ASM Handbook®

Comprehensive Index

Prepared under the direction of the
ASM International Handbook Committee

ASM INTERNATIONAL®
The Materials
Information Society

First printing, April 1994

This book is a collective effort involving hundreds of technical specialists. It brings together a wealth of information from worldwide sources to help scientists, engineers, and technicians solve current and longrange problems.

Great care is taken in the compilation and production of this Volume, but it should be made clear that NO WARRANTIES, EXPRESS OR IMPLIED, INCLUDING, WITHOUT LIMITATION, WARRANTIES OF MERCHANTABILITY OR FITNESS FOR A PARTICULAR PURPOSE, ARE GIVEN IN CONNECTION WITH THIS PUBLICATION. Although this information is believed to be accurate by ASM, ASM cannot guarantee that favorable results will be obtained from the use of this publication alone. This publication is intended for use by persons having technical skill, at their sole discretion and risk. Since the conditions of product or material use are outside of ASM's control, ASM assumes no liability or obligation in connection with any use of this information. No claim of any kind, whether as to products or information in this publication, and whether or not based on negligence, shall be greater in amount than the purchase price of this product or publication in respect of which damages are claimed. THE REMEDY HEREBY PROVIDED SHALL BE THE EXCLUSIVE AND SOLE REMEDY OF BUYER, AND IN NO EVENT SHALL EITHER PARTY BE LIABLE FOR SPECIAL, INDIRECT OR CONSEQUENTIAL DAMAGES WHETHER OR NOT CAUSED BY OR RESULTING FROM THE NEGLIGENCE OF SUCH PARTY. As with any material, evaluation of the material under enduse conditions prior to specification is essential. Therefore, specific testing under actual conditions is recommended.

Nothing contained in this book shall be construed as a grant of any right of manufacture, sale, use, or reproduction, in connection with any method, process, apparatus, product, composition, or system, whether or not covered by letters patent, copyright, or trademark, and nothing contained in this book shall be construed as a defense against any alleged infringement of letters patent, copyright, or trademark, or as a defense against liability for such infringement.

Comments, criticisms, and suggestions are invited, and should be forwarded to ASM International.

Library of Congress Cataloging Card Number: 94-070831
ISBN: 0-87170-383-1
SAN: 204-7586

ASM International®
Materials Park, OH 440730002

Printed in the United States of America

Preface

This *Comprehensive Index* is a convenient, single-volume compilation of indexes to 28 handbook volumes published by ASM International. It features indexes to 17 volumes of the current *ASM Handbook* (formerly *Metals Handbook*), and—as a service to owners of a complete set of the 9th Edition *Metals Handbook*—it includes indexes for the six 9th Edition volumes that have been, or will soon be, superseded by revised *ASM Handbook* volumes. In addition, the *Comprehensive Index* includes indexes to the four-volume *Engineered Materials Handbook* as well as the *Electronic Materials Handbook*, Volume 1, *Packaging*. A complete list of the handbooks covered in this Volume is given in the table below.

The format for the *Comprehensive Index* has been designed for ease of use. The letter and number code in boldface following a subject entry indicates the handbook series and volume number. The letter designations are **A**, *ASM Handbook*, Volumes 1-4, 6-18 (Volume 5 of *ASM Handbook* is scheduled for publication in Autumn 1994 and is not included in this index); **M**, *Metals Handbook*, 9th Edition, Volumes 1-6; **EM**, *Engineered Materials Handbook*, Volumes 1-4; and **EL**, *Electronic Materials Handbook*, Volume 1. Owners of 9th Edition *Metals Handbook* volumes should note that Volumes 7 through 17 of that series have been folded into the *ASM Handbook* (and have been re-covered in the signature green *ASM Handbook* cover). Thus, index entries for these volumes are coded with **A**; however, these entries are also valid for the red-covered *Metals Handbook* volumes.

Following the boldface series and volume indicator for each entry are the page numbers that indicate the location of information on the indicated subject within that particular volume. For example, **A4**: 253-257, 655 directs the reader to *ASM Handbook*, Volume 4, pages 253 to 257 and 655. *ASM Handbook*, Volume 3, *Alloy Phase Diagrams* is numbered by sections; an example of an entry is **A3**: 1•24, which directs the reader to section 1, page 24 of that volume.

Because the separate indexes combined in the *Comprehensive Index* vary considerably in style and approach, the user of this Volume may want to try a number of strategies for finding information on a given subject. For example, to find information about heat treating of a particular alloy, the user should look not only under "Heat treating" but also under the name/designation of the particular material as well as under the names of specific heat treating processes. An electronic version of this index that facilitates searching for key terms is also available from ASM International.

Handbook volumes included in the *Comprehensive Index*

Volume No.	Index code	Title	Year of publication
ASM Handbook			
1	A1	*Properties and Selection: Irons, Steels, and High-Performance Alloys*(a)	1990
2	A2	*Properties and Selection: Nonferrous Alloys and Special-Purpose Materials*(a)	1990
3	A3	*Alloy Phase Diagrams*	1992
4	A4	*Heat Treating*	1991
6	A6	*Welding, Brazing, and Soldering*	1993
7	A7	*Powder Metallurgy*(b)	1984
8	A8	*Mechanical Testing*(b)	1985
9	A9	*Metallography and Microstructures*(b)	1985
10	A10	*Materials Characterization*(b)	1986
11	A11	*Failure Analysis and Prevention*(b)	1986
12	A12	*Fractography*(b)	1987
13	A13	*Corrosion*(b)	1987
14	A14	*Forming and Forging*(b)	1988
15	A15	*Casting*(b)	1988
16	A16	*Machining*(b)	1989
17	A17	*Nondestructive Testing and Quality Control*(b)	1989
18	A18	*Friction, Lubrication, and Wear Technology*	1992
Metals Handbook			
1	M1	*Properties and Selection: Irons and Steels*	1978
2	M2	*Properties and Selection: Nonferrous Alloys and Pure Metals*	1979
3	M3	*Properties and Selection: Stainless Steels, Tool Materials, and Special-Purpose Materials*	1980
4	M4	*Heat Treating*	1981
5	M5	*Surface Cleaning, Finishing, and Coating*	1982
6	M6	*Welding, Brazing, and Soldering*	1983
Engineered Materials Handbook			
1	EM1	*Composites*	1987
2	EM2	*Engineering Plastics*	1988
3	EM3	*Adhesives and Sealants*	1990
4	EM4	*Ceramics and Glasses*	1991
Electronic Materials Handbook			
1	EL1	*Packaging*	1989

(a) Originally released as *Metals Handbook*, 10th Ed. (b) Originally released as *Metals Handbook*, 9th Ed.

Numbered Entries

0.6-0.9C-10Cr-Mo alloy
abrasive wear volume **A18:** 805
2θ geometry, variable
RDF analysis **A10:** 396
4-(2-pyridylazo)-resorcino(PAR) reagent
use in ion chromatography **A10:** 661
5-2½ alloy *See* Titanium alloys, specific types, Ti-5Al-2.5Sn
5-end satin weave *See* Satin (crowfoot) weave
6/4 alloy Ti
sawing **A16:** 360
7-14CuMo
composition **M6:** 354
7-Mo Plus *See* Stainless steels, specific types, S32950
8-8-2-3 beta alloy *See* Titanium alloys, specific types, Ti-8Mo-8V-2Fe-3Al
8-end satin weave *See* Satin (crowfoot) weave
8-hydroxyquinoline
as precipitant **A10:** 170
as solvent extractant **A10:** 170
8-quinolinol
as precipitant **A10:** 169
10 alloy *See* Copper alloys, specific types, C17500
13-11-3 *See* Titanium alloys, specific types, Ti-13V-11Cr-3Al
14-MeV fast neutron activation analysis
elemental concentrations in NBS fly ash determined using **A10:** 239
14CrMoV69 steel
fatigue curves **M1:** 541
15-5 PH *See* Stainless steels
15-5PH *See* Stainless steels, specific types, S15500
15-7PH *See* Stainless steels, specific types, S15700
15-15LC *See* Stainless steels, specific types, S21300
16-25-6
18% maraging steel, constitutional liquation in multicomponent systems **A6:** 568
arc welding **M6:** 364
broaching **A16:** 744, 745
composition **A6:** 564 **A16:** 736 **M6:** 354
electron-beam welding **A6:** 869
machining **A16:** 738, 741-743, 746-747, 749-758
17-4PH *See* Stainless steels, specific types, S17400
17-14CuMo
composition **A16:** 736
17-22 AS
composition **M1:** 649
17-22 AV
composition **M1:** 649
17-22A
broaching **A16:** 203, 209
drilling **A16:** 750
18-2FM *See* Stainless steels, specific types, S18200
18-2Mn *See* Stainless steels, specific types, S24100
18-18 Plus *See* Stainless steels, specific types, S28200
18Ni steels *See* Maraging steels
19 alloy/20 alloy thermocouple *See* Thermocouples, materials, nonstandard
19-9 DL
composition **A4:** 771, 794
19-9DL
annealing **M4:** 655
arc welding **M6:** 364
composition **A16:** 736 **M4:** 651-652 **M6:** 354
drilling **A16:** 750
flash welding **M6:** 557
machining **A16:** 738, 741-743, 746-747, 749-758
stress relieving **M4:** 655
19-9DX
flash welding **M6:** 557
20Cb-3
filler metal for stainless steel casting alloys **A6:** 496
21-6-9LC *See* Stainless steels, specific types, S21904
22 40 steel
composition and heat treatment **M1:** 542
22-13-5 *See* Stainless steels, specific types, S20910
25 alloy *See* Copper alloys, specific types, C17200
30 CD 12 steel
composition and heat treatment **M1:** 542
31CrMoV9 steel
composition and heat treatment **M1:** 542
32 point groups *See* Crystal classes
38-6-44 *See* Titanium alloys, specific types, Ti-3Al-8V-6Cr-4Zr-4Mo
50 alloy *See* Copper alloys, specific types, C17600
60 metal *See* Tantalum alloys, specific types, Ta-10U
61 metal *See* Tantalum alloys, specific types, "61" metal
63 metal *See* Tantalum alloys, specific types, "63" metal
100-AR
composition and mechanical properties **M1:** 621
135M steel *See* Nitralloy 135M
165 alloy *See* Copper alloys, specific types, C17000
263
composition **A6:** 573
300M
composition **M1:** 422
heat treatment **M1:** 427
mechanical properties **M1:** 427-429
notches, effect on fatigue behavior **M1:** 668, 670
processing **M1:** 427
885 °F embrittlement **A6:** 848

A

55% Aluminum-zinc alloy coating **M5:** 348-350
cooling **M5:** 349
corrosion resistance **M5:** 348
heat treating **M5:** 349-350
microstructure **M5:** 348-350
precleaning **M5:** 349
process **M5:** 348-349
steel sheet and wire **M5:** 348-350
A *See* Absorbance; Crack length; Crack size; Crystal lattice length along the a axis; Stress ratio **EM3:** 3
defined **EM2:** 2
A basis
of design values **A8:** 662
α- and β-fibers
in rolled copper **A10:** 363
A-40 *See* Titanium alloys, specific types, Ti grade 2
A-44 *See* Titanium alloys, specific types, Ti grade 3
A-70 *See* Titanium alloys, specific types, Ti grade 4
A-110AT *See* Titanium alloys, specific types, Ti-5Al-2.5Sn
A-286
aging **A4:** 796, 800, 801
aging cycle **M4:** 656, 657
aging, effect on properties **M4:** 659, 664
aging precipitates **A4:** 796
annealing **M4:** 655
band sawing **A16:** 738, 756
broachability constant **A16:** 200
broaching **A16:** 203-206, 209, 743-746
carbon pickup **A4:** 798
cold working effect on aging **A4:** 801
composition **A4:** 794 **A16:** 736 **M4:** 651, 652 **M6:** 354
contour band sawing **A16:** 363
cooling rate, effect on properties **M4:** 661
double-aging **A4:** 796
drilling **A16:** 738, 739
electrochemical grinding **A16:** 547
electrochemical machining **A16:** 534, 539, 541
end milling **A16:** 539, 738, 739
face milling **A16:** 738, 739
fixtures **A4:** 799
flash welding **M6:** 557
gas nitriding **A4:** 387, 401
gas tungsten arc welding **M6:** 365-366
grain growth **A4:** 799, 800
grinding **A16:** 547, 759, 760
machinability **A16:** 737
machining **A16:** 738, 741-743, 746-747, 749-758
machining characteristics compared in table **A16:** 738, 739
mechanical properties, effect of heat treatment **A4:** 800
milling **A16:** 547
nickel content and alloy classification **A4:** 800
precipitation strengthening and grain growth **A4:** 799
reaming **A16:** 738, 739
sawing **A16:** 360
A-286 (continued)
solution heat treatment **A4:** 796, 801
solution treating **M4:** 656, 657
springs, strip for **M1:** 286
springs, wire for **M1:** 285
straddle milling **A16:** 738, 739
stress relieving **M4:** 655
tapping **A16:** 738, 739
threading **A16:** 738, 739
turning **A16:** 738, 739, 740
α-Al_2O_3 (white ceramic) **A16:** 98
A-allowable *See* A-basis
α-β alloy, SCC environment
AES analysis **A10:** 563-564
A-basis *See also* B-basis; S-basis; Typical basis; Typical-basis **EM3:** 3
defined **EM1:** 3 **EM2:** 2
α-brass alloy
Auger analysis **A10:** 563
a-carbon
plasma-assisted physical vapor deposition **A18:** 848
A-frame core knockout machine **A15:** 504-505
A-fritting **A18:** 682, 683
α-phase alloy, SCC environment
AES analysis **A10:** 563-564
A-porosity in cemented carbides **A9:** 274
A-scan display modes
applications **A17:** 242
data interpretation **A17:** 244-246
data, optical recording of **A17:** 228
display **A17:** 242, 244-246
optical coherent signal processor for **A17:** 227
pulse-echo ultrasonic inspection **A17:** 241-242
scanning acoustical holography **A17:** 443
signal display **A17:** 242
system setup **A17:** 242
***a*-spacing**
layer lattice solid lubricants **A18:** 113
A-spot
defined **A18:** 1
A-stage *See also* B-stage; C-stage **EM3:** 5
defined **EM1:** 4 **EM2:** 5
A.O. Smith iron powder
functions of annealing **A7:** 182
A.W.G. system *See* American Wire Gage system
A/W glass ceramic **EM4:** 1008
bonding to bone **EM4:** 1010, 1011
in vertebral surgery **EM4:** 1011
A2, A4, A5, A6, etc *See* Toot steels, specific types
A10 + SiO_4 + chromium oxide
coating for gas-lubricated bearings **A18:** 532
A15 superconductors *See also* Superconducting materials; Superconductivity; Superconductors
alloying, with third element additions **A2:** 1062-1063
applications **A2:** 1070-1074
assembly techniques **A2:** 1065-1067
bronze tape conductors **A2:** 1065
cable and winding **A2:** 1069-1070
chloride deposition **A2:** 1065
commercial magnets **A2:** 1070
conductor alloy **A2:** 12
critical current density **A2:** 1063-1064
defined **A2:** 1060
deformation **A2:** 1067
development **A2:** 1060-1062
fusion application **A2:** 1071-1072
high-energy physics **A2:** 1071
history **A2:** 1028
jelly roll method **A2:** 1067
layer growth **A2:** 1063
liquid quenching **A2:** 1065
matrix materials **A2:** 1064-1065
modified jelly roll process **A2:** 1066
multifilamentary wire assembly **A2:** 1065-1067
niobium tube process **A2:** 1066-1067
phase diagrams **A2:** 1062
powder metallurgy **A2:** 1067
power generation applications **A2:** 1070-1071
processing **A2:** 1065-1070
properties **A2:** 1065-1070
reaction heat treatments **A2:** 1068-1069
rod process **A2:** 1065-1066
surface diffusion **A2:** 1065
tape conductor assembly **A2:** 1065

A16-SG alumina
rheological behavior in injection molding ... EM4: 174-175
A286 *See* Stainless steels, specific types, S66286 (AISI A286)
AAS *See* Atomic absorption spectrometry
ABA copolymers ... EM3: 3
defined ... EM2: 2
ABAQUS computer program for structural analysis ... EM1: 268, 26
ABAQUS Finite-element analysis code ... EM3: 480, 486
Abbe number
dispersion property of glass ... EM4: 565
Abbé offset ... A18: 337
Abbé principle in dimensional measurement ... A18: 337
Abbe's criterion
for microscope magnification ... A12: 80
Abbott-Firestone curve *See also* Bearing area
defined ... A18: 1
Abbreviations ... A4: 968-970
and symbols ... A8: 724-726 A10: 689-692 A11: 796-798 A12: 492-494 A13: 1375-1377 A14: 944-945 A15: 896-897 A17: 758-760 EL1: 1166-1168 EM1: 948-950 EM2: 850-852 EM3: 852-853
Abbreviations, symbols, and tradenames ... A1: 1038-1041 A2: 1273-1277
Aberration
chromatic, defined ... A10: 670
defined ... A9: 1 A10: 668
spherical ... A10: 682
Aberration, spherical
in SEM imaging ... A12: 167-168
ABEX wet abrasion test ... A18: 189
Abhesive ... EM3: 3
defined ... EM2: 2
Abietic acid ... A6: 129
ABL bottle
defined ... EM1: 3 EM2: 2
Ablation ... EM3: 3
defined ... EM1: 3 EM2: 2
Ablation, laser
for solid sample analysis ... A10: 36
Ablative plastic ... EM3: 3
defined ... EM1: 3 EM2: 2
Abnormal grain growth *See also* Grain growth ... A9: 689-690
Abnormal zinc homeostasis
biologic indicators ... A2: 1255
Aborescent powder
defined ... A7: 1
Abradable seals
from composite powders ... A7: 175
Abraded ribbons
FMR study of ... A10: 274
Abraded surfaces *See also* Abrasion artifacts; Abrasion damage
flatness ... A9: 39-40
plastic deformation ... A9: 39
Abrasion ... A8: 1
for adhesive bonding ... EM1: 68
carbon fiber ... EM1: 5
cobalt-base alloys ... A13: 663
and crushing failure, steel wire rope ... A11: 519
data, cobalt-base alloys ... A2: 450
defined ... A9: 1, 35 A11: 1
definition ... A16: 40
edge retention of sample during ... A9: 44-45
effect, gas/oil wells ... A13: 479-480
failures, in pharmaceutical production ... A13: 1229
as fatigue crack origin ... A12: 263
fuel pump failure by ... A11: 465
of glass fibers ... EM1: 4
high-carbon steels ... A12: 285
of lead ... A9: 41
range of wear coefficients ... A8: 601-602
resistance, carbon and tungsten effects ... A2: 450
rub marks by ... A11: 27
Abrasion (continued)
surface, from mechanical damage ... A11: 342
of thermoplastic mounting materials ... A9: 30
in XPS samples ... A10: 575
Abrasion (abrasive wear) *See also* Abrasive erosion, Abrasive wear
defined ... A18: 1
Abrasion, aluminum alloys
soldering ... A6: 628
Abrasion artifacts *See also* Polishing artifacts; Tempering artifacts ... A9: 37-39
in austenitic stainless steel ... A9: 34
in austenitic steels ... A9: 37
in brass ... A9: 33, 37
defined ... A9: 1
deformation etch markings ... A9: 37
in ferrite steels ... A9: 35, 38
in metals with noncubic crystal structure ... A9: 37-38
in pearlitic steels ... A9: 35, 38
in plain carbon steels ... A9: 36
in surface oxide layers ... A9: 46
in very soft materials ... A9: 46-47
in zinc ... A9: 34, 37-38
Abrasion damage *See also* Polishing damage ... A9: 37-39
effect on hardness ... A9: 39
effect on transmission electron microscopy samples ... A9: 39
in gray iron ... A9: 36, 38-39
relationship of hardness to depth of ... A9: 37
Abrasion fluid
defined ... A9: 1
"Abrasion level" concept ... A18: 758
Abrasion, of dies *See also* Abrasives; Adhesion; Galling ... A14: 47, 56, 505
Abrasion process
defined ... A9: 1
Abrasion protection
thermal spray coatings ... A6: 1007
Abrasion rate
defined ... A9: 1
Abrasion resistance *See also* Fuzz ... M3: 582, 583
austenitic manganese steel ... M3: 580-583
chromate conversion coating ... A13: 392-393
of coatings ... A13: 395
comparisons ... EM2: 167
of core blowing machines ... A15: 191
defined ... EL1: 1133
galvanized steels ... A13: 438
of metal patterns ... A15: 195
polyamides (PA) ... EM2: 126
polyurethanes (PUR) ... EM2: 259
superhard tool materials ... M3: 453, 455, 456, 457-458, 461, 464
thermoplastic polyurethanes (TPUR) ... EM2: 206
ultrahigh molecular weight polyethylenes (UHMWPE) ... EM2: 167
Abrasion resistance test
magnesium alloy finishes ... M5: 638
Abrasion soldering
definition ... M6: 1
Abrasion surface ... EM3: 35
Abrasion testing
austenitic manganese steel ... M3: 582
Abrasion wear *See* Abrasive wear
Abrasion-resistant (AR) steels ... A18: 649
wear rates for test plates in drag conveyor bottoms ... A18: 720
Abrasion-resistant cast iron ... A9: 245
as-cast against a chill ... A9: 254
high-chromium, as-cast ... A9: 255
Abrasion-resistant cast irons *See* Alloy cast irons
abrasion resistance ... M1: 81, 87-88
alloying elements, effects on depth of chill ... M1: 76-80
applications ... M1: 81
characteristics ... M1: 75-76, 81
compositions ... M1: 76, 81, 82
Abrasion-resistant cast irons (continued)
heat treatment ... M1: 81-83
mechanical properties ... M1: 86-87
microstructure ... M1: 76, 83-86
physical properties ... M1: 83, 87, 88
production ... M1: 81
Abrasive
defined ... A9: 1
definition ... M6: 1
Abrasive belt grinding
Al alloys ... A16: 770, 801
cast Irons ... A16: 663
Abrasive belt polishing *See* Belt polishing, abrasive
Abrasive blast cleaning ... A13: 414-415, 912, 1143
normalizing scale removal method ... A4: 40
of stainless steel forgings ... A14: 230
Abrasive blasting *See also* Grit blasting; specific blasting methods by name ... M5: 83-96
abrasive types used ... M5: 83-86, 91, 93-95
aluminum and aluminum alloys ... M5: 91, 571-573
applications ... M5: 83, 91-93
bronze ... M5: 91
cast iron ... M5: 91
castings ... M5: 86, 91-92, 614
ceramic coating processes ... M5: 537-539
copper and copper alloys ... M5: 614
for deburring ... A7: 458
definition ... M6: 1
dry *See* Dry abrasive blasting ... A17: 81, 82
electropolishing processes ... M5: 304, 306, 308
environmental control ... M5: 88-90
equipment ... M5: 86-91, 95-96
heat-resistant alloys ... M5: 563-565
hot dip galvanized coating process ... M5: 326, 329, 332
hot dip tin coating process ... M5: 353
iron ... M5: 91
magnesium alloys ... M5: 628-629
mechanism of action ... M5: 4
molybdenum and tungsten ... M5: 659
nickel and nickel alloys ... M5: 669-670
nonmetallic materials ... M5: 91
painting pretreatment ... M5: 332
painting process, wire brushing compared with ... M5: 476
porcelain enameling process ... M5: 515
refractory metals ... M5: 652-653, 659-663, 667
rust and scale removal by ... M5: 11-14
safety precautions ... M5: 21, 96
stainless steel ... M5: 552-553
steel ... M5: 91, 94
tantalum and niobium ... M5: 663
titanium and titanium alloys ... M5: 652-653
weldments ... M5: 91
wet *See* Wet abrasive blasting
zirconium and hafnium alloys ... M5: 667
Abrasive cleaning
after investment casting ... A15: 263
grit, zirconium castings ... A15: 838
of Replicast products ... A15: 271
Abrasive cloth wear testing ... M1: 603
Abrasive cutoff machine
for optical metallography specimen preparation ... A10: 300
Abrasive cutoff sawing
Al alloys ... A16: 800
Cu alloys ... A16: 818
refractory metals ... A16: 867, 868
Ti alloys ... A16: 846
Abrasive cutoff wheel cutting
specimen ... A12: 76, 92
Abrasive cutoff wheels
hot upset forging ... A14: 86
for nitrided steels ... A9: 218
used for tool steels ... A9: 256
Abrasive cutter ... A9: 23
Abrasive cutting
rhenium ... A2: 562

SUBJECTS OF THE INDEXED VOLUMES: ASM Handbook (designated by the letter "A"): **A1:** Properties and Selection: Irons, Steels, and High-Performance Alloys (1990); **A2:** Properties and Selection: Nonferrous Alloys and Special-Purpose Materials (1990); **A3:** Alloy Phase Diagrams (1992); **A4:** Heat Treating (1991); **A6:** Welding, Brazing, and Soldering (1993); **A7:** Powder Metallurgy (1984); **A8:** Mechanical Testing (1985); **A9:** Metallography and Microstructures (1985); **A10:** Materials Characterization (1986); **A11:** Failure Analysis and Prevention (1986); **A12:** Fractography (1987); **A13:** Corrosion (1987); **A14:** Forming and Forging (1988); **A15:** Casting (1988); **A16:** Machining (1989); **A17:** Nondestructive Testing and Quality Control (1989); **A18:** Friction, Lubrication, and Wear Technology (1992). **Metals Handbook, 9th Edition** (designated by the letter "M"): **M1:** Properties and Selection: Irons and Steels (1978); **M2:** Properties and Selection: Nonferrous Alloys and Pure Metals (1979); **M3:** Properties and Selection: Stainless Steels, Tool Materials and Special-Purpose Materials (1980); **M4:** Heat Treating (1981); **M5:** Surface Cleaning, Finishing, and Coating (1982); **M6:** Welding, Brazing, and Soldering (1983). **Engineered Materials Handbook** (designated by the letters "EM"): **EM1:** Composites (1987); **EM2:** Engineering Plastics (1988); **EM3:** Adhesives and Sealants (1990); **EM4:** Ceramics and Glasses (1991); **Electronic Materials Handbook** (designated by the letters "EL"): **EL1:** Packaging (1989).

Abrasive cutting used for sectioning A9: 24-26
solutions to problems encountered A9: 24
Abrasive disk grinding
Al alloys A16: 770
Ti alloys A16: 846
Abrasive disks
for surface preparation A17: 52
Abrasive erosion *See also* Erosion
in boilers and steam equipment A11: 623
defined A18: 1
Abrasive fillers
two-component flexible epoxies EL1: 818
Abrasive flow A16: 19
Abrasive flow machining (AFM) A16: 509, 514-519
abrasive grains A16: 517
machines A16: 515-516
media A16: 516-517
and postprocessing A16: 35
process applications A16: 518-519
process capabilities A16: 517-518
process characteristic flow rates A16: 514-515, 516
stock removal A16: 518
surface finish A16: 518
tooling A16: 516
Abrasive fluid jet cutting EM4: 313, 314
Abrasive fluid jet machining EM4: 363-366
abrasive slurry jets EM4: 365
abrasive waterjet cutting fundamentals EM4: 363, 364
advantages EM4: 363, 364
limitations EM4: 364
process capabilities EM4: 363-364
system components EM4: 363, 364
applications EM4: 365, 366
cutting parameters' effect on surface quality EM4: 364-365
abrasives used EM4: 365
cut surface properties EM4: 365
cutting speeds EM4: 365
future outlook EM4: 366
interactions at the grinding zone EM4: 315-317
abrasive grain size effect EM4: 316, 317, 325
grinding direction effect EM4: 316, 317
surface finish EM4: 316, 317, 318, 325, 327
principles of EM4: 363
recommended cutting speeds for selected ceramics EM4: 366
Abrasive fraction A18: 431
Abrasive grain shape A18: 185
Abrasive grinding
of investment castings A15: 264
Abrasive jet machining (AJM) A16: 509, 511-513
advantages and disadvantages A16: 512
applications A16: 511
compared to sandblasting A16: 511
material removal abrasive powders A16: 512
material removal flow rates A16: 512
material removal nozzle tip distance A16: 513
stainless steels A16: 706
system components A16: 511
tolerance and finish A16: 513
Abrasive machining EM4: 313, 314
defined A9: 1
Abrasive machining methods
and milling operation A16: 329
surface alterations produced A16: 23-24
Abrasive machining, principles of EM4: 315-327
grindability of ceramics versus metals EM4: 315
elastic deformation EM4: 315
strength EM4: 315
thermal conductivity EM4: 315
material removal mechanism in the grinding of ceramics EM4: 317-320
chip formation and surface generation EM4: 319-320
chip formation model for precision grinding of ceramics EM4: 318-320
ductile regime grinding model EM4: 317-318
indentation fracture mechanism EM4: 317, 319
plastic deformation EM4: 319, 320
systems approach EM4: 320-327
machine tool parameters EM4: 320, 321-324
operational factors EM4: 320, 326-327
wheel specification EM4: 320, 324
work material properties EM4: 320, 325-326
Abrasive minerals
hardness of M1: 89
Abrasive paint stripping method M5: 18-19
Abrasive papers
flatness obtained compared to fixed-abrasive lap A9: 39
Abrasive processes A16: 19
Abrasive processing A16: 32-33
Abrasive removal
nickel-titanium shape memory effect (SME) alloys A2: 899
Abrasive slurry
used with wire saws A9: 26
Abrasive tumbling A16: 27, 35
as mechanical cleaning method A17: 82
Abrasive water-jet cutting EM1: 673-67
applications EM1: 674-67
benefits/problems EM1: 673-67
cutting characteristics EM1: 674-67
equipment tools EM1: 67
future trends EM1: 67
materials cut by EM1: 675
Abrasive waterjet cutting A14: 743-755
abrasives A14: 747-748
advantages/limitations A14: 743
applications A14: 752-755
cut quality A14: 751-752
cutting principle A14: 743-744
defined A14: 743
as new metalworking process A14: 18-19
safety A14: 755
surface finish A14: 746-748
system components A14: 743-747
waterjet speeds, calculation A14: 748-751
Abrasive waterjet machining (AWJM)
See also Waterjet/abrasive waterjet machining A16: 509
carbon and alloy steels A16: 677
compared to friction band sawing A16: 365
MMCs A16: 893-894, 896, 897
stainless steels A16: 704, 706
Abrasive wear *See also* Adhesive wear; Fretting; Fretting corrosion; Oxidative wear; specific type by name, such as Scratching abrasion A8: 1, 602 A18: 184-190, 613
abrasive particle size M1: 601, 602
aluminum-silicon alloys A18: 788
bearing steels A18: 732-733
of bearings A11: 494-495
categories A18: 184-190
of cemented carbides A7: 779 A18: 797-798, 799
ceramics A18: 814
of cobalt-base wear-resistant alloys A2: 447
cobalt-base wrought alloys A18: 767, 768
composite restorative materials (dental), studies and testing A18: 670
correlations between dissimilar tests M1: 600-602
damage dominated by chip formation A18: 179-180
defined A9: 1 A11: 1 A18: 1, 184
dental amalgam abrasion test methods A18: 669
dental cement testing A18: 673
denture acrylics A18: 674
die material, material loss (dental) A18: 675
effect of material properties on A11: 158-159
electroplated coating applications A18: 835
embeddability of soft metals M1: 609-610
environmental effect A18: 187-189
abrasive A18: 188
corrosive effects A18: 189
humidity A18: 188-189
load A18: 188
speed of contact A18: 188
temperature A18: 188
failures A11: 146-148
gray cast iron M1: 24
hardfacing alloys A18: 758, 759, 760-761, 763, 764, 765
hardfacing for A7: 823
hardness of abrasive M1: 601-603
hardness of metal M1: 603-605
internal combustion engine parts A18: 555, 558, 559
ion implantation A18: 855-856, 857, 858
jet engine components A18: 588, 590
laboratory vs. field tests M1: 600
laser-hardened gray iron A18: 864
Abrasive wear (continued)
lubricant analysis case history A18: 308
mass loss measures of wear A18: 362
material properties, effects of A18: 186-187
abrasive grain size A18: 187
alloying A18: 186
crystal structure and orientation A18: 186, 187
fracture toughness A18: 186
hardness correlation with abrasion rate A18: 186
modulus of elasticity A18: 187
second phase size A18: 186-187
solidus temperature A18: 186
material selection for A13: 333
materials A18: 189-190
ceramics A18: 189
metals A18: 189-190
plastics A18: 190
mechanism M1: 599
mechanisms for material removal A18: 184
cutting A18: 184, 185
microcracking A18: 184, 185-186
microfatigue A18: 184
plowing A18: 184-185, 186
wedge formation A18: 184
metal-matrix composites A18: 804-805, 806, 807, 809, 810
mining and mineral industries A18: 649-650, 651, 652
types of A18: 649
nickel, electroless A18: 837
nitrided surfaces A18: 879, 880, 881, 882
on gear teeth A11: 595-596
processes A18: 184, 185
pump sleeve failure by A11: 159
pumps A18: 595, 597-598, 599
resistance, cemented carbides A2: 958-959
rubbing wear particles in lubricant analysis A18: 302
semiconductors A18: 685
in shafts A11: 465
of shell liner A11: 375-377
in sliding bearings A11: 488 A18: 742, 743
solid particle erosion A18: 199, 203-204
stainless steels A18: 713-714, 716, 717, 718-720, 722-723
surface texture applications A18: 343
tester A8: 605
testing M1: 599, 600-603, 618
theory A11: 146-148 A18: 189
thermal spray coating applications A18: 831, 832-833
performance factors A18: 833
thermoplastic composites A18: 821
third-particle A11: 494
tool steels A18: 736-737, 738
toothbrush and dentifrice prophylactic wear of human dental tissues A18: 665
wear studies A18: 667-669
types M1: 597, 599
under lubrication A11: 150
volume rate per unit length of sliding, symbol for A11: 798
wear resistance tables M1: 617, 621
wear resistance versus hardness, annealed unalloyed metals A18: 707, 708
Abrasive wear factor
defined A18: 1
Abrasive wear, of dies *See also* Abrasion; Wear A14: 47
Abrasive wear resistance A18: 490-491
of cemented carbides A7: 778
in ferrous P/M materials A7: 464
Abrasive wheel cutting
of carbon and alloy steels A9: 165-166
Abrasive wheel grinding
Al alloys A16: 770
Abrasive wheels
as copper-based powder application A7: 733
metal bonded A7: 797
nonconsumable A9: 25
powders used for A7: 572
Abrasive wheels, consumable
coolants for use with A9: 24-25
edge wear used to determine suitability A9: 25
selection of A9: 24

Abrasive wheels, consumable (continued)
shelf life ... **A9:** 25
speeds ... **A9:** 25
Abrasive wood/plastic composites
diamond for machining ... **A16:** 105
Abrasive(s)
for blast cleaning ... **A15:** 510-511
distribution/angle of impact ... **A15:** 517-518
flow rates ... **A15:** 516-517
operating mix ... **A15:** 511
parameters of ... **A15:** 518-520
steel shot/grit as ... **A15:** 506
Abrasives *See also* Abrasion; Abrasive waterjet cutting ... **EM4:** 329-335
ceramic machining guidelines ... **EM4:** 333-335
grinding of glass ... **EM4:** 333-334
grinding of high-alumina ceramics ... **EM4:** 333
grinding of technical ceramics ... **EM4:** 334-335
coatings to improve bond properties ... **EM4:** 332-333
nickel metal coatings ... **EM4:** 333
controlling abrasive properties ... **EM4:** 331-332
bondability ... **EM4:** 332
friability ... **EM4:** 332
toughness index ... **EM4:** 332
diamond grinding applications ... **EM4:** 333
drilling ... **A16:** 229
garnet ... **A14:** 746-748
for grinding ceramics and glasses ... **EM4:** 331
grit size ... **A14:** 752
matching abrasive and bond properties ... **EM4:** 331
nomenclature ... **EM4:** 333
grinding ratio ... **EM4:** 333
grit size ... **EM4:** 333
specific grinding energy ... **EM4:** 333
specific grinding ratio ... **EM4:** 333
performance characteristics ... **A13:** 415
physical properties of engineered materials ... **EM4:** 329
hardness ... **EM4:** 329, 330
modulus of resilience ... **EM4:** 329, 330
position of abrasives on the hardness/MOR map ... **EM4:** 329-331
properties ... **A13:** 921
silica ... **A14:** 747-748
Abrasives for aluminum alloys ... **A9:** 352-353
for aluminum-silicon alloys ... **A9:** 40
for beryllium ... **A9:** 389
for beryllium-copper alloys ... **A9:** 392-393
for beryllium-nickel alloys ... **A9:** 392-393
for carbon and alloy steels ... **A9:** 168-169
for carbon steel casting specimens ... **A9:** 230
for cast irons ... **A9:** 243
for copper and copper alloys ... **A9:** 400
effect on flatness ... **A9:** 40
for electrical contact materials ... **A9:** 550
for electrogalvanized sheet steel ... **A9:** 197
embedding in specimens ... **A9:** 39
for ferrites and garnets ... **A9:** 533
for fiber composites, grinding ... **A9:** 588-589
for fiber composites, polishing ... **A9:** 589-591
for hafnium ... **A9:** 497
for hand polishing ... **A9:** 35
for heat-resistant casting alloys ... **A9:** 330
for hot-dip galvanized sheet steel ... **A9:** 197
for hot-dip zinc-aluminum coated sheet steel ... **A9:** 197
for iron-cobalt and iron-nickel alloys ... **A9:** 532-533
for low-alloy steel casting samples ... **A9:** 230
for powder metallurgy materials ... **A9:** 505-506
for preservation of nonmetallic inclusions ... **A9:** 39
for refractory metals ... **A9:** 439
for sleeve bearing materials ... **A9:** 565-567
for tin and tin alloy coatings ... **A9:** 450-451
for tin and tin alloys ... **A9:** 449
for titanium and titanium alloys ... **A9:** 458-459
Abrasives for aluminum alloys (continued)
for transmission electron microscopy specimens ... **A9:** 104
for use with tool steels ... **A9:** 257
for wire sawing ... **A9:** 26
for wrought heat-resistant alloys ... **A9:** 305-307
for wrought stainless steel ... **A9:** 279
for zinc and zinc alloys ... **A9:** 488
for zirconium and zirconium alloys ... **A9:** 497
Abrasives, polishing
types used ... **M5:** 108
Abrasivity
defined ... **A18:** 1
ABS *See* Acrylonitrile-butadiene-styrene; Acrylonitrile-butadiene-styrenes
Absolute coil arrangement
eddy current inspection ... **A17:** 175-176
Absolute density
by computed tomography (CT) ... **A17:** 361
defined ... **A7:** 1
Absolute depth scale
by FIM/AP ... **A10:** 593
Absolute humidity ... **EM3:** 3
defined ... **EM2:** 2
Absolute impact velocity *See* Impact velocity
Absolute magnetic permeability ... **A6:** 365
Absolute methanol, and bromine
to isolate inclusions in steel ... **A10:** 176
Absolute pore size
defined ... **A7:** 1
Absolute probes
eddy current inspection ... **A17:** 180-181
Absolute viscosity *See* Viscosity ... **EM3:** 3
defined ... **EM2:** 2
Absorbance
abbreviation for ... **A10:** 689
defined ... **A10:** 668
as function of wavelength ... **A10:** 63
in IR quantitative analysis ... **A10:** 117
in IR spectra ... **A10:** 110
optimum, UV/VIS ... **A10:** 68
UV/VIS, as function of sample concentration ... **A10:** 62-64
vs radiation energy ... **A10:** 85-86
Absorbance-subtraction techniques
as IR qualitative analysis ... **A10:** 116
for polymer curing reactions ... **A10:** 120
Absorbed dose
SI derived unit and symbol for ... **A10:** 685
SI unit/symbol ... **A8:** 721
Absorbed moisture *See* Absorption; Moisture absorption
Absorbed specimen current detection
in scanning electron microscopy ... **A9:** 90
Absorbed water
and permittivity ... **EL1:** 600-601
Absorber
ultraviolet ... **EM2:** 1
Absorption *See also* Adsorption; Moisture absorption; Radiographic absorption; Water absorption ... **A13:** 1, 147, 329-333 **EM3:** 3
of a photon ... **A10:** 61
broad-beam ... **A17:** 310
characteristics, compared with neutron and x-ray scattering ... **A10:** 421
and chemical susceptibility ... **EM2:** 572
coefficients ... **A17:** 310
contrast, AEM ... **A10:** 444-445
correction, EPMA ... **A10:** 524
cross section, Mössbauer effect ... **A10:** 288
cross section, Mössbauer spectroscopy ... **A10:** 288
curve, for uranium, as function of wavelength ... **A10:** 85
defined ... **A10:** 84, 668 **EM1:** 3 **EM2:** 2
edges ... **A10:** 85-86
edges, defined ... **A17:** 309
effect in AAS ... **A10:** 43
effect of low neutron ... **A10:** 423
Absorption (continued)
effective, of x-rays ... **A17:** 310-311
-emission, model approximations of ... **A10:** 97
of energy, by honeycomb structures ... **EM1:** 728
enhancement effects, interelement ... **A10:** 97
in ferromagnetic resonance ... **A10:** 267
and fluorescence spectra, N-phenyl carbazole ... **A10:** 75
of hydrogen ... **A12:** 124
jump ... **A10:** 85-86
as leakage ... **A17:** 58
of light, effect of sample thickness ... **A10:** 61
lineshapes, NMR ... **A10:** 280
and Lorentz polarization, in surface stress measurement ... **A10:** 385-386
matrix, as XRPD source of error ... **A10:** 341
measured as function of applied magnetic field, FMR ... **A10:** 267
micro-, as XRPD source of error ... **A10:** 341
of microwaves ... **A17:** 204
molecular, and de-excitation processes, MFS Jablonsky diagram for ... **A10:** 73
molecular, of UV/VIS radiation, as requirement for fluorescence ... **A10:** 73
negative ... **A10:** 98
neutron and x-ray ... **A17:** 387
neutrons, process ... **A17:** 390
particle, as XRPD source of error ... **A10:** 341
and photoelectric effect ... **A10:** 97
photoelectric, in EXAFS ... **A10:** 409
and porosity, wave effects ... **A17:** 212
probability, fluorescence intensity as measure of ... **A10:** 411
spectra, ESR ... **A10:** 260
spectra, Mössbauer spectroscopy ... **A10:** 294
spectrum, K-edge, of krypton gas ... **A10:** 410
total, above absorption edge ... **A10:** 409
total, electromagnetic radiation attenuation ... **A17:** 309
tracer gas, and system responses ... **A17:** 69
of ultrasonic energy, ultrasonic beams ... **A17:** 238
ultrasonic waves, attenuation by ... **A17:** 231
x-ray, as cause of interelement effects ... **A10:** 97
x-ray, effect in AEM-EDS microanalysis ... **A10:** 448
x-ray, in XRS ... **A10:** 84
Absorption coefficients ... **A18:** 463, 464, 466
Absorption contrast
AEM ... **A10:** 444-445
defined ... **A10:** 668
Absorption correction (A)
in EPMA analysis ... **A10:** 524
Absorption diffraction method
XRPD analysis ... **A10:** 339-340
Absorption edge
defined ... **A10:** 668
Absorption of high-energy electrons ... **A9:** 111
Absorption spectroscopy
defined ... **A10:** 668
Absorptive lens
definition ... **M6:** 1
Absorptivity ... **A6:** 265-266
defined ... **A10:** 668
molar ... **A10:** 62-63
in UV/VIS ... **A10:** 62
Abundance, natural
and atomic mass, for naturally occurring isotopes ... **A10:** 643
Abuse, electrical *See* Electrical abuse
Abusive drilling ... **A16:** 29
Abusive final grinding
cracking from ... **A12:** 335
Abusive grinding ... **A16:** 26
of tool steel parts ... **A11:** 567, 569
AC *See* Acetal (AC) copolymers; Acetal (AC) homopolymers; Acetal (AC) resins; Acetal copolymers; Alternating current; Homopolymer and copolymer acetals (AC)

AC corona
defined ... **EM2:** 461
ac noncapacitive arc
defined ... **A10:** 668
Acc cooling (AcC) ... **A4:** 58
Accelerated aging *See* Artificial weathering
Accelerated corrosion
by stray current ... **A13:** 87
Accelerated corrosion tests ... **A11:** 174 **A13:** 194
defined ... **A13:** 1
of intergranular corrosion ... **A13:** 239
laboratory, for exfoliation corrosion ... **A13:** 242
of magnesium/magnesium alloys ... **A13:** 745
of pitting corrosion ... **A13:** 231-233
of uniform corrosion ... **A13:** 229
Accelerated cost recovery system ... **A13:** 372
Accelerated fatigue reliability testing ... **EL1:** 741, 747-751
Accelerated life prediction
accelerated testing/analysis ... **EM2:** 789-790
analytical plan development ... **EM2:** 793-794
compliance model ... **EM2:** 790-791
curing/aging/environment, effects of ... **EM2:** 788-789
durability prediction synopsis ... **EM2:** 794
failure model ... **EM2:** 791-792
laminate model ... **EM2:** 792-793
Accelerated stress-corrosion crack testing ... **A8:** 496-501
for alloy systems ... **A8:** 522-532
interpretation of results ... **A8:** 500-501
medium, correlation of ... **A8:** 522-523
purposes ... **A8:** 495
test environment for ... **A8:** 521-522
Accelerated test
defined ... **EM1:** 3
Accelerated testing *See also* Testing
acceleration factor ... **EL1:** 889
for aging and solderability ... **EL1:** 631
cautions ... **EL1:** 893
defined ... **EL1:** 887, 1133
failure kinetics, VLSI mechanisms ... **EL1:** 889-893
failure rate ... **EL1:** 889
fatigue reliability ... **EL1:** 741, 747-751
pur-pose ... **EL1:** 887
time-to-failure modeling ... **EL1:** 887-888
time-to-failure statistics ... **EL1:** 888-889
Accelerated thermal cycle testing
life cycle ... **EL1:** 136-139
Accelerated-life test *See also* Artificial aging ... **EM3:** 3
defined ... **EM2:** 2
Accelerating and storage complex (UNK)
as niobium-titanium superconducting material application ... **A2:** 1055
Accelerating potential
defined ... **A9:** 1
definition ... **M6:** 1
Accelerating voltage
defined ... **A10:** 668
SEM imaging ... **A12:** 167
Acceleration *See also* Angular acceleration
angular, SI derived unit and symbol for ... **A10:** 685
of cracking ... **A11:** 744
factor, in accelerated testing ... **EL1:** 889
factors, humidity testing ... **EL1:** 497
and inertia, strain rate testing ... **A8:** 40
nonuniform ... **EL1:** 893
potential, electron, effect in x-ray emission ... **A10:** 84
SI defined unit and symbol for ... **A10:** 685
SI unit/symbol for ... **A8:** 721
transform, fatigue ... **EL1:** 741, 745
wear testing ... **A8:** 604, 606
Acceleration, gravitational
solidification effects ... **A15:** 147-158
Acceleration period
defined ... **A18:** 1
Accelerator *See also* Catalyst; Promoter ... **EM3:** 3
defined ... **EM2:** 2
Accelerator pedals
economy in manufacture ... **M3:** 848
Accelerator test
phosphate coating solutions ... **M5:** 442
Accelerators ... **A13:** 384, 393
defined ... **EM1:** 3
Accelerators (continued)
for epoxy curing ... **EM1:** 137
high-energy x-ray machines ... **A17:** 388-389
low-voltage ... **A17:** 388
for machine guns ... **A7:** 685
as neutron source ... **A17:** 388-389
phosphate coating process ... **M5:** 435, 442
Van de Graaff ... **A17:** 389
Accelerators, combustion
use in high-temperature combustion ... **A10:** 221-222
Acceptable quality level
defined ... **EL1:** 1133
Acceptable quality level (AQL)
polished components ... **EM4:** 469
Acceptable quality levels (AQL) ... **EM3:** 785
Acceptable weld
definition ... **M6:** 1
Acceptance *See also* Acceptance or rejection
conditional ... **A17:** 675
criteria, NDI ... **A17:** 663
and NDE response ... **A17:** 675
standards, liquid penetrant inspection ... **A17:** 88
standards, radiography ... **A17:** 347
threshold criterion ... **A17:** 676
Acceptance or rejection
by coordinate measuring machines ... **A17:** 18
criteria, adhesive-bonded joints ... **A17:** 633
measurement, by dimensional laser sorting methods, acoustic emission inspection ... **A17:** 278
weld ... **A17:** 590
Acceptance test ... **EM3:** 3
Acceptance tests ... **A13:** 193, 207, 239, 240
mechanical ... **A11:** 19
Acceptance, user
of corrosion test results ... **A13:** 316-317
Accepted reference value
defined ... **A8:** 1
Access control
to radiation facilities ... **A17:** 302
Accessibility
to corrosives ... **A13:** 340
Accessory gear
engine M-1 Abrams tank ... **A7:** 688
Accessory seal
defined ... **A18:** 2
Accidents *See* Safety
Accumulating-type multiblock continuous wire-drawing machine ... **A14:** 333-334
Accumulation period *See* Acceleration period
Accumulator
defined ... **EM2:** 2
for hydraulic torsional system ... **A8:** 216
Accumulator ring
hydrogen embrittlement fracture ... **A11:** 337, 338
Accumulator-drive presses
for hot extrusion ... **A14:** 319
Accuracies
defined, in welding ... **A17:** 590
diffraction pattern technique ... **A17:** 13
human vs. machine vision ... **A17:** 30
laser triangulation sensors ... **A17:** 13
of photodiode array imaging ... **A17:** 12-13
of scanning laser gage ... **A17:** 12
volumetric, coordinate measuring machines ... **A17:** 26
Accuracy *See also* Allowances; Dimensional accuracy; Tolerances ... **A8:** 1
of activity coefficient ... **A15:** 55
of blanking operations ... **A14:** 456-457
characteristics, forming machines ... **A14:** 16
component placement, factors affecting ... **EL1:** 732
of contour roll forming ... **A14:** 633
defined ... **A10:** 668
of failure analysis and fracture mechanics ... **A11:** 55-57
in HERF processing ... **A14:** 105
of hot forming titanium alloys ... **A14:** 842
locational, technological capabilities ... **EL1:** 508
of low-acceleration fatigue tests ... **EL1:** 741
of mechanical presses ... **A14:** 39-40
of microanalysis, standards for ... **A10:** 530
number of samples and ... **A8:** 623-624
parameter ... **A8:** 387
in piercing ... **A14:** 466-467
and precision, compared ... **A10:** 525
Accuracy (continued)
of press forming ... **A14:** 552
of presses ... **A14:** 39-40, 495
of radioanalysis ... **A10:** 246-247
in shearing ... **A14:** 706, 714
of single-crystal analysis ... **A10:** 352
of stretch forming ... **A14:** 596-597
of temperature control, precision forging ... **A14:** 165
translation/rotational, for component placement ... **EL1:** 732
in UV/VIS ... **A10:** 70
wire forming ... **A14:** 695
Aceramic Neolithic period
metalworking in ... **A15:** 15
Acetal
critical surface tension ... **EM3:** 180
friction coefficient data ... **A18:** 73
Acetal (AC) copolymers ... **EM3:** 3
Celcon ... **EM3:** 278
defined ... **EM2:** 2
primer ... **EM3:** 278
surface preparation ... **EM3:** 278, 291
Acetal (AC) homopolymers ... **EM3:** 3
defined ... **EM2:** 2
Delrin ... **EM3:** 278
surface preparation ... **EM3:** 278
Acetal (AC) resins *See also* Homopolymer and copolymer acetals (AC); Polyoxymethylene (POM) ... **EM3:** 3
chemistry ... **EM2:** 65, 100
defined ... **EM2:** 2, 100
as engineering thermoplastics ... **EM2:** 448
as structural plastic ... **EM2:** 65
Acetaldehyde
physical properties ... **EM3:** 104
Acetate chemical group
an polymer naming ... **EM2:** 56
Acetate solution
copper/copper alloy SCC ... **A13:** 633
Acetate tape
for plastic replicas ... **A17:** 53
Acetates
as salt precursors ... **EM4:** 113
Acetic acid ... **A13:** 1158-1159
in an aqueous solution as an etchant for magnesium alloys ... **A9:** 426
boiling glacial, austenitic stainless steel corrosion ... **A13:** 556
boiling, nickel-base alloy corrosion ... **A13:** 648
as chemical cleaning solution ... **A13:** 1141
copper/copper alloy corrosion in ... **A13:** 629
corrosion of stainless steels in ... **M3:** 78-81
as ferrous cleaning agent ... **A12:** 75
nickel-base alloy resistance ... **A13:** 646, 648
nitric acid and glycerol as an etchant for tin-lead alloys ... **A9:** 450
and nitric acid as an etchant for stainless steels welded to carbon or low alloy steels ... **A9:** 203
in petroleum refining and petrochemical operations ... **A13:** 1268
stainless steel corrosion in ... **A13:** 558
Acetic acid-salt spray (fog) test ... **A13:** 225
Acetic anhydride
copper/copper alloy resistance ... **A13:** 629
Acetic anhydride and perchloric acid ... **A9:** 51
Acetic nitrate-pickling for macroetching magnesium alloys ... **A9:** 426
Acetic-nitrate pickling
magnesium alloys ... **M5:** 630-631, 640-641
Acetic-picral etchants used for magnesium alloys ... **A9:** 426
Acetone
for cleaning electron gun components and workpiece parts ... **A6:** 257
cleaning with ... **M5:** 40-41
deep immersion in ... **A8:** 36
effect on bearing strength, aluminum alloy sheet ... **A8:** 60
epoxy resin removal by ... **EM1:** 153
milling of WC ... **A16:** 72
surface tension ... **EM3:** 181
Acetone (C_3H_6O)
as solvent used in ceramics processing ... **EM4:** 117
Acetylacetone
as solvent extractant ... **A10:** 161

Acetylene
chemical bonding ... **M6:** 900
chemisorption and solid friction ... **A18:** 28
cutting gas for oxyfuel gas cutting ... **M6:** 899-901
fuel gas for flame spraying of cast irons ... **A6:** 720
fuel gas for oxyfuel gas cutting ... **A6:** 1156, 1157, 1158, 1161, 1162
fuel gas for torch brazing ... **M6:** 950
in high-velocity oxyfuel powder spray process ... **A18:** 830
for oxyfuel gas cutting ... **A14:** 722-723
oxyfuel gas welding fuel gas ... **A6:** 281, 282, 283, 284, 285, 287-288, 290
underwater cutting ... **M6:** 923
use in oxyfuel gas welding ... **M6:** 584
valve thread connections for compressed gas cylinders ... **A6:** 1197
cost ... **A6:** 1157
heat content ... **A6:** 1157
heat-affected zone ... **A6:** 1157
properties ... **A6:** 1157
safety hazards ... **A6:** 1157
Acetylene (C_2H_2)
fuel gas for torch brazing ... **A6:** 328
Acetylene addition polyimides ... **EM1:** 85
Acetylene end-capped oligomer imides ... **EM1:** 84
Acetylene end-capped polyimide resin ... **EM1:** 78, 80
Acetylene-terminated thermosetting polymers ... **EM2:** 631
Acheson process ... **EM4:** 49
Achromat objectives ... **A9:** 72-73
effect of lens defects in using ... **A9:** 75
Achromatic
defined ... **A9:** 1 **A10:** 668
Achromatic lens
defined ... **A10:** 668
Achromatic objective
defined ... **A9:** 1
ACI casting alloys *See* Heat-resistant alloys, ACI specific types; Stainless steels, ACI specific types
ACI specifications *See* Heat-resistant alloys, ACI specific types; Stainless steels, ACI specific types
Acicular alpha
defined ... **A9:** 1
Acicular constituents
in iron-chromium-nickel heat-resistant casting alloys ... **A9:** 332
Acicular eutectic microstructure ... **A3:** 1•19, 20
Acicular ferrite
defined ... **A13:** 1
Acicular ferrite (AF) ... **A6:** 76, 77, 78, 79, 94, 99, 1011
Acicular ferrite, classification of
in weldments ... **A9:** 581
Acicular ferrite steels ... **A1:** 148, 400, 404-405
Acicular needles
martensite ... **A12:** 328
Acicular powders
defined ... **A7:** 1
particle shapes ... **A7:** 233, 234
Acid
defined ... **A15:** 1
Acid anhydrides ... **EM1:** 70, 132
Acid attack
in qualitative classical wet methods ... **A10:** 168
Acid bath plating
copper *See* Copper plating, acid process
gold ... **M5:** 282
tin ... **M5:** 271-272
zinc *See* Zinc acid chloride plating
Acid cleaning ... **M5:** 59-67
acid attack during ... **M5:** 63
acid pickling compared to ... **M5:** 59-60
additives used in ... **M5:** 60
agitation used in ... **M5:** 64
alkaline cleaning combined with ... **M5:** 61-62
Acid cleaning (continued)
alligatoring from ... **A12:** 351
aluminum and aluminum alloys ... **M5:** 578-579
analysis, cleaner ... **M5:** 64
antifoaming agents used in ... **M5:** 60
applications ... **M5:** 61-62
barrel process ... **M5:** 60-64
chemical brightening *See* Chemical brightening
chemical etching ... **A12:** 75-76
chromic acid process ... **M5:** 59-60
cleaner compositions and operating conditions ... **M5:** 59-60, 64-65, 579
cold acid cleaners ... **M5:** 15
corrosion resistance enhanced by ... **M5:** 59-60
cutting fluids removed by ... **M5:** 7-8
drying process ... **M5:** 63
electrolytic process ... **M5:** 60-65
equipment ... **M5:** 62-64
hydrogen embrittlement by ... **A12:** 22
immersion process ... **M5:** 60-65
inhibitors used in ... **M5:** 60
iron ... **M5:** 59-67
for liquid penetrant inspection ... **A17:** 81, 82
maintenance schedules ... **M5:** 65
mechanism of action ... **M5:** 4
mineral acid *See* Mineral acid cleaning
molybdenum ... **M5:** 659
niobium ... **M5:** 663
organic acid *See* Organic acid cleaning
phosphoric acid process ... **M5:** 59-65
pigmented drawing compounds removed by ... **M5:** 7-8
polishing and buffing compounds removed by ... **M5:** 11
process control ... **M5:** 64-65
process types *See also* specific processes by name ... **M5:** 60-65
selection criteria ... **M5:** 61-62
rinsing process ... **M5:** 63-64
rust and scale removal by ... **M5:** 13
safety precautions ... **M5:** 21, 65
sludge formation, control of ... **M5:** 63-65
spray process ... **M5:** 7-9, 60-65
of stainless steel forgings ... **A14:** 230
steel ... **M5:** 59-67
surfactants used in ... **M5:** 59-60, 64
tantalum ... **M5:** 663
temperature ... **M5:** 63-64
tungsten ... **M5:** 659
unpigmented oils and greases removed by ... **M5:** 5, 8-9
waste treatment ... **M5:** 65, 313
wiping process ... **M5:** 60-65
Acid concentration
effect in copper powders ... **A7:** 111-113
effect on iron corrosion rate ... **A11:** 175
Acid contamination
of penetrants ... **A17:** 85
Acid digestion
bomb ... **A10:** 165
oxidizing or nonoxidizing ... **A10:** 165-166
for residue isolation ... **A10:** 176
Acid dipping
aluminum and aluminum alloys ... **M5:** 603-604, 606
brass die castings ... **M5:** 29
cadmium plating systems ... **M5:** 263
copper and copper alloys ... **M5:** 619-621
stainless steel ... **M5:** 561
steel ... **M5:** 16-17, 36
zinc and zinc alloy die castings ... **M5:** 677
Acid electroless nickel plating ... **M5:** 220-222
Acid electrolytic brightening
aluminum and aluminum alloys ... **M5:** 582
Acid electropolishing *See also* Electrolytic polishing ... **M5:** 303-305, 308
Acid etching ... **A7:** 462 **EM3:** 42
alkaline etching used with ... **M5:** 585
Acid etching (continued)
aluminum and aluminum alloys ... **M5:** 582-586
anodic *See* Anodic acid etching
equipment and operating procedures ... **M5:** 585-586
magnesium alloys ... **M5:** 638-639, 645
nickel alloys ... **M5:** 564
solution compositions and operating conditions ... **M5:** 584-586
stainless steel ... **M5:** 560-561
Acid extraction
defined ... **A9:** 1
Acid, flexural effects
unsaturated polyesters ... **EM2:** 247
Acid fluorides
in Group VI electrolytes ... **A9:** 54
Acid gases
as samples in gas analysis by mass spectroscopy ... **A10:** 152
Acid gases corrosion
gas/oil wells ... **A13:** 482
Acid insoluble test ... **A7:** 247
Acid melting practice ... **A15:** 363-365
melting heats and raw materials ... **A15:** 364-365
steelmaking ... **A15:** 363-364
Acid percolator
for integrated circuit analyses ... **A11:** 767, 768
Acid pickling *See also* Pickling; specific systems by name
of cold extruded parts ... **A14:** 304
corrosion inhibitors in ... **A13:** 524
enameling ... **EM3:** 303
of investment castings ... **A15:** 264
normalizing scale removal method ... **A4:** 40, 41
for precoat cleaning ... **A15:** 561
of stainless steel forgings ... **A14:** 230
of titanium alloy forgings ... **A14:** 280-281
Acid pickling baths
acidity-basicity measured ... **A10:** 172
wet chemical analysis of ... **A10:** 165
Acid plating *See* Acid bath plating
Acid process
defined ... **A15:** 1
Acid rain
defined ... **A13:** 1
Acid ratio test
phosphate coating solutions ... **M5:** 443
Acid refractory *See also* Basic refractory
defined ... **A15:** 1
Acid resistance
chlorendics ... **EM1:** 93
Acid retardation (acid purification) process
plating waste treatment ... **M5:** 317-318
Acid split
for water content determination ... **A14:** 516
Acid spot test
porcelain enamel ... **M5:** 527-529
Acid stain resistance ... **A7:** 594
Acid surface activating solution ... **M5:** 642, 644, 646
Acid washing
of tungsten powders ... **A7:** 152
Acid(s) *See also* Chemical processing industry; specific acids
cemented carbides corrosion in ... **A13:** 856
in chromate conversion coatings ... **A13:** 393
cleaners, for surface oxides ... **A13:** 381
copper/copper alloy corrosion ... **A13:** 627-629
defined ... **A13:** 1
deposition, corrosion testing ... **A13:** 204
electroplated chromium deposit resistance to ... **A13:** 873
embrittlement, defined ... **A13:** 1
environments, inhibitors ... **A13:** 524-525
fatty, for corrosion inhibitors ... **A13:** 481
-gas corrosion ... **A13:** 482
gold corrosion in ... **A13:** 798
magnesium/magnesium alloys in ... **A13:** 742
mineral, stainless steel corrosion ... **A13:** 557-558
mixtures ... **A13:** 646-647, 727

SUBJECTS OF THE INDEXED VOLUMES: ASM Handbook (designated by the letter "A"): **A1:** Properties and Selection: Irons, Steels, and High-Performance Alloys (1990); **A2:** Properties and Selection: Nonferrous Alloys and Special-Purpose Materials (1990); **A3:** Alloy Phase Diagrams (1992); **A4:** Heat Treating (1991); **A6:** Welding, Brazing, and Soldering (1993); **A7:** Powder Metallurgy (1984); **A8:** Mechanical Testing (1985); **A9:** Metallography and Microstructures (1985); **A10:** Materials Characterization (1986); **A11:** Failure Analysis and Prevention (1986); **A12:** Fractography (1987); **A13:** Corrosion (1987); **A14:** Forming and Forging (1988); **A15:** Casting (1988); **A16:** Machining (1989); **A17:** Nondestructive Testing and Quality Control (1989); **A18:** Friction, Lubrication, and Wear Technology (1992). **Metals Handbook, 9th Edition** (designated by the letter "M"): **M1:** Properties and Selection: Irons and Steels (1978); **M2:** Properties and Selection: Nonferrous Alloys and Pure Metals (1979); **M3:** Properties and Selection: Stainless Steels, Tool Materials and Special-Purpose Materials (1980); **M4:** Heat Treating (1981); **M5:** Surface Cleaning, Finishing, and Coating (1982); **M6:** Welding, Brazing, and Soldering (1983). **Engineered Materials Handbook** (designated by the letters "EM"): **EM1:** Composites (1987); **EM2:** Engineering Plastics (1988); **EM3:** Adhesives and Sealants (1990); **EM4:** Ceramics and Glasses (1991); **Electronic Materials Handbook** (designated by the letters "EL"): **EL1:** Packaging (1989).

Acid(s) (continued)
naphthenic ... **A13:** 1271-1273
nickel-base alloy corrosion resistance ... **A13:** 643
organic ... **A13:** 544, 558-559
osmium corrosion in ... **A13:** 806
palladium corrosion in ... **A13:** 804
in phosphate baths ... **A13:** 384
platinum corrosion in ... **A13:** 801
porcelain enamels resistance ... **A13:** 446, 449, 451
production, in biological corrosion ... **A13:** 43
pure, corrosion rates in ... **A13:** 645, 647
pure tin in ... **A13:** 771-772
reducing ... **A13:** 678
rhodium corrosion in ... **A13:** 805
solutions ... **A13:** 29, 722
strong, effect in oil/gas production ... **A13:** 1233
sulfuric, alloy steel corrosion ... **A13:** 544
tantalum resistance to ... **A13:** 725-727, 729-730
uranium/uranium alloys in ... **A13:** 815-816
zinc corrosion in ... **A13:** 763
Acid-acceptor ... **EM3:** 3
defined ... **EM2:** 2
Acid-base indicators
common ... **A10:** 172
Acid-base titration
alkaline cleaners ... **M5:** 25
Acid-base titrations
equilibrium in ... **A10:** 163
of industrial materials ... **A10:** 172-173
Acid-catalyzed alkoxide gels ... **EM4:** 450
Acid-consumed value test
phosphate coating solutions ... **M5:** 443
Acidic solutions
as corrosive environment ... **A12:** 24
Acidic water
effect on aluminum alloys ... **A9:** 353
Acidified chloride fatigue testing
at ambient and elevated temperatures ... **A8:** 418-420
continuous bubbling ... **A8:** 419
control of oxidizing nature of environment ... **A8:** 418
electrochemical potential ... **A8:** 419
electrodes ... **A8:** 419
flowing test solution ... **A8:** 419
fracture surface analysis ... **A8:** 420
high-pressure pumps ... **A8:** 419
material selection ... **A8:** 418
post-test analysis ... **A8:** 420
potentiostat for imposed potential ... **A8:** 419-420
seals ... **A8:** 418-419
solution containment ... **A8:** 418-419
specimens ... **A8:** 419-420
standard hydrogen electrode scale ... **A8:** 419
temperature uniformity ... **A8:** 419
Acidity *See also* pH ... **A18:** 84
defined ... **A15:** 1
effects, aqueous corrosion ... **A13:** 37
in iron corrosion ... **A13:** 37-38
porcelain enamels resistance to ... **A13:** 446, 449, 451
Acids *See* Aqueous acids
analytic methods for ... **A10:** 7
-base indicators ... **A10:** 172
-base solutions, conductometric titration used in ... **A10:** 203
-base titrations ... **A10:** 172-173
concentrated, analysis of solutions in ... **A10:** 35
corrosive (inorganic) soldering fluxes ... **A6:** 628
determined ... **A10:** 215-216
equivalent weight of an unknown ... **A10:** 216
functional group analysis of ... **A10:** 215-216
intermediate (organic) soldering fluxes ... **A6:** 628
isolation, as second-phase test method ... **A10:** 177
mixtures, for sample dissolutions ... **A10:** 165, 166
purity of ... **A10:** 216
Acids, as corrosive
copper casting alloys ... **A2:** 352
Acme screws
economy in manufacture ... **M3:** 854
ACO
as synchrotron radiation source ... **A10:** 413
Acoustic Barkhausen noise *See* Barkhausen noise
Acoustic detectors
used in thermal-wave imaging ... **A9:** 90
Acoustic electron spin resonance
as ESR supplemental technique ... **A10:** 258
Acoustic emission
adhesion measurements ... **A6:** 143
Acoustic emission (AE) technique
defect detection ... **EM3:** 751
for measuring fiber fragment lengths ... **EM3:** 397, 398, 401
Acoustic emission (AE) testing ... **EM1:** 3, 777
Acoustic emission inspection *See also* Acoustic emission; NDE reliability ... **A11:** 18
A17: 278-294
acoustic emission waves/propagation ... **A17:** 279-280
of adhesive-bonded joints ... **A17:** 627
codes, boilers/pressure vessels ... **A17:** 642
data displays ... **A17:** 283-284
data evaluation procedures ... **A17:** 290-291
emission counts ... **A17:** 281-282
hit-driven systems ... **A17:** 282
in-service, tubular products ... **A17:** 574
instrumentation principles ... **A17:** 281-284
load control/repeated loadings ... **A17:** 286
in materials studies ... **A17:** 286-289
measurements, typical ... **A17:** 280
multichannel ... **A17:** 283
noise, precautions against ... **A17:** 285-286
with other NDT methods, combined ... **A17:** 278
preamplifiers ... **A17:** 280-281
of pressure vessels ... **A17:** 654-656
in production quality control ... **A17:** 289-290
range of applicability ... **A17:** 278-279
sensors ... **A17:** 280-281
signal detection ... **A17:** 281-282
signal measurement parameters ... **A17:** 282-283
special purpose ... **A17:** 284
structural test applications ... **A17:** 290-292
of weldments ... **A17:** 598-602
Acoustic emission testing ... **A18:** 435
Acoustic emission testing (AE) ... **A6:** 1081, 1084
brazed joints ... **A6:** 1119, 1123
Acoustic emission(s) *See also* Acoustic emission inspection
counts ... **A17:** 281-284
defined ... **A17:** 278
defined, powder metallurgy parts ... **A17:** 540-541
from crack growth ... **A17:** 287
scales of ... **A17:** 278
sensor, for Barkhausen noise ... **A17:** 160
sources, mechanisms of ... **A17:** 287
waveform transmission ... **A17:** 280
waves and propagation ... **A17:** 279-280
Acoustic emissions
inspection for weld discontinuities ... **M6:** 849-850
laser beam weld inspection ... **M6:** 668
Acoustic horn
for amplifying converter output ... **A8:** 246
extension, resonance properties ... **A8:** 245
for ultrasonic fatigue testing ... **A8:** 240, 244-245
Acoustic impedance *See also* Impedance
effects, ultrasonic beams ... **A17:** 238
ultrasonic inspection ... **A17:** 234-235
of various materials ... **A17:** 476
Acoustic inspection
of solder joints ... **EL1:** 942
Acoustic leak testing
of pressure systems
Acoustic lens focused ultrasonic search units ... **A17:** 259-260
Acoustic material signatures (AMS) ... **A18:** 407
Acoustic microscopes
for optical imaging ... **EL1:** 1069-1071
Acoustic microscopy *See also* Acoustic ultrasonic inspection; C-mode scanning acoustic microscopy (C-SAM); Scanning acoustic microscopy (SAM); Scanning laser acoustic microscopy (SLAM) ... **A17:** 465-482
applications ... **A17:** 468-481
bearing steel inclusion imaging and length verification ... **A18:** 726
of ceramic materials ... **A17:** 469-472
color images by ... **A17:** 485, 487
of composite materials ... **A17:** 469
of integrated circuits (ICs) ... **A17:** 474-480
of metals ... **A17:** 472-473
methods, compared ... **A17:** 468
methods, fundamentals of ... **A17:** 465-468
of microelectronic components ... **A17:** 473-481
microstructure effect ... **A17:** 51
Acoustic microscopy (continued)
as nondestructive testing ... **EL1:** 369-371
of soldered joints ... **A17:** 607
techniques, compared ... **A17:** 470
Acoustic noise *See also* Noise
in acoustic emission inspection ... **A17:** 285
Acoustic pressure
and angle of reflection/refraction ... **A17:** 237
Acoustic properties
of metals/nonmetals ... **A17:** 235
Acoustic reflection coefficient ... **A6:** 143
Acoustic stressing
for optical holographic interferometry ... **A17:** 408-409
Acoustic ultrasonic inspection *See also* Acoustic microscopy; Ultrasonic inspection
of powder metallurgy parts ... **A17:** 539-541
Acoustic wave train ... **A8:** 243-244
Acoustic waveguide sensors
of nuclear waste ... **A17:** 281
Acoustic-ultrasonic test technique ... **EM1:** 776-777
Acoustical holography *See also* Holography; Optical holographic interferometry; Optical holography; Ultrasonic inspection
of adhesive-bonded joints ... **A17:** 630-631
applications ... **A17:** 445-447
calibration ... **A17:** 444-445
defined ... **A17:** 438
interpretation of results ... **A17:** 445
liquid-surface ... **A17:** 438-440
liquid-surface and scanning systems compared ... **A17:** 443-444
liquid-surface equipment, commercial ... **A17:** 441
readout methods ... **A17:** 444
scanning ... **A17:** 440-441
scanning equipment, commercial ... **A17:** 441-443
and ultrasonic inspection ... **A17:** 240
of welds ... **A17:** 445-446
Acoustical microphone elements
powder used ... **A7:** 573
Acoustical microscopy
and ultrasonic inspection ... **A17:** 240
Acoustical plastics
powders used ... **A7:** 574
Acoustical-control knockout machines ... **A15:** 505
Acoustophoresis-based techniques ... **EM4:** 74
Acquisition of data
forging process design ... **A14:** 439-442
Acrawax
burn off ... **A7:** 191, 351, 352
as lubricant ... **A7:** 190-193
Acronyms
for analytical techniques ... **A10:** 689
Acrylamate polymer *See also* Urethane hybrids
characteristics ... **EM2:** 268
prepreg composites ... **EM2:** 270
Acrylamide copolymer
as flocculant ... **EM4:** 92
Acrylate coatings *See also* Acrylates; Coatings
development ... **EL1:** 785
equipment ... **EL1:** 787
masking ... **EL1:** 787
materials ... **EL1:** 785-786
processing techniques ... **EL1:** 786-787
UV future ... **EL1:** 788
Acrylate pressure-sensitive adhesives
automotive decorative trim ... **EM3:** 552
Acrylate resins *See* Acrylic resins ... **EM3:** 3
Acrylate(s) *See also* Acrylate coatings
caprolactone ... **EL1:** 862
as coatings/encapsulants ... **EL1:** 759
difunctional/multifunctional, in radiation-cure systems ... **EL1:** 857
epoxy, photochemistry of ... **EL1:** 854-866
UV-curable, for solder masking ... **EL1:** 555
Acrylated silicone urethanes ... **EM3:** 90
Acrylated silicones ... **EM3:** 90
as ultraviolet-curable ... **EL1:** 824
Acrylated urethanes ... **EM3:** 90
for electronic general component bonding ... **EM3:** 573
Acrylates ... **EM3:** 594
advantages and limitations ... **EM3:** 85
for automotive electronics bonding ... **EM3:** 553
for automotive structural bonding ... **EM3:** 554
bonding substrates and applications ... **EM3:** 85
for electrically conductive bonding ... **EM3:** 572

Acrylates (continued)
for electronic general component bonding **EM3:** 573
temperature limits **EM3:** 621
for window assembly attachments **EM3:** 553
Acrylic
laser cutting of **A14:** 742
Acrylic acid polymers
as binders **EM4:** 474
Acrylic adhesives *See also* Acrylic coatings; Acrylics **EM1:** 684, 686
for flexible printed boards **EL1:** 582
for surface mounting **EL1:** 671
Acrylic casting resin method
for plastic replicas **A17:** 53
Acrylic caulks with silicone
plumbing sealant **EM3:** 608
Acrylic coatings *See also* Acrylic adhesives; Acrylics
as conformal coatings **EL1:** 763
and urethane, silicone, epoxy, compared **EL1:** 775
Acrylic emulsion
applications **EM3:** 56
Acrylic in methyl ethyl ketone
batch weight of formulation when used in nonoxidizing sintering atmospheres **EM4:** 163
Acrylic lacquers
as preservative **A12:** 73
Acrylic latex
characteristics **EM3:** 53
chemistry **EM3:** 50
E, water-base PSA properties **EM3:** 88
G, water-base PSA properties **EM3:** 88
G/rosin (50/50), water-base PSA properties **EM3:** 88
properties **EM3:** 50
for recreational vehicle sealing **EM3:** 58
Acrylic latexes
as binders **EM4:** 474
Acrylic plastic *See also* Acrylics; Polymethyl methacrylate; Polymethyl methacrylate (PMMA) **EM3:** 3
defined **EM1:** 3 **EM2:** 2
environmental effects **EM2:** 427-428
Acrylic plastics
tools for shaped tube electrolytic machining **A16:** 555
Acrylic resin mounting materials **A9:** 30-31
for electropolishing **A9:** 49
for tin and tin alloys **A9:** 449
Acrylic resins **EM3:** 3
defined **EM2:** 2
as media for decorating **EM4:** 475
rubber as toughener **EM3:** 185
Acrylic resins and coatings **M5:** 473-479, 495, 498, 500, 504
Acrylic rubber phenolics **EM3:** 104
suppliers **EM3:** 104
Acrylic urethanes **A13:** 409
Acrylic vinyl emulsions
residential applications **EM3:** 675
Acrylic wax binders
as binders for spray drying before dry pressing **EM4:** 146
Acrylic-based silk screen processes **EL1:** 115
Acrylic-styrene-acrylonitriles (ASA) *See* Styrene-acrylonitriles
Acrylic/epoxy and sulfur dioxide cold box process *See also* Cold box processes; Cold box resin binder processes; Coremaking
as coremaking system **A15:** 238
Acrylics *See also* Acrylic adhesives; Acrylic plastic; Acrylic resin; Acrylic- coatings; Thermoplastic resins **A13:** 1, 404-406 **EM3:** 44, 75, 92, 119-125, 594
for acrylonitrile-butadiene-styrene (ABS) bonding **EM3:** 121
advantages and limitations **EM3:** 90

Acrylics (continued)
aerospace application **EM3:** 559
for aluminum aircraft repair and primary bonding **EM3:** 125
for aluminum boats **EM3:** 125
for aluminum bonding **EM3:** 121, 122
for aluminum construction use **EM3:** 125
for aluminum window and door bonding **EM3:** 119
applications **EM2:** 104 **EM3:** 44
automotive applications **EM3:** 45-46
blow molding **EM2:** 107
for boat construction **EM3:** 125
for bonding **EM3:** 60
bonding composites to composites **EM3:** 293
for bronze bonding **EM3:** 122
butyl methacrylate **EM3:** 119
cast sheet **EM2:** 103
for caulking **EM3:** 607-608
chain reactions **EM3:** 120
characteristics **EM2:** 105-108 **EM3:** 90
chemical resistance properties **EM3:** 639
chemistry **EM3:** 50, 119-120
for circuit protection **EM3:** 592
for coating/encapsulation **EL1:** 242
cold-cured (single-part) toughened shear properties **EM3:** 321
commercial formulations **EM2:** 103
compared to epoxies **EM3:** 98, 119, 120, 121, 124
compared to polysulfides **EM3:** 195
compared to urethanes **EM3:** 119, 120, 121, 124
competing with anaerobics **EM3:** 116
competing with silicones **EM3:** 134
competitive materials **EM2:** 104
conformal overcoat **EM3:** 592
for construction applications **EM3:** 675
for copper bonding **EM3:** 122
cost factors **EM3:** 124-125
cross-linking **EM3:** 90
curing methods **EM3:** 114-121, 123, 124
degradation **EM3:** 679
design considerations **EM2:** 106
development **EM3:** 119
DuPont Hypalon (DH) acrylics **EM3:** 121, 122
electrical properties **EL1:** 822
elevated-temperature resistance **EM3:** 123
emulsion, properties **EM3:** 677
as engineering adhesives family **EM3:** 567
equal-mix (no-mix, honeymoon type) **EM3:** 123
for equipment enclosures and housings **EM3:** 125
for exterior seals in construction **EM3:** 56
for faying surfaces **EM3:** 604
fillers **EM3:** 122
for fillets **EM3:** 604
for flexible printed wiring boards (PMBs) **EM3:** 45
future trends **EM3:** 124-125
for galvanized steel bonding **EM3:** 60, 121, 123-124
as gap-filling materials **EM3:** 125
generations of adhesives **EM3:** 121
glass transition temperature **EM3:** 119
as glassy polymers **EM3:** 617
grades **EM2:** 103
heating and air conditioning duct sealing and bonding **EM3:** 610
high-impact (HI) **EM3:** 122
high-performance (HP) **EM3:** 119, 122
hybrid epoxy-acrylic **EM3:** 122-123
impact-modified **EM2:** 103, 105, 107
for in-house glazing **EM3:** 58
injection molding **EM2:** 106-107
for interior seals in window systems **EM3:** 57
latex E **EM3:** 90
latex G **EM3:** 90
for lead bonding **EM3:** 122
for magnesium bonding **EM3:** 122
for metal building construction **EM3:** 57, 58

Acrylics (continued)
methyl methacrylate **EM3:** 119, 120
as modifiers to improve surface wetting **EM3:** 181
molecule **EL1:** 671
for motorcycle construction **EM3:** 125
for nickel bonding **EM3:** 122
odor **EM3:** 119
patents **EM3:** 122, 124
performance **EM3:** 674
phase structure **EM3:** 413
photocurable **EM3:** 124
for plastic appliance component bonding **EM3:** 125
for plastics bonding **EM3:** 125
for polycarbonate lens bonding **EM3:** 124
for polystyrene bonding **EM3:** 121
for polyvinyl chloride (PVC) bonding **EM3:** 121
for printed wiring board manufacture **EM3:** 591
processing **EM2:** 106-107
producers **EM2:** 108
product properties **EM2:** 105-106
production volume **EM2:** 104
properties **EM3:** 92, 121, 122
for protecting electronic components **EM3:** 59
reacting with acrylonitrile-butadiene rubber **EM3:** 148
resin, as conformal coating **EL1:** 761
resin compound types **EM2:** 108
resin pricing **EM2:** 103-104
resistant to many aggressive materials **EM3:** 637
for rivets **EM3:** 604
for rubber bonding **EM3:** 121
shear strength of mild steel joints **EM3:** 670
for sheet molding compound (SMC) bonding **EM3:** 46
shelf life **EM3:** 119, 120, 124
silane coupling agents **EM3:** 182
as solder resists **A6:** 133
solution
applications **EM3:** 56
properties **EM3:** 677
solvent cements **EM3:** 567
for sporting equipment bonding **EM3:** 125
for stainless steel bonding **EM3:** 60
for steel bonding **EM3:** 121
steel joints wedge tested **EM3:** 668
as structural plastic **EM2:** 65
substrate cure rate and bond strength for cyanoacrylates **EM3:** 129
suppliers **EM3:** 58, 124
surface-activated (SAA) **EM3:** 123-124
for surface-mount technology bonding **EM3:** 570
tackifiers for **EM3:** 183
thermal properties **EM3:** 52
thermoforming **EM2:** 107
for thermoplastics bonding **EM3:** 121
as top coats **EM3:** 640
tougheners **EM3:** 183
water-base *See also* Latex **EM3:** 210-214
for wood bonding **EM3:** 121
for zinc bonding **EM3:** 60, 124
Acrylics as sealants **EM3:** 188, 191, 675
component terminal sealant **EM3:** 612
dielectric sealants **EM3:** 611
resident-al construction applications **EM3:** 188
tensile strength affected by temperature **EM3:** 189
Acrylics, for denture teeth
studies **A18:** 673-674
Acrylonitrile **EM3:** 3
defined **EM2:** 3
Acrylonitrile rubber
maskant material for chemical milling of Al alloys **A16:** 803
Acrylonitrile-butadiene rubber (NBR) *See also* Nitrile rubber
additives and modifiers **EM3:** 149
for auto brake and clutch linings **EM3:** 148

Acrylonitrile-butadiene rubber (NBR) (continued)
for bonding gasketing ... **EM3:** 148
chemistry ... **EM3:** 148
commercial forms ... **EM3:** 148
cross-linking ... **EM3:** 149
exposure in electrochemically inert conditions ... **EM3:** 629
for fabrication of aircraft honeycomb structures ... **EM3:** 148
hose and belt manufacture ... **EM3:** 148
markets ... **EM3:** 148-149
for printed wiring boards ... **EM3:** 148
properties ... **EM3:** 148
shelf life ... **EM3:** 148
for shoe sole attachment ... **EM3:** 148
Acrylonitrile-butadiene-styrene (ABS)
critical surface tension ... **EM3:** 180
cross-link density ... **EM3:** 418
extrusion effects ... **EM1:** 100
mechanical keying ... **EM3:** 416
medical applications ... **EM3:** 576
rubber-toughened ... **EM3:** 413
solvent cements ... **EM3:** 567
substrate cure rate and bond strength, for cyanoacrylates ... **EM3:** 129
surface preparation ... **EM3:** 278-279, 291
Acrylonitrile-butadiene-styrene (ABS) resins ... **EM3:** 3
in flexible epoxies ... **EL1:** 818
Acrylonitrile-butadiene-styrenes (ABS) *See also* Thermoplastic resins
alloys and blends ... **EM2:** 109
applications ... **EM2:** 110-111
characteristics ... **EM2:** 111-114
commercial forms ... **EM2:** 109
competitive materials ... **EM2:** 111
costs and production volume ... **EM2:** 109-110
defined ... **EM2:** 3
design considerations ... **EM2:** 111-112
future trends ... **EM2:** 111
processing ... **EM2:** 112-113
product properties ... **EM2:** 111
resin compound types ... **EM2:** 114
suppliers ... **EM2:** 114
vs polycarbonates (PC) ... **EM2:** 151
ACS *See* American Carbon Society; American Ceramic Society
ACSR wire ... **M1:** 264
Actinide carbonyls ... **A7:** 135
Actinide metals *See also* Pure metals; Transplutonium metals
actinium, properties ... **A2:** 1189
neptunium, properties ... **A2:** 1190
plutonium, properties ... **A2:** 1192
protactinium, properties ... **A2:** 1194
thorium, properties ... **A2:** 1195
ultrapurification by electrotransport process ... **A2:** 1094-1095
uranium, properties ... **A2:** 1197
Actinides
ICP-MS analysis of ... **A10:** 40
laser-induced resonance ionization mass spectrometry for ... **A10:** 142
in periodic table ... **A10:** 688
Actinium *See also* Actinide metals
as actinide metal, properties ... **A2:** 1189
pure ... **M2:** 714, 832-833
Actinolite
chemical composition ... **A6:** 60
Actions, as application programs
computer-aided design ... **EL1:** 129-130, 133
Activated carbon
Amoco PX-21, diffraction data ... **A10:** 399
RDF analysis of ... **A10:** 399-400
Activated charcoal
Raman analysis ... **A10:** 132
Activated Diffusion Bonding (ADB) *See* Diffusion brazing
Activated Diffusion Healing (ADH) *See* Diffusion brazing
Activated reactive evaporation ... **EM4:** 218
Activated reactive evaporation (ARE) ... **A18:** 840, 844, 849
compounds and synthesized deposition rates ... **A18:** 848
titanium alloys ... **A18:** 780
Activated rosin flux
definition ... **A6:** 1206 **M6:** 1
Activated sintering ... **A7:** 316-319
chemical additions ... **A7:** 319
and conventional sintering compared ... **A7:** 318
criteria for activators ... **A7:** 318
defined ... **A7:** 1
kinetics ... **A7:** 318
phase diagram ... **A7:** 319
tungsten ... **A7:** 318
of tungsten nickel powders ... **A7:** 308
Activation ... **EM3:** 3, 35
control, aqueous corrosion ... **A13:** 30-32
defined ... **A7:** 1 **A13:** 1 **EM1:** 3 **EM2:** 3
energy, in aqueous solutions ... **A13:** 17
Activation analysis *See also* Prompt gamma activation analysis (PGAA)
chemical elements measured by ... **A10:** 243
defined ... **A10:** 668
Activation barrier
for nucleation ... **A15:** 103
Activation energy
for aluminum and aluminum alloys ... **A11:** 772
for crack propagation in SMIE ... **A11:** 244
defined ... **A7:** 1
determination ... **EL1:** 890
ESR analysis ... **A10:** 266
RBS analysis ... **A10:** 628
for steel-cadmium embrittlement ... **A11:** 240, 244
Activation energy, creep and self-diffusion related ... **A8:** 309
Activation, surface *See* Surface activation
Activator *See also* Accelerator; Accelerators ... **EM3:** 3
in polyester resin reaction ... **EM1:** 133
Activator base in computer printers
soft magnetic application ... **A7:** 641
Activator key
spring-material test apparatus ... **A8:** 134-135
Activators
criteria in activated sintering ... **A7:** 318
defined ... **A7:** 1
as flux component ... **EL1:** 644
palladium as ... **A7:** 318
Active
defined ... **A13:** 1
Active analog components *See also* Active digital components
active devices, bipolar IC technology ... **EL1:** 145-146
bipolar transistor analysis ... **EL1:** 150-154
field effect transistor (FET) analysis ... **EL1:** 154-159 -150
gallium arsenide technology ... **EL1:** 148
high-performance active devices, bipolar technology ... **EL1:** 146-147
monolithic diodes ... **EL1:** 144
monolithic resistors/capacitors ... **EL1:** 144
MOS integrated circuit fabrication ... **EL1:** 147-148
npn bipolar junction transistor IC technology ... **EL1:** 144-145
Active circuit technology ... **EL1:** 7, 143
Active clearance control
blade tips of jet engines ... **A18:** 589
Active component fabrication *See also* Active devices; Fabrication
bipolar junction transistor technology ... **EL1:** 195-196
device -solation techniques ... **EL1:** 199
diffusion of impurities ... **EL1:** 194-195
digital devices ... **EL1:** 201
epitaxial growth ... **EL1:** 192-193
gallium arsenide technology ... **EL1:** 199-200
metal-oxide semiconductor integrated circuit ... **EL1:** 196-198
new directions ... **EL1:** 200-201
oxidation ... **EL1:** 195
packaging considerations ... **EL1:** 199
photolithographic process ... **EL1:** 193-194
wafer preparation ... **EL1:** 191-192
yield considerations ... **EL1:** 198-199
Active components *See also* Active analog components; Active component fabrication; Active digital components
defined ... **EL1:** 1133
Active devices
active element ... **EL1:** 1013-1017
Active devices (continued)
in bipolar IC technology ... **EL1:** 145-146
defined ... **EL1:** 1133
failure mechanisms ... **EL1:** 1006-1017
high-performance, with bipolar technology ... **EL1:** 146-147
in npn BJT technology ... **EL1:** 196
zone 2, package exterior ... **EL1:** 1006-1008
zone 2, package interior ... **EL1:** 1008-1013
Active digital components *See also* Active analog components
complementary metal-oxide semiconductor (CMOS) ... **EL1:** 162-163
emitter-coupled logic (ECL) ... **EL1:** 163-165
future trends ... **EL1:** 176-177
introduction ... **EL1:** 160-161
large-scale integration (LSI), types ... **EL1:** 166--168
packaging considerations ... **EL1:** 172-174
signal transmission ... **EL1:** 168-172
technologies, compared ... **EL1:** 165-166
thermal management ... **EL1:** 174-176
transistor-transistor logic (TTL) ... **EL1:** 161-162
Active layer ... **A18:** 177
Active metal
defined ... **A13:** 1
Active path corrosion
abbreviation ... **A8:** 724
SCC testing ... **A8:** 532
Active potential
defined ... **A13:** 1
Activity
defined ... **A13:** 1
Activity (of radionuclides)
SI derived unit and symbol for ... **A10:** 685
SI unit/symbol ... **A8:** 721
Activity (thermodynamic)
alloy components, calculated ... **A15:** 61-70
Darken formalism for ... **A15:** 62-63
defined ... **A15:** 51
of iron and nickel, binary solutions ... **A15:** 51
liquid aluminum-magnesium alloys ... **A15:** 57
of liquid aluminum-silicon alloys ... **A15:** 57
liquid copper-aluminum alloys ... **A15:** 58
liquid copper-nickel alloys ... **A15:** 58
liquid copper-tin alloys ... **A15:** 58
liquid copper-zinc alloys ... **A15:** 59
and thermal properties, aluminum/copper alloys ... **A15:** 55
Activity coefficient
accuracy ... **A15:** 55
of aluminum-magnesium alloys ... **A15:** 57
of aluminum-silicon alloys ... **A15:** 57
at infinite dilution, liquid metals ... **A15:** 60
carbon, in iron melt ... **A15:** 62
defined ... **A13:** 1
as departure from ideality ... **A15:** 51-52
liquid copper-aluminum alloys ... **A15:** 58
liquid copper-nickel alloys ... **A15:** 58
liquid copper-tin alloys ... **A15:** 58
liquid copper-zinc alloys ... **A15:** 59
silicon, in iron melt ... **A15:** 62
Activity quotient ... **A6:** 56
Actual contact area (real area of contact)
defined ... **A18:** 2
Actual slip *See* Macroslip
Actual throat
definition ... **A6:** 1206
Actual tire load ... **A18:** 578
Actuator
hydraulic rotary ... **A8:** 157-158
linear ... **A8:** 160
torsional hydraulic, finite-element model of ... **A8:** 217
Acute angles
by press-brake forming ... **A14:** 536
Acute berylliosis
from beryllium powder/dust exposure ... **A2:** 687
Acute pulmonary disease (chemical pneumonitis)
from beryllium exposure ... **A2:** 1239
Acute toxicity *See* Toxicity
Adapter bending
in tooling design ... **A7:** 337
Adapter grip
for axial fatigue testing ... **A8:** 369
Adaptive control (AC) ... **A16:** 618-626
adaptive control with constraints (ACC) system ... **A16:** 619, 620, 622-624, 625

Adaptive control (AC) (continued)
adaptive control with optimization (ACO) system **A16:** 619-620, 621-622, 624
CAD/CAM/CIM systems integration **A16:** 624
compared to CNC **A16:** 618
computer-integrated manufacturing (CIM) **A16:** 624
control systems **A16:** 618-619
geometric adaptive control (GAC) system **A16:** 619, 621, 624
tool life **A16:** 618, 621, 622, 623, 624
variable-gain ACC systems **A16:** 624, 625
Adaptive mask selection **A18:** 353
Adatoms **A18:** 848
ADC *See* Analog-to-digital converter
ADCL/DCRF nomograph
as die casting analysis method **A15:** 290
Add-on hybrid components
advantages/limitations **EL1:** 258-259
Add-on SIMS instrument
for qualitative analysis **A10:** 613
Addition agent
defined **A13:** 1 **A15:** 1
Addition curing
of room-temperature vulcanizing (RTV) silicones **EL1:** 823
Addition polyimides *See also* Polyimide resins
acetylene **EM1:** 85
acetylene end-capped **EM1:** 80
for aerospace prepregs **EM1:** 141
bismaleimides **EM1:** 78-79
chemistry **EM1:** 78-80
Reverse Diels-Alder (RDA) **EM1:** 79-80
Addition polymerization *See also* Polymerization **EM3:** 3
defined **EM1:** 3 **EM2:** 3
mechanism **EM1:** 752
Addition techniques
potentiometric membrane electrodes **A10:** 183
Additive *See also* Filler
for carbon-graphite materials **A18:** 816
corrosivity **A18:** 84
defined **A18:** 2 **EM1:** 3
filterability **A18:** 85
gear oils **A18:** 86
high-vacuum polymer lubricant applications **A18:** 154
lubricating oils for valve train assembly **A18:** 558-559
polymers in misting oils **A18:** 87
pultrusion **EM1:** 539
sliding bearing lubrication **A18:** 519, 520
stability of liquid lubricants **A18:** 84
for thermoplastic composites **A18:** 820
to inhibit corrosion of internal combustion engine parts **A18:** 555, 558
Additive laminates
rigid printed wiring boards **EL1:** 548
Additive manufacturing sequences
rigid printed wiring boards **EL1:** 549
Additive processing
of rigid printed wiring boards **EL1:** 547
Additive rigid printed wiring boards
processing **EL1:** 540
Additive technology
for printed wiring boards **EL1:** 505
Additives *See also* Alloying elements **EM2:** 493-507 **EM3:** 3
antioxidants **EM2:** 67
antistatic agents **EM2:** 501-502
blowing agents **EM2:** 503
colorants **EM2:** 500-501
compounding problems **EM2:** 493-494
conductive, selection **EM2:** 473-474
defined **A15:** 1 **EM2:** 3
effect on particle size **A7:** 54
effects, chemical susceptibility **EM2:** 572-573
electrical breakdown from **EM2:** 466
electrical conductivity **EM2:** 474-475

Additives (continued)
fiber reinforcements **EM2:** 504-506
fillers and extenders **EM2:** 497-500
flame retardants **EM2:** 67, 503-504
for fluxes **EL1:** 644
GC/MS analysis of **A10:** 639
for glazes **EM3:** 308
heat stabilizers **EM2:** 67, 494-495
high-density polyethylenes (HDPE) **EM2:** 166
impact modifiers **EM2:** 497
iron **A7:** 614
light stabilizers **EM2:** 495
in lubricants **A14:** 514-515 **EM2:** 496-497
metal powder, ultrahigh molecular weight polyethylenes (UHMWPE) **EM2:** 170
nucleating agents **EM2:** 502-503
plasticizers, and polymer miscibility **EM2:** 496
properties effect **EM2:** 424-425
properties modification by **EM2:** 493-507
pultrusion **EM2:** 395
rigid epoxies **EL1:** 813
rotational molding **EM2:** 362
solvent leaching of **EM2:** 774
ultraviolet stabilizers **EM2:** 67
unmelted and undissolved **A7:** 486, 487
wax, for investment casting **A15:** 254
Additives, corrosion control
for steam equipment **A11:** 620
Adduct **EM3:** 3
defined **EM2:** 3
Adhere **EM3:** 4
Adherence *See also* Adhesion (adhesive force); Adhesion, mechanical
defined **A18:** 2
of porcelain enamels **A13:** 450-451
Adherence, particle
as nonrelevant indication **A17:** 105
Adherence, sand
as casting defect **A15:** 549
Adherences
glass-to-metal seals **EL1:** 455
Adherend **EM3:** 4
defined **EM1:** 3 **EM2:** 3
definition **EM3:** 39
Adherend defects
adhesive-bonded joints **A17:** 612-613
Adherend preparation **EM3:** 4
Adherent packing material
as casting defect **A11:** 385
Adhesion *See also* Abrasion; Abrasives; Adhesive wear; Galling **EM3:** 4, 39-43
of anodized coatings **A13:** 397
basic concepts **EM3:** 39-40
bonding mechanisms **EM3:** 40
electrostatic attraction **EM3:** 40
formation of covalent chemical bonds **EM3:** 40
mechanical interlocking **EM3:** 40
classification of adhesives **EM3:** 8-9, 40-41
elastomers **EM3:** 41
hybrids **EM3:** 41
laminating adhesives **EM3:** 41
natural glues **EM3:** 41
polyolefins **EM3:** 41
pressure-sensitive adhesives **EM3:** 41
rubbery thermosets **EM3:** 41
structural adhesives **EM3:** 41
thermoplastics **EM3:** 41
thermoset polymers **EM3:** 41
of coatings **A13:** 395
columnar dendritic structure by **A15:** 132
composite joints **EM3:** 777
conductor, rigid printed wiring boards **EL1:** 548
defined **A7:** 1 **EL1:** 1133 **EM1:** 3 **EM2:** 3
detection **EL1:** 1091
and die life **A14:** 505
during milling **A7:** 57
effect in metal-plastic composites **A7:** 612

Adhesion (continued)
failures, caused by thin-film contaminants **A11:** 43
of gold-palladium powders **A7:** 151
hardfacing for **A7:** 823
high-performance adhesives **EM3:** 41
of hot dip galvanized steel **A13:** 438
initial, as fretting **A11:** 148
joint design **EM3:** 42-43
and matrix strength in composite plastic-filler mixtures **A7:** 612
metal, and nonlubricated wear **A11:** 154-155
as nomenclature processing problem **EL1:** 559
in organic-coated steels **A14:** 565
paint **A13:** 442
particle size, filler, and elongation at break **A7:** 612, 613
performance **EM3:** 41
of plated coatings **A13:** 426
plating, as PTH failure mechanism **EL1:** 1025
of polyimides **EL1:** 769
polymer-metal **A7:** 606
printed board coupons **EL1:** 576
promoter, defined **EM2:** 3
promoters, thin-film hybrids **EL1:** 326
promotion, substrates **EL1:** 624
of rigid epoxies **EL1:** 810
solder mask **EL1:** 554
solder mask/protective coat, materials and processes selection **EL1:** 115
and surface chemical analysis **A7:** 250
surface parameters **EM3:** 41-42
testing **A13:** 418
theory of friction **A11:** 149
of thick-film conductor inks **EL1:** 207
time-temperature superposition **EM3:** 40
to metal, in extrusion **EM2:** 386
of ultrahigh molecular weight polyethylenes (UHMWPE) **EM2:** 170
and wear **A8:** 601
work of .... **A18:** 400 , 403 , 404, 405, 435
Adhesion (adhesive force) **A18:** 31
defined **A18:** 2
Adhesion 1-Adhesion 13 .... **EM3:** 69
Adhesion and Adhesives **EM3:** 67, 71
Adhesion and Bonding **EM3:** 70
Adhesion and the Formulation of Adhesives **EM3:** 70
Adhesion coating
coated carbide tools **A2:** 960
Adhesion coefficient *See* Coefficient of adhesion
Adhesion energy **A6:** 145
Adhesion factor **EM3:** 397
Adhesion, mechanical *See also* Adherence; Mechanical adhesion
defined **A18:** 2
Adhesion promoters **EM1:** 3, 122-123 **EM3:** 4, 49, 52, 254-258, 674
for anaerobics **EM3:** 113
for butyls **EM3:** 202
for polyimides **EM3:** 157
for polysulfides **EM3:** 196-197
silane coupling agents **EM3:** 255-256, 257
Adhesion Science and Technology **EM3:** 70
Adhesion tests **A18:** 404
Adhesion theory of friction
AES analyses **A10:** 565
Adhesion, thermally aged *See* Thermally aged adhesion
Adhesion—Fundamentals and Practice **EM3:** 69
Adhesive *See also* Adherend; Adhesives; Anaerobic adhesive; Cold-setting adhesive; Contact adhesive; Gap-filling adhesive; Heat-activated adhesive; Heat-sealing adhesive; Hot-melt adhesive; Intermediate temperature setting adhesive; Press re-sensitive adhesive; Structural adhesive
amino resin **EM2:** 628
bonding **EM2:** 725
bonds **EM1:** 481-484

Adhesive (continued)
defined **EM1:** 3 **EM2:** 3
failure, defined **EM2:** 3
film, defined *See* Film adhesive
shear model, elastic-plastic **EM1:** 484
shearing **EM1:** 482
strength, defined **EM1:** 3 **EM2:** 3
Adhesive Abstracts **EM3:** 65
Adhesive, anaerobic *See* Anaerobic adhesive
Adhesive and Sealant Compound Formulations **EM3:** 68
Adhesive assembly **EM3:** 4
Adhesive bond
definition **M6:** 1
Adhesive bond strength classifier algorithm **A17:** 611
Adhesive bonding **A7:** 457 **EM1:** 484
of aircraft components, neutron radiography of **A17:** 392-393
aluminum **M2:** 201-202
beryllium-copper alloys **A2:** 414-415
cereal as sand addition for **A15:** 211
copper metals **M2:** 456
definition **A6:** 1206 **M6:** 1
of ductile iron **A15:** 665
of epoxy materials **EL1:** 832-833
for honeycomb sandwich structures **EM1:** 727
magnesium **M2:** 547-549, 550
of magnesium and magnesium alloys **A2:** 474, 477
protective film **A17:** 614
safety precautions **A6:** 1204-1205
and sinter bonding **A7:** 457
in space and low-gravity environments **A6:** 1023
surface preparation **EM1:** 681-682
of thermoplastic composite **EM1:** 552
ultrasonic inspection **A17:** 232
with zinc alloys **A15:** 795
Adhesive Bonding ALCOA Aluminum **EM3:** 69
Adhesive Bonding Aluminum **EM3:** 69
Adhesive Bonding of Aluminum Alloys **EM3:** 69
Adhesive Bonding of Wood **EM3:** 68
Adhesive Bonding—Techniques and Applications **EM3:** 70
Adhesive, cold-setting *See* Cold-setting adhesive
Adhesive, contact *See* Contact adhesive
Adhesive dispersion **EM3:** 4
Adhesive dot height
as process control measure **EL1:** 673
Adhesive failure **EM3:** 4
defined **EM1:** 3
Adhesive film *See also* Films **EM3:** 4
defined **EM1:** 3 **EM2:** 3
Adhesive flash
adhesive-bonded joints **A17:** 612
Adhesive friction component **A18:** 432
Adhesive, gap-filling *See* Gap-filling adhesive
Adhesive, heat-activated *See* Heat-activated adhesive
Adhesive, heat-sealing *See* Heat-sealing adhesive
Adhesive, hot-setting *See* Hot-setting adhesive
Adhesive, intermediate temperature setting *See* Intermediate temperature setting adhesive
Adhesive joint *See also* Joint(s)
defined **EM1:** 3 **EM2:** 3
Adhesive joints **EM3:** 4, 33-34
shear testing standards **A8:** 62
testing **EM3:** 37
type of **EM3:** 36
Adhesive modifiers **EM3:** 175-186
adhesion promoters **EM3:** 180-182
chemical bonding **EM3:** 181-182
polarity and hydrogen bonding **EM3:** 181
surface wetting **EM3:** 180-181
cost factors **EM3:** 185-186
filler considerations **EM3:** 179-180
processing considerations **EM3:** 177-179
electrical conductivity improvements **EM3:** 178
electrical property enhancement **EM3:** 178-179
flame retardance **EM3:** 179
pigmentation **EM3:** 179
rheology control **EM3:** 177-178
smoke suppression **EM3:** 179
thermal conductivity improvement **EM3:** 178
tackifiers **EM3:** 182-183
Adhesive modifiers (continued)
tougheners **EM3:** 183-186
plasticization **EM3:** 184
single-phase toughening **EM3:** 184-185
two-phase toughening **EM3:** 185-186
Adhesive plumbum materials
applications **A2:** 555
Adhesive polymers
solid-film **EL1:** 220-221
Adhesive, pressure-sensitive *See* Pressure-sensitive adhesive
Adhesive sheet
brazing filler metals available in this form **A6:** 119
Adhesive strength **EM3:** 4
defined **EM1:** 3
Adhesive, structural *See* Structural adhesive
Adhesive system **EM3:** 4
Adhesive theory of friction **A18:** 31
Adhesive wear *See also* Abrasive wear; Galling; Scoring; Scuffing; Seizing
aircraft brakes **A18:** 583
in ball bearings **A11:** 496, 498
in band printer **A8:** 607
bearing steels **A18:** 732, 733
defined .... **A8:** 1 , 601-602 **A9:** 1 **A11:** 1 **A18:** 1
effect of material properties on **A11:** 158
electroplated coating applications **A18:** 835
failures **A11:** 145-146
from wire wooling **A11:** 466
gray cast iron **M1:** 24
internal combustion engine parts **A18:** 555, 559
ion implantation **A18:** 855-856, 857, 858
jet engine components **A18:** 588, 590
lubrication **A8:** 602
metal-matrix composites **A18:** 806, 807-808, 809, 810
mining and mineral industries **A18:** 651
nitrided surfaces **A18:** 879, 880, 881, 882
off shafts **A11:** 466
on gear teeth **A11:** 596
in inner cone, roller bearing **A11:** 158
rolling contact wear **A18:** 257
shear fracture damage **A18:** 179
sliding **A8:** 601
sliding bearings **A18:** 742, 743
stainless steels **A18:** 715, 716, 717, 720-721
in steel gears **A11:** 596
surface property effect **A18:** 342, 343
theory **A11:** 145-146
thermal spray coating applications **A18:** 831-832, 833
performance factors **A18:** 832
thermoplastic composites **A18:** 821
tool steels **A18:** 736, 737-738
Adhesive(s)
application methods **EL1:** 1046
application technology **EL1:** 671-672
card/planar assembly **EL1:** 117
classification **EL1:** 671
conductive, flexible printed boards **EL1:** 590
curing technology **EL1:** 672
dot height **EL1:** 673
failure mechanisms **EL1:** 1046-1047
for flexible printed boards **EL1:** 582
flexible printed wiring, compared **EL1:** 582
for microelectronic applications **EL1:** 110
organic, as component attachment **EL1:** 348-349
organic, as die attachment **EL1:** 213
postsolder deposits **EL1:** 674
process control **EL1:** 672-674
properties **EL1:** 1046
requirements **EL1:** 670
for surface mounting **EL1:** 631, 670-674
for surface-mount soldering **EL1:** 700
troubleshooting guide **EL1:** 673
types **EL1:** 670-671
Adhesive-bonded aluminum sheet
thermal inspection **A17:** 403
Adhesive-bonded joints *See also* Joint(s) **A17:** 610-640
defects, description of **A17:** 610-616
defects during fabrication **A17:** 612
glue-line thickness, by eddy current inspection **A17:** 188-189
inspection results, evaluation **A17:** 636-639
NDT, in product cycle **A17:** 632-636
Adhesive-bonded joints (continued)
NDT method, applications and limitations **A17:** 616-632
NDT method, selection **A17:** 631-632
tap test **A17:** 626-627
thermal stress **EL1:** 57-58
Adhesiveless
defined **EL1:** 582
Adhesively Bonded Joints: Testing, Analysis, and Design **EM3:** 67
Adhesives *See also* Adhesive
acrylic **EM1:** 684, 686
adhesive joint design **EM1:** 683
alternatives to **EM1:** 687
anaerobic **EM3:** 4
auxiliary materials/products, specifications for **EM1:** 695-696
bismaleimide (BMI) **EM1:** 684
cold-curing **EM3:** 665
cold-setting **EM3:** 4
compared to sealants **EM3:** 56
contact **EM3:** 4
corrosivity **A13:** 342
curing **EM1:** 685
definition **EM3:** 4, 33, 39
epoxy **EM1:** 684, 685-686
evaluation data **EM1:** 329
with fasteners **EM1:** 687
forms for application **EM3:** 40
functions **EM3:** 33
gap-filling **EM3:** 4
heat-activated **EM3:** 4
heat-sealing **EM3:** 4
hot-curing **EM3:** 665
hot-melt **EM1:** 13, 684, 687 **EM3:** 4
hot-setting **EM3:** 4
intermediate-temperature-setting **EM3:** 4
material property specifications **EM1:** 690-695
phenolic-based **EM1:** 684
plasticization of **EM3:** 617
polyimide **EM1:** 684
polysulfide **EM1:** 687
pressure-sensitive **EM3:** 4
processing specifications **EM1:** 689-690
reactive rubber **EM1:** 686-687
rubber-base **EM1:** 686-687
selection **EM1:** 683-688
silicone **EM1:** 684, 687
specifications/standards **EM1:** 688-701
structural **EM3:** 4
surface tensions **EM3:** 181
test methods **EM1:** 696-699
void formation with **EM1:** 687
Adhesives, Adherends, Adhesion **EM3:** 70
Adhesives Age **EM3:** 65
Adhesives and Glue—How to Choose and Use Them **EM3:** 69
Adhesives and Sealants **EM3:** 71
Adhesives and sealants, specific types
3501 (epoxide), moisture ingression on joint performance **EM3:** 361-362
3501-6 (epoxy)
fracture process **EM3:** 509
fracture toughness **EM3:** 510
5208 (epoxy resin)
diffusion coefficient changes **EM3:** 362
fracture process **EM3:** 509
fracture toughness **EM3:** 510
acrylic latex E **EM3:** 88
acrylic latex G **EM3:** 88
acrylic latex G/rosin **EM3:** 88
AF 42 (epoxy) **EM3:** 657
AF 55
adhesive defect detected **EM3:** 750
variable-quality standard **EM3:** 770, 771
x-ray opaque adhesive **EM3:** 749-750, 759
AF 126 .... **EM3:** 364
surface preparation evaluation **EM3:** 802, 803
AF 126-2, shear modulus and proportionality **EM3:** 665, 666
AF 147, weak bond inspection **EM3:** 530, 531
AF 147/FM 300, storage time **EM3:** 522
AF 163
carrier cloth effect **EM3:** 515
high-temperature structural **EM3:** 508
Aflas (fluoroelastomer sealant) **EM3:** 226
Agomet U3 (polyester) **EM3:** 657

Adhesives and sealants, specific types (continued)
Alodine 1200 (primer)
effect on bond stability EM3: 802, 803
surface preparation EM3: 806
Apical (polyimide) EM3: 160
Araldite (epoxy), for use with polysulfides EM3: 139
Araldite 508 (epoxy) EM3: 184
Araldite 6010 (epoxy) EM3: 184
Ardel (polyarylate resin) EM3: 279
Astrel (polyaryl sulfone resins) EM3: 279
AW 106 (epoxy) EM3: 657
AY 103 .... EM3: 482, 483
BR 34B-18 (primer) EM3: 294
BR 127 (primer) EM3: 364, 644, 775
wedge testing of epoxy FM EM3: 667
BSL 312 (epoxy), shear properties EM3: 321
Capcure (polysulfide) EM3: 139, 141
Celanex (polyester) EM3: 279
Chemlock 205 (primer) EM3: 630
Chemlock 205/220 (primer) EM3: 629
Chlorobutyl 1066 .... EM3: 199
DC 1-2577 .... EM3: 597
DEN 438 (epoxy resin) EM3: 286
Durimid 100 (polyimide) EM3: 268
EA 934 (epoxy) EM3: 644
thickness effect on debond load EM3: 481
to bond metallic fitting to graphite-epoxy composite tube EM3: 496
EA 934NA (epoxy)
dynamic mechanical analysis results EM3: 646-647, 649
tension lap-shear test results EM3: 646-647
EA 9309, lap-shear and peel test results EM3: 809
EA 9309.3NA (epoxy)
dynamic mechanical analysis results EM3: 647, 649
tension lap-shear test results EM3: 647
EA 9320NA (epoxy)
dynamic mechanical analysis results EM3: 647-648, 649
tension lap-shear test results EM3: 647-648
EA 9334 (epoxy)
dynamic mechanical analysis results EM3: 648, 649
tension lap-shear test results EM3: 648
EA 9394 (epoxy, Hysol), structural schematic EM3: 493
EA 9396 (epoxy)
dynamic mechanical analysis results EM3: 648-649
tension lap-shear test results EM3: 648-649
EA 9628
adhesive defect detected EM3: 750
EA 9628NW (epoxy, Dexter-Hysol) EM3: 363, 364
lap-shear and peel test results EM3: 809
water absorption EM3: 363, 364
EA 9649
carrier cloth effect EM3: 515
effect of viscoelasticity EM3: 513
high-temperature structural EM3: 508
EA 9649R, weak bond inspection EM3: 530, 531
EC 2320 (primer), to bond wedge specimens EM3: 802, 803
EC 3445 (epoxy)
carrier cloth effect EM3: 515
fracture process EM3: 509
fracture toughness EM3: 510
EC 3448, graphite-epoxy substrates EM3: 524
EC 3924/AF EM3: 669
EH 330 (polysulfide) EM3: 139
ELP 3 (polysulfide) EM3: 138, 140, 141, 142
Epon (epoxy) for use with polysulfides EM3: 139

Adhesives and sealants, specific types (continued)
Epon 828 (epoxy resin) EM3: 285, 286
adhesion strength of AS4 carbon fibers to polyethylene (UHMW-PE) EM3: 393, 395-396, 402
coating on carbon/graphite EM3: 288
Epon 834 (epoxy resin) EM3: 184
epoxy B EM3: 595
ERL-4221 (epoxy) EM3: 96
Exxon Butyl 0-65 .... EM3: 199
Exxon Butyl 268 .... EM3: 199
F-185 (epoxy)
fracture process EM3: 509
fracture toughness EM3: 510
Fluorel (fluoroelastomer sealant) EM3: 226
FM 34B-18 (polyimide) EM3: 294
FM 37 (epoxy), incipient defects in bonded structures EM3: 524
FM 47, fatigue testing EM3: 502
FM 73 (epoxy) EM3: 355-357, 359
aluminum with BR 127 wedge tested EM3: 668
bond line porosities produced by shimming EM3: 523-524, 525
effect of high temperatures EM3: 513
lap-shear and peel test results EM3: 809
lap-shear strength versus adhesive thickness plotted EM3: 773, 774, 775
mechanical properties EM3: 360
moisture ingression EM3: 364
overlap edge stress concentration for single-lap joints EM3: 360-361
pulse-echo ultrasonic method for evaluation EM3: 778
shear modulus and proportionality EM3: 665, 666
shear properties EM3: 321
stress-strain curves in shear EM3: 474
wedge testing involving BR 127 primer EM3: 667
FM 73M (epoxy), peel and lap-shear test results EM3: 466
FM 300 (epoxy) EM3: 355-356, 357, 358-359
carrier cloth effect EM3: 515
direct compliance method of fracture analysis EM3: 340
effect of high temperatures EM3: 513
fracture process EM3: 509
fracture toughness EM3: 510
fracture toughness envelopes EM3: 446
fracture toughness values EM3: 447, 448
incipient defects bonded structures EM3: 524
mechanical properties EM3: 360
overlap edge stress concentration for single-lap joints EM3: 360-361
FM 300K (epoxy)
moisture ingression EM3: 364
peel and lap-shear test results EM3: 466
weak bond detection EM3: 530
weak bond inspection EM3: 531
FM 300M (epoxy)
dynamic mechanical analysis EM3: 648-649
tension lap-shear test results EM3: 9
Ti-6Al-4V adherend EM3: 270
FM 400 (epoxy)
for advanced composite structures EM3: 822, 824-825, 827
fatigue testing EM3: 468
peel and lap-shear test results EM3: 466
x-ray opaque adhesive EM3: 749-750, 759
FM 1000 (Cyanamid, epoxy)
effect of temperature on strength EM3: 619
shear modulus and proportionality EM3: 665, 666
tensile mechanical properties EM3: 619
FM series (polyimide) EM3: 160
FMS 1013 type 1B EM3: 528, 529
GORE-TEX (polytetrafluoroethylene) EM3: 225
Gylon (polytetrafluoroethylene) EM3: 225
HDS-100 (polysulfide) EM3: 139

Adhesives and sealants, specific types (continued)
HT 939 .... EM3: 96
HX 205 .... EM3: 205
fracture process EM3: 509
fracture toughness EM3: 510
HY 940 .... EM3: 96
Hydrite 121 (polysulfide) EM3: 139
Hypalon
applications EM3: 56
sealant characteristics (wet seals) EM3: 57
Hysol EA 934, shear stress peaking factor EM3: 483
IM 6/3501-6 .... EM3: 495
Kalar (butyl) EM3: 199
Kalene (butyl) EM3: 199
Kalrez (fluoroelastomer sealant) EM3: 226, 227
Kapton (polyimide) EM3: 159, 160, 165
Kynar EM3: 279, 454
LARC TPI (polyimide) EM3: 266-267, 270
LARC-13 .... EM3: 154
latex A/phthalate ester (40/60) (SBR), water-base PSA properties EM3: 88
latex D/low molecular weight polystyrene (60/40) (SBR), water-base PSA properties EM3: 88
latex H/phthalate ester (40/60) (SBR), water-base PSA properties EM3: 88
latex J/rosin (50/50) (SBR), water-base PSA properties EM3: 88
latex/rosin (40/60) (natural rubber), water-base PSA properties EM3: 88
latex/terpene (40/60) (natural rubber), water-base PSA properties EM3: 88
LP (liquid polysulfide) EM3: 138, 139
LP-2 .... EM3: 140
LP-3 .... EM3: 139, 140, 142
mechanical properties EM3: 141
physical properties EM3: 141
LP-12 .... EM3: 140
LP-31 .... EM3: 140
LP-32 .... EM3: 140
LP-33 .... EM3: 140
MB 329 .... EM3: 821
Metlbond 329 (M-329) EM3: 834
Metlbond 1113 .... EM3: 502, 503, 505, 506, 510
carrier cloth effect EM3: 515
catastrophic threshold values EM3: 516
crack growth behavior EM3: 509
crack propagation under mixed-mode cyclic loading EM3: 513
dynamic fatigue EM3: 367
experiment synopsis EM3: 517-518
fatigue testing EM3: 350
onset threshold values EM3: 516
opening load versus in-plane shear load for ILMMS EM3: 517
plane-strain conditions EM3: 511
relaxation behavior in bulk EM3: 365
stress-whitening EM3: 514, 515
Metlbond 1113-2 .... EM3: 503, 510
carrier cloth effect EM3: 515
catastrophic threshold values EM3: 516
crack growth behavior EM3: 509
crack propagation under mixed-mode cyclic loading EM3: 513
dynamic fatigue EM3: 367
experiment synopsis EM3: 517-518
onset threshold values EM3: 516
opening load versus in-plane shear load for ILMMS EM3: 517
plastic deformation zone EM3: 511
stress-whitening EM3: 514, 515
MY 0500 (epoxy) EM3: 94
MY 720 (epoxy) EM3: 94-95
MY 750 .... EM3: 482, 483
neoprene latex (carboxylated)/rosin (60/40), water-base PSA properties EM3: 88

Adhesives and sealants, specific types (continued)
neoprene latex (noncarboxylated)/ rosin (60/40), water-base PSA properties ... **EM3:** 88
Nomex ... **EM3:** 807
NR/C205-220 (natural rubber) ... **EM3:** 629
Oppanol (butyl) ... **EM3:** 199
P-1700 (polysulfone) ... **EM3:** 364
PAG 200 ... **EM3:** 669
Parylene C ... **EM3:** 600
Parylene D ... **EM3:** 600
Parylene N ... **EM3:** 600
PasaJell ... **EM3:** 105, 799
nontank procedure peel strength evaluation ... **EM3:** 803, 804, 805, 806
PasaJell 107 ... **EM3:** 799
Permabond ESP110 ... **EM3:** 515
Permapol P-2 (polysulfide) ... **EM3:** 141
Permapol P-2/P-3 (polysulfide) ... **EM3:** 138-139
Permapol P-3 (polysulfide) ... **EM3:** 141
PMR-15 (polyimide) ... **EM3:** 154, 155, 158-159, 160
RB-7116, mechanically induced artificial delamination ... **EM3:** 523, 524
Redux 319 (epoxy), moisture ingression ... **EM3:** 364
Redux 775 (phenolic), corrosion resistance ... **EM3:** 671
Redux 775RN (phenolic), shear properties ... **EM3:** 321
Reliabond 398
for lap-shear specimen fabrication ... **EM3:** 744
storage time ... **EM3:** 522
RTV-I ... **EM3:** 597, 598
RTV-II ... **EM3:** 597-598
Ryton (polyphenylene sulfide (PPS)) ... **EM3:** 280
SF 340 (bonded aileron) ... **EM3:** 562-563
Skybond 700 series (polyimide lacquer) ... **EM3:** 158
Solithane 60 ... **EM3:** 283
Syntorg IP200-705 (polyphenylquinox-aline) ... **EM3:** 164
Syntorg IP200-710 (polyphenylquinox-aline) ... **EM3:** 164
Syntorg IP200-715 (polyphenylquinox-aline) ... **EM3:** 164
Tenite (polyester) ... **EM3:** 279
Thermid AL grade (polyimide) ... **EM3:** 158
Thermid IP-6001 ... **EM3:** 158
Thermid LR series (polyimide) ... **EM3:** 158
Toluene (polysulfide) ... **EM3:** 139, 181
Udel (polysulfone resin) ... **EM3:** 279
Ultem (polyetherimide) ... **EM3:** 280
Upilex (polyimide) ... **EM3:** 160
Valox (polyester) ... **EM3:** 279
Valox (polyester), lap-shear strength ... **EM3:** 276
Vespel (polyimide) ... **EM3:** 277-278
Victrex (polyether sulfone (PESV)) ... **EM3:** 280
Vistanex L-100, molecular weight ... **EM3:** 199
Vistanex L-140, molecular weight ... **EM3:** 199
Vistanex LM-MH, molecular weight ... **EM3:** 199
Vistanex LM-MS, molecular weight ... **EM3:** 199
Viton (fluoroelastomer sealant) ... **EM3:** 226
Adhesives, Annual Book of ASTM Standards ... **EM3:** 67
Adhesives for Metals: Theory and Technology ... **EM3:** 70
Adhesives for Structural Applications ... **EM3:** 68
Adhesives for the Composite Wood Panel Industry ... **EM3:** 68
Adhesives for Wood-Research Applications, and Needs ... **EM3:** 68
Adhesives from Renewable Resources ... **EM3:** 71
Adhesives Handbook ... **EM3:** 69
Adhesives in Modern Manufacturing ... **EM3:** 69
Adhesives, Sealants, and Coatings for the Electronics Industry ... **EM3:** 68
Adhesives, Sealants, and Gaskets—A Survey ... **EM3:** 68
Adhesives, Sealants and Primers ... **EM3:** 69
Adhesives specifications ... **EM1:** 689-701
Adhesives Technology Handbook ... **EM3:** 67
Adhesives Technology—Developments Since 1979 ... **EM3:** 68
Adhesives, wheel polishing
types used ... **M5:** 208
Adhesives/sealants
defined ... **EM3:** 61
ADI *See* Austempered ductile iron
Adiabatic *See also* Extrusion ... **EM3:** 4
conditions, in high strain rate regime ... **A8:** 191
defined ... **EM2:** 3
heating, in processing map ... **A8:** 572
shear band, as flow localizations ... **A8:** 155
shear instabilities, in torsion testing ... **A8:** 217-218
Adiabatic diesel engines ... **EM4:** 987-994
adiabatic engine designs ... **EM4:** 987-989
applications
ceramic ... **EM4:** 960
summary ... **EM4:** 990
design considerations ... **EM4:** 993-994
materials ... **EM4:** 989-993
tests to qualify coatings for engine applications ... **EM4:** 992
thermal properties of engine wall insulator lining materials ... **EM4:** 992
Adiabatic fast passage
defined as ESR supplemental technique ... **A10:** 258
Adiabatic shear
bands, in titanium alloy ... **A12:** 43
defined ... **A12:** 31
strain rates for ... **A12:** 31-33, 43
Adiabatic stability
in superconductors ... **A2:** 1038
ADINA
finite-element analysis code ... **EM3:** 479, 480, 486
ADINA computer program for structural analysis ... **EM1:** 268, 269
ADINAT thermal analysis software ... **EL1:** 446-447
Adjustable light spot analyzers ... **A7:** 229
Adjustable reamers ... **A16:** 241, 242-243
Adjustable-bed presses
for piercing ... **A14:** 463
Adjustable-iris diaphragm in optical microscopes ... **A9:** 72
Adjustments
elimination of ... **EL1:** 123
Admiralty alloy, SCC environment
AES analysis of ... **A10:** 563-564
Admiralty brass
brazing ... **A6:** 629-630
dezincification of ... **A13:** 128, 132
weldability ... **A6:** 753
Admiralty metal
applications and properties ... **A2:** 318-319
electrolytic etching ... **A9:** 401
Admiralty metals
antimonial *See* Copper alloys, specific types, C44400
arsenical *See* Copper alloys, specific types, C44300
arsenical, impingement attack ... **A11:** 634-635
B brass, intergranular attack ... **A11:** 182
brass heat-exchanger tubes, failed ... **A11:** 637
inhibited *See* Copper alloys, specific types, C44300, C44400 and C44500
phosphorized *See* Copper alloys, specific types, C44500
SCC in ... **A11:** 635
Admittance matrices
transmission line characteristic impendance and ... **EL1:** 35
Admixture ... **EM3:** 4
defined ... **EM1:** 3 **EM2:** 3
ADONE
as synchrotron radiation source ... **A10:** 413
Adsorbed contaminants ... **EM3:** 41
Adsorbed surfactants
IR analysis of ... **A10:** 109
Adsorption *See also* Absorption ... **EM3:** 4
as contamination of gravimetric samples ... **A10:** 163
defined ... **A13:** 1 **EM1:** 3 **EM2:** 3
-enhanced plasticity ... **A13:** 160
of hydrogen ... **A12:** 124
-induced brittle fracture, crack propagation ... **A13:** 161
inhibitors ... **A13:** 1141
as initial oxidation, gaseous corrosion ... **A13:** 67
isotherms, for surface species, Raman analysis as probe for ... **A10:** 134
LEED analysis of ... **A10:** 536
of molecular species, IR identification of ... **A10:** 109
Adsorption (continued)
pyridine, Raman studies ... **A10:** 134
surface, and hydrogen damage ... **A13:** 164
surface, as SCC parameter ... **A13:** 147
surface, FIM/AP study of ... **A10:** 583
tracer gas ... **A17:** 69
Adsorption chromatography *See also* Liquid-solid chromatography
defined ... **A10:** 668
Adsorption isotherms ... **EM4:** 70, 155
Adsorption, water
in molding clays ... **A15:** 210-212
Adsorption-induced embrittlement
in axles ... **A11:** 718, 719
ADV-pH test
for sand reclamation ... **A15:** 355
Advance separation
crack path after joining of ... **A8:** 443
Advanced (loose powder) filling
systems ... **A7:** 431
Advanced aluminum materials *See also* Aluminum; Aluminum alloys
aluminum-base discontinuous metal-matrix composites ... **A14:** 251
aluminum-lithium alloys ... **A14:** 250
forging of ... **A14:** 249-251
prealloyed P/M alloys ... **A14:** 250-251
"Advanced" ceramics ... **EM4:** 45-49
additives to aid ceramic processing ... **EM4:** 49
applications ... **EM4:** 1-2, 16, 47, 48
electrical discharge machining ... **EM4:** 375-376
electrical/electronic ... **EM4:** 1105-1106
carbides ... **EM4:** 47
centerless grinding ... **EM4:** 341
development ... **EM4:** 16-17
estimated market in U.S. ... **EM4:** 47
fracture toughness lacking ... **EM4:** 16
grinding capabilities for varying cycle times ... **EM4:** 341
materials included ... **EM4:** 16
mixed oxide ceramics ... **EM4:** 47
nitrides ... **EM4:** 47
oxides ... **EM4:** 47
processing innovations ... **EM4:** 47-49
properties ... **EM4:** 1, 2, 16, 48, 1105
raw materials ... **EM4:** 32-33, 48
shaping and finishing ... **EM4:** 313
strength versus temperature properties ... **EM4:** 1001
stress-rupture characteristics ... **EM4:** 1001
subcritical crack growth ... **EM4:** 16
substrate formulations ... **EM4:** 1105
testing ... **EM4:** 547
ultrasonic machining ... **EM4:** 360
workpieces finished by production grinding techniques ... **EM4:** 339
Advanced ceramics, application
internal combustion engine parts ... **A18:** 561
Advanced CMOS logic (ACL)
development ... **EL1:** 160
Advanced composite packaging materials ... **EL1:** 1117-1118
Advanced composite structures *See* Composite structures, advanced
Advanced composites *See also* Aerospace applications; Aerospace composite structure fabrication; Composites ... **EM3:** 4
in aircraft industry ... **EM1:** 801-809
for aircraft manufacturing industry ... **EM1:** 34
applications ... **EM1:** 799-847
axial tensile strength ... **EM1:** 192
chemical structures ... **EM1:** 101
damping properties analysis ... **EM1:** 206-217
defined ... **EM2:** 3
matrix resins ... **EM1:** 73, 545
testing requirements ... **EM1:** 283
thermoplastic ... **EM1:** 544-553
thermoplastic resins ... **EM2:** 621-622
tooling, quality control ... **EM1:** 738-739
Advanced cure monitor (ACM) 101 instrumentation ... **EM1:** 762
Advanced failure analysis techniques
for corrosion failure ... **EL1:** 1102-1116
Advanced forging process modeling ... **A14:** 409-416

Advanced gas turbine (AGT) program, DOE Office of Transportation Systems **EM4:** 716-720
ceramic gasifier turbine scroll assembly for AGT-100 engine **EM4:** 717, 718, 720
design/manufacturing trade-offs **EM4:** 720
radial inflow turbine rotor burst **EM4:** 718-719
Advanced gas turbines **EM4:** 995-1001
applications **EM4:** 995
nonautomotive **EM4:** 995
background **EM4:** 995
ceramic gas turbine engine development in Europe **EM4:** 1000
ceramic materials **EM4:** 1000-1001
ceramic turbine engine development in Japan **EM4:** 999-1000
ceramics versus superalloys for turbine nozzles **EM4:** 995
challenges **EM4:** 995-996
the DOE-ATTAP program **EM4:** 996-998, 999, 1001
U.S. development for auxiliary power units **EM4:** 998-999, 1000
versus conventional internal combustion engines **EM4:** 995
Advanced inertial Reference Sphere for ICBM, for beryllium P/M parts **A7:** 761
Advanced Materials & Processes Magazine **EM3:** 65
Advanced Materials and Processes
trade magazine **EM2:** 92
Advanced statistical concepts of fracture in brittle materials **EM4:** 709-715
distributional models **EM4:** 709-710
multiaxial failure **EM4:** 710
Weibull distribution **EM4:** 709
Griffith/Orowan criteria **EM4:** 709
Weibull estimators for combined data **EM4:** 710-715
Bartlett correction factor **EM4:** 714
bootstrap techniques for confidence and tolerance **EM4:** 712-713
classes of problem **EM4:** 710
confidence and tolerance bounds **EM4:** 712
example of Class IV strength data **EM4:** 710, 711
future needs and directions **EM4:** 714-715
likelihood techniques for confidence and tolerance **EM4:** 713-714
maximum likelihood estimator **EM4:** 710-712, 714
probabilities of failure **EM4:** 711-712
profile log likelihood **EM4:** 713
Advanced structural ceramics
properties needed in these design areas **EM4:** 690, 691
Advanced thermal management design
tape automated bonding (TAB) **EL1:** 287
Advanced titanium materials *See also* Titanium; Titanium alloys
forging of **A14:** 282-283
titanium aluminides **A14:** 283
titanium metal-matrix composites **A14:** 283
titanium powder metallurgy materials **A14:** 283
Advanced titanium-base alloys **A6:** 524-527
alpha-beta alloys **A6:** 524, 526
electron-beam welding **A6:** 526
families **A6:** 524
fusion zone **A6:** 526
gas-tungsten arc welding **A6:** 526
heat-affected zone **A6:** 526, 527
heat-affected zone cracking **A6:** 526
intermetallic alloys **A6:** 525, 526
intermetallics **A6:** 525
laser-beam welding **A6:** 526
metastable-beta alloys **A6:** 524, 526
microstructure **A6:** 524, 525-526
near-beta alloys **A6:** 524, 526
physical metallurgy **A6:** 524-525
Advanced titanium-base alloys (continued)
postweld heat treatment **A6:** 526, 527
titanium-aluminum phase diagram **A6:** 525
titanium-matrix composites **A6:** 524, 525, 527
weldability **A6:** 525-527
Advanced titanium-base alloys, specific types
Alpha-2
chemical composition **A6:** 525
electron-beam welding **A6:** 526
mechanical properties **A6:** 525
weldability **A6:** 525, 526
Beta 21S
chemical composition **A6:** 524
mechanical properties **A6:** 524, 525
weldability **A6:** 524, 526
Beta C
chemical composition **A6:** 524-525
gas-tungsten arc welding **A6:** 526
mechanical properties **A6:** 524-525
weldability **A6:** 524, 526
Corona 5
chemical composition **A6:** 524
mechanical properties **A6:** 524, 525
weldability **A6:** 524, 526
Gamma
chemical composition **A6:** 525
mechanical properties **A6:** 525
microstructures **A6:** 525
weldability **A6:** 525, 526
IMI 834
chemical composition **A6:** 524
mechanical properties **A6:** 524, 525
weldability **A6:** 524, 526
Orthorhombic
mechanical properties **A6:** 525
weldability **A6:** 525, 527
Ti-10V-2Fe-3Al
chemical composition **A6:** 524
mechanical properties **A6:** 524-525
weldability **A6:** 524, 526
Ti-1100
chemical composition **A6:** 524
mechanical properties **A6:** 524, 525
weldability **A6:** 524, 526
Advanced tool motions **A7:** 327-328
Advanced turbine technology application project (ATTAP) **EM4:** 716
ceramic-to-ceramic joints deficient **EM4:** 720
design/manufacturing trade-offs **EM4:** 720
Garrett radial inflow turbine redesign and foreign object damage **EM4:** 719
Advanced waveform analysis package
acoustic emission inspection **A17:** 286
Advances in Adhesives: Applications, Materials and Safety **EM3:** 69
Advances in Polymer Technology
as information source **EM2:** 93
Advantage
Fellgett's or multiplex **A10:** 129
Advantages of adhesive joining **EM3:** 33, 34
AE inspection *See* Acoustic emission inspection
AEM *See* Analytical electron microscopy
Aerate
defined **A15:** 1
Aerated concrete
powders used **A7:** 527
Aerated conditions
high-temperature pure water **A8:** 420-422
Aeration *See also* Deaeration
in chemical processing plant **A13:** 1135
corrosion effect **A13:** 1162
defined **A13:** 1
differential **A13:** 339, 785, 788
effect, nickel-base alloy corrosion in HF **A13:** 648
effect, zinc corrosion in distilled water **A13:** 760
in foam fluxers **EL1:** 682
of solution, total immersion tests **A13:** 221-222
water, corrosion effects **A13:** 207
Aeration cell *See* Differential aeration cell
Aerial piping
ductile iron **M1:** 99, 100
Aerobic bacteria
corrosion by **A11:** 190
Aerobic corrosion
of iron and steels **A13:** 117
Aerodynamic lubrication *See* Gas lubrication
Aerogels **EM4:** 449
alkoxide-derived gels **EM4:** 210-211, 212, 213
sol-gel process for formation **EM4:** 62
Aerosol pyrolysis technique
high-temperature superconductors **A2:** 1086
Aerosol samples
atmospheric, PIXE particle size analysis of **A10:** 102
collection **A10:** 94
Aerospace **A6:** 385-388
brazeability and solderability applications **A6:** 617-618
ceramic applications **EM4:** 960
Aerospace alloys
CBN tools for machining **A16:** 639
design values for bearing and tensile properties **A8:** 61
fracture mechanics data **A11:** 54
high-speed tool steels used **A16:** 59
Aerospace and military applications **EM4:** 1016-1020
frangible glasses **EM4:** 1020
glass fibers **EM4:** 1029
infrared glasses **EM4:** 1019-1020
mirrors for space **EM4:** 1016, 1017, 1018
missile nose cones (radomes) **EM4:** 1017-1019
solar cell covers **EM4:** 1019
spacecraft windows **EM4:** 1016-1017, 1018
Aerospace applications **A7:** 646-656 **EM4:** 1003-1006
aircraft propulsion **EM4:** 1003-1005
benefits of ceramics in aerospace systems **EM4:** 1004
benefits of thermostructural ceramics **EM4:** 1003
of beryllium powders **A7:** 169, 759-760
bulk molding compounds **EM1:** 162-163
as casting market **A15:** 34
constructional steels for elevated temperature use **M1:** 647
of copper-based powder metals **A7:** 733
factors **EM4:** 1003
High Temperature Engine Materials Program (NASA) **EM4:** 1004
of hybrids **EL1:** 254
investment casting **A15:** 265
key technical issues **EM4:** 1004
limitations of thermostructural ceramics **EM4:** 1003
liquid crystal polymers (LCP) **EM2:** 180
matrices for **EM1:** 32-33
nickel alloy powders for **A7:** 142
of parts design **EM2:** 616
polybenzimidazoles (PBI) **EM2:** 147
polyether sulfones (PES, PESV) **EM2:** 159
potential applications in aircraft propulsion **EM4:** 1004
powders used **A7:** 572
of prepreg resins **EM1:** 139-141
refractory metals for **A7:** 765
of semisolid metal casting/forging **A15:** 327
Small Engine Component Technology Study **EM4:** 1004
space and missile systems **EM1:** 816-822
for stainless steel alloys **A7:** 730
Stirling Engine **EM4:** 1005
technologies **A7:** 18
thermoplastic fluoropolymers **EM2:** 117
thermoplastic polyimides (TPI) **EM2:** 177
of titanium powder **A7:** 164
turbopump components **EM4:** 1005
ultra-clean powders for **A7:** 36

SUBJECTS OF THE INDEXED VOLUMES: ASM Handbook (designated by the letter "A"): **A1:** Properties and Selection: Irons, Steels, and High-Performance Alloys (1990); **A2:** Properties and Selection: Nonferrous Alloys and Special-Purpose Materials (1990); **A3:** Alloy Phase Diagrams (1992); **A4:** Heat Treating (1991); **A6:** Welding, Brazing, and Soldering (1993); **A7:** Powder Metallurgy (1984); **A8:** Mechanical Testing (1985); **A9:** Metallography and Microstructures (1985); **A10:** Materials Characterization (1986); **A11:** Failure Analysis and Prevention (1986); **A12:** Fractography (1987); **A13:** Corrosion (1987); **A14:** Forming and Forging (1988); **A15:** Casting (1988); **A16:** Machining (1989); **A17:** Nondestructive Testing and Quality Control (1989); **A18:** Friction, Lubrication, and Wear Technology (1992). **Metals Handbook, 9th Edition** (designated by the letter "M"): **M1:** Properties and Selection: Irons and Steels (1978); **M2:** Properties and Selection: Nonferrous Alloys and Pure Metals (1979); **M3:** Properties and Selection: Stainless Steels, Tool Materials and Special-Purpose Materials (1980); **M4:** Heat Treating (1981); **M5:** Surface Cleaning, Finishing, and Coating (1982); **M6:** Welding, Brazing, and Soldering (1983). **Engineered Materials Handbook** (designated by the letters "EM"): **EM1:** Composites (1987); **EM2:** Engineering Plastics (1988); **EM3:** Adhesives and Sealants (1990); **EM4:** Ceramics and Glasses (1991); **Electronic Materials Handbook** (designated by the letters "EL"): **EL1:** Packaging (1989).

Aerospace applications (continued)
unidirectional/two-directional fabrics **EM1:** 126
urethane hybrids **EM2:** 268
of vacuum induction melting **A15:** 393
Aerospace applications for adhesives **EM3:** 558-566
adhesive bonding of aircraft canopies and windshields **EM3:** 559, 563-564
application methods **EM3:** 559, 560-561
bonding honeycomb to primary surfaces **EM3:** 559, 562-563
classes **EM3:** 559
F-15 composite speedbrake **EM3:** 563
honeycomb core construction **EM3:** 559-560
honeycomb core sandwich construction **EM3:** 559, 560-561
honeycomb structure damage **EM3:** 559, 566
laminate repair **EM3:** 559, 565
lightning strike applications **EM3:** 559, 564-565
missile radome-bonded joint **EM3:** 559, 563, 564
noise-suppression systems for engine nacelles **EM3:** 559, 561-562
product forms **EM3:** 559
repair of adhesive disbonds **EM3:** 559, 566
secondary honeycomb sandwich bonding **EM3:** 559, 561
shelf life **EM3:** 559
Aerospace composite structure fabrication *See also* Advanced composites; Aerospace applications **EM1:** 73, 575-577
automated/mechanized lay-up **EM1:** 577
design requirements **EM1:** 575
labor-intensive lay-up **EM1:** 575-577
ply shapes **EM1:** 575-576
process requirements **EM1:** 575
Aerospace construction
codes governing **M6:** 824-825
joining processes **M6:** 56-57
Aerospace industry **A13:** 1058-1106
corrosion in **A13:** 1058-1106
manned spacecraft, corrosion of **A13:** 1058-1101
P/M superalloy applications **A13:** 836
pin bearing testing in **A8:** 59
space boosters and space satellites **A13:** 1101-1105
Aerospace industry applications *See also* Aircraft industry applications
acoustic emission inspection **A17:** 290-292
aluminum and aluminum alloys **A2:** 12
aluminum P/M alloys **A2:** 200
aluminum-lithium alloys **A2:** 178, 182
beryllium **A2:** 683-687
computed tomography (CT) **A17:** 363
of NDE reliability models **A17:** 702-715
neutron radiography **A17:** 391-395
of precious metals **A2:** 693
refractory metals and alloys **A2:** 557-559
rocket motors, computed tomography **A17:** 363
space shuttle program, as NDE reliability case study **A17:** 685-686
titanium and titanium alloy castings **A2:** 634, 644-645
of titanium and titanium alloys **A2:** 587-588
titanium P/M products **A2:** 655, 657-658
Aerospace industry market **EM3:** 58, 604-605
Aerospace manufacturing processes *See also* Aerospace composite structure fabrication **EM1:** 575-663
autoclave cure systems **EM1:** 645-648
automated integrated system **EM1:** 636-638
automated ply lamination **EM1:** 639-641
computer-controlled ply cutting labeling **EM1:** 619-623
computerized autoclave cure control **EM1:** 649-653
contoured tape laying **EM1:** 631-635
curing BMI resins **EM1:** 657-661
curing epoxy resins **EM1:** 654-656
curing polyimide resins **EM1:** 662-663
elastomeric tooling **EM1:** 590-601
electroformed nickel tooling **EM1:** 582-585
flat tape laying **EM1:** 624-630
graphite-epoxy tooling **EM1:** 586-589
manual lay-up **EM1:** 602-604
mechanically assisted lay-up **EM1:** 605-607
overview and basic operations **EM1:** 575-577
Aerospace manufacturing processes (continued)
preparation for cure **EM1:** 642-644
tooling for autoclave molding **EM1:** 578-581
ultrasonic ply cutting **EM1:** 615-618
Aerospace Material Specification (of SAE) **A8:** 724
Aerospace Material Specification, 4779 standard
powders used for flame spraying **A6:** 715
Aerospace Material Specifications (AMS) **EM1:** 41
Aerospace material specifications (SAE AMS) **EM2:** 91
Aerospace Materials Specifications **A13:** 322
Aerospace Materials Specifications (AMS) **EM3:** 62
Aerostatic lubrication *See* Pressurized gas lubrication
AES *See* Atomic emission spectrometry; Auger electron spectroscopy
AF2-lDA
thread grinding **A16:** 275
AF1410 steel **A1:** 431, 446-447
heat treatment for **A1:** 447
properties of **A1:** 445, 446, 447
AFBMA
defined **A18:** 2
AFC *See* Automatic frequency control
AFNOR (French) standards for steels **A1:** 158
compositions of **A1:** 186-189
cross-referenced to SAE-AISI steels **A1:** 166-174
AFNOR penetrameters
as step wedges **A17:** 341
Africa
metalworking in **A15:** 19
AFS *See* Atomic fluorescence spectrometry
AFS 50-70 test sand
defined **A18:** 2
AFS clay test
for sand reclamation **A15:** 355
AFS grain fineness number **A15:** 209
After-fabrication galvanizing
as zinc coating **A2:** 527-528
Afterbake *See* Postcure **EM3:** 4
Afterglow
defined **A17:** 383
AFWAL *See* Air Force Wright Aeronautical Laboratories
Ag-Al (Phase Diagram) .... **A3:** 2 • 25
Ag-As (Phase Diagram) .... **A3:** 2 • 25
Ag-Au (Phase Diagram) .... **A3:** 2 • 25
Ag-Au-Cu (Phase Diagram) .... **A3:** 3 • 5
Ag-Be (Phase Diagram) .... **A3:** 2 • 26
Ag-Bi (Phase Diagram) .... **A3:** 2 • 26
Ag-Ca (Phase Diagram) .... **A3:** 2 • 26
Ag-Cd (Phase Diagram) .... **A3:** 2 • 27
Ag-Cd-Cu (Phase Diagram) .... **A3:** 3 • 5-6
Ag-Cd-Zn (Phase Diagram) .... **A3:** 3 • 6-7
Ag-Ce (Phase Diagram) .... **A3:** 2 • 27
Ag-Co (Phase Diagram) .... **A3:** 2 • 27
Ag-Cu (Phase Diagram) .... **A3:** 2 • 28
Ag-Cu-Zn (Phase Diagram) .... **A3:** 3 • 7
Ag-Dy (Phase Diagram) .... **A3:** 2 • 28
Ag-Er (Phase Diagram) .... **A3:** 2 • 28
Ag-Eu (Phase Diagram) .... **A3:** 2 • 29
Ag-Fe (Phase Diagram) .... **A3:** 2 • 29
Ag-Ga (Phase Diagram) .... **A3:** 2 • 29
Ag-Gd (Phase Diagram) .... **A3:** 2 • 30
Ag-Ge (Phase Diagram) .... **A3:** 2 • 30
Ag-Hg (Phase Diagram) .... **A3:** 2 • 30
Ag-Ho (Phase Diagram) .... **A3:** 2 • 31
Ag-In (Phase Diagram) .... **A3:** 2 • 31
Ag-La (Phase Diagram) .... **A3:** 2 • 31
Ag-Li (Phase Diagram) .... **A3:** 2 • 32
Ag-Mg (Phase Diagram) .... **A3:** 2 • 32
Ag-Mo (Phase Diagram) .... **A3:** 2 • 32
Ag-Na (Phase Diagram) .... **A3:** 2 • 33
Ag-Nd (Phase Diagram) .... **A3:** 2 • 33
Ag-Ni (Phase Diagram) .... **A3:** 2 • 33
Ag-P (Phase Diagram) .... **A3:** 2 • 34
Ag-Pb (Phase Diagram) .... **A3:** 2 • 34
Ag-Pb-Sn (Phase Diagram) .... **A3:** 3 • 7-8
Ag-Pd (Phase Diagram) .... **A3:** 2 • 34
Ag-Pr (Phase Diagram) .... **A3:** 2 • 35
Ag-Pt (Phase Diagram) .... **A3:** 2 • 35
Ag-S (Phase Diagram) .... **A3:** 2 • 35
Ag-Sb (Phase Diagram) .... **A3:** 2 • 35
Ag-Sc (Phase Diagram) .... **A3:** 2 • 36
Ag-Se (Phase Diagram) .... **A3:** 2 • 36
Ag-Si (Phase Diagram) .... **A3:** 2 • 37
Ag-Sm (Phase Diagram) .... **A3:** 2 • 37
Ag-Sn (Phase Diagram) .... **A3:** 2 • 37
Ag-Sr (Phase Diagram) .... **A3:** 2 • 38
Ag-Te (Phase Diagram) .... **A3:** 2 • 38
Ag-Ti (Phase Diagram) .... **A3:** 2 • 38
Ag-Tl (Phase Diagram) .... **A3:** 2 • 39
Ag-Y (Phase Diagram) .... **A3:** 2 • 39
Ag-Yb (Phase Diagram) .... **A3:** 2 • 39
Ag-Zn (Phase Diagram) .... **A3:** 2 • 40
Ag-Zr (Phase Diagram) .... **A3:** 2 • 40
Agate
honing stone selection **A16:** 476
Age hardenable alloys
microstructural features of precipitation **A9:** 651
Age hardening *See also* Precipitation hardening
aluminum alloys **M2:** 30-32, 33, 36-38, 40-42, 43
beryllium-copper alloys **A2:** 405-408, 421-422
copper alloys **A2:** 236 **M2:** 256
defined **A9:** 1 **A13:** 1
development of .... **A3:** 1 • 25
maraging steels **M1:** 445-446, 448-449
mechanically alloyed oxide alloys **A2:** 947
dispersion-strengthened (MA ODS)
nickel and nickel alloys **A4:** 907, 911-912 **M4:** 757, 758-759
process .... **A3:** 1 • 22
of uranium alloys **A2:** 674
Age hardening (of grease)
defined **A18:** 2
Age hardening treatment **A9:** 646
Age-hardenable alloys
effects of solution heat treatment **A11:** 122
Age-hardenable nickel-base wrought heat- resistant alloys **A9:** 309
Age-hardenable stainless steels
thermal expansion coefficient **A6:** 907
Age-life history **EM3:** 735-736
Aged, and equalized
heat treatment **A13:** 934
Agency-related properties
data sheets **EM2:** 409
Agents *See* Antistatic agents; Coupling agents; Foaming agents; Mold release agent; Release agent
AGGIE computer program for structural analysis **EM1:** 268, 269
Agglomerate
definition **EM4:** 632
Agglomerate (noun)
defined **A7:** 1
Agglomerate (verb)
defined **A7:** 1
Agglomerate size distribution
in atomizing systems **A7:** 76
in spray drying **A7:** 73-74
Agglomerated particle pattern **A7:** 186
Agglomeration *See also* Deagglomeration; Thermal agglomeration
avoidance in silver powders **A7:** 147
by spray drying **A7:** 73-74, 76
chemical reactions in **A7:** 58
and grain refinement **A15:** 106
as milling process **A7:** 62
in oxide reduction **A7:** 52
of point defects and interstitials **A9:** 116
of precious metal powders **A7:** 149
sieves and **A7:** 216
Aggregate **EM3:** 4
defined **EM1:** 3 **EM2:** 3
Aggregate (noun)
defined **A7:** 1
Aggregate (verb)
defined **A7:** 1
Aggregate molding materials *See also* Clays; Molding aggregates; Molding materials Plastic materials; Sand mixes; Sand(s) **A15:** 208-211
aluminum silicates **A15:** 209-210
bentonites **A15:** 210
chromite **A15:** 209
clays **A15:** 210-211
fireclay **A15:** 210-211
olivine **A15:** 209
plastics **A15:** 211

Aggregate molding materials (continued)
sand mixes, additions to **A15:** 211
sands **A15:** 208-210
silica sands **A15:** 208-209
Southern bentonite **A15:** 210
Western bentonite **A15:** 210
zircon **A15:** 209
Aggregate properties approach *See also* Design; Properties
computer data bases **EM2:** 411
data sheet alternatives **EM2:** 411
data sheet limitations **EM2:** 410-411
data sheets **EM2:** 407-411
design information **EM2:** 407
Aggregated two-phase structures
defined **A9:** 604
Aggregates
molecular, in complexometric titrations **A10:** 164
polycrystalline, x-ray topographic analysis **A10:** 365
x-ray powder diffraction analysis of **A10:** 333-343
Aggregation, particle
effects **A15:** 144
Aggressive environments
and fatigue resistance **A1:** 677, 681
Aggressive tack **EM3:** 4
Aging *See also* Age hardening; Aging temperature; Artificial aging; Artificial weathering; Environmental effects; Heat aging; Natural aging; Overaging; Precipitation; Weather resistance **A1:** 641-642, 946-947 **EM3:** 4
accelerated **EL1:** 677-678
Alnico alloys **A9:** 539
aluminum alloys **A15:** 761
artificial, defined (under artificial aging) **A13:** 2
artificial, wrought aluminum alloy **A2:** 40
and cooling stress, effects **EM2:** 751
and deep drawing **A14:** 575
defined **A8:** 1 **A9:** 1 **A13:** 1 **EM1:** 3 **EM2:** 3
dilation during, cast copper alloys **A2:** 360
double **A12:** 34, 47
effect, long-term reliability **EM2:** 788-789
effect, mechanical properties **EM2:** 756
effect on carbide precipitation in wrought heat-resistant alloys **A9:** 311
effect on embrittlement **A12:** 34
effect on iron fracture surfaces **A12:** 458-459
effect on microstructure of titanium alloys **A9:** 461
effect on toughness of austenitic stainless steel **A1:** 947
effects, corrosion-resistant high-alloy **A15:** 728
effects in iron-chromium-cobalt alloy autocorrelograms for **A10:** 599, 600
effects, rheological plot **EL1:** 835
in electrical resistance alloys **A2:** 822
and elevated-temperature failures **A11:** 267
of epoxy resin matrices **EM1:** 77
of glass-polyester composites **EM1:** 93, 95
as hardening, defined **A2:** 762
and high-strain behavior **EM2:** 757
isothermal **EM2:** 566
long-term, polymer die attach **EL1:** 219
magnesium alloys **M4:** 745, 746, 747
materials, FIM/AP study of nucleation growth, and coarsening of precipitates in **A10:** 583
natural and artificial **A11:** 87
natural, wrought aluminum alloy **A2:** 39-40
nickel alloys **A12:** 397
nonferrous high-temperature materials **A6:** 572-574
over-, kinetics of **A10:** 317
physical, in polymers **A11:** 758
and physical properties **EM2:** 756
polymer, liquid chromatography monitoring of stability during **A10:** 649
Aging (continued)
of polymers, and failure analysis **EM2:** 732
precipitates, in gravimetric sample preparation **A10:** 163
quench, as embrittlement **A12:** 129-130
of shape memory alloys **A2:** 900
silicon steels, as magnetically soft materials **A2:** 769
stainless steel casting alloys **A6:** 497, 498
strain **A14:** 12, 547
strain, as embrittlement **A12:** 129-130
styrene-maleic anhydrides (S/MA) **EM2:** 219
TEM for **A12:** 129
temperature **EM2:** 569, 751
time, effect on zinc alloy tensile strength **A2:** 532
of titanium alloys **A14:** 842
and transition behavior **EM2:** 757
in Unicast process **A15:** 252
weather **EM2:** 575-580
wrought titanium alloys **A2:** 615, 619-620
Aging temperature *See also* Fabrication characteristics; Temperature(s)
aluminum casting alloys **A2:** 157-177
beryllium-copper alloys **A2:** 407
cast copper alloys **A2:** 356-391
uranium alloys **A2:** 680
wrought aluminum and aluminum alloys **A2:** 62-122
Aging temperatures for precipitation-hardenable stainless steels **A9:** 285
Aging time
effect on constructional steels **M1:** 656-658
Agitation
acid cleaning **M5:** 64
air *See* Air agitation
alkaline cleaning process **M5:** 11, 578
anodizing process **M5:** 593
bath, effect on ladle desulfurization **A15:** 78
chemical brightening process **M5:** 580
chromium plating process **M5:** 178
cleaning, for urethane coatings **EL1:** 777
copper plating process **M5:** 162-166
electroless nickel plating **M5:** 236
electropolishing processes **M5:** 305
emulsion cleaning process **M5:** 35
mechanical, for grain refinement **A15:** 476-477
mechanism of action **M5:** 4
melting bath, vacuum induction furnace **A15:** 397
nickel plating process **M5:** 206-207, 212, 214
paint dipping process **M5:** 481
pickling process **M5:** 73
radiographic film processing **A17:** 352
solvent cleaning processes **M5:** 41-42, 44
tin-lead plating process **M5:** 277-278
ultrasonic cleaning process **M5:** 4
in wave soldering **EL1:** 689
zinc alloy **A15:** 787
zinc cyanide plating process **M5:** 248, 252
Agitation during solidification
effect on dendritic structures in copper alloy ingots **A9:** 637-638
Agitation, effect
zinc corrosion in distilled water **A13:** 760
Agitation of
quenching media **M4:** 33, 47, 48, 49, 60-61
factors controlling agitation **M4:** 60-61
measurement of velocity **M4:** 61
molten salt **M4:** 49, 60
oil flow **M4:** 47, 48, 49, 60
turbulent agitation **M4:** 61
variables affecting agitation **M4:** 61, 62, 64-65
water and brine **M4:** 60
Agitator
defined **A7:** 1
Agree life
tests **EL1:** 494, 499
Agricultural applications
of copper-based powder metals **A7:** 733
homopolymer/copolymer acetals **EM2:** 101
powders used **A7:** 572
spraying, atomization mechanism of **A7:** 27
for stainless steels **A7:** 730
thermoplastic polyurethanes (TPUR) **EM2:** 205
urethane hybrids **EM2:** 268
Agricultural equipment
ductile iron **M1:** 36, 47
hardenable steels **M1:** 458-459
implements, wear testing of **M1:** 600
Agricultural machinery components
hardened steel for **A1:** 456
Agricultural mass finishing media **M5:** 135
Agricultural materials *See also* Food products
use of ICP-AES for **A10:** 31
Agricultural products
arsenic toxicity of **A2:** 1237
AGT *See* Advanced gas turbine program, DOE Office of Transportation Systems
Aileron
bonded **EM3:** 562-563
Air *See also* Atmospheres; Atmospheric corrosion; Oxygen
analytic methods for **A10:** 8
assay, for toxic elements, NAA for **A10:** 233
beryllium corrosion in **A13:** 808-809
conditions, intermittent immersion tests **A13:** 223
cooling, ultrasonic testing **A8:** 247
crack growth in **A11:** 54
as cutting fluid for tool steels **A18:** 738
direct sampling by AAS **A10:** 43
dry, as converter gas **A15:** 426
dry, for fracture preservation **A12:** 73
environment, effect on torsional fatigue testing **A8:** 152
explosivity of **A7:** 194
fatigue crack growth rate of nickel-base superalloy in **A8:** 412-413
humid, fracture effects **A12:** 72
inlet, annular, in Whiting cupola **A15:** 30
introduction, to cupola **A15:** 29
maximum service temperatures, stainless steels **A13:** 558
mean free path **A17:** 59
mean free path value at atmospheric conditions **A18:** 525
microwaves propagated through **A17:** 202
nickel alloy cracking in **A12:** 396
ovens, temperature control in **A8:** 36
as particle medium **A17:** 101
physical properties as atmosphere **A7:** 341
pollutants, GFAAS analysis **A10:** 58
tantalum corrosion rate in **A13:** 736
turbulence, effect in optical holographic interferometry **A17:** 413
uranium oxidation rate in **A13:** 813
UV/VIS trace analysis of **A10:** 60
and vacuum, fatigue fractures in **A12:** 48, 55
vs. vacuum environment at elevated temperatures **A8:** 412-413
Air acetylene welding
definition **M6:** 1
Air agitation
copper plating process **M5:** 162, 164-166
Air agitation cleaning
disadvantages of **M5:** 578, 580
Air aspiration
of atomized aluminum powder **A7:** 125
Air atomization *See also* Atomization
of aluminum powders **A7:** 130
of tin powders **A7:** 123-124
Air bearing *See also* Gas lubrication; Pressurized gas lubrication
defined **A18:** 2
Air bearings
coordinate measuring machines **A17:** 24

Air bend die
defined **A14:** 1
Air bending
for press-brake forming **A14:** 536
Air blast (pressure) abrasive blasting systems **M5:** 87, 90, 92-93
Air blast cleaning
of fractures **A12:** 74
Air bubbles *See also* Bubbles
in iron castings **A11:** 357
Air cap
definition **M6:** 1
Air carbon arc cutting
carbon absorption **A14:** 734
electrodes **A14:** 733
pipe fabrication **A14:** 732
power/air supplies **A14:** 732-733
rough cutting **A14:** 732
technique **A14:** 733-734
Air carbon arc cutting and gouging **M6:** 918-920
absorption of carbon **M6:** 920
air supply **M6:** 918-919
cutting action **M6:** 918
definition **M6:** 1401
electrodes **M6:** 919
power supply **M6:** 918
rough cutting **M6:** 918
technique **M6:** 919-920
Air carbon arc process
cast irons **A6:** 712
Air chambers
of cupolas **A15:** 29
Air channel
defined **A15:** 1
Air classification
defined **A7:** 1
versus weighting factor **EM4:** 85
Air conditioners
powders used **A7:** 572
Air conditioning
as casting market **A15:** 34
Air contaminant control
schematic **EL1:** 781
Air coolers
finned tubing for **A11:** 628
Air cooling *See also* Cooling
with fans **EL1:** 309-310
vs liquid cooling **EL1:** 50
Air core inductors
passive devices **EL1:** 1005
Air cutting gun (gas metal arc cutting)
definition **A6:** 1206
Air damping
effects **EM1:** 212
Air dried
defined **A15:** 1
Air dry enamel **M5:** 501
Air drying
of rammed graphite molds **A15:** 273
of thick-film circuits **EL1:** 249
Air feed
definition **M6:** 1
Air filters
GFAAS atomizers as **A10:** 58
PIXE analysis of **A10:** 102
powders used **A7:** 572
Air flow directionality
through honeycomb structures **EM1:** 728
Air Force
airplane damage tolerance requirements (MIL-A-83444) **EM3:** 504
contract to Boeing for surface preparation qualitative procedure evaluation **EM3:** 802, 803
Large Area Composite Structure Repair (LACOSR) program **EM3:** 821, 826
primer development efforts **EM3:** 254
repair procedures in contracts F33615-73-C-5171 and F33615-76-C-3137 ..... **EM3:** 807
repair test program (advanced composites structures article) **EM3:** 844
Air Force Wright Aeronautical Laboratories **A8:** 724
Air furnace *See also* Reverberatory furnace
for commercially pure iron **A2:** 764
defined **A15:** 1
Air furnace (continued)
pig iron, early use **A15:** 25
Air gap
defined **EM2:** 3
Air hammer
defined **A18:** 2
Air hammers *See also* Air-lift hammers; Hammers
for drop hammer forming **A14:** 654
for heat-resistant alloys **A14:** 234
Air hardenability test **A1:** 465-466 **M1:** 473
Air hardening of low-alloy steels **A1:** 644-645, 646
Air heaters
flue-gas corrosion in **A11:** 619
Air heaters, high-temperature
corrosion of **A13:** 998-999
Air hole
defined **A15:** 1
Air jet
for aramid fiber **EM1:** 115
for texturized fiberglass fabric **EM1:** 111
Air knives *See also* Hot air solder leveling (HASL)
debridging hot **EL1:** 691-693
defined **EL1:** 691
wave soldering **EL1:** 702
wipe-off **EL1:** 683
Air lances
for ash removal **A11:** 619
Air melting **A7:** 25
ultrahigh-strength steels **M1:** 426-429, 437, 439-441
Air permeater (Blaine permeater) **A7:** 264
Air plasma spraying (APS)
molten particle deposition **EM4:** 204, 205
Air ring
defined **EM2:** 3
Air setting
defined **A15:** 1
Air shotted copper **A7:** 106, 107
Air spray devices
for lubricant application **A14:** 515
Air tool exhaust muffler
powders used **A7:** 573
Air tools
for knockout **A15:** 504-505
Air unloading
of presses **A14:** 500
Air vent **EM1:** 4, 168
defined **EM2:** 3
Air venting *See also* Venting
die casting **A15:** 291
Air vibrators *See also* Vibrators
development **A15:** 28
Air-acetylene flame atomizer **A10:** 48
Air-assist forming
defined **EM2:** 3
Air-bubble void *See also* Voids **EM3:** 4
defined **EM1:** 3-4 **EM2:** 3
Air-carbon arc cutting (CAC-A) **A6:** 1104, 1105, 1172, 1177
air supply requirements (minimum) **A6:** 1174
aluminum **A6:** 1172, 1173, 1176
aluminum alloys **A6:** 1176
aluminum bronze **A6:** 1176
aluminum-nickel-bronze **A6:** 1176
applications **A6:** 1172
automatic air-carbon arc U-shaped groove operating data **A6:** 1176
carbon steels **A6:** 1172, 1175, 1176
cast iron **A6:** 1176
copper alloys **A6:** 1172, 1176
current ranges for various electrode sizes **A6:** 1175
definition **A6:** 1206
description of process **A6:** 1172
ductile iron **A6:** 1172, 1176
electrodes **A6:** 1172, 1173-1174, 1175, 1176, 1177
equipment selection **A6:** 1174
automatic systems **A6:** 1174
gouging torches **A6:** 1174
power sources **A6:** 1174
vacuum gouging **A6:** 1174
gray iron **A6:** 1172
heat-affected zone **A6:** 1176
high-carbon steels **A6:** 1176
low-alloy steels **A6:** 1175
magnesium alloys **A6:** 1176
malleable iron **A6:** 1172, 1176
Air-carbon arc cutting (CAC-A) (continued)
nickel **A6:** 1172, 1176
noise level and welding safety **A6:** 1192
nonferrous metals **A6:** 1172, 1173-1174
operating techniques **A6:** 1173-1174
beveling **A6:** 1174
gouging with manual torches **A6:** 1173, 1174
severing techniques **A6:** 1173-1174
washing **A6:** 1174
principles of operation **A6:** 1172-1173
carbon electrode **A6:** 1172, 1173, 1175
compressed air **A6:** 1172-1173
control of automatic gouging torches **A6:** 1173
gouging torch **A6:** 1173, 1175
power sources **A6:** 1172
process variables of importance **A6:** 1174-1176
safety and health **A6:** 1176-1177
electrical power **A6:** 1176
electrodes **A6:** 1176
fire and burn hazards **A6:** 1177
personal and protective equipment and clothing **A6:** 1176-1177
torches **A6:** 1176
ventilation hazards **A6:** 1177
stainless steels **A6:** 1172, 1176
suggested viewing filter plates **A6:** 1191
Air-cooled integral heat exchanger
as thermal control **EL1:** 47, 54-55
Air-dried strength
defined **A15:** 1
Air-drying lacquers
use of **M5:** 626-627
Air-film-metal interference effect **A9:** 136
Air-fired fracture surface
XPS survey **A10:** 577
Air-hardening alloy
microstructure **A7:** 488
Air-hardening medium-alloy cold-work steels
composition limits **A18:** 735
Air-hardening medium-alloy tool steels
for hot-forging dies **A18:** 623, 624
Air-hardening steels
service temperature of die materials in forging **A18:** 625
Air-hardening steels, medium alloy
cold work tool steels **A1:** 763-765
Air-hardening steels, medium-alloy *See* Medium-alloy air-hardening steels
Air-knife terne coating system **M5:** 359-360
Air-lift hammer *See also* Drop hammer; Gravity hammer **A14:** 1, 25
Air-melted alloys
counter-gravity low-pressure casting **A15:** 317-319
Air-mounted
punches **A7:** 323
Air-moving systems
explosion proof **A7:** 197
Air-operated molding machines
development **A15:** 29
Air-oxidizing coatings **M5:** 500-501
Air-slip forming
defined **EM2:** 3
Air-slip thermoforming **EM2:** 401
Air-water mist spray
of steel castings **A15:** 312
Airblasting *See* Abrasive cleaning; Blast cleaning; Blasting
Airborne dust
as contaminant **EL1:** 661
Airborne particulates, effects
coordinate measuring machines **A17:** 27
Airborne radars
and microwave holography **A17:** 228
Aircraft
accelerometer **A13:** 1121
airframes **A13:** 1019-1036
bilges, corrosion in **A13:** 1032
bonded airframe structures **A13:** 1034-1035
case histories/failures **A13:** 1045-1054
catapult-hook attachment fitting **A11:** 88, 91
codes governing **M6:** 824-825
corrosion fatigue **A13:** 1031
corrosion-related failures **A13:** 1022-1035
crevice corrosion **A13:** 1025-1026
deck plate, service failure by fatigue cracking **A11:** 311-312

Aircraft (continued)
engine air-intake assembly, fatigue fracture ... **A11:** 310-311
engine governor, fatigue fracture ... **A11:** 308-309
engines, reciprocating, fatigue cracking ... **A11:** 477
erosion-corrosion ... **A13:** 1034-1035
fighter, stress-corrosion failure of clamp for ... **A11:** 309
filiform corrosion ... **A13:** 107, 1028-1030
fretting corrosion ... **A13:** 1030
fretting wear ... **A18:** 242
fuel-tank floors, fatigue factors of ... **A11:** 126-127
galvanic corrosion ... **A13:** 1022-1025
hot-salt SCC ... **A13:** 1039
intergranular corrosion ... **A13:** 1028
jet-impingement experiments, corrosive wear ... **A18:** 274-275
landing gear component, SCC failure of ... **A11:** 213
material/process corrosion solutions ... **A13:** 1020-1022
microbial growth ... **A13:** 120
microbiological corrosion ... **A13:** 120, 1031-1032
powerplants, corrosion ... **A13:** 1037-1045
pressure cabin, cracking ... **A13:** 1020
propellor blade, fatigue fracture ... **A11:** 125
shaft, steel, corrosion-fatigue cracking ... **A11:** 260-261
stress-corrosion cracking ... **A13:** 1026-1028
structural corrosion failures ... **A13:** 1046-1054
wheel half, fatigue fracture from subsurface defect ... **A11:** 330-331
wheel half, forged, fatigue cracking ... **A11:** 323
wing clamp, forging failure from burning ... **A11:** 334-335
wing nut, intergranular fracture ... **A11:** 29
wing slat track, bending distortion of ... **A11:** 140-141
wing-attachment bolt, cracked along Seam ... **A11:** 530, 532
Aircraft accelerometer
corrosion failure analysis of ... **EL1:** 1111-1112
Aircraft alloys
thermal coefficient of expansion for ... **EM1:** 716
Aircraft applications
bearings, steels for ... **M1:** 606
cold-finished steel bars ... **M1:** 221
investment casting ... **A15:** 265
liquid crystal polymers (LCP) ... **EM2:** 180
polyether-imides (PEI) ... **EM2:** 156
ultrahigh-strength steels ... **M1:** 424, 427, 429, 434, 441
Aircraft brakes, friction and wear of ... **A18:** 582-586
aircraft friction materials ... **A18:** 582
brake characteristics ... **A18:** 582
carbon brakes ... **A18:** 582, 583, 584-586
brake wear ... **A18:** 585
coefficient of friction ... **A18:** 585
friction coefficient variability ... **A18:** 586
mechanical properties ... **A18:** 584-585
moisture problems ... **A18:** 585
oxidation ... **A18:** 585, 586
physical properties ... **A18:** 584-585
processing ... **A18:** 584
raw materials ... **A18:** 584
thermal properties ... **A18:** 584-585
vibration ... **A18:** 586
wear debris analysis ... **A18:** 585
wear mechanism ... **A18:** 585
steel brakes ... **A18:** 582-584
balance ... **A18:** 583
brake design ... **A18:** 583
brake wear ... **A18:** 583
chemistry ... **A18:** 582-583
coefficient of friction ... **A18:** 583-584
friction material selection ... **A18:** 583
processing ... **A18:** 583
service life ... **A18:** 584
spalling ... **A18:** 584
Aircraft brakes, friction and wear of (continued)
thermal properties ... **A18:** 583
vibration ... **A18:** 584
wear mechanisms ... **A18:** 583
wear rates ... **A18:** 583
testing and its requirements ... **A18:** 586
Aircraft composite applications ... **EM1:** 801-809
air-frame components ... **EM1:** 34
carbon-carbon composite ... **EM1:** 922-924
composite components used ... **EM1:** 801-809
current production ... **EM1:** 802-807
early commercial ... **EM1:** 801-802
military ... **EM1:** 804-809
thermoset/thermoplastic trade-offs for ... **EM1:** 100
of towpregs ... **EM1:** 152
transport, flight service evaluations ... **EM1:** 826-831
Aircraft construction
joining processes ... **M6:** 56-57
Aircraft cord wire ... **A1:** 286 **M1:** 269
Aircraft drift measurements
powder used ... **A7:** 572
Aircraft engine components
surface finish requirements ... **A16:** 22
Aircraft engines
components ... **A7:** 647, 649
Aircraft gas turbine
development of ... **A1:** 995
Aircraft industry ... **A13:** 1019-1057
composite applications ... **EM1:** 801-809, 922-924
composite fabrication for ... **EM1:** 34
cost savings, from composites ... **EM1:** 97
damage tolerance requirements ... **EM1:** 259-267
full-scale tests for ... **EM1:** 346-351
Aircraft industry applications *See also* Aerospace industry applications
acoustic emission inspection ... **A17:** 290-292
adhesive-bonded aluminum honeycomb structures, neutron radiography ... **A17:** 392-393
airframe, titanium and titanium alloy castings ... **A2:** 587, 634, 644
airframes, and NDE reliability ... **A17:** 664, 680-681
aluminum and aluminum alloys ... **A2:** 12
aluminum-lithium alloys ... **A2:** 182
Clevis/Lua attachments, eddy current bushing inspection ... **A17:** 192-193
components, neutron radiography of ... **A17:** 392
components, ultrasonic inspection ... **A17:** 232
of electric current perturbation ... **A17:** 138-140
engine components, eddy current inspection ... **A17:** 189-194
engine structural maintenance plan ... **A17:** 666
fasteners ... **A17:** 191-192
fluorescent magnetic particle inspection ... **A17:** 103
fracture control/damage tolerance ... **A17:** 666-673
galvanic exfoliation corrosion of aluminum wing skins ... **A17:** 191
machine vision ... **A17:** 44
of magnetic rubber inspection ... **A17:** 123, 125
microwave holography, for concealed weapons ... **A17:** 226
nickel alloys ... **A2:** 430
on-aircraft eddy current inspection ... **A17:** 191-194
splice joints, eddy current inspection ... **A17:** 193
structural, eddy current inspection ... **A17:** 189-194
subassemblies, eddy current inspection ... **A17:** 190-191
Aircraft quality ... **A1:** 254
of low-alloy steel ... **A1:** 209
Aircraft quality alloy steel
bars ... **M1:** 209
sheet and strip ... **M1:** 164
wire rod ... **M1:** 256
Aircraft quality plates ... **A1:** 237
Aircraft sealants
suppliers ... **EM3:** 59
Aircraft structural assemblies
adhesive bonded joints ... **EM3:** 743-745
Aircraft structural quality
of low-alloy steel ... **A1:** 209
Aircraft structures
fracture mechanics of ... **A8:** 459
Aircraft structures, repair of advanced composite commercial ... **EM3:** 829-837
damage types ... **EM3:** 829, 830
inspecton techniques used ... **EM3:** 829, 830
repair development ... **EM3:** 829-833
graphite-polyimide composite materials ... **EM3:** 832-833
materials graphite-epoxy composite ... **EM3:** 830-832
scarf repairs ... **EM3:** 829-830
repair durability ... **EM3:** 833-837
baseline results ... **EM3:** 835
exposure and test plan ... **EM3:** 834
exposure results ... **EM3:** 835-836
outdoor exposure test setup ... **EM3:** 834-835
program synopses ... **EM3:** 836-837
tabbed laminate specimen ... **EM3:** 834
Aircraft wings
compression testing of ... **A8:** 55
AiResist 13
composition ... **A4:** 795 **A6:** 929 **A16:** 737
machining ... **A16:** 738, 741-743, 746-758
AiResist 213
composition ... **A4:** 795 **A6:** 929
machining ... **A16:** 738, 741-743, 746-747, 749-758
AiResist 215
composition ... **A4:** 795 **A6:** 929 **A16:** 737
machining ... **A16:** 738, 741-743, 746-758
Airfoil
microstructure ... **A7:** 563
Airfoils
electrochemical machining ... **A16:** 540
Airframe components
titanium and titanium alloy ... **A2:** 587, 634, 644
Airframe industry
advanced composites for ... **EM1:** 34
Airframe structures ... **A14:** 150, 246
Airframes
aircraft ... **A13:** 1019-1036
and NDE reliability ... **A17:** 664, 680-681
Airless abrasive blast wheel systems ... **M5:** 86-87, 90, 92-93
Airless spray
as coating application technique ... **A13:** 416
Airless spray devices
for lubricant application ... **A14:** 515
Airless spraying
paint ... **M5:** 478, 494
Airport runways
microwave holographic visualization ... **A17:** 226-227
Airwash separator
use in dry blasting ... **M5:** 88-89
AISC *See* American Institute of Steel Construction
AISI *See* American Iron and Steel Institute
AISI carbon and alloy steels
compositions ... **A9:** 177
AISI designations *See* SAE-AISI designations
AISI specifications *See also* AISI-SAE specifications; Steels, AISI specific types
constructional steels for elevated temperature use ... **M1:** 647, 649
AISI tube steels
compositions ... **A9:** 211
AISI-SAE specifications *See also* Steels, AISI-SAE specific types
alloy steels, discussion of ... **M1:** 126-127
carbon steels, discussion of ... **M1:** 125-126
composition ranges and limits ... **M1:** 125-132
designation system ... **M1:** 124-127
hot rolled bars, equivalent AMS specifications ... **M1:** 209
hot rolled bars, equivalent ASTM specifications ... **M1:** 208
ultrahigh-strength steels ... **M1:** 421-428, 431-434

SUBJECTS OF THE INDEXED VOLUMES: ASM Handbook (designated by the letter "A"): **A1:** Properties and Selection: Irons, Steels, and High-Performance Alloys (1990); **A2:** Properties and Selection: Nonferrous Alloys and Special-Purpose Materials (1990); **A3:** Alloy Phase Diagrams (1992); **A4:** Heat Treating (1991); **A6:** Welding, Brazing, and Soldering (1993); **A7:** Powder Metallurgy (1984); **A8:** Mechanical Testing (1985); **A9:** Metallography and Microstructures (1985); **A10:** Materials Characterization (1986); **A11:** Failure Analysis and Prevention (1986); **A12:** Fractography (1987); **A13:** Corrosion (1987); **A14:** Forming and Forging (1988); **A15:** Casting (1988); **A16:** Machining (1989); **A17:** Nondestructive Testing and Quality Control (1989); **A18:** Friction, Lubrication, and Wear Technology (1992). **Metals Handbook, 9th Edition** (designated by the letter "M"): **M1:** Properties and Selection: Irons and Steels (1978); **M2:** Properties and Selection: Nonferrous Alloys and Pure Metals (1979); **M3:** Properties and Selection: Stainless Steels, Tool Materials and Special-Purpose Materials (1980); **M4:** Heat Treating (1981); **M5:** Surface Cleaning, Finishing, and Coating (1982); **M6:** Welding, Brazing, and Soldering (1983). **Engineered Materials Handbook** (designated by the letters "EM"): **EM1:** Composites (1987); **EM2:** Engineering Plastics (1988); **EM3:** Adhesives and Sealants (1990); **EM4:** Ceramics and Glasses (1991); **Electronic Materials Handbook** (designated by the letters "EL"): **EL1:** Packaging (1989).

AISI-SAE steels *See* Steels, AISI-SAE
AISI/SAE alloy steels *See also* AISI/SAE alloy steels, specific types; Steel(s)
austenitization effects **A12:** 339
bolts, spontaneous rupture **A12:** 299
Charpy impact fracture **A12:** 338
drill pipe, corrosion fatigue fracture **A12:** 291
fractal analysis **A12:** 212-214
fractographs **A12:** 291-344
fracture surface, pure tensile fatigue **A12:** 342
fracture/failure causes illustrated **A12:** 216
high-strength low-alloy, fracture surfaces **A12:** 344
hydrogen flaking **A12:** 125
quasi-cleavage facets, dimples, and voids **A12:** 330
spontaneous sulfide-SCC fracture **A12:** 299
temper embrittlement **A12:** 134
AISI/SAE alloy steels, specific types *See also* AISI/SAE alloy steels; Steel(s)
AISI 304 (SUS 304), effect of frequency and wave form effect on fatigue properties **A12:** 62
AISI 508 B60, torsional fatigue fracture **A12:** 323
AISI 1040 bolts, quench cracks **A12:** 149
AISI 1040 bolts, SCC failure **A12:** 151
AISI 1070, transverse fracture **A12:** 142, 163
AISI 1085, cathodic cleaning **A12:** 75
AISI 1085, ultrasonic cleaning **A12:** 74
AISI 1340, corrosion fatigue fracture **A12:** 291
AISI 4130, brittle fracture **A12:** 291
AISI 4130, effect of stress intensity factor range on fatigue crack growth rate **A12:** 57
AISI 4130, fatigue fracture surface **A12:** 293
AISI 4130, frequency and wave form effects on fatigue properties **A12:** 59
AISI 4130, hydrogen-embrittled **A12:** 31
AISI 4130, metallographic study, hitchpost failure **A12:** 292
AISI 4140, ductile fracture **A12:** 298
AISI 4140, embrittlement by liquid cadmium **A12:** 30, 39
AISI 4140, fatigue fracture surface **A12:** 295
AISI 4140, fracture surface, near weld toe **A12:** 294
AISI 4140, improper heat treatment **A12:** 298
AISI 4140, microstructures, with temper embrittlement **A12:** 153
AISI 4140, service fracture **A12:** 297
AISI 4140, tire tracks **A12:** 23
AISI 4142, splitting **A12:** 106
AISI 4146, improper induction hardening **A12:** 300
AISI 4150, star and beach marks **A12:** 301
AISI 4315, hydrogen embrittlement **A12:** 301
AISI 4315, tension overload fracture **A12:** 301
AISI 4320, fatigue failure **A12:** 111, 120
AISI 4340, Charpy impact fractures **A12:** 314
AISI 4340, dimples, SEM fractograph **A12:** 207
AISI 4340, dimples with inclusions **A12:** 65, 67
AISI 4340, effect of frequency and wave form on fatigue properties **A12:** 59
AISI 4340, effect of lead on fracture morphology **A12:** 30, 38
AISI 4340, effect of stress intensity factor range on fatigue crack growth rate **A12:** 57
AISI 4340, effects of decreasing stress **A12:** 315
AISI 4340, embrittlement **A12:** 214
AISI 4340, fatigue fracture surface **A12:** 303
AISI 4340, fractal analyses **A12:** 212-215
AISI 4340, fractal dimensions **A12:** 213
AISI 4340, fractographic analysis **A12:** 302
AISI 4340, fracture appearance, impact energy vs test temperature **A12:** 109
AISI 4340, fretting wear **A12:** 308
AISI 4340, hammer blow mechanical failure **A12:** 305
AISI 4340, hydrogen damage **A12:** 302
AISI 4340, hydrogen embrittlement **A12:** 306
AISI 4340, improper heat treatment **A12:** 309
AISI 4340, low-cycle fatigue fracture **A12:** 308
AISI 4340, mating fracture surface **A12:** 310
AISI 4340, mating segments, fatigue fracture **A12:** 315-316

AISI/SAE alloy steels, specific types (continued)
AISI 4340, pre-existing crack as fracture origin **A12:** 65
AISI 4340, profile angular distributions **A12:** 203
AISI 4340, quasi-cleavage in hydrogen-embrittled **A12:** 31
AISI 4340, radial fracture **A12:** 312
AISI 4340, roughness parameters **A12:** 213
AISI 4340, service failure **A12:** 317
AISI 4340, stringers on fracture surface **A12:** 67
AISI 4340, tensile fracture **A12:** 103
AISI 4340, tension overload fracture **A12:** 304, 311-313
AISI 4340, true profile length values **A12:** 200
AISI 4340, unfavorable grain flow **A12:** 67-68
AISI 4615, high-cycle bending fatigue fracture **A12:** 321
AISI 4817, fatigue fracture surface **A12:** 322
AISI 4817, rotating bending fatigue
AISI 4817, subcase fatigue cracking **A12:** 322
AISI 5046, fatigue zone, subcase fatigue cracking **A12:** 323
AISI 5132, mating fracture surface **A12:** 323
AISI 5140H, effect of strain rate on fracture appearance **A12:** 31, 41
AISI 5160, ribbonlike inclusions **A12:** 326
AISI 5160 wire spring, fracture from seam **A12:** 63, 64
AISI 5160H, fracture from seam **A12:** 326
AISI 6150, fatigue fracture surface **A12:** 327
AISI 8617, bending fatigue fracture **A12:** 329
AISI 8620, bending fatigue fracture **A12:** 330
AISI 8620, fatigue striations **A12:** 331
AISI 8620, spalling fatigue fracture **A12:** 329
AISI 8620, torsional overload fracture **A12:** 330
AISI 8640, beach marks and final fast fracture **A12:** 331
AISI 8640, service fracture surface **A12:** 331
AISI 8645, fatigue fracture surface **A12:** 333
AISI 8740, decohesive rupture **A12:** 24
AISI 8740, tensile-overload fracture **A12:** 334
AISI 9254, brittle intergranular fracture **A12:** 335
AISI 9310, fatigue fracture, reversed cyclic bending **A12:** 120
AISI 9310, inclusion in service fracture surface **A12:** 66
AISI 52100, effect of rapid heating in austenitizing **A12:** 328
AMS 6434, impact fracture **A12:** 319
AMS 6434 steel sheet, tension overload fracture **A12:** 319
AMS 6434, stress-corrosion cracking **A12:** 320
AMS 6434, tension overload fracture **A12:** 319
Cr-V alloy, high-cycle fatigue fracture **A12:** 337
D6B, fracture surface **A12:** 343
fracture **A12:** 321
SAE 21-4N (EV 8) steel, photo-illumination effects **A12:** 87
SAE 51 B60 railroad spring, torsion failure **A12:** 121
SAE 81 B45, fatigue fracture surface **A12:** 328
SAE 4150, overtempering **A12:** 301
SAE 4150, reversed torsional fatigue fracture **A12:** 301
SAE 5160, impact fracture, with mating surface **A12:** 324-325
Ajax Metal Company (Philadelphia) **A15:** 32
AKS-doped tungsten
properties **A2:** 578
Al-2.5% Mg
weld microstructures **A6:** 53
AL-905XL
composition **A6:** 1037
furnace brazing **A6:** 1040
AL-9052
composition **A6:** 1037
furnace brazing **A6:** 1040
Al-As (Phase Diagram) ..... **A3:** 2 • 40
Al-Au (Phase Diagram) ..... **A3:** 2 • 41
Al-Ba (Phase Diagram) ..... **A3:** 2 • 41
Al-Be (Phase Diagram) ..... **A3:** 2 • 41
Al-Bi (Phase Diagram) ..... **A3:** 2 • 42
Al-Ca (Phase Diagram) ..... **A3:** 2 • 42
Al-Cd (Phase Diagram) ..... **A3:** 2 • 42
Al-Ce (Phase Diagram) ..... **A3:** 2 • 43
Al-Co (Phase Diagram) ..... **A3:** 2 • 43
Al-Cr (Phase Diagram) ..... **A3:** 2 • 43

Al-Cr-Fe (Phase Diagram) ..... **A3:** 3 • 8
Al-Cr-Mg (Phase Diagram) ..... **A3:** 3 • 8-9
Al-Cr-Mn (Phase Diagram) ..... **A3:** 3 • 9
Al-Cr-Ni (Phase Diagram) ..... **A3:** 3 • 9
Al-Cr-Ti (Phase Diagram) ..... **A3:** 3 • 9
Al-Cu (Phase Diagram) ..... **A3:** 2 • 44
Al-Cu-Fe (Phase Diagram) ..... **A3:** 3 • 9-10
Al-Cu-Mn (Phase Diagram) ..... **A3:** 3 • 10-11
Al-Cu-Ni (Phase Diagram) ..... **A3:** 3 • 11-12
Al-Cu-Si (Phase Diagram) ..... **A3:** 3 • 12
Al-Cu-Zn (Phase Diagram) ..... **A3:** 3 • 12-13
Al-Er (Phase Diagram) ..... **A3:** 2 • 44
Al-Fe (Phase Diagram) ..... **A3:** 2 • 44
Al-Fe-Mn (Phase Diagram) ..... **A3:** 3 • 13-14
Al-Fe-Ni (Phase Diagram) ..... **A3:** 3 • 14-15
Al-Fe-Si (Phase Diagram) ..... **A3:** 3 • 15-16
Al-Fe-Zn (Phase Diagram) ..... **A3:** 3 • 16
Al-Ga (Phase Diagram) ..... **A3:** 2 • 45
Al-Gd (Phase Diagram) ..... **A3:** 2 • 45
Al-Ge (Phase Diagram) ..... **A3:** 2 • 45
Al-H (Phase Diagram) ..... **A3:** 2 • 46
Al-Hg (Phase Diagram) ..... **A3:** 2 • 46
Al-Ho (Phase Diagram) ..... **A3:** 2 • 46
Al-In (Phase Diagram) ..... **A3:** 2 • 47
Al-La (Phase Diagram) ..... **A3:** 2 • 47
Al-Li (Phase Diagram) ..... **A3:** 2 • 47
Al-Mg (Phase Diagram) ..... **A3:** 2 • 48
Al-Mg-Mn (Phase Diagram) ..... **A3:** 3 • 17
Al-Mg-Si (Phase Diagram) ..... **A3:** 3 • 17-18
Al-Mg-Zn (Phase Diagram) ..... **A3:** 3 • 18-19
Al-Mn (Phase Diagram) ..... **A3:** 2 • 48
Al-Mn-Si (Phase Diagram) ..... **A3:** 3 • 19
Al-Mo-Ni (Phase Diagram) ..... **A3:** 3 • 20
Al-Mo-Ti (Phase Diagram) ..... **A3:** 3 • 20
Al-Nb (Phase Diagram) ..... **A3:** 2 • 48
Al-Nd (Phase Diagram) ..... **A3:** 2 • 49
Al-Ni (Phase Diagram) ..... **A3:** 2 • 49
Al-Ni-Ti (Phase Diagram) ..... **A3:** 3 • 20-21
Al-Pb (Phase Diagram) ..... **A3:** 2 • 49
Al-Pd (Phase Diagram) ..... **A3:** 2 • 50
Al-Pr (Phase Diagram) ..... **A3:** 2 • 50
Al-Pt (Phase Diagram) ..... **A3:** 2 • 50
Al-S (Phase Diagram) ..... **A3:** 2 • 51
Al-Sb (Phase Diagram) ..... **A3:** 2 • 51
Al-Se (Phase Diagram) ..... **A3:** 2 • 51
Al-Si (Phase Diagram) ..... **A3:** 2 • 52
Al-Si-Zn (Phase Diagram) ..... **A3:** 3 • 21-22
Al-Sn (Phase Diagram) ..... **A3:** 2 • 52
Al-Sr (Phase Diagram) ..... **A3:** 2 • 52
Al-Ta (Phase Diagram) ..... **A3:** 2 • 53
Al-Te (Phase Diagram) ..... **A3:** 2 • 53
Al-Th (Phase Diagram) ..... **A3:** 2 • 53
Al-Ti (Phase Diagram) ..... **A3:** 2 • 54
Al-Ti-V (Phase Diagram) ..... **A3:** 3 • 22
Al-U (Phase Diagram) ..... **A3:** 2 • 54
Al-V (Phase Diagram) ..... **A3:** 2 • 54
Al-W (Phase Diagram) ..... **A3:** 2 • 55
Al-Y (Phase Diagram) ..... **A3:** 2 • 55
Al-Yb (Phase Diagram) ..... **A3:** 2 • 55
Al-Zn (Phase Diagram) ..... **A3:** 2 • 56
Al-Zr (Phase Diagram) ..... **A3:** 2 • 56
Al_2O_3
solderable and protective finishes for substrate materials **A6:** 979
Al_3Ti
as grain refinement compound **A15:** 105-108
Alarms, light and sound
eddy current readout **A17:** 178
Albite (Na_2O-Al_2O_3-$6SiO_2$) **EM4:** 6
crystal structure **EM4:** 882
framework structure **EM4:** 759
purpose for use in glass manufacture **EM4:** 381
volume expansion coefficient **EM4:** 761
Alcanodox anodizing process
aluminum and aluminum alloys **M5:** 592
Alceram **EM4:** 1095
Alclad
chemical milling and scribing **A16:** 580
electrode potential in NaCl-H_2O_2 solution **A6:** 730
Alclad 3003
electrode potential in NaCl-H_2O_2 solution **A6:** 730
relative rating of filler alloys for welding **A6:** 731, 732, 733, 734, 735
weldability **A6:** 534

Alclad 3004
relative rating of filler alloys for welding ... **A6:** 731, 732, 733, 734, 735
weldability ... **A6:** 534
Alclad 6061
electrode potential in NaCl-H_2O_2 solution ... **A6:** 730
Alclad alloys
resistance welding ... **A6:** 848
Alclad products
alloying effects ... **A2:** 44
core and cladding combinations ... **M2:** 211
corrosion resistance ... **M2:** 210-211
Alclad wrought aluminum alloys
applications and properties ... **A2:** 82-86, 102, 115, 119
Alclad/alclad products ... **A13:** 1, 588
Alcoa
statistical technique for tensile properties ... **A8:** 662
Alcoa process
aluminum powder production ... **A7:** 127-129
for demagging aluminum alloys ... **A15:** 474
Alcogel
sol-gel transition role ... **EM4:** 210
Alcohol
defined ... **EM2:** 3
and perchloric acid (Group I electrolytes) ... **A9:** 52-54
role in etchants for wrought stainless steels ... **A9:** 281
Alcohol cleaners ... **M5:** 40-41, 44
Alcoholic ferric chloride as an etchant for tin ... **A9:** 450
Alcohols ... **EM3:** 4
as cleaning solvents ... **EL1:** 663
copper/copper alloy corrosion in ... **A13:** 635
determined by EFG ... **A10:** 216-217
as organic cleaning solvents ... **A12:** 74
for polyesters ... **EM1:** 132
and SCC in titanium and titanium alloys ... **A11:** 224
as solvent used in ceramics processing ... **EM4:** 117
titanium/titanium alloy SCC ... **A13:** 687-688
to control dusting ... **A18:** 684
Alcohols used in etchants ... **A9:** 67-68
nominal compositions ... **A9:** 68
Alcology *See* Copper alloys, specific types, C68800
Alcoloy
applications and properties ... **A2:** 336-337
Alconox
as ferrous and aluminum detergent ... **A12:** 74-75
Alcop (aluminum-copper bronze)
cage material for rolling-element bearings ... **A18:** 503
Aldehydes ... **EM3:** 4
copper/copper alloy corrosion in ... **A13:** 634
defined ... **EM2:** 3
functional group analysis of ... **A10:** 217
and ketones, determined ... **A10:** 217
physical properties ... **EM3:** 104
Alexandrine ... **EM4:** 18
ALEXTR CAD/CAM program for hot extrusion ... **A14:** 323-325
Alfalfa seeds
magnetically cleaned ... **A7:** 589
Alfenol
photochemical machining etchant ... **A16:** 590
Alfesil
roll welding ... **A6:** 314
Alfesil, unsuitability as cladding ... **M6:** 691
aluminum alloys ... **M6:** 1030
Alginates
removal ... **EM4:** 137
Algorithmic test generation methods
system-level ... **EL1:** 375
Algorithms
adhesive bond strength classifier ... **A17:** 611
CT image reconstruction ... **A17:** 359-360
Algorithms (continued)
defined ... **A17:** 383
for EIC count, in design ... **EL1:** 513
Hightower ... **EL1:** 531
Lee ... **EL1:** 529-531
pattern-fit ... **EL1:** 531-532
routing ... **EL1:** 529
Aliasing ... **A18:** 294-295
defined ... **A17:** 383
of image artifacts ... **A17:** 375-376
of variable effects ... **A17:** 747
Aligned discontinuous fibers *See also* Discontinuous fiber composites ... **EM1:** 153-156
Aligning bearing
defined ... **A18:** 2
Alignment *See also* Misalignment ... **A18:** 351
changes, effect on fatigue cracking ... **A11:** 475
defined ... **A9:** 1
fiber, in composites ... **A11:** 731
in metal casting ... **A15:** 40
mold ... **A15:** 190
shaft, and bearing failure ... **A11:** 507
and specimen buckling, axial compression testing ... **A8:** 55
tooling ... **A14:** 160
vertical, for Scleroscope hardness testing ... **A8:** 105
Alignment, measurement
by interferometer ... **A17:** 14-15
Alignment of silicon boule
for cutting along crystallographic planes ... **A10:** 342
Alignment techniques
for recovered carbon fiber ... **EM1:** 153-155
Aliphatic amines, and curing agents
epoxies ... **EL1:** 827
Aliphatic hydrocarbons ... **EM3:** 4
defined ... **EM2:** 3
to control dusting ... **A18:** 684
Aliphatic petroleum cleaners ... **M5:** 40-41
Aliphatic polyester adducts ... **EM3:** 100
Aliphatic polyol epoxies/epoxy esters ... **EM2:** 272
Aliquot
defined ... **A10:** 668
Alkali alumina borosilicate glass
Corning glass code 8111 derived ... **EM4:** 463
Alkali aluminosilicate glass, applications
electronic processing ... **EM4:** 1059
laboratory and process ... **EM4:** 1087
Alkali and alkaline earth silicate glasses
thermal conductivity ... **EM1:** 47
Alkali catalyst
phenolics ... **EM3:** 104
Alkali halides
ESR studied ... **A10:** 263
as sample in Raman analysis of metal oxides ... **A10:** 131
Alkali hydroxides
for polishing amphoretic metals ... **A9:** 54
Alkali metal
addition to fluxes affecting ionization process ... **A6:** 57
Alkali metal removal
by flux injection ... **A15:** 453
from aluminum melts ... **A15:** 79-80
from cast iron and steels ... **A15:** 74
processes for ... **A15:** 470-471
Alkali metals
in alkaline cleaners ... **M5:** 23-24
defined ... **A13:** 1
eluent suppression technique for ... **A10:** 660
fire ... **A13:** 96
flame source emissions for trace analyses ... **A10:** 29-30
optical emission spectroscopy ... **A10:** 21, 29-30
reaction with oxygen ... **A13:** 94
solvent extractants for ... **A10:** 170
Alkali oxides
role in glazes ... **EM4:** 1062
Alkali silicate coatings
water-soluble ... **M5:** 533-534
Alkali silicate frits
melting/fining ... **EM4:** 392
Alkali silicate glasses
density ... **EM4:** 846
electrical properties ... **EM4:** 851, 852
heat capacity ... **EM4:** 847-848
modifier effect ... **EM4:** 845
optical properties ... **EM4:** 854
self-diffusion coefficients of alkali ions ... **EM4:** 461
thermal expansion ... **EM4:** 847
viscosity ... **EM4:** 848
Alkali zinc borosilicate glass, properties
non-CRT applications ... **EM4:** 1048-1049
Alkali zinc silicate glass
applications, information display ... **EM4:** 1049
properties, non-CRT applications ... **EM4:** 1048-1049
ceram ... **EM4:** 1048-1049
opal ... **EM4:** 1048-1049
Alkali-borate glasses
chemical properties ... **EM4:** 855
optical properties ... **EM4:** 854
properties ... **EM4:** 846, 847, 848, 851
Alkali-borosilicate glasses
ground coat enamels ... **EM4:** 1065-1066
leachable, composition ... **EM4:** 427
leachable, processing ... **EM4:** 427
properties ... **EM4:** 1057
properties, non-CRT applications ... **EM4:** 1048-1049
Alkali-free dielectric glass, properties
non-CRT applications ... **EM4:** 1048-1049
Alkali-free phosphate glasses
electrical properties ... **EM4:** 852
Alkali-germanate glasses
chemical properties ... **EM4:** 855, 856
electrical properties ... **EM4:** 852
mechanical properties ... **EM4:** 846, 847, 848-849
optical properties ... **EM4:** 854
Alkali-lime-silicate glass, applications
optical glass ... **EM4:** 1080
Alkali-metal aluminoborosilicate glasses
enameling ... **EM3:** 303
Alkali-metal containing glasses, applications
thick film circuits ... **EM4:** 1141
Alkali-oxide-containing glasses
ion exchange ... **EM4:** 460
Alkalies ... **A16:** 105
alloy steel corrosion ... **A13:** 544
copper/copper alloy resistance ... **A13:** 629-630
electrolytes for electrochemical machining ... **A16:** 536
magnesium/magnesium alloys in ... **A13:** 742
nickel-base alloy corrosion resistance in ... **A13:** 647
porcelain enamels resistance ... **A13:** 449, 451
stainless steel corrosion ... **A13:** 559
tantalum resistance to ... **A13:** 727-728
titanium/titanium alloy resistance ... **A13:** 680
zirconium/zirconium alloy resistance ... **A13:** 716
Alkaline
defined ... **A13:** 1
Alkaline aluminosilicate
as an addition to doped tungsten ... **A9:** 442
Alkaline batteries ... **A13:** 1318
Alkaline boilout solutions ... **A13:** 1141
Alkaline cleaners
for surface oxides ... **A13:** 381
Alkaline cleaning ... **M5:** 22-39
acid cleaning combined with ... **M5:** 61-62
acid-base titration ... **M5:** 25
agitation during, effects of ... **M5:** 11, 578
aluminum and aluminum alloys ... **M5:** 15, 24, 577-578, 590-591
carbonate cleaners ... **M5:** 23-24, 28, 35
chelating agents used in ... **M5:** 71
cleaner composition and operating additives, use in ... **M5:** 24
builders, use in ... **M5:** 23-24, 28, 35

Alkaline cleaning (continued)
conditions **M5:** 4, 23-25, 28, 35, 70-71, 578, 590-591, 603-604, 606
surfactants, use in **M5:** 24, 28, 35, 577
cleaning cycles **M5:** 7, 29, 36
cleaning mechanisms **M5:** 4, 23
dispersion **M5:** 23
emulsification **M5:** 23
saponification **M5:** 23
cold alkaline cleaners **M5:** 15
in cold extrusion **A14:** 304
copper and copper alloys **M5:** 617-620
cutting fluids removed by **M5:** 10
derusters in *See also* Alkaline descaling and derusting **M5:** 22
electrolytic *See* Electrolytic cleaning, alkaline
electropolishing processes **M5:** 304
equipment **M5:** 25, 30-31, 37-38
etching cleaners **M5:** 577-578
composition and operating conditions **M5:** 578
flow process **M5:** 23
free alkalinity titration **M5:** 25
hot dip galvanized coating process **M5:** 325
hydroxide cleaners **M5:** 24
immersion process
cleaner composition and operating conditions **M5:** 24
equipment **M5:** 25
pigmented drawing compounds removed by **M5:** 5-7
process **M5:** 22
rust and scale removal **M5:** 12
steel stampings, wire fabrications, and fasteners **M5:** 29
inhibited cleaners **M5:** 7, 10, 577
for liquid penetrant inspection **A17:** 81, 82
magnesium alloys ... **M5:** 630-631, 633, 635, 637-639, 645, 647
niobium **M5:** 663
nonetching cleaners **M5:** 577-578
compositions and operating conditions **M5:** 578
nonsilicated cleaners **M5:** 577
oils and grease removed by **M5:** 9, 577
phosphate cleaners **M5:** 23-24, 28
phosphate coating process **M5:** 439, 450-451, 453-454
pickling process, precleaning **M5:** 70-71, 81
pigmented drawing compounds removed by **M5:** 6-7
plating process precleaning **M5:** 5, 16-18
polishing and buffing compounds removed by **M5:** 10-11
porcelain enameling process **M5:** 514-516
process selection **M5:** 4-7, 9-13, 15-18, 21
rinsing process **M5:** 7, 9, 18, 23, 578
safety precautions **M5:** 21, 453-454
silicated *See* Silicate alkaline cleaners
soak process
cadmium plating systems **M5:** 263
cleaner compositions and operating conditions **M5:** 619, 630-631
copper and copper alloys **M5:** 618-619
electropolishing processes **M5:** 304
magnesium alloys **M5:** 630-631, 639, 642
oils and greases removed by **M5:** 9
pigmented drawing compounds removed by **M5:** 7
plating process precleaning **M5:** 17-18
polishing and buffing compounds removed by **M5:** 10-12
zinc alloy die castings **M5:** 677
solution composition and operating conditions **M5:** 325, 618-620
solution control and testing **M5:** 25, 29-30, 36-37
solution strength, effects of **M5:** 325
spray process **M5:** 5-7, 11-12, 22-25
cleaner composition and operating conditions **M5:** 24
equipment **M5:** 25
pigmented drawing compounds removed by **M5:** 5-7
polishing and buffing compounds removed by **M5:** 11-12
stainless steel **M5:** 561
steam process **M5:** 23

Alkaline cleaning (continued)
steel **M5:** 16-18, 24, 70-71, 81
surfactants used in **M5:** 24, 28, 35, 577
tanks, construction and equipment ... **M5:** 25, 30-31, 37-38
tantalum **M5:** 663
total alkalinity titration **M5:** 25
uninhibited cleaners **M5:** 7
wastewater treatment **M5:** 312-313
zinc **M5:** 24
Alkaline cleaning, of aluminum alloys
fatigue fracture from **A11:** 126-127
Alkaline descaling and derusting **M5:** 12-13
electrolytic **M5:** 12
Alkaline earth aluminoborosilicate glass, applications
information display **EM4:** 1045
Alkaline earth aluminosilicate cladding glass **EM4:** 1101
Alkaline earth aluminosilicate glass **EM4:** 1102
applications **EM4:** 1045, 1095
chemical corrosion **EM4:** 1047
properties, non-CRT applications.... **EM4:** 1048-1049
Alkaline earth borate glasses
thermal expansion **EM4:** 847
Alkaline earth boroaluminosilicate glass
chemical corrosion **EM4:** 1047
properties, non-CRT applications.... **EM4:** 1048-1049
Alkaline earth elements *See also* Rare earths
purification, from aluminum melts **A15:** 79-80
removal, from cast iron and steels **A15:** 74
Alkaline earth metal hydroxide catalyst
phenolics **EM3:** 104
Alkaline earth metals
complexometric titrations for **A10:** 164
eluent suppression technique for **A10:** 660
extractants for **A10:** 170
optical emission spectroscopy for **A10:** 21
Alkaline earth silicate
coefficient of thermal expansion **EM4:** 1102
composition **EM4:** 1102
softening point **EM4:** 1102
Alkaline earths
role in glazes **EM4:** 1062
specific properties imparted in CRT tubes **EM4:** 1039
Alkaline electroless nickel plating **M5:** 220-221
Alkaline electrolytes (Group VII electrolytes) **A9:** 53-54
Alkaline electrolytic brightening
aluminum and aluminum alloys **M5:** 582
Alkaline electrolytic stripping
chromium plate **M5:** 198
Alkaline environments
corrosion of nickel and cobalt in aqueous **A10:** 135
oxidation of silver electrodes in **A10:** 135
Alkaline etching
acid etching used with **M5:** 585
aluminum and aluminum alloys **M5:** 582-586, 590
bleed-out in **M5:** 584
desmutting process **M5:** 584
dimensional change in **M5:** 584-585
equipment and operating procedures **M5:** 583-584
rinsing process **M5:** 584
sequestrants used in **M5:** 583
sodium hydroxide process **M5:** 583
solution composition and control **M5:** 583
waste treatment **M5:** 583-584
Alkaline ferricyanide reagents *See also* Murakami's reagent
as an etchant for wrought stainless steels **A9:** 281
Alkaline hypochlorite solution **EM3:** 35
Alkaline oxide
chromate conversion coating **A13:** 394
Alkaline plating
copper *See* Copperplating, alkaline process
tin **M5:** 270-271
zinc *See* Zinc alkaline noncyanide plating
Alkaline potassium ferricyanide as an etchant for type 304 stainless steel **A9:** 65
Alkaline scale-conditioning
heat-resistant alloys **M5:** 564-566

Alkaline sodium picrate as an etchant for
austenitic manganese steel casting specimens **A9:** 239
carbon and alloy steels **A9:** 166
silicon iron alloys **A9:** 531
Alkaline solution cleaning
refractory metals and alloys **A2:** 563
Alkaline solutions **A13:** 722, 1141
cast iron resistance **A13:** 570
in chemical etching cleaning **A12:** 75
storage of **A9:** 51
Alkaline solutions composition **A7:** 459
Alkaline surface activating solutions **M5:** 642, 644, 646
Alkaline-etch cleaning process
pitting from **A11:** 127
Alkaline/phenolic/ester no-bake processes *See also* Coremaking; No-bake processes
as coremaking system **A15:** 238
Alkalinity **A18:** 84
of lead **A13:** 788
Alkalis
specific properties imparte in CTV tubes **EM4:** 1039
Alkalis, corrosion of nickel alloys in
See also Caustic solutions **M3:** 173-174
Alkenes
determined **A10:** 219
Alkyd modified DGEBA diacrylate
properties **EM3:** 92
Alkyd plastic **EM3:** 4
defined **EM1:** 4 **EM2:** 3-4
as injection-moldable **EM2:** 321
Alkyd resins **A13:** 400-403
as media for screening and stamping processes **EM4:** 475
Alkyd resins and coatings **M5:** 473-475, 498-499, 501, 505
modified **M5:** 475, 498-499
Alkyd-urethane no-bake resins **A15:** 216
Alkyds
defined **A13:** 1
topcoat, marine corrosion coatings **A13:** 914
as urethane coating **A13:** 410
Alkyl amines
as solvent extractants **A10:** 170
Alkyl aromatic
properties **A18:** 81
Alkyl mercury, as biologic indicator
mercury toxicity **A2:** 1249
Alkylation
defined **A13:** 1
Alkylcyanoacrylate adhesives
moisture ingression on joint performance **EM3:** 361
Alkyls catalysts
powder used **A7:** 572
All-ceramic mold casting
procedure **A15:** 249
All-chloride nickel plating **M5:** 202-203, 217
All-sulfate nickel plating **M5:** 201-202
All-weld-metal tensile test **A6:** 103
All-weld-metal tension tests **A6:** 103-104, 105
All-weld-metal test specimen
defined **A8:** 1
definition **M6:** 1
Allanophanate **EM3:** 4
formation of **EM3:** 203-204
Allen, Ethan
as early founder **A15:** 26
Allen, William
as early founder **A15:** 25
Allen's "golden rule" of powder sampling **EM4:** 83
Allentheses
orthopedic implants as **A11:** 670
Allergenicity *See also* Toxicity
of complex salts of platinum **A2:** 1258
Allergic hypersensitive reactions
dental alloys **A13:** 1337
Alligator shears
for bar **A14:** 714-715
for plate and flat sheet **A14:** 702
Alligator skin *See also* Orange peel
as casting defect **A11:** 384
Alligatoring **A6:** 240 **A13:** 1, 406
austenitic stainless steels **A12:** 351

Alligatoring (continued)
defined ... **A11:** 1
in rolled slab ... **A14:** 358
Alligatoring, defect *See also* Orange peel ... **A8:** 1, 595
Allison C-4
friction performance requirements for hydraulic fluids ... **A18:** 99
Allophanate
defined ... **EM2:** 4
Alloprene ... **EM3:** 4
defined ... **EM2:** 4
Allotriomorphic alpha (GB α) phase
titanium alloys ... **A6:** 84
Allotriomorphic crystal
defined ... **A9:** 1
Allotriomorphic ferrite ... **A6:** 77, 78-79
Allotropes
iron ... **A13:** 46, 48
Allotropic modification
pure tin ... **A13:** 770
Allotropic transformation
of zirconium and zirconium alloys ... **A2:** 665-666
Allotropic transformations
in pure metals ... **A9:** 655
Allotropy *See also* Graphite ... **A3:** 1 • 1 **EM3:** 4
defined ... **A9:** 1 **EM1:** 4 **EM2:** 4
Allowable depth of wear for sliding bearings ... **A18:** 515
Allowable stress (SA) ... **A6:** 389
Allowables *See* Design allowables; Lamina allowables
Allowance
defined ... **A15:** 1
Allowances *See also* Accuracy; Dimensional accuracy; Machining allowance; Tolerances
bend, for bar ... **A14:** 663
defined ... **A14:** 73
dimensional ... **A15:** 614-623
distortion ... **A15:** 193
for finish ... **A15:** 193, 621
machining, centrifugal casting molds ... **A15:** 302
machining, in ring rolling ... **A14:** 125-126
pattern ... **A15:** 192-193
radial forging ... **A14:** 148
for seamless rolled rings ... **A14:** 126-127
for shaving ... **A14:** 457, 470
for shift, three-roll forming ... **A14:** 619
shrinkage, for patterns ... **A15:** 192-193
shrinkage, magnesium alloys ... **A15:** 805
shrinkage, metals/alloys ... **A15:** 303
and tolerances, open-die forging ... **A14:** 71-73
Allowed (diagram) lines
x-radiation ... **A10:** 86
Alloy *See also* Alloy designation systems; Alloy selection; Alloying; Specific metals and alloys ... **EL1:** 42 **EM3:** 4
design, in ordered intermetallics ... **A2:** 913-914
development, aluminum-lithium alloys ... **A2:** 180-184
formation, rare earth metals ... **A2:** 726-727
glass frit attachment ... **EL1:** 734
as lead frame material ... **EL1:** 731
lead frames, for CERDIP packages ... **EL1:** 203-204
preparation, of niobium-titanium superconductors ... **A2:** 1043-1044
with shape memory effect ... **A2:** 897
special, low-expansion ... **A2:** 895
Alloy 25
composition ... **A6:** 598
Alloy 42
ceramic/metal seals ... **EM4:** 506
properties ... **EM4:** 503
Alloy 49
glass/metal seals ... **EM4:** 494
Alloy 52 (52Ni-Fe)
glass/metal seals ... **EM4:** 536-537
Alloy 150
composition ... **A6:** 598
Alloy 214
heat-affected-zone cracks ... **A6:** 93
solidification cracking ... **A6:** 88, 91
Alloy 625
composition ... **A6:** 573
mill annealing temperature range ... **A6:** 573
solidification cracking, weldability study ... **A6:** 89, 90
solution annealing temperature range ... **A6:** 573
Alloy 718
heat-affected-zone cracks ... **A6:** 93
Alloy additions *See also* Additives; Alloying elements; Composition control; Inoculation
acid steelmaking ... **A15:** 365
analysis, for electric furnace ... **A15:** 365
kinetics of ... **A15:** 71-74
liquid at melt temperature ... **A15:** 71-72
solid at melt temperature ... **A15:** 72-74
Alloy brazing ... **M6:** 962
Alloy carburizing steels
machinability ratings ... **M1:** 568, 579-580
Alloy cast iron
use for deep drawing dies ... **M3:** 496, 498
Alloy cast iron, austenitic
alloying elements, effects of ... **M1:** 75-80
automotive applications ... **M1:** 96
compositions ... **M1:** 76, 82
corrosion resistance in acids and alkalies ... **M1:** 90
inoculants, effects of ... **M1:** 80
mechanical properties ... **M1:** 85-87, 89, 92, 94, 95
oxidation ... **M1:** 92-94
patternmakers' rules for ... **M1:** 33
physical properties ... **M1:** 88
Alloy cast irons *See also* specific types of cast iron ... **A1:** 11, 85-104
abrasion-resistant cast irons ... **A1:** 11, 90-98
abrasion resistance ... **A1:** 96-98
annealing, effect of ... **A1:** 92
compositions ... **A1:** 85, 91
hardness, conversions ... **A1:** 95
hardness, of microconstituents ... **A1:** 97
heat treatment ... **A1:** 91-92
mechanical properties ... **A1:** 94
microstructure ... **A1:** 92-94, 95
physical properties ... **A1:** 96
relative toughness ... **A1:** 96
tensile strength ... **A1:** 95-96
transverse strength ... **A1:** 96
wear rates ... **A1:** 97-98
alloying elements, effects of ... **A1:** 11, 86-90
carbon ... **A1:** 86, 88
chromiumium ... **A1:** 86, 88-90, 100
copper ... **A1:** 89
manganese ... **A1:** 87
molybdenum ... **A1:** 89, 90
nickel ... **A1:** 89
on depth of chill ... **A1:** 87
on rate of growth ... **A1:** 101, 103
on scaling ... **A1:** 101-102, 103
phosphorus ... **A1:** 87-88
silicon ... **A1:** 86-87, 88, 100
sulfur ... **A1:** 87
vanadium ... **A1:** 89-90
arc welding ... **M6:** 316
arc welding of ... **A15:** 528-529
for automotive service ... **A1:** 103-104
classification of ... **A1:** 5
corrosion-resistant irons ... **A1:** 86
heat-resistant irons ... **A1:** 86
white cast irons ... **A1:** 85-86
corrosion-resistant cast irons ... **A1:** 11, 98-100
alloying elements, effects of ... **A1:** 98
compositions ... **A1:** 85
high-chromium irons ... **A1:** 99-100
high-nickel irons ... **A1:** 100
high-silicon irons ... **A1:** 98-99
mechanical properties ... **A1:** 99
Alloy cast irons (continued)
physical properties ... **A1:** 96
cutting tool material selection based on machining operation ... **A18:** 617
heat-resistant cast irons ... **A1:** 11, 100-104
alloy ductile irons ... **A1:** 103
alloying elements, effects of ... **A1:** 100
composition ... **A1:** 85
creep ... **A1:** 102
growth ... **A1:** 100-101
high-aluminum irons ... **A1:** 103
high-chromium irons ... **A1:** 103
high-nickel irons ... **A1:** 100, 103
high-silicon irons ... **A1:** 102-103
high-temperature strength ... **A1:** 102
mechanical properties ... **A1:** 99
physical properties ... **A1:** 96
scaling ... **A1:** 100, 101-102
inoculants, effects of ... **A1:** 90
oxidation of ... **A1:** 101
Alloy cast steel
die material for sheet metal forming ... **A18:** 628
Alloy casting *See* Cast irons; Cast steels
Alloy Casting Institute ... **A15:** 34
Alloy Casting Institute designations
for iron-chromium-nickel heat-resistant casting alloys ... **A9:** 330
for stainless steel ... **A9:** 298
Alloy composition *See also* Composition
single-phase alloys ... **A15:** 114
Alloy constructional steel
hardness conversion tables ... **A8:** 109-113
Alloy crown
(Au-Ag-Cu) proper-ties ... **A18:** 666
simplified composition or microstructure ... **A18:** 666
Alloy design use of phase diagrams in ... **A3:** 1 • 25-26
Alloy designation systems *See also* Alloy; Specific metals and alloys; Temper designation system
aluminum and aluminum alloys ... **A2:** 4-5, 15-16, 29, 32-33
aluminum casting alloys ... **A2:** 22-25, 123-126
aluminum-lithium alloys ... **A2:** 178, 180-184
ISO and Aluminum Association International, equivalents ... **A2:** 26
for magnesium and magnesium alloys ... **A2:** 1401-1402
magnetically soft materials ... **A2:** 763-778
for wrought unalloyed aluminum and wrought aluminum alloys ... **A2:** 17-21
Alloy development
aluminum alloys ... **A15:** 762
stress-corrosion cracking in ... **A8:** 495, 508
Alloy ductile irons ... **A1:** 103
Alloy loss *See* Alloying elements
Alloy M25 *See* Copper alloys, specific types, C17300
Alloy MA 754
mechanically alloyed oxide dispersion-strengthened (MA ODS) alloy ... **A2:** 944-945
Alloy MA 758
mechanically alloyed oxide dispersion-strengthened (MA ODS) alloy ... **A2:** 945-946
Alloy MA 760
mechanically alloyed oxide dispersion-strengthened (MA ODS) alloy ... **A2:** 947
Alloy MA 956
mechanically alloyed oxide dispersion-strengthened (MA ODS) alloy ... **A2:** 946
Alloy MA 6000
mechanically alloyed oxide dispersion-strengthened (MA ODS) alloy ... **A2:** 946-947
Alloy mechanical coatings ... **M5:** 300-301
Alloy oxidation ... **A13:** 73, 76

Alloy phase constitution
thermodynamic analysis ... **A15:** 101
Alloy phases that possess superlattices ... **A9:** 681
Alloy plating
defined ... **A13:** 1
Alloy powder, alloyed powder
defined ... **A7:** 1
Alloy segregation *See* Segregation
Alloy selection *See also* Alloy; Materials selection
beryllium-copper alloys ... **A2:** 416-421
for buried metals ... **A11:** 192
effect on SCC in aluminum alloys ... **A11:** 219
effect on SCC in marine-air environment ... **A11:** 309-310
for investment castings ... **A15:** 265
for niobium-titanium superconductors ... **A2:** 1043
and steel casting failure ... **A11:** 391
Alloy sensitization
evaluation ... **A13:** 218-219
Alloy sheet and strip
powders used ... **A7:** 574
Alloy steel *See also* Low-alloy steel
alloying elements in ... **A1:** 144-147, 456-457
bulk formability of ... **A1:** 581-590
composition-on ... **A1:** 152-153
definition of ... **A1:** 149
distortion in heat treatment ... **A1:** 369-370
embrittlement
aluminum nitride ... **A1:** 694-696
blue brittleness ... **A1:** 692
graphitization ... **A1:** 696-697
quench-age ... **A1:** 692-693
strain-age ... **A1:** 693-694, 695
fabrication of parts and assemblies ... **A1:** 463
hardenable alloy steels ... **A1:** 453
induction and flame hardening ... **A1:** 463
machinability of through-hardening ... **A1:** 600-601
mechanical properties ... **A1:** 457-458
temper embrittlement in ... **A1:** 698
tempering ... **A1:** 458-459, 462
weldability of ... **A1:** 609
Alloy steel bars *See* Cold finished bars, Hot rolled bars ... **A1:** 245-246
aircraft quality and magnaflux quality ... **A1:** 246
axle shaft quality ... **A1:** 246
ball and roller bearing quality and bearing quality ... **A1:** 246
cold-shearing quality ... **A1:** 246
cold-working quality ... **A1:** 246
quality descriptors ... **A1:** 253-254
regular quality ... **A1:** 246
structural quality ... **A1:** 246
Alloy steel billet
bleeding ... **A9:** 175
butt tears ... **A9:** 175
carbon spots ... **A9:** 175
center segregation ... **A9:** 174
flakes ... **A9:** 174
flute marks ... **A9:** 174
splash ... **A9:** 174
Alloy steel forging
center burst ... **A9:** 176
Alloy steel piling *See* Piling
Alloy steel plate *See also* Plate ... **M1:** 181, 183-184
ASTM specifications ... **M1:** 183-184, 186-189
mechanical properties ... **M1:** 190-193
Alloy steel rod ... **A1:** 275
qualities and commodities for ... **A1:** 275
special requirements for ... **A1:** 275-276
Alloy steel spring wire ... **A1:** 287
Alloy steel wire
fabrication characteristics ... **M1:** 587-593
oil-tempered, tensile strength ranges ... **M1:** 269
types ... **M1:** 269-270
Alloy steel wire rod
decarburization limits ... **M1:** 256-257
grain size ... **M1:** 257
hardenability ... **M1:** 257
heat-analysis limits ... **M1:** 256
inspection and testing ... **M1:** 257
qualities and commodities ... **M1:** 256
special requirements ... **M1:** 256-257
Alloy steels *See also* AISI/SAE alloy steels; Alloy(s); Alloying; ASTM/ASME alloy steels; Carbon and alloy steels; Heat-resistant alloy steels; High-alloy steels; High-strength low-alloy steels; High-strength medium-carbon quenched and tempered steels; Linepipe steels; Plate steels; Quenched and tempered high-strength alloy steels; specific types by designation or trade name; Steel(s) ... **A9:** 165-196 **A13:** 531-546 **A14:** 215-221 **A18:** 693 **EM4:** 968, 969, 971, 972
4000, 5000, 6000, 8000, and 9000 series ... **A16:** 93
abrasion resistance ... **M1:** 600, 602-605, 617-618, 620-621 **M3:** 582, 583
AISI compositions ... **A9:** 177
alloying at elevated temperatures ... **A13:** 538-539
aluminum coating of ... **M5:** 335, 342-343, 345-346
applications ... **M1:** 455, 470
in aqueous chloride solution ... **A8:** 405
arc welding ... **M6:** 247-306
atmospheric corrosion resistance ... **A13:** 532-533
bar ... **A14:** 148
bar and tube, die materials for drawing ... **M3:** 525
bar, hydrogen flaking in ... **A11:** 316
bearing applications compositions ... **M1:** 609-610
blanking, die materials for ... **M3:** 485-487
blue, brittleness of ... **M1:** 684
bolt applications, hardenability and cost comparison ... **M1:** 275-276
capacitor discharge stud welding ... **M6:** 730
carbon content effect on warm workability ... **A14:** 174
carbon content, effect on weldability ... **M1:** 561
and carbon steel galvanic couple ... **A13:** 544
carburized, heat treatment of ... **M1:** 532, 533, 539
carburized, toughness ... **M1:** 534-536
case hardenability ... **M1:** 534-535
case hardened, hardness profiles ... **M1:** 633-634
for case hardening, compositions of ... **A9:** 219
cast, patternmakers' rules for ... **M1:** 31, 33
cermet tools for milling ... **A16:** 96
cermets and grooving ... **A16:** 95
cermets and threading ... **A16:** 95
in chemical-processing industry ... **A13:** 544-545
classifications and designations ... **M1:** 117-143
quality descriptors ... **M1:** 118-119
cold extrusion, tool materials for ... **M3:** 515-517
cold heading of ... **A14:** 291
composition ranges and limits
AISI-SAE standard grades ... **M1:** 127-131
bars ... **M1:** 121, 127-129
billets ... **M1:** 121, 127-129
blooms ... **M1:** 121, 127-129
grades formerly listed by SAE ... **M1:** 133-134
H-steels ... **M1:** 129-131
plate ... **M1:** 122, 129, 136 140
SAE experimental grades ... **M1:** 131-132
slabs ... **M1:** 121, 127-129
compositions, AMS grades ... **M1:** 141-143
contour band sawing ... **A16:** 364
core hardenability ... **M1:** 534-535
corrosion, in specific end-use environments ... **A13:** 533-542
corrosion of ... **A13:** 531-546
corrosive environments ... **A13:** 531-532
cutting fluids used ... **A16:** 125
cutting tool material selection based on machining operation ... **A18:** 617
die forging, materials for ... **M3:** 529, 530, 532
dies ... **A14:** 86
distortion in heat treatment ... **M1:** 469-470
drilling ... **A16:** 231
electrogas welding ... **M6:** 239
embrittlement of ... **M1:** 684-686
etchants ... **A9:** 169-175
explosion welding ... **A6:** 896 **M6:** 710
fabric knives ... **M1:** 625
as fastener material ... **A11:** 530
for fine-edge blanking and piercing ... **A14:** 472
flakes in ... **A11:** 121
flame hardening ... **M1:** 532
flash welding ... **M6:** 557
flux cored arc welding ... **M6:** 298-299
for forging ... **A14:** 218
forging costs ... **M1:** 350, 351
forging effects on properties ... **A14:** 217
forging lubricants ... **A14:** 217-218
forging of ... **A14:** 215-221
forging temperatures ... **A14:** 81
friction welding ... **A6:** 153 **M6:** 721
gas metal arc welding ... **M6:** 299-301
Alloy steels (continued)
gas tungsten arc welding ... **M6:** 182, 203, 303-304
gas-metal arc welding shielding gases ... **A6:** 67
gas-tungsten arc welding ... **A6:** 190
gear shaving with high-speed steel tools ... **A16:** 342-343
granular bainite ... **A9:** 665
graphitization of ... **M1:** 686
grinding ... **A9:** 168
grinding by CBN wheels ... **A16:** 455
hardenability ... **M1:** 471-525
hardenability equivalence table ... **M1:** 484-488
hardenable ... **M1:** 455-470
hardening of ... **M1:** 459-460
hardness conversion tables for ... **A8:** 109-113
heat treatment ... **A14:** 218-219
high frequency resistance welding ... **M6:** 760
high-speed machining ... **A16:** 598
hobbing with high-speed steel tools ... **A16:** 345-346
honing ... **A16:** 477
hot dip tin coating ... **M5:** 351
for hot forging ... **A14:** 43, 215-217
hot swaging of ... **A14:** 142
for hot upset forging ... **A14:** 83, 86
hydrochloric acid corrosion ... **A13:** 1162
hydrogen embrittlement of ... **M1:** 687
ion plating of ... **M5:** 421
Jominy equivalent hardenability ... **M1:** 537
laser cutting ... **A14:** 741
lift pin, fatigue fracture in ... **A11:** 77
lifting-fork an-n, fracture of ... **A11:** 325-326
machinability ... **M1:** 239, 568, 574, 579-583
machinability test matrix ... **A16:** 639-640
machining ... **A2:** 967
macroetching ... **A9:** 170-177
marine applications ... **A13:** 539, 542-545
mechanical properties ... **A14:** 165
microalloyed forging ... **A14:** 219-221
microetching ... **A9:** 169
microstructure ... **M1:** 455
microstructures ... **A9:** 177-179
milling ... **A16:** 312, 313, 314
miscellaneous, figure numbers for ... **A12:** 216
miscellaneous, fracture causes illustrated ... **A12:** 216
mounting ... **A9:** 166-168
nickel plating of ... **M5:** 215-216, 218
nitrided ... **M1:** 540-542
notch toughness ... **M1:** 690, 693-698, 704-707, 709
oxyacetylene pressure welding ... **M6:** 595
oxyfuel gas cutting ... **A6:** 1159 **A14:** 724 **M6:** 904
photochemical machining ... **A16:** 591
physical properties ... **M1:** 145-151
pickling of ... **M5:** 68-70, 80
piston pins, wear influenced by surface finish ... **M1:** 635
plasma arc welding ... **A6:** 197 **M6:** 304-306
polishing ... **A9:** 168-169
polishing and buffing of ... **M5:** 108, 120
postweld heat treatments ... **M6:** 259, 296, 301
AISI-SAE steel bars ... **M6:** 291
quenched and tempered steels ... **M6:** 288-289
reduction of residual stresses ... **M6:** 890-891
precoated before soldering ... **A6:** 131
preheating ... **M6:** 259-261, 287, 296, 301
AISI-SAE steel bars ... **M6:** 291
effect on weldability ... **M6:** 259-261
electrode selection ... **M6:** 260-261
recommended heat inputs ... **M6:** 288-289
reduction of residual stresses ... **M6:** 890
product composition tolerances ... **M1:** 123
properties, historical study ... **A12:** 2
protection from corrosion ... **M1:** 751-759
recommended shielding gas selection for gas-metal arc welding ... **A6:** 66
resistance seam welding ... **M6:** 494
rough and finish cutting of bevel gears ... **A16:** 349
Schaeffler diagrams used to predict microstructures in dissimilar-metal welded joints ... **A9:** 582
in seawater, corrosion factors ... **A13:** 539
seawater, corrosion in ... **M1:** 739-746
sectioning ... **A9:** 165-166
selected grades, tensile and fatigue properties ... **M1:** 680
selection for hardenability ... **M1:** 482-491

Alloy steels (continued)
selection guide based on hardenability **M1:** 464-466
shaping gears with high-speed steel tools **A16:** 347-348
sheet and strip **M1:** 163-165
ASTM specifications **M1:** 164
composition **M1:** 163
mechanical properties **M1:** 165
mill heat treatment **M1:** 164-165
production of **M1:** 163
quality descriptors **M1:** 163-164
tolerances **M1:** 163
shielded metal arc welding **M6:** 297-298
soil corrosion **M1:** 725-731
specimen preparation **A9:** 165-169, 171
specimen preservation **A9:** 172
springs
grades for *See also* specific alloy designations **M1:** 284-285, 311-312
hardenability requirements **M1:** 297, 301-303, 311-312
steam service applications **M1:** 747
stud arc welding...... **A6:** 210, 211, 212, 213, 214 **M6:** 730
stud material **M6:** 730
submerged arc welding..... **A6:** 202, 203 **M6:** 301-303
temper embrittlement of **M1:** 684-685
thread grinding **A16:** 274
timing chain components, wear compared to carbon steel **M1:** 628
tubular products *See* Steel pipe; Steel tubes; Steel tubing
turning **A16:** 144, 146, 147
valve springs, fatigue fracture of **A11:** 551
weld overlay material **M6:** 807
weldability **M1:** 561-564
wire, die materials for drawing **M3:** 522
for wire-drawing dies **A14:** 336

Alloy steels (cast)
thermal expansion coefficient **A6:** 907

Alloy steels, specific types *See also* HSLA steels; SAE specific types; Steels, AISI-SAE specific types; Steels, ASTM specific types; Tool steels, specific types
0.03C-2Mn, lath martensite **A9:** 670
1.2C-0.5Cr-0.9Mo-0.2V, bar, hot-rolled **A9:** 194
1Cr-$^1/_2$Mo, thermal expansion **M1:** 653
2$^1/_4$Cr-1Mo, elevated temperature behavior **M1:** 639-641, 653-662
9Cr-1Mo, elevated temperature properties **M1:** 649, 651-653
12H2N4A gear, ion carburized and diffused **A9:** 225-226
51B60, bar, hot-rolled, different heat treatments compared **A9:** 195
2340, austenitized, isothermally transformed, upper bainite plates **A9:** 663
3310 bar, austenitized, cooled and tempered **A9:** 225
3310, gas carburized **A9:** 221
3310, pack carburized and cooled in the pot **A9:** 224
3310H, gas carburized, different etches compared **A9:** 222-223
4118, gas carburized, with decarburized surface layer **A9:** 223
4118, gas carburized, with grain boundary oxidation **A9:** 222
4118, hot rolled, annealed and gas nitrided **A9:** 228
4130, bar, hot-rolled **A9:** 191
4130, normalized by austenitizing **A9:** 191
4140, bar, different heat treatments compared **A9:** 191
4140, flow lines in a forged hook **A9:** 176
4140, gas nitrided **A9:** 228
4140, ion nitrided **A9:** 229
4140, isothermal transformation **A9:** 191

Alloy steels, specific types (continued)
4140, oxide inclusions **A9:** 191
4140, resulfurized, forging **A9:** 191
4320, gas carburized **A9:** 222
4340, total dilation vs. transformation temperature during isothermal formation of bainite **A9:** 662
4350, bar, partial isothermal transformation **A9:** 191-192
4360, austenitized, isothermally transformed, lower bainite **A9:** 665
4360, austenitized, isothermally transformed, upper bainite **A9:** 663
4620, gas carburized and hardened **A9:** 223
4620, gas carburized at 1.00% carbon potential **A9:** 220-221
4620, pack carburized and cooled in the pot **A9:** 224
5132, forging, austenitized and water quenched **A9:** 192
8617 bar, annealed by austenitizing **A9:** 224
8617, carbonitrided **A9:** 226
8620 bar, carbonitrided **A9:** 227
8620 bar, normalized by austenitizing and cooled in still air **A9:** 225
8620, gas carburized, butterfly alterations at microcracks **A9:** 223
8620, gas carburized, with grain boundary oxidation **A9:** 222
8620H, gas carburized **A9:** 222
8620H, ion carburized **A9:** 225
8620H tubing, gas carburized, hardened and tempered **A9:** 224
8620H, vacuum carburized **A9:** 225
8720, hot-rolled, gas carburized **A9:** 221-222
8822H bar, austenitized and cooled **A9:** 225
8822H, gas carburized **A9:** 222
8822H, gas carburized, roller for contact- fatigue test **A9:** 223
9310 bar, normalized by austenitizing and cooled **A9:** 225
9310, gas carburized, different surface carbon contents compared **A9:** 220
52100, bar, different heat treatments compared **A9:** 195-196
52100, bar, different magnifications compared **A9:** 195
52100, damaged by an abrasive cutoff wheel **A9:** 196
52100, rod, austenitized and slack quenched in oil **A9:** 196
52100, roller, crack from a seam in bar stock **A9:** 196
AF 1410, lath martensite under different illuminations **A9:** 81
AMS 6419, different heat treatments compared **A9:** 192
AMS 6470, gas nitrided, different processing procedures compared **A9:** 227-228
API 5L-X60, pipe, resistance weld **A9:** 178
ASTM A517, grade J, arc butt weld **A9:** 178
C-$^1/_2$Mo, thermal expansion **M1:** 653
Fe-5Cr-1Mo-2Cu-0.5P-2.5C, pressed and sintered **A9:** 530
Fe-16Ni, lath martensite **A9:** 670
Fe-20Ni, lath martensite **A9:** 670
Fe-32Ni, plate martensite **A9:** 671
Fe-33.5Ni, plate martensite in an austenitic single crystal **A9:** 671

Alloy system
defined **A9:** 1

Alloy wire **A1:** 286-287

Alloy(s) *See also* Alloy steels; Alloying
and alloy surfaces **A13:** 46
binary, oxide scale growth **A13:** 75-76
blending **A7:** 69
chemistry, as metallurgically influenced corrosion **A13:** 123

Alloy(s) (continued)
coated composite powders **A7:** 174
"psuedo" **A7:** 147
defined **A13:** 46
development, doping principle **A13:** 73
with high SCC resistance **A13:** 1103
hydrogen embrittlement relative resistance **A13:** 1104
hydrogen susceptibilities **A13:** 288
with low SCC resistance **A13:** 1104
metal couples, compatibility **A13:** 1040
with moderate SCC resistance **A13:** 1103
oxidation: doping principle **A13:** 73
passive, crevice corrosion of **A13:** 303
purity **A7:** 179
single phase and polyphase **A13:** 46
steels, sintering of *See also* Stainless steels; Steels **A7:** 366
as surgical implants **A13:** 1325
system, and homogenization **A7:** 315
used in electronics systems **A13:** 1108

Alloy-environment systems
stress-corrosion cracking **A13:** 146, 326

Alloy-junction transistor
development and packaging **EL1:** 958

Alloy-tin couple test
for tinplate **A13:** 781-782

Alloyable coatings
as surface preparation **EL1:** 679

Alloyed steel
solderability **A6:** 978

Alloyed tungsten carbides **A7:** 773
properties **A7:** 776
selection of **A7:** 777

Alloying *See also* specific alloying elements
in A15 superconducting materials **A2:** 1062-1063
aluminum coatings **A13:** 434
at elevated temperatures **A13:** 538-539
for atmospheric corrosion prevention **A13:** 83, 512-515
as austenite stabilizers **A13:** 47
carbon steels, for atmospheric corrosion **A13:** 512-515
and compressibility, iron powder **A14:** 190
content, in cold extrusion **A14:** 301
and corrosion resistance **A13:** 47-48, 323, 693
development, for powder forging **A14:** 189
effect, formability of aluminum alloys **A14:** 791
effect of, on hardenability and tempering of steel **A1:** 392-394, 395, 396, 468-469
effect, on anodic polarization of uranium **A13:** 816
effect on fatigue strength **A11:** 119
effect on magnetic properties **A2:** 762
effect on SCC in copper alloys **A11:** 220-221
effect on SCC in magnesium alloys **A11:** 223
elements, volatilizaton **A13:** 344
equipment **A7:** 723
excessive, and brazed joint failures **A11:** 452
as ferrite stabilizers **A13:** 47
galvanized steel **A13:** 433
and impurity specifications, aluminum alloy ingot **A2:** 16
laser surface **A13:** 501-504
liquid-metal corrosion by **A13:** 56, 59, 92
in marine atmospheres **A13:** 540, 906
mechanical **A7:** 23, 723
nickel aluminides **A2:** 914-915, 920
with noble metal, as selective oxidation **A13:** 73
palladium **A13:** 799-800
platinum **A13:** 798
primary phase or matrix, defined **A13:** 46
problems **A13:** 48
rare earth metals **A2:** 727-729
resistance to cavitation erosion **A18:** 217
secondary phase, or precipitate, defined **A13:** 46
of silver **A13:** 794-795

SUBJECTS OF THE INDEXED VOLUMES: ASM Handbook (designated by the letter "A"): **A1:** Properties and Selection: Irons, Steels, and High-Performance Alloys (1990); **A2:** Properties and Selection: Nonferrous Alloys and Special-Purpose Materials (1990); **A3:** Alloy Phase Diagrams (1992); **A4:** Heat Treating (1991); **A6:** Welding, Brazing, and Soldering (1993); **A7:** Powder Metallurgy (1984); **A8:** Mechanical Testing (1985); **A9:** Metallography and Microstructures (1985); **A10:** Materials Characterization (1986); **A11:** Failure Analysis and Prevention (1986); **A12:** Fractography (1987); **A13:** Corrosion (1987); **A14:** Forming and Forging (1988); **A15:** Casting (1988); **A16:** Machining (1989); **A17:** Nondestructive Testing and Quality Control (1989); **A18:** Friction, Lubrication, and Wear Technology (1992). **Metals Handbook, 9th Edition** (designated by the letter "M"): **M1:** Properties and Selection: Irons and Steels (1978); **M2:** Properties and Selection: Nonferrous Alloys and Pure Metals (1979); **M3:** Properties and Selection: Stainless Steels, Tool Materials and Special-Purpose Materials (1980); **M4:** Heat Treating (1981); **M5:** Surface Cleaning, Finishing, and Coating (1982); **M6:** Welding, Brazing, and Soldering (1983). **Engineered Materials Handbook** (designated by the letters "EM"): **EM1:** Composites (1987); **EM2:** Engineering Plastics (1988); **EM3:** Adhesives and Sealants (1990); **EM4:** Ceramics and Glasses (1991); **Electronic Materials Handbook** (designated by the letters "EL"): **EL1:** Packaging (1989).

Alloying (continued)
for surface stability of superalloys........ **A1:** 953, 955, 956-957, 958-959
titanium/titanium alloys **A13:** 693, 695
to modify as-cast properties in gray iron.. **A1:** 28-29
uranium/uranium alloys....................... **A13:** 814-815
wrought aluminum alloy................. **A2:** 36-37, 44-57
wrought copper and copper alloy products... **A2:** 241-242
zinc/zinc alloys and coatings **A13:** 759
Alloying additions
to austenitic manganese steel castings effects on microstructure....................... **A9:** 239
to modify eutectic phase chemistries and volume fractions............................ **A9:** 621
Alloying element
defined **A9:** 1
Alloying elements *See also* Alloy additions; individual elements by name; Inoculation; Minor elements
400 to 500 °C embrittlement, effect on susceptibility to..................................... **M1:** 686
500 °F embrittlement, effect on susceptibility to... **M1:** 685
activity in solution, calculated....................... **A15:** 55
aluminum alloys...................................... **A15:** 743-746
atmospheric corrosion resistance, effect on... **M1:** 721-722
cast iron.. **M1:** 76, 80
cast steel, effect on **M1:** 386-389, 393-399
concentration, thermodynamic effects...... **A15:** 55-60
constructional steels for elevated temperature use, effect on **M1:** 647-650
corrosion protection.. **M1:** 751
for ductile iron............... **A15:** 648-649 **M1:** 34-42, 45
effect, carbon solubility **A15:** 68
effect, hydrogen solubility in copper **A15:** 466
flame AAS analysis in steels.......................... **A10:** 56
general effects **M1:** 114-115
gray cast iron **M1:** 26-27, 28-30
in gray iron... **A15:** 639
hardenability affected by **M1:** 476-477
hardenable steels **M1:** 455-456, 459-460, 466-469
HSLA steels **M1:** 410-411
hydrogen solubility in steels, influence on.. **M1:** 687
influence, carbon activity.......................... **A15:** 61-70
interaction coefficients in FE-C-X alloys.. **A15:** 62
in liquid copper alloys, interaction coefficients **A15:** 60
loss, acid steelmaking.................................. **A15:** 365
loss, basic steelmaking **A15:** 367
low-alloy steel.. **A15:** 715-716
machinability affected by....................... **M1:** 580, 582
magnesium alloys **A15:** 802
maraging steels.. **M1:** 445
minor, in aluminum melts............................ **A15:** 79
P/M materials.................................. **M1:** 333, 334-335
particle buoyancy effects **A15:** 72
partitioning of, FIM/AP analysis................ **A10:** 583
pearlitic steel, effect on transition temperature **M1:** 417
rare earth, as grain refiners **A15:** 481
sampling trainload of metal pipe for percentage of... **A10:** 15
selective evaporation **A15:** 396
solubility effects, iron-carbon systems **A15:** 68-69
steel sheet, effects on formability........... **M1:** 554-557
temper embrittlement, effect on **M1:** 684-685
zinc alloys **A15:** 787-788
Alloying, localized
as failure mechanism.......................... **EL1:** 1013-1014
Alloying materials
as coatings **A7:** 174, 817
effect on compressibility **A7:** 286
refractory metals as... **A7:** 765
Alloying process, laser *See* Laser processing techniques, laser alloying
Alloys *See also* Aircraft alloys; Metals; Metals and alloys, characterization of; specific alloys by name
addition effect for three heats................. **A8:** 479-481
additions, and hydrogen embrittlement **A8:** 487

Alloys (continued)
additions, and stress-corrosion cracking.. **A8:** 487, 489
aerospace, fracture mechanics data **A11:** 54
analytic methods for.. **A10:** 4
in aqueous environments **A8:** 406-407
arc-welded, heat-resisting..................... **A11:** 433-434
and blends, applications **EM2:** 491
change to prevent corrosion.................. **A11:** 193-194
characterized **A10:** 1
chemical analysis by controlled-potential coulometry............... **A10:** 207
chemical vapor deposition coating of **M5:** 382-383
commercial, hydrochloric acid for sample dissolution of.......................... **A10:** 165
crack nucleation in.. **A8:** 366
defined **EM1:** 4 **EM2:** 4
effect on toughness and crack growth **A11:** 54
effects of composition on mass absorption **A10:** 97
engineering, classes...................................... **EM2:** 632
of engineering plastics......................... **EM2:** 632-637
ferrous, LME in.. **A11:** 234
FIM images of... **A10:** 589-590
FIM/AP study of point defects in **A10:** 583
formation, RBS analysis **A10:** 628
galvanic series.. **M5:** 431
gold plating, uses in **M5:** 282-283
hardenable, heat treatment of...................... **A11:** 140
heat-resisting, corrosion in **A11:** 200-201
high-purity, voltammetric analysis of......... **A10:** 188
high-temperature, dissolution mediums... **A10:** 166
ion-implanted, AEM microstructural analysis.. **A10:** 484-487
light, intergranular corrosion of.................. **A11:** 182
liquid, NMR analysis of electronic structure of.. **A10:** 284
metals in, quantitative determination by electrogravimetry........................... **A10:** 197
multicomponent, phase separation analysis by SAXS/SANS/SAS **A10:** 402
mutual solubility in **A11:** 452
nonferrous, liquid-metal embrittlement of.. **A11:** 233-234
ordered, effect of antiphase boundaries on FIM images.............................. **A10:** 589
ordered, ladder diagrams of **A10:** 594
perchloric acid as dissolution medium....... **A10:** 166
phosphate coating of **M5:** 437-438
polymer, environmental effects **EM2:** 431
prompt gamma activation analysis (PGAA) of **A10:** 240
segregation, banding and flakes from......... **A11:** 121
spark source excitation for elemental analysis of **A10:** 29
spot test kits for.. **A10:** 168
surfaces, FIM imaged **A10:** 590
and temperature effect on torsional ductility **A8:** 164-166
trace impurities detected **A10:** 31, 43
two-phase, SEM atomic number contrast analysis of **A10:** 508
unsuitable, for pressure vessels............ **A11:** 644-646
verification of inorganic solids, applicable methods.. **A10:** 4-6
workability behaviors...................................... **A8:** 165
Alloys and metals characterization *See* Metals and alloys, characterization of
Alloys, high-temperature
cemented carbide machining **A16:** 88
and surface integrity....................................... **A16:** 22
Allyl
uses and properties..................................... **EM3:** 126
Allyl glycidyl ether (AGE)
as epoxy diluent **EM1:** 67, 70
Allyl plastic
defined **EM1:** 4
Allyl resins... **EM3:** 4
defined **EM2:** 4
Allylics *See* Allyl resins; Allyls
Allyls (DAP, DAIP) *See also* Thermosetting resins
applications **EM2:** 226-227
characteristics... **EM2:** 227-229
commercial forms... **EM2:** 226
competitive materials **EM2:** 227

Allyls (DAP, DAIP) (continued)
costs **EM2:** 226
as low-temperature resin system **EM2:** 440
processing **EM2:** 228-229
product forms **EM2:** 229
production volume **EM2:** 226
suppliers.. **EM2:** 229
thermoset **EM2:** 630
Alnico
honing stone selection **A16:** 476
Alnico alloy
fracture topography **A12:** 461
Alnico alloys *See also* Cast Alnico alloys; Magnetic materials; Magnetic materials, specific types; Permanent magnet materials, specific types
applications **A2:** 793
as commercial permanent magnet materials **A2:** 785-787
crystallographic texture developed to achieve desired properties **A9:** 701
microstructure... **A9:** 538-539
specimen preparation **A9:** 533
spinodal decomposition **A9:** 654
Alnico magnets, sintering
time and temperature... **M4:** 796
Alnico permanent magnets **A7:** 641-642
produced by Osprey atomizing....................... **A7:** 531
Alpha
defined **A9:** 1
Alpha + beta alloys
titanium.. **A2:** 586, 602
Alpha alloy *See* Titanium alloys, specific types, Ti-5Al-2.5Sn
Alpha alloys
titanium.. **A2:** 586, 600-601
Alpha alumina............................. **EM4:** 6, 111, 112, 113
applications
dental **EM4:** 1008
medical.. **EM4:** 1009
canisters, HIP method used **EM4:** 196
engineering properties **EM4:** 752
properties... **EM4:** 759
thermal etching.. **EM4:** 575
Alpha brass powder
sintering **A7:** 308
Alpha bronze *See also* Bronze.......................... **A7:** 464
microstructure, bearing alloy **A7:** 707
structure, sintered bronze clutch.................... **A7:** 737
Alpha case
defined **A9:** 1
in titanium and titanium alloys....................... **A9:** 460
Alpha case removal
titanium alloys **A15:** 264
Alpha cellulose... **EM3:** 4
defined **EM2:** 4
Alpha double prime
defined **A9:** 1
Alpha double-prime phase in titanium alloys... **A9:** 461, 475
Alpha eucryptite
specialty refractory................................ **EM4:** 907-908
Alpha ferrite **A6:** 457
defined **A13:** 1
Alpha iron
defined **A9:** 1 **A13:** 1
high-stacking fault energy material.............. **A8:** 173
Alpha loss peak **EM3:** 4
defined **EM2:** 4
Alpha Model 1
defined **A18:** 2
Alpha nickel-aluminum bronze
properties and applications............................ **A2:** 386
Alpha parameter
temperature and strain rate effect on............ **A8:** 172
Alpha particle
defined **EL1:** 965-966
Alpha particle induced failures
silicon *p-n* junctions **A11:** 782-784
Alpha particles
as radiation.. **A17:** 295
Alpha phase particles
mean free distance **A9:** 129
Alpha phase unalloyed uranium
fabrication techniques....................................... **A2:** 671
Alpha phases
in Alnico alloys.. **A9:** 539
of cemented carbides **A9:** 274

Alpha phases (continued)
in cobalt-base heat-resistant casting alloys **A9:** 334
in iron-cobalt-vanadium alloys **A9:** 538
in massive transformations **A9:** 656
of titanium alloys **A9:** 458, 460-461
in uranium and uranium alloys **A9:** 476-487
Alpha plate colonies
titanium and titanium alloy castings **A2:** 638
Alpha prime
defined **A9:** 1
Alpha prime phase
in Alnico alloys **A9:** 539
in titanium alloys **A9:** 461
Alpha process *See also* Shell molding
defined **A15:** 1
Alpha radiation, effect
E-glass fibers **EM1:** 47
Alpha radiation from depleted uranium **A9:** 477
Alpha ray barriers
polyimides as **EL1:** 770
Alpha segregation defects in titanium and titanium alloys **A9:** 459
Alpha stabilizer
defined **A9:** 1
Alpha stabilizers
titanium alloys **A2:** 598
zirconium alloys **A2:** 665
Alpha stabilizers in titanium ..... **A3:** 1 • 23
Alpha structures
in hafnium **A9:** 497-499, 501-502
wrought titanium alloys **A2:** 606
in zirconium and zirconium alloys **A9:** 497-501
Alpha titanium alloys
brazeability **M6:** 1049
weldability **M6:** 446
Alpha transus
defined **A9:** 1
Alpha-2 alloys *See also* Ordered intermetallics; Titanium aluminides
crystal structure and deformation **A2:** 926
material processing **A2:** 926
mechanical/metallurgical properties **A2:** 926-927
Alpha-beta alloys
interface fracture **A8:** 487
workability **A8:** 575
Alpha-beta brass
effects of different amounts of NH40H in the suspending liquid in vibratory polishing **A9:** 44
Alpha-beta forging of titanium alloys **M3:** 368, 369
Alpha-beta structure
defined **A9:** 1
Alpha-beta titanium alloys **A9:** 458 **A14:** 267-268, 271, 839
beta flecks in **A9:** 459-460
brazeability **M6:** 1049-1050
strengthening of flash welds **M6:** 579
torsion for flow softening study **A8:** 177
weldability **M6:** 446
Widmanstattten alpha microstructures in **A8:** 178
workability **A8:** 165, 575
Alpha-brass
crystallographic planes parallel to the rolling plane **A9:** 686
Cu-30Zn, cold worked and annealed **A9:** 156
defined **A9:** 1
deformation modes **A9:** 686
line etched grains **A9:** 63
line etching **A9:** 62
microstructural deformation modes as a function of strain **A9:** 686
Alpha-bronze
line etching **A9:** 62
Alpha-martensite
formation in 300 series stainless steels **A9:** 66
Alpha-particle emission
as radioactive decay mode **A10:** 244-245
Alpha-particle radiation
contamination by **EL1:** 45
Alpha-prime embrittlement **A6:** 444
Alpha-prime martensite
in austenitic stainless steels **A9:** 283
Alpha-stabilized voids
in titanium alloy forgings **A17:** 497-498
Alpha-titanium *See also* Optically anisotropic metals
etching by polarized light **A9:** 59
Alpha-titanium alloys
plastic response **A8:** 231
Alpha-uranium
etching by polarized light **A9:** 59
Alpha-zirconium
etching by polarized light **A9:** 59
Alpha/near-alpha titanium alloys **A14:** 267-268, 839
Alphabenzoinoxime
as narrow-range precipitant **A10:** 169
Alphabet, Greek
symbols for **A10:** 692 **A11:** 798
ALPID *See* Analysis of large plastic incremental deformation
Alrok oxide conversion coating process **M5:** 598-599
Alternate current (ac) mode
fluxes for welding **A6:** 57
Alternate extension wires
defined **A2:** 977
Alternate immersion
apparatus, lift-type **A13:** 223
test, defined **A13:** 1
Alternate immersion test
for aluminum alloys **A8:** 522-523
following heat treatment to decrease SCC susceptibility **A8:** 508
Alternate polarity operation
definition **M6:** 1
Alternate-blow swaging **A14:** 131
Alternating bending
and fatigue-crack propagation **A11:** 108
Alternating copolymer **EM3:** 4
Alternating copolymers **EM2:** 4, 58
Alternating current
abbreviation **A8:** 724
abbreviation for **A10:** 689
impedance measurement **A13:** 200
magnetization by **A7:** 576
stray-current corrosion by **A13:** 87
Alternating current (ac)
characteristics, WSI **EL1:** 356-357
demagnetization **A17:** 93, 121
effect, electrical contact materials **A2:** 840
forms **A17:** 92
full-wave rectified single-phase, defined **A17:** 91
injection, electric current perturbation **A17:** 136
losses, in superconductors **A2:** 1039-1040
in magnetic particle inspection **A17:** 91-92
magnetic properties **A2:** 773, 777
permeability, in magabsorption **A17:** 146
rectification **A17:** 91
ripple failures **EL1:** 997
skin effect **A17:** 91
solid ferromagnetic conductor carrying **A17:** 96
vs. direct current, magnetic particle inspection **A17:** 98-110
Alternating grinding (AG)
ceramics and wear studies **A18:** 409, 410
Alternating stress *See also* Stress **EM3:** 4
amplitude, defined **EM1:** 4
defined **EM1:** 4 **EM2:** 4
effect on fatigue cracking **A12:** 15
Alternating stress amplitude **EM3:** 4
defined **EM2:** 4
symbol **A8:** 726
symbol for **A11:** 797
Alternating stresses
in shafts **A11:** 461
Alternating torsion
fatigue test specimens **A8:** 368
Alternating twisting moment (torque)
in shafts **A11:** 461
Alternative hypotheses
one-sided **A8:** 626
and probability **A8:** 624
two-sided **A8:** 626
Alternative materials
for high-speed digital PWBs **EL1:** 604-608
Alternator contacts
powders used **A7:** 572
Alternator pole pieces
powders used **A7:** 572
Alternator regulator
powders used **A7:** 572
Alum solution decolorizing
powder used **A7:** 574
Alumilite anodizing processes
aluminum and aluminum alloys **M5:** 587, 592
Alumina *See also* Aluminum oxide; Molding sands; Sand(s) ... **A16:** 666-680 **EM3:** 33
as a reinforcing filler for mounting materials for epoxy-matrix composites **A9:** 588
AES spectra with peakshifts and plasmon loss peak structures **A10:** 552
in austenitic manganese steel castings **A9:** 239
background fluorescence of **A10:** 205
ceramic filler polyimide-base adhesives **EM3:** 159
ceramics **A2:** 1021
chromia catalysts, ESR analysis of **A10:** 265
copper bonding in gas-metal eutectic method **EM3:** 305
defined **A15:** 1
diluent for boriding **A4:** 441
as filler for conductive adhesives **EM3:** 76
fusion fluxes for **A10:** 167
grinding wheels for W **A16:** 859
for improved thermal performance **EM3:** 584
lapping **A16:** 492-493, 497, 505
laser cladding material **M6:** 797-798
as lining material, induction furnaces **A15:** 372
as mold refractory, investment casting **A15:** 258
silicon carbide whisker-reinforced **A2:** 1023-1024
sulfuric acid as dissolution medium **A10:** 165
transition metals as source of fluorescence on **A10:** 130
ultrasonic machining **A16:** 530, 532
unfused abrasive for lapping **A16:** 493
zirconia-toughened **A2:** 1022-1023
Alumina (79%)
refractory physical properties **EM4:** 897, 898, 899
Alumina (86%) ceramics
brazing with Cusil-ABA in Ar-O_2 atmosphere **A6:** 957
ceramic envelope used in vacuum device industry **A6:** 331
chemical composition **A6:** 60
fluxes used for SAW applications **A6:** 62
friction surfacing inclusion **A6:** 323
friction welding **A6:** 317
overlayed on cast irons **A6:** 721
properties **A6:** 992
sintering and diffusion bonding **A6:** 159
solid-state-welded interlayers **A6:** 169
Alumina (90%)
characteristics **EM4:** 976
properties **EM4:** 976
refractory physical properties **EM4:** 897, 898, 899
substrate properties **EM4:** 1108
Alumina (99.5%)
substrate properties **EM4:** 1108
Alumina (Al_2O_3) *See also* Aluminum oxide
$2Al_2O_3{\cdot}2MgO{\cdot}5SiO_2$ *See* Cordierite
$3Al_2O_3{\cdot}2SiO_2$ *See* Mullite

Alumina (Al_2O_3) (continued)
abnormal grain growth development **EM4:** 307
abrasive machining **EM4:** 320, 321, 322, 326, 327
abrasive wear **A18:** 189
as additive for pressure densification **EM4:** 298-299
additive to Si_3N_4 .... **EM4:** 226
Al_2O_3-Al_2TiO_5, composite **EM4:** 19
Al_2O_3-Ti composite
grain size affected by electrical discharge machining **EM4:** 376
properties **EM4:** 191
Al_2O_3-TiO_2, for substrates **EM4:** 17
Al_2O_3-ZrO_2 .... **EM4:** 2
applications **EM4:** 48
key product properties **EM4:** 48
properties **EM4:** 512
raw materials **EM4:** 48
Al_2O_3·MgO *See* Spinel
as an embedding agent **EM4:** 572
application as a coating **EM4:** 208
applications **A18:** 812 **EM4:** 46, 47, 48, 752, 963, 964
aerospace **EM4:** 1005
heat exchangers **EM4:** 982
medical and dental **EM4:** 1008-1009
wear **EM4:** 974, 975, 976, 977
Auger electron spectroscopy spectrum **A18:** 454
bioceramic physical characteristics **EM4:** 1009
brazing with glasses **EM4:** 519, 520
ceramic coatings for dies **A18:** 643
ceramic substrates **EM4:** 1107-1109
in ceramic tiles **EM4:** 926
ceramic/ceramic joining **EM4:** 480
ceramic/metal joints **EM4:** 515-516
ceramic/metal seals **EM4:** 502, 503, 505, 506, 507, 508, 509
chemical integrity of seals **EM4:** 540
chemical properties **EM4:** 753
for coarse and fine polishing before microstructural analysis **EM4:** 573
coating formation in molten particle deposition **EM4:** 206, 207
coatings for cutting tool materials **A18:** 614
coefficient of thermal expansion **EM4:** 503, 685
component in photochromic ophthalmic and flat glass composition **EM4:** 442
component in photosensitive glass composition **EM4:** 440
composition **EM4:** 46
in composition of glass-ceramics **EM4:** 499
in composition of leachable alkali-borosilicate glasses **EM4:** 428
in composition of textile products **EM4:** 403
in composition of wool products **EM4:** 403
crystal structure **EM4:** 30
for cutting tools **EM4:** 959
damage dominated by brittle fracture **A18:** 180
densified, applications, medical and dental **EM4:** 1007
in dental polishing pastes **A18:** 666
direct bonding **EM4:** 480
in drinkware compositions **EM4:** 1102
effect of adding a second phase in ceramic matrix composites **EM4:** 862-863
effect on chemical properties of glass **EM4:** 857
effect on elastic modulus **EM4:** 849
effect on glass hardness **EM4:** 851
electrical properties **EM4:** 752-753
electrical/electronic applications **EM4:** 1105
engineering properties **EM4:** 752
erodent particles in metals **A18:** 202, 203, 204
erosion in ceramics **A18:** 205, 206
erosion rate testing particles **A18:** 200
failure strength of bars analyzed using CARES program **EM4:** 706-707
as filler **EM4:** 6
fracture mechanics and R-curves **EM4:** 647
fracture surface **EM4:** 641
freeze drying **EM4:** 62
fretting wear **A18:** 250
friction coefficient data **A18:** 72
fused **EM4:** 50
fusion-cast **EM4:** 391
gas pressure sintering for pressure densification **EM4:** 299

Alumina (Al_2O_3) (continued)
for gas-lubricated bearings **A18:** 532
glass/metal seal in high-pressure sodium vapor lamp **EM4:** 498
in glaze composition for tableware **EM4:** 1102
grain-growth inhibitor **EM4:** 188
as grinding abrasive before microstructural analysis **EM4:** 572
high-purity **EM4:** 19
hot pressed for pressure densification **EM4:** 296
hot pressing **EM4:** 191
humidity sensor basis **EM4:** 1148
hydrated, as dentifrice abrasive **A18:** 665
impurity found in gypsum **EM4:** 380
ion-exchange **EM4:** 462
joining method and wear application **EM4:** 974
key product properties **EM4:** 48
lapping abrasive **EM4:** 351, 352
laser beam machining **EM4:** 361, 367, 368, 369
liquid-phase sintering **EM4:** 286, 287, 288, 289
load-bearing lifetimes for orthopedic prostheses **EM4:** 1008-1009
material to which crystallizing solder glass seal is applied **EM4:** 1070
matrix for reinforced composites in directed metal oxidation **EM4:** 233-235
matrix material for ceramic-matrix composites **EM4:** 840
mechanical integrity of seals **EM4:** 533, 534, 535, 538
mechanical properties **A18:** 774, 813 **EM4:** 331, 751, 753, 973, 974
melting/fining **EM4:** 391
metal brazing **EM4:** 490
opaque properties **EM4:** 1110
optical properties **EM4:** 753, 754
in ovenware compositions **EM4:** 1103
phase analysis **EM4:** 25
physical properties **A18:** 192, 801, 813, 814 **EM4:** 752
as plasma spray coating for titanium alloys **A18:** 780
plasma-sprayed
properties, adiabatic engine use **EM4:** 980
scuffing temperatures and coefficients of friction between ring and cylinder liner materials **EM4:** 991
polycrystalline fracture surface **EM4:** 639-641
pressure densification
pressure **EM4:** 301
technique **EM4:** 301
temperatures **EM4:** 301
primary sources in U.S. **EM4:** 379
processing defects **EM4:** 643
produced by calcination **EM4:** 111-112
properties **A6:** 949 **EM4:** 30, 191, 330, 424, 503, 512, 761
adiabatic engine use **EM4:** 990
property data of composite components **EM4:** 863
protective coating for solid-electrolyte sensors **EM4:** 1137
raw materials **EM4:** 48
refractory compositions **EM4:** 895, 896
as refractory filler **EM4:** 1072
reinforced for cutting tools **EM4:** 959
role in glazes **EM4:** 1062
scanning acoustic microscopy for wear studies **A18:** 409
scanning laser acoustic microscopy **EM4:** 625
sintered creep data **EM4:** 753
sintered for handling and processing equipment **EM4:** 959
sintered, fretting wear **A18:** 248, 249
sintered, strength **EM4:** 753
sintering **EM4:** 262, 263, 266, 511
sintering aid **EM4:** 188
sliding wear **A18:** 389
solid particle erosion **A18:** 210
solid particle impingement erosion of cobalt-base wrought alloys **A18:** 768, 769, 770
solid-state joining **EM4:** 487
solid-state sintering **EM4:** 273, 276-277, 278, 279, 280
specialty calcined **EM4:** 50

Alumina (Al_2O_3) (continued)
specific properties imparted in CTV tubes **EM4:** 1039
spherical media composition for wet milling **EM4:** 78
structure **EM4:** 752
for substrates **EM4:** 17
substrates for thick-film circuits **EM4:** 1142
superplasticity **EM4:** 301
supply sources **EM4:** 46
in tableware compositions **EM4:** 1101
tabular **EM4:** 50
thermal properties **EM4:** 30, 191, 331, 503, 685, 974
thermal shock parameter **EM4:** 191
as tooling for pressure calcintering **EM4:** 300
as tooling for uniaxial hot pressing **EM4:** 298
transition **EM4:** 111
translucent tubes produced for high-pressure sodium vapor lamps **EM4:** 150
in typical ceramic body compositions **EM4:** 5
ultrasonic machining **EM4:** 361
water slurry abrasive effect on Al-13Si alloy **A18:** 195
wear rates, medical prostheses **EM4:** 1009
Y-TZP composites **EM4:** 516
Alumina abrasives
for ferrous metals **A9:** 24
for hand polishing **A9:** 35
Alumina AD-998
brazing with Cusil-ABA in Ar-O_2 atmosphere **A6:** 957
Alumina, amorphous
formed during active metal brazing **EM4:** 524
Alumina borosilicate glass
composition, laboratory glassware **EM4:** 1088
properties, laboratory glassware **EM4:** 1088
Alumina brick, applications
refractory **EM4:** 900, 901-902, 903
Alumina ceramic
diffusion bonded joint **A9:** 62
Alumina ceramic coatings **M5:** 532, 534-536, 540-542
Alumina ceramics
applications and properties **A2:** 1021
green machining feeds and speeds **EM4:** 183
thermal expansion coefficient **A6:** 907
Alumina cermets
thermal expansion coefficient **A6:** 907
Alumina fiber
high, as synthetic reinforcement **EM1:** 117-118
importance **EM1:** 43
properties **EM1:** 60, 62, 118
Alumina Fiber FP (DuPont) **EM4:** 223
Alumina fiber metal matrix composites **EM1:** 889-890
Alumina grit blast
surface treatment **EM3:** 294
Alumina, high-fired
ceramic substrate for soldering **A6:** 132
Alumina inclusions in steel **A9:** 185
Alumina oxide coating
molybdenum **M5:** 662
Alumina phosphate
as coremaking system **A15:** 238
no-bake resin binder process **A15:** 217
Alumina powder
applications **EM4:** 196
rheological behavior in injection molding **EM4:** 174-175
Alumina silicon carbide
hot pressing **EM4:** 191
Alumina silicon carbide whisker-reinforced **EM4:** 548
direct brazing **EM4:** 519
fracture toughness **EM4:** 191, 586, 973
joining oxide ceramics **EM4:** 512
properties **EM4:** 191
Alumina substrates **EL1:** 104-106, 336-337
Alumina titanate
as filler for solder glass **EM4:** 1072
Alumina titanium oxide
applications as a coating **EM4:** 208
coating formation in molten particle deposition **EM4:** 206

Alumina trihydrate
as filler (flame retardant) **EM3:** 179
Alumina zirconia
abrasive **A16:** 98, 99, 432, 434, 436, 440
Alumina zirconium silicate (AZS)
applications **EM4:** 963, 964
property comparison, mineral processing **EM4:** 962
refractory applications **EM4:** 904-905, 907
Alumina-base inserts
high removal rate machining **A16:** 608
Alumina-boria-silica fibers **EM1:** 43, 60-61
Alumina-containing ceramics
chemical etching **EM4:** 575
Alumina-ethylene-vinyl acetate (Alumina-EVA)
viscosity in injection molding **EM4:** 174
Alumina-glass composites
crack-second-phase interactions **EM4:** 865
effect of volume fraction and size of dispersed phase **EM4:** 866
fracture strength data comparison **EM4:** 866
fracture toughness **EM4:** 867
hot-pressed, hardness **EM4:** 967
Alumina-modified silica sols
charge reversal **EM4:** 447
Alumina-molybdenum-manganese
metallizing process **EM4:** 490-491
Alumina-silica fibers **EM1:** 60-61
Alumina-silica hybrid multidirectional composite
design and application **EM1:** 939-940
Alumina-silicate fibers **EM1:** 117-118
Alumina-titanium carbide (black ceramic composite) **A16:** 98, 99
laser-enhanced etching **A16:** 576
Alumina-zirconia ($Al_2O_3 \bullet ZrO_2$)
properties **A6:** 949
Alumina/Al matrix
directed metal oxidation **EM4:** 232
Alumina/titania
thermal spray coating material **A18:** 832
Alumina/titanium carbide
machining example of irons and steels for cutting tools **EM4:** 968
Alumina/zirconia
flexural strength aging time **EM4:** 773
flexural strength dependence on alumina content **EM4:** 772
properties **EM4:** 759, 770
Aluminate
carburizing affected by content in steels **A18:** 875
Aluminate basic
fluxes used for SAW applications **A6:** 62
Aluminate glasses
electrical properties **EM4:** 851
Aluminates
in austenitic manganese steel castings **A9:** 239
formation in steel **A9:** 625
Aluminide coatings
superalloys and refractory metals **M5:** 376-380, 661-666
Aluminides
for intermetallic matrix composites **A2:** 910
iron **A2:** 920-925
nickel **A2:** 914-920
silicides **A2:** 933-935
titanium **A2:** 925-929
trialuminides **A2:** 929-933
types, ordered intermetallics **A2:** 914
Aluminized explosives **A7:** 600-601
Aluminized steels *See also* Aluminum coatings **A13:** 434-435, 458, 527, 528, 1014
Aluminized wire **A1:** 281
Aluminizing *See* Aluminum coatings
defined **A13:** 1
Aluminoborate glasses
electrical properties **EM4:** 951
Aluminoborosilicate glass **EM4:** 1056
applications
electronic processing **EM4:** 1055
laboratory and process **EM4:** 1087
Aluminogermanate glasses
electrical properties **EM4:** 851
Aluminon method
analysis for aluminum in ferrosilicon/ferroboron by **A10:** 68
analysis for beryllium in copper-beryllium alloys by **A10:** 65
Aluminosilicate
for accelerating adhesive cure **EM3:** 179
filler for urethane sealants **EM3:** 205
glass-to-metal seals **EM3:** 302
Aluminosilicate fiber
applications **EM4:** 46
composition **EM4:** 46
supply sources **EM4:** 46
Aluminosilicate glass-ceramics
melting/fining **EM4:** 392
Aluminosilicate glasses
applications **EM4:** 960
aerospace **EM4:** 1017
electronic processing **EM4:** 1056, 1058
laboratory and process **EM4:** 1089
lighting **EM4:** 1032, 1034, 1036, 1037
composition **EM4:** 742
when used in lamps **EM4:** 1033
electrical properties **EM4:** 851
island structure **EM4:** 758
maximum operating temperature **EM4:** 1035
properties **EM4:** 742, 849, 1033, 1034, 1057
tempered, aerospace applications **EM4:** 1017, 1018
uses **EM4:** 742
Aluminosilicates
melting/fining **EM4:** 391
Aluminothermic reaction **M6:** 692-693
Aluminothermic reduction
niobium powders **A7:** 162
Aluminothermic steel
compensation for short preheats **M6:** 695-696
fusion thermit welding **M6:** 694
transformation in thermit welding **M6:** 697
Aluminous fireclay
chamotte as **A15:** 248-249
Aluminum *See also* Advanced aluminum materials; Aluminum alloys; Aluminum alloys specific types; Aluminum alloys, specific types; Aluminum alloys, Specific types; Aluminum alloys, specific types; Aluminum alloys,specific types; Aluminum casting alloys; Aluminum casting alloys, specific types; Aluminum foundry products; Aluminum melts; Aluminum metallization; Aluminum mill and engineered products; Aluminum nitride; Aluminum P/M alloys; Aluminum P/M parts; Aluminum P/M processing; Aluminum powders, specific types; Aluminum recycling; Aluminum, specific types; Aluminum toxicity; Aluminum wire bonding; Aluminum-lithium alloys; Aluminum-lithium alloys, specific types; Aluminum-magnesium alloys; Aluminum-silicon alloys; Arc-welded aluminum alloys; Atomized aluminum powders; Cast Alnico alloys; High-purity aluminum; High-strength aluminum alloys; Liquid aluminum; Liquid aluminum alloys; P/M aluminum alloys; Prealloyed aluminum powders; Pure aluminum; Pure metals; Wrought aluminum alloys; Wrought aluminum alloys, specific types; Wrought aluminum and aluminum alloys; Wrought aluminum and aluminum alloys, specific types **A13:** 583-609
(liquid), contact angles on beryllium at various test temperatures in argon and vacuum atmospheres **A6:** 116
as a conductive coating for scanning electron microscopy specimens **A9:** 97
Aluminum (continued)
abradable seal material **A18:** 589
abrasive blasting of **M5:** 571-573
abrasive wear **A18:** 188
absorptivity **A6:** 265
acid dipping of **M5:** 603-604, 606
activation energy of **A11:** 772
activation energy of creep for **A8:** 309
addition to epoxy adhesives **EM3:** 515
addition to filler metals for ceramic materials **A6:** 951
addition to precipitation-hardenable nickel alloys **A6:** 576
addition to solid-solution nickel alloys **A6:** 575
addition to strengthen chromium equivalent **A6:** 100
addition to superalloys to resist oxidation **A4:** 798
adherends
compared to graphite-epoxy adherends **EM3:** 506
direct compliance method of fracture analysis **EM3:** 340
durability **EM3:** 261-264
plastic deformation and Moire interferometry **EM3:** 451
surface preparation **EM3:** 259-264
adhesion measurement of fcc metals **A6:** 144
adhesion promoters used for urethane sealants **EM3:** 206
adhesion-dominated durability **EM3:** 665
adhesive bond causes of failure **EM3:** 249
AES spectra with peakshifts and plasmon loss peak structures **A10:** 552
age-hardened, FIM/AP study of precipitates in **A10:** 583
aiding surface hardening **A4:** 263
air-carbon arc cutting **A6:** 1172, 1173, 1176
aircraft structure repair by adhesives **EM3:** 801-819
alkaline cleaning of **M5:** 577-578, 590-591
alloying effect in titanium alloys **A6:** 508, 509
alloying effect on copper alloys **M6:** 400
alloying effect on nickel-base alloys **A6:** 589
alloying effects, electrolytic-solution potential **A13:** 584
alloying, in microalloyed uranium **A2:** 677
alloying, in niobium-titanium superconducting alloys **A2:** 1045
alloying, in wrought copper and copper alloys **A2:** 242
alloying, in wrought titanium alloys **A2:** 599
alloying, of nickel-base alloys **A13:** 641
and aluminum alloys **A15:** 743-770
as an addition to austenitic manganese steel castings **A9:** 239
as an alloying addition to nickel-base wrought heat-resistant alloys **A9:** 309-311
as an alloying addition to precipitation- hardening stainless steels **A9:** 285
as an alpha stabilizer in titanium alloys **A9:** 458
analysis in ferrosilicon/ferroboron, by aluminon method **A10:** 68
anodes **A13:** 920-921
anodic coatings **M5:** 586, 589-591, 606-607, 608-610
anodized **EM3:** 416
anodizing process **M5:** 572, 580-598, 601, 603-610
corrosion resistance, anodized aluminum **M5:** 594-596
limitations, factors causing **M5:** 589-591
anodizing to study grain structure **A9:** 142
applications **A7:** 205
applications and properties **A2:** 63-64
applications, internal combustion engine parts **A18:** 553
applications, sheet metals **A6:** 399
architectural finishes **M5:** 609-610

SUBJECTS OF THE INDEXED VOLUMES: ASM Handbook (designated by the letter "A"): **A1:** Properties and Selection: Irons, Steels, and High-Performance Alloys (1990); **A2:** Properties and Selection: Nonferrous Alloys and Special-Purpose Materials (1990); **A3:** Alloy Phase Diagrams (1992); **A4:** Heat Treating (1991); **A6:** Welding, Brazing, and Soldering (1993); **A7:** Powder Metallurgy (1984); **A8:** Mechanical Testing (1985); **A9:** Metallography and Microstructures (1985); **A10:** Materials Characterization (1986); **A11:** Failure Analysis and Prevention (1986); **A12:** Fractography (1987); **A13:** Corrosion (1987); **A14:** Forming and Forging (1988); **A15:** Casting (1988); **A16:** Machining (1989); **A17:** Nondestructive Testing and Quality Control (1989); **A18:** Friction, Lubrication, and Wear Technology (1992). **Metals Handbook, 9th Edition** (designated by the letter "M"): **M1:** Properties and Selection: Irons and Steels (1978); **M2:** Properties and Selection: Nonferrous Alloys and Pure Metals (1979); **M3:** Properties and Selection: Stainless Steels, Tool Materials and Special-Purpose Materials (1980); **M4:** Heat Treating (1981); **M5:** Surface Cleaning, Finishing, and Coating (1982); **M6:** Welding, Brazing, and Soldering (1983). **Engineered Materials Handbook** (designated by the letters "EM"): **EM1:** Composites (1987); **EM2:** Engineering Plastics (1988); **EM3:** Adhesives and Sealants (1990); **EM4:** Ceramics and Glasses (1991); **Electronic Materials Handbook** (designated by the letters "EL"): **EL1:** Packaging (1989).

Aluminum (continued)
atmospheric corrosion **A13:** 82
atmospheric etch **M5:** 578
atomic interaction descriptions **A6:** 144
atomized, effect of particle size on apparent density **A7:** 273
atomized, shipment tonnage **A7:** 24
Auger electron microscopy map of a gallium arsenide field-effect transistor **EM3:** 241
Auger electron spectroscopy spectra when oxidized **EM3:** 240
Auger electron spectroscopy spectrum **A18:** 454
and austenitic grain growth **A1:** 227
in austenitic stainless steels **A6:** 468
backing bars **M6:** 382-383
barrel finishing of **M5:** 572-574
base metal solderability **EL1:** 677
biological corrosion **A13:** 118-119
blanking-shear tests for **A8:** 64
bonded by polyamides and polyesters **EM3:** 82
bonded structure with incipient defects **EM3:** 524
bonding fixtures **EM3:** 707
brass condenser tube, failed **A11:** 632
brass plating of **M5:** 603-605
bright dipping of **M5:** 579-582
bright finishing of **M5:** 574, 576
brightening of ... **M5:** 579-582, 591, 596-597, 607-608, 610
chemical **M5:** 579-582, 591, 596, 607-608, 610
electrolytic **M5:** 580-582, 596-597, 607-608
Brinell test block for **A8:** 88
buffing *See* Aluminum, polishing and buffing of
butt welding **M6:** 674
cadmium plating of **M5:** 605
calculated electron range for **EL1:** 1095
capacitor discharge stud welding **A6:** 221, 222
carburizing container coating **A4:** 327
care of **A13:** 602-603
cast, history of **A15:** 22
as cast in plaster molds **A15:** 243
in cast iron **A1:** 5, 6, 8
in cast iron, gas porosity from **A15:** 82
as casting material **A15:** 35
castings, cleaning and finishing of **M5:** 571-574, 576, 583, 588, 590-591, 601-603, 606
castings, markets for **A15:** 42
cathodic corrosion **A13:** 748
ceramic coating of **M5:** 610
ceramic cutting tool cost effectiveness **EM4:** 967
characterized **A13:** 583
chemical brightening of **M5:** 579-582, 590-591, 603-604, 606
process selection, factors affecting **M5:** 581-582
chemical cleaning of *See also* specific processes by name **M5:** 576-579, 607, 610
chemical conversion coating of **M5:** 597-600, 606-607, 609-610
chemical finishes, designation system **M5:** 609-610
chemical plating of **M5:** 604-606
chemical resistance **M5:** 4, 7-10
chemical state information by AES when oxidized **EM3:** 244
chlorine corrosion **A13:** 1170
chromate conversion coating of **M5:** 457-458, 599-600
chromating process sequence **A13:** 390
chromium plating of **M5:** 172, 180, 184-185, 609-610
hard chrome plating **M5:** 172, 180, 184-185
removal of plate **M5:** 184-185
cladding material for brazing **A6:** 347
cleaning fluxes, ternary phase diagram **A15:** 446
cleaning processes **M5:** 571-586, 590-591, 603-604, 606-608
cleaning solutions for substrate materials **A6:** 978
cleanliness, degree of, testing **M5:** 576
coated, filiform corrosion **A13:** 107
coating for resistance seam welding **M6:** 502
as coating for silica fiber to resist fatigue **EM4:** 744
coating for valve train assembly components **A18:** 559

Aluminum (continued)
for coating surfaces before scanning transmission electron microscopy **EM3:** 242
coatings **A13:** 527
coatings, for fasteners **A11:** 542
coextrusion welding **A6:** 311
cold welding **A6:** 307-308, 309
commercial-purity, cold rolled **A9:** 687-696
in compacted graphite iron **A1:** 56
composite processing map for **A14:** 365
conductor inks **EL1:** 208
contact angle with mercury **A7:** 269
content effect in beryllium alloys, electron-beam welding **A6:** 872
content in magnesium alloys **M6:** 427
convergent-beam electron diffraction (CBED) pattern **A18:** 387
conversion **A13:** 396
in copper alloys **A6:** 752
copper plating of **M5:** 160-161, 163, 168-169, 603-605
corroded honeycomb core **EM3:** 845
corrosion defect detection **EM3:** 751
corrosion in **A11:** 201
corrosion of **A13:** 583-609
corrosion resistance **A13:** 583
corrosion resistance, anodized aluminum **M5:** 581, 594-596
crack interception of particle in aluminum in glass **EM4:** 863
creep rupture testing of **A8:** 302
critical angles for cutting **A18:** 185
critical surface tensions **EM3:** 180
crystal formation, in master alloy processing **A15:** 107
crystallization and coarsening stages, x-ray topography **A10:** 376
cutting tool material selection based on machining operation **A18:** 617
cutting tool materials and cutting speed relationship **A18:** 616
for cylinder blocks of automobile internal combustion engines **A18:** 553
damping capacity **M1:** 32
debris effect on wear **A18:** 249
deformation mechanisms map **A8:** 310
densities **A2:** 47
deoxidization, of low-alloy steels **A15:** 715
as deoxidizer **A15:** 91
deoxidizing, copper and copper alloys **A2:** 236
deoxidizing process **M5:** 8, 12-13
depth profiling of **A7:** 258
designation systems **A2:** 15-28
determined in plant tissues **A10:** 41
difficult to inspect by infrared methods **EM3:** 763
diffusion bonding **A6:** 145
diffusivity of silicon in **A11:** 777
dip brazing **A6:** 336
distortion **A6:** 1098-1099
dominant texture orientations **A10:** 359
in ductile iron **A15:** 649
dwell pressures, cold isostatic pressing **A7:** 449
dynamic yield stress vs. strain rate in **A8:** 41
EDTA titration **A10:** 173
effect, CG irons **A15:** 673
effect of impact angle on **A11:** 155
effect of, on notch toughness **A1:** 741
effect of, on steel **A1:** 146, 577
effect on case depth in carbonitriding **A4:** 376
effect on SCC of copper **A11:** 221
effect on thermal conductivity **EM3:** 620
effect on tin-base alloys **A18:** 748
effects, cartridge brass **A2:** 300-301
as electrical contact materials **A2:** 849-850
electrical contacts, use in **M3:** 672
electrical resistance applications **M3:** 641
electrical resistivity **A15:** 756
in electrical steels **A9:** 537
as electrically conductive filler **EM3:** 178
electrodes, nylon liners **A6:** 184
electrogas welding **A6:** 275, 278 **M6:** 239
electroless nickel plating **M5:** 219, 221, 228
electrolytic cleaning of **M5:** 578
electromigration **EL1:** 964
electron-beam welding **A6:** 851, 857
electronic applications **A6:** 990

Aluminum (continued)
in electronic nickel, photometric method for **A10:** 65
electroplating of **M5:** 600-610
electropolishing of **M5:** 305-306, 308
electroslag welding, reactions **A6:** 273, 274
embrittlement **A13:** 178
embrittlement by **A11:** 234 **A13:** 179-180
emulsion cleaning of **M5:** 4-5, 36, 577
in enamel cover coats **EM3:** 304
enamel frits for **A13:** 447
in enameling ground coat **EM3:** 304
engine block microporosity sealants **EM3:** 609
engineered products **A2:** 5-6
epoxy and stress analysis **EM3:** 478
epoxy joint system primer application **EM3:** 626
epoxy lap-shear strength to **EM3:** 99
erosion mechanisms **A18:** 202
erosive attack of melt on die surface **A18:** 630
etching of **M5:** 8-9, 582-586, 590, 608-610
acid **M5:** 582-586
explosion welding **A6:** 162, 163, 303, 304 **M6:** 710
electrical applications **M6:** 713-714
marine applications **M6:** 714-715
for explosively loaded torsional Kolsky bar **A8:** 224
explosivity **A7:** 194-196, 199, 601
exposure limits and toxicity **A7:** 205-206
extended solubility of iron in **A10:** 294-295
extruded, static recrystallization **A9:** 691
extruded, subgrains in **A9:** 690
extrusion welding **M6:** 714
extrusions, ultrasonic inspection **A17:** 271
fabrication characteristics **A2:** 7-9
FASIL adhesion excellent **EM3:** 678
as fastener material **EM1:** 716-718
fatigue strength, anodic coatings affecting **M5:** 597
faying surface sealing with **EM1:** 720
in ferrite **A1:** 404, 408
ferrite formation **M6:** 346
fibers as fillers **EM3:** 178
field evaporation of **A10:** 586, 587
as filler for conductive adhesives **EM3:** 76
filler for polymers **A7:** 606
as filler for polyphenylquinoxalines **EM3:** 167
in filler metals for active metal brazing **EM4:** 523, 524, 525, 526
in filler metals for direct brazing **EM4:** 518-519
fine-grained, superplasticity **A8:** 553
finishes **M5:** 606-607, 609-610
designation system **M5:** 609-610
standards for **M5:** 606-607
finishing processes **M5:** 573-577, 585-610
fixturing for induction brazing **M6:** 971-972
in flame atomizers **A10:** 48
flash welding **A6:** 247
flat-rolled products (plate, sheet, foil) **A2:** 5
flow stress deformed in torsion **A8:** 178-179
flow stress during rolling **A8:** 179
as flux for stud arc welding of carbon steels **A6:** 660
foil, for flexible printed boards **EL1:** 581
foil, solid-state welding **A6:** 169
Forest Products Laboratory procedure, compared to steel **EM3:** 271
forgings, cleaning and finishing **M5:** 590-591
forming by stainless tools **A18:** 633
forming, lubricants in **A14:** 519
friction coefficient data **A18:** 71, 72
friction surfacing **A6:** 321
friction welding **A6:** 152
fume generation from shielding gases **A6:** 68
functions in FCAW electrodes **A6:** 188
furnace, heat transfer in **A15:** 454
gages, combination, use for **M3:** 556
galvanic corrosion **A13:** 84
gas metal arc welding **M6:** 153
gas removal ratio **A15:** 85
gas-metal arc welding **A6:** 24
gas-tungsten arc welding **A6:** 190, 191
gas-tungsten arc welding shielding gas selection **A6:** 67
gases in **A15:** 85-86
gasket material **A18:** 550
in glass electrodes **EM4:** 1089

Aluminum (continued)
globular-to-spray transition currents
for electrodes ... A6: 182
gold plating of ... M5: 605
Hall-Heroult process ... A2: 3
in hardfacing alloys ... A18: 765
in heat-resistant alloys ... A4: 512
high fatigue strength ... A18: 743
high frequency resistance welding ... M6: 760
high-frequency welding ... A6: 252
high-temperature solid-state welding ... A6: 298, 299
history ... A2: 3
honeycomb ... EM1: 722-723, 728
honeycomb components, neutron radiography ... A17: 392-393
honeycomb core ... EM3: 807
hot dip galvanized coating, use as alloying element in ... M5: 324
hot extrusion of ... A14: 321
hydrogen embrittlement ... A12: 23-24, 124
hydrogen solubility ... A15: 85, 456
immersion plating of ... M5: 601, 603-607
impurity in solders ... M6: 1071
induction heating energy requirements for metalworking ... A4: 189
induction heating temperatures for metalworking processes ... A4: 188
induction soldering, physical properties ... A6: 364
ingot, macrograph of ... A10: 302
inoculants for ... A6: 53
inorganic fluxes for soldering ... A6: 980
introduction to ... A2: 3-14
ion implantation and oxidation resistance ... A18: 856
ion removal from ... A10: 200
ion sputtering and oxidized ... EM3: 244
ion-induced Auger yields ... A10: 550
iron in, extended solubility of ... A10: 294-295
in iron-base alloys, flame AAS analysis for ... A10: 56
lacquering of ... M5: 572, 583
laminated coatings ... M5: 609-610
laminates, fatigue levels ... EM3: 504, 505
as lap plate material ... EM4: 353
lap welding ... M6: 673
lap-shear strength, storage, and weathering effect ... EM3: 658
laser alloying ... A18: 866
laser beam welding ... M6: 647, 661
weld properties ... M6: 661
laser cladding ... A18: 867
laser melt/particle inspection ... M6: 802
laser-beam welding ... A6: 263, 874, 876
as lead additive ... A2: 545
LEISS spectra ... A10: 604
liftoff effect ... A17: 223
liquid erosion resistance ... A11: 167
liquid, Gibbs free energies, element Solution ... A15: 59
liquid-metal embrittlement of ... A11: 233 M1: 688
load-displacement curves ... A8: 229, 231
low-temperature solid-state welding ... A6: 300, 301
lubricant indicators and range of sensitivities ... A18: 301
machining of ... A2: 966
in magnesium alloys ... A9: 427
in malleable iron ... A1: 10
manufactured forms ... A2: 5-7
in maraging steels ... A4: 220, 222, 224
mass finishing of ... M5: 129-130, 134
matrix, elastic properties ... EM1: 188
maximum strain level ... A8: 551
mechanical filter ... A8: 224, 227
mechanical finishes, designation system ... M5: 609-610
mechanical properties ... EM4: 316
in medical therapy, toxic effects ... A2: 1256

Aluminum (continued)
for metal core molding ... EM3: 591
metal forming lubricants ... A18: 147
metallization corrosion ... A11: 770, 771, 775
for metallization in wafer processing ... EM3: 581
microalloying of ... A14: 220
in moist chlorine ... A13: 1173
molten, effect on carbon fibers ... EM1: 52
molten-salt dip brazing ... A6: 338
molybdenum-implanted ... A10: 484, 486
Monte Carlo electron trajectories in ... A12: 167
NDT methods used with skins and cores of ... EM3: 767
neutron and x-ray scattering, and absorption compared ... A10: 421
in Ni-Al/Ni-Cr-B-SiC, thermal spray coating material ... A18: 832
nickel plating bath contamination by ... M5: 208, 210
nickel plating of ... M5: 203, 206, 216, 218, 232, 604-605
electroless ... M5: 232
nickel undercoat, for chrome plating ... M5: 180
pretreatment for ... M5: 457
in nickel-chromium coatings ... A7: 174
nickel-implanted ... A10: 485
nitric acid corrosion ... A13: 1156
nitride-forming element ... A18: 878
nitrocellulose coated, flaws ... A13: 108
nitrogen-based sintering atmospheres ... A7: 345
oil coat minimizing good adhesion of urethane sealants ... EM3: 205
oxide breakage and bond formation model ... A6: 145
oxide spectrum by x-ray photoelectron spectroscopy (XPS) ... EM3: 237, 238
oxyacetylene welding ... A6: 281
oxyfuel gas cutting ... A6: 1158
oxygen cutting, effect on ... M6: 898
oxygen produced by ... A15: 78
painting of ... M5: 17, 19, 457, 606-607, 609
stripping ... M5: 19
pattern equipment ... A15: 195
perchloric acid as electrolyte for ... A9: 48
phosphate coating of ... M5: 438, 597-599
phosphoric acid anodizing ... EM3: 42, 249-250, 845
-phosphorus-oxygen SBD ... EM3: 250, 251
photometric analysis methods ... A10: 64
as physical modeling material ... A14: 432-433
physical properties ... EM4: 316
physical properties related to thermal stresses ... A4: 605
pin bearing testing ... A8: 59-60
plain carbon steel resistance to ... A13: 515
plasma and shielding gas compositions ... A6: 197
plasma arc cutting ... A6: 1167, 1169, 1170, 1171 M6: 916-917
plasma arc welding ... A6: 197
plasma-MIG welding ... A6: 224
plastic deformation, sequence of ... A9: 685
polishing and buffing of ... M5: 573-576, 581-582, 609-610
and polybenzimidazoles ... EM3: 170
polycrystalline, bright-field image of dislocations in ... A10: 444
polycrystalline, liquid mercury embrittlement of ... A11: 226
polycrystalline, ring pattern from ... A10: 437
as polymer contaminant ... A12: 479
porcelain enameling of ... M5: 509-511, 514-516, 519-521, 524-525, 527-529, 606-607, 609-610
design parameters ... M5: 524-525
evaluation of enameled surfaces ... M5: 527-529
frits, composition ... M5: 511
methods ... M5: 519-521
selecting, factors in ... M5: 514

Aluminum (continued)
surface preparation for ... M5: 515-516
porcelain enamels ... EM4: 937
porosity in ... M6: 839-840
powder as pigment ... EM3: 179
powder metallurgy materials, etching ... A9: 509
powder metallurgy materials, polishing ... A9: 507
power brushing of ... M5: 153
prebond treatment ... EM3: 35
precipitate stability and grain boundary pinning ... A6: 73
in precipitation-hardening steels ... M6: 350
precoated before soldering ... A6: 131
preformed butyl tapes press-in-place application with glass ... EM3: 190
processing map ... A8: 572
product classifications ... A2: 9-14
production ... A2: 3-4, 6
products, cross-referencing system ... A2: 16
projection welding ... A6: 233 M6: 506
properties ... A2: 3 A6: 629, 992 EM4: 677
protective film ... A13: 82
pure *See* Pure aluminum ... M2: 714-715
pure, integrated circuit defects ... A12: 481
pure, K-M CBEDP from ... A10: 441
pure, properties ... A2: 1099-1100
as pyrophoric ... A7: 194-196, 199, 601
radiographic absorption equivalence ... A17: 311
radiographic film selection ... A17: 328
rare earth alloy additives ... A2: 728-729
as reactive material, properties ... A7: 597
recommended guidelines for selecting PAW shielding gases ... A6: 67
recommended impurity limits of solders ... A6: 986
recommended shielding gas selection for gas-metal arc welding ... A6: 66
recovery from selected electrode coverings ... A6: 60
reductant of metal oxides ... M6: 692, 694
reflectance values, cleaning affecting ... M5: 596-598
and finishing processes
relationship to graphitization ... M6: 834
relative solderability ... A6: 134
relative solderability as a function of flux type ... A6: 129
relative weldability ratings, resistance spot welding ... A6: 834
resinous coatings ... M5: 609-610
resistance brazing ... A6: 340 M6: 976
resistance of, to liquid-metal corrosion ... A1: 635-636
resistance seam welding ... A6: 245 M6: 494
resistance soldering ... A6: 357
Rockwell scale for ... A8: 76
roll bonding processes ... A18: 755-756
roll welding ... A6: 312-314, 317 M6: 676
rolling, metalworking lubricant functions and requirements ... A18: 139
room-temperature bend strength of silicon nitride metal joints ... EM4: 526
sacrificial anodes ... A13: 469
sacrificial corrosion ... A13: 587
sampling and chemical analysis ... A7: 248
satin finishing of ... M5: 574-575, 577
SCC of ... A12: 28
SCC prevention ... A13: 328
seal adhesive wear ... A18: 549
segregation and solid friction ... A18: 28
selected-area diffraction patterns for single-crystallite ... A18: 386
sensitive tint used to examine ... A9: 138
for shallow forming dies ... A18: 633
shear stress ... EM3: 402
shielded metal arc welding ... A6: 176
shielding, buried telephone cable, galvanic corrosion ... A13: 85

Aluminum (continued)
shielding gas purity ... **A6:** 65
shot peening of ... **M5:** 141-142, 145
-silicon interdiffusion ... **A11:** 777-778
as silicon modifier ... **A15:** 161
silver plating of ... **M5:** 605-606
single-crystal specimen Kolsky bar ... **A8:** 222, 224
single-crystal, spot diffraction pattern from ... **A10:** 437
sintering, time and temperature ... **M4:** 796
slide welding ... **M6:** 674
smut removal ... **M5:** 580-581, 584-585, 590-591
as solder impurity ... **EL1:** 637
solderability ... **A6:** 971, 978
solderable and protective finishes for substrate materials ... **A6:** 979
soldering ... **A6:** 628, 631, 632 **M6:** 1072, 1075
solubility in magnesium ... **M2:** 525
solvent cleaning of ... **M5:** 576-577
species weighed in gravimetry ... **A10:** 172
spectrometric metals analysis ... **A18:** 300
specular finishes ... **M5:** 574, 576, 580-581, 609-610
for sporting goods manufacturing ... **EM3:** 576
spray material for oxyfuel wire spray process ... **A18:** 829
stacking-fault energy ... **A18:** 715
stainless steel-clad ... **A13:** 889-890
in stainless steels ... **A18:** 712, 716
standardized products ... **A2:** 5
in steel ropes, fretting wear ... **A18:** 243
in steels, inclusion-forming ... **A15:** 90
strain rate sensitivity ... **A8:** 38, 40, 230-231
strain-age cracking in precipitation-strengthened alloys ... **A6:** 573
strengths of ultrasonic welds ... **M6:** 752
stress corrosion, microwave inspection ... **A17:** 215
stripping methods ... **M5:** 19
structural bonding with titanium, primers ... **EM3:** 254
structure of porous anodic film ... **EM3:** 416
stud arc welding ... **A6:** 210, 211, 212, 213, 214, 217, 218-219
stud material ... **M6:** 731, 735
submerged arc welding effect on cracking ... **M6:** 129
substrate cure rate and bond strength for cyanoacrylates ... **EM3:** 129
as substrate for polyimides ... **EM3:** 161
as substrate for polysulfides ... **EM3:** 141
suitability for cladding combinations ... **M6:** 691
and sulfur, ductility/impact strength effects ... **A15:** 93
super-purity, solid solution effects ... **A2:** 38
in superalloys, solution treating ... **A4:** 799
superpure, ram speed vs temperature extrusion ... **A14:** 318
surface preparation ... **EM3:** 558
surface preparation for finishing processes ... **M5:** 180, 586, 597-598, 601
surface preparation for porcelain enameling ... **A13:** 447
surface preparation methods ... **EM3:** 259-264
surfaces ... **A2:** 3
susceptibility to hydrogen damage ... **A11:** 249
tap density ... **A7:** 277
temperature dependence of strength of lap joints ... **EM3:** 619
test conditions effect on interfacial energies ... **A6:** 117
thermal diffusivity from 20 to 100 °C ... **A6:** 4
thermal expansion coefficient ... **A6:** 907
thermal properties ... **A18:** 42
thermal properties when used as engine wall insulator lining ... **EM4:** 992
thermal radiation reflection, anodic coatings affecting ... **M5:** 596-598
for thermal spray coatings ... **A6:** 1004-1009 **A13:** 460-461
thermal stressing unsatisfactory ... **EM3:** 761
in thermite, AAS analysis for ... **A10:** 56
thermocompression welding ... **M6:** 674
threshold stress intensity ... **A8:** 256
tin plating of ... **M5:** 604-605
as tin solder impurity ... **A2:** 520
TNAA detection limits ... **A10:** 237
to deoxidize carburizing steels, for grain size control ... **A4:** 366-367

Aluminum (continued)
to fabricate integrated circuits ... **EM3:** 378
-to-aluminum joints ... **EM3:** 330
-to-aluminum lap joints ... **EM3:** 327
-to-aluminum specimens for bond testing ... **EM3:** 531
in tool steels ... **A18:** 739
torch brazing ... **A6:** 328
torch soldering ... **A6:** 351
torsion flow curves, plane-strain compression tests on ... **A8:** 162-164
torsion tests at hot working temperatures ... **A8:** 163
torsional ductility ... **A8:** 166-167
for torsional Kolsky bar strain rate testing ... **A8:** 227
as trace element, cupolas ... **A15:** 388
transistor base lead, fatigue failure ... **A12:** 483
for tubing in carbonitriding ... **A4:** 383
TWA limits for particulates ... **A6:** 984
twist reversal after deformation and strain ... **A8:** 174
U.S. shipments ... **M2:** 4, 17
ultimate shear stress ... **A8:** 148
ultrapure, by zone-refining technique ... **A2:** 1094
ultrasonic welding ... **A6:** 324, 326, 327
unalloyed, compositions ... **A2:** 22-25
under static, dynamic, incremental strain rate loading in shear ... **A8:** 224, 226
as unknown particle ... **A10:** 457
use in flux cored electrodes ... **M6:** 103
use of direct current electrode positive ... **M6:** 185-186
used to microalloy beryllium ... **A9:** 390
vacuum-deposited electrodes ... **EM3:** 429
vacuum-deposited, electronic defects ... **A12:** 484, 486-487
vapor degreasing of ... **M5:** 45-46, 53-55
vapor pressure, relation to temperature ... **A4:** 495 **M4:** 309, 310
Vickers and Knoop microindentation hardness numbers ... **A18:** 416
vitreous coatings ... **M5:** 609-610
voltage-specific corrosion of ... **A11:** 771
volumetric procedures for ... **A10:** 175
wavelength dispersive x-ray spectroscopy map ... **EL1:** 1100
wavy slip lines in a single crystal ... **A9:** 689
weakened or distorted by welding ... **EM3:** 33
wedge testing of sheets ... **EM3:** 666, 667, 668, 669
weighed as the phosphate ... **A10:** 171
welding electrodes ... **A6:** 176
wettability ... **A6:** 115
wetting by gallium, LME by ... **A11:** 719
for wire bonding ... **EM3:** 585
wire connections, EPMA failure analysis ... **A10:** 531-532
wire, rod, and bar ... **A2:** 5
work material for ion implantation ... **A18:** 858
world production ... **M2:** 4
wrought, cleaning and finishing ... **M5:** 574, 576, 583, 597, 603-606
wrought unalloyed, compositions ... **A2:** 17-21
x-ray characterization of surface wear results for various microstructures ... **A18:** 469
yield strength vs. fracture toughness ... **A6:** 1017
in zinc alloys ... **A9:** 489 **A15:** 788
zinc plating of ... **M5:** 601-606
in zinc/zinc alloys and coatings ... **A13:** 759
zincating process ... **M5:** 590-591, 601-606
zone refined, impurity concentration ... **M2:** 713

Aluminum (liquid)
contact angles on beryllium at various test temperatures in argon and vacuum atmospheres ... **A6:** 116

Aluminum aircraft structures
2% hydrofluoric (HF) acid method, peel testing ... **EM3:** 806
aerodynamically flush honeycomb core-aluminum skin plug repair ... **EM3:** 814, 816
aerodynamically flush patch line maintenance repair ... **EM3:** 811, 812
Alclad dissolution and disbond ... **EM3:** 818
autoclave repair techniques ... **EM3:** 808, 817-818

Aluminum aircraft structures (continued)
chromate conversion coating surface preparation, peel testing ... **EM3:** 806-807
damage assessment, water contamination and environmental deterioration ... **EM3:** 801-802
damage tolerance ... **EM3:** 801
FPL etch surface preparation, peel test results ... **EM3:** 138, 803, 805
fully repaired or rebuilt bonded structures ... **EM3:** 808, 818-819
honeycomb core plug line maintenance repair ... **EM3:** 811
life-limited repairs ... **EM3:** 808, 813-814
metal sheet repair materials ... **EM3:** 807
nonaerodynamic patch line maintenance repairs ... **EM3:** 811-812
partial repairs ... **EM3:** 808, 814-817
PasaJell 105 nontank procedure peel testing ... **EM3:** 803, 804, 805, 806
phosphoric acid nontank anodizing ... **EM3:** 805-806
potted core technique line maintenance repairs ... **EM3:** 811
prefabricated aerodynamically flush plug line maintenance repair ... **EM3:** 813
preimpregnated cloth method ... **EM3:** 811
puncture and gouge line maintenance repairs ... **EM3:** 809
riveted line maintenance repairs ... **EM3:** 808, 809, 810
selection of repair materials ... **EM3:** 807
skin and honeycomb core damage requiring wet lay-up ... **EM3:** 811
skin damage or delamination line maintenance repairs ... **EM3:** 809-811
small-area line maintenance work repair ... **EM3:** 808, 809-813
solvent wipe and abrade surface preparation, peel testing ... **EM3:** 806
surface dent line maintenance repairs ... **EM3:** 809
surface preparation nontank procedures ... **EM3:** 804-807
surface preparation qualitative procedure comparison ... **EM3:** 802-803
surface preparation relationship to service life ... **EM3:** 802
trailing-edge damage line maintenance repairs ... **EM3:** 812-813
type of repairs ... **EM3:** 808-819

Aluminum alkyls
powder used ... **A7:** 574

Aluminum alloy extruded parts ... **A14:** 307-310

Aluminum alloy filler metals
compositions of ... **A9:** 359

Aluminum alloy filler metals, specific types
4047, brazed joint between 6063-O sheets ... **A9:** 387
4245, brazed joint between 7004-O sheets ... **A9:** 387
4343, brazed joint in 12-O brazing sheets ... **A9:** 387
ER2319, electron beam weld in 2219-T37 sheet ... **A9:** 384
ER2319, gas tungsten arc weld in 2219-T37 sheet ... **A9:** 384
ER4043, gas tungsten arc weld, 6061-T6 tubing to A356-T6 casting ... **A9:** 382-383
ER4043, gas tungsten arc weld, butt joint ... **A9:** 381
ER5356, gas tungsten arc fillet weld ... **A9:** 381
R-SG70A, repair weld in 356-F casting ... **A9:** 383

Aluminum alloy ingots
dendrite arm spacing in ... **A9:** 629
grain refining inoculants for use in ... **A9:** 630
grain structures in ... **A9:** 629-631
homogenization of ... **A9:** 632
hydrogen porosity in ... **A9:** 633
macrosegregation in ... **A9:** 633
microsegregation in ... **A9:** 631-632
solidification structures of ... **A9:** 629-636
surface defects ... **A9:** 633-634

Aluminum alloy tubing
as tube stock ... **A14:** 671

Aluminum alloy(s) *See also* Aluminum; Aluminum alloys, specific types; Aluminum foundry products; Aluminum mill and engineered products;

Aluminum alloy(s) (continued)
Aluminum-lithium alloys; High-strength aluminum P/M products; Wrought aluminum alloys; Wrought aluminum alloys, specific types
beryllium in **A2:** 426
building and construction applications **A2:** 9-10
casting compositions, defined **A2:** 4-5
castings and ingot, compositions **A2:** 22-25
castings, production **A2:** 5-7
chemical milling **A2:** 8
consumer durable applications **A2:** 13
containers and packaging applications **A2:** 10
copper-alloyed, designation system **A2:** 15
densities **A2:** 47
designation systems **A2:** 15-28
electrical applications **A2:** 12-13
engineered products **A2:** 5-7
fabrication characteristics **A2:** 7-9
flat-rolled products (plate, sheet, foil) **A2:** 5
forgeability **A2:** 8-9
forging **A2:** 6, 34
formability **A2:** 8
ground/machined with superabrasives/ ultrahard tool materials **A2:** 1013
impacts **A2:** 6
ingot, designation system **A2:** 15-16
joining .... **A2:** q
machinability **A2:** 7-8
machinery and equipment applications **A2:** 13-14
magnesium-alloyed, designation system **A2:** 15
manufactured forms **A2:** 5-7
mechanical properties **A2:** 49-51
metal-matrix composites (MMCs) **A2:** 7
phases, wrought aluminum alloys **A2:** 36-37
physical properties **A2:** 45-46
powder metallurgy (P/M) parts **A2:** 6-7
product classifications **A2:** 9-14
production **A2:** 3-4
products, cross-referencing system **A2:** 16
properties **A2:** 3
silicon-alloyed, designation system **A2:** 15
standardized products **A2:** 5
strength improvement, methods of **A2:** 36-41
tin-alloyed, designation system **A2:** 15
transportation applications **A2:** 10-12
wire, rod, and bar **A2:** 5
wrought, applications and properties **A2:** 62-122
wrought, compositions **A2:** 4, 17-21
zinc-alloyed, designation system **A2:** 15
Aluminum alloys *See also* Aluminum; Aluminum alloys, specific types; Aluminum alloys, specific-types; Aluminum bronzes; Aluminum casting alloys; Aluminum P/M parts; Aluminum powders; Aluminum powders, specific types; Aluminum-silicon alloys; Aluminum-titanium alloys; Arc-welded aluminum alloys; Atomized aluminum powders; Cast aluminum alloys; High-strength aluminum alloys; Liquid aluminum alloys; P/M aluminum alloys; Prealloyed aluminum powders; Wrought aluminum alloys; Wrought aluminum alloys, specific types **A9:** 351-388 **A13:** 583-609 **A14:** 241-254, 791-804 **A16:** 761-804 **A18:** 693
1100, corrosion resistance **EM3:** 671
2000 series, not susceptible to corrosion if alclad **EM3:** 751
2024
aircraft structure repairs **EM3:** 801
bond line corrosion **EM3:** 671
chromic acid anodization of aluminum adherends **EM3:** 261
FPL etching procedure **EM3:** 260
mechanically induced artificial delamination **EM3:** 523, 524
phosphoric acid anodization **EM3:** 261
wedge testing **EM3:** 667-668

Aluminum alloys (continued)
wet peel test **EM3:** 669
2024-T3
for aerospace adhesive testing **EM3:** 733
bond line corrosion **EM3:** 671
bond line porosities produced by shimming **EM3:** 525
bonded to composite **EM3:** 644
dynamic fatigue **EM3:** 367
epoxy joint hydration of oxide layers **EM3:** 624
fatigue specimen **EM3:** 468
and FM 47 adhesive **EM3:** 502
and FM 73 with variation in lap-shear strength **EM3:** 774
high-performance acrylic adhesive bonding when oiled **EM3:** 122
for lap-shear testing **EM3:** 460
metal sheet as repair materials **EM3:** 807
metal-to-metal bond test specimens **EM3:** 530
primer-coated sheet for life-limited repairs **EM3:** 813-814
thickness influence on shear strength of adhesives **EM3:** 473
wedge test used to evaluate surface preparation **EM3:** 803
wedge testing **EM3:** 668
2024-T3B, surface preparation methods using wedge testing **EM3:** 804
2024-T6, test specimens for comparison of NDI techniques **EM3:** 528, 529
2024-T351, moisture ingression at joints **EM3:** 364
5052
honeycomb core **EM3:** 733
repaired with 5056 aluminum honeycomb core **EM3:** 807
5052-O, high-performance acrylic adhesive bonding when oiled **EM3:** 122
5056
corrosion resistance **EM3:** 671
honeycomb core as repair materials **EM3:** 807
6061-T **EM3:** 6
end fitting for a small-diameter graphite-epoxy tube **EM3:** 495, 498
high-performance acrylic adhesive bonding when oiled **EM3:** 122
7000 series, corrosion attack in the bond line **EM3:** 751
7075
aircraft structure repairs **EM3:** 801
bond line corrosion **EM3:** 671
7075-T6
inverse skin-doubler coupon **EM3:** 472
lap-shear coupon test **EM3:** 473
peel test results after surface preparation **EM3:** 805
as repair material **EM3:** 807
7075-T6C, surface preparation methods using wedge testing **EM3:** 804
7075-T73, metal fitting with graphite-epoxy composite tube **EM3:** 496
abrasive belt grinding **A16:** 801
abrasive blasting of **M5:** 571-573
abrasive cutoff sawing **A16:** 800
abrasive flow machining **A16:** 517
acid dipping of **M5:** 603-604, 606
activation energy **A11:** 772
adhesive bonding **M2:** 201-202
advanced, forging of **A14:** 249-251
aerospace applications **EM3:** 559-560
air bottle, intergranular cracking failure **A11:** 436
air-carbon arc cutting **A6:** 1176
alkaline cleaning of **M5:** 15, 577-578, 590-591
alloy designations **A6:** 528
alloying elements, effects **A15:** 743-746
alternate immersion test **A8:** 523
aluminum filler alloys **A6:** 724

Aluminum alloys (continued)
aluminum melting and reverberatory furnaces **EM4:** 903
analysis for boron, by Can-nine method **A10:** 68
annealing **M2:** 28-29, 30
anodic coatings **M5:** 586, 589-591, 594-598, 606-607, 609-610
anodic films, features revealed with **A9:** 351, 354
anodized, resistance of **A13:** 599-600
anodizing process **M5:** 572, 580-598, 601, 603-610
limitations, factors causing **M5:** 589-591
Antioch process for **A15:** 247
applications **A6:** 727, 736 **A15:** 743, 768-769 **A18:** 753, 790, 791 **M2:** 16-23
aerospace **A6:** 386, 387
building and construction **M2:** 16-17
consumer **M2:** 21-22
container and packaging **M2:** 17-18
electrical **M2:** 20-21
internal combustion engine parts **A18:** 561
machinery and equipment **M2:** 22
pistons for internal combustion engines **A18:** 553, 555
sheet metals **A6:** 399
transportation **M2:** 18-20
arc welding *See* Arc welding of aluminum alloys
arc-welded **A11:** 434-437
architectural application **A15:** 22
architectural finishes **M5:** 609-610
assembly, fatigue failure at spot welds **A11:** 310-311
atmospheric corrosion **A13:** 596
bacteria-produced tubercule **A13:** 120
bar and tube, die materials for drawing **M3:** 525
bar vs. tubing **A16:** 764
barrel finishing of **M5:** 572-574
beading **A14:** 804
as bearing alloys **A18:** 748, 752-753
advantages **A18:** 752
applications **A18:** 752-753
compositions **A18:** 752, 753
designations **A18:** 752, 753
mechanical properties **A18:** 752-753
microstructural features **A18:** 752
bearing material systems **A18:** 745, 746, 747
applications **A18:** 746
bearing performance characteristics **A18:** 746
compositions used **A18:** 746
load capacity rating **A18:** 746
bend radii for **A8:** 129-130
in bimetal bearing material systems **A18:** 747
binary, relative potency factors **A6:** 89
blanking, die materials for **M3:** 485, 486, 487
blanking of **A14:** 793-794
and blended powders **A7:** 125
boring **A16:** 162, 768, 771-772, 777-778, 791, 797
brass plating of **M5:** 603-606
brazing *See* Brazing of aluminum alloys **A6:** 828
dip brazing **M2:** 199-200
filler metals **M2:** 199-200
sheet for **M2:** 29, 201
summary of procedures **M2:** 200
torch brazing **M2:** 199-200
vacuum furnace brazing **M2:** 199-200
brazing and soldering characteristics **A6:** 627-628
brazing with clad brazing materials **A6:** 347
bright dipping of **M5:** 579-582
bright finishing of **M5:** 574, 576
brightening of ... **M5:** 579-582, 591, 596-597, 607-608, 610
chemical **M5:** 578-582, 591, 596, 607-609, 610
electrolytic **M5:** 580-582, 596-597, 607-608
Brinell test load for **A8:** 84
broachability constant **A16:** 200
broaching **A16:** 203-206, 208, 769, 774-775, 778-780

SUBJECTS OF THE INDEXED VOLUMES: ASM Handbook (designated by the letter "A"): **A1:** Properties and Selection: Irons, Steels, and High-Performance Alloys (1990); **A2:** Properties and Selection: Nonferrous Alloys and Special-Purpose Materials (1990); **A3:** Alloy Phase Diagrams (1992); **A4:** Heat Treating (1991); **A6:** Welding, Brazing, and Soldering (1993); **A7:** Powder Metallurgy (1984); **A8:** Mechanical Testing (1985); **A9:** Metallography and Microstructures (1985); **A10:** Materials Characterization (1986); **A11:** Failure Analysis and Prevention (1986); **A12:** Fractography (1987); **A13:** Corrosion (1987); **A14:** Forming and Forging (1988); **A15:** Casting (1988); **A16:** Machining (1989); **A17:** Nondestructive Testing and Quality Control (1989); **A18:** Friction, Lubrication, and Wear Technology (1992). **Metals Handbook, 9th Edition** (designated by the letter "M"): **M1:** Properties and Selection: Irons and Steels (1978); **M2:** Properties and Selection: Nonferrous Alloys and Pure Metals (1979); **M3:** Properties and Selection: Stainless Steels, Tool Materials and Special-Purpose Materials (1980); **M4:** Heat Treating (1981); **M5:** Surface Cleaning, Finishing, and Coating (1982); **M6:** Welding, Brazing, and Soldering (1983). **Engineered Materials Handbook** (designated by the letters "EM"): **EM1:** Composites (1987); **EM2:** Engineering Plastics (1988); **EM3:** Adhesives and Sealants (1990); **EM4:** Ceramics and Glasses (1991); **Electronic Materials Handbook** (designated by the letters "EL"): **EL1:** Packaging (1989).

Aluminum alloys (continued)
buffing ... **M5:** 573-576, 581-582, 610
bulging ... **A14:** 804
cadmium plating of ... **M5:** 605
canning ... **A18:** 738
capacitor discharge stud welding ... **M6:** 730, 736
care of ... **A13:** 602-603
castability, ratings ... **A15:** 766
casting alloys
alloy systems ... **M2:** 140-141
casting processes ... **M2:** 143-148
characteristics ... **M2:** 143, 144, 145
designation system ... **M2:** 141-143, 144, 145
mechanical properties ... **M2:** 148-151
modification ... **M2:** 149, 150
quality of castings ... **M2:** 148
casting alloys, properties ... **A15:** 763-769
casting processes ... **A15:** 746
castings ... **M2:** 9-10
castings, cleaning and finishing of ... **M5:** 571-574, 576, 583, 588, 590-591, 601-603, 606
castings, gas-tungsten arc welding ... **A6:** 192
castings, inspection of ... **A15:** 556-557
castings, inspection/quality control ... **A17:** 532-534
cavitation erosion ... **A18:** 216
cemented carbide tool life ... **A16:** 76
ceramic coating of ... **M5:** 610
ceramic cutting tools applied ... **A16:** 103
cermet tools applied ... **A16:** 92
chemical brightening of ... **M5:** 579-582, 590-591, 603-604, 606
chemical cleaning of *See also* specific process by name ... **M5:** 576-579, 607, 610
chemical compositions ... **A15:** 743
chemical conversion coatings ... **M5:** 597-603, 606-607, 609-610
chemical finishes, designation system ... **M5:** 609-610
chemical milling ... **A16:** 579, 581-586, 802-804
chemical plating of ... **M5:** 604-606
chemical resistance ... **M5:** 4, 7-10 **EM3:** 639
chip formation ... **A16:** 761, 765, 769-770, 780, 783, 787, 791-792
chips for machinability ratings ... **A16:** 761
chromate conversion coating ... **M5:** 599-600
chromating process sequence ... **A13:** 390
chromium plating of ... **M5:** 184-185, 609-610
removal of plate ... **M5:** 184-185
circular sawing ... **A16:** 365, 794, 800
classes of ... **A9:** 357
classification, system ... **A16:** 761-763
cleaning ... **A15:** 762-763
cleaning processes ... **M5:** 571-586, 590-591, 603-604, 606-608
coated carbides for machining ... **A16:** 80, 81, 83
coated, filiform corrosion ... **A13:** 107
coatings, for fasteners ... **A11:** 542
coextrusion welding ... **A6:** 311
coining ... **A14:** 804
cold extrusion, tool materials for ... **M3:** 515, 516, 517
cold swaging ... **A14:** 128
compared to gray iron ... **A16:** 797
comparison of P/M and I/M ... **A7:** 747
composite graph for Gill-Goldhoff correlation for ... **A8:** 337
composition control ... **A15:** 79-81
composition, effects on anodizing ... **M5:** 590
composition/microstructure, corrosion effects ... **A13:** 585-587
compositions ... **A7:** 741 **A18:** 753
compositions of ... **A9:** 359
constitutional liquation ... **A6:** 75
contact with foods, pharmaceuticals, and chemicals ... **A13:** 602
containing copper ... **A13:** 592-593
containing copper, line etching ... **A9:** 62
continuous immersion test ... **A8:** 523
contour band sawing ... **A16:** 363, 800-801
contour milling ... **A16:** 604, 797
contour roll forming ... **A14:** 634, 795
contrasting by interference layers ... **A9:** 60
copper plating of ... **M5:** 603-605
copper-free ... **A13:** 593
corrosion fatigue ... **A13:** 595
corrosion fatigue behavior ... **A8:** 408
corrosion in ... **A11:** 201

Aluminum alloys (continued)
corrosion of ... **A13:** 583-609
corrosion ratings ... **A13:** 586-588
corrosion resistance ... **A6:** 729-730
alclad products ... **M2:** 210-211
anodized products ... **M2:** 225-226, 229, 232
atmospheric corrosion ... **M2:** 219-228
cathodic protection ... **M2:** 210-211
chemical products, packaging ... **M2:** 228-229, 231-233
composition, effect of ... **M2:** 206-209
corrosion fatigue ... **M2:** 219, 220
deposition corrosion ... **M2:** 211-212
erosion-corrosion ... **M2:** 219
exfoliation corrosion ... **M2:** 218-220
food, packaging ... **M2:** 228-229, 231
galvanic corrosion ... **M2:** 207, 209-210
high purity waters ... **M2:** 222-223
intergranular corrosion ... **M2:** 212
microstructure, effect of ... **M2:** 206-209
natural waters ... **M2:** 223
nonmetallic building materials ... **M2:** 226-228
oxide film protection ... **M2:** 204-205
pharmaceuticals, packaging ... **M2:** 228-229, 231-233
pitting ... **M2:** 204-206
ratings ... **M2:** 209-211, 213
seawater ... **M2:** 223-225, 228-232
soil ... **M2:** 22
solution potentials ... **M2:** 206-207
stress-corrosion cracking ... **M2:** 210-211, 212-218
corrosion resistance, ratings ... **A15:** 766
counterboring ... **A16:** 766-767
coupling nut, SCC cracked in marine atmosphere ... **A11:** 209
curling ... **A14:** 804
cutting, effects of ... **A9:** 351
cutting fluids ... **A16:** 125, 128, 761, 765-766, 769-795, 800-801
cutting force and power ... **A16:** 763-764
cutting speed ... **A16:** 765, 769, 770, 775, 779
cutting, tools for ... **M3:** 477
deep drawing ... **A14:** 795-797
deep drawing, tool materials for ... **M3:** 494, 495, 496
defects in cold extrusion of ... **A8:** 591-592, 595
defects, permanent mold casting ... **A15:** 285
deformation bonding ... **A6:** 157
degassing of ... **A15:** 456-462
demagging of ... **A15:** 471-474
dendrite arm spacing vs. cooling rate ... **A7:** 33, 36
deoxidizing process ... **M5:** 8, 12-13
deposition corrosion ... **A13:** 589
designations ... **A18:** 753
diamond tools for machining ... **A16:** 105
die casting alloys of ... **A15:** 286
die casting compositions ... **A15:** 755
die cutting speeds ... **A16:** 301
die design ... **A14:** 246-247
die forging, tool materials for ... **M3:** 529, 530, 532
die manufacture ... **A14:** 247
die material ... **A14:** 246
die threading ... **A16:** 791
diffraction contrast ... **A18:** 387
diffraction techniques, elastic constants, and bulk values for ... **A10:** 382
diffusion bonding ... **A6:** 157
diffusion brazing ... **A6:** 343
diffusion welding ... **A6:** 884-885, 886
dip brazing ... **A6:** 338
direct-chill continuous casting ... **A15:** 313-314
discontinuous metal-matrix composites ... **A14:** 251
dispersoid control ... **A8:** 484
distortion ... **A6:** 727, 728
distortion and dimensional variation ... **A16:** 770-772
drilling ... **A16:** 220-231, 237, 766-769, 775-785, 791
drilling in automatic bar and chucking machines ... **A16:** 782-784
drilling/countersinking ... **A16:** 899
drop hammer forging ... **A14:** 803-804
drop hammer forming ... **A14:** 656-657
dye penetrant testing, preparation for ... **A9:** 351
effect of copper solute content on secondary dendrite arm spacing ... **A9:** 629

Aluminum alloys (continued)
effects, nonmetallic building materials ... **A13:** 600-602
electric current perturbation inspection ... **A17:** 136
electrical discharge machining ... **A16:** 558, 560
electrochemical grinding ... **A16:** 543, 547
electrochemical machining ... **A16:** 533, 534, 535
electroforming in EDM ... **A16:** 560
electrohydraulic forming ... **A14:** 802
electrolytic cleaning of ... **M5:** 578
electromagnetic forming ... **A14:** 802-803
electron beam machining ... **A16:** 570
electron beam welding ... **M6:** 641-642
electron-beam welding ... **A6:** 739, 828, 855, 859, 871-872
electronic applications ... **A6:** 990, 998
electroplating of ... **M5:** 600-610
electroslag welding ... **A6:** 738
elements implanted to improve wear and friction properties ... **A18:** 858
embossing ... **A14:** 804
embrittlement by low-melting alloys ... **A12:** 29
emulsion cleaning of ... **M5:** 577
enamels ... **EM3:** 303
end milling ... **A16:** 325, 766-767, 772-773, 784-787, 790-795
environments that cause stress-corrosion cracking ... **A6:** 1101
erosion-corrosion ... **A13:** 595-596
erosive attack on die surfaces ... **A18:** 630
etchants ... **A9:** 351-357
etching ... **A9:** 351-357
etching of ... **M5:** 8, 582-586, 590
acid ... **M5:** 582-586
alkaline ... **M5:** 582-585, 590
exfoliation corrosion ... **A13:** 334, 594-595, 1032
exothermic brazing ... **A6:** 345
expanding ... **A14:** 804
explosion welding ... **A6:** 739, 896
explosive forming ... **A14:** 800-802
extruded parts ... **A14:** 307-310
extrusion rate vs flow stress ... **A14:** 317
extrusion welding ... **M6:** 677
extrusions ... **M2:** 5-6
extrusions, mechanical properties ... **A7:** 574
fabrication characteristics ... **M2:** 14-16
face milling ... **A16:** 788-789, 791, 793, 795, 797
fasteners ... **M2:** 202-203
fasteners, use in ... **M3:** 184, 185
fatigue at subzero temperatures ... **M3:** 732, 733, 746
fatigue strength, anodic coatings affecting ... **M5:** 597
fatigue striations in ... **A12:** 176
ferrographic application to identify wear particles ... **A18:** 305
filiform corrosion ... **A13:** 596-597, 1034-1034
filler alloys for sustained elevated-temperature service ... **A6:** 729
fillet ratio ... **A16:** 804
FIM sample preparation of ... **A10:** 586
for fine-edge blanking and piercing ... **A14:** 472
finishes ... **M5:** 606-607, 609-610
designation system ... **M5:** 609-610
standards for ... **M5:** 606-607
finishing processes ... **M5:** 573-577, 585-610
flaking during rolling ... **A16:** 281
flash welding ... **M6:** 558
flow stress and workability as function of temperature ... **A14:** 169
fluxing of ... **A15:** 445-447
FM-73, peel test results after surface preparation ... **EM3:** 805
foamed plaster molding of ... **A15:** 247
forge welding ... **A6:** 306 **M6:** 676
forgeability ... **A14:** 241-242 **M2:** 6
forging equipment ... **A14:** 244-245
forging methods ... **A14:** 242-244
forging of ... **A14:** 241-254
forging process ... **A14:** 248
forging, SCC fracture by decohesion ... **A12:** 18, 25
forging temperatures ... **A14:** 242
forgings ... **M2:** 5-9
forgings, cleaning and finishing ... **M5:** 590-591
forgings, flaws and inspection methods ... **A17:** –497
forgings, processing of ... **A14:** 247-249

Aluminum alloys (continued)
formability ... **A14:** 791-792
forming ... **M2:** 14-16
forming equipment and tools ... **A14:** 792-793
forming limit diagrams for ... **A14:** 20
forming of ... **A14:** 791-804
foundry practice for specialty castings ... **A15:** 755-757
foundry products ... **M2:** 140-151
fracture surfaces, preservation of ... **A9:** 351
fracture toughness ... **M3:** 728, 732, 746
fracture toughness testing of ... **A8:** 458-462
fretting wear ... **A18:** 248, 250, 252
friction band sawing ... **A16:** 365
friction surfacing ... **A6:** 321
friction welding ... **A6:** 152, 153, 739, 890 **M6:** 722
furnace brazing ... **A6:** 330
fusion welding to steels ... **A6:** 828
fusion zone ... **A6:** 727
future metalworking of ... **A14:** 20-21
galvanic corrosion ... **A6:** 729 **A13:** 84, 587-589
galvanic couples ... **A6:** 1065, 1066
galvanic series ... **A13:** 587
gas metal arc welding ... **M6:** 153
gas tungsten arc welding ... **M6:** 182, 203
gas-metal arc welding ... **A6:** 180, 722, 723, 724, 726, 729, 730, 731-735, 737-738, 739
gas-metal arc welding shielding gases ... **A6:** 66-67
gas-tungsten arc welding ... **A6:** 725, 729, 730, 731-735, 736, 737, 738, 739, 871
gating and risering ... **A15:** 754-755
general machining conditions ... **A16:** 764-765
gold plating of ... **M5:** 605
grain boundaries ... **A13:** 156
grain refinement ... **A15:** 476-480
grinding ... **A9:** 352 **A16:** 547, 774, 783, 792, 798-802
grinding, effects of ... **A9:** 351
grinding fluids ... **A16:** 801-802
grinding wheel core material ... **A16:** 456
groove joints, tensile strength after welding ... **A6:** 729
gun drilling ... **A16:** 769, 778-779, 782, 784, 791
heat and temperature effects on strength retention ... **A18:** 745
heat treatable ... **A16:** 761
heat treatment ... **A15:** 757-762 **M2:** 28-43
  aging ... **M2:** 29-33, 36-38, 40-42, 43
  castings ... **M2:** 32, 33
  cold work, effect after quenching ... **M2:** 38, 39
  corrosion resistance, effect of quenching ... **M2:** 32-35
  dimensional change during ... **M2:** 39-42, 43
  dimensional stability ... **M2:** 42-43
  precipitation hardening ... **M2:** 29-33, 36-38, 40-42, 43
  quality control ... **M2:** 32
  quenching ... **M2:** 32-35, 40-41, 43
  refrigeration, effect on aging ... **M2:** 35-38
  solution heat treatment ... **M2:** 31, 32, 35, 38-40
  wrought alloys ... **M2:** 30-32
heat-affected zone ... **A6:** 725, 726, 727, 729
heat-treatable cast ... **A6:** 724
heat-treatable commercial wrought, solution potentials ... **A13:** 584
heat-treatable wrought ... **A6:** 723, 728
heat-treated high-strength, intergranular corrosion ... **A13:** 241
HERF forgeability ... **A14:** 104
high removal rate machining ... **A16:** 609
high silicon-content alloys *See* Aluminum alloys, specific types (380, 390)
high-frequency resistance welding ... **M6:** 760
high-speed machining ... **A16:** 597, 598, 600-604
high-speed tool steels used ... **A16:** 57, 58, 59
high-strength / high temperature, forging and rolling ... **A7:** 522
high-strength, grain-boundary separation ... **A12:** 174
high-strength, SCC failure in ... **A11:** 27, 28

Aluminum alloys (continued)
hole flanging ... **A14:** 804
honing ... **A16:** 472, 477, 484, 775, 801, 802
honing stone selection ... **A16:** 476
horns for ultrasonic impact grinding machines ... **A16:** 529
hot extrusion, billet temperatures for ... **M3:** 537
hot extrusion of ... **A14:** 321
hot extrusion, press capacities ... **M3:** 538
hot extrusion, tool materials for dies ... **M3:** 538
hot forging ... **A18:** 625
hot isostatic pressing, effects ... **A15:** 541
hot pressing products ... **A7:** 509
hydraulic forming ... **A14:** 803
hydrogen damage ... **A13:** 169-170
hydrogen effects ... **A15:** 457
hydrogen solubility, effects, and removal ... **A15:** 747-749
hydrogen-embrittled, types ... **A12:** 23-24
ideal oxide configuration for adhesion ... **EM3:** 744
identification of phases ... **A9:** 355-357
identification of temper ... **A9:** 358
immersion plating ... **M5:** 601-607
impacts ... **M2:** 8-10, 13
inclusions in ... **A15:** 488
intergranular corrosion ... **A12:** 126 **A13:** 130, 240-241, 589-590
in irons, for elevated-temperature oxidation resistance ... **A15:** 701
joining ... **M2:** 16, 191-202
laminated coatings ... **M5:** 609-610
lap welding ... **M6:** 673
lapping ... **A16:** 492, 499, 802-803
laser beam machining ... **A16:** 574, 575
laser beam welding ... **M6:** 647, 661
  weld properties ... **M6:** 661
laser cladding with alumina ... **M6:** 798
laser cutting ... **A14:** 742
laser melt/particle injection ... **A18:** 870-871
laser-beam welding ... **A6:** 263, 739
life-limiting factors for die-casting dies ... **A18:** 629
lifting sling, fractured ... **A11:** 528
liquation cracking ... **A6:** 75, 83
liquid erosion resistance ... **A11:** 167
liquid-metal embrittlement ... **A11:** 27-28
lithium and corrosion resistance ... **EM3:** 671
localized attack measurement ... **A13:** 194
low tin, in trimetal bearing material systems ... **A18:** 748
lubricant effect on bearing strength ... **A8:** 60
lubricants for forming ... **A14:** 519, 793
lubrication of tool steels ... **A18:** 738
lug, fracture surfaces in ... **A11:** 31
machinability ... **A16:** 645
machinability grouping ... **A16:** 761-763
machinability, ratings ... **A15:** 766 **A16:** 762, 763
machining ... **M2:** 5, 6, 13, 15, 187-190
  chip characteristics ... **M2:** 189-190
  machinability ratings ... **M2:** 188-189
  surface finish ... **M2:** 190
  tool wear ... **M2:** 187-189
machining problems, sources of ... **A16:** 773-774
macroexamination ... **A9:** 353-354
for match plate pattern plaster mold casting ... **A15:** 245
material for die-casting dies ... **A18:** 629
material for jet engine components ... **A18:** 588
as matrix material ... **EL1:** 1120
mechanical cutting ... **A6:** 1179, 1180
mechanical finishes, designation system ... **M5:** 609-610
mechanical properties ... **A7:** 468, 474 **A15:** 767
melt refining of ... **A15:** 470-471
melting and metal treatment ... **A15:** 746-747
melting heat for ... **A15:** 376
melting point ... **A16:** 601
melting practice, reverberatory furnace ... **A15:** 376-380
melts, forced convection ... **A15:** 453-456

Aluminum alloys (continued)
metal preparation ... **A15:** 753
metal removal rate ... **A16:** 764
in metal-matrix composites ... **A12:** 466 **A13:** 587
metallurgical corrosion effects ... **A13:** 130
microexamination ... **A9:** 354-357
microstructures ... **A9:** 357-360
mill products ... **M2:** 4-5, 44-62
milling ... **A16:** 307, 312-313, 326-329, 547, 766-769, 784, 791-804
modification and refinement ... **A15:** 751-753
modulus of elasticity at different temperatures ... **A8:** 23
mold coatings for ... **A15:** 282
mounting ... **A9:** 352
mounting, effects of ... **A9:** 351
multiple-operation machining ... **A16:** 761, 764, 783, 793, 797
n value ... **A8:** 550
NC machining operations ... **A16:** 774
necking ... **A14:** 804
nickel plating of ... **M5:** 604-605
nitrogen-sintered, properties ... **A7:** 742
nominal compositions ... **A7:** 381
non-heat-treatable cast ... **A6:** 723, 726
non-heat-treatable wrought ... **A6:** 727, 728
nondestructive testing ... **A6:** 1086
nonheat-treatable commercial wrought, solution potentials ... **A13:** 584
nonmetallic inclusions in ... **A11:** 316
not readily diffusion bondable ... **A6:** 156
overheating ... **A16:** 769, 771-774, 784
oxides as inclusions ... **A15:** 95
oxyfuel gas welding ... **A6:** 281, 282, 285, 738-739 **M6:** 583
P/M chips versus processed wrought chips ... **A16:** 890
painting of ... **M5:** 17, 606-609
as part of bimetal bearings ... **A9:** 567
parts, cold extrusion ... **A14:** 307-310
PCD tooling application ... **A16:** 109, 110
peck drilling ... **A16:** 899
peel test use ... **EM3:** 332
percussion welding ... **M6:** 740
peripheral milling ... **A16:** 324, 786-787, 793
in petroleum refining and petrochemical operations ... **A13:** 1263
phase designations ... **A9:** 356-357
phase identification ... **A9:** 358
phosphate coating of ... **M5:** 597-599
photochemical machining ... **A16:** 588, 589, 590, 591
physical properties ... **A15:** 764-765 **A18:** 192
piercing of ... **A14:** 793-794
pitting corrosion ... **A13:** 583-584
plane-strain fracture toughness ... **A8:** 451
planing ... **A16:** 184, 773, 778
plasma and shielding gas compositions ... **A6:** 197
plasma arc cutting ... **A14:** 731 **M6:** 916
plasma arc welding ... **A6:** 195, 197, 199, 735, 736-737 **M6:** 214
plasma-MIG welding ... **A6:** 223, 224
plate, double-shear test results ... **A8:** 63
plate, flat-face tensile fracture in ... **A11:** 76
polishing ... **A9:** 351-353 **M5:** 573-574, 609
polysulfides as sealants ... **EM3:** 196
pop-in precracking ... **A8:** 517
porcelain enameling of *See* Aluminum porcelain enameling of
postweld heat treatments ... **A6:** 83, 726-727, 728
pouring ... **A15:** 754
pouring temperatures ... **A15:** 238, 283
powder compact, die contact surface cracking ... **A14:** 402
powder metallurgy materials microstructures ... **A9:** 511
powder metallurgy parts ... **A9:** 358 **M2:** 10-13
power band sawing ... **A16:** 795, 797, 800
power hacksawing ... **A16:** 796, 800, 801

Aluminum alloys (continued)
power requirements **A16:** 764-765, 791-792
prealloyed P/M **A14:** 250
precipitation temperatures **A9:** 351
precision forgings **A14:** 251-254
press-brake forming of **A14:** 794-795
press-formed parts, materials for forming tools **M3:** 492, 493
pressure welding **A6:** 739
primary testing direction **A8:** 667
process selection, factors affecting **M5:** 581-582
product forms **A18:** 753 **M2:** 4-14
projection welding **A6:** 233 **M6:** 503
propeller blade, fatigue fracture **A11:** 125
properties *See also* data compilations for specific alloys **M2:** 3-4
pure Al **A16:** 761
quality control **A15:** 762
r value **A8:** 550
R-curve for **A11:** 64
rapidly solidified **A9:** 615
ratio-analysis diagrams **A11:** 62-63
reaming **A16:** 767, 780-782, 785-787
recommended hot extrusion tool steels and hardnesses **A18:** 627
recovery temperatures **A9:** 351
refining, with reactive gases **A15:** 80
reflectance values, cleaning and finishing processes affecting **M5:** 596-598
removal from permanent molds **A15:** 284
resinous coatings **M5:** 609-610
resistance brazing **A6:** 342
resistance seam welding **A6:** 241 **M6:** 494
resistance soldering **A6:** 357
resistance spot welding **A6:** 229 **M6:** 479-480
resistance welding *See* Resistance welding of aluminum alloys **A6:** 833, 834, 840, 841, 847, 848-849
Rockwell scale for **A8:** 76
roll welding **A6:** 312 **M6:** 676
roller burnishing **A16:** 253, 787
for rolling **A14:** 343, 355
rubber-modified epoxy **EM3:** 329
rubber-pad forming **A14:** 799-800
salt corrosion **A13:** 598-599
sampling and chemical analysis of **A7:** 248
satin finishing of **M5:** 574-577
sawing **A16:** 357, 358, 800-801
SCC of **A12:** 18, 25, 28, 133
SCC protection systems **A13:** 607
SCC resistance **A13:** 264-268, 1023
SCC testing of **A8:** 498-499, 519, 523-525
seawater effects **A13:** 603-607
second-phase constituents, solution potentials **A13:** 585
sectioning **A9:** 351-352
selection of alloy and temper **A16:** 764
semisolid **A15:** 327, 334-336
semisolid forged, mechanical properties **A15:** 333
SENB specimens **EM3:** 447
shaping **A16:** 191, 778
sheet, axial-stress fatigue strength **A13:** 595
sheet, blanking-shear and single-shear tests compared **A8:** 65
sheet, stamping **M2:** 13, 14, 180-186
  biaxial stretching **M2:** 180
  characteristics **M2:** 183-184
  deep drawing **M2:** 180
  flanging **M2:** 180, 186
  forming-limit diagrams **M2:** 182-183
  material properties **M2:** 180-181
  pure bending **M2:** 180
  shape analysis **M2:** 184, 185
  stretch bending **M2:** 180, 181, 186
  stretch/draw **M2:** 184-186
  tests **M2:** 180-182
sheet, thick, weathering data **A13:** 598-599
shielded metal arc welding **A6:** 738 **M6:** 75
shot peening **A14:** 803
shot peening of **M5:** 141-142, 145
shrinkage allowances **A15:** 303
silver plating of **M5:** 605-606
simultaneous honing of Al and gray iron **A16:** 802
sintered in dissociated ammonia **A7:** 384
sintered in vacuum **A7:** 384

Aluminum alloys (continued)
sintering **A7:** 381-385
skin milling **A16:** 795
slide welding **M6:** 674
slotting **A16:** 790
smut removal **M5:** 580-581, 584-585, 590-591
for soft metal bearings **A11:** 483, 484
soft spots **A16:** 772-773
solderable and protective finishes for substrate materials **A6:** 979
soldering **A6:** 628, 631, 632, 739 **M2:** 200-201
solid-state phase transformations in welded joints **A9:** 481
solid-state transformations in weldments **A6:** 83
solidification structures in welded joints **A9:** 479
solution heat treatment **A6:** 83 **A11:** 122
solution potentials **A13:** 584-585
solvent cleaning of **M5:** 576-577
sources of porosity **M6:** 839-840
for space boosters/satellites **A13:** 1101
spade drilling **A16:** 225, 777, 782
spar mill defect **A16:** 766, 772
specific power **A16:** 18
specimen preparation **A9:** 351-353
specular finishes **M5:** 574, 576, 580-581, 609-610
spinning of **A14:** 797-798
springback in **A8:** 552, 564
squeeze casting of **A15:** 323
stability **A15:** 762
stamping **A14:** 804
stampings **M2:** 4, 13-14
standard designations **A15:** 743
storage time and temperature **EM3:** 522
strain rate regimes for **A8:** 519
strain-hardenable alloys **A16:** 761
strain-hardened 5xxx, intergranular corrosion **A13:** 240-241, 264-268, 590-594, 607, 1023
stress-corrosion cracking **A6:** 727
stress-corrosion cracking in **A11:** 218-220
stress-corrosion cracking resistance **A8:** 522
stretch forming **A14:** 798-799
stripping of **M5:** 218
structure control **A15:** 749-751
stud arc welding **M6:** 730, 733
stud material **M6:** 730
suitability for cladding combinations **M6:** 691
and sulfur compounds **A16:** 35
superplastic forming **A14:** 800
surface parameters **EM3:** 41
surface preparation **EM3:** 521
susceptibility to hydrogen damage **A11:** 249
susceptibility to porosity **M6:** 44
tapping **A16:** 261
tapping, cold form **A16:** 266, 267
temper designations **M2:** 24-27
temperature, and extrusion **A14:** 318
and tempers, corrosion ratings **A13:** 587
tensile properties at subzero temperatures **M3:** 722-730, 733-746
testing with precracked specimens **A8:** 524
thermal energy method of deburring **A16:** 577-578
thermal expansion coefficient **A6:** 907
thermal properties **A6:** 17
for thermal spray coatings **A13:** 460-461
thermal treatments **A7:** 381
thermodynamic properties **A15:** 55-60
thread grinding **A16:** 270
thread milling **A16:** 269
thread rolling **A16:** 288, 290, 291, 292-293
threading **A16:** 766-767, 769, 791
threading using radial chasers **A16:** 297
Ti-6Al-4V
  aerospace applications **EM3:** 264
  anodization procedures **EM3:** 265, 266
  chromic acid anodization **EM3:** 266, 267, 269, 270
  wedge-crack propagation test **EM3:** 268, 269
tin free, in trimetal bearing material systems **A18:** 748
tin plating of **M5:** 604-605
tool design **A16:** 761, 766, 769-771, 773-775, 787
tool life **A16:** 761, 764-765, 768-772, 775-785, 797
tool material **A16:** 766-768

Aluminum alloys (continued)
tube stock **A14:** 671
turning **A16:** 94, 110, 135-136, 380-381, 766-770, 774-777
ultrasonic fatigue testing **A8:** 252
ultrasonic welding **A6:** 739, 894, 895 **M6:** 746
Unicast process for **A15:** 251
upset welding **A6:** 249
vacuum effects **A12:** 46
vacuum induction melting **A15:** 396
versus steel properties **A18:** 712-713
vitreous coatings **M5:** 609-610
water corrosion **A13:** 597-598
weld microstructures **A6:** 51
weld model **A6:** 1133
weld-crack resistance in **A11:** 435
weldability, ratings **A15:** 766
weldbonding **M2:** 202
welding **A15:** 763
  filler metals **M2:** 193-194
  finishing **M2:** 195-196
  joint preparation **M2:** 193-195
  joint types **M2:** 193-194
  processes **M2:** 196-199
  weld strength **M2:** 195, 197-199
  weldability **M2:** 192-193
welding electrodes **A6:** 176
weldments **A13:** 344-345
wire, bonding **EL1:** 224-226
wire, die materials for drawing **M3:** 522
for wire-drawing dies **A14:** 336
workability **A8:** 165, 575
workability test for **A7:** 411
wrought alloys
  bar **M2:** 51-52
  designation system **M2:** 44-51
  elevated-temperature properties **M2:** 56, 57-58, 62
  extrudability **M2:** 54
  extrusions, interconnecting **M2:** 55, 57-58
  flat rolled products **M2:** 51
  low-temperature properties **M2:** 62
  mechanical properties **M2:** 55, 58, 59-62
  physical properties **M2:** 53-54, 58
  rod **M2:** 51-52
  shapes **M2:** 52-57
  tubular products **M2:** 52
  wire **M2:** 51-52
wrought, cleaning and finishing **M5:** 574, 576, 583, 588, 597, 603-604
wrought commercial, SCC resistance ratings **A11:** 220
YAG lot
zinc plating of **M5:** 601-606
zincating process **M5:** 601-606

Aluminum alloys, annealing
castings **M4:** 709
controlled-atmosphere **M4:** 709
full annealing **M4:** 707-708
partial annealing **M4:** 708-709
stress-relief **M4:** 709
temperature control **M4:** 709

Aluminum alloys, brazing sheet, specific types, annealing
11 ..... **M4:** 708
12 ..... **M4:** 708
21 ..... **M4:** 708
22 ..... **M4:** 708
23 ..... **M4:** 708
24 ..... **M4:** 708

Aluminum alloys, cast, specific types
319.0, composition **A6:** 529
355.0, composition **A6:** 529
356.0, composition **A6:** 529, 724
357.0, composition **A6:** 529

Aluminum alloys, casting, specific types *See* Aluminum casting alloys, specific types

Aluminum alloys, heat treating *See also* Aluminum alloys, quenching; Aluminum alloys, solution heat treating **A4:** 841-879
aging, natural **A4:** 841, 842, 843, 844, 856, 859-866, 869 **M4:** 697-700
alloy systems **A4:** 841-842
alloying additions **A4:** 842-843
annealing **A4:** 869-871, 872
applications **A4:** 865
brazing sheet, annealing **A4:** 871

Aluminum alloys, heat treating (continued)
cold work strain hardening **A4:** 865, 866, 870–871, 879 **M4:** 699
corrosion behavior **A4:** 851, 856, 857-858, 873, 876, 877, 878
dimensional changes **A4:** 875-876 **M4:** 713-714
dimensional stability **A4:** 876 **M4:** 714
distortion and its control after heat treatment **A4:** 617
electrical conductivity **A4:** 843-844, 856, 858, 877-878 **M4:** 715
equipment maintenance **A4:** 872-873, 876 **M4:** 714
forming **A4:** 841, 869 **M4:** 695
fracture toughness **A4:** 856, 860, 867, 870, 876, 878 **M4:** 717
furnaces, air **A4:** 848, 872-873, 874 **M4:** 711
grain growth **A4:** 842, 843, 856, 865, 869, 871-872, 877 **M4:** 709-710
Guinier-Preston (GP) zone solvus line **A4:** 841-842, 860, 862
hardness tests **A4:** 877, 878 **M4:** 715, 716
induction heating **A4:** 872-873
intergranular-corrosion test **A4:** 852, 857, 858, 877, 878 **M4:** 715
Lüders lines formation **A4:** 856
with magnesium alloys in same furnace **A4:** 902
precipitate-free zones (PFZs) **A4:** 843
precipitation treatments **A4:** 841-844, 845-847, 859-866, 867, 868, 874, 875, 876, 877, 879 **M4:** 676, 677-684, 685-686, 700, 713-714
property development **A4:** 867
quality assurance **A4:** 876-878 **M4:** 714-717
reheating **A4:** 853, 865, 868-869, 870 **M4:** 701-703, 706, 707
residual stresses **A4:** 608-609
safety precautions **A4:** 844, 850, 873, 874 **M4:** 710-711
salt baths **A4:** 848, 872-873, 874 **M4:** 684, 710-712
soak time **A4:** 848, 862, 863, 865-867, 869, 872-873 **M4:** 704
solubility-temperature relationship **A4:** 841, 842 **M4:** 675-676, 677-683, 684, 685-686
stress relief, mechanical **A4:** 854, 867-868, 876, 879 **M4:** 694, 695-696
stress relief, thermal treatments **A4:** 854, 855, 867-868, 870-871, 876, 879 **M4:** 696-697
sulfur dioxide atmosphere harmful **A4:** 902
temper designations **A4:** 878-879 **M4:** 717-718
temper(s) **A4:** 844-847, 860-862, 864-872, 875-879
temperature control **A4:** 841, 844, 865, 871, 873-875 **M4:** 704
instrument calibration **A4:** 874, 875 **M4:** 712-713
probe checks **A4:** 874 **M4:** 711
radiation effects **A4:** 865, 874-875 **M4:** 712
sensing elements **A4:** 874, 875 **M4:** 711
uniformity surveys **A4:** 874-875 **M4:** 711-712
tempers **M4:** 700-704, 705
tensile strength **A4:** 867
tensile tests **A4:** 852, 860, 876-877 **M4:** 714-715
transverse-flux heating **A4:** 873
yield strength **A4:** 862-863, 864, 867, 868, 870

Aluminum alloys, powder metallurgy, specific types *See also* Aluminum alloys, wrought, specific types; Aluminum casting alloys, specific types
201 AB
as-sintered properties **M2:** 13
fatigue curves **M2:** 15
tensile strength, effects of density and thermal condition on **M2:** 14
601 AB
as-sintered properties **M2:** 11
fatigue curves **M2:** 14
tensile strength, effects of density and thermal condition on **M2:** 13

Aluminum alloys, quenching *See also* Aluminum alloys, heat treating; Aluminum alloys, solution heat treating **A4:** 844, 848, 851-859, 867-870
air-blast **A4:** 858
castings **A4:** 866-867, 871 **M4:** 688, 689
corrosion behavior **A4:** 851, 856, 857-858 **M4:** 691, 692
delay **A4:** 851-852 **M4:** 684, 688-689
dimensional changes **A4:** 851, 875 **M4:** 713-714
Lüders lines formation **A4:** 856
mechanical properties **A4:** 851, 852, 858
quench severity (Grossmann numbers) **A4:** 853, 854, 859
quench-factor analysis **A4:** 851, 852, 856-859 **M4:** 690-691
residual stress **A4:** 851, 854-855, 856, 868 **M4:** 692-695
spray **A4:** 851, 852-853, 855 **M4:** 690
tensile strength **A4:** 852, 853
warpage **A4:** 854-855, 857, 867, 868, 869, 875 **M4:** 692-695
water immersion **A4:** 851, 852, 853, 854, 858
water-immersion **M4:** 689-690
wrought **A4:** 853, 854, 856, 866-867 **M4:** 688
yield strength **A4:** 851, 852, 853, 858-859, 866-867 **M4:** 691-692, 693

Aluminum alloys, solution heat treating *See also* Aluminum alloys, heat treating; Aluminum alloys, quenching **A4:** 841, 844-851, 867, 872, 878-879
dimensional changes **A4:** 848, 850-851, 867, 875 **M4:** 713
high-temperature oxidation **A4:** 848-851 **M4:** 687-688
nonequilibrium melting **A4:** 844 **M4:** 684
overheating **A4:** 844 **M4:** 683-684
treating time **A4:** 844, 848 **M4:** 684-687
underheating **A4:** 844-845
vacuum heat-treating support fixture material **A4:** 503

Aluminum alloys, specific types *See also* Aluminum; Aluminum alloys; Aluminum alloys, specific types; Aluminum P/M parts; Aluminum powders; Aluminum powders, specific types; Atomized aluminum powders; High-strength aluminum alloys; Prealloyed aluminum powders; Sleeve bearing alloys, specific types; Wrought aluminum alloys
1*xxx* series
base alloy, filler alloy for welding, for sustained elevated-temperature service **A6:** 729
capacitor discharge stud welding **A6:** 222
weldability **A6:** 725
welding, for sustained elevated-temperature service **A6:** 729
1*xxx* series, insoluble particles in **A9:** 358
2*xxx* series
corrosion resistance **A6:** 729
elevated-temperature properties **A6:** 729
weldability **A6:** 725
2*xxx* series, macroetching **A9:** 354
2*xxx* series, soluble phases **A9:** 358
3*xxx* series, capacitor discharge stud welding **A6:** 222
3*xxx* series, homogenization of **A9:** 632
3*xxx* series, intermetallic phases **A9:** 358
5*xxx* series
brazeability **A6:** 937
capacitor discharge stud welding **A6:** 222
corrosion resistance **A6:** 729-730
ductility **A6:** 728
filler alloy choices **A6:** 730
gas-metal arc welding **A6:** 738
hydrogen solubility **A6:** 722
not recommended for use at sustained temperatures **A6:** 729
stud arc welding **A6:** 215

Aluminum alloys, specific types (continued)
surface preparation **A6:** 736
5*xxx* series, insoluble particles **A9:** 358
6*xxx* series
capacitor discharge stud welding **A6:** 222
electron-beam welding **A6:** 739
weldability **A6:** 725
6*xxx* series, macroetching **A9:** 354
6*xxx* series, soluble phases **A9:** 358
6*xxx*-T4 series, groove weld strength **A6:** 727
7*xxx* series
corrosion resistance **A6:** 729
ductility **A6:** 728
weldability **A6:** 725-726
7*xxx* series, macroetching **A9:** 354
7*xxx* series, soluble phases **A9:** 358
43, porcelain enameling of **M5:** 513
82Al-9Pd-9Ga, brazing **A6:** 944
98Zn-2Al **A6:** 351
99.99%, seawater pitting **A13:** 907
201, graphite-aluminum composite **A9:** 594
201, machining **A16:** 771-774, 776, 777, 779, 781, 782, 786, 788-790, 794-796, 798-800
201-F, as premium quality cast, with shrinkage cavities **A9:** 377
201-T7, premium quality cast, solution heat treated and stabilized **A9:** 377
201.0 applications and properties **A2:** 152-153
201AB, effect of vacuum level on tensile strength **A7:** 384
201AB, electrical and thermal conductivity **A7:** 742
201AB, mechanical properties **A7:** 474
201AB, powder material, microstructure **A9:** 511
201AB, properties **A7:** 742
202AB, compacts, properties **A7:** 742
204.0, applications and properties **A2:** 154
206.0, applications and properties **A2:** 154-155
208, cleaning and finishing **M5:** 604, 606
208, machining **A16:** 763, 771-774, 776, 777, 779, 781, 782, 786, 788-790, 794-796, 798-800
208.0
composition **A6:** 723
properties **A6:** 723
weldability **A6:** 723
208.0, applications and properties **A2:** 155-156
212, contour band sawing **A16:** 363
213, machining **A16:** 763, 771-774, 776, 777, 779, 781, 782, 786, 788-790, 794-796, 798-800
222, machining **A16:** 763, 771-774, 776, 777, 779, 781, 782, 786, 788-790, 794-796, 798-800
222-T61, sand cast, solution heat treated and artificially aged **A9:** 372
222.0
composition **A6:** 724
physical properties **A6:** 724
weldability **A6:** 724
welding, for sustained elevated-temperature service **A6:** 729
224, close-tolerance alloy **A15:** 769
224, machining **A16:** 771-774, 776, 777, 779, 781, 782, 786, 788-790, 794-796, 798-800
224-F, as premium quality cast **A9:** 378
224-T7, premium quality cast, solution heat treated and stabilized **A9:** 378
238, machinability rating **A16:** 763
238-F, as permanent mold cast **A9:** 372
238.0
composition **A6:** 723
properties **A6:** 723
weldability **A6:** 723
238.0, applications and properties **A2:** 156-157
238.0-F, resistance spot welding **A6:** 848
238.0-F, resistance welding **M6:** 536
240.0
composition **A6:** 724
physical properties **A6:** 724

SUBJECTS OF THE INDEXED VOLUMES: ASM Handbook (designated by the letter "A"): **A1:** Properties and Selection: Irons, Steels, and High-Performance Alloys (1990); **A2:** Properties and Selection: Nonferrous Alloys and Special-Purpose Materials (1990); **A3:** Alloy Phase Diagrams (1992); **A4:** Heat Treating (1991); **A6:** Welding, Brazing, and Soldering (1993); **A7:** Powder Metallurgy (1984); **A8:** Mechanical Testing (1985); **A9:** Metallography and Microstructures (1985); **A10:** Materials Characterization (1986); **A11:** Failure Analysis and Prevention (1986); **A12:** Fractography (1987); **A13:** Corrosion (1987); **A14:** Forming and Forging (1988); **A15:** Casting (1988); **A16:** Machining (1989); **A17:** Nondestructive Testing and Quality Control (1989); **A18:** Friction, Lubrication, and Wear Technology (1992). **Metals Handbook, 9th Edition** (designated by the letter "M"): **M1:** Properties and Selection: Irons and Steels (1978); **M2:** Properties and Selection: Nonferrous Alloys and Pure Metals (1979); **M3:** Properties and Selection: Stainless Steels, Tool Materials and Special-Purpose Materials (1980); **M4:** Heat Treating (1981); **M5:** Surface Cleaning, Finishing, and Coating (1982); **M6:** Welding, Brazing, and Soldering (1983). **Engineered Materials Handbook** (designated by the letters "EM"): **EM1:** Composites (1987); **EM2:** Engineering Plastics (1988); **EM3:** Adhesives and Sealants (1990); **EM4:** Ceramics and Glasses (1991); **Electronic Materials Handbook** (designated by the letters "EL"): **EL1:** Packaging (1989).

Aluminum alloys, specific types (continued)
weldability ... **A6:** 724
242, machining ... **A16:** 763, 771-774, 776, 777, 779, 781, 782, 786, 788-790, 794-796, 798-800
242-F, as permanent mold cast ... **A9:** 373
242-T77, sand cast and heat treated ... **A9:** 373
242-T571, permanent mold cast and artificially aged ... **A9:** 373
242.0
composition ... **A6:** 529, 724
physical properties ... **A6:** 724
weldability ... **A6:** 724
242.0, applications and properties ... **A2:** 157
295, cleaning and finishing ... **M5:** 604, 606
295, machining ... **A16:** 763, 771-774, 776, 777, 779, 781, 782, 786, 788-790, 794-796, 798-800
295-T6 investment casting, electron beam weld ... **A9:** 385
295.0
composition ... **A6:** 724
physical properties ... **A6:** 724
weldability ... **A6:** 724
welding, for sustained elevated-temperature service ... **A6:** 729
295.0, applications and properties ... **A2:** 157-159
296.0, applications and properties ... **A2:** 159-160
300M, recommended upsetting pressures for flash welding ... **A6:** 843
308, machining ... **A16:** 763, 771-774, 776, 777, 779, 781, 782, 786, 788-790, 794-796, 798-800
308-F, as permanent mold cast ... **A9:** 376
308.0, applications and properties ... **A2:** 160
308.0-F, resistance spot welding ... **A6:** 848
308.0-F, resistance welding ... **M6:** 536
312, contour band sawing ... **A16:** 363
316, FMR probes for ... **A17:** 221
319, automotive application ... **A15:** 769
319, cleaning and finishing ... **M5:** 603-604, 606
319, machining ... **A16:** 763, 768, 771-774, 776, 777, 779, 781, 782, 786, 788-790, 794-796, 798-800
319, purge gas efficiency ... **A15:** 462
319-F, as permanent mold cast ... **A9:** 376
319-T6, permanent mold cast, solution heat treated, artificially aged ... **A9:** 376
319.0
composition ... **A6:** 724
electrode potential in NaCl-H_2O_2 solution ... **A6:** 730
physical properties ... **A6:** 724
relative rating of filler alloys for welding ... **A6:** 731
weldability ... **A6:** 534, 724
welding, for sustained elevated-temperature service ... **A6:** 729
319.0, applications and properties ... **A2:** 160-161
328, machining ... **A16:** 771-774, 776, 777, 779, 781, 782, 786, 788-790, 794-796, 798-800
332.0
composition ... **A6:** 724
physical properties ... **A6:** 724
weldability ... **A6:** 724
332.0, applications and properties ... **A2:** 161
333, machining ... **A16:** 763, 771-774, 776, 777, 779, 781, 782, 786, 788-790, 794-796, 798-800
333.0
composition ... **A6:** 724
electrode potential in NaCl-H_2O_2 solution ... **A6:** 730
physical properties ... **A6:** 724
relative rating of filler alloys for welding ... **A6:** 731
weldability ... **A6:** 534, 724
welding, for sustained elevated-temperature service ... **A6:** 729
333.0-T6, resistance spot welding ... **A6:** 848
333.0-T6, resistance welding ... **M6:** 536
335.0, applications and properties ... **A2:** 161-162
336.0
composition ... **A6:** 724
physical properties ... **A6:** 724
weldability ... **A6:** 724
336.0, plug gaging of, wear of gage materials ... **M3:** 556
339.0, applications ... **A2:** 162
354, machining ... **A16:** 763, 771-774, 776, 777, 779, 781, 782, 786, 788-790, 794-796, 798-800

Aluminum alloys, specific types (continued)
354.0
composition ... **A6:** 724
physical properties ... **A6:** 724
relative rating of filler alloys for welding ... **A6:** 731
weldability ... **A6:** 534, 724
welding, for sustained elevated-temperature service ... **A6:** 729
354.0, applications and properties ... **A2:** 162-163
355, cleaning and finishing ... **M5:** 604, 606
355, for match plate pattern plaster mold casting ... **A15:** 245
355, machining ... **A16:** 763, 771-774, 776, 777, 779, 781, 782, 786, 788-790, 794-796, 798-800
355-F, as investment cast ... **A9:** 374
355-F, modified with Al-IOSr, as investment cast ... **A9:** 374
355-T6, permanent mold cast, solution heat treated and artificially aged ... **A9:** 374
355.0
composition ... **A6:** 724
physical properties ... **A6:** 724
relative rating of filler alloys for welding ... **A6:** 731
weldability ... **A6:** 534, 724
welding, for sustained elevated-temperature service ... **A6:** 729
355.0, applications and properties ... **A2:** 163-164
355.0, composition/characteristics ... **A15:** 159
355.0, use for combination gages ... **M3:** 556
355.0-T6, electrode potential in NaCl-H_2O_2 solution ... **A6:** 730
356 and 356-T-6, cleaning and finishing ... **M5:** 515, 597, 599, 603, 605
356, gas effect on tensile/yield strengths ... **A15:** 457
356, machining ... **A16:** 763, 771-774, 776, 777, 779, 781, 782, 786, 788-790, 794-796, 798-800
356, permanent mold castings ... **A15:** 275
356, purge gas efficiency ... **A15:** 462
356-F, as investment cast with sodium-
356-F, as investment cast with sodium-modified ingot ... **A9:** 374
356-F, as sand cast ... **A9:** 374
356-F, modified with 0.025% Na, as sand cast ... **A9:** 374
356-T4, modified with 0.025% Na, sand cast and heat treated ... **A9:** 374
356-T4, sand cast, solution heat treated and quenched ... **A9:** 374
356-T6, investment cast in a hot mold ... **A9:** 374
356-T6, investment cast with sodium-modified ingot, solution heat treated, artificially aged ... **A9:** 374
356-T6, permanent mold casting, hydrogen porosity ... **A9:** 375
356-T7, modified with sodium, sand cast solution heat treated, and stabilized ... **A9:** 375
356-T51, power requirements ... **A16:** 765
356-T51, sand cast, artificially aged ... **A9:** 375
356.0
brazeability ... **A6:** 937
filler alloys for best color match ... **A6:** 730
melting range ... **A6:** 937
physical properties ... **A6:** 724
relative rating of filler alloys for welding ... **A6:** 731
resistance brazing ... **A6:** 342
weldability ... **A6:** 534, 724, 728
welding, for sustained elevated-temperature service ... **A6:** 729
356.0, applications and properties ... **A2:** 164-165
356.0, use for combination gages ... **M3:** 556
356.0-T6
electrode potential in NaCl-H_2O_2 solution ... **A6:** 730
resistance spot welding ... **A6:** 848
356.0-T6, resistance welding ... **M6:** 536
357, dendritic/nondendritic microstructures compared ... **A15:** 327
357, machining ... **A16:** 763, 771-774, 776, 777, 779, 781, 782, 786, 788-790, 794-796, 798-800
357, magnetohydrodynamically cast ... **A15:** 329
357, semisolid automotive forg-ng ... **A15:** 335

Aluminum alloys, specific types (continued)
357, strain impact on microstructure SIMA processed ... **A15:** 330
357.0, applications and properties ... **A2:** 166
357.0, composition ... **A6:** 529
359, machining ... **A16:** 763, 771-774, 776, 777, 779, 781, 782, 786, 788-790, 794-796, 798-800
359.0
composition ... **A6:** 724
physical properties ... **A6:** 724
relative rating of filler alloys for welding ... **A6:** 731
weldability ... **A6:** 534, 724
welding, for sustained elevated-temperature service ... **A6:** 729
359.0, applications and properties ... **A2:** 166-167
360, machining ... **A16:** 763, 771-774, 776, 777, 779, 781, 782, 786, 788-790, 794-796, 798-800
360.0
composition ... **A6:** 723
electrode potential in NaCl-H_2O_2 solution ... **A6:** 730
properties ... **A6:** 723
weldability ... **A6:** 723
360.0, applications and properties ... **A2:** 167-168
360.0, composition ... **A15:** 159
364, machinability rating ... **A16:** 763
380 and A380, cleaning and finishing ... **M5:** 573-574, 584, 603
380, automotive application ... **A15:** 769
380, machining ... **A16:** 363, 640, 763, 768, 771-774, 776, 777, 779, 781, 782, 786, 788-791, 793-800
380-F, die casting ... **A9:** 378-379
380.0
composition ... **A6:** 723
electrode potential in NaCl-H_2O_2 solution ... **A6:** 730
properties ... **A6:** 723
relative rating of filler alloys for welding ... **A6:** 731
weldability ... **A6:** 534, 723
380.0, applications and properties ... **A2:** 168-169
380.0, composition ... **A15:** 159
383, machining ... **A16:** 771-774, 776, 777, 779, 781, 782, 786, 788-790, 794-796, 798-800
383.0, applications and properties ... **A2:** 169-170
384, machinability rating ... **A16:** 763
384, sludge crystals from ... **A9:** 380
384-F, die casting ... **A9:** 379-380
384.0, applications and properties ... **A2:** 170-171
390, laser cladding ... **M6:** 799
390, machining ... **A16:** 640, 763, 765, 768-777, 779-783, 786-800
390.0, applications and properties ... **A2:** 171
390.0, composition ... **A15:** 159
392, machining ... **A16:** 771-774, 776, 777, 779, 781, 782, 786, 788-790, 794-796, 798-800
392-F, as permanent mold cast ... **A9:** 376
392-F, as permanent mold cast with phosphorus added ... **A9:** 376
413, machining ... **A16:** 763, 771-774, 776, 777, 779, 781, 782, 786, 788-790, 794-796, 798-800
413-F, as die cast ... **A9:** 375
413-F, die castings, different defects from the gate area ... **A9:** 380-381
413.0
composition ... **A6:** 723
electrode potential in NaCl-H_2O_2 solution ... **A6:** 730
properties ... **A6:** 723
relative rating of filler alloys for welding ... **A6:** 731
weldability ... **A6:** 534, 723
welding, for sustained elevated-temperature service ... **A6:** 729
413.0, applications and properties ... **A2:** 171-172
413.0, composition ... **A15:** 159
413.0-F, resistance spot welding ... **A6:** 848
413.0-F, resistance welding ... **M6:** 536
443, machinability rating ... **A16:** 763
443-F, as sand cast ... **A9:** 376
443.0
brazeability ... **A6:** 937
composition ... **A6:** 723
electrode potential in NaCl-H_2O_2 solution ... **A6:** 730
filler alloys for best color match ... **A6:** 730

Aluminum alloys, specific types (continued)
melting range ... A6: 937
properties ... A6: 723
relative rating of filler alloys for welding ... A6: 731
resistance brazing ... A6: 342
weldability ... A6: 534, 723
welding, for sustained elevated-temperature service ... A6: 729
443.0, applications and properties ... A2: 172-173
443.0-F, resistance spot welding ... A6: 848
443.0-F, resistance welding ... M6: 536
444.0, relative rating of filler alloys for welding ... A6: 731
511.0
composition ... A6: 723
filler alloys for best color match ... A6: 730
properties ... A6: 723
weldability ... A6: 534, 723
512.0
composition ... A6: 723
properties ... A6: 723
weldability ... A6: 534, 723
513.0
composition ... A6: 723
properties ... A6: 723
weldability ... A6: 534, 723
513.0-F, resistance spot welding ... A6: 848
513.0-F, resistance welding ... M6: 536
514, machining ... A16: 763, 771-774, 776, 777, 779, 781, 782, 786, 788-790, 794-796, 798-800
514.0
composition ... A6: 723
electrode potential in NaCl-H_2O_2 solution ... A6: 730
filler alloys for best color match ... A6: 730
properties ... A6: 723
relative rating of filler alloys for welding ... A6: 731, 732, 733
weldability ... A6: 534, 723
514.0, applications and properties ... A2: 173
518, as die casting, composition ... A15: 286
518, machining ... A16: 763, 771, 772, 776, 779, 781, 782, 786, 788-790, 794, 796, 800
518.0
composition ... A6: 723
properties ... A6: 723
weldability ... A6: 723
518.0, applications and properties ... A2: 173-174
520, machining ... A16: 763, 771-774, 776, 777, 779, 781, 782, 786, 788-790, 794-796, 798-800
520-F, as sand cast, effect of solution heat treatment ... A9: 377
520.0
composition ... A6: 724
physical properties ... A6: 724
weldability ... A6: 724
520.0, applications and properties ... A2: 174
520.0-T4, resistance spot welding ... A6: 848
520.0-T4, resistance welding ... M6: 536
535, machining ... A16: 763, 771-774, 776, 777, 779, 781, 782, 786, 788-790, 794-796, 798-800
535.0
composition ... A6: 723
filler alloys for best color match ... A6: 730
properties ... A6: 723
weldability ... A6: 723
535.0, applications and properties ... A2: 174-175
535.0, hot tear sensitivity ... A15: 617
601AB, criteria on of deformation ... A7: 411
601AB, effect of vacuum level on tensile strength ... A7: 384
601AB, electrical and thermal conductivity ... A7: 742
601AB, mechanical properties ... A7: 474
601AB, powder metallurgy materials microstructure ... A9: 511
601AB, properties ... A7: 742
601AB, workability test ... A7: 411

Aluminum alloys, specific types (continued)
601AC, atomized powder ... A9: 529
602AB, electrical and thermal conductivity ... A7: 742
602AB, properties ... A7: 742
705, machining ... A16: 763, 771-774, 776, 777, 779, 781, 782, 786, 788-790, 794-796, 798-800
707, machining ... A16: 763, 771-774, 776, 777, 779, 781, 782, 786, 788-790, 794-796, 798-800
710.0
brazeability ... A6: 937
composition ... A6: 723
melting range ... A6: 937
properties ... A6: 723
weldability ... A6: 534, 723
711.0
brazeability ... A6: 937
composition ... A6: 723
melting range ... A6: 937
properties ... A6: 723
weldability ... A6: 534, 723
712-F, resistance spot welding ... A6: 848
712-F, resistance welding ... M6: 536
712.0
composition ... A6: 723
properties ... A6: 723
weldability ... A6: 534, 723
712.0, applications and properties ... A2: 175
713, machining ... A16: 763, 771-774, 776, 777, 779, 781, 782, 786, 788-790, 794-796, 798-800
713.0, applications and properties ... A2: 175-176
750, contour band sawing ... A16: 363
771, machining ... A16: 771-774, 776, 777, 779, 781, 782, 786, 788-790, 794-796, 798-800
771.0, applications and properties ... A2: 176
850, machining ... A16: 763, 771-774, 776, 777, 779, 781, 782, 786, 788-790, 794-796, 798-800
850-F, as permanent mold cast, with hot tear ... A9: 377
850.0, applications and properties ... A2: 176-177
1050, composition ... A6: 538
1060
composition ... A6: 538, 722
mechanical properties, gas-shielded arc welded butt joints ... A6: 727
physical properties ... A6: 722
relative rating of filler alloys for welding ... A6: 731, 732, 733, 734, 735
weldability ... A6: 534, 722
1060, extrudability ... A2: 35
1060, heat exchanger tube alloy ... A2: 33
1060, machining ... A16: 762, 771-774, 776, 777, 779, 781, 782, 786, 788-790, 794-796, 798-800
1060, resistance welding ... M6: 536-539
1060-0, strain rate sensitivity ... A8: 38, 40
1060-0, uniaxial stress/strain/strain rate data for ... A8: 40
1060-H18, resistance spot welding ... A6: 848
1070, weldability ... A6: 534
1080, weldability ... A6: 534
1100
applications ... A6: 537
Auger electron spectroscopy, electron micrographs ... A18: 454, 455
brazeability ... A6: 937
cleaning and finishing of ... M5: 572, 580, 583, 599, 603-604, 606
composition ... A6: 533, 538, 722, 724 M3: 723
corrosion resistance ... M6: 536
electrode potential in NaCl-H_2O_2 solution ... A6: 730
electron-beam welding ... A6: 873
erosion ... A18: 200
fasteners, use for ... M3: 184, 185
filler alloy ... A6: 537-538
filler alloy, electrode potential in NaCl-H_2O_2 solution ... A6: 730
filler alloy for welding, for sustained elevated temperature service ... A6: 729

Aluminum alloys, specific types (continued)
filler alloy, relative rating for fillet welding or butt welding two component base alloys ... A6: 734, 735
filler alloy, ultimate tensile strength at selected temperatures of GSAW groove joints ... A6: 729
filler alloy, weldability ... A6: 534
filler alloys for best color match ... A6: 730
friction welding ... A6: 152, 153, 154 M6: 722
gages, combination, use for ... M3: 556
hot dip coating with ... M5: 339
mechanical properties, gas-shielded arc welded butt joints ... A6: 727
melting range ... A6: 724, 937
minimum shear strengths of fillet welds ... A6: 728
physical properties ... A6: 722
porcelain enameling of ... M5: 513
press forming, tool materials for ... M3: 492, 493
relative rating of filler alloys for welding ... A6: 731, 732, 733, 734, 735
resistance welding ... M6: 536, 539
roll welding ... A6: 313
thermal diffusivity from 20 to 100 °C ... A6: 4
ultrasonic welding ... A6: 894
ultrasonic welding power requirements ... M6: 750
weldability ... A6: 534, 722, 725
weldments, tensile properties at subzero temperatures ... M3: 723
1100, anodic polarization curve ... A13: 583-584
1100, as plate alloy ... A2: 33
1100, bending specimen thickness ... A8: 130
1100, center cracking ... A9: 634
1100, columnar grains in ... A9: 630
1100, corrosion depth/tensile strength loss ... A13: 596
1100, distribution of porosity and hydrogen in as-cast ingot ... A9: 633
1100, dynamically recovered ... A10: 470
1100, effect of grain refiner on ... A9: 630-631
1100, effect of metal-feed location on ... A9: 631
1100, extrudability ... A2: 35
1100, fastener coating ... A11: 542
1100, feather crystals in ... A9: 631
1100, forging alloy ... A2: 34
1100, low-temperature alloy ... A2: 59
1100, machining ... A16: 15, 762, 764, 771-774, 776, 777, 779, 781, 782, 786, 788-790, 794-796, 798-800
1100, minimum bend radii ... A8: 129
1100, nomograph, hydrogen content ... A15: 460
1100, oxide stringer inclusion ... A9: 635
1100, pitting as function of coating thickness ... A13: 607
1100, shear testing ... A8: 65
1100, tube alloy ... A2: 33
1100, weathering data ... A13: 597
1100, wet chlorine attack ... A13: 1173
1100-H12, typical chips for machinability rating ... A16: 761
1100-H14, chemical solution corrosion ... A13: 608
1100-H14, ultrasonic welding ... A6: 894
1100-H18, cold rolled sheet ... A9: 360
1100-H18, resistance spot welding ... A6: 848
1100-O, analog records ... A8: 228
1100-O, effective true strain contour maps ... A14: 436
1100-O, high strain rate pressure-shear tests ... A8: 236
1100-O, metal flow simulation ... A14: 435
1100-O, sheet, cold rolled and annealed ... A9: 360
1100-O, stress-strain curves ... A8: 236-237
1100-O, tension and torsion effective fracture strains ... A8: 168
1100-O, weldment properties ... A6: 539
1145, composition ... A6: 538

Aluminum alloys, specific types (continued)
1145, machining..... **A16:** 771-774, 776, 777, 779, 781, 782, 786, 788-790, 794-796, 798-800
1175, composition........ **A6:** 538
1175, machining..... **A16:** 771-774, 776, 777, 779, 781, 782, 786, 788-790, 794-796, 798-800
1188
composition........ **A6:** 533, 538, 724
corrosion resistance........ **A6:** 730
filler alloy........ **A6:** 537-538
filler alloy for welding, for sustained elevated temperature service........ **A6:** 729
filler alloy, relative rating for fillet welding or butt welding two component base alloys........ **A6:** 735
filler alloy, weldability........ **A6:** 534
filler for best color match to 1100, 3003, 5005, and 5050........ **A6:** 730
melting range........ **A6:** 724
1199, chloride ion effect on pitting potential........ **A13:** 584
1200, composition........ **A6:** 538
1230
composition........ **A6:** 538
resistance welding........ **A6:** 848
1230, clad to 2024-T3........ **A9:** 264
1230, clad to 2024-T4, resistance spot weld........ **A9:** 386
1230, resistance welding........ **M6:** 535
1235, composition........ **A6:** 538
1235, machining..... **A16:** 771-774, 776, 777, 779, 781, 782, 786, 788-790, 794-796, 798-800
1345, composition........ **A6:** 538
1350
brazeability........ **A6:** 937
composition........ **A6:** 538, 722
electrode potential in NaCl-H_2O_2 solution........ **A6:** 730
mechanical properties, gas shielded arc welded butt joints........ **A6:** 727
melting range........ **A6:** 937
physical properties........ **A6:** 722
relative rating of filler alloys for welding........ **A6:** 731, 732, 733, 734, 735
weldability........ **A6:** 534, 722
1350, electrical conductor alloy........ **A2:** 33
1350, extrudability........ **A2:** 35
1350, resistance welding........ **M6:** 536
1350-H19, resistance spot welding........ **A6:** 848
01420-T6, tensile properties........ **A6:** 550
2011, applications and properties........ **A2:** 66-67
2011, bar, rod, wire alloy........ **A2:** 33
2011, cleaning and finishing........ **M5:** 604, 606
2011, effect of homogenization on structure........ **A9:** 632
2011, extrudability........ **A2:** 35
2011, machining..... **A16:** 762, 764, 771-774, 776, 777, 779, 781, 782, 786, 788-790, 794-796, 798-800
2011-T3, machining........ **A16:** 761, 764-765, 778, 782, 784
2014
applications........ **A6:** 530
button diameter of welds........ **M6:** 542
composition........ **A6:** 529, 723 **M3:** 723
crack sensitivity ratings of base alloy/filler alloy combinations........ **A6:** 725
electron-beam welding........ **A6:** 739
fatigue life........ **M3:** 746
flash welding........ **M6:** 558
fracture toughness........ **M3:** 746
physical properties........ **A6:** 723
Poisson's ratio........ **M3:** 725
relative rating of filler alloys for welding........ **A6:** 731, 732, 733, 734, 735
resistance welding........ **M6:** 536, 542
tensile properties at subzero temperatures........ **M3:** 724, 726
thermal diffusivity from 20 to 100 °C........ **A6:** 4
weld microstructures........ **A6:** 54
weldability........ **A6:** 534, 723, 726
welding, for sustained elevated-temperature service........ **A6:** 729
Young's modulus........ **M3:** 725
2014 Alclad, applications and properties........ **A2:** 67-68
2014, applications and properties........ **A2:** 67-68
2014, elevated-temperature behavior........ **A2:** 59

Aluminum alloys, specific types (continued)
2014, extrudability........ **A2:** 35
2014, flow stress vs strain rate........ **A14:** 242
2014, forging alloy........ **A2:** 34
2014, fracture toughness........ **A2:** 59
2014, low-temperature alloy........ **A2:** 59
2014, machining..... **A16:** 762, 771-774, 776, 777, 779, 781, 782, 786, 788-790, 794-796, 798-800
2014, minimum bend radii........ **A8:** 129
2014 plate, exfoliation corrosion........ **A11:** 201
2014, tube alloy........ **A2:** 33
2014-f6, aircraft component fatigue failure........ **A12:** 175
2014-T-6, shot peening........ **M5:** 141
2014-T4, closed-die forging, solution heat treated and quenched........ **A9:** 363
2014-T4, electrode potential in NaCl-H_2O_2 solution........ **A6:** 730
2014-T6
electrode potential in NaCl-H_2O_2 solution........ **A6:** 730
resistance spot welding........ **A6:** 848
ultrasonic welding........ **A6:** 327
2014-T6 (base alloy), mechanical properties, gas-shielded arc welded butt joints........ **A6:** 728
2014-T6 actuator barrel lug, SCC failed...... **A11:** 219
2014-T6, aircraft part, exfoliation........ **A13:** 1022
2014-T6 aircraft wheel half, fatigue cracking........ **A11:** 323, 325
2014-T6, bar, pressure weld........ **A9:** 387
2014-T6, broaching........ **A16:** 209
2014-T6 catapult-hook attachment fitting, fracture from straightening process........ **A11:** 88, 91
2014-T6, chemical milling........ **A16:** 585
2014-T6, closed-die forging, hydrogen porosity........ **A9:** 363
2014-T6, closed-die forging, rosettes from eutectic melting........ **A9:** 363
2014-T6, closed-die forging, solution heat treated, aged and overaged........ **A9:** 363
2014-T6, diffraction techniques, elastic constants, and bulk values for........ **A10:** 382
2014-T6, flange failure........ **A13:** 1052
2014-T6 hinge bracket, stress-corrosion cracking in........ **A11:** 219
2014-T6, knobbly structure........ **A12:** 33
2014-T6, pitting corrosion, space shuttle orbiter........ **A13:** 1065
2014-T6, shear testing compared........ **A8:** 65
2014-T61, closed-die forging, hydrogen porosity........ **A9:** 264
2014-T651, double-shear test........ **A8:** 63
2014-T651, SCC performance........ **A8:** 524
2014-T651, shear strength........ **A8:** 64
2014-T652 turning........ **A16:** 598
2017, applications and properties........ **A2:** 68, 70
2017, cleaning and finishing........ **M5:** 604, 606
2017, flash welding........ **M6:** 558
2017, machining..... **A16:** 282, 762, 764, 771-774, 776, 777, 779, 781, 782, 786, 788-790, 794-796, 798-800
2017, use for combination gages........ **M3:** 556
2017-T4, machining........ **A16:** 763, 764, 780, 782-784
2018, machining..... **A16:** 771-774, 776, 777, 779, 781, 782, 786, 788-790, 794-796, 798-800
2020, ultrasonic welding........ **A6:** 894
2021, machining..... **A16:** 771-774, 776, 777, 779, 781, 782, 786, 788-790, 794-796, 798-800
2024
button diameter of welds........ **M6:** 542
composition........ **A6:** 529, 723 **M3:** 723
electrical conductivity........ **M6:** 536
electrode potential in NaCl-H_2O_2 solution........ **A6:** 730
electron-beam welding........ **A6:** 739
fasteners, use for........ **M3:** 184, 185
flash welding........ **M6:** 558
fracture toughness........ **M3:** 746
friction welding........ **A6:** 153, 154
gages, combination, use for........ **M3:** 556
physical properties........ **A6:** 723
properties........ **A6:** 529-530
resistance welding........ **M6:** 536
shrinkage during cooling........ **M6:** 536
solidification cracking........ **A6:** 531
strength of ultrasonic welds........ **M6:** 752

Aluminum alloys, specific types (continued)
tensile properties at subzero temperatures........ **M3:** 727
ultrasonic welding power requirements........ **M6:** 750
upset welding........ **A6:** 249
weldability........ **A6:** 723
2024, aircraft alloy........ **A2:** 33
2024, aircraft part, intergranular corrosion........ **A13:** 1033
2024 Alclad, applications and properties........ **A2:** 70-71
2024, applications and properties........ **A2:** 70-71
2024, cleaning and finishing........ **M5:** 579, 583, 594, 597, 603-604, 606
2024, dendritic structures........ **A9:** 634
2024, elevated-temperature behavior........ **A2:** 59
2024, extrudability........ **A2:** 35
2024, fracture toughness........ **A2:** 50
2024, fracture-limit line for........ **A8:** 583
2024, low-temperature alloy........ **A2:** 59
2024, machining..... **A16:** 282, 762, 764, 771-774, 776, 777, 779, 781, 782, 786, 788-790, 794-796, 798-800
2024, minimum bend radii........ **A8:** 129
2024, nondestructive dimple profiles........ **A12:** 199
2024, radiographic absorption........ **A17:** 311
2024, sheet, bend testing........ **A8:** 127-128
2024, sheet, specimen thickness........ **A8:** 130
2024, tube alloy........ **A2:** 33
2024, with oxide inclusion........ **A9:** 635
2024, workability criteria for centerbursting in........ **A8:** 577-758 **A14:** 370
2024-0, plate, hot rolled and annealed........ **A9:** 365
2024-0, sheet........ **A9:** 365
2024-351, power requirements........ **A16:** 765
2024-T-4, shot peening........ **M5:** 141
2024-T-6, shot peening........ **M5:** 145
2024-T3
electrode potential in NaCl-H_2O_2 solution........ **A6:** 730
ultrasonic welding........ **A6:** 327, 894
2024-T3, chemical milling........ **A16:** 583, 584
2024-T3, ductile striations........ **A8:** 481, 484
2024-T3, fatigue behavior........ **A2:** 43-44
2024-T3, fatigue fracture stages........ **A11:** 104
2024-T3, fatigue striations........ **A12:** 19, 20
2024-T3, shear testing........ **A8:** 65
2024-T3, sheet clad with alloy 1230
2024-T3, sheet, solution heat treated, different quenches compared........ **A9:** 264
2024-T4, aircraft part, exfoliation corrosion........ **A13:** 1022
2024-T4, aircraft part, pitting corrosion.... **A13:** 1025
2024-T4, alclad sheet........ **A9:** 385
2024-T4, bolts, screws alloy........ **A2:** 33
2024-T4, electrode potential in NaCl-H_2O_2 solution........ **A6:** 730
2024-T4, exposure time and temperature effects on tensile properties........ **A8:** 37
2024-T4, extruded bar, section through cold rolled threads........ **A9:** 388
2024-T4, extruded bar, section through machined threads........ **A9:** 388
2024-T4, impact wear........ **A18:** 264
2024-T4, machining........ **A16:** 15, 761, 764, 782, 784, 787, 789
2024-T4 plates, liquid mercury embrittlement of........ **A11:** 79
2024-T4, ringing in........ **A8:** 40, 44
2024-T4, sheet clad with 1230, resistance spot weld........ **A9:** 386
2024-T6, for notched-pins........ **A8:** 221
2024-T6, lubricant effect on bearing strength of........ **A8:** 60
2024-T6, sheet, stretched from 2% to 20%........ **A9:** 264
2024-T35, compression tests........ **A14:** 391
2024-T62, pitting corrosion, space shuttle orbiter........ **A13:** 1066
2024-T81, galvanic corrosion, space shuttle orbiter........ **A13:** 1067
2024-T351, corroded aircraft part........ **A13:** 1020
2024-T351, diffraction techniques, elastic constants, and bulk values for...... **A10:** 382
2024-T351, double shear tests........ **A8:** 63
2024-T351, fracture loci in upset test specimens........ **A8:** 580-581

Aluminum alloys, specific types (continued)
2024-T351, fracture locus **A14:** 392
2024-T351, fracture strain lines **A14:** 397
2024-T351, intergranular cracking **A13:** 1049
2024-T351, SCC resistance **A8:** 522
2024-T351, shear strength **A8:** 64
2024-T361, resistance spot welding **A6:** 848
2024-T851, crevice corrosion, space shuttle orbiter **A13:** 1069
2024-T851, plate, cold rolled, solution heat treated, stretched, artificially aged, different sections compared **A9:** 365
2024-T851, plate, hot rolled, solution heat treated, stretched, artificially aged **A9:** 365
2024-T851, SCC resistance **A8:** 522
2024-T3511, chemical milling **A16:** 584
2024T, drilling **A16:** 237
2025, forging alloy **A2:** 34
2025, machining **A16:** 762, 771-774, 776, 777, 779, 781, 782, 786, 788-790, 794-796, 798-800
2025-T6, closed-die forging, solution heat treated and artificially aged **A9:** 366
2036
laser beam welding **M6:** 661
relative rating of filler alloys for welding **A6:** 731, 732, 733, 734, 735
resistance welding **M6:** 535-536
ultrasonic welding **A6:** 894
weldability **A6:** 534
2036, applications and properties **A2:** 71-72
2036, minimum bend radii **A8:** 129
2036-T4, resistance spot welding **A6:** 848
2048, applications and properties **A2:** 74
2090
composition **A6:** 529, 550, 723
electron-beam welding **A6:** 551
gas-tungsten arc welding **A6:** 551
laser-beam welding **A6:** 551
physical properties **A6:** 723
properties **A6:** 549, 550, 551
weldability **A6:** 550, 551, 552, 723, 726
2090, microstructure **A9:** 357
2090-T8, tensile properties **A6:** 550
2091
composition **A6:** 550
properties **A6:** 549, 550
weldability **A6:** 550
2091-T8, tensile properties **A6:** 550
2094
composition **A6:** 550
electron-beam welding **A6:** 551
properties **A6:** 549, 550, 551
weldability **A6:** 550, 551
2095
composition **A6:** 550
properties **A6:** 549, 550
weldability **A6:** 551, 726
2117, machining **A16:** 762, 771-774, 776, 777, 779, 781, 782, 786, 788-790, 794-796, 798-800
2117, rivet, fittings alloy **A2:** 33
2117-T4, cold upset rivet, solution heat treated and quenched **A9:** 366
2124, aircraft alloy **A2:** 33
2124, applications and properties **A2:** 74-75
2124, copper and magnesium micro-segregation in **A9:** 631
2124, exfoliation corrosion **A13:** 242
2124, fracture toughness **A2:** 60
2124, fracture toughness of plate **M3:** 746
2124, second-phase constituents **A2:** 42
2124, variation in copper concentration **A9:** 633
2124-T851, aircraft alloy **A2:** 59
2124-UT, hydrogen-embrittled **A12:** 33
2125, fracture toughness, fatigue behavior **A2:** 42
2195
composition **A6:** 529, 550

Aluminum alloys, specific types (continued)
properties **A6:** 549, 550
2195-T8, tensile properties **A6:** 550
2214, fracture toughness **A2:** 60
2218, applications and properties **A2:** 75, 77-78
2218, machining **A16:** 762, 771-774, 776, 777, 779, 781, 782, 786, 788-790, 794-796, 798-800
2218-T61, closed-die forging, solution heat treated and artificially aged **A9:** 366
2219
applications **A6:** 530
composition **A6:** 529, 723 **M3:** 723
crack sensitivity ratings of base alloy/filler alloy combinations **A6:** 725
cryogenic applications **A6:** 535
electron-beam welding **A6:** 257, 871, 872
fatigue life **M3:** 746
fracture toughness **M3:** 746
gas tungsten arc welding **M6:** 396
gas-tungsten arc welding **A6:** 871, 872
laser beam welding **M6:** 661
laser cladding of alumina **M6:** 800
laser-beam welding **A6:** 264
physical properties **A6:** 723
Poisson's ratio **M3:** 725
properties **A6:** 539
relative rating of filler alloys for welding **A6:** 731, 732, 733, 734, 735
resistance welding **M6:** 536
tensile properties at subzero temperatures **M3:** 722, 729, 731, 732
ultimate tensile strength at selected temperatures for GSAW groove joints **A6:** 729
weldability ... **A6:** 534, 535, 551, 723, 725, 726, 727, 728
welding, for sustained elevated-temperature service **A6:** 729
Young's modulus **M3:** 725
2219, aircraft alloy **A2:** 33
2219 Alclad, applications and properties **A2:** 79-80
2219, applications and properties **A2:** 79-80
2219, forging alloy **A2:** 34
2219, fracture toughness **A2:** 59
2219, gas metal arc weld **A9:** 584
2219, machining **A16:** 585, 762, 771-774, 776, 777, 779, 781, 782, 786, 788-790, 794-796, 798-800
2219-T3, electrode potential in $NaCl-H_2O_2$ solution **A6:** 730
2219-T4, electrode potential in $NaCl-H_2O_2$ solution **A6:** 730
2219-T6, closed-die forging, solution heat treated and artificially aged **A9:** 366
2219-T6, electrode potential in $NaCl-H_2O_2$ solution **A6:** 730
2219-T6, filiform corrosion, space shuttle orbiter **A13:** 1066
2219-T8, electrode potential in $NaCl-H_2O_2$ solution **A6:** 730
2219-T31 (base alloy), mechanical properties, gas-shielded arc welded butt joints **A6:** 728
2219-T37
resistance spot welding **A6:** 848
ultimate tensile strength at selected temperatures for GSAW groove joints **A6:** 729
2219-T37 (base alloy), mechanical properties, gas-shielded arc welded butt joints **A6:** 728
2219-T37, chemical milling **A16:** 584
2219-T37, sheet **A9:** 383
2219-T37, sheet, electron beam weld with ER2319 ..... **A9:** 384
2219-T37, sheet, gas tungsten arc weld with ER2319 ..... **A9:** 384
2219-T62, shear strength **A8:** 64

Aluminum alloys, specific types (continued)
2219-T81 (base alloy), mechanical properties, gas-shielded arc welded butt joints **A6:** 728
2219-T87
corrosion of weldments, galvanic couples **A6:** 1065, 1066
corrosion resistance **A6:** 534, 535
differences in space-based (Skylab) and earth-based weld samples **A6:** 1024
electron-beam welding in a space environment **A6:** 1023- 1025
gas-tungsten arc welding **A6:** 532, 533, 534, 729
2219-T87 (base alloy)
mechanical properties, gas-shielded arc welded butt joints **A6:** 728
properties and compositions studied in M512 melting experiments **A6:** 1024
2219-T87, chemical milling **A16:** 584
2219-T87, fracture toughness **A2:** 60
2219-T87, SCC performance **A8:** 522, 524
2219-T87, weldment corrosion **A13:** 345
2219-T851, cubic spline curve fit to fatigue crack growth **A8:** 681
2219-T851, *K*-gradient effect on near-threshold fatigue crack growth **A8:** 379-380
2219-T851, stress-life data and best-fit curves **A8:** 697
2219-T851, water vapor corrosion fatigue **A13:** 143
2219-T851, water vapor effect on crack propagation rate **A12:** 40, 52
2319
composition **A6:** 533, 724
corrosion of weldments **A6:** 1065, 1066
crack sensitivity ratings of base alloy/filler alloy combinations **A6:** 725
filler alloy, electrode potential in $NaCl-H_2O_2$ solution **A6:** 730
filler alloy for welding, for sustained elevated temperature service **A6:** 729
filler alloy, relative rating for fillet welding or butt welding two component base alloys **A6:** 731-735
filler alloy, weldability **A6:** 534
filler metal for aluminum-lithium alloys **A6:** 550, 551, 552
as filler metals **A6:** 533, 534, 535
gas-tungsten arc welding **A6:** 729
melting range **A6:** 724
minimum shear strengths of fillet welds **A6:** 728
weldability, filler metal **A6:** 725, 726, 728
2319 (filler alloy)
mechanical properties, gas-shielded arc welded butt joints **A6:** 728
ultimate tensile strength at selected temperatures for GSAW groove joints **A6:** 729
2319, applications and properties **A2:** 80-81
2319, weld filler metal, corrosion **A13:** 345
2419, cold work effects **A2:** 48
2419, fracture toughness **A2:** 60
2419-T851, aircraft alloy **A2:** 59
2519
applications **A6:** 530, 540
composition **A6:** 529
weldability **A6:** 534
2618
composition **A6:** 723
physical properties **A6:** 723
weldability **A6:** 723
2618, applications and properties **A2:** 81-82
2618, forging alloy **A2:** 34
2618, machining **A16:** 762, 771-774, 776, 777, 779, 781, 782, 786, 788-790, 794-796, 798-800
2618-T4, closed-die forging, solution heat treated and quenched **A9:** 366

SUBJECTS OF THE INDEXED VOLUMES: ASM Handbook (designated by the letter "A"): **A1:** Properties and Selection: Irons, Steels, and High-Performance Alloys (1990); **A2:** Properties and Selection: Nonferrous Alloys and Special-Purpose Materials (1990); **A3:** Alloy Phase Diagrams (1992); **A4:** Heat Treating (1991); **A6:** Welding, Brazing, and Soldering (1993); **A7:** Powder Metallurgy (1984); **A8:** Mechanical Testing (1985); **A9:** Metallography and Microstructures (1985); **A10:** Materials Characterization (1986); **A11:** Failure Analysis and Prevention (1986); **A12:** Fractography (1987); **A13:** Corrosion (1987); **A14:** Forming and Forging (1988); **A15:** Casting (1988); **A16:** Machining (1989); **A17:** Nondestructive Testing and Quality Control (1989); **A18:** Friction, Lubrication, and Wear Technology (1992). **Metals Handbook, 9th Edition** (designated by the letter "M"): **M1:** Properties and Selection: Irons and Steels (1978); **M2:** Properties and Selection: Nonferrous Alloys and Pure Metals (1979); **M3:** Properties and Selection: Stainless Steels, Tool Materials and Special-Purpose Materials (1980); **M4:** Heat Treating (1981); **M5:** Surface Cleaning, Finishing, and Coating (1982); **M6:** Welding, Brazing, and Soldering (1983). **Engineered Materials Handbook** (designated by the letters "EM"): **EM1:** Composites (1987); **EM2:** Engineering Plastics (1988); **EM3:** Adhesives and Sealants (1990); **EM4:** Ceramics and Glasses (1991); **Electronic Materials Handbook** (designated by the letters "EL"): **EL1:** Packaging (1989).

Aluminum alloys, specific types (continued)
2618-T4, forging, solution heat treated and cooled in air **A9:** 366
2618-T61, forging, solution heat treated cooled in still air, aged, and stabilized **A9:** 366
2618-T61, forging, solution heat treated quenched and stabilized **A9:** 366
3002, machinability rating **A16:** 762
3003
brazeability **A6:** 937
brazing **A6:** 944
composition **A6:** 538, 722 **M3:** 723
electrode potential in $NaCl-H_2O_2$ solution **A6:** 730
explosion welding, wave morphology of trilayer **A6:** 162
fatigue-crack-growth rates **M3:** 732, 733
filler alloys for best color match **A6:** 730
mechanical properties, gas-shielded arc welded butt joints **A6:** 727
melting range **A6:** 937
physical properties **A6:** 722
relative rating of filler alloys for welding **A6:** 731, 732, 733, 734, 735
resistance to corrosion **M6:** 536, 1032
resistance welding **M6:** 535-536, 539
tensile properties at subzero temperatures **M3:** 734
ultimate tensile strength at selected temperatures for GSAW groove joints **A6:** 729
weldability **A6:** 534, 722, 725
welding, for sustained elevated-temperature service **A6:** 729
3003 Alclad, applications and properties **A2:** 82-84
3003 Alclad, heat exchanger tube **A2:** 33
3003, applications and properties **A2:** 82-84
3003, as plate alloy **A2:** 33
3003, bending specimen thickness **A8:** 129-130
3003, bleed bands in **A9:** 634
3003, clad with 4343 brazing filler metal **A9:** 387
3003, cleaning and finishing **M5:** 513, 580, 583, 599, 603-604, 605
3003, corrosion depth and tensile strength loss **A13:** 596
3003, dendrite arm spacing for different solidification rates **A9:** 630
3003, extrudability **A2:** 35
3003, feather crystals in **A9:** 631
3003, foil alloy **A2:** 33
3003, forging alloy **A2:** 34
3003, heat exchanger tube alloy **A2:** 33
3003, low-temperature alloy **A2:** 59
3003, machining **A16:** 762, 764, 771-774, 776, 777, 779, 781, 782, 786, 788-790, 794-796, 798-800
3003, pipe alloy **A2:** 33
3003, shear testing **A8:** 65
3003, tensile properties **A8:** 555
3003, tube alloy **A2:** 33
3003, weathering data **A13:** 597
3003-0, annealed sheet **A9:** 361
3003-F, extruded tube **A9:** 360
3003-F, hot rolled sheet **A9:** 361
3003-H14, atmospheric corrosion **A13:** 597
3003-H14 clad with 7072, eddy current inspection **A17:** 572-573
3003-H18
resistance spot welding **A6:** 848
weldment properties **A6:** 539
3003-O, weldment properties **A6:** 539
3004
applications **A6:** 537
brazeability **A6:** 937
composition **A6:** 538, 722
electrode potential in $NaCl-H_2O_2$ solution **A6:** 730
melting range **A6:** 937
physical properties **A6:** 722
relative rating of filler alloys for welding **A6:** 731, 732, 733, 734, 735
resistance to corrosion **M6:** 1032
resistance welding **M6:** 535

Aluminum alloys, specific types (continued)
weldability **A6:** 534, 722
3004 Alclad, applications and properties **A2:** 84-86
3004, applications and properties **A2:** 84-86
3004, bending specimen thickness **A8:** 129-130
3004, cleaning and finishing **M5:** 599, 604, 606
3004, corrosion depth and tensile strength loss **A13:** 596
3004, electromagnetic cast **A9:** 634
3004, machining **A16:** 762, 771-774, 776, 777, 779, 781, 782, 786, 788-790, 794-796, 798-800
3004, shear testing **A8:** 65
3004, weathering data **A13:** 597
3004-H14, weight loss measurements **A13:** 583
3004-H32, typical chips for machinability rating **A16:** 761
3004-H38, resistance spot welding **A6:** 848
3005, composition **A6:** 538
3005, machining **A16:** 771-774, 776, 777, 779, 781, 782, 786, 788-790, 794-796, 798-800
3105, applications and properties **A2:** 87
3105, composition **A6:** 538
3105, minimum bend radii **A8:** 244
4004
brazing filler metal **A6:** 627
fluxless vacuum brazing **A6:** 627
4009
composition **A6:** 724
filler alloy for welding, for sustained elevated temperature service **A6:** 729
melting range **A6:** 724
4010
composition **A6:** 724
filler alloy for welding, for sustained elevated temperature service **A6:** 729
filler for best color match to 356.0, A356.0, A357.0, and 443.0 ..... **A6:** 730
melting range **A6:** 724
4011
composition **A6:** 724
filler alloy for welding, for sustained elevated temperature service **A6:** 729
melting range **A6:** 724
4032, applications and properties **A2:** 87-88
4032, composition **A6:** 538
4032, forging alloy **A2:** 34
4032, machining **A16:** 762, 764, 771-774, 776, 777, 779, 781, 782, 786, 788-790, 794-796, 798-800
4043
brazing **A6:** 944
composition **A6:** 533, 538, 724
corrosion resistance **A6:** 729
crack sensitivity ratings of base alloy/filler alloy combinations **A6:** 725
filler alloy **A6:** 537-538
filler alloy, electrode potential in $NaCl-H_2O_2$ solution **A6:** 730
filler alloy for welding, for sustained elevated temperature service **A6:** 729
filler alloy, relative rating for fillet welding or butt welding two component base alloys **A6:** 731-735
filler alloy, weldability **A6:** 534
filler for best color match to 356.0, A356.0, A357.0, and 443.0 ..... **A6:** 730
filler metal for aluminum-lithium alloys **A6:** 550, 551, 552
as filler metals **A6:** 533, 534
fillet weld strength **A6:** 727
melting range **A6:** 724
minimum shear strengths of fillet welds **A6:** 728
weldability **A6:** 539
weldability, filler metal **A6:** 725, 727-728
4043 (filler alloy)
mechanical properties, gas-shielded arc welded butt joints **A6:** 728
ultimate tensile strength at selected temperatures for GSAW groove joints **A6:** 729
4043, applications and properties **A2:** 88-89
4043 filler, brittle fracture **A11:** 527
4045, composition **A6:** 538
4047
brazing filler metal **A6:** 627
composition **A6:** 533, 538, 724

Aluminum alloys, specific types (continued)
filler alloy **A6:** 537-538
filler alloy, electrode potential in $NaCl-H_2O_2$ solution **A6:** 730
filler alloy for welding for sustained elevated temperature service **A6:** 729
filler for best color match to 356.0, A356.0, A357.0, and 443.0 ..... **A6:** 730
filler metal for aluminum metal-matrix composites **A6:** 556
filler metal for aluminum-lithium alloys **A6:** 551
melting range **A6:** 724
weldability **A6:** 539
weldability, filler metals **A6:** 725
4104, fluxless vacuum brazing **A6:** 627
4145
composition **A6:** 533, 538, 724
crack sensitivity ratings of base alloy/filler alloy combinations **A6:** 725
filler alloy, electrode potential in $NaCl-H_2O_2$ solution **A6:** 730
filler alloy for welding, for sustained elevated temperature service **A6:** 729
filler alloy, relative rating for fillet welding or butt welding two component base alloys **A6:** 731, 732, 734
filler alloy, weldability **A6:** 534
melting range **A6:** 724
weldability, filler metal **A6:** 725, 728
4153, broaching **A16:** 199
4343
brazing filler metal **A6:** 627
composition **A6:** 538
4643
composition **A6:** 533, 538, 724
filler alloy for welding, for sustained elevated temperature service **A6:** 729
melting range **A6:** 724
minimum shear strengths of fillet welds **A6:** 728
weldability, filler metal **A6:** 727
5005
brazeability **A6:** 937
composition **A6:** 538, 722
electrode potential in $NaCl-H_2O_2$ solution **A6:** 730
filler alloys for best color match **A6:** 730
mechanical properties, gas-shielded arc welded butt joints **A6:** 727
melting range **A6:** 937
physical properties **A6:** 722
relative rating of filler alloys for welding **A6:** 731, 732, 733, 734
weldability **A6:** 534, 722
welding, for sustained elevated-temperature service **A6:** 729
5005, applications and properties **A2:** 89
5005, bending specimen thickness **A8:** 129-130
5005, cleaning and finishing **M5:** 580, 583, 594, 599
5005, machining **A16:** 762, 771-774, 776, 777, 779, 781, 782, 786, 788-790, 794-796, 798-800
5005, resistance welding **M6:** 535
5005-H38, resistance spot welding **A6:** 848
5050
composition **A6:** 538, 722
electrode potential in $NaCl-H_2O_2$ solution **A6:** 730
filler alloys for best color match **A6:** 730
mechanical properties, gas-shielded arc welded butt joints **A6:** 727
melting range **A6:** 937
physical properties **A6:** 722
properties **A6:** 539
relative rating of filler alloys for welding **A6:** 731, 732, 733, 734
weldability **A6:** 534, 722
welding, for sustained elevated-temperature service **A6:** 729
5050, applications and properties **A2:** 89-90
5050, bending specimen thickness **A8:** 129-130
5050, machining **A16:** 762, 771-774, 776, 777, 779, 781, 782, 786, 788-790, 794-796, 798-800
5050, resistance welding **M6:** 534-535
5050, shear testing **A8:** 65
5050, tube alloy **A2:** 33

Aluminum alloys, specific types (continued)
5050-H32, weldment properties **A6:** 539
5050-H38
resistance spot welding **A6:** 848
weldment properties **A6:** 539
5050-O, weldment properties **A6:** 539
5052
brazeability **A6:** 937
composition **A6:** 538, 722
corrosion resistance **M6:** 535
crack sensitivity ratings of base alloy/filler alloy combinations **A6:** 725
electrode potential in NaCl-H_2O_2 solution **A6:** 730
explosion welding, interface failure of trilayer **A6:** 163, 164
filler alloys for best color match **A6:** 730
laser melt/particle inspection **M6:** 802
mechanical properties, gas-shielded arc welded butt joints **A6:** 727
melting range **A6:** 937
physical properties **A6:** 722
relative rating of filler alloys for welding **A6:** 731, 732, 733, 734
resistance welding **M6:** 534-535, 539, 542
roll welding **A6:** 313
thermal diffusivity from 20 to 100 °C **A6:** 4
ultimate tensile strength at selected temperatures for GSAW groove joints **A6:** 729
weldability **A6:** 534, 722
welding, for sustained elevated-temperature service **A6:** 729
5052 aircraft part, corrosion fatigue **A13:** 1038
5052, applications and properties **A2:** 90-91
5052, as plate alloy **A2:** 33
5052, ceramic/metal joints **EM4:** 515
5052, cleaning and finishing **M5:** 580, 583, 599, 603-604, 606
5052, effect of homogenization on **A9:** 632
5052, explosively bonded to tantalum **A9:** 445
5052, foil alloy **A2:** 33
5052, hydrogen porosity in **A9:** 632
5052, laser melt/particle injection **A18:** 869
5052, machining **A16:** 762, 764, 771-774, 776, 777, 779, 781, 782, 786, 788-790, 794-796, 798-800
5052, minimum bend radii **A8:** 129
5052, seawater pitting **A13:** 907
5052-H38, resistance spot welding **A6:** 848
5052-O, GTA fillet weld in sheet, different sections **A9:** 381
5056 Alclad alloy **A2:** 33
5056 Alclad, applications and properties **A2:** 91-92
5056, applications and properties **A2:** 91-92
5056, cleaning and finishing **M5:** 599-600
5056, composition **A6:** 538
5056, foil alloy **A2:** 33
5056, machining **A16:** 762, 771-774, 776, 777, 779, 781, 782, 786, 788-790, 794-796, 798-800
5056 rivet material, composite applications **A11:** 530
5056, zipper alloy **A2:** 33
5056-H38, chips difficult to control **A16:** 764
5056-O, flash welding **M6:** 575-576
5056-O, surface stress measurement **A10:** 383
5083
applications **A6:** 537, 540
composition **A6:** 538, 722 **M3:** 723
crack sensitivity ratings of base alloy/filler alloy combinations **A6:** 725
cryogenic applications **A6:** 383
electrode potential in NaCl-H_2O_2 solution **A6:** 730
electron-beam welding **A6:** 538-539
explosion welding, interface failure of trilayer **A6:** 163, 164
explosion welding, wave morphology of trilayer **A6:** 162

Aluminum alloys, specific types (continued)
fatigue life **M3:** 746
fatigue-crack-growth rates **M3:** 733
filler alloys for best color match **A6:** 730
filler metal for aluminum metal-matrix composites **A6:** 556
flash welding **M6:** 558
fracture toughness **M3:** 746
friction welding **A6:** 153
gas tungsten arc welding **M6:** 396
gas-tungsten arc welding **A6:** 736
laser-beam welding **A6:** 264, 878
mechanical properties, gas-shielded arc welded butt joints **A6:** 727
physical properties **A6:** 722
Poisson's ratio **M3:** 728
properties **A6:** 539
relative rating of filler alloys for welding **A6:** 731, 732, 733, 734
resistance welding **M6:** 536
tensile properties at subzero temperatures **M3:** 734, 735
ultimate tensile strength at selected temperatures for GSAW groove joints **A6:** 729
weldability **A6:** 534, 722, 728, 729
Young's modulus **M3:** 727
5083, applications and properties **A2:** 92-93
5083, cold rolled plate **A9:** 362
5083, extrudability **A2:** 35
5083, for marine, cryogenics, pressure vessels **A2:** 33
5083, forging alloy **A2:** 34
5083, low-temperature alloy **A2:** 59
5083, machining **A16:** 762, 771-774, 776, 777, 779, 781, 782, 786, 788-790, 794-796, 798-800
5083, minimum bend radii **A8:** 129
5083 plate, hot short weld cracks **A11:** 435
5083-H32, weldment properties **A6:** 539
5083-H112, cold rolled plate **A9:** 362
5083-H116, gas-metal arc welding **A6:** 726
5083-H131 weldment, mercury-cracked **A13:** 589
5083-H321, resistance spot welding **A6:** 848
5083-O
weldability **A6:** 729
weldment properties **A6:** 539
5083-O, cryogenic alloy **A2:** 59
5083-O, fracture toughness **A2:** 60
5083-O plate, microstructures **A13:** 594
5086
composition **A6:** 538, 722
electrode potential in NaCl-H_2O_2 solution **A6:** 730
filler alloys for best color match **A6:** 730
laser beam welding **M6:** 661
mechanical properties, gas-shielded arc welded butt joints **A6:** 727
physical properties **A6:** 722
relative rating of filler alloys for welding **A6:** 731, 732, 733, 734
resistance welding **M6:** 536
stud arc welding **A6:** 211
ultimate tensile strength at selected temperatures for GSAW groove joints **A6:** 729
weldability **A6:** 534, 722
5086 Alclad, applications and properties **A2:** 93-94
5086, applications and properties **A2:** 93-94
5086, bending specimen thickness **A8:** 129-130
5086, extrudability **A2:** 35
5086, for marine, cryogenics, pressure vessels **A2:** 33
5086, machining **A16:** 762, 771-774, 776, 777, 779, 781, 782, 786, 788-790, 794-796, 798-800
5086, shear testing **A8:** 65
5086-H34, plate, cold rolled and stabilized **A9:** 362
5086-H34, resistance spot welding **A6:** 848

Aluminum alloys, specific types (continued)
5154
composition **A6:** 538, 722
crack sensitivity ratings of base alloy/filler alloy combinations **A6:** 725
electrode potential in NaCl-H_2O_2 solution **A6:** 730
filler alloys for best color match **A6:** 730
mechanical properties, gas-shielded arc welded butt joints **A6:** 727
physical properties **A6:** 722
relative rating of filler alloys for welding **A6:** 731, 732, 733
weldability **A6:** 534, 722
5154, applications and properties **A2:** 94-95
5154, bending specimen thickness **A8:** 130
5154, machining **A16:** 762, 771-774, 776, 777, 779, 781, 782, 786, 788-790, 794-796, 798-800
5154, resistance welding **M6:** 536
5154, shear testing **A8:** 65
5154-H38, resistance spot welding **A6:** 848
5182
laser beam welding **M6:** 661
resistance welding **M6:** 535-536
5182, applications and properties **A2:** 95
5182-O, resistance spot welding **A6:** 848
5183
composition **A6:** 533, 538, 724
corrosion of weldments **A6:** 1065, 1066
filler alloy **A6:** 537-538
filler alloy, electrode potential in NaCl-H_2O_2 solution **A6:** 730
filler alloy, relative rating for fillet welding or butt welding two component base alloys **A6:** 731-735
filler alloy, weldability **A6:** 534
filler for best color match to 5083, 5086, 5454, and 5456 ..... **A6:** 730
as filler metal **A6:** 534, 535
gas-tungsten arc welding **A6:** 729
melting range **A6:** 724
minimum shear strengths of fillet welds **A6:** 728
properties **A6:** 539
stud arc welding **A6:** 211
weldability, filler metals **A6:** 725, 727, 728, 729
5183 (filler alloys), ultimate tensile strength at selected temperatures for GSAW groove joints **A6:** 729
5252, applications and properties **A2:** 95-96
5252, bright finishing alloy **A2:** 37
5252, machining **A16:** 762, 771-774, 776, 777, 779, 781, 782, 786, 788-790, 794-796, 798-800
5252, minimum bend radii **A8:** 130
5254
composition **A6:** 538, 722
corrosion resistance **A6:** 730
electrode potential in NaCl-H_2O_2 solution **A6:** 730
physical properties **A6:** 722
relative rating of filler alloys for welding **A6:** 731, 732, 733
weldability **A6:** 534, 722
5254, applications and properties **A2:** 96-97
5254, machining **A16:** 762, 771-774, 776, 777, 779, 781, 782, 786, 788-790, 794-796, 798-800
5254, minimum bend radii **A8:** 130
5257, machinability rating **A16:** 762
5356
composition **A6:** 533, 538, 724
corrosion resistance **A6:** 729
crack sensitivity ratings of base alloy/filler alloy combinations **A6:** 725
filler alloy **A6:** 537-538
filler alloy, electrode potential in NaCl-H_2O_2 solution **A6:** 730
filler alloy, relative rating for fillet welding or butt welding two component base alloys **A6:** 731-735

Aluminum alloys, specific types (continued)
filler alloy, weldability ... A6: 534
filler for best color match to 6061, 6063, 511.O, 514.O, and 535.O ... A6: 730
as filler metal ... A6: 531, 533
filler metal for aluminum-lithium alloys ... A6: 550
melting range ... A6: 724
minimum shear strengths of fillet welds ... A6: 728
relative rating of filler alloys for welding ... A6: 731, 732, 733, 734
stud arc welding ... A6: 211
weldability, filler metals ... A6: 725, 727, 728
5356 (filler alloy)
mechanical properties, gas-shielded arc welded butt joints ... A6: 728
ultimate tensile strength at selected temperatures for GSAW groove joints ... A6: 729
5356, applications and properties ... A2: 97
5356 weld filler metal, seawater corrosion ... A13: 345
5356-H12, microstructure, SCC susceptibility ... A13: 591
5357 and 5357-H-32, cleaning and finishing ... M5: 580, 588, 594
5357, machinability rating ... A16: 762
5386, relative rating of filler alloys for welding ... A6: 731, 732, 733
5454
applications ... A6: 537
composition ... A6: 538, 722
crack sensitivity ratings of base alloy/filler alloy combinations ... A6: 725
electrode potential in NaCl-H_2O_2 solution ... A6: 730
filler alloys for best color match ... A6: 730
mechanical properties, gas-shielded arc welded butt joints ... A6: 727
physical properties ... A6: 722
relative rating of filler alloys for welding ... A6: 731, 732
ultimate tensile strength at selected temperatures for GSAW groove joints ... A6: 729
weldability ... A6: 534, 722, 729
welding, for sustained elevated-temperature service ... A6: 729
5454, applications and properties ... A2: 97-98
5454, hot-rolled slab, with oxide stringer ... A9: 362
5454, machining ... A16: 762, 771-774, 776, 777, 779, 781, 782, 786, 788-790, 794-796, 798-800
5454, resistance welding ... M6: 536
5454, shear testing ... A8: 65
5454-H34, resistance spot welding ... A6: 848
5456
composition ... A6: 538, 722 M3: 723
crack sensitivity ratings of base alloy/filler alloy combinations ... A6: 725
electrode potential in NaCl-H_2O_2 solution ... A6: 730
filler alloys for best color match ... A6: 730
gas-tungsten arc welding ... A6: 736
laser beam welding ... M6: 661
laser-beam welding ... A6: 264
mechanical properties, gas-shielded arc welded butt joints ... A6: 727
physical properties ... A6: 722
properties ... A6: 539
relative rating of filler alloys for welding ... A6: 731, 732, 733, 734
resistance welding ... M6: 536
stud arc welding ... A6: 211
tensile properties at subzero temperatures ... M3: 736, 737
ultimate tensile strength at selected temperatures for GSAW groove joints ... A6: 729
weldability ... A6: 534, 722
5456, applications and properties ... A2: 98-99
5456, for marine, cryogenics, pressure vessels ... A2: 33
5456, hot rolled plate, dynamic recrystallization ... A9: 363
5456, low-temperature alloy ... A2: 59
5456, machining ... A16: 762, 771-774, 776, 777, 779, 781, 782, 786, 788-790, 794-796, 798-800
5456, minimum bend radii ... A8: 130
5456, plate, cold rolled and stress relieved ... A9: 363
5456, shear testing ... A8: 64-65
5456-H116, gas-tungsten arc welding ... A6: 532, 533, 534
5456-H321
corrosion of weldments, galvanic couples ... A6: 1065, 1066
corrosion resistance ... A6: 534, 535
gas-metal arc welding ... A6: 729
resistance spot welding ... A6: 848
5456-H321, double shear tests ... A8: 63
5456-H321, plate, electron beam weld ... A9: 384
5456-H321, weldment corrosion ... A13: 345
5456-O, plate, hot rolled and annealed ... A9: 363
5457, applications and properties ... A2: 99-100
5457, bending specimen thickness ... A8: 130
5457, cleaning and finishing ... M5: 580, 583, 596-597
5457, composition ... A6: 538
5457, machining ... A16: 762, 771-774, 776, 777, 779, 781, 782, 786, 788-790, 794-796, 798-800
5457-F, extrusion ... A9: 361
5457-F, plate ... A9: 361
5457-O, plate, effect of cold rolling ... A9: 361
5554
composition ... A6: 533, 538, 724
crack sensitivity ratings of base alloy/filler alloy combinations ... A6: 725
filler alloy ... A6: 537-538
filler alloy, electrode potential in NaCl-H_2O_2 solution ... A6: 730
filler alloy for welding, for sustained elevated temperature service ... A6: 729
filler alloy, relative rating for fillet welding or butt welding two component base alloys ... A6: 731-735
filler alloy, weldability ... A6: 534
as filler metal ... A6: 535
melting range ... A6: 724
minimum shear strengths of fillet welds ... A6: 728
weldability, filler metals ... A6: 729
5556
composition ... A6: 533, 538, 724
corrosion of weldments ... A6: 1065, 1066
crack sensitivity ratings of base alloy/filler alloy combinations ... A6: 725
filler alloy ... A6: 537-538
filler alloy, electrode potential in NaCl-H_2O_2 solution ... A6: 730
filler alloy, relative rating for fillet welding or butt welding two component base alloys ... A6: 731-735
filler alloy, weldability ... A6: 534
filler for best color match to 5083, 5086, 5454,and 5456 ... A6: 730
as filler metal ... A6: 535
fillet weld strength ... A6: 727
gas-metal arc welding ... A6: 729
melting range ... A6: 724
minimum shear strengths of fillet welds ... A6: 728
stud arc welding ... A6: 211
weldability, filler metals ... A6: 725, 727-728
5556 (filler alloy)
mechanical properties, gas-shielded arc welded butt joints ... A6: 728
ultimate tensile strength at selected temperatures for GSAW groove joints ... A6: 729
5556 weld filler metal, corrosion ... A13: 345
5557, machinability rating ... A16: 762
5652
composition ... A6: 538, 722
corrosion resistance ... A6: 730
physical properties ... A6: 722
relative rating of filler alloys for welding ... A6: 731, 732, 733, 734
weldability ... A6: 534, 722
5652, applications and properties ... A2: 100
5652, machining ... A16: 762, 771-774, 776, 777, 779, 781, 782, 786, 788-790, 794-796, 798-800
5652, minimum bend radii ... A8: 130
5654
composition ... A6: 533, 538, 724
corrosion resistance ... A6: 730
crack sensitivity ratings of base alloy/filler alloy combinations ... A6: 725
filler alloy ... A6: 537-538
filler alloy, electrode potential in NaCl-H_2O_2 solution ... A6: 730
filler alloy, relative rating for fillet welding or butt welding two component base alloys ... A6: 731, 732, 733, 734
filler alloy, weldability ... A6: 534
filler for best color match to 5052 and 5154 ... A6: 730
melting range ... A6: 724
minimum shear strengths of fillet welds ... A6: 728
weldability, filler metals ... A6: 727
5657, applications and properties ... A2: 100
5657, bright finishing alloy ... A2: 37
5657, composition ... A6: 538
5657, grain growth in ... A9: 632
5657, grain structure in ... A9: 632
5657, ingot ... A9: 362
5657, machining ... A16: 762, 771-774, 776, 777, 779, 781, 782, 786, 788-790, 794-796, 798-800
5657, minimum bend radii ... A8: 130
5657, sheet, with banding from dendritic segregation ... A9: 362
5657-F, cold rolled, stress relieved, and annealed ... A9: 362
6003, resistance welding ... A6: 848
6005
composition ... A6: 529
relative rating of filler alloys for welding ... A6: 731, 732, 733
6005, applications and properties ... A2: 100-101
6005, machinability rating ... A16: 762
6009
composition ... A6: 529, 723
laser beam welding ... M6: 661
physical properties ... A6: 723
resistance welding ... M6: 535-536
weldability ... A6: 723
6009, applications and properties ... A2: 101
6009-T4 (base alloy), mechanical properties, gas-shielded arc welded butt joints ... A6: 728
6009-T4, resistance spot welding ... A6: 848
6009-T4, tensile properties ... A8: 555
6010, applications and properties ... A2: 101-102
6010, composition ... A6: 529
6010, resistance welding ... M6: 535-536
6010-T4, resistance spot welding ... A6: 848
6013
composition ... A6: 529, 723
physical properties ... A6: 723
weldability ... A6: 723
6013-T6
gas-tungsten arc welding ... A6: 531
liquation cracking ... A6: 531
6033, flash welding ... M6: 558
6053
brazeability ... A6: 937
melting range ... A6: 937
resistance welding ... A6: 848
6053, machining ... A16: 771-774, 776, 777, 779, 781, 782, 786, 788-790, 794-796, 798-800
6053, resistance welding ... M6: 535
6053, rivet, fittings alloy ... A2: 33
6061
adhesive wear resistance ... A18: 721
applications ... A6: 530
brazeability ... A6: 937
button diameter of welds ... M6: 542
composition ... A6: 529, 723 M3: 723
corrosion resistance ... A6: 729 M6: 535
corrosive wear ... A18: 719
crack sensitivity ratings of base alloy/filler alloy combinations ... A6: 725
electrical resistivity ... A18: 713
electron-beam welding ... A6: 860
fasteners, use for ... M3: 184, 185
fatigue life ... M3: 746

Aluminum alloys, specific types (continued)
fiber for reinforcement ... A18: 803
filler alloys for best color match ... A6: 730
flash welding ... M6: 558
fracture toughness ... M3: 746
friction welding ... A6: 153 M6: 722
gas-tungsten arc welding, weld microstructures ... A6: 51
low-heat electron beam welding ... M6: 625-626
matrix hardness ... A18: 789
melting range ... A6: 937
physical properties ... A6: 723
postbraze heat treatment ... A6: 940
processing technique ... A18: 803
properties ... A6: 530, 539 A18: 713, 714, 803
relative rating of filler alloys for welding ... A6: 731, 732
resistance welding ... M6: 535-536, 539, 542
tensile properties at subzero temperatures ... M3: 738, 739
thermal diffusivity from 20 to 100 °C ... A6: 4
weldability ... A6: 534, 723, 727-728
welding, for sustained elevated-temperature service ... A6: 729
6061, 6061-S, 6061-T-4, and 6061-T-6 cleaning and finishing ... M5: 513, 580, 583, 594, 597, 599
6061 Alclad, applications and properties ... A2: 102-103
6061, applications and properties ... A2: 102-103
6061, artificial aging ... A2: 40
6061, as plate alloy ... A2: 33
6061, bending specimen thickness ... A8: 130
6061, boron-aluminum composite ... A9: 595, 597
6061, flow stress vs strain rate ... A14: 242
6061, forging alloy ... A2: 34
6061, graphite-aluminum composite ... A9: 594, 597
6061, halide-flux inclusion ... A12: 65, 67
6061, low-temperature alloy ... A2: 59
6061, machining ... A16: 762, 764, 771-774, 776, 777, 779, 781, 782, 786, 788-790, 794-796, 798-800, 895
6061, pipe alloy ... A2: 33
6061, rotary forged ... A14: 179
6061, structural shape alloy ... A2: 34
6061, tube alloy ... A2: 33
6061-F, plate, as hot rolled ... A9: 367
6061-F, sheet, hot rolled ... A9: 367
6061-O, effective true strain contour maps ... A14: 436
6061-O, metal flow simulation ... A14: 435
6061-T4 (base alloy)
electrode potential in NaCl-H_2O_2 solution ... A6: 730
mechanical properties, gas-shielded arc welded butt joints ... A6: 727
postweld heat treatments ... A6: 532-533
welding condition effect on weld strength ... A6: 728
6061-T6
adhesive wear resistance ... A18: 721
corrosion resistance ... A6: 729
electrode potential in NaCl-H_2O_2 solution ... A6: 730
erosion mechanisms ... A18: 203
friction coefficient data ... A18: 71, 73, 74
gas-tungsten arc welding ... A6: 532, 533, 534
heat-affected zone microstructures ... A6: 727
mechanical properties, gas-shielded arc welded butt joints ... A6: 728
postweld heat treatments ... A6: 532-533
resistance spot welding ... A6: 848
ultimate tensile strength at selected temperatures for GSAW groove joints ... A6: 729
weldability ... A6: 727
welding condition effect on weld strength ... A6: 728
6061-T6 combustion chamber, cavitation erosion ... A11: 168-169
6061-T6 connector tube, failed ... A11: 312-313
6061-T6, crevice corrosion, space shuttle orbiter ... A13: 1070
6061-T6, extruded tube, with gas tungsten arc fillet weld to A356-T6 investment casting ... A9: 382-383
6061-T6 extrusions, ductile overload fracture ... A11: 86-87, 91
6061-T6, for notched pins ... A8: 221
6061-T6 forged and formed truck wheels ... A14: 248
6061-T6, high-cycle fatigue striations in ... A11: 78
6061-T6, lubricant effect on bearing strength ... A8: 60
6061-T6, machining ... A16: 15, 58, 600, 761, 764
6061-T6, pitting/corrosion products ... A13: 1047
6061-T6, plate ... A9: 382
6061-T6, plate, AC and DC welds compared ... A9: 382
6061-T6, plate with welded butt joint ... A9: 381
6061-T6, pressure-shear waves for ... A8: 231
6061-T6, shear testing ... A8: 65
6061-T6, sheet ... A9: 382
6061-T6, sheet, AC and DC welds compared ... A9: 382
6061-T6, sheet, electron beam weld ... A9: 385
6061-T6, sheet with welded butt joint ... A9: 381
6061-T6, striations on fatigue crack fronts ... A12: 23
6061-T651, double shear tests ... A8: 63
6061-T651, fracture toughness ... A2: 60
6061-T651, friction welding ... A6: 558
6061-T651, power requirements ... A16: 765
6061-T651, SCC resistance ... A8: 522
6061-T651, shear strength ... A8: 64
6061-T651, stress vs. rupture life ... A8: 332
6061T, drilling ... A16: 237
6063
brazeability ... A6: 937
composition ... A6: 529, 723
corrosion resistance ... A6: 729 M6: 535
electrode potential in NaCl-H_2O_2 solution ... A6: 730
electron-beam welding ... A6: 860
filler alloys for best color match ... A6: 730
flash welding ... M6: 558
friction surfacing ... A6: 323
melting range ... A6: 937
physical properties ... A6: 723
relative rating of filler alloys for welding ... A6: 731, 732, 733
resistance welding ... M6: 535
weldability ... A6: 534, 723, 728
welding, for sustained elevated-temperature service ... A6: 729
6063 and 6063-T-6, cleaning and finishing ... M5: 580, 583, 594, 599
6063, angular interdendritic porosity ... A9: 633
6063, applications and properties ... A2: 103-104
6063, columnar grains in ... A9: 630
6063, effect of grain refiner on ... A9: 630
6063, equiaxed grains in ... A9: 630
6063, extrudability ... A2: 35
6063, hydrogen porosity in ... A9: 633
6063, machining ... A16: 762, 771-774, 776, 777, 779, 781, 782, 786, 788-790, 794-796, 798-800
6063, pipe alloy ... A2: 33
6063, tube alloy ... A2: 33
6063-O, sheets, brazed joint with 4047 filler ... A9: 387
6063-T5, extrusion ... A9: 367
6063-T6 (base alloy), mechanical properties, gas-shielded arc welded butt joints ... A6: 728
6063-T6 extension ladder side-rail, buckling and plastic deformation of ... A11: 137
6063-T6, resistance spot welding ... A6: 848
6066, applications and properties ... A2: 104-105
6066, electron-beam welding ... A6: 860
6066, machining ... A16: 762, 771-774, 776, 777, 779, 781, 782, 786, 788-790, 794-796, 798-800
6070
relative rating of filler alloys for welding ... A6: 731, 732
weldability ... A6: 534
6070, applications and properties ... A2: 105
6070, machining ... A16: 762, 771-774, 776, 777, 779, 781, 782, 786, 788-790, 794-796, 798-800
6101
composition ... A6: 723
physical properties ... A6: 723
relative rating of filler alloys for welding ... A6: 731, 732, 733
weldability ... A6: 534, 723
6101, applications and properties ... A2: 105-106
6101, electrical conductor alloy ... A2: 33
6101, machining ... A16: 771-774, 776, 777, 779, 781, 782, 786, 788-790, 794-796, 798-800
6101, resistance welding ... M6: 536
6101-T6, resistance spot welding ... A6: 848
6151
relative rating of filler alloys for welding ... A6: 731, 732, 733
resistance brazing ... A6: 342
weldability ... A6: 534
6151, applications and properties ... A2: 106
6151, machining ... A16: 762, 771-774, 776, 777, 779, 781, 782, 786, 788-790, 794-796, 798-800
6151-T6, closed-die forging ... A9: 367
6201
relative rating of filler alloys for welding ... A6: 731, 732, 733
weldability ... A6: 534
6201, applications and properties ... A2: 106-107
6201, electrical conductor alloy ... A2: 33
6205, applications and properties ... A2: 107
6253, machining ... A16: 771-774, 776, 777, 779, 781, 782, 786, 788-790, 794-796, 798-800
6262
composition ... A6: 529, 723
physical properties ... A6: 723
weldability ... A6: 723
6262, applications and properties ... A2: 107
6262, machining ... A16: 762, 764, 771-774, 776, 777, 779, 781, 782, 786, 788-790, 794-796, 798-800
6262, wire, bar, and rod alloy ... A2: 33
6262-T9, machinability of hexagonal nut ... A16: 764
6351
composition ... A6: 723
physical properties ... A6: 723
relative rating of filler alloys for welding ... A6: 731, 732, 733
weldability ... A6: 534, 723
6351, applications and properties ... A2: 107-108
6351-T6, extruded tube ... A9: 367
6463, applications and properties ... A2: 108
6463, machining ... A16: 762, 771-774, 776, 777, 779, 781, 782, 786, 788-790, 794-796, 798-800
6951
brazeability ... A6: 937
composition ... A6: 723
melting range ... A6: 937
physical properties ... A6: 723
relative rating of filler alloys for welding ... A6: 731, 732, 733
weldability ... A6: 534, 723
6951, machining ... A16: 762, 771-774, 776, 777, 779, 781, 782, 786, 788-790, 794-796, 798-800
7001, machining ... A16: 762, 771-774, 776, 777, 779, 781, 782, 786, 788-790, 794-796, 798-800
7004 extrusion, with speed cracks ... A11: 91
7004, machining ... A16: 771-774, 776, 777, 779, 781, 782, 786, 788-790, 794-796, 798-800

Aluminum alloys, specific types (continued)
7004, properties ... **A6:** 530
7004-O, sheets, brazed joint with 4245 filler ... **A9:** 387
7005
brazeability ... **A6:** 937
composition ... **A6:** 529, 723 **M3:** 723
corrosion of weldments ... **A6:** 1065
crack sensitivity ratings of base alloy/filler alloy combinations ... **A6:** 725
electron-beam welding ... **A6:** 871
mechanical properties ... **M6:** 1032
melting range ... **A6:** 937
not brazed ... **A6:** 627
physical properties ... **A6:** 723
precipitation hardening after brazing ... **M6:** 1031
properties ... **A6:** 530
relative rating of filler alloys for welding ... **A6:** 731, 732
tensile properties at subzero temperatures ... **M3:** 740
weldability ... **A6:** 534, 723, 725, 726, 728 **M6:** 373
welding, for sustained elevated-temperature service ... **A6:** 729
7005, applications and properties ... **A2:** 108-109
7005, low-temperature alloy ... **A2:** 59
7005, machining ... **A16:** 762, 771-774, 776, 777, 779, 781, 782, 786, 788-790, 794-796, 798-800
7005, seawater corrosion ... **A13:** 345
7005-T6
corrosion resistance ... **A6:** 729
electrode potential in NaCl-H_2O_2 solution ... **A6:** 730
7005-T53 (base alloy), mechanical properties, gas-shielded arc welded butt joints ... **A6:** 728
7010, diffusion welding ... **A6:** 885
7010, forging alloy ... **A2:** 34
7021, relative rating of filler alloys for welding ... **A6:** 731, 732
7039
applications ... **A6:** 540
composition ... **A6:** 529, 723 **M3:** 723
corrosion of weldments ... **A6:** 1065
crack sensitivity ratings of base alloy/filler alloy combinations ... **A6:** 725
electron-beam welding ... **A6:** 871
fatigue life ... **M3:** 746
fracture toughness ... **M3:** 746
gas tungsten arc welding ... **M6:** 396
physical properties ... **A6:** 723
properties ... **A6:** 530
relative rating of filler alloys for welding ... **A6:** 731, 732
tensile properties at subzero temperatures ... **M3:** 741-743
weldability ... **A6:** 534, 535, 723, 725, 726, 728 **M6:** 373
7039, applications and properties ... **A2:** 109-111
7039, forging alloy ... **A2:** 34
7039, fracture toughness ... **A2:** 60
7039, ingot ... **A9:** 368
7039, low-temperature alloy ... **A2:** 59
7039, machining ... **A16:** 771-774, 776, 777, 779, 781, 782, 786, 788-790, 794-796, 798-800
7039, plate, as hot rolled, reduced 50% and 83% ... **A9:** 368
7039-T6 C-ring, stress-corrosion cracking in ... **A11:** 78-79
7039-T6, electrode potential in NaCl-H_2O_2 solution ... **A6:** 730
7039-T61 (base alloy), mechanical properties, gas-shielded arc welded butt joints ... **A6:** 727
7039-T63, plate, electron beam weld ... **A9:** 385
7039-T64, industrial atmospheric SCC ... **A13:** 266
7039-T651
corrosion of weldments, galvanic couples ... **A6:** 1065, 1066
corrosion resistance ... **A6:** 534, 535
gas-tungsten arc welding ... **A6:** 729
7039-T651, weldment corrosion ... **A13:** 345
7046, relative rating of filler alloys for welding ... **A6:** 731, 732
7049, applications and properties ... **A2:** 111-113
7049, forging alloy ... **A2:** 34
7049, machining ... **A16:** 771-774, 776, 777, 779, 781, 782, 786, 788-790, 794-796, 798-800
7050, aircraft alloy ... **A2:** 33
7050, applications and properties ... **A2:** 113-114
7050, composition ... **A6:** 529
7050, forging alloy ... **A2:** 34
7050, fracture toughness ... **A2:** 60
7050, fracture toughness, fatigue behavior ... **A2:** 42
7050 low copper, hydrogen-embrittled ... **A12:** 33
7050, machining ... **A16:** 771-774, 776, 777, 779, 781, 782, 786, 788-790, 794-796, 798-800
7050, SCC propagation rates ... **A13:** 269
7050, second-phase constituents ... **A2:** 42
7050-T6, diffraction techniques, elastic constants, and bulk values for ... **A10:** 382
7057-T6, high-speed machining ... **A16:** 603
7070-T651, in potassium iodide solution electrode potential effect ... **A8:** 407-408
7072
brazeability ... **A6:** 937
cladding material only ... **A6:** 627
electrode potential in NaCl-H_2O_2 solution ... **A6:** 730
7072, applications and properties ... **A2:** 114-115
7072, chemical milling ... **A16:** 803-804
7072, clad to 7075-T6 ... **A9:** 369
7072, clad to 7178-T76, sacrificially corroded ... **A9:** 369
7072, minimum bend radii ... **A8:** 130
7072, resistance welding ... **M6:** 535
7075
button diameter of welds ... **M6:** 542
composition ... **A6:** 529, 723 **M3:** 723
electron-beam welding ... **A6:** 739, 872
fatigue life ... **M3:** 746
flash welding ... **M6:** 558
fracture toughness ... **M3:** 746
physical properties ... **A6:** 723
properties ... **A6:** 530
resistance welding ... **M6:** 535-536
shrinkage during cooling ... **M6:** 536
solidification cracking ... **A6:** 531
tensile properties at subzero temperatures ... **M3:** 744
thermal diffusivity from 20 to 100 °C ... **A6:** 4
weldability ... **A6:** 723, 725-726, 728
7075, aircraft alloy ... **A2:** 33
7075 Alclad, applications and properties ... **A2:** 115-116
7075, applications and properties ... **A2:** 115-116
7075, bending specimen thickness ... **A8:** 130
7075, cleaning and finishing ... **M5:** 579, 583, 597, 599, 603
7075, effect of solidification rate on secondary dendrite arm spacing ... **A9:** 629
7075, effect of temperature on strength and ductility ... **A8:** 36
7075, extrudability ... **A2:** 35
7075 extrusion, temper effects on exfoliation ... **A13:** 595
7075, flow stresses and deformation ... **A14:** 250
7075, forging alloy ... **A2:** 34
7075, fracture toughness ... **A2:** 59
7075, low-temperature alloy ... **A2:** 59
7075, machining ... **A16:** 776, 777, 779, 781, 782, 786, 788-790, 794-796, 798-800, 803
7075, SCC relative susceptibility ... **A13:** 267
7075, SCC testing ... **A13:** 251
7075, segregation in ... **A9:** 633
7075, stress-corrosion cracking ... **A13:** 159
7075, temper effect on SCC ... **A13:** 251
7075, thermal treatment to decrease SCC susceptibility ... **A8:** 508
7075, tube alloy ... **A2:** 33
7075, with CrAI7 inclusion ... **A9:** 635
7075-O, sheet, annealed, different cooling rates compared ... **A9:** 368
7075-T-6, shot peening ... **M5:** 141, 145
7075-T6
electrode potential in NaCl-H_2O_2 solution ... **A6:** 730
properties ... **A6:** 537
resistance spot welding ... **A6:** 848
ultrasonic welding ... **A6:** 327
7075-T6 aircraft landing gear component, SCC failure ... **A11:** 213
7075-T6, aircraft part, galvanic corrosion ... **A13:** 1024
7075-T6, alclad sheet, brittle fracture ... **A9:** 372
7075-T6, alclad sheet, ductile fracture ... **A9:** 372
7075-T6, compressed with orange peel effect ... **A8:** 57
7075-T6, crack tip stress intensity control fatigue cracking ... **A13:** 297
7075-T6, diffraction techniques, elastic constants, and bulk values for ... **A10:** 382
7075-T6, effect of corrosion on fatigue strength ... **A12:** 43, 54
7075-T6, extruded bar, intergranular stress- corrosion cracks ... **A9:** 372
7075-T6, extrusion, exfoliation-type corrosion ... **A9:** 372
7075-T6, extrusion, fractures ... **A9:** 371
7075-T6, extrusion, intergranular corrosion ... **A9:** 371
7075-T6, extrusion, pitting-type corrosion ... **A9:** 371
7075-T6 fasteners, fabrication failure ... **A11:** 530, 533
7075-T6, fatigue behavior ... **A2:** 43-44
7075-T6, fatigue crack propagation ... **A8:** 364-365, 404
7075-T6, fatigue fracture, resistance spot weld ... **A12:** 66, 67
7075-T6 fatigue-fracture surfaces ... **A11:** 112, 253
7075-T6, forging, shrinkage cavities and internal cracks ... **A9:** 370-371
7075-T6, forging, stress corrosion cracking ... **A9:** 370
7075-T6, forging, with fold at machined fillet ... **A9:** 369-370
7075-T6, forging, with parting-plane fracture ... **A9:** 369
7075-T6 forgings, thermal treatment effects ... **A11:** 340
7075-T6, lubricant effect on bearing strength ... **A8:** 60
7075-T6, machining ... **A16:** 583, 584, 585, 604, 764, 768, 797
7075-T6, mixed-mode fracture in ... **A11:** 84
7075-T6, mud crack pattern ... **A13:** 1049
7075-T6 plate, fatigue-fracture surface ... **A11:** 104
7075-T6 plate, grain structure ... **A13:** 592
7075-T6 plates, fatigue-fracture zones ... **A11:** 110
7075-T6, residual stresses ... **A13:** 591
7075-T6 rifle receivers, exfoliation failure ... **A11:** 338-342
7075-T6, *S-N* curve ... **A8:** 364
7075-T6, SEM and TEM fractographs, compared ... **A12:** 187-188
7075-T6, sequential erosion ... **A18:** 202, 203
7075-T6, shear testing ... **A8:** 65
7075-T6, sheet clad with 7072 ... **A9:** 369
7075-T6 sheet, eddy current inspection ... **A17:** 188
7075-T6, stress-corrosion fracture ... **A12:** 28, 35
7075-T6, surface appearance ... **A13:** 1047
7075-T6, tension and torsion effective fracture strains ... **A8:** 168
7075-T73, electrode potential in NaCl-H_2O_2 solution ... **A6:** 730
7075-T73 landing gear torque arm, fatigue fracture design ... **A11:** 114
7075-T73 very large precision forging section ... **A14:** 254
7075-T651, anodic polarization curves ... **A8:** 532
7075-T651, anodic polarization curves, SCC testing ... **A13:** 265
7075-T651, directionality effect on SCC in ... **A8:** 501
7075-T651, double shear test ... **A8:** 63
7075-T651, electron-beam welding ... **A6:** 872
7075-T651, fayed to 4130 steel, surface fretting ... **A9:** 387
7075-T651, milling ... **A16:** 767, 769
7075-T651, plate, surfaces of electrochemically machined hole ... **A9:** 388
7075-T651, SCC cracking ... **A13:** 1045
7075-T651, SCC resistance ... **A8:** 522
7075-T651, shear strength ... **A8:** 64

Aluminum alloys, specific types (continued)
7075-T651, sheet, effect of saturation peening ... A9: 387
7075-T651, stress-corrosion cracking ... A13: 591
7075-T7351, SCC performance ... A8: 522, 524
7075-T7352, forging, solution heat treated cold reduced, artificially aged ... A9: 368
7075T, coatings and tool life ... A16: 58
7076, applications and properties ... A2: 116-118
7079
composition ... A6: 723
physical properties ... A6: 723
weldability ... A6: 723
7079, forging alloy ... A2: 34
7079, fracture toughness ... A2: 59
7079, machining ... A16: 762, 771-774, 776, 777, 779, 781-782, 786, 788-790, 794-796, 798-800
7079-T-6, shot peening ... M5: 141
7079-T6, clamshell marks ... A13: 1048
7079-T6, forging, solution heat treated artificially aged, reduced from 40% to 85% ... A9: 367
7079-T6, stress-corrosion cracking ... A13: 1027-1029
7079-T6, stress-corrosion failure ... A13: 1102
7079-T651, corrosion fatigue behavior ... A13: 300
7079-T651, corrosive environment effect on SCC ... A13: 268
7079-T651, double shear test ... A8: 63
7079-T651, effect of corrosive environment on SCC in ... A8: 499-500
7079-T651, relative SCC susceptibility ... A13: 267
7079-T651, SCC performance ... A8: 524
7079-T651, shear strength ... A8: 64
7090, flow stresses and deformation ... A14: 250
7090, microstructure ... A9: 358
7090, P/M aerospace forgings ... A7: 746
7091, flow stresses and deformation ... A14: 250
7091, microstructure ... A9: 358
7146, relative rating of filler alloys for welding ... A6: 731, 732
7150, aircraft alloy ... A2: 33
7150-T651, autographic bearing load vs. bearing deformation curves for ... A8: 61
7175, applications and properties ... A2: 118
7175, machining ... A16: 771-774, 776, 777, 779, 781, 782, 786, 788-790, 794-796, 798-800
7178
composition ... A6: 529, 723
crack sensitivity ratings of base alloy/filler alloy combinations ... A6: 725
physical properties ... A6: 723
weldability ... A6: 723, 725-726, 728
7178, aircraft alloy ... A2: 33
7178 Alclad, applications and properties ... A2: 119
7178, applications and properties ... A2: 119
7178, extrudability ... A2: 35
7178, machining ... A16: 762, 771-774, 776, 777, 779, 781, 782, 786, 788-790, 794-796, 798-800
7178, minimum bend radii ... A8: 130
7178-T6 aircraft fuel-tank floors, fatigue fracture ... A11: 126-127
7178-T6, fatigue cracking ... A11: 311-312
7178-T6, pitting/cracking ... A13: 1046
7178-T76, clad with 7072 ... A9: 369
7178-T76, exposed to a salt fog ... A9: 369
7178-T651, double shear test ... A8: 63
7178-T651, exfoliation corrosion ... A13: 595
7178-T651, SCC performance ... A8: 524
7178-T651, shear strength ... A8: 64
7475, aircraft alloy ... A2: 33
7475, applications and properties ... A2: 119, 121-122
7475, diffusion welding ... A6: 885
7475, effect of stress intensity factor range on fatigue crack growth rate ... A12: 57
7475, fracture toughness ... A2: 60
7475, fracture toughness, fatigue behavior ... A2: 42
7475, second-phase constituents ... A2: 42
7475, superplasticity in ... A14: 800
7475-T7351, fracture toughness ... A8: 461
7475-T7651, fatigue striation spacing ... A12: 22
8079, foil alloy ... A2: 33
8090
composition ... A6: 529, 550
diffusion welding ... A6: 885
electron-beam welding ... A6: 551
properties ... A6: 549, 550
8090, microstructure ... A9: 357
8090-T6, tensile properties ... A6: 550
8111, foil alloy ... A2: 33
8280, machinability rating ... A16: 762
A 201.0
composition ... A6: 724
physical properties ... A6: 724
weldability ... A6: 724
A 242.0
composition ... A6: 724
physical properties ... A6: 724
weldability ... A6: 724
A 356.0
composition ... A6: 533, 724
filler alloys for best color match ... A6: 730
melting range ... A6: 724
physical properties ... A6: 724
relative rating of filler alloys for welding ... A6: 731
weldability ... A6: 534, 724
welding, for sustained elevated-temperature service ... A6: 729
A 357.0
composition ... A6: 533, 724
filler alloys for best color match ... A6: 730
melting range ... A6: 724
physical properties ... A6: 724
relative rating of filler alloys for welding ... A6: 731
weldability ... A6: 534, 724
welding, for sustained elevated-temperature service ... A6: 729
A 444.0
composition ... A6: 723
electrode potential in NaCl-H_2O_2 solution ... A6: 730
filler alloys for best color match ... A6: 730
properties ... A6: 723
weldability ... A6: 534, 723
welding, for sustained elevated-temperature service ... A6: 729
A 514.0, relative rating of filler alloys for welding ... A6: 731, 732, 733
A 712.0
electrode potential in NaCl-H_2O_2 solution ... A6: 730
relative rating of filler alloys for welding ... A6: 731, 732
A201-T7, hot isostatic pressing ... A15: 540, 541
A240, machinability rating ... A16: 763
A240-F, as investment cast, with shrinkage voids ... A9: 372
A242, machinability rating ... A16: 763
A332, machining ... A16: 763, 771-774, 776, 777, 779, 781, 782, 786, 788-790, 794-796, 798-800
A332-F, as investment cast ... A9: 373
A332-T65, sand cast, solution heat treated and artificially aged ... A9: 373
A332-T551, sand cast and artificially aged ... A9: 373
A354-F, as investment cast ... A9: 373
A354-T4, investment casting with fusion voids ... A9: 373
A356, as widely used ... A15: 159
A356, grain size/modification effects ... A15: 752
A356, machining ... A16: 763, 771-774, 776, 777, 779, 00
A356, mechanical properties ... A15: 166
A356, porosity effect, hot isostatic pressing ... A15: 540
A356, strontium effects ... A15: 166, 167
A356, volume fraction porosity ... A15: 165
A356-F, sand casting ... A9: 375
A356-F, sand casting, with grain refiner added ... A9: 375
A356-T6, hot isostatic pressing ... A15: 541
A356-T6, investment casting welded to 6061-T6 tube ... A9: 382-383
A356-T62, dendrite cell size and tensile properties ... A15: 167
A356-T62, tensile properties vs dendrite cell size ... A15: 749
A356.0, applications and properties ... A2: 164-165
A356.0, composition/characteristics ... A15: 159
A356,0, minimum tensile properties ... A15: 160
A357, helicopter application ... A15: 769
A357, machinability rating ... A16: 763
A357-F, as premium quality cast ... A9: 378
A357-T6, hot isostatic pressing ... A15: 541
A357-T6, premium quality cast, solution heat treated and artificially aged ... A9: 378
A357-T61, permanent mold cast, solution heat treated, quenched and aged ... A9: 375
A357.0, applications and properties ... A2: 166
A357.0, composition/characteristics ... A15: 159
A360, as die cast'ng, composition ... A15: 286
A360, machining ... A16: 763, 771-774, 776, 777, 779, 781, 782, 786, 788-790, 794-796, 798-800
A360.0, applications and properties ... A2: 167-168
A380, as die casting, composition ... A15: 286
A380, machining ... A16: 763, 771-774, 776, 777, 779, 781, 782, 786, 788-790, 794-796, 798-800
A380.0, applications and properties ... A2: 168-169
A383, as die casting, composition ... A15: 286
A384, as die casting, composition ... A15: 286
A384, machining ... A16: 771-774, 776, 777, 779, 781, 782, 786, 788-790, 794-796, 798-800
A384.0, applications and properties ... A2: 170-171
A390, machining ... A16: 643, 763, 895
A390.0, applications and properties ... A2: 171
A413, as die cast'ng, composition ... A15: 286
A413, machining ... A16: 763, 771-774, 776, 777, 779, , 798-800
A413.0, applications and properties ... A2: 171-172
A444, machinability rating ... A16: 763
A444.0, applications and properties ... A2: 172-173
A514, machining ... A16: 763, 771-774, 776, 777, 779, , 798-800
A535, machinability rating ... A16: 763
A535.0, applications and properties ... A2: 174-175
A712, machining ... A16: 763, 771-774, 776, 777, 779, 781, 782, 786, 788-790, 794-796, 798-800
A850, machining ... A16: 763, 771-774, 776, 777, 779, 781, 782, 786, 788-790, 794-796, 798-800
A1050, non-oxide ceramic joining ... EM4: 480
A2010 (Al-4.7Cu)
fiber for reinforcement ... A18: 803
processing technique ... A18: 803
properties ... A18: 803
A2024, sliding wear in metal-matrix composites with hard particles ... A18: 809
A12014, contact conditions effect on wear rate ... A18: 808
AA8009, friction welding ... A6: 546, 547
AC-8A, non-oxide ceramic joining ... EM4: 480
ADC 12
fiber for reinforcement ... A18: 803
processing technique ... A18: 803
properties ... A18: 803
Al-Zn-Mg alloy, fretting wear ... A18: 250

Aluminum alloys, specific types (continued)
Al-0.12Cu-1.2Mn (3003), reactive metal brazing, filler metal ... **A6:** 945
Al-1.7Cu, information from FIM image of ... **A10:** 589-590
Al-1.91Mg, deformed 0.62%, grain-boundary sliding ... **A9:** 691
Al-1Mg-0.6Si-0.379Cu, pressed and sintered ... **A9:** 525-526
Al-4.4Cu-0.8Si-0.4Mg, pressed and sintered ... **A9:** 525
Al-4.5 Cu, atomized powder ... **A9:** 616
Al-4.5Mg alloy, specific wear rate ... **A18:** 805
Al-4.7Cu, diffusion-induced grain-boundary migration ... **A10:** 461-464
Al-4.7Cu, typical CBEDP for ... **A10:** 464
Al-4Cu, diffraction patterns ... **A9:** 118
Al-4Cu, fracture surface measurements ... **A12:** 205-206
Al-4Cu, fractured, SEM projected facets ... **A12:** 195
Al-4Cu, fractured, serial sectioning profile ... **A12:** 198
Al-4Cu, friction surfacing ... **A6:** 322, 323
Al-4Cu, SEM fractograph, calculation of features ... **A12:** 206-207
Al-4Cu, structure-factor contrast ... **A9:** 118
Al-4Cu, true profile length values ... **A12:** 200
Al-4Mg and Al-5Si, yield strength vs. temperature ... **A6:** 1016
Al-5.2Si (4043), reactive metal brazing, filler metal ... **A6:** 945
Al-5Cu, thermal analysis of ... **A15:** 184
Al-6Ag, structure-factor contrast ... **A9:** 118
Al-6Si, electrohydrodynamic atomized powder ... **A9:** 616
Al-6Ti, addition to aluminum-silicon alloys ... **A18:** 787
Al-6U, peritectic structures ... **A9:** 676
Al-7Si, dendrite arm spacing ... **A15:** 164
Al-7Si ingots, grain refinement effects ... **A15:** 750
Al-8.4Fe-3.6Ce,for niobium-titanium superconducting materials ... **A2:** 1045
Al-8.4Fe-3.7Ce
chemical composition ... **A6:** 542
microstructure ... **A6:** 542
Al-8.5Fe-1.3V-1.7Si (AA 8009 alloy)
chemical composition ... **A6:** 542
microstructure ... **A6:** 542
Al-8.7Fe-2.8Mo-1V, chemical composition ... **A6:** 542
Al-8Fe, vacuum-atomized ... **A9:** 617
Al-8Fe-1.7Ni, chemical composition ... **A6:** 542
Al-8Fe-2.3Mo, chemical composition ... **A6:** 542
Al-8Fe-2Mo
electron-beam welding ... **A6:** 544, 545
laser-beam welding ... **A6:** 544, 545, 546
Al-8Fe-4Ce
capacitor-discharge welding ... **A6:** 544
microstructure ... **A6:** 542
Al-9Fe-3Mo-1V
friction welding ... **A6:** 546
microstructure ... **A6:** 542
Al-9Fe-4Ce (AA 8019 alloy)
chemical composition ... **A6:** 542
friction welding ... **A6:** 546
Al-9Fe-7Ce, chemical composition ... **A6:** 542
Al-10Fe-5Ce
chemical composition ... **A6:** 542
gas-tungsten arc welding ... **A6:** 543, 544
Al-10Si, dendrite arm spacing ... **A15:** 164
Al-11.7Fe-1.2V-2.4Si (FVS 1212 alloy), chemical composition ... **A6:** 542
Al-11.7Si-0.3Fe, backscattered electron images and secondary electron images compared ... **A9:** 93
Al-11.7Si-1Co-1Mg-1Ni-0.3Fe, cast ... **A9:** 98
Al-12Mn, melt spun, alpha-aluminum ... **A9:** 615
Al-12Si, dendrite arm spacing ... **A15:** 164
Al-12Si-50Al_2O_3, wear rate ... **A18:** 806
Al-13Si alloy, polishing wear ... **A18:** 195
Al-13Si, graphite-aluminum composite ... **A9:** 594
Al-15Ag, Guinier-Preston zones ... **A9:** 651
Al-15Mn, melt spun ... **A9:** 615
Al-18Ag, cellular precipitation colonies ... **A9:** 651
Al-18Ag, intragranular Widmanstatten precipitation ... **A9:** 648

Aluminum alloys, specific types (continued)
Al-20Ag, Widmanstatten precipitation ... **A9:** 649
Al-20SiC alloy
adhesive wear conditions ... **A18:** 808
erosive wear rate ... **A18:** 806
Al-22Si-1Ni-1Cu, phosphorus refinement ... **A15:** 753
Al-Mg-Si: 97% Al, 1% Mg, 1% Si, thermal properties ... **A18:** 42
Al-Sn alloys, bearing material microstructures ... **A18:** 743, 744
Al-Ti-B, addition to aluminum-silicon alloys ... **A18:** 787
Al-Zn-Mg-Mn-Cr, SCC responses with bending vs. tension ... **A8:** 503
Al40, machining ... **A16:** 771-774, 776, 777, 779, 781, 798-800
Alcan B54S-O rivet, SCC failure after heating ... **A11:** 544-545
Alclad 2014, bending specimen thickness ... **A8:** 130
Alclad 2014-T6, shear testing ... **A8:** 65
Alclad 2024, shear testing ... **A8:** 65
Alclad 6061-T6, shear testing ... **A8:** 65
Alclad 7075, chemical milling ... **A16:** 585
Alclad 7075-T6, shear testing ... **A8:** 65
B 514.0, relative rating of filler alloys for welding ... **A6:** 731, 732, 733
B295, machining ... **A16:** 763, 771-774, 776, 777, 779, 781, 782, 786, 788-790, 794-796, 798-800
B358, machinability rating ... **A16:** 763
B390, as die casting, composition ... **A15:** 286
B443, machining ... **A16:** 771-774, 776, 777, 779, 781, 798-800
B443-F, as permanent mold cast ... **A9:** 376
B443.0, applications and properties ... **A2:** 172-173
B443.0, composition/characteristics ... **A15:** 159
B514, machining ... **A16:** 763, 771-774, 776, 777, 779, 781, 782, 786, 788-790, 794-796, 798-800
B535, machinability rating ... **A16:** 763
B535.0, applications and properties ... **A2:** 174-175
B850, machining ... **A16:** 763, 771-774, 776, 777, 779, 00
C 355.0
composition ... **A6:** 533, 724
filler alloys for welding, for sustained elevated temperature service ... **A6:** 729
melting range ... **A6:** 724
physical properties ... **A6:** 724
relative rating of filler alloys for welding ... **A6:** 731, 732
weldability ... **A6:** 534, 724
welding, for sustained elevated-temperature service ... **A6:** 729
C 355.0-T61, resistance spot welding ... **A6:** 848
C335, reaming ... **A16:** 781
C355, machining ... **A16:** 763, 771-774, 776, 777, 779, 781, 782, 786, 788-790, 794-796, 798-800
C355.0, applications and properties ... **A2:** 163-164
C355.0, characteristics ... **A15:** 159
C355.0-T61, resistance welding ... **M6:** 536
C443, machining ... **A16:** 771-774, 776, 777, 779, 781, 798-800
C443-F, as die cast ... **A9:** 376
C443.0, applications and properties ... **A2:** 172-173
C712, machinability rating ... **A16:** 763
CW67, flow stresses and deformation ... **A14:** 250
D712, machining ... **A16:** 763, 771-774, 776, 777, 779, 781, 782, 786, 788-790, 794-796, 798-800
D712-F, as investment cast, intergranular fusion voids ... **A9:** 377
D712-F, as sand cast ... **A9:** 377
DTD 3064, tension and torsion effective fracture strains ... **A8:** 168
EC, machining ... **A16:** 323-325, 771-774, 776, 777, 779, 781, 782, 786, 788-790, 794, 795, 798-800
F 514.0, relative rating of filler alloys for welding ... **A6:** 731, 732, 733
F132-T5, power requirements ... **A16:** 765
F332, machining ... **A16:** 763, 771-774, 776, 777, 779, 781, 782, 786, 788-790, 794-796, 798-800
F514, machinability rating ... **A16:** 763
FVS1212, friction welding ... **A6:** 546, 547
Hiduminium RR -350 machining ... **A16:** 771-774, 776, 777, 779, 781, 782, 786, 788-790, 794-796, 798-800

Aluminum alloys, specific types (continued)
I/M 2024-T351, friction welding ... **A6:** 546, 547
IN9021, flow stresses and deformation ... **A14:** 250
L514, machinability rating ... **A16:** 763
modified ingot, with aluminum-oxide inclusions ... **A9:** 375
Pure Al
alloy designations ... **A6:** 528
gas-tungsten arc welding, weld microstructures ... **A6:** 51
SIMA 6262, semisolid forging ... **A15:** 334
solution heat treated ... **A9:** 264
X2020-T6, ultrasonic welding ... **A6:** 326
X7064, flow stresses and deformation ... **A14:** 250
X7090, mechanical properties ... **A7:** 474
X7091, mechanical properties ... **A7:** 474
X7091, true profile length values ... **A12:** 200
XAP001, mechanical properties ... **A7:** 475
XAP002, mechanical properties ... **A7:** 475
XAP004, mechanical properties ... **A7:** 475
Aluminum alloys, use for bearings *See also* Bearings, sliding ... **M3:** 817-820
Aluminum alloys, wrought, specific types *See also* Aluminum alloys, powder metallurgy, specific types; Aluminum casting alloys, specific types
1050 ... **M2:** 63-64
composition ... **M2:** 45
mechanical properties ... **M2:** 59
physical properties ... **M2:** 53
product forms ... **M2:** 45
1060 ... **M2:** 64-65
annealing temperature ... **M2:** 29
applications ... **M2:** 46
composition ... **M2:** 45
corrosion resistance ... **M2:** 46, 209
fabrication characteristics ... **M2:** 46
mechanical properties ... **M2:** 59
physical properties ... **M2:** 53
product forms ... **M2:** 45
solution potential ... **M2:** 207
stress-corrosion cracking resistance ... **M2:** 209
weldability ... **M2:** 193
1060, annealing ... **A4:** 871 **M4:** 708
1100 ... **M2:** 65-66, 67
annealing ... **A4:** 870, 871, 872
annealing curves ... **M2:** 28, 30
annealing temperature ... **M2:** 29
anodic polarization curve ... **M2:** 205
applications ... **M2:** 46
atmospheric corrosion resistance ... **M2:** 221, 222
composition ... **M2:** 45
corrosion in chemical solutions ... **M2:** 233
corrosion pit density vs anodic coating thickness ... **M2:** 229
corrosion resistance ... **M2:** 46, 209, 223, 224, 230, 231, 232
fabrication characteristics ... **M2:** 46
forming-limit diagram ... **M2:** 183
grain growth ... **A4:** 872
impacts, mechanical properties ... **M2:** 10
impacts, minimum wall thickness ... **M2:** 9
machinability rating ... **M2:** 188
mechanical properties ... **M2:** 59
physical properties ... **M2:** 53
product forms ... **M2:** 45
properties, expected for welds ... **M2:** 197
properties, gas metal-arc welded plate ... **M2:** 198
solution potential ... **M2:** 207
stress-corrosion cracking resistance ... **M2:** 209
weldability ... **M2:** 193
1100, annealing ... **M4:** 708, 709
1135 atmospheric corrosion resistance ... **M2:** 224
1145 ... **M2:** 66-67
composition ... **M2:** 45
physical properties ... **M2:** 53
product forms ... **M2:** 45
1188
atmospheric corrosion resistance ... **M2:** 224
1199 ... **M2:** 67-68
atmospheric corrosion resistance ... **M2:** 224
composition ... **M2:** 45
physical properties ... **M2:** 53
pitting potential, effect of chloride-ion activity ... **M2:** 206
product forms ... **M2:** 45

Aluminum alloys, wrought, specific types (continued)
seawater corrosion resistance **M2:** 230
1350 ..... **M2:** 68-69, 70
annealing temperature **M2:** 29
applications **M2:** 46
composition **M2:** 45
corrosion resistance **M2:** 46, 209
fabrication characteristics **M2:** 46
machinability rating **M2:** 188
mechanical properties **M2:** 59
physical properties **M2:** 53
product forms **M2:** 45
stress-corrosion cracking resistance **M2:** 209
weldability **M2:** 193
wire conductors **M2:** 20
1350, annealing **A4:** 871 **M4:** 708
01429 precipitation heat treatment **A4:** 843
2008
precipitation heat treatment **A4:** 845
solution heat treatment **A4:** 845
2011 ..... **M2:** 69-70, 71
age hardening **A4:** 865
applications **M2:** 46
composition **M2:** 45
corrosion resistance **M2:** 46, 209
fabrication characteristics **M2:** 46
machinability rating **M2:** 188
mechanical properties **M2:** 59
physical properties **M2:** 53
precipitation heat treatment **A4:** 845 **M4:** 677-683
product forms **M2:** 45
solution heat treatment **A4:** 845 **M4:** 677-683
stress-corrosion cracking resistance **M2:** 209, 213
weldability **M2:** 193
2014 ..... **M2:** 70-73
age hardening **A4:** 864, 865
aging **M4:** 701-703
aging characteristics at low aging temperatures **M2:** 36-38
annealing **A4:** 871 **M4:** 708
annealing temperature **M2:** 29
applications **M2:** 46
composition **M2:** 45
corrosion resistance **M2:** 46, 209, 224, 225
fabrication characteristics **M2:** 46
fatigue curves **M2:** 15
forgeability, relative **M2:** 6
forging, relative cost vs forging weight **M2:** 5
fracture toughness **A4:** 878
hardness **A4:** 877
hardness values **M4:** 716
impacts, mechanical properties **M2:** 10
impacts, minimum wall thickness **M2:** 9
intergranular corrosion **A4:** 877
machinability rating **M2:** 188
mechanical properties **M2:** 59
physical properties **M2:** 53
precipitation heat treatment **A4:** 845, 876 **M4:** 677-683
product forms **M2:** 45
quenching **A4:** 856, 869, 875 **M4:** 695, 713
quenching stresses **M2:** 40, 43
reheating **M4:** 707
reheating schedules **A4:** 869
solution heat treatment **A4:** 844, 845, 854 **M4:** 677-683
solution potential **M2:** 207
stress-corrosion cracking resistance **M2:** 209, 213
weldability **M2:** 193
2017
annealing **A4:** 871 **M4:** 708
annealing temperature **M2:** 29
corrosion resistance **M2:** 209, 223
galvanic corrosion with magnesium **M2:** 607

Aluminum alloys, wrought, specific types (continued)
machinability rating **M2:** 188
precipitation heat treatment **A4:** 845 **M4:** 677-683
quenching **A4:** 856 **M4:** 695
solution heat treatment **A4:** 844, 845, 854 **M4:** 677-683
stress-corrosion cracking resistance **M2:** 209
2018
corrosion resistance **M2:** 209
precipitation heat treatment **A4:** 845 **M4:** 677-683
solution heat treatment **A4:** 845 **M4:** 677-683
stress corrosion cracking resistance **M2:** 209
2020 precipitation heat treatment **A4:** 843
2024 ..... **M2:** 72-75, 76-78
age hardening **A4:** 860, 864, 865, 866
aging **M4:** 701-703
aging characteristics at low aging temperatures **M2:** 31, 36-38
aging time and temperature effect on mechanical properties **A4:** 835-836
annealing **A4:** 871 **M4:** 713
annealing temperature **M2:** 29
applications **M2:** 46
composition **M2:** 45
corrosion **M4:** 691, 692
corrosion behavior **A4:** 858, 859
corrosion rate affected by quenching rate **M2:** 34
corrosion resistance **M2:** 46, 209, 224, 225, 226, 232
dimensional changes **A4:** 876
fabrication characteristics **M2:** 46
fatigue characteristics **A4:** 858 **M2:** 35 **M4:** 695
fracture toughness **A4:** 867, 878
galvanic corrosion with magnesium **M2:** 607
hardness **A4:** 842, 877
hardness values **M4:** 716
intergranular corrosion **A4:** 877
machinability rating **M2:** 188
mechanical properties **M2:** 59
natural aging curve **M2:** 31
physical properties **M2:** 53
precipitation heat treatment **A4:** 845, 874, 876 **M4:** 677-683
product forms **M2:** 45
properties, elevated temperatures **M2:** 56, 62
quenching **A4:** 856, 857, 858, 859 **M4:** 691, 695
quenching rate, effect on yield strength after aging **M2:** 34
reheating **A4:** 869, 870 **M4:** 707
solution heat treatment **A4:** 844, 845, 848, 852, 854 **M4:** 677-683, 687
solution potential **M2:** 207
stress-corrosion cracking resistance **M2:** 209, 213
tensile properties affected by cold work before aging **M2:** 39
tensile properties, effect of cold work **M4:** 705
weldability **M2:** 193
yield strength **A4:** 867 **M4:** 689
2025
corrosion resistance **M2:** 209
forging, relative cost vs forging weight **M2:** 5
precipitation heat treatment **A4:** 845 **M4:** 677-683
solution heat treatment **A4:** 845 **M4:** 677-683
stress-corrosion cracking resistance **M2:** 209
2036 ..... **M2:** 75-76
annealing **A4:** 871 **M4:** 708
annealing temperature **M2:** 29
applications **M2:** 46
composition **M2:** 45
corrosion resistance **M2:** 46, 209
fabrication characteristics **M2:** 46

Aluminum alloys, wrought, specific types (continued)
machinability rating **M2:** 188
mechanical properties **M2:** 59
mechanical properties, sheet **M2:** 181
physical properties **M2:** 53
precipitation heat treatment **A4:** 845 **M4:** 677-683
product forms **M2:** 45
solution heat treatment **A4:** 845 **M4:** 677-683
solution potential **M2:** 207
stress-corrosion cracking resistance **M2:** 209
2038
precipitation heat treatment **A4:** 845
solution heat treatment **A4:** 845
2048 ..... **M2:** 77-78, 80-82
composition **M2:** 45
mechanical properties **M2:** 59
physical properties **M2:** 53
product forms **M2:** 45
stress-corrosion cracking resistance **M2:** 213
2048, fracture toughness **A4:** 878
2090
fracture toughness **A4:** 878
precipitation heat treatment **A4:** 843, 845
solution heat treatment **A4:** 845, 852
tensile properties **A4:** 852
2091
fracture toughness **A4:** 878
precipitation heat treatment **A4:** 843, 845
solution heat treatment **A4:** 845
2117
annealing **A4:** 871 **M4:** 708
annealing temperature **M2:** 29
corrosion resistance **M2:** 209
precipitation heat treatment **A4:** 845 **M4:** 677-683
quenching **A4:** 856 **M4:** 695
solution heat treatment **A4:** 845, 854 **M4:** 677-683
stress-corrosion cracking resistance **M2:** 209
2124 ..... **M2:** 78-79, 82-84
age hardening **A4:** 864, 865
annealing **A4:** 871
annealing temperature **M2:** 29
composition **M2:** 45
fracture toughness **A4:** 878
mechanical properties **M2:** 59
physical properties **M2:** 53
product forms **M2:** 45
stress-corrosion cracking resistance **M2:** 213
2124, annealing **M4:** 708
2218 ..... **M2:** 79, 85
applications **M2:** 46
composition **M2:** 45
corrosion resistance **M2:** 46, 209
fabrication characteristics **M2:** 46
mechanical properties **M2:** 59
physical properties **M2:** 53
precipitation heat treatment **A4:** 845 **M4:** 677-683
product forms **M2:** 45
solution heat treatment **A4:** 845 **M4:** 677-683
stress-corrosion cracking resistance **M2:** 209
weldability **M2:** 193
2219 ..... **M2:** 86-88, 89, 91
age hardening **A4:** 864, 865
annealing **A4:** 871 **M4:** 708
annealing temperature **M2:** 29
applications **M2:** 46
composition **M2:** 45
corrosion resistance **M2:** 46, 209
fabrication characteristics **M2:** 46
fracture toughness **A4:** 878
hardness **A4:** 842
intergranular corrosion **A4:** 877
machinability rating **M2:** 188
mechanical properties **M2:** 59
physical properties **M2:** 53

Aluminum alloys, wrought, specific types (continued)
precipitation heat treatment........ **A4:** 845, 876 **M4:** 677-683
product forms........ **M2:** 45
quenching........ **A4:** 856, 875 **M4:** 695
solution heat treatment........ **A4:** 844, 845, 875 **M4:** 677-683
solution potential........ **M2:** 207
stress-corrosion cracking resistance........ **M2:** 209, 213
weldability........ **M2:** 193
2224 fracture toughness........ **A4:** 878
2319
physical properties........ **M2:** 53
product forms........ **M2:** 45
2324
age hardening........ **A4:** 860-861, 865
fracture toughness........ **A4:** 878
2419
age hardening........ **A4:** 865
fracture toughness........ **A4:** 878
2618........ **M2:** 88-90, 92, 93
applications........ **M2:** 47
composition........ **M2:** 45
corrosion resistance........ **M2:** 47, 209
fabrication characteristics........ **M2:** 47
forgeability, relative........ **M2:** 6
machinability rating........ **M2:** 188
mechanical properties........ **M2:** 59
physical properties........ **M2:** 53
precipitation heat treatment........ **A4:** 845 **M4:** 677-683
product forms........ **M2:** 45
solution heat treatment........ **A4:** 845 **M4:** 677-683
stress-corrosion cracking resistance........ **M2:** 209
weldability........ **M2:** 193
3003........ **M2:** 90-92, 94, 95
annealing temperature........ **M2:** 29
applications........ **M2:** 47
atmospheric corrosion resistance........ **M2:** 221, 222
composition........ **M2:** 45
corrosion resistance........ **M2:** 47, 209, 223, 224, 225, 230, 231, 232
fabrication characteristics........ **M2:** 47
fatigue characteristics, weldments........ **M2:** 195, 199
galvanic corrosion with magnesium........ **M2:** 607
impacts, mechanical properties of........ **M2:** 10
machinability rating........ **M2:** 188
mechanical properties........ **M2:** 59
mechanical properties, sheet........ **M2:** 181
physical properties........ **M2:** 53
product forms........ **M2:** 45
properties, expected for welds........ **M2:** 197
properties, gas metal-arc welded plate........ **M2:** 197
solution potential........ **M2:** 207
stress-corrosion cracking resistance........ **M2:** 209
weldability........ **M2:** 193
3003, annealing........ **A4:** 869, 871 **M4:** 708
3004........ **M2:** 92-94, 96, 97
annealing temperature........ **M2:** 29
applications........ **M2:** 47
atmospheric corrosion resistance........ **M2:** 221, 222
composition........ **M2:** 45
corrosion in distilled water........ **M2:** 205
corrosion resistance........ **M2:** 47, 209, 224, 225, 232
fabrication characteristics........ **M2:** 47
machinability rating........ **M2:** 188
mechanical properties........ **M2:** 60
physical properties........ **M2:** 53
product forms........ **M2:** 45
properties, expected for welds........ **M2:** 197
solution potential........ **M2:** 207
stress-corrosion cracking resistance........ **M2:** 209
weldability........ **M2:** 193
3004, annealing........ **A4:** 871 **M4:** 708
3105........ **M2:** 94-95, 97
annealing temperature........ **M2:** 29
applications........ **M2:** 47
composition........ **M2:** 45
corrosion resistance........ **M2:** 47, 209
fabrication characteristics........ **M2:** 47
mechanical properties........ **M2:** 60
physical properties........ **M2:** 53
product forms........ **M2:** 45
stress-corrosion cracking resistance........ **M2:** 209

Aluminum alloys, wrought, specific types (continued)
weldability........ **M2:** 193
3105, annealing........ **A4:** 871 **M4:** 708
4032........ **M2:** 95-96, 98
applications........ **M2:** 47
composition........ **M2:** 45
corrosion resistance........ **M2:** 209
fabrication characteristics........ **M2:** 47
forgeability, relative........ **M2:** 6
mechanical properties........ **M2:** 60
physical properties........ **M2:** 53
precipitation heat treatment........ **A4:** 845 **M4:** 677-683
product forms........ **M2:** 45
solution heat treatment........ **A4:** 845 **M4:** 677-683
stress-corrosion cracking resistance........ **M2:** 209
weldability........ **M2:** 193
4043........ **M2:** 97-98
atmospheric corrosion resistance........ **M2:** 204
composition........ **M2:** 45
mechanical properties........ **M2:** 60
physical properties........ **M2:** 53
product forms........ **M2:** 45
5005........ **M2:** 98-99
annealing temperature........ **M2:** 29
applications........ **M2:** 47
composition........ **M2:** 45
corrosion resistance........ **M2:** 47, 209, 224
fabrication characteristics........ **M2:** 47
machinability rating........ **M2:** 188
mechanical properties........ **M2:** 60
physical properties........ **M2:** 53
product forms........ **M2:** 45
properties, expected for welds........ **M2:** 197
stress-corrosion cracking resistance........ **M2:** 209
weldability........ **M2:** 193
5005, annealing........ **A4:** 871 **M4:** 708
5050........ **M2:** 99-100
annealing temperature........ **M2:** 29
applications........ **M2:** 47
composition........ **M2:** 45
corrosion resistance........ **M2:** 47, 209, 224, 225
fabrication characteristics........ **M2:** 47
machinability rating........ **M2:** 188
mechanical properties........ **M2:** 60
physical properties........ **M2:** 53
product forms........ **M2:** 45
properties, expected for welds........ **M2:** 197
properties, gas metal-arc welded plate........ **M2:** 197
solution potential........ **M2:** 207
stress-corrosion cracking resistance........ **M2:** 209
weldability........ **M2:** 193
5050, annealing........ **A4:** 871 **M4:** 708
5052........ **M2:** 101-102
annealing curves........ **M2:** 28, 30
annealing temperature........ **M2:** 29
applications........ **M2:** 47
composition........ **M2:** 45
corrosion resistance........ **M2:** 47, 209, 224, 225, 230, 231
fabrication characteristics........ **M2:** 47
galvanic corrosion with magnesium........ **M2:** 607
machinability rating........ **M2:** 188
mechanical properties........ **M2:** 60
mechanical properties, sheet........ **M2:** 181
physical properties........ **M2:** 53
product forms........ **M2:** 47
properties, expected for welds........ **M2:** 197
properties, gas metal-arc welded plate........ **M2:** 197
solution potential........ **M2:** 207
stress-corrosion cracking resistance........ **M2:** 209
weldability........ **M2:** 193
5052, annealing........ **A4:** 870, 871, 872 **M4:** 708, 709
5056........ **M2:** 102-103
annealing temperature........ **M2:** 29
applications........ **M2:** 47
composition........ **M2:** 45
corrosion resistance........ **M2:** 47, 209, 230, 231
fabrication characteristics........ **M2:** 47
galvanic corrosion with magnesium........ **M2:** 607
machinability rating........ **M2:** 188
mechanical properties........ **M2:** 60
mechanical properties, sheet........ **M2:** 181
physical properties........ **M2:** 53

Aluminum alloys, wrought, specific types (continued)
product forms........ **M2:** 45
solution potential........ **M2:** 207
stress-corrosion cracking resistance........ **M2:** 209
weldability........ **M2:** 193
5056, annealing........ **A4:** 871 **M4:** 708
5083........ **M2:** 103, 104
annealing temperature........ **M2:** 29
applications........ **M2:** 47
composition........ **M2:** 45
corrosion resistance........ **M2:** 47, 209, 224, 230, 232
cruciform weldment, mercury cracking of........ **M2:** 212
fabrication characteristics........ **M2:** 47
forgeability, relative........ **M2:** 6
machinability rating........ **M2:** 188
mechanical properties........ **M2:** 60
physical properties........ **M2:** 53
product forms........ **M2:** 45
properties, expected for welds........ **M2:** 197
properties, gas metal-arc welded plate........ **M2:** 197
solution potentials........ **M2:** 207
stress-corrosion cracking resistance........ **M2:** 209, 216-217
weldability........ **M2:** 193
5083, annealing........ **A4:** 871 **M4:** 708
5086........ **M2:** 104-105
annealing temperature........ **M2:** 29
applications........ **M2:** 48
composition........ **M2:** 45
corrosion resistance........ **M2:** 48, 209, 224, 232
fabrication characteristics........ **M2:** 48
machinability ratings........ **M2:** 188
mechanical properties........ **M2:** 60
mechanical properties, sheet........ **M2:** 181
physical properties........ **M2:** 53
product forms........ **M2:** 45
properties, expected for welds........ **M2:** 197
properties, gas metal-arc welded plate........ **M2:** 197
solution potential........ **M2:** 207
stress-corrosion cracking resistance........ **M2:** 209
weldability........ **M2:** 193
5086, annealing........ **A4:** 871 **M4:** 708
5154........ **M2:** 105, 106
annealing temperature........ **M2:** 29
applications........ **M2:** 48
composition........ **M2:** 45
corrosion resistance........ **M2:** 48, 209, 224, 230, 232
fabrication characteristics........ **M2:** 48
fatigue characteristics, weldments........ **M2:** 195, 199
machinability rating........ **M2:** 188
mechanical properties........ **M2:** 60
physical properties........ **M2:** 53
product forms........ **M2:** 45
properties, gas metal-arc welded plate........ **M2:** 197
solution potential........ **M2:** 207
stress-corrosion cracking resistance........ **M2:** 209
weldability........ **M2:** 193
5154, annealing........ **A4:** 871 **M4:** 708
5182........ **M2:** 106-107
annealing temperature........ **M2:** 29
composition........ **M2:** 45
machinability rating........ **M2:** 188
mechanical properties........ **M2:** 60
mechanical properties, sheet........ **M2:** 181
physical properties........ **M2:** 53
product forms........ **M2:** 45
solution potential........ **M2:** 207
5182, annealing........ **A4:** 871 **M4:** 708
5252........ **M2:** 107
applications........ **M2:** 48
composition........ **M2:** 45
corrosion resistance........ **M2:** 48, 209
fabrication characteristics........ **M2:** 48
mechanical properties........ **M2:** 60
mechanical properties, sheet........ **M2:** 181
physical properties........ **M2:** 53
product forms........ **M2:** 45
stress-corrosion cracking resistance........ **M2:** 209
weldability........ **M2:** 193
5254........ **M2:** 108-109
annealing temperature........ **M2:** 29
applications........ **M2:** 48

Aluminum alloys, wrought, specific types (continued)
composition **M2:** 45
corrosion resistance **M2:** 48, 209
fabrication characteristics **M2:** 48
mechanical properties **M2:** 60-61
physical properties **M2:** 53
product forms **M2:** 45
stress-corrosion cracking resistance **M2:** 209
weldability **M2:** 193
5254, annealing **A4:** 871 **M4:** 708
5356 .... **M2:** 109
composition **M2:** 45
microstructure, effect on susceptibility to stress-corrosion cracking **M2:** 213-215
physical properties **M2:** 53
product forms **M2:** 45
5357 atmospheric corrosion resistance **M2:** 224
5454 .... **M2:** 109-110
annealing temperature **M2:** 29
applications **M2:** 48
composition **M2:** 45
corrosion resistance **M2:** 48, 209, 224, 232
fabrication characteristics **M2:** 48
machinability rating **M2:** 188
mechanical properties **M2:** 61
physical properties **M2:** 53
product forms **M2:** 45
properties, expected for welds **M2:** 197
solution potential **M2:** 207
stress-corrosion cracking resistance **M2:** 209
weldability **M2:** 193
5454, annealing **A4:** 871 **M4:** 708
5456 .... **M2:** 110-111
annealing temperature **M2:** 29
applications **M2:** 48
composition **M2:** 45
corrosion resistance **M2:** 48, 209, 224, 232
fabrication characteristics **M2:** 48
machinability rating **M2:** 188
mechanical properties **M2:** 61
physical properties **M2:** 48
product forms **M2:** 45
properties, expected for welds **M2:** 197
solution potential **M2:** 207
stress-corrosion cracking resistance **M2:** 209
weldability **M2:** 193
5456, annealing **A4:** 871 **M4:** 708
5457 .... **M2:** 111-112
annealing temperature **M2:** 29
applications **M2:** 48
composition **M2:** 45
corrosion resistance **M2:** 48, 209, 232
fabrication characteristics **M2:** 48
machinability rating **M2:** 188-189
mechanical properties **M2:** 61
physical properties **M2:** 53
product forms **M2:** 45
stress-corrosion cracking resistance **M2:** 209
weldability **M2:** 193
5457, annealing **A4:** 871 **M4:** 708
5652 .... **M2:** 112-113
annealing temperature **M2:** 29
applications **M2:** 48
composition **M2:** 45
corrosion resistance **M2:** 48, 209
fabrication characteristics **M2:** 48
mechanical properties **M2:** 61
physical properties **M2:** 53
product forms **M2:** 45
stress-corrosion cracking resistance **M2:** 209
weldability **M2:** 193
5652, annealing **A4:** 871 **M4:** 708
5657 .... **M2:** 113
applications **M2:** 48
composition **M2:** 45
corrosion resistance **M2:** 48, 209
fabrication characteristics **M2:** 48

Aluminum alloys, wrought, specific types (continued)
machinability rating **M2:** 189
mechanical properties **M2:** 61
physical properties **M2:** 53
product forms **M2:** 45
stress-corrosion cracking resistance **M2:** 209
weldability **M2:** 193
6005 .... **M2:** 113-114
annealing **A4:** 871 **M4:** 708
annealing temperature **M2:** 29
applications **M2:** 48
composition **M2:** 45
corrosion resistance **M2:** 48
mechanical properties **M2:** 61
physical properties **M2:** 53
precipitation heat treatment **A4:** 846 **M4:** 677-683
product forms **M2:** 45
solution heat treatment **A4:** 846 **M4:** 677-683
6009 .... **M2:** 114-115
annealing **A4:** 871 **M4:** 708
annealing temperature **M2:** 29
composition **M2:** 45
machinability rating **M2:** 189
mechanical properties **M2:** 61
mechanical properties, sheet **M2:** 181
physical properties **M2:** 54
precipitation heat treatment **A4:** 846 **M4:** 677-683
product forms **M2:** 45
solution heat treatment **A4:** 846 **M4:** 677-683
solution potential **M2:** 207
6010 .... **M2:** 115
annealing **M4:** 708
annealing temperature **M2:** 29
composition **M2:** 45
machinability rating **M2:** 189
mechanical properties **M2:** 61
mechanical properties, sheet **M2:** 181
physical properties **M2:** 54
precipitation heat treatment **M4:** 677-683
product forms **M2:** 45
solution heat treatment **M4:** 677-683
solution potential **M2:** 207
6010, annealing **A4:** 871
6013
precipitation heat treatment **A4:** 846
solution heat treatment **A4:** 846
6051 corrosion resistance **M2:** 223, 230, 231
6053
annealing **A4:** 871 **M4:** 708
annealing temperature **M2:** 29
corrosion resistance **M2:** 209, 224, 231, 232
hardness **A4:** 877
hardness values **M4:** 716
precipitation heat treatment **A4:** 846 **M4:** 677-683
solution heat treatment **A4:** 846 **M4:** 677-683
stress-corrosion cracking resistance **M2:** 209
6061 .... **M2:** 115-117
age hardening **A4:** 865
aging **M4:** 701-703
aging characteristics at low aging temperatures **M2:** 31, 36-38
annealing **A4:** 871 **M4:** 708
annealing temperature **M2:** 29
applications **M2:** 48
composition **M2:** 45
corrosion resistance **M2:** 48, 209, 224, 225, 230, 231, 232
dimensional changes **A4:** 876
fabrication characteristics **M2:** 48
fatigue characteristics, weldments **M2:** 195, 199
fatigue curves **M2:** 14
forgeability, relative **M2:** 6
forging, relative cost vs forging weight **M2:** 5

Aluminum alloys, wrought, specific types (continued)
forming, change from alloy 5052 to eliminate cracking during **M2:** 16
forming-limit diagram **M2:** 183
galvanic corrosion with magnesium **M2:** 607
hardness **A4:** 877
hardness values **M4:** 716
impacts, mechanical properties of **M2:** 10
impacts, minimum wall thickness **M2:** 9
machinability rating **M2:** 189
mechanical properties **M2:** 61
mechanical properties, sheet **M2:** 181
natural aging curve **M2:** 31
physical properties **M2:** 54
precipitation heat treatment **A4:** 846, 851 **M4:** 677-683
product forms **M2:** 45
properties, expected for welds **M2:** 197
properties, gas metal-arc welded plate **M2:** 197
quenching **A4:** 856, 858 **M4:** 694, 695
quenching rate, effect on yield strength after aging **M2:** 34
reheating **M4:** 707
reheating schedules **A4:** 869
solution heat treatment **A4:** 844, 846, 855 **M4:** 677-683
solution potential **M2:** 207
stress-corrosion cracking resistance **M2:** 209, 213
weldability **M2:** 193
yield strength **M4:** 689
6062, reheating **M4:** 707
6062, reheating schedules **A4:** 869
6063 .... **M2:** 117-118
annealing **A4:** 871 **M4:** 708
annealing temperature **M2:** 29
applications **M2:** 49
composition **M2:** 45
corrosion resistance **M2:** 49, 209, 226, 231, 232
fabrication characteristics **M2:** 49
galvanic corrosion with magnesium **M2:** 607
hardness **A4:** 877
hardness values **M4:** 716
machinability rating **M2:** 189
mechanical properties **M2:** 61
physical properties **M2:** 54
precipitation heat treatment **A4:** 846, 851 **M4:** 677-683
product forms **M2:** 45
properties, expected for welds **M2:** 197
reheating **M4:** 707
reheating schedules **A4:** 869
solution heat treatment **A4:** 846, 851 **M4:** 677-683
solution potential **M2:** 207
stress-corrosion cracking resistance **M2:** 209
weldability **M2:** 193
6066 .... **M2:** 118-119
annealing **A4:** 871 **M4:** 708
annealing temperature **M2:** 29
applications **M2:** 49
composition **M2:** 45
corrosion resistance **M2:** 49, 209
fabrication characteristics **M2:** 49
mechanical properties **M2:** 61
physical properties **M2:** 54
precipitation heat treatment **A4:** 846 **M4:** 677-683
product forms **M2:** 45
solution heat treatment **A4:** 846 **M4:** 677-683
stress-corrosion cracking resistance **M2:** 209
weldability **M2:** 193
6070 .... **M2:** 119
applications **M2:** 49
composition **M2:** 45
corrosion resistance **M2:** 49, 209, 232
fabrication characteristics **M2:** 49

Aluminum alloys, wrought, specific types (continued)
mechanical properties **M2:** 61
physical properties **M2:** 54
precipitation heat treatment........ **A4:** 846 **M4:** 677-683
product forms **M2:** 45
solution heat treatment........ **A4:** 846 **M4:** 677-683
stress-corrosion cracking resistance........ **M2:** 209
weldability **M2:** 193
6101 .... **M2:** 119-120
application **M2:** 49
composition **M2:** 45
corrosion resistance **M2:** 49, 209
fabrication characteristics **M2:** 49
mechanical properties **M2:** 61
physical properties **M2:** 54
product forms **M2:** 45
stress-corrosion cracking resistance........ **M2:** 209
weldability **M2:** 193
6101, precipitation heat treatment........ **A4:** 843-844
6111
precipitation heat treatment........ **A4:** 846
solution heat treatment........ **A4:** 846
6151 .... **M2:** 120-121
applications **M2:** 49
composition **M2:** 45
corrosion resistance **M2:** 209
fabrication characteristics **M2:** 49
forgeability, relative........ **M2:** 6
forging, relative cost vs forging weight........ **M2:** 5
hardness........ **A4:** 877
hardness values........ **M4:** 716
mechanical properties **M2:** 61
mechanical properties, sheet........ **M2:** 181
physical properties **M2:** 54
precipitation heat treatment........ **A4:** 846 **M4:** 677-683
product forms........ **M2:** 45
quenching........ **A4:** 857, 875
quenching stress........ **M2:** 40, 41, 43
solution heat treatment........ **A4:** 846 **M4:** 677-683
solution potential **M2:** 207
stress-corrosion cracking resistance........ **M2:** 209
6201
applications **M2:** 49
composition........ **M2:** 45
corrosion resistance **M2:** 49, 209
fabrication characteristics **M2:** 49
mechanical properties **M2:** 61
physical properties **M2:** 54
product forms........ **M2:** 45
stress-corrosion cracking resistance........ **M2:** 209
weldability **M2:** 193
6201, precipitation heat treatment........ **A4:** 843-844
6205 .... **M2:** 121-122
composition........ **M2:** 45
mechanical properties **M2:** 61
physical properties **M2:** 54
product forms........ **M2:** 45
6262 .... **M2:** 122
applications **M2:** 49
composition........ **M2:** 45
corrosion resistance **M2:** 49, 209
fabrication characteristics **M2:** 49
machinability rating **M2:** 189
mechanical properties **M2:** 61
physical properties **M2:** 54
precipitation heat treatment........ **A4:** 846 **M4:** 677-683
product forms........ **M2:** 45
solution heat treatment........ **A4:** 846 **M4:** 677-683
stress-corrosion cracking resistance........ **M2:** 209
weldability **M2:** 193
6351 .... **M2:** 122-123
composition........ **M2:** 45
impacts, mechanical properties **M2:** 10
mechanical properties **M2:** 62
physical properties **M2:** 54
product forms........ **M2:** 45
properties, expected for welds **M2:** 197
seawater corrosion resistance **M2:** 232
solution potential **M2:** 207
6463 .... **M2:** 123
applications **M2:** 49
composition........ **M2:** 45

Aluminum alloys, wrought, specific types (continued)
corrosion resistance **M2:** 49, 209
fabrication characteristics **M2:** 49
machinability rating **M2:** 189
mechanical properties **M2:** 62
physical properties **M2:** 54
precipitation heat treatment........ **A4:** 846, 851 **M4:** 677-683
product forms........ **M2:** 45
solution heat treatment........ **A4:** 846 **M4:** 677-683
stress-corrosion cracking resistance........ **M2:** 209
weldability **M2:** 193
6951
precipitation heat treatment........ **A4:** 846 **M4:** 677-683
solution heat treatment........ **A4:** 846 **M4:** 677-683
7001
annealing **A4:** 871 **M4:** 708
annealing temperature **M2:** 29
corrosion resistance **M2:** 209
precipitation heat treatment........ **A4:** 847 **M4:** 677-683
solution heat treatment........ **A4:** 847 **M4:** 677-683
stress-corrosion cracking resistance........ **M2:** 209
7005 .... **M2:** 123-125
annealing **A4:** 871 **M4:** 708
annealing temperature **M2:** 29
composition........ **M2:** 45
mechanical properties **M2:** 62
physical properties **M2:** 54
precipitation heat treatment........ **A4:** 847, 851 **M4:** 677-683
product forms........ **M2:** 45
solution heat treatment........ **A4:** 847, 851 **M4:** 677-683
solution potential **M2:** 207
stress-corrosion cracking resistance........ **M2:** 213
weldability **M2:** 193
7021, mechanical properties, sheet........ **M2:** 181
7029, mechanical properties, sheet........ **M2:** 181
7039
seawater corrosion resistance **M2:** 232
stress-corrosion cracking resistance........ **M2:** 213, 232
7049 .... **M2:** 125-126
annealing **A4:** 871
annealing temperature **M2:** 29
composition........ **M2:** 45
fracture toughness **A4:** 878
machinability rating **M2:** 189
mechanical properties **M2:** 62
physical properties **M2:** 54
precipitation heat treatment........ **A4:** 862
product forms........ **M2:** 45
solution potential **M2:** 207
stress-corrosion cracking resistance........ **M2:** 213
7049, annealing **M4:** 708
7050 .... **M2:** 126-128
age hardening **A4:** 863
aging characteristics at room temperature........ **M2:** 36-38
annealing **A4:** 871 **M4:** 708
annealing temperature **M2:** 29
composition........ **M2:** 45
fracture toughness **A4:** 878
machinability rating **M2:** 189
mechanical properties **M2:** 62
physical properties **M2:** 54
precipitation heat treatment........ **A4:** 847, 862, 876 **M4:** 677-683
product forms........ **M2:** 45
quenching rate, effect on yield strength after aging **M2:** 34
solution heat treatment........ **A4:** 847 **M4:** 677-683
solution potential **M2:** 207
stress-corrosion cracking resistance........ **M2:** 213
yield strength........ **A4:** 866, 867 **M4:** 689, 704
7072 .... **M2:** 128-129
composition........ **M2:** 45
mechanical properties **M2:** 62
physical properties **M2:** 54
product forms........ **M2:** 45
seawater corrosion resistance **M2:** 230
solution potential **M2:** 207
7075 .... **M2:** 129-132
age hardening **A4:** 860

Aluminum alloys, wrought, specific types (continued)
aging characteristics at low aging temperatures **M2:** 36-38
annealing **A4:** 869, 871 **M4:** 708
annealing temperature **M2:** 29
annealing,effect on ductility **M4:** 708
applications **M2:** 49
composition........ **M2:** 45
cooling curves........ **M4:** 692
corrosion rate affected by quenching rate **M2:** 34
corrosion resistance **M2:** 49, 209, 224, 225, 226, 230, 232
exfoliation corrosion resistance **M2:** 219
fabrication characteristics **M2:** 49
forgeability, relative........ **M2:** 6
forged part, mechanical properties........ **M2:** 13
forging, relative cost vs forging weight........ **M2:** 5
fracture toughness **A4:** 878
galvanic corrosion with magnesium **M2:** 607
hardness........ **A4:** 877
hardness values........ **M4:** 716
impacts, mechanical properties **M2:** 10, 13
impacts, minimum wall thickness **M2:** 9
machinability rating **M2:** 189
mechanical properties **M2:** 62
natural aging curve **M2:** 31
physical properties **M2:** 54
precipitation heat treatment........ **A4:** 847, 862, 876 **M4:** 677-683
precipitation-hardening curves **M2:** 31, 36-38, 43
product forms........ **M2:** 45
properties, elevated temperatures **M2:** 56, 58, 62
quality assurance **A4:** 876
quenching........ **A4:** 856, 857, 859, 875 **M4:** 690, 693, 695
quenching rate, effect on yield strength after aging **M2:** 34
reheating........ **A4:** 869 **M4:** 707
solution heat treatment........ **A4:** 847, 851, 852, 853, 854, 855, 875 **M4:** 677-683
solution potential **M2:** 207
stress-corrosion cracking resistance........ **M2:** 209, 213-216
tempers **A4:** 862
weldability **M2:** 193
yield strength........ **A4:** 851, 860, 866 **M4:** 689, 693, 704, 714
7079
annealing **A4:** 871 **M4:** 708
annealing temperature **M2:** 29
atmospheric corrosion resistance **M2:** 224, 232
forged parts, variation in mechanical properties........ **M2:** 8, 9
forging, relative cost vs forging weight........ **M2:** 5
fracture toughness **A4:** 878
hardness........ **A4:** 877
hardness values........ **M4:** 716
quenching........ **A4:** 856 **M4:** 695
stress-corrosion cracking resistance........ **M2:** 213
7146, mechanical properties, sheet........ **M2:** 181
7149, fracture toughness **A4:** 878
7149, stress corrosion cracking resistance........ **M2:** 213
7150
fracture toughness **A4:** 878
solution heat treatment........ **A4:** 852
7175 .... **M2:** 131-134
applications **M2:** 49
composition........ **M2:** 45
corrosion resistance **M2:** 49
fabrication characteristics **M2:** 49
fracture toughness **A4:** 878
mechanical properties **M2:** 10, 62
physical properties **M2:** 54
precipitation heat treatment........ **A4:** 847, 862 **M4:** 677-683
product forms........ **M2:** 45
quenching........ **A4:** 856 **M4:** 695
solution heat treatment........ **A4:** 847 **M4:** 677-683

Aluminum alloys, wrought, specific types (continued)
stress-corrosion cracking resistance **M2:** 214
7178 .... **M2:** 134-135
annealing **A4:** 871 **M4:** 708
annealing temperature **M2:** 29
applications **M2:** 49
composition **M2:** 45
corrosion resistance **M2:** 49, 209
fabrication characteristics **M2:** 49
hardness **A4:** 842, 877
hardness values **M4:** 716
machinability rating **M2:** 189
product forms **M2:** 45
quenching **A4:** 856 **M4:** 695
reheating **M4:** 707
reheating schedules **A4:** 869
solution heat treatment **A4:** 854
solution potential **M2:** 207
stress-corrosion cracking resistance **M2:** 209, 214
weldability **M2:** 193
7475 .... **M2:** 135-139
annealing **A4:** 871 **M4:** 708
annealing temperature **M2:** 29
composition **M2:** 45
fracture toughness **A4:** 878
machinability rating **M2:** 489
mechanical properties **M2:** 62
physical properties **M2:** 54
precipitation heat treatment **A4:** 847, 862 **M4:** 677-683
product forms **M2:** 45
solution heat treatment **A4:** 847, 852 **M4:** 677-683
solution potential **M2:** 207
stress-corrosion cracking resistance **M2:** 214
7475 Alclad
precipitation heat treatment **M4:** 677-683
solution heat treatment **M4:** 677-683
8090
fracture toughness **A4:** 878
mechanical properties **A4:** 853
precipitation heat treatment **A4:** 843, 845
solution heat treatment **A4:** 845, 852, 853
Al-4.6Cu, precipitate formation **A4:** 834, 835
Al-4Cu
hardness as function of aging time **A4:** 836
precipitation hardening **A4:** 834, 835
Al-5Cu
aging effect on phase diagram **A4:** 834
microstructure, precipitate formation **A4:** 834, 835
alclad 2024 .... **M2:** 72-75, 76-77
galvanic corrosion with magnesium **M2:** 607
alclad 3003 .... **M2:** 90-92, 94
alclad 3004 .... **M2:** 92-94
alclad 5056 .... **M2:** 102-103
alclad 5086 .... **M2:** 104-105
alclad 6061 .... **M2:** 115-117
alclad 7075 .... **M2:** 129-132
galvanic corrosion with magnesium **M2:** 607
mechanical properties
alclad 7178 .... **M2:** 134-135
CP276
precipitation heat treatment **A4:** 843, 845
solution heat treatment **A4:** 845
X7016, solution potential **M2:** 207
X7021, solution potential **M2:** 207
X7029, solution potential **M2:** 207
X7146, solution potential **M2:** 207
Aluminum anodizing **A13:** 396-398
Aluminum Association **A7:** 130 **A15:** 743
Aluminum Association International Alloy designations
cross-referencing system **A2:** 16, 17-25
ISO equivalents for **A2:** 26
Aluminum bearing alloys
corrosion resistance **A18:** 744
Aluminum brass
resistance spot welding **A6:** 850
thermal expansion coefficient **A6:** 907
Aluminum brasses
corrosion resistance **A13:** 611
Aluminum brazing alloys
compositions **A9:** 359
Aluminum brazing alloys, specific types
4047, brazed joint between 6063-O sheets **A9:** 387
4245, brazed joint between 7004-O sheets **A9:** 387
4343, brazed joint in 12-0 brazing sheets **A9:** 387
Aluminum bronze *See* Copper- aluminum alloys, heat treating; Copper-aluminum alloys, heat treating
5%
electrical resistivity **A18:** 713
thermal conductivity **A18:** 713
thermal expansion **A18:** 713
10.5%, adhesive wear resistance **A18:** 721
95% Cu, 5% Al, thermal properties **A18:** 42
cavitation resistance **A18:** 600
for deep-drawing dies **A18:** 634
die material for sheet metal forming **A18:** 628
electrolytic etching **A9:** 401
graphite-aluminum bronze composite **A9:** 594
hardfacing alloys **A18:** 762
laser melt/particle injection **A18:** 869
microstructures of **A9:** 551
pickling of **M5:** 612
properties **A18:** 713
for shallow forming dies **A18:** 633
thermal conductivity **A18:** 713
thermal expansion **A18:** 713
thermal spray coating material **A18:** 832
Aluminum bronze alloys
corrosion in **A11:** 201
Aluminum bronze, specific types
ASTM B148, Grade 9C, heat treated **A9:** 156
Cu-11.8A1, different illuminations compared **A9:** 79
Cu-IOAI-5Ni-4Fe-2Mn, corroded surface **A9:** 99
Aluminum bronze/nickel-graphite
abradable seal material **A18:** 589
Aluminum bronzes *See also* Aluminum casting alloys; Cast copper alloys; specific types, under Bronzes; Wear-resistant alloys, nonferrous **A6:** 752 **A13:** 461, 611, 627
air-carbon arc cutting **A6:** 1176
alpha aluminum bronzes, heat treating **M2:** 259
applications **A2:** 227, 325-334, 348, 350, 355, 383-387 **A15:** 784
brazeability **M6:** 1034
brazing **A6:** 931
composition and properties **M6:** 401
corrosion in various media **M2:** 390-391, 468-469
corrosion ratings **A2:** 353-354
deep drawing dies, use for **M3:** 496
drosses as **A15:** 96
electrodes for shielded metal arc welding **A6:** 765-766
filler metals **A6:** 756, 765
foundry properties for sand casting **A2:** 348
freezing range **A2:** 348
gas metal arc welding **M6:** 420-422
gas tungsten arc welding **M6:** 412-413
gas-metal arc butt welding **A6:** 760
gas-metal arc welding **A6:** 754, 755, 765
gas-tungsten arc welding **A6:** 754, 764-765
to high-carbon steel **A6:** 827, 828
to low-alloy steel **A6:** 827, 828
to medium-carbon steel **A6:** 827, 828
to stainless steel **A6:** 827, 828
hardening **A2:** 236
hardfacing **A6:** 795
hardfacing coating **A6:** 808
Aluminum bronzes (continued)
heat treatment **A15:** 782
as high-shrinkage foundry alloy **A2:** 346
iron content for grain structure refinement **A2:** 348
laser hardfacing **A6:** 806
lubrication **M3:** 591-592
mechanical properties **A2:** 348, 350
melt treatment **A15:** 774-775
nominal compositions **A2:** 347
plasma arc welding **A6:** 754
precoated before soldering **A6:** 131
press forming dies, use for **M3:** 490
properties **A2:** 227, 325-334, 383-387
quenching/tempering **A2:** 350
recycling **A2:** 1214
relative solderability **A6:** 134
relative solderability as a function of flux type **A6:** 129
resistance spot welding **A6:** 850
shielded metal arc welding **A6:** 754, 755, 763, 765-766 **M6:** 425
shielding gas **A6:** 764
shrinkage allowance **A15:** 303
solderability **A6:** 978
solidification range **A2:** 348
thermal expansion coefficient **A6:** 907
wear applications **M3:** 591
wear tests **M3:** 591, 592
weld cladding **A6:** 822
weld overlay for hardfacing alloys **A6:** 820
weld overlay material **M6:** 816
weldability **A6:** 753
Aluminum buffing compounds **M5:** 117
Aluminum casting alloys *See also* Aluminum; Aluminum alloys; Aluminum alloys, specific types; Aluminum casting alloys; Aluminum casting alloys, specific types; Aluminum foundry products; Aluminum-silicon alloys; Cast aluminum alloys **A13:** 587, 593-594
alloy systems **A2:** 123-126
alloying elements and effects **A2:** 131-133
aluminum-base metal-matrix composites (MMCs) **A2:** 126
aluminum-copper-silicon alloys **A2:** 125
aluminum-lithium alloys **A2:** 126
aluminum-magnesium alloys **A2:** 125
aluminum-silicon alloys **A2:** 125
aluminum-tin alloys **A2:** 126
aluminum-zinc-magnesium alloys **A2:** 125-126
appearance **A15:** 762-763
bearing alloys **A2:** 128, 131
by Cosworth process **A15:** 38
castability ratings **A2:** 129 **A15:** 766
casting processes for **A2:** 136-145
categories of **A2:** 126-131
commercial Duralumin alloys **A2:** 129-130
compositions **A2:** 22-25 **A9:** 359
compressive yield strength **A2:** 146
corrosion resistance ratings **A2:** 129
designation system **A2:** 15-16, 123-126
fatigue strength **A2:** 146
general characteristics **A2:** 123
hardness values **A2:** 146
heat treatments **A2:** 137-138
machinability **A2:** 129
physical properties **A2:** 127-128 **A15:** 764-765
piston/elevated-temperature alloys **A2:** 130
premium quality alloys **A2:** 130, 147
production processes **A2:** 5
properties **A15:** 763-769
properties of **A2:** 145-150
registered, compositions **A15:** 744-745
rotor castings **A2:** 128-129
selection, for foundry products **A2:** 126-131
shear strength values **A2:** 146
specifications **A2:** 125
standard general-purpose alloys **A2:** 130-131

SUBJECTS OF THE INDEXED VOLUMES: ASM Handbook (designated by the letter "A"): **A1:** Properties and Selection: Irons, Steels, and High-Performance Alloys (1990); **A2:** Properties and Selection: Nonferrous Alloys and Special-Purpose Materials (1990); **A3:** Alloy Phase Diagrams (1992); **A4:** Heat Treating (1991); **A6:** Welding, Brazing, and Soldering (1993); **A7:** Powder Metallurgy (1984); **A8:** Mechanical Testing (1985); **A9:** Metallography and Microstructures (1985); **A10:** Materials Characterization (1986); **A11:** Failure Analysis and Prevention (1986); **A12:** Fractography (1987); **A13:** Corrosion (1987); **A14:** Forming and Forging (1988); **A15:** Casting (1988); **A16:** Machining (1989); **A17:** Nondestructive Testing and Quality Control (1989); **A18:** Friction, Lubrication, and Wear Technology (1992). **Metals Handbook, 9th Edition** (designated by the letter "M"): **M1:** Properties and Selection: Irons and Steels (1978); **M2:** Properties and Selection: Nonferrous Alloys and Pure Metals (1979); **M3:** Properties and Selection: Stainless Steels, Tool Materials and Special-Purpose Materials (1980); **M4:** Heat Treating (1981); **M5:** Surface Cleaning, Finishing, and Coating (1982); **M6:** Welding, Brazing, and Soldering (1983). **Engineered Materials Handbook** (designated by the letters "EM"): **EM1:** Composites (1987); **EM2:** Engineering Plastics (1988); **EM3:** Adhesives and Sealants (1990); **EM4:** Ceramics and Glasses (1991); **Electronic Materials Handbook** (designated by the letters "EL"): **EL1:** Packaging (1989).

Aluminum casting alloys (continued)
structure control, for foundry products **A2:** 133-136
tensile properties **A2:** 143
weldability ratings **A2:** 129
Aluminum casting alloys, former ASTM designations
C4A *See* Aluminum casting alloys, specific types, 295.0
CN42A *See* Aluminum casting alloys, specific types, 242.0
CS43A *See* Aluminum casting alloys, specific types, 208.0
G4A *See* Aluminum casting alloys, specific types, 514.0
G8A *See* Aluminum casting alloys, specific types, 518.0
G10A *See* Aluminum casting alloys, specific types, 500.0
GH70B *See* Aluminum casting alloys, specific types, 535.0
S5A *See* Aluminum casting alloys, specific types, B443.0
S5B *See* Aluminum casting alloys, specific types, 443.0
S5C *See* Aluminum casting alloys, specific types, C443.0
S12A *See* Aluminum casting alloys, specific types, A413.0
S12B *See* Aluminum casting alloys, specific types, 413.0
SC51A *See* Aluminum casting alloys, specific types, 355.0
SC51B *See* Aluminum casting alloys, specific types, C355.0
SC64A *See* Aluminum casting alloys, specific types, 308.0
SC64D *See* Aluminum casting alloys, specific types, 319.0
SC84A *See* Aluminum casting alloys, specific types, A380.0
SC84B *See* Aluminum casting alloys, specific types, 380.0
SC92A *See* Aluminum casting alloys, specific types, 354.0
SC102A *See* Aluminum casting alloys, specific types, 383.0
SC114A *See* Aluminum casting alloys, specific types, 384.0
SG70A *See* Aluminum casting alloys, specific types, 356.0
SG70B *See* Aluminum casting alloys, specific types, A356.0
SG91A *See* Aluminum casting alloys, specific types, 359.0
SG100A *See* Aluminum casting alloys, specific types, A360.0
SG100B *See* Aluminum casting alloys, specific types, 360.0
SN122A *See* Aluminum casting alloys, specific types, 336.0
ZC81A.B *See* Aluminum casting alloys, specific types, 713.0
ZG61A *See* Aluminum casting alloys, specific types, 712.0
ZG71B *See* Aluminum casting alloys, specific types, 771.0
Aluminum casting alloys, heat treating
annealing **M4:** 709
hardening **M4:** 685-686, 704-706
Aluminum casting alloys, specific types *See also* Aluminum; Aluminum alloys; Aluminum alloys, powder metallurgy, specific types; Aluminum alloys, specific types; Aluminum alloys, wrought, specific types; Aluminum casting alloys
13 *See* Aluminum casting alloys, specific types, 413.0
40E *See* Aluminum casting alloys, specific types, 712.0
43 *See* Aluminum casting alloys, specific types, 443.0, A443.0, B443.0
100.1
corrosion resistance **M2:** 211
stress-corrosion cracking resistance **M2:** 211
108 *See* Aluminum casting alloys, specific types, 208.0

Aluminum casting alloys, specific types (continued)
142 *See* Aluminum casting alloys, specific types, 242.0
150.1
corrosion resistance **M2:** 211
stress-corrosion cracking resistance **M2:** 211
170.1
corrosion resistance **M2:** 211
stress-corrosion cracking resistance **M2:** 211
174.0
aging **M4:** 685-686
solution heat treatment **M4:** 685-686
195 *See* Aluminum casting alloys, specific types, 295.0
201.0 .... **M2:** 152-154
aging **A4:** 849 **M4:** 685-686
composition **M2:** 142
solution heat treatment **A4:** 849 **M4:** 685-686
tensile properties, test bars **M2:** 149
201.0, applications and properties **A2:** 152-153
204.0
aging **A4:** 849 **M4:** 685-686
solution heat treatment **A4:** 849 **M4:** 685-686
204.0, applications and properties **A2:** 154
206.0 .... **M2:** 154-155
aging **A4:** 849
composition **M2:** 142
solution heat treatment **A4:** 849
tensile properties, test bars **M2:** 149
206.0, applications and properties **A2:** 154-155
208.0 .... **M2:** 155
aging **A4:** 849 **M4:** 685-686
characteristics **M2:** 144
composition **M2:** 142
corrosion resistance **M2:** 210, 227
machinability rating **M2:** 188
solution heat treatment **A4:** 849 **M4:** 685-686
solution potential **M2:** 207
stress-corrosion cracking resistance **M2:** 210
tensile properties, test bars **M2:** 149
weldability **M2:** 193
208.0, applications and properties **A2:** 155-156
213.0
characteristics **M2:** 144
weldability **M2:** 193
214 *See* Aluminum casting alloys, specific types, 514.0
218 *See* Aluminum casting alloys, specific types, 518.0
220 *See* Aluminum casting alloys, specific types, 520.0
222.0
aging **A4:** 849 **M4:** 685-686
characteristics **M2:** 144
solution heat treatment **A4:** 849 **M4:** 685-686
weldability **M2:** 193
224.0
corrosion resistance **M2:** 210
stress-corrosion cracking resistance **M2:** 210
238.0, applications and properties **A2:** 156-157
238.0 solution potential **M2:** 207
240.0
corrosion resistance **M2:** 210
stress-corrosion cracking resistance **M2:** 210
242.0 .... **M2:** 155-157
aging **A4:** 849 **M4:** 685-686
characteristics **M2:** 144
composition **M2:** 142
corrosion resistance **M2:** 210, 211
machinability rating **M2:** 188
solution heat treatment **A4:** 849 **M4:** 685-686
stress-corrosion cracking resistance **M2:** 210, 211
tensile properties, test bars **M2:** 149
weldability **M2:** 193
242.0, applications and properties **A2:** 157
249.0
corrosion resistance **M2:** 210
stress-corrosion cracking resistance **M2:** 210
282.0
aging **M4:** 685-686
solution heat treatment **M4:** 685-686
295.0 .... **M2:** 157-158
aging **A4:** 849 **M4:** 685-686
characteristics **M2:** 144
composition **M2:** 142
corrosion resistance **M2:** 210, 227

Aluminum casting alloys, specific types (continued)
machinability rating **M2:** 188
solution heat treatment **A4:** 849 **M4:** 685-686
solution potential **M2:** 207
stress-corrosion cracking resistance **M2:** 210
tensile properties, test bars **M2:** 149
weldability **M2:** 193
295.0, applications and properties **A2:** 157-159
296.0 .... **M2:** 158-159
aging **A4:** 849 **M4:** 685-686
characteristics **M2:** 144
composition **M2:** 142
solution heat treatment **A4:** 849 **M4:** 685-686
solution potential **M2:** 207
tensile properties, test bars **M2:** 149
weldability **M2:** 193
296.0, applications and properties **A2:** 159-160
299.0
aging **M4:** 685-686
solution heat treatment **M4:** 685-686
308.0 .... **M2:** 159
characteristics **M2:** 144
composition **M2:** 142
corrosion resistance **M2:** 211
machinability rating **M2:** 188
solution potential **M2:** 207
stress-corrosion cracking resistance **M2:** 211
tensile properties, test bars **M2:** 149
weldability **M2:** 193
308.0, applications and properties **A2:** 160
319 Allcast *See* Aluminum casting alloys, specific types, 319.0
319.0 .... **M2:** 159-160
aging **A4:** 849 **M4:** 685-686
characteristics **M2:** 144
composition **M2:** 142
corrosion resistance **M2:** 210, 211, 227
machinability rating **M2:** 188
solution heat treatment **A4:** 849 **M4:** 685-686
solution potential **M2:** 207
stress-corrosion cracking resistance **M2:** 210, 211
tensile properties, test bars **M2:** 149
weldability **M2:** 193
319.0, applications and properties **A2:** 160-161
328.0
aging **A4:** 849 **M4:** 685-686
characteristics **M2:** 144
solution heat treatment **A4:** 849 **M4:** 685-686
weldability **M2:** 193
332.0
aging **A4:** 849 **M4:** 685-686
characteristics **M2:** 144
corrosion resistance **M2:** 211
solution heat treatment **A4:** 849 **M4:** 685-686
stress-corrosion cracking resistance **M2:** 211
weldability **M2:** 193
332.0, applications and properties **A2:** 161
333.0
aging **A4:** 849 **M4:** 685-686
characteristics **M2:** 144
solution heat treatment **A4:** 849 **M4:** 685-686
weldability **M2:** 193
336.0 .... **M2:** 160-161
aging **A4:** 849 **M4:** 685-686
characteristics **M2:** 144
composition **M2:** 142
corrosion resistance **M2:** 211
solution heat treatment **A4:** 849 **M4:** 685-686
stress-corrosion cracking resistance **M2:** 211
tensile properties, test bars **M2:** 149
weldability **M2:** 193
336.0, applications and properties **A2:** 161-162
339.0, applications **A2:** 162
354.0 .... **M2:** 161, 162, 163
aging **A4:** 849 **M4:** 685-686
characteristics **M2:** 144
composition **M2:** 142
corrosion resistance **M2:** 211
machinability rating **M2:** 188
solution heat treatment **A4:** 849 **M4:** 685-686
stress-corrosion cracking resistance **M2:** 211
tensile properties, test bars **M2:** 149
weldability **M2:** 193
354.0, applications and properties **A2:** 162-163
355
composition **M3:** 723

Aluminum casting alloys, specific types (continued)
tensile properties at subzero temperatures ... M3: 744
355.0 ... M2: 161-165
aging ... A4: 849 M4: 685-686
applications ... M2: 145
composition ... M2: 142
corrosion resistance ... M2: 210, 211, 227
machinability rating ... M2: 188
solution heat treatment ... A4: 849 M4: 685-686
solution potential ... M2: 207
stress-corrosion cracking resistance ... M2: 210-211
tensile properties, test bars ... M2: 149
weldability ... M2: 193
355.0, applications and properties ... A2: 163-164
355.0, characteristics ... A2: 131
356
composition ... M3: 723
tensile properties at subzero temperatures ... M3: 744
356.0 ... M2: 164-167
aging ... A4: 849 M4: 685-686
applications ... M2: 145
characteristics ... M2: 144
composition ... M2: 142
corrosion resistance ... M2: 210, 211, 227
dimensional changes ... A4: 876
machinability rating ... M2: 188
microstructure, effect of sodium modification ... M2: 150
precipitation hardening ... M2: 32, 33
precipitation heat treatment ... A4: 868
solution heat treatment ... A4: 849 M4: 685-686
solution potential ... M2: 207
stress-corrosion cracking resistance ... M2: 210, 211
tensile properties, test bars ... M2: 149
weldability ... M2: 193
356.0, applications and properties ... A2: 164-165
356.0, characteristics ... A2: 130-131
357.0 ... M2: 167
aging ... A4: 850 M4: 685-686
applications ... M2: 146
characteristics ... M2: 144
composition ... M2: 142
machinability rating ... M2: 188
solution heat treatment ... A4: 850 M4: 685-686
tensile properties, test bars ... M2: 149
weldability ... M2: 193
357.0, applications and properties ... A2: 166
358.0
corrosion resistance ... M2: 211
stress-corrosion resistance ... M2: 211
359.0 ... M2: 167-168, 169
aging ... A4: 850 M4: 685-686
characteristics ... M2: 144
composition ... M2: 142
corrosion resistance ... M2: 211
machinability rating ... M2: 188
solution heat treatment ... A4: 850 M4: 685-686
stress-corrosion cracking resistance ... M2: 211
tensile properties, test bars ... M2: 149
weldability ... M2: 193
359.0, applications and properties ... A2: 166-167
360.0 ... M2: 168-169
applications ... M2: 145
characteristics ... M2: 145
composition ... M2: 142
corrosion resistance ... M2: 211
machinability rating ... M2: 188
stress-corrosion cracking resistance ... M2: 211
tensile properties, test bars ... M2: 149
360.0, applications and properties ... A2: 167-168
364.0
corrosion resistance ... M2: 211
stress-corrosion cracking resistance ... M2: 211
380.0 ... M2: 169, 170
applications ... M2: 145

Aluminum casting alloys, specific types (continued)
characteristics ... M2: 145
composition ... M2: 142
corrosion resistance ... M2: 211
machinability rating ... M2: 188
stress-corrosion cracking resistance ... M2: 211
tensile properties, test bars ... M2: 149
380.0, applications and properties ... A2: 168-169
383.0 ... M2: 170
characteristics ... M2: 145
composition ... M2: 142
corrosion resistance ... M2: 211
stress-corrosion cracking resistance ... M2: 211
tensile properties, test bars ... M2: 149
383.0, applications and properties ... A2: 169-170
384.0 ... M2: 170-171
characteristics ... M2: 145
composition ... M2: 142
corrosion resistance ... M2: 211
stress-corrosion cracking resistance ... M2: 211
tensile properties, test bars ... M2: 149
384.0, applications and properties ... A2: 170-171
390.0 ... M2: 171-172
composition ... M2: 142
corrosion resistance ... M2: 211
machinability rating ... M2: 188
stress-corrosion cracking resistance ... M2: 211
tensile properties, test bars ... M2: 149
390.0, applications and properties ... A2: 171
390.0, characteristics ... A2: 131
392.0
corrosion resistance ... M2: 211
stress-corrosion cracking resistance ... M2: 211
413.0 ... M2: 172-173
applications ... M2: 145
characteristics ... M2: 145
composition ... M2: 142
corrosion resistance ... M2: 211
machinability rating ... M2: 188
stress-corrosion cracking resistance ... M2: 211
tensile properties, test bars ... M2: 149
weldability ... M2: 193
413.0, applications ... A2: 130
413.0, applications and properties ... A2: 171-172
441.0
aging ... M4: 685-686
solution heat treatment ... M4: 685-686
443, applications ... A2: 130
443.0 ... M2: 173-174
composition ... M2: 142
corrosion resistance ... M2: 210, 227
machinability rating ... M2: 188
solution potential ... M2: 207
stress-corrosion cracking resistance ... M2: 210
tensile properties, test bars ... M2: 150
weldability ... M2: 193
443.0, applications and properties ... A2: 172-173
444, applications ... A2: 130
512.0
characteristics ... M2: 144
corrosion resistance ... M2: 210
stress-corrosion cracking resistance ... M2: 210
weldability ... M2: 193
513.0
applications ... M2: 145
characteristics ... M2: 144
corrosion resistance ... M2: 210, 211
stress-corrosion cracking resistance ... M2: 210, 211
weldability ... M2: 193
514.0 ... M2: 174
characteristics ... M2: 144
composition ... M2: 142
corrosion resistance ... M2: 210
machinability rating ... M2: 188
solution potential ... M2: 207
stress corrosion cracking resistance ... M2: 210
tensile properties, test bars ... M2: 150

Aluminum casting alloys, specific types (continued)
weldability ... M2: 193
514.0, applications and properties ... A2: 173
518.0 ... M2: 174-175
applications ... M2: 145
characteristics ... M2: 145
composition ... M2: 142
corrosion resistance ... M2: 211
machinability rating ... M2: 188
stress-corrosion cracking resistance ... M2: 211
tensile properties, test bars ... M2: 150
518.0, applications and properties ... A2: 173-174
519.0
aging ... M4: 685-686
solution heat treatment ... M4: 685-686
520.0 ... M2: 175
aging ... A4: 850 M4: 685-686
applications ... M2: 146
characteristics ... M2: 144
composition ... M2: 142
corrosion resistance ... M2: 210, 227, 232
machinability rating ... M2: 188
solution heat treatment ... A4: 850 M4: 685-686
solution potential ... M2: 207
stress-corrosion cracking resistance ... M2: 210
tensile properties, test bars ... M2: 150
weldability ... M2: 193
520.0, applications and properties ... A2: 174
535.0 ... M2: 175-176
aging ... A4: 850 M4: 685-686
characteristics ... M2: 144
composition ... M2: 142
corrosion resistance ... M2: 210
solution heat treatment ... A4: 850 M4: 685-686
stress-corrosion resistance ... M2: 210
tensile properties, test bars ... M2: 150
weldability ... M2: 193
535.0, applications and properties ... A2: 174-175
592.0
aging ... M4: 685-686
solution heat treatment ... M4: 685-686
613 Tenzaloy *See* Aluminum casting alloys, specific types, 713.0
621.0
aging ... M4: 685-686
solution heat treatment ... M4: 685-686
630.0
aging ... M4: 685-686
solution heat treatment ... M4: 685-686
639.0
aging ... M4: 685-686
solution heat treatment ... M4: 685-686
648.0
aging ... M4: 685-686
solution heat treatment ... M4: 685-686
657.0
aging ... M4: 685-686
solution heat treatment ... M4: 685-686
665.0
aging ... M4: 685-686
solution heat treatment ... M4: 685-686
674.0
aging ... M4: 685-686
solution heat treatment ... M4: 685-686
692.0
aging ... M4: 685-686
solution heat treatment ... M4: 685-686
705.0
aging ... A4: 850 M4: 685-686
characteristics ... M2: 144
corrosion resistance ... M2: 210, 227
solution heat treatment ... A4: 850 M4: 685-686
stress-corrosion cracking resistance ... M2: 210
weldability ... M2: 193
707.0
aging ... A4: 850 M4: 685-686
characteristics ... M2: 144
corrosion resistance ... M2: 210, 211, 227
solution heat treatment ... A4: 850 M4: 685-686

Aluminum casting alloys, specific types (continued)
stress-corrosion cracking resistance **M2:** 210, 211
weldability **M2:** 193
710.0
aging **A4:** 850 **M4:** 685-686
characteristics **M2:** 144
corrosion resistance **M2:** 210, 227
solution heat treatment **A4:** 850 **M4:** 685-686
stress-corrosion cracking resistance **M2:** 210
weldability **M2:** 193
711.0
aging **A4:** 850 **M4:** 685-686
characteristics **M2:** 144
corrosion resistance **M2:** 211, 227
solution heat treatment **A4:** 850 **M4:** 685-686
stress-corrosion cracking resistance **M2:** 211, 227
weldability **M2:** 193
712.0 .... **M2:** 176-177
aging **A4:** 850 **M4:** 685-686
composition **M2:** 142
corrosion resistance **M2:** 210, 227
machinability rating **M2:** 188
solution heat treatment **A4:** 850 **M4:** 685-686
stress-corrosion cracking resistance **M2:** 210
tensile properties, test bars **M2:** 150
weldability **M2:** 193
712.0, applications and properties **A2:** 175
713.0 .... **M2:** 177-178
aging **A4:** 850 **M4:** 685-686
applications **M2:** 146
characteristics **M2:** 144
composition **M2:** 142
corrosion resistance **M2:** 210, 211, 227
machinability rating **M2:** 188
solution heat treatment **A4:** 850 **M4:** 685-686
stress-corrosion cracking resistance **M2:** 210, 211
tensile properties, test bars **M2:** 150
weldability **M2:** 193
713.0, applications and properties **A2:** 175-176
718.0
aging **M4:** 685-686
solution heat treatment **M4:** 685-686
726.0
aging **M4:** 685-686
solution heat treatment **M4:** 685-686
734.0
aging **M4:** 685-686
solution heat treatment **M4:** 685-686
745.0
aging **M4:** 685-686
solution heat treatment **M4:** 685-686
750 *See* Aluminum casting alloys, specific types, 850.0
757.0
aging **M4:** 685-686
solution heat treatment **M4:** 685-686
765.0
aging **M4:** 685-686
solution heat treatment **M4:** 685-686
771.0 .... **M2:** 178
aging **A4:** 850 **M4:** 685-686
characteristics **M2:** 144
composition **M2:** 142
corrosion resistance **M2:** 210
solution heat treatment **A4:** 850 **M4:** 685-686
stress-corrosion cracking resistance **M2:** 210
tensile properties, test bars **M2:** 150
weldability **M2:** 193
771.0, applications and properties **A2:** 176
781.0
aging **M4:** 685-686
solution heat treatment **M4:** 685-686
850.0 .... **M2:** 178-179
aging **A4:** 850 **M4:** 685-686
characteristics **M2:** 144
composition **M2:** 142
corrosion resistance **M2:** 210, 211
machinability rating **M2:** 188
solution heat treatment **A4:** 850 **M4:** 685-686
stress-corrosion cracking resistance **M2:** 210, 211
tensile properties, test bars **M2:** 150
850.0, applications and properties **A2:** 176-177

Aluminum casting alloys, specific types (continued)
851.0
aging **A4:** 850 **M4:** 685-686
characteristics **M2:** 144
corrosion resistance **M2:** 210
solution heat treatment **A4:** 850 **M4:** 685-686
stress-corrosion cracking resistance **M2:** 210
852.0
aging **A4:** 850 **M4:** 685-686
characteristics **M2:** 144
corrosion resistance **M2:** 210
solution heat treatment **A4:** 850 **M4:** 685-686
stress-corrosion cracking resistance **M2:** 210
A13 *See* Aluminum casting alloys, specific types, A413.0
A43 *See* Aluminum casting alloys, specific types, C443.0
A108 *See* Aluminum casting alloys, specific types, 308.0
A132 *See* Aluminum casting alloys, specific types, 336.0
A206.0 .... **M2:** 154-155
composition **M2:** 142
tensile properties, test bars **M2:** 149
A218 *See* Aluminum casting alloys, specific types, A535.0
A242.0
corrosion resistance **M2:** 210
stress-corrosion cracking resistance **M2:** 210
A332.0 *See* Aluminum casting alloys, specific types, 336.0
A356
composition **M3:** 723
tensile properties at subzero temperatures **M3:** 744
A356.0 .... **M2:** 164-167
aging **A4:** 849 **M4:** 685-686
applications **M2:** 145, 146
characteristics **M2:** 144
composition **M2:** 142
corrosion resistance **M2:** 211
machinability rating **M2:** 188
microstructure, effect of sodium modification **M2:** 150
solution heat treatment **A4:** 849 **M4:** 685-686
stress-corrosion cracking resistance **M2:** 211
weldability **M2:** 193
A356.0, applications and properties **A2:** 164-165
A357.0 .... **M2:** 167, 168
aging **A4:** 850 **M4:** 685-686
applications **M2:** 145
characteristics **M2:** 144
composition **M2:** 142
corrosion resistance **M2:** 211
machinability rating **M2:** 188
solution heat treatment **A4:** 850 **M4:** 685-686
stress-corrosion cracking resistance **M2:** 211
tensile properties, test bars **M2:** 149
weldability **M2:** 193
A357.0, applications and properties **A2:** 166
A360.0 .... **M2:** 168-169
characteristics **M2:** 145
composition **M2:** 142
corrosion resistance **M2:** 211
machinability ratings **M2:** 188
stress-corrosion cracking resistance **M2:** 211
tensile properties, test bars **M2:** 149
A360.0, applications and properties **A2:** 167-168
A380.0 .... **M2:** 170
characteristics **M2:** 145
composition **M2:** 142
corrosion resistance **M2:** 211
machinability rating **M2:** 188
stress-corrosion cracking resistance **M2:** 211
tensile properties, test bars **M2:** 149
A380.0, applications and properties **A2:** 168-169
A384.0 .... **M2:** 170-171
tensile properties, test bars **M2:** 149
A384.0, applications and properties **A2:** 170-171
A390.0 .... **M2:** 171-172
composition **M2:** 142
machinability rating **M2:** 188
tensile properties, test bars **M2:** 149
A390.0, applications and properties **A2:** 171
A413.0 .... **M2:** 172-173
characteristics **M2:** 145
composition **M2:** 142

Aluminum casting alloys, specific types (continued)
corrosion resistance **M2:** 211
stress-corrosion cracking resistance **M2:** 211
tensile properties, test bars **M2:** 149
A413.0, applications **A2:** 130
A413.0, applications and properties **A2:** 171-172
A443.0 .... **M2:** 173-174
composition **M2:** 142
weldability **M2:** 193
A443.0, applications and properties **A2:** 172-173
A444, applications **A2:** 130
A444.0
aging **A4:** 850 **M4:** 685-686
corrosion resistance **M2:** 211
solution heat treatment **A4:** 850 **M4:** 685-686
stress-corrosion cracking resistance **M2:** 211
A535.0 .... **M2:** 175-176
composition **M2:** 142
A535.0, applications and properties **A2:** 174-175
Almag 25 *See* Aluminum casting alloys, specific types, 535.0
B195 *See* Aluminum casting alloys, specific types, 296.0
B218 *See* Aluminum casting alloys, specific types, B535.0
B295.0 *See* Aluminum casting alloys, specific types, 296.0
B443.0 .... **M2:** 174-174
applications **M2:** 145, 146
characteristics **M2:** 144
composition **M2:** 142
corrosion resistance **M2:** 211
stress-corrosion cracking resistance **M2:** 211
tensile properties, test bars **M2:** 150
weldability **M2:** 193
B443.0, applications and properties **A2:** 172-173
B535.0 .... **M2:** 175-176
composition **M2:** 142
corrosion resistance **M2:** 210
machinability rating **M2:** 188
stress-corrosion cracking resistance **M2:** 210
B535.0, applications and properties **A2:** 174-175
C355
composition **M3:** 723
tensile properties at subzero temperatures **M3:** 744
C355.0 .... **M2:** 161-165
aging **A4:** 849 **M4:** 685-686
applications **M2:** 145, 146
characteristics **M2:** 144
corrosion resistance **M2:** 210, 211
machinability rating **M2:** 188
mechanical properties **M2:** 151
solution heat treatment **A4:** 849 **M4:** 685-686
stress-corrosion cracking resistance **M2:** 210, 211
tensile properties, test bars **M2:** 149
weldability **M2:** 193
C355.0, applications and properties **A2:** 163-164
C443.0 .... **M2:** 173-174
characteristics **M2:** 145
composition **M2:** 142
corrosion resistance **M2:** 211
stress-corrosion cracking resistance **M2:** 211
tensile properties, test bars **M2:** 150
C443.0, applications and properties **A2:** 172-173
D612 *See* Aluminum casting alloys, specific types, 712.0
D712.0 *See* Aluminum casting alloys, specific types, 712.0
F356.0
corrosion resistance **M2:** 211
stress-corrosion cracking resistance **M2:** 211
Precedent 71A *See* Aluminum casting alloys, specific types, 771.0
Aluminum chloride
in petroleum refining and petrochemical operations **A13:** 1268
Aluminum chloride, anhydrous
description **A9:** 68
safety hazards **A9:** 69
Aluminum coated steel
soldering **A6:** 631
Aluminum coating **M5:** 333, 347
applications **M5:** 333-335, 337, 339-342, 345
atmospheric exposure **M5:** 333-334
elevated-temperature **M5:** 334-335, 342, 345

Aluminum coating (continued)
high-production parts M5: 339-341
base metal and formability A1: 219
cladding by rolling M5: 345-346
cobalt-base alloys M5: 340, 342-343
composition, effects of M5: 338, 345
composition of base metal effects of M5: 338, 342-343
corrosion protection M5: 333-335, 337, 344-347
corrosion resistance A1: 219
diffusion processes M5: 334-335, 339-346
temperature and time effects M5: 343-347
electrophoresis process M5: 347
electroplating process M5: 347
equipment
continuous hot dip method M5: 336-337
ion vapor deposition process M5: 346
pack diffusion processes M5: 340-341
slurry processes M5: 342-343
fabricability, effects on M5: 336-337
fused-salt fluxing method M5: 337-339
handling and storage A1: 220
heat reflection A1: 219-220
heat resistance A1: 219
heat-resistant alloys M5: 566
high-stress, high temperature applications M5: 335
hot dip process M5: 333-339
batch process M5: 333-334, 337, 339
continuous process M5: 335-337
immersion time, effects of M5: 338
limitations M5: 338-339
ion plating method M5: 420-421
iron-aluminum interfacial layer characteristics and effects of thickness M5: 333, 335-336, 338, 345-347, 338, 345
iron-base alloys M5: 341
mechanical properties A1: 218, 219
molybdenum and tungsten M5: 661-662
nickel-base alloys M5: 340-343
oxidation and corrosion resistance M5: 333-335, 337, 344-347
oxidation protective types *See* Oxidation protection coatings
pack diffusion process M5: 340-341
painting A1: 220
procedures and control
batch hot dip method M5: 337-338
continuous hot dip method M5: 336
high-production part coating processes M5: 339-341
pack diffusion process M5: 340-341
slurry processes M5: 342-343
spray coating processes M5: 343-345
sealing process, spray coats M5: 344-345
silicon affecting M5: 334, 336, 338, 345
slurry processes M5: 341, 343
composition of slurries M5: 342-343
procedures and equipment M5: 342-343
spray process M5: 343-346
procedures M5: 343-346
stainless steel M5: 342-343, 345
steel M5: 333-347
strength and hardness of base metal affected by M5: 336-338, 341-344
structure, effects of time and temperature on M5: 343-344
superalloys M5: 335, 339, 342
surface preparation for M5: 333, 336-337, 339-340, 342-346
temperature effects M5: 334-335, 338, 342-346
diffusion process M5: 343-345
elevated temperatures M5: 334-335, 342, 345
tensile strength affected by M5: 336-337, 342
thickness effects M5: 335, 337-338, 342-346
interfacial layer M5: 338, 345
for threaded steel fasteners A1: 295
time effects M5: 338-339, 343-345
weldability A1: 220

Aluminum coatings
aqueous corrosion resistance A13: 435
atmospheric corrosion resistance A13: 434-435
cast irons M1: 102-104
corrosion protection by M1: 752-753
microstructure A13: 434
protection A13: 434
steel fence wire M1: 271
steel sheet M1: 171-173
applications M1: 172, 173
coating weights M1: 172
corrosion resistance M1: 172
formability M1: 172
handling and storage M1: 173
heat reflection M1: 172-173
heat resistance M1: 172
mechanical properties M1: 171, 172
painting M1: 173
weldability M1: 173
steel wire M1: 263, 264, 269
for steels A13: 458
Aluminum complex soap A18: 126, 129
Aluminum conductor steel reinforced wire A1: 283
Aluminum conductors, steel reinforced (ACSR)
construction A2: 12
Aluminum die casting alloys
characteristics A2: 131
Aluminum die casting mold inserts A7: 796
Aluminum die forgings
T-grain direction A8: 668
Aluminum electrolytic capacitors
failure mechanism EL1: 179, 972
Aluminum engineered wrought products *See* Aluminum mill and engineered wrought products
Aluminum eutectic A9: 619
Aluminum flake pigments A7: 594-595
Aluminum flake powder
properties A7: 600
Aluminum flakes
as conductive filler EM2: 470-471
Aluminum flame spraying
marine corrosion A13: 919
Aluminum fibers
as intentional inclusions A15: 88
Aluminum fluxing compounds
binary phase diagrams A15: 446
Aluminum foil
leak detection with A17: 66
Aluminum foundry products
alloy selection, casting A2: 126-131
alloy systems A2: 123-126
alloying elements and effects A2: 131-133
aluminum casting alloys, properties of A2: 145-150
aluminum-base metal-matrix composite (MMC) A2: 126, 904-907
aluminum-copper alloys A2: 124-125
aluminum-copper-silicon alloys A2: 125
aluminum-lithium casting alloys A2: 126
aluminum-magnesium casting alloys A2: 125
aluminum-silicon alloys A2: 125
aluminum-tin alloys A2: 126
aluminum-zinc-magnesium alloys A2: 125-126
casting process A2: 136-145
centrifugal casting A2: 141
commercial Duralumin alloys A2: 129-130
composite-mold casting A2: 141
compositional groupings A2: 123-126
continuous casting A2: 141
designations A2: 123-124
die casting A2: 131, 136-139
evaporative (lost-foam) pattern casting A2: 140
fluidity effects A2: 145-146
hot isostatic pressing A2: 141
hybrid permanent mold processes A2: 141-145
investment casting A2: 140-141
mechanical properties A2: 147-150

Aluminum foundry products (continued)
microstructural effects on mechanical properties A2: 133-136
permanent mold (gravity die) casting A2: 139
physical properties A2: 145-147
piston/elevated-temperature alloys A2: 130
premium-quality castings A2: 130
rotor castings A2: 128-129
sand casting A2: 139-140
semisolid-metal casting A2: 142
shell mold casting A2: 140
shrinkage effects A2: 146-147
solidification A2: 146
squeeze casting (liquid metal forging) A2: 141-145
standard general-purpose aluminum casting alloys A2: 130-131
structure control A2: 133-136
Aluminum fracture
cleavage and void growth A8: 478-479
Aluminum housings eliminating cracks in .... A3: 1 • 28
Aluminum hydroxide
used in chemical treatment before disposal A16: 131
Aluminum hydroxide ($Al(OH)_3$)
pressure densification EM4: 300
Aluminum in cast iron *See also* High aluminum cast irons M1: 76
Aluminum in steel M1: 115, 411, 417
500 °F embrittlement, effect on M1: 685
effect on formability M1: 555
graphitization, effect on M1: 686
hydrogen solubility, effect on M1: 687
nitriding, effect on M1: 540-541
notch toughness, effect on M1: 693, 695
strain-age embrittlement, effect on susceptibility to M1: 684
Aluminum in titanium M3: 356
Aluminum matrix composites
with alumina fibers polishing A9: 590
with boron fibers A9: 591-592
with graphite fibers A9: 590-592
Aluminum melts
electrochemical refining of A15: 80-81
inoculation practice in A15: 105
purification of A15: 79
refining, by evaporation treatment A15: 80
refining, with reactive gases A15: 80
Aluminum metal-matrix composites A6: 554-558
aluminum oxide-reinforced A6: 554, 556-558
capacitor discharge welding A6: 555, 558
chemical reactions A6: 554-555
coefficient of thermal expansion A6: 555
electron-beam welding A6: 555, 557
filler metals A6: 555
flash-butt welding A6: 558
forge welding A6: 558
friction welding A6: 555, 558
fusion-welding processes A6: 555-558
gas-metal arc welding A6: 555, 556, 557
gas-tungsten arc welding A6: 555-556
general joining considerations A6: 554-555
interaction effects on solidification A6: 555
laser-beam welding A6: 555, 556-557
microstructure A6: 555, 556
resistance welding A6: 555, 558
shielding gases A6: 555
silicon carbide-reinforced A6: 554, 556, 557, 558
solidification cracking A6: 555
transient liquid-phase bonding A6: 555, 557-558
viscosity effects A6: 554
weld preparation A6: 555
Aluminum metal-matrix composites, specific types
2014/SiC/15p, electron-beam welding A6: 557
6061 (as-welded), gas-metal arc welding A6: 555, 557
6061, gas-tungsten arc welding A6: 555, 556
6061-T6, gas-metal arc welding A6: 557

SUBJECTS OF THE INDEXED VOLUMES: ASM Handbook (designated by the letter "A"): **A1:** Properties and Selection: Irons, Steels, and High-Performance Alloys (1990); **A2:** Properties and Selection: Nonferrous Alloys and Special-Purpose Materials (1990); **A3:** Alloy Phase Diagrams (1992); **A4:** Heat Treating (1991); **A6:** Welding, Brazing, and Soldering (1993); **A7:** Powder Metallurgy (1984); **A8:** Mechanical Testing (1985); **A9:** Metallography and Microstructures (1985); **A10:** Materials Characterization (1986); **A11:** Failure Analysis and Prevention (1986); **A12:** Fractography (1987); **A13:** Corrosion (1987); **A14:** Forming and Forging (1988); **A15:** Casting (1988); **A16:** Machining (1989); **A17:** Nondestructive Testing and Quality Control (1989); **A18:** Friction, Lubrication, and Wear Technology (1992). **Metals Handbook, 9th Edition** (designated by the letter "M"): **M1:** Properties and Selection: Irons and Steels (1978); **M2:** Properties and Selection: Nonferrous Alloys and Pure Metals (1979); **M3:** Properties and Selection: Stainless Steels, Tool Materials and Special-Purpose Materials (1980); **M4:** Heat Treating (1981); **M5:** Surface Cleaning, Finishing, and Coating (1982); **M6:** Welding, Brazing, and Soldering (1983). **Engineered Materials Handbook** (designated by the letters "EM"): **EM1:** Composites (1987); **EM2:** Engineering Plastics (1988); **EM3:** Adhesives and Sealants (1990); **EM4:** Ceramics and Glasses (1991); **Electronic Materials Handbook** (designated by the letters "EL"): **EL1:** Packaging (1989).

Aluminum metal-matrix composites, specific types (continued)
6061/Al_2O_3/10p, gas-tungsten arc welding ... **A6:** 556
6061/Al_2O_3/15p, transient liquid-phase bonding ... **A6:** 557, 558
6061/Al_2O_3/20p
gas-metal arc welding ... **A6:** 557
gas-tungsten arc welding ... **A6:** 556
laser-beam welding ... **A6:** 556
6061/Al_2O_3/20p as-welded, gas-metal arc welding ... **A6:** 557
6061/Al_2O_3/20p-T6, gas-metal arc welding ... **A6:** 557
6061/SiC/25p, transient liquid-phase bonding ... **A6:** 557
7005, gas-tungsten arc welding ... **A6:** 556
7005/Al_2O_3/10p, gas-tungsten arc welding ... **A6:** 556
A-356/SiC/15p
electron-beam welding ... **A6:** 557
laser-beam welding ... **A6:** 557
A-356/SiC/20p, friction welding ... **A6:** 558
A359/SiC/10p, resistance welding ... **A6:** 558
A359/SiC/20p, friction welding ... **A6:** 558

Aluminum metallization
bipolar junction transistor technology ... **EL1:** 195
chemical effects ... **EL1:** 965
corrosion ... **A11:** 770
cracked ... **A11:** 775
H-WSI systems ... **EL1:** 88
phosphorus-related corrosion by ... **A11:** 771
temperature-sensitive failures with ... **EL1:** 960

Aluminum mill and engineered wrought products
See also Aluminum; Aluminum alloys; Aluminum alloys, specific types; Wrought aluminum alloys
1*xxx* series, characteristics ... **A2:** 29
2*xxx* series, characteristics ... **A2:** 29
3*xxx* series, characteristics ... **A2:** 29
4*xxx* series, characteristics ... **A2:** 69
5*xxx* series, characteristics ... **A2:** 29, 32
6*xxx* series, characteristics ... **A2:** 32
7*xxx* series, characteristics ... **A2:** 32-33
alloying, general effects ... **A2:** 44-46
alloying, specific effects ... **A2:** 46-57
antimony alloying ... **A2:** 46
applications ... **A2:** 29-32
arsenic alloying ... **A2:** 46
bend properties ... **A2:** 58
beryllium alloying ... **A2:** 46
bismuth alloying ... **A2:** 46
boron alloying ... **A2:** 46-47
cadmium alloying ... **A2:** 47
calcium alloying ... **A2:** 47
carbon alloying ... **A2:** 47
chromium alloying ... **A2:** 47
cobalt alloying ... **A2:** 47
conventional high-strength alloys ... **A2:** 59
copper alloying ... **A2:** 47-51
copper-magnesium alloying ... **A2:** 48
defined ... **A2:** 29
design of shapes ... **A2:** 34-36
elevated-temperature properties ... **A2:** 59
fatigue behavior ... **A2:** 42-44, 59
fatigue crack growth ... **A2:** 59
formability ... **A2:** 41-42
fracture toughness ... **A2:** 42-44, 58-59
gallium alloying ... **A2:** 11, 11-11, 51
heat-treatable, strengthening ... **A2:** 39-41
hydrogen alloying ... **A2:** 51
indium alloying ... **A2:** 51-52
iron alloying ... **A2:** 52
lead alloying ... **A2:** 52
lithium alloying ... **A2:** 52
low-temperature properties ... **A2:** 59-60
magnesium alloying ... **A2:** 52
magnesium-manganese alloying ... **A2:** 52
magnesium-silicide alloying ... **A2:** 52-53
manganese alloying ... **A2:** 53-54
mechanical properties, limits ... **A2:** 57
mercury alloying ... **A2:** 54
mill products, types ... **A2:** 33-34
molybdenum alloying ... **A2:** 54
niobium alloying ... **A2:** 54
non-heat-treatable, strengthening ... **A2:** 37-39
phases in aluminum alloys ... **A2:** 36-37

Aluminum mill and engineered wrought products (continued)
phosphorus alloying ... **A2:** 54
physical metallurgy ... **A2:** 36-57
properties ... **A2:** 57-60
silicon alloying ... **A2:** 54-55
silver alloying ... **A2:** 55
strengthening mechanisms ... **A2:** 37-41
strontium alloying ... **A2:** 55
sulfur alloying ... **A2:** 55
tensile property limits ... **A2:** 58
tin alloying ... **A2:** 55
values, typical ... **A2:** 57
vanadium alloying ... **A2:** 55
wrought alloy series ... **A2:** 29, 32-33
zinc-magnesium alloying ... **A2:** 55-56
zinc-magnesium-copper alloying ... **A2:** 56
zirconium alloying ... **A2:** 56-57

Aluminum monohydrate
calcination ... **EM4:** 111-112

Aluminum nitride
ceramic filler for polyimide-base adhesives ... **EM3:** 159
as ceramic substrate ... **EL1:** 337
as CTE-matched material ... **EL1:** 306
as filler ... **EM3:** 596
as heat sink ... **EL1:** 1129
metallization ... **EL1:** 306-307

Aluminum nitride (AIN) ... **EM4:** 32, 47
additive to Si_3N_4 ... **A16:** 100
adiabatic temperatures ... **EM4:** 229
applications ... **EM4:** 48, 230
as ceramic substrate ... **EM4:** 1107, 1110
chemical vapor deposition ... **EM4:** 217
electrical/electronic applications ... **EM4:** 1105
grain growth inhibitor ... **EM4:** 188
hot pressing ... **EM4:** 191, 192
in joining non-oxide ceramics ... **EM4:** 528-529
key product properties ... **EM4:** 48
non-oxide ceramic joining ... **EM4:** 480
pressure densification ... **EM4:** 298
properties ... **EM4:** 191
raw materials ... **EM4:** 48
as replacement for alumina as substrate in microelectronic devices ... **EM4:** 480
sintering aid ... **EM4:** 188
substrate properties ... **EM4:** 1108
for substrates ... **EM4:** 17
substrates for thick films circuits ... **EM4:** 1142
synthesized by SHS process ... **EM4:** 229, 230
thermal shock resistance ... **EM4:** 1003
thick film circuit substrates ... **EM4:** 1144

Aluminum nitride (AlN)
properties ... **A6:** 629

Aluminum nitride embrittlement ... **A1:** 694-696
intergranular fracture in castings ... **A1:** 694-695
panel cracking ... **A1:** 695
reduced hot ductility ... **A1:** 695-696

Aluminum nitrides
in electrical steels ... **A9:** 537
in plate steels, examination for ... **A9:** 203

Aluminum nitrides, as particles in stainless steel tubing
EDS/EELS spectra ... **A10:** 461

Aluminum oxide ... **A13:** 381, 396-398, 860-861 **EM3:** 235, 592
bearing fatigue life effect ... **A18:** 726
in binary phosphate glasses ... **A10:** 131
blasting with ... **M5:** 93-94
bonding to aluminum ... **EM3:** 624, 625
ceramic coatings for dies ... **A18:** 644
chemical vapor deposition of ... **M5:** 383-384
coating for gas-lubricated bearings ... **A18:** 532
coating for seals ... **A18:** 551
for gas-lubricated bearings ... **A18:** 532
hardness, true and Vickers ... **M5:** 546-547
high-purity, implant material for total replacement synovial joints ... **A18:** 657, 658
ion sputtering effect ... **EM3:** 245
laser-induced CVD for synthesis ... **A18:** 848
load and sliding distance effect on wear versus that of steel ... **A18:** 365-366
platens, elevated-temperature compression testing ... **A8:** 196
polishing with ... **M5:** 548
polycrystalline, thermal properties ... **A18:** 42
pressure bars ... **A8:** 202

Aluminum oxide (continued)
production considerations ... **EM3:** 264
properties ... **A6:** 722
removal of ... **M5:** 612
rocks ... **EL1:** 1038
sapphire, thermal properties ... **A18:** 42
as sintered to ceramic substrates ... **EL1:** 386
surface behavior diagrams to study surface composition ... **EM3:** 247-248
surface preparation ... **EM3:** 259, 260, 261
thermal spray coating material ... **A18:** 832
to improve thermal conductivity ... **EM3:** 178, 620, 621
in tungsten powder production ... **A7:** 153
Vickers and Knoop microindentation hardness numbers ... **A18:** 416

Aluminum oxide (Al_2O_3) *See also* Alumina ... **A16:** 29, 98, 99, 101 **EM4:** 13, 32, 33
abrasive cutoff wheel machining of MMCs ... **A16:** 895
and abrasive flow machining ... **A16:** 517
abrasive for cast irons ... **A16:** 661
abrasive for ECG ... **A16:** 545
abrasive for lapping ... **A16:** 493, 497, 503
as abrasive grains or cutting tool tips for grinding or machining ... **EM4:** 329
and abrasive jet machining ... **A16:** 512, 513
abrasive property control ... **EM4:** 332
abrasives ... **A16:** 34, 440, 448, 453-456, 463, 467-468
abrasives and tool grinding ... **A16:** 450
additive to Si_3N_4 ... **A16:** 100
applications ... **EM4:** 331, 960
bond type ... **EM4:** 331
bondability ... **EM4:** 332
calcination ... **EM4:** 110, 111-112
ceramic dies ... **EM4:** 187
ceramic powder, batch weight of formulation when used in oxidizing sintering atmospheres ... **EM4:** 163
coating for carbide tools ... **A16:** 639, 656
coating for cemented carbides ... **A16:** 652
compared to CBN abrasive grinding ... **A16:** 462, 464, 465
composition ... **EM4:** 14
corrosion resistance of refractories ... **EM4:** 391
default material for structural applications ... **EM4:** 29
and gear manufacture ... **A16:** 350, 351, 353, 354
and gray iron metal removal rates ... **A16:** 652
grinding ... **A16:** 421
grinding wheels for Al alloys ... **A16:** 801
grinding wheels for carbon and alloy steels ... **A16:** 676
grinding wheels for Cu alloys ... **A16:** 818, 819
grinding wheels for heat-resistant alloys ... **A16:** 758, 760
grinding wheels for Ni alloys ... **A16:** 843
grinding wheels for refractory metals ... **A16:** 869
grinding wheels for stainless steels ... **A16:** 705
grinding wheels for thread grinding ... **A16:** 271
grinding wheels for Ti alloys ... **A16:** 848-849, 850
grinding wheels for tool steels ... **A16:** 727, 728, 732
grinding wheels for tools machining Mg alloys ... **A16:** 821
grinding wheels for Zr ... **A16:** 854-855, 856
and high-speed machining ... **A16:** 602
honing stones ... **A16:** 475, 476, 477, 478, 490
inclusion for stainless steels ... **A16:** 688
manufacturing processes ... **EM4:** 331
mechanical properties ... **EM4:** 316
medical applications ... **EM4:** 960
and Mg alloys ... **A16:** 827
physical properties ... **EM4:** 316
primary applications ... **A16:** 639
properties ... **EM4:** 14, 330
softening point ... **A16:** 601
thermal properties ... **EM4:** 316
tool material for machining cast irons ... **A16:** 651, 652, 656
tool material properties ... **A16:** 107
ultrasonic machining ... **EM4:** 359, 360, 361
ultrasonic machining abrasive slurry ... **A16:** 529, 530

Aluminum oxide (Al_2O_3)(sapphire)
properties ... **A6:** 629

Aluminum oxide and aluminum metal-matrix
composites ... **A2:** 906, 907

Aluminum oxide and aluminum metal-matrix (continued)
effect, in aluminum joining A2: 9
MMC reinforcements A2: 7
Aluminum oxide cermets
applications and properties A2: 992-993
Aluminum oxide chromium cermets
temperature effects A2: 994
Aluminum oxide fibers *See also* Continuous aluminum oxide fiber MMCS EM1: 31, 60, 874
Aluminum oxide product
acoustic microscopy inspection A17: 471-472
Aluminum oxide wheels
grinding A7: 462
Aluminum oxide-based cermets A7: 802-803
Aluminum oxide-chromium cermet products A7: 803
Aluminum oxide-containing cermets composition and properties A7: 804
effect of temperature on strength of A7: 805
Aluminum oxide-containing cermets, specific types
aluminum oxide, composition A7: 804
chromium, composition A7: 804
molybdenum, composition A7: 804
titanium oxide, composition A7: 804
Aluminum oxide-silicon carbide (Al_2O_3-SiC) A16: 98, 99-101
Aluminum oxide-silicon carbide (whisker-reinforced Al_2O_3-Si_wC) A16: 99-100, 102-103
machinability A16: 639, 640, 646
turning heat-resistant alloys A16: 740
Aluminum oxide-titanium carbide (Al_2O_3-TiC) A16: 99, 100, 101, 106
coatings on ceramics A16: 103
gray iron metal removal rates A16: 652
ground by diamond wheels A16: 462
machinability A16: 639, 640, 646
tool material for high removal rate machining A16: 608
tool material for machining cast irons A16: 652, 656
tool material properties A16: 107
turning heat-resistant alloys A16: 740
Aluminum oxide-titanium oxide (Al_2O_3-TiO)
ceramic A16: 98
Aluminum oxide-zirconium oxide (Al_2O_3-ZrO_2) A16: 101, 102
ceramic A16: 98, 99
grinding wheels for heat-resistant alloys A16: 760
Aluminum oxide/aluminum composites A13: 860-861
Aluminum oxynitride (ALON) EM4: 18
Aluminum P/M alloys *See also* Aluminum alloys; Aluminum P/M processing; P/M aluminum alloys
alloy design research A2: 204-210
aluminum P/M processing A2: 201-204
ambient-temperature strength A2: 204-206
can vacuum degassing A2: 202-203
conventionally pressed and sintered alloys A2: 210-213
corrosion resistance A2: 204-206
dipurative degassing A2: 203
direct powder forming A2: 203
dynamic compaction A2: 204
elevated-temperature properties A2: 206-208
high-modulus and/or low-density alloys A2: 209-210
high-strength A2: 200-215
hot isostatic pressing (HIP) A2: 203
intermetallics A2: 210
mechanical attrition process A2: 202
mechanically attrited alloys A2: 205, 207-208
metal-matrix composites A2: 209-210
part processing A2: 210-213
powder degassing and consolidation A2: 202-204
Aluminum P/M alloys (continued)
powder production A2: 201-202
rapid omnidirectional consolidation A2: 203-204
rapid solidification alloys A2: 204-207
strengthening features A2: 200, 202
stress-corrosion cracking A2: 204-206
superplastic forming (SPF) A2: 210
technology, advantages A2: 200-201
vacuum degassing in reusable chamber A2: 203
wear resistant RS alloys A2: 205
Aluminum P/M parts *See also* Aluminum; Aluminum alloys; Aluminum alloys, specific types; Aluminum powders; Aluminum powders, specific types; Atomized aluminum powders; High-strength aluminum alloys; Prealloyed aluminum powders; Wrought aluminum alloys; Wrought aluminum alloys, specific types
applications A7: 744-745
dimensional changes during sintering A7: 384-385
heating cycles A7: 382, 743
material properties A7: 741-743
mechanical properties A7: 742
microstructure A7: 385
processing A7: 743-744
sintering A7: 381-385
technology A7: 741-748
Aluminum P/M processing *See also* Aluminum alloys; Aluminum P/M alloys; Powder metallurgy (P/M) processing
atomization A2: 201
can vacuum degassing A2: 202-203
dipurative degassing A2: 203
direct powder forming A2: 203
dynamic compaction A2: 204
hot isostatic pressing (HIP) A2: 203
mechanical alloying A2: 202
mechanical attrition A2: 202
melt-spinning techniques A2: 202
powder degassing and consolidation A2: 202-204
powder production A2: 201-202
rapid omnidirectional consolidation A2: 203-204
reaction milling A2: 202
sinter-aluminum-pulver (SAP) technology A2: 202
splat cooling A2: 201-202
vacuum degassing in reusable chamber A2: 203
Aluminum paints
metallic flake pigments for A7: 594
Aluminum, paste flux for
inorganic fluxes for A6: 980
Aluminum phase diagrams, discussion of
aluminum-bismuth ..... A3: 1 • 28
aluminum-copper ..... A3: 1 • 22, 23
aluminum-gold ..... A3: 1 • 28, 29
aluminum-iron ..... A3: 1 • 26, 27
aluminum-lead ..... A3: 1 • 28
Aluminum powder alloys, specific types
Clevite 66, properties A7: 408
Aluminum powders *See also* Aluminum; Aluminum alloys; Aluminum alloys, specific types; Aluminum P/M parts; Aluminum powders, specific types; Atomized aluminum powders; High-strength aluminum alloys; Prealloyed aluminum powders; Wrought aluminum alloys; Wrought aluminum alloys, specific types
aerospace applications A7: 652-653
Alcoa process for A7: 127-129
aluminum as A7: 699
apparent density A7: 297
ASTM standards for A7: 601
chemical requirements A7: 601
compressibility curve A7: 287
consolidation and hot working A7: 526-527
degassing A7: 526
densities with high-energy compacting A7: 305
dimensional change during sintering A7: 384-385
Aluminum powders (continued)
effect of presintering atmosphere A7: 384
explosibility A7: 125, 127, 130
for explosives, classification and apparent density A7: 600
explosives containing A7: 597, 599-600
for flame cutting A7: 842
flow rate through Hall and Carney funnels A7: 279
grades A7: 125
heating cycles A7: 382
HIP temperatures and process times A7: 437
as incendiary A7: 603
oxide film formation A7: 248
oxide thickness A7: 129-130
physical and chemical properties A7: 129-130
porous A7: 699
preparation and uses A7: 600
pressure-density relationships A7: 299
production A7: 125-130
in propellants, pyrotechnics, and explosives A7: 597, 599-600
reactivity and combustibility A7: 130
relative density, highest A7: 305
sintering A7: 381-385
sintering atmospheres A7: 383-384
size and shape A7: 129
synthetic organic waxes with A7: 192
tap and apparent densities compared A7: 297
Aluminum powders, specific types *See also* Aluminum; Aluminum alloys; Aluminum alloys, specific types; Aluminum P/M parts; Aluminum powders; Atomized aluminum powders; High-strength aluminum alloys; Prealloyed aluminum powders; Wrought aluminum alloys; Wrought aluminum alloys, specific types
201 AB, mechanical properties A7: 468, 469
601AB, mechanical properties A7: 468, 474
7075, density with high-energy compacting A7: 305
7090, tensile properties A7: 653
7091, roll compacted A7: 408
7091, SEM micrographs A7: 235
7091, tensile properties A7: 653
Aluminum, pure
no phase transformation A6: 84
Aluminum recovery
as secondary aluminum A2: 46
Aluminum recycling
aluminum recyclability A2: 1205
automobile scrap recycling A2: 1211-1213
continuous melting A2: 1211
delacquering A2: 1209-1210
melting, preparation, casting A2: 1210-1211
process developments A2: 1211
recycling loop A2: 1205-1206
scrap streams, developing A2: 1207-1208
skin formation A2: 1211
technological aspects A2: 1208-1211
trends A2: 1206-1207
Aluminum see-butoxide EM4: 209
Aluminum sheet
pin bearing test fixture for A8: 60
thermal inspection of A17: 403
Aluminum sheet and copper
explosively welded A9: 386
Aluminum sheet and steel
explosively welded A9: 386
Aluminum shot, as carrier core
copier powders A7: 584-585
Aluminum silicate (AS)
regenerator cores for advanced gas turbines EM4: 1001
regenerator disks for gas turbine engines EM4: 720-721
Aluminum silicates *See also* Mullite; Sand(s); Silica; Silica sands
as mold refractories, investment casting A15: 258

Aluminum silicates (continued)
as molding sands, characteristics **A15:** 209-210
as refractory, core coatings **A15:** 240
Aluminum smelter flue dusts
gallium recovery from **A2:** 742-743
Aluminum soap **A18:** 126
Aluminum, specific types *See also* Aluminum; Aluminum alloys; Aluminum alloys, specific types; Cast aluminum alloys, specific types; Wrought aluminum alloys, specific types
1050, applications and properties **A2:** 62
1060, applications and properties **A2:** 62-63
1100, applications and properties **A2:** 63-64
1145, applications and properties **A2:** 64
1199, applications and properties **A2:** 64-65
1350, applications and properties **A2:** 65-66
Aluminum stearate
lubricant for hot forging of tool steels **A18:** 738
Aluminum stub substrates
SEM specimens **A12:** 172
Aluminum sulfate
used in chemical treatment before disposal **A16:** 131
Aluminum titanate **EM4:** 49, 60-61, 676
applications and properties **A2:** 1021-1022
key features **EM4:** 676
properties **EM4:** 677
Aluminum trihydrate
calcination **EM4:** 111
as flame retardant **EM2:** 504
Aluminum windings
applications **A2:** 13
Aluminum wing skins
galvanic exfoliation corrosion inspection **A17:** 191
Aluminum wire bonding
physical characteristics **EL1:** 110
wire in **EL1:** 224-226
wire selection **EL1:** 110-111
Aluminum wrought alloy series, 1*xxx* through 7*xxx*
characteristics **A2:** 29, 32-33
Aluminum wrought alloys *See* Aluminum mill and engineered wrought products; Wrought aluminum and aluminum alloys; Wrought aluminum and aluminum alloys, specific types
Aluminum-alloy microstructures
aluminum-18% silver .... **A3:** 1 • 22
aluminum-33% copper .... **A3:** 1 • 19
aluminum-silicon .... **A3:** 1 • 19, 20
Aluminum-aluminum oxide
cold rolled and annealed **A9:** 696
Aluminum-base discontinuous metal-matrix composites **A14:** 251
Aluminum-base filler alloys
brazing corrosion resistance **A13:** 884
Aluminum-base mechanically alloyed oxide dispersion-strengthened (MA ODS) alloys **A2:** 943
Aluminum-base metal-matrix composites **A13:** 859-861
for casting **A2:** 126, 904-907
Aluminum-boron carbide cermets
applications and properties **A2:** 1002-1003
Aluminum-coated (aluminized) steel **A13:** 434-435, 458, 527-528, 1014
Aluminum-coated fence wire **A1:** 285
Aluminum-coated sheet steel
specimen preparation **A9:** 197
Aluminum-coated sheet steel, specific types
1008, with type 1 hot-dip coating **A9:** 200
1008, with type 2 hot-dip coating **A9:** 200
Aluminum-coated steels
brazing **M6:** 1030
press forming of **A14:** 561-562
resistance spot welding **M6:** 480, 491-492
Aluminum-coated strand wire **A1:** 282
Aluminum-copper alloy system
equation for relationship between dendrite arm spacing and solidification time **A9:** 629
Aluminum-copper alloys *See also* Aluminum alloys, specific types; Wrought aluminum alloys, specific types
as aluminum casting alloys **A2:** 124-125
applications and properties **A2:** 62-82
binary phase diagram **A15:** 760
cavitation erosion **A18:** 216
Aluminum-copper alloys (continued)
CBEDP for **A10:** 462
characteristics **A2:** 29
comparison of measured and calculated grain size **A9:** 131
corrosion resistance **A6:** 729
electromigration-induced failures **A11:** 772-773
Guinier-Preston zone, FIM images of **A10:** 589-590
liquidus temperature, determined **A15:** 185
natural aging curves **A2:** 40
recommended guidelines for selecting PAW shielding gases **A6:** 67
solidification crack sensitivity **A6:** 726
strain contrast **A9:** 7
weldability **A6:** 725, 728, 753
Aluminum-copper alloys, facets
SEM projection **A12:** 195
Aluminum-copper eutectic alloy
interfaces between phases **A9:** 126
lamellae **A9:** 129
Aluminum-copper phase diagram **A9:** 614 **M6:** 22
Aluminum-copper system .... **A3:** 1 • 22-23
Aluminum-copper-lithium alloys
plasma arc welding **A6:** 736-737
Aluminum-copper-silicon alloys
as aluminum casting alloys **A2:** 125
weld-crack resistance **A11:** 435
Aluminum-epoxy adhesive bonds **EM3:** 235
Aluminum-gold intermetallics
in wire bonding **EL1:** 228-229
Aluminum-iron alloys
rapidly solidified **A9:** 615
Aluminum-iron-cerium alloys
homogenization by high-energy milling **A7:** 67
Aluminum-killed steel
strain measurements and forming limit diagram for **A8:** 566
Aluminum-killed steel, porcelain enameling of **M5:** 513-514
vacuum process **M5:** 388, 390-395, 399-406, 408
vapor deposition process **M5:** 346-347
Aluminum-killed steels *See also* Special-killed low-carbon steel sheet and strip **A1:** 6, 578
anisotropy **M1:** 549
box annealed *See* Box-annealed aluminum-killed steel
cold extrusion applications **M1:** 591
cold rolled *See* Cold rolled aluminum killed steel
forming limit diagram **M1:** 551-552
grain structure **M1:** 558-559
sheet and strip, for severe drawing applications **M1:** 155
sheet, mechanical properties **M1:** 552
typical load-extension curve **M1:** 548
Aluminum-lead alloy strip
and unsintered powder rolled **A7:** 408
Aluminum-lithium alloys *See also* Aluminum alloys; Aluminum alloys, specific types; Aluminum-lithium alloys, specific types **A6:** 549-552 **EM1:** 35
aging **A6:** 551
alloy development **A2:** 180-184
as aluminum casting alloys **A2:** 126
applications **A2:** 178, 181-182 **A6:** 549-550
classification **A6:** 549
commercial alloys **A2:** 184-197 **A6:** 549-550
compositions of **A9:** 357
corrosion **A6:** 552
double arc contrast **A10:** 467
ductility **A12:** 440
electron-beam welding **A6:** 551, 552
explosion potential **A2:** 182-183
fatigue **A6:** 551
filler metals **A6:** 550-551
gas-metal arc welding **A6:** 551, 552
gas-tungsten arc welding **A6:** 551, 552
heat treating **A2:** 183
heat-affected zone **A6:** 551, 552
hot cracking **A6:** 550
ingot quality **A2:** 182
laser-beam welding **A6:** 551
macroetching to reveal grain structure **A9:** 354
manufacturing **A2:** 182-184
metalworking of **A14:** 20, 250
Aluminum-lithium alloys (continued)
microstructure **A6:** 551-552 **A9:** 357
physical metallurgy **A2:** 179-180
plasma arc welding **A6:** 551, 552
porosity **A6:** 550
postweld heat treatment **A6:** 551
properties **A6:** 549, 550, 551
Pumphrey-Moore cracking index **A6:** 550
recycling **A2:** 183-184
solidification crack sensitivity **A6:** 726
specimen preparation **A9:** 357
strengthening phases and designations **A6:** 549
ternary alloys **A2:** 179-180
thermomechanical effects **A2:** 180
trans-Varestraint test for weldability **A6:** 551
types **A2:** 178-179
warm-worked and annealed, recovered subgrain structure **A10:** 470
weld characterization **A6:** 551
weldability **A6:** 549, 550-551, 726
Aluminum-lithium alloys, specific types *See also* Aluminum-lithium alloys
01420
composition **A6:** 550
properties **A6:** 549, 550
weldability **A6:** 550, 551
01420-T6
gas-tungsten arc welding **A6:** 552
tensile properties **A6:** 552
2090-T8
electron-beam welding **A6:** 552
gas-metal arc welding **A6:** 552
gas-tungsten arc welding **A6:** 552
plasma arc welding **A6:** 552
tensile properties **A6:** 552
2094-T8
gas-metal arc welding **A6:** 552
gas-tungsten arc welding **A6:** 552
plasma arc welding **A6:** 552
tensile properties **A6:** 552
8090-T6
electron-beam welding **A6:** 552
tensile properties **A6:** 552
8090-T8
gas-metal arc welding **A6:** 552
gas-tungsten arc welding **A6:** 552
plasma arc welding **A6:** 552
tensile properties **A6:** 552
Alloy 2090, corrosion **A2:** 187
Alloy 2090, design considerations **A2:** 186-187
Alloy 2090, fatigue **A2:** 187-188
Alloy 2090, finishing characteristics **A2:** 189-190
Alloy 2090, forming **A2:** 188
Alloy 2090, nominal composition **A2:** 178
Alloy 2090, strength and toughness **A2:** 185-186
Alloy 2091, applications **A2:** 190
Alloy 2091, corrosion **A2:** 190-191
Alloy 2091, fatigue **A2:** 191
Alloy 2091, finishing **A2:** 191
Alloy 2091, forming **A2:** 11, 191
Alloy 2091, nominal composition **A2:** 178
Alloy 8090, applications **A2:** 191-192
Alloy 8090, corrosion performance **A2:** 194-195
Alloy 8090, design considerations **A2:** 193-194
Alloy 8090, fatigue **A2:** 195-196
Alloy 8090, finishing characteristics **A2:** 197
Alloy 8090, forming **A2:** 196
Alloy 8090, nominal composition **A2:** 178
Alloy 8090, strength and toughness **A2:** 192-193
Alloy 8090, welding **A2:** 197
Weldalite 049, applications and characteristics **A2:** 184-185
Weldalite 049, nominal composition **A2:** 178
Weldalite CP276, nominal composition **A2:** 178
Aluminum-lithium investment castings
as materials development **A15:** 38-39
Aluminum-magnesium alloys *See also* Aluminum alloys, specific types; Wrought aluminum alloys, specific types
as aluminum casting alloys **A2:** 125
applications and properties **A2:** 87, 89-100, 102-103, 105-108
cavitation erosion **A18:** 216
characteristics **A2:** 32, 39
corrosion **A6:** 622
for foamed plaster molding **A15:** 247
hydrogen solubility **A6:** 722

Aluminum-magnesium alloys (continued)
liquid, integral thermal properties ... **A15:** 57
Lüders bands in ... **A8:** 548, 553
second-phase precipitation ... **A6:** 622
solidification crack sensitivity ... **A6:** 726
tension and torsion effective fracture strains ... **A8:** 168
thermodynamic properties ... **A15:** 57
torsional ductility ... **A8:** 166-167
weldability ... **A6:** 725, 728
zincating process ... **M5:** 601-604
Aluminum-magnesium composite, with carbon fibers
cross section ... **A9:** 162
Aluminum-magnesium silicide
solidification crack sensitivity ... **A6:** 726
Aluminum-magnesium-manganese alloys
properties ... **A2:** 39
Aluminum-magnesium-silicon alloys
corrosion resistance ... **A6:** 729
weld-crack resistance ... **A11:** 435
Aluminum-manganese alloys *See also* Aluminum alloys, specific types
applications and properties ... **A2:** 84-87
characteristics ... **A2:** 29
Aluminum-matrix composites
continuous fiber ... **A2:** 904-906
discontinuous fiber ... **A2:** 906-907
Aluminum-nickel alloy
diffraction pattern ... **A9:** 109
Aluminum-nickel-bronze
air-carbon arc cutting ... **A6:** 1176
Aluminum-nickel-copper-cobalt-iron alloys *See* Cast Alnico alloys
Aluminum-plated steel
resistance spot welding ... **M6:** 480
Aluminum-silicon alloy
comparison of abrasives for ... **A9:** 40
Aluminum-silicon alloys ... **A18:** 785-791
as aluminum casting alloys ... **A2:** 125
applications ... **A18:** 753, 785, 789-791
aerospace components ... **A18:** 791
Australian-3HA alloy automotive applications ... **A18:** 789
automotive components ... **A18:** 789-791
consumer electronics components ... **A18:** 791
French alloy automotive applications ... **A18:** 789, 790
pistons for heavy-duty engines ... **A18:** 555
applications and properties ... **A2:** 87-89, 100-102, 104-107
brazing, available product forms of filler metals ... **A6:** 119
brazing, joining temperatures ... **A6:** 118
breakthroughs in aluminum-silicon wear-resistant materials ... **A18:** 791
alloys from foreign countries ... **A18:** 789-791
coatings/surface treatments ... **A18:** 791
metal/matrix composites ... **A18:** 791
powder metallurgy ... **A18:** 791
spray casting ... **A18:** 791
characteristics ... **A2:** 29
coefficient of thermal expansion ... **A18:** 785
common, compositions of ... **A15:** 159
compositions ... **A18:** 556, 753, 786
for use in bearing applications ... **A18:** 790
cross-reference to equivalent wear-resistant aluminum-silicon alloys ... **A18:** 786
dendrite structure ... **A15:** 165-166
designations ... **A18:** 753
development ... **A18:** 785
as eutectic ... **A15:** 159
grain structure ... **A15:** 159-161
hypereutectic, refinement of ... **A15:** 753
hypoeutectic, modification of ... **A2:** 134
inoculant for ... **A15:** 105
intermetallic phases ... **A15:** 166
laser melting ... **A18:** 865
Aluminum-silicon alloys (continued)
liquid, integral thermal properties ... **A15:** 57
low-temperature solid-state welding ... **A6:** 300
metallurgy ... **A18:** 785-786
binary system ... **A18:** 785-786
minor element control in ... **A15:** 79
modification ... **A15:** 481-486, 751-753
modifiers, effects of ... **A15:** 162-165
percent shaped castings manufactured ... **A15:** 159
permanent mold, characteristics of ... **A15:** 159
product forms ... **A18:** 753
properties ... **A18:** 556, 785-788, 789
heat treatment ... **A18:** 787
properties, structural effects on ... **A15:** 167-168
quench modification ... **A15:** 162
recommended gap for braze filler metals ... **A6:** 120
refinement ... **A15:** 751-753
resistance brazing filler metals ... **A6:** 342
silicon modif-ication ... **A15:** 161-162
silicon particles, primary ... **A15:** 165-166
solidification crack sensitivity ... **A6:** 729
solidification of ... **A15:** 159-168
structural assessment ... **A15:** 166
structure ... **A18:** 786-788
principles of microstructural control ... **A18:** 787-788
thermodynamic properties ... **A15:** 57
wear behavior ... **A18:** 788-789
abrasive wear ... **A18:** 788
intermetallic constituents ... **A18:** 789
matrix hardness ... **A18:** 789
silicon particles ... **A18:** 788-789
sliding wear ... **A18:** 788
Aluminum-silicon alloys, for brazing
composition and properties ... **A7:** 839
Aluminum-silicon alloys, specific types
332.0-T5, application, passenger car pistons ... **A18:** 555
6061, graphite fiber precursor effect ... **A18:** 809
A201.0, graphite fiber precursor effect ... **A18:** 809
A206.0, abrasive wear conditions ... **A18:** 808
A356.0, titanium addition effect on microstructure ... **A18:** 787
A357.0, microstructure ... **A18:** 785, 786
A390.0 application, as piston cylinder liner material ... **A18:** 556
microstructure ... **A18:** 785, 786
Al 356, contact conditions effect on wear rate ... **A18:** 808
Al-9.9Si-0.8Fe, erosive attack of aluminum melt on the die surface ... **A18:** 630
aluminum-silicon-lead alloys, mixed bearing microstructure ... **A18:** 744
aluminum-silicon-tin alloys, mixed bearing microstructure ... **A18:** 744
aluminum-silicon/polyester, abradable seal material ... **A18:** 589
Aluminum-silicon brazing alloy
used to braze a titanium alloy ... **A9:** 158
Aluminum-silicon bronze *See also* Cast copper alloys
properties and applications ... **A2:** 386
Aluminum-silicon carbide cermets
applications and properties ... **A2:** 1002
Aluminum-silicon cast alloys
tension and torsion effective fracture strains ... **A8:** 168
Aluminum-silicon eutectic ... **A9:** 620, 622
Aluminum-silicon interdiffusion
alloy penetration pits formed by ... **A11:** 777
epitaxial mesas formed by ... **A11:** 778
Aluminum-silicon-copper alloys
weldability ... **A6:** 725
weldability, filler metals ... **A6:** 725
Aluminum-silicon-magnesium alloys
characteristics ... **A2:** 32
Aluminum-silver alloys
silver precipitates in ... **A9:** 117
Aluminum-tin alloy *See* Titanium alloys, specific types, Ti-5Al-2.5Sn
Aluminum-tin alloys
as aluminum casting alloys ... **A2:** 126
Aluminum-tin bearing alloy ... **M1:** 609
Aluminum-tin bearing alloys
composition ... **A2:** 524
Aluminum-tin bearings
etching ... **A9:** 451
Aluminum-titanium alloys
binary phase diagram ... **A15:** 160
as inoculant ... **A15:** 105
Aluminum-titanium-boron alloys
as melt addition ... **A15:** 105
Aluminum-to-aluminum structures
eddy current inspection ... **A17:** 187-188
Aluminum-zinc alloy coatings ... **A1:** 220-221 **A13:** 435-436
Aluminum-zinc alloys
applications and properties ... **A2:** 108-122
characteristics ... **A2:** 32-33
corrosion resistance ... **A6:** 729
gravity castings ... **A2:** 530
zinc solubility ... **A2:** 46
Aluminum-zinc coated sheet steel
analysis ... **A13:** 391-392
Aluminum-zinc-magnesium alloys
as casting alloys ... **A2:** 125-126
fretting damage ... **A13:** 141
SCC failure in ... **A11:** 27
weldability ... **A6:** 725, 728
Aluminum/air batteries ... **A13:** 1319-1320
Aluminum/aluminum
acrylic properties ... **EM3:** 122
Aluminum/brass
part material for ion implantation ... **A18:** 858
Aluminum/brass interface, aluminum wire connections
EPMA and SEM images of ... **A10:** 531-532
Aluminum/iron interface, aluminum wire connections
EPMA and SEM images of ... **A10:** 531-532
Aluminum/titanium alloys
age hardening and strain-age cracking ... **A6:** 563
Alundum (200 mesh)
Miller numbers ... **A18:** 235
Alundum (400 mesh)
Miller numbers ... **A18:** 235
Alundum HS (Al_2O_3)
properties ... **A18:** 548
Alusil (Al-Si; 80% Al; 20-22% Si)
thermal properties ... **A18:** 42
AM 100A
contour band sawing ... **A16:** 363
AM-355
broaching ... **A16:** 206
Amaigam gilding ... **A15:** 20-21
Amalgam
defined ... **A13:** 1
dental ... **A13:** 1359
Amalgam formation
in controlled-potential electrolysis ... **A10:** 208
Amalgamation
aluminum alloys ... **A13:** 589
and surface modification, compared ... **A13:** 498
Amalgams
admixed
high-copper, friction as described by various properties ... **A18:** 669
properties ... **A18:** 666
simplified composition or microstructure ... **A18:** 666
dental ... **A7:** 661-662
lathe-cut, low-copper, friction described by various properties ... **A18:** 669
spherical, low-copper, friction described by various properties ... **A18:** 669
unicompositional
properties ... **A18:** 666

Amalgams (continued)
simplified composition or microstructure ... **A18:** 666
AMAX-LP copper *See* Copper alloys, specific types, C10800
applications and properties ... **A2:** 268-269
Amber glass ... **EM4:** 1082
Ambient
defined ... **A11:** 1 **EM1:** 4
Ambient illumination
for optical holography ... **A17:** 413
Ambient spin testing ... **EM4:** 718
Ambient temperature *See also* Room temperature
acidified chloride at ... **A8:** 418-420
aqueous solution fatigue testing ... **A8:** 415-417
load application in ... **A8:** 417
strength, aluminum P/M alloys ... **A2:** 204-206
symmetric rod impact test at ... **A8:** 204-205
vacuum and gaseous environments at ... **A8:** 410-412
Ambient viscosity of lubricant
nomenclature for lubrication regimes ... **A18:** 90
Ambient-temperature cure system
for vinyl esters ... **EM2:** 274
American blackheart malleable iron
development ... **A15:** 30-31
American Carbon Society (ACS)
as information source ... **EM1:** 40
American Ceramic Society ... **EM4:** 38-39, 40, 116, 691, 692
American Ceramic Society (ACS)
as information source ... **EM1:** 40, 42
American Conference of Governmental and Industrial Hygienists ... **A7:** 136
American Dental Association, Specification No. 37
abrasivity test for prophylactic paste that uses radiotracer method ... **A18:** 668
American foundries *See* Early American foundries
American Foundry Equipment Company *See* Wheelabrator-Frye Co
American Foundrymen's Society ... **A15:** 34, 35, 209
American Gear Manufacturers Association (AGMA)
gear failure modes (20) classified ... **A18:** 535
service requirements of industrial gear oils ... **A18:** 99
viscosity grades of lubricants ... **A18:** 85
American Institute of Steel Construction ... **A8:** 724
American Iron and Steel Institute ... **A8:** 724
American Iron and Steel Institute (AISI)
classification system for high-speed tool steels ... **A16:** 51
American Malleable Casting Association ... **A15:** 34
American National standard taper pipe threads (NPT) ... **A16:** 301
American National Standards Institute ... **EM4:** 40
expected performance of eye and face protecting devices (Z87.1) ... **EM4:** 1074
guidelines for expected optical and mechanical performance of ophthalmic glass products (80.1) ... **EM4:** 1074
impact resistance tests and guidelines for sunglass lenses (Z80.3) ... **EM4:** 1074
Inc ... **A8:** 724
notation and terminology ... **A13:** 369-370
specifications ... **A13:** 322
specifications for ceramic tile including definitions (A 137.1) ... **EM4:** 925
terminology and some recommendations for low-vision aids of ophthalmic glass products (Z80.9) ... **EM4:** 1074
American National Standards Institute (ANSI) ... **EL1:** 734 **EM2:** 462
designation systems ... **A2:** 15-28
friction test standards ... **A18:** 53
American Petroleum Institute ... **A8:** 724
specifications ... **A13:** 322
American Petroleum Institute (API)
engine oil service classifications ... **A18:** 162-163
Engine Service Classification System ... **A18:** 86, 163
gravity ... **A18:** 82
performance specifications for engine lubricants ... **A18:** 98
performance testing of engine oils ... **A18:** 170
service categories for gasoline engine oils ... **A18:** 164-165
American Petroleum Institute (API) (continued)
service classifications ... **A18:** 98, 99
service symbol ... **A18:** 163, 164
Standard 618, guidelines on loading limits of rider rings in reciprocating compressors ... **A18:** 602
American Petroleum Institute (API), 1104
pipeline welding specification ... **A6:** 98
American Powder Metallurgy Institute ... **A7:** 19
American Revolution
foundries in ... **A15:** 26-27
American Society for Metals ... **A7:** 19
software by ... **A11:** 55
American Society for Testing and Materials *See* ASTM ... **A8:** 724
graded series for center segregation ... **A9:** 173-174
American Society for Testing and Materials (ASTM) *See also* ASTM Special Technical Publications; ASTM standards; ASTM test methods; Specifications; Standards ... **A7:** 19, 463 **EM2:** 90, 334, 425, 461 **EM3:** 61 **EM4:** 40
adhesives, applicable test methods ... **EM3:** 643
AST M grade HX, composition ... **A16:** 737
ASTM ... **A16:** 738, 741-743, 746-747, 749-757
ASTM A ... **A16:** 738, 741-743, 746-747, 749-757
ASTM Book of Standards ... **EM3:** 72
ASTM Committee C24 on building seals and sealants ... **EM3:** 61-62
ASTM Committee D14 on adhesives ... **EM3:** 61-62, 642
ASTM F 15 alloy
bonding application to borosilicate glass ... **EM3:** 300
glass-to-metal seals ... **EM3:** 301, 302
ASTM grade HW, composition ... **A16:** 737
as information source ... **EM1:** 40, 701
photometric tests approved ... **A10:** 64
requirements for high-speed tool steels ... **A16:** 51
sampling procedures ... **A7:** 212
Specification A ... **A16:** 51
standards for metal cutting and grinding fluids ... **A16:** 127, 128
American Society of Mechanical Engineers ... **A8:** 724
American Society of Lubrication Engineers (ASLE) ... **A8:** 601
American Steel Foundries (St. Louis) ... **A15:** 34
American Tyler Series
sieve aperture variety ... **EM4:** 66
American Water Jet Conferences ... **A18:** 222
American Welding Society ... **A15:** 532
specifications ... **A13:** 322
American Welding Society (AWS)
classification and identification of welding electrode types ... **A6:** 176
American wire gage (AWG)
defined ... **EL1:** 1133
American wire gage system ... **M1:** 259
Americium *See also* Transplutonium actinide metals
applications and properties ... **A2:** 1198-1201
pure ... **M2:** 715, 832-833
AMGO 54 computer program for structural analysis ... **EM1:** 268, 269
AMGO 72 computer program for structural analysis ... **EM1:** 268, 269
Amides
intermediate (organic) soldering fluxes ... **A6:** 628
Amine adduct ... **EM3:** 4
defined ... **EM2:** 4
Amine curing agents, EH-330
for use with polysulfides ... **EM3:** 139
Amine-epoxide reaction
in epoxy composite curing ... **EM1:** 67-71
Amines ... **A13:** 633, 1269-1270
aliphatic ... **EM1:** 74, 75
aromatic, in epoxy curing ... **EM1:** 70
as curing agent, epoxies ... **EL1:** 827-828
determined ... **A10:** 217-218
as hardeners, molding compounds ... **EL1:** 804-805
intermediate (organic) soldering fluxes ... **A6:** 628
for prepregs, reaction ... **EM1:** 140
as samples in gas analysis by mass spectrometry ... **A10:** 52
simple, eluent suppression technique for ... **A10:** 660
synergists, with photoinitiators ... **EL1:** 855
Amines, tertiary
for core curing ... **A15:** 240
Amino ... **EM3:** 4
defined ... **EM2:** 4
Amino acids
corrosion of stainless steels in ... **M3:** 81
Amino acids, irradiated
ESR analysis of ... **A10:** 266
Amino resins *See also* Amino; Aminos; Resins ... **EM3:** 4
chemistry ... **EM2:** 65
defined ... **EM2:** 4
Amino-borane electroless nickel plating process ... **M5:** 221-224, 229-231
coatings, properties of ... **M5:** 229-231
solution composition and operating conditions ... **M5:** 221-224
Aminoethylpiperazine (AEP) ... **EM1:** 69-70
Aminos *See also* Amino; Amino resins; Thermosetting resins
applications ... **EM2:** 230, 628
characteristics ... **EM2:** 230-231
costs and production volume ... **EM2:** 230
environmental effects ... **EM2:** 428
as low-temperature resin system ... **EM2:** 439
molding compound properties ... **EM2:** 231
processing ... **EM2:** 230-231
as structural plastics ... **EM2:** 65
suppliers ... **EM2:** 231
as thermoset resins ... **EM2:** 627-628
Aminosilanes
adhesion promoter for polyimides ... **EM3:** 157
Ammeter
defined ... **A13:** 1
for electrogravimetry ... **A10:** 200
in ultrasonic hardness tester ... **A8:** 101
Ammonia ... **A13:** 1180-1182
for accelerated SCC testing of copper and copper alloys ... **A8:** 525-526
agricultural, structural steel SCC from ... **A11:** 204
atmosphere, effect of grain size on SCC resistance ... **A11:** 208
atmospheric effect helping control dusting ... **A18:** 684
boiler, space shuttle orbiter ... **A13:** 1080
-caused SCC, copper/copper alloys ... **A13:** 615, 633-634
conductometric titration of ... **A10:** 203
corrosion ... **A13:** 1181-1182
corrosion of stainless steels in ... **M3:** 81
as corrosive environment ... **A12:** 24
cracking in ... **A11:** 215
dissociator ... **A7:** 344
dissolved, in water ... **A13:** 489
effect in carbonitriding ... **A7:** 454
eluent suppression technique for ... **A10:** 660
explosive range ... **A7:** 348
leach process, Sherritt ... **A7:** 139
physical properties as atmosphere ... **A7:** 341
poisoning ... **A7:** 349
as sample in gas analysis by mass spectrometry ... **A10:** 152
SCC effects produced in copper-zinc alloys ... **A11:** 635
solution, effect on critical strain rate ... **A8:** 519
undissociated, effect in high-carbon steels ... **A12:** 282
use in brass plating ... **M5:** 286-287
Ammonia dissociator ... **A7:** 344
Ammonia reactions
as gas/leak detection devices
Ammonium alum
calcination ... **EM4:** 111, 112
Ammonium bifluoride
as chemical cleaning solution ... **A13:** 1140
chemical milling etchant ... **A16:** 873
use in color etching ... **A9:** 142
Ammonium bisulfite
in oil/gas production ... **A13:** 1244-1245
Ammonium bisulfite as an etchant for carbon and alloy steel weldments ... **A9:** 175-176
Ammonium chloride
photochemical machining etchant ... **A16:** 591
Ammonium chloride zinc plating system ... **M5:** 251-252

Ammonium citrate
as ferrous cleaning agent **A12:** 75
Ammonium compound-caused SCC
copper/copper alloys **A13:** 615
Ammonium dihydrogen phosphate
as common analyzing crystal, x-ray spectroscopy **A10:** 88
Ammonium fluoroborate
for flame retardance **EM3:** 179
Ammonium hexachloroplatinate
as allergen **A2:** 1258
Ammonium hexacyanoferrate (II)
decomposition temperatures **EM4:** 55
Ammonium hexacyanoferrate (III)
decomposition temperatures **EM4:** 55
Ammonium hydroxide
and ammonium persulfate and water as an etchant for tin and tin alloy coatings **A9:** 451
as an electrolytic reagent for wrought stainless steels **A9:** 281
copper/copper alloy corrosion **A13:** 630
and hydrogen peroxide and water as an etchant for copper and copper alloys **A9:** 400
and hydrogen peroxide as an etchant for copper-base alloys **A9:** 551
and hydrogen peroxide as an etchant for copper-base powder metallurgy materials **A9:** 509
and hydrogen peroxide as an etchant for electrical contact materials **A9:** 551
and hydrogen peroxide as an etchant for silver -base alloys **A9:** 551
identification labelling **A9:** 67
as precipitant **A10:** 168
used to electrolytically etch heat-resistant casting alloys **A9:** 331-333
used to neutralize etching acids **A9:** 172
in water with hydrogen peroxide as an etchant for tin and tin alloy coatings **A9:** 451
Ammonium hydroxide, as corrosive
copper casting alloys **A2:** 352
Ammonium ion
organic precipitant for **A10:** 169
Ammonium iron (II) sulfate
decomposition temperatures **EM4:** 55
Ammonium iron (III) citrate
decomposition temperatures **EM4:** 55
Ammonium iron (III) sulfate
decomposition temperatures **EM4:** 55
Ammonium molydate
description **A9:** 68
and nitric acid as an etchant for lead and lead alloys **A9:** 415
Ammonium nitrate
in solid propellants **A7:** 598
Ammonium nitrate, apparent threshold stress values
low-carbon steels **A8:** 526
Ammonium nitrate, as sample modifier
GFAAS analysis **A10:** 55
Ammonium orthophosphate
in flame-retardant compounds **EM3:** 179
Ammonium oxalate solutions
as ferrous cleaning agents **A12:** 75
Ammonium paratungstate
in tungsten powder production **A7:** 152
Ammonium perchlorate propellants **A7:** 598, 600
Ammonium persulfate
as a macroetchant for carbon and alloy steels **A9:** 174
and ammonium hydroxide and water as an etchant for tin and tin alloy coatings **A9:** 451
as an etchant for welded joints in plate steels **A9:** 203
photochemical machining etchant **A16:** 591, 593
Ammonium persulfate (continued)
and potassium cyanide as an etchant for palladium and palladium alloys **A9:** 551
Ammonium persulfate hydroxide
as an etchant for beryllium-copper alloys **A9:** 394
Ammonium persulfate solution
and tartaric and solution as an etchant for lead-lead-antimony-tin alloys **A9:** 417
Ammonium phosphate, and magnesium nitrate
as sample modifiers, GFAAS analysis **A10:** 55
Ammonium phosphomolybdate
as precipitate **A10:** 173
Ammonium polyelectrolyte
application or function optimizing powder treatment and green forming **EM4:** 49
Ammonium polyphosphate
in flame-retardant compounds **EM3:** 179
Ammonium purpureate
as metallochromic indicator **A10:** 174
Ammonium pyrrolidinedithio-carbamate
as solvent extractant **A10:** 170
Ammonium salts
photochemical machining etchant **A16:** 589
Ammonium sulfate
and citric acid as an electrolyte for wrought heat-resistant alloys **A9:** 308
corrosion of stainless steels in **M3:** 81
and tartaric acid as an electrolyte for wrought heat-resistant alloys **A9:** 308
Ammonium trisoxalaftoferrate (III)
decomposition temperatures **EM4:** 55
Ammonolysis
aramid fibers **EM3:** 285
AMMRC *See* Army Materials and Mechanics Research Center
Ammunition
copper and copper alloys **A14:** 823-824
lead and lead alloy **A2:** 554
powder used **A7:** 573
Amonton coefficient of friction **A18:** 60, 66
Amontons' laws **A18:** 30-31, 432
asperity deformation **A18:** 34
defined **A18:** 2
Amorphous carbon
oxidized in high-temperature combustion resistance furnaces **A10:** 224
RDF structural ordering in **A10:** 393
Amorphous iron cut with a wire saw **A9:** 25
Amorphous materials
amorphous superconductors **A2:** 816-817
applications
atomic diffusion **A2:** 813
brazing materials **A2:** 819
chemical properties **A2:** 817-818
coatings **A2:** 819
crystallization **A2:** 812
defined **EL1:** 93
electronic properties **A2:** 815
EXAFS analysis **A10:** 407, 410
future developments **A2:** 820
glass transition **A2:** 812
heat capacity **A2:** 812-813
historical background **A2:** 804
magnetic properties **A2:** 815-816
mechanical properties **A2:** 813-814
RDF determined interatomic distance distributions and coordination numbers **A10:** 393
reinforcing fibers **A2:** 819
soft magnetic materials **A2:** 818-819
solid state amorphization **A2:** 807-809
structural models, diffraction experiments crystallization **A2:** 809
structure **A2:** 809-812
synthesis and processing **A2:** 806-807
technology **A2:** 818-820
Amorphous materials (continued)
thermal transport **A2:** 11, 813
thermodynamic properties **A2:** 812-813
Amorphous metals **A13:** 864-870
applications **A13:** 869
chromium effects **A13:** 866
corrosion behavior **A13:** 864-867
and crystalline metals, compared **A13:** 864
hydrogen embrittlement **A13:** 869
localized corrosion **A13:** 867-868
pitting **A13:** 867-868
production **A13:** 864
stress-corrosion cracking **A13:** 868-869
transition metal-metal binary alloys **A13:** 866
transition metal-metalloid alloys **A13:** 866-867
Amorphous oxides
texture **A13:** 66
Amorphous plastic **EM3:** 4
defined **EM2:** 4
Amorphous plastic (amorphous phase)
defined **EM1:** 4
Amorphous polymers *See also* Amorphous plastic; Amorphous resins; Polymer(s)
chemistry of **EM2:** 64
defined **A11:** 758
high-modulus graphite fibers in **EM2:** 758-759
immiscible blends **EM2:** 633-635
miscible blends **EM2:** 635-636
Amorphous polypropylene **EM3:** 80
Amorphous powder metals **A7:** 1, 794-797
Amorphous resins *See also* Amorphous polymers; Polymers; Resins
immiscible blends **EM2:** 632-633
miscible blends **EM2:** 636-637
thermoplastic **EM2:** 620-621
Amorphous silica *See* Aluminum silicates; Silica; Silica sands
Amorphous solar cell formation, as application
electron spectroscopy **EL1:** 1080-1081
Amorphous solid *See also* Metallic glass
defined **A10:** 668 **A13:** 1
Amount of phase
as x-ray diffraction analysis **A10:** 325
Amount of substance
SI base unit and SI symbol for **A10:** 685
AMP/HEDP nonheavy-metal inhibitor pair **A13:** 496
Ampco 21 *See* Copper alloys, specific types, C62500
Ampco A1 *See* Copper alloys, specific types, C95200
Ampco alloys *See* Cast copper alloys, specific types
Ampco B2 *See* Copper alloys, specific types, C95300
Ampco C3 *See* Copper alloys, specific types, C95400
Ampco D4 *See* Copper alloys, specific types, C95500
Ampcoloy 495 *See* Cast copper alloys, specific types; Copper alloys, specific types, C95700
Amperage
in magnetic particle applications **A7:** 576-577
Amperage, effect
electrical contact materials **A2:** 840
Ampere
symbol for **A8:** 724 **A11:** 796
Amperometric gas sensors
capabilities **A10:** 181
Amperometric titration
as electrometric **A10:** 204
Amperometry
capabilities, compared with voltammetry **A10:** 188
and conductometry, compared **A10:** 204
defined **A10:** 668
Amphiboles
chain structure **EM4:** 759, 760
Amphoretic metals
electropolishing with alkali hydroxides **A9:** 54
Amphoteric **A18:** 107
conductors **A13:** 65-66
defined **A13:** 1
metals, stray-current corrosion **A13:** 87
oxides, defect structure **A13:** 66

Amphoteric surfactants
use in alkaline cleaners **M5:** 24
Amplification
as electronic function **EL1:** 89
Amplification, ac
microwave inspection **A17:** 217
Amplifier
defined **A9:** 1
in ultrasonic hardness tester **A8:** 101
Amplifier hybrid microcircuits
corrosion failure analysis **EL1:** 1115
Amplifiers
microwave **A17:** 209
and preamplifiers, in x-ray spectrometer detectors **A10:** 91
stabilized operational, in electrogravimetry **A10:** 199
ultrasonic inspection **A17:** 253
Amplifying system
drift in direct current electrical potential method **A8:** 389
Amplitude
constant, *S-N* curves **A8:** 363
detection devices **A8:** 245-246
effect, in erosion/cavitation testing **A13:** 313
effect, in fretting **A13:** 139
in explosively loaded torsional Kolsky bar **A8:** 224
stress **A13:** 150, 295
ultrasonic ply cutting **EM1:** 616-617
Amplitude, constant
and sinusoidal loading, ***S-N*** curves **A11:** 103
Amplitude detection
ultrasonic testing **A8:** 245-246
Amplitude modulation
in magabsorption **A17:** 143
Amplitude reflection factor
microwave inspection **A17:** 204
Amplitude signals
magabsorption **A17:** 148
remote-field eddy current inspection **A17:** 198-199
Amplitude(s) *See also* Amplitude signals; Signal amplitude
in acoustic emission inspection **A17:** 282
distribution function, acoustic emission inspection **A17:** 284
echo, ultrasonic inspection **A17:** 245-246
microwave, reflected and transmitted **A17:** 204
modulation, in magabsorption **A17:** 143
reflection factor **A17:** 204
ultrasonic inspection
vibration, in ultrasonic inspection **A17:** 231
AMS *See* Aerospace Material Specification (of SAE); Aerospace Material Specifications **A16:** 547
AMS 4925 Ti
broaching **A16:** 203
AMS 6302C (forging)
for steel aircraft brakes **A18:** 584
AMS 6385C (plate)
for steel aircraft brakes **A18:** 584
AMS 6508
electron-beam welding **A6:** 865
AMS 6545
electron-beam welding **A6:** 868
AMS designations of carbon and alloy steels **A1:** 153-154, 159, 160-162
AMS Index of Specifications **EM3:** 62
AMS specifications *See also* listings in data compilations for individual alloys; Nonferrous alloys, AMS specific types; Steels AMS specific types
alloy steel bars **M1:** 209
carbon contents, carbon steels **M1:** 140
constructional steels for elevated temperature use **M1:** 647, 649
copper tubular products **M2:** 264
nominal compositions, alloy steels **M1:** 141-143
titanium **M3:** 358
AMSIL copper *See* Copper alloys, specific types, C10400, C10500 and C10700
applications and properties **A2:** 267-268
Amsler
double-shear tool **A8:** 62-63
resonant fatigue tester **A8:** 393
Amsler wear machine
defined **A18:** 2
amu *See* Atomic mass unit
Amyl acetate (98.8%)/1.2% nitrocellulose
as vehicle for solder glass powder **EM4:** 1069-1070
Amyl-nital as an etchant
for hot-dip galvanized sheet steels **A9:** 197
Amylaceous **EM3:** 5
Amzirc Brand copper *See* Copper alloys, specific types, C15000
applications and properties **A2:** 280-281
Anaerobic
corrosion **A13:** 43, 116-117
defined **A13:** 1
Anaerobic acrylates
for thermally conductive bonding **EM3:** 571, 572
Anaerobic acrylics
properties compared **EM3:** 92
shear strength of mild steel joints **EM3:** 670
Anaerobic adhesive *See also* Adhesive
defined **EM1:** 4 **EM2:** 4
Anaerobic adhesives **A7:** 457 **EM3:** 5, 48
Anaerobic bacteria
corrosion by **A11:** 190
Anaerobic sulfides, production
aqueous corrosion in **A13:** 43
Anaerobics **EM3:** 44, 48, 74, 75, 113-118
accelerator component **EM3:** 113, 114
for acetal applications **EM3:** 79
additives and modifiers **EM3:** 117
advantages and limitations **EM3:** 77
applications **EM3:** 79
automotive applications **EM3:** 45-46, 609
for automotive gasketing **EM3:** 553
for automotive threadlocking **EM3:** 553
chemical gasketing applications **EM3:** 609, 610
chemical properties **EM3:** 52
chemistry **EM3:** 50-51, 79, 113-114
commercial forms **EM3:** 115
compared to epoxies **EM3:** 98
competitive adhesives **EM3:** 116-117
cost factors **EM3:** 116
cure properties **EM3:** 51
cured, properties **EM3:** 117
curing mechanisms **EM3:** 50-51, 79, 113-114, 116, 118, 553
for cylindrical parts, fitting **EM3:** 115, 116
for cylindrical parts retaining **EM3:** 115
design **EM3:** 53
dispensing equipment **EM3:** 117, 118
for electronic general component bonding **EM3:** 573
for engine sealing **EM3:** 609
as engineering adhesives family **EM3:** 567
for flange sealing **EM3:** 116
form modifier component **EM3:** 113, 114, 117
functional types **EM3:** 114-115
flange sealants **EM3:** 114, 115
porosity sealants **EM3:** 114, 115
retaining adhesives **EM3:** 114, 115
structural adhesives **EM3:** 114, 115
threadlocking adhesives **EM3:** 114
threadsealing adhesives **EM3:** 114
threadsealing anaerobic products **EM3:** 114, 115
UV-curing adhesives **EM3:** 114, 115, 116, 117
for gasketing **EM3:** 54
initiator component **EM3:** 113-114
for lock washer replacements **EM3:** 51
markets **EM3:** 115-116
monomer (resin) component **EM3:** 113
for nylon **EM3:** 79
packaging of **EM3:** 114, 118
performance modifier component **EM3:** 113, 114, 117
performance of **EM3:** 124
photoinitiators added **EM3:** 114
for pipe flanges **EM3:** 547
for polyolefins **EM3:** 79
for polyvinylidene chloride **EM3:** 79
for porosity sealing **EM3:** 115-116
for porosity sealing for engine blocks **EM3:** 57
pot life **EM3:** 117-118
predicted 1992 scales **EM3:** 77
primers or activators used **EM3:** 118
product design **EM3:** 117
properties **EM3:** 50-51, 78, 116, 117
sealant formulations **EM3:** 677
Anaerobics (continued)
for shaft seal collars **EM3:** 547
shelf life **EM3:** 114, 117-118
slip fit and press and shrink fit values **EM3:** 116
stabilizer component **EM3:** 113, 114
for structural bonding **EM3:** 116
suppliers **EM3:** 58, 79, 116
tackifiers **EM3:** 183
for Teflon **EM3:** 79
thermal properties **EM3:** 52
for threadlocking **EM3:** 115
for threadsealing **EM3:** 115
tougheners **EM3:** 183
uncured, properties **EM3:** 117
for welded wheel rim seals for tubeless tires **EM3:** 116
Anaglyphe method of stereomicroscopy **A9:** 97
Analog dot mapping *See also* Dot mapping; Mapping; X-ray maps
and digital compositional mapping **A10:** 528
as elemental-distribution mapping **A10:** 525-528
Analog imaging, and digital imaging
of iron in aluminum matrix **A10:** 448
Analog instrumentation
eddy current inspection **A17:** 175, 178
Analog integrated circuits
of gallium compounds **A2:** 740
Analog oscilloscope **A8:** 228
Analog parasitic effects
in modeling/ simulation **EL1:** 77-78
Analog patterns from magnetic etching **A9:** 65
Analog-to-digital (A/d) converter **A18:** 294, 351
Analog-to-digital converter
defined **A10:** 668
in ICP systems **A10:** 39
Analog-to-digital recorder
for high strain rate data acquisition **A8:** 192
for split Hopkinson bar test data acquisition **A8:** 202
Analysis *See also* Analysis of large plastic incremental deformation; Analytical process modeling; Detail process analysis tools; Modeling; Process modeling; Simulation; Statistical analysis
chemical, and sampling **A10:** 15
of creep and creep-rupture methods of **A8:** 689-693
defined **A10:** 668
detail, finite-element method for **A14:** 409
of experiments **A8:** 639, 641, 653-661
of fatigue crack growth **A8:** 678-679
general, of organic gases **A10:** 11
metallographic, of powder forged parts **A14:** 204
methods, data-specific **A8:** 623
nondestructive, MOLE as **A10:** 130
on-line, PGAA as **A10:** 239-240
remote, electrometric titration as **A10:** 202
rough, KBES for **A14:** 409
stress, and die design **A14:** 414
of variance **A8:** 669, 677
Analysis and Testing of Adhesive Bonds **EM3:** 67
Analysis of large plastic incremental deformation
computational scheme **A14:** 426
as detail analysis tool **A14:** 411-412
example applications **A14:** 428-431
finite-element problem formulation **A14:** 425-426
flow chart for **A14:** 426
results, field variables and flow lines **A14:** 427-428
versions of **A14:** 426-427
Analysis of structure *See* Structural analysis
Analysis of surfaces *See* Surface; Surface analysis; Surface analysis and characterization
Analysis of variance
data analysis method **EM2:** 602-603
Analysis, off
as cause of casting failure **A11:** 391
Analytes
in complexometric titrations **A10:** 164
concentration in sample for ICP-AES **A10:** 33
concentration yielded by back-titration **A10:** 173
defined **A10:** 668
in gravimetric analysis **A10:** 170
ion association in solvent extraction **A10:** 170
moieties, weighing as the, gravimetric analysis **A10:** 171
separation **A10:** 164

Analytes (continued)
submicrogram amounts determined by electrometric titration ... **A10:** 202
titer technique for ... **A10:** 172
Analytic modeling *See also* Model(s); Modeling; Simulation
cost/performance ... **EL1:** 14-15
interconnection network simulator ... **EL1:** 14
of packaging density ... **EL1:** 13-14
of solder shear fatigue ... **EL1:** 743-746
thermal analysis ... **EL1:** 14
Analytical chemistry
complexometric titrations ... **A10:** 164
defined ... **A10:** 668
gravimetric ... **A10:** 163
ion exchange separation ... **A10:** 164-165
oxidation-reduction reactions ... **A10:** 163-164
solvent extraction ... **A10:** 164
Analytical curve
defined ... **A10:** 668
Analytical electron microscope (AEM) ... **A18:** 380
Analytical electron microscopes
conventional transmission electron ... **A10:** 430-431
convergent-beam electron-diffraction pattern ... **A10:** 431
electron microdiffraction ... **A10:** 431
electron optical column ... **A10:** 431
imaging in the ... **A10:** 440-446
modem ... **A10:** 431-432
scanning electron ... **A10:** 430-431
signal detector positioning ... **A10:** 434
x-ray microanalysis in the ... **A10:** 446-449
Analytical electron microscopy *See also* Analytical transmission electron microscopy
absorption contrast ... **A10:** 444-445
capabilities ... **A10:** 516
contrast mechanisms ... **A10:** 444-445
defined ... **A10:** 668
dislocation cell structure analysis by ... **A10:** 470-473
effect of lenses in ... **A10:** 432
and energy-dispersive x-ray spectroscopy, as electron-beam microanalytical technique ... **A10:** 446
imaging in TEM mode ... **A10:** 440-442
phase contrast ... **A10:** 445
special techniques ... **A10:** 445-446
TEM and STEM images, relationship between ... **A10:** 442-444
in the STEM mode ... **A10:** 442
Analytical electron microscopy (AEM)
for microstructural analysis ... **EM4:** 25
Analytical gap
defined ... **A10:** 668
Analytical line
defined ... **A10:** 668
Analytical methods *See also* Life tests; Test methods
for germanium and germanium compounds ... **A2:** 736
of thermal durability modeling ... **EL1:** 51
Analytical process modeling *See also* Analysis; Modeling; Process modeling
approximate and closed form solutions ... **A14:** 425
finite-element methods ... **A14:** 425
simulation methods ... **A14:** 913-919
Analytical transmission electron microscopy ... **A10:** 429-489
analytical electron microscope ... **A10:** 430-432
applications ... **A10:** 429, 453, 473-487
applications, techniques and procedures for ... **A10:** 429, 453-473
Bragg's law ... **A10:** 436
bright-field imaging ... **A10:** 441-446
capabilities ... **A10:** 365, 402, 490, 516, 536
capabilities, compared with optical metallography ... **A10:** 299
capabilities, compared with Rutherford backscattering spectrometry ... **A10:** 628
Analytical transmission electron microscopy (continued)
dark-field imaging ... **A10:** 441-446
defect analysis by TEM/STEM ... **A10:** 464-468
deformation, recovery, and recrystallization structures by TEM ... **A10:** 468-470
diffusion measurements by AEM ... **A10:** 476-478
diffusion-induced grain-boundary migration analysis by EDS/CBED ... **A10:** 461-464
dislocation cell analysis by AEM ... **A10:** 470-473
electrochemical thinning vs electrojet thinning ... **A10:** 451
electron beam/specimen interactions ... **A10:** 432-434
electron diffraction ... **A10:** 436-440
electron energy loss ... **A10:** 432, 435
electron energy loss spectroscopy and ... **A10:** 449-450
electron optics ... **A10:** 432
estimated analysis time ... **A10:** 429
general uses ... **A10:** 429
grain-boundary segregation analysis by AEM ... **A10:** 481-484
image morphology ... **A10:** 309
imaging, analytic electron microscope ... **A10:** 440-446
introduction ... **A10:** 430
ion beam milling ... **A10:** 451-452
Kikuchi line patterns ... **A10:** 437-438
lattice imaging ... **A10:** 445-446
light-element analysis by EDS/UTW-EDS/ EELS ... **A10:** 459-461
limitations ... **A10:** 429
microtomes, for sample sectioning ... **A10:** 452
orientation relationships and habit plane determination by ... **A10:** 453-455
phase diagrams determined by AEM ... **A10:** 473-476
related techniques ... **A10:** 429
replicas, extraction ... **A10:** 452
sample preparation ... **A10:** 450-453
samples ... **A10:** 429, 432-434, 450-453
selected-area diffraction ... **A10:** 436
signal detectors ... **A10:** 434-436
unknown phase identification by electron diffraction/EDS ... **A10:** 455-459
weld metal microstructure, by AEM ... **A10:** 478-481
x-ray microanalysis in ... **A10:** 446-449
Analytical wavelength
defined ... **A10:** 668
Analyzer, optical microscope ... **A9:** 76
defined ... **A9:** 1
path difference of reflected light ... **A9:** 79
Analyzers
cylindrical mirror ... **A10:** 554, 571
defined ... **A10:** 668
double-pass cylindrical mirror ... **A10:** 571
electron energy, or velocity ... **A10:** 570-571
electrostatic LEISS ... **A10:** 607
hemispherical, for AES analysis ... **A10:** 554
for image analysis ... **A10:** 310-313
retarding field, for AES analysis ... **A10:** 554
sector, for AES analysis ... **A10:** 554
semiautomatic digital image ... **A10:** 310
systems, x-ray spectrometers ... **A10:** 91
velocity ... **A10:** 570-571
Analyzers, atmospheres *See* Surface carbon content control, analyzers
Analyzing crystals
common ... **A10:** 88
in wavelength-dispersive x-ray spectrometers ... **A10:** 88
Anchor Hocking tempered borosilicate glass
coefficient of thermal expansion ... **EM4:** 1103
composition ... **EM4:** 1103
Anchor link
fatigue failure of ... **A11:** 397
Anchor pattern ... **EM3:** 5
Anchorage
defined ... **EM2:** 4
Anchorite
defined ... **A13:** 1
Anchors, and ties
in masonry walls ... **A13:** 1302, 1309-1310
Ancorsteel *See* Iron Powders, specific types
Andalusite
as aluminum silicate molding sand ... **A15:** 209
applications ... **EM4:** 46
composition ... **EM4:** 46
island structure ... **EM4:** 758
supply sources ... **EM4:** 46
Anderson and Stuart model
activation energy for conduction ... **EM4:** 852
Anderson-Darling (A-D) goodness-of-fit tests ... **EM4:** 701, 702, 704, 705, 707
Anderson-Darling (AD) goodness-of-fit test ... **EM1:** 302-305
Andesite
stone molds of ... **A15:** 15
Andrade creep equation ... **A8:** 318
Andreasen's Pipette
to analyze ceramic powder particle size ... **EM4:** 67
Anelastic deformation ... **EM3:** 5
Anelasticity
defined ... **EM1:** 4
Anemia
use of iron powders to combat/prevent ... **A7:** 614
Anemias
from cobalt deficiencies ... **A2:** 1251
refractory, and iron toxicity ... **A2:** 1252
Angle
of bend ... **A8:** 1, 125, 127
between normal to the diffracting lattice for ... **A10:** 692
planes and the sample surface, symbol
conversion factors ... **A8:** 722 **A10:** 686
of incidence ... **A10:** 114, 668
plane and solid, SI unit/symbol for ... **A8:** 721
symbol for ... **A10:** 692
Angle, displacement *See* Displacement angle
Angle, draft *See* Draft angle
Angle interlock weave
for multidirectionally reinforced fabrics and preforms ... **EM1:** 130
Angle of attack
defined ... **A18:** 2
Angle of bend ... **A8:** 1, 125, 127
Angle of bite
defined ... **A14:** 1
nomenclature for lubrication regimes ... **A18:** 90
Angle of contact
defined ... **A18:** 2
Angle of impact
of abrasives ... **A15:** 517-518
Angle of incidence
defined ... **A18:** 2
microwave effects ... **A17:** 204
ultrasonic inspection ... **A17:** 235-236
Angle of reflection
defined ... **A9:** 1
Angle of repose ... **A7:** 1, 282-285
drained, measurement methods ... **A7:** 282, 283
dynamic, measurement method ... **A7:** 282
factors affecting ... **A7:** 284
versus mixture composition ... **A7:** 284
Angle press
defined ... **EM2:** 4
Angle, slip *See* Slip angle
Angle, wind *See* Wind angle
Angle wrap
defined ... **EM1:** 4 **EM2:** 4
Angle-beam techniques
contact-type ultrasonic search units ... **A17:** 257
polar backscattering ... **A17:** 248

Angle-beam techniques (continued)
standard reference blocks, miniature **A17:** 264-265
surface wave **A17:** 247
ultrasonic inspection **A17:** 246-248
ultrasonic, of rolled plate **A17:** 270
Angle-dispersive diffractometry
used to construct pole figures **A9:** 695
Angle-ply laminate
defined **EM1:** 4
Angled blade plate
fatigue failure of **A11:** 683, 687
Angles, acute
by press brake forming **A14:** 536
Angstrom **A10:** 669, 691
symbol for **A8:** 724
Angstrom unit
defined **A9:** 1
Angular acceleration
SI derived unit and symbol for **A10:** 685
SI unit/symbol for **A8:** 721
Angular aperture
defined **A9:** 2
Angular change
butt welds **M6:** 876, 878
distortion caused by **M6:** 876-880
fillet welds **M6:** 876-879
Angular deflection scale
for cantilever beam bend test **A8:** 132-133
Angular distribution
fractal analysis **A12:** 212
profile, for fracture surface area **A12:** 202-204
Angular distribution functions
nuclear transitions of multipolarity **A10:** 290
Angular measure
abbreviation for **A11:** 798
symbol for **A10:** 691
Angular momentum quantum number **A18:** 450, 451
Angular movement
symbol for **A8:** 726
Angular scan, channeling
lattice strain measurement by **A10:** 635
Angular strain *See* Shear strain
Angular velocity
SI derived unit and symbol for **A10:** 685
SI unit/symbol for **A8:** 721
Angular velocity of cylindrical contact
nomenclature for lubrication regimes **A18:** 90
Angular velocity of gear rad/s
symbol and units **A18:** 544
Angular velocity of pinion
symbol and units **A18:** 544
Angular-contact ball bearings **A18:** 500, 506
applications **A11:** 490
basic load rating **A18:** 505
fretting failure **A11:** 498
f_v factors for lubrication method **A18:** 511
z and *y* factors **A18:** 511
Angular-contact bearings
defined **A18:** 2
Angular-contact groove ball bearings **A18:** 509
Angularity
defined **A14:** 1 **A15:** 1
of sand grains, molding effects **A15:** 208
Anhydride
defined **EM2:** 4
Anhydride/epoxide reaction
in epoxy composite curing **EM1:** 67-71
Anhydrides **EM3:** 5
commercial, list of **EM3:** 99
for curing epoxies **EM3:** 95
reaction with epoxies **EM3:** 99
Anhydrides, as curing agents
epoxies **EL1:** 828-829
Anhydrides, boric acid
as flux **A10:** 167
Anhydrides, types
epoxy curing **EM1:** 70-71
Anhydrous aluminum chloride in Group Vi electrolytes **A9:** 54
Anhydrous ammonia **A13:** 328, 544, 630
Anhydrous borax ($Na_2O{\cdot}2B_2O_3$)
purpose for use in glass manufacture **EM4:** 381
Anhydrous dibasic calcium phosphate
dentifrice abrasive **A18:** 665
Anhydrous ethyl alcohol (solvent)
batch weight of formulation when used in oxidizing sintering atmospheres **EM4:** 163
Anhydrous hydrogen fluoride **A13:** 1166-1170
Aniline **EM3:** 5
defined **EM2:** 4
Aniline point
defined **A18:** 2
Aniline-formaldehyde resins **EM3:** 5
defined **EM2:** 4
Animal feed
powder used **A7:** 572
Animal glue
applications **EM3:** 45
characteristics **EM3:** 45
for packaging **EM3:** 45
Animal medication
powder used **A7:** 572
Animal tissue
AAS analysis of trace metals **A10:** 55
NAA analysis of retention of toxic elements in **A10:** 233
voltammetric detection of herbicide/pesticide residues in **A10:** 188
Anion *See also* Cation; Ion **A13:** 1, 18-19, 65-66, 71, 330
Anion-exchange column
simulated pressurized water reactor water system **A8:** 423-424
Anionic detergents
for surface cleaning **A13:** 380
Anionic surfactants
use in alkaline cleaners **M5:** 24
Anions
defined **A10:** 669
determined on contaminated surfaces, by ion chromatography **A10:** 658
exchange resins, use in ion chromatography **A10:** 659
green and yellow, as determined by single- crystal analysis **A10:** 354
influence in stress-corrosion cracking **A8:** 499
inorganic, determined by ion chromatography **A10:** 663
as negatively charged ions **A10:** 659
potentiometric membrane electrodes quantification **A10:** 181
qualitative and quantitative analysis by ion chromatography **A10:** 658-667
Anisotropic *See* Optically anisotropic **EM3:** 5
defined **EM1:** 4 **EM2:** 4
laminate, defined **EM1:** 4
Anisotropic alloys (Alnico) **A15:** 738
Anisotropic effects
ESR studied **A10:** 256
Anisotropic hyperfine coupling constants
electron spin resonance **A10:** 261
Anisotropic laminate *See also* Laminate(s)
defined **EM2:** 4
Anisotropic materials
high-temperature solid-state welding **A6:** 299
Anisotropic mechanical properties
and mechanical fibering **A9:** 686
Anisotropic metals
polarized light optical color metallography **A9:** 138
Anisotropic thermal expansion coefficients
XRPD analysis **A10:** 333
Anisotropic thermal motions
and crystal structure determination **A10:** 352, 353
Anisotropy *See also* Directionality; Planar anisotropy **A8:** 1
average normal, and *r* value **A8:** 550
and birefringence **EM2:** 596-597
constants, FMR for obtaining **A10:** 272, 273
defined **A9:** 2
design requirements for **EM1:** 181
effect, deep drawing **A14:** 584
effect of, on notch toughness **A1:** 744-745
in forging **A11:** 316, 319
forgings **M1:** 356-360
of fracturing, wrought products **A11:** 317
in high-strength steel **A1:** 343, 344, 345
high-temperature superconductors **A2:** 1088
of laminates, defined **EM1:** 4
magnetic, determined **A10:** 272-273
Anisotropy (continued)
material, properties effects **EM2:** 405
microstructural, effect on fatigue **M1:** 681
microwave measurement **A17:** 215
notch toughness of steels **M1:** 695-696, 700
phenomenological theory **A8:** 143
planar, and earing **A14:** 576
planar, and *r* value **A8:** 550
in plastic torsion **A8:** 143
plate, instability of **EM1:** 446
steel sheet **M1:** 549
of unidirectional composite materials **EM1:** 218
unidirectional composites **EM1:** 218
in wrought alloys **A14:** 367
zirconium **A2:** 667
Anisotropy effect
inclusion-forming **A15:** 91
Anisotropy of P/M high-speed tool steels **A16:** 62
Anneal
definition **EM4:** 632
Anneal to temper
defined **A9:** 2
Anneal-resistant electrolytic copper *See* Copper alloys, specific types, C11100
applications and properties **A2:** 272-274
Annealed and tempered glass **EM4:** 453-459
applications **EM4:** 453-454
automotive glass forming **EM4:** 457-459
measurement of stress in glass **EM4:** 456-457
Babinet Compensator **EM4:** 457
differential surface refractometer **EM4:** 457
polarimeter **EM4:** 457
scattered light polarimeter **EM4:** 457
strength of glass **EM4:** 456
break pattern of tempered glass **EM4:** 456
tempering process **EM4:** 454-456
equipment **EM4:** 454-455
physics of process **EM4:** 455-456
types of tempering systems **EM4:** 454
Annealed brass strip
thickness for Scleroscope testing **A8:** 105
Annealed glass
seal design techniques **EM4:** 534
Annealed low-carbon manufacturers' wire **A1:** 282 **M1:** 264
Annealed microstructures
in tool steels **A9:** 258
Annealed powder
defined **A7:** 1
Annealed spring wire
characteristics of **A1:** 307-308
Annealed steel
press forming of **A14:** 558
Annealed temper **A1:** 850
Annealed wire
springs wound from **M1:** 289-290
Annealing *See also* Annealing temperature; Decarburization; Fabrication characteristics; Heat treatment; Stress relieving; Stress-relief anneal; Stress-relief annealing **A1:** 122-123, 132-133, 272 **EM3:** 5
(growth) twin, rutile images of **A10:** 443
abrasion resistance of steels **A18:** 490
alloy steel sheet and strip **M1:** 164
alloy steels **A4:** 37, 39
aluminum alloys **A15:** 761 **M2:** 28, 29 **M4:** 707-708, 709
applied to a cold-worked metal microstructural stages **A9:** 692-699
atmosphere, magnetically soft materials **A2:** 763
austenitic ductile irons **A15:** 700
Austenitizing temperatures, carbon-iron compacts **A7:** 456
batch **A1:** 122
beehive furnace, early **A15:** 31
black **A1:** 280
bright **A1:** 280
carbon steels **A4:** 37, 39
cast irons **A6:** 714
cast steels, effect on mechanical properties **M1:** 383, 384, 385, 386-387
chilled cast iron, effect on hardness and combined carbon **M1:** 82
chilled-iron railroad car wheels **A15:** 30
cold finished bars **M1:** 234

Annealing (continued)
cold rolled low-carbon steel sheet and strip **M1:** 156-157
of cold-rolled steel products **A1:** 132-133
continuous **A1:** 122-123
control of malleable iron **A15:** 688-690
copper alloys **M4:** 719-724
copper and copper alloys **A2:** 216
copper metals **M2:** 253-255
copper wire **M2:** 272
decarburization in QMP iron powder **A7:** 86-87
and deep drawing **A14:** 575
defined **A7:** 182 **A9:** 2 **A13:** 1 **EM1:** 4 **EM2:** 4
definition **EM4:** 632
and diffusion bonding **A6:** 157
dimensional change during **A7:** 481
distortion during **A7:** 480
Domfer iron powder process **A7:** 90-91
of dual-phase steels **A1:** 424-425
of ductile iron **A1:** 41 **A15:** 657 **M1:** 37
effect on crystallographic texture **A9:** 700-701
effect on ductile iron microstructure **A9:** 245
effect on green strength **A7:** 302
effect on tool steel microstructure **A9:** 258
electrolytic iron powder **A7:** 94
enameling **EM3:** 303
enhanced by accelerated cooling **A4:** 167
flame **M4:** 506
formation of metal-silicon contacts **EM3:** 581
functions in iron powder processes **A7:** 182-183
for gas cutting **A14:** 724
gear materials **A18:** 261
gold and gold alloys **A4:** 941, 942-944
grain-coarsening, mechanically alloyed oxide alloys **A2:** 944
dispersion-strengthened (MA ODS)
gray cast iron **M1:** 23
of gray iron **A1:** 23-24 **A15:** 642-643 **M4:** 529-531
hafnium **A2:** 663
high-chromium white irons **A15:** 684
of hot-rolled steel bars **A1:** 241
improper, defect depth distribution from **A10:** 628
with induction heating **A4:** 193-194
intermediate, for heat-resistant alloys **A14:** 779
intermediate, in three-roll forming **A14:** 620-621
of Invar **A2:** 890--891
iridium **A4:** 945, 946-947
killed steel to avoid **A14:** 548
lime bright **A1:** 280
of low-alloy steel sheet/strip **A1:** 209
magnetically y so materials **A2:** 762-763
malleable cast iron **M1:** 57-63
of malleable iron **A1:** 72-73
metal powders **A7:** 182-185
molybdenum alloys **A6:** 581
molybdenum-implanted aluminum **A10:** 486
nickel and nickel alloys **A4:** 907-911 **M4:** 754-757
nickel plate **M5:** 217-218
nickel strip **A7:** 401
nickel-base corrosion-resistant alloys containing molybdenum **A6:** 596
niobium-titanium ingot **A2:** 1044
nonferrous high-temperature materials **A6:** 572
notch toughness of steels, effect on **M1:** 706, 708
palladium **A2:** 716 **A14:** 850
palladium and palladium alloys ... **A4:** 944, 945, 946
physical aging **EM3:** 422
plain carbon steels **A15:** 713
platinum **A2:** 709 **A14:** 850
platinum alloys **A14:** 851
platinum and platinum alloys **A4:** 944, 945-946
post-treatment after atomic deposition method for metallizing **EM4:** 542
of powder metallurgy high-speed tool steels **A1:** 783
as powder treatment **A7:** 24
precious metals **M4:** 760-761, 762
prehistoric **A15:** 15

Annealing (continued)
of press formed parts **A14:** 548
process, effect on cold heading properties **A14:** 293
of pure copper, as electrical contact material **A2:** 843
Raman analysis of **A10:** 133
recrystallization, copper/copper alloys **A13:** 615
regenerative burner fuel savings from furnace **A4:** 522
of rhodium **A4:** 945, 946-947 **A14:** 850
ring forging cracked after **A11:** 574, 580
ruthenium **A4:** 946-947
salt **A1:** 280
SAS techniques for **A10:** 405
as secondary operation **A7:** 456
of shape memory effect (SME) alloys **A2:** 899-900
silver and silver alloys **A4:** 939-941
of sintered high-speed steels **A7:** 374
solution annealing, austenitic stainless steels **A1:** 898-899, 912, 945
solution, beryllium-copper alloys **A2:** 405-406
in specialty P/M strip production **A7:** 403
spheroidize **A1:** 209, 280
spherulite enlargement **EM3:** 410
stainless steel **M4:** 624, 625-626, 628, 633-634, 639 **M5:** 102
steel **M4:** 14-27
of steel wire **A1:** 280 **M1:** 262
steel wire, during fabrication **M1:** 590-591
steel wire rod **M1:** 253
strand **A1:** 280
stress-relief, for fracture resistance **A11:** 125
structures of hypoeutectic bearing caps after **A11:** 350
studies, by NMR **A10:** 277
titanium **M4:** 765
of titanium alloy forgings **A14:** 281
to develop equiaxed alpha grains in titanium and titanium alloys **A9:** 460
to reveal as-cast solidification structures in steel **A9:** 624
of tool and die steels **A14:** 53-54
tool steels **A4:** 734-737, 739, 743, 750, 757, 758, 759 **M4:** 563-564, 565
ultrahigh-strength steels **M1:** 424, 430, 431, 432, 433, 434, 435, 438, 441
wrought copper and copper alloy wiredrawing and wire stranding **A2:** 256
wrought copper and copper alloys **A2:** 245-247
wrought titanium alloys **A2:** 619
and x-ray diffraction results **A18:** 469
zirconium **A2:** 2803
Annealing behavior of cold-worked metals **A9:** 692-699
Annealing carbon *See* Temper carbon
Annealing out of point defects during recovery **A9:** 693
Annealing point **EM4:** 424
defined **EM1:** 47
Annealing temperature *See also* Annealing; Fabrication characteristics
aluminum casting alloys **A2:** 157
cast copper alloys **A2:** 367-368, 383-387
effect on recrystallization **A9:** 686
wrought aluminum and aluminum alloys **A2:** 63-122
Annealing, thermal
as failure mechanism **EL1:** 1012
Annealing twin
defined **A9:** 2 **A11:** 1
in steel **A9:** 178
Annealing twin bands *See also* Twin bands **A9:** 2
Annual Book of ASTM Standards **EM3:** 61
Annual interest compounding **A13:** 370-371
Annular air inlet
in Whiting cupola **A15:** 30

Annular bearing
defined **A18:** 2
Annular fracture surface
metal-matrix composites **A12:** 466
Annular free fall nozzle designs **A7:** 26, 29
Annular gas jets
confined atomization **A7:** 26, 29
Annular nozzle
atomizing technique **A7:** 123
Annular ring
as plated-through hole failure **EL1:** 1022
Annular snap joints **EM2:** 719-720
Anodal anodizing process
aluminum and aluminum alloys **M5:** 587
Anode *See also* Cathode
in aqueous corrosion **A13:** 29
area, ratio to cathode area, galvanic corrosion **A11:** 186
corrosion, defined **A13:** 1
corrosion efficiency, defined **A13:** 1
defined **A11:** 1 **A13:** 1
effect, defined **A13:** 1
efficiency, defined **A13:** 1
film, defined **A13:** 2
materials, cathodic protection **A13:** 468-469, 920-922
single, formulas **A13:** 470
splines, in metal-processing equipment **A13:** 1316
Anode aperture
defined **A9:** 2
Anode materials
x-ray tubes **A10:** 89
Anode melting rate
atomization **A7:** 41, 43
Anode polarization *See* Polarization
Anodes
aluminum and aluminum alloys **A2:** 14
and antenna covers, microwave inspection **A17:** 202
for electropolishing **A9:** 49
hooded **A17:** 305
lead and lead alloy **A2:** 555
reflection technique, microwave inspection **A17:** 206
silver **A7:** 148
transmission technique, microwave inspection **A17:** 205
x-ray tubes **A17:** 302
Antenna(s)
Anodes, dual
in electrogravimetry **A10:** 199
Anodic acid etching *See also* Anodic etching
steel **M5:** 16-18
Anodic back series
in equine serum **A13:** 1330
Anodic breakdown pitting
titanium/titanium alloys **A13:** 683-684
Anodic cleaning
defined **A13:** 2
Anodic coating **A13:** 1, 424, 811
Anodic coatings *See also* Anodizing
abrasion resistance **M5:** 596
aluminum and aluminum alloys **M5:** 586, 589-591, 594-598, 606-607, 609-610
color anodizing **M5:** 595-596, 609-610
integral process **M5:** 595-596, 609-610
two-step (electrolytic) process **M5:** 596
evaluation of **M5:** 596
fatigue strength affected by **M5:** 597
lightfastness **M5:** 596
magnesium alloys **M5:** 629, 632-638, 640-641, 643-644, 647
problems and corrections **M5:** 636-638, 642-643
process control of **M5:** 635-636, 640-641
repair of **M5:** 635
surface preparation **M5:** 632
reflectance values affected by **M5:** 596-598
sealing processes **M5:** 590-591, 594-596, 603-604, 606

Anodic coatings (continued)
standards ... **M5:** 606-607
surface and mechanical properties affected by ... **M5:** 596-598
thermal radiation reflectance affected by ... **M5:** 596-598
thickness ... **M5:** 596-601
weight, as function of time ... **M5:** 588, 590-591, 595-596
Anodic dissolution
effect on cleavage ... **A12:** 42
in electrochemical principles ... **A9:** 144
as SCC mechanism ... **A12:** 25
to extract phases from wrought heat-resistant alloys ... **A9:** 308
Anodic electrocleaning
aluminum and aluminum alloys ... **M5:** 578
copper and copper alloys ... **M5:** 618-620
magnesium alloys ... **M5:** 631
periodic reverse system ... **M5:** 34-35
processes and materials ... **M5:** 33-35, 618-620
stainless steel ... **M5:** 561
steel ... **M5:** 16-18
zinc alloy die castings ... **M5:** 677
Anodic electroplating
molybdenum and tungsten ... **M5:** 660-661
solution compositions and operating conditions ... **M5:** 660
Anodic etching *See also* Anodic acid etching ... **A9:** 61
defined
hard chromium plating pretreatment by ... **M5:** 180, 183
steel ... **M5:** 180
Anodic film replicas
used to examine aluminum alloys ... **A9:** 351
Anodic films on aluminum alloys
features revealed with reflected plane-polarized light ... **A9:** 351, 354
Anodic hard coating
postforging defects from ... **A11:** 333
Anodic inhibitors *See also* Cathodic-inhibitors; Inhibitors ... **A13:** 1141
chromates ... **A13:** 494
as corrosion control ... **A11:** 197
defined ... **A13:** 2
molybdates ... **A13:** 494
nitrites ... **A13:** 494
water-recirculating systems ... **A13:** 494-495
Anodic metal dissolution
at dislocations ... **A13:** 46
Anodic overvoltage
in electrogravimetry ... **A10:** 198
Anodic oxidation of isotropic metals and alloys
polarized light etching ... **A9:** 59
Anodic oxides
as barrier protection ... **A13:** 377-378
Anodic passivation at heterogeneities in electrolytic polishing ... **A9:** 48
Anodic polarization *See also* Polarization
behavior, intergranular corrosion ... **A13:** 123
behavior, typical ... **A13:** 217
curve, aluminum alloy ... **A13:** 265, 584
curve, schematic ... **A13:** 464
curves, aluminum alloy ... **A8:** 532
curves, for zirconium in phosphoric acid ... **A13:** 717
curves, for zirconium in sulfuric acid ... **A13:** 708-709
defined ... **A13:** 2
potentiostatic, measurement ... **A13:** 1333
potentiostatic passive ... **A13:** 218
stress-corrosion cracking under ... **A8:** 537
studies, apparatus ... **A13:** 464
uranium, alloying effects ... **A13:** 816
Anodic potential ranges ... **A9:** 144
Anodic protection *See also* Cathodic-protection ... **A13:** 2, 463-465, 694
for corrosion control ... **A11:** 196
Anodic reaction
aqueous corrosion rate control by ... **A13:** 32
defined ... **A13:** 2
noble metals ... **A13:** 807
Anodic reaction products
in electropolishing ... **A9:** 49-50
Anodic reactions in electrochemical etching ... **A9:** 60-61
Anodic reactivation polarization testing ... **A13:** 220
Anodic Tafel slope ... **A18:** 274
Anodic-cathodic pickling system
iron and steel ... **M5:** 76
Anodization
of thin-film hybrids ... **EL1:** 313
Anodize
chromic ... **A13:** 396
classification (MIL-A-8625) ... **A13:** 396
sulfuric ... **A13:** 396-397
Anodized aluminum ... **A13:** 219, 599-600, 607, 1081
coating for seals ... **A18:** 551
corrosion resistance ... **M2:** 225-226, 229-232
for resistant to scuffing ... **A18:** 538
ears
sensitive tint used to examine ... **A9:** 138
Anodized coatings ... **A13:** 397
Anodizing *See also* Anodic coatings ... **EM3:** 416-417
agitation used in ... **M5:** 593
alloy composition affecting ... **M5:** 590
aluminum and aluminum alloys ... **M5:** 572, 580-598, 601, 603-610
limitations, factors causing ... **M5:** 589-591
of aluminum and aluminum alloys ... **A11:** 195
aluminum casting alloys ... **A2:** 175
bleaching prior to, aluminum and aluminum alloys ... **M5:** 572
bulk processing ... **M5:** 594
ceramic coating for adiabatic diesel engines ... **EM4:** 992
chromic *See* Chromic anodizing
color ... **M5:** 595-596, 609-610
commercial processes, solution composition and operating conditions ... **M5:** 586-587
defined ... **A13:** 599
electrolytic polishing pretreatment ... **M5:** 305
electroplating pretreatment ... **M5:** 601-603
equipment and process control ... **M5:** 591-594
racks, design and materials for ... **M5:** 593-594
equipment, corrosion of ... **A13:** 1314-1316
hafnium alloys ... **M5:** 667-668
hard *See* Hard anodizing
heat treatment affecting ... **M5:** 591
inserts, handling ... **M5:** 591
molybdenum ... **M5:** 661
niobium ... **M5:** 663
operating procedures ... **M5:** 586-589, 591-594
power requirements ... **M5:** 593
prior processing affecting ... **M5:** 591-592
problems and corrections for examples ... **M5:** 594
process types *See also* specific processes by name ... **M5:** 586-589, 592
special ... **M5:** 589, 592
reasons for ... **M5:** 585-586
rough finishing, affecting ... **M5:** 591
selective, masking for ... **M5:** 593
solution composition and operating conditions ... **M5:** 587-592, 596-597
reflectance values affected by ... **M5:** 596-597
sulfuric *See* Sulfuric anodizing
surface finish, effects on ... **M5:** 590-591
surface preparation ... **M5:** 586
tantalum ... **M5:** 663
temper affecting ... **M5:** 591
temperature control ... **M5:** 593
tungsten ... **M5:** 667-668
used to enhance contrast in hafnium ... **A9:** 497-499
used to enhance contrast in zirconium and zirconium alloys ... **A9:** 497-499
used to produce interference films of oxides ... **A9:** 137, 142-143
used to render isotropic metals optically active ... **A9:** 78
of zinc alloy castings ... **A15:** 797
zinc alloys ... **A2:** 530
zinc-coated steel ... **M1:** 169
zirconium alloys ... **M5:** 667-668
Anodizing solutions
for uranium and uranium alloys ... **A9:** 478
Anolyte
defined ... **A13:** 2
Anomalies, structural
as casting defects ... **A11:** 387-388
Anomalous absorption of high-energy electrons ... **A9:** 111
Anomalous service conditions
of continuous fiber-reinforced composites ... **A11:** 733
Anomalous transmission
divergent beam topography ... **A10:** 370
in x-ray topography ... **A10:** 367
Anorthic crystal system ... **A3:** 1 • 10, 15 **A9:** 706
Anorthite ($CaO \cdot Al_2O_3 \cdot 2SiO_2$) ... **EM4:** 6
crystal structure ... **EM4:** 882
purpose for use in glass manufacture ... **EM4:** 381
Anoxal anodizing process
aluminum and aluminum alloys ... **M5:** 587
Anschultz, George
as early founder ... **A15:** 26
ANSI *See* American National Standards Institute
ANSI/ASTM D 2270
reference oil viscosities tabulated and formula for viscosity index numbers over ... **A18:** 83
ANSYS
finite-element analysis code ... **EM3:** 479, 480
ANSYS computer program ... **EM4:** 700-702
ANSYS REVISION 4.0+ computer program for structural analysis ... **EM1:** 268, 270
ANSYS software program
for heat transfer problems ... **A15:** 30
Antacids
as bismuth application ... **A2:** 1256
magnesium in ... **A2:** 1259
Antechamber
defined ... **A7:** 1
Antenna marker beacon ... **A13:** 1119-1120
Antenna marker beacons
corrosion failure analysis ... **EL1:** 1108
Anthracene, and naphthalene
total luminescence spectrum (EEM) of mixture ... **A10:** 78
Anti-acid metal *See* Copper alloys, specific types, C93800
properties and applications ... **A2:** 380-382
Anti-fouling
defined ... **A13:** 2
topcoats, organic coatings ... **A13:** 918
Anti-fouling paints
powders used ... **A7:** 572
Anti-fouling ship paints
powders used ... **A7:** 574
Anti-galling pipe joint compound lubricant
powders used ... **A7:** 573
Anti-personnel bombs
powders used ... **A7:** 573
Anti-segregation process
for tool steels ... **A7:** 784-787
Anti-segregation process (ASP) ... **A1:** 780
high-speed tool steel bend testing ... **A16:** 62
steel grade ... **A16:** 61, 62
steel grindability ... **A16:** 61, 62, 63
Anti-static surfaces ... **A7:** 610-611
by specialty polymers ... **A7:** 606
Anti-Stokes lines ... **EM4:** 56
Anti-Stokes Raman line
defined ... **A10:** 669
Anti-Stokes scattering
energy-level diagram ... **A10:** 127
in Raman spectroscopy ... **A10:** 126-128
Antichills *See also* Chills
in permanent mold casting ... **A15:** 283
Anticlastic curvature
in elastic bending ... **A8:** 118
Anticorrosion agents
grease additives ... **A18:** 124
Anticorrosive additive
defined ... **A18:** 3
Antiextrusion ring
defined ... **A18:** 3
Antiferromagnetic materials
defined ... **A9:** 63
ESR identification of magnetic states in ... **A10:** 253
variable temperature studies of ... **A10:** 257
Antiferromagnets
domains ... **A9:** 602
Antifoam additive
defined ... **A18:** 3
hydraulic oils ... **A18:** 86
for metalworking lubricants ... **A18:** 141-142, 144, 147

Antifoam additive (continued)
in rust and oxidation (R&O) oils ... **A18:** 133
Antifoam compounds
for ceramic shell investment molds ... **A15:** 259
Antifoaming agents
as lubricant additives ... **A14:** 515
for spray drying ... **A7:** 75
use in acid cleaners ... **M5:** 60
Antifriction ... **A18:** 499
Antifriction bearing *See also* Roller bearing; Rolling-element bearing; Self-lubricating bearing
defined ... **A18:** 3
wear failure of ... **A11:** 764-765
Antifriction material
defined ... **A18:** 3
Antihalation films
camera lens ... **A12:** 84
Antihistamines
coulometric titration of ... **A10:** 205
Antilock braking systems (ABS)
as automotive hybrid application ... **EL1:** 382
Antimicrobial agents
as lubricant additives ... **A14:** 515
for metalworking lubricants ... **A18:** 142, 143-144
Antimigration barriers ... **A18:** 151
Antimisting agents
as lubricant additives ... **A14:** 515
Antimonial admiralty metal
applications and properties ... **A2:** 318-319
Antimonial naval brass
applications and properties ... **A2:** 319-320
Antimonial-tin solder
applications and compositions ... **A2:** 521
Antimony
as addition to aluminum-silicon alloys ... **A18:** 788
alloying, aluminum casting alloys ... **A2:** 130
alloying, wrought aluminum alloy ... **A2:** 46
in aluminum alloys ... **A15:** 743
as an addition to electrical steels ... **A9:** 537
applications ... **A2:** 1258
as-cleaved, twins, river patterns, and cracks ... **A9:** 159
in carbon-graphite materials ... **A18:** 816
in cast iron ... **A1:** 5, 8
cast polycrystalline, early TEM ... **A12:** 6
cause of temper embrittlement ... **A4:** 124, 135
compatibility in bearing materials ... **A18:** 743
in composition, effect on ductile iron ... **A4:** 686
in composition, effect on gray irons ... **A4:** 671
in copper alloys, determined by iodoantimonite method ... **A10:** 68
determined by controlled-potential coulometry ... **A10:** 209
-doped ASTM/ASME alloy steels ... **A12:** 350
-doped gold contact wires ... **EL1:** 958
effect in copper alloys ... **A11:** 635
effect, lead-acid battery corrosion ... **A13:** 1317
effects of, on notch toughness ... **A1:** 742
embrittlement by ... **A11:** 234-235 **A13:** 180
epithermal neutron activation analysis (ENAA) ... **A10:** 239
etching by polarized light ... **A9:** 59
as eutectic refiner ... **A15:** 164
fire refining effect ... **A15:** 453
fractured ingots, history ... **A12:** 1
gaseous hydride, for ICP sample introduction ... **A10:** 36
as halogen trap material ... **A10:** 224
heat-affected zone fissuring in nickel-base alloys ... **A6:** 588
impurity in solders ... **M6:** 1072
as lead additive ... **A2:** 545
in lead-base alloys ... **A18:** 749
lubricant indicators and range of sensitivities ... **A18:** 301
as minor element, ductile iron ... **A15:** 648
as minor toxic metal, biologic effects ... **A2:** 1258-1259
as modifier addition ... **A15:** 484
Antimony (continued)
photometric analysis methods ... **A10:** 64
plain carbon steel resistance to ... **A13:** 515
price per pound ... **A6:** 964
pure ... **M2:** 715, 716
pure, properties ... **A2:** 1100
quartz tube atomizers for ... **A10:** 49
recommended impurity limits of solders ... **A6:** 986
resistance of, to liquid-metal corrosion ... **A1:** 635
safety standards for soldering ... **M6:** 1099
segregation to grain boundaries ... **A12:** 350
as silicon modifier ... **A15:** 79
as solder impurity ... **EL1:** 637
species weighed in gravimetry ... **A10:** 172
spectrometric metals analysis ... **A18:** 300
in steel weldments ... **A6:** 420
thermal diffusivity from 20 to 100 °C ... **A6:** 4
as tin solder impurity ... **A2:** 520
TNAA detection limits ... **A10:** 237, 238
toxicity ... **A6:** 1195
as trace element ... **A15:** 394
as tramp element ... **A8:** 476
TWA limits for particulates ... **A6:** 984
vapor pressure ... **A6:** 621
vapor pressure, relation to temperature ... **A4:** 495
volatilizing ... **A10:** 166
volumetric procedures for ... **A10:** 175
weighed as the sulfide ... **A10:** 171
Antimony alloys, Sb-14Ni
peritectic transformations ... **A9:** 678
Antimony alloys, specific types
95Sn-5Sb ... **A6:** 351
Antimony in fusible alloys ... **M3:** 799
Antimony in steel
notch toughness, effect on ... **M1:** 694
temper embrittlement, role in ... **M1:** 684-685
Antimony oxide
cost-effective replacements for ... **EM3:** 179
as filler (flame retardant) ... **EM3:** 179
Antimony oxides
examination under polarized light ... **A9:** 400
Antimony precipitate in lead-antimony alloys ... **A9:** 417
Antimony trioxide
as flame retardant ... **EM2:** 504
Antimony trioxide (Sb_2O_3)
component in photosensitive glass composition ... **EM4:** 440
as fining agent ... **EM4:** 380
in ovenware compositions ... **EM4:** 1103
purpose for use in glass manufacture ... **EM4:** 381
specific properties imparted in CTV tubes ... **EM4:** 1039
in tableware compositions ... **EM4:** 1101
volatilization losses in melting ... **EM4:** 389
Antimony, vapor pressure
relation to temperature ... **M4:** 310
Antimony-opacified porcelain enamel
composition of ... **M5:** 510
Antimony/antimony alloys
chemical analysis and sampling ... **A7:** 248
Antinode strain
cooling in ultrasonic testing ... **A8:** 247
Antioch process
dehydration/rehydration ... **A15:** 246
drying temperature ... **A15:** 246-247
metals cast ... **A15:** 247
mold assembly ... **A15:** 247
molds, composition ... **A15:** 246
pouring practice ... **A15:** 247
sequence of operations ... **A15:** 246
Antioxidant
defined ... **A18:** 3 **EM1:** 4
grease additive ... **A18:** 124
for metalworking lubricants ... **A18:** 141, 143
Antioxidants ... **EM3:** 5
added to elastomeric adhesives ... **EM3:** 143
Antioxidants (continued)
as additive, effects ... **EM2:** 425
defined ... **EM2:** 4
effect, chemical susceptibility ... **EM2:** 572
GC/MS analysis of ... **A10:** 639
as polymer additive ... **EM2:** 67
ultrahigh molecular weight polyethylenes (UHMWPE) ... **EM2:** 170
Antiozonants, as additive
effects ... **EM2:** 425
Antiphase boundaries ... **A9:** 681-683
defined ... **A9:** 681
dislocation generated ... **A9:** 682-683
schematic representation of edge dislocation ... **A9:** 682
transmission electron microscopy ... **A9:** 118-119
Antiphase boundaries, in ordered alloys
effect in FIM images ... **A10:** 589
Antiphase domain boundaries ... **A9:** 601
Antiphase domain structures
in nonferrous martensite ... **A9:** 672-673
Antipitting agent
defined ... **A13:** 2
Antipitting agents
use in nickel plating ... **M5:** 200-203, 206, 209
Antique green brass coloring solution ... **M5:** 586
Antirust additives
high-vacuum lubricant applications ... **A18:** 157
Antiscuffing lubricant
defined ... **A18:** 3
Antiseize additives
in nonengine lubricant formulations ... **A18:** 111
Antiseizure coating
for steel bolts ... **A11:** 536
Antiseizure property
defined ... **A18:** 3
Antistatic
agents ... **EM2:** 4, 501-502
compounds, types ... **EM2:** 468-469
Antistatic agents ... **EM3:** 5
defined ... **EM1:** 4
Antiwear additive
defined ... **A18:** 3
high-vacuum liquid lubricants ... **A18:** 157
hydraulic oils ... **A18:** 86
for metalworking lubricants ... **A18:** 141
mineral oils ... **A18:** 155
Antiwear and extreme-pressure (EP) agents ... **A18:** 99, 101-103
antiseize additives ... **A18:** 102
applications ... **A18:** 101, 102
ASTM sequence IIIE and VE engine tests ... **A18:** 101
dithiophosphoric acid zinc salts ... **A18:** 102
formation of ... **A18:** 102-103
in engine lubricant formulations ... **A18:** 111
for internal combustion engine lubricants ... **A18:** 162
load factor ... **A18:** 101
in nonengine lubficant formulations ... **A18:** 111
in rust and oxidation (R&O) oils ... **A18:** 133
testing ... **A18:** 101
Antiwear films
AES characterized ... **A10:** 566
Antiwear number (AWN) *See also* Archard wear law
defined ... **A18:** 3
Antiweld characteristic *See* Antiseizure property
Anvil
in Charpy V-notch test ... **A8:** 263
defined ... **A14:** 1
effect in Brinell test workpiece ... **A8:** 85-86, 88
effect in Rockwell hardness testing ... **A8:** 76-77
free-surface transverse velocity ... **A8:** 235-236
hammers, types ... **A14:** 41-42
in open-die forging ... **A14:** 61
for Rockwell hardness testers ... **A8:** 80
support, cylindrical workpieces ... **A8:** 83
Anvil cap *See* Sow block

Anvil die closure
for helical gear compaction **A7:** 324
Anvil effect **A18:** 417
Anvil plate
before lapping **A8:** 235
normal stress-particle velocity **A8:** 232
pressure-shear impact testing **A8:** 231-235
properties **A8:** 235
AOD *See* Argon oxygen decarburization
AP *See* Atom probe; Atom probe microanalysis
Apartment buildings
corrosion in **A13:** 1299
Apatite **EM4:** 1010, 1012
hardness **A18:** 433
on Mohs scale **A8:** 108
APB *See* Antiphase boundaries
APC *See* Active path corrosion
Apeizon oil
defined **A18:** 3
Aperture
camera lens, selection **A12:** 80-81
defined **A9:** 2
number to *f*/number conversions for Macro-Nikkor lenses **A12:** 81
optimum, problem of **A12:** 80-81
size effects in SEM imaging **A12:** 168
Aperture, collimator
defined **A17:** 383
Aperture diaphragm in optical microscopes **A9:** 72
Aperture size
defined **A7:** 1
secondary electron imaging (SEI) **EL1:** 1096-1097
Aperture size, effective
computed tomography (CT) **A17:** 373
API *See* American Petroleum Institute
API GL-4
extreme pressure performance requirements for hydraulic fluids **A18:** 99
API gravity (API degree)
defined **A18:** 3
API pipe steels *See also* Steel pipe, specific types
compositions **A9:** 211
API Specification for line pipe **A9:** 210
API specifications
for steel tubular products **A1:** 328, 329, 330, 332
API Specifications for oil country tubulars **A9:** 210
API specifications, steel pipe **M1:** 317, 319, 321, 322
Aplanatic
defined **A9:** 2 **A10:** 669
Aplite (K, Na, Ca, Mg, alumina silicate) **EM4:** 379
purpose for use in glass manufacture **EM4:** 381
Apochromatic lens
defined **A10:** 669
Apochromatic objective
defined **A9:** 2
Apochromatic objective lenses **A9:** 73
Apparatus
recommended practices **A7:** 249
Apparent (effective) permeability
in magnetic particle inspection **A17:** 99
Apparent area of contact *See also* Hertzian contact area; Nominal contact area
defined **A8:** 1 **A18:** 3
Apparent creep modulus
phenolics **EM2:** 244
Apparent density **A7:** 272-275
of atomized aluminum powder **A7:** 129
change with particle shape **A7:** 188, 189
change with particle size in packing **A7:** 296
control by high-energy milling **A7:** 69
of copper powders, effect of acid concentration **A7:** 113
defined **A7:** 1 **A10:** 669
effect of particle size distribution **A7:** 297
effect of production method **A7:** 297
effect of stainless steel mixture on **A7:** 273
effect on atomized copper powder **A7:** 118
effect on powder compact **A7:** 211
of electrolytic copper powders **A7:** 113, 114
factors affecting **A7:** 272-273
and flow rate **A7:** 273
and green strength **A7:** 288, 289
of low-alloy steel powder **A7:** 102
of magnesium powder **A7:** 131, 132
Apparent density (continued)
measurement equipment **A7:** 273-275
in milling of single particles **A7:** 59
as packed density **A7:** 297
and tap density, compare **A7:** 297
theoretical, and flow rate **A7:** 280
tin powders **A7:** 123, 124
Apparent hardness *See also* Hardness
of copper-based P/M materials **A7:** 470
defined **A7:** 1
effect of sintering temperature on **A7:** 367
of ferrous P/M materials **A7:** 465, 466
and Microhardness **A7:** 489
Apparent pore volume
defined **A7:** 1
Apparent porosity
defined **EL1:** 1134
Apparent resistance *See* Resistance, apparent
Apparent signal line impedance
defined **EL1:** 37
Apparent viscosity, defined *See* under Viscosity. Applications, composite **EM1:** 799-847
aircraft industry **EM1:** 801-809
automotive **EM1:** 832-836
commercial **EM1:** 832-836, 845-847
continuous SiC fiber MMCs **EM1:** 864-865
and experience **EM1:** 799-847
flight service **EM1:** 823-831
glass fibers **EM1:** 107
glass rovings **EM1:** 109
high-temperature **EM1:** 810-815
introduction **EM1:** 799
long-term environmental effects in **EM1:** 823-831
marine **EM1:** 837-844
space and missile systems **EM1:** 816-822
sports and recreational equipment **EM1:** 845-847
Appearance
aluminum and aluminum alloys **A2:** 3
of conformal coatings **EL1:** 765
of metal castings **A15:** 40, 762-763
Appearance standards
for weldments **A17:** 591
Appliance applications
acrylonitrile-butadiene-styrenes (ABS) **EM2:** 111
blow molding **EM2:** 359
high-impact polystyrenes (PS, HIPS) **EM2:** 195
homopolymer/copolymer acetals **EM2:** 100-101
of part design **EM2:** 616
phenolics **EM2:** 243
polycarbonates (PC) **EM2:** 151
polyphenylene ether blends (PPE PPO) **EM2:** 183
polyphenylene sulfides (PPS) **EM2:** 186
polyurethanes (PUR) **EM2:** 259
reinforced polypropylenes (PP) **EM2:** 192-193
styrene-acrylonitriles (SAN, OSA ASA) **EM2:** 215
Appliances
P/M parts for **A7:** 622-623, 736
P/M self-lubricating bearings in **A7:** 705
use of stainless steels **A7:** 730-732
Applicable load **A18:** 511
Application methods **EM3:** 4, 36
adhesives **EL1:** 1046
fluoropolymer coatings **EL1:** 783-784
of parylene coatings **EL1:** 797-800
Application pressure ratio **A18:** 605
Application(s) *See* specific metals and alloys
bottom-brazed flatpacks **EL1:** 992
of composites in packages **EL1:** 1126-1128
environmental stress screening **EL1:** 876-877
-related failure mechanisms **EL1:** 974-975
of welding processes **EL1:** 1041
Application-specific integrated circuits (ASICS)
ceramic hybrids using **EL1:** 297, 298
as new digital IC class **EL1:** 161
Applications *See also* End uses; specific metals and alloys **EM3:** 44-47, 76-77
ablative barriers, sealants **EM3:** 605
acetal , anaerobics **EM3:** 79
acrylic windows bonding, cyanoacrylates **EM3:** 128
acrylics **EM2:** 104
acrylonitrile-butadiene-styrene (ABS) bonding acrylics **EM3:** 121
acrylonitrile-butadiene-styrenes (ABS) **EM2:** 110-111
Applications (continued)
adhesion promoter in water-base coatings, urethanes **EM3:** 111
adhesive bonding of aircraft canopies and
silicones **EM3:** 563-564
thermoplastics **EM3:** 563
windshields
advanced aircraft, polybenzimidazoles **EM3:** 171
aerodynamic smoothing compounds
polysulfides **EM3:** 194
sealants **EM3:** 58, 677
silicones **EM3:** 59
aerospace **EM3:** 558-566
epoxies **EM3:** 97
polyether silicones **EM3:** 230
silicone sealants **EM3:** 192
aerospace and automotive gaskets, fluoroelastomer sealants **EM3:** 227
aerospace and automotive packing materials, fluoroelastomer sealants **EM3:** 227
aerospace and automotive seals, fluoroelastomer sealants **EM3:** 227
aerospace honeycomb core construction
epoxies **EM3:** 560
phenolics **EM3:** 560
aerospace honeycomb sandwich construction, epoxies **EM3:** 560
aerospace industry
phenolics **EM3:** 105
polysulfides **EM3:** 193, 194, 196
primers **EM3:** 255
silicones **EM3:** 218, 220
syntactics **EM3:** 176
aerospace missile radome-bonded joint, epoxy **EM3:** 564
aerospace protective coating, potential in polyphenylquinoxalines **EM3:** 164
aerospace secondary honeycomb sandwich bonding **EM3:** 561
aerospace structural parts adhesive, polyimides **EM3:** 151
aerospace window sealant silicones (RTV) **EM3:** 220
air conditioner sealing
asphalt **EM3:** 59
polypropylenes **EM3:** 59
solvent acrylics **EM3:** 209
air conditioning heater components, EPDM **EM3:** 57
air filter gasket formation, plastisols **EM3:** 46
aircraft
adhesive-bonded joints **EM3:** 743-745
epoxy-phenolics **EM3:** 80
integral fuel tanks **EM3:** 50
aircraft and aerospace industries, high-temperature adhesives **EM3:** 80
aircraft and device encapsulation, silicones **EM3:** 677
aircraft assembly **EM3:** 41, 52
phenolics **EM3:** 79
aircraft construction with improved heat, chemical, and solvent resistance, fluoroelastomer sealants **EM3:** 226, 227
aircraft engines, epoxies **EM3:** 97
aircraft fuel tanks, Buna-N **EM3:** 604
aircraft honeycomb acoustic panels **EM3:** 751
aircraft inspection plates
fluorosilicones **EM3:** 604
polysulfides **EM3:** 604
aircraft interior assembly
epoxies **EM3:** 97
neoprenes **EM3:** 44
nitriles **EM3:** 44
aircraft, investment casting **A15:** 265
aircraft skins, phenolics **EM3:** 105
aircraft structural parts and repair, polyimides **EM3:** 160
aircraft wing **EM3:** 36
aircraft/aerospace industry, structural adhesives **EM3:** 44
airfield joint sealants, asphalts and coal tar resins **EM3:** 51
for alloy cast irons **A1:** 103-104
alloy steels **M1:** 455, 470
alloys and blends **EM2:** 491
allyls (DAP, DAIP) **EM2:** 226-227

Applications (continued)
alpha-particle memory chip protection, polyimides EM3: 161
aluminum aircraft repair and primary bonding, acrylics EM3: 125
aluminum aircraft structure repairs ... EM3: 801-819
aluminum alloys A15: 743, 768-769 M2: 46-49
aluminum boats, acrylics EM3: 125
aluminum bonding
acrylics EM3: 121, 122
polysulfides EM3: 138
aluminum construction use, acrylics EM3: 125
aluminum skin bonding to aircraft bodies, epoxies EM3: 97
aluminum window and door bonding, acrylics EM3: 119
of aluminum-silicon castings A15: 159
aminos EM2: 230, 628
appliance seal and gasket applications, polypropylenes EM3: 51
appliance sealing
butyl rubber EM3: 146
butyls EM3: 200
silicones EM3: 192
appliances, water-base adhesives EM3: 87
architectural A15: 18, 20-22
armature coil sealing and bonding epoxies EM3: 612
art material cement, natural rubber EM3: 145
assembly bonding, hot-melt adhesives EM3: 82
assembly of electronic connectors and circuit boards EM3: 707
assistance bonding, hot-melt adhesives EM3: 82
attaching stone, metal, or glass to a building side EM3: 188
austempered steel M4: 105, 107-109
for austenitic manganese steels A1: 822 A15: 735
for austenitic stainless steels A1: 948-949
auto battery casings, polypropylene EM3: 58
auto body sealant
butyl EM3: 57
nitrile sealants EM3: 57
styrene-butadiene rubber (SBR) EM3: 57
auto body sealants and glazing materials
gilsonites EM3: 57
plastisols EM3: 57
auto brake and clutch linings, acrylonitrile-butadiene rubber EM3: 148
auto glass seal, butyl rubber EM3: 146
auto headliners, styrene-butadiene rubber EM3: 147
auto tire tread cracking, viscoelastic structural adhesives EM3: 513
auto transmission blades, phenolics EM3: 105
automobile interior seam sealing, plastisol sealants EM3: 720
automobile valve stem seals, fluoroelastomer sealants EM3: 227
automobile windshield sealing, urethane sealants EM3: 204
automotive A15: 32-34, 44, 327, 334-336, 691 EM3: 551-557
polyether silicones EM3: 230
polyurethanes EM3: 111
pressure-sensitive adhesives EM3: 83, 84
UV/EB-cured adhesives EM3: 91
automotive body assembly bonding, epoxies EM3: 609
automotive body sheet EM3: 41
automotive bonding and sealing
hot melts EM3: 609
vinyl plastisols EM3: 609
automotive brakeshoes, phenolics EM3: 79
automotive carpet bonding, hot melts EM3: 46
automotive circuit devices
epoxies EM3: 610
urethanes EM3: 610
automotive crash damage repair, epoxies EM3: 556-557

Applications (continued)
automotive decorative trim EM3: 551-552
acrylate pressure-sensitive adhesives EM3: 552
elastomeric adhesives EM3: 552
poly(butyl acrylate) adhesives EM3: 552
polychloroprene rubber EM3: 149
automotive electronic tacking and sealing EM3: 610
automotive electronics bonding
acrylates EM3: 553
epoxy resins EM3: 553
silicones EM3: 553
urethanes EM3: 553
automotive fabric adhesive EM3: 551-552
automotive filter element bonding, urethanes EM3: 110
automotive gasketing EM3: 547
anaerobics EM3: 553
silicones (RTV) EM3: 220
automotive industry
butyls EM3: 200
carbon black EM3: 178
phenolics EM3: 105
polybenzimidazoles EM3: 171
primers EM3: 255
silicone sealants EM3: 192
automotive interior trim bonding, natural rubber EM3: 145
automotive padding and insulation, styrene-butadiene rubber EM3: 147
automotive repair, urethane sealants EM3: 204
automotive sound absorption, elastomeric adhesives EM3: 555-556
automotive structural bonding
acrylates EM3: 554
epoxies EM3: 554-555
urethanes EM3: 554
automotive threadlocking, anaerobics EM3: 553
automotive under-hood elastomers EM3: 52
automotive valve sealing, elastomeric sealants EM3: 547
automotive windshield bonding, urethane EM3: 723-724
back lights (auto)
butyls EM3: 50
polyurethanes EM3: 554
battery construction, epoxies EM3: 45
battery separators, polybenzimidazoles EM3: 169
belt lamination, polychloroprene rubber EM3: 149
belting, natural rubber EM3: 145
binding electrical, industrial, and decorative laminates EM3: 105
biomedical/dental area
acrylics EM3: 76
cyanoacrylates EM3: 76
inorganics EM3: 76
modified epoxies EM3: 76
boat construction, acrylics EM3: 125
body assembly, epoxies EM3: 553, 554
epoxies EM3: 553, 554
reactive plastisol adhesives EM3: 554
urethanes EM3: 554
body assembly sealant, epoxies EM3: 608
epoxies EM3: 608
polysulfides EM3: 609
silicones (RTV) EM3: 609
urethanes EM3: 608, 609
body sealants and glazing materials
butyl EM3: 57
ethylene-vinyl acetate EM3: 57
nitrile mastics EM3: 57
polyurethanes EM3: 57
styrene-butadiene rubber (SBR) EM3: 57
body seam (auto) sealant
hot melts EM3: 46
vinyl plastisols EM3: 51
bonded aircraft, aluminum EM3: 42

Applications (continued)
bonded auto lens assembly, thermoset polyester EM3: 575
bonding aluminum oxide and silicon carbide, phenolics EM3: 105
bonding coated abrasives, phenolics (resols) EM3: 105
bonding composites, polyimides EM3: 160
bonding decorative wall panels, latex EM3: 577
bonding gasketing, acrylonitrile-butadiene rubber EM3: 148
bonding honeycomb to primary surfaces, epoxy EM3: 563
bonding of electrical wires and devices
conductive adhesives EM3: 45
semiconductive adhesives EM3: 45
bonding to paper of wood, metals, plastics, and metal, phenolics EM3: 105
bonding together galvanized steel sheets EM3: 577
book binding
hot-melt adhesives EM3: 82
styrene-butadiene rubber EM3: 147
brake blocks, phenolics EM3: 105
brake linings, phenolics EM3: 105, 639
brake shoe bonding EM3: 551, 552
brick and stone masonry pointing, solvent acrylics EM3: 209
bronze bonding, acrylics EM3: 122
building and highway expansion joints, silicones (RTV) EM3: 218-219
building construction, epoxies EM3: 97
cabin pressure sealing
polyurethanes EM3: 59
sealants EM3: 58
canopy sealing, sealants EM3: 605
for carbon and low-alloy steel sheet and strip A1: 200
carbon bonding, polysulfides EM3: 138
carbon steels M1: 455, 457-459, 470
carburized steels M1: 626
carton closing, butyl rubber EM3: 146
catheters bonding, cyanoacrylates EM3: 128
caulking
acrylics EM3: 607-608
butyl EM3: 607-608
silicones EM3: 607
urethanes EM3: 607-608
cell, automated A15: 568-569
ceramic and tile bonding compounds, styrene-butadiene rubber EM3: 147
ceramic bonding, polysulfides EM3: 138
ceramic molding A15: 248
channel sealing, sealants EM3: 677
chemical gasketing
anaerobics EM3: 609, 610
silicones (RTV) EM3: 609
chipboard construction, phenolics EM3: 105
circuit protection
acrylics EM3: 592
epoxies EM3: 592
glass-base systems EM3: 592
parylene EM3: 592
polyimides EM3: 592
silicones EM3: 592
UV-curable acrylics EM3: 592
circulation pumps A15: 456
civil engineering, polysulfides EM3: 194
CLA process A15: 317-318
CLAS process A15: 319
cloth bonding, phenolics EM3: 105
clutch disks, phenolics EM3: 79
clutch facings, phenolics EM3: 105, 639
coated and bonded abrasives, phenolics EM3: 105
coating and encapsulation
benzocyclobutene EM3: 580
block copolymers EM3: 580
epoxy EM3: 580

SUBJECTS OF THE INDEXED VOLUMES: ASM Handbook (designated by the letter "A"): **A1:** Properties and Selection: Irons, Steels, and High-Performance Alloys (1990); **A2:** Properties and Selection: Nonferrous Alloys and Special-Purpose Materials (1990); **A3:** Alloy Phase Diagrams (1992); **A4:** Heat Treating (1991); **A6:** Welding, Brazing, and Soldering (1993); **A7:** Powder Metallurgy (1984); **A8:** Mechanical Testing (1985); **A9:** Metallography and Microstructures (1985); **A10:** Materials Characterization (1986); **A11:** Failure Analysis and Prevention (1986); **A12:** Fractography (1987); **A13:** Corrosion (1987); **A14:** Forming and Forging (1988); **A15:** Casting (1988); **A16:** Machining (1989); **A17:** Nondestructive Testing and Quality Control (1989); **A18:** Friction, Lubrication, and Wear Technology (1992). **Metals Handbook, 9th Edition** (designated by the letter "M"): **M1:** Properties and Selection: Irons and Steels (1978); **M2:** Properties and Selection: Nonferrous Alloys and Pure Metals (1979); **M3:** Properties and Selection: Stainless Steels, Tool Materials and Special-Purpose Materials (1980); **M4:** Heat Treating (1981); **M5:** Surface Cleaning, Finishing, and Coating (1982); **M6:** Welding, Brazing, and Soldering (1983). **Engineered Materials Handbook** (designated by the letters "EM"): **EM1:** Composites (1987); **EM2:** Engineering Plastics (1988); **EM3:** Adhesives and Sealants (1990); **EM4:** Ceramics and Glasses (1991); **Electronic Materials Handbook** (designated by the letters "EL"): **EL1:** Packaging (1989).

Applications (continued)
parylenes ... **EM3:** 580
polyimides ... **EM3:** 580
silicone-polyimides ... **EM3:** 580
silicones ... **EM3:** 580
sol-gels ... **EM3:** 580
urethanes ... **EM3:** 580
for cobalt-base superalloys ... **A1:** 965, 967-968
cold finished bars ... **M1:** 220, 221
commercial tape foundation, natural rubber ... **EM3:** 145
for compacted graphite iron ... **A1:** 70
component carrier tapes, silicones ... **EM3:** 134
component covers
fluorosilicones ... **EM3:** 611
silicones ... **EM3:** 611
component terminal sealant
acrylics ... **EM3:** 612
silicones ... **EM3:** 612
urethanes ... **EM3:** 612
composite panel assembly or repair, epoxies ... **EM3:** 98, 100
composite primary aircraft components, MB 329 adhesive ... **EM3:** 821
composite repair, epoxies ... **EM3:** 97
compression packings, fluorocarbon sealants ... **EM3:** 223, 225-226
computer, cupola ... **A15:** 384
concrete and mortar patch repair, polysulfides ... **EM3:** 38
concrete bonding, epoxies ... **EM3:** 96, 101
concrete crack repair, polysulfides ... **EM3:** 138
conductive sealants ... **EM3:** 604-605
construction industry ... **EM3:** 76
polyether silicones ... **EM3:** 229
polysulfides ... **EM3:** 193-194, 196
pressure-sensitive adhesives ... **EM3:** 83, 84
sealants ... **EM3:** 605-608
silicone sealants ... **EM3:** 192
silicones ... **EM3:** 218-220
styrene-butadiene rubber ... **EM3:** 147
urethanes ... **EM3:** 606, 607
water-base adhesives ... **EM3:** 87, 88
construction joints sealants ... **EM3:** 605
consumer do-it-yourself silicone sealants ... **EM3:** 192
consumer goods, water-base adhesives ... **EM3:** 88
contact bonds, polychloroprene rubber ... **EM3:** 149
containers, hot-melt adhesives ... **EM3:** 82
contour pattern flocking, urethanes ... **EM3:** 110
contraction joints, sealants ... **EM3:** 605
control joints, solvent acrylics ... **EM3:** 209
copper alloy castings ... **A15:** 771, 782-785
copper bonding acrylics ... **EM3:** 122
copper metals ... **M2:** 466
copper tubular products ... **M2:** 261, 262
of core coatings ... **A15:** 240-241
cork and rubber gasket sealing, rosin sealant ... **EM3:** 57-58
corrosion protection
pressure-sensitive adhesives ... **EM3:** 83, 84
sealants ... **EM3:** 604
corrosion-inhibiting sealants ... **EM3:** 194
cure-in-place gasketing (CIPG) sealants ... **EM3:** 548
curtain wall construction
polysulfide ... **EM3:** 549
polyurethane ... **EM3:** 549
silicone ... **EM3:** 549
CV process ... **A15:** 318
cyanates ... **EM2:** 232-234
cylindrical parts fitting, anaerobics ... **EM3:** 115, 116
cylindrical parts retaining, anaerobics ... **EM3:** 115
decals ... **EM3:** 41
deck sealing, plastisols ... **EM3:** 46
design guidelines ... **EM2:** 710
dialysis membranes bonding, cyanoacrylates ... **EM3:** 128
diaper construction, hot-melt adhesives ... **EM3:** 47
die attach adhesives, polyimides ... **EM3:** 159, 161
die attach in device packaging
epoxies ... **EM3:** 584
polyimides ... **EM3:** 584
thermoplastics ... **EM3:** 584

Applications (continued)
thermosets ... **EM3:** 584
die attachment and interconnection
epoxies ... **EM3:** 580
polyimides ... **EM3:** 580
thermoplastics ... **EM3:** 580
thermosets ... **EM3:** 580
dielectric interlayer adhesion, polyimides ... **EM3:** 161
dielectric sealants
acrylics ... **EM3:** 611
silicones ... **EM3:** 611
silicones (RTV) ... **EM3:** 611
urethanes ... **EM3:** 611
dielectrics, polymers ... **EM3:** 378
differential cover sealing, elastomeric sealants ... **EM3:** 547
dip coating ... **EM3:** 36
disk pads, phenolics ... **EM3:** 105
display panels bonding, cyanoacrylates ... **EM3:** 28
domestic households, polybenzimidazoles ... **EM3:** 171
drip rail steel molding urethane sealants ... **EM3:** 204
for ductile iron ... **A1:** 35-38 **M1:** 34-35, 36, 47
of ductile iron castings ... **A15:** 665
electrical ... **EM3:** 36
electrical contact assemblies
epoxies ... **EM3:** 611
polyesters ... **EM3:** 611
silicones ... **EM3:** 611
urethanes ... **EM3:** 611
electrical insulator protection
epoxies ... **EM3:** 612
epoxy urethanes ... **EM3:** 612
silicones ... **EM3:** 612
electrical, pressure-sensitive adhesives ... **EM3:** 83, 84
electrical/electronics
acrylics ... **EM3:** 44
cyanoacrylates ... **EM3:** 44
epoxies ... **EM3:** 44
polyamide hot melts ... **EM3:** 44
polyimides ... **EM3:** 44
polyurethanes ... **EM3:** 44
silicones ... **EM3:** 44
electrically conductive bonding
acrylates ... **EM3:** 572
epoxies ... **EM3:** 572, 573
silicones ... **EM3:** 572
urethanes ... **EM3:** 572
electrically conductive sealants, polysulfides ... **EM3:** 194
electron beam melting ... **A15:** 414
electronic components potting and encapsulation ... **EM3:** 48
electronic general component bonding
acrylated-urethanes ... **EM3:** 573
acrylates ... **EM3:** 573
anaerobics ... **EM3:** 573
cyanoacrylates ... **EM3:** 573
epoxies ... **EM3:** 573
methacrylates ... **EM3:** 573
silicones ... **EM3:** 573
thermoplastic hot melts ... **EM3:** 573
electronic insulation urethanes ... **EM3:** 110
electronics
butyls ... **EM3:** 201
polybenzimidazoles ... **EM3:** 171
UV/EB-cured adhesives ... **EM3:** 91
water-base adhesives ... **EM3:** 87
electronics industry, silicones ... **EM3:** 218, 220
electronics market ... **EM3:** 76
electronics/microelectronics end-uses, polyimides ... **EM3:** 151
for electrostatic flocking of plastics, textiles, or rubbers, urethanes ... **EM3:** 110, 111
elevated-temperature, CG irons ... **A15:** 675-676
encapsulation of electronic and electrical components ... **EM3:** 50, 52
epoxies ... **EM3:** 51
engine gasketing, RTV silicones ... **EM3:** 553
engine sealant
anaerobic polymers ... **EM3:** 609

Applications (continued)
sealants ... **EM3:** 609
engineering change order (ECO) wire tacking, cyanoacrylates ... **EM3:** 569, 572
of engineering plastics ... **EM2:** 1, 68-73
environmental barriers, sealants ... **EM3:** 608, 609
environmental seals, sealants ... **EM3:** 604
epoxies ... **EM2:** 240-241
equipment cabinet ... **EM3:** 36
equipment enclosures and housings, acrylics ... **EM3:** 125
expansion joint sealing, polyurethanes ... **EM3:** 204
expansion joints, sealants ... **EM3:** 605, 607, 608
exterior horizontal and vertical control and expansion joints, sealants ... **EM3:** 56
exterior horizontal paving or traffic-bearing joints ... **EM3:** 56
exterior joints in water/chemical retainment structures ... **EM3:** 56
exterior mirrors bonding, silicones ... **EM3:** 553
exterior mullion and curtain wall panel joints ... **EM3:** 56
exterior panel joints, solvent acrylics ... **EM3:** 209
exterior perimeter joints ... **EM3:** 56
exterior pointing of brick and stone masonry ... **EM3:** 56
exterior roof joints (parapets, reglets, flashings, pipes, ducts, vent openings) ... **EM3:** 56
exterior sealants in buildings ... **EM3:** 49
exterior sealing of glass into window frames ... **EM3:** 56
exterior seals around windows, doors, and
acrylics ... **EM3:** 56
latex caulks ... **EM3:** 56
sidings
silicones ... **EM3:** 56
exterior seals for roofing penetrations
acrylics ... **EM3:** 56
latex caulks ... **EM3:** 56
silicones ... **EM3:** 56
exterior seals on aluminum and vinyl siding, butyl caulks ... **EM3:** 56
exterior use in internal seals in window framing systems ... **EM3:** 56
extruded gaskets
EDPM ... **EM3:** 58
neoprenes ... **EM3:** 58
F-111 flaps bonding ... **EM3:** 762
fabricated structures, polybenzimidazoles ... **EM3:** 169
fabrication of aircraft honeycomb structures, acrylonitrile-butadiene rubber ... **EM3:** 148
faying surfaces ... **EM3:** 48
acrylics ... **EM3:** 604
fluorosilicones ... **EM3:** 604
polysulfides ... **EM3:** 604
polyurethanes ... **EM3:** 604
rubber compounds ... **EM3:** 604
sealants ... **EM3:** 604
silicones (RTV) ... **EM3:** 604
ferrite cores fixturing, cyanoacrylates ... **EM3:** 128
fiber optics, UV/EB-cured adhesives ... **EM3:** 91
fiber-reinforced plastic (FRP) bonding, polysulfides ... **EM3:** 141, 142
fiberglass insulation bonding, hot melts ... **EM3:** 45
fiberglass molds potting, cyanoacrylates ... **EM3:** 128
fillets
acrylics ... **EM3:** 604
fluorosilicones ... **EM3:** 604
polysulfides ... **EM3:** 604
polyurethanes ... **EM3:** 604
rubber compounds ... **EM3:** 604
silicones (RTV) ... **EM3:** 604
filling unitized steel shells, polyurethanes ... **EM3:** 577
filter caps fixturing, cyanoacrylates ... **EM3:** 128
filter sealants and bonding agents, plastisols ... **EM3:** 46
fingerprint development, cyanoacrylates ... **EM3:** 127
fire stops
latex sealants ... **EM3:** 177
silicone ... **EM3:** 177

Applications (continued)
solvent-base sealants **EM3:** 177
urethanes **EM3:** 177
fire-resistant aircraft seats, polybenzimidazoles **EM3:** 169
flame hardening **M4:** 484
flame-retardant fire stop sealants, silicones (RTV) **EM3:** 219-220
flange sealing, anaerobics **EM3:** 116
flanges, fluorocarbon sealants **EM3:** 223, 224
flexible packaging, urethanes **EM3:** 110
flexible printed wiring board substrate material, polyimides **EM3:** 161
flexible printed wiring boards (PWBs)
acrylics **EM3:** 45
polyester film **EM3:** 45
polyimides **EM3:** 45
flight suits, polybenzimidazoles **EM3:** 169
flow brush **EM3:** 36
flow gun **EM3:** 36
fluorosilicones **EM3:** 50
foam and fabric laminations, urethanes **EM3:** 110
foam sponges, urethanes **EM3:** 110
foams, polybenzimidazoles **EM3:** 169
footwear, natural rubber **EM3:** 145
forgings **M1:** 349, 350, 354, 357, 369, 371-373
form-in-place gasketing (FIPG) **EM3:** 547-548
formed-in-place gaskets silicone rubber **EM3:** 57
foundry and shell moldings, phenolics **EM3:** 105
foundry sand patterns, phenolic resin binders **EM3:** 47
freezer coils bonding, hot melts **EM3:** 45
friction materials, phenolics **EM3:** 105
fuel containment, silicone sealants **EM3:** 192
fuel tank sealants
polysulfides **EM3:** 194
sealants **EM3:** 677
fuel tank structures (airframe) bonding, polysulfides **EM3:** 195
furniture **EM3:** 41
furniture assembly **EM3:** 41
polychloroprene rubber **EM3:** 149
structural adhesives **EM3:** 46
galvanized steel bonding
acrylics **EM3:** 60, 121, 123-124
polysulfides **EM3:** 141
galvanized steel G-60 bonding, polysulfides **EM3:** 142
gas tank seam sealant
Buna-N **EM3:** 610
fluorosilicones **EM3:** 610
polysulfides **EM3:** 610
urethanes **EM3:** 610
gasket bonding **EM3:** 707
gasket dressings **EM3:** 51
gasket fixturing, cyanoacrylates **EM3:** 128
gasket replacements, silicones **EM3:** 547
gasketing **EM3:** 51
RTV silicones **EM3:** 54
gears to shaft bonding, cyanoacrylates **EM3:** 128
gearshift indicators bonding, cyanoacrylates **EM3:** 128
general glazing for weatherproofing, silicones (RTV) **EM3:** 219
general industrial sealant, silicones (RTV) **EM3:** 220-221
glass bonding
epoxies **EM3:** 96, 101
polysulfides **EM3:** 138, 141, 142
glass unit insulation, polysulfides **EM3:** 194-195, 196
glass-phenolic laminate, phenolics **EM3:** 107
glass-to-mullion joints, solvent acrylics **EM3:** 209
glazing of insulating glass panels, solvent acrylics **EM3:** 209
glazing, sealants **EM3:** 606
gold in dentistry **M2:** 684-687

Applications (continued)
grain refinement **A15:** 477-478
granite construction, polyurethanes **EM3:** 607
gray cast iron **M1:** 11, 16-19, 22-24, 26, 30
for gray iron **A1:** 12
of gray irons **A15:** 644-645
grouting compounds, polysulfides **EM3:** 138
gutter repair, urethane sealants **EM3:** 204
gyroscope bearings **EM3:** 37
handle assemblies bonding
epoxies **EM3:** 45
polyurethanes **EM3:** 45
hardboard construction, phenolics **EM3:** 105
hardenable steels **M1:** 457-459, 470
hardware to printed circuit boards, cyanoacrylates **EM3:** 128
headliner adhesives, thermoplastic hot-melt adhesives **EM3:** 552
heat sinks fixturing, cyanoacrylates **EM3:** 128
heat-protective apparel, polybenzimida-zoles **EM3:** 169
heating and air conditioning duct sealing and
acrylics **EM3:** 610
bonding
butyls **EM3:** 610
silicones (RTV) **EM3:** 610
urethanes **EM3:** 610
heel filler beads for glazing, solvent acrylics **EM3:** 209
helicopter blade bonding **EM3:** 762
helicopter rotor blades **EM3:** 33
hem flange bonding of car door **EM3:** 48
hemmed flange bonding, vinyl adhesives **EM3:** 553-554
hermetic packaging, epoxies **EM3:** 587
high-chromium white irons **A15:** 684-685
high-density polyethylenes (HDPE) **EM2:** 163-165
high-impact polystyrenes (PS HIPS) **EM2:** 194-196
high-performance military aircraft, polysulfides **EM3:** 195
high-silicon irons **A15:** 699, 701
high-speed application machinery **EM3:** 35
for high-temperature bearing steels **A1:** 384-386
high-temperature, dynamic-type sealing, fluoroelastomer sealants **EM3:** 226
high-temperature insulation and seals for plasma spray masking, silicones **EM3:** 134
high-temperature sealing of aircraft assemblies, silicones **EM3:** 808
high-temperature wire insulation, potential in polyphenylquinoxalines **EM3:** 164
highway construction
epoxies **EM3:** 97
silicone sealants **EM3:** 192
highway construction joints
asphalts and coal tar resins **EM3:** 51, 57
neoprene rubber **EM3:** 57
silicones **EM3:** 57
tar **EM3:** 57
urethanes **EM3:** 57
homopolymer/copolymer acetals **EM2:** 100-101
honing stones bonding, cyanoacrylates **EM3:** 128
hood sealing, plastisols **EM3:** 46
horizontal centrifugal casting **A15:** 300
horizontal joints, urethane sealants **EM3:** 204
horizontal stabilizer stub box test, graphite composite **EM3:** 539
horizontal stabilizers for aircraft, composites **EM3:** 533, 535
hose and belt manufacture, acrylonitrile-butadiene rubber **EM3:** 148
hose and belting production, polychloroprene rubber **EM3:** 149
hoses, natural rubber **EM3:** 145
hospital/first aid, pressure-sensitive adhesives **EM3:** 83, 84

Applications (continued)
hot pressing weather-resistant plywood, phenolics
hot rolled bars and shapes **M1:** 206-209, 211-213
household cookware coatings **EM3:** 172
household, epoxies
housewares, water-base adhesives **EM3:** 87
housing assemblies bonding
epoxies **EM3:** 45
polyurethanes **EM3:** 45
for HSLA steels **A1:** 399, 415-423 **M1:** 403, 405-406
hybrid circuit encapsulants, polyimides **EM3:** 159
hybrid circuit protective overcoat, polyimides **EM3:** 161
in-house glazing
acrylics **EM3:** 58
silicones **EM3:** 58
industrial bonding, hot-melt adhesives **EM3:** 82
industrial gaskets, fluorocarbon sealants **EM3:** 223, 224, 225, 226
industrial sealants
butyls **EM3:** 58
fluorocarbons **EM3:** 223
thermoplastics **EM3:** 58
insulated double-pane window construction
polysulfides **EM3:** 46
polyurethanes **EM3:** 46
silicones **EM3:** 46
thermoplastics **EM3:** 46
insulated glass construction
butyls **EM3:** 58
polyisobutylenes **EM3:** 58
polysulfides **EM3:** 58
polyurethanes **EM3:** 58
silicones **EM3:** 58
insulated glass window sealant **EM3:** 50
butyl rubber **EM3:** 146
butyls **EM3:** 190, 200, 201
insulating tile bonding to space shuttle **EM3:** 41
insulation materials, phenolics **EM3:** 105
insulation of glass windows and doors, butyls **EM3:** 675
integral fuel tanks in aircraft **EM3:** 50
integral fuel tanks, sealants **EM3:** 58
integrated circuit dielectric films, polyimides **EM3:** 161
interior duct, pipe, electrical penetration joints **EM3:** 56
interior fire stops **EM3:** 56
interior floor joints **EM3:** 56
interior horizontal and vertical control and expansion joints **EM3:** 56
interior perimeter joints **EM3:** 56
interior plaster and trim joints **EM3:** 56
interior sanitary fixture sealants **EM3:** 56
interlevel insulators, polymers **EM3:** 378
internal appliance gap sealing, silicones **EM3:** 59
internal seals of window and curtain wall
acrylics **EM3:** 177
butyls **EM3:** 177
silicones **EM3:** 177
systems
urethanes **EM3:** 177
of investment castings **A15:** 264-266
ionomers **EM2:** 120
jet engine firewall sealing, silicones **EM3:** 58, 59
jet engine lip seals, fluoroelastomer sealants **EM3:** 227
jet turbine blades, sealants **EM3:** 605
jet window sealing, silicones **EM3:** 59
joint sealing, butyls **EM3:** 675
joints
auto **EM3:** 51
building **EM3:** 51
joints between roadway concrete slabs **EM3:** 188
jumper wires fixturing, cyanoacrylates **EM3:** 128
label attachment **EM3:** 41, 707

SUBJECTS OF THE INDEXED VOLUMES: ASM Handbook (designated by the letter "A"): **A1:** Properties and Selection: Irons, Steels, and High-Performance Alloys (1990); **A2:** Properties and Selection: Nonferrous Alloys and Special-Purpose Materials (1990); **A3:** Alloy Phase Diagrams (1992); **A4:** Heat Treating (1991); **A6:** Welding, Brazing, and Soldering (1993); **A7:** Powder Metallurgy (1984); **A8:** Mechanical Testing (1985); **A9:** Metallography and Microstructures (1985); **A10:** Materials Characterization (1986); **A11:** Failure Analysis and Prevention (1986); **A12:** Fractography (1987); **A13:** Corrosion (1987); **A14:** Forming and Forging (1988); **A15:** Casting (1988); **A16:** Machining (1989); **A17:** Nondestructive Testing and Quality Control (1989); **A18:** Friction, Lubrication, and Wear Technology (1992). **Metals Handbook, 9th Edition** (designated by the letter "M"): **M1:** Properties and Selection: Irons and Steels (1978); **M2:** Properties and Selection: Nonferrous Alloys and Pure Metals (1979); **M3:** Properties and Selection: Stainless Steels, Tool Materials and Special-Purpose Materials (1980); **M4:** Heat Treating (1981); **M5:** Surface Cleaning, Finishing, and Coating (1982); **M6:** Welding, Brazing, and Soldering (1983). **Engineered Materials Handbook** (designated by the letters "EM"): **EM1:** Composites (1987); **EM2:** Engineering Plastics (1988); **EM3:** Adhesives and Sealants (1990); **EM4:** Ceramics and Glasses (1991); **Electronic Materials Handbook** (designated by the letters "EL"): **EL1:** Packaging (1989).

Applications (continued)
labels, pressure-sensitive adhesives **EM3:** 83, 84
laminate bonding, contact cements **EM3:** 577
laminated film packaging, urethanes **EM3:** 110
laminates
nylon-epoxies **EM3:** 78
phenolics **EM3:** 103, 105
polyimides **EM3:** 161
styrene-butadiene rubber **EM3:** 148
lamination of building panels, urethanes **EM3:** 110
laminations for furniture construction, urethanes **EM3:** 110
latex compounds in paper, natural rubber **EM3:** 145
lead **M2:** 495-498
lead bonding, acrylics **EM3:** 122
lead-free automotive body solder, polysulfides **EM3:** 138
leather bonding, phenolics **EM3:** 105
leather goods bonding, urethanes **EM3:** 110
leather shoe sole bonding
natural rubber **EM3:** 145
polychloroprene rubber **EM3:** 149
lens bonding
epoxies **EM3:** 575
methacrylates **EM3:** 575
polyurethanes **EM3:** 575
lighting subcomponent bonding
hot-melt thermoplastics **EM3:** 552
room-temperature-vulcanizing silicones **EM3:** 552
two-part urethanes **EM3:** 552
ultraviolet (UV)-curable acrylates **EM3:** 552
lightning strike, epoxy resins **EM3:** 565
liquid carburizing **M4:** 245-246, 247
liquid crystal display potting and sealing, UV-curable epoxies **EM3:** 612
liquid crystal polymers (LCP) **EM2:** 180-181
liquid membranes or sheet goods between concrete slabs, urethane sealants **EM3:** 204
localized metal reinforcement, epoxies **EM3:** 555
lock washer replacements, anaerobics **EM3:** 51
loud speaker assembly
cyanoacrylates **EM3:** 575
methacrylates **EM3:** 575
low-stress construction, butyls **EM3:** 577
machinery, water-base adhesives **EM3:** 87
magnesium **M2:** 525
magnesium alloy sand casting **A15:** 806
magnesium bonding, acrylics **EM3:** 122
magnet bonding for motors and speakers, free-radical reactive adhesives **EM3:** 45
magnetic media, UV/EB-cured adhesives **EM3:** 91
malleable iron **A15:** 690-691 **M1:** 57, 63, 70, 71, 73
for malleable irons **A1:** 73, 74, 83, 84
maraging steels **M1:** 451
marine, butyl rubber **EM3:** 146
butyl rubber **EM3:** 146
epoxies **EM3:** 97
martempered steel parts **M4:** 96, 97, 98
masking of printed circuit boards, silicones **EM3:** 134
medical **EM3:** 576
UV/EB-cured adhesives **EM3:** 91
medical adhesive to close wounds, cyanoacrylates **EM3:** 127
medical bonding (class IV approval)
cyanoacrylates **EM3:** 576
silicones **EM3:** 576
UV-curing methacrylates **EM3:** 576
UV-curing urethane-acrylates **EM3:** 576
membranes, polybenzimidazoles **EM3:** 169
metal bonding, phenolics **EM3:** 79
metal building construction
acrylics **EM3:** 57, 58
butyls **EM3:** 58, 177
latex **EM3:** 57
silicones **EM3:** 57
urethanes **EM3:** 57
metal casting **A15:** 37-45
metal finishing, UV/EB-cured adhesives **EM3:** 91

Applications (continued)
metal-to-metal bonding
acrylics **EM3:** 46
epoxies **EM3:** 46, 101
evaporation adhesives **EM3:** 75
polyimides **EM3:** 160
polyurethanes **EM3:** 46
microelectronics industry protective-coating, polyimides **EM3:** 160
microelectronics, polybenzimidazoles **EM3:** 172
mirror assemblies, silicones **EM3:** 553
missile assembly
epoxies **EM3:** 97
polybenzimidazoles **EM3:** 171-172
mobile home manufacture, butyl rubber **EM3:** 146
motor magnet bonding
methacrylates **EM3:** 574, 575
urethane-acrylates **EM3:** 574, 575
motorcycle construction, acrylics **EM3:** 125
mounting wires and circuitry, silicones **EM3:** 134
multichip modules, polyimides **EM3:** 160
multilayer packaging **EM3:** 41
nameplates bonding, cyanoacrylates **EM3:** 128
nickel bonding, acrylics **EM3:** 122
nickel-chromium white irons **A15:** 681
nonrigid bonding, water-base adhesives **EM3:** 87, 88
nonwoven goods binder, styrene-butadiene rubber **EM3:** 147-148
nylon, anaerobics **EM3:** 79
O-rings bonding **EM3:** 707
cyanoacrylates **EM3:** 128
O-rings sealing, fluoroelastomer sealants **EM3:** 227
off-the-road tires, natural rubber **EM3:** 145
office/graphic arts, pressure-sensitive adhesives **EM3:** 83, 84
oil recovery systems, polybenzimidazoles **EM3:** 171
opera window sealing, urethane sealants **EM3:** 204
orbital vehicles, polybenzimidazoles **EM3:** 172
organic junction coatings, silicones **EM3:** 587
oriented strand board (OSB) construction, phenolics **EM3:** 105
orthopedic devices, styrene-butadiene rubber **EM3:** 147
overwraps for plumbing repairs, polyester cloth **EM3:** 608
P/M materials **M1:** 327, 339
pacers fixturing, cyanoacrylates **EM3:** 128
pack carburizing **M4:** 105
package lamination, UV/EB-cured adhesives **EM3:** 91
packaging
animal glues **EM3:** 45
casein **EM3:** 45
dextrines **EM3:** 45
hot-melt adhesives **EM3:** 45, 82
isocyanates **EM3:** 45
polyvinyl and acrylic resin emulsions **EM3:** 45
pressure-sensitive adhesives **EM3:** 83, 84
rubber lattices **EM3:** 45
silicates **EM3:** 45
solvent cements **EM3:** 45
starch conversions **EM3:** 45
styrene-butadiene rubber **EM3:** 147
urethanes **EM3:** 110
water-base adhesives **EM3:** 87, 88
packaging large appliances in protective containers, hot melts **EM3:** 45
panel sealing, butyls **EM3:** 675
parking deck sealing, urethane sealing **EM3:** 204, 206
particle board construction
phenol formaldehyde **EM3:** 106
phenolics **EM3:** 105
of parts design **EM2:** 616-617
paving joint sealing, polyurethanes **EM3:** 204
phenolics **EM2:** 242-243
photoresists, polymers **EM3:** 378
photovoltaic cell construction, methacrylates **EM3:** 578
pipe flanges, anaerobics **EM3:** 547
plain carbon steels **A15:** 714

Applications (continued)
of plaster molding **A15:** 242
plastic appliance component bonding, acrylics **EM3:** 125
plastic film lamination, urethanes **EM3:** 110
plastic plumbing joint sealant, vinyl plastisols **EM3:** 51
plastics bonding
acrylics **EM3:** 125
epoxies **EM3:** 96
phenolics **EM3:** 105
polysulfides **EM3:** 138
plate **M1:** 181
platen seals
nitrile rubber **EM3:** 694
polytetrafluoroethylene (PTFE)-encapsulated silicone rubber **EM3:** 694
silicone rubber **EM3:** 694
plumbing repairs, epoxies **EM3:** 608
plumbing sealants, polyvinyl acetate **EM3:** 608
plywood **EM3:** 41
phenolics **EM3:** 79, 105
polysulfides **EM3:** 141, 142
polyamide- imides (PAI) **EM2:** 128
polyamides (PA) **EM2:** 125-126
polyarylates (PAR) **EM2:** 138-139
polyaryletherketones (PAEK, PEK PEEK, PEKK) **EM2:** 142
polybenzimidazoles (PBI) **EM2:** 147-148
polybutylene terephthalates (PBT) **EM2:** 153-154
polycarbonate lens bonding, acrylics **EM3:** 124
polycarbonates (PC) **EM2:** 151
polyether silicones (s-m RTV) **EM3:** 231-232
polyether sulfones (PES, PESV) **EM2:** 159-160
polyether-imides (PEI) **EM2:** 156
polyethylene terephthalates (PET) **EM2:** 172
polymer thick film, polyimides **EM3:** 161
polyolefins, anaerobics **EM3:** 79
polyphenylene ether blends (PPE PPO) **EM2:** 183-184
polyphenylene sulfides (PPS) **EM2:** 186
polystyrene bonding, acrylics **EM3:** 121
for polystyrene patterns, investment casting **A15:** 255
polysulfones (PSU) **EM2:** 200
polyurethanes (PUR) **EM2:** 258-260
polyvinyl chloride (PVC) bonding, acrylics **EM3:** 121
polyvinylidene chloride, anaerobics **EM3:** 79
porosity sealants for engine blocks
anaerobics **EM3:** 57
unsaturated polyesters **EM3:** 57
porosity sealing, anaerobics **EM3:** 115-116
porosity sealing in castings
sodium silicates **EM3:** 51
unsaturated polyesters **EM3:** 51
potting
epoxies **EM3:** 585
polyurethanes **EM3:** 585
silicones **EM3:** 585
for precipitation-hardening semiaustenitic stainless steels **A1:** 944
prefabricated construction, butyls **EM3:** 200
prefabricated metal buildings, butyl rubber **EM3:** 146
pressure sealants, polysulfides **EM3:** 194
printed board manufacture
acrylics **EM3:** 591
epoxies **EM3:** 591
epoxy-glass **EM3:** 589, 590
polyimide-glass **EM3:** 589, 590
printed circuit boards, polybenzimida-zoles **EM3:** 169
printed circuits **EM3:** 36
printed wiring boards
acrylonitrile-butadiene rubber **EM3:** 148
polyimide resins **EM3:** 569
product, squeeze casting **A15:** 323-327
protection of terminals and wiring connections, PVC liquid coating **EM3:** 611
pumps, fluorocarbon sealants **EM3:** 223
pushbuttons bonding, cyanoacrylates **EM3:** 128
for quenched and tempered martensitic stainless steels **A1:** 940-942
quick-repair sealants, polysulfides **EM3:** 194
radiation masks, polymers **EM3:** 378

Applications (continued)
reaction vessels, fluorocarbon sealants EM3: 223
recreational vehicle modifications
silicones (RTV) EM3: 610, 611
urethanes EM3: 610, 611
recreational vehicle sealant
acrylic latex EM3: 58
butyls EM3: 58
polybutene EM3: 58
refrigerator and freezer cabinet sealing
asphalt EM3: 59
polybutene EM3: 59
polypropylene EM3: 59
polyvinyl chloride (PVC) plastisol EM3: 59
silicones EM3: 59
refrigerator and freezer sealing EM3: 45
polybutene caulk EM3: 59
refrigerator cabinet seams, foam stop hot-melt adhesive EM3: 45
refrigerator component sealing, PVC plastisol EM3: 59
reinforced polypropylenes (PP) EM2: 192-193
release coatings, UV/EB-cured adhesives EM3: 91
release-coated paper, hot-melt adhesives EM3: 82
replacement glass sealant, butyls EM3: 200
Replicast process A15: 37, 271-272
reservoirs and canals, sealants EM3: 188
residential construction
acrylic sealants EM3: 188
skinning butyls EM3: 188
retread and tire patch compounds, natural rubber EM3: 145
rigid bonding, water-base adhesives EM3: 88
rivet heads, sealants EM3: 604
rivet holes in an aluminum fuselage structure EM3: 762
rivets
acrylics EM3: 604
fluorosilicones EM3: 604
polyurethanes EM3: 604
rubber compounds EM3: 604
silicones (RTV) EM3: 604
robotic foundry A15: 566-568
rocket foam-to-tank bond lines EM3: 762
rocket motor nozzles, phenolics EM3: 105-106
roofing panel sealing, plastisols EM3: 46
rubber athletic flooring, urethanes EM3: 110
rubber battery sealant
asphalt EM3: 58
epoxies EM3: 58
rubber bonding, acrylics EM3: 121
rubber bumpers bonding, cyanoacrylates EM3: 128
rubber coating bonding onto aluminum EM3: 762
rubber feet bonding, cyanoacrylates EM3: 128
rubber roof installation, butyl rubber EM3: 146
rubbers bonding, phenolics EM3: 105
sanitary mildew-resistant sealant, silicones (RTV) EM3: 219, 220
satellites ,polybenzimidazoles EM3: 171, 172
sealed relays EM3: 37
sealing aircraft assemblies, polysulfides EM3: 808
sealing construction, polyurethanes EM3: 606
self-sealing tires
butyl sealants EM3: 58
natural rubber EM3: 58
semiconductor component coating, elastomeric sealants EM3: 612
of semisolid metal casting/forging A15: 327
shaft seal collars, anaerobics EM3: 547
sheet molding compound (SMC) bonding
acrylics EM3: 46
epoxies EM3: 46
hot-melt adhesives EM3: 46

Applications (continued)
polyurethanes EM3: 46
sheet molding compound bonding to a metal frame, epoxies EM3: 97
shell molding, phenolics EM3: 105
shelving and window construction, silicones EM3: 45
shoe sole attachment
acrylonitrile-butadiene rubber EM3: 148
urethanes EM3: 110
shunt plates fixturing, cyanoacrylates EM3: 128
sidewalk repair, urethane sealants EM3: 204
silicones EM3: 50, 57-58
silicones (SI) EM2: 266-267
silicones, insulated glass assembly EM3: 50
slotted screwheads potting, cyanoacrylates EM3: 128
solar panel construction, methacrylates/silicones EM3: 578
solid rocket space booster joints, fluoroelastomer sealants EM3: 227
sound-deadening material bonding
hot melts EM3: 45
plastisols EM3: 46
space shuttle orbiter thermal protection tiles bonding EM3: 762
space vehicles, polybenzimidazoles EM3: 171
speaker cones bonding, cyanoacrylates EM3: 128
sporting goods manufacturing
acrylics EM3: 125
cyanoacrylates EM3: 576, 577
epoxy EM3: 576
methacrylates EM3: 577
urethanes EM3: 576-577
stainless steel bonding, acrylics EM3: 60
static gasketing, fluoroelastomer sealants EM3: 226
steel bonding
acrylics EM3: 121
polysulfides EM3: 138
for steel plate A1: 226
stepped-lap joints for aircraft wings and tails EM3: 471
strain gages bonding, cyanoacrylates EM3: 128
structural bonding, anaerobics EM3: 116
structural glazing, silicones EM3: 192, 219, 606, 607
structural integrity EM3: 604
structural wood bonding, phenolics (resorcinols) EM3: 105
styrene-acrylonitriles (SAN, OSA ASA) EM2: 215
styrene-maleic anhydrides (S/MA) EM2: 217-219
subsonic aircraft, epoxy adhesives EM3: 808
substrate attach adhesive, polyimides EM3: 161
for superalloys A1: 950
supersonic aircraft repair, epoxy adhesives EM3: 808
surface-mount technology bonding
acrylics EM3: 570
epoxies EM3: 570
pressure-sensitive adhesives EM3: 570
silicones EM3: 570
thermoplastic hot melts EM3: 570
urethane-acrylates EM3: 570
surgical plaster foundation, natural rubber EM3: 145
T-top roof sealing, urethane sealants EM3: 204
tack coat for concrete, polysulfides EM3: 138
taillight and headlight bonding, polyurethanes EM3: 46
taillight sealing EM3: 11
hot melts EM3: 46
urethanes EM3: 204
tape automated bonding (TAB), polyimides EM3: 161
tapes
butyl rubber EM3: 146
pressure-sensitive adhesives EM3: 83, 84

Applications (continued)
water-base adhesives EM3: 88
tapes (sealant), fluorocarbon sealants EM3: 223
tapes and labels, thermoplastic elastomers EM3: 82
Teflon, anaerobics EM3: 79
tennis racket parts, fixturing, cyanoacrylates EM3: 128
textile binding, natural rubber (vulcanized latex) EM3: 145
thermal barriers, sealants EM3: 605
thermally conductive bonding
anaerobic acrylates EM3: 571
anaerobic/aerobics EM3: 572
epoxies EM3: 571
methacrylates EM3: 571, 572
silicones EM3: 571
Thermex process press tooling EM3: 711
thermoplastic fluoropolymers EM2: 117-118
thermoplastic polyimides (TPI) EM2: 177
thermoplastic polyurethanes (TPUR) EM2: 205-206
thermoplastic resins EM2: 623-625
thermoplastics bonding
acrylics EM3: 121
epoxies EM3: 96
thermosetting resins EM2: 223-224
thermostat housings, sealants EM3: 548
threaded fittings EM3: 51
threadlocking, anaerobics EM3: 115
threadsealing, anaerobics EM3: 115
tin powder M2: 616
tire cord adhesion, phenolics EM3: 105
tire cord coating, styrene-butadiene rubber EM3: 147
titanium alloys A15: 824-825, 833-834
titanium bonding with advanced composites, polyphenylquinoxalines EM3: 163
to fill surface dents in aluminum aircraft structures, epoxy adhesives EM3: 809
to mate the decks and hulls of small fiberglass boats EM3: 707
transistors potting, cyanoacrylates EM3: 128
transportation, water-base adhesives EM3: 88
trim bonding, urethanes EM3: 110
trowel EM3: 36
truck and aircraft tire assembly EM3: 761
truck and trailer roof and floor bonding, polychloroprene rubber EM3: 149
truck trailer joints
neoprenes EM3: 58
polysulfides EM3: 58
polyurethanes EM3: 58
truss networks EM3: 495
tube perimeter repair, urethane sealants EM3: 204
tubing, fluorocarbon sealants EM3: 223
ultrahigh molecular weight polyethylenes (UHMWPE) EM2: 167-168
ultrahigh-strength steels M1: 423, 424, 427, 429, 431-434, 437, 441
undersea cable splicing
epoxies EM3: 612
polyesters EM3: 612
silicones EM3: 56
urethanes EM3: 56
vacuum bag sealants, butyls EM3: 59
valves, fluorocarbon sealants EM3: 223
vertical joints, urethane sealants EM3: 204
video game cartridges bonding, cyanoacrylates EM3: 128
vinyl repair, urethanes EM3: 110
wafer board construction, phenolics EM3: 105
wall construction, silicones EM3: 56
unsaturated polyesters EM2: 246-247
urethane hybrids EM2: 268
vacuum arc remelting A15: 414
vacuum induction melting A15: 396
vinyl esters EM2: 272-273

Applications (continued)
water deflector sealant, polypropylene **EM3:** 57
water tank sealing, sealants **EM3:** 58
water-deflector films (auto), butyls **EM3:** 50
wave solder masking, silicones **EM3:** 134
weather seals (auto), polypropylenes **EM3:** 51
weather seals (roofing), polypropylenes **EM3:** 51
weather strip sealants, hot melts **EM3:** 46
welded wheel rim seals for tubeless tires, anaerobics **EM3:** 116
wet systems in constructing butyl tapes **EM3:** 56
wind noise baffles (auto), butyls **EM3:** 50
window (auto) glazing, butyls **EM3:** 50
window assembly attachments **EM3:** 553
acrylates **EM3:** 553
epoxies **EM3:** 553
window glazing
butyl tapes **EM3:** 56
ethylene-propylene-diene monomer (EPDM) rubber **EM3:** 56
neoprene **EM3:** 56
preformed vinyl **EM3:** 56
sealants **EM3:** 545
silicones **EM3:** 56
window glazing and mounting silicones **EM3:** 577
window glazing and wall panel taping, butyls **EM3:** 199-200
window repair, urethane sealants **EM3:** 204
windshield (auto) installation **EM3:** 50
urethanes **EM3:** 53
windshield bonding, urethanes **EM3:** 554
windshield glass bonding, polyether silicones **EM3:** 229, 231
windshield glazing
polyurethanes **EM3:** 57
urethanes **EM3:** 57
windshield installation, polyurethanes **EM3:** 554
windshield sealants **EM3:** 605
polysulfides **EM3:** 194, 608
polyurethanes **EM3:** 608
urethanes **EM3:** 46
wiper blades bonding, cyanoacrylates **EM3:** 128
wire splice protection, rubber **EM3:** 612
wire tacking on motor assemblies, UV-curable resins **EM3:** 612
wire-winding operation adhesive
epoxies **EM3:** 574
UV-curing adhesives **EM3:** 573, 574
wood bonding
acrylics **EM3:** 121
phenolics **EM3:** 103, 106
polysulfides **EM3:** 138
wood, epoxies **EM3:** 101
wood, fibrous and granulated, phenolics **EM3:** 105
wood veneer plywood production, phenolics **EM3:** 107
woodworking
epoxies/acrylates **EM3:** 46
hot-melt adhesives **EM3:** 46
solvent cement/weld **EM3:** 46
structural adhesives **EM3:** 46
water-base adhesives **EM3:** 46
wrapping and bundling of wires, silicones **EM3:** 134
wrapping transformers, silicones **EM3:** 134
zinc **M2:** 629-637
zinc bonding, acrylics **EM3:** 60, 124

Applications and examples
(111) pole figures from Cu tubing **A10:** 363
300/400 series stainless steels, analysis of **A10:** 100
AAS instruments as detectors for other analytical equipment **A10:** 55
acids, determined **A10:** 215-216
activated carbon, analysis of **A10:** 399-400
AES application to tribology **A10:** 566
Al-killed steel, analysis of **A10:** 231
alcohols, determined **A10:** 216-217
aldehydes and ketones determined **A10:** 217
alignment of silicon boule for cutting along crystallographic planes **A10:** 342
alkenes, determined **A10:** 219
amines, determined **A10:** 217-218

Applications and examples (continued)
anisotropy, magnetic, determined **A10:** 272-273
archaeological samples, copper determined in **A10:** 186
aromatic hydrocarbons, determined **A10:** 218
atmospheric particles, analysis of **A10:** 106
atomic number contrast, in analysis of two-phase alloys **A10:** 508
baby lotion, analysis of parabens in **A10:** 655-656
biological materials, powdered, analysis of **A10:** 106-107
blocking and channeling surface structure study by **A10:** 633
borosilicate glass, determination of B and F in **A10:** 179
brine analysis, geological **A10:** 665
bromine, indirect determined of **A10:** 70
bulk samples, composition of **A10:** 631
calcite, quantitative analysis of ZnO in **A10:** 342
carbon, activated, analysis of **A10:** 399-400
carbon determined by combustion **A10:** 214
carbon, surface, determination of **A10:** 224
catalysts, chromia alumina **A10:** 265
cellular decomposition of martensite in uranium alloy, kinetics of **A10:** 316-318
cement, analysis of **A10:** 99-100
ceramic nuclear waste form simulant study of **A10:** 532-535
ceramics, SAXS/SANS analysis of **A10:** 405
channeling and blocking, surface structure study by **A10:** 633
chemical reaction products, analysis of **A10:** 656-657
chloride ions, nickel determined in samples containing **A10:** 201
chromia alumina catalysts **A10:** 265
chromium depletion in weld zone **A10:** 179
coal fly ash, determination of Cu and Pb in **A10:** 147-148
cobalt compounds, trace nickel determined in **A10:** 194-195
cold-rolled steel, corrosion resistance **A10:** 556-557
combustion, determination of carbon hydrogen, and nitrogen by **A10:** 214
complex, number of ligands determined in **A10:** 70
composition fluctuations, local, in ternary 3:5 semiconductor **A10:** 601
composition of bulk samples **A10:** 631
composition, of mixtures, determined **A10:** 213
composition vs depth, passive film on Sn-Ni substrate **A10:** 608-609
compounds, organic, identification of **A10:** 213
concentrations, high, and well-resolved peaks gold-copper alloy analysis with **A10:** 530
coordination geometry, determination of **A10:** 354
copper determined in archaeological samples **A10:** 186
copper, hydrogen determined in **A10:** 231-232
copper oxidation using ^{18}O **A10:** 609
copper tubing, (111) pole figures from **A10:** 363
copper-manganese alloy, copper content estimated **A10:** 201
corrosion and surface strains, connector **A10:** 607
corrosion in pyrotechnic actuators **A10:** 510-511
corrosion of Ni and Co in aqueous alkaline media **A10:** 135
corrosion on lead surfaces **A10:** 135
corrosion on metals, analysis of **A10:** 134-135
corrosion resistance of cold-rolled steel **A10:** 556-557
crystal growth and electronic device material studies **A10:** 375-376
crystal kinetics and material transformation **A10:** 376
crystalline phase, unknown, use of unit cells to identify **A10:** 353
crystallographic planes, alignment of silicon boule for cutting along **A10:** 342
curing mechanism of a polyimide resin **A10:** 285
dc arc excitation to determine trace metal impurities in $CaWO_4$ **A10:** 29

Applications and examples (continued)
deer hair, analysis of sulfur in **A10:** 224
defects, analysis of **A10:** 464-468
deformation analysis **A10:** 376-378, 468-470
depth profiles, heavy element impurities **A10:** 632-633
depth profiling, granular sample using ATR DRS, and PAS **A10:** 120
detection of phase changes **A10:** 282-283
detection of surface phase **A10:** 293
determination of BTU and ash content in coal **A10:** 100
determination of Cr, Ni, and Mn in stainless steel **A10:** 146-147
determination of trace Sn and Cr in HO_2, by GFAAS **A10:** 57-58
determination of ultratrace uranium by laser-induced fluorescence spectroscopy **A10:** 80
diamond, synthetic, characterization of metal impurities in **A10:** 417
diffusion measurements **A10:** 476-478
diffusion of plutonium into thorium **A10:** 249
diffusion phenomena, investigation of **A10:** 576-577
diffusion-induced grain-boundary migration analysis of **A10:** 461-464
direct determination of benzo(a)pyrene **A10:** 79
direct solid-sample AAS analysis **A10:** 58
dislocation cell structure analysis **A10:** 470-473
dopant atoms, lattice location of **A10:** 633-634
electrical contacts, surface films on **A10:** 578-579
electrochemically based corrosion systems in aqueous environments **A10:** 134-135
electrolytically generated radical ions **A10:** 265
electronic device material studies, and crystal growth **A10:** 375-376
electronic structure, liquid metals and alloys **A10:** 284
elemental analysis, Schöniger flask method for **A10:** 215
elemental mapping of high-temperature solder **A10:** 532
elements, light, analyses of **A10:** 459-461, 558-559
empirical formulas, determination of **A10:** 213
enzymatic determination of glucose using oxygen quenching of fluorescence **A10:** 79-80
enzyme activity, assay of **A10:** 70
esters, determined **A10:** 218
estimation of Cu in a Cu-Mn alloy **A10:** 201
exchange stiffness **A10:** 275
exotic effects, studied in disordered magnetic material **A10:** 276
explosive actuator, determination of oxygen isotopes in **A10:** 625-626
extended solubility of iron in aluminum **A10:** 294-295
factor analysis and curve fitting applied to polymer blend system **A10:** 118
failure of Al wire connections **A10:** 531-532
failure, overload, of steel threaded rod **A10:** 511-513
fcc materials, features of rolling textures in **A10:** 363-364
ferromagnetic alloys, order-disorder in **A10:** 284-285
films, grain size of silver **A10:** 543
films, thick, analysis of **A10:** 561
fingerprinting, multielement **A10:** 195
flame AAS for elemental analysis **A10:** 55-56
flame analysis determination of alloying elements in steels **A10:** 56
formulas, empirical determination of **A10:** 213
fracture behavior and deformation **A10:** 376-378
fracture surfaces, SCC **A10:** 562-564
free lime in portland cement, determination of **A10:** 179
geological brine analysis of **A10:** 665
geometric and elemental analysis, particles produced by explosive detonation **A10:** 318-320
GFAAS determination of Bi in Ni **A10:** 57
glass microballoons, analysis of **A10:** 665-667
glass surface layers, analysis of **A10:** 624-625
glasses, SAXS/SANS analysis of **A10:** 405

Applications and examples (continued)
glow charge to determine C, P, and S in low-alloy steels and cast iron **A10:** 29
gold determined **A10:** 210-211
gold-copper alloys, analysis with well-resolved peaks and high concentrations for **A10:** 530
grain size, Ag film grown on mica **A10:** 543-544
grain-boundary chemistry, study of **A10:** 561-562
grain-boundary migration, diffusion-induced, analysis of **A10:** 461-464
grain-boundary segregation, analysis of **A10:** 481-484
graphites, analysis of **A10:** 132-133
graphites, wettability of **A10:** 543
grinding, local variations in residual stress produced by **A10:** 390
habit plane and orientation relationships **A10:** 453-455
heavy element impurities, depth profiles of **A10:** 632-633
hexamethyldisiloxane, structure and degradation of plasma-polymerized **A10:** 285-286
high resolution, analysis of Jominy jar using **A10:** 508
high-temperature solder, elemental mapping of **A10:** 532
historical objects, analysis of **A10:** 107
hydrided TiFe, phase analysis of **A10:** 293-294
hydrocarbons, aromatic, determined **A10:** 218
hydrogen determined by combustion **A10:** 214
hydrogen determined in copper **A10:** 231
hydroxyl and boron content in glass, quantitative analysis **A10:** 121-122
impurities, heavy element, depth profiles of **A10:** 632-633
impurities in nickel, determined **A10:** 240
impurities in UO_2, determination of **A10:** 149-150
impurity analysis in LPCVD thin films, quantitative **A10:** 624
Inconel 600 tubing, residual stress and percent cold work distribution **A10:** 390
indirect determination of bromine **A10:** 70
inhomogeneities, magnetic, determined **A10:** 274-275
integrated circuit problems, SEM analysis for **A10:** 513-514
interatomic bond lengths and physical properties **A10:** 355
interface/superlattice studies **A10:** 634-635
interfacial segregation studies in Mo **A10:** 599-601
intermetallic compounds, sublattice ordering in **A10:** 283-284
internal electrolysis, separation of Cd and Pb by **A10:** 201
ion-implantation profile in silicon, phosphorus, quantitative analysis of **A10:** 623-624
ion-implanted ions, microstructural analysis of **A10:** 484-487
iridium anomaly at Cretaceous-Tertiary boundary **A10:** 240-241
iron in copper, analysis of phases of **A10:** 294
iron oxide films, composition determined **A10:** 135
iron-base magnet alloy , spinodal decomposition of **A10:** 598-599
isotopes, oxygen, determined in explosive actuator **A10:** 625-626
Jominy bar, use of high resolution in analysis of **A10:** 508
Karl Fischer method, water determination **A10:** 219
ketones and aldehydes determined **A10:** 217
kinetics, crystal, and material transformations **A10:** 376

Applications and examples (continued)
kinetics of cellular decomposition of martensite in uranium alloy **A10:** 316-318
kinetics of radical production and subsequent decay **A10:** 265-266
Kjeldahl method, to determine nitrogen **A10:** 214-215
Knight shift measurements on metallic glass **A10:** 284
Kovar-glass seals, shear fracture studies of **A10:** 577-578
laser treatment, of stainless steel, surface composition effects during **A10:** 622-623
lattice location of dopant atoms **A10:** 633-634
lead surfaces, corrosion on **A10:** 135
ligands in complex, numbers determined **A10:** 70
light element analysis **A10:** 459-461, 558-559
liquid metals and alloys, electronic structure of **A10:** 284
local composition fluctuations in ternary 3:5 semiconductor **A10:** 601-602
local variations in residual stress produced by surface grinding **A10:** 390-391
longitudinal residual stress distribution in welded railroad rail **A10:** 391-392
LPCVD thin films, quantitative impurity analysis of **A10:** 624
machining, magnitude and direction of by **A10:** 392
maximum residual stress produced
magnetic anisotropy, determined **A10:** 272-273
magnetic disordered materials, exotic effects in **A10:** 276
magnetic inhomogeneities determined **A10:** 274-275
magnetization, determined **A10:** 271
major constituent analysis with peak overlap **A10:** 530
material transformations and crystal kinetics **A10:** 376
maximum residual stress, magnitude and direction produced by machining **A10:** 392
mercury switch, analysis of surface films on electrical contacts in **A10:** 578-579
metal oxide systems, analysis of **A10:** 130-131
metallic glass, Knight shift measurements on **A10:** 284
metals, analysis by SAXS and SANS **A10:** 405
metatorbernite, analysis of **A10:** 265
microballoons, glass, analysis of **A10:** 665-666
microcircuit process gas analysis **A10:** 156-157
microstructure, ion-implanted alloys determined **A10:** 484-487
microstructure, weld metal **A10:** 478-481
mixtures, composition determined **A10:** 213
molecular orientation in drawn polymer films determined **A10:** 120
molybdenum, interfacial segregation studies in **A10:** 599-601
monitoring polymer-curing reactions using ATR **A10:** 120-121
monolayers adsorbed on metal surfaces examined **A10:** 118-120
multielement fingerprinting and approximate quantification in effluent samples **A10:** 195
natural waters, analysis of **A10:** 41
$Nd_2(CoFe_{0.9})_{14}B$, Rietveld analysis of **A10:** 425
near-surface defects in single-crystals, study of **A10:** 633
nickel determined in samples containing chloride ions **A10:** 201
nickel, impurities determined in **A10:** 240
nickel-base superalloy, phase chemistry and phase stability of **A10:** 598

Applications and examples (continued)
nickel-phosphorus film, coverage determined on platinum substrate **A10:** 608
nitrogen determined by combustion **A10:** 214
nitrogen determined by Kjeldahl method **A10:** 214-215
nontransparent samples, surfaces characterized **A10:** 70-71
nuclear waste form simulant, ceramic, study of **A10:** 532-535
optical microscopy, complementarity of SEM and SAM in **A10:** 509-510
order-disorder in ferromagnetic alloys **A10:** 284-285
organic compounds identified **A10:** 213
orientation relationships and habit plane **A10:** 453-455
overload failure, quench-cracked steel threaded rod **A10:** 511-513
oxidation of Ag electrodes in alkaline environments **A10:** 135
oxidation of copper using ^{18}O, study of **A10:** 609
oxygen in silicon wafers, quantitative analysis of **A10:** 122-123
oxygen in Ti, determined **A10:** 231
oxygen isotopes determined in explosive actuator **A10:** 625-626
parabens, analysis in baby lotion **A10:** 655-656
particles produced by explosive detonation geometric and elemental analysis **A10:** 318-320
passive film on Sn-Ni substrate, composition vs depth of **A10:** 608
passive films, study of thin **A10:** 557-558
peak overlap, analysis of major constituents with **A10:** 530
peaks, well resolved, and high concentrations gold-copper alloy analysis with **A10:** 530
pearlite growth, use of SACP to understand **A10:** 508-509
peroxides, determined **A10:** 218
petroleum products, analysis of **A10:** 100-101
phase analysis of hydrided TiFe **A10:** 293-294
phase analysis of iron in copper **A10:** 294
phase changes, detection of **A10:** 282-283
phase chemistry and phase stability of Ni-base superalloy **A10:** 598
phase diagrams, determination of **A10:** 473-476
phases, surface, on silicon, qualitative analysis of **A10:** 341-342
phases, unknown crystalline, unit cell information to identify **A10:** 353
phases, unknown, identification of **A10:** 455-459
phenols, determined **A10:** 218
phosphorus ion-implantation profile in silicon quantitative analysis of **A10:** 623-624
PIXE and proton microprobe **A10:** 107
plant tissues, analysis of **A10:** 41
plasma-polymerized hexamethyldisiloxane structure and degradation of **A10:** 285-286
platinum substrate, determined coverage of Ni-P film on **A10:** 608
plutonium diffusion into thorium **A10:** 249
pole figures (111), from copper tubing **A10:** 363
polyimide resin, curing mechanism of **A10:** 285
polymer and plasticizer materials in vinyl film, identification of **A10:** 123-124
polymers, analyses of **A10:** 131-132, 405, 647-648
porous graphite atomizers as filters for air particulates **A10:** 58
powdered biological materials, analysis of **A10:** 106-107
praseodymium and neodymium, separation of **A10:** 249-250

SUBJECTS OF THE INDEXED VOLUMES: ASM Handbook (designated by the letter "A"): **A1:** Properties and Selection: Irons, Steels, and High-Performance Alloys (1990); **A2:** Properties and Selection: Nonferrous Alloys and Special-Purpose Materials (1990); **A3:** Alloy Phase Diagrams (1992); **A4:** Heat Treating (1991); **A6:** Welding, Brazing, and Soldering (1993); **A7:** Powder Metallurgy (1984); **A8:** Mechanical Testing (1985); **A9:** Metallography and Microstructures (1985); **A10:** Materials Characterization (1986); **A11:** Failure Analysis and Prevention (1986); **A12:** Fractography (1987); **A13:** Corrosion (1987); **A14:** Forming and Forging (1988); **A15:** Casting (1988); **A16:** Machining (1989); **A17:** Nondestructive Testing and Quality Control (1989); **A18:** Friction, Lubrication, and Wear Technology (1992). **Metals Handbook, 9th Edition** (designated by the letter "M"): **M1:** Properties and Selection: Irons and Steels (1978); **M2:** Properties and Selection: Nonferrous Alloys and Pure Metals (1979); **M3:** Properties and Selection: Stainless Steels, Tool Materials and Special-Purpose Materials (1980); **M4:** Heat Treating (1981); **M5:** Surface Cleaning, Finishing, and Coating (1982); **M6:** Welding, Brazing, and Soldering (1983). **Engineered Materials Handbook** (designated by the letters "EM"): **EM1:** Composites (1987); **EM2:** Engineering Plastics (1988); **EM3:** Adhesives and Sealants (1990); **EM4:** Ceramics and Glasses (1991); **Electronic Materials Handbook** (designated by the letters "EL"): **EL1:** Packaging (1989).

Applications and examples (continued)
preferred crystallographic growth, use of SACP to establish **A10:** 509
pyrotechnic actuators, corrosion in **A10:** 510-511
radical ions, electrolytically generated **A10:** 265
radical production and subsequent decay, kinetics of **A10:** 265
recovery, analysis of **A10:** 468-470
recrystallization structure analysis **A10:** 468-470
relay weld integrity, determined **A10:** 156
residual stress analyses **A10:** 390, 392, 425-426
Rietveld analysis **A10:** 425
rod, overload failure of **A10:** 511-513
rolling textures in fcc materials, features of **A10:** 363-364
SACP **A10:** 508-509
sampling a trainload of metal pipe for percentage of alloying element **A10:** 15
SAXS/SANS analyses **A10:** 405
Schöniger flask method for elemental determinations **A10:** 215
seals, shear fracture studies of Kovar-glass **A10:** 577-578
segregation, analysis of grain-boundary **A10:** 481-484
segregation, interfacial, in molybdenum **A10:** 599-601
segregation, of Pb to surface of Sn-Pb solder **A10:** 607-608
segregation, surface **A10:** 564-566
SEM analysis of integrated circuit problems **A10:** 513-514
SEM and SAM, complementary contributions of optical microscopy **A10:** 509-510
semiconductor, local composition fluctuations in **A10:** 601-602
separation of Cd and Pb by internal electrolysis **A10:** 201
separation of praseodymium and neodymium **A10:** 249-250
shear fracture studies of Kovar-glass seals **A10:** 577-578
silica glass, x-ray diffraction pattern **A10:** 398-399
silicon, analysis of phosphorus ion-implantation profile in **A10:** 623-624
silicon boule, alignment for cutting along crystallographic planes **A10:** 342
silicon, qualitative analysis of surface phase **A10:** 341-342
silver film grown on mica, grain size in **A10:** 543-544
silver scrap metal, analysis of **A10:** 41
single crystals, study of near-surface defects in **A10:** 633
solder, analyses of **A10:** 179, 532, 607-608
solubility, extended, of iron in aluminum **A10:** 294-295
spark source excitation for elemental analysis of metal or alloy **A10:** 29
spectrophotometric titrations **A10:** 70
spin relaxation rates, determined **A10:** 275-276
spin wave resonances **A10:** 275
spinodal decomposition of Fe-base magnet alloy **A10:** 598-599
stable free radical hydrazyl **A10:** 265
stainless steel alloy, "umpire" analysis of **A10:** 178-179
stainless steels, surface composition effects during laser treatment of **A10:** 622-623
strains, measurement of **A10:** 275
strains, surface, and corrosion products, connector **A10:** 607
stress-corrosion crack fracture surfaces **A10:** 563-564
structure and degradation of plasma-polymerized hexamethyldisiloxane **A10:** 285-286
sublattice ordering in intermetallic compounds **A10:** 283-284
subsurface residual stress and hardness distributions in induction-hardened steel shaft **A10:** 389-390
sulfur in deer hair, analysis of **A10:** 224
superlattice/interface studies **A10:** 634-635
sur-face-phase detection **A10:** 293
surface carbon, determined **A10:** 224

Applications and examples (continued)
surface composition effects during laser treatment of stainless steels **A10:** 622-623
surface effects, determined **A10:** 273
surface films on electrical contacts, mercury switch, analysis of **A10:** 578-579
surface layers, glass, analysis of **A10:** 624-625
surface phase on silicon, qualitative analysis **A10:** 341-342
surface segregation **A10:** 564-566
surface species on nonmetals, analysis of **A10:** 134
surface strains and corrosion products on connector, identification of **A10:** 607
surface structure, analysis of **A10:** 133-134
surface structure study, by channeling and blocking **A10:** 633
surfaces of nontransparent samples characterization of **A10:** 70-71
surfactant molecules in water, examination of structural changes in **A10:** 118
synthetic diamond, metal impurities in **A10:** 417
synthetic substance, analysis of new **A10:** 353-354
texture, measurement and analysis of **A10:** 425
thermite, flame AAS analysis of **A10:** 56-57
thick films, analysis of **A10:** 561
thickness, of thin films, determination of **A10:** 100, 631-632
thin film composition and layer thickness **A10:** 631-632
thin films, passive **A10:** 557-558
thin films, quantitative impurity analysis in LPCVD **A10:** 624
tin-nickel substrate, composition vs depth of passive film on **A10:** 608
titrations, spectrophotometric **A10:** 70
toxic trace elements in ground water measurement of **A10:** 148
trace amounts of Ni in Co compounds **A10:** 194-195
transformations, material, and crystal kinetics **A10:** 376
transformer cores, analysis **A10:** 224
tribology, AES application to **A10:** 566
turquoise, analysis of **A10:** 265
two-phase alloys, atomic number contrast in analysis of **A10:** 508
umpire analysis, stainless steel alloy **A10:** 178-179
unit cells, use to identify unknown crystalline phase **A10:** 353
unknown phases , identification of **A10:** 455-459
unsaturation, determined **A10:** 219
uranium, determined **A10:** 211
vanadium determined in compounds **A10:** 206
water determination, Karl Fischer method for **A10:** 219
weld metal microstructure, interpretation of **A10:** 478-481
welded railroad rail, longitudinal residual stress distribution in **A10:** 391-392
wettability, of graphite **A10:** 543
wire connections, aluminum, failure of **A10:** 531-532
x-ray diffraction pattern, of silica glass **A10:** 398-399
ZnO in calcite, quantitative analysis of **A10:** 342
Applications engineering
with machine vision **A17:** 42-43
Applications of metal powders **A7:** 569-574
Applied bending moment
reversing in shafts **A11:** 462
Applied current, magnitude
magnetic particle inspection **A17:** 111
Applied electromotive force (emf)
in electrogravimetry **A10:** 197
Applied fracture mechanics **EM4:** 645-650
fracture mechanics analysis of ceramics **EM4:** 646-649
elliptical crack **EM4:** 647
flaw shape parameter **EM4:** 648
inclusions **EM4:** 648
irregularly shaped two-dimensional cracks **EM4:** 647-648
multiple flaws **EM4:** 648-649
residual stress effects **EM4:** 648
semielliptical crack **EM4:** 647, 648

Applied fracture mechanics (continued)
stress-intensity factor **EM4:** 647-649
voids **EM4:** 648, 649
linear elastic fracture mechanics **EM4:** 645-646
fracture toughness **EM4:** 646
stress field of mode I loading **EM4:** 645-646
stress-intensity factor for mode I loading **EM4:** 645, 646
Applied load **A18:** 508
abbreviation for **A11:** 797
and distortion failures **A11:** 136-138
effect in composites **A11:** 733
ion implantation **A18:** 779
symbol for **A8:** 725
and thickness, effect on buckling and plastic deformation **A11:** 137
Applied normal force **A18:** 434
Applied Polymer Symposia **EM3:** 68
Applied potential
changes for corrosion control **A11:** 198
effect on electrolytic polishing **A9:** 105
Applied pressure, and density *See also* Density; Pressure; Relative density
Applied statistics **A8:** 623-627
Applied stress **A13:** 277, 292, 615
of castings **A11:** 346
defined **A11:** 102
effect, embrittlement **A12:** 29
effect, striation spacing **A12:** 120
examining **A12:** 92
in fatigue cracking **A11:** 102
in fatigue-crack initiation **A11:** 102
in fracture mechanics **A11:** 48
intensity, and fatigue-crack growth rate **A11:** 107
periodic interruptions, wrought aluminum alloys **A12:** 418
as permanent distortion **A11:** 467
pulsation from **A12:** 304
and residual stress distribution **A8:** 124
vs case depth, in gears **A11:** 594
Applied stresses
effect on magnetoresistance **A17:** 144
magabsorption measurement of **A17:** 154-155
Applied thrust load **A18:** 508
Approximate-failure theories
laminates **EM1:** 235
Approximation
circuit **EL1:** 30-31
transmission line **EL1:** 31-32
Approximation, single-scattering
in EXAFS **A10:** 409
Aqua regia as an etchant for
heat-resistant casting alloys **A9:** 330
wrought stainless steels **A9:** 281
Aqueous
defined **A13:** 2
Aqueous acids and chemicals used in etchants **A9:** 67-68
Aqueous caustic molten salt bath descaling process
refractory metals **M5:** 654
Aqueous cleaning
materials for **EL1:** 663
Aqueous corrosion *See also* Aqueous corrosion resistance
activation control **A13:** 30-32
aluminized steels **A13:** 527
of beryllium **A13:** 809-810
biological effects **A13:** 41-43
of carbon steels **A13:** 512
cemented carbides **A13:** 850-855
of cobalt-base corrosion-resistance alloys **A2:** 453
defined **A13:** 29
effects, environmental variables **A13:** 37-44
effects, metallurgical variables **A13:** 45-49
electrode potentials **A13:** 19-21
electrode processes **A13:** 18-19
exposure types **A13:** 516
of galvanized steel **A13:** 433-434, 526-527
and gaseous corrosion, compared **A13:** 61
kinetics **A13:** 17, 29-36
and liquid-metal corrosion, compared **A13:** 17
mass transport control **A13:** 33-35
mechanism **A13:** 433
niobium **A13:** 723
oxiding power (potential) **A13:** 39

Aqueous corrosion (continued)
passivation **A13:** 35-36
pH, effect **A13:** 37-39
potential measurements, reference electrodes **A13:** 21-24
potential vs. pH (Pourbaix) diagrams **A13:** 24-28
rate, measurement of **A13:** 32-33
temperature and heat transfer **A13:** 39-40
testing, copper/copper alloys **A13:** 636-638
thermodynamics **A13:** 18-28, 32
titanium/titanium alloy SCC **A13:** 689
velocity/fluid flow rate **A13:** 40
Aqueous corrosion resistance *See also* Aqueous corrosion
aluminum coatings **A13:** 435
aluminum-zinc alloy coatings **A13:** 435-436
galvanized steel **A13:** 433-434
Aqueous electroplating
oxidation-resistant coatings **M5:** 664-666
Aqueous environment synthesis **A8:** 416
Aqueous environments **A13:** 143, 314
cracking from hydrogen charging in **A11:** 245-247
electrochemically based corrosion systems determined in **A10:** 134
Aqueous hydrofluoric acid **A13:** 1166-1170
Aqueous media
Raman analyses of polymers in **A10:** 131
Aqueous nitric acid as a macroetchant for tool steels **A9:** 256
Aqueous samples
optical emission spectroscopy **A10:** 21
Aqueous slip (slurry)
in rigid tool compaction **A7:** 327
Aqueous solution
metal electrode potential in **A8:** 416-417
SCC testing of nickel alloys in **A8:** 531
Aqueous solutions
at ambient temperature **A8:** 415-417
containing nickel and cobalt ions, spectra compared **A10:** 65
ionic, use in ion chromatography **A10:** 658-664
corrosion in **A13:** 17
corrosion protection in **A13:** 377-379
SCC testing **A13:** 274
SCC testing of titanium alloys in **A8:** 531
Uranium/uranium alloys in **A13:** 814
AR 213
composition **A16:** 736
Ar-5CO_2-4O_2, effect on carbon manganese, and silicon losses and on weld strength values **A6:** 68
Ar-10CO_2 shielding gas, effect on carbon manganese, and silicon losses and on weld strength values **A6:** 68
Ar-12O_2 shielding gas, effect on carbon manganese, and silicon losses and on weld strength values **A6:** 68
Ar-18CO_2 shielding gas, effect on carbon manganese, and silicon losses and on weld strength values **A6:** 68
Ar-25CO_2 shielding gas, effect on carbon manganese, and silicon losses and on weld strength values **A6:** 68
AR-213
composition **A6:** 564 **M6:** 354
Aragonite
Miller numbers **A18:** 235
Aragonite ($CaCO_3$) **EM4:** 379
purpose for use in glass manufacture **EM4:** 381
Aramid *See also* Aramid composites; Aramid fibers; Aramid fibers, specific types; Kevlar **EM3:** 5
aramid fibers **EM1:** 4
defined **EM2:** 4
fibers, as reinforcements **EM2:** 506
as thixotrope **EM3:** 178
Aramid composites
drilling of **EM1:** 668-669
Aramid composites (continued)
fabric, properties **EM1:** 149
shedding, by cutting **EM1:** 667
Aramid fiber reinforced plastics, and aluminum weights compared **EM1:** 35
Aramid fibers *See also* Aramid composites; Aramid fibers, specific types; Fiber properties analysis; Fibers; Kevlar aramid fibers; p-aramid fibers; Para-aramid fibers **EM1:** 114-116 **EM3:** 283-286
continuous filament **EM1:** 114-115
damping in **EM1:** 208
discontinuous filament forms **EM1:** 115
effect, vinyl ester resins **EM1:** 92
elastic buckling stress **EM1:** 197
fabrics **EM1:** 114, 115
felts **EM1:** 115
forms **EM1:** 360
impact properties **EM1:** 36
introduction **EM1:** 30
Kevlar, fibrillar structure **EM1:** 55
laser cutting of **A14:** 742
locking leno pattern yams **EM1:** 127-128
moisture absorption in **EM1:** 190
papers **EM1:** 115
with polyester resins **EM1:** 92
properties **EM1:** 58, 361
pulp **EM1:** 115
as reinforcement **EL1:** 535, 615-618
rovings **EM1:** 114
for short-fiber reinforced composites **EM1:** 120
in space and missile applications **EM1:** 817
spun yams **EM1:** 115
spunlaced sheets **EM1:** 115
staple/spun yams **EM1:** 115
structure **EL1:** 605
tensile properties **EM3:** 285
textured **EM1:** 115
thermal expansion properties **EL1:** 615-618
unidirectional, reinforced epoxies, strength of **EM1:** 35
vs. glass fibers, cost **EM1:** 105
for woven materials **EM1:** 125-128
woven rovings **EM1:** 114-115
yarns **EM1:** 114, 115
Aramid fibers, specific types
HM-50, properties **EM1:** 56
Kevlar 29, creep **EM1:** 55
Kevlar 29, fiber properties **EL1:** 615
Kevlar 29, properties **EM1:** 55
Kevlar 29, yam and roving sizes **EM1:** 114
Kevlar 49, baseline STEB design **EM1:** 514
Kevlar 49, creep **EM1:** 55
Kevlar 49, electron radiation effects **EM1:** 56
Kevlar 49, fabric/woven roving specifications **EM1:** 114
Kevlar 49, properties **EL1:** 535 **EM1:** 55, 175
Kevlar 49, temperature effect on tensile strength/modulus **EM1:** 55
Kevlar 49, tensile modulus vs. temperature **EM1:** 362
Kevlar 49, tensile strength retention **EM1:** 55
Kevlar 49, ultimate tensile strength vs temperature **EM1:** 362
Kevlar 49, yam and roving sizes **EM1:** 114
Kevlar 149, properties **EL1:** 615 **EM1:** 55
Nomex, as honeycomb core material **EM1:** 723
Nomex, as paper **EM1:** 115
Technora HM-50, properties **EL1:** 535
Aramid glass
signal bandwidth effects **EL1:** 82
Aramid paper
for flexible printed boards **EL1:** 583
Aramid papers **EM1:** 115
Aramid printed wiring boards **EL1:** 616-618
Aramid pulp **EM1:** 115
Aramid-epoxy
single-filament interfacial bond strength **EM3:** 284
Aramid-epoxy composites
applications **A9:** 592
fabric, microstructure of **EM1:** 768
local fiber failure mechanisms **EM1:** 198
polishing **A9:** 590
unidirectional, with voids **A9:** 593
Aramid-graphite-epoxy hybrid composites, fastener holes
techniques/tools for **EM1:** 715
Aramid-polystyrene composites
properties **EM3:** 286
Arbitrary body and coordinate system
for crack-tip stresses **A11:** 47
Arbitration bar
defined **A8:** 1 **A15:** 1
Arbor
defined **A15:** 1
Arbor-type punches
for press-brake forming **A14:** 537
Arc
direct-current **A10:** 25
initiation for gas tungsten arc welding **M6:** 183-184
systems for gas tungsten arc welding **M6:** 184
Arc blow
backward **M6:** 87-88
definition **A6:** 1206 **M6:** 1, 87
forward **M6:** 87-88
shielded metal arc welding **M6:** 78-79, 87-88
alternating current **M6:** 79
direct current **M6:** 78
stud arc welding **M6:** 733
Arc blow in
shielded metal arc welds of nickel alloys **M6:** 441
submerged arc welds **M6:** 127-128
Arc brazing
definition **M6:** 1
Arc brazing (AB)
definition **A6:** 1206
Arc burns **A6:** 1073
in weldments **A17:** 582
Arc butt weld
optical macrograph **A10:** 303
Arc column **A6:** 67
Arc cutting
defined **A14:** 720
definition **M6:** 1
types **A14:** 729-734
Arc cutting (AC)
definition **A6:** 1206
Arc deposition
and drilling **A16:** 219
Arc emission spectroscopy
to analyze the bulk chemical composition of starting powders **EM4:** 72
Arc energy input **A6:** 412
Arc erosion
as function of cadmium oxide in dispersion-strengthened silver **A7:** 717
Arc force
definition **A6:** 1206 **M6:** 1
Arc furnace
defined **A15:** 1
Arc furnace method
to make compacts of silicon **EM4:** 237
Arc gouging
definition **A6:** 1206 **M6:** 1
Arc, immersed
vacuum arc degassing **A15:** 437
Arc length
electrogas welding **M6:** 242-243
shielded metal arc welding **M6:** 86-87
Arc light sources for microscopes **A9:** 72
Arc melting **A7:** 25
defined **A15:** 1
Arc oxygen cutting
definition **A6:** 1206 **M6:** 1
Arc plasma **A6:** 64

Arc resistance ... **EM3:** 5
defined ... **EM1:** 4 **EM2:** 4, 460, 467
Arc seam weld
definition ... **A6:** 1206 **M6:** 1
Arc sources
applications ... **A10:** 29
compared ... **A10:** 27
for optical emission spectroscopy ... **A10:** 25
Arc spot weld
definition ... **A6:** 1206 **M6:** 1
Arc spraying (ASP)
cast irons ... **A6:** 720
definition ... **A6:** 1206
Arc stabilizers
as coating material ... **A7:** 817
Arc starting
gas tungsten arc welding of austenitic stainless steels ... **M6:** 339
percussion welding ... **M6:** 741
shielded metal arc welding ... **M6:** 78-79
alternating current ... **M6:** 79
direct current ... **M6:** 78
submerged arc welding ... **M6:** 134-135
Arc strike
arc burns ... **A6:** 1073
definition ... **A6:** 1206 **M6:** 1
Arc strikes
damage by ... **A11:** 413, 414
fatigue crack initiation at ... **A12:** 261
in fracture surfaces, medium-carbon steels ... **A12:** 255
hard spot caused by ... **A11:** 97
in weldments ... **A17:** 582
Arc stud welding *See* Stud arc welding
Arc time
definition ... **M6:** 1
percussion welding ... **M6:** 740-741
Arc tracking
defined ... **EM2:** 590
resistance ... **EM2:** 586-588
Arc voltage
definition ... **M6:** 1
electrogas welding ... **M6:** 242-243
Arc weld
gas tungsten ... **A11:** 438
Arc welding *See also* Arc welds; Weld(s); Welding; Weldments; Weldments, failures of
adaptive robotic ... **A17:** 43
aluminum coated steel ... **M1:** 173
of aluminum metal-matrix composites ... **A6:** 555-556
of aluminum-lithium alloys ... **A6:** 551, 552
and cast iron microstructure ... **A15:** 522
of cast irons ... **A15:** 520-529
cast steels ... **M1:** 401
comparison to gas welding oxyacetylene ... **M6:** 601
comparison to projection welding ... **M6:** 522-523
consumables ... **A15:** 523-524
of corrosion- and heat-resistant cast irons ... **A15:** 529
definition ... **M6:** 1
of ductile irons ... **A15:** 527
electrodes, selection of ... **M1:** 562-564
flux-cored ... **A11:** 414
fusion zone ... **A15:** 523
gas metal ... **A11:** 414-415
gas tungsten ... **A11:** 415
of gray irons ... **A15:** 526-527
hardfacing ... **M6:** 783-787
hardfacing deposition in mining and mineral industries ... **A18:** 653
heat-affected zone, cast iron ... **A15:** 522
high-deposition submerged ... **A7:** 821-822
hydrogen embrittlement of steel during ... **M1:** 687
of Invar ... **A2:** 892-893
low-carbon steel electrodes, function and composition ... **A7:** 817
of malleable irons ... **A15:** 528
maraging steel ... **M1:** 563
martensite tempered by ... **M1:** 564
medium-carbon low-alloy steel ... **M1:** 561
metallography of joints ... **A9:** 577-586
metallurgy ... **A15:** 520-522
partially melted region ... **A15:** 522-523
plasma ... **A11:** 415
Arc welding (continued)
postweld heat treatment ... **A15:** 526
preheat/interpass temperature ... **A15:** 525-526
preparation ... **A15:** 524-525
processes, types ... **A15:** 523
as secondary operation ... **A7:** 456
submerged ... **A11:** 415
techniques ... **A15:** 526
of thermocouple thermometers ... **A2:** 871
vs. laser-beam welding ... **A6:** 262
of white/alloy cast irons ... **A15:** 528-529
wire *See* Welding wire
Arc welding (AW)
definition ... **A6:** 1206
Arc welding and cutting
safety precautions ... **A6:** 1191
Arc welding electrode
definition ... **A6:** 1206 **M6:** 1
Arc welding gun
definition ... **A6:** 1206 **M6:** 1
Arc welding of
alloy steels ... **M6:** 247-306
hardenable carbon steels ... **M6:** 247-306
heat-resistant low-alloy steels ... **M6:** 247, 292-294
high-strength alloy steels ... **M6:** 291-292
high-strength medium-carbon quenched and tempered steels ... **M6:** 247
line pipe ... **M6:** 278-280
tool steels ... **M6:** 294-297
Arc welding of aluminum alloys ... **A6:** 722-739
aluminum filler alloys ... **A6:** 724
applications ... **A6:** 727, 736
base metals ... **M6:** 373
corrosion resistance ... **A6:** 724, 729-730
cracking ... **M6:** 386-387
distortion ... **A6:** 727, 728
ductility ... **A6:** 724, 728
edge preparation ... **M6:** 375
electrical conductivity ... **A6:** 723
electrodes ... **A6:** 722-723 **M6:** 381-382, 390
filler alloy choices ... **A6:** 730
filler alloy, selection criteria ... **A6:** 724-730
filler alloys for sustained elevated-temperature service ... **A6:** 729
filler metals ... **A6:** 722, 724-727, 729, 730-737, 739 **M6:** 373-374, 377-378, 391-392
fillet weld strength ... **A6:** 727-728
fixtures ... **M6:** 379-380
forms of aluminum ... **A6:** 724
fusion zone ... **A6:** 727
galvanic corrosion ... **A6:** 729
gas metal arc welding ... **M6:** 380-390, 397-398
gas tungsten arc welding ... **M6:** 390-398
groove joints, tensile strength after welding ... **A6:** 729
groove weld strength ... **A6:** 726-727
heat-affected zone ... **A6:** 725, 726, 727, 729
heat-treatable cast aluminum alloys ... **A6:** 724
heat-treatable wrought aluminum alloys ... **A6:** 723, 727, 728
hydrogen solubility ... **A6:** 722
joining aluminum to other metals ... **A6:** 739
joint design ... **M6:** 374-375
joint designs ... **A6:** 730-736
non-heat-treatable cast aluminum alloys ... **A6:** 723, 726
oxide ... **A6:** 722
percussion welding ... **M6:** 399
porosity ... **M6:** 386
postweld heat treatments ... **A6:** 726-727, 728
powder metallurgy parts ... **A6:** 724
power supplies ... **M6:** 383-385, 392-395
preheating ... **M6:** 378-379
preparation for welding ... **A6:** 730-736
preweld cleaning ... **M6:** 375, 378
process selection and comparsion ... **M6:** 397-399
properties affecting welding ... **A6:** 722
properties of aluminum ... **A6:** 722-724
quality control requirements for filler rods and electrodes ... **A6:** 730
sensitivity to weld cracking ... **A6:** 724, 725-726
shielded metal arc welding ... **M6:** 398-399
shielding gases ... **M6:** 380, 390-391
soundness of welds ... **M6:** 385-388
spray transfer arc ... **M6:** 381
stress-corrosion cracking ... **A6:** 727
stud welding ... **M6:** 399
Arc welding of aluminum alloys (continued)
surface preparation ... **A6:** 736
temperature vs. performance ... **A6:** 724, 729
thermal characteristics ... **A6:** 723-724
weld and base metal color match ... **A6:** 730
weld backing ... **M6:** 382-383 , 392
weldability ... **M6:** 373
casting alloys ... **M6:** 373
wrought alloys ... **M6:** 373
welding processes ... **A6:** 736-739
electrogas welding ... **A6:** 738
electron-beam welding ... **A6:** 739
electroslag welding ... **A6:** 738
gas-metal arc welding ... **A6:** 722, 723, 726, 729, 730, 731-735, 737-738, 739
gas-tungsten arc welding ... **A6:** 725, 729, 730, 731-735, 736, 737, 738, 739
laser-beam welding ... **A6:** 739
oxyfuel gas welding ... **A6:** 738-739
plasma arc welding ... **A6:** 735, 736-737
shielded metal arc welding ... **A6:** 738
Arc welding of beryllium
joint design ... **M6:** 462
processes and procedures ... **M6:** 461-462
safety ... **M6:** 462
shielding gases ... **M6:** 462
surface preparation ... **M6:** 461
weld repair ... **M6:** 462
Arc welding of carbon steels ... **A6:** 641-660
definition of carbon steels ... **A6:** 64
electrodes ... **A6:** 641, 642, 643, 652
electrogas welding ... **A6:** 652, 653, 658-659, 660
electroslag welding ... **A6:** 652, 653, 659
flux-cored arc welding ... **A6:** 643, 647, 652, 653, 654, 657-658
gas-metal arc welding ... **A6:** 647, 652, 653, 654, 655, 657
gas-tungsten arc welding ... **A6:** 652, 653, 654, 655-656, 658
heat affected zone ... **A6:** 641, 642, 647, 648, 649, 652, 658, 659, 660
hydrogen-induced cracking ... **A6:** 641, 642-649
factors causing ... **A6:** 642
hydrogen sources ... **A6:** 643
prevention of ... **A6:** 644-647
temperature range ... **A6:** 644
tensile stresses ... **A6:** 643-644
inspection methods for crack detection ... **A6:** 642-643
lamellar tearing ... **A6:** 651, 652
plasma arc welding ... **A6:** 652, 653, 654, 658
porosity ... **A6:** 641, 651-652
killed steels ... **A6:** 652
postweld heat treatment ... **A6:** 641, 645-647, 648, 649
process selection ... **A6:** 652-655
shielded metal arc welding ... **A6:** 651, 652, 653, 654, 656-657
solidification cracking ... **A6:** 649-651, 657
stud arc welding ... **A6:** 652, 653, 656-660
submerged arc welding ... **A6:** 642, 643, 647, 648, 649, 650, 652, 653, 654, 658
weldability considerations ... **A6:** 642-652
welding consumable selection and procedure development ... **A6:** 655-660
Arc welding of cast irons
applications ... **M6:** 307
consumables ... **M6:** 310-311
copper-based electrodes ... **M6:** 311
filler metals for flux cored arc welding ... **M6:** 311
filler metals for other processes ... **M6:** 311
filler metals for shielded metal arc welding ... **M6:** 310-311
nickel-based electrodes ... **M6:** 311
corrosion- and heat-resistant cast irons ... **M6:** 316
design requirements ... **M6:** 307
ductile irons ... **M6:** 315-316
electrodes ... **M6:** 310-311
filler metals ... **M6:** 310-311
gray irons ... **M6:** 314-315
malleable irons ... **M6:** 315-316
microstructures ... **M6:** 308-310
fusion zone ... **M6:** 310
heat-affected zone ... **M6:** 309-310
partially melted region ... **M6:** 309-310
postweld heat treatment ... **M6:** 314

Arc welding of cast irons (continued)
practical applications **M6:** 316-318
preparation for welding **M6:** 311-312
procedures and processes **M6:** 310
flux cored arc welding **M6:** 310
gas metal arc welding **M6:** 310
gas tungsten arc welding **M6:** 310
shielded metal arc welding **M6:** 310
submerged arc welding **M6:** 310
repair welding **M6:** 316-317
temperature, preheat
and interpass **M6:** 312-313
martensite behavior **M6:** 313
preheating **M6:** 312-313
welding metallurgy of **M6:** 307-308
compacted or vermicular graphite
irons **M6:** 308
ductile cast iron **M6:** 308
gray cast irons **M6:** 308
malleable iron **M6:** 308
white iron **M6:** 307-308
welding practices **M6:** 314
welding techniques **M6:** 313
backstep technique **M6:** 313
block sequence **M6:** 313
cascade sequence **M6:** 313
for heat input control **M6:** 313
peening **M6:** 313
stringer and weave beads **M6:** 313
white and alloy cast irons **M6:** 316
Arc welding of copper and copper alloys **M6:** 400-426
developments in welding
fine-wire gas metal arc welding **M6:** 426
plasma arc welding **M6:** 426
processes **M6:** 426
pulsed-current gas metal arc
welding **M6:** 426
pulsed-current gas tungsten arc
welding **M6:** 426
submerged arc welding **M6:** 426
effects of alloying elements **M6:** 400-402
factors affecting weldability **M6:** 402-404
hot cracking **M6:** 402
joint design **M6:** 402
porosity **M6:** 402
precipitation-hardenable alloys **M6:** 402
shielding gas **M6:** 402
surface condition **M6:** 404
thermal conductivity **M6:** 402
welding position **M6:** 402
gas metal arc welding **M6:** 415-424
aluminum bronzes **M6:** 420-421
beryllium copper, high-conductivity **M6:** 418-419
beryllium copper, high-strength **M6:** 419-420
brasses **M6:** 420
copper alloys to dissimilar metals **M6:** 424
copper nickels **M6:** 421-422
coppers **M6:** 417-418
dissimilar copper alloys **M6:** 422-424
electrode wires **M6:** 416-417
filler metals **M6:** 415, 417
phosphor bronzes **M6:** 420
silicon bronzes **M6:** 421
welding conditions **M6:** 417
welding position **M6:** 417
gas tungsten arc welding **M6:** 400, 404-415
aluminum bronzes **M6:** 412-413
beryllium copper, high-conductivity **M6:** 408
beryllium copper, high-strength **M6:** 408-409
cadmium and chromium coppers **M6:** 409
copper nickels **M6:** 414
copper-zinc alloys **M6:** 409-410
coppers **M6:** 405-408
dissimilar metals **M6:** 414-415
electrodes **M6:** 404
filler metals **M6:** 404
mechanized applications **M6:** 404
Arc welding of copper and copper alloys (continued)
nickel silver **M6:** 409
phosphor bronze **M6:** 410-412
silicon bronzes **M6:** 413-414
type of current **M6:** 404
shielded metal arc welding **M6:** 424-426
aluminum bronzes **M6:** 425
brasses **M6:** 425
copper nickels **M6:** 426
copper to aluminum **M6:** 425
coppers **M6:** 425
phosphor bronzes **M6:** 425
silicon bronzes **M6:** 425-426
Arc welding of heat-resistant alloys **M6:** 353-370
alloy composition **M6:** 354
cobalt-based alloys **M6:** 367-370
cleaning **M6:** 368
gas metal arc welding **M6:** 370
gas tungsten arc welding **M6:** 368-370
shielded metal arc welding **M6:** 370
weld defects **M6:** 368
electrode composition **M6:** 359
filler metal composition **M6:** 359
fixtures **M6:** 354-355
backing bars **M6:** 353
for gas tungsten arc welding **M6:** 355
gas metal arc welding **M6:** 362, 366-367, 370
joint designs **M6:** 362
welding techniques **M6:** 362
gas tungsten arc welding **M6:** 358-361, 365-366, 368-370
crack repair **M6:** 369-370
filler metals **M6:** 358-360, 365
joint preparation and fit-up **M6:** 365-366
resistance seam welding **M6:** 360-361
shielding gas **M6:** 358, 365
tack welding **M6:** 366
welding techniques **M6:** 361
iron-nickel-chromium and
iron-chromium-nickel alloys **M6:** 364-367
gas metal arc welding **M6:** 366-367
gas tungsten arc welding **M6:** 365-366
joint design **M6:** 365
shielded metal arc welding **M6:** 367
submerged arc welding **M6:** 367
treatments, pre- and postweld **M6:** 365
nickel-based alloys **M6:** 355-364
cold work effect **M6:** 358
gas metal arc welding **M6:** 362
gas tungsten arc welding **M6:** 358-362
joint design **M6:** 356-357
overaging **M6:** 358
shielded metal arc welding **M6:** 362-363
treatments, pre- and postweld **M6:** 357-358
weld defects **M6:** 363-365
shielded metal arc welding **M6:** 362-363, 367, 370
electrodes **M6:** 363, 367
welding conditions **M6:** 363
submerged arc welding **M6:** 367
weld defects **M6:** 363-365, 368
cold shuts and surface pits **M6:** 364
cracks and fissures **M6:** 363
notches **M6:** 364
porosity and inclusions **M6:** 363
strain-age cracking **M6:** 363-364
void at the root joint **M6:** 364
workpiece cleaning **M6:** 353
Arc welding of magnesium alloys **M6:** 427-435
filler metals **M6:** 427-428
gas metal arc welding **M6:** 429-430
metal transfer **M6:** 429
operating conditions **M6:** 429-430
power supplies **M6:** 429
welding positions **M6:** 429
gas tungsten arc welding **M6:** 429-432
automatic welding **M6:** 432-433
manual welding **M6:** 430-432
power supplies **M6:** 430
Arc welding of magnesium alloys (continued)
short-run production **M6:** 431
joint design **M6:** 428-429
edges **M6:** 428
fit-up **M6:** 428
grooves in backing bars **M6:** 428-429
welding fixtures **M6:** 428-429
postweld heat treatment **M6:** 435
preheating **M6:** 429
repair welding of castings **M6:** 432-435
shielding gases **M6:** 428
surface preparation **M6:** 429
weldability **M6:** 427
Arc welding of molybdenum and tungsten **M6:** 462-465
filler metals **M6:** 464
gas tungsten arc welding **M6:** 464
interstitial contamination **M6:** 463
preheating **M6:** 464
stress relieving **M6:** 463-464
surface cleaning **M6:** 463
alkaline-acid procedure **M6:** 463
dual-acid procedure **M6:** 463
welding conditions **M6:** 464
welding heat input **M6:** 464
Arc welding of nickel alloys **M6:** 436-445
cast nickel alloys **M6:** 438
cleaning of workpieces **M6:** 437
fixtures for welding **M6:** 437-438
backing bars **M6:** 437
clamping and restraint **M6:** 438
gas metal arc welding **M6:** 439-440
electrodes **M6:** 439
filler metals **M6:** 440
joint design **M6:** 440
shielding gas **M6:** 439-440
welding current **M6:** 439
welding techniques **M6:** 440
gas tungsten arc welding **M6:** 438-439
electrodes **M6:** 438-439
filler metals **M6:** 439
joint design **M6:** 439
shielding gas **M6:** 438
welding techniques **M6:** 439
joining of dissimilar metals **M6:** 443-445
dilution of weld metal **M6:** 444
filler metal selection **M6:** 444-445
joint design **M6:** 437
beveled joints **M6:** 437
corner and lap joints **M6:** 437
design considerations **M6:** 437
plasma arc welding **M6:** 440
postweld treatment **M6:** 436
precipitation-hardenable alloys **M6:** 438
general welding procedures **M6:** 438
susceptibility to cracking **M6:** 438
treatments, pre- and postweld **M6:** 438
preweld heating and heat treating **M6:** 436
shielded metal arc welding **M6:** 440-442
cleaning the weld bead **M6:** 442
electrodes **M6:** 440-441
joint design **M6:** 441-442
welding current **M6:** 441
welding position **M6:** 441
welding techniques **M6:** 441-442
submerged arc welding **M6:** 442
bead deposition **M6:** 442
electrodes **M6:** 439, 442
fluxes **M6:** 442
joint design **M6:** 442-443
welding current **M6:** 442
weld defects **M6:** 442-443
cracking **M6:** 443
effect of slag on weld metal **M6:** 443
porosity **M6:** 442-443
Arc welding of niobium
gas tungsten arc welding **M6:** 460
surface preparation **M6:** 460
weldability **M6:** 459

Arc welding of niobium (continued)
welding processes **M6:** 459-460
Arc welding of stainless steels
alloying elements **M6:** 320
austenitic stainless steels **M6:** 320-344
carbide precipitation **M6:** 321-323
contamination of welds **M6:** 321-322
electrogas welding **M6:** 344
electroslag welding **M6:** 344
ferrite estimation **M6:** 322-323
flux cored arc welding **M6:** 326-327
gas metal arc welding **M6:** 329-333
gas tungsten arc welding **M6:** 333-342
heat of welding **M6:** 321-322
microfissuring **M6:** 321-323
plasma arc welding **M6:** 342-344
porosity **M6:** 324
postweld stress relieving **M6:** 323-324
preheating **M6:** 323
shielded metal arc welding **M6:** 324-326
solution annealing **M6:** 321
stabilized steels **M6:** 321
submerged arc welding **M6:** 326-329
underbead cracking **M6:** 323
weld characteristics **M6:** 320
austenitic stainless steels, nitrogen-strengthened **M6:** 344-345
electrogas welding **M6:** 345
electroslag welding **M6:** 345
gas metal arc welding **M6:** 345
gas tungsten arc welding **M6:** 345
plasma arc welding **M6:** 345
shielded metal arc welding **M6:** 345
submerged arc welding **M6:** 345
electrodes
flux cored arc welding **M6:** 326-327, 349
gas metal arc welding **M6:** 331-332, 349
gas tungsten arc welding **M6:** 333, 349
shielded metal arc welding **M6:** 324-325, 345
submerged arc welding **M6:** 327-328, 349
ferritic stainless steels **M6:** 346-348
corrosion resistance **M6:** 347
ductility **M6:** 346
electrogas welding **M6:** 348
electroslag welding **M6:** 348
filler metal selection **M6:** 346-347
flux cored arc welding **M6:** 347
gas metal arc welding **M6:** 348
gas tungsten arc welding **M6:** 347-348
grain size **M6:** 346
plasma arc welding **M6:** 348
postweld annealing **M6:** 346
preheating **M6:** 346
shielded metal arc welding **M6:** 347
submerged arc welding **M6:** 347
temperature, effect on notch toughness **M6:** 346
welding heat, effects of **M6:** 346
martensitic stainless steels **M6:** 348-349
flux cored arc welding **M6:** 349
gas tungsten arc welding **M6:** 349
postweld heat treating **M6:** 348-349
preheating **M6:** 348-349
shielded metal arc welding **M6:** 349
submerged arc welding **M6:** 349
precipitation-hardening stainless steels **M6:** 349-352
austenitic **M6:** 351-352
compositions **M6:** 350
martensitic steels **M6:** 349-350
semiaustenitic steels **M6:** 350-351
shielding gas
gas metal arc welding **M6:** 332
gas tungsten arc welding **M6:** 334
welding current
flux cored arc welding **M6:** 327
gas metal arc welding **M6:** 329
gas tungsten arc welding **M6:** 333-335
Arc welding of tantalum **M6:** 460-461
applications **M6:** 461
gas tungsten arc welding **M6:** 460-461
surface preparation **M6:** 461
weldability **M6:** 460
welding processes **M6:** 460-461
Arc welding of titanium and titanium alloys **M6:** 446-456
cleaning **M6:** 448-449
degreasing **M6:** 448-449
oxide removal **M6:** 449
filler metals **M6:** 447
composition **M6:** 447
preparation **M6:** 447
gas metal arc welding **M6:** 446, 455-456
gas tungsten arc welding **M6:** 446, 453-455
equipment **M6:** 453-455
hot wire process **M6:** 455
procedures **M6:** 454-455
joint preparation **M6:** 448
plasma arc welding **M6:** 446, 456
repair welding **M6:** 456
setups for welding **M6:** 447
shielding gas **M6:** 447-448
stress relieving **M6:** 456
weldability **M6:** 446
alpha alloys **M6:** 446
alpha-beta alloys **M6:** 446
beta alloys **M6:** 446
unalloyed titanium **M6:** 446
welding in chambers **M6:** 449-451
metal chambers **M6:** 449-450
plastic chambers **M6:** 451
welding out of chambers **M6:** 451-453
Arc welding of zirconium and hafnium **M6:** 456-459
filler metals **M6:** 457
shielding gases **M6:** 458
weldability **M6:** 456-457
welding process **M6:** 458-459
Arc welding processes
application **EL1:** 238
Arc welds
contours **A11:** 412
discontinuities in **A17:** 582-585
double-submerged **A11:** 698
failure origins **A11:** 412-415
gas tungsten, voids in **A11:** 93
plasma, defective **A11:** 415
radiographic inspection **A17:** 334-335
Arc-gouged drain groove
dendritic crack structure in **A11:** 647
Arc-outs **A6:** 860
Arc-rib mist hackle **EM4:** 639, 640
Arc-sprayed coatings
resistance to cavitation erosion **A18:** 217
Arc-welded alloy steel
failures in **A11:** 423-426
Arc-welded aluminum alloys
distortion in **A11:** 436
failures in **A11:** 434-437
gas porosity **A11:** 434
inclusions **A11:** 435-436
incomplete fusion in **A11:** 435
undercuts in **A11:** 435
weld cracks in **A11:** 435
Arc-welded hardenable carbon steel
failures in **A11:** 422-423
Arc-welded heat-resisting alloys
failures in **A11:** 433-434
Arc-welded low-carbon steel
failures in **A11:** 415-422
hot cracking in **A11:** 416
inclusions in **A11:** 416
Arc-welded nonmagnetic ferrous tubular products
inspection **A17:** 566-567
Arc-welded stainless steel
austenitic **A11:** 427-428
failures in **A11:** 426-433
ferritic **A11:** 428
martensitic **A11:** 427
precipitation-hardening **A11:** 428
Arc-welded titanium and titanium alloys
failures in **A11:** 437-439
Archaeological samples
of books and artifacts, PIXE analysis for **A10:** 102
ion selective electrode to determine copper in **A10:** 186
as NAA application **A10:** 234
Archard wear law *See also* Wear coefficient; Wear constant; Wear factor **A18:** 282
defined **A18:** 3
Archard's equation **A18:** 185, 188, 189, 571
Archimedes experiment
to determine density of ceramic powders **EM4:** 27
Architectural Aluminum Manufacturers Association (AAMA) peel adhesion test
for latex **EM3:** 213
Architectural applications
of cast iron **A15:** 18
of electroplating **A15:** 22
of lead **A15:** 20-21
Architectural bronze
applications and properties **A2:** 312-313
Architectural glass **EM4:** 1021
cooling energy versus electric lighting requirements **EM4:** 1022
laminated glass **EM4:** 1021
parameters that can be controlled by specific residential and commercial glazing **EM4:** 1022
regulation of heat and light **EM4:** 1021
tempered glass **EM4:** 1021
total transmission versus light transmission **EM4:** 1022
wired glass **EM4:** 1021
Architectural panels, porcelain enameled
steel sheets for **M1:** 180
Architecture
design, introduction **EL1:** 1
evaluated by modeling **EL1:** 15
innovations **EL1:** 10
machine, engineering design system for **EL1:** 127
system design **EL1:** 2
system, WSI, testing **EL1:** 376-377
VSLI-optimized **EL1:** 2
Architecture applications
titanium and titanium alloy **A2:** 589-590
Arcing **A6:** 365
brush/slip ring assembly failure from **A12:** 488
Arcing contacts *See also* Electrical contact materials
defined **A2:** 840
as failure factor **A2:** 841
property requirements for make-break contacts **A2:** 841
and sliding contacts, compared **A2:** 841
Arcing damage
in electrical contact materials **A9:** 563-564
Arctic pipeline steel X-80
laser-beam welding **A6:** 264
ARE (BARE) process **A18:** 845
Area
as basic convex figure quantity **A12:** 194
of closed figure **A12:** 195
conversion factors **A8:** 722 **A10:** 686
of cross section, symbol **A8:** 724
effect in galvanic corrosion **A13:** 83
fractal analysis **A12:** 214-215
fracture surface **A12:** 201-205
of fracture surface, importance **A12:** 193
of interconnections **EL1:** 20
of irregular fracture surfaces **A12:** 211-215
parametric relationships **A12:** 204-205
planar circuit, interconnection requirements **EL1:** 6
profile angular distributions for **A12:** 202-204
ratios, for partially oriented surfaces **A12:** 201
reduction, and strain **A8:** 575
reduction of **A11:** 8, 338
SI derived unit and symbol for **A10:** 685
SI unit/symbol for **A8:** 721
stereological relationships **A12:** 196
and surface parameters **A12:** 200
total facet surface **A12:** 202
triangular elements for **A12:** 201-202
true, and true length **A12:** 204
true fracture surface, importance **A12:** 211
true mean facet **A12:** 208
true, vertical sectioning for **A12:** 198-199, 211-212
vs lead count **EL1:** 211
Area (array) tape automated bonding (TAB)
defined **EL1:** 275
Area amplitude *See also* Ultrasonic inspection
blocks **A17:** 264
blocks, standard reference **A17:** 264
curves, determined, ultrasonic inspection **A17:** 265-266

Area channeling analysis *See also* Channeling
crystallographic texture measurement and analysis by **A10:** 357
Area cooling
for thermal inspection **A17:** 398
Area fraction **A18:** 464
effects of varying in image analysis **A10:** 311-312
Area heating
for thermal inspection **A17:** 398
Area of contact
defined **A18:** 3
Area reduction *See also* Reduction **A14:** 368, 378
Area scan
EMPA analog mapping as **A10:** 525
Areal ratio
equality of volume fraction to **A9:** 125
Areal weight **EM1:** 4, 125, 286, 737
defined **EM2:** 4
Argentometric titration
to analyze the bulk chemical composition of starting powders **EM4:** 72
Argon
active metal brazing atmosphere **EM4:** 525
adsorbed, effects in FIM **A10:** 588
as an ion bombardment etchant for fiber composites **A9:** 591
atomized aluminum powder **A7:** 130
atomized nickel-based superalloy powder, Auger composition-depth profile **A7:** 254
atomized powders, Auger profile of **A7:** 254, 255
atomized stainless steel powders **A7:** 101-103
atomized superalloys, composition-depth profile **A7:** 254
bubbling, nitrogen removal by **A15:** 84
by residual gas analysis (RGA) **EL1:** 1065
in CAP process **A7:** 533
characteristics in a blend **A6:** 65
contamination **A7:** 434
continuous-wave gas lasers **A10:** 128
as converter gas **A15:** 426
crack propagation in **A1:** 719, 720
as cutting fluid for tool steels **A18:** 738
duplex stainless steel welding backing gas **A6:** 474
electrogas welding shielding **A6:** 275
electron-beam welding atmosphere **A6:** 857
electron-beam welding, high-strength alloy steels **A6:** 867
flux-cored arc welding shielding gas **A6:** 187, 189
fume generation from shielding gases **A6:** 68
gas atomization **A7:** 27
gas-metal arc welding shielding gas **A6:** 181, 185, 489
for aluminum alloys **A6:** 738
for copper alloys **A6:** 755
for nickel alloys **A6:** 743-744, 745, 746
gas-tungsten arc welding shielding gas **A6:** 1022
for aluminum alloys **A6:** 736
for copper alloys **A6:** 756
ferritic stainless steels **A6:** 445, 452, 453
hot isostatic pressing atmosphere **A6:** 884
ionization potential **A6:** 64
low-heat-input welding of low-alloy steels **A6:** 662
magnesium alloy shielding gas **A6:** 772
plasma-MIG welding **A6:** 224
plasma-MIG welding shielding gas **A6:** 224
shielding gas for aluminum metal-matrix composites **A6:** 555
shielding gas for arc welding of low-alloy steels **A6:** 66
shielding gas for FCAW **A6:** 67
shielding gas for GMAW **A6:** 66, 67
shielding gas for GMAW of cast irons **A6:** 718, 719
shielding gas for GTAW **A6:** 32, 33, 67
shielding gas for GTAW of cast irons **A6:** 720
Argon (continued)
shielding gas for plasma arc welding **A6:** 197
shielding gas for zirconium alloys **A6:** 787-788
shielding gas properties **A6:** 64
shielding gas purity and moisture **A6:** 65
steel weldment soundness in gas-shielded processes **A6:** 408, 409
stud arc welding shielding gas **A6:** 211, 214
thermal conductivity **A6:** 64
torch gas for gas metal arc welding **A6:** 23
valve thread connections for compressed gas cylinders **A6:** 1197
for high-temperature torsion testing **A8:** 159
ion etching, use in XPS **A10:** 575
ion-sputtering, XPS analysis **A10:** 575, 576
ionization potentials and imaging fields for **A10:** 586
leakage flow rates **A7:** 433, 434
liquid, for hot isostatic pressing **A7:** 421, 422
for low-current plasma arc welding **A6:** 68
mean free path **A17:** 59
-oxygen mixture **A15:** 426
in plasma arc powder spraying process **A18:** 830
for plasma arc spraying **A6:** 811
for plasma arc welding **A6:** 197
plasma, use in ICP-AES **A10:** 31
purging system, vacuum induction furnace **A15:** 397
purified, for sintering powder-rolled titanium strip **A7:** 394
recirculation and purification system **A7:** 27, 30
removal **A7:** 180-181
sigma values for ionization **A17:** 68
in sputter deposition process **A18:** 840
in stainless steels **A1:** 930
used for ion-beam thinning of transmission electron microscopy specimens **A9:** 107
in weld relay, gas mass spectrometry of **A10:** 156
working gas for ion implantation **A18:** 857
Argon dry box welding **A7:** 431
Argon gas
in insulating glass unit production **EM3:** 196
Argon ion lasers
for optical holographic interferometry **A17:** 417
Argon National Laboratory
Intense Pulsed Neutron Source **A10:** 424
Argon oxygen decarburization **A14:** 222
Argon shielding
in plasma melting/casting **A15:** 420
Argon shielding gas
comparison to helium **M6:** 197-198
effect on weld metallurgy **M6:** 41
furnace atmosphere for brazing **M6:** 1008-1009
removal of oxygen **M6:** 199
Argon shielding gases for
arc welding of
coppers **M6:** 402
magnesium alloys **M6:** 428
molybdenum and tungsten **M6:** 462-464
stainless steels, austenitic **M6:** 331-332, 334
titanium and titanium alloys **M6:** 447-448
electron beam welding **M6:** 623
gas metal arc welding **M6:** 163-164
aluminum alloys **M6:** 380
nickel alloys **M6:** 439-440
nickel-based heat-resistant alloys **M6:** 362
gas tungsten arc welding **M6:** 197-199
aluminum alloys **M6:** 390-391
iron-nickel-chromium and iron-chromium-nickel heat-resistant alloys **M6:** 365
nickel alloys **M6:** 438
nickel-based heat-resistant alloys **M6:** 358-360
plasma arc welding **M6:** 217
Argon stirring
direct current arc furnace **A15:** 368
Argon-carbon dioxide
flux-cored arc welding, low-alloy steels **A6:** 662
flux-cored arc welding shielding gas **A6:** 189
gas-metal arc welding, low-alloy steels **A6:** 662
Argon-helium
flux-cored arc welding shielding gas **A6:** 189
gas-metal arc welding shielding gas **A6:** 181, 185
gas-tungsten arc welding shielding gas for aluminum alloys **A6:** 736
shielding gas for GMAW **A6:** 66-67
shielding gas for GTAW **A6:** 67
shielding gas for plasma arc welding **A6:** 197
Argon-hydrogen
plasma arc welding of aluminum alloys **A6:** 735
shielding gas for GTAW **A6:** 67-68
shielding gas for plasma arc cutting **A6:** 1167, 1168, 1170
thermal conductivity **A6:** 64
Argon-oxygen
flux-cored arc welding, low-alloy steels **A6:** 662
gas-metal arc welding, low-alloy steels **A6:** 662
plasma-MIG welding shielding gas **A6:** 224
shielding gas for GMAW **A6:** 66, 67
surface tension **A6:** 65
Argon-oxygen decarburization
development **A15:** 36
equipment **A15:** 427
fundamentals **A15:** 426-427
nickel alloys **A15:** 820
oxygen top and bottom blowing, as extension **A15:** 428-429
of plain carbon steels **A15:** 709
processing **A15:** 427-428
as secondary refining **A15:** 426
vessel, schematic **A15:** 427
Argon-oxygen decarburization (AOD)
ultrahigh-strength steels **A4:** 207
Argon-oxygen decarburization (AOD) process **A1:** 841, 970 **A6:** 593
ferritic stainless steels **A6:** 443, 444
shielding gas for GMAW **A6:** 66, 67
Argon-oxygen deoxidation (AOD) **A1:** 930
Argon/carbon dioxide/hydrogen
shielding gas for GMAW **A6:** 67
Argon/oxygen/carbon dioxide
shielding gas for GMAW **A6:** 67
ARGUS computer program for structural analysis **EM1:** 268, 270
Argyria, local
as silver poisoning **A7:** 205
Arimax **EM3:** 294
Arithmetic average surface roughness **A18:** 436
Arithmetic mean *See* Sample average
Arm
definition **M6:** 1
Arm bearing
compaction of iron-based **A7:** 706
Armature binding wire **A1:** 851-852
Armature, dc motor
fabrication failure of **A11:** 421
Armature-type sensitive relays
recommended microcontact materials **A2:** 866
Armco iron **A18:** 878
stress-strain curves **A8:** 174
Armco iron friction welded to carbon steel **A9:** 156
Armor plate
drilling **A16:** 219
Hy 80, milling **A16:** 317
Hy 100, milling **A16:** 317
wrought, milling **A16:** 312, 313
Armor-piercing cores
powders used **A7:** 573
Arms
for quick-release clamp **A8:** 221

SUBJECTS OF THE INDEXED VOLUMES: ASM Handbook (designated by the letter "A"): **A1:** Properties and Selection: Irons, Steels, and High-Performance Alloys (1990); **A2:** Properties and Selection: Nonferrous Alloys and Special-Purpose Materials (1990); **A3:** Alloy Phase Diagrams (1992); **A4:** Heat Treating (1991); **A6:** Welding, Brazing, and Soldering (1993); **A7:** Powder Metallurgy (1984); **A8:** Mechanical Testing (1985); **A9:** Metallography and Microstructures (1985); **A10:** Materials Characterization (1986); **A11:** Failure Analysis and Prevention (1986); **A12:** Fractography (1987); **A13:** Corrosion (1987); **A14:** Forming and Forging (1988); **A15:** Casting (1988); **A16:** Machining (1989); **A17:** Nondestructive Testing and Quality Control (1989); **A18:** Friction, Lubrication, and Wear Technology (1992). **Metals Handbook, 9th Edition** (designated by the letter "M"): **M1:** Properties and Selection: Irons and Steels (1978); **M2:** Properties and Selection: Nonferrous Alloys and Pure Metals (1979); **M3:** Properties and Selection: Stainless Steels, Tool Materials and Special-Purpose Materials (1980); **M4:** Heat Treating (1981); **M5:** Surface Cleaning, Finishing, and Coating (1982); **M6:** Welding, Brazing, and Soldering (1983). **Engineered Materials Handbook** (designated by the letters "EM"): **EM1:** Composites (1987); **EM2:** Engineering Plastics (1988); **EM3:** Adhesives and Sealants (1990); **EM4:** Ceramics and Glasses (1991); **Electronic Materials Handbook** (designated by the letters "EL"): **EL1:** Packaging (1989).

Army Materials and Mechanics Research Center (AMMRC) **A8:** 724
verification of Charpy impact test apparatus **A8:** 263
Arnold meters **A7:** 274, 275
Aroma components
IR determination of **A10:** 109
Aromatic
defined **A10:** 669 **EM1:** 4 **EM2:** 4
Aromatic amines **EM1:** 70
Aromatic amines, as curing agents
epoxies **EL1:** 828
Aromatic co-polyester **EM3:** 601
Aromatic dianhydrides
for polyimide coatings **EL1:** 767
Aromatic divinyl compounds
bismaleimide reaction with **EM2:** 255
Aromatic ethers
as engineering thermoplastic **EM2:** 449
Aromatic hydrocarbons
determined **A10:** 218
Aromatic isocyanates
in polyurethanes **EM2:** 257
Aromatic polyamide
typical properties **EM3:** 83
Aromatic polyamide fibers *See also* Aramid fibers; p-aramid fibers; Para-aramid fibers **EM1:** 30
Aromatic polyarylates (PARs) *See also* Polyarylates (PAR)
chemistry **EM2:** 66
Aromatic polyester *See also* Polyesters **EM3:** 5
defined **EM2:** 4
Aromatic polysulfones (PSU) chemistry **EM2:** 66
properties **EM2:** 450
Aromatic rings, polymer
chemical structure **EM2:** 52
Aromatic silicones
Raman analysis **A10:** 132
Aromatic solvents
as cleaning agents **EL1:** 663
Aromatic sulfones *See also* Aromatic polysulfones; Polysulfones
as engineering thermoplastics **EM2:** 449
Aromatic thermoplastic polyesters **EM2:** 65-66
Aromatic thermoplastic polyimides (TPIs) **EM2:** 177-178
Aromatics **EM3:** 5
high-temperature resistant **EM3:** 76
Arrangement of atoms
as x-ray diffraction analysis **A10:** 325
Array
-based system functions **EL1:** 8-9
physical, types **EL1:** 9
Arrays
data, defined **A17:** 383
detector, defined **A17:** 383
orthogonal **A17:** 748-750
processor, defined **A17:** 383
types of **A7:** 296
Arrest
of pipeline fractures **A11:** 704-706
Arrest line *See also* Beach marks
in ceramics **A11:** 747
definition **EM4:** 632
Arrest mark
definition **EM4:** 632
Arrest marks *See* Beach marks
Arrhenius collision coefficient **A18:** 571
Arrhenius constant
tribological **A18:** 280, 283
Arrhenius equation **EM4:** 460
Arrhenius expression
simplified **EM4:** 55-56
Arrhenius function
for liquid diffusivity **A15:** 103
Arrhenius plots
delayed failure by SMIE **A13:** 186
Arrhenius plots of delayed failure **A11:** 240, 244
Arrowhead defects
1008 steel **A9:** 183
Arsenic *See also* Arsenic toxicity
alloying, wrought aluminum alloy **A2:** 46
alloying, wrought copper and copper alloys **A2:** 242
in cast iron **A1:** 5, 8
cause of temper embrittlement **A4:** 124, 135
Arsenic (continued)
in compacted graphite iron **A1:** 59
determined by controlled-potential coulometry **A10:** 209
-doped gold contact wires **EL1:** 958
effect in copper alloys **A11:** 221, 635
effect of addition on lead-antimony-tin microstructures **A9:** 417
effect on tin-base alloys **A18:** 748
effects, cartridge brass **A2:** 301
effects, electrolytic tough pitch copper **A2:** 270
effects of, on notch toughness **A1:** 742
epithermal neutron activation analysis of **A10:** 239
evaporation fields for **A10:** 587
extending epoxy lifetimes **EM3:** 170
as fining agent **EM4:** 380
fire refining effect **A15:** 453
gaseous hydride for ICP sample introduction **A10:** 36
in glass, K-edge EXAFS spectra **A10:** 411
high-purity, production **A2:** 747
ICP-determined in natural waters **A10:** 41
impurity in solders **M6:** 1072
as inoculant **A15:** 105
iodimetric titrations of **A10:** 174
as lead additive **A2:** 545
in lead-base alloys **A18:** 749-750
as major toxic metal with multiple effects **A2:** 1237-1239
as minor element, ductile iron **A15:** 648
photometric analysis methods **A10:** 64
pure **M2:** 716
pure, properties **A2:** 1101
quartz tube atomizers for **A10:** 49
recommended impurity limits of solders **A6:** 986
safety standards for soldering **M6:** 1098
sample modification, for GFAAS analysis **A10:** 55
in silicon on ion-implanted silicon samples **A10:** 632
as solder impurity **EL1:** 637
species weighed in gravimetry **A10:** 172
in steel weldments **A6:** 420
as tin solder impurity **A2:** 520
TNAA detection limits **A10:** 237
toxicity **A2:** 1237-1238 **A6:** 1195
as trace element **A15:** 394
vapor pressure, relation to temperature **A4:** 495
volatilization losses in melting **EM4:** 389
volatilizing **A10:** 166
Volhard titration for **A10:** 173
volumetric procedures for **A10:** 175
weighed as the sulfide **A10:** 171
Arsenic as platinum alloy **A7:** 15
Arsenic chloride
distillation **A10:** 169
Arsenic in lead, age-hardening
effect on **M4:** 741
Arsenic in steel
notch toughness, effect on **M1:** 694
temper embrittlement, role in **M1:** 684-685
Arsenic oxide (AS_2O_3)
glass forming ability **EM4:** 494
melting point **EM4:** 484
in ovenware compositions **EM4:** 1103
purpose for use in glass manufacture **EM4:** 381
specific properties imparted in CTV tubes **EM4:** 1039
in tableware compositions **EM4:** 1101
viscosity at melting point **EM4:** 494
Arsenic oxides
examination under polarized light **A9:** 400
Arsenic toxicity
biologic indicators **A2:** 1238
biotransformation **A2:** 1237
carcinogenicity **A2:** 1238
cellular effects **A2:** 1237
disposition **A2:** 1237
reproductive effects and teratogenicity **A2:** 1238
toxicology **A2:** 1237-1238
treatment **A2:** 1238
Arsenic, vapor pressure
relation to temperature **M4:** 310
Arsenical admiralty metal
applications and properties **A2:** 318-319
Arsenical babbitts
applications **A2:** 553
Arsenical copper
in Bronze Age **A15:** 15-16
Arsenical leaded Muntz metal
applications and properties **A2:** 311
Arsenical naval brass
applications and properties **A2:** 319-320
Arsine gas
formation and toxicity **A2:** 1238
Art founding
classical sculpture **A15:** 20-21
colossal statues **A15:** 21
gilding **A15:** 21
modem statuary **A15:** 21-22
Arthrodeses
as corrective orthopedic surgery **A11:** 671
Arthroplasty **A18:** 656-658
Artifact
defined **A17:** 383
Artifacts
in aluminum alloys as a result of mechanical polishing **A9:** 353
in carbon and alloy steels **A9:** 165-166, 168
defined **A9:** 2
distorting SIMS depth profiles **A10:** 619
in infiltrated powder metallurgy materials **A9:** 504
peaks, as AEM-EDS microanalytic limitation **A10:** 448
in replicas **A12:** 184-185
spectral, sum and escape peaks as **A10:** 520
structures **A9:** 36-37
in wavelength-dispersive spectrum **A10:** 520
Artificial accelerated aging *See* Accelerated aging
Artificial aging *See also* Aging **EM3:** 5
defined **A9:** 2 **A13:** 2 **EM2:** 4
wrought aluminum alloy **A2:** 40
Artificial graphite, electrodes
resistance brazing **A6:** 341
Artificial intelligence **A14:** 247, 409
and NDE reliability models **A17:** 713
Artificial joints
types of **A11:** 670-671
Artificial lift wells **A13:** 1247-1248
Artificial twist boundaries in gold
transmission electron microscopy **A9:** 120-121
Artificial weathering *See also* Aging; Weather resistance **EM3:** 5
defined **EM1:** 4 **EM2:** 4-5
Artificially aged tempers and solution-heat- treated tempers in aluminum alloys
etchants for distinguishing **A9:** 355
Artware
glazes **EM4:** 1061
Artwork *See also* Master drawing
compensation **EL1:** 627
flexible printed boards **EL1:** 593
layer, and substrate compensation **EL1:** 623-624
master, in final design package **EL1:** 524-525
quality, and computer-aided design **EL1:** 527, 529
rigid printed wiring boards **EL1:** 548-549
Aryldiazonium salt,
cationic curing with **EL1:** 859
As-Au (Phase Diagram) .... **A3:** 2•56
As-Bi (Phase Diagram) .... **A3:** 2•57
As-brazed
definition **M6:** 1
As-cast condition
defined **A15:** 1
As-casting
gear materials **A18:** 261
As-Cd (Phase Diagram) .... **A3:** 2•57
As-Co (Phase Diagram) .... **A3:** 2•58
As-Cu (Phase Diagram) .... **A3:** 2•58
As-Fe (Phase Diagram) .... **A3:** 2•58
As-Ga (Phase Diagram) .... **A3:** 2•59
As-Ge (Phase Diagram) .... **A3:** 2•59
As-In (Phase Diagram) .... **A3:** 2•59
As-K (Phase Diagram) .... **A3:** 2•60
As-Mn (Phase Diagram) .... **A3:** 2•60
As-Nd (Phase Diagram) .... **A3:** 2•60
As-Ni (Phase Diagram) .... **A3:** 2•61
As-P (Phase Diagram) .... **A3:** 2•61
As-Pb (Phase Diagram) .... **A3:** 2•61

As-Pd (Phase Diagram) .... A3: 2 • 62
As-quenched hardness A1: 471, 476 M1: 478, 481
As-rolled pearlitic steels A1: 399, 404
for steel tubular products A1: 328, 329, 330, 331, 332, 333, 334
As-rolled structural steels *See also* Heat treated HSLA steels; Microalloyed HSLA steels M1: 403, 404
compositions, typical M1: 404
mechanical property distributions M1: 412-416
As-S (Phase Diagram) .... A3: 2 • 62
As-Sb (Phase Diagram) .... A3: 2 • 62
As-Se (Phase Diagram) .... A3: 2 • 63
As-Si (Phase Diagram) .... A3: 2 • 63
As-Sn (Phase Diagram) .... A3: 2 • 63
As-Te (Phase Diagram) .... A3: 2 • 64
As-Tl (Phase Diagram) .... A3: 2 • 64
As-welded
definition A6: 1206 M6: 1
As-Yb (Phase Diagram) .... A3: 2 • 64
As-Zn (Phase Diagram) .... A3: 2 • 65
ASA *See* Styrene-acrylonitriles
ASAAS II computer program for structural analysis EM1: 268, 270
Asahi
solar-cell cover glass product EM4: 1019
Asbestos EM1: 60, 115 EM3: 175
for compression packings EM3: 226
crystal structure EM4: 882
drilling A16: 229, 230
as filler EM3: 177, 178
function and composition for mild steel SMAW electrode coatings A6: 60
for gaskets EM3: 224, 225
no longer a slag former A6: 61
for organic brake linings A18: 569, 570
toxicity of brake wear debris A18: 574
Asbestos board
waterjet machining A16: 522
Asbestosis A7: 202
ASEA presses
Quintus fluid forming A14: 614-615
rubber-pad forming A14: 608
ASEA Quintus fluid-cell process A14: 610-611
ASEA-STORA process A7: 784-787 A16: 60
ASF-1307, crystallizing solder glass designation
commercially available EM4: 1070
ASF-1307B, crystallizing solder glass designation
commercially available EM4: 1070
Ash A18: 84
and BTU determined in coal A10: 100
deposits, superheater A13: 1201
-handling systems, fossil fuel power plants A13: 1007-1008
Miller numbers A18: 235
Ash content EM3: 5
defined EM1: 4 EM2: 5
Ash removal
steam equipment A11: 619-620
Ashby-Brown contrast *See also* Black-white contrast
of second-phase precipitates A9: 117
of stacking-fault tetrahedra A9: 117
Ashby-Frost deformation maps A14: 421
Ashing
of organic liquids and solutions, for analysis A10: 10
of organic solids, for analysis A10: 9
oxygen plasma dry A10: 167
of samples, GFAAS analysis A10: 55
in second-phase testing A10: 177
ASKA computer program for structural analysis EM1: 268, 270
ASLE *See* American Society of Lubrication Engineers
ASM INTERNATIONAL EM4: 38, 39, 40
as information source EM1: 40
ASME *See* American Society of Mechanical Engineers
ASME specifications *See also* listings in data compilations for individual alloys; Steels, ASME specific types M1: 132
constructional steels for elevated temperature use M1: 647-648, 654
copper tube and pipe M2: 264
Asp fasteners
materials and composite applications for A11: 530
ASP steels (anti-segregation process steels) A7: 784-787
austenitizing A7: 786
grades, compositions, applications A7: 785
heat treatment of A7: 785-787
properties A7: 784-785
Aspect ratio A18: 580, 802, 803 EM1: 4, 120 EM3: 5
at fracture, forged disk A7: 411
conductive filler EM2: 474
defined EM2: 5
effect on free-surface strain A8: 580
effects A14: 66
of particle shape A7: 242
Asperities
in adhesive wear A11: 145
defined A11: 1 A18: 3
deformation A11: 149
in delamination theory of wear A11: 148
in initial adhesion A11: 148
Asperity A18: 46
defined A8: 1 A18: 3
Asperity contact theory A18: 46
Asperity deformation behavior A18: 33, 59, 60
Asperity film thickness or microelastohydrodynamic film thickness A18: 94
nomenclature for lubrication regimes A18: 90
Asperity lubrication modes A18: 94-96
micro-EHL and friction polymer films A18: 94-95
oxide film A18: 94, 96
physically adsorbed and other surface films A18: 94, 95-96
Asphalt
appliance market applications EM3: 59
applications EM3: 56
as carbon mold addition A15: 211
cure properties EM3: 51
extender for urethane sealants EM3: 205
fracture/failure causes illustrated A12: 217
for highway construction joints EM3: 57
for refrigerator and freezer cabinet sealing EM3: 59
for rubber battery sealing EM3: 58
sulfur, fracture surface A12: 473
suppliers EM3: 59
Asphalt reclamation
cemented carbide tools for A2: 973
Asphalt roof coating
powder used A7: 572
Asphalt tile mixer gate
materials for wear resistance M1: 622
Asphalts
as bituminous coatings A13: 406
Asphalts and coal tar resins
chemistry EM3: 50, 51
for highway and airfield joint sealing EM3: 51
properties EM3: 50, 51
Asphyxiation
dangers of sintering atmospheres A7: 349
Aspiration atomization A7: 125
Assay
defined A10: 669
Assembled structures
computed tomography (CT) inspection A17: 364
Assembly *See also* Assembly and manufacture; Flip chip assembly EM3: 5
and assembly forms EM1: 665, 681-737
board, lead frame materials EL1: 488
board, through-hole packages EL1: 970
of boilers/pressure vessels A17: 646
Assembly (continued)
Boothroyd-Dewhurst design for, method EL1: 124-125
by electromagnetic forming A14: 644
by spinning A14: 604
ceramic multilayer, technologies compared EL1: 297-298
CERDIPs EL1: 487
costs EM2: 649
defined EL1: 1134
design requirements for EM1: 182-183
for differentiation EL1: 438-441
drawing, in final design package EL1: 524
of dual-in line package, plastic postmolded EL1: 487
equipment, selection EL1: 732
failures EL1: 943, 1058
final, cost of EM2: 650
fixtures A16: 404, 406
hermetic/nonhermetic EL1: 484-485
high-density, on substrates EL1: 439
levels, environmental stress screening EL1: 877-878, 884
and manufacture, design for EL1: 119-126
methods EM2: 711-725
NC implemented A16: 616
for optimum chip performance EL1: 438
package, lead frame EL1: 484-485
package requirements, lead frame EL1: 487-488
of plastic packages EL1: 471-479
of plastic pin-grid arrays EL1: 475-476
purpose EL1: 449
for quality EL1: 441-448
as stress source A11: 205
techniques, thick/thin-film technology EL1: 255
through-hole and surface mount EL1: 437-438
time, defined EM1: 4 EM2: 5
ultrasonic EM2: 721-722
uniaxis EL1: 121-122
wafer-scale, as hybrids EL1: 250
Assembly adhesive EM3: 5
Assembly and assembly forms *See also* Machining; specific assembly techniques EM1: 665, 681-737
adhesive bonding surface preparation EM1: 681-682
adhesives selection EM1: 683-688
adhesives specifications EM1: 689-701
blind fastening EM1: 709-711
bonding cure considerations EM1: 702-705
dissimilar material separation EM1: 716-718
fastener hole considerations EM1: 712-715
faying surface sealing EM1: 719-720
fiber properties analysis EM1: 731-735
honeycomb structure EM1: 721-728
mechanical fastener selection EM1: 706-708
quality control EM1: 729-730
resin properties analysis EM1: 736-737
Assembly and manufacture *See also* Assembly
design for EL1: 119-126
design for manufacturability EL1: 121-125
early manufacturing involvement (EMI) EL1: 125
future directions EL1: 125-126
historical background EL1: 119-120
value engineering EL1: 120-121
Assembly methods
heat welding and sealing EM2: 724-725
inserts EM2: 722-724
mechanical fastening EM2: 711-713
press and snap fits EM2: 713-721
solvent and adhesive bonding EM2: 725
ultrasonic assembly EM2: 721-722
Assembly, mold *See* Mold assembly
Assembly techniques
for A15 tape superconductors A2: 1065-1067
for multifilamentary wire A15 superconductors A2: 1065-1067
Assembly time EM3: 5

Assessment
of corrosion damage **A13:** 194-195
Asset depreciation range **A13:** 372
Assumption of randomness
in projected images **A12:** 194-195
ASTAP simulation **EL1:** 33-34, 38
Asthma
as toxic reaction to cobalt **A7:** 204
Astigmatism
defined **A9:** 2
in SEM illuminating/imaging system **A12:** 167
ASTM *See* American Society for Testing and Materials; American Society, for Testing and Materials; American Society for Testing and Materials (ASTM)
ASTM 163 etchant for
niobium **A9:** 440
tantalum **A9:** 440
ASTM A 27 .... **A9:** 230
ASTM A 47 .... **A9:** 245
ASTM A 48 .... **A9:** 245
ASTM A 148 .... **A9:** 230-231
ASTM A 216 .... **A9:** 230-231
ASTM A 220 .... **A9:** 245
ASTM A 247 .... **A9:** 161
ASTM A 319 .... **A9:** 161
ASTM A 352 .... **A9:** 231
ASTM A 436 .... **A9:** 245
ASTM A 439 .... **A9:** 245
ASTM A 487 .... **A9:** 231
ASTM A 518 .... **A9:** 245
ASTM A 532 .... **A9:** 245
ASTM A 536 .... **A9:** 245
ASTM A 602 .... **A9:** 245
ASTM A 677
silicon iron electrical steels **A9:** 537
ASTM A 683
silicon iron electrical steels **A9:** 537
ASTM A 763, Practice Z
for ferritic stainless steels **A13:** 125-126
ASTM B 215
for powder metallurgy materials **A9:** 505
ASTM B 276
for apparent porosity of cemented carbides **A9:** 274
ASTM B 328
for powder metallurgy materials **A9:** 503-504
ASTM B 390
for grain size of cemented carbides **A9:** 274
ASTM B 657
for microstructure of cemented carbides **A9:** 274
ASTM compositions for plate steels **A9:** 202
ASTM E 112 .... **A9:** 129
application to austenitic manganese steel castings **A9:** 238
applied to grain size measurement of aluminum alloys **A9:** 357
dendrite arm spacing measurements in zinc **A9:** 490
ASTM E 381
center segregation **A9:** 173-174
ASTM F 746
pitting corrosion test method **A13:** 231
ASTM G 48
pitting corrosion test method **A13:** 231
ASTM G 61
pitting corrosion test method **A13:** 231
ASTM grain-size number **A9:** 129
for austenitic manganese steel castings obtaining **A9:** 238
nomograph for **A9:** 130
ASTM microcontact tester
life test using **A2:** 858
ASTM pipe steels
compositions **A9:** 211
ASTM Special Technical Publications
wear tests for ceramics (STP 1010) **EM4:** 605, 608
wear tests for coatings (STP 769) **EM4:** 605
wear tests for metals (STP 615) **EM4:** 605
wear tests for plastics (STP 701) **EM4:** 605
ASTM specifications *See also* listings in data compilations for individual alloys; Nonferrous alloys, ASTM specific types; Steels ASTM specific types **A1:** 150, 154, 156, 162, 163-164, 334
abrasion-resistant cast irons **M1:** 82, 86
ASTM specifications (continued)
for alloy steel pressure pipe and pressure tubes **A1:** 331, 334
alloy steel sheet and strip **M1:** 164
alloy steel, use at subzero temperatures **M3:** 739-740, 754
aluminum casting alloys, former designations *See* Aluminum casting alloys, former ASTM designations
for aluminum coatings **A1:** 218-219
for aluminum-zinc alloy coatings **A1:** 220
for austenitic grain size **A1:** 274
austenitic manganese steel **M3:** 568, 574
bearing materials **M3:** 813, 814
of carbon and alloy steel pipe **A1:** 332
for carbon and alloy steel pressure tubes **A1:** 333
for carbon and alloy steel structural and mechanical tubing **A1:** 335
for carbon and low-alloy steels for elevated-temperature services **A1:** 618
for carbon steel rod **A1:** 274
for carburizing steels **A1:** 483
cast steels **M1:** 377-378, 401
for chromate passivation **A1:** 215
coatings for steel sheet **M1:** 168-174
aluminum coatings **M1:** 172
chromate passivation, testing of **M1:** 169
hot dip galvanized **M1:** 170-171
terne coatings **M1:** 174
tin coatings **M1:** 173
zinc coatings, tests for **M1:** 168
for cold-finished steel bars **A1:** 251
composition ranges and limits **M1:** 135-140
for concrete reinforcement rod **A1:** 274
constructional steels for elevated temperature use **M1:** 647, 648
copper casting alloys **M2:** 384-386
copper tube and pipe **M2:** 264
copper wire **M2:** 266-273
creep and stress-rupture tests **M3:** 229
for deformation rates **A8:** 39-40
description **M1:** 119, 132, 134
for ductile iron **A1:** 34, 36, 40, 41 **M1:** 34, 35, 36
fastener materials **M3:** 184, 185
for flake graphite **A1:** 13
for fracture toughness **A1:** 341
generic designations **M1:** 134-135
gray cast iron **M1:** 16-17
for gray iron **A1:** 15, 16, 17, 22
heat-resistant castings **M3:** 269
for high-carbon steels **A1:** 483
for high-strength carbon and low-alloy steels **A1:** 390
hot rolled steel bars and shapes **M1:** 204-212
for hot-rolled steel bars and shapes **A1:** 240, 241, 242, 243, 244, 245, 246, 247
HSLA steels **M1:** 403, 405-407, 409-410
for low-alloy steel **A1:** 208, 209
low-carbon steel sheet and strip **M1:** 154, 155
for machinability testing for screw machines **A1:** 593
magnesium alloys designations **M2:** 525-526, 527, 528
for nonmetallic inclusion testing **A1:** 274
for notch toughness **A1:** 753
on hardenability **A1:** 464
P/M materials **M1:** 330, 332, 333
pipe, steel **M1:** 317, 318, 320-321, 322, 324, 325
plate **M1:** 183-184, 186-189
plate, discussion of **M1:** 134-136
pressure-vessel plate, discussion of **M1:** 136
sheet products, discussion of **M1:** 134-135
for stainless steel products **A1:** 932
for steel castings **A1:** 364, 365, 366, 368, 370, 377, 378, 379
for steel tubular products **A1:** 335
for structural quality steel plate **A1:** 230, 236
structural shapes, discussion of **M1:** 135-136
for structural steel **A1:** 664
temper designations, copper metals **M2:** 248-251
for terne coatings **A1:** 221
testing of superhard tool materials **M3:** 452, 453
for threaded fasteners **A1:** 289, 290, 294, 295, 296
for tin coatings **A1:** 221
titanium **M3:** 357, 360
for tool steels **A1:** 757, 759
ASTM specifications (continued)
tubing, steel **M1:** 321, 323, 324, 325, 326
for weathering steels **A1:** 399
for wrought tool steels **A1:** 757
for zinc coatings **A1:** 213-215
ASTM standards
air-entraining admixtures for concrete (C 260) **EM4:** 921
air-entraining cements designated (C 175) **EM4:** 12
annealing grade measurement of glass containers (C 148) **EM4:** 1085
annealing point of glass determined (C 336) **EM4:** 567
annealing point of glass determined by midpoint deflection of a glass beam (C 598) **EM4:** 567
bend strength tests on ceramics (C 1161) **EM4:** 710
blended cement manufacture (C 595) **EM4:** 918
block-on-ring test (G 77) **EM4:** 606
bond-strength measurement test on coated specimens (C 633) **EM4:** 992
ceramic tile property measurements **EM4:** 927
chemical admixtures for concrete (C 494) **EM4:** 921
chemical durability rating basis (C 225) **EM4:** 876, 877
chipping extent of brick allowable (C 216) **EM4:** 946
compressive strength as an index of structural clay product durability (C 67) **EM4:** 946
emittance **EM4:** 611
"Flow Rate and Tap Density of Electrical Grade Magnesium Oxide for Use in Sheathed-Type Electric Heating Elements" (D 3347-86) **EM4:** 71
flowability measurement procedure (B 213) **EM4:** 107
fly ash and raw or calcined natural pozzolan for use as a mineral admixture in concrete (C 618) **EM4:** 921
ground blast-furnace slag for use in concrete and mortars (C 989) **EM4:** 921
industrial flooring prescribed types (C 410) **EM4:** 950
lead and cadmium extractions, test methods for (C 738, C 895, C 927, C 1034) **EM4:** 1065
lead and cadmium release from porcelain enamel surfaces (C 872) **EM4:** 1065
for material selection **A13:** 322
"Method for Tension and Vacuum Testing Metallized Ceramic Seals " (F 19) **EM4:** 513, 515, 516
mortar specifications (C 270) **EM4:** 947
on plastics **EM2:** 90
on strength during overpressurization or thermal shock (C 47-62, C-149-50) **EM4:** 743
paving brick specification, discontinued in 1980 (C 7) **EM4:** 949
pedestrian and light traffic paving brick specifications (C 902) **EM4:** 949
portland cement manufacture (C 150) **EM4:** 918
powder bulk density measurement procedure (B 212) **EM4:** 107
sieve aperture variety (E 11-61, E 11-87) **EM4:** 66
size distribution measurement procedure (B 214) **EM4:** 107
sliding wear of ceramics test (pin-on-disk) (G 99) (1990) **EM4:** 605, 606
softening point temperature and viscosity of glass related (C 338) **EM4:** 567
specific heat capacity **EM4:** 611
specifications on strength during overpressurization or thermal shock (C 47-62, C 149-50) **EM4:** 743
standard definitions of terms relating to thermophysical properties (E 1142) **EM4:** 611
"Standard Test Method for the Specific Gravity of Soils" (D 854-83) **EM4:** 71
strain point determined in glass (C 336, C 598) **EM4:** 567

ASTM standards (continued)
strength measurement technique for whiteware products (C 674) **EM4:** 567
test for photoelasticity of glass (C 770) **EM4:** 565
test measuring the viscosity of glass above the softening point (C 965) **EM4:** 567
for testing waters/materials in water **A13:** 208
thermal conductivity/diffusivity **EM4:** 611
thermal expansion **EM4:** 611
thermal expansion measurement of glass (C 372, E 228) **EM4:** 568
three tests to measure the durability of glass containers (C 225) **EM4:** 566
to determine density achieved by compacting spray-dried powder at a given pressure (B 331) **EM4:** 107
ultrasonic attenuation determination using immersion technique (E 664) **EM4:** 623
ultrasonic velocity measurement in materials (E 494) **EM4:** 621

ASTM Standards (American Society for Testing and Materials)
abrasive wear testing of TiN coated tool steels (G 65-10 modified) **A18:** 739
API gravity definition and formula (D 287) **A18:** 82
"Apparent Viscosity of Lubricating Greases" (D 1092) **A18:** 127, 128
apparent viscosity, pour-point depressants, and cold cranking simulator (D 2602) **A18:** 83, 108
ash measurement of oils (D 482) **A18:** 84
block-on-ring machine and friction testing (D2 Committee) **A18:** 49
borderline pumping temperature determination (D 3892) **A18:** 163
borderline pumping temperature of engine oil (D 4684) **A18:** 84
borderline pumping test, pour-point depressants (D 3829) **A18:** 108
Brookfield viscometer, detection of presence of glycol coolants (D 2983) **A18:** 83, 300
"Calibration and Operation of Alpha Model LFW-1 Friction and Wear Testing Machine" (D 2714; D-2 Committee on lubricants) **A18:** 50
capillary viscometer to measure viscosity **A18:** 84
carbon/graphite types of materials in coefficient of friction tests (C-5 Committee) **A18:** 49
"Carburizing Steels for Anti-Friction Bearings," inclusion standards (A 534) **A18:** 875
cavitation damage resistance assessment, vibratory tests (G 32) **A18:** 217, 218, 226, 600, 762, 763, 769
"Classification of and Specifications for Automotive Service Greases" (D 4950) **A18:** 125
classification of porosity into 3 types (B 276 procedure) **A18:** 797
Cleveland open cup procedure to determine flash and fire points (G 92) **A18:** 84
coefficient of friction and wear definitions (G 40) **A18:** 47, 48, 228, 367
"Coefficient of Friction and Wear of Sintered Metal," inclined plane method (B 526; B-9 Committee on metal powders) **A18:** 50
"Coefficient of Friction, Yam to Yam," extension of D 3108 test (D 3412; D-13 Committee on textiles) **A18:** 49, 51
"Coefficient of Static Friction of Corrugated and Solid Fiberboard" **A18:** 49, 51
inclined plane method (D 3248; D-6 Committee on paper) **A18:** 49, 51

ASTM Standards (American Society for Testing and Materials) (continued)
sled testing (D 3247; D-6 Committee on paper) **A18:** 49, 51
color matching with standards for lubricants **A18:** 82
color-indicator methods for oils, acidity, and alkalinity **A18:** 84, 300
"Compatibility of Lubricating Grease with Elastomers" (D 4289) **A18:** 129
composite friction materials for clutches and brakes (B-9 Committee) **A18:** 48
concentration of water in lubricants measured (D 1744) **A18:** 300
Cone Penetration Test (NGLI specification IIE, grease testing) for classifying consistency of greases (D 217) **A18:** 13, 125, 127, 136
Conradson method for measurement of carbon residue (D 187) **A18:** 84
conversion factors between current and previously used viscosity units (D 2161) **A18:** 82-83
crossed-cylinder wear test (G 83) **A18:** 721
detection of presence of glycol coolants (D 2982, D 2983, D 2984) **A18:** 300
determination of a lubricant's boiling point range using temperature-programmed gas chromatography (D 2887) **A18:** 84
determination of the Miller number, a measure of abrasivity of a slurry (G 75) **A18:** 1, 17, 234-235, 597
determination of the precision of a test method by way of interlaboratory study (E 691) **A18:** 486, 487
"Determining Pavement Surface Frictional and Polishing Characteristics using a Small Torque Device" (E 510; E-17 Committee on traveled surfaces) **A18:** 52, 53
development of a standard format for friction databases, maximum erosion rate and incubation period equations (G-2 Committee) **A18:** 58, 228
dropping point test, test description and results (D 566) **A18:** 127
dry sand-rubber wheel abrasive wear test (G 65) **A18:** 2, 7, 362, 759, 761-762, 768, 774, 804, 805, 807
dry sand/rubber wheel abrasion tests (G 65-B) **A18:** 808
"Dynamic Coefficient of Friction and Wear of Sintered Metal Friction Materials under Dry Conditions" (B 460; B-9 Committee on metal powders) **A18:** 50
"Evaluation of Test Data Obtained by Using the Horizontal Slipmeter or the James Machine for Measurement of Static Slip Resistance of Footwear, Sole, Heel, or Related Material" (F 695-81; F-13 Committee on footwear) **A18:** 53
"Evaporation Loss of Grease," evaporation loss of lubricating greases at temperatures between 95 and 315 °C (D 2595) **A18:** 127
evaporation test procedure used for motor oils and other oils (D 972) **A18:** 84
foaming tendency assessment of a lubricant **A18:** 85, 108
friction and floor finishes (D-21 Committee) **A18:** 49
friction coefficient of flooring and sole leather (D-7 Committee) **A18:** 49
friction test for plastic films versus other solids (D-20 Committee) **A18:** 49
friction tests on footwear (F-13 Committee) **A18:** 49
"Frictional Characteristics of Sintered Metal Friction Materials Run in Lubricants" (B 461; B-9 Committee on metal powders) **A18:** 50

ASTM Standards (American Society for Testing and Materials) (continued)
frictionometer used to determine coefficient of friction of plastic solids and sheeting (D 3028; D-20 Committee on plastics) **A18:** 49, 51
graphite defined by shape, size, and distribution (A 247-47 and A 247-67) **A18:** 698
Jernkontoret system, resistance to rolling contact fatigue (spalling) and carburizing (E 45) **A18:** 875
kinematic viscosity measurement of lubricants (D 445) **A18:** 82, 134, 140, 163, 300
Knoop (microindentation) hardness number determination test method **A18:** 12, 417
liquid impingement erosion testing and erosion resistance number definition **A18:** 226-227, 228, 229-230
"Low Temperature Torque of Ball Bearing Greases" (D 1478) **A18:** 127
"Low Temperature Torque of Grease Lubricated Wheel Bearings" (D 4693) **A18:** 127
low-temperature viscosity determination (D 2802, modified) **A18:** 163
measurement of carbon residue by Ramsbottom method (D 524) **A18:** 84
measurement of evaporation tendency of lubricants (D 2715) **A18:** 84
"Measuring Surface Frictional Properties using the British Pendulum Tester" (E E-17 Committee on traveled surfaces) **A18:** 49, 52
method for calculating viscosity index from kinematic viscosities at two temperatures **A18:** 134
nomenclature and grading of graphite (A 247-47; A 247-67) **A18:** 698
nominal compositions (A 485-1; A 485-3) **A18:** 725
oxidative stability test method (D 943; D 2272, D 2893, D 4742) **A18:** 84, 105
performance specifications for engine lubrication **A18:** 98
permanent viscosity loss measurement (D 3945) **A18:** 84
pin-on-disk sliding wear testing (G 99) **A18:** 363, 364, 365
potentiometric method to measure acidity and alkalinity (D 664) **A18:** 84, 300
pour point testing of lubricants (D 97) **A18:** 83, 108, 127
"Preparation of Substrate Surfaces for Coefficient of Friction Testing" (D 4103, D-21 Committee on polishes) **A18:** 49, 52
"Rating of Static Coefficient of Shoe Sole and Heel Materials as Measured by the James Machine" (F 489; F-13 Committee on footwear) **A18:** 52, 53
"Reciprocating Pin-on-Flat Evaluation of Friction and Wear Properties of Polymeric Materials for Use in Total Joint Prostheses" (F F-4 on medical and surgical materials) **A18:** 53
reference oil viscosities tabulated and formula for viscosity index numbers over 100 (D 2270) **A18:** 83
"Reporting Friction and Wear Test Results of Manufactured Carbon and Graphite Bearing and Seal Materials" (C 808; C-5 Committee on carbon/graphite) **A18:** 50
Rockwell hardness testing procedure (E 40) **A18:** 16
rotational tapered plug tests (D 4741) **A18:** 84
rubber wheel abrasion tests (B 611) **A18:** 804, 805, 807
sequence dynamometer engine tests **A18:** 100
sequence IID engine test for engine oils corrosion inhibition **A18:** 106

ASTM Standards (American Society for Testing and Materials) (continued)
sequence IIIE and VE engine tests for antiwear agent effectiveness **A18:** 101, 105
shear stability test, test description and results (D 217A) **A18:** 127
shell roller test, test description and results (D 1831) **A18:** 127
"Side Force Friction on Paved Surfaces using the Mu-Meter" (E 670; E-17 Committee on traveled surfaces) **A18:** 52, 53
"Simulated Service Testing of Wood and Wood-Base Finish Flooring" (D 2394; D-7 Committee on Wood) **A18:** 50
"Skid Resistance of Paved Surfaces using the North Carolina State University Variable Speed Friction Tester" (E 707; E-17 Committee on traveled surfaces) **A18:** 52, 53
sled and inclined plane tests (D-6 Committee) **A18:** 49
slurry/steel wheel test for high-stress abrasion (B 611) **A18:** 768, 797, 799
slurry/steel wheel test for high-stress abrasion resistance (B 611, modified) **A18:** 759
solid-particle gas-jet impingement erosion test (G 76) **A18:** 767, 768, 769, 805
"Standard Method for Extreme-Pressure Properties of Lubricating Grease (Four-Ball Method)" (D 2596) **A18:** 127
"Standard Method for Measurement of Extreme-Pressure Properties of Lubricating Grease (Timken Method)" (D 2509) **A18:** 127
"Static and Kinetic Coefficients of Friction of Plastic Films and Sheeting" (D 1894) **A18:** 50
"Static Coefficient of Friction of Polish Coated Floor Surfaces as Measured by the James Machine" (D 2047; D-21 Committee on polishes) **A18:** 49, 50
sulfated ash measurement of unused oils with metal-containing additives (D 874) **A18:** 84
surveillance panels, performance testing of engine oils **A18:** 170
tapered bearing simulator to measure viscosity (D 4683) **A18:** 84
"Test Method for Static Slip Resistance of Footwear, Sole, Heel, or Related Materials by Horizontal Pull Slipmeter (HPS)" (F 609) **A18:** 53
"Testing of Fabrics Woven from Polyolefin Monofilaments," static coefficient of friction test using inclined planes (D 3334; D-13 Committee on textiles) **A18:** 49, 51
tin-base bearing alloy compositions (B 23) **A18:** 748, 749
to measure lubricant rust prevention properties (D 665, D 3603) **A18:** 84
to test greases at high speeds (D 3336) **A18:** 127
total base number measurement (D 2896) **A18:** 310
U.S. Steel, bleeding and evaporation testing, test description and results **A18:** 127
Vickers microindentation hardness test (E 284) **A18:** 20
viscosity grades of lubricants (D 2422) **A18:** 85
viscosity-temperature equation, defined **A18:** 3
viscosity-temperature relation (MacCoull-Walther) equation (D 341) **A18:** 83
wear rate determination of self-lubricated materials, thrust washer test procedure (D 3702) **A18:** 820, 821, 822
wear testing using the block-on-ring machine (G 77) **A18:** 2, 49
wheel bearing test, test description and results (D 1263) **A18:** 127
yarn versus solid materials test method using the capstan formula for measuring friction (D 3108; D-13 Committee on textiles) **A18:** 49, 51
STM standards for qualitative metallography of cemented carbides **A9:** 274

ASTM test methods **EM2:** 334
ASTM test methods, P-W (C 225)
durability of glasses **EM4:** 540
ASTM tube steels
compositions **A9:** 211
ASTM/ASME alloy steels *See also* ASTM/ASME alloy steels, specific types; Steels(s)
fractographs **A12:** 345-350
fracture/failure causes illustrated **A12:** 216
hydrogen flaking **A12:** 125
temper embrittlement **A12:** 134
ASTM/ASME alloy steels, specific types *See also* ASTM/ASME alloy steels; Steel(s)
ASME SA213, creep failure **A12:** 346
ASTM 533B pressure vessel, hydrogen effects on fracture appearance **A12:** 37, 51
ASTM A325 bolt, SCC in **A12:** 133
ASTM A372, solidification cracking, laser-beam welds **A12:** 345
ASTM A490, bolt specimens **A12:** 103, 104
ASTM A508 class II, overheating **A12:** 146
ASTM A508, fracture by overpressurization **A12:** 345
ASTM A514F, effect of inclusions on fatigue crack propagation **A12:** 346
ASTM A517H, brittle fracture **A12:** 347
ASTM A533B, cavitated intergranular fracture, hydrogen attack **A12:** 349
ASTM A533B, effect of inclusions on fatigue crack propagation **A12:** 347, 348
Cr-Mo-V, elevated-temperature fracture surface **A12:** 349
Cr-Mo-V, phosphorus effect on cavities **A12:** 349
Cr-Mo-V, phosphorus effect on ductility **A12:** 349
Ni-Cr antimony-doped, hydrogen effects **A12:** 350
Astralloy-V
adhesive wear resistance **A18:** 721
corrosive wear **A18:** 719
Astrology
application of station function approach to **A8:** 334, 336
Astroloy *See* Nickel-base superalloys, specific types
aging **A4:** 796
aging cycle **M4:** 656
aging precipitates **A4:** 796
annealing **M4:** 655
applications **A4:** 807
composition **A4:** 794 **A6:** 573 **A16:** 736 **M4:** 651-652
drilling **A16:** 746, 747
heat treatments **A4:** 807-808
machining **A16:** 738, 741-743, 746-747, 749-758
sawing **A16:** 360
solution treating **A4:** 796 **M4:** 656
stress relieving **M4:** 655
thread grinding **A16:** 275
Astrophysical research
qualitative spectral analysis for **A10:** 43
Astroquartz II fibers **EM1:** 61
Asymmetric populations **A8:** 632
Asymmetric reflections
in double-crystal spectrometry **A10:** 371
Asymmetric rod impact test **A8:** 204-205
Asymmetric stripline properties
impedance models **EL1:** 602-603
Asymmetric transmission
Guinier camera arrangement **A10:** 335
Asymmetric waves
in wave soldering **EL1:** 688
Asymmetrical diffraction optics **A18:** 467-468
Asymptotic continuum mechanics
with fracture mechanics **A8:** 465
Asymptotic curvature
fractal **A12:** 211-212
AT&T interactive systems analysis model **EL1:** 13-15
Atactic polypropylene
rheological behavior in injection molding **EM4:** 174-175
shear modulus **EM4:** 176
Atactic stereoisomerism *See also* Isotactic stereoisomerism; Syndiotactic stereoisomerism **EM3:** 5
defined **EM2:** 5

ATE electrical testing
for failure verification/fault isolation **EL1:** 1060
ATEM *See* Analytical transmission electron microscopy
Athermal
defined **A9:** 2
Athermal martensite in titanium alloys **A9:** 461
ATL *See* Automatic tape layers
Atlantic Ocean
carbon steel/wrought iron corrosion **A13:** 898-899
Atmosphere
abbreviation for **A8:** 724 **A11:** 796
of creep-rupture testing **A8:** 303
for elevated/low temperature tension testing **A8:** 37
steel corrosion protection in **M1:** 751, 752
Atmosphere control **A4:** 568-572
dew point instrument **A4:** 568, 569, 570, 571
endothermic **A4:** 568, 569-572 **M4:** 362-364
equipment adjustment **A4:** 570, 571-572 **M4:** 364
exothermic **A4:** 568, 569-572 **M4:** 362-364
infrared analyzer **A4:** 568, 569, 570, 571
oxygen analysis **A4:** 569, 571
oxygen probes **A4:** 568-569, 570-571 **M4:** 365-366
purpose **A4:** 568 **M4:** 361-362
shim analysis **A4:** 568
systems **A4:** 570, 572 **M4:** 361-362
total combustibles analyzer (catalytic type) **A4:** 571
Atmosphere heat treating
high-speed tool steels **A16:** 55-56
Atmospheres *See also* Atmospheric corrosion; Industrial atmospheres; Marine atmospheres; Reducing atmosphere; Rural atmospheres; Sintering atmospheres specific sintering atmospheres; Urban atmospheres
analyzers **A4:** 577-586
annealing of steel, furnaces **A4:** 45-46, 50
brazeability and solderablity considerations **A6:** 621-622, 623, 626
brazing **A7:** 457
brazing, clad brazing material applications **A6:** 963
brazing of ceramic and ceramic-to-metal joints **A6:** 955- 956, 957
brazing of cobalt-base alloys **A6:** 928
brazing of reactive metals **A6:** 941
brazing of refractory metals **A6:** 941
brazing, safety precautions **A6:** 1202
brazing types per AWS specification B2.2 .... **A6:** 622, 628
carbon restoration **A4:** 598-599
carburizing atmospheres **A2:** 834
composition ,effect on sintered iron graphite powders **A7:** 363-364
composition for sintering steel **A7:** 339
contamination **A2:** 835
controlled, for brazing **M6:** 1016
inert gases **M6:** 1017
pure dry hydrogen **M6:** 1016-1017
vacuum **M6:** 1017
cooling by **A7:** 341
copper or copper alloy brazing **A6:** 931, 933, 934
corrosion rates by **A13:** 510-511
corrosion testing in **A13:** 204-206
corrosivity of **A11:** 192
crucible furnace **A15:** 383
dewpoints **A7:** 340, 341
diffusion welding **A6:** 884
of dissociated ammonia **A7:** 344
effect, electrical contact materials **A2:** 859-860
effect on sintered aluminum P/M parts **A7:** 385
effect on stress-corrosion cracking **A11:** 208-209
effects of composition **A7:** 367
of endothermic gas **A7:** 341-343
of exothermic gas **A7:** 343-344
explosive ranges of **A7:** 348
functions **A7:** 339
furnace, for austenitizing **A7:** 453
furnace, improper control of **A11:** 573
gas carburizing **M4:** 143, 161
gaseous, physical properties **A7:** 341
for heat-resistant alloys, brazing of **A6:** 927
for heating element materials **A2:** 833-835

Atmospheres (continued)
high-temperature solid-state welding **A6:** 297, 298
for hot pressing **A7:** 503-504
of hydrogen **A7:** 344-345
indoor **A13:** 746-747
industrial **A11:** 192-193
liquid, physical properties **A7:** 342
magnesium alloy heat treatments **A4:** 901-902
marine and marine-air **A11:** 193, 309-310
molybdenum high-temperature behavior in gases **A4:** 818
neutral, carbon potential **A7:** 340
nickel alloy heat treating **A4:** 909
nickel and nickel alloys **M4:** 756, 757
nickel heat treating **A4:** 909
for nickel-base alloys, brazing of **A6:** 928
for nitrocarburizing **A7:** 455
nitrogen-based **A7:** 345-346
oxidizing **A2:** 833-834
pressure and flow rate in **A7:** 339
production sintering **A7:** 339-350
protective, for brazing of steels **M6:** 934-935
brazing with silver alloy filler metal **M6:** 944-945
protective, for martensitic stainless steels **A4:** 781
protective, for sintering **A7:** 295, 380
protective, furnace brazing of copper **M6:** 1037
reactive metal brazing **A6:** 947
reducing atmospheres **A2:** 834-835
reduction-oxidation function of **A7:** 343
refractory metal brazing **A6:** 947
repair welding of magnesium alloy castings **A6:** 780
rural **A11:** 193
seacoast, SCC in aluminum alloys **A13:** 265
simulated, testing by **A13:** 226
sintering **A7:** 341-346, 360-361 **M4:** 794-796
for sintering brasses and nickel silvers **A7:** 380
sintering, effect in stainless steel powders **A7:** 729
sintering, in activated sintering **A7:** 319
for sintering tungsten and molybdenum **A7:** 389
stabilized austenitic alloys heat treatment **A4:** 769
stainless steel brazeability **A6:** 911
steel furnaces **M4:** 20-21
titanium, furnace **M4:** 771-772
tool steels, heat treating **A4:** 728-729 **M4:** 579-580
for tungsten heavy alloys **A7:** 392
for tungsten-rhenium thermocouples **A2:** 876
types **A13:** 510
types, for simulated service testing **A13:** 204
ultradry hydrogen **A11:** 450
vacuum **A7:** 341, 345
vacuum arc remelting **A15:** 407
vacuum, for brazing **A11:** 450
zoned **A7:** 346-348
Atmospheres for
brazing of molybdenum **M6:** 1058
brazing of nickel-based alloys **M6:** 1019
brazing of niobium **M6:** 1057
brazing of reactive metals **M6:** 1053
brazing of steel **M6:** 931
brazing of tungsten **M6:** 1059
furnace brazing of stainless steel **M6:** 1004
air **M6:** 1009
argon **M6:** 1008-1009
dissociated ammonia **M6:** 1007-1008
dry hydrogen **M6:** 1004-1007
induction brazing of stainless steel **M6:** 1011
laser brazing **M6:** 1065
Atmospheric
aerosol sample collection **A10:** 94
aerosols, PIXE analysis by particle size **A10:** 102, 106
physics and chemistry, PIXE studies in **A10:** 106
Atmospheric (continued)
pressure, abbreviation for **A10:** 691
Atmospheric control
for hot-die/isothermal forging dies **A14:** 154-155
Atmospheric corrosion *See also* Rust **A13:** 80-83 **M1:** 717-723
aluminized steels **A13:** 527
aluminum coatings protecting against **M5:** 333-334
aluminum-zinc alloy coatings **A13:** 435
aluminum/aluminum alloys **A13:** 434-435, 596
aluminum/nonferrous metals, rates for **A13:** 601
atmospheric factors **A13:** 511-512
atmospheric variables **A13:** 81-82
austenitic stainless steels **A13:** 554
cadmium plate protecting against **M5:** 256, 263-264, 266, 269
carbon steels **A13:** 510-512
cast aluminum alloys **A13:** 601
cast carbon/low-alloy steels **A13:** 573-574
cast irons **A13:** 570
cast steels **A13:** 573-575, 577
chromate conversion coatings protecting against **M5:** 457-458
chromium content, effect of **M1:** 717, 721-722
composition, effect of **M1:** 721-723
contaminants **A13:** 81
of copper casting alloys **A2:** 352
copper content, effect of **M1:** 717, 721-723
copper/copper alloys **A13:** 616-618, 634, 637-638
corrosion rates **M1:** 719-722
corrosion-product films **A11:** 192-193
corrosivity of **A11:** 192
defined **A13:** 2
defined and described **M5:** 431
dew, effects of **M1:** 717-720
effects, on SCC **A13:** 266
from moisture **A13:** 17
galvanic **A13:** 237-238
galvanic couples **M1:** 718
galvanized steel **A13:** 432-433
geographic and meteorological factors **M1:** 717-718
hot dip galvanized coatings protecting against **M5:** 323-24, 331-32
industrial environment **M1:** 721, 723
inert dusts **M1:** 717, 720
kinetics **A13:** 512
lead/lead alloys **A13:** 787
magnesium alloys **M5:** 628, 637, 646
magnesium/magnesium alloys **A13:** 742
marine exposure **M1:** 720, 721, 723
marine, world test sites **A13:** 917-918
mechanical coatings **M5:** 300-302
nickel content, effect of **M1:** 717, 721-722
nickel plating **M5:** 199-200, 207
nickel-base alloy applications for **A13:** 654
oxygen, role in **M1:** 718
paint protecting against **M5:** 474-475, 491, 504
passivity **A11:** 193
penetration, measurement of **M1:** 719, 722
phosphorus content, effect of **M1:** 721-722
of porcelain enamels **A13:** 449, 451
prevention **A13:** 82-83
preventive measures **M1:** 722
pure tin **A13:** 770
rates of **A11:** 192
rates, various metals **A13:** 82
relative humidity **M1:** 718
resistance, nickel alloys **A2:** 429
rural environment **M1:** 723
and salinity **A13:** 908
silicon content, effect of **M1:** 717, 721-722
soft solder **A13:** 774
space shuttle orbiter **A13:** 1061
of specific systems **A13:** 82
stainless steels **A13:** 555 **M3:** 65-70
steel, weight loss from **A11:** 193
Atmospheric corrosion (continued)
of structures **A13:** 1303-1308
sulfur oxides, role in **M1:** 717, 718
test sites, marine environments **A13:** 906
testing **A13:** 204-206, 237-238, 757-758 **M1:** 718-719
tests **A13:** 204-206, 638, 906
thermal spray coatings for **A13:** 460-461
in threaded fasteners **A11:** 535
of tin-coated steel **A13:** 777
tin/tin alloy coatings **A13:** 775-776
tin/tin alloys **A13:** 774-775
tropical environments **M1:** 721, 723
types **A13:** 80-81
vs. time, industrial environment **A13:** 531
wrought aluminum alloys **A13:** 600
zinc coatings protecting steel from **M5:** 253-255
zinc/zinc alloys and coatings **A13:** 756-758
Atmospheric corrosion testing **A13:** 204-206, 237-238, 757-758
Atmospheric etch
aluminum surfaces **M5:** 578
Atmospheric evaporation
plating waste recovery process **M5:** 316
Atmospheric galvanic corrosion
evaluation **A13:** 237-238
Atmospheric pollution **A13:** 511-512
Atmospheric pressure
consolidation by **A7:** 533-536
defined in leak testing **A17:** 58
Atmospheric pressure sintering **A7:** 376
Atmospheric riser *See also* Blind riser
defined **A15:** 1
Atmospheric zone
marine structures **A13:** 542-544
Atmospheric-corrosion resistance *See also* Weather resistance
of engineering plastics **EM2:** 1
Atom positions in unit cells **A9:** 708
for simple metallic crystals **A9:** 716-718
Atom probe
abbreviation **A10:** 689
defined **A10:** 584, 669
principle of **A10:** 591
voltage-pulsed and laser-pulsed compared **A10:** 597
Atom probe microanalysis *See also* Field ion microscopy **A10:** 583-602
alternate data representation **A10:** 594
calibration **A10:** 592
complete system **A10:** 591-592
composition profiles **A10:** 593-594
field evaporation as basis of **A10:** 587
field ion microscopy and **A10:** 583-602
high-resolution energy-compensated atom probe **A10:** 592, 597
imaging atom probe **A10:** 596-597
instrument design and operation **A10:** 591
mass resolution, ECAP effect **A10:** 597
mass spectra and interpretation **A10:** 591-593
mass spectra of γ matrix, IN 939, and primary γ' precipitates **A10:** 598, 599
principles **A10:** 591
pulsed laser atom probe **A10:** 597-598
quantitative analysis **A10:** 594-595
single-layer depth resolution of **A10:** 595
spatial resolution, factors limiting **A10:** 595-596
ATOMET iron powders
by QMP process **A7:** 87-89
ATOMET iron powders, specific types *See also* ATOMET iron powders; Iron powders
ATOMET 25 .... **A7:** 87-89
ATOMET 28 .... **A7:** 87, 89
ATOMET 30 .... **A7:** 87-89
ATOMET 67 .... **A7:** 89
ATOMET 68 .... **A7:** 89
ATOMET 602 .... **A7:** 89
ATOMET 664 .... **A7:** 89
ATOMET 669 .... **A7:** 89

Atomic absorption (AA)
spectrometric metals analysis **A18:** 300
Atomic absorption (AA), for trace element analysis **EM4:** 24
Atomic absorption analysis
for trace elements **A2:** 1095
Atomic absorption coefficient
radiography **A17:** 309
Atomic absorption spectrometers
alternative designs **A10:** 53, 54
double-beam **A10:** 50
multielement **A10:** 52
Atomic absorption spectrometry **A10:** 43-59
accessory equipment **A10:** 55
advantages **A10:** 46
analytical sensitivities, hydride-generation AAS systems **A10:** 50
applications **A10:** 43, 55-58
atomizers **A10:** 46-50
Boltzman equation **A10:** 44
capabilities **A10:** 31, 102, 181, 233, 333
capabilities, compared with ion chromatography **A10:** 658
capabilities, compared with molecular fluorescence spectroscopy **A10:** 72
capabilities, compared with UV/VIS absorption spectroscopy **A10:** 60
capabilities, compared with x-ray spectrometry **A10:** 82
continuum-source **A10:** 52-53
direct solid-sample analysis **A10:** 58
discrete atomic line-source lamps, effects of **A10:** 52
electrothermal vaporization technique for **A10:** 36
estimated analysis time **A10:** 43
flame atomizer versions **A10:** 44-46
hydride-generation system **A10:** 50
of inorganic liquids and solutions **A10:** 7
of inorganic solids **A10:** 4-6
instruments, as detectors **A10:** 55
introduction **A10:** 43-44
limitations **A10:** 43
neutron activation analysis and compared **A10:** 233
and optical emission spectroscopy **A10:** 21
of organic solids **A10:** 9
principles and instrumentation **A10:** 44-46
quantitative elemental analysis by **A10:** 43
related techniques **A10:** 43, 44-46
research and future trends **A10:** 52-55
sampling **A10:** 43, 47, 54-58
sensitivities **A10:** 46
signal error in **A10:** 45
spectrometers **A10:** 50-52
Atomic absorption spectrophotometry
capabilities, compared with classical wet analytical chemistry **A10:** 161
capabilities, compared with electrometric titration **A10:** 197
residue analysis by **A10:** 177
Atomic absorption spectroscopy **A13:** 391-392, 1115
as advanced failure analysis technique **EL1:** 129
Atomic absorption spectroscopy (AAS)
for chemical analysis **EM4:** 552-553, 555
chemical analysis of glass-quality sand **EM4:** 378
for metallizing ceramics **EM4:** 542
to analyze feldspars and nepheline syenite **EM4:** 379
to analyze salt cake **EM4:** 380
to analyze soda ash **EM4:** 380
to analyze the bulk chemical composition of starting powders **EM4:** 72
Atomic absorption tests **EM1:** 737
Atomic bonding
as milling process **A7:** 62
Atomic clusters
and surface atoms **A7:** 259
Atomic coordinates
for defining crystal structure **A10:** 348
effect of Rietveld method on **A10:** 423
Atomic diffusion
amorphous materials and metallic glasses **A2:** 813
Atomic diffusion at elevated temperatures and low strain rates **A9:** 690
Atomic displacement profile **A18:** 851
Atomic emission
optical systems for **A10:** 23-24
Atomic emission spectrometry
atomic absorption spectrometry compared **A10:** 44-46
capabilities **A10:** 102, 181, 333
energy-level transitions **A10:** 44
principles and instrumentation **A10:** 44-45
Atomic emission spectroscopy (AES)
spectrometric metals analysis **A18:** 300
Atomic fluorescence spectrometry
and atomic absorption spectrometry compared **A10:** 44-46
effort sources **A10:** 46
energy-level transitions **A10:** 44
principles and instrumentation **A10:** 45-46
Atomic FM **A6:** 145
Atomic force microscope (AFM) ... **A18:** 397, 401, 402
Atomic force microscopy (AFM) **A6:** 144
Atomic hydrogen welding
definition **M6:** 1
Atomic mass unit **A10:** 669, 689
Atomic mobility
and crystal 109 growth/solidification
Atomic number
abbreviation for **A10:** 691 **A11:** 798
backscattering coefficient and secondary electron yield as a function of **A9:** 92
and characteristic x-ray wavelength relationship between **A10:** 433
contrast, two-phase alloys **A10:** 508
correction (Z) **A10:** 524
defined **A10:** 669
effect on depth of information in secondary electron imaging **A9:** 95
of elements **A10:** 688
imaging, defined **A10:** 669
Atomic number contrast, scanning electron microscopy **A9:** 93-94
of wrought stainless steels **A9:** 282
Atomic number correction
in EPMA analysis **A10:** 524
Atomic number imaging
defined **A10:** 669
electrons and uses for **A12:** 168
Atomic order
as fine structure effect **A10:** 438
Atomic ordering
in $AlFe_3$ phase **A9:** 682
Atomic oxygen **A13:** 1099-1100
effect on silver solar cell interconnect **A12:** 481
Atomic percent
abbreviation for **A10:** 689
symbol for **A8:** 724
Atomic replica **A9:** 2
Atomic scattering factor **A9:** 2
defined **A10:** 329
Atomic sensitivity factors
x-ray photoelectron spectroscopy **A10:** 574
Atomic spectroscopy *See also* Spectroscopy
energy-level transitions **A10:** 44
Fourier transform spectrometers in **A10:** 39
Atomic structure
defined **A10:** 669
and electronic phenomena **EL1:** 90-92
of lanthanum-nickel-platinum alloy **A10:** 284
neutron diffraction analysis for **A10:** 420
symmetry related to crystal symmetry and crystal systems **A10:** 348
Atomic theory
basic **A10:** 33
and optical emission spectroscopy **A10:** 21-23
Atomic transport
in low-gravity eutectic alloys **A15:** 151-152
Atomic volume
of rare earth elements **A2:** 722
Atomic wear
defined **A18:** 3
Atomic weight
defined **A10:** 669
Atomic weight of specimens
effect on depth of information in x-ray scanning electron microscopy **A9:** 93
Atomization *See also* Air atomization- Gas atomization; Water atomization
advanced, oxide reduction by **A7:** 256
aluminum P/M alloys **A2:** 201
of Astroloy powder **A7:** 428
of beryllium **A2:** 684-685
for brazing and soldering powders **A7:** 837
centrifugal **A2:** 685 **A7:** 25, 26, 49
of cobalt powders **A7:** 146
of composite bearings **A7:** 407
confined **A7:** 26, 29
of copper alloy powder, flowchart **A7:** 121
copper powders **A2:** 392-393 **A7:** 106, 116-118, 121, 734
cross-jet **A7:** 123
defined **A7:** 1 **A10:** 669
definition **M6:** 2
effect of particle collisions **A7:** 29-30, 34, 254
efficiency **A7:** 30
engineering properties **A7:** 25
estimation of powder yields **A7:** 33, 35
flame, typical system **A10:** 48
free-fall **A7:** 26, 29
iron powder produced by **A7:** 23
mechanism **A7:** 27-30
nickel powders produced by **A7:** 134, 142
nozzle designs **A7:** 28, 123
and particle size distribution **A7:** 47, 75-76
process, schematic **A7:** 26
reductions in particle collision during **A7:** 29-30, 34, 254
and roll compacting, effects **A7:** 401
rotating electrode process **A7:** 39-42
silicon diffusion during **A7:** 252
of silver powders **A7:** 148
spray drying systems **A7:** 73-76
stages **A7:** 28, 39-44
systems, AAS sample introduction **A10:** 54
techniques for **A7:** 75-76
of tin powder **A7:** 123-124
of titanium powder **A7:** 167
two-fluid **A7:** 25, 27
ultrasonic **A7:** 25, 26, 28
vacuum **A7:** 25, 26, 43-45
-vaporization interferences **A10:** 33, 34
variables **A7:** 27
and volatilization, in graphite furnace atomizers **A10:** 53
water, of low-carbon iron **A7:** 83-86
Atomized aluminum powders *See also* Aluminum; Aluminum powders
air-atomized **A7:** 129
apparent density **A7:** 297
chemical analysis **A7:** 129-130
compressibility curve **A7:** 287
effect of stearate coating on explosivity **A7:** 195
effect of tapping on loose powder density **A7:** 297
explosibility **A7:** 125, 127, 130, 195
for metal cutting **A7:** 843
Micromerograph particle size **A7:** 129
oxygen content and surface area **A7:** 130
particle size and shape **A7:** 129
powder shipments **A7:** 24
production process **A7:** 125, 126
Atomized cobalt-based hardfacing powders **A7:** 146
Atomized copper powders *See also* Copper powders
effect of additions on apparent density **A7:** 34, 37
pressure density relationships **A7:** 299
Atomized iron powders
compressibility curve **A7:** 287
different amounts of carbon compared **A9:** 517-518
green density and green strength **A7:** 289
mercury porosimetry for **A7:** 268
microstructures **A9:** 509
for welding **A7:** 818
Atomized magnesium particles **A7:** 132
Atomized nickel-based alloys **A7:** 255
Atomized nickel-based hardfacing powders **A7:** 142
Atomized particles
SEM examination **A7:** 238

Atomized powders *See also* Atomization; specific elemental powders
for brazing and soldering ... **A7:** 837
chemical composition and microstructure of ... **A7:** 32-34
consolidation by atmospheric pressure (CAP process) ... **A7:** 533-536
microstructure ... **A7:** 36-39
properties ... **A7:** 25
solidification structures ... **A9:** 616-617
Atomized stainless steel powders
apparent density ... **A7:** 297
Atomized steel powders
surface composition ... **A7:** 256
Atomizer test cleaning process efficiency ... **M5:** 20
Atomizers ... **A7:** 46
air-acetylene flame, characteristics ... **A10:** 48
for atomic absorption spectrometry ... **A10:** 46-50
centrifugal (rotating disk) ... **A7:** 74-76
defined ... **A10:** 669
electrical plasma ... **A10:** 53-54
flame ... **A10:** 47-49
furnaces ... **A10:** 48, 49
graphite furnace ... **A10:** 48, 49
Langmuir torch ... **A10:** 54
nitrous oxide-acetylene flame ... **A10:** 48
nozzle ... **A7:** 74-76
quartz tube for ... **A10:** 48
Atomizing nozzle technology ... **A7:** 125, 127
Atoms ... **A13:** 46
absolute configurations ... **A10:** 344
arrangement of ... **A10:** 287, 325
black ... **A10:** 348-349
chemisorbed, EXAFS for geometry of ... **A10:** 407
coordination numbers and geometries ... **A10:** 344
defined ... **A10:** 669
dopant, lattice location of ... **A10:** 633
electronic structure by UV/VIS ... **A10:** 60
energy-level diagram of ... **A10:** 570
field-evaporated, effects of ... **A10:** 590, 602
fluorescent ... **A10:** 72-74
geometry around ... **A10:** 345, 352
hydrogen ... **A10:** 410, 420
inorganic, MFS analysis ... **A10:** 74
interstitial, FIM images of ... **A10:** 588
irradiated, decay rate equals production rate ... **A10:** 235
kinetics of ... **A15:** 50
layered, in FIM images ... **A10:** 590
location ... **A10:** 344, 349, 350
near-neighbor environment, analysis in solids by NMR ... **A10:** 277
phase difference in scattering from different electrons within ... **A10:** 328
scattering factor ... **A10:** 349
shells, photoejection of electrons from ... **A10:** 85
single-layer depth analysis ... **A10:** 595
sputter removal of ... **A10:** 554
surface positions of ... **A10:** 536
wave from, mathematical form of ... **A10:** 349
white ... **A10:** 348-349
x-rays generated from disturbance of electron orbitals of ... **A10:** 83
Atoms n atomic structure ... **EL1:** 90-92
and conductivity ... **EL1:** 93-94
in crystals ... **EL1:** 93
ATR *See* Attenuated total reflectance spectroscopy
Attachment
alloys, millimeter/microwave applications ... **EL1:** 755-757
compliancy, SM solder joint attachments ... **EL1:** 741-743
component, technologies ... **EL1:** 347-351
component, trends ... **EL1:** 387-388
configurations, flexible printed boards ... **EL1:** 590
leadless solder ... **EL1:** 744
methods, die ... **EL1:** 213-223
millimeter/microwave applications ... **EL1:** 755-756
Attachment (continued)
process, selection ... **EL1:** 734
reliability, dies ... **EL1:** 61-62
rigidity effects ... **EL1:** 740
solder joint ... **EL1:** 740-742
solder, leaded ... **EL1:** 744-745
surface mounting ... **EL1:** 631
Attachment alloys
microwave applications ... **EL1:** 756-757
Attachment fitting, aluminum alloy catapult-hook
fractured ... **A11:** 88, 91
Attachment method
sliding electrical contacts ... **A2:** 842
Attachments, pressure vessels
failures of ... **A11:** 644
Attack polishing *See* Chemical-mechanical polishing; Polish-etching
Attack-polishing
beryllium ... **A9:** 389
defined ... **A9:** 2
of uranium and uranium alloys ... **A9:** 478, 480
ATTAP *See* Advanced turbine technology applications project
Attapulgite ... **A18:** 569
Attenuated total reflectance spectroscopy ... **A10:** 113-114
depth profiling granular sample by ... **A10:** 120
for monolayer adsorption on metal surfaces ... **A10:** 119
and photoacoustic spectroscopy compared ... **A10:** 115
of polymer curing reactions ... **A10:** 120-121
Wilks' ATR attachment ... **A10:** 120-121
Attenuation *See also* Damping ... **EM3:** 5
by absorption and scattering ... **A17:** 231
Compton scattering ... **A17:** 309
defined ... **A10:** 669 **EL1:** 1134 **EM1:** 4, 776 **EM2:** 5
differential, radiographic inspection ... **A17:** 309
effective absorption, x-rays ... **A17:** 310-311
of electromagnetic radiation ... **A17:** 309-311
gating, forms ... **EM2:** 841-842
high-frequency digital systems ... **EL1:** 80
measurement, as ultrasonic test technique ... **EM1:** 776
as measurement, microwave inspection ... **A17:** 205
method, microstructural ultrasonic inspection ... **A17:** 274
of neutron beams ... **A17:** 390
of neutrons and x-rays, compared ... **A17:** 387
of optical systems ... **EL1:** 15
overall, ultrasonic beams ... **A17:** 240
porosity effects ... **A17:** 212
rate, glass fiber production ... **EM1:** 108
signal, design considerations ... **EL1:** 41-42
sound, in transmission method, ultrasonic inspection ... **A17:** 240
of ultrasonic beams ... **A17:** 238-240
ultrasonic inspection, welding ... **A17:** 597
ultrasonic wave ... **A17:** 232-233
variation, in neutron radiography ... **A17:** 387
of x-rays, defined ... **A17:** 383
Attenuation formula ... **A18:** 325
Attenuation period *See* Deceleration period
Attenuators
microwave inspection ... **A17:** 202
Attitude (attitude angle) ... **A18:** 525-526
defined ... **A18:** 3
Attribute data
in Shewhart control charts ... **A17:** 734
Attrition
compaction ... **A7:** 56
defined ... **A18:** 3
mechanical, for copier powders ... **A7:** 587
Attrition reclaimers
of sands ... **A15:** 227, 351-352
Attrition wear
as cemented carbide tool wear mechanism ... **A2:** 954
Attritioned beryllium powders ... **A7:** 170-171
Attritioned mills *See also* Ball milling; Ball mills
ball ... **A7:** 23, 68
mechanism ... **A7:** 68-69
of silver powders ... **A7:** 148
Attritioning process
beryllium ... **A2:** 684
Attritor
ball milling ... **A7:** 23
defined ... **A7:** 1
grinding, defined ... **A7:** 1
Au-Be (Phase Diagram) ... **A3:** 2 • 65
Au-Bi (Phase Diagram) ... **A3:** 2 • 65
Au-Ca (Phase Diagram) ... **A3:** 2 • 66
Au-Cd (Phase Diagram) ... **A3:** 2 • 66
Au-Ce (Phase Diagram) ... **A3:** 2 • 67
Au-Co (Phase Diagram) ... **A3:** 2 • 67
Au-Cr (Phase Diagram) ... **A3:** 2 • 67
Au-Cu (Phase Diagram) ... **A3:** 2 • 68
Au-Cu-Ni (Phase Diagram) ... **A3:** 3 • 22-23
Au-Dy (Phase Diagram) ... **A3:** 2 • 68
Au-Eu (Phase Diagram) ... **A3:** 2 • 68
Au-Fe (Phase Diagram) ... **A3:** 2 • 69
Au-Ga (Phase Diagram) ... **A3:** 2 • 69
Au-Ge (Phase Diagram) ... **A3:** 2 • 69
Au-Hg (Phase Diagram) ... **A3:** 2 • 70
Au-In (Phase Diagram) ... **A3:** 2 • 70
Au-K (Phase Diagram) ... **A3:** 2 • 70
Au-La (Phase Diagram) ... **A3:** 2 • 71
Au-Li (Phase Diagram) ... **A3:** 2 • 71
Au-Mg (Phase Diagram) ... **A3:** 2 • 71
Au-Mn (Phase Diagram) ... **A3:** 2 • 72
Au-Na (Phase Diagram) ... **A3:** 2 • 72
Au-Nb (Phase Diagram) ... **A3:** 2 • 73
Au-Ni (Phase Diagram) ... **A3:** 2 • 73
Au-Pb (Phase Diagram) ... **A3:** 2 • 73
Au-Pd (Phase Diagram) ... **A3:** 2 • 74
Au-Pr (Phase Diagram) ... **A3:** 2 • 74
Au-Pt (Phase Diagram) ... **A3:** 2 • 74
Au-Pu (Phase Diagram) ... **A3:** 2 • 75
Au-Rb (Phase Diagram) ... **A3:** 2 • 75
Au-Sb (Phase Diagram) ... **A3:** 2 • 75
Au-Se (Phase Diagram) ... **A3:** 2 • 76
Au-Si (Phase Diagram) ... **A3:** 2 • 76
Au-Sn (Phase Diagram) ... **A3:** 2 • 76
Au-Sr (Phase Diagram) ... **A3:** 2 • 77
Au-Te (Phase Diagram) ... **A3:** 2 • 77
Au-Th (Phase Diagram) ... **A3:** 2 • 77
Au-Ti (Phase Diagram) ... **A3:** 2 • 78
Au-Tl (Phase Diagram) ... **A3:** 2 • 78
Au-U (Phase Diagram) ... **A3:** 2 • 78
Au-V (Phase Diagram) ... **A3:** 2 • 79
Au-Yb (Phase Diagram) ... **A3:** 2 • 79
Au-Zn (Phase Diagram) ... **A3:** 2 • 79
Au-Zr (Phase Diagram) ... **A3:** 2 • 80
Audio equipment
commercial hybrid applications ... **EL1:** 385
Audit process control
solder joint inspection as ... **EL1:** 735
Auditing
of design ... **EL1:** 127
Audrey
defined ... **EM1:** 4
Auger
bevel gears ... **A7:** 675, 676
electron emission ... **A7:** 250
spectra ... **A7:** 251-255
transition ... **A7:** 250, 251
Auger analysis
cutting fluids and Ti alloys ... **A16:** 846
of thermal embrittlement of maraging steels ... **A1:** 697-698
Auger analysis used to study the fiber-matrix interface of fiber composites ... **A9:** 592
Auger chemical shift
defined ... **A10:** 669
Auger composition depth profile
P/M stainless steel ... **A13:** 830

Auger electron microscopy
compared to scanning electron microscopy **A9:** 90
Auger electron spectroscopy **A7:** 250-255 **A10:** 549-567 **A13:** 391, 830
applications **A7:** 252-255 **A10:** 549, 556-566
capabilities **A10:** 333, 490, 516, 603, 610
capabilities, and FIM/AP **A10:** 583
capabilities, compared with classical wet analytical chemistry **A10:** 161
capabilities, compared with x-ray photoelectron spectroscopy **A10:** 568
chemical effects **A10:** 552
compared with XPS and x-ray analysis **A10:** 569
data acquisition **A10:** 555
defined **A10:** 669
electron beam artifacts **A10:** 556
elemental detection sensitivity **A10:** 556
equipment **A7:** 250-252
estimated analysis time **A10:** 549
experimental methods **A10:** 554-556
general uses **A10:** 549
high vapor pressure samples **A10:** 556
of inorganic solids, types of information from **A10:** 4-6
instrumentation **A10:** 554
introduction **A10:** 550
limitations **A10:** 549, 556
point analysis **A10:** 557-558
principles **A10:** 550-554
quantitative analysis **A10:** 553
related techniques **A10:** 549
sample charging **A10:** 556
samples **A10:** 549, 556-566
and scanning Auger microscopy **A7:** 254
sensitivity **A10:** 550, 556
spectral peak overlap **A10:** 556
sputtering artifacts **A10:** 556
and x-ray photoelectron spectroscopy **A7:** 250
Auger electron spectroscopy (AES) **A1:** 689 **A18:** 445, 446-447, 449, 450-456 **EM1:** 285 **EM2:** 816-817 **EM3:** 237, 238-240
advantages and limitations **EM3:** 239
alloy identification **A18:** 305
applications **A18:** 454-455
chemical state information **EM3:** 244
combined with ion sputtering **EM3:** 244, 245
for crystallographic damage (cratering) **EL1:** 1043
data analysis **A18:** 452-453
depth profiling by ball cratering **EM3:** 245
depth profiling by inelastic scattering ratio **EM3:** 247
depth profiling by multiple anodes/different transitions **EM3:** 247
development of **A11:** 33
effects of chemical environment on AES spectrum **A18:** 454
electron escape depth **A18:** 451-452
elemental depth profiling **EL1:** 1079-1080
elemental mapping **EL1:** 1079
and EPMA failure analysis **A11:** 39-41
equipment **A18:** 452
failure analysis **EM3:** 248-249
fundamentals **A18:** 450-451
instrumental resolution **EL1:** 1078-1079
instrumentation **EL1:** 1077-1078
ion etching (sputtering) **A18:** 453-454, 455
limitations **A18:** 455-456
maps of gallium arsenide field-effect gallium **EM3:** 241
transistor and gold, aluminum, and
material interactions and resolution **EL1:** 1079
microstructural identification of carburized steels **A4:** 368
photoemission in **EL1:** 1075
platinized titania application **A18:** 449-450, 451
process **EM3:** 238-240
quantification **EM3:** 242-243
spectrum **A18:** 452-453
sputter depth profiles of PAA samples **EM3:** 250
sputter-depth profiles of microelectronic bonding pad **EM3:** 245
for surface analysis **EM4:** 25
surface behavior diagrams **EM3:** 247
as surface failure analysis **EL1:** 1107
to observe embedding of erodent fragments in metals **A18:** 203
Auger electron spectroscopy (AES) (continued)
to observe seal wear **A18:** 552
uses **A11:** 37
versus SIMS **A18:** 458, 459
as wafer-level physical test method **EL1:** 922-924
Auger electron yield
defined **A10:** 669
Auger electrons *See also* Electrons
chemical effects **A10:** 552-553
defined **A10:** 669
energy, principal **A10:** 550, 551
escape depths, and x-ray emission depths compared **A10:** 552
inner-shell ionization and de-excitation by **A10:** 433
kinetic energy of **A10:** 550, 551
peaks **A10:** 551
produced **A10:** 550
spectra **A10:** 550-551
yield, defined **A10:** 669
Auger emissions
and light-element sensitivity **A10:** 550
probabilities, and qualitative analysis **A10:** 550
Auger extrusion **EM4:** 9
Auger imaging
of integrated circuit **A10:** 555
Auger map
defined **A10:** 669
Auger matrix effects
defined **A10:** 669
Auger microprobe analysis
capabilities **A10:** 516
Auger parameter
in XPS analysis **A10:** 572
Auger process
defined **A10:** 669
Auger spectroscopy (AES) **A6:** 144
Auger transition designations
defined **A10:** 669
Augers
as sampling tools **A10:** 16
Ault-Wald-Bertolo Charpy fracture
toughness correlation **A8:** 265
Aus-bay quenching **A4:** 212
Austempered ductile iron (ADI) **A1:** 34, 35, 37-38
advantages **A1:** 37-38
effect of austempering temperature on strength and ductility **A1:** 40
heat treatment for **A1:** 40, 42
mechanical property requirements **A1:** 34
properties **A1:** 42
specifications **A1:** 36
Austempered ductile irons *See also* Ductile iron
development of **A15:** 35, 38-39
specifications **A15:** 655
Austempering **A1:** 455, 457
6150 steel **M1:** 431
defined **A9:** 2
ductile iron **A15:** 659
hardenable steels **M1:** 460
properties achieved **A15:** 660
Austempering of ductile cast iron **A4:** 682, 685, 688-689, 691, 692
Austempering of steel **A4:** 152-163
advantages **A4:** 152
applications **A4:** 155-156, 157 **M4:** 105, 107-109
austenitizing temperature **A4:** 153, 154, 156, 159-160 **M4:** 105-106
dimensional changes **A4:** 160-161, 162 **M4:** 114, 115
equipment **A4:** 156-159 **M4:** 110, 111-112
modified **A4:** 156, 162-163 **M4:** 115-116
problems **A4:** 160, 161-162 **M4:** 113-115
process control **A4:** 152, 156-160 **M4:** 110-111
quenching media **A4:** 152-153, 157, 158, 159 **M4:** 104-105
racking **A4:** 158-159 **M4:** 110
safety precautions **A4:** 160, 163 **M4:** 116
salt baths **A4:** 152-153, 156, 157, 158, 160 **M4:** 109, 110-113
section thickness limitations **A4:** 154-155 **M4:** 106-107
steel selection **A4:** 153-154 **M4:** 105, 106
titration procedure for chloride determination **A4:** 158
washing and drying **A4:** 156, 159, 163 **M4:** 110, 116
Austempering of steel, furnaces
batch **A4:** 156, 158, 159
belt **A4:** 156, 157 **M4:** 109
carboaustempering systems **A4:** 158
direct-fired tunnel-type **A4:** 157 **M4:** 109-110
fluidized bed **A4:** 158
gantry **A4:** 158
induction heating systems **A4:** 158
multiple-quench batch-type **A4:** 158
pusher **A4:** 156, 158
rotary retort **M4:** 109
rotary-retort **A4:** 156-157, 160
salt pot **A4:** 159
shaker-hearth **A4:** 156, 157, 159 **M4:** 109
Austenite *See also* Retained austenite **A3:** 1 • 23
abrasion resistance **M1:** 614
in abrasion-resistant cast irons **A1:** 93-94
bright-field image and diffraction pattern **A10:** 440
carbon metastable, cleavage **A8:** 481
carbon solubility in **A15:** 66
in cast iron **A15:** 82
defined **A9:** 2 **A13:** 2 **A15:** 1
in duplex alloys **A13:** 127
in ferritic stainless steels **A13:** 127
formation during compact sintering **A7:** 314
formation, from cast iron solidification **A15:** 82
formation of, during intercritical annealing **A1:** 425
formation, symbol for temperature at **A8:** 724
grain size, effect on notch toughness **M1:** 699, 701
grain size, in bearing materials **A11:** 509
-graphite eutectic, graphite structure **A15:** 169
-graphite growth curves, in cast iron **A15:** 170
gray cast iron **M1:** 13, 28, 29
hydrogen solubility in **A15:** 82
inhibition of cleavage fracture **M1:** 701
as iron allotrope **A13:** 46, 48
in maraging steels **A1:** 794
peritectic transformation from ferrite **A9:** 679-680
primary, in cast iron **A15:** 173-174
retained in steels, Mössbauer analysis of **A10:** 287
role in nucleation and growth of pearlite **A9:** 659-661
stabilizers **A3:** 1 • 25 **A13:** 47
in steel **A9:** 177-178
temperature at which cementite completes solution in, symbol for **A11:** 796
temperature at which formation begins symbol for **A11:** 796
transformation of, after intercritical annealing **A1:** 425
transformation temperature of ferrite to symbol **A11:** 796
Austenite dendrites
in cast iron **M1:** 5, 6
Austenite grains
influence of size of, on hardenability **A1:** 392-393
size of, and steel plate production **A1:** 227-228
transformation of, to ferrite **A1:** 586
Austenite, retained *See* Retained austenite
contact fatigue resistance **A18:** 260
retained percent as function of carbon content **A18:** 874
Austenite stabilizer **A6:** 100
Austenite-finish temperature (A_f) **A6:** 438
Austenite-flake graphite eutectic
graphite structure **A15:** 169-170
nucleation of **A15:** 170-172
Austenite-iron carbide eutectic
growth, multidirectional solidification **A15:** 179-180
iron carbide structure **A15:** 173
Austenite-spheroidal graphite eutectic
nucleation of **A15:** 172-173
Austenite-stabilizing elements in wrought stainless steels **A9:** 283, 285
Austenite-start temperature (A_s) **A6:** 438
Austenitic alloy irons
advantages **A6:** 797
applications **A6:** 797
Austenitic alloy irons, specific types
Cr-Mo, advantages and applications of materials for surfacing, build-up, and hardfacing **A18:** 650

Austenitic alloy irons, specific types (continued)
Ni-Cr, advantages and applications of materials for surfacing, build-up, and hardfacing ... **A18:** 650
Austenitic alloys ... **A18:** 649
carburization detection in ... **A11:** 272
as high-alloy steels ... **A15:** 722-723, 731-732
Austenitic cast iron impellers
shrinkage porosity in ... **A11:** 355-356
Austenitic cast irons
compositions ... **M1:** 76
for corrosion resistance ... **M1:** 91
growth at high temperature ... **M1:** 93
for heat resistance ... **M1:** 93-96
mechanical properties ... **M1:** 89, 92
oxidation at high temperature ... **M1:** 93-94
Austenitic cast irons, corrosion-resistant
composition ... **M4:** 556
dimensional stabilizing ... **M4:** 557
high-temperature stabilizing ... **M4:** 557
mechanical properties ... **M4:** 557
solution treating ... **M4:** 557
spheroidize annealing ... **M4:** 557
stress relieving ... **M4:** 556-557
Austenitic cast steels
corrosion fatigue ... **A13:** 581
general corrosion ... **A13:** 577-578
intergranular corrosion ... **A13:** 578-580
localized corrosion ... **A13:** 580-581
Austenitic ductile irons *See also* Ductile irons; High-alloy graphitic irons ... **A9:** 245
heat treatment ... **A15:** 700-701
melting ... **A15:** 700
molding and casting ... **A15:** 700
shakeout ... **A15:** 700
Austenitic ferritic chromium steel
by STAMP process ... **A7:** 248
Austenitic grades
of corrosion-resistant steel castings ... **A1:** 913
Austenitic grain size
defined ... **A9:** 2
Austenitic gray irons, as corrosion resistant
elevated-temperature service alloys ... **A15:** 699-700
Austenitic iron
classification and composition of hardfacing alloys ... **A18:** 652
Austenitic manganese steel ... **M3:** 568-588
abrasion resistance ... **M3:** 580-583
abrasion testing ... **M3:** 582
alloy modifications ... **M3:** 568-572, 584-585
applications ... **A18:** 702 **M3:** 568
classification and composition of hardfacing alloys ... **A18:** 652
composition ... **M3:** 568-572, 573, 584-585
corrosion resistance ... **M3:** 583
hardness ... **M3:** 573, 580, 582, 583, 587
hardness measurement ... **M3:** 578
heat treatment ... **M3:** 573-575, 578-560
machinable grade ... **M3:** 587, 588
machining ... **M3:** 586-587, 588
magnetic properties ... **M3:** 584, 587
mechanical properties ... **M3:** 568-573, 575-576, 578, 579, 580, 583, 584, 585, 586, 587
microstructure ... **M3:** 568-575, 577-580
physical properties ... **M3:** 574, 583, 584, 587
properties ... **A18:** 702
reheating ... **M3:** 578-580
stress-strain curve ... **M3:** 576
temperature effects ... **M3:** 583-584
tensile properties ... **M3:** 568-573, 575, 576-578, 580, 583, 585, 586, 587
toughness ... **M3:** 575, 576, 578, 580, 583, 584
welding ... **M3:** 586
work hardening ... **M3:** 576-578
Austenitic manganese steel castings ... **A9:** 237-241
effect of cooling rate on microstructure ... **A9:** 237
effect of superheat on grain size ... **A9:** 237
grain size ... **A9:** 238
macroetching ... **A9:** 238
Austenitic manganese steel castings (continued)
macroexamination ... **A9:** 238
microexamination ... **A9:** 238-239
microstructure ... **A9:** 239
mounting ... **A9:** 237
polishing ... **A9:** 237-238
sectioning ... **A9:** 237
specimen preparation ... **A9:** 237-238
Austenitic manganese steels ... **A1:** 822-840
abrasion resistance vs. carbon content ... **M1:** 608
abrasion resistance vs. toughness ... **M1:** 607
applications ... **A1:** 822 **A15:** 735
as-cast properties ... **A1:** 828, 829
commercial use of castings ... **A1:** 828, 829
heavy sections ... **A1:** 829
brittle fracture in ... **A11:** 393-395
composition of ... **A1:** 822
bismuth ... **A1:** 825-826
carbon ... **A1:** 822-824
common alloy modifications ... **A1:** 824-825
copper ... **A1:** 825
manganese ... **A1:** 822-824, 825
phosphorus ... **A1:** 822, 824
silicon ... **A1:** 822, 824
sulfur ... **A1:** 826
titanium ... **A1:** 826
vanadium ... **A1:** 825
compositions ... **A9:** 239 **A15:** 733
corrosion of ... **A1:** 836
effect of temperature on ... **A1:** 836-837
arc welding ... **A1:** 838
magnetic properties ... **A1:** 837
precautions ... **A1:** 838
welding ... **A1:** 837-838
heat treatment for precautions ... **A1:** 830, 831
procedures ... **A1:** 822, 830
as high-alloy ... **A15:** 733-735
higher manganese content steels ... **A1:** 826-827
fabrication ... **A1:** 828
oxidation ... **A1:** 828
physical properties ... **A1:** 44-45
machinability ... **A15:** 734-735
machining ... **A1:** 838-839
machinable grade ... **A1:** 839
procedures ... **A1:** 839
mechanical properties ... **A15:** 734
mechanical properties after heat treatment ... **A1:** 830-831
melt practice ... **A1:** 828
microstructural wear ... **A11:** 161
reheating ... **A1:** 833
wear in grinding mills ... **M1:** 622-624
wear resistance ... **A1:** 834
abrasion testing ... **A1:** 835-836
metal-to-metal contact ... **A1:** 834-835
weldability ... **A15:** 735
work hardening ... **A1:** 831-832
determination of rate ... **A1:** 832
methods of ... **A1:** 832-833
service limitations ... **A1:** 832
Austenitic manganese steels, applications
railroad equipment ... **A6:** 398
Austenitic manganese steels, specific types
ASTM A128, as-cast ... **A9:** 240
ASTM A128, cast and solutionized ... **A9:** 241
ASTM A128, cast, heat treated and quenched ... **A9:** 240-241
ASTM A128 grade A, as-cast ... **A9:** 240
ASTM A128 grade A, cast and overheated ... **A9:** 241
ASTM A128 grade A, cast, heat treated and quenched ... **A9:** 241
ASTM A128 grade C, cast, heat treated and quenched ... **A9:** 241
ASTM A128 grade D, cast, heat treated and quenched ... **A9:** 241
ASTM A128, grade D, tensile test specimen ... **A9:** 158
Austenitic manganese steels, specific types (continued)
ASTM A128 grade E2, air cooled and partially solutionized ... **A9:** 241
ASTM A128, heat treated and quenched ... **A9:** 240
experimental alloy, as-cast ... **A9:** 240
experimental alloys, heat treated and quenched ... **A9:** 240
Fe-15Mn, sheet martensite ... **A9:** 672
Fe-19Mn, sheet martensite ... **A9:** 672
Austenitic nickel-alloy irons
applications, and types ... **A15:** 699-701
for high-temperature service ... **A15:** 699
Austenitic nitrocarburizing ... **A4:** 264, 430-431
advantages ... **A4:** 430
Alpha Plus industrial process ... **A4:** 431, 432
Beta industrial process ... **A4:** 431, 432
compound layer formation ... **A4:** 430-431
gaseous, physical metallurgy ... **A4:** 430-431
industrial techniques ... **A4:** 431, 432
Nitrotec C process ... **A4:** 431
production applications ... **A4:** 432
Austenitic nodular irons, corrosion-resistant
composition ... **M4:** 557
heat treating procedures
mechanical properties ... **M4:** 557
Austenitic precipitation-hardenable stainless steels
See also Wrought stainless steels; Wrought stainless steels, specific types ... **A9:** 285
Austenitic stainless steel *See also* Stainless steel; Stainless steel, austenitic
for autoclaves in high-temperature aerated water testing ... **A8:** 420
in chloride solution, electrode potential control ... **A8:** 407
constant-stress creep curve ... **A8:** 320-321
deformation processing map for ... **A8:** 154
environments for stress-corrosion cracking ... **A8:** 527
fatigue crack propagation in liquid sodium ... **A8:** 426, 428
hardness conversion tables ... **A8:** 109
high-energy-rate-forged extrusion ... **A8:** 573
intergranular stress-corrosion cracking ... **A8:** 420-422
for liquid metal systems ... **A8:** 425
mill finishes ... **M5:** 552, 561
modulus of elasticity at different temperatures ... **A8:** 23
negative creep ... **A8:** 331
pickling of ... **M5:** 72-73
porcelain enameling of ... **M5:** 513
sheet, combined creep-fatigue data in reversed bending ... **A8:** 356
stress-strain curve ... **A8:** 177
susceptibility to chloride cracking ... **A8:** 528-529
true stress/true strain curve for ... **A8:** 24-25
Austenitic stainless steel powders
annealing ... **A7:** 185
for encapsulation ... **A7:** 428
properties ... **A7:** 100, 729
sintering ... **A7:** 308
Austenitic stainless steel tubing
ultrasonic inspection ... **A17:** 569-570
Austenitic stainless steels *See also* Austenite; Austenitic alloys; Austenitic stainless steels, specific types; Austenitic steels; Cast stainless steels; Stainless steel(s); Stainless steels; Stainless steels, austenitic grades; Stainless steels, specific types; Steel(s); Steels; Wrought stainless steel ... **A14:** 224-226
abrasion artifacts in ... **A9:** 34
advantages ... **A6:** 797
aging ... **A1:** 946-947
alloying effects at elevated temperatures ... **A13:** 538-539
applications ... **A1:** 948-949 **A6:** 378, 383, 797
arc welding *See* Arc welding of stainless steels

Austenitic stainless steels (continued)
arc-welded ... **A11:** 427-428
in boiling glacial acetic acid ... **A13:** 556
brazing ... **A6:** 913
second-phase precipitation ... **A6:** 622, 625
brazing and soldering characteristics ... **A6:** 625-626
in carbonate melts ... **A13:** 91
cast, SCC prevention ... **A13:** 327
categories of ... **A1:** 931, 944-945
caustic (SCC) embrittlement, welds ... **A13:** 353
characterized ... **A13:** 549-550
clad to carbon, welding of ... **A6:** 501-502
clad to low-alloy steels, welding of ... **A6:** 501-502
compositions ... **M6:** 526
compositions of ... **A1:** 843, 847-848
constitutional liquation ... **A6:** 75
consumables for ferritic stainless steel welding ... **A6:** 449, 450
corrosion resistance effects, heat-tint oxides ... **A13:** 351-353
crevice corrosion ... **A13:** 323
cryogenic service ... **A6:** 1016
dislocation interaction ... **A10:** 469
dissimilar metal joining ... **A6:** 827
dissimilar metal joining, buttering ... **A6:** 825-826
ductility loss, hydrogen damage ... **A13:** 171
electrodes for arc welding of high-carbon steels ... **A6:** 64
electrogas welding ... **M6:** 239
electron beam welding ... **M6:** 638
electron-beam brazing ... **A6:** 922-923
electron-beam welding ... **A6:** 868-869
electroslag welding ... **A6:** 278
elevated-temperature properties ... **A1:** 944-949
embrittlement by zinc ... **A11:** 236-237
eutectic joining ... **EM4:** 526
extralow-carbon ... **A14:** 225
fatigue crack growth ... **A1:** 948, 949
fatigue properties ... **A1:** 947-948
fatigue striations ... **A12:** 21
ferritic iron-aluminum, brittleness of ... **A12:** 365
forgeability of ... **A1:** 891-893
forging of ... **A14:** 224-226
forging procedure ... **A14:** 225
formability of ... **A1:** 888-889
forming operations, suitability ... **A14:** 759
fractographs ... **A12:** 351-365
fracture/failure causes illustrated ... **A12:** 217
friction surfacing ... **A6:** 321, 323
friction welding ... **A6:** 152, 153, 154 **M6:** 721
furnace brazing ... **A6:** 918-919
galvanic corrosion ... **A6:** 625
gas-metal arc welding shielding gases ... **A6:** 67
gas-tungsten arc welding ... **A6:** 1018
Charpy V-notch absorbed energy vs. strength ... **A6:** 1018
glass-metal seals ... **EM4:** 875
H grades ... **A1:** 931, 945
hardfacing ... **A6:** 790, 791, 798
high-energy-rate-forged extrusion ... **A14:** 365
high-temperature solid-state welding ... **A6:** 298
hot cracking ... **A6:** 677, 687, 688, 694
hydrochloric acid corrosion ... **A13:** 1162
hydrogen charging ... **A13:** 329
hydrogen effects ... **A12:** 39
hydrogen embrittlement in ... **A1:** 715
hydrogen fluoride/hydrofluoric acid corrosion ... **A13:** 1165-1168
hydrogen mitigation, underwater welding ... **A6:** 1011-1012, 1014
impact toughness ... **A1:** 947
intergranular corrosion ... **A12:** 126, 142 **A13:** 239, 325 **M6:** 48
intergranular corrosion of ... **A11:** 180
irradiation embrittlement ... **A12:** 127
knife-line attack ... **A13:** 124
laser surface alloying ... **A13:** 504
laser-beam brazing ... **A6:** 922
laser-beam welding ... **A6:** 877
liquid erosion resistance ... **A11:** 167
localized biological corrosion ... **A13:** 117
localized corrosion resistance ... **A13:** 562-563
machinability of ... **A1:** 894-896
machining ... **A2:** 966
as magnetically soft materials ... **A2:** 776-777
metallurgical corrosion effects ... **A13:** 124-125
microbiologically influenced corrosion ... **A6:** 1068

Austenitic stainless steels (continued)
molten zinc effects ... **A13:** 334
Nelson diagram ... **A6:** 380
nitric acid corrosion ... **A13:** 1155
nitrogen pickup, oxygen partial pressure effect ... **A15:** 443
nitrogen-strengthened ... **A14:** 225-226
oxyfuel gas welding ... **A6:** 284
in pharmaceutical production facilities ... **A13:** 1226
physical properties ... **M6:** 527
plasma and shielding gas compositions ... **A6:** 197
plasma-MiG welding ... **A6:** 224
preferential attack, weld metal precipitates ... **A13:** 347-348
press forming ... **A14:** 763-764
primary austenite solidification ... **A6:** 686-695
primary ferrite solidification ... **A6:** 686
principal ASTM specifications for weldable steel sheet ... **A6:** 399
projection welding ... **M6:** 506
repair welding ... **A6:** 1105-1106
resistance of, to stress-corrosion cracking ... **A1:** 726-728
resistance welding ... **A6:** 847-848 **M6:** 527-530
SCC behavior ... **A13:** 272
SCC failures of ... **A12:** 133
sensitization of, to intergranular corrosion ... **A1:** 706-707
sheet metals ... **A6:** 399, 400
shielded metal arc welding ... **A6:** 1018
shielded metal arc welding, Charpy V-notch absorbed energy vs. strength ... **A6:** 1018
sigma phase embrittlement in ... **A1:** 709-711
sigma-phase embrittlement ... **A12:** 132
sintered, corrosion resistance ... **A13:** 831
solid solubility of carbon in ... **A13:** 827
solid-state transformations in weldments ... **A6:** 82
in sour gas environments ... **A11:** 300
stress-corrosion cracking in ... **A11:** 27, 215-217, 624, 635
stud arc welding ... **A6:** 211, 213 **M6:** 733
sulfuric acid corrosion ... **A13:** 1150-1151
susceptibility to hydrogen damage ... **A11:** 249
temperature effects ... **A12:** 50-52
tensile properties ... **A1:** 934, 945-946
thermal expansion coefficient ... **A6:** 907
thermal properties ... **A6:** 17
for thermal spray coatings ... **A13:** 461
void swelling in ... **A1:** 655-656
weight loss as function of absorbed nitrogen ... **A13:** 829
weld cladding ... **A6:** 817, 818
weld decay, and prevention ... **A13:** 351
weldability of ... **A1:** 897-905
welding consumables for field repair of heavy machinery ... **A6:** 1065-1066, 1067
welding to carbon steels ... **A6:** 500-501, 502
welding to dissimilar austenitic stainless steels ... **A6:** 500
welding to low-alloy steels ... **A6:** 500-501
weldments ... **A13:** 125, 347-355
zinc embrittlement ... **A13:** 184

Austenitic stainless steels, new alloy development ... **A3:** 1 • 26

Austenitic stainless steels, nitrogen-strengthened
arc welding *See* Arc welding of stainless steels
composition ... **M6:** 526
resistance welding ... **M6:** 527

Austenitic stainless steels, specific types *See also* Austenitic stainless steels
AISI 301, fatigue fracture surface ... **A12:** 351
AISI 301, hydrogen-embrittled ... **A12:** 31, 39, 52
AISI 302, alligatoring ... **A12:** 351
AISI 302, high-cycle fatigue fracture ... **A12:** 352
AISI 302, hydrogen effect ... **A12:** 39, 52
AISI 302, rock-candy fracture ... **A12:** 351
AISI 304, chloride SCC ... **A12:** 354
AISI 304, effect of strain rate on creep crack propagation ... **A12:** 354
AISI 304, hydrogen embrittlement ... **A12:** 355, 357
AISI 304, polythionic acid SCC ... **A12:** 354
AISI 304, SCC facets ... **A12:** 354
AISI 304L, hydrogen damaged ... **A12:** 356
AISI 316, channel fracture ... **A12:** 365

Austenitic stainless steels, specific types (continued)
AISI 316, chloride SCC ... **A12:** 357
AISI 316, fatigue fracture appearance ... **A12:** 51, 56
AISI 316, intergranular SCC ... **A12:** 357
AISI 316L, fatigue fracture, orthopedic implant ... **A12:** 359-364
P/M 316L, brittle fracture ... **A12:** 358
SIS 2343, corrosion pit cracking ... **A12:** 358

Austenitic stainless steels, wrought ... **A6:** 456-469
alloy types ... **A6:** 456
alloying effects ... **A6:** 459-461
carbide precipitation ... **A6:** 465-466, 469
coefficient of thermal expansion ... **A6:** 456, 469
composition ... **A6:** 456-457
copper contamination cracking ... **A6:** 465
corrosion behavior ... **A6:** 465-467
crevice corrosion ... **A6:** 467
distortion ... **A6:** 469
ductility dip cracking ... **A6:** 465
electron-beam welding ... **A6:** 459, 462, 464
gas-tungsten arc welding ... **A6:** 462-463, 464, 465, 466, 468
heat-affected zone ... **A6:** 456, 464, 465, 466, 467, 468
heat-affected zone liquation cracking ... **A6:** 464-465
high-energy density weld solidification behavior and microstructure ... **A6:** 462-463
intergranular attack ... **A6:** 465-466
lack-of-fusion defects ... **A6:** 456
laser-beam welding ... **A6:** 459, 463, 464
liquation cracking ... **A6:** 456, 464-465
measurement of weld-metal ferrite ... **A6:** 461-462, 465
mechanical properties ... **A6:** 467-468
microbiologically influenced corrosion ... **A6:** 467
microstructural development ... **A6:** 456, 457-458, 463
nitrogen-strengthened stainless steels ... **A6:** 462
pitting ... **A6:** 467
postweld heat treatment ... **A6:** 466, 467, 469
shielded metal arc welding ... **A6:** 461
sigma phase ... **A6:** 465
sigma-phase embrittlement ... **A6:** 465
solid-state transformations and ferrite morphologies ... **A6:** 459-461
solidification behavior ... **A6:** 458-459
solidification cracking ... **A6:** 456, 458-459, 461, 462, 463-464, 467
superaustenitic stainless steels ... **A6:** 467
weld defect formation ... **A6:** 463-465
weld penetration characteristics ... **A6:** 468
weld porosity ... **A6:** 465
weld thermal treatments ... **A6:** 468-469
weld-metal liquation cracking ... **A6:** 465
welding characteristics ... **A6:** 456-457

Austenitic steels *See also* Austenite; Austenitic alloys; Austenitic stainless steels; Steel(s)
abrasion artifacts in ... **A9:** 37
casting failure ... **A11:** 393
dissimilar-metal welds with ... **A11:** 620
as hardfacing alloys ... **A7:** 828-829
solidification structures in welded joints ... **A9:** 479
thermal expansion coefficients ... **A11:** 620

Austenitic-ferritic (duplex) alloys *See* Duplex stainless steels

Austenitization
aging after, effect in iron alloy ... **A12:** 458
AISI/SAE alloy steels ... **A12:** 298
effect on fracture characteristics ... **A12:** 339
effect on fracture toughness ... **A12:** 340
of high-Cr white irons ... **A15:** 684
incomplete, AISI/SAE alloy steels ... **A12:** 291
phosphorus segregation during, tool steels ... **A12:** 375
rapid heating effects in ... **A12:** 328

Austenitizing
ASP steels ... **A7:** 786
constructional steels for elevated temperature use ... **M1:** 654
defined ... **A9:** 2 **A13:** 2
die-casting dies ... **A18:** 632
furnaces and furnace atmospheres for ... **A7:** 453
nonuniform, effect in heat-treated steel ... **A11:** 140-141
over-, of tools and dies ... **A11:** 569-571

Austenitizing (continued)
as secondary operation **A7:** 453
of sintered high-speed steels **A7:** 374
surface hardening of steels **M1:** 528-529, 531, 539, 542
temperature control **A11:** 571
temperature of, effect on notch toughness **M1:** 699
temperatures, and hardness **A7:** 451, 452
of tool and die steels **A14:** 54
Austenitizing of steel
to reveal as-cast solidification structures **A9:** 624
Austenitizing temperatures for hardening
carbon steel **M4:** 28, 29, 30
low-alloy steel **M4:** 28, 29, 30
Austenitizing temperatures of tool steels
effects on microstructure **A9:** 258-259
Auto-oxidative cross-linked resins **A13:** 400
Autocatalytic pitting corrosion **A13:** 112-113
Autocatalytic process
for plated-through hole (PTH) **EL1:** 114
Autoclave *See also* Autoclave cure; Autoclave molding **EM3:** 5
defined **EM1:** 4 **EM2:** 5
heat, pressure, control systems **EM1:** 704
heat rate **EM1:** 747
lamination capabilities **EL1:** 510
loading **EM1:** 747
unbiased, as humidity test **EL1:** 495
Autoclave cure
control, computerized **EM1:** 649-653
control systems **EM1:** 647
gas stream heating and circulation sources **EM1:** 645-647
gas stream pressurizing systems **EM1:** 647
lay-up preparation for **EM1:** 642-644
loading system **EM1:** 647
management synergisms **EM1:** 649
materials processed **EM1:** 645
modified autoclaves **EM1:** 647-648
of polyimide resins **EM1:** 662
pressure vessel **EM1:** 645
process modeling of **EM1:** 500-501
safety and installation **EM1:** 648
system control logic **EM1:** 651-652
systems **EM1:** 645-648
vacuum systems **EM1:** 647
Autoclave cure control
computerized **EM1:** 649-653
dynamics **EM1:** 650-652
system programming dynamics **EM1:** 652-653
utility support-program concepts **EM1:** 653
Autoclave dewaxing, as pattern removal
investment casting **A15:** 262
Autoclave molding *See also* Autoclave **EM2:** 5, 338, 340-341 **EM3:** 5
defined **EM1:** 4
of epoxy **EM1:** 71
of thermoplastic resin composites **EM1:** 549
tooling for **EM1:** 578-581
Autoclave processing **EM4:** 224
Autoclaving
zirconium and hafnium alloys **M5:** 667-668
Autocorrelation analysis
of atom probe composition profiles **A10:** 594
Autocorrelation function (ACF) **A18:** 335-336, 337
Autocorrelograms
for iron-chromium-cobalt alloy, effects of aging **A10:** 599
of spinodally decomposing iron-chromium- cobalt permanent-magnet alloy **A10:** 594, 600
Autofrettage
pressure vessel fracture during **A11:** 327-328
Autogenous GTA welding
preferential attack **A13:** 359-360
Autogenous ignition **A7:** 194, 198-199
Autogenous weld
definition **A6:** 1206 **M6:** 2
Autographic recorder
curve vs. bearing deformation curve, pin bearing testing **A8:** 61
with drop tower compression test **A8:** 197
for strain or deflection **A8:** 58
Autohesion **A7:** 62, 63
Automated batch-manufacturing systems **A16:** 309
Automated bonding, tape *See* Type automated bonding (TAB) technology
Automated defect evaluation systems
real-time radiography **A17:** 320-321
Automated die-closing swaging machine **A14:** 135
Automated equipment *See also* Equipment
for liquid penetrant inspection **A17:** 79-80
specific applications, by magnetic particle inspection **A17:** 116-120
Automated forging design
as knowledge-based expert system **A14:** 410
Automated image analysis *See* Image analysis
Automated integrated manufacturing system **EM1:** 636-638
integrated laminating center **EM1:** 636
modules of operation **EM1:** 636-638
Automated logic diagram (ALD) **EL1:** 128
Automated pick-and-place equipment
selection **EL1:** 732
Automated ply lamination **EM1:** 639-641
Automated swaging machines **A14:** 134
Automated ultrasonic inspection *See also* Ultrasonic inspection
of boilers/pressure vessels **A17:** 649
Automated uniaxial pressing **EM4:** 126
Automated weaving machines **EM1:** 129-131
Automatic bar and chucking machines **A16:** 371-379
Al alloy drilling **A16:** 778, 782-784
Al alloy reaming **A16:** 782, 787
boring **A16:** 168
Cu alloy machining **A16:** 815
die threading **A16:** 296
Monel R-405 for high production rates **A16:** 836
multiple-spindle bar and chucking machines **A16:** 376-378
reamers **A16:** 242
roller burnishing **A16:** 252
single-spindle automatic bar and chucking machines **A16:** 371-374
thread rolling **A16:** 284, 285, 286
vertical multiple-spindle automatic chucking machines **A16:** 378-379
Automatic brazing
definition **M6:** 2
Automatic chargers for filling **EM4:** 384
Automatic counting **EM4:** 66
Automatic electromagnetic forming **A14:** 646
Automatic exposure devices
used in photomicroscopy **A9:** 84-85
Automatic frequency control
abbreviation **A10:** 689
Automatic gas cutting
definition **M6:** 2
Automatic guided vehicles
foundry **A15:** 570-571
Automatic lathes **A16:** 153-158, 367-393
single-spindle automatic lathes **A16:** 367-369
Automatic mold
defined **EM1:** 4 **EM2:** 5
Automatic optical inspection (AOI)
as component/board-level physical test method **EL1:** 941-942
defect types **EL1:** 568-569
drivers **EL1:** 569
and electrical testing **EL1:** 565, 568-571
illumination alternatives **EL1:** 570
method description **EL1:** 568
Automatic optical inspection (AOI) (continued)
optical system selection **EL1:** 570-571
for quality control, printed wiring boards **EL1:** 873-874
system technology **EL1:** 569-570
Automatic oxygen cutting
definition **M6:** 2
Automatic pouring systems *See also* Pouring
benefits **A15:** 497-498
bottom-pour **A15:** 570
control schemes **A15:** 500-501
electric heating of pouring vessels **A15:** 498-499
electrically heated pouring furnaces **A15:** 499-500
foundry **A15:** 569-570
laser level measurement **A15:** 570
methods **A15:** 498
pouring control parameters **A15:** 500
robotic **A15:** 570
Automatic press
defined **A7:** 1 **A14:** 1 **EM1:** 4 **EM2:** 5
Automatic press roll straightening **A14:** 687-688
Automatic press stop
defined **A14:** 1
Automatic radial-axial multi-ple-mandrel ring mills **A14:** 114-115
Automatic radiographic film processing **A17:** 353-355
Automatic tape layers (ATL)
for tape prepreg **EM1:** 145
Automatic tool lifters
shaping **A16:** 190
Automatic trace routing
channel routers **EL1:** 532
gridless routing **EL1:** 533
Hightower algorithm **EL1:** 531
Lee algorithm **EL1:** 529-531
net list sorting **EL1:** 533
pattern-fit algorithm **EL1:** 531-532
rip-up routers **EL1:** 532-533
routing algorithms **EL1:** 529
Automatic transmission parts
automobile **A7:** 617, 619
Automatic welding
definition **A6:** 1206 **M6:** 2
Automatically programmed tool (APT) language
See Numerical control
Automation *See also* Computer-aided design/computer-aided manufacture; Computer-aided engineering; Modeling **EM3:** 699-702, 705, 716-725
advancements in dispensing technology **EM3:** 719-720
applications **EM3:** 716-725
automotive body shop robotic sealing **EM3:** 723-724
automotive door bonding **EM3:** 721-722
automotive interior seam sealing **EM3:** 720-721
automotive windshield bonding **EM3:** 723-724
blow molding, costs **EM2:** 299
compression molding, and costs **EM2:** 297
of continuous casting **A15:** 315
costs, permanent mold casting **A15:** 285
developing a robotic system **EM3:** 724-725
dispensing equipment for robotic applications **EM3:** 716-719
bead management methods **EM3:** 718-719
dispensing gun or valve **EM3:** 716, 717-718
header system **EM3:** 716, 717
pumping system **EM3:** 716-717
of electroslag remelting **A15:** 404
of feed metal availability **A15:** 584-585
of filament winding **EM1:** 135
of forging process design **A14:** 409-416
of foundries **A15:** 33-36, 566-573
injection molding, costs **EM2:** 295
of pattern assembly, investment casting **A15:** 257
of patternmaking **A15:** 198
of permanent mold casting **A15:** 276
and process selection **EM2:** 278

Automation (continued)
process, vacuum induction remelting and shape casting ... **A15:** 399-400
resin transfer molding, costs ... **EM2:** 301
of semisolid metal casting/forging ... **A15:** 327, 333
of solid graphite mold casting ... **A15:** 285
of squeeze casting ... **A15:** 323
of vacuum induction melting ... **A15:** 397-399
Automobile
exhaust extract, fluorescence spectrum of liquid-chromatographic fraction of ... **A10:** 79
paint, NAA forensic studies of ... **A10:** 233
Automobile axle shafts
fatigue life of ... **M1:** 675
Automobile radiators
as copper recycling scrap ... **A2:** 1214
Automobile scrap recycling *See also* Recycling
gravity separation ... **A2:** 1212
jigging system ... **A2:** 1213
low-temperature separation ... **A2:** 1212
technology ... **A2:** 1211-1213
Automobile stub axles
slag inclusions in ... **A11:** 322-323
Automobile transmission stick-shift
failure of ... **A11:** 763
Automobiles ... **A6:** 393-395
ceramic applications ... **EM4:** 960
material composition changes ... **EM3:** 551
Automotive and truck drive trains, friction and wear of ... **A18:** 563-568
automatic transmission components ... **A18:** 563-568
automatic transmission fluid (ATF) issues ... **A18:** 563-564
clutch bands and plates ... **A18:** 564
engine components ... **A18:** 567
manual transmission components ... **A18:** 564-565
cluster gear wear ... **A18:** 565
synchronizer wear ... **A18:** 554-565
transfer cases ... **A18:** 565-567
sprocket and chain wear ... **A18:** 565-566
thrust surface wear ... **A18:** 566-567
wheel end components ... **A18:** 567-568
Automotive applications *See also* Automotive industry ... **A7:** 617-621 **EM1:** 832-836
acrylonitrile-butadiene-styrenes (ABS) ... **EM2:** 111
of alloy cast irons ... **A1:** 103-104 **M1:** 96
aluminum flake coatings ... **A7:** 594-595
of aluminum P/M parts ... **A7:** 744-745
blow molding ... **EM2:** 359
of bronze P/M parts ... **A7:** 736
of bulk molding compounds ... **EM1:** 163
bus transmission gears, service life vs processing ... **M1:** 635
cast alloy steel ... **A15:** 32
cast steels for ... **M1:** 383-384
as casting market ... **A15:** 34
CG iron ... **M1:** 7-8
for comfort, hybrids ... **EL1:** 382
commercial hybrids ... **EL1:** 381-382, 385
compacted and sintered parts ... **A7:** 617-618
composite materials and processes ... **EM1:** 832-834
of copper-based powder metals ... **A7:** 733
corrosion protection ... **M1:** 755, 757
critical properties ... **EM2:** 458
ductile iron ... **M1:** 35, 36
engine blocks and liners, materials for ... **M1:** 604
engine parts ... **A7:** 617, 619
engine parts, corrosive wear ... **M1:** 636-637
environmental testing ... **EL1:** 500
future trends ... **EL1:** 393
of glass roving applications ... **EM1:** 109
gray cast iron ... **M1:** 11, 16, 18-19, 24-26
of gray cast irons ... **A1:** 19
high-impact polystyrenes (PS, HIPS) ... **EM2:** 195
of high-strength low-alloy steels ... **A1:** 417
homopolymer/copolymer acetals ... **EM2:** 100
hot formed parts ... **A7:** 618, 620
hot rolled bars ... **M1:** 206, 208, 209
of hybrids ... **EL1:** 254
malleable cast irons ... **M1:** 63, 70, 71, 73
malleable iron castings ... **A15:** 691
metal casting trends ... **A15:** 44
part design ... **EM2:** 616
phenolics ... **EM2:** 243
phosphate coating to reduce wear in torque converter ... **M1:** 631, 637

Automotive applications (continued)
piston pins, wear influenced by surface finish ... **M1:** 635
piston rings, wear affected by operating variables ... **M1:** 601, 602, 636
polyarylates (PAR) ... **EM2:** 138
polybutylene terephthalates (PBT) ... **EM2:** 153
polyether sulfones (PES, PESV) ... **EM2:** 159
polyether-imides (PEI) ... **EM2:** 156
polyethylene terephthalates (PET) ... **EM2:** 172
polyphenylene ether blends (PPE PPO) ... **EM2:** 183
polyphenylene sulfides (PPS) ... **EM2:** 186
polyurethanes (PUR) ... **EM2:** 258-259
powders used ... **A7:** 572
reinforced polypropylenes (PP) ... **EM2:** 192
self-lubricating bearings ... **A7:** 704
semisolid aluminum ... **A15:** 334-336
of semisolid metal casting/forging ... **A15:** 327
of stainless steels ... **A1:** 881 **A7:** 730
styrene-acrylonitriles (SAN, OSA ASA) ... **EM2:** 215
styrene-maleic anhydrides (S/MA) ... **EM2:** 217-218
surface-hardened steel parts ... **M1:** 527, 537
thermoplastic fluoropolymers ... **EM2:** 117
thermoplastic polyimides (TPI) ... **EM2:** 177
thermoplastic polyurethanes (TPUR) ... **EM2:** 205
timing chain, wear affected by lubricant contamination ... **M1:** 636
torque converter seal rings, materials for ... **M1:** 624
ultrahigh-strength steels ... **M1:** 423, 431, 432
urethane hybrids ... **EM2:** 268
vacuum induction shape casting ... **A15:** 401
Automotive bearing materials
fatigue life ... **A11:** 487
fretting failure ... **A11:** 498
Automotive body panels
stretching of ... **A8:** 547
Automotive brakes, friction and wear of ... **A18:** 569-577
automotive brake frictional characteristics ... **A18:** 574-575
brake design basis ... **A18:** 574
design factors ... **A18:** 574-575
effectiveness ... **A18:** 574, 576
self-actuation ... **A18:** 574
automotive brake linings ... **A18:** 569-570
carbon-based brake linings ... **A18:** 570
metallic brake linings ... **A18:** 570
organic friction materials ... **A18:** 569-570
brake drum and disk wear ... **A18:** 572-574
brake lining chemistry effects ... **A18:** 572
external abrasive effects ... **A18:** 572-574
graphite morphology effects ... **A18:** 572
local cast iron wear ... **A18:** 572
normal cast iron wear ... **A18:** 572
transfer coatings ... **A18:** 573-574
brake frictional performance ... **A18:** 575-576
blister fade ... **A18:** 575
burnished effectiveness ... **A18:** 575
contamination fade ... **A18:** 575
delayed fade ... **A18:** 575
effectiveness drift ... **A18:** 576
environmental sensitivity ... **A18:** 575
fade recovery ... **A18:** 575
fade resistance ... **A18:** 575
flash fade ... **A18:** 575
green effectiveness ... **A18:** 575
moisture sensitivity ... **A18:** 575-576
rust effects ... **A18:** 576
speed sensitivity ... **A18:** 575
wet friction ... **A18:** 575
brake lining wear ... **A18:** 570-571
aftermarket (AM) friction materials ... **A18:** 571
brake rubbing speed effects ... **A18:** 570
brake temperature effects ... **A18:** 570
brake torque effects on wear ... **A18:** 570
brake usage severity effects ... **A18:** 570-571
break-in wear ... **A18:** 570
heavy-duty (HD) friction materials ... **A18:** 570
original equipment (OE) ... **A18:** 571
specific wear rate ... **A18:** 570
brake lining wear modeling ... **A18:** 571-572
interfacial temperature effects ... **A18:** 571-572

Automotive brakes, friction and wear of (continued)
lining cure effect ... **A18:** 571
laboratory and vehicle brake evaluation ... **A18:** 576-577
correlation of laboratory and vehicle test results ... **A18:** 576-577
federal braking requirements and other brake tests ... **A18:** 577
friction assessment screening test (FAST) machine ... **A18:** 576
friction material performance ... **A18:** 577
friction materials test machine ... **A18:** 576
full brake inertia dynamometer ... **A18:** 576
SAE-recommended practices ... **A18:** 577
semimet frictional behavior ... **A18:** 576
environmental sensitivity ... **A18:** 576
thermal fade ... **A18:** 576
water sensitivity ... **A18:** 576
toxicity of brake wear debris ... **A18:** 574
Automotive construction
joining processes ... **M6:** 57
Automotive engine, friction and wear of *See* Internal combustion engine parts, friction and wear of
Automotive industry *See also* Automotive applications ... **A13:** 1011-1018
composite applications ... **EM1:** 834-835
corrosion forms ... **A13:** 1011
corrosion testing ... **A13:** 1016-1017
design ... **A13:** 1016
design for simultaneous engineering ... **EM1:** 835-836
inhibitor applications ... **A13:** 525
P/M production ... **A7:** 17-18, 569
paint systems ... **A13:** 1015-1017
precoated steels ... **A13:** 1011-1015
structural vs. appearance composite requirements ... **EM1:** 832
Automotive industry applications *See also* Parts; Shafts; specific automotive parts
aluminum and aluminum alloys ... **A2:** 10
borescopes ... **A17:** 7
cobalt-base wear-resistant alloys ... **A2:** 451
copper and copper alloys ... **A2:** 239
of holography ... **A17:** 16
machine vision ... **A17:** 38, 43
powder metallurgy parts ... **A17:** 542-543
of structural ceramics ... **A2:** 1019
titanium and titanium alloy ... **A2:** 589
of titanium P/M products ... **A2:** 657
Automotive parts
closed-die forging ... **A14:** 82
driveshafts, electromagnetic forming ... **A14:** 648
precoated steel sheet for ... **M1:** 167, 169, 172-176
transmission shaft, by radial forging ... **A14:** 147
Automotive poppet valves, steel
aluminum coating process ... **M5:** 335, 339-341
Automotive turbocharger wheels
design practices for structural ceramics ... **EM4:** 722-726
Automotive valve spring
distortion failure of ... **A11:** 138-139
Autophoretic paints ... **M5:** 472
Autoradiography
in neutron radiography ... **A17:** 387
to reveal as-cast solidification structures in steel ... **A9:** 624
Autorouting technology
Lee algorithm in ... **EL1:** 529-531
Autospectrum ... **A18:** 294
Auxiliary anode
defined ... **A13:** 2
Auxiliary brighteners
use in nickel plating ... **M5:** 205
Auxiliary cooling
permanent mold casting ... **A15:** 283
Auxiliary electrode
defined ... **A13:** 2
Auxiliary equipment *See also* Equipment; Tooling; Tools
for explosive forming ... **A14:** 638
for open-die forging ... **A14:** 61-63
for power spinning ... **A14:** 603
sheet metal forming ... **A14:** 489, 499-503
Auxiliary heating
for component removal ... **EL1:** 723-724

Auxiliary magnifier
definition ... **M6:** 2
Auxiliary metals
in tungsten carbide powder production ... **A7:** 157
Avalanche photodiode (APD)
defined ... **EL1:** 1134
Average (bulk) asperity contact pressure
nomenclature for lubrication regimes ... **A18:** 90
Average (bulk) hydrodynamic pressure
nomenclature for lubrication regimes ... **A18:** 90
Average (bulk) surface temperature rise
nomenclature for lubrication regimes ... **A18:** 90
Average cornering force ... **A18:** 579
Average density
defined ... **A7:** 1
Average energy flux (AEF) ... **A18:** 583
Average erosion rate
defined ... **A18:** 3
Average film ... **A18:** 94
Average flash temperature ... **A18:** 41, 43
Average frictional power ... **A18:** 478
Average grain diameter
defined ... **A9:** 2
Average grain dislocation density ... **A10:** 358
Average grain orientation
texture as measure of ... **A10:** 358
Average linear strain *See* Engineering strain
Average lubricant film thickness
nomenclature for hydrostatic bearings with orifice or capillary restrictor ... **A18:** 92
nomenclature for lubrication regimes ... **A18:** 90
Average lubricant flow
nomenclature for hydrostatic bearings with orifice or capillary restrictor ... **A18:** 92
Average lubricant shear stress
nomenclature for lubrication regimes ... **A18:** 90
Average molecular weight ... **EM3:** 5
defined ... **EM2:** 5
Average normal anisotropy
and *r* value ... **A8:** 550
of sheet metals ... **A8:** 555-556
Average of the sample ... **EM3:** 786
Average pressure
nomenclature for hydrostatic bearings with orifice or capillary restrictor ... **A18:** 92
Average root-mean-square (rms) surface roughness
symbol and units ... **A18:** 544
Average solid composition ... **A6:** 47
Average stiffness
nomenclature for hydrostatic bearings with orifice or capillary restrictors ... **A18:** 92
Average stress criterion
of failure ... **EM1:** 235, 254-255
Average stress method ... **EM3:** 481
Average surface roughness ... **A18:** 340, 341-342
Average total heat flux (q_{av} ... **A18:** 40
Average true stress
in necking ... **A8:** 25
Averages, lot *See* Lot averages
Averaging extensometer
dial-type ... **A8:** 616
for elevated/low temperature tension testing ... **A8:** 36
LVDT ... **A8:** 618
for strain measurement ... **A8:** 49
Aviation
as casting market ... **A15:** 34
Avimid K-III
as condensation polyimide ... **EM1:** 78
Avimid N (NR-150B2)
temperature capabilities ... **EM1:** 78
Avionics
space shuttle orbiter ... **A13:** 1074
Avogadro's number
defined ... **A10:** 162, 669
AX-140 filler metal
hydrogen-induced cracking ... **A6:** 413
Axes
defined ... **EM1:** 357, 359
Axial
compression, in shafts ... **A11:** 461
defined ... **A11:** 1 **EM1:** 275
fatigue, in shafts ... **A11:** 461
Axial closed-die rolling process
principle ... **A14:** 112
Axial compression testing ... **A8:** 55-58
Axial compressive strength
analysis of ... **EM1:** 196-197
Axial compressive stress
at equator of upset cylinder ... **A8:** 578
Axial displacement ... **A6:** 315
Axial effects
and alternative analysis methods ... **A8:** 180-184
Axial fatigue
of P/M forged steel ... **A7:** 301, 416, 468, 469
Axial fatigue testing *See also* Axial fatigue testing machine
constant-amplitude ... **A8:** 149
grips ... **A8:** 369
loading, notch-sensitivity for steel ... **A8:** 373
machine ... **A8:** 369, 371
universal open-front holders ... **A8:** 369
Axial fatigue tests
reliability ... **M1:** 676
Axial flow blast cleaning machine ... **A15:** 509
Axial force
of body-centered cubic metals during torsion testing ... **A8:** 181, 183
of face-centered cubic metals during torsion testing ... **A8:** 181-182
Axial line
application on gage section surface ... **A8:** 157
Axial load bearing *See* Thrust bearing
Axial loading
defined ... **A7:** 1
effect on ball or roller-path patterns ... **A11:** 492
electrohydraulic testing machine ... **A8:** 159-160
fatigue test specimen ... **A8:** 371
fatigue testing by ... **A11:** 102
and fatigue-crack propagation ... **A11:** 109
pure, fatigue failures ... **A11:** 109
rolling-contact fatigue from ... **A11:** 503-504
Axial pulse attenuator ... **A8:** 227
Axial radial loads
ball bearing ... **A11:** 491-492
Axial ratio
defined ... **A9:** 2
Axial rolls ... **A14:** 1, 114
Axial seal *See* Face seal
Axial shear failure
matrix ... **EM1:** 198
Axial shrinkage
as casting defect ... **A11:** 382
Axial strain ... **A8:** 1 **EM3:** 5
in cold upset testing ... **A8:** 579-580
defined ... **A11:** 1 **EM1:** 4 **EM2:** 5
Axial stress
dependence on temperature in copper ... **A8:** 181
during torsion testing ... **A8:** 180-181
fatigue test specimens ... **A8:** 368
Axial stress notch fatigue strength ... **A7:** 301, 416, 468, 469
Axial structure
carbon fibers ... **EM1:** 50
Axial tensile strength
analysis of ... **EM1:** 192-194
Axial tests ... **A8:** 351-352
Axial upset ... **A6:** 315
Axial winding *See also* Polar winding; Winding
defined ... **EM1:** 4 **EM2:** 5
Axial-load fatigue test ... **A1:** 861
Axicell Mirror Fusion Test Facility
thermonuclear fusion containment ... **A2:** 1057
Axis (crystal)
defined ... **A9:** 2
Axis of a weld
definition ... **M6:** 2
Axis of revolution
horizontal centrifugal castings with ... **A15:** 296
Axis of symmetry
assumed ... **A12:** 202-203
Axle grease
as fracture preservative ... **A12:** 73
Axle, hollow
by radial forging ... **A14:** 147
Axle ratio ... **A18:** 566
Axle shaft
roll forging of ... **A14:** 98
Axle shaft quality steel ... **A1:** 253
Axles *See also* Locomotive axles, failures of
highway trailer, weld failure ... **A11:** 419
housing, fatigue fracture ... **A11:** 397
housing, fracture surface ... **A11:** 389
locomotive, failure analysis studies ... **A11:** 723-724
normal microstructure of ... **A11:** 723
steel drive, fatigue fracture ... **A11:** 117
stub, slag inclusions in ... **A11:** 322-323
torsion tests for ... **A8:** 139
tractor, U-bolt fitting failure ... **A11:** 533-535
ultrasonic inspection ... **A17:** 232
AZ 31B-92A
contour band sawing ... **A16:** 363
AZS
glass-contact refractories ... **EM4:** 392
melting/fining ... **EM4:** 391, 392
AND gate
defined ... **EL1:** 1134
The Adhesives & Sealants Newsletter ... **EM3:** 65
The Adhesives and Sealants Newsletter ... **EM3:** 71

B

1,3-bis(3-aminophenoxy)benzene (APB) ... **EM3:** 155
B *See* Burger's vector; Crystal lattice; Magnetic flux density ... **EM3:** 5
defined ... **EM2:** 5
B 1900
composition ... **M4:** 653
***B* basis**
of design values ... **A8:** 662
β- and α-fibers
in rolled copper ... **A10:** 363
B-1 aircraft
as NDE reliability case study ... **A17:** 680-681
B-66
welding conditions effect ... **A6:** 581
B-120 *See* Titanium alloys, specific types, Ti-13V-11Cr-3Al
B-1900
aging cycle ... **A4:** 812
composition ... **A4:** 795 **A16:** 737
machining ... **A16:** 738, 741-743, 746-758
B-1900 + Hf
aging cycle ... **A4:** 812
B-basis *See also* A-basis; A-basis,- Design allowables; S-basis; Typical basis; Typical-basis ... **EM3:** 6
defined ... **EM1:** 5 **EM2:** 6
normal method ... **EM1:** 304
statistical analysis for ... **EM1:** 302-307
Weibull method ... **EM1:** 304
B-C (Phase Diagram) ... **A3:** 2 • 80
B-C-Fe (Phase Diagram) ... **A3:** 3 • 23-24
B-Co (Phase Diagram) ... **A3:** 2 • 80
B-Cr (Phase Diagram) ... **A3:** 2 • 81
B-Cu (Phase Diagram) ... **A3:** 2 • 81
β-diketones
as extractant ... **A10:** 170
B-Fe (Phase Diagram) ... **A3:** 2 • 81
B-fritting ... **A18:** 682
B-H characteristics *See also* Magnetic field testing
magabsorption measured ... **A17:** 145
B-Mn (Phase Diagram) ... **A3:** 2 • 82

B-Mo (Phase Diagram) .... **A3:** 2 • 82
B-Nb (Phase Diagram) .... **A3:** 2 • 82
B-Ni (Phase Diagram) .... **A3:** 2 • 83
B-Pd (Phase Diagram) .... **A3:** 2 • 83
B-porosity in cemented carbides **A9:** 274
B-Pt (Phase Diagram) .... **A3:** 2 • 83
B-Re (Phase Diagram) .... **A3:** 2 • 84
B-Ru (Phase Diagram) .... **A3:** 2 • 84
B-Sc (Phase Diagram) .... **A3:** 2 • 84
B-scan display modes
applications **A17:** 243
display **A17:** 242-243
pulse-echo ultrasonic inspection **A17:** 242
scanning acoustical holography **A17:** 443
signal display **A17:** 243
signal processor for **A17:** 227-228
system setup **A17:** 243
B-Si (Phase Diagram) .... **A3:** 2 • 85
B-stage *See also* A-stage; C-stage **EM3:** 6
defined **EM1:** 5-6 **EM2:** 7
filament winding process **EM1:** 507
B-Ta (Phase Diagram) .... **A3:** 2 • 85
B-Ti (Phase Diagram) .... **A3:** 2 • 85
B-V (Phase Diagram) .... **A3:** 2 • 86
B-W (Phase Diagram) .... **A3:** 2 • 86
B-Whitockite **EM4:** 1010
B-Y (Phase Diagram) .... **A3:** 2 • 86
B-Zr (Phase Diagram) .... **A3:** 2 • 87
B. & S. G. system *See* Brown and Sharp Gage system
B. W. G. system *See* Birmingham Wire Gage system
B_{10} life *See* Rating life
Ba-Ca (Phase Diagram) .... **A3:** 2 • 87
Ba-Cd (Phase Diagram) .... **A3:** 2 • 87
Ba-Cu (Phase Diagram) .... **A3:** 2 • 88
Ba-Ga (Phase Diagram) .... **A3:** 2 • 88
Ba-Ge (Phase Diagram) .... **A3:** 2 • 88
Ba-H (Phase Diagram) .... **A3:** 2 • 89
Ba-Hg (Phase Diagram) .... **A3:** 2 • 89
Ba-In (Phase Diagram) .... **A3:** 2 • 89
Ba-Li (Phase Diagram) .... **A3:** 2 • 90
Ba-Mg (Phase Diagram) .... **A3:** 2 • 90
Ba-Na (Phase Diagram) .... **A3:** 2 • 90
Ba-P (Phase Diagram) .... **A3:** 2 • 91
Ba-Pb (Phase Diagram) .... **A3:** 2 • 91
Ba-Se (Phase Diagram) .... **A3:** 2 • 91
Ba-Si (Phase Diagram) .... **A3:** 2 • 92
Ba-Te (Phase Diagram) .... **A3:** 2 • 92
Ba-Tl (Phase Diagram) .... **A3:** 2 • 92
Ba-Zn (Phase Diagram) .... **A3:** 2 • 93
Ba_2FeO_x **EM4:** 58
$Ba_2P_4O_7$/Ti **EM4:** 18
Babbitt *See also* Babbitt metal; Lead babbitt; Tin babbitt **A9:** 419
500 °F embrittlement, susceptibility to **M1:** 685
ability to embed abrasives **M1:** 609
and broaching **A16:** 204
compatibility with steel journals **M1:** 606-607
neutron embrittlement, susceptibility to **M1:** 686
notch toughness, effect on **M1:** 699, 701, 702, 704, 706
spray material for oxyfuel wire spray process **A18:** 829
thermal spray coating material **A18:** 832
wear resistance compared with martensite **M1:** 613
Babbitt bearing alloys **A13:** 774
Babbitt metal **A18:** 693
bearings for reciprocating pumps **A18:** 597
compatibility in bearing materials **A18:** 743
defined **A18:** 3
in friction bearings **A11:** 715
seal adhesive wear **A18:** 549
sliding bearings (steel backed) **A18:** 516, 518, 520
tin and lead base **A11:** 483
Babbitt metals
as bearing alloy **A2:** 523-524
lead-base **A2:** 523-524, 553-554
part of bimetal bearings **A9:** 567
recycling **A2:** 1219
Babbitted bearings *See also* Tin and tin alloy coatings
etching **A9:** 451
grinding **A9:** 451
Babbitting **M5:** 356-357
applications **M5:** 356
Babbitting (continued)
bearing shells **M5:** 356-357
cast iron **M5:** 356
centrifugal process **M5:** 304
cleaning and degreasing methods **M5:** 356
mechanical bonding process **M5:** 356
metal-spray process **M5:** 357
rotating speeds **M5:** 357
solidification of babbitt **M5:** 357
static process **M5:** 356-357
Babington nebulizers
for ICP sample introduction **A10:** 36
Baby lotion
liquid chromatographic analysis of parabens in **A10:** 655-656
Back bead
definition **A6:** 1206
Back contacts
large area **EL1:** 958
Back draft
defined **A15:** 1
Back extrusion punches
cemented carbide **A2:** 970-971
Back gages
straight-knife shearing **A14:** 703
Back gouging
definition **A6:** 1206 **M6:** 2, 68
in gas metal arc welding of aluminum alloys **M6:** 383
Back pressure
defined **EM1:** 4 **EM2:** 5
Back pressuring
defined as pressure testing **A17:** 65
Back rake angles **A16:** 18, 143, 162, 175, 192
Back reflection *See also* Percentage of back reflection technique
defined **A9:** 2
intensity, various metals **A17:** 238
loss, ultrasonic inspection **A17:** 246
method, for forgings **A17:** 505
technique, ultrasonic inspection **A17:** 263, 505
Back ring
defined **A18:** 3
Back rolls **A14:** 96-97
Back scatter *See also* Backscattering
protection, radiographic inspection **A17:** 344
shadow formation, radiography **A17:** 313-314
Back scattering **A18:** 448, 449
Back screen
radiography **A17:** 315
Back taper *See also* Undercut
defined **EM2:** 5
Back weld
definition **M6:** 2
Back-pressure relief port
defined **EM2:** 5
Back-propagation **A18:** 412
Back-reflection Laue method
x-ray diffraction **A10:** 329, 330
Back-slagging pit
electric arc furnace **A15:** 359
Back-titrations
in nitrogen determination **A10:** 173
yield of analyte concentration **A10:** 173
Back-to-back ring seal
defined **A18:** 3
Back-up bars
electrogas welding **M6:** 242
Backer cams
designs of **A14:** 131
Backer rod
for urethane sealants **EM3:** 205
Backers
for swaging **A14:** 131
Backfill
defined **A13:** 2
Backfire
definition **A6:** 1206 **M6:** 2
Backfires **A6:** 1201
Background
correction systems, atomic absorption spectrometry **A10:** 51-52
defined **A10:** 669
fluorescence **A10:** 130
intensity, abbreviation for **A10:** 690
parameters defining, RDF analysis **A10:** 396
removal, in EXAFS analysis **A10:** 412
Background (continued)
spectral, defined **A10:** 682
Background, as signal
defined **A17:** 678
Background noise
effect on scanning electron microscopy images **A9:** 90
Background papers
fractographic **A12:** 78
Background variables (factors) **A8:** 639
Backhand welding
definition **A6:** 1206 **M6:** 2
oxyfuel gas welding **M6:** 589
Backhand welding technique **A6:** 183
Backing
defined **A18:** 3
definition **A6:** 1206 **M6:** 2
for oxyfuel gas repair welding **M6:** 592
Backing bars
arc welding of
heat-resistant alloys **M6:** 353
magnesium alloys **M6:** 428-429
nickel alloys **M6:** 437
stainless steels **M6:** 327-328
edge preparations **M6:** 68
electroslag welding **M6:** 227-228
gas metal arc welding of
aluminum alloys **M6:** 382-383
of coppers **M6:** 406
gas tungsten arc welding **M6:** 197
keyhole welding **M6:** 218-220
submerged arc welds **M6:** 134
Backing bead
definition **A6:** 1207 **M6:** 2
Backing board *See also* Bottom board
defined **A15:** 1
Backing filler metal
definition **A6:** 1207
Backing film
defined **A9:** 2
Backing pass
definition **A6:** 1207 **M6:** 2
Backing piece
in welding process **A17:** 582
Backing piece left on **A6:** 1073
Backing plate *See* Backing board
defined **EM2:** 5
Backing plates
metallographic examination of welds made with **A9:** 578
Backing ring
definition **A6:** 1207
Backing rings
arc welding of austenitic stainless steels **M6:** 329
definition **M6:** 2
submerged arc welds **M6:** 134
Backing rings, weld
crevice corrosion **A13:** 350-351
Backing shoe
definition **A6:** 1207
Backing strips
definition **M6:** 2
for gas metal arc welding **M6:** 169
Backing weld
definition **A6:** 1207 **M6:** 2
Backlash
in Rockwell hardness testing **A8:** 74
n tension testing machine **A8:** 47
Backlighting
for automatic optical Inspection **EL1:** 942
Backoff angle
broaching **A16:** 195
Backplane
placement and level **EL1:** 76
Backprojection, defined
computed tomography (CT) **A17:** 381-383
Backscatter, beta
as plating thickness testing **EL1:** 943
Backscattered electron **A10:** 669, 689
Backscattered electron (BSE)
imaging **EL1:** 1094, 1096-1098
micrographs **EL1:** 1099-1100
Backscattered electron detectors
contrast with **A10:** 502-504
ring geometry **A10:** 503
used for wrought stainless steels **A9:** 282

Backscattered electron image
compared to secondary electron image **A9:** 92
Backscattered electron imaging **A9:** 95
of solder and solder joints **A9:** 451
Backscattered electron intensity
used in electronic image analysis **A9:** 152
Backscattered electron microscopy
used to study titanium alloy subgrain boundaries **A9:** 461
Backscattered electrons **A18:** 377-378, 379, 380
energy **A9:** 92
fractographs, sulfur concrete fracture surfaces **A12:** 472
in magnetic contrast **A9:** 95
production of **A9:** 91-92
in scanning transmission electron microscopy **A9:** 104
and secondary electrons, compared **A12:** 168
SEM illuminating/imaging system **A12:** 167-168
Backscattered scanning electron microscopy **A9:** 91-92
depth of information as a function of acceleration voltage of primary electron beam **A9:** 91
voltages **A9:** 91-92
Backscattering *See also* Rutherford backscattering spectrometry; Scattering
analysis and signal processing **A10:** 631
in EPMA quantitative analysis **A10:** 524
fine structure from **A10:** 408
and industrial computed tomography compared **A17:** 362
microwave inspection **A17:** 220
polar, as angle-beam ultrasonic inspection **A17:** 248
in Rutherford backscattering spectrometry **A10:** 629
as two-body elastic collision process **A10:** 629
Backscattering coefficient and secondary electron yield as a function of atomic number **A9:** 92
topographic contrast **A9:** 93
Backstep sequence
definition **A6:** 1207 **M6:** 2
Backstep welding of cast irons **M6:** 313
Backstep welding technique
cast irons **A15:** 526
Backstop tongs
in hot upset forging **A14:** 87
Backup
definition **M6:** 2
Backup coat
defined **A15:** 1
Backup refractories
Shaw process **A15:** 249
Backup roll, steel
fracture in shipping **A11:** 97-98
Backward coupled noise, saturated **EL1:** 35
Backward extrusion *See also* Extrusion
of cuplike parts **A14:** 305
hot **A14:** 316
load vs displacement curve **A14:** 37
of titanium alloys **A14:** 273
Backward tube spinning **A14:** 576-676
Backward wave oscillator (BWO) tube
microwave inspection **A17:** 209
Backward-coupling coefficients
three coupled lines **EL1:** 40
Bacteria
anaerobic biological corrosion by **A13:** 116
films **A13:** 42, 88, 900-901
-induced corrosion, gas/oil wells **A13:** 482-483
in pipelines **A13:** 1288-1289
rod-shaped **A13:** 88
sulfate-reducing **A13:** 116-117, 482-483
sulfur cycle, biological corrosion **A13:** 41-42
tubercule formed by **A13:** 120
Bacteria, thiobacillus
as corrosive **A12:** 245
Bacteriacides **A13:** 483
Bacterial corrosion *See also* Biological corrosion; Localized biological corrosion; Microbiological corrosion
on metals **A11:** 190-191
telephone cables **A13:** 1130
Bacterial film
in seawater **A13:** 900-901
Bacterial resistance
and weather aging **EM2:** 580
Baddeleyite (ZrO_2) *See also* Zirconia; Zirconium oxide **EM4:** 45, 50
radioactivity **EM4:** 50
$BaFe_{12}O_{19}$
applications **EM4:** 48
key product properties **EM4:** 48
raw materials **EM4:** 48
Baffle **EM3:** 5
defined **EM2:** 5
Baffles
effect in powder mixing **A7:** 189
for solder wave configurations **EL1:** 688
Bag molding *See also* Vacuum bag molding **EM3:** 5
defined **EM1:** 4 **EM2:** 5
Bag side
defined **EM1:** 4 **EM2:** 5
Bagaryatski orientation relationship
in pearlite **A9:** 658
in upper bainite **A9:** 664
Baggage, aircraft
microwave holography inspection **A17:** 226
Bagging *See* also Vacuum bag; Vacuum bagging **EM3:** 5
defined **EM1:** 4, 703 **EM2:** 5
lay-up, sequence for **EM1:** 703
quality control **EM1:** 755
Baghouse collectors **A7:** 73
Bailey-Orowan
equation, for creep **A8:** 309-310
model **A8:** 301
Bainite **A1:** 128, 129 **A9:** 662-667 **A13:** 2, 566
Bainite in carbon and alloy steels **A9:** 179
defined **A9:** 2
etching to reveal **A9:** 170
nonferrous **A9:** 665-666
Bainite packet
for cleavage fracture **A8:** 466-467
Bainite structure
ductile iron **A15:** 35
Bainitic microstructure
brittleness from **A11:** 325-326
low-alloy steel **A11:** 393
Bainitic transformation **A9:** 655
Bake
defined **A15:** 1
Bake (verb)
defined **A7:** 1
Bake sand
defined **A15:** 1
Bake-hardening steels **A4:** 61
Bake-out cycle
for hydrogen removal **A13:** 330
Bake-out temperature
for environmental test chamber **A8:** 411
Baked core
defined **A15:** 1
Baked sand molding
Alnico alloys **A15:** 736
Bakelite **A7:** 606
defined **EM2:** 5
drilling **A16:** 230
Bakelite as a mounting material *See also* Mounting materials **A9:** 26
for carbon and alloy steels **A9:** 166-167
for carbonitrided and carburized steels **A9:** 217
copper and copper alloys **A9:** 399
for nitrided steels **A9:** 218
powder metallurgy materials **A9:** 504
Bakelite as a mounting material (continued)
titanium and titanium alloys **A9:** 458
Bakelite mounting
for optical metallography sample **A10:** 300
Bakelite premolds **A9:** 26
Baking
in all-ceramic mold casting **A15:** 249
hard chromium plating **M5:** 186
maraging steels **M1:** 448
material, rigid printed wiring boards **EL1:** 541
nickel plating process **M5:** 217-218
ovens, for core production **A15:** 32
paint *See* Paint and painting, curing methods
postlamination **EL1:** 544
prelamination inner layer **EL1:** 543
of rammed graphite molds **A15:** 273
ultrahigh-strength steels **M1:** 425, 436
Baking enamel **M5:** 501
Baking mold expansion as casting defect **A11:** 386
Baking, postplate
high-carbon steels **A12:** 284
Balance
abbreviation for **A8:** 724
Balance beam
for constant-stress testing **A8:** 319-320
leveling motor, for creep test stand **A8:** 311-312
Balance construction
defined **EM2:** 5
Balance displacement pendulum
weighing system **A8:** 613
Balanced biaxial stretching
with Marciniak test **A8:** 558
in sheet metal forming **A8:** 547
Balanced construction
defined **EM1:** 4
Balanced design
defined **EM1:** 4 **EM2:** 5
Balanced incomplete block experiment plan **A8:** 646-649
Balanced laminate *See also* Laminate(s); Symmetrical laminate
defined **EM1:** 5 **EM2:** 5
Balanced twist
defined **EM1:** 5 **EM2:** 5
Balanced-in-plane contour
defined **EM1:** 4-5 **EM2:** 5
Balancing cam
in constant-stress testing **A8:** 320
Baling wire **A1:** 282 **M1:** 264
Ball and roller bearing quality and bearing quality **A1:** 253-254
Ball and roller bearings
alloy steel wire for **A1:** 286-287
measurement of residual stress and hardness of raceway of **A10:** 380
Ball bar
defined **A17:** 18
Ball bearing *See also* Rolling-element bearings, friction and wear of
defined **A18:** 3
fatigue spalling life **A18:** 258
scanning acoustic microscopy for quality assurance of ceramics **A18:** 408
wedging film action of hydrodynamic lubrication **A18:** 89
Ball bearing cup and race **A7:** 570
Ball bearings *See also* Bearings; Rolling-element bearings
alloy steel wire for **M1:** 269
ball path patterns **A11:** 491-492
deformation of raceway **A11:** 510
and roller bearings, compared **A11:** 490
rolling-contact fatigue fracture, from subsurface inclusions **A11:** 504
stainless steel, pitting failure of **A11:** 495, 497
types of **A11:** 490
Ball bonding *See* Thermocompression welding
gold wire **EL1:** 350

Ball bonding (continued)
methods of ... **EL1:** 224-226
Ball clay ... **EM4:** 32, 44
in typical ceramic body compositions ... **EM4:** 5
Ball complement
defined ... **A18:** 1
Ball cratering
for AES sputtering problems ... **A10:** 556
Ball indented bearing
defined ... **A18:** 3
Ball indenter
Brinell ... **A8:** 86-88
hardened steel ... **A8:** 84
in Rockwell hardness testing ... **A8:** 77
tungsten carbide ... **A8:** 84
verification ... **A8:** 88
Ball joints
wedging film action of hydrodynamic lubrication ... **A18:** 89
Ball mandrels
for bending tube ... **A14:** 667
Ball mill
fractured linings from ... **A11:** 377-378
ore samples crushed in ... **A10:** 165
Ball mill components *See* Grinding balls, Liners
Ball milling *See also* Ball mills ... **A7:** 56-70 **A16:** 100-101
beryllium ... **A2:** 684
cobalt powders ... **A7:** 145
cobalt-covered tungsten carbide ... **A7:** 173
defined ... **A7:** 1
grinding elements ... **A7:** 60
of magnesium powders ... **A7:** 131
platelet tantalum particle shapes ... **A7:** 161
in Pyron process ... **A7:** 82
in QMP process ... **A7:** 86
of silver powders ... **A7:** 148
Ball mills *See also* Attrition mills; Ball milling
brittle cathode process ... **A7:** 72
charge parameters ... **A7:** 66
defined ... **A7:** 1
high-energy ... **A7:** 723
for metallic flake pigments ... **A7:** 593
powders, for brazing and soldering ... **A7:** 837
production, of aluminum powder ... **A7:** 125
Ball path patterns
by axial and unidirectional radial loads ... **A11:** 491-492
effect of tilt and bearing clearance ... **A11:** 492
Ball punch tests
as simulative stretching test ... **A8:** 561
Ball sizing ... **A16:** 210
Ball-and-wedge bonds
cycle, steps ... **EL1:** 226
formation, plastic packages ... **EL1:** 472
Ball-on-ball impact fracture(s)
AISI/SAE alloy steels ... **A12:** 336
Ball-plane wear test
for print band ... **A8:** 607
reciprocating ... **A8:** 603
stress and strokes combined ... **A8:** 603
Ballast resistors
of electrical resistance alloys ... **A2:** 823
Balling, pitch
in rammed graphite molds ... **A15:** 274
Balling up
definition ... **A6:** 1207 **M6:** 2
Balls, steel
magnetic particle inspection ... **A17:** 98-99
BAMACAST software program ... **A15:** 888
Banbury ... **EM3:** 5
defined ... **EM2:** 5
Band density ... **EM1:** 5, 508
defined ... **EM2:** 5
Band head
in molecular emission ... **A10:** 23
Band printer ... **A8:** 607
Band saw cutting
for macroscopic examination ... **A12:** 92
Band sawing *See* Sawing
Band sawing, of titanium alloys *See also* Sawing ... **A14:** 841
Band saws used in sectioning ... **A9:** 23
Band theory
quantum mechanical (solid state) ... **EL1:** 96-103
Band thickness
defined ... **EM1:** 5 **EM2:** 5
Band width
defined ... **EM1:** 5 **EM2:** 5
Band-aid joints ... **EM3:** 550
Banded structure
defined ... **A11:** 1 **A13:** 2
from alloy segregation ... **A11:** 121
Banding *See* Bands
defined ... **A9:** 2
segregation ... **A15:** 306
in stainless steel as a result of hot rolling ... **A9:** 627
in steel as a result of hot rolling ... **A9:** 626
Bandpass
acoustic emission inspection ... **A17:** 281
Bands
AISI/SAE alloy steels ... **A12:** 333
as alloy segregation, effect on fatigue strength ... **A11:** 121
deformation, butterflies as ... **A12:** 115, 134
deformation, defined ... **A11:** 3
ferrite-pearlite ... **A11:** 316
microstructural, from chemical segregation and mechanical working ... **A11:** 315
microstructural, in forged product ... **A11:** 315
shear ... **A12:** 32, 42
shear, defined ... **A11:** 9
slip, in fatigue cracking ... **A11:** 102
in steel bearing ring ... **A11:** 505
in steel plate ... **A11:** 320
transformed ... **A12:** 32
twin, defined ... **A11:** 11
Bands, molecular
in emission spectroscopy ... **A10:** 23
Bandwidth
defined ... **A17:** 255
of load-measuring system ... **A8:** 193
Bandwidths
distribution of available ... **EL1:** 5
maximum , limits ... **EL1:** 6
of optical systems ... **EL1:** 16
reduction, as future development ... **EL1:** 10
Bank sand *See also* Sand
defined ... **A15:** 1
Banking concept
of radiation safety doses ... **A17:** 301
Bar *See also* Bar bending; Bar drawing; Bar sections; Bars; Round Bar
beryllium ... **A2:** 683
beryllium-copper alloys ... **A2:** 403, 411
double-shear tests for ... **A8:** 62-63
extruded, and shapes, of wrought magnesium alloys ... **A2:** 459
materials ... **A2:** 769
mechanically alloyed oxide dispersion-strengthened (MA ODS) alloys ... **A2:** 948-949
mover ... **A8:** 201
primary testing direction ... **A8:** 667
rolled steel, specimen fracture surface ... **A8:** 278, 279
rolling, of nickel-titanium shape memory effect (SME) alloys ... **A2:** 899
silicon iron, as magnetically soft
round, for workability ... **A8:** 156
selection for torsional Kolsky bar strain rate ... **A8:** 227
solid, dimensional changes in torsion ... **A8:** 143
specimen locations for ... **A8:** 60
in split Hopkinson pressure bar test ... **A8:** 200-202
stainless steel *See* Stainless steel, bar
steel *See* Steel, bar
stopper ... **A8:** 201
stress intensity ranges, ultrasonic testing ... **A8:** 252
for tensile tests ... **A8:** 155
for torsion testing ... **A8:** 139, 143
for ultrasonic testing ... **A8:** 242-243, 250
wrought aluminum alloy ... **A2:** 33
wrought beryllium-copper alloys ... **A2:** 409
wrought titanium alloys ... **A2:** 610-611
Bar (screw) machines *See also* Automatic bar and chucking machines ... **A16:** 160, 256
Bar and tubing
economy in manufacture ... **M3:** 849
Bar bending ... **A14:** 661-664
Bar compound application buffing system ... **M5:** 116, 127
Bar drawing ... **A14:** 330, 334-337
Bar drawing dies
cemented carbide ... **M3:** 523-525
chromium plating ... **M3:** 525
diamond ... **M3:** 521, 523, 525
polishing ... **M3:** 525
sectional, adjustable ... **M3:** 524
tool breakage ... **M3:** 525
tool steels ... **M3:** 523, 524, 525
Bar magnet
magnetic fields of ... **A17:** 90
Bar rolling *See also* Bars
fracture prediction ... **A14:** 397-399
pass, finite-element computer modeling ... **A14:** 350-351
Bar sections
bending of ... **A14:** 661-664
shearing of ... **A14:** 714-719
straightening of ... **A14:** 680-689
Bar, steel *See* Alloy steel bars; Carbon steel bars; Cold finished steel bars; Hot-rolled steel bars and shapes
annealing ... **A4:** 39 **M4:** 19, 25-26
normalizing ... **A4:** 39-40
Bar stock
ultrasonic inspection ... **A17:** 267-268
Bar(s) *See also* Barstock; Steel bar
cylindrical, frequency selection, eddy current inspection ... **A17:** 174
eddy current inspection system ... **A17:** 166
magabsorption measurement of ... **A17:** 155
magnetized, defined ... **A17:** 80
radiographic methods ... **A17:** 296
solid cylindrical, impedance of ... **A17:** 171-172
steel, magabsorption measurement ... **A17:** 155
steel, ultrasonic inspection ... **A17:** 271-272
Bar-shaped steel compacts
full density ... **A7:** 505
Barba's law
for ductility measurement ... **A8:** 26
Barbed wire fence ... **M1:** 271
Barcol hardness *See also* Hardness ... **EM3:** 5
defined ... **EM1:** 5 **EM2:** 5
polyester resins ... **EM1:** 91, 92
Bare copper circuit boards *See* Circuit boards
Bare electrode
definition ... **M6:** 2
Bare glass
defined ... **EM1:** 5 **EM2:** 5
Bare metal arc welding
definition ... **M6:** 2
Bare silicon circuit board (SCB)
testing ... **EL1:** 362-363
Barite ... **A18:** 572 **EM3:** 175
Barite/barytes ($BaSO_4$)
purpose for use in glass manufacture ... **EM4:** 381
Barium
applications ... **A2:** 1259
in enamel cover coats ... **EM3:** 304
in enamel ground coat ... **EM3:** 304
glass-to-metal seals ... **EM3:** 302
gravimetric finishes ... **A10:** 171
lubricant indicators and range of sensitivities ... **A18:** 301
as minor toxic metal, biologic effects ... **A2:** 1259
pure ... **M2:** 716-717
pure, properties ... **A2:** 1101
as silicon modifier ... **A15:** 161
species weighed in gravimetry ... **A10:** 172
sulfate ion separation ... **A10:** 169
sulfuric acid as dissolution medium ... **A10:** 165
TNAA detection limits ... **A10:** 237, 238
toxicity ... **A6:** 1195
ultrapure, by distillation process ... **A2:** 1094
used to make detergents ... **A18:** 100
weighed as chromate ... **A10:** 171
weighed as sulfate ... **A10:** 171
Barium aluminoborosilicate
chemical corrosion ... **EM4:** 1047
properties, non-CRT applications ... **EM4:** 1048-1049
Barium carbonate
carburizing role ... **A4:** 325
Barium carbonate ($BaCO_3$)
decomposition ... **EM4:** 110
purpose for use in glass manufacture ... **EM4:** 381

Barium carbonate and iron oxide mixture
mixing index affected by time and heat treatment EM4: 97
Barium chloride
molten A11: 277
Barium ferrite A9: 539, 549 EM4: 56
comminution milling types EM4: 78
hexagonal EM4: 59
thermal etching EM4: 575
Barium fluoride
rolling-element bearing lubricant A18: 138
to control dusting A18: 684
Barium glass
replacement for quartz filler in dental clinical studies A18: 671
Barium hexaferrite EM4: 97
Barium lead borosilicate glass
properties EM4: 1057
Barium metaborate
for flame retardance EM3: 179
Barium osmullite
maximum use temperature EM4: 875
Barium oxide
in binary phosphate glasses A10: 131
Barium oxide (BaO)
in composition of textile products EM4: 403
in composition of wool products EM4: 403
in drinkware compositions EM4: 1102
in glaze composition for tableware EM4: 1102
in ovenware compositions EM4: 1103
properties EM4: 424
specific properties imparted in CTV tubes EM4: 1039
in tableware compositions EM4: 1101
Barium porcelain enamels
composition of M5: 510-511
Barium salts
toxic effects A2: 1259
Barium soap A18: 126, 129
Barium stearate
lubricants A7: 191
Barium sulfate
biologic inhalation effects A2: 1259
Barium titanate
for accelerating adhesive cure EM3: 179
as transducer element A17: 255
Barium titanate (Ba_2TiO_3) *See also* Barium titanium oxide EM4: 55
application EM4: 542
for capacitors EM4: 17
ceramic powder, batch weight of formulation when used in non-oxidizing sintering atmospheres EM4: 163
chemical etching EM4: 575
discovery EM4: 16
piezoelectric property EM4: 16
precipitation process EM4: 60
production process EM4: 109
rare earth doped EM4: 58
used in Langevin-type piezoelectric vibrators EM4: 1119
Barium titanium oxide (Ba_2TiO_3) *See also* Barium titanate
anisotropic dielectric constants as function of temperature EM4: 771
applications EM4: 48, 300
electrical/electronic applications EM4: 1106
gas pressure sintering for pressure densification EM4: 299
hot pressing EM4: 192
key product properties EM4: 48
pressure densification
pressure EM4: 301
technique EM4: 301
temperature EM4: 301
raw materials EM4: 48
Barium titanium oxide carbonate ($BaTiO(C_2O_4)_2$)
pressure densification EM4: 300
Barium, vapor pressure
relation to temperature A4: 495 M4: 310
Barkhausen effect A9: 534
Barkhausen jumps
defined A17: 159
Barkhausen magnetic methods
capabilities of A10: 380
Barkhausen noise *See also* Noise
acoustic A17: 160
capabilities and limitations A17: 160
as decarburization measure A17: 134
instrumentation A17: 160
as magnetic material characterization A17: 129
measurement, magnetic A17: 132
residual stress measurement by A17: 159-160
stress dependence A17: 160
Barn A10: 669, 691
Barnacles
as biofouling organisms .... A13: 88, ill, 114
Barrel *See* Extruder
defined A7: 1
Barrel acid cleaning M5: 60-64
Barrel cracking A6: 993
Barrel cracks
printed board coupons EL1: 576
Barrel finishing
aluminum and aluminum alloys M5: 572-574
applications M5: 134-135
centrifugal *See* Centrifugal barrel finishing
dry *See* Dry barrel finishing
magnesium alloys M5: 629-630
mass finishing *See* Mass finishing
media M5: 134-136
self-tumbling *See* Self-tumbling
of stainless steel forgings A14: 230
wet *See* Wet barrel finishing
zinc alloys M5: 676
Barrel image distortion A9: 77
Barrel plating
brass M5: 285-286
bronze M5: 288
cadmium M5: 257-261
chromium M5: 179-180
copper M5: 162-163, 167
gold M5: 282
lead and lead alloys M5: 275
nickel *See* Nickel plating, barrel process
rhodium M5: 290-291
Barrel, shotgun
distortion in A11: 139-140
Barreling
in compression testing A8: 55-58, 196-197
and compressive strength A8: 58
of cylindrical specimens A8: 56-57
and friction A8: 56
in nonlubricated, nonisothermal hot forging A8: 582
Barreling, defined *See also* Compression test A14: 1
Barretters
microwave detection by A17: 208
Barrier
as cure processing material EM1: 644
platings, as surface preparation EL1: 679
properties, parylene coatings EL1: 794-795
as vitreous dielectric application EL1: 109
Barrier coat *See also* Coatings
defined EM1: 5 EM2: 5
Barrier coatings *See also* Barrier protection M5: 362, 433
electroplated hard chromium A13: 871-875
for galvanic corrosion A13: 86-87
for marine corrosion A13: 916
Barrier film *See also* Films
defined EM1: 5 EM2: 5
Barrier oxide film *See* Barrier coatings; Film; Oxide film; Oxides
Barrier plastics EM3: 6
defined EM2: 5-6
Barrier protection
anodic oxides A13: 377-378
ceramic coatings A13: 378
conversion coatings A13: 378-379
corrosion inhibitors A13: 378
organic coatings A13: 378
Bars *See also* Bar bending; Bar drawing; Bar rolling; Bar sections; Cold finished bars, Hot rolled bars; Rolling
bending A14: 661-664
by radial forging A14: 145
computing ovality in A14: 133
cutting and fullering A14: 63
cutting, for closed-die forging A14: 80-81
dies and die materials for drawing A14: 337
drawbenches for A14: 334-335
drawing of A14: 330, 334-337
flash welding M6: 558, 577
forged, allowances and tolerances A14: 72
forged, bursts in A11: 317-318
forming A14: 622, 836
forming, nickel-base alloy A14: 836
friction welding M6: 720-721, 726
in-line drawing and straightening machine for A14: 333
infiltration-brazed butted iron- copper A7: 558
metal flow during swaging A14: 130
oxyfuel gas cutting M6: 913
porter A14: 63
reinforcement thermit welding M6: 702-703
rotary swaging of A14: 128-144
round steel, quench crack in A11: 94
shearing of A14: 714-719
springs, hot wound, use for M1: 297, 301
straightening of A14: 680-689
tool materials for drawing A14: 336
yeild strength, relation to hardness M1: 301
Bars, steel *See* Cold finished steel bars, Hot rolled steel bars; Steel bars
Barsom Charpy/fracture toughness correlation A8: 265
Barsom-Rolfe Charpy/fracture toughness correlation A8: 265
Barus equation A18: 83
Basal plane
defined A9: 2
Basalt
applications EM4: 963, 964
high-level waste disposal in A13: 975
Basalt fused cast, property comparison
mineral processing EM4: 962
Basalt stoneware EM4: 3, 4
Base diffusion
bipolar junction transistor technology EL1: 195
Base material
definition A6: 1207 M6: 2
Base materials *See* also phase metals
materials and processes selection EL1: 113-114
Base metal
definition A6: 1207 M6: 2
Base metal erosion
in brazed joints A17: 603
Base metal test specimen
definition M6: 2
Base metals *See also* Metals
common cleaners for EL1: 678
as dental alloys A13: 1351
PFM alloys A13: 1356
and solderability EL1: 676
Base plate
defined A7: 1
Base sands *See also* Reclamation; Sands
reclamation effects A15: 355
Base SI units
guide for A10: 685
Base units
Système International d'Unités (SI) A11: 793
Base-centered space lattice .... A3: 1 • 15

Base-line technique
defined ... **A10:** 669
Base-metal cracking
in laser beam welds ... **A11:** 449
Base-metal thermocouple
for creep test temperature control ... **A8:** 314
Base-metal thermocouples *See* Thermocouple materials; Thermocouple(s)
Base/bed construction
coordinate measuring machines ... **A17:** 24
Bases
and acids, indicators ... **A10:** 172
and acids, titrations ... **A10:** 172-173
analytic methods for ... **A10:** 7
aqueous corrosion in ... **A13:** 38-39
defined ... **A13:** 2, 1140
pure tin in ... **A13:** 772
uranium/uranium alloys in ... **A13:** 815-816
zinc corrosion in ... **A13:** 763
BASF-A wax
additive to atactic polypropylene in injection molding ... **EM4:** 174, 176
$BaSi_2O_5$/Pb ... **EM4:** 18
Basic brick
applications ... **EM4:** 903, 913
Basic chemical equilibria
and analytical chemistry ... **A10:** 162-165
Basic design language for structure (BDL/S)
in computer aided design ... **EL1:** 128-129
defined ... **EL1:** 128
Basic dynamic capacity ... **A18:** 505
Basic dynamic load capacity *See also* Basic load rating
defined ... **A18:** 3
Basic dynamic load rating ... **A18:** 505
Basic fluoride
fluxes used for SAW applications ... **A6:** 62
Basic helix angle
symbol and units ... **A18:** 544
Basic load rating *See also* Basic dynamic load capacity; Dynamic load ... **A18:** 505
defined ... **A18:** 3
Basic melting practice
steelmaking ... **A15:** 366
Basic NMR frequency
defined ... **A10:** 669
Basic oxygen furnace ... **M1:** 109-110, 112
Basic oxygen process (BOP) ... **A1:** 110, 111, 112
Kawasaki basic oxygen process ... **A1:** 111
quick-quiet basic oxygen process ... **A1:** 112
Basic oxygen steelmaking (BOS) ... **EM4:** 44
Basic plumbum materials
applications ... **A2:** 555
Basic refractory *See also* Acid refractory
defined ... **A15:** 1
Basic solutions
as corrosive environment ... **A12:** 24
Basic static load rating
defined ... **A18:** 3
Basic theory of solid friction ... **A18:** 27-37
basic mechanisms of friction ... **A18:** 30
ceramics, friction of ... **A18:** 36
composition ... **A18:** 28-29
chemical compound formation ... **A18:** 29
chemisorption ... **A18:** 28-29
mechanical compound formation ... **A18:** 29
reconstruction ... **A18:** 28
segregation ... **A18:** 28
definition of friction ... **A18:** 27
definition of solid friction ... **A18:** 27
elastomers, friction of ... **A18:** 36
friction under lubricated conditions ... **A18:** 29-30
future outlook ... **A18:** 37
graphite, friction of ... **A18:** 36
history ... **A18:** 30-31
English School ... **A18:** 31
French School ... **A18:** 30-31
ice, friction of ... **A18:** 36
metals, friction of ... **A18:** 31-35
adhesion ... **A18:** 31-33
asperity deformation ... **A18:** 33-34
deformation energy ... **A18:** 34-35
third-body effects ... **A18:** 35
molybdenum disulfide, friction of ... **A18:** 36
nature of surfaces ... **A18:** 27
polymers, friction of ... **A18:** 35-36
deformation zone friction ... **A18:** 36
Basic theory of solid friction (continued)
interfacial zone shear ... **A18:** 36
rolling friction ... **A18:** 37
anelastic hysteresis losses ... **A18:** 37
microslip at the interface ... **A18:** 37
surface roughness ... **A18:** 37
subsurface microstructure ... **A18:** 29
topography ... **A18:** 27-28
asperity, distribution model ... **A18:** 28
macrodeviations ... **A18:** 27, 28
microroughness ... **A18:** 27-28
roughness ... **A18:** 27
roughness measurement ... **A18:** 28
waviness ... **A18:** 27
Basicity ... **A18:** 84
index ... **M6:** 125
effect of oxygen ... **M6:** 41
relationship to weld metallurgy ... **M6:** 41
relationship to weld-metal oxygen content ... **M6:** 125
Basicity index (BI) ... **A6:** 204
Basis metals
electroplated hard chromium ... **A13:** 872
Basis values
defined and computed ... **EM1:** 302-303
Basket weave ... **EM1:** 111, 125, 148
defined ... **EM2:** 6
Basketweave
defined ... **A9:** 2
Basquin-Coffin-Manson
strain-life relationship ... **A8:** 698
Batch *See also* Lot ... **EM3:** 6
defined ... **A7:** 1 **A8:** 1 **A15:** 1 **EM1:** 5 **EM2:** 6
fabric prepreg, defined *See* Fabric prepreg batch
glass, melting/forming ... **EM1:** 107-108
mixers, for coremaking ... **A15:** 239
operation, induction furnaces ... **A15:** 373-374
Batch attrition mills ... **A7:** 69
Batch furnace
furnace brazing ... **A6:** 121, 122
porcelain enameling ... **M5:** 519-521
Batch hot dip aluminum coating process
steel ... **M5:** 333-334, 337, 339
Batch hot dip galvanized coating *See* Hot dip galvanized coating
batch process
Batch life
emulsions ... **A18:** 143
Batch mixers for seed separation ... **A7:** 589
Batch mixing
SMC resin pastes ... **EM1:** 159
Batch ovens, paint curing
direct- and indirect-fired ... **M5:** 486-487, 505
Batch pickling process ... **M5:** 68-69, 71
Batch process
hot dip galvanizing by ... **A13:** 436-444
Batch reactors
precious metal powders ... **A7:** 149
Batch sintering ... **A7:** 1
Batch vacuum coating process ... **M5:** 397-398
Batch vertical process equipment
condensation (vapor phase) soldering ... **EL1:** 703
Batch-carburizing furnace
radiant tube failure in ... **A11:** 292-294
Batch-tumbling barrel blast cleaning equipment ... **A15:** 506-507
Batch-type furnaces ... **A7:** 356-357
for Al powders ... **A7:** 381, 743
bell and elevator ... **A7:** 356-357
for carbonitriding ... **A7:** 454
sintering ... **A7:** 743
vacuum sintering ... **A7:** 357-359
Batch-type muller
green sand preparation ... **A15:** 344-345
Batched production
fixturing ... **A16:** 410
Batches, process streams and
as corrosive ... **A11:** 210
Batching ... **EM4:** 95
masterbatching ... **EM4:** 95
objectives ... **EM4:** 95
pickup ... **EM4:** 95, 97
safety precautions ... **EM4:** 95
whitewares ... **EM4:** 95
Batdorf fracture theory method
prediction of fast-fracture reliability of ceramics ... **EM4:** 700, 701, 703, 706, 707
Bath *See also* Hardening bath
additions, as contamination ... **EL1:** 679
agitation, effect, ladle desulfurization ... **A15:** 78
chemical analysis, as bath control ... **EL1:** 680
defined ... **A15:** 1
liquid metal, kinetic paths for melting ... **A15:** 71
stratification, reverberatory furnace ... **A15:** 378
Baths
chromate coating ... **A13:** 389-390
chromium plating ... **A13:** 871
cleaner, control of ... **A13:** 382
phosphate, testing of ... **A13:** 384-385
stop, for radiographic film ... **A17:** 353
strength, for magnetic particles ... **A17:** 102
wet developer ... **A17:** 79
zinc plating ... **A13:** 767
Baths, chemical
UV/VIS analysis ... **A10:** 233
Bathtub curve
device failure rate ... **EL1:** 887
hermeticity, passivation considerations ... **EL1:** 244-245
of integrated circuit failure rates ... **A11:** 766
phases of ... **EL1:** 740
reliability ... **EL1:** 897-899
"Bathtub" failure rate curve ... **A18:** 494-495
Bathtub-type plug-in package ... **EL1:** 452
Bathythermograph
as parylene application ... **EL1:** 800
Batt
defined ... **EM2:** 6
Battelle Columbus Laboratories
run-arrest experiments ... **A8:** 285
Battelle Memorial Institute
DWTT test ... **A11:** 61-62
Batteries
from P/M porous parts ... **A7:** 700
powders used ... **A7:** 573
Batteries, lead-acid
cycle life and failure mechanisms in ... **A10:** 135
Battery corrosion ... **A13:** 1317-1323
alkaline, sintered nickel electrode in ... **A13:** 1318
aluminum/air batteries ... **A13:** 1319-1320
fuel cells ... **A13:** 1320-1321
lead-acid ... **A13:** 1317-1318
lithium ambient-temperature batteries ... **A13:** 1318-1319
lithium/sulfur dioxide batteries ... **A13:** 1319
sodium/sulfur batteries ... **A13:** 1320
Battery grid alloys *See also* Lead; Lead alloys
compositions ... **A2:** 544, 551
containing tin ... **A2:** 526
as lead application ... **A2:** 548-549
Battery-recycling chain
recent changes ... **A2:** 1221-1222
Batts ... **EM1:** 5, 62
Bauer-Vogel (MBV) oxide conversion coating process ... **M5:** 598-599
Baumé hydrometer ... **M5:** 173
Baumé syrup
for rammed graphite molds ... **A15:** 273-274
Bauschinger effect ... **A8:** 1
in low-cycle fatigue testing ... **A8:** 367
in titanium alloys ... **A14:** 839
wrought titanium alloys ... **A2:** 614
Bauxite ... **EM4:** 45, 49-50, 895, 896, 903
applications ... **EM4:** 46
composition ... **EM4:** 46
gallium recovery from ... **A2:** 739, 741
refractory material composition ... **EM4:** 896
supply sources ... **EM4:** 46
Bauxite (Al_2O_3)
methods used for synthesis ... **A18:** 802
Miller numbers ... **A18:** 235
Bayer aluminum process ... **EM4:** 56, 111
followed by powder washing ... **EM4:** 92-93
Bayer process
of gallium recovery from bauxite ... **A2:** 742
Bayer-Ku zero-sliding-wear theory ... **A18:** 265, 266
Bayerite ... **EM3:** 262, 264 **EM4:** 111, 112
bcc *See* Body centered cubic (bcc) materials; Body-centered cubic lattice; Body-centered cubic materials
BE *See* Backscattered electron
BE P/M technology *See* Blended elemental titanium P/M compacts/products

Be-38Al
physical properties ... **A6:** 941
Be-Co (Phase Diagram) ... **A3:** 2•93
Be-Cr (Phase Diagram) ... **A3:** 2•93
Be-Cu (Phase Diagram) ... **A3:** 2•94
Be-Fe (Phase Diagram) ... **A3:** 2•94
Be-Hf (Phase Diagram) ... **A3:** 2•95
Be-Nb (Phase Diagram) ... **A3:** 2•95
Be-Ni (Phase Diagram) ... **A3:** 2•95
Be-Pd (Phase Diagram) ... **A3:** 2•96
Be-Si (Phase Diagram) ... **A3:** 2•96
Be-Th (Phase Diagram) ... **A3:** 2•96
Be-Ti (Phase Diagram) ... **A3:** 2•97
Be-W (Phase Diagram) ... **A3:** 2•97
Be-Zr (Phase Diagram) ... **A3:** 2•97
Beach marks *See also* Striation
AISI/SAE alloy steels ... **A12:** 301, 322, 331, 332
austenitic stainless steels ... **A12:** 358, 359
cast aluminum alloys ... **A12:** 408
circular ... **A12:** 273
and circular spall ... **A12:** 114, 125-126
defined ... **A8:** 1 **A11:** 1 **A13:** 2
in diesel truck crankshaft ... **A11:** 77, 78
ductile iron crankshaft ... **A12:** 228
in fatigue fracture ... **A12:** 111-112
as fatigue striations ... **A12:** 175
in forged components ... **A11:** 321
fracture mechanics of ... **A11:** 57
in fracture surface fatigue regions ... **A11:** 104
from advancing fatigue-crack front ... **A11:** 26
from stable crack growth ... **A11:** 87
high-carbon steels ... **A12:** 281, 285
in integral coupling and gear ... **A11:** 129
macroscopy of ... **A11:** 104
martensitic stainless steels ... **A12:** 369
in medium-carbon steels ... **A12:** 260, 267, 273-276
on aircraft fuel-tank floors ... **A11:** 126
oval ... **A12:** 273
in spring failures ... **A11:** 554
in steel knuckle pins ... **A11:** 129
superalloys ... **A12:** 391
titanium alloys ... **A12:** 446, 452
tool steels ... **A12:** 377
in unidirectional-bending fatigue, shafts ... **A11:** 461
wrought aluminum alloys ... **A12:** 415, 416, 421, 427
Bead
defined ... **A15:** 1
Bead weld
definition ... **A6:** 1207
Bead(s)
defined ... **A14:** 1
draw ... **A14:** 582
formation ,during radial-axial rolling ... **A14:** 115
press forming of ... **A14:** 552
Beaded flange
defined ... **A14:** 1
Beading
aluminum alloy ... **A14:** 804
dies, for press-brake forming ... **A14:** 537-538
of titanium alloys ... **A14:** 846
Beam
balance, in torsional testing machine ... **A8:** 146
grid deformations in longitudinal and cross ... **A8:** 118-120
sections, bend tests for ... **A8:** 117
skeletal points ... **A8:** 326-327
springback in simple bending ... **A8:** 552
torsion tests for ... **A8:** 139
Beam bending
formulas for ... **EM2:** 653
Beam breaks
inner/outer tape automated bonding (TAB) ... **EL1:** 287
Beam diagrams for transmission electron microscopy ... **A9:** 104
Beam, electron
in SEM imaging ... **A12:** 167-168
Beam hardening ... **A17:** 376, 383
Beam intensity
ultrasonic inspection ... **A17:** 237-238
Beam leads
bonding, as component attachment ... **EL1:** 351
defined ... **EL1:** 1135
Beam splitters
for optical holography ... **A17:** 418
Beam stops, integrated
as optical structure ... **EL1:** 10
Beam tape constructions
types ... **EL1:** 275-277
Beam theory ... **EM3:** 0
Beam(s)
angle, calibration, ultrasonic inspection ... **A17:** 266
diameter, ultrasonic ... **A17:** 240
high-energy x-ray ... **A17:** 307
model, radiographic inspection ... **A17:** 710
models, ultrasonic inspection ... **A17:** 705
neutron, attenuation ... **A17:** 390
P/S polarization, intetferometers ... **A17:** 14
scanning light ... **A17:** 12
splitters, optical holographic interferometry ... **A17:** 418
spread, calibration, ultrasonic inspection ... **A17:** 266
spreading, ultrasonic ... **A17:** 240
ultrasonic, attenuation of ... **A17:** 238-240
Beam-broadening effect
electron-beam welding ... **A6:** 854
Beam-condensing optics
use in IR diamond- anvil cells ... **A10:** 113
Beam-induced damage
from surface analytical techniques ... **A7:** 251, 255
Beam-lead sealed junction (BLSJ) chip ... **EL1:** 961
Beam-to-chip separation
tape automated bonding (TAB) ... **EL1:** 287
Beam-to-substrate separation
tape automated bonding (TAB) ... **EL1:** 287
Beams *See* Electron beams
damping analysis of ... **EM1:** 209-210
equalizer, fracture of ... **A11:** 388-390
laser cutting of ... **EM1:** 678
small-rotation assumption and ... **EM2:** 692-694
steel cantilever, distortion and stress ratios ... **A11:** 137
Beardsley, Elmer
as early founder ... **A15:** 28
Bearing
area, defined ... **A8:** 1 **EM1:** 5
defined ... **A18:** 3
deformation curves, vs. autographic bearing load, pin bearing testing ... **A8:** 61
load, in pin bearing testing ... **A8:** 59
metals, Rockwell scales for ... **A8:** 76
strain, defined ... **EM1:** 5
stress, defined ... **EM1:** 5
stress, pin bearing testing for ... **A8:** 59
yield, calculated in pin bearing testing ... **A8:** 61
Bearing alloys ... **A13:** 183, 774
aluminum base ... **A2:** 128, 131
fatigue resistance ... **A2:** 554
lead ... **A2:** 544, 553-554
tin ... **A2:** 522-523
Bearing alloys, specific types
52100 steel bar, different heat treatments compared ... **A9:** 195-196
52100 steel bar, different magnifications compared ... **A9:** 195
52100 steel, damaged by an abrasive cutoff wheel ... **A9:** 196
52100 steel rod, austenitized and slack quenched in oil ... **A9:** 196
52100 steel roller, crack from a seam in bar stock ... **A9:** 196
Bearing applications
copper alloy castings ... **M2:** 392-393
hot rolled bars ... **M1:** 208
journal bearing alloys, compatibility with shafting alloys ... **M1:** 606
Bearing applications (continued)
journal bearings, wire wooling selection to avoid ... **M1:** 606
rolling contact bearings, steels for ... **M1:** 606
compositions ... **M1:** 609-610
Bearing area
defined ... **A18:** 3-4 **EM2:** 6
Bearing bronze
applications and properties ... **A2:** 325, 378-380
Bearing bronzes (lead bronzes) ... **A18:** 693
bearing material microstructures ... **A18:** 743, 744
in bimetal bearing material systems ... **A18:** 747
casting processes with tin in alloy ... **A18:** 754, 755
corrosion resistance ... **A18:** 744
defined ... **A18:** 4
sliding bearings ... **A18:** 516
in trimetal bearing material systems ... **A18:** 748
Bearing cap
failed, by stress raiser and low-strength microstructure ... **A11:** 347-350
structure, hypereutectic gray iron ... **A11:** 349
structure, hypoeutectic cast iron ... **A11:** 349
Bearing cap bolts
sulfide SCC failure ... **A12:** 299
Bearing capacity ... **A18:** 507
Bearing characteristic number *See also* Capacity number; Sommerfeld number
defined ... **A18:** 4
Bearing diameter
nomenclature for Raimondi-Boyd design chart ... **A18:** 91
Bearing endurance ... **A18:** 508
Bearing fraction
defined ... **A18:** 4
Bearing friction torque ... **A18:** 510-511
Bearing internal speeds ... **A18:** 513
Bearing land thickness
nomenclature for hydrostatic bearings with orifice or capillary restrictor ... **A18:** 92
Bearing length ... **A18:** 63
nomenclature for Raimondi-Boyd design chart ... **A18:** 91
Bearing life ... **A18:** 507
life adjustment factors ... **A18:** 507-508
Bearing materials
effect of free lead on boundary lubrication ... **A11:** 161
embrittlement by ... **A11:** 236
peak stresses for normal fatigue life ... **A11:** 487
rolling-element ... **A11:** 490
sliding ... **A11:** 483-484
steel, fabrication ... **A11:** 490
steels, heat treatment variations ... **A11:** 509
tin alloy ... **M2:** 614-615
Bearing materials, ASTM specific types
B23
composition ... **M3:** 813, 814
properties ... **M3:** 813, 814
Bearing materials, SAE specific types
AAR M501, composition ... **M3:** 814
SAE 11, composition ... **M3:** 813
SAE 12
bearing life, effect of babbitt thickness ... **M3:** 806, 813
composition ... **M3:** 813
SAE 16, composition ... **M3:** 814
SAE 19, composition ... **M3:** 814
SAE 190, composition ... **M3:** 814
Bearing materials, sleeve *See* Sleeve bearing materials
Bearing number ... **A18:** 522-523, 524, 525
Bearing pitch diameter ... **A18:** 505, 511
Bearing pocket diameter
nomenclature for hydrostatic bearings with orifice or capillary restrictor ... **A18:** 92
Bearing properties
aluminum casting alloys ... **A2:** 154
wrought aluminum and aluminum alloys ... **A2:** 111, 113

Bearing quality
alloy steel sheet and strip **M1:** 164
alloy steel wire rod **M1:** 256
of low-alloy steel **A1:** 209
Bearing races
Barkhausen noise measurement **A17:** 160
flux leakage inspection **A17:** 133
microhoned **A16:** 491
Bearing radius
nomenclature for Raimondi- Boyd design chart **A18:** 91
Bearing shells
babbitting processes **M5:** 356-357
Bearing speed (rev/s)
nomenclature for Raimondi-Boyd design chart **A18:** 91
Bearing steels **A1:** 149, 380-388
carburizing **A1:** 381-382, 383
composition
carburizing steels **A1:** 382
corrosion-resistant steels **A1:** 388
high-carbon steels **A1:** 381
high-temperature steels **A1:** 387
deformation resistance vs rolling temperature **A14:** 119
heat treatment, effect of **A1:** 383, 384, 385
high- or low-carbon steels for **A1:** 24-25, 380-381
high-carbon **A1:** 381, 382
induction-hardened **A1:** 380, 381
mechanical properties
hardness **A1:** 381, 387
impact strength **A1:** 381
nonmetallic inclusion rating **A1:** 385
tensile strength **A1:** 381
microstructure characteristics **A1:** 381-382, 383
carburizing **A1:** 381-382, 383
high-carbon **A1:** 381, 382
quality of **A1:** 382-384, 385
rolling-contact fatigue **A12:** 115, 134
special-purpose **A1:** 384-388
Bearing steels, friction and wear of **A18:** 693, 725-733
abrasive wear **A18:** 732-733
adhesive wear **A18:** 732, 733
application, internal combustion engine parts **A18:** 556
bearing life **A18:** 728, 729, 730
composition **A18:** 725-727
carburizing steel advantages **A18:** 725-726
cleanness **A18:** 726-727
cost **A18:** 726
fatigue life **A18:** 726, 727, 728, 730
high-carbon steel advantages **A18:** 725
properties **A18:** 726
vacuum arc remelting (VAR) ... **A18:** 726, 727, 731
vacuum induction melted/vacuum arc remelted (VIM/VAR) **A18:** 726, 727, 731
concentrated contacts **A18:** 727-729
asperity slope **A18:** 729
bearing life **A18:** 728, 729
fatigue life **A18:** 727-729
friction coefficient **A18:** 728
load effect on bearing life **A18:** 728
damage classification **A18:** 728
electroslag remelting (ESR) **A18:** 731
friction coefficient data **A18:** 74
lambda ratio and modes of wear **A18:** 729-732
application of environmental condition factor a_3 .... **A18:** 729, 731-732
fatigue spall criteria **A18:** 731
friction coefficient **A18:** 730
life adjustment factors **A18:** 730-731
load-life equation **A18:** 730, 731
material factor a_2 .... **A18:** 731
reliability factor a_1 .... **A18:** 731
rolling contact fatigue **A18:** 260
wear mode **A18:** 728
Bearing strain
defined **A8:** 1 **EM2:** 6
Bearing strength **A8:** 1
defined **EM1:** 5
and grain direction, pin bearing testing **A8:** 61
lubricant effect in aluminum alloys **A8:** 60
pin and bolt **EM1:** 314-316
pin bearing testing for **A8:** 59
Bearing stress
defined **A8:** 1 **EM2:** 6
Bearing test
defined **A8:** 1
Bearing ultimate strength
computation of derived **A8:** 667
effect of lubricants and cleaners **A8:** 60
pin bearing testing for **A8:** 59
symbols and unit **A8:** 662
Bearing yield strength **A8:** 1
effect of lubricants and cleaners **A8:** 60
pin bearing testing for **A8:** 59
symbols and unit **A8:** 662
Bearing(s) *See also* Bearing alloys; Bearing bronze; Bearing properties
aluminum and aluminum alloys **A2:** 11
applications, beryllium-copper alloys **A2:** 418
cemented carbide **A2:** 973
construction, coordinate measuring machines **A17:** 24
of copper casting alloys **A2:** 352, 354-355
engine, silver in **A2:** 691
indium plating **A2:** 635
life, self-lubricating sintered bronze bearings **A2:** 395-396
races **A17:** 133, 160
rings, magnetic particle inspection **A17:** 117
rollers, magnetic particle inspection methods **A17:** 116-117
rolling-element antifriction, flux leakage inspection **A17:** 133
strength, of magnesium alloys **A2:** 460-461
Bearing-load ratings **A11:** 490-491
Bearings *See also* Ball bearings; Gas-lubricated bearings; Rolling contact wear of; Rolling- element bearings, friction and wear of; Rolling-element bearings; Rolling-element bearings, failures of; Sliding bearings; Sliding bearings, failures of; Tin and tin alloy coatings **A7:** 704-709
alloy, performance characteristics **A7:** 408
alloy steel wire for **M1:** 269
antifriction, wear of **A11:** 764-765
for appliances **A7:** 623
assembly, fretting damage **A11:** 341
automotive front-wheel, fretting failure **A11:** 498
babbitts **A11:** 483
ball and roller, compared **A11:** 490
bronze P/M **A7:** 736
butterflies in **A12:** 115, 134
caps, failed **A11:** 347-350
carbon fiber reinforced nylon **EM1:** 35
composite and sleeve, powder-rolled **A7:** 406-408
copper-lead alloy, deleading failure **A11:** 488
cylindrical-roller **A11:** 490
deformation, as fastener failure **A11:** 531
design using solid lubrication **A11:** 153
early P/M techniques for **A7:** 16
effect of porosity **A7:** 451
examination of failed **A11:** 491
failures **A11:** 486-489
friction, overheated, failure of locomotive axles from **A11:** 715-727
grease and oil lubrication guides **A11:** 511
halves, fatigue failure **A11:** 489
high-performance **A11:** 483
industrial, powders used **A7:** 573
infiltration use **A7:** 565
length **A7:** 709
lubricant viscosities **A11:** 511
materials for **A11:** 490
misalignment of long and short **A11:** 489
oilite **A8:** 201
overheated traction-motor support, locomotive axle failures from **A11:** 715-727
overloading, effects of **A11:** 501
pillow-block, misaligned **A11:** 475
porous **A7:** 17, 706
porous, lubricant for **A7:** 706
powders used **A7:** 572
rolling contact fatigue tester **A8:** 370
rolling of strip for **A7:** 406, 407
self-lubricating **A7:** 16, 18
and shaft failures **A11:** 466
sleeve, powder-rolled **A7:** 407-408
sliding **A11:** 483-489
spalling fatigue in **A12:** 114-115
Bearings (continued)
spherical, corrosion fatigue failure **A11:** 488
-surface failure, of rivets **A11:** 544
surfaces, grinding bums **A11:** 89
for three-roll forming machines **A14:** 619
thrust, electrical wear **A11:** 487
tin powders for **A7:** 123-124
trimetal, distortion failure **A11:** 489
wear, and shaft failure **A11:** 459
Bearings, sliding
aluminum alloys **M3:** 806-807, 818-820
bearing life, effect of babbitt thickness **M3:** 806, 813
bimetal systems **M3:** 807-809
casting of **M3:** 808-811
cemented carbides **M3:** 820-821
classification **M3:** 802-803
configuration **M3:** 803
copper alloys **M3:** 806, 816-818
corrosion **M3:** 804, 806
electroplating **M3:** 809, 812
gray cast irons **M3:** 820
lead alloys **M3:** 814-815
manufacturing method **M3:** 803
microstructures **M3:** 805
nonmetallic materials for **M3:** 821-822
operating conditions **M3:** 803-804
overlays **M3:** 815-816
powder metallurgy **M3:** 811-812
roll bonding **M3:** 808, 812
silver alloys **M3:** 820
single-metal systems **M3:** 806, 807
size **M3:** 803
structural characteristics **M3:** 803
testing **M3:** 805
tin alloys **M3:** 813-814
trimetal systems **M3:** 809, 810
Becquerel, as SI derived unit
symbol for **A10:** 685
Bed
defined **A14:** 1
Bed filters
for particle filtration **A15:** 490
Bed-of-nails handlers
for electrical testing **EL1:** 566-567
Bedding
defined **A15:** 1
Bedding a core
defined **A15:** 1
Beer
barrels, corrosion in **A13:** 1223
copper/copper alloy resistance **A13:** 631
pure tin resistance **A13:** 772
Beer fermentation process
powder used **A7:** 574
Beer's law
as basis for IR quantitative analysis **A10:** 117
defined **A10:** 669-670
deviations from **A10:** 70
in ion chromatography **A10:** 665
in UV/VIS absorption spectroscopy **A10:** 61-63, 70
Beer-Lambert law **EM4:** 1042
Beeswax
for investment casting **A15:** 253-254
Beet-sugar solution
copper alloy corrosion in **A13:** 635
Begley-Logsdon three-point Charpy fracture toughness
correlation **A8:** 265
Behavior
as form of CAD model definition **EL1:** 1104
Behavioral model
as rules **EL1:** 1104
Beilby layer
defined **A8:** 1 **A18:** 4
Beilby theory of polishing **A18:** 197
Beja process
of gallium recovery from bauxite **A2:** 742
Belite **EM4:** 11
Bell
defined **A7:** 1
Bell and elevator batch-type furnaces **A7:** 356-357
Bell jar
carbon **A12:** 173
Bell OH-58 helicopter **A7:** 760

Bell process
for demagging aluminum alloys **A15:** 473
Bell-and-spigot joint
welding cracks **A11:** 424
Bell-type furnace
defined **A7:** 1
Bellcrank assemblies
economy in manufacture **M3:** 848-849
Belleek china *See* Frit china
Belleville spring
in manned spacecraft **A13:** 1091
Belleville washers
heat-treatment distortion **A11:** 140
Bellows
in vacuum fatigue test chamber **A8:** 412, 414
Bellows expansion joint
intergranular fatigue cracking in **A11:** 131-133
Bellows grips
with elevated-temperature compression testing **A8:** 196
Bellows length
photomacrography **A12:** 79
Bellows liners, welded
fatigue fracture in **A11:** 118
Bellows seal
defined **A18:** 4
Bells
by Paul Revere **A15:** 26
cast, history of **A15:** 19-20
cast steel, German **A15:** 31
Liberty, history of **A15:** 27
Belt grinding
belt life, stainless steel processes **M5:** 556-557
mechanized, stainless steel **M5:** 556
safety precautions **M5:** 557
stainless steel **M5:** 556-557
titanium and titanium alloys **M5:** 652
Belt polishing
abrasive **M5:** 109-110, 115, 676
applicability **M5:** 110, 115
construction, coated abrasives **M5:** 109-110
contact wheels, types, characteristics, and uses of **M5:** 110-111, 115, 125-126
flat part polishing machines **M5:** 122-124, 127
grades, abrasive **M5:** 110
grit size and belt speeds **M5:** 113, 115
process selection factors **M5:** 110-111
surface finishes, range of **M5:** 110
aluminum and aluminum alloys **M5:** 573-574
copper and copper alloys **M5:** 616
lubricants used **M5:** 115
magnesium alloys **M5:** 632
stainless steel **M5:** 557
titanium and titanium alloys **M5:** 655-656
zinc alloys **M5:** 676
Belt-drop hammer **A14:** 42
Belt-type continuous furnace
for carbonitriding **A7:** 455
Beltless sheet molding compound machines **EM1:** 159-160
Beltrami maximum strain energy **A8:** 344
Bench lathes **A16:** 153
Bench micrometer
laser **A17:** 13
Bench molding
defined **A15:** 1
Bench units
for circular magnetization **A17:** 97
for magnetic particle inspection **A17:** 111
Bench-mounted microhardness tester **A8:** 91-92
Bend *See also* Bar bending; Defect; Twist
allowance, for bar **A14:** 663
defined **A14:** 1
holes close to **A14:** 553
Bend allowance
steel wire fabrication **M1:** 588
Bend angle
defined **A14:** 1
Bend bar
modulus of rupture tests **EM4:** 547
Bend contours **A9:** 111, 113
pattern, for strain fields in crystals **A10:** 368
polycrystalline molybdenum-rhenium alloy **A10:** 445
Bend deflection
P/M and ingot metallurgy tool steels **A7:** 471, 472
Bend formability
beryllium-copper alloys **A2:** 411
Bend forming
lead frame materials **EL1:** 487
Bend fracture strength
CPM alloys **A16:** 64-65
Bend fracture stress
P/M and ingot metallurgy tool steels **A7:** 471, 472
Bend loading vs pure tensile loading **A11:** 746
Bend radii
steel wire fabrication **M1:** 587-588
Bend radius **A8:** 1, 125 **A14:** 1, 523
Bend test **A1:** 582, 583, 610 **A8:** 1 **EM3:** 6
cadmium plate adhesion **M5:** 269
defined **EM2:** 6
pass-fail results **A8:** 117
for welds **A6:** 101
Bend test data
alloy steel forgings **M1:** 358-360, 374
Bend test, plate
for weldability **M1:** 198
Bend test requirements
hot dip galvanized steel sheet **M1:** 170, 171
hot rolled bars **M1:** 205, 206, 212
low-carbon steel sheet and strip **M1:** 155
Bend testing **A16:** 61, 62
for bulk workability assessment **A8:** 577-578
and linear elastic behavior **A8:** 117
simple **A8:** 560
simulative **A8:** 560-561
Bend tests
Lehigh **A11:** 59
of rocket-motor case fracture **A11:** 96
simple **A11:** 18
Bendability
of steels **A14:** 523
Benders
as impression dies **A14:** 44
Bending *See also* Bend; Bend testing; Bending ductility tests; Bending stress; Deformation; Elastic bending; Elastic plastic bending; Press bending; Pure plastic bending **A14:** 665-672
air, for press-brake forming **A14:** 536
alternating **A11:** 108
analysis **A14:** 915-918
of bars **A14:** 661-664
behavior, laminates **EM1:** 224
of beryllium **A14:** 805-807
cold draw, minimum radii **A14:** 665
cold, vs hot bending **A14:** 671
compression, of bar **A14:** 661
in contour roll forming **A14:** 624
of copper and copper alloys **A14:** 812-814
of curved flanges **A14:** 530
cylindrical **A8:** 118-119
cylindrical parts **A14:** 529
defined **A14:** 1
dies, defined **A14:** 1
distortion, of aircraft wing slat track **A11:** 140-141
draw, of bar **A14:** 661
ductility, tests for **A8:** 117, 125-131
edge **A14:** 529
elastic **A8:** 118-119, 552
elastic, below yield stress **A14:** 881
elastic-plastic **A8:** 119-120
fatigue fracture, of alloy steel pushrod **A11:** 469
fatigue fracture, steel pump shaft **A11:** 109
fatigue machine **A8:** 369
Bending (continued)
forced-displacement system **A8:** 392
forced-vibration system **A8:** 392
and forming, of tubing **A14:** 665-674
free **A14:** 532
hand vs power **A14:** 665
high-cycle fatigue, alloy steels **A12:** 296
historical studies **A12:** 3
hot **A14:** 669-671
in hot upset forging **A14:** 83
HSLA steels **M1:** 406, 408, 419
lateral, from thin-lip ruptures **A11:** 606
load vs displacement curve **A14:** 37
loaded elements **EM1:** 325
loaded subcomponents **EM1:** 334
lubrication for **A14:** 672-673
machines **A14:** 668-669, 671
of magnesium and magnesium alloy parts **A2:** 476-477
manual, of wire **A14:** 695-696
method, selection **A14:** 665
modulus of rupture in **A11:** 7
moment **A11:** 462
multifunction machining **A16:** 387, 388
in multiple-slide forming **A14:** 569
of nickel-base alloys **A14:** 834-837
noncylindrical **A8:** 119
notch-sensitivity with notch radius for steels in **A8:** 373
of organic-coated steels **A14:** 565
orientation **A14:** 524-525
in pipe **A11:** 704
plane strain **A8:** 120-121
plastic **A8:** 118, 120-122
plate **M1:** 194
power, of wire **A14:** 695-696
press-brake forming **A8:** 119
proof strength **A8:** 132-135
proof stress **A8:** 134
properties, wrought aluminum alloy **A2:** 58
pure, bent beam for **A8:** 505
pure plastic **A8:** 120-122
residual stress and springback **A8:** 122-124
resonance system **A8:** 392
reversed, alloy steel lift pin **A11:** 77
roll **A8:** 119
roll, of bar **A14:** 661
rotational **A11:** 108-109
rotational bending system **A8:** 392
servomechanical system **A8:** 392
severe, tool materials for **A14:** 539
sheet **M1:** 552-555
as sheet metal forming **A8:** 547
sheet, process modeling **A14:** 913-914
simple **A8:** 552 **A14:** 881-882
simple, tool materials for **A14:** 539
as springback **A8:** 552, 565
of stainless steel tubing **A14:** 777
steel wire fabrication **M1:** 587-588
strain curvature **A8:** 118
strength, selected structural metals **A2:** 478
strength test **A8:** 132-136
as stress, in fatigue fracture **A11:** 75
as stress on shafts **A11:** 461
stress-moment equations **A8:** 118
stress-strain curve **A8:** 132
stress-strain relationships **A8:** 118-124
stretch, of bar **A14:** 661
and stretching **A8:** 552-553
tests **A14:** 889-890
thin-wall tubes **A14:** 671-672
three-point **A8:** 118-119
of titanium alloys **A14:** 848
tools **A14:** 665-666, 671-672
tube stock for **A14:** 671
of tubing **A14:** 665-673
tubing, with mandrel **A14:** 666-668
tubing, without mandrel **A14:** 668
unidirectional **A11:** 108

Bending (continued)
universal testing machines for **A8:** 612
unsymmetrical **A8:** 119
vs. uniaxial tension, in SCC testing **A8:** 503
in wing slat track, from service stresses **A11:** 140-141
Bending brake
defined **A14:** 1
Bending ductility tests *See also* Bend testing; Bending **A8:** 117, 125-131
apparatus **A8:** 125-126
characteristics **A8:** 128-129
devices for **A8:** 125-126
effect of bending method in **A8:** 127
specimens **A8:** 126-127, 130
strain distributions **A8:** 128
terms used **A8:** 125
test method and interpretation **A8:** 127-129
Bending fatigue
failure, steel wire hoisting rope **A11:** 518
fracture, of shaft assembly, from misalignment **A11:** 475
fracture, steel pump shaft **A11:** 109
in shafts **A11:** 109, 461
testing **A11:** 102
Bending fatigue fracture(s)
AISI/SAE alloy steels **A12:** 296, 321, 329-330, 332
alloy steel gear teeth **A12:** 329-330
iron **A12:** 220
rotating, alloy steels **A12:** 321
surfaces, tool steels **A12:** 376
Bending fatigue machine **A8:** 369
Bending impact fracture
low-carbon steel **A12:** 242
Bending Lam waves
ultrasonic inspection **A17:** 234
Bending machines and presses
for bar **A14:** 661-663
bending presses **A14:** 668-669
powered rotary benders **A14:** 668
roll benders **A14:** 669
for tube **A14:** 668-669
Bending, modulus of rupture in *See* Modulus of rupture, in bending
Bending moment
conversion factors **A8:** 722 **A10:** 686
-deflection, for spring-tempered C77000 copper alloy strip **A8:** 134
sign convention for **A8:** 119-120
symbol for **A8:** 725
Bending overload fracture, classic
medium- carbon steels **A12:** 272
Bending proof strength
in three- and four-point bend tests **A8:** 132-135
Bending proof stress
in three- and four-point bend test **A8:** 134
Bending rolls
defined **A14:** 1
Bending strength test **A8:** 132-136
Bending stress
definition **EM4:** 632
magnesium structures **M2:** 552
Bending stress, defined *See also* Bending **A14:** 1
Bending tests **A13:** 247-248, 286-288 **A14:** 376-377, 792, 889-890
aluminum alloys **A14:** 792
three-point **A14:** 376
for workability **A14:** 377
Bending-twisting coupling
defined **EM1:** 5 **EM2:** 6
Benefit-cost ratios method
economic analysis **A13:** 370
Bengough-Stewart anodizing process
aluminum and aluminum alloys **M5:** 598
Benin bronzes
of Nigeria **A15:** 19
Bent-beam specimens
SCC testing **A8:** 503-504 **A13:** 248-249
Benton (clay) **A18:** 126, 129
Bentonite
as binder **EM4:** 120
chemical composition **A6:** 60
as suspending agent for ceramic coatings **EM4:** 955
Bentonite, modified
as filler **EM3:** 178
Bentonites
as binder **A15:** 29
defined **A15:** 1
as molding clay, characteristics/types **A15:** 210
sodium and calcium, blending effects **A15:** 210
types, for green sand molding **A15:** 341
Benzaldehyde (C_6H_5CHO)
as solvent used in ceramics processing **EM4:** 117
Benzalkonium chloride
description **A9:** 68
Benzene
copper/copper alloy resistance **A13:** 631
Benzene, adsorbed
SERS analyses for **A10:** 136
Benzene ring **EM3:** 6
defined **EM2:** 6
Benzine (C_6H_6)
as solvent used in ceramics processing **EM4:** 117
Benzo(a)pyrene
direct determination of **A10:** 79
Benzocyclobutene **EM3:** 161
for coating and encapsulation **EM3:** 580
for multichip structures **EL1:** 302-303
for printed board material systems **EM3:** 592
Benzoic acids
as ion chromatography eluents **A10:** 660
Benzol
copper/copper alloy resistance **A13:** 631
Benzotriazole **A18:** 141
Benzoyl peroxide
added to acrylic adhesives **EM3:** 120
formulation **EM3:** 123
Benzoyl peroxide (BPO) **EM1:** 133
as vinyl ester cure **EM2:** 274
Benzyl alcohol (C_7H_7OH)
as solvent used in ceramics processing **EM4:** 117
Beraha's reagent
as a color etchant for carbon and low alloy steels **A9:** 142
Beraha's tint etchant
used with carbon and alloy steels **A9:** 170
Berg clip lead
thermal expansion mismatch **EL1:** 611
Berg-Barrett reflection topography method
applicability **A10:** 368
camera for **A10:** 369
topographs of shadowing due to cleavage steps **A10:** 369
Berkelium *See also* Transplutonium metals
applications and properties **A2:** 1198-1201
pure **M2:** 717, 832-833
Bernal-Finney DRPHS model
of amorphous materials and metallic glasses **A2:** 810
Bernoulli equation **A6:** 161
Bernoulli's equation **A18:** 594, 600
Bernoulli's theorem
in gating design **A15:** 590-591
Bertrandite *See also* Beryllium
mining and refining **A2:** 684
Beryl *See also* Beryllium
crystal structure **EM4:** 881
fusion flux for **A10:** 167
isolated group structure **EM4:** 758
mining and refining **A2:** 684
sintering agent for **A10:** 166
Berylco 10 *See* Copper alloys, specific types, C17500
Berylco 165 *See* Copper alloys, specific types, C17000
Berylco alloys
applications and properties **A2:** 284
Beryllia
as ceramic substrate **EL1:** 106, 337
as filler **EM3:** 596
as filler for conductive adhesives **EM3:** 76
as heat sink **EL1:** 1129
for improved thermal performance **EM3:** 584
thermal expansion coefficient **A6:** 907
Beryllia (99.5%) ceramics
properties **A6:** 992
Beryllia (BeO) *See also* Beryllium oxide **EM4:** 13-14
as chemical substate **EM4:** 1110
freeze drying **EM4:** 62
properties **EM4:** 503
substrate properties **EM4:** 1108
thermal shock resistance **EM4:** 1003
Beryllia insulators
for thermocouples **A2:** 883
Beryllia-silicon carbide **EM4:** 191
Beryllide phase
in beryllium-copper alloys **A9:** 395-397
Beryllides in beryllium-copper alloys
as revealed by etching **A9:** 394
Berylliosis
acute and chronic **A2:** 687, 1239
Berylliosis (chronic beryllium disease)
description and symptoms **A7:** 202
Beryllium *See also* Beryllium grades, specific types; Beryllium oxide; Beryllium powders; Beryllium toxicity; Beryllium-copper alloys; Beryllium-nickel alloys; Optically anisotropic metals **A9:** 389-391 **A13:** 808-812 **A14:** 805-808
in air **A13:** 808-809
aircraft break, pitted **A13:** 1026
alloying, aluminum casting alloys **A2:** 131
alloying effect on copper alloys **M6:** 400
alloying, wrought aluminum alloy **A2:** 46
in alloys, oxyfuel gas cutting **A6:** 1165
in aluminum alloys **A2:** 426 **A15:** 744
applications **A2:** 683 **A7:** 202 **M6:** 1052
aqueous corrosion **A13:** 809-810
arc welding *See* Arc welding of beryllium
atomic interaction descriptions **A6:** 144
atomization **A2:** 684-685
attack-polishing **A9:** 389
blocks, vacuum hot pressing **A7:** 508
brazing **M6:** 1052
applications **M6:** 1054
brazeability **M6:** 1052
filler metals **M6:** 1052
preparation **M6:** 1053
safety **M6:** 1052
brazing to aluminum **M6:** 1031
chemical analysis and sampling **A7:** 248
chips **A7:** 170
color differences under polarized light **A9:** 389
commercial grades, chemistry **A2:** 686
consolidated part, secondary ion mass spectroscopy of **A7:** 258
contact angles of liquid metals on **A6:** 116
-containing alloys, safe handling **A2:** 426-427
in copper alloys **A6:** 752
corrosion protection **A13:** 81
current industrial practices **A2:** 684
deep drawing of **A14:** 806-807
deoxidizing, copper and copper alloys **A2:** 236
dies and workpieces, heating of **A14:** 806
effect on maraging steels **A4:** 222, 224
elasticity and density **A7:** 169
electron-beam welding **A6:** 872-873
autogenous welds **A6:** 872-873
braze welds **A6:** 873
preheat effect on weldability **A6:** 873
equipment and tooling **A14:** 805-806
etchants for **A9:** 390
etching **A9:** 390
etching by polarized light **A9:** 59
evaporation fields for **A10:** 587
extrusions **A7:** 514, 759
in filler metal used for direct brazing **EM4:** 519
flash welding **M6:** 558
forgings **A7:** 759
formability **A14:** 805
gas tungsten arc welding **M6:** 206
gas-tungsten arc welding **A6:** 192
grades and their designations **A2:** 686-687
grain size control **A9:** 390
grinding **A9:** 389
handling and storage **A13:** 810
hot pressing **A7:** 172, 513
in-process corrosion **A13:** 810
infrared reflectivity **A2:** 684
ingot, reduction to chips **A7:** 170
instrument grades **A2:** 686-687
joining techniques **A2:** 683
low-temperature solid-state welding **A6:** 300, 301
lubrication for **A14:** 806
macroexamination **A9:** 389
in magnesium systems **A2:** 426
as major toxic metal with multiple effects **A2:** 1238-1239
mechanical properties **M6:** 461

Beryllium (continued)
microalloying **A9:** 390
microexamination **A9:** 389-390
microstructures **A9:** 390
microyield strength **A2:** 684
mining and refining **A2:** 684
mounting **A9:** 389
near-net shape processes **A2:** 685-686
nuclear applications **A7:** 664
P/M technology **A7:** 755-762
photometric analysis methods **A10:** 64
physical properties **A2:** 683 **A6:** 941 **M6:** 461
polishing **A9:** 389
powder characteristics, commercial **A7:** 171
powder consolidation methods **A2:** 685-686
powder metallurgy (P/M) production **A2:** 683-686
powder production operations **A2:** 684-685
precipitation **A10:** 169
production and consolidation **A7:** 757-758
properties **A7:** 756
properties of importance **A2:** 683-684
pure **M2:** 717-718
pure, properties **A2:** 1102
for pyrotechnics **A7:** 597
relative solderability **A6:** 134
relative solderability as a function of flux type **A6:** 129
roll welding **A6:** 314
safety **A9:** 389
safety practice **A14:** 806
safety standards for soldering **M6:** 1098-1099
sample, copper impurities detected **A7:** 258, 259
scanning acoustic microscopy for machining damage **A18:** 409
sectioning **A9:** 389
sheet metal envelope, hot pressing of **A7:** 509
solderability **A6:** 978
species weighed in gravimetry **A10:** 172
specimen preparation **A9:** 389
spinning of **A14:** 807-808
stress relieving of **A14:** 806
stretch forming of **A14:** 807
strip, temperature profiles **A14:** 357
structural grades **A2:** 686
structural members, Hughes satellite **A7:** 759
stylus shanks **A7:** 762
substrate with internal cavities HIP **A7:** 427
surface treatments/coatings **A13:** 811
tensile properties **A7:** 756
thermal diffusivity from 20 to 100 °C **A6:** 4
thermal expansion coefficient **A6:** 907
three-roll bending of **A14:** 807
-to-Monel brazed joint, embrittlement **A13:** 879
toxicity **A6:** 1195, 1196 **A7:** 202 **A9:** 389
toxicity, health, and safety **A2:** 687
unsuitability for cladding combinations **M6:** 691
UV/VIS analysis for chromium, by diphenylcarbazide method **A10:** 68
V(*z*)-curves of surfaces **A18:** 407
vacuum hot pressing **A2:** 685
vacuum hot-pressed, compositions **A13:** 808
vacuum hot-pressed grades **A9:** 390
vapor pressure, relation to temperature **A4:** 495
wettability **A6:** 115-116
window **A10:** 519, 670
workability **A8:** 165, 575
wrought product forms **A7:** 758-759
wrought products and fabrication **A2:** 687

Beryllium alloys **A16:** 870-873
applications **A6:** 945
brazing **A6:** 947
chemical milling **A16:** 871, 872-873
cold isostatic pressing **A16:** 870
cutting fluids **A16:** 159, 872
diffusion welding **A6:** 885, 886

Beryllium alloys (continued)
ductile-to-brittle transition temperature **A6:** 945
electrochemical grinding **A16:** 543, 547
electrochemical machining removal rates **A16:** 534
elements implanted to improve wear and friction properties **A18:** 858
etching and heat treating **A16:** 871-872
filler metals **A6:** 945-946
flash welding **M6:** 558
fluxes **A6:** 946
gas tungsten arc welding **M6:** 182
grinding **A16:** 547
health concerns **A16:** 870
hot isostatic pressing **A16:** 870
material considerations **A16:** 873
milling **A16:** 547, 870
photochemical machining **A16:** 588, 590, 872-873
properties affecting its handling **A16:** 870
sawing **A16:** 364
sintering **A16:** 870
surface damage **A16:** 870-872
surface finish **A16:** 872
thermal expansion coefficient **A6:** 907
tool life **A16:** 872
tools **A16:** 872
transverse tensile properties **A16:** 871
trepanning **A16:** 176
turning **A16:** 872
vacuum heat treating **A16:** 871
vacuum hot pressing **A16:** 870

Beryllium alloys, specific types
Be-Cu **A16:** 476, 588, 590
Be-Ni, milling **A16:** 313
S65 Be, composition **A16:** 870
S65 Grade 1319A, transverse tensile properties **A16:** 871
S65 Grade S200E, transverse tensile properties **A16:** 871
Select S65 Be, composition **A16:** 870

Beryllium carbide
thermal expansion coefficient **A6:** 907

Beryllium copper *See also* Copper alloys, specific types, C17000, C17200, C17300 and C17600; Copper alloys specific types, C17200; Wear-resistant alloys, nonferrous **A6:** 752
and brass, barrier coatings for **EL1:** 679
cleaning solutions for substrate materials **A6:** 978
composition and properties **M6:** 401
electrical conductivity **M6:** 551
electronic applications **A6:** 998
foil, for flex printed boards **EL1:** 581
gas metal arc welding **M6:** 418-420
high-conductivity grades **M6:** 418-419
high-strength grades **M6:** 419-420
gas tungsten arc welding **M6:** 408-409
high-conductivity grades **M6:** 408
high-strength grades **M6:** 408-409
high-conductivity, gas-tungsten arc welding **A6:** 761, 762
high-strength, gas-metal arc welding **A6:** 761-762
high-strength, gas-tungsten arc welding **A6:** 760, 761, 762
molds for plastics and rubber, use in **M3:** 548, 549
nonconsumable wire guide tubes **M6:** 227
relative hydrogen susceptibility **A8:** 542
relative solderability **A6:** 134
relative solderability as a function of flux type **A6:** 129
resistance welding *See* Resistance welding of copper and copper alloys **A6:** 849
Rockwell scale for **A8:** 76
self-lubrication **M3:** 594
soldering **M6:** 1075
springs, strip for **M1:** 286

Beryllium copper (continued)
springs, wire for **M1:** 284
thermal expansion coefficient **A6:** 907
wear-resistance techniques **M3:** 593, 594
weldability **A6:** 753

Beryllium copper alloys
precipitation hardening of **A14:** 809

Beryllium copper nickel 72C *See also* Cast copper alloys
properties and applications **A2:** 391

Beryllium coppers
10C *See* Copper alloys, specific types, C82000
20C *See* Copper alloys, specific types, C82500
30C *See* Copper alloys, specific types, C82200
50C *See* Copper alloys, specific types, C81800
70C *See* Copper alloys, specific types, C81400
165C *See* Copper alloys, specific types, C82400
245C *See* Copper alloys, specific types, C82600
275C *See* Copper alloys, specific types, C82800
Be-modified chrome copper *See* Copper alloys, specific types, C81400
casting alloy 10C *See* Copper alloys, specific types, C82000
casting alloy 30C *See* Copper alloys, specific types, C82200
casting alloy 35C *See* Copper alloys, specific types, C82200
casting alloy 53B *See* Copper alloys, specific types, C82200
casting alloy 165C *See* Copper alloys, specific types, C82400
casting alloy 245C *See* Copper alloys, specific types, C82600
casting alloy 275C *See* Copper alloys, specific types, C82800
grain-refined casting alloy 21C *See* Copper alloys, specific types, beryllium copper 21C
heat treating **M2:** 256-257, 258
heat treatment **A15:** 782
melt treatment **A15:** 775
standard casting alloy *See* Copper alloys, specific types, C82500

Beryllium cupro-nickel *See* Copper alloys, specific types, C96600 and Beryllium copper nickel 72C

Beryllium cupro-nickel alloy
properties and applications **A2:** 388

Beryllium fluoride
molten **A13:** 52-54

Beryllium grades, specific types
I-70, for optical components **A2:** 686
I-220, ductility and microyield strength **A2:** 687
I-400, microyield strength **A2:** 687
O-50, infrared reflectivity grade **A2:** 687
S-65, aerospace grade **A2:** 686
S-200E, attritioned powder **A2:** 686
S-200F, as most commonly used grade **A2:** 686
S-200FH, HIP consolidated **A2:** 686

Beryllium nitride **A16:** 100

Beryllium oxide **A16:** 100, 530
based cermets **A7:** 803
in binary phosphate glasses **A10:** 131
effect on tensile elongation of high-purity beryllium powders **A7:** 758
to improve thermal conductivity **EM3:** 178
to increase thermal conductivity **EM3:** 620, 621
vacuum heat-treating support fixture material **A4:** 503

Beryllium oxide (BeO) *See also* Beryllia
applications **EM4:** 48
composition **EM4:** 14
content effect, electron-beam welding **A6:** 872
electrical/electronic applications **EM4:** 1105
grain-growth inhibitor **EM4:** 188
key product properties **EM4:** 48
properties **A6:** 629 **EM4:** 14
raw materials **EM4:** 48
sintering aid **EM4:** 188

Beryllium oxide (BeO) (continued)
solderable and protective finishes for substrate materials **A6:** 979
solid-state sintering **EM4:** 273
for substrates **EM4:** 17
ultrasonic machining **EM4:** 359
Beryllium oxide as a constituent of vacuum hot-pressed beryllium **A9:** 390
appearance under polarized light **A9:** 390
effect on ductility **A9:** 390
effect on grain size **A9:** 390
effect on strength **A9:** 390
Beryllium oxide cermets
applications and properties **A2:** 993
Beryllium oxide, effects
beryllium powder metallurgy **A2:** 686
Beryllium powders *See also* Beryllium; Beryllium oxide
applications **A7:** 169, 759-762
as-cast grain size **A7:** 169
block impact ground powder **A7:** 170
comminution, pole density, and ductility **A7:** 171
compacting pressure and density **A7:** 171
composition **A7:** 171
consolidation **A7:** 172
consumer applications **A7:** 761-762
high-temperature applications **A7:** 761
hot pressed block **A7:** 172
impact attrition mill **A7:** 757
instrument applications **A7:** 761
manufacture **A7:** 170-172
nuclear and structural grades **A7:** 758
particle size, oxide content, and grain size **A7:** 171
physical and mechanical properties **A7:** 169, 170, 756-757
production **A7:** 169-172
purity improvement **A7:** 757
safety exposure limits **A7:** 202
strength and grain size **A7:** 171
Beryllium, specific grades
1-220, consolidated from impact-ground powder **A9:** 391
1-400, consolidated from ball-milled powder **A9:** 391
S-65B, consolidated from impact-ground powder **A9:** 391
S-200F, consolidated from impact-ground powder **A9:** 391
SR-200, sheet, rolled from S-200E **A9:** 391
Beryllium thrust tube
Japanese CS-2 satellite **A7:** 759
Beryllium toxicity
disposition **A2:** 1238
in metallic wire **EM1:** 118
pulmonary effects **A2:** 1239
skin effects **A2:** 1239
Beryllium, vapor pressure
relation to temperature **M4:** 309, 310
Beryllium window **A10:** 519, 670
with EDX detectors **A11:** 38-39
for x-ray tubes **A17:** 306
Beryllium-aluminum alloys
correlation between microstructure and mechanical properties **A9:** 30
elastic modulus as a function of complexity index **A9:** 31
interpenetrating two-phase **A9:** 31
yield strength as a function of complexity index **A9:** 31
Beryllium-copper alloys *See also* Cast beryllium-copper alloys; Copper-beryllium alloys; Wrought beryllium-copper alloys **A9:** 392-398
adhesive bonding **A2:** 414-415
age hardening **A2:** 405-408
applications **A2:** 284-290, 356-363, 403
cast products **A2:** 422-423
casting, properties and applications **A2:** 358-363
cleaning **A2:** 414
cleaning and finishing **M5:** 612, 614, 620
cold working and age hardening **A2:** 421-422
compositions **A9:** 395
corrosion resistance **A2:** 420-421
cryogenic temperature thermal/electrical conductivity **A2:** 420
design and alloy selection **A2:** 416-421
Beryllium-copper alloys (continued)
dimensional change (age hardening) **A2:** 412-414
elastic springback **A2:** 412
electrolytic etching **A9:** 401
electropolishing **A9:** 393
etchants for **A9:** 394
etching **A9:** 394
extrusions, properties **A2:** 412
fabrication characteristics **A2:** 411-416
fatigue strength **A2:** 419
fatigue strength and resilience **A2:** 417-418
forgings, properties **A2:** 412
formability **A2:** 411
galling stress **A2:** 418
general corrosion behavior **A2:** 361
grinding of **A9:** 392-393
heat treatment **A2:** 405-408
high-conductivity wrought alloys, age hardening **A2:** 406-407
high-strength wrought alloys, age hardening **A2:** 406
hot-working processes **A2:** 415
machining **A2:** 415-416
macroetching **A9:** 393-394
macroexamination **A9:** 393-394
magnetic susceptibility **A2:** 418-419
mechanical properties **A2:** 409-411
melting, casting, hot working **A2:** 421
metallographic AES study **A10:** 558-559
microexamination **A9:** 394
microstructure **A2:** 404-405
microstructures **A9:** 394-395
mounting **A9:** 392-393
overaging **A2:** 407-408
peak-age treatments **A2:** 407-408
phase diagram **A2:** 403-404
physical metallurgy **A2:** 403-404
physical properties **A2:** 408-409
polishing **A9:** 392-393
precipitation hardening **A2:** 403
production metallurgy **A2:** 421-423
properties **A2:** 284-290, 356-363, 390-391, 408-410
quenching **A2:** 406
resilience **A2:** 416
resistance to specific agents **A2:** 361
safety precautions **A9:** 392
secondary electron micrograph **A10:** 559
sectioning **A9:** 392
soldering **A2:** 414
solution annealing **A2:** 405-406
specimen preparation **A9:** 392-393
spinodal decomposition hardening **A2:** 236
strength and electrical conductivity **A2:** 417
strip, temper designations and properties **A2:** 409-410
tempering **A2:** 410-411
thermal conductivity **A2:** 419
thermal stability of spring properties **A2:** 416-417
underage treatments **A2:** 407-408
welding **A2:** 414
wire, mechanical and electrical properties **A2:** 411
wrought, age hardening **A2:** 236
Beryllium-copper alloys, specific types
C17000 ..... **A9:** 395
C17000, wrought high-strength, composition **A2:** 403
C17200 ..... **A9:** 395
C17200, alloy plate, homogenized and hot worked **A9:** 397
C17200, alloy strip, AM temper **A9:** 396
C17200, alloy strip, heated and quenched 398
C17200, alloy strip, solution annealed and age hardened **A9:** 398
C17200, alloy strip, XHMS temper **A9:** 396
C17200, solution annealed and cold rolled **A9:** 397
C17200, solution annealed and overaged **A9:** 397
C17200, solution annealed and quenched **A9:** 396
C17200, solution annealed, cold rolled and precipitation hardened **A9:** 397
C17200, solution annealed, quenched and precipitation hardened **A9:** 396
C17200, wrought high-strength, composition **A2:** 403-404
Beryllium-copper alloys, specific types (continued)
C17300 ..... **A9:** 395
C17500 ..... **A9:** 395
C17500, alloy strip, solution annealed, cold rolled, and precipitation hardened **A9:** 397
C17500, alloy strip, solution annealed quenched and precipitation hardened **A9:** 397
C17510 ..... **A9:** 395
C17510, alloy rod, solution annealed and aged **A9:** 398
C17510, alloy strip, solution annealed, cold rolled, and precipitation hardened **A9:** 397
C17510, alloy strip, solution annealed quenched and precipitation hardened **A9:** 397
C82000 ..... **A9:** 395
C82200 ..... **A9:** 395
C82200, as-cast, beryllide phase **A9:** 396
C82400 ..... **A9:** 395
C82400, high-strength casting, composition **A2:** 403-404
C82500 ..... **A9:** 395
C82500, high-strength casting, composition **A2:** 403-404
C82500, solution annealed and aged **A9:** 396
C82510 ..... **A9:** 395
C82600 ..... **A9:** 395
C82600, high-strength casting, composition **A2:** 403-404
C82800 ..... **A9:** 395
C82800, high-strength casting, composition **A2:** 403-404
Beryllium-modified chrome copper
properties and applications **A2:** 356-357
Beryllium-nickel alloy, UNS 36000
alloy strip, solution annealed, quenched, and aged **A9:** 397
Beryllium-nickel alloys **A9:** 392-398
cleaning **A2:** 424
compositions **A2:** 423 **A9:** 396
etchants **A2:** 423
etchants for **A9:** 394
etching **A9:** 394
fatigue behavior **A2:** 425
grinding of **A9:** 392-393
heat treatment **A2:** 423
joining **A2:** 424
macroetching **A9:** 393-394
macroexamination **A9:** 393-394
mechanical properties **A2:** 423-424
melting and casting (foundry products) **A2:** 425
microexamination **A9:** 394
microstructures **A9:** 395-396
mounting **A9:** 392-393
physical and electrical properties **A2:** 426
physical metallurgy **A2:** 423
polishing **A9:** 392-393
production metallurgy **A2:** 425-426
safety precautions **A9:** 392
sectioning **A9:** 392
specimen preparation **A9:** 392-393
wrought, mechanical and physical properties **A2:** 423-424
Beryllium-window x-ray tube
for radiographing adhesive-bonded structures **EM3:** 759
Berzelius, Johns Jakob
as chemist **A15:** 29
Bessel function
modified **A6:** 10
Bessemer
Sir Henry **A15:** 31-32, 34
Bessemer converter
development **A15:** 31-32
Bessemer dry stamping process **A7:** 125
BESSY
as synchrotron radiation source **A10:** 413
Best Best quality telephone and telegraph wire **M1:** 264-265
Best candidate system description
leadless packaging **EL1:** 986-987
Best efficiency point (BEP) **A18:** 594
Best image voltage **A10:** 588, 689

Best line fit
x-ray spectrometry **A10:** 97
Best-fit curve *See also* Goodness of fit
for aluminum alloy at different stress ratios **A8:** 697
comparison using stress and life as dependent variables **A8:** 698
BET *See* Braunauer-Emmett-Teller
BET method *See* Brunauer-Emmet-Teller (BET) method
Beta
defined **A9:** 2
Beta alloys
titanium.......... **A2:** 586-587, 602 **A14:** 267-268, 271-272, 839
Beta anneal
cycle and microstructure.......... **A6:** 510
Beta annealing
wrought titanium alloys **A2:** 619, 620
Beta backscatter
as plating thickness inspection **EL1:** 943
Beta brass
martensitic structures **A9:** 672-673
Beta C alloy *See* Titanium alloys, specific types, Ti-3Al-8V-6Cr-4Zr-4Mo
Beta emissions from depleted uranium **A9:** 477
Beta eucryptite ($Li_2O{\cdot}Al_2O_3{\cdot}SiO_2$)
chemical system.......... **EM4:** 870
as filler for solder glass **EM4:** 1072
glass/metal seals **EM4:** 499
as refractory filler **EM4:** 1072
thermal expansion coefficients.......... **EM4:** 499
Beta eutectoid group of alloying elements for titanium alloys **A9:** 458
Beta eutectoid stabilizer
defined **A9:** 2
Beta flecks
defined **A9:** 2
macroscopic appearance **A9:** 472
in titanium and titanium alloys.......... **A9:** 459
Beta forging
of titanium.......... **A2:** 612-613
of titanium alloys **A14:** 271-272
Beta forging of titanium alloys **M3:** 368-369
Beta gage.......... **EM3:** 6
defined **EM2:** 6
Beta grain size
titanium and titanium alloy castings.......... **A2:** 638
Beta III *See* Titanium alloys, specific types, Ti-11.5Mo-6Zr-4.5Sn
Beta isomorphous group of alloying elements for titanium alloys **A9:** 458
Beta isomorphous stabilizer
defined **A9:** 2
Beta loss peak.......... **EM3:** 6
defined **EM2:** 6
Beta particles
defined **EL1:** 966
in neutron radiography.......... **A17:** 391
as radiation.......... **A17:** 295
Beta phase
in beryllium-copper alloys **A9:** 394-395
in brass, electrolytic etching to reveal.......... **A9:** 401
of cemented carbides.......... **A9:** 274
in copper alloys **A9:** 639
in silver-aluminum alloy, single crystal.......... **A9:** 657
of titanium and titanium alloys.......... **A9:** 458, 461
in uranium and uranium alloys **A9:** 476-487
Beta phase brass alloys
massive transformations in **A9:** 655-656
Beta phase unalloyed uranium
fabrication techniques.......... **A2:** 671
Beta prime phase in titanium alloys.......... **A9:** 461, 475
Beta quench
cycle and microstructure.......... **A6:** 510
Beta radiation, effect
E-glass fibers **EM1:** 47
Beta segregation defects in titanium and titanium alloys.......... **A9:** 459
Beta stabilizers in titanium .... **A3:** 1 • 23
Beta structure
defined **A9:** 2
Beta structures
wrought titanium alloys **A2:** 607-608
Beta titanium alloys
brazeability.......... **M6:** 1049
weldability.......... **M6:** 446
workability **A8:** 575
Beta transus
defined **A9:** 2
Beta transus temperature of titanium and titanium alloys.......... **A9:** 458
effect of beta flecks on.......... **A9:** 459
Beta transus temperatures
wrought titanium alloys **A2:** 623
Beta-alumina
ionic conductor.......... **EM4:** 18
solid-state sintering.......... **EM4:** 272, 279
thermal etching.......... **EM4:** 575
Beta-aluminum-nickel alloy
diffraction pattern **A9:** 109
Beta-particle emission
as radioactive decay mode **A10:** 245
Beta-quartz
coefficient of thermal expansion **EM4:** 1103
composition **EM4:** 1103
primary phase glass-ceramics based on.......... **EM4:** 870, 872
solid solution, aluminosilicate glass ceramics.......... **EM4:** 435
volume expansion coefficient.......... **EM4:** 761
Beta-ray emission, radiochemical
destructive TNAA.......... **A10:** 238
Beta-ray gage *See* Beta gage
Beta-spodumene ($Li_2O{\cdot}Al_2O_3{\cdot}4SiO_2$)
coefficient of thermal expansion **EM4:** 1103
composition.......... **EM4:** 1103
glass/metal seals **EM4:** 499
maximum use temperature **EM4:** 875
primary phase glass-ceramic based on **EM4:** 870, 871
solid solution, aluminosilicate glass ceramics.......... **EM4:** 435
specialty factory.......... **EM4:** 908
thermal expansion coefficient **EM4:** 499
volume expansion coefficient.......... **EM4:** 761
Beta-stabilizers
titanium alloys **A2:** 598-599
zirconium alloys **A2:** 665
Beta-titanium
erosion test results **A18:** 200
orthodontic wires **A18:** 666, 675-676
Beta/metastable beta titanium alloys.......... **A14:** 267-268
Betaketone.......... **EM3:** 626
Bethe formula
ion stopping power.......... **A18:** 323
Bevel
definition.......... **A6:** 1207 **M6:** 2
Bevel (extruding) angle
thread rolling **A16:** 282, 284, 287, 294
Bevel angle
definition.......... **A6:** 1207 **M6:** 2
Bevel cut
for plating thickness inspection.......... **EL1:** 943
Bevel cutting.......... **M6:** 911-912, 917
by plasma arc.......... **A14:** 731
Bevel gears
blanks, production methods.......... **A14:** 124-126
differential, by precision forming **A14:** 163
hypoid **A11:** 587, 588
spiral **A11:** 587, 588
spiral, precision forming of.......... **A14:** 173-175
straight **A11:** 587
tooth-gear contact in.......... **A11:** 588
zerol **A11:** 587
Bevel groove weld
definition.......... **A6:** 1207
Bevel groove welds
definition, illustration.......... **M6:** 2, 60-61
double, applications of.......... **M6:** 61
gas tungsten arc welding.......... **M6:** 360
of heat-resistant alloys **M6:** 356-357, 360
oxyfuel gas welding.......... **M6:** 590-591
preparation **M6:** 67-68
single, applications of.......... **M6:** 61
Bevel joints
radiographic inspection.......... **A17:** 334
Beveling
multiple-operation machining **A16:** 376
Beverage cans
aluminum alloy **A2:** 10
Beverage industry
stainless steel corrosion.......... **A13:** 559-560
BG42
nominal compositions **A18:** 726
Bi-Ca (Phase Diagram) .... **A3:** 2 • 98
Bi-Cd (Phase Diagram).......... **A3:** 2 • 98
Bi-Cs (Phase Diagram) .... **A3:** 2 • 98
Bi-Cu (Phase Diagram).......... **A3:** 2 • 99
Bi-Ga (Phase Diagram).......... **A3:** 2 • 99
Bi-Ge (Phase Diagram).......... **A3:** 2 • 99
Bi-Hg (Phase Diagram).......... **A3:** 2 • 100
Bi-In (Phase Diagram) .... **A3:** 2 • 100
Bi-K (Phase Diagram) .... **A3:** 2 • 100
Bi-La (Phase Diagram).......... **A3:** 2 • 101
Bi-Li (Phase Diagram) .... **A3:** 2 • 101
Bi-lubricant systems.......... **A7:** 192
Bi-Mg (Phase Diagram).......... **A3:** 2 • 101
Bi-Mn (Phase Diagram).......... **A3:** 2 • 102
Bi-Na (Phase Diagram).......... **A3:** 2 • 102
Bi-Nd (Phase Diagram) .... **A3:** 2 • 102
Bi-Ni (Phase Diagram) .... **A3:** 2 • 103
Bi-Pb (Phase Diagram) .... **A3:** 2 • 103
Bi-Pd (Phase Diagram) .... **A3:** 2 • 103
Bi-Pt (Phase Diagram) .... **A3:** 2 • 104
Bi-Rb (Phase Diagram).......... **A3:** 2 • 104
Bi-S (Phase Diagram).......... **A3:** 2 • 104
Bi-Sb (Phase Diagram) .... **A3:** 2 • 105
Bi-Se (Phase Diagram).......... **A3:** 2 • 105
Bi-Sm (Phase Diagram) .... **A3:** 2 • 106
Bi-Sn (Phase Diagram) .... **A3:** 2 • 106
Bi-Sr (Phase Diagram) .... **A3:** 2 • 106
Bi-Te (Phase Diagram) .... **A3:** 2 • 107
Bi-Tl (Phase Diagram) .... **A3:** 2 • 107
Bi-U (Phase Diagram) .... **A3:** 2 • 107
Bi-Y (Phase Diagram) .... **A3:** 2 • 108
Bi-Yb (Phase Diagram).......... **A3:** 2 • 108
Bi-Zn (Phase Diagram).......... **A3:** 2 • 108
Bi-Zr (Phase Diagram).......... **A3:** 2 • 109
$Bi_2Sr_2Ca_2Cu_3O_{10}$.......... **EM4:** 47
$Bi_2Sr_2CaCu_2O_8$.......... **EM4:** 47
applications **EM4:** 48
key product properties.......... **EM4:** 48
raw materials **EM4:** 48
solid-state sintering.......... **EM4:** 274
Biamperometric titration.......... **A10:** 204
Bias
in corrosion testing **A13:** 195-196
defined **A8:** 1 **A10:** 670
oxide contamination effect with/without **EL1:** 959
of test methodologies, and sampling **A10:** 12, 13
Bias buffs.......... **M5:** 118, 125
Bias fabric
defined **EM1:** 5 **EM2:** 6
Biased autoclave test
for humidity-induced stress.......... **EL1:** 495
Biaxial compressive stress system **A8:** 576
Biaxial contact
defined in ceramics.......... **A11:** 755
Biaxial fatigue test
specimen **A8:** 370
Biaxial load **EM3:** 6
defined **EM1:** 5 **EM2:** 6
Biaxial stress
at equator of upset cylinder.......... **A8:** 578

Biaxial stress (continued)
correction factor for cracks in **A11:** 124
effect on dimple rupture **A12:** 31, 39
Biaxial stress, defined (under Principal stress normal) **A13:** 10-11
Biaxial stretch testing **A8:** 558-559 **A14:** 887-888
Biaxial tensile stress
and compressive hydrostatic stress **A8:** 559
system **A8:** 576
Biaxial tension
in fatigue fracture **A11:** 75
Biaxial tension-uniaxial compression system **A8:** 576
Biaxial winding *See also* Winding
defined **EM1:** 5 **EM2:** 6
BICMOS *See also* Bipolar complementary metal-oxide semiconductor
defined **EL1:** 177
Bidentate ligands
role in metal toxicity **A2:** 1235
Bidirectional laminate *See also* also Laminate(s); Laminate(s); Unidirectional laminate
defined **EM1:** 5 **EM2:** 6
Bidirectional seal
defined **A18:** 4
Bidirectional waves
in wave soldering **EL1:** 688
Bifilar eyepiece
defined **A9:** 2
Bifurcated fiber optic light source
fractographic **A12:** 79
Bifurcated fiber optic tubes
dynamic notched round bar testing **A8:** 277
Bifurcation
definition **EM4:** 632
Big bang mechanism
of equiaxed grain growth **A15:** 131
Big-end bearing (bottom-end bearing, crankpin bearing, large-end bearing) *See also* Little-end bearing
defined **A18:** 4
BIGHT
defined **EL1:** 1135
Bilinear RAMOD-2 equation **EM3:** 515
Billet
cleanliness, niobium-titanium superconducting materials **A2:** 1046
defined **EM1:** 5
for wrought titanium alloys **A2:** 610
Billet method
hot extrusion of powder mixtures **A2:** 988
Billet separation equipment
precision forging **A14:** 162-164
Billet shape
effect on stress state in forging **A8:** 588
Billet stacking method
superconductor manufacture **A14:** 338-341
Billet, steel
pickling of **M5:** 69
Billets *See also* Ingots
copper and copper alloy **A14:** 257
cutting, in ring rolling **A14:** 123
defined **A14:** 343
eddy current inspection **A17:** 559-560
extrusion, defined **A14:** 5
flow lines in **A14:** 428
forging, defined **A14:** 6
heating **A14:** 164
magnetic particle inspection **A17:** 115-116, 119-120, 558-559
for mechanical testing, forging modes **A14:** 197
molybdenum **A14:** 237-238
niobium **A14:** 237
preparation, for hot extrusion **A14:** 322
processing flaws **A17:** 492-493
shape, and enclosure **A14:** 279
of stainless steel **A14:** 223
steel, surface discontinuities **A17:** 116
surface preparation **A17:** 558
temperatures, for hot extrusion **A14:** 315
ultrasonic inspection **A17:** 267-268
Billets CAP process for **A7:** 533
in cold isostatic pressing **A7:** 448
defined **A7:** 1
filled, extrusion process **A7:** 518
forging and rolling **A7:** 522-529
Billets CAP process for (continued)
for hot working into mill shapes CAP process for **A7:** 533
Billets, mill annealed
hardness of **M1:** 203
Bimetal *See also* Centrifugal casting
defined **A15:** 1
Bimetal bearing
defined **A18:** 4
Bimetal bearing alloys **A7:** 408
Bimetal bearings **A9:** 567
Bimetal casting *See* Dual-metal centrifugal casting
defined **A15:** 5
Bimetal electrical contact tape **A9:** 563
Bimetal foil laminates
XPS analyses of **A11:** 44
Bimetal tubes
for heat exchangers/condensers **A13:** 627
Bimetallic parts
processing by rapid omnidirectional compaction **A7:** 544-546
Bimetallic strip, for rod bearings
roll compacting in **A7:** 406
Bimetallic tubes
horizontal centrifugal casting **A15:** 299-300
Bimodal particle size distribution **A7:** 41, 43
Binary alkali silicate glasses
ion exchange **EM4:** 460
Binary alloy
defined **A9:** 2
Binary alloy phase diagrams .... **A3:** 2 • 25-383
Binary alloys **A13:** 75-76
Binary alloys diagram
Au-Al phases **EL1:** 1013
Binary alloys index .... **A3:** 2 • 5-21
Binary aluminum alloys
relative potency factors **A6:** 89
Binary collisions
low-energy ion-scattering spectroscopy **A10:** 604
Binary eutectic alloys
schematic phase diagram **A9:** 618
Binary gold-copper alloys
EPMA analysis of **A10:** 530
Binary image **A18:** 346, 350, 352
Binary iron-base systems *See also* Iron-base alloys
Fe-C, thermodynamics of **A15:** 61-62
Fe-Si, thermodynamics of **A15:** 62
phase diagram **A15:** 61
Binary iron-chromium equilibrium phase diagram **A6:** 678, 681
Binary iron, relative potency factors **A6:** 89
Binary isomorphous phase diagram **A6:** 46, 47
Binary metal infiltration systems **A7:** 554
Binary nickel-base alloys
relative potency factors **A6:** 89
Binary nonheat-treatable alloys
aluminum-silicon alloys as **A15:** 159
Binary optics
use in laser surface transformation hardening **A4:** 292
Binary phase diagrams **A13:** 46-47
Binary phosphate glasses
cations bonding in **A10:** 131
Binary polymer blends
types **EM2:** 632
Binary system
defined **A9:** 2
in machine vision **A17:** 33
Binary system or diagram description .... **A3:** 1 • 2-4
Binder
content, cermet systems **EL1:** 340
defined **EM1:** 5 **EM2:** 6
phase, thick-film formulations **EL1:** 249
system, rigid epoxy encapsulants **EL1:** 812
Binder, cobalt
in cemented carbides **A13:** 846
Binder mean free path of cemented carbides
determination of **A9:** 275
Binder metal
defined **A7:** 2
Binder phase
defined **A7:** 2
Binder phase of cemented carbides **A9:** 274
Binder resins for brake linings **A18:** 569
Binder systems *See also* Binders; Bonded sand molds
chemical, aluminum alloys **A15:** 203
cold box, properties **A15:** 215
Binder systems (continued)
of core coatings **A15:** 240
heat-cured, properties **A15:** 215
no-bake **A15:** 214-217
resin, market status **A15:** 215
Binder volume fraction of cemented carbides
determination of **A9:** 275
Binders *See also* Binder systems; Bonded sand molds; Organic binders; Resin binder processes **A7:** 1 **EM3:** 6
for butyls **EM3:** 202
ceramic shell molds, investment casting **A15:** 258-259
as coating material **A7:** 817
core-oil **A15:** 218-219
for cores, types **A15:** 238
for decorating materials **EM4:** 474-475
defined **A15:** 1
hot box **A15:** 218
hybrid, investment casting **A15:** 259
inorganic and organic, in spray drying **A7:** 74, 77
in metal casting, plastics for **A15:** 211
organic, development of **A15:** 35
plastic **A15:** 211
resin, processes of **A15:** 214-221
warm box **A15:** 218
Binding agents
x-ray spectrometry **A10:** 94
Binding energies **A18:** 445, 446, 447, 448
Bingham rheology **EM4:** 33
Bingham solid
defined **A18:** 4
Bingham-plastic flow behavior **EM4:** 156
Binocular vision *See* Stereo vision; Three-dimensional
Binodal curve
defined **A9:** 2
Binomial distribution **A8:** 628, 635-636
cumulative distribution function **A8:** 629, 636
mean **A8:** 629, 636
parameter estimates **A8:** 629, 636
percentile **A8:** 629
probability density function **A8:** 629
probability mass function **A8:** 635-646
variance **A8:** 629, 636
Bio-fouling corrosion
control of **A11:** 191
on metals **A11:** 190-191
Bioavailability, of iron powders
food enrichment **A7:** 614-615
Biochemical mixtures
liquid chromatography of **A10:** 649
Biocides **A18:** 110 **EM3:** 674
applications **A18:** 110
and cutting fluids **A16:** 131
functions **A18:** 110
for metalworking lubricants **A18:** 142, 143, 144
in nonengine lubricant formulations **A18:** 110
Biocompatibility
of implant materials **A11:** 672
of metallic implants **A13:** 1328-1329
of shape memory alloys **A2:** 901
Biodegradation
defined **EM2:** 784
measured **EM2:** 784-785
mechanisms **EM2:** 783-784
Biodeterioration
defined/measured **EM2:** 784-785
Biodisintegration
of plastic-starch blends **EM2:** 786
Biofilm formation
in water **A13:** 492, 900-902
Biofouling *See also* Anti-fouling; Biological corrosion
copper-base paint **A13:** 907
copper/copper alloys **A13:** 610, 625-626
on aluminum alloys **A13:** 598
organisms **A13:** 88
in water **A13:** 492-494
Bioglass **EM4:** 1008, 1096
45S5, endosseous ridge maintenance **EM4:** 1011
stainless steel fiber reinforced **EM4:** 1088
Biologic indicators *See also* Indicator tissues
of abnormal zinc homeostasis **A2:** 1255
of arsenic **A2:** 1238
of cadmium **A2:** 1241
of lead toxicity **A2:** 1246
of mercury toxicity **A2:** 1249

Biologic indicators (continued)
of nickel toxicity ... **A2:** 1250
Biological corrosion *See also* General biological corrosion; Localized biological corrosion; Microbiological corrosion
of aluminum ... **A13:** 118-119
aqueous ... **A13:** 17, 41-43
cell ... **A13:** 42-43
of copper alloys ... **A13:** 119
defined ... **A13:** 2
of iron and steel ... **A13:** 116-117
localized ... **A13:** 114-120
mechanisms ... **A13:** 41-43
oil/gas production ... **A13:** 1234
organisms, characteristics ... **A13:** 41
in seawater ... **A13:** 900-902
of stainless steel ... **A13:** 117-118
telephone cables ... **A13:** 1130
tuberculation ... **A13:** 119-120
Biological materials
characterization of ... **A10:** 1
ESR analysis of ... **A10:** 264
fluid, potentiometric membrane electrode analysis ... **A10:** 181
freeze-dried, as x-ray spectrometric samples ... **A10:** 93
GC/MS analysis of ... **A10:** 639
MFS analysis of carcinogenic polynuclear aromatic compounds in ... **A10:** 72
NAA determination of toxic elements in ... **A10:** 233
PIXE analysis of ... **A10:** 106-107
powdered, analysis of ... **A10:** 106-107
use of ICP-AES for ... **A10:** 31
voltammetric monitoring of metals and nonmetals in ... **A10:** 188
Biological testing
of lubricants ... **A14:** 517
Biomechanical implant failures ... **A11:** 681-687
Biomechanical stability
of internal fixation devices ... **A11:** 671
Biomedical applications
investment castings ... **A15:** 266
Biomedical prosthetic devices
corrosion of ... **A13:** 1324-1335
Biomolecular Strauss coupling reaction
polyimides ... **EM3:** 157
Biomolecules
EXAFS analysis of ... **A10:** 407
Biopsy
GFAAS detection of metals in ... **A10:** 55
Biotransformation, of arsenic
as toxic metal ... **A2:** 1237
BiO_x-Au glass thin-film system
Auger electron spectroscopy ... **A18:** 453, 454
secondary ion mass spectrometry (SIMS) spectrum, depth-profile data ... **A18:** 459, 460, 461
Bipolar
defined ... **EL1:** 1135
Bipolar complementary metal-oxide semiconductor (BICMOS) ... **EL1:** 177, 390
Bipolar drivers ... **EL1:** 8, 83
Bipolar electrode
defined ... **A13:** 2
Bipolar emitter-coupled logic (ECL)
as future trend ... **EL1:** 390
Bipolar junction transistor (BJT) technology
active devices in ... **EL1:** 145-146
active-component fabrication ... **EL1:** 191, 195-196
current components ... **EL1:** 150
fabrication steps ... **EL1:** 197
high-performance active devices with ... **EL1:** 146-147
IC technology, npn ... **EL1:** 144-145
Bipolar transistor analysis
current gain ... **EL1:** 151-152
Ebers-Moll equations ... **EL1:** 151-152
Bipolar transistor analysis (continued)
frequency response-small signal model ... **EL1:** 153-154
mechanism ... **EL1:** 150-151
Bipolar transistor(s)
defined ... **EL1:** 1135
failure mechanisms ... **EL1:** 975
Bipotentiometric titration
as electrometric ... **A10:** 204
Bird, William
as early founder ... **A15:** 26
Birefringence ... **A10:** 115, 129 **A18:** 304
of aluminum alloy phases ... **A9:** 356-357, 360
defined ... **A9:** 3
in optical testing ... **EM2:** 596-597
Birefringent crystal
defined ... **A10:** 670
Birmingham wire gage (BWG) system ... **A1:** 277
Birmingham Wire Gage system ... **M1:** 259
Birotational seal
defined ... **A18:** 4
Bis(4-maleimidodiphenyl) methane ... **EM2:** 252-253
Biscuit *See* Cull; Preform
Bismaleimide (BMI) *See also* Bismaleimide resins (BMIs)
defined ... **EM1:** 5
Bismaleimide 5245C resin ... **EM1:** 83, 88
Bismaleimide Matrimid 5292 ... **EM1:** 81-82, 86
Bismaleimide resins (BMIs)
addition-type ... **EM1:** 78-90
adhesives ... **EM1:** 684
for aerospace application ... **EM1:** 32
commercial ... **EM1:** 80
constituent properties ... **EM1:** 83, 88
cure cycles ... **EM1:** 658-660
cure time/temperature relationships ... **EM1:** 659-660
curing ... **EM1:** 657-661
descriptions/types ... **EM1:** 80-83
and fiber-resin composites ... **EM1:** 373-380
as high-temperature thermoset ... **EM1:** 373-380
hot/wet in-service temperatures ... **EM1:** 33
Michael addition ... **EM1:** 79, 80
modified, constituent properties ... **EM1:** 83, 88
-olefin copolymers ... **EM1:** 80
processing parameters ... **EM1:** 81
properties ... **EM1:** 289, 291, 373-380
property optimization ... **EM1:** 657-658
for resin transfer molding ... **EM1:** 169
sample chemical reactions ... **EM1:** 751-753
sizings for ... **EM1:** 123
tests for ... **EM1:** 289, 291
-triazine ... **EM1:** 81
two-component BMI-olefin resins ... **EM1:** 81
Bismaleimide Rhone-Poulenc Kerimid ... **EM1:** 80-81, 86
Bismaleimide triazine/epoxy resin *See also* BT resins ... **EL1:** 1670
properties ... **EL1:** 534-535
Bismaleimide-triazine (BT) resins ... **EM1:** 81-83, 87
Bismaleimides (BMI) *See also* Thermosets; Thermosetting resins ... **EM3:** 6, 44, 153
adhesive prepreg material ... **EM3:** 426
aerospace application ... **EM3:** 559
chemical reaction mechanism ... **EM3:** 423
chemistry ... **EM2:** 252-253
commercial forms ... **EM2:** 253-354
defined ... **EM2:** 6
as high-temperature resin system ... **EM2:** 252, 444
high-temperature stability ... **EM3:** 320
homopolymerization of ... **EM2:** 254
in Kerimid ... **EM2:** 252
Kerimid 601 ... **EM3:** 424
mechanical properties ... **EM2:** 256
preparation ... **EM2:** 252
processing ... **EM2:** 254-255
properties ... **EM2:** 254-256
suppliers ... **EM2:** 253-254
Bismaleimides (BMI) (continued)
surface contamination of carbon laminate ... **EM3:** 845-846, 847
thermoplastic additives for toughness ... **EM3:** 185
Bismuth *See also* Liquid bismuth ... **A13:** 56, 180, 515
as addition to low-melting fusible alloy solders ... **A6:** 968
additive for improved machinability of stainless steels ... **A16:** 685, 687, 688
additive for improved machinability of steels ... **A16:** 125, 673-677, 679
in alloy cast irons ... **A1:** 90
alloying, aluminum casting alloys ... **A2:** 132
alloying, wrought aluminum alloys ... **A2:** 46
in aluminum alloys ... **A15:** 744
as an addition to permanent magnets ... **A9:** 539
in austenitic manganese steel ... **A1:** 825-826
bismuth-base solders ... **A2:** 756-757
cast, deformation twins ... **A9:** 59
in cast iron ... **A1:** 5, 8
compatibility in bearing materials ... **A18:** 743
content additions to P/M materials ... **A16:** 885
in copper alloys ... **A6:** 753
determined by controlled-potential coulometry ... **A10:** 209
effect, gas dissociation ... **A15:** 83
effect, nitrogen solubility, iron-base alloys ... **A15:** 83
effect on tin-base alloys ... **A18:** 748
embrittlement by ... **A11:** 235
etching by polarized light ... **A9:** 59
fire refining effect ... **A15:** 453
in free-machining metals ... **A16:** 389
fusible alloys ... **A2:** 755-756 **EL1:** 636
gaseous hydride, for ICP sample introduction ... **A10:** 36
grain boundary adhesion ... **A6:** 144
gravimetric finishes ... **A10:** 171
heat-affected zone fissuring in nickel-base alloys ... **A6:** 588
history ... **A2:** 753
impurity in solders ... **M6:** 1072
and indium ... **A2:** 750
in malleable iron ... **A1:** 10
in medical therapy, toxic effects ... **A2:** 1256-1257
as minor element, ductile iron ... **A15:** 648
in nickel, determination by GFAAS ... **A10:** 57
occurrence ... **A2:** 753
photometric analysis methods ... **A10:** 64
price per pound ... **A6:** 964
pricing history ... **A2:** 754
properties ... **A2:** 754-755
pure, properties ... **A2:** 1103
quartz tube atomizers with ... **A10:** 49
recommended impurity limits of solders ... **A6:** 986
recovery methods ... **A2:** 753-754
removal, from lead alloys ... **A15:** 476
resistance of, to liquid-metal corrosion ... **A1:** 636
safety standards for soldering ... **M6:** 1099
separated from copper ... **A10:** 200
as solder impurity ... **EL1:** 637
as solid lubricant inclusion for stainless steels ... **A18:** 716
species weighed in gravimetry ... **A10:** 172
thermal diffusivity from 20 to 100 °C ... **A6:** 4
as tin solder impurity ... **A2:** 520
toxicity, effects and treatment ... **A2:** 1256-1257
as trace element ... **A15:** 388, 394
as trace metal in iron-based steel, AAS analysis ... **A10:** 55
TWA limits for particulates ... **A6:** 984
ultrapure, by zone refining technique ... **A2:** 1094
volatilization losses in melting ... **EM4:** 389
weighed as the phosphate ... **A10:** 171
wetting of copper, LME by ... **A11:** 719
Bismuth containing fusible mounting alloys
for electropolishing ... **A9:** 49

Bismuth germanate
detector configuration **A18:** 325
Bismuth germanium oxide crystals
production **A2:** 743
Bismuth glycolyarsanilate
as medicinal application **A2:** 1256
Bismuth in cast iron
carbide-inducing inoculant, use as **M1:** 80
Bismuth in fusible alloys **M3:** 799
Bismuth in iron
malleable cast irons **M1:** 58
Bismuth oxide
in binary phosphate glasses **A10:** 131
metallizing by thick-film compound adhesion **EM4:** 544
Bismuth, pure **M2:** 718-719
solution potential **M2:** 207
Bismuth salts
catalyst for urethane sealants **EM3:** 204
Bismuth, vapor pressure
relation to temperature **A4:** 495 **M4:** 310
Bismuth-base solder alloys
applications **A2:** 756-757
Bisnadimides (BNI) *See also* Thermosets **EM3:** 153
Bisphenol A (BPA) **EM3:** 96, 590, 594-595, 617
for forming epoxies **EM3:** 94
physical properties **EM3:** 104
Bisphenol A (BPA) fumarate resins *See also* Polyester resins
application/preparation **EM1:** 90
basic solution resistance **EM1:** 93
clear casting mechanical properties **EM1:** 91
electrical properties **EM1:** 94
in fiberglass-polyester composites **EM1:** 91
glass content effects **EM1:** 91
mechanical properties **EM1:** 90
thermal stability **EM1:** 93
Bisphenol A (BPA) fumarates *See also* Unsaturated polyesters
electrical properties **EM2:** 250
preparation, properties **EM2:** 246
Bisphenol A vinyl ester resin **EM2:** 272
Bisphenol F **EM3:** 97, 594, 595
epoxidation **EM3:** 95
Bisphenol F (DGEBF)
as epoxy resin **EL1:** 826-827
Bisque (porcelain enameling) **M5:** 518, 530
Bisque firing **EM4:** 181
BISRA *See* British Iron and Steel Research Association
Bisulfate chemi-mechanical pulping digesters **A13:** 1218
Bit
definition **A6:** 1207 **M6:** 2
Bithermal weld **A6:** 605, 606
Bitter patterns
magnetic etching **A9:** 63
Bitter technique for studying magnetic domains **A9:** 534-535
Bitumastic sealers
aluminum coatings **M5:** 345
Bitumen
defined **EM2:** 6
Bituminous coal
for hot-blast furnace **A15:** 30
Bituminous coating
in cast iron pipes **M1:** 97-98, 100
Bituminous coatings
asphalts **A13:** 406
Bituminous-base caulks **EM3:** 188
Biuret
formation of **EM3:** 204
BIV *See* Best image voltage
Bivalve molds
historic use **A15:** 16-17
Bivariant equilibrium .... **A3:** 1 • 2
defined **A9:** 3
Black annealing **A1:** 280
defined **A9:** 3
Black anodizing copper coloring solution **M5:** 625
Black body
in thermal image analysis **EL1:** 368
Black boxes **A13:** 1074-1075, 1107-1108, 1111
Black copper coloring solutions **M5:** 625-626
Black coring **EM4:** 258
Black iron oxide
pressure-tightness by **A7:** 464
Black light *See* Ultraviolet light
Black liquor
defined **A13:** 2
processing equipment **A13:** 1213-1214
Black magnetic-powder indications
cold shut **A17:** 101
Black marking
defined **EM2:** 6
Black mix **A7:** 156
Black nickel plating **M5:** 119, 204-205, 623
Black oxide
defined **A13:** 2
Black oxide, coating for taps
Ti alloys **A16:** 847
Black spots
as casting defect **A11:** 388
Black-and-white films for photomicroscopy **A9:** 84-86
Black-and-white imaging compared to color **A9:** 135
Black-and-white photography
filters for **A9:** 72
Black-level suppression for image modification in scanning electron microscopy **A9:** 95
Black-white contrast
of dislocation loops **A9:** 116-117
of second-phase precipitates **A9:** 117
of stacking-fault tetrahedra **A9:** 117
Blackbodies, effect
electromagnetic radiation **A17:** 396
Blackbody
defined **A9:** 3
in emission spectroscopy **A10:** 115
spectral radiance **EM4:** 615-616
Blackburn equation
in creep analysis **A8:** 689
Blackening *See* Density
line **A10:** 142
as optical density **A10:** 143
photoplate **A10:** 143
Blackheart malleable cast iron **M1:** 57
Blackheart malleable iron **A1:** 74
American **A15:** 30-31
Blacking
defined **A15:** 1
Blacking inclusions
as casting defect **A11:** 387
Blacking scab
as casting defect **A11:** 385
Blacksmith welding
definition **A6:** 1207
Bladder
defined **EM1:** 5 **EM2:** 6
Blade slots, fan-disk
electric current perturbation inspection **A17:** 138
Blade-and-fork connections
and controlled impedance **EL1:** 87
Blades *See* Shear blades
aircraft propeller, repair deformation **A11:** 125
helicopter, spindle fatigue fracture **A11:** 126
jet-engine turbine, high-temperature fatigue fracture **A11:** 131
turbine **A13:** 1000-1001
Blaine air permeability apparatus **A7:** 264
Blaine apparatus **EM4:** 70
Blank
defined **A7:** 2
Blank preparation
for drop hammer forming **A14:** 655
HERF processing **A14:** 104
for ring rolling **A14:** 122-123
in three-roll forming **A14:** 619
of titanium alloys **A14:** 840-841
Blank(s) *See also* Blank preparation; Blankholders; Blanking; Cutoff; Slug
beryllium, development of **A14:** 806-807
bevel gear, production methods **A14:** 124-126
carburizing, in powder forging **A14:** 202
cutting off, multiple-slide forming **A14:** 569
deburring, in press forming **A14:** 548
defined **A14:** 445
design, effect in ring rolling **A14:** 116
design, for fine-edge blanking and piercing **A14:** 473
Blank(s) (continued)
developed, vs trimming, deep drawing **A14:** 589
distorted by punch/blank diameter **A14:** 119
feeding of **A14:** 500
finished ring profile from **A14:** 119
forged and pierced, shapes of **A14:** 61
forging and piercing of **A14:** 69
heated, explosive forming of **A14:** 643
holding **A14:** 571
large, welded **A14:** 450
layout of **A14:** 449-450
manufacture **A14:** 120
non-nesting **A14:** 551
preformed, refractory metals **A14:** 788
preheating, for drop hammer forming **A14:** 657
preparation **A14:** 119
processing, factors **A14:** 448-449
refractory metal **A14:** 788
round **A14:** 449-450
short, multiple-slide forming **A14:** 570
size, for deep drawing **A14:** 578
transfer, in multiple-slide forming **A14:** 570
welded **A14:** 450
Blanked edges
characteristics **A14:** 447
Blanket
defined **EM1:** 5 **EM2:** 6
Blanket epitaxy
as GaAs-silicon wafer production technique **A2:** 747
Blankholder ring
in sheet metal forming **A8:** 547
Blankholders
deep-drawing, materials for **A14:** 508-509
defined **A14:** 1
design, for explosive forming **A14:** 639
force, deep drawing **A14:** 582
magnesium alloy, pressures **A14:** 828
materials for **A14:** 511
and metal flow restraint, deep drawing **A14:** 581-582
shaped, in press forming **A14:** 549
types, deep drawing **A14:** 582
Blankholding
in drawing presses **A14:** 578
Blanking *See also* Blank preparation; Blank(s); Blankholders; Blanking dies; Blanking tools; Fine-edge blanking; Piercing; Punching; Shearing
accuracy **A14:** 456-457
aluminum alloy **A14:** 519, 793-794
auxiliary equipment **A14:** 477-478
blank layout **A14:** 449-450
blanked edges, characteristics **A14:** 447
burr removal after **A14:** 458
by Guerin process **A14:** 607
by multiple slide forming **A14:** 569-570
of carbon/low-alloy steels, lubricants for **A14:** 518
chemical **A14:** 458
conventional dies **A14:** 453-456
of copper and copper alloys **A14:** 811
defined **A14:** 1, 445-446 **EM2:** 6
die clearance **A14:** 447
ductile and brittle fractures from **A11:** 88
of electrical steel sheet **A14:** 476-482
fine-edge **A14:** 458
force requirements, calculation **A14:** 448
fracture in **A8:** 548
heat-resistant alloys, lubricant for **A14:** 519
high-carbon steel **A14:** 556
load vs displacement curve **A14:** 37
of low-carbon steel **A14:** 445-458
methods, in presses **A14:** 445-447
of nickel-base alloys **A14:** 520, 832
operating conditions **A14:** 456
of organic-coated steels **A14:** 565
and piercing, compared **A14:** 459
and piercing, with compound dies **A14:** 456-457
in press brake **A14:** 538
presses **A14:** 451, 458
process, for ceramic multilayer packages **EL1:** 463
processing factors **A14:** 448-449
refractory metals and alloys **A2:** 561
rigid printed wiring boards **EL1:** 547

Blanking (continued)
safety ... **A14:** 458
short-run dies for ... **A14:** 451-453
of stainless steels ... **A14:** 759, 761-762
tool materials for ... **A14:** 539
tools, for ring rolling ... **A14:** 121, 124
welded blanks ... **A14:** 450-451
work metal form selection ... **A14:** 449
work metal thickness, effect ... **A14:** 456
wrought copper and copper alloys ... **A2:** 248
Blanking and piercing dies ... **M3:** 484-488
Blanking dies ... **A14:** 451-456
applications ... **A14:** 485-486
material selection for ... **A14:** 483-486
tool materials ... **A14:** 484-485
Blanking fracture
effect in high-carbon steels ... **A12:** 285
Blanking lines *See* Cut-to-length lines
Blanking-shear test ... **A8:** 64-65
Blanks
applying circle grids to ... **A8:** 567
defined ... **A10:** 222, 670
preparation, in UV/VIS analysis ... **A10:** 68
sampling error reduction by ... **A10:** 12
Blanks, quartz
defects in ... **EL1:** 979
Blast cleaning *See also* Abrasive cleaning; Abraslue blasting; Blasting; Grit Blasting; Sandblasting; specific methods by name ... **A13:** 414-415, 521
abrasive flow rates ... **A15:** 516-517
abrasive parameters ... **A15:** 518-520
abrasives ... **A15:** 510-511
angle of impact, abrasive ... **A15:** 517-518
of castings ... **A15:** 506-520
centrifugal wheels, types ... **A15:** 511-516
defined ... **A15:** 1, 506
equipment, selection ... **A15:** 506-510
of internal surfaces ... **A15:** 520
Blast furnace ... **A1:** 107-108 **M1:** 109-110
current technology for ... **A1:** 108-109
Blast meter
cupola ... **A15:** 30
Blast pattern
defined ... **A15:** 517
Blast residues, effect
magnesium/magnesium alloys ... **A13:** 741
Blast tubes
of cupolas ... **A15:** 29
Blasting *See also* Abrasive cleaning; Blast cleaning; Grit; Sandblasting
for core buildup removal ... **A15:** 240
defined ... **A15:** 1
definition ... **M6:** 2
Blasting-tumbling machines ... **M5:** 88-89, 93, 95
Bleaching
prior to anodizing aluminum ... **M5:** 572
Bleed ... **EM3:** 6
defined ... **A15:** 1 **EM2:** 6
Bleed bands in aluminum alloy ingots ... **A9:** 633
in alloy 3003 ... **A9:** 634
Bleed hole, hydraulic-oil
corrosion of ... **A11:** 37
Bleed out
defined ... **A7:** 2
Bleed-out ... **EM3:** 6
Bleedback, penetrant
as inspection method ... **A17:** 74
Bleeder
as cure processing material ... **EM1:** 643-644
Bleeder cloth ... **EM3:** 6
defined ... **EM1:** 5 **EM2:** 6
Bleeding ... **EM3:** 6
additive ... **EM2:** 493
colorant ... **EM2:** 501
defined ... **A18:** 4 **EM1:** 5 **EM2:** 6
plasticizer ... **EM2:** 496
Bleeding revealed by macroetching ... **A9:** 173-174
in alloy steel billet ... **A9:** 175
Bleedout
defined ... **EM1:** 5 **EM2:** 6
Blend (noun)
defined ... **A7:** 2
Blend, blending (verb)
defined ... **A7:** 2
Blend draft
forgings ... **M1:** 362, 363
Blended elemental (BE) titanium compacts *See* also Titanium P/M products
blended elemental Ti-6AI-4V ... **A2:** 648-649
fracture toughness ... **A2:** 648-650
mechanical properties ... **A2:** 647-651
tensile properties ... **A2:** 648-650
Blended elemental alloy production ... **A7:** 654, 655
compaction, nuts and parts produced by ... **A7:** 753
powders for shapemaking ... **A7:** 749-750
titanium and titanium alloy powders ... **A7:** 164-165, 167, 681, 748-749
Blended elemental Ti-6AI-4V compacts
mechanical properties ... **A2:** 648-649
Blended elemental titanium P/M products
types and processes ... **A2:** 654-655
Blended sand *See also* Sand(s)
defined ... **A15:** 2
Blenders ... **A7:** 189, 213
Blending *See also* Reblending ... **A7:** 24, 186-289
of core coatings ... **A15:** 240-241
defined ... **A18:** 4
degree of uniformity ... **A7:** 186, 188
equipment ... **A7:** 189
of glass ... **EM1:** 107-108
and grinding, of samples ... **A10:** 17
of molybdenum powders ... **A7:** 155-156
P/M alloys, consolidation and ... **A7:** 308
of P/M ferrous powders ... **A7:** 683
P/M materials ... **M1:** 331
as P/M oxide-dispersion technique ... **A7:** 711-718
powder characteristics and variables ... **A7:** 187-189
in QMP powder process ... **A7:** 87
reclaimed and new sands ... **A15:** 353
and roll compacting ... **A7:** 402
and sampling ... **A10:** 16
statistical analysis ... **A7:** 187
of tungsten powders ... **A7:** 154
Blends
and alloys, applications ... **EM2:** 491
amorphous-crystalline, properties ... **EM2:** 490
binary polymer ... **EM2:** 632
chemistry ... **EM2:** 66-67
components ... **EM2:** 490
of engineering plastics ... **EM2:** 632-637
immiscible, amorphous resins ... **EM2:** 632-633
immiscible, amorphous/semicrystalline polymers ... **EM2:** 633-635
isomorphic ... **EM2:** 632
mechanical properties ... **EM2:** 634-636
miscible, amorphous polymers ... **EM2:** 635-636
miscible, amorphous/semicrystalline resins ... **EM2:** 636-637
miscible, semicrystalline polymers ... **EM2:** 637
plastic-starch ... **EM2:** 784-787
polymer, environmental effects ... **EM2:** 431
Blind bolts
for advanced composites ... **A11:** 530
Blind fasteners
defined ... **A11:** 529
failures in ... **A11:** 545
types of ... **A11:** 545-546
Blind fastening *See also* Fastener holes; Fasteners; Mechanical fastening ... **EM1:** 709-711
galvanic compatibility ... **EM1:** 709
installation effects ... **EM1:** 710
joint strength ... **EM1:** 710
robotics installation ... **EM1:** 711
sensitivity to hold quality ... **EM1:** 711
sheet take-up ... **EM1:** 710-711
Blind hole
defined ... **EM2:** 6
Blind hole clamps ... **EM1:** 711
Blind holes *See also* Hole(s)
magnetic rubber inspection methods for ... **A17:** 123
Blind pores
measured with surface area ... **A7:** 264-265
Blind riser *See also* Atmospheric riser
defined ... **A15:** 2
Blind rivets
material and composite applications ... **A11:** 530
Blind sample
defined ... **A10:** 670
Blind shrinkage
as casting defect ... **A11:** 382
Blind stagger
as selenium poisoning ... **A2:** 1254-1255
Blind vias *See also* Vias
geometry and alternatives ... **EL1:** 112-113
Blinded sieves ... **A7:** 216
Blindjoint
definition ... **M6:** 2
Blisks ... **A18:** 592
Blister ... **EM3:** 6
defined ... **A13:** 2 **A15:** 1 **EM2:** 6
Blister cracking *See* Hydrogen-induced cracking (HIC)
Blister, solder *See* Solder blister
Blister test ... **A18:** 404
Blister tests ... **EM3:** 384-386, 389
axisymmetric fracture mechanics ... **EM3:** 385-386, 389
Blistering *See also* Hydrogen blistering ... **A13:** 164, 242, 331, 1277-1278
in BMI laminate curing ... **EM1:** 661
copper plate ... **M5:** 168
defect, squeeze casting ... **A15:** 325-326
defined ... **A7:** 2
during pickling, effects of ... **M5:** 80-81
as hydrogen damage ... **A12:** 124, 125, 142
hydrogen-induced ... **A11:** 5, 247-248
paint ... **M5:** 489, 491, 493, 495
of porcelain enameled steel ... **M1:** 179
radiographic inspection methods ... **A17:** 296
in semisolid casting and forging ... **A15:** 336-337
in steel bar and wire ... **A17:** 549
in tubular products ... **A17:** 567
Blisters
carbon-graphite materials ... **A18:** 818
from rolling heating practice ... **A14:** 358-359
Bloch equations
T_1, and T_2 ... **A10:** 280
Bloch walls ... **A9:** 534
Bloch waves
bright-dark image oscillations ... **A9:** 112
dynamical diffraction theory for imperfect crystals ... **A9:** 112
equation ... **A9:** 111
Block *See* Pin or Mandrel
Block and cascade welding technique
cast irons ... **A15:** 526
Block and finish
defined ... **A14:** 1
Block brazing
definition ... **M6:** 2
Block copolymers ... **EM2:** 6, 58 **EM3:** 6, 80, 147, 148
for coating and encapsulation ... **EM3:** 580
predicted 1992 sales ... **EM3:** 81
sealant formulations ... **EM3:** 677
Block diagrams
acoustic emission inspection ... **A17:** 282-283
reliability ... **EL1:** 899
Block experimental designs ... **A8:** 643-650
balanced incomplete ... **A8:** 646, 648-649
chain ... **A8:** 646-647
estimating and testing block effects ... **A8:** 657, 661
incomplete ... **A8:** 644-646
Latin squares ... **A8:** 647-650

Block experimental designs (continued)
and randomized design **A8:** 643-650
select factorial **A8:** 645
and stress amplitude **A8:** 374
to minimize nuisance variables **A8:** 641
types of **A8:** 640
Youden square **A8:** 650
Block, first, second, and finish
defined **A14:** 1
Block grease
defined **A18:** 4
Block polymers
SAS applications **A10:** 405
Block sequence
cast irons, welding of **A6:** 715
definition **A6:** 1207 **M6:** 2
welding of cast irons **M6:** 313
Block terpolymers
properties **EM3:** 82
Blocked curing agent **EM3:** 6
Blocked factorial experimental designs **A8:** 642-643
Blocked tee configuration **A18:** 199
Blocker dies
defined **A14:** 2
design **A14:** 77, 410
Blocker initial design
as knowledge-based expert system **A14:** 410
Blocker-type forgings
aluminum alloy **A14:** 243
by closed-die forging **A14:** 76
defined **A14:** 2
Blockers, as impression dies *See also* Preforms **A14:** 44
Blocking **EM3:** 6
defined **A14:** 2 **EM2:** 6
impression, defined **A14:** 2
in open-die forging **A14:** 64
Blocking (experimental)
defined **EM2:** 600
Blocking and channeling
for surface structure **A10:** 633
Blocks **A14:** 1, 665-666 **A18:** 569
Blocks, gray iron cylinder
cracking in **A11:** 345-346
Blocks, test *See* Test blocks
Blocky alpha
defined **A9:** 3
in zirconium and zirconium alloys **A9:** 501
Blocky impact ground beryllium powder **A7:** 170, 172
Blok's contact temperature theory **A18:** 539-540
Blok's critical temperature theory **A18:** 538
Blok's equation **A18:** 541
Blok's flash temperature equation (AGMA 2001-B88) **A18:** 539
Blok-Jaeger method
calculation of maximum surface contact temperature **A18:** 93
Blood lead levels
national estimates **A2:** 1243
Blood plasma
ionic composition of **A11:** 672
Bloom **EM3:** 6
defined **A14:** 2, 343 **EM1:** 5 **EM2:** 6-7
product, of stainless steel **A14:** 223
Blooming *See* Blushing
of additives **EM2:** 493
colorant **EM2:** 501
dot mapping and **A10:** 526-527
Blooming mill
defined **A14:** 2
Blooms
ultrasonic inspection **A17:** 267-268
Blow
defined **A15:** 2
Blow down
defined **A13:** 2
Blow, forging
rapidity and intensity effects **A14:** 57
Blow forming
as superplastic forming process **A14:** 18, 857
Blow holes
defined **A15:** 2
from rolling **A14:** 358
as gray iron defect **A15:** 641
Blow molding **EM2:** 352-359
acrylics **EM2:** 107
conventional, properties effects **EM2:** 285
cost summary **EM2:** 299
defined **EM2:** 7
economic factors **EM2:** 298-299
engineering thermoplastics advantages **EM2:** 356-357
high molecular weight **EM2:** 164
injection, properties effects **EM2:** 285
large part, of high-density polyethylenes (HDPE) **EM2:** 164
machinery and processing methods **EM2:** 352-353
market opportunities **EM2:** 359
material distribution **EM2:** 357-358
molded-in color **EM2:** 306
molds and associated parameters **EM2:** 353-356
part design **EM2:** 357
piece-part cost analysis **EM2:** 358-359
reinforced polypropylenes (PP) **EM2:** 193
size and shape effects **EM2:** 290
textured surfaces **EM2:** 305
Blow pin
defined **EM2:** 7
Blow pressure
defined **EM2:** 7
Blow rate
defined **EM2:** 7
Blow, surface *See* Surface blow
Blow-up ratio
defined **EM2:** 7
Blowby **A18:** 98
Blower gears
economy in manufacture **M3:** 855, 856
Blowers
cupolas **A15:** 385-386
development of **A15:** 27
types, illustrated **A15:** 27
Blowhole **EM3:** 6
definition **A6:** 1207
Blowholes
in austenitic manganese steel castings
causes of **A9:** 238
as casting defects **A11:** 382
defined **A9:** 3
in iron **A12:** 221
as preheating defect **EL1:** 687
Blowing
oxygen, for carbon and silicon removal **A15:** 78
sand, with hydraulic squeeze **A15:** 29
Blowing agents
as additives **EM2:** 503
chemical, defined **EM2:** 9
defined **EM2:** 7
physical **EM2:** 30, 503
structural foams **EM2:** 510-512
Blown core
adhesive-bonded joints **A17:** 613-614
Blown linseed oil **M5:** 498-499
Blown oil
defined **A18:** 4
Blown tubing
defined **EM2:** 7
Blown-film extrusion
products **EM2:** 383-384
Blowoff, mold
as green sand mold finishing **A15:** 347
Blowout
of aluminum alloy connector tubes **A11:** 312
Blowpipe
definition **M6:** 2
Blue annealing
defined **A9:** 3
Blue brittleness **A1:** 692 **M1:** 684
defined **A13:** 2
notch toughness, effect on **M1:** 701-703
of steels **A11:** 98
Blue coring **EM4:** 258
Blue porcelain enamel
composition of **M5:** 510
Blue, thymol and bromthymol
as acid-base indicators **A10:** 172
Blue-black copper coloring solution **M5:** 626
Blue-black porcelain enamel
composition of **M5:** 510
Blueing
defined **EM2:** 7
finish for nails **M1:** 271
Blum's slice model **EM4:** 602
Blunting
at crack tip **A12:** 15, 21
crack-tip **A11:** 56
Blunting line
in elastic-plastic fracture toughness **A8:** 455
Blur circles
photographic **A12:** 84
Blushing **EM3:** 6
defined **A13:** 2
BMC *See* Bulk molding compound; Bulk molding compound
BMI *See* Bismaleimide; Bismaleimide resins (BMIs); Bismaleimides
Board failure *See* Board-level physical test methods; Circuit(s); Failure(s)
Board(s) *See also* Circuit boards,- Integrated circuit boards (ICs)
assembly, through-hole packages **EL1:** 970
fibers, as contaminants **EL1:** 661
-level cooling **EL1:** 47
particles, as contaminants **EL1:** 661
single-sided, double-sided **EL1:** 711-712
solderability of leaded and leadless surface-mount joints **EL1:** 731
thickness, defined **EL1:** 1135
-to-board interconnections, line segments **EL1:** 3
trade-offs **EL1:** 21
wiring density, design and manufacture effects **EL1:** 129
Board-drop hammers *See also* Drop hammer **A14:** 2, 25-26, 41-42
Board-level physical test methods
automatic optical inspection **EL1:** 941-942
capabilities **EL1:** 941
environmental testing **EL1:** 944
for plating thickness **EL1:** 942-943
for solderability **EL1:** 943-944
solderjoint inspection **EL1:** 942
Boat
defined **A7:** 2
Boat-growth horizontal Bridgeman (HB) GaAs single-crystal growth method **A2:** 744
Boating *See also* Marine applications
composite material applications for **EM1:** 847
Bobbin
on fly-shuttle loom **EM1:** 127
Bobbing
nickel alloys **M5:** 674-675
Bochumer Verein Company (Germany) **A15:** 31
Body **EM3:** 6
defined **A18:** 4 **EL1:** 1135
Body environment
biochemical **A11:** 673
biochemical attack **A11:** 676-677
bone healing in **A11:** 673-674
bone resorption in **A11:** 674-676
dynamic **A11:** 673, 676-677
mechanical properties of bone **A11:** 676
Body fluids and tissues
tantalum resistance to **A13:** 728
Body fluids, dried
as corrosive **A12:** 361
Body putty
defined **EM1:** 5
Body solders
for plastic reinforcing, powders used **A7:** 574
powders used **A7:** 572
Body-centered
defined **A9:** 3
Body-centered cubic (bcc) crystal structures
abrasive wear **A18:** 186
adhesive wear **A18:** 179
cavitation erosion **A18:** 216
Body-centered cubic (bcc) materials
ductile-to-brittle transition **A11:** 84-85
flow strength **A11:** 138
fracture transition **A11:** 66
fractures in **A11:** 75
hydrogen damage in **A11:** 338
transgranular brittle fracture in **A11:** 22

Body-centered cubic (bcc) metals
abbreviation ... **A8:** 724
cyclic deformation of ... **A8:** 256
cyclic stress-strain curve ... **A8:** 256
ductile-to-brittle transition temperature ... **A8:** 34, 36
effects of strain aging in ... **A8:** 256
epitaxial growth direction ... **A6:** 50-51
fatigue limits of ... **A8:** 253
flow stress ... **A8:** 224
mechanical twinning in ... **A8:** 34-35
texture components and axial forces during torsion testing ... **A8:** 181, 183
Body-centered cubic array
density and coordination number ... **A7:** 296
Body-centered cubic forms
in uranium ... **A9:** 476
Body-centered cubic lattice
slip planes ... **A9:** 684
Body-centered cubic materials
abbreviation for ... **A10:** 689
and fcc materials, Kurdjumov-Sachs orientation relationship ... **A10:** 439
formation of dislocation loops ... **A9:** 116
iron-nickel alloy as, EXAFS analysis ... **A10:** 416-417
Body-centered cubic metals
compacts of ... **A7:** 310
defined ... **A13:** 45
Body-centered cubic wires
curly grain structure in ... **A9:** 687
Body-centered space-lattice ... **A3:** 1 • 15
Body-centered-cubic metals
embrittlement ... **A12:** 123
hydrogen embrittlement ... **A12:** 22
intergranular fracture in ... **A12:** 123
iron, slip lines ... **A12:** 219
in low temperatures, effect on fracture mode ... **A12:** 33
state of stress effects on ... **A12:** 31
strain rate effect on ... **A12:** 31
"Body-in-white" blank automotive bodies
laser-beam welding ... **A6:** 264
Boehmite ... **EM3:** 250, 262, 264, 624 **EM4:** 111, 112
pressure calcintering for study of compaction kinetics ... **EM4:** 300
sol-gel, solid-state sintering ... **EM4:** 276
Boeing Company
BAC 5555 proprietary process ... **EM3:** 802
environmental attack of a bonded surface postulated ... **EM3:** 802
surface preparation qualitative procedure evaluation ... **EM3:** 802, 803
Boeing wedge-crack test ... **EM3:** 472
Bog ore
early casting with ... **A15:** 24
Bohr magneton
in electron spin resonance ... **A10:** 254
Bohr model of atom ... **A12:** 168
Boil scab
as casting defect ... **A11:** 385
Boiled linseed oil ... **M5:** 498
"Boiler and Pressure Vessel Code" ... **A6:** 677
Boiler drum, utility
failure during hydrotesting ... **A11:** 647
Boiler plate
creep-induced failure in ... **A11:** 30
Boiler tube steels ... **A1:** 616, 937
maximum-use temperature of ... **A1:** 617
Boiler tubes
ASTM specifications for ... **M1:** 323
sigma-phase embrittlement of ... **M1:** 686
Boiler water embrittlement detector test
for carbon/low-alloy steels ... **A13:** 270
Boilers *See also* Boilers and related equipment, failures of; Pressure vessels; Steam equipment; Weld(s); Weldments
acoustic emission inspection ... **A17:** 642
burning municipal solid waste ... **A13:** 997-998
caustic cracking in ... **A11:** 214
Boilers (continued)
cleaning ... **A13:** 1138-1139
coal-fired ... **A13:** 995-996
codes governing ... **M6:** 823
corrosion control during idle periods ... **A11:** 616
corrosion protection ... **A11:** 615-616
design and construction ... **A11:** 619
dew point corrosion ... **A13:** 1001-1004
dissimilar-metal welds in ... **A11:** 620-621
eddy current inspection ... **A17:** 642
fire-side corrosion ... **A11:** 616-620
forgings, inspection ... **A17:** 644-645
fossil fuel, remote-field eddy current inspection ... **A17:** 200-201
generating bank, corrosion in ... **A13:** 1202
in-service quantitative evaluation ... **A17:** 653-654
inspection codes ... **A17:** 4
inspection during fabrication ... **A17:** 645-646
joining processes ... **M6:** 56
liquid penetrant inspection ... **A17:** 642
magnetic particle inspection ... **A17:** 642
oil-fired ... **A13:** 995-996
plate, inspection ... **A17:** 644-645
radiographic inspection ... **A17:** 641-642
recovery, wood pulp industry ... **A13:** 1198-1202
replication microscopy inspection ... **A17:** 642-644
service corrosion of carbon steels ... **A13:** 513-514
steam/water-side ... **A13:** 990-993
tubes, corrosion-fatigue cracks in ... **A11:** 79
tubes, inspection ... **A17:** 644-645
ultrasonic inspection ... **A17:** 642
water, and steam chemistry ... **A13:** 992-993
water-side corrosion ... **A11:** 614-616
Boilers and related equipment, failures of ... **A11:** 602-627
by corrosion or scaling ... **A11:** 614-621
by erosion ... **A11:** 623-624
by fatigue ... **A11:** 621-623
by overheating ruptures ... **A11:** 603-614
by stress-corrosion cracking ... **A11:** 624-626
causes, in steam equipment ... **A11:** 602-603
failure analysis procedures ... **A11:** 602
multiple-mode ... **A11:** 626
with sudden tube rupture ... **A11:** 603
Boiling
cooling, temperature drop for ... **EL1:** 364
of organic solvent cleaners ... **EL1:** 663
Boiling acid test
porcelain enamel ... **M5:** 525-528
Boiling liquid-warm liquid-vapor cleaning techniques ... **A7:** 459
Boiling liquid-warm liquid-vapor degreasing system ... **M5:** 46-47, 51, 54-55
Boiling point
of rare earth metals ... **A2:** 723-724
in reduction reactions ... **A7:** 53
Boiling water
with contaminants, fatigue crack growth testing ... **A8:** 426-430
embrittlement detector testing, low-carbon steels ... **A8:** 526
Boiling water reactors ... **A13:** 927-937
alloy X-750 jet pump beams, SCC of ... **A13:** 933-935
corrosion fatigue in ... **A13:** 928, 937
feedwater nozzles, corrosion fatigue ... **A13:** 937
irradiation-assisted SCC ... **A13:** 935-936
and light water reactors ... **A13:** 948
nitrided stainless steel, SCC in ... **A13:** 933
piping, intergranular SCC ... **A13:** 928-933
radiation fields, corrosion influence ... **A13:** 949-951
stress-corrosion cracking in ... **A13:** 927-928
Bolometers (barretters)
microwave detection by ... **A17:** 208
Bolster
defined ... **EM2:** 7
Bolster plates ... **A14:** 2, 62, 497
Bolt
modified Charpy V-notch testing of ... **A8:** 542
Bolt (continued)
-on attachment, crack-opening displacement transducer ... **A8:** 384
Bolt, and pin
bearing strength ... **EM1:** 314-316
Bolt(s)
cold-forged high-tensile, eddy current inspection ... **A17:** 554
explosive, neutron radiography of ... **A17:** 394
threads, diffraction pattern measurement ... **A17:** 13
Bolted connectors
recommended contact materials ... **A2:** 861
Bolted joints *See also* Joint(s) ... **EM1:** 479, 488-492
crevice corrosion in ... **A11:** 184
dissimilar material separation ... **EM1:** 716-717
elements, testing ... **EM1:** 325-326
fastener selection for ... **EM1:** 716-717
long-term exposure testing ... **EM1:** 825-826
subcomponents, testing ... **EM1:** 336-337
Bolted material
pin bearing testing for ... **A8:** 59-61
Bolting
galvanized members ... **A13:** 441
weathering steel ... **A13:** 519
Boltmaking machines
for cold heading ... **A14:** 292
Bolts *See also* Threaded fasteners; Threaded steel fasteners for elevated-temperature service ... **A1:** 296, 620, 631
AISI/SAE alloy steels, spontaneous rupture ... **A12:** 299
alloy steel, shear-face tensile fracture in ... **A11:** 76
bearing cap, sulfide SCC in ... **A12:** 299
blind, material and composite applications ... **A11:** 530
breaking strength ... **M1:** 277
bronze, preferential corrosion ... **A12:** 403
chevrons from fracture origin, SEM fractographs ... **A12:** 170
clamping forces ... **M1:** 280-282
coatings ... **M1:** 280
cold heading ... **M1:** 274, 280
delayed failure from hydrogen embrittlement ... **A11:** 539-540
elevated temperatures, steels recommended for ... **M1:** 280
fabrication of ... **M1:** 274, 280
failures, fractographic study ... **A12:** 248
fatigue strength ... **M1:** 275-276, 279-280, 282
fractured, SEM fractographs of radial marks ... **A12:** 169
hardened, cap screw substitution for ... **A11:** 142
hardness ... **M1:** 278-281
holes, stress concentration cracking at ... **A11:** 346-347
hot heading ... **M1:** 274, 275, 280
low hardness wear failure of ... **A11:** 160-161
mechanical fastening of ... **EM2:** 711
mechanical properties ... **M1:** 274, 278-282
preloading, and fastener performance ... **A11:** 542
proof stress ... **M1:** 273, 274, 277-278
quench cracks ... **A12:** 131
relaxation tests ... **A1:** 624
selection of steel for ... **A1:** 292, 293, 295 **M1:** 273-276
as spring ... **A11:** 530-531, 533
stainless steel, hydrogen embrittlement of ... **A11:** 249
stainless steel, SCC of ... **A11:** 536-537
steel wire for ... **M1:** 265-266
strength grades and property classes ... **M1:** 273-277
strength-hardness relations ... **M1:** 278-281
stress-corrosion tests for antiseizure coating ... **A11:** 536
structural, reversed-bending fatigue failure ... **A11:** 322
T-, SCC of ... **A11:** 538

Bolts (continued)
U-, fatigue fracture ... **A11:** 533-535
unitemp 212, hydrogen-induced delayed cracking ... **A11:** 249
wing-attachment, cracked along seam ... **A11:** 530, 532
Bolts, round-headed
economy in manufacture ... **M3:** 852, 853
Boltzmann
distribution, defined ... **A10:** 670
equation ... **A10:** 24, 44
Boltzmann constant ... **A18:** 441
Boltzmann superposition principle ... **EM2:** 412
Boltzmann's constant ... **EL1:** 98
Bomb
acid digestion ... **A10:** 165
Bomb calorimeter ignition
for sample dissolution ... **A10:** 167
Bombs, smoke *See* Smoke bombs
Bond ... **EM3:** 6
definition ... **A6:** 1207
Bond angle ... **EM3:** 6
defined ... **EM2:** 7
Bond breakers ... **EM3:** 549-550
Bond bridge ... **A18:** 236
Bond clay *See also* Bonding agent; Clays
defined ... **A15:** 2
Bond coat
definition ... **M6:** 2
Bond coat (thermal spraying)
definition ... **A6:** 1207
Bond coats
for thermal spray coatings ... **A13:** 460
Bond distances
determined by single-crystal analysis ... **A10:** 354
Bond energies
polymers ... **EM2:** 57
Bond face ... **EM3:** 6
Bond failure ... **EM3:** 629-635
Bond length ... **EM3:** 6
Bond lengths, interatomic
and physical properties ... **A10:** 355
Bond line ... **EM3:** 6
definition ... **A6:** 1207 **M6:** 2
Bond line corrosion ... **EM3:** 670-671
Bond strength *See also* also Bonding; Bond(s); Bonding agent; Peel strength ... **EM3:** 6
defined ... **A15:** 1 **EM1:** 5 **EM2:** 7
Bond strength development
water-base organic-solvent-base adhesives properties ... **EM3:** 86
Bond systems
for bonded-abrasive grains
Bond testers, ultrasonic *See* Ultrasonic bond testers
Bond testing
thermal spray coatings ... **M5:** 372
Bond zone formation ... **M6:** 706
Bond(s)
bonded-abrasive grains ... **A2:** 1015
brazing, ultrasonic inspection ... **A17:** 232
clay and water, types ... **A15:** 210-212
electroplated bond systems ... **A2:** 1015
formed, in molding aggregates ... **A15:** 212-213
lack as planar flaw ... **A17:** 50
metal bond systems ... **A2:** 1015
organic ... **A15:** 213
phenol-aralkyl bonds ... **A2:** 1014
phosphoric acid ... **A15:** 213
resin bonds ... **A2:** 1014
silica-based ... **A15:** 212-213
thermoplastic resins ... **A2:** 1014
vitreous bond systems ... **A2:** 1014-1015
weak, adhesive-bonded joints ... **A17:** 616
ond-related failures, zone 2
package interior ... **EL1:** 1011-1013
ondascope 2100 ... **EM3:** 757-758, 778
for adhesive-bonded joints ... **A17:** 622
onded asbestos brake linings
powders used ... **A7:** 573
onded film lubricant *See* Bonded solid lubricant
onded honeycomb structures
microwave inspection ... **A17:** 202
onded joints *See also* Joint(s) ... **EM1:** 480-488
adhesive shear stresses ... **EM1:** 683
dissimilar material separation ... **EM1:** 717
elements, testing ... **EM1:** 326
Bonded joints (continued)
subcomponents, testing ... **EM1:** 337-339
surface preparation ... **EM1:** 681-682
ultrasonic inspection ... **A17:** 272-273
Bonded lead
intermetallic formation around ... **A11:** 776
Bonded metal-metal laminary composite systems
clad metals ... **A13:** 887
Bonded particle filters
foundry and die casting ... **A15:** 490
Bonded plumbum materials
applications ... **A2:** 555
Bonded resistance-strain gage *See also* Strain gage
for displacement measurement ... **A8:** 193
Bonded sand molds *See also* Binder systems; Binders; Resin binder systems
dry sand molding ... **A15:** 228
green sand molds ... **A15:** 222-228
loam molding ... **A15:** 228-229
phosphate molds ... **A15:** 229-230
silicate molds ... **A15:** 229-230
skin-dried molds ... **A15:** 228
Bonded solid lubricant
defined ... **A18:** 4
Bonded structure repair
introduction and overview ... **EM3:** 799-800
Bonded-abrasive grains
bonds ... **A2:** 1014
coatings ... **A2:** 1015
in grinding wheels ... **A2:** 1013
Bonded-phase chromatography
defined ... **A10:** 670
stationary phases ... **A10:** 652
Bonding *See also* Tape automated bonding (TAB); Tape automated bonding (TAB) technology; Ultrasonic bonding
adhesive ... **EM2:** 725
adhesive, of epoxy materials ... **EL1:** 832-833
adhesive, surface preparation for ... **EM1:** 681-682
Auger spectral lineshapes showing ... **A10:** 552
automation ... **EL1:** 274
and bonding distance, EXAFS determined ... **A10:** 407, 415
by welding, during milling ... **A7:** 57
cermet ... **A7:** 801-802
of cermets ... **A2:** 990-991
of clad metals ... **A13:** 887-888
cure ... **EM1:** 702-705
defined ... **A7:** 2
eutectic, as die attachment method ... **EL1:** 213, 349
explosive, refractory metals and alloys ... **A2:** 559
failures
gang (mass) ... **EL1:** 278
infinite nature, and single-crystal analysis ... **A10:** 345
inner lead ... **EL1:** 278-281
integrity polymer die attach ... **EL1:** 217-218
interatomic bond-lengths, and physical properties ... **A10:** 355
interfacial, carbon fibers ... **EM1:** 52
in metal-matrix composites ... **A12:** 466
outer lead ... **EL1:** 283-286
pi, defined ... **A10:** 679
polymer ... **EM2:** 63-64
reaction, of structural ceramics ... **A2:** 1020-1021
secondary *See* Secondary bonding ... **EM2:** 59
secondary, cast irons ... **A15:** 238
sigma, defined ... **A10:** 681
silver-glass ... **EL1:** 349
single-point ... **EL1:** 278
solid-state, metal-matrix composites ... **A2:** 903
solvent ... **EM2:** 725
for stray-current corrosion ... **A13:** 87
substance differences due to ... **A10:** 345
substrate ... **EL1:** 627
sulfur-cement ... **A12:** 472
and surface chemistry ... **A7:** 260
in ternary molybdenum chalcogenides (chevrel phases) ... **A2:** 1078
thermocompression ... **EL1:** 278-279, 734
to frame, as cooling ... **EL1:** 310
topologies, RDF analysis and ... **A10:** 393, 398
types, in aggregate molding ... **A15:** 212-213
ultrasonic, aluminum transistor base lead ... **A12:** 483
void distributions ... **EL1:** 214
void formation during ... **EM1:** 687
Bonding (continued)
wires (chip and wire assembly) ... **EL1:** 110
within mer ... **EM2:** 57
zone 2, package interior ... **EL1:** 1011-1013
Bonding agent *See also* Bond clay; Bond strength
defined ... **A15:** 2
Bonding force
definition ... **A6:** 1207 **M6:** 2
Bonding preparation, application, and tooling ... **EM3:** 703-708
adhesive application ... **EM3:** 704-705
film ... **EM3:** 705
liquid ... **EM3:** 705
new application techniques ... **EM3:** 705
paste ... **EM3:** 705
bonding preparations ... **EM3:** 703-704
bond line thickness control ... **EM3:** 703-704
prefit evaluation ... **EM3:** 704
prefitting of adherends ... **EM3:** 703
selection ... **EM3:** 703
surface treatment prior to adhesive application ... **EM3:** 704
tooling ... **EM3:** 705-708
autoclave bonding fixtures ... **EM3:** 706-707
bonding fixtures ... **EM3:** 705-707
curing equipment ... **EM3:** 708
fixture design ... **EM3:** 707
fixtureless bonding ... **EM3:** 707
press bonding fixtures ... **EM3:** 705-706
pressure applicators ... **EM3:** 707-708
pressure bag fixtures ... **EM3:** 706
Bonding success
factors ... **EM3:** 43
Bonding systems *See also* Aggregate molding materials; Binders; Binding systems; Bonded sand molds
plastic ... **A15:** 211
Bonding, tape automated *See* Tape automated bonding (TAB) technology
Bonding wire(s)
physical characteristics ... **EL1:** 110
in plastic packages ... **EL1:** 211
types ... **EL1:** 227
Bone *See also* Bone fractures; Bone plate; Bone screw; Implants; Metallic orthopedic implants, failures of; Skeletal system
breakage, as implant failure ... **A11:** 672
cement degradation, effect on hip prosthesis ... **A11:** 690-693
cortical, microradiography from thin section of ... **A11:** 672
grafting, primary ... **A11:** 673
heating ... **A11:** 672-674
mechanical properties of ... **A11:** 676
plates ... **A11:** 671
resorption, as implant failure ... **A11:** 672, 674-676
resorption, effects on hip prosthesis ... **A11:** 690-693
Bone ash ... **EM4:** 44
in ceramic tiles ... **EM4:** 926
in typical ceramic body compositions ... **EM4:** 5
Bone china ... **EM4:** 4
absorption ... **EM4:** 4
composition ... **EM4:** 5, 45
imports ... **EM4:** 935
physical properties ... **EM4:** 934
products ... **EM4:** 4
properties of fired ware ... **EM4:** 45
Bone fractures *See also* Bone; Metallic orthopedic implants, failures of
bone resorption in ... **A11:** 674-676
combined dynamic and biochemical attack ... **A11:** 676-677
corrective surgery and treatment for ... **A11:** 671
healing of ... **A11:** 673-674
total hip joint prostheses ... **A11:** 692-693
Bone oil
defined ... **A18:** 4
Bone plate *See also* Bone
corrosion and wear products transported in tissue ... **A11:** 688, 692
crack initiation on ... **A11:** 680, 685-686
with fatigue crack and broken screw ... **A11:** 681, 686-687
fracture surfaces of failed ... **A11:** 682, 686-687
straight, fatigue initiation ... **A11:** 679, 680, 684-686
surface, with fatigue cracks ... **A11:** 680, 684

Bone screw *See also* Bone
broken, and fatigue cracked bone plate..... **A11:** 681, 686-687
cancellous........ **A11:** 671
cortical........ **A11:** 671
fatigue failure........ **A11:** 679, 682
fracture surfaces of failed........ **A11:** 682, 686-687
head, titanium, wear and fretting at plate hole........ **A11:** 689, 692
hole, with fretting and fretting corrosion........ **A11:** 688, 691-692
pitting corrosion in........ **A11:** 687, 691
pure titanium, shearing fracture........ **A11:** 677, 682
retrieved, from cobalt-chromium alloy with casting defects........ **A11:** 674, 680
sheared-off........ **A11:** 678, 682
stainless steel, fatigue failure........ **A11:** 679, 682
stainless steel, shearing fracture........ **A11:** 676, 682
Boniszewski basicity index........ **A6:** 204
Book mold casting
wrought copper and copper alloys........ **A2:** 242
Boolean substitutions
in design layout........ **EL1:** 513, 516
Boosters, space
corrosion of........ **A13:** 1101-1105
Boothroyd-Dewhurst design for assembly (DFA) method........ **EL1:** 124-125
Boots/Technochemie Compimide resins........ **EM1:** 82-83
BOPACE 3d version 6 computer program for structural analysis........ **EM1:** 268, 270
Borate alkaline cleaners........ **M5:** 24
Borate glasses
electrical properties........ **EM4:** 851, 853
heat capacity........ **EM4:** 847
thermal expansion........ **EM4:** 847
typical oxide compositions of raw materials........ **EM4:** 550
Borate materials
applications........ **EM4:** 380
in fiberglass........ **EM4:** 380
mining techniques........ **EM4:** 380
properties........ **EM4:** 380
purpose for use in glass manufacture........ **EM4:** 381
sources........ **EM4:** 380
Borates
effects on electroless nickel plating........ **M5:** 223
in torch brazing flux........ **M6:** 962
Borax
for flame retardance........ **EM3:** 179
as flux........ **A10:** 167
flux removal after torch brazing........ **M6:** 963
fluxing agent for dip brazing........ **M6:** 991
Borax bead tests
as qualitative wet analyses........ **A10:** 168
Borax coating
steel wire........ **M1:** 262, 266
steel wire rod........ **M1:** 253
Borax frits and glazes
typical oxide compositions........ **EM4:** 550
Borazon
abrasive for ECG........ **A16:** 545
Bordie particles
as inoculant........ **A15:** 106-107
Bore holes
gray iron crankcase failure from........ **A11:** 363
Bore scope
in split Hopkinson pressure bar........ **A8:** 201
Bore seal
defined........ **A18:** 4
Borehole logging
PGAA use in........ **A10:** 240
Bores, splined and tapered
in gears and gear trains........ **A11:** 589-590
Borescopes........ **A17:** 3-10
applications........ **A17:** 9-10
direction of view........ **A17:** 9
extendable........ **A17:** 5
field of view........ **A17:** 9

Borescopes (continued)
flexible........ **A17:** 5-10
flexible fiberscopes........ **A17:** 5
focusing and resolution........ **A17:** 8
hybrid........ **A17:** 5
illumination........ **A17:** 8-9
measuring........ **A17:** 8
miniborescopes........ **A17:** 4
mirror sheaths........ **A17:** 5
for residual core detection........ **A17:** 393
rigid........ **A17:** 4-5
scanning........ **A17:** 5
selection........ **A17:** 8-9
videoscopes with CCD probes........ **A17:** 5-8
in visual leak detection........ **A17:** 66
working length........ **A17:** 9
Boric acid
in corrosion fatigue tests........ **A8:** 423
for flame retardance........ **EM3:** 179
flux constituent for torch brazing........ **M6:** 962
flux removal after torch brazing........ **M6:** 963
nickel plating, use in........ **M5:** 199-200, 204-205, 207, 209
for pH level reduction........ **A13:** 944
tin-lead plating process using........ **M5:** 276-278
Boric acid ($B_2O_3 \cdot 3H_2O$)
purpose for use in glass manufacture........ **EM4:** 381
Boric acid, as binding agent for samples
x-ray spectrometry........ **A10:** 94
Boric acid as injection molding binder........ **A7:** 498
Boric oxide
role in glazes........ **EM4:** 1062
Boride cermets *See also* Cermets........ **A7:** 811-813
application and properties........ **A2:** 1003
chromium boride cermets........ **A2:** 1004
defined........ **A2:** 979
molybdenum boride cermets........ **A2:** 1004
titanium boride cermets........ **A2:** 1004
zirconium boride cermets........ **A2:** 1003
Boride-based cermets........ **A7:** 789-799, 811-813
and metal borides, properties........ **A7:** 812
Boride-containing nickel-base alloys
hardfacing........ **A6:** 790, 794-795, 796
Borides........ **A1:** 952, 955-956
applications........ **EM4:** 203
chemical vapor deposition of........ **M5:** 381
maximum service temperature........ **EM4:** 203
plasma spray material........ **EM4:** 203
properties........ **EM4:** 203
in structural ceramics........ **A2:** 1019, 1021-1024
Borides in wrought heat-resistant alloys........ **A9:** 309, 311-312
anodic dissolution to extract........ **A9:** 308
Boriding........ **A4:** 437-446
advantages........ **A4:** 437, 443-444, 446
alloying elements and their effects........ **A4:** 441
alternative nonthermochemical surface-coating processes........ **A4:** 437
applications........ **A4:** 446
thermochemical........ **A4:** 444-445
boroaluminizing........ **A4:** 444
borochromizing........ **A4:** 444, 445
borochromtitanizing........ **A4:** 444, 445
borochromvanadized steels........ **A4:** 444
borosiliconizing........ **A4:** 444
borovanadized steels........ **A4:** 444
Borudif process........ **A4:** 442
case depths........ **A4:** 440, 442
characteristic features of layers........ **A4:** 437-438
chemical vapor deposition (CVD)........ **A4:** 445-446
diamond lapping........ **A4:** 438
disadvantages........ **A4:** 437-438, 442, 443, 444
egg shell effect........ **A4:** 440
electroless salt bath........ **A4:** 442
electrolytic salt bath........ **A4:** 442-443, 444
ferrous materials........ **A4:** 438-441
fluidized bed boriding........ **A4:** 443-444
gas boriding........ **A4:** 443
growth........ **A4:** 438

Boriding (continued)
heat treatment after........ **A4:** 441
liquid boriding........ **A4:** 442-443
multicomponent boriding........ **A4:** 444
nonferrous materials........ **A4:** 441
pack boriding........ **A4:** 441-442, 444
paste boriding........ **A4:** 442, 444
plasma boriding........ **A4:** 443
process description........ **A4:** 437
rolling contact fatigue properties........ **A4:** 438
safety precautions........ **A4:** 443
titanium alloys........ **A18:** 780-781
tool steels........ **A18:** 641, 642, 739
wear testing........ **A4:** 437, 438, 439, 441
Boring........ **A16:** 160-174
accuracy of form and diameter........ **A16:** 170-171
adapters........ **A16:** 381, 382
aircraft engine components, surface finish requirements........ **A16:** 22
Al alloys........ **A16:** 768, 771-772, 777-778, 791, 797
bars........ **A16:** 170, 171
carbon and alloy steels........ **A16:** 668
cast irons........ **A16:** 655, 658
cemented carbides used........ **A16:** 75
ceramics used........ **A16:** 101
close tolerance........ **A16:** 171
compared to broaching........ **A16:** 194, 196
compared to grinding........ **A16:** 426, 427
compared to honing........ **A16:** 473
compared to reaming........ **A16:** 239
composition and hardness of workpiece........ **A16:** 169
in conjunction with broaching........ **A16:** 194, 195, 209-210, 211
in conjunction with drilling........ **A16:** 214, 216, 218, 219, 221, 235
in conjunction with milling........ **A16:** 304, 306, 329
in conjunction with tapping........ **A16:** 256
in conjunction with turning........ **A16:** 135, 139, 140-142
control of vibration and chatter........ **A16:** 171-172
Cu alloys........ **A16:** 810, 813
cutting fluids........ **A16:** 170
cylinder block........ **A16:** 115
equipment use for other operations........ **A16:** 173-174
finish machining........ **A16:** 33
fixtures........ **A16:** 404
heat-resistant alloys........ **A16:** 743, 751-752
high-speed tool steels........ **A16:** 152
inertia-disk dampers........ **A16:** 172
jig machines and milling........ **A16:** 329
machines........ **A16:** 1, 160-161, 212, 223, 252
in machining centers........ **A16:** 393
as machining process........ **A7:** 461
Mg alloys........ **A16:** 820, 821-823
mills........ **A16:** 170, 240, 473
MMCs........ **A16:** 896
monitoring systems........ **A16:** 414, 416
multifunction machining........ **A16:** 366-368, 371, 378-380, 384
NC implemented........ **A16:** 613, 614
Ni alloys........ **A16:** 837
number of operations........ **A16:** 168-169
P/M materials........ **A16:** 881, 882, 889
pilots and supports........ **A16:** 162-163
plug dampers........ **A16:** 171-172
production quantity........ **A16:** 170
refractory metals........ **A16:** 859, 862, 863
speed and feed........ **A16:** 163
stainless steels........ **A16:** 154, 691, 692
surface finish........ **A16:** 173
taper........ **A16:** 387
tool design........ **A16:** 162, 164, 170, 173
tool, for porous bronze bearings........ **A7:** 462
tool life........ **A16:** 162-163, 166, 168, 169, 172, 173
tool materials........ **A16:** 162
tool steels........ **A16:** 716, 727
tools........ **A16:** 161-162, 163, 166
and transfer machines........ **A16:** 394, 397

Boring (continued)
workpiece configuration **A16:** 166-168
workpiece size **A16:** 164-166
zirconium **A16:** 853, 856
Zn alloys **A16:** 831-832
Boring bars
cemented carbide **A2:** 971
Boring bars, machine-tool
economy in manufacture **M3:** 854
Boring mill
development of **A15:** 20
Boring plungers
cemented carbide **A2:** 971
Boring, tunnel and shaft
cemented carbide tools **A2:** 973
Borland's concepts
solidification cracking **A6:** 89, 90
Bormioli alkaline earth silicate
coefficient of thermal expansion **EM4:** 1102
composition **EM4:** 1102
softening point **EM4:** 1102
Borocementite **A4:** 440
Borohydride-reduced electroless nickel plating *See* Sodium borohydride
electroless nickel plating process
Boron *See also* Cubic boron nitride **A13:** 728, 860
AAS analysis of **A10:** 46
acid-base titrations **A10:** 172
as addition to brazing filler metals **A6:** 904
as addition to carbon-graphite materials **A18:** 816
addition to high-temperature alloys **A6:** 563
addition to improve hardenability of low-carbon steels **A4:** 366
addition to inhibit deleterious effect of AIN particles **A4:** 60, 61
as addition to nickel aluminide alloys **A18:** 772, 774, 775
addition to underwater welds, microstructural development **A6:** 1011
additive to stainless steels improving machinability **A16:** 683, 687-688
alloying, aluminum casting alloys **A2:** 132
alloying effect on nickel-base alloys **A6:** 589-590
alloying, ordered intermetallics **A2:** 913
alloying, wrought aluminum alloy **A2:** 46-47
in aluminum alloys **A15:** 744
at elevated-temperature service **A1:** 641
atomic interactions and adhesion **A6:** 144
in austenitic stainless steels **A6:** 458, 463
cemented carbides **A16:** 81
chemical vapor deposition of **M5:** 381
content effect on heat-affected zone cracks **A6:** 93
content effect on hot-ductility response **A6:** 91, 93
content effect on solidification cracking **A6:** 90
content, in glass, IR analysis **A10:** 121-122
continuous, as reinforcements **A2:** 7
deoxidizing, copper and copper alloys **A2:** 236
determined in borosilicate glass **A10:** 179
detrimental to welding of alloy systems **A6:** 89
diffusion, neutron radiography of **A17:** 391
in ductile iron **A15:** 649
effect of, on hardenability of steels **A1:** 395, 469-470
effect of, on notch toughness of steels **A1:** 741
effect on cobalt-base corrosion-resistant alloys **A6:** 598
effect on precipitation-hardening stainless steels **A6:** 490
in electroless nickel **A18:** 837
electroslag welding, reactions **A6:** 278
in enamel cover coats **EM3:** 304
in enameling ground coat **EM3:** 304
filaments, microwave inspection **A17:** 215
in flux for submerged arc welding **M6:** 124
forming lower melting point eutectics **A6:** 588
for grain refinement, inclusion-forming **A15:** 95
in hardfacing alloys **A18:** 763-764, 765
heat-affected zone fissuring in nickel-base alloys **A6:** 588
in heat-resistant alloys **A4:** 512
in high-strength low-alloy steels **A1:** 408
Boron (continued)
impurity in diamond **A16:** 454
as inoculant **A15:** 105
in interlayer metal for joining non-oxide ceramics **EM4:** 528
for laser alloying **A18:** 866
lubricant indicators and range of sensitivities **A18:** 301
in malleable iron **A1:** 10
in Mo/Ni-Cr-B-Si blend, thermal spray coating material **A18:** 832
in Mo/Ni-Cr-B-SiC, thermal spray coating material **A18:** 832
in Ni-Al/Ni-Cr-B-SiC, thermal spray coating material **A18:** 832
in Ni-Cr-B-SiC (fused), thermal spray coating material **A18:** 832
in Ni-Cr-B-SiC (unfused), thermal spray coating material **A18:** 832
in Ni-Cr-B-SiC-Al-Mo, thermal spray coating material **A18:** 832
in Ni-Cr-B-SiC/WC (fused), thermal spray coating material **A18:** 832
nickel-chromium-boron filler metal **A6:** 344
as paint for laser-alloyed stainless steel **A18:** 866
photometric analysis methods **A10:** 64
powders fused after flame spraying of cast irons **A6:** 720
prompt gamma activation analysis of **A10:** 240
pure **M2:** 719-720
pure, properties **A2:** 1103
recovery from selected electrode coverings **A6:** 60
in seawater, determined by electrometric titration **A10:** 205
segregation and solid friction **A18:** 28
separation by distillation **A10:** 169
-shielded samples for epithermal neutron bombardment **A10:** 234
spectrometric metals analysis **A18:** 300
in steel **A1:** 145
in superalloys **A1:** 954, 984
in superalloys, solution treating **A4:** 799
UV/VIS analysis in aluminum alloys Carmine method **A10:** 68
vapor pressure, relation to temperature **A4:** 495
volatilization losses in melting **EM4:** 389
volumetric procedures for **A10:** 175
in WC/Ni-Cr-B-SiC (fused), thermal spray coating material **A18:** 832
in WC/Ni-Cr-B-SiC (unfused), thermal spray coating material **A18:** 832
Boron as an addition to
cobalt-base heat-resistant casting alloys **A9:** 334
electrical steels **A9:** 537
nickel-base heat-resistant casting alloys **A9:** 334
wrought heat-resistant alloys **A9:** 311-312
Boron carbide
in abrasive flow machining operation **A16:** 517
abrasive for lapping **A16:** 493, 505
in abrasive slurry for ultrasonic machining **A16:** 529, 531
friction coefficient data **A18:** 72
for gas-lubricated bearings **A18:** 532
honing stone selection **A16:** 476
metallic binder phase, effects **A2:** 1008
pellets, swelling of **A7:** 666
as structural ceramic, applications and properties **A2:** 1022
as superhard material **A2:** 1008
for truing of CBN grinding wheels **A16:** 468
use in nuclear control rods and shielding **A7:** 666
Vickers and Knoop microindentation hardness numbers **A18:** 416
Boron carbide (B_4C) **EM4:** 47
as abrasive for ultrasonic machining **EM4:** 360
adiabatic temperatures **EM4:** 229
applications **EM4:** 1, 48, 200, 230, 806
electrical properties **EM4:** 806
fabrication **EM4:** 805
grain-growth inhibitor **EM4:** 188
hardness **EM4:** 351
hot pressing **EM4:** 191
Boron carbide (B_4C) (continued)
key product properties **EM4:** 48
manufacture **EM4:** 804-805
mechanical properties versus product condition **EM4:** 807
nuclear properties **EM4:** 806
phase diagram **EM4:** 805
pressure densification **EM4:** 298
properties **A6:** 629 **EM4:** 191, 806-806
raw materials **EM4:** 48
sintering aid **EM4:** 188
stoichiometry **EM4:** 804
structure **EM4:** 804, 805
synthesized by SHS process **EM4:** 229, 230
thermal expansion coefficient **A6:** 907
uses **EM4:** 806
Boron carbide-alumina **EM4:** 191
Boron content
effect in sintering **A7:** 373
Boron deoxidation
of copper alloys **A15:** 469
Boron fiber reinforced aluminum (B-Al)
applications **EL1:** 1122-1125
Boron fiber titanium composites
production **EM1:** 851
Boron fibers *See also* Continuous boron fiber MMCs **EM1:** 58
brittleness, as testing, problem **EM1:** 732
defined **EM1:** 5
and epoxy resins **EM1:** 75-76
fabrication of **A9:** 592
filament modifications **EM1:** 852
galvanic corrosion of **EM1:** 717
importance **EM1:** 43
introduction **EM1:** 31
manufacture **EM1:** 851-853
mean/range strengths **EM1:** 193
mechanical testing **EM1:** 731-732
in metal matrix composites(MMCs) **EM1:** 31
microstructure **EM1:** 59
properties **EM1:** 58, 118, 175, 851
as reinforcements **EM2:** 505
surface treatment vs. sizing of **EM1:** 122
Boron in cast iron
depth of chill, effect on **M1:** 77
Boron in iron
malleable cast iron **M1:** 58, 60
Boron in steel **M1:** 115, 411
hardenability affected by **M1:** 477
modified low-carbon steels **M1:** 162
notch toughness, effect on **M1:** 692
Boron metalloid carbide cermets
application and properties **A2:** 1002
Boron nitride *See* Tool materials, superhard
as filler for conductive adhesives **EM3:** 76
sulfuric acid as dissolution medium for **A10:** 165
thermal expansion coefficient **A6:** 907
Boron nitride (BN)
adiabatic temperatures **EM4:** 229
applications **EM4:** 230, 820
chemical vapor deposition **EM4:** 217
grain-growth inhibitor **EM4:** 188
for laser cladding **A18:** 868
non-oxide ceramic joining **EM4:** 480
properties **EM4:** 820
refractory material **EM4:** 14
sintering aid **EM4:** 188
structure **EM4:** 820
synthesized by SHS process **EM4:** 229
Boron nitride crucibles
vacuum coating **M5:** 390, 392, 400
Boron nitride wheels
used to section tool steels **A9:** 256
Boron oxide (B_2O_3)
component in photochromic ophthalmic and flat glass composition **EM4:** 442
in composition of glass-ceramics **EM4:** 499
in composition of leachable alkali-borosilicate glasses **EM4:** 428
in composition of textile products **EM4:** 403
in composition of wool products **EM4:** 403
direct evaporation **A18:** 844
in drinkware compositions **EM4:** 1102
glass-forming ability **EM4:** 494
in glaze composition for tableware **EM4:** 1102

Boron oxide (B_2O_3) (continued)
melting point **EM4:** 494
in ovenware compositions **EM4:** 1103
properties **EM4:** 424
in tableware compositions **EM4:** 1101
Boron phosphide
protective coating against liquid impingement erosion **A18:** 222
Boron powder
as filler for polyphenylquinoxalines **EM3:** 167
Boron steels **A1:** 208
threaded fasteners **M1:** 276
Boron tetrafluoride
electrode **A10:** 184
Boron trichloride
for extinguishing magnesium fires **A4:** 906
Boron trifluoride
complexing of **EM1:** 140
for extinguishing magnesium fires **A4:** 906
Boron trifluoride complexes **EM3:** 95-96
Boron trioxide
as flux **A10:** 167
Boron tungsten matrix composites **EM1:** 117
Boron type igniters **A7:** 604
Boron, vapor pressure
relation to temperature **M4:** 309, 310
Boron-aluminum composites **EM1:** 851, 854-856
applications **A9:** 592
diffusion bonded to Ti-6Al-4V **A9:** 595
fiber on foil, diffusion bonded **A9:** 595, 597
grinding **A9:** 589
liquid impingement erosion protection applications **A18:** 222
polishing **A9:** 590
Boron-aluminum continuous fiber metal-matrix composites **A2:** 904
Boron-carbon matrix composites **EM1:** 117
Boron-copper woven tape
continuous reentrant fill **EM1:** 127
Boron-epoxy 27-ply multidirectional laminate **EM3:** 821
Boron-epoxy composites **A9:** 592
application **EM1:** 31
failures, tension-loaded **EM1:** 200
Boronizing *See* Boriding
Borophosphosilicate glasses (BPSG)
applications electronic processing **EM4:** 1056
Borosilicate cladding glass
composition **EM4:** 1101
properties **EM4:** 1101
Borosilicate glass
applications
biomedical **EM4:** 19
dental **EM4:** 1093
electronic processing **EM4:** 1056, 1058
glass containers **EM4:** 1082, 1083, 1084
laboratory and process **EM4:** 1087, 1088, 1089
lighting **EM4:** 1032, 1034, 1036, 1037
optical glass products **EM4:** 1076
chemical properties **EM4:** 857
composition **EM4:** 566, 741, 742, 1033, 1083, 1088
determination of B and F in **A10:** 179
E-glass properties **EM4:** 1057
electrical properties **EM4:** 851
in glass enamel **EM4:** 1066
glass-contact and fused AZS refractories **EM4:** 904
heat transfer coefficients compared **EM4:** 1090
maximum operating temperature **EM4:** 1035
melting/fining **EM4:** 392
not strengthened by ion-exchange **EM4:** 462
for ovenware **EM4:** 1103
porcelain enamels a variation of **EM4:** 937
properties **EM4:** 566, 742, 849, 863, 1033, 1083
laboratory glassware **EM4:** 1088
non-CRT applications **EM4:** 1048-1049
refractive index **EM4:** 566
regenerative heat exchanger refractory applications **EM4:** 906-907
Borosilicate glass (continued)
S-glass **EM4:** 1057
softening point **EM4:** 1166
uses **EM4:** 742
volatilization and devitrification **EM4:** 389
Borosilicate glasses
glass-to-metal seals **EM3:** 302
types and properties **EM1:** 45-47
Borosilicate specialty glasses
applications **EM4:** 380
borate materials in composition **EM4:** 380
Borosilicates
as bases for photochromic glasses **EM4:** 441
Boroxine-polycarbosilane-derived coatings
time- dependent weight loss **EM4:** 226
Borrmann effect *See also* Anomalous transmission
diffraction geometry for **A10:** 370
in x-ray diffraction **A10:** 367
Borrmann fan
effect of crystal thickness on **A10:** 367
BOSOR
finite difference method code **EM3:** 480
BOSOR 4 computer program for structural analysis **EM1:** 268, 270
BOSOR 5 computer program for structural analysis **EM1:** 268, 270
Boss
defined **A15:** 2
Bosses *See also* Hub **A14:** 2, 552
by metal casting **A15:** 40
defined **EM2:** 7
design of **EM2:** 615
forgings **M1:** 362, 364
polyamide-imides (PAI) **EM2:** 131
Boston round
defined **EM2:** 7
Bottle
definition **A6:** 1207
Bottle, ABL *See* ABL bottle
Bottleneck-shaped forgings **A14:** 71
Bottles, steel
by radial forging **A14:** 145
Bottom blow
defined **EM2:** 7
Bottom board
defined **A15:** 2
Bottom draft
defined **A14:** 2
Bottom filling technique
in Cosworth and FM processes **A15:** 38
Bottom gating
permanent mold casting **A15:** 279
Bottom plate
defined **EM2:** 7
Bottom pouring *See* Bottom running
Bottom punch
defined **A7:** 2
Bottom running
defined **A15:** 2
Bottom setting *See* Coining
Bottom-braze packages **EL1:** 77, 992-993
Bottom-end bearing *See* Big-end bearing
Bottom-pour ladle *See also* Ladles; Nozzle
automatic **A15:** 497-498
defined **A15:** 2
for plain carbon steels **A15:** 710
Bottoming bending
defined **A14:** 2
Boule, silicon
alignment for cutting along crystallographic planes **A10:** 342
Boundaries *See also* High-angle boundaries
incoherent, in massive transformations **A9:** 655
resulting from plastic deformation **A9:** 693
Boundaries, rugged
fractals as descriptors **A7:** 243-244
Boundary additives
for metalworking lubricants **A18:** 140-141, 142, 143
Boundary conditions
laminate **EM1:** 229-230
Boundary element analysis
for high-frequency digital system design **EL1:** 81
Boundary energies in pure metals **A9:** 610
Boundary grain
defined **A9:** 3
Boundary integral method
of magnetic flaw characterization **A17:** 131
Boundary layer model
mass transfer limited kinetics **A15:** 53
Boundary layer thickness
in carbon diffusion in iron-carbon melts **A15:** 72-74
Boundary lubricant
defined **A18:** 4
Boundary lubricants *See also* Lubricant(s); Lubricantforms; Lubrication
for sheet metal forming **A14:** 512
for wire forming **A14:** 696
Boundary lubrication *See also* Elastohydrodynamic lubrication; Extreme-pressure lubrication; Lubricants; Lubrication; Thin-film lubrication **A18:** 80, 89, 94, 96
by free lead in bearing alloy **A11:** 161
chemically reacted surface films **A18:** 96
chemisorption **A18:** 96
defined **A18:** 4
friction and wear in **A11:** 150
and hydrodynamic lubrication **A11:** 484
physisorption **A18:** 96
straight-chain fatty-acid molecules in **A11:** 152
valve train assembly of internal combustion engine **A18:** 558
and wear **A11:** 151
zone **A18:** 29
Boundary precipitates
quantitative metallography of **A9:** 29
Boundary representation
as geometric modeler **A15:** 858
Boundary structure
and crystals, compared **A10:** 358
EPMA analysis of compositional gradients in **A10:** 516
Boundary-element calculations
remote-field eddy current model **A17:** 198
Bourdon tube gauge **A7:** 423
Bourdon tube hydraulic test gage
with load measuring system **A8:** 613
Bow
defined **A14:** 2 **EM2:** 7
Bowden-Tabor model
friction coefficient **A18:** 46
Bowl vibratory finishing **M5:** 131-132
Box annealing
defined **A9:** 3
Box beam tests **EM3:** 555
Box furnace
defined **A7:** 2
Box milling
in conjunction with milling **A16:** 322
Box skin tooling design **EM1:** 597-599
Box-annealed aluminum-killed steel
strain-age embrittlement of **M1:** 684
Box-annealed rimmed steel
strain-age embrittlement of **M1:** 684
Box-Behnken experimental design **EM2:** 601
Box-forming dies
for press-brake forming **A14:** 537
Box-girder webs (bridge)
penetrations through **A11:** 710-711
Box-Wilson experimental design **EM2:** 601-602
Boxing
definition **A6:** 1207 **M6:** 2
of tubing **A11:** 630
Boyden, Seth
as metallurgist **A15:** 31, 33

BPA epoxy resins *See also* Epoxy resins **EM1:** 66-77
BPA fumarate *See* Bisphenol A (BPA) fumarates
Brackish water
corrosion of steels in **M1:** 739, 744
defined **A13:** 2
Bradelloy (Hastelloy X honeycomb + braze/nickel aluminum)
abradable seal material **A18:** 589
Bragg angle **A18:** 463, 464
defined **A9:** 3
XRPD analysis **A10:** 337
Bragg case
reflection topography **A10:** 366
Bragg conditions
in electron-channeling patterns **A9:** 94
Bragg diffraction **A18:** 388
Bragg equation *See also* Bragg's law
defined **A9:** 3
derivation of **A10:** 327
Bragg equation, relationship between angle of incidence
wavelength, and interplanar spacing **EM4:** 557-558, 559
Bragg method
defined **A9:** 3
Bragg peaks **A18:** 464, 465, 468
Bragg reflections
excited by high-energy electron diffraction **A9:** 111
Bragg's equation **A18:** 386
Bragg's law **A6:** 1150 **A18:** 464, 465, 468
defined **A10:** 670
in determining pole figures **A10:** 360
diffraction defined by **A10:** 381
effect on inelastically scattered electrons **A9:** 109
electron diffraction patterns in TEM **A10:** 436
electron probe x-ray microanalysis **A10:** 520-521
in single-crystal analysis **A10:** 348-349
in x-ray diffraction **A10:** 329
in x-ray powder diffraction **A10:** 337
and x-ray spectrometers **A10:** 87
Bragg-Brentano diffractometers
geometry of **A10:** 337
Bragg-Brentano geometry
XRPD analysis **A10:** 337
Bragg-Brentano x-ray diffractometer **EM4:** 73
Braided composites *See also* Braiding
cost, comparative **EM1:** 519
properties **EM1:** 525-527
two-/three-dimensional **EM1:** 525-527
Braiding **EM1:** 519-528
application, automotive industry **EM1:** 833-835
in ceramic-ceramic composites **EM1:** 934
classifications **EM1:** 520-521
computer-aided **EM1:** 522-523
defined **EM1:** 5, 519-520 **EM2:** 7
and filament winding, compared **EM1:** 519
three-dimensional **EM1:** 523-527
two-dimensional **EM1:** 521-523, 525
Brake asperity "flash" temperatures **A18:** 570
Brake bands
powders used **A7:** 572
Brake drum, ductile iron
brittle fracture of **A11:** 370-371
Brake drums and discs
alloy cast iron for **M1:** 96
Brake effectiveness **A18:** 574, 576
Brake heat sink **A18:** 582
Brake life **A18:** 583
Brake lining **A18:** 569
Brake linings
joined by welding **A7:** 457
powders used **A7:** 572
rake-shoe components
economy in manufacture **M3:** 850
raking systems, antilock (ABS)
as automotive hybrid application **EL1:** 382
ale
defined **A8:** 2
ale indenter
defined **A18:** 4
amson's formula (emissivity) **A6:** 265
anch lines
wafer-scale integration **EL1:** 361
anched polymer **EM3:** 6
Branched polymers **EM1:** 5, 751 **EM2:** 7, 63
Branching **EM3:** 6
defined **EM2:** 7
in eutectic structures **A9:** 620
mer, and melt properties **EM2:** 62
as molecular structure **EM2:** 58
Brass *See also* Cartridge brass; Copper alloys, specific types; Copper and copper alloys; Copper-base alloys, specific types; Copper-zinc alloys
abrasion artifacts in **A9:** 33, 37
adhesive wear versus tool steel **A18:** 237-238
bainitic-like microstructures **A9:** 666
base metal solderability **EL1:** 677
bearing material systems **A18:** 745, 747
and beryllium copper, barrier platings for **EL1:** 679
bonded by polyamides and polyesters **EM3:** 82
brazeability **M6:** 1033
Brinell test block for **A8:** 88
broaching **A16:** 203, 204, 206
buffing *See* Brass, polishing and buffing of
cage material for rolling-element bearings **A18:** 503
capacitor discharge stud welding **M6:** 738
carbides for machining **A16:** 75, 108
cartridge, EDS and WDS analysis of **A10:** 530
castings, cleaning and finishing **M5:** 615-616
cermet tools applied **A16:** 92
chromium plating, hard **M5:** 171
cleaning and finishing processes **M5:** 611-621, 625-626
coloring solutions **M5:** 625-626
composition and properties **M6:** 401
compositions of various types **M6:** 546
contact bridge when sliding on silicon carbide surface **A18:** 236
cutting tool material selection based on machining operations **A18:** 617
damage dominated by shear fracture **A18:** 179
diamond abrasive for honing **A16:** 476
die castings
electrolytic cleaning of **M5:** 33, 35-36
emulsion cleaning of **M5:** 35
polishing and buffing compounds removed from **M5:** 10-11
polishing and buffing of **M5:** 112-114
die materials for blanking **M3:** 487
drilling **A16:** 220, 221, 227, 229
EDG wheels **A16:** 565
EDM electrode polarity **A16:** 558, 559, 561
electrochemical machining **A16:** 540
electrochemical machining tool **A16:** 533, 536, 537, 541
electroplated onto sleeve bearing liners **A9:** 567
electropolishing of **M5:** 305-08
emulsion cleaning of **M5:** 35
enamels **EM3:** 303
erosive attack of melt on die surface **A18:** 630, 631
fixturing for induction brazing **M6:** 971
flash welding **M6:** 558
forgings, cleaning and finishing **M5:** 611-613
form turning **A16:** 381
free-cutting **A16:** 297-299, 301
friction coefficient data **A18:** 71
gas metal arc welding **M6:** 416, 420
gas tungsten arc welding **M6:** 408-410
hardness and density of P/M materials **A16:** 882
honing stone selection **A16:** 476
induction heating energy requirements for metalworking **A4:** 189
induction heating temperatures for metalworking processes **A4:** 188
isolation of lead in **A10:** 173
lapping process **A16:** 499
laser cladding **A18:** 867
Brass, 60Cu-40Zn
lead and sulfur content and flaking **A16:** 281
lead-free, for flame head for oxy-fuel gas flame heating **A4:** 274
leaded free-cutting, ultrasonic inspection **A17:** 274-275
martensitic structures **A9:** 672-673
maximum strain level **A8:** 551
milling with PCD tooling **A16:** 110
Brass (continued)
multipoint cutting tools used **A16:** 59
nickel plating of **M5:** 20, 215-216, 238-240
electroless **M5:** 238-240
photochemical machining **A16:** 588
physical properties **M6:** 546
planing **A16:** 184
polishing and buffing of **M5:** 108, 112, 123-124
polishing damage in **A9:** 41
powder metallurgy materials, etching **A9:** 509
powder metallurgy parts, preparation for plating **M5:** 620-621
radial tangential turning **A16:** 380
radiographic absorption **A17:** 311
reaming **A16:** 239, 247, 248
recovery, by microelectrogravimetry **A10:** 200
relative hydrogen susceptibility **A8:** 542
resistance welding *See* Resistance welding of copper and copper alloys
Rockwell C and B scales for **A8:** 74
rod, cleaning and finishing **M5:** 612-614, 618
rubber bonding cross-link density **EM3:** 418
sawing **A16:** 362
for shallow forming dies **A18:** 633
shaping **A16:** 191
shear stresses and HP **A16:** 15
shielded metal arc welding **M6:** 425
shim, in torsional impact machine **A8:** 217
shot peening of **M5:** 145-146
spade drilling **A16:** 225
specific energy factors **A16:** 18
stud material **M6:** 735
substrate cure rate and bond strength for cyanoacrylates **EM3:** 129
tapping **A16:** 259
tapping, cold form **A16:** 266
tension and torsion effective fracture strain **A8:** 168
texture orientations **A10:** 360
thermal energy method of deburring **A16:** 577-578
thread milling **A16:** 269
thread rolling **A16:** 282, 288, 290-293
tool bit tool steels used **A16:** 57
tool life **A16:** 299
tool steel alloys used **A16:** 57, 58
tubing, cleaning and finishing **M5:** 613-614
turning **A16:** 135, 380, 381
turning operation with cermet tools **A16:** 94
turning with PCD tooling **A16:** 110
ultrasonic impact grinding machine cutting tools **A16:** 529
vapor degreasing of **M5:** 45, 53-54
wire for traveling-wire EDM
yellow, ultimate shear stress **A8:** 148
zinc, low and high **M6:** 554-556
Brass, 70Cu-30Zn
thermal properties **A18:** 42
Brass alloys
lubrication for tool steels **A18:** 738
Brass alloys, beta phase
massive transformations in **A9:** 655-656
"Brass chills" copper poisoning **A7:** 205
Brass coatings
steel wire **M1:** 263
Brass electrical contact material **A9:** 553
Brass mill
sheet and strip manufacturing process **A2:** 241-248
Brass plating **M5:** 285-287
aluminum and aluminum alloys **M5:** 603-605
ammonia used in **M5:** 286-287
anodes **M5:** 287
applications **M5:** 285
barrel (bulk) process **M5:** 285-286
carbonate used in **M5:** 286
color, alloy composition determining **M5:** 285
copper content **M5:** 285-286
current densities **M5:** 286
cyanide process **M5:** 285-287
cyanide-to-zinc ratio **M5:** 285-286
decorative **M5:** 285
efficiency of **M5:** 286
engineering **M5:** 285
equipment **M5:** 287
free cyanide **M5:** 285
gold-colored **M5:** 285-286

Brass plating (continued)
high-speed process **M5:** 286-287
impurities **M5:** 286
noncyanide process **M5:** 285
pH control **M5:** 286
proprietary additions **M5:** 286
solution compositions and operating conditions **M5:** 285-287
steel **M5:** 285
temperature **M5:** 286-287
yellow brass **M5:** 285
zinc content **M5:** 285-286
Brass powder alloys, specific types *See also* Brasses
B-126, flow rate through Hall and Carney funnels **A7:** 279
CZP-0010, mechanical properties **A7:** 470
CZP-0020, mechanical properties **A7:** 470
CZP-0030, mechanical properties **A7:** 470
CZP-0210, mechanical properties **A7:** 470
CZP-0230, mechanical properties **A7:** 470
Brass rack guide
rack and pinion steering column **A7:** 738
Brass, sintering
time and temperature **M4:** 796
Brass, specific types
70-30, relative hydrogen susceptibility **A8:** 542
70-30, rolled 60%, shear bands **A9:** 686
70-30, strain to produce deformation in cold rolled **A9:** 685-686
70-30, strain-hardening exponent and true stress values for **A8:** 24
70-30, tensile properties **A8:** 555
alpha-brass, deformation modes **A9:** 686
Brass striking
aluminum and aluminum alloys **M5:** 603-604
Brass(es) *See also* Copper-zinc alloys **A6:** 752
brazing **A6:** 931, 932, 934
cleaning solutions for substrate materials **A6:** 978
electronic applications **A6:** 998
flash welding **A6:** 247
friction welding **A6:** 152
gas-tungsten arc welding **A6:** 192
high-frequency welding **A6:** 252
induction brazing **A6:** 333, 335
inorganic fluxes for **A6:** 980
leaded, thermal expansion coefficient **A6:** 907
oxyacetylene welding **A6:** 281
plain, thermal expansion coefficient **A6:** 907
plasma arc cutting **A6:** 1170
precoated before soldering **A6:** 131
relative solderability **A6:** 134
relative solderability as a function of flux type **A6:** 129
relative weldability ratings, resistance spot welding **A6:** 834
resistance welding **A6:** 847
shielded metal arc welding **A6:** 755
solderability **A6:** 978
stud arc welding **A6:** 210, 218-219
torch brazing **A6:** 328
torch soldering **A6:** 351
ultrasonic welding **A6:** 326
Brass-plated wire **A1:** 281
Brasses *See also* Aluminum brasses; Bronzes; Copper; Copper alloy powders; Copper alloys; Copper alloys, specific types; Nickel silvers; Tin brasses; Yellow brasses .... **A3:** 1 • 22
56-2-10-12 *See* Copper alloys, specific types, C97300
63-1-1-35 *See* Copper alloys, specific types, C85700 and C85800
67-1-3-29 *See* Copper alloys, specific types, C85400
70-30 *See* Copper alloys, specific types, C26000
72-1-3-24 *See* Copper alloys, specific types, C85200
76-2$^1/_2$-6$^1/_2$-15 *See* Copper alloys, specific types, C84800
81-3-7-9 *See* Copper alloys, specific types, C84400

Brasses (continued)
82-4-14 *See* Copper alloys, specific types, C87500 and C87800
85-5-5-5 *See* Copper alloys, specific types, C83600
Admiralty brass *See* Copper alloys, specific types, C44300, C44400 and C44500
bushings **A11:** 470
cartridge, applications and properties **A2:** 300-302
cartridge, as solid-solution copper alloy **A2:** 234
cartridge brass, 70% *See* Copper alloys, specific types, C26000
cast, corrosion ratings **M2:** 390-391
castings in, African **A15:** 19
clock brass *See* Copper alloys, specific types, C34200 and C35300
clock brass, applications and properties **A2:** 308-309
as copper alloy powder **A7:** 121-122
copper-base structural parts from **A2:** 397
corrosion pitting fracture **A12:** 404
corrosion resistance **A13:** 610
dezincification **A13:** 131
dezincification of **A11:** 633
die castings, size **A2:** 346
dimensional change **A7:** 292
effect of hot pressing temperature and pressure on density **A7:** 504
effect of lithium stearate lubrication **A7:** 191
electrical and thermal conductivity **A7:** 742
engraver's brass *See* Copper alloys, specific types, C34200 and C35300
engraver's brass, applications and properties **A2:** 308-309
extra quality brass *See* Copper alloys, specific types, C26000
extra-high leaded brass *See* Copper alloys, specific types, C35600
extra-high-leaded brass, applications and properties **A2:** 310
failure mode domains **A13:** 153
flakes, particle size measurement **A7:** 225
forgeability and application **A14:** 255
forging brass *See* Copper alloys, specific types, C37700
forging brass, applications and properties **A2:** 312
forming limit diagrams for **A14:** 20
free-cutting ,applications and properties **A2:** 310-311
free-cutting brass *See* Copper alloys, specific types, C36000
free-cutting tube brass *See* Copper alloys, specific types, C33200
free-cutting yellow brass *See* Copper alloys, specific types, C36000
free-turning brass *See* Copper alloys, specific types, C36000
galvanic corrosion with magnesium **M2:** 607
gold bronze as **A7:** 593
heavy-leaded, applications and properties **A2:** 308-309
heavy-leaded brass *See* Copper alloys, specific types, C34200 and C35300
high, applications and properties **A2:** 306
high brass *See* Copper alloys, specific types, C33000
high copper yellow brass *See* Copper alloys, specific types, C85200
high strength yellow brass *See* Copper alloys, specific types, C86100, C86200, C86300 and C86500
high-leaded brass *See* Copper alloys, specific types, C34200, C35300 and C36000
high-leaded brass (tube) *See* Copper alloys, specific types, C33200
high-leaded naval brass, applications and properties **A2:** 321

Brasses (continued)
high-zinc, susceptibility to dezincification **A11:** 633
horizontal centrifugal casting **A15:** 296
laser cutting **A14:** 742
leaded high strength yellow brass *See* Copper alloys, specific types, C86400
leaded naval, applications and properties **A2:** 320-321
leaded nickel brass *See* Copper alloys, specific types, C97300
leaded red brass *See* Copper alloys, specific types, C83600
leaded semi-red brass *See* Copper alloys, specific types, C84400 and C84800
leaded, susceptibility to hot cracking **A11:** 450
leaded yellow brass *See* Copper alloys, specific types, C85200, C85400, C85700 and C85800
limiting draw ratios **A14:** 575
liquid embrittlement in **A11:** 27
in liquid mercury, crack propagation rate **A11:** 227, 232
low brass, 80% *See* Copper alloys, specific types, C24000
low brass, applications and properties **A2:** 299-300
low-leaded **A2:** 306
low-leaded brass *See* Copper alloys, specific types, C33500
low-leaded brass (tube) *See* Copper alloys, specific types, C33000
mechanical properties **A7:** 738
medium-leaded, applications and properties **A2:** 307-308, 309
medium-leaded brass, 62% *See* Copper alloys, specific types, C35000
medium-leaded brass, 64.5% *See* Copper alloys, specific types, C34000
medium-leaded naval brass, applications and properties **A2:** 320-321
microstructural analysis **A7:** 488-489
microstructures of **A9:** 551
naval, applications and properties **A2:** 319-322
naval brass *See* Copper alloys, specific types, C46400, C46500, C46600 and C46700
naval brass, antimonial *See* Copper alloys, specific types, C46600
naval brass, arsenical *See* Copper alloys, specific types, C46500
naval brass, high leaded *See* Copper alloys, specific types, C48500
naval brass, inhibited *See* Copper alloys, specific types, C46500, C46600 and C46700
naval brass, leaded *See* Copper alloys, specific types, C48200 and C48500
naval brass, medium leaded *See* Copper alloys, specific types, C48200
naval brass, phosphorized *See* Copper alloys, specific types, C46700
naval brass, uninhibited *See* Copper alloys, specific types, C46400
No. 1 yellow brass *See* Copper alloys, specific types, C85400
P/M parts **A7:** 737-739
plumbing goods brass *See* Copper alloys, specific types, C84800
as pressed and sintered prealloyed powders **A7:** 464
properties **A7:** 122
red brass, 85% *See* Copper alloys, specific types, C23000
red brass, applications and properties **A2:** 298-299
SCC of **A12:** 28, 36
SCC resistance **A13:** 615
semisolid application **A15:** 336
silicon brass *See* Copper alloys, specific types, C87500 and C87800

Brasses (continued)
silicon red brass *See* Copper alloys, specific types, C69400
sintering **A7:** 378-381
spinning brass *See* Copper alloys, specific types, C26000
spinning brass, applications and properties **A2:** 300-302
spring brass *See* Copper alloys, specific types, C26000
spring brass, applications and properties **A2:** 300-302
stress-corrosion cracking **A2:** 216
temper designations **M2:** 248-249
tin brass *See* Copper alloys, specific types, C41900
tin brass, applications and properties **A2:** 315
uninhibited naval brass, applications and properties **A2:** 319-320
white manganese brass *See* Copper alloys, specific types, C99700
yellow, applications and properties **A2:** 302-304
yellow brass *See* Copper alloys, specific types, C26800, C27000 and C33000
yellow brass, 65% *See* Copper alloys, specific types, C27000
yellow brass, 66% *See* Copper alloys, specific types, C26800
yellow, intermetallic inclusions in **A15:** 96
Brasses, specific types
33 wt% Zn, properties **A6:** 992
65-35, capacitor discharge stud welding **A6:** 222
70-30, capacitor discharge stud welding **A6:** 222
(hard), ultrasonic welding **A6:** 326
properties **A6:** 629
Braunauer-Emmett-Teller (BET)
equation **EM4:** 70
surface area given by adsorption isotherm **EM4:** 213
for surface area measurements of ceramic powders **EM4:** 27
method **EM4:** 70, 428
surface area analysis **EM4:** 272
Bravais lattice *See also* Crystal structure; Lattice **A3:** 1 • 10
defined **A9:** 706
Bravais lattices
defined **EL1:** 93, 95
Braycoat 815Z polymer **A18:** 156
Braze
definition **A6:** 1207 **M6:** 3
Braze 071
brazing, composition **A6:** 117
wettability indices on stainless steel base metals **A6:** 118
Braze 580
brazing, composition **A6:** 117
wettability indices on stainless steel base metals **A6:** 118
Braze 630
brazing, composition **A6:** 117
wettability indices on stainless steel base metals **A6:** 118
Braze 655
brazing, composition **A6:** 117
wettability indices on stainless steel base metals **A6:** 118
Braze 852
brazing, composition **A6:** 117
wettability indices on stainless steel base metals **A6:** 118
Braze cladding **M6:** 804
Braze filler metals
gold, platinum, and palladium **A11:** 450
nickel- and cobalt-base alloys **A11:** 450
phosphide embroiling in **A11:** 452
Braze interface
definition **A6:** 1207
Braze runoff
by liquid penetrant inspection **A17:** 86
Braze strength
first-level package **EL1:** 991-992
Braze welding **A6:** 124-125
advantages **A6:** 715-716
cast irons **A6:** 715-716
Braze welding (continued)
composition and properties of rods and electrodes used with cast irons **A6:** 716
copper alloys to dissimilar metals **M6:** 424
definition **A6:** 1207 **M6:** 3
equipment **A6:** 716
filler metals used **A6:** 716
limitations **A6:** 716
oxyacetylene **M6:** 604
techniques **A6:** 716
to solve problems in joining thin sections by oxyfuel gas welding **A6:** 288
Brazeability *See also* Brazing
aluminum and aluminum alloys **A2:** 13
definition **M6:** 3
materials and properties **A11:** 450
wrought aluminum alloys **A2:** 30-32
Brazeability and solderability of engineering materials **A6:** 617-636
aluminum alloys **A6:** 627-628
brazing **A6:** 627-628
soldering **A6:** 628, 631, 632
applications **A6:** 617-619
atmospheres selection **A6:** 622, 623, 626
austenitic stainless steels **A6:** 625-626
carbides **A6:** 635-636
cast irons **A6:** 626-627
ductile iron **A6:** 626
gray iron **A6:** 626-627
malleable iron **A6:** 626
ceramic materials **A6:** 635-636
characteristics of engineering materials **A6:** 623-636
cobalt-base alloys **A6:** 634
copper **A6:** 628-631
copper alloys **A6:** 628-631
dissimilar material joints **A6:** 619
duplex stainless steels **A6:** 626
ferritic stainless steels **A6:** 626
fluxes **A6:** 621-622, 625
graphite **A6:** 635
heat-resistant alloys **A6:** 632-633
heating method effect **A6:** 624, 629
interfacial reactions **A6:** 619-620, 621
joint clearance **A6:** 620, 621, 623
liquid filler flowability **A6:** 620-621
low-carbon steels **A6:** 624
martensitic stainless steels **A6:** 626
materials selection **A6:** 619-623
metallugical considerations **A6:** 622-623
molybdenum **A6:** 634
mutual dissolution and erosion **A6:** 621, 624, 625
nickel-base alloys **A6:** 631-632
niobium **A6:** 634
nitride ceramics **A6:** 636
oxide ceramics **A6:** 636
precipitation-hardening stainless steels **A6:** 626
refractory metals **A6:** 634-635
requirements **A6:** 617-619
spreading **A6:** 619, 620, 622, 624, 625, 626, 628
stainless steels **A6:** 625-626
tantalum **A6:** 634
titanium **A6:** 633-634
titanium alloys **A6:** 633-634
tool steels **A6:** 624-625
tungsten **A6:** 634-635
vapor pressure **A6:** 621, 625
wetting **A6:** 617, 619, 620, 624, 625, 626, 627-628, 629, 631, 632, 635, 636
Brazed assemblies *See also* Weldment(s)
flaw types **A17:** 602-603
inspection methods **A17:** 603
joint integrity **A17:** 603
liquid penetrant inspection **A17:** 604
pressure testing **A17:** 604
proof testing **A17:** 604
radiographic inspection **A17:** 604
thermally quenched phosphor inspection **A17:** 604-605
ultrasonic inspection **A17:** 604
visual inspection **A17:** 603-604
Brazed honeycomb panels *See also* Honeycomb structures
magnetic printing inspection **A17:** 126
Brazed joints *See also* Brazing; Joints **A13:** 876-886
beryllium-to-Monel, embrittlement **A13:** 879
Brazed joints (continued)
chemical etching **A9:** 401
contrasting by interference layers **A9:** 60
electrolytic polishing **A9:** 400
flaws in **A17:** 602-603
niobium, shear test data **A13:** 885
silver-copper-palladium filler alloy **A13:** 880
as source, nonrelevant indications **A17:** 106
stainless steel to stainless steel
vacuum deposition of interference films **A9:** 148
Brazed joints, evaluation and quality control of **A6:** 1117-1123
brazing process planning and control **A6:** 1118-1120
case studies **A6:** 1121-1123
design testing, evaluation, and feedback **A6:** 1120-1121
overall quality system **A6:** 1117-1118
quality standards for brazing and brazing processes **A6:** 1118
specifications for overall quality systems **A6:** 1117-1118
Brazed joints, failures of **A11:** 450-455
examples of **A11:** 453-455
major defects **A11:** 451-453
testing and inspection **A11:** 451
Brazement
definition **A6:** 1207 **M6:** 3
Brazer
definition **A6:** 1207 **M6:** 3
Brazing *See also* Brazeability; Brazing filler metals; Joining; specific processes; Welding **A6:** 109-110 **A7:** 837-841
advantages **A6:** 109
alloy powders, composition and properties **A7:** 838-839
aluminum **M2:** 199-201
aluminum alloys **A6:** 828, 937-940
with amorphous materials and metallic glasses **A2:** 819
applications
aerospace **A6:** 387
automotive **A6:** 393, 395
atmosphere types per AWS specification B2.2 **A6:** 622, 628
atmospheres **A7:** 457
as attachment method, sliding contacts **A2:** 842
automation and mass production **A6:** 110
basic requirements **A6:** 109
of beryllium **A2:** 683
bonds, ultrasonic inspection **A17:** 232
chemical alloy powders, analysis and sampling **A7:** 249
conditions **A11:** 450-451
copper alloys **M2:** 449-453
corrosion forms **A13:** 876-879
crack nucleation **A6:** 110
debond, as planar flaw **A17:** 50
defined **A11:** 450 **A13:** 876
definition **A6:** 1207 **M6:** 3
dip brazing **A6:** 110
dispersion-strengthened aluminum alloys **A6:** 543
dissimilar metal joining **A6:** 822
dissimilar metals, embrittlement **A6:** 622, 623, 629
distortion **A6:** 110
ductile and brittle fractures from **A11:** 92-94
ductile iron castings **A15:** 664-665
of electrical contact materials **A2:** 841
electrical resistance alloys **A2:** 822
failure mechanisms **EL1:** 1045
ferritic malleable iron **A15:** 693
filler metal **A6:** 109
filler metal atomized powders **A7:** 837
-flux residues, SCC failures produced by **A11:** 453
flux types per AWS specification B2.2 **A6:** 622, 627
fluxes **A7:** 840
fluxless **A11:** 451
furnace brazing **A6:** 110
furnaces **A7:** 457
galvanic corrosion **A13:** 876
heat-affected zone **A6:** 110
history and development **A6:** 109

Brazing (continued)
induction brazing **A6:** 110
joint strength **A6:** 109
with Kovar materials **EL1:** 734
leaded and leadless surface-mount joints **EL1:** 734
limitations **A6:** 110
malleable cast irons **M1:** 66-67, 71
of malleable iron **A1:** 76, 83-84
mechanical properties of base metals **A6:** 110
mechanics of **A6:** 110
metal powders for **A7:** 837-841
methods **A11:** 450 **A13:** 880
molybdenum **A2:** 564
as package sealing method **EL1:** 237-239
for particulate depositions in metallizing **EM4:** 543
pearlitic/martensitic malleable iron **A15:** 696-697
powder types **A7:** 837
preferential attack **A13:** 876
procedures **A7:** 457
process selection **A13:** 879-880
processes, failure mechanisms **EL1:** 1045
refractory metals and alloys **A2:** 564
repair, distortion from **A11:** 141-142
safety precautions **A6:** 1191, 1202 **M6:** 58
as secondary operation **A7:** 457
sheet metals **A6:** 398-399
silver metallization onto alumina **EM4:** 544
in sintering process **A7:** 340
in space and low-gravity environments **A6:** 1023
steps **A6:** 110
tantalum alloys **A6:** 580
temperatures, corrosion tests, wettability Zircaloy-2 sheet **A13:** 881-883
titanium-matrix composites **A6:** 527
to solve problems in joining thin sections by oxyfuel gas welding **A6:** 288
torch brazing **A6:** 110
tungsten alloys **A6:** 581
use of fluxes or salts **A11:** 451
vs. other welding processes **A6:** 110
vs. soldering **A6:** 109, 110
vs. welding **A6:** 109
Brazing alloy
definition **A6:** 1207
Brazing alloy powders, specific types
aluminum-silicon alloys, compositions and properties **A7:** 839
cobalt alloys, compositions and properties **A7:** 838
copper, compositions and properties **A7:** 839
copper-phosphorus alloys, compositions and properties **A7:** 839
gold alloys, compositions and properties **A7:** 839
nickel alloys, compositions and properties **A7:** 838
silver alloys, compositions and properties **A7:** 838
Brazing alloys **EM3:** 40 **EM4:** 489-490, 491
Brazing alloys, aluminum
compositions of **A9:** 359
Brazing consumables, selection criteria **A6:** 903-905
filler metals **A6:** 904-905
joint considerations **A6:** 903-904
process stages **A6:** 903
product forms **A6:** 903-904
rapid solidification (RS) technology **A6:** 904
Brazing filler metal
definition **A6:** 1207 **M6:** 3
Brazing filler metals
refractory metals and alloys **A2:** 564
silver-base, properties **A2:** 702
Brazing, fundamentals of **A6:** 114-125
base-metal characteristics **A6:** 116, 117
braze welding **A6:** 124-125
Brazing, fundamentals of (continued)
capillary attraction **A6:** 114
definition **A6:** 114
developments **A6:** 114
dip **A6:** 122-123
electron-beam brazing **A6:** 123-124
elements of the brazing process **A6:** 116-120
exothermic **A6:** 123
filler-metal characteristics **A6:** 117-119
filler-metal flow **A6:** 116-117
furnace **A6:** 121
heating methods **A6:** 120-125
induction brazing **A6:** 121-122
infrared (quartz) brazing **A6:** 123, 124
joint design and clearance **A6:** 116, 119-120
laser brazing **A6:** 123-124
manual torch **A6:** 121, 123
microwave brazing **A6:** 124
physical principles **A6:** 114-116
protection by an atmosphere or flux **A6:** 116
rate and source of heating **A6:** 116
resistance **A6:** 123
salt-bath **A6:** 121, 122
surface preparation **A6:** 116, 119
temperature and time **A6:** 116, 117-118, 120
torch, manual **A6:** 121, 122, 123
wetting **A6:** 114-116
Brazing joints
mechanical properties **A7:** 841
powders used **A7:** 573
Brazing of aluminum alloys **A6:** 937-940 **M6:** 1022-1032
alloy brazing **M6:** 1030
assembly **A6:** 938-939 **M6:** 1026
base metals **A6:** 937 **M6:** 1022
brazing sheet **A6:** 937 **M6:** 1023-1024
brazing to other metals **M6:** 1030-1031
copper **M6:** 1030
ferrous metals **M6:** 1030
nonferrous metals **M6:** 1031
corrosion, resistance to **M6:** 1032
dip brazing **A6:** 939 **M6:** 1026-1027
equipment **M6:** 1026
modifications **M6:** 1030
technique **M6:** 1026-1027
filler metals **A6:** 937, 938, 939 **M6:** 1022-1023
finishing **M6:** 1031-1032
flux removal **M6:** 1031
flux removal techniques **A6:** 939
fluxes **A6:** 937-938 **M6:** 1023-1025
stopoffs **M6:** 1024-1025
fluxless vacuum brazing **A6:** 939
furnace brazing **A6:** 939 **M6:** 1027-1029
joint design **A6:** 938 **M6:** 1025
mechanical properties **M6:** 1032
motion brazing **M6:** 1030
postbraze heat treatment **A6:** 939-940
prebraze cleaning **A6:** 938 **M6:** 1025
resistance brazing **M6:** 1030
safety **M6:** 1032
safety precautions **A6:** 940
silicon diffusion **M6:** 1023
specialized processes **M6:** 1030
to copper **A6:** 627
to ferrous alloys **A6:** 627
to other nonferrous metals **A6:** 627-628
torch brazing **A6:** 939 **M6:** 1029-1030
equipment **M6:** 1029
technique **M6:** 1029-1030
vacuum brazing, fluxless **M6:** 1029
equipment **M6:** 1029
technique **M6:** 1029
Brazing of carbon and graphite **M6:** 1061-1063
applications **M6:** 1062
brazing characteristics **M6:** 1062
brazing to dissimilar metal **M6:** 1062
thermal expansion **M6:** 1062
wettability **M6:** 1062
filler metals **M6:** 1062-1063
Brazing of carbon and graphite (continued)
heating methods **M6:** 1063
material production **M6:** 1061-1062
Brazing of carbon steels **A6:** 906-910
base-metal brazeability **A6:** 906
cleaning procedures **A6:** 908-909
dissimilar metals **A6:** 906
filler metals **A6:** 906-908
fixturing procedures **A6:** 909
flux/atmosphere procedures **A6:** 909
furnace brazing **A6:** 910
heat-treatment requirements **A6:** 908
heating methods **A6:** 909-910
induction brazing **A6:** 909-910
preforms **A6:** 909
salt-bath brazing **A6:** 908, 910
torch brazing **A6:** 909
Brazing of cast irons **A6:** 906-910 **M6:** 996-1000
applicability **M6:** 996
base-metal brazeability **A6:** 906
brazeability **M6:** 996
ductile iron **M6:** 996
gray iron **M6:** 996
malleable iron **M6:** 996
cleaning procedures **A6:** 908-909
dip brazing in fused salt bath **M6:** 999-1000
dissimilar metals **A6:** 906
filler metal **M6:** 996
filler metals **A6:** 906-908, 909
fixturing procedures **A6:** 909
flux **M6:** 996
flux/atmosphere procedures **A6:** 909
furnace brazing **A6:** 909, 910
fused salt cleaning **M6:** 996-997
heat-treatment requirements **A6:** 908
heating methods **A6:** 909-910
induction brazing **A6:** 909-910
preforms **A6:** 909
preheating **M6:** 997
preparation of castings **M6:** 996
joint designs **M6:** 996
surface preparation **M6:** 996
production applications **M6:** 997-999
salt-bath brazing **A6:** 910
strength, retention of **M6:** 999
torch brazing **A6:** 909
Brazing of ceramic and ceramic-to-metal joints **A6:** 948-958
applications **A6:** 953
brazing parameters **A6:** 954-956
ceramic materials **A6:** 948-950
categories **A6:** 948
processing **A6:** 949-950
ceramic-to-metal joints **A6:** 956-958
direct brazing of ceramics with metallic filler metals **A6:** 951-952
direct brazing of graphitic materials **A6:** 957-958
direct brazing with nonmetallic glasses **A6:** 952-953
eutectic brazing **A6:** 953
filler metals **A6:** 949, 950, 951-952, 953, 954, 956, 957-958
gas-metal eutectic brazing **A6:** 953
graphitic materials **A6:** 950
chemical-vapor infiltration (CVI) **A6:** 950
liquid impregnation **A6:** 950
indirect brazing **A6:** 950-951
molybdenum-manganese (Mo-Mn) process **A6:** 951
partial transient liquid-phase joining **A6:** 953
procedure development **A6:** 950-956
reaction bonding **A6:** 949
reaction sintering **A6:** 949
surface preparation **A6:** 953-954
Brazing of copper and copper alloys **A6:** 628-630, 931-935 **M6:** 1033-1048
applications **A6:** 933-934, 935
atmospheres **A6:** 931, 933, 934
brazeability **M6:** 1033-1034

Brazing of copper and copper alloys (continued)
brazing fluxes **A6:** 932 **M6:** 1035
brazing processes **A6:** 932-935
dip brazing **M6:** 1048
minimizing distortion **M6:** 1048
filler metals **A6:** 931, 932, 933, 934 **M6:** 1034-1035
furnace brazing **A6:** 933 **M6:** 1035-1037
accelerated heating **M6:** 1036-1037
advantages **M6:** 1036
assembly **M6:** 1037
furnace atmosphere **M6:** 1037
furnaces **M6:** 1036
limitations **M6:** 1036
temperatures **M6:** 1036
venting **M6:** 1037
induction brazing **A6:** 934-935 **M6:** 1042-1045
advantages **M6:** 1042
avoiding flux use **M6:** 1045
cost of brazing **M6:** 1045
design of inductors **M6:** 1043
limitations **M6:** 1042-1043
mass production **M6:** 1043-1044
power supplies **M6:** 1043
use of fluxes **M6:** 1044-1045
joint clearance **A6:** 932 **M6:** 1035
joint design **A6:** 932
process selection **M6:** 1035
resistance brazing **A6:** 935 **M6:** 1045-1048
high production **M6:** 1048
leads to commutator bars **M6:** 1046-1047
multiple-strand copper wire **M6:** 1045-1046
portable machines **M6:** 1047-1048
salt-bath dip brazing **A6:** 935
torch brazing **A6:** 933-934 **M6:** 1037-1042
applications **M6:** 1037
filler metals **M6:** 1038-1039
fluxes **M6:** 1036
fuel gases **M6:** 1039-1042
joint design **M6:** 1041
manual brazing **M6:** 1037
mechanized and automatic **M6:** 1040
precision torch brazing **M6:** 1041
process selection **M6:** 1037
Brazing of heat-resistant alloys **A6:** 924-929
brazing filler metals **A6:** 924
chemical cleaning methods **A6:** 925-926
cobalt-base alloys **A6:** 928-929
cobalt-based alloys **M6:** 1021
controlled atmospheres **A6:** 927 **M6:** 1016-1018
inert gases **M6:** 1017
pure dry hydrogen **M6:** 1016
vacuum **M6:** 1017
filler metals **M6:** 1014
product forms **M6:** 1014-1015
fixturing **A6:** 926-927 **M6:** 1016
hydrogen fluoride cleaning **A6:** 926
mechanical cleaning **A6:** 926
microstructure of filler metals **A6:** 924, 925, 926
nickel flashing **A6:** 926
nickel-base alloys **A6:** 927-928
nickel-based alloys **M6:** 1018-1020
oxide dispersion-strengthened alloys **M6:** 1020-1021
oxide-dispersion-strengthened (ODS) alloys **A6:** 924, 928
powder metallurgy (P/M) products **A6:** 924, 925
product forms **A6:** 924-925
surface cleaning and
chemical cleaning methods **M6:** 1015-1016
mechanical cleaning **M6:** 1016
nickel flashing **M6:** 1016
preparation **M6:** 1015-1016
surface cleaning and preparation **A6:** 925-926
razing of low-alloy steels **A6:** 624, 924, 929-930
atmospheres **A6:** 930
brazing filler metals **A6:** 929-930
fixturing **A6:** 930
fluxes **A6:** 930
precleaning **A6:** 930
razing of low-carbon steels **A6:** 624
razing of precious metals **A6:** 931, 935-936
applications **A6:** 936
atmospheres **A6:** 936
brazing processes **A6:** 936
filler metals **A6:** 935-936
for gold jewelry brazing applications **A6:** 936
fluxes **A6:** 936
Brazing of precious metals (continued)
furnace brazing **A6:** 936
induction brazing **A6:** 936
material composition **A6:** 935
resistance brazing **A6:** 936
torch brazing **A6:** 936
Brazing of stainless steels **A6:** 911-923
applicability **A6:** 911 **M6:** 1001
applications **A6:** 920
atmosphere and dewpoint **A6:** 622
austenitic, second-phase precipitation **A6:** 622, 625
brazeability **A6:** 911
brazing filler metal **A6:** 911-913, 914, 915, 919, 920, 922
cobalt **A6:** 913
copper **A6:** 911, 913, 917, 920
gold **A6:** 911, 913, 915-916, 920
nickel **A6:** 911, 913, 915, 917, 920
silver **A6:** 911-913, 914, 915, 920
carbide precipitation **A6:** 913
chromium oxide formation **A6:** 911
dip brazing in a salt bath **A6:** 911, 921-922
dip brazing in salt bath **M6:** 1012
electron beam brazing **M6:** 1012-1013
applications **M6:** 1012-1013
filler metal brazing and service temperatures **A6:** 631
filler metals **M6:** 1001-1004
fluxes **A6:** 913-914 **M6:** 1004
furnace atmospheres **M6:** 1004
furnace brazing **A6:** 911, 913, 914-918 **M6:** 1004-1010
in air atmosphere **A6:** 919-920 **M6:** 1009
in argon **A6:** 915, 919 **M6:** 1008-1009
in dissociated ammonia **A6:** 915, 918-919 **M6:** 1007-1008
furnace brazing in dry hydrogen **A6:** 915, 916, 917
in hydrogen **M6:** 1004-1007
in vacuum **M6:** 1009-1010
in vacuum atmosphere **A6:** 920-921
high-energy-beam brazing **A6:** 922-923
induction brazing **A6:** 911, 921, 922 **M6:** 1011-1012
atmospheres **M6:** 1011
in vacuum **M6:** 1011-1012
process fundamentals **M6:** 1001
inclusions and surface contaminants **M6:** 1001
torch brazing **A6:** 911, 914 **M6:** 1010-1012
filler metals **M6:** 1010
flame adjustment **M6:** 1010
flux **M6:** 1010
vacuum brazing **M6:** 1009-1010
effect of filler-metal composition **M6:** 1010
Brazing of tool steels **A6:** 624-625, 924, 929-930
atmospheres **A6:** 930
brazing filler metals **A6:** 929-930
fixturing **A6:** 930
fluxes **A6:** 930
precleaning **A6:** 930
Brazing operator
definition **A6:** 1207 **M6:** 3
Brazing procedure
definition **M6:** 13
Brazing salts **A6:** 336-337, 338
Brazing sheet
definition **M6:** 3
Brazing technique
definition **M6:** 3
Brazing temperature
definition **M6:** 3
Brazing temperature range
definition **M6:** 3
Brazing with clad brazing materials **A6:** 347-348
aluminum alloys **A6:** 347
applications **A6:** 347, 348
cladding materials **A6:** 347-348
advantages **A6:** 348
fabrication of **A6:** 347-348
copper **A6:** 347
definition of clad brazing materials **A6:** 347
design and manufacturing considerations **A6:** 348
filler metals **A6:** 347, 348
fluxes **A6:** 347, 348
formation of clad brazing material **A6:** 347
Brazing with clad brazing materials (continued)
stainless steels **A6:** 347
steel **A6:** 347
titanium **A6:** 347
"Break-in" coatings **A18:** 875
Break-in cycle *See* Wear-in
Breakage
cemented carbides **A7:** 779
Breakage, cold
as casting defects **A11:** 383
Breakage, die
in drawing **A14:** 337
Breakaway oxidation
scales **A13:** 72
Breakaway torque *See* Starting torque
Breakdown
defined **A14:** 2
ingot, for stainless steel forging **A14:** 222-223
of nickel-base alloys **A14:** 261
Breakdown potential
defined **A13:** 2
Breakdown voltage *See also* Arc resistance; Dielectric breakdown voltage; Dielectric strength; Electrical breakdown
defined **EM2:** 7
Breaker cores
and riser necks **A15:** 587-588
Breaker plate
defined **EM2:** 7
Breaking
extension, defined **EM1:** 5
factor, defined **EM1:** 5
length, defined **EM1:** 5
Breaking extension **EM3:** 6
defined **EM2:** 7
Breaking factor **EM3:** 6
defined **EM2:** 7
Breaking, final
of specimens **A12:** 77
Breaking length
defined **EM2:** 7
Breaking load
defined **A8:** 2
Breaking radiation *See* Bremsstrahlung radiation
Breaking strength (stress) **EM3:** 33
average **EM3:** 34
Breaking stress *See* Rupture stress
Breakout *See also* Fiber breakout **EM3:** 6
defined **EM1:** 5 **EM2:** 7
Brearley, H
as metallurgist **A15:** 32
Breather **EM1:** 5, 644
defined **EM2:** 7
Breather cloth *See* Vent cloth
Breathers
powder used **A7:** 573
Breathing *See also* Permeability
defined **EM1:** 5 **EM2:** 7
Brehmsstrahlung radiation **A18:** 446
Brehmsstrahlung x-rays **A18:** 378
Bremmstrahlung radiation *See also* Continuum; Radiation
defined **A10:** 83, 325-326
as inelastic scattering process **A10:** 433
K_{α} aluminum or magnesium x-ray lines as filter for **A10:** 570
and synchrotron radiation, compared **A10:** 411
as x-ray source for EXAFS **A10:** 411
Bremsstrahlung **EM4:** 558, 579
in computed tomography **EM4:** 620
Bremsstrahlung rays, defined *See also* X-rays **A17:** 298
Bremsstrahlung x-ray spectrum **A9:** 92
Brenner nickel-phosphorus alloy plating bath **M5:** 204
Brewery industry **A13:** 1221-1225
corrosion control methods **A13:** 1221-1223
equipment **A13:** 1223-1224
plant structures **A13:** 1224-1225
Brewster angle
effect in microwave inspection **A17:** 204
Brick
friction coefficient data **A18:** 75
as natural fiber reinforced matrix composite **EM1:** 117
uniaxial strength **EM4:** 591

Brick Institute of America (BIA) **EM4:** 947, 948, 951
Brick linings **A13:** 455, 1153
Brick plumbum materials
applications **A2:** 555
Bridge (dental) alloys
of precious metals **A2:** 696
Bridge box girder
cracked web **A11:** 710-712
Bridge comparator for microscopes **A9:** 83, 85
Bridge components *See also* Bridge components, failures of
box-girder webs **A11:** 710-712
cracks and defects in **A11:** 708-710
floor-beam-girder connection plates **A11:** 712
large initial defects and cracks in **A11:** 708-710
low fatigue strength of **A11:** 707-708
materials for **A11:** 515
multiple-girder diaphragm **A11:** 712-714
tied-arch floor beams **A11:** 714
Bridge components, failures of **A11:** 707-714
cracking categories **A11:** 707
details and defects **A11:** 707-711
out-of-plane distortion **A11:** 711-714
Bridge formation
in electrical contact materials **A2:** 841
Bridge reamers **A16:** 241, 245
Bridge sites
number with cracking **A11:** 707
Bridge steels
Charpy toughness requirements for **A8:** 265
crack arrest toughness of **A8:** 284-286
impact *K* testing of **A8:** 453
toughness criteria **A8:** 264-265, 453
Bridge unbalance system
eddy current inspection **A17:** 177-178
Bridge wheel, steel crane
fracture of **A11:** 527-528
Bridge wire **M1:** 272
Bridge wire, galvanized
coating weight **M1:** 263
description **M1:** 264
mechanical properties **M1:** 265
Bridge wires, exploding
with symmetric rod impact test **A8:** 204
Bridge-type coordinate measuring machines **A17:** 21
Bridgeman method
ferrite processing **EM4:** 1163
Bridges *See also* Impedance bridge **A6:** 366, 375-377
acoustic emission inspection **A17:** 290
aluminum and aluminum alloys **A2:** 9
codes governing **M6:** 824
corrosion in **A13:** 1299
detector, magabsorption **A17:** 149-150
impedance, typical **A17:** 176
induction, eddy current inspection **A17:** 178
unbalance system, eddy current inspection **A17:** 177-178
Wheatstone, strain gage **A17:** 450
Bridges, Robert
as early founder **A15:** 24
Bridgesize
definition **M6:** 3
Bridgewire
SEM micrograph of corrosion and corrosion product in **A10:** 511
Bridging *See also* Dimensional control; Solidification; Void
as conformal coating application **EL1:** 762
defined **A7:** 2 **A15:** 2 **EM1:** 5 **EM2:** 7
and fluxes **EL1:** 647
from excess joint solder **EL1:** 691
interconnections as **EL1:** 12
laser/IR inspected **EL1:** 942
in loose powder compaction **A7:** 298
in top-poured ingots **A11:** 315
Bridgman correction factor
assumptions of **A8:** 25-26
effect on true stress/true strain curve **A8:** 26
relationship to true tensile strain **A8:** 26
Bridgman method of growing single crystals *See also* Unidirectional solidification **A9:** 607
Bright acid tin
from codeposited organics **EL1:** 679
Bright annealing **A1:** 280
defined **A9:** 3
steel wire **M1:** 262
Bright bronze plating solution **M5:** 289
Bright dipping
aluminum and aluminum alloys **M5:** 579-582
cadmium plating **M5:** 263, 269
copper and copper alloys **M5:** 611-613, 619-620
nickel alloys **M5:** 670-671
solution compositions and operating conditions **M5:** 611-612, 620
Bright dips
copper and copper alloy forgings **A14:** 258
Bright electrodeposited copper
applications **M5:** 159
Bright field
defined **EL1:** 1067
Bright finish
steel wire **M1:** 261
Bright finishing
aluminum and aluminum alloys **M5:** 574, 576
Bright flake
wrought aluminum alloys **A12:** 415
Bright flake in aluminum alloys **A9:** 358
Bright nickel plating **M5:** 199, 204-206
Bright rolling
copper and copper alloys **M5:** 615
Bright soft wire
definition **M1:** 262
Bright stock
defined **A18:** 4
Bright-dark oscillations in transmission electron microscopy **A9:** 112
Bright-field illumination **A9:** 76
coarse-grain iron alloy **A12:** 93-94
and color etching **A9:** 136
and dark-field, SEM images, compared **A12:** 92
defined **A9:** 3
for slip **A12:** 121
used for porcelain enameled sheet steel **A9:** 198
Bright-field images
of annealing twin in rutile **A10:** 443
in ceramic containing crystalline and amorphous phases **A10:** 445
fcc matrix (austenite) **A10:** 440
for image analyzer microscopes **A10:** 310
iron-base superalloy **A10:** 442
of polycrystalline aluminum **A10:** 444
precipitates on grain boundary, iron-base superalloy **A10:** 447
transmission electron microscopy **A10:** 441-446, 457
of unknown phase/particle **A10:** 457
Bright-field transmission electron microscopy **A9:** 103
beam diagram **A9:** 104
of dislocations **A9:** 113
intensities calculated using the dynamical theory **A9:** 111
two-beam thickness contours **A9:** 112
Bright-finishing wrought aluminum alloy **A2:** 37
Brightener
defined **A13:** 2
Brightener-levelers
use in nickel plating **M5:** 205
Brighteners
auxiliary **M5:** 205
cadmium plating using **M5:** 256-257
Brighteners (continued)
copper plating using **M5:** 168
nickel plating process **M5:** 204-206
organic **A11:** 45
zinc plating using **M5:** 246, 250
Brightening
aluminum and aluminum alloys **M5:** 579-582, 590-591, 596-597, 603-604, 606-608, 610
chemical *See* Chemical brightening
electrolytic *See* Electrolytic brightening
process selection **M5:** 581-582
Brightness
of image, machine vision process **A17:** 33
of response, inspection materials **A17:** 678
Brine **A13:** 2, 1233-1234
geological analysis **A10:** 665-667
as ion chromatography solution **A10:** 658
Brine quenching
advantages and disadvantages **M4:** 36, 41
contamination **M4:** 43
cooling rates **M4:** 41-42
temperature **M4:** 36, 37, 42-43
Brine-heater shell
fracture at welds **A11:** 637
Brinell hardness **A7:** 312 **A16:** 15
abbreviation **A8:** 724
abbreviation for **A10:** 690
Al alloys **A16:** 774
of gray iron **A1:** 16-17
number (HB), defined **A11:** 1
PCBN tools **A16:** 113
symbol for **A11:** 796
test, defined **A11:** 1
Brinell hardness balls
tolerances **A8:** 86
Brinell hardness number *See also* Hardness number **A8:** 2, 86
equivalent hardness numbers, steel **A8:** 111
equivalent Rockwell B hardness numbers steel **A8:** 109-110
maximum range **A8:** 84
Rockwell C hardness conversions for steel **A8:** 110
Vickers hardness conversions, steel **A8:** 112-113
Brinell hardness number (HB) **A1:** 40 **A18:** 4
Brinell hardness test
for casting alloys **A17:** 521
defined **A18:** 4
Brinell hardness testing *See also* Brinell hardness number (HB); Hardness testing; Knoop hardness test; Rockwell hardness test; Scleroscope hardness test; Vickers hardness test **A8:** 84-89
applications **A8:** 89, 102
defined **A8:** 2, 84
deformed grid pattern **A8:** 72
depth **A8:** 102
indentation process **A8:** 85
indenters **A8:** 84, 102
load selection **A8:** 84
method of measurement **A8:** 102
minimum thickness requirements **A8:** 86
of nonferrous metals **A8:** 80
precautions and limitations **A8:** 85-86
and Rockwell hardness test **A8:** 74
as static indentation test **A8:** 71
surface preparation **A8:** 102
techniques compared **A8:** 102
testing machines **A8:** 86-88
verification of loads, indenters, and microscopes **A8:** 88
Brinell indentation diameter (BID) **A1:** 40
Brinell pressure **A18:** 682
Brinell test
for ductile iron **A1:** 40
for gray iron **A1:** 18-19, 30
Brinelling *See also* False Brinelling
in bearings **A11:** 490
defined **A8:** 2 **A11:** 1 **A18:** 4
false, defined **A11:** 4

Brinelling (continued)
true and false, compared in bearings **A11:** 499-500
Briquet roll
grinding cracks in **A11:** 362
Briquet(te)
defined **A7:** 2
nickel powder **A7:** 141-142
Briquets
as samples **A10:** 93-94
Britannia metal *See* Tin alloys, specific types, pewter
British Anti Lewisite (BAL)
as chelator **A2:** 1235-1236
British Iron and Steel Research Association (BISRA)
cam plastometer **A8:** 194
British standards (BS) for steels **A1:** 158
compositions of **A1:** 182-186
cross-referenced to SAE-AISI steels **A1:** 166-174
British Standards Institution **A8:** 724
British thermal unit **A8:** 724
Brittle *See also* Brittlefractures
defined **A11:** 1 **A12:** 173
materials, erosion and wear failure in **A11:** 156
Brittle cathode process **A7:** 72
Brittle coating-drilling technique **A6:** 1095
Brittle corrosion fatigue cracking **A8:** 408
Brittle crack propagation
defined **A8:** 2 **A11:** 1
Brittle erosion behavior
defined **A8:** 2 **A11:** 1
Brittle fracture *See also* Embrittlement; Fracture; Premature fracture **A7:** 58-59 **A8:** 2 **A13:** 2, 161 **M6:** 881-883, 885
in creep tests **A8:** 353-354
defined **A9:** 3
effect of residual stress **M6:** 881-882, 886
effect of stress relieving **M6:** 882-883, 886
of FIM samples **A10:** 587
in micro-fracture mechanics **A8:** 465-466
in repeated tension test **A8:** 353-354
and stress-corrosion cracking **A8:** 495
Brittle fracture failure
thermal **EL1:** 56
Brittle fracture transition
HSLA steels **M1:** 415, 417-418
Brittle fracture(s) *See also* Brittle; Brittle intergranular fracture(s); Brittleness
AISI/SAE alloy steels **A12:** 291, 335
ASTM/ASME alloy steels **A12:** 347
austenitic stainless steels **A12:** 354, 356, 358
by pure tensile fatigue **A12:** 342
cast aluminum alloys **A12:** 405-408, 410
cemented carbides **A12:** 470
ductile irons **A12:** 227, 231-232, 235-237
effect of grain size **A12:** 106-107
granular, macrograph **A12:** 103
historical study **A12:** 5
in-service, tool steels **A12:** 376
intergranular **A12:** 174-175, 335, 354
interpretation of **A12:** 105-111
low-carbon steels **A12:** 243, 249
macroscopic characteristics **A12:** 107
magnesium matrix, metal-matrix composites **A12:** 464
malleable iron **A12:** 238
materials illustrated in **A12:** 217
matrix, in composites **A12:** 468
medium-carbon steels **A12:** 258, 270, 273
metal-matrix composites **A12:** 466-468
microscopic characteristics **A12:** 109
polymer **A12:** 479
with river patterns, ductile iron **A12:** 230
SEM characterized **A12:** 174-175
splines, alloy steels **A12:** 330
titanium alloys **A12:** 449, 454
tool steels **A12:** 375
wrought aluminum alloys **A12:** 424
Brittle fractures
of alloy steel chain links **A11:** 522
of alloy steel fasteners **A11:** 540-541
aluminum alloy lifting-sling member **A11:** 527
appearance **A11:** 82-83
by liquid erosion **A11:** 164-166, 225, 227, 231
by liquid-metal embrittlement **A11:** 227, 231
by temper embrittlement **A11:** 75
Brittle fractures (continued)
by transgranular or intergranular cracking **A11:** 82
of cast austenitic manganese steel chain link **A11:** 393-395
of cast low-alloy steel jaws **A11:** 389-391
causes of **A11:** 85
in cemented carbide **A11:** 26
characteristics of **M1:** 689
characterized **A11:** 76-77
of clamp-strap assembly **A11:** 69-70
of clapper weldment for disk valve, improper filler metal **A11:** 645-646
in composites **A11:** 734
defined **A11:** 1
and ductile fractures **A11:** 82-101
of ductile iron brake drum **A11:** 370-371
ductile-to-brittle transition **A11:** 84-85
failures, from residual stresses **A11:** 97-98
from coarse grain size **A11:** 70
from steel embrittlement **A11:** 98-101
of gray iron nut **A11:** 369-370
identification chart for **A11:** 80
improper electroplating, failures from **A11:** 97
improper fabrication, failures from **A11:** 87-94
improper thermal treatment, failures from **A11:** 94-97
intergranular **A11:** 22, 25-26, 76-77, 536-537
intergranular and transgranular facets **A11:** 76-77
intergranular, in temper-embrittled steel **A11:** 22
of iron casting oil-pump gear **A11:** 344-345
in large steel component **A11:** 84
lead inclusion in steels, effect on **A11:** 242
liquid lead induced **A11:** 225
of locking collar, from fibering or banding **A11:** 320
of locomotive axles **A11:** 717
of polymers **A11:** 761
of pressure vessels **A11:** 663-666
propagation modes **A11:** 696
of rehardened high-speed steels **A11:** 574
of rephosphorized, resulfurized steel check-valve poppet **A11:** 70-71
of rimmed steel tube, after strain aging by cold swaging **A11:** 648
of roadarm weldment **A11:** 391-392
of roll-assembly sleeve, from microstructure **A11:** 327
service failure from **A11:** 69-71
of shafts **A11:** 459, 466
of splines on rotor shafts **A11:** 478
in stainless steel bolts **A11:** 536-537
of steel stop-block guide, crane **A11:** 526-527
topography of **A11:** 75
transgranular **A11:** 22, 25
transgranular fracture path **A11:** 75
Brittle intergranular fracture
in acoustic emission inspection **A17:** 287
of ordered intermetallics **A2:** 913
Brittle intergranular fracture(s) *See also* Brittle fracture(s); Brittleness
austenitic stainless steels **A12:** 354
low-carbon steel **A12:** 245
precipitation-hardening stainless steels **A12:** 374
steel alloy **A12:** 30, 38
Brittle intermetallics
electronic industry **A13:** 1110
Brittle materials
advanced statistical concepts of fracture **EM4:** 709-715
bend tests for **A8:** 117
chevron-notched specimens for fracture toughness testing **A8:** 469
impact response curves for **A8:** 269
stress-strain curve for **A8:** 23
Brittle materials milling of **A7:** 56-70
mill
ultrafine grinding of **A7:** 59
Brittle nugget
definition **A6:** 1207
Brittle phase
mechanical properties **A17:** 289
Brittle reaction zones, composites
acoustic emission inspection **A17:** 288
Brittle striations
formation **A12:** 35
Brittle surface fracture **A18:** 183
Brittle-coating method
of stress analysis **A17:** 51, 453
Brittlelike fracture
of polymers **A11:** 761
Brittleness *See also* Brittle fracture(s); Brittle intergranular fracture(s); Tough-brittle transition
in bismaleimides (BMI) **EM2:** 253
blue **A11:** 98
of boron fibers **EM1:** 732
caused by carbides, iron castings **A11:** 361
of composites **EM1:** 259-260
defined **A9:** 3 **A11:** 1
examining **A12:** 92
explosion characteristics **A7:** 196
of ferritic iron-aluminum alloys **A12:** 365
from coarse grain size **A11:** 70
in iron-base alloys **A12:** 460
as laminate fracture cause **EM1:** 234
ordered intermetallics **A2:** 914
phenolics **EM2:** 242
refractory metals and alloy welds **A2:** 564
of single fibers, as testing problem **EM1:** 732
temperature, supplier data sheets **EM2:** 642
as temperature-dependent **A12:** 106
testing for **EM2:** 738-739
in thermocouples **A2:** 882
thermoplastics **EM1:** 101
of thermosetting resins **EM2:** 225
of urethane hybrids **EM2:** 268
weld metals **A12:** 375
Broaches **A16:** 194-196, 205
burnishing **A16:** 210
coatings and increased tool life **A16:** 58
high-speed tool steels used **A16:** 59
TIN coatings **A16:** 57, 58
types of **A16:** 199-202, 203
Broaching **A16:** 194-211
aircraft engine components, surface finish requirements **A16:** 22
Al alloys **A16:** 769, 774, 775, 778-780
applicability **A16:** 195-196
applications of P/M high-speed tool steel for **A1:** 785-786
broach breakage causes and prevention **A16:** 210
broach length selection **A16:** 206
broach life affected by work metal and hardness **A16:** 208
broach repair **A16:** 211
broachability constants **A16:** 200
burnishing **A16:** 210
cast irons **A16:** 650, 652, 655, 656, 658
cemented carbides used **A16:** 75
chip breakers **A16:** 205
compared to shaping and slotting **A16:** 187, 192, 193
in conjunction with boring **A16:** 168
in conjunction with drilling **A16:** 223
in conjunction with turning **A16:** 140
Cu alloys **A16:** 811, 813
cutting fluids **A16:** 125, 127, 206-208
dimensional accuracy **A16:** 208-209
fixtures **A16:** 205
and gear manufacture **A16:** 330, 333-335, 344
heat-resistant alloys **A16:** 743-746
horizontal broaching machines **A16:** 196, 197-198, 208
of investment castings **A15:** 264
machines **A16:** 1, 196-200, 202, 203, 206
and machining of internal gears **A16:** 339-340
Mg alloys **A16:** 823
and milling **A16:** 329
monitoring systems **A16:** 414, 415, 417
and multifunction machining **A16:** 368, 377
Ni alloys **A16:** 837
notch root radius **A8:** 382
P/M high-speed tool steels applied **A16:** 66-67, 68
as P/M tool steel applications **A7:** 791
pot broaching machines **A16:** 199, 200
power requirement determination **A16:** 199
selection of broach length **A16:** 206
selection of stroke speed **A16:** 206
stainless steels **A16:** 693-695, 700, 701, 704
strip broaching **A16:** 199
stroke speed selection **A16:** 206

Broaching (continued)
surface alterations produced **A16:** 23
tool design **A16:** 203-205
tool life **A16:** 199-200, 202-203, 204, 208, 209
tool steels **A16:** 717
tools **A16:** 196-197, 199, 200-203
types of broaches **A16:** 199-202, 203
vertical broaching machines **A16:** 197, 198, 208
vs. alternative processes **A16:** 209-210
vs. planing **A16:** 186
Broaching allowance
forgings **M1:** 368
Broad glass **EM4:** 395
Broad goods **EM1:** 5, 125-127
defined **EM2:** 7
Broad-beam absorption/geometry
radiography **A17:** 310
Broad-beam SIMS instrument
for qualitative analysis **A10:** 613
Broadband tunable dye lasers **A10:** 128
Broadcasting
from root node **EL1:** 5
Broadening *See also* Line broadening
dipole-dipole, in ESR spectra **A10:** 255
separation, x-ray diffraction residual stress techniques **A10:** 386
Broken-up structure (BUS) treatment *See also* Titanium P/M products
prealloyed titanium P/M compacts **A2:** 653-654
Bromcresol green
as acid-base indicator **A10:** 172
Bromide
component in photochromic ophthalmic and flat glass composition **EM4:** 442
as fining agents **EM4:** 380
Bromides
anions, separation by ion chromatography **A10:** 659
determined by precipitation titration **A10:** 164
as electrodes **A10:** 184
ion chromatography analysis of geological waters for **A10:** 665-666
ion chromatography calibration curves for **A10:** 666
titanium/titanium alloy resistance to **A13:** 683
Brominated epoxy resins
manufacture **EM1:** 66
Bromination
aramid fibers **EM3:** 285
Bromine
epithermal neutron activation analysis **A10:** 239
etchant for laser-enhanced etching **A16:** 576
in graphites, Raman analysis **A10:** 133
indirect determination of **A10:** 70
and methanol, for isolating inclusions from steel **A10:** 176
methyl acetate, second-phase test method **A10:** 177
safety hazards **A9:** 69
species weighed in gravimetry **A10:** 172
tantalum resistance to **A13:** 731
titration with **A10:** 205
TNAA detection limits **A10:** 237
volatilization losses in melting **EM4:** 389
Bromine-containing compounds
as flame retardants **EM2:** 504
Bromoform
corrosion of stainless steels in **M3:** 81
Bromthymol blue
as acid-base indicator **A10:** 172
Bronchitis
from occupational vanadium exposure **A2:** 1262
Bronchitis, chronic
as cobalt toxic reaction **A7:** 204
Bronchopneumonia
from occupational vanadium exposure **A2:** 1262
Bronsted acids
cycloaliphatic epoxide polymerization with **EL1:** 861
Bronze *See also* Copper alloys, specific types; Copper and copper alloys powder metallurgy materials, etching; Copper-base alloys, specific types **A9:** 509
abrasive blasting of **M5:** 91
bearing material systems **A18:** 746
applications **A18:** 746
bearing performance characteristics **A18:** 746
load capacity rating **A18:** 746
bearing materials **A11:** 484
bearing shells, babbitting of **M5:** 357
Benin, of Nigeria **A15:** 19
bonded to aluminum-silicon-tin or aluminum-silicon-lead alloys **A18:** 744
for brake linings **A18:** 570
broaching **A16:** 200, 204, 206
cage material for rolling-element bearings **A18:** 503
cast, ritual vessels, historic **A15:** 17
casting, in Renaissance **A15:** 21-22
casting processes **A18:** 754, 755
cermet tools applied **A16:** 92
compositions for various types **M6:** 546-547
contamination **M6:** 321
cylinder, friction bearings as **A11:** 715
diamond as abrasive for honing **A16:** 476
drilling **A16:** 227, 229, 231
ECM tool **A16:** 533, 536, 537
enamels **EM3:** 302, 303
friction bearings **A11:** 715
friction welding **A6:** 152
gasket material **A18:** 550
grinding wheel core material **A16:** 456
guide shoe material for honing **A16:** 478
hardfacing **A6:** 807
hardfacing alloys **A18:** 758, 765
high-speed machining **A16:** 597, 598
hone forming **A16:** 488
honing stone selection **A16:** 476
inorganic fluxes for **A6:** 980
lapping **A16:** 499
for large cast bells **A15:** 19
laser cladding **A18:** 867
leaded, galling resistance with various material combinations **A18:** 596
low-lead, bearing material microstructures **A18:** 743, 744
oxyacetylene welding **A6:** 281
P/M materials, hardness and density **A16:** 882
phosphor, modulus of elasticity and proof strength in bending **A8:** 136
phosphor, Rockwell scale for **A8:** 76
physical properties **M6:** 546-547
pickling solutions contaminated by **M5:** 80
planing **A16:** 185
polishing and buffing of **M5:** 112-114
porous
sliding bearings **A18:** 516
sliding bearings that are steel-backed and impregnated with PTFE and lead **A18:** 516
sliding bearings with graphite impregnation **A18:** 516
properties **A6:** 992
pump impeller, cavitation damage failure **A11:** 167-168
radiographic film selection **A17:** 328
relative solderability **A6:** 134
resistance welding *See* Resistance welding of copper and copper alloys **A6:** 850
roller burnishing **A16:** 253, 254
sawing **A16:** 362
seal adhesive wear **A18:** 549
shaping **A16:** 191
Bronze (continued)
sliding bearings used in small electric motors **A18:** 516
solderability **A6:** 978
solid graphite mold casting of **A15:** 285
spade drilling **A16:** 225
spray material for oxyfuel wire spray process **A18:** 829
spring-tempered phosphor, load-deflection plot **A8:** 136
superabrasive wheel bonds **A16:** 433-434
thermal properties (75Cu-25Sn) **A18:** 42
tool bit tool steels used **A16:** 57
turning with cermet tools **A16:** 94
white *See* Cast zinc
Bronze age
casting in **A15:** 15-17
Bronze alloys
corrosion in **A11:** 201
friction bearing composition **A11:** 715
Bronze bearings **A7:** 705
copper powder **A7:** 105
microstructure diluted **A7:** 706
microstructure produced by transient liquid-phase sintering **A7:** 319, 320
one-way, overrunning clutch **A7:** 737
properties **A2:** 325, 394-396
self-lubricating, chemical composition **A7:** 705
self-lubricating, permissible loads **A7:** 707
spherical compaction of **A7:** 707
tin powders for **A7:** 123-124
Bronze, commercial, bearing material systems **A18:** 745, 746, 747
applications **A18:** 746
bearing performance characteristics **A18:** 746
load capacity rating **A18:** 746
Bronze, leaded
suitability for journal bearings **M1:** 610
Bronze plating **M5:** 288-289
alloy composition, control of **M5:** 288
anodes **M5:** 288
applications **M5:** 288
barrel (bulk) process **M5:** 288
bright bronze plating solution **M5:** 289
corrosion resistance **M5:** 288
current density **M5:** 289
equipment **M5:** 288
Rochelle salt addition **M5:** 288-289
solution compositions and operating conditions **M5:** 288-289
speculum plating **M5:** 288-289
standard bronze plating solution **M5:** 289
steel **M5:** 288
temperature **M5:** 288-289
Bronze premix blends
lubricating **A7:** 191, 192
Bronze process
P/M superconducting materials **A7:** 636-637
wire, superconducting **A7:** 638
Bronze, sintering
time and temperature **M4:** 795
Bronze welding
definition **A6:** 1207 **M6:** 3
Bronze-coated wire **A1:** 281
Bronze-graphite
bearing material systems **A18:** 746-747
Bronzes *See also* Alpha bronze; Aluminum bronzes; Bronze bearings; Copper; Copper alloy powders; Copper alloys; Nickel silvers **A13:** 614, 774-775
64-4-4-8-20 *See* Copper alloys, specific types, C97600
66-5-2-2-25 *See* Copper alloys, specific types, C97800
70-5-25 *See* Copper alloys, specific types, C94300
75-3-8-2-12 manganese aluminum bronze *See* Copper alloys, specific types, C95700
78-7-15 *See* Copper alloys, specific types, C93800
79-6-15 *See* Copper alloys, specific types, C93900

Bronzes (continued)
80-10-10 *See* Copper alloys, specific types, C93700
81-4-4-11 aluminum bronze *See* Copper alloys, specific types, C95500
83-4-6-7 *See* Copper alloys, specific types, C83800
83-7-7-3 *See* Copper alloys, specific types, C93200
84-10-2$^1/_2$-0-3$^1/_2$ *See* Copper alloys, specific types, C92900
85-4-11 aluminum bronze *See* Copper alloys, specific types, C95400
85-5-9-1 *See* Copper alloys, specific types, C93500
86$^1/_2$-12-0-0-1$^1/_2$ *See* Copper alloys, specific types, C91700
87-8-1-4 *See* Copper alloys, specific types, C92300
87-11-1-0-1 *See* Copper alloys, specific types, C92500
88-3-9 aluminum bronze *See* Copper alloys, specific types, C95200
88-6-1$^1/_2$-4$^1/_2$ *See* Copper alloys, specific types, C92200
88-8-0-4 *See* Copper alloys, specific types, C90300
88-10-0-2 *See* Copper alloys, specific types, C90500
88-10-2-0 *See* Copper alloys, specific types, C92700
89-1-10 aluminum bronze *See* Copper alloys, specific types, C95300
89-6-5 *See* Copper alloys, specific types, C87200
92-4-4 *See* Copper alloys, specific types, C87200
95-1-4 *See* Copper alloys, specific types, C87200
444 bronze *See* Copper alloys, specific types, C54400
alpha nickel aluminum bronze *See* Copper alloys, specific types, C95800
aluminium bronze, 7% *See* Copper alloys, specific types, C61300 and C61400
aluminum, applications and properties **A2:** 325-334
aluminum bronze, 5% *See* Copper alloys, specific types, C60600 and C60800
aluminum bronze, 8% *See* Copper alloys, specific types, C61000
aluminum bronze, 9% *See* Copper alloys, specific types, C62300
aluminum bronze 9A *See* Copper alloys, specific types, C95200
aluminum bronze 9B *See* Copper alloys, specific types, C95300
aluminum bronze 9C *See* Copper alloys, specific types, C95400
aluminum bronze 9D *See* Copper alloys, specific types, C95500
aluminum bronze, 11 % *See* Copper alloys, specific types, C62400
aluminum bronze A *See* Copper alloys, specific types, C60600
aluminum bronze D *See* Copper alloys, specific types, C61400
aluminum bronze E *See* Copper alloys, specific types, C63000
architectural, applications and properties **A2:** 312-313
architectural bronze *See* Copper alloys, specific types, C38500
bearing, applications and properties **A2:** 325, 394-396
bearing bronze *See* Copper alloys, specific types, C54400
bearing bronze 660 *See* Copper alloys, specific types, C93200
bushing and bearing bronze *See* Copper alloys, specific types, C93700
cast, corrosion ratings **M2:** 390-391
commercial, applications and properties **A2:** 296-297
commercial bronze, 90% *See* Copper alloys, specific types, C22000
as copper alloy powder **A7:** 121, 122
copper-base structural parts from **A2:** 396
couch roll shell, primary crack **A12:** 403
density-pressure relationships **A7:** 300
dimensional change **A7:** 292, 481
electrical and thermal conductivity **A7:** 742
filters **A7:** 308, 699
flow rate **A7:** 279
"G"-bronze *See* Copper alloys, specific types, C90300

Bronzes (continued)
high leaded tin bronze *See* Copper alloys, specific types, C93200, C93566, C93700, C93800, C93900 and C94300
high-conductivity, applications and properties **A2:** 313
high-conductivity bronze *See* Copper alloys, specific types, C40500
high-silicon bronze *See* Copper alloys, specific types, C65500
high-silicon bronze A *See* Copper alloys, specific types, C65500
history **A7:** 14
hydraulic bronze *See* Copper alloys, specific types, C83800
isolation of lead in **A10:** 173
jewelry, applications and properties **A2:** 297-298
jewelry bronze, 87$^1/_2$% *See* Copper alloys, specific types, C22600
laser cutting **A14:** 742
leaded commercial, applications and properties **A2:** 305-309
leaded commercial bronze *See* Copper alloys, specific types, C31400
leaded commercial bronze, nickel-bearing *See* Copper alloys, specific types, C31600
leaded Navy "G" bronze *See* Copper alloys, specific types, C92300
leaded nickel bronze *See* Copper alloys, specific types, C97600 and C97800
leaded nickel-tin bronze *See* Copper alloys, specific types, C92900
leaded tin bronze *See* Copper alloys, specific types, C92300, C92500, C92600 and C92700
low-silicon bronze *See* Copper alloys, specific types, C65100
low-silicon bronze B *See* Copper alloys, specific types, C65100
manganese aluminum bronze *See* Copper alloys, specific types, C95700
manganese bronze (60 000 psi) *See* Copper alloys, specific types, C86400
manganese bronze (65 000 psi) *See* Copper alloys, specific types, C86500
manganese bronze (90 000 psi) *See* Copper alloys, specific types, C86100 and C86200
manganese bronze (100 00 psi) *See* Copper alloys, specific types, C86300
mechanical properties **A7:** 464, 470
medium bronze *See* Copper alloys, specific types, C94500
microstructural analysis **A7:** 488
Navy "M" bronze *See* Copper alloys, specific types, C92200
nickel alloying, history **A2:** 429
nickel aluminum bronze *See* Copper alloys, specific types, C63000 and C63200
nickel gear bronze *See* Copper alloys, specific types, C91700
nickel-aluminum, applications and properties **A2:** 331-333
P/M parts **A7:** 736-737
penny, applications and properties **A2:** 313
penny bronze *See* Copper alloys, specific types, C40500
phosphor bronze, 1.25% E *See* Copper alloys, specific types, C50500
phosphor bronze, 5%. A *See* Copper alloys, specific types, C51000
phosphor bronze, 8% C *See* Copper alloys, specific types, C52100
phosphor bronze, 10% D *See* Copper alloys, specific types, C52400
phosphor bronze B-2 *See* Copper alloys, specific types, C54400
phosphor bronze, free-cutting *See* Copper alloys, specific types, C54400
phosphor gear bronze *See* Copper alloys, specific types, C90700
as porous materials **A7:** 698
premixed, lubricants for **A7:** 191, 192
propeller bronze *See* Copper alloys, specific types, C95800
properties of alloy compositions **A7:** 122
shell, SCC microstructure **A12:** 403
silicon, applications and properties **A2:** 334-335

Bronzes (continued)
silicon bronze *See* Copper alloys, specific types, C87200
sintering of **A7:** 377-378
sleeve, counterbalance mechanism for copier machines **A7:** 737
soft bronze *See* Copper alloys, specific types, C94300
steam bronze *See* Copper alloys, specific types, C92200
stem manganese bronze *See* Copper alloys, specific types, C86400
structural parts, sintered, properties of **A7:** 737
tin bronze *See* Copper alloys, specific types, C90300 and C90500
tin bronze, 65 *See* Copper alloys, specific types, C90700
Brookfield viscosity
for molecular weight **EM2:** 533-534
Broust, Louis J
as chemist **A15:** 29
Brown and Sharp Gage system **M1:** 259
Brown copper and brass coloring solutions **M5:** 626
Brownian motion
to analyze ceramic powder particle sizes **EM4:** 67
Brownian movement
in magnetic etching fluids **A9:** 64
Brucite ($Mg(OH)_2$) **EM4:** 6, 113
calcination **EM4:** 111
Brunauer-Emmet-Teller (BET) method
of measuring specific surface area **A7:** 262, 263
Brush blasting
before painting **M5:** 332
Brush cleaning
of fractures **A12:** 74
Brush coating *See also* Coatings
of conformal coatings **EL1:** 764
of silicone conformal coatings **EL1:** 774
of urethanes **EL1:** 779
Brush fluxers
stationary **EL1:** 682
Brush materials
for sliding contacts **A2:** 842, 862-863
Brush plating
selective plating compared to **M5:** 292
Brush seams, in billets
magnetic particle inspection **A17:** 115
Brush, stationary
as fluxer type **EL1:** 682
Brush Wellman beryllium extraction process **A2:** 684
Brush-on paint stripping method **M5:** 18
Brushes
metal-graphite **A7:** 634-636
powders used **A7:** 573
Brushing
of magnetic paint **A17:** 126-128
nickel alloys **M5:** 674-675
paint application **A13:** 415
porcelain enameling process **M5:** 519
power *See* Power brushing
scratch *See* Scratch brushing
Tampico *See* Tampico brushing
wire *See* Wire brushing
zinc alloys **A2:** 530
BSCCO superconducting materials *See also* High-temperature superconductors
properties
BT resins **EM1:** 79, 81, 87
for printed wiring boards **EL1:** 607
BT-9
erosion test results **A18:** 200
BT-10
erosion test results **A18:** 200
BT-11
erosion test results **A18:** 200
BT-12
erosion test results **A18:** 200
BT-24
erosion test results **A18:** 200
Btu *See* British thermal unit
BTU, and ash content
in coat **A10:** 100
Bubble **EM3:** 6
defined **A7:** 2

Bubble mass
in foam fluxers **EL1:** 682
Bubble memory device
gallium gadolinium garnet **A2:** 740-741
Bubble memory devices **EL1:** 818, 820
Bubble testing
of pressure systems **A17:** 60
Bubble tube
as leak detection method **A17:** 61
Bubble-forming solutions
as pressure system leak detectors **A17:** 60
Bubbler
defined **EM2:** 7
Bubbler molding cooling
defined **EM2:** 7
Bubbler techniques
ultrasonic inspection **A17:** 248
Bubbles *See also* Cavities; Microvoids
air, in iron castings **A11:** 357
aluminum chemical purification by **A15:** 79-80
behavior in liquid erosion **A11:** 163, 167
caps, crevice corrosion pitting in **A11:** 183
gas, in heat-exchanger pitting **A11:** 632
gas-filled, in liquid erosion **A11:** 163
steam, in iron castings **A11:** 357
in tungsten-rhenium thermocouples **A2:** 876
Bubbles, air
in cured epoxy systems **EL1:** 818
Bubbly oil
defined **A18:** 4
Bucket shackle, dragline
failure of **A11:** 399-400
Bucket tooth dragline
hydrogen failure in **A11:** 410
Bucket trucks
acoustic emission inspection **A17:** 290-292
Buckhorn plantain **A7:** 589, 591
Buckle *See also* Buckling; Crush; Dip coat; Rattail
as casting defect **A11:** 384
defined **A11:** 1 **A15:** 1
in iron castings **A11:** 353
Buckled plate test **A6:** 147
Buckling **A8:** 2 **EM3:** 6
analytical model **EM1:** 197
boron-epoxy laminate **EM1:** 332, 335
in composite compressive fracture **A11:** 739
control, in drop hammer forming **A14:** 655
defined **A11:** 1 **A14:** 2 **EM1:** 6 **EM2:** 7
and distortion failure **A11:** 137
effects in axial compression testing **A8:** 55-58
elastic **A11:** 143
failure, in riveted joint **A11:** 544
as failure mode **EM1:** 196-197
in fibers **A11:** 739
as formability problem **A8:** 548
impact damage by **EM1:** 263
inelastic cyclic, and distortion failure **A11:** 144
instability, in graphite-epoxy test structure **A11:** 742-743
in iron castings **A11:** 353
and material properties **A8:** 551
in pin bearing testing **A8:** 59
in pipe **A11:** 704
post-, of plates **EM1:** 447-449
prevention , by limit analysis **A11:** 137
in shafts **A11:** 467
sheet metal **A14:** 878
in sheet metal forming **A8:** 548
side-slip, schematic **A8:** 56
in single-shear tests **A8:** 63-64
specimen, in axial compression testing **A8:** 55-56
stress, axial compression testing **A8:** 55
tests **A14:** 892-893
tests for **A8:** 563-564
in torsion specimen **A8:** 156
Buckling distortion **M6:** 879-880
compressive loading **M6:** 884-887
columns **M6:** 884-885
plate **M6:** 885-886

Buckyballs **EM4:** 829
Buddha, the Great
cast statue **A15:** 21
Buffalo Steel Company *See* Pratt and Letchworth Company
Buffer
defined **A9:** 3
Buffer gas
defined **A7:** 2
Buffer layers
of superconducting thin-film materials **A2:** 1081
Buffer materials
in design **EL1:** 517
Buffering agent
use in surface cleaning **A13:** 381
Buffers
defined **A10:** 670
effects in dc arc sources **A10:** 25
in nonengine lubricant formulations **A18:** 111
potential, use in electrogravimetry **A10:** 200
Buffing *See also* Polishing **A16:** 19 **M5:** 107-108, 115-127
aluminum *See* Aluminum, polishing and buffing of
applications **M5:** 107-108, 120, 123
automatic systems **M5:** 112-114, 119-121, 127, 575-576
rotary machines, types used **M5:** 120-122, 127
brass *See* Brass, polishing and buffing of
chemical and electrolytic brightening compared to **M5:** 581-582
color *See* Color buffing
compound systems **M5:** 116-118
compounds
problems with **M5:** 125
selection of **M5:** 116-117
types used **M5:** 116-118
compounds, removal of
aluminum and aluminum alloys **M5:** 576-577
copper and copper alloys **M5:** 619-620
processes **M5:** 5, 10-12, 56-57
contact, process **M5:** 115-116
contact time, determining **M5:** 125
copper *See* Copper, polishing and buffing of
copper plating process **M5:** 168
equipment **M5:** 107, 119-125
glossary of terms **M5:** 125-127
hard **M5:** 115, 557
heads **M5:** 119-120, 126
high-luster, aluminum parts **M5:** 574-575
liquid compounds **M5:** 116-118
magnesium alloys **M5:** 631-632
materials used **M5:** 107-108
mush, process **M5:** 116, 119
nickel *See* Nickel, polishing and buffing of
problems **M5:** 124-125
process types, described **M5:** 115-116
refractory metals **M5:** 655-656
rotary automatic systems **M5:** 120-122
safety precautions **M5:** 648-649
semiautomatic systems **M5:** 120, 122
stainless steel *See* Stainless steel, polishing and buffing of
steel *See* Steel, polishing and buffing of
straight-line machines
camming arrangements **M5:** 124
noncircular parts **M5:** 124
tooling for **M5:** 124
types used **M5:** 121-124
terminology, glossary of **M5:** 125-127
titanium and titanium alloys **M5:** 655-656
tooling for **M5:** 124
wheel *See* Wheel buffing
work-holding mechanisms **M5:** 119-120, 122
zinc *See* Zinc, polishing and buffing of
Buffing, mechanical
zinc alloys **A2:** 530
Build-up alloys (hardfacing)
abrasion resistance **A18:** 759
applications **A18:** 759

Build-up alloys (hardfacing) (continued)
compositions **A18:** 759
impact resistance **A18:** 759
properties **A18:** 759
Builders
use in alkaline cleaners **M5:** 23-24, 28, 35
Building and construction applications **A7:** 733
of copper-based powder metals powders used **A7:** 572
for stainless steels **A7:** 730
Building applications *See also* Construction applications
aluminum and aluminum alloys **A2:** 9-10
copper and copper alloys **A2:** 239-240
corrosion protection **M1:** 755
Building materials
precoated steel sheet **M1:** 167, 169, 172, 175, 176
Building Seals and Sealants; Fire Standards; Building Constructions, Annual Book of ASTM Standards **EM3:** 67
Buildings **A6:** 375-377
codes governing **M6:** 824
Buildup
definition **A6:** 1207
Buildup of metal
due to electrical arcing **A9:** 563
Built-in self-test (BIST)
system-level **EL1:** 374-376
Built-up edge
as cemented carbide tool wear mechanism **A2:** 954
Built-up edge (BUE) *See also* Wedge formation **A18:** 610, 613, 617
defined **A18:** 4
prow behavior during polishing **A18:** 195
Built-up laminated wood **EM3:** 6
Bulbous joint
fatigue life of **EL1:** 642
Bulge formation
affecting microhardness readings **A8:** 96
Bulge forming
refractory metals and alloys **A2:** 562
Bulge test **A18:** 421
Bulging
aluminum alloy **A14:** 804
by rubber-pad forming **A14:** 673
defined **A14:** 2
as distortion in shotgun barrel **A11:** 139-140
localized strains on **A14:** 389
plane-strain compression test for **A14:** 377-379
punches, fluid forming **A14:** 614
Bulk adherend **EM3:** 6
Bulk adhesive **EM3:** 6
Bulk analysis *See also* Bulk characterization; Macroanalysis
of inorganic gases, analytic methods for **A10:** 8
of inorganic liquids and solutions, applicable methods **A10:** 7
of inorganic solids, applicable analytical methods **A10:** 4-6
of organic solids and liquids, techniques for **A10:** 9, 10
Bulk characterization *See also* Bulk analysis
atomic absorption spectrometry **A10:** 43-59
classical wet analytical chemistry **A10:** 161-180
controlled-potential coulometry **A10:** 207-211
crystallographic texture measurement and analysis **A10:** 357-364
electrochemical analysis **A10:** 181-211
electrogravimetry **A10:** 197-201
electrometric titration **A10:** 202-206
electron spin resonance **A10:** 253-266
elemental and functional group analysis **A10:** 212-220
extended x-ray absorption fine structure **A10:** 407-419
ferromagnetic resonance **A10:** 267-276

Bulk characterization (continued)
gas chromatography/mass spectrometry **A10:** 639-648
inductively coupled plasma atomic emission spectroscopy **A10:** 31-42
infrared spectroscopy **A10:** 109-125
ion chromatography **A10:** 658-667
liquid chromatography **A10:** 649-657
molecular fluorescence spectrometry **A10:** 72-81
Mössbauer spectroscopy **A10:** 287-295
neutron activation analysis **A10:** 233-242
neutron diffraction **A10:** 420-426
nuclear magnetic resonance **A10:** 277-286
optical emission spectroscopy **A10:** 21-30
optical metallography **A10:** 299-308
particle-induced x-ray emission **A10:** 102-108
potentiometric membrane electrodes **A10:** 181-187
radial distribution function analysis **A10:** 393-401
radioanalysis **A10:** 243-250
Raman spectroscopy **A10:** 126-138
Rutherford backscattering spectrometry **A10:** 628-636
single-crystal x-ray diffraction **A10:** 344-356
small-angle x-ray and neutron scattering **A10:** 402-406
spark source mass spectrometry **A10:** 141-150
ultraviolet/visible absorption spectroscopy **A10:** 60-71
voltammetry **A10:** 188-196
x-ray diffraction **A10:** 325-332
x-ray diffraction residual stress techniques **A10:** 380-392
x-ray powder diffraction **A10:** 333-343
x-ray spectrometry **A10:** 82-101
x-ray topography **A10:** 365-379
Bulk chemical analysis **A7:** 246-249
acid insoluble test **A7:** 247
hydrogen loss testing **A7:** 246-247
metallographic examination **A7:** 247
sampling method **A7:** 246
of titanium sponge **A7:** 254
Bulk conductivity **EM3:** 433, 436
Bulk deformation, processes **A8:** 571
forging and rolling **A7:** 522
Bulk density *See also* Density
of copier powders **A7:** 585
defined **A7:** 2 **EM1:** 6 **EM2:** 7
lubricant, in iron premixes **A7:** 190
and mix flow, with zinc stearate lubricant **A7:** 190
in spray drying **A7:** 73-74
and stripping pressure, lubricant effect in iron powders **A7:** 190
Bulk diffusion
as SCC parameter **A13:** 147
Bulk dispensing technique
general solder paste parameters **A6:** 988
Bulk factor **EM1:** 6, 508, 509
defined **EM2:** 8
Bulk formability of steels **A1:** 581-590
characteristics of
bulk versus sheet formability **A1:** 581
of carbon and alloy steels **A1:** 581
tests for **A1:** 581
flow localization **A1:** 584-585
flow stress and forging pressure **A1:** 585
formability characteristics **A1:** 581
formability tests **A1:** 581-584
bend test **A1:** 582, 583
compression test **A1:** 582
ductility testing **A1:** 582
hot twist testing **A1:** 583, 584
nonisothermal upset test **A1:** 584
notched-bar upset test **A1:** 584
partial-width indentation test **A1:** 583
plane-strain compression test **A1:** 582-583
ring compression test **A1:** 583
secondary-tension test **A1:** 583
sidepressing test **A1:** 583-584
tension test **A1:** 581-582
torsion test **A1:** 582
truncated-cone indentation test **A1:** 584
wedge-forging test **A1:** 583, 584
microalloyed steels **A1:** 585-586
comparison of microalloyed plate and bar products **A1:** 588, 589
Bulk formability of steels (continued)
processing of microalloyed bars **A1:** 587-588
processing of microalloyed forging steels **A1:** 588-589, 590
processing of microalloyed plate steels **A1:** 586-587
stainless steels **A1:** 889-894
Bulk forming *See also* Sheetforming
defined **A14:** 2, 15-16
processes **A14:** 16
workability theory and application in **A14:** 388-404
Bulk fused silica
strength **EM4:** 755
thermal properties **EM4:** 754
Bulk materials
chemical analysis of **A11:** 30
sampling of **A10:** 12-18
Bulk metal powders
angle of repose **A7:** 282-285
properties **A7:** 211
Bulk metallic glasses
technology and applications **A2:** 819-820
Bulk meters **EM3:** 693-695, 696, 697, 698, 700
Bulk modulus *See* Bulk modulus of elasticity **EM3:** 316
defined **EM1:** 6 **EM2:** 8
isentropic secant **A18:** 82
isothermal secant **A18:** 82
Bulk modulus of elasticity **EM3:** 6
defined **A8:** 2
Bulk modulus of elasticity (*K*)
defined **A11:** 1
Bulk molding compound (BMC) **EM3:** 6
Bulk molding compounds (BMC) *See also* Molding compounds; Premix; Sheet molding compound; Sheet molding compounds (SMC) **EM1:** 161-163
applications **EM1:** 162-163
chopped glass in **EM1:** 110
compound preparation **EM1:** 161
defined **EM1:** 6 **EM2:** 8
discontinuous fiber matrix for **EM1:** 33
formulation **EM1:** 161
as injection-moldable **EM2:** 321-322
markets **EM1:** 162-163
molded-in color **EM2:** 306
molding methods **EM1:** 161
processing **EM1:** 161
properties **EM1:** 161-162
properties effects **EM2:** 286
short fibers for **EM1:** 121
size and shape effects **EM2:** 291-292
Bulk plastic
nondestructive testing **A6:** 1086
Bulk plating *See* Barrel plating
Bulk samples
composition of **A10:** 562, 631
defined as gross sample **A10:** 674
DRS analysis for **A10:** 114
electron beam spreading in **A10:** 434
preparation for AEM analysis **A10:** 450
surface, SEM chemical composition **A10:** 490
Bulk sampling programs
design of **A10:** 12
Bulk solids
dust collection of **A13:** 1369
Bulk temperature **A18:** 39, 40, 41
gears **A18:** 539, 540
symbols and units **A18:** 544
Bulk unloaders **EM3:** 716-717
Bulk volume
defined **A7:** 2
Bulk volumetric temperature **A18:** 438
Bulk, weight of
and flow rate **A7:** 280
Bulk workability *See also* Bulk workability testing; Workability **A8:** 577-578
Bulk workability testing *See also* Workability **A8:** 571-597
cold upset testing **A8:** 578-581
in drawing **A8:** 591-593
evaluating workability **A8:** 571-578
in extrusion **A8:** 591-593
for forging **A8:** 587-591
hot compression testing **A8:** 581-584
hot tension testing **A8:** 586-587
Bulk workability testing (continued)
partial-width indentation test **A8:** 584-585
plane-strain compression test **A8:** 583
ring compression test **A8:** 585-586
rolling **A8:** 593-596
Bulkiness
of mer **EM2:** 57, 62
Bulkwelding process
for hardfacing **A7:** 835
Bull blocks **A14:** 2, 333-334
Bull ladles *See also* Ladles
early usage **A15:** 33
Bull's-eye structure **M1:** 7-8
malleable cast iron **M1:** 60, 61
Bull's-eye structure in ductile iron **A9:** 245
Bulldozer
defined **A14:** 2
Bulldozing of springs *See* Presetting of springs
Bump check
definition **EM4:** 632
Bump contacts
defined **EL1:** 1136
Bump foil strip **A18:** 531
Bump-contact chip
development **EL1:** 961
Bumped tape automated bonding **EL1:** 275, 479
Bumper *See also* Jolt ramming
defined **A15:** 2
Bumper assembly, automotive
economy in manufacture **M3:** 852, 853
Bumping *See* Breathing
Buna-N *See also* Acrylonitrile-butadiene rubber; Nitrile rubber
for aircraft fuel tanks **EM3:** 604
gas tank seam sealant **EM3:** 610
Bunch stranded copper conductors **M2:** 266, 272
Bundle *See also* Fiber(s); Filament(s)
defined **EM1:** 6 **EM2:** 8
Bundles
in composite fractures **A11:** 738
Bunge formalism
ODF coefficient determined **A10:** 362
and Roe formalism, compared **A10:** 362
in specifying orientation in crystallographic measurement **A10:** 359-361
Buoyancy
alloy, effect in alloy additions **A15:** 72
-driven interdendritic solute transport **A15:** 154
Burden (overhead)
defined **EM2:** 86
as piece cost component **EM2:** 82
types **EM2:** 86
Bureau of Mines
explosion testing chamber **A7:** 197
Buret
defined **A10:** 670
titrimetry and coulometry **A10:** 205
walls, wetting of **A10:** 172
Burger's vector
abbreviation for **A10:** 689
one-dimensional defect analysis **A10:** 465
orientation, effect on FIM contrast from dislocations **A10:** 588-589
Burgers vector **A18:** 388, 468
of body-centered cubic metals **A9:** 684
contrast profiles of single perfect dislocations **A9:** 114
defined **A9:** 719
dislocation displacement field **A9:** 113-114
and dislocation loops **A9:** 116-117
in dislocation pairs **A9:** 114-115
in dislocations, determination **A9:** 115-116
of face-centered cubic metals **A9:** 684-685
of grain boundary dislocations **A9:** 120
of hexagonal close-packed metals **A9:** 684
of molybdenum and molybdenum carbide interface **A9:** 122
of partial dislocations **A9:** 685
relationship to slip **A9:** 684
Burial zone
marine structures **A13:** 543
Buried metals
corrosion of **A11:** 191-192
Buried piping *See* Underground installations
Buried plants **A13:** 1128, 1133
Buried via(s) *See also* Via(s)
inner layers **EL1:** 543

Buried via(s) (continued)
rigid printed wiring boards EL1: 550
Buried-layer diffusion
bipolar junction transistor technology EL1: 195
Burn in
as casting defect A11: 384
Burn on
as casting defect A11: 384
Burn through
definition A6: 1207
Burn through weld
definition A6: 1207
Burn-in
on tape-on-chip EL1: 281-282
as temperature-induced stress test EL1: 497-498
Burn-in defect
core, coatings for A15: 240
mold porosity effect A15: 209
Burn-off *See also* Mold stabilization
in all-ceramic mold casting A15: 249
defined A15: 252
in Shaw process A15: 252
Burn-off failures
in locomotive axles A11: 714, 715
Burn-through
definition M6: 3
Burn-through weld
definition M6: 3
Burnback A6: 27
Burned EM3: 6
defined EM1: 6 EM2: 8
Burned adhesive
adhesive-bonded joints A17: 612
Burned sand
defined A15: 2
Burned-in sand
blast cleaning of A15: 506
Burned-on sand *See also* Metal penetration
cores, coatings for A15: 240
defined A15: 2
Burner
definition A6: 1207
Burners
crucible furnaces A15: 383
flame hardening M4: 489-492
as OES flame source A10: 28
reverberatory furnaces A15: 375-376
Burning *See also* Flammability; Metallurgical bum; Overheating; Oxidation ..... A3: 1 • 19
alloy steels A12: 300
bridge web fracture from A11: 528
defined A9: 3 A13: 2 A18: 4
definition A6: 1207
effect on fatigue strength A11: 120
of forged steel rocker arm A11: 119-120
in forging A17: 493
forging failures from A11: 332
for macroscopic examination A12: 92
in medium-carbon steels A12: 259
polyester resistance to EM1: 96
in steels A12: 127
in steels, color etching to reveal A9: 142
tool and die failure from A11: 573
Burning rate EM3: 7
defined EM2: 8
Burnish
defined A18: 4
Burnishing A7: 462
aluminum and aluminum alloys, barrel finishing process M5: 272-274
compared to reaming A16: 239
in conjunction with thread rolling A16: 280
defined A9: 3
macroexamination of A12: 72
mass finishing process M5: 135
media for M5: 135
of shafts A11: 459
thermal spray-coated materials M5: 370
Burnishing (continued)
to improve surface integrity A16: 35
and transfer machines A16: 397
Burnishing, and shaving
of pierced holes A14: 470
Burnishing surface modification technique A18: 178
Burnoff A6: 844
defined A7: 2
lubricant A7: 191
zone in furnaces A7: 351
Burnout
of substrates EL1: 465
Burnout, and mold firing
investment casting A15: 262
Burnout method of purging M6: 948-949
Burnup, with plastic flow
roller bearing A11: 500-501
Burr *See also* Deburring
Burr tolerance
forgings M1: 366, 367
Burring
multifunction machining A16: 374
Burrow's formula
impedance change A17: 173
Burrs *See also* Deburring
by blanking A11: 88
in cantilever beam bend test specimens A8: 132
defined A14: 2
detected by diffraction pattern technique A17: 13
fatigue fractures from A11: 123
height, in electrical steel sheet processing A14: 481
in hole, spherical bearing A11: 489
liquid penetrant inspection A17: 86
microwave inspection A17: 215
on slit edges A14: 708
on upset butt welded steel wire A11: 443
in press forming A14: 548
removal, after blanking A14: 458
and shear testing A8: 68
Burrs, removal of *See* Deburring
Burst strength
defined EM1: 6 EM2: 8
of three-directional hybridized fiber fabric EM1: 127
Burst-type signals
acoustic emission inspection A17: 281
Bursts *See also* Chevron patterns
brittle fracture by A11: 185
in electron beam welds A11: 446-447
of forged bar A11: 317-318
as forging defect A11: 85, 317, 327
as forging process flaws A17: 493
intemal, as subsurface discontinuities A11: 121
intemal, effects on cold-formed part failure A11: 307
internal, radiographic methods A17: 296
ultrasonic inspection A17: 232
Bursts, revealed by macroetching A9: 174
in alloy steel forging A9: 176
Bus bar
defined EL1: 116
Bus bar conductors
aluminum alloy A2: 12-13
Bus bars
in metal-processing equipment A13: 1316
BUS circulating fluid-bed process
for zinc recycling A2: 1225
BUS treatment *See* Broken-up structure (BUS) treatment
Buses
aluminum and aluminum alloys A2: 11
Bush bearing
defined A18: 4
Bushing *See also* Glass filament bushing
ball-bearing, in ultrasonic hardness tester A8: 101
defined EM1: 6 EM2: 8
for hydraulic torsional system A8: 216
Bushing and bearing bronze
properties and applications A2: 379-380
Bushings
brass or steel A11: 470
pilot-valve, fatigue fracture of A11: 121
powders used A7: 572
steel A11: 470
Bushings (industrial)
powders used A7: 573
Bushings, nonmetallic
eddy current inspection A17: 192-193
Business machine applications
of aluminum P/M parts A7: 744-745
of bronze P/M parts A7: 736
high-impact polystyrenes (PS, HIPS) EM2: 195
P/M parts for A7: 667-670
polyphenylene ether blends (PPE, PPO) EM2: 183
powders used A7: 573
self-lubricating bearings in A7: 705
of stainless steels A7: 732
Business machine component
brittle fracture in A11: 90-92
Buster
defined A14: 2
Butadiene EM3: 7
defined EM2: 8
polymerization of A10: 132
Butadiene-acrylonitrile elastomers
as sizing EM1: 124
Butadiene-acrylonitrile rubber EM3: 82-83
Butadiene-co-maleic anhydride (BMA)
electrodeposition of carbon/graphite fibers EM3: 287
Butadiene-styrene plastic EM3: 7
Butadiene-styrene-plastics
defined EM2: 8
Butadiene-styrenes EM3: 594
Butane
valve thread connections for compressed gas cylinders A6: 1197
Butler finishing
stainless steel M5: 559
Butler-Volmer equation A13: 30
Butt fusion
defined EM2: 8
Butt joint EM3: 454, 545
brazing A6: 120
defined EM1: 6 EM2: 8
definition A6: 1207
electron-beam welding A6: 260
flash welding A6: 247
heat-treatable wrought aluminum alloys, gas-shielded arc welded, properties A6: 728
high-strength low-alloy quench and tempered structural steels, heat input A6: 666
hydrogen-induced cold cracking A6: 436
incomplete fusion A6: 408
lamellar tearing A6: 95-96
laser-beam welding A6: 264, 879, 880
nickel-base alloys A6: 591
non-heat-treatable aluminum alloys, gas-shielded arc welded, properties A6: 727
oxyfuel gas welding A6: 286, 287, 288
plasma arc welding A6: 197, 198
soldering A6: 130
wrought martensitic stainless steel A6: 438
Butt joint test A18: 404 EM3: 321-322
Butt joints
arc welding of nickel alloys M6: 437
brazing of aluminum alloys M6: 1025

Butt joints (continued)
definition, illustration **M6:** 3, 60-61
electron beam welds **M6:** 615-618
electroslag welding **M6:** 225-226
explosion welding **M6:** 712
furnace brazing of steels **M6:** 943
gas metal arc welding **M6:** 165, 166-167
of aluminum alloys **M6:** 374, 379, 384
of commercial coppers **M6:** 403
of coppers and copper alloys **M6:** 416
gas tungsten arc welding **M6:** 201
of aluminum alloys **M6:** 374, 379, 395-396, 398
of magnesium alloys **M6:** 431-432
nickel-based heat-resistant alloys **M6:** 356, 363
oxyacetylene braze welding **M6:** 597
oxyfuel gas welding **M6:** 589-590
radiographic inspection **A17:** 334
recommended grooves for arc welding **M6:** 69, 71
resistance brazing **M6:** 983
submerged arc welding **M6:** 114
tolerances for laser beam welds **M6:** 663-664
Butt seam joints **A6:** 239
Butt tears, revealed by macroetching **A9:** 174
in alloy steel billet **A9:** 175
Butt weld
definition **A6:** 1207
Butt weld detail, vertical
in bridges **A11:** 709
Butt welded cold finished mechanical tubing **M1:** 324
Butt welding
clamp design **M6:** 673-674
clamping procedure **M6:** 673-674
high frequency welds **M6:** 759
machines **M6:** 674
metals welded **M6:** 674
seam welding **M6:** 501-502
thermoplastic tape **EM1:** 551
ultrasonic welding **M6:** 747
Butt welding, fluid flow phenomena
GTAW **A6:** 21
Butt welds
acoustic emission inspection **A17:** 599-600
angular change **M6:** 876, 878
effect of back chipping **M6:** 876
effect of groove shape **M6:** 876
arc welding of austenitic stainless steels **M6:** 328-329, 334, 338
definition **M6:** 3
effect of restraint on transverse shrinkage **M6:** 873-875
effect of tensile loading on residual stress **M6:** 881, 884
effect of welding sequence on transverse shrinkage **M6:** 875-876
flux cored arc welding **M6:** 109
longitudinal distortion **M6:** 880, 882
longitudinal shrinkage **M6:** 876
magnetic particle inspection **A17:** 109, 114
non-heat-treatable aluminum alloys **A6:** 540
nonrelevant indications in **A17:** 106-107
plasma arc welding **M6:** 218
rotational distortion **M6:** 873
submerged arc welds **M6:** 129
transverse shrinkage **M6:** 870-875, 880
Butt welds, upset
failure origins **A11:** 443-444
Butterflies
in bearings **A12:** 115, 134
Butterflies, stress
in steel bearing ring **A11:** 505
Butterfly alterations
in gas carburized steels **A9:** 223
Butterfly cone
definition **EM4:** 632
Butterfly wings
as submicrostructure **A11:** 593
Buttering **A6:** 498, 789
cast irons **A6:** 712, 716, 717
definition **A6:** 1207 **M6:** 3
dissimilar copper alloys **A6:** 770
dissimilar metal joining **A6:** 823-824, 825, 827
lamellar tearing **A6:** 96
Button
defined **A8:** 2
definition **A6:** 1207 **M6:** 3
Button defects *See* Scabs
Button melting
electron beam **A15:** 412
Buttonhead
grips **A8:** 51
specimen, in constant-load testing **A8:** 314
Butyl
as bag material **A7:** 447
Butyl acetate/1.2% nitrocellulose
as vehicle for solder glass powder **EM4:** 1069-1070
Butyl benzyl phthalate (plasticizer)
batch weight of formulation when used in non-oxidizing sintering atmospheres **EM4:** 163
Butyl carbitol, description **A9:** 68
Butyl caulks
for exterior seals on aluminum and vinyl siding **EM3:** 56
Butyl cellosolve
description **A9:** 68
Butyl glycidyl ether
as epoxy diluent **EM1:** 67, 70
Butyl phenol **EM3:** 103
Butyl rubber **EM3:** 76, 82-83
additives and modifiers **EM3:** 146-147
for appliance seating **EM3:** 146
for auto glass seal **EM3:** 146
for carton closing **EM3:** 146
chemistry **EM3:** 146, 198
commercial forms **EM3:** 146
cross-linking **EM3:** 146
for insulated glass window sealant **EM3:** 146
for marine uses **EM3:** 146
markets **EM3:** 146
maskant material for chemical milling **A16:** 803
for mobile home manufactures **EM3:** 146
performance **EM3:** 674
for prefabricated metal buildings **EM3:** 146
properties **EM3:** 146, 198
for rubber roof installation **EM3:** 146
sealant applications **EM3:** 198
as substitute for fluoropolymers **EM3:** 678
for tapes **EM3:** 146
Butyl rubber insulation
wrought copper and copper alloy products **A2:** 258
Butyl skinning sealants **EM3:** 191
Butyl stearate
as plasticizer not affecting viscosity in injection molding **EM4:** 174
Butyl tapes
for wet systems in construction **EM3:** 56
for window glazing **EM3:** 56
Butylene glycol diglycidyl ether (BGDGE)
epoxy resin **EM1:** 67, 69
Butylene plastics **EM3:** 7
defined **EM2:** 8
Butyls *See also* Butyl rubber; Polyisobutylene **EM3:** 198-202
additives **EM3:** 202
advantages and limitations **EM3:** 675
for appliance sealing tapes **EM3:** 200
applications **EM3:** 56, 201
for auto back lights **EM3:** 50
for auto body sealing **EM3:** 57
automotive applications **EM3:** 50, 200
binders used **EM3:** 200
for body sealants and glazing materials **EM3:** 57
for caulking **EM3:** 607-608
characteristics **EM3:** 53
chemistry **EM3:** 50, 198
construction market applications **EM3:** 199-200
cure capability **EM3:** 199
cure mechanism **EM3:** 200-201
cure properties **EM3:** 51
electronics applications **EM3:** 201
features **EM3:** 675
fillers used **EM3:** 200
Food and Drug Administration regulations **EM3:** 201
forms **EM3:** 50, 199, 200
for glazing windows and wall panels **EM3:** 199-200
heating and air conditioning duct sealing and bonding **EM3:** 610
Butyls (continued)
high hydrocarbon-content sealants **EM3:** 201
as hot-applied sealants **EM3:** 200
for industrial sealants **EM3:** 58
insulated glass applications **EM3:** 200, 201
for insulated glass construction **EM3:** 58
for interior seals in window systems **EM3:** 57
life expectancy of tape **EM3:** 200-201
for low-stress construction applications **EM3:** 577
for metal building market **EM3:** 58
methods of application **EM3:** 199, 200
mixing equipment used **EM3:** 198
polymer grades **EM3:** 199
for prefabricated construction **EM3:** 200
properties **EM3:** 50, 201, 202
for recreational vehicle sealing **EM3:** 58
replaced by polyurethanes in automotive industry **EM3:** 50
as replacement glass sealant **EM3:** 200
sealant characteristics (wet seals) **EM3:** 57
sealant formulation **EM3:** 198-199
for self-sealing tires **EM3:** 58
shelf life of tapes **EM3:** 199
silane coupling agent **EM3:** 182
substrate cure rate and bond strength for cyanoacrylates **EM3:** 129
suppliers **EM3:** 58, 199, 202
thermal properties **EM3:** 52
uses and properties **EM3:** 126
for vacuum bag sealing **EM3:** 59
Butyls as sealants **EM3:** 188, 190
for insulating glass windows and doors **EM3:** 675
for joint sealing **EM3:** 675
for panel sealing **EM3:** 675
properties **EM3:** 677
service life **EM3:** 190
BWRA penetrameters
as step wedges **A17:** 341
By-pass sample station
for analyzing liquid metal purity **A8:** 426
n-Butanol
surface tension **EM3:** 181
n-Butyl acetate
surface tension **EM3:** 181
n-Butyl n-butyrate
as solvent used in ceramics processing **EM4:** 117
n-Butyraldehyde
physical properties **EM3:** 104

C

c *See* Crystal lattice; Velocity
C glass, effect
mechanically alloyed oxide dispersion-strengthened (MA ODS) alloys **A2:** 947
C-103 *See* Niobium alloys, specific types
C-110M *See* Titanium alloys, specific types, Ti-8Mn
C-120AV *See* Titanium alloys, specific types, Ti-6Al-4V
C-129Y *See* Niobium alloys, specific types
C-135AMo *See* Titanium alloys, specific types, Ti-7Al-4Mo
C-263
composition **A4:** 794
C-bend
as lead formation **EL1:** 733-734
c-chart *See also* Control charts; Quality; Quality control
for number of defects **A17:** 736-737
C-Co (Phase Diagram) .... **A3:** 2 • 109
C-Cr (Phase Diagram) .... **A3:** 2 • 109
C-Cr-Fe (Phase Diagram) .... **A3:** 3 • 24-25
C-Cr-Mo (Phase Diagram) .... **A3:** 3 • 25-26
C-Cr-N (Phase Diagram) .... **A3:** 3 • 26
C-Cr-V (Phase Diagram) .... **A3:** 3 • 26-27
C-Cr-W (Phase Diagram) .... **A3:** 3 • 27
C-Cu (Phase Diagram) .... **A3:** 2 • 110
C-Cu-Fe (Phase Diagram) .... **A3:** 3 • 27-28
C-Fe (Phase Diagram) .... **A3:** 2 • 110
C-Fe-Mn (Phase Diagram) .... **A3:** 3 • 28-30
C-Fe-Mo (Phase Diagram) .... **A3:** 3 • 30-31
C-Fe-N (Phase Diagram) .... **A3:** 3 • 31-32
C-Fe-Ni (Phase Diagram) .... **A3:** 3 • 32

C-Fe-Si (Phase Diagram) .... **A3:** 3 • 33-34
C-Fe-V (Phase Diagram) .... **A3:** 3 • 34
C-Fe-W (Phase Diagram) .... **A3:** 3 • 35
C-glass *See also* Glass, fibers
chemical resistance **EM1:** 107
composition and use **EM1:** 45
defined **EM1:** 6 **EM2:** 9
specific heat **EM1:** 47
C-grade system
cemented carbides **A2:** 953
C-HEMP
to determine dynamic flow curve **A8:** 205
C-Hf (Phase Diagram) .... **A3:** 2 • 111
C-hooks
design and materials for **A11:** 522-523
C-La (Phase Diagram) .... **A3:** 2 • 111
C-Mn (Phase Diagram) .... **A3:** 2 • 111
C-Mo (Phase Diagram) .... **A3:** 2 • 112
C-mode scanning acoustic microscope (C-SAM) **EL1:** 1070-1071
C-mode scanning acoustic microscopy (C-SAM)
color image **A17:** 487
defined **A17:** 465
operating principles **A17:** 466-467
parameters/techniques, compared **A17:** 470
of soldered joints **A17:** 607
C-Ni (Phase Diagram) .... **A3:** 2 • 112
C-porosity
in cemented carbides **A9:** 274
C-Pr (Phase Diagram) .... **A3:** 2 • 112
C-profiled ring
rolling production stages **A14:** 117
C-ring, aluminum alloy
fracture of **A11:** 78, 79
C-ring specimens
SCC testing **A13:** 249
C-SAM *See* C-mode scanning acoustic microscopy (C-SAM)
C-Sc (Phase Diagram) .... **A3:** 2 • 113
C-scan **EM1:** 8, 262, 776 **EM3:** 9
defined **EM2:** 12
ultrasonic **EM2:** 838-845
C-Scan Acoustic Microscopy (C-SAM) **EM4:** 624-625
C-scan display modes
display **A17:** 243-244
gating **A17:** 243-244
microwave inspection **A17:** 213
of polar backscattering **A17:** 249
pulse-echo ultrasonic inspection **A17:** 242
pulsed leaky Lamb wave testing **A17:** 253
scanning acoustical holography **A17:** 443
system setup **A17:** 243
of titanium-matrix composite **A17:** 250-251
ultrasonic imaging, of powder metallurgy parts **A17:** 540
C-scan ultrasonic imaging
of soldered joints **A17:** 607
C-Si (Phase Diagram) .... **A3:** 2 • 113
***c*-spacing**
layer lattice solid lubricants **A18:** 113
C-stage *See also* A-stage; B-stage **EM3:** 9
defined **EM1:** 8 **EM2:** 12
C-Ta (Phase Diagram) .... **A3:** 2 • 113
C-Th (Phase Diagram) .... **A3:** 2 • 114
C-Ti (Phase Diagram) .... **A3:** 2 • 114
C-U (Phase Diagram) .... **A3:** 2 • 114
C-V (Phase Diagram) .... **A3:** 2 • 115
C-W (Phase Diagram) .... **A3:** 2 • 115
C-Y (Phase Diagram) .... **A3:** 2 • 115
C-Zr (Phase Diagram) .... **A3:** 2 • 116
C1, C2, C3, etc *See* Carbide tools, specific types
Ca (Phase Diagram) .... **A3:** 2 • 120
Ca-Cd (Phase Diagram) .... **A3:** 2 • 116
Ca-Cu (Phase Diagram) .... **A3:** 2 • 116
Ca-Ga (Phase Diagram) .... **A3:** 2 • 117
Ca-Ge (Phase Diagram) .... **A3:** 2 • 117
Ca-Hg (Phase Diagram) .... **A3:** 2 • 117
Ca-In (Phase Diagram) .... **A3:** 2 • 118
Ca-Li (Phase Diagram) .... **A3:** 2 • 118
Ca-Mg (Phase Diagram) .... **A3:** 2 • 118
Ca-Na (Phase Diagram) .... **A3:** 2 • 119
Ca-Nd (Phase Diagram) .... **A3:** 2 • 119
Ca-Ni (Phase Diagram) .... **A3:** 2 • 119
Ca-Pb (Phase Diagram) .... **A3:** 2 • 120
Ca-Pd (Phase Diagram) .... **A3:** 2 • 120
Ca-Pt (Phase Diagram) .... **A3:** 2 • 121
Ca-Sb (Phase Diagram) .... **A3:** 2 • 121
Ca-Si (Phase Diagram) .... **A3:** 2 • 121
Ca-Sr (Phase Diagram) .... **A3:** 2 • 122
Ca-Tl (Phase Diagram) .... **A3:** 2 • 122
Ca-Yb (Phase Diagram) .... **A3:** 2 • 122
Ca-Zn (Phase Diagram) .... **A3:** 2 • 123
CA811 to CA879 *See* Cast copper alloys, specific types
Cabinet blasting machines **M5:** 88-89, 95-96
Cabinets
salt spray (fog) testing **A13:** 225-226
Cable *See also* Cable sheathing; Cabling; Thermocouple wire; Wire; Wire(s)
of A15 conductors **A2:** 1069-1070
classifications, wrought copper and copper alloys **A2:** 251-253
copper **M2:** 265-274
extruded **EM2:** 385
jacketing, polyamide **EM2:** 125
thermocouple extension wire, color codes **A2:** 879
thermoplastic polyurethanes (TPUR) **EM2:** 205
and wire, wrought copper and copper alloys **A2:** 250-260
Cable sheath
lead **A9:** 418
Cable sheathing
aluminum and aluminum alloys **A2:** 12
lead and lead alloy **A2:** 550-551, 544
Cable-sheathing **A2:** 544
Cables
bridge **M1:** 272
for direct contact magnetization **A17:** 94
electrical system pipe-type **A13:** 1292
flux leakage inspection method **A17:** 133
for magnetizing large forgings/castings **A17:** 112
power, galvanic corrosion **A13:** 85
surface roughness, optical sensors for **A17:** 10
telephone, corrosion of **A13:** 85, 1127-1133
Cables, elevator
fatigue fracture **A11:** 520-521
Cabling
of niobium-titanium superconducting materials **A2:** 1051-1052
Cabling system
as level 4 components **EL1:** 76
Cabot alloy 214
heat-affected-zone cracks **A6:** 91, 93
solidification cracking **A6:** 90, 91
Cabrera-Mott (thin-film) theory **A13:** 67
CAD/CAM applications **A16:** 627-636
advantages **A16:** 633-634
analysis benefit **A16:** 630
analysis capabilities function **A16:** 628
Automatically Programmed Tools (APT) part program **A16:** 627, 629, 631, 633, 635
benefits **A16:** 630-631
bill of material **A16:** 627
CAD to CAM interfaces **A16:** 633-634
computer aided process planning (CAPP) systems **A16:** 628-629
computer-aided design (CAD) ... **A16:** 627, 629, 635, 636
computer-aided manufacturing (CAM) **A16:** 627
computer-aided numerical control (CNC) **A16:** 634-635, 636
cutter location file (CL) **A16:** 629, 633, 635
data base **A16:** 633
CAD/CAM applications (continued)
direct numerical control (DNC) **A16:** 634, 635, 636
distributed numerical control **A16:** 635-636
drafting function **A16:** 628
finite-element model analysis **A16:** 628
function of a CAD/CAM system **A16:** 627-630
hardware components **A16:** 631-633
improved accuracy benefit **A16:** 631
inspection function **A16:** 630
Machine Control Data (MCD) codes **A16:** 630
Machine Control Unit (MCU) memory **A16:** 630
machines programmed **A16:** 629
Materials Requirement Planning (MRP) **A16:** 636
minimum hardware requirements in NC machining **A16:** 633
NC machines **A16:** 634
NC part programming **A16:** 634
NC part programming benefit **A16:** 631
part machining function **A16:** 630
part programming function **A16:** 629
postprocessor **A16:** 634
process planning function **A16:** 628
program verification function **A16:** 629-630
tool design benefit **A16:** 630
understandable drawings benefit **A16:** 630-631
CAD/CAM integrated composite manufacturing center **EM1:** 621-622
CAD/CAM techniques *See* Computer-aided design; Computer-aided design and manufacture; Computer-aided manufacture
CAD/CAM/CIM systems **A16:** 624-625
Cadmium *See also* Solid cadmium **EL1:** 636, 637
as addition to aluminum alloys **A4:** 843
addition to aluminum-base bearing alloys **A18:** 752
as addition to brazing filler metals **A6:** 904-905
air-acetylene flame atomizer for **A10:** 48
alloying, aluminum casting alloys **A2:** 132
alloying effects on copper alloys **M6:** 402
alloying, wrought aluminum alloy **A2:** 47
alloying, wrought copper and copper alloys **A2:** 242
in alloys, oxyfuel gas cutting **A6:** 1165
in aluminum alloys **A15:** 745
Arrhenius plots of delayed steel failure in **A11:** 240, 244
in blood, GFAAS analysis **A10:** 55
in cast iron **A1:** 8
chemical analysis and sampling **A7:** 248
chemical resistance **M5:** 4
coatings, for fasteners **A11:** 542
as colorant **EM4:** 380
compatibility in bearing materials **A18:** 743
constant-current electrolysis **A10:** 200
in copper alloys **A6:** 753
dangers in dip brazing **M6:** 995
determined by controlled-potential coulometry **A10:** 209
effects, cartridge brass **A2:** 301
effects, electrolytic tough pitch copper **A2:** 270
embrittlement by **A11:** 230-235
embrittlement, by mercury-indium **A11:** 230, 232
in enameling ground coat **EM3:** 304
friction coefficient data **A18:** 71
in furnace atomizers **A10:** 49
gravimetric finishes **A10:** 171
hydrogen damage to **A11:** 126
ICP-determined in plant tissues **A10:** 41
impurity in solders **M6:** 1072
lap welding **M6:** 673
and lead, separated by internal electrolysis **A10:** 201
liquid, low-alloy steel embrittlement by **A12:** 30, 39
liquid-metal embrittlement in **A11:** 234
as low-melting embrittler **A12:** 29

Cadmium (continued)
as major toxic metal with multiple effects ... **A2:** 1239-1242
microelectrogravimetry of ... **A10:** 200
molten, as embrittler of titanium ... **A11:** 226, 228
nickel plating bath contamination by ... **M5:** 209-210
as pigment ... **EM3:** 179
-plated alloy steel bolts, hydrogen damage to ... **A11:** 540-541
-plated steel aircraft wing clamp, forging failure from burning ... **A11:** 332-334
-plated steel, embrittlement in ... **A11:** 242
-plated steel, intergranular fracture ... **A11:** 29
-plated steel nut, hydrogen embrittlement failure ... **A11:** 246-247
-plated steel plate, arc striking fracture at hard spot in ... **A11:** 97
plating for tool steels ... **A18:** 739
precoating ... **A6:** 131
prompt gamma activation analysis of ... **A10:** 240
pure, properties ... **A2:** 1104
recommended impurity limits of solders ... **A6:** 986
reductant of metal oxides ... **M6:** 692, 694
relative solderability ... **A6:** 134
relative solderability as a function of flux type ... **A6:** 129
release into glass housewares limited ... **EM4:** 1100
resistance of, to liquid-metal corrosion ... **A1:** 635
safety standards for soldering ... **M6:** 1098
and SCC in titanium and titanium alloys ... **A11:** 223
shielded metal arc welding ... **A6:** 179
-shielded samples for epithermal neutron bombardment ... **A10:** 234
soldering ... **A6:** 631
solid, effect on crack depth in titanium alloys ... **A11:** 239-241
species weighed in gravimetry ... **A10:** 172
thermal diffusivity from 20 to 100 °C ... **A6:** 4
as tin solder impurity ... **A2:** 520
toxicity ... **A2:** 1240-1241 **A6:** 1195, 1196 **M5:** 267
as trace element, cupolas ... **A15:** 388
TWA limits for particulates ... **A6:** 984
urinary concentration, and renal dysfunction ... **A2:** 1241
vapor pressure ... **A6:** 621
vapor pressure, relation to temperature ... **A4:** 495
Vickers and Knoop microindentation hardness numbers ... **A18:** 416
volatilization losses in melting ... **EM4:** 389
volumetric procedures for ... **A10:** 175
weighed as the phosphate ... **A10:** 171
in zinc alloys ... **A15:** 788
Cadmium alloy, Cd-25Ni
peritectic envelopes ... **A9:** 679
Cadmium boride
glass forming ability ... **EM4:** 494
melting point ... **EM4:** 494
viscosity at melting point ... **EM4:** 494
Cadmium coating
hydrogen embrittlement testing in ... **A8:** 542
Cadmium coatings
cast irons ... **M1:** 102
forgings ... **M1:** 356
springs, steel ... **M1:** 291
threaded fasteners ... **M1:** 279
for threaded steel fasteners ... **A1:** 295
Cadmium copper *See* Copper alloys, specific types, C16200 ... **A9:** 553
Cadmium copper, deoxidized *See* Copper alloys, specific types, C14300
Cadmium coppers ... **A6:** 762
Cadmium cyanide plating ... **M5:** 256-269
analytical procedures plating baths ... **M5:** 265-267
cyanide-to-cadmium ratio ... **M5:** 256-257
equipment ... **M5:** 258-261
maintenance ... **M5:** 260
solution compositions and
cadmium metal content ... **M5:** 257, 265-266
operating conditions ... **M5:** 256-259, 261-263, 265-268
sodium carbonate content ... **M5:** 257, 266-267
sodium cyanide content ... **M5:** 257, 266
sodium hydroxide content ... **M5:** 257, 266
temperature ... **M5:** 257, 259, 262, 264
Cadmium cyanide plating (continued)
throwing power ... **M5:** 256-257, 259, 261-262
Cadmium electrodeposited coatings ... **A13:** 426, 911-912
embrittlement ... **A13:** 179
embrittlement by ... **A13:** 180, 335
plain carbon steel resistance to ... **A13:** 515
sacrificial corrosion ... **A13:** 587
Cadmium etching by polarized light ... **A9:** 59
slip planes ... **A9:** 684
in zinc alloys ... **A9:** 489
Cadmium ferrites ... **A9:** 538
Cadmium fluoborate plating ... **M5:** 256-258, 264
Cadmium fluoride
added to silicone vapor to control dusting ... **A18:** 684
Cadmium in copper ... **M2:** 241-243
Cadmium in fusible alloys ... **M3:** 799
Cadmium mechanical coatings ... **M5:** 300-302
Cadmium noncyanide plating ... **M5:** 256-257
Cadmium oxide
as a constituent of silver-base electrical contact materials ... **A9:** 551-552, 555-557
in binary phosphate glasses ... **A10:** 131
as glaze ... **EM4:** 1063
Cadmium oxide-graphite
layer lattice solid lubricant ... **A18:** 115
Cadmium plating ... **M5:** 256-269
acid dipping in ... **M5:** 263
adhesion, testing ... **M5:** 269
alkaline soak cleaning in ... **M5:** 263
aluminum and aluminum alloys ... **M5:** 605
anodes ... **M5:** 258, 262-263, 265-266
composition ... **M5:** 258
conforming ... **M5:** 262-263, 265-266
isoluble ... **M5:** 258
applications ... **M5:** 256, 260-266
automatic systems ... **M5:** 257, 259-261
barrel systems ... **M5:** 257-261
bright dipping ... **M5:** 263, 269
brighteners, use of ... **M5:** 256-257
carbonate content, effects of ... **M5:** 257-258, 266-267
cast iron ... **M5:** 256-257, 262, 264
chromate conversion coating of ... **M5:** 269
cleaning process ... **M5:** 261-263
composite process (nickel and cadmium) ... **M5:** 264
copper and copper alloys ... **M5:** 622
corrosion protection ... **M5:** 256, 263-264, 266, 269
corrosion protection by ... **M1:** 753
current density ... **M5:** 257-259
current efficiency ... **M5:** 258-259
cyanide dipping in ... **M5:** 263
cyanide process *See* Cadmium cyanide plating
deposition rates ... **M5:** 259
diffused coatings ... **M5:** 264
discoloration of ... **M5:** 269
drying process ... **M5:** 268
electrolytic cleaning process ... **M5:** 263
equipment ... **M5:** 258-261
maintenance ... **M5:** 260
fasteners ... **M5:** 36, 261
filtration and purification ... **M5:** 258
fluoborate process ... **M5:** 256-258, 264
hardness, base metal, effects of ... **M5:** 263
heat-resisting alloys ... **M5:** 264
hydrogen embrittlement ... **M5:** 256, 263-264, 268-269
magnesium alloys, stripping of ... **M5:** 647
maintenance schedules ... **M5:** 260
method, selection of ... **M5:** 260-261
molybdenum coating ... **M5:** 269
nickel alloys ... **M5:** 264
noncyanide process ... **M5:** 256-257
phosphate treatment ... **M5:** 269
postplating processes ... **M5:** 261-263, 268-269
rectifiers used in ... **M5:** 259
rinsing process ... **M5:** 260-263, 268
rough or pitted deposits causes of ... **M5:** 257-258
safety precautions ... **M5:** 267
selective ... **M5:** 298
shapes requiring conforming anodes ... **M5:** 262, 265-266
solderability ... **M5:** 264
solution compositions and operating
cadmium metal content ... **M5:** 257, 265-266
conditions ... **M5:** 256-259, 261-263, 265-269, 622
Cadmium plating (continued)
sodium carbonate content ... **M5:** 257, 266-267
sodium cyanide content ... **M5:** 257, 266
sodium hydroxide content ... **M5:** 257, 266
springs ... **M5:** 261, 263-264
stainless steel ... **M5:** 264
steel ... **M5:** 29, 256, 261-264, 266-268
still tank systems ... **M5:** 256-259
stripping of ... **M5:** 267-268, 647
temperature ... **M5:** 257, 259, 262, 264
operating ... **M5:** 257, 259, 262
service ... **M5:** 264
thickness ... **M5:** 261-267
measuring ... **M5:** 267
normal variations ... **M5:** 263, 266
throwing power ... **M5:** 256-257, 259, 261-262
valve bodies ... **M5:** 260-261
zinc plating compared to ... **M5:** 253-254, 264, 266
Cadmium, pure ... **M2:** 720-721
solution potential ... **M2:** 207
Cadmium sulfide (CdS), chemical vapor
deposition ... **EM4:** 217
Cadmium sulfide interference film
formation ... **A9:** 142
Cadmium toxicity
biologic indicators ... **A2:** 1241
carcinogenicity ... **A2:** 1241
chronic pulmonary disease ... **A2:** 1240
critical concentration ... **A2:** 1240
disposition ... **A2:** 1239-1240
hypertension and cardiovascular disease ... **A2:** 1241
metallothionein, role in ... **A2:** 1240
skeletal system effects ... **A2:** 1241
of the kidney ... **A2:** 1240
treatment ... **A2:** 1241-1242
Cadmium, vapor pressure
relation to temperature ... **M4:** 309, 310
Cadmium-coated steels
resistance spot welding ... **M6:** 480
Cadmium-copper alloys *See also* Wrought coppers and copper alloys
applications and properties ... **A2:** 277, 283-284
work hardening ... **A2:** 230
Cadmium-plated alloys
galling ... **A18:** 715
Cadmium-tin eutectic ... **A9:** 622
CAE *See* Computer-aided engineering
Caesium, vapor pressure
relation to temperature ... **A4:** 495 **M4:** 310
Cage *See also* Separator ... **A18:** 499, 503
defined ... **A18:** 4
materials ... **A18:** 503
Cake
defined ... **A7:** 2
Caking
of sampling materials ... **A10:** 16
Cal *See* Calorie
CAL (continuous-anneal technology) ... **A4:** 58
Calcareous deposition
in seawater ... **A13:** 897-898
Calcia ceramics
chemical etching ... **EM4:** 575
Calcination ... **EM4:** 109-114
agglomeration ... **EM4:** 112-113
aluminum oxide ... **EM4:** 110-112
effect of process variables ... **EM4:** 113-114
nature of the precursor ... **EM4:** 113
temperature ... **EM4:** 113-114
time ... **EM4:** 113-114
kinetics ... **EM4:** 109
magnesium oxide ... **EM4:** 110-112
processes ... **EM4:** 109
production of fine particles ... **EM4:** 110
purpose ... **EM4:** 42
solid-state sintering ... **EM4:** 271-272
surface area reduction as temperature increased ... **EM4:** 111
thermodynamics ... **EM4:** 109
Calcined alumina
applications ... **EM4:** 47
composition ... **EM4:** 47
hardness ... **EM4:** 351
supply sources ... **EM4:** 47
Calcined magnesium silicate
abrasive in commercial prophylactic paste ... **A18:** 666, 668

Calcining
of aluminum silicates ... **A15:** 209
granulated powders as feedstock ... **EM4:** 100
thermal, for sand reclamation ... **A15:** 227
Calcintering
pressure densification ... **EM4:** 300
Calcite ... **EM4:** 379
in ceramic tiles ... **EM4:** 926
chemical composition ... **A6:** 60
hardness ... **A18:** 433
on Mohs scale ... **A8:** 108
typical oxide compositions of raw materials ... **EM4:** 550
XRDP analysis of ZnO in ... **A10:** 342
Calcitite ... **EM4:** 1008
Calcium
added to nuclear waste, EPMA analysis ... **A10:** 532-535
alloying, aluminum casting alloys ... **A2:** 132
alloying, wrought aluminum alloy ... **A2:** 47
in aluminum alloys
in cast iron ... **A1:** 5
cations important for water quality of cutting fluids ... **A16:** 128
in cement, optical emission spectroscopy for ... **A10:** 21
in compacted graphite iron ... **A1:** 56
deoxidation of stainless steel ... **A16:** 688
deoxidation treatment of turned steels ... **A16:** 673
deoxidizing, copper and copper alloys ... **A2:** 236
determination in paint, absorption and enhancement effects ... **A10:** 98
dietary, and lead toxicity ... **A2:** 1246
EDTA titration ... **A10:** 173
effect of, on machinability of carbon steels ... **A1:** 599
effects of, on notch toughness ... **A1:** 742
as electrode ... **A10:** 184
in enamel cover coats ... **EM3:** 304
forms, as desulfurization reagents ... **A15:** 75
gravimetric finishes ... **A10:** 171
gray cast iron content ... **A16:** 654
ICP-determined in plant tissues ... **A10:** 41
ions, exchanged in water softeners ... **A10:** 658-659
as lead additive ... **A2:** 545
liquid, ultrapure, by external gettering ... **A2:** 1094
lubricant indicators and range of sensitivities ... **A18:** 301
as modifier addition ... **A15:** 161, 484
nickel plating bath contamination by ... **M5:** 208, 210
pure ... **M2:** 721-722
pure, properties ... **A2:** 1105
as pyrophoric ... **A7:** 199
species weighed in gravimetry ... **A10:** 172
spectrometric metals analysis ... **A18:** 300
sulfate ion separation ... **A10:** 169
sulfuric acid as dissolution medium ... **A10:** 165
tantalum corrosion by ... **A13:** 733
TNAA detection limits ... **A10:** 237
ultrapure, by distillation ion ... **A2:** 1094
use in flux cored electrodes ... **M6:** 103
used to make detergents ... **A18:** 100
vapor pressure ... **A6:** 621
volumetric procedures for ... **A10:** 175
weighed as the fluoride ... **A10:** 171
Calcium 12-hydroxystearate ... **A18:** 129
Calcium aluminate ... **EM4:** 45
Calcium aluminate cements
applications ... **EM4:** 47
composition ... **EM4:** 47
supply sources ... **EM4:** 47
Calcium aluminate glasses
applications aerospace ... **EM4:** 1020
optical glass products ... **EM4:** 1074
electrical properties ... **EM4:** 851
Calcium aluminate inclusions in low carbon steel ... **A9:** 628
Calcium aluminates
carburizing affected by content in steels ... **A18:** 875
Calcium aluminoborate glass
electrical conductivity ... **EM4:** 566
Calcium aluminoborosilicate
as E-glass composition ... **EM1:** 45
Calcium aluminoferrite ... **EM4:** 11
Calcium aluminosilicates
aerospace applications ... **EM4:** 1020
Calcium as an addition to steel to reduce sulfur
levels ... **A9:** 628
solubility in lead-calcium alloys ... **A9:** 417
Calcium bentonites *See also* Bentonites; Southern bentonite
and sodium bentonites, blending effects ... **A15:** 210
Calcium boroaluminate glasses (CABAL glasses)
chemical integrity of seals ... **EM4:** 540
Calcium carbide ... **A16:** 71
as flux addition ... **A15:** 389
Calcium carbonate ... **EM3:** 175-176
carburizing role ... **A4:** 325
in ceramic tiles ... **EM4:** 926
decomposition ... **EM4:** 56
dentifrice abrasive ... **A18:** 665, 668
for explosion prevention ... **A7:** 197
as filler ... **EM3:** 179
as filler for polysulfides ... **EM3:** 139
as filler for sealants ... **EM3:** 674
filler for urethane sealants ... **EM3:** 205
function and composition for mild steel SMAW electrode coatings ... **A6:** 60
Miller numbers ... **A18:** 235
and water quality of cutting fluids ... **A16:** 128
Calcium carbonate, as filler/extender
polypropylenes (PP) ... **EM2:** 192
Calcium carbonate, scale
water-formed ... **A13:** 490-491
Calcium carbonates
as sheet molding compound filler ... **EM1:** 158
Calcium chloride
additive causing corrosion in reinforced concrete ... **EM4:** 921
effect on vaporization interferences ... **A10:** 29
Calcium chloride, pitting
stainless steel ... **A13:** 113
Calcium complex soap ... **A18:** 126, 129
Calcium EDTA
as chelator ... **A2:** 1236
Calcium, effects on inclusions and cracking
low-carbon steels ... **A12:** 247
Calcium fluoride
lubricating behavior for thermoplastic composites ... **A18:** 823
rolling element bearing lubricant ... **A18:** 138
thermogravimetric analysis ... **A18:** 823
Calcium fluoride (CaF_2) ... **EM4:** 18
internal origin in a bend specimen ... **EM4:** 640
opal core glass
composition ... **EM4:** 1101
properties ... **EM4:** 1101
opal glass
composition ... **EM4:** 1101
properties ... **EM4:** 1101
Calcium hydroxides
as sheet molding compound thickener ... **EM1:** 158
Calcium hypochlorite ... **A13:** 1180
Calcium in steel ... **M1:** 115
machinability improved by ... **M1:** 576
notch toughness, effect on ... **M1:** 694
Calcium molybdate/zinc phosphate blend
formulation ... **EM3:** 122
Calcium nitrate, apparent threshold stress values
low-carbon steels ... **A8:** 526
Calcium oxide
in binary phosphate glasses ... **A10:** 131
uncombined, analysis in Portland cement ... **A10:** 179
Calcium oxide (CaO)
in composition of textile products ... **EM4:** 403
in composition of wool products ... **EM4:** 403
in drinkware compositions ... **EM4:** 1102
in glaze composition for tableware ... **EM4:** 1102
in ovenware compositions ... **EM4:** 1103
properties ... **EM4:** 424
in tableware compositions ... **EM4:** 1101
Calcium oxides
as sheet molding compound thickener ... **EM1:** 158
Calcium phosphate
thermal conductivity of ... **A11:** 604
and vaporization interferences ... **A10:** 29
Calcium phosphate glass
applications
dental ... **EM4:** 1094
medical ... **EM4:** 1009
Calcium phosphate, scale
water-formed ... **A13:** 491
Calcium pyrophosphate
dentifrice abrasive ... **A18:** 665
Calcium silicate
as filler ... **EM3:** 179
Calcium silicate-low silica
fluxes used for SAW applications ... **A6:** 62
Calcium silicate-neutral
fluxes used for SAW applications ... **A6:** 62
Calcium silicon
defined ... **A15:** 2
Calcium soap ... **A18:** 126, 129
Calcium stearate
as mold release agent ... **EM1:** 158
Calcium stearates
as lubricant ... **A7:** 191
Calcium sulfate
dentifrice abrasive ... **A18:** 665
in plaster molds/cores ... **A15:** 242-243
thermal conductivity of ... **A11:** 604
Calcium sulfate dihydrate (gypsum) ... **A18:** 235, 433
die material for denture teeth ... **A18:** 675
properties ... **A18:** 666
Calcium sulfate, scale
water-formed ... **A13:** 491
Calcium sulfide
precipitates and turning of carbon steels ... **A16:** 673
Calcium sulfide inclusions in low carbon steel ... **A9:** 628
Calcium sulfonate ... **A18:** 126
Calcium titanate ($CaTiO_3$)
hot pressing ... **EM4:** 192
Calcium tungstate ($CaWO_4$) ... **EM4:** 18
Calcium, vapor pressure
relation to temperature ... **A4:** 495 **M4:** 310
Calcium wire injection
direct current arc furnaces ... **A15:** 368
Calcium-aluminum-silicate/calumite slag
purpose for use in glass manufacture ... **EM4:** 381
Calcium-high silica
fluxes used for SAW applications ... **A6:** 62
Calcium-magnesium (Ca-Mg) ... **EM4:** 22
Calcium-modified zinc phosphate ... **A13:** 386
Calcium/magnesium (water hardness) as electrode ... **A10:** 184
Calculated quantities
relationship to measured quantities ... **A9:** 124
Calculation matrix *See also* Experimental design; Orthogonal arrays
variable choice ... **A17:** 750
Calculator, programmable
composite material analysis by ... **EM1:** 277-279
Calculators
desktop ... **A10:** 310
Calender
defined ... **EM1:** 6 **EM2:** 8
Caliber rolling ... **A14:** 347
Calibrate
defined ... **A8:** 2

SUBJECTS OF THE INDEXED VOLUMES: ASM Handbook (designated by the letter "A"): **A1:** Properties and Selection: Irons, Steels, and High-Performance Alloys (1990); **A2:** Properties and Selection: Nonferrous Alloys and Special-Purpose Materials (1990); **A3:** Alloy Phase Diagrams (1992); **A4:** Heat Treating (1991); **A6:** Welding, Brazing, and Soldering (1993); **A7:** Powder Metallurgy (1984); **A8:** Mechanical Testing (1985); **A9:** Metallography and Microstructures (1985); **A10:** Materials Characterization (1986); **A11:** Failure Analysis and Prevention (1986); **A12:** Fractography (1987); **A13:** Corrosion (1987); **A14:** Forming and Forging (1988); **A15:** Casting (1988); **A16:** Machining (1989); **A17:** Nondestructive Testing and Quality Control (1989); **A18:** Friction, Lubrication, and Wear Technology (1992). **Metals Handbook, 9th Edition** (designated by the letter "M"): **M1:** Properties and Selection: Irons and Steels (1978); **M2:** Properties and Selection: Nonferrous Alloys and Pure Metals (1979); **M3:** Properties and Selection: Stainless Steels, Tool Materials and Special-Purpose Materials (1980); **M4:** Heat Treating (1981); **M5:** Surface Cleaning, Finishing, and Coating (1982); **M6:** Welding, Brazing, and Soldering (1983). **Engineered Materials Handbook** (designated by the letters "EM"): **EM1:** Composites (1987); **EM2:** Engineering Plastics (1988); **EM3:** Adhesives and Sealants (1990); **EM4:** Ceramics and Glasses (1991); **Electronic Materials Handbook** (designated by the letters "EL"): **EL1:** Packaging (1989).

Calibrate (verb)
defined ... **A7:** 2
Calibrated leaks
for leak rate measurement ... **A17:** 70
Calibration
accuracy tolerances ... **A8:** 612
acoustical holography ... **A17:** 444-445
agency requirements of ... **A8:** 611
automatic, in CMMs ... **A17:** 20
by laser interferometer ... **A17:** 15
of Charpy pendulum impact machine ... **A8:** 266
classical wet analyses for ... **A10:** 162
and crack-extension force ... **A8:** 441-443
curves, for quantitative x-ray spectrometry ... **A10:** 97-98
and data reduction of dynamic tests torsional Kolsky bar ... **A8:** 228
defined ... **A8:** 2, 611
durometer ... **A8:** 107
in dynamic notched round bar testing ... **A8:** 279
elastic devices for ... **A8:** 614-615
energy, in spectrum-fitting programs ... **A10:** 91
error, in direct current electrical potential method ... **A8:** 389
of extensometers ... **A8:** 616-619
for ICP-AES ... **A10:** 34
load cells ... **A8:** 615-616
of optical measuring device, round bar testing ... **A8:** 280-281
samples, by RBS ... **A10:** 628
of Scleroscope hardness test ... **A8:** 105-106
Scleroscope HFRSc/HFRSd ... **A8:** 104
sensitivity, acoustic emission inspection ... **A17:** 280
temperature, for thermal inspection ... **A17:** 400
of testing equipment ... **A8:** 611-619
thermocouple ... **A2:** 878-881
thermocouple, changes during service ... **A2:** 881-882
for TNAA ... **A10:** 236
ultrasonic inspection equipment ... **A17:** 266-267
ultrasonic inspection of forgings ... **A17:** 505-506
and verification, of testing equipment ... **A8:** 611
wavelength, MFS analysis ... **A10:** 77
of wavenumber, F-F-IR spectroscopy ... **A10:** 112
for XPS qualitative analysis ... **A10:** 572
Calibration curve
for compact-type specimen ... **A8:** 386-387
and electrical potential measurement accuracy ... **A8:** 386
of single-edge notched specimen ... **A8:** 386
Calibrator
for mechanical/electrical extensometers and load-elongation recorders ... **A8:** 618
requirements of ... **A8:** 611
Californium *See also* Transplutonium actinide metals
applications and properties ... **A2:** 1198-1201
pure ... **M2:** 832-833
Californium-252 neutron sources
use in borehole logging ... **A10:** 240
Caliper survey
gas/oil production monitoring ... **A13:** 1251
Calipers
recording ... **A7:** 229
Callable graphics libraries
for process automation ... **A14:** 410
Calomel electrode *See also* Electrode potential; Reference electrode; Saturated calomel electrode
for acidified chloride solutions ... **A8:** 419
defined ... **A10:** 670 **A13:** 2
Calorie ... **A8:** 724
abbreviation for ... **A10:** 691
Calorimeter ... **EM3:** 7
defined ... **EM2:** 8
Calorizing
defined ... **A13:** 2
Calrod preheaters
wave soldering systems ... **EL1:** 684
Calumite slag
as fining agent ... **EM4:** 380
Cam backer
designs of ... **A14:** 131
Cam lobe
antiwear film analyzed for ... **A10:** 565-566
lubricated wear in ... **A11:** 361
Cam plastometer ... **A8:** 193-196
Cam press
defined ... **A14:** 2
Cam-actuated flanging dies
for press bending ... **A14:** 526-527
Cam-driven compacting press ... **A7:** 330
Cam-driven dies
for press-brake forming ... **A14:** 538
Cam-lever
constant-stress ... **A8:** 319-320
Camber ... **EL1:** 468, 483
defined ... **A14:** 2
in electrical steel sheet ... **A14:** 480-481
interference microscope measurement ... **A17:** 17
slitting, coiled metals ... **A14:** 709-710
in straightening ... **A14:** 680
wrought copper and copper alloys ... **A2:** 247-248
Camber angle ... **A18:** 578
Cambridge Crystallographic Data File ... **A10:** 355
Camcorders
thick-film hybrid applications ... **EL1:** 385
Cameras *See also* Television cameras
35-mm single-lens-reflex ... **A12:** 78-79
back-reflection pinhole, schematic ... **A10:** 334
Debye-Scherrer, XRPD analysis ... **A10:** 335
Gandolfi, XRPD analysis ... **A10:** 335
glancing-angle, XRPD analysis ... **A10:** 336
Guinier, in asymmetric transmission arrangement ... **A10:** 335
Guinier, XRPD analysis ... **A10:** 335-336
Huber Guinier ... **A10:** 336
human eye vs. vidicon ... **A17:** 30
improved resolution, machine vision ... **A17:** 43
Laue, XRPD analysis ... **A10:** 334-335
magnification ... **A12:** 80
micro-, XRPD analysis ... **A10:** 336
pinhole, XRPD analysis ... **A10:** 334-335
Read, XRPD analysis ... **A10:** 336
reflection ... **A10:** 369
solid-state, vision machine ... **A17:** 32-33
transmission pinhole, schematic ... **A10:** 334
vidicon, vision machine ... **A17:** 31-32
view ... **A12:** 78-79
Cams and camshafts ... **A7:** 616-621
Cams, cast iron
coatings for ... **M1:** 104
Camshaft, carburized
wear compared to induction hardened ... **M1:** 629, 630
Camshafts
economy in manufacture ... **M3:** 847
magnetizing ... **A17:** 94
Can (pack)
for rolling titanium and nickel-base alloys ... **A14:** 356
Can ironing press ... **A14:** 335-336
Can, metal
as package ... **EL1:** 958
Can seaming
of metal strip ... **A14:** 572
Can solder ... **A9:** 422
Can vacuum degassing
aluminum P/M alloys ... **A2:** 202-203
Can/cast fluid dies ... **A7:** 543, 544
Canadian Department of Mines, Energy
and Resources, cam plastometer at ... **A8:** 194
Canadian Department of Mines, Energy, and Resources ... **A8:** 194
for high strain rate compression testing ... **A8:** 187
for hot compression testing ... **A8:** 582
for medium-rate compression testing ... **A8:** 190
Canadian heavy-water-moderated (CANDU) nuclear reactors ... **A7:** 664
Canadian industries Limited flow test *See* CIL flow test
Canadian Plastics
as information source ... **EM2:** 93
Canadian Standards Association ... **EM2:** 461
Canasite
polishing ... **EM4:** 469
Cancellous bone screw
as internal fixation device ... **A11:** 671
Cancer
beryllium-caused ... **A7:** 202
from nickel toxins ... **A7:** 203
lung, copper toxicity ... **A7:** 205
Candela
as investment casting wax ... **A15:** 253
Candela, as SI base unit
symbol for ... **A10:** 685
Candescent lamp filaments ... **A7:** 16
Candles, smoke *See* Smoke candles
Canning ... **A18:** 738
defined ... **A14:** 2
Cannon tubes
tests for ... **A11:** 281
Cannula tubes
abrasive flow machining ... **A16:** 518, 519
Cans *See also* Containers
beverage, aluminum alloy ... **A2:** 10
as cemented carbide application ... **A2:** 971
defined ... **A7:** 2
leak-free, for powder encapsulation ... **A7:** 433
shape and dimensional control by design of ... **A7:** 432
sheet metal powder, rectangular and cylindrical ... **A7:** 431
tantalum ... **A2:** 559
tin ... **A13:** 778-779
Canted-vane wheels
blast cleaning ... **A15:** 515
Cantilever
beams, distortion and stress ratios ... **A11:** 137
curl, in ceramics ... **A11:** 747
curl, in four-point bend specimens ... **A11:** 746
loading, fractures in ... **A11:** 108
Cantilever beam bend test *See also* Cantilever beam test ... **A8:** 132-134
Cantilever beam clip gage
for displacement measurement ... **A8:** 383
Cantilever beam machine
for fatigue testing ... **A8:** 369
Cantilever beam specimen
stress-corrosion cracking in ... **A8:** 500
Cantilever beam strip specimen
stress-relaxation bend testing ... **A8:** 326
Cantilever beam test
and contoured double-cantilever beam test compared ... **A8:** 538
for hydrogen embrittlement ... **A8:** 537-538 **A13:** 284
procedure ... **A8:** 538
and wedge-opening load test compared ... **A8:** 538-539
Cantilever bend specimens
constant-curvature, stress-relaxation bend testing ... **A8:** 326
SCC testing ... **A8:** 511 **A13:** 254
Cantilever reverse bending
fatigue test specimen ... **A8:** 371
Cantilever rolls ... **A18:** 57
Cantilever snap joints ... **EM2:** 714-718
Cantilever spring leaves
characteristics of ... **M1:** 309-311
Cantilever-type coordinate measuring machines ... **A17:** 20-21
Canton flannel buffs ... **M5:** 118-119, 125
Canvas awnings protective coatings
powders used ... **A7:** 572
CaO(calcium oxide)-TiO_2-SiO_2 (CTS) glass ... **A6:** 952
CAP process (consolidation of atomized powders) ... **A7:** 533-536
elements ... **A7:** 533-534
materials ... **A7:** 535-536
sequence and advantages ... **A7:** 533-535
Cap screws *See also* Bolts
commercial, service distortion of ... **A11:** 142
fatigue fracture of ... **A11:** 533, 535
selection of steel for ... **M1:** 275-276
strength distribution ... **M1:** 279
Capacitance
defined ... **EL1:** 90, 417-418 **EM2:** 8
distributed, flexible printed boards ... **EL1:** 587-588
effects in pulse polarography ... **A10:** 193
formulas for ... **EL1:** 29
SI derived unit and symbol ... **A10:** 685
SI unit/symbol for ... **A8:** 721
variation with voltage ... **EL1:** 158
Capacitance gages
amplitude detection, ultrasonic testing ... **A8:** 245-246
for Hugoniot elastic limit measurement ... **A8:** 211

Capacitance gages (continued)
as strain gage ... **A8:** 618
Capacitance manometer
in environmental test chamber ... **A8:** 411
Capacitive environments
vs transmission line environments ... **EL1:** 601
Capacitive loading ... **EL1:** 26, 37-39
Capacitive ratio test
for plastic package hermeticity ... **EL1:** 953
Capacitor
for strain measurement ... **A8:** 202
Capacitor banks
for electromagnetic forming ... **A14:** 650
Capacitor dielectrics, ceramic *See* Ceramic capacitor dielectrics
Capacitor discharge (CD) stud welding ... **A6:** 210, 221-222
advantages ... **A6:** 221
aluminum ... **A6:** 221, 222
applications ... **A6:** 221-222
definition ... **A6:** 221
disadvantages ... **A6:** 221
drawn-arc mode ... **A6:** 221, 222
equipment ... **A6:** 222
heat-affected zone (HAZ) ... **A6:** 221
initial-contact mode ... **A6:** 221-222
initial-gap mode ... **A6:** 221-222
personnel responsibilities ... **A6:** 222
stainless steel ... **A6:** 221
Capacitor discharge stud welding of
alloy steel ... **M6:** 730
aluminum alloys ... **M6:** 730, 736
brass ... **M6:** 738
carbon steel ... **M6:** 730
copper ... **M6:** 738
copper alloys ... **M6:** 730
low-carbon steel ... **M6:** 738
stainless steel ... **M6:** 730, 736
titanium and titanium alloys ... **M6:** 738
zinc alloys ... **M6:** 738
Capacitor discharge welding (CDW)
aluminum metal-matrix composites ... **A6:** 555, 558
dispersion-strengthened aluminum alloys ... **A6:** 543, 544
Capacitor-discharge resistance brazing ... **M6:** 985-987
Capacitor-discharge welding ... **M6:** 740-743
control ... **M6:** 742
current ... **M6:** 742-743
design and size of workpieces ... **M6:** 739-740
displacement ... **M6:** 742-743
high-voltage welding ... **M6:** 744-745
applications ... **M6:** 744
machines ... **M6:** 744-745
low-voltage welding ... **M6:** 743-744
high-frequency-start machines ... **M6:** 743-744
nib-starter machines ... **M6:** 743
preparation of workpieces ... **M6:** 742
sequence of steps ... **M6:** 742
stud welding *See* Stud welding
voltage ... **M6:** 742-743
Capacitors *See also* Tantalum capacitors
aluminum alloy ... **A2:** 13
barium-titanate, and lead germanate ... **A2:** 743
decoupling, placement ... **EL1:** 28
defined ... **EL1:** 90
devitrifying dielectrics for ... **EL1:** 109
dielectrics, thick-film pastes ... **EL1:** 342-343
dipped mica, solderability defects ... **EL1:** 1036-1037
electrolytic, refractory metals and alloys ... **A2:** 557, 559
electronic ... **A7:** 160-163
fabrication ... **EL1:** 185-187
failure mechanisms ... **EL1:** 971-973
high-voltage, percussion welding ... **M6:** 740
as IC modification ... **EL1:** 249
implementation at microwave frequency ... **EL1:** 178
low-voltage, percussion welding ... **M6:** 740
Capacitors (continued)
materials selection ... **EL1:** 182
miscellaneous ... **EL1:** 998
monolithic, active analog components ... **EL1:** 144
MOS, structures ... **EL1:** 156
multilayer ceramic ... **A7:** 151
parylene coatings ... **EL1:** 799
in passive components ... **EL1:** 178-179
passive devices, failure mechanisms ... **EL1:** 994-999
removal methods ... **EL1:** 724-727
resistance brazing ... **M6:** 977
thin-film ... **EL1:** 320-321
types ... **EL1:** 178-179
wound-film ... **EL1:** 999
Capacity
defined ... **A8:** 2
Capacity number (C_n)
defined ... **A18:** 4
Capillarity
and component removal ... **EL1:** 715
effects, eutectic growth ... **A15:** 124
model, of nucleation during solidification ... **A15:** 103
Capillary action
definition ... **M6:** 3
Capillary attraction
defined ... **A7:** 2
Capillary balance test
for solderability ... **EL1:** 944
Capillary condensation ... **A18:** 400
Capillary drilling (CD) ... **A16:** 509, 551-553
acid electrolytes ... **A16:** 551, 552, 553
advantages ... **A16:** 551
applications ... **A16:** 551
CNC machines ... **A16:** 552
equipment and tooling ... **A16:** 552-553
limitations ... **A16:** 551
process capabilities ... **A16:** 551
process parameters ... **A16:** 553
Capillary infiltration methods ... **A7:** 552, 553
Capillary protocol
for molecular weight ... **EM2:** 535
Capillary rheometer (viscometer) ... **EM3:** 322, 323
Capillary rheometry
for rheological behavior measurement ... **EM4:** 174
Capillary rise
in liquid penetrant inspection ... **A17:** 71-73
Capillary spaces
residue cleaning from ... **EL1:** 666
Capillary tubing
used in mounting wire specimens ... **A9:** 31
Capital equipment
economic analysis ... **EM2:** 294
Capped steels ... **A1:** 141, 143 **M1:** 112, 123
sheet strain-age embrittlement of ... **M1:** 683
wire rod ... **M1:** 255, 257
Capping
defined ... **A17:** 383
Capping of abrasive particles, defined ... **A9:** 3
Carbide, defined ... **A9:** 3
Carbide-formation rate in CF stainless steel casting alloys ... **A9:** 298
Caprolactam ... **EM3:** 7
defined ... **EM2:** 8
Capsil 9, brazing
composition ... **A6:** 117
Captive hybrid markets ... **EL1:** 253
Capture cross section, defined
neutron radiography ... **A17:** 390
Capture efficiency *See* Collection efficiency
Carbanion ion ... **EM3:** 7
Carbanium ion
defined ... **EM2:** 8
Carbide
abrasive wear materials ... **A18:** 189
coating finishing ... **A18:** 831
coating thickness limitations ... **A18:** 831
effect on cast iron ... **M6:** 999
Carbide (continued)
extension of tool life, via ion implantation, examples ... **A18:** 643
hardfacing material ... **M6:** 776-777
methods used for synthesis ... **A18:** 802
plasma-assisted physical vapor deposition process ... **A18:** 848
spray material for oxyfuel powder spray method ... **A18:** 830
Carbide ceramic coatings ... **M5:** 535-536, 541, 546
hardness ... **M5:** 546
melting points ... **M5:** 534-535
Carbide ceramics
substrate for thermoreactive deposition/diffusion process ... **A4:** 449
Carbide cermets *See also* Carbides; Cermets ... **A7:** 799, 804-811
aluminum-boron carbide cermets ... **A2:** 1002
aluminum-silicon carbide cermets ... **A2:** 1002
and carbonitride cermets ... **A2:** 995-1003
chromium carbide cermets ... **A2:** 1000-1001
defined ... **A2:** 979
hafnium carbide cermets ... **A2:** 1001
nickel-bonded titanium carbide cermets ... **A2:** 995
niobium carbide cermets ... **A2:** 1001
steel-bonded titanium carbide cermets ... **A2:** 996-998
steel-bonded tungsten ... **A2:** 1000
tantalum carbide cermets ... **A2:** 1001
zirconium carbide cermets ... **A2:** 1001
Carbide, chemical vapor
deposition of ... **M5:** 381
Carbide composition
as remaining life indicator ... **A17:** 55
Carbide cutting tools, eliminating brittleness of ... **A3:** 1•28
Carbide degeneration ... **A1:** 985
Carbide fibers ... **EM1:** 63-64
commercially available types ... **EM1:** 61
development ... **EM1:** 60
and oxide fibers, compared ... **EM1:** 64
Carbide grain size in cemented carbides
determination of ... **A9:** 275
Carbide hardfacing powders ... **A7:** 827-828
Carbide oxidation-resistant coating ... **M5:** 665-666
Carbide particles
delineation of, in heat-resistant casting alloys ... **A9:** 330-331
effect on high-temperature strength in iron-alloys ... **A9:** 333
chromium-nickel heat-resistant casting
formation of, in cobalt-base heat-resistant casting alloys ... **A9:** 334
formation of, in iron-chromium-nickel heat- resistant casting alloys ... **A9:** 333
in high speed steels, revealed by differential interference contrast ... **A9:** 59
in high-speed steel, contrast enhancement for scanning electron microscopy ... **A9:** 98-99
identification of, in heat-resistant casting alloys ... **A9:** 332
in roller bearing steels, revealed by differential interference contrast ... **A9:** 59
Carbide phase in ferritic chromium steel
detection by phase contrast etching ... **A9:** 59
Carbide precipitates in
austenitic manganese steel castings ... **A9:** 239
Carbide precipitation
in austenitic stainless steels ... **A9:** 283-284
in bainite ... **A9:** 662-664
brazing and ... **A6:** 117
ductility discontinuities from ... **A8:** 34
in stainless steel casting alloys, effect on intergranular corrosion ... **A9:** 297-298
Carbide strengthening
of nickel and nickel alloys ... **A2:** 429-430

SUBJECTS OF THE INDEXED VOLUMES: ASM Handbook (designated by the letter "A"): **A1:** Properties and Selection: Irons, Steels, and High-Performance Alloys (1990); **A2:** Properties and Selection: Nonferrous Alloys and Special-Purpose Materials (1990); **A3:** Alloy Phase Diagrams (1992); **A4:** Heat Treating (1991); **A6:** Welding, Brazing, and Soldering (1993); **A7:** Powder Metallurgy (1984); **A8:** Mechanical Testing (1985); **A9:** Metallography and Microstructures (1985); **A10:** Materials Characterization (1986); **A11:** Failure Analysis and Prevention (1986); **A12:** Fractography (1987); **A13:** Corrosion (1987); **A14:** Forming and Forging (1988); **A15:** Casting (1988); **A16:** Machining (1989); **A17:** Nondestructive Testing and Quality Control (1989); **A18:** Friction, Lubrication, and Wear Technology (1992). **Metals Handbook, 9th Edition** (designated by the letter "M"): **M1:** Properties and Selection: Irons and Steels (1978); **M2:** Properties and Selection: Nonferrous Alloys and Pure Metals (1979); **M3:** Properties and Selection: Stainless Steels, Tool Materials and Special-Purpose Materials (1980); **M4:** Heat Treating (1981); **M5:** Surface Cleaning, Finishing, and Coating (1982); **M6:** Welding, Brazing, and Soldering (1983). **Engineered Materials Handbook** (designated by the letters "EM"): **EM1:** Composites (1987); **EM2:** Engineering Plastics (1988); **EM3:** Adhesives and Sealants (1990); **EM4:** Ceramics and Glasses (1991); **Electronic Materials Handbook** (designated by the letters "EL"): **EL1:** Packaging (1989).

Carbide tool inserts
vapor forming ... **M5:** 382-383
Carbide tools
boring tools ... **A16:** 162-164, 166, 168-169, 171-173
boring tools for Al alloys ... **A16:** 771, 777, 778
boring tools for Cu ... **A16:** 810, 813
boring tools for heat-resistant alloys ... **A16:** 742
boring tools for refractory metals ... **A16:** 859, 863
boring tools for tool steels ... **A16:** 716
broaching ... **A16:** 197, 200-202, 203, 206, 207
broaching tools for Al alloys ... **A16:** 775, 779
broaching tools for Cu alloys ... **A16:** 813
circular saws for Al alloys ... **A16:** 794, 800
circular saws for Cu alloys ... **A16:** 817
circular saws for Mg alloys ... **A16:** 827, 828
coated ... **A2:** 959-962
compared to cast Co alloy tools ... **A16:** 70
compared to TiN-coated tool steels ... **A16:** 58
counterboring tools for cast irons ... **A16:** 660
counterboring tools for Mg alloys ... **A16:** 825
counterboring tools for refractory metals ... **A16:** 860
disposable tools ... **A16:** 150, 155
drills ... **A16:** 43, 217, 218-220, 233-237
drills for Al alloys ... **A16:** 781, 784, 785
drills for cast irons ... **A16:** 658
drills for Cu alloys ... **A16:** 814
drills for heat-resistant alloys ... **A16:** 747, 748
drills for MMCs ... **A16:** 896, 897
drills for P/M materials ... **A16:** 890
drills for refractory metals ... **A16:** 860, 861, 865
drills for Ti alloys ... **A16:** 847
drills for tool steels ... **A16:** 718
drills for uranium alloys ... **A16:** 875
drills for Zn alloys ... **A16:** 832
electrochemical grinding ... **A16:** 543, 544
end milling ... **A16:** 43
end milling tools for cast irons ... **A16:** 663
end milling tools for Cu alloys ... **A16:** 817
end milling tools for heat-resistant alloys ... **A16:** 754
end milling tools for Mg alloys ... **A16:** 827
end milling tools for Ti alloys ... **A16:** 849
end milling tools for tool steels ... **A16:** 722
end milling tools forAl alloys ... **A16:** 768-787
end milling-slotting tools for refractory metals ... **A16:** 866, 867
face milling tools ... **A16:** 43
face milling tools for Al alloys ... **A16:** 788-789
face milling tools for cast irons ... **A16:** 662
face milling tools for Cu alloys ... **A16:** 816
face milling tools for heat-resistant alloys ... **A16:** 753
face milling tools for Mg alloys ... **A16:** 827
face milling tools for refractory metals ... **A16:** 859
face milling tools for Ti alloys ... **A16:** 850
friction coefficient and specific power ... **A16:** 18
gear hand finishing ... **A16:** 343
ground and shaped with diamond wheels ... **A16:** 455, 460-461
gun drills for Al alloys ... **A16:** 779
gun drills for Mg alloys ... **A16:** 824
high removal rate machining ... **A16:** 608
high-speed machining ... **A16:** 601, 603
hollow milling tools for refractory metals ... **A16:** 862
indexable inserts ... **A16:** 456
machining parts tested when made from high-strength steel grades ... **A16:** 679
milling ... **A16:** 311-315, 317-318, 321-323, 325-327, 329
milling Al alloys ... **A16:** 769, 785, 792, 793, 797
milling cast irons ... **A16:** 660, 661
milling cutters for Cu alloys ... **A16:** 816
milling cutters for Hf ... **A16:** 856
milling cutters for MMCs ... **A16:** 898
milling cutters for refractory metals ... **A16:** 867
milling cutters for stainless steels ... **A16:** 703
milling cutters for Ti alloys ... **A16:** 846
milling cutters for tool steels ... **A16:** 726
multiple-operation machining tools ... **A16:** 369, 380, 381, 386, 388-389
peripheral end mill tools for refractory metals ... **A16:** 866, 867
peripheral end mill tools for Ti alloys ... **A16:** 849
peripheral milling of Mg alloys ... **A16:** 827
planing tools ... **A16:** 184, 185-186

Carbide tools (continued)
planing tools for Al alloys ... **A16:** 773, 778
planing tools for cast irons ... **A16:** 657, 660
planing tools for Cu ... **A16:** 811, 813
planing tools for heat-resistant alloys ... **A16:** 743
planing tools for Hf ... **A16:** 856
power band saws for MMCs ... **A16:** 897
reamers ... **A16:** 240-241, 243-247, 456
reamers for Al alloys ... **A16:** 781
reamers for cast irons ... **A16:** 659, 660
reamers for heat-resistant alloys ... **A16:** 750
reamers for Mg alloys ... **A16:** 823, 825
reamers for P/M materials ... **A16:** 890
reamers for refractory metals ... **A16:** 862-865, 867
reamers for stainless steels ... **A16:** 702-703, 705
reamers for Ti alloys ... **A16:** 847, 852
reamers for tool steels ... **A16:** 718
reamers for Zn alloys ... **A16:** 832, 833
shaping and slotting operations ... **A16:** 190
shaping tools for Al alloys ... **A16:** 778
shaping tools for Hf ... **A16:** 856
spade drills ... **A16:** 223-225
spotfacing tools for cast irons ... **A16:** 660
spotfacing tools for refractory metals ... **A16:** 860
taps ... **A16:** 256, 259
tip drills ... **A16:** 229
tool grinding ... **A16:** 450
tools compared to TiN-coated tool steels ... **A16:** 58
tools for adaptive control ... **A16:** 618
trepanning tools for refractory metals ... **A16:** 860
turning tools ... **A16:** 43, 145, 148, 150-152, 154, 156-159
turning tools for Al alloys ... **A16:** 770, 774, 775, 776
turning tools for Be alloys ... **A16:** 872
turning tools for carbon and alloy steels ... **A16:** 670, 673
turning tools for cast irons ... **A16:** 653, 654, 658
turning tools for Cu alloys ... **A16:** 809, 811, 812
turning tools for heat-resistant alloys ... **A16:** 739, 741
turning tools for Hf ... **A16:** 856
turning tools for MMCs ... **A16:** 898
turning tools for Ni alloys ... **A16:** 837
turning tools for refractory metals ... **A16:** 858
turning tools for stainless steels ... **A16:** 692, 693, 696, 697
turning tools for Ti alloys ... **A16:** 846, 847
turning tools for tool steels ... **A16:** 708-710, 713
turning tools for uranium alloys ... **A16:** 875
turning tools for Zn alloys ... **A16:** 831
twist drills ... **A16:** 456
wear pads ... **A16:** 222-223

Carbide tools, specific types
C-1, counterboring tools for refractory metals ... **A16:** 860
C-1, drills ... **A16:** 233, 234
C-1, planing tools for cast irons ... **A16:** 657
C-1, spotfacing tools for refractory metals ... **A16:** 860
C-1, tools for cast iron machining ... **A16:** 656, 657
C-1, tools for Hf ... **A16:** 855
C-1, tools for refractory metals ... **A16:** 860
C-1, tools for Zr ... **A16:** 855
C-1, turning tools for Zr ... **A16:** 853
C-2, boring tools for Al alloys ... **A16:** 771
C-2, boring tools for Cu alloys ... **A16:** 813
C-2, boring tools for heat-resistant alloys ... **A16:** 742
C-2, boring tools for refractory metals ... **A16:** 859
C-2, circular saws for Al alloys ... **A16:** 794
C-2, counterboring tools for carbon steels ... **A16:** 251
C-2, counterboring tools for cast irons ... **A16:** 660
C-2, counterboring tools for heat-resistant alloys ... **A16:** 752
C-2, counterboring tools for refractory metals ... **A16:** 860
C-2, drills ... **A16:** 233, 234, 236, 237
C-2, drills for cast irons ... **A16:** 658
C-2, drills for heat-resistant alloys ... **A16:** 747, 749
C-2, drills for refractory metals ... **A16:** 860
C-2, drills for tool steels ... **A16:** 718
C-2, end milling tools ... **A16:** 325, 326, 722
C-2, end Milling tools for cast irons ... **A16:** 663
C-2, end milling tools for heat-resistant alloys ... **A16:** 754

Carbide tools, specific types (continued)
C-2, face milling tools ... **A16:** 323
C-2, face milling tools for Al alloys ... **A16:** 788-789
C-2, face milling tools for cast irons ... **A16:** 662
C-2, face milling tools for heat-resistant alloys ... **A16:** 753
C-2, face milling tools for refractory metals ... **A16:** 863
C-2, face milling tools for Ti alloys ... **A16:** 848, 850
C-2, gun drills for Al alloys ... **A16:** 779
C-2, hollow milling tools for refractory metals ... **A16:** 862
C-2, milling cutters for Al alloys ... **A16:** 797
C-2, milling cutters for refractory metals ... **A16:** 864
C-2, milling cutters for stainless steels ... **A16:** 699
C-2, milling cutters for Ti alloys ... **A16:** 845
C-2, oil hole or pressurized coolant drills for refractory metals ... **A16:** 861
C-2, peripheral end milling tools for Al alloys ... **A16:** 786-787
C-2, peripheral end milling tools for refractory metals ... **A16:** 866
C-2, peripheral end milling tools for Ti alloys ... **A16:** 849
C-2, planing tools for Al alloys ... **A16:** 773
C-2, planing tools for cast irons ... **A16:** 657
C-2, reamers for Al alloys ... **A16:** 781
C-2, reamers for cast irons ... **A16:** 659
C-2, reamers for heat-resistant alloys ... **A16:** 750, 751
C-2, reamers for Ni alloys ... **A16:** 839
C-2, reamers for refractory metals ... **A16:** 862
C-2, reamers for stainless steels ... **A16:** 702-703
C-2, reamers for Ti alloys ... **A16:** 852
C-2, reamers for tool steels ... **A16:** 719
C-2, spade and gun drills for Ni alloys ... **A16:** 839
C-2, spade drilling tools for Al alloys ... **A16:** 777
C-2, spade drills for refractory metals ... **A16:** 861
C-2, spotfacing tools for carbon steels ... **A16:** 251
C-2, spotfacing tools for cast irons ... **A16:** 660
C-2, spotfacing tools for heat-resistant alloys ... **A16:** 752
C-2, spotfacing tools for refractory metals ... **A16:** 860
C-2, tools for Al alloys ... **A16:** 767
C-2, tools for cast irons ... **A16:** 656, 657
C-2, tools for Hf ... **A16:** 855
C-2, tools for plastic molds ... **A16:** 723
C-2, tools for refractory metals ... **A16:** 860
C-2, tools for Ti alloys ... **A16:** 844
C-2, tools for Zr ... **A16:** 855
C-2, trepanning tools ... **A16:** 179
C-2, trepanning tools for refractory metals ... **A16:** 860
C-2, turning tools ... **A16:** 145, 147
C-2, turning tools for Al alloys ... **A16:** 770
C-2, turning tools for cast irons ... **A16:** 651, 652, 653
C-2, turning tools for Cu alloys ... **A16:** 811
C-2, turning tools for heat-resistant alloys ... **A16:** 739, 741
C-2, turning tools for MMCs ... **A16:** 897, 898
C-2, turning tools for Ni alloys ... **A16:** 838
C-2, turning tools for refractory metals ... **A16:** 858
C-2, turning tools for Ti alloys ... **A16:** 845, 847
C-2, turning tools for Zr ... **A16:** 853
C-3, boring tools for Al alloys ... **A16:** 771
C-3, boring tools for Cu alloys ... **A16:** 813
C-3, boring tools for heat-resistant alloys ... **A16:** 742
C-3, drills ... **A16:** 233, 234
C-3, end milling-slotting tools ... **A16:** 866
C-3, face milling tools for refractory metals ... **A16:** 863
C-3, planing tools for cast irons ... **A16:** 657
C-3, tools for Al alloys ... **A16:** 767, 769
C-3, tools for cast irons ... **A16:** 656, 657
C-3, tools for Ti alloys ... **A16:** 844
C-3, turning tools for Al alloys ... **A16:** 770
C-3, turning tools for cast irons ... **A16:** 653
C-3, turning tools for Cu alloys ... **A16:** 811
C-3, turning tools for heat-resistant alloys ... **A16:** 739, 741
C-3, turning tools for Ti alloys ... **A16:** 847
C-3, turning tools for uranium alloys ... **A16:** 875
C-3, turning tools for Zr ... **A16:** 853

Carbide tools, specific types (continued)
C-4, boring tools for refractory metals **A16:** 859
C-4, grade Carbaloy **A16:** 872
C-4, grade Teledyne HF **A16:** 872
C-4, tools for cast irons **A16:** 656, 657
C-4, tools for refractory metals **A16:** 860
C-4, turning tools for Be alloys **A16:** 872
C-4, turning tools for refractory metals **A16:** 858
C-4, turning tools for uranium alloys **A16:** 875
C-4, turning tools for Zr **A16:** 853
C-5, climb milling tools for heat-resistant alloys **A16:** 755
C-5, drills **A16:** 236, 237
C-5, electrochemical discharge grinding tools
C-5, end milling tools **A16:** 325, 328
C-5, end milling tools for castirons **A16:** 663, 664
C-5, end milling tools for heat-resistant alloys **A16:** 754
C-5, face milling tools **A16:** 325, 328
C-5, machinability testing **A16:** 646
C-5, planing tools for cast irons **A16:** 657
C-5, slab milling tools **A16:** 328
C-5, tools for cast irons **A16:** 656, 657
C-5, turning tools for tools for uranium alloys **A16:** 875
C-6, boring tools for heat-resistant alloys **A16:** 742
C-6, electrochemical discharge grinding **A16:** 550
C-6, face milling tools **A16:** 323, 714, 715
C-6, face milling tools for cast irons **A16:** 662
C-6, milling tools for stainless steels **A16:** 699
C-6, planing tools for cast irons **A16:** 657
C-6, planing tools for heat-resistant alloys **A16:** 743
C-6, reaming tools for Ni alloys **A16:** 839
C-6, spade and gun drilling tools for Ni alloys **A16:** 839
C-6, tools for cast irons **A16:** 656, 657
C-6, trepanning tools **A16:** 179
C-6, turning tools **A16:** 145, 147
C-6, turning tools for cast irons **A16:** 653
C-6, turning tools for heat-resistant alloys **A16:** 741
C-6, turning tools for Ni alloys **A16:** 839
C-6, turning tools for uranium alloys **A16:** 875
C-7, boring operation **A16:** 164
C-7, tools for cast irons **A16:** 656, 657
C-7, turning tools **A16:** 145
C-7, turning tools for cast irons **A16:** 653
C-7, turning tools for heat-resistant alloys **A16:** 741
C-7, turning tools for Ni alloys **A16:** 838
C-7, turning tools for uranium alloys **A16:** 875
C-8, boring tools **A16:** 164
C-8, boring tools for heat-resistant alloys **A16:** 742
C-8, tools for cast irons **A16:** 656, 657
C-8, turning tools for heat-resistant alloys **A16:** 739, 741
C-8, turning tools for Ni alloys **A16:** 838
C-10, tools for Hf **A16:** 855
C-10, tools for Zn **A16:** 855
C-11, tools for Hf **A16:** 855
C-11, tools for Zn **A16:** 855
Carbide(s)
brazing and soldering characteristics **A6:** 635-636
hardfacing **A6:** 790, 792, 793, 794, 796
inclusions formed in fluxes **A6:** 56
T-111
electron-beam welding **A6:** 871
T-222
electron-beam welding **A6:** 871
torch brazing **A6:** 328
Carbide-base and refractory metal composites
as electrical contact materials **A2:** 854-855
Carbide-boride grain refinement model
kinetics of **A15:** 105-106
Carbide-forming elements
in cobalt-base heat-resistant casting alloys **A9:** 334
in steel **A9:** 178
in steel, effect on pearlite growth **A9:** 661
Carbide-forming elements (CFE)
thermoreactive deposition/diffusion process **A4:** 449
Carbide-inducing inoculants
alloy cast iron **M1:** 80
Carbide-metal cermets
substrate for thermoreactive deposition/diffusion process **A4:** 449
Carbide-tipped cutters
for milling procedures **A7:** 462
Carbide/malleable iron application
piston ring materials **A18:** 557
Carbides *See also* Carbide cermets; Carbide hardfacing; Carbide tools; Cemented carbides **A1:** 952, 954-955
abrasive flow machining **A16:** 517, 519
alloying, nickel-base alloys **A13:** 641-642
applications **EM4:** 203
in austenitic stainless steels **A1:** 946-947
austenitizing for surface hardening effect on **M1:** 531
in carbon steels, etching **A9:** 170
in cast iron **M1:** 3-5
in cast irons **A16:** 650, 651
in cast irons, magnifications to resolve **A9:** 245
chevron-notched specimens of **A8:** 470
chloride salt corrosion **A13:** 90
in cobalt-base alloys **A1:** 986
compared to ceramics **A16:** 101
containing steels, microstructural wear effects **A11:** 161
defined **A7:** 2
diamond as abrasive for honing **A16:** 476
in ductile iron, alloying **A15:** 649
electrical discharge grinding **A16:** 565, 566, 567
electrochemical discharge grinding **A16:** 548, 549, 550
embrittling effect, AISI/SAE alloy steels **A12:** 341
as embrittling intergranular networks **A11:** 359
enlargement, annealing-caused **A7:** 185
films, grain-boundary **A11:** 267, 405
formation, from cast iron inoculation **A15:** 170
formation of in $2^1/_4$Cr-1Mo steel **A1:** 632-633, 638, 641-642
formation, ternary iron-base alloys **A15:** 68
formers **A7:** 373, 429
grains, dispersion, in cermets **A2:** 991
gray cast iron **M1:** 12, 13, 14, 26
grinding by superabrasives **A16:** 432, 433, 437
ground by diamond wheels **A16:** 455, 461, 462
hardness of **A1:** 394
in high-speed tool steel **A16:** 55, 56
honing grit size selection **A16:** 478
as inclusions **A10:** 176
infiltration **A7:** 556-557
interdendritic, in high-alloy graphitic iron **A15:** 698
intergranular, cracking in **A11:** 407
locating points for shaped tube electrolytic machining **A16:** 555
maximum service temperature **EM4:** 203
mixed, in cemented carbides **A9:** 274
morphology, tool and die failure and **A11:** 575
networks, carburizing for **A11:** 121
in nickel-chromium irons **A15:** 681-682
particles, brittle fracture from **A11:** 327
particles of, atom probe mass spectrum **A10:** 592
particles, weld-interface **A11:** 444
plasma spray material **EM4:** 203
in plate steels, examination **A9:** 203
platelets, after annealing **A7:** 184
powders used **A7:** 572
Carbides (continued)
precipitation **A13:** 349, 551
precipitation, and brazed 'joint defects **A11:** 451
properties **EM4:** 203
reactions in elevated-temperature failures **A11:** 267
rigid tool **A7:** 322-328
sectioning by fracturing **A9:** 23
segregation, tool and die failure due to **A11:** 575
solubility in austenite **A1:** 407
spheroidization **A1:** 642
spheroidization, ASTM/ASME alloy steels **A12:** 346
in steel
formability influenced by **M1:** 558-559
wear resistance influenced by ... **M1:** 608, 610-612, 614, 622
in structural ceramics **A2:** 1019, 1021-1024
in tool steels **A9:** 258-259
tools, for turning or boring **A7:** 461
tungsten-titanium-tantalum (niobium) **A2:** 950-951
vacuum sintering atmospheres for **A7:** 345
in wrought heat-resistant alloys **A9:** 308-309, 311
in wrought stainless steels, etching **A9:** 281-282
in wrought stainless steels, in austenitic grades **A9:** 283-284
in wrought stainless steels, in ferritic grades **A9:** 285
Carbides, coated **A16:** 79-83
boring heat-resistant alloys **A16:** 742
cast iron machining **A16:** 653, 656
heat-resistant alloy machining **A16:** 739, 740, 742
machinability **A16:** 639-646
P/M material machining **A16:** 882
threading **A16:** 95
Ti alloy machining **A16:** 844
tool life **A16:** 79, 80
Carbides in steels .... **A3:** 1 • 24
Carbides, uncoated
machinability **A16:** 639-641, 643-646
P/M materials machining **A16:** 881, 889
primary applications **A16:** 639
Carbitol
description **A9:** 68
Carbofrax D
erosion test results **A18:** 200
Carboloy 883
cutting tool material for refractory metals **A16:** 863
Carbon *See also* Binary iron-base systems; Carbon content; Carbon equivalent; Carbon fiber; Carbon fiber reinforced composites; Carbon solubility; Combined carbon; Iron-base alloys **EM3:** 7
as a conductive coating for scanning electron microscopy specimen **A9:** 97-98
absorption, air carbon arc cutting **A14:** 734
absorption in air carbon arc cutting **M6:** 920
activated, RDF analysis of **A10:** 399-400
activity, effect, alloying elements **A15:** 61-70
addition to solid-solution-strengthened superalloys **A4:** 809
AES analysis of surface chemistry **A10:** 552-553
in alloy cast irons **A1:** 86, 88
alloying effect on nickel-base alloys **A6:** 589
alloying, in wrought titanium alloys **A2:** 599-600
alloying, nickel-base alloys **A13:** 641-642
alloying, stainless steels **A13:** 550
alloying, wrought aluminum alloy **A2:** 47
amorphous, EELS edge shapes for **A10:** 460
as an addition to austenitic manganese steel castings **A9:** 239
as an addition to beryllium-nickel alloys **A9:** 395
as an addition to cobalt-base heat-resistant casting alloys **A9:** 334
as an addition to nickel-base heat-resistant casting alloys **A9:** 334

SUBJECTS OF THE INDEXED VOLUMES: ASM Handbook (designated by the letter "A"): **A1:** Properties and Selection: Irons, Steels, and High-Performance Alloys (1990); **A2:** Properties and Selection: Nonferrous Alloys and Special-Purpose Materials (1990); **A3:** Alloy Phase Diagrams (1992); **A4:** Heat Treating (1991); **A6:** Welding, Brazing, and Soldering (1993); **A7:** Powder Metallurgy (1984); **A8:** Mechanical Testing (1985); **A9:** Metallography and Microstructures (1985); **A10:** Materials Characterization (1986); **A11:** Failure Analysis and Prevention (1986); **A12:** Fractography (1987); **A13:** Corrosion (1987); **A14:** Forming and Forging (1988); **A15:** Casting (1988); **A16:** Machining (1989); **A17:** Nondestructive Testing and Quality Control (1989); **A18:** Friction, Lubrication, and Wear Technology (1992). **Metals Handbook, 9th Edition** (designated by the letter "M"): **M1:** Properties and Selection: Irons and Steels (1978); **M2:** Properties and Selection: Nonferrous Alloys and Pure Metals (1979); **M3:** Properties and Selection: Stainless Steels, Tool Materials and Special-Purpose Materials (1980); **M4:** Heat Treating (1981); **M5:** Surface Cleaning, Finishing, and Coating (1982); **M6:** Welding, Brazing, and Soldering (1983). **Engineered Materials Handbook** (designated by the letters "EM"): **EM1:** Composites (1987); **EM2:** Engineering Plastics (1988); **EM3:** Adhesives and Sealants (1990); **EM4:** Ceramics and Glasses (1991); **Electronic Materials Handbook** (designated by the letters "EL"): **EL1:** Packaging (1989).

Carbon (continued)
as an alpha stabilizer in titanium alloys ... **A9:** 459
as an austenite-stabilizing element in steel ... **A9:** 177
as an austenite-stabilizing element in wrought stainless steels ... **A9:** 283
as an embedding agent ... **EM4:** 572
as an interference film ... **A9:** 147
applications ... **EM4:** 46
at elevated-temperature service ... **A1:** 640
atoms, in diamond/graphite ... **A2:** 1010
Auger chemical map for ... **A10:** 557
as austenite stabilizer ... **A13:** 47
in austenitic manganese steel ... **A1:** 822-824
in austenitic stainless steels.... **A6:** 457, 458, 465, 468
batch size ... **EM4:** 382
in bearing steels ... **A18:** 726
brazing of ... **M6:** 1061-1063
burn off ... **A7:** 191
in cast Co alloys ... **A16:** 69
in cast iron ... **A1:** 5
CDJ, properties ... **A18:** 549, 551
in cemented carbides ... **A9:** 273-275
chain polymers, chemical structure ... **EM2:** 49
cladding dilution ... **M6:** 809-810
coating for SEM specimens ... **A18:** 380
in cobalt-base wrought alloys ... **A18:** 766, 768
codeposited with chromium electroplating ... **A18:** 835
combined ... **A13:** 3
combined, defined ... **A15:** 3
combined, effects in sintering ... **A7:** 362-363
combustion method for elemental analysis ... **A10:** 214
in commercial CPM tool steel compositions ... **A16:** 63, 64
compatibility with steel ... **A18:** 743
composition ... **EM4:** 46
in composition, effect on dimensional changes in heat treatment ... **A4:** 612
in composition, effect on ductile iron ... **A4:** 687, 689, 690, 692
in composition, effect on flame hardening ... **A4:** 277
in composition, effect on gray irons **A4:** 671, 672, 673, 675, 677, 678, 679
composition-depth profiles ... **A7:** 256
concentration, cupolas ... **A15:** 380
contamination ... **M6:** 321
content affecting broach life ... **A16:** 208
content, and liquation cracking **A6:** 568-569, 570
content effect, electron-beam welding ... **A6:** 867
content effect on alloy solidification cracking ... **A6:** 89-90
content effect on minimum sensitization time in 300-series stainless steels ... **A6:** 1067
content effect on quench cracking **A4:** 77, 78, 79, 80
content in carbon steels ... **A16:** 149-150, 358
content in heat-treatable low-alloy (HTLA) steels ... **A6:** 670
content in HSLA and Q&T steels ... **A6:** 665
content in martensitic stainless steels.... **M6:** 348-349
content in nickel-base and cobalt-base high-temperature alloys ... **A6:** 573
content in P/M materials **A16:** 884, 887, 888, 889
content in stainless steels...... **A16:** 682-683, 684, 688, 689, 690 **M6:** 320, 322
content in tool steels ... **A16:** 708, 726, 727
content in ultrahigh-strength low-alloy steels ... **A6:** 673
content, microstructures, and properties of steels ... **A1:** 127-128, 144, 576
content of weld deposits ... **A6:** 675
content related to flaking ... **A16:** 281
content, sink/float density separations for ... **A10:** 177
control agents in nitrogen-based atmospheres ... **A7:** 346
control, ductile iron ... **A15:** 647
cracking sensitivity in stainless steel casting alloys ... **A6:** 497
CVD *See* Chemical vapor deposited carbon
in CVD process ... **A16:** 80
defined ... **EM1:** 6 **EM2:** 8

Carbon (continued)
deoxidizing, copper and copper alloys ... **A2:** 236
determination by high-temperature combustion ... **A10:** 221-225
diffusion coefficient in nitrocarburizing ... **A18:** 878
dissolution, in cast iron ... **A15:** 72-73
dissolution, powder forging ... **A14:** 192
double carbides (η phase) ... **A16:** 73, 74
in ductile iron ... **A1:** 40, 43
in duplex stainless steels ... **A6:** 471
effect, gas disassociation ... **A15:** 83
effect, hydrogen/nitrogen solubility ... **A15:** 82
effect in cast iron, discovered ... **A15:** 29
effect in iron, and copper diffusion ... **A7:** 480
effect of, on cast stainless steel corrosion resistance ... **A1:** 912, 913
effect of, on hardenability........ **A1:** 392, 465, 467-468
effect of, on notch toughness ... **A1:** 739
effect on ferrite formation in heat-resistant casting alloys ... **A9:** 333
effect on hardenability ... **A4:** 25
effect on hardfacing spraying ... **M6:** 789
effect on shielding gas ... **M6:** 103
effect on weldability in stainless steels ... **M6:** 525
electrode content and arc welding of low-alloy steels ... **A6:** 662
electrode systems based on ... **A10:** 191
electroslag welding, reactions ... **A6:** 273, 274
equivalent, calculated ... **A15:** 68-69
-FeO reaction in iron ... **A12:** 221
in ferrite ... **A1:** 406
ferritic stainless steel content ... **M6:** 346
as filler material ... **EM2:** 499
filler to gain electrical conductivity ... **EM3:** 572
flux, defined ... **A15:** 74
glow discharge to determine ... **A10:** 29
graphitic, determined by selective combustion ... **A10:** 223-224
graphitic, neutron, x-ray scattering and absorption characteristics ... **A10:** 421
in hardfacing alloys **A18:** 758, 760, 761, 762, 763, 764
hardness of steels affected by content ... **A4:** 185-186
in heat-resistant alloys ... **A4:** 510, 511, 512
in high-alloy white irons ... **A15:** 679
in high-speed tool steels ... **A16:** 51, 52, 53
historical studies ... **A12:** 3
impregnation effects on typical graphite base material ... **A18:** 817
as impurity in uranium alloys ... **A9:** 477
as impurity, magnetic effects ... **A2:** 762
induction hardening cracking tendency ... **A4:** 202
infrared detection, high temperature combustion ... **A10:** 223
as inoculant ... **A15:** 105
in inorganic solids, applicable analytical methods ... **A10:** 4, 6
interaction coefficient, ternary iron-base alloys ... **A15:** 62
interstitial contamination ... **M6:** 463
interstitial, content in silicon wafers ... **A10:** 123
ion implantation of titanium alloys..... **A18:** 779, 780
KVV lineshapes, effect on quantitative analysis ... **A10:** 553
for laser alloying ... **A18:** 864
loss effect on welding parameters ... **A6:** 68
as major element, gray iron ... **A15:** 629-630
in metal carbide, EELS edge shape for ... **A10:** 460
metallurgical effects, projection welding ... **M6:** 506
microalloying of ... **A14:** 219-220
Miller numbers ... **A18:** 235
mobile, determined in iron/steels ... **A10:** 178
nearest neighbors, liquid iron-carbon alloys ... **A15:** 168
in nickel-base superalloys ... **A1:** 984
oxygen cutting, effect on ... **M6:** 898
in P/M alloys ... **A1:** 809
in P/M high-speed tool steels ... **A16:** 61
in P/M stainless steels ... **A13:** 827-829
paste or plate for backing ... **M6:** 592
penetration from porosity in carbonitriding ... **A7:** 454, 455

Carbon (continued)
pickup in milling of electrolytic iron powders ... **A7:** 64, 65
potential in protective atmospheres ... **M6:** 934
in precipitation-hardening steels ... **M6:** 350
presence in cast irons ... **M6:** 307-308
prompt gamma activation analysis of ... **A10:** 240
properties ... **A18:** 817
recovery from, selected electrode coverings ... **A6:** 60
reductant of metal oxides ... **M6:** 694
as refractory, for core coatings ... **A15:** 240
relationship to hot cracking ... **M6:** 38
removal, by oxygen blowing ... **A15:** 78
removal, by solid-state refining techniques ... **A2:** 1094
removal from organic matrix ... **A10:** 167
removal of, salt bath descaling process ... **M5:** 99-100
removal rates, degassing processes compared ... **A15:** 430
residuals, effect iron powders ... **A7:** 183
segregation and solid friction ... **A18:** 28
segregation, electroslag remelting effects ... **A15:** 405
in sintered austenitic stainless steels ... **A13:** 831
solid solubility in Fe-Cr-Ni alloy ... **A13:** 929
solubility ... **A15:** 66-68, 465
solubility, cast irons and carbon steels **A13:** 46-47
solubility in ferrite vs temperature ... **A1:** 132
in stainless steels.... **A6:** 678, 682 **A18:** 710, 712, 713, 716
in steel weldments ... **A6:** 416, 418, 419, 420
substrate for thermoreactive deposition/diffusion process ... **A4:** 449
and sulfur effects, ultimate tensile strength ... **A14:** 200
sulfuric acid corrosion ... **A13:** 1154
supply sources ... **EM4:** 46
surface, cleaning of ... **A13:** 380
surface, determined ... **A10:** 223-224
tantalum corrosion at elevated temperatures ... **A13:** 728
thermal diffusivity from 20 to 100 °C ... **A6:** 4
thermal expansion coefficient ... **A6:** 907
in thermal spray coating materials ... **A18:** 832
thin films ... **A12:** 173
tin-lead plating baths treated with ... **M5:** 278
in tool steels **A18:** 734, 735-736, 737, 738, 739
trace element analysis, by combustion technique ... **A2:** 1095
use in oxide reduction ... **A7:** 52, 53
vacuum heat-treating support fixture material ... **A4:** 503
vapor pressure ... **A4:** 493, 494
vapor pressure, relation to temperature ... **A4:** 495
Vickers and Knoop microindentation hardness numbers ... **A18:** 416
vitreous, Raman analysis ... **A10:** 132
weld-metal content, underwater welding ... **A6:** 1010, 1011, 1012
in wrought stainless steels ... **A1:** 872

Carbon (AS-4) fibers
bonded to polyethylene (UHMW-PE) by epoxy resin ... **EM3:** 393, 395-396, 402

Carbon (HMU) fibers
microindentation test ... **EM3:** 400

Carbon alloys
13Cr-4Ni-0.05C, hydrogen-induced cold cracking resistance ... **A6:** 438
electron-beam welding ... **A6:** 581
Fe-0.2C-12Cr-1Mo
Charpy V-notch data for weld joints of EB welded alloy ... **A6:** 440
microhardness traverse test results ... **A6:** 441

Carbon and alloy steels, friction and wear of ... **A18:** 702-708
carbon content ... **A18:** 707
correlation to relative wear content ... **A18:** 707
heat-affected zone (HAZ) ... **A18:** 707
depth of hardened regions ... **A18:** 706
relation of hardness to microstructure ... **A18:** 707-708
steel metallurgy ... **A18:** 702-704
microstructures ... **A18:** 702-704, 705

Carbon and alloy steels, friction and wear of (continued)
properties influenced by microstructure ... **A18:** 704
steel selection based on relative costs ... **A18:** 706
steel transformation diagram ... **A18:** 704-705
toughness ... **A18:** 706-707
wear properties of carbon steel ... **A18:** 705-706
corrosion resistance improved by altering microstructure ... **A18:** 706
hardness as a function of carbon content ... **A18:** 705
improving wear properties of mild steels ... **A18:** 705
Carbon and low-alloy steel plate *See* Steel plate
Carbon and low-alloy steels *See also* Alloy steel; Carbon steel; Low-alloy steel
alloy designations and specifications for elevated-temperature service ... **A1:** 617-618
alloying elements, effects of ... **A1:** 144-147
aluminum ... **A1:** 146
boron ... **A1:** 145
carbon ... **A1:** 144
chromium ... **A1:** 145-146
copper ... **A1:** 145
lead ... **A1:** 145
manganese ... **A1:** 144
molybdenum ... **A1:** 146
nickel ... **A1:** 146
niobium ... **A1:** 146
phosphorus ... **A1:** 144
silicon ... **A1:** 145
sulfur ... **A1:** 144-145
titanium ... **A1:** 146
zirconium ... **A1:** 147
chemical analysis ... **A1:** 141-142
heat and product analysis ... **A1:** 141
residual elements ... **A1:** 141
silicon content ... **A1:** 141
classification of ... **A1:** 140-141
deoxidation practice ... **A1:** 142-143
capped steel ... **A1:** 143
killed steel ... **A1:** 142
rimmed steel ... **A1:** 143
semikilled steel ... **A1:** 142-143
quality descriptors ... **A1:** 143-144
spheroidization and graphitization in ... **A1:** 644
Carbon and low-alloy steels, selection of ... **A6:** 405-407
chromium-molybdenum steels ... **A6:** 405, 406, 407
classification of steels ... **A6:** 405-406
composition and carbon equivalent of selected steels ... **A6:** 406
heat-affected zone ... **A6:** 405, 406, 407
heat-treatable low-alloy (HTLA) steels ... **A6:** 405, 406
high-strength low-alloy (HSLA) steels ... **A6:** 405, 406, 407
low-carbon steels ... **A6:** 405, 406, 407
microalloyed steels ... **A6:** 405-406
mild steels ... **A6:** 405-406
quenched-and-tempered steels ... **A6:** 405, 406, 407
relative susceptibility of steels to hydrogen-assisted cold cracking ... **A6:** 407
thermal-mechanical-controlled processing (TMCP) steels ... **A6:** 405, 406, 407
weldability ... **A6:** 405
fabrication weldability ... **A6:** 405
service weldability ... **A6:** 405
Carbon arc brazing (CAB)
definition ... **A6:** 1207
Carbon arc cutting
definition ... **M6:** 3
Carbon arc cutting (CAC)
definition ... **A6:** 1207
power source selected ... **A6:** 37
Carbon arc welding
definition ... **M6:** 3
Carbon arc welding (CAW) ... **A6:** 124-125, 200-201
applications ... **A6:** 201
of cast irons ... **A6:** 201
of copper ... **A6:** 201
definition ... **A6:** 200, 1207
electrodes ... **A6:** 200-201
of galvanized steel ... **A6:** 201
operation ... **A6:** 200-201
single-electrode ... **A6:** 200
twin-electrode ... **A6:** 200-201
of steels ... **A6:** 200
Carbon austenite fracture
transmission electron micrograph ... **A8:** 483
Carbon black ... **EM2:** 8, 470, 501 **EM3:** 7
additive for urethane sealants ... **EM3:** 205
automotive industry applications ... **EM3:** 178
extender ... **EM3:** 176
filler for elastomeric adhesives ... **EM3:** 150
pigment ... **EM3:** 179
to enhance electrical conductivity in adhesives ... **EM3:** 178
"Carbon boil" ... **A6:** 595
Carbon boiling
during porcelain enameling ... **M1:** 177-179
Carbon brushes
as copper powder application ... **A7:** 105
Carbon coating
for extraction replicas ... **A9:** 108-109
Carbon composition resistors
failure mechanisms ... **EL1:** 971, 1002-1003
Carbon content *See also* Carbon
400 to 500 °C embrittlement effect on ... **M1:** 686
alloy cast irons ... **M1:** 76, 77, 78, 82
alloy steel, effect on weldability ... **M1:** 561
at austenitizing temperatures ... **A7:** 451, 452
cast iron, effect on weldability ... **M1:** 563-564
CG iron, optimum ... **A15:** 668
in cold extrusion ... **A14:** 300
concentration profiles, iron-carbon melt ... **A15:** 73
control during sintering ... **A7:** 370, 386, 390
distribution, in steel plate ... **M1:** 189, 195-196
effect, cast iron ... **A15:** 29
effect, hydrogen/nitrogen solubility ... **A15:** 82
effect of sintering atmosphere on ... **A7:** 340, 341
effect on formation of upper bainite ... **A9:** 663
effect on martensitic start temperature ... **A9:** 669
effect on sintering ... **A7:** 372
effect on torsional ductility in steels and aluminum alloys ... **A8:** 166-167
effect on warm workability ... **A14:** 174
equivalent, effect on strength of cold extruded steel ... **M1:** 592
eutectic, in high-alloy graphitic irons ... **A15:** 698
fatigue limit of alloy steel effect on ... **M1:** 675, 676
of ferritic stainless steels ... **A9:** 284-285
and fracture toughness ... **A8:** 481, 484
hydrogen solubility, effect on ... **M1:** 687
of martensitic stainless steels ... **A9:** 285
quench-age embrittlement effect on ... **M1:** 684
in stainless steel casting alloys, effects of ... **A9:** 298
in stainless steels ... **A15:** 431
of steel, and swageability ... **A14:** 128
of steel, effect on martensite ... **A9:** 178
steel plate, effect on mechanical properties ... **M1:** 194, 197
temper embrittlement, effect on ... **M1:** 684, 685
and tempering, effects on hardness ... **A14:** 198
of tool steels ... **A9:** 258-259
vs mechanical properties, iron-carbon alloys ... **A14:** 200
white cast irons ... **M1:** 75, 76, 77, 78
Carbon control, evaluation *See also* Surface carbon content control ... **A4:** 587-600
carbon gradients ... **A4:** 588, 592-598, 599 **M4:** 438-440, 441, 442
carbon restoration ... **A4:** 598-599 **M4:** 446, 447
Carbon control, evaluation (continued)
case properties, effect of carbon gradient ... **A4:** 594 **M4:** 440, 442
case-depth variation ... **A4:** 590-592, 593 **M4:** 437
coercive-force testing ... **A4:** 590 **M4:** 436
consecutive cut analysis ... **A4:** 588-589 **M4:** 433-434
electromagnetic testing ... **A4:** 590 **M4:** 436
hardness testing ... **A4:** 587, 588 **M4:** 432-433
low surface carbon ... **A4:** 595 **M4:** 442-443
magnetic-comparator testing ... **A4:** 590 **M4:** 436
microscopic examination ... **A4:** 587-588 **M4:** 433
quality control ... **A4:** 599-600 **M4:** 447-448
rejected parts, disposition ... **A4:** 600 **M4:** 448
rolled wire analysis ... **A4:** 590 **M4:** 436
shim stock analysis ... **A4:** 589-590 **M4:** 434-435
spectrographic analysis ... **A4:** 590 **M4:** 436
surface carbon content ... **A4:** 594 **M4:** 440, 442
surface carbon variability ... **A4:** 595-598 **M4:** 443, 444, 445, 446
test results ... **A4:** 591-592 **M4:** 438, 439
Carbon correction *See* Carbon restoration
Carbon deficiency in cemented carbides ... **A9:** 274-275
on a fracture surface ... **A9:** 276
Carbon deposition
in extraction replicas ... **A17:** 54
Carbon depth profile
thin-film hybrids ... **EL1:** 319
Carbon diffusion, in dissimilar-metal welds ... **A11:** 620
equivalent ... **A11:** 391-392, 796
films, lustrous, as casting defect ... **A11:** 388
groups, causing stress-corrosion cracking ... **A11:** 207
nitrogen interaction, in elevated-temperature failures ... **A11:** 273
pickup, metallographic sectioning ... **A11:** 24
-plus-nitrogen, effect on stress rupture life ... **A11:** 268
saturation, symbol for ... **A11:** 797
total content, abbreviation for ... **A11:** 798
Carbon dioxide
atmospheric effect helping control dusting ... **A18:** 684
by residual gas analysis ... **EL1:** 1065
cause of porosity in nickel alloy welds ... **M6:** 442-443
characteristics in a blend ... **A6:** 65
content, endothermic gas ... **A7:** 343
as converter gas ... **A15:** 426
copper/copper alloy resistance ... **A13:** 632
core curing with ... **A15:** 240
corrosion, gas/oil production ... **A13:** 1233, 1247
corrosion rate, fresh water effect on ... **M1:** 733
as corrosive, copper casting alloys ... **A2:** 352
as cutting fluid for tool steels ... **A18:** 738
dissociation ... **A6:** 66
dissociation and recombination ... **A6:** 64
dissolved, corrosive effects in seawater ... **A13:** 896-898
dissolved, in water ... **A13:** 489
effect, zinc corrosion in distilled water ... **A13:** 760
electrogas welding shielding gas ... **A6:** 270, 275
and explosivity ... **A7:** 195
as extraction media for binder removal in injection molding ... **EM4:** 179
flux-cored arc welding shielding gas ... **A6:** 189
fume generation from shielding gases ... **A6:** 68
gas mass analysis of ... **A10:** 155
as SFC solvent ... **A10:** 116
gas-metal arc welding shielding gas ... **A6:** 181, 183, 185
injection, gas/oil production ... **A13:** 1253-1254
ionization potential ... **A6:** 66
laser-beam welding shielding gas ... **A6:** 878
lasers, as germanium applications ... **A2:** 743
low-heat-input welding of low-alloy steels ... **A6:** 662
in marine atmospheres ... **A13:** 904

Carbon dioxide (continued)
mean free path ... **A17:** 59
partial pressure, AOD, VODC, and VOD compared ... **A15:** 430
partial pressure, effect on corrosion rate alloy steels ... **A13:** 536
process, defined ... **A15:** 2
reactivity/oxidation potential ... **A6:** 64
in Shaw process ... **A15:** 249
shielding gas for FCAW ... **A6:** 67
shielding gas for GMAW ... **A6:** 66, 67
shielding gas for GMAW of cast irons ... **A6:** 718, 719
shielding gas from fluxes ... **A6:** 58
shielding gas properties ... **A6:** 64
shielding gas purity and moisture content ... **A6:** 65
sigma values for ionization ... **A17:** 68
solubility in water ... **M1:** 733, 734
steam system corrosion effects ... **M1:** 733
steel weldment soundness in gas-shielded processes ... **A6:** 408, 409
in superheaters ... **A13:** 1201
tantalum corrosion by ... **A13:** 731
use to determine carbon ... **A10:** 221-225
valve thread connections for compressed gas cylinders ... **A6:** 1197
in weld relay, gas mass spectroscopy of ... **A10:** 156
Carbon dioxide (dc excited) continuous wave lasers, parameters
laser-beam welding applications ... **A6:** 263
Carbon dioxide (rf excited) continuous wave lasers, parameters
laser-beam welding applications ... **A6:** 263
Carbon dioxide gas
metalworking laser for laser surface transformation hardening ... **A4:** 290-291
Carbon dioxide gas dynamic continuous wave lasers, parameters
laser-beam welding applications ... **A6:** 263
Carbon dioxide lasers *See* Laser beam machining
characteristics and advantages ... **M6:** 656
comparison of weld results ... **M6:** 658
laser beam transport ... **M6:** 665
suitability for welding ... **M6:** 651-652
Carbon dioxide pulsed lasers, parameters
laser-beam welding applications ... **A6:** 263
Carbon dioxide shielding gas
characteristics ... **M6:** 103
containers ... **M6:** 103-104
effect on weld metallurgy ... **M6:** 41
flow rate ... **M6:** 104
purity ... **M6:** 104
Carbon dioxide shielding gas for
arc welding of austenitic stainless steels ... **M6:** 331-332
electrogas welding ... **M6:** 241
flux cored arc welding ... **M6:** 103-104
gas metal arc welding ... **M6:** 164
Carbon dioxide silicate molding
of Alnico alloys ... **A15:** 737
Carbon dioxide welding *See* Gas metal arc welding
Carbon electrode
definition ... **M6:** 3
Carbon equivalent ... **M6:** 250, 262
abbreviation ... **A8:** 724
calculated ... **A15:** 68-69
cast iron ... **M1:** 4
code requirements ... **M6:** 250
compacted graphite irons ... **A15:** 667
ductile iron ... **A15:** 648
effect on weldability ... **M6:** 250
as solidification impact ... **A15:** 629
weldability of iron and steel effect on ... **M1:** 561
Carbon equivalent (CE) ... **A6:** 94
carbon steels ... **A6:** 648, 649
cast irons ... **A6:** 710, 711
related to hydrogen-assisted cold cracking ... **A6:** 407
Carbon equivalent (CE) value ... **A18:** 695-696, 698
Carbon equivalent (C_{eq}) formula ... **A6:** 379, 381
Carbon fabric reinforced phenolic resins
as medium-temperature thermoset matrix composite ... **EM1:** 382
Carbon fiber *See also* Carbon fiber reinforced composites; Fiber(s); Graphite; Graphite fiber; Pyrolysis ... **EM3:** 7
abrasive waterjet machining ... **A16:** 527
as conductive reinforcements ... **EM2:** 471-472
defined ... **EM2:** 8
as filler/extender, polypropylenes (PP) ... **EM2:** 192
PCD tooling ... **A16:** 110
rovings, pultrusions ... **EM2:** 393
Carbon fiber ceramic composites ... **EM1:** 929-930
Carbon fiber reinforced aluminum (C-Al)
applications ... **EL1:** 1122-1125
Carbon fiber reinforced composites
with cyanates ... **EM2:** 235-236
short-beam shear ... **EM2:** 237
Carbon fiber reinforced copper(C-Cu)
applications ... **EL1:** 1122-1125
Carbon fiber reinforced epoxy (C-Ep)
applications ... **EL1:** 1122
Carbon fiber reinforced epoxy composites ... **EM3:** 293
Carbon fiber reinforced epoxy panels
surface contamination ... **EM3:** 846
Carbon fiber reinforced nylon
for bearings ... **EM1:** 35
Carbon fiber reinforced plastic
nondestructive testing ... **A6:** 1086
Carbon fiber reinforced plastic (CFRP)
with aligned discontinuous fibers ... **EM1:** 153-155
and aluminum, weights compared ... **EM1:** 35
beams, damping in ... **EM1:** 210, 213
compression properties ... **EM1:** 155
electrical waveguides ... **EM1:** 36
interlaminar shear strength ... **EM1:** 156
reduced energy dissipation in ... **EM1:** 36
scrap, recovering/recycling of ... **EM1:** 153-156
tensile properties ... **EM1:** 154, 155
Carbon fiber reinforced plastic adherends
mechanical properties ... **EM3:** 333
Carbon fiber reinforced plastics (CFRP)
adherend shear strains ... **EM3:** 326
Carbon fiber reinforced polymers
friction and wear properties ... **EM1:** 36
Carbon fiber(s) *See also* Fibers; Pit(-h-base carbon fibers
reinforcement ... **EL1:** 1126-1128
vapor-grown ... **EL1:** 1117-1118
Carbon fiber-epoxy matrix systems ... **EM3:** 402-403
microindentation test ... **EM3:** 400, 401
Carbon fibers *See also* Continuous graphite fiber MMCs; Graphite fibers ... **EM1:** 49-53, 112-113
aerospace application ... **EM1:** 139
assay, methods ... **EM1:** 285
axial structure ... **EM1:** 50
bulk properties ... **EM1:** 51-52
in carbon/carbon composites ... **EM1:** 916
classes ... **EM1:** 51
commercially available, mechanical properties ... **EM1:** 113
conversion processes ... **EM1:** 112-113
critical lengths ... **EM1:** 120
defined ... **EM1:** 6
diameters ... **EM1:** 51-52
elastic buckling stress ... **EM1:** 197
elastic properties ... **EM1:** 188
environmental interaction ... **EM1:** 52
and epoxy resins ... **EM1:** 75
fastener corrosion with ... **EM1:** 706, 709
feedstock ... **EM1:** 112
forms ... **EM1:** 360
and graphite fibers, compared ... **EM1:** 867
importance ... **EM1:** 43
incorporation in thermoplastic composites ... **A18:** 820, 823-824
interfacial bonding ... **EM1:** 52
introduction ... **EM1:** 29-30
laminates, wet/dry properties ... **EM1:** 76
manufacture, for continuous graphite fiber MMCs ... **EM1:** 867-868
mesophase pitch precursor fibers ... **EM1:** 51
microstructures ... **EM1:** 50-52
modulus, determined ... **EM1:** 50
nickel-coated ... **EM1:** 36
with polyacrylonitrile precursor fibers properties ... **EM1:** 49
precursors for ... **EM1:** 49, 868
Carbon fibers (continued)
prepreg ... **EM1:** 139
prepreg tow, for filament winding ... **EM1:** 138
processes ... **EM1:** 49-50
properties ... **EM1:** 51-52, 113, 361, 867
radial structure ... **EM1:** 51
rapier loom for ... **EM1:** 127
reinforced epoxy resin composites ... **EM1:** 400, 410-412
reinforcement, ceramic composites ... **EM1:** 929
scrap, recyling of ... **EM1:** 153-156
for short fiber reinforced composites ... **EM1:** 120
and silica fibers, cylinder showing ... **EM1:** 129
sizings ... **EM1:** 122, 868
in space and missile applications ... **EM1:** 817
spools ... **EM1:** 112
structure, types/effects ... **EM1:** 50-52
surface treatment vs. sizing of ... **EM1:** 122
surfaces ... **EM1:** 868
tensile strength/modulus, compared ... **EM1:** 113
-thermoset resins, interlaminar fracture toughness ... **EM1:** 98
three-dimensional structure ... **EM1:** 51
tow sizes ... **EM1:** 105
types, commercially available ... **EM1:** 113
unidirectional, reinforced epoxies, strength of ... **EM1:** 35
vs. glass fibers, cost ... **EM1:** 105
for woven materials ... **EM1:** 125-128
Carbon fibers, specific types
AS-4, properties ... **EM1:** 58
T1000, specific tensile strength and specific tensile modulus ... **EM1:** 28
Carbon filament precursors for boron fibers ... **A9:** 592
Carbon flotation
as casting intemal discontinuity ... **A11:** 354
defined ... **A11:** 1
improper surface finish caused by ... **A11:** 358-359
in iron castings ... **A11:** 358
Carbon free-cutting steels
thermal expansion coefficient ... **A6:** 907
Carbon gages
for Hugoniot elastic limit measurement ... **A8:** 211
Carbon gradients *See* Carbon control, evaluation
Carbon graphite
anodic oxidation ... **EM3:** 287, 290
AS-4 fibers ... **EM3:** 287
Celion 6000 fibers ... **EM3:** 287
electrodeposition ... **EM3:** 287
epoxy finish layer ... **EM3:** 288
fast atom bombardment ... **EM3:** 287
flame treatment ... **EM3:** 286
Fortafil 5T ... **EM3:** 287
interfacial shear stress and shear strength ... **EM3:** 288
nitric acid oxidation ... **EM3:** 286
oxidation of fibers ... **EM3:** 286
plasma treatments ... **EM3:** 288
properties ... **EM3:** 287
surface preparation ... **EM3:** 286-290
testing problems ... **EM3:** 333
wet chemical treatment ... **EM3:** 286
Carbon in cast iron *See also* Graphite; Temper carbon ... **M1:** 3-5, 6, 9
ductile iron ... **M1:** 36-37, 38, 40-41, 47, 49
gray iron ... **M1:** 11-12, 21-23, 29, 31-32
malleable cast irons ... **M1:** 58-64, 73
Carbon in P/M materials ... **M1:** 336-337, 340, 342
Carbon in steel ... **M1:** 114-115, 410, 417
abrasion resistance ... **M1:** 600, 605, 617, 620
alloy steels ... **M1:** 459-460
carburizing and carbonitriding ... **M1:** 533
castings, effect in ... **M1:** 384, 385, 386-389, 393, 394-395, 399, 400
compressive yield strength hardened 86*xx* steels ... **M1:** 536
constructional steels for elevated temperature use, effect on ... **M1:** 647
depth of hardening for induction hardened bars ... **M1:** 530
fabric knives ... **M1:** 625
formability affected by ... **M1:** 553
hardenability ... **M1:** 456-457, 470, 474-476
hardening temperature and minimum hardness ... **M1:** 529

Carbon in steel (continued)
hardness affected by **M1:** 472, 478, 480
hardness vs. carbon content and percent martensite **M1:** 457, 458, 529, 530, 561, 607, 608
machinability affected by **M1:** 572-573, 577
maraging steels **M1:** 445-447, 449
notch toughness, effect on **M1:** 692, 697, 707, 709
relation to strength and ductility **M1:** 463, 470
wear resistance vs. carbon content **M1:** 607, 608
Carbon iron powders
annealing **A7:** 183
Brinell hardness of compacts **A7:** 312
density from vibratory compacting **A7:** 306
density with high-energy compacting **A7:** 305
for food enrichment **A7:** 615
Carbon manganese boiler plate
tension and torsion effective fracture strains **A8:** 168
Carbon matrix composites, ceramic or glassy, microstructure **A9:** 592
polishing **A9:** 591
Carbon, max
chemical compositions per ASTM specification B 550-92 ..... **A6:** 787
Carbon molds *See* Mold(s)
Carbon monofluoride
as layer lattice solid lubricant **A18:** 116
Carbon monoxide **A13:** 632, 731, 1200-1201
electrometric titration for **A10:** 205
explosive range **A7:** 348
fume generation from arc welding **A6:** 68
measured on nickel powder surface **A7:** 258
poisoning **A7:** 349
pressure in carbonyl vapormetallurgy processing **A7:** 92
reaction **A6:** 58
reaction, underwater welding **A6:** 1010-1011, 1012
reaction with nickel **A7:** 134
Carbon monoxide, evolution
in aluminum **A15:** 82
during porcelain enameling **M1:** 179
Carbon monoxide reduction
of copper oxide **A7:** 107, 109
of iron powders for food enrichment **A7:** 615
of oxides **A7:** 52, 53
Carbon pearlitic alloy
temperature effect on torsional flow curve for **A8:** 176
Carbon pickup
Replicast process for **A15:** 270
Carbon planchets
for SEM specimens **A12:** 172
Carbon potential
defined **A9:** 3
Carbon powders
in laser cladding material **A18:** 867
Carbon refractory
defined **A15:** 2
Carbon replicas
techniques, TEM **A12:** 7
two-stage **A12:** 182
Carbon resistors
types and construction **EL1:** 178
Carbon restoration **A1:** 261, 263-264 **M1:** 235 **M4:** 446, 447
defined **A9:** 3
Carbon solubility
in copper alloys **A15:** 465
in multicomponent systems **A15:** 68
ternary iron-base systems **A15:** 66-68
Carbon spots
in alloy steel billet **A9:** 175
macroetching to reveal **A9:** 173
Carbon steel *See also* Hardenable steels; High-carbon steel; Low-carbon steel; Medium-carbon steel
adhesive wear (low, 12% C) **A18:** 238-239
alloying elements in **A1:** 144-147, 456-457
alloys, hardenability **A7:** 451, 452

Carbon steel (continued)
aluminum coating of **M5:** 334-335, 343-344
AMS designations **A1:** 153-154, 159, 160-162
application, internal combustion engine parts **A18:** 556
applications **M1:** 455, 457-459, 470
ASTM specifications **A1:** 150, 154, 156, 162, 163-164
atmospheric corrosion of **M1:** 717-723
for autoclave construction **A8:** 424
bars *See* Cold finished bars; Hot rolled bars
blue brittleness of **M1:** 684
bulk formability of **A1:** 581
carbon contents, AMS grades **M1:** 140
carbon contents of **A1:** 148, 454-456
cast, carbon-content classifications **M1:** 377-378
cast, patternmakers rules **M1:** 31, 33
castings and **A1:** 363
cavitation erosion rate **A18:** 774
classification and composition of hardfacing alloys **A18:** 652, 653
classification of **A1:** 147
classifications and
designations **M1:** 117-140, 457-459
quality descriptors **M1:** 118-119
cold finished bars **M1:** 215-251
combined effect of strain rate and temperature in **A8:** 38, 40
compacts, austenitizing temperatures **A7:** 453
composition of **A1:** 147, 149-151, 363
composition ranges and limits
15*xx* grades **M1:** 126
AISI-SAE standard grades **M1:** 125-127
cold finished bars **M1:** 120, 125, 126
grades formerly listed by SAE **M1:** 132
H-steels **M1:** 127
hot rolled bars **M1:** 120, 125, 126
merchant-quality grades **M1:** 126
plate **M1:** 120, 125, 136, 138
resulfurized and rephosphorized grades **M1:** 126
seamless tubing ..... **M1:** 120 , 125, 126
semifinished products for forging **M1:** 120, 125, 126
sheet and strip **M1:** 120, 125, 135
structural shapes **M1:** 120, 125, 136
welded tubing **M1:** 120, 125
wire rod **M1:** 120, 125, 126
corrosion in river water **M1:** 737-738
corrosion in seawater **M1:** 739-746
corrosion protection **M1:** 751-759
corrosive wear **A18:** 723
creep damage from service exposures **A8:** 338
cutting tool material selection based on machining operation **A18:** 617
damage by plastic deformation **A18:** 178
damage dominated by extrusion **A18:** 179
definition of **A1:** 147
deoxidation practice **A1:** 148
die material for sheet metal forming **A18:** 628
distortion in heat treatment **A1:** 369-370 **M1:** 469-470
ductility measurement from hot torsion tests **A8:** 165-166
electropolishing of **M5:** 305-08
elevated temperature properties **M1:** 639, 647, 649-653
elevated-temperature properties **A1:** 618, 619
embrittlement
aluminum nitride **A1:** 694-696
blue brittleness **A1:** 692
graphitization **A1:** 696-697
quench-age **A1:** 692-693
strain-age **A1:** 693-694
endodontic instruments manufactured from **A18:** 666, 675
endothermic gas sintering atmosphere **A7:** 341

Carbon steel (continued)
fabrication of parts and assemblies **A1:** 463
hardenability **A1:** 451-454
flow curves via torsion and tension testing for **A8:** 163, 165
fluid die, processing steps **A7:** 543
forging costs **M1:** 350, 351
forming by aluminum bronze tools **A18:** 633
fretting corrosion seen in light microscopy **A18:** 372
fretting wear **A18:** 248, 249
friction coefficient data **A18:** 73
galvanic effect in seawater **M1:** 740-741
graphitization of **M1:** 686
hardenability **M1:** 471-497
selection for **M1:** 482-491
hardenability equivalence table **M1:** 484-488
hardenable **M1:** 455-470
hardness conversion tables for **A8:** 109-113
heat treatment **A7:** 559
heat-treated **A18:** 693
hot dip tin coating of **M5:** 351
impact resistance and abrasion resistance properties **A18:** 759
impact wear **A18:** 268
induction and flame hardening **A1:** 463
induction hardening temperature **M4:** 452, 455
international designations and
British (BS) steel compositions ... **A1:** 158, 166-174, 182-184
French (AFNOR) steel compositions **A1:** 158, 166-174, 186-188
German (DIN) steel compositions **A1:** 157, 166-174, 175-178
Italian (UNI) steel compositions **A1:** 159, 166-174, 190-191
Japanese (JIS) steel compositions **A1:** 157-158, 166-174, 180
specifications for **A1:** 156-159, 166-194
Swedish (SS) steel compositions **A1:** 159, 166-174, 193
laser melting **A18:** 864
low-strength, oxygen contamination effect **A8:** 408
machinability **M1:** 236-239, 572-579
machinability of **A1:** 595-597
with calcium **A1:** 599
with lead **A1:** 599-600
with nitrogen **A1:** 599
with phosphorus **A1:** 599
resulfurized **A1:** 597-599
with selenium **A1:** 599
with tellurium **A1:** 599
mechanical properties **A1:** 202, 205, 206, 457-458
mechanical properties hardenable grades **M1:** 460-463
microstructure **A7:** 487, 488
modulus of elasticity at different temperatures **A8:** 23
nickel plating of **M5:** 215-258, 230-231
electroless **M5:** 230-231
notch toughness **M1:** 691-701, 704-707, 708-709
NR/C **EM3:** 629
nylon adhesive wear rate **A18:** 240
P/M parts, quenching **A7:** 453
parameter Z dependence on grain size in **A8:** 175
physical properties **A1:** 195-199 **M1:** 145-151
pickling of **M5:** 68-73
piling *See* Piling
pipeline, fatigue check extension in **A8:** 405
plate *See also* Plate **M1:** 181, 183
ASTM specifications **M1:** 183-185
mechanical properties **M1:** 188-190, 192, 194-198
polishing and buffing **M5:** 112-114, 120-121
in pressure vessel fabrication **A1:** 618
product composition tolerances **M1:** 122
quality descriptors **A1:** 201, 203

Carbon steel (continued)
quench-age embrittlement ... **M1:** 684
rust and scale removal ... **M5:** 12-13
SAE-AISI designations ... **A1:** 149-151
seawater exposure effect on adhesives ... **EM3:** 632
selected grades, tensile and fatigue properties ... **M1:** 680
shafting, selection for ... **M1:** 606
for shallow forming dies ... **A18:** 633
sheet and strip *See* Low-carbon steel, sheet and strip
sheet formability of ... **A1:** 573-579
sheet, minimum bend radii ... **M1:** 554
sheet, precoated *See* Steel sheet, precoated
sheet, terne coating of ... **M1:** 173-174
shot peening of ... **M5:** 141-142
single-phase austenite, stress-strain curves ... **A8:** 174
soil corrosion ... **M1:** 725-731
specimen size effect on fatigue limit reversed bending ... **A8:** 372
spring steel, abrasion resistance ... **M1:** 603
springs
grades for *See also* specific alloy designations ... **M1:** 284-285
hardenability requirements ... **M1:** 297, 301
steam service applications ... **M1:** 747
strain aging ... **A8:** 179
strain-age embrittlement of ... **M1:** 683-684
strain-hardening exponent and true stress values for ... **A8:** 24
stress ratio effect on corrosion fatigue crack propagation ... **A8:** 406-407
temper embrittlement of ... **M1:** 684-685
tempering ... **A1:** 458-459, 462 **M1:** 463, 466-469
threaded fasteners ... **M1:** 275, 277
timing chain components, wear compared to alloy steel ... **M1:** 628
for tooling in HIP units ... **A7:** 423
torsion for flow softening study ... **A8:** 177
torsional flow stress data, Zener-Hollomon parameter ... **A8:** 162-163
tubular products *See* Steel pipe; Steel tubes; Steel tubing
UNS designations ... **A1:** 151, 153
versus stainless steel properties ... **A18:** 713
weldability ... **M1:** 561-564
weldability of ... **A1:** 608-609
wire
fabrication characteristics ... **M1:** 587-593
manufacture of ... **M1:** 259-262
products *See also* Fasteners; Fence; Rope; Springs ... **M1:** 271-272
standard size tolerances ... **M1:** 261
wire rod
compositions ... **M1:** 254
decarburization limits ... **M1:** 254-256
grain size ... **M1:** 255
heat treatment ... **M1:** 255-256
inspection and testing ... **M1:** 255
manganese content, effect on tensile strength ... **M1:** 257
mechanical properties ... **M1:** 256, 257
qualities and commodities ... **M1:** 254-255
special requirements for ... **M1:** 255-256
tensile strength ... **M1:** 256, 257
workability ... **A8:** 165, 575
Carbon steel bars, for specific applications *See also* Cold-finished steel bars; Hot-rolled steel bars and shapes
axle shaft quality ... **A1:** 245
cold-shearing quality ... **A1:** 245
cold-working quality ... **A1:** 245
structural quality ... **A1:** 245
Carbon steel castings *See also* Austenitic manganese steel castings; Low-alloy steel castings ... **A9:** 230-235
abrasives for ... **A9:** 230
compositions of ... **A9:** 230-231
etchants ... **A9:** 230
etching ... **A9:** 230
grinding ... **A9:** 230
mechanical properties ... **A9:** 230
microstructures ... **A9:** 230-231
mounting ... **A9:** 230
polishing ... **A9:** 230
sectioning ... **A9:** 230
Carbon steel plate *See also* Steel plate ... **A1:** 226, 227, 232-233, 235
explosive-bonded to zirconium ... **A9:** 155
Carbon steel rod ... **A1:** 272
mechanical properties of ... **A1:** 275-276
qualities and commodities of ... **A1:** 272-274
special requirements for ... **A1:** 274
Carbon steel sheet and strip ... **A1:** 200-208
application of ... **A1:** 200
control of flatness ... **A1:** 205-208
direct casting methods ... **A1:** 211
mechanical properties of carbon steels ... **A1:** 202, 205, 206
mill heat treatment of cold-rolled steel products ... **A1:** 202-204
modified low-carbon steel sheet and strip ... **A1:** 206, 207
production of ... **A1:** 200-201, 203, 204
quality descriptors for carbon
commercial quality ... **A1:** 201
drawing quality ... **A1:** 201-202
steels ... **A1:** 201-202, 203
structural quality ... **A1:** 202
surface characteristics ... **A1:** 204-205
strain aging ... **A1:** 204-205
stretcher strains ... **A1:** 204
Carbon steel spring wire
characteristics of ... **A1:** 307
Carbon steels *See also* Carbon steels, specific types; Hardenable carbon steels; High-carbon steel; High-carbon steels; Low-alloy steels; Low-carbon steels; Medium carbon steels; Medium-carbon steels; Plain carbon steel; Plain carbon steels; Plate steels; specific types; Steel(s); Steels; Steels, AISI- SAE; Steels, specific types ... **A9:** 165-196 **A14:** 215-221
abrasion artifacts in ... **A9:** 36
air-carbon arc cutting ... **A6:** 1172, 1175, 1176
aircraft part, fatigue cracking ... **A13:** 1024
AISI compositions ... **A9:** 177
and alloy steel galvanic couple ... **A13:** 544
ammonium nitrate induced SCC in ... **A11:** 211
annealing ... **A4:** 37, 39
annual steel production ... **A13:** 509
applications ... **A6:** 641
automotive ... **A6:** 395
sheet metals ... **A6:** 399
applications, austempered parts ... **A4:** 155, 157, 161-162, 163
applications, protection tubes and wells ... **A4:** 533
aqueous corrosion ... **A13:** 512
arc-welded hardenable, failures in ... **A11:** 422-423
atmospheric corrosion ... **A13:** 510-512
austenitic-stainless-clad, welding of ... **A6:** 501-502
back reflection intensity ... **A17:** 238
backing bars ... **M6:** 382-383
bar and tube, die materials for drawing ... **M3:** 525
blanking, die materials for ... **M3:** 485, 486, 487
boiler service corrosion ... **A13:** 513-514
boiler tube, hydrogen damage and decarburization ... **A11:** 612
boriding ... **A4:** 437
brazing properties ... **M6:** 966
brazing temperature effect on hardness ... **A6:** 908
brine-heater shell, failed ... **A11:** 638
capacitor discharge stud welding ... **M6:** 730
carbon content effect on warm workability ... **A14:** 174
carbon solubility ... **A13:** 46
for case hardening, compositions of ... **A9:** 219
cast, weldability ... **A15:** 532-534
casting ... **A13:** 573-574
clad stainless steel, seawater corrosion ... **A13:** 889
cladding of austenitic stainless steel to ... **A6:** 502-504
classified ... **A15:** 702
cold extrusion, tool materials for ... **M3:** 515, 516, 517
cold heading of ... **A14:** 291
cold reduction effect, hydrogen absorption ... **A13:** 329
columnar bainite ... **A9:** 665
composition of tool and die steel groups ... **A6:** 674
Carbon steels (continued)
compositions of electrodes ... **M6:** 162-163
connecting rod, forging fold-fractured ... **A11:** 328, 330
corrosion by hydrogen sulfide ... **A11:** 631
corrosion in concrete ... **A13:** 513
corrosion losses, chemical plant atmospheres ... **A13:** 533
corrosion of ... **A11:** 199, 253, 300, 631 **A13:** 509-530
corrosion prevention guide, various environments ... **A13:** 523
corrosion protection ... **A13:** 521-525
corrosion, types of ... **A13:** 509-515
corrosion-fatigue cracks in ... **A11:** 253
counterbalance spring, fatigue failure ... **A11:** 558
covering for welding electrodes ... **A6:** 177
deaerator tanks, weld corrosion ... **A13:** 366
decarburization bands when held in a fluidized bed ... **A4:** 486
deep drawing, tool materials for ... **M3:** 494, 495, 496
deformation resistance vs rolling temperature ... **A14:** 119
degassing procedures ... **A15:** 428
description of ferrite/carbide microconstituents in low-carbon steel welds ... **A6:** 101
die forging, tool materials for ... **M3:** 529, 530, 532, 534
diffusion welding ... **A6:** 884
dip brazing ... **A6:** 336-338
discharge line, failure of ... **A11:** 639
dissimilar metal joining ... **A6:** 821, 824, 825
distortion in heat treatment ... **A4:** 612, 614, 615
drop hammer forming of ... **A14:** 655-656
effect of austenite grain size on ferrite and pearlite ... **A9:** 179
effect of bushings on fatigue strength ... **A11:** 470
electrodes, chemical composition of ... **M6:** 241
electrodes for flux-cored arc welding ... **A6:** 188
electrodes for shielded metal arc welding ... **M6:** 81-82
electrodes, submerged arc welding ... **A6:** 204, 205
electrogas welding ... **A6:** 652, 653, 658-659, 660
electron beam welding, preheating ... **M6:** 613
electron-beam welding ... **A6:** 259, 828, 860
electroslag welding ... **A6:** 273, 274, 276-277, 652, 653, 659
etchants ... **A9:** 169-175
explosion welding ... **A6:** 303, 896 **M6:** 706-707, 710, 713
electrical applications ... **M6:** 714
for fine-edge blanking and piercing ... **A14:** 472
flash welding ... **M6:** 557
flux-cored arc welding ... **A6:** 186, 187, 188, 643, 647, 652, 653, 654, 657-658
forge welding ... **A6:** 306 **M6:** 676
for forging ... **A14:** 218
forging effects on properties ... **A14:** 217
forging lubricants ... **A14:** 217-218
forging of ... **A14:** 215-221
forging temperatures ... **A14:** 81
friction surfacing ... **A6:** 321, 323
friction welded to Armco iron ... **A9:** 156
friction welding ... **A6:** 889 **M6:** 721
fusion welding to stainless steels ... **A6:** 826, 827
galvanizing vat, failure of ... **A11:** 273-274
gas metal arc welding ... **M6:** 153
gas tungsten arc welding ... **M6:** 182, 203
gas-metal arc welding ... **A6:** 68, 180, 467, 652, 653, 654, 655, 657
gas-metal arc welding shielding gases ... **A6:** 66, 67
gas-tungsten arc welding ... **A6:** 190, 652, 653, 654, 655-656, 658
for gastight shell of cold-wall vacuum furnace ... **A4:** 498
graphitization ... **M6:** 834
grinding ... **A9:** 168
grooving corrosion ... **A13:** 130-131
hardfacing ... **A6:** 789
hardfacing alloys for ... **A6:** 791
heat treatment ... **A14:** 218-219
HERF forgeability ... **A14:** 104
high frequency welding ... **M6:** 760
high-frequency welding ... **A6:** 252
hook, brittle fracture of ... **A11:** 332-333
hot forging behavior ... **A14:** 215-217

Carbon steels (continued)
for hot upset forging **A14:** 83
hot-twist testing **A14:** 216
humidity and atmospheric pollutant effects **A13:** 511-512
hydrochloric acid corrosion **A13:** 1162
hydrogen attack **A13:** 332
hydrogen fluoride/hydrofluoric acid corrosion **A13:** 1166-1167
hydrogen-induced cracking **A6:** 94, 642-649
induction heating energy requirements for metalworking **A4:** 189
induction heating temperatures for metalworking processes **A4:** 188
inhibitors for **A13:** 524
JIS S45C, friction welding **A6:** 441
lamellar tearing **A6:** 651, 652
laser beam welding **M6:** 647
laser cutting **A14:** 741
laser surface hardening **A4:** 286
liquid erosion resistance **A11:** 167
liquid nitriding **A4:** 419
liquid-metal corrosion **A13:** 514-515
liquid-metal embrittlement **A11:** 28
locomotive axles **A11:** 715-727
longitudinal properties **A14:** 219
macroetching **A9:** 170-177
mass, effect on hardness **A4:** 39
mechanical properties **A14:** 164
mechanical properties, austempered parts **A4:** 155
mechanical properties of electrodes **M6:** 162-163
in metal-matrix composites **A12:** 466
metallic coated **A13:** 526-527
microalloyed forging **A14:** 219-221
microetching **A9:** 169
microstructure **A6:** 642-643, 648
microstructure of annealed material **M6:** 22, 24
microstructures **A9:** 177-179
in mineral acids **A13:** 575
minimum bend radius **A14:** 523-524
mounting **A9:** 166-168
nonrelevant indications **A17:** 106-108
normalizing **A4:** 35-37, 38, 39
ocean corrosion rates **A13:** 898-899
organic coated **A13:** 528-529
overlay material **M6:** 602
oxide stability **A11:** 452
oxy/methylacetylene-propadiene- stabilized gas cutting **M6:** 904
oxyacetylene gas cutting **M6:** 901-902
oxyacetylene pressure welding **M6:** 595
oxyfuel gas cutting **A6:** 1156, 1159-1160 **M6:** 897
oxyfuel gas cutting, manual OFC-A **A6:** 1157
oxyfuel gas welding **A6:** 281, 286 **M6:** 583
oxynatural gas cutting **M6:** 903
oxypropane gas cutting **M6:** 904
oxypropylene gas cutting **M6:** 905
parts, casting failure **A11:** 392
in petroleum refining and petrochemical operations **A13:** 1262-1263
phases **A13:** 48
pipe, cavitation damage **A13:** 333
pipe, hydrogen embrittlement in **A11:** 645
pipe, weld inspection **A17:** 112
pipe, with uniform corrosion **A11:** 300
plasma (ion) nitriding **A4:** 423
plasma arc cutting **A6:** 1168, 1169-1170 **A14:** 731 **M6:** 916-917
plasma arc welding **A6:** 197, 652, 653, 654, 658 **M6:** 214
polished, monolayers adsorbed on **A10:** 118-120
polishing **A9:** 168-169
porosity **A6:** 641, 651-652
killed steels **A6:** 652
precoated before soldering **A6:** 131
preferential HAZ corrosion **A13:** 363
press forming dies, use for **M3:** 490

Carbon steels (continued)
protection needed for vacuum chamber constructions **A4:** 503, 504
recommended guidelines for selecting PAW shielding gases **A6:** 67
recommended machining specifications for turning with HIP metal-oxide ceramic insert cutting tools **EM4:** 969
recommended shielding gas selection for gas-metal arc welding **A6:** 66
repair welding **A6:** 1105
repair welding and corrosion of weldments **A6:** 1066
Replicast castings of **A15:** 272
residual stress distribution in **A10:** 389
residual stresses **A6:** 643, 651
resistance seam welding **A6:** 241
resistance soldering **A6:** 357
resistance spot welding **M6:** 478, 486
roll welding **A6:** 313, 314
SCC failures **A12:** 133
SCC in **A11:** 635
SCC susceptibility, temperature/concentration limits **A13:** 328
SCC testing **A13:** 270
Schaeffler diagrams used to predict joints **A9:** 582
microstructures in dissimilar-metal welded
in seawater **A13:** 539, 1256
sectioning **A9:** 165-166
shielded metal arc welding **A6:** 651, 652, 653, 654, 656-657 **M6:** 75, 83
shielding gas purity **A6:** 65
shrinkage allowances **A15:** 303
slow strain rate testing **A13:** 263
soil corrosion **A13:** 512-513
solidification cracking **A6:** 641, 649-651, 657
in sour gas environments **A11:** 300
specimen preparation **A9:** 165-169, 171
specimen preservation **A9:** 172
spraying for hardfacing **M6:** 789
springs, failure of **A11:** 555, 556
stainless steel clad, corrosion control **A13:** 888
and stainless steels, formabilities compared **A14:** 760
steam tube, corroded inner surface **A11:** 176
structure, principles **A13:** 46-47
stud arc welding **A6:** 210, 211, 212, 213, 214, 216, 652, 653, 659-660 **M6:** 730, 733
stud material **M6:** 730
submerged arc welding **A6:** 202, 203, 204, 642, 643, 647, 648, 649, 650, 652, 653, 654, 658 **M6:** 115
effect on cracking **M6:** 128
electrodes for **M6:** 120
suitability for cladding combinations **M6:** 691
sulfur dioxide concentration effect on corrosion rate **A13:** 909
sulfuric acid corrosion **A13:** 1148
superheater tube, pitting corrosion **A11:** 615
surface effects of grinding analyzed by Mössbauer **A10:** 287
susceptibility to hydrogen damage **A11:** 126, 249
swageability of **A14:** 128-129
swaging cold reduction effects **A14:** 129
thermal properties **A6:** 17
thermite welding **A6:** 292
thermophysical constants for laser surface hardening **A4:** 289, 290
thermoreactive deposition/diffusion process **A4:** 452
tool, concrete roughers, bending failure **A11:** 564-565
tool, hydrogen flaking **A11:** 574
torch brazing *See* Torch brazing of steels
tubing, remote-field eddy current inspection **A17:** 200-201
type SA-516-70 thermal cycling effects **M6:** 712
ultrasonic welding **A6:** 893
weathering steels **A13:** 515-521

Carbon steels (continued)
welding to austenitic stainless steels **A6:** 500-501, 502
welding to ferritic stainless steels **A6:** 501
welding to martensitic stainless steels **A6:** 501
weldment, brittle fracture **A11:** 93-94
weldment properties **A6:** 417, 424
weldments **A13:** 362-367
wetting by copper, LME from **A11:** 719
wire, die materials for drawing **M3:** 522
for wire-drawing dies **A14:** 336
wires, biological corrosion **A13:** 116

Carbon steels and alloy steels **A16:** 666-680
abrasive waterjet cutting **A16:** 677
annealing process **A16:** 672, 674, 676
automatic screw machine test **A16:** 677
bainite **A16:** 667, 669, 671-672
bevel gear rough and finish cutting **A16:** 349
boring **A16:** 163, 164, 668
Ca treatment for deoxidation **A16:** 673, 675
cementite **A16:** 666-667, 670-671
cermet tools **A16:** 92, 93, 95, 96
chip formation **A16:** 668, 669, 670, 674
classification into 3 groups **A16:** 666, 667
coating of tools **A16:** 57
cold form tapping **A16:** 266
composition percentages **A16:** 667
contour band sawing **A16:** 361, 364
cutoff band sawing **A16:** 360
cutting fluids **A16:** 125, 677, 691
cutting speeds **A16:** 668, 678, 679
deformation zones **A16:** 669
drill eccentricity test **A16:** 679
drill force test **A16:** 678-679
drilling **A16:** 222, 225-226, 229-232, 237, 673-675
ductile fracture **A16:** 669
electrochemical grinding **A16:** 677
end milling **A16:** 325
feed and cutting forces compared **A16:** 764
feed and metal removal rates compared **A16:** 764
ferrites **A16:** 666-667, 669, 671-672, 674, 675
gear shaping **A16:** 347
gear shaving **A16:** 342
grinding **A16:** 676-677
heat treatment **A16:** 671, 672, 674, 676
hobbing **A16:** 345
honing **A16:** 477
laser cutting **A16:** 677
leaded or resulfurized **A16:** 149
machinability **A16:** 668, 669-670, 671
machinability test matrix **A16:** 639-640
machinability testing **A16:** 677-680
martensite **A16:** 667, 669, 671-672, 674
mechanical properties compared **A16:** 763, 764
microcrack formation **A16:** 669, 672, 674-675, 676
microhoning **A16:** 490
microstructures **A16:** 666-669, 675-676
microvoid coalescence **A16:** 669, 672, 674-675
milling **A16:** 312, 313, 315, 675-676
multifunction machining **A16:** 386, 392
normalizing **A16:** 674
pearlite **A16:** 666-667, 669-672, 675-677
photochemical machining **A16:** 588, 591
planing **A16:** 676
plunge test **A16:** 677-678
power requirements **A16:** 690
rephosphorized steels **A16:** 672
resulfurization **A16:** 672-673, 675
saw bands **A16:** 357, 358, 359, 362, 364
sawing **A16:** 359, 360
shaping **A16:** 347, 676
shear formation **A16:** 671
slab milling **A16:** 324
spheroidizing **A16:** 672, 674, 676
surface finish **A16:** 670, 672, 673, 675
susceptibility to seaming **A16:** 282
tapping **A16:** 256, 260, 262, 263, 676
thread grinding **A16:** 274

Carbon steels and alloy steels (continued)
thread rolling **A16:** 282, 288, 290, 292-293
threading **A16:** 300
tool life **A16:** 669-673, 675, 677, 679
turning........ **A16:** 144, 146-147, 149, 668-673, 675-676
twist drill geometries **A16:** 674
Carbon steels, specific types *See also* Carbon steels
0.20% C, water quenched **A9:** 185
0.20C-1.0Mn, as-quenched **A9:** 185
10B35, different heat treatments
compared **A9:** 186-187
1008, aluminum-killed, different
anneal temperatures **A9:** 182
1008, aluminum-killed, hot-rolled,
open skin lamination **A9:** 183
1008, aluminum-killed, hot-rolled
sheet arrowhead defects **A9:** 183
1008, capped, finished hot, coiled cold, hot
90% **A9:** 180-181
rolled, reductions from 10% to
1008, cold-rolled, longitudinal streaks **A9:**
183-184
1008, cold-rolled sheet, ingot scab
sliver **A9:** 183
1008, cold-rolled sheet, mill scale
defect **A9:** 183
1008, cold-rolled sheet, surface pits
from rolled-in sand **A9:** 184
1008, rimmed, coiled, cold rolled,
effects of different process
temperatures **A9:** 182
1008, rimmed, orange peel **A9:** 182
1008, rimmed, stretcher strains **A9:** 182
1010, as hot rolled, effects of insuffi-
cient grinding **A9:** 167
1010, carbonitrided **A9:** 227
1010, decarburized, effects of improper
polishing **A9:** 169
1010, hot-rolled, different sectioning
techniques **A9:** 166
1010, liquid nitrided **A9:** 229
1012 modified, cold-rolled strip
carbonitrided **A9:** 226
1018 bar, austenitized and furnace
cooled **A9:** 224
1018 bar, carbontrided **A9:** 226
1018, gas carburized, different times
and surface carbon contents
compared **A9:** 219-220
1020, carbonitrided **A9:** 227
1020, carburized, prior austenite grain
boundaries **A9:** 185
1020, cyanided **A9:** 227
1025, normalized by austenitizing **A9:** 186
1030, isothermal transformation of
austenite **A9:** 186
1035 & 10B35, effect of boron on
hardenability **A9:** 187
1035 modified, salt bath nitrided **A9:** 229
1038, bar, as-forged, secondary pipe **A9:** 187
1038, bar, as-forged, severely
overheated **A9:** 187
1039, gas carburized roller for contact-
fatigue tests, with butterfly
alterations **A9:** 223
1040, bar, different heat treatments **A9:** 187
1040, isothermal transformation **A9:** 187
1045, bar, different heat treatments
compared **A9:** 188
1045, bar stock, different heat treat-
ments compared **A9:** 188
1045, forging, different heat treatments
compared **A9:** 188
1045 modified, salt bath nitrided **A9:** 229
1045, partial isothermal transformation **A9:** 188
1045, sheet, normalized by
austenitizing **A9:** 188
1050, different heat treatments
compared **A9:** 189
1052, forging, various inclusions **A9:** 190
1055, rod, patented by austenitizing **A9:** 192
1055, wire, patented by austenitizing **A9:** 192
1060, decarburized **A9:** 193
1060, rod, different heat treatments
compared **A9:** 192
1060, wire, air patented by
austenitizing **A9:** 193

Carbon steels, specific types (continued)
1064, strip, cold-rolled **A9:** 193
1065, wire, patented by austenitizing **A9:** 193
1070, valve-spring wire, 80% reduction **A9:** 193
1074, sheet, cold-rolled, different heat
treatments compared **A9:** 193
1080, bar, hot-rolled, different cooling
rates compared **A9:** 193-194
1095, different heat treatments
compared **A9:** 194
1095, wire, different heat treatments
compared **A9:** 194
1541, forged, different heat treatments **A9:** 190
1541, forging lap **A9:** 190
AISI 1042, temperature effect on frac-
ture mode **A12:** 33, 45
AISI 1060, knobbly structure **A12:** 33
AISI 1080, temperature effect on frac-
ture mode **A12:** 33, 44
AISI 1085, cathodic cleaning **A12:** 75-76
ASTM A27, annealed by austenitizing **A9:** 231
ASTM A27, as-cast **A9:** 231
ASTM A27, grade 70-36, as-cast **A9:** 232
ASTM A27, grade 70-36, normalized
by austenitizing **A9:** 232
ASTM A27, grade 70-36, quenched
and tempered **A9:** 232
ASTM A27, quenched and tempered **A9:** 231
ASTM A148, annealed by austenitizing **A9:** 233
ASTM A148, grade 90-60, as-cast **A9:** 232
ASTM A148, grade 90-60, normalized
and tempered **A9:** 233
ASTM A148, grade 90-60, normalized
by austenitizing **A9:** 233
ASTM A148, grade 105-85, quenched
and tempered **A9:** 233
ASTM A148, quenched and tempered **A9:** 233
ASTM A216, grade WCA, annealed by
austenitizing **A9:** 233-234
ASTM A216, grade WCA, as-cast **A9:** 233
ASTM A216, grade WCA, normalized
and tempered **A9:** 233-234
ASTM A216, grade WCA, quenched
and tempered **A9:** 234
ASTM A216, grade WCB, annealed by
austenitizing **A9:** 234
ASTM A216, grade WCB, as-
quenched **A9:** 234-235
ASTM A216, grade WCB, normalized
by austenitizing **A9:** 234-235
Fe-0.57C, plate martensite **A9:** 671
Fe-0.69C, lower bainite **A9:** 665
Fe-1.2C, plate martensite **A9:** 671
Fe-1.34C, inverse bainite **A9:** 665
Carbon treating
as bath control **EL1:** 680
Carbon tube furnace
defined **A7:** 2
Carbon vane pump
in environmental test chamber **A8:** 411
Carbon, vapor pressure
relation to temperature **M4:** 309, 310
Carbon-arc microscope illumination
systems **A9:** 72
Carbon-boron-nitrogen-silicon composition
tetrahedron
as superhard material **A2:** 1008
Carbon-carbon
defined **EM1:** 6
Carbon-carbon composites **EM1:** 911-924 **EM4:** 20
in aircraft brakes **A18:** 582
applications **A9:** 592 **EM1:** 922-924 **EM4:** 20
for brake linings **A18:** 570
for carbon aircraft brakes **A18:** 584
chemical vapor infiltration process **EM1:** 911-912
coatings, for oxidation resistance **EM1:** 920
continuous carbon fiber reinforced **EM1:** 911-914
etching **A9:** 591
fabrication processes **A9:** 591
heat sink properties **EM1:** 36
high-performance, properties **EM1:** 920
liquid impregnation process **EM1:** 911-912
manufacture **EM1:** 924
mounting and mounting materials **A9:** 588
multidirectionally reinforced **EM1:** 129, 915-919
oxidation-resistant **EM1:** 920-921
polarized light rnicroscopy **A9:** 592

Carbon-carbon composites (continued)
polishing **A9:** 590-591
process **EM1:** 911-912
properties **EM4:** 20
redensified with pyrolytic graphite **A9:** 595-596
scanning electron microscopy **A9:** 592
structurally reinforced **EM1:** 922-924
tape-wound, shrinkage cracks and
voids **A9:** 595
thermal conductivity **EM1:** 924
unidirectional properties **EM1:** 912-914
Carbon-carbon deposits
manufacturing methods **A18:** 816, 817
Carbon-carbon engineering properties **EM4:**
835-838, 842-843
Carbon-carbon friction materials
for carbon aircraft brakes **A18:** 584
Carbon-chromium equilibrium curves **A15:** 426
Carbon-containing cermets **A7:** 799
defined **A2:** 979
Carbon-epoxy composites
applications **A9:** 592
fiber volume fraction effect on thermal
expansion **EM1:** 190
flexural damping variation **EM1:** 208
grinding **A9:** 588
ply angle effects on damping **EM1:** 210
polishing **A9:** 590
S-N curves **EM1:** 201
Carbon-equivalent number (CEN) **A6:** 416
Carbon-filled epoxy
volume resistivity and conductivity **EM3:** 45
Carbon-film resistors *See also* Resistors
construction **EL1:** 178
failure mechanism **EL1:** 971, 999-1001
Carbon-graphite **A18:** 693
electrodes for resistance brazing **A6:** 341
electrodes, resistance brazing **A6:** 340-341
filler for seals **A18:** 551
G-14 (grade), properties **A18:** 549, 551
hard, oxidation resistant, electrodes,
resistance brazing **A6:** 340-341
P-658RC, properties **A18:** 549, 551
properties (70%-30%) **A18:** 817
seal material **A18:** 551
soft, electrodes, resistance brazing **A6:** 340-341
Carbon-graphite composites
single-lap shear, fatigue failure in fas-
tened joints **A11:** 549
static tensile failures in **A11:** 548-549
Carbon-graphite materials, friction and
wear of **A18:** 816-819
applications **A18:** 816, 819
component design **A18:** 819
corrosive wear **A18:** 816
electrical carbons **A18:** 816
electrographite **A18:** 818, 819
"film control" **A18:** 818
friction and wear **A18:** 817-819
impregnation effects on typical car-
bon-graphite base material **A18:** 817
impregnation effects on typical graph-
ite-base material **A18:** 817
lubrication **A18:** 818
manufacturing methods **A18:** 816-817
mechanical carbons **A18:** 816
microstructure **A18:** 816
physical and mechanical properties **A18:** 816,
817, 818, 819
rubbing speed **A18:** 818
Carbon-manganese
submerged arc welding **M6:** 118-119
Carbon-manganese alloys
slow-strain rate testing **A13:** 263
Carbon-manganese cast steels **A1:** 373
as low-alloy **A15:** 715
Carbon-manganese steel **A8:** 499
Carbon-manganese steels
applications, machinery and
equipment **A6:** 390
composition of **A1:** 151
hot ductility tests on **A1:** 696
principal ASTM specifications for
weldable sheet steels **A6:** 399
sheet and strip **A1:** 206, 208
weld microstructure **A6:** 53
weldability **A6:** 420, 421, 422, 423, 424, 426-427

Carbon-manganese structural steels
hot-rolled ... **A1:** 390, 391
Carbon-molybdenum
resistance to graphitization ... **M6:** 834
submerged arc welding electrodes for ... **M6:** 121
Carbon-molybdenum desulfurizer welds
cracking in ... **A11:** 663
Carbon-molybdenum steel
creep damage from service exposures ... **A8:** 338
Carbon-molybdenum steel, flux-cored arc welding
designator ... **A6:** 189
Carbon-molybdenum steels
boiler applications ... **M1:** 747
graphitization in ... **A11:** 100
Carbon-phenolic prepregs
as flame resistant ... **EM1:** 141
Carbon-platinum replicas ... **A9:** 108
Carbon-resin composites
as advanced composite ... **EM1:** 28
property data ... **EM1:** 294
Carbon-silicon steels ... **A1:** 208
Carbon-tungsten special-purpose tool steel *See* Tool steels, carbon-tungsten, special purpose
Carbon-tungsten special-purpose tool steels
forging temperatures ... **A14:** 81
Carbon/graphite matrix composites
applications ... **EM1:** 918
densification processing ... **EM1:** 917-918
multidirectionally reinforced ... **EM1:** 915-919
preforms, woven ... **EM1:** 915-917
properties ... **EM1:** 918
Carbonaceous
additions, to molding sand mixes ... **A15:** 211
defined ... **A15:** 2
Carbonate
brass plating, use in ... **M5:** 286, 571, 573
cadmium cyanide plating bath content effects of ... **M5:** 257-258, 266-267
copper plating, formation during, effects and removal of ... **M5:** 163-164
zinc cyanide plating bath content ... **M5:** 248
Carbonate alkaline cleaners ... **M5:** 23-24, 28, 35
Carbonates
as molten salt ... **A13:** 50, 91
as salt precursors ... **EM4:** 113
Carbonic acid
conductometric titration of ... **A10:** 203
Carbonitride cermets *See also* Carbide cermets ... **A16:** 91-92
applications and properties ... **A2:** 1004-1005
defined ... **A2:** 979
titanium carbonitride cermets ... **A2:** 998-1000
Carbonitrided steels ... **A9:** 217-229
electron microscopy ... **A9:** 217
etchants for ... **A9:** 217
etching ... **A9:** 217
grinding ... **A9:** 217
microstructures ... **A9:** 218
mounting ... **A9:** 217
polishing ... **A9:** 217
sectioning ... **A9:** 217
specimen preparation ... **A9:** 217
Carbonitrides
appearance of, in iron-chromium-nickel heat-resistant casting alloys ... **A9:** 332
for fatigue resistance ... **A11:** 121
identification of ... **A11:** 39
as inclusions ... **A10:** 176
Carbonitrides, precipitation of
in steel ... **A1:** 115-116
Carbonitriding ... **A4:** 264, 266, 376-386 **M4:** 176-190
applications ... **A4:** 379-380, 385 **M4:** 176-178
atmosphere constituents ... **A4:** 380 **M4:** 181
atmosphere control ... **A4:** 380-382 **M4:** 182-183, 184
atmospheres, batch furnace ... **A4:** 382-383 **M4:** 183-184
atmospheres, continuous-furnace ... **A4:** 383 **M4:** 185
Carbonitriding (continued)
case and core hardenabilities ... **M1:** 533-538
case composition ... **A4:** 376-377, 382 **M4:** 178, 179, 180
case depth ... **A4:** 376, 377-379, 381, 382, 383-384, 385-386 **M4:** 180-181, 182, 183
case hardenability ... **A4:** 376, 379, 382, 386 **M4:** 180, 182
case hardening by ... **A7:** 454
case properties ... **M1:** 533, 536, 538
cermets ... **A7:** 813
core properties ... **M1:** 534-535
defined ... **A9:** 3 **A13:** 2
and drilling operations ... **A16:** 219
equipment and techniques ... **A7:** 454
examples ... **A4:** 377, 378-379, 380, 382, 383-384
in fluidized beds ... **A4:** 486, 490
furnaces ... **A4:** 380-383 **M4:** 181
gas quenching ... **A4:** 376, 385 **M4:** 188
hardenability affected by ... **M1:** 493-495
hardness gradients ... **A4:** 379, 380 **M4:** 179-180, 181, 182
hardness profiles developed by ... **M1:** 632-633
hardness testing ... **A4:** 385-386 **A7:** 454 **M4:** 189
methods ... **M1:** 533, 539-540
as modified gas carburizing ... **A7:** 454
oil quenching ... **A4:** 376, 378, 383, 386 **M4:** 188
P/M drive gears ... **A7:** 667
P/M materials ... **M1:** 339, 343
powder metallurgy parts ... **A4:** 383, 386 **M4:** 189-190
ferrous ... **A4:** 386 **M4:** 799
quenching media ... **A4:** 376, 384-385 **M4:** 177
refractory cermets ... **A7:** 813
retained austenite ... **A18:** 260
retained austenite, control of ... **A4:** 384 **M4:** 187
safety precautions ... **A4:** 383 **M4:** 185
selection of steel for ... **M1:** 535-538
temperature selection ... **A4:** 383-384 **M4:** 186-187
tempering ... **A4:** 379, 385, 386 **M4:** 189-190
threaded fasteners ... **M1:** 277, 279
time and temperature, effect of ... **A4:** 376, 378, 379, 382, 383 **M4:** 180-181, 182, 183
titanium alloys ... **A18:** 866
uses ... **A4:** 376 **M4:** 176
void formation ... **A4:** 379, 382 **M4:** 179, 181
water quenching ... **A4:** 385 **M4:** 187
Carbonium ion ... **EM3:** 7
defined ... **EM2:** 8
Carbonization ... **EM1:** 6, 112 **EM3:** 7
defined ... **A18:** 4
Carbonizing flame
definition ... **A6:** 1207
Carbontetrachloride
as carcinogenic cleaning agent ... **A12:** 74
Carbonyl iron powders
microstructures ... **A9:** 509
Carbonyl nickel powder
density of compacts ... **A7:** 310
microstructures ... **A7:** 396
production ... **A7:** 134-138
properties ... **A7:** 396
shrinkage of iron mixes with ... **A7:** 480, 481
Carbonyl powders
cluster ... **A7:** 135
defined ... **A7:** 2
metal ... **A7:** 135
in thermal decomposition ... **A7:** 54-55
vapor ... **A7:** 137-138
Carbonyl vapormetallurgy processing ... **A7:** 92-93, 134-138
Carborundum (220 mesh)
Miller numbers ... **A18:** 235
Carborundum's sintered alpha (SA) silicon carbide ... **EM4:** 980
Carboxy DGEBA diacrylate
properties ... **EM3:** 92
Carboxyl-terminated butadiene acrylonitrile adhesive
principal strains ... **EM3:** 329
Carboxyl-terminated butadiene acrylonitrile epoxy
compared to epoxy-urethane ... **EM3:** 185
Carboxyl-terminated butadiene acrylonitrile epoxy system
strain-energy release rates ... **EM3:** 365
Carboxyl-terminated butadiene acrylonitrile liquid polymers
properties ... **EM3:** 185
Carboxyl-terminated polybutadiene acrylonitrile rubber
as polyimide toughening agent ... **EM3:** 161
Carboxylated butadiene acrylonitrile
used as modifiers ... **EM3:** 121
Carboxymethylcellulose
removal ... **EM4:** 137
Carburization *See also* Decarburization ... **A16:** 71
at elevated-temperature service ... **A1:** 643
by macroetching ... **A9:** 174-175
in carbon and alloy steels, revealed of iron-alloys, effect of silicon on ... **A9:** 333
chromium-nickel heat-resistant casting
case hardening by ... **A7:** 453-454
-caused brittle fracture, austenitic stainless steels ... **A12:** 358
effect in tool steels ... **A11:** 571-572
effect on chromium depletion and magnetic permeability ... **A11:** 272
effect on stainless steel microstructure ... **A11:** 271
effect on stainless steels ... **A11:** 271-273
excessive, of tools and dies ... **A11:** 571-573
of fatigue surface ... **A8:** 373
faulty, effect on distortion ... **A11:** 141
in forging ... **A11:** 335-336 **A17:** 493
gas ... **A7:** 453
growth or shrinkage during ... **A7:** 480
of heat-resistant cast alloys ... **A13:** 576
in heat-treating components ... **A13:** 1311-1313
high-temperature ... **A13:** 99-101
historical study ... **A12:** 1-2
improper, steel gear and pinion failure from ... **A11:** 336
in iron casting ... **A11:** 361-362
mechanically alloyed oxide dispersion-strengthened (MA ODS) alloys ... **A2:** 947
nickel-base alloys and welding considerations ... **A6:** 591-592
nickel-base corrosion-resistant alloys containing molybdenum ... **A6:** 596
of P/M superalloys ... **A13:** 839
pack ... **A7:** 453
postforging defects from ... **A11:** 333
of powder metallurgy materials ... **A9:** 503, 512
of stainless steels ... **A13:** 559
in steel castings ... **A11:** 396, 406
of steel, microstructure ... **A11:** 326-327
surface ... **A11:** 406
tool steels, high-speed ... **A16:** 55
of tungsten carbide powder, furnaces for ... **A7:** 157
of tungsten carbide powder production ... **A7:** 156-157
Carburized cases
microhardness testing ... **A8:** 96
Carburized hardenability test ... **A1:** 464-465, 466 **M1:** 472-473
Carburized steel
welding factor ... **A18:** 541
Carburized steel parts
surface hardening of ... **M1:** 532
Carburized steels ... **A9:** 217-229
characteristics of ... **A1:** 380-381
electron microscopy ... **A9:** 217
etchants for ... **A9:** 217
etching ... **A9:** 217
grinding ... **A9:** 217

Carburized steels (continued)
machinability of ... **A1:** 600
microstructures ... **A9:** 218
mounting ... **A9:** 217
polishing ... **A9:** 217
sectioning ... **A9:** 217
specimen preparation ... **A9:** 217
thermoreactive deposition/diffusion process ... **A4:** 448
Carburizing ... **A18:** 873-876
applications ... **A18:** 873, 874
carburizing process ... **A18:** 873
pack carburizing ... **A18:** 873
salt bath carburizing ... **A18:** 873
vacuum and plasma carburizing ... **A18:** 873
carburizing steels ... **A18:** 875-876
cleanliness ... **A18:** 875
fabricability ... **A18:** 876
hardenability ... **A18:** 875
preoxidation prior to carburizing ... **A18:** 876
secondary hardening alloys ... **A18:** 876
case and core hardenabilities ... **M1:** 533-538
case properties ... **M1:** 532-536
characteristics of carburized surfaces ... **A18:** 873-875
diffusion of carbon ... **A18:** 873
microstructure ... **A18:** 873-874
residual stress ... **A18:** 874-875
surface oxidation ... **A18:** 875
transformation of retained austenite ... **A18:** 875
copper brazed assemblies ... **M6:** 936
core properties ... **M1:** 537-538
defined ... **A9:** 3 **A13:** 2
and drilling ... **A16:** 219
gear materials ... **A18:** 261
history and development ... **A18:** 873
mainshaft bearings of jet engines ... **A18:** 590
methods ... **M1:** 533, 538-539
notch toughness of steels, effect on ... **M1:** 704-705, 706
in powder forging ... **A14:** 201-202
process/materials selection for wear resistance ... **A18:** 876
retained austenite ... **A18:** 260
stainless steels ... **A18:** 715, 716, 723
steels for ... **M1:** 535-538, 610
as surface treatment in wrought tool steels ... **A1:** 779
threaded fasteners ... **M1:** 277, 279
titanium alloys ... **A18:** 780-781
tool steels ... **A18:** 642, 739
tungsten carbide ... **A18:** 795
typical hardness profiles ... **M1:** 634
for valve train assembly components ... **A18:** 559
Carburizing and cyaniding salts
dip brazing ... **A6:** 337, 338
Carburizing atmospheres
heating-element materials ... **A2:** 834
Carburizing bearing steels ... **A1:** 381-382, 383
Carburizing potential
of cemented carbides in hydrogen atmosphere ... **A7:** 386-387
neutral ... **A7:** 388
Carburizing salts
use in dip brazing ... **M6:** 991
Carburizing steels
compressive layer and hardness produced by ... **A10:** 380
machinability ... **M1:** 568, 578-581
CARCA *See* Computer-assisted rocking curve analysis
Carcass foundation stiffness ... **A18:** 580
Carcinogenesis
and dental alloys ... **A13:** 1339
of lead toxicity ... **A2:** 1245-1246
and metal toxicity ... **A2:** 1235
of nickel toxicity ... **A2:** 1250
Carcinogenic organic solvents
cleaning ... **A12:** 74
Carcinogenic polynuclear aromatic compounds
MFS analysis of ... **A10:** 72
Carcinogenicity
of arsenic ... **A2:** 1238
of cadmium ... **A2:** 1241
of metals, chronology of observations ... **A2:** 1236
of platinum complexes ... **A2:** 1258
Carcinogens
in metallic wires ... **EM1:** 118
Card *See* Printed board; Printed wiring board (PWB)
Card assembly materials
and processes selection ... **EL1:** 116-117
Card edge
substrate ... **EL1:** 624
Card-on-board
defined ... **EL1:** 1136
Card-on-mother-board package *See* Card-on-board; Third-level package
Cardboard
carbides for machining ... **A16:** 75
Carded glass fibers ... **EM1:** 11
Carding process
defined ... **EM1:** 111
Cardiomyopathy
as cobalt toxicity ... **A7:** 204
from cobalt toxicity ... **A2:** 1251
Cardiovascular disease
from cadmium exposure ... **A2:** 1241
Care
of fractures ... **A12:** 72
CARES computer program *See also* Probabilistic design of ceramic components, NASA/CARES computer program ... **EM4:** 700, 744
Cariogenesis ... **A13:** 1337
Carmen-Kozeny equation
permeation of viscous flow along tortuous channels through many small capillaries ... **EM4:** 70
Carmine method
for boron in Al alloys ... **A10:** 68
Carnauba
as investment casting wax ... **A15:** 253
Carney funnel
for determining apparent density ... **A7:** 273-274
for determining powder rate ... **A7:** 279
Carnotite ore (uranium)
toxicity of ... **A2:** 1261-1262
Carousel-type rig
for thermal fatigue testing ... **A11:** 278-279
Carpenter Hampden steel
properties ... **A8:** 234
Carpenter Stentor steel
properties ... **A8:** 234
Carpenters' saws
wear of ... **M1:** 624
Carpet plot *See also* Fracture surface map
Carpet plots ... **EM1:** 233, 310-312
fractured titanium alloy ... **A12:** 172
stereo imaging ... **A12:** 171
titanium alloy, by stereophotogrammetry ... **A12:** 198
Carriage
cam plastometer ... **A8:** 195-196
in quick-release clamp ... **A8:** 221
Carrier
defined ... **A10:** 670
effects in dc arc sources ... **A10:** 25
gas, in analytic ICP systems ... **A10:** 34
Carrier core composition
in copier powders ... **A7:** 584
Carrier density
as function of position ... **EL1:** 152
Carrier gas
definition ... **M6:** 3
Carriers *See* Chip Carriers
in core castings ... **A15:** 240
liquid, wet abrasive blasting process ... **M5:** 94-95
nickel plating, use in ... **M5:** 204-205
Carriers (bases)
of lubricants ... **A14:** 514
Carrousel sample holder
in Auger spectrometer ... **A10:** 554
Cartridge brass *See also* Brasses; Copper alloys; Copper alloys, specific types; Copper alloys, specific types, C26000; Wrought coppers and copper alloys
applications and properties ... **A2:** 300-302
EDS and WDS analysis of ... **A10:** 530
electrolytic etching ... **A9:** 401
hardness conversion tables ... **A8:** 109
resistance spot welding ... **A6:** 850
SCC failures in ... **A11:** 27
as solid-solution copper alloy ... **A2:** 234
weldability ... **A6:** 753
Cartridge cases
20-mm ... **A7:** 680
Cartridge filters
as rigid filter ... **A15:** 490
Cascade ... **A18:** 851
Cascade cleaning ... **A13:** 1139
Cascade development
of copier powders ... **A7:** 582
Cascade impaction
to analyze ceramic powder particle sizes ... **EM4:** 67
Cascade method, cast irons
welding of ... **A6:** 715
Cascade separator
defined ... **A7:** 2
Cascade sequence
definition ... **M6:** 3
welding of cast irons ... **M6:** 313
Cascade tests
for gas-turbine components ... **A11:** 280-281
Cascade welding sequence
cast irons ... **A15:** 526
Cascades
of peritectic reaction ... **A15:** 128
Case
carburizing, for fatigue resistance ... **A11:** 121
and core, low ductility of ... **A11:** 389-391
crushing, gear failure from ... **A11:** 595
defined ... **A9:** 3
hardness of gear tooth ... **A11:** 600
Case carburizing
in powder forging ... **A14:** 201-202
Case crushing
defined ... **A18:** 4
Case debonding
as squeeze casting defect ... **A15:** 326
Case depth *See also* Case depth, measurement
case depth, control of
measurement ... **M4:** 155, 156, 158-159, 172-174
variation ... **M4:** 153-154, 154-155
and Knoop hardness readings ... **A8:** 96, 101
metallographic sections for ... **A11:** 24
mounting for examination of, in cast irons ... **A9:** 243
of powder metallurgy materials
measuring ... **A9:** 508
preparation of specimens for
examining ... **A9:** 217-218
vs applied stress, in gears ... **A11:** 594
Case depth, measurement ... **A4:** 454-461
chemical method ... **A4:** 454-456 **M4:** 276-277
procedure for carburized cases ... **A4:** 454-455 **M4:** 276, 277
spectrographic analysis ... **A4:** 455-456 **M4:** 276-277, 278
destructive methods ... **A4:** 460-461
macroscopic visual procedures ... **A4:** 457 **M4:** 278-279
mechanical method ... **A4:** 456 **M4:** 277-278
cross-section procedure ... **A4:** 456 **M4:** 277-278
step-grind procedure ... **A4:** 456 **M4:** 278
taper-grind procedure ... **A4:** 456 **M4:** 278
microhardness testing ... **A4:** 459-460
microscopic visual procedures ... **A4:** 457-459 **M4:** 278-279
annealed condition ... **A4:** 457-458 **M4:** 279-280
carbonitrided cases ... **A4:** 458 **M4:** 280
carburized cases ... **A4:** 457-458
cyanided cases ... **A4:** 458-459 **M4:** 280
hardened condition ... **A4:** 457-459 **M4:** 279
nitrided cases ... **A4:** 459 **M4:** 281
selectively hardened cases ... **A4:** 459 **M4:** 281
nondestructive methods ... **A4:** 460-461
process parameter contributions to variations ... **A4:** 624, 625
specifications ... **A4:** 454
Case depth requirements
surface hardened steels ... **M1:** 528, 538
Case hardened steel
Brinell testing of ... **A8:** 89
Rockwell hardness testing of ... **A8:** 83
shallow, Rockwell scale for ... **A8:** 76
Case hardening ... **M1:** 457, 470, 473, 491-496
brittle fracture and ... **A11:** 90-92
by carbonitriding ... **A7:** 454
by carburizing ... **A7:** 453-454
by gas nitriding ... **A7:** 455

Case hardening (continued)
chemical surface studies of ... **A10:** 177
defined ... **A9:** 3 **A13:** 2
depths ... **A7:** 454
effect on fatigue behavior ... **M1:** 673-675
effective case depth ... **A4:** 454, 459 **M4:** 276
faulty, effect on distortion ... **A11:** 141
hardness profiles developed by ... **M1:** 632, 633
hardness surveys ... **A4:** 454 **M4:** 275
of iron-carbon P/M drive gears ... **A7:** 667
of parts ... **A11:** 94
as secondary operation ... **A7:** 453-455
of shaft, high-cycle fatigue in ... **A11:** 106
steels for ... **M1:** 491-496
total case depth ... **A4:** 454, 458, 459 **M4:** 276
Case hardening, effects
visual examination ... **A12:** 72
Case hardening steels ... **A9:** 217-229
carbonitrided, specimen preparation ... **A9:** 217
carburized, specimen preparation ... **A9:** 217
compositions of ... **A9:** 219
microstructures ... **A9:** 218
nitrided, specimen preparation ... **A9:** 217-218
Case properties
carburized and carbonitrided steels ... **M1:** 533-537
nitrided steels ... **M1:** 540
Case-hardening
gear materials ... **A18:** 261
mainshaft bearings of jet engines ... **A18:** 590
rolling-element bearings ... **A18:** 262
Case-hardening steels, specific types
805A 17, residual stresses ... **A4:** 609
805A20, residual stresses ... **A4:** 609
832M 13, residual stresses ... **A4:** 609
897M39, residual stresses ... **A4:** 609
905M39, residual stresses ... **A4:** 609
Casein ... **EM3:** 7
applications ... **EM3:** 45
characteristics ... **EM3:** 45
defined ... **EM2:** 8
for packaging ... **EM3:** 45
Casein adhesive ... **EM3:** 7
Casing *See* Steel tubular products ... **M1:** 320
extrusion press ... **A11:** 37
fractured, after bisection of bleed hole ... **A11:** 38
CASS test ... **A13:** 4, 225
CASSE computer program for structural analysis ... **EM1:** 268, 270
Cassette ... **EM3:** 65
Cassiterite ... **A7:** 160
fusion flux for ... **A10:** 167
as stream tin ... **A15:** 16
Cassiterite (SnO_2)
applications ... **EM4:** 48
electrical/electronic applications ... **EM4:** 1106
key product properties ... **EM4:** 48
raw materials ... **EM4:** 48
Cast
defined ... **EM2:** 8
Cast alloy steels *See also* Casting alloys; Ferrous casting alloys
development of ... **A15:** 32
Cast alloy steels, specific types
14% Cr alloy (VM), development ... **A15:** 32
20Cr-7Ni (VA), development of ... **A15:** 32
80% Ni 20% Cr, development of ... **A15:** 32
Cast alloy tools
cast iron machining ... **A16:** 656
Ni alloy machining ... **A16:** 837
turning Ti alloys ... **A16:** 846
Cast alloys *See also* Zinc and zinc alloys
scanning electron microscopy used to determine microstructural morphology ... **A9:** 101
Scheil equation for the solute distribution characteristic ... **A9:** 631
Cast Alnico alloys
chemical analysis and magnetic property control ... **A15:** 737
Cast Alnico alloys (continued)
foundry practice ... **A15:** 736-738
grinding ... **A15:** 738-739
heat treatment ... **A15:** 738
history ... **A15:** 736
inspection and testing ... **A15:** 738-739
melting and casting ... **A15:** 737-738
molding ... **A15:** 736-737
patterns ... **A15:** 736
structure and properties ... **A15:** 739
Cast aluminum *See also* Aluminum; Aluminum casting alloys
designation system ... **A2:** 15-16
elongation measurements of ... **A8:** 655
Cast aluminum alloys *See also* Aluminum; Aluminum casting alloys; Cast aluminum alloys, specific types
atmospheric corrosion ... **A13:** 596, 601
chemical analysis ... **A12:** 405
composition and microstructure effects on corrosion ... **A13:** 585-587
designation system ... **A2:** 15-16
effects of freezing and heat treatment ... **A12:** 409
experimental, transverse fracture surface ... **A12:** 413
foundry products ... **A2:** 123-151
fractographs ... **A12:** 405-413
fracture/failure causes illustrated ... **A12:** 217
nomenclatures ... **A2:** 4-5
properties ... **A2:** 152-177
shrinkage cavities ... **A12:** 409
solution potentials ... **A13:** 585
Cast aluminum alloys, specific types *See also* Cast aluminum alloys
356.0-T6, brittle fracture ... **A12:** 405-406
356.0-T6, fatigue fracture ... **A12:** 407-408
356.0-T6, service fracture ... **A12:** 409-410
380.0, brittle fracture ... **A12:** 410
518.0, overload fracture in service ... **A12:** 411-412
A357 blade, porosity in ... **A12:** 66, 67
A357-T6 gear housing, shrinkage void ... **A12:** 66, 67
A357-T6, inclusion in fracture surface ... **A12:** 65, 66
A357-T6, shrinkage void ... **A12:** 67
Cast aluminum and aluminum alloys
cleaning and finishing of ... **M5:** 571-574, 576, 583, 588, 590-591, 597, 601-603, 606
Cast aluminum bronze
dealuminification ... **A13:** 130
Cast aluminum molds
rotational molding ... **EM2:** 366
Cast aluminum-magnesium alloy
nose-splitting in ... **A8:** 595-596
Cast austenitic stainless steels
ferrite in ... **A1:** 909, 910-911
SCC prevention ... **A13:** 327
Cast austenitic-manganese steels, specific types
12Mn, nominal compositions and applications ... **A18:** 703
12Mn-1Mo, nominal compositions and applications ... **A18:** 703
12Mn-1Mo-Ti, nominal compositions and applications ... **A18:** 703
C-Mn, nominal compositions and applications ... **A18:** 703
Cr-Ni-Mo, nominal compositions and applications ... **A18:** 703
Mn-Cr-Mo, nominal compositions and applications ... **A18:** 703
Cast beryllium-copper alloys *See also* Beryllium-copper alloys
mechanical properties ... **A2:** 412
microstructure ... **A2:** 404-405
Cast beryllium-nickel alloys
melting and casting ... **A2:** 403
Cast billet
beryllium-copper alloys ... **A2:** 403
Cast chromium steels
mineral acid corrosion ... **A13:** 575
Cast Co-Cr-W alloys
for planer tools ... **A16:** 183
Cast cobalt alloys
for cutting tool materials ... **A18:** 614, 615
hot hardness values compared ... **A16:** 69
machining of Zn alloys ... **A16:** 831
processing ... **A16:** 69
properties and applications ... **A16:** 69
Tantung 144 ... **A16:** 69-70
Tantung G ... **A16:** 69-70
tools ... **A16:** 69, 70, 154, 654
Cast cobalt-base superalloys *See also* Polycrystalline cast superalloys
compositions of ... **A1:** 983
design of ... **A1:** 985-986
physical properties ... **A1:** 983
stress-rupture properties ... **A1:** 985-987
tensile properties ... **A1:** 984
Cast cobalt-chromium-molybdenum endoprosthesis ... **A13:** 663
Cast copper alloys *See also* Cast copper alloys, specific types; Copper; Copper alloys; Copper alloys, specific types ... **A13:** 619-620, 624
applications ... **A2:** 224-228
availability ... **A2:** 216
corrosion ratings ... **A2:** 231
properties ... **A2:** 224-228, 356-390
Cast copper alloys, specific types *See also* Cast copper alloys
beryllium copper 21C, properties and applications ... **A2:** 390-391
C81100, composition, applications properties, fabrication ... **A2:** 356
C81300, properties and applications ... **A2:** 356
C81400, properties and applications ... **A2:** 356-357
C81500, properties and applications ... **A2:** 357
C81800, properties and applications ... **A2:** 357-358
C82000, properties and applications ... **A2:** 358-359
C82200, properties and applications ... **A2:** 359
C82400, properties and applications ... **A2:** 359-360
C82500, properties and applications (standard) ... **A2:** 360-362
C82600, properties and applications ... **A2:** 362
C82800, properties and applications ... **A2:** 362-363
C83300, properties and applications ... **A2:** 363-364
C83600, properties and applications ... **A2:** 364
C83800, properties and applications ... **A2:** 365
C84400, properties and applications ... **A2:** 365
C84800, properties and applications ... **A2:** 365-366
C85200, properties and applications ... **A2:** 366
C85400, properties and applications ... **A2:** 366
C85700, properties and applications ... **A2:** 366-367
C85800, properties and applications ... **A2:** 366-367
C86100, properties and applications ... **A2:** 367
C86200, properties and applications ... **A2:** 367
C86300, properties and applications ... **A2:** 367-368
C86400, properties and applications ... **A2:** 368-369
C86500, properties and applications ... **A2:** 369
C86700, properties and applications ... **A2:** 370
C86800, properties and applications ... **A2:** 370-371
C87300, properties and applications (formerly C87200) ... **A2:** 371-372
C87500, properties and applications ... **A2:** 372-373
C87600, properties and applications ... **A2:** 372
C87610, properties and applications ... **A2:** 372
C87800, properties and applications ... **A2:** 372-373
C87900, properties and applications ... **A2:** 373
C90300, properties and applications ... **A2:** 374
C90500, properties and applications ... **A2:** 374
C90700, properties and applications ... **A2:** 374-375
C91700, properties and applications ... **A2:** 375
C92200, properties and applications ... **A2:** 375-376
C92300, properties and applications ... **A2:** 376
C92500, properties and applications ... **A2:** 376-377
C92600, properties and applications ... **A2:** 377-378
C92700, properties and applications ... **A2:** 378
C92900, properties and applications ... **A2:** 378
C93200, properties and applications ... **A2:** 378-379
C93400, properties and applications ... **A2:** 379

Cast copper alloys, specific types (continued)
C93500, properties and applications **A2:** 379
C93700, properties and applications **A2:** 379-380
C93800, properties and applications **A2:** 380-382
C93900, properties and applications **A2:** 382
C94300, properties and applications **A2:** 382
C94500, properties and applications **A2:** 382
C95200, properties and applications **A2:** 382-383
C95300, properties and applications **A2:** 383-384
C95400, properties and applications **A2:** 384-385
C95500, properties and applications **A2:** 385
C95600, properties and applications **A2:** 386
C95700, properties and applications **A2:** 386
C95800, properties and applications **A2:** 386-387
C96200, properties and applications **A2:** 387
C96400, properties and applications **A2:** 387-388
C96600, properties and applications **A2:** 388
C97300, properties and applications **A2:** 388
C97600, properties and applications **A2:** 388-389
C97800, properties and applications **A2:** 389
C99400, properties and applications **A2:** 389
C99500, properties and applications **A2:** 389-390
C99700, properties and applications **A2:** 390
Cast dies **A14:** 53
Cast film *See also* Films
defined **EM2:** 8
Cast film extrusion *See* Chill roll extrusion
Cast grit
blasting with **M5:** 83-84, 86
Cast iron *See also* Alloy cast irons; Compacted graphite iron; Ductile iron; Gray iron; Iron castings; Malleable iron; specific type, such as Alloy cast iron; White iron **A1:** 3-104
abrasion resistance **M3:** 582, 583
abrasive blasting of **M5:** 91
acid cleaning of **M5:** 59, 62
air-carbon arc cutting **A6:** 1176
alloy production, powders used **A7:** 572
alloyed **M1:** 4-5, 6, 7, 75-96
alloying elements, graphitization potential of **A1:** 6
architectural uses **A15:** 18
austenite dendrites **M1:** 5, 6
austenitic, impellers, shrinkage porosity damage **A11:** 355-356
automotive applications of gray **A1:** 19
babbitting of **M5:** 356
basic metallurgy of **A1:** 3-11
bearing, cap, hypoeutectic **A11:** 349-350
brazing and soldering characteristics **A6:** 626-627
Brinell test application **A8:** 84, 89
cadmium plating of **M5:** 256-257, 262, 264
carbon content of **A1:** 3, 5
carbon dissolution in **A15:** 72-73
carbon effects, discovered **A15:** 29
carbon equivalent **M1:** 4
carbon in **M1:** 3-5, 6, 9
cementite in **M1:** 3-5, 6
chemical analysis and sampling **A7:** 249
chemical composition **A15:** 185, 522
chemical pipe sealants for plumbing **EM3:** 608
chilled
abrasion resistance **M1:** 81, 87-88
annealing, effect on hardness and combined carbon **M1:** 82
compositions **M1:** 76
depth of chill, effect of alloying elements **M1:** 76-80
mechanical properties **M1:** 85-87
structure **M1:** 75-76, 82-85
chrome plating, hard **M5:** 171-172
classification of **A1:** 3-11
classified **A15:** 627-628
cleaning solutions for substrate materials **A6:** 978
common **A1:** 3
compacted graphite irons **A1:** 3, 8-9
composition ranges **A1:** 5, 6
Connellsville coke for **A15:** 30
constitutional liquation **A6:** 75
cooling curve analysis **A15:** 180
corrosion in near-neutral soil **M1:** 726-727
corrosion of **A11:** 199-200
coupled zone in **A15:** 174
cutting, tools for **M3:** 477
defined **A15:** 2
definition of **A1:** 3, 12
Cast iron (continued)
desulfurization **A15:** 75-76
development of **A15:** 29
dip brazing **A6:** 338
ductile *See* Ductile cast iron
ductile iron **A1:** 3, 7-8
ductile, stress concentration cracking **A11:** 346-347
electroslag welding **A6:** 278
emulsion cleaning of **M5:** 35
enamel application **EM3:** 301, 303
as eutectic alloy **A15:** 168-181
ferrite in **M1:** 6-9
ferrosilicon as inoculant for **A15:** 105
fluxing of **M5:** 353
fresh water corrosion, effect of H_2S **M1:** 733
friction welding **A6:** 152
gages, use for **M3:** 556
gas porosity in **A15:** 82
gas-tungsten arc welding **A6:** 192
gases in **A15:** 82-85
graphite in **M1:** 3-9
with graphite particles, Rockwell hardness testing of **A8:** 8o
graphite shape **A1:** 3, 6
graphitic, postweld heat treatment **A15:** 527
gray *See* Gray cast iron **A11:** 199
gray iron **A1:** 3, 4-7
gray, paper-roll driers, failures of **A11:** 653-654
gray, yield strength of **A8:** 21
growth of eutectic in **A15:** 174-180
hardened steel ball indenters for **A8:** 74
hardfacing **A6:** 798, 807
hardness testing **A7:** 452
heat treatment **A1:** 7
high-nickel **A11:** 200
high-silicon **A11:** 200
history of **A15:** 17-18
hot dip tin coating of **M5:** 351-355
hydrogen removal **A15:** 84-85
hydrogen solubility in **A15:** 82
hypoeutectic, structure of bearing cap cast from **A11:** 349
inclusions in **A15:** 94-95
inert gas flushing of **A15:** 84-85
inoculants for **A15:** 105
inorganic fluxes for **A6:** 980
iron-iron carbide-silicon system **M1:** 3-4
kinetics of gas-liquid reactions **A15:** 82-83
ladle desulfurization **A15:** 77
as lap plate material **EM4:** 352-353
liquid iron-carbon alloys, structure of **A15:** 168-169
liquid treatment of **A1:** 7
malleable *See* Malleable cast iron
malleable and ductile **A11:** 199
malleable irons **A1:** 3, 9-11
martensite in **M1:** 5-7
matrix **A1:** 3
melts, purification of **A15:** 75-79
metallurgy of **M1:** 3-9
metalworking rolls, use for **M3:** 504-506
microhardness testing **A8:** 97
microstructural wear **A11:** 161
microstructure **M1:** 3-9
microstructure, and weldability **A15:** 522
microstructures and processing for obtaining common commercial **A1:** 4
mottled, defined *See* Mottled cast iron
mottled iron **A1:** 3
nickel plating, electroless **M5:** 238-240
nitrogen removal **A15:** 84-85
nitrogen solubility in **A15:** 82-83
nucleation of eutectic in **A15:** 169-174
oxyacetylene welding of **A15:** 529-531
oxyfuel gas cutting **A6:** 1155
oxyfuel gas welding **A6:** 281
oxygen removal **A15:** 74
paper-roll dryer, journal -to-head failure **A11:** 655
pearlite in **M1:** 4, 6-9
phosphate coating of **M5:** 436-438, 441, 448
pickling of **M5:** 72
polishing and buffing **M5:** 112-114
porcelain enameling of.... **M5:** 509-513, 515, 519-522, 524-525, 527-528
design parameters **M5:** 524-525
Cast iron (continued)
evaluation of enameled surfaces **M5:** 527-528
frits, composition **M5:** 510
methods **M5:** 519-520
selecting, factors in **M5:** 512-513
surface preparation for **M5:** 515
precoated before soldering **A6:** 131
press forming. dies, use for **M3:** 490, 493
principles of metallurgy of **A1:** 3-4
properties **EM4:** 677, 990, 992
pump impeller, graphitic corrosion **A11:** 374-375
relative solderability as a function of flux type **A6:** 129
repair welding **A6:** 1105
repair welding and corrosion of weldments **A6:** 1066
Rockwell hardness testing of **A8:** 82
Rockwell scale for **A8:** 76
rust and scale removal **M5:** 13-14
salt bath descaling of **M5:** 98-100, 102
scrap, as charge **A15:** 388
scuffing temperatures and coefficients between ring and cylinder- liner materials **EM4:** 991
shielded metal arc welding **A6:** 176
silicon in **M1:** 3-5
solderability **A6:** 971
solderable and protective finishes for substrate materials **A6:** 979
soldering **A6:** 631
solidification of **A15:** 83-84, 168-181
special **A1:** 3, 11
spheroidal graphite iron **A1:** 7-8
structural diagrams for **A15:** 68-70
sulfur removal **A15:** 74
tensile strength **A1:** 6, 7
ternary, third element effects **A15:** 65-68
tooling for pressed ware **EM4:** 398
turning and milling recommended ceramic grade inserts for cutting tools **EM4:** 972
types **M1:** 3-9
urea effect **A15:** 238
vapor degreasing of **M5:** 54
weld repair with high-nickel alloys **A6:** 1066
weldability **M1:** 563-564
welding of **A15:** 520-531
white iron **A1:** 3
white irons, hardfacing **A6:** 790, 791
Cast iron cylinder compressed air blast
early use **A15:** 25
Cast iron, glow discharge to determine C
P, and S in **A10:** 29
Cast iron, induction hardening
flake-graphite gray iron **M4:** 470, 476-477
applications **M4:** 477
distortion **M4:** 476
ferrite **M4:** 476
pearlite **M4:** 476
nodular iron **M4:** 470, 477-479
as cast **M4:** 477
normalized and tempered **M4:** 478
prior treatments **M4:** 477-478
quenched and tempered **M4:** 478-479
pearlitic malleable iron **M4:** 470, 479-480
Cast iron pipe *See also* Ductile iron pipe; Gray iron pipe **M1:** 97-100
applications **M1:** 97-100
coatings and linings **M1:** 97-98, 100
joints and fittings **M1:** 97-100
specifications for **M1:** 97-98, 100
standard laying conditions **M1:** 98, 99
Cast iron rolls
applications and types **A14:** 352-353
Cast iron shot peening **M5:** 141
Cast irons *See also* Ductile irons; Gray iron; Gray irons; Iron; Malleable iron Nodular iron; Malleable irons ...... **A3:** 1 • 23-24, 26 **A9:** 242-255 **A13:** 566-572 **A16:** 184-185, 648-665
abrasion damage in **A9:** 38-39
abrasive belt grinding **A16:** 663-664
abrasive machining **A16:** 648
acicular structures **A16:** 9, 650
alloying effects **A13:** 556-567
alloying element effect on machining characteristics **A16:** 652-654
anaerobic biological corrosion **A13:** 116

Cast irons (continued)
annealing treatments A16: 651
applications, protection tubes and wells A4: 533
arc welding *See* Arc welding of cast irons
austenite A16: 649
bainite A16: 648
basic metallurgy A13: 566
boriding A4: 440
boring A16: 162, 655
brazing *See* Brazing of cast irons
Brinell hardness indicator A16: 648
broaching A16: 197, 200, 203-204, 206, 208, 656
carbides A16: 649
carbon solubility A13: 46-47
for casings on air-fuel gas burners A4: 274
cemented carbide tools A16: 86
centerless roll lapping A16: 496
ceramic cutting tools A16: 98, 101-103
cermet tools applied A16: 92
chemical compositions M6: 309
chill A16: 649, 650
chill cast iron rolls A16: 98
classification, by carbon form/shape A13: 566
coated carbide tools A16: 79
coatings A13: 570-571
color etching A9: 141-142
commercially available A13: 567-568
compacted graphite A16: 648
containing tin A2: 526
corrosion forms A13: 468-569
corrosion-resistant, hydrochloric acid corrosion A13: 1163
counterboring A16: 660
cutting fluids A16: 125, 651-652, 654, 665
cutting speed related to tool life A16: 651
cylindrical grinding A16: 664
diamond as abrasive for honing A16: 476
drilling A16: 218, 219, 226-230, 236-237, 658
dry machining A16: 392
ductile, fracture modes A12: 230
ductile iron machining characteristics A16: 651
electrochemical grinding A16: 544, 547
electrochemical machining A16: 538, 539
end milling A16: 663-664
etchants A9: 244-246
face milling A16: 648, 651, 652
ferrite A16: 648, 649, 650
flame hardening A4: 284
friction band sawing A16: 365
friction welding M6: 722
frits for A13: 446-447
gas metal arc welding M6: 153
granular brittle fractures A12: 103
graphite nodules in, revealed by different illuminations A9: 81
graphite retention A9: 243-244
graphitic corrosion A13: 568
gray cast iron metal removal rates compared A16: 652
gray iron machining characteristics A16: 649-651
grinding A9: 243 A16: 112, 547, 661-664
ground/machined with superabrasives/ ultrahard tool materials A2: 1013
guide shoe materials for honing A16: 478
hardness of microconstituents A16: 648
hardness test results as function of testing scale used A4: 623
high-Ni irons A16: 656
high-Si irons A16: 661
highly alloyed A13: 567
honing A16: 476, 484, 664-665
induction hardening A4: 668
iron-iron carbide-silicon ternary phase diagram A13: 566
lap material A16: 498, 502, 503, 504
lapping A16: 492, 494, 499, 503, 664-665
laser surface transformation hardening A4: 265, 286, 287, 290, 293, 294-295

Cast irons (continued)
liquid nitriding A4: 419
low/moderately alloyed A13: 567
machinability test matrix A16: 639-640
machine lapping between plates A16: 494-495, 497
machining A2: 966
malleable iron machining characteristics A16: 652
martensite A16: 9
matrix microstructure effect on tool life A16: 650
microhoning A16: 40
microstructure A13: 567
microstructure effect on machinability A16: 648-649, 650
microstructure, effect on properties A9: 242
microstructures A9: 245-246
milling A16: 319, 328, 547, 660-661
mold stress relieving A4: 668
mounting A9: 243
nickel alloying, history A2: 429
nodular ferritic, salt bath nitrided A9: 229
oxyacetylene braze welding *See* Oxyacetylene braze welding of steel and cast irons
oxyacetylene welding M6: 601-605
applications M6: 601-602, 605
comparison to arc welding M6: 601
corrosion-resistant cast irons M6: 605
ductile iron M6: 604
ductility M6: 601
filler metals M6: 603
fluxes M6: 603
gray iron M6: 603-604
hardfacing M6: 602
malleable and white iron M6: 604-605
porosity M6: 601
postweld heat treatment M6: 602
preheating M6: 602
preparation of castings M6: 602
repair and reclamation M6: 601-602
welding rods M6: 602-603
oxyfuel gas cutting M6: 897, 112-113
oxyfuel gas welding M6: 583
PCBN tooling used A16: 111, 112, 113, 115
pearlite A16: 648, 649, 650, 652
in petroleum refining and petrochemical operations A13: 1263
planing A16: 184, 185, 657, 659-660
plasma (ion) nitriding A4: 423
polishing A9: 243-244
reaming A16: 239, 245, 659
resistance, to corrosive environments A13: 569-570
roller burnishing A16: 252
sampling A9: 242
sand content A16: 650
sectioning A9: 242-243
selection A13: 571
shaping A16: 190, 659-660
shielded metal arc welding M6: 75
shifting or swelling of castings A16: 650
shrinks A16: 650
silicon carbide for abrasive on honing stones A16: 476
spade drilling A16: 225
specimen preparation A9: 242-245
spheroidite A16: 649, 650
steadite A16: 649, 652
structure, principles A13: 46-47
substrate for thermoreactive deposition/diffusion process A4: 449
sulfuric acid corrosion A13: 1149-1150
surface grinding A16: 664
surface preparation, porcelain enameling A13: 447
tapping A16: 259, 263, 264, 266, 660, 661
tests performed during specimen preparation A9: 242

Cast irons (continued)
thermal energy method of deburring A16: 578
thread milling A16: 269
threading A16: 298
tool bit tool steels use A16: 57
tool life A16: 112, 648-652, 654, 656
tool materials A16: 654-656
tool steel alloys used A16: 57
turning A16: 112, 135, 158, 159, 653, 654
types of A4: 667
unalloyed A13: 567
WC-Co tools A16: 74, 75
white iron A16: 112, 115, 535, 648-649, 652, 656

Cast irons, friction and wear of A18: 693, 695-701
adhesive wear A18: 241
alloy A18: 649
for deep-drawing dies A18: 634
applications A18: 701
internal combustion engine parts A18: 553, 556, 557, 561
laser transformation A18: 863
for brake drums and disk brake rotors A18: 572, 573
for brake linings A18: 570
carbon contents and steel metallurgy A18: 702
cast iron
carbon equivalent value A18: 694-696
constitution A18: 695-696
cooling rate A18: 696
section sensitivity A18: 696
cavitation resistance A18: 600
cutting tool material selection based on machining operation A18: 617
cutting tool materials and cutting speed relationship A18: 616
die material for sheet metal forming A18: 628
diesel engine wear, lubricant analysis case history A18: 308-309
ferrographic analysis A18: 306
galling resistance with various material combinations A18: 596
gear materials, surface treatment and minimum surface hardness A18: 261
graphite A18: 698-700
flake A18: 698-699
nodular A18: 699
gray iron A18: 695, 697-698, 699
applications A18: 695, 701
microstructure A18: 695, 697-698, 699, 700-701
properties A18: 695
section thickness A18: 698
tensile strength A18: 696, 698
grinding media, mining industry A18: 654
high-resolution electron microscopy to study sliding wear A18: 389
laser alloying A18: 866
laser melted gray cast iron A18: 864
laser melting A18: 864
laser transformation hardening A18: 862, 863, 864
light microscopy A18: 372
malleable iron A18: 695-701
microstructure A18: 695, 700-701
properties A18: 695
mechanical properties and supplementary information A18: 700
mottled iron A18: 695
nodular or spherical graphite iron A18: 695, 698, 699-700
applications A18: 695, 700, 701
microstructure A18: 699-701
properties A18: 695
sliding bearings A18: 516
for sliding vane rotary compressor cylinder A18: 606
surface replica versus SEM micrograph of gear tooth A18: 374
tensile properties A18: 696-697
section effect on strength A18: 696-697
thermal spray coating applications A18: 832, 833

Cast irons, friction and wear of (continued)
typical microstructures **A18:** 700-701
acicular transformation structures **A18:** 701
austenite **A18:** 701
bainitic transformation structures **A18:** 701
cementite **A18:** 700
ferrite **A18:** 700
martensite **A18:** 701
pearlite **A18:** 700
phosphide eutectic **A18:** 700-701
versus aluminum-silicon alloy A390.0 for piston cylinder liners **A18:** 556
white iron (chilled iron) **A18:** 695, 696
applications **A18:** 695, 701
composition **A18:** 698
hardness specification **A18:** 696, 698
ledeburite formation **A18:** 697
microstructure **A18:** 695, 697, 698, 700-701
Cast irons, heat treating **A4:** 667-669
equipment **A4:** 668 **M4:** 526-527
hardness measurement **A4:** 668, 669
hardness measurements **M4:** 525-526, 527
heating media **A4:** 668 **M4:** 528
heating rate **A4:** 667 **M4:** 527
proven applications for borided ferrous materials **A4:** 445
quenching media **A4:** 668 **M4:** 528
quenching temperature, effect on carbon content **A4:** 669 **M4:** 525
temperature control **A4:** 667, 668 **M4:** 527-528
Cast irons, high-alloy *See* High-alloy cast irons, heat treating
Cast irons, welding of **A6:** 708-721
air-carbon arc process **A6:** 712
alloying additions **A6:** 709
annealing **A6:** 714
arc spraying **A6:** 720
arc-welding processes **A6:** 716-721
austempered ductile iron **A6:** 708
block sequence **A6:** 715
braze welding **A6:** 715-716
cascade method **A6:** 715
classification by commercial designation, microstructure, and fracture **A6:** 708
compacted graphite iron **A6:** 708, 710
contaminants **A6:** 711, 712
defect removal **A6:** 711, 712
ductile iron **A6:** 708, 709, 712
braze welding **A6:** 716
gas-metal arc welding **A6:** 719
postweld heat treatment **A6:** 714
shielded metal arc welding **A6:** 717
electron-beam welding **A6:** 720
flame spraying **A6:** 715, 720
flux-cored arc welding **A6:** 716, 719-720
gas-metal arc welding **A6:** 716, 718-719, 720
globular transfer mode **A6:** 718
short circuiting transfer mode **A6:** 718, 719
spray transfer mode **A6:** 718
gas-tungsten arc welding **A6:** 716, 720
gray iron **A6:** 708, 712
braze welding **A6:** 716
flame spraying **A6:** 715
hot cracking **A6:** 715
oxyfuel welding **A6:** 714
postweld heat treatment **A6:** 714
shielded metal arc welding **A6:** 717
groove face grooving **A6:** 712, 713
heat-affected zone ... **A6:** 709, 710, 712, 713-714, 717, 718, 720
joint design modifications **A6:** 712, 713
laser-beam welding **A6:** 720
malleable iron **A6:** 708, 709
braze welding **A6:** 716
microstructures **A6:** 709, 711
minimizing dilution of the base iron casting **A6:** 721
mottled iron **A6:** 708
nodularizer additives **A6:** 709
overlaying **A6:** 720-721
oxyfuel gas welding **A6:** 720
oxyfuel welding **A6:** 714-715
peening **A6:** 712, 713
plasma spraying **A6:** 720
postwelding treatment **A6:** 713-714
processing steps **A6:** 709
production of **A6:** 708
Cast irons, welding of (continued)
shielded metal arc welding **A6:** 716-718
stress relieving **A6:** 714
stringer bead welding **A6:** 710
studding **A6:** 712-713
submerged arc welding **A6:** 716, 720, 721
surfacing **A6:** 720-721
surfacing materials **A6:** 720-721
cast iron alloys **A6:** 721
ceramic materials **A6:** 721
copper alloys **A6:** 720-721
hardfacing alloys **A6:** 721
high-nickel alloys **A6:** 721
stainless steels **A6:** 721
temperature zone schematic representation in a typical welding **A6:** 710
thermal spraying **A6:** 720
weldability **A6:** 710-713
base-metal preparation **A6:** 711-712
of castings **A6:** 710
fusion zone **A6:** 710
heat-affected zone **A6:** 709, 710
identification of iron casting **A6:** 710
partially melted region **A6:** 710
preheat **A6:** 710-711
preweld testing **A6:** 711
special techniques **A6:** 712-713
welding processes and consumables **A6:** 714-716
white iron **A6:** 708
Cast irons, white
superplasticity **A14:** 869-871
Cast lean manganese-austenitic steels, specific types
6Mn-5Cr-1Mo, nominal compositions and applications **A18:** 703
9Mn-1Mo-Ti, nominal compositions and applications **A18:** 703
Cast magnesium alloys *See also* Cast magnesium alloys, specific types; Magnesium; Magnesium alloys; Magnesium alloys, cast
properties of **A2:** 491-516
Cast magnesium alloys, specific types *See also* Magnesium alloys
AM60A, properties **A2:** 491-492
AM60B, properties **A2:** 491-492
AMIOOA, properties **A2:** 492
AS41XB, properties **A2:** 492-493
AZ63A, properties **A2:** 493-494
AZ81 A, properties **A2:** 494-496
AZ91 D, properties **A2:** 496-497
AZ91 E, properties **A2:** 496-497
AZ91A, properties **A2:** 496-497
AZ91C, properties **A2:** 496-497
AZ92A, properties **A2:** 497-498
EQ21, properties **A2:** 499-501
EZ33A, properties **A2:** 501
HK31A, properties **A2:** 501-503
HZ32A, properties **A2:** 503-504
K1A, properties **A2:** 504-505
QE22A, properties **A2:** 505-506
QH21A, properties **A2:** 506-507
WE43, properties **A2:** 507-508
WE54, properties **A2:** 508-510
ZC63, properties **A2:** 510-511
ZE41A, properties **A2:** 511
ZE63A, properties **A2:** 511-512
ZH62A, properties **A2:** 513-514
ZK51A, properties **A2:** 514-515
ZK61A, properties **A2:** 515-516
Cast materials
corrosion testing **A13:** 193-194
Cast metal
quasi-static torsional testing of **A8:** 145
structure, grain size **A8:** 573
and wrought metal workability at varied temperatures **A8:** 574
Cast metal-matrix composites *See also* Metal-matrix composites
applications/properties **A15:** 849, 851-852
casting techniques **A15:** 842-848
development **A15:** 36
fiber-metal wettability, casting effects **A15:** 840-842
fluidity, of composites **A15:** 849-850
matrix-dispersoid combinations **A15:** 841
microstructures **A15:** 850-851
remelting/degassing effects **A15:** 848-849
Cast metal-matrix composites (continued)
structure **A15:** 840
Cast metals
workabilities **A14:** 366
Cast microstructure
of TI-6AI-4V **A2:** 637
Cast nickel *See also* Nickel; Nickel alloys; Nickel alloys, cast, specific types, CZ-100; Nickel alloys, specific types
applications **A15:** 823
heat treatment **A15:** 822
mechanical properties **A15:** 817
welding **A15:** 822
Cast nickel-base superalloys *See also* Polycrystalline cast superalloys
compositions of **A1:** 982
design of **A1:** 983-985
directionally solidified alloys **A1:** 996-998
castability **A1:** 996-997
compositions Of **A1:** 996
heat treatment **A1:** 997-998
stress-rupture properties **A1:** 998
heat treatment **A1:** 993
hot isostatic pressing **A1:** 993-994
effect on fatigue properties **A1:** 991, 992, 994
investment casting **A1:** 989-990
melting practice **A1:** 986-988
microstructures **A1:** 990-992
carbides ..... **A1:** 990-99 I
dendrites **A1:** 990
eutectic segregation ..... **A1:** 99 I
grain size **A1:** 992
porosity **A1:** 991-992
single-crystal alloys **A1:** 998-1006
castability **A1:** 998
compositions **A1:** 996, 997
fatigue properties **A1:** 1002, 1004, 1005
heat treatment **A1:** 1000-1002
microstructure **A1:** 1000-1002
oxidation of **A1:** 1004, 1006
stress-rupture properties **A1:** 1000, 1002, 1004
stress-rupture properties **A1:** 985, 986, 987, 991, 993
tensile properties **A1:** 984
Cast products
beryllium-copper alloys **A2:** 422-423
zinc applications **A2:** 530
Cast replica
defined **A9:** 3
Cast sheet
acrylic **EM2:** 103
Cast shot
blasting with **M5:** 83-84, 86
Cast stainless steel lever
vibration fatigue fracture **A11:** 113-114
Cast stainless steels *See also* High-alloy steels; Stainless steels **A1:** 908-929
C-type alloys, corrosion behavior **A13:** 576-582
composition **A13:** 574-575
composition and microstructure of **A1:** 909, 910
ferrite control **A1:** 911
ferrite in cast austenitic stainless steels **A1:** 909, 910-911
corrosion-resistant steel castings **A1:** 909, 912
compositions **A1:** 909, 912-913
corrosion characteristics **A1:** 915-917
mechanical properties **A1:** 909, 914, 917-920
microstructures **A1:** 909, 913-915
grade designations and compositions **A1:** 908
C-type (corrosion-resistant) steel castings **A1:** 908-910
H-type (heat-resistant) steel castings **A1:** 910
H-type alloys, corrosion behavior **A13:** 575-576
heat treatment of **A1:** 911
homogenization **A1:** 911
sensitization and solution annealing of austenitic alloys **A1:** 912
heat-resistant cast steels **A1:** 920
galling **A1:** 928-929
general properties **A1:** 909, 919, 920-921
iron-chromium-nickel **A1:** 922-925
iron-nickel-chromium **A1:** 921, 923, 925-927
magnetic properties **A1:** 929
manufacturing characteristics **A1:** 919, 928, 929
metallurgical structures **A1:** 921-922
properties of heat-resistant alloys **A1:** 920, 921, 924, 925, 926, 927-928

Cast stainless steels (continued)
straight chromium heat-resistant castings ... **A1:** 922
intergranular corrosion ... **A13:** 581
microstructure ... **A13:** 574-575
research alloys, composition ... **A13:** 581
stress-corrosion cracking ... **A13:** 581-582
sulfuric acid corrosion ... **A13:** 1151
weldability ... **A15:** 535-537
Cast stainless steels, selection of ... **A6:** 495-498
aging ... **A6:** 497, 498
"buttering" ... **A6:** 498
compositions and typical microstructures of corrosion-resistant stainless steel casting alloys ... **A6:** 496
electron-beam welding ... **A6:** 496
electroslag welding ... **A6:** 496
fusion welding ... **A6:** 495-496, 498
fusion zone ... **A6:** 497, 498
gas-metal arc welding ... **A6:** 496
gas-tungsten arc welding ... **A6:** 496
heat-affected zone ... **A6:** 497, 498
hot cracking ... **A6:** 497-498
laser-beam welding ... **A6:** 496
plasma arc welding ... **A6:** 496
postweld heat treatment ... **A6:** 497, 498
postweld solutionizing ... **A6:** 498
shielded metal arc welding ... **A6:** 496
stainless steel casting alloys ... **A6:** 495
categories ... **A6:** 495
nomenclature ... **A6:** 495
properties ... **A6:** 495
"upgrading" ... **A6:** 495
welding and weldability ... **A6:** 495-498
defects ... **A6:** 495, 497-498
martensitic stainless steel castings ... **A6:** 497
metallurgical considerations ... **A6:** 496, 497
parameters ... **A6:** 498
welding processes ... **A6:** 495-496
Cast stainless steels, specific types
A27
composition ... **A6:** 642
mechanical properties ... **A6:** 642
A216
composition ... **A6:** 642
mechanical properties ... **A6:** 642
SAE J435c
composition ... **A6:** 642
mechanical properties ... **A6:** 642
Cast statuary
history ... **A15:** 20-22
Cast steel *See also* Steel castings; Steels, AMS; Steels, ASTM ... **M1:** 377-402
abrasion resistant ... **M1:** 616, 618
alloying elements, effect of ... **M1:** 388-389, 394-395
aluminum coating of ... **M5:** 339
annealing, effect on mechanical properties ... **M1:** 383, 384, 385, 386-388
applications ... **M1:** 383-384, 386-388, 393-399
ASTM specifications ... **M1:** 377-378, 401
carbon content, effect of ... **M1:** 384, 385, 386-388, 393, 394-395, 399, 400
chromium additions, effect of ... **M1:** 388
composition ... **M1:** 377-378, 379, 382, 383
contraction of ... **M1:** 400
corrosion resistance ... **M1:** 400
Cr-Mo hold times in push-pull loading test ... **A8:** 351
Cr-Mo-V, low-cycle fatigue data in hold periods in tension ... **A8:** 351
Cr-Mo-V, reversed bend ... **A8:** 352
Cr-Mo-V, static creep, repeated tension reversed cyclic creep tests with ... **A8:** 353-354
elastic constants ... **M1:** 393
engineering properties ... **M1:** 400-401
fatigue properties ... **M1:** 383, 389, 397
hardenability ... **M1:** 377, 380, 496
heat resistance ... **M1:** 400
Cast steel (continued)
heat treatments ... **M1:** 379, 381, 382, 388-389, 392, 398
impact properties ... **M1:** 378-381, 382, 389, 390, 392-393, 398, 399
low carbon, short-time tensile properties ... **M1:** 52
machinability ... **M1:** 54-56, 400, 567-568
magnetic properties ... **M1:** 399
manganese content, effect of ... **M1:** 384, 386-391, 394, 399
mass, effect of ... **M1:** 384-386, 392, 393, 398
mechanical properties ... **M1:** 378-399
metalworking rolls, use for ... **M3:** 506
microstructure, control of ... **M1:** 386, 389
Mo, hold times in push-pull loading test ... **A8:** 351
Mo push-pull fatigue ... **A8:** 352
molybdenum content, effect of ... **M1:** 388, 394-395
nickel content, effect of ... **M1:** 388, 394-395
nondestructive inspection ... **M1:** 401-402
normalizing, effect on mechanical properties ... **M1:** 384, 385, 386-388
notch toughness ... **M1:** 705-707, 708, 709
oxidation resistance ... **M1:** 46
patternmakers' rules ... **M1:** 31
phosphorus content, effect of ... **M1:** 399
physical properties ... **M1:** 392, 393, 399-400
pickling of ... **M5:** 72
press forming dies, use for ... **M3:** 490
S-N curves ... **M1:** 389, 397
SAE specifications ... **M1:** 377-378, 379
section size and thickness, effect of ... **M1:** 379, 381, 383, 392, 398
silicon content, effect of ... **M1:** 378, 399
specifications ... **M1:** 377-378, 379, 380, 400-401
sulfur content, effect of ... **M1:** 399
tempering ... **M1:** 379, 382, 383, 384, 385, 386-389
testing ... **M1:** 378, 379, 380-381, 400-402
U.S. Government specifications ... **M1:** 377, 378
vanadium content, effect of ... **M1:** 388
volumetric changes, effect of ... **M1:** 400
weak resistance ... **M1:** 400
wear vs. toughness ... **M1:** 607
weldability ... **M1:** 400-401, 563
Cast steel rolls ... **A14:** 353
Cast steel shot peening ... **M5:** 140-141
Cast steels *See also* Cast alloy steels; Cast iron; Steel castings; Steels; Steels, specific types ... **A1:** 363-379 **A13:** 573-582
atmospheric corrosion ... **A13:** 573-575
ball and rod grinding media, mining industry ... **A18:** 654
carbon/low-alloy corrosion ... **A13:** 573-574
cermet tools for milling ... **A16:** 96
composition, categories ... **A13:** 573
corrosion rates ... **A13:** 574
dendritic structure revealed by macroetching ... **A9:** 173
development of ... **A15:** 19, 31-32
die material for sheet metal forming ... **A18:** 628
high-speed tool steels used ... **A16:** 58
microalloyed steel castings, applications compositions ... **A1:** 420
nominal compositions and applications ... **A18:** 703
notch toughness of ... **A1:** 746-747
stainless, corrosion ... **A13:** 574-582
tool bit tool steels used ... **A16:** 57, 58
tool steel alloys used ... **A16:** 57
weldability ... **A15:** 532, 535
welding of ... **A15:** 531-537
welding processes ... **A15:** 531
Cast structure
defined ... **A9:** 3 **A15:** 2
Cast tin bronzes
destannification ... **A13:** 133
Cast tooling, of patterns
investment casting ... **A15:** 256-257
Cast-in inserts
in magnesium alloy parts ... **A2:** 466
Castability
aluminum alloys ... **A15:** 766
aluminum casting alloys ... **A2:** 153-177
aluminum-silicon alloys ... **A15:** 159, 167
of compacted graphite iron ... **A1:** 57
compacted graphite irons ... **A15:** 671
of copper casting alloys ... **A2:** 346, 348
defined ... **A2:** 346 **A15:** 2
of gray iron ... **A1:** 12-13
vs fluidity, copper casting alloys ... **A2:** 346
Castability of gray iron ... **M1:** 11-12
Castable
defined ... **A15:** 2
Castable mounting materials
for carbon and alloy steels ... **A9:** 167
Castable plastics as mounting materials ... **A9:** 30-31
for aluminum alloys ... **A9:** 352
Caster oil, hydrogenated
as investment casting wax ... **A15:** 254
Casting *See also* Cast products; Casting processes; Casting temperatures; Metal casting; specific casting processes ... **EM3:** 585
aluminum and aluminum alloys ... **A2:** 4-5
art ... **A15:** 20-22
with beryllium ... **A2:** 683
beryllium-copper alloys ... **A2:** 421-423
bore, and lubrication hole, fatigue cracking from ... **A11:** 346
brittle fracture from ... **A11:** 85
carbon steel ... **A11:** 392
centrifugal, defined *See* Centrifugal casting
centrifugal, development of ... **A15:** 34
computer applications/modeling ... **A15:** 855-891
continuous, defined *See* Continuous casting
in continuous flow melting ... **A15:** 415
defects, testing and inspection ... **A15:** 544-561
defined ... **A15:** 2
design, titanium and titanium alloy castings ... **A2:** 640-642
development of crystallographic texture during ... **A9:** 700-701
die, development of ... **A15:** 35
of discontinuous ceramic fiber MMCs ... **EM1:** 905
of ductile iron ... **A15:** 651-652
end uses ... **A15:** 42-43
of epoxies ... **EL1:** 831-832
fatigue strength ... **A11:** 120
gas porosity in ... **A15:** 82
of graphite-reinforced MMCs ... **EM1:** 872
history of ... **A15:** 15-23
incomplete, as casting defect ... **A11:** 385-386
industry, markets for ... **A15:** 41-42
integrated system, automated ... **A15:** 570
load bearing ability ... **A11:** 346
low-alloy steel ... **A11:** 392-393
magnesium alloy, product form selection ... **A2:** 462-463
magnesium alloys ... **A2:** 456-459
markets, development ... **A15:** 34
materials, corrosion failures in ... **A11:** 401-405
materials, developments in ... **A15:** 38-39
and melting, in investment casting ... **A15:** 262-263
as metalworking ... **A14:** 15
operations, process developments in ... **A15:** 38
parameters, macrostructural effects ... **A15:** 130
permanent mold ... **A15:** 34-35
processes, aluminum casting alloys ... **A2:** 136-145
production, flow diagram for ... **A15:** 203
radiographic inspection of ... **A11:** 17
refurbishment, by hot isostatic pressing ... **A15:** 544
removal, from permanent molds ... **A15:** 283-284
shrinkage, defined ... **A15:** 2
slush, development ... **A15:** 34-35
stresses, cracking in gray iron cylinder blocks by ... **A11:** 345-346

Casting (continued)
stresses, failed gray iron crankcase by **A11:** 362-365
surface roughness arithmetic average extremes **A18:** 340
temperature, cast copper alloys **A2:** 357
tire-mold, surface defect **A11:** 358
tungsten-fiber matrix **EM1:** 885
types, copper and copper alloys **A2:** 224-228
unalloyed and alloyed aluminum **A2:** 22-25
US birthplace of **A15:** 24
variables, aluminum casting alloys **A2:** 148-149
zirconium **A2:** 663-664
Casting alloys *See also* Cast alloy steels; Ferrous casting alloys; Nonferrous casting alloys
advances in **A15:** 29-32
cast alloy steels **A15:** 32
cast steel **A15:** 31-32
chilled iron **A15:** 30
Connellsville coke **A15:** 30
cupola iron **A15:** 29-30
malleable iron **A15:** 30-31
Casting alloys, stainless steel *See* Stainless steel casting alloys
Casting cold shut
as planar flaw **A17:** 50
Casting conditions
effect on microstructure of nickel-base heat-resistant casting alloys **A9:** 334
Casting defects *See also* Castings; Defects; Discontinuities; Flaws; Gas defects; Inclusions
cavities **A11:** 382 **A17:** 512, 514
classification **A11:** 380 **A15:** 545-553 **A17:** 512
control of **A11:** 380, 388
defective surface **A17:** 515-517
defective surface as **A11:** 383-385
defined **A15:** 2
discontinuities **A11:** 383 **A17:** 512, 515
flaws, radiographic appearance **A17:** 348-349
inclusions **A17:** 512, 519-520
inclusions or structural anomalies as **A11:** 387-388
incomplete casting **A17:** 517-518
incomplete casting as **A11:** 385-386
incorrect dimension/shape **A17:** 512, 518-519
incorrect dimensions or shape as **A11:** 386-387
internal discontinuities **A15:** 544-545
metallic projections **A11:** 381 **A17:** 512-513
in paper-drier head **A11:** 352-354
shrinkage as **A11:** 355, 357
steel **A11:** 380-391
testing and inspection **A15:** 544-561
Casting design *See also* Design; Design considerations **A15:** 598-613
austenitic ductile irons **A15:** 700
changing thermal shape **A15:** 606-610
computer-aided **A15:** 610-611
economical, rules for **A15:** 602-604, 611-612
feeding **A15:** 606
high-chromium white irons **A15:** 683
high-silicon irons **A15:** 699-701
and mechanical properties **A15:** 765
mold complexity **A15:** 611-613
nickel-chromium white irons **A15:** 680
solidification **A15:** 598-599
solidification sequence **A15:** 599-606
titanium alloys **A15:** 829-831
Casting ejection *See* Ejection
Casting equipment *See also* Equipment
advances, history **A15:** 27-29, 33-36
history of **A15:** 24-36
two-handed scythe **A15:** 24
Casting industry
structure **A15:** 41
Casting Industry Supplier's Association **A15:** 34
Casting ingot
beryllium-copper alloys **A2:** 403
Casting markets
for casting industry **A15:** 42
and end uses **A15:** 41-45
shipment tonnages **A15:** 41-42
trends in **A15:** 43-45
Casting methods
direct **A1:** 211
Casting molds
effect on pure metal solidification structures **A9:** 608-610
Casting on
historic use **A15:** 17
Casting operations
process development in **A15:** 38
Casting plants
jobbing **A15:** 28
Casting processes *See also* Molding processes
aggregate molding materials **A15:** 208-211
alloy selection **A2:** 136
aluminum alloys **A15:** 746
aluminum casting alloys **A2:** 136-145
blast cleaning **A15:** 506-520
capabilities **A15:** 615
centrifugal casting **A2:** 141 **A15:** 296-307
ceramic molding **A15:** 248-252
classification **A15:** 203-207
coating **A15:** 561-565
composite-mold casting **A2:** 141
continuous casting **A2:** 141 **A15:** 308-316
core knockout **A15:** 502-506
coremaking **A15:** 238-241
die casting **A2:** 136-139 **A15:** 285-295
evaporative (lost-foam) pattern casting (EPC) **A2:** 140
flow charts of **A15:** 203-207
hot isostatic pressing **A2:** 141 **A15:** 538-544
investment casting **A15:** 253-269
investment casting, aluminum casting alloys **A2:** 140-141
molding aggregates, bonds formed in **A15:** 212-213
new and emerging processes **A15:** 317-338
permanent mold (gravity die) casting **A2:** 139
permanent mold casting **A15:** 275-285
plaster molding **A15:** 242-247
rammed graphite molds **A15:** 273-274
Replicast process **A15:** 270-272
resin binder processes **A15:** 214-221
sand casting **A2:** 139-140
sand molding **A15:** 222-237
selection, factors affecting **A15:** 614
shakeout **A15:** 502-506
shell mold casting **A2:** 140
testing and inspection, of defects **A15:** 544-561
welding, cast irons and steels **A15:** 520-537
Casting, rotational *See* Rotational casting
Casting section thickness
by FM processes **A15:** 38
defined **A15:** 2
Casting shrinkage *See* Liquid shrinkage; Shrinkage cavity; Solid shrinkage; Solidification shrinkage
Casting size, effects of
gray cast iron **M1:** 14-16
Casting speed
effect on center cracking of aluminum alloy ingots **A9:** 634-635
effect on grain structures in copper alloy ingots **A9:** 642
effect on macrosegregation of aluminum alloy ingots **A9:** 633
Casting stresses
defined **A15:** 2
Casting temperature *See also* Casting; Fabrication characteristics; Temperature(s)
aluminum casting alloys **A2:** 157-177
cast copper alloys **A2:** 357
effect on grain size in continuous cast copper alloy wirebars **A9:** 642
of lead and lead alloys **A2:** 547-548
Casting thickness
defined **A15:** 2
Casting volume
defined **A15:** 2
Casting yield
defined **A15:** 2
Castings *See also* Casting defects; Die castings; specific metals and alloys; Steel castings **A17:** 512-535
abrasive blasting of **M5:** 86, 91-92
aluminum *See* Cast aluminum
aluminum alloy, inspection **A17:** 532-534
brass *See* Brass, castings
casting defects **A17:** 512-520
coating of **A15:** 561-565
Castings (continued)
copper alloy *See* Copper alloys, castings
copper and copper alloy, inspection **A17:** 534-535
copper, annealing **M4:** 724
damaged, welding repair of **A15:** 529-530
dimensional inspection, computer-aided **A17:** 521-524
ductile iron **A17:** 532
eddy current inspection **A17:** 525-526
fine-grain, hot isostatic pressing **A15:** 545
grain growth in pure metal **A9:** 608-610
gray iron, inspection **A17:** 531
half-wave current, magnetic particle inspection **A17:** 91
heat-affected-zone cracks **A6:** 91
hollow, by slush casting **A15:** 35
hot isostatic pressing **A15:** 538-544
imported, market effects **A15:** 44
incomplete, as defect **A17:** 517-518
inspection categories **A17:** 512
inspection procedures **A17:** 512, 520-521
intergranular fractures in **A1:** 694-695
leak testing **A17:** 531
liquid penetrant inspection **A17:** 524-525
magnetic particle inspection of **A17:** 112-114, 525
magnetizing **A17:** 94
magnetizing cable **A17:** 112
malleable iron **A17:** 531-532
mangesium alloy *See* Magnesium alloys, cast
nickel- and cobalt-based alloys **A7:** 468
nonferrous, inspection of **A17:** 531-532
pickling of **M5:** 72, 326
precision, computed tomography (CT) **A17:** 363
preferred crystallographic orientations in **A10:** 358
products, liquid penetrant inspection of **A17:** 71
quality control tests **A12:** 141-142
radiographic methods **A17:** 296, 526-529
salt bath descaling of **M5:** 98-102
small, from crucible steel **A15:** 31
small, magnetic particle inspection methods **A17:** 117
steel *See* Steel, castings
steel, normalizing **M4:** 12
structural evaluation **A17:** 530-531
ultrasonic inspection **A17:** 267, 529-531
vapor degreasing of **M5:** 54, 56
of varying thicknesses **A15:** 581-582
worn, repair of **A15:** 529-530
Castings, stainless steel *See* Stainless steel castings
Castings, steel
normalizing **A4:** 36, 38, 40
CAT scanning *See* Computed tomography
Catalan forge
historic use **A15:** 31
Catalysis
heterogeneous, FIM/AP study of **A10:** 583
surface, AES analysis for **A10:** 549
surface analysis technique **A7:** 250
Catalyst *See also* Accelerator, Curing agent; Hardener; Inhibitor; Promoter **EM3:** 7, 674
defined **EM2:** 8
Catalysts *See also* Accelerator; Curing agent; Hardener; Inhibitor; Promoter
analytic methods applicable **A10:** 6
BET analysis of specific surface area **A7:** 262
chlorine in, determined by electrometric titration **A10:** 205
chromia alumina **A10:** 265
defined **EM1:** 6
effects, neutron diffraction **A10:** 420
electron spin resonance for **A10:** 263
for epoxy resins **EM1:** 139, 140
EXAFS structural analysis of **A10:** 407
influence of surface structure on **A10:** 536
metal oxide, Raman analysis **A10:** 133
polyester resin **EM1:** 133
for promoting synthesis of water powder used **A7:** 574
prompt gamma activation analysis of **A10:** 239
properties analysis of **EM1:** 736
for sheet molding compounds **EM1:** 141, 157-158
for silicone-based coatings **EL1:** 773
surfaces of, studies **A10:** 114, 253

Catalytic techniques, and electrometric titration
compared **A10:** 202
Catalyzing
in magnetic rubber inspection **A17:** 122-123
Catapult-hook attachment fitting
fracture of **A11:** 88, 91
Catastrophic crack propagation **EM3:** 513
threshold values, epoxy adhesives **EM3:** 516-517, 518
Catastrophic failure *See also* Failure
alloy steels **A12:** 336
compressive strength defined by **A8:** 57
fracture equation for **A7:** 58
graphite-aluminum metal matrix composite **A13:** 860
stress-corrosion fracture **A13:** 148
Catastrophic failures *See also* Failure
defined **EM1:** 6
Catastrophic period
defined **A18:** 4
Catastrophic thermal failure
defined **EL1:** 46
Catastrophic threshold **EM3:** 507
Catastrophic wear
defined **A8:** 2 **A11:** 1 **A18:** 5
Catatectic reaction .... **A3:** 1 • 5
Catchlights
on fracture surfaces **A12:** 83-84
Catchment efficiency *See* Collection efficiency
CATE program *See* Ceramic applications in turbine engines (CATE) program
Categorization *See also* Classification; Nomenclature; Terminology
of plastics **EM2:** 68
of polymers **EM2:** 63-64
Catenary **EM1:** 6, 109
defined **EM2:** 8
Caterpillar 1K/lH2
single-cylinder engine tests **A18:** 101
Caterpillar TO-2/TO-4
friction performance requirements for hydraulic fluids **A18:** 99
Cathedral mist hackle **EM4:** 639
Cathetometer
for creep testing **A8:** 303
for measuring deflection **A8:** 134
Cathode *See also* Anode
for anodic protection **A13:** 464
bombardment **A10:** 26
brittle, process of **A7:** 72
defined **A13:** 2-3
deposition **A7:** 71-72, 93-94
efficiency, defined **A13:** 3
film, defined **A13:** 3
galvanic, anodic protection of **A13:** 465
mercury, in electrogravimetry **A10:** 199
metals adherence, in electrogravimetry **A10:** 198
shielding alternatives **A7:** 612
silver grown on **A7:** 148
sputtering **A10:** 27
Cathode area, ratio to anode area
galvanic corrosion **A11:** 186
Cathode cleaning action **A6:** 31
Cathode materials
for interference films, optical constants **A9:** 149
for reactive sputtering of interference layers **A9:** 60
Cathode probes for in situ local electropolishing **A9:** 55
Cathode ray tube
as synchronized with SEM imaging system **A12:** 169-171
in x-ray analyses **A12:** 168
Cathode sputtering *See* Cathodic sputtering; Reactive sputtering
Cathode sputtering system **M5:** 413-414
Cathode-ray oscilloscope **A8:** 724
for split-Hopkinson bar test **A8:** 213
Cathode-ray tube **A10:** 670, 690
abbreviation **A8:** 724
with scanning electron microscope **EL1:** 1094
Cathode-ray tubes *See* CRTs and TV picture tubes
powders used **A7:** 573
Cathode-type gage
for vacuum pumping system **A8:** 414
Cathodes
for electropolishing **A9:** 49-50
Cathodic
breakdown test, electrochemicals **A13:** 219
charging, titanium/titanium alloys **A13:** 673
cleaning, defined **A13:** 3
coatings, corrosion prevention mechanisms **A13:** 424-425
corrosion, defined **A13:** 3
disbondment, defined **A13:** 3
hydrogen uptake, titanium/titanium alloys **A13:** 685-686
pickling, defined **A13:** 3
polarization behavior **A13:** 217
Cathodic attack
of titanium **A11:** 202
Cathodic cleaning
of fractures **A12:** 75
Cathodic electrocleaning
aluminum and aluminum alloys **M5:** 578
contamination by hexavalent chromium **M5:** 34, 618
copper and copper alloys **M5:** 618-620
hydrogen embrittlement by **M5:** 27-28, 34-35
magnesium alloys **M5:** 631, 639-640
periodic reverse system **M5:** 27-28, 34-35
processes and materials **M5:** 33-35, 619-620
Cathodic electrocoating systems
painting **M5:** 474, 484
Cathodic etching *See* Ion etching
Cathodic inhibitors *See also* Anodic inhibitors; Inhibitors **A13:** 1141
as corrosion control **A11:** 197
defined **A13:** 3
multicomponent systems **A13:** 495-496
phosphonates **A13:** 495
polyphosphates **A13:** 495
precipitating **A13:** 495
in water recirculating systems **A13:** 495-497
zinc ions **A13:** 495
Cathodic polarization
effect on ultrasonic and conventional-frequency corrosion fatigue **A8:** 254
hydrogen stress cracking under **A8:** 537
Cathodic potential ranges **A9:** 144
Cathodic protection *See also* Anodic protection; Corrosion protection **A13:** 466-477 **M1:** 751, 757-759
against corrosion **M5:** 433
of aluminum alloys **A13:** 588-589
anode materials **A13:** 468-469, 920-922
and anodic protection **A13:** 463
in aqueous corrosion **A13:** 29
in breweries **A13:** 1223-1224
of carbon steels **A13:** 525
of cast steels **A13:** 581
for chloride SCC **A13:** 327
for corrosion control **A11:** 195-197
criteria **A13:** 467-468, 920
defined **A13:** 3
design/power sources **A13:** 470
example resistance calculations **A13:** 470-477
fresh water systems **M1:** 738
from hydrogen damage **A11:** 246-247
fundamentals **A13:** 466-467
galvanic corrosion **A13:** 84, 87
gas/oil production equipment **A13:** 1237-1240
impressed-current **A13:** 467-469, 922
impressed-current and sacrificial anode systems **A13:** 922
as kinetic **A13:** 377
maraging steels **M1:** 451
Cathodic protection (continued)
for marine corrosion **A13:** 919-924
of offshore structures **A13:** 922-924
of pipelines **A13:** 922, 1291
sacrificial anode **A13:** 467-469, 922
seawater corrosion **M1:** 743
of ship hulls **A13:** 914
soil corrosion prevented by **M1:** 731
space boosters/satellites **A13:** 1103
system, for buried steel tank **A11:** 196
system, schematic **A13:** 466
of tantalum/tantalum alloys **A13:** 735, 736
types **A13:** 467
zinc anodes **M2:** 654-655
zinc/zinc alloys and coatings **A13:** 764
Cathodic reaction **A13:** 3, 29
aqueous corrosion rate control by **A13:** 32
Cathodic sputtering *See also* Reactive sputtering **A9:** 62
Cathodic Tafel slope **A18:** 274
Cathodic vacuum etching **A9:** 62
Cathodoluminescence **A10:** 507, 670, 689 **A18:** 378, 385, 391
in scanning electron microscopy **A9:** 90
used in electronic image analysis **A9:** 152
Catholyte
defined **A13:** 3
Cation *See also* Anion; Ion **A13:** 3, 18-19, 65-66, 71 **EM3:** 7
defined **EM2:** 8
-exchange column, simulated pressurized water reactor water system **A8:** 423-424
influence in stress-corrosion cracking **A8:** 499
Cation exchange capacity (CEC) **EM4:** 117
Cationic alumina plus organic flocculant
application or function optimizing powder treatment and green forming **EM4:** 49
Cationic cure systems *See also* Coating(s); Cure; Curing **EL1:** 859-865
conformal coatings **EL1:** 864
dual cure **EL1:** 863-864
formulations **EL1:** 861-863
photoinitiators **EL1:** 859-861
Cationic detergents
for surface cleaning **A13:** 380
Cationic surfactants
use in alkaline cleaners **M5:** 24
Cations
defined **A10:** 670
inorganic, determined by ion chromatography **A10:** 663
ion chromatography analyses **A10:** 658-667
as positively charged ions **A10:** 659
potentiometric membrane electrode quantitative analysis **A10:** 181
Cauchy peak location method
XRD residual stress techniques **A10:** 386
Caul **EM3:** 7
Caul plates **EM1:** 6, 581, 593
defined **EM2:** 8
Caulk weld
definition **A6:** 1207
Caulking compound
powder used **A7:** 572
Caulks
compared to sealants **EM3:** 56
Caustic
corrosion, steam/water-side boilers **A13:** 991
defined **A13:** 3
dip, defined **A13:** 3
environments, nickel-base alloy SCC in **A13:** 650
in petroleum refining and petrochemical operations **A13:** 1269
Caustic cracking *See also* Caustic embrittlement; Embrittlement **A8:** 2
in boilers **A11:** 214

Caustic cracking (continued)
defined ... **A11:** 2
failure, low-carbon steels ... **A8:** 526
failures, in wrought carbon and low-alloy steels ... **A11:** 214-215
Caustic embrittlement *See also* Caustic cracking; Embrittlement ... **A13:** 3, 353-354, 650
of low carbon steels, by potassium hydroxide ... **A11:** 658-660
in pressure vessels ... **A11:** 658
Caustic environments
stress-corrosion cracking in ... **A11:** 217
Caustic etchant used for aluminum alloys ... **A9:** 354
Caustic fusion
use in ion chromatography ... **A10:** 664
Caustic permanganate
as chemical cleaning solution ... **A13:** 1141
Caustic SCC *See* Caustic embrittlement
Caustic soda ... **A13:** 328, 1174-1178
etchant for chemical milling of aluminum alloys ... **A16:** 803
Caustic soda/sodium hydroxide (NaOH)
purpose for use in glass manufacture ... **EM4:** 381
Caustic solutions, corrosion of zirconium in *See also* Alkalis ... **M3:** 785-786
Caustic-permanganate pickling process
iron and steel ... **M5:** 73
Cautions *See* Safety
Cavitating disk apparatus
defined ... **A18:** 5
Cavitation *See also* Cavitation damage; Cavitation erosion; Erosion ... **A6:** 374 **A8:** 571 **A18:** 272
aircraft powerplants ... **A13:** 1044-1045
at high temperatures ... **A8:** 154, 572
by oxide scale ... **A13:** 61
carbon steel pipe ... **A13:** 333
collapse pressures for ... **A11:** 163-164
copper/copper alloys ... **A13:** 613
creep, as decohesive rupture ... **A12:** 20
damage ... **A8:** 2 **A13:** 334
damage, cylinder lining diesel motor ... **A11:** 377-378
data, test-service correlations ... **A13:** 313
defined ... **A8:** 154 **A11:** 2 **A13:** 3 **A14:** 19 **A18:** 5
effect on corrosion of titanium ... **M3:** 415
erosion ... **A11:** 190
evaluation of ... **A13:** 311-313
formation ... **A14:** 364
formation by methane gas bubbles ... **A12:** 37, 51
from liquid-erosion ... **A11:** 163-164
from superplastic forming ... **A14:** 800
grain-boundary ... **A8:** 574 **A12:** 19, 26, 219, 349
intergranular creep rupture by ... **A12:** 19, 26
in iron ... **A12:** 219
in iron castings ... **A11:** 377
local parameter of ... **A11:** 167
material selection for ... **A13:** 333-334
in mining/mill application ... **A13:** 1295-1296
in oil/gas production ... **A13:** 1234
resistance, materials rating for ... **A13:** 1297
seals ... **A18:** 549-550
shock waves ... **A11:** 165
in sliding bearings ... **A11:** 488 **A18:** 742
and slip lines, iron ... **A12:** 219
spheres as wear particles ... **A18:** 303
stainless steels ... **A18:** 715
of superplastic metals ... **A14:** 867
surface property effect ... **A18:** 342
testing ... **A13:** 311-313
used to obtain sonogels ... **EM4:** 211
Cavitation cloud
defined ... **A18:** 5
Cavitation damage *See also* Cavitation
in brazed joints ... **A11:** 451
of bronze pump impeller ... **A11:** 167-168
components and structures sustaining ... **A11:** 163
defined ... **A8:** 2 **A11:** 2, 163
in impeller specimen ... **A11:** 357
ultrasonic cleaning of specimens to prevent ... **A9:** 28
Cavitation erosion *See also* Cavitation; Cavitation damage; Corrosion; Crevice corrosion; Erosion ... **A18:** 214-219
of aluminum alloy combustion chamber ... **A11:** 168-169
Cavitation erosion (continued)
in boiler and steam equipment ... **A11:** 614
of cast iron suction bell ... **A11:** 624
cavity clusters ... **A18:** 214-215
of cobalt-base wear resistant alloys ... **A2:** 450-451
cobalt-base wrought alloys ... **A18:** 768, 769
collapse velocity ... **A18:** 214
combined effects of cavitation erosion and corrosion ... **A18:** 217-218
cavitation effect on the corrosion process ... **A18:** 217-218
corrosion effect on cavitation process ... **A18:** 218
components and structures sustaining ... **A11:** 163
conclusions ... **A18:** 219
defined ... **A11:** 163 **A18:** 5, 214
energy dissipation ... **A18:** 214, 215
factors ... **A18:** 214-215
hardfacing alloys ... **A18:** 762, 763, 765
hardfacing for ... **A7:** 823
iron-base alloys ... **A18:** 768, 769
materials factors ... **A18:** 215-217
erosion of metals and alloys ... **A18:** 215-217
localized loading ... **A18:** 215
surface coatings and treatments ... **A18:** 217
means of combating erosion ... **A18:** 218-219
air injection ... **A18:** 218-219
coatings ... **A18:** 218
control of operating temperature or pressure ... **A18:** 219
materials selection and development ... **A18:** 218
surface treatments ... **A18:** 218
system design ... **A18:** 218
nickel-base alloys ... **A18:** 768, 769
pumps ... **A18:** 593, 597, 599-600
resistance ratings, in seawater ... **A11:** 190
sliding bearing surface ... **A11:** 488
sliding bearings ... **A18:** 520
testing ... **A18:** 218, 230
cavitating jet ... **A18:** 218
flow channels ... **A18:** 218
rotating disk test equipment ... **A18:** 218
to screen materials for service under liquid impingement conditions ... **A18:** 230
vibratory (ultrasonic) equipment ... **A18:** 218
versus liquid impingement erosion ... **A18:** 222, 225-226
in water ... **A11:** 190
of water pump impeller ... **A11:** 167-168
Cavitation number
defined ... **A18:** 5
Cavitation tunnel
defined ... **A18:** 5
Cavitation-erosion ... **A13:** 3, 142
Cavities *See also* Bubbles; Defects; Shrinkage cavities; Shrinkage cavity; Voids
absorption ... **A17:** 216
in alloy steel ... **A12:** 349
as casting defects ... **A11:** 382 **A15:** 547 **A17:** 512, 514
circular ... **A17:** 217
die, EDM failures in ... **A11:** 566-567
and dimples, compared ... **A12:** 20, 220
elongated, titanium alloys ... **A12:** 450
fatigue fracture from ... **A12:** 419
gas, cast aluminum alloys ... **A12:** 405-406
gas-filled, in liquid erosion ... **A11:** 163
grain-boundary ... **A12:** 219, 349
as gray iron defect ... **A15:** 640-641
in intergranular iron fracture ... **A12:** 219
in iron ... **A12:** 219-220
mechanics of growth, collapse and rebound ... **A11:** 163
microwave inspection ... **A17:** 202
microwave-resonant ... **A10:** 256
nodule-bearing, in ductile iron ... **A12:** 229
nonsymmetrical collapse, stages of ... **A11:** 164
nucleation ... **A12:** 122
r-type ... **A12:** 122, 140
separation, copper alloys ... **A12:** 402
shrinkage ... **A12:** 140, 160 **A15:** 640-641, 838
transmission electron microscopy ... **A9:** 117
types, austenitic stainless steels ... **A12:** 364
Cavities, formation of
during irradiation ... **A1:** 654
Cavities, internal
for copper castings ... **A2:** 355
Cavity
defined ... **A15:** 2 **EM1:** 6 **EM2:** 8
development, in creep damage ... **A8:** 344
due to creep deformation ... **A8:** 306
Cavity clusters ... **A18:** 214-215, 222
Cavity packages
as hybrid package form ... **EL1:** 452
Innovative solutions for ... **EL1:** 448
Cavity retainer plates *See also* Force retainer plates
defined ... **EM2:** 8-9
Cavity-less expanded polystyrene casting process *See* Lost foam casting
CB-752 *See* Niobium alloys, specific types
electrical discharge machining ... **A16:** 868
CBED *See* Convergent-beam electron diffraction
CBEDP *See* Convergent-beam electron- diffraction pattern
CBN *See* Cubic boron nitride
CBS-600
nominal compositions ... **A18:** 726
CBS-1000M
nominal compositions ... **A18:** 726
CCT *See* Center-cracked-tension, or Continuous cooling transformation
CCT diagrams *See* Continous cooling transformation diagrams
Cd-Cu (Phase Diagram) ... **A3:** 2 • 123
Cd-Eu (Phase Diagram) ... **A3:** 2 • 123
Cd-Ga (Phase Diagram) ... **A3:** 2 • 124
Cd-Gd (Phase Diagram) ... **A3:** 2 • 124
Cd-Ge (Phase Diagram) ... **A3:** 2 • 124
Cd-Hg (Phase Diagram) ... **A3:** 2 • 125
Cd-In (Phase Diagram) ... **A3:** 2 • 125
Cd-La (Phase Diagram) ... **A3:** 2 • 125
Cd-Li (Phase Diagram) ... **A3:** 2 • 126
Cd-Mg (Phase Diagram) ... **A3:** 2 • 126
Cd-Na (Phase Diagram) ... **A3:** 2 • 126
Cd-Ni (Phase Diagram) ... **A3:** 2 • 127
Cd-P (Phase Diagram) ... **A3:** 2 • 127
Cd-Pb (Phase Diagram) ... **A3:** 2 • 127
Cd-Sb (Phase Diagram) ... **A3:** 2 • 128
Cd-Sb-Sn (Phase Diagram) ... **A3:** 3 • 35-36
Cd-Se (Phase Diagram) ... **A3:** 2 • 128
Cd-Sm (Phase Diagram) ... **A3:** 2 • 128
Cd-Sn (Phase Diagram) ... **A3:** 2 • 129
Cd-Sr (Phase Diagram) ... **A3:** 2 • 129
Cd-Te (Phase Diagram) ... **A3:** 2 • 129
Cd-Th (Phase Diagram) ... **A3:** 2 • 130
Cd-Tl (Phase Diagram) ... **A3:** 2 • 130
Cd-Y (Phase Diagram) ... **A3:** 2 • 130
Cd-Yb (Phase Diagram) ... **A3:** 2 • 131
Cd-Zn (Phase Diagram) ... **A3:** 2 • 131
CDA *See* Copper Development Association
CDF *See* Centered dark-field (image)
CDJ (unfilled carbon)
properties ... **A18:** 549, 551
CE *See* Carbon equivalent; Carbon Equivalent
Ce (Phase Diagram) ... **A3:** 2 • 135
Ce-Co (Phase Diagram) ... **A3:** 2 • 131
Ce-Cu (Phase Diagram) ... **A3:** 2 • 132
Ce-Fe (Phase Diagram) ... **A3:** 2 • 132
Ce-Ga (Phase Diagram) ... **A3:** 2 • 133
Ce-Ge (Phase Diagram) ... **A3:** 2 • 133
Ce-In (Phase Diagram) ... **A3:** 2 • 133
Ce-Ir (Phase Diagram) ... **A3:** 2 • 134
Ce-Mg (Phase Diagram) ... **A3:** 2 • 134
Ce-Mn (Phase Diagram) ... **A3:** 2 • 134
Ce-Ni (Phase Diagram) ... **A3:** 2 • 135
Ce-Pd (Phase Diagram) ... **A3:** 2 • 135
Ce-Pu (Phase Diagram) ... **A3:** 2 • 136
Ce-S (Phase Diagram) ... **A3:** 2 • 136
Ce-Si (Phase Diagram) ... **A3:** 2 • 136
Ce-Sn (Phase Diagram) ... **A3:** 2 • 137
Ce-Te (Phase Diagram) ... **A3:** 2 • 137
Ce-Ti (Phase Diagram) ... **A3:** 2 • 137
Ce-Tl (Phase Diagram) ... **A3:** 2 • 138
Ce-Zn (Phase Diagram) ... **A3:** 2 • 138
Celion 6000
properties ... **A18:** 803
Cell *See also* Cellular plastic; Electrochemical cell ... **EM3:** 7
aeration/differential aeration/oxygen *See* Differential aeration cell
biological corrosion ... **A13:** 42-43
defined ... **A13:** 3 **EM1:** 6 **EM2:** 9
electrochemical ... **A13:** 20
electrochemical polarization ... **A13:** 214

Cell (continued)
filiform corrosion, diagrams **A13:** 106
formation in creep deformation **A8:** 310
potentials, and electromotive force series **A13:** 20-21
profiles for symmetric rod impact test **A8:** 205-206
size, defined **EM1:** 6
Cell applications
automated **A15:** 568-569
Cell configuration
as electrolytic inclusion and phase isolation **A10:** 176
Cell size *See* Secondary dendrite arm spacing
Cell structure *See also* Cellular structures
formed by dislocation tangles **A9:** 685, 688
use of thin-foil specimens to 'Observe **A9:** 693
Cell voltage as a function of anode current density for electropolishing copper **A9:** 48
Cell-based circuits
defined **EL1:** 168
Cellosolve
description **A9:** 68
and hydrochloric acid as an electrolyte for magnesium alloys **A9:** 426
Cells
classical types **A10:** 199
constant-current **A10:** 199-200
controlled-potential **A10:** 199-200
in dendritic growth **A15:** 116-119
dendritic structures, single-phase alloys **A15:** 116
and electrodes, for electrogravimetry **A10:** 199-200
electrolysis **A10:** 199, 207-208
electrolytic, to decrease concentration polarization **A10:** 200
half-, volumetric analysis of **A10:** 163
for internal electrolysis **A10:** 199
kinetics of martensite decomposition in **A10:** 316-318
size, and eutectic structure aluminum-silicon alloys **A15:** 167-168
small, rotation about axes **A10:** 473
spacing, velocity effect **A15:** 117
structure, dislocation analysis of **A10:** 470-473
structure factor equation for **A10:** 329
in the deformed state **A9:** 693
typical dual anode **A10:** 199
Cellular
effects, of arsenic, as toxic metal **A2:** 1237
growth, low-gravity **A15:** 153
interface, particle behavior at **A15:** 144-145
metabolism, of mercury, as toxin **A2:** 1248
structures, single-phase alloys **A15:** 116
Cellular adhesive **EM3:** 7
Cellular growth as a result of impurities in high-purity tin **A9:** 607
Cellular logic arrays **EL1:** 8
Cellular plastic *See also* Cell; Foamed plastic; Syntactic cellular plastics **EM3:** 7
defined **EM2:** 9
Cellular precipitation **A9:** 647-649
in beryllium-copper alloys **A9:** 395
Cellular structures *See also* Cell structure
formation **A9:** 601
in rapidly solidified alloys **A9:** 615-616
solidification **A9:** 612-613
Cellulose
chemical composition **A6:** 60
critical surface tension **EM3:** 180
function and composition for mild steel SMAW electrode coatings **A6:** 60
powdered, as binding agent for x-ray spectrometry samples **A10:** 94
properties of **EM2:** 450
removal of sulfur trioxide, in high-temperature combustion **A10:** 222
Cellulose (continued)
surface preparation **EM3:** 291
as x-ray tube filter **A10:** 90
Cellulose acetate **EM3:** 7
defined **EM2:** 9
surface preparation **EM3:** 291
Cellulose acetate butyrate **EM3:** 7
defined **EM2:** 9
surface preparation **EM3:** 291
Cellulose acetate replica(s)
with acetone **A12:** 180
for SEM imaging **A12:** 171-172
tape, as fracture preservative **A12:** 73
tape, for light microscopy **A12:** 94-95, 99
Cellulose acetate rust-preventive compound
use of **M5:** 465
Cellulose additions
to molding sand mixes **A15:** 211
Cellulose ester **EM3:** 7
Cellulose ethers
as binders **EM4:** 475, 955
Cellulose nitrate **EM3:** 7
defined **EM2:** 9
surface preparation **EM3:** 291
Cellulose nitrate, with amyl-
ethyl-, or methyl acetate, for replicas **A12:** 180
Cellulose propionate **EM3:** 7
Cellulose propionate, defined **EM2:** 9
Cellulose-base insulating lacquers as mounting materials for electropolishing **A9:** 49
Cellulosic plastics **EM3:** 7
Cellulosics **EM2:** 9, 450
solvent cements **EM3:** 567
surface preparation **EM3:** 279
Celsian
in glass-ceramics **EM4:** 1102
Celsius, vs. Fahrenheit degrees
in creep and creep-rupture analyses **A8:** 685
Cement **EM3:** 7
analyses of **A10:** 99-100
calcium in, optical emission spectroscopy for **A10:** 21
natural fiber reinforcement of **EM1:** 117
Portland **A10:** 30, 179
Cement coatings nails **M1:** 271
Cement copper
chemical analysis **A7:** 106, 119
as hydrometallurgical copper powder **A7:** 118-119
Cement grinding balls **A7:** 58
Cement lining
in cast iron pipe **M1:** 97-98, 100
Cement mixer
for lacquer coating **A7:** 588
Cement-asbestos
in oil/gas production **A13:** 1243-1244
Cement-mill equipment
elevated-temperature failures in **A11:** 294
Cementation
copper production by **A7:** 105, 119
metal precipitation by **A7:** 54
Cementation processes
ceramic coating **M5:** 542-545
fluidized-bed process *See* Fluidized- bed cementation process
pack cementation *See* Pack cementation
vapor streaming, ceramic coatings **M5:** 545
Cemented carbides *See also* Carbides; Cemented carbides, specific types; Cermets; Tool materials, superhard **A2:** 950-977 **A7:** 17, 773-783 **A9:** 273-278 **A13:** 846-858 **A16:** 70-90, 113 **EM4:** 330, 808-810
abrasion resistance **M3:** 453, 455, 456, 457, 464, 582
Al_2O_3-coated tool material for machining cast irons **A16:** 652
applications **A7:** 73, 76, 156, 777-779 **A13:** 847 **M3:** 449, 450, 451, 452, 460 **EM4:** 808
Cemented carbides (continued)
aqueous corrosion **A13:** 850-855
batch-type vacuum furnace for sintering **A7:** 359
bearings, valve seats, valve stems **A2:** 973
blanking and piercing dies, use for **M3:** 485, 487, 488
boring bars and plungers **A2:** 971
brittle fracture **A11:** 26
carbon deficiency **A9:** 274-275
classification **A2:** 953-954, 968 **A7:** 773
classification of **A16:** 74, 75
Co-bonded, properties **A16:** 73
coated carbide tools **A2:** 959-962 **A16:** 79-83
coated carbides **M3:** 456-458, 459
coatings **A13:** 856-857
coatings to improve bond properties **EM4:** 332
cobalt powder as binder **A7:** 144, 145
cobalt-bonded carbides **M3:** 452, 457-458
cobalt-bonded, properties **EM4:** 809
cobalt/carbon effect on phases **A13:** 849
for coining **A14:** 183
cold heading tools, use for **M3:** 512-513
cold-forming applications **A2:** 970-971
compared with cermets **A16:** 93-95
composition **M3:** 459, 460
composition effect on properties **A13:** 846-847
compositions **A2:** 951-953, 968 **EM4:** 808, 809
compositions and microstructures **A16:** 72-74
compressive properties **A16:** 78
containerless **A7:** 441, 443
contrasting by interference layers **A9:** 60
for corrosion applications **A13:** 848-850
corrosion resistance **A13:** 848-855 **M3:** 456
cutting tools **A16:** 41
deep drawing dies, materials for **M3:** 499
defined **A7:** 2 **A9:** 273
density property **A16:** 79
diamond indenter for **A8:** 74
drawing dies **A2:** 969
elemental cobalt alloying **A2:** 446
eta phase on a fracture surface **A9:** 276
eta phases **A9:** 274-275
etchants for **A9:** 274
fluid-handling components **A2:** 972-973
fractographs **A12:** 470
fracture toughness **A12:** 470
fracture toughness property **A16:** 78-79
fracture/failure causes illustrated **A12:** 217
free carbon in **A9:** 276
friction welding **A6:** 152
gages, use for **M3:** 554-556
galvanic corrosion **A13:** 855, 857
grain size determination **A9:** 274
grinding **A9:** 273 **EM4:** 334
grooving operation **A16:** 97
ground with superabrasives **A2:** 1013
hardness **M3:** 453, 455, 456, 457-458, 459, 460, 464
hardness property **A16:** 77-78
for high temperature use, microexamination of **A9:** 274
high-pressure dies and punches **A2:** 972
hot isostatic pressing **M3:** 451-452
hot pressing **A7:** 441, 443
imaging methods for **A9:** 275
indexable carbide inserts **A16:** 84-85, 86, 88
inserts, in deep-drawing dies **A14:** 511
inserts, indexable **M3:** 451
machinability **A16:** 639
machined by ultrahard tool materials **A2:** 1013
machining **A7:** 777
for machining applications **A2:** 965-968 **A16:** 86-88
macroexamination **A9:** 273-274
manufacture **M3:** 451-452
manufacture of **A2:** 950-951 **A16:** 71-72
mechanical properties **A7:** 469, 476 **M3:** 453, 456, 457-458, 459, 460
metalforming applications **A2:** 968-971

SUBJECTS OF THE INDEXED VOLUMES: ASM Handbook (designated by the letter "A"): **A1:** Properties and Selection: Irons, Steels, and High-Performance Alloys (1990); **A2:** Properties and Selection: Nonferrous Alloys and Special-Purpose Materials (1990); **A3:** Alloy Phase Diagrams (1992); **A4:** Heat Treating (1991); **A6:** Welding, Brazing, and Soldering (1993); **A7:** Powder Metallurgy (1984); **A8:** Mechanical Testing (1985); **A9:** Metallography and Microstructures (1985); **A10:** Materials Characterization (1986); **A11:** Failure Analysis and Prevention (1986); **A12:** Fractography (1987); **A13:** Corrosion (1987); **A14:** Forming and Forging (1988); **A15:** Casting (1988); **A16:** Machining (1989); **A17:** Nondestructive Testing and Quality Control (1989); **A18:** Friction, Lubrication, and Wear Technology (1992). **Metals Handbook, 9th Edition** (designated by the letter "M"): **M1:** Properties and Selection: Irons and Steels (1978); **M2:** Properties and Selection: Nonferrous Alloys and Pure Metals (1979); **M3:** Properties and Selection: Stainless Steels, Tool Materials and Special-Purpose Metals (1980); **M4:** Heat Treating (1981); **M5:** Surface Cleaning, Finishing, and Coating (1982); **M6:** Welding, Brazing, and Soldering (1983). **Engineered Materials Handbook** (designated by the letters "EM"): **EM1:** Composites (1987); **EM2:** Engineering Plastics (1988); **EM3:** Adhesives and Sealants (1990); **EM4:** Ceramics and Glasses (1991); **Electronic Materials Handbook** (designated by the letters "EL"): **EL1:** Packaging (1989).

Cemented carbides (continued)
microexamination **A9:** 274-275
microstructure **M3:** 452-453, 454-455, 459, 460
microstructures **A2:** 951-953 **A7:** 387-388, 780-783 **A13:** 847
milling **A16:** 79
mining and oil and gas drilling **A2:** 974-977
mounting **A9:** 273
nickel-bonded titanium carbide **M3:** 459, 460
for nonmachining applications **A2:** 968-977
nozzles **A2:** 973
oxidation resistance **A13:** 855-856
phases **A9:** 274
physical properties **A7:** 780-783 **M3:** 453, 455, 456, 457-458, 459, 460
polishing **A9:** 273
porosity determination **A9:** 274
powder compacting dies and punches **A2:** 971
powders used **A7:** 572
preparation of specimens **A9:** 273
primary applications **A16:** 639
production of **A7:** 156-158
production processes **EM4:** 808-810
properties **EM4:** 330, 808
properties of **A2:** 955-959
properties of hot pressed **A7:** 515
property test methods **A16:** 77
proprietary designations **EM4:** 809
qualitative metallography **A9:** 274-275
quantitative metallography **A9:** 275
reaction rates base on Murakami's reagent **A9:** 274
rebar rolls **A2:** 970
refractory-metal carbides **M3:** 453
Rockwell hardness testing of **A8:** 83
Rockwell scale for **A8:** 76
rod mill rolls **A2:** 969
saw tips and corrosion **A13:** 856
scanning electron microscopy of fractures **A9:** 99
seal rings **A2:** 972
sectioning **A9:** 273
selection of **A7:** 776-777
Sendzimir mill rolls **A2:** 969-970
sintering **A7:** 308-309, 385-389
sliding bearings, use in **M3:** 820-821
slitter knives **A2:** 970
specifications **M3:** 453
spray drying **A7:** 73, 76
stamping punches and dies **A2:** 971
steel-bonded carbide **M3:** 459-461
strength and toughness in **A7:** 153
structural components **A2:** 971-972
structural components, use for **M3:** 558
substrated for thermoreactive deposition/diffusion process **A4:** 450, 452
surface treatments **A13:** 857
test methods for determining properties of **M3:** 2, 453, 455-456
thermal shock resistance **A16:** 79
tool holding **A16:** 83-86
tool life **A16:** 75-76, 81-82, 85, 87, 88, 110, 113
tool wear mechanisms **A2:** 954-955 **A16:** 75-77
tools and toolholding **A2:** 962-965
toughness **M3:** 456, 457-458, 459
transportation and construction applications **A2:** 973-974
transverse rupture strength **A2:** 961, 989 **A16:** 78
tungsten, for drawing dies **A14:** 336
vacuum deposition of interference films **A9:** 148
vibratory compacting of **A7:** 306
for wear applications **A7:** 777-780
wire flattening rolls **A2:** 970
Cemented carbides, friction and wear of **A18:** 693, 795-800
applications **A18:** 795
as bearing alloys **A18:** 748, 754
applications **A18:** 754
composition **A18:** 754
mechanical properties **A18:** 754
chemical vapor deposition **A18:** 849
compatibility with steel **A18:** 743
corrosive wear **A18:** 795
for cutting tool materials **A18:** 616
damage dominated by brittle fracture **A18:** 180
Cemented carbides, friction and wear of (continued)
damage dominated by dissolution or diffusion **A18:** 181
damage dominated by plastic deformation **A18:** 178
die material for sheet metal forming **A18:** 628
elements implanted to improve wear and friction properties **A18:** 858
erosion resistance (WC) **A18:** 204
for hot-forging dies **A18:** 625, 627
ion plating **A18:** 849
laboratory testing methods for solid friction **A18:** 57
lubrication **A18:** 796
magnetron sputtering **A18:** 849
manufacturing methods **A18:** 795-796
finishing operations **A18:** 796
grade powders **A18:** 795-796
physical or chemical vapor deposition (PVD or CVD) **A18:** 796
preforming or shaping operations **A18:** 796
pressing or powder consolidation **A18:** 796
sintering operations **A18:** 796
properties **A18:** 796-797, 798, 799, 800
applications **A18:** 797
binder content **A18:** 797
microstructures **A18:** 797, 799
nominal composition **A18:** 796, 797
relative abrasion resistance **A18:** 796
raw materials **A18:** 795
chromium carbide **A18:** 795
cobalt **A18:** 795
nickel **A18:** 795
tantalum/titanium/niobium carbides **A18:** 795
tungsten carbide **A18:** 795
wear properties **A18:** 797-800
Cemented carbides, specific types
5WC-8Mo-79TiC-8Ni **A9:** 276
43WC-50(Ta,Ti,Nb,W)C-6Co **A9:** 278
73WC-21(Ta,Ti,Nb,W)C-6Co **A9:** 278
75WC-25Co, coarse grain structure **A9:** 278
76WC-16(Ta,Ti,Nb,W)C-8Co **A9:** 278
78WC-15(Ta,Ti,Nb,W)C-7Co **A9:** 278
79WC-14(Ta,Ti,Nb,W)C-7Co **A9:** 276
80WC-13(Ta,Ti,Nb,W)C-7Co, etch series **A9:** 276
83WC-10(Ta,Ti,Nb,W)C-8Co **A9:** 278
85WC-8(Ta,Ti,Nb,W)C-7Co, eta phase **A9:** 277
85WC-9(Ta,Ti,Nb,W)C-6Co, with coatings **A9:** 277
85WC-15Co, coarse grain structure **A9:** 278
86WC-8(Ta,Ti,Nb,W)C-6Co, with C-porosity **A9:** 276
86WC-8(Ta,Ti,Nb,W)C-6Co, with CVD coating **A9:** 278
89WC-10(Ta,W)C-ICo, progressive etching **A9:** 277
89WC-11Co, medium size grain structure **A9:** 278
90WC-10CO **A9:** 277
92WC-2(Ta,W)C-6Co **A9:** 277
94WC-6Co, eta-phase, fracture surface **A12:** 470
94WC-6Co, mating fracture analysis **A12:** 470
97WC-3Co, brittle fractures **A12:** 470
97WC-3Co, type 3-F microstructure **A9:** 277
ISO P20 ..... **A9:** 274
WC-12Co, dot map **A9:** 91
WC-12Co, fracture toughness test specimen **A9:** 91
WC-12Co, orientation contrast **A9:** 91
WC-12Co, worn drill **A9:** 91
WC-Co, ZnSe interference film for color **A9:** 158
Cemented tungsten carbide
liquid-phase sintering **A7:** 320
Cemented tungsten powders
particle size and distribution for **A7:** 154
Cementite *See also* Carbides; Eutectic carbide; Pearlite ..... **A3:** 1 • 23 **A6:** 708 **A13:** 3, 47
500 °F embrittlement, role in **M1:** 685
annealing with **A7:** 182, 185
in carbon and alloy steels **A9:** 178-179
cast iron **M1:** 3-5, 6, 12
defined **A9:** 3 **A15:** 2
etching to reveal **A9:** 170
from cast iron solidification **A15:** 82
Cementite (continued)
in gray iron **A15:** 632
in lower bainite **A9:** 664
proeutectoid **A1:** 127, 129-130
solubility of nitrogen in **A15:** 82
stability of **A15:** 61
in tool steels **A18:** 734
in upper bainite **A9:** 663
Cementitious materials
steel in **A13:** 1306-1308
Cements for repairing castings and metal parts
powders used **A7:** 574
Cementum
abrasion of dentifrices **A18:** 668
Centane number
defined **A18:** 5
Center bead cracks
by liquid penetrant inspection **A17:** 86
Center burst in alloy steel forging **A9:** 176
Center bursts
cold extruded steel **M1:** 591, 592
Center crack, panel
plan views **A8:** 452
Center cracking in aluminum alloy ingots
in alloy 1100 ..... **A9:** 634
effect of casting speed on **A9:** 634-635
effect of grain refiners on **A9:** 630
effect of ingot diameter on **A9:** 635
enter cracks in copper alloy ingots **A9:** 642
Center cracks
fracture analysis of **EM1:** 252-257
Center defects *See also* Defects
in cold-formed parts **A11:** 307
in friction welds **A11:** 444
Center for Professional Development (Brunswick NJ) **EM2:** 95
Center heating **EM4:** 630
Center upsetting, in cold heading
complex workpieces **A14:** 294-295
Center-cracked tension specimen *See also* Compact specimens
for fatigue crack growth analysis **A8:** 377-379, 678
geometries for **A8:** 251
gripping arrangements **A8:** 382
precracking **A8:** 382
size **A8:** 380-381
stress-intensity factor solutions **A8:** 379-380
thickness **A8:** 381
for vacuum and gaseous fatigue testing **A8:** 411
for vacuum and oxidizing fatigue testing **A8:** 414-415
Center-cracked-tension
abbreviation **A8:** 724
Center-gated mold
defined **EM2:** 9
Centerburst
in aluminum alloys, workability criteria **A8:** 577-578
at center of extruded or drawn products **A8:** 592, 595
fracture **A8:** 573-574
Centerbursting *See* Central burst
Centered dark-field image **A10:** 689
Centering
multifunction machining **A16:** 375
Centerless grinding *See* Grinding
by scanning laser gage **A17:** 12
nuclear fuels **A7:** 665
Centerless wire-filled brushes **M5:** 156
Centerline **EM3:** 791-796
segregation, squeeze casting **A15:** 325
shrinkage, defined **A15:** 2
symbol for **A8:** 726
Centerline average (CLA) **A18:** 475
Centerline cracking **A6:** 51
Centerline cracks **A6:** 409
low-carbon steel **A12:** 244
Centerline shrinkage **A17:** 349, 391
as casting defect **A11:** 382
defined **A11:** 2
forging **A11:** 315
in ingots **A11:** 315
Centimeter
abbreviation **A8:** 724
Centistoke **A18:** 140

Central burst
during extrusion **A14:** 399-400
in forgings **A14:** 401-402
prediction, wire drawing **A14:** 395
workability criteria **A14:** 370
Central bursts *See* Bursts; Chevron patterns
Central composite designs
experimental **EM2:** 601-602
Central conductors
applications **A17:** 94
for bearing rings **A17:** 117
circular magnetization by **A17:** 96
for cylinders **A17:** 112
defined **A17:** 95
magnetizing by **A17:** 94, 130
offset **A17:** 96-97
solid ferromagnetic, ac/dc current **A17:** 96
solid nonmagnetic, dc current **A17:** 96
within hollow ferromagnetic cylinder **A17:** 96-97
Central crack
crack tip strip zone model for **A8:** 449
vector lines from crack tip and crack center for **A8:** 444
Central electronic processor
image analyzers **A10:** 310
Central fibrous region
alloy steels **A12:** 334
Central film thickness **A18:** 539
Central processing unit (CPU)
in design process **EL1:** 128
Central tendency
measures of **A8:** 624-625
Centrifugal (rotating disk) atomizer **A7:** 74-76
Centrifugal atomization ... **A1:** 972-973 **A7:** 25, 26, 49, 75-77
beryllium powder **A2:** 685
in titanium powder production **A7:** 167
Centrifugal babbitting **M5:** 357
Centrifugal barrel finishing **M5:** 133-134
chemically accelerated **M5:** 134
fatigue strength improved by **M5:** 134
Centrifugal casting *See also* Bimetal; Castings; Centrifuge casting; Foundry products; Horizontal centrifugal casting; Pressure casting; Vertical centrifugal casting
aluminum alloys **M2:** 146-147
aluminum casting alloys **A2:** 141
applications **A15:** 299-300
of copper alloys **A2:** 346, 348 **M2:** 384
defects in **A15:** 306-307
defined **EM1:** 6 **EM2:** 9
dual-metal, defined *See* Dual-metal centrifugal casting
equipment **A15:** 296-297, 307
horizontal **A15:** 296-300
machining allowances **A15:** 302
of metal-matrix composites **A15:** 844
molds **A15:** 296-297, 300-304
as permanent mold process **A15:** 34, 276-277
process details **A15:** 297-298, 304-306
processes **A15:** 34, 37, 296, 300
and static casting, compared **A15:** 301
vertical **A15:** 300-307
Centrifugal disk finishing **M5:** 133
Centrifugal finishing **A7:** 459
Centrifugal high-energy deburring **A7:** 459
Centrifugal pressure impregnation in infiltration **A7:** 554
Centrifugal screen **A7:** 177
Centrifugal sedimentation
to analyze ceramic powder particle sizes **EM4:** 67
Centrifugal shrinkage
defined **A15:** 2
Centrifugal wheels
performance **A15:** 516-517
types **A15:** 511-516
Centrifugally atomized powders **A13:** 833
Centrifugally atomized specialty powders
rigid tool compaction of **A7:** 322
Centrifuge
as physical testing **EL1:** 944
Centrifuge casting *See also* Centrifugal casting
defined **A15:** 2
Centrifuge centrifugal casting
defined **A15:** 300
Centrifuge testing method
hard chromium plating bath composition **M5:** 173
Centroid aspect ratio (CAR), as particle shape factor *See also* Aspect ratio
Centrosymmetry
determining effect on ODF coefficient **A10:** 362
CERABULL (database) **EM4:** 40, 692
Ceracon process **A7:** 537-541
applications and properties **A7:** 540-541
flowchart **A7:** 537
mechanical properties **A7:** 540
part geometries **A7:** 540-541
pin location for lateral pressure measurement in **A7:** 539
processing sequence **A7:** 537-538, 540
Ceramers
alkoxide-derived gels **EM4:** 210, 211
Ceramic *See also* Ceramic composites; Ceramic fibers; Sintering
defined **EL1:** 1136 **EM1:** 6
for RTM tooling **EM1:** 168-169
Ceramic (96% Al)
recommended waterjet cutting speeds **EM4:** 366
Ceramic Abstracts On-Line (data base) **EM4:** 40
Ceramic alumina
applications **EM4:** 331
bond type **EM4:** 331
Ceramic and glass bonding
ceramic-ceramic bonding **EM3:** 309
glass-glass bonding **EM3:** 309
glazes **EM3:** 308-309
compositions **EM3:** 309
opacifying **EM3:** 310
Ceramic applications in turbine engines (CATE) program **EM4:** 716, 719, 720
design/manufacturing trade-offs **EM4:** 720
Ceramic body preparation
structural ceramics **A2:** 1019-1020
Ceramic capacitators
acoustic microscopy of **A17:** 479-480
Ceramic capacitor dielectrics **EM4:** 1112-1117
barium titanate-based dielectrics **EM4:** 1112-1114
barium-neodymium-titanate **EM4:** 1114-1115
calcium titanate **EM4:** 1114
commercial capacity requirements **EM4:** 1113
dielectric compositions with high lead content **EM4:** 1115-1116
dielectric constants at 25 °C **EM4:** 1113
integrated capacitors **EM4:** 1117
magnesium titanate **EM4:** 1114
multilayer ceramic capacitors **EM4:** 1112
processing
disks and tubulars **EM4:** 1116
multilayer ceramic capacitors **EM4:** 1116-1117
thick-film and thin-film capacitors **EM4:** 1117
strontium titanate **EM4:** 1114
Ceramic capacitors
failure mechanisms **EL1:** 972, 994-995
Ceramic coating **M5:** 532-547
abrasive blasting processes **M5:** 537-539
alumina coatings **M5:** 532, 534-536, 540-542
aluminum and aluminum alloys **M5:** 609-610
applicability **M5:** 538-539, 544-545
applications **M5:** 532-538, 540-542, 544-545
applying, methods of *See also* specific processes by name **M5:** 532-546
bond strength, testing **M5:** 547
carbide coatings **M5:** 535-536, 541, 546
hardness **M5:** 546
Ceramic coating (continued)
melting points **M5:** 534-535
cementation processes **M5:** 542-545
fluidized-bed cementation **M5:** 542, 544-545
pack cementation **M5:** 542-544
vapor streaming cementation **M5:** 545
cermet coatings **M5:** 536-537
chemical cleaning processes **M5:** 537-538, 542-543
compatability, chemical and mechanical **M5:** 532
continuity of, effects of sharp and round corners **M5:** 542-543
crystallized glass coatings **M5:** 533
dipping process **M5:** 536-538
electrophoretic coating process **M5:** 545-546
elevated-temperature
conditions **M5:** 533-534, 536-537, 540, 546
high-temperature test **M5:** 546
equipment **M5:** 538-543
flame spraying process **M5:** 534-536, 538-541
combustion system **M5:** 539-542, 546
detonation gun system **M5:** 542, 546
plasma-arc system **M5:** 535, 541-542, 546-547
flow coating process **M5:** 538-539
hardness, true and Vickers **M5:** 546-547
heat-resisting alloys **M5:** 537-538, 566
impact strength, testing **M5:** 546-547
materials used, types and characteristics *See also* specific types by name **M5:** 533-537
melting points **M5:** 534-536
molybdenum and molybdenum alloys **M5:** 543-545, 662
oxidation resistance **M5:** 535-537, 543, 546
oxide coatings **M5:** 534-536, 546-547
hardness **M5:** 546-547
melting points **M5:** 534-535
phosphate-bonded coatings **M5:** 535-537
densities and maximum service temperatures **M5:** 536-537
process steps **M5:** 537-546
protection mechanisms **M5:** 532
quality control **M5:** 533, 546-547
refractory metals **M5:** 532-533, 535, 537, 542
repairability **M5:** 533
selection factors **M5:** 532-533
service environment, effects of **M5:** 532
silicate coatings **M5:** 533-534
silicide coatings **M5:** 535, 537, 542-545
spray process **M5:** 533-542, 546
stainless steel **M5:** 537-538
structure, testing **M5:** 546-547
surface preparation **M5:** 537-539, 542-545
temperature **M5:** 533-534, 536-537, 539-546
elevated temperatures **M5:** 533-534, 536-537, 539-540, 546
thermal barrier coatings **M5:** 541-542
thickness **M5:** 534-537, 541-545
control of **M5:** 541-542
time-and-temperature effects **M5:** 541, 543-545
trowel coating process **M5:** 545
tungsten **M5:** 662
vapor deposition process **M5:** 535
wear properties, testing **M5:** 547
zirconia coatings **M5:** 534-536, 540-542
Ceramic coatings *See also* Ceramics; Coatings **EM4:** 953-958
applications **EM4:** 953
as barrier protection **A13:** 378
for carbon steels **A13:** 524
coating fit **EM4:** 957
crazing **EM4:** 957
shivering **EM4:** 957
dry application techniques **EM4:** 956
dry-powder cast iron enameling **EM4:** 956
electrostatic dry-powder coatings **EM4:** 956
flame spraying **EM4:** 956
firing **EM4:** 956-957
box furnaces **EM4:** 956-957

SUBJECTS OF THE INDEXED VOLUMES: ASM Handbook (designated by the letter "A"): **A1:** Properties and Selection: Irons, Steels, and High-Performance Alloys (1990); **A2:** Properties and Selection: Nonferrous Alloys and Special-Purpose Materials (1990); **A3:** Alloy Phase Diagrams (1992); **A4:** Heat Treating (1991); **A6:** Welding, Brazing, and Soldering (1993); **A7:** Powder Metallurgy (1984); **A8:** Mechanical Testing (1985); **A9:** Metallography and Microstructures (1985); **A10:** Materials Characterization (1986); **A11:** Failure Analysis and Prevention (1986); **A12:** Fractography (1987); **A13:** Corrosion (1987); **A14:** Forming and Forging (1988); **A15:** Casting (1988); **A16:** Machining (1989); **A17:** Nondestructive Testing and Quality Control (1989); **A18:** Friction, Lubrication, and Wear Technology (1992). **Metals Handbook, 9th Edition** (designated by the letter "M"): **M1:** Properties and Selection: Irons and Steels (1978); **M2:** Properties and Selection: Nonferrous Alloys and Pure Metals (1979); **M3:** Properties and Selection: Stainless Steels, Tool Materials and Special-Purpose Materials (1980); **M4:** Heat Treating (1981); **M5:** Surface Cleaning, Finishing, and Coating (1982); **M6:** Welding, Brazing, and Soldering (1983). **Engineered Materials Handbook** (designated by the letters "EM"): **EM1:** Composites (1987); **EM2:** Engineering Plastics (1988); **EM3:** Adhesives and Sealants (1990); **EM4:** Ceramics and Glasses (1991); **Electronic Materials Handbook** (designated by the letters "EL"): **EL1:** Packaging (1989).

Ceramic coatings (continued)
continuous furnaces **EM4:** 957
fuel costs **EM4:** 957
single-fire processing **EM4:** 957
two-fire processing **EM4:** 957
frit-melting furnaces **EM4:** 953-954
minimizing coating defects **EM4:** 957-958
bubbles **EM4:** 957-958
crawling **EM4:** 958
crazing **EM4:** 957
metal marking **EM4:** 958
peeling **EM4:** 957
shivering **EM4:** 957
specking **EM4:** 958
raw materials **EM4:** 953, 954
wet application processes **EM4:** 954-956
application techniques **EM4:** 955-956
bactericides **EM4:** 955
binders **EM4:** 955
dipping **EM4:** 955
doctor blade method **EM4:** 956
electrolytes **EM4:** 955
electrostatic spray coating **EM4:** 955-956
mill additives **EM4:** 954-955
organic color-code dyes **EM4:** 955
painting and brushing **EM4:** 956
silk screen process **EM4:** 956
slip preparation **EM4:** 954
spraying **EM4:** 955-956
suspending agents **EM4:** 955
wetting agents **EM4:** 955
Ceramic collars
for insulating specimen from machine **A8:** 389
Ceramic composites *See also* Ceramic; Ceramic fibers; Ceramic-ceramic composites; Composites; Metal-matrix composites; Multidirectionally, reinforced ceramics **EM1:** 925-944
fiber reinforcement of **EM1:** 59, 925-926
future directions and problems **A2:** 1024
metal-ceramic composites **A2:** 1024
multidirectionally reinforced **EM1:** 933-940
structural **EM1:** 925-932
system characteristics **EM1:** 927-931
types **A2:** 1023-1024
ultrasonic machining **EM4:** 359
whisker-reinforced **EM1:** 941-944
Ceramic cutting tools **EM4:** 966-972
cost effectiveness of machining with ceramics **EM4:** 967
difficult-to-machine materials **EM4:** 970-972
high-productivity machining of cast iron with Si_3N_4-based ceramics **EM4:** 969-970
history **EM4:** 966-967
machining of irons and steels with oxide-based ceramic inserts **EM4:** 967-969, 970
properties distinguishing them from traditional steel and tungsten-carbide cutting materials **EM4:** 966
Ceramic design and process engineering **EM4:** 29-36
ceramic material design **EM4:** 29
ceramic processing methods **EM4:** 32-36
firing **EM4:** 35-36
forming processes **EM4:** 33-35
preparation for forming **EM4:** 33
raw materials for advanced ceramics **EM4:** 32-33
raw materials for traditional ceramics **EM4:** 32
testing and evaluation **EM4:** 36
design flow chart **EM4:** 31-32
design methodology **EM4:** 29-30
deterministic **EM4:** 29, 30
empirical **EM4:** 29, 30
probabilistic **EM4:** 29-30
design process overview **EM4:** 678-688
material selection **EM4:** 29
proof testing **EM4:** 31
Weibull statistics **EM4:** 30-31
Ceramic design process overview **EM4:** 676-688
conceptual design **EM4:** 677-678
ceramic precombustion chambers **EM4:** 677
indirect-injection diesel engines **EM4:** 677
sample applications **EM4:** 677
Ceramic design process overview (continued)
turbochargers **EM4:** 677-678
detailed design **EM4:** 678-688
effective volume **EM4:** 680-682
fast fracture reliability **EM4:** 678-679
finite-element models for reliability **EM4:** 683-684
joints, attachments, and interfaces **EM4:** 685-686
lifetime reliability **EM4:** 685
modulus of rupture **EM4:** 680, 683
probability of failure **EM4:** 682, 683, 684, 686
proof testing **EM4:** 686-688
reliability prediction **EM4:** 679, 683
reliability selection **EM4:** 684
reliability sensitivity to various parameters **EM4:** 680-683
risk of rupture **EM4:** 679, 680-681, 684, 687
simplified structural ceramic design technique **EM4:** 683
thermal shock considerations **EM4:** 686
Weibull material property generation **EM4:** 680-687
Weibull theory **EM4:** 679-680, 683-687
key structural ceramic characteristics **EM4:** 676
modulus of rupture **EM4:** 676
major phases of design **EM4:** 676
Ceramic dual-in-line package (CERDIP)
assembly sequence **EL1:** 487
defined **EL1:** 1137
die attachments **EL1:** 213
humidity in **EL1:** 962
nuclear radiation induced device failure **EL1:** 1056
package outline **EL1:** 203
residual gas analysis results for **EL1:** 1066
substrates and **EL1:** 203-204
world market **EL1:** 460
Ceramic facing, preparation
Shaw process **A15:** 249
Ceramic fibers *See also* Ceramic composites; Discontinuous ceramic fiber MMCs **EM1:** 60-65
alumina-silica/alumina-boria-silica **EM1:** 60-61
aluminum oxide **EM1:** 60
continuous oxide **EM1:** 60-61
continuous silicon carbide **EM1:** 63-64
development **EM1:** 60
discontinuous, for metal matrix composites **EM1:** 903-910
discontinuous oxide **EM1:** 62-63
discontinuous silicon carbide/silicon nitride whiskers **EM1:** 64
fused-silica **EM1:** 61
leached-glass **EM1:** 61
nonoxide **EM1:** 63-64
-reinforced piston for high-performance diesel engines **A2:** 922
as thermocouple wire insulation **A2:** 882-883
zirconia-silica **EM1:** 61
Ceramic film
defined **EL1:** 1136
Ceramic films and coatings on metals
chemical vapor deposition **EM3:** 307, 308
coating techniques **EM3:** 307
evaporation **EM3:** 307
ion implantation **EM3:** 308
plasma spraying **EM3:** 307-308
sol-gel coatings **EM3:** 308
sputtering **EM3:** 307
uses **EM3:** 307
Ceramic filters *See also* Ceramics; Filtration
foam, as inclusion control **A15:** 90
foam, for nonferrous casting **A15:** 490-491
in gating design **A15:** 594-597
Ceramic heat exchangers **EM4:** 960
applications **EM4:** 960
Ceramic investment mold method
of encapsulation **A7:** 428, 429
Ceramic joining technologies, material types and uses **EM4:** 478-480
ceramic/ceramic joining **EM4:** 480
liquid-phase active-metal techniques **EM4:** 480
solid-state brazing **EM4:** 480
Ceramic joining technologies, material types and uses (continued)
solid-state pressure/diffusion bonding **EM4:** 480
ceramic/metal joining **EM4:** 479-480
active metal brazing **EM4:** 479
applications **EM4:** 479
disadvantages **EM4:** 479
glass bonding of ceramics to metals **EM4:** 479
liquid-phase joining **EM4:** 479
moly-manganese (Mo-Mn) process **EM4:** 479
precious metal brazes **EM4:** 479
refractory metal brazing **EM4:** 479
solid-state joining **EM4:** 479-480
tungsten brazes **EM4:** 479
glass-ceramic/metal sealing **EM4:** 479
applications **EM4:** 479
properties **EM4:** 479
glass/metal seals **EM4:** 478-479
applications **EM4:** 478
field-assisted bonding **EM4:** 478-479
glass seal classification scheme **EM4:** 478
metals and metallic alloys **EM4:** 478
non-oxide ceramic joining **EM4:** 480
applications **EM4:** 480
brazing of Si_3N_4 ...... **EM4:** 480
composite interlayer bonding **EM4:** 480
direct brazing SiC and Si_3N_4 ...... **EM4:** 480
liquid-phase joining techniques **EM4:** 480
properties **EM4:** 480
solid-state joining techniques **EM4:** 480
Ceramic leadless chip carrier (LCC)
heat sinks **EL1:** 1129-1131
Ceramic magnet materials *See* Ferrites
Ceramic magnets *See also* Permanent magnet materials
as ferrimagnetic **A2:** 782
as hard ferrites **A2:** 788-790
Ceramic mass finishing media **M5:** 135
Ceramic materials
acoustic microscopy methods **A17:** 469-472
engineering, computed tomography (CT) of **A17:** 364
liquid penetrant inspection **A17:** 71
mechanical properties **EL1:** 1120
microwave inspection **A17:** 202
multichip structures, unsuitable **EL1:** 305
polarized, as transducer elements **A17:** 255
standard conditions for sliding **A18:** 236
thermal etching **A9:** 62
thermal, mechanical, electrical properties **EL1:** 335-336
thermal properties **EL1:** 335-336, 1120
Ceramic matrix composites **A9:** 592
with silicon-carbide fibers **A9:** 596-597
Ceramic metallization and joining **EM3:** 304-306
active metal process **EM3:** 305
direct bonding **EM3:** 305-306
electroforming **EM3:** 306
gas-metal eutectic (direct) method **EM3:** 305-306
graded-powder process **EM3:** 306
liquid-phase metallizing **EM3:** 306
metal powder-glass frit method **EM3:** 305
moly-manganese paste process **EM3:** 304-305
nonmetallic fusion process **EM3:** 306
pressed diffusion joining **EM3:** 306
sintered metal powder (SMP) process **EM3:** 304-305
vapor-phase ceramic coatings **EM3:** 306
vapor-phase metallizing **EM3:** 306
Ceramic microspheres
as extender **EM3:** 176
Ceramic microwave resonators
solid-state sintering **EM4:** 281
Ceramic mold process in hot isostatic pressing **A7:** 425, 426, 428
manufacturing sequence **A7:** 751
with superalloys **A7:** 440
Ceramic molding *See also* Ceramic(s); Cope; Drag **A15:** 248-252
all-ceramic mold casting, procedure **A15:** 249
applications **A15:** 248
defined **A15:** 2
and investment molding, compared **A15:** 248
Shaw process **A15:** 248-250
Unicast process **A15:** 250-252

Ceramic multilayer assemblies
Hybrid, as VLSI packaging approach **EL1:** 270
technologies compared **EL1:** 297-298
Ceramic multilayer package fabrication **EL1:** 460-469
ceramic covered **EL1:** 460
layer personalization **EL1:** 463-464
market **EL1:** 460
materials preparation **EL1:** 460-463
physical properties **EL1:** 467-468
reliability **EL1:** 468
substrate fabrication **EL1:** 464--467
substrate layers **EL1:** 460-462
thick-film metallization **EL1:** 462-463
Ceramic nuclear waste forms
EPMA study of simulant **A10:** 532-535
Ceramic oxides **EM4:** 17
Ceramic packages *See also* Ceramic dual-in-line package (CERDIP); Ceramic multilayer package fabrication; CERPA K; Encapstilation; Pressed ceramic packages **EM3:** 585, 588
defined **EL1:** 454
failure mechanisms **EL1:** 961-962
glass-sealed **EL1:** 203-204
substrates **EL1:** 203-206
thermal performance of **EL1:** 409-410
Ceramic powders *See also* Ceracon process; Ceramic investment mold method,- Ceramic mold process; Cermets
as contaminants **A7:** 178
firing temperatures **A7:** 151
grain characteristics **A7:** 539
for hardfacing **A7:** 830
high densities for packed **A7:** 297
hot pressed, products **A7:** 514
multilayer capacitors **A7:** 151
particles **A7:** 537-541
precious metal powders in **A7:** 151
tapping or vibrating **A7:** 297
Ceramic powders and processing
applications **EM4:** 41
characterization methods **EM4:** 41
introduction **EM4:** 41-42
mixing methods **EM4:** 42
starting materials **EM4:** 41
characteristics **EM4:** 42
Ceramic powders, characterization **EM4:** 65-74
chemical
bulk composition **EM4:** 66
phases **EM4:** 66
surface composition **EM4:** 66
chemical composition **EM4:** 72-73
arc emission spectroscopy **EM4:** 72
argentometric titration **EM4:** 72
atomic absorption spectroscopy **EM4:** 72
atomic emission spectroscopy **EM4:** 72
combustion **EM4:** 72
coulometry **EM4:** 72
direct current plasma emission spectroscopy **EM4:** 72
electrochemical techniques **EM4:** 72
gravimetry **EM4:** 72
inductively coupled plasma emission spectroscopy **EM4:** 72
Kjeldahl **EM4:** 72
mass spectrometry **EM4:** 72, 73
methods for bulk chemical analysis **EM4:** 72
neutron activation analysis **EM4:** 72-73
potentiometric titration **EM4:** 72
selective-ion potentiometry **EM4:** 72
x-ray fluorescence spectroscopy **EM4:** 72
density **EM4:** 71
bulk **EM4:** 71
tap **EM4:** 71
theoretical **EM4:** 71
methods of analysis **EM4:** 66-69
centrifugal techniques **EM4:** 68
Ceramic powders, characterization (continued)
specific gravity balance **EM4:** 68
morphological analysis **EM4:** 69
definition **EM4:** 69
fractal analysis **EM4:** 69
Luerkens equations **EM4:** 69
particle shape definitions **EM4:** 69
nominal size ranges of particles **EM4:** 67
particle size **EM4:** 65
particle size analysis methods **EM4:** 67
phase composition **EM4:** 73
nuclear magnetic resonance spectroscopy **EM4:** 73
Rietveld refinement methods **EM4:** 73
x-ray powder diffraction **EM4:** 73
physical
agglomerates **EM4:** 66
grains **EM4:** 66
porosity **EM4:** 71-72
gas adsorption **EM4:** 71-72
mercury porosimetry **EM4:** 71
nuclear magnetic resonance **EM4:** 71, 72
porosimetry curve characteristics **EM4:** 71, 72
research methods **EM4:** 72
small angle neutron scattering **EM4:** 71, 72
size analysis steps **EM4:** 65-66
data collection **EM4:** 66
dispersion **EM4:** 65-66
error, total overall estimation **EM4:** 65
interpretation **EM4:** 66
measurement **EM4:** 66
sampling from powder lot **EM4:** 65
wetting **EM4:** 65-66
size distribution **EM4:** 65
specific surface area **EM4:** 69-71
equivalent spherical diameter of a particle **EM4:** 69
gas adsorption **EM4:** 69-70
Harkins-Jura methods **EM4:** 70-71
multipoint method **EM4:** 70
permeametry **EM4:** 70, 71
single point method **EM4:** 70
total surface area **EM4:** 70
surface composition **EM4:** 73-74
electrokinetic properties **EM4:** 74
electron spin resonance **EM4:** 73-74
Fourier transform infrared spectroscopy **EM4:** 73
INMR spectroscopy **EM4:** 73
Raman spectroscopy **EM4:** 73
x-ray photoelectron spectroscopy **EM4:** 73
Ceramic printed wiring boards *See also* Boards; Printed wiring boards; Thick-film ceramic wiring boards
circuit construction **EL1:** 387
component attachment **EL1:** 388
defined **EL1:** 505
for high-bandwidth digital systems **EL1:** 76
materials **EL1:** 388-389
Ceramic properties data base systems **EM4:** 690-692
ceramics property information **EM4:** 690-691
assessment criteria **EM4:** 691
data reliability considerations **EM4:** 690-691
evaluated data **EM4:** 691
important properties **EM4:** 690, 691
metadata **EM4:** 690
existing compilations ceramic property data **EM4:** 691
computerized data bases **EM4:** 691-692
printed compilations **EM4:** 691
issues for future developments **EM4:** 692
linking of materials property data bases **EM4:** 692
structured query language **EM4:** 692
Ceramic rod flame spraying
definition **M6:** 3
Ceramic substrates *See also* Ceramic; Substrates **EM4:** 1107-1111
advanced **EL1:** 8
alumina substrates **EM4:** 1107-1110
aluminum nitride **EM4:** 1110
beryllia substrates **EM4:** 1110
with cermets, medical and military applications **EL1:** 386, 388
fabrication **EL1:** 464-467
glass-ceramic materials **EM4:** 1110-1111
large-aspect ratio **EL1:** 8
layers **EL1:** 460-462
materials **EL1:** 336-338
for thick-film circuits **EL1:** 249
thick-film hybrids **EL1:** 334-338
world market **EL1:** 460
Ceramic Technology for Advanced Heat Engines program (Oak Ridge National Laboratory (ORNL) **EM4:** 692
Ceramic tiles *See* Tile whiteware **A18:** 649
Ceramic wall tile
recommended waterjet cutting speeds **EM4:** 366
Ceramic(s) *See also* Ceramic filters; Ceramic molding; Filtration; Preformed ceramic core; Structural ceramics
bivalve mold **A15:** 17
body preparation **A2:** 1019-1020
and cermets, compared **A2:** 978
coating and firing, Replicast process **A15:** 271
cores, manufacture of **A15:** 261
defined **A2:** 1019 **A15:** 2
as ferrites for high-frequency applications **A2:** 776
market effects **A15:** 44
metal-matrix composites, low gravity effects **A15:** 152-153
mixed-phase, as superconducting materials **A2:** 1028
in Neolithic period **A15:** 15
of precious metals **A2:** 693
processing future and problems **A2:** 1024
shell molds, manufacture of **A15:** 257-261
structural **A2:** 1019-1024
thermal sprayed, as coating **A15:** 563
as thermocouple protection **A2:** 882-884
toughened **A2:** 1022-1023
Ceramic-ceramic composites **EM1:** 118, 930-931, 933-940 **EM4:** 47
glass-encapsulated HIP processed **EM4:** 200
Ceramic-coated carbides
cutting speed and work material relationship **A18:** 616
Ceramic-lined die assembly **A7:** 506
Ceramic-matrix composites **EM4:** 20, 29
acrospace applications **EM4:** 1004-1005
application in future jet engine components **A18:** 592
chemical vapor deposition **EM4:** 215
chemical vapor infiltration **EM4:** 215
electrical discharge machining.... **EM4:** 371, 374, 376
properties **EM4:** 20, 835, 383-843
reaction sintering **EM4:** 291
Ceramic-matrix fiber-reinforced composites
fabrication processes **EM4:** 35
Ceramic/metal seals **EM4:** 502-509, 513
active brazing process considerations **EM4:** 504-509
active brazed joints **EM4:** 509
active brazing filler metals **EM4:** 505
process mechanism **EM4:** 505
processing steps **EM4:** 505
screenable paste versus foil preform-n for edge brazing **EM4:** 509
techniques **EM4:** 504
materials **EM4:** 502
coefficent of thermal expansion **EM4:** 502
static fatigue **EM4:** 502

Ceramic/metal seals (continued)
typical materials used and their properties **EM4:** 503
moly-manganese process **EM4:** 502-504, 506
process mechanism **EM4:** 502-503
processing steps **EM4:** 503-504
process evolution **EM4:** 502-503
seal design finite-element stress-analytic techniques **EM4:** 538

Ceramics *See also* Ceracon process; Ceramic investment mold method; Ceramic mold process; Ceramics, characterization of; Ceramics, specific types; Cermets **A16:** 2, 98-104
abrasive flow machining **A16:** 517, 519
abrasive jet machining **A16:** 511, 512
additives to optimize powder treatment and green forming **EM4:** 49
aerospace applications **A6:** 617-618
Al_2O_3-$10ZrO_2$ **A9:** 94
analytic methods applicable **A10:** 5
applications **A6:** 948 **A16:** 101-103, 639 **EM4:** 959-960
applications as a coating **EM4:** 208
at fracture **A12:** 471
boring **A16:** 101, 162
brazing **A6:** 948-950
brazing and soldering characteristics **A6:** 635-636
carbides for machining **A16:** 75
cast iron machining **A16:** 656, 658
Cermets
characterized **A10:** 1
chemical vapor deposition **EM3:** 308
chevron-notched specimens of **A8:** 470
coated **A16:** 103
compared to high-speed tool steels **A16:** 54
containing crystalline and amorphous phases **A10:** 445
cooling effects of cutting fluids **A16:** 122
crack propagation **EM4:** 694-698
crystalline materials **A10:** 381
crystallographic texture measurement and analysis **A10:** 357
cutters, high removal rate machining **A16:** 607, 608
cutting fluids **A16:** 125
determining causes **A11:** 749-757
diamond as abrasive for honing **A16:** 476
drilling **A16:** 102
edge preparation and machinability **A16:** 646-647
electromagnetic forming with **A14:** 649-650
electron beam machining **A16:** 570
electronic applications **A6:** 991
extensometers, elevated-temperature testing with **A8:** 36
failure analysis of **A11:** 744-757
fractographs **A12:** 471
fracture surface analysis by instrumented stereometry **A9:** 96
fracture toughness testing **A8:** 469
fracture/failure causes illustrated **A12:** 217
friction welding **A6:** 152, 154, 891
glass precoats, for titanium alloy forgings **A14:** 279
grinding **A16:** 102, 432, 433
grinding wheel core material **A16:** 456
ground by CBN wheels **A16:** 455
ground by diamond wheels **A16:** 455, 460, 461, 462, 463
high-palladium **A13:** 1361
high-speed machining **A16:** 604
high-temperature solid-state welding **A6:** 298, 299
honing **A16:** 472, 476, 477, 478
impressed-current anodes **A13:** 469, 921-922
indexable-insert milling cutters **A16:** 315
joining of **A6:** 617, 618, 619
lapping **A16:** 492, 494, 499
laser cutting of **A14:** 742
laser hardfacing **A6:** 806
laser-enhanced etching **A16:** 576
linings, chemical-setting **A13:** 453-455
liquid-metal corrosion **A13:** 59
location of fracture origin **A11:** 744-747
metallized, electron diffraction/EDS method to identify unknown phase in **A10:** 457-458
microstructural changes **A10:** 366

Ceramics (continued)
microstructures **A6:** 953-954
microwave brazing **A6:** 124
milling **A16:** 102, 327
multiphase, IA quantitative determination of second-phase phenomena **A10:** 309
multiphase, SIMS phase distribution analysis in **A10:** 610
nitric acid corrosion **A13:** 1156
nuclear applications **A6:** 618, 619
PCD tooling **A16:** 110
Permalloy film thickness, x-ray spectrometry for **A10:** 100-101
phase separation analysis by SAXS/ SANS/ SAS **A10:** 402, 405
polycrystalline, fracture mirrors **A11:** 745
polycrystalline, fracture of **A11:** 26
powder, XRPD analysis of crystalline phases in **A10:** 333
prebond treatment **EM3:** 35
production process **A16:** 98
properties **A6:** 948 **A16:** 101
SAS applications **A10:** 405
sectioning by fracturing **A9:** 23
Si_3N_4-base tool materials **A16:** 100
SIMS analysis of surface layers **A10:** 610
single-phase, image analysis of **A10:** 309
solderable and protective finishes for substrate materials **A6:** 979
techniques of fractography **A11:** 747-749
TEM bright-field images and diffraction patterns **A10:** 445
thermal shock **A16:** 102
thread grinding **A16:** 271
for three-dimensional grains and particles **A9:** 132
tool geometries **A16:** 101
tool steels **A16:** 714
tools and machinability **A16:** 642, 646
tools for boring **A16:** 714
tools for turning **A16:** 708-710
truing of CBN grinding wheels **A16:** 468
turning **A16:** 101, 146, 150, 151
turning of Al alloys **A16:** 777
turning of heat-resistant alloys **A16:** 739, 740
turning operation **A16:** 146, 154
for two-dimensional planar figures **A9:** 131
ultrasonic fatigue testing of **A8:** 240
ultrasonic machining **A16:** 530, 531
vacuum deposition of interference films **A9:** 148, 158
Cesium chloride, diffraction patterns **A9:** 109
Chalkley equations
vapor-phase soldering **A6:** 369
waterjet machining **A16:** 520, 527
wear mechanisms **A16:** 40
wear resistance **A16:** 108
weld overlay material **M6:** 807
whisker-reinforced **A16:** 99

Ceramics, and thermosetting resins
compared **EM2:** 222

Ceramics, characterization of *See also* Ceramics
analytical transmission electron microscopy **A10:** 429-489
atomic absorption spectrometry **A10:** 43-59
Auger electron spectroscopy **A10:** 549-567
classical wet analytical chemistry **A10:** 161-180
controlled-potential coulometry **A10:** 207-211
electrochemical analysis **A10:** 181-211
electrogravimetry **A10:** 197-201
electrometric titration **A10:** 202-206
electron probe x-ray microanalysis **A10:** 516-535
electron spin resonance **A10:** 253-266
extended x-ray absorption fine structure **A10:** 407-419
inductively coupled plasma atomic emission spectroscopy **A10:** 31-42
infrared spectroscopy **A10:** 109-125
ion chromatography **A10:** 658-667
low-energy electron diffraction **A10:** 536-545
low-energy ion-scattering spectroscopy **A10:** 603-609
neutron activation analysis **A10:** 233-242
neutron diffraction **A10:** 420-426
optical emission spectroscopy **A10:** 21-30
particle-induced x-ray emission **A10:** 102-108

Ceramics, characterization of (continued)
potentiometric membrane electrodes **A10:** 181-187
radial distribution function analysis **A10:** 393-401
Raman spectroscopy **A10:** 126-138
Rutherford backscattering spectrometry **A10:** 628-636
scanning electron microscopy **A10:** 490-515
secondary ion mass spectroscopy **A10:** 610-627
single-crystal x-ray diffraction **A10:** 344-356
small-angle x-ray and neutron scattering **A10:** 402-406
spark source mass spectrometry **A10:** 141-150
ultraviolet/visible absorption spectroscopy **A10:** 60-71
voltammetry **A10:** 188-196
x-ray diffraction **A10:** 325-332
x-ray diffraction residual stress techniques **A10:** 380-392
x-ray photoelectron spectroscopy **A10:** 568-580
x-ray powder diffraction **A10:** 333-343
x-ray spectrometry **A10:** 82-101
x-ray topography **A10:** 365-379

Ceramics Correspondence Institute **EM4:** 40

Ceramics, friction and wear of **A18:** 158, 693, 812-815
abrasion resistance **A18:** 490
abrasive wear materials **A18:** 186
applications **A18:** 812
in future jet engine components **A18:** 592
internal combustion engine parts **A18:** 554, 561
rolling-element bearings **A18:** 261
cermets **A18:** 812
applications **A18:** 812
definition **A18:** 812
coating finishing **A18:** 831
coating thickness limitations **A18:** 831
coating to improve abrasive wear resistance **A18:** 639
compatibility with steel **A18:** 743
cutting speed and work material relationship **A18:** 616
for cutting tool insert materials **A18:** 616-617
damage dominated by brittle fracture **A18:** 180
elements implanted to improve wear and friction properties **A18:** 858
erosion of **A18:** 199, 204-206
particle hardness **A18:** 205-206
for gas-lubricated bearings **A18:** 532
heat-treatable ceramics **A18:** 814-815
material parameters that should be documented to ensure repeatability when testing tribosystems **A18:** 55
mechanical and physical properties **A18:** 812-814
microfracture **A18:** 186
particle size effect of erosion rate **A18:** 200
phase contrast imaging **A18:** 389
plasma-spray coating for pistons **A18:** 556, 561
product lines **A18:** 812
relative erosion factors **A18:** 201
rolling contact bearings **A18:** 815
rolling contact fatigue **A18:** 260-261
rolling contact wear **A18:** 260
roughness measurement and surface texture **A18:** 340-341
scanning acoustic microscopy for wear studies **A18:** 409
scanning acoustic microscopy to study machining damage **A18:** 409, 410
sliding and adhesive wear **A18:** 237, 240-241
spray material for oxyfuel powder spray method **A18:** 830
spray material for plasma arc powder spraying **A18:** 830
thermal spray coating recommended **A18:** 832
tool steel coatings **A18:** 643-644
wear properties **A18:** 814
abrasive wear **A18:** 814
erosive wear **A18:** 814

Ceramics joined to glasses
surface considerations **EM3:** 298-310

Ceramics, specific types
Al_2O_{32} + 3 glass, true profile length **A12:** 200
Al_2O_{32}-glass, area/length parametric relation **A12:** 204
alpha-SiC, corrosion pitting fracture **A12:** 471

Ceramics technology readiness development (CTRD) program **EM4:** 716
Ceramics, traditional *See* Traditional ceramics
Ceramography
for microstructural analysis **EM4:** 25-26
CeraPerl, applications
dental **EM4:** 1095
Ceravital **EM4:** 1008, 1010
bonding to bone **EM4:** 1010, 1011
for middle ear surgery **EM4:** 1011
Cerclage stainless steel wire
intercrystalline corrosion on **A11:** 676, 681
Cercor
composition **EM4:** 871
properties **EM4:** 871
CERDIP *See* Ceramic- dual-in-line package
Cereals
in molding sand mixes **A15:** 211
Ceresin, as wax
investment casting **A15:** 254
Cerestore **EM4:** 1095, 1096
Ceria
thermionic emission production **A6:** 30
Cerium *See also* Rare earth metals
in cast iron **A1:** 5
Ce_2O_3 ..... **A16:** 100
CeO_2 ..... **A16:** 100
classification in tungsten alloy electrodes for GTAW **A6:** 191
in compacted graphite iron **A1:** 56
in ductile iron **A15:** 648
effect of, on steel composition and formability **A1:** 577
effect on ductile iron welds **M6:** 604
in ferrite **A1:** 408
gray cast iron content **A16:** 654
in malleable iron **A1:** 10
in nodular graphite composition **A18:** 699
oxide, abrasive for lapping **A16:** 493
pure **M2:** 722-723
as pyrophoric **A7:** 199, 597
as rare earth metal, properties **A2:** 720, 1178
redox titration **A10:** 175
SiCeON **A16:** 100
as silicon modifier **A15:** 161
TNAA detection limits **A10:** 238
volumetric procedures for **A10:** 175
Cerium (Ce)
in heat-resistant alloys **A4:** 512
vapor pressure, relation to temperature **A4:** 495
Cerium in steel **M1:** 115, 556
Cerium, nodulizing agent
ductile iron **M1:** 6, 37
Cerium oxide (CeO_2)
as additive for pressure densification **EM4:** 298-299
for coarse and fine polishing before microstructural analysis **EM4:** 573
component in photosensitive glass composition **EM4:** 440
crystal structure **EM4:** 30
properties **EM4:** 30
purpose for use in glass manufacture **EM4:** 381
sol-gel processing **EM4:** 447
specific properties, imparted in CTV tubes **EM4:** 1039
Cerium, vapor pressure
relation to temperature **M4:** 310
Cerium-doped glasses, applications
solar cell covers **EM4:** 1019
Cerium-sulfur ratio, HSLA steels
effect on toughness **M1:** 418
Cermet
defined **EM1:** 6
Cermet billets *See also* Cermets
hot extrusion of **A2:** 987-988
Cermet ceramic coatings **M5:** 536-537
Cermet films
PVD applications **EM4:** 219
Cermet forming techniques, specific types
cold hydrostatic pressing **A7:** 800
extrusion **A7:** 800
hot isostatic pressing **A7:** 800
infiltration **A7:** 800
plasma spraying **A7:** 800
slip casting **A7:** 800
static cold pressing **A7:** 800
static hot pressing **A7:** 800
Cermet paste systems
conductors **EL1:** 339-341
dielectrics and encapsulants **EL1:** 341-343
resistors **EL1:** 343-345
Cermet powder mixtures
warm extrusion of **A2:** 982-983
Cermet(s)
in medical and military applications **EL1:** 386, 388
Cermets *See also* Cemented carbides; Ceramics; Cermet forming techniques, specific types; Metal-matrix composites
aluminum oxide cermets **A2:** 992-993
aluminum-boron carbide cermets **A2:** 1002-1003
aluminum-silicon carbide cermets **A2:** 1002
application **A2:** 978-979
applications **A16:** 95-97, 639 **A18:** 812 **EM4:** 203, 810
applications as a coating **EM4:** 208
beryllium oxide cermets **A2:** 993
bonding **A2:** 990-992
bonding and microstructure **A7:** 801-802
boride cermets **A2:** 1003-1004
bucket, stress-rupture properties of infiltrated **A7:** 562
carbide **A7:** 804-811
carbide and carbonitride cermets **A2:** 995-1003
carbonitride- and nitride-based cermets **A2:** 1004-1005
cast iron machining **A16:** 656
chemical processing **A7:** 55
chromium boride cermets **A2:** 1004
chromium carbide cermets **A2:** 1000-1001
classification **A2:** 979 **A7:** 798-799
cold hydrostatic pressing **A2:** 981-982
compared with cemented carbides **A16:** 93-95
composition and microstructure **A16:** 90-92
compositions **A18:** 806
cutting speed and work material relationship **A18:** 616
for cutting tool materials **A18:** 616-617
defined **A2:** 978 **A7:** 2, 798 **A18:** 812
detonation gun used for molten particle deposition **EM4:** 204, 206
electrical discharge machining **EM4:** 371, 374, 376
fabrication techniques **A2:** 979-990 **A7:** 799-801
forming techniques **A2:** 981
future directions and problems **A2:** 1024
for gas-lubricated bearings **A18:** 532
graphite- and diamond-containing cermets **A2:** 1005
grooving **A16:** 95, 97
hafnium carbide cermets **A2:** 1001
history **A2:** 978
hot extrusion of cermet billets **A2:** 987-988
hot isostatic pressing (HIP) **A2:** 986-987
infiltration process **A2:** 989-990
magnesium oxide cermets **A2:** 993
malleable cast iron machining **A16:** 653, 654
maximum service temperature **EM4:** 203
metal-matrix high-temperature superconductor cermet **A2:** 995
microstructure **A2:** 990-992
milling **A16:** 96, 97
molybdenum boride cermets **A2:** 1004
nickel-bonded titanium carbide cermets **A2:** 995
niobium carbide cermets **A2:** 1001
Cermets (continued)
oxide cermets
P/M injection molding (MIM) process **A2:** 984-985
phase contrast imaging **A18:** 389
plasma spray material **EM4:** 203
plasma-spray coating for pistons **A18:** 556
powder preparation **A2:** 979-980
powder rolling (roll compacting) **A2:** 983-984
powders **A7:** 55
product development and marketing **A2:** 978
properties **EM4:** 203
properties and grade selection **A16:** 92-95
silicide cermets **A2:** 1005
silicon oxide cermets **A2:** 992
sintering **A2:** 985-986
sintering-compacting combination **A2:** 988-989
slip casting **A2:** 984
solid particle erosion **A18:** 207
solubility **A2:** 990-991
spray material for oxyfuel powder spray method **A18:** 830
static cold pressing **A2:** 980-981
steel-bonded titanium carbide cermets **A2:** 996-998
steel-bonded tungsten carbide cermets **A2:** 1000
tantalum carbide cermets **A2:** 1001
thorium oxide cermets **A2:** 993
threading **A16:** 95, 96, 97
titanium boride cermets **A2:** 1004
titanium carbonitride cermets **A2:** 998-1000
tool life **A16:** 92, 95, 97
turbined blade, graded **A7:** 562, 563
turning **A16:** 95-97
uranium carbide cermets **A2:** 1002
uranium oxide cermets **A2:** 993-994
warm extrusion of cermet powder mixtures **A2:** 982-983
wetting **A2:** 991
zirconium boride cermets **A2:** 1003
zirconium carbide cermets **A2:** 1001
Ceroxides
as casting defect **A11:** 387
Cerpacks
and substrates **EL1:** 204
Cerquads
and substrates **EL1:** 204
Certification
of liquid penetrant inspection personnel **A17:** 85-86
operators, and NDE reliability **A17:** 663, 677, 678
titanium and titanium alloy castings **A2:** 645
Cervit
composition **EM4:** 871
properties **EM4:** 871
Cesium **A13:** 92, 94-95, 733
-cadmium-induced cleavage fracture **A11:** 235
cations, in glasses, Raman analysis **A10:** 131
epithermal neutron activation analysis **A10:** 239
explosive reactivity in moisture **A7:** 194
flooding, in secondary ion mass spectroscopy **A7:** 258
liquid, Zircaloy claddings embrittled by **A11:** 230
organic precipitant for **A10:** 169
species weighed in gravimetry **A10:** 172
pure **M2:** 723-724
as pyrophoric **A7:** 199
TNAA detection limits **A10:** 238
use with flame emission sources **A10:** 30
vapor pressure **A6:** 621
Cesium iodide (CsI)
radiographic screens of **A17:** 318
Cesium oxide-silicon dioxide (Cs_2O-SiO_2)
self-diffusion coefficients of alkali ions **EM4:** 461
Cesium, pure
properties **A2:** 1107
CESR
as synchrotron radiation source **A10:** 413

CG iron *See* Compacted graphite cast iron; Compacted graphite irons
CG Nicalon **EM4:** 224
composition **EM4:** 225
mechanical properties, room-temperature **EM4:** 225
room-temperature properties **EM4:** 226
CHA *See* Concentric hemispherical analyzer
Chafing *See* Fretting
defined **A18:** 5
Chafing fatigue *See also* Fretting
defined **A8:** 2 **A11:** 2
Chain belt compaction
sheet molding compound machine **EM1:** 160
Chain block experimental plan **A8:** 646-647
Chain ditchers
of cemented carbides **A2:** 974
Chain entanglements
chemistry of **EM2:** 64
Chain extenders
as thermoplastic polyurethanes (TPUR) **EM2:** 204, 257
Chain fittings and hooks
materials for **A11:** 515
Chain intermittent weld
definition **A6:** 1207
Chain intermittent welds
definition **M6:** 3
Chain length **EM3:** 7
defined **EM1:** 6 **EM2:** 9
Chain link *See also* Chains
cast conveyor, fabrication weld fracture **A11:** 400
failures of **A11:** 521-522
steel, brittle fracture in **A11:** 393-395
steel, fatigue failure **A11:** 398
steel, fracture surface **A11:** 409
Chain link fence
wire **M1:** 269, 271
Chain link fence wire **A1:** 285
Chain links, welded
magnetic particle inspection **A17:** 116
Chain quality rod **A1:** 273
Chain scission process **EM3:** 654, 655, 678
Chain transfer agent **EM3:** 7
defined **EM2:** 9
Chain-quality carbon steel wire rod **M1:** 254
Chains *See also* Chain link
failures of **A11:** 521-522
materials for **A11:** 515
sling, steel hook failure on **A11:** 524
weld defect fracture **A11:** 521-522
Chains, welded
alloy steel wire for **M1:** 270
Chair, office
failure of roller on **A11:** 763-764
Chalcide glass, applications
optical glass products **EM4:** 1074-1075
Chalcogenide glasses
chemical properties **EM4:** 855-856
properties **EM4:** 846, 847, 849, 850, 851
structural role of components **EM4:** 845
structures **EM4:** 846
Chalcogenides **EM4:** 22
applications **EM4:** 22
production processes **EM4:** 22
two-state (switching) behavior **EM4:** 22
Chalcolithic period
metalworking in **A15:** 15
Chalking **EM2:** 9, 494 **EM3:** 7
defined **A13:** 3 **EM1:** 6
Chalkley method
for determining the surface- to-volume ratio of discrete particles **A9:** 125
Chalmers method, of growing single crystals **A9:** 607
Charge density, effect on potentiostatic etching **A9:** 146
Charging effects in scanning electron microscopy specimens, prevention **A9:** 97-98
Chamber
for fatigue crack growth testing in high- pressure or pressurized-water **A8:** 427, 429
liquid metal environmental **A8:** 427
Chamber furnace
defined **A7:** 2
Chambers
arc welding of titanium and titanium alloys **M6:** 449-450
inert atmosphere for arc welding **M6:** 464
Chamberscopes, rigid *See also* Borescopes **A17:** 5
Chamfer
defined **A14:** 2
definition **A6:** 1207
Chamfer angles
and die threading **A16:** 300
thread rolling **A16:** 292
Chamfering **A16:** 33
in conjunction with boring **A16:** 168, 169
in conjunction with drilling **A16:** 215, 216-217, 221, 222, 235
in conjunction with turning **A16:** 135, 138, 157, 158
multifunction machining **A16:** 375
Chammotte **EM4:** 45
applications **EM4:** 46
supply sources **EM4:** 46
Chamotte
in composite ceramic molds **A15:** 248-249
CHAMPION 3D computer program for structural analysis **EM1:** 268, 270
Change in quantity
symbol for **A10:** 692
Change of state
as x-ray diffraction analysis **A10:** 325
Channel
number of lines **EL1:** 19
number per centimeter **EL1:** 19-20
routers, as automatic trace routing **EL1:** 532
stop, ion-implantation for **EL1:** 197-198
Channel dies
for press-brake forming **A14:** 536
Channel fracture
austenitic stainless steels **A12:** 365
Channel induction
furnace **A15:** 368, 636
heating, of pouring vessels **A15:** 499
Channel injection process
for demagging aluminum alloys **A15:** 473-474
Channel plate multiplier
for x-ray photoelectron spectroscopy **A10:** 571
Channel plots
acoustic emission inspection **A17:** 283-284
Channel segregation **A15:** 140-141, 156
Channel widening **EM4:** 372
Channeling *See also* Electron channeling
angular scan, lattice strain measurement by **A10:** 635
and blocking, surface structure study by **A10:** 633
in conjunction with milling **A16:** 308
contrast, source of **A10:** 504
and dechanneling **A10:** 634
defined **A18:** 5
effect, RBS analysis **A10:** 630-631
electron, capabilities **A10:** 365
electron, patterns and contrast **A10:** 504-506
ion scattering to study surface structure by **A10:** 633
for lattice location of solute atoms **A10:** 633
patterns **A10:** 504-506, 670
selected-area patterns **A10:** 505-506
spectrum **A10:** 632-633
Channeling angular scan
lattice strain measurement by **A10:** 635
Channeling patterns **A10:** 504-506, 670
Channels
wrought aluminum alloy **A2:** 34
Chaplet **A15:** 2, 17
Chaplet, unfused
as casting defect **A11:** 383
Chaplets, unfused
radiographic appearance **A17:** 349
Characteristic
defined **A8:** 2
Characteristic curves
x-ray film **A17:** 324
Characteristic electron energy loss phenomena
defined **A10:** 670
Characteristic function
in optical holographic interferometry **A17:** 415
Characteristic impedance
flexible printed boards **EL1:** 588
Characteristic K_α peaks
and bremsstrahlung radiation **A10:** 571
Characteristic radiation
defined **A10:** 670
spectra **A10:** 326
Characteristic spectrum **EM4:** 558
Characteristic x-ray analysis **A13:** 1117
as advanced failure analysis technique **EL1:** 1106
Characteristic x-rays
defined **A12:** 168
Characterization **EM4:** 24-28, 547-548
ceramic powder characterization **EM4:** 26-27
density **EM4:** 27
particle size **EM4:** 26
porosity **EM4:** 27
rheometry **EM4:** 27
surface area **EM4:** 27
surface properties **EM4:** 26-27
surface roughness **EM4:** 27
chemical analysis **EM4:** 24-25
additives and their effects **EM4:** 24
bulk chemistry analysis methods **EM4:** 24
bulk chemistry analyzed **EM4:** 24
microchemical analysis **EM4:** 25
surface or interfaces **EM4:** 25
failure analysis **EM4:** 28
microstructural analysis **EM4:** 25-26
ceramography **EM4:** 25
image analysis **EM4:** 26
optical microscopy **EM4:** 26
scanning electron microscopy **EM4:** 26
transmission electron microscopy **EM4:** 26
phase analysis **EM4:** 25
crystal diffraction **EM4:** 25
electron diffraction **EM4:** 25
error-causing factors **EM4:** 25
neutron diffraction **EM4:** 25
x-ray diffraction **EM4:** 25
x-ray powder diffraction **EM4:** 25
properties of glasses **EM4:** 25
spectrochemical absorption methods **EM4:** 25
sources of potential frustration **EM4:** 24
structure of glasses **EM4:** 25
techniques, acronyms for **A10:** 689
testing **EM4:** 27-28
hardness and wear **EM4:** 27
nondestructive evaluation **EM4:** 27
proof testing **EM4:** 27
strength **EM4:** 27
thermophysical properties **EM4:** 27-28
toughness **EM4:** 27
of thin films **A10:** 559-561
Characterization of ceramic powders *See* Ceramic powders, characterization
Characterization of ceramics *See* Ceramics, characterization of
Characterization of corrosion products *See* Corrosion products, characterization of
Characterization of gases *See* Gases, characterization of
Characterization of geologic samples *See* Geologic samples, characterization of
Characterization of glasses *See* Glasses, characterization of
Characterization of inorganic materials *See* Inorganic materials, characterization of
Characterization of liquids *See* Liquids, characterization of
Characterization of metals and alloys *See* Metals and alloys, characterization of
Characterization of minerals *See* Minerals, characterization of
Characterization of organic materials *See* Organic materials, characterization of
Characterization of solids *See* Solids, characterization of
Characterization of surfaces *See* Surface; Surface analysis and characterization
Characterization of surfaces by acoustic imaging techniques **A18:** 406-412
high-frequency acoustic imaging (HAIM) **A18:** 406, 409-410, 411
applications **A18:** 410

Characterization of surfaces by acoustic imaging techniques (continued)
principles ... **A18:** 409-410
scanning acoustic microscopy (SAM) ... **A18:** 406-409
applications ... **A18:** 408-409
evaluation of surface conditions caused by machining ... **A18:** 408-409
thin-film thickness measurements ... **A18:** 408
scanning laser acoustic microscopy (SLAM) ... **A18:** 406, 410-412
principles ... **A18:** 410-411
reconstruction of images by holography ... **A18:** 411-412
summary ... **A18:** 412
Charcoal
as fuel ... **A15:** 15, 26
Charcoal, activated
Raman analysis ... **A10:** 132
Charcoal-base atmospheres
composition ... **M4:** 394, 411-412
Charge *See also* Static charge
acid steelmaking ... **A15:** 364
basic steelmaking ... **A15:** 366-367
calculations, cupolas ... **A15:** 388
crucible furnace ... **A15:** 383
defined ... **A7:** 2 **A15:** 112 **EM2:** 9
equipment for hot pressing ... **A7:** 502
material, dry, induction furnaces ... **A15:** 374
materials, cupolas ... **A15:** 387-389
metal ... **A15:** 388
and physical characteristics of precious metal powders ... **A7:** 149
polarity, of copier powders ... **A7:** 583-584
tap-and-, induction furnaces ... **A15:** 374
techniques, reverberatory furnaces ... **A15:** 379
water dissipators, effect in blending and premixing ... **A7:** 188
Charge bucket
electric arc furnace ... **A15:** 361
Charge, electrical
defined ... **EL1:** 89-92
Charge injection
silicon oxide interface failures ... **A11:** 782
Charge transfer ... **A13:** 30, 32
schematic of ... **EL1:** 1076
Charge trapping
in silicon oxide failures ... **A11:** 780-781
Charge-coupled device
abbreviation for ... **A11:** 796
Charge-coupled device (CCD)
as image sensor, borescopes ... **A17:** 4
in machine vision ... **A17:** 32
optical sensors with ... **A17:** 10
videoscopes with ... **A17:** 54
Charge-up time
capacitive load ... **EL1:** 26-27
for receiver circuits ... **EL1:** 30-34
Charged device model (CDM)
defined ... **EL1:** 966
Charged particle beam
in x-ray spectrometry ... **A10:** 82
Charged particle detectors
for x-ray diffraction ... **A10:** 245-246
Charges
for explosive forming ... **A14:** 636, 641, 642
Charging
in chemical testing ... **EM1:** 285
Charpy C-notch impact
CPM alloys ... **A16:** 65
Charpy C-notch impact energy
CPM Rex 20 compared to M ... **A16:** 64
Charpy impact test *See also* Impact strength; Impact test ... **EM3:** 7
defined ... **EM1:** 6 **EM2:** 9
Charpy test *See also* Charpy V-notch impact test; Impact properties, Notch toughness; Izod test ... **M1:** 689-691
defined ... **A8:** 2 **A11:** 2
Charpy test (continued)
for notch toughness ... **A11:** 57-60
Charpy three-point bend specimen ... **A8:** 262
Charpy V-notch
abbreviation ... **A8:** 724
cast copper alloys ... **A2:** 357-391
Charpy V-notch impact
and bend fracture strengths, CPM alloys ... **A7:** 789
P/M and ingot metallurgy tool steels ... **A7:** 471, 472
P/M forged low-alloy steel powders ... **A7:** 470
response in fully dense iron powder ... **A7:** 415
Charpy V-notch impact energy test
low-alloy steels ... **A15:** 717
plain carbon steels ... **A15:** 702
Charpy V-notch impact test *See also* Impact energy; Impact test; Izod test ... **A6:** 101, 103, 104, 374, 376-377, 384
abbreviation for ... **A11:** 796
and alternative dynamic bend tests ... **A8:** 259
change in properties ... **A8:** 262
and concept of impact response curves ... **A8:** 259
of ductile and brittle fractures ... **A11:** 84-85
for dynamic fracture testing ... **A8:** 259, 261-268
and dynamic notched round bar testing compared ... **A8:** 276
effect of neutron irradiation ... **A11:** 69
effect of specimen orientation on ... **A11:** 68
electrodes for SMAW of HSLA steels ... **A6:** 663
ferritic stainless steels ... **A6:** 452, 453, 454
fracture-transition data, steel ... **A11:** 67
fractures ... **A12:** 106, 108-110, 341
for high strain rate fracture toughness testing ... **A8:** 187
instrumented ... **A8:** 264-267
and keyhole Charpy test, compared ... **A11:** 57-58
precracked ... **A8:** 259, 267-268
shear dimples in shear-lip zone of fracture ... **A11:** 76
solid-state transformations in weldments ... **A6:** 78
specimen, Type A ... **A8:** 263
standard ... **A8:** 262-264
and tests using inertial loading ... **A8:** 259
as toughness control in bridge steels ... **A8:** 265, 453
and toughness test, compared ... **A11:** 55
transition temperatures from ... **A11:** 67
wrought martensitic stainless steels ... **A6:** 440
Charpy V-notch specimen
for hydrogen embrittlement testing ... **A8:** 539-540
modified, for testing of bolts ... **A8:** 542
time-to-fracture measurement ... **A8:** 270
Charpy V-notch test ... **A1:** 610-611, 737, 753
correlation of, to fracture mechanics ... **A1:** 753
hydrogen embrittlement ... **A13:** 287
for steel castings ... **A1:** 367
variability of results ... **A1:** 749-753
Charred carbon
ESR studied ... **A10:** 263
Charring
defined ... **EM1:** 6 **EM2:** 9
Chart recorders, strip
eddy current inspection ... **A17:** 179
Chase
defined ... **EM2:** 9
Chasers *See* Thread chasers
Chatter
defined ... **A18:** 5
shear cracks from ... **A11:** 464
Chatter sleek
definition ... **EM4:** 632
Check
defined ... **A14:** 2 **A15:** 2
Check marks *See* Checks
Check-valve poppet
brittle fracture and redesign of ... **A11:** 70-71
Checker board test procedures ... **EL1:** 375
Checking *See* Craze cracking
defined ... **A13:** 3
in precision forging ... **A14:** 162
resistance, of die materials ... **A14:** 47
Checking fixtures
for part shape measurement ... **A8:** 549
Checking wear
solid graphite molds ... **A15:** 285
Checks
defined ... **A13:** 3
in torsional testing equipment ... **A8:** 146
Chelants
as chemical cleaning solution ... **A13:** 1141
Chelatable lead
toxicity of ... **A2:** 1246
Chelate *See also* Complexation ... **EM3:** 8
defined ... **A10:** 670 **A13:** 3 **EM2:** 9
used in formulating anaerobics ... **EM3:** 114
Chelate fertilizers
powder used ... **A7:** 572
Chelating agent
defined ... **A13:** 3
Chelation *See also* Chelators
of aluminum ... **A2:** 1256
defined ... **A2:** 1235 **A13:** 3
and metal toxicities ... **A2:** 1235-1237
therapy, for bismuth toxicity ... **A2:** 1257
Chelators *See also* Chelation
BAL (British Anti Lewisite) ... **A2:** 1235-1236
calcium EDTA ... **A2:** 1236
desferrioxamine ... **A2:** 1236
dithiocarb ... **A2:** 1236-1237
DMPS (2,3-dimercapto-1-propanesulfonic acid) ... **A2:** 1236
penicillamine ... **A2:** 1236
Chelometric titration ... **A10:** 164, 173, 174
Chelons
formation ... **A10:** 164
Chem milling *See* Chemical milling
Chemcor process ... **EM4:** 1059
Chemi-thermomechanical pulping equipment ... **A13:** 1217-1218
Chemical activators
use in spray drying ... **A7:** 75
Chemical additives
in activated sintering ... **A7:** 319
effects on explosivity ... **A7:** 194, 196
Chemical analysis *See also* Chemical susceptibility; Evaluation; Inspection; Testing ... **EM4:** 549-555
advanced failure analysis techniques ... **EL1:** 1103-1106
analytical process steps ... **EM4:** 549-550
evaluation of data and report preparation ... **EM4:** 549
measurement (including calibration graph) ... **EM4:** 549, 550
problem identification and method selection ... **EM4:** 549
sample preparation according to selected method ... **EM4:** 549-550
sampling ... **EM4:** 549
analytical technique selection ... **EM4:** 554-555
bath ... **EL1:** 680
bulk ... **A7:** 246-249
of bulk materials ... **A11:** 30
cast aluminum alloys ... **A12:** 405, 407
certified reference materials sources ... **EM4:** 551
for classifying steels ... **A1:** 141
continuous, and corrosion rates ... **A11:** 199
electron spectroscopy for ... **A7:** 251, 255-257
electron spectroscopy for (ESCA) ... **A11:** 35
of failed locomotive axles ... **A11:** 724
as failure analysis ... **A11:** 29-31
heat exchanger failed parts ... **A11:** 629
hydrogen loss testing ... **A7:** 246-247
of investment castings ... **A15:** 264

Chemical analysis (continued)
major analytic techniques **EM4:** 550-554
atomic absorption spectrophotometry **EM4:** 552-553, 555
direct current plasma-emission spectrometry **EM4:** 553
energy dispersion spectrometry....... **EM4:** 550-551
inductively coupled radio frequency **EM4:** 553, 554, 555
plasma-emission spectrophotometry...... **EM4:** 553, 554
ultraviolet/visible spectrophotometry **EM4:** 553-554
wavelength dispersion XRFS **EM4:** 550-551
x-ray fluorescence spectrometry **EM4:** 550-552, 553, 554, 555
metallographic examination **A7:** 247
for microbiological corrosion **A13:** 314
for pitting.. **A11:** 177
precision and accuracy................................ **EM4:** 555
purposes... **EM4:** 549
recommended practices................................. **A7:** 249
relation to designations **M1:** 120-124
of rocket-motor case fracture **A11:** 96
and sampling, ASTM standards............. **A7:** 248-249
sampling method **A7:** 246
for SCC **A11:** 213
of shafts.. **A11:** 460
of surfaces and deposits................................. **A11:** 30
of test spots **A11:** 30-31
thermoplastic resins **EM2:** 533-543
trace elements in pure metals **M2:** 711
types of materials.. **EM4:** 549
of worn parts **A11:** 158
Chemical analysis, and microstructure
of precipitates... **A17:** 55
Chemical analysis of microstructural elements
by scanning electron microscopy **A9:** 90-93
Chemical analysis of thermoset resins
chromatography **EM2:** 517-522
composition characterization **EM2:** 517-522
gel permeation chromatography (GPC)............ **EM2:** 518-519
high-performance liquid chromatography (HPLC) **EM2:** 517-518
infrared (IR) spectroscopy **EM2:** 521-522
liquid-solid chromatography (LSC) **EM2:** 519-520
processing characterization **EM2:** 522-528
thin-layer chromatography (TLC) **EM2:** 520-521
Chemical analysis techniques
advanced failure analysis **EL1:** 1103
atomic absorption spectroscopy (AAS)..... **EL1:** 1104
emission spectroscopy **EL1:** 1104
gas and liquid chromatography **EL1:** 1104-1105
infrared, visible ultraviolet spectroscopies............................ **EL1:** 1103-1104
mass spectroscopy............................... **EL1:** 1105-1106
Raman spectroscopy **EL1:** 1104
wet chemistry and microelemental analysis **EL1:** 1106
x-ray diffraction.. **EL1:** 1104
Chemical and Engineering News **EM2:** 95
Chemical applications
of copper-based powder metals **A7:** 733
of metal powders, powders used.................. **A7:** 572
for stainless steels... **A7:** 731
Chemical binding processes *See also* Binding; Binding systems
hardening and compaction of....................... **A15:** 203
Chemical blanking *See* Photochemical machining **A14:** 458
refractory metals and alloys **A2:** 561
Chemical blowing agent *See also* Blowing agents
as additives... **EM2:** 9, 503
Chemical bonding
defined **A10:** 670
effects in EXAFS **A10:** 415
types **EL1:** 92-93
Chemical brightening
agitation used in.. **M5:** 580
aluminum and aluminum alloys **M5:** 579-582, 591, 596, 607-608, 610
process selection, factors affecting...... **M5:** 581-582
buffing compared to **M5:** 581-582
copper used in **M5:** 579
dragout, effects of **M5:** 579-580
electrolytic brightening vs **M5:** 581
Chemical brightening (continued)
phosphoric and phosphoric- sulfuric acid baths **M5:** 580
phosphoric-nitric acid baths................... **M5:** 579-580
process selection **M5:** 581-582
surfactants used in **M5:** 579-580
Chemical cleaning *See also* Cleaning; specific processes by name **A13:** 1139-1143
for coatings **A15:** 561
methods **A13:** 1139
methods, liquid penetrant inspection........... **A17:** 81
procedures **A13:** 1141-1143
of process equipment **A13:** 1137-1143
for solderability **EL1:** 678
solutions **A13:** 1140-1141
Chemical coatings
for corrosion control **A11:** 195
Chemical compatibility
of plastics **EM2:** 1
Chemical complexes
ESR studied **A10:** 263
Chemical composition *See* Composition
aluminum casting alloys................. **A2:** 148, 152-177
cast copper alloys **A2:** 356-391
cast magnesium alloys **A2:** 491-516
of dielectric materials, microwave inspection **A17:** 215
effect on corrosion-fatigue **A11:** 256
electrical resistance alloys......................... **A2:** 835-839
heat tinting to identify **A9:** 136
of iron castings **A11:** 363
and magnetic characterization **A17:** 131
magnetic effects **A17:** 132
as NDE area **A17:** 49
pewter.. **A2:** 522
pure metals.. **A2:** 1100-1178
as surface parameter **EM3:** 41
wrought aluminum and aluminum alloys **A2:** 62-122
wrought copper and copper alloys......... **A2:** 265-345
wrought magnesium alloys..................... **A2:** 480-491
of zinc alloys **A2:** 532-542
Chemical composition, effect of
on weldability **A1:** 606, 609
Chemical contouring *See* Chemical milling
Chemical control
of inclusion-forming reactions **A15:** 90
of reaction rate **A15:** 83
Chemical conversion coating *See also* Coatings; Conversion coatings **EM3:** 42
aluminum and aluminum alloys........... **M5:** 597-600, 606-610
chromate *See* Chromate conversion coating
for corrosion control **A11:** 195
defined **A13:** 3 **A18:** 5
magnesium alloys **M5:** 629, 632-638, 640-641, 643-644, 647
oxide *See* Oxide conversion coating processes
phosphate *See* Phosphate coating process
procedure and equipment **M5:** 598-600
solution compositions and operating conditions......................... **M5:** 598-600, 656-657
standards.. **M5:** 606-607
surface preparation **M5:** 598
thickness.. **M5:** 657
titanium and titanium alloys **M5:** 656-658
wear testing **M5:** 656-658
Chemical conversion coatings
atmospheric corrosion resistance enhanced by... **M1:** 722
cast irons **A15:** 563
Chemical corrosion *See also* Chemical processing industry **M5:** 430-431
aircraft powerplants...................................... **A13:** 1045
of aluminum/aluminum alloys **A13:** 602
of lead/lead alloys **A13:** 780-781
polymer susceptibilities to.................. **EM2:** 573-574
prevention... **A11:** 19
of soldered joints **A17:** 609
of stainless steel **A13:** 556
Chemical drop test
cadmium plate thickness measuring **M5:** 267
Chemical effects on adhesive joints..... **EM3:** 637-643
chemical agents often encountered..... **EM3:** 637-639
chemical resistance of adhesives by chemical type **EM3:** 639
Chemical effects on adhesive joints (continued)
chemical resistance of common adherends.. **EM3:** 639
chemical resistance test methods **EM3:** 642-643
composite surface contamination and pretreatment on joints effect.............. **EM3:** 638
elastomeric sealants **EM3:** 641-642
methods of bonded joint protection ... **EM3:** 639-642
moisture absorbed into the bond line **EM3:** 638
paint systems **EM3:** 640-641
water-displacing corrosion inhibitors......... **EM3:** 641
Chemical environments
effect on corrosion-fatigue **A11:** 255
Chemical equilibrium............................ **A15:** 50-52, 56
defined **A10:** 163
Chemical equipment
cast iron, coatings for **M1:** 104-106
organic coatings for **M1:** 755
Chemical etches, for surface preparation
SCC testing **A8:** 510
Chemical etching *See also* Color etching; Electrochemical etching; Etching; Macroetching **A9:** 61 **A10:** 144-145
aluminum and aluminum alloys.......... **M5:** 582-586, 609
before heat tinting **A9:** 136
of casting surfaces, as inspection **A17:** 512
copper and copper alloys **A9:** 400-401
of electrical contact materials.................. **A9:** 550-551
as fracture cleaning technique **A12:** 75-76
for liquid penetrant inspection **A17:** 81
magnesium alloys **M5:** 638-639, 645
post treatments **M5:** 583
precleaning **M5:** 583
selected examples of ceramics **EM4:** 575
titanium alloys before ultrasonic welding **A6:** 894
to remove surface oxide layer for solid-state welding **A6:** 165
wet **EL1:** 327
Chemical exposure
as environmental factor........................... **EM2:** 70-71
Chemical extraction
of wrought stainless steel second phases **A9:** 283
Chemical finishing *See also* Finishes; specific processes by name
of aluminum alloys **A15:** 762
zinc alloys **A2:** 530
Chemical flux cutting **A14:** 728-729
definition... **M6:** 3
Chemical flux cutting (FOC)
definition.. **A6:** 1207
Chemical industry applications *See also* Process industry applications
bismuth **A2:** 1256
lead pipe and traps **A2:** 552
liquid crystal polymers (LCP)...................... **EM2:** 180
nickel alloys **A2:** 430
polybenzimidazoles (PBI) **EM2:** 147
polyphenylene sulfides (PPS) **EM2:** 186
polysulfones (PSU)... **EM2:** 200
of precious metals **A2:** 693
titanium and titanium alloy castings........... **A2:** 634, 644-645
of titanium and titanium alloys...................... **A2:** 588
titanium P/M products........................... **A2:** 655-656
vinyl esters **EM2:** 272
Chemical inertness
of structural ceramics **A2:** 1019
Chemical ion plating **EM4:** 218
Chemical kinetics
in composite curing **EM1:** 748-751
interphase mass transport **A15:** 52-53
nucleation **A15:** 52-53
Chemical layer theory.................................... **EM3:** 300
Chemical lead *See also* Lead
composition **A2:** 543-545
corrosion resistance.. **A2:** 551
Chemical lead sheet
micrograph of **A9:** 418
Chemical machining **EM4:** 313
Chemical machining (CM) *See also* Chemical milling **A16:** 34
4340 steel surface characteristics **A16:** 33
and fatigue strength... **A16:** 25
and postprocessing **A16:** 35

Chemical machining (CM) (continued)
surface alterations produced ... **A16:** 24, 25
surface integrity effects in material removal processes ... **A16:** 28
Chemical Manufacturers Association
performance testing of engine oils ... **A18:** 170
Chemical microanalysis
determination of nature of surface films ... **A18:** 369
FIM/AP ... **A10:** 583
Chemical milling ... **EM3:** 772
aluminum alloy ... **A15:** 763
aluminum and aluminum alloys ... **A2:** 8
of investment castings ... **A15:** 264
refractory metals and alloys ... **A2:** 562
titanium alloys ... **A15:** 832
titanium and titanium alloy castings ... **A2:** 401
Chemical milling (CHM) *See also* Chemical machining ... **A16:** 509, 579-586
advantages and disadvantages ... **A16:** 586
Al alloys ... **A16:** 802, 803-804
applications ... **A16:** 579, 584-586
Be alloys ... **A16:** 871, 872-873
chem-mill quality ... **A16:** 582
chemical leaching ... **A16:** 586
computer-guided lasers used ... **A16:** 580
controls ... **A16:** 581
design features ... **A16:** 582
equipment ... **A16:** 579-581
etchants ... **A16:** 579, 581-586
etching ... **A16:** 580-581
historical development ... **A16:** 579
hydrogen absorption ... **A16:** 584
masking ... **A16:** 579-580, 581, 586
MCAIR process ... **A16:** 584
mechanical properties of machined parts ... **A16:** 583-584
military specifications ... **A16:** 581
minimum residual stress (MRS) ... **A16:** 582
MMCs ... **A16:** 896-897
Ni- and Co-base alloys ... **A16:** 802
precleaning ... **A16:** 579
process characteristics ... **A16:** 579-581
process defects ... **A16:** 582, 583
refractory metals ... **A16:** 859, 868
scribing ... **A16:** 580
steels ... **A16:** 802
surface finish ... **A16:** 579, 581-583, 584, 585, 586
Ti alloys ... **A16:** 802, 846, 852, 853
triethanolamine (TEA) addition to Al etchant ... **A16:** 583, 584, 585
Chemical modification
aluminum-silicon alloys ... **A15:** 162
of core coatings ... **A15:** 240
Chemical names
of polymers ... **EM2:** 53
Chemical pitting
nickel alloys ... **A2:** 432
Chemical plating *See* Electroless plating
and drilling operations ... **A16:** 219
Chemical pneumonitis (acute pulmonary disease)
from beryllium exposure ... **A2:** 1239
Chemical polishing *See also* Bright dipping; Chemical brightening
aluminum alloys ... **A9:** 353
aluminum and aluminum alloys ... **M5:** 579-582, 596
copper and copper alloys ... **M5:** 623-624
defined ... **A9:** 3
of magnesium alloys ... **A9:** 426
process selection factors ... **M5:** 624
stainless steel ... **M5:** 559
of transmission electron microscopy specimens ... **A9:** 105
Chemical polishing solutions
for aluminum alloys ... **A9:** 353
for electrical steels ... **A9:** 533
Chemical potential
defined ... **A13:** 3
Chemical preparations
radioanalysis ... **A10:** 247
Chemical process fluids
lead/lead alloys in ... **A13:** 789-790
Chemical processing industry ... **A13:** 1134-1185
alloy steel corrosion in ... **A13:** 544-545
ammonia corrosion ... **A13:** 1180-1182
chemical cleaning, process equipment ... **A13:** 1137-1143
chlorine corrosion ... **A13:** 1170-1180
cobalt-base alloy applications ... **A13:** 667
corrosion under thermal insulation ... **A13:** 1144-1147
failure causes in ... **A13:** 338
hydrogen chloride/hydrochloric acid corrosion ... **A13:** 1160-1166
hydrogen fluoride/hydrofluoric acid corrosion ... **A13:** 1166-1170
nickel-base alloys ... **A13:** 653
niobium applications ... **A13:** 723
nitric acid corrosion ... **A13:** 1154-1156
organic acid corrosion ... **A13:** 1157-1160
plants, waste systems ... **A13:** 1369-1370
process/environmental variables ... **A13:** 1134-1137
protective linings for ... **A13:** 455
sulfuric acid corrosion ... **A13:** 1148-1154
zirconium/zirconium alloy applications ... **A13:** 719-720
Chemical processing industry applications
cobalt-base wear-resistant alloys ... **A2:** 451
Chemical processing quality
checked by wedge-crack extension test ... **EM3:** 738
Chemical properties *See also* Chemical resistance; Chemical structure; Chemical susceptibility; Chemistry; Corrosion
actinide metals ... **A2:** 1189-1198
amorphous materials and metallic glasses ... **A2:** 817-818
commercially pure tin ... **A2:** 518-519
copper-clad E-glass laminates ... **EL1:** 536-537
effect on aluminum metallization ... **EL1:** 965
electrical resistance alloys ... **A2:** 836-839
fluoropolymer coatings ... **EL1:** 782-783
germanium and germanium compounds ... **A2:** 733
lead frame materials ... **EL1:** 489
of make-break arcing contacts ... **A2:** 841
niobium alloys ... **A2:** 567-571
palladium and palladium alloys ... **A2:** 715-718
pewter ... **A2:** 522
platinum and platinum alloys ... **A2:** .846
polyimides ... **EL1:** 324-325
of polymers ... **EM2:** 61-62
pure metals ... **A2:** 1100-1178
of rare earth metals ... **A2:** 725, 1178-1189
silver and silver alloys ... **A2:** 699-704
tantalum alloys ... **A2:** 573-574
titanium and titanium alloy castings ... **A2:** 637
transplutonium actinide metals ... **A2:** 1199
wrought aluminum and aluminum alloys ... **A2:** 62-122
wrought copper and copper alloys ... **A2:** 265-345
of zinc alloys ... **A2:** 532-542
Chemical Propulsion Information Agency (CPIA)
as information source ... **EM1:** 40
Chemical purification *See also* Melt purification
for hydrogen, in aluminum melts ... **A15:** 79-80
Chemical reactions
gelation/curing ... **EL1:** 850-852
products, liquid chromatography analysis of ... **A10:** 656-657
rates of, measured ... **A10:** 60, 243
scanning electron microscopy study of specimens ... **A9:** 97
surfaces (chemisorbed layers), LEED analysis ... **A10:** 536
Chemical reactor filters
powder used ... **A7:** 574
Chemical reagents *See also* Reagents
analytic methods for ... **A10:** 6-10
GFAAS analysis of ... **A10:** 58
inorganic, analytic methods for ... **A10:** 6, 7
organic, analytic methods for ... **A10:** 9, 10
Chemical resistance
ASTM test methods ... **EM2:** 334
of castable resins ... **A9:** 30
and degradation ... **EM2:** 424
of ECR-glass ... **EM1:** 107
of engineering plastics ... **EM2:** 1
epoxies ... **EM2:** 241
of glass fibers ... **EM1:** 46
high-impact polystyrenes (PS, HIPS) ... **EM2:** 197
measured ... **EM2:** 614
measurement ... **EL1:** 536-537
metals vs plastic ... **EM2:** 77
polyamide-imides (PAI) ... **EM2:** 130, 134
polyamides (PA) ... **EM2:** 126-127
polyarylates (PAR) ... **EM2:** 140
polyaryletherketones (PAEK, PEK PEEK, PEKK) ... **EM2:** 142, 144
polybenzimidazoles (PBI) ... **EM2:** 149-150
of polyester resins ... **EM1:** 93-94
polyether sulfones (PES, PESV) ... **EM2:** 160, 161
of polymers ... **EM2:** 62
polyphenylene sulfides (PPS) ... **EM2:** 186
and process selection ... **EM2:** 277
of pultruded composites ... **EM1:** 541-542
of pultrusions ... **EM2:** 396
resin systems ... **EL1:** 534-535
of rigid epoxies ... **EL1:** 810
as selection criterion, electrical contact materials ... **A2:** 840
silicone conformal coatings ... **EL1:** 774
silicones (SI) ... **EM2:** 267
structural foams ... **EM2:** 509
styrene-maleic anhydrides (S/MA) ... **EM2:** 219
of substrates ... **EL1:** 105
tests for ... **EM2:** 425-426
of thermoplastic resins ... **A9:** 30 **EM2:** 618
of thermoplastics ... **EM1:** 293-294
of thermosetting resins ... **A9:** 29 **EM2:** 223
tungsten alloys ... **A2:** 579
ultrahigh molecular weight polyethylenes (UHMWPE) ... **EM2:** 169-170
unsaturated polyesters ... **EM2:** 247
vinyl esters ... **EM2:** 272-273
Chemical segregation
banding from ... **A11:** 315
in cast alloys, and forging failure ... **A11:** 314-315
forging failures from ... **A11:** 323-325
ingot ... **A17:** 491
Chemical shifts
in XPS analysis ... **A10:** 572
Chemical solutions
used with wire sawing ... **A9:** 26
Chemical spinodal ... **A9:** 652
Chemical spot tests ... **A1:** 1030, 1031
substrate ... **A13:** 421
Chemical stability
lead frame materials ... **EL1:** 489
Chemical stoneware
composition ... **EM4:** 5
Chemical stripping ... **EM3:** 42
Chemical structure
heterochain thermoplastic polymers ... **EM2:** 53
hydrocarbon thermoplastic polymers ... **EM2:** 50
nonhydrocarbon carbon-chain thermoplastic polymers ... **EM2:** 51
of polymers ... **EM2:** 48-52
thermosets ... **EM2:** 55
Chemical surface studies
classical wet chemistry ... **A10:** 177-178
Chemical susceptibility *See also* Chemical analysis; Chemical properties; Chemical structure; Chemistry
absorption and transport ... **EM2:** 572
additive effects ... **EM2:** 572-573

Chemical susceptibility (continued)
chemical corrosion **EM2:** 573-574
degradation detection **EM2:** 574
environmental corrosion **EM2:** 573
photooxidative degradation **EM2:** 573
polymer structure **EM2:** 571-572
thermal degradation **EM2:** 573
thermal oxidative degradation **EM2:** 573
Chemical synthesis **EM4:** 52-62
gas phase reactions **EM4:** 62
high-intensity arcs **EM4:** 62
lasers **EM4:** 62
plasma jets **EM4:** 62
preparation from solution **EM4:** 56-62
coated colloids or slurries **EM4:** 60
collection of powder **EM4:** 61
combustion **EM4:** 61
emulsion methods **EM4:** 61
ferricyanides **EM4:** 58
ferrocyanides **EM4:** 58
freeze drying **EM4:** 61
hydroxides **EM4:** 59-60
metal organics **EM4:** 60
mixed reagents **EM4:** 60
overall process **EM4:** 56
oxalate ion **EM4:** 58
precipitation process **EM4:** 56-58
sol-gel process **EM4:** 62
spray drying **EM4:** 61
spray roasting **EM4:** 61
thermal evaporation **EM4:** 61
thermal methods **EM4:** 61
thermal decomposition principles **EM4:** 52-56
decomposition temperatures of iron-lithium mixtures **EM4:** 55
mechanisms **EM4:** 53-56
particle size **EM4:** 56
rate law expression **EM4:** 55
rates **EM4:** 53-56
reactivity **EM4:** 53-56
stoichiometry **EM4:** 52-53
techniques **EM4:** 52-53
Chemical thermodynamics *See also* Thermodynamics
enthalpy and heat capacity **A15:** 50
Gibbs free energy **A15:** 50-51
phase diagrams **A15:** 52
Chemical thinning
of transmission electron microscopy specimens **A9:** 105
Chemical toxicity of depleted uranium **A9:** 477
Chemical Treatment No. 9
magnesium alloys **M5:** 632, 636-637, 640-641, 643
Chemical Treatment No. 17 magnesium alloys **M5:** 632, 636-638, 640-641, 643
Chemical vapor deposited (CVD) carbon
defined **EM1:** 6
Chemical vapor deposition **A13:** 3, 457 **M5:** 381-385
abbreviation for **A11:** 796
alloys coated by **M5:** 382-383
aluminum oxide coatings **M5:** 383-384
applications **M5:** 382-384
chromium coatings **M5:** 383
coated materials characteristics of **M5:** 385
coating structure **M5:** 385
coating types **M5:** 382-384
compared to thermoreactive deposition/diffusion process **A4:** 448
defined **A7:** 2
deposition pressure **M5:** 385
deposition temperature **M5:** 384-385
fluidized bed process **M5:** 385
grain structure, control of **M5:** 385
limitations and advantages of **M5:** 381-382
nickel coatings **M5:** 382
open-tube process, chromium coatings **M5:** 383
oxidation-resistant coating **M5:** 664-666
for passivation in wafer processing **EM3:** 582
passivation preparation **EM3:** 593
process control **M5:** 384-385
of pyrolytic graphite to redensify a carbon- carbon composite **A9:** 595-596
reactant concentration **M5:** 384
silicon nitride grown **EM3:** 583
steel coated by **M5:** 382-384
surface preparation for **M5:** 384
Chemical vapor deposition (continued)
for thin-film hybrids **EL1:** 313
titanium coatings **M5:** 382-384
tools coated by **M5:** 382-384
tungsten alloy welds **A6:** 582
tungsten coatings **M5:** 382-383, 385
used to make boron fibers **A9:** 592
wear resistance, coating **M5:** 382-383
of wear resistant coatings **M1:** 635
Chemical vapor deposition (CVD) *See also* PVD and CVD coatings **A18:** 840, 841, 846-851 **EM4:** 32, 124
advanced ceramics **EM4:** 47
advanced techniques **A18:** 848
anti-ice and antifog protection to aircraft window glass **EM4:** 1024
applications **A18:** 846
for boron fiber production **EM1:** 58, 851-852
carbon aircraft brakes **A18:** 584
for carbon/carbon densification **EM1:** 918
of carbon/graphite fibers **EM1:** 868
cemented carbides **A18:** 796
ceramic coatings for adiabatic diesel engines **EM4:** 992
classification of reactions **A18:** 846
coated carbide tools **A2:** 959-960
coating method for valve train assembly components **A18:** 559
comparison with polymer-derived coatings **EM4:** 225
complex reactions **A18:** 846
conventional (CCVD) **A18:** 840, 846-847
corrosive by-products **EM4:** 445
defined **EM1:** 6
definition **A18:** 846
development **EM4:** 377
diamond coatings **EM4:** 826-827
diamond synthesis **EM4:** 822
and flaw identification **EM4:** 666
hot-filament CVD **A18:** 848
infiltration unit **EM1:** 868
laser-induced (LCVD) **A18:** 846, 848
low-pressure (LPCVD) **A18:** 847
metal-organic (MOCVD) **A18:** 846
microwave excitation **A18:** 840, 849
mullite ultrafine powder **EM4:** 763-764
nonoxide fibers by **EM1:** 63-64
organo-metallic (OMCVD) **A18:** 846
parameters **A18:** 841
photon excitation **A18:** 840
plasma-assisted **A18:** 840, 846, 847-848
processes **A18:** 840, 841, 846-848
rate-limiting steps **A18:** 846
in reaction sintering **EM4:** 294
reactors **A18:** 846-847
refractory metals and alloys **A2:** 563
rf excitation **A18:** 840
in SiC fiber production **EM1:** 59, 63, 858
silicon carbide manufacture **EM4:** 806
silicon carbide monofilament fibers by **EM1:** 58
thermal **A18:** 840, 846-847
titanium carbide coatings **A18:** 158, 645
critical normal force versus substrate hardness **A18:** 436
to make ferrite films **EM4:** 1163
to make garnet films **EM4:** 1163
tool steels **A18:** 641, 643, 645, 646, 739
tungsten carbide coatings, comparison of coatings for cold upsetting **A18:** 645
ultralow-expansion glass production **EM4:** 1016
as ultrapurification process **A2:** 1094
Chemical vapor deposition (CVD) process **A16:** 57
application of coatings to cemented carbides **A16:** 71
cemented carbide coatings **A16:** 81-82, 83-85, 86, 87
cemented carbides **A16:** 80
and drilling **A16:** 219
Chemical vapor deposition coatings on cemented carbides
effect on eta phases **A9:** 275
preparation for examination of **A9:** 273
Chemical vapor deposition films
crystallographic texture in **A9:** 700
Chemical vapor deposition for purifying metals **M2:** 711
Chemical vapor deposition/chemical vapor infiltration
ceramic prototypes of tubular heat exchangers **EM4:** 983
Chemical vapor impregnations
carbon-carbon composites **EM4:** 835
Chemical vapor infiltration (CVI) **EM4:** 35
for advanced ceramics **EM4:** 47
carbon aircraft brakes **A18:** 584
of ceramic matrix composites **EM4:** 224
ceramic-matrix composites **EM4:** 840
fiber-reinforced for aerospace applications **EM4:** 1004-1005
fiber-reinforced composites **EM4:** 215, 219-220
of matrix phase **EM4:** 35
in reaction sintering **EM4:** 293-294
silicon carbide manufacture **EM4:** 806
to fabricate ceramic-matrix composites **EM4:** 20
Chemical washing
as surface preparation **A13:** 413-414
Chemical wear *See* Corrosive wear
Chemical(s)
coatings, explosion characteristics **A7:** 196
composition **A7:** 246-249
decomposition, as ignition source **A7:** 197
decomposition, defined **A7:** 2
deposition, defined **A7:** 2
embrittlement **A7:** 52, 55
industry, refractory metals in **A7:** 17
methods of powder production **A7:** 52-55
precipitation, of tin powders **A7:** 123
reduction metal precipitation by **A7:** 54
refractory metals, use in **A7:** 765
Chemical-element segregation
flaking from **A11:** 121
Chemical-mechanical polishing *See also* Polish-etching **A9:** 39
etch attack in **A9:** 42
of hafnium **A9:** 497-498
in skid polishing **A9:** 42
in vibratory polishing **A9:** 42
of zirconium and zirconium alloys **A9:** 497-498
Chemical-processing equipment
ultrasonic inspection **A17:** 232
Chemical-setting ceramic linings
inorganic **A13:** 453-455
Chemically accelerated centrifugal
barrel finishing **M5:** 134
Chemically bonded sand molding
and green sand molding, compared **A15:** 341
self-setting **A15:** 37
Chemically clean surfaces **A9:** 28
Chemically deposited coatings for contrast enhancement in scanning electron
microscopy **A9:** 99
Chemically precipitated powder
defined **A7:** 2
Chemically reactive adhesives **EM3:** 74-75
shelf life **EM3:** 75
Chemically reactive paints **M5:** 500-501
Chemicals
analysis **A10:** 72, 177
commercial, voltammetric characterization of metals in **A10:** 188
surface species identified **A10:** 568
systems, analyzed **A10:** 263
wastes assay for toxic elements, NAA for **A10:** 233
Chemicals used in etchants **A9:** 66-68
disposal **A9:** 69
labelling **A9:** 67
purity grades **A9:** 67
storage containers **A9:** 69
Chemietching *See* Chemical milling
Chemiluminescence
monitoring **EM2:** 426
Chemisorption *See also* Physisorption
defined **A10:** 670 **A13:** 3
Chemistry *See also* Analytical chemistry; Chemical analysis; Chemical properties; Classical wet analytical chemistry; Photochemistry; Physical chemistry
atmospheric, PIXE studies in **A10:** 106
of bismaleimides (BMI) **EM2:** 252-253
cationic systems **EL1:** 859-864
of cermet systems **EL1:** 340-341

Chemistry (continued)
of elements at surfaces, AES lineshapes showing **A10:** 552-553
of epoxy resins/hardeners **EL1:** 474, 825-827
free radical systems
of grain boundaries **A10:** 561-562
irradiation **A10:** 235
liquid crystal polymers (LCP) **EM2:** 179
metallurgical **A15:** 27
of PMDA-ODA polyimide **EL1:** 324
polycarbonates (PC) **EM2:** 151
polyimide **EL1:** 324-326, 767
polymer **EM2:** 63-67
and single-crystal castability **A1:** 998-1003
Chemlock 205 .... **EM3:** 630
Chemlock 205/220 .... **EM3:** 629
Chert
abrasive wear **A18:** 188
Chevron cracking
submerged arc welds **M6:** 129-130
Chevron marks *See* Chevron pattern
AISI/SAE alloy steels **A12:** 306, 307
ASTM/ASME alloy steels **A12:** 347
as brittle **A12:** 107-108, 111, 173, 258
cast aluminum alloys **A12:** 413
causes **A12:** 107-108
fracture, brittleness and ductility of **A12:** 108
fractures, crack front in **A12:** 107
from fracture origin, SEM fractographs **A12:** 170
medium-carbon steels **A12:** 255-258, 260-261, 269
on cleavage fracture surface **A12:** 13
polymers **A12:** 480
visual examination **A12:** 107, 111-113
wrought aluminum alloys **A12:** 429, 436
Chevron notch
alternative **A13:** 259
Chevron pattern
at crack origin **A11:** 397
in brittle fracture **A11:** 82
defined **A8:** 2 **A11:** 2 **A13:** 3
fatigue fracture **A11:** 116
and fracture origin **A11:** 80
in low-alloy steel ship-plate **A11:** 77
on fracture surface, steel axle shaft **A11:** 21
and river marks, contrasted **A11:** 25
in steel plate punched hole **A11:** 90
from tensile brittle fracture **A11:** 76-77
Chevron slot
angle for curved **A8:** 472
geometries **A8:** 470, 473
thickness **A8:** 471
Chevron-notched specimens
advantages **A8:** 469
calibration constant for **A8:** 471-472
data analysis with **A8:** 473-474
elastic plastic behavior with **A8:** 471-473
for fracture toughness testing **A8:** 469-475
geometry of **A8:** 470
linear elastic fracture mechanics test with **A8:** 470-471
short rod/bar, for fracture toughness testing **A8:** 461
single **A8:** 469
and test equipment, fracture toughness testing **A8:** 469-470, 472
Chevrons
in steel bar and wire **A17:** 550
ultrasonic inspection for **A17:** 271
Chi phase
in austenitic stainless steels **A9:** 284
in iron-chromium-nickel heat-resistant casting alloys **A9:** 332-333
in wrought heat-resistant alloys **A9:** 309
Chi-square (χ^2) distribution **A18:** 482, 485
Chi-square distribution
and reliability prediction **EL1:** 902
Chi-square test
to determine normal distribution **A8:** 663-664, 673
Chill
casting, irons **A12:** 219
defined **A11:** 2
formation, macroetching of cast iron for **A11:** 344
inverse, ductile iron fracture from **A12:** 227
inverse, in iron castings **A11:** 362
tests, applications **A12:** 141
to control shrinkage **A11:** 354
use to accelerate cooling **A11:** 345
Chill casting
used to obtain random orientation **A9:** 701
Chill coating
defined **A15:** 3
Chill, correction of
gray cast iron **M1:** 22
Chill ring
definition **A6:** 1207
Chill roll
defined **EM2:** 9
Chill test
for cupolas **A15:** 390
Chill tests
in ductile iron **A1:** 39
Chill(s) *See also* Antichills; Chilled iron; Insulating pads and sleeves; Inverse chill
copper alloy casting **A15:** 779, 781
as dendritic structure growth mechanism **A15:** 118
effect, solidification sequence **A15:** 606-608
of high-alloy graphitic irons **A15:** 698
magnesium alloy **A15:** 806
in permanent mold casting **A15:** 283
water-cooled copper, DS/SC furnaces **A15:** 400-401
Chill-block melt-spinning **A7:** 48, 49
Chilled car wheel iron
development of **A15:** 30
Chilled iron *See also* Cast irons, chilled; Chill; Inverse chill
arc welding of **A15:** 528-529
defined **A15:** 3
development of **A15:** 30
Chilled iron rolls **A14:** 353
Chilled iron shot peening **M5:** 140-141
Chilled zone in ductile iron **A9:** 251
Chilling tendency
CG irons **A15:** 671-673
of compacted graphite iron **A1:** 57
Chills
shielded metal arc welding **M6:** 91
Chills, for directional solidification
copper alloy casting **A2:** 348
China
absorption **EM4:** 3
composition **EM4:** 4
early metalworking in **A15:** 16-18
glazing **EM4:** 3
properties **EM4:** 3-4
subclassifications based on use or quality **EM4:** 4
China clay **A6:** 60 **EM4:** 32, 44
pressure calcintering for study of compaction kinetics **EM4:** 300
in typical ceramic body compositions **EM4:** 5
China clay (kaolin)
as filler material **EM2:** 499-500
Chinese script .... **A3:** 1 • 19, 20
in aluminum alloys **A9:** 359
beryllide phase in beryllium-copper alloys **A9:** 395
eutectic, defined **A9:** 3
in eutectics **A9:** 621
"Chinese script" eutectic **A18:** 785
Chinese script-type eutectic
irregular **A15:** 120
Chip attach
ceramic multilayer packages **EL1:** 466-467
materials, in plastic packages **EL1:** 211
Chip attach (continued)
plastic pin-grid array **EL1:** 476
Chip breakers *See also* Chip control; Chip formation **A16:** 143, 148
boring **A16:** 166, 167
and broaching **A16:** 204, 205
cast irons **A16:** 649
drilling **A16:** 220, 227
milling **A16:** 316
Ni alloys **A16:** 837
stainless steels **A16:** 681, 696, 697, 704
trepanning **A16:** 178
Chip bumping, tape automated bonding
plastic packages **EL1:** 477
Chip carrier (CC) *See also* Chip(s); Leadless chip carrier (LCC); Plastic leaded chip carrier (PLCC)
conductive polymer interconnection for **EL1:** 15
conformal coatings/encapsulants, introduction **EL1:** 759-760
CTE mismatch problem **EL1:** 661
defined **EL1:** 1137
dimensional tolerances **EL1:** 734
leaded and leadless **EL1:** 730-734
as package family **EL1:** 404
routability **EL1:** 15
in surface-mount technology **EL1:** 76
Chip characteristics
in machining aluminum **M2:** 189-190
relation to machinability **M1:** 574
Chip control *See also* Chip breakers; Chip formation **A16:** 19
adaptive control implementation **A16:** 624
boring **A16:** 162, 164
broaching **A16:** 195, 196, 197, 205, 206, 208
cast irons **A16:** 654
cemented carbides **A16:** 84-85, 86
drilling **A16:** 219, 221, 222, 224, 229, 236-238
electrical discharge machining **A16:** 560
and fixturing **A16:** 404, 405
grinding and abrasives **A16:** 433, 434, 437, 442, 444
and grinding of steel gears **A16:** 354
high removal rate machining **A16:** 608
honing **A16:** 488
milling **A16:** 307, 318, 319, 321, 327-328
multifunction machining **A16:** 392, 396
Ni alloys **A16:** 837, 840
planing **A16:** 182, 184, 186
reaming **A16:** 58
sawing **A16:** 358, 365
tapping **A16:** 256, 257, 259, 261, 262, 267
thread grinding **A16:** 278
trepanning **A16:** 177, 178, 180
and truing and dressing of grinding wheels **A16:** 466, 467
turning **A16:** 159
Chip cutting **A16:** 23
Chip failures
corrosion, plastic packages **EL1:** 479
integrated circuits **A11:** 766
mechanical stresses **EL1:** 480
Chip formation *See also* Chip breakers; Chip control **A16:** 7-12
Al alloys **A16:** 761, 765, 769-770, 780, 783, 787, 791-792
analysis **A16:** 13-18
broaching of heat-resistant alloys **A16:** 743, 744, 745
carbon and alloy steels **A16:** 668, 669, 670
carbon steels and drilling **A16:** 674-675
characteristic types **A16:** 12
Cu alloys **A16:** 805, 808-812, 815, 817
cutting parameters **A16:** 10-11
energy **A16:** 14
friction coefficient **A16:** 10
high-speed machining **A16:** 597, 598-600, 603
and machinability **A16:** 642, 646
Mg alloys **A16:** 820, 821-822, 824, 825, 828

Chip formation (continued)
P/M materials ... **A16:** 883, 885
P/M tool steels ... **A16:** 735
power consumption in production processes ... **A16:** 17, 18
properties and their effect ... **A16:** 10
reaming ... **A16:** 239
stainless steels ... **A16:** 681, 696
stress distributions -n metal cutting ... **A16:** 14-17
types ... **A2:** 964, 966
uranium alloys ... **A16:** 874, 875, 876
Zn alloys ... **A16:** 831, 832, 834
Zr reactive metals ... **A16:** 853, 855
Chip inductors
types ... **EL1:** 179
Chip interconnection *See also* Chip(s); Interconnection; Interconnects
direct ... **EL1:** 231-232
laser pantography ... **EL1:** 232-233
pressure contacts ... **EL1:** 232
schemes ... **EL1:** 232
for very-large-scale integration ... **EL1:** 231
Chip packages from DIPS, evaluated ... **EL1:** 15
epoxies and molding processes ... **EL1:** 760-761
Chip passivation computer modeling
technology ... **EL1:** 244
Chip removal rate
volumetric ... **A18:** 611
Chip resistors tools and removal methods ... **EL1:** 724-727
types ... **EL1:** 178
Chip size DIP, effects ... **EL1:** 53
thermal resistance effect ... **EL1:** 411
Chip velocity ... **A18:** 610, 611
Chip(s) *See also* Chip-to-chip; On-(-hip; Semiconductor chips
beam-lead design, development ... **EL1:** 961
bump-contact design ... **EL1:** 961
discrete, mounting technologies ... **EL1:** 143
dynamic random access memory (DRAM), as future trend ... **EL1:** 390
edges ... **EL1:** 7
fabrication vs packaging ... **EL1:** 397
failures ... **EL1:** 479-480
gate counts per, growth of ... **EL1:** 416
instrumentation/testing
in interconnection hierarchy ... **EL1:** 4
load, defined ... **EL1:** 1137
metallization, computer modeling of
mount pad, size ... **EL1:** 411
mounting, bipolar junction transistor technology ... **EL1:** 195-196
passivation ... **EL1:** 244
power dissipation vs time ... **EL1:** 46
-related failure mechanisms, through-hole packages ... **EL1:** 974
semiconductor, packaging ... **EL1:** 397
silicon, thermomechanical design considerations ... **EL1:** 415-416
single vs multichip module ... **EL1:** 252
size, constant ... **EL1:** 143
size, effects ... **EL1:** 53
soldering methods ... **EL1:** 180
tape bonded ... **EL1:** 484
technologies, mix of ... **EL1:** 443
temperature, thermal management ... **EL1:** 45
test, as design tools ... **EL1:** 419
thermal parameters ... **EL1:** 47
-type components, fabrication ... **EL1:** 178-190
and wire assembly, physical characteristics ... **EL1:** 110
yield, in WSI ... **EL1:** 354
Chip-level packaging
current level 1 packages ... **EL1:** 403-405
elimination of ... **EL1:** 407
integrated circuit trends ... **EL1:** 399-401
level 1 package trends ... **EL1:** 405-407
overview ... **EL1:** 398-407
package design considerations ... **EL1:** 401-403
Chip-on-board defined ... **EL1:** 1137
technology, as cavityless hybrid packaging ... **EL1:** 452
Chip-to-bump separation
as tape automated bonding (TAB) mechanical failure ... **EL1:** 287
Chip-to-chip issues
physical performance ... **EL1:** 7-8
Chipboard
milling with PCD tooling ... **A16:** 110
Chipbreaking
carbide metal cutting tools ... **A2:** 963-964
Chipping
of cemented carbides ... **A7:** 779
gear-tooth ... **A11:** 595
for magnesium powder production ... **A7:** 131
multiple, in ceramics ... **A11:** 754
scoring damage by ... **A11:** 157
Chipping, porcelain enamel
resistance to ... **M5:** 527, 529
Chipping resistance
porcelain enamels ... **A13:** 451
Chips
contamination by, vapor degreasing solvents ... **M5:** 48
continuous ... **A16:** 122
discontinuous ... **A16:** 122
reduction of beryllium ingot to ... **A7:** 170
removal, from steel parts ... **M5:** 5, 9-10
Chips, steel
AAS analysis ... **A10:** 55
Chisel steels
early fractographs ... **A12:** 5
Chisel test
explosion welds ... **M6:** 711
resistance spot welds ... **M6:** 487
Chisel-point fractures
fcc metals ... **A12:** 100
Chlorate, sodium
electrolyte for electrochemical machining ... **A16:** 533, 535, 536
Chlorate solutions
copper/copper alloy SCC in ... **A13:** 634
Chlorendics *See also* Polyester resins
acid resistance ... **EM1:** 93
clear casting mechanical properties ... **EM1:** 91
in fiberglass-polyester resin composites ... **EM1:** 91
glass content effect ... **EM1:** 91
mechanical properties ... **EM1:** 90
preparation/application ... **EM1:** 90
Chloride
anions, separation by ion chromatography ... **A10:** 659
contamination, in austenitic stainless steels ... **A11:** 635
content, high-purity oxygenated water ... **A8:** 420
corrosion fatigue test specification ... **A8:** 423
determined by precipitation titration ... **A10:** 164
as electrode ... **A10:** 184
electrolyte for Ni alloy electrochemical machining ... **A16:** 843
ferric, analysis in graphite ... **A10:** 133
hot, SSC of pressure vessels from ... **A11:** 660
integral-finned stainless steel tube cracked by ... **A11:** 636
ion pitting, of Kovar lead material ... **A11:** 770
ions ... **A10:** 169, 201
ions, pitting of heat-exchanger tubes by ... **A11:** 630
Mohr titration for ... **A10:** 173
pitting, U-bend heat-exchanger tubes failure from ... **A11:** 637
pressurized water reactor specification ... **A8:** 423
quantitative determination of metals in presence of, electrogravimetry ... **A10:** 197
salts, hot dry, and SCC ... **A11:** 223
SCC in stainless steel shaft from ... **A11:** 660
SCC of nuclear steam-generator vessel by ... **A11:** 656-657
SCC tube failure from ... **A11:** 635
solutions, effect on critical strain rate ... **A8:** 519
weighing as the, gravimetry ... **A10:** 171
Chloride cracking tests
of stainless steels ... **A1:** 725
Chloride deposition
for A15 superconductor assembly ... **A2:** 1065
Chloride salts
for SCC testing of titanium alloys ... **A8:** 531
Chloride(s)
aqueous, crevice corrosion in titanium ... **A13:** 672
dissolved, corrosion resistance effect nickel-base alloys ... **A13:** 646
effect, pulp bleach plants ... **A13:** 1193
ion activity, aluminum alloy ... **A13:** 584
Chloride(s) (continued)
ions, as cause of steel corrosion in concrete ... **A13:** 513
level, for pitting/crevice corrosion ... **A13:** 1367
magnesium/magnesium alloys in ... **A13:** 743
in marine atmospheres ... **A13:** 903
organic ... **A13:** 1268
salts, molten ... **A13:** 90
SCC, prevention ... **A13:** 327
solution, copper/copper alloy SCC in ... **A13:** 634
titanium/titanium alloy resistance ... **A13:** 682-683
zirconium/zirconium alloy corrosion ... **A13:** 717
Chloride-sulfate nickel plating ... **M5:** 201-202
Chlorides
effect in alloy steels ... **A12:** 291
as fining agents ... **EM4:** 380
role in freshwater corrosion ... **M1:** 733, 735
solution, as corrosive environment ... **A12:** 24
stress-corrosion cracking ... **A12:** 357
in torch brazing flux ... **M6:** 962
Chlorimet alloys *See* Nickel alloys, cast, specific types
Chlorinated diphenyl
for SCC testing of titanium alloys ... **A8:** 531
Chlorinated hydrocarbon ... **EM3:** 8
Chlorinated hydrocarbon solvent cleaners ... **M5:** 40-42, 57
Chlorinated hydrocarbons
safety precautions ... **A6:** 1196
Chlorinated lubricant *See also* Extreme- pressure lubricant; Sulfochlorinated lubricant; Sulfurized lubricant
defined ... **A18:** 5
Chlorinated rubber coatings ... **A13:** 404, 405, 914
Chlorinated rubber resins and coatings ... **M5:** 473-474, 495, 498, 500-502, 504-505
Chlorinated solvents ... **M5:** 617
for chemical cleaning ... **A15:** 561
corrosion of stainless steels in ... **M3:** 81-82
Chlorination
in blended elemental titanium alloys ... **A7:** 164
inclusion testing by ... **A10:** 176
for iron and manganese carbides and sulfides in residues ... **A10:** 177
and microbiological corrosion ... **A13:** 314
niobium extraction ... **A7:** 160
as second-phase test method ... **A10:** 177
Chlorine ... **A13:** 1170-1180
alloy steel corrosion ... **A13:** 544-545
in coal or catalysts, determined by electrometric titration ... **A10:** 205
cold form tapping ... **A16:** 266, 267
compounds, oxidizing, titanium/titanium alloy resistance ... **A13:** 677
contamination ... **A13:** 1162
corrosion of corrosion-resistant steel castings from ... **A1:** 913
corrosion of nickel alloys in ... **M3:** 174
degassing, of magnesium alloys ... **A15:** 462-463
for demagging aluminum alloys ... **A15:** 472-474
detected by Auger electron spectroscopy ... **A7:** 251, 254
dry, tantalum resistance to ... **A13:** 731
effect, copper/copper alloys in seawater ... **A13:** 624-625
in EP additives ... **A16:** 123, 125, 126
gas, hafnium-zirconium composition in ... **A13:** 720
and grinding operation ... **A16:** 437, 438
ionic composition, and metal implants ... **A11:** 672, 673
lubricant indicators and range of sensitivities ... **A18:** 301
measured in lubricants ... **A14:** 516
moist ... **A13:** 1172-1173
as NAA sample contaminant ... **A10:** 236
production ... **A7:** 134
as reagent, aluminum melts ... **A15:** 80
refrigerated liquid ... **A13:** 1171-1172
SCC in titanium and titanium alloys by ... **A11:** 223
for SCC testing of titanium alloys ... **A8:** 531
species weighed in gravimetry ... **A10:** 172
spectrometric metals analysis ... **A18:** 300
in thread grinding oils ... **A16:** 273
TNAA detection limits ... **A10:** 237
volumetric procedures for ... **A10:** 175

Chlorine (continued)
-water corrosion **A13:** 1173-1174
Chlorine (Cl)
component in photochromic ophthalmic and flat glass composition **EM4:** 442
volatilization losses in melting **EM4:** 389
Chlorine compounds
and stress corrosion **A16:** 35
Chlorine dioxide
tantalum resistance to **A13:** 727
Chlorine extraction
defined **A9:** 3
Chlorine-containing compounds
as flame retardants **EM2:** 504
Chlorinity
seawater **A13:** 894
Chlorite **EM4:** 6
Chloro-alkali chemical processing plants
pollution control **A13:** 1369
Chlorobutyl 1066
molecular weight **EM3:** 199
Chlorocarbon(s) blends, physical properties **EL1:** 664
molecular structure of **EL1:** 662
solvents, as organic cleaners **EL1:** 662
Chlorofluorocarbon (CFC) solvents
alternatives for cleaning PWB assemblies **A6:** 354
as organic cleaners **EL1:** 662-663, 667
Chlorofluorocarbon cleaners **M5:** 40-41
Chlorofluorocarbon plastics **EM3:** 8
Chlorofluorocarbons (CFCs) **A6:** 112
Chlorofluorohydrocarbon plastics **EM3:** 8
Chloroform
extractants for **A10:** 170
Chloroplasts
ESR studied **A10:** 264
Chloroprene *See also* Neoprene **EM3:** 51
Chloroprene rubber
exposure in electrochemically inert conditions **EM3:** 629
Chlorosulfonated polyethylene
used as modifier **EM3:** 121
Chlorosulfonic acid
corrosion of stainless steels in **M3:** 82
Chlorotrifluoroethylene *See also* Polychlorotrifluoroethylene (CTFE)
as fluoropolymer **EM2:** 116
Choked flow
in leaks **A17:** 58
Chokes
copper alloy casting **A15:** 776-777
gating system effects **A15:** 590-591
Cholesteric crystals
for optical imaging **EL1:** 1072
Cholesteric liquid crystal
use in integrated circuit failure analysis **A11:** 767-768
Cholesteric liquid crystals
as temperature sensors **A17:** 399
Chopped fiber
as toughener **EM3:** 185
Chopped fibers *See also* Chopped glass; Discontinuous,fibers; Short fiber; Staple fiber
length, composite effect **EM1:** 162
for sheet molding compounds **EM1:** 141
for spray lay-up technique **EM1:** 132
Chopped glass **EM1:** 109-110
Chopped mat *See* Mats
Chopped-strand glass fiber products **EM1:** 109-110
Chopped-strand mats *See* Mats
Chopper dies **A14:** 132
Chord measurements
image analysis **A10:** 316
Chord modulus *See also* Modulus of elasticity,; Modulus of elasticity; Modulus of elasticity (E) **A8:** 2
defined **A11:** 2 **A14:** 2
Christmas trees
oil/gas production **A13:** 1237
Chromadizing *See also* Chromating; Chromizing
defined **A13:** 3
Chromalloy
broaching **A16:** 203, 204
Chromalloy steel
flash welding **M6:** 557
Chromate *See also* Chromate conversion coatings
as anodic inhibitor **A13:** 494
finishes, for zinc castings **A15:** 796
as refractory, core coatings **A15:** 240
rinses, after phosphate coating **A13:** 387
sands, reclamation of **A15:** 355
Chromate coating
hot dip galvanized products **M5:** 332
mechanical coating process using **M5:** 301-302
phosphate coating resisted by **M5:** 438
Chromate concentration
phosphate coating solutions **M5:** 443
Chromate conversion coating **M5:** 457-458
aluminum and aluminum alloys **M5:** 457
cadmium plate **M5:** 269
chromium chromate type **M5:** 457-458
chromium phosphate type **M5:** 457
corrosion resistance **M5:** 457-458
equipment **M5:** 599-600
mechanism of action **M5:** 457
properties **M5:** 457-458, 599
quality control **M5:** 600
solution compositions and operating conditions **M5:** 599-600
solution control **M5:** 599-600
zinc plated parts **M5:** 254-255
Chromate conversion coatings **A13:** 389-395 **M1:** 752, 754
alkaline oxide **A13:** 394
applications **A13:** 389
as barrier protection, aqueous solutions **A13:** 379
for beryllium **A13:** 811
for carbon steel **A13:** 523-524
cast irons **M1:** 104
chromate-type **A13:** 394
chromium phosphate **A13:** 394
control and testing **A13:** 393-394
corrosion inhibition with **M1:** 169, 175
equipment and application **A13:** 390
micrographs **A13:** 392
no-rinse processes **A13:** 394
on aluminum **A13:** 394
on cadmium **A13:** 395
on magnesium **A13:** 395
on steels **A13:** 395
on zinc and galvanized steels **A13:** 394-395
passivation, zinc-coated surfaces **M1:** 168-169
processes **A13:** 389-390
properties **A13:** 390-393
safety and waste treatment **A13:** 395
salt spray data **A13:** 393
standard practices/specifications for **A13:** 394
testing methods and standards **A13:** 395
Chromate films
deposited by color etching **A9:** 141
Chromate passivation **A1:** 214-215
Chromate treatment
defined **A13:** 3
Chromate, weighing as the
gravimetric analysis **A10:** 171
Chromates
as corrosion inhibitors **A18:** 277
Chromatic aberration **EM3:** 8
defined **A10:** 670
Chromatic aberrations
defined **A9:** 3
Chromatic aberrations (continued)
effect on secondary electron imaging **A9:** 95
Chromatic contrast by an interference film **A9:** 158
Chromating *See also* Chromadizing
defined **A13:** 3
process **A13:** 389-390
zinc alloys **A2:** 530
Chromatizing *See* Chromadizing
Chromatograms
common anions separated by single-column ion chromatography **A10:** 661
defined **A10:** 670
eluent-suppressed ion **A10:** 660
fiber-suppressed ion **A10:** 660
for geological brines **A10:** 666
for glass microballoons **A10:** 667
for ion chromatography separation and detection of alkali and transition metals **A10:** 660-661
ion-exchange, of radioactive alkali metals **A10:** 653
ion-pair, of napthylamine sulfonic acids **A10:** 653
normal phase **A10:** 652
reverse-phase, of an organic mixture **A10:** 653
reverse-phase, of parabens and baby lotion extract **A10:** 655
single-ion **A10:** 645
size-exclusion **A10:** 654
typical total ion **A10:** 644
Chromatographic effluents
IR identification of **A10:** 109
Chromatographs
high-performance liquid **A10:** 665
liquid **A10:** 650-651, 665
liquid, essential components **A10:** 650
Chromatographs, gas
for leak detection **A17:** 64
Chromatography *See also* High-performance liquid chromatography (HLPC); Testing; Thin-layer chromatography (TLC) **EM3:** 8
bonded-phase **A10:** 652
defined **A10:** 670
gas chromatography/mass spectroscopy **A10:** 639-648
ion **A10:** 658-667
ion-exchange **A10:** 168, 653
liquid **A10:** 649-657
liquid-liquid **A10:** 652
liquid-solid **A10:** 651-652
normal-phase **A10:** 652
paper **A10:** 168
potentiometric membrane electrodes as detectors for **A10:** 181
preparative liquid **A10:** 654
reversed-phase **A10:** 652-653
silica-based supports **EM4:** 1088-1089
size-exclusion **A10:** 654
of thermoplastic resins **EM2:** 539-540
of thermoset resins **EM2:** 517-522
versus weighting factor **EM4:** 85
Chromatography for epoxies **EL1:** 833
gas and liquid, as failure analyses **EL1:** 1104-1105
Chromatopyrogram
failed nitrile sheath **A10:** 648
intact neoprene sheath **A10:** 648
Chrome
alloying, magnetically soft materials **A2:** 762
coating for seals **A18:** 551
copper alloys, properties and applications **A2:** 357
refractory physical properties **EM4:** 897, 898, 899
for valve springs for reciprocating compressors **A18:** 604
Chrome brick **EM4:** 896
applications, refractory **EM4:** 901-902, 903, 906
Chrome compounds
as colorants **EM4:** 380

Chrome conversion coating EM3: 42
Chrome copper *See* Copper alloys, specific types, C81500
Chrome Copper 999 *See* Copper alloys, specific types, C18200
Chrome ore EM4: 45
applications EM4: 46
composition EM4: 46
supply sources EM4: 46
Chrome pickle
defined A13: 3
Chrome pickling, magnesium alloys M5: 629, 632-634, 636, 640-641, 643, 647-648
modified M5: 632, 636, 640-641
sealed M5: 632, 634, 636-637, 640-642
stripping of chrome pickle M5: 648
Chrome plating *See* Chromium plating; Decorative chromium plating; Hard chromium plating
Chrome steels
6Cr-1Mo, abrasive wear data A18: 705
thermal properties A18: 42
Chrome yellows
as pigment EM3: 179
Chromed steel
glass-metal seals EM4: 875
Chromel A
composition A6: 573
Chromel-Alumel thermocouples A8: 330
Chromia
melting/fining EM4: 391
Chromia (Cr_2O_3)
plasma spray coating for titanium alloys A18: 780
Chromia alumina catalysts
ESR analysis of A10: 265
Chromia-AZS
melting/fining EM4: 391
Chromia/alumina
melting/fining EM4: 391
Chromic acid A13: 677, 1141
as an electrolyte for zinc and zinc alloys A9: 489
as an etchant for carbon-carbon composites A9: 591
as an etchant for chromized sheet steel A9: 198
as an etchant for copper and copper alloys A9: 401
as an etchant for stainless-clad sheet steel A9: 198
as cleaning agent A12: 75
corrosion of stainless steels in M3: 82
decorative chromium plating processes M5: 189-191
description A9: 68
in Group VI electrolytes A9: 54
passivation with M1: 169
phosphate coating processes M5: 439, 442, 444, 448, 454
photochemical machining etchant A16: 593
safety hazards A9: 69
safety precautions in handling M5: 454
with sodium sulfate as an etchant for zinc and zinc alloys A9: 488
used to electrolytically etch heat-resistant casting alloys A9: 331
in water (Group V electrolytes) A9: 52-54
in water as an electrolyte for refractory metals A9: 440
in water with sodium sulfate as an etchant for electrogalvanized sheet steel A9: 197
Chromic acid anodization (CAA) process
bond line corrosion EM3: 670-671
of polyphenylquinoxalines EM3: 166, 167
pretreatment before wet peel testing of aluminum EM3: 668
surface preparation, processing quality control EM3: 738
Chromic acid cleaning process M5: 9-10
cast iron and stainless steel M5: 59-60
Chromic acid electropolishing solutions M5: 303, 305, 308
Chromic acid etching bath M5: 180
Chromic acid hard chromium plating baths M5: 172-175, 182, 186-187
Chromic acid pickling
magnesium alloys M5: 630-631, 635-637, 640-642, 647
Chromic acid purification
plating wastes M5: 317-318
Chromic anodize A13: 396
Chromic anodizing, aluminum and aluminum alloys M5: 586-589, 591-595
pH control M5: 586, 589
Chromic oxide
chemical composition A6: 60
description A9: 68
Chromic-acetic acid as electrolyte
current- voltage relation A9: 48-49
Chromic-nitric acid pickling
magnesium alloys M5: 638, 645, 647
Chromic-sulfuric acid pickling
magnesium alloys M5: 630, 640-641
Chromindur ductile permanent magnets
FIM/AP analysis of A10: 598-599
Chromite
erosion test results A18: 200
Chromite ($FeO-Cr_2O_3$)
purpose for use in glass manufacture EM4: 381
refractory material composition EM4: 896
solid-state sintering EM4: 278-279
Chromite, as molding sand
characteristics A15: 209
Chromite ores
sample dissolution mediums A10: 166
sulfuric acid as dissolution medium A10: 165
Chromium *See also* Chromium toxicity; Chromium-copper alloys; Electrical resistance alloys
as a beta stabilizer in titanium alloys A9: 459
as a carbide former in steel A9: 178, 661
as a ferrite-stabilizing element in wrought stainless steels A9: 283
as a substitute for manganese in sulfides A9: 279, 284
in abrasion-resistant cast irons A1: 115
as addition to cemented carbides A18: 800
addition to cylinder liner materials for strength A18: 556
addition to ferritic stainless steels A6: 444
addition to low-alloy steels for pressure vessels and piping A6: 667
addition to solid-solution nickel alloys A6: 575
addition to superalloys to resist oxidation A4: 798
as adhesion layer for polyimides EM3: 158
in alloy cast irons A1: 86, 88-89, 100
alloyed with Ni A16: 835
alloying, aluminum casting alloys A2: 132
alloying effect in titanium alloys A6: 508
alloying effect on copper alloys M6: 402
alloying effect on nickel-base alloys A6: 589
alloying effects A13: 47-48
alloying, magnetic property effect A2: 762
alloying, of cast irons A13: 567
alloying, of nickel-base alloys A13: 641
alloying, of stainless steels A13: 550
alloying, wrought copper and copper alloys A2: 242
in aluminum alloys A15: 745
in aluminum-silicon alloys A18: 788
in amorphous metals A13: 865-868
as an addition to austenitic manganese steel castings A9: 239
as an addition to beryllium-nickel alloys A9: 395
as an addition to nickel-iron alloys A9: 538
as an addition to permanent magnet alloys A9: 538
as an addition to wrought stainless steels A9: 284-285
as an addition to zirconium A9: 497
at elevated-temperature service A1: 640
in austenitic manganese steel A1: 824, 825
in austenitic stainless steels A6: 457, 458, 459, 461
in cast Co alloys A16: 69
in cast iron A1: 11-111-6, 28
cast iron content A16: 649
cause of temper embrittlement A4: 135
coarse primary carbides produced by carburizing of alloys A18: 874
coating for dental feldspathic porcelain and ceramics A18: 674
Chromium (continued)
coating for TEM specimens A18: 382
coating to improve abrasive wear resistance A18: 639
coatings for dies A18: 641
coefficient of friction A18: 836, 837
in commercial CPM tool steel compositions A16: 63
in composition, effect on ductile iron A4: 686
in composition, effect on gray irons A4: 671, 672, 673, 676, 678, 680
concentration, in stainless steels A15: 431
containing low-alloy steel powders A7: 101
contaminant of optical fibers EM4: 413-414
content, effect on scale flaking or spalling A18: 210
content in heat-treatable low-alloy (HTLA) steels A6: 670
content in HSLA Q & T steels A6: 665
content in nickel-base and cobalt-base high-temperature alloys A6: 573
content in saw bands A16: 358
content in stainless steels A16: 681, 682-684, 685, 686 M6: 320
content in tool and die steels A6: 674
content in tool steels affecting grindability A16: 727, 729, 732
content in ultrahigh-strength low-alloy steels A6: 673
content of weld deposits A6: 675
in copper alloys A6: 753
corrosion films on A7: 259
and crack growth A8: 487
crevice corrosion A13: 110
depletion, austenitic stainless steels A12: 51
depletion, by molten salt corrosion A13: 89
depletion, effect of carburization on A11: 272
depletion, in a weld zone A10: 179
depletion, in wrought heat-resisting alloys A11: 277
deposits, microcrack structure A13: 871-872
determined by controlled-potential coulometry A10: 209
determined in stainless steel A10: 146
in dichromate ion, analysis for A10: 70
distribution in pearlite A9: 661
as dopant for tungsten carbide A18: 795
dual or duplex *See* Microcracked chromium plating
in ductile iron A15: 649
in duplex stainless steels A6: 471-473, 478
effect of, on corrosion resistance A1: 912, 913
effect of, on hardenability A1: 395, 468
effect of, on notch toughness A1: 741
effect on activity coefficient in gas carburizing A4: 315
effect on base metal color matching in aluminum alloys A6: 730
effect on cast iron microstructure A18: 701
effect on equilibrium temperature, cast irons A15: 65
effect on iron borides A4: 441
effect on maraging steels A4: 222
effect, oxidation resistance, cast steels A13: 578
effects, cartridge brass A2: 301
effects of thermoreactive deposition/diffusion process A4: 449, 451
electrochemical grinding A16: 543
electrodeposited coatings A13: 426-427
in electrodes, weld metal hydrogen vs. oxygen content A6: 59
in electroplated coatings A18: 835-836, 838
applications A18: 835-836, 838
electroplating of dies A18: 644
electroslag welding, reactions A6: 273, 274
embrittlement sources A12: 123
erosion resistance A18: 228
evaporation fields for A10: 587
in ferrite A1: 408
as ferrite stabilizer A13: 47
in ferritic stainless steels A6: 450, 451
frequency-distribution curves A10: 601
friction coefficient A18: 71, 72
functions in FCAW electrodes A6: 188
glass/metal seals EM4: 1037
in gray iron A1: 22
as gray iron alloying element A15: 639

Chromium (continued)
as hard plating material, and fatigue strength **A11:** 126
hardfacing **A6:** 807
in hardfacing alloys **A18:** 759-760, 763-764, 765
in heat-resistant alloys **A4:** 510, 511, 512, 514
in heat/corrosion-resistant casting alloys **A13:** 574-581
hexavalent *See* Hexavalent chromium
in high-alloy white irons **A15:** 680
in high-speed tool steels **A16:** 52
in hydrogen peroxide, GFAAS analysis **A10:** 57-58
ICP-determined in plant tissues **A10:** 41
ICP-determined in silver scrap metal **A10:** 41
impurity concentrations **A2:** 1097
ion implantation and oxidation resistance **A18:** 856
in iron-base alloys, flame AAS analysis **A10:** 56
iron-chromium phase diagram **A6:** 447
isotope composition and intensity **A10:** 146
for laser alloying **A18:** 866
laser cladding **A18:** 867
in limestone **EM4:** 379
loss, analysis at weld zone **A10:** 179
in low-alloy steels **A16:** 150
lubricant indicators and range of sensitivities **A18:** 301
as major toxic metal with multiple effects **A2:** 1242
metal-to-metal oxide equilibria **A7:** 340
molybdenum steel and sawing **A16:** 363
neutron and x-ray scattering, and absorption compared **A10:** 421
nickel plating bath contamination by **M5:** 200, 208
in nickel-base superalloys **A1:** 984
in nickel-chromium white irons **A4:** 700-702 **A15:** 680
nickel-chromium-boron filler metal **A6:** 344
nitride-forming element **A18:** 878
oxygen cutting, effect on **M6:** 898
in P/M alloys **A1:** 810
in P/M high-speed tool steels **A16:** 61
partitioning oxidation states in **A10:** 178
permanganate titration for **A10:** 176
phosphoric acid as dissolution medium **A10:** 165
photometric analysis methods **A10:** 64
piston ring liners for cast iron rings only **A18:** 556
pitting effect, amorphous metals **A13:** 867-868
plating **A13:** 871-875
plating baths **A13:** 871
plating, effect in AISI/SAE alloy steel fracture **A12:** 297
plating for deep-drawing dies **A18:** 635
plating for piston rings **A18:** 556
plating for tool steels **A18:** 739
plating of drills **A16:** 219, 847
plating of pilot boring tools **A16:** 163
plating of taps **A16:** 259
for plating of tool steels to prevent galling **A18:** 633
pure, oxide scale formation **A13:** 97
pure, properties **A2:** 1107
as pyrophoric **A7:** 199
qualitative tests to identify **A10:** 168
recovery from selected electrode coverings **A6:** 60
redox titrations **A10:** 175
relative solderability **A6:** 134
relative solderability as a function of flux type **A6:** 129
selective oxidation of **A13:** 134
sensitization, Inconel 600 .... **A10:** 483
solderability **A6:** 978

Chromium (continued)
specific properties imparted in CTV tubes **EM4:** 1040, 1042
spectrometric metals analysis **A18:** 300
spraying for hardfacing **M6:** 789
in stainless steels **A18:** 710, 712, 716, 719, 721
in steel **A1:** 145-146, 577
steel, feather markings **A12:** 18
steel, general corrosion of **A11:** 674, 680
in steel weldments **A6:** 417
in Stellite alloys **A13:** 658
stripping of **M5:** 218
submerged arc welding **A6:** 200
tap density **A7:** 277
in thermal spray coating materials **A18:** 832
TNAA detection limits **A10:** 238
to enhance case hardness **A4:** 263
to form chemical bonds with alumina substrate in metallizing **EM4:** 545
to improve hardenability in carburized steels **A4:** 366, 367
to promote hardness **A4:** 124, 128-129
in tool steels **A16:** 53 **A18:** 734, 735-736, 737, 739
toxicity **A6:** 1195, 1196
as trace element, cupolas **A15:** 388
trace, in hydrogen peroxide, GFAAS analysis for **A10:** 57
trace levels in H_2O_2, GFAAS determined **A10:** 57-58
trivalent *See* Trivalent chromium
TWA limits for particulates **A6:** 984
ultrapure, by iodide/chemical vapor deposition **A2:** 1094
use in flux cored electrodes **M6:** 103
UV/VIS analysis in beryllium, by diphenylcarbazide method **A10:** 68
vapor pressure **A4:** 493
vapor pressure, relation to temperature **A4:** 495
in vapor-phase metallizing **EM3:** 306
varying, effect on stress-rupture life **A11:** 268
volumetric procedures for **A10:** 175
wear resistance of die material **A18:** 635-636
in wrought heat-resistant alloys **A9:** 310-311
in wrought stainless steels **A1:** 871 **A9:** 283-285
x-ray characterization of surface wear results for various microstructures **A18:** 469
in zinc alloys **A15:** 788
in zinc/zinc alloys and coatings **A13:** 759
Chromium alloys
eutectic joining and joint properties **EM4:** 526
fretting wear **A18:** 248, 250
oxygen domination of Auger electron spectroscopy spectrum **A18:** 456
Chromium alloys, containing rhenium
calculation of compressive strain **A9:** 127
Chromium alloys, specific types
2.25Cr-1Mo steel, time versus mass-based E/C rates **A18:** 209
9Cr-1Mo steel, erosion-enhanced corrosion **A18:** 208
9Cr-1Mo steel, scale spalling **A18:** 210
12Cr-Mo-0.3V (HT9)
applications **A6:** 433
filler metals, specific welding recommendations **A6:** 440
gas-tungsten arc welding **A6:** 435, 836
laser welding **A6:** 441
microstructure **A6:** 435
orientation and PWHT effect **A6:** 437
tempering behavior **A6:** 440
13Cr-4Ni-0.05C, hydrogen-induced cold cracking resistance **A6:** 438
18Cr-8Ni-Fe, refractory metal brazing, filler metal **A6:** 942
21Cr-6Ni-9Mn
cryogenic service **A6:** 1017

Chromium alloys, specific types (continued)
fatigue strength for gas-tungsten arc welds **A6:** 1018
25Cr-20Ni, hot cracking **A6:** 497
25Cr-20Ni-Fe, refractory metal brazing, filler metal **A6:** 942
105Cr6, nominal compositions **A18:** 725
Cr-Ni-Co-Fe superalloys, thermal expansion coefficient **A6:** 907
Cr-Ni-Fe superalloys, thermal expansion coefficient **A6:** 907
Fe-0.2C-12Cr-1Mo
Charpy V-notch data for weld joints of EB welded alloy **A6:** 440
microhardness traverse test results **A6:** 441
Chromium boride cermets
application and properties **A2:** 1004
Chromium boride-based cermets **A7:** 812
Chromium buffing compounds **M5:** 117
Chromium carbide **A16:** 72, 73, 74
in cast irons **A16:** 649
in cemented carbides **A18:** 795
chemical vapor deposition process **M5:** 383
coating for dies **A18:** 643
coating for jet engine components **A18:** 592
formation **A13:** 48
for gas-lubricated bearings **A18:** 532
and hard-phase Ni alloys **A16:** 835
honing stone selection **A16:** 476
in laser cladding material **A18:** 867
and machinability of stainless steels **A16:** 689
oxidation-resistant coating process **M5:** 665-666
properties **A18:** 795
thermal spray coating material **A18:** 832
Vickers and Knoop microindentation hardness numbers **A18:** 416
Chromium carbide (Cr_3C_2)
crystal structure **EM4:** 30
properties **EM4:** 30
Chromium carbide cermet
thermal expansion coefficient **A6:** 907
Chromium carbide cermets
applications and properties **A2:** 1000-1001
Chromium carbide-based cermets **A7:** 805-806
properties **A7:** 806, 807
Chromium carbide-based cermets, specific types
chromium carbide, composition **A7:** 806
nickel, composition **A7:** 806
tungsten, composition **A7:** 806
Chromium carbide-molybdenum
plasma-spray coating for pistons **A18:** 556
Chromium carbide-nickel
plasma-spray coating for pistons **A18:** 556
Chromium carbide-nickel chromide (Cr_3C_2-NiCr)
application as a coating **EM4:** 208
Chromium carbides, appearance of, in iron- chromium-nickel heat-resistant casting
alloys **A9:** 332
Chromium chromate coatings **M5:** 457-458
Chromium coating
chemical vapor deposition of **M5:** 383
molybdenum and tungsten **M5:** 661
pack cementation process **M5:** 383
Chromium content of heat-resistant casting alloys, effect on oxidation and sulfidation
resistance **A9:** 333-334
Chromium copper *See* Copper alloys, specific types, C18200, C18400, C18500 and C81500 .... **A9:** 553
Chromium copper alloys
heat treatment **A15:** 782
melt treatment **A15:** 774
Chromium coppers **A6:** 762
brazing **A6:** 931
heat treating **M2:** 257, 259
induction brazing **A6:** 935
resistance brazing **A6:** 339
tongs for manual resistance brazing **A6:** 340

SUBJECTS OF THE INDEXED VOLUMES: ASM Handbook (designated by the letter "A"): **A1:** Properties and Selection: Irons, Steels, and High-Performance Alloys (1990); **A2:** Properties and Selection: Nonferrous Alloys and Special-Purpose Materials (1990); **A3:** Alloy Phase Diagrams (1992); **A4:** Heat Treating (1991); **A6:** Welding, Brazing, and Soldering (1993); **A7:** Powder Metallurgy (1984); **A8:** Mechanical Testing (1985); **A9:** Metallography and Microstructures (1985); **A10:** Materials Characterization (1986); **A11:** Failure Analysis and Prevention (1986); **A12:** Fractography (1987); **A13:** Corrosion (1987); **A14:** Forming and Forging (1988); **A15:** Casting (1988); **A16:** Machining (1989); **A17:** Nondestructive Testing and Quality Control (1989); **A18:** Friction, Lubrication, and Wear Technology (1992). **Metals Handbook, 9th Edition** (designated by the letter "M"): **M1:** Properties and Selection: Irons and Steels (1978); **M2:** Properties and Selection: Nonferrous Alloys and Pure Metals (1979); **M3:** Properties and Selection: Stainless Steels, Tool Materials and Special-Purpose Materials (1980); **M4:** Heat Treating (1981); **M5:** Surface Cleaning, Finishing, and Coating (1982); **M6:** Welding, Brazing, and Soldering (1983). **Engineered Materials Handbook** (designated by the letters "EM"): **EM1:** Composites (1987); **EM2:** Engineering Plastics (1988); **EM3:** Adhesives and Sealants (1990); **EM4:** Ceramics and Glasses (1991); **Electronic Materials Handbook** (designated by the letters "EL"): **EL1:** Packaging (1989).

Chromium equivalence (Cr_{eq}) **A6:** 457, 459-461, 462, 463, 464, 483, 503
Chromium equivalent **A6:** 817-818, 819, 825
Chromium heat-resistant castings **A1:** 922
Chromium hot-work steels **A1:** 762
catastrophic die failure/plastic deformation **A18:** 641
ceramic coatings for dies **A18:** 643
composition limits... **A18:** 735
for hot extrusion tools **A18:** 627
for hot-forging dies **A18:** 623, 624, 625
resistance to abrasive wear **A18:** 638
service temperature of die materials in forging **A18:** 625
thermal fatigue in dies **A18:** 639
Chromium hydroxide
sol-gel processing **EM4:** 447
Chromium in cast iron *See also* High-chromium cast irons; Nickel-chromium cast irons **M1:** 77, 78-79
ductile iron, effect on magnetic properties **M1:** 54
gray iron **M1:** 21, 26, 27, 28, 29, 30
Chromium in steel **M1:** 115, 411, 417
400 to 500 °C embrittlement role in............. **M1:** 686
500 °F embrittlement, role in **M1:** 685
atmospheric corrosion resistance **M1:** 717, 721-722
castings, effect in **M1:** 388
constructional steels for elevated temperature use effect on **M1:** 647, 649-650
formability reduced by.................................. **M1:** 554
hardenability affected by **M1:** 477
nitriding, effect on.................................... **M1:** 540-541
notch toughness, effect on **M1:** 693
seawater corrosion, effect on.......... **M1:** 741-742, 745
sigma-phase embrittlement, role in **M1:** 686
soil corrosion, effect on **M1:** 730
temper embrittlement, role in **M1:** 684, 703
Chromium iron *See also* Cast irons.......... **A9:** 245-246
Chromium iron, 25Cr iron
erosion test results **A18:** 200
Chromium iron rolls
PCBN cutting tools **A16:** 115
Chromium metallization
thin-film hybrids **EL1:** 326
Chromium nitride
coating for jet engine components **A18:** 592
in fracture surface **A7:** 253, 254
Chromium nitride sensitization
P/M stainless steels **A13:** 828, 830
Chromium nitrides, appearance of
in iron-chromium-nickel heat-resistant casting alloys **A9:** 332
in wrought stainless steels............................. **A9:** 284
Chromium oxide
metal-to-metal oxide equilibria...................... **A7:** 340
overlayed on cast irons **A6:** 721
Chromium oxide (Cr_2O_3)
applications **EM4:** 47
applications as a coating.............................. **EM4:** 208
ceramic coatings for dies **A18:** 643-644
coating for gas-lubricated bearings............ **A18:** 532
coating for seals.. **A18:** 551
coating formation in molten particle deposition... **EM4:** 206
composition **EM4:** 47
corrosion resistance of refractories **EM4:** 391
crystal structure **EM4:** 30
densified coating, properties, adiabatic engine use **EM4:** 990
densified, scuffing temperatures and coefficients of friction between ring and cylinder liner material.. **EM4:** 991
for gas-lubricated bearings **A18:** 532
plasma-spray coating for pistons **A18:** 556
plasma-sprayed, properties **EM4:** 990
plasma-sprayed, scuffing temperatures and coefficients of friction between ring and cylinder liner material **EM4:** 991
properties **EM4:** 30
supply sources **EM4:** 47
thermal spray coating material.................... **A18:** 832
Chromium oxide coating
molybdenum **M5:** 662
Chromium oxide scale
on alloy 800 .... **A13:** 97
Chromium oxide-silicon oxide (Cr_2O_3-SiO_2)
applications as a coating............................ **EM4:** 208
Chromium phosphate
chromate conversion coating **A13:** 394
Chromium phosphate coatings...................... **M5:** 457
Chromium plate
broaching **A16:** 203
diamond as abrasive for honing **A16:** 476
friction coefficient data.................................... **A18:** 74
galling resistance with various material combinations **A18:** 596
hone forming... **A16:** 488
honing **A16:** 476, 477
micronhoning **A16:** 490
Chromium plating
atmospheric corrosion resistance enhanced by... **M1:** 722
cast irons **M1:** 102
corrosion protection **M1:** 753, 754
maraging steels **M1:** 448
surface condition for....................................... **M1:** 157
threaded fasteners **M1:** 279
for wear resistance **M1:** 606, 638
Chromium plating aluminum and aluminum alloys **M5:** 608-610
applications **M5:** 170-171, 179-180, 182-183, 188, 190-195
copper and copper alloys **M5:** 622
decorative *See* Decorative chromium plating
equipment.......................... **M5:** 173, 177-180, 192-195
hafnium alloys **M5:** 668
hard *See* Hard chromium plating
heat-resistant alloys **M5:** 566
hexavalent process **M5:** 188-189
magnesium alloys, copper-nickel-chromium and decorative chromium systems **M5:** 646-647
stripping of **M5:** 646
maintenance schedules.......................... **M5:** 179, 195
microcracked *See* Microcracked chromium plating
molybdenum **M5:** 660
nickel plating process using.................. **M5:** 205, 207
niobium **M5:** 663-664
rinsewater recovery **M5:** 317-318
selective **M5:** 194
solution compositions and operating conditions......... **M5:** 172-177, 181-183, 189-191, 663-664, 668
stripping of.. **M5:** 198, 646
tantalum **M5:** 663-664
trivalent process **M5:** 196-198
tungsten... **M5:** 660
zirconium alloys **M5:** 668
Chromium plating of specimens for edge retention **A9:** 32
Chromium, pure...................................... **M2:** 724-725
solution potential.. **M2:** 207
Chromium reduction process
plating waste disposal............................ **M5:** 311-312
Chromium stainless steel
composition of **A1:** 912
Chromium stainless steels, aluminum coating
effects of variables on..................................... **M5:** 343
Chromium steels
composition of tool and die steel groups... **A6:** 674
corrosion in nitric acid **M3:** 85-86
dissimilar metal joining.................................. **A6:** 827
SCC effect of nickel alloying **A13:** 273
Chromium sulfides in austenitic stainless steels **A9:** 284
Chromium toxicity... **A2:** 1242
Chromium trioxide
as an etchant for wrought stainless steels... **A9:** 282
grades **A9:** 67
in Group V electrolytes **A9:** 54
Chromium vacuum coating............. **M5:** 389-390, 392
Chromium, vapor pressure
relation to temperature **M4:** 309, 310
Chromium white irons
hardness and good abrasive resistance....... **A18:** 188
Chromium-22, treatment
magnesium alloys **M5:** 634-635, 640-641, 644
Chromium-antimony alloys
peritectic transformations **A9:** 678
Chromium-base hot-work tool steels
compositions **A14:** 43
forging temperatures **A14:** 81
heat treating **A14:** 54
tempering temperature effects **A14:** 55
Chromium-bearing copper nickel alloys
applications and properties............................ **A2:** 341
Chromium-carbide/nickel-chromium
coating compositions for jet engine components.................................... **A18:** 590, 591
Chromium-coated steels
resistance spot brazing **M6:** 479
Chromium-cobalt-iron alloys *See* Magnetic materials
Chromium-containing alloys, specific types
0.32C-3Cr-1Mo-0.3V, plasma nitriding.......... **A4:** 405
1.2C-1.5Cr, distortion in heat treatment **A4:** 614
2C-12Cr, distortion in heat treatment........... **A4:** 612, 615
18Cr-9Ni, boriding **A4:** 439
35Ni-15Cr
for cast element material in heat-treating furnaces.......................... **A4:** 472
heat-resistant alloy applications......... **A4:** 515, 516, 517
35Ni-18Cr, recommended for parts and fixtures for salt baths **A4:** 514
35Ni-18Cr-44Fe, ribbon material in heat-treating furnaces **A4:** 472
35Ni-20Cr, heat-resistant alloy applications.. **A4:** 515
68Ni-20Cr, for element strip material in heat-treating furnaces **A4:** 472
80Ni-20Cr
for element strip material in heat-treating furnaces.......................... **A4:** 472
heat-resistant alloy applications................. **A4:** 516
Cr-Ni-V steel
applications, austempered parts **A4:** 157
austenitizing.. **A4:** 162
Ni-Cr-Mo, distortion in heat treatment......... **A4:** 614
Chromium-copper alloys
age hardenable.. **A2:** 236
applications and properties...................... **A2:** 290-291
Chromium-cracked panels
for liquid penetrant inspection **A17:** 88
Chromium-iron-niobide phase (Z phase)
in austenitic stainless steels............................ **A9:** 284
Chromium-molybdenum
alloy composition and abrasion resistance **A18:** 189
coating for titanium alloys **A18:** 781, 782
Chromium-molybdenum alloy steels
horizontal centrifugal casting **A15:** 299
as low-alloy **A15:** 716
Chromium-molybdenum cast steels................ **A1:** 374
Chromium-molybdenum heat-resistant steels.. **A1:** 619-630
ASTM specifications **A1:** 157
compositions of.. **A1:** 158
definition of... **A1:** 149
Chromium-molybdenum steel
creep curve **A8:** 331
creep damage from service exposures **A8:** 338
log stress vs. log rupture life curves **A8:** 333
rupture strength **A8:** 340
Chromium-molybdenum steel alloys
fatigue fracture in.......................... **A11:** 129, 395-396
pinion fracture **A11:** 395-396
superheater tubes, overheating rupture of... **A11:** 609
Chromium-molybdenum steels................ **A1:** 149-150, 618-620
0.5Mo steel... **A1:** 619-620
1.0Cr-0.5Mo steel **A1:** 620
2.25Cr-1Mo steel **A1:** 620, 645-647
9Cr-1Mo steel...................... **A1:** 620, 622, 623, 625, 937
allowable stresses **A1:** 625
boiler applications **M1:** 747
classification and group description..... **A6:** 405, 406, 407
compositions **A1:** 618
corrosion resistance in steam systems.......... **M1:** 751
creep embrittlement **A12:** 124
creep strengths... **A1:** 620
creep-rupture strength **A1:** 622, 937

Chromium-molybdenum steels (continued)
damaged classification vs. expended creep-life fraction **A6:** 1115
diffusion welding **A6:** 884
dissimilar metal joining **A6:** 823
electroslag welding **A6:** 276, 277, 278, 279
flux-cored arc welding, designator **A6:** 189
high-frequency welding **A6:** 252
modified chromium-molybdenum steels **A1:** 621, 939
room-temperature tensile properties **A1:** 618
stress-rupture strength variation **A6:** 1113
tensile properties at elevated temperatures **A1:** 624
weldability **A6:** 420-421
Chromium-molybdenum steels, specific types
2.5Cr-1Mo, trace element impurity effect on GTA weld penetration **A6:** 20
9Cr-1Mo, transformation effect on transient weld stresses **A6:** 80
12CrMo, transformation effect on transient weld stresses **A6:** 80
26Cr-1Mo, weldability **A6:** 452, 453
A 217, composition and carbon content **A6:** 406
A 387, composition and carbon content **A6:** 406
Chromium-molybdenum-vanadium steels
for elevated-temperature service **A1:** 619, 620-621, 624, 937, 939
Chromium-nickel alloys *See* Nickel-chromium alloys; Nickel-iron- chromium alloys
17Cr-4Ni precipitation-hardening stainless steel **A18:** 222
milling **A16:** 313
Chromium-nickel stainless steel
principal ASTM specifications for weldable steel sheet **A6:** 399
Chromium-nickel steels
solderability **A6:** 971
Chromium-nickel-molybdenum alloy steel
effect of bushings on fatigue strength of **A11:** 470
Chromium-plated steel
resistance spot welding **M6:** 491
Chromium-plated steels
deformation resistance vs rolling temperature **A14:** 119
for draw rings **A14:** 510
press forming of **A14:** 563-564
Chromium-silicon monoxide (Cr-SiO)
hot pressing **EM4:** 191
Chromium-silicon steel *See also* Steel, ASTM specific types, A401; Steels, AISI-SAE specific types, 9254
modulus of rigidity **M1:** 300
spring wire, cost **M1:** 305
spring wire, stress relieving **M1:** 291
Chromium-silicon steel spring wire and strip
characteristics of **A1:** 306-307
Chromium-silicon steel VSQ wire
characteristics of **A1:** 307
Chromium-to-nickel ratio in heat-resistant casting alloys, effect on ferrite
formation **A9:** 333
Chromium-tungsten-cobalt alloys
advantages **A6:** 797
applications **A6:** 797
Chromium-tungsten-cobalt alloys, advantages and applications of materials for surfacing build-up
and hardfacing **A18:** 650
Chromium-vanadium
spray for hardfacing **M6:** 789
Chromium-vanadium steel *See also* Steels, AMS specific types; Steels, ASTM specific types A231, A232
electropolishing of **M5:** 308
modulus of rigidity **M1:** 300
Chromium-vanadium steel (continued)
spring wire, cost **M1:** 305
spring wire, stress relieving **M1:** 291
Chromium-vanadium steel spring wire and strip
characteristics of **A1:** 306-307
Chromium-vanadium steel wire
characteristics of **A1:** 307
Chromized 1006 sheet steel **A9:** 200
Chromized sheet steels
color etched **A9:** 156
specimen preparation **A9:** 197-198
Chromizing *See also* Chromadizing
defined **A13:** 3
Chromodizing *See* Chromadizing
Chromophores, isolated
in UV/VIS analyses **A10:** 63
Chromosil steel **M1:** 624
Chronic berylliosis
from beryllium powder/dust exposure **A2:** 687
Chronic granulomatous pulmonary disease (berylliosis) **A2:** 1239
Chronic interstitial nephropathy
as lead toxicity effect **A2:** 1244
Chronic manganese poisoning (manganism) **A2:** 1253
Chronic pulmonary disease
from cadmium exposure **A2:** 1240
Chrysler Corporation
flow rate test **A7:** 280
Chrysotile
as filler **EM3:** 177
Chucking **A16:** 156, 157
stainless steel **A16:** 154
Chucking machines (chuckers) **A16:** 136, 140, 141, 286
boring **A16:** 160
die threading **A16:** 296
reamers **A16:** 242
roller burnishing **A16:** 252
tapping **A16:** 256
Church shape factor **A7:** 239
Churchill two-line method
SSMS calibration curves **A10:** 143
Chute
defined **A7:** 2
rifflers, particle sizing sampling technique **A7:** 226
Chute feeds
blanks **A14:** 500
Chvorinov's rule
solidification times **A15:** 601, 779-780, 860
CIDI cervit
elastic constant **EM4:** 875
Cigarette lighter flint
powder used **A7:** 578
CIL flow test **EM3:** 8
defined **EM2:** 9
CIM *See* Compression injection molding
Cinchonine
as narrow-range precipitant **A10:** 169
Cinders
steel corrosion in **M1:** 729
CIP *See* Cold isostatic pressing
Circ winding *See* Circumferential ("circ") winding
CIRCLE *See* Cylindrical internal reflection cell
Circle arc elongation test
for sheet metals **A8:** 556
Circle grid
analysis **A8:** 567 **A14:** 2, 895-896
defined **A14:** 2
uniaxial tensile testing **A8:** 554
Circle grid analysis
for steel sheet **A1:** 575-576
Circle shearing *See* Rotary shearing
Circles, as gage marks
sheet metal forming **A8:** 549
Circuit
boards, defined **EM1:** 6
defined **EM1:** 6 **EM2:** 9
Circuit areas, planar
interconnections **EL1:** 6
Circuit board *See also* Printed circuit board; Printed wiring board **EM3:** 8
defined **EM2:** 9-10
Circuit boards *See also* Boards; Circuitry; Circuits; FR 4 glass-epoxy boards; Printed wiring assemblies; Printed wiring boards
bare copper **EL1:** 561
solderability, inspection methods **EL1:** 944
wire-wrapped **EL1:** 7
Circuit breakers **A13:** 1118-1119
life tests in **A2:** 859
recommended contact materials **A2:** 863
Circuit effects extraction
high-frequency digital systems **EL1:** 86
Circuit voltage, effect
electrical contact materials **A2:** 840
Circuit wiring demand *See* Wiring demand
Circuit(s) applicable technologies **EL1:** 161
approximation **EL1:** 30-31
breakers **EL1:** 1108
cell-based **EL1:** 168
configuration, commercial hybrids **EL1:** 381
design, flexible printed boards **EL1:** 586-588
-related failure mechanisms **EL1:** 974-975
speed as determined by dielectric constant **EL1:** 1
types **EL1:** 160-161
Circuit-pack level
of interconnection **EL1:** 13
Circuitization
materials and processes selection **EL1:** 115
sequential process **EL1:** 133-134
yield loss, and cost **EL1:** 112
Circuitry *See also* Circuit(s); Circuitization
characteristics, electrical contact materials
configuration, commercial hybrids **EL1:** 381
defective, replacement classes **EL1:** 9
direction of current, defined **EL1:** 95
eddy current inspection **A17:** 167
gating **A17:** 253
magabsorption **A17:** 148
marginal oscillator **A17:** 151
modfied Villard/Greinacher **A17:** 305
network logic **EL1:** 2
pulsar, ultrasonic inspection **A17:** 252
receiver-amplifier, ultrasonic inspection **A17:** 253
thermocouple extension wires **A2:** 876
Villard, radiography **A17:** 305
Circuits
electrical, EPMA failure analysis for **A10:** 531
for electrolysis **A10:** 199
polarographic **A10:** 189
Circular end brushes **M5:** 155
Circular faceplates
in torsional testing equipment **A8:** 145-146
Circular fluorescent-light tubes
photographic **A12:** 83
Circular holes *See also* Hole
fracture analysis **EM1:** 252-255
specimens/experimental data **EM1:** 255-257
Circular magnetization *See also* Magnetization
by central conductors **A17:** 96
current strength **A17:** 105
defined **A17:** 90-91
electrical **A7:** 576
Circular plaque design
of penetrameters **A17:** 339
Circular sawing *See* Sawing
Circular spall
in steels **A12:** 113-115, 123-128
Circular test grids
for Hilliard's grain size measurement **A9:** 130
superimposed on a micrograph **A9:** 127
used in quantitative metallography **A9:** 124
Circular-step bearing *See also* Step bearing
defined **A18:** 5

Circulating fluid systems **A18:** 133
Circulating oil lubrication systems **A18:** 133
Circulating oils
pour-point depressants **A18:** 108
Circulating systems
for liquid metal systems **A13:** 95-96
Circulation
aluminum melt **A15:** 453-456
of chemical cleaners **A13:** 1139
forced, advantages **A15:** 454-456
Circulation loop
for liquid metal systems **A13:** 95-96
Circulation pump *See also* Molten metal pump
aluminum furnace **A15:** 455
electric centrifugal **A15:** 456
for hydrogen removal **A15:** 461
Circulation-type attrition mills **A7:** 69
Circumferential ("circ") winding
defined **EM1:** 6 **EM2:** 10
as hoop patterns **EM1:** 509-510, 514
Circumferential annulus
hydrostatic gas-lubricated bearings **A18:** 528
Circumferential corrosion-fatigue cracks
low-alloy steel superheater tube **A11:** 79
Circumferential strain
in upset testing **A8:** 579-580
Circumferential stress
during bending **A8:** 121, 123
Circumferential weaving
of fabrics and preforms **EM1:** 129, 131
Circumferential welding
electroslag welds **M6:** 226, 233
gas tungsten arc welds **M6:** 210
in stainless steels **M6:** 339-340
plasma arc welds **M6:** 221
in stainless steels **M6:** 344
submerged arc welds **M6:** 139-140
in stainless steels **M6:** 329
Cis isomers
chemistry **EM2:** 64
CIS stereoisomer **EM3:** 8
defined **EM2:** 10
Cis-1,4-polyisoprene *See* Rubber, natural
Citrate solutions
copper/copper alloy SCC in **A13:** 634
Citric acid
as chemical cleaning solution **A13:** 1141
corrosion of stainless steels in **M3:** 82
Citric acid spot test
porcelain enamel **M5:** 527-529
CL *See* Cathodoluminescence
Cl-Cs (Phase Diagram) .... **A3:** 2 • 138
Cl-Ga (Phase Diagram) .... **A3:** 2 • 139
Cl-Hg (Phase Diagram) .... **A3:** 2 • 139
Cl-In (Phase Diagram) .... **A3:** 2 • 139
Cl-Na (Phase Diagram) .... **A3:** 2 • 140
CLA process *See* Counter-gravity low-pressure casting
CLA2D laminate analysis computer program **EM1:** 274
Clad brazing materials, application of **A6:** 961-963
advantages **A6:** 963
assembly **A6:** 962-963
atmosphere for brazing **A6:** 963
brazing parameters **A6:** 963
cleaning **A6:** 962-963
cold roll bonding process steps **A6:** 962
copper cladding amount **A6:** 962
definition of clad brazing material **A6:** 961
design considerations **A6:** 961
embrittlement **A6:** 961
fabrication of **A6:** 961
galvanic coupling, corrosion due to **A6:** 961
low-carbon steel base metals **A6:** 961
material selection **A6:** 961
precautions during brazing cycle **A6:** 961
stainless steel base metals **A6:** 961
stamping **A6:** 962-963
thickness ratio **A6:** 962
three-layer **A6:** 963
two-layer **A6:** 962
Clad brazing materials, brazing with **A6:** 347-348
aluminum alloys **A6:** 347
applications **A6:** 347, 348
cladding materials **A6:** 347-348
advantages **A6:** 348
Clad brazing materials, brazing with (continued)
fabrication of **A6:** 347-348
copper **A6:** 347
definition of clad brazing material **A6:** 347
design and manufacturing considerations **A6:** 348
filler metals, 5W-1 .... **A6:** 348
fluxes **A6:** 347, 348
formation of clad brazing material **A6:** 347
stainless steels **A6:** 347
steel **A6:** 347
titanium **A6:** 347
Clad brazing sheet
definition **A6:** 1207 **M6:** 3
Clad metal
definition **M6:** 3
Clad metal combinations
high purity nickel strip for **A7:** 403
Clad metal composites
as heat sinks **EL1:** 1130-1131
Clad metals **A13:** 887-890
for carbon steel **A13:** 523
defined **A13:** 3
Clad overlays
precious metal **A2:** 848
Clad plate
refractory metals and alloys **A2:** 559
Clad tube
wrought aluminum alloy **A2:** 33
Clad vessels
flaws and inspection methods **A17:** 646, 654
Cladding
aluminum-lead strip **A7:** 408
angle arrangements **M6:** 705
brazing filler metals available in this form **A6:** 119
by roll welding **M6:** 689-691
corrosion-resistant **A13:** 652, 931
defined **A18:** 5
definition **M6:** 3
of electrical contacts **A2:** 848
laser process *See* Laser cladding; Laser processing techniques
metal combinations **M6:** 690-691
as metallic coating for corrosion control **A11:** 195
nickel-base alloys and welding considerations **A6:** 591
oxidation-resistant coatings **M5:** 665-666
parallel arrangements **M6:** 705
plasma-MIG welding **A6:** 224
procedure **M6:** 689-691
resistance to cavitation erosion **A18:** 217
steel sheet with aluminum rolling process **M5:** 345-346
strip roll welding **A6:** 314
of uranium dioxide fuel rods **A7:** 664
and weld repair, nuclear reactors **A13:** 970-971
Cladding materials for reactors
neutron embrittlement of **M1:** 686
Cladding thickness of aluminum alloys
etchants for examination **A9:** 355
Cladless hot isostatic pressing **EM4:** 194, 196-197, 199
Cladosporium
biological corrosion by **A13:** 118-119
Clamer, Dr. G.H
as inventor **A15:** 32
Clamp mounting
of porcelain enameled sheet steel **A9:** 198
zinc and zinc alloys **A9:** 488
Clamp mounts **A9:** 28-29, 167-168
Clamp spring
stress-relaxation test setup **A8:** 327
Clamp-off
as casting defect **A11:** 384
Clamp-up
of mechanical fasteners **EM1:** 706-707
Clamp/joint friction grip
for fatigue test specimen **A8:** 371
Clamping blocks
for bending **A14:** 665-666
Clamping dies
flash welding **M6:** 529-532
Clamping forces
of threaded steel fasteners **A1:** 300-301
Clamping of carburized steel specimens **A9:** 217
Clamping pressure
defined **EM1:** 6 **EM2:** 10
Clamps
blind hole **EM1:** 711
for direct contact magnetization **A17:** 94
hold-down, distortion in **A11:** 140
ring, brittle fracture from burning of **A11:** 332-333
for stored-torque Kolsky bar **A8:** 219-221
-strap assembly, brittle fracture of **A11:** 69-70
strap-type, stress-corrosion failure **A11:** 309
as tension source for stress-corrosion cracking **A8:** 502
in torsional testing equipment **A8:** 146
Clamshell markings *See* Beach marks; Fatigue striations
Clamshell marks *See* Beach marks
CLAP laminate analysis computer program **EM1:** 274
Clapeyron, Benoit .... **A3:** 1 • 8
Clarification process, plating waste disposal **M5:** 312-313
retention time **M5:** 312-313
sludge blanket in **M5:** 312-313
Clarity
of polycarbonates (PC) **EM2:** 151
CLAS process *See* Counter-gravity low-pressure casting
Clash **A18:** 564
Class A extensometers **A8:** 618
Class B-1 extensometers **A8:** 618-619
Class B-2 extensometers **A8:** 619
CLASS computer program for laminate analysis **EM1:** 269, 274, 275-281, 452-454
Class I alloys
creep in **A8:** 308
Class I through Class IV
P/M parts **A7:** 332, 333, 463
Classic Moore hip endoprosthesis **A11:** 670
Classic Taylor test *See* Rod impact (Taylor) test
Classical Bagby compression bone plate
as internal fixation device **A11:** 671
Classical design
and distortion failures **A11:** 136
Classical diffusion theory **EM3:** 631-632
Classical, electrochemical, and radiochemical analysis
capabilities **A10:** 333
classical wet analytical chemistry **A10:** 161-180
controlled-potential coulometry **A10:** 207-211
electrogravimetry **A10:** 198-201
electrometric titration **A10:** 202-206
elemental and functional group analysis **A10:** 212-220
high-temperature combustion **A10:** 221-225
inert gas fusion **A10:** 226-232
neutron activation analysis **A10:** 233-242
potentiometric membrane electrodes **A10:** 181-187
radioanalysis **A10:** 243-250
voltammetry **A10:** 188-196
Classical gravimetric analysis **A10:** 170-171
Classical nucleation theory **EM3:** 408, 409
Classical sculpture
metalworking of **A15:** 20-21
Classical Sherman bone plate
as internal fixation device **A11:** 671
Classical wet analytical chemistry **A10:** 161-180
applications **A10:** 161, 178-179
appropriateness of methods **A10:** 162
basic chemical equilibria and analytical chemistry of **A10:** 162-165
as basis for spectrographic calibration **A10:** 162
buffer solution **A10:** 163
capabilities compared with ion chromatography **A10:** 501
capabilities, compared with voltammetry **A10:** 188
chemical equilibrium **A10:** 162-165
chemical surface studies **A10:** 177-178
common ion effect **A10:** 163
coprecipitation **A10:** 163
estimated analysis time **A10:** 161
general uses **A10:** 161
gravimetry **A10:** 170-171

Classical wet analytical chemistry (continued)
half-cell reactions **A10:** 163-164
inclusion and second-phase testing **A10:** 176-177
introduction **A10:** 162
Jones reductor **A10:** 175-176
Kjeldahl determination **A10:** 172-173
limitations **A10:** 161
partitioning oxidation states **A10:** 178
qualitative methods **A10:** 167-168
reduction-oxidation reactions **A10:** 163-164
related techniques **A10:** 161
sample dissolution **A10:** 165-167
samples **A10:** 161, 165, 178-179
separation techniques **A10:** 168-170
solubility products constant **A10:** 163
techniques for subdividing solids in **A10:** 165
titrations, acid-base **A10:** 172-173
titrimetry **A10:** 171-176
Classical wet chemistry *See* Classical wet analytical chemistry
Classical wet chemistry analysis
for phosphorus detected in glassivation layers **A11:** 41
Classical wet methods
appropriateness of **A10:** 162
Classification *See also* Categorization; Nomenclature; Terminology
of carbides **A2:** 968
of casting defects, international **A15:** 545-553
of casting processes **A15:** 203-207
of cemented carbides, machining applications **A2:** 953-954
of cermets **A2:** 979
of coremaking processes **A15:** 138
of ferrous casting alloys **A15:** 627-628
of gear failures **A11:** 590-597
of hybrid packages **EL1:** 451-454
international, of casting defects **A11:** 380
of- sliding bearings **A11:** 483
performance range, of electronic packages **EL1:** 25
of resin binder processes **A15:** 214
of stress **EM2:** 751-752
Classification, classifying
classify defined **A7:** 2
Classification of cast iron **A1:** 3
Classification of steel **A1:** 140-141
Clausius, Rudolf .... **A3:** 1 • 7, 8
Clauslus-Clapeyron equation .... **A3:** 1 • 8, 10
Clay
chemical composition **A6:** 60
function and composition for mild steel SMAW electrode coatings **A6:** 60
as grease thickener **A18:** 126, 129
high level waste disposal in **A13:** 980
Miller numbers **A18:** 235
as sheet molding compound filler **EM1:** 158
typical oxide compositions of raw materials **EM4:** 550
Clay molds
historic use **A15:** 19
Clay products **EM4:** 1
structural, testing **EM4:** 547
Clay(s)
bentonites **A15:** 210, 341
bonding efficiency **A15:** 225
defined **A15:** 3
fireclay **A15:** 210-211
for green sand molding **A15:** 224-225, 341
investment molds **A15:** 16
properties, controlling **A15:** 224-225
Southern bentonite **A15:** 210, 341
Western bentonite **A15:** 210, 341
Clay-bonded sand
ramming of **A15:** 203
reclamation of **A15:** 354-355
Clay-water bonds *See also* Aggregate moldin material; Bonds; Clay(s); Sand(s)
characteristics **A15:** 212
Clays
in ceramic tiles **EM4:** 926, 928
Clean bright wire finish
for steel wire **A1:** 279 **M1:** 261
Clean rooms
pressure for **EM1:** 144
Clean surface **EM3:** 8
Clean-room microscopy **A9:** 83
Cleanability, flux
defined **EL1:** 644
Cleaner bath
control of **A13:** 382
Cleaners **EM3:** 52
acid **A13:** 381
alkaline **A13:** 381
for base metals **EL1:** 678
and coaters, iron phosphating **A13:** 386
formulation **A13:** 382
high- and low-temperature **A13:** 382
liquid **A13:** 382
organic solvent **EL1:** 662-663
powder **A13:** 382
Cleaning *See also* Cleaning materials; Cleaning stations; Cleanliness; "No-clean" applications; Postcleaning; Precleaning; Surface preparation
abrasive **A13:** 414, 1143
abrasive blast **A13:** 414 **A14:** 230
abrasive blast, of Replicast castings **A15:** 271
after soldering/interconnection **EL1:** 117
alkaline **A13:** 304
alkaline, effect on fatigue fracture **A11:** 126-127
of aluminum alloy forgings **A14:** 248-249
aluminum alloys **A15:** 762-763
of aluminum/aluminum alloys **A13:** 603
aqueous **EL1:** 663-666
beryllium-copper alloys **A2:** 414
beryllium-nickel alloys **A2:** 424
chemical, for solderability **EL1:** 678
chemical, of boiler tubes **A11:** 616
chemical, of process equipment **A13:** 1137-1143
chemical, of- heat-exchanger tubing **A11:** 630
for chromate conversion coating **A13:** 389
of copper and copper alloy forgings **A14:** 258
of crucible furnaces **A15:** 383
in deep drawing **A14:** 589
effect on fatigue strength **A11:** 126
effect, optical properties **EM2:** 485-486
effects on bearing strength in aluminum alloys **A8:** 60
effects, stainless steel corrosion **A13:** 551
electrolytic alkaline **A7:** 459
electrolytic, postforging defects from **A11:** 333
emulsion **EL1:** 666
of encapsulated powders **A7:** 430
of equipment, design effects **A13:** 340
equipment, for surface conversion **A13:** 381-382
excessive, as casting detect **A11:** 385
in extrusion **EM2:** 386
of failed parts **A11:** 173
flexible printed boards **EL1:** 590
fluxes **A15:** 446
of fracture surfaces **A11:** 19
of friction welds **A11:** 444
grades, abrasive blasting **A13:** 414
hand tool **A13:** 414
health and safety regulations **A17:** 81-82
of heat-exchanger tubes **A11:** 629, 630
hydrogen embrittlement of bolts from **A11:** 540-541
for imaging, rigid printed wiring boards **EL1:** 542
immersion specimens **A13:** 221
magnesium alloy forgings **A14:** 260
of magnesium alloys **A13:** 750
for marine organic coatings **A13:** 912
materials **EL1:** 661-666
materials selection **EL1:** 668
methods **A7:** 459
methods, liquid penetrant inspection **A17:** 80-81
Cleaning (continued)
of military board designs **EL1:** 517
and mold life **A15:** 281
oil/gas production pipe **A13:** 1258-1259
oils, surface **A13:** 380
on-line **A13:** 1143
operations, robotic **A15:** 468-569
for passivation **A13:** 552
plain carbon steels **A15:** 712-713
postweld, associated corrosion **A13:** 350
power tool **A13:** 414
preoperational, chemical processing **A13:** 1138-1139
of press formed parts **A14:** 548
printed wiring boards, for conformal coatings **EL1:** 763
procedures, pin bearing testing **A8:** 59
process, flow chart **EL1:** 777
refractory metals and alloys **A2:** 563
and removal methods, for conversion **A13:** 380-381
rigid printed wiring boards **EL1:** 547
sample, uniform corrosion testing **A13:** 230
as secondary operation **A7:** 451, 459
and simultaneous deburring **A7:** 458
in solder masking **EL1:** 554
of soldered joints **A17:** 609
solutions, hydrogen embrittlement testing for **A8:** 541
solvent **A13:** 413-414
stainless steels **A14:** 230
steam **A13:** 414
substrate, for urethane conformal coating **EL1:** 776-778
surface -mount assemblies **EL1:** 666-667
surface, before coating **A15:** 561
for surface conversion **A13:** 380-382
thermal **A13:** 1143
for thermal spray coatings **A13:** 460
for thin-film hybrids **EL1:** 319
time, and surfactant concentration **A13:** 380
of titanium alloys **A14:** 280-281, 839-840
typical shapes requiring **A7:** 459
water **A13:** 414, 1143
of welds, for inspection **A17:** 591
wrought copper and copper alloys **A2:** 247
Cleaning and finishing
stainless steel **M3:** 52-55
Cleaning compounds
safety precautions **A6:** 1196
Cleaning fluxes, aluminum
ternary phase diagram **A15:** 446
Cleaning for
arc welding of
cobalt-based heat-resistant alloys **M6:** 368
titanium and titanium alloys **M6:** 448-449
brazing of aluminum alloys **M6:** 1025
electron beam welding **M6:** 612
furnace brazing of steels **M6:** 938
resistance brazing **M6:** 982-983
resistance welding of
aluminum alloys **M6:** 539
copper and copper alloys **M6:** 548
Cleaning index
determining **M5:** 20
Cleaning materials *See also* Cleaning **EL1:** 661-666
aqueous cleaning **EL1:** 663-666
contaminant types **EL1:** 658-661
contaminants/residues, sources of **EL1:** 658
emulsion cleaning **EL1:** 666
environmental considerations **EL1:** 667
materials selection parameters **EL1:** 668
measurement of cleanliness **EL1:** 667-668
no cleaning **EL1:** 666
organic solvent cleaners **EL1:** 662-663
for surface-mount assemblies **EL1:** 666-667
Cleaning of specimens to be mounted **A9:** 28
Cleaning process, selection of **M5:** 03-21
cleanliness, degree of *See* Cleanliness, degree of

Cleaning process, selection of (continued)
mechanism of action **M5:** 3-4
pollution control and resource recovery **M5:** 20
safety considerations **M5:** 3-20
soil types, effects of **M5:** 3-15
substrate considerations **M5:** 3-14
surface preparation procedures **M5:** 5, 16-19
Cleaning solvents *See also* Cleaning,- Solvents
resistance, of flexible epoxies **EL1:** 821
Cleaning stations
liquid penetrant inspection **A17:** 78, 80-82, 84
Cleaning techniques
air blast **A12:** 74
brush **A12:** 74
cathodic **A12:** 75
chemical etching **A12:** 75-76
of fracture surfaces **A12:** 73-77, 179-183
organic solvents **A12:** 74
replica-stripping **A12:** 74
water-base detergent **A12:** 74
cleanliness
and fatigue resistance **A1:** 678-679, 681, 682
measurement of **EL1:** 667-668
in soldering process **EL1:** 631
wafer-surface **EL1:** 192
Cleanliness, billet
for niobium-titanium superconducting materials **A2:** 1046
Clear acrylic lacquers
as fracture preservatives **A12:** 73
Clear chromate conversion coating
cadmium plate **M5:** 269
Clear openings
screening **A7:** 176
Clearance *See also* Die clearance
blanking/piercing, stainless steels **A14:** 761-762
in deep drawing **A14:** 575
defined **A7:** 2
gib, and press accuracy **A14:** 495
large, in piercing **A14:** 461-462
punch-to-die, deep drawing effects **A14:** 581
side, die design with **A14:** 133
slitting knives **A14:** 708-709
small, in piercing **A14:** 461
in straight-knife shearing **A14:** 705
and tool size, piercing **A14:** 462-463
Clearance ratio
defined **A18:** 5
Cleavage *See also* Cleavage facets; Cleavage fractures; Cleavage steps; Intergranular fracture; Intergranular fractures; Transgranular cleavage fractures **A8:** 2
alloy steels **A12:** 338
Alnico alloy **A12:** 461
of alpha on alpha-beta phase field **A8:** 487
austenitic stainless steels **A12:** 352
brittle fracture by **A11:** 82
in brittle materials **A11:** 75
by high deformation rates **A8:** 251
of carbon metastable austenite **A8:** 481
crack, defined **A11:** 2
-crack nucleation, fracture surface **A11:** 23
crack path, low-carbon steel **A12:** 117
cyclic **A8:** 484-485, 487
defined **A9:** 4 **A11:** 2 **A12:** 13-14 **A13:** 3 **EM2:** 10
effect of anodic dissolution **A12:** 42
effect on measurement point in structural steels **A8:** 458
facets, of transgranular brittle fracture **A11:** 22
fracture, and aluminides **A2:** 930-931
fracture, defined **A13:** 3
fracture mechanics and **A11:** 47
fracture mechanics of **A8:** 439
from brittle ordered intermetallics **A2:** 914
historical study **A12:** 4
intergranular and transgranular, compared **A12:** 290
iron alloys **A12:** 223, 459
local, tool steels **A12:** 382
low-carbon steel **A12:** 174
materials illustrated in **A12:** 217
and mechanical twinning **A12:** 4
metal-matrix composites **A12:** 467
overload fracture **A8:** 479-481
plane, defined **A11:** 2
plane, step-wise growth on Fe-Ni **A8:** 484-485

Cleavage (continued)
planes **A12:** 13
and ripples, compared **A12:** 453
river patterns on **A12:** 252, 424
and slip, in hcp and bcc metals **A11:** 75
steps **A12:** 223, 352
stress-intensity factor range effect **A8:** 485-486
in subcritical fracture mechanics (SCFM) **A11:** 47
surface features **A12:** 13-18
tools steels **A12:** 382
transcrystalline, iron **A12:** 222
transgranular **A12:** 175, 461, 467
as transgranular fracture mode, SEM defined **A12:** 175
and trialuminides **A2:** 930-931
and void growth, aluminum fracture **A8:** 478-479
wrought aluminum alloys **A12:** 418
Cleavage crack
defined **A9:** 4
definition **EM4:** 632
Cleavage facets
AISI/SAE alloy steels **A12:** 319
and dimples, sizes of **A12:** 328
formation **A12:** 339
light fractographs **A12:** 93-95
molybdenum alloy **A12:** 464
nickel alloys **A12:** 396-397
titanium alloys **A12:** 453
Cleavage fracture **A8:** 2
bainite packet grain size **A8:** 467
of cadmium monocrystals, by LME **A11:** 225
carbide density variations in **A8:** 458
crack arrest testing for **A8:** 453-455
defined **A9:** 4 **A11:** 2
and fracture toughness evaluation **A8:** 450
in low-carbon steels **A8:** 466
in notched impact steel **A11:** 22
RKR critical stress model for **A8:** 466-467
run-arrest **A8:** 285
second phase, of Ti-6Al-4V **A8:** 485, 488
stress-controlled modet for **A8:** 466
Cleavage fractures
AISI/SAE alloy steels **A12:** 302-303, 319
in Armco iron, shear step **A12:** 224
at three different magnifications **A12:** 175
brittle, ductile iron **A12:** 227
by early TEM study **A12:** 6
by mercury vapor embrittlement **A12:** 30, 38
by SCC, titanium alloys **A12:** 453
of corrosion products **A12:** 29
defined **A12:** 13
ductile irons **A12:** 232, 236, 237
effect of subgrain and grain boundaries **A12:** 17
etch pits on **A12:** 101
feather pattern, steps **A12:** 18
flat, high-purity iron **A12:** 219
formation **A12:** 13, 17
in hydrogen-embrittled stainless steel **A12:** 31
iron-aluminum alloys **A12:** 365
iron-base alloys **A12:** 224, 365, 457, 460
light fractographs **A12:** 94, 99
low-carbon steels **A12:** 249, 252
low-melting metals **A12:** 30, 38
nickel alloys **A12:** 297
in polycrystalline metals **A12:** 252
with river patterns, tongues, grain boundary **A12:** 224
surfaces **A12:** 13, 17-18
titanium alloys **A12:** 30, 38, 453
transgranular **A12:** 302, 460
transition to dimple rupture **A12:** 33, 45
woody, alloy steels **A12:** 319
wrought aluminium alloys **A12:** 424, 430
Cleavage plane
defined **A9:** 4
Cleavage steps **A12:** 13, 263
AISI/SAE alloy steels **A12:** 301
Alnico alloy **A12:** 461
in Armco iron **A12:** 18
austenitic stainless steels **A12:** 352-353
giant, titanium alloys **A12:** 450
in iron **A12:** 17, 457
medium-carbon steels **A12:** 263
precipitation-hardening stainless steels **A12:** 370
terraced facets with **A12:** 448

Cleavage steps (continued)
titanium alloys **A12:** 448, 450
wrought aluminum alloys **A12:** 417, 432
Cleavage strength **EM3:** 8
Cleavage surfaces
examination by phase contrast etching **A9:** 59
Cleavage, transgranular
acoustic emission inspection **A17:** 287
Cleaved alpha grains
in hydrogen-charged Ti-6Al-6V-2Sn microstructure **A8:** 490
Cleland clamp design
for torsional Kolsky bar **A8:** 196
Cleveland open cup procedure
to determine flash and fire points **A18:** 84
Clevis/Lua attachments
eddy current bushing inspection **A17:** 192-193
Cliff-Lorimer (standardless ratio) technique
microanalysis **A10:** 447
Cliffs
wrought aluminum alloys **A12:** 418
CLIMAT test
galvanic corrosion **A13:** 238
Climate effect
atmospheric corrosion **A13:** 81
Climb dislocation
pure metals **A8:** 308
Climb milling
carbon and alloy steels **A16:** 675
MMCs **A16:** 900
refractory metals **A16:** 861, 867
Ti alloys **A16:** 845, 846, 848
zirconium **A16:** 854
Zn alloys **A16:** 834
Climbing temperature program (CTP) **EM4:** 189-190
Clinched lead attachment
flexible printed boards **EL1:** 590
Clinging, electrostatic
of sample materials **A10:** 16
Clinoenstatite
chemical system **EM4:** 870
Clip gage
for crack arrest testing **A8:** 454
to monitor crack extension in corrosive environments **A8:** 428
Clip leads
thermal expansion mismatch problem **EL1:** 611
Clips, spring
failures in **A11:** 548-549
Clock brass
applications and properties **A2:** 308-309
Clock skew **EL1:** 7
Clock springs *See* Power springs
Clock(s)
as external-to-internal global communication **EL1:** 5
frequencies **EL1:** 76
global, in WSI technology **EL1:** 9
signals, on-chip **EL1:** 7
Clocking, optical
for clock skew **EL1:** 7
Close tolerance
defined **A14:** 307
Close tolerances
of composites **EM1:** 36
Close-packed
defined **A9:** 4
Close-to-finish factor
forgings **M1:** 349
Close-tolerance forging *See also* Precision forging **A14:** 2, 77
Closed assembly time **EM3:** 8
Closed circuit TV systems used in optical microscopy **A9:** 83
Closed climate **EM3:** 656-657, 661
Closed dies **A14:** 2, 101
aluminum and aluminum alloys **A2:** 6
Closed pass
defined **A14:** 2
Closed pore
defined **A7:** 2
Closed porosity
in encapsulated hot isostatic pressing **A7:** 435
Closed thermodynamic system .... **A3:** 1•5
Closed-cell cellular plastics **EM3:** 8
defined **EM2:** 10

Closed-cell foam *See* under Open-cell foam
Closed-die axial rolling machines A14: 114
Closed-die forging *See also* Die block; Forging; Impression dies; Impression-dieforging A8: 587, 590-591
of aluminum alloys A14: 243-244
blocker die design A14: 77-78
CAD/CAM of dies A14: 80
classifications A14: 76-77
cooling practice A14: 82
of copper and copper alloys A14: 255
defects in A14: 385-386
dies for A14: 43-45, 77-80, 101
equipment for A14: 80-81
flash design A14: 78-79
forging pressure, prediction A14: 79-80
friction and lubrication in A14: 76
in hammers and presses A14: 75-82
of heat-resistant alloys A14: 231
hot trimming punches for A14: 230
hot/warm, as precision forging A14: 158
load versus displacement curve A14: 37
load-stroke curves A14: 79
materials for A14: 75-76
metal flow in A14: 79
multiple-impression dies for A14: 44
process capabilities A14: 75
and ring rolling, combined A14: 112, 125-126
and ring rolling, compared A14: 122-123
sequence, typical A14: 82
shape complexity in A14: 77
of stainless steels A14: 222-223
temperatures for A14: 81-82
of titanium alloys A14: 272-273, 276
tolerances for A14: 243
trimming method A14: 82
wrought aluminum alloy A2: 34
Closed-die forging tools
die failure, causes M3: 526-527
forging shapes, effect on die life M3: 530-531
hammer-forging dies M3: 528, 529-530
hardness ranges M3: 527, 528-530
plug-type inserts M3: 528, 529-530
press forging dies M3: 529-530
selection of materials for M3: 526-532
trimming tools M3: 531-532
Closed-die forgings A1: 337-357
allowance for machining A1: 352
decarburization A1: 352
design for tooling economy A1: 352-353
design of hot extrusion forgings A1: 354, 355, 356
machining allowance A1: 357
mechanical properties A1: 356, 357
mismatch tolerances A1: 356-35
design of hot upset forgings A1: 353
design of specific parts A1: 353-354, 35
machining stock allowances A1: 353, 35
tolerances A1: 353
design stress calculations A1: 342, 343-346, 34
fundamentals of hammer and press
draft A1: 346, 347, 348
fillets and radii A1: 347-348, 34
forgings A1: 34
holes and cavities A1: 348
lightening holes in webs A1: 348-349
minimum web thickness A1: 348, 349
parting line A1: 346, 347, 348
ribs and bosses A1: 346-347
scale control A1: 349
inspection techniques A17: 495
material control A1: 338-339
combined specifications A1: 339
critical forging A1: 339
ductility and amount of forging reduction A1: 340-341
end-grain exposure A1: 342
fatigue strength A1: 341, 342
fracture toughness A1: 341-342
grain flow A1: 341
Closed-die forgings (continued)
grain size and microconstituents A1: 341
identification A1: 339
material specification A1: 339
quality assurance and quality control A1: 339
residual stress A1: 342
routine production A1: 339
test plans A1: 339, 340
tests and test coupons A1: 339-340
wrought structure and ductility A1: 340
mechanical properties A1: 342
anisotropy in high-strength steel A1: 343, 344, 345
grain flow and anisotropy A1: 342-343
selection of steel for A1: 337
cost A1: 338
design requirements A1: 338
forgeability A1: 338
microalloyed high-strength low-alloy (HSLA) steels A1: 337
precipitation-hardenable stainless steels A1: 337-338
tolerances A1: 349, 350
broaching allowance A1: 352
die wear A1: 351
draft A1: 351
flash A1: 351
hot shearing A1: 351
length A1: 349-350, 351
piercing A1: 351-352
shift or mismatch tolerance A1: 349, 350
trimming A1: 351
types A14: 76-77
types of A1: 337
Closed-die steel forgings M1: 349-375
anisotropy M1: 356-359
applications M1: 349, 350, 354, 357, 369, 371-373
cost considerations M1: 350, 351, 367, 369
design considerations M1: 350, 359-361, 369-375
design stress calculations M1: 359-361
fatigue data M1: 354-355, 359-361
fundamentals M1: 360-364
grain flow M1: 354
machining allowance M1: 368-371, 375
material control M1: 350-356
mechanical properties M1: 354-359, 374, 375
test plans M1: 352-353
selection of steel for M1: 349-350
testing M1: 351-359
tolerances M1: 362, 364-368, 370-371, 373-375
Closed-end protection tubes
for thermocouples A2: 883-884
Closed-form theoretical calibration equations
for crack advance A8: 391
Closed-loop
low-cycle torsional fatigue testing A8: 151
Closed-loop cure EM1: 761-763
Closed-loop electrohydraulic machine
loading tests on A8: 717-718
Closed-loop resonant fatigue tester
components A8: 393
Closed-loop servo-controlled testing machine
for strain rate A8: 582
Closed-loop servohydraulic testing machines
speed of A8: 43
Closed-loop servomechanical tester A8: 394-395
with furnace chamber and electronic controls A8: 395-396
Closing crack, in fatigue *See also* Cracks A12: 15
Closure
defined EM1: 6 EM2: 10
Cloth *See* Fabric; Nonwoven fabric; Polishing cloth; Roving cloth; Vent cloth; Woven fabric
defined A7: 2
of fiber reinforcements EM2: 506
Cloud *See* Dust cloud
Cloud chamber
powder coating M5: 503
Cloud point
defined A18: 5
Clover seeds A7: 589
Cluster carbonyl A7: 135
Cluster gears
economy in manufacture M3: 855-856
Cluster mill *See also* Four-high mill; Two-high mill
defined A14: 2
rolling A14: 351-352
Cluster porosity *See also* Porosity
in arc welds A11: 413
in weldments A17: 583
Cluster rolling mills
for wrought copper and copper alloys A2: 244
Clustering
atom probe composition profile of A10: 593
Clusters
design, investment casting A15: 257
energetics of A15: 103
flux, kinetics A15: 103
and pattern assembly, investment casting A15: 257
preparation, investment casting A15: 260
of solids during solidification A15: 103-105
Clutch
lever bearing A7: 617, 619
magnets, powders used A7: 573
plate, from P/M friction materials A7: 701
Clutch-drive assembly
tapered pin fatigue fracture A11: 545-546
Clutches
eddy current A14: 497
friction A14: 497
positive A14: 496-497
CLV process *See* Counter-gravity low-pressure processes
CM-X
preparation and properties EM1: 102-103
CMA *See* Cylindrical mirror analyzers
CMM *See* Coordinate measuring machines
CMOD *See* Crack mouth opening displacement
CMOS *See* Complementary metal-oxide semiconductors
CMSX-2
aging cycle A4: 812
composition A4: 795
CNC direct die sinking
CAM-driven A14: 247
Co-carburization, tungsten carbide/titanium
carbide A7: 158
Co-Cr (Phase Diagram) A3: 2 • 140
Co-Cr-Fe (Phase Diagram) A3: 3 • 36-37
Co-Cr-Ni (Phase Diagram) A3: 3 • 37
Co-Cr-Ti (Phase Diagram) A3: 3 • 38
Co-Cr-W (Phase Diagram) A3: 3 • 38
Co-Cu (Phase Diagram) A3: 2 • 140
Co-curing *See also* Cure; Secondary bonding EM3: 8
defined EM1: 7 EM2: 10
Co-Dy (Phase Diagram) A3: 2 • 141
Co-Er (Phase Diagram) A3: 2 • 141
Co-Fe (Phase Diagram) A3: 2 • 141
Co-Fe-Mo (Phase Diagram) A3: 3 • 38-39
Co-Fe-Ni (Phase Diagram) A3: 3 • 39-40
Co-Fe-W (Phase Diagram) A3: 3 • 40-41
Co-Ga (Phase Diagram) A3: 2 • 142
Co-Gd (Phase Diagram) A3: 2 • 142
Co-Ge (Phase Diagram) A3: 2 • 142
Co-Hf (Phase Diagram) A3: 2 • 143
Co-Ho (Phase Diagram) A3: 2 • 143
Co-Mn (Phase Diagram) A3: 2 • 143
Co-Mo (Phase Diagram) A3: 2 • 144
Co-Mo-Ni (Phase Diagram) A3: 3 • 41
Co-Nb (Phase Diagram) A3: 2 • 144
Co-Nd (Phase Diagram) A3: 2 • 144
Co-Ni (Phase Diagram) A3: 2 • 145
Co-Ni-Ti (Phase Diagram) A3: 3 • 41
Co-P (Phase Diagram) A3: 2 • 145
Co-Pd (Phase Diagram) A3: 2 • 145

SUBJECTS OF THE INDEXED VOLUMES: ASM Handbook (designated by the letter "A"): **A1:** Properties and Selection: Irons, Steels, and High-Performance Alloys (1990); **A2:** Properties and Selection: Nonferrous Alloys and Special-Purpose Materials (1990); **A3:** Alloy Phase Diagrams (1992); **A4:** Heat Treating (1991); **A6:** Welding, Brazing, and Soldering (1993); **A7:** Powder Metallurgy (1984); **A8:** Mechanical Testing (1985); **A9:** Metallography and Microstructures (1985); **A10:** Materials Characterization (1986); **A11:** Failure Analysis and Prevention (1986); **A12:** Fractography (1987); **A13:** Corrosion (1987); **A14:** Forming and Forging (1988); **A15:** Casting (1988); **A16:** Machining (1989); **A17:** Nondestructive Testing and Quality Control (1989); **A18:** Friction, Lubrication, and Wear Technology (1992). **Metals Handbook, 9th Edition** (designated by the letter "M"): **M1:** Properties and Selection: Irons and Steels (1978); **M2:** Properties and Selection: Nonferrous Alloys and Pure Metals (1979); **M3:** Properties and Selection: Stainless Steels, Tool Materials and Special-Purpose Materials (1980); **M4:** Heat Treating (1981); **M5:** Surface Cleaning, Finishing, and Coating (1982); **M6:** Welding, Brazing, and Soldering (1983). **Engineered Materials Handbook** (designated by the letters "EM"): **EM1:** Composites (1987); **EM2:** Engineering Plastics (1988); **EM3:** Adhesives and Sealants (1990); **EM4:** Ceramics and Glasses (1991); **Electronic Materials Handbook** (designated by the letters "EL"): **EL1:** Packaging (1989).

Co-Pr (Phase Diagram) ... A3: 2 • 146
Co-precipitation ... A7: 54, 55
as P/M oxide-dispersion technique ... A7: 718-719
Co-Pt (Phase Diagram) ... A3: 2 • 146
Co-Pu (Phase Diagram) ... A3: 2 • 146
Co-Re (Phase Diagram) ... A3: 2 • 147
Co-S (Phase Diagram) ... A3: 2 • 147
Co-Sb (Phase Diagram) ... A3: 2 • 147
Co-Se (Phase Diagram) ... A3: 2 • 148
Co-Si (Phase Diagram) ... A3: 2 • 148
Co-Sm (Phase Diagram) ... A3: 2 • 148
Co-Sn (Phase Diagram) ... A3: 2 • 149
Co-spray roasting
of ferrites ... EM4: 1163
Co-Ta (Phase Diagram) ... A3: 2 • 149
Co-Tb (Phase Diagram) ... A3: 2 • 149
Co-Te (Phase Diagram) ... A3: 2 • 150
Co-Th (Phase Diagram) ... A3: 2 • 150
Co-Ti (Phase Diagram) ... A3: 2 • 150
Co-V (Phase Diagram) ... A3: 2 • 151
Co-W (Phase Diagram) ... A3: 2 • 151
Co-Y (Phase Diagram) ... A3: 2 • 151
Co-Zn (Phase Diagram) ... A3: 2 • 152
CO_2 process *See* Carbon dioxide process
Co_3Ti trialuminide alloys
properties ... A2: 929-930
Co_3V trialuminide alloy
properties ... A2: 929-930
Coagulation ... EM3: 8
defined ... EM2: 10
Coal ... EM3: 175
analysis, PGAA for ... A10: 240
analytic methods for ... A10: 9
analyzed by x-ray spectrometry ... A10: 100
ash, XRPD study of phases in ... A10: 333
BTU and ash content in ... A10: 100
chlorine in, determined by electrometric titration ... A10: 205
combustion, beryllium toxicity from ... A2: 1238
derivatives, analytic methods for ... A10: 9
fly ash, gallium recovery from ... A2: 742-743
fly ash, SSMS analysis ... A10: 147-148
as fuel ... A15: 30
gasification and liquefaction products, GC/MS analysis of volatile compounds in ... A10: 639
and germanium ... A2: 733
hydrosulfurization, Raman surface analysis of molybdenum oxide catalysts used for ... A10: 133
local structure of trace impurities, EXAFS determined ... A10: 407
Miller numbers ... A18: 235
service life of coal-handling equipment ... A18: 719
sintering agents for ... A10: 166
stainless steels for coal-handling equipment ... A18: 722
x-ray spectrometric results ... A10: 100
Coal ash
alternative fuels and ... A11: 616
composition of ... A11: 616
corrosion, boiler tubes and steam equipment ... A11: 617-618
Coal fly ash
copper and lead determined in ... A10: 147-148
Coal gasification
corrosive wear ... A18: 271
nickel alloy applications ... A2: 430
Coal mine
machinery, aluminum and aluminum alloys ... A2: 14
tools, cemented carbide ... A2: 975-976
Coal pulverizer shaft, steel
fatigue failure ... A11: 468
Coal tar
cure properties ... EM3: 51
enamels ... A13: 406
epoxies, for marine corrosion ... A13: 916-918
extender for urethane sealants ... EM3: 205
Coal tar distillate
as toughener ... EM3: 184
Coal tar resins and coatings ... M5: 496, 500-502, 504-505
Coal-tar coatings
for corrosion protection ... M1: 757
Coal-tar-based laminate
in culvert pipe ... M1: 176
Coal-tar-epoxy lining
in cast iron pipe ... M1: 100
Coalescence *See also* Microvoid coalescence
of aluminum alloy phases ... A9: 359
defined ... A9: 4
definition ... A6: 1207 M6: 3
ductile fracture by ... A8: 572 A11: 82
microcrack-to-macrocrack ... A8: 57
particle growth by ... A15: 79
of particles in atomization ... A7: 25, 28, 34
Coarse fraction
defined ... A7: 2
Coarse grains
defined ... A9: 4
Coarse grit
for abrasive blasting ... A7: 458
Coarse powders
effect on packed density ... A7: 296
Coarse shot
for abrasive blasting ... A7: 458
Coarse-grained region (CGR) ... A6: 81
Coarsening *See also* Grain growth ... A9: 697-698
alloy steels ... A12: 292
dendrite ... A15: 138, 154-155
during precipitation reactions ... A9: 647
nucleation effects ... A15: 101
secondary arm ... A15: 117
Coarsening, and crystallization
aluminum, by x-ray topography ... A10: 376
Coast Metal 64
coating compositions for LPT blade interlocks ... A18: 590
Coated abrasive product
defined ... A9: 4
Coated aluminum
alloys, types ... A13: 107
filiform corrosion ... A13: 107
foil ... A13: 105
Coated atomized powders ... A7: 125
Coated carbide tools *See also* Cemented carbides ... A16: 79-83
for boring ... A16: 716
chemical vapor-deposited coatings ... A2: 959
cobalt enrichment ... A2: 960-961
compositions of CVD coatings ... A2: 960
diffusion wear ... A2: 960
hardness and tool life ... A2: 960
for high-speed machining ... A16: 601
laminated coatings ... A2: 959
physical vapor deposition (PVD) ... A2: 961-962
thermal expansion and coating adhesion ... A2: 960-961
for turning ... A16: 708-710, 714
Coated electrode
definition ... A6: 1207
Coated electrodes for arc welding
powders used ... A7: 573
Coated lenses ... A12: 84
Coated magnesium ... A13: 107
Coated metals
polarized light used to examine ... A9: 79
projection welding ... M6: 506
Coated microelectronic devices
failure mechanisms ... EL1: 1049-1057
Coated microstrip
impedance models ... EL1: 602
Coated sheet steel ... A9: 197
Coated spherical developer bead
copier powder ... A7: 580
Coated steel
characteristics ... EM4: 976
properties ... EM4: 976
Coated steels
aluminum ... M6: 480, 491-492
brazing to aluminum ... M6: 1030
cadmium ... M6: 480
characteristics ... A13: 104-105
chromium ... M6: 479
filiform corrosion ... A13: 104-107
gas tungsten arc welding ... M6: 183
lacquered can lid, filiform corrosion ... A13: 104
press forming of ... A14: 560-566
projection welding ... M6: 520-521
resistance seam welding ... M6: 494, 502
resistance spot
lead-tin alloy ... M6: 491
Coated steels (continued)
welding ... M6: 479-480, 491
terne metal ... M6: 479-480, 491
tin ... M6: 479-480, 491-493
types, filiform corrosion ... A13: 107
zinc ... M6: 479-480, 491-492
Coated superabrasive grains
types ... A2: 1015
Coated surfaces *See also* Coating(s); Surface(s)
magnetic rubber inspection of ... A17: 123-124
Coating
bend tests for ... A8: 117
for correction of spangles ... A8: 548
definition ... M6: 3
disk-pressure test for hydrogen embrittlement in ... A8: 540-541
protective, for high-temperature torsion test specimen ... A8: 159
Coating adhesion
coated carbide tools ... A2: 960
Coating density
definition ... A6: 1207 M6: 4
Coating failure mechanisms *See also* Coating(s)
corrosion ... EL1: 1049-1052
interconnect ... EL1: 1054-1056
nuclear radiation induced device failure ... EL1: 1056
stress-related ... EL1: 1052-1054
Coating intensity method
for thin-film sample preparation ... A10: 95
Coating methods
conformal coatings ... EL1: 763-764
urethane coatings ... EL1: 775, 778-779
Coating(s) *See also* Acrylate coatings; Application methods; Coating failure mechanisms; Coating methods; Conformal coatings; Fluoropolymer coatings; Parylene coatings; Polyimide coatings; Protective coatings; Silicon-based coatings; Urethane coatings
acrylate ... EL1: 785-788
for adhesive-bonded joints ... A17: 629
alloyable, as surface preparation ... EL1: 679
amorphous materials and metallic glasses ... A2: 819
bonded-abrasive grains ... A2: 1015
chemical vapor-deposited (CVD), coated carbide tools ... A2: 959-660
codeposited organics in ... EL1: 679-680
conformal, types ... EL1: 762-763
cure formulation ... EL1: 856-859, 861-863
diamond, spiked nickel ... A2: 1014
effect, solder masking ... EL1: 556
effects, nondestructive evaluation ... A17: 123
ferromagnetic, magabsorption measurement ... A17: 152
flexible epoxy junction ... EL1: 819
fluoropolymer ... EL1: 782-784
function ... EL1: 822
heat-sensitive, thermal inspection ... A17: 399
high-gloss, radiation cure formulation ... EL1: 858
laminated, coated carbide tools ... A2: 959
methods, conformal coatings ... EL1: 763-764
organic, as preservation ... EL1: 563
organic, for magnetic printing ... A17: 125
organic, in solderable systems ... EL1: 680
parylene ... EL1: 789-801
passive device, failure mechanisms ... EL1: 1000
plastic, in magnetic printing ... A17: 125
plumbum, uses ... A2: 555-556
polyimide ... EL1: 767-772
polyimide, as thermocouple wire insulation ... A2: 882
polymer, for package sealing ... EL1: 239-243
precious metals ... A2: 695
properties, typical ... EL1: 783
protective, materials and processes selection ... EL1: 115
radiation curing ... EL1: 854
refractory metals and alloys ... A2: 564-565
rhodium, for sterling silver ... A2: 691
silicone conformal coating ... EL1: 822-824
silver ... A2: 691
superhard, low-pressure synthesis is ... A2: 1009
terne ... A2: 554-555
thickness, eddy current inspection ... A17: 164
thickness inspection ... EL1: 943
types, conformal ... EL1: 761

Coating(s) (continued)
urethane EL1: 775-781
UV-curable EL1: 785-788
wire, wrought copper and copper alloys A2: 256-257
zinc, types A2: 527-528
Coatings *See also* Antiseizing coatings; Barrier coatings; Chemical coatings; Chemical conversion coatings; Conversion coatings; Curtain coating; Elastomeric coatings; Electroplated coatings; Fluidized-bed coating; Get coat; Metal coatings; Metallic coatings; Platings, Linings, Finishes and specific coating types by name; Protective coatings; specific coatings; Surface coatings A13: 522-524
acrylic A13: 404-405
AES in-depth compositional analysis A10: 549
aluminum A13: 527
aluminum anodizing A13: 396-398
aluminum, for fasteners A11: 542
anodic A13: 811
anodic hard, postforging defects from A11: 333
as anodic/cathodic, to substrate A13: 419
anodized A13: 396-398
antiseizure, for steel bolts A11: 536
for appliances M1: 105, 755
for atmospheric corrosion A13: 83
atmospheric corrosion resistance enhanced by M1: 722
barrier A13: 86-87
for beryllium A13: 811
bituminous A13: 406
cadmium A13: 426, 571
cadmium, for fasteners A11: 542
carbon/graphite EM3: 287-288
carrier, for copier powders A7: 585-586
cast iron pipe M1: 97-98, 100
for cast irons *See also* specific coating types by name A13: 570-571 M1: 101-106
of castings A15: 561-565
cathodic damage to A13: 748
for cemented carbides A13: 856-857
ceramic A13: 378, 524 EM4: 20
ceramic glass precoat A14: 279
ceramic, Replicast process A15: 271
chemical conversion A11: 195
chemical conversion, aluminum alloys A15: 763
chemically bonded media A15: 341
chlorinated rubber A13: 404
chromate conversion A13: 389-395, 523-524
chromium A13: 426-427
clad metals A13: 523
clusters, investment casting A15: 260-261
cobalt A7: 174
continuity A13: 451
conversion A13: 378-379 A15: 563
conversion, for shafts A11: 482
copper A13: 427
copper metals, corrosion protection M2: 465
copper wire M2: 272
of copper-based powder metals A7: 733
for copper/copper alloys A13: 636
for cores A15: 240-241
corrosion protection M1: 751-755
CVD/PVD A13: 456-458
defective, as casting defect A11: 385
diffusion A15: 563
for dry sand molding A15: 228
effect of hydrogen reduction on A7: 173
effect on formability A14: 563
elastomeric A11: 170-171
electrochemical evaluation A13: 219-220
electrogalvanized A13: 766-767
electrolytic and chemical, for corrosion control A11: 195
electroplated A13: 419-431, 523
electroplated hard chromium A13: 871-875
electroplated, intergranular cracks from A11: 337

Coatings (continued)
electroplating A15: 561-562
electroplating as A7: 460
for elevated temperatures A1: 296
external, for photolytic protection EM2: 782
extrusion EM2: 385-386
for fasteners A11: 530, 542
finish effects, permanent molds A15: 285
fluoride, for beryllium A13: 811
fresh water corrosion protection from M1: 738
fused dry-resin A15: 565
galfan A14: 561
for gas wells A13: 1249
gel coat EM1: 134
glass frit A14: 237
glass, tantalum forgings A14: 238
glasses and aluminides, for tantalum A14: 238
for graphite A13: 458
for graphite grains, rammed graphite molds A15: 273
green sand A15: 341
hot dip A13: 522-523 A15: 562
hot rolled bars and shapes M1: 200
for hydrogen damage A11: 251
in-mold EM2: 306
iron castings M1: 101-106
lead A13: 427, 571
leaf springs, steel M1: 313
life, permanent molds A15: 282
for lost foam casting A15: 232-233
lubricant, and explosivity A7: 194, 196
as lubricant form A14: 514
lubricating, for fasteners A11: 542
magnesium/magnesium alloys A13: 749-753
measuring thickness of A7: 259
mechanical A7: 459
metal matrix composites A13: 861-862
metal, measurement A10: 177
metallic, cast irons A13: 570-571
metallic, corrosivity A13: 342-343
metallic, for carbon steels A13: 526-527
metallic, for corrosion control A11: 195
metallic, galvanic corrosion A13: 84
metallic, nickel-base alloys A14: 832
metallic zinc, economics A13: 755
methods, for cast irons A13: 571
microstructural wear and A11: 161
microstructure A13: 526
mixed-oxide A13: 379
mold, for specific alloys A15: 282
and mold life A15: 281
in multiple-slide forming A14: 567
nickel electrodeposited A13: 426
nickel-based hardfacing A7: 142
nickel-chromium A13: 427-428
nickel-phosphorus A13: 426
nitric acid corrosion A13: 1156-1157
nonmetallic A13: 524
notch toughness of steels, effect on M1: 705
for oil/gas production equipment A13: 1236-1237, 1259
on nonferrous metals A13: 776
and optical properties EM2: 484-485
organic A13: 378, 399-418, 524, 528-529, 912-918 A14: 564-565 A15: 564-565
organic, measurement A10: 177
painting as A7: 459-460
passivation A13: 381
pattern A15: 195-196
permanent molds A15: 281-282
pharmaceutical production materials A13: 1231
phosphate A14: 304
phosphate conversion A13: 383-388, 523-524
photoelastic A11: 134
photoelastic, for crack length measurements EM3: 342
PIXE analysis of A10: 102
plastic, for zinc castings A15: 796
for polyaramid EM3: 286

Coatings (continued)
porcelain enamel A13: 446-452
porcelain enameling A15: 563
porous surface, for prosthesis systems A11: 671
processes, copier powders A7: 587-588
products of, x-ray spectrometric analysis A10: 100
protective (dielectric), for pipeline A13: 1259, 1289-1290
protective, for springs A11: 560
protective, tests for EM2: 425
quality, thermal spray A13: 462
refractory ceramic, for oxidation resistance EM1: 921
refractory materials for A15: 240
requirements, permanent molds A15: 281
resin-bonded sand systems A15: 213
resinous and inorganic base, for corrosion A11: 194
resins, amino EM2: 628
resins, principal A13: 401-402
robotic A15: 568
seawater corrosion, protection from M1: 745
as secondary operation A7: 451, 459-461
selection A13: 529
silicon carbide (SiC), for oxidation resistance EM1: 920
soft metal, for wear applications M1: 634
of solid graphite molds A15: 285
solvent, for woven fabric prepregs EM1: 149
for specific temperatures A13: 461
sputter A12: 173
steel sheet M1: 167-176
steel wire M1: 262-265
steel wire products M1: 271-272
steel wire rod M1: 253
for stray-current corrosion A13: 87
stress A11: 134
for structural corrosion A13: 1303
sulfuric acid corrosion A13: 1154
surface A12: 72-73, 83
surface preparation EM1: 681-682
system, diffusion phenomena in A10: 576-577
tallow-base, gas chromatographic measurement A10: 177
techniques, cast irons A13: 571
testing methods/standards for A13: 395
tests and designations A1: 212-214, 215
thermal evaporation A12: 173
thermal spray A13: 459-462, 523
thickness, and aluminum alloy pitting A13: 607
thickness, and mold temperature A15: 282
threaded fasteners, effect on fatigue strength M1: 279
tin A13: 427, 571, 780-782 M2: 613-614
tin-lead A13: 571
tin/tin alloys A13: 775, 780-782
to improve glass strength EM4: 743-744
cold-end EM4: 743
hot-end EM4: 743
shrink-wrapped EM4: 744
to prevent production problems associated with automotive welds A6: 395
to protect printed boards from chlorine and sulfur EM3: 592
for tungsten A14: 238
use in breweries A13: 1222
used to improve image quality in scanning electron microscopy A9: 97-99
vapor-deposited A13: 456-458, 523
wear resistant M1: 631-632, 634-635
weight A1: 218-219 A13: 385, 391, 395
for wood/wood laminate patterns A15: 194
zinc A13: 410-412, 426, 526-527, 571, 756-759, 765-769
zinc dust/zinc oxide A13: 768-769
zinc, for fasteners A11: 542
zinc on iron and steel M2: 651-653
zinc phosphate, for wire A14: 697

Coatings (continued)
zinc-phosphate, Auger imaging **A10:** 558
zinc-phosphate, on steels **EM3:** 272-273
zinc-rich..... **A13:** 410-412, 768-769
zinc-rich, for corrosion control **A11:** 194
Coatings, curing of
induction heating energy requirements **A4:** 189
induction heating temperatures **A4:** 188
Coaxial cables **A13:** 1127
Coaxial capacitor/condenser
for strain measurement **A8:** 199, 202
Coaxing
defined **A8:** 705
Cobalt *See also* Cobalt alloy powders; Cobalt powder strip; Cobalt powders; Cobalt toxicity; Cobalt-base alloys; Cobalt-base alloys, specific types; Cobalt-base corrosion-resistant alloys; Cobalt-base high-temperature alloys; Cobalt-base wear-resistant alloys; Elemental cobalt
as a beta stabilizer in titanium alloys **A9:** 459
addition to complex metal carbonitrides **A16:** 91-92
addition to solid-solution nickel alloys **A6:** 575
additions to gold electroplating solution **A18:** 837
adhesion and solid friction **A18:** 32-33
in age hardening in maraging steels **A1:** 794
alloying effect in titanium alloys **A6:** 508
alloying, magnetic property effect **A2:** 762
alloying, nickel-base alloys **A13:** 641
alloying, wrought aluminum alloy **A2:** 47
alloys as substrate for thermoreactive deposition/diffusion process **A4:** 449, 452
as an addition to beryllium-copper alloys **A9:** 394-395
applications **A7:** 144
in austenitic stainless steels **A9:** 284
binder, cemented carbides **A13:** 846
as binder for gas-lubricated bearings **A18:** 532
binder for WC **A16:** 71, 72
biologic effects and toxicity **A2:** 1251
cavitation erosion **A18:** 216
cemented carbide machining applications **A16:** 86
cemented carbide thermal properties **A16:** 81-82
in cemented carbides **A18:** 795, 796, 797, 798, 799
in ceramics and cermets **A18:** 812, 813
chemical analysis and sampling **A7:** 248
coating to improve abrasive wear resistance **A18:** 639
coatings **A7:** 174
coatings, bonded-abrasive grains **A2:** 1015
coatings for dies **A18:** 641
and cobalt alloys **A2:** 446-454
as colorant **EM4:** 380
in commercial CPM tool steel compositions **A16:** 63
compounds, trace nickel determined in **A10:** 194-195
content effect on tungsten carbide properties **A18:** 813
content in nickel-base and cobalt-base high-temperature alloys **A6:** 573
content in tool and die steels **A6:** 674
content in tool steels **A16:** 708
content in ultrahigh-strength low-alloy steels **A6:** 673
content in WC, PCD tooling **A16:** 110
in CPM Rex **A16:** 63
damage dominated by plastic deformation in cemented carbide **A18:** 178
deficiencies, biologic effects **A2:** 1251
determined by controlled-potential coulometry **A10:** 209
determined in samples containing chloride ions **A10:** 201
diffusion brazing **A6:** 343
E-pH diagram **A13:** 27
effect on catalytic activity and disbondment **EM3:** 634
effect on hardenability **A4:** 25
effect on tool steel grindability **A16:** 726
effect on tool steel tempering **A4:** 722
electrochemical grinding **A16:** 543

Cobalt (continued)
electroplated coating material **A18:** 838
electroplating of dies **A18:** 644
elemental **A2:** 446
in enameling ground coat **EM3:** 303, 304
enrichment, coated carbide tools **A2:** 960-961
enrichment of tools **A16:** 82-83
erosion resistance **A18:** 228
as essential metal **A2:** 1251
evaporation fields for **A10:** 587
ferromagnetism **A9:** 533
friction coefficient data **A18:** 71
friction welding **A6:** 154
gamma spectrum, radionuclide **A18:** 326
in hardfacing alloys **A18:** 761-762
in heat-resistant alloys **A4:** 512
hydrometallurgical processing **A7:** 118
as inoculant **A15:** 105
ion-irradiated, analysis of small dislocation loops **A9:** 117
in iron-base alloys, flame AAS analysis of **A10:** 56
iron-nickel-cobalt ASTM F 15 alloy **EM3:** 301
isolation in high-temperature alloys **A10:** 174
low explosivity class **A7:** 196-197
magnetic contrast **A9:** 94
in MAR-M 247 ..... **A1:** 1016-1018
in maraging steel composition **A4:** 219, 220, 221-222, 223, 224
martensitic structures **A9:** 672-673
in milling cutters **A16:** 314
milling with PCBN tools **A16:** 114
neutron and x-ray scattering, and absorption compared **A10:** 421
and nickel, corrosion in aqueous alkaline media **A10:** 135
in nickel-base superalloys **A1:** 984
oxygen cutting, effect on **M6:** 898
in P/M high-speed tool steels **A16:** 61
particle-impact-induced brittle fracture **A18:** 182
particles **A7:** 145
in permanent magnets **A9:** 538-539
photometric analysis methods **A10:** 64
physical metallurgy **A13:** 658
projected U.S. supply and demand of **A7:** 144
properties **A18:** 795
pure **M2:** 725-726
pure, properties **A2:** 1109
as pyrophoric **A7:** 199
qualitative tests to identify **A10:** 168
refining process, Sherritt Gordon **A7:** 144
separation, by phenylthiohydantoic acid **A10:** 169
solvent/catalyst of diamond **A16:** 105
species weighed in gravimetry **A10:** 172
specific properties imparted in CTV tubes **EM4:** 1042
in stainless steel brazing filler metals **A6:** 913
in stainless steels **A18:** 723
submerged arc welding **A6:** 206
suitability for cladding combinations **M6:** 691
in T15 and M42 high-speed tool steels **A16:** 63-64
tap density **A7:** 277
thermal diffusivity from 20 to 100 °C **A6:** 4
thermal expansion coefficient **A6:** 907
in thermal spray coating materials **A18:** 832
TNAA detection limits **A10:** 238
in tool material for drilling **A16:** 234
in tool steels **A18:** 734, 735-736
toxicity **A6:** 1195, 1196
toxicity, diseases, exposure limits **A7:** 204
in Udimet 700 ..... **A1:** 1014-1016
vapor pressure, relation to temperature **A4:** 495
Vickers and Knoop microindentation hardness numbers **A18:** 416
volumetric procedures for **A10:** 175
in Waspaloy **A1:** 1014-1016
Cobalt alloy, MAR-M509
effect of temperature on strength and ductility **A8:** 36
Cobalt alloy powders *See also* Cobalt; Cobalt powder strip; Cobalt powders; Cobalt-based hardfacing alloys
changes in powder particle morphology **A7:** 62

Cobalt alloy powders (continued)
effect of milling time on density and flowability **A7:** 59
for hardfacing **A7:** 145, 825-828
mechanical properties **A7:** 468, 472
production **A7:** 144-146
Cobalt alloys *See also* Cast cobalt alloys; Cobalt alloys, specific types; Heat-resistant alloys
aluminum coating of **M5:** 341-343
brazing, available product forms of filler metals **A6:** 119
brazing, joining temperatures **A6:** 118
explosion welding **A6:** 896
fractographs **A12:** 398
fracture/failure causes illustrated **A12:** 217
general welding characteristics **A6:** 563
in metal-matrix composites, ductile dimpled rupture **A12:** 470
radiographic inspection **A17:** 308
solid-state phase transformation in welded joints **A9:** 581
Cobalt alloys, high temperature *See* High temperature cobalt alloys
Cobalt alloys, specific types *See also* Superalloys, cobalt-base, specific types; Wear-resistant alloys, nonferrous
ASTM F75 cast, fatigue fracture **A12:** 398
ASTM F75, stage I fatigue fracture appearance **A12:** 16
Co-8Fe, deformation bands in a single crystal deformed 44% **A9:** 689
Co-8Fe, slip bands in a plastically deformed single crystal **A9:** 689
Co-12Fe-6TI, coherent metastable precipitate **A9:** 650
Co-12Fe-6TI, general and grain boundary precipitation **A9:** 648
Stellite 6, wear data **M3:** 590
Stellite 6B **M3:** 590, 591
composition **M3:** 210, 265, 590
erosion shield, use for **M3:** 591
physical properties **M3:** 217
property data **M3:** 264, 265
wear data **M3:** 590
Stellite 6K **M3:** 590
composition **M3:** 265, 590
property data **M3:** 265-266
wear data **M3:** 590
Tribaloy T-400 ..... **M3:** 590-591
composition **M3:** 590
wear data **M3:** 590
Tribaloy T-800 ..... **M3:** 590-591
composition **M3:** 590
wear data **M3:** 590
Vitallium, fatigue fracture **A12:** 398
Cobalt alloys, specific types L-605
aluminum coating, tensile strength affected by **M5:** 342
Cobalt and its alloys
eutectic joining **EM4:** 526
medical applications **EM4:** 1009
semifinish turning component example using ceramic insert cutting tools **EM4:** 972
Cobalt binder phase in cemented carbides **A9:** 274
Cobalt electroplating **M1:** 102, 753-754
Cobalt ferrites **A9:** 538
Cobalt, gold plating
use in **M5:** 282-283
Cobalt in steel **M1:** 445-447
Cobalt nitrate
dc and differential pulse polarograms of nickel in **A10:** 194-195
Cobalt oxide (CoO)
(Co_2O_3·CoO), purpose for use in glass manufacture **EM4:** 381
as colorant **EM4:** 380
pressure densification
/titania stain **EM4:** 474
pressure **EM4:** 391
technique **EM4:** 301
temperature **EM4:** 301
Cobalt oxide, thermal
decomposition of **A7:** 145-146
Cobalt oxtacarbonyl **A7:** 135-136

Cobalt powder strip *See also* Cobalt; Cobalt alloy powders; Cobalt powders
properties, and production **A7:** 402

Cobalt powders *See also* Cobalt; Cobalt alloy powders; Cobalt powder strip; Cobalt-based hardfacing alloys
applications **A7:** 144
hydrometallurgical processed properties **A7:** 145
need for substitute **A7:** 144
production **A7:** 144-146
reduced **A7:** 145-146

Cobalt sheet and strip
powder used **A7:** 574

Cobalt silicide in Cu-2.5Co-1.2Cd-0.5Si **A9:** 553

Cobalt toxicity
biologic effects **A2:** 1251

Cobalt tungsten-carbide coatings (WC-Co)
adhesion pressure of molten particle deposition **EM4:** 204
applications as a coating **EM4:** 208

Cobalt, vapor pressure
relation to temperature **M4:** 310

Cobalt-12% iron-6% titanium alloy, microstructure of .... **A3:** 1 • 22

Cobalt-aluminum oxidation protective coating
superalloys **M5:** 376

Cobalt-base alloy
design for **A1:** 983, 985-986
heat treatment for **A1:** 993

Cobalt-base alloys *See also* Cobalt; Cobalt-base alloys, specific types **A13:** 658-668
in acid media, alloying effects **A13:** 660
air melting practices **A15:** 812-813
applications **A13:** 665-667 **A15:** 811
brazing **A6:** 928-929
brazing and soldering characteristics **A6:** 634
as carrier, sliding electrical contacts **A2:** 842
cast, fatigue fracture **A11:** 284
cavitation erosion **A18:** 217
composition **A6:** 929
compositions **A13:** 659
corrosion of **A13:** 658-668
corrosion-resistant **A2:** 403, 448, 453-454
in corrosive environments, behavior **A13:** 658-661
deslagging **A15:** 813
development **A15:** 811
diffusion brazing **A6:** 344
diffusion welding **A6:** 885
electron-beam welding **A6:** 869
environmental embrittlement **A13:** 661-662
erosion resistance for pump components **A18:** 598
explosive forming **A14:** 783-784
fabrication **A13:** 662-665
fatigue strengths **A13:** 666
forge welding **A6:** 306
forging temperatures and forgeability **A14:** 232
forming practice **A14:** 783
fretting **A13:** 138
fusion welding to steels **A6:** 828
galling **A18:** 715
galvanic corrosion **A13:** 85
general corrosion **A13:** 658-661
hardfacing **A6:** 789, 790, 792, 793-794, 796, 797, 799, 803
hardfacing and wear **A13:** 663-664
heat treatment **A15:** 813-814
heat-resistant **A2:** 446-448, 451-453
heat-resistant, forging of **A14:** 233-234
hydrochloric acid corrosion **A13:** 661
hydrogen embrittlement **A13:** 661-662
hydrogen peroxide dissolution medium **A10:** 166
laser cladding components and techniques **A18:** 869
localized corrosion **A13:** 661
machining, ultrahard materials for **A2:** 1010

Cobalt-base alloys (continued)
master ingots, manufacture and remelting **A15:** 811-812
material for jet engine components **A18:** 588, 591
melting ranges **A15:** 812
in nitric acid **A13:** 661
nominal compositions **A15:** 811
not readily diffusion bondable **A6:** 156
in nuclear reactor systems **A13:** 951
physical metallurgy **A13:** 658
preheat and pouring **A15:** 813-814
and rare earth alloys, as permanent magnet materials **A2:** 787-788
special metallurgical welding considerations **A6:** 575-579
creep-resistant secondary carbide-strengthened alloys welded with nickel alloys **A6:** 577
overaging **A6:** 575
special welded product conditions **A6:** 577-579
stacking faults in fcc phase **A10:** 466
superalloys, stress rupture data **A2:** 452
as surgical implants **A13:** 1326-1327
thermal spray coating recommended **A18:** 832
Unicast process for **A15:** 251
wear-resistant **A2:** 446-451
weldability **A13:** 662-665
welded overlays for resistance to cavitation erosion **A18:** 217
wrought, as implant materials **A11:** 672
yttrium segregation in **A10:** 483

Cobalt-base alloys, carbide type
abrasive wear **A18:** 761
applications **A18:** 758
coefficient of friction **A18:** 764
composition **A18:** 762
composition of alternate alloy replacements **A18:** 764
properties **A18:** 758, 761, 762, 764
sliding wear **A18:** 762
wear resistance **A18:** 758

Cobalt-base alloys, heat treating *See also* Heat-resistant alloys, heat treating
aging **M4:** 669-670
annealing **M4:** 669
atmospheres **M4:** 670
temperatures **M4:** 655, 656, 657, 669-670

Cobalt-base alloys, specific types
17Co7Cr3W, laser alloying and abrasive wear resistance **A18:** 866
19-9DL, recommended upsetting pressures for flash welding **A6:** 843
52Co22Cr9W, laser alloying and abrasive wear resistance **A18:** 866
54Co-27Cr-6Mo-5Ni, refractory metal brazing, filler metal **A6:** 942
55Co-20Ni-15W-10Ni, refractory metal brazing, filler metal **A6:** 942
AiResist 215, vacuum casting **A15:** 812
Alloy 6
abrasive wear **A18:** 767
erosive wear **A18:** 768
Alloy 6B
abrasive wear **A18:** 767, 770
annealed hardness **A18:** 768
applications **A18:** 770
carbide volume fraction **A18:** 766
erosive wear **A18:** 767, 768
hot hardness **A18:** 770
microstructure after etching **A18:** 767
nominal composition **A18:** 766
Alloy 6K
abrasive wear **A18:** 770
applications **A18:** 770
carbide volume fraction **A18:** 766
microstructure after etching **A18:** 767
nominal composition **A18:** 766
Alloy 20CB-3
annealed hardness **A18:** 768

Cobalt-base alloys, specific types (continued)
erosive wear **A18:** 768
Alloy 25
abrasive wear **A18:** 767
annealed hardness **A18:** 768
applications **A18:** 770
carbide volume fraction **A18:** 766
erosive wear **A18:** 768
hot hardness **A18:** 770
metastability **A18:** 767
microstructure after etching **A18:** 767
nominal composition **A18:** 766
Alloy 31, friction coefficient **A18:** 770
Alloy 188
carbide volume fraction **A18:** 766
erosive wear **A18:** 767
hot hardness **A18:** 770
metastability **A18:** 766
microstructure after etching **A18:** 767
nominal composition **A18:** 766
physical properties **A18:** 766
Alloy 255
abrasive wear **A18:** 767
annealed hardness **A18:** 768
erosive wear **A18:** 767, 768
Alloy 625
annealed hardness **A18:** 768
erosive wear **A18:** 768
Alloy 718, erosive wear **A18:** 767-768
alloy 1233, pitting resistance **A2:** 453-454
Alloy C-276
annealed hardness **A18:** 768
erosive wear **A18:** 767-768
sliding wear **A18:** 769
Co-10Cr, multilayer oxide scale **A13:** 98
F75 cobalt-chromium-molybdenum, cast broken hip prosthesis **A11:** 690-693
F563 wrought, fatigue-fracture structures showing toughness of **A11:** 685, 689
H-31, electron-beam welding **A6:** 869
Havar, corrosion at boiling temperatures **A13:** 661
Havar, nitric acid corrosion **A13:** 661
Haynes 188, solid particle erosion **A2:** 450
Haynes 188, sulfidation data **A2:** 453
Haynes alloy 6B, product forms and microstructure **A2:** 449
Haynes alloy 25, solid particle erosion **A2:** 450
Haynes alloy 25, sulfidation data **A2:** 453
Haynes alloys, boiling temperature environments **A13:** 661
Haynes alloys, sulfuric acid corrosion **A13:** 660
HS-21, air casting **A15:** 812
HS-21, electron-beam welding **A6:** 869
HS-25, air casting **A15:** 812
MAR-M 302, vacuum pouring **A15:** 811-813
MAR-M 509, vacuum pouring **A15:** 811-812
MP35N
nominal composition **A18:** 766
physical properties and carbon content **A18:** 766
strength levels controlled by cold reduction and aging **A18:** 766
MP35N, corrosion at boiling temperatures **A13:** 661
MP35N multiphase alloy, mechanical properties **A2:** 454
MP35N, nitric acid corrosion **A13:** 661
MP159
nominal composition **A18:** 766
physical properties and carbon content **A18:** 766
strength levels controlled by cold reduction and aging **A18:** 766
physical properties **A18:** 766
S-816, electron-beam welding **A6:** 869
Stellite 1, overlay microstructures **A13:** 659
Stellite 6, overlay microstructures **A13:** 660
Stellite 6 sheet, liquid-metal corrosion **A13:** 91

SUBJECTS OF THE INDEXED VOLUMES: ASM Handbook (designated by the letter "A"): **A1:** Properties and Selection: Irons, Steels, and High-Performance Alloys (1990); **A2:** Properties and Selection: Nonferrous Alloys and Special-Purpose Materials (1990); **A3:** Alloy Phase Diagrams (1992); **A4:** Heat Treating (1991); **A6:** Welding, Brazing, and Soldering (1993); **A7:** Powder Metallurgy (1984); **A8:** Mechanical Testing (1985); **A9:** Metallography and Microstructures (1985); **A10:** Materials Characterization (1986); **A11:** Failure Analysis and Prevention (1986); **A12:** Fractography (1987); **A13:** Corrosion (1987); **A14:** Forming and Forging (1988); **A15:** Casting (1988); **A16:** Machining (1989); **A17:** Nondestructive Testing and Quality Control (1989); **A18:** Friction, Lubrication, and Wear Technology (1992). **Metals Handbook, 9th Edition** (designated by the letter "M"): **M1:** Properties and Selection: Irons and Steels (1978); **M2:** Properties and Selection: Nonferrous Alloys and Pure Metals (1979); **M3:** Properties and Selection: Stainless Steels, Tool Materials and Special-Purpose Materials (1980); **M4:** Heat Treating (1981); **M5:** Surface Cleaning, Finishing, and Coating (1982); **M6:** Welding, Brazing, and Soldering (1983). **Engineered Materials Handbook** (designated by the letters "EM"): **EM1:** Composites (1987); **EM2:** Engineering Plastics (1988); **EM3:** Adhesives and Sealants (1990); **EM4:** Ceramics and Glasses (1991); **Electronic Materials Handbook** (designated by the letters "EL"): **EL1:** Packaging (1989).

Cobalt-base alloys, specific types (continued)
Stellite alloy 1, for castings/overlays **A2:** 449
Stellite alloy 1, microstructure **A2:** 449
Stellite alloy 6, application **A2:** 449
Stellite alloy 6, microstructure **A2:** 449
Stellite alloy 12, application **A2:** 449
Stellite alloy 12, microstructure **A2:** 449
Stellite alloy 21, as wear-resistant **A2:** 449
Stellite, chromium effects **A13:** 658
Tribaloy alloy (T-800), Laves precipitates **A2:** 449
Ultimet
abrasive wear **A18:** 767
annealed hardness **A18:** 768
applications **A18:** 770
erosive wear **A18:** 768
hot hardness **A18:** 770
microstructure after etching **A18:** 767
no grain boundary carbide precipitation **A18:** 766
nominal composition **A18:** 766
physical properties and carbon content **A18:** 766
sliding wear **A18:** 769, 770
WI-52, air casting **A15:** 812
X-40, electron-beam welding **A6:** 869
Cobalt-base corrosion-resistant alloys **A6:** 585
alloy compositions and product forms **A2:** 453
applications **A2:** 454
corrosion properties **A2:** 453-454
mechanical properties **A2:** 454
nominal compositions **A2:** 448
postweld heat treatment **A6:** 599
types of aqueous corrosion **A2:** 453
Cobalt-base corrosion-resistant alloys, selection of **A6:** 598-599
applications **A6:** 598
composition of selected alloys **A6:** 598
gas-metal arc welding **A6:** 598
heat-affected zone, cracking **A6:** 598
microstructure **A6:** 598
weldability characteristics **A6:** 598-599
Cobalt-base hardfacing alloys
laser cladding materials **A18:** 866-867
Cobalt-base heat-resistant casting alloys
compositions of **A9:** 331
microstructures **A9:** 334
Cobalt-base heat-resistant casting alloys, specific types
98M2 Stellite, as investment cast, different magnifications compared **A9:** 348
Haynes 21, as-cast and aged **A9:** 347
Haynes 31, as-cast and aged, thin and thick sections compared **A9:** 347
Haynes 151, as-cast and aged **A9:** 347
MAR-M 302, as-cast **A9:** 348
MAR-M 509, as-cast and aged **A9:** 349
WI-52, as-cast **A9:** 348
Cobalt-base high-temperature alloys
alloy compositions and product forms **A2:** 451-452
applications **A2:** 452-453
nominal compositions **A2:** 448
oxidation/sulfidation resistances **A2:** 452
Cobalt-base superalloys *See also* specific types; Super- alloys, cobalt-base, specific types; Wrought heat-resistant alloys
cast cobalt-base superalloys **A1:** 983, 985-987
extreme-temperature solid lubricants **A18:** 118
phases in **A15:** 812
powder metallurgy (P/M) cobalt-base alloys **A1:** 977-980
properties **A15:** 812, 814
thermal expansion coefficient **A6:** 907
VIM melt protocol **A15:** 394
wrought cobalt-base superalloys **A1:** 950, 962-968
Cobalt-base superalloys, arc-welded
failures in **A11:** 433-434
Cobalt-base wear-resistant alloys
abrasive wear **A2:** 447
alloy compositions and product forms **A2:** 448-449
applications **A2:** 451
erosive wear **A2:** 448
mechanical and physical properties **A2:** 451
nominal compositions **A2:** 448
physical/mechanical properties **A2:** 451
Cobalt-base wear-resistant alloys (continued)
sliding wear **A2:** 447-448
wear data **A2:** 449-451
Cobalt-base wrought alloys, friction and wear of **A18:** 766-770
annealed hardnesses of specific alloys **A18:** 768
categories **A18:** 766
hot hardnesses of specific alloys **A18:** 770
mechanical properties **A18:** 770
wear properties **A18:** 767-770
wear-related applications **A18:** 770
Cobalt-based alloys
brazing **M6:** 1021
compositions, characteristics and applications **M6:** 1019
electron beam welding **M6:** 638
forge welding **M6:** 676
friction welding **M6:** 722
hardfacing material **M6:** 774
plasma arc welding **M6:** 214
Tribaloy alloys laser cladding **M6:** 796-797
weld overlay material **M6:** 806
Cobalt-based hardfacing alloys **A7:** 145, 825-828
carbide percentages **A7:** 827
compositions and hardness **A7:** 145
corrosion rates **A7:** 828
wear data **A7:** 828
Cobalt-based heat-resistant alloys *See* specific types and Heat-resistant alloys
Cobalt-bonded tungsten carbide
chemical vapor deposition **EM4:** 217
Cobalt-bonded tungsten carbide, and heat-treatable steel-bonded carbides
compared **A2:** 996-997
Cobalt-chromium alloys
broken adjustable Moore pins from **A11:** 675, 681
cold-worked, implant failure of **A11:** 675, 681
dental **A13:** 1361
as implant materials **A11:** 672
orthodontic wires **A18:** 666, 675-676
part material for ion implantation **A18:** 858
retrieved bone screw from **A11:** 674, 680
Cobalt-chromium-aluminum- yttrium oxidation protective coating, superalloys **M5:** 376-378
structure **M5:** 376-377
thermal mechanical fatigue behavior and ductility **M5:** 376-378
Cobalt-chromium-molybdenum alloys
applications, femoral components of hip and knee replacements **A18:** 657, 658
bone screw, sheared-off **A11:** 678, 682
cast, as implant materials **A11:** 672
composition **A18:** 658
cycle rotating beam fatigue limit **A7:** 660
fretting wear **A18:** 250
orthopedic implants **A7:** 657
physical and mechanical properties **A18:** 659
tensile properties **A7:** 660
Cobalt-chromium-tungsten alloys
classification and composition of hardfacing alloys **A18:** 652
for cutting tool materials **A18:** 615
Cobalt-covered tungsten carbide **A7:** 173
Cobalt-iron powder alloys
properties of **A7:** 402
Cobalt-nickel-phosphorus
electroplated coatings **A18:** 838
Cobalt-phosphorus
electroplated coatings **A18:** 838
Cobalt-rare earth alloys *See* Magnetic materials
Cobalt-samarium permanent magnets **A7:** 643
Cobalt-tin alloys
peritectic transformations **A9:** 678
Cobalt-tungsten-carbon phase diagram .... **A3:** 1 • 29
Cobalt/rare earth permanent magnets **A7:** 642-643
Coble diffusional creep **EM4:** 296
Cobron *See* Copper alloys, specific types, C66400
applications and properties **A2:** 335-336
Cockcroft and Latham criterion **A8:** 168-169
Cockcroft model
of fracture **A14:** 394
Cocking
of mechanical fasteners **EM1:** 706
Cocoa (red mud)
defined **A18:** 5
Cocurrent drying **A7:** 73, 74
COD *See* Crack opening displacement
COD method
measurement of crack closure load **EM3:** 510-511
Codeposited organics
effects in plated coatings **EL1:** 679-680
Codeposition
in constant-current methods of electrogravimetry **A10:** 198
Codes
for boiler/pressure vessel inspection methods **A17:** 641-644
Codes for
aircraft and spacecraft **M6:** 824-825
boilers and pressure vessels **M6:** 823
bridges, buildings similar structures **M6:** 824
concrete requirements **M6:** 702
field-welded storage tanks **M6:** 824
flash welding **M6:** 580
industrial machinery **M6:** 825
industrial pipelines and piping **M6:** 824
liquid-penetrant inspection **M6:** 826-827
magnetic-particle inspection **M6:** 827
nuclear reactors **M6:** 823-824
pressure piping **M6:** 824
radiographic inspection **M6:** 827
railroad rolling stock **M6:** 824
shipbuilding **M6:** 824
ultrasonic inspection **M6:** 827
Coding, color
of thermocouple wires and extension wires **A2:** 878
Coding of levels (experimental)
defined **EM2:** 600
COE *See* Cube-on-edge texture
Coefficient of adhesion **A6:** 144 **A18:** 475
defined **A18:** 5
Coefficient of elasticity *See also* also Compliance, Young's modulus; Compliance **EM3:** 8
defined **EM1:** 7 **EM2:** 10
Coefficient of expansion **EM2:** 1, 10 **EM3:** 8
defined **EM1:** 7
electrical resistance alloys **A2:** 822
Coefficient of expansion effect
fastener performance at elevated temperatures **A11:** 542
Coefficient of friction *See also* Friction **A6:** 144 **A8:** 2 **A18:** 27, 28, 31-33, 35, 40, 432-436 **EM3:** 8
of acetals **EM2:** 100
acrylic versus porcelain denture teeth **A18:** 673, 674
automotive
brakes **A18:** 574-575
sprocket and chain wear **A18:** 566
bearing steels **A18:** 730
boundary lubrication **A18:** 96
ceramics **A18:** 814, 815
chromium **A18:** 836, 837
chromium electroplating **A18:** 835
cobalt-base wrought alloys **A18:** 770
crankshaft bearings in internal combustion engines **A18:** 559-561
decreased with sliding velocity **A18:** 43
defined **A18:** 5 **EM1:** 7 **EM2:** 10
dental tissue (human) and laboratory studies **A18:** 667
fretting wear in a vacuum **A18:** 249
gallium arsenide, as a function of temperature and doping **A18:** 688-689
gray iron **A18:** 695
historical development **A18:** 45-46
metal-matrix composites **A18:** 803, 804, 805, 806, 807, 808-809, 810
of nickel aluminide alloys **A18:** 772-777
nickel, electroless **A18:** 837
nickel, electroplated **A18:** 836
orthodontic wires **A18:** 676
polyether etherketone **A18:** 824, 825, 826
polytetrafluoroethylene **A18:** 824, 825
pump materials **A18:** 595
seals **A18:** 547
silicon (*n*-type), as a function of temperature and doping **A18:** 688-689
sliding bearings **A18:** 515, 516, 518, 519
and Sommerfeld number **A11:** 485
stainless steels **A18:** 717
strip rolling **A18:** 66

Coefficient of friction (continued)
surface texture of magnetic storage tapes ... **A18:** 343
symbol for ... **A8:** 726 **A11:** 796
synovial joints, major load-bearing natural ... **A18:** 656
thermoplastic composites ... **A18:** 820, 821, 822, 824, 825, 826
titanium alloys ... **A18:** 778, 779, 780, 781, 782
tool steels ... **A18:** 737, 738, 739
tribotest example ... **A18:** 483, 484, 485
ultrahigh molecular weight polyethylenes (UHMWPE) ... **EM2:** 167
VAMAS round-robin sliding wear tests ... **A18:** 487
Coefficient of friction testing ... **A13:** 962
Coefficient of linear thermal expansion *See also* Coefficient of thermal expansion; Thermal expansion
aluminum and aluminum alloys ... **A2:** 9
cast copper alloys ... **A2:** 356-391
defined ... **A2:** 9
temperature dependence ... **EL1:** 814
Coefficient of thermal expansion *See also* Coefficient of linear thermal expansion; Thermal expansion
aluminum casting alloys ... **A2:** 153-177
of beryllium ... **A2:** 683
of gray iron ... **A1:** 31
iron-nickel low-expansion alloys ... **A2:** 893
variability in ... **A8:** 623
wrought aluminum and aluminum alloys ... **A2:** 62-122
Coefficient of thermal expansion (CTE) *See also* CTE-matched materials; Linear expansion; Temperature(s); Thermal ... **A6:** 433 **EM3:** 8, 401, 444, 575, 582-583, 595
of aircraft alloys ... **EM1:** 716
aluminum metal-matrix composites ... **A6:** 555
aluminum-silicon alloys ... **A18:** 785
austenitic stainless steels, 8D-1 ... **A6:** 456, 469
bonding fixture ... **EM3:** 707
calculated ... **EM1:** 358-359
carbon fiber/fabric reinforced epoxy resin ... **EM1:** 411
carbon fibers ... **EM1:** 52
ceramic material brazing ... **A6:** 635
ceramic materials ... **A6:** 949, 950, 952, 957
ceramic multilayer packages ... **EL1:** 468
ceramics ... **A18:** 814 **EL1:** 336
of common packaging materials ... **EL1:** 454
of composite materials ... **EM1:** 716
composite packaging materials ... **EL1:** 1122
of composite tooling ... **EM1:** 580-581, 586
control, in PWBs ... **EL1:** 77
cryogenic service ... **A6:** 1018
data sheet information ... **EM2:** 410
defined ... **EM1:** 7 **EM2:** 10
differences causing thermal stresses ... **EM3:** 617
dissimilar metal joining ... **A6:** 824, 825, 826
dissimilar metals joined to carbon steels and cast irons ... **A6:** 906
effect on cryogenic service ... **A6:** 1018
and elasticity ... **EM1:** 188-190
of engineering plastics ... **EM2:** 69
epoxy resin system composites ... **EM1:** 402, 406, 408, 411, 413
glass addition effect ... **EM2:** 70
glass fiber reinforced epoxy resin ... **EM1:** 406
glass fibers ... **EM1:** 47 , 107
graphite fiber reinforced epoxy resin ... **EM1:** 413
graphitic materials ... **A6:** 950
heat sink effects ... **EL1:** 1129-1131
heat-treatable aluminum alloys ... **A6:** 530
high-temperature thermoset matrix composites ... **EM1:** 376, 380
of homogeneous solids, defined ... **EM1:** 189
hydrogen mitigation, underwater welding ... **A6:** 1011-1012
importance, hybrid packages ... **EL1:** 451
Coefficient of thermal expansion (CTE) (continued)
Kevlar 49 fiber/fabric reinforced epoxy resin ... **EM1:** 408
laminate ... **EM1:** 225-226
lead frame alloys ... **EL1:** 491
linear, for various materials ... **EM2:** 752
longitudinal, for graphite laminate ... **EM1:** 184
low-temperature thermoset matrix composites ... **EM1:** 396
and material selection ... **EM1:** 38
matrix materials ... **EL1:** 1120-1121, 1127-1128
medium-temperature thermoset matrix composites ... **EM1:** 384, 387, 390, 391
metal matrix composites ... **EL1:** 1127-1128
mismatch between solder and substrate ... **A6:** 964
mismatch problem ... **EL1:** 77, 611
nickel alloys ... **A6:** 750
for package materials ... **EL1:** 415
for packaging materials ... **EL1:** 58
polyamide-imides (PAI) ... **EM2:** 129
polyimide ... **EM3:** 597
of polymer matrix composites ... **EL1:** 1117
of reinforcements ... **EL1:** 535
and reliability, electronic interconnect ... **EL1:** 730
resin effects ... **EL1:** 534-535
rigid epoxies ... **EL1:** 811
sheet metals ... **A6:** 399
sheet molding compounds ... **EM1:** 158
silicon carbide particle reinforced aluminum ... **EL1:** 1124
silicone conformal coatings ... **EL1:** 822
slag vs. weld differences ... **A6:** 61
in SMT design ... **EL1:** 733-734
soldered joint mismatches ... **A6:** 1128
soldering in electronic applications ... **A6:** 992-993, 996, 997
and specific heat ... **EM2:** 455-456
stainless steel dissimilar welds ... **A6:** 501
stainless steels ... **A6:** 847
and strain, vs temperature, copper-Invar-copper ... **EL1:** 622
for substrate materials, thin-film hybrids ... **EL1:** 318
substrates ... **EL1:** 318, 612
surface-mount solder joints, effect on ... **EL1:** 633
tailoring ... **EL1:** 614-628, 741-743
target, for substrates ... **EL1:** 612-613
and thermal loading ... **EL1:** 57
and thermal stresses ... **EM2:** 751
thermoplastic matrix composites ... **EM1:** 366, 369, 371
thermoplastic polyimides (TPI) ... **EM2:** 177
ultrahigh molecular weight polyethylenes (UHMWPE) ... **EM2:** 170
values, glasses and metals ... **EL1:** 457
vs. temperature, Kevlar ... **EM1:** 362
Coefficient of wear *See* Archard wear law; Wear coefficient; Wear constant; Wear factor
Coefficients
mass absorption ... **A10:** 85
Coefficients of swelling
and elasticity ... **EM1:** 188-190
Coefficients of thermal expansion (CTE)
in drinkware compositions ... **EM4:** 1102
in ovenware compositions ... **EM4:** 1103
in tableware compositions ... **EM4:** 1101
uniaxial hot pressing ... **EM4:** 187
Coercive force
defined ... **A17:** 100
and hardness, steel ... **A17:** 134
of magnetic materials ... **A2:** 761
Coercivity
magnetic ... **A7:** 643
Coercivity in permanent magnets ... **A9:** 538
Coextrusion
as process ... **EM2:** 387
Coextrusion welding
definition ... **M6:** 4
Coextrusion welding (CEW) ... **A6:** 311
advantages ... **A6:** 311
aluminum ... **A6:** 311
aluminum alloys ... **A6:** 311
applications ... **A6:** 311
copper ... **A6:** 311
definition ... **A6:** 311, 1207
low-carbon steel ... **A6:** 311
nickel ... **A6:** 311
nickel-base alloys ... **A6:** 311
niobium ... **A6:** 311
steel ... **A6:** 311
tantalum ... **A6:** 311
titanium ... **A6:** 311
zirconium ... **A6:** 311
Coffee-bean contrast in copper-cobalt alloys ... **A9:** 117
Coffin-Manson relationship
low-cycle fatigue testing ... **A8:** 367
Cofired ceramic multilayer packages *See also* Ceramic multilayer package, fabrication
physical characteristics ... **EL1:** 106
process flow ... **EL1:** 461
as substrate material ... **EL1:** 106
thermal expansion properties ... **EL1:** 615
Cofired tape technology
as thick film modification ... **EL1:** 249
Cofiring
defined ... **EL1:** 1137
Cogging ... **A1:** 971
defined ... **A14:** 2
of ingot products ... **A11:** 327
Cohen-Grest model ... **EM4:** 849
Coherence
in the illuminating beam of a transmission electron microscope ... **A9:** 103
Coherent atomic scattering intensity
abbreviation for ... **A10:** 690
Coherent Bragg diffraction
analytical transmission electron microscopy ... **A10:** 436
Coherent diffraction
of inelastically scattered electrons ... **A9:** 109-110
Coherent interface
between matrix and precipitate ... **A9:** 648
defined ... **A9:** 604, 647
Coherent phase transformations
revealed by differential interference contrast ... **A9:** 59
Coherent precipitates
defined ... **A9:** 4
structure-factor contrast ... **A9:** 112-113
Coherent scattering *See* Rayleigh scattering
defined ... **A9:** 4
Cohesion ... **EM3:** 8
composite joints ... **EM3:** 777
defined ... **EM1:** 7 **EM2:** 10
work of ... **A18:** 399, 400, 403
Cohesion coefficient ... **A6:** 144
Cohesive
failure, defined ... **EM1:** 7
strength, defined ... **EM1:** 7
Cohesive blocking ... **EM3:** 8
Cohesive failure ... **EM3:** 8
defined ... **EM2:** 10
Cohesive force *See* Adhesion
Cohesive powders
angle of repose for ... **A7:** 284
mobility and angle of repose ... **A7:** 285
Cohesive strength ... **EM3:** 8
defined ... **EM2:** 10
Cohesive-matrix failure
in composites ... **A11:** 734
Coil annealing
wrought copper and copper alloys ... **A2:** 246
Coil cradles ... **A14:** 501
Coil hooks
fatigue fracture ... **A11:** 523-524
materials for ... **A11:** 515, 522

Coil impedance *See also* Impedance
components **A17:** 166-167
in eddy current inspection **A17:** 166-167
phasor representation of sinusoids **A17:** 167
Coil lift-off locus
defined **A17:** 172
Coil springs *See* Helical springs
Coil, steel
pickling of **M5:** 71, 76, 80
Coil stock
for blanking **A14:** 449
for electromagnetic forming **A14:** 650
feeds for **A14:** 499
Coil termination adhesives **EM3:** 573, 574
Coil with support
definition **M6:** 4
Coil without support
definition **M6:** 4
Coil(s) *See also* Coil impedance; Conductor(s); Eddy current inspection; Probes; Sensors
applications **A17:** 94
arrangement, eddy current inspection **A17:** 174-175, 185
encircling **A17:** 176, 183
encircling tubing, impedance of **A17:** 169-170
exciter, in eddy current vs. electric current perturbation methods **A17:** 136-137
flux leakage measurement with **A17:** 130
impedance **A17:** 166-167, 169-170
inspection, eddy current inspection **A17:** 175-177
internal **A17:** 183
for longitudinal magnetization **A17:** 95
magnetizing, advantages/limitations **A17:** 94
multiple **A17:** 176, 196
multiple sector, remote-field eddy current inspection **A17:** 196
probe **A17:** 176
for reflection method, eddy current inspection **A17:** 183
single, and resistor **A17:** 177
sizes and shapes **A17:** 176-177
for transmission method, eddy current inspection **A17:** 183
types, eddy current inspection **A17:** 165, 176
Coil-handling equipment **A14:** 501-502
Coil-type electrical magnetization equipment longitudinal magnetization by **A7:** 576
Coiled sheet *See also* Sheet forming; Sheet metals
flatteners and levelers **A14:** 713
slitting **A14:** 708-711
Coiled strip *See also* Strip
flatteners and levelers **A14:** 713
multiple-slide forming of **A14:** 567
shearing (cut-to-length lines) **A14:** 711-713
slitting of **A14:** 708-711
Coiled wire
multiple-slide forming of **A14:** 567
Coils
superconducting **A2:** 1056-1057
Coin
production of **A14:** 184, 823
straightening, defined **A14:** 2
Coin test **EM3:** 8
defined **EM1:** 7 **EM2:** 10
Coin(s), coining
of copper-based powder metals **A7:** 733
defined **A7:** 2
early technology for **A7:** 16
effect on copper powder conductivity **A7:** 116
powders used **A7:** 573
as secondary pressing operation **A7:** 337-338
Coinability
of metals **A14:** 183
Coinage
as copper and copper alloy application **A2:** 239-240
copper and copper alloys **A14:** 823
silver, properties **A2:** 700-702
Coincidence boundaries in pure metals **A9:** 610
Coincidence-site model for grain boundaries **A9:** 119
Coining **A14:** 180-187 **A15:** 3, 264
aluminum alloy **A14:** 804
aluminum and aluminum alloys, defined **A2:** 6
applicability **A14:** 180
Coining (continued)
beryllium-copper alloys **A2:** 411
by multiple-slide forming **A14:** 567
capacity **A14:** 181
of carbon/low-alloy steels, lubricants for **A14:** 518
compressive residual stresses by **A11:** 125
of copper and copper alloys **A14:** 818
decorative **A14:** 180-182 **M3:** 508-510
defined **A14:** 2
die materials **A14:** 181-183
dies, defined **A14:** 2
dimension control **A14:** 186-187
drop hammer **A14:** 655
finish control **A14:** 186-187
in green machining **EM4:** 183
hammers and presses **A14:** 180-181
load vs displacement curve **A14:** 37
lubricants **A14:** 181
metals, coinability **A14:** 183
process steps **A14:** 180
production practice **A14:** 183-186
in progressive dies **A14:** 182
as sheet metal forming **A8:** 548
of sheet metals **A14:** 877-878
of stainless steels **A14:** 759
vs machining **A14:** 185
weight control **A14:** 186-187
working hardnesses **A14:** 182
Coining dies
cemented carbides, use for **M3:** 511
decorative coining **M3:** 508-510
gears **M3:** 510
hubbed dies **M3:** 509
machined dies **M3:** 509
P/M steels, use for **M3:** 510-511
progressive forming **M3:** 510
silverware **M3:** 509-510
working hardnesses **M3:** 510
Coke
bed, cupolas **A15:** 389
as cupola fuel **A15:** 30
defined **A15:** 3 **EM1:** 7 **EM2:** 10
fine, injection **A15:** 384
furnace, defined **A15:** 3
high-temperature gasification **A15:** 53
as reducing agent in Hoeganaes process **A7:** 79-82
specifications, cupolas **A15:** 387-388
Coke bed *See also* Flask
defined **A15:** 3
Coke breeze
defined **A15:** 3
Coke pig iron *See also* Pig iron
early usage **A15:** 30
Coke-oven car wheels
fatigue fracture of **A11:** 130
Cokeless cupolas **A15:** 392
Cokemaking **A1:** 107
Colburn process **EM4:** 399
Cold acid cleaners **M5:** 15
Cold alkaline cleaners **M5:** 15
Cold box processes *See also* Cold box resin binder processes
as coremaking systems **A15:** 238
defined **A15:** 3
release agents for **A15:** 240
Cold box resin binder processes *See also* Cold box processes
free radical cure process **A15:** 220
phenolic ester cold box process **A15:** 220-221
phenolic urethane cold box **A15:** 219
SO_2 process (Furan/SO_2) **A15:** 219-221
sodium silicate/CO_2 system **A15:** 221
Cold break failures
as copper penetration failure in locomotive axles **A11:** 715
Cold breakage
as casting defect **A11:** 383
Cold cathode **A6:** 30
Cold cell attachments for optical microscopes **A9:** 82
Cold chamber machine *See also* Hot chamber machine; Plunger; Port
defined **A15:** 3
Cold chamber pressure casting *See* Pressure casting
Cold chamber process
as die casting method **A15:** 286
gating system **A15:** 290
schematic **A15:** 287
Cold cleaning (solvent) **M5:** 40-44
Cold coined forging
defined **A14:** 2
Cold compacting
defined **A7:** 2
and rapid omnidirectional compacting **A7:** 543-544
Cold compacts
effect of metal powders in polymer **A7:** 607
Cold corrosion
aircraft powerplants **A13:** 1041-1045
Cold crack
definition **A6:** 1207
Cold cracking *See also* Hot cracking; Lamettar tearing; Stress-relief cracking **A6:** 410
as casting defect **A15:** 548
defined **A13:** 3 **A15:** 3
examination/interpretation **A12:** 137-138
high-strength low-alloy steels **A6:** 73
stainless steels **A6:** 677
Cold cracks *See also* Crack(s); Cracking; Cracks; Mechanical cracks
in arc welds **A11:** 413
in base metals **A11:** 440
in Electroslag welds **A11:** 440
in flash welds **A11:** 443
in iron castings **A11:** 353
radiographic appearance **A17:** 349
Cold die filling
in hot pressing **A7:** 502-503
Cold die quenching **M4:** 66
Cold dies
failure of **A14:** 56
Cold drawing *See* Drawing
defined **EM2:** 10
for sizing of tubular products **M2:** 264
Cold drawing, of Invar **A2:** 891
wrought copper and copper alloys **A2:** 250
Cold drawn steel
hardening of **M1:** 460, 461
machinability **M1:** 581-582, 585
spring wire grades **M1:** 284-285
wire, for concrete reinforcement **M1:** 271
Cold dynamic degassing **A7:** 180, 181
Cold equipment
as thermal inspection application **A17:** 402
Cold etching
defined **A9:** 4
of tools and dies **A11:** 563
Cold extrusion *See also* Extrusion **A14:** 299
of aluminum alloy parts **A14:** 307
composition effects **A14:** 300-301
of copper and copper alloy parts **A14:** 310-311
dimensional accuracy **A14:** 307
equipment **A14:** 301-302
extrusion ratio **A14:** 300
and impact extrusion, magnesium alloys **A14:** 311-312
of nickel-base alloys **A14:** 836-837
problems/causes **A14:** 307
procedure selection **A14:** 304-307
quality **A14:** 301
quality carbon steel wire rod **M1:** 254, 255
slug preparation **A14:** 303-304
of steel, condition effects **A14:** 300-301
steel, lubricants for **A14:** 304
steel wire fabrication **M1:** 591-592
for stepped shafts **A14:** 305-306
surface defects **A8:** 591-592, 595
tool materials **A14:** 303
tooling **A14:** 302-303
vs alternative processes **A14:** 300
vs hot upset forging **A14:** 95
Cold extrusion, forward
chevrons from **A11:** 88
Cold extrusion tools **M3:** 514-520
air-hardening tool steel **M3:** 518
backward extrusion **M3:** 514-516, 517, 518
cemented carbides **M3:** 516, 519-520
dies **M3:** 514-520
drawing tools **M3:** 515-516
forward extrusion **M3:** 514, 516-517

Cold extrusion tools (continued)
high speed steels ... **M3:** 518-519
high-carbon, high-chromium tool steels ... **M3:** 518
lubrication ... **M3:** 515
oil-hardening tool steels ... **M3:** 518
punches ... **M3:** 515-517, 518, 519
secondary components ... **M3:** 516, 517-518
shock-resisting tool steels ... **M3:** 518
wear, effect of ... **M3:** 515
Cold finger
defined ... **A10:** 670
Cold finished bars ... **M1:** 215-251
applications ... **M1:** 220, 221
bar types ... **M1:** 215
carbon restoration ... **M1:** 235
diameter tolerances ... **M1:** 217-219
elevated temperature drawing ... **M1:** 246-249
fatigue strength ... **M1:** 227-231
grades available ... **M1:** 215-216
heat treatment ... **M1:** 219, 232-235
heavy draft drawing ... **M1:** 247-249
impact properties ... **M1:** 225, 227-229, 231, 244, 247, 250
machinability ... **M1:** 216, 236-239
mechanical properties ... **M1:** 221-234, 241-251
microstructure, control of ... **M1:** 234, 235
product types ... **M1:** 215-219
quality descriptors ... **M1:** 219-221
residual stress ... **M1:** 225-226, 232, 234-235
special die drawing ... **M1:** 245-251
straightening ... **M1:** 225, 235
straightness tolerances ... **M1:** 219
stress relieving ... **M1:** 225-226, 232, 234
stress-strain curves ... **M1:** 221, 244, 249
tolerances ... **M1:** 217-219
transition temperature ... **M1:** 228, 230, 231, 232
Cold finishing
quality carbon steel wire rod ... **M1:** 254
for steel tubular products ... **A1:** 328-329 **M1:** 316-317, 324-325
Cold finishing quality rod ... **A1:** 273
Cold flow *See also* Creep; Deformation under load ... **EM3:** 8
of acetals ... **EM2:** 100
defined ... **EM1:** 7 **EM2:** 10
Cold forging
alloy steel wire for ... **M1:** 269-270
by HERF processing ... **A14:** 104
classification of cracks ... **A8:** 590, 592
cold heading as ... **A14:** 291
cracking in ... **A14:** 385
and powder forging, compared ... **A14:** 196
precision forging as ... **A14:** 158
workability tests for ... **A8:** 589-591
Cold formed
defined ... **A11:** 307
Cold forming *See also* Cold working
with cemented carbides ... **A2:** 970-971
of heat-resistant alloys ... **A14:** 779
and hot forming, combined ... **A14:** 620
of magnesium alloys ... **A14:** 825
of titanium alloys ... **A14:** 841
vs hot forming, three-roll forming ... **A14:** 619
wrought titanium alloys ... **A2:** 615
Cold heading *See also* Heading; Upsetting ... **A14:** 291-298
alloy steel wire for ... **M1:** 269-270
bolts ... **M1:** 274, 280
carbon and alloy steels ... **A14:** 291
complex workpieces ... **A14:** 294-295
defined ... **A14:** 2
dimensional accuracy ... **A14:** 295-296
economy in ... **A14:** 295
equipment ... **A14:** 291-292
and extrusion, combined ... **A14:** 296-297, 306
lubrication ... **A14:** 294
and machining, compared ... **A14:** 291
materials for ... **A14:** 291
Cold heading (continued)
of nickel-base alloys ... **A14:** 836-837
quality steel wire rod ... **M1:** 255, 256
steel wire fabrication ... **M1:** 589-592
steels, microalloyed ... **A14:** 221
straightening stainless steel for ... **A14:** 687
surface finish ... **A14:** 296
tool materials ... **A14:** 293
as tool steel application ... **A7:** 792
tools ... **A14:** 292-293
vs hot upset forging ... **A14:** 95
and warm heading ... **A14:** 297-298
work metal, preparation ... **A14:** 293-294
Cold heading quality alloy steel rod ... **A1:** 275
Cold heading tools
selection of materials for ... **M3:** 512-513
Cold hearth
melting, plasma ... **A15:** 424
refining process, electron beam ... **A15:** 414-415
Cold hydrostatic extrusion ... **A14:** 328
Cold hydrostatic pressing
advantages/disadvantages ... **A2:** 982
as cermet forming technique ... **A7:** 800
of cermets
dry-bag pressing ... **A2:** 982
wet-bag method ... **A2:** 982
Cold isostatic pressing ... **A16:** 100-101
beryllium ... **A16:** 870
cemented carbides ... **A18:** 796
FULDENS process ... **A16:** 64-65, 66
in production of cemented carbides ... **A16:** 72
Cold isostatic pressing (CIP) *See also* Cold pressing; Hot isostatic pressing (HIP); Hot pressing ... **EM4:** 124, 147-151, 188-189
advantages ... **A7:** 444-445 **EM4:** 147, 151
applications ... **A7:** 449-450 **EM4:** 150
beryllium powder ... **A2:** 685-686
of blended elemental Ti-6Al-4V ... **A2:** 649
constituents of powder formulation ... **EM4:** 126
defined ... **A7:** 2
dry-bag isostatic pressing ... **EM4:** 123, 147-148, 149, 150, 151
advantages ... **EM4:** 147-148
applications ... **EM4:** 151
automated lines ... **EM4:** 151
disadvantages ... **EM4:** 148
dwell pressures ... **A7:** 449
encapsulation of silicon nitride powder ... **EM4:** 197
equipment ... **A7:** 445-446 **EM4:** 149-150
flowchart ... **A7:** 448
formation of rod from titanium boride ... **EM4:** 199
history of process ... **EM4:** 147
limitations ... **EM4:** 147
materials ... **EM4:** 148-149
green body properties ... **EM4:** 148-149
subsystems ... **EM4:** 150
mechanical consolidation ... **EM4:** 125, 126
of metal powders ... **A7:** 444-450
parameters ... **A7:** 448-449
and powder properties ... **A7:** 448
schematic ... **A7:** 446
of titanium-based alloys ... **A7:** 164, 438-439
tooling ... **EM4:** 149
latex molds ... **EM4:** 149
natural rubber molds ... **EM4:** 149
neoprene rubber molds ... **EM4:** 149
nitrile rubber molds ... **EM4:** 149
polysulfide molds ... **EM4:** 149
polyurethane molds ... **EM4:** 149
polyvinyl chloride molds ... **EM4:** 149
silicone molds ... **EM4:** 149
units ... **A7:** 446
warm temperatures ... **EM4:** 150
wet bag isostatic pressing ... **EM4:** 123, 147, 148, 149, 150, 151
advantages ... **EM4:** 147
applications ... **EM4:** 151
Cold isostatic pressing (CIP) (continued)
automated line ... **EM4:** 151
disadvantages ... **EM4:** 147
Cold lap *See also* Cold shut
as casting defect ... **A11:** 383
as defect, squeeze casting ... **A15:** 326
defined ... **A14:** 2-3 **A15:** 3
Cold melting
malleable cast irons ... **M1:** 57
Cold molding
defined ... **EM2:** 10
Cold mounting
of carburized and carbonitrided steels ... **A9:** 217
Cold nickel plating ... **M5:** 202-203
Cold parison blow molding
defined ... **EM2:** 10
"Cold" plasma treatment ... **EM3:** 35
Cold plastic flow
rolling element bearings failure by ... **A11:** 499
Cold plates *See* Heat sinks
Cold press molding
defined ... **EM2:** 10
properties effects ... **EM2:** 287
size and shape effects ... **EM2:** 291
Cold pressing ... **EM3:** 8
of beryllium powders ... **A7:** 171, 172
ceramics ... **A16:** 98
cermets ... **A7:** 799
coins ... **A7:** 172
compared to rapid omnidirectional compaction ... **A7:** 545
defined ... **A7:** 2 **EM2:** 10
and dimensional change from tooling ... **A7:** 480
nuclear fuels ... **A7:** 665
and sintering ... **A7:** 172, 522, 545, 665
Cold pressure welding *See also* Cold welding
definition ... **A6:** 1207
Cold processing
ternary molybdenum chalcogenides (chevrel phases) ... **A2:** 1079
Cold proof tests ... **EM4:** 718
Cold re-pressing
elastic springback during ... **A7:** 480
Cold reduction
beryllium-copper alloys ... **A2:** 406
Cold resistance ... **EM3:** 52
water-base versus organic solvent-base adhesives properties ... **EM3:** 86
Cold roll bonding
of clad metals ... **A13:** 887
Cold rolled aluminum-killed steel
mechanical properties ... **M1:** 178
for porcelain enameling ... **M1:** 177
Cold rolled rimmed steel
mechanical properties ... **M1:** 178
porcelain enameling of ... **M1:** 177-178
sag resistance ... **M1:** 179-180
Cold rolled sheet and strip
annealing ... **M1:** 156-157
ASTM specifications ... **M1:** 154, 155
bend limitation for ... **M1:** 555
characteristics ... **M1:** 154, 556
classes of ... **M1:** 153-154
flatness ... **M1:** 157, 160-161
leveling ... **M1:** 157, 160-161
mechanical properties ... **M1:** 155-156, 160-161, 178
mechanical properties related to formability ... **M1:** 547-549
mill heat treatment ... **M1:** 156-157
minimum bend radii, selected grades ... **M1:** 554
modified low-carbon steels ... **M1:** 161-162
normalizing ... **M1:** 157
Olsen ductility ... **M1:** 156, 161
porcelain enameling of ... **M1:** 177-178
production of ... **M1:** 153-154, 178-179
quality descriptors ... **M1:** 154-155
sag resistance, enameling steel ... **M1:** 179-180
spring strip grades ... **M1:** 285-286
standard sizes ... **M1:** 154

Cold rolled sheet and strip (continued)
strain aging **M1:** 154, 157, 162
stretcher strains, in annealed stock **M1:** 157
surface characteristics **M1:** 157
thickness **M1:** 153, 154, 161
width range **M1:** 153, 154
Cold rolled steel
porcelain enameling of **M5:** 512, 517, 527
Cold rolling
edge cracking in **A8:** 594
green strip densification by **A7:** 406, 407
hafnium **A2:** 663
metalworking lubricants and friction **A18:** 147
nickel strip **A7:** 401
in processing of solid steel **A1:** 121-122, 123
reduction, in tension testing **A8:** 595-596
workability limits **A8:** 594
wrought copper and copper alloys **A2:** 244-245
zirconium **A2:** 663
Cold rolling quality carbon steel wire rod **M1:** 254
Cold sawing, for stock preparation
hot upset forging **A14:** 86
Cold setting of springs *See* Presetting of springs
Cold shearing, for stock preparation
hot upset forging **A14:** 86
Cold sheet welding *See* Lap welding
Cold shortness
sulfur effects **A15:** 29
Cold shot
as casting defect **A11:** 387
defined **A11:** 2 **A15:** 3
in iron castings **A11:** 353
Cold shut *See also* Cold lap
as casting defect **A11:** 282
in closed-die forgings **A14:** 385
defined **A11:** 2 **A14:** 3
insert **A11:** 383
in iron castings **A11:** 352
preventing, in hot upset forging **A14:** 88
stainless steel fuel-control lever fractured at **A11:** 388
as surface discontinuities, iron castings **A11:** 352-354
Cold shut defect
in closed-die forging **A8:** 590
Cold shuts
arc welds in heat-resistant alloys **M6:** 364
black magnetic-power indications **A17:** 101
in copper alloy ingots **A9:** 642
defined **A15:** 3
as die casting defect **A15:** 294
as discontinuities, defined **A12:** 64-65
as forging flaws **A17:** 494
magnetic particle inspection **A17:** 108
in permanent mold castings **A15:** 285
as planar flaw **A17:** 50
and pouring temperature **A15:** 283
radiographic appearance **A17:** 349
radiographic inspection methods **A17:** 296
in semisolid metal casting and forging **A15:** 336-337
tool steels **A12:** 378
wrought aluminum alloys **A12:** 435
zirconium alloys **A15:** 838
Cold slug
defined **EM2:** 10
Cold soldered joint
definition **A6:** 1207 **M6:** 4
Cold stamping marks
quench cracking from **A11:** 94
Cold static degassing **A7:** 180-181
Cold straightening
effect on fatigue fracture **A11:** 125
Cold stretch
defined **EM2:** 10
Cold swaging *See also* Swagers; Swaging
reduction by **A14:** 128
rolls and backers for **A14:** 131
Cold swaging, of rimmed steel tube
brittle fracture after **A11:** 648
Cold tearing
as casting defect **A11:** 383
Cold test
defined **A18:** 5
Cold trap
with gas chromatographs **A17:** 68
Cold treating
steel **A4:** 203-204 **M4:** 117-118
Cold trimming **A14:** 3, 82
forgings **M1:** 367
Cold upset testing **A8:** 579-581
Cold upset welding *See* Butt welding
Cold water *See also* Water
zinc corrosion in **A13:** 761
Cold water rinsing
of cold extruded parts **A14:** 304
Cold weld
defined **A17:** 562
Cold welding *See also* Welding **A6:** 307-309 **M6:** 673-674
aluminum **A6:** 307-308, 309
butt welding **M6:** 673-674
cold pressure butt welding **A6:** 308-309
cold pressure lap welding **A6:** 307-308
copper **A6:** 307-308, 309
defined **A7:** 2
definition **A6:** 1207 **M6:** 4
in drawing processes **A6:** 309
equipment **A6:** 307, 308, 309
from fretting **A13:** 138
lap welding **M6:** 673
mechanically alloyed oxide alloys **A2:** 943
dispersion-strengthened (MA ODS)
as mechanism of green strength **A7:** 303-304
multiple-step upsetting method **A6:** 308
and nonlubricated wear **A11:** 154-155
single-step upsetting method **A6:** 308-309
slide welding **M6:** 674
surface extension parameter **A6:** 308
variations **A6:** 307-309
Cold work **A9:** 684
percent, measured **A10:** 380, 390
plastic strain, topographic methods for **A10:** 368
Cold worked structure
defined **A9:** 4
Cold working *See also* Cold extrusion; Cold forging; Cold forming; Cold heading; Deformation, Forming; Hot working; Warm working **A13:** 3, 48-49, 741
in aluminum alloys, etchants for examination of **A9:** 355
aluminum alloys, relation to heat treatment **M2:** 32, 38, 40-42
beryllium-copper alloys **A2:** 421-422
copper and copper alloys **A2:** 219, 223
copper and copper alloys, formability effects **A14:** 809-810
copper metals **M2:** 241
damage from sectioning **A9:** 23
defined **A14:** 3
effect on cellular precipitation **A9:** 649
effect on magnetic properties **A9:** 539
effect on maraging steel **A11:** 218
effect on microhardness **A8:** 96
effect on sigma phase formation in austenitic stainless steels **A9:** 284
effect on sigma phase formation in duplex stainless steels **A9:** 286
effect on SSC resistance **A11:** 299
hardness as measure of **A7:** 61
of iridium **A14:** 851
of magnesium alloys during sectioning **A9:** 425
of maraging steels **A1:** 795 **M1:** 447
of palladium **A14:** 850
of platinum **A14:** 849-850
platinum alloys **A14:** 851
of precipitation-hardenable stainless steels to produce martensite **A9:** 285
of rhodium **A14:** 850
strain-age embrittlement, effect on **M1:** 683
sulfide stress cracking, effect on **M1:** 687
unidirectional, Brinell indentation in **A8:** 85
and upset forging **A7:** 690
working methods in **A14:** 383
zirconium **A2:** 666
Cold working alloys
ion implantation strengthening mechanisms **A18:** 855, 858
wear resistance versus hardness of materials **A18:** 708
Cold working of stainless steels **M3:** 43-44
Cold working temperature
defined **A8:** 575
ductile fracture at **A8:** 154, 573-574
flow stress **A8:** 575
Cold wound springs ... **M1:** 283-286, 288-300, 303-313
steels for **M1:** 283-285
Cold-drawn bar
eddy current inspection **A17:** 553
hexagonal, inspection **A17:** 552-554
ultrasonic inspection **A17:** 551-552
Cold-drawn steel
machinability of **A1:** 601
Cold-drawn wire
flaw detection **A17:** 552, 554
Cold-extruded steel parts
ultrasonic inspection **A17:** 271
Cold-finished steel bars **A1:** 248-271
bar sizes **A1:** 248
classifications **A1:** 248
commercial grades **A1:** 248-249
heat treatment **A1:** 260
carbon restoration **A1:** 261, 263-264
machinability **A1:** 264-265
mechanical properties **A1:** 254, 257
hardness **A1:** 258
impact properties **A1:** 259
tensile and yield strengths **A1:** 257-258
product quality descriptors **A1:** 252
alloy steel quality descriptors **A1:** 253-254
carbon steel quality descriptors **A1:** 252-253
product types **A1:** 248
cold-drawn bars **A1:** 250-251
machined bars **A1:** 249-250
turning versus cold drawing **A1:** 251
residual stresses **A1:** 259
straightening **A1:** 259
stress relieving **A1:** 259-261
special die drawing **A1:** 268-269
drawing at elevated temperatures **A1:** 269-271
heavy drafts **A1:** 269
strength considerations **A1:** 265-266, 268
Cold-formed parts, failures of **A11:** 307-313
by materials defects **A11:** 307
design problems **A11:** 307-308
prevention of **A11:** 308-313
process problems **A11:** 307-308
Cold-forming
ductile or brittle fractures from **A11:** 88
Cold-forming strip
high-strength low-alloy steels for **A1:** 418-419
Cold-heading applications
cemented carbides **A2:** 971
Cold-heading wire **A1:** 851
Cold-manifold molding *See* Warm-runner molding
Cold-mounting epoxies
as mounting material for tool steels **A9:** 257
Cold-mounting materials **A9:** 30-31
for resin-matrix composites **A9:** 588
Cold-mounting resins
used for powder metallurgy materials **A9:** 504
Cold-pressure welding
of aluminum/copper to-packages **EL1:** 239
Cold-rolled high-strength low-alloy steels **A1:** 420-421
Cold-rolled sheet
defined **A14:** 3
Cold-rolled steel
AES analysis of corrosion resistance in **A10:** 556-557
effects of steelmaking practices on formability of **A1:** 577-578
for glass-to-metal seals **EL1:** 455
tensile properties and formability factors of **A1:** 398, 420
Cold-rolled steel products
mill heat treatment of **A1:** 202-204
Cold-rolled steels
enamels **EM3:** 303
etching procedure **EM3:** 272
Cold-runner molding *See also* Warm-runner molding
defined **EM2:** 10
Cold-set phenolic urethane molds
Alnico alloys **A15:** 736-737
Cold-setting adhesive
defined **EM1:** 7 **EM2:** 10
Cold-setting adhesives **EM3:** 8

Cold-setting epoxy
as a mounting material for wrought stainless steels ... **A9:** 279
Cold-setting process *See also* No-bake binder
defined ... **A15:** 3
Cold-slug well
defined ... **EM2:** 10
Cold-thermal-wave excitation
for thermal inspection ... **A17:** 398, 403
Cold-welding ... **A18:** 367
"Cold-welding" phenomenon ... **A18:** 150
Cold-work finishing processes ... **A16:** 26
Cold-work tool steels powder metallurgy ... **A1:** 786-789
wrought ... **A1:** 763-766
air-hardening, medium-alloy ... **A1:** 763-765
high-carbon, high-chromium ... **A1:** 765
oil-hardening ... **A1:** 765
Cold-work tools
ASP steel grades application ... **A16:** 61
Cold-worked metals
nucleation sites ... **A9:** 694
stages of annealing ... **A9:** 692
Cold-worked steels
macroetching ... **A9:** 172
Coldstream impact process milling ... **A7:** 2, 69
Colemanite ($Ca_2B_6O_{11}{\cdot}5H_2O$)
purpose for use in glass manufacture ... **EM4:** 381
Collapse
defined ... **EM2:** 10
of liquid-erosion bubbles or cavities ... **A11:** 163-164
pressures, from cavitation ... **A11:** 163-164
Collapsibility *See also* Core filler
defined ... **A15:** 3
of polyol urethane and phenolic urethane compared ... **A15:** 217
Collapsible cores
permanent mold casting ... **A15:** 279-280
Collapsible tool
defined ... **A7:** 2
Collapsible tubes
tin-base alloy ... **A2:** 525
Collar
definition ... **M6:** 4
in split-Hopkinson bar tension test ... **A8:** 212-213
Collar oiler
defined ... **A18:** 5
Collaring
definition ... **M6:** 4
Collation, of layers
ceramic packages ... **EL1:** 464
Collection efficiency
defined ... **A18:** 5
Collection of indexable data (CID)
of engineering design system ... **EL1:** 129
Collection optics ... **A10:** 24, 128
Collector slit
in gas mass spectrometers ... **A10:** 153, 154
Collet
defined ... **EM1:** 7 **EM2:** 10
Collet grip
for fatigue testing ... **A8:** 368
Collet retainer tube
SCC in ... **A13:** 933
Collet-type machines
die threading operations ... **A16:** 296
Colliau cupola
development of ... **A15:** 29-30
Colligative properties ... **EM3:** 8
defined ... **EM2:** 10
Collimate
defined ... **A10:** 670
Collimated ... **EM3:** 8
defined ... **EM1:** 7 **EM2:** 10
Collimated electron beam
SEM imaging ... **A12:** 167
Collimated roving *See also* Roving
defined ... **EM1:** 7 **EM2:** 10
Collimation
defined ... **A9:** 4 **A10:** 670
neutron diffraction ... **A10:** 422
in neutron radiography ... **A17:** 388-389
x-ray, basic methods ... **A10:** 403
Collimator ... **EM3:** 8
defined ... **EM2:** 10
Collimators
aperture, defined ... **A17:** 383
defined ... **A17:** 383
for scattered radiation ... **A17:** 344
Soller ... **A10:** 87
x-ray, computed tomography (CT) ... **A17:** 368-369
Collision efficiency *See* Collection efficiency
Collision kinematics
RBS analysis ... **A10:** 629
Collision-activated dissociation mass spectra
gas chromatography/mass spectrometry ... **A10:** 647
Collision-point velocity (Vc) ... **A6:** 160, 161, 162
Collisional line broadening
in emission spectroscopy ... **A10:** 22
Collisions
elastic and inelastic ... **A12:** 168
Collisions (particle)
effect on particle shape ... **A7:** 32
effects on quench rates ... **A7:** 38, 39
force or energy in grinding ... **A7:** 60
kinematics, as simple elastic binary collision ... **A7:** 259
milling mechanisms of ... **A7:** 57, 65-70
of particles, in atomization ... **A7:** 29-30, 34
Collisions, binary
LEISS analysis ... **A10:** 604
Collodian replica
defined ... **A9:** 4
Colloidal ... **EM3:** 8
defined ... **EM1:** 7 **EM2:** 10
Colloidal silica
as binder, investment casting ... **A15:** 258
bonds, characteristics ... **A15:** 212
as filler ... **EM3:** 178
for hand polishing ... **A9:** 35
used for very soft materials ... **A9:** 47
used in mechanical polishing ... **A9:** 43
Colloidal suspensions of magnetic particles ... **A9:** 63-64
Colmoloy
laser cladding components and techniques ... **A18:** 869
Colmonoy
honing stone selection ... **A16:** 476
Colonies
defined ... **A9:** 4
formation by precipitation ... **A9:** 647-649
in pearlite ... **A9:** 658
Colony size of acicular alpha in titanium and titanium alloys
effect on properties ... **A9:** 460
Colony structures, eutectic ... **A9:** 619-620
aluminum and Mg_2Al_3 ... **A9:** 619
$CuAl_2$-Al ... **A9:** 620
niobium-carbide rods in nickel matrix ... **A9:** 619
Colophony ... **EM3:** 8
Color
as an aluminum alloy phase identifier ... **A9:** 359-360
brass plating, alloy composition determining ... **M5:** 285
change, gas/leak detection by ... **A17:** 61
of chromate conversion coatings ... **A13:** 390, 392
coding, thermocouple wires and extension wires ... **A2:** 878
colorants, as additives ... **EM2:** 500-501
copper and copper alloys ... **A2:** 216, 219
for degradation detection ... **EM2:** 574
development and stability, in UV/VIS analysis ... **A10:** 68
images, by various NDE methods ... **A17:** 483-488
Color (continued)
inherent, of engineering plastics ... **EM2:** 1
of lead compounds ... **A2:** 548
in milling of single particles ... **A7:** 59
models, digital image enhancement ... **A17:** 458
molded-in, processes for ... **EM2:** 305-306
or tint etching, of image analysis samples ... **A10:** 313
oxide ... **EM2:** 501
paint *See* Paint and painting, color
palladium-silver alloys ... **A2:** 716
of penetrants ... **A17:** 75
of polyamide-imides (PAI) ... **EM2:** 130
porcelain enamel *See* Porcelain enameling, color
of porcelain enamels ... **A13:** 448, 451
sensing, by machine vision ... **A17:** 43
styrene-acrylonitriles (SAN, OSA ASA) ... **EM2:** 216
styrene-maleic anhydrides (S/MA) ... **EM2:** 219
surface, optical testing ... **EM2:** 598
of thermoplastic polyurethanes (TPUR) ... **EM2:** 206
use, in digital image enhancement ... **A17:** 458
of welded gray iron ... **A15:** 527
Color anodizing
aluminum and aluminum alloys ... **M5:** 595-596, 609-610
Color background
for optical testing ... **EL1:** 571
Color buffing
compound ... **M5:** 117
copper and copper alloys ... **M5:** 616
process ... **M5:** 115-116, 126
silver ... **M5:** 306
stainless steel ... **M5:** 557-558
Color centers
defined ... **A10:** 670
detected by ESR ... **A10:** 263
Color comparison
kits ... **A10:** 66-67
in UV/VIS absorption analysis ... **A10:** 66
Color concentrate ... **EM3:** 8
defined ... **EM2:** 10
Color contrast
in potentiostatic etching ... **A9:** 145-147
in wrought stainless steels ... **A9:** 281
Color differences
in beryllium under polarized light ... **A9:** 389
Color dipping
copper and copper alloys ... **M5:** 611-612
Color etchants ... **A9:** 136-137, 139-142
Color etching to obtain interference films ... **A9:** 136
principles ... **A9:** 139-142
Color films
for photomicroscopy ... **A9:** 84-86
Color filter
defined ... **A9:** 4
Color filter nomograph for photography ... **A9:** 140
Color images
by acoustic microscopy ... **A17:** 485, 487
by computed tomography ... **A17:** 483, 486
by digital radiography ... **A17:** 485
by stress analysis ... **A17:** 488
by ultrasonic inspection ... **A17:** 484-486, 488
under polarized light ... **A9:** 78
Color metallography ... **A9:** 135-162
advantages ... **A9:** 135
interference film deposition *See also* Anodizing; Color etching; Heat tinting; Potentiostatic etching; Reactive sputtering,, Vacuum deposition ... **A9:** 135-138
methods ... **A9:** 135-139
Color photograpahy ... **A9:** 139-140, 142
effect of lens defects on ... **A9:** 75
Color temperature
defined ... **A9:** 4
of light source in optical microscopes balancing to film ... **A9:** 72

Color TV glasses
material to which crystallizing solder glass seal is applied **EM4:** 1070
material to which vitreous solder glass seal is applied **EM4:** 1070
Color TV panel
defect inclusion levels **EM4:** 392
Color-producing groups *See* Chromophores
Colorants **EM3:** 304 **EM4:** 380
as additives **EM2:** 500-501
for epoxies **EM3:** 99
Colored container glass
iron content **EM4:** 378
Colorimeter
paint color testing **M5:** 491
Colorimetry
schematic of Nessler tube **A10:** 66
Coloring process **M5:** 624-626
copper and copper alloys **M5:** 624-626
nickel alloys **M5:** 674-675
procedures **M5:** 624-626
solution compositions and operating conditions **M5:** 625-626
Colorizing
pack diffusion processes **M5:** 340
Colors
fatigue area, austenitic stainless steels **A12:** 352
formed by precipitation etching **A9:** 61
of fracture, crack growth measurement by **A12:** 120
of interference films, effect of thickness on **A9:** 136, 141, 143, 145-146
temper, of cracks **A12:** 65
Colossal statues
metalworking of **A15:** 21
Colt-Crucible ceramic mold process containerization, system for **A7:** 751
parts produced by **A7:** 754
titanium powder production **A7:** 167, 755
Columbite **A7:** 160
niobium pentoxide recovery fro m **A2:** 1043
Columbium *See* Niobium
Columbium, vapor pressure, relation to temperature *See also* Niobium **M4:** 310
Columbium-stabilized steel
porcelain enameling of **M5:** 512-513
Column hydrodynamic chromatography
to analyze ceramic powder particle sizes **EM4:** 67
Column type
coordinate measuring machines **A17:** 21
Columnar bainite **A9:** 665
Columnar fractures **A12:** 2
Columnar front
described **A15:** 132-133
Columnar grain structure
in aluminum alloy 1100 .... **A9:** 630
in aluminum alloy 6063 .... **A9:** 630
in aluminum alloy ingots **A9:** 629-631
in pure metals **A9:** 610
in titanium alloy welded joints **A9:** 579
Columnar grains
in castings, crystallographic texture in **A9:** 700-701
in DHP copper **A9:** 641
formation in Alnico alloys **A9:** 539
in steel macrostructure **A9:** 623
Columnar growth *See* Twinned columnar growth
Columnar structure
defined **A11:** 2
Columnar structure defined **A9:** 4
formation of **A9:** 603
Columnar structures
defined **A15:** 3
from directional solidification **A15:** 319-320
growth, vs equiaxed grain growth **A15:** 135
modeling of **A15:** 884-885
Columnar to equiaxed transition **A15:** 130-135
casting parameters, effect **A15:** 130
columnar vs equiaxed grain growth **A15:** 135
equiaxed grains, growth of **A15:** 132-135
equiaxed nuclei, origin of **A15:** 130-132
Columnar zone
in ferrous alloy welded joints **A9:** 581
Columns
cleaning of **A13:** 1138
Coma
defined **A9:** 4 **A10:** 670
Combination die *See* Compound die
defined **A15:** 3
Combination mechanical coatings **M5:** 300-302
Combination mold *See* Family mold
Combination polishing wheels .... **M5:** log
Combination rotary automatic polishing and buffing machines **M5:** 121
Combined carbon *See also* Carbon; Free carbon
defined **A9:** 4 **A13:** 3 **A15:** 3
effect in sintering iron-graphite powder **A7:** 362-363
effect on microstructure **A7:** 363
effect on tensile strength of wrought (rolled) steel **A7:** 362
effect on transverse-rupture strength of sintered steel **A7:** 362
Combined cyclic stress
analysis **EM1:** 202-203
Combined environments reliability test **A13:** 1114
Combined environments reliability testing (CERT)
as advanced failure analysis technique **EL1:** 1095
Combined Industry Standards and Military Specifications **EM3:** 72
Combined roughness **A18:** 146
Combined temperature-moisture-mechanical effects on adhesive joints **EM3:** 651-655
evaluation parameters **EM3:** 651
model selection **EM3:** 652-653
multiple stresses **EM3:** 653-654
selecting the number of experiments **EM3:** 652
service life tests **EM3:** 651-653, 654
temperature, moisture, and mechanical stress **EM3:** 654-655
Combined wear
mechanisms of **A11:** 159
Combines
P/M parts for **A7:** 675-676
Combing
defined **EM2:** 10
Combusted fuel gas
brazing atmosphere source **A6:** 628
Combustible-gas detectors
for leak testing **A17:** 62
Combustion *See also* Flame retardants; Flammability
accelerators **A10:** 221-222
chamber, aluminum alloy, cavitation erosion **A11:** 168
of composites **EM1:** 35
control, steam equipment **A11:** 619
determination of carbon, hydrogen, and nitrogen by **A10:** 214
furnaces, high-temperature combustion **A10:** 221-225
high-frequency **A10:** 221-225
high-frequency, typical configuration **A10:** 222
mechanisms of metals in oxygen **A7:** 597
natural gas **M6:** 588
oxyacetylene **M6:** 587-588
oxyhydrogen **M6:** 588
of polyaryl sulfones (PAS) **EM2:** 146
of polyarylates (PAR) **EM2:** 140
polyvinyl chlorides (PVC) **EM2:** 209
products, detection in high-temperature combustion **A10:** 222
propane **M6:** 588
in space heater catalysts, powder used **A7:** 572
to analyze the bulk chemical composition of starting powders **EM4:** 72
total and selective, sample preparation **A10:** 223-224
Combustion chamber liners **A2:** 922
Combustion equipment
cast iron coatings for **M1:** 104
Combustion flame spraying
ceramic coatings **M5:** 539-542, 546
equipment **M5:** 540-542
gravity-feed powder system **M5:** 540-542
pressure-feed powder system **M5:** 540
process steps **M5:** 539-540
rod spray system **M5:** 540-542
surface preparation for **M5:** 539
Combustion method, for elemental analysis of carbon
hydrogen, and nitrogen **A10:** 214
Combustion ratio for fuel gases **M6:** 900
Combustion reactions, zones
cupolas **A15:** 389
Combustion synthesis (CS) **EM4:** 228
Combustion technique, for trace element analysis
in carbon **A2:** 1095
Combustion turbines
corrosion of **A13:** 999-1001
Combustion wire process *See* Oxyfuel wire spray process
Comet tails
defined **A9:** 4
in magnesium alloys **A9:** 425
Commands
initialization **EL1:** 5
Commercial alloys
cast copper alloys **A2:** 356-391
mechanically alloyed oxide alloys **A2:** 944-947
dispersion-strengthened (MA ODS) with rare earth metals **A2:** 720, 730
Commercial applications *See also* Marine applications; Sports and recreational equipment
automotive electronics **EL1:** 381-382
consumer electronics **EL1:** 385
environmental testing for **EL1:** 493-503
hybrid characteristics **EL1:** 381
matrices for **EM1:** 31-32
of structural composites **EM1:** 832-836
telecommunications **EL1:** 382-383
Commercial bronze *See* Copper alloys, specific types, C22000 .... **A6:** 752
applications **A18:** 750, 751
applications and properties **A2:** 296-297
composition **A18:** 751
designations **A18:** 751
mechanical properties **A18:** 750, 752
product form **A18:** 751, 752
weldability **A6:** 753
Commercial cast irons
types **A13:** 567-568
Commercial clays **EM4:** 6
properties **EM4:** 7
Commercial coppers
gas-metal arc butt welding **A6:** 760
Commercial Duralumin alloys
as aluminum casting alloys **A2:** 126-127, 129-130
Commercial ferrous forging **A7:** 415-417
Commercial fine gold
properties **A2:** 704-705
Commercial heat-treatable aluminum alloys
types **A2:** 40-41
Commercial hybrids *See* Hybrids
Commercial iron (III) oxide
decomposition temperatures **EM4:** 55
Commercial items descriptions (CIDS) **EM3:** 63
Commercial names
of polymers **EM2:** 53-56
Commercial nitrogen-base atmospheres
advantages **M4:** 403-404
applications **M4:** 406-408
blending equipment **M4:** 404, 405-406
carbon-controlled **M4:** 403
components **M4:** 404-405, 406
protective atmospheres **M4:** 403
reactive atmospheres **M4:** 403
Commercial permeametry **A7:** 263
Commercial powder metallurgy **A7:** 16-18
Commercial prepreg, defined *See* under Prepreg
Commercial purity
of metals **A2:** 1093
Commercial pycnometers **A7:** 265
Commercial quality
of carbon steels **A1:** 201
Commercial quality sheet and strip
low-carbon steel **M1:** 154, 155
Commercial rolled zinc alloys
properties **A2:** 539-540
Commercial salt bath hardening
high-speed tool steels **A16:** 55, 56
Commercial shape memory effect (SME) alloys
copper-base shape memory alloys **A2:** 899-900
nickel-titanium alloys **A2:** 899
Commercial suppliers
of UV-cured conformal coating material **EL1:** 786

Commercial vacuum hot pressed beryllium products **A7:** 172
Commercial x-ray photoelectron spectroscopy systems **A7:** 255
Commercially pure copper
applications, properties, types **A2:** 223, 230, 234
Commercially pure iron
air-furnace melted, magnetically soft **A2:** 764-765
vacuum-induction melted, magnetically soft **A2:** 764
Commercially pure lead **A9:** 418
Commercially pure nickel
applications and characteristics **A2:** 435, 437, 441
Commercially pure palladium
properties **A2:** 714-716
Commercially pure platinum
properties **A2:** 707-709
Commercially pure silver
properties **A2:** 699-700
Commercially pure tin
properties **A2:** 518-519
Commercially pure titanium *See* Titanium; Titanium alloys, specific types, unalloyed Ti
vacuum effects on fatigue **A12:** 48-49
Comminution **EM4:** 75-81
additives and their effect **EM4:** 77
alteration of powder properties **EM4:** 77
of beryllium powder **A7:** 170
compacting by **A7:** 58
control of milling systems **EM4:** 77
defined **A7:** 2
definition **EM4:** 75
dry-milling technique **EM4:** 75, 76
effect of chemical reactions **A7:** 58
equipment **EM4:** 78-81
basic mechanisms **EM4:** 78
dry-milling methods **EM4:** 78
types **EM4:** 78
wet-milling, agitation ball **EM4:** 78, 80-81
wet-milling, fluid energy **EM4:** 80, 81
wet-milling, methods **EM4:** 78
wet-milling, particle size distribution **EM4:** 79
wet-milling, planetary ball **EM4:** 78
wet-milling, tube **EM4:** 78
wet-milling, vibratory ball **EM4:** 78, 79-80
for hard metals and oxide powders **A7:** 56
of magnesium **A7:** 131, 132
mechanical, of precious metal powders **A7:** 149
mechanochemical effects **EM4:** 76-77
objectives **EM4:** 75
advantages **EM4:** 75
breakage phenomenon **EM4:** 75
disadvantages **EM4:** 75
for prealloyed titanium powder production **A7:** 165-167
process **EM4:** 75-76
models **EM4:** 76, 77
purpose **EM4:** 42
slip casting **EM4:** 157
of sliver powders **A7:** 148
tape casting **EM4:** 161
uniaxial hot pressing **EM4:** 188
wet-milling technique **EM4:** 75, 76
Committee C-24 (ASTM) on Building Seals and Sealants **EM3:** 71
Committee D-14 (ASTM) on Adhesives **EM3:** 71
Committee D-30 on High Modulus Fibers and Their Composites (of ASTM) **EM1:** 40
Committee for Acoustic Emission in Reinforced Plastics (CARP) **A17:** 291
Committee on Characterization of Materials, Materials Advisory Board
National Research Council **A10:** 1
Common cause **EM3:** 785, 795
Common desilverized lead *See* Leads and lead alloys, specific types, corroding lead
Common dry drawn finish
steel wire **M1:** 261
Common emitter current gain
bipolar transistor analysis **EL1:** 152-153
Common hardness scales **A7:** 489
Common ion effect
in gravimetric analysis **A10:** 163
Common lead *See also* Lead
composition **A2:** 543, 545
Common logarithm (base 10)
abbreviation for **A10:** 690
Common manufacturing information system (CMIS) **EL1:** 130
Common modes
modification factor **EL1:** 38-39
in parallel lines **EL1:** 37-38
parameters **EL1:** 39
Common release processing system (CRPS) **EL1:** 130
Communication
issues and types **EL1:** 3-5
Communications applications *See also* Telecommunications applications
acrylonitrile-butadiene-styrenes (ABS) **EM2:** 111
for hybrids **EL1:** 254
urethane hybrids **EM2:** 268
Communications equipment **A13:** 1113-1126
analysis techniques **A13:** 1113-1118
chemical analysis **A13:** 1114-1117
electron optics **A13:** 1117
electronic/electrical corrosion, examples **A13:** 1118-1126
Commutator **A6:** 39, 40
Commutator-controlled welding
definition **M6:** 4
Commutators **A7:** 715
Compact
defined **A14:** 3
Compact specimen *See also* Center-cracked tension specimens
abbreviation **A8:** 724
calibration curve **A8:** 386-387
with copper current input connections and potential probes **A8:** 389
for corrosive environmental testing **A8:** 427, 429
crack measurement intervals **A8:** 378
and current input and potential probe locations **A8:** 386
equipotential distribution from electrical analog patterns **A8:** 387
for fatigue crack growth analysis **A8:** 379, 678-679
for fatigue crack propagation testing **A8:** 377-378
grip **A8:** 382
non-dimensional voltage with lead position on **A8:** 388
for plane-strain fracture toughness testing **A8:** 450-451
precracking **A8:** 382
predicted and experimental compliance **A8:** 385
size **A8:** 380-381
stress-intensity factor solutions **A8:** 379-380
thickness **A8:** 381
for vacuum and gaseous fatigue testing **A8:** 411
Compact tension *See also* Tension
abbreviation for **A11:** 796
specimen, crack growth in **A11:** 63
Compact testing
system-level **EL1:** 375
Compact(s) *See also* Green compacts
of body-centered cubic metals **A7:** 310
carbonyl nickel powder, density **A7:** 310
changes in mechanical properties **A7:** 311
compact (noun), defined **A7:** 2
compact (verb), defined **A7:** 2
dimensional change **A7:** 290-292
electrolytic copper, density of **A7:** 310
green **A7:** 288-289, 311
green density **A7:** 310
Compact(s) (continued)
hardness and sintering temperature **A7:** 312
hot pressing fully dense **A7:** 501-521
Kirkendall porosity formation and coarsening in nickel blend **A7:** 314
mean composition, effect on densification and expansion during homogenization **A7:** 315
microstructure of copper particle blend **A7:** 314
puffed **A7:** 9
shrinkage as function of particle size **A7:** 309
shrinkage as function of sintering temperature **A7:** 191
sintering **A7:** 309-314
sintering, of homogeneous metal powders **A7:** 302-312
variation of sintered density during homogenization in nickel-copper **A7:** 314
Compacted graphite cast iron **M1:** 7-9
applications **M1:** 8
foundry practices **M1:** 8
graphite shape **M1:** 7-9
microstructure **M1:** 7-9
relation of properties to structure **M1:** 8-9
Compacted graphite iron **A1:** 3, 8-9, 56-70
advantages **A1:** 70
applications **A1:** 70
castability
chilling tendency **A1:** 57
fluidity **A1:** 57
shrinkage characteristics **A1:** 57
composition of **A1:** 5, 8-9, 56-57
cooling rate **A1:** 9
corrosion resistance **A1:** 68, 69
damping capacity **A1:** 69-70
ferritization tendency **A1:** 57
graphite morphology **A1:** 56
heat treatment for **A1:** 9
liquid treatment of **A1:** 9
machinability **A1:** 68-69
mechanical properties at elevated
growth and scaling **A1:** 63, 66
temperature **A1:** 63-64
tensile properties **A1:** 63, 65
thermal fatigue **A1:** 63-64, 66, 67
mechanical properties at room
compressive properties **A1:** 58, 60-61, 62, 63, 64
fatigue strength **A1:** 58, 62-63, 65
impact properties **A1:** 61-62, 63, 64
modules of elasticity **A1:** 58, 61, 62, 63
shear properties **A1:** 61
temperature **A1:** 57-63
tensile properties and hardness **A1:** 57, 58, 59-60, 62
thermal conductivity **A1:** 58
microstructure **A1:** 56, 61
physical properties **A1:** 66-69
damping capacity **A1:** 69-70
sonic and ultrasonic properties **A1:** 67-68, 69
thermal conductivity **A1:** 58, 64, 66-68, 69
thermal expansion **A1:** 67, 68
Compacted graphite irons *See also* Cast iron; High-alloy graphitic irons
applications **A15:** 675-676
castability **A15:** 671
chemical composition **A15:** 667-668
chilling tendency **A15:** 671
corrosion resistance **A15:** 675
damping capacity **A15:** 675
defined **A15:** 13, 667
dross formation **A15:** 671
elevated-temperature properties **A15:** 673-675
fluidity **A15:** 671
linear expansion **A15:** 675
machinability **A15:** 675
mechanical properties **A15:** 671-673
melt treatment **A15:** 668-670
molding materials **A15:** 671

Compacted graphite irons (continued)
physical properties ... **A15:** 675
process control ... **A15:** 670-671
production techniques ... **A15:** 667-671
quality control ... **A15:** 671
shrinkage ... **A15:** 671
tensile strength and hardness ... **A15:** 672-673
thermal conductivity ... **A15:** 675
welding metallurgy ... **A15:** 521-522
Compacted/vermicular graphite eutectic, growth
in multidirectional solidification ... **A15:** 178-179
Compactibility
and compressibility ... **A7:** 286
defined ... **A7:** 2
in milling of single particles ... **A7:** 59
of stainless steels ... **A7:** 184
Compacting *See also* Compact(s); Compaction
aluminum P/M alloys ... **A2:** 210-211
of aluminum P/M parts ... **A7:** 743
automatic, and stroke capacity ... **A7:** 325
crack, defined ... **A7:** 2
defined ... **A7:** 2
effect on copper powder tensile strength ... **A7:** 116
ferrous P/M materials ... **M1:** 329, 331
force, defined ... **A7:** 2
methods for higher density ... **A7:** 304-307
physical fundamentals of ... **A7:** 308-321
-sintering combination, of cermets ... **A2:** 988-989
sliding and plastic deformation in ... **A7:** 298
as tool steel application ... **A7:** 792
vibratory ... **A7:** 306
Compacting crack
defined ... **A7:** 2
Compacting force
defined ... **A7:** 2
Compacting lubricants
commonly used ... **A7:** 352
Compacting mill *See also* Compacting presses
Compacting powders
equiaxed shaped, apparent density ... **A7:** 272
green strength ... **A7:** 288-289
lubricant effects ... **A7:** 192
Compacting presses
adjustable die filling ... **A7:** 329
deflection analysis ... **A7:** 336
designing ... **A7:** 335-337
double-action tooling ... **A7:** 333
double-reduction gearing systems for ... **A7:** 330
floating die tooling ... **A7:** 333-334
flow rate through ... **A7:** 278
gearing systems ... **A7:** 329-330
hydraulic ... **A7:** 330-332
requirements ... **A7:** 329
secondary operations ... **A7:** 337-338
selection ... **A7:** 331-332
single-reduction gearing systems for ... **A7:** 330
tooling systems for ... **A7:** 332-334
types of ... **A7:** 334-335
withdrawal tooling systems ... **A7:** 334
Compacting pressure
for beryllium powders ... **A7:** 171
defined ... **A7:** 2
and density, isostatic pressing ... **A7:** 299
density of carbonyl nickel powder compacts as function of ... **A7:** 310
effect on densification and expansion during homogenization ... **A7:** 315
electrolytic copper powder ... **A7:** 115
and lubricant mixing time ... **A7:** 191
in rigid tool compaction ... **A7:** 325
Compacting tool set
defined ... **A7:** 2
Compacting tools
for free-flowing powder ... **A7:** 278
rectangular test specimens ... **A7:** 286
Compaction *See also* Compact(s); Compactibility; Compacting; Compacting presses; Compacting pressure; Compacting tools; Isostatic compaction; Loose powder compaction ... **EM3:** 8
by Wollaston press ... **A7:** 15
copper-base structural parts ... **A2:** 397
defined ... **A7:** 2 **EM1:** 7 **EM2:** 10
force, in molding machines ... **A15:** 345
high-speed, effect on green strength ... **A7:** 302
history ... **A7:** 14
of iron-based arm bearing ... **A7:** 706
Compaction (continued)
isostatic, and rigid tool compaction ... **A7:** 323
of loose powder, stages ... **A7:** 297-298
and lubrication ... **A7:** 190
in manual lay-up ... **EM1:** 604
of P/M ferrous powders ... **A7:** 683
of polymers ... **A7:** 607
posts, titanium P/M treatments ... **A2:** 653
presses ... **A7:** 295
pressure ... **A7:** 289
properties affecting ... **A7:** 211
rapid omnidirectional ... **A7:** 542-546
in rigid dies, pressure and relative density of powders ... **A7:** 298
rigid tool, and shape attainment *See also* Rigid tool compaction
roll, and rigid tool compaction ... **A7:** 323
sand, coremaking ... **A15:** 239-240
second stage ... **A7:** 58
self-lubricating sintered bronze bearings ... **A2:** 394
sheet molding compound, types ... **EM1:** 160
split die ... **A7:** 326, 327
third stage ... **A7:** 58
triaxial chamber for ... **A7:** 304
wet magnet ... **A7:** 327-328
Compacts, sintered
tin and tin alloy ... **A2:** 519-520
Comparative experiments *See also* Experiment ... **A8:** 639-642
block designs for ... **A8:** 643-650
determining optimum levels ... **A8:** 650-652
factorial ... **A8:** 641-643
nature of ... **A8:** 639-641
planning, introduction ... **A8:** 623
precision of ... **A8:** 640
randomized designs for ... **A8:** 643-650
requisites and tools ... **A8:** 640
Comparators
in acoustic emission inspection ... **A17:** 281
microwave inspection ... **A17:** 205
optical ... **A17:** 10-11
Comparison method
of thermocouple calibration ... **A2:** 881
Comparison microscopes ... **A9:** 83-84
Comparison standard
defined ... **A9:** 4
Compatibility ... **EM3:** 8
agents ... **EM2:** 488-489
defined ... **EM1:** 7 **EM2:** 10
of dissimilar materials ... **A13:** 340-343
of polymer-polymer mixtures ... **EM2:** 487-488
Compatibility (frictional)
defined ... **A18:** 5
Compatibility (lubricant)
defined ... **A18:** 5-6
Compatibility (metallurgical)
defined ... **A18:** 6
Compensating extension wires
defined ... **A2:** 877
Compensating eyepieces
defined ... **A10:** 671
Compensating microscope eyepieces ... **A9:** 73
defined ... **A9:** 4
effect on image distortion ... **A9:** 77
Compensation
defined ... **A18:** 6
Competition
and quality design and control ... **A17:** 719
Compimide bismaleimides ... **EM1:** 81-83, 87
Complementary error function
defined ... **EL1:** 194
Complementary metal-oxide semiconductor (CMOS) ... **EM3:** 593
active-component fabrication ... **EL1:** 196-198
as ASIC technology ... **EL1:** 161
as circuit interface ... **EL1:** 160-161
defined ... **EL1:** 1138
development ... **EL1:** 160
for digital ICs ... **EL1:** 162-163
and ECL, TTL, compared ... **EL1:** 165-166
failure mechanisms ... **EL1:** 978
as future trend ... **EL1:** 390
IC trend lines ... **EL1:** 399-400
ICs, heat flux dissipated ... **EL1:** 402
logic circuits ... **EL1:** 76
Complementary metal-oxide semiconductor (CMOS) (continued)
very-large-scale integration (VLSI) devices, as future trend ... **EL1:** 390
Complementary pnp bipolar junction transistors
active devices ... **EL1:** 145-146
Complete fusion
definition ... **A6:** 1207 **M6:** 4
Complete joint penetration
definition ... **A6:** 1207 **M6:** 4
Complete penetration
definition ... **A6:** 1207
Completely randomized experimental plans ... **A8:** 643-644, 646
Complex
numbers of ligands determined in ... **A10:** 70
Complex castings
by ceramic molding ... **A15:** 248
by permanent mold casting ... **A15:** 275
Complex center upset
production of ... **A14:** 295
Complex composite contacts
properties ... **A2:** 853
Complex dielectric constant *See also* Dielectric constant
defined ... **EM1:** 7 **EM2:** 10
Complex failures
determined ... **A11:** 29
Complex inclusions
EPMA detected ... **A11:** 39
in steels ... **A15:** 92-93
Complex modulus ... **EM2:** 11, 551 **EM3:** 8
Complex modulus test (Oberst Bar ASTM E 756) ... **EM3:** 556
Complex multicomponent glass
EDS and WDS x-ray spectra of ... **A10:** 521
Complex multilayer systems
for cladding ... **A13:** 889-890
Complex oxides *See also* Oxides
as inclusions, aluminum alloys ... **A15:** 95
Complex parts
as integral unit ... **A15:** 40
metal casting advantages ... **A15:** 39
Complex permittivity
microwave inspection ... **A17:** 205
Complex relative permittivity ... **EM3:** 428, 429
Complex reliability block diagram ... **EL1:** 900
Complex shapes *See also* Shapes; Workpieces
forged, ultrasonic inspection ... **A17:** 504
radiographic inspection ... **A17:** 333-334
thermal inspection ... **A17:** 396
tool materials for drawing ... **A14:** 336
Complex shear modulus
defined ... **EM1:** 7
Complex silicate inclusions
defined ... **A9:** 4
Complex stresses
effect on fatigue strength ... **A11:** 112
Complex subsystems
reliability prediction in ... **EL1:** 900
Complex viscosity
for closed-loop cure ... **EM1:** 761
Complex workpieces *See also* Shapes; Workpieces
cold heading of ... **A14:** 294
Complex Young's modulus *See* Complex dielectric constant; Young's modulus ... **EM3:** 8
defined ... **EM1:** 7
Complex-shaped parts
CAP process for ... **A7:** 533
Complexation *See also* Cheitite; Coordination compound; Ligand
defined ... **A10:** 671 **A13:** 3
in internal electrogravimetry, effects of ... **A10:** 200
organic complexing agents and metals determined fluorimetrically by ... **A10:** 74
selected, UV/VIS ... **A10:** 65-66
separation, for interferences ... **A10:** 65
in voltammetry ... **A10:** 193-194
Complexing agents
defined ... **A13:** 1140
effect on decomposition potentials ... **A10:** 198
effect on UV/VIS analysis ... **A10:** 64
organic, determined fluorimetrically ... **A10:** 74
Complexity
of integrated circuits ... **EL1:** 399-401
of rigid printed wiring boards ... **EL1:** 551-552

Complexity index
for assessing the effects of heat treatment ... **A9:** 130
Complexity, mold
and design ... **A15:** 611-613
Complexometric titrations
classical wet chemical analysis ... **A10:** 164
in internal electrolysis ... **A10:** 201
Compliance *See also* Coefficient of elasticity ... **EM3:** 8, 504
and crack length ... **A8:** 383
defined ... **EM1:** 7 **EM2:** 11
indirect, for crack growth in aqueous solutions ... **A8:** 417
measurement mathematics ... **A8:** 385-386
measurements ... **EM3:** 318-319, 445-446, 447, 448
model, accelerated life prediction ... **EM2:** 790-791
normalized, computing ... **A8:** 385
predicted and experimental for compact-type fatigue specimen ... **A8:** 385
and tensile load ... **A8:** 383-386
Compliance calibration
Rice J integral procedure as ... **A8:** 448
Compliance method
for crack extension measurement ... **A8:** 382-386, 412
and displacement ... **A8:** 383
laboratory practice ... **A8:** 383-384
measurement mathematics ... **A8:** 385-386
normalized, computing ... **A8:** 385
system components ... **A8:** 383
to monitor crack extension in corrosive environments ... **A8:** 428
Compliant surface approach
as interconnection option ... **EL1:** 985
Compliant surface bearings ... **A18:** 530-531
advantages ... **A18:** 529
applications ... **A18:** 531
bending-dominated continuous foil bearings ... **A18:** 530, 531
bending-dominated segmented foil bearings ... **A18:** 530-531
design analysis ... **A18:** 531
modified bending-dominated continuous foil bearings ... **A18:** 531
pressurized-membrane bearings ... **A18:** 531
surface coatings ... **A18:** 532
Compocast *See also* Metal-matrix composites
defined ... **A15:** 338
of metal-matrix composites ... **A15:** 844-847
Component
defined ... **A7:** 2
Component and discrete chip mounting technologies ... **EL1:** 143
Component attachment technology
beam lead bonding ... **EL1:** 351
eutectic bonding ... **EL1:** 349
organic adhesives ... **EL1:** 348-349
silver-glass bond ... **EL1:** 349
soldering ... **EL1:** 347-348
thick-film hybrids ... **EL1:** 347-351
wire bonding ... **EL1:** 349-350
Component design *See also* Complex parts; Design; Parts
metal casting advantages ... **A15:** 39-41
Component lead materials/finishes
electronics industry ... **A13:** 1109
Component of variance
defined ... **A8:** 2
Component removal
factors affecting ... **EL1:** 713-714
methods ... **EL1:** 715-718, 724-727
surface-mount technology ... **EL1:** 722-724
tools ... **EL1:** 724-727
Component testing ... **EM3:** 540-541
aircraft aft fuselage
aircraft graphite composite horizontal stabilizer stub box test ... **EM3:** 539, 540
damage tolerance testing ... **EM3:** 540
Component testing (continued)
environmental tests ... **EM3:** 540
multirib wing box test fixture ... **EM3:** 540, 541
to evaluate galvanic corrosion ... **A13:** 234
Component(s) *See also* Active analog components; Active component fabrication; Active digital components; Component attachment technology; Component removal; Component-level physical testing methods; Passive component fabrication
active digital ... **EL1:** 160-177
attachment, trends ... **EL1:** 387-388
burn-in, TAB advantages ... **EL1:** 281
defined ... **EL1:** 1138
design/layout ... **EL1:** 513-516
digital logic ... **EL1:** 160-161
dimensions, process control ... **EL1:** 672-673
dynamic characteristics ... **EL1:** 65
electronic, dynamic response functions, microcircuitry vs conventional ... **EL1:** 89
hybrid, integral and add-on ... **EL1:** 258-259
insertion problems, through-hole packages ... **EL1:** 970
leadless, joint failures ... **EL1:** 735
masking of ... **EL1:** 764
miniature electrical ... **EL1:** 800
physical testing methods ... **EL1:** 941-944
placement, in surface-mount soldering ... **EL1:** 700-701
preparation, surface-mount joints ... **EL1:** 731
re working processes ... **EL1:** 712-715
realignment ... **EL1:** 289
removal and replacement ... **EL1:** 288-289
repair ... **EL1:** 289
selection or design ... **EL1:** 747
solderability ... **EL1:** 731, 944
surface-mount, number of I/Os per ... **EL1:** 730
testing, tape automated bonding ... **EL1:** 281
urethane-coated, removal ... **EL1:** 780
Component-level cooling
as thermal control ... **EL1:** 47
Component-level physical test methods
automatic optical inspection ... **EL1:** 941-942
capabilities ... **EL1:** 941
environmental testing ... **EL1:** 944
for plating thickness ... **EL1:** 942-943
solder joint inspection ... **EL1:** 942
for solderability ... **EL1:** 943-944
Components *See also* Part(s); Parts
aircraft ... **EM1:** 801-809
by compression molding ... **EM1:** 560-561
customer requirements ... **EM1:** 38
material selection for ... **EM1:** 38-39
thin plastic, design/analysis techniques ... **EM2:** 691-700
transport aircraft, flight service evaluations ... **EM1:** 826-831
winding ... **EM1:** 507
Components of a system ... **A3:** 1 • 2
Composi-Lok fasteners ... **EM1:** 710-711
Composite B
properties ... **A18:** 548
Composite bearing material
defined ... **A18:** 6
Composite bearings
performance characteristics ... **A7:** 408
powder melting ... **A7:** 406-407
production setup ... **A7:** 406, 407
as roll compacting specialty applications ... **A7:** 406-408
Composite camshaft assembly as nonconventional automobile application ... **A7:** 620
Composite cements
material loss on abrasion (dental) ... **A18:** 673
Composite ceramics
silicon carbide whisker-reinforced alumina ... **A2:** 1023-1024
Composite coating
defined ... **A7:** 2
Composite compact
defined ... **A7:** 3
Composite cure control ... **EM1:** 649-653
Composite cylinder assemblage (CCA) ... **EM1:** 187-188
Composite die construction
specifications ... **EM1:** 611
Composite dust
part rejection for ... **EM1:** 36
Composite electrical contacts *See also* Electrical contact materials
categories of materials ... **A2:** 850
hybrid consolidation ... **A2:** 857-858
internal oxidation ... **A2:** 857
manufacturing methods ... **A2:** 856-858
properties ... **A2:** 850-853
properties, for electrical make-break contacts ... **A2:** 851-853
refractory metal and carbide-base composites ... **A2:** 854-855
silver-base composites ... **A2:** 855-856
Composite electrode
definition ... **M6:** 4
Composite external scales
gaseous corrosion ... **A13:** 74-76
Composite fiber-matrix bond tests ... **EM3:** 391-405
Composite fluid dies ... **A7:** 544
Composite friction material applications of cement copper ... **A7:** 119
Composite image building
digital image enhancement ... **A17:** 457
Composite joint
definition ... **M6:** 4
Composite material *See* Composites ... **EM3:** 8
defined ... **A7:** 3
Composite material age ... **A8:** 716
Composite material(s) *See also* Composite material analysis and design; Composite(s); Dissimilar materials; Material properties; Material properties analysis ... **EM1:** 176-179
advantages ... **EM1:** 105
analysis, programmable calculator programs for ... **EM1:** 277-279
analysis, software for ... **EM1:** 269, 274, 275-281
anisotropic properties, and design ... **EM1:** 38
applications ... **EM1:** 799-847
basic elements/axis systems ... **EM1:** 176
cost drivers in
defined ... **EM1:** 7, 27
description ... **EM1:** 355-359
design cycle for composite structures ... **EM1:** 179
directional nature ... **EM1:** 177
evaluation of ... **EM1:** 38-39
fabrication processes ... **EM1:** 179
failure causes ... **EM1:** 767
fibers, form and properties ... **EM1:** 175, 179, 360-362
flight service evaluation ... **EM1:** 823, 826-830
forms ... **EM1:** 353-415
general use considerations ... **EM1:** 35-37
high-performance, applications ... **EM1:** 206
long-term environmental effects ... **EM1:** 823-826
matrix, forms ... **EM1:** 175-176
microstructures of ... **EM1:** 177, 768
prebond surface treatment ... **EM1:** 681-682
process modeling of ... **EM1:** 499-502
properties ... **EM1:** 173, 177-179, 353-415
quality control ... **EM1:** 729-763
raw materials ... **EM1:** 105-171
reaction control, as quality control ... **EM1:** 729
requirements ... **EM1:** 38
selection of ... **EM1:** 38-39
service characteristics ... **EM1:** 179-180
and single joint data ... **EM1:** 313-319
software for analysis of ... **EM1:** 269, 274, 275-281
technology, information sources for ... **EM1:** 40-42

Composite material(s) (continued)
ten-year worldwide ground-based exposure tests ... **EM1:** 823-825
terminology ... **EM1:** 176-177
thermal coefficient of expansions for ... **EM1:** 716
thermoplastic matrix ... **EM1:** 363-372
thermoset, high-strength medium-temperature ... **EM1:** 399-415
thermoset, high-temperature ... **EM1:** 373-380
thermoset, low-temperature ... **EM1:** 392-398
thermoset, medium temperature ... **EM1:** 381-391
Composite materials *See also* Fiber composites; Metal matrix composites
acoustic emission inspection ... **A17:** 287-288
acoustic microscopy methods ... **A17:** 469
analysis requirements for ... **A8:** 713-718
C-SAM image ... **A17:** 469
computed tomography (CT) ... **A17:** 363
failure mechanisms ... **A8:** 714
fiber-matrix, microwave inspection ... **A17:** 215
interface debonding in ... **A8:** 714
laminate code for ... **A8:** 714
metallic, eddy current inspection ... **A17:** 190-191
microwave inspection ... **A17:** 202, 215
optical holography of ... **A17:** 429
screens, radiography ... **A17:** 316
SLAM images ... **A17:** 469
titanium matrix, ultrasonic inspection ... **A17:** 250
use in electrical contacts ... **M3:** 665, 672-681, 682
Composite materials analysis and design ... **EM1:** 173-281
computer programs for structural analysis ... **EM1:** 268-274
damage tolerance ... **EM1:** 259-267
damping properties analysis ... **EM1:** 206-217
design requirements ... **EM1:** 181-184
failure analysis, of laminates ... **EM1:** 236-251
fatigue analyses, of laminates ... **EM1:** 236-251
fracture analysis, of laminates ... **EM1:** 252-258
laminate properties analysis ... **EM1:** 218-235
material properties analysis ... **EM1:** 185-205
overview ... **EM1:** 173, 175-180
software for composite materials analysis ... **EM1:** 275-281
strength analysis, of laminates ... **EM1:** 236
Composite metal particles
mechanically alloyed oxide dispersion-strengthened (MA ODS) alloys ... **A2:** 943
Composite mold
Shaw process ... **A15:** 250
Unicast process ... **A15:** 251
Composite packaging materials *See also* Advance composite packaging materials
as heat sinks, types ... **EL1:** 1122
requirements for ... **EL1:** 1122
Composite parts *See also* Part(s)
defined ... **EM1:** 422-423
Composite patterns ... **A15:** 193, 195
Composite powders
alloy-coated ... **A7:** 174
applications ... **A7:** 174-175
core and coating ... **A7:** 174
defined ... **A7:** 3, 173
extrusion die ... **A7:** 442, 443
with graphite ... **A7:** 174
high-energy milling ... **A7:** 69
hot isostatic pressing of ... **A7:** 442-443
nickel-coated, production ... **A7:** 173-174
Osprey process for ... **A7:** 530
P/M history ... **A7:** 17
production ... **A7:** 173-175
properties ... **A7:** 174
ratio of components ... **A7:** 173
roll compaction for ... **A7:** 406
Composite processing map
for aluminum ... **A14:** 365
Composite rolls ... **A14:** 353
Composite roughness ... **A18:** 30
Composite sample
defined ... **A10:** 13
Composite structural analysis ... **EM1:** 458-462
applications ... **EM1:** 460-462
classical lamination theory ... **EM1:** 458-460
numerical ... **EM1:** 463-478
numerical codes for ... **EM1:** 469
Composite structure
defined ... **A7:** 3
Composite structures *See also* Composite structural analysis; Structural
analysis ... **EM1:** 458-462
analysis and design ... **EM1:** 417-495
cost drivers, in design/manufacture of ... **EM1:** 419-427
failure of ... **EM1:** 432-435
fatigue strength of ... **EM1:** 436-444
instability considerations in ... **EM1:** 445-449
integrated manufacturing center for ... **EM1:** 621-622
interfaces, design/tooling/ manufacturing of ... **EM1:** 428-431
joint design ... **EM1:** 479-495
laminate ranking tool for ... **EM1:** 450-457
laminate sizing for ... **EM1:** 450-457
numerical design and analysis ... **EM1:** 463-478
static strength ... **EM1:** 432-435
Composite structures, advanced
blind-side bonded-scarf repair ... **EM3:** 824
blind-side sandwich repair ... **EM3:** 824-825
bonded, external patch repair ... **EM3:** 825, 826
bonded-scarf joint flush repair ... **EM3:** 822-823, 825, 826
double-scarf joint flush repair ... **EM3:** 823-824
load-transfer analysis ... **EM3:** 827-828
panel repair, 100 mm diameter holes ... **EM3:** 825-826
repair concepts ... **EM3:** 821-828
Composite strut
design of ... **EM1:** 183-184
Composite surface roughness ... **A18:** 539
Composite target sputtering
as thin-film deposition technique ... **A2:** 1081-1082
Composite testing
graphite composite outboard elevator test ... **EM3:** 540
Composite tools *See also* Electroformed nickel tooling; Tooling
graphite-epoxy ... **EM1:** 586-587
prepreg tool fabrication ... **EM1:** 739
wet lay-up fabrication ... **EM1:** 739
Composite x-ray target
refractory metals and alloys ... **A2:** 559
Composite(s) *See also* also Advanced composites; Composite material(s); Composite packaging materials; Composite structural analysis; Composite structures; Constituent materials; Continuous fiber reinforced composite; Continuous fiber reinforced composites; Continuous fibers; Discontinuous fiber composites; Fiber-reinforced composites; High performance composites; High-modulus composites; Metal-matrix composites; Particulate-reinforced composites Short fiber composites; Reinforced composites; specific composites; Subcomposites; Tensile strength
adhesive bonding, surface preparation ... **EM1:** 681-682
advantages ... **EM1:** 36-37
aerospace, epoxy ... **EM1:** 73
aircraft cost savings by ... **EM1:** 97
application, in packages ... **EL1:** 1126-1128
autoclave curing of ... **EM1:** 645-653
chop length, effect of ... **EM1:** 164
commercial aircraft ... **EM1:** 100
components ... **EM1:** 355
composition ... **EM1:** 289
for consumer products ... **EM1:** 554
CTE relationships ... **EL1:** 611
curing process ... **EM1:** 702-705
damage tolerance of ... **EM1:** 259-267
damping properties analysis of ... **EM1:** 206-217
design ... **EM1:** 33
discontinuous phase ... **EM1:** 27
evaluation of ... **EM1:** 38-39
fabric weave types ... **EM1:** 256, 355
fabrication ... **EM1:** 33-34
failure causes ... **EM1:** 767
fiber-resin applications ... **EM1:** 355, 356
fibers ... **EM1:** 29-31
forms ... **EM1:** 355
general use considerations ... **EM1:** 35-37
as heat sinks ... **EL1:** 1130-1131
high-performance consumer ... **EM1:** 554
Composite(s) (continued)
inorganic-inorganic, laser cutting of ... **EM1:** 679-680
inorganic-organic, laser cutting of ... **EM1:** 679
interlaminar fracture toughness ... **EM1:** 99
introduction to ... **EM1:** 27-34
low-performance consumer ... **EM1:** 554
material forms ... **EM1:** 33
matrices ... **EM1:** 31-33
metal-matrix ... **EL1:** 1126-1128
and metals, compared ... **EM1:** 35, 216, 259-260
molded, properties ... **EM1:** 161
organic-organic, laser cutting of ... **EM1:** 678-679
particulate, defined *See* Particulate composite, defined
polymer matrix ... **EL1:** 1117-1118
principle, for packaging ... **EL1:** 1117
reinforced, flexible printed boards ... **EL1:** 583
resin matrix effect ... **EM1:** 162
selection of ... **EM1:** 38-39
short fiber reinforced ... **EM1:** 119-121
solvent impregnation/solvent resistance problem ... **EM1:** 102
structural, polyimides for ... **EM1:** 78
structural testing ... **EM1:** 313-345
types of ... **EM1:** 768-769
versatility ... **EM1:** 35
woven fibrous, damping analysis ... **EM1:** 213
Composite-mold casting *See also* Castings; Foundry products
aluminum alloys ... **M2:** 147
aluminum casting alloys ... **A2:** 141
Composite-to-metal joining ... **A6:** 1041-1047
amorphous bonding ... **A6:** 1043, 1044, 1045
bolted joints ... **A6:** 1045, 1046
bonded joints ... **A6:** 1042-1046
composite-to-metal bolted joint concept ... **A6:** 1045
galvanic corrosion ... **A6:** 1041-1042
joint designs ... **A6:** 1041, 1042-1044, 1045-1046
Man-Rated Demonstration Article (MRDA) ... **A6:** 1045, 1046
organic-matrix composites ... **A6:** 1041-1047
STEPLAP computer procedure ... **A6:** 1041, 1042-1043, 1044
submersible thick composite-to-metal joint ... **A6:** 1046-1047
thermoplastic composites ... **A6:** 1042, 1044-1045
thermoset composites ... **A6:** 1042, 1044
Ultem film ... **A6:** 1044, 1045
Composite-to-metal joints
filament wound ... **EM1:** 511, 514
Composites *See also* Advanced composites; Cast metal-matrix composites; Fiberglass-polyester composites; Matrix alloys; Metal-matrix composites; Resin- matrix composites ... **EM3:** 71
acrylamate prepreg, physical properties ... **EM2:** 270
advanced, fasteners for ... **A11:** 530
advanced, thermoplastic resins ... **EM2:** 621-622
aluminum alloy ... **A13:** 587
aluminum-base discontinuous metal-matrix ... **A14:** 251
aluminum-base metal-matrix, for casting ... **A2:** 126
analytic methods applicable ... **A10:** 6
BMI, mechanical properties ... **EM2:** 256
bonding fixtures ... **EM3:** 707
carbon fiber reinforced ... **EM2:** 235-236
ceramic, applications and properties ... **A2:** 1023-1024
chopped fiber reinforced acrylamate ... **EM2:** 270
clad metals as ... **A13:** 887
coinability of ... **A14:** 183
compression fracture from buckling in ... **A11:** 742-743
continuous fiber ... **A11:** 731
continuous fiber reinforced magnesium alloys ... **A2:** 460
corrosion-resistant, welding ... **A13:** 652
cutting tool material selection based on machining operation ... **A18:** 617
defined ... **A10:** 671 **A11:** 2 **EM2:** 98
deformation processed copper refractory metals ... **A2:** 922
die materials ... **A14:** 639

Composites (continued)
for electrical make-break contacts, properties of ... **A2:** 851-853
electromagnetic forming with ... **A14:** 649-650
engineering properties ... **A11:** 731
fiber, deformation of ... **A11:** 761
fiberglass-polyester resin ... **EM2:** 248-249
Fiberite T300 carbon fibers ... **EM3:** 644
fracture modes in ... **A11:** 733
friction as described by various properties ... **A18:** 669
glass reinforced acrylamate, physical properties ... **EM2:** 269
graphite fiber reinforced copper matrix, for space power radiator panels ... **A2:** 922
ground by diamond wheels ... **A16:** 455, 460
hybrid
properties ... **A18:** 666
simplified composition of microstructure ... **A18:** 666
hypoeutectic alloys as ... **A15:** 167-168
load states in ... **A11:** 735
machinability test matrix ... **A16:** 639-640
material, defined ... **EM2:** 11
material parameters that should be documented to ensure repeatability when testing tribosystems ... **A18:** 55
mechanical fasteners for ... **A11:** 529
metal-ceramic ... **A2:** 1024
microfilled
properties ... **A18:** 666
simplified composition on microstructure ... **A18:** 666
microstructural changes studied by x-ray topography ... **A10:** 366
microstructures ... **A11:** 732
moisture-induced failure in ... **EM2:** 765-766
with molybdenum skeletons, as electrical contact materials ... **A2:** 855
multifilamentary NbTi superconducting ... **A2:** 1043-1052
organic, analytic methods for ... **A10:** 9
phase contrast imaging ... **A18:** 389
phenolics for adhesive ... **EM3:** 105
plastic, environmental effects ... **EM2:** 428-429
polyphenylene sulfides (PPS) ... **EM2:** 190
for pressure vessels, welding of ... **A11:** 654-656
reaming of Ni alloys ... **A16:** 839
refractory metal fiber reinforced ... **A2:** 582-584
SiC whisker-reinforced aluminum metal-matrix ... **A14:** 20
silicon-nitride matrix ... **A2:** 1024
solidification, low-gravity eutectic alloys ... **A15:** 150-153
space shuttle orbiter ... **A13:** 1066
structural, cyanates ... **EM2:** 232-233
structure, high-impact polystyrenes (PS HIPS) ... **EM2:** 195
surface preparation for adhesive-bonded repair ... **EM3:** 840-844
T300/934 ... **EM3:** 644
tailorability ... **A11:** 731
titanium metal-matrix ... **A14:** 283
with tungsten carbide skeletons, as electrical contact materials ... **A2:** 855
types of ... **A11:** 731-732
vinyl ester ... **EM2:** 274
vs metal, compared ... **EM2:** 371-373
wear resistant ... **M1:** 635
wire, superconductors as ... **A14:** 338-342
Composites Technology (CMPS) standardization area *See* Department of Defense (DoD)
Composition *See also* Alloy steel; Carbon steel; Gray cast iron; HSLA steel; Nominal composition
alloy steel sheet and strip ... **M1:** 163
aluminum casting alloys ... **A15:** 744-745
of aluminum oxide-containing cermets ... **A7:** 804
Composition (continued)
atomic-scale, FIM/AP for ... **A10:** 584
of bulk garnets ... **A10:** 628
bulk, of nickel alloys ... **A10:** 562
of bulk samples ... **A10:** 631
bulk, XRS analysis ... **A10:** 83
of cast iron pump parts, wear from ... **A11:** 365-367
cemented carbides ... **A2:** 968
of cemented carbides, machining applications ... **A2:** 951-953
changes, during solidification ... **A15:** 101-103
chemical ... **A11:** 256, 363
chemical, aluminum alloys ... **A15:** 743
chemical, determined by atom probe microanalysis ... **A10:** 591
chemical, of surfaces, AES analysis for ... **A10:** 549
of cobalt-base alloys ... **A2:** 448-449
cobalt-base corrosion-resistant alloys ... **A2:** 453-454
cobalt-base high-temperature alloys ... **A2:** 451-452
cobalt-base wear-resistant alloys ... **A2:** 448-449
of common aluminum-silicon alloys ... **A15:** 159
of copper casting alloys ... **A2:** 347 **M2:** 383-384
of copper-based P/M materials ... **A7:** 470
of CVD coatings ... **A2:** 960
diagrams, vs free energy ... **A15:** 102
ductile iron ... **A15:** 652, 656-657
effect, columnar grain growth ... **A15:** 135
effect on dislocation distribution ... **A9:** 693
effect on maraging steels ... **A11:** 218
effect on metal properties ... **A11:** 325
effect on toughness and crack growth ... **A11:** 54
effects, aluminum/aluminum alloys ... **A13:** 585-587
effects on ductile-to-brittle transition temperature, structural steels ... **A11:** 68
effects on expansion coefficient, of Invar ... **A2:** 889-890
effects on stress-corrosion cracking ... **A11:** 206
effects, stainless steels ... **A13:** 550
of gray iron ... **A15:** 629-630
heat-resistant high-alloy steels ... **A15:** 724
heat-treated copper casting alloys ... **A2:** 355
homogeneity, nucleation effects ... **A15:** 101
influence, iron-carbon-silicon alloys ... **A15:** 172
of iron oxide films, determined ... **A10:** 135
and layer thickness, of thin films ... **A10:** 631-632
limits, HSLA steels (ASTM grades) ... **M1:** 407
of lunar surface, NAA application for ... **A10:** 234
of magnesium alloys ... **A2:** 457
material selection ... **A11:** 391
of materials, and materials characterization I
of matrix, in cermets ... **A2:** 991
mean, effect on densification and expansion during homogenization ... **A7:** 315
metal, testing ... **A13:** 193-194
microanalysis, electron probe x-ray microanalysis ... **A10:** 517-518
of mixtures, EFG analysis ... **A10:** 213
nominal, cobalt-base alloys ... **A2:** 448
nondestructive XPS ... **A10:** 568
off analysis of ... **A11:** 391
of P/M stainless steels ... **A7:** 468
P/M steels ... **M1:** 333, 337-339
in phase diagrams ... **A15:** 57
and purity, NAA analyis for ... **A10:** 233
relation to designations ... **M1:** 120-124
of shafts ... **A11:** 478
of silica sand ... **A15:** 208
single-phase alloys ... **A15:** 114
SSMS verification of alloy ... **A10:** 141
in stainless steels ... **A11:** 391
steel casting failures due to ... **A11:** 391-392
temperature, and sintering atmosphere ... **A7:** 246
of unalloyed and alloyed aluminum castings and ingots ... **A2:** 22-25
unspecified and trace elements ... **A11:** 392
vs depth, passive film on tin-nickel substrate ... **A10:** 608-609
Composition (continued)
wrought aluminum and aluminum alloys ... **A2:** 62-122
of wrought unalloyed aluminum and wrought aluminum alloys ... **A2:** 17-21
Composition control *See also* Alloy additions; Melt purification; Presolidification ... **A15:** 71-81
alloy additions, kinetics of ... **A15:** 71-74
ductile iron ... **A15:** 647
mechanisms of ... **A15:** 71
purification of metals ... **A15:** 74-81
of scrap ... **A15:** 71
Composition conversion ... **A3:** 1 • 18
Composition gradient ... **A6:** 47, 51, 52
Composition gradients
effect on microsegregation ... **A9:** 614
Composition limits
cast copper alloys ... **A2:** 356-391
Composition metal *See* Copper alloys, specific types, C83600
Composition metal foil
characteristics and composition ... **A2:** 555
Composition of specimens
semiquantitative analysis by scanning electron microscopy ... **A9:** 92
Composition profiles *See also* Depth profiles; Depth profiling
depth, atom probe microanalysis ... **A10:** 593
depth, gold-nickel-copper metallization system ... **A10:** 560
depth, LEISS analyses for ... **A10:** 603
of elemental distribution in thin films, by XPS ... **A10:** 568
of FIM/AP analyzed ductile magnets ... **A10:** 599
for iron-chromium-cobalt alloy ... **A10:** 600
low-level dopants ... **A10:** 610
x-ray microanalysis, Al-4.7Cu ... **A10:** 462
Composition scales ... **A3:** 1 • 18
Composition-depth profiles ... **A7:** 37, 252-254, 256
of atomized steel powder ... **A7:** 256
of green sintered compacts ... **A7:** 252, 253
ion gun for ... **A7:** 255
P/M type 316L stainless steel ... **A7:** 252, 253
of stainless steel alloys ... **A7:** 252, 253
Compositional depth profile *See also* Composition; Composition profiles; Compositional mapping
defined ... **A10:** 671
Compositional gradients
microbeam analysis ... **A10:** 530
Compositional mapping
by electron probe x-ray microanalysis ... **A10:** 525-529
digital ... **A10:** 528-529
of heterogeneous specimens ... **A10:** 516
Compound ... **EM3:** 8
defined ... **EM2:** 11
Compound analysis *See also* Compound or phase identification; Compounds
chemical, flame AAS for major and minor component analysis ... **A10:** 56
gas analysis by mass spectroscopy ... **A10:** 151-157
of inorganics ... **A10:** 7, 8
of organics ... **A10:** 9, 10
Compound buffing systems *See* Buffing, compound systems
Compound dies ... **A14:** 456-457
for blanking ... **A14:** 454
defined ... **A14:** 3
flanging and hemming, for press bending ... **A14:** 527
for piercing ... **A14:** 465-466
for press bending ... **A14:** 527
for press forming ... **A14:** 546
for sheet metal drawing ... **A14:** 579
for stainless steels ... **A14:** 765-766
Compound impact wear ... **A18:** 263
Compound interest
annual vs. continuous ... **A13:** 370-371

Compound interest (continued)
factors, functional forms **A13:** 370
Compound or phase identification
analytical transmission electron microscopy **A10:** 429-489
electron probe x-ray microanalysis **A10:** 516-535
elemental and functional group analysis **A10:** 212-220
field ion microscopy **A10:** 583-602
gas analysis by mass spectrometry **A10:** 151-157
gas chromatography/mass spectrometry **A10:** 639-648
infrared spectroscopy **A10:** 109-125
liquid chromatography **A10:** 649-657
molecular fluorescence spectrometry **A10:** 72-81
Mössbauer spectroscopy **A10:** 287-295
neutron diffraction **A10:** 420-426
nuclear magnetic resonance **A10:** 277-286
optical metallography **A10:** 299-308
Raman spectroscopy **A10:** 126-138
single-crystal x-ray diffraction **A10:** 344-356
small-angle x-ray and neutron scattering **A10:** 402-406
x-ray diffraction **A10:** 380-392
x-ray powder diffraction **A10:** 333-343
Compound reaction
as liquid-metal corrosion **A13:** 92
Compound reduction, liquid-metal corrosion by **A13:** 56, 59
Compound types *See* Resin compounds
Compound(s) *See also* Unsaturated compounds
bulk molding, composition **EM1:** 161
defined **EM1:** 7
ordered metallic **A2:** 913-942
rare earth metal **A2:** 726
Compounders, independent *See* Suppliers
Compounding
of additives **EM2:** 493-494
of powder metal-filled plastics **A7:** 606
of sands **A15:** 32
Compounds *See also* Compound analysis; Compound or phase identification; Organic compounds
as aluminum alloy inoculants, types **A15:** 105
chemical analysis by controlled-potential coutometry **A10:** 207
chemical, flame AAS for major and minor component analysis **A10:** 56
cobalt, voltammetric analysis of trace nickel in **A10:** 194-195
extended x-ray absorption fine structure for **A10:** 407
fluorescent organic, in volumetric analysis **A10:** 164
formation in arc welds **A11:** 413
identification of, as x-ray diffraction analysis **A10:** 325
inorganic, applicable analytic methods **A10:** 6, 151
inorganic, gas analysis of **A10:** 151
inorganic, molecular requirements for fluorescence in **A10:** 73-74
inorganic solid, bulk analyses methods for **A10:** 6
for molecular structure, dynamics and environment of **A10:** 109
NMR analysis **A10:** 277
nucleant **A15:** 105-108
organic **A13:** 558-559, 570, 631
organic, analysis of **A10:** 60, 73-74, 151, 213
purified, liquid chromatography isolation for synthetic purposes **A10:** 649
sublattice ordering in intermetallic, NMR analysis **A10:** 283
vanadium determined in **A10:** 206
weighing as the, gravimetric analysis **A10:** 171
Compressed air
for tin powder atomization **A7:** 123
Compressed hydrogen gas
as atmosphere **A7:** 344
Compressed powdered iron
low-carbon steels **A2:** 765
Compressibility *See also* Green strength **A7:** 286-287
and compactibility **A7:** 286
defined **A7:** 3, 211
effect of internal pore size **A7:** 269, 270
Compressibility (continued)
effect of lubrication **A7:** 191-192, 288, 302
ferrous P/M materials **M1:** 327-330
and green strength **A7:** 85, 108, 110, 288, 289
and green strength, P/F vs P/M applications **A14:** 189
high-carbon iron, residual carbon effect **A7:** 183
influencing factors **A7:** 286
of iron powder **A14:** 190
of iron powders **A7:** 23, 84, 93-95
of loose powders **A7:** 298
of mercury **A7:** 268-269
stainless steel powder **A7:** 102
Compressibility curve
defined **A7:** 3, 287
Compressibility number **A18:** 522-523, 525, 526, 527, 528, 529
defined **A18:** 6
Compressibility test
defined **A7:** 3
Compression *See also* Dry and baked compression test; Stress(es)
axial, in shafts **A11:** 461
CG iron **A15:** 673
computer-code simulation, steel **A8:** 57
cracking **A8:** 57
ductile iron **A15:** 660
effect, magabsorption measurement **A17:** 154
effective stress-strain curves, 304L stainless steel **A8:** 162, 164
effects in resin-matrix composites **A12:** 478
elastic proving rings for **A8:** 614
elongation percent, thermoset matrix composites **EM1:** 396
in fatigue fracture **A11:** 75
flow stress in **A1:** 585 **A14:** 375-376
in forced-displacement system **A8:** 392
forced-vibration system **A8:** 392
forming **A14:** 594-595
fracture **A8:** 57-58 **EM1:** 792
fracture, composites, delamination and interlocking **A11:** 739
fracture, graphite-epoxy test structure **A11:** 742-743
of gray iron **A15:** 643-644
as in-plane failure mode **EM1:** 781-782
jig, for sheet **A8:** 56
lip, in ceramics **A11:** 746, 747
load requirements for **A8:** 58
-loaded elements **EM1:** 323-324
-loaded subcomponents **EM1:** 331
loading, of shafts **A11:** 460-461
longitudinal, and damping **EM1:** 207-208
malleable iron, pearlitic/martensitic **A15:** 696
measurement, by Barkhausen noise **A17:** 160
milling **A7:** 56
molding of polymers **A7:** 606
process, analysis **A14:** 376-377
pure, elastic-stress distribution **A11:** 461
resonance system **A8:** 392
rotational bending system **A8:** 392
servomechanical system **A8:** 392
strains between tapered dies **A7:** 412
strength **A7:** 318
strength, in composites **A11:** 731
as stress on shafts **A11:** 461
surface, effect on fatigue strength **A11:** 125-126
test *See* Compression test **EM1:** 298-299
test fixture **A8:** 198
translaminar **EM1:** 792
transverse, ply **EM1:** 238
uniaxial, ply **EM1:** 137
universal testing machines for **A8:** 612
vertical, deformation under **A14:** 389
Compression, and uniaxial flow
rheology **EL1:** 839, 844
Compression bending
of bar **A14:** 661
Compression bone plate with glide holes
as intemal fixation device **A11:** 671
Compression crack **A7:** 3
Compression forming **A14:** 594-595
Compression injection molding (CIM) **EM2:** 323
Compression load
applications **A8:** 55
Compression modulus *See* Bulk modulus cf elasticity (*K*); Bulk modulus of elasticity
Compression molding *See also* Autoclave molding **EM1:** 559-563 **EM4:** 224
of bulk molding compounds **EM1:** 161
of cast irons **A9:** 243
for coating/ encapsulation **EL1:** 240
components, processing/manufacture **EM1:** 560-561
computer-aided design (CAD) **EM2:** 335-336
cost comparison **EM2:** 333-335
defined **EM1:** 7 **EM2:** 11, 302-303
design considerations **EM2:** 326-333
for discontinuous fibers **EM1:** 120-121
economic factors **EM2:** 296-298
of epoxy composites **EM1:** 71
finite-element analysis (FEA) **EM2:** 336-337
materials and methods **EM2:** 324-326
materials, for sheet manufacture **EM1:** 559-560
mechanical properties **EM2:** 333
molded-in color **EM2:** 306
of polyimide resins **EM1:** 663
properties effects **EM2:** 285-287
sheet molding compounds by **EM1:** 157-160
size and shape effects **EM2:** 290
structural **EM1:** 561-562
surface finish **EM2:** 303
textured surfaces **EM2:** 305
thermoplastic **EM1:** 562-563
unsaturated polyesters **EM2:** 249-250
Compression mount packages
two-terminal **EL1:** 432
Compression ratio **A7:** 3, 286, 297
defined **EM2:** 11
Compression set **EM3:** 51-52
defined **A18:** 6
Compression springs **A1:** 302, 319-320, 322
design **M1:** 303, 306-307
fatigue properties **M1:** 291-296, 297
relaxation **M1:** 296-297, 298-300
residual stresses **M1:** 290-291
wire for **M1:** 288-289
Compression stamping *See also* Stamping
computer-aided design (CAD) **EM2:** 335-336
cost comparison **EM2:** 333-335
defined **EM2:** 324
design considerations **EM2:** 326-333
finite-element analysis (FEA) **EM2:** 336-337
mechanical properties **EM2:** 333
size and shape effects **EM2:** 290
textured surfaces **EM2:** 305
Compression strength *See* Compressive strength
Compression test
defined **A14:** 3
free surface combinations for **A14:** 392
on aluminum alloys **A14:** 391
plane-strain **A14:** 377-379
specimens **A14:** 391
for workability **A14:** 374-376
Compression test specimen
cylindrical **A8:** 578-580
flanged **A8:** 578-579
lubricated and non-lubricated **A8:** 195-197
orientation from hot rolled steel **A8:** 580-581
tapered **A8:** 578-579
and tension test specimen orientation **A8:** 581
Compression testing **A8:** 2
advantage over tension testing, high strain rates **A8:** 191
as-machined specimen **A8:** 197
axial **A8:** 55-58
barreling **A8:** 196-197
for bulk workability assessment **A8:** 577-578
disadvantage **A8:** 57-58
elevated-temperature **A8:** 196
friction in **A8:** 192
fully reversed, ultrasonics **A8:** 248
glass frits **A8:** 195
grips **A8:** 191
high strain rate **A8:** 190-207
hot **A8:** 581-584
isothermal hot, Ti-6242Si specimen **A8:** 172
lubrication during **A8:** 192, 195-196
medium strain rate, cam plastometer **A8:** 193-196
medium strain rate, with conventional load frames **A8:** 192-193
necking **A8:** 577

Compression testing (continued)
plane-strain, room-temperature torsion flow curves on copper and aluminum ... **A8:** 162-164
powdered tungsten disulfide ... **A8:** 195
room-temperature ... **A8:** 195
strain rate ranges for ... **A8:** 40
stress-relaxation ... **A8:** 325-328
stress-strain curves for Armco iron ... **A8:** 174
tungsten carbide bearing blocks for ... **A8:** 57
Compression waves *See* Longitudinal waves
Compression-after-impact test
for damage tolerance ... **EM1:** 97, 98, 100
Compression-hold-only test
saturation effect ... **A8:** 348-349
and symmetrical-hold-only tests ... **A8:** 347
Compression-ignition engines ... **A18:** 553
Compression-mounting epoxies ... **A9:** 29
used as a mounting material for wrought stainless steels ... **A9:** 279
Compression-mounting epoxy resins *See also* Epoxy resins
used to mount tool steels ... **A9:** 256-257
Compression-mounting materials ... **A9:** 29-30
for carbonitrided and carburized steels ... **A9:** 217
Compression/deflection measurements (tests)
butyl tapes ... **EM3:** 202
Compressive
defined ... **A11:** 2 **A13:** 3
strength, defined ... **A13:** 3-4
stress, defined ... **A13:** 4
Compressive failure
continuous fiber composites ... **EM1:** 792
discontinuous fiber composites ... **EM1:** 796-797
fiber mode ... **EM1:** 200
matrix mode ... **EM1:** 200
Compressive flow stress curve
with strain softening ... **A8:** 583
Compressive hoop strain
true ... **A8:** 564
Compressive hoop stress
Barkhausen noise measurement ... **A17:** 160
Compressive linear strain *See* Linear strain
Compressive loading
end effects correction ... **A14:** 376
extrapolation method for and effects in ... **A8:** 583
Compressive modulus ... **EM3:** 8
defined ... **EM1:** 7 **EM2:** 11
elastic, Kevlar fiber/fabric reinforced epoxy resin ... **EM1:** 408
glass fabric reinforced epoxy resin ... **EM1:** 404
Compressive properties *See also* Compressive field strength; Compressive strength; Mechanical properties
cast copper alloys ... **A2:** 365, 367, 372, 378
cemented carbides ... **A2:** 955-956
of compacted graphite iron ... **A1:** 58, 60-61, 62, 63, 64
of ductile iron ... **A1:** 42, 45 **M1:** 38, 41
surface hardened 86*xx* steels ... **M1:** 536
wrought magnesium alloys ... **A2:** 483
Compressive residual stresses
by heat treatment ... **A11:** 97
surface, and fatigue strength ... **A11:** 112
Compressive residual surface stresses
and fatigue strength ... **A8:** 374
Compressive seals ... **EM3:** 301
Compressive strain, in cast chromium-base alloys containing rhenium, calculation
of ... **A9:** 127
Compressive strength *See also* Mechanical properties; Strength; Ultimate compressive strength ... **EM3:** 8, 51-52
aluminum casting alloys ... **A2:** 154-177
analytical model ... **EM1:** 196-197
axial, analysis ... **EM1:** 196-197
as damage tolerance property ... **EM1:** 99
defect effects ... **EM1:** 261
Compressive strength (continued)
defined ... **A8:** 2 **A14:** 3 **EM1:** 7 **EM2:** 11
detained ... **A11:** 2
engineering plastics ... **EM2:** 245
of epoxy resin matrices ... **EM1:** 73
of gray iron ... **A1:** 17-18, 19
long-term exposure testing ... **EM1:** 825
of magnesium alloys ... **A2:** 460
of malleable irons ... **A1:** 82
and material selection ... **EM1:** 38
p-aramid fibers ... **EM1:** 55
phenolics ... **EM2:** 244
polyester resins ... **EM1:** 91, 92
residual, thermosetting/thermoplastic systems ... **EM1:** 98
resin shear modulus influence ... **EM1:** 262
short fiber based CFRP ... **EM1:** 155
test ... **EM1:** 298-299
testing, for damage tolerances ... **EM1:** 264
Compressive stress *See also* Stress(es); Tensile stress ... **A8:** 2 **EM3:** 9
defined ... **A11:** 2 **A14:** 3 **EM1:** 7 **EM2:** 11
in fatigue tests ... **A11:** 102
residual ... **A11:** 97, 112
system ... **A8:** 576
systems, as controlling workability ... **A14:** 369
Compressive yield strength *See also* Mechanical properties; Yield strength ... **A8:** 662, 667
aluminum casting alloys ... **A2:** 154-177
of aluminum oxide-containing cermets ... **A7:** 804
of ASP steels ... **A7:** 785
chromium carbide-based cermets ... **A7:** 806
copper casting alloys ... **A2:** 349
of copper-based P/M materials ... **A7:** 470
of titanium carbide-based cermets ... **A7:** 808
wrought aluminum and aluminum alloys ... **A2:** 85, 90, 92-94, 96
Compressive yield strength, of powder forged materials *See also* Yield strength ... **A14:** 201-202
Compressometer ... **A8:** 2-3, 56
Compressor blades and disks
corrosion of ... **A13:** 999-1000
Compressor disks
aircraft engines ... **A7:** 647
Compressor shaft
peeling-type fatigue cracking ... **A11:** 471
Compressor stator vane
Ti alloy powder ... **A7:** 681-682
Compressors ... **A13:** 1139, 1223
in HIP processing ... **A7:** 421-422
Compressors, friction and wear of ... **A18:** 602-608
centrifugal compressors ... **A18:** 606-608
bearings ... **A18:** 607-608
lubrication ... **A18:** 607, 608
seals ... **A18:** 607-608
types ... **A18:** 606-607
wear considerations ... **A18:** 608
compressor comparison ... **A18:** 608
reciprocating compressors ... **A18:** 602-605, 608
crosshead ... **A18:** 604
crosshead-type ... **A18:** 603, 604
cycle ... **A18:** 602
cylinder ... **A18:** 602-603
frame portion ... **A18:** 604-605
overview ... **A18:** 605
piston rod packing ... **A18:** 604
piston-ring materials ... **A18:** 603
trunk-type ... **A18:** 602, 604
valves ... **A18:** 603-604
rotary compressors ... **A18:** 605-606, 608
dry helical lobe ... **A18:** 605-606, 608
flooded helical lobe ... **A18:** 606, 608
lobe type ... **A18:** 604-606
Lysholm compressor ... **A18:** 605
seals used ... **A18:** 606
sliding vane and its components ... **A18:** 606
SRM compressor ... **A18:** 605
Compton edge ... **A18:** 325
Compton equation ... **A18:** 325
Compton imaging, and industrial computed tomography
compared ... **A17:** 362
Compton scatter/scattering
defined ... **A10:** 671
from x-ray absorption ... **A10:** 84
intensity, effect of decreasing mass absorption on ... **A10:** 99
and Rayleigh scatter, x-rays ... **A10:** 85
for rhodium tube ... **A10:** 99
Compton scattering ... **A18:** 324, 325
in electromagnetic radiation attenuation ... **A17:** 309
Compton wavelength
defined ... **A10:** 84
Computation
as electronic function ... **EL1:** 89
Computation models, numerical
for material properties ... **EM1:** 187
Computed tomography (CT) *See also* Industrial computed tomography
of castings ... **A17:** 528-529
defined ... **A17:** 383
of powder metallurgy parts ... **A17:** 538
Computer *See also* Data base; Information
control, test stands ... **A8:** 312-313
-controlled resonant fatigue tester ... **A8:** 394
data acquisition systems, for high strain rates ... **A8:** 192
data banks, and materials selection ... **EM2:** 1
data-acquisition, for load-displacement data ... **A8:** 385
tension testing machine, block diagram ... **A8:** 51
use of IBM PC-AT-compatible for microdebonding indentation system ... **EM3:** 401
used with Bondascope 2100 to get PortaScan (portable ultrasonic color scan imaging) ... **EM3:** 758
Computer analysis and repair system (CARS) ... **EL1:** 131
Computer applications *See also* Computer-aided design and manufacture Modeling; Computer-aided design/computer-aided manufacture; Computer-aided engineering; Modeling; Process modeling; Simulation; Software
of closed-die forging deformation ... **A14:** 80
contour roll forming ... **A14:** 634-635
for die and die materials ... **A14:** 57-58
hot forging ... **A14:** 57-58
introduction ... **A15:** 857
in metal casting ... **A15:** 855-891
modeling of combined fluid flow and heat/mass transfer ... **A15:** 877-882
modeling of fluid flow ... **A15:** 867-876
modeling of microstructural evolution ... **A15:** 883-891
modeling of solidification heat transfer ... **A15:** 858-866
for rolling ring mill ... **A14:** 117
Computer control *See also* Automated; Automation; Computers
through-hole soldering ... **EL1:** 695-696
Computer control of scanning electron microscopy
for electronic image analysis ... **A9:** 152
Computer disks
scanning laser gages for ... **A17:** 12
Computer disks, as samples
XRS ... **A10:** 95
Computer graphic nesting systems ... **EM1:** 620-621
Computer hardware *See also* Automation; Computers; Software
coordinate measuring machines ... **A17:** 25-26
Computer modeling *See* Computer applications; Modeling
galvanic corrosion evaluation by ... **A13:** 234

Computer Numerical Control **A14:** 464, 501
Computer numerical control (CNC)
electrical discharge machining technology **EM4:** 371
systems **EM4:** 141
tools used in green machining ... **EM4:** 181, 182, 183
for ultrasonic machining processing equipment **EM4:** 361
Computer numerical control (CNC) shape-cutting machines **A6:** 1169
Computer programs
evaluations **EM1:** 269-273
for fiber fragment analysis **EM3:** 397
for laminate analysis...... **EM1:** 269, 274-281, 451-454
Levenberg-Marquardt routine **EM3:** 512
for resin bleeder schedule **EM1:** 756
for structural analysis **EM1:** 268-274
subroutine ZXSSQ **EM3:** 512
to predict ceramic component reliability **EM4:** 700
Computer simulation *See also* Modeling
of solidification **A15:** 36, 863-865
Computer software *See* Software
ADINA **EM4:** 737-738
ANSYS **EM4:** 737-738, 739
CARES, failure probabilities of ceramics **EM4:** 724
cold welded butt-welded joints **A6:** 308
finite-element stress analytic techniques for seal design **EM4:** 536-537
JAC2D, finite-element stress-analytic techniques **EM4:** 538
MARC **EM4:** 737-738, 739
MSC/NASTRAN **EM4:** 737-738, 739
SCARE (later CARES), failure probabilities of ceramics **EM4:** 724
TCARES algorithm **EM4:** 733, 737-738, 739
Computer vision *See* Machine vision
Computer(s)
applications, for hybrids **EL1:** 254
data handling, in final design package **EL1:** 525-526
mainframe, multichip assemblies in **EL1:** 298
midrange, mechanical package for **EL1:** 22
Military, ceramic PWB for **EL1:** 387
personal, for electrical testing **EL1:** 567
thermal control **EL1:** 48-49
Computer-aided analysis **EL1:** 132-135
Computer-aided design *See also* Computer-aided design and manufacture; Computer-aided manufacture **A12:** 421
application **A14:** 905-906
costs/advantages **A14:** 905
equipment **A14:** 904-905
as geometry representation tool **A14:** 410
part geometry for **A14:** 409
Computer-aided design (CAD)
automatic trace routing **EL1:** 529-533
-based methodology, for high-frequency digital systems **EL1:** 81
compression molding/stamping **EM2:** 335-336
as computer-aided analysis **EL1:** 132
design verification **EL1:** 129
development **EL1:** 127
for final design package **EL1:** 523-526
functional objectives **EL1:** 127-128
for layout **EL1:** 513
logic design **EL1:** 129
model simulation **EL1:** 129
physical design **EL1:** 129
physical partitioning **EL1:** 128-129
of printed wiring boards (PWBS) **EL1:** 505
process flow **EL1:** 128
release data flow **EL1:** 128
routing algorithms **EL1:** 529-533
system objectives **EL1:** 527-528
technology rules **EL1:** 128
as tool **EL1:** 419
for VLSI **EL1:** 8
work flow through design system **EL1:** 528-529
Computer-aided design (CAD) system **A6:** 1057-1058
Computer-aided design and computer-aided manufacturing *See* CAD/CAM applications in machining
Computer-aided design and manufacture
for aluminum alloy forgings **A14:** 246-247, 253-254
applications for flat dies **A14:** 323-324
for closed-die forging **A14:** 80
for die and die materials **A14:** 57-58
future of **A14:** 326, 909-910
for hot extrusion **A14:** 321, 323-326
for hot forging **A14:** 57-58
information flow **A14:** 909
for laser cutting **A14:** 740-741
for lubricated extrusion **A14:** 325-326
in sheet forming **A14:** 903-910
system selection **A14:** 904
Computer-aided design/computer-aided manufacture *See also* Automation Computer-aided engineering; Computer applications; Modeling; Software
advantages **A15:** 857
as automatic sorting and inspection **A15:** 572
casting **A15:** 610-611
cooling analyses, die casting **A15:** 293
finite-difference method **A15:** 610-611
finite-element method **A15:** 610-611
for geometric modeling **A15:** 858-859
heat transfer analysis **A15:** 610
of metal casting **A15:** 36
for patternmaking **A15:** 198-199
sand casting tolerances **A15:** 618
vision inspection system, schematic **A15:** 572
Computer-aided design/computer-aided manufacture (CAD/CAM)
with coordinate measuring machines **A17:** 20
with mathematical modeling, machine vision process **A17:** 37
and NDE reliability models **A17:** 702-703
ultrasonic inspection model with **A17:** 712-713
Computer-aided dimensional inspection
of castings **A17:** 521-524
Computer-aided engineering **A14:** 252, 409, 909-910
advantages **A15:** 857
Computer-aided engineering (CAE)
for high-bandwidth systems **EL1:** 85
for layout **EL1:** 513
method, preferred **EL1:** 86
software tools **EL1:** 81
Computer-aided machining (CAM) technology
in green machining **EM4:** 181
Computer-aided manufacture *See also* Computer-aided design; Computer-aided design and manufacture **A14:** 907-909
computer requirements **A14:** 907
costs/advantages **A14:** 908
equipment **A14:** 907
machining with **A14:** 908-909
NC machine tools **A14:** 907-908
use **A14:** 908
Computer-aided manufacturing (CAM)
and computer-aided design **EL1:** 129
data generation **EL1:** 130
process **EL1:** 129-132
process flow **EL1:** 128
Computer-aided process design **A14:** 21
acquisition of data **A14:** 439-442
for bulk forming **A14:** 407-442
forging process design **A14:** 409-416
introduction **A14:** 407-408
modeling techniques **A14:** 417-438
Computer-aided roll pass designs
for shape rolling **A14:** 347-350
Computer-aided testing *See* Life cycle testing; Testing
Computer-assisted design **A7:** 569-570
Computer-assisted rocking curve analysis
abbreviation **A10:** 689
with position-sensitive detector **A10:** 372
Computer-control
electron-beam heat treating **M4:** 520
Computer-controlled ply cutting and labeling **EM1:** 619-623
Computer-integrated manufacturing (CIM)
in hybrid facility **EL1:** 256-257
Computer-integrated manufacturing process and control (CIMPAC) system **EL1:** 130
Computer-integrated manufacturing system
Replicast **A15:** 569
Computer-related materials
XRS analysis of **A10:** 82, 95
Computerized axial tomography (CAT scanning)
of castings **A17:** 528-529
Computerized modeling *See also* Model(s); Modeling; Simulation
of assembly/packaging **EL1:** 442-444
Computerized numerical control (CNC), automatic joint tracking
electron-beam welding **A6:** 863
Computerized properties-prediction and technology planning **A4:** 638--654
computer simulation objectives **A4:** 638-639
gas carburizing process parameters **A4:** 650-654
property-prediction systems (PPS) **A4:** 641-650
applications examples **A4:** 646-650
austenite grain size computation **A4:** 642-643
austenitization time and temperature calculation **A4:** 642
continuous cooling transformation modelling **A4:** 644-645
cooling curve computation **A4:** 644
design of a heat-treatment method **A4:** 648
grain size computation **A4:** 642-643
hardness calculation **A4:** 645, 652
heating curve computation **A4:** 642
kinetic functions for isothermal conditions **A4:** 644
mechanical property estimates **A4:** 646, 647, 649
microstructural transformations determinations **A4:** 645, 647-648, 649
microstructure calculation **A4:** 641
model description **A4:** 641
property calculation **A4:** 641
selection procedure for hardenable steel grades **A4:** 648-650, 651
tempering computer model **A4:** 645-648
TTT diagram calculation **A4:** 643-644
simulation softwares **A4:** 639-641
case hardening simulation **A4:** 641, 652-654
data base systems **A4:** 639, 640
development trends **A4:** 639
distortion analysis **A4:** 641
dynamic models **A4:** 639
examples of heat-treatment softwares **A4:** 639, 640, 652, 653-654
hardenability prediction **A4:** 641, 648-650, 651, 652-654
heat-treatment process selection **A4:** 641
material selection **A4:** 641
property prediction **A4:** 639, 641
residual stress analysis **A4:** 641
static models **A4:** 639, 641
Computerized systems for heat treating
advantages **M4:** 375
applications **M4:** 369-370
atmospheres control, carburizing furnace **M4:** 370-371
carbon potential, monitoring **M4:** 370
components **M4:** 370-371
computer basics **M4:** 367-368
computer control **M4:** 374
glossary, computer terms **M4:** 375-377
limitations **M4:** 370, 371
management applications **M4:** 375
memory systems **M4:** 368-369
monitoring **M4:** 367, 370
operation **M4:** 374, 375
software **M4:** 369
time vs temperature program **M4:** 371-373, 374
Computers *See also* Microcomputers **A13:** 317, 483
in acoustic emission inspection development **A17:** 282
in computed tomography (CT) **A17:** 372
-controlled gas mass spectrometer **A10:** 152
for ICP-AES systems **A10:** 39
for MFS analysis **A10:** 77
as readout, eddy current inspection **A17:** 179
for vision systems **A17:** 44
x-ray powder diffraction **A10:** 340
Computers (nuclear engineering)
powder used **A7:** 573

Computers, used in transmission electron microscopy **A9:** 113
of precipitates **A9:** 117
using dynamical theory of electron diffraction to study grain boundaries **A9:** 120
Concave fillet weld
definition **M6:** 4
Concave grating
defined **A10:** 671
Concave root surface
definition **M6:** 4
Concave surfaces
electropolishing of **A9:** 55
Concavity
definition **A6:** 1207 **M6:** 4
Concentrated load
symbol for **A8:** 725
Concentration *See also* Species concentration
absorbance as function of **A10:** 63, 64
of amount of substance, SI defined unit and symbol for **A10:** 685
of analyte elements, and intensity in x-ray spectrometry **A10:** 99
atomic, of surface elements **A10:** 603
component, constant, maintained by electrometric titration **A10:** 202
constituent, aqueous corrosion **A13:** 40-41
defined **A10:** 671
and diffusion current **A10:** 190
effect on corrosion rate **A11:** 175
exponential decay of radiant power as function of **A10:** 62
gradual, of corrosive substances **A11:** 211
high, and well-resolved peaks, gold-copper alloy analysis with **A10:** 530
profiles **A10:** 475, 610
spin wave stiffness as function of **A10:** 273
UV/VIS measured **A10:** 63-66
vs sputter etching time, Inconel **A10:** 558, 625
weight fractions as, EPMA quantitative analysis **A10:** 524
Concentration cell *See also* Differential aeration cell
defined **A13:** 4
effects, telephone cables **A13:** 1129-1130
oxygen/chemical **A13:** 42-43
Concentration, of amount of substance
SI unit and symbol for **A8:** 721
Concentration polarization **A13:** 4, 34, 214
Concentration profiles
elemental, by SIMS **A10:** 610
hydrogen **A10:** 610
molybdenum, in Ni-Cr-Mo alloy **A10:** 475
oxide surfaces, SIMS **A10:** 610
Concentration-cell corrosion *See also* Crevice corrosion; Galvanic corrosion **M1:** 713
of metals **A11:** 183
pitting by **A11:** 631-632
Concentration-depth profile
by sputter analysis **A7:** 251
Concentration-distance profile
aqueous corrosion **A13:** 33
Concentric hemispherical analyzer
abbreviation **A10:** 689
Concentric nebulizers
for analytic ICP systems **A10:** 34-35
Concentric-lay stranded copper conductors **M2:** 266, 268
Concentricity
of a forging **A14:** 73
Conchoidal fracture
as casting defect **A11:** 383
low-alloy steel **A11:** 392
Conchoidal marks *See* Beach marks
Concrete **A13:** 454, 455, 513, 1303
diamond for machining **A16:** 105
epoxy bonding **EM3:** 96
fracture/failure causes illustrated **A12:** 217
Concrete (continued)
galvanic corrosion of metals embedded in **A11:** 186-187
ground by diamond wheels **A16:** 455
roughers, bending failure of **A11:** 564-565
sulfur, fracture surfaces **A12:** 472-473
thermal expansion rate **A7:** 611
waterproofing, powders used **A7:** 574
Concrete reinforcement
steel wire for **M1:** 271
wire rod for **M1:** 255
Concrete reinforcing bars **M1:** 211-212
Concrete reinforcing wire
coating weight **M1:** 263
description **M1:** 264
Concrete-reinforcing bars **A1:** 246-247
Concrete-reinforcing wire **A1:** 282-283
Concurrent heating
definition **M6:** 4
Condensate chemistry
nuclear reactors **A13:** 958
Condensate corrosion **A13:** 622, 989
Condensates
as ion chromatography solutions **A10:** 658
Condensation **EM3:** 9
on high-impedance circuits **EL1:** 762
on PWB assemblies **EL1:** 761
Condensation (vapor phase) soldering **EL1:** 702-704
Condensation and evaporation
as material transport systems in sintering of compacts **A7:** 313
Condensation polyamic acid-polyimide
synthesizing **EL1:** 767
Condensation polyimides *See also* Polyimide resins
for aerospace prepregs **EM1:** 141
chemistry, formation, solvents **EM1:** 78, 79
as high temperature resistant **EM1:** 810-815
types **EM1:** 79
Condensation polymerization *See also* Polymerization
defined **EM1:** 7 **EM2:** 11
mechanism **EM1:** 752
Condensation resin
defined **EM2:** 11
Condensation review **EM3:** 9
Condensed phase equilibrium, and activity
defined **A15:** 51
Condenser
defined **A10:** 671
lens, defined **A10:** 671
Condenser aperture
defined **A9:** 4
Condenser lens
defined **A9:** 4
Condenser of an optical microscope **A9:** 72
defined **A9:** 4
Condenser tubes
ASTM specifications for **M1:** 323
Condensers **A13:** 637, 986-989
corrosion of **A11:** 615
tube, failed aluminum brass **A11:** 632
Conditional probability
in NDE reliability **A17:** 675
Conditioners
magnetic particle **A17:** 101
Conditioning **EM3:** 9
defined **EM1:** 7 **EM2:** 11
Conditioning time **EM3:** 9
Conditioning treatments
hardenable steels **M1:** 457
Condominium structures
corrosion of **A13:** 1299
Conductance
defined **A10:** 671 **EM2:** 590, 592
mho, as original ion chromatography unit of **A10:** 659
SI derived unit and symbol for **A10:** 685
SI unit/symbol for **A8:** 721
Conductimetric sensors *See* Oxygen sensors, semiconductor sensors
Conducting materials
for rigid printed wiring boards **EL1:** 538-539
Conduction *See also* Conductivity; Eletrical condictivity; Thermal conductivity
and cooling techniques **EL1:** 413-414
extrinsic, defined **EL1:** 99
formulas for **EL1:** 51-52
four-point probes **EM3:** 434-435
intrinsic, defined **EL1:** 99
microwave **A17:** 204
parallel plate measurements **EM3:** 434
in thermal inspection **A17:** 396
Conduction band
defined **EL1:** 97-98
Conduction electrons
ESR study **A10:** 261-263
excitation leading to secondary-electron (low-energy) emission **A10:** 433-434
Conduction enhanced circuit pack
for PWBs **EL1:** 47
Conduction equation **A6:** 8
Conduction heat transfer
in sintering **A7:** 341
Conduction, in boiler tubes
heat transfer by **A11:** 603
Conduction-mode electron-beam welds **A6:** 20, 21
Conduction-mode welding **A6:** 264
Conductive
adhesives **A7:** 607
fillers as shielding alternatives **A7:** 612
paints, powders used **A7:** 105, 572
plastics, powders used **A7:** 572
processes, atomization **A7:** 47-48
rapid quenching techniques **A7:** 47-48
silver inks **A7:** 147
Conductive (hot bar) soldering **EL1:** 705-706
Conductive adhesives **EM3:** 76
for bonding electrical wires and devices **EM3:** 45
flexible printed boards **EL1:** 590
suppliers **EM3:** 76
used to mount scanning electron microscopy specimens **A9:** 97-98
Conductive and non-spark flooring
powder used **A7:** 572
Conductive belt soldering **EL1:** 706
Conductive coatings used in scanning electron microscopy **A9:** 97
Conductive films
of indium/indium-tin oxides **A2:** 752
Conductive heating
for component removal **EL1:** 723
Conductive inks
for thick-film screen printing **EL1:** 207-208
Conductive materials
physical characteristics **EL1:** 106-107
thick-film, types **EL1:** 249
Conductive mounting materials **A9:** 29, 32
Conductive plastic materials **EM2:** 467-478
antistatic compounds **EM2:** 468-469
electrical conductivity, factors influencing **EM2:** 473-475
electrical resistivity testing **EM2:** 475
electromagnetic interference (EMI) shielding **EM2:** 476-478
fillers and reinforcements **EM2:** 469-473
static elimination testing **EM2:** 475-476
Conductive plastics
defined **EM2:** 461
Conductive polymer film interconnections
function **EL1:** 15
Conductive resins
as electropolishing mounting materials **A9:** 49
Conductive solids nebulizer
abbreviation for **A10:** 690
as solid-sampling device **A10:** 36

Conductive-thermal detection
inert gas fusion **A10:** 229-230
Conductivity *See also* Conduction; Electrical; Electrical conductivity; Electrical properties; Superconductive materials; Superconductivity; Thermal conductivity **A13:** 4, 17 **EM3:** 9
band theory of **EL1:** 97
of carbon fibers **EM1:** 52
common metals and alloys **A17:** 168
of composite laminates **EM1:** 36
copper casting alloys **M2:** 393
corrosion fatigue test specification **A8:** 423
curve, impedance-plane diagram **A17:** 168
defined **EL1:** 89 **EM1:** 7 **EM2:** 11
detection, ion chromatography **A10:** 659-661, 663, 665
eddy current inspection **A17:** 164, 167
electrical **EL1:** 93-96
electrical, as ion chromatography detector **A10:** 659
and hardness, eddy current inspection **A17:** 168
of heat sinks **EL1:** 1130
high-purity oxygenated water **A8:** 420
high-temperature, in rare earth cuprates **A2:** 1027
ionic **EL1:** 93-96
lower, PLAP analysis of **A10:** 597
and microwave inspection **A17:** 202-203
and moisture diffusion, in UDCs **EM1:** 191-192
monitoring at ambient temperature **A8:** 421
of polymers **EM2:** 62
specific, pressurized water reactor specification **A8:** 423
temperature effect **EL1:** 98
thermal, conversion factors **A10:** 686
thermal, SI derived unit and symbol for **A10:** 685
transients, and crack growth rate in stainless steel/water system **A8:** 421-422
Conductivity detection
eluent-suppressed **A10:** 659-660
of inorganic anions and cations determined by ion chromatography **A10:** 663
of inorganic ions determined by ion chromatography **A10:** 663
single-column ion chromatography with **A10:** 660-661
Conductivity detectors
ion chromatography **A10:** 665
Conductivity, specimen
for SEM imaging **A12:** 171
Conductometric oxidation-reduction titrations (redox)
rarity of **A10:** 203
Conductometry
and amperometry, compared **A10:** 204
capabilities, compared with voltammetry **A10:** 188
Conductor accessories
aluminum and aluminum alloys **A2:** 12
Conductor films, deposition of
vacuum coating process **M5:** 408
Conductor spacing
selection criteria **EL1:** 518-519, 587
Conductor(s)
adhesion, rigid printed wiring boards **EL1:** 548
cermet thick-film paste systems **EL1:** 339-341
corners, 90-degree **EL1:** 76
design, flexible printed boards **EL1:** 586-587
design spacing/design width, defined **EL1:** 1140
ideal materials for **EL1:** 112
inks, characteristics **EL1:** 208
materials **EL1:** 1041
multilevel **EL1:** 324
printed board coupons **EL1:** 576-577
selection criteria **EL1:** 518
spacing **EL1:** 518-519, 587
thickness, design effects **EL1:** 517
thin-film **EL1:** 316-320, 324
Conductors *See also* A15 superconductors; High-temperature superconductors; Niobium-titanium superconductors; Superconducting materials; Superconductivity; Superconductors; Ternary molybdenum chalcogenides (chevrel phases); Thin-film materials
high-conductivity, wrought copper and copper alloys **A2:** 251
Conductors (continued)
monofilamentary **A2:** 1046-1047
multifilamentary **A2:** 1047-1049
one-dimensional, nonstoichiometric salts as **A10:** 355
use of glow discharges with **A10:** 28
Conductors, central *See also* Central conductors; Coils **A17:** 95-97
Conductors, nonmetallic
galvanic corrosion **A13:** 84
Conduit pipe **A1:** 331 **M1:** 318
Condylar angle blade plate
as internal fixation device **A11:** 671
Cone
and cup samplers **A7:** 226
definition **A6:** 1207 **M6:** 4
and quartering samples **A7:** 226-227
Cone and plate test **EM3:** 443, 444
Cone and plate viscometer **EM3:** 322-323
Cone angle **A18:** 60, 61, 63, 66
Cone blenders
for sampling **A10:** 17
Cone crack **EM4:** 319
Cone geometries
in melt rheology **EM2:** 535-540
Cone indentation test
truncated **A14:** 384
Cone resistance value *See also* Penetration (of a grease)
defined **A18:** 6
Cone tests **EM3:** 387, 388
Cones
deficiency and diffraction **A10:** 327, 371
formation, AES analysis **A10:** 556
spinning of **A14:** 599-602
truncated, forming **A14:** 621-622
Confidence bands
bearing steel test groups **A18:** 728
Confidence interval **A18:** 482
defined **A8:** 626
width, with increasing sample size **A8:** 706
Confidence intervals
hit/miss data **A17:** 696
for POD(A) function **A17:** 695
signal response analysis **A17:** 698
as statistical method **A17:** 746
Confidence level **A18:** 481-482, 484, 485
defined **A8:** 626 **A10:** 671
and reliability levels **A8:** 627
Student's t value **A8:** 700
Confidence limits
ninety-five percent computation, on response curves **A8:** 703
on least squares parameters **A8:** 702
on mean fatigue curve **A8:** 700
and probability **A8:** 624
for Probit test data **A8:** 703
in sampling bulk materials **A10:** 13
for simple linear regression line **A8:** 700
statistical **A8:** 626
for stress **A8:** 703
unknown distribution **A8:** 666
Configuration
as design category **A11:** 115
SAS techniques for **A10:** 405
Configuration, or reconfiguration
of defective circuitry **EL1:** 9
Configuration parameter
surface roughness **A12:** 200
Configurationally frozen liquid *See* Amorphous materials; Metallic glasses
Configurations **EM3:** 9
defined **EM2:** 11
Confined atomization **A7:** 26, 29
Confocal microscopy **A18:** 357-361
applications **A18:** 359-361
development **A18:** 357
equipment **A18:** 357
experimental techniques **A18:** 358-359
computerized image processing **A18:** 359
image acquisition **A18:** 359
microscope configurations **A18:** 358-359
specimen requirements and limitations **A18:** 359
hardware configurations for confocal microscopes **A18:** 358
Confocal microscopy (continued)
image processing techniques for optical sections **A18:** 360
principles of the process **A18:** 357-358
Confocal scanning laser microscope
block diagram **EL1:** 367
Conforma clad process
hardfacing **A7:** 836
Conformability
defined **A18:** 6
Conformal coating **EM3:** 9
Conformal coatings *See also* Acrylics; Coatings; Epoxies; Epoxy; Parylene coatings; Polyurethane; Protective coatings; Urethane coatings; Urethanes
acrylic **EL1:** 763
applications **EL1:** 761-762
cationic cure systems **EL1:** 864
characteristics **EL1:** 765
coating methods **EL1:** 763-764
coating types **EL1:** 762-763
cure types **EL1:** 764-765
defined **EL1:** 761-762, 1138
in design **EL1:** 517
electrical properties **EL1:** 822
epoxy **EL1:** 763
fluoropolymer **EL1:** 782-784
introduction **EL1:** 759-760
manufacturing variables **EL1:** 762
non-UV curable **EL1:** 785
operating variables **EL1:** 763
overview **EL1:** 761-766
parylene **EL1:** 763
polyurethane **EL1:** 763
properties **EL1:** 762
radiation curing **EL1:** 854
silicone **EL1:** 763, 773-774, 822-824
specific energies **EL1:** 783
urethane **EL1:** 775-781
UV curable, properties **EL1:** 785-786
vessication, by ionic residues **EL1:** 660-661
Conformal surfaces
defined **A18:** 6
Conformance
solder masks **EL1:** 554
Conformation
effects, polymer blend system IR analysis **A10:** 118
molecular, and stereochemistry, IR determination of **A10:** 109
Conformations **EM3:** 9
defined **EM2:** 11
Conforming contact
in sliding contact wear tests **A8:** 605-606
Conforming shear blades **A14:** 717-718
Confound/alias (experimental)
defined **EM2:** 600
Confounding, statistical
in experimental design **A17:** 747, 749
Congruent phase change **A3:** 1•4
Congruent phase transformation **A3:** 1•4, 10
Congruent point **A3:** 1•10
Conical cap wrinkling test
for sheet metals **A8:** 563-564
Conical joints
friction welds **M6:** 726-727
Conjugate phases **A3:** 1•3
defined **A9:** 4
Conjugate planes
defined **A9:** 4
Connecting of three points method
of object orientation **A17:** 34
Connecting rod
cap, fatigue fracture **A11:** 119-120
fracture, from forging fold **A11:** 328, 330
and shafts, failures of **A11:** 459
Connecting rod, truck engine
fatigue fracture from forging lap **A11:** 328-329
Connecting rods
by Ceracon process **A7:** 541
cap, by precision forging **A14:** 163
forging data **A14:** 368
hot formed from P/M preform **A7:** 620
magnetic particle inspection methods **A17:** 117-119
Connection plates
floor-beam-girder **A11:** 712

Connection plates (continued)
multiple-girder diaphragm **A11:** 712-714
Connection rods
economy in manufacture **M3:** 850
Connectivity
as roughness parameter **A12:** 201
Connector alloys
beryllium-copper **A2:** 416
Connector link arm
engine **A7:** 750
Connectors
copper and copper alloys **A14:** 821
eye, sand-cast low-alloy steel **A11:** 390
failure of aluminum wire **A10:** 531-532
LEISS identification of surface stains and corrosion products on **A10:** 607
stained, LEISS analysis **A10:** 607
steel, tensile fracture in **A11:** 289-390
surface strains and corrosion products on **A10:** 607
tubes, aluminum alloy, flowout failure **A11:** 312-313
Connectors, bolted
recommended contact materials **A2:** 861
Connectors(s) *See also* Fundamental interconnection issues; Interconnection(s); Interconnects
effect, switching speeds **EL1:** 76
electrical failure analysis of **EL1:** 1112
failure mechanisms **EL1:** 981
high-density **EL1:** 394
as level 3 components **EL1:** 76
low-voltage, corrosion failure analysis **EL1:** 1112
as mechanical support **EL1:** 21
metallurgy **EL1:** 22
rack-and-panel, defined **EL1:** 1154
reliability of **EL1:** 21-23
system, requirements **EL1:** 86
Connellsville coke
development of **A15:** 30
Connes's advantage
in FT-IR spectrometers **A10:** 112
Conostan C-20
EDS determination of sulfur in **A10:** 101
Conradson method **A18:** 84
Conservation of Strategic Aerospace Materials (COSAM) program **A1:** 1009
Considere's construction
point of maximum load **A8:** 25
Consistency *See also* Cone resistance value; Penetration hardness number **EM3:** 9
defined **A18:** 6
Consolidation *See also* Powder consolidation **A7:** 295
alloy system variables **A7:** 315
applications **A7:** 295
beryllium powder **A2:** 685-696
of beryllium powders **A7:** 172
by atmospheric pressure (CAP process) **A7:** 533-536
by curing **EM1:** 655
of composite bearings **A7:** 407-408
of copper powders **A7:** 734
defined **A7:** 295 **EM1:** 7
deformation during, Ceracon process **A7:** 538
homogenization variables **A7:** 315
laminate level of, as quality-control variable **EM1:** 730
mechanical fundamentals of **A7:** 296-307
mechanically alloyed oxide alloys **A2:** 943-944
dispersion-strengthened (MA ODS)
and mechanics of metal powder aggregates **A7:** 295
methods, aluminum and aluminum alloys **A2:** 7
parts **EM1:** 36
physical fundamentals of **A7:** 308-321
powder, cemented carbides **A2:** 951
Consolidation (continued)
powder, high-strength aluminum P/M alloys **A2:** 203-204
powder variables **A7:** 315
processes, in roll compacting **A7:** 407
processes, special and developing **A7:** 295
and production **A7:** 23-24
Rapi-Press **EM1:** 872
rapidly solidified, permanent magnet materials **A2:** 791-792
and sintering **A7:** 295
techniques, P/M **A7:** 719-720
of titanium powders **A7:** 748-749
Consolidation at atmospheric pressure (CAP process) **A1:** 780, 973 **A16:** 60
Consolidation, parts
and costs **EM2:** 85
Constancy of volume
for plastic deformation **A8:** 22
Constant
definition **A6:** 38
symbol and units **A18:** 544
Constant acceleration
package-level testing **EL1:** 937-938
Constant amplitude, and sinusoidal loading
S-N curves **A11:** 103
Constant amplitude tests
low-alloy steels **A15:** 717
plain carbon steels **A15:** 703-704
Constant cell potential
abbreviation for **A10:** 690
Constant current imaging **A18:** 393
Constant head flow coating method
porcelain enameling **M5:** 516
Constant *K* specimens
SCC testing **A8:** 515 **A13:** 256
Constant load amplitude fatigue crack growth **A1:** 370, 371, 372
Constant maximum shear stress **A8:** 72
Constant rate of extension testing
machines **A8:** 47
Constant strain amplitude
low-cycle fatigue tests **A8:** 367
Constant strain rate
for titanium alloy **A8:** 171
Constant stress
cam-lever apparatus **A8:** 319-320
creep curve for austenitic stainless steel **A8:** 320-321
curves **A8:** 311, 333, 335
vs. constant load, in creep testing **A8:** 305
Constant stress creep **EL1:** 840, 846--847
Constant tensile load testing **EM2:** 802
Constant volume
as material behavior **A8:** 343
Constant-amplitude fatigue tests
ASTM E 466 in **A8:** 149
for fatigue crack growth rate **A8:** 678
Constant-amplitude stress cycle
vs. crack growth **A8:** 379, 678-679
Constant-amplitude tests of smooth bars **A1:** 369-370
Constant-current cells
for electrogravimetry **A10:** 199-200
Constant-current electrolysis
separation and analysis of metal ions by **A10:** 200
Constant-current methods
of electrogravimetry **A10:** 198
Constant-curvatuve cantilever specimen
stress-relaxation bend testing **A8:** 326
Constant-deflection
for high-cycle fatigue tests .... **A8:** 1 -367
Constant-displacement specimens
K-decreasing tests **A8:** 517-518
Constant-force springs **A1:** 302
Constant-level pouring
permanent mold method **A15:** 276
Constant-life diagram
temperature behavior of S-816 alloy **A11:** 131
use of **A11:** 111
Constant-life fatigue diagram **A8:** 3
Goodman **A8:** 712-713
Constant-lifetime diagram
and fatigue resistance **A1:** 675
Constant-load testing
amplitude, for high-cycle fatigue tests **A8:** 367
bending, SCC testing **A8:** 503
and constant-stress testing, lead wire **A8:** 319-320
creep curve **A8:** 311
for creep, stress rupture, stress relaxation **A8:** 311, 313-318
data presentation **A8:** 315
interruptions **A8:** 315
K-increasing **A8:** 517
notched-specimen testing **A8:** 315-318
rate tension machines **A8:** 47
specimen loading **A8:** 314
strain rate and time for creep **A8:** 331
for stress-corrosion **A8:** 496, 502
temperature control **A8:** 314-315
threshold stresses by **A8:** 499
vs. constant stress, in creep testing **A8:** 305
Constant-load tests
SCC evaluation **A13:** 246-247
Constant-load vs constant-strain testing **EM2:** 801-802
Constant-load wheel, with balancing cam
constant-stress testing **A8:** 320
Constant-radii fillets
ultrasonic fatigue testing specimens **A8:** 250-251
Constant-rate drying period **EM4:** 105
Constant-strain cycling
stress-strain loop for **A8:** 367
Constant-strain testing **EM2:** 802
Constant-strain tests
effect of low strain rate in **A8:** 499
effects of changing stress on crack growth in **A8:** 502
SCC evaluation **A13:** 246-247
of stress corrosion cracking **A8:** 496, 502
threshold stresses by **A8:** 499
vs constant-load test, in SCC testing **A8:** 502
Constant-stress testing **A8:** 318-321
with balancing cam and constant-load wheel **A8:** 320
compression **A8:** 320-32J
and constant-load testing, lead wire **A8:** 319-320
creep curve **A8:** 311
equipment and methods **A8:** 311, 318-321
hyperbolic-weight apparatus **A8:** 318-319
system and furnace **A8:** 321-322
test methods and equipment **A8:** 318-321
weight pan knife edge **A8:** 321-322
Constant-stress testing equipment
cam-lever apparatus **A8:** 319-320
hyperbolic-weight apparatus **A8:** 318-319
Constant-time curves **A8:** 333, 335
Constant-voltage electrogravimetry **A10:** 199
Constantan *See* Electrical resistance alloys; Electrical resistance alloys, specific types; Thermocouple materials
thermal properties **A18:** 42
Constantin
photochemical machining etchant **A16:** 590
Constituent **EM3:** 9
defined **A7:** 3 **A9:** 4 **EM1:** 7 **EM2:** 11
Constituent material forms **EM1:** 105-171
aramid fibers **EM1:** 114-116
bulk molding compounds **EM1:** 161-163
carbon fibers **EM1:** 112-113
fiber sizing **EM1:** 122-124
filament-winding resins **EM1:** 135-138
glass fibers **EM1:** 107-111
injection molding compounds **EM1:** 164-167
multidirectional tape prepregs **EM1:** 146-147

Constituent material forms (continued)
multidirectionally reinforced fabrics
preforms **EM1:** 129-131
other continuous fibers **EM1:** 117-118
other discontinuous fibers **EM1:** 119-121
prepreg resins **EM1:** 139-142
recycling carbon fiber scrap **EM1:** 153-156
resin transfer molding (RTM)
materials **EM1:** 168-171
sheet molding compounds **EM1:** 157-160
two-directional fabrics **EM1:** 125
unidirectional fabrics **EM1:** 125-128
unidirectional tape prepregs **EM1:** 143-145
wet lay-up resins **EM1:** 132-134
woven fabric prepregs **EM1:** 148-150
Constituent materials
boron and silicon carbide fibers **EM1:** 58-59
carbon/graphite fibers **EM1:** 49-53
ceramic fibers **EM1:** 60-65
epoxy resins **EM1:** 66-77
glass fibers **EM1:** 45-48
introduction **EM1:** 43
organic fibers **EM1:** 54-57
polyester resins **EM1:** 90-96
polyimide resins **EM1:** 78-89
properties of **EM1:** 43-104
thermoplastic resins **EM1:** 97-104
Constitution diagram
stainless steel weld metal **M3:** 51
Constitution diagrams **A6:** 127, 677, 678, 679, 680, 681, 682
Constitutional diagram *See* Phase diagram ..... **A3:** 1 • 2
Constitutional liquation **A6:** 74-75, 91
nickel-base alloys **A6:** 588
Constitutional supercooling **A6:** 46-48, 49, 51, 52, 53 **A9:** 611-612
effect on dendritic structures **A9:** 613
equation, simplified **A9:** 621
and equiaxed growth **A15:** 130-131
Constitutional undercooling
effect on eutectic structures **A9:** 619
Constitutive equations
for material modeling **A14:** 417-420
Constrained fiber printed wiring board **EL1:** 984
Constrained flow stress **A8:** 577
Constrained recovery
of shape memory alloys **A2:** 900
Constrained technique
as interconnection system option **EL1:** 984-985
Constraint
defined **A8:** 3 **A15:** 3
Constricted arc
definition **M6:** 4
Constricted arc (plasma arc welding and cutting)
definition **A6:** 1207
Constricting nozzle
definition **M6:** 4
Constricting orifice
definition **M6:** 4
Constriction resistance *See also* A-spot; Contact resistance; Film resistance
defined **A18:** 6
Construction *See also* Fabrication
fabric pattern as **EM1:** 125
of passive components **EL1:** 178-179
reliability effects, passive components **EL1:** 180-181
Construction and Structural Adhesives and Sealants—An Industrial Guide **EM3:** 67
Construction applications *See also* Building applications; Plumbing applications
acrylonitrile-butadiene-styrenes (ABS) **EM2:** 111
aluminum and aluminum alloys **A2:** 9-10
architectural covercoat enamels **EM4:** 1066
cemented carbides **A2:** 973-974
critical properties **EM2:** 458
exterior, polyarylates (PAR) **EM2:** 139
polybutylene terephthalates (PBT) **EM2:** 153
polyurethanes (PUR) **EM2:** 259
styrene-acrylonitriles (SAN, OSA ASA) **EM2:** 215
unsaturated polyesters **EM2:** 246
Construction glass **EM4:** 1024-1025
cellular glass **EM4:** 1025
ceramic building cladding **EM4:** 1025
foam glass **EM4:** 1025
glass block **EM4:** 1025
Construction glass (continued)
glass ceramic versus marble and granite **EM4:** 1025
transparent materials properties **EM4:** 1024
Construction joint sealants **EM3:** 52-53
design and categorization **EM3:** 53
Construction Sealants and Adhesives **EM3:** 70
Construction sections
bend tests for **A8:** 117
Constructional steel *See* Alloy steel; Carbon steel; Low-alloy steel
Constructional steels *See* Structural steels
Constructive solid geometry
as geometric modeler **A15:** 858
Consumable electrode remelting **M1:** 111
Consumable electrode vacuum arc remelting (VAR) *See also* Vacuum arc remelting
for niobium-titanium superconducting materials **A2:** 1044
Consumable electrode vacuum melting *See* Vacuum arc remelting
Consumable guide electroslag welding **M6:** 227
definition **M6:** 4
Consumable insert
definition **A6:** 1207
Consumable inserts
definition **M6:** 4
gas tungsten arc welding **M6:** 202
Consumable vacuum-melted (CVM)
rolling contact component material **A18:** 503
Consumable-abrasive cutting **A9:** 24-25
Consumable-electrode remelting *See also* Electroslag remelting; Vacuum arc remelting
defined **A15:** 3
Consumable-wheel abrasive cutting **A9:** 24-25
Consumables
in cast iron welding **A15:** 523-524
Consumables for
arc welding of cast irons **M6:** 310-311
high-strength low-alloy
flux cored arc welding **M6:** 278
gas metal arc welding **M6:** 274-278
shielded metal arc welding **M6:** 274-275
steels **M6:** 271-278
submerged arc welding **M6:** 278
submerged arc welding **M6:** 119-127
Consumer electronics
self-lubricating bearings in **A7:** 705
Consumer housewares applications **EM4:** 1100-1103
categories of glass houseware **EM4:** 1100
drinkware **EM4:** 1102
classifications **EM4:** 1102
compositions used commercially **EM4:** 1102
durability **EM4:** 1102
material requirements **EM4:** 1102
properties **EM4:** 1102
types **EM4:** 1102
ground coat enamels **EM4:** 1066
history **EM4:** 1100
material requirements **EM4:** 1100
decoration durability **EM4:** 1100
durability **EM4:** 1100
loss tangent and microwave heating **EM4:** 1100
strength **EM4:** 1100
thermal shock resistance **EM4:** 1100
ovenware **EM4:** 1102-1103
compositions **EM4:** 1103
durability **EM4:** 1103
glass **EM4:** 1103
glass-ceramic **EM4:** 1103
properties **EM4:** 1102-1103
safety and health **EM4:** 1100
Food and Drug Administration Standards **EM4:** 1100
tableware
alternate materials **EM4:** 1102
compositions **EM4:** 1101
glass-ceramics **EM4:** 1101-1102
glaze compositions **EM4:** 1102
laminates **EM4:** 1101
opals (opaque) **EM4:** 1101
properties **EM4:** 1101
soda-lime dinnerware **EM4:** 1101
top-of-stove ware **EM4:** 1103
aluminosilicate glass compositions **EM4:** 1103
beta-quartz glass ceramics **EM4:** 1103
Consumer hybrid applications
types **EL1:** 255, 385
Consumer product applications
blow molding **EM2:** 359
copper and copper alloys **A2:** 239
durables, aluminum and aluminum alloys **A2:** 13
high-impact polystyrenes (PS, HIPS) **EM2:** 195
homopolymer/copolymer acetals **EM2:** 101
liquid crystal polymers (LCP) **EM2:** 180
polyamides (PA) **EM2:** 125
polybutylene terephthalates (PBT) **EM2:** 153
polyether sulfones (PES, PESV) **EM2:** 159
polysulfones (PSU) **EM2:** 200
polyurethanes (PUR) **EM2:** 259
reinforced polypropylenes (PP) **EM2:** 193
silicones (SI) **EM2:** 266
styrene-acrylonitriles (SAN, OSA ASA) **EM2:** 215
titanium and titanium alloys **A2:** 590
Consumer product manufacturing processes **EM1:** 554-574
compression molding **EM1:** 554, 559-563
injection molding **EM1:** 554, 555-558
resin transfer molding **EM1:** 554, 564-568
tube rolling **EM1:** 554, 569-574
Consumer product market **EM3:** 47
Consumer's risk **EM4:** 86
Contact
cracking **A11:** 754
and fracture mechanics **A11:** 57
stresses, in fatigue fracture **A11:** 75
wear, spiral bevel gear teeth **A11:** 596
windows, electromigration effects at **A11:** 778
Contact adhesive
defined **EM1:** 7 **EM2:** 13
Contact adhesives **EM3:** 9
phenolics **EM3:** 105
Contact angle
defined **A18:** 6
Contact angle (in a bearing) *See* Angle of contact
Contact angles
liquid metals on beryllium **A6:** 116
wetting and brazing process **A6:** 115, 116
Contact bond adhesives **EM3:** 9, 36
Contact bridge **A18:** 236
Contact buffing **M5:** 115-116
Contact cements
for laminate bonding **EM3:** 577
Contact compliance **A18:** 423
Contact corrosion
defined **A13:** 4
Contact dermatitis
from beryllium **A2:** 1239
Contact fatigue *See also* Faitigue; Spalling **A18:** 242
defined **A11:** 2, 465
failure **A11:** 133-134
in gears **A11:** 594
rolling **A11:** 500-505
in shafts **A11:** 465
subsurface **A11:** 134
surface pitting as **A11:** 134
testing, surface pitting from **A11:** 133-134
Contact fatigue wear
tool steels **A18:** 736, 737
Contact force distribution **EM3:** 450
Contact load
in fretting **A13:** 139
Contact lubrication
defined **A18:** 6
Contact marks, as flaws
defined **A17:** 562
Contact metal
properties and applications **A2:** 363-364
Contact method probe, for aircraft subassemblies
eddy current inspection **A17:** 190
Contact molding *See also* Spray lay-up; Spray molding
defined **EM1:** 7 **EM2:** 11
processes **EM2:** 338-339
Contact pad metallurgy **EL1:** 116
Contact patch **A18:** 438
Contact potential
in adhesive-bonded joints **A17:** 611
defined **A13:** 4
Contact pressure **A18:** 476

Contact pressure resins ... **EM3:** 9
defined ... **EM1:** 7 **EM2:** 11
Contact printing of micrographs ... **A9:** 85
Contact probing
as instrumentation/testing ... **EL1:** 371
Contact radius ... **A18:** 43
Contact resistance ... **A18:** 682-683
defined ... **A18:** 6
Contact resistance (resistance welding)
definition ... **A6:** 1207
Contact stress ... **A18:** 506
defined ... **A18:** 6
maximum normal ... **A18:** 506
Contact supports
oxide-dispersion-strengthened copper ... **A2:** 401
Contact surface fracture
prediction ... **A14:** 399
Contact surface temperature ... **A18:** 40
Contact temperature ... **A18:** 438-439, 441
symbol and units ... **A18:** 544
Contact through transmission inspection
of adhesive-bonded joints ... **A17:** 617
Contact tube
definition ... **A6:** 1208 **M6:** 4
Contact ultrasonic ringing
defect detection ... **EM3:** 751
Contact wheels
polishing and buffing ... **M5:** 111, 125-126
Contact window alloy spikes
as failure mechanism ... **EL1:** 1015-1016
Contact(s)
area, defined ... **A7:** 3
composite metals for ... **A7:** 17
density of ... **EL1:** 440
design, and connector reliability ... **EL1:** 22-23
electromigration effects ... **EL1:** 964
failure, VLSI mechanisms ... **EL1:** 890-891
infiltration ... **A7:** 3, 553
injecting ... **EL1:** 958
large-area back ... **EL1:** 958
material, defined ... **A7:** 3
ohmic ... **EL1:** 958
pitting, semiconductor chip ... **EL1:** 964
powders used ... **A7:** 572
transfer method of painting ... **A7:** 460
Contact-point compressometer ... **A8:** 56
Contact-pressure laminates *See* Laminate(s)
Contact-type search ultrasonic units *See also* Search units
angle-beam ... **A17:** 257
delay-tip ... **A17:** 258
dual-element ... **A17:** 257-258
paintbrush transducers ... **A17:** 258
straight-beam ... **A17:** 257
Contacting needle
in spring-material test apparatus ... **A8:** 134-135
Contacting ring seal
defined ... **A18:** 6
Contactless probing
as instrumentation/ testing ... **EL1:** 371-372
Contactors
resistance spot welding ... **M6:** 470-471
Contacts, electrical
XPS analysis of surface films on ... **A10:** 578-579
Container applications
steel wire ... **M1:** 264
Container disposal ... **EM3:** 693
Container glass
alumina content ... **EM4:** 379
applications ... **EM4:** 1015
composition ranges ... **EM4:** 382
melters ... **EM4:** 391-392
melting furnace ... **EM4:** 389, 390
strength a key factor ... **EM4:** 741
Container materials ... **A13:** 50, 971-980
Container selection ... **EM3:** 693-694, 717
Containerless hot isostatic pressing ... **A7:** 436, 441
Containers *See also* Cans
aluminum and aluminum alloys ... **A2:** 10
Containers (continued)
assembly ... **A7:** 430-431
can designs ... **A7:** 430
conditions and explosivity ... **A7:** 195
corner designs, in hot isostatic pressing ... **A7:** 430
deep drawing of ... **A14:** 575
defect inclusion levels ... **EM4:** 392
design, and encapsulation techniques ... **A7:** 429-430
fabrication of ... **A7:** 430
fill tubes for ... **A7:** 430, 431
filling practices ... **A7:** 431, 433
leak testing ... **A7:** 431-434
packaged, sampling from ... **A7:** 213
plugs, during evacuation and gassing ... **A7:** 430, 431
precoated steel sheet for ... **M1:** 172, 173
rectangular, from sheet metal ... **A7:** 430
sheet metal powder ... **A7:** 430, 431
welding, in containerized hot isostatic pressing ... **A7:** 431
Containers for chemicals ... **A9:** 69
Containers/tableware glass
applications ... **EM4:** 379
Contaminant ... **EM3:** 9
defined ... **EM1:** 7 **EM2:** 11
Contaminants
defined ... **A13:** 380
effect on compressibility and sintering ... **A7:** 211
effects, types, removal ... **A7:** 178-181
process and handling operations with ... **EL1:** 658
solder-bath ... **EL1:** 638
surface ... **A7:** 260
trapping, by particle impact noise detection ... **EL1:** 954
types ... **EL1:** 658-661
Contamination
adhesion failures caused by thin-film ... **A11:** 43
air, control of ... **EL1:** 781
airborne, in marine atmospheres ... **A13:** 903-905
of aluminum melts ... **A15:** 79-81
analysis, on grain-boundary fractures ... **A11:** 41-42
of apparatus or reagents, controlling for ... **A10:** 12
argon, and leakage flow in compacts ... **A7:** 434
in as-sintered, high-speed steels ... **A7:** 379
atmospheres, heating-element materials ... **A2:** 835
in bearing lubricated systems ... **A11:** 485, 486
by nonvolatile organics, in XPS samples ... **A10:** 575
control, parylene coatings ... **EL1:** 800
coupling, microwave inspection ... **A17:** 202
defects, defined ... **EL1:** 978-979
of developers ... **A17:** 85
effect, fatigue strength titanium P/M compact ... **A2:** 653
effect, in magnesium ... **A13:** 741
effects and cleaning of ... **A7:** 178-181
of emulsifiers ... **A17:** 85
of encapsulated powders ... **A7:** 430
environmental, influencing corrosion fatigue crack propagation ... **A8:** 405, 407-408
flux, control of ... **EL1:** 649
hydrogen, in aluminum-silicon melts ... **A15:** 164-165
identification by secondary ion mass spectroscopy ... **A7:** 258
ionic ... **EL1:** 1026-1027
iron powder ... **A14:** 190-191
lubricant failure from ... **A11:** 153
of melts ... **A15:** 74, 79-81
metal, feeding aids ... **A15:** 586
in microanalytical samples ... **A11:** 36
mobile ion ... **EL1:** 34
with modifiers ... **A15:** 484-485
organic ... **EL1:** 1027-1030
organic surface, SIMS analysis ... **EL1:** 1086-1087
oxide ... **EL1:** 959-960
Contamination (continued)
particle, identification of ... **EL1:** 1089
of particle surfaces, and green strength ... **A7:** 288-289
of penetrants ... **A17:** 85
of personnel, in nuclear applications ... **A7:** 666
prevention, ultrahigh vacuum for ... **A7:** 251
of process fluid, and material selection ... **A13:** 323
radioactive ... **A13:** 950
radioactive, in NAA samples ... **A10:** 236
removal and yield of cleaned red clover seeds ... **A7:** 591
of rolling-element bearings ... **A11:** 511
sampling, quality assurance and ... **A10:** 17
in seeds, magnetic separation for ... **A7:** 589
silicon inversion due to ... **A11:** 781
in sintered tool steel microstructures ... **A7:** 376
in steam or boiling water, fatigue crack growth testing ... **A8:** 426-430
in steam treating ... **A7:** 453
surface, blast cleaning for ... **A15:** 506
surface, RBS analysis ... **A10:** 628
surface, trace analysis ... **A10:** 177
in tungsten oxide reduction ... **A7:** 153
type, in liquid penetrant inspection ... **A17:** 81
under lead plating ... **EL1:** 990
variable resistor ... **EL1:** 1001-1002
visual inspection ... **A17:** 3
Contamination line
nondestructive profiles ... **A12:** 199
Contamination, surface
optical effects ... **EM2:** 597-598
Contiguity in cemented carbides
determination of ... **A9:** 275
Continental dies *See* Steel-rule dies
Contingency costs
of plastics program ... **EM2:** 87
Continuity
law of ... **A15:** 591
Continuity bond
defined ... **A13:** 4
Continuous aluminum oxide fiber MMCs *See also* Aluminum oxide, fibers ... **EM1:** 874-877
applications ... **EM1:** 876
constituent materials ... **EM1:** 874
fabrication ... **EM1:** 874-875
properties ... **EM1:** 875-876
Continuous annealing furnace
as shoving furnace ... **A15:** 31
Continuous attrition mills ... **A7:** 69
Continuous boron fiber MMCs *See also* Boron fibers ... **EM1:** 851-857
applications ... **EM1:** 856-857
boron fiber production ... **EM1:** 851-853
composite processing ... **EM1:** 854
composite properties ... **EM1:** 855
Continuous boron MMC reinforcements ... **A2:** 7
Continuous bubbling
in acidified chloride solutions ... **A8:** 419
Continuous butt-welded steel pipe
inspection of ... **A17:** 567
Continuous carbon fiber reinforced carbon matrix composites ... **EM1:** 911-914
Continuous cast copper alloy ingots
freezing front ... **A9:** 641
grain structures ... **A9:** 640-642
Continuous casting *See also* Castings, Foundry products; Shape casting processes; Strand casting ... **A15:** 308-316 **EM4:** 44
aluminum alloys ... **M2:** 147
aluminum casting alloys ... **A2:** 141
defined ... **A15:** 3
effect on Lüders lines ... **A8:** 553
history ... **A15:** 308
horizontal ... **A15:** 313
machines ... **A15:** 308-315

Continuous casting (continued)
of metallic glasses ... **A2:** 806
of nonferrous alloys ... **A15:** 313-315
as permanent mold method ... **A15:** 277
plant layout ... **A15:** 309-310
sequence of operations ... **A15:** 308
of steel ... **A15:** 308-313
steel, effect on gray iron casting ... **A15:** 43-44
types ... **A15:** 309
wire rod ... **A2:** 254
wrought copper and copper alloys ... **A2:** 243
Continuous casting machines ... **A15:** 308-312
Continuous compaction
defined ... **A7:** 3
Continuous cooling *See also* Cooling
diffusional growth in peritectic transformations ... **A9:** 677
peritectic transformation during ... **A15:** 127
Continuous cooling transformation
abbreviation ... **A8:** 724
white cast iron ... **M1:** 84
Continuous cooling transformation (CCT) diagram
steel weldments ... **A6:** 416, 417, 418, 419
Continuous cooling transformation curve
low carbon-manganese steel plate weld metal ... **A9:** 580
Continuous cooling transformation diagram
constructional steels for elevated temperature use ... **M1:** 654
Continuous cooling transformation diagrams ... **M6:** 25-26, 39-40
used to describe nonequilibrium phases in welded joints ... **A9:** 580
Continuous cycling, mode
sonic converters ... **A8:** 243
Continuous depth recording (CDR) ... **A18:** 419-420, 421, 423, 426
Continuous distribution function
gradients in x-ray characterization of surface wear ... **A18:** 465
Continuous dry abrasive blasting systems ... **M5:** 88-89, 92-93
Continuous electron beam accelerator facility (CEBAF)
as niobium-titanium superconducting material application ... **A2:** 1056
Continuous emission *See also* Acoustic emission inspection; Acoustic emission(s)
acoustic emission inspection of ... **A17:** 281-284
and burst-type, compared ... **A17:** 287
Continuous fiber aluminum metal-matrix composites ... **A2:** 7, 904-906
Continuous fiber placement method
ceramic-ceramic composites ... **EM1:** 934
Continuous fiber reinforced composites *See also* Composites, Fiber composites Fiber-reinforced composites
basic failure modes ... **EM1:** 781-785
defined ... **EM1:** 27
failure analysis ... **EM1:** 768-769
fibers in ... **EM1:** 29
fractography for ... **EM1:** 786-793
general considerations ... **EM1:** 768-769
multidirectional ... **EM1:** 933-934
optical micrograph of ... **EM1:** 769
as packaging materials ... **EL1:** 1122-1125
strength tests ... **EM4:** 595-596
Continuous fiber reinforced composites, failure analysis of ... **A11:** 731-743
causes of failure ... **A11:** 732-733
fracture modes in composites ... **A11:** 733-739
procedures for ... **A11:** 739-743
types of composites ... **A11:** 731-739
Continuous fiber reinforced magnesium composites ... **A2:** 460
Continuous fiber-reinforced composites *See also* Continuous fiber-reinforced composites, failure analysis of
compression fracture from buckling ... **A11:** 742-743
defined ... **A11:** 731
failure modes ... **A11:** 735, 736
fracture modes in ... **A11:** 733-739
inclined microcracks in ... **A11:** 736
matrix feathering in ... **A11:** 736
optical micrograph, laminated construction ... **A11:** 732
planes of separation in ... **A11:** 734
Continuous fiber-reinforced composites (continued)
types of ... **A11:** 731-739
Continuous fibers *See also* fiber(s)
boron (B) filaments ... **EM1:** 117
carbon/graphite, properties ... **EM1:** 867
effect, fluidity ... **A15:** 849-850
high-alumina ... **EM1:** 117-118
in metal-matrix composites ... **A15:** 840
metallic wire ... **EM1:** 118
natural ... **EM1:** 117
product forms ... **EM1:** 33
as reinforcement ... **EL1:** 1119-1121
silicon nitride ... **EM1:** 118
synthetic ... **EM1:** 117-118
vs. discontinuous fibers, cost ... **EM1:** 105
Continuous filament *See also* Continuous fibers; Fiber(s); Filaments
aramid fiber ... **EM1:** 114-115
Continuous filament yarn *See also* Fiber(s); Filaments; Yarn; Yarn(s)
defined ... **EM1:** 7 **EM2:** 11
Continuous flow electron beam melting
equipment ... **A15:** 415-416
melted materials, characteristics ... **A15:** 416-417
principles ... **A15:** 414-415
vs drip method ... **A15:** 415
Continuous flow melting
by electron beam ... **A15:** 414-415
historic ... **A15:** 27
Continuous furnace
furnace brazing ... **A6:** 121, 122
porcelain enameling ... **M5:** 520
Continuous furnace(s) ... **A7:** 3
for Al powders ... **A7:** 381-382
humpback ... **A7:** 351, 353-354, 355, 356
mesh-belt conveyor ... **A7:** 351-353
production nomograph for ... **A7:** 353
pusher ... **A7:** 351, 354, 356
roller-hearth ... **A7:** 351, 354, 357
for sintering P/M compacts ... **A7:** 351-359
vacuum ... **A7:** 357-358
walking-beam ... **A7:** 354-358
Continuous galvanized strip
chromating ... **A13:** 390
Continuous glass fibers
types ... **EM1:** 107
Continuous grain growth *See also* Grain growth ... **A9:** 697
grain shape distribution ... **A9:** 697
grain size distribution ... **A9:** 697
Continuous graphite fiber MMCs ... **EM1:** 867-873
carbon/graphite fiber manufacture ... **EM1:** 867-868
casting ... **EM1:** 872-873
diffusion bonding ... **EM1:** 869-870
direct-metal infiltration processing ... **EM1:** 872
precursors ... **EM1:** 868-869
pultrusion ... **EM1:** 870-872
Rapi-Press consolidation ... **EM1:** 872
Continuous graphite/copper metal-matrix composites ... **A2:** 909
Continuous heating transformation (CHT) diagrams ... **A6:** 73
Continuous hot dip aluminum coating process
mill products ... **M5:** 335-337
Continuous immersion test
for SCC susceptibility ... **A8:** 523
Continuous jet impingement
versus liquid impingement erosion ... **A18:** 222
Continuous ladle-desulfurization processes ... **A15:** 77-78
Continuous magnetism *See also* Magnetization
magnetic fields ... **A17:** 91
as magnetic particle inspection method ... **A17:** 110
Continuous mixers
for coremaking ... **A15:** 239
Continuous mixing
SMC resin pastes ... **EM1:** 159
Continuous muller
green sand preparation ... **A15:** 344-345
Continuous multifrequency techniques
eddy current inspection ... **A17:** 174
Continuous on-line oil monitoring ... **A18:** 299
Continuous ovens, convection and radiant
paint curing process ... **M5:** 487-488
Continuous oxide fibers
alumina-silica/alumina-boria-silica ... **EM1:** 60-61
Continuous oxide fibers (continued)
aluminum ... **EM1:** 60
commercially available types ... **EM1:** 60
fused-silica ... **EM1:** 61
leached-glass ... **EM1:** 61
properties/types ... **EM1:** 62
zirconia-silica ... **EM1:** 61
Continuous phase
of composites ... **EM1:** 27
defined ... **A9:** 4
Continuous phase separation *See* Spinodal decomposition
Continuous pickling process ... **M5:** 68-69, 71-72
Continuous precipitation ... **A9:** 647
Continuous processing
polyurethanes (PUR) ... **EM2:** 263-264
Continuous quenching
of steel ... **M4:** 58, 63
Continuous random network model
amorphous materials and metallic glasses ... **A2:** 809-810
Continuous reinforcement *See also* Fillers; Reinforcements
pultrusions ... **EM2:** 393
Continuous rotary automatic polishing and buffing machines ... **M5:** 121-122
Continuous second-phase networks
and forging failure ... **A11:** 327
Continuous sequence
definition ... **M6:** 4
Continuous service motors and generators
as magnetically soft material application ... **A2:** 779
Continuous silicon carbide fiber MMCs *See also* Silicon carbide (SiC) fibers ... **EM1:** 858-866
applications ... **EM1:** 864-865
chemical vapor deposition ... **EM1:** 858
composite processing ... **EM1:** 859-862
fiber properties ... **EM1:** 63-64
future trends ... **EM1:** 865
properties ... **EM1:** 862-863
SiC fiber production ... **EM1:** 858-859
Continuous silicon carbide fibers ... **EM1:** 63-64
Continuous sintering ... **A7:** 3
vacuum, of iron and steel compacts ... **A7:** 359
Continuous solid solution ... **A3:** 1 • 2, 18
Continuous spectrum
defined ... **A9:** 4
Continuous strand annealing
of wrought copper and copper alloys ... **A2:** 246-247
Continuous strand rovings
tests for ... **EM1:** 291
Continuous tumbling barrel blast cleaning ... **A15:** 507-508
Continuous tungsten fiber MMCs ... **EM1:** 878-888
creep resistance ... **EM1:** 880-881
design ... **EM1:** 884-885
fabrication techniques ... **EM1:** 885-887
fiber-matrix compatibility ... **EM1:** 879-880
hot corrosion ... **EM1:** 882-883
impact strength ... **EM1:** 883-884
oxidation ... **EM1:** 882-883
stress rupture strength ... **EM1:** 880
thermal conductivity ... **EM1:** 884
thermal fatigue ... **EM1:** 881-882
Continuous tungsten fiber reinforced copper composites ... **A2:** 908-909
Continuous vacuum coating ... **M5:** 392, 397, 404-405
Continuous variables
in fractional factorial design ... **A17:** 746
Continuous wave (CW) lasers ... **A18:** 861
Continuous wave laser beam welding ... **M6:** 657-658
Continuous waves
microwave inspection ... **A17:** 202, 205-206
solid state devices ... **A17:** 209
Continuous wear ... **A18:** 737
Continuous weld
definition ... **M6:** 4
Continuous x-rays
production of ... **A17:** 298
Continuous yielding
tensile load vs. elongation ... **A9:** 684
Continuous-anneal process line (CAPL) technology ... **A4:** 58

Continuous-beam testing
ultrasonic inspection **A17:** 249
Continuous-belt sheet molding compound machines **EM1:** 159-160
Continuous-drive welding
titanium alloys **A6:** 522
Continuous-fiber composites, coatings for **A13:** 859, 861-862
Continuous-filament fiber composites *See* Fiber composites
Continuous-monorail hanger blast cleaning machine **A15:** 508
Continuous-rim resin-bonded wheels **A9:** 25
Continuous-strand mat **EM1:** 109, 169
Continuous-wave (CW) carbon dioxide (CO_2) lasers **A6:** 262, 263, 265, 266, 267
Continuous-wave (CW) lasers
optical holographic interferometry **A17:** 407
Continuous-wave gas lasers
Raman spectroscopy **A10:** 128
Continuous-wave NMR spectrometer
with field sweep and crossed coil detector **A10:** 283
Continuous-wave optical holographic interferometry **A17:** 410
Continuous-wave reflectometers
microwave inspection **A17:** 212-213
Continuous-wave spectrometers ... **A10:** 258, 283, 690
Continuous-welded cold-finished mechanical tubing **A1:** 335
Continuum
defined **A10:** 671
effects, electron probe x-ray microanalysis **A10:** 527-528
emission, intensity of **A10:** 83-84
overlap, as spectral interference in ICP- AES **A10:** 34
-source background correction, atomic absorption spectrometry **A10:** 51
-source systems, atomic absorption spectrometry **A10:** 52-53
x-rays, as inelastic scattering process **A10:** 433
Continuum mechanics **EM3:** 328
and multiaxial creep **A8:** 343
Continuum radiation
defined **A10:** 83
Contour band sawing
and blanking **A14:** 458
metal-matrix composites **A16:** 895
stainless steels **A16:** 705
Contour boring machines **A16:** 174
Contour dies
for rotary swaging **A14:** 132
Contour forging
of double bottleneck-shaped workpiece **A14:** 73
open-die **A14:** 71
shapes of **A14:** 61
Contour forming *See* Rollforming; Stretch forming; Tangent bending; Wiper forming
Contour, geodesic-isotensoid *See* Geodesic isotensoid contour
Contour maps
aluminum alloys **A14:** 436
Contour measurement
by coordinate measuring machines **A17:** 19
Contour plots
fractured titanium alloy **A12:** 172
stereo imaging **A12:** 171
Contour ring rolling **A14:** 117-120
Contour roll forming **A14:** 624-635
accuracy **A14:** 633
aluminum alloys **A14:** 795
auxiliary equipment **A14:** 627-628
auxiliary operations **A14:** 624
bending in **A14:** 624
categories **A14:** 624
computer-aided **A14:** 634-635
of copper and copper alloys **A14:** 818
Contour roll forming (continued)
defined **A14:** 624
lubricants for **A14:** 625
machines **A14:** 625-627
materials **A14:** 624
postcut method **A14:** 624-625
power and speed **A14:** 625
process variables **A14:** 625
quality **A14:** 633
stainless steels **A14:** 775-776
straightness **A14:** 632-633
surface finish **A14:** 633-634
and three-roll forming **A14:** 623
of titanium alloys **A14:** 846
tolerances **A14:** 632-634
tooling **A14:** 624, 628-630
of tube and pipe **A14:** 630-632, 673
Contour roll forming machines **A14:** 625-627
drive systems **A14:** 627
selection **A14:** 627
spindle support **A14:** 626
station configuration **A14:** 626-627
Contour rolling *See* Contour roll forming; Profile rolling
Contoured double-cantilever beam test
compared with wedge-opening load test and cantilever beam test **A8:** 538
for hydrogen embrittlement **A8:** 538-539
Contoured double-cantilevered beam test
for hydrogen embrittlement **A13:** 285-286
Contoured polishing wheels **M5:** 111
Contoured ring *See also* Ring rolling
production stages **A14:** 109
Contoured tape laying **EM1:** 631-635
automated machine development **EM1:** 631
future technology **EM1:** 634-635
machine features **EM1:** 631-633
machine programming **EM1:** 633-634
Contouring
in conjunction with drilling **A16:** 235
holographic **A17:** 408, 425-429
with image enhancement, digital **A17:** 457
in machining centers **A16:** 393
multifunction machining **A16:** 375
of surfaces, by acoustical holography **A17:** 447
Contours, curved
reflex, by metal casting **A15:** 40
Contracting
aluminum alloy **A14:** 804
Contracting geometry
rate law expression **EM4:** 55
Contraction
at crack tip **A11:** 51
hindered or irregular, as casting defect **A11:** 386
Contraction (solidification)
defined **A15:** 3
liquid-liquid **A15:** 598
liquid-solid **A15:** 599
in sand casting **A15:** 618
solid-solid **A15:** 598-599
Contraction of cast steels
Contrast
absorption, analytical electron microscopy **A10:** 444-445
atomic number, use in two-phase alloy analysis **A10:** 508
with backscattered electron detector **A10:** 502-504
between phases, revealed by scanning electron microscopy **A9:** 94
calculations for grain boundaries **A9:** 120
channeling, source of **A10:** 504
diffraction, analytical electron microscopy **A10:** 444-445
of dislocation pairs **A9:** 114-115
electron channeling **A10:** 504-506
emission, scanning electron microscopy **A10:** 502
image, scanning electron microscopy **A10:** 500-504
Contrast (continued)
of interfaces **A9:** 118-119
magnetic, as SEM special technique **A10:** 506
with magnetic particles **A17:** 100
mechanisms, analytical electron microscopy **A10:** 444-445
phase, analytical electron microscopy **A10:** 445
radiographic **A17:** 298-299
sample material influence on **A10:** 497
of single perfect dislocations **A9:** 114
voltage, as SEM special technique **A10:** 506-507
Contrast enhancement
by coating in scanning electron microscopy **A9:** 98-99
defined **A9:** 4
Contrast enhancement techniques
color images **A17:** 485
Contrast fliter
defined **A9:** 4
Contrast oscillations in transmission electron microscopy **A9:** 112
Contrast perception
defined **A9:** 4
Contrast scale
defined **A17:** 383
Contrast sensitivity
by image intensifiers, real-time radiography **A17:** 318-319
defined **A17:** 383
defined, radiography **A17:** 299
Contrast simulations
high resolution electron microscopy **A9:** 121
Contrast stretching
digital image enhancement **A17:** 456
in machine vision process **A17:** 33-34
Contrast-detail-dose diagram (CDD) **A17:** 375, 383
Contrasting by interference layers **A9:** 59
Contrasting chamber for sputtering of interference layers **A9:** 59-60
Control *See also* Production control; Quality- control **A8:** 392
degree of, in robust experimental design **A17:** 750
factors, quality design **A17:** 722
of inspection materials, and NDE reliability **A17:** 678
limits, and tolerances, in quality control assessments **A17:** 739
of parts, by machine vision **A17:** 40-41
in wear tests **A8:** 604, 606
Control arm ball joint ball **A7:** 617, 619
Control charts *See also* Quality control **EM3:** 792-797 **EM4:** 85-86, 87
c-chart, for number of defects **A17:** 736-737
for castings **A17:** 523
combined usage **A17:** 728
examples **A17:** 733-734, 736-737
exponentially weighted moving average (EWMA) **A17:** 732
for individual measurements **A17:** 732-734
initial, interpretation **A17:** 728
moving range **A17:** 732-734
p-charts, for fraction defective **A17:** 735-737
R-control charts, construction and interpretation **A17:** 725
revised, interpretation **A17:** 728
sample means, construction interpretation **A17:** 725
Shewhart model **A17:** 725-728
tests for **A17:** 731-732
u-chart, for number of defects per unit **A17:** 737
x control charts **A17:** 732-734
zone rules, for analysis **A17:** 730-732
Control equipment
for rolling mills **A14:** 354-355
Control modes
in cyclic torsional testing **A8:** 149
in fatigue experiments **A8:** 696
load/torsional moment **A8:** 149

Control modes (continued)
rotational angle **A8:** 149
strain control **A8:** 149
Control, process or production *See* Process control
Control, quality *See* Quality control; Quality design; Statistical methods
Control systems *See also* Process control
autoclave **EM1:** 704
for cold isostatic pressing **A7:** 445
temperature **M4:** 345-360
ultrasonic inspection **A17:** 254
Control valves
in servo-hydraulic test frames **A8:** 192
Control-rolled steels **A1:** 148
mechanical properties of **A1:** 409
Controlled atmosphere **A7:** 3
Controlled cooling **A1:** 272
steel wire rod **M1:** 253, 256, 257
Controlled energy flow forging machines **A14:** 29, 101
Controlled etching
defined **A9:** 4
Controlled expansion alloys
applications and properties **A2:** 441
nickel-base, applications and properties **A2:** 440
types **A2:** 443
Controlled impact test
abrasive strength **EM4:** 332
Controlled impedance *See also* Impedance; Impedance models
connector description **EL1:** 86
connector systems, high-frequency **EL1:** 87
high I/O connector, performance requirements **EL1:** 86
multilayer lamination **EL1:** 510
Controlled pore glass (CPG) **EM4:** 429-430
Controlled rolling **A1:** 115, 117-118, 131, 408-409, 587-588
conventional controlled rolling **A1:** 117, 409
defined **A9:** 4
dynamic recrystallization controlled rolling **A1:** 117-118, 409
mechanical properties of control-rolled steel **A1:** 409
of microalloyed bar **A1:** 587-588
of microalloyed plate **A1:** 586-587
recrystallization controlled rolling **A1:** 117, 409
temperature-time schedules **A1:** 130, 131
Controlled slow cooling
ductile iron **A15:** 657
Controlled spray deposition (CSD) **A16:** 60
Controlled spray deposition and hot working (CSD and Osprey) **A1:** 780
Controlled spray deposition process **A7:** 531-532
Controlled thermal severity test **A1:** 612-613
Controlled tilting experiment
for unknown phase/particle confirmation **A10:** 458
Controlled waveform spark
optical emission spectroscopy **A10:** 25, 26
Controlled-potential cells
for electrogravimetry **A10:** 199-200
Controlled-potential coulometry **A10:** 207-211
apparatus for **A10:** 208
applications **A10:** 207, 210-211
capabilities **A10:** 188, 197
and controlled-potential electrolysis **A10:** 208-210
defined **A10:** 671
electrometric titration and, compared **A10:** 202
estimated analysis time **A10:** 207
general uses **A10:** 207
introduction **A10:** 207, 210
limitations **A10:** 207
related techniques **A10:** 207
reversible processes **A10:** 208
samples **A10:** 207
technique **A10:** 210
Controlled-potential electrogravimetry
automatic potentiostat for **A10:** 200
Controlled-potential electrolysis
capabilities **A10:** 207, 208-210
as electrogravimetnc method **A10:** 199
Controlled-speed drawing machines **A14:** 334
Controlled-toughness high-strength alloys
fracture toughness of **A8:** 458
Controls
corrosion test **A13:** 195
and measurement electronics **EL1:** 567
wave soldering **EL1:** 702
Convection
in alloy additions **A15:** 71
and big bang theory **A15:** 131
cooling **EL1:** 522
defined **A15:** 3
and dendrite detachment **A15:** 131-132
effects in voltammetry **A10:** 189
forced, aluminum melt circulation **A15:** 453
formulas for **EL1:** 51-52
heat exchange, modeling of **A15:** 857
heat transfer coefficient, defined **A17:** 396
as heat transfer mode in boilers **A11:** 603
level, and insoluble particles **A15:** 146
level, liquid **A15:** 144
and liquid melt temperature field **A15:** 147-148
postfilling buoyant **A15:** 880-881
in reflow soldering **EL1:** 694
and solute redistribution **A15:** 148-149
in thermal inspection **A17:** 396
Convection coefficient **A6:** 1133
Convection continuous oven
paint curing process **M5:** 487-488
Convection during solidification
effect on grain growth in copper alloy ingots **A9:** 641
Convection heat transfer
in sintering **A7:** 341
Convection losses **A6:** 611
Convection processes
atomization **A7:** 48-49
Convection-dominant reflow soldering **A6:** 353
Convective flow
for on-eutectic growth **A15:** 150-151
Convective heat transfer gages **A4:** 508
Convective heating
for component removal **EL1:** 723
Convective mixing
statistical analysis **A7:** 187
Convenience applications
automotive hybrids **EL1:** 382
Conventional (structural) reinforcement
corrosion of **A13:** 1309
Conventional abrasives applications **EM4:** 331
bond type **EM4:** 331
hardness **EM4:** 331
mechanical properties **EM4:** 331
modulus of resilience **EM4:** 331
not for use in metal bonds **EM4:** 331
thermal properties **EM4:** 331
Conventional blanking dies *See also* Blanking
compound **A14:** 454
multiple **A14:** 455-456
progressive **A14:** 454-455
single-operation **A14:** 453-454
tool materials **A14:** 453
transfer **A14:** 455
Conventional controlled rolling (CCR) **A1:** 117, 408-409
Conventional cut gasket
design **EM3:** 54
Conventional forming
refractory metals and alloys **A2:** 562
Conventional hot extrusion *See also* Extrusion **A14:** 315-326
Conventional hybrids *See also* Hybrids
test procedures **EL1:** 372
Conventional IC packaging *See also* Integrated circuits (IC); Packaging
types/advantages **EL1:** 7
Conventional interconnection environments *See also* Connections; Fundamental interconnection issues; Interconnections
communication issues **EL1:** 4
physical interconnection hierarchy **EL1:** 2-4
Conventional load frames
crosshead movement **A8:** 192
grip design **A8:** 192-193
for high strain rate compression testing **A8:** 187
for high strain rate tension testing **A8:** 187
for medium strain rate compression testing **A8:** 192-193
screw-driven machines **A8:** 192
Conventional load frames (continued)
servo-hydraulic test frames **A8:** 192
Conventional logic
minimum device size for **EL1:** 2
Conventional memory integrated circuits **EL1:** 8
Conventional molding processes
plaster, sequence of operations **A15:** 243-245
types **A15:** 37
Conventional packaging *See* Packaging
Conventional pinch-type three-roll forming machines **A14:** 616-617
Conventional radiography *See* Radiography
Conventional strain *See* Engineering strain
Conventional stress *See* Engineering stress
Conventional strip galvanizing
as zinc coating **A2:** 527
Conventional transmission electron microscope (CTEM) **A18:** 380, 390-391
Convergent magnetic lenses
SEM **A12:** 167
Convergent-beam diffraction in transmission electron microscopy **A9:** 109-110
Convergent-beam electron diffraction
analytical transmission electron microscopy **A10:** 438-440
defined **A10:** 671
in diffraction-induced grain boundary migration **A10:** 462-464
patterns **A10:** 441, 439, 689
for phase diagram determination **A10:** 474
structural analysis by **A10:** 461-464
Convergent-beam electron diffraction (CBED) pattern **A18:** 386-387
Convergent-beam electron-diffraction pattern
abbreviation **A10:** 689
with analytical electron microscopy **A10:** 431, 439
Conversion coating *See* Chemical conversion coating; Chemical conversion coatings; specific types by name
Conversion coatings *See also* Chromate conversion coating; Chromate treatment; Coatings; Phosphate conversion coating; Phosphating; Protective coatings; specific type by name
aluminum anodizing **A13:** 396-398
as barrier protection **A13:** 378-379
for cast irons **A13:** 571 **M1:** 104
casting **A15:** 563
chemical, aluminum alloys **A15:** 763
chromate **A13:** 389-395
cleaning for **A13:** 380-382
corrosion protection **M1:** 754
defined **A13:** 4
magnesium/magnesium alloys **A13:** 751
phosphate compounds in **A13:** 383
porcelain enamel **A13:** 446-452
for shafts **A11:** 482
types **A13:** 378-379
Conversion electron Mössbauer scattering
austenite result of **A10:** 294
Conversion factors
common uniform corrosion rate units **A13:** 229
for leak testing **A17:** 57
metric conversion guide **A11:** 794
Conversion guide
metric **A1:** 1035-1037 **A2:** 1270-1272 **A11:** 793-795 **A14:** 941-943 **A15:** 893-895 **A17:** 755-757 **EL1:** 1163-1165 **EM1:** 945-947 **EM2:** 847-849
Conversion oxide film replicas
formation **A12:** 181
Conversion processes
carbon fiber **EM1:** 112-113
Conversion screens
for neutron radiography **A17:** 387, 391
Conversion tables
hardness **A8:** 109-113
Converter
current/voltage **A10:** 199
magnetostrictive **A8:** 244
microchannel plate-image, field ion microscopy **A10:** 584
piezoelectric, in ultrasonic hardness tester **A8:** 101
Converters
Bessemer **A15:** 32-33
metallurgy of **A15:** 426-431
steel production by **A15:** 31

Converters (continued)
Tropenas **A15:** 33
vacuum oxygen decarburization **A15:** 430
Converters, hybrid digital-to-analog
corrosion failure analysis **EL1:** 1115
Convex figure
basic quantities for **A12:** 194
Convex fillet weld
definition **M6:** 4
Convex particles *See also* Particles
properties of **A9:** 133
relationships between in their projection **A9:** 134
in space **A9:** 134
in their sections **A9:** 134
Convex root surface
definition **M6:** 4
Convex surfaces
electropolishing of **A9:** 55
Convexity
definition **A6:** 1208 **M6:** 4
Conveyor chains
corrosive wear of **M1:** 637
Conveyor loaders
as transfer equipment **A14:** 501
Convolution(s)
computed tomography (CT) of **A17:** 383
and fibers, filtered-backprojection technique **A17:** 380
filters, for image enhancement **A17:** 459
Convolutional integral
ESR line and **A10:** 261
Cook-Norteman process
zinc-base coatings **A13:** 526
Cookware **EM4:** 4
absorption **EM4:** 4
applications **EM4:** 4
composition **EM4:** 5
glazing **EM4:** 4
process **EM4:** 4
products **EM4:** 4
properties **EM4:** 4
Cookware, cast iron
coatings for **M1:** 106
Cool components
gas-turbine **A11:** 284
Cool time
definition **M6:** 4
Coolant leakage
by shrinkage porosity **A11:** 355, 357
Coolant-system assembly, radar
brazed, joint failure **A11:** 452-453
Coolants *See also* Cutting fluids; Superconducting materials
for consumable abrasive wheel cutting **A9:** 24
for diamond wheel cutting **A9:** 25
for machining of magnesium and magnesium alloys **A2:** 475
for power spinning **A14:** 604
for sawing **A9:** 23
Coolants, liquid-vapor metal
corrosion effects **A13:** 93
Coolers, air
hydrogen, and oil, finned tubing for **A11:** 628
Coolidge process
incandescent lamp filaments **A7:** 16-17
Coolidge x-ray tubes
in wavelength-dispersive x-ray spectrometers **A10:** 88
Cooling *See also* Continuous cooling; Cooling curve analysis; Cooling curves; Cooling rate; Cryogenic cooling; Heat removal; Immersion cooling; Slow cooling; Spiral mold cooling; Supercooling; Undercooling
analysis, thermoplastic injection molding **EM2:** 312
auxiliary, permanent mold casting **A15:** 283
blow molding **EM2:** 355-356
board-level **EL1:** 47

Cooling (continued)
boiling **EL1:** 364
in cemented carbide sintering **A7:** 388-389
centerline rates, uranium alloys **A2:** 679
channels, NTT **EL1:** 310
chills for **A11:** 345
component-level **EL1:** 47
continuous **A15:** 127
continuous, effect on flow stress vacuum-melted iron **A8:** 177
controlled slow **A15:** 242, 657
convection **EL1:** 522
design trade-offs
deviation, from equilibrium solidification **A15:** 183
devices, for sand
in die casting, analysis **A15:** 293-294
die, in hot upset forging **A14:** 87
of dry sand castings **A15:** 228
effect on solidification structures of steel **A9:** 623-624
of forged heat-resistant alloys **A14:** 235
galling caused by **A11:** 366
heat sinks used in **EL1:** 414
heat transfer coefficients **EL1:** 24
importance during deformation **A8:** 178
induction furnace **A15:** 371
intermetallic inclusions during **A15:** 95
interrupted, ductile iron **A15:** 657
liquid, ultrasonic testing **A8:** 247-248
in lost foam casting **A15:** 231
magnesium alloy forgings **A14:** 260
methods, effect on mold life **A15:** 281
of mounting presses **A9:** 30
multichip technology **EL1:** 307-309
nickel-base forgings **A14:** 263
of nuclear waste, acoustic emission inspection **A17:** 281
practice, closed-die forging **A14:** 82
rate, iron-copper-carbon alloys **A14:** 200
rate, stainless steels **A14:** 229
of sand **A15:** 348-350
scanning electron microscopy study of specimens **A9:** 97
scheme, selection **EL1:** 50
selective forced, for stress and distortion **A15:** 616
slow, of irons **A11:** 360
splat, aluminum P/M alloys **A2:** 201-202
stresses **EM2:** 751
surfaces **EL1:** 310
symbol for temperature at which ferrite transforms upon **A8:** 724
systems, conventional and ultrasonic fatigue testing **A8:** 246-248
techniques **EL1:** 23-24
techniques, for thermal design **EL1:** 413-414
in thermoforming **EM2:** 400
in thermoplastic extrusion **EM2:** 382
titanium alloys **A12:** 454
in tungsten and molybdenum sintering **A7:** 391
in tungsten heavy alloy sintering **A7:** 393
under-, effect on bearing cap **A11:** 350
uneven, shape distortions from **A11:** 266
wafer-scale integration (WSI) **EL1:** 363-364
water, electric arc furnace **A15:** 359
water, permanent steel molds **A15:** 304
x-ray tubes **A17:** 306
Cooling and coiling system
precipitation **A1:** 119-120, 123
in processing of solid steel **A1:** 118-119, 121
Cooling channels
defined **EM2:** 11
Cooling cracks *See also* Flakes; Thermal cracks
revealed by macroetching **A9:** 173
Cooling curve
defined **A9:** 4

Cooling curve analysis *See also* Cooling curves; Thermal analysis
cast iron **A15:** 180
interpretation and use **A15:** 182-185
Cooling curves *See also* Cooling curve analysis; Thermal analysis .... **A3:** 1 • 15, 16, 17
effect of oxidation **M4:** 50, 62
interpretation and use **A15:** 182-185
for master alloy processing **A15:** 108
and phase diagrams, relationship **A15:** 182
quenching of steel **M4:** 32-34
stages **M4:** 33
Cooling fixture
defined **EM2:** 11
Cooling, rapid
martensite formation **A13:** 47
Cooling rate *See also* Heating rate
as a beta stabilizer in titanium alloys **A9:** 459
as a conductive coating for scanning electron microscopy specimens **A9:** 97
as a reactive sputtering cathode material **A9:** 60
in aluminum powder metallurgy alloys **A9:** 511
as an addition to nickel-iron alloys **A9:** 538
as an addition to permanent magnets **A9:** 538
as an addition to precipitation-hardenable stainless steels **A9:** 285
as an addition to tin-zinc alloys, effects on microstructure **A9:** 452
bearing alloys **A2:** 553
bright-field bend contours **A9:** 113
cell voltage as a function of anode current density in electropolishing of **A9:** 48
chills and antichills, effects **A15:** 283
color etching **A9:** 141-142
of compacted graphite irons **A1:** 9
current-voltage relation in electropolishing of **A9:** 48
defined **A9:** 4
diffraction contrast of a single dislocation **A9:** 113
in diffusion-alloyed steel powder metallurgy materials **A9:** 510
dislocation loops **A9:** 116
dislocation pairs **A9:** 115
dislocations in single crystal **A9:** 689
of ductile iron **A1:** 8
effect of addition on lead-antimony-tin microstructures **A9:** 417
effect on carbon steel casting microstructures **A9:** 231
effect on dendritic structure in copper alloy ingots **A9:** 637-640
effect on gray iron microstructure **A9:** 245
effect on massive transformation **A9:** 655
effect on microstructure of austenitic manganese steel castings **A9:** 237
effect on microstructure of titanium and titanium alloys **A9:** 460-461
effect on nonequilibrium constituents in aluminum alloy ingots **A9:** 634
effect on peritectic reactions **A9:** 676
effect on precipitation strengthening **A1:** 402
effects, copper casting alloys **A2:** 350
electrical resistivity of, changes during isothermal recovery **A9:** 693
electroplated onto sleeve bearing liners **A9:** 567
etch pits due to dislocation lines **A9:** 127
fatigued, deformation marks **A9:** 100
grain-nucleating particles in lead alloy **A9:** 418
of gray iron **A1:** 6-7 **A15:** 634
influence, Maurer diagram **A15:** 69
lead and lead alloys **A2:** 545
line etching **A9:** 62
magnetically soft materials **A2:** 763
of malleable iron **A1:** 10

Cooling rate (continued)
microstructural deformation modes as a function of strain **A9:** 686
in nickel steel powder metallurgy materials **A9:** 510
oriented dislocation arrays in thin foil **A9:** 128
plastic deformation, modes of **A9:** 686
plastic deformation, sequence of **A9:** 693
polishing pure powder materials **A9:** 507
powder metallurgy materials, etching **A9:** 509
relationship of ductility and volume fraction of dispersions in **A9:** 125
single crystal sphere, oxidized **A9:** 137
stacking-fault tetrahedra **A9:** 117
thickness contours, bright field and dark field compared **A9:** 112
in zinc alloys **A9:** 489
Cooling rates
effect on iron-graphite powders **A7:** 365, 366
in forced convection cooling **A7:** 46
hardenability related to **M1:** 471, 481-488, 491, 492
in sintering **A7:** 370
Cooling stress *See also* Thermal stress
and aging effects **EM2:** 751
Cooling stresses
defined **A15:** 3
Cooling water
analysis, nuclear reactors **A13:** 958
jackets **A13:** 1139
polluted, copper/copper alloys in **A13:** 625
systems **A13:** 487, 1134-1135
Cooling-rate equation **A6:** 10, 14-15
Cooperative Test Program **A8:** 285
Coordinate measuring machines (CMMS) **A17:** 18-28
applications **A17:** 20
bridge-type **A17:** 21
cantilever-type **A17:** 20-21
components **A17:** 23-26
error characterization by interferometer **A17:** 15
fixed-table type **A17:** 23
gantry CMMs **A17:** 21
horizontal CMMs **A17:** 21-23
implementation **A17:** 27-28
measurement techniques **A17:** 19
moving-ram type **A17:** 23
moving-table type **A17:** 23
operating principles **A17:** 18-20
performance factors **A17:** 26-27
specifications **A17:** 21
types **A17:** 20-23
vertical CMMs **A17:** 21
Coordinate systems
coordinate measuring machines **A17:** 19
Coordinate tolerancing
of parts **A15:** 622-623
Coordinates
transformation of **EM1:** 459
Coordinating European Council
performance testing of engine oils **A18:** 170
Coordination catalysis *See also* Ziegler-Natta catalysts **EM3:** 9
defined **EM2:** 11
Coordination compound *See also* Chelate; Complexation; Ligand
defined **A10:** 671 **A13:** 4
Coordination geometry
determination of **A10:** 354
effects on XANES spectrum **A10:** 415
Coordination number
defined **A7:** 296 **A10:** 671
Coors Si/SiC, SC-2 (RBSC)
properties **EM4:** 240
Cope *See also* Ceramic molding; Drag; Flask; Mold; Pattern
defined **A15:** 189, 203
and drag molding machines **A15:** 342-343
and drag patterns **A15:** 190
Cope defect
casting **A11:** 384
Cope spall
as casting defect **A11:** 385
Cope-side defect
ductile irons **A15:** 94
Copier machine parts of stainless steel powders **A7:** 732
Copier powders **A7:** 580-588
carrier core composition **A7:** 584-585
charge polarity **A7:** 583-584
coating processes **A7:** 587-588
development processes **A7:** 582-584
electrostatic history **A7:** 581
evolution **A7:** 580-582
irregularly shaped **A7:** 582
manufacturing process **A7:** 586-587
P/M parts for **A7:** 667
properties **A7:** 585
recycling **A7:** 582
spherical, estimated consumption **A7:** 582
Coplanarity
lead **EL1:** 731
Copolymer *See also* Polymer(s) **EM3:** 9
bismaleimides as **EM1:** 78
blending, BT resins **EM1:** 79
block, microphase separation by SAXS/ SANS/SAS **A10:** 402
defined **EM1:** 7
formation **A13:** 404
ratios, NMR determination of **A10:** 277
Copolymer(s) *See also* Acetals; Copolymer acetals; Polymers; Styrene copolymers
chemistry of **EM2:** 63
compatibilities, styrene-acrylonitriles (SAN, OSA, ASA) **EM2:** 216
defined **EM2:** 11
environmental effects **EM2:** 429
polyphenylene ether blends (PPE, PPO) **EM2:** 183
styrene-acrylonitriles **EM2:** 214-216
styrene-maleic anhydride (SMA) **EM2:** 66
types **EM2:** 58
Copolymerization *See* also **EM3:** 9
bismaleimides **EM1:** 78-81
molecular **EM2:** 58
Copper *See also* Atomized copper powders; Brasses; Bronzes; Copper alloy castings; Copper alloy strip; Copper alloys; Copper alloys, Copper alloys, specific types; Copper alloys, specific types; Copper contact alloys; Copper P/M products; Copper powders; Copper recycling; Copper-base structural parts; Copper-based powder metals; Copper-matrix composites; Copper-nickel alloys; Copper-zinc alloys; Electrolytic tough pitch (ETP) copper; Nickel silvers; Nickel silvers Tin brasses; Pure copper; specific copper alloys; Tumbaga; Wrought copper and copper alloys **A13:** 610-640 **A14:** 809-824 **M2:** 239-247
in abrasion-resistant cast irons **A1:** 115
abrasive blasting **M5:** 614
abrasive wear **A18:** 188, 189
in active metal process **EM3:** 305
addition to aluminum-base alloys **A18:** 752
addition to cylinder liner materials for strength **A18:** 556
addition to epoxy adhesives **EM3:** 515
additions, effect in P/M stainless steel **A13:** 832
adherends
alkaline-chlorite etch **EM3:** 269-270
Ebonol C etch **EM3:** 269-270
ferric chloride-nitric acid solution ... **EM3:** 269-270
adhesion and solid friction **A18:** 32, 33
adhesion, in copper-clad E-glass laminates **EL1:** 535-536
air-shotted **A7:** 106, 107
alkaline cleaning of **M5:** 617-620
in alloy cast irons **A1:** 89
alloy strip, stress-strain curve for **A8:** 134
alloying, aluminum casting alloys **A2:** 132
alloying in aluminum alloys **M6:** 373
alloying, in cast irons **A13:** 567
alloying, magnetically soft materials **A2:** 762
alloying, nickel-base alloys **A13:** 641
alloying, wrought aluminum alloy **A2:** 47-48
in aluminum alloys **A15:** 745
aluminum alloys containing **A13:** 592
in aluminum-silicon alloys **A18:** 786, 787, 789
analysis in presence of lead, cell for **A10:** 199
anneal-resistant **M2:** 241, 242, 243
annealing **M2:** 241
furnace atmospheres **A4:** 549, 550
Copper (continued)
with precious metals **A4:** 939, 940-941, 946
anodized **EM3:** 417
applications **A2:** 239-240 **A7:** 205
arc welding *See* Arc welding of copper and copper alloys
in archaeological samples **A10:** 186
arsenical, in Bronze Age **A15:** 15-16
atmospheric corrosion **A13:** 82, 621
Auger electron spectroscopy application **A18:** 452, 453
in austenitic manganese steel **A1:** 825
axial force dependence on temperature in **A8:** 181
backing bars for aluminum alloys **M6:** 382-383
bare, solder mask over **EL1:** 550
base metal solderability **EL1:** 677
bearing lubricants, nuclear reactors **A13:** 964
-bearing steel **A13:** 516
beryllium, Rockwell scale for **A8:** 76
biologic effects and toxicity **A2:** 1251-1252
bonded by polyamides and polyesters **EM3:** 82
brass plating bath content **M5:** 285-286
brazeability **M6:** 1033
brazeability of **A11:** 450
brazing *See* Brazing of copper and copper alloys
brazing, furnace atmosphere **A4:** 548, 552
for brazing in ceramic/matrix seals **EM4:** 504, 505, 506
brazing properties **M6:** 966-967
Brinell test load for **A8:** 84
bronze plating bath content **M5:** 288-289
buffing *See* Copper, polishing and buffing of
butt welding **M6:** 674
cadmium plating of **M5:** 622
calculated electron range **EL1:** 1095
capacitor discharge stud welding **M6:** 738
in carbon-graphite materials **A18:** 816
cast, applications and properties **A2:** 224-228
as cast in plaster molds **A15:** 243
in cast iron **A1:** 6, 28
catalytic effect on oxidation of polyolefins **EM3:** 418
cementation **A7:** 54
characteristics **M6:** 400
chemical analysis **A7:** 248
chemical and electrolytic cleaning of **M5:** 617-619
chemical and electrolytic polishing of **M5:** 623-624
chemical brightening baths, use in, effect of **M5:** 579-580
chemical pipe sealants for plumbing **EM3:** 608
chemical resistance **M5:** 4
chip combustion accelerators **A10:** 222
chlorine corrosion **A13:** 1170-1171
chromium plating baths contaminated by **M5:** 173-174
chromium plating of **M5:** 171, 622
cleaning processes *See also* specific processes by name **M5:** 611-621
for coating in ceramic/metal seals **EM4:** 504
coatings, bonded-abrasive grains **A2:** 1015
coil liners for inductor coils **A4:** 174
coinability of **A14:** 183
cold extrusion of **A14:** 310
cold working **A2:** 219, 223
color **A2:** 219
as colorant **EM4:** 380
coloring process **M5:** 611-612, 624-626
in compacted graphite iron **A1:** 57, 59
in composite metals, history **A7:** 17
in composition, effect on ductile iron **A4:** 686, 688, 689
in composition, effect on gray irons **A4:** 671, 673, 681
compositions and properties **M6:** 401
compositions for various types **M6:** 546-547
as conductive, thick-film material **EL1:** 249
conductor inks **EL1:** 208
for conductors **EL1:** 114
conductors on printed boards **EM3:** 591
constant-current electrolysis **A10:** 200
contact angle with mercury **A7:** 269
contamination **M6:** 321
contamination, in low-carbon steel **A12:** 249
content in stainless steels **M6:** 320

Copper (continued)
content in steels, atmospheric corrosion effects **A13:** 514
controlled-potential electrogravimetry of **A10:** 200
cooling in ultrasonic testing **A8:** 247
and copper alloys **A15:** 771-785
copper metals, production **A2:** 237-239
corrosion by some epoxy adhesives **EM3:** 37
corrosion fatigue test specification **A8:** 423
corrosion, freshwater **A13:** 621
corrosion in **A11:** 201
corrosion ratings **A2:** 353-354
corrosion resistance **A2:** 216 **A13:** 610 **M2:** 239-240
cracks, from low-stress sliding **A8:** 603
critical angles for cutting **A18:** 185
critical relative humidity **A13:** 82
critical surface tensions **EM3:** 180
cross-wire projection welding **M6:** 518
in Cu-Ni, thermal spray coating material **A18:** 832
in Cu-Ni-In, thermal spray coating material **A18:** 832
damage dominated by chip formation **A18:** 179, 180
debris effect on wear **A18:** 249
decorative colors and finishes **M2:** 240
in dental amalgam **A18:** 669
deoxiders **A2:** 236-237
deposition **A10:** 198, 199
deposition in gas metal welding arc **A11:** 721
determination using internal electrolysis **A10:** 200
determined by controlled-potential coulometry **A10:** 209
determined in coal fly ash **A10:** 147
diffusion welding **M6:** 677-678
dimple formation **A12:** 173
dispersion-strengthened **A7:** 711-716
dominant texture orientations **A10:** 359
in ductile iron **A1:** 44
dwell pressures for cold isostatic pressing **A7:** 449
dynamically recrystallized grain size Zener-Hollomon parameter **A8:** 175
effect of arsenic, phosphorus, antimony, and silicon on SCC of **A11:** 221
effect of, on hardenability **A1:** 393, 395
effect of, on notch toughness **A1:** 741
effect of particle size on apparent density **A7:** 273
effect on hot workability **A11:** 722
effect on maraging steels **A4:** 222
effect on thermal conductivity **EM3:** 620, 621
effect, weathering steels **A13:** 515
effect, wrought/cast aluminum alloys **A13:** 586
effects of aluminum, nickel, tin, and zinc on SCC of **A11:** 221
electrical conductivity **A2:** 219 **EM3:** 596
electrical contacts, use in **M3:** 665-666
electrical coppers **A2:** 223, 230, 234
electrical resistance applications **M3:** 641
as electrically conductive filler **EM3:** 178
as electrode **A10:** 184
electrodeposited coatings **A13:** 427
in electroforming **EM3:** 306
electroless, as additive process **EL1:** 548
electrolytic cleaning of **M5:** 618-620
electrolytically deposited/rolled-annealed **EL1:** 581
electron beam welding **M6:** 642
in electroplated coatings **A18:** 838
electroplated on surface for sintered metal powder process **EM3:** 305
electroplating for SEM specimen preparation **A18:** 380-381
electroplating of bearing materials **A18:** 756
electropolishing of **M5:** 305, 308

Copper (continued)
embrittlement **A13:** 178-179
embrittlement by **A11:** 235, 721 **A13:** 180
emulsion cleaning of **M5:** 618
in enameling ground coat **EM3:** 303, 304
enamels **EM3:** 302, 303
end-use applications **A2:** 239
erosion mechanisms **A18:** 202, 203
erosion-corrosion in **A11:** 201-202
as essential metal **A2:** 1251-1252
estimated in copper-manganese alloy **A10:** 201
ethylene-vinyl acetate (EVA) coatings for **EM3:** 411
evaporation fields for **A10:** 587
excitation and emission for photoejection of electrons **A10:** 85, 87
explosion welding **M6:** 713
electrical applications **M6:** 714
exposure limits **A7:** 205
extruded parts **A14:** 310-311
extrusion welding **M6:** 676-677
fabrication, ease of **A2:** 219
fabricators **A2:** 238
fatigue life as function of stress amplitude in **A8:** 253
in ferrite **A1:** 406-407
fiber for reinforcement **A18:** 803
fibers as filler **EM3:** 178
filler for conductive adhesives **EM3:** 76
filler for polymers **A7:** 606
in filler metals for active metal brazing **EM4:** 523-524, 526, 529, 530
in filler metals used for direct brazing **EM4:** 517-518, 519
films, LEISS study of oxidation of **A10:** 609
finished, cracking problems analyzed by inert gas fusion **A10:** 231-232
finishing processes *See also* specific processes by name **M5:** 621-627
fixturing for induction brazing **M6:** 971-972
for flame head for oxy-fuel gas flame heating **A4:** 274
flow stress, strain-rate history effect **A8:** 179
forging of **A14:** 255-258
forgings, burning of **A11:** 332, 334
forming, lubricants for **A14:** 519
forming of **A14:** 809-S24
-free aluminum alloys **A13:** 593
free machining **A12:** 401
fretting wear **A18:** 248
friction coefficient data **A18:** 71, 72
galvanic corrosion with magnesium **M2:** 607
gas metal arc welding **M6:** 153, 417-418
of deoxidized coppers **M6:** 418
of oxygen-free coppers **M6:** 418
to aluminum bronze **M6:** 422
to copper nickels **M6:** 422
of tough pitch coppers **M6:** 418
gas tungsten arc welding **M6:** 405-408
of deoxidized coppers **M6:** 406
of oxygen-free coppers **M6:** 407
of tough pitch coppers **M6:** 407
in gas-metal eutectic method **EM3:** 305
gases in **A15:** 86
gasket material **A18:** 550
gaskets, for fatigue test chamber **A8:** 412
glass coatings referred to as enamels **EM3:** 301-302
glass-ceramic/metal seals **EM4:** 499
glass-to-metal seals **EL1:** 455 **EM3:** 302
as gold alloy **A2:** 690
gold plating of **M5:** 621-623
gold plating, uses in **M5:** 283
grain-boundary embrittlement, low-carbon steel **A12:** 246
gravimetric finishes **A10:** 171
as gray iron alloying element **A15:** 639
ground, reduction of oxidized **A7:** 107-110
hardness conversion tables **A8:** 109

Copper (continued)
heat treatment **A2:** 223
high fatigue strength **A18:** 743
high frequency resistance welding **M6:** 760
high-conductivity, degassing **A15:** 469
high-purity **A12:** 399-400
high-purity, SSMS analysis **A10:** 144
high-resolution electron microscopy to study sliding wear **A18:** 389
hot extrusion of **A14:** 322
hot working **A2:** 223
hydrochloric acid corrosion **A13:** 1163
hydrogen determined in **A10:** 231-232
hydrogen embrittlement **M2:** 239-240
hydrogen solubility **A15:** 466
ICP-determined in plant tissues **A10:** 41
ICP-determined in silver scrap metal **A10:** 41
impurity in solders **M6:** 1072
-induced LME in steels **A11:** 232, 233
induction heating energy requirements for metalworking **A4:** 189
induction heating temperatures for metalworking processes **A4:** 188
industry structure **A2:** 238-239
infiltration **A7:** 457
ingot, spider cracks **A10:** 302
inhibitors **A13:** 497
injection nozzle material for laser melt/particle injection **A18:** 868
inspection methods **A17:** 534
introduction to **A2:** 216-240
ion-scattering spectra from **A10:** 604
iron in, phase analysis **A10:** 294
in iron-base alloys, flame AAS analysis **A10:** 56
lamination background color, optical testing **EL1:** 571
lap welding **M6:** 673
laser beam welding **M6:** 647
as lead frame material **EL1:** 731
in lead-base alloys **A18:** 750
LEISS study of oxidation, using oxygen **A10:** 609
liquid, effect of hydrogen and steam pressures **A7:** 117
liquid, standard Gibbs free energies for solution in **A15:** 60
liquid-metal embrittlement of **A11:** 233-234
long-term corrosion, testing **A13:** 195
lubricant indicators and range of sensitivities **A18:** 301
in magnesium alloys, by hydrobromic acid-phosphoric acid method **A10:** 65
malleability **A15:** 26
mass absorption coefficient vs x-ray energy **A10:** 85, 87
mass finishing of **M5:** 614-616
material to which vitreous solder glass seal is applied **EM4:** 1070
materials for conductors **EM4:** 1142
as matrix material, A15 superconductors **A2:** 1064
matrix materials, niobium-titanium superconducting materials **A2:** 1045
mechanical working **A2:** 219, 223 **M2:** 241
melting of **A7:** 106
for metal core molding **EM3:** 591
in metal powder-glass frit method **EM3:** 305
in metal-matrix composites **A18:** 808-809, 810
metallization, for multichip structures **EL1:** 303-304
for metallizing **EM4:** 542, 54
metallochrome color, use in titration **A10:** 174
microhardness traverses **A8:** 230
microstructure of compacted blend **A7:** 314
mills, prompt gamma activation analysis use in **A10:** 240
mining, prompt gamma activation analysis use in **A10:** 240

SUBJECTS OF THE INDEXED VOLUMES: ASM Handbook (designated by the letter "A"): **A1:** Properties and Selection: Irons, Steels, and High-Performance Alloys (1990); **A2:** Properties and Selection: Nonferrous Alloys and Special-Purpose Materials (1990); **A3:** Alloy Phase Diagrams (1992); **A4:** Heat Treating (1991); **A6:** Welding, Brazing, and Soldering (1993); **A7:** Powder Metallurgy (1984); **A8:** Mechanical Testing (1985); **A9:** Metallography and Microstructures (1985); **A10:** Materials Characterization (1986); **A11:** Failure Analysis and Prevention (1986); **A12:** Fractography (1987); **A13:** Corrosion (1987); **A14:** Forming and Forging (1988); **A15:** Casting (1988); **A16:** Machining (1989); **A17:** Nondestructive Testing and Quality Control (1989); **A18:** Friction, Lubrication, and Wear Technology (1992). **Metals Handbook, 9th Edition** (designated by the letter "M"): **M1:** Properties and Selection: Irons and Steels (1978); **M2:** Properties and Selection: Nonferrous Alloys and Pure Metals (1979); **M3:** Properties and Selection: Stainless Steels, Tool Materials and Special-Purpose Materials (1980); **M4:** Heat Treating (1981); **M5:** Surface Cleaning, Finishing, and Coating (1982); **M6:** Welding, Brazing, and Soldering (1983). **Engineered Materials Handbook** (designated by the letters "EM"): **EM1:** Composites (1987); **EM2:** Engineering Plastics (1988); **EM3:** Adhesives and Sealants (1990); **EM4:** Ceramics and Glasses (1991); **Electronic Materials Handbook** (designated by the letters "EL"): **EL1:** Packaging (1989).

Copper (continued)
as minor element, ductile iron ... **A15:** 648
in moist chlorine ... **A13:** 1173
molds, horizontal centrifugal casting ... **A15:** 296
mutual solubility of ... **A11:** 452
nickel plating bath contamination by ... **M5:** 208-210
nickel plating of ... **M5:** 207, 215-216
in nickel-chromium white irons ... **A15:** 680
OFHC, friction coefficient data ... **A18:** 71
OFHC, shear stress-strain curve ... **A8:** 216-217
organic coating of ... **M5:** 621, 626-627
oxidation using ^{18}O ... **A10:** 609
oxygen cutting, effect on ... **M6:** 898
oxygen effect on corrosion rates ... **A13:** 627
oxygen-free high-conductivity ... **A12:** 401
oxygen-free, high-conductivity (OFHC)
for coil construction ... **A4:** 180
cold working effect on mechanical properties ... **A4:** 827
oxygen-free, impurities effect on conductivity ... **A7:** 106
in P/M alloys ... **A1:** 809
P/M parts lubrication ... **A7:** 192, 193
particles, partially oxidized and reduced ... **A7:** 108, 110
passivation of ... **M5:** 624
penetration failures, in locomotive axles ... **A11:** 715-727
penetration, x-ray elemental dot map of ... **A11:** 720
percussion welding ... **M6:** 740, 745
phase structure influence on microstructure ... **EM3:** 418
photoejection of electrons in ... **A10:** 85
photometric analysis methods ... **A10:** 64
physical properties ... **M6:** 546-547
physical properties related to thermal stresses ... **A4:** 605
pickling and bright dipping ... **M5:** 611-615, 619-620
pickling solutions contaminated by ... **M5:** 80
pigments ... **A7:** 595-596
pipe, erosion pitting from river water ... **A11:** 189
plasma arc cutting ... **M6:** 915
plating procedures ... **M5:** 619-624
electroless process ... **M5:** 621-622
electroplating ... **M5:** 622-624
immersion process ... **M5:** 622
surface preparation ... **M5:** 619-621
plating, uses ... **EL1:** 679
polishing and buffing of ... **M5:** 615-617
polycrystalline, crack growth data ... **A8:** 255
polyethylene adhesion ... **EM3:** 414
polyethylene coatings ... **EM3:** 412
polyimide adhesion ... **EM3:** 158
Pourbaix (potential-pH) diagram ... **A13:** 28
prebond treatment ... **EM3:** 35
precipitation strengthening with ... **A1:** 411
in precipitation-hardening steels ... **M6:** 350
prehistoric use ... **A15:** 15
in pressed diffusion joining ... **EM3:** 306
pressurized water reactor specification ... **A8:** 423
processing technique ... **A18:** 803
properties ... **A18:** 803
properties of importance ... **A2:** 216, 219
protective film ... **A13:** 82
pure ... **M2:** 726-733
pure, direct-chill casting ... **A15:** 314
pure, melt treatment ... **A15:** 774
pure, properties ... **A2:** 1110
pure, structural parts application ... **A7:** 105
qualitative tests to identify ... **A10:** 168
radiation ... **A10:** 326
radiographic absorption ... **A17:** 311
rare earth alloy additives ... **A2:** 729
recovery in copper foil ... **A10:** 200
recycling ... **A2:** 1213-1216
resistance brazing ... **M6:** 976
resistance seam welding ... **M6:** 494
resistance welding *See* Resistance welding of copper and copper alloys
rhodium plating of ... **M5:** 623
roll welding ... **M6:** 676
rolled, Euler space ... **A10:** 363
room-temperature torsion flow curves and plane-strain compression tests on ... **A8:** 162-164
rubber bonding cross-link density ... **EM3:** 418

Copper (continued)
sampling practices ... **A7:** 249
scratch brushing of ... **M5:** 616
screens, radiography ... **A17:** 316
seal adhesive wear ... **A18:** 549
segregation and solid friction ... **A18:** 28
separation, from zinc ... **A10:** 198
serpentine glide formation ... **A12:** 17
as SERS metal ... **A10:** 136
shear stress-strain curves ... **A8:** 229, 231
shielded metal arc welding to aluminum ... **M6:** 425
silver plating of ... **M5:** 617, 621-623
sintering time and temperature ... **M4:** 796
slide welding ... **M6:** 674
slugs for correcting flux patterns of inductor coils ... **A4:** 176, 177
soft, yield strength of ... **A8:** 21
as solder impurity ... **EL1:** 637-638
soldering ... **M6:** 1075
solid particle erosion ... **A18:** 388-389
solid-state sintering ... **EM4:** 273
solution potential ... **M2:** 207
solvent cleaning of ... **M5:** 617-618
sources ... **A7:** 205
species weighed in gravimetry ... **A10:** 172
spectrometric metals analysis ... **A18:** 300
spraying for hardfacing ... **M6:** 789
spring-tempered alloy strip, bending moment-deflection data ... **A8:** 134
stacking fault energy ... **A18:** 715
in stainless steels ... **A18:** 712, 713
in steel ... **A1:** 145, 577
strain-hardening exponent and true stress values for ... **A8:** 24
stress-corrosion cracking ... **M2:** 239-240
stress-corrosion cracking in ... **A11:** 220-223
strip combustion accelerators ... **A10:** 222
substrates ... **EL1:** 307
suitability for cladding combinations ... **M6:** 691
sulfuric acid corrosion ... **A13:** 1153
superplasticity of ... **A8:** 553
supply and reserves ... **A2:** 240
surface fatigue damage in ... **A8:** 603
surface, oxidized, LEISS spectra from ... **A10:** 609
surface preparation ... **EM3:** 259
tarnish removal ... **M5:** 613, 619
temper designations ... **A2:** 223 **M2:** 248-251
temperature effect, corrosion rate ... **A13:** 910
tension and torsion flow curves at room temperature ... **A8:** 162-164
texture dependence on temperature ... **A8:** 181-182
thermal conductivity ... **A2:** 219
thermal properties ... **A18:** 42
thermocompression bonding with ... **EL1:** 734
thermocompression welding ... **M6:** 674-675
threshold stress intensity ... **A8:** 256
tilt pins, in plate impact testing ... **A8:** 234
as tin solder impurity ... **A2:** 520
tin, tin-lead, and tin-copper alloy plating of ... **M5:** 621-624
in tin-base alloys ... **A18:** 749
TNAA detection limits ... **A10:** 237
to reduce electromigration ... **EM3:** 581
in torsion, effect of increasing or decreasing strain rate on flow stress ... **A8:** 177-178
torsionally prestrained, tensile fracture in ... **A8:** 156
toxicity ... **A7:** 204-205
as trace element ... **A15:** 388, 394
transition diagram for ... **A10:** 87
transition for K lines of ... **A10:** 86, 87
triggering anaerobic curing mechanism ... **EM3:** 113, 118
tubing ... **A10:** 361, 363
tubing, (111) pole figures from ... **A10:** 363
tubing for master work coils ... **A4:** 177, 180, 181
ultrapure, by zone-refining technique ... **A2:** 1094
ultrasonic cleaning of ... **M5:** 619
ultrasonic welding power requirements ... **M6:** 750
UNS C10100
coefficient of thermal expansion ... **EM4:** 503
elastic modulus ... **EM4:** 503
elongation modulus ... **EM4:** 503

Copper (continued)
temperature range ... **EM4:** 503
tensile strength ... **EM4:** 503
yield strength ... **EM4:** 503
usage, PWB manufacturing ... **EL1:** 510
use in breweries ... **A13:** 1222
use in thick-film technology ... **A7:** 151
vapor degreasing of ... **M5:** 617-618
vapor pressure ... **A4:** 493
relation to temperature ... **A4:** 495
vapor pressure, relation to temperature ... **M4:** 309, 310
in vapor-phase metallizing ... **EM3:** 306
Vickers and Knoop microindentation hardness numbers ... **A18:** 416
volume resistivity and conductivity ... **EM3:** 45
volumetric procedures for ... **A10:** 175
weak beam imaging ... **A18:** 388
wetting by bismuth, LME by ... **A11:** 719
for wire-drawing dies ... **A14:** 336
in zinc alloys ... **A15:** 788
zinc at grain boundaries of ... **A10:** 527
and zinc specifications, cast copper alloys ... **A2:** 370
zinc-containing, digital composition map ... **A10:** 528
in zinc/zinc alloys and coatings ... **A13:** 759

Copper (commercially pure)
friction welding ... **A6:** 153

Copper adhesion
in copper-clad E-glass laminates ... **EL1:** 535-536

Copper alloy castings
alloy selection ... **A2:** 346-355
applications ... **A2:** 346-355
bearing and wear properties ... **A2:** 352, 354-355
castability ... **A2:** 346, 348
for corrosion service ... **A2:** 352
cost considerations
dimensional tolerances ... **A2:** 350
electrical and thermal conductivity ... **A2:** 355
general-purpose alloys ... **A2:** 351-352
heat-treated, composition/typical properties ... **A2:** 355
machinability ... **A2:** 351
mechanical properties ... **A2:** 348, 350
nominal compositions ... **A2:** 347
solidification, control of ... **A2:** 348
types of copper ... **A2:** 346

Copper alloy extruded parts ... **A14:** 310-311

Copper alloy filler metals, specific types
BAg-1, joining C12200 ... **A9:** 409
BAg-8a, joining C10100 ... **A9:** 408
BCuP-5, joining C10100 ... **A9:** 408
BCuP-5, joining C12200 ... **A9:** 409

Copper alloy powders *See also* Atomized copper powders; Brasses; Bronzes; Copper; Copper powders; Copper-based powder metals; Nickel silvers; Pure copper
applications ... **A7:** 733-734
for brazing, composition and properties ... **A7:** 839
compacts, mechanical properties ... **A7:** 510
effect of infiltration with precipitation-hardened ... **A7:** 628
friction materials, nominal compositions ... **A7:** 702
HIP temperatures and process times for ... **A7:** 437
hot pressed, products ... **A7:** 509-510
mechanical properties ... **A7:** 464, 470
prealloyed, lubrication of ... **A7:** 191
processing of dispersion-strengthened ... **A7:** 528
production of ... **A7:** 121-122, 734
sampling practices ... **A7:** 249
shipments of ... **A7:** 24, 571
sintering of ... **A7:** 376-381

Copper alloy strip
bending of ... **A8:** 134

Copper alloys *See also* Bronze, commercial; Cast copper alloys; Copper; Copper alloy castings; Copper alloys, specific types; Copper contact alloys; Copper P/M products; Copper-base alloys, specific types; Electrical resistance alloys; Electrical resistance alloys, specific types; High-purity copper; Oxygen- free high-conductivity copper; Tin bronze; Tumbaga;

Copper alloys (continued)
Wrought copper and copper alloy products; Wrought copper and copper alloys ... **A6:** 752-771 **A13:** 610-640 **A14:** 809-824 **A16:** 805-819 **M2:** 239-274
abrasive blasting ... **M5:** 614
abrasive cutoff ... **A16:** 818, 819
acid corrosion ... **A13:** 627-629
addition affecting stainless steel machinability ... **A16:** 689, 690
admiralty brass, weldability ... **A6:** 753
advantages ... **A6:** 752, 797
advantages and applications of materials for surfacing, build-up, and hardfacing ... **A18:** 650
advantages of rapid solidification ... **A9:** 640
AEM analysis in ... **A10:** 471-473
age hardenable ... **A2:** 236
age-hardenable ... **M2:** 242
air-carbon arc cutting ... **A6:** 1172, 1176
alkali corrosion ... **A13:** 629-630
alkaline cleaning ... **M5:** 617-620
alloy families ... **M2:** 241-242
alloy metallurgy ... **A6:** 752-753
alloy ranges, major ... **A15:** 772
alloy systems ... **A2:** 238
alloyed with Ni ... **A16:** 835
alloying element effect on machinability ... **A16:** 805-808
aluminum brass, arsenical, weldability ... **A6:** 753
aluminum brasses ... **A13:** 611
aluminum bronzes ... **A2:** 236 **A6:** 754, 764-766 **A13:** 611
gas-metal arc welding ... **A6:** 755, 770
gas-tungsten arc welding ... **A6:** 769
weldability ... **A6:** 753
analysis for nickel, by dimethylgloxime method ... **A10:** 66
annealed, Rockwell scale for ... **A8:** 76
Antioch process for ... **A15:** 247
applications ... **A2:** 216, 220-223, 239-240 **A6:** 752, 797 **A13:** 610 **A15:** 771, 782-785 **A18:** 750-752
sheet metals ... **A6:** 400
arc welding *See* Arc welding of copper and copper alloys
arc welding processes ... **A6:** 754-755
arc welding with nickel alloys ... **M6:** 443
atmospheric exposure ... **A13:** 616-618
back extrusion of ... **A14:** 398-399
bainitic-like structures ... **A9:** 666
bar and tube, die materials for drawing ... **M3:** 525
bearing material systems ... **A18:** 745, 746, 747
applications ... **A18:** 746
bend ductility data ... **A8:** 131
bending of ... **A14:** 812-813
beryllium copper
cutoff band sawing with bimetal blades ... **A6:** 1184
thermal diffusivity from 20 to 100 °C ... **A6:** 4
weldability ... **A6:** 753
beryllium-containing, cleaning and finishing ... **M5:** 612, 614, 620
beta phase ... **A9:** 639
billets/slugs, heating of ... **A14:** 257
in bimetal bearing material systems ... **A18:** 747
biofouling ... **A13:** 625-626
biological corrosion ... **A13:** 87, 119
blanking, die materials for ... **M3:** 485, 486
blanking of ... **A14:** 811-812
boring ... **A16:** 810, 813
brasses ... **A13:** 610
brasses, gas-metal arc welding ... **A6:** 755
brazeability of ... **A11:** 450
brazing *See* Brazing of copper and copper alloys
brazing and soldering characteristics ... **A6:** 631
brazing, available product forms of filler metals ... **A6:** 119
brazing, joining temperatures ... **A6:** 118

Copper alloys (continued)
brazing with cast iron ... **M6:** 996
broaching ... **A16:** 200, 204, 208, 811, 813
cadmium coppers ... **A6:** 762
cadmium plating of ... **M5:** 622
capacitor discharge stud welding ... **M6:** 730
carbides for machining ... **A16:** 75
cartridge brass, weldability ... **A6:** 753
cast, corrosion ratings, various media ... **A13:** 619-620
cast grain structures ... **A9:** 640-642
castings, cleaning and finishing ... **M5:** 614-616, 619-620
centerless grinding ... **A16:** 818
cermet tools applied ... **A16:** 92
characteristics ... **M6:** 400
chemical and electrochemical cleaning of ... **M5:** 617-619
chemical and electrolytic polishing of ... **M5:** 623-624
chemical resistance ... **M5:** 7
chip formation ... **A16:** 805, 808-812, 817
chromium coppers ... **A6:** 762
chromium plating of ... **M5:** 622
circular sawing ... **A16:** 817-818
classification, generic ... **A2:** 236
classification into 3 subgroups ... **A16:** 805
classification system ... **A14:** 809
cleaning ... **A14:** 258
cleaning processes *See also* specific processes by name ... **M5:** 611-621
cleaning solutions for substrate materials ... **A6:** 978
closed-die forging ... **A14:** 255
coextrusion welding ... **A6:** 311
coining of ... **A14:** 184-185, 818
cold extrusion of ... **A14:** 310-311
cold form tapping ... **A16:** 266, 267
cold reduction swaging ... **A14:** 128-129
cold work effect ... **A16:** 808
cold working ... **A2:** 219, 223
color ... **A2:** 219
color etching ... **A9:** 141-142
coloring process ... **M5:** 611-612, 624-626
commercial bronze
thermal diffusivity from 20 to 100 °C ... **A6:** 4
weldability ... **A6:** 753
compositions ... **A18:** 751, 752
compositions and properties ... **M6:** 401, 546
contact tube, gas-metal arc welding ... **A6:** 183
contaminated naphtha ... **A13:** 635
content additions to P/M materials ... **A16:** 885
contour band sawing ... **A16:** 363
contour roll forming of ... **A14:** 818
copper fabricators ... **A2:** 238
copper nickels ... **A13:** 611
copper-nickel alloys ... **A6:** 754, 766-769
gas-metal arc welding ... **A6:** 755, 770
gas-tungsten arc welding ... **A6:** 769
recommended shielding gas selection for gas-metal arc welding ... **A6:** 66
weldability ... **A6:** 753
copper-phosphorus alloys
brazing, available product forms of filler metals ... **A6:** 119
brazing, joining temperatures ... **A6:** 118
recommended gap for braze filler metals ... **A6:** 120
copper-silicon alloys ... **A13:** 611
copper-silver-phosphorus alloys
brazing, available product forms of filler metals ... **A6:** 119
brazing, joining temperatures ... **A6:** 118
copper-tin alloys
brazing, available product forms of filler metals ... **A6:** 119

Copper alloys (continued)
brazing, joining temperatures ... **A6:** 118
copper-zinc alloys ... **A6:** 762-763
brazing, available product forms of filler metals ... **A6:** 119
brazing, joining temperatures ... **A6:** 118
recommended gap for braze filler metals ... **A6:** 120
coppers and high-copper alloys, corrosion resistance ... **A13:** 610
coppers, gas-metal arc welding ... **A6:** 755
corrosion guide ... **A13:** 612
corrosion in ... **A11:** 201-202
corrosion in gases ... **A13:** 632-633
corrosion in organic compounds ... **A13:** 631
corrosion in salts ... **A13:** 630
corrosion, in specific environments ... **A13:** 616-633
corrosion resistance ... **A2:** 216 **A13:** 610
crevice corrosion in ... **A11:** 632
critical strain rate for SCC in ... **A8:** 519
crystallographic directions ... **A9:** 638
Cu graphite as electrode material for EDM ... **A16:** 559, 560
Cu-infiltrated iron ... **A16:** 882
Cu-Te and EDM ... **A16:** 559
Cu-W and EDM ... **A16:** 558, 559, 560
Cu-W, tool for electrochemical machining ... **A16:** 536
cutoff band sawing ... **A16:** 360
cutting fluids ... **A16:** 125, 159, 808, 811, 814-815, 817
cutting, tools for ... **M3:** 477
cylindrical grinding ... **A16:** 818
decorative colors and finishes ... **M2:** 240
deep drawing, tool materials for ... **M3:** 494, 495, 496
defects, permanent mold casting ... **A15:** 285
degassing of ... **A15:** 86, 464-468
deoxidation of ... **A15:** 468-470
deoxidized coppers, weldability ... **A6:** 753
deoxidizers ... **A2:** 236-237 **M2:** 243
designations ... **A18:** 751
diamond for machining ... **A16:** 105
die casting types ... **A15:** 286
die forging, tool materials for ... **M3:** 529, 530
dies ... **A14:** 256
dissimilar metals ... **A6:** 769-770
gas-metal arc welding ... **A6:** 769-770
gas-tungsten arc welding ... **A6:** 769
drawing of ... **A14:** 814-818
drilling ... **A16:** 226, 229, 230, 237, 811-816
dry machining ... **A16:** 392
ductile irons, gas-metal arc welding ... **A6:** 770
effect of alloying elements on SCC ... **A11:** 220-221
effect of casting temperature ... **A9:** 642
effects, alloy compositions ... **A13:** 610-611
electrical conductivity ... **A2:** 219
as electrical contact materials ... **A2:** 842-843
electrical discharge machining ... **A16:** 558, 559, 560, 561
electrochemical grinding ... **A16:** 543, 545, 547
electrochemical machining ... **A16:** 535, 537
electrochemical machining removal rates ... **A16:** 534
electrodes ... **A6:** 754, 764-765
electrolytic cleaning of ... **M5:** 618-620
electrolytic tough pitch (ETP) copper, weldability ... **A6:** 753
electrolytic, ultrasonic welding ... **A6:** 326
electromagnetic forming ... **A14:** 819
electron beam drilling ... **A16:** 570
electron beam welding ... **M6:** 642
electron-beam welding ... **A6:** 855, 872
electronic applications ... **A6:** 988-990, 998
elements implanted to improve wear and friction properties ... **A18:** 858
embossing of ... **A14:** 819
emulsion cleaning of ... **M5:** 618
end milling ... **A16:** 325, 816-817

SUBJECTS OF THE INDEXED VOLUMES: ASM Handbook (designated by the letter "A"): **A1:** Properties and Selection: Irons, Steels, and High-Performance Alloys (1990); **A2:** Properties and Selection: Nonferrous Alloys and Special-Purpose Materials (1990); **A3:** Alloy Phase Diagrams (1992); **A4:** Heat Treating (1991); **A6:** Welding, Brazing, and Soldering (1993); **A7:** Powder Metallurgy (1984); **A8:** Mechanical Testing (1985); **A9:** Metallography and Microstructures (1985); **A10:** Materials Characterization (1986); **A11:** Failure Analysis and Prevention (1986); **A12:** Fractography (1987); **A13:** Corrosion (1987); **A14:** Forming and Forging (1988); **A15:** Casting (1988); **A16:** Machining (1989); **A17:** Nondestructive Testing and Quality Control (1989); **A18:** Friction, Lubrication, and Wear Technology (1992). **Metals Handbook, 9th Edition** (designated by the letter "M"): **M1:** Properties and Selection: Irons and Steels (1978); **M2:** Properties and Selection: Nonferrous Alloys and Pure Metals (1979); **M3:** Properties and Selection: Stainless Steels, Tool Materials and Special-Purpose Materials (1980); **M4:** Heat Treating (1981); **M5:** Surface Cleaning, Finishing, and Coating (1982); **M6:** Welding, Brazing, and Soldering (1983). **Engineered Materials Handbook** (designated by the letters "EM"): **EM1:** Composites (1987); **EM2:** Engineering Plastics (1988); **EM3:** Adhesives and Sealants (1990); **EM4:** Ceramics and Glasses (1991); **Electronic Materials Handbook** (designated by the letters "EL"): **EL1:** Packaging (1989).

Copper alloys (continued)
end-use applications **A2:** 239
environments that cause
stress-corrosion cracking **A6:** 1101
erosion-corrosion in **A11:** 201-202
erosive attack on die surfaces **A18:** 630
erosive wear during die casting **A18:** 629
in ester solutions, corrosion **A13:** 633
explosion welding **A6:** 896 **M6:** 710
extruded parts **A14:** 310-311
extrusion welding **M6:** 677
fabrication characteristics **A2:** 219-223
face milling **A16:** 816
factors affecting weldability **A6:** 753-754
hot cracking **A6:** 754
porosity **A6:** 754
precipitation-hardenable alloys **A6:** 754
surface condition **A6:** 754
thermal conductivity **A6:** 754
welding position **A6:** 754
fasteners, use in **M3:** 184
fatigue strength at subzero
temperatures **M3:** 733-734
feeding **A15:** 778-782
ferrographic application to identify
wear particles **A18:** 305
filler metals **A6:** 755-756, 763, 764, 765, 766, 768
FIM sample preparation of **A10:** 586
for fine-edge blanking and piercing **A14:** 472-473
finishing processes *See also* specific
processes by name **M5:** 621-627
fire refining effects **A15:** 453
flash welding **M6:** 558
flow-through water-cooled **A15:** 311
fluxes **A6:** 755
fluxing of **A15:** 448-451
as forged **A14:** 256
forged, appearance of **A14:** 258
forging alloys **A14:** 255
forging brass, forgeability **A14:** 255
forging machines **A14:** 256
forging of **A14:** 255-258
forging severities **A14:** 257
forging temperatures **A14:** 257
forgings, cleaning and finishing **M5:** 611, 613-616
forming limit analysis **A14:** 820-821
forming of **A14:** 809-824
forming operations **A14:** 809
foundry properties, for sand casting **A2:** 348
fractographs **A12:** 399-404
fracture/failure causes illustrated **A12:** 217
free machining **A2:** 216
free-cutting alloys **A16:** 805, 808, 811-818
free-cutting, die cutting speeds **A16:** 301
freezing front in continuous casting **A9:** 641
in freshwater **A13:** 621-622
fretting wear **A18:** 248, 250
friction band sawing **A16:** 365
friction welding **M6:** 722
galling in shallow forming dies **A18:** 633
galvanic corrosion **A13:** 85
gas metal arc welding **M6:** 153
to dissimilar metals **M6:** 424
gas porosity, inspection of **A15:** 557-558
gas tungsten arc welding **M6:** 182, 206
gas-metal arc welding **A6:** 180, 752, 754-755
shielding gases **A6:** 66-67
gas-tungsten arc welding **A6:** 192, 752, 754, 755, 756, 764-765, 766
gases in **A15:** 86, 466
gating **A15:** 776-778
general biological corrosion **A13:** 87
general-purpose, casting **A2:** 351-352
generic classification **A13:** 611
gilding, weldability **A6:** 753
gold plating of **M5:** 621-623
grain refining of **A15:** 481
gray irons, gas-metal arc welding **A6:** 770
grinding **A16:** 547, 818-819
grinding fluids **A16:** 819
group I **A15:** 771, 774-775, 778-781
group II **A15:** 771-772, 775-776, 778-781
group III **A15:** 771-772, 776, 781-782
groups divided into **A6:** 752
hardfacing material **M6:** 777
in heat exchangers and condensers **A13:** 626-627
heat treatment **A2:** 223 **A15:** 782

Copper alloys (continued)
heat-affected zone **A6:** 754
heating of dies **A14:** 257
high frequency resistance welding **M6:** 760
high-carbon steel, gas-metal arc
welding **A6:** 770
high-carbon steels, gas-tungsten arc
welding **A6:** 769
high-silicon bronze, cutoff band sawing with bimetal blades **A6:** 1184
high-speed machining **A16:** 597
high-zinc brasses
gas-metal arc welding **A6:** 770
weldability **A6:** 753
hone forming **A16:** 488
honing **A16:** 476, 477, 819
hot extrusion
billet temperatures **M3:** 537
die life **M3:** 539-540
die materials **M3:** 538
hot extrusion of **A14:** 322
hot working **A2:** 223
hydraulic forming of **A14:** 819
hydrochloric acid corrosion **A13:** 1163
hydrogen damage **A13:** 170
impingement attack **A13:** 624
inclusions in **A15:** 96, 488
industry structure **A2:** 238-239
ingot solidification structures **A9:** 637-645
inhibited alloys **A13:** 611
inoculants for **A15:** 105
insoluble alloying elements **A2:** 236
insoluble elements, effect on
machinability **M2:** 242-243
inspection methods **A17:** 534-535
inspection of **A15:** 557-558
intergranular corrosion evaluation **A13:** 241
internal grinding **A16:** 818
knurling **A16:** 815-816
lapping **A16:** 494
laser beam welding **M6:** 647
laser cutting **A14:** 742
lead effect on machinability **A16:** 805, 808
lead frame strip, manufacture **EL1:** 483
leaded, electroless nickel plating **M5:** 232-233
life-limiting factors for die-casting dies **A18:** 629
line etching **A9:** 62
liquid-erosion resistance **A11:** 167
liquid-metal embrittlement of **A11:** 28
low brass, weldability **A6:** 753
low-alloy steels
gas-metal arc welding **A6:** 770
gas-tungsten arc welding **A6:** 769
low-carbon steel
gas-metal arc welding **A6:** 770
gas-tungsten arc welding **A6:** 769
low-zinc brasses
gas-metal arc welding **A6:** 770
gas-tungsten arc welding **A6:** 769
weldability **A6:** 753
lubricants **A14:** 257, 519
machinability ratings and variations **A16:** 805
malleable irons, gas-metal arc welding **A6:** 770
manganese bronze A
cutoff band sawing with bimetal
blades **A6:** 1184
weldability **A6:** 753
manganese bronzes, gas-metal arc
welding **A6:** 755
manganese-nickel-aluminum bronzes,
gas-metal arc welding **A6:** 755
marine pitting **A13:** 906
martensitic structures **A9:** 672-673
mass finishing of **M5:** 614-616
materials for die-casting dies **A18:** 629
as matrix material **EL1:** 1120
mechanical working **A2:** 219, 223
medium-carbon steel
gas-metal arc welding **A6:** 770
gas-tungsten arc welding **A6:** 769
melt refining **A15:** 449-450
melt treatment **A15:** 774-782
melting heat for **A15:** 376
melting practice **A15:** 772-774
metal forming lubricants **A18:** 147
microdrilling **A16:** 238
microstructures of **A9:** 551

Copper alloys (continued)
milling **A16:** 110, 312, 313, 314, 327, 547, 816-817
minimum-draft forgings **A14:** 258
modified solid-solution alloys **A2:** 234-235
in moist chlorine **A13:** 1173
mold coatings for **A15:** 282
multiple-operation machining **A16:** 815
muntz metal, weldability **A6:** 753
naval brass
thermal diffusivity from 20 to 100 °C **A6:** 4
weldability **A6:** 753
nickel alloys
gas-metal arc welding **A6:** 770
gas-tungsten arc welding **A6:** 769
nickel plating of **M5:** 199-200, 215, 218, 232-233
electroless **M5:** 232-233
nickel silvers **A13:** 611
nickel silvers, weldability **A6:** 753
nickel-aluminum bronzes, gas-metal
arc welding **A6:** 755
nickel-chromium alloys
gas-metal arc welding **A6:** 770
gas-tungsten arc welding **A6:** 769
nickel-chromium-iron alloys
gas-metal arc welding **A6:** 770
gas-tungsten arc welding **A6:** 769
nickel-copper alloys
gas-metal arc welding **A6:** 770
gas-tungsten arc welding **A6:** 769
nickel-iron alloys
gas-metal arc welding **A6:** 770
gas-tungsten arc welding **A6:** 769
no resistance seam welding **A6:** 241-242
nondestructive testing **A6:** 1086
nonmetallic inclusions in **A11:** 316
organic coating of **M5:** 621, 626-627
oxyfuel gas welding **A6:** 281, 285 **M6:** 583
oxyfuel welding **A6:** 756
oxygen-free copper (OFC), weldability **A6:** 753
oxygen-free high conductivity **A12:** 401
passivation of **M5:** 624
PCD tooling application **A16:** 109, 110
percussion welding **M6:** 740
peripheral milling **A16:** 815, 816
peritectic transformations **A9:** 677-678
permanent mold casting **A15:** 275
petroleum refining and petrochemical
operations **A13:** 1263
phosphor bronzes **A13:** 611
gas-metal arc welding **A6:** 755, 770
gas-tungsten arc welding **A6:** 769
weldability **A6:** 753
phosphorus as inclusion forming **A15:** 90
phosphorus deoxidation **A15:** 468-469
photochemical machining **A16:** 587, 588, 591, 593
photochemical machining etchant **A16:** 590, 591
physical properties **M6:** 546-547
pickling and bright dipping **M5:** 611-615, 619-620
piercing of **A14:** 811-812
planing **A16:** 810-811, 813
plasma and shielding gas
compositions **A6:** 197
plasma arc cutting **A6:** 1169, 1170
plasma arc welding **A6:** 197, 752, 754, 755, 756 **M6:** 214
plating procedures **M5:** 619-624
electroless process **M5:** 621-622
electroplating **M5:** 622-624
immersion process **M5:** 622
surface preparation **M5:** 619-621
polishing and buffing **M5:** 615-617
in polluted cooling waters **A13:** 625
porosity in **A15:** 86, 557-558
pouring **A15:** 776-778
pouring temperatures **A15:** 283
power band sawing **A16:** 818
power hacksawing **A16:** 818
precoated before soldering **A6:** 131
press forming, tool materials for **M3:** 492, 493
principal ASTM specifications for
weldable nonferrous sheet metals **A6:** 400
product forms **A18:** 751
production of **A2:** 237-238
products, ceramic molded **A15:** 248
projection welding **M6:** 503
properties **A18:** 750-752
properties of **A2:** 216, 219

Copper alloys (continued)
property requirements, for formed products ... **A14:** 821-824
protective coatings ... **A13:** 636
reaming ... **A16:** 248, 812-813, 814, 815
recessing ... **A16:** 815-816
recommended shielding gas selection for gas-metal arc welding ... **A6:** 66
red brass, weldability ... **A6:** 753
relative hydrogen susceptibility ... **A8:** 542
removal, from permanent molds ... **A15:** 284
resistance brazing ... **A6:** 339, 342 **M6:** 976
resistance seam welding ... **M6:** 494
resistance soldering ... **A6:** 357
resistance spot welding ... **M6:** 479-480
resistance welding *See* Resistance welding of copper and copper alloys ... **A6:** 834, 841, 847, 849-850
rhodium plating of ... **M5:** 623
ring rolling ... **A14:** 255-256
Rockwell scale for ... **A8:** 76
roll welding ... **A6:** 312 **M6:** 676
roller burnishing ... **A16:** 254, 813
rolling of ... **A14:** 343, 355-356
rotary forging of ... **A14:** 179
rubber-pad forming of ... **A14:** 818-819
safe welding practices ... **A6:** 770-771
in salt water ... **A13:** 622-625
sawing ... **A14:** 257 **A16:** 362
SCC failures ... **A12:** 133
SCC, in specific environments ... **A13:** 633-636
SCC testing ... **A13:** 268-270
SCC testing of ... **A8:** 525-526
scratch brushing of ... **M5:** 616
in seawater, critical surface shear stress ... **A13:** 624
selective leaching ... **A13:** 334
semisolid metal casting and forging ... **A15:** 327
shaping ... **A16:** 191
shear stresses and horsepowers ... **A16:** 15
shearing ... **A14:** 256-257
shielded metal arc welding ... **A6:** 176, 179, 752, 754, 755 **M6:** 75
shrinkage voids in ... **A15:** 558
silicon bronzes ... **A6:** 754, 766
  gas-metal arc welding ... **A6:** 755, 770
  gas-tungsten arc welding ... **A6:** 769
  weldability ... **A6:** 753
silicon-containing, pickling ... **M5:** 612-613
silver plating of ... **M5:** 621-623
slab milling ... **A16:** 324
slitting ... **A16:** 817-818
Sn effect on machinability ... **A16:** 808
in soils and groundwater ... **A13:** 618-621
soldering ... **A6:** 630-631
solid solution ... **M2:** 242
solid-solution alloys ... **A2:** 1019
solvent cleaning of ... **M5:** 617-618
spade drilling ... **A16:** 225
special alloys ... **A6:** 752
special brasses, gas-metal arc welding ... **A6:** 770
specific power ... **A16:** 18
spinning of ... **A14:** 818
spring materials, forming of ... **A14:** 819-820
springback in ... **A14:** 820
squeeze casting of ... **A15:** 323
stainless steels
  gas-metal arc welding ... **A6:** 770
  gas-tungsten arc welding ... **A6:** 769
stampings, cleaning and finishing ... **M5:** 614-615, 620
standardized system of identification ... **A6:** 752
steam as SCC source ... **A13:** 328
in steam condensate ... **A13:** 622
stock preparation ... **A14:** 256-257, 671
stress-corrosion cracking in ... **A11:** 27, 220-223
stress-corrosion testing ... **A13:** 636-639
stress-relieving parameters ... **A13:** 615
stretch forming of ... **A14:** 814-818

Copper alloys (continued)
stripping of ... **M5:** 218
stud arc welding ... **M6:** 730
stud material ... **M6:** 730
submerged arc welding ... **A6:** 752, 755
suction shell ... **A13:** 1204
sulfur effect on machinability ... **A16:** 805-808
sulfuric acid corrosion ... **A13:** 1153
supply and reserves ... **A2:** 240
surface activation of ... **M5:** 232-233
surface finish ... **A16:** 805, 808, 812, 815, 817, 819
surface grinding ... **A16:** 818
swaging of ... **A14:** 128-129, 819
tapping ... **A16:** 263, 813-815, 816
tarnish removal ... **M5:** 613, 619
Te effect on machinability ... **A16:** 805-808
temper designations ... **A2:** 223
temperature distribution and cutting of Al bronzes ... **A16:** 808
tensile properties at subzero temperatures ... **M3:** 732-733, 747
testing in Mattsson's solution ... **A8:** 525
testing mediums ... **A8:** 525-526
thermal conductivity ... **A2:** 219
thermodynamic properties ... **A15:** 55-60
threading ... **A16:** 813-815
threading operation and circular chasers ... **A16:** 297, 298
tin brasses ... **A13:** 610-611
tin brasses, weldability ... **A6:** 753
tin, gas-metal arc welding ... **A6:** 770
tin, tin-lead, and tin-copper plating of ... **M5:** 621-624
tin-containing ... **A2:** 526
tin-copper as ... **A13:** 774
tool design ... **A16:** 809-810, 811, 812
tool for electrochemical machining ... **A16:** 533, 535, 536, 540
tool life ... **A16:** 815
torch brazing ... **A6:** 328
torch soldering ... **A6:** 351, 352
trimming ... **A14:** 257
tube stock ... **A14:** 671
turning ... **A16:** 135, 809-810, 811, 812
turning operation with cermet tools ... **A16:** 94
turning operation with PCD tooling ... **A16:** 110
types ... **A15:** 771-772
types of ... **A2:** 346-347
types of attack ... **A13:** 612-616
ultrasonic cleaning of ... **M5:** 619
ultrasonic welding ... **A6:** 894-895 **M6:** 746
upset forging ... **A14:** 255
UV/VIS analysis for antimony, by iodoantimonite method ... **A10:** 68
vacuum induction melting ... **A15:** 396
vapor degreasing of ... **M5:** 617-618
versus stainless steel properties ... **A18:** 712-713
in water ... **A13:** 621-627
water-base cutting or grinding fluids ... **A16:** 128
weld overlay material ... **M6:** 806, 816
weldability ... **A6:** 752-753
welding electrodes ... **A6:** 176
welding of coppers ... **A6:** 756-760
welding of high-strength beryllium coppers ... **A6:** 760-762
wire, die materials for drawing ... **M3:** 522
for wire-drawing dies ... **A14:** 336
wrought, corrosion ratings, various media ... **A13:** 617-618
yellow brass, weldability ... **A6:** 753
yield strengths, plastic packages ... **EL1:** 490
Zn effect on machinability ... **A16:** 808

Copper alloys, annealing ... **A4:** 881-884, 889 **M4:** 722-724
castings ... **A4:** 884 **M4:** 724
cooling ... **A4:** 884
fire cracking ... **A4:** 881, 884 **M4:** 723
hydrogen embrittlement ... **A4:** 884 **M4:** 723
impurities ... **A4:** 884 **M4:** 723

Copper alloys, annealing (continued)
loading ... **A4:** 884 **M4:** 723
lubricants, effect of ... **A4:** 884 **M4:** 723
oxidation ... **A4:** 884 **M4:** 723
pretreatment, effect of ... **A4:** 884 **M4:** 723
sulfur stains ... **A4:** 884
sulphur stains ... **M4:** 723-724
testing ... **A4:** 884 **M4:** 723
thermal shock ... **A4:** 884 **M4:** 723
time, effect of ... **A4:** 884 **M4:** 723
wrought products ... **A4:** 881-883 **M4:** 719-721

Copper alloys, cast, specific types
C81300, stress-relieving temperature ... **A4:** 885
C81500, stress-relieving temperature ... **A4:** 885
C81540, stress-relieving temperature ... **A4:** 885
C81700, stress-relieving temperature ... **A4:** 885
C81800, stress-relieving temperature ... **A4:** 885
C82200, stress-relieving temperature ... **A4:** 885
C82400, stress-relieving temperature ... **A4:** 885
C82500, stress-relieving temperature ... **A4:** 885
C82600, stress-relieving temperature ... **A4:** 885
C82700, stress-relieving temperature ... **A4:** 885
C82800, stress-relieving temperature ... **A4:** 885
C83300, stress-relieving temperature ... **A4:** 885
C84800, stress-relieving temperature ... **A4:** 885
C95200, stress-relieving temperature ... **A4:** 885
C95400, stress-relieving temperature ... **A4:** 885
C95500, stress-relieving temperature ... **A4:** 885
C95800, stress-relieving temperature ... **A4:** 885
C96600, stress-relieving temperature ... **A4:** 885
C97800, stress-relieving temperature ... **A4:** 885
C99300, stress-relieving temperature ... **A4:** 885

Copper alloys, castings
applications ... **M2:** 387, 389-390
ASTM specifications ... **M2:** 384, 386
centrifugal casting ... **M2:** 384
composition ... **M2:** 383-384
conductivity ... **M2:** 393
corrosion resistance ... **M2:** 390-391
cost considerations ... **M2:** 393-394
dimensional tolerances ... **M2:** 388-389
machinability ... **M2:** 388
mechanical properties ... **M2:** 385-387
permanent mold casting ... **M2:** 384
plaster mold casting ... **M2:** 384-385
sand casting ... **M2:** 384
solidification control ... **M2:** 385

Copper alloys, corrosion environments
acetic acid ... **M2:** 473, 475-476
alkalis ... **M2:** 477
ammonium hydroxide ... **M2:** 477
anhydrous ammonia ... **M2:** 477
atmospheric ... **M2:** 467, 470
beer ... **M2:** 480
benzol ... **M2:** 480
biofouling ... **M2:** 472
carbon dioxide ... **M2:** 480-481
carbon monoxide ... **M2:** 480-481
creosote ... **M2:** 479
dry oxygen ... **M2:** 482-483
fatty acids ... **M2:** 476
fresh water ... **M2:** 470-471
  copper aluminum alloys ... **M2:** 471
  copper nickels ... **M2:** 471
  copper-silicon alloys ... **M2:** 471
  copper-zinc alloys ... **M2:** 470
gasoline ... **M2:** 479, 483
halogen gases ... **M2:** 475, 481
hydrochloric acid ... **M2:** 473-675
hydrocyanic acid ... **M2:** 476-477
hydrofluoric acid ... **M2:** 475
hydrogen ... **M2:** 481-483
hydrogen sulfide ... **M2:** 481
linseed oil ... **M2:** 480
oleic acid ... **M2:** 476
organic compounds ... **M2:** 478, 482
oxidizing salts ... **M2:** 478-479
phosphoric acids ... **M2:** 473-474
salt water ... **M2:** 471-472

Copper alloys, corrosion environments (continued)
salts ... **M2:** 477-479
selection for specific environment ... **M2:** 466-467, 468-469
soil ... **M2:** 467, 470
steam ... **M2:** 471
stearic acid ... **M2:** 476-477
sugar ... **M2:** 480, 483
sulfur compounds ... **M2:** 480
sulfur dioxide ... **M2:** 481, 483
sulfuric acid ... **M2:** 473, 474, 475
tartaric acid ... **M2:** 477
Copper alloys, corrosion resistance
aluminum brasses ... **M2:** 464
aluminum bronzes ... **M2:** 465
brasses ... **M2:** 461, 463, 464
copper nickels ... **M2:** 464-465
copper-silicon ... **M2:** 464, 465
inhibited brasses ... **M2:** 464
nickel silvers ... **M2:** 464, 465
phosphor bronzes ... **M2:** 464, 465
tin brasses ... **M2:** 463-464
Copper alloys, corrosion service ... **M2:** 466-483
tubes, condenser ... **M2:** 472
tubes, heat exchange ... **M2:** 472
Copper alloys, heat treating ... **A4:** 880-898
aluminum bronzes ... **M2:** 259
annealing *See* Copper alloys, annealing
annealing temperatures ... **M2:** 256
atmospheres, protective ... **A4:** 887, 888-889, 894 **M4:** 727-728
beryllium coppers ... **M2:** 256-257, 258
aging ... **M2:** 257, 258
solution treating ... **M2:** 256-257
chromium coppers ... **M2:** 257-258, 259
copper-nickel-phosphorus alloys ... **M2:** 257
hardening ... **A4:** 886 **M4:** 724-726
homogenizing ... **A4:** 880-881 **M4:** 719
low-temperature hardening alloys ... **A4:** 886-887 **M4:** 724
order-hardening alloys ... **A4:** 886-887 **M4:** 725-726
precipitation-hardening alloys ... **A4:** 886, 894, 896 **M4:** 724-725
quench hardening ... **A4:** 887 **M4:** 726
spinodal-hardening alloys ... **A4:** 886, 894-897 **M4:** 725, 736
microduplexing ... **A4:** 897
stress relieving ... **A4:** 884-885, 886 **M4:** 722-723, 724
tempering ... **A4:** 887 **M4:** 726
zirconium copper ... **M2:** 258-259
Copper alloys, heat treating equipment
aging ... **M4:** 727
furnaces ... **M4:** 726
salt baths ... **M4:** 726-727
stress-relieving ... **M4:** 727
Copper alloys, heat-treating equipment
aging ... **A4:** 887-888, 889
furnaces ... **A4:** 887, 888
salt baths ... **A4:** 887, 888
stress-relieving ... **A4:** 34, 887-888
Copper alloys, microstructure of specific types
C23000 ... **A3:** 1 • 22, 23
C24000 ... **A3:** 1 • 22, 23
C26000 ... **A3:** 1 • 22, 23
C27000 ... **A3:** 1 • 22, 23
C28000 ... **A3:** 1 • 22, 23
C71500 ... **A3:** 1 • 18
Copper alloys, specific types *See also* Copper; Copper alloys
4Cu-1Mg, precipitation hardening ... **A4:** 836
64Cu-27Ni-9Fe, spinodal decomposition ... **A12:** 402
64Cu-27Ni-9Fe, tensile overload fracture ... **A12:** 402
70Cu-30Ni, crevice corrosion ... **A13:** 110
70Cu-30Zn brass, dealloying corrosion ... **A13:** 128, 132
70Cu-30Zn cartridge brass, SCC failure in ... **A11:** 27
75-5-20, bearing wear ... **M3:** 592
75Cu-25Ni, roll welding ... **A6:** 314
80Cu-20Zn brass, corrosion pitting fracture ... **A12:** 404
90-10 Cu-Ni, photochemical machining ... **A16:** 588
90Cu-10Ni tube, biological pitting corrosion ... **A13:** 120

Copper alloys, specific types (continued)
90Cu-10Sn-lPb, stress-corrosion cracking ... **A12:** 403
95Cu-2.5Co-2.5Be, wear data ... **M3:** 593
95Cu-5Al (Al bronze), properties ... **A18:** 714
97% aluminum bronze, thermal diffusivity from 20 to 100 °C ... **A6:** 4
170
springs, strip for ... **M1:** 286
springs, wire for ... **M1:** 284
184, resistance brazing ... **M6:** 976
260 (cartridge brass)
friction welding ... **A6:** 153
thermal diffusivity from 20 to 100 °C ... **A6:** 4
353, resistance brazing ... **M6:** 976
360, resistance brazing ... **M6:** 976
400, crevice corrosion ... **A13:** 110
510
springs, strip for ... **M1:** 286
springs, wire for ... **M1:** 284
793 (SAE), used as a surface layer for bushings ... **A18:** 750
10100 tubing, (111) pole figure from inside wall ... **A10:** 363
10100 tubing, ODF using Euler plots ... **A10:** 361
10100 tubing, pole figure, from midwall ... **A10:** 362
10300, roll welding ... **A6:** 313
11000, roll welding ... **A6:** 313
11600, stress relaxation ... **M2:** 486, 488
26000 (cartridge brass, 70%), microstructure from several directions ... **A10:** 305
Al-4Cu, friction surfacing ... **A6:** 322, 323
aluminum bronze D, cutoff band sawing with bimetal blades ... **A6:** 1184
AMS 4881, wear-test results ... **M3:** 591
beryllium copper 21C ... **M2:** 438
beryllium copper nickel 72C ... **M2:** 438-439
beryllium copper, use in die-casting dies ... **M3:** 543
C10100 (ETP Cu) ... **A16:** 809, 810, 811
C10100, composition and machinability rating ... **A16:** 806
C10100, material for ECM tool s ... **A16:** 537
C10100, oxygen-free, characteristics ... **A2:** 230
C10100-10800, composition and machinability rating ... **A16:** 806
C10100-C10800 (oxygen-free Cu) ... **A16:** 809, 810, 811, 813
C10200 ... **A16:** 363
composition ... **M3:** 746
stress relaxation ... **M2:** 484-487
tensile properties at subzero temperatures ... **M3:** 747
tubes, applications ... **M2:** 262
tubes, mechanical properties ... **M2:** 263
C10200, brazing with phosphor bronze cladding ... **A6:** 347
C10200 forging, effects of overheating and burning ... **A11:** 332, 334
C10200, oxygen-free, characteristics ... **A2:** 230
C11000
acetic acid, corrosion in ... **M2:** 476
acetic acid-acetic enhydride, corrosion in ... **M2:** 475
alcohols, corrosion in ... **M2:** 482
aldehydes, corrosion in ... **M2:** 481
amine-system, corrosion in ... **M2:** 478
ammonia, corrosion in ... **M2:** 477
atmospheric corrosion ... **M2:** 470
beet-sugar, corrosion in ... **M2:** 483
$CaCl_2$ refrigeration brine, corrosion in ... **M2:** 478
composition ... **M2:** 467
contaminated naphtha, corrosion in ... **M2:** 483
ester solutions, corrosion in ... **M2:** 479-480
ethers, corrosion in ... **M2:** 481
ethylene glycol, corrosion in ... **M2:** 482
fasteners ... **M2:** 443
hydrocyanic acid, corrosion in ... **M2:** 477
hydrogen, corrosion in ... **M2:** 482, 483
hydrogen cyanide, corrosion in ... **M2:** 476
isopropyl ether-acetic acid, corrosion in ... **M2:** 476
ketones, corrosion in ... **M2:** 481
sodium chloride brine, corrosion in ... **M2:** 478
stearic acid corrosion in ... **M2:** 476-477

Copper alloys, specific types (continued)
stress relaxation ... **M2:** 484-487
sulfuric acid, corrosion in ... **M2:** 474
tubes, piercing temperature ... **M2:** 263
C11000 (tough pitch copper), electron-beam welding ... **A6:** 860, 872
C11000, characteristics ... **A2:** 230
C11000, use for fasteners ... **M3:** 184
C11300-C11600 (Ag-bearing ETP Cu) composition and machinability rating ... **A16:** 806
C11400, softening characteristics ... **A2:** 234
C12000
atmospheric corrosion ... **M2:** 470
ketones, corrosion in ... **M2:** 481
steam condensate, corrosion in ... **M2:** 471
stress relaxation ... **M2:** 485, 488
tubes, used for ... **M2:** 471
C12200
clad to 304 L stainless steel, clad brazing ... **A6:** 347
clad to 409 stainless steel, clad brazing ... **A6:** 347
clad to 1008 steel, clad brazing ... **A6:** 347
composition ... **M2:** 467 **M3:** 746
tensile properties at subzero temperatures ... **M3:** 747
tubes, applications ... **M2:** 262
tubes, mechanical properties ... **M2:** 263
tubes, piercing temperature ... **M2:** 263
C12200 (phosphorus-deoxidized Cu) composition and machinability rating ... **A16:** 806
C12300, tubes ... **M2:** 472
C12500, characteristics ... **A2:** 223
C12500-C13000 (fire-refined ETP Cu) composition and machinability rating ... **A16:** 806
C13400, stress relaxation ... **M2:** 486, 488
C13700, stress relaxation ... **M2:** 485, 488
C14200
steam condensate, corrosion in ... **M2:** 471
sulfuric acid, corrosion in ... **M2:** 474
C14300, cadmium-copper, cold rolling ... **A2:** 234
C14300, softening characteristics ... **A2:** 234
C14500 (Te-Cu) ... **A16:** 323, 324, 325, 815
C14500, composition and machinability rating ... **A16:** 806
C14500, electrochemical machining ... **A16:** 537
C14500, tellurium-bearing, machinability ... **A2:** 230
C14700 ... **A16:** 323, 324, 325
C14700, sulfur-bearing, machinability ... **A2:** 230
C15000, as age hardenable ... **A2:** 236
C15100, as age hardenable ... **A2:** 236
C15100, copper-zirconium, conductivity ... **A2:** 234
C15500, copper-silver-magnesium-phosphorus ... **A2:** 234
C15710 ... **M2:** 298
C15720 ... **M2:** 299
C15735 ... **M2:** 299, 300
C16200 ... **M2:** 300
stress relaxation ... **M2:** 486, 488
C16200 (Cd-Cu), composition and machinability rating ... **A16:** 806
C16500, composition and machinability rating ... **A16:** 806
C17000 ... **A16:** 361, 363 **M2:** 301-302
aging, proper-ties corresponding to ... **M2:** 258
C17000, as gold alloy ... **A2:** 236
C17200 ... **M2:** 303-304, 305
aging, properties corresponding to ... **M2:** 258
composition ... **M3:** 746
fatigue life ... **M3:** 748
stress relaxation ... **M2:** 487
tensile properties at subzero temperatures ... **M3:** 747
wear data ... **M3:** 593
C17200 (Be-Cu), composition and machinability rating ... **A16:** 806
C17200, as gold alloy ... **A2:** 236
C17300 ... **M2:** 303-304, 305
C17300 (Be-Cu) ... **A16:** 323, 324, 325
C17300, age hardenable ... **A2:** 236

Copper alloys, specific types (continued)
C17300, composition and machinability rating ... A16: 806
C17300, wear data ... M3: 593
C17500 ... M2: 306-307
aging, properties corresponding to ... M2: 258
stress relaxation ... M2: 488
C17500 (low-Be Cu), composition and machinability rating ... A16: 806
C17500, age hardenable ... A2: 236
C17500, wear data ... M3: 593
C17510, age hardening ... A2: 236
C17600 ... M2: 308
C18200 ... M2: 309
C18200, as age hardenable ... A2: 236
C18200, resistance brazing ... A6: 339
C18200-C18500 (chrome Cu), composition and machinability rating ... A16: 806
C18400 ... M2: 309
C18500 ... M2: 309
C18500, as age hardenable ... A2: 236
C18700 ... M2: 310
C18700 (leaded Cu) ... A16: 323, 324, 325
C18700, composition and machinability rating ... A16: 806
C18700, electrochemical machining ... A16: 537
C18700, end milling ... A16: 325
C18700, face milling ... A16: 323
C18700, slab milling ... A16: 324
C19000 ... M2: 488
C19000 (high-Cu alloy), composition and machinability rating ... A16: 806
C19000, as age hardenable ... A2: 236
C19100 ... A16: 323, 324, 325
C19100, as age hardenable ... A2: 236
C19200 ... M2: 311
tubes, applications ... M2: 262
tubes, mechanical properties ... M2: 263
C19400 ... M2: 311-312, 313
C19500 ... M2: 313
C19500 (Strescon), composition and machinability rating ... A16: 806
C19500, copper-iron-cobalt-tin-phosphorus ... A2: 234
C21000 ... M2: 256, 316
C21000 (gilding, 95%), composition and machinability rating ... A16: 806
C22000 ... M2: 317, 318
composition ... M2: 467 M3: 746
stress-relieving temperatures ... M2: 256
tensile properties at subzero temperatures ... M3: 747
tubes, piercing temperature
C22000 (commercial bronze, 90%) ... A16: 363
C22000, composition and machinability rating ... A16: 806
C22000, electrochemical machining ... A16: 537
C22600 ... M2: 318, 319
C22600 jewelry bronze), composition and machinability rating ... A16: 806
C23000 ... M2: 320-321, 322
alcohols, corrosion in ... M2: 482
atmospheric corrosion ... M2: 470
composition ... M2: 467
contaminated naphtha, corrosion in ... M2: 483
galvanic corrosion
steam condensate, corrosion in ... M2: 471
stress-relieving temperatures ... M2: 256
sulfur compounds, corrosion in ... M2: 480
tartaric acid, corrosion in ... M2: 477
tubes, applications ... M2: 262
tubes, mechanical properties ... M2: 263
tubes, piercing temperature ... M2: 263
C23000 (red brass, 85%), composition and machinability rating ... A16: 806
C24000 ... M2: 256, 322, 323
C24000 (low brass, 80%), composition and machinability rating ... A16: 806

Copper alloys, specific types (continued)
C24000, susceptibility to dezincification ... A11: 633
C26000 ... M2: 323-326, 327, 481
ammonia, corrosion in ... M2: 477
atmospheric corrosion ... M2: 470
composition ... M2: 467 M3: 746
ethylene glycol, corrosion in ... M2: 482
fasteners ... M2: 444
fatigue life ... M3: 748
ketones, corrosion in ... M2: 481
oleic acid, corrosion in ... M2: 476
stearic acid, corrosion in ... M2: 476-477
stress-relieving temperatures ... M2: 256
sulfuric acid, corrosion in ... M2: 474
tartaric acid, corrosion in ... M2: 477
tensile properties at subzero temperatures ... M3: 747
tubes, applications ... M2: 262
tubes, extrusion ... M2: 261-262
tubes, mechanical properties ... M2: 263
tubes, piercing temperature ... M2: 263
C26000 (cartridge brass, 70%), composition and machinability rating ... A16: 806
C26000 cartridge brass, dezincification of ... A11: 633
C26000 cartridge brass pipe, dezincification differences from domestic water supply ... A11: 178
C26000, electrochemical machining ... A16: 537
C26000, intergranular corrosion of ... A11: 182
C26000, SCC from amines ... A13: 633
C26000 steam-turbine condenser tube, dezincification failure of ... A11: 634
C26800 ... M2: 327-328
C26800-C27000 (yellow brass) ... A16: 808
C26800-C27000, composition and machinability rating ... A16: 806
C27000 ... M2: 327-328
dezincification ... M2: 461
fasteners ... M2: 443
stress-relieving temperatures ... M2: 256
C27000, use for fasteners ... M3: 184
C27000 yellow brass air-compressor innercooler tube, dezincification failure ... A11: 179
C28000 ... M2: 328, 329-330
composition ... M2: 467
contaminated naphtha, corrosion in ... M2: 483
fasteners ... M2: 444
stress-relieving temperatures ... M2: 256
sulfur compounds, corrosion in ... M2: 480
tubes, piercing temperature ... M2: 263
C28000 (Muntz metal, 60/40 brass) ... A16: 808, 809, 811
C28000, composition and machinability rating ... A16: 806
C31400 ... M2: 330
C31400 (leaded commercial bronze) ... A16: 323, 324, 325, 363, 808
C31400, composition and machinability rating ... A16: 806
C31600 ... M2: 331-331
C31600 (leaded commercial bronze-Ni) ... A16: 323, 324, 325
C31600, composition and machinability rating ... A16: 806
C33000 ... M2: 331
tubes, applications ... M2: 262
tubes, mechanical properties ... M2: 263
C33000 (low-leaded brass, tube) ... A16: 323, 324, 325
C33000, composition and machinability rating ... A16: 806
C33200 ... M2: 332
C33200 (high-leaded brass tube) ... A16: 323, 324, 325
C33200, composition and machinability rating ... A16: 806

Copper alloys, specific types (continued)
C33500 ... M2: 333
C33500 (low-leaded brass) ... A16: 323, 324, 325
C33500, composition and machinability rating ... A16: 806
C34000 ... M2: 333-334
C34000 (medium-leaded brass) ... A16: 323, 324, 325
C34000, composition and machinability rating ... A16: 806
C34200 ... M2: 333, 334
C34200 (high-leaded brass) ... A16: 323, 324, 325, 815, 817
C34200, composition and machinability rating ... A16: 806
C34200, use for fasteners ... M3: 184
C34900 ... M2: 335, 336
C34900 (brass) ... A16: 323, 324, 325
C34900, composition and machinability rating ... A16: 806
C35000 ... A16: 323, 324, 325 M2: 336
C35300 ... A16: 323, 324, 325 M2: 334, 335
C35600 ... M2: 336-337
C35600 (extrahigh-leaded brass) ... A16: 323, 324, 325
C35600, composition and machinability rating ... A16: 806
C36000 ... M2: 337-338
composition ... M2: 467
fasteners ... M2: 444
stress-relieving temperatures ... M2: 256
tubes, applications ... M2: 262
tubes, extrusion ... M2: 261-262
C36000 (free-cutting brass) ... A16: 323, 324, 325, 363
C36000, composition and machinability rating ... A16: 806
C36000, electrochemical machining ... A16: 537
C36000, free cutting brass, semisolid application ... A15: 336
C36000, use for fasteners ... M3: 184
C36500 ... M2: 338-339
C36500-C36800 (leaded Muntz metal) ... A16: 323, 324, 325
C36500-C36800, composition and machinability rating ... A16: 806
C36600 ... M2: 338-339
C36700 ... M2: 338-339
C36800 ... M2: 338-339
C37000 ... M2: 339
C37000 (free-cutting Muntz metal) ... A16: 323, 324, 325
C37000, composition and machinability rating ... A16: 806
C37700 ... M2: 340-341
C37700 (forging brass) ... A16: 323, 324, 325, 814
C37700, composition and machinability rating ... A16: 806
C37700, high-zinc brass ... A2: 236
C38500 (architectural bronze) ... A16: 323, 324, 325
C38500, composition ... M2: 342, 467
C38500, composition and machinability rating ... A16: 806
C40500 ... M2: 342
C40800 ... M2: 343
C41100 ... M2: 343, 344
C41100 (Lubaloy), composition and machinability rating ... A16: 806
C41500 ... M2: 344-345
C41900 ... M2: 345
C42200 ... M2: 345
C42500 ... M2: 346
C42500-C43500 (Sn brass), composition and machinability rating ... A16: 806
C43000 ... M2: 346, 347
C43400 ... M2: 347
C43500 ... M2: 347-348
tubes, applications ... M2: 262

Copper alloys, specific types (continued)
tubes, mechanical properties **M2:** 263
C44200
amine-system, corrosion in **M2:** 478
atmospheric corrosion **M2:** 470
contaminated naphtha, corrosion in **M2:** 483
ethylene glycol, corrosion in **M2:** 482
C44300 ..... **M2:** 348-349
beet-sugar, corrosion in **M2:** 483
composition **M2:** 467
gasoline, corrosion in **M2:** 483
impingement attack **M2:** 460
steam condensate, corrosion in **M2:** 471
stress-relieving temperatures **M2:** 256
tubes **M2:** 472
tubes, applications **M2:** 262
tubes, extrusion **M2:** 261-262
tubes, mechanical properties **M2:** 263
C44300, gasoline corrosion **A13:** 635
C44300 heat-exchanger tube, impingement corrosion failure **A11:** 635
C44300, impingement attack **A11:** 634-635
C44300, SCC in **A11:** 635
C44300-C44500 (inhibited admiralty) composition and machinability rating **A16:** 806
C44400 ..... **M2:** 348-349
alcohols, corrosion in **M2:** 482
beet-sugar, corrosion in **M2:** 483
composition **M2:** 467
steam condensate, corrosion in **M2:** 471
tubes **M2:** 472, 480
tubes, applications **M2:** 262
tubes, mechanical properties **M2:** 263
C44400, SCC in **A11:** 635
C44500 ..... **M2:** 348-349
beet-sugar, corrosion in **M2:** 483
composition **M2:** 467
intercrystalline corrosion **M2:** 461
steam condensate, corrosion in **M2:** 471
tubes **M2:** 472, 480
tubes, applications **M2:** 262
tubes, mechanical properties **M2:** 263
C44500, SCC in **A11:** 635
C46200 fasteners **M2:** 443
C46200, use for fasteners **M3:** 184
C46400 ..... **M2:** 349-351
composition **M2:** 467
contaminated naphtha, corrosion in **M2:** 483
fasteners **M2:** 443, 444
tubes, applications **M2:** 262
tubes, mechanical properties **M2:** 263
tubes, piercing temperature **M2:** 263
C46400 (naval brass), composition and machinability rating **A16:** 806
C46400, electrochemical machining **A16:** 537
C46400, use for fasteners **M3:** 184
C46500 ..... **M2:** 349-351
composition **M2:** 467
fasteners **M2:** 444
tubes, applications **M2:** 262
tubes, mechanical properties **M2:** 263
C46500-C46700 (naval brass), composition and machinability rating **A16:** 806
C46600 ..... **M2:** 349-351
composition **M2:** 467
fasteners **M2:** 444
tubes, applications **M2:** 262
tubes, mechanical properties **M2:** 263
C46700 ..... **M2:** 349-351
composition **M2:** 467
fasteners **M2:** 444
tubes, applications **M2:** 262
tubes, mechanical properties **M2:** 263
C48200 ..... **M2:** 351-352
C48200 (medium-leaded naval brass) **A16:** 323, 324, 325
C48200, composition and machinability rating **A16:** 806
C48500 ..... **M2:** 352-353
fasteners **M2:** 444
C48500 (leaded naval brass) **A16:** 323, 324, 325
C48500, composition and machinability rating **A16:** 806
C50500 ..... **M2:** 353-354

Copper alloys, specific types (continued)
C50500-C52400 (phosphor bronze) composition and machinability rating **A16:** 806
C51000 ..... **M2:** 354-355
aldehydes, corrosion in **M2:** 481
composition **M2:** 467 **M3:** 746
ester solutions, corrosion in **M2:** 480
ethylene glycol, corrosion in **M2:** 482
fasteners **M2:** 443
oleic acid, corrosion in **M2:** 476
paper-mill vapor, corrosion **M2:** 483
stress relaxation **M2:** 489
stress-relieving temperatures **M2:** 256
sulfuric acid, corrosion in **M2:** 474
tensile properties at subzero temperatures **M3:** 747
C51000 (phosphor bronze) **A16:** 361, 814, 815
C51100 ..... **M2:** 355
C52100 ..... **M2:** 355-356
atmospheric corrosion **M2:** 470
composition **M2:** 467
paper-mill vapor, corrosion in **M2:** 483
C52400 ..... **M2:** 356
C52400, clad to phosphor bronze C10200 copper, clad bronzing **A6:** 347
C54400 ..... **M2:** 356-357
C54400 (free-cutting phosphor bronze) **A16:** 323, 324, 325, 363
C54400, composition and machinability rating **A16:** 806
C60600 ..... **M2:** 357-358
C60800 ..... **M2:** 358
ester solutions, corrosion in **M2:** 479
ethylene glycol, corrosion in **M2:** 482
tubes, applications **M2:** 262
tubes, mechanical properties **M2:** 263
C60800-C61000 (Al bronze) **A16:** 813, 814
C60800-C62400, composition and machinability rating **A16:** 806
C61000 ..... **M2:** 358-359
atmospheric corrosion **M2:** 470
C61300 ..... **M2:** 359-360, 361
composition **M2:** 467
oleic acid, corrosion in **M2:** 476
wear applications **M3:** 592
wear-test results **M3:** 591
C61400 (Al bronze, D) **A16:** 361, 363
C61400, fasteners **M2:** 359, 360-361, 443
C61400, wear applications **M3:** 592
C61500 ..... **M2:** 362
C61800
ethylene glycol, corrosion in **M2:** 482
paper-mill vapor, corrosion in **M2:** 483
C62300 ..... **M2:** 362-363
ester solutions, corrosion in **M2:** 479
C62300-C62400 (Al bronze) **A16:** 323, 324, 325
C62400 ..... **M2:** 364
wear applications **M3:** 592
wear-test results **M3:** 591
C62500 ..... **M2:** 364-365
wear applications **M3:** 592
wear-test results **M3:** 591
C62500 (Ampco 21, Wearite 4-14) composition and machinability rating **A16:** 806
C63000 ..... **M2:** 365-366
bearing wear **M3:** 593
ethylene glycol, corrosion in **M2:** 482
fasteners **M2:** 443
wear applications **M3:** 592
C63000 (NI-AL bronze), composition and machinability rating **A16:** 806
C63000 aluminum bronze, fatigue fracture **A11:** 114
C63200 ..... **M2:** 366, 367
C63600 ..... **M2:** 367
ester solutions, corrosion in **M2:** 479
C63700, composition **M2:** 467
C63800 ..... **M2:** 367-369
C63800 (Coronze) **A16:** 323, 324, 325
C63800, composition and machinability rating **A16:** 806
C63800, high-strength **A2:** 235
C63900 (Al-Si bronze), electrochemical machining **A16:** 537
C64200 (Al bronze) **A16:** 323, 324, 325

Copper alloys, specific types (continued)
C64200, composition and machinability rating **A16:** 806
C64200, fasteners **M2:** 443
C64200, use for fasteners **M3:** 184
C64700, age hardenable **A2:** 236
C65100 ..... **M2:** 369
composition **M2:** 467
fasteners **M2:** 443, 444
tubes, applications **M2:** 262
tubes, mechanical properties **M2:** 263
C65100 (low-Si bronze, B), composition and machinability rating **A16:** 806
C65100, use for fasteners **M3:** 184
C65400, high-strength alloy **A2:** 236
C65500 ..... **M2:** 369-370
acetic acid, corrosion in **M2:** 476
acetic acid-acetic enhydride, corrosion in **M2:** 475
alcohols, corrosion in **M2:** 482
aldehydes, corrosion in **M2:** 481
amine-system, corrosion in **M2:** 478
atmospheric corrosion **M2:** 470
composition **M2:** 467
ester solution, corrosion in **M2:** 479
ethers, corrosion in **M2:** 481
ethylene glycol, corrosion in **M2:** 482
fasteners **M2:** 443, 444
hydrocyanic acid, corrosion in **M2:** 477
hydrogen cyanide, corrosion in **M2:** 476
ketones, corrosion in **M2:** 481
oleic acid, corrosion in **M2:** 476
stearic acids, corrosion in **M2:** 476-477
sulfuric acid, corrosion in **M2:** 473
tubes, applications **M2:** 262
tubes, mechanical properties **M2:** 263
C65500 (high-Si bronze, A), composition and machinability rating **A16:** 806
C65500, in sulfuric acid **A13:** 627
C65500, use for fasteners **M3:** 184
C65600 (high-Si bronze, D) **A16:** 361
C65800
hydrochloric acid, corrosion in **M2:** 475
paper-mill vapor, corrosion in **M2:** 483
C65800, hydrochloric acid corrosion **A13:** 628-629
C66100 (leaded Si bronze D), composition and machinability rating **A16:** 806
C66100 fasteners **M2:** 443
C66400 ..... **M2:** 370, 371
C66400, low-zinc brass **A2:** 236
C67500 (Mn bronze, A) **A16:** 361
C67500, composition and machinability rating **A16:** 806
C67500 fasteners **M2:** 443
C68700
composition **M2:** 467
tubes, applications **M2:** 262
tubes, mechanical properties **M2:** 263
tubes, used for **M2:** 472
C68700 (Al brass), composition and machinability rating **A16:** 806
C68700, effect of arsenic in **A11:** 635
C68800 ..... **M2:** 370-371
C68800 (Alcoloy), composition and machinability rating **A16:** 806
C68800, high-strength modified aluminum brass **A2:** 235
C69000 ..... **M2:** 372
C69400 ..... **M2:** 372
C69400 (Si red brass) **A16:** 363
C69400, composition and machinability rating **A16:** 806
C70250, age hardenable **A2:** 236
C70400 ..... **M2:** 373
C70600 ..... **M2:** 373-374
composition **M2:** 467 **M3:** 746
ethylene glycol, corrosion in **M2:** 482
tensile properties at subzero temperatures **M3:** 747
tubes, applications **M2:** 262
tubes, mechanical properties **M2:** 263
tubes, used for **M2:** 472
Young's modulus **M3:** 733
C70600 (Cu-Ni, 10%), composition and machinability rating **A16:** 806
C70600, chlorination effects **A13:** 626

Copper alloys, specific types (continued)
C70600, corrosion rate as function of seawater velocity and sulfide content A13: 625
C70600, electrochemical machining A16: 537
C70600, fouling rates as function of seawater velocity A13: 626
C70600 heat-exchanger tubing A11: 633
C70600, in seawater, corrosion rates with sulfide additions A13: 625
C70600, iron effect on seawater corrosion A13: 623
C70600, salt water corrosion A13: 623
C70600 tube, hydraulic oil cooler A11: 633
C70600/C71500, galvanic couple data, in flowing seawater A13: 624
C71000 .... M2: 374-375
beet-sugar, corrosion in M2: 483
fasteners M2: 443
steam condensate, corrosion in M2: 471
tartaric acid, corrosion in M2: 477
C71000 (Cu-Ni), composition and machinability rating A16: 806
C71300 tartaric acid, corrosion in M2: 477
C71500 .... M2: 375-376
amine-system, corrosion in M2: 478
beet-sugar, corrosion in M2: 483
composition M2: 467 M3: 746
ester solutions, corrosion in M2: 479
ethylene glycol, corrosion in M2: 482
fasteners M2: 443
gasoline, corrosion in M2: 483
hydrofluoric, corrosion in M2: 475
stress relieving temperatures M2: 256
tensile properties at subzero temperatures M3: 747
tubes, applications M2: 262
tubes, mechanical properties M2: 263
tubes, used for M2: 472
Young's modulus M3: 733
C71500 (Cu-Ni, 30%), composition and machinability rating A16: 806
C71500, gasoline corrosion A13: 635
C71500, salt water corrosion A13: 623
C71640, iron effects on seawater corrosion A13: 623
C71900 .... M2: 376-377
C71900, spinodal decomposition hardening A2: 236
C72200 .... M2: 377-378
C72200, iron effects on seawater corrosion A13: 623
C72500 (Cu-Ni, tin-bearing), composition and machinability rating A16: 806
C72500 stress relaxation M2: 489
C73200 paper-mill vapor, corrosion in M2: 483
C74500 .... M2: 378-379
fasteners M2: 444
C74500 (Ni-Ag, 65-10), composition and machinability rating A16: 806
C75200 .... M2: 379-380
composition M2: 467
paper-mill vapor, corrosion in M2: 483
stress-relieving temperatures M2: 256
C75200 (Ni-Ag, 65-18), composition and machinability rating A16: 806
C75400 .... M2: 380
C75400 (Ni-Ag, 65-15), composition and machinability rating A16: 806
C75700 .... M2: 380-381
C75700 (Ni-Ag, 65-12) A16: 363
C75700, composition and machinability rating A16: 806
C77000 .... M2: 381
paper-mill vapor, corrosion in M2: 483
C77000 (Ni-Ag), composition and machinability rating A16: 806
C78200 .... M2: 382
C78200 (leaded Ni-AG) A16: 323, 324, 325

Copper alloys, specific types (continued)
C78200, composition and machinability rating A16: 806
C80100-C81100 (Cu), composition and machinability rating A16: 807
C81100 .... M2: 395
C81300-C82800 (Be-Cu), composition and machinability rating A16: 807
C81400 .... M2: 395-396
composition M2: 393
mechanical properties M2: 393
C81400, as age hardenable A2: 236
C81500 .... M2: 396
composition M2: 393
mechanical properties M2: 393
C81800 .... M2: 396-397
composition M2: 393
mechanical properties M2: 393
C82000 .... M2: 397-398
composition M2: 393
mechanical properties M2: 393
C82200 .... M2: 398-399
composition M2: 393
mechanical properties M2: 393
C82400 .... M2: 399-400
C82500 .... M2: 400-402
composition M2: 393
mechanical properties M2: 393
C82500, use in molds for plastics and rubber M3: 549
C82600 .... M2: 402-403
C82600, use in molds for plastics and rubber M3: 549
C82800 .... M2: 403-405
composition M2: 393
mechanical properties M2: 393
C82800, use in molds for plastics and rubber M3: 549
C83300-C83800 (leaded red brass) composition and machinability rating A16: 807
C83420, bearing material systems A18: 746
C83520, bearing material systems A18: 746
C83600 .... M2: 404-406
applications M2: 392
composition M2: 384
foundry properties, sand casting M2: 385
machinability M2: 389
mechanical properties M2: 386
working stress, ASME Code castings M2: 390
C83600, allowable working stresses A2: 352
C83600, as general-purpose copper casting alloy A2: 351
C83800 .... M2: 406-407
applications M2: 392
composition M2: 384
machinability M2: 389
mechanical properties M2: 386
C83800, as general-purpose copper casting alloy A2: 351
C84200-C84800 (leaded semi-red brass) composition and machinability rating A16: 807
C84400 .... M2: 407
applications M2: 392
composition M2: 384
foundry properties, sand casting M2: 385
machinability M2: 389
mechanical properties M2: 386
C84400, as general-purpose copper casting alloy A2: 351
C84800 .... M2: 407-408
applications M2: 392
composition M2: 384
foundry properties, sand casting M2: 385
machinability M2: 389
mechanical properties M2: 386
C84800, as general-purpose copper casting alloy A2: 351-352

Copper alloys, specific types (continued)
C85200 .... M2: 408
applications M2: 392
composition M2: 384
machinability M2: 389
mechanical properties M2: 386
C85200, as general-purpose copper casting alloy A2: 352
C85200-C85800 (leaded yellow brass) composition and machinability rating A16: 807
C85400 .... M2: 408-409
applications M2: 392
composition M2: 384
foundry properties, sand casting M2: 385
machinability M2: 389
mechanical properties M2: 386
C85700 .... M2: 409
composition M2: 384
mechanical properties M2: 386
C85700, as general-purpose copper casting lloy A2: 352
C85800 .... M2: 409
composition M2: 384
foundry properties, sand casting M2: 385
mechanical properties M2: 386
C85800, as die casting, composition A15: 286
C86100 .... M2: 409-410
C86100-C86800 (Mn bronze), composition and machinability rating A16: 807
C86200 .... M2: 409-410
compositions M2: 384
mechanical properties M2: 386
C86300 .... M2: 410-411
composition M2: 384
foundry properties, sand casting M2: 385
machinability M2: 389
mechanical properties M2: 386
C86300, wear-test results M3: 591
C86400 .... M2: 411-412
composition M2: 384
machinability M2: 389
mechanical properties M2: 386
C86500 .... M2: 412-415
composition M2: 384
foundry properties, sand casting M2: 385
machinability M2: 389
mechanical properties M2: 386
C86700
composition M2: 384
mechanical properties M2: 386
C87200 .... M2: 416
composition M2: 384
foundry properties, sand casting M2: 385
mechanical properties M2: 386
C87200-C87400 (Si bronze), composition and machinability rating A16: 807
C87400
composition M2: 384
mechanical properties M2: 386
C87500 .... M2: 416-417
composition M2: 384
foundry properties, sand casting M2: 385
mechanical properties M2: 386
C87500-C87900 (Si brass), composition and machinability rating A16: 807
C87600
composition M2: 384
mechanical properties M2: 386
C87800 .... M2: 416-417
composition M2: 384
mechanical properties M2: 386
C87800, as die casting, composition A15: 286
C87900
composition M2: 384
mechanical properties M2: 386
C87900, as die casting, composition A15: 286
C90200-C91700 (Sn bronze), composition and machinability rating A16: 807

Copper alloys, specific types (continued)
C90300 ... **M2:** 417-418
composition ... **M2:** 384
foundry properties, sand casting ... **M2:** 385
machinability ... **M2:** 389
mechanical properties ... **M2:** 386
C90300, as general-purpose copper casting alloy ... **A2:** 352
C90500 ... **M2:** 418
composition ... **M2:** 384
machinability ... **M2:** 389
mechanical properties ... **M2:** 386
C90500, as general-purpose copper casting alloy ... **A2:** 352
C90500, bearing wear ... **M3:** 592
C90700 ... **M2:** 418
C90700, phosphor bronze, for gear applications ... **A2:** 352
C91100
applications ... **M2:** 394
composition ... **M2:** 384
C91100, bridge turntable application ... **A2:** 352
C91300
applications ... **M2:** 394
composition ... **M2:** 384
C91300, bridge turntable application ... **A2:** 352
C91700 ... **M2:** 418-419
C91700, wear-test results ... **M3:** 591
C92200 ... **M2:** 419-421
composition ... **M2:** 384
foundry properties, sand casting ... **M2:** 385
machinability ... **M2:** 389
mechanical properties ... **M2:** 386
working stress, ASME Code castings ... **M2:** 390
C92200 (Navy M bronze), composition and machinability rating ... **A16:** 807
C92200, allowable working stresses ... **A2:** 352
C92200, corrosion resistance ... **A2:** 352
C92300 ... **M2:** 422
composition ... **M2:** 384
machinability ... **M2:** 389
mechanical properties ... **M2:** 386
C92300 (Navy G bronze), composition and machinability rating ... **A16:** 807
C92500 ... **M2:** 422
C92500-C92800 (leaded Sn bronze) composition and machinability rating ... **A16:** 807
C92600 ... **M2:** 422-423
C92700 ... **M2:** 423
C92900 ... **M2:** 423
C93200 ... **M2:** 424
applications ... **M2:** 394
composition ... **M2:** 384
machinability ... **M2:** 389
mechanical properties ... **M2:** 386
C93200, bearing wear ... **M3:** 592
C93200-C93900 (high-leaded Sn bronze) composition and machinability rating ... **A16:** 807
C93500 ... **M2:** 424
applications ... **M2:** 394
composition ... **M2:** 384
machinability ... **M2:** 389
mechanical properties ... **M2:** 386
C93700 ... **M2:** 424-427
applications ... **M2:** 394
composition ... **M2:** 384
foundry properties, sand casting ... **M2:** 385
machinability ... **M2:** 389
mechanical properties ... **M2:** 386
C93800 ... **M2:** 427-428
applications ... **M2:** 394
composition ... **M2:** 384
machinability ... **M2:** 389
mechanical properties ... **M2:** 386
C93900 ... **M2:** 428
C94100 composition ... **M2:** 384
C94300 ... **M2:** 428
applications ... **M2:** 394
composition ... **M2:** 384
foundry properties, sand casting ... **M2:** 385
machinability ... **M2:** 389
mechanical properties ... **M2:** 386
C94400, composition ... **M2:** 384

Copper alloys, specific types (continued)
C94500 ... **M2:** 429
composition ... **M2:** 384
C94700
composition ... **M2:** 384
mechanical properties ... **M2:** 386
C94700-C94800 (Ni-Sn bronze), composition and machinability rating ... **A16:** 807
C94800
composition ... **M2:** 384
mechanical properties ... **M2:** 386
C94900
composition ... **M2:** 384
mechanical properties ... **M2:** 386
C95200 ... **M2:** 429-431
composition ... **M2:** 384
machinability ... **M2:** 389
mechanical properties ... **M2:** 386
C95200-C95500 (Al bronze), composition and machinability rating ... **A16:** 807
C95300 ... **M2:** 430-431
composition ... **M2:** 384
foundry properties, sand casting ... **M2:** 385
machinability ... **M2:** 389
mechanical properties ... **M2:** 386
C95400 ... **M2:** 431-433
bearing wear ... **M3:** 592
composition ... **M2:** 384
machinability ... **M2:** 389
mechanical properties ... **M2:** 386
wear applications ... **M3:** 592
wear-test results ... **M3:** 591
C95500 ... **M2:** 433-434
composition ... **M2:** 384
machinability ... **M2:** 389
mechanical properties ... **M2:** 386
C95500, wear-test results ... **M3:** 591
C95600 (Si-Al bronze) ... **A16:** 817
C95600, composition and machinability rating ... **A16:** 807
C95600, machinability ... **M2:** 389
C95700 ... **M2:** 434
C95700 (Mn-Al bronze), composition and machinability rating ... **A16:** 807
C95800 ... **M2:** 434-435
foundry properties, sand casting ... **M2:** 385
C95800 (propeller bronze), composition and machinability rating ... **A16:** 807
C96200-C96400 (Cu Ni), composition and machinability rating ... **A16:** 807
C96400 ... **M2:** 435
C96600 ... **M2:** 435-436
C96600 (Be cupro-Ni), composition and machinability rating ... **A16:** 807
C97300 ... **M2:** 436
composition ... **M2:** 384
machinability ... **M2:** 389
mechanical properties ... **M2:** 386
C97300-C97400 (leaded Ni brass) composition and machinability rating ... **A16:** 807
C97600 ... **M2:** 436-437
composition ... **M2:** 384
foundry properties, sand casting ... **M2:** 385
mechanical properties ... **M2:** 386
C97600-C97800 (leaded Ni bronze) composition and machinability rating ... **A16:** 807
C97800 ... **M2:** 437
composition ... **M2:** 384
foundry properties, sand casting ... **M2:** 385
mechanical properties ... **M2:** 386
C99300 (Incramant 900), composition and machinability rating ... **A16:** 807
C99400 ... **M2:** 437
C99700 ... **M2:** 437-438
Cu-1-10
fiber for reinforcement ... **A18:** 803
processing technique ... **A18:** 803
properties ... **A18:** 803
Cu-2.5Be, fully aged, strength ... **A12:** 402
Cu-2.5Be, underaged fracture-toughness test ... **A12:** 402
Cu-3Zn, expected pole orientations of preferred orientations ... **A10:** 360
Cu-3Zn, grain orientations ... **A10:** 360
Cu-3Zn, measure (111) pole figure for ... **A10:** 360

Copper alloys, specific types (continued)
Cu-5Sn-5Pb-4Zn, corrosion fatigue failure ... **A12:** 403
Cu-5Zn
annealing time and temperature effect on hardness ... **A4:** 828
annealing time and temperature effect on microstructure ... **A4:** 829
annealing time effect on annealing process ... **A4:** 830
microstructure showing deformation and bent annealing twins ... **A4:** 828
recrystallization ... **A4:** 829
Cu-8Al, intergranular fatigue cracking ... **A11:** 254
CU-10MoS$_2$ alloys, sliding wear ... **A18:** 810
Cu-10Ni *See* C70600
Cu-10Sn (tin-bronze), preferential corrosion ... **A12:** 403
Cu-20MoS$_2$ alloys, sliding wear ... **A18:** 810
Cu-20Pd-3In, brazing ... **A6:** 945
Cu-30Ni *See* C71500
Cu-30Zn brass, diamond abrasive polishing wear ... **A18:** 195
Cu-30Zn brass, fracture stress as function of liquid mercury exposure ... **A11:** 226
Cu-38Zn-lPb, erosive attack of brass melt on the die surface ... **A18:** 631
Cu-40Zn
heat-treatment effect on hardness ... **A4:** 837, 838
microstructure after quenching ... **A4:** 838
two-phase structure development ... **A4:** 837, 838
Cu-42Zn, two-phase structure development ... **A4:** 838, 839
Cu-43Zn, two-phase structure development ... **A4:** 838, 839
Cu-Ag-P-Mg
fiber for reinforcement ... **A18:** 803
processing technique ... **A18:** 803
properties ... **A18:** 803
Cu-Zn, plastic deformation ... **A4:** 827
Cu$_3$Au, atomic interaction descriptions ... **A6:** 144
ETP, dislocation cell structure, by cold rolling ... **A10:** 469
OFHC, dislocation cell structure analysis ... **A10:** 472-473
phosphor bronze 5% A, cutoff band sawing with bimetal blades ... **A6:** 1184
phosphor bronze C (C52100) wire cloth, fatigue fracture from warp tension ... **A12:** 404
Copper alloys, use for bearings *See also* Bearings, sliding ... **M3:** 816-818
Copper alloys with zinc
solderable and protective finishes for substrate materials ... **A6:** 979
Copper alloys, wrought, specific types
78200, annealing temperature ... **A4:** 882
C10100, annealing temperature ... **A4:** 882
C10200, annealing temperature ... **A4:** 882
C10300, annealing temperature ... **A4:** 882
C10400, annealing temperature ... **A4:** 882
C10500, annealing temperature ... **A4:** 882
C10600, annealing temperature ... **A4:** 882
C10700, annealing temperature ... **A4:** 882
C10800, annealing temperature ... **A4:** 882
C11100
annealing temperature ... **A4:** 882
stress-relieving temperature ... **A4:** 885
C11300, annealing temperature ... **A4:** 882
C11400, annealing temperature ... **A4:** 882
C11500, annealing temperature ... **A4:** 882
C11600, annealing temperature ... **A4:** 882
C12000
annealing temperature ... **A4:** 882
stress-relieving temperature ... **A4:** 885
C12200
annealing temperature ... **A4:** 882
stress-relieving temperature ... **A4:** 885
C12500, annealing temperature ... **A4:** 882
C12700, annealing temperature ... **A4:** 882
C13000, annealing temperature ... **A4:** 882
C14200, stress-relieving temperature ... **A4:** 885
C14500, annealing temperature ... **A4:** 882
C14700, annealing temperature ... **A4:** 882
C15000
aging temperature ... **M4:** 725
hardness ... **A4:** 886

Copper alloys, wrought, specific types (continued)
heat treatments ... A4: 896
properties, heat treated ... M4: 725, 735
properties, heat-treated ... A4: 886
solution-treating temperature ... M4: 725
C15500, annealing temperature ... A4: 882
C16200, annealing temperature ... A4: 882
C17000
aging temperature ... A4: 889 M4: 725, 728
annealing temperature ... A4: 882 M4: 720
precipitation treatment ... M4: 728, 730-731
precipitation treatments ... A4: 891
properties, heat treated ... M4: 725, 730-731
properties, heat-treated ... A4: 886
solution-treating temperature ... A4: 889 M4: 725, 728
C17200
aging ... A4: 888, 890, 894
aging temperature ... A4: 889, 892 M4: 725, 728, 729
annealing temperature ... A4: 882 M4: 720
hardness ... A4: 894, 895
hardness measurement ... M4: 733
hardness variation ... M4: 734
precipitation treatment ... M4: 728, 730-731, 732
precipitation treatments ... A4: 891, 892
properties, heat treated ... M4: 725, 730-731, 732
properties, heat-treated ... A4: 886, 892
solution-treating ... A4: 890
solution-treating temperature ... A4: 889 M4: 725, 728, 729
tensile strength ... A4: 894
tensile strength variation ... M4: 734
C17300
aging temperature ... A4: 889 M4: 725, 728
properties, heat treated ... M4: 725
properties, heat-treated ... A4: 886
solution-treating temperature ... A4: 889 M4: 725, 728
C17500
aging temperature ... A4: 889, 892 M4: 725, 728
annealing temperature ... A4: 882 M4: 720
precipitation treatment ... A4: 891, 892 M4: 728, 730-731, 732
properties, heat treated ... M4: 725, 730-731, 732
properties, heat-treated ... A4: 886, 892
solution-treating ... A4: 890
solution-treating temperature ... A4: 889 M4: 725, 728, 729
C17510
aging temperature ... A4: 889 M4: 728
precipitation treatment ... A4: 891 M4: 728, 730-731
properties, heat treated ... M4: 730-731
solution-treating temperature ... A4: 889 M4: 728
C17600
aging temperature ... M4: 725
properties, heat treated ... M4: 725
solution-treating temperature ... M4: 725
C17600, properties, heat-treated ... A4: 886
C18000
aging temperature ... M4: 725
hardness and electrical conductivity ... A4: 886
properties, heat treated ... M4: 725
properties, heat-treated ... A4: 886
solution-treating temperature ... M4: 725
C18200
aging temperature ... M4: 725
hardness ... A4: 886
heat treatments ... A4: 895
properties, heat treated ... M4: 725, 735
properties, heat-treated ... A4: 886
solution-treating temperature ... M4: 725
C18400
aging temperature ... M4: 725
properties, heat treated ... M4: 725
solution-treating temperature ... M4: 725
C18400, properties, heat-treated ... A4: 886

Copper alloys, wrought, specific types (continued)
C18500
aging temperature ... M4: 725
properties, heat treated ... M4: 725
solution-treating temperature ... M4: 725
C18500, properties, heat-treated ... A4: 886
C19000, precipitation heat treatment ... A4: 894
C19100, precipitation heat treatment ... A4: 894
C19200, annealing temperature ... A4: 882
C19400, annealing temperature ... A4: 882
C19500, annealing temperature ... A4: 882
C21000
annealing temperature ... A4: 882 M4: 720
stress-relieving temperature ... A4: 885 M4: 724
C22000
annealing temperature ... A4: 882 M4: 720
stress-relieving temperature ... A4: 885 M4: 724
C22600
annealing temperature ... A4: 882
stress-relieving temperature ... A4: 885
C22600, annealing temperature ... M4: 720
C23000
annealing temperature ... A4: 882 M4: 720
stress-relieving temperature ... A4: 885 M4: 724
C24000, annealing temperature ... A4: 882 M4: 720
C26000
annealing ... A4: 881, 883
annealing data ... M4: 722-723
annealing temperature ... A4: 882 M4: 720
properties, annealed ... A4: 884 M4: 723
stress-relieving temperature ... A4: 885 M4: 724
C26800, annealing temperature ... A4: 882 M4: 720
C27000
annealing ... A4: 881
annealing data ... M4: 721
annealing temperature ... A4: 882 M4: 720
stress-relieving temperature ... A4: 885 M4: 724
C27400, annealing temperature ... A4: 882 M4: 720
C28000
annealing temperature ... M4: 720
stress-relieving temperature ... M4: 724
C28000, annealing temperature ... A4: 882
C31400
annealing temperature ... A4: 882
stress-relieving temperature ... A4: 885
C31400, annealing temperature ... M4: 720
C31600, annealing temperature ... A4: 882
C33000
annealing temperature ... A4: 882
stress-relieving temperature ... A4: 885
C33000, annealing temperature ... M4: 720
C33200
annealing temperature ... A4: 882
stress-relieving temperature ... A4: 885
C33200, annealing temperature ... M4: 720
C33500
annealing temperature ... A4: 882
stress-relieving temperature ... A4: 885
C33500, annealing temperature ... M4: 720
C34000
annealing temperature ... A4: 882
stress-relieving temperature ... A4: 885
C34000, annealing temperature ... M4: 720
C34200, annealing temperature ... A4: 882 M4: 720
C35000
annealing temperature ... A4: 882
stress-relieving temperature ... A4: 885
C35000, annealing temperature ... M4: 720
C35300
annealing temperature ... A4: 882
stress-relieving temperature ... A4: 885
C35300, annealing temperature ... M4: 720
C35600
annealing temperature ... A4: 882
stress-relieving temperature ... A4: 885
C35600, annealing temperature ... M4: 720
C36000
annealing temperature ... A4: 882 M4: 720

Copper alloys, wrought, specific types (continued)
stress-relieving temperature ... A4: 885 M4: 724
C36500, annealing temperature ... A4: 882 M4: 720
C36600, annealing temperature ... A4: 882 M4: 720
C36700, annealing temperature ... A4: 882 M4: 720
C36800, annealing temperature ... A4: 882 M4: 720
C37000, annealing temperature ... A4: 882 M4: 720
C37700
annealing temperature ... A4: 882
stress-relieving temperature ... A4: 885
C37700, annealing temperature ... M4: 720
C38500, annealing temperature ... A4: 882 M4: 720
C41100, annealing temperature ... A4: 882
C41300, annealing temperature ... A4: 882
C42500, annealing temperature ... A4: 882
C43000, stress-relieving temperature ... A4: 885
C43400, stress-relieving temperature ... A4: 885
C44300
annealing temperature ... M4: 720
stress-relieving temperature ... M4: 724
C44330
annealing temperature ... A4: 882
stress-relieving temperature ... A4: 885
C44400
annealing temperature ... A4: 882 M4: 720
stress-relieving temperature ... A4: 885 M4: 724
C44500
annealing temperature ... A4: 882 M4: 720
stress-relieving temperature ... A4: 885 M4: 724
C46200
annealing temperature ... A4: 882
stress-relieving temperature ... A4: 885
C46200, annealing temperature ... M4: 720
C46400-C46700
annealing temperature ... A4: 882
stress-relieving temperature ... A4: 885
C48200, annealing temperature ... A4: 882 M4: 720
C48500, annealing temperature ... A4: 882 M4: 720
C50500, annealing temperature ... A4: 882 M4: 720
C51000
annealing temperature ... A4: 882 M4: 720
stress-relieving temperature ... A4: 885 M4: 724
C51100, properties, annealed ... A4: 884 M4: 723
C52100
annealing temperature ... A4: 882 M4: 720
homogenization ... A4: 880, 881
stress-relieving temperature ... A4: 885 M4: 724
C52400
annealing temperature ... A4: 882
homogenization ... A4: 881
C53200
annealing temperature ... A4: 882 M4: 720
properties, annealed ... A4: 884 M4: 723
C53400
annealing temperature ... A4: 882 M4: 720
properties, annealed ... A4: 884 M4: 723
C54200, annealing temperature ... M4: 720
C54400
annealing temperature ... A4: 882 M4: 720
properties, annealed ... A4: 884 M4: 723
stress-relieving temperature ... A4: 885
C60600
annealing temperature ... A4: 882
heat treatments ... A4: 898
C60600, annealing temperature ... M4: 720
C60800, annealing temperature ... A4: 882 M4: 720
C61000
annealing temperature ... A4: 882
heat treatments ... A4: 898
C61000, annealing temperature ... M4: 720
C61300
annealing temperature ... A4: 882 M4: 720
heat treatments ... A4: 897, 898
stress-relieving temperature ... M4: 724
C61400
annealing temperature ... A4: 882 M4: 720
heat treatments ... A4: 897, 898
stress-relieving temperature ... M4: 724
C61500, annealing ... A4: 886

Copper alloys, wrought, specific types (continued)
C61800, annealing temperature **A4:** 882 **M4:** 720
C61900, annealing temperature **A4:** 882 **M4:** 720
C62300, annealing temperature **A4:** 882
C62400
annealing temperature **A4:** 882 **M4:** 720
properties, heat treated **A4:** 898 **M4:** 737
C62500, annealing temperature **A4:** 882
C63000
annealing temperature **A4:** 882 **M4:** 720
properties, heat treated **M4:** 737
properties, heat-treated **A4:** 898
quenching **A4:** 898
C63200, annealing temperature **A4:** 882 **M4:** 720
C63800
annealing **A4:** 886
annealing temperature **A4:** 882
C64200, annealing temperature **A4:** 882 **M4:** 720
C65100
annealing temperature **A4:** 882
stress-relieving temperature **A4:** 885
C65100, annealing temperature **M4:** 720
C65500
annealing temperature **A4:** 882 **M4:** 720
stress-relieving temperature **A4:** 885 **M4:** 724
C66700, annealing temperature **A4:** 882
C67000, annealing temperature **A4:** 882 **M4:** 720
C67400, annealing temperature **A4:** 882
C67500, annealing temperature **A4:** 882 **M4:** 720
C68700
annealing temperature **A4:** 882
stress-relieving temperature **A4:** 885
C68700, annealing temperature **M4:** 720
C68800
annealing **A4:** 886
annealing temperature **A4:** 882
C69000, annealing **A4:** 886
C69700, stress-relieving temperature **A4:** 885
C70600
annealing temperature **A4:** 882 **M4:** 720
stress-relieving temperature **A4:** 885 **M4:** 724
C71000, annealing temperature **A4:** 882
C71500
annealing temperature **A4:** 882 **M4:** 720
stress-relieving temperature **A4:** 885 **M4:** 724
C71900
aging **A4:** 897
aging temperature **M4:** 725
homogenization applications **A4:** 880-881
properties, heat treated **M4:** 725
properties, heat-treated **A4:** 886
solution-treating temperature **A4:** 896 **M4:** 725
times for spinodal alloys **A4:** 896
C72500, annealing temperature **A4:** 882
C72600
aging times **A4:** 897
solution-treating temperature **A4:** 896
strengths **A4:** 897
times for spinodal alloys **A4:** 896
C72700
aging times **A4:** 897
solution-treating temperature **A4:** 896
strengths **A4:** 897
times for spinodal alloys **A4:** 896
C72800
aging temperature **M4:** 725
aging times **A4:** 897
properties, heat treated **A4:** 886 **M4:** 725
solution-treating temperature **A4:** 896 **M4:** 725
strengths **A4:** 897
times for spinodal alloys **A4:** 896
C72900
aging times **A4:** 897
solution-treating temperature **A4:** 896
strengths **A4:** 897
times for spinodal alloys **A4:** 896
C73500, stress-relieving temperature **A4:** 885
C74500
annealing temperature **A4:** 882
stress-relieving temperature **A4:** 885
C75200
annealing temperature **A4:** 882 **M4:** 720
properties, annealed **A4:** 884
stress-relieving temperature **A4:** 885 **M4:** 724
C75400
annealing temperature **A4:** 882

Copper alloys, wrought, specific types (continued)
stress-relieving temperature **A4:** 885
C75700
annealing temperature **A4:** 882
stress-relieving temperature **A4:** 885
C75700, annealing temperature **M4:** 720
C76200, properties, annealed **M4:** 723
C77000
annealing temperature **A4:** 882
stress-relieving temperature **A4:** 885
C77000, annealing temperature **M4:** 720
C81300
aging temperature **A4:** 892 **M4:** 733
properties, heat treated **M4:** 733
properties, heat-treated **A4:** 892
solution-treating temperature **A4:** 892 **M4:** 733
C81500
aging temperature **M4:** 725
heat treatment **A4:** 895
properties, heat treated **M4:** 729, 735
properties, heat-treated **A4:** 886
solution-treating temperature **M4:** 725
C81540
aging temperature **M4:** 725
properties, heat treated **M4:** 725
solution-treating temperature **M4:** 725
C81540, properties, heat-treated **A4:** 886
C81700
aging temperature **A4:** 892 **M4:** 733
properties, heat treated **M4:** 733
properties, heat-treated **A4:** 892
solution-treating temperature **A4:** 892 **M4:** 733
C81800
aging temperature **A4:** 892 **M4:** 733
properties, heat treated **M4:** 733
properties, heat-treated **A4:** 892
solution-treating temperature **A4:** 892 **M4:** 733
C82000
aging temperature **A4:** 892 **M4:** 733
properties, heat treated **M4:** 733
properties, heat-treated **A4:** 892
solution-treating temperature **A4:** 892 **M4:** 733
C82100
aging temperature **A4:** 892 **M4:** 733
properties, heat treated **M4:** 733
properties, heat-treated **A4:** 892
solution-treating temperature **A4:** 892 **M4:** 733
C82200
aging temperature **A4:** 892 **M4:** 733
properties, heat treated **M4:** 733
properties, heat-treated **A4:** 892
solution-treating temperature **A4:** 892 **M4:** 733
C82400
aging temperature **A4:** 892 **M4:** 733
properties, heat treated **M4:** 733
properties, heat-treated **A4:** 892
solution-treating temperature **A4:** 892 **M4:** 733
C82500
aging temperature **A4:** 892 **M4:** 733
properties, heat treated **M4:** 733
properties, heat-treated **A4:** 892
solution-treating temperature **A4:** 892 **M4:** 733
C82600
aging temperature **A4:** 892 **M4:** 733
properties, heat treated **M4:** 733
properties, heat-treated **A4:** 892
solution-treating temperature **A4:** 892 **M4:** 733
C82700
aging temperature **A4:** 892 **M4:** 733
properties, heat treated **M4:** 733
properties, heat-treated **A4:** 892
solution-treating temperature **A4:** 892 **M4:** 733
C82800
aging temperature **A4:** 892 **M4:** 733
properties, heat treated **M4:** 733
properties, heat-treated **A4:** 892
solution-treating temperature **A4:** 892 **M4:** 733
C94700
aging **A4:** 894
aging temperature **M4:** 725
properties, heat treated **M4:** 725, 735
properties, heat-treated **A4:** 886, 896
solution-treating temperature **A4:** 896 **M4:** 725
C94800
aging **A4:** 894
properties, heat-treated **A4:** 896

Copper alloys, wrought, specific types (continued)
solution-treating temperature **A4:** 896
C94800, properties, heat treated **M4:** 735
C95300
annealing temperature **A4:** 882
properties, heat-treated **A4:** 898
C95400
annealing temperature **A4:** 882
properties, heat-treated **A4:** 898
C95400, properties, heat treated **M4:** 737
C95500
annealing temperature **A4:** 882
properties, heat-treated **A4:** 898
C95500, properties, heat treated **M4:** 737
C95800 annealing temperature **A4:** 882
C96300, properties, heat treated **M4:** 737
C96400, homogenization **A4:** 881
C96600
aging **A4:** 894
properties, heat-treated **A4:** 896
solution-treating temperature **A4:** 896
C96600, properties, heat treated **M4:** 735
C99400
aging **A4:** 894
aging temperature **M4:** 725
properties, heat treated **M4:** 725, 735
properties, heat-treated **A4:** 886, 896
solution-treating temperature **A4:** 896 **M4:** 725
C99500
aging **A4:** 894
properties, heat-treated **A4:** 896
solution-treating temperature **A4:** 896
C99500, properties, heat treated **M4:** 735
Copper ammonium chloride
ferric chloride and hydrochloric acid as an etchant for stainless steels welded to carbon or low alloy steels **A9:** 203
Copper and aluminum sheet
explosively welded **A9:** 386
Copper and copper alloys **A9:** 399-414
compositions **A9:** 402
etchants for **A9:** 401
eutectic alloys, solidification structures of welded joints **A9:** 580
eutectics **A9:** 620
grinding **A9:** 399-400
microstructures **A9:** 401
mounting **A9:** 399
polishing **A9:** 400
specimen preparation **A9:** 399-401
stress relaxation **M2:** 484-490
stress relaxation, mechanical components **M2:** 487-489
Copper bearing alloys
weldability **A6:** 725
Copper bearing cast steels
as low-alloy **A15:** 716
Copper brazing
definition **A6:** 1208 **M6:** 4
Copper brazing in furnaces **A7:** 457
Copper cable **M2:** 265-274
Copper carbonate hydroxide, as corrosion deposit
copper alloys **A11:** 633
Copper cementation
powder used **A7:** 573
Copper chip combustion accelerator **A10:** 222
Copper chloride
catalyst for polychloroprene rubber **EM3:** 149
Copper chromium powders **A7:** 625
Copper clad
Invar, as heat sinks **EL1:** 1129-1131
laminates, PWB technologies **EL1:** 505
laminates, resins and reinforcements **EL1:** 534-537
molybdenum, as heat sink material **EL1:** 1129-1131
Copper Cliff nickel refinery **A7:** 137
Copper coated tungsten spheres
wetting during liquid phase sintering **A9:** 100
Copper concentrate
Miller numbers **A18:** 235
Copper contact alloys *See also* Copper alloys
applications **A2:** 843
as electrical contact material **A2:** 842-843
Copper content of tin-antimony-copper alloys
effect on microstructure **A9:** 452

Copper current input connections
screw-threaded ... **A8:** 389
Copper cyanide plating ... **M5:** 159-169
dilute process ... **M5:** 159, 161-162, 167
cleaning action of ... **M5:** 159, 161
equipment ... **M5:** 167
high-efficiency sodium and potassium processes ... **M5:** 159-162, 164-165, 167-169
Rochelle process ... **M5:** 159, 161-163, 169
cleaning action of ... **M5:** 159, 161
pH, adjusting ... **M5:** 163
solution composition and operating conditions ... **M5:** 159-165, 168-169
surface preparation for ... **M5:** 161
Copper deficiencies
biologic effects ... **A2:** 1251
Copper deoxidizers
testing and effectiveness ... **A15:** 470
Copper Development Association ... **A8:** 724
Copper diallyl phthalate as a mounting material ... **A9:** 32
Copper displacement test
electrolytic cleaning ... **M5:** 37
Copper drossing
for lead refining ... **A15:** 474-475
Copper electrodes
dipersion-strengthened ... **A7:** 624-626
Copper electroplating
corrosion protection ... **M1:** 753
on cast iron ... **M1:** 102
Copper ferrites ... **A9:** 538
Copper flashing
mechanical coating process ... **M5:** 302
nickel alloy surfaces, prevention and removal of ... **M5:** 673
Copper fluoborate plating ... **M5:** 160-161, 166-167
Copper foil
for flexible printed boards ... **EL1:** 581
Copper graphite
electrical contact applications ... **A7:** 631
Copper in cast iron ... **M1:** 77, 79
ductile iron ... **M1:** 40, 41, 47, 49
gray iron ... **M1:** 15, 28
malleable iron ... **M1:** 58, 66
Copper in steel ... **M1:** 115, 410-411, 417
atmospheric corrosion resistance ... **M1:** 717, 721-723
formability affected by ... **M1:** 554-555
modified low-carbon steels ... **M1:** 162
neutron embrittlement, effect on susceptibility to ... **M1:** 686
notch toughness, effect on ... **M1:** 694
P/M alloys ... **M1:** 337, 340
seawater corrosion, effect on ... **M1:** 739-742, 744, 745
soil corrosion., effect on ... **M1:** 730
Copper ions, in fresh water
effect on corrosion rate ... **M1:** 735
Copper iron powders
composition ... **A7:** 464
Copper lubricants
powder used ... **A7:** 573
Copper metals ... **M2:** 243-246
applications ... **M2:** 239, 240-241, 246-247, 466
corrosion fatigue ... **M2:** 459, 463
corrosion potentials ... **M2:** 458-459, 460
crevice corrosion ... **M2:** 459, 460
dealloying ... **M2:** 459, 461
deposit attack ... **M2:** 458, 459
fabricated mill products ... **M2:** 245-246
fretting ... **M2:** 459, 460-461
galvanic corrosion ... **M2:** 458-459
general corrosion ... **M2:** 458, 459
impingement attack ... **M2:** 459, 460-461
intercrystalline corrosion ... **M2:** 459, 461-462
mining and refining ... **M2:** 243-245
pitting ... **M2:** 459-461
stress-corrosion cracking ... **M2:** 462-463
sulfur corrosion ... **M2:** 463, 464
Copper metals (continued)
water-line attack ... **M2:** 459, 460
Copper metals, annealing
continuous strand ... **M2:** 255
grain size ... **M2:** 253, 254, 255
grain-size stabilized alloys ... **M2:** 253
hydrogen embrittlement ... **M2:** 255
mechanical properties, correlation with ... **M2:** 253-254
pretreatment, effect of ... **M2:** 255
recrystallization ... **M2:** 253, 255
temperatures ... **M2:** 253
testing ... **M2:** 254
time, effect of ... **M2:** 255
Copper metals, brazing
atmospheres ... **M2:** 450-452
beryllium coppers ... **M2:** 450
cadmium-bearing copper ... **M2:** 450
chromium copper ... **M2:** 450
copper-aluminum alloys ... **M2:** 450
copper-nickel-zinc alloys ... **M2:** 450
copper-silicon alloys ... **M2:** 450
copper-tin alloys ... **M2:** 450
copper-zinc alloys ... **M2:** 450
deoxidized coppers ... **M2:** 449
dissimilar metals ... **M2:** 450, 451
filler metals ... **M2:** 450, 451
fluxes ... **M2:** 450-452
joint clearance ... **M2:** 452
oxygen-free coppers ... **M2:** 449
postbraze treatment ... **M2:** 453
processes ... **M2:** 452-453
surface preparation ... **M2:** 452
tough pitch coppers ... **M2:** 449
zirconium coppers ... **M2:** 450
Copper metals, cast
corrosion ratings ... **M2:** 390-391
Copper metals, heat treating
age hardening ... **M2:** 256
homogenizing ... **M2:** 252-253
martensitic transformation ... **M2:** 259
precipitation hardening ... **M2:** 256
spinodal decomposition ... **M2:** 259-260
stress relieving ... **M2:** 255-256
Copper metals, joining *See also* Copper metals, brazing; Copper metals, soldering; Copper metals, welding
adhesive bonding ... **M2:** 456
diffusion bonding ... **M2:** 456
electroplating ... **M2:** 457
mechanical joining ... **M2:** 440-443, 444
process selection ... **M2:** 440, 441, 442
roll bonding ... **M2:** 457
Copper metals, soldering
advantages ... **M2:** 443, 445
coated copper alloys ... **M2:** 448
fluxes ... **M2:** 446-447
mechanical properties, joints ... **M2:** 448-449, 450
methods ... **M2:** 447-448
solders ... **M2:** 444-446
surface preparation ... **M2:** 447
testing ... **M2:** 448
Copper metals, welding
aluminum bronzes ... **M2:** 455
beryllium coppers ... **M2:** 453, 454
brasses ... **M2:** 454-455
cadmium-coppers ... **M2:** 453, 454
chromium-copper ... **M2:** 453, 454
copper-nickels ... **M2:** 456
deoxidized coppers ... **M2:** 453
electron beam ... **M2:** 457
filler metals ... **M2:** 453, 454
friction ... **M2:** 457
gas metal-arc ... **M2:** 453, 454
gas tungsten-arc ... **M2:** 453, 454
laser ... **M2:** 457
nickel silvers ... **M2:** 456
oxygen-free coppers ... **M2:** 453
resistance ... **M2:** 453, 455
Copper metals, welding (continued)
silicon bronzes ... **M2:** 455-456
tin brasses ... **M2:** 455
tin bronzes ... **M2:** 455
tough pitch coppers ... **M2:** 453
ultrasonic ... **M2:** 457
zirconium coppers ... **M2:** 453, 454
Copper mirror test ... **A6:** 130
Copper nickel *See* Copper alloys, specific types, C70600, C71500
Copper nickel, 30%, microstructure of ... **A3:** 1 • 18
Copper nickels
10% *See* Copper alloys, specific types, C70600
20% *See* Copper alloys, specific types, C71000
30% *See* Copper alloys, specific types, C71500
70-30 *See* Copper alloys, specific types, C96400
brazeability ... **M6:** 1034
brazing ... **A6:** 630, 931
chromium-bearing *See* Copper alloys, specific types, C71900 and C72200
composition and properties ... **M6:** 401
compositions of various types ... **M6:** 546
corrosion in various media ... **M2:** 468-469
explosion welding ... **A6:** 303
friction welding ... **A6:** 152
gas metal arc welding ... **M6:** 421-422
gas tungsten arc welding ... **M6:** 414
gas-metal arc butt welding ... **A6:** 760
gas-metal arc welding
to high-carbon steel ... **A6:** 828
to low-alloy steels ... **A6:** 828
to low-carbon steels ... **A6:** 828
to medium-carbon steel ... **A6:** 828
to stainless steels ... **A6:** 828
gas-tungsten arc welding
to high-carbon steels ... **A6:** 827
to low-alloy steel ... **A6:** 827
to low-carbon steel ... **A6:** 827
to medium-carbon steel ... **A6:** 827
to stainless steel ... **A6:** 827
hot cracking ... **A6:** 754
oxyacetylene welding ... **A6:** 281
physical properties ... **M6:** 546
plasma arc cutting ... **M6:** 916
relative solderability as a function of flux type ... **A6:** 129
resistance welding *See* Resistance welding of copper and copper alloys ... **A6:** 850
shielded metal arc welding ... **M6:** 426
tin-bearing *See* Copper alloys, specific types, C72500
weld overlay for hardfacing alloys ... **A6:** 820
weldability ... **A6:** 753
welding (bonding) ... **A6:** 145
Copper oxide
examination under polarized light ... **A9:** 400
grinding of ... **A7:** 107, 108
as lubricant for hot forging of tool steels ... **A18:** 738
nonwetting ... **A6:** 129
particles ... **A7:** 108, 109, 111
removal of ... **M5:** 619-620
Copper oxide (CuO)
component in photochromic ophthalmic and flat glass composition ... **EM4:** 442
Copper oxide cake ... **A7:** 111
Copper oxide reduction ... **A7:** 52-53
copper powder produced by ... **A7:** 105-110
free energies and heats of reaction ... **A7:** 107, 109
temperatures ... **A7:** 108, 109
Copper P/M products
atomization ... **A2:** 392-393
beta stabilizers ... **A2:** 599
consumption ... **A2:** 392
copper-base structural parts ... **A2:** 396-398
electrolysis ... **A2:** 393
friction materials ... **A2:** 398-400

Copper P/M products (continued)
hydrometallurgy ... **A2:** 393-394
oxide-dispersion-strengthened copper ... **A2:** 400-401
porous bronze filters ... **A2:** 401-402
powder production ... **A2:** 392-394
reduction of oxide ... **A2:** 393
self-lubricating sintered bronze bearings ... **A2:** 394-396
Copper penetration failures
causes ... **A11:** 716-717
of- locomotive axles ... **A11:** 715-727
x-ray elemental composition maps of ... **A11:** 725
Copper peritectic alloys *See also* Beryllium copper; Brass; Silicon bronze
solidification structures in welded joints ... **A9:** 580
Copper plating ... **M5:** 159-169 **EL1:** 545
acid process ... **M5:** 159-162, 165-167
equipment for ... **M5:** 167
solution composition and operating conditions ... **M5:** 160-163, 165-166
surface preparation for ... **M5:** 161
adhesion of ... **M5:** 168
agitation ... **M5:** 162-166
alkaline process ... **M5:** 159-168
equipment for ... **M5:** 167
solution composition and operation conditions ... **M5:** 159-165, 168
surface preparation for ... **M5:** 161
aluminum and aluminum alloys ... **M5:** 160-161, 163, 168-169
anode bags ... **M5:** 164, 167
anodes ... **M5:** 167
applications ... **M5:** 159
barrel plating process ... **M5:** 162-163, 167
blistering of ... **M5:** 168
brightness, achieving ... **M5:** 168
buffing and electropolishing ... **M5:** 168
carbonate formation during ... **M5:** 163-164
characteristics of ... **M5:** 168-169
cleaning processes ... **M5:** 161
contamination factors ... **M5:** 162-166
copper striking in ... **M5:** 159-161, 168-169
cost factors ... **M5:** 169
current density ... **M5:** 161-166
current interruption cycle process ... **M5:** 159-160, 164, 168-169
decorative chromium plating processes ... **M5:** 193-194
die castings ... **M5:** 160-163, 168
equipment ... **M5:** 167
filtration processes ... **M5:** 162-163, 165
fluoborate processes ... **M5:** 160-161, 166-167
hafnium alloys ... **M5:** 668
hardness ... **M5:** 168-169
heat-resisting alloys ... **M5:** 566-568
stripping method ... **M5:** 567-568
high-efficiency cyanide systems ... **M5:** 159-165, 167-169
leveling ... **M5:** 164-165, 169
magnesium ... **M5:** 160-161, 163, 168-169
multiplate systems ... **M5:** 161, 169
nickel striking in ... **M5:** 160
niobium ... **M5:** 663-664
organic contamination ... **M5:** 162-163, 165-166
orthophosphate formation during ... **M5:** 165
periodic reversal process ... **M5:** 159-160, 162, 164-165, 168
cycle efficiency ... **M5:** 164-165
porosity of ... **M5:** 168
pyrophosphate process ... **M5:** 160-162, 165, 167
racks ... **M5:** 167
rinsewater recovery ... **M5:** 318
roughness of ... **M5:** 168
solderability of ... **M5:** 168
solution composition and operating conditions ... **M5:** 159-165, 168-169, 603-605, 663-664, 668
specifications and practices for ... **M5:** 161
stainless steel ... **M5:** 163, 168, 561-562
steel ... **M5:** 160-161, 163
stresses in ... **M5:** 163
stripping of ... **M5:** 567-568, 646-647
sulfate process ... **M5:** 160, 165-167
surface preparation for ... **M5:** 161
tanks ... **M5:** 167
Copper plating (continued)
tantalum ... **M5:** 663-664
temperature ... **M5:** 162-164, 166
thickness ... **M5:** 159-162, 165, 168
throwing power ... **M5:** 261-262, 265
titanium and titanium alloys ... **M5:** 658
ultrasonic vibration process ... **M5:** 162
wastewater control and treatment ... **M5:** 166-167
water, purity of ... **M5:** 162
zinc die castings ... **M5:** 160-162, 168
zirconium alloys ... **M5:** 668
Copper plating of specimens for edge retention ... **A9:** 32
Copper powder compacts *See also* Copper; Copper alloy powders; Copper powders
effect of sintering temperature and time on densification ... **A7:** 735
effects of lubricant on density and strength of ... **A7:** 289
expansion ... **A7:** 310
mechanical properties of hot pressed ... **A7:** 510
microstructure ... **A7:** 311
shrinkage of ... **A7:** 309
Copper powder spheres
progress of sintering ... **A9:** 99
Copper powders *See also* Atomized copper powders; Brasses; Bronzes; Copper alloy powders; Copper powders, specific types; Copper-based powder metals; Electrolytic copper powders; Nickel silver; Pure copper
air- and water-atomized ... **A7:** 106, 107
apparent and tap densities ... **A7:** 297
atomization ... **A2:** 392
atomized, alloying additions ... **A7:** 117-118
automotive applications ... **A7:** 17-18
chemical analysis and sampling ... **A7:** 247, 248
commercial, characteristics ... **A2:** 392
commercial grades, properties ... **A2:** 393
conductivity ... **A7:** 106
dendritic, density with high-energy compacting ... **A7:** 305
density and kinetic energy of projectile ... **A7:** 305
density increase for three types ... **A7:** 276
effect of compacting pressure on compact porosities ... **A7:** 269
effect of copper concentration ... **A7:** 112
effect of hot pressing temperature and pressure on density ... **A7:** 504
effect of nozzle diameter and pouring temperature ... **A7:** 32, 36
effect of oxide films on green strength ... **A7:** 303
effect of particle shape ... **A7:** 189, 276
effects of electrolyte composition ... **A7:** 111-112
explosivity ... **A7:** 196-197
formation, effect of acid concentration ... **A7:** 111-112
hot pressed, products ... **A7:** 509-510
hydrometallurgical processing ... **A7:** 118-120
infiltrated parts ... **A7:** 740
Kirkendall porosity ... **A7:** 314
loss of green strength and electrical conductivity ... **A7:** 304
measurement of particle size ... **A7:** 223-225
mechanical properties of specific types ... **A7:** 470
oxidation ... **A7:** 106-109
P/M ... **A7:** 733-740
particle size distribution ... **A7:** 188
particle size/shape ... **A2:** 392
porous, I-pore and V-pore shrinkage ... **A7:** 299
prealloyed atomized bronze ... **A2:** 392-393
prealloyed, of brass and nickel silver ... **A2:** 392
pressure and green density ... **A7:** 298
processing ... **A7:** 734-735
production of ... **A7:** 105-120
properties of commercial grades ... **A7:** 111
pyrophoricity ... **A7:** 199
reduction of oxide ... **A2:** 393
shipments ... **A7:** 24, 571
Copper powders, specific types *See also* Copper; Copper alloy powders; Copper powders
C15715, mechanical properties ... **A7:** 713
C15715 strip, mechanical properties ... **A7:** 714
C15760, dispersion-strengthened ... **A7:** 625
RWMA class 2 standard ... **A7:** 625
SAE 792, alloy properties ... **A7:** 408
SAE 794, alloy properties ... **A7:** 408
Copper pyrophosphate plating ... **M5:** 160-162, 165, 167
Copper recycling
melt refining practices ... **A2:** 1216
scrap classification ... **A2:** 1213-1215
scrap metals ... **A2:** 1213
technology ... **A2:** 1215-1216
Copper refractory metals
electrical and magnetic applications ... **A7:** 633-634
Copper removal
Sherritt nickel powder production ... **A7:** 139-140
Copper resources ... **M2:** 246, 247
Copper salts
acute poisoning by ... **A2:** 1252
Copper solute content in aluminum alloys
effect on secondary dendrite arm spacing ... **A9:** 635
Copper steel powders, composition ... **A7:** 464
heat treatment effects ... **A7:** 558
microstructural analysis ... **A7:** 487
Copper steels ... **A1:** 208
Copper striking
aluminum and aluminum alloys ... **M5:** 603-605
copper plating process ... **M5:** 159-161, 168-169
decorative chromium plating processes ... **M5:** 193-194
magnesium alloys ... **M5:** 639, 645-646, 647
stripping of ... **M5:** 646
zinc alloys ... **M5:** 677
Copper strip combustion accelerators ... **A10:** 222
Copper substrates
multichip structures ... **EL1:** 307
Copper sulfate, formation
in high-temperature combustion ... **A10:** 222
Copper sulfate plating ... **M5:** 160, 165-167
Copper tensile specimen, fracture surfaces
in sodium nitrite solution ... **A8:** 490
Copper toxicity
biologic effects ... **A2:** 1251-1252
Copper tube shells, production
extrusion ... **M2:** 261-262
rotary piercing ... **M2:** 262, 263
Copper tubes, production
cold drawing ... **M2:** 264
reducing ... **M2:** 264
Copper tubing
ODF using Euler plots ... **A10:** 361
pole figures from ... **A10:** 363
Copper, tubular products ... **M2:** 261-264
applications ... **M2:** 261, 262
joints ... **M2:** 261
mechanical properties ... **M2:** 261, 263
Copper wire ... **M2:** 265-274
characteristics ... **M2:** 267-273
coating ... **M2:** 272
rectangular wire ... **M2:** 266, 272
round wire ... **M2:** 266, 267
square wire ... **M2:** 266
stranded wire ... **M2:** 266-273
tensile stress-relaxation curve ... **A8:** 325
tin coated ... **M2:** 266, 273
Copper wire, drawing
annealing ... **M2:** 272
flat wire ... **M2:** 272
processes ... **M2:** 271-272
rod preparation ... **M2:** 271
Copper wire, materials
electrical bronzes ... **M2:** 265-266
high-conductivity ... **M2:** 265-266
high-copper alloys ... **M2:** 265-266
Copper wire rod, fabrication
continuous casting ... **M2:** 269-270
Hazelett process ... **M2:** 271
Outokumpu process ... **M2:** 271
Properzi system ... **M2:** 270
rolling ... **M2:** 266-269
Southwire system ... **M2:** 270
wirebar ... **M2:** 266
Copper(s) ... **A6:** 752
absorptivity ... **A6:** 265
as addition to brazing filler metals ... **A6:** 905
addition to low-alloy steels for pressure vessels and piping ... **A6:** 667
addition to strengthen nickel equivalent ... **A6:** 100
as addition to tin-lead solders ... **A6:** 966, 967, 969
adhesion measurement of fcc metals ... **A6:** 144

Copper(s) (continued)
adhesion to nickel ... **A6:** 144
alloying addition to heat-treatable aluminum alloys ... **A6:** 528, 529, 530, 531, 532
alloying effect in titanium alloys ... **A6:** 508
alloying effect on nickel-base alloys ... **A6:** 588-589
applications, sheet metals ... **A6:** 400
atomic interaction descriptions ... **A6:** 144
brazing and soldering characteristics ... **A6:** 631
brazing with clad brazing materials ... **A6:** 347
cladding for composite laminates, soldering ... **A6:** 132
cladding material for brazing ... **A6:** 347
cleaning solutions for substrate materials ... **A6:** 978
coextrusion welding ... **A6:** 311
cold welding ... **A6:** 307-308, 309
contact angles on beryllium at various test temperatures in argon and vacuum atmospheres ... **A6:** 116
contact tube, gas-metal arc welding ... **A6:** 183
contamination source for niobium electron-beam welding ... **A6:** 871
content in HSLA Q & T steels ... **A6:** 665
diffusion bonding ... **A6:** 159
in diffusion welding welds ... **A6:** 884-885, 886
dip brazing ... **A6:** 336
distortion ... **A6:** 757-758
in duplex stainless steels ... **A6:** 471
electrodes, nylon liners ... **A6:** 183-184
electron-beam welding ... **A6:** 851, 872
in eutectic alloys ... **A6:** 127
explosion welding ... **A6:** 162, 163, 303, 304
filler metals ... **A6:** 757, 761, 762
flash welding ... **A6:** 247
foil, solid-state welding ... **A6:** 169
friction welding ... **A6:** 152, 153
gas-metal arc welding ... **A6:** 759-760
  to high-carbon steel ... **A6:** 828
  to low-alloy steel ... **A6:** 828
  to low-carbon steel ... **A6:** 828
  to medium-carbon steel ... **A6:** 828
  to stainless steel ... **A6:** 828
gas-tungsten arc welding ... **A6:** 190, 192, 756-759
  to high-carbon steel ... **A6:** 827
  to low-alloy steels ... **A6:** 827
  to low-carbon steel ... **A6:** 827
  to medium-carbon steels ... **A6:** 827
  to stainless steels ... **A6:** 827
heat-affected zone ... **A6:** 759
high-frequency welding ... **A6:** 252
induction brazing ... **A6:** 333
induction soldering, physical properties ... **A6:** 364
inorganic fluxes for ... **A6:** 980
in intermetallic compounds ... **A6:** 127
joined to aluminum alloys ... **A6:** 739
low-temperature solid-state welding ... **A6:** 300
mechanical cleaning not recommended ... **A6:** 131
mechanical cutting ... **A6:** 1178
molten-salt dip brazing ... **A6:** 338
no resistance seam welding ... **A6:** 241-242
oxyacetylene welding ... **A6:** 281
oxyfuel gas welding ... **A6:** 285
plasma and shielding gas compositions ... **A6:** 197
postweld heat treatment ... **A6:** 761
precoated before soldering ... **A6:** 131
precoating ... **A6:** 131
price per pound ... **A6:** 964
projection welding ... **A6:** 233
properties ... **A6:** 629, 992
recommended gap for braze filler metals ... **A6:** 120
recommended guidelines for selecting PAW shielding gases ... **A6:** 67
recommended impurity limits of solders ... **A6:** 986

Copper(s) (continued)
recovery from selected electrode coverings ... **A6:** 60
relative solderability ... **A6:** 134
relative solderability as a function of flux type ... **A6:** 129
relative weldability ratings, resistance spot welding ... **A6:** 834
resistance brazing ... **A6:** 339, 340, 342
resistance soldering ... **A6:** 357
resistance welding ... **A6:** 833, 849-850
roll welding ... **A6:** 312-314
rosin flux use ... **A6:** 129
shielded metal arc welding ... **A6:** 176, 179, 755, 760
shielding gas purity ... **A6:** 65
solderability ... **A6:** 978
solderable and protective finishes for substrate materials ... **A6:** 979
soldering ... **A6:** 130-131, 630-631
in stainless steel brazing filler metals ... **A6:** 911, 913, 917, 920
thermal conductivity value ... **A6:** 587
thermal diffusivity from 20 to 100 °C ... **A6:** 4
thermal expansion coefficient ... **A6:** 907
torch brazing ... **A6:** 328
torch soldering ... **A6:** 351
toxicity ... **A6:** 1195, 1196
TWA limits for particulates ... **A6:** 984
ultrasonic welding ... **A6:** 324, 326
weld discontinuities, plasma arc welding ... **A6:** 1078
wettability ... **A6:** 115

Copper-accelerated acetic acid-salt spray (fog) test ... **A13:** 4, 225

Copper-aluminum alloys *See* Aluminum bronzes ... **A6:** 752
dealuminification ... **A13:** 133
dental ... **A13:** 1352, 1363
freshwater corrosion ... **A13:** 622

Copper-aluminum alloys, heat treating ... **A4:** 842, 843, 844, 845, 898
alpha aluminum bronzes ... **A4:** 898
alpha-beta aluminum bronzes ... **A4:** 898

Copper-aluminum alloys, heat treatings ... **M4:** 738-739
alpha aluminum bronzes ... **M4:** 734, 739
alpha-beta aluminum bronzes ... **M4:** 737, 739

Copper-aluminum alloys, liquid
thermodynamic properties ... **A15:** 58

Copper-aluminum dispersion alloys
strain rate as a function of mean free distance between particles ... **A9:** 131

Copper-aluminum powder depth profiling ... **A7:** 258

Copper-aluminum-nickel alloys
as shape memory effect (SME) alloys ... **A2:** 899-900

Copper-aluminum-nickel-manganese alloys
as shape memory effect (SME) alloys ... **A2:** 899-900

Copper-ammonia-water system potential
pH diagram ... **A7:** 54

Copper-base alloys ... **A18:** 748, 750-752
bearing alloys ... **A18:** 750-752
  classifications ... **A18:** 750
  commercial bronze ... **A18:** 750, 751, 752
  copper-lead alloys ... **A18:** 750-752
  high-lead tin bronzes ... **A18:** 750, 751, 752
  low-lead tin bronzes ... **A18:** 750, 751, 752
  mechanical properties ... **A18:** 752
  medium-lead tin bronzes ... **A18:** 750, 751, 752
  tin bronzes ... **A18:** 750, 751, 752
  unleaded tin bronze ... **A18:** 750, 751, 752
composition ... **M4:** 797
dissimilar metal joining ... **A6:** 824
fusion welding to steels ... **A6:** 828
hardfacing ... **A6:** 789, 795-796
heat and temperature effects on strength retention ... **A18:** 745
lubrication of tool steels ... **A18:** 737, 738
projection welding ... **A6:** 233

Copper-base alloys (continued)
sintering temperatures ... **M4:** 797
torch brazing filler metals ... **A6:** 328

Copper-base alloys, specific types *See also* Copper alloy filler metals, specific types; Copper base powder metallurgy materials, specific types; Sleeve bearing materials, specific types
51.5Cu-33.5Ni-15Fe, spinodal microstructure ... **A9:** 653
66.3Cu-30Ni-2.8Cr (wt%), spinodal microstructure ... **A9:** 654
92Cu-8Sn, atomized filter powder ... **A9:** 529
100 RXM, powder ... **A9:** 529
10100 (oxygen-free electronic), bar, electron beam welded ... **A9:** 408
10100 (oxygen-free electronic), brazed with BAg-8a filler metal ... **A9:** 408
10100 (oxygen-free electronic), brazed with BCu-5 filler metal ... **A9:** 408
10100 (oxygen-free electronic), rolled and annealed ... **A9:** 410
10200 (oxygen-free), different heat treatments compared ... **A9:** 406
10200 (oxygen-free), test rod ... **A9:** 410
11000 (electrolytic tough pitch), cold-rolled bar, annealed, tungsten arc welded ... **A9:** 408
11000 (electrolytic tough pitch), extruded compared ... **A9:** 406-407
rod, different heat treatments
11000 (electrolytic tough pitch), hot-rolled rod ... **A9:** 406
11000 (electrolytic tough pitch), static cast ... **A9:** 403
11000 (electrolytic tough pitch), test bar ... **A9:** 410
11000 (ETP copper), wirebar ... **A9:** 642
12200 (deoxidized high phosphorus), brazed with BAG-1 ... **A9:** 409
12200 (deoxidized high phosphorus), brazed with BCuP-5 ... **A9:** 409
12200 (deoxidized high phosphorus), drawn condenser tube, intergranular stress-corrosion cracks ... **A9:** 409
12200 (deoxidized high phosphorus) internal oxidation ... **A9:** 407
12200 (deoxidized high phosphorus), lap defects in condenser tubes ... **A9:** 407
12200 (deoxidized high phosphorus), static cast ... **A9:** 403
12200 (DHP copper), continuous cast different sections compared ... **A9:** 641
12200 (DHP copper), effect of solidification conditions on dendrite arm spacing ... **A9:** 637
12200 (DHP copper), variation in dendrite arm spacing in continuously cast ingot ... **A9:** 638
12500 (fire-refined tough pitch), hot-rolled ... **A9:** 407
14520, hot-rolled and drawn rod ... **A9:** 407
14700, rod, cold worked, 50% reduction ... **A9:** 407
17200 (beryllium copper), heat treated, and cold rolled ... **A9:** 407-408
18200, solutionized ... **A9:** 405
19400 (aluminum bronze), direct-chill cast ... **A9:** 640
19400 (aluminum bronze), electromagnetic casting ... **A9:** 639
26000 (cartridge brass), annealed ... **A9:** 404
26000 (cartridge brass), cast, cooled quenched ... **A9:** 404
26000 (cartridge brass), drawn cup ... **A9:** 409
26000 (cartridge brass), formation of nonequilibrium beta phase ... **A9:** 639
26000 (cartridge brass), hot rolled, annealed cold rolled ... **A9:** 410

Copper-base alloys, specific types (continued)
26000 (cartridge brass), local dezincification **A9:** 411
26000 (cartridge brass), processed to obtain grain sizes from 5 μm to 200 μm **A9:** 411
26000 (cartridge brass), semicontinuous cast cold shuts **A9:** 643
26000 (cartridge brass), transgranular corrosion crack **A9:** 410
26000 (cartridge brass), tube, drawn annealed, cold-reduced 5% **A9:** 409
28000 (Muntz metal), as-cast ingot **A9:** 411
28000 (Muntz metal), hot-rolled plate **A9:** 412
36000 (free-cutting brass), as-cast **A9:** 403
36000 (free-cutting brass), extrusion internal cracks **A9:** 645
36000 (free-cutting brass) ,phase formation **A9:** 639
36000 (free-cutting brass), semi-solid processed plumbing fitting **A9:** 414
36000 (free-cutting brass), semicontinuous cast, zinc-rich smudges from condensation **A9:** 644
36000 (free-cutting brass), semicontinuous cast, cold shuts **A9:** 643
36000 (free-cutting brass), semicontinuous cast, different sections of ingot **A9:** 639
36000 (free-cutting brass), semicontinuous cast, intergranular cracks **A9:** 643
36000 (free-cutting brass), semicontinuous cast, longitudinal crack ... **A9:** 642, 644
36000 (free-cutting brass), static cast **A9:** 642
44300 (arsenical admiralty), drawn tube **A9:** 412
46400 (uninhibited naval brass), as-cast **A9:** 404
46400 (uninhibited naval brass), extruded drawn and annealed **A9:** 405
51000 (phosphor bronze), rod, extruded cold drawn, annealed **A9:** 412
63800, ingot, hot rolled, 80% reduction **A9:** 645
64700, solution treated and aged **A9:** 412
67500, extruded rod **A9:** 412
68700 (arsenical aluminum brass) annealed **A9:** 404
68700 (arsenical aluminum brass), as-cast **A9:** 404
70600, grain-boundary cracks **A9:** 412
70600, laser welded to 1020 steel **A9:** 408
70600, semicontinuous cast **A9:** 405
71500, as-cast **A9:** 404
71500, coring in the columnar region **A9:** 639
71500, direct-chill, semicontinuous cast **A9:** 640
71500, effect of solidification conditions on dendrite arm spacing **A9:** 637
71500, semicontinuous cast, shrinkage porosity **A9:** 643
71500, variation in dendrite arm spacing in semicontinuous-cast ingot **A9:** 638
74500 (nickel silver), cold-rolled sheet annealed **A9:** 412
86200, as sand cast **A9:** 413
86300, as sand cast **A9:** 413
95400 (aluminum bronze), dealuminized, as sand cast **A9:** 413
95400 (aluminum bronze), heat treated **A9:** 413
95500, as sand cast **A9:** 413
97800, as sand cast **A9:** 413
ASTM B 148, Grade 9C, heat treated **A9:** 156
C69000 ..... **A9:** 552
Copper-base powder metallurgy materials microstructures **A9:** 511
Cu-0.1Al, oxidized single crystal sphere **A9:** 137
Cu-2.1Be-0.4Ni, precipitation at grain boundaries and slip planes **A9:** 649
Cu-2.5Co-0.5Be **A9:** 553
Cu-2.5Co-1.2Cd-0.5Si **A9:** 553
Cu-3.1Co, aged, metastable precipitate **A9:** 650
Cu-3Ti, early stages of cellular reaction **A9:** 650
Cu-3Ti ,Widmanstatten precipitation **A9:** 649
Cu-4Ti (wt%), spinodally decomposed electron diffraction pattern **A9:** 653
Cu-4Ti, cellular reaction **A9:** 650

Copper-base alloys, specific types (continued)
Cu-5Ni-2.5Ti, hot extruded, heat treated **A9:** 413-414
Cu-8.9P sand cast alloy, different illuminations compared **A9:** 79
Cu-8Sn, filter powder, gravity sintered **A9:** 530
Cu-10Al, intrinsic stacking fault, bright field and dark field images compared **A9:** 120
Cu-10Cd, peritectic transformations **A9:** 678
Cu-10Co, cobalt solid solution dendrites **A9:** 614
Cu-10Sn, pressed, sintered and sized **A9:** 524-525
Cu-11.8Al, heat treated, different illuminations compared **A9:** 79
Cu-12Al, martensitic structures **A9:** 673
Cu-14Al, martensitic structures **A9:** 673
Cu-18.4Ga-5Ge (at.%), massive transformation **A9:** 630
Cu-19.3Al (at.%), beta-to-alpha massive transformation **A9:** 630
Cu-20Sn, directionally solidified, peritectic reaction **A9:** 677
Cu-20Zn-2Pb, pressed, sintered re-pressed **A9:** 525
Cu-21.5Ga (at.%), massive transformation **A9:** 630
Cu-21Ga-1.5Ge, massive transformation **A9:** 630
Cu-27.0Sn, alpha phase with interlath precipitation **A9:** 666
Cu-27.5Zn-1.05Sn, tube, stress-corrosion crack **A9:** 412
Cu-27Sn, nonlamellar eutectoid structure **A9:** 659
Cu-30Zn, cold worked and annealed **A9:** 156
Cu-30Zn, crystallite orientation distribution function **A9:** 705
Cu-37.7Zn (at.%), partial massive transformation **A9:** 630
Cu-37Zn-2Al-2Fe, iron-rich precipitates **A9:** 644
Cu-39Zn, martensitic structures **A9:** 672
Cu-41.4Zn, bainite plates **A9:** 666
Cu-44.1Zn, surface relief from formation of bainitic plates **A9:** 662
Cu-50Zr, glass transition temperature **A9:** 640
Cu-70Sn, directionally solidified, peritectic reaction **A9:** 677
low-oxygen, high-purity, hot tears **A9:** 643
phosphorus-deoxidized, explosively bonded to tantalum **A9:** 445
Copper-base castings
markets for **A15:** 42
Copper-base electrodes
as cast iron welding consumable **A15:** 524
Copper-base filler alloys
brazing corrosion resistance **A13:** 883
Copper-base metal matrix composites **A13:** 861
Copper-base powder metallurgy materials, specific types
70Cu-30W **A9:** 554
75Cu-25W **A9:** 554
Copper-base structural parts
applications **A2:** 396
P/M, from brass, nickel silver, or bronze **A2:** 396-397
pure copper P/M parts **A2:** 397-398
Copper-based powder metals *See also* Atomized copper powders; Brasses; Bronzes; Copper; Copper alloy powders; Copper powders; Nickel silvers; Pure copper **A7:** 376-381, 733-740
Copper-bearing cast steels **A1:** 374
Copper-bearing lead *See also* Lead **A9:** 418
composition **A2:** 545
Copper-bearing steel
corrosion resistance **M1:** 735, 737-738, 751
as weathering steel **A13:** 516
Copper-beryllium
microstructure of **A9:** 551
Copper-beryllium alloys
analysis for copper by aluminon method **A10:** 65
analysis for iron, by thiocyanate method **A10:** 68
Unicast process for **A15:** 251

Copper-beryllium alloys, heat treating **A4:** 889-894
aging **A4:** 889, 890, 892, 893 **M4:** 729-734
electrical conductivity **A4:** 892
examples **A4:** 893-894
hardness **A4:** 893-894
inspection **A4:** 893-894
mechanical properties **A4:** 891, 892
oxidation **A4:** 890
precipitation hardening **A4:** 890-894
quality control **A4:** 893-894
quenching **A4:** 890 **M4:** 729
solution treating **M4:** 728, 729
solution-treating **A4:** 889-890, 892-893
Copper-bismuth alloys
occurrence of SMIE in **A11:** 243
Copper-boron carbide as nuclear reactor shielding material **A7:** 666
Copper-cadmium alloys
peritectic transformations **A9:** 678
Copper-cadmium-zirconium
electrodes for resistance spot welding **M6:** 480
Copper-chromium
microstructure of **A9:** 551
relative solderability as a function of flux type **A6:** 129
Copper-chromium alloys
heat treating **A4:** 894, 895
Copper-chromium alloys, heat treating **M4:** 734, 735
C11000 ..... **M4:** 720
Copper-clad alloys
glass-to-metal alloys **EM3:** 302
Copper-clad nickel
glass-to-metal seals **EM3:** 302
Copper-cobalt alloys
transmission electron microscopy of precipitates **A9:** 117
Copper-cobalt systems liquid-phase sintering **A7:** 319
Copper-gallium alloys *See also* Copper alloys, specific types
feathery structures in **A9:** 630
Copper-graphite
microexamination of **A9:** 550
pressed and sintered **A9:** 553
Copper-hardened rolled zinc alloy
properties **A2:** 540-541
Copper-infiltrated iron **A7:** 558
Copper-infiltrated steels, powder metallurgy materials
microstructures **A9:** 510
pressed and infiltrated **A9:** 519
Copper-Invar-copper
for metal core molding **EM3:** 591
Copper-Invar-copper (CIC) **EL1:** 620-625, 627-628
Copper-iron alloys
Mössbauer absorption spectrum **A10:** 294
phase analysis, Mössbauer spectroscopy **A10:** 294
Copper-iron compact microstructure **A7:** 559
Copper-iron-cobalt-tin-phosphorus alloy
characteristics **A2:** 234
Copper-lead alloys **A18:** 750-752
applications **A18:** 750, 751, 752
bearing material microstructures **A18:** 743, 744
casting processes **A18:** 754, 755
composition **A18:** 750, 751
corrosion resistance **A18:** 744
designations **A18:** 751
mechanical properties **A18:** 750-752
microstructures **A18:** 750
as part of bimetal bearings **A9:** 567
product form **A18:** 751, 752
in trimetal bearing material systems **A18:** 748
Copper-lead babbitt
bearing material microstructures **A18:** 744
Copper-lead bearing alloys
materials for **A11:** 483-484
Copper-lead P/M parts **A7:** 739-740
Copper-magnesium alloying
wrought aluminum alloy **A2:** 48
Copper-manganese alloys
copper content estimated **A10:** 201
estimation of copper in **A10:** 201
for niobium-titanium superconducting materials **A2:** 1046

Copper-manganese-aluminum
liquid impingement erosion of stator vanes ... **A18:** 223
Copper-manganese-nickel resistance alloys *See also* Manganins; Resistance alloys
properties and application ... **A2:** 823, 825
Copper-matrix composites
continuous graphite-copper MMCs ... **A2:** 909
continuous tungsten fiber reinforced ... **A2:** 908-909
Copper-molybdenum-copper (CMC)
for metal core construction ... **EL1:** 620
Copper-nickel
base metal solderability ... **EL1:** 677
Copper-nickel alloys *See* Nickel-copper alloys ... **A6:** 752, 754, 766-769
applications ... **A2:** 228, 338-342
corrosion resistance ... **A13:** 611
denickelification ... **A13:** 133
denickelification of ... **A11:** 633, 634
electrolytic etching ... **A9:** 401
filler metals ... **A6:** 756
fluxes ... **A6:** 755
freshwater corrosion ... **A13:** 621
gas-metal arc welding ... **A6:** 754, 755, 768
gas-tungsten arc welding ... **A6:** 754, 766-768
for heat exchangers/condensers ... **A13:** 627
iron alloying, seawater corrosion effects ... **A13:** 623
liquid, integral thermal properties ... **A15:** 58
melt treatment ... **A15:** 775-776
for niobium-titanium superconducting materials ... **A2:** 1045
oxidation ... **A18:** 591
plasma arc welding ... **A6:** 754
properties ... **A2:** 228, 338-342
recommended shielding gas selection for gas-metal arc welding ... **A6:** 66
shielded metal arc welding ... **A6:** 755, 768-769
shrinkage allowance ... **A15:** 303
surface condition ... **A6:** 754
thermal conductivity ... **A6:** 754
weldability ... **A6:** 753
Copper-nickel P/M parts ... **A7:** 739
Copper-nickel phase diagram ... **M6:** 21-22
Copper-nickel plating systems
magnesium alloys ... **M5:** 646
Copper-nickel resistance alloys *See also* Electrical resistance alloys properties and applications
Copper-nickel-chromium plating systems ... **M5:** 193
Copper-nickel-indium alloys
oxidation ... **A18:** 591
Copper-nickel-phosphorus alloys
age hardenable ... **A2:** 236
heat treating ... **M2:** 242, 257
Copper-nickel-silicon alloys ... **M2:** 242
age hardenable ... **A2:** 236
Copper-nickel-zinc alloy powder ... **A7:** 122
Copper-oxygen alloys, as-cast
effect of oxygen content on microstructure ... **A9:** 405-406
Copper-oxygen phase diagram ... **A15:** 466
Copper-phosphorus alloys
for brazing ... **A7:** 839
brazing, available product forms of filler metals ... **A6:** 119
brazing, joining temperatures ... **A6:** 118
liquid-phase sintering ... **A7:** 319
recommended gap for braze filler metals ... **A6:** 120
resistance brazing filler metals ... **A6:** 342
Copper-plated steel
welding factor ... **A18:** 541
Copper-refractory metal composites
deformation processed ... **A2:** 922
Copper-silicon
fretting wear ... **A18:** 248
relative solderability as a function of flux type ... **A6:** 129
Copper-silicon alloys ... **A6:** 752, 753 **A13:** 611, 621-622
stacking fault energies and deformation ... **A9:** 686
Copper-silver alloys
metal brazing ... **EM4:** 489-490, 491
relationship between composition and dendrite arm spacing ... **A9:** 638
Copper-silver system
wetting and spreading ... **EM4:** 485
Copper-silver-magnesium-phosphorus alloy
characteristics ... **A2:** 234
Copper-silver-phosphorus alloys
brazing, available product forms of filler metals ... **A6:** 119
brazing, joining temperatures ... **A6:** 118
Copper-tin
base metal solderability ... **EL1:** 677
hot cracking ... **A6:** 754
Copper-tin alloys *See also* Phosphor bronzes; Tin-copper alloy ... **A15:** 58-59
brazing, available product forms of filler metals ... **A6:** 119
brazing, joining temperatures ... **A6:** 118
peritectic reactions ... **A9:** 677
Copper-tin plating systems
magnesium alloys ... **M5:** 646
Copper-tin powders
air atomized ... **A7:** 122
compacts, effect of density on strength ... **A7:** 736
lubricant systems ... **A7:** 186
premixes ... **A7:** 480, 481
pressing characteristics ... **A7:** 736
systems, liquid-phase sintering ... **A7:** 319, 320
Copper-titanium alloys
high-resolution mass scan for ... **A10:** 616
Copper-tungsten
microexamination of ... **A9:** 550
percussion welding ... **M6:** 740
resistance brazing ... **M6:** 965
tongs for manual resistance brazing ... **A6:** 340
Copper-tungsten powders
applications ... **A7:** 631
Copper-zinc alloys *See also* Brasses ... **A6:** 752
brazing, available product forms of filler metals ... **A6:** 119
brazing, joining temperatures ... **A6:** 118
composition ... **A6:** 762
composition and properties ... **M6:** 401
corrosion in ... **A11:** 201
dealloying ... **A2:** 216
dezincification ... **A13:** 131-133, 614
failure by SCC and dezincification ... **A11:** 222, 635
as filler metal for dip brazing ... **A6:** 338
filler metals ... **A6:** 756
fracture history ... **A12:** 1
freshwater corrosion ... **A13:** 621
gas tungsten arc welding ... **M6:** 409-410
gas-metal arc welding ... **A6:** 762, 763
gas-tungsten arc welding ... **A6:** 762, 763
grain orientation in ... **A10:** 360
liquid, thermodynamic properties ... **A15:** 59
shielded metal arc welding ... **A6:** 762, 763
zinc flaring in ... **A15:** 466
Copper-zinc alloys, specific types
RBCuZn-A, dip brazing filler metal ... **A6:** 337
Copper-zinc phase diagram ... **A3:** 1 • 22
Copper-zinc powders ... **A7:** 121, 292, 380, 570
Copper-zinc system ... **A3:** 1 • 22
Copper-zinc-aluminum alloys
as shape memory effect (SME) alloys ... **A2:** 899-900
Copper-zinc-aluminum-manganese alloys
as shape memory effect (SME) alloys ... **A2:** 899-900
Copper-zinc-nickel alloys ... **A6:** 752
Copper-zinc-tin alloys
SCC and dezincification ... **A13:** 132
Copper-zirconium
electrodes for resistance spot welding ... **M6:** 480
Copper-zirconium (Cu-Zr)
fiber for reinforcement ... **A18:** 803
processing technique ... **A18:** 803
properties ... **A18:** 803
Copper-zirconium alloys
heat treating ... **A4:** 894, 896 **M4:** 734-736
Copper-zirconium powder metallurgy product ... **A9:** 414
Coppered finish
steel wire ... **M1:** 262
Coppered finish and liquor finishes
for steel wire ... **A1:** 279
Copperheads
in enameled steel sheet ... **M1:** 179
Coppers
corrosion in various media ... **M2:** 468-469
electrical ... **M2:** 240-241
Coppers, specific types
C10100 ... **M2:** 275-278
C10200 ... **M2:** 275-278
C10300 ... **M2:** 279
C10400 ... **M2:** 280-281
C10500 ... **M2:** 280-281
C10700 ... **M2:** 280-281
C10800 ... **M2:** 281-282
C11000 ... **M2:** 282-290
C11100 ... **M2:** 291
C11300 ... **M2:** 291-292, 293
C11400 ... **M2:** 291-292, 293
C11500 ... **M2:** 291-292, 293
C11600 ... **M2:** 291-292, 293
C12500 ... **M2:** 292-293, 294
C12700 ... **M2:** 292-293, 294
C12800 ... **M2:** 292-293, 294
C12900 ... **M2:** 292-293, 294
C13000 ... **M2:** 292-293, 294
C14300 ... **M2:** 294, 295
C14310 ... **M2:** 294, 295
C14500 ... **M2:** 295
C14700 ... **M2:** 295-296
C15000 ... **M2:** 296-297, 298
stress relaxation ... **M2:** 485-488
Coppers, wrought, specific types, annealing temperature
C10200 ... **M4:** 720
C11300 ... **M4:** 720
C11400 ... **M4:** 720
C11500 ... **M4:** 720
C11600 ... **M4:** 720
C12000 ... **M4:** 720
C12200 ... **M4:** 720
C14500 ... **M4:** 720
Coprecipitation
of ferrite ... **EM4:** 1163
in gravimetric analysis ... **A10:** 163
Coprecipitation, as manufacturing process
composite contact materials ... **A2:** 857-858
Copy milling
aluminum alloy forging dies ... **A14:** 247
Copy turning
PCBN cutting tools ... **A16:** 115, 116
Copying machine parts, powders used
See also Copier powders ... **A7:** 573
Cord ... **EM4:** 387, 388, 389, 391, 392
Cord-filled brushes ... **M5:** 165
Cordierite
affecting slag removal in welds ... **A6:** 61
Cordierite ($2Al_2O_3{\cdot}MgO{\cdot}5SiO_2$)
applications ... **EM4:** 759
ceramic corrosion in the presence of combustion products ... **EM4:** 982
chemical system ... **EM4:** 870, 872
composition ... **EM4:** 759
crystal structure ... **EM4:** 881
as filler for solder glass ... **EM4:** 1072
in heat exchangers ... **EM4:** 980
isolated group structure ... **EM4:** 758-759
maximum use temperature ... **EM4:** 875
mullite-cordierite composite ... **EM4:** 859, 860, 861

SUBJECTS OF THE INDEXED VOLUMES: ASM Handbook (designated by the letter "A"): **A1:** Properties and Selection: Irons, Steels, and High-Performance Alloys (1990); **A2:** Properties and Selection: Nonferrous Alloys and Special-Purpose Materials (1990); **A3:** Alloy Phase Diagrams (1992); **A4:** Heat Treating (1991); **A6:** Welding, Brazing, and Soldering (1993); **A7:** Powder Metallurgy (1984); **A8:** Mechanical Testing (1985); **A9:** Metallography and Microstructures (1985); **A10:** Materials Characterization (1986); **A11:** Failure Analysis and Prevention (1986); **A12:** Fractography (1987); **A13:** Corrosion (1987); **A14:** Forming and Forging (1988); **A15:** Casting (1988); **A16:** Machining (1989); **A17:** Nondestructive Testing and Quality Control (1989); **A18:** Friction, Lubrication, and Wear Technology (1992). **Metals Handbook, 9th Edition** (designated by the letter "M"): **M1:** Properties and Selection: Irons and Steels (1978); **M2:** Properties and Selection: Nonferrous Alloys and Pure Metals (1979); **M3:** Properties and Selection: Stainless Steels, Tool Materials and Special-Purpose Materials (1980); **M4:** Heat Treating (1981); **M5:** Surface Cleaning, Finishing, and Coating (1982); **M6:** Welding, Brazing, and Soldering (1983). **Engineered Materials Handbook** (designated by the letters "EM"): **EM1:** Composites (1987); **EM2:** Engineering Plastics (1988); **EM3:** Adhesives and Sealants (1990); **EM4:** Ceramics and Glasses (1991); **Electronic Materials Handbook** (designated by the letters "EL"): **EL1:** Packaging (1989).

Cordierite ($2Al_2O_3 \cdot MgO \cdot 5SiO_2$) (continued)
mullite-cordierite fracture toughness **EM4:** 865
properties **EM4:** 512, 759-760, 761
as refractory filler **EM4:** 1072
refractory material **EM4:** 907
solid-state sintering **EM4:** 279
structure **EM4:** 759-760
thermal expansion curves for low- and high- temperature forms **EM4:** 761
Cordierite glass-ceramics
aluminosilicate glass- ceramics **EM4:** 435-436
"Cordierite-type" glass **EM4:** 872
Cordless electric toothbrush and razor
powders used **A7:** 573
Core *See also* Force plug
crush, defined **EM1:** 7
defined **EM1:** 7 **EM2:** 11
depression, defined **EM1:** 7
fluted *See* Fluted core
magnetic, of magnetically soft materials **A2:** 780
RTM and SRIM, compared **EM2:** 347
separation, defined **EM1:** 8
size, copper alloys casting **A2:** 351
splicing, defined **EM1:** 8
vs coreless cavity design, copper alloy casting **A2:** 355
Core assembly
defined **A15:** 3
Core bake-out method
knockout **A15:** 506
Core binder
defined **A15:** 3
Core blow
defined **A15:** 3
Core blower
defined **A15:** 3
Core blowing machines
abrasive wear by **A15:** 191
usage **A15:** 239-240
Core box *See also* Blow holes
CAD/CAM processing **A15:** 618
defined **A15:** 3, 191
full split aluminum **A15:** 191
for plaster molding **A15:** 243
Core, broken or crushed
as casting defect **A11:** 381
Core, defined **A7:** 3
hardness tested by magnetic bridge sorting **A7:** 491
Core distortion
from directional solidification **A15:** 321
Core drilling
compared to reaming **A16:** 239
Core dryers
defined **A15:** 3
Core excitation
effect on Auger electrons **A10:** 551
Core filler *See also* Collapsibility
defined **A15:** 3
Core hardness **A1:** 481-482
Core, honeycomb
adhesive-bonded joints **A17:** 613-615
Core knockout **A15:** 503-506
defined **A15:** 503
equipment **A15:** 504-505
high-frequency drive machine **A15:** 506
Core knockout machine
defined **A15:** 3
Core losses in electrical steels
factors affecting **A9:** 537-538
Core materials
Antioch process **A15:** 246
selection, permanent mold casting **A15:** 280
Core oil
cores, release agents for **A15:** 240
defined **A15:** 3
hot box presses **A15:** 238
process, defined **A15:** 218
types of **A15:** 218
Core pin *See also* Cot-e
defined **EM2:** 11
Core pin plate
defined **EM2:** 11
Core plates
defined **A15:** 3
Core plating *See also* Insulation
of electrical steel sheet **A14:** 482
Core prints **A15:** 4, 192
Core properties, carburized and carbonitrided steels **M1:** 491-492, 534-535, 537-538
Core rod, defined **A7:** 3
materials **A7:** 337
mounting **A7:** 336
steps **A7:** 325
Core rods *See* Core wires
Core sand
defined **A15:** 4
Core setting
as green sand mold finishing **A15:** 347
Core shift
defined **A15:** 4
radiographic methods **A17:** 296, 349
Core shrinkage
as casting defect **A11:** 382
Core vents
defined **A15:** 4
Core wash
defined **A15:** 4
Core wires
defined **A15:** 4
Core(s) *See also* Collapsible cores; Green sand core; Preformed ceramic core
breaker, and riser necks **A15:** 587-588
ceramic, manufacture of **A15:** 261
defects, power inductors **EL1:** 1004
defined **A15:** 3
for die casting **A15:** 287
distortion **A15:** 321-322
effect, solidification sequence **A15:** 606-608
erosion of **A15:** 589
for gray iron **A15:** 640
large permanent, materials for **A15:** 280
passive device failure mechanisms **EL1:** 1000
patterns with **A15:** 195
permanent mold casting **A15:** 279-280
preformed **A15:** 261
processes, development of **A15:** 35
produced by shell process **A15:** 217-218
removal, as postcasting operation **A15:** 263
self-formed **A15:** 261
strainer (choke) **A15:** 596
system, selection of **A15:** 238
use **A15:** 191
Core-level ionization **A18:** 450
Cored bar
defined **A7:** 3
Cored mold
defined **EM1:** 7 **EM2:** 11
Cored solder
definition **A6:** 1208 **M6:** 4
Coreless induction furnaces **A15:** 3, 368-369, 500, 636
Coreless induction heating
of pouring vessels **A15:** 499
Coremaking **A15:** 238-241
with baking ovens **A15:** 32
by cold box process, and products **A15:** 239
coating **A15:** 240-241
compaction **A15:** 239-240
core system selection **A15:** 238
curing **A15:** 240
defined **A15:** 238
development of **A15:** 32-33
magnesium alloy casting **A15:** 805
mechanization of **A15:** 32
mixing **A15:** 238-239
release agents **A15:** 240
Corers
as sampling tools **A10:** 16
Cores (electronic)
powders used **A7:** 572
Cores, transformer
analysis of **A10:** 224
Coring *See also* Extrusion pipe; Liquation .... **A3:** 1•18, 19
in cast structures **A9:** 611-617
copper alloy ingots **A9:** 638-640
defined **A9:** 4 **A14:** 3 **A15:** 3 **EM2:** 11
polyamide-imides (PAI) **EM2:** 130-131
Scheil equation for **A9:** 631
Coring operation
in conjunction with broaching **A16:** 194, 195
Cork **EM3:** 49
Corner blowholes
as casting defects **A11:** 382
Corner castings, steel
aluminum coating of **M5:** 339
Corner joint
definition **A6:** 1208
electron-beam welding **A6:** 260
lamellar tearing **A6:** 95
non-heat-treatable aluminum alloys **A6:** 540
Corner joints
arc welding of
heat-resistant alloys **M6:** 356-357, 360, 363
nickel alloys **M6:** 437
stainless steels, austenitic **M6:** 334
definition **M6:** 4
design, illustration **M6:** 60-61
electron beam welds **M6:** 616-617
electroslag welding **M6:** 225
flux cored arc welding **M6:** 100
gas metal arc welding of aluminum alloys **M6:** 384
gas tungsten arc welding **M6:** 201
of magnesium alloys **M6:** 431-432
of silicon bronzes **M6:** 413
laser beam welding **M6:** 664
oxyfuel gas welding **M6:** 589-590
recommended grooves **M6:** 70
shielded metal arc welding of heat-resistant alloys **M6:** 356, 363
Corner radii, steel forgings **M1:** 362-364, 370, 371
Corner radius
computing of **A14:** 133
Corner scab
as casting defect **A11:** 381
Corner setting *See* Coining
Corner shrinkage
as casting defect **A11:** 382
Corner thinning
in cold-formed parts **A11:** 308
Corner weld
laser-beam welding **A6:** 879
Corner-edge joints
oxyfuel gas welding **A6:** 286, 287
Corner-flange weld
definition **A6:** 1208 **M6:** 4
Cornering stiffness **A18:** 578, 579-580
Corners
joint, lamellar tearing in HAZ of **A11:** 92
sharp, as notches **A11:** 85
sharp, as stress concentrators **A11:** 318
sharp internal, bending-fatigue fracture from **A11:** 469
sharp, oil quenching cracks from **A11:** 565
Corning Corelle Core Glass (Borosilicate cladding glass) **EM4:** 1101
Corning Corelle Skin Glass (CaF_2 opal core glass) **EM4:** 1101
Corning Corning Ware Beta-spodumene glass-ceramic **EM4:** 1103
Corning glass codes
0010 .... **EM4:** 497
0080 .... **EM4:** 497
0110 .... **EM4:** 497
0120 .... **EM4:** 497
0129 .... **EM4:** 497
0213 .... **EM4:** 1019, 1056
0281 .... **EM4:** 1103
0313 .... **EM4:** 463, 1020
0315 .... **EM4:** 463
0319 .... **EM4:** 463
0335 .... **EM4:** 877
0336 .... **EM4:** 871, 872
1720 .... **EM4:** 497
1723 .... **EM4:** 497, 1017
1990 .... **EM4:** 497
1991 .... **EM4:** 497
3320 .... **EM4:** 497
6720 .... **EM4:** 1101
6810 .... **EM4:** 863
7040 .... **EM4:** 497
7050 .... **EM4:** 497
7052 .... **EM4:** 497, 498
7056 .... **EM4:** 497
7251 .... **EM4:** 1103

Corning glass codes (continued)
7570 ... **EM4:** 497
7583 ... **EM4:** 875
7720 ... **EM4:** 497
7740 ... **EM4:** 863
7750 ... **EM4:** 497
7971 ... **EM4:** 1016, 1018
8111 ... **EM4:** 463
8361 ... **EM4:** 463
9010 ... **EM4:** 497
9013 ... **EM4:** 536-537
9019 ... **EM4:** 497
9455 ... **EM4:** 871, 877
9606 ... **EM4:** 872, 874, 875, 876, 877
9608 ... **EM4:** 871, 875, 876, 877
9617 ... **EM4:** 871, 877
9623 ... **EM4:** 872
9741 ... **EM4:** 1056
9753 ... **EM4:** 1020
9754 ... **EM4:** 1020
Corning Inc. photosensitive glass products ... **EM4:** 440, 442
Corning multiform process ... **EM4:** 1056
Corning Pyroceram Code
0308 (K-richterite glass-ceramic)
9609 (Nepheline ceramic)
Corning Visions beta-quartz glass-ceramic
coefficient of thermal expansion ... **EM4:** 1103
composition ... **EM4:** 1103
Cornish stone
flux composition ... **EM4:** 932
Corona
definition ... **M6:** 4
Corona (AC)
defined ... **EM2:** 461
Corona (DC)
defined ... **EM2:** 461
Corona (resistance welding)
definition ... **A6:** 1208
Corona 5 alloy
annealing ... **A4:** 915
Corona extinguishing voltage (CEV)
defined ... **EM2:** 461
Corona resistance ... **EM3:** 9
defined ... **EM2:** 11
Corona starting voltage (CSV)
defined ... **EM2:** 461
Corona treatment ... **EM3:** 35, 42
Coronal plane
defined ... **A17:** 383
Coronze *See* Copper alloys, specific types, C63800
applications and properties ... **A2:** 333-334
Corporate Average Fuel Economy (CAFE) standards ... **A18:** 554
Correction
defined ... **A8:** 3
Correction factors
cracks in biaxial stress ... **A11:** 124
for fatigue test data ... **A11:** 115
Corrective lens
definition ... **M6:** 4
Corrective orthopedic surgery
internal fixation devices as ... **A11:** 671
Correlated fluctuations
SAS techniques for ... **A10:** 405
Correlation ... **A8:** 623
Correlation distance ... **A18:** 469
Correlation parameter ... **A18:** 465
Corroded surfaces
microscopic examination ... **A11:** 173-174
Corrodents, atmospheric
types encountered ... **M5:** 333
Corroding lead *See also* Lead-, Pure lead
composition ... **A2:** 543-544
Corrodkote test
defined ... **A13:** 4
Corrosion *See also* Atmospheric corrosion; Corrosion economics; Corrosion failure analysis; Corrosion failure(s); Corrosion fatigue; Corrosion fatigue evaluation; Corrosion fatigue fracture(s); Corrosion pitting; Corrosion prevention; Corrosion product protection; Corrosion products; Corrosion products, characterization of; Corrosion rate; Corrosion resistance; Corrosion testing; Corrosion thinning; Corrosion-resistant alloys; Corrosive environments; Corrosivity; Crevice corrosion; Denickelification; Dezincification; Electrical corrosion; Electrolytic corrosion; Electronic corrosion; Erosion- corrosion; Exfoliation; Exfoliation corrosion Galvanic exfoliation corrosion Intergranular corrosion; Filiform corrosion; Freshwater corrosion; Fretting corrosion; Galvanic corrosion; General corrosion; Graphitic corrosion; Impingement attack; Interdendritic corrosion; Intergranular corrosion; Internal oxidation; Liquid corrosion; Oxidation; Parting; Pitting; Pitting corrosion; Poultice corrosion; Rust; Seawater corrosion; Selective leaching; Soil corrosion; specific corroding agent or specific type of metal; specific corrosion types; Stray- current corrosion; Stress corrosion; Stress-corrosion cracking; Stress-cracking corrosion (SCC); Sulfide stress cracking; Uniform corrosion ... **A6:** 374 **EM3:** 48, 628-636, 661
accelerated testing ... **EL1:** 891
in adhesive-bonded aluminum, neutron radiography of ... **A17:** 392-393
adhesive-bonded joints ... **A17:** 615-616
of adhesives ... **EL1:** 674
AES identification of chemical-reaction products in ... **A10:** 549
in AISI/SAE alloy steels ... **A12:** 318
and alloying for surface stability ... **A1:** 953, 956-957, 958-959
aluminum alloys ... **A6:** 729-730 **A15:** 765
of aluminum alloys containing copper as a result of using magnesium oxide ... **A9:** 353
aluminum metallization ... **A11:** 770
aluminum, neutron radiography of ... **A17:** 392-393
aluminum-lithium alloys ... **A6:** 552
aluminum-magnesium alloys ... **A6:** 622
atmospheric ... **A11:** 192-193, 535
of austenitic manganese steel ... **A1:** 836
bacterial and bio-fouling ... **A11:** 190-191
in bearing failures ... **A11:** 493
bearing materials ... **M3:** 804, 806, 807, 808-809
behavior, cast copper alloys ... **A2:** 383-385, 388, 391
behavior, commercially pure tin ... **A2:** 518-519
behavior, testing ... **A13:** 193-196
in boilers and steam equipment ... **A11:** 614-621
in bolts ... **A12:** 248
by marine organisms ... **A11:** 191
cells ... **A11:** 770
of cemented carbides ... **A7:** 779 **M3:** 456
in ceramic packages ... **EL1:** 962
chloride-induced bond-pad ... **EL1:** 891
coal-ash, steam equipment ... **A11:** 617
of cobalt-base alloys ... **A1:** 968
cobalt-base corrosion-resistant alloys ... **A2:** 453-454
composite-to-metal joining ... **A6:** 1041-1042
of composites, and cost ... **EM1:** 35
of condensers and feedwater heaters ... **A11:** 615
conditions and forms ... **A13:** 17
control, in steam-generator tubes ... **A11:** 615
control, parylene coatings ... **EL1:** 800
copper alloys ... **A12:** 403
of copper alloys, as electrical contact materials ... **A2:** 843
copper metals ... **M2:** 458-465
corrective and preventive measures for ... **A11:** 193-199
and corrosion products, SEM analysis of bridgewire ... **A10:** 511
costs, U.S ... **A13:** 755, 1039
crevice ... **A8:** 500 **A11:** 535, 631-632
data, and steel composition ... **A11:** 193

Corrosion (continued)
defined ... **A11:** 2 **A13:** 4
design changes for ... **A11:** 197
dezincification ... **A6:** 753
dissimilar metal joining ... **A6:** 826-827
in dissimilar-metal welds ... **A11:** 620-621
due to galvanization being burned off ... **EM3:** 724
duplex stainless steels ... **A6:** 626
economics ... **A13:** 369-374
effect of boundary precipitation and solute segregation on ... **A10:** 549
effect on liquid-erosion failures ... **A11:** 167
effects, corrective, in fatigue failures ... **A11:** 259-260
effects, electrical resistance alloys ... **A2:** 824
effects in creep testing ... **A8:** 303
electrochemically based, determined in aqueous environments ... **A10:** 134
electrochemically based, Raman studies ... **A10:** 135
electrolytic ... **EL1:** 493
electrolytic, of gold ... **A11:** 771
electronic materials ... **A12:** 482
erosion ... **A11:** 189
and erosion-corrosion ... **A11:** 189, 268-271
examination techniques ... **EL1:** 1043
exfoliation ... **A11:** 338-342
failure mechanisms ... **EL1:** 1043, 1049-1052
fatigue, and SCC, unified theory ... **A12:** 42
fatigue, defined ... **A11:** 2
fatigue, ductile iron ... **A15:** 662
fire-side ... **A11:** 616-620
fissure, environmentally assisted fracture at ... **A8:** 499
flux, defined ... **EL1:** 643-644
in forging ... **A11:** 338
fracture, cleaning of ... **A12:** 73-76
of fracture surfaces ... **A11:** 212
fractures, identification chart for ... **A11:** 80
fretting, defined ... **A11:** 5
from quenching ... **A7:** 453
furnace wall ... **A11:** 618
galvanic ... **A11:** 5, 185-186, 535-536 **EL1:** 493
gas-phase, Raman analysis ... **A10:** 135
general ... **A11:** 630 **A12:** 41
general, as cause of premature cracking ... **A8:** 605
gold metallization ... **A11:** 770
graphitic, defined ... **A11:** 5
hardfacing as preventative measure ... **A6:** 789
hardfacing for ... **A7:** 823
of heat exchangers ... **A11:** 630-635
heat-resistant alloys ... **M3:** 196, 207-209, 218-219, 270, 319-320
heat-treatable aluminum alloys ... **A6:** 534-535
high-silicon stainless steels ... **A6:** 795
high-temperature ... **A11:** 130-133
hot ... **A11:** 200, 269-271 **A12:** 391
hot gas ... **A15:** 730
impingement ... **A11:** 189, 634-635
of implants, dissolution as ... **A11:** 672
-induced, hydrogen-assisted fracture, low-carbon steels ... **A12:** 248
inhibitors ... **A14:** 515
inhibitors, for lubricant t-failure ... **A11:** 154
in integrated circuits ... **A11:** 769-770
intergranular ... **A11:** 180-182, 403-404
intergranular, austenitic stainless steels ... **A12:** 364
introduction ... **A13:** 17
of iron, in water and dilute aqueous solutions ... **A11:** 198
of iron-base alloys ... **A11:** 632
kinetics ... **A12:** 41
and liquid erosion ... **A11:** 167
liquid-immersion, of threaded fasteners ... **A11:** 535
localized, and premature cracking ... **A8:** 496
in low-carbon steel ... **A12:** 243, 248
low-temperature, steam equipment ... **A11:** 618-619
mechanically assisted, defined ... **A13:** 79

Corrosion (continued)
metal, Raman analysis **A10:** 126, 134-135
of metallic materials, from ionic residues **EL1:** 660
metallurgically influenced, defined **A13:** 79
in milling environment **A7:** 63
monitoring processing variables to control **A11:** 198-199
monitoring, ultrasonic inspection **A17:** 275-276
Mössbauer analysis of **A10:** 287
nickel alloys **M3:** 171-174
of nickel and cobalt in aqueous alkaline media **A10:** 135
of nickel and cobalt, Raman studies **A10:** 135
non-heat-treatable aluminum alloys **A6:** 540
in nuclear waste containers **A17:** 275-276
of- continuous aluminum oxide fiber MMCs **EM1:** 876
of- glass fiber **EM1:** 46
oil-ash, steam equipment **A11:** 618
on metals, analysis of **A10:** 134-135
organic inhibitor additives **EM3:** 641
-oxidation, in ceramics **A11:** 755-757
passive **A12:** 41-42
passive devices **EL1:** 1000
and pH, in boiler tubes **A11:** 612-613
phosphorus related, by aluminum metallization **A11:** 771
pitting **A11:** 176-177, 467
in plastic packages **EL1:** 479
potential effect on cracking in elevated-temperature water **A8:** 422
potential, of lubricants **A14:** 516
poultice, defined **A11:** 8
preferential, from urban atmosphere **A12:** 403
prevention, surface coatings for **A12:** 73
protection, for fasteners **A11:** 541-542
in pyrotechnic actuators **A10:** 510
rate, effect of acid concentration **A11:** 175
rates **A11:** 174, 175, 188, 612-613
rates for cobalt- and nickel-based hardfacing alloys **A7:** 828
ratings, cast copper alloys **A2:** 231-233
ratings, copper and copper alloys **A2:** 229-230
research **A13:** 193
resistance from chromic acid anodizing **EM3:** 738
resistance of maraging steels to **A1:** 799-800
-resistant prealloyed P/M aluminum alloy forgings **A14:** 251
rolling-element bearings failures by **A11:** 498-499
service, of copper casting alloys **A2:** 352
in shafts **A11:** 467
sites, from dot maps **A12:** 168
in sliding bearings **A11:** 486-487
of soldered joints **A17:** 608
soldering in electronic applications **A6:** 990, 991
as source, acoustic emissions **A17:** 287
special test procedures **EL1:** 953
specific types **M6:** 805
spring failures caused by **A11:** 559-560
stainless steel **M3:** 6, 56-93
stainless steel castings **M3:** 94-103
of stainless steels **A1:** 869-884, 935-936
stainless steels and brazing **A6:** 622
of steam equipment **A11:** 602, 619
steel weldments **A6:** 424-425
of steel wire rope **A11:** 518
strain amplitude and stress effects on **A12:** 418
subsurface, eddy current inspection **A17:** 193
surface, AES analysis for **A10:** 549
surface chemical analysis of **A7:** 250
and surface stains, LEISS identified on connector **A10:** 607
system, defined **A13:** 4
tantalum **A2:** 573
technology for cyclic crack growth studies **A8:** 422-423
and temperature, effects on wear failure **A11:** 156
testing **A11:** 174
tests **A14:** 516-517
thermal stability in elevated temperatures for aerospace applications **A6:** 385
thin-film chip resistors **EL1:** 1003
titanium alloys **A6:** 509 **M3:** 354, 373, 413-417
of titanium and titanium alloys **A2:** 588-589

Corrosion (continued)
types **A11:** 174 **A13:** 79
of unalloyed uranium **A2:** 672
under thermal insulation **A11:** 184
under-deposit, copper alloys **A12:** 403
uniform **A11:** 174-176, 300, 402
uranium **M3:** 778, 779
visual inspection **A17:** 3
voltage-specific, of aluminum **A11:** 771
in water-containing fuels **A11:** 191
water-side **A11:** 614-616
weight-loss **A11:** 300
weld overlays to overcome **M6:** 805
welding **EL1:** 1043
of welds **A11:** 400-401
zirconium **M3:** 784-791
zone 2, package interior **EL1:** 1008-1009
Corrosion analysis specimens
mounting **A9:** 31
Corrosion barrier **A13:** 889
Corrosion cells
electrolytic, caused by applied bias **A11:** 770
galvanic, caused by dissimilar metals **A11:** 770
Corrosion current **A18:** 274
Corrosion current density **A18:** 274
Corrosion de trepidation *See also* Fretting wear **A18:** 242
Corrosion economics
analysis methods **A13:** 370
of anodic protection **A13:** 465
depreciation **A13:** 371-372
examples/applications **A13:** 374
generalized equations **A13:** 372-374
hot dip galvanizing **A13:** 443
money/time and **A13:** 369
notation and terminology **A13:** 369-370
Corrosion embrittlement
defined **A13:** 4
Corrosion environments, zinc
behavior in various **M2:** 648-650
Corrosion failure analysis *See also* Advanced failure analysis; Corrosion; Corrosion failure analysis; Failure analysis; Failure mechanisms
advanced techniques for **EL1:** 1102-1116
of aircraft accelerometers **EL1:** 1111-1112
analysis techniques **EL1:** 1102-1107
of antenna marker beacon **EL1:** 1107
chemical analysis techniques **EL1:** 1103-1106
of circuit breakers **EL1:** 1108
of disk recorder heads **EL1:** 1112
of electrical connectors **EL1:** 1112
electron optics **EL1:** 1106
electronic/electrical corrosion, examples **EL1:** 1108-1116
environmental testing **EL1:** 1102-1103
of fuses **EL1:** 1110-1111
of hybrid microcircuits **EL1:** 1115
of integrated circuits **EL1:** 1114-1115
of klystron electron tubes **EL1:** 1115-1116
of microwave detectors **EL1:** 1112-1113
of nickel/boron-plated panels **EL1:** 1113-1114
package moisture content analysis **EL1:** 1106-1107
of printed wiring boards **EL1:** 1109-1110
of steering potentiometer **EL1:** 1111
of stepper motors **EL1:** 1111
surface analysis **EL1:** 1107
of tin whiskers **EL1:** 1116
of vacuum tubes **EL1:** 1110
Corrosion failures *See also* Corrosion **A11:** 172-202
analysis of **A11:** 172-174
atmospheric **A11:** 191-192
bacterial and bio-fouling **A11:** 190-191
of buried metals **A11:** 191-192
of cast materials **A11:** 401-405
concentration-cell **A11:** 183
corrective and preventive measures for **A11:** 193-199
crevice **A11:** 183-184
differential-temperature cells **A11:** 184-185
factors that influence **A11:** 172
galvanic **A11:** 185-188
of heat-exchanger tee fitting **A11:** 630-631
inclusions, selective attack on **A11:** 182-183
intergranular **A11:** 180-182
pitting **A11:** 176-177
selective leaching **A11:** 178-180

Corrosion failures (continued)
of specific metals and alloys **A11:** 199-202
uniform **A11:** 174-176
velocity-affected, in water **A11:** 188-190
Corrosion fatigue *See also* Corrosion; Corrosion fatigue crack growth rate; Fatigue; Fatigue strength **A8:** 3, 374-375, 403-410 **A13:** 142-144
aircraft **A13:** 1031, 1045
aluminum alloys **M2:** 219, 220
of aluminum alloys in aqueous halide solutions **A8:** 407-408
aluminum/aluminum alloys **A13:** 595
applied stress parameters **A13:** 292
in boilers and steam equipment **A11:** 623
in boiling water reactors **A13:** 928-933
as cause of premature fracture **A8:** 496
circumferential cracks from **A11:** 79
control and monitoring bulk water chemistry **A8:** 415
copper metals **M2:** 459, 463
copper/copper alloys **A13:** 614
of corrosion-resistant casting alloys **A13:** 581
of corrosion-resistant steel castings **A1:** 918, 920
crack closure effects **A8:** 408-409
crack growth, modeled effect **A13:** 298
crack growth rate **A12:** 41
crack initiation **A11:** 252-253 **A13:** 142-143
crack propagation **A11:** 81, 253-254 **A13:** 143-144
crack propagation rate **A13:** 297
cracking, gray iron cylinder inserts **A11:** 371-372
defined **A8:** 374 **A11:** 2 **A12:** 36 **A13:** 4
effect of cathodic polarization at ultrasonic/conventional frequencies **A8:** 254
effect of frequency **A8:** 374-375
electrochemical potential **A8:** 415
electrode potential **A8:** 405-407
elevated-temperature **A8:** 405-406
endurance data, steels **A13:** 296
environment **A8:** 403, 405, 407-408
environmental effects on crack initiation **A8:** 375
evaluation of **A13:** 291-302
experimentation **A8:** 409-410
failure **A11:** 134, 637
failure analysis **A11:** 256, 260-261
fracture mechanics **A13:** 295-297
fracture mechanics approach **A8:** 403-405, 417
frequency effect **A8:** 405-406
of gray iron component **A11:** 372
in heat exchangers **A11:** 636-637
high-cycle **A8:** 254
and hydrogen embrittlement **A13:** 143-144
in iron castings **A11:** 371
in low-carbon steel **A12:** 245, 250
metal matrix composite **A13:** 861
metallurgical variables **A8:** 408
microstructure influence **A8:** 405, 408
multiple cracking from **A11:** 24, 79
nickel alloys **A2:** 432
in nuclear feedwater nozzles **A13:** 937
oil/gas production **A13:** 1235
in petroleum refining and petrochemical operations **A13:** 1280-1281
portable machine, with ultrasonics **A8:** 243
predicting **A8:** 403
rate-limiting crack tip electromechanical reactions **A8:** 404
in shafts **A11:** 467
short-crack **A11:** 254
in sliding bearings **A11:** 487
in steam generators **A13:** 945
in steam turbines **A13:** 993
of steel in hydrogen **A8:** 409-410
of steel in moist air **A8:** 403
of steel in seawater **A8:** 407
and steels **A13:** 764
stress amplitude effect **A8:** 375
and stress corrosion/hydrogen embrittlement compared **A13:** 291
stress intensity range **A13:** 143
stress ratio **A8:** 405-407
and stress-corrosion cracking **A8:** 495, 499 **A13:** 143-144
and stress-intensity range **A8:** 405
sustained-load failure **A8:** 486
in telephone cables **A13:** 1130, 1132

Corrosion fatigue (continued)
test cell ... **A8:** 415
test specifications ... **A8:** 423
testing ... **A13:** 291, 1204-1205
testing, environment supply system for ... **A8:** 248
testing methods for measuring crack extension ... **A8:** 428
testing specimens, aqueous solutions ... **A8:** 417
in Ti-6Al-4V in aqueous sodium chloride ... **A8:** 405-406
titanium/titanium alloys ... **A13:** 676
in U-bend heat-exchanger tubes ... **A11:** 637
in ultrasonic fatigue testing ... **A8:** 241, 253-254
variables influencing ... **A8:** 405-409 **A13:** 297-300
yield strength influence ... **A8:** 405, 408
zinc/zinc alloys and coatings ... **A13:** 763-764
Corrosion fatigue crack growth rate
acidified chloride solution test specimen ... **A8:** 420
in alloy steels in aqueous chloride ... **A8:** 405
by applied stress-intensity range ... **A8:** 403
and cathodic potential ... **A8:** 417
cycle-dependent ... **A8:** 405
electrochemical potential ... **A8:** 419
elevated-temperature ... **A8:** 405-406
environment in acidified chloride solution testing ... **A8:** 419
loading frequency effect in salt water ... **A8:** 404
monitoring crack length ... **A8:** 420-422
near-threshold, oxygen content effect ... **A8:** 427, 430
pH effect ... **A8:** 427, 430
in pressurized water reactor ... **A8:** 425
purity effect ... **A8:** 411
in seawater ... **A8:** 406-407
in steam or boiling water with contaminants ... **A8:** 426-430
temperature effect ... **A8:** 411
time-dependent ... **A8:** 405
in ultrahigh-strength steel ... **A8:** 405-406
vs. stress intensity ... **A8:** 403-404
Corrosion fatigue crack growth testing
initial analysis frequency ... **A8:** 425
in pressurized-water ... **A8:** 423
Corrosion fatigue crack propagation ... **A8:** 403-410
acidified chloride at ambient and elevated temperatures ... **A8:** 418-420
electrode potential ... **A8:** 406-407
environment composition control ... **A8:** 410
of steel in moist air and steam ... **A8:** 409
stress ratio effect, carbon steel in pressurized nuclear reactor water ... **A8:** 406-407
testing in ambient temperature, aqueous solutions ... **A8:** 410
Corrosion fatigue cracking ... **A13:** 291, 299-300
Corrosion fatigue evaluation ... **A13:** 291-302
crack propagation tests ... **A13:** 295-300
cycles to failure tests ... **A13:** 291-295
data presentation ... **A13:** 292
effect, specimen size ... **A13:** 293
effect, stress concentration ... **A13:** 293-294
environmental effects ... **A13:** 295
standards/practices ... **A13:** 291
surface effects ... **A13:** 294-295
testing regimes ... **A13:** 292
Corrosion fatigue experimentation
containing environment about crack ... **A8:** 410
specimen thickness ... **A8:** 410
Corrosion fatigue fracture(s)
copper alloys ... **A12:** 403
crack propagation, P/M aluminum alloys ... **A12:** 440
low-carbon steel ... **A12:** 250
transgranular ... **A12:** 438
wrought aluminum alloys ... **A12:** 432, 438
Corrosion fatigue of stainless steel ... **M3:** 64-65
Corrosion fatigue test specification ... **A8:** 423
Corrosion film *See* Film

Corrosion in water
aluminum alloys ... **M2:** 220-222, 205, 208
Corrosion inhibitors *See* Inhibitors ... **EM3:** 41, 52
use in aluminum coating processes ... **M5:** 342-343
Corrosion inhibitors/metal deactivators
in engine lubricant formulations ... **A18:** 111
for metalworking lubricants ... **A18:** 141, 143, 144, 147
in nonengine lubricant formulations ... **A18:** 111
Corrosion of iron and steel
in alkali metals ... **M1:** 715
in atmospheric environments ... **M1:** 717-723
in fresh water ... **M1:** 731-736
in seawater ... **M1:** 737-743
in soils ... **M1:** 725-731
in steam systems ... **M1:** 745-748
Corrosion of weldments ... **A6:** 1065-1069
chevron cracking ... **A6:** 1068
forms of weld corrosion ... **A6:** 1065-1068
delta ferrite role in stainless steel weld deposits ... **A6:** 1066-1067, 1068
galvanic couples ... **A6:** 1065-1066
heat-tint oxide formation ... **A6:** 1068
hydrogen damage ... **A6:** 1068
microbiologically influenced corrosion (MIC) ... **A6:** 1068
pitting ... **A6:** 1067, 1068
stainless steel weld decay ... **A6:** 1066
stress-corrosion cracking ... **A6:** 1066, 1067
microsegregation ... **A6:** 1065
microstructural gradients ... **A6:** 1065
microstructurally distinct regions (five) ... **A6:** 1065
base metal ... **A6:** 1065
fusion zone ... **A6:** 1065, 1066, 1068
heat-affected zone ... **A6:** 1065, 1066, 1067, 1068
partially melted region ... **A6:** 1065
unmixed region ... **A6:** 1065
underbead cracking ... **A6:** 1068
welding practices to minimize corrosion ... **A6:** 1068-1069
Corrosion performance *See* Corrosion resistance
Corrosion pits *See* Pits; Pitting
Corrosion pitting *See also* Pitting; Pitting corrosion
AISI/SAE alloy steels ... **A12:** 291, 318
austenitic stainless steels ... **A12:** 358
in ceramic ... **A12:** 471
copper alloys ... **A12:** 404
radiographic methods ... **A17:** 296
steel tubing, magnetic characterization ... **A17:** 131
as volumetric flaw ... **A17:** 50
wrought aluminum alloys ... **A12:** 414, 415, 419
Corrosion potential ... **A13:** 4, 17, 200, 344-345
Corrosion prevention *See also* Anodic protection; Cathodic protection; Coatings; Conversion coatings; Corrosion protection; Corrosion resistance; Corrosion testing; Protective coatings; specific corrosion types ... **A13:** 523, 989, 1130-1131
Corrosion product
black, biological ... **A13:** 119
in brazed joints ... **A13:** 880
defined ... **A13:** 4
effect, electroplated coatings ... **A13:** 419
effect, immersion tests ... **A13:** 221
formation, by copper in groundwater ... **A13:** 621
formation, in liquid metals ... **A13:** 57-58
general form reaction ... **A13:** 57
and plaque ... **A13:** 1347
precipitation, manned spacecraft ... **A13:** 1097-1099
process stream testing for ... **A13:** 201
and time of wetness ... **A13:** 82
wedging, effect on SCC growth ... **A13:** 268
Corrosion products
AISI/SAE alloy steels ... **A12:** 299, 306
austenitic stainless steels ... **A12:** 37, 361
cadmium plate vs zinc plate ... **M5:** 265
defined ... **A12:** 29
effect on crack propagation ... **A12:** 133
effect on dimple rupture ... **A12:** 29
electronic materials ... **A12:** 482

Corrosion products (continued)
hot dip galvanized coating process producing ... **M5:** 331-332
hydroxide ... **A12:** 372
IR determination of molecular structure and orientation in ... **A10:** 109
layers, on low-carbon steel gas-dryer piping ... **A11:** 631
LEISS determined ... **A10:** 603
in low-carbon steel ... **A12:** 245
magnesium alloy ... **A12:** 456
magnesium alloys ... **M5:** 636-637
niobium alloy ... **A12:** 37
on hip implant ... **A12:** 37
on intergranular fracture surface ... **A12:** 37
on reheat steam pipe fracture surface ... **A11:** 653
on steel, Mössbauer analysis ... **A10:** 287
P/M aluminum alloys ... **A12:** 440
precipitation-hardening stainless steels ... **A12:** 372
in SCC fractures ... **A12:** 27
XRPD analysis of crystalline phases in ... **A10:** 333
Corrosion products, characterization of
analytical transmission electron microscopy ... **A10:** 429-489
atomic absorption spectrometry ... **A10:** 43-59
Auger electron spectroscopy ... **A10:** 549-567
classical wet analytical chemistry ... **A10:** 161-180
controlled-potential coutometry ... **A10:** 207-211
electrochemical analysis ... **A10:** 181-211
electrogravimetry ... **A10:** 197-201
electrometric titration ... **A10:** 202-206
electron probe x-ray microanalysis ... **A10:** 516-535
inductively coupled plasma atomic emission spectroscopy ... **A10:** 31-42
infrared spectroscopy ... **A10:** 109-125
ion chromatography ... **A10:** 658-667
low-energy ion-scattering spectroscopy ... **A10:** 603-609
Mössbauer spectroscopy ... **A10:** 287-295
optical emission spectroscopy ... **A10:** 21-30
particle-induced x-ray emission ... **A10:** 102-108
potentiometric membrane electrodes ... **A10:** 181-187
Raman spectroscopy ... **A10:** 126-138
scanning electron microscopy ... **A10:** 490-515
secondary ion mass spectroscopy ... **A10:** 610-627
spark source mass spectrometry ... **A10:** 141-150
voltammetry ... **A10:** 188-196
x-ray photoelectron spectroscopy ... **A10:** 568-580
x-ray spectrometry ... **A10:** 82-101
Corrosion properties ... **A6:** 97
Corrosion protection *See also* Anodic protection; Cathodic protection; Coatings; Corrosion resistance; Inhibitors; Protective coatings; specific coatings ... **M5:** 429-433
50% aluminum-zinc alloy coatings ... **M5:** 348
acid cleaning enhancing ... **M5:** 59-60
aluminum coating ... **M5:** 333-335, 337, 344-347
aluminum/aluminum alloys ... **A13:** 587-589
anodizing ... **M5:** 594-596
in aqueous solutions ... **A13:** 377-379
atmospheric corrosion ... **M5:** 431
barrier ... **A13:** 377-379
barrier coatings ... **M5:** 433
beryllium ... **A13:** 811
bronze plating ... **M5:** 288
by environmental control ... **A13:** 379
by metallurgical design ... **A13:** 379
by paint films ... **A13:** 528
by reducing reactivity of reacting phases ... **A13:** 521
by separating reacting phases ... **A13:** 521
by structural design ... **A13:** 379
cadmium plating ... **M5:** 256, 263-264, 266, 269
carbon steels ... **A13:** 521-525
cathodic ... **M5:** 433
ceramic coatings ... **A13:** 524
chemical corrosion ... **M5:** 430-431

SUBJECTS OF THE INDEXED VOLUMES: ASM Handbook (designated by the letter "A"): **A1:** Properties and Selection: Irons, Steels, and High-Performance Alloys (1990); **A2:** Properties and Selection: Nonferrous Alloys and Special-Purpose Materials (1990); **A3:** Alloy Phase Diagrams (1992); **A4:** Heat Treating (1991); **A6:** Welding, Brazing, and Soldering (1993); **A7:** Powder Metallurgy (1984); **A8:** Mechanical Testing (1985); **A9:** Metallography and Microstructures (1985); **A10:** Materials Characterization (1986); **A11:** Failure Analysis and Prevention (1986); **A12:** Fractography (1987); **A13:** Corrosion (1987); **A14:** Forming and Forging (1988); **A15:** Casting (1988); **A16:** Machining (1989); **A17:** Nondestructive Testing and Quality Control (1989); **A18:** Friction, Lubrication, and Wear Technology (1992). **Metals Handbook, 9th Edition** (designated by the letter "M"): **M1:** Properties and Selection: Irons and Steels (1978); **M2:** Properties and Selection: Nonferrous Alloys and Pure Metals (1979); **M3:** Properties and Selection: Stainless Steels, Tool Materials and Special-Purpose Materials (1980); **M4:** Heat Treating (1981); **M5:** Surface Cleaning, Finishing, and Coating (1982); **M6:** Welding, Brazing, and Soldering (1983). **Engineered Materials Handbook** (designated by the letters "EM"): **EM1:** Composites (1987); **EM2:** Engineering Plastics (1988); **EM3:** Adhesives and Sealants (1990); **EM4:** Ceramics and Glasses (1991); **Electronic Materials Handbook** (designated by the letters "EL"): **EL1:** Packaging (1989).

Corrosion protection (continued)
chromate conversion coatings **A13:** 392 **M5:** 457-458
chromium plating, decorative **M5:** 189, 198
clad metals **A13:** 523
coatings, for carbon steels **A13:** 522-524
copper metals **M2:** 459, 465
corrosion inhibitors **M5:** 432
corrosion theory **M5:** 429-433
basic mechanisms **M5:** 429-430
corrosion rate, factors influencing **M5:** 430
corrosive conditions **M5:** 430-431
galvanic effect **M5:** 430-432
types of corrosion *See also* specific types by name **M5:** 430-432
defined **A13:** 4
electroplated coatings **A13:** 523
electropolishing **M5:** 306-07
extent of determining **M5:** 460
from test data **A13:** 316-317
galvanic corrosion **M5:** 431-432
hot dip coating processes **A13:** 522-523
hot dip galvanized coating **M5:** 323-24, 331-32
ion implantation coatings **M5:** 425-426
ion plating **M5:** 420-421
lead plating **M5:** 275
localized corrosion, aluminum surfaces **M5:** 578
magnesium alloys **M2:** 602, 603 **M5:** 628-629, 631-633, 634, 636-638, 646
mechanical coatings **M5:** 300-302
nickel plating **M5:** 199-201, 204, 207-208, 229-231, 237-240
electroless **M5:** 229-231, 237-240
nonmetallic coatings for *See also* specific coatings by name **M5:** 432-433
on kinetic basis **A13:** 377
on thermodynamic basis **A13:** 377
organic coatings **A13:** 524
oxidation protective coatings **M5:** 375-376, 379
paint **M5:** 474-476, 491, 500-501, 504
passivity and **M5:** 431-432
phosphate coatings **M5:** 435, 442, 453, 455-456
phosphate or chromate conversion coatings **A13:** 523-524
pitting **M5:** 432-433
porcelain enamel **M5:** 525-526
sacrificial coatings **M5:** 432-433
shot peening improving **M5:** 145-146
soil corrosion **M5:** 430-431
stray current corrosion **M5:** 432
surface preparation for **A13:** 522
temporary **A13:** 522
thermal spray coating **M5:** 361
thermal spray coatings **A13:** 523
for threaded steel fasteners **A1:** 291, 296-297
vacuum coatings **M5:** 395, 397, 402
vapor-deposited coatings **A13:** 523
water corrosion **M5:** 430, 457-458
with zinc anodes **A13:** 764
zinc plating **M5:** 253-255
of zirconium **A13:** 718
Corrosion protection of steel
cathodic **M1:** 751, 757-759
coatings for *See also* Coatings **M1:** 751-754
inhibitors **M1:** 751, 754-757
various environments, protection guide **M1:** 751
Corrosion rate and anodic coatings **A13:** 419
in aqueous solutions **A13:** 17
atmospheric, measurement **A13:** 205
by ion concentration **A13:** 229
calcium concentration effect **A13:** 490
calculated, applied cathodic potential **A13:** 378
cast steels **A13:** 574
change, as function of time **A13:** 911
in chemical cleaning **A13:** 1142
constant extension, notched tensile tests **A13:** 963
control, aqueous solutions **A13:** 32
conversion of electrochemical current data to **A13:** 213
and critical humidity level **A13:** 511-512
data interpretation for **A13:** 316
defined **A13:** 4
effect of concrete **A13:** 513
and film formation **A13:** 195
of high-silicon cast irons **A13:** 567
Corrosion rate and anodic coatings (continued)
iron-base alloys, molten salts **A13:** 51
maximum achievable **A13:** 34
measurement **A13:** 32-33
metallic glasses and crystalline metals compared **A13:** 865
and relative humidity **A13:** 511
short-term, copper alloy in saline groundwater **A13:** 621
sulfide inclusion effect **A13:** 49
test coupons **A13:** 197-198
testing **A13:** 195, 197-198, 962-964
tin/tin alloys, in alkalies **A13:** 772
units, relationships among **A13:** 229
Corrosion rates
atmospheric corrosion **M1:** 719-721
cast steel **M1:** 401
seawater **M1:** 739-744
basic rate **M1:** 739, 741-742
galvanic couples **M1:** 740-74
tropical waters **M1:** 742
zones **M1:** 740, 742-744
Corrosion resistance *See also* Corrosion; General corrosion; specific types of corrosion **A1:** 581, 584 **EM3:** 9
alloying effects **A13:** 47-48, 323
aluminum **A2:** 3 **A13:** 583, 586-588 **M2:** 204-236
aluminum alloys **A15:** 766
aluminum alloys, effect of quenching **M2:** 32-35
aluminum casting alloys **A2:** 150
of aluminum coatings **A1:** 219 **M1:** 172
aluminum P/M alloys **A2:** 204-206
of aluminum P/M parts **A7:** 741-742
aluminum-lithium alloys **A2:** 187, 190-191, 194-195
aluminum-silicon alloys **A15:** 159, 167
of anodized aluminum **A13:** 599-600
of anodized coatings **A13:** 397
atmospheric, of engineering plastics **EM2:** 1
beryllium-copper alloys **A2:** 420-421
of brazed joints **A13:** 880-885
of carbon fiber reinforced composites (CFRP) **EM1:** 709
cast magnesium alloys **A2:** 492-516
of cast steels **A1:** 376 **M1:** 400
cemented carbides **A13:** 851-852
of cemented carbides for wear applications **A7:** 778
of cermets **A2:** 978
CG irons **A15:** 675
of chemical-setting ceramic linings **A13:** 455
coating cast iron for **M1:** 102, 106
of cold-rolled steel **A10:** 556-557
of compacted graphite iron **A1:** 68, 69
of composites **EM1:** 36
copper **M2:** 239-240
copper alloys **M2:** 463-465
copper and copper alloys **A2:** 216
of copper casting alloys **A2:** 352 **M2:** 390-391
critical pitting temperature and **A13:** 347
defined **A13:** 4 **EM1:** 8 **EM2:** 11
ductile iron **A15:** 663
effect, metal form **A13:** 193-194
effects of carbon on **A1:** 912, 913
effects of chromium on **A1:** 912, 913
effects of molybdenum on **A1:** 912
effects of silicon on **A1:** 912, 913
electrical resistance alloys **A2:** 822
and elimination, cyanates **EM2:** 232
environmental parameters for selecting alloys **A6:** 585
of ferritic malleable iron **A1:** 76 **A15:** 692-693
ferritic stainless steels **A2:** 778 **A9:** 285
fiber-reinforced vinyl ester resins **EM2:** 275
fiberglass-polyester resin composites **EM2:** 249
of gold **A13:** 796 **M2:** 669-670
gray cast iron, alloying elements for **M1:** 28
hard chromium plated steel, in salt spray **A13:** 871-872
Hastelloy nickel alloy series **A2:** 429
in heat exchangers, types of **A11:** 628
heat treatment effects **A13:** 48-49
high-alloy steels **A15:** 722
high-chromium white irons **A15:** 685
of high-phosphorus weathering steel **A1:** 400
HSLA steels **M1:** 405-406, 409
Corrosion resistance (continued)
of implant materials **A11:** 672
inconel, in molten salts **A13:** 51
of Invar **A2:** 891
iridium **M2:** 669
iron alloys **A2:** 778
of laser-processed materials **A13:** 503
of lead and lead alloys **A2:** 547
low-alloy steels **A13:** 82 **A15:** 720
low-melting temperature indium-base solders **A2:** 752
magnesium alloys **M2:** 596-609
of magnetically soft materials **A2:** 778
malleable cast irons **M1:** 66
maraging steels **M1:** 450-451
of mechanical fasteners **EM1:** 706, 709
mechanically alloyed oxide dispersion-strengthened (MA ODS) alloys **A2:** 943
mechanism, stainless steels **A13:** 550
metal composition and **A11:** 315
of microcrystalline alloys **A7:** 797
molybdenum and molybdenum alloys **A2:** 575
of nebulizers **A10:** 35
nickel alloys **A2:** 431-433
nickel-base alloy, acid media **A13:** 643-647
nickel-iron alloys **A2:** 778
nitrided carbon and low-alloy steels **M1:** 540
of- glass fiber-polyester resin composites **EM1:** 93, 94
osmium **A13:** 807
package level testing of **EL1:** 935-936
painted **A13:** 385
painted, cold-rolled steel **A10:** 556
palladium **A13:** 799 **M2:** 669
petroleum refining and petrochemical operations **A13:** 1265-1266
pewter **A2:** 522
of phosphate coatings **A13:** 385
of platinum **A13:** 798 **M2:** 668-669
of polyester resins **EM1:** 93
of pultruded composites **EM1:** 541-542
of pultrusions **EM2:** 396
of rare earths **A2:** 731
refractory metals and alloys **A2:** 558
rhenium **A2:** 581-582
of rhodium **A13:** 801-802 **M2:** 669
in sealed cans **A13:** 779
silicon steels **A2:** 778
of silver **A13:** 793 **M2:** 670
of sintered stainless steels **A13:** 825-832
of spray chambers **A10:** 35
in stainless steel casting alloys **A9:** 298
of stainless steels, to crevice corrosion **A13:** 110
of STAMP products **A7:** 549
structural ceramics **A2:** 1019
tanks, coatings, linings for metal-finishing shop **A13:** 1315
tantalum, mechanism **A13:** 725
terne coatings **M1:** 173, 174
of tin-containing P/M stainless steels **A7:** 254
tin/tin-nickel alloys **A13:** 778
titanium **A2:** 586
titanium/titanium alloys expanding/enhancing **A13:** 693-696
to molten glass, mechanically alloyed oxide dispersion-strengthened (MA ODS) alloys **A2:** 947
to pitting **A13:** 114
in tubing **A11:** 628
tungsten alloys **A2:** 579
of uranium alloys **A2:** 672
vinyl ester composites **EM2:** 274
vinyl esters **EM2:** 272-273
welds **A6:** 100
wrought aluminum, alloying effects **A2:** 44
wrought aluminum and aluminum alloys **A2:** 70, 111
X-ray photoelectron spectroscopy for **A7:** 256
zinc **M2:** 646-655
zinc coatings **M1:** 167-168, 170
zinc, in soils **A13:** 762
zirconia grain-stabilized (ZGS) platinum and platinum alloys **A2:** 713-714
zirconium/zirconium alloys **A13:** 707-717, 946
Corrosion resistance function **EM3:** 33, 36

Corrosion surfaces
scanning electron microscopy used to study ... **A9:** 99
Corrosion testing *See also* Evaluation; Laboratoy corrosion testing
assessment, corrosion damage ... **A13:** 194-195
ASTM standard methods/practices ... **A13:** 230
automotive industry ... **A13:** 1016-1017
brazed joint corrosion, Zircaloy specimens ... **A13:** 885
brazed joints ... **A13:** 881-883
controls ... **A13:** 195
corrosive solutions, for immersion tests ... **A13:** 220-222
environmental, electronics/communication equipment ... **A13:** 1113-1114
implant metals ... **A13:** 1332-1333
laboratory ... **A13:** 212-228
magnesium alloys ... **M2:** 603, 604
marine atmospheric, world test sites ... **A13:** 917
materials, procurement ... **A13:** 193-194
media selection ... **A13:** 194
for microbiological corrosion ... **A13:** 314-315
molten salt/liquid-metal ... **A13:** 17
objectives ... **A13:** 194
planning and preparation ... **A13:** 193-196, 316-317
results, accelerated ... **A13:** 194
results, interpretation and use ... **A13:** 316-317
results, reliability ... **A13:** 195-196
short-term, effects ... **A13:** 195
simulated service ... **A13:** 208-210
specimens ... **A13:** 194
stainless steels ... **A13:** 465
in the atmosphere ... **A13:** 204-206
tin coatings ... **A13:** 780-782
types of ... **A11:** 174
unexpected trends ... **A13:** 196
in water ... **A13:** 207-208
Corrosion testing of stainless steel ... **M3:** 65
Corrosion thinning
tube, remote-field eddy current inspection ... **A17:** 195
as volumetric flaw ... **A17:** 50
Corrosion tunnel model ... **A13:** 160, 162
Corrosion-erosion *See* Erosion-corrosion
Corrosion-fatigue failures ... **A11:** 134, 252-262
analysis of ... **A11:** 256-259
corrective measures ... **A11:** 259-260
crack initiation ... **A11:** 252-253
crack propagation ... **A11:** 253-254
environment, effect of ... **A11:** 255-256
failure analysis ... **A11:** 260-261
fatigue strength, effect of environment on ... **A11:** 252
loading parameters, effect of ... **A11:** 254-255
metallurgical parameters, effect of ... **A11:** 256
short-crack ... **A11:** 254
Corrosion-fatigue strength
defined ... **A8:** 375
Corrosion-product films
from atmospheric corrosion ... **A11:** 192-193
Corrosion-resistance metals
body assembly sealants used ... **EM3:** 608
Corrosion-resistant alloys
explosion welding ... **A6:** 303
Corrosion-resistant cast irons *See* Alloy cast irons ... **A9:** 245
alloying elements ... **M1:** 76, 88-89
arc welding of ... **A15:** 529
compositions ... **M1:** 76
high silicon irons ... **M1:** 89, 90
high-alloy, as-cast ... **A9:** 255
high-chromium irons ... **M1:** 90-91
high-nickel irons ... **M1:** 91
high-silicon, as-cast ... **A9:** 255
mechanical properties ... **M1:** 89
physical properties ... **M1:** 88
Corrosion-resistant cast steels
mechanical properties of ... **A1:** 909, 914, 917-920
Corrosion-resistant casting alloys
as C-type alloys ... **A13:** 576
composition ... **A13:** 576
corrosion behavior ... **A13:** 576-582
fully austenitic, general corrosion ... **A13:** 578
martensitic, general corrosion ... **A13:** 576-577
Corrosion-resistant high-alloy steels
austenitic grades ... **A15:** 722-723
austenitic-ferritic ... **A15:** 723
duplex ... **A15:** 723
ferritic grades ... **A15:** 723
iron-chromium-nickel ... **A15:** 724, 733
iron-nickel-chromium ... **A15:** 724, 733
martensitic grades ... **A15:** 722
mechanical properties ... **A15:** 726-728
precipitation hardening grades ... **A15:** 723
weldability ... **A15:** 730
Corrosion-resistant nickel alloy castings ... **M3:** 175-178
Corrosion-resistant steel castings *See also* Stainless steels, ACI specific types ... **A1:** 908-910, 912 **M3:** 94-103
austenitic grades, ferrite in ... **M3:** 96-98
constitution diagram ... **M3:** 98
corrosion resistance ... **M3:** 98-99, 102
galling ... **M3:** 102
heat treatment ... **M3:** 96, 99-101
iron-chromium-nickel alloys ... **M3:** 96
iron-nickel-chromium alloys ... **M3:** 96
machining ... **M3:** 100, 102
mechanical properties ... **M3:** 96-98, 103
precipitation-hardening alloys ... **M3:** 95-96
welding ... **M3:** 101-103
Corrosion-resisting paints
powders used ... **A7:** 572
Corrosive environment *See also* Stress-corrosion cracking
effect on dimple rupture ... **A12:** 24-29
liquid, effects and types ... **A12:** 36, 41-46
types ... **A12:** 24
Corrosive environments *See also* Atmospheres; Environment; Environments
alkali metals ... **M1:** 715
aqueous substances ... **M1:** 713-714
boiling water with contaminants ... **A8:** 426-430
corroding mediums, described ... **M1:** 713
liquid fertilizers ... **M1:** 714, 715
liquid metals ... **M1:** 714-715
nickel-base alloy behavior in ... **A13:** 643-647
Corrosive fluids
bearing damage by ... **A11:** 494, 496
Corrosive flux
definition ... **M6:** 4
Corrosive pitting ... **A18:** 349
Corrosive wear *See also* Oxidative wear ... **A18:** 271-277 **M1:** 637-638
abrasive wear ... **A18:** 189
carbon steel ... **A18:** 723
carbon-graphite materials ... **A18:** 816
cemented carbides ... **A18:** 795, 799-800
corrosive wear particles in lubricant analysis ... **A18:** 304
defined ... **A8:** 3 **A11:** 2 **A18:** 6, 271
electroplated coating applications ... **A18:** 834, 837
electroplating with lead-tin alloys ... **A18:** 838
environmental factors' effect on corrosive wear ... **A18:** 272-274
grinding wear: impact and three-body abrasive-corrosive wear ... **A18:** 273-274
slurry particle impingement on two-body corrosive wear ... **A18:** 272-273
experimental measurement of corrosion-wear synergism ... **A18:** 274-276
closed-loop pipeline experiments ... **A18:** 275
grinding wear systems ... **A18:** 275-276
jet-impingement experiments ... **A18:** 274-275
rotating ball-on electrode experiments ... **A18:** 274-276
Corrosive wear (continued)
rotating cylinder/anvil experiments ... **A18:** 275
slurry particle impingement systems ... **A18:** 274-275
slurry pot experiments ... **A18:** 274, 275
gold electroplating ... **A18:** 837
gray cast iron ... **M1:** 24
hardfacing alloys ... **A18:** 759, 765
internal combustion engine parts ... **A18:** 555, 556, 558
ion implantation ... **A18:** 856, 857, 858
mechanisms affecting aqueous corrosion ... **A18:** 856
mechanisms affecting oxidation resistance ... **A18:** 856
lubricant analysis case history ... **A18:** 309-310
means for combating corrosive wear ... **A18:** 276
cathodic protection ... **A18:** 277
design of pumps, valves, elbows ... **A18:** 277
materials selection ... **A18:** 277
modification of the materials handling environment ... **A18:** 277
slurry parameters ... **A18:** 277
surface treatment ... **A18:** 277
use of corrosion inhibitors ... **A18:** 277
mechanisms of wear/corrosion synergism in abrasive and impact wear ... **A18:** 276
abrasion ... **A18:** 276
corrosion ... **A18:** 276
impact ... **A18:** 276
mining and mineral industries ... **A18:** 649, 650- 651
nickel
electroless ... **A18:** 837
electroplated ... **A18:** 837
nitrided surfaces ... **A18:** 882
occurrences in practice ... **A18:** 271-272
crushing ... **A18:** 271
grinding ... **A18:** 271
high-temperature processes ... **A18:** 271
power-generation plants ... **A18:** 271-272
sliding wear ... **A18:** 271
slurry handling ... **A18:** 271
pumps ... **A18:** 593
seals ... **A18:** 549-550
sliding bearing materials ... **A18:** 742, 743, 744- 745
stainless steels ... **A18:** 715, 716, 718-720, 722-723
surface property effect ... **A18:** 342
thermal spray coating applications ... **A18:** 833
thermoplastic composites ... **A18:** 820
tool steels ... **A18:** 736, 737
use of transmission electron microscopy for surface studies ... **A18:** 381
Corrosive wear rate
abrasive particle dependence on ... **A18:** 272-273
corrosion products and the mass transfer of oxygen ... **A18:** 273
hydrodynamics, dependence on ... **A18:** 273
slurry, dependence on ... **A18:** 273
Corrosives
storage and handling ... **EM3:** 686
Corrosivity
defined ... **A13:** 4
rapid changes, and coupon testing ... **A13:** 197
relative, simulated service testing ... **A13:** 204
site ... **A13:** 205
Corrugated board
waterjet machining ... **A16:** 522, 525
Corrugated manufacturing technique
for honeycomb ... **EM1:** 722
Corrugated sheet
press-brake forming of ... **A14:** 541
Corrugated steel pipe
natural water corrosion ... **A13:** 433
Corrugating *See also* Corrugations; Crimping ... **A14:** 3
Corrugation
in thermoplastic extrusion ... **EM2:** 382-383
Corrugations ... **A14:** 3, 277

Corten A
corrosion resistance of ... **A1:** 400
Cortical bone screw
as internal fixation device ... **A11:** 671
Corundum ... **EM4:** 6, 49, 111
abrasive for lapping ... **A16:** 493
defined ... **A9:** 4 **A15:** 4
fusion with acidic fluxes ... **A10:** 167
hardness ... **A18:** 433
on Mohs scale ... **A8:** 108
polishing with ... **M5:** 108
Corundum grinding balls ... **A7:** 58
COSAM program approach ... **A1:** 1013
advanced processing ... **A1:** 1018
alternate materials ... **A1:** 1018-1020
results ... **A1:** 1020-1021
substitution ... **A1:** 1014-1018
Cosine wave
in elastic pressure bar ... **A8:** 199
Cosmetic pass ... **A6:** 259-260
definition ... **M6:** 4
Cosmetics
powders used ... **A7:** 573
Cosmochemical research
use of NAA in ... **A10:** 233
Cost *See also* Economics
of aluminum alloy precision forgings ... **A14:** 253-254
aluminum-lithium alloys ... **A14:** 250
of extrusion vs brazed assembly ... **A14:** 311
factor in selection for wear resistance ... **M1:** 606-607
heat sink assembly methods ... **EM3:** 572
of hot-die/isothermal forging ... **A14:** 150, 155-157
and machine size selection ... **A14:** 85
of precision forging ... **A14:** 158-159, 286
tolerances, effect on ... **A14:** 94-95
water-base versus organic-solvent-base adhesives properties ... **EM3:** 86
Cost considerations *See also* Cost(s)
alternative metals ... **EM2:** 85
burden (overhead) ... **EM2:** 86
contingency costs ... **EM2:** 87
cost control ... **EM2:** 87-88
cost optimizations ... **EM2:** 87
final design ... **EM2:** 88
part costing breakdown ... **EM2:** 83-85
of plastic program ... **EM2:** 82-88
plastic vs metal ... **EM2:** 83
product cost components ... **EM2:** 83
project level ... **EM2:** 82
quoting prices ... **EM2:** 86-87
system level ... **EM2:** 82-83
Cost control
guidelines ... **EM2:** 88
Cost drivers
in design/manufacture of composite structures ... **EM1:** 419-427
effects (CDE) ... **EM1:** 421-425
and lot size ... **EM1:** 420-421
material ... **EM1:** 420
of mechanical system ... **EM1:** 419
related to development ... **EM1:** 419
test, inspection, evaluation ... **EM1:** 421
Cost reduction *See also* Cost(s)
assembly/packaging for ... **EL1:** 448-449
as driving force ... **EL1:** 438
Cost(s) *See also* Cost considerations; Cost drivers; Economic process selection factors; Economics; Material costs; Pricing history
acrylics ... **EM2:** 103-104
acrylonitrile-butadiene-styrenes (ABS) ... **EM2:** 109-110
allyls (DAP, DAIP) ... **EM2:** 226
aminos ... **EM2:** 230
analysis, resin transfer molding ... **EM1:** 170-171
analytic modeling ... **EL1:** 14-15
assembly ... **EM2:** 649
of braided composites ... **EM1:** 519
of commercial products ... **EL1:** 1
of composite structures ... **EM1:** 417, 419-427
of conductive plastics ... **EM2:** 478
connector ... **EL1:** 23
considerations ... **EM2:** 82-88
contingency ... **EM2:** 87
control, guidelines ... **EM2:** 88
of copper casting alloy selection ... **A2:** 355

Cost(s) (continued)
cyanates ... **EM2:** 232
of damage tolerance ... **EM1:** 259
design-to ... **EM1:** 419
designer's worksheet for ... **EM1:** 425-427
and diffusion technology ... **EL1:** 961
distributed plane construction ... **EL1:** 628
drill bits vs drill diameter ... **EL1:** 508
elastomeric tooling ... **EM1:** 595-596
of electrical contact materials ... **A2:** 868
electrical testing ... **EL1:** 372
of engineering plastics ... **EM2:** 73
of engineering thermoplastics ... **EM2:** 99
of environmental stress screening ... **EL1:** 876
epoxies ... **EM2:** 240
estimating ... **EM1:** 424-425
extrusion ... **EM2:** 386
fibers ... **EL1:** 1120-1121
of filament winding, comparative ... **EM1:** 503
final assembly ... **EM2:** 650
finishing ... **EM2:** 650
of flexible printed boards ... **EL1:** 579-580, 592-594
forming/processing ... **EM2:** 647-648
guidelines, material selection ... **EM1:** 105
hermetic packages ... **EL1:** 468
high-density polyethylenes (HDPE) ... **EM2:** 163
high-impact polystyrenes (PS, HIPS) ... **EM2:** 194
homopolymer/copolymer acetals ... **EM2:** 100
as hybrid technology criterion ... **EL1:** 250-252
and hybridization ... **EM1:** 35
impact of decisions on ... **EM1:** 420
for injection molds ... **EM2:** 296
learning curve (LC) for ... **EM1:** 424-425
of level 1 packages ... **EL1:** 403
liquid crystal polymers (LCP) ... **EM2:** 180
low, filament winding ... **EM2:** 368
maintenance, of aircraft composites ... **EM1:** 259
manufacturing, of surface-mount components ... **EL1:** 730
manufacturing-to ... **EM1:** 419
material ... **EM2:** 82, 648-649
material, of composites ... **EM1:** 35
as material selection parameter ... **EM1:** 38-39, 105
and materials selection ... **EM2:** 1
of non-water-based sizing ... **EM1:** 123
optimization ... **EM2:** 87
packaging ... **EM2:** 650
parts, estimating ... **EM2:** 82, 647, 709-710
per system function, system-level products ... **EL1:** 12
permanent magnet materials ... **A2:** 792-793
phenolics ... **EM2:** 242
piece, components ... **EM2:** 82
polyamide-imides (PAI) ... **EM2:** 128, 130
polyamides (PA) ... **EM2:** 125
polybenzimidazoles (PBI) ... **EM2:** 147
polybutylene terephthalates (PBT) ... **EM2:** 153
polycarbonates (PC) ... **EM2:** 151
polyethylene terephthalates (PET) ... **EM2:** 172
of polyimide resins ... **EM1:** 43
polyphenylene ether blends (PPE, PPO) ... **EM2:** 183
polyurethanes (PUR) ... **EM2:** 258
polyvinyl chlorides (PVC) ... **EM2:** 209
prefinishing ... **EM2:** 50
of PWB materials ... **EL1:** 608
-quality relationships, magnesium alloys ... **A2:** 463
reduction, assembly/packaging for ... **EL1:** 448-449
rough order of magnitude (40M) ... **EL1:** 585
of shape memory alloys ... **A2:** 901
SiC fibers ... **EM1:** 59, 859
silicones (SI) ... **EM2:** 265
solder alloys ... **EL1:** 633, 642
solder masking ... **EL1:** 558
styrene-acrylonitriles (SAN, OSA ASA) ... **EM2:** 214
styrene-maleic anhydrides (S/MA) ... **EM2:** 217
substrate ... **EL1:** 612
summary, blow molding ... **EM2:** 299
of thermoplastic composite ... **EM1:** 552-553
thermoplastic fluoropolymers ... **EM2:** 117
thermoplastic polyurethanes (TPUR) ... **EM2:** 204-205
of thermoplastic resin composites ... **EM1:** 97
of thermoplastic resins ... **EM2:** 622
thermosetting engineering plastics ... **EM2:** 222
titanium and titanium alloy castings ... **A2:** 634

Cost(s) (continued)
of titanium P/M products ... **A2:** 647
-to-price translations ... **EM2:** 86
total system, and design ... **EM1:** 181
ultrahigh molecular weight polyethylenes (UHMWPE) ... **EM2:** 167
unit volume prices, engineering plastics ... **EM2:** 79
unsaturated polyesters ... **EM2:** 246
urethane hybrids ... **EM2:** 268
Cost-effective material selection ... **A13:** 335
Costs *See also* Economical casting design rules; Equipment costs; Operating costs
of cobalt-base alloys ... **A15:** 811
and design, of castings ... **A15:** 598
energy, of continuous casting ... **A15:** 310
equipment, inspection, tubular products ... **A17:** 561-562
factors affecting ... **A15:** 612
gating removal ... **A15:** 589
green sand molding ... **A15:** 341
human vs. machine vision ... **A17:** 30
of liquid penetrant methods ... **A17:** 77-78
of machine vision ... **A17:** 42
machining, mold cavity ... **A15:** 280
of melt purification ... **A15:** 74
of metal casting ... **A15:** 40-41
molding, vertical centrifugal casting ... **A15:** 301
operating, of inspection ... **A17:** 562
permanent mold casting ... **A15:** 285
of plaster molding ... **A15:** 242
slag, least cost mix, cupolas ... **A15:** 388
and tolerances ... **A15:** 614
Cosworth process
of aluminum casting ... **A15:** 38
Cotangent
abbreviation for ... **A10:** 690
Cotter pin wire ... **A1:** 852
Cotton
as natural fiber ... **EM1:** 117
Cotton pickers ... **A7:** 672, 673
Cotton thread
friction coefficient data ... **A18:** 75
Cottoning ... **EM3:** 9
Coulomb
symbol and abbreviation for ... **A10:** 685, 691
Coulomb coefficient of friction ... **A8:** 582 **A18:** 60, 66
Coulomb friction ... **A18:** 31
defined ... **A18:** 6
Coulomb's law
of friction ... **A8:** 576
Coulomb's laws *See* Amontons' laws
Coulomb-Mohr maximum pressure reduced shear stress ... **A8:** 344
Coulombic sliding friction ... **A16:** 15, 16
Coulometers
electronic ... **A10:** 203
Coulometric titration
capabilities, compared with electrometric titration ... **A10:** 197
cell, with platinum generator electrode ... **A10:** 205
as electrometric ... **A10:** 204-205
measured electric quantities in ... **A10:** 202
Coulometry *See also* Controlled-potential coulometry
capabilities, compared with voltammetry ... **A10:** 188
defined ... **A10:** 671
to analyze the bulk chemical composition of starting powders ... **EM4:** 72
Coulter Counter analysis
of melts ... **A15:** 493
Coulter counter for aluminum powder measurement ... **A7:** 129
Coulter principle ... **EM4:** 67
Count *See also* Pick count; Strand count; Thread count
defined ... **EM1:** 8
Counter top plastics
powders used ... **A7:** 572
Counter-gravity low-pressure casting
of air-melted alloys ... **A15:** 317-318
air-melted, sand casting ... **A15:** 317, 319
check valve casting ... **A15:** 317, 318
of vacuum-melted alloys ... **A15:** 317, 318
Counterbalance spring
hydrogen damage to ... **A11:** 558

Counterbalances
in press slides ... **A14:** 497
Counterblow equipment
defined ... **A14:** 3
Counterblow forging equipment
defined ... **A14:** 3
Counterblow hammers ... **A14:** 3, 28, 42
Counterboring ... **A16:** 249-251
Al alloys ... **A16:** 766-767
cast irons ... **A16:** 660
in conjunction with boring ... **A16:** 168, 169
in conjunction with drilling ... **A16:** 214, 219, 221, 222, 228
heat-resistant alloys ... **A16:** 751, 752
Mg alloys ... **A16:** 823, 825
refractory metals ... **A16:** 860, 863
speed and feed ... **A16:** 251
tools ... **A16:** 250
Countercurrent fountain spray drying ... **A7:** 73, 74
Counterdrilling
in conjunction with drilling ... **A16:** 219
Counterelectrode *See* Auxiliary electrode
for aqueous solutions ... **A8:** 416
defined ... **A10:** 671
x-ray diffraction ... **A10:** 246
Counterformal surfaces
defined ... **A18:** 6
Counterlocks ... **A14:** 3, 48-49
Countershaft pinion
fatigue fracture ... **A11:** 396
Countersinking ... **A16:** 249-250 **EM1:** 552, 671-672
in conjunction with drilling ... **A16:** 219, 222
speed and feed ... **A16:** 249-250
tools ... **A16:** 249-250
Counterweighting
and turning ... **A16:** 135
Counterweights
composite metals for ... **A7:** 17
powders used ... **A7:** 572
Counts
effect of image analysis on ... **A10:** 309
in image analysis ... **A10:** 312
per second, abbreviation for ... **A10:** 690
pulse, Auger spectrometer ... **A10:** 554
x-ray diffraction ... **A10:** 246
Couper-Gorman curves
and elevated-temperature service ... **A1:** 632
Couplants
ultrasonic inspection ... **A17:** 231, 256
Coupled lines
nomographs for ... **EL1:** 39-41
Coupled noise *See* Cross talk
Coupled two-phase eutectic structures ... **A9:** 620-621
Coupled zone
in cast iron ... **A15:** 174
diagram, aluminum-silicon alloys ... **A15:** 162, 164
directionally grown iron-carbon eutectic ... **A15:** 171
of eutectics ... **A15:** 123-124
Couplers ... **A18:** 110
in engine and nonengine lubricant formulations ... **A18:** 111
function ... **A18:** 110
Couples *See also* Galvanic corrosion; Galvanic couples
diffusion ... **A10:** 477-478
dissimilar-metal, compatible ... **A13:** 1061
Couples, galvanic *See* Galvanic couples
Coupling
extensional-bending, defined *See* Extensional-bending coupling
extensional-shear, defined *See* Extensional shear coupling
laminate ... **EM1:** 219, 220-222
magnetic particle inspection methods ... **A17:** 108, 117
Coupling agent(s)
defined ... **EM1:** 8
for injection molding compounds ... **EM1:** 164
Coupling agent(s) (continued)
for moisture resistance ... **EM1:** 123
sizing as ... **EM1:** 122
Coupling agents ... **EM3:** 9, 42, 181-182, 254-257, 674
as additives ... **EM2:** 499-500
chemistry ... **EM3:** 255, 256
combined with primer ... **EM3:** 256-257
defined ... **EM2:** 12
effect, reinforced polypropylenes (PP) ... **EM2:** 193
for epoxies ... **EM3:** 99-100
examples ... **EM3:** 256
for glass ... **EM3:** 281-283
in metalworking lubricants ... **A18:** 143, 147
for organic bonding of ceramics and glasses with metals ... **EM3:** 309
organo-titanates ... **EM3:** 256
for polyaramid ... **EM3:** 286
silane ... **EM3:** 257
zircoaluminates ... **EM3:** 256
Coupling, bending-twisting *See* Bending-twisting coupling
Couplings
by radial forging ... **A14:** 145
Coupon *See also* Testing ... **EM3:** 9
defined ... **A8:** 3 **A15:** 4 **EM1:** 8 **EM2:** 12
Coupon(s)
printed board ... **EL1:** 572-577
registration ... **EL1:** 870
Coupon(s), testing
bleach plant, crevice corrosion ... **A13:** 349
cleaning and evaluation ... **A13:** 199
corrosion rates ... **A13:** 197-198
crevice corrosion ... **A13:** 198
design ... **A13:** 197
finish ... **A13:** 198
galvanic corrosion ... **A13:** 198
gas/oil production ... **A13:** 1250
for immersion tests ... **A13:** 221
marine corrosion ... **A13:** 920
options ... **A13:** 198
retractable, holder for ... **A13:** 199
sensitized ... **A13:** 199
stress-corrosion ... **A13:** 198
test rack ... **A13:** 199
testing, direct ... **A13:** 197-199
for uniform corrosion rate testing ... **A13:** 229-230
unpainted coated steel ... **A13:** 1014
welded ... **A13:** 198-199
Covalent bond
defined ... **A10:** 671
definition ... **A6:** 1208 **M6:** 4
Covalent bonding
chemical ... **EL1:** 92
Cover coats
flexible printed boards ... **EL1:** 584
Cover core *See also* Parting line
defined ... **A15:** 4
Cover fluxes
aluminum alloys ... **A15:** 446
neutral, for copper alloys ... **A15:** 449
Cover layers
flexible printed boards ... **EL1:** 583-584
Cover lens
definition ... **M6:** 4
Cover plate
bridge, cracked girder at ... **A11:** 707
bridge, typical large crack at weld toe ... **A11:** 707
cracked groove weld in ... **A11:** 710
definition ... **M6:** 4
polished-and-etched section, showing lack of fusion ... **A11:** 710
Cover-coat porcelain enamels *See* Porcelain enameling, cover-coat enamels
Covered electrode
definition ... **A6:** 1208 **M6:** 4
Covered electrode welding *See* Shielded metal arc welding
Covering power *See also* Throwing power
defined ... **A13:** 4
CPG-85 computer program
for contoured tape laying machines ... **EM1:** 634
CPIA *See* Chemical Propulsion Information Agency
CPM *See* Crucible Particle Metallurgy process
CPM (Crucible particle metallurgy)process
alloy development ... **A7:** 787-788
processed tool steels compositions ... **A7:** 788
temper resistance of ... **A7:** 788
CPM alloys
Charpy C-notch impact ... **A16:** 65
hot hardness ... **A16:** 65
lathe tool test results ... **A16:** 65
temper resistance ... **A16:** 65
Cr-Cu (Phase Diagram) ... **A3:** 2 • 152
Cr-Fe (Phase Diagram) ... **A3:** 2 • 152
Cr-Fe-Mo (Phase Diagram) ... **A3:** 3 • 42
Cr-Fe-N (Phase Diagram) ... **A3:** 3 • 43
Cr-Fe-Ni (Phase Diagram) ... **A3:** 3 • 43-44
Cr-Fe-W (Phase Diagram) ... **A3:** 3 • 45
Cr-Ga (Phase Diagram) ... **A3:** 2 • 153
Cr-Ge (Phase Diagram) ... **A3:** 2 • 153
Cr-Hf (Phase Diagram) ... **A3:** 2 • 153
Cr-Ir (Phase Diagram) ... **A3:** 2 • 154
Cr-Lu (Phase Diagram) ... **A3:** 2 • 154
Cr-Mn (Phase Diagram) ... **A3:** 2 • 154
Cr-Mo (Phase Diagram) ... **A3:** 2 • 155
Cr-Mo-Ni (Phase Diagram) ... **A3:** 3 • 45
Cr-Mo-V steel, nitrided
torsional fatigue strength ... **M1:** 541
Cr-Mo-W (Phase Diagram) ... **A3:** 3 • 46
Cr-Nb (Phase Diagram) ... **A3:** 2 • 155
Cr-Nb-Ni (Phase Diagram) ... **A3:** 3 • 46-47
Cr-Nb-W (Phase Diagram) ... **A3:** 3 • 47
Cr-Ni (Phase Diagram) ... **A3:** 2 • 155
Cr-Ni-Ti (Phase Diagram) ... **A3:** 3 • 47-48
Cr-Ni-W (Phase Diagram) ... **A3:** 3 • 48
Cr-O (Phase Diagram) ... **A3:** 2 • 156
Cr-Os (Phase Diagram) ... **A3:** 2 • 156
Cr-Pd (Phase Diagram) ... **A3:** 2 • 156
Cr-Pt (Phase Diagram) ... **A3:** 2 • 157
Cr-Re (Phase Diagram) ... **A3:** 2 • 157
Cr-Rh (Phase Diagram) ... **A3:** 2 • 157
Cr-Ru (Phase Diagram) ... **A3:** 2 • 158
Cr-S (Phase Diagram) ... **A3:** 2 • 158
Cr-Sb (Phase Diagram) ... **A3:** 2 • 158
Cr-Sc (Phase Diagram) ... **A3:** 2 • 159
Cr-Se (Phase Diagram) ... **A3:** 2 • 159
Cr-Si (Phase Diagram) ... **A3:** 2 • 160
Cr-Sn (Phase Diagram) ... **A3:** 2 • 160
Cr-Ta (Phase Diagram) ... **A3:** 2 • 160
Cr-Te (Phase Diagram) ... **A3:** 2 • 161
Cr-Ti (Phase Diagram) ... **A3:** 2 • 161
Cr-Ti-W (Phase Diagram) ... **A3:** 3 • 49
Cr-U (Phase Diagram) ... **A3:** 2 • 161
Cr-V (Phase Diagram) ... **A3:** 2 • 162
Cr-W (Phase Diagram) ... **A3:** 2 • 162
Cr-Zr (Phase Diagram) ... **A3:** 2 • 162
Crab-form graphite
in cast iron ... **M1:** 6, 7
Crack *See also* Crack arrest; Crack arrest marks; Crack closure; Crack extension; Crack front; Crack growth; Crack initiation; Crack length; Crack nucleation; Crack opening; Crack opening displacement; Crack origin; Crack path; Crack propagation; Crack size; Crack speed; Crack tip; Crack tip opening displacement; Cracking; Cracks; Fatigue crack growth; Fatigue crack growth rate; Fatigue crack propagation; Fatigue striation spacings; Fatigue striations ... **A6:** 1073, 1075-1076, 1079
behavior, electrical potential method for measuring ... **A8:** 390-391
blunting ... **A8:** 499
central, vector lines from crack tips and crack center for ... **A8:** 444
classification of cold forging ... **A8:** 590, 592
closure ... **A8:** 391, 408-409, 449, 682
in composite materials ... **A8:** 716

SUBJECTS OF THE INDEXED VOLUMES: ASM Handbook (designated by the letter "A"): **A1:** Properties and Selection: Irons, Steels, and High-Performance Alloys (1990); **A2:** Properties and Selection: Nonferrous Alloys and Special-Purpose Materials (1990); **A3:** Alloy Phase Diagrams (1992); **A4:** Heat Treating (1991); **A6:** Welding, Brazing, and Soldering (1993); **A7:** Powder Metallurgy (1984); **A8:** Mechanical Testing (1985); **A9:** Metallography and Microstructures (1985); **A10:** Materials Characterization (1986); **A11:** Failure Analysis and Prevention (1986); **A12:** Fractography (1987); **A13:** Corrosion (1987); **A14:** Forming and Forging (1988); **A15:** Casting (1988); **A16:** Machining (1989); **A17:** Nondestructive Testing and Quality Control (1989); **A18:** Friction, Lubrication, and Wear Technology (1992). **Metals Handbook, 9th Edition** (designated by the letter "M"): **M1:** Properties and Selection: Irons and Steels (1978); **M2:** Properties and Selection: Nonferrous Alloys and Pure Metals (1979); **M3:** Properties and Selection: Stainless Steels, Tool Materials and Special-Purpose Materials (1980); **M4:** Heat Treating (1981); **M5:** Surface Cleaning, Finishing, and Coating (1982); **M6:** Welding, Brazing, and Soldering (1983). **Engineered Materials Handbook** (designated by the letters "EM"): **EM1:** Composites (1987); **EM2:** Engineering Plastics (1988); **EM3:** Adhesives and Sealants (1990); **EM4:** Ceramics and Glasses (1991); **Electronic Materials Handbook** (designated by the letters "EL"): **EL1:** Packaging (1989).

Crack (continued)
configuration, precracked SCC testing specimens **A8:** 516
and crack growth mechanism characteristics **A8:** 590, 593
definition **A6:** 1208 **EM4:** 632
deformation, three modes of **A8:** 441-443
-density limit, of composite materials **A8:** 714
detection in cold upset testing **A8:** 579
development and extension, forces causing **A8:** 439-464
driving tendency, methods representing **A8:** 439-464
-extension force **A8:** 439-443
formation at hot and warm temperatures **A8:** 572
front, in fracture mechanics **A8:** 439-464
ideal model **A8:** 440
instability curve, short-lived stress pulses **A8:** 284
jump **A8:** 454-455, 474
measurement, precracked specimens **A8:** 518
mechanical driving force, measured **A8:** 497
nucleation **A8:** 363, 366, 537
part-through surface, shape of **A8:** 444
path, after joining out-of-plane advance separation **A8:** 443
path tortuosity **A8:** 477, 480
plane, direction of **A8:** 440
problems, three dimensional, *K* for **A8:** 444
profile, in direct current electrical potential method **A8:** 389
running **A8:** 284, 453
in shear band during high-energy-rate forging **A8:** 573
as smooth, continuous surface **A8:** 439
stationary **A8:** 259, 439
subsurface, in copper crystal from low-stress sliding **A8:** 603
surfaces, displacement pattern **A8:** 444
tendency of stress-strain field to drive **A8:** 439
velocity **A8:** 284
visible, defining **A8:** 128
Crack arrest
and crack propagation **A8:** 284
fracture toughness specimens **A8:** 288-293
marking and measuring **A8:** 291-292
mechanism of **A12:** 15
value of *K* at **A8:** 286
Crack arrest fracture toughness **A8:** 288-293
Crack arrest marks *See also* Fatigue striations
in hydrogen-embrittled titanium alloys **A12:** 23, 32
wrought aluminum alloys **A12:** 414
Crack arrest testing
for dynamic fracture **A8:** 284-286
in fracture mechanics **A8:** 453-455
for high strain rate fracture toughness **A8:** 187
round-robin test program **A8:** 286, 288-293
specimens for **A8:** 290, 454
static post-arrest stress intensity in **A8:** 285
Crack arrest toughness
defined **A8:** 284
from run-arrest tests **A8:** 286
measurement of **A8:** 453-455
symbol for **A8:** 725
Crack branching
definition **EM4:** 632
information from **A11:** 744
in thermal shock fracture **A11:** 744
Crack closure **A8:** 391, 408-409, 449 **A12:** 15 **EM3:** 508
by partial slip reversal **A12:** 21
decarburization softening-induced enhancement **A12:** 38
as mechanical damage **A12:** 72
Crack closure, effects
corrosion fatigue **A13:** 299
Crack closure load level **EM3:** 509
Crack deflection
model **A12:** 206
Crack depth *See also* Crack direction; Crack extension; Crack growth; Crack length; Crack(s)
aspect ratio as function of **A11:** 124
effect of solid cadmium on **A11:** 239-241
Crack detection **A7:** 483-484, 577
aluminum sheets, eddy current inspection **A17:** 187-188
by magnetic printing **A17:** 126
in failure analysis **A17:** 54
microwave system **A17:** 219
slow sweep mode, eddy current inspection **A17:** 190
trigger pulse mode, eddy current inspection **A17:** 190
Crack direction
in composites **A11:** 735-738
information from **A11:** 744
Crack extension *See also* Crack depth; Crack length; Crack-extension force; Effective crack size; Original crack size; Physical crack size **A8:** 3
by fracture **A11:** 49-50
by fracture, *J* as function of **A11:** 63
control during precracking **A8:** 382
and crack tip stress **A8:** 442
defined **A11:** 2
determined by SEM **A8:** 542
dynamic impulse effect on **A8:** 454
effect of crack speeds on **A8:** 440
increasing, effect on singularity field size **A8:** 447
nominal crack tip opening displacement as measure of **A8:** 446
resistance to, measuring **A8:** 456-457
and stress-intensity factor *(K)*, *R*-curves from **A8:** 450
surface cracks role in **A8:** 450
values of real **A8:** 451
Crack formation
in line etching **A9:** 62
thermal-wave imaging used to study **A9:** 91
Crack front
advancing, mechanical damage **A12:** 72
defined **A12:** 176
direction, wrought aluminum alloys **A12:** 430
fracture mechanics of **A8:** 439-464
in ideal crack model **A8:** 440
located by beach marks **A12:** 112
plane strain, and fracture toughness **A8:** 450
plastic zone, in progressive fracturing **A8:** 440
primary, ductile irons **A12:** 233
relation to fracture process zone **A8:** 440
schematic **A8:** 440
segment, simplified for stress field analysis **A8:** 442
stresses and dynamic characterization **A8:** 444-445
striations on **A12:** 23
Crack growth *See also* Crack growth rate; Fatigue crack growth **EM1:** 8, 261 **EM3:** 9
acoustic emissions from **A17:** 287
and additions of dispersoid-forming elements **A8:** 487
analysis, software for **A11:** 55
by microvoid coalescence **A11:** 227, 230
change in cleaved intermetallic compound **A8:** 482
creep, in P/M superalloys **A13:** 838
as creep-rupture phase **A8:** 344
criterion, and tearing instability **A8:** 446
curves, comparing materials by **A8:** 518-519
cyclic **A8:** 376, 378-379, 422-423
direction, titanium alloys **A12:** 441-442
effect of vibration **A12:** 109
effects of changing stresses on **A8:** 502
electrical potential method for measuring **A8:** 390
fatigue **A12:** 14-18, 420
in fatigue, examination by in situ electropolishing **A9:** 55-56
and fatigue striation spacings **A12:** 205
fatigue-striation counts for **A11:** 56
fracture mechanics for **A11:** 47, 56-57
fracture mechanics models **A17:** 287
in fracture toughness specimens electropolishing **A9:** 55-56
examination by in situ
from surface flaw **A12:** 420
in graphite-epoxy composites **A11:** 737
in high toughness materials **A8:** 461
life curve **A8:** 681-682
Crack growth (continued)
life, prediction of **A8:** 678
linear elastic fracture mechanics (LEFM) for **A11:** 52
measurement, and acoustic emission inspection **A17:** 286
measuring systems, ultrasonic testing **A8:** 246
mechanism, and crack characteristics **A8:** 590, 593
modes, medium-carbon steels **A12:** 273
OFHC copper **A12:** 401
in pipeline steel **A11:** 54
in polyethylene **A12:** 480
prediction **A11:** 52
rate, and applied stress intensity **A11:** 107
rates **A11:** 53, 103, 107
resistance **A11:** 64
in sintered ceramic **A11:** 756
slow, in ceramics **A11:** 757
stable **A11:** 75, 87
stable and subcritical, compared **A11:** 63
stable, mechanical damage in **A12:** 72
stages, schematic **A11:** 710
stress-corrosion cracking and **A11:** 53
structure sensitivity, titanium alloys **A12:** 442-443
studied by bend or tension loading **A8:** 517
subcritical **A13:** 283, 1085
surface roughness from **A12:** 276
and toughness, parameters for **A11:** 54
ultrasonic testing of **A8:** 252
unstable, and fracture **A11:** 75
vs. constant-amplitude stress cycle **A8:** 379
Crack growth energy release rate
symbol for **A11:** 797
Crack growth per cycle *(da/dN)*
wrought aluminum alloy **A2:** 43
Crack growth rate *See also* Crack propagation rate; Fatigue crack growth rate
and applied stress intensity **A11:** 107
calculation **A8:** 518-519, 679-680
calculation, SCC testing **A13:** 260
copper **A8:** 255
cyclic **A8:** 376, 378-379
determined **A11:** 53
and fracture **A11:** 75
from hydrogen embrittlement **A8:** 537
LME, effect of- loading on **A11:** 726
in maraging steels **A11:** 218
methods to determine **A8:** 365
model of **A8:** 678
modeling of **A8:** 680-681
in precracked specimens **A8:** 518-519
schematic of typical **A8:** 242
steam turbine materials **A13:** 957-958
and stress-intensity factor **A8:** 242, 365
test, curve, and stress ratio **A11:** 53
Crack growth rate data
constructional steels for elevated temperature use **M1:** 657-658, 660-661
Crack initiation
acoustic emissions from **A17:** 287
AISI/SAE alloy steels **A12:** 302
at corrosion pits **A13:** 149
at surface discontinuity **A13:** 148-149
in axial test **A8:** 351-352
by discontinuities **A11:** 106
by intergranular corrosion **A13:** 149
by slip dissolution **A13:** 149
by slip-plane fracture **A11:** 104
by tool marks and corrosion pits **A12:** 419
comparison of measured fracture toughness data for **A8:** 467
corrosion fatigue **A13:** 142
corrosion-fatigue **A11:** 252-253
and crack propagation, compared **A13:** 150
as creep-rupture phase **A8:** 344
defined **A8:** 366
in elastic-plastic fracture toughness tests **A8:** 390
electrical potential technique **A8:** 389-390
environmental effects on **A8:** 375
fatigue **A8:** 363-364, 366-375 **A12:** 112
as fatigue stage **A11:** 102, 104
as fatigue stage I **A12:** 175
in fracture mechanics **A8:** 465
in high strain rate fracture testing **A8:** 259
hydrogen flaking and **A11:** 316

Crack initiation (continued)
incubation time ... **A11:** 242, 244
in inert/aggressive environments, compared ... **A13:** 142
in low-cycle torsional fatigue testing ... **A8:** 152
malleable iron ... **A12:** 238
mechanisms ... **A13:** 149-150
multiple, ratchet marks from ... **A11:** 77
on stainless steel dynamic compression plate ... **A11:** 679, 683-686
P/M aluminum alloys ... **A12:** 440
point, in one-point bend tests ... **A8:** 275
processes, phenomenology ... **A13:** 148-150
relation to environment ... **A11:** 106
shear, in punch-loading Kolsky bar ... **A8:** 229-230
sites ... **A8:** 366
and stress-corrosion cracking ... **A11:** 203
study, by magnetic rubber inspection ... **A17:** 125
subsurface ... **A11:** 79
tests ... **A13:** 291-295
and topography of crack surfaces ... **A11:** 212
toughness ... **A8:** 284, 466-467, 725
in welded structures ... **A8:** 366
Crack initiation testing
fatigue ... **A8:** 363
Crack initiation toughness ... **A8:** 284, 466-467, 725
Crack jumps ... **A8:** 454-455, 474
Crack length *See also* Crack depth; Crack length measurement; Crack length monitoring; Crack size; Cracking ... **A8:** 3
and compliance ... **A8:** 383
and crack path preference ... **A12:** 201
defined ... **A11:** 2
five-point average ... **A8:** 415
fracture band width as function of, polymer pipe ... **A11:** 762
Krak-Gages for ... **A8:** 391
mean, calculated ... **A12:** 207
measuring ... **A12:** 77
normalized, and laboratory fatigue failure compared ... **A11:** 762
symbol for ... **A8:** 724 **A11:** 796
true, calculated ... **A12:** 207
vs. elapsed-cycle, fatigue crack propagation ... **A8:** 377
vs. load drop, dynamic notched round bar testing ... **A8:** 278, 279
Crack length measurement
alternating-current electric potential method ... **A8:** 417
compliance method ... **A8:** 383-386
direct-current electric potential method for ... **A8:** 417
displacement measurement hardware attachment ... **A8:** 384-385
effects of random error in ... **A8:** 679
electric potential ... **A8:** 386-391
in high temperature vacuum fatigue crack growth rate ... **A8:** 415
for high-temperature aerated water testing ... **A8:** 422
interval, for compact-type specimens ... **A8:** 378
in liquid metals environments ... **A8:** 426
optical ... **A8:** 382-383
techniques ... **A8:** 382-391
Crack length monitoring
in acidified chloride solutions ... **A8:** 420
aqueous solutions ... **A8:** 417
electric potential ... **A8:** 386-391
in steam or boiling water with contaminants ... **A8:** 428
in vacuum and gaseous fatigue testing ... **A8:** 412
Crack mouth opening displacement (CMOD) *See also* Crack opening displacement; Crack opening displacement (COD)
abbreviation ... **A8:** 724
load as function of ... **A8:** 451
Crack nucleation ... **A8:** 363, 366, 537
as fatigue stage ... **A11:** 102, 104
Crack opening
by slip ... **A12:** 21
displacement, measurement ... **A17:** 286
microwave sensitivity ... **A17:** 203
relation to *K*, crack-extension force, *J*-integral ... **A8:** 440
secondary ... **A12:** 77
size near physical crack tip ... **A8:** 446
as source, acoustic emissions ... **A17:** 287
Crack opening displacement (COD) *See also* Mode
abbreviation ... **A8:** 724
behind crack front in progressive fracturing ... **A8:** 439
defined ... **A11:** 2
factors affecting measurement accuracy ... **A8:** 384-385
in fracture mechanics ... **A8:** 439
test, for fracture toughness ... **A11:** 62
transducer, bolt-on attachment ... **A8:** 384-385
Crack opening displacements (COD) ... **EM3:** 507
Crack origin *See also* Fracture origin
at abraded surface area ... **A12:** 263
and chevron patterns ... **A11:** 76
in composites ... **A11:** 742
determined by beach marks ... **A12:** 112
fractured medum-carbon steel shell ... **A12:** 257
in hydrogen damage ... **A11:** 250
lap as ... **A12:** 64, 65
medium-carbon steels ... **A12:** 261, 272
multiple ... **A12:** 272
superalloys ... **A12:** 390
wrought aluminum alloys ... **A12:** 415
Crack path *See also* Fracture path
ductile and brittle ... **A12:** 102
in low-carbon steel ... **A12:** 245
in SCC ... **A12:** 133
tortuosity ... **A12:** 38, 206
transgranular, SCC caused ... **A12:** 133-134, 152
in white iron ... **A12:** 239
Crack path tortuosity ... **A8:** 477, 480
Crack plane orientation
defined ... **A8:** 3 **A11:** 2
Crack profile measurements ... **EM3:** 452-453
Crack propagating toughness
velocity dependence of ... **A8:** 284
Crack propagation *See also* Crack propagation rates; Fatigue crack propagation
activation energy, in SMIE ... **A11:** 244
along grain boundaries, surfaces of ... **A11:** 76, 77
applied stress/flow depth effects on ... **A13:** 277
in argon and liquid lithium ... **A1:** 720
in bearings ... **A11:** 502
blended elemental titanium P/M compacts ... **A2:** 650-651
in brittle fracture(s) ... **A11:** 76
by alternate slip ... **A12:** 15, 21
by atom displacement at crack tip ... **A11:** 226, 230
by multilayer self-diffusion of embrittlers ... **A11:** 244
calculated by SCFM ... **A11:** 55-57
corrosion fatigue ... **A13:** 143-144
corrosion-fatigue ... **A11:** 253-254
crack enlargement ... **A11:** 106
and crack initiation, compared ... **A13:** 150
direction, and fatigue striation spacings ... **A12:** 205
direction and origin, composites ... **A11:** 742
direction, and river patterns ... **A11:** 77
ductile iron ... **A12:** 228
effect of hardness on ... **A11:** 475
effect of liquid in ... **A11:** 227
effect of twin boundaries, austenitic stainless steels ... **A12:** 351
elliptical ... **A11:** 123-124
environmental effects ... **A12:** 35
environmental factors ... **A13:** 151-155
factors controlling ... **A8:** 365
fatigue ... **A8:** 376-402, 403-435 **A11:** 103
and fatigue life prediction ... **EM1:** 246
Crack propagation (continued)
fatigue, localized failures in steel bridge components by ... **A11:** 707
as fatigue stage ... **A11:** 102, 104
as fatigue stage II ... **A12:** 175
fatigue, testing for ... **A11:** 103
fatigue, wrought titanium alloys ... **A2:** 624
final ... **A11:** 106-107
historical studies ... **A12:** 3
initial ... **A11:** 106
in iron ... **A12:** 222-223
liquid role, LME ... **A13:** 174
LME and SMIE, compared ... **A11:** 240
in low-carbon iron ... **A12:** 222
malleable iron ... **A12:** 238, 239
material chemistry ... **A13:** 155-158
mechanical factors ... **A13:** 158-159
mechanical fracture models ... **A13:** 160-162
mechanisms ... **A13:** 159-162
microstructure ... **A13:** 155-158
in monotonic fracture ... **A12:** 229
for nonlinear and plastic deformation ... **A8:** 445
in polymers ... **A11:** 762
prealloyed titanium P/M compacts ... **A2:** 652-653
processes ... **A13:** 150-162
rapid, in brittle fracture ... **A11:** 8-
rate, effect of grain-boundary cavitation on ... **A12:** 38
rate in liquid-metal embrittlement ... **A11:** 719
rate, liquid in ... **A11:** 227, 232
schematic ... **A13:** 161
SEM studies ... **A12:** 169
slow, in ceramics ... **A11:** 757
of steel in indium ... **A11:** 244
in stress-corrosion cracking ... **A8:** 496 **A13:** 148
study, by magnetic rubber inspection ... **A17:** 125
testing ... **A11:** 102
testing, fracture mechanics methods for ... **A8:** 363
tests, corrosion fatigue ... **A13:** 295
tests, for dynamic fracture testing ... **A8:** 284
types ... **A17:** 54
unstable ... **A11:** 60
vs. stress intensity, visual fit model ... **A8:** 681
Crack propagation in ceramics ... **EM4:** 694-698
environmentally enhanced crack growth ... **EM4:** 695-696
chemical wedge ... **EM4:** 695
cyclic loading ... **EM4:** 696
Griffith energy condition ... **EM4:** 695, 696
static loading ... **EM4:** 695-696
threshold ... **EM4:** 696
fast fracture ... **EM4:** 694-695
fractography ... **EM4:** 694-695
strength-controlling flaws ... **EM4:** 694
future trends ... **EM4:** 698
high-temperature crack growth ... **EM4:** 696, 697
interfacial crack propagation ... **EM4:** 698
delamination ... **EM4:** 698
four-point flexure specimen ... **EM4:** 698
toughening ceramics ... **EM4:** 696-698
crack-growth resistance curve (R-curve) ... **EM4:** 697
Weibull modulus ... **EM4:** 697-698
Crack propagation laws ... **EM3:** 506-508
Crack propagation rate *See also* Crack growth rate; Crack propagation; Fatigue crack growth; Fatigue crack growth rate; Fatigue testing
average, and oxidation ... **A13:** 155
average, in stainless steel, in water ... **A13:** 154
calculation methods ... **A8:** 678-680
corrosion fatigue ... **A13:** 297
factors influencing ... **A13:** 148
as function of crack tip stress intensity ... **A13:** 147
modeling of ... **A8:** 680-681
over entire stress intensity range ... **A8:** 681
and plasticity ... **A8:** 684
sigmoidal log-log plot ... **A8:** 681
and strain rate, stainless steel ... **A13:** 160

Crack propagation rate (continued)
subcritical ... **A13:** 158
vs. temperature, intergranular SCC ... **A13:** 154
Crack propagation testing ... **A8:** 363
Crack size *See also* Crack length; Physical crack size ... **A8:** 3
and beach marks ... **A11:** 57
defined ... **A11:** 2
determined ... **A11:** 52
final, fracture mechanics for ... **A11:** 56
and fracture stress ... **A11:** 56
load vs. displacement curves for ... **A8:** 448
plasticity adjustment ... **A8:** 452
strength as decreasing with ... **A11:** 55
symbol for ... **A8:** 724
Crack speed
and crack front stresses ... **A8:** 445
and crack tip stress-intensity factor Homolite 100 ... **A8:** 445, 446
dynamic ... **A8:** 444-445
effect on crack extension ... **A8:** 440
elastic analysis limit for ... **A8:** 445
relation to *K* in semi-brittle material ... **A8:** 441
to *K*, as toughness evaluation ... **A8:** 450
Crack stress field ... **A8:** 443-444
Crack tip
alternate slip ... **A12:** 15, 21
blunting ... **A11:** 56 **A12:** 15, 21
in bridge components, TEM fractograph ... **A11:** 708
characterization, by H-R-R singularity field ... **A8:** 446
chemistry, effect on corrosion-fatigue ... **A11:** 256
contraction at ... **A11:** 51
corrosion rate, crack depth effect ... **A13:** 150
cracking at ... **A11:** 47
diffraction, flaw sizing by ... **A17:** 654
dislocation nucleation ... **A12:** 30
displacement, by COD test ... **A11:** 62
displacement of atoms at ... **A11:** 230
environmentally-assisted fracture at ... **A8:** 499
film formation, alloy-environment systems for ... **A13:** 146
fracture at ... **A11:** 47
plastic zone ... **A12:** 15, 16
plastic zone, acoustic emissions ... **A17:** 287
plastic zone, determined ... **A11:** 49
processes, schematic ... **A13:** 148
protective films at, and stress-corrosion cracking ... **A8:** 499
resharpening ... **A12:** 21
Rice *J*-integral contour around ... **A8:** 447
size of crack opening near physical analyses for ... **A8:** 446
slip at ... **A12:** 15, 21
stress intensity ... **A13:** 147, 297
stress-intensity factor (*K*) ... **A8:** 497
stress-intensity factor, and crack speed Homolite 100 ... **A8:** 446
stresses ... **A11:** 47-49, 51
Crack tip opening angle
abbreviation ... **A8:** 724
Crack tip opening displacement *See also* Plane stress
abbreviation ... **A8:** 724
as measure of crack extension tendency ... **A8:** 446
in microvoid coalescence ... **A8:** 466
obtained by Rice *J*-integral ... **A8:** 449
symbol for ... **A8:** 726
use in *R*-curve format ... **A8:** 457
in wedge-opening load test for hydrogen embrittlement ... **A8:** 538
Crack tip opening displacement (CTOD) *See also* Crack opening and displacement ... **A6:** 81
tests ... **A6:** 104
and toughness ... **A11:** 56
Crack tip opening displacement (CTOD) tests ... **A1:** 611, 662-663
Crack tip plastic zone ... **EM3:** 508
Crack tip stress
analysis, for prediction of stress corrosion ... **A8:** 497
fields, and stress-intensity factor ... **A8:** 441-443
in fracture mechanics ... **A8:** 465
-intensity, control of fatigue crack propagation in aluminum alloy ... **A8:** 404
Crack tip stress (continued)
-intensity, dynamic vs. static ... **A8:** 282
Crack tip stress-whitening zones (CTSWZ) ... **EM3:** 514
Crack tip strip zone model
for central crack ... **A8:** 449
Crack width
in truncated cone identification test ... **A14:** 380
Crack(s) *See also* Crack growth; Crack initiation; Crack opening Crack propagation; Crack tip; Cracking; Crazing; Defects Discontinuities; Fatigue cracks; Flaw(s) Flaw detection; Inclusions; Internal shrinkage cracks; Microcracking; Microcracks; Microcracks Nonmetallic inclusions; Stress crazing; Subsurface crack(s); Subsurface flaws; Surface crack(s); Surface defects; Surface flaws Weldments
advancing fatigue front of ... **A11:** 26
analysis, by replication ... **A17:** 54
angle of forking, information from ... **A11:** 744
base-metal ... **A11:** 449
in bearing materials ... **A11:** 504-506
in billets, magnetic particle inspection ... **A17:** 115
branching ... **A11:** 81, 702, 744
in bridge components ... **A11:** 708-710
center ... **A11:** 61
characterization, by electric current perturbation ... **A17:** 140-141
circumferential corrosion-fatigue ... **A11:** 79
cleavage, defined ... **A11:** 2
closure, as source, acoustic emissions ... **A17:** 287
cold, in arc welds ... **A11:** 413
cold, radiographic appearance ... **A17:** 349
crater ... **A17:** 86, 582, 585
creep ... **A17:** 54-55
defined ... **EM2:** 12
depth ... **A11:** 124, 239-241
detection ... **A7:** 483-484
development ... **EM2:** 807-810
direction ... **A11:** 81, 735-738, 744
double-edge ... **A11:** 61
during quenching ... **A7:** 453
eddy current inspection of ... **A17:** 164-166
elliptical, schematic in origin region ... **A11:** 124
enlargement ... **A11:** 106
extension ... **A11:** 20 **EM2:** 805
face friction, acoustic emission inspection ... **A17:** 287
fastener hole ... **A17:** 139-140
fatigue, as planar flaw ... **A17:** 50
in forging ... **A11:** 327
formation ... **EM2:** 806
frequency of branching, information from ... **A11:** 744
front, acoustic emissions from ... **A17:** 287
front tunneling ... **A11:** 20
grinding, as planar flaw ... **A17:** 50
growth, defined ... **EM2:** 12
growth rates, in maraging steels ... **A11:** 218
heat treatment, as planar flaw ... **A17:** 50
Hertzian cone, ceramics ... **A11:** 753
hook ... **A11:** 449
hot, in arc welds ... **A11:** 413
hot-shortness ... **A11:** 444
-inducing substances ... **A11:** 210
initiation ... **A11:** 106, 203
internal and surface ... **A11:** 108
jumps as source, acoustic emissions ... **A17:** 287
length as criterion, NDE reliability ... **A17:** 664
liquid penetrant inspection ... **A17:** 71, 86
location, by microwave inspection ... **A17:** 203
magnetic field testing detection ... **A17:** 129
magnetic particle detection ... **A7:** 484, 577
and microlaminations ... **A7:** 486
microwave inspection ... **A17:** 212
monitoring, by ultrasonic inspection ... **A17:** 273
morphology, in hydrogen damage ... **A11:** 250
multiple-origin ... **A11:** 24
nucleation ... **A11:** 23, 106
origin, in hydrogen damage ... **A11:** 250
part-through ... **A11:** 51
patterns, and SCC ... **A11:** 212
peeling-type, in shafts ... **A11:** 471
plating, as planar flaw ... **A17:** 50
quench ... **A11:** 122
radiographic appearance ... **A17:** 350
radiographic methods ... **A17:** 296
Crack(s) (continued)
in resistance welds ... **A11:** 441
restraint, in friction welds ... **A11:** 444
root ... **A11:** 92, 93
secondary ... **A11:** 19-20, 79
short-, corrosion-fatigue behavior ... **A11:** 254
single-edge ... **A11:** 61
speed ... **A11:** 87, 91
in steel bar and wire ... **A17:** 549-550
stress corrosion, as planar flaw ... **A17:** 50
surface, laser-detected ... **A17:** 17
surfaces, topography of ... **A11:** 212
thermal, as forging defects ... **A11:** 317
through-thickness ... **A11:** 51
transgranular, transverse section of ... **A11:** 28
and unbonded particles, detected by metallography ... **A7:** 484
weld, in arc-welded aluminum alloys ... **A11:** 435
in weldments ... **A17:** 50, 582, 584-585
Crack-arrest lines *See* Beach marks
Crack-closure method
for energy release rate ... **EM1:** 248-250
Crack-extension force ... **A8:** 3, 439-443
Crack-extension resistance
defined ... **A8:** 3
Crack-tip plane strain
defined ... **A8:** 3 **A11:** 3
Crack-tip stress-intensity factor (K)
in fatigue- crack propagation ... **A11:** 103, 107
Crack-tip stresses
in a hole ... **A11:** 48
by linear clastic fracture mechanics (LEFM) ... **A11:** 47
coordinate system for ... **A11:** 47
distribution for ... **A11:** 49
in plane strain ... **A11:** 51
and toughness ... **A11:** 48
Cracked gas ... **A7:** 3
Cracked lap-shear (CLS) specimens ... **EM3:** 508, 509, 512
Cracking *See also* Crack; Crack propagation; Crack propagation rate; Crack(s); Delayed cracking; Fatigue crack growth; Fatigue crack growth rate; Fracture; Hot cracking; Macrocrack; Microcrack; Stress-corrosion cracking (SCC) ... **A6:** 88-96 **A14:** 19, 364, 380, 385
as a result of electric discharge machining ... **A9:** 27
acceleration rates ... **A11:** 744
in ammonia ... **A11:** 215
at crack tip ... **A11:** 47
base-metal cracking ... **M6:** 93
base-metal, in laser beam welds ... **A11:** 449
in bearing failures ... **A11:** 494
in bending ductility tests ... **A8:** 117
brazing and ... **A6:** 110
of bridge components ... **A11:** 707
by hydrogen embrittlement, in titanium alloys ... **A8:** 522
calculation of susceptibility ... **M6:** 45
as casting defect ... **A15:** 548
causes ... **M6:** 834-835
causes of ... **A11:** 96-97
caustic, defined ... **A11:** 2
caustic, in boilers ... **A11:** 214
centerline cracking ... **M6:** 93
chevron cracking ... **M6:** 48
of coating, defined ... **A13:** 4
cold ... **A12:** 137-138
cold cracking ... **A6:** 93-95 **M6:** 44-46
shielded metal arc welding ... **M6:** 93
cold, defined ... **A15:** 3
compression ... **A8:** 57
consequences ... **M6:** 843-844
contact, in ceramic ... **A11:** 754
corrosion-fatigue ... **A11:** 371-372
crater cracking ... **A6:** 88, 90, 91
creep, elastic-plastic fracture mechanics (EPFM) for ... **A11:** 47
defined ... **A18:** 6
defined for stress corrosion ... **A8:** 495
delayed ... **A11:** 95
delayed, uranium alloys ... **A2:** 675-676
of drive-gear assembly ... **A11:** 143
ductility-dip cracking ... **M6:** 48
due to creep deformation ... **A8:** 306
effect of restraint ... **M6:** 251-252

Cracking (continued)
electron-beam welding...... **A6:** 866-867, 870, 871-873
electron-beam welds **A6:** 866-867
environmentally induced **A13:** 145, 217-219
fast **A11:** 75
fatigue **A8:** 366 **A11:** 105-107
ferrite vein cracking **M6:** 46
fiber **EM1:** 195
in fiber composites **A11:** 761
fisheye **M6:** 831-832, 834-835
from extrusion speed **A11:** 87, 91
from gaseous hydrogen **A11:** 247
from hydride formation **A11:** 248-249
from hydrogen charging, aqueous environments **A11:** 245-247
from precipitation of internal hydrogen **A11:** 248
general, SCC **A13:** 247
graphitization **M6:** 834
hairline **A11:** 28
heat-affected zone *See* Heat affected zone
heat-affected-zone cracks **A6:** 90-93
hot **A12:** 138
hot cracking..... **A6:** 88-90, 91, 92 **M6:** 47-48, 832-833
arc welds of coppers **M6:** 402
beryllium susceptibility **M6:** 461-462
causes **M6:** 47-48, 833
ductility-dip cracking **M6:** 832
liquation cracking **M6:** 832
prevention **M6:** 833
reheat cracking **M6:** 832-833
shielded metal arc welds **M6:** 92-93
solidification cracking **M6:** 832
stress-relief cracking **M6:** 832-833
susceptibility **M6:** 47
in hydrogen damage **A13:** 163-171
hydrogen-assisted **A15:** 532-534
hydrogen-induced cold
alloy and carbon steels **M6:** 248-249, 883
cracking **M6:** 44-46, 831-832
root cracking **M6:** 830-831
toe cracking **M6:** 830-831
underbead cracking **M6:** 830
hydrogen-induced cracking **A6:** 93-95
causes and cures **A6:** 95
intergranular, and creep deformation **A11:** 29
interlaminar **EM1:** 241-244
kinetics of **A8:** 449
lamellar tearing **A6:** 95-96 **M6:** 46, 832
layer **EM1:** 436-437
layer, computer program for **EM1:** 277
in liquid-metal embrittlement **A13:** 171-184
localized (SCC) **A13:** 247
matrix, energy method **EM1:** 241
measurement, SCC testing **A13:** 259
mechanism **EM2:** 805-807
microfissures **M6:** 832
microporosity **A11:** 355-357
multiple, from corrosion fatigue **A11:** 79
in nitrate solutions **A11:** 214-215
of nut, as fastener failure **A11:** 531
of oxide **A13:** 72
path, in maraging steels **A11:** 218
path, magnesium alloys **A11:** 223
in permanent mold castings **A15:** 285
in polythionic acid **A8:** 528
pre **A12:** 75, 236-237, 397
prevention **M6:** 45, 834-835
prevention in electron
beam welds **M6:** 630
of hardenable steels **M6:** 637
quench, factors controlling **A11:** 94-95
reheat cracking **A6:** 92, 93 **M6:** 46-47
retardation effect **A8:** 682
season, defined **A11:** 9
sensitivity **A6:** 94
shatter **A13:** 164
slow **A11:** 75
slow, stable, and crack enlargement **A8:** 440

Cracking (continued)
solid graphite molds **A15:** 285
in solid-metal induced embrittlement **A13:** 184-187
solidification cracking...... **A6:** 88-90, 91, 92 **M6:** 249, 254-255
stepwise, low-strength steel **A13:** 170
strain-age cracking **A6:** 92, 93
strain-age cracking, nickel- based heat-resistant alloys **M6:** 363-364
stress corrosion cracking **M6:** 833
in carbon steels **M6:** 883-884
in nickel alloy welds **M6:** 443
stress-corrosion **A13:** 145-163, 247, 259, 275
stress-relief **A12:** 139-140
stress-relief cracking **A6:** 92, 93
stress-relief cracking in alloy steels **M6:** 249
subcritical fracture mechanics of **A11:** 47
submerged arc welding **A6:** 202-203
subsurface **A11:** 134
surface corrosion from **A12:** 72
sustained-load, SCC testing **A13:** 275
tests, environmental **A11:** 301-302
thermal stresses **M6:** 251
types in bridge components **A11:** 707
types of weld discontinuities **M6:** 829-835
underbead cracking in shielded metal arc welds **M6:** 92-93
weld **A12:** 137-140
of welds **A13:** 344
Cracking, control of **M1:** 457, 460, 470
Cracking in
arc welds of nickel alloys **M6:** 443
nickel-based heat-resistant alloys **M6:** 363-364
electroslag welds **M6:** 233-234
flash welds **M6:** 579
gas metal arc welds **M6:** 172
of aluminum alloys **M6:** 386-387
resistance welds of aluminum alloys **M6:** 543-544
resistance welds of stainless steels **M6:** 533
shielded metal arc welds **M6:** 92-93
base-metal cracking **M6:** 93
centerline cracking **M6:** 92-93
cold cracking **M6:** 93
hot cracking **M6:** 93
underbead cracking **M6:** 92-93
submerged arc welds **M6:** 128-130
weld overlays **M6:** 817
Cracking, in finished copper
IGF analysis **A10:** 231-232
Cracking, package
and stress **EL1:** 480
Cracks *See also* Grinding cracks **EM3:** 9
in broken/unbroken aluminum alloy compared **A12:** 121, 137, 138
center, fracture analysis **EM1:** 252-257
corrosion of **A12:** 72
defined **EM1:** 8
definition **M6:** 4
direction, effect, interlaminar/intralaminar fracture **EM1:** 787-790
in fiber composites, mounting to prevent fiber tearout **A9:** 588
first, onset **EM1:** 243
high-speed, effects in low-carbon steel **A12:** 252
as hydrogen damage **A12:** 124
laser/IR inspected **EL1:** 942
length, and flaw size **EM1:** 255
in microetched carbon and alloy steels **A9:** 173
multiple, laminar **EM1:** 244
nucleation, austenitic stainless steels **A12:** 360
part-through **A12:** 72
in powder metallurgy materials **A9:** 512
pre-existing, as fracture origin **A12:** 65
primary, opening **A12:** 77
secondary, opening **A12:** 77
separations, measuring **A12:** 77
size, effect, laminate failure **EM1:** 255
through-thickness, composite effects **EM1:** 261

Cracks (continued)
transverse, laminar **EM1:** 242-244
Crane
hooks, magnetic particle inspection of **A17:** 107, 112-113
overhead, magnetic particle inspection **A17:** 115
Crane and vehicle applications
high-strength low-alloy steels for **A1:** 419
Crane hooks
materials for **A11:** 522
Crane ladle *See also* Ladles
development of **A15:** 33
early ladle movement by **A15:** 27
jib, historic **A15:** 33
Cranes
-bridge wheel, fracture of **A11:** 527-528
brittle fracture of stop-block guide **A11:** 526-527
electric overhead traveling **A14:** 63
for explosive forming **A14:** 637
failures of **A11:** 525-528
fatigue fracture of **A11:** 524-525
fatigue fracture of alloy steel lift pin from **A11:** 77
materials for **A11:** 515
and related members **A11:** 525-528
steel drive axle wheel, fatigue fracture **A11:** 116-117
Crank
defined **A14:** 3
Crank and lever testing machine **A8:** 369-370
Crank presses **A14:** 3, 40
Crankcase
failed gray iron **A11:** 362-365
Crankpin bearing *See* Big-end bearing
Crankshaft and camshaft sprocket gears **A7:** 617, 619
Crankshaft bearings **A18:** 559-561
Crankshaft drive
mechanical presses **A14:** 493-494
Crankshaft lathes **A16:** 153
Crankshafts *See also* Shafts **A16:** 111
diesel-engine, fatigue failure from subsurface inclusions **A11:** 323
ductile iron automotive, fatigue-fracture surface **A11:** 104
fatigue cracking from segregation of nonmetallic inclusions **A11:** 477-478
fatigue failure analysis of **A11:** 123-125
fatigue failure from stress raisers **A11:** 472
fatigue fracture, from metal spraying **A11:** 480
fatigue limits of **M1:** 674
magnetic particle inspection methods **A17:** 117
magnetizing **A17:** 94
main-bearing journals **A11:** 358-359
misalignment of **A11:** 475
residual stress, magabsorption measurement **A17:** 157-158
Crater crack
definition **M6:** 4
Crater cracks
liquid penetrant inspection **A17:** 86
in weldments **A17:** 582, 585
Crater fill current
definition **M6:** 4
Crater fill time
definition **M6:** 4
Crater fill voltage
definition **M6:** 5
Crater filling **A6:** 747
Crater wall effects **EM3:** 241
Crater wear **A18:** 609, 610
as cemented carbide tool wear mechanism **A2:** 954
defined **A18:** 6
Cratering **EM3:** 9
of cermets **A2:** 979
defined **EM2:** 12
as failure mechanism **EL1:** 1043
TC **A16:** 74

Cratering (continued)
titanium carbide for resistance to **A7:** 158
Cratering due to electrical arcing.......... **A9:** 563
Craters *See also* Bubbles; Cavities.......... **A6:** 703
definition.......... **A6:** 1208 **M6:** 4
formed by liquid erosion.......... **A11:** 165
in shielded metal arc welds **M6:** 93
Crawler tractor track pins, nitrided
wear of **M1:** 628, 630
CRAY 2
immersion cooling of.......... **EL1:** 49
Craze cracking.......... **A18:** 621-622
defined **A18:** 6
pumps.......... **A18:** 595
Craze fibrils
defined **A11:** 760
Crazes, structure in glassy polymers
by SAXS/SANS/SAS **A10:** 402
Crazing *See also* Corrosion resistance; Crack, Stress; Crackling; Hairline craze; Star craze; Stress crack; Stress crazing **EM3:** 9, 655, 683
brittleness testing **EM2:** 738-739
craze growth **EM2:** 737
defined **A13:** 4 **EM1:** 8 **EM2:** 12
ductile-brittle transition.......... **EM2:** 735
effect on toughness **EM2:** 737
environmental effects **EM2:** 736
environmental stress **EM2:** 796-804
failure analysis of.......... **EM2:** 734-740
formation **EM2:** 657
fracture toughness testing **EM2:** 739-740
initiation criteria **EM2:** 736-737
of matrices **EM1:** 31
pattern, thermal fatigue of steel tube **A11:** 623
polymeric behavior **EM2:** 734
in polymers.......... **A11:** 759-760
as PTH failure mechanism **EL1:** 1025
as SCC **A11:** 451
in thin polystyrene film **A11:** 758
CRB-7
nominal composition.......... **A18:** 726
CRC L-37 and CRC L-42
axle tests for EP agent performance **A18:** 101
CRC L-38 testing
oxidation inhibitors.......... **A18:** 105
CRC L-60 test
oxidation inhibitors.......... **A18:** 105
Creams *See* Solder (-reams
Creams, topical
as bismuth application **A2:** 1256
Credibility
of corrosion test results.......... **A13:** 316
Creel
defined **EM1:** 8 **EM2:** 12
Creel warp supply spools.......... **EM1:** 127, 128
Creep *See also* Cold flow; Creep characteristics; Creep curve; Creep forming; Creep modulus; Creep properties; Creep rate; Creep recovery; Creep resistance; Creep rupture; Creep rupture properties; Creep rupture strength; Creep rupture testing; Creep strength; Creep testing; Creep-fatigue interaction; Deformation under load; Deformation under load, Flow; Elevated-temperature properties; Elevated-temperature service; Flow; Plastic flow; Tensile creep; Yield.......... **A6:** 374 **EM3:** 9, 51-52, 353-359
activation energy vs. self-diffusion activation energy **A8:** 309
analysis, matrix for **A8:** 685-686
ASTM test methods **EM2:** 334
at application temperature, plastics.......... **EM2:** 1
behavior, negative.......... **A8:** 331
-brittle materials **A8:** 344
bulk specimen behavior.......... **EM3:** 364-365
of cast stainless steels **A1:** 920, 924, 927
classical behavior.......... **A1:** 629
compliance.......... **EM3:** 367-368
compliance and relaxation modulus, as viscoelasticity.......... **EM1:** 190-191
components of.......... **A8:** 301
constants as function of stress **A8:** 318-319
crack formation, schematic.......... **A17:** 55
crack growth, fracture mechanics of.......... **A11:** 52
crack growth, P/M superalloys.......... **A13:** 838
as crack propagation.......... **A17:** 54

Creep (continued)
crack propagation, austenitic stainless steels.......... **A12:** 354
cracking, elastic-plastic fracture mechanics (EPFM) for.......... **A11:** 47
creep damage in superalloys **M3:** 225-226
and creep rupture, forms.......... **EM2:** 672-673
creep strain/time relationships **A8:** 686
curves **A8:** 304-305, 308-309
cyclic **A8:** 353-354
damage.......... **A8:** 337-339, 344 **A11:** 290, 605 **A17:** 54-55
damage, as indicator of fracture mode **A12:** 62-63
data analysis.......... **A8:** 685-694
data, analysis of.......... **EM2:** 667-672
data presentation.......... **A8:** 303-305, 315
data, thermoplastic resins.......... **EM2:** 621
defects, from stress and thermal load **A17:** 54
defined **A8:** 3, 301 **A11:** 3 **A13:** 4 **EM1:** 8, 190 **EM2:** 12
definition of.......... **A1:** 622, 927, 932
deformation.......... **A7:** 664 **A8:** 306, 308-310, 331
deformation and intergranular cracking.......... **A11:** 29
design of overlap to prevent.......... **EM3:** 474
dislocation, mechanisms of **A8:** 309-310
distortion failure from.......... **A11:** 138
ductility, data analyses of.......... **A8:** 693
ductility, in steels **A11:** 98
effect on fracture **A12:** 59
elevated temperature **A8:** 302
in elevated-temperature failures.......... **A11:** 263-264
embrittlement, in steels **A11:** 98
embrittlement, of chromium-molybdenum steels **A12:** 124
embrittlement, petroleum refining and petrochemical operations **A13:** 1265
equations.......... **A8:** 686-689
experiments **A8:** 302-303
failure **A11:** 406-407, 610-612
failure analysis of.......... **EM2:** 728-730
failure, ASTM/ASME alloy steels.......... **A12:** 346
failure, by wedge cracking.......... **A12:** 364
failures.......... **EL1:** 56, 632
-fatigue interaction.......... **A8:** 346-360
fatigue interaction, in elevated-temperature failures.......... **A11:** 266
fatigue life prediction by **A12:** 123
fissures, ultrasonic inspection **A17:** 652
fissuring **A11:** 272
fits to experimental curves **A8:** 688
fractures, identification chart for.......... **A11:** 80
full densification of powder compact by.......... **A7:** 502
general behavior during **A8:** 301
of gray iron.......... **A1:** 27, 102
in heat-resistant alloys.......... **A1:** 920, 924, 927 **A15:** 729
-induced failure, boiler plate.......... **A11:** 30
influence of multiaxial stressing on 343-345 I
interaction with fatigue.......... **M3:** 234-235
as joint design factor.......... **EM3:** 43
Kevlar aramid fibers.......... **EM1:** 55
life, and ductility, effect of multiaxial stresses.......... **A8:** 343
life dependence on deformation and fracture **A8:** 344
logarithmic.......... **A8:** 308
low-temperature/high-temperature **A8:** 301
matrix **A12:** 349
measurement.......... **A8:** 339
measurement device **EM3:** 665
microstructural crystallinity **EM3:** 411
microstructure during **A8:** 305-306
microvoid linking by **A12:** 364
models of **EM2:** 659-666
modified.......... **A11:** 264
moisture effects.......... **EM2:** 763-765
Nicalon.......... **EM1:** 64
in nickel-base alloy, cracking from **A11:** 284
nonclassical behavior.......... **A8:** 331-332
para-aramid fibers.......... **EM1:** 55
parameters **A8:** 689-690
power law.......... **A8:** 303-304, 310
and prediction of deformation under load.......... **EM2:** 673-678
in pressure vessels **A11:** 666-668
primary and tertiary **A8:** 308 **A11:** 263, 264

Creep (continued)
primary, defined.......... **A11:** 8
properties of niobium alloys **A7:** 772
pure titanium **A2:** 596
rate **A11:** 3, 264
rate of, defined.......... **EM1:** 8 **EM2:** 12
rate, zirconium alloys **A2:** 668
relationship with hardness **A1:** 640
relaxation **A8:** 347
and relaxation, compared **A11:** 542
resistance.......... **A7:** 144, 439 **A11:** 131
resistance, of nickel alloys **A2:** 429
rotational, in rolling-element bearings **A11:** 492, 496
rupture, defined.......... **A12:** 18-20
-rupture embrittlement.......... **A12:** 123-124
-rupture embrittlement, defined.......... **A13:** 4
and rupture relationship, with STAMP process **A7:** 549
rupture strength, defined.......... **A11:** 3 **A13:** 4
schematic of.......... **A8:** 301
secondary **A11:** 263
in shafts **A11:** 459, 467
in shear/compression, polyether sulfones (PES, PESV) **EM2:** 161
as solder joint failure mechanism **EL1:** 1031
stages **A8:** 311, 331 **A12:** 19, 25
stages of **A11:** 264
steam equipment failure by **A11:** 602
step excitation test methods **EM2:** 549-551
strain **A8:** 337
strain, defined **A11:** 3
strain/time relationships **A8:** 686-687, 691
strength, defined **A11:** 3
strength for steels **A8:** 330
strength, of boiler tubes **A11:** 603
stress, defined **A11:** 3
and stress relaxation.......... **A8:** 306-307 **A11:** 144 **EM2:** 659-678
and stress-rupture failures.......... **A11:** 29
in superalloys.......... **A7:** 473, 652 **A8:** 331
in tension, polyether sulfones (PES, PESV) **EM2:** 160-161
tertiary.......... **A11:** 263-264
test planning for **A8:** 685
transient **A8:** 308
testing, spin-test rig and pit for.......... **A11:** 280
tests **EM2:** 334, 435, 549-551
tests, for high-temperature effects.......... **A12:** 121
tests, medium-density polyethylene **A12:** 480
thermal **EL1:** 56
in thermoplastic fluoropolymers **EM2:** 118-119
of thermoplastics **EM1:** 97, 100, 293
thermoplastics compared to epoxies **EM3:** 98
as time-dependent plastic distortion **A11:** 75
as time-temperature relation in polymers.......... **EM3:** 421
tungsten-reinforced MMCs.......... **EM1:** 880-881
ultrahigh molecular weight polyethylenes (UHMWPE) **EM2:** 169
uniaxial tensile.......... **EM2:** 666-667
and uniform loading.......... **EM2:** 660
and viscoelasticity **EM2:** 414
voids, in steam tube walls.......... **A11:** 605
of wrought stainless steels.......... **A1:** 932
Creep (microslip)
defined **A18:** 6
Creep behavior *See also* Elevated- temperature properties, Stress rupture properties
alloy cast irons.......... **M1:** 92-93
constructional steels for elevated temperature use **M1:** 640, 642-647, 649, 650, 655-656
ductile iron **M1:** 47, 48, 49
gray cast iron **M1:** 26, 27
Creep characteristics
commercially pure tin **A2:** 518
lead alloys.......... **A2:** 551
Creep compliance **EM3:** 318
Creep curve **A8:** 304-305, 308-309
of alloy 2V tested in argon and air.......... **A8:** 332
of chromium-molybdenum steel with nonclassical early stage **A8:** 331
for constant-load test.......... **A8:** 311
constant-stress, austenitic stainless steel **A8:** 320-321
for constant-stress test.......... **A8:** 311

Creep curve (continued)
creep stages ... **A8:** 311
effect of oxide strengthening ... **A8:** 332
and temperature and stress effect ... **A8:** 318-319
Creep damage ... **A8:** 337-339, 344
Creep deformation ... **A8:** 139-310, 306, 331
Creep embrittlement ... **A12:** 123-124
and elevated-temperature service ... **A1:** 626
Creep forming
of titanium alloys ... **A14:** 846-847
Creep modulus
apparent, phenolics ... **EM2:** 244
determination ... **EM2:** 75
Creep properties
sand cast magnesium alloys ... **A2:** 496-516
wrought magnesium alloy ... **A2:** 485
Creep rate ... **A8:** 3
curve ... **A8:** 308
dependence on stress and temperature ... **A8:** 302, 308
equivalent, in tensile and compressive directions ... **A8:** 343
ferritic steels ... **A8:** 331
measurement, after service ... **A8:** 339
minimum, effect of frequency on ... **A8:** 347
-time curve, for silver-lead alloy cable sheath ... **A8:** 321, 323
Creep ratio *See also* Creep
defined ... **A18:** 6
Creep recovery ... **EM3:** 9
defined ... **A8:** 3 **EM2:** 12
polyether sulfones (PES, PESV) ... **EM2:** 161
Creep resistance
of acetals ... **EM2:** 100
allyls (DAP, DAIP) ... **EM2:** 227
phenolics ... **EM2:** 244
polyether sulfones (PES, PESV) ... **EM2:** 160
Creep resistance in tungsten alloys
effect of doping on ... **A9:** 442
Creep rupture *See also* Creep-rupture properties; Mechanical properties
aluminum casting alloys ... **A2:** 153-177
analysis, by computer ... **A8:** 690
characteristics, cast copper alloys ... **A2:** 366, 368-369, 375-376, 383, 386, 389
creep strain/time relationships ... **A8:** 686
damage, by cyclic stress ... **A8:** 354-355
data analysis ... **A8:** 329-330, 685-694 **EM2:** 668-670
data extrapolation ... **A8:** 332-337
equations for ... **A8:** 686-689
forms ... **EM2:** 672-673
influence of multiaxial stressing on ... **A8:** 343-345
life, static, prior fatigue effect on ... **A8:** 353, 355
mixed criteria ... **A8:** 344
phases ... **A8:** 344
properties ... **A8:** 329-342
scatter ... **A8:** 329-330
step excitation ... **EM2:** 552-553
strength ... **A8:** 317, 319
testing ... **A8:** 301-306, 334-337, 685
white metal ... **A2:** 525
wrought aluminum and aluminum alloys ... **A2:** 74-79, 81, 88, 113, 119, 122
Creep rupture, in stainless steels
ultrasonic inspection ... **A17:** 650-652
Creep rupture strength
defined ... **EM2:** 12
plastic ... **EM2:** 75
Creep rupture(s)
defined ... **A12:** 18-20
embrittlement, interpreting ... **A12:** 123-124
high-purity copper ... **A12:** 399
intergranular, austenitic stainless steels ... **A12:** 364
strain rates for ... **A12:** 31
superalloys ... **A12:** 389, 393, 395
Creep strain ... **A8:** 3, 337
Creep strength ... **A6:** 374
of $2^1/_4$Cr-1Mo steel ... **A1:** 635, 636, 647
of carbon steel ... **A1:** 629
Creep strength (continued)
cast copper alloys ... **A2:** 364
chromium and molybdenum, effect of ... **A1:** 643
of chromium-molybdenum steels ... **A1:** 620, 623, 629
compared with hardness ... **A1:** 640
defined ... **A8:** 3
of ductile iron ... **A1:** 48, 49, 50
effects of chromium on ... **A1:** 643
maximum-use temperatures based on ... **A1:** 617
of wrought stainless steels ... **A1:** 932-933
wrought titanium alloys ... **A2:** 626
Creep strength of iron-chromium-nickel heat- resistant casting alloys
influence of microstructure ... **A9:** 333
Creep strengthening
effect of segregation in ... **A10:** 598
Creep stress
defined ... **A8:** 3
Creep testing ... **A8:** 3, 311-328 **EM3:** 316-318, 319
closed-loop servomechanical system ... **A8:** 396
constant-load ... **A8:** 313-318
constant-stress ... **A8:** 318-321
design and analysis of adhesive bonding ... **EM3:** 467, 468
equipment ... **A8:** 311-313
polybenzimidazoles ... **EM3:** 170
specimen, thermocouple ... **A8:** 312-313
step-down ... **A8:** 324
strain rate ranges for ... **A8:** 40
temperature control and measurement in furnace ... **A8:** 312-313
torsional ... **A8:** 147
vibration and shock load effect ... **A8:** 312
Creep testing equipment ... **A8:** 311-313
Creep tests ... **EM4:** 36
constructional steels for elevated temperature use ... **M1:** 640, 642
malleable cast irons ... **M1:** 66, 72
Creep theory ... **EM4:** 189
Creep-fatigue interaction ... **A8:** 346-360
10% rule ... **A8:** 354
and chromium-molybdenum steels ... **A1:** 625, 632, 633
damage rate analysis ... **A8:** 358
diagrams ... **A8:** 355-356
effect of ductility on ... **A1:** 633, 935
effects ... **A8:** 346-354
fractional damage equation ... **A8:** 355-356
frequency effect ... **A8:** 346-347
frequency-modified fatigue equation ... **A8:** 356-357
historical perspective ... **A8:** 346
hold periods ... **A8:** 346-347
linear damage rule ... **A8:** 355
material behavior in ... **A8:** 357
partitioned strain range/cycles to failure in ... **A8:** 357
plastic-strain fatigue resistance of stainless steel ... **A8:** 348
plot for hold-time data ... **A8:** 355-356
prediction techniques ... **A8:** 354-358
strain-range partitioning ... **A8:** 357-358
t-n diagram analysis ... **A8:** 358
tensile hysteresis energy approach ... **A8:** 358
test environment effect on ... **A8:** 354
under mean load and high temperatures ... **A8:** 254
and wrought stainless steels ... **A1:** 934-935
wrought titanium alloys ... **A2:** 626-627
Creep-feed grinding ... **A16:** 443-444
Creep-feed grinding (CFG)
ceramics and wear studies ... **A18:** 409, 410
Creep-resistant steels
reheat cracking ... **A6:** 92
Creep-rupture data extrapolation ... **A8:** 332-337
Creep-rupture properties
assessment and use of ... **A8:** 329-342
estimation of required properties from insufficient data ... **A8:** 335-337
Creep-rupture properties (continued)
evaluating creep damage and remaining service life ... **A8:** 337-339
extrapolation procedures ... **A8:** 332-335
interpolation procedures ... **A8:** 332-335
measuring rupture properties after service ... **A8:** 338
Monkman-Grant relationship ... **A8:** 335-337
nonclassical creep behavior ... **A8:** 331-332
planning a test program ... **A8:** 339-340
scatter ... **A8:** 329-330
Creep-rupture strength *See also* Stress rupture ... **EM3:** 9
of 1Cr-1Mo-0.25V steel ... **A1:** 629
of $2^1/_4$Cr-1Mo steel ... **A1:** 622, 623, 636, 647
of 9Cr-1Mo steel ... **A1:** 622, 623
austenitic stainless steels ... **A1:** 934
304 stainless steel ... **A1:** 622
effect of solution annealing temperature on ... **A1:** 945
carbon steel ... **A1:** 629
chromium-molybdenum steels ... **A1:** 622
chromium-molybdenum-vanadium steels ... **A1:** 619
defined ... **A8:** 3
effect of heat treatment on
$2^1/_4$Cr- 1 Mo ... **A1:** 641, 642
austenitic stainless steel ... **A1:** 945
effect of microstructure on ... **A1:** 638
effect of spheroidization on ... **A1:** 644
ferritic steels ... **A1:** 622, 939, 941
maximum-use temperatures based on ... **A1:** 617
Creep-rupture strength, of iron-chromium- nickel heat-resistant casting alloys
influence of microstructure ... **A9:** 333
Creep-rupture testing ... **A8:** 3, 301-306, 334-337
Creeping-spindle swaging ... **A14:** 131
Creosote
copper/copper alloy resistance ... **A13:** 631
Crescent crack
definition ... **EM4:** 632
Crescents *See* Beach marks
Cresol novolacs ... **EM3:** 96, 594, 595
epoxidized ... **EM3:** 94
as epoxy resins ... **EL1:** 826-827
Cresol resins ... **EM3:** 105
Cresol-base epoxy-novolac resins ... **EM3:** 104
suppliers ... **EM3:** 104
Cresols ... **EM3:** 103
Cretaceous-Tertiary boundary
iridium determined by NAA at ... **A10:** 240-241
Creusot-Loire system ... **A4:** 645
Crevice corrosion *See also* Cavitation erosion; Concentration-cell corrosion; Corrosion; Crevice corrosion evaluation; Pitting corrosion ... **A13:** 108-113 **M5:** 432-433
aircraft ... **A13:** 1025-1026
assembled test specimen ... **A13:** 564
austenitic stainless steel weldments ... **A13:** 348-349
austenitic stainless steels ... **A6:** 467
automotive industry ... **A13:** 1011
in brazed joints ... **A11:** 451
in brazing ... **A13:** 877
by dirt in river water ... **A11:** 633
cast irons ... **A13:** 568
characteristics of ... **A11:** 632
in chloride-containing natural waters ... **A13:** 112
in copper/copper alloys ... **A13:** 612-613
coupons ... **A13:** 198
crevice geometry ... **A13:** 111-112
defined ... **A11:** 3 **A13:** 4, 303
and design ... **A13:** 339
electrochemical testing methods ... **A13:** 216-217
evaluation of ... **A13:** 303-310
indexes, stainless steels ... **A13:** 556
initiation ... **A13:** 305-309
in manned spacecraft ... **A13:** 1080-1082
in marine atmospheres ... **A13:** 112
material selection to avoid/minimize ... **A13:** 323

SUBJECTS OF THE INDEXED VOLUMES: ASM Handbook (designated by the letter "A"): **A1:** Properties and Selection: Irons, Steels, and High-Performance Alloys (1990); **A2:** Properties and Selection: Nonferrous Alloys and Special-Purpose Materials (1990); **A3:** Alloy Phase Diagrams (1992); **A4:** Heat Treating (1991); **A6:** Welding, Brazing, and Soldering (1993); **A7:** Powder Metallurgy (1984); **A8:** Mechanical Testing (1985); **A9:** Metallography and Microstructures (1985); **A10:** Materials Characterization (1986); **A11:** Failure Analysis and Prevention (1986); **A12:** Fractography (1987); **A13:** Corrosion (1987); **A14:** Forming and Forging (1988); **A15:** Casting (1988); **A16:** Machining (1989); **A17:** Nondestructive Testing and Quality Control (1989); **A18:** Friction, Lubrication, and Wear Technology (1992). **Metals Handbook, 9th Edition** (designated by the letter "M"): **M1:** Properties and Selection: Irons and Steels (1978); **M2:** Properties and Selection: Nonferrous Alloys and Pure Metals (1979); **M3:** Properties and Selection: Stainless Steels, Tool Materials and Special-Purpose Materials (1980); **M4:** Heat Treating (1981); **M5:** Surface Cleaning, Finishing, and Coating (1982); **M6:** Welding, Brazing, and Soldering (1983). **Engineered Materials Handbook** (designated by the letters "EM"): **EM1:** Composites (1987); **EM2:** Engineering Plastics (1988); **EM3:** Adhesives and Sealants (1990); **EM4:** Ceramics and Glasses (1991); **Electronic Materials Handbook** (designated by the letters "EL"): **EL1:** Packaging (1989).

revice corrosion (continued)
mechanisms A13: 100, 1012
of metal surfaces A11: 183-184
in mining/mill applications A13: 1295
oil/gas production A13: 1234
paper machine A13: 1189-1190
phases, defined A13: 303
pitting, stainless steel bubble caps A11: 183
prevention A13: 112-113
propagation A13: 308
resistance, amorphous metals A13: 868
resistance, casting alloys A13: 582
resistance, titanium/titanium alloys A13: 694-696
in seawater A13: 108-112
space shuttle orbiter A13: 1069
of stainless steels A13: 303, 465, 554
steam surface condensers A13: 987-988
and stress corrosion cracking A8: 500
in threaded fasteners A11: 535
in titanium A11: 202
in titanium/titanium alloys A13: 672-673, 681-683
tubesheet annular A13: 944
of tubing in hydraulic-oil cooler A11: 632-633
types
under residual slag A13: 349
in water, heat exchangers A11: 631-632
water-recirculating systems A13: 488
in zirconium/zirconium alloys A13: 717
revice corrosion evaluation *See also* Crevice corrosion A13: 303-310
aspects/guidelines A13: 303
electrochemical tests A13: 308-309
ferric chloride tests A13: 304
immersion tests A13: 303-308
Materials Technology Institute tests A13: 304-305
multiple-crevice assembly testing A13: 305-308
spool specimen test racks A13: 304
revices A13: 111-112, 303
rimp
defined EM1: 8 EM2: 12
rimped wire radial brushes M5: 151-152
rimping *See also* Corrugating
defined A14: 3
steel wire fabrication M1: 587-588
wrought aluminum alloy A2: 36
rimping assembly
furnace brazing of steels M6: 940
ristobalite
chemical system EM4: 870-871
crystal structure EM4: 879-880, 881
framework structure EM4: 759
in glass-ceramics EM4: 1102
as quartz transition A15: 208
result of pyrophyllite phase transformation EM4: 761
thermal expansion coefficients EM4: 499
itical angle for jetting A6: 162
itical anodic current density
defined A13: 4
itical contact angle
liquid impact erosion A18: 224
itical cooling rate
defined A9: 4
itical crevice temperature
cast/wrought alloys A13: 581
itical current density *See also* Current density
in A15 superconductors A2: 1063-1064
ternary molybdenum chalcogenides (chevrel phases) A2: 1079
itical curve
defined A9: 5
itical damping EM3: 9
defined EM2: 12
itical diameter M1: 474, 476
itical dimension
defined A15: 4
itical failure load A18: 491, 492
itical fiber length EM3: 368, 396, 397-398
defined EM1: 119
itical film thickness A18: 63
itical flaw size
defined A13: 4
itical flow transition velocity A6: 162
itical heat input index A6: 1100
itical humidity
defined A13: 4

Critical illumination
defined A9: 5
Critical laminate strain
determining EM1: 242
Critical length
defined EM1: 8 EM2: 12
Critical load A18: 491
Critical load effect A18: 423
Critical longitudinal stress
defined EM1: 8 EM2: 12
Critical micelle concentration A10: 118, 671, 690
Critical normal force A18: 434- 435, 436
Critical ordering temperature
ordered intermetallics A2: 913-914
Critical pitting potential
defined A13: 4
Critical pitting temperature (CPT) A6: 697
Critical point .... A3: 1 • 2
defined A9: 5
Critical pressure
defined A9: 5
Critical quality test
after etching EL1: 872
Critical rake angle
defined A9: 5
Critical relative humidity
defined A13: 82
Critical resolution point *See also* Fractal plot A12: 211
Critical sliding velocity
ceramics A18: 814
Critical state theory (Schofield and Wroth) EM4: 274
Critical strain
defined A9: 5 EM1: 8 EM2: 12
in recrystallization A9: 696
Critical strain rate A8: 519
Critical strain rate, regimes
SCC A13: 262
Critical stress intensity factor A7: 3 EM1: 255-256
cemented carbides A16: 78-79
Critical stress-intensity factor A18: 421 EM4: 36
effects of thickness on A11: 54
Critical surface
defined A9: 5
Critical surface tensions EM3: 180
Critical temperature A6: 375
defined A9: 5
Critical temperatures A1: 115-116, 126-127, 130
Critical-point phenomena
Mössbauer analysis of A10: 287
CRO *See* Cathode-ray oscilloscope
Crochet's delayed failure equation EM3: 354-359, 365
Crocus (iron oxide)
natural abrasive A16: 434
Croning process *See also* Shell molding process
defined A15: 4
organic binders A15: 35
Cross direction *See* Transverse direction
Cross drilling
in conjunction with turning A16: 135
Cross fittings
use in pipe welding M6: 591
Cross laminate *See also* Laminate(s); Parallel laminate EM3: 9
defined EM2: 12
Cross laminates, under fatigue loading
axial cracking in A8: 714
Cross linking EM3: 9
between molecules EM2: 58-59
defined EM2: 12
microstructural analysis EM3: 412-413
polyaramid EM3: 286
polyimides EM3: 154, 155
in polyurethanes (PUR) EM2: 257
related to glass transition temperature EM3: 320
in thermoset resins EM2: 626-627
ultrahigh molecular weight polyethylenes (UHMWPE) EM2: 170
Cross rolling
defined A9: 5
Cross section
as plating thickness inspection EL1: 943
Cross talk
among coupled transmission lines EL1: 35-37
coplanar backward EL1: 83-84

Cross talk (continued)
data rate effect EL1: 7
defined EL1: 34, 417, 603-604
as design consideration EL1: 518-519
near-end, defined EL1: 36
optical interconnections EL1: 9, 16
in preheating process, wave soldering EL1: 686
in signal transmission EL1: 172
in VHSIC technology EL1: 76
wafer-scale integration
Cross talk noise *See* Cross talk
Cross wire weld
definition M6: 5
Cross wire welding of
aluminum alloys M6: 543
copper M6: 518
low-carbon steel M6: 518
nickel-based alloys M6: 518
stainless steel M6: 518, 531-532
Cross wire welding, projection welding M6: 517-519
electrode design M6: 518
electrode force M6: 518
metals welded M6: 518
weld time M6: 518
welding current M6: 518
Cross-check defect
precipitation-hardening stainless steels A12: 374
Cross-checking A18: 654
Cross-flow blenders A7: 189
Cross-jet atomization for tin powders A7: 123
Cross-linked polyethylene (XLPE) insulation
for copper and copper alloy products A2: 258
Cross-linked polyimides EM1: 78
Cross-linked polymers
formation/application EM1: 752
Cross-linked rubber
ductility dependence on strain rate for A8: 39, 43
Cross-linked thermosetting coatings A13: 406-410
Cross-linking
defined EM1: 8
degree of, defined EM1: 8
of polyester resins EM1: 90
of resin cure, filament winding EM1: 135
in thermoplastic matrices EM1: 33
Cross-plied tapes *See also* Multidirectional type prepregs
properties EM1: 147
Cross-ply laminate *See also* Laminate(s)
defined EM1: 8 EM2: 12
Cross-section transmission electron microscopy
capabilities A10: 628
Cross-sectional area
defined A15: 4
Cross-slip of dislocations
and formation of cell walls A9: 693
Cross-travel shaft
fatigue fracture of A11: 525
Crosscutting
waterjet machining A16: 525-526
Crossed-axes helical gears
described A11: 586
Crossed-cylinder apparatus A18: 400
Crossed-field amplifier (CFA)
microwave inspection A17: 209
Crossed-polarized light in optical microscopy A9: 72
Crossflow nebulizers
for analytic ICP systems A10: 34-35
Crosshead
displacement, defined as relative displacement A8: 41, 45
displacement, strain measurement based on A8: 35
with drop tower compression system A8: 196
movement, in conventional load frames A8: 192
specific rate of A8: 41
speed A8: 39, 43-44
Crossheading, plastics
in extrusion EM2: 383
Crosslink
of UV conformal coatings EL1: 786
Crossovers
devitrifying dielectrics for EL1: 109
dielectrics, thick-film EL1: 341-342
flexible printed boards EL1: 586

Crosstalk, detector
defined ... **A17:** 383
Crosswise direction ... **EM3:** 9
defined ... **EM1:** 8 **EM2:** 12
Crow's feet
as casting defect ... **A11:** 384
Crowdion diffusion ... **A13:** 68
Crowfoot
defined ... **EM2:** 12
Crowfoot satin *See* Four-harness satin
Crowfoot satin weave *See* Satin (crowfoot) weave
Crowfoot weave *See* Satin (crowfoot) weave
Crown
defined ... **A14:** 3
interference microscope measurement ... **A17:** 17
Crown (dental) alloys
of precious metals ... **A2:** 696
Crown glass process ... **EM4:** 395
Crown polyethers
as solvent extractant ... **A10:** 170
Crowning, of rolls
three-roll forming machines ... **A14:** 618
Crows feet cracking ... **A8:** 591, 594
CRT *See* Cathode ray tube; Cathode-ray tube
CRT displays
digital image enhancement ... **A17:** 461-462
CRTs and TV picture tubes ... **EM4:** 1038-1044
B&W picture tubes ... **EM4:** 1039, 1040
color ... **EM4:** 1041-1042
CRTs ... **EM4:** 1038
direct view CRT ... **EM4:** 1038-1039, 1040, 1041, 1042-1043
future trends ... **EM4:** 1044
projection CTV tube ... **EM4:** 1039
properties of glasses ... **EM4:** 1042-1044
specific properties imparted by various oxides used in CTV tubes ... **EM4:** 1039-1041
transmittance ... **EM4:** 1041
x-ray radiation limits for TV tubes ... **EM4:** 1041
Crucible B-120VCA *See* Titanium alloys, specific types, Ti-13V-11Cr-3Al
Crucible furnaces ... **A7:** 3 **A15:** 374, 381-383
defined ... **A15:** 4
design considerations ... **A15:** 383
early steel production ... **A15:** 31
movable/pouring ladies ... **A15:** 382-383
stationary ... **A15:** 381-382
tilting ... **A15:** 382
types ... **A15:** 381-383
Crucible Particle Metallurgy (CPM) process ... **A1:** 780 **A7:** 787-788
bend fracture strengths ... **A16:** 65
Crucible refractories
developments in ... **A15:** 31
Crucible Steel Company (New York) ... **A15:** 32
Crucible wall scrapers
induction furnaces ... **A15:** 373
Crucible(s)
cobalt-base alloys ... **A15:** 812-813
defined ... **A15:** 4
historical excavation of ... **A15:** 16
magnesium alloy ... **A15:** 800
materials, vacuum induction furnace ... **A15:** 394
and mold assembly, African ... **A15:** 19
plasma cold, casting ... **A15:** 424-425
steel, development of ... **A15:** 31
vacuum arc skull casting ... **A15:** 409-410
VIM, refractory linings ... **A15:** 394
Crucible-type electric resistance heated furnaces ... **A15:** 500
Crucibles
elevated-temperature failures in ... **A11:** 296
graphite, for IGF analysis ... **A10:** 227
precious metal applications ... **A2:** 693
for sinters/fusions ... **A10:** 166-167
tungsten, nuclear applications ... **A2:** 558
Crud
in steam generators ... **A13:** 948, 950
Crude oil ... **A18:** 123
Crude oil refineries
corrosion inhibitors for ... **A13:** 485-486
Cruise Missile engine radial compressor rotor
by Colt-Crucible ceramic mold process ... **A7:** 754
Crush *See also* Buckle
as casting defect ... **A11:** 381
defined ... **A11:** 3 **A15:** 4
Crush (nip)
defined ... **A18:** 6
Crush bead *See* Crush strip
Crush rolls for thread forms
thread grinding applications ... **A16:** 278
Crush strip
defined ... **A15:** 4
Crushers
for sampling ... **A10:** 16
Crushing
in compression testing ... **A8:** 57
defined ... **A7:** 3
of drive-gear assembly ... **A11:** 143
in QMP iron powder process ... **A7:** 87
test for tantalum capacitor powder anodes ... **A7:** 163
as wet chemical technique for subdividing solids ... **A10:** 165
Crushing test
defined ... **A8:** 3
Cryogels
alkoxide-derived gels ... **EM4:** 210, 211
Cryogenic *See also* Cryogenic applications Low temperature
stability, in superconductors ... **A2:** 1037-1038
temperature behavior, elongation, wrought aluminum alloy ... **A2:** 56-58
Cryogenic applications *See also* Low-temperature properties
aluminum-lithium alloys ... **A2:** 184
beryllium-copper alloys ... **A2:** 420
materials for ... **M3:** 721-772
niobium-titanium superconductors ... **A2:** 1043
thermocouples ... **A2:** 872-873, 882, 885
Cryogenic cooling ... **EL1:** 49-50
with spraying techniques in molten particle deposition ... **EM4:** 207
Cryogenic ESR studies ... **A10:** 257
Cryogenic grinding
for samples ... **A10:** 16
Cryogenic service, welding for ... **A6:** 1016-1018
applications ... **A6:** 1017
effects on properties ... **A6:** 1016-1018
Charpy V-notch impact energy at 77 K ... **A6:** 1017-1018
coefficient of thermal expansion ... **A6:** 1018
electron-beam welding ... **A6:** 1017
fatigue strength ... **A6:** 1018
fracture toughness at 4 K ... **A6:** 1016-1017
gas-metal arc welding ... **A6:** 1017
gas-tungsten arc welding ... **A6:** 1017
inclusion content ... **A6:** 1018
shielded metal arc welding ... **A6:** 1017
strength ... **A6:** 1016
submerged arc welding ... **A6:** 1017
titanium alloys ... **A6:** 1017
ultrahigh-strength alloys ... **A6:** 1017
Cryogenic temperatures
effect on fatigue ... **A12:** 52-53
polybenzimidazole behavior ... **EM3:** 170
Cryogenic tensile shear strengths
polybenzimidazoles ... **EM3:** 171
Cryogenic trap
in environmental test chamber ... **A8:** 411
Cryogenic treatment of steels ... **A4:** 203, 204-206
Cryogenics
austenitic stainless steels ... **A6:** 468, 686, 689
furnace brazing of stainless steels in a vacuum atmosphere ... **A6:** 920-921
heat-affected zone ... **A6:** 1017
heat-treatable aluminum alloys ... **A6:** 535
Cryogenics (continued)
materials for cryogenic applications ... **A6:** 383
stainless steels ... **A6:** 677
tin-lead solders, mechanical properties ... **A6:** 967
Cryolite
chemical composition ... **A6:** 60
in dip brazing flux ... **M6:** 991
Cryolite (Na_3AlF_6)
attack resisted by RBSN ... **EM4:** 238-239
corrosion resistance of silicon oxynitride for containers ... **EM4:** 239
purpose for use in glass manufacture ... **EM4:** 381
Cryomicroscopy methods ... **A18:** 376
Cryopump
defined ... **A10:** 671
in vacuum pumping system ... **A8:** 414
Cryptands
as general-use reagent ... **A10:** 170
Crystal
defined ... **A9:** 5
description ... **A3:** 1 • 10
dimensions ... **A3:** 1 • 10-15
ordering ... **A3:** 1 • 10
orientation ... **A8:** 188, 553
properties, use in phase-diagram determination ... **A3:** 1 • 17-18
single, in torsional Kolsky bar test ... **A8:** 222
single, Kolsky bar testing for ... **A8:** 219
structure ... **A3:** 1 • 10-17
structure, effect on deformation ... **A8:** 34
systems ... **A3:** 1 • 10
thermal conductivity of ... **EM1:** 47
Crystal analysis
defined ... **A9:** 5
Crystal classes
described ... **A10:** 346-347
Crystal Data
data base ... **A10:** 326, 355
Crystal defects
types ... **A13:** 45-46
Crystal defects from plastic deformation
determination of ... **A9:** 686
Crystal diffraction ... **A10:** 346
for phase analysis ... **EM4:** 25
Crystal geometry
as x-ray diffraction analysis ... **A10:** 325
Crystal growth *See also* Crystallization; Crystals; Growth; Needle crystals; Spherical crystals
of bulk GaAs material ... **EL1:** 199-200
equiaxed ... **A15:** 115
GaAs ... **EL1:** 200
liquid/solid states ... **A15:** 109-110
mass and heat transport ... **A15:** 111-113
nucleation effects ... **A15:** 101
packages, defect types ... **EL1:** 979
single, Czochralski process ... **EL1:** 191
solid/liquid interface ... **A15:** 110-111
and solidification ... **A15:** 109-113
Crystal growth, gallium
methods ... **A2:** 744
Crystal imperfections ... **A9:** 601
Crystal lattice *See also* Lattices; Superlattices
length along the *a* axis ... **A10:** 689
length along the *b* axis ... **A10:** 689
length along the *c* axis ... **A10:** 689
strain measured, for residual stress calculation ... **A10:** 381
Crystal lattice imperfections
corrosion from ... **A17:** 215
Crystal lattice length along a axis
symbol for ... **A8:** 724
Crystal multiplication
in copper alloy ingots ... **A9:** 638
effect of dendrite structure on ... **A9:** 641
Crystal orientation
in a steel ingot ... **A9:** 623
dependence of mechanical and physical properties on ... **A9:** 701
determination of ... **A9:** 701-706

Crystal orientation (continued)
determined from diffraction patterns **A9:** 109-110
in eutectoid compositions **A9:** 658-659
in pure metals **A9:** 610
as x-ray diffraction analysis **A10:** 325
Crystal perfection
as x-ray diffraction analysis **A10:** 325
Crystal pulling
defined **EL1:** 958
Crystal quartz sensitive tint plate
placement **A9:** 138
Crystal structure *See also* Body-centered cubic metals; Face-centered cubic metals; Hexagonal close-packed metals
A15 superconductors **A2:** 1060-1061
analysis, assumptions of **A10:** 352
atom positions **A9:** 708, 716-718
beryllium-copper alloy **A2:** 286-290
and boundary structure, compared **A10:** 358
by neutron diffraction **A10:** 420
calculated density of unit cell **A9:** 708
calculating intensities from **A10:** 349
cartridge brass **A2:** 302
cast copper alloys **A2:** 360, 362, 384-385, 387
"cat" layer crystallized in monoclinic space group *P2* **A10:** 347, 349
commercial bronze **A2:** 297
commercially pure titanium **A2:** 592-594
of cubic boron nitride (CBN) **A2:** 1010
defects **A9:** 710, 719-720
defined **A9:** 706
defining **A10:** 348
determined **A10:** 345
of diamond **A2:** 1009-1010
of diamond and graphite **A10:** 345, 355
effect, iron-base alloy corrosion **A13:** 126
effect of alloy composition **A9:** 708
effect of massive transformation on **A9:** 655-657
effect on dislocations **A9:** 684
extraction replicas used to study **A9:** 108
of ferrous martensite **A9:** 669, 672
in gaseous corrosion **A13:** 61
illustrated by Oak Ridge Thermal Ellipsoid Program **A10:** 352, 354
initial guessing procedure **A10:** 350
of inorganic solids, applicable analytical methods **A10:** 4-6
least squares refinement of atomic positions to determine **A10:** 351
long-range ordered, intermetallic **A2:** 913-914
of organic solids, methods for analysis **A10:** 9
plane designation **A9:** 708-710
point groups **A9:** 708
of pure metals **A13:** 62-63
of selected oxides **A13:** 64
structure prototypes **A9:** 707-708, 711-718
symbols for identifying **A9:** 706-707, 716-718
of the elements **A9:** 709-710
titanium aluminides **A2:** 926-928
transplutonium actinide metals **A2:** 1198
of two carbon forms **A10:** 345
uranium, effect on mechanical properties **A2:** 671
Crystal structures
of aluminum alloy phases **A9:** 359
$AuCu_3$ **A9:** 681-682
AX structures **EM4:** 879-882
BiF_3 ..... **A9:** 682-683
CsCl **A9:** 682-683
summary of ceramic crystal structures **EM4:** 880
of titanium and titanium alloys **A9:** 459, 460-461
unit cells **A9:** 681
Crystal supports
powders used **A7:** 573
Crystal symmetry **A10:** 346
determined from diffraction patterns **A9:** 109-110
Crystal systems **A10:** 347, 348
defined **A9:** 5
relationships of edge lengths and interaxial angles **A9:** 706
types **A9:** 706
Crystal zone
defined **A9:** 710
Crystal(s) *See also* Crystal growth; Crystallization; Needle crystals; Spherical crystals
affecting grain refinement nucleation **A15:** 105-108
Crystal(s) (continued)
beta, nucleation and growth **A15:** 125-126
distribution of atoms **EL1:** 93
formation, nucleant action **A15:** 105
for optical imaging **EL1:** 1071-1072
structure, and electrical properties **EL1:** 93
through-hole packages **EL1:** 979
Crystal-figure etching
defined **A9:** 5
Crystal-particle statistics
as XRPD source of error **A10:** 340
Crystal-structure
nomenclature ..... **A3:** 1 • 15-16
prototypes ..... **A3:** 1 • 16
Crystalline defects
effects **EL1:** 93, 979
Crystalline fracture *See also* Fibrous fracture; Fracture; Granular fracture; Silky fracture
defined **A8:** 3 **A9:** 5 **A11:** 3
Crystalline fractures **A12:** 2
Crystalline interfaces
idealized constructions **A9:** 119
Crystalline lattice, preferential alignment in a polycrystalline aggregate *See* Texture, crystallographic
rotation during compression **A9:** 700
rotation to a stable orientation **A9:** 700
Crystalline materials
ferromagnetic properties **A2:** 761-763
membrane electrodes, solid **A10:** 182
powder compacts, preferred orientations in **A10:** 358
solids, ESR detection of color centers and defects in **A10:** 253-266
x-ray diffraction residual stress techniques for **A10:** 381
Crystalline phases
from crystal structures **A10:** 345
identified by XRPD **A10:** 333
unknown, unit cell identification **A10:** 353
use of cell information to identify unknown **A10:** 353
Crystalline plastic *See also* Semicrystalline **EM3:** 9
defined **EM1:** 8 **EM2:** 12
Crystalline polymers
chemistry of **EM2:** 64
Crystalline rock
high level waste disposal in **A13:** 974-976
Crystalline segments
as toughener **EM3:** 185
Crystalline solids
progressive fracturing in **A8:** 439
Crystalline state
defined **A10:** 326
Crystallinity *See also* Secondary crystallization; Stress-induced crystallization **EM3:** 9, 654
defined **EL1:** 93 **EM2:** 12
effect, elongation/toughness **EM1:** 101
effect, environmental stress crazing **EM2:** 800
failure analysis of **EM2:** 731-732
of homopolymer/copolymer acetals **EM2:** 101
initiators, injection molding **EM1:** 167
of linear thermoplastics **EM1:** 98
of parylene coatings **EL1:** 796
polymer **EM2:** 59, 64
polyphenylene sulfides (PPS) **EM2:** 189
properties effects **EM2:** 437
thermoplastic resins **EM2:** 619-620
of thermoplastics **EM1:** 100
Crystallinity, of fossil fuels
ESR determined **A10:** 253
Crystallite
defined **A9:** 5
Crystallite orientation distribution function **A9:** 706
of Cu-30Zn **A9:** 705
Crystallite size
EPMA analysis **A10:** 516-535
field ion microscopy **A10:** 583-602
measured, scanning electron microscopy **A10:** 490-515
and subgrain size/shape, measured by x-ray topography **A10:** 365
x-ray diffraction **A10:** 325-332
Crystallization
amorphous materials and metallic glasses **A2:** 809-813
heterogeneous nucleation effect **A15:** 104
homogeneous nucleation effect **A15:** 103
impurity effects **A15:** 103
liquid resistance to **A15:** 103
Crystallization, and coarsening
in aluminum x-ray topographs **A10:** 376
Crystallization, fractional *See* Fractional crystallization
Crystallographic anisotropy
in plastic torsion **A8:** 143
Crystallographic cleavage
defined **A8:** 3
Crystallographic damage
in bond failure **EL1:** 1043
Crystallographic features revealed by electron- channeling patterns **A9:** 94
Crystallographic growth *See* Preferred crystallographic growth
Crystallographic information *See also* Crystallographic texture measurement and analysis; Texture
analytical transmission electron microscopy **A10:** 429-489
crystallographic texture measurement and analysis **A10:** 357-364
electron spin resonance **A10:** 253-266
extended x-ray absorption fine structure **A10:** 407-419
ferromagnetic resonance **A10:** 267-276
field ion microscopy **A10:** 583-602
infrared spectroscopy **A10:** 109-125
low-energy electron diffraction **A10:** 536-545
Mössbauer spectroscopy **A10:** 287-295
neutron diffraction **A10:** 420-426
nuclear magnetic resonance **A10:** 277-286
radial distribution function analysis **A10:** 393-401
Raman spectroscopy **A10:** 126-138
scanning electron microscopy **A10:** 490-515
single-crystal x-ray diffraction **A10:** 344-356
small-angle x-ray and neutron scattering **A10:** 402-406
x-ray diffraction **A10:** 325-332
x-ray powder diffraction **A10:** 333-343
x-ray topography **A10:** 365-379
Crystallographic orientation
effect on massive transformations **A9:** 657
etch pits used to determine **A9:** 62
heat tinting to reveal **A9:** 136
line etching used to determine **A9:** 62
scanning electron microscopy used to study **A9:** 101
in sheet metal rolling **A8:** 553
Crystallographic phase
and magnetic characterization **A17:** 131
Crystallographic phase transitions
by ESR analysis **A10:** 257
Crystallographic planes
alignment for cutting along **A10:** 333, 342
stress-corrosion cracking along **A8:** 501
Crystallographic planes of slip **A9:** 684
density of, parallel to the rolling plane of alpha-brass **A9:** 686
Crystallographic preferred orientations
measurement and analysis **A10:** 357-364
Crystallographic relations
between precipitate and parent phases **A9:** 647
Crystallographic texture *See also* Crystallographic information; Crystallographic texture measurement and analysis; Texture
anisotropic Young's modulus **A10:** 358
measurement and analysis **A10:** 357-364
topographic methods for **A10:** 368
of torsion tested material **A8:** 155
Crystallographic texture, in electrical steels **A9:** 537
preparation of specimens **A9:** 531
Crystallographic texture measurement and analysis *See also* Crystallographic texture; Orientation distribution function; Preferred orientation; Texture
applications **A10:** 357, 363-364
Bragg's law **A10:** 360
descriptions of preferred orientation **A10:** 358-361
estimated time analysis **A10:** 357

Crystallographic texture measurement and analysis (continued)
introduction **A10:** 358
limitations ..... **A10:** 357 , 362-363
pole figure **A10:** 360
related techniques **A10:** 357
samples **A10:** 357
series method of ODF analysis **A10:** 361-363
texture measurements **A10:** 358
Crystallographic transformation
types of invariant point **EM4:** 883
Crystallography
and electronic phenomena **EL1:** 93
of shape memory alloys **A2:** 898
surface, LEED analysis **A10:** 536
surface, vocabulary of **A10:** 537-538
Crystals *See also* Liquid crystals; Single crystals
analyzing, by x-ray spectrometry **A10:** 87, 88
axes, stereographic projection of **A10:** 359
and boundary structure, compared **A10:** 358
classes of **A10:** 346-347
defect intensities, measured by x-ray topography **A10:** 365
diffraction in **A9:** 110-113
diffraction of **A10:** 346
effect of cooling on analysis **A10:** 352
effect of thickness on topographic methods **A10:** 367-370
effects of rotation **A10:** 368
EXAFS analysis of **A10:** 407
fluorescent, radiography **A17:** 317
forces, Raman analyses lattice vibrations to obtain **A10:** 130
fracture surface of **A10:** 376-377
growth **A10:** 365, 375-376
ideal **A10:** 351
ideally imperfect **A10:** 351
imperfections **A10:** 332
ionic, ESR studied **A10:** 263
as isotropic **A10:** 358
kinetics of **A10:** 376
local environment around transition ions characterized by ESR **A10:** 253-266
mosiac **A10:** 351
near-perfect, diffraction in **A10:** 365, 367
next-nearest neighbors EXAFS determined **A10:** 407
perfect **A10:** 325, 351
point-group symmetry of **A10:** 346
polar, LEISS identification of faces **A10:** 603
quartz, as piezoelectric elements **A17:** 254-255
silicon, spin-dependent recombination analysis of **A10:** 258
single **A10:** 129, 256
size of, as x-ray diffraction analysis **A10:** 325
stored energy as a result of cold working **A9:** 692
structural information by EXAFS **A10:** 407
structure definition **A10:** 348
structure, effect on magnetic properties **A17:** 131-132
symmetry of **A10:** 346
systems **A10:** 347
Cs-Ge (Phase Diagram) ..... **A3:** 2 • 163
Cs-Hg (Phase Diagram) ..... **A3:** 2 • 163
Cs-In (Phase Diagram) ..... **A3:** 2 • 163
Cs-K (Phase Diagram) ..... **A3:** 2 • 164
Cs-Na (Phase Diagram) ..... **A3:** 2 • 164
Cs-O (Phase Diagram) ..... **A3:** 2 • 164
Cs-Rb (Phase Diagram) ..... **A3:** 2 • 165
Cs-S (Phase Diagram) ..... **A3:** 2 • 165
Cs-Sb (Phase Diagram) ..... **A3:** 2 • 165
Cs-Se (Phase Diagram) ..... **A3:** 2 • 166
Cs-Sn (Phase Diagram) ..... **A3:** 2 • 166
Cs-Te (Phase Diagram) ..... **A3:** 2 • 166
Cs-Tl (Phase Diagram) ..... **A3:** 2 • 167
CSA specifications
of carbon and alloy steel pipe **A1:** 332
of steel tubular products **A1:** 328, 329, 330, 332
CT *See* Compact specimen
CT numbers
defined **A17:** 383
CTBN-ATBN
as toughener **EM3:** 185
CTE *See* Coefficient of thermal expansion; CTE-matched materials
CTE tailoring *See* Coefficient of thermal expansion (CTE)
CTE-matched materials
aluminum nitride **EL1:** 306
heat sink effects **EL1:** 1129-1131
leadless packaging **EL1:** 985
silicon carbide **EL1:** 306
CTFE *See* Polychlorotrifluoroethylene
CTOD *See* Crack tip opening displacement
Cu (Phase Diagram) ..... **A3:** 2 • 174
Cu-Dy (Phase Diagram) ..... **A3:** 2 • 167
Cu-Er (Phase Diagram) ..... **A3:** 2 • 167
Cu-Eu (Phase Diagram) ..... **A3:** 2 • 168
Cu-Fe (Phase Diagram) ..... **A3:** 2 • 168
Cu-Fe-Ni (Phase Diagram) ..... **A3:** 3 • 49-50
Cu-Ga (Phase Diagram) ..... **A3:** 2 • 168
Cu-Gd (Phase Diagram) ..... **A3:** 2 • 169
Cu-Ge (Phase Diagram) ..... **A3:** 2 • 169
Cu-H (Phase Diagram) ..... **A3:** 2 • 169
Cu-Hf (Phase Diagram) ..... **A3:** 2 • 170
Cu-Hg (Phase Diagram) ..... **A3:** 2 • 170
Cu-In (Phase Diagram) ..... **A3:** 2 • 170
Cu-Ir (Phase Diagram) ..... **A3:** 2 • 171
Cu-La (Phase Diagram) ..... **A3:** 2 • 171
Cu-Li (Phase Diagram) ..... **A3:** 2 • 171
Cu-Mg (Phase Diagram) ..... **A3:** 2 • 172
Cu-Mn (Phase Diagram) ..... **A3:** 2 • 172
Cu-Nb (Phase Diagram) ..... **A3:** 2 • 172
Cu-Nd (Phase Diagram) ..... **A3:** 2 • 173
Cu-Ni (Phase Diagram) ..... **A3:** 2 • 173
Cu-Ni-Sn (Phase Diagram) ..... **A3:** 3 • 50
Cu-Ni-Zn (Phase Diagram) ..... **A3:** 3 • 51
Cu-P (Phase Diagram) ..... **A3:** 2 • 174
Cu-Pb (Phase Diagram) ..... **A3:** 2 • 175
Cu-Pb-Zn (Phase Diagram) ..... **A3:** 3 • 51-52
Cu-Pd (Phase Diagram) ..... **A3:** 2 • 175
Cu-Pt (Phase Diagram) ..... **A3:** 2 • 175
Cu-Pu (Phase Diagram) ..... **A3:** 2 • 176
Cu-Rh (Phase Diagram) ..... **A3:** 2 • 176
Cu-S (Phase Diagram) ..... **A3:** 2 • 176
Cu-Sb (Phase Diagram) ..... **A3:** 2 • 177
Cu-Sb-Sn (Phase Diagram) ..... **A3:** 3 • 52
Cu-Se (Phase Diagram) ..... **A3:** 2 • 178
Cu-Si (Phase Diagram) ..... **A3:** 2 • 178
Cu-Sn (Phase Diagram) ..... **A3:** 2 • 178
Cu-Sn-Zn (Phase Diagram) ..... **A3:** 3 • 52
Cu-Sr (Phase Diagram) ..... **A3:** 2 • 179
Cu-Te (Phase Diagram) ..... **A3:** 2 • 179
Cu-Th (Phase Diagram) ..... **A3:** 2 • 180
Cu-Ti (Phase Diagram) ..... **A3:** 2 • 181
Cu-V (Phase Diagram) ..... **A3:** 2 • 181
Cu-Yb (Phase Diagram) ..... **A3:** 2 • 181
Cu-Zn (Phase Diagram) ..... **A3:** 2 • 182
Cu-Zr (Phase Diagram) ..... **A3:** 2 • 182
Cube texture
defined **A9:** 5
in electrical steels **A9:** 537
in nickel-iron alloys **A9:** 538
Cube-on-edge (COE) texture **A9:** 537
Cubex **A9:** 537
Cubic
defined **A9:** 5
Cubic boron nitride *See* Tool materials, superhard **EM4:** 329
abrasive property control **EM4:** 332
applications **EM4:** 481
bond type **EM4:** 331
bondability **EM4:** 332
for cutting tools **EM4:** 959
grinding of technical ceramics **EM4:** 334, 335
hardness **EM4:** 330, 331, 351, 806
key product properties **EM4:** 48
Cubic boron nitride (continued)
mechanical properties **EM4:** 331
modulus of resilience **EM4:** 330, 331
nickel metal coatings **EM4:** 333
raw materials **EM4:** 48
thermal properties **EM4:** 331
Cubic boron nitride (CBN) *See also* Sintered polycrystalline cubic boron nitride **A16:** 2, 105-117, 453-471
abrasive applications **A16:** 455, 456, 465, 467
abrasive for cast irons **A16:** 661
advantages of abrasives in precision production grinding **A16:** 457
applications **A16:** 639
atom arrangement **A16:** 106
Borazon (trade name) **A16:** 454
Borazon grinding wheels for machining CPM 10V tool steel **A16:** 735
boring heat-resistant alloys **A16:** 742
boring of P/M materials **A16:** 889
compared to Al_2O_3 abrasive grinding **A16:** 462, 464, 465
compared to diamond **A16:** 105
concentration affecting tool life of grinding wheels **A16:** 464
conditioning **A16:** 469-470
cutting speed and work material relationship **A18:** 616
cutting tool material **A18:** 614, 617
dressing methods **A16:** 468-469
equilibrium diagram **A2:** 1009
erosion test results **A18:** 200
finish machining of hardened steels and irons **A16:** 116
and gear manufacture **A16:** 350, 351, 353, 354
grinding **A16:** 421, 428, 429
grinding of 52100 bearing steel **A16:** 466
grinding wheels **A16:** 462-465
grinding wheels for heat-resistant alloys **A16:** 758
grinding wheels for tool steels **A16:** 724, 728, 730
high-speed machining **A16:** 602, 604
honing of P/M materials **A16:** 891
in honing stones **A16:** 476, 477, 478
insert for turning fiber FP Al metal-matrix composite **A16:** 898
laser beam machining **A16:** 574
machining of P/M materials **A16:** 881
machining parameters **A16:** 112
milling **A16:** 112
milling operation eliminated **A16:** 460
PCBN and copy turning **A16:** 115, 116
PCBN and facing **A16:** 115, 116
PCBN and grinding **A16:** 116
PCBN and grooving **A16:** 115, 116
PCBN and milling **A16:** 116
PCBN and threading **A16:** 115, 116
PCBN rough machining applications **A16:** 112-113
PCBN tool life **A16:** 112, 113, 115
PCBN wear resistance **A16:** 112
physical properties **A18:** 192
plasma-assisted physical vapor deposition **A18:** 848
polycrystalline abrasion resistance **A16:** 109
polycrystalline cubic boron nitride (PCBN) **A16:** 105, 106, 107
polycrystalline cutting tools **A16:** 111
properties of **A2:** 1010-1011
rough machining applications **A16:** 113-116
softening point **A16:** 601
superabrasive **A16:** 4, 430-432, 450, 453-455
as superhard material **A2:** 1008
synthesis **A16:** 105
thermal conductivity **A2:** 1010
threading **A16:** 114
tool fabrication **A16:** 111
tool geometries **A16:** 109-110, 112

Cubic boron nitride (CBN) (continued)
tool life and grit sizes of grinding wheels ... **A16:** 457, 462, 469
tool material for machining cast irons ... **A16:** 651, 654, 656
toolholders for solid PCBN inserts ... **A16:** 111
tools for high removal rate machining ... **A16:** 608
tools for turning ... **A16:** 708
truing methods ... **A16:** 466-467, 468
truing parameters ... **A16:** 468
turning of Be alloys ... **A16:** 872
turning of heat-resistant alloys ... **A16:** 739, 740, 742
turning of P/M materials ... **A16:** 889
wear resistance ... **A16:** 108-109, 111, 112
wear resistance of PCBN ... **A16:** 115
Cubic carbide *See also* Carbide particles
appearance of, in iron-chromium-nickel heat-resistant casting alloys ... **A9:** 332
Cubic crystal system ... **A3:** 1•10, 15 **A9:** 706
Cubic crystals
transmission electron microscopy contrast studies ... **A9:** 121
Cubic metals
appearance under polarized light ... **A9:** 58
Cubic phases
in aluminum alloys ... **A9:** 359-360
Cubic unit cells ... **A10:** 346-348
Cull
defined ... **EM2:** 12
Cullet ... **EM4:** 381
composition ... **EM4:** 381
melt stability ... **EM4:** 381
mixing ... **EM4:** 381
particle size ... **EM4:** 381
purpose ... **EM4:** 381
Culvert stock
galvanized sheet steel ... **A9:** 199
Cumulative damage
analysis of ... **EM1:** 203-204
Cumulative distribution function
binomial distribution ... **A8:** 636
normal distribution ... **A8:** 631
Poisson distribution ... **A8:** 637
of statistical distributions ... **A8:** 628-629
Weibull distribution ... **A8:** 632-633
Cumulative erosion-time curve
defined ... **A18:** 6-7
Cumulative group mode failure
analysis ... **EM1:** 195
Cumulative linear damage
theories of ... **A11:** 112
Cumulative material transfer
in rolling-element bearings ... **A11:** 496
Cumulative normal distribution ... **A8:** 631
Cumulative probability curve
increasing variability with increasing mean life ... **A8:** 699
Cumulative sum (CuSum)-chart ... **EM3:** 795
Cumulative sum charts
in control chart quality control ... **A17:** 732-733
Cumulative weakening failure
analysis ... **EM1:** 194-195
Cunico *See* Magnetic materials, specific types; Permanent magnet materials, specific types
Cunife *See also* Magnetic materials; Magnetic materials, specific types; Permanent magnet materials, specific types ... **A9:** 538
as commercial permanent magnet material ... **A2:** 785
Cup *See also* Cupping; Cupping test
defined ... **A14:** 3
forming, analysis of ... **A14:** 924
fracture, defined ... **A14:** 3
-shaped workpieces, by deep drawing ... **A14:** 575
and tubes, drawing with moving mandrel ... **A14:** 335-336
Cup and cone samplers ... **A7:** 226
Cup drawing
r value in ... **A8:** 550
Cup, drawn
with ears ... **A8:** 550
Cup for mounting powder metallurgy materials ... **A9:** 505
Cup fracture
defined ... **A8:** 3 **A11:** 3
Cup-and-cone fracture *See also* Cup fracture; Cup, fracture
ductile and brittle ... **A11:** 82-83
fractographs ... **A11:** 83
Cup-and-cone fracture(s)
in alloy steels ... **A12:** 302
examining ... **A12:** 98
in fcc metals ... **A12:** 100
medium-carbon steels ... **A12:** 253
precipitation-hardening stainless steels ... **A12:** 370
studies ... **A12:** 3
tensile ... **A12:** 3, 98, 100, 102, 253
titanium alloys ... **A12:** 451
tool steels ... **A12:** 377
zones ... **A12:** 102
Cup-type brushes ... **M5:** 152, 154
Cupferron
as precipitant ... **A10:** 169
as solvent extractant ... **A10:** 170
Cuplike parts
cold extruded ... **A14:** 305
Cupola iron
desulfurization ... **A15:** 75-76
development of ... **A15:** 29-30
Cupola malleable cast iron ... **M1:** 57
Cupolas ... **A15:** 383-392
automation of ... **A15:** 35
basic slag composition effects ... **A15:** 390-391
charge materials ... **A15:** 387-389
combustion reactions/zones ... **A15:** 389
computer use ... **A15:** 391-392
construction/operation ... **A15:** 384-387
control principles ... **A15:** 389-390
control tests and analyses ... **A15:** 390
defined ... **A15:** 4
desulfurization methods ... **A15:** 391
drop bottom ... **A15:** 29
equipment, refinements in ... **A15:** 384
gray iron melting ... **A15:** 636
plasma arc ... **A15:** 36
size ranges ... **A15:** 385
specialized ... **A15:** 392
straight-sided ... **A15:** 30
Cupping *See also* Deep drawing ... **A8:** 3 **A14:** 3
defined ... **A9:** 5 **A11:** 3 **A17:** 383
radiographic methods ... **A17:** 296
test, defined ... **A8:** 3-4
Cupping and drawing
for steel tubular products ... **A1:** 328 **M1:** 316
Cupping test *See also* Cup, fracture; Erichsen test; Olsen ductility test ... **A14:** 3
Cuppy-core fractures ... **A8:** 592
Cupric ammonium chloride
description ... **A9:** 68
Cupric chloride ... **A16:** 69-70
photochemical machining etchant ... **A16:** 589, 591, 593
Cupric chloride, hydrochloric acid and ethanol as an etchant for stainless steels welded
to carbon or low alloy steels ... **A9:** 203
Cupric salts
as impurities ... **A13:** 1161-1162
Cupro-nickel alloys
pickling of ... **M5:** 612
Cupronickel *See* Copper-nickel alloys ... **A6:** 313, 314
thermal expansion coefficient ... **A6:** 907
Cupronickel powders
applications ... **A2:** 402
Cupronickels *See* Copper nickels
Cuprous oxide
examination under polarized light ... **A9:** 400
Curatives *See also* Cure; Curing agents
for epoxy resins ... **EM1:** 134, 136, 139-141
for prepreg resins ... **EM1:** 140
Cure *See also* Co-curing; Cure cycle; Cure monitoring, electrical; Cure quality control; Cure stress; Curing; Curing agents; Degree of cure; Post cure; Postcure; Precure; Set (mechanical); Undercure ... **EM3:** 9
ambient-temperature system ... **EM2:** 274
autoclave ... **EM1:** 500-501, 642-648, 704
of BMI resins ... **EM1:** 657-661
bonding ... **EM1:** 702-705
closed-loop ... **EM1:** 761-763
composite, management synergisms ... **EM1:** 649
computer modeling of ... **EM1:** 500-501, 704-705, 758-759
Cure (continued)
defined ... **EM1:** 8 **EM2:** 12
degree of, solder masks ... **EL1:** 554
dual ... **EL1:** 859, 863-864
effect, part dimensions, polyurethanes (PUR) ... **EM2:** 264
of epoxy adhesives ... **EM1:** 685
of epoxy resins ... **EM1:** 67-73, 134-141, 654-656
in filament winding ... **EM1:** 507
final schedule, thin-film hybrids ... **EL1:** 328-329
and gelling, chemical ... **EM1:** 132
hold temperatures ... **EM1:** 655-656
in-process control of material during ... **EM1:** 744
modeling ... **EM1:** 500-501, 704-705, 758-759
monitoring ... **EM2:** 12, 528
of organic coatings ... **A13:** 418
ply thickness for ... **EM1:** 761
of polyester resins ... **EM1:** 133
of polyimide resins ... **EM1:** 662-663
preparation for ... **EM1:** 642-644
processing materials ... **EM1:** 642-644
quality control ... **EM1:** 745-760
rapid, polymers for ... **EM1:** 121
reverse Diels-Alder reaction during ... **EM1:** 83
rigid epoxies ... **EL1:** 810
shrinkage, rigid epoxies ... **EL1:** 810
stress residuals ... **EM1:** 761
temperature/time effects on viscosity during ... **EM1:** 649
of thermoset molding compounds ... **EM1:** 167
time/temperature relationships, BMIs ... **EM1:** 659-660
tube rolling ... **EM1:** 573-574
types, conformal coatings ... **EL1:** 764-765
variables in ... **EM1:** 759-760
in wet lay-up ... **EM1:** 132-134
Cure acclerators ... **EM3:** 52
Cure cell management computer system ... **EM3:** 714
Cure cycle *See also* Cure
defined ... **EM1:** 8
developing ... **EM1:** 704
with diffusion control ... **EM1:** 758
epoxies ... **EM1:** 140-141
flow in ... **EM1:** 753-755
PMR-15 polyimide ... **EM1:** 141
Cure, degree of
microwave inspection ... **A17:** 215
Cure modeling ... **EM1:** 500-501, 704-705, 758-759
Cure monitoring, electrical *See also* Cure
defined ... **EM1:** 8
Cure quality control ... **EM1:** 745-760
chemical kinetics ... **EM1:** 748-751
cure modeling ... **EM1:** 758-759
curing variables ... **EM1:** 759-760
diffusion control ... **EM1:** 757-758
fluid hydrostatic pressure ... **EM1:** 755-757
heat transfer theory ... **EM1:** 746-748
now ... **EM1:** 753-755
sample chemical reactions ... **EM1:** 751-753
Cure rate index ... **EM3:** 9
Cure stress *See also* Cure ... **EM3:** 9
defined ... **EM1:** 8 **EM2:** 12
Cure time
measurement of ... **EM3:** 189
Cured strength
as process control measure ... **EL1:** 674
Curie and Neel point measurements
by Mössbauer spectroscopy ... **A10:** 287
Curie point *See* Curie temperature ... **A6:** 365
for ferromagnetic materials ... **A17:** 89
in magnetic hysteresis ... **A17:** 99-100
microstructural dependence ... **A17:** 131-132
Curie temperature ... **A10:** 671, 691
Curie temperatures
magnetically soft materials ... **A2:** 761
permanent magnet materials ... **A2:** 791
Curing *See also* Chemistry; Cure; Cure strength; Drying; Shadow cure; Ultraviolet-curable silicone coatings
agent, for flexible epoxies ... **EL1:** 818
agents, selection ... **EL1:** 827-831
allyls (DAP, DAIP) ... **EM2:** 229
anaerobics ... **EM3:** 50-51
autoclave heat, pressure, control systems ... **EM1:** 704
autoclave, process modeling of ... **EM1:** 500-501

Curing (continued)
bagging operation **EM1:** 703-704
by cyclotrimerization **EM2:** 233
chemistry, cyanates **EM2:** 232
coatings/encapsulants **EL1:** 242
computer modeling of **EM1:** 500-501, 704-705, 758-759
of cores **A15:** 240
dielectric, defined *See* Dielectric curing
effect, long-term reliability **EM2:** 788-789
effect on rheological properties **EM3:** 323
electronic general component adhesives **EM3:** 573
epoxies **EM3:** 95-96, 101-102, 617-618
epoxy materials **EL1:** 825, 827-831
epoxy, reaction mechanisms **EL1:** 828
final, Unicast process **A15:** 252
future development **EM1:** 705
heat **EL1:** 773
industrial applications for adhesives **EM3:** 568
of leaded and leadless surface-mount joints **EL1:** 733
mechanism, of polymide resin **A10:** 285
methods **A15:** 238
parameters **EM3:** 37
of polyamide-imides (PAI) **EM2:** 137
polyimide coatings **EL1:** 771-772
of polymers, ATR monitoring of **A10:** 120-121
of polyphenylene sulfides (PPS) **EM2:** 187
printed board coupons **EL1:** 573
processing quality control **EM3:** 739-742
radiation, types **EL1:** 854
radio frequency (RF) method **EM3:** 577
reactions, of epoxy resins **EM1:** 67-71
rheological behavior **EL1:** 850-852
shadow **EL1:** 824
sheet molding compounds **EM1:** 142
of silicone conformal coatings **EL1:** 823-824
SMT adhesives **EL1:** 672
of solder paste **EL1:** 733
in surface-mount soldering **EL1:** 700
surface-mount technology adhesives **EM3:** 570-571
temperatures, wet lay-up **EM1:** 132
theoretical aspects **EM1:** 704-705
time (no bake), defined **A15:** 4
time, defined **EM2:** 12
ultraviolet (UV) **EL1:** 785-788, 854
variables **EM1:** 759-760
viscosity, role of **EM1:** 702
vs imidization **EL1:** 772
Curing agent *See also* Catalyst; Latent curing agent **EM3:** 10
defined **EM2:** 12
Curing agents *See also* Cure; Hardener
aliphatic amine **EM1:** 753
amine **EM1:** 134, 753
anhydride **EM1:** 753
aromatic amine **EM1:** 753
defined **EM1:** 8
for epoxy resins **EM1:** 67, 134, 139, 140
for filament winding **EM1:** 505
properties analysis of **EM1:** 736
resin injection **EM1:** 531-532
Curing and cure controls **EM3:** 709-715
autoclave processing system **EM3:** 709-710
bagless autoclave processing **EM3:** 710
mechanical restraint methods **EM3:** 710
press curing **EM3:** 710
computer-controlled cure systems **EM3:** 711-715
evolution of control methods **EM3:** 711
future computer control techniques **EM3:** 714-715
gaseous pressurizing systems **EM3:** 713
mechanical press (pressure control) **EM3:** 713
pressure control **EM3:** 713
reporting functions **EM3:** 714
temperature controls **EM3:** 712-713
Curing and cure controls (continued)
temperature sensor **EM3:** 711-712
thermocouple failure detection **EM3:** 712
vacuum control **EM3:** 713-714
water pressurizing systems **EM3:** 713
ovens **EM3:** 709
convection oven/autoclave **EM3:** 709
microwave **EM3:** 709
radiant **EM3:** 709
press tooling **EM3:** 710-711
matched metal tooling **EM3:** 710
repair using adhesives **EM3:** 711
roll consolidations **EM3:** 710
single-sided tooling **EM3:** 710
Thermex process **EM3:** 711
Curing, paint *See* Paint and painting, curing methods
Curing temperature **EM3:** 10
Curing time **EM3:** 10
Curing times of castable resins **A9:** 30-31
Curium *See also* Transplutonium actinide metals
applications and properties **A2:** 1198-1201
pure **M2:** 733, 832-833
Curling
aluminum alloy **A14:** 804
dies, for press-brake forming **A14:** 537
in TEM replicas **A12:** 181
Curly grain structure **A9:** 686-687
in iron wire **A9:** 688
Curly lamellar structure
in pearlitic steel wire **A9:** 688
Current *See* Alternating current (ac); Direct current (dc); Electric current; specific process
components, bipolar-junction transistor **EL1:** 150
defined **A13:** 4
direction, defined **EL1:** 95
efficiency, defined **A13:** 4
electric, SI base unit and symbol for **A10:** 685
exchange **A13:** 6
flowing, effect on electrolysis **A10:** 197-198
gain, bipolar transistor analysis **EL1:** 151-152
-potential relationship, aqueous corrosion **A13:** 30-31
requirements, in design process **EL1:** 128
Current and wire feed speed
effect on weld attributes **A6:** 182
Current densities
critical, A15 superconductors **A2:** 1063-1064
critical, ternary molybdenum chalcogenides (chevrel phases) **A2:** 1079
for electrolytic etching of zinc and zinc alloys **A9:** 489
for electropolishing nickel alloys **A9:** 439
for electropolishing nickel-copper alloys **A9:** 439
in potentiostatic etching **A9:** 144-147
and single-electrode potential for electrolytes possessing polishing action **A9:** 48
superconducting magnets **A2:** 1028
in superconductors **A2:** 1033-1034
Current density
conversion factors **A8:** 722 **A10:** 686
defined **A13:** 4
effect on median time-to-failure of aluminum and aluminum alloys **A11:** 772
as electrolytic inclusion and phase isolation **A10:** 176
exchange **A13:** 6
exponent determination **EL1:** 890
measured **A18:** 274
reference electrodes **A13:** 24
SI derived unit and symbol for **A10:** 685
SI unit/symbol for **A8:** 721
vs life **EL1:** 963-964
Current density-voltage curve
for electropolishing **A17:** 52
Current detection
in scanning electron microscopy specimens **A9:** 90
Current efficiency
in electrolytic copper powders **A7:** 113
Current flow *See also* Electric current; Magnetizing current
direction, in magnetic fields **A17:** 90
Current input
leads, high-current cable **A8:** 389
locations optimized **A8:** 388
and potential measurement probe locations, test specimen geometries **A8:** 386
Current interruption cycle copper plating **M5:** 159-160, 164, 168-169
Current interruption technique
reference electrodes **A13:** 24
Current markets *See* Applications
Current rise
rate of **A6:** 38
Current stability
in direct current electrical potential method **A8:** 389
Current-decay method
of demagnetization **A17:** 93
Current-sampled polarography
as improved voltammetry **A10:** 193
Current-voltage curves for electropolishing **A9:** 105
Current-voltage relationships
voltammetry and dc polarograms for **A10:** 189, 190
Curtain coating
defined **EM2:** 12
paint **M5:** 482-483, 494
Curvature of field
defined **A9:** 5
Curvature, particle
in solidification **A15:** 102-103
Curvature, radius of
abbreviation for **A10:** 691
Curve fitting, and factor analysis
IR application to polymer blend system **A10:** 117-118
Curve tracer testing
in failure verification/fault isolation **EL1:** 1059-1060
Curve-fitting equations
in fracture mechanics **A11:** 53
Curved plates
radiographic inspection **A17:** 332
Curved-vane wheels
blast cleaning **A15:** 513-515
Cushion bearing **A18:** 662
Cusil
brazing, composition **A6:** 117
wettability indices on stainless steel base metals **A6:** 118
Cusil-ABA
ceramic/metal seals **EM4:** 506
wetting behavior and joining **EM4:** 491-492
Cusiltin 5
brazing, composition **A6:** 117
wettability indices on stainless steel base metals **A6:** 118
Cusiltin 10
brazing, composition **A6:** 117
wettability indices on stainless steel base metals **A6:** 118
Cusin-1-ABA
ceramic/metal seals **EM4:** 507, 509
Custom Age 625 PLUS
aging **A4:** 796
composition **A4:** 794
solution-treating **A4:** 796
Custom assembly
discrete two-terminal devices **EL1:** 432

Custom reference electrode
for acidified chloride solutions over boiling ... **A8:** 419
Customary names
of polymers ... **EM2:** 53
Customer requirements
and material selection ... **EM1:** 38
Customized packaging
for differentiation ... **EL1:** 440-441
CuSum charts
in control chart quality control method ... **A17:** 732-733
Cut
defined ... **A11:** 3 **A15:** 4 **A18:** 7
Cut layers
defined ... **EM2:** 13
Cut off
defined ... **A15:** 4
as postcasting operation, investment casting ... **A15:** 263
Cut-off
defined ... **EM2:** 13
Cut-stay fence ... **M1:** 271
Cut-to-length lines
blanking lines ... **A14:** 712
capacity ... **A14:** 712
dimensional accuracy ... **A14:** 712-713
edge-trim slitters ... **A14:** 712
flying-die or rocker shear ... **A14:** 711
rotary drum shears ... **A14:** 711-712
of sheet ... **A14:** 711-713
stationary shears ... **A14:** 711
stop/start ... **A14:** 711
Cutoff *See also* Blank; Multiple
bar shearing, method selection ... **A14:** 714
for blanks ... **A14:** 445
carbide metal cutting tools ... **A2:** 965
defined ... **A14:** 3
equipment, contour roll forming ... **A14:** 628
impact cutoff machines, for shearing ... **A14:** 714, 718-719
unit, multiple-slide machines ... **A14:** 568
Cutoff tools
coatings and increased tool life ... **A16:** 58
multifunction machining ... **A16:** 366, 375, 376
Cutting *See also* Cutting machines; Sectioning; Shearing ... **A18:** 184, 185
abrasive water-jet ... **EM1:** 673-675
abrasive waterjet ... **A14:** 18-19, 743-755
abrasive wheel ... **A12:** 92
air carbon arc ... **A14:** 732-734
alignment of silicon boule for ... **A10:** 342
band saw ... **A12:** 92
band saw vs. ultrasonic ... **EM1:** 615
of bar stock, closed-die forging ... **A14:** 80-81
bevel ... **A14:** 731
of billets, ring rolling ... **A14:** 123
blanking as ... **A14:** 445
of blanks, three-roll forming ... **A14:** 619
by diamond tooling ... **EM1:** 295
by ultramicrotome ... **A12:** 199
chemical flux ... **A14:** 728-729
of chromized sheet steel ... **A9:** 198
double, of angle sections ... **A14:** 718
dry ... **A11:** 19
of electrogalvanized sheet steel ... **A9:** 197
flame ... **A11:** 19
force, in blanking ... **A14:** 448
gas, and blanking ... **A14:** 458
gouging ... **A14:** 732-734
of hot-dip galvanized sheet steel ... **A9:** 197
of hot-dip zinc-aluminum coated sheet steel ... **A9:** 197
laser ... **A14:** 18, 735-742 **EM1:** 676-680
mechanical and flame, compared for steam equipment ... **A11:** 602
mechanical, and thermal, compared ... **A14:** 720
metal powder ... **A14:** 728
oxyfuel gas ... **A14:** 720-728
pattern, for tube rolling ... **EM1:** 573
plasma arc ... **A14:** 729-732
of porcelain enameled sheet steel ... **A9:** 198
refractory metals and alloys ... **A2:** 560-562
residual stress distributions from ... **A10:** 392
of samples ... **A10:** 16
specimen ... **A12:** 76-77
speed, bar shearing ... **A14:** 714

Cutting (continued)
of stainless-clad sheet steel ... **A9:** 198
of substrates ... **EL1:** 465
thermal ... **A14:** 720-734
of thermoplastic composite ... **EM1:** 552
tips, oxyfuel gas cutting ... **A14:** 726
to length, and cold heading ... **A14:** 294
tool steels ... **A18:** 738-739
tools ... **EM1:** 667-670
torches, oxyfuel gas cutting ... **A14:** 726
ultrasonic ... **EM1:** 615-618
underwater, oxygen arc ... **A14:** 734
water-jet ... **EM1:** 673-675
wrought copper and copper alloys ... **A2:** 247-248
Cutting attachment
definition ... **A6:** 1208 **M6:** 5
Cutting blowpipe
definition ... **A6:** 1208
Cutting edge wear ... **A18:** 610
conventional and P/M tool steels ... **A7:** 786
Cutting fluids *See also* Coolants
affecting die threading accuracy and finish ... **A16:** 300
Al alloys ... **A16:** 761, 765-766, 769-773, 775, 778-782, 784-790, 792
Be alloys ... **A16:** 872
boring ... **A16:** 165-172
broaching ... **A16:** 205, 206-208
and built-up edges ... **A16:** 40, 121, 122
carbon and alloy steels ... **A16:** 677
cast irons ... **A16:** 651-652, 654
and chip formation ... **A16:** 121
chlorine prohibited when machining Ti ... **A16:** 28
contour band sawing ... **A16:** 362, 363
cooling ... **A16:** 122
Cu alloys ... **A16:** 808-809, 811, 814, 815, 817
drilling ... **A16:** 212-213, 216, 220-224, 228-230, 232, 234, 238
end milling ... **A16:** 174
filtering ... **A16:** 302
functions ... **A16:** 121-132
and gear cutting ... **A16:** 343, 344-346
grinding ... **A16:** 423-424
gun drilling ... **A16:** 173
heat-resistant alloys ... **A16:** 747, 750-751, 753, 755-757
Hf machining ... **A16:** 856
high-speed machining ... **A16:** 602
influencing output ... **A16:** 299
and lubrication ... **A16:** 121-122
and machinability ratings ... **A16:** 643
for machining of magnesium and magnesium alloys ... **A2:** 475
Mg alloys ... **A16:** 820-823, 825, 826, 828, 829
milling ... **A16:** 320, 321, 322, 327-328
multifunction machining ... **A16:** 384, 386, 392, 393
Ni alloys ... **A16:** 836-837, 838, 839, 840, 841, 842-843
pipe threading ... **A16:** 302
and planing ... **A16:** 186
reaming ... **A16:** 244, 245, 246, 247, 248
removal from steel parts ... **M5:** 5, 9-10
removal to avoid stress corrosion ... **A16:** 35
roller burnishing ... **A16:** 253
sawing systems ... **A16:** 358, 360, 361, 362, 363
shaping ... **A16:** 191, 192
soluble oil emulsion ... **A16:** 301
stainless steels ... **A16:** 691-698, 700, 701, 703-704, 705
sulfurated cutting oil ... **A16:** 301
sulfurized oils ... **A16:** 299
and surface integrity ... **A16:** 31-32
tapping ... **A16:** 257, 259-265, 267
thread rolling ... **A16:** 288
threading ... **A16:** 301
Ti alloys ... **A16:** 844-848, 850, 851, 853, 854
tool steel machining ... **A16:** 716
trepanning ... **A16:** 175, 176, 178, 180
turning
uranium alloys ... **A16:** 874, 875, 876, 877
Zn alloys ... **A16:** 831, 832, 833, 834
Zr machining ... **A16:** 855
Cutting fluids for machining operations ... **A7:** 461
Cutting head
definition ... **M6:** 5
Cutting lubricants ... **EM3:** 41

Cutting machines *See also* Cutting; Shearing
directions ... **A14:** 727-728
portable ... **A14:** 727
stationary ... **A14:** 727
tape control ... **A14:** 728
Cutting methods *See also* Sectioning
depth of deformation in different metals ... **A9:** 23
Cutting nozzle
definition ... **A6:** 1208
Cutting process
definition ... **M6:** 5
Cutting speed ... **A1:** 591-592
Cutting speed, machining of steel ... **M1:** 566, 568, 571
interrelation with machining costs ... **M1:** 582-585
interrelation with tool life ... **M1:** 566, 572, 574, 581, 583
machinability ratings based on ... **M1:** 570
recommended cutting speeds selected steels ... **M1:** 569
Cutting tip
definition ... **A6:** 1208
Cutting tips
definition ... **M6:** 5
for oxyfuel gas cutting ... **M6:** 906-907
Cutting tool
AES thick films analysis of ... **A10:** 561
Cutting tool applications
cermets for ... **A2:** 978
Cutting tool grades
coated carbide grade steels ... **A2:** 967-968
uncoated alloyed carbide grade steels ... **A2:** 967
uncoated straight WC-CO grade steels ... **A2:** 967
Cutting tools
drills ... **M3:** 472-473
end mills ... **M3:** 474-475
hobs ... **M3:** 477
material selection, general requirements ... **M3:** 470
materials for ... **M3:** 470-477
milling cutters ... **M3:** 475-476
powders used ... **A7:** 573
reamers ... **M3:** 473
single-point tools ... **M3:** 470-472
taps ... **M3:** 474
tool life ... **M3:** 472-473, 474-475
Cutting tools and cutting tool materials, friction and wear of ... **A18:** 609-619
cutting fluid systems ... **A18:** 617-619
application mechanism ... **A18:** 618, 619
cutting fluid handling system ... **A18:** 618, 619
cutting fluids ... **A18:** 618-619
cutting tool failure modes ... **A18:** 618
cutting tool material selection guidelines ... **A18:** 614-617
cast cobalt alloys ... **A18:** 614, 615
cemented carbides ... **A18:** 614, 616
ceramic inserts ... **A18:** 614, 616-617
high-speed steel (HSS) ... **A18:** 614, 615
micrograin high-speed steels ... **A18:** 614, 615-616
ultrahard tool materials ... **A18:** 614, 617
cutting tool testing ... **A18:** 617
machine, cutting tool, and tool wear interactions ... **A18:** 613-614
temperatures in the wear zones ... **A18:** 611-612
wear environment ... **A18:** 609
wear mechanisms ... **A18:** 612-613
classes ... **A18:** 612
initial ... **A18:** 612-613
steady-state ... **A18:** 613
tertiary ... **A18:** 613
wear surfaces ... **A18:** 609-611
motion along the wear surfaces ... **A18:** 611
stresses ... **A18:** 610-611, 612
Cutting tools, eliminating brittleness of carbide ... **A3:** 1 • 28
Cutting torch
definition ... **M6:** 5
Cutting torch (arc)
definition ... **A6:** 1208
Cutting torch (oxyfuel gas)
definition ... **A6:** 1208
Cutting torches
oxyfuel gas cutting ... **A14:** 726
for samples ... **A10:** 16

Cutting torches for
oxyfuel gas cutting **M6:** 906
plasma arc cutting **M6:** 915-916
Cutting velocity **A18:** 609, 610, 611, 614, 615, 618
Cutting wear *See* Abrasive wear
Cutting wheels *See also* Abrasive cutting wheels
for cemented carbides **A9:** 273
Cutting-down operation
nickel alloys **M5:** 674-675
CV process *See* Counter-gravity low-pressure casting
CVD *See* Chemical vapor deposition
CVD carbon *See* Chemical vapor deposited (CVD) carbon
CVD coatings *See* PVD and CVD coatings
CVN *See* Charpy V-notch
CY-10C
properties **A18:** 549
CY-179 (epoxy
Ciba-Geigy Corporation) **EM3:** 96
Cyanate
cyanide reduced to **M5:** 312
Cyanate ester resins **EL1:** 534-535, 606-607
Cyanate esters *See* Cyanates **EM1:** 290
Cyanate resins **EM3:** 10
defined **EM2:** 13
Cyanates *See also* Thermosetting resins
applications **EM2:** 232-234
characteristics **EM2:** 234-238
competitive materials **EM2:** 233-234
cost/volume **EM2:** 232
curing chemistry **EM2:** 232
and epoxies, co-reactivity **EM2:** 234
fiberglass laminates **EM2:** 235
prepolymers **EM2:** 232
processing **EM2:** 236-238
properties **EM2:** 236
resin forms **EM2:** 236
suppliers **EM2:** 238
unreinforced **EM2:** 234-235
Cyanic esters *See* Cyanates
Cyanide
in alkaline electrolytes **A9:** 54
as an etchant for beryllium-containing alloys **A9:** 394
cleaning up spills **A9:** 54
free **M5:** 285
safety hazards **A9:** 69
safety precautions **M6:** 995
Cyanide bath plating
brass **M5:** 285-287
bronze **M5:** 288-289
cadmium *See* cadmium cyanide plating
copper *See* copper cyanide plating
gold *See* Hot cyanide gold bath plating
silver **M5:** 279-280
zinc *See* Zinc cyanide plating
Cyanide dipping
cadmium plating process **M5:** 263
copper and copper alloys **M5:** 619, 621
Cyanide ions
ion chromatographic analysis of **A10:** 661
Cyanide oxidation process
plating waste disposal **M5:** 312
Cyanide peroxide hydroxide as an etchant for beryllium-containing alloys **A9:** 394
Cyanide salts
procedure for use in brazing **M6:** 994
use in dip brazing **M6:** 991
Cyanide tarnish removal solutions
use of and safety precautions **M5:** 613
Cyanided 1020 steel **A9:** 227
Cyanided steels
microstructures **A9:** 218
specimen preparation **A9:** 217
Cyanides
as electrodes **A10:** 185
nickel titration using **A10:** 174
Cyaniding *See* Liquid carburizing **M1:** 539
defined **A9:** 5
hardness profiles developed by **M1:** 626
Cyanoacrylate Resins—The Instant Adhesives **EM3:** 68
Cyanoacrylates **EM3:** 10, 35, 44, 74-75, 126-132
for acrylic windows bonding **EM3:** 128
additives **EM3:** 130
advantages and limitations **EM3:** 77, 79
aerospace applications **EM3:** 559
applications **EM3:** 128-129
automotive applications **EM3:** 45-46
for catheters bonding **EM3:** 128
chemical resistance properties **EM3:** 639
chemistry **EM3:** 78-79, 126-127
compared to epoxies **EM3:** 98
consumer product **EM3:** 47
cost factors **EM3:** 128
curing mechanisms **EM3:** 78, 126-127, 128
degradation **EM3:** 679
design considerations **EM3:** 130-131
for dialysis membranes bonding **EM3:** 128
for display panels bonding **EM3:** 128
durability **EM3:** 670
electrical/electronics applications **EM3:** 44
for electronic general component bonding **EM3:** 573
engineering adhesives family **EM3:** 567
for engineering change order (ECO) wire tacking **EM3:** 11-1-569, 572
fastest sales growth **EM3:** 77
for ferrite cores fixturing **EM3:** 128
fiberglass molds potting **EM3:** 128
for filter caps fixturing **EM3:** 128
for fingerprint development **EM3:** 127
for gaskets fixturing **EM3:** 128
for gears to shaft bonding **EM3:** 128
for gearshift indicators bonding **EM3:** 128
for hardware to printed circuit boards **EM3:** 128
health and safety considerations **EM3:** 131
for heat sinks fixturing **EM3:** 128
for honing stones bonding **EM3:** 128
hydrolytic reactions **EM3:** 127, 128
for jumper wires fixturing **EM3:** 128
for loud speaker assembly **EM3:** 575
markets **EM3:** 127
for medical adhesives to close wounds **EM3:** 127
medical applications **EM3:** 127, 576
for medical bonding (class IV approval) **EM3:** 576
moisture cured **EM3:** 74
for nameplates bonding **EM3:** 128
not approved for medical applications by FDA **EM3:** 127
for O-rings bonding **EM3:** 128
performance of **EM3:** 124
predicted 1992 sales **EM3:** 77
processing parameters **EM3:** 131
properties **EM3:** 78, 128-130
for pushbuttons bonding **EM3:** 128
for rubber bumpers bonding **EM3:** 128
for rubber feet bonding **EM3:** 128
shear strength of mild steel joints **EM3:** 670
shelf life **EM3:** 131
for shunt plates fixturing **EM3:** 128
for slotted screwheads potting **EM3:** 128
for spacers fixturing **EM3:** 128
for speaker cones bonding **EM3:** 128
for sporting goods manufacturing **EM3:** 576, 577
for strain gages bonding **EM3:** 128
suppliers **EM3:** 79, 127-128, 129
surface activators **EM3:** 127
synthesis of cyanoacrylate monomers **EM3:** 126
temperature limits **EM3:** 621
for tennis racket parts fixturing **EM3:** 128
test methods **EM3:** 130
time-weighted average (TWA) concentrations recommended by OSHA **EM3:** 131
tougheners **EM3:** 183
Cyanoacrylates (continued)
transistors potting **EM3:** 128
for video game cartridges bonding **EM3:** 128
for wiper blades bonding **EM3:** 128
for wire-tacking **EM3:** 574
Cyanogen
abbreviation for **A10:** 690
emission molecule **A10:** 25
Cyanogen (CN)
produced by graphite reaction with nitrogen in furnaces **EM4:** 246
Cyanosilicones **EM3:** 677
advantages and disadvantages **EM3:** 675
for severe environments **EM3:** 673
Cyanosiloxanes
sealant formulations **EM3:** 677-678
Cycle **A8:** 4
-by-cycle approach, damage analysis **A8:** 682
defined **EM2:** 13
-dependent corrosion fatigue cracking **A8:** 405-406
frequency and number, composite material age **A8:** 716
life **A8:** 351-352, 358
time vs. cycles to fracture **A8:** 353
Cycle (*W*)
defined **A11:** 3
Cycle annealing **A1:** 261
cold finished bars **M1:** 232
defined **A9:** 5
Cycle frequency, and mold temperature
permanent molds **A15:** 282
Cycle period
vs. cycles to fracture fatigue behavior **A8:** 351
Cycle-time reduction
and costs **EL1:** 448
Cycled humidity-temperature test
under bias **EL1:** 495
Cycled THB life test **EL1:** 495
Cycles
to failure, vs. plastic strain range **A8:** 346-347
to fracture **A8:** 348-352
-to-crack-initiation, in fatigue testing ductile materials **A8:** 696
-to-failure **A8:** 364, 696
Cycles, number of
effect in fretting **A13:** 140
Cycles per second
abbreviation for **A10:** 690
Cycles to failure tests
corrosion fatigue **A13:** 291-295
Cyclic acidified salt spray tests **A13:** 242-243
Cyclic cleavage **A8:** 484-485, 487
Cyclic crack growth **A8:** 376, 422-423
fracture mechanics knowledge for **A8:** 422-423
rate **A8:** 376, 378-379
water chemistry knowledge for **A8:** 422-423
Cyclic creep, reversed
test results **A8:** 353-354
Cyclic ductile decohesion
fatigue failure **A8:** 482-484
load ratio effect **A8:** 484
Cyclic flow curve
grain coarsening **A8:** 175
Cyclic frequency, effect on fatigue cracking *See also* Frequency **A12:** 15
Cyclic hydrocarbons **EM3:** 10
defined **EM2:** 13
Cyclic intergranular process
fatigue failure **A8:** 484
Cyclic loading *See also* Frequency; Loading
in corrosion fatigue **A13:** 298
defined **A11:** 3
effect on fatigue cracking **A11:** 102
effects, produced fluids **A13:** 479
fatigue fractures by **A11:** 77
of forged components **A11:** 321
machine parts fatigue under **A11:** 371
machinery, mining **A13:** 1297

Cyclic loading (continued)
retainer spring failure due to **A11:** 561-562
in shafts, torsional-fatigue fracture **A11:** 464
weld crack under **A11:** 117
Cyclic loading, effect on fatigue *See also* Loading **A12:** 53-54, 111
Cyclic loads **A8:** 3, 377
metal vs plastic **EM2:** 76
Cyclic microvoid process
D-6 AC steel **A8:** 484
Cyclic oxidation *See also* Oxidation
mechanically alloyed oxide
alloys **A2:** 947
dispersion-strengthened (MA ODS)
Cyclic potentiodynamic polarization **A13:** 217-218
Cyclic strain accumulation *See* Ratcheting
Cyclic strain softening
defined **A11:** 144
Cyclic straining *See also* Ultrasonic fatigue testing
bar **A8:** 242
Cyclic stress **A13:** 292, 614
combined **EM1:** 202-203
creep-rupture damage by **A8:** 354-355
metal vs. composites **EM1:** 261–62
single **EM1:** 201-202
test parameters **A8:** 364
in ultrasonic fatigue testing **A8:** 240
Cyclic stresses
in corrosion fatigue, brasses and stainless steels **A11:** 636
effect on corrosion-fatigue **A11:** 254
fatigue from **A11:** 102
and fracture mechanics accuracy **A11:** 55
mechanical **A11:** 636-637
peeling-type fatigue cracking from **A11:** 471
in stress-corrosion cracking **A11:** 206
testing, schematic with test parameters
for **A11:** 102
thermally generated **A11:** 636
Cyclic stresses, wrought aluminum alloys *See also* Stress **A12:** 421, 430
Cyclic tension-tension fatigue
loading **A8:** 717-718
Cyclic torsional testing **A8:** 149-153
Cyclic variation in the stress intensity factor (ΔK) **A6:** 1013
Cyclic voltammetry
fundamentals **A10:** 192
Cyclic-load dominant fracture tests **EM3:** 507
Cycling *See* Power cycling; Temperat recycling-sensitive
frequency effect in low-cycle fatigue
behavior of lead **A8:** 347
Cycloaliphatic **EM1:** 753
Cycloaliphatic amines
as curing agents **EL1:** 827-828
Cycloaliphatic epoxides **EL1:** 854-866 **EM3:** 594, 595
polymerization, with Lewis acids **EL1:** 860
and products, commercial **EL1:** 862
specific types, properties **EL1:** 864
Cycloaliphatic resins **EM3:** 94, 96
Cycloconverter power sources **A6:** 39
Cyclodextrin
defined **A10:** 671
Cyclohexanol
dendrites in **A9:** 608
Cyclohexanon (homogenizer)
batch weight of formulation when
used in non-oxidizing sintering
atmospheres **EM4:** 163
Cyclohexonone cements with PVC fillers
medical applications **EM3:** 576
Cyclone
abatement chambers **A7:** 127
defined **A7:** 3
separator, in spray drying **A7:** 73
for tin powders **A7:** 123-124
Cyclone blowers
development of **A15:** 27
Cyclotetramethylene tetranitramine (HMX)
friction coefficient data **A18:** 75
Cyclotrimerization *See also* Cyanates
of aryl dicyanate **EL1:** 607
curing by **EM2:** 233
rates **EM2:** 234
Cyclotrimethylene trinitramine (RDX)
friction coefficient data **A18:** 75
Cyclotron resonance principle
for microwave amplifiers **A17:** 209
Cylinder
hollow **A8:** 143
for measuring displacement **A8:** 193
Rockwell hardness testing correction
factors for **A8:** 82
specimen, torque vs. angle of twist **A8:** 327
upset, consequences of friction **A8:** 573
Cylinder compressed air blast
first cast iron **A15:** 25
Cylinder manifold
definition **A6:** 1208
Cylinder(s)
as basic casting shape **A15:** 599
plate intersections, as basic casting
shapes **A15:** 599
solidification and design **A15:** 606
Cylinder-head exhaust port
failed **A11:** 355, 357
Cylinders *See also* Cylindrical; Cylindrical shapes; Shapes; Solid cylindrical bar; Tubes; Tubing; Tubular products; Workpiece(s)
blocks, gray iron cracking in **A11:** 345-346
castings, door-closer **A11:** 363-365
definition **M6:** 5
forging of **A14:** 67
hollow, closed at one end, inspection **A17:** 112
hollow ferromagnetic, central conductor within **A17:** 96
hollow, magnetic particle inspection of **A17:** 111-112
inelastic cycle buckling of **A11:** 144
inserts, corrosion fatigue cracking in **A11:** 371-372
large, forming **A14:** 621
lining, diesel motor, cavitation damage **A11:** 377-378
for oxyfuel gas welding **M6:** 584
penetrameters/identification markers
radiographic inspection **A17:** 343
small, forming **A14:** 621
solid, eddy current inspection **A17:** 184-185
solid, radiographic inspection **A17:** 332-333
solid, tubes on **A17:** 185
steel and aluminum powders **A7:** 411
steel, carburization distortion **A11:** 141-142
upset, stresses on bulge surface **A7:** 410
Cylinders, hollow
of unalloyed uranium **A2:** 671
Cylindrical
bars, frequency selection, eddy current
inspection **A17:** 274
Cylindrical bending
requirements **A8:** 119
Cylindrical billets **A7:** 424-425
typical can shrinkage **A7:** 425, 428
Cylindrical charge
for explosive forming **A14:** 636
Cylindrical chip
inductors, fabrication **EL1:** 188
resistors, construction **EL1:** 178
Cylindrical compression test specimen **A8:** 579-580
Cylindrical cup
drawing r value in **A8:** 550
Cylindrical flux and unidirectional flux
in sintering **A7:** 315
Cylindrical grinding *See* Grinding
Cylindrical internal reflection cell **A10:** 118, 689
Cylindrical magnetron technique
sputtering **A18:** 841, 842
Cylindrical mirror analyzer **A7:** 251
Cylindrical mirror analyzers **A10:** 554, 689
Cylindrical parts *See also* Cylinder(s); Oil well tubing; Part(s); Tube; Tubes; Tubing; Tubular products
centrifugal casting of **A15:** 296
economy in manufacture **M3:** 850, 851
Cylindrical roller bearings
basic load rating **A18:** 505
f_1 factors **A18:** 511
f_v factors for lubrication method **A18:** 511
Cylindrical roller thrust bearings
basic load rating **A18:** 505
Cylindrical sag
interference microscope measurement **A17:** 17
Cylindrical shapes *See also* Cylinders; Shapes; Workpiece(s)
bending of **A14:** 529
by manual spinning **A14:** 599
deep drawn **A14:** 575
drawing of **A14:** 585
large, forming **A14:** 621
multiple-slide forming **A14:** 572
small , forming **A14:** 621
Cylindrical shells, welded
economy in manufacture **M3:** 853
Cylindrical specimen *See also* Cylindrical compression test specimen; Specimen
barreling of **A8:** 56-57
and compression fracture **A8:** 57
fracture loci **A8:** 580-581
stress-relaxation compression testing **A8:** 326
upset, with reduced gage section **A8:** 589
Cylindrical specimens
fatigue corrosion testing **A13:** 293
Cylindrical-roller bearings
deformation of raceway by **A11:** 510
described **A11:** 490
Cysteine hydrochloride staining test
for tinplate **A13:** 782
Czerny-Turner monochromators **A10:** 23, 38
Czerny-Turner spectrometers
with Triplemate device **A10:** 129
Czochralski crystal growth process
silicon **EL1:** 191
Czochralski crystal-growing technique
See also Unidirectional
solidification **A9:** 607
Czochralski method
ferrite processing **EM4:** 1163
d, as symbol
defined **A8:** 724
enter porosity, in carbon and alloy steels
revealed by macroetching **A9:** 173
enter segregation, ASTM graded series **A9:** 173-174
Centerline chill in ductile iron casting **A9:** 251
Ceramic carbon matrix composites,
polishing **A9:** 591
Ceramic-filled plastics, as mounting
materials **A9:** 45
Ceramic layers, interference films
used to improve contrast **A9:** 59
The Composites & Adhesives Newsletter

D

4,4′-diphenyl methane diisocyanate (MDI) **EM3:** 203
D log E curve
radiography **A17:** 324
D-6 AC
flash welding **M6:** 557
plasma arc welding **M6:** 220
D-6a
composition **A4:** 207 **M1:** 422 **M4:** 120
hardness **A4:** 212
heat treatment **M1:** 429-430
heat treatments **A4:** 211-212
heat-treatment temperatures **A4:** 208
mechanical properties **A4:** 212 **M1:** 430-432 **M4:** 124
processing **M1:** 429
tempering temperature **M4:** 124
D-6a/6ac steel **A1:** 436
heat treatment for **A1:** 436
properties of **A1:** 436-437, 438
D-6AC *See* D-6a
fracture toughness **A4:** 211, 212
heat treatments **A4:** 211-212
heat-treatment temperatures **A4:** 208
plane-strain fracture toughness **A4:** 213
tensile properties **A4:** 212
D-6ac low-alloy ultrahigh-strength steel
laser-beam welding **A6:** 263
D-6ac steel
laser-beam welding **A6:** 264
D-6ac, tensile properties *See also* D-6a **M4:** 124

D-43 unmodified
welding conditions effect ... **A6:** 581
D-43Y
welding conditions effect ... **A6:** 581
D-979
broaching ... **A16:** 203, 209
composition ... **A6:** 573 **M4:** 651-652
D-glass *See also* Glass fibers
defined ... **EM1:** 9 **EM2:** 13
dielectric constant of ... **EM1:** 107
as reinforcements ... **EL1:** 535, 604-605
***d*-spacing gradient** ... **A18:** 465, 468
***d*-spacings**
formulas for ... **A10:** 328
use in unknown phase/particle identification ... **A10:** 455-456
D-stage washer corrosion
pulp bleach plants ... **A13:** 1193-1194
d.f *See* Degrees of freedom
D1, D2, D3, D4, etc *See* Tool steels, specific types
da/dN *See* Crack growth rate; Fatigue crack growth rate
Dacron paint rollers ... **M5:** 503
Daimler-Benz AGI
ceramic components for an automotive research gas turbine ... **EM4:** 717
DAIP *See* Allyls
Dairy equipment
metal coatings for ... **M1:** 102
tin coated steel sheet for ... **M1:** 173
Dairy metal *See* Copper alloys, specific types, C97600
properties and applications ... **A2:** 388-389
Daisy chain routing
for VHSIC systems ... **EL1:** 83
Dalton's Law ... **A6:** 58
Daly collector
in gas mass spectrometers ... **A10:** 154-155
Dam
in cure processing ... **EM1:** 8, 644
defined ... **EM2:** 13
Damage
cumulative, analysis ... **EM1:** 203-204
dent depth as ... **EM1:** 265
from electric discharge machining ... **A9:** 27
from sectioning ... **A9:** 23
modes, exact ... **EM1:** 247
and static strength ... **EM1:** 434-435
visible, and tolerances ... **EM1:** 266
Damage accumulation
as fatigue failure mechanism ... **EM1:** 202
fatigue prediction models from ... **EM1:** 244-247
Damage analysis
of creep-fatigue interaction ... **A8:** 358
cycle-by-cycle approach ... **A8:** 682
equivalent constant-amplitude stress approach ... **A8:** 682
in fatigue crack growth ... **A8:** 681-682
Damage factors ... **A18:** 266
Damage fraction ... **EM3:** 517
Damage growth
model and analysis ... **EM1:** 201-202
Damage, impact *See* Impact damage
Damage nucleation maps ... **A14:** 421
Damage resistance
of metals ... **A11:** 166-167
Damage tolerance *See also* NDE reliability; Tolerance ... **EM3:** 10
of composites ... **EM1:** 180, 233, 259-267
in composites and metals compared ... **EM1:** 259-260
constituent properties for improved ... **EM1:** 99
control plan ... **A17:** 669-670
cyclic loading ... **EM1:** 261-262
damage detectability ... **EM1:** 265
defect sensitivity and ... **EM1:** 261
defined ... **A17:** 668 **EM1:** 8, 259 **EM2:** 13
design ... **A17:** 664, 702
designing for ... **EM1:** 182
Damage tolerance (continued)
fracture control philosophy as ... **A17:** 666
full-scale tests ... **EM1:** 351
laminate ... **EM1:** 233
material effects on ... **EM1:** 262-264
of natural fiber ... **EM1:** 117
requirements, aircraft ... **EM1:** 264-266
requirements, thermoplastic resins ... **EM1:** 97-98
residual strength effects ... **EM1:** 265
safety aspects ... **EM1:** 259
and structural efficiency ... **EM1:** 259
tests ... **EM1:** 97-98, 264, 293, 343-344, 351
of thermoplastics ... **EM1:** 293
Damage tolerance testing ... **EM3:** 540, 543
Damaged castings
repair of ... **A15:** 529-530
Damp corrosion ... **A13:** 80
Damping *See also* Attenuation; Damping properties analysis; Hysteresis ... **EM3:** 10
aerodynamic ... **EM1:** 206
air ... **EM1:** 212
Debye-Waller and mean-free- path ... **A10:** 410
defined ... **EM1:** 8 **EM2:** 13
features, of composites ... **EM1:** 35-36
fiber volume fraction effect ... **EM1:** 207
flexural, variation with fiber volume fraction ... **EM1:** 208
internal, sources of ... **EM1:** 06
measuring ... **EM1:** 207
ply angle effects ... **EM1:** 210
properties, analysis of ... **EM1:** 206-217
and strength ... **EM1:** 216
temperature dependence of ... **EM1:** 215-216
vibration, as beneficial ... **EM1:** 190
as vibration property, defined ... **EM1:** 206
Damping capacity
1018 steel ... **M1:** 41, 45
aluminum ... **M1:** 32
CG iron castings ... **A15:** 675
of compacted graphite iron ... **A1:** 69-70
of ductile iron ... **A1:** 45-46 **M1:** 32, 41, 45
eutectoid steel ... **M1:** 32
of gray iron ... **A1:** 31-32 **A15:** 644 **M1:** 32, 41, 45
of magnesium alloys ... **A2:** 462
of malleable iron ... **A1:** 82, 84 **M1:** 32
pure iron ... **M1:** 32
steels, selected grades ... **M1:** 145
white cast iron ... **M1:** 32
Damping devices
P/M porous parts ... **A7:** 700
Damping properties analysis
of beams cut from laminated plates ... **EM1:** 209-210
of laminated plates ... **EM1:** 210-213
of sandwich laminates ... **EM1:** 214
and strength ... **EM1:** 216
temperature effects ... **EM1:** 215-216
of unidirectional composites ... **EM1:** 207-209
of woven fibrous composites ... **EM1:** 213
Damping temperature
effects ... **EL1:** 45
Dams
acoustic emission inspection ... **A17:** 290
Dams for
electrogas welding ... **M6:** 239-240
electroslag welding ... **M6:** 227-228
Dancer roll ... **A18:** 57
Danner process ... **EM4:** 399-400, 1032, 1038
DAP *See* Allyls
for coating/encapsulation ... **EL1:** 242
Darcy's Law
permeation of a fluid through a packed bed of a powder ... **EM4:** 70
Dark field
defined ... **EL1:** 1067
Dark reaction ... **EM3:** 10
Dark red copper coloring solutions ... **M5:** 625
Dark-field illumination ... **A9:** 76 **A10:** 690
defined ... **A9:** 5
optical etching ... **A9:** 58
Dark-field illumination (continued)
principles of ... **A9:** 58
used for porcelain enameled sheet steel ... **A9:** 199
Dark-field illumination fractographs
coarse-grain iron alloy ... **A12:** 93-94
light-field, and SEM, compared ... **A12:** 92-100
Dark-field images
of annealing twin in rutile ... **A10:** 443
band of precipitate particles ... **A10:** 460
iron-base superalloy ... **A10:** 442
nitrogen-implanted Ti alloy ... **A10:** 485
transmission electron microscopy ... **A10:** 441-446, 460
Dark-field transmission electron microscopy ... **A9:** 103
beam diagram ... **A9:** 104
intensities calculated using dynamical theory ... **A9:** 111
two-beam thickness contours ... **A9:** 112
weak-beam images ... **A9:** 111
Darken formalism
for alloy component activity ... **A15:** 62-63
Darkroom photographic procedures used in photomicroscopy ... **A9:** 84-85
Darling-Saunder extensometer ... **EM2:** 657
Darting
of woven fabric prepregs ... **EM1:** 150
Dash pot
defined ... **EM2:** 13
Data
acquisition, uniform corrosion testing ... **A13:** 230
analysis and utilization, personal computer ... **A13:** 317
bases, on aerospace alloys ... **A11:** 54
corrosion, qualitative ... **A13:** 780
corrosion test, misapplication ... **A13:** 316
fatigue, presentation ... **A13:** 292
quantitative corrosion, chemicals ... **A13:** 780
scatter, in corrosion testing ... **A13:** 195
stress-corrosion cracking, precision of ... **A13:** 278
and tests, fracture toughness ... **A11:** 53-55
Data acquisition *See also* Analysis; Modeling; Process modeling
by hot compression testing ... **A14:** 439
for forging process design ... **A14:** 439-442
interface data ... **A14:** 441-442
of workability ... **A14:** 439-441
Data acquisition system (DAS)
defined ... **A17:** 383
Data analysis *See also* NDE reliability data analysis ... **EM2:** 602-606
dynamic notched round bar testing ... **A8:** 281-282
of fatigue crack growth ... **A8:** 678-684
fracture toughness using chevron-notched specimens ... **A8:** 473
graphical analysis ... **EM2:** 603-604
of hit/miss data ... **A17:** 689-691, 696-697
introduction ... **A8:** 623-627
least squares/analysis of variance ... **EM2:** 602-603
model building ... **EM2:** 602
mold shrinkage, as example ... **EM2:** 604-605
NDE reliability
sensitivity analysis, as example ... **EM2:** 605-606
software ... **EM2:** 607-608
for split Hopkinson pressure bar test ... **A8:** 202
Data base *See also* Computer applications; Simulation
for composite cures ... **EM1:** 705
design ... **EM1:** 181-182
for forging software tools ... **A14:** 412-413
fracture toughness ... **A8:** 462
on solidification ... **A15:** 862-863
relational, for KI SHELL ... **A14:** 412
workpiece and die thermophysical properties ... **A14:** 412
Data base management systems
for process design ... **A14:** 409-412

Data base systems *See* Ceramic properties data base systems
Data bases **EM4:** 40
computer, of application-specific properties **EM2:** 411
infrared spectra **A10:** 116
on materials information **EM2:** 95
in single-crystal analysis **A10:** 355-356
x-ray energy **A10:** 522-523
Data displays
acoustic emission inspection **A17:** 283-284
Data integrity
in computer-aided design **EL1:** 528-529
Data logging
through-hole soldering **EL1:** 695-696
Data processing
as casting market **A15:** 34
Data reduction
options in stainless steels, x-ray spectrometry for **A10:** 100
RDF analysis **A10:** 396-398
for silica glass **A10:** 397
Data reduction processing
machine vision process **A17:** 33-34
Data sheets *See also* Aggregate properties approach; Information; Information sources; Supplier data sheets
alternatives to **EM2:** 411
electrical properties **EM2:** 409
introduction **EM2:** 405
mechanical properties **EM2:** 408
physical properties **EM2:** 407
supplier, interpreting **EM2:** 638-645
thermal properties **EM2:** 409
Data signals, SEM
origin and detection **A12:** 168
Data storage
for coordinate measuring machines **A17:** 20
Data-processing equipment
ultrasonic inspection **A17:** 253-254
Dating of prehistoric materials
radioanalysis of **A10:** 243
Daubing
defined **A15:** 4
Daughter ion scans
gas chromatography/mass spectrometry **A10:** 646
Daylight *See also* Shut height **A14:** 3
defined **EM1:** 8 **EM2:** 13
DBX explosive
aluminum powder containing **A7:** 601
dc *See* Direct current (dc)
dc arc source
for optical emission spectroscopy **A10:** 25
DC corona
defined **EM2:** 461
dc intermittent noncapacitive arc
defined **A10:** 671
dc plasma excitation
defined **A10:** 671
dc polarography *See* Polarography
DC surface resistance
defined **EM2:** 461
DC surface resistivity
defined **EM2:** 461
DCAP computer program for structural analysis **EM1:** 268, 270
DCB *See* Double-cantilever beam
DCI
as synchrotron radiation source **A10:** 413
DCP *See* Direct-current plasma
de la Breteque process
of gallium recovery from bauxite **A2:** 742
***De La Pirotechnia* (V. Biringuccio)** **A12:** 1
de Maresquelle, Louis (Louis Ansort)
as early founder **A15:** 26
de-excitation
atomic fluorescence spectrometry **A10:** 46
Auger electron **A10:** 433, 550
by x-ray photon emission **A10:** 433
in neutron activation analysis **A10:** 234
processes, and molecular absorption Jablonsky diagram **A10:** 73
Deactivation **A13:** 4, 603
Dead time
correcting, EDS dot mapping **A10:** 528-529
defined **A10:** 519, 671
Dead time (continued)
practical effect of **A10:** 519-520
wavelength-dispersive spectrometer **A10:** 520
x-ray spectrometry **A10:** 92
Dead weight
as tension source for stress-corrosion cracking **A8:** 502
Dead weight Brinell hardness tester **A8:** 87
Dead zone, calibration
ultrasonic inspection **A17:** 267
Dead-burned
defined **A15:** 4
Dead-burned dolomite *See* Dolomite brick
Dead-load tests
SCC testing **A8:** 502
Dead-metal zones
fracture **A8:** 574
Dead-stop end-point titration *See* Biamperometric titration
Deaerated water
high-temperature pure **A8:** 422-425
Deaerating feedwater heaters
failures of **A11:** 657-658
Deaeration
vacuum **A13:** 1244
Deaerators
corrosion forms of **A13:** 990
Deagglomeration *See also* Agglomeration
Dealloying *See also* Chromium depletion; Dealloying corrosion; Dealuminification; Decarburization Decobtiltization; Denickelification; Dezincification; Graphic corrosion; Selective leaching
alloys/environments subject to **A13:** 130
copper and copper alloys **A2:** 216
copper metals **M2:** 459, 461
defined **A13:** 4
from molten salt corrosion **A13:** 89
material selection for **A13:** 333-334
mechanisms **A13:** 131
in mining/mill applications **A13:** 1295
of noble metal systems **A13:** 133
as SCC mechanism **A12:** 25
of steam surface condensers **A13:** 987
tin effects **A13:** 132
Dealloying corrosion
in aqueous environments **A13:** 131-133
in brazing **A13:** 877
of copper/copper alloys **A13:** 614
evaluation **A13:** 134
as metallurgically influenced corrosion **A13:** 131-134
Dealuminification *See also* Dealloying **A13:** 130, 133
Debinding
in injection molding **A7:** 497
Debond *See also* Delamination; Disbond **EM3:** 10
defined **EM1:** 8 **EM2:** 13
of matrices **EM1:** 31
Debonding *See also* Interface debonding
case/extrusion **A15:** 326
die attach **EL1:** 371
in fiber composites **A11:** 761
interfacial, by axial cracking **EM1:** 195
interlaminar **EM1:** 234
optical holography of **A17:** 405
of precipitates, as acoustic emission source **A17:** 287
ultrasonic inspection of **A17:** 232
Deborah number **A6:** 1138
Deboronization
chemical surface studies **A10:** 177
Debossed
defined **EM2:** 13
Debridging hot air knives **EL1:** 691-693
Debris *See also* Wear debris **A18:** 376
in bearing lubricated systems **A11:** 486
from field fractures **A12:** 92
generation, as fretting **A11:** 148
wear failure from **A11:** 156
Debulking
defined **EM1:** 8 **EM2:** 13
Deburring *See also* Burrs **A16:** 33 **M5:** 151-156, 614-616
abrasive flow machining **A16:** 514, 517, 518-519
aluminum and aluminum alloys barrel finishing process **M5:** 572-574
of blanks, in press forming **A14:** 548
Deburring (continued)
by blast cleaning **A15:** 506
compounds **M5:** 12, 152
and electrochemical machining **A16:** 533
electrolytic *See* Electrolytic deburring
mass finishing processes **M5:** 128, 134, 614-616
media for **M5:** 134
methods **A7:** 458
power brush process **M5:** 151-155
rigid printed wiring boards **EL1:** 544, 548
as secondary operation **A7:** 458-459
and simultaneous cleaning **A7:** 458
of stainless steels **A14:** 762
thermal energy method (TEM) **A16:** 577-578
typical shapes requiring **A7:** 459
Deburring drum
fatigue failure of **A11:** 346-348
Debye rings
defined **A9:** 5
as indicators of crystal orientation **A9:** 701
Debye-Scherrer
camera, for XRPD analysis **A10:** 335
diffracted x-rays, imaging polycrystalline substructure by **A10:** 374
powder method, schematic **A10:** 335
Debye-Scherrer method
defined **A9:** 5
Debye-Waller damping
EXAFS analysis **A10:** 410
Decapping
in integrated circuit failure analysis **A11:** 767
Decapsulation
as coating/encapsulant removal **EL1:** 243
Decarburization *See also* Annealing; Argon oxygen decarburization; Carburization; Dealloying; White zone **A8:** 4 **A13:** 4, 134
allowance, forging **M1:** 369
alloy steels **A12:** 300, 333
and annealing in QMP iron powder process **A7:** 86-87
in annealing of metal powders **A7:** 182-185
argon-oxygen, development **A15:** 36
at elevated-temperature service **A1:** 642-643
in carbon and alloy steels, revealed by macroetching **A9:** 174-175
of carbon steel pipe **A11:** 645
chemical surface studies of **A10:** 177
in coarse-grain pearlitic steel, effects **A11:** 77-78
defined **A9:** 5 **A11:** 3, 122
in Domfer iron powder process **A7:** 90
effect in Rockwell hardness testing **A8:** 83
effect, in VOD process **A15:** 435
effect of hydrogen **A12:** 37-38, 51
effect on fatigue strength **A11:** 122
and fatigue resistance **A1:** 679-680, 681
fatigue resistance, effect on **M1:** 673-674
of fatigue surface **A8:** 373
in forging **A11:** 335-336
high-carbon steels **A12:** 288
hot rolled bars **M1:** 200
in hot-rolled steel bars **A1:** 241
in iron casting **A11:** 361-362
in low-alloy steels **A15:** 428
malleable iron, effect on machinability **M1:** 66
in martensitic stainless steels **A9:** 285
metallographic sectioning **A11:** 24
microhardness testing for **A8:** 83
and notch toughness in wrought steels **A1:** 746
notch toughness of steels effect on **M1:** 705, 707
partial, alloy steels **A12:** 327
in quench cracking **A11:** 94
Reaumur's **A15:** 31
in sintering process **A7:** 340
springs **M1:** 290, 301
stainless steel **A14:** 222
stainless steel pan failure from **A11:** 274
in stainless steels **A15:** 428
steel plate **M1:** 182
and steel plate imperfections **A1:** 230
of steel springs **A1:** 308-309
steel wire rod, limits for **M1:** 254-257
of surface, bending-fatigue fracture from **A11:** 469
surface, powder forged parts **A14:** 204
TiC coating **A16:** 80
tool steels, high-speed **A16:** 55
of tools and dies **A11:** 572

Decarburization (continued)
ultrahigh-strength steels ... **A4:** 208
use of reducing atmosphere ... **A7:** 182
Decarburization limits
for alloy steel rod ... **A1:** 274
for carbon steel rod ... **A1:** 274
Decarburization, magnetic measurement
steels ... **A17:** 134
Decarburized enameling steel
mechanical properties ... **M1:** 178
porcelain enameling of ... **M1:** 178
production ... **M1:** 179
Decarburized steel
porcelain enameling of ... **M5:** 509-510, 512-515, 521
Decarburizing reactions
effects on eta phases of cemented carbides ... **A9:** 275
Decay
constant (λ), defined ... **A10:** 671
kinetics of radical production and ... **A10:** 265-266
spin lattice relaxation time and ... **A10:** 257
Deceleration
atomic number correction for ... **A10:** 524
Deceleration period
defined ... **A18:** 7
Dechanneling
lattice strain measurement by ... **A10:** 634-635
Decibel
abbreviation ... **A8:** 724
Decinary system or diagram ... **A3:** 1 • 2
Decision-tree guide
for leak testing methods ... **A17:** 69
Deck protective coatings
powders used ... **A7:** 572
Deck-type shakeouts
green sand molding ... **A15:** 347
Deckle rod
defined ... **EM2:** 13
Decobaltification *See also* Dealloying; Selective leaching
defined ... **A13:** 4
Decohesion *See also* Decohesive rupture(s); Intergranular decohesion fracture(s) ... **A13:** 72, 164
along grain boundaries, elongated grains ... **A12:** 23
along grain boundaries, equiaxed grains ... **A12:** 23
austenitic stainless steels ... **A12:** 364
glide-plane ... **A12:** 391
intergranular, band formation by hydrogen ... **A12:** 37, 51
intergranular, effect of strain rate ... **A12:** 41
intergranular, SCC fracture by ... **A12:** 27
nodule, ductile irons ... **A12:** 236
particle-mix ... **A11:** 82-83
tearing, in welds ... **A12:** 139
through weak grain-boundary phase ... **A12:** 23
Decohesive rupture(s) *See also* Decohesion; Grain-boundary separation; Intergranular brittle fracture
by SEM imaging ... **A12:** 174-175
creep rupture ... **A12:** 18-20
defined ... **A12:** 18-20
intergranular, steels ... **A12:** 31
materials illustrated in ... **A12:** 217
mechanisms ... **A12:** 18
in precipitation-hardenable stainless steel ... **A12:** 18, 24
Decomposition
cellular, image analysis kinetic analysis ... **A10:** 316-318
defined ... **A7:** 3
Freiberger ... **A10:** 167
kinetics of reactions ... **EM4:** 110
polyester resin, at elevated temperatures ... **EM1:** 93
potential ... **A10:** 198, 201
of sample, in Raman analyses ... **A10:** 130
Decomposition (continued)
spinodal ... **A10:** 583, 593
of thermoplastics ... **EM1:** 101
Decomposition of organic precursors
of ferrites ... **EM4:** 1163
Decomposition potential
defined ... **A13:** 4
Decomposition, thermal
of polymers ... **EM2:** 60
Decomposition voltage *See* Decomposition potential
Deconvolution
based on multiplet splitting ... **A10:** 573
Decorating ... **EM4:** 471-475
direct methods ... **EM4:** 471-472
acid etching ... **EM4:** 472
chemical frosting ... **EM4:** 472
curtain and roll coating ... **EM4:** 471
laser etching ... **EM4:** 472
sand blasting ... **EM4:** 472
sand carving ... **EM4:** 472
spraying process ... **EM4:** 471
water-jet cutting ... **EM4:** 472
history of the art ... **EM4:** 471
indirect methods ... **EM4:** 471, 472-474
decalomania (decals) ... **EM4:** 473
double offset ... **EM4:** 473-474
offset screen printing ... **EM4:** 474
silicone pad printers ... **EM4:** 472-473, 475
materials ... **EM4:** 474-475
glass enamels ... **EM4:** 474
lusters ... **EM4:** 474
media ... **EM4:** 474-475
metallics ... **EM4:** 474
organic colors and processes ... **EM4:** 475
stains ... **EM4:** 474
photosensitive glass ... **EM4:** 474
process comparison for multicolored designs ... **EM4:** 474
processes ... **EM4:** 471
rubber stamping ... **EM4:** 472
screen printing ... **EM4:** 472
stenciling ... **EM4:** 472
semi-direct methods ... **EM4:** 472
flexography ... **EM4:** 472
Decoration aging technique, for studying recrystallization in beta titanium
alloys ... **A9:** 461, 475
Decoration of dislocations
defined ... **A9:** 5
Decorative applications
by coining ... **A14:** 180
copper and copper alloys ... **A14:** 822
Decorative brass plating ... **M5:** 285
Decorative chromium plating ... **M5:** 188-198
adhesion of ... **M5:** 188
anodes ... **M5:** 191-194
auxiliary ... **M5:** 192-194
bipolar ... **M5:** 194
nonconforming ... **M5:** 192
appearance ... **M5:** 198
applications ... **M5:** 188, 190-195
catalysts in ... **M5:** 189-190
chromic acid concentration ... **M5:** 189-191
chromic acid to sulfate ratio ... **M5:** 189, 191
chromic anhydride to sulfate ratio ... **M5:** 189-191
copper plating and copper striking processes ... **M5:** 193-194
copper-nickel-chromium combinations ... **M5:** 193
corrosion resistance and testing ... **M5:** 189, 198
cracking of ... **M5:** 188-189, 197
crack pattern ... **M5:** 189
current density ... **M5:** 189-191, 194, 196
current distribution control of ... **M5:** 192-193
current shields ... **M5:** 192
design-cost requirements ... **M5:** 194-197
die castings ... **M5:** 189-192, 194-197
difficult-to-plate parts ... **M5:** 192-193, 196
ductility ... **M5:** 198
durability ... **M5:** 198
Decorative chromium plating (continued)
equipment ... **M5:** 192-195
maintenance of ... **M5:** 195
hard chromium plating differing from ... **M5:** 188
hexavalent process ... **M5:** 188-189
leveling ... **M5:** 198
magnesium alloys ... **M5:** 646
maintenance schedules ... **M5:** 195
metallic impurities, effects of ... **M5:** 189
microcracked *See* Microcracked chromium plating
mischromes, causes and correction ... **M5:** 190, 195
mixed catalyst baths, compositions and operating conditions ... **M5:** 190
nickel plating process ... **M5:** 193-194
physical properties ... **M5:** 188-189
plating time ... **M5:** 190-196
costs affected by ... **M5:** 195-196
porosity of ... **M5:** 188-89, 197
problems and corrections ... **M5:** 194-195, 197-198
racks ... **M5:** 192-193
selective ... **M5:** 194
solution compositions and operating conditions ... **M5:** 189-191
solution control ... **M5:** 191
steel ... **M5:** 189-191, 196-197
stop-offs ... **M5:** 194
striking process ... **M5:** 192-194
stripping of ... **M5:** 198
tanks ... **M5:** 194
temperature ... **M5:** 190-191, 194-196
heating ... **M5:** 194-195
thickness ... **M5:** 188-192, 194, 198
porosity affected by ... **M5:** 188-189
thieves or robbers ... **M5:** 192
trivalent process ... **M5:** 196-198
zinc ... **M5:** 189-191, 194-195
Decorative coatings
cast iron ... **M1:** 102, 104, 105
Decorative gold plating ... **M5:** 284
Decorative nickel plating ... **M5:** 199, 205, 218
Decorative paints
powders used ... **A7:** 572
Decorative parts
malleable iron for ... **M1:** 9
Decorative rhodium plating ... **M5:** 290-291
Decorative vacuum coating ... **M5:** 392-394, 396-403, 407-408
Decoring *See* Core knockout
Decoupled modes
propagation of ... **EL1:** 36-37
Decoupling capacitors
placement ... **EL1:** 28
Decrepitation ... **EM4:** 379
Dedicated SIMS instrument
for probe imaging ... **A10:** 613
Dedicated test heads
for quality control ... **EL1:** 873
Dedusted aluminum powders ... **A7:** 125
Deep drawability
Swift cup test for ... **A8:** 562-563
Deep drawing *See also* Drawability; Drawing ... **A14:** 575-590
aluminum alloys ... **A14:** 519, 795-797
of beryllium ... **A14:** 806-807
beryllium-copper alloys ... **A2:** 411
of boxlike shells ... **A14:** 585
cleaning, of workpieces ... **A14:** 589
combined operations ... **A14:** 510
of copper and copper alloys ... **A14:** 519, 816-817
defined ... **A14:** 3, 575
developed blanks vs final trimming ... **A14:** 589
dies ... **A14:** 579-580
dimensional accuracy ... **A14:** 589-590
direct redrawing ... **A14:** 584-585
drawability ... **A14:** 576-577
drawing fundamentals ... **A14:** 476
finite-element analysis programs ... **A14:** 923-924
with fluid-forming presses ... **A14:** 587
formability, magnesium alloys ... **A2:** 468-469

Deep drawing (continued)
of hemispheres ... **A14:** 586
lubricants for ... **A14:** 519-520
magnesium ... **M2:** 543, 545
of magnesium alloys ... **A14:** 827-828
maraging steels ... **M1:** 447
materials for ... **A14:** 583-584
multiple-step ... **A14:** 816-817
of nickel-base alloys ... **A14:** 520, 833
organic-coated steels ... **A14:** 565
presses ... **A14:** 577-579
of pressure vessels ... **A14:** 587
process, defined ... **A14:** 508
process variables, effects ... **A14:** 580-583
redrawing, tooling for ... **A14:** 585
reducing drawn shells ... **A14:** 586
reverse redrawing ... **A14:** 585
safety ... **A14:** 590
of stainless steels ... **A14:** 519, 759, 767-771
strain rates, of beryllium ... **A14:** 807
of superplastic alloys ... **A14:** 857, 859
technique, ASEA-Quintus ... **A14:** 615
and three-roll forming ... **A14:** 623
tin/terne-coated steels, and press forming ... **A14:** 563
of titanium alloys ... **A14:** 520, 844-845
and trimming ... **A14:** 588-589
workpiece ejection ... **A14:** 587-588
workpieces, expanding ... **A14:** 586-587
of workpieces with flanges ... **A14:** 585-586
zinc alloys, properties ... **A2:** 539
Deep drawing dies
cemented carbide die parts ... **M3:** 499
chromium plating ... **M3:** 498
draw rings ... **M3:** 495-497
galling ... **M3:** 496, 497, 498
lubrication ... **M3:** 494, 495
plastic die parts ... **M3:** 499
punches ... **M3:** 497, 498, 499
tools for ... **M3:** 494, 495
wear ... **M3:** 496, 497-498, 499
zinc alloy die parts ... **M3:** 499
Deep drawing zinc *See* Zinc alloys, specific types, commercial rolled zincs
Deep etching *See also* Macroetching
copper and copper alloys ... **A9:** 399
defined ... **A9:** 5
Deep eutectic alloys
as metallic glasses ... **A2:** 805
Deep groundbed *See also* Groundbed
defined ... **A13:** 4
Deep-draw mold
defined ... **EM1:** 8 **EM2:** 13
Deep-drawing dies
materials for ... **A14:** 508-511
performance ... **A14:** 508
service problems ... **A14:** 509-511
Deep-drawn steel sheet
Lüders lines in ... **A9:** 687
Deep-field microscopy
of fractures ... **A12:** 96
Deep-groove ball bearings
applicable load ... **A18:** 511
f_v factors for lubrication method ... **A18:** 511
y and z factors ... **A18:** 511
Deep-lying discontinuities
magnetic particle inspection ... **A17:** 105
Deep-penetration electron-beam and laser welding
fluid flow in the keyhole ... **A6:** 22
instability in keyhole fluid flow ... **A6:** 22-23
keyhole formation ... **A6:** 22
Deep-penetration-mode welding ... **A6:** 264-265, 266
Deep-submergence vessel float structure
magnetic particle inspection ... **A17:** 114
Deer hair
analysis of sulfur in ... **A10:** 224
Defect *See also* Weld defects
in cold extrusion of aluminum alloys ... **A8:** 591-592, 595
definition ... **A6:** 1081, 1208 **M6:** 5
detection by Marciniak biaxial stretching test ... **A8:** 558
in explosion welds ... **M6:** 707-708
as flow localization source ... **A8:** 170
in rolling ... **A8:** 593-595
Defect analysis *See also* Defects; Defects, analytical methods
analytical transmission electron microscopy ... **A10:** 464-468
by nuclear magnetic resonance ... **A10:** 277
by SAXS/SANS/SAS ... **A10:** 402
by scanning electron microscopy ... **A10:** 490
electron spin resonance for ... **A10:** 254
experimental parameters ... **A10:** 464-465
of inorganic solids, applicable analytical methods ... **A10:** 4-6
one-dimensional ... **A10:** 465-466
three-dimensional ... **A10:** 466
two-dimensional ... **A10:** 466
weak-beam microscopy ... **A10:** 466-467
Defect and fault tolerance
system-level ... **EL1:** 377
Defect centers
ESR detected ... **A10:** 253
Defect evaluation systems
automated ... **A17:** 320-321
Defect growth of graphite
theory of ... **A15:** 177
Defect imaging ... **A10:** 367-368, 370
Defect leakage fields
origin ... **A17:** 129
Defect verification process ... **EL1:** 872-874
Defect(s) *See also* Casting defects; Inclusion-forming reactions; Inclusions; Presolidification, specific defect types
analysis, solderability ... **EL1:** 1035
automatic detection, solder joints ... **EL1:** 739
burn-in ... **A15:** 208-209
in centrifugal casting ... **A15:** 306-307
classification, international ... **A15:** 545-553
cold lap, defined *See* Cold lap
cold shuts ... **A15:** 294
cope-side ... **A15:** 94
as crack initiation sites ... **A13:** 148
defined ... **A15:** 4
in die castings ... **A15:** 294-295
in directional solidification ... **A15:** 319-321
drilling ... **EL1:** 870, 1021
etching ... **EL1:** 1019
expansion ... **A15:** 346
and fault tolerance ... **EL1:** 377
flow lines ... **A15:** 337
flux, from low specific gravity ... **EL1:** 684
freckles ... **A15:** 407
in gray iron ... **A15:** 640-642
hard, detecting ... **EL1:** 568
heat check fins ... **A15:** 295
importance, electronic materials ... **EL1:** 93
ingot ... **A15:** 404, 407
joint, and solder impurities ... **EL1:** 642
laminate ... **EL1:** 1022
latent, defined ... **EL1:** 244, 867-868
lustrous carbon ... **A15:** 270
magnesium castings ... **A15:** 809-810
misoriented grains ... **A15:** 321
mold wall movement ... **A15:** 29
near, detecting ... **EL1:** 568
non-fill ... **A15:** 336-337
permanent mold casting ... **A15:** 285
plating ... **EL1:** 1027
polycrystalline, effects ... **EL1:** 93
porosity ... **A15:** 325
raining ... **A15:** 306-307
registration ... **EL1:** 1022
sand expansion ... **A15:** 210
segregation banding ... **A15:** 306
in semisolid casting and forging ... **A15:** 336-337
shrinkage ... **A15:** 321
shrinkage porosity ... **A15:** 294
solder joint ... **EL1:** 738-739
in solder joint inspection ... **EL1:** 735
soldering ... **A15:** 294-295
soldering, types ... **EL1:** 1034
structure, oxides ... **A13:** 65-66
surface, hot isostatic pressing effects ... **A15:** 542
surface, investment castings ... **A15:** 264
testing and inspection of ... **A15:** 544-561
tree ring patterns ... **A15:** 407
verification process ... **EL1:** 872-874
vibration ... **A15:** 307
welding of ... **A15:** 529-530
white spots ... **A15:** 407
Defect(s) (continued)
zirconium castings, weld repair ... **A15:** 838
Defective
defined ... **A15:** 4
Defective coating
as casting defect ... **A11:** 385
Defective surfaces
as casting defects ... **A11:** 383-385 **A17:** 515-517
Defective weld
definition ... **M6:** 5
Defects *See also* Bend; Casting defects; Center defects; Cracks; Defect analysis; Defects, analytical methods; Discontinuities; Edge defects; Extrusion pipe; Flaw detection Fraction defective; Flaws; Forging defects; Inclusions; Interior defects; Internal defects; Material flaws; Nonmetallic inclusions Subsurface cracks; Pipe; Postforging defects; specific defects; Subsurface defects; Surface cracks; Surface defects Weldments; Twist; Weld defects ... **EM1:** 261
accumulations of, characterized by x-ray topography ... **A10:** 365
in adhesive-bonded joints ... **A17:** 610-616
alligatoring ... **A14:** 358
blister ... **A14:** 358-359
blow holes ... **A14:** 358
buried distribution of ... **A10:** 632
burrs ... **A17:** 13
-caused brittle fracture ... **A11:** 85
causing pipeline failures ... **A11:** 697
center, cold-formed parts ... **A11:** 307
centers of ... **A10:** 253
in crystalline solids, ESR detection of ... **A10:** 253
defined ... **A17:** 49, 103
and details, of bridge components ... **A11:** 707
distribution depth profile, in single crystal ... **A10:** 511
effect on fatigue cracking ... **A11:** 102
electrical breakdown from ... **EM2:** 465
in extrusion ... **EM2:** 386
FIM images, in pure metals ... **A10:** 588-589
finning ... **A14:** 359
fir tree ... **A14:** 403
fishtail ... **A14:** 359
forging ... **A11:** 317, 327-331 **A14:** 385-386
from postforging processes ... **A11:** 331-333
imaging of ... **A10:** 367-368
in imperfect crystals ... **A10:** 365
initial, and crack growth analysis ... **A11:** 57
interfacial, imaged by x-ray topography ... **A10:** 365
large, in bridge components ... **A11:** 708-710
longitudinal weld, in pipe ... **A11:** 704
macrodefects, titanium alloy forgings ... **A17:** 498
materials, in cold-formed parts ... **A11:** 307
near-surface, in single crystals ... **A10:** 633
number of, c-chart for ... **A17:** 736-737
number, per unit, u-chart for ... **A17:** 737
observable using optical microscopy ... **A10:** 307
overlap ... **A14:** 359
pipe ... **A14:** 358
in pipe body ... **A11:** 697, 699-701
point ... **A10:** 358, 556-559, 583, 588
in powder metallurgy parts ... **A17:** 536-537
reference standard unavailability ... **A17:** 676
-related dielectric breakdown, integrated circuits ... **A11:** 778-779
in rolling ... **A14:** 358-359
scale ... **A14:** 358-359
seams ... **A14:** 358
size, by NDE methods ... **A17:** 677
slivers ... **A14:** 358
spiking ... **A11:** 350-352
strain-induced, as fine structure effect ... **A10:** 438, 440
subsurface ... **A7:** 576, 578 **A11:** 330-331
subsurface, from welding ... **A11:** 127
surface, cold-formed parts ... **A11:** 307
surface, from welding ... **A11:** 127
in tantalum and tantalum alloys ... **A14:** 238
weld, fatigue failure from ... **A11:** 127-128
welding ... **A11:** 400
Defects, analytical methods *See also* Defect analysis; Defects
analytical transmission electron microscopy ... **A10:** 429-489
electron spin resonance ... **A10:** 253-266

Defects, analytical methods (continued)
field ion microscopy **A10:** 583-602
Rutherford backscattering spectrometry **A10:** 628-636
scanning electron microscopy **A10:** 490-515
x-ray powder diffraction **A10:** 325-332
x-ray topography **A10:** 365-379
Defects and distortion in heat-treated parts *See* Tool steels, defects and distortion in heat-treated parts
Defects, bubbles
and chemical integrity of seals **EM4:** 541
Defense Advanced Research Projects Agency (DARPA) **EM4:** 716
gallium research **A2:** 747
Defense, Department of (DoD)
adopted standards **EM3:** 61, 62
Index of Specifications and Standards (DoDISS) **EM3:** 63-64
Defense standardization and specification program **EM2:** 89
Defensive missile systems
composite components **EM1:** 816-817
Deficiency cones
in divergent-beam topographs **A10:** 371
Define (x-rays)
defined **A9:** 5
Definition
defined **A9:** 5
Definitions *See also* Glossary of terms; Nomenclature
glossary of **EL1:** 1133-1162
of high-level model **EL1:** 129
magnetic particle inspection **A17:** 103
operational, control charts **A17:** 734
of quality **A17:** 720
of samples, statistical **A17:** 728-730
and terms **A8:** 1-15
Deflagration explosives **A7:** 600
Deflashing
defined **EM1:** 8 **EM2:** 13
of molded plastic packages **EL1:** 475
Deflection
autographic recording of **A8:** 58
in cantilever beam bend test **A8:** 133
cathetometer for measuring **A8:** 134
defined **A14:** 3
deflectometer for measuring **A8:** 134
machine, in tube spinning **A14:** 678
in three- and four-point bend tests **A8:** 134
vs. load data, three- and four-point bending **A8:** 135-136
vs. strain **A8:** 58
Deflection analysis
in tooling design **A7:** 336-337
Deflection model
crack path tortuosity by **A12:** 206
Deflection temperature under load
defined **EM1:** 8
Deflection temperature under load (DTUL) **EM3:** 10
defined **EM2:** 13
Deflectometer
for measuring deflection **A8:** 134
Deflocculants
as electrolytes for ceramic coatings **EM4:** 955
Deflocculating agents
for slurry in spray drying **A7:** 75
Defluxing
of surface-mount soldering **EL1:** 707-708
Defoamants
for lubricant failures **A11:** 154
Defoamers
as aqueous cleaners **EL1:** 665
for metalworking lubricants **A18:** 141-142, 144, 147
Deformability *See* Conformability
Deformation *See also* Bending; Cold work, Forming; Deformation mechanism maps; Deformation processing maps; Deformation rate; Elastic deformation; Localized deformation; Necking; Plastic deformation; Viscous deformation **A8:** 4 **A13:** 72, 137
in A15 superconductor assembly **A2:** 1067
acoustic emission inspection of **A17:** 286-289
amorphous materials and metallic glasses **A2:** 813
analysis of **A10:** 468-470
and annealing structures **A9:** 602-603
austenitic bar, martensitic strain-induced transformation from **A8:** 481-482
in austenitic manganese steel castings **A9:** 237
bands, butterflies as **A12:** 115, 134
bands, defined **A11:** 3
bearing, as fastener failure **A11:** 531
bending **M1:** 587-588
bending, in sheet metal forming **A8:** 547
blue brittleness, effect on **M1:** 684
by forging machines **A14:** 25
by liquid erosion **A11:** 164-166
by slip **A8:** 34, 188
by stress wave **A8:** 40, 44
caused by sectioning in carbon and alloy steels **A9:** 166
of Ceracon preforms **A7:** 538
characteristics **EM2:** 58
Charpy impact testing of **A8:** 262
cold extrusion **M1:** 591-592
cold heading **M1:** 589-592
cold, substructure due to **A10:** 468-469
in composite structural analysis **EM1:** 459-460
compressive, cam plastometer **A8:** 194
contours, symmetric rod impact test **A8:** 206
cooling importance **A8:** 178
copper and copper alloys **A2:** 219
creep, and intergranular cracking **A11:** 29
and creep life **A8:** 344
crystallographic texture of **A10:** 357-364
cyclic, in fcc metals **A8:** 256
defined **A7:** 3 **A11:** 3 **EM2:** 412
of dendritic pattern in steel, effect of hot rolling on **A9:** 626-627
depth of, in different metals due to cutting method **A9:** 23
differences in titanium alloys **A8:** 583-584
double-barreled, in rolling **A8:** 593-594
and ductility **A7:** 298
during cold upset testing **A8:** 579-580
dynamic recovery and recrystallization in single-phase materials **A8:** 173-177
effect on ductility **A8:** 164-165
effect on green strength **A7:** 288-289
elastic, defined *See* Elastic deformation
elastic, magnetic printing detection **A17:** 126
elastic/elastoplastic/ thermoelastic **EL1:** 55
end and side, in rolling **A8:** 593, 595
as equal to crosshead displacement **A8:** 41
failure modes in **A8:** 573-574
fatigue fracture of steering knuckle by **A11:** 342
of fiber composites **A11:** 761
fields, in fracture mechanics **A8:** 465
flow localization alpha parameter application to **A8:** 171-172
flow, medium-carbon steels **A12:** 258
forging **A7:** 410-411, 413
and fracture behavior **A10:** 365, 376-378
with fracture, iron alloy **A12:** 458-459
and fracture mechanisms, of polymers **A11:** 758-761
friction during **A8:** 575-576
galling **A8:** 576
in gas cutting **A14:** 725
of gas-nitrided drive-gear assembly **A11:** 142-143
geometry for plastic straining of hollow cylinder **A8:** 143
geometry, pure plastic bending **A8:** 120, 122
grain boundary sliding **A8:** 188
grain-boundary void and crack formation at hot and warm temperatures **A8:** 572
hardness as resistance to **A8:** 71
heat transfer, and compression flow stress **A8:** 171
heating, in torsion testing **A8:** 161
hot, substructure due to **A10:** 469-470
hot, wrought titanium alloys **A2:** 612
impact forces **A7:** 60
in-plane, speckle metrology measurement **A17:** 432-434
inhomogeneity **A8:** 573
kinematic analysis **A12:** 166
limit, defined **A14:** 3
localized, tests for **A14:** 384-385
long-term, polyether sulfones (PES PESV) **EM2:** 160-161
in loose powder compaction **A7:** 298
measurement of **A14:** 878-879
measurement, sheet metal forming **A8:** 548-549
measuring restraining force due to **A8:** 567
mechanisms **A9:** 602
mechanisms, process modeling **A14:** 420-421
microstructure development during **A8:** 173-178
mode change due to strain rate **A8:** 188
mode I **A11:** 60-61
modeling, of open-die forging **A14:** 65-67
nonhomogeneous **A8:** 44
nonlinear, in progressive fracturing **A8:** 440
in nonlubricated, nonisothermal hot forging **A8:** 582
opening mode of *See* Stress-intensity factor
in ordered alloys **A2:** 913-914
out-of-plane, speckle metrology measurement **A17:** 434
patterns, nonisothermal forging **A14:** 375
in pin bearing testing **A8:** 59
plane stress **A8:** 576
plastic, effect on fringe patterns **A10:** 368
plastic, magnetic printing detection **A17:** 126
polymer, loading rate effects **EM2:** 680-681
prediction, under load **EM2:** 673-678
-processed copper-refractory metal composites **A2:** 922
processes, high-rate, examples **A8:** 190
processes, primary and secondary **A7:** 522
processing, ductile fracture in **A14:** 363-364
processing of preforms **A7:** 531
rate, and yield strength, in niobium **A8:** 38, 39
rates, ASTM specifications for **A8:** 39-40
recovery, and recrystallization structure analysis **A10:** 468-470
redundant work of **A14:** 331
resistance, complexing of cermet compositions for **A2:** 991
resistance, universal testing machine to determine **A8:** 612
resistance, vs temperature, carbon/ alloy steels **A14:** 216
of roller and ball bearing raceway **A11:** 510
seizing **A8:** 575-576
simple shear, inclination effect of mean burgers vector on length change **A8:** 180-181
of single crystals in shear, Kolsky bar **A8:** 219
specimen temperature during **A8:** 191
squeezing, rolling as **A14:** 343
steel wire **M1:** 587-593
strain **A8:** 574-575
strain rate control **A8:** 178
strain rate influence on **A8:** 188
and stress, defined **A8:** 308
stress states **A8:** 575
swaging **M1:** 593
temperature **A8:** 188, 575
temperature, iron powder preforms **A14:** 194

Deformation (continued)
-temperature-time schedule and flow behavior of aluminum in torsion ... **A8:** 178-179
-temperature-time sequence, for microalloyed steels ... **A8:** 179
and tensile stress ... **A8:** 168-169
texture, due to mechanical processing, PST for ... **A10:** 374
thermoplastic composites ... **A18:** 821
time-dependent, in pressure vessels ... **A11:** 666-668
titanium aluminides ... **A2:** 926-928
torsional ... **A8:** 139, 179
in tungsten alloys, effect of doping on ... **A9:** 442
twins, in threshold regime Inconel ... **A8:** 718, 488
twist reversal on aluminum after strain and ... **A8:** 174
uniform ... **A8:** 305
under horizontal tension and vertical compression ... **A14:** 389
under tension, effect of crystal structure ... **A8:** 34
wedge, simulation of ... **A14:** 429
work of ... **A8:** 191
workpiece, rotary forging ... **A14:** 178
zone parameter, in rolling ... **A8:** 593
Deformation bands
as a result of tensile load ... **A9:** 684
in Co-8Fe single crystal deformed 44% ... **A9:** 689
defined ... **A8:** 4 **A9:** 5, 685, 687
Deformation bonding ... **A6:** 145
aluminum alloys ... **A6:** 157
Deformation cells
effect of recovery on ... **A9:** 694
Deformation curve *See* Stress-strain diagram
Deformation lines
defined ... **A9:** 5
Deformation maps
Ashby-Frost ... **A14:** 421
damage nucleation ... **A14:** 421
process modeling of ... **A14:** 420-421
Rao-Raj ... **A14:** 421
Deformation mechanism maps ... **A8:** 310
Deformation processing
failure modes ... **A14:** 365-366
fracture mechanisms ... **A14:** 363-364
map ... **A14:** 365, 370-371
Deformation processing map ... **A8:** 154
Deformation rate
aluminum alloy, and forgeability ... **A14:** 242
ASTM specifications for ... **A8:** 39-40
effect on strength, quantifying ... **A8:** 39-40
and fracture mechanisms ... **A8:** 154
and strain rate, defined ... **A8:** 44
titanium alloys ... **A14:** 269-270
Deformation strength
steel wire fabrication ... **M1:** 592
Deformation studies by scanning electron microscopy ... **A9:** 100-101
Deformation texture ... **A9:** 700-701
Deformation theory of friction (Bikerman) ... **A18:** 46
Deformation twinning
in beryllium as a result of grinding ... **A9:** 389
Deformation twins *See also* Twinning ... **A9:** 686-688
in 70-30 brass, strain to produce ... **A9:** 685
in cast bismuth ... **A9:** 59
in metals with noncubic crystal structure ... **A9:** 37-38
in silver ... **A9:** 554-555
in zinc ... **A9:** 37-38
Deformation under load *See also* Cold flow; Creep ... **EM3:** 10
defined ... **EM1:** 8 **EM2:** 13
Deformation wear
defined ... **A18:** 7
Deformation welding
applications ... **M6:** 686-691
cold welding of aluminum tubing ... **M6:** 691
metal cladding by strip roll welding ... **M6:** 689-691
roll welded heat exchangers ... **M6:** 689
seal welds ... **M6:** 686-689
cold welding ... **M6:** 673-674
butt welding ... **M6:** 673-674
lap welding ... **M6:** 673
Deformation welding (continued)
slide welding ... **M6:** 674
differentiation from diffusion welding ... **M6:** 672
extrusion welding ... **M6:** 676-677
mating of surfaces ... **M6:** 678
pressure ... **M6:** 679
solid-state sintering ... **M6:** 678-679
surface contaminants ... **M6:** 678-679
surface extension ... **M6:** 678
surface roughness ... **M6:** 679
thennocompression welding ... **M6:** 674-675
Deformed grain structure
formation of ... **A9:** 603
Deformed metals
polarized light used to examine ... **A9:** 79
Deformed mold, as casting
defect ... **A11:** 387
Deformed pattern
as casting defect ... **A11:** 387
Deformed state ... **A9:** 692-693
polycrystalline specimens ... **A9:** 693
Degasification *See* Degassing
Degasifier
defined ... **A15:** 4
Degassing *See also* Breathing; Degassing processes; Fluxes; Gases; Outgassing; Vacuum degassing ... **EM3:** 10
in CAP process ... **A7:** 533
and capsule filling station ... **A7:** 431, 433
of copper alloys ... **A15:** 86, 467-468
defined ... **A7:** 3 **A15:** 4
end point for ... **A7:** 434
flux injection ... **A15:** 453
hexachloroethane ... **A15:** 460, 462-463
of high-performance aluminum powder ... **A7:** 526
of Invar ... **A2:** 890
metal-matrix composites ... **A15:** 848-849
as powder cleaning ... **A7:** 180-181
powder, high-strength aluminum P/M alloys ... **A2:** 202-203
solid ... **A15:** 467-468
of vacuum induction furnace ... **A15:** 395
in vacuum melting ultrapurification ... **A2:** 1094
Degassing for purifying metals ... **M2:** 710
Degassing processes *See also* Degassing; Gases; Secondary metallurgy
for aluminum alloys ... **A15:** 456-462
compared ... **A15:** 439, 440
converter metallurgy ... **A15:** 426-431
for copper alloys ... **A15:** 464-468
defined ... **A15:** 426
effect on modification ... **A15:** 485-486
gases used ... **A15:** 426
hexachloroethane ... **A15:** 460, 462-463
hydrogen removal ... **A15:** 459-462
ladle metallurgy ... **A15:** 432-444
for magnesium ... **A15:** 462-464
oxidation-deoxidation, copper alloys ... **A15:** 466
porous plug ... **A15:** 461-462
rotary ... **A15:** 460-461
vacuum oxygen decarburization, ladle ... **A15:** 434-435
Degradation ... **EM3:** 10
abrasive, of glass filaments ... **EM1:** 45
in atmosphere ... **A13:** 204
by light and heat, of polymers ... **A11:** 761
cavitation erosion ... **A13:** 142
corrosion fatigue ... **A13:** 142-144
defined ... **EM1:** 8 **EM2:** 13
detection ... **EM2:** 574
erosion ... **A13:** 136-138
in flow properties, as hydrogen damage ... **A13:** 164
fretting corrosion ... **A13:** 138-140
fretting fatigue ... **A13:** 141
hydrogen, classification ... **A13:** 165
measuring ... **A13:** 195
mechanical ... **EM2:** 424
mechanically assisted, types ... **A13:** 79
microbial ... **EM2:** 424, 783-787
of orthopedic implants ... **A11:** 672, 689-692
photolytic ... **EM2:** 776-782
photooxidative ... **EM2:** 573
of polymers, Raman analyses ... **A10:** 131
preferential dissolution as ... **A13:** 17
radiation ... **EM2:** 424
Degradation (continued)
and structure ... **A10:** 285-286
thermal ... **EM2:** 423-424, 568-570, 573
thermal, of thermoplastic matrices ... **EM1:** 33
thermal oxidative ... **EM2:** 573
types ... **EM2:** 424
water drop impingement ... **A13:** 142
WSI system ... **EL1:** 9
Degradation of microstructures due to morphological changes, scanning electron
microscopy used to study ... **A9:** 101
Degrease ... **EM3:** 10
Degreasing *See* Grease, removal of; Vapor degreasing
for thermal spray coatings ... **A13:** 460
ultrasonic cleaning operation ... **A7:** 459
Degree of cure *See also* Cure; Curing
for close-loop cure ... **EM1:** 761
epoxy resin lamination ... **EM1:** 74-75
as quality-control variable ... **EM1:** 730
solder masks ... **EL1:** 554
Degree of hardness (for water)
symbol for ... **A11:** 798
Degree of overbasing ... **A18:** 100
Degree of polymerization ... **EM3:** 10
defined ... **EM1:** 8 **EM2:** 13
Degree of saturation ... **EM3:** 10
defined ... **EM2:** 13
Degree of wear ... **A18:** 431
Degrees of freedom ... **A3:** 1 • 2
abbreviation ... **A8:** 724
defined ... **A8:** 625-626 **A9:** 5
percentiles ... **A8:** 707, 709
percentiles of the *t* distribution for ... **A8:** 655
use in determining design allowables ... **A8:** 665, 668, 672-677
Degussa OX-50 ... **EM4:** 447, 449
Degussit ... **A16:** 98
Dehydrated Kaolin ... **EM3:** 175
Dehydration
Antioch process ... **A15:** 246
Delamination *See also* Debond; Disbond; Interlaminar fractures ... **A8:** 714 **EL1:** 1003, 1023-1024 **EM3:** 10
at laminate edges ... **EM1:** 230
at rivet hole ... **A11:** 551-553
of coatings during specimen preparation of sheet steels ... **A9:** 197-198
in composite compression fracture ... **A11:** 739
defined ... **EM1:** 8 **EM2:** 13
detection of ... **A18:** 421
failure analysis case histories ... **EM2:** 817-822
as fatigue failure ... **EM1:** 437-438
in fiber composites, as a result of manufacture ... **A9:** 591
in fiber composites, as a result of specimen mounting ... **A9:** 588
preparation, prevention during
free edge, interlaminar cracking by ... **EM1:** 241-242
growth ... **EM1:** 784
as impact damage ... **EM1:** 259-260
interfaces, laminate ... **EM1:** 242
and interlocking, translaminar fracture ... **EM1:** 792
internal, by machining and drilling ... **EM1:** 668
internal, damage tolerances ... **EM1:** 262
mechanically induced artificial ... **EM3:** 522-523
as out-of-plane failure mode ... **EM1:** 781, 783-784
pipe-wall, hook crack in ... **A11:** 448
resistance, as damage tolerance property ... **EM1:** 99
resistance, testing for ... **EM1:** 264
surface, by cutting ... **EM1:** 667
theory, of wear ... **A11:** 148
wear, under lubrication ... **A11:** 150
Delamination test ... **A18:** 404
Delamination wear
defined ... **A18:** 7
Delaminations *See also* Splitting
by computed tomography (CT) ... **A17:** 361
causes for ... **A12:** 105
defined ... **A12:** 104
microwave inspection ... **A17:** 202, 212
superalloys ... **A12:** 393
thermal inspection ... **A17:** 402
ultrasonic inspection ... **A17:** 250
wrought aluminum alloys ... **A12:** 418
Delay cartridges ... **A7:** 600, 602, 603

Delay formulations
burning times ... **A7:** 603-604
Delay intrinsic device ... **EL1:** 2
RLC lines ... **EL1:** 359-361
signal, and local physical performance ... **EL1:** 5
speed-of-light, as communication limit ... **EL1:** 2
time, interconnections ... **EL1:** 20
total, defined ... **EL1:** 2
transmission line ... **EL1:** 6
Delay lines, computers
powders used ... **A7:** 573
Delay-tip contact-type ultrasonic search unit ... **A17:** 258
Delayed cracking *See also* Crack propagation; Cracking ... **A6:** 410
ductile and brittle fracture from ... **A11:** 95
hydrogen-induced, in bolts ... **A11:** 249
quench ... **A11:** 95-96, 122
uranium alloys ... **A2:** 675-676
Delayed failure *See also* Hydrogen embrittlement ... **EM3:** 353-355
by solid-metal embrittlement ... **A13:** 185-187
of electroplated steel bolt, hydrogen embrittlement ... **A11:** 539-540
hydrogen effect ... **A13:** 329-330, 535-536
in hydrogen-embrittled steels ... **A11:** 28-29
in nickel-base alloys ... **A13:** 650-652
in SMIE and LME systems ... **A11:** 242-244
Delayed fluorescence
in molecular fluorescence spectroscopy ... **A10:** 73
Delayed fracture
martensitic low-carbon steel ... **A12:** 248
Delayed hydride cracking (DHC) ... **A6:** 788
Delayed solidification
in iron casting ... **A11:** 350
Delayed-neutron counting ... **A10:** 238, 690
Delayed-tack adhesives ... **EM3:** 75
Deleading failure
copper-lead alloy bearings ... **A11:** 488
Delineating etchants for heat-resistant casting alloys ... **A9:** 330-332
Deliquescence ... **EM3:** 10
defined ... **EM2:** 13
DeLong diagrams ... **A6:** 82, 457, 458, 462, 501, 503, 677-678, 679 **M6:** 40
constitution diagram ... **A6:** 810
weld cladding prediction ... **A6:** 818, 819
Deloro 60
laser cladding components and techniques ... **A18:** 869
Delta ferrite *See* Ferrite ... **A6:** 457
in 18-8 stainless steel, polarization curve for potentiostatic etching ... **A9:** 145
in austenitic steel welded joints ... **A9:** 579
defined ... **A9:** 5
effect on sigma in austenitic stainless steel ... **A9:** 136-137
peritectic formation of austenite from ... **A9:** 679-680
in type 304 stainless steel, revealed by magnetic etching ... **A9:** 65-66
Delta iron
defined ... **A9:** 5
Delta phase
in wrought heat-resistant alloys ... **A9:** 309-310
Delta-ferrite in wrought stainless steels
in austenitic grades ... **A9:** 283-284
etchants compared ... **A9:** 289
etching ... **A9:** 282
magnetic etching ... **A9:** 282
in martensitic grades, prevention ... **A9:** 285
in precipitation-hardenable grades ... **A9:** 285
Delta-glaze 347m/349m
as lubricant for compression testing ... **A8:** 195
Delta-I noise ... **EL1:** 27, 28
Delube
defined ... **A7:** 3
Delubrication
effects of particle size distribution and mixture fluctuations ... **A7:** 186
growth and shrinkage during ... **A7:** 480
in sintering atmosphere ... **A7:** 340
Delubrication zone
furnaces ... **A7:** 351
Demagging
of aluminum alloys ... **A15:** 471-474
Demagnetization *See also* Magnetic hysteresis; Magnetism; Magnetization ... **M3:** 616-617, 637, 639
after magnetic particle inspection ... **A17:** 120-122
with alternating current (ac) ... **A17:** 121
by mobile equipment ... **A17:** 93
current-decay method ... **A17:** 93
curve for permanent magnet alloy ... **A7:** 639
curves ... **A2:** 785-790
with direct current (dc) ... **A17:** 121
resistance, of permanent magnets ... **A2:** 782, 784
Demagnetization curve ... **EM4:** 1161
Demagnetizing field ... **A10:** 690
Demarest process ... **A14:** 3, 610
Dementia
from aluminum toxicity ... **A2:** 1256
Deming's fourteen points
for quality control ... **A17:** 719-720
Demixed particle pattern ... **A7:** 186
Demixing ... **A7:** 3, 188
Demodulation methods
eddy current inspection ... **A17:** 174
Demulsifiers ... **A18:** 106-107
applications ... **A18:** 107
formation of ... **A18:** 107
moieties ... **A18:** 106
structures ... **A18:** 107
Denatured alcohols
use in etchants ... **A9:** 67-68
Dendrite
defined ... **A8:** 4 **A9:** 5 **A11:** 3 **A13:** 4
Dendrite arm spacing *See also* Primary dendrite arm spacing; Secondary dendrite arm spacing
in aluminum alloy 3003 ... **A9:** 630
in aluminum alloy ingots ... **A9:** 629
as an indicator of solidification history and its effect on postsolidification processing ... **A9:** 625
in cast aluminum alloys ... **A9:** 357
in copper alloy ingots, effect of cooling rate on ... **A9:** 637, 639
and relationship to solidification time equation for ... **A9:** 629
in steel ... **A9:** 623-625
in zinc alloys, measuring ... **A9:** 490
Dendrite arm spacing, vs. cooling rate
Al alloys ... **A7:** 33, 36
Dendrite arms
maraging steels ... **A12:** 384
Dendrite cell size, in cast aluminum alloys used to determine solidification rate and strength ... **A9:** 357
Dendrite formation
and tin whisker growth ... **EL1:** 969-970
zone 2, package interior ... **EL1:** 1009
Dendrite(s) *See also* Dendritic; Dendritic structure
arm spacing, aluminum alloys ... **A15:** 749-750
coarsening of ... **A15:** 138, 154-155
columnar ... **A15:** 130-135
defined ... **A15:** 4
detachment, as equiaxed grain growth mechanism ... **A15:** 131-132
equiaxed ... **A15:** 130-135
and eutectics, competitive growth of ... **A15:** 122-124
as grains, after solidification ... **A15:** 118
growth, and low gravity ... **A15:** 153-156
instabilities ... **A15:** 122-123
length, calculated ... **A15:** 117
morphology, in microsegregation ... **A15:** 137
Dendrite(s) (continued)
secondary ... **A15:** 117
spacing, low gravity effect ... **A15:** 150
tip, equiaxed ... **A15:** 135
Dendrite-arm spacing
aluminum casting alloys ... **A2:** 133
Dendrites ... **A3:** 1 • 19
as a result of supercooling in pure metals ... **A9:** 609
in aluminum alloy 2024 ... **A9:** 634
in aluminum alloy ingots ... **A9:** 629
in copper alloy ingots ... **A9:** 637-641
in cyclohexanol ... **A9:** 608
in rapidly solidified alloys ... **A9:** 615-617
redistribution of solutes during solidification ... **A9:** 614
SEM analysis ... **A10:** 490
solidification ... **A9:** 612-614
Dendritic
microstructure ... **A3:** 1 • 20
segregation ... **A3:** 1 • 18, 19
Dendritic coarsening ... **A15:** 138, 154-155
Dendritic crack structure
arc-gouged drain groove ... **A11:** 647
Dendritic growth
from ionic residues ... **EL1:** 660
Dendritic growth in steel ... **A9:** 623-628
segregation during ... **A9:** 625-626
Dendritic interface
particle behavior at ... **A15:** 144-145
Dendritic macrostructure in aluminum alloy ingots ... **A9:** 629
relationship to grain size ... **A9:** 631
Dendritic particle shapes ... **A7:** 233, 234
Dendritic particles
apparent density ... **A7:** 272
Dendritic pattern in steel
effect of hot rolling on ... **A9:** 627-628
Dendritic powder
defined ... **A7:** 3
Dendritic segregation
defined ... **A9:** 5
Dendritic solidification
microscopic solids in ... **A15:** 102
particle behavior in ... **A15:** 145-146
Dendritic solidification structure
Ni-5Ce alloy ... **A10:** 307
Dendritic structure
aluminum-silicon alloys ... **A15:** 165
in carbon and alloy steels, revealed by macroetching ... **A9:** 173
electrolytic copper powder ... **A7:** 71
electrolytic iron sheet ... **A7:** 72
growth, in mold ... **A15:** 118-119
high-alloy steels ... **A15:** 731
in low alloy steel ... **A9:** 623
and semisolid cast microstructure compared ... **A15:** 327
single-phase alloys ... **A15:** 116-119
velocity effect ... **A15:** 116-119
Dendritic theory ... **EM3:** 300
Denickelification *See also* Corrosion; Dealloying; Selective leaching ... **A13:** 4, 130, 133
defined ... **A11:** 3
failure, copper-nickel alloy heat-exchanger tubes ... **A11:** 634
as selective leaching ... **A11:** 178, 633
Denier ... **EM1:** 8-9, 114
defined ... **EM2:** 13
Denitriding
as annealing of metal powders ... **A7:** 182
Denitriding, surface
chemical studies ... **A10:** 177
Dense random packing of hard spheres (DRPHS)
in amorphous materials and metallic glasses ... **A2:** 810
Dense silicon nitride ... **EM4:** 812-183
additives ... **EM4:** 812
fabrication ... **EM4:** 812

SUBJECTS OF THE INDEXED VOLUMES: ASM Handbook (designated by the letter "A"): **A1:** Properties and Selection: Irons, Steels, and High-Performance Alloys (1990); **A2:** Properties and Selection: Nonferrous Alloys and Special-Purpose Materials (1990); **A3:** Alloy Phase Diagrams (1992); **A4:** Heat Treating (1991); **A6:** Welding, Brazing, and Soldering (1993); **A7:** Powder Metallurgy (1984); **A8:** Mechanical Testing (1985); **A9:** Metallography and Microstructures (1985); **A10:** Materials Characterization (1986); **A11:** Failure Analysis and Prevention (1986); **A12:** Fractography (1987); **A13:** Corrosion (1987); **A14:** Forming and Forging (1988); **A15:** Casting (1988); **A16:** Machining (1989); **A17:** Nondestructive Testing and Quality Control (1989); **A18:** Friction, Lubrication, and Wear Technology (1992). **Metals Handbook, 9th Edition** (designated by the letter "M"): **M1:** Properties and Selection: Irons and Steels (1978); **M2:** Properties and Selection: Nonferrous Alloys and Pure Metals (1979); **M3:** Properties and Selection: Stainless Steels, Tool Materials and Special-Purpose Materials (1980); **M4:** Heat Treating (1981); **M5:** Surface Cleaning, Finishing, and Coating (1982); **M6:** Welding, Brazing, and Soldering (1983). **Engineered Materials Handbook** (designated by the letters "EM"): **EM1:** Composites (1987); **EM2:** Engineering Plastics (1988); **EM3:** Adhesives and Sealants (1990); **EM4:** Ceramics and Glasses (1991); **Electronic Materials Handbook** (designated by the letters "EL"): **EL1:** Packaging (1989).

Dense silicon nitride (continued)
flexural strength **EM4:** 816
sintering **EM4:** 812
Densification
by cold rolling **A7:** 406, 407
by hydrostatic extrusion, stages **A7:** 300
in carbon/carbon composites **EM1:** 917-918
ceramic-ceramic composites **EM1:** 935-937
and chemical composition **A7:** 246-249
chemical vapor deposition (CVD) for **EM1:** 918
of copper powder compacts **A7:** 735
during cemented carbide sintering **A7:** 386
effect on electrical conductivity and tensile properties of P/M copper **A7:** 736
end point, effects of contained argon **A7:** 434
factors affecting during homogenization **A7:** 315
and flow stress **A7:** 300
forging **A7:** 410-411
full, of powder compact **A7:** 502
as function of compact green density **A7:** 310
of green strip **A7:** 405
of hollow sphere under hydrostatic pressure **A7:** 298
incomplete sintering, effects **EM1:** 937-938
of loose powders, plastic flow stage **A7:** 298
model for ductile metal powders **A7:** 299-300
in nuclear fuel pellet fabrication **A7:** 665
process, defined **EM1:** 9
SAS techniques for **A10:** 405
and shrinkage during sintering **A7:** 310
sintering **EM4:** 260-265, 266, 267-268
and sintering in CAP process **A7:** 533
and sol-gel processing **EM4:** 449-450
solid-state sintering **EM4:** 275-276, 279-280
Densification, of castings
by hot isostatic pressing **A15:** 263
Densification process
defined **EM2:** 13
Densification/coarsening ratio **EM4:** 297
Density *See also* Bulk density; Current density; Densification; Density distribution; Electric charge density; Electric flux density; Energy density; Green density; Heat flux density; Mass characteristics; Packed density; Porosity; Specific gravity **EM4:** 424, 582-583
absolute **A7:** 3
absolute, defined **A8:** 4
active circuitry **EL1:** 7
alloy cast iron **M1:** 88
of aluminum alloys **A7:** 474
aluminum and aluminum alloys **A2:** 47
aluminum casting alloys **A2:** 153-177
of aluminum oxide-containing cermets **A7:** 804
available, hybrids **EL1:** 251
band, in filament winding **EM1:** 508
of beryllium **A2:** 683-684
beryllium-copper alloys **A2:** 409
board wiring, design/manufacture effect **EL1:** 129
bulk **EM4:** 582
bulk, defined *See* Bulk density
carbon fibers **EM1:** 51
of carbonyl nickel powder compacts **A7:** 310
cast copper alloys **A2:** 356-391
of cast steel **A1:** 374
cast steels **M1:** 392, 393
of cemented carbides **A2:** 957
change during aging, cast copper alloys **A2:** 360, 362
of chromium carbide-based cermets **A7:** 806
compact, and fatigue strength, titanium P/M **A2:** 652
compacted, of tungsten powders **A7:** 390
compaction pressure for prealloyed steel powder **A7:** 683
of compacts in sintering **A7:** 309
composite, for filament winding **EM1:** 508, 509
of composite tools **EM1:** 586-587
of contacts **EL1:** 440
copper-based P/M materials **A7:** 470
crystal defect, measured by x-ray topography **A10:** 365
current **A10:** 685
defined **A10:** 671
defined, thermal inspection **A17:** 396
dry **A7:** 3
Density (continued)
of ductile iron **A1:** 50 **A15:** 663 **M1:** 49
eddy current, and depth **A17:** 169
eddy current, effect, remote-field eddy current inspection **A17:** 195
effect, environmental stress crazing **EM2:** 800
effect in blending and premixing **A7:** 188
effect of lubrication in rigid dies **A7:** 302
effect of packing method in lead shot **A7:** 296
effect on elastic modulus, Poisson's ratio **A7:** 466
effect on hardness testing **A7:** 452
effect on tensile/impact properties **A14:** 202-203
effect, paper radiographs **A17:** 314
effect, PUR foams **EM2:** 262
effect, radiographic inspection **A17:** 295
effective **EM4:** 583
effects from steam treating ferrous P/M materials **A7:** 466
effects in welding **A7:** 456
electric charge **A10:** 685
of electrolytic copper powder compacts **A7:** 310
of electrostatic copier powders **A7:** 585
energy **A10:** 685
of ferrous P/M materials **A7:** 466
and fill ratio, in rigid tool compaction **A7:** 321-325
final **A7:** 463
flux **EL1:** 683
forging and ejection pressures **A7:** 417
as function of energy in explosive compacting **A7:** 305
as function of hardenability **A7:** 451
as function of pressure in cold isostatic pressing **A7:** 449
function, of variables **A8:** 624
functional, of printed wiring boards **EL1:** 505
of gases, defined **A13:** 4
of glass fibers **EM1:** 46
of gray iron **A1:** 31 **A15:** 644 **M1:** 31
increase by tapping **A7:** 296
increase by vibration **A7:** 296
increased, methods in powder compacts **A7:** 304-307
of injection molded P/M materials **A7:** 471
interconnection **EL1:** 7, 10
of iron powder compacts **A7:** 511
lead and lead alloys **A2:** 545
of loosely packed nickel powder **A7:** 397
low, of engineering plastics **EM2:** 1
of magnesium, and casting weight **A15:** 808
malleable iron **M1:** 67
maraging steels **M1:** 450, 451
mass **A10:** 685
mass, SI units/symbol for **A8:** 721
and material selection **EM1:** 38
and materials selection **A7:** 262
measurement and quality control **A7:** 482-483
measurement of **EM1:** 285-286
mechanical fundamentals **A7:** 296
and mechanical properties, powder forging **A14:** 189
of metal borides and boride-based cermets **A7:** 812
metals/metal oxides, for gating systems **A15:** 593
methods to determine **A7:** 262-271
mold aggregate, and tolerances **A15:** 618
of molybdenum and molybdenum alloys **A7:** 476
of nickel-powder compacts, sintered **A7:** 314
optical, blackening as **A10:** 143
of P/M and I/M alloys **A7:** 747
P/M and wrought titanium and alloys **A7:** 475
P/M forged low-alloy steel powders **A7:** 470
of P/M stainless steels **A7:** 468 **A13:** 830-832
P/M steels, designations and ranges **M1:** 333
packaging, factors **EL1:** 438-440
packaging, model of **EL1:** 13-14
packaging, through-hole vs surface mount technologies **EL1:** 730
packing, grain size distribution effect **A15:** 208
palladium **A2:** 715
polyamide-imides (PAI) **EM2:** 133
polyaryl sulfones (PAS) **EM2:** 146
of polyester resins **EM1:** 91
Density (continued)
of powder forged parts **A14:** 204
of powders, empirical determination **A7:** 296-297
and pressure relationships, Ceracon process **A7:** 538
and pressure relationships, powder compacts **A7:** 298-300
profile, polyurethanes (PUR) **EM2:** 260
pycnometry as measure of **A7:** 262, 265-266
of rare earth elements **A2:** 722-723
ratio, defined **A7:** 3
reflection, radiography **A17:** 314, 323
relative, measurement by computed tomography (CT) **A17:** 361
of semiconductor ICs **EL1:** 297
sheet molding compounds **EM1:** 158
signal line, VHSIC interconnects **EL1:** 389
of solids and liquids, defined **A13:** 4
of STAMP-processed materials **A7:** 548
steels, selected grades **M1:** 145
structural ceramics **A2:** 1019
symbol and units **A18:** 544
symbol for **A8:** 726 **A10:** 692
of synthetic sands **A15:** 29
terminations, comparison **EL1:** 730
theoretical **A7:** 280, 287 **EM4:** 582-583
of titanium carbide-based cermets **A7:** 808
of titanium carbide-steel cermets **A7:** 810
transmission, radiography **A17:** 323
true **EM4:** 582-583
and types of arrays **A7:** 296
uniform **A7:** 298, 300-301
of uranium and uranium alloys **A2:** 670
variability in **A8:** 623
versus sintering time, liquid-phase sintering **A7:** 320
vs costs, hybrids **EL1:** 250-252
and weighing errors **A7:** 483
wet **A7:** 3
white cast iron **M1:** 31
wiring, and thermal expansion **EL1:** 613
wiring, metal cores **EL1:** 620-621
wrought aluminum and aluminum alloys **A2:** 62-122
of x-ray film **A17:** 323-324
Density distribution *See also* Densification; Green density; Packed density; Porosity
in compacts **A7:** 301
in dies with sidearm **A7:** 300
improved with top and bottom pressure **A7:** 301
measuring **A7:** 301
and stress **A7:** 300-302
Density, material properties
flyer plate and explosion welding **A6:** 161
Density of powder metallurgy materials **A9:** 503
Density resolution
defined **A17:** 383
Dent depth *See* Damage
Dental alloys *See also* Dental amalgam; Dental fillings; Sliver alloys **A13:** 1336-1366
beryllium-nickel **A2:** 426
classification/characterization **A13:** 1348-1359
compositions/properties **A13:** 1336-1337
containing tin **A2:** 526
crown and bridge alloys **A2:** 696
crown, bridge, and partial **A13:** 1350
direct filling alloys **A2:** 696
with gallium **A2:** 741
implant alloys **A2:** 696
implant rejection **A13:** 1339
interstitial vs. oral fluid environments and artificial solutions **A13:** 1340-1341
intraoral surface **A13:** 1346-1347
intraoral vs. simulated exposures **A13:** 1347-1348
oral corrosion pathways/electrochemical properties **A13:** 1342-1344
oral corrosion process **A13:** 1344-1346
partial denture alloys **A2:** 696
porcelain fused to metal alloys **A2:** 696
properties **A2:** 698
saliva composition effects **A13:** 1341-1342
soldering alloys **A2:** 696-697
tarnish/corrosion resistance **A13:** 1337-1340
tarnish/corrosion under simulated/accelerated conditions **A13:** 1359-1363

Dental alloys (continued)
trade practices ... **A2:** 691
wrought orthopedic wires ... **A2:** 696
Dental alloys, tin
use in ... **M2:** 615
Dental amalgam *See also* Dental alloys; Silver alloys ... **M2:** 678
properties ... **A2:** 703-704
Dental amalgams
liquid mercury with silver-based alloy powders ... **A7:** 661-662
powders used ... **A7:** 573
Dental applications ... **A7:** 657-663 **EM4:** 1091-1097
dental glass in composite resin materials ... **EM4:** 1091-1092
glass and glass-ceramic crown and bridge materials ... **EM4:** 1093-1096
glass implant materials ... **EM4:** 1096
glasses in dental ceramics ... **EM4:** 1092-1093
health and safety regulations ... **EM4:** 1096-1097
Dental composites
single-pass sliding with amalgam material ... **A18:** 669
Dental fillings
germanium gold alloys for ... **A2:** 743
Dental impression rubber method
for plastic replicas ... **A17:** 53
Dental inlays
investment cast ... **A15:** 35
Dental materials, friction and wear of ... **A18:** 665-676
composite restorative materials ... **A18:** 669-672
abrasion tests ... **A18:** 670
aging and chemical softening ... **A18:** 670, 671
clinical studies ... **A18:** 671-672
correlating abrasion with hardness and tensile data ... **A18:** 670
double-pass sliding studies ... **A18:** 670
fracture toughness studies ... **A18:** 670
fundamental laboratory studies ... **A18:** 669-670
Leinfelder technique ... **A18:** 671
simulation studies ... **A18:** 670-671
single-pass sliding studies ... **A18:** 670, 671
dental amalgam ... **A18:** 669
abrasion tests ... **A18:** 669
clinical studies ... **A18:** 669
fracture toughness tests ... **A18:** 669
friction ... **A18:** 669
simulation studies ... **A18:** 669
single-pass sliding studies ... **A18:** 669
dental cements, studies ... **A18:** 673
material loss on abrasion ... **A18:** 673
dental feldspathic porcelain and ceramics, studies ... **A18:** 674
friction and wear properties for single-pass sliding ... **A18:** 674
denture acrylics, studies ... **A18:** 666, 674
die materials (stone, resin, and metal), studies ... **A18:** 666, 675
material loss on abrasion ... **A18:** 675
endodontic instruments, studies ... **A18:** 666, 675
human dental tissues ... **A18:** 665-666
noble and base metal alloys, studies ... **A18:** 673
orthodontic wires ... **A18:** 675-676
periodontal instruments, studies ... **A18:** 675
pits and fissure sealants, studies ... **A18:** 672
porcelain and plastic denture teeth, studies ... **A18:** 673-674
friction and wear properties for single-pass sliding ... **A18:** 674
interpenetrating polymer network (IPN) ... **A18:** 674
loss of material on opposing restoration ... **A18:** 674
wear studies ... **A18:** 666-669
classification of wear situations in dentistry ... **A18:** 666
clinical studies ... **A18:** 668-669
failure classification scale (lab) ... **A18:** 667
Dental materials, friction and wear of (continued)
fundamental laboratory studies ... **A18:** 667-668
simulation studies ... **A18:** 668
Dentin
abrasion of dentifrices ... **A18:** 668
human teeth, properties ... **A18:** 666
Denting *See also* Brinelling
resistance, in heat-exchanger tubes ... **A11:** 628
in steam generators ... **A13:** 940
Dentistry
precious metals in ... **A2:** 695-698
use of gold ... **M2:** 684-687
Dents
adhesive-bonded joints ... **A17:** 613
Dents, by dirt
slivers in die ... **A8:** 548
Denture acrylics
properties ... **A18:** 666
simplified composition on microstructure ... **A18:** 666
Deoxidation *See also* Oxygen; Oxygen removal
in acid steelmaking ... **A15:** 364, 365
basic steelmaking ... **A15:** 367
brittle fracture from ... **A11:** 85
by aluminum ... **A15:** 91
of copper alloys ... **A15:** 468-470
copper and copper alloys ... **A2:** 236-237
defined ... **A15:** 4
of ferrous melts ... **A15:** 74, 78-79
of Invar ... **A2:** 890
low-alloy steels ... **A15:** 721
of plain carbon steels ... **A15:** 708-709
single-element ... **A15:** 78
systems ... **A15:** 78
and temper annealing, ferrous alloys ... **A7:** 185
vacuum, in water-atomized tool steel powders ... **A7:** 104
vacuum induction furnace ... **A15:** 395
Deoxidation of steel
effect on macrostructure ... **A9:** 623
Deoxidation practice ... **M1:** 111-112, 114, 121, 123-124
capped steel ... **A1:** 143
influence of, on graphitization ... **A1:** 696
killed steel ... **A1:** 142
notch toughness, effect on ... **M1:** 694-695, 698
rimmed steel ... **A1:** 143
semikilled steel ... **A1:** 142-143
steel plate production ... **A1:** 226-227
Deoxidation products
defined ... **A9:** 5
Deoxidized copper
filler metals ... **A6:** 755
gas-metal arc welding ... **A6:** 759
gas-tungsten arc welding ... **A6:** 192, 756, 757
globular-to-spray transition current for electrodes ... **A6:** 182
roll welding ... **A6:** 313
weldability ... **A6:** 753
Deoxidized steel
pipe revealed by macroetching ... **A9:** 174
Deoxidizer
addition, testing for ... **A15:** 470
defined ... **A15:** 4
Deoxidizers, aluminum processing *See also* Oxide, removal of ... **M5:** 8, 12-13
Deoxidizing
defined ... **A13:** 4
Deoxidizing elements ... **A6:** 753
Department of Defense (DoD) ... **EL1:** 458, 906-916
as information source ... **EM1:** 40, 42
Department of Defense Ceramics Analysis Information Center (CINDA) ... **EM4:** 692
Department of Defense index of Specifications and Standards (DoDISS) ... **EL1:** 906 **EM1:** 701 **EM2:** 89-90
Department of Defense Index of Standards and Specifications ... **EM4:** 40
Departure from nucleate boiling (DNB) ... **A11:** 1, 605, 796
Departure side pinning
superalloys ... **A12:** 393
Dependability
of metal castings ... **A15:** 40
Dependent variables
in fatigue testing ... **A8:** 697-698
Dephosphorization
defined ... **A15:** 4
Depleted uranium *See* Uranium
Depleted uranium (DU), heat treating ... **A4:** 928-938
aging ... **A4:** 931, 932, 933, 934-935, 936, 937-938
analysis, chemical and gas ... **A4:** 928
annealing ... **A4:** 929-930, 931
applications ... **A4:** 928
atmospheres, furnace ... **A4:** 932, 936, 937
cast ... **A4:** 930-931
cleaning ... **A4:** 937
cold working ... **A4:** 929, 936
cooling rates ... **A4:** 929, 931, 932, 933
corrosion in molten salts ... **A4:** 937
density ... **A4:** 928-929, 930
dilute alloys ... **A4:** 928, 931-936
fixtures ... **A4:** 937
furnaces ... **A4:** 936, 937, 938
grain size ... **A4:** 928-929, 930, 936
hardness data ... **A4:** 929, 934, 935, 938
mechanical properties ... **A4:** 929, 930, 935, 936, 938
metallurgical characteristics ... **A4:** 928
metastable high alloys ... **A4:** 936
orientation control ... **A4:** 928-929
procedures, examples ... **A4:** 938
quenching ... **A4:** 931, 932-934, 936, 937, 938
safety requirements ... **A4:** 938
salt baths ... **A4:** 932, 936, 937-938
solution treatment ... **A4:** 931, 932, 933, 936, 937, 938
stress leveling ... **A4:** 934, 935
tensile properties ... **A4:** 931, 932, 936, 938
vacuum ... **A4:** 936-937, 938
vapor pressure of uranium, relation to temperature ... **A4:** 495
wrought ... **A4:** 928, 932, 937
Depleted uranium, heat treating
aging ... **M4:** 782, 784, 785, 786
analysis, chemical and gas ... **M4:** 778
annealing ... **M4:** 778-779, 780, 781
atmospheres, furnace ... **M4:** 784
cast ... **M4:** 779-781, 782
cleaning ... **M4:** 784
cold working ... **M4:** 778, 779
cooling rates ... **M4:** 782
corrosion in molten salts ... **M4:** 784
density ... **M4:** 777
dilute alloys ... **M4:** 781-782, 783
fixtures ... **M4:** 785, 787
furnaces ... **M4:** 784, 786
grain size ... **M4:** 778
hardness data ... **M4:** 779
mechanical properties ... **M4:** 779, 781
metallurgical characteristics ... **M4:** 777, 778
metastable high alloys ... **M4:** 783
procedures, examples ... **M4:** 785-786
quenching ... **M4:** 782-783
safety requirements ... **M4:** 786
salt baths ... **M4:** 785
solution treatment ... **M4:** 782
tensile properties ... **M4:** 782
vacuum ... **M4:** 784, 786
Depleted-uranium screens
radiography ... **A17:** 316
Depletion *See also* Chromium; Selective leaching
chromium, in a weld zone ... **A10:** 179
defined ... **A9:** 5 **A11:** 3
rate, in calibration of gas mass spectrometers ... **A10:** 155
Depletion gilding ... **A15:** 19, 21
Depletion mode
MOSFET ... **EL1:** 158-159

Deply testing technique EM1: 765-766, 774
Depolarization
defined A13: 5
Depolarizer
defined A13: 5
Deposit (thermal spraying)
definition A6: 1208
Deposit attack *See* Deposit corrosion; Poultice corrosion
Deposit corrosion *See also* Poultice corrosion
of aluminum/aluminum alloys A13: 589
chemical, identification A13: 1138
control, gas/oil production A13: 1247
defined A13: 5
Deposit etching A9: 61
and line etching A9: 62
of silicon steel transformer sheets A9: 62-63
Deposit on sequence
definition M6: 5
Deposited metal
definition A6: 1208 M6: 5
Deposition *See also* Electrodeposition EM3: 10
of arsenic, as toxic metal A2: 1237
conditions for complete electrogravimetric A10: 198
of copper on platinum cathode, as electrogravimetry example A10: 198
defined EM1: 9 EM2: 13
effect of medium A10: 198
electrolytic, piezoelectric effect in A10: 198
and electrometric titration A10: 203
incomplete, use in electrogravimetry A10: 198
matrix, processes A2: 903
methods, spray thermal coatings A13: 459-460
physical properties of deposits A10: 198
reversal A10: 198
thin-film, superconducting materials A2: 1081-1083
vapor A13: 456-458
Deposition, copper
by sputtering EL1: 303
Deposition corrosion
aluminum alloys M2: 211-212
Deposition, direct
electrolytic A7: 71-72
Deposition efficiency
definition M6: 5
Deposition efficiency (arc welding)
definition A6: 1208
Deposition efficiency (thermal spraying)
definition A6: 1208
Deposition, film
by shadowing A12: 172
Deposition process (electroslag welding) A18: 644
Deposition rate
definition M6: 5
flux cored arc welding M6: 104
Deposition sequence
definition A6: 1208
Deposits *See also* Scaling
chemical analysis A11: 30
on boilers and steam equipment, effects A11: 608
on fracture surface, in hydrogen damage A11: 250
solid, effects on crevice corrosion A11: 184
Deposits, to improve wear resistance *See also* Coatings, Electroplating; Hardfacing M1: 634-635
Depreciation
in corrosion economics A13: 371-372
Depressurization rate
of cold isostatic pressing A7: 449
Depressurization systems
for cold isostatic pressing A7: 445
Depth EM3: 10
defined EM2: 13
three-dimensional measurement A12: 207
vs composition, passive film on tin-nickel substrate A10: 608-609
vs elements, semiquantitative PIXE analysis for A10: 102
Depth analysis *See* Depth profiles; Depth profiling
Depth dose EM3: 10
defined EM2: 13
Depth filtration
inclusion removal A15: 489
Depth measurement
Brinell test A8: 85
Depth of cut A18: 609
Depth of field A10: 497, 671
defined A9: 5
for macrophotography A9: 87
miniborescopes A17: 4
of optical microscopes A9: 76
of optical microscopes, relationship among light A9: 78
numerical aperture and wavelength of
Depth of focus A6: 877-878
definition A6: 1208
of scanning electron microscopes A9: 89
Depth of fusion
definition M6: 5
Depth of hardening M1: 478-480
Depth of penetration *See also* Penetration A10: 113, 671
dependence, microwave inspection A17: 202
and frequencies, eddy current inspection A17: 169
Depth profile analysis
AES A11: 33
schematic of effects A11: 37
for surface sodium detection A11: 43
time taken by A11: 36
Depth profiles *See also* Composition profiles; Depth profiling
Auger, antiwear films analysis A10: 565-566
Auger elemental A10: 554
compositional 304 stainless steel A10: 555
compositional, AES A10: 549
from ATR spectra A10: 113
of granular sample, using ATR, DRS, and PAS A10: 120
of heavy-element impurities A10: 632-633
negative, LPCVD thin films A10: 624
organometallic silicate film A10: 617
phosphorus, for ion-implanted silicon substrate A10: 624
RBS defect distribution, in single crystals A10: 628
secondary ion mass spectroscopy A10: 617-620, 623-626
of solid samples, infrared spectroscopy A10: 109, 115
spark source mass spectrometry A10: 142
x-ray photoelectron spectroscopy A10: 573-574
Depth profiling *See also* Composition profiles; Depth profile analysis; Depth profiles; Surface analysis techniques EL1: 1079-1080, 1090
Auger electron spectroscopy A10: 549-567
electron probe x-ray microanalysis A10: 516-535
field ion microscopy A10: 583-602
granular sample using ATR, DRS, and PAS A10: 120
low-energy ion-scattering spectroscopy A10: 603-609
neutron diffraction A10: '420-426
particle-induced x-ray emission A10: 102-108
photoacoustic spectroscopy A10: 115
Rutherford backscattering spectrometry A10: 628-636
scanning electron microscopy A10: 490-515
secondary ion mass spectroscopy A10: 610-627
surface analytical technique A7: 251
vs sputter analysis A7: 251
Depth, seawater
effect on copper/copper alloys A13: 624
Depth-contour mappings
by optical holography A17: 405
Depth-dose profiles *See* Depth dose
Depth-of-cut notching
as cemented carbide tool wear mechanism A2: 955
Depth-of-field
fractographic effects A12: 78, 80-81, 87-88
optimum aperture for A12: 80
SEM A12: 166
Depth-to-width (D/W) ratio A6: 608
Derakane vinyl ester resins EM1: 33, 137-138
Derbies
defined A9: 476
metallic uranium ingots as A2: 670
Derham process
for demagging aluminum alloys A15: 473
Derivative thermogravimetric curves (DTG) EM4: 52, 53
Derived property
defined A8: 667
Derived SI units
guide for A10: 685
Derived units
Système International d'Unités (SI) A11: 793
Derjaguin approximation A18: 400, 401
Dermatitis
contact, from beryllium A2: 1239
from gold medical therapy A2: 1257
from nickel toxicity A2: 1250
from platinum A2: 1258
worker EM1: 36
Dermatitis (nickel itch) A7: 203
Derrick hooks
materials for A11: 522
Derusters, alkaline cleaners *See also* Alkaline descaling and derusting
Desalination plant equipment
seawater corrosion of M1: 744-745
Descaling *See also* Alkaline descaling and derusting; Defects; Salt bath descaling; Scale, removal of; Scaling
after cold heading A14: 294
of annealed stainless steel forgings A14: 230
defined A13: 5 A15: 4
high-pressure waterjets A14: 87
in hot upset forging A14: 87
mechanical methods A14: 87
molten-salt A14: 280
in open-die forging A14: 64
salt-bath A14: 230
in vacuum degassing A7: 435
Descriptive fractography EM4: 629, 635-644
fracture markings EM4: 635-638
arrest line EM4: 637, 638, 639
gull wings EM4: 637, 641
mirror, mist, and velocity hackle EM4: 635-636, 637, 638, 639, 640, 641, 643
rib mark EM4: 637
scarps EM4: 637-638
tail EM4: 637
twist hackle EM4: 636-637, 638, 642
wake hackle EM4: 637, 641
Wallner lines EM4: 636, 637, 638, 639, 640, 642
fracture origins EM4: 641-644
impact sites EM4: 641-643
indentation sites EM4: 641-643
machining flaws EM4: 643
processing defects EM4: 643-644
fracture surfaces in materials EM4: 638-641
glass EM4: 638
polycrystalline ceramics EM4: 639-640
single crystals EM4: 638-639
indentation and impact sites EM4: 642-643
half-penny crack EM4: 642
scanning electron microscopy EM4: 640, 641
significance EM4: 635
techniques EM4: 635
uses EM4: 635
velocity hackle EM4: 640
Descriptive statistics A8: 624
Desferrioxamine
as chelator A2: 1236
Desiccant EM3: 10
defined EM2: 13
Design *See also* Balanced design; Casting design; Computer-aided design (CAD); Computer-aided design/computer-aided manufacture; Computer-aided engineering; Design allowables; Design considerations; Design corrosion control; Design detail; Design for manufacturability (DFM); Design guidelines; Design requirements; Design trade-offs; Die design; Electrical design; Experimental design; Forging process; Joint design and quality; Life-cycle optimization; Mechanical design; Part design; Physical design; Process design; Quality control; Quality design; Structural analysis and design EL1: 513-526
adhesive joint EM1: 683
aircraft A13: 1023
allowables EM1: 9, 308-312
of aluminum alloy precision forgings A14: 251

Design (continued)
analysis costs ... **EM2:** 83
anodic protection system ... **A13:** 464-465
application ... **EL1:** 516-517
application example ... **EM1:** 183-184
approach, engineering plastics ... **EM2:** 74-81
for assembly and manufacture ... **EL1:** 119-126
automation ... **EL1:** 127-129
automotive ... **A13:** 1016
beryllium-copper alloys ... **A2:** 416
of blanks, effect in ring rolling ... **A14:** 116
of blocker (preform) dies ... **A14:** 77-78
for blow molding ... **EM2:** 357
of boilers ... **A11:** 603, 619
Boothroyd-Dewhurst, method for assembly ... **EL1:** 124-125
of borescopes ... **A17:** 3
bread-boarded ... **EL1:** 505
capability, error-free ... **EL1:** 85-86
cascading effects ... **EM1:** 37
casting ... **A11:** 345
casting, surface finish effect ... **A15:** 285
of cathodic protection systems ... **A13:** 470
changes, for corrosion control ... **A11:** 196-197
charts, for laminate selection ... **EM1:** 233
of check-valve poppet ... **A11:** 70
classical, and distortion failure ... **A11:** 136
of cluster, investment casting ... **A15:** 257
of cold heading tools ... **A14:** 293
of cold-formed parts ... **A11:** 307-308
component, metal casting advantages ... **A15:** 39-41
of composite structure, cost drivers in ... **EM1:** 419-427
with composites ... **EM1:** 33
configuration, effect, damage tolerance ... **EM1:** 262-264
configurations, evaluating ... **A8:** 683
considerations, aluminum-lithium alloys ... **A2:** 186-187, 193-194
considerations, permanent magnet materials ... **A2:** 799-802
constraint, for parts ... **EM2:** 78-79
continuous casting machine ... **A15:** 310-312
of continuous fiber-reinforced composites ... **A11:** 731-732
costs ... **EM2:** 82
of crack-tolerant structures, fracture mechanics as ... **A11:** 47
criteria, aluminum alloy precision forgings ... **A14:** 251-252
of crucible furnaces ... **A15:** 383
cycle, for composite structures ... **EM1:** 179
damage tolerance ... **A17:** 702
data base ... **EM1:** 181-182
data, fatigue loading ... **EM2:** 706
defects, steam equipment failure by ... **A11:** 602-603
for deposition corrosion, aluminum alloys ... **A13:** 589
die, defects from ... **A15:** 294-295
die, for heat-resistant alloys ... **A14:** 234
for ease of orientation ... **EL1:** 123
effect of composite anisotropy on ... **EM1:** 38
effect of dynamic fracture toughness on ... **A8:** 259
effects, stainless steel corrosion ... **A13:** 551
elastomeric mandrels ... **EM1:** 593-594
electrical ... **EL1:** 25-44, 402-403, 517-523
end plate wedge and pin, metal molds ... **A15:** 303
engineer, information for
environmental methodologies ... **EL1:** 45-46, 65-67
error, and spring failures ... **A11:** 551
errors, as failure cause ... **EM1:** 767
errors, continuous fiber-reinforced composites ... **A11:** 732-733
of ESR plant ... **A15:** 403
of experiments ... **A14:** 928, 937-939
of experiments, factorial designs ... **A17:** 740-750
factors, for forging failures ... **A11:** 318
and failure, in forging ... **A11:** 317-319

Design (continued)
fastener ... **A11:** 542
fatigue data for ... **EM2:** 705-706
fatigue failure from ... **A11:** 396-397
faulty, of drive-gear assembly ... **A11:** 143
of fiber-reinforced composites ... **EM1:** 769
of filament winding applications ... **EM1:** 508-510
final design package ... **EL1:** 523-526
of fine-edge blanking and piercing tools ... **A14:** 474
and fit-up, of joints ... **A11:** 450
flash, closed-die forging ... **A14:** 78-79
of flash gutter ... **A14:** 50
for flexibility ... **EL1:** 588-589
forging failure frequency from ... **A11:** 342
forging, minimum-draft ... **A14:** 258
forging sequence ... **A14:** 413-416
function, defined ... **A13:** 338
furnace, with water cooling ... **A15:** 360
for galvanic corrosion ... **A13:** 87, 749
of gating ... **A15:** 289, 589-597
guidelines, general ... **EM2:** 707-710
guidelines, in RTM material selection ... **EM1:** 168
guidelines, leaded and leadless surface-mount joints ... **EL1:** 733-734
guidelines/methodology
high-frequency digital systems ... **EL1:** 82-85
of heat exchangers ... **A11:** 636-639
of high-frequency digital systems ... **EL1:** 76-88
high-level, defined ... **EL1:** 129
with homopolymer/copolymer acetals ... **EM2:** 101
of honeycomb sandwich structures ... **EM1:** 727-728
for inclusion control ... **A15:** 91
influence on fatigue strength ... **A11:** 115-118
influence on tool and die failure ... **A11:** 564-566
information ... **EM2:** 410
of internal fixation devices ... **A11:** 677-680
of investment castings ... **A15:** 264-266
of iron castings ... **A11:** 344-352
of joints ... **A11:** 116, 639
layout ... **EL1:** 514-516
level 1 package ... **EL1:** 1401-403
for life cycle optimization ... **EL1:** 127-141
low-level, defined ... **EL1:** 129
of magnesium and magnesium alloy parts ... **A2:** 476-479
magnesium/magnesium alloys ... **A13:** 749
and material ... **EM2:** 612-617
and materials selection ... **A13:** 321
mechanical, and lubricant failure ... **A11:** 154
mechanical and structural, categories of ... **A11:** 115
mechanical, costs ... **EM2:** 83
mechanical methodologies ... **EL1:** 45-46, 55-65
mechanical parameters ... **EL1:** 517-523
metal matrix composite corrosion prevention ... **A13:** 862-863
metal vs plastics ... **EM2:** 78
metallurgical, corrosion protection by ... **A13:** 379
methodologies, for durability ... **EL1:** 1, 45-75, 401
for microcircuit performance ... **EL1:** 260-261
and mismatch ... **A14:** 49
mold, effect on mold life ... **A15:** 281
mold, permanent mold casting ... **A15:** 277-278
of molds, for resin transfer molding ... **EM1:** 168-169
mounting, effect on bearing material failure ... **A11:** 506
of NDE reliability experiments ... **A17:** 692-694
numerical methods ... **EM1:** 463-478
objectives ... **EM1:** 39
optimization, example ... **A17:** 750-752
optimization, parameter ... **A17:** 751
parameters, shafts ... **A11:** 459
part ... **EM2:** 78-81, 357, 615-616
of parts, and paint films ... **A13:** 528-529
of parts, effect on SCC in marine-air environment ... **A11:** 309-310
in patternmaking ... **A15:** 191, 193, 198-199

Design (continued)
and performance, testing ... **EL1:** 954-955
of permanent mold castings ... **A15:** 284
personnel, interfaces with tool/manufacturing personnel ... **EM1:** 428-430
phase, environmental stress screening ... **EL1:** 876
philosophy, linear elastic fracture mechanics (LEFM) ... **A17:** 664
of pipeline ... **A13:** 1291
of plastic parts ... **EM2:** 78-81
for plating ... **A13:** 422
power distribution, WSI ... **EL1:** 354
practice, and laminate complexities ... **EM1:** 311-312
of precision tooling ... **A14:** 159-160
of preforms ... **A14:** 50-51
probability of detection (POD) functions for ... **A17:** 663
problems, in failure analysis ... **A11:** 747
process, and material selection ... **EM1:** 38
process, hot-die/isothermal forging ... **A14:** 153-155
process, tooling/manufacturing effects on ... **EM1:** 428-431
product, for die casting ... **A15:** 286-288
product, thermoplastic resins ... **EM2:** 622-623
properties, low- and elevated-temperature ... **A8:** 670-671
PWB, future ... **EL1:** 506
and quality control, statistical ... **A17:** 719-753
-related discontinuities, welding ... **A17:** 582
for reliability tool ... **EL1:** 741, 746-748
requirements ... **EM1:** 181-184
requirements, aggregate properties approach ... **EM2:** 407-411
of risers
robust, implementing ... **A17:** 750-752
of rolls ... **A14:** 352
rule, for integrated circuits ... **EL1:** 161
for simultaneous engineering, automotive industry ... **EM1:** 835-836
of solder masks ... **EL1:** 558-559
for space and missile hardware ... **EM1:** 816-822
stainless steels ... **A13:** 552-553
of steam turbines ... **A13:** 995
and stress raisers, iron castings ... **A11:** 346
for structural corrosion ... **A13:** 379, 1303
structural, requirements ... **EM1:** 313, 314
tension testing and ... **A8:** 19
for testability ... **EL1:** 374
thermal methodologies ... **EL1:** 45-55, 287, 402
of thermoplastic extruders ... **EM2:** 379-383
thin-film resistors and conductors ... **EL1:** 316-320
time-frame, for new products ... **EL1:** 390
of titanium alloy aircraft part ... **A14:** 274
to minimize corrosion ... **A13:** 338-343
tool, in powder forging ... **A14:** 197
of tooling, aluminum alloy precision forgings ... **A14:** 252-253
tools ... **EL1:** 419
trade-off studies ... **EM1:** 426-427
trade-offs, flexible epoxies ... **EL1:** 821
transmission line, WSI ... **EL1:** 354
of tungsten-reinforced composites ... **EM1:** 884-885
of two-directional fabrics ... **EM1:** 125-127
of unidirectional fabrics ... **EM1:** 125-127
upper limit, defined ... **A11:** 136
vacuum induction furnace ... **A15:** 397
values, minimum ... **A8:** 662
vehicle structural ... **EM1:** 346-351
with weathering steels ... **A13:** 518
weld, stainless steels ... **A13:** 551-552
weld-joint, incomplete fusion voids and ... **A11:** 92
worksheet, for costs ... **EM1:** 425-427

Design allowables *See also* A-basis; B-basis; S-basis; Typical basis ... **EM3:** 10
computation of derived properties ... **A8:** 667-668
defined ... **EM1:** 9 **EM2:** 13
determining by regression analysis ... **A8:** 668-670
determining distribution form ... **A8:** 663-664

Design allowables (continued)
direct computation for normal and unknown distribution **A8:** 664-666
examples of computational procedures **A8:** 672-677
general procedures **A8:** 662-663
generation of **EM1:** 308-312
for low- and elevated-temperature properties **A8:** 670-671
for static metallic material properties **A8:** 662-667
Design and analysis of adhesive bonding **EM3:** 459-469
crack propagation tests **EM3:** 460, 462-463
cleavage test **EM3:** 462
design of adhesive joints **EM3:** 468-469
fatigue and creep allowable stress testing **EM3:** 467-468
lap-shear test **EM3:** 460-461
model joint for stress analysis **EM3:** 459-460
adhesive shear stiffness **EM3:** 459, 460
skin doubler specimen **EM3:** 459-460, 467, 468, 469
peel tests (metal-to-metal) **EM3:** 460, 461-462
bell **EM3:** 461
climbing drum **EM3:** 461
"T" peel **EM3:** 461
shear stiffness testing **EM3:** 463-465
skin-doubler adhesive strain in the nonlinear range **EM3:** 465-467
skin-doubler analysis verification **EM3:** 465
Design and analysis of experiments **EM3:** 797
Design capabilities
P/M products **A7:** 295
Design considerations *See also* Casting design; Design; Design trade-offs **EL1:** 408-421 **EM2:** 48-95
acrylics **EM2:** 106
acrylonitrile-butadiene-styrenes (ABS) **EM2:** 111-112
casting **A15:** 598-613
compression molding/stamping **EM2:** 326-333
costs **EM2:** 82-88
design tools **EL1:** 419
dimensional tolerances and allowances **A15:** 614-623
driving forces/trade-offs **EL1:** 408
electrical considerations **EL1:** 416-419
engineering plastics, characteristics **EM2:** 68-73
engineering plastics, design approach ... **EM2:** 74-81
flexible printed boards **EL1:** 584-592
gating **A15:** 589-597
general, introduction **EM2:** 1
homopolymer/copolymer acetals **EM2:** 101
information sources **EM2:** 92-95
introduction **EM4:** 675
ionomers **EM2:** 121-122
of polyamide-imides (PAI) **EM2:** 130-132
polyamides (PA) **EM2:** 127
polybenzimidazoles (PBI) **EM2:** 150
polybutylene terephthalates (PBT) **EM2:** 154
polycarbonates (PC) **EM2:** 151-152
polyether sulfones (PES, PESV) **EM2:** 160-162
polyethylene terephthalates (PET) **EM2:** 172-175
polymer chemistry, overview **EM2:** 63-67
polymer science for engineers **EM2:** 48-62
polyphenylene ether blends (PPE PPO) **EM2:** 185
polysulfones (PSU) **EM2:** 201
risers **A15:** 577-588
specifications and standards **EM2:** 89-91
styrene-acrylonitriles (SAN, OSA ASA) **EM2:** 216
styrene-maleic anhydrides (S/MA) **EM2:** 219-220
thermal considerations **EL1:** 408-414
thermomechanical considerations **EL1:** 414-416
thermoplastic fluoropolymers **EM2:** 118
thermoplastic injection molding **EM2:** 308
Design constraints
common **EM2:** 78-79
Design corrosion control *See also* Design
and compatibility **A13:** 340-343
function **A13:** 338
and material component failure **A13:** 338
and materials selection **A13:** 338-339, 343
mechanical factors **A13:** 343
and shape **A13:** 339-340
of surfaces **A13:** 343
Design detail *See also* Design
effect, rotational molding **EM2:** 363-364
filament winding **EM2:** 371-373
as process selection factor **EM2:** 288-292
RTM/SRIM, compared **EM2:** 349
to minimize corrosion **A13:** 338-343
Design, die casting
zinc **M2:** 633-634
Design engineering, original
costs **EM2:** 83
Design enhancement plating equalizers **EL1:** 872
Design, experimental *See* Designed experiments; Experimental design
Design for manufacturability (DFM) *See also* Design; Manufacturability
and early manufacturing involvement (EMI) **EL1:** 125
quantitative measure **EL1:** 124-125
rules **EL1:** 121
uniaxis assembly **EL1:** 121-122
Design guidelines *See also* Design
cost estimating plastics parts **EM2:** 709-710
end-use requirements, defining **EM2:** 707
part geometry **EM2:** 707-709
strength of plastics **EM2:** 709
structure/properties/processing/applications **EM2:** 710
Design life **EM3:** 36
Design limit load (DLL) failure **A6:** 1046
Design lubricant film thickness
nomenclature for hydrostatic bearings with orifice or capillary restrictor **A18:** 92
Design of experiments **A14:** 928, 937-939
Design of forgings **M1:** 350, 357-360, 369-375
Design of friction and wear experiments **A18:** 480-488
calculations of elastic contact dimensions and stresses **A18:** 487-488
α and β parameter constants **A18:** 487, 488
formulas **A18:** 488
nomenclature **A18:** 487
categories of tribotests **A18:** 480-481
conditions of tribotests **A18:** 480-481
design of experiments **A18:** 482-485
one-at-a-time approach **A18:** 482-483
simple complete factorials approach **A18:** 483-485
evaluation of tribotests **A18:** 481-482
laboratory friction and wear tests and simulative tribotesting **A18:** 481, 482
round-robin tests **A18:** 485
case study, international round-robin sliding wear tests **A18:** 486-487
general rules **A18:** 486
terminology **A18:** 485-486
Design optimization problem
example **A17:** 750-752
Design package
final **EL1:** 523-526
Design practices for glass and glass fibers
applications **EM4:** 741-745
forming methods **EM4:** 741-742
history of processing and raw materials used **EM4:** 741
intrinsic strength of glass **EM4:** 742
lessons learned **EM4:** 744-745
material composition **EM4:** 741
mechanisms of strength reduction **EM4:** 742-743
component shape **EM4:** 742
internal flaws **EM4:** 742
residual stresses **EM4:** 743
surface flaws **EM4:** 742-743
methods for improving glass strength **EM4:** 743-744
acid etching **EM4:** 743
annealing **EM4:** 743
coatings **EM4:** 743-744
proof testing **EM4:** 744
tempering **EM4:** 743
properties of glass **EM4:** 741
reliability analysis **EM4:** 744
flaw statistics **EM4:** 744
proof testing **EM4:** 744, 745
time-dependent effects **EM4:** 744
time-to-failure equation **EM4:** 744
strength testing techniques **EM4:** 743
Design practices for structural ceramics in automotive turbocharger wheels **EM4:** 722-726
ceramic advantages **EM4:** 722
component developments **EM4:** 726
burst speed **EM4:** 726
vehicle testing **EM4:** 726
cost considerations **EM4:** 726
design methodology **EM4:** 723-725
brazed joint analyses **EM4:** 724-725
joining silicon nitride to metals **EM4:** 724
mechanical attachment **EM4:** 725
minimizing failure probability **EM4:** 723-724
minimizing stress concentrations **EM4:** 723
resisting foreign object damage **EM4:** 724
history **EM4:** 722-723
material selection and design trade-offs **EM4:** 725-726
nondestructive evaluation **EM4:** 726
turbocharger operation **EM4:** 722
Design practices for structural ceramics in gas turbine engines **EM4:** 716-721
design methodology **EM4:** 717, 718
lessons learned **EM4:** 721
close coordination **EM4:** 721
difficulty of large components **EM4:** 721
thermally induced stresses **EM4:** 721
thorough analysis **EM4:** 721
materials selection, design trade-offs, and component development **EM4:** 717-719
component evaluation **EM4:** 718-719
design trade-offs **EM4:** 718-719
manufacturability **EM4:** 718
material selection **EM4:** 717-718
program history **EM4:** 716-717
successful ceramic components **EM4:** 719-721
cyclic durability and time dependence **EM4:** 719-720
design/manufacturing trade-offs **EM4:** 719, 720-721
foreign object damage **EM4:** 719
high temperatures and speeds **EM4:** 719-720
high-speed blade rubs **EM4:** 719
time-dependent and cyclic-durability data base **EM4:** 719, 720
Design practices for structural ceramics in gasoline engines **EM4:** 728-732
ceramic engine components under commercial production in Japan **EM4:** 728
ceramic material properties compared **EM4:** 729
ceramic valve design **EM4:** 730-731
revolution limit **EM4:** 730
stress distribution **EM4:** 730-731
wear resistance **EM4:** 730, 731
weight ratio basis **EM4:** 730
design procedure **EM4:** 728-730
finite-element method analysis **EM4:** 729, 731
proof testing **EM4:** 729, 730
material selection **EM4:** 728
partially stabilized zirconia **EM4:** 728
silicon carbide **EM4:** 728
silicon nitride **EM4:** 728, 729, 730, 731
sliding parts **EM4:** 731
rocker arms engine components **EM4:** 731
turbine wheels **EM4:** 731-732
Design practices for whisker-toughened ceramic components **EM4:** 733-740
estimating thermoelastic properties **EM4:** 733-736
examples **EM4:** 738-740
noninteractive reliablity models **EM4:** 736-737, 740
TCARES algorithm **EM4:** 733, 737-378
Design proof testing **EM3:** 533-543
applications **EM3:** 533-535
building block test program approach **EM3:** 534-535
component testing **EM3:** 539-541
coupon testing **EM3:** 536-537
environmental considerations **EM3:** 535-536
moisture analysis **EM3:** 535-536
thermal analysis **EM3:** 535-536
full-scale tests **EM3:** 541-543
advanced composite elevator test **EM3:** 542-543
composite stabilizer ground test **EM3:** 542-543

Design proof testing (continued)
composite stabilizer test **EM3:** 542-543
subcomponent testing **EM3:** 537-539
bonded-stiffener runout detail **EM3:** 537, 538
bonded-stiffener tension damage tolerance panel **EM3:** 539
composite-to-titanium step-lap bonded joint **EM3:** 537, 538
honeycomb panel attachment detail **EM3:** 537, 538
pressure restraint test detail **EM3:** 537, 538
short-column crippling test specimen **EM3:** 537, 538-539
YC-14 honeycomb panel shear test **EM3:** 538, 539
YC-14 honeycomb panel stability test **EM3:** 538, 539
Design, quality *See* Quality control; Quality design; Statistical methods
Design requirements
anisotropy **EM1:** 181
application example **EM1:** 183-184
assembly **EM1:** 182-183
damage tolerance **EM1:** 182
design data base **EM1:** 181-182
design process **EM1:** 183
environmental effects **EM1:** 182
inspection **EM1:** 183
manufacturing and quality control **EM1:** 182
repair **EM1:** 183
total system cost **EM1:** 181
Design rules **EL1:** 15, 161
Design spacing of conductors
defined **EL1:** 1140
Design stress **A1:** 316-317, 319
Design tradeoffs *See also* Design
connector reliability **EL1:** 21-23
cooling **EL1:** 23-24
in electronic packaging **EL1:** 18-24
interconnections **EL1:** 18-21
mechanical support **EL1:** 21
Design verification, semiconductors
SEM for **A10:** 490
Design width of conductor
defined **EL1:** 1140
Design-for-reliability
as design tool **EL1:** 741, 746-748
Design-stress calculations
forgings **M1:** 357-360
Design-to-cost (DTC) process **EM1:** 419, 422-423
Designation of sheet texture **A9:** 701
Designations
as applied to solder mask **EL1:** 559
Designed experiments *See also* Experimental design
analysis of **A8:** 623, 653-661
factorial **A8:** 653-654
fractional factorial **A8:** 654-656
incomplete block **A8:** 657-661
randomized block **A8:** 656-657
Designing adhesive joints **EM3:** 457-458
design proof testing **EM3:** 457
Desiliconification
of silicon bronzes **A13:** 133
Desilvering
as lead refining **A15:** 475-476
Desizing
defined **EM1:** 9 **EM2:** 13
Deslagging *See also* Slag
cobalt-base alloys **A15:** 813
Desludging
for inclusion control **A15:** 96
Desmearing
parameters for **A10:** 403
Desmutting *See* Smut, removal of
Desoldering
defined **EL1:** 719-722
heat rate recognition **EL1:** 714-715
large thermal mass joints **EL1:** 721
planar-mounted components **EL1:** 722
Desoldering (continued)
of terminals and sockets **EL1:** 721-722
for touch up and rework **EL1:** 722
unclinching of leads by **EL1:** 722
Desorption *See also* Absorption; Adsorption **EM3:** 10
of adsorbed layers, LEISS analysis of **A10:** 603
defined **EM1:** 9 **EM2:** 13
images, gated **A10:** 596, 600-601
as leakage **A17:** 58
Dessicator
for fracture preservation **A12:** 73
Destannification
of cast tin bronzes **A13:** 133
Destaticization **EM3:** 10
defined **EM2:** 13
Destructive cross-section metallography
detection of subsurface cracking **A18:** 369
Destructive etching **A9:** 60-62
Destructive interference
x-ray spectrometers **A10:** 88
Destructive testing
ceramic coatings **M5:** 546-547
Destructive testing of
explosion welds **M6:** 711
high frequency welds **M6:** 766-767
soldered joints **M6:** 1090-1091
Destructive tests **EM3:** 37
correlated with nondestructive test results **EM3:** 772
grading of adhesive defects **EM3:** 526
inspection of final preparation **EM3:** 737
for joint design **EM3:** 43
skin peeler construction **EM3:** 739
types **EM1:** 774
Desulfomonas
biological corrosion by **A13:** 116
Desulfotomaculum
biological corrosion by **A13:** 116
Desulfovibrio
biological corrosion by **A13:** 116, 118-119
Desulfurization *See also* Sulfur; Sulfur removal **A1:** 109, 110, 228 **A13:** 561, 1001-1006
in argon oxygen decarburization **A15:** 428
cupolas **A15:** 391
of ferrous melts **A15:** 74-78
in low-carbon steel **A12:** 247
plain carbon steels **A15:** 709-710
ratio **A15:** 75-78
requirements **A15:** 75-78
systems **A15:** 75-76
Desulfurization, hot metals
magnesium powders **A7:** 131
Desulfurization of steel plate **M1:** 181
Desulfurization ratio **A15:** 75-78
Desulfurization reactor, naphtha
cracks from hydrogen damage **A11:** 664
Desulfurizer welds, carbon-molybdenum
cracking in **A11:** 663
Desulfurizing *See also* Sulfur
defined **A15:** 4
Detail (micro) testing methods
defined **EM1:** 774
Detail fracture(s)
high-carbon steels **A12:** 288
as rail fracture mode **A12:** 117, 136
rail head **A12:** 289
Detail perceptibility, of images
radiography **A17:** 300
Detail process analysis tools *See also* Analysis
ALPID system **A14:** 411-412
NIKE as FEM stress analysis program **A14:** 412
TOPAZ **A14:** 412
Detectability
low contrast, defined **A17:** 383
modeling of **A17:** 710
Detectability, damage
and tolerance requirements **EM1:** 265-266
Detected area fraction
effects of varying in iron-carbon alloys **A10:** 309
Detected signals, used for imaging and analysis in scanning electron
microscopy **A9:** 90
Detection *See also* Bridge detector; Crack detection; Detector probe Discontinuities; Detectors; Gas detection; Leak rate; Leak testing Probability of detection (POD); Leak(s) Leakage
automated, electrometric titration for **A10:** 202
by magnetic particle inspection **A17:** 103-105
characteristics, of discontinuities **A17:** 104
of combustion products in high-temperature combustion **A10:** 222-223
crack depth effects **A17:** 104
electronics and interface, inductively coupled plasma **A10:** 39
of ESR spectrometers **A10:** 256-257
feature, by image analysis **A10:** 310
of flaws **A17:** 49-50
fluorescence, EXAFS analysis **A10:** 412
infrared **A10:** 223, 230
magabsorption **A17:** 148-152
minimum, electron spin resonance **A10:** 259
modes, ion chromatography **A10:** 659-662
NDE, for volumetric flaws **A17:** 50
neutron, methods of **A17:** 390-391
of phase changes **A10:** 282-283
preferential, image analysis of AISI 416 stainless steel **A10:** 311
of radioactivity **A10:** 245-246
redox endpoint **A10:** 164
setting, effects on area fraction detected **A10:** 312
spectrophotometric, with ion chromatography **A10:** 661
surface-phase **A10:** 293
techniques, EXAFS analysis **A10:** 418
thermal-conductive **A10:** 223, 229-230
transmission mode, EXAFS analysis **A10:** 412
of x-rays **A10:** 326
XRPD methods of **A10:** 331
Detection limits
Auger electron spectroscopy **A10:** 556
defined **A10:** 671
elemental, AEM-EDS **A10:** 449
fluorescence analysis **A10:** 76
gas chromatography/mass spectrometry **A10:** 645
for ICP-AES analysis **A10:** 33
of minor elements in oil **A10:** 101
PIXE vs XRF analysis **A10:** 106
radioanalysis **A10:** 246-247
single-element interference free, for TNAA **A10:** 238
thermal neutron activation analysis **A10:** 237-238
UV/VIS absorption spectroscopy **A10:** 70
varying **A10:** 96
Detective quantum efficiency (DQE)
defined **A17:** 298, 383
and quantum noise (mottle) **A17:** 371
Detector aperture
defined **A17:** 383
Detector probe(s)
accumulation technique, with tracer gases **A17:** 65
modes of **A17:** 5
technique, specific-gas **A17:** 64-65
Detector(s)
cleanup time **A17:** 69
x-ray, defined **A17:** 383
Detectors
AAS instruments as **A10:** 55
backscatter, effect on SEM **A10:** 490
backscattered electron, contrast with **A10:** 502-504
backscattered electron, ring geometry of **A10:** 503
charged particle **A10:** 245-246

Detectors (continued)
early gas x-ray ... **A10:** 83
electron multiplier, x-ray photoelectron spectroscopy ... **A10:** 571
electron probe x-ray microanalysis, resolution (peak broadening) ... **A10:** 519
electron-optical, UV/VIS absorption spectroscopy ... **A10:** 67
energy-dispersive spectroscopy, cross section ... **A10:** 435
of energy-dispersive x-ray spectrometer ... **A10:** 519
flow proportional ... **A10:** 521
for gamma-rays ... **A10:** 235
gas-filled ... **A10:** 88
germanium, and x-ray spectrometry ... **A10:** 83
glowing-gas proportional ... **A10:** 88
and image formation, SEM microscopes ... **A10:** 493-494
imaging, for retrofitting spark ... **A10:** 145
lithium-doped silicon ... **A10:** 83, 519
mercuric iodide ... **A10:** 95
molecular fluorescence spectroscopy ... **A10:** 77
multichannel ... **A10:** 128, 137
photographic film ... **A10:** 326
photon ... **A10:** 246, 326
photoplate ... **A10:** 143
position-sensitive ... **A10:** 326, 372, 422
preamplifier and amplifier, x-ray spectrometers ... **A10:** 91
scintillation ... **A10:** 88-89
secondary electron ... **A10:** 495, 554
signal, analytical transmission electron microscopy ... **A10:** 434-436
solid-state lithium-drifted silicon ... **A10:** 89, 90
specimen current ... **A10:** 506
used for imaging and analysis in scanning electron microscopy ... **A9:** 90
used for x-ray scanning electron microscopy ... **A9:** 92-93
vidicon and diode array, use in Raman spectroscopy ... **A10:** 129
for x-ray spectrometry ... **A10:** 88-91
Detergency
engine oils ... **A18:** 169
Detergent additive
defined ... **A18:** 7
Detergent oil
defined ... **A18:** 7
Detergent/dispersant additives
hydraulic oils ... **A18:** 86
Detergents ... **A18:** 99, 100-101, 102
analytic methods for ... **A10:** 9
applications ... **A18:** 101
bases ... **A18:** 100, 101
degree of overbasing ... **A18:** 100
detection by infrared spectroscopy ... **A18:** 301
emulsifiers and ... **A18:** 106
in engine lubricant formulations ... **A18:** 111
formation of ... **A18:** 100-101
for lubricant failure ... **A11:** 154
metals used in formation of ... **A18:** 100
Miller numbers ... **A18:** 235
multifunctional nature ... **A18:** 111
neutral, idealized structures ... **A18:** 102
in nonengine lubricant formulations ... **A18:** 111
powder, as binding agents for samples ... **A10:** 94
as rust and corrosion inhibitors ... **A18:** 106
in soluble by-products ... **A18:** 101
for surface cleaning ... **A13:** 380
total base number ... **A18:** 100
Deterioration *See also* Embrittlement
by temper embrittlement ... **A11:** 69
Determination
defined ... **A10:** 671
Determination of structure *See* Structure determinations
Deterministic surface ... **A18:** 346, 348
Detonation
explosive, dynamic fracture by ... **A8:** 259
of explosively loaded torsional Kolsky bar ... **A8:** 224, 227
explosives ... **A7:** 600
gun spray process for hardfacing ... **A7:** 835
Detonation circuit
for explosive forming ... **A14:** 637-638
Detonation flame spraying
definition ... **M6:** 5
Detonation gun (D-gun) ... **EM4:** 203, 204, 206, 207
Detonation gun (D-gun) spraying *See also* High-velocity oxyfuel powder spray process ... **A18:** 644
coatings for jet engine components ... **A18:** 592
Detonation gun spraying
ceramic coatings for adiabatic diesel engines ... **EM4:** 992
Detonation gun systems
ceramic coating ... **M5:** 542, 546
oxidation-resistant coating ... **M5:** 665-666
Detonation velocity ... **A6:** 160, 161, 897
Detritus *See* Wear debris
Deuterium lamp
for UV/VIS analysis ... **A10:** 66
Deuteron
defined ... **A10:** 671
Deutsche Deramische Gesellschaft ... **EM4:** 38
Developed blank *See also* Blanks ... **A14:** 4
Developers *See also* Liquid penetrant inspection; Penetrant(s)
application of ... **A17:** 83
contamination of ... **A17:** 85
forms ... **A17:** 77
purpose/properties ... **A17:** 76
solution, activity, radiographic film processing ... **A17:** 353-354
stations for ... **A17:** 78-79
Developing
as multilayer inner layer process ... **EL1:** 542
Development
film, arresting ... **A17:** 353
procedure, radiographic film ... **A17:** 351-352
Development and production records system (DPRS) ... **EL1:** 130-132
Developmental carriers
with metal-matrix composites (MMCS) ... **EL1:** 1126-1128
Developments in Adhesives 1 and 2 ... **EM3:** 70
Deviation (x-ray)
defined ... **A9:** 5
Device current
use with integrated circuits ... **A11:** 768-769
Device delay
defined ... **EL1:** 2
Device operating failure
abbreviation for ... **A11:** 796
Device(s)
density, future trends ... **EL1:** 390
failure analysis ... **EL1:** 917-918
future trends ... **EL1:** 390-391
gate-level ... **EL1:** 2
history, determination ... **EL1:** 1058
isolation techniques, active component fabrication ... **EL1:** 199
memory bit-cell ... **EL1:** 2
minimum size, for conventional logic ... **EL1:** 2
package, trends ... **EL1:** 416
passive ... **EL1:** 994-1005
speed, maximum usable intrinsic ... **EL1:** 2
Devitrification
defined ... **A9:** 5 **EM1:** 9 **EM2:** 13
Devitrifying dielectric compositions
applications ... **EL1:** 109
Devitrifying solder glass (for soda-lime), properties
non-CRT applications ... **EM4:** 1048-1049
Dew formation
in accelerated corrosion ... **A13:** 82
Dew point ... **EM3:** 10
corrosion, fossil fuel plants ... **A13:** 1001-1004
defined ... **EM2:** 13
nomograph ... **EL1:** 1065
for sintered P/M stainless steels ... **A13:** 828
Dew-point analyzers
atmospheres ... **M4:** 426-428
Dewar flask
defined ... **A10:** 671
Dewaxing ... **A15:** 4, 262
defined ... **A7:** 3
Dewaxing, low-pressure
of cermets ... **A2:** 989
Dewetting
as cleaning defect ... **EL1:** 777
copper alloys ... **A6:** 631
defined ... **EL1:** 642
Dewetting (continued)
from sulfur impurities ... **EL1:** 642
plated-through hole ... **EL1:** 1026
as solderability mechanism ... **EL1:** 676, 1032-1034
vs wetting ... **EL1:** 990
Dewpoint
defined ... **A7:** 3
in exothermic gas ... **A7:** 344
and metal/metal oxide equilibria of hydrogen ... **A7:** 498
of sintering atmospheres ... **A7:** 340, 341
and water vapor content of sintering atmospheres ... **A7:** 361
Dexter, Thomas
as early founder ... **A15:** 24
Dextrines
applications ... **EM3:** 45
characteristics ... **EM3:** 45
for packaging ... **EM3:** 45
Dezincification *See also* Corrosion; Dealloying; Selective leaching
of admiralty brass ... **A13:** 128, 132
in aqueous environments ... **A13:** 131-133
of brass, as selective leaching ... **A11:** 628
in brasses ... **A13:** 614
in copper alloy (cartridge brass) pipe ... **A11:** 178
copper-zinc alloy failure by ... **A11:** 222
of copper-zinc alloys ... **A13:** 131-133
cracking ... **A13:** 129, 132
defined ... **A7:** 3 **A8:** 4 **A9:** 5 **A11:** 3 **A13:** 5
definition and mechanics of ... **A12:** 26
inhibitors ... **A13:** 132
plug and layer types ... **A11:** 633
plug-type ... **A13:** 128, 132
as selective leaching in brasses ... **A11:** 633
yellow brass failure by ... **A11:** 179
Dezincification resistance
in manganese bronze ... **A2:** 348
Df *See* Dilution factor
DGEBA *See also* Diglycidyl ether of bisphenol A; Epoxies (EP)
as diglycidyl ether of bisphenol A ... **EM2:** 240
properties ... **EM2:** 241
structure ... **EM2:** 240
DGV *See* Distributed gage volume
Di-(2-ethylhexyl)phosphoric acid
as solvent extractant ... **A10:** 169-170
Di-amine polyimides ... **EM1:** 78
Di-tungsten carbide ... **A9:** 274-275
Diadic polyamide ... **EM3:** 10
defined ... **EM2:** 13
Dial
in spring-material test apparatus ... **A8:** 134-135
Dial comparators (micrometers) paint film thickness tests ... **M5:** 491-492
DIAL computer program for structural analysis ... **EM1:** 268, 270-271
Dial feeds
of blanks ... **A14:** 500
Diallyl chlorendate (DAC) *See* Allyls
Diallyl isophthalate (DAIP) *See* Allyls
Diallyl maleate (DAM) *See* Allyls
Diallyl phthalate
surface preparation ... **EM3:** 278
Diallyl phthalate (DAP) *See* Allyls
in polyester reaction ... **EM1:** 132-133
Diallyl phthalate as a mounting material ... **A9:** 29
for cemented carbides ... **A9:** 273
for copper and copper alloys ... **A9:** 399
for hafnium ... **A9:** 497
for titanium and titanium alloys ... **A9:** 458
for zirconium and zirconium alloys ... **A9:** 497
Diallyl phthalate resins *See also* Allyls (DAP, DAIP)
characteristics ... **EM2:** 227-229
electrical properties ... **EM2:** 228
as injection-moldable ... **EM2:** 321
physical properties ... **EM2:** 228
suppliers ... **EM2:** 229
Diamagnetic glasses ... **EM4:** 855
Diamagnetic materials
defined ... **A9:** 63
Diamagnetism
of superconducting materials ... **A2:** 1030
Diameter ... **A10:** 198, 690, 692
Diameter strain
and forging modes ... **A7:** 414

Diameter to thickness ratio (*D/t*)
in pin bearing testing **A8:** 59
Diameter tolerances
cold finished bars **M1:** 217-219
Diameter(s) *See also* Dimensional measurement; Tube(s); Tubing
changes, tube, remote-field eddy current inspection **A17:** 195
measurement, by lasers **A17:** 12-16
tube, eddy current method selection by **A17:** 179-184
ultrasonic beams **A17:** 240
Diameters *See* Fiber diameter
cylinder, in three-roll forming **A14:** 621
reducing, by necking/nosing **A14:** 586
rolls, three-roll forming machines **A14:** 618
for three-roll forming **A14:** 616
Diametral expansion
dependence on overgrind **A14:** 142
Diametrical strength
defined **A7:** 3
Diamond *See also* Diamond abrasive grains; Diamond drill oil bits; Diamond grit; Diamond-containing cermets; Polycrystalline diamond; Sintered polycrystalline diamond; Superabrasives; Synthetic diamond; Tool materials, superhard; Ultrahard tool materials **A16:** 2, 105-117
as abrasive **EM4:** 324
abrasive applications **A16:** 454, 456, 457, 458
abrasive bond advantages **EM4:** 325
abrasive bond limitations **EM4:** 325
in abrasive flow machining **A16:** 517
abrasive for Al-Si alloys **A16:** 402
abrasive for ECG **A16:** 545
abrasive for honing stones **A16:** 476
abrasive for lapping **A16:** 493, 495, 503, 505
abrasive for truing **EM4:** 347
as abrasive material **EM4:** 329
abrasive property control **EM4:** 332
abrasives for electronic ceramics grinding **EM4:** 336, 337-339
applications **A16:** 110 **EM4:** 331, 825-827
atom arrangement **A16:** 106
bond type **EM4:** 331
bondability **EM4:** 332
bonded, in honing stones **A16:** 475, 477
centerless grinding of advanced ceramics **EM4:** 341
chemical inertness **EM4:** 822-824
for coarse and fine polishing before microstructural analysis **EM4:** 573, 574
compared to CBN **A16:** 105
core drills for ultrasonic machining **A16:** 528-529, 530, 531
cutting tool material **A18:** 617
cutting tools, tungsten carbide use in **A2:** 976
CVD coatings **EM4:** 826-827
in dental polishing pastes **A18:** 666
dressing methods **EM4:** 347-348
dry abrasive effect on Al-13Si alloy **A18:** 195
electrical discharge grinding **A16:** 107-108
electrical properties **EM4:** 824
engineering properties **EM4:** 821-831
fracture toughness **A18:** 192
friability of abrasives **EM4:** 332
friction coefficient data **A18:** 75
and graphite, crystal structures of **A10:** 345, 355
and graphite, theoretical properties **EM1:** 49
grinding **A16:** 421-422, 423, 424, 425
as grinding abrasive before microstructural analysis **EM4:** 572, 574
grinding methods and applications based on abrasive grain size **EM4:** 343
grinding of 52100 bearing steel **A16:** 465, 466
grinding of electronic ceramics **A16:** 462
and grinding of gears **A16:** 351, 353
grinding of technical ceramics **EM4:** 334-335
grinding wheel construction **EM4:** 337-339, 343
Diamond (continued)
grinding wheel tolerance specifications **EM4:** 344, 806, 821, 823
grinding wheels and their selection **A16:** 460-461
grit, synthesis of **A2:** 1008-1009
hardness **A18:** 433 **EM4:** 330, 351
high-alumina ceramics grinding **EM4:** 333
high-strength metal bonds with abrasives **EM4:** 331
in honing stones **A16:** 476, 477, 478, 479
impurities **EM4:** 822
indenter **A8:** 74-75, 83, 90
indexable-insert milling cutters **A16:** 315, 317
lapping abrasive **EM4:** 351, 352
local structure of trace impurities, EXAFS determined **A10:** 407
machining parameters **A16:** 110
mechanical properties **EM4:** 331, 823, 824
modulus of resilience **EM4:** 330, 333
natural (type la), thermal properties **A18:** 42
nickel metal coatings **EM4:** 333
on Mohs scale **A8:** 108
optical properties **EM4:** 825, 827
for orifice of waterjet nozzles **A16:** 521, 522
parameters affecting abrasive and wheel bond type selection for machining ceramics **EM4:** 343
paste **A8:** 234
PCD (polycrystalline diamond) **A16:** 105, 106
PCD abrasion resistance **A16:** 109
PCD and boring, Al alloys **A16:** 772, 778
PCD and face milling, Al alloys **A16:** 789
PCD and grinding, Al alloys **A16:** 770
PCD and grinding wheels, Mg alloys **A16:** 821
PCD and gun drilling, Al alloys **A16:** 791
PCD and machining, Al alloys **A16:** 765, 766, 767, 792, 793, 797-800
PCD and machining, Cu alloys **A16:** 809, 811, 813, 816
PCD and machining, Mg alloys **A16:** 821, 827
PCD and machining, Zn alloys **A16:** 832
PCD and sawing **A16:** 110
PCD and turning, Al alloys **A16:** 767-769, 774, 775, 776, 777
PCD applications **A16:** 639
PCD cutting tools **A16:** 106-107, 111
PCD, grinding of **A16:** 107
PCD machinability **A16:** 639, 640
PCD routing **A16:** 110
PCD Syndite microstructures, 4 different grades **A16:** 108
PCD tool fabrication **A16:** 111
physical properties **A18:** 192 **EM4:** 316
plasma-assisted physical vapor deposition **A18:** 848
polytype parameters **EM4:** 823
polytypes **EM4:** 822, 823
properties **EM4:** 822-825
properties of **A2:** 1009-1010
protective coating against liquid impingement erosion **A18:** 222
relationship between particle size and particles per carat **EM4:** 467
single-crystal tools **A16:** 107, 110
softening point **A16:** 601
with spiked nickel coating **A2:** 1014
structure **EM4:** 821
superabrasive **A16:** 4, 102, 430-432, 439, 441-444, 453-454
superabrasive grinding wheels standard marking system **EM4:** 466
superabrasive tool grinding **A16:** 450
as superhard material **A2:** 1008
synthesis **A16:** 105
synthesis of **A2:** 1008-1009
synthesized **EM4:** 822
synthetic **A16:** 453-454
synthetic (polycrystalline), thermal properties **A18:** 42
Diamond (continued)
synthetic, EXAFS characterization of metal impurities in **A10:** 417
thermal properties **EM4:** 316, 331, 824-825, 826
thin Type II, for diamond-anvil cell **A10:** 113
thread grinding wheels **A16:** 271, 272-273
tool geometries **A16:** 109-110
tool life **A16:** 108-109, 110
tooling **EM1:** 295
toughness **EM4:** 332, 333
toughness index **EM4:** 332
truing device in thread grinding **A16:** 272
truing methods **A16:** 466-467
truing methods for production grinding **EM4:** 346, 348
truing parameters **A16:** 468
wear resistance **A16:** 108-109, 111, 112
wheels for electrochemical grinding **A16:** 542
wire guide, EDM **A16:** 562
Diamond abrasive grains
shapes **A2:** 1010-1011
sizes **A2:** 1010-1011
Diamond abrasives
ability to delineate abrasion damage **A9:** 37
for cemented carbides **A9:** 273
for hand polishing **A9:** 35
material removal rates in polishing **A9:** 37
polycrystalline **A9:** 43
results with napless cloths **A9:** 40
used for tool steels **A9:** 257
Diamond cutting wheels **A9:** 25
used to section tool steels **A9:** 256
Diamond dies
for drawing **A14:** 336
Diamond film *See also* Diamond like film
defined **A18:** 7
Diamond grit
synthesis of **A2:** 1008-1009
Diamond insulating films
electrical/electronic applications **EM4:** 1105
Diamond like carbon (DLC)
coating for titanium alloys **A18:** 778, 781-782
Diamond like film *See also* Diamond film
defined **A18:** 7
Diamond like hydrogenated carbon electroplated coatings **A18:** 838
Diamond oil drill bits
application **A2:** 976
Diamond pyramid hardness
abbreviation **A8:** 724
abbreviation for **A11:** 796
Diamond pyramid hardness (DPH) **A18:** 415
Diamond pyramid hardness indentation **A7:** 61
Diamond pyramid hardness numbers *See also* Vickers hardness numbers
conversion tables for **A8:** 109-110
Diamond pyramid hardness test *See also* Vickers hardness test; Vickers microhardness test **A8:** 4
Diamond pyramid indentation
barrel-shaped **A8:** 100, 102
distortion due to elastic effects **A8:** 102
pincushion **A8:** 100, 102
for Vickers test **A8:** 91
Diamond saw cuts of honeycombs **A9:** 24
Diamond saws **A9:** 24-25
Diamond tools
grinding, Ti alloys **A16:** 848
honing, P/M materials **A16:** 891
milling **A16:** 327
sawing, honeycomb structures **A16:** 900-901
Diamond wheels *See* Diamond cutting wheels
Diamond-abrasive leadfoil lap, used in abrading materials with a surface oxide layer **A9:** 46
Diamond-anvil cell **A10:** 113, 116
Diamond-containing cermets **A7:** 813-814
application and properties **A2:** 1005

SUBJECTS OF THE INDEXED VOLUMES: ASM Handbook (designated by the letter "A"): **A1:** Properties and Selection: Irons, Steels, and High-Performance Alloys (1990); **A2:** Properties and Selection: Nonferrous Alloys and Special-Purpose Materials (1990); **A3:** Alloy Phase Diagrams (1992); **A4:** Heat Treating (1991); **A6:** Welding, Brazing, and Soldering (1993); **A7:** Powder Metallurgy (1984); **A8:** Mechanical Testing (1985); **A9:** Metallography and Microstructures (1985); **A10:** Materials Characterization (1986); **A11:** Failure Analysis and Prevention (1986); **A12:** Fractography (1987); **A13:** Corrosion (1987); **A14:** Forming and Forging (1988); **A15:** Casting (1988); **A16:** Machining (1989); **A17:** Nondestructive Testing and Quality Control (1989); **A18:** Friction, Lubrication, and Wear Technology (1992). **Metals Handbook, 9th Edition** (designated by the letter "M"): **M1:** Properties and Selection: Irons and Steels (1978); **M2:** Properties and Selection: Nonferrous Alloys and Pure Metals (1979); **M3:** Properties and Selection: Stainless Steels, Tool Materials and Special-Purpose Materials (1980); **M4:** Heat Treating (1981); **M5:** Surface Cleaning, Finishing, and Coating (1982); **M6:** Welding, Brazing, and Soldering (1983). **Engineered Materials Handbook** (designated by the letters "EM"): **EM1:** Composites (1987); **EM2:** Engineering Plastics (1988); **EM3:** Adhesives and Sealants (1990); **EM4:** Ceramics and Glasses (1991); **Electronic Materials Handbook** (designated by the letters "EL"): **EL1:** Packaging (1989).

Diamond-impregnated wires for sawing A9: 26
Diamond-stop lapping fixture, modified
for plate specimen thickness A8: 234-235
Diapers, baby
waterjet machining A16: 526
Diaphragm
defined A10: 671
for measuring displacement A8: 193
Diaphragm compressors
in HIP processing A7: 421-422
Diaphragm forming
of thermoplastic resin composite EM1: 549-551
Diaphragm gate
defined EM2: 13
Diaphragms
for scattered radiation A17: 344
Diaspore EM4: 49, 111
Diatomaceous silicon dioxide
abrasive in commercial prophylactic paste A18: 666
Diatomite
in alloy-coated composite powders A7: 174
Dibasic acid esters
lubricants for rolling-element bearings A18: 134-135
properties A18: 81
Dibasic calcium phosphate dihydrate
dentifrice abrasive A18: 665
Diborides, as inclusions
aluminum alloys A15: 95
Dicalcium silicate
in composition of portland cement EM4: 11, 12
Dichromate coating, magnesium alloys M5: 629, 632, 634-637, 640-641, 643, 647-648
stripping of M5: 648
Dichromate sealing process
anodic coatings M5: 595
Dichromate treatment
defined A13: 5
Dichromate-bifluoride post treatment
magnesium alloys M5: 633-634, 639
Dicing A18: 686-688
Dickite EM4: 5, 6
Dicor, applications
dental EM4: 1094, 1095, 1096
Dicumene chromium
chemical vapor deposition process M5: 383
Dicyandiamide (DICY) EM3: 95
as curing agent EM3: 102, 590, 594
Dicyandiamide, as curing agent
epoxies EL1: 830-831
Die
cast zinc alloy A13: 241
chilling simulation, torsion test for A8: 180
circle grid analysis for A8: 566-567
definition M6: 5
forging, aluminum, transverse grain direction A8: 668
forging, primary testing direction, various alloys A8: 667
punch and, sheet metal forming A8: 547
-type wipe bending devices A8: 125
Die adapter
defined EM2: 14
Die angle A18: 63
Die applications M1: 429, 437
Die assembly
defined A14: 4
Die attach *See also* Die attachment methods; Die(s)
debonding EL1: 371
lead frame assembly EL1: 487
molded plastic packages EL1: 471-472
tests, package-level EL1: 931-934
thermal design considerations EL1: 411-412
thermomechanical design considerations EL1: 415
zone 2, package interior EL1: 1010-1011
Die attach adhesives EM3: 76
Die attachment methods *See also* Die attach; Die(s)
eutectic die attach EL1: 213-215
glass die attach EL1: 215
gold-silicon eutectic die attach EL1: 217
for hermetic packages EL1: 213-217
plastic-encapsulated devices EL1: 217-221
polymer die attach EL1: 217-220
silver-glass die attach EL1: 215-216
Die attachment methods (continued)
solder die attach EL1: 216-217, 221
solid-film polymers EL1: 220-221
Die block *See also* Closed-die forging A14: 4
Die block steels A14: 230
Die blocks
ultrasonic inspection A17: 232
Die burns A6: 844
Die casting *See also* Castings, Foundry products; Pressure casting A15: 286-295
alloying element and impurity specifications A2: 16
alloys, compositions A15: 286
aluminum alloy A15: 753-755
aluminum alloys M2: 140, 143, 145, 147
aluminum casting alloys A2: 136-139
cold chamber process A15: 286
of copper alloys A2: 346
copper casting alloys M2: 384
cycles, preparation for A15: 294
defects A15: 294-295
defined A15: 4
development of A15: 35
direct injection process A15: 286
ejection, casting A15: 294
gating system A15: 288-292, 755
heat flow paths A15: 295
heat removal A15: 292-294
hot chamber process A15: 286
low-pressure A15: 276
magnesium alloy A15: 799, 807-810
of metal-matrix composites A15: 844
as permanent mold process A15: 34
postcasting operations A15: 295
process control in A15: 286
product design for A15: 286-288
single-cavity, components A15: 288
temperature, aluminum casting alloys A2: 168, 172
tolerances A15: 289, 619-620
Die casting alloys
intergranular corrosion evaluation A13: 241
Die casting machines
zinc alloys A15: 789
Die castings
aluminum, polishing and buffing M5: 573-574, 576
brass *See* Brass, die castings
chromium plating, decorative M5: 189, 192, 194-197
cleaning and finishing processes M5: 677
copper plating of M5: 160-161, 163, 168
magnesium alloy *See* Magnesium alloys, die castings
polishing and buffing of M5: 10-11, 108, 112-114, 123, 676
zinc *See* Zinc and zinc alloy die castings M2: 630-634
Die cavity A14: 4, 159
Die check
defined A14: 4
Die clearance *See also* Clearance
in blanking A14: 447
deep drawing A14: 581
defined A14: 4
rotary swaging A14: 132
selection, for piercing A14: 459-460
Die closure
defined A14: 4
speeds, of forming machines A14: 100
Die cone angle A18: 60, 61, 63, 66
Die contact surface fracture
prediction A14: 402-403
Die cracking EL1: 215, 1045-1046
Die cushions *See also* Die(s) A14: 497-499
defined A14: 4
hydropneumatic A14: 498-499
pneumatic A14: 498
Die cutting *See also* Blanking
cutting die EM1: 610-611
defined EM2: 14
die storage/retrieval EM1: 612-613
pad EM1: 609-610
ply EM1: 608-614
presses EM1: 608-609
Die design *See also* Die(s)
for aluminum alloys A14: 246-247
Die design (continued)
with CAD A14: 906-907
cold extrusion A14: 302-303
for explosive forming A14: 639
for heat-resistant alloys A14: 234
for high-velocity pneumatic-mechanical forging A14: 101-102
hot extrusion A14: 321
minimum-draft forging A14: 258
segmented, for physical modeling A14: 435
with side clearance A14: 133
and stress analysis, computer-aided A14: 414
for titanium alloys A14: 276-277
Die erosion
as casting defect A11: 384
Die failures
analytical approach A11: 563
causes of A11: 564-577
design and A11: 564-566
heat treatment, influence of A11: 564, 567-574
machining, influence of A11: 564, 566-567
service conditions A11: 575-577
steel quality and grade A11: 564, 566, 574-575
types of A11: 564
Die forger hammers A14: 28
Die forging
defined A14: 4
Die forging of titanium alloys M3: 366-368, 369
Die forming
defined A14: 4
Die hardness *See also* Hardness
for stainless steels A14: 228
Die heating *See also* Die temperature; Die(s); Heat treatment; Heating; Temperature(s) A14: 235, 248
aluminum alloy A14: 248
for beryllium forming A14: 806
for copper and copper alloy forging A14: 257
for heat-resistant alloys A14: 235
for magnesium alloy forming A14: 259, 828
in precision forging A14: 162
in pultrusion EM2: 390-391
for stainless steels A14: 229
for titanium alloys A14: 278-279
Die heating/dies
pultrusion EM1: 534-536
Die height
defined A14: 4
Die holder *See also* Sub-sow block A14: 4
Die impression
defined A14: 4
Die inserts
defined A14: 4
gripper A14: 48
in hot upset forging A14: 86
types A14: 47-48
Die life A14: 57
and abrasion A14: 505
blanking/piercing high-carbon steels A14: 556
defined A14: 4
fine-edge blanking and piercing A14: 474-475
forging severity effects A14: 228
hardness effects A14: 99
in high-energy-rate forging A14: 102
impact cutoff machines A14: 719
of press forming dies A14: 505-507
of roll dies A14: 99
in squeeze casting A15: 323
stainless steels A14: 228
steel and carbide, compared A14: 485
titanium alloy forging A14: 277
of wire-drawing dies A14: 336
Die line *See also* Galling; Pickup; Scoring A14: 4
Die locks *See also* Counterlocks; Locks
for mismatch A14: 49
Die lubricant *See also* Drawing compound; Lubricants; Lubrication
defined A14: 4
for stainless steels A14: 229
Die lubrication A14: 87, 229, 508
Die making
for aluminum alloys A14: 245, 247
CAD/CAM applied to A14: 903-910
for heat-resistant alloys A14: 235
for titanium alloys A14: 277

Die match
defined ... **A14:** 4
Die materials *See also* Tool materials; Work materials ... **A14:** 45
for aluminum alloys ... **A14:** 245-246
for bar bending ... **A14:** 663
for blanking and piercing ... **A14:** 479, 483-486, 761
for coining ... **A14:** 181
common ... **A14:** 45
compositions ... **A14:** 43
computer-aided design and manufacturing of ... **A14:** 57-58
for contour forging ... **A14:** 71
for drawing ... **A14:** 336-337, 580
for electrical steel sheet ... **A14:** 479
for explosive forming ... **A14:** 638-639
for fine-edge blanking and piercing ... **A14:** 474
and hardness, die life effects ... **A14:** 57
heat treating, types ... **A14:** 53-55
for heat-resistant alloys ... **A14:** 234-235
for HERF processing ... **A14:** 102
for hot forging ... **A14:** 43-58
hot-die/isothermal forging ... **A14:** 154
life of ... **A14:** 157
for magnesium alloy forming ... **A14:** 828
for nickel-base alloys ... **A14:** 261
piercing ... **A14:** 479, 761
for press bending ... **A14:** 527
for roll forging ... **A14:** 97
safety with ... **A14:** 58
selection of ... **A14:** 45-47
for sheet metals ... **A14:** 479, 580
stainless steel ... **A14:** 227-228, 761
for titanium alloys ... **A14:** 275-276
for trimming and punching ... **A14:** 55-56
for wire drawing ... **A14:** 336
Die pad
defined ... **A14:** 4
Die plates
materials for ... **A14:** 484
Die presses ... **A14:** 502
Die pressing
advanced ceramics ... **EM4:** 49
at elevated pressure ... **EM4:** 188-189
in reaction sintering ... **EM4:** 292
Die proof *See also* Proof ... **A14:** 4
Die pull
defined ... **A15:** 4
Die radius
defined ... **A14:** 4
Die run
and burr height ... **A14:** 481
Die separation
defined ... **A15:** 4
Die set
defined ... **A14:** 4
Die shaft
defined ... **A14:** 4
Die shape
for rotary swaging ... **A14:** 132
simulation ... **A14:** 428
Die shoes
defined ... **A14:** 4
Die sinking ... **A14:** 4, 247
Die space ... **A14:** 4, 84
Die springs ... **A1:** 302
Die stamping
defined ... **A14:** 4
Die steels
hardfacing ... **A6:** 798
Die swell ratio
defined ... **EM2:** 14
Die systems *See also* Die(s); Dies, specific types
aluminum precision forging ... **A14:** 253
for explosive forming ... **A14:** 638-639
for hot-die/isothermal forging ... **A14:** 154
Die taper angle
for swaging ... **A14:** 135, 139
Die temperature *See also* Die heating; Temperature(s)
aluminum alloy forging ... **A14:** 242
control of ... **A14:** 81-82
effects, on forging pressure ... **A14:** 153
hot-die/isothermal forging ... **A14:** 153
ranges, aluminum alloy forging ... **A14:** 242
for titanium alloys ... **A14:** 270
Die threading *See also* Dies
accuracy and finish ... **A16:** 300
Acme threads ... **A16:** 299
Al alloys ... **A16:** 783, 791
automatic turret lathes ... **A16:** 296
by a single-point tool ... **A16:** 299
by grinding ... **A16:** 299
chamfer angles ... **A16:** 297, 298
chip clearance ... **A16:** 298
collet-type machines ... **A16:** 296
compared to thread rolling ... **A16:** 295
composition and hardness of work metal ... **A16:** 299
factors that influence output ... **A16:** 299
lead control ... **A16:** 296, 300-301
Mg alloys ... **A16:** 825-826
MMCs ... **A16:** 896
pipe threading speed ... **A16:** 302
selection of machine ... **A16:** 296
single-chaser threading ... **A16:** 298
single-point tool ... **A16:** 302
solid dies ... **A16:** 38-39, 296-297
stainless steels ... **A16:** 694, 698, 700-701
taper threading of pipe ... **A16:** 301-302
thread size ... **A16:** 300
threading to a shoulder ... **A16:** 300
tool life ... **A16:** 301
tools for pipe threading ... **A16:** 302
turret lathes ... **A16:** 296, 297
using bar and chucking machines ... **A16:** 296, 298
Zn alloys ... **A16:** 833-834
Die trimming
in die casting ... **A15:** 295
Die wall friction ... **EM4:** 126, 127
Die wear ... **M1:** 364-367
Die(s) *See also* Attachment; Die assembly; Die attach; Die attachment methods; Die block; Die clearance; Die cushions; Die design; Die heating; Die life; Die lubrication; Die making; Die materials; Die systems; Die temperature; Dies, specific types; Hermetic- packages; Plastic-encapsulated devices
for aluminum and aluminum alloys ... **A2:** 6
angle and friction, top surface strains ... **A7:** 412
arrangement, rotary forging ... **A14:** 17
attach reliability ... **EL1:** 61-62
back side preparation, eutectic die attach ... **EL1:** 213-214
barrel ... **A7:** 3, 304
beading ... **A14:** 537-538
for bending ... **A14:** 666
for blanking ... **A14:** 451-456, 478-479, 484, 569
body ... **A7:** 3
bolster ... **A7:** 3
box-forming ... **A14:** 537
breakage, in drawing ... **A14:** 337
breakthrough ... **A7:** 3
cam-actuated flanging ... **A14:** 526-527
cam-driven ... **A14:** 526-527, 538
cast ... **A14:** 53
cavity ... **A7:** 3
channel ... **A14:** 536
chokes ... **A7:** 325
chopper ... **A14:** 132
for closed-die (impression die) forging ... **A14:** 43-45
for coining silverware ... **A14:** 182
cold ... **A14:** 56
for cold extrusion ... **A14:** 308-309
combination, defined *See* Combination die
composite ... **A14:** 639
Die(s) (continued)
compound ... **A14:** 454, 465-466, 527, 546
compound flanging and hemming ... **A14:** 527
computer modeling ... **A14:** 903-910
computer-aided design and manufacture ... **A14:** 57-58
construction, for press bending ... **A14:** 525-527
contour ... **A14:** 132
for copper and copper alloys ... **A14:** 256
cracking ... **EL1:** 215, 1045-1046
cracks, types of ... **EL1:** 215
curling ... **A14:** 537
cushions for ... **A14:** 497-499
cutting force, blanking ... **A14:** 448
for decorative coining ... **A14:** 181-182
deep drawing ... **A14:** 508-511, 579-580
defined ... **A7:** 3 **A14:** 4
design ... **A7:** 335
for drawing ... **A14:** 336-337
drawing, cemented carbide ... **A2:** 969
for electrical steel sheet ... **A14:** 478-479
extrusion ... **EM2:** 381
fabrication, HERF processing ... **A14:** 102
failure, causes of ... **A14:** 55-56
failure mechanisms, through-hole packages ... **EL1:** 974
fill ... **A7:** 3
filling ... **A7:** 297, 329
forging, for SIMA process ... **A15:** 332
forging machine ... **A14:** 45
forgings, tensile properties ... **A7:** 653
four-piece, ovality in ... **A14:** 133
fully cylindrical ... **A14:** 97
heat treating ... **A14:** 53-55
for heat-resistant alloys ... **A14:** 234-235
heating, in pultrusions ... **EM2:** 390-391
high-energy-rate forging ... **A14:** 101-104
high-pressure, cemented carbide ... **A2:** 972
for hot forging ... **A14:** 43-58
for hot pressing ... **A7:** 502
for hydrostatic extrusion ... **A14:** 328
insert ... **A7:** 3, 337
life of ... **A14:** 57
liner ... **A7:** 3
lubricant ... **A7:** 3
lubricants for ... **A14:** 279
lubrication of ... **A14:** 87, 229, 508
and machines, interaction of fluid flow ... **A15:** 292
for magnesium alloys ... **A14:** 259
for mandrel swaging ... **A14:** 137-138
materials for ... **A14:** 45-47
mismatch ... **A14:** 49
for multiple-slide forming ... **A14:** 569
number, for rotary swaging ... **A14:** 131
for open-die forging ... **A14:** 61, 62
opening ... **A7:** 4
and parting lines ... **A14:** 48
performance, deep-drawing ... **A14:** 508
for piercing, electrical steel sheet ... **A14:** 478-479
plate ... **A7:** 4
powder compacting, cemented carbide ... **A2:** 971
for precision forging ... **A14:** 51-52
press forming ... **A14:** 504-507, 545-546
for press-brake forming ... **A14:** 535-539
primary, pultrusion ... **EM2:** 392
punching ... **A14:** 55
radii, deep drawing effects ... **A14:** 580-581
as reusable reverse patterns ... **A15:** 192
rigid, pressing of powders ... **A7:** 297
role, types ... **A14:** 97-99
in rotary forging ... **A14:** 178
rubber *See* Rubber-pad forming
safety with ... **A14:** 58
set ... **A7:** 4, 331-332
shape ... **A14:** 132, 428
shaped, for hot extrusion ... **A14:** 320, 325
shaving ... **A14:** 458
with shear, cutting force ... **A14:** 448
with side clearance, design ... **A14:** 133

Die(s) (continued)
with sidearm, direction of powder flow ... **A7:** 300
silicon microelectronic, silicone conformal ... **EL1:** 22
size, and IC complexity ... **EL1:** 400
for stainless steels ... **A14:** 227-228
stamping, cemented carbide ... **A2:** 971
swaging ... **A14:** 132-133
tapered, strains in compression between ... **A7:** 412
temperature, in die casting ... **A15:** 289
temperature, wrought titanium alloys ... **A2:** 614
thermophysical properties, data base ... **A14:** 412
for titanium alloys ... **A14:** 274-277
volume ... **A7:** 4
wall, friction, during compacting ... **A7:** 302
for zinc alloy casting ... **A15:** 789-790
Die-cast zinc alloy nut
SCC of ... **A11:** 538-539
Die-cast zinc alloys
capacitor discharge stud welding ... **A6:** 222
Die-casting dies
injection components ... **M3:** 542, 543
selection of materials for ... **M3:** 542-543
slides, guides, cores and pins ... **M3:** 542-543
trim dies ... **M3:** 543
Die-closing swagers ... **A14:** 131
Die-parting line
defined ... **EM2:** 14
Die-workpiece
contact area (footprint), rotary forging ... **A14:** 17
interface, data base for ... **A14:** 413
Die/isostatic pressing
for gas turbine component fabrication ... **EM4:** 718
Dieing machines ... **A14:** 502
Dielectric *See also* Complex dielectric constant; Dielectric constant ... **EM3:** 10
breakdown ... **EM2:** 581-582
breakdown voltage, of glass fibers ... **EM1:** 46-47
curing, defined ... **EM1:** 9
defined ... **EM1:** 9, 359 **EM2:** 14, 460
heating, defined ... **EM1:** 9
loss, defined ... **EM1:** 9
monitoring, defined ... **EM1:** 9 **EM2:** 14
properties, of polymers ... **EM2:** 62
Dielectric absorption
defined ... **EL1:** 1141
Dielectric breakdown
copper-clad E-glass laminates ... **EL1:** 537
defect-related ... **A11:** 778-779
defined ... **EL1:** 1141
as failure mechanism for MOS and CMOS ... **A11:** 778-779
glass capacitors ... **EL1:** 998
intrinsic, integrated circuits ... **A11:** 779-780
VLSI, and accelerated testing ... **EL1:** 892
wet-slug tantalum capacitors ... **EL1:** 997-998
Dielectric constant *See also* Complex dielectric constant; Dielectric ... **EM3:** 10, 428, 429
at temperature, and material selection ... **EM1:** 38
ceramic multilayer packages ... **EL1:** 467-468
of ceramics ... **EL1:** 336
copper-clad E-glass laminates ... **EL1:** 534-536
defined ... **EL1:** 99, 1141 **EM1:** 9 **EM2:** 14, 461, 467
as determining circuit speed ... **EL1:** 1
and dissipation factor ... **EM2:** 582-585
of engineering plastics ... **EM2:** 73
epoxy resin system composites ... **EM1:** 403, 405, 409, 415
as function of frequency ... **EL1:** 82
glass fabric reinforced epoxy resin ... **EM1:** 405
glass fibers ... **EM1:** 107
high-temperature thermoset matrix composites ... **EM1:** 376
Kevlar 49 fiber/fabric reinforced epoxy resin ... **EM1:** 409
low-temperature thermoset matrix composites ... **EM1:** 397
medium-temperature thermoset matrix composites ... **EM1:** 385, 388
microwave inspection ... **A17:** 204
of packaging materials ... **EL1:** 21
polyether-imides (PEI) ... **EM2:** 158
and propagation time ... **EL1:** 20-21
PWB substrate, effect ... **EL1:** 505

Dielectric constant (continued)
quartz fabric reinforced epoxy resin ... **EM1:** 415
reduced, high-performance electronic systems ... **EL1:** 15
sheet molding compounds ... **EM1:** 158
thermoplastic matrix composites ... **EM1:** 367, 369, 372
of thermoplastics/thermosets ... **EM2:** 470
Dielectric curing ... **EM3:** 10
Dielectric films, deposition of
vacuum coating process ... **M5:** 408-409
Dielectric fluid
for electric discharge machining ... **A9:** 26
Dielectric heating ... **EM3:** 10
defined ... **EM2:** 14
Dielectric inks
thick-film ... **EL1:** 208, 346
Dielectric loss ... **EM3:** 10
angle, defined ... **EM2:** 14
cyanates ... **EM2:** 232
defined ... **EM2:** 14
of engineering plastics ... **EM2:** 73
factor, defined ... **EM2:** 14, 461
Dielectric loss angle ... **EM3:** 10
Dielectric loss factor ... **EM3:** 10
Dielectric materials *See also* Dielectrics; Substrates
chemical composition, by microwave inspection ... **A17:** 215
constraining, CTE tailoring ... **EL1:** 614
effects ... **EL1:** 83
microwave thickness gaging ... **A17:** 212
for multichip structures ... **EL1:** 301-303
nonmetallic, dielectric properties ... **A17:** 205
polarization ... **EL1:** 99-100
polyimide glass ... **EL1:** 83
for printed wiring boards ... **EL1:** 597-610
quantum mechanical band theory (solid state) ... **EL1:** 99-100
for rigid printed wiring boards ... **EL1:** 538-539
silicon dioxide ... **EL1:** 88
Dielectric monitoring ... **EM3:** 10
Dielectric permittivity ... **EM3:** 428
Dielectric phase angle ... **EM3:** 10
Dielectric power factor ... **EM3:** 10
defined ... **EM2:** 14
Dielectric process
and organic binders ... **A15:** 35
Dielectric properties
of adhesives ... **EL1:** 674
defined ... **EL1:** 90
electrical properties ... **EL1:** 597-601
and high-speed electrical performance issues ... **EL1:** 597-610
materials properties ... **EL1:** 604-608
printed board manufacturability ... **EL1:** 608-609
as properties change monitor ... **EL1:** 836
recommendations ... **EL1:** 608
Dielectric shield
defined ... **A13:** 5
Dielectric strength ... **EM3:** 10
defined ... **EL1:** 90, 336, 1141 **EM1:** 9, 359 **EM2:** 14, 460, 467
and dielectric breakdown ... **EM2:** 581-582
of engineering plastics ... **EM2:** 73
epoxy resin system composites ... **EM1:** 403, 409
high-temperature thermoset matrix composites ... **EM1:** 377
Kevlar 49 fiber/fabric reinforced epoxy resin ... **EM1:** 409
low-temperature thermoset matrix composites ... **EM1:** 398
and mean free path ... **A17:** 59
medium-temperature thermoset matrix composites ... **EM1:** 385
retention, glass-polyester composites ... **EM1:** 95
thermoplastic matrix composites ... **EM1:** 367, 370, 372
unsaturated polyesters ... **EM2:** 250
Dielectric(s) *See also* Dielectric breakdown; Dielectric constant; Dielectric materials; Dielectric properties; Dielectric strength; Dynamic dielectric analysis (DDA); Substrates
capacitor ... **EL1:** 342-343
chemistry, thick-film pastes ... **EL1:** 342
composition, physical characteristics ... **EL1:** 108-109
crossover and multilayer ... **EL1:** 341-342
defects, ceramic capacitors ... **EL1:** 994-995

Dielectric(s) (continued)
defined ... **EL1:** 109
devitrifying/vitreous ... **EL1:** 109
as insulators, defined ... **EL1:** 90
isolation techniques ... **EL1:** 199
losses, and signal attenuation ... **EL1:** 603
multichip, comparative tests ... **EL1:** 303
multilayer thin-film hybrid ... **EL1:** 323-324
reinforcements, thermal expansion properties ... **EL1:** 615-619
reliability, tests ... **EL1:** 468
strength ... **EL1:** 90, 336, 1141
substrates, flexible printed boards ... **EL1:** 582-583
thick-film ... **EL1:** 341
thickness, of PWBs ... **EL1:** 82
thin, organic ... **EL1:** 299
Dielectric-phase angle
defined ... **EM2:** 14
Dielectrics/piezoelectrics ... **EM4:** 17
Dielectrometry ... **EM3:** 10
defined ... **EM1:** 9 **EM2:** 14
Dieless drawing
as superplastic forming ... **A14:** 857
Diels-Aider reaction ... **EM1:** 79-83
Diels-Alder reaction
of polyphenylquinoxalines ... **EM3:** 163
Dies *See also* Coining dies; Deep drawing dies; Die failures; Die threading; Thread-rolling dies; Tools; Tools and dies, failures of
adjustable solid ... **A16:** 302
adjusting screw type ... **A16:** 296
cavities, EDM failures in ... **A11:** 566-567
characteristics ... **A11:** 563
chromium-plated blanking, failed ... **A11:** 575
cold heading ... **A16:** 519
drawing ... **A16:** 519
erosion ... **A11:** 384
extrusion ... **A16:** 515, 516, 519
forging, fracture by segregation ... **A11:** 324-326
nonadjustable solid ... **A16:** 302
one-piece nonadjustable ... **A16:** 296
operation, and failure ... **A11:** 564
plastic mold, white-etching in ... **A11:** 567-568
pultrusion ... **EM1:** 534-536
rehardened high-speed steel ... **A11:** 574
self-opening ... **A16:** 298-299, 300
self-opening and tools for pipe threading ... **A16:** 302
self-opening revolving ... **A16:** 297
self-opening stationary ... **A16:** 297
setup, and failure ... **A11:** 564
solid ... **A16:** 300
solid vs. self-opening ... **A16:** 298
spring-type adjustable ... **A16:** 296-297
tools and, failures of ... **A11:** 563-585
Dies and die materials, friction and wear ... **A18:** 621-646
die materials for cold extrusion ... **A18:** 627-628
die materials for cold heading ... **A18:** 627
die materials for hot extrusion ... **A18:** 627
die materials for hot forging ... **A18:** 622-627
alloy steels for hot-forging dies ... **A18:** 622-623
cast steels ... **A18:** 625-626
comparison of die steels for hot forging ... **A18:** 623
maraging steels ... **A18:** 625
process variables effect ... **A18:** 623-625
rating of die steels for hot forging ... **A18:** 623
selection factors ... **A18:** 622
shock-resisting tool steels ... **A18:** 625
superalloys ... **A18:** 626
workpiece properties effect ... **A18:** 623-625
die materials for sheet metal forming ... **A18:** 628
die wear and failure mechanisms ... **A18:** 621
abrasive wear ... **A18:** 621
adhesive wear ... **A18:** 621
corrosive wear ... **A18:** 621
erosive wear ... **A18:** 621
gross cracking ... **A18:** 621, 622
mechanical fatigue ... **A18:** 621, 622
plastic deformation ... **A18:** 621, 622
thermal fatigue (heat checking) ... **A18:** 621-622
thermal shock ... **A18:** 621, 622
die wear in hot forging dies ... **A18:** 635-641
abrasive wear characterization factors ... **A18:** 635
abrasive wear factors ... **A18:** 635-638

Dies and die materials, friction and wear (continued)
catastrophic die failure/plastic deformation ... **A18:** 641
mechanical fatigue ... **A18:** 640-641
methods of improving resistance to abrasive wear ... **A18:** 638-639
thermal fatigue ... **A18:** 639-640
factors influencing wear and failure in die-casting dies ... **A18:** 631-632
die material considerations ... **A18:** 631-632
geometric factors ... **A18:** 631
heat checking ... **A18:** 632
interface considerations ... **A18:** 632
processing conditions ... **A18:** 632
thermal fatigue ... **A18:** 632
heat treatment effect ... **A18:** 621
lubrication effect ... **A18:** 621
materials for die-casting dies ... **A18:** 628-629
materials for dies and molds ... **A18:** 622
surface treatments and coatings ... **A18:** 641-646
boriding ... **A18:** 641, 642
carburizing ... **A18:** 642
ceramic coatings ... **A18:** 643-644
chemical vapor deposition (CVD) ... **A18:** 641, 643, 645, 646
for cold upsetting ... **A18:** 646
electroplating ... **A18:** 644
hardcoatings for cold extrusion ... **A18:** 645-646
hardfacing ... **A18:** 644-645
ion implantation ... **A18:** 642-643, 645
ion nitriding ... **A18:** 641
MetalLife treatment ... **A18:** 643
nitriding ... **A18:** 641-642, 645
nitrocarburizing ... **A18:** 645
PVD ... **A18:** 645
vanadizing ... **A18:** 645
wear and failure modes in die-casting dies ... **A18:** 629-631
chemical attack ... **A18:** 629
corrosion ... **A18:** 629
erosion and abrasive wear ... **A18:** 629-630
erosion due to liquid metal attack ... **A18:** 629
gross cracking ... **A18:** 629
soldering ... **A18:** 629
test for erosion and corrosion evaluation ... **A18:** 629-630
thermal fatigue ... **A18:** 629, 630-631
thermal shock ... **A18:** 629, 631
wear and failure modes in die casting ... **A18:** 629-631
wear reduction methods ... **A18:** 629
wear in sheet metal forming dies ... **A18:** 632-635
deep-drawing dies ... **A18:** 633-635
shallow forming dies ... **A18:** 632-633
Dies, axisymmetric
for wires and rods ... **A10:** 359
Dies, forging *See* Closed-die forging tools; Hot upset forging tools
Dies, specific types *See also* Die systems; Die(s)
compound ... **A14:** 579
diamond, for drawing ... **A14:** 336
double-action ... **A14:** 579
double-extension ... **A14:** 132
double-taper ... **A14:** 132
drop-through ... **A14:** 453
flat ... **A14:** 67, 323-324
flat-back ... **A14:** 97
flattening ... **A14:** 536
FM (free from Mannesmann Effect) ... **A14:** 61
impression ... **A14:** 43-45, 52-53
inverted ... **A14:** 453
lock-seam ... **A14:** 537
locked and counterlocked ... **A14:** 48-49
long-taper ... **A14:** 132
multiple ... **A14:** 455-456
multiple, with transfer mechanism ... **A14:** 579-580
multiple-part ... **A14:** 51
multiple-slide ... **A14:** 569
Dies, specific types (continued)
offset ... **A14:** 536
one-stroke hemming ... **A14:** 537
open ... **A14:** 43
piloted ... **A14:** 132
pipe-forming ... **A14:** 537
pressure ... **A14:** 666
progressive ... **A14:** 182, 454-455, 466, 478-479, 527-528, 546, 569, 579
resinking ... **A14:** 53
return ... **A14:** 453-454
rocker-type ... **A14:** 537
rotary-bending ... **A14:** 526
segmented ... **A14:** 587
semicylindrical ... **A14:** 97-98
short-run, for blanking ... **A14:** 451-453
single-action ... **A14:** 579
single-extension ... **A14:** 132
single-operation ... **A14:** 453-454, 465, 527, 546, 579
single-station ... **A14:** 478
single-taper ... **A14:** 132
sliding, upsetting with ... **A14:** 90-91
solid ... **A14:** 292-293, 639
steel-rule ... **A14:** 451-452
subpress ... **A14:** 452-453
taper-point ... **A14:** 132
template ... **A14:** 452
three-pass ,for gear blanks ... **A14:** 45
transfer ... **A14:** 455, 466, 528
trimming ... **A14:** 55-56, 257
tube-forming ... **A14:** 537
two-piece ... **A14:** 132-133
U-bending ... **A14:** 526
V ... **A14:** 525
vertical four-station hot upset forging ... **A14:** 84
wing ... **A14:** 527
wiper ... **A14:** 666
wiping ... **A14:** 525-526
wrap, for precision forging ... **A14:** 52
Diesel burn
in injection molding ... **EM1:** 167
Diesel engines ... **A7:** 621
crankshaft, fatigue fracture from subsurface inclusions ... **A11:** 323
cylinder lining, cavitation damage ... **A11:** 377-378
locomotive, corrosion-fatigue cracking ... **A11:** 371-372
rocker levers, malleable, fatigue failure ... **A11:** 350-352
Diesel truck crankshaft
beach marks from fatigue fracture in ... **A11:** 77, 78
Dietary iron source
elemental iron as ... **A7:** 614
Diethylene glycol
description ... **A9:** 68
Diethylene triamine (DETA) ... **EM3:** 96
reaction with phthalic anhydride ... **EM3:** 96
Difference image ... **A18:** 350, 351-352
Difference mode
parameters ... **EL1:** 39
Differential aeration *See also* Crevice corrosion
aluminum alloy exfoliation from ... **A13:** 1022
cell ... **A13:** 5, 43
and design ... **A13:** 339
Differential aeration cell ... **A13:** 5, 43
in lead ... **A13:** 785, 788
in pipeline corrosion ... **A13:** 1288
pitting by ... **A11:** 631-632
Differential attenuation
in radiographic inspection ... **A17:** 309
Differential bevel gear
by precision forming ... **A14:** 163
Differential coil arrangement *See also* Coils
eddy current inspection ... **A17:** 174
multiple coils ... **A17:** 176
Differential expansion ... **EM3:** 37
Differential index of plasticity ... **A18:** 424, 425
Differential interference contrast
after Nomarski, principles of ... **A9:** 59
Differential interference contrast (continued)
color metallography ... **A9:** 138, 150-152
etching ... **A9:** 59
used for precious metals and alloys ... **A9:** 552
Differential interference contrast microscopy
for as-polished samples ... **A12:** 95, 100
titanium alloys ... **A12:** 445
Differential interference contrast-Nomarski system
for imaging ... **EL1:** 1068
Differential interference-contrast illumination ... **A9:** 79
defined ... **A9:** 5
Differential probes
eddy current inspection ... **A17:** 180-181
Differential pulse polarography ... **A10:** 193, 194
Differential pulse stripping voltammetry
principles ... **A10:** 193
Differential scanning calorimeter ... **A7:** 197
Differential scanning calorimeter (DSC)
for shape memory alloys ... **A2:** 898
Differential scanning calorimetry (DSC) ... **EM2:** 523-525, 540, 825, 830-832 **EM3:** 10, 323 **EM4:** 52
defined ... **EM2:** 14
of epoxies ... **EL1:** 835
kinetic thermal analysis ... **EM3:** 424-425
for phase analysis ... **EM4:** 562
for temperature resistance ... **EM2:** 559-564
thin-film hybrids ... **EL1:** 324
to measure scanning calorimetry relative energy difference ... **EM4:** 27-28
Differential scanning calorimetry (DSC) analysis ... **EM1:** 9, 704, 780
Differential shear ... **EM3:** 478
Differential sintering ... **EM4:** 125
Differential sputtering
as artifact in AES analysis ... **A10:** 556
Differential testing
limitations ... **A8:** 183
for shear stress ... **A8:** 182-183
Differential thermal analysis *See also* Cooling curve analysis; Thermal analysis
equipment, set-up ... **A15:** 184
for second- phase testing ... **A10:** 177
simplified ... **A15:** 183-184
standard ... **A15:** 184
Differential thermal analysis (DTA) *See also* Thermal analysis ... **EM2:** 825, 830-832 **EM3:** 10 **EM4:** 52, 53
defined ... **EM1:** 9 **EM2:** 14
melting temperature ranges measurement ... **A6:** 89
for phase analysis ... **EM4:** 561-562
of polyphenylene sulfides (PPS) ... **EM2:** 187
for temperature-time control in polymer removal techniques ... **EM4:** 137
to measure heat capacity ... **EM4:** 615
to measure temperature difference relative to a reference ... **EM4:** 27-28
Differential thermat analysis ... **A3:** 1 • 17
Differential thermogravimetric analysis (DTG)
for phase analysis ... **EM4:** 561
Differential-pressure method
of quantity loss leak testing ... **A17:** 65-66
Differential-temperature cells
corrosion of ... **A11:** 184-185
Differentiation
assembly/packaging for ... **EL1:** 438-441
Diffracted beams
intensities of ... **A10:** 328-329
Diffracted intensities
resolving of ... **A10:** 424
Diffraction *See also* X-ray diffraction
angle, diffraction x-ray beam at ... **A10:** 381
angle, symbol for ... **A10:** 692
by scanning laser gages ... **A17:** 12
coherent Bragg ... **A10:** 436
cones ... **A10:** 327, 371
contrast ... **A10:** 444, 671

Diffraction (continued)
convergent-beam electron **A10:** 438-440
crystal **A10:** 346
defined **A9:** 5
dynamical theory of **A10:** 366-367
electron **A10:** 410, 436-440
experiments, two types of single crystal **A10:** 330
from surfaces, principles of **A10:** 538-539
geometry of **A10:** 326-327
grating, defined **A10:** 672
intensities, in single-crystal x-ray diffraction **A10:** 348-349
intensities, kinematic and dynamic effects in **A10:** 366-367
kinematical theory of **A10:** 366
of light by line gratings **A10:** 345
line intensity **A10:** 328
of monochromatic x-ray beams at high diffraction angle **A10:** 381
in near-perfect crystal **A10:** 367
pattern, to determine atom locations within unit cells **A10:** 345-346
peaks, in surface stress measurement **A10:** 385
of scattered radiation **A17:** 345
selected-area **A10:** 436-438
single-crystal methods **A10:** 329-331
single-crystal, principles of **A10:** 350
spots **A10:** 345-346, 351, 366
techniques, recommended for ferrous and nonferrous alloys **A10:** 382
of ultrasonic beams **A17:** 239
vector **A10:** 690
wave theory of **A10:** 83
and x-ray topography **A10:** 366-367
Diffraction angle
symbol for **A10:** 692
Diffraction cones **A10:** 327, 371
Diffraction contrast **A10:** 444, 671
of an edge dislocation **A9:** 114
of defects, theory of **A9:** 110-113
of dislocation pairs in copper and silver **A9:** 115
in imperfect crystals **A9:** 111-113
in perfect crystals **A9:** 110-111
of single perfect dislocations **A9:** 114
Of translation interfaces, calculated by the matrix method **A9:** 118-119
Diffraction contrast studies, of grain boundaries by transmission electron microscopy **A9:** 119-120
Diffraction experiments
amorphous materials and metallic glasses **A2:** 809
Diffraction grating
defined **A9:** 5-6 **A10:** 672
photoresist for applying **A8:** 234
in plate impact testing **A8:** 233
Diffraction intensities
calculation **A9:** 111
Diffraction limited focal spot size **A6:** 875
Diffraction lines
created by electron- channeling patterns **A9:** 94
Diffraction methods
crystallographic texture measurement and analysis **A10:** 357-364
extended x-ray absorption fine structure **A10:** 407-419
neutron diffraction **A10:** 420-426
polycrystalline **A10:** 331-332
radial distribution function analysis **A10:** 393-401
single-crystal **A10:** 329-331
small-angle x-ray and neutron scattering **A10:** 402-406
x-ray diffraction **A10:** 325-332
x-ray diffraction residual stress techniques **A10:** 380-392
x-ray powder diffraction **A10:** 333-343
x-ray topography **A10:** 365-379
Diffraction pattern (x-rays)
defined **A9:** 6
Diffraction pattern technique
of laser inspection **A17:** 13
Diffraction patterns
defined **A10:** 345, 672
from superlattice **A10:** 540
Diffraction patterns (continued)
GaAs, LEED analysis **A10:** 541
oriented pyrolytic graphite **A10:** 543
point-group symmetry from **A10:** 346
precession photographs showing three layers of **A10:** 346
for single crystals **A10:** 345, 351
as three-dimensional **A10:** 346
Diffraction patterns in transmission electron microscopy **A9:** 109-110
effect of second phase precipitates on **A9:** 118
effects caused by spinodal decomposition **A9:** 653
from different modes **A9:** 104
Kikuchi lines **A9:** 109-110
wrought heat-resistant alloys **A9:** 308
Diffraction ring
defined **A9:** 6
Diffraction spots
diffraction patterns at **A10:** 345-346
identification, for single-crystal analysis **A10:** 351
intensity variation within **A10:** 366
Diffraction studies of grain boundaries **A9:** 120-121
Diffraction vector **A10:** 690
Diffraction vectors
different images used to distinguish moiré patterns **A9:** 110
Diffraction-peak location
methods, compared **A10:** 385-386
x-ray diffraction residual stress techniques **A10:** 385-386
Diffractometer technique
used to determine crystallographic texture **A9:** 702
Diffractometers
automated single-crystal **A10:** 351
Bragg-Brentano **A10:** 337
conventional four-circle **A10:** 351
conventional x-ray powder diffraction **A10:** 337
double-crystal **A10:** 371-372
Eulerian cradle mounted on **A10:** 360
Guinier **A10:** 337
horizontal laboratory **A10:** 389
low-energy electron diffraction **A10:** 540
micro **A10:** 337-338
neutron powder **A10:** 422
Seeman-Bohlin **A10:** 337
θ-θ **A10:** 337
thin film **A10:** 337
time-of-flight powder **A10:** 422
time-of-flight single-crystal, at pulsed neutron source **A10:** 424
use in x-ray diffraction residual stress techniques **A10:** 387
Diffractometry
angle-dispersive, used to construct pole figures **A9:** 695
energy-dispersive, used to construct pole figures **A9:** 695-706
Diffuse reflectance spectroscopy **A10:** 114, 120 **EM4:** 52
Diffuse slip
in high-cycle corrosion **A8:** 254
Diffuse transmittance
defined **A10:** 672
Diffused germanium
temperature effect **EL1:** 958
Diffused silicon
temperature effect **EL1:** 958
Diffused strain gages **EL1:** 445
Diffusion *See also* Interface diffusion **A13:** 5, 61, 67-69, 147 **EM3:** 11 **EM4:** 136
at very high solidification rates **A15:** 125
and atomic transport, in low gravity eutectic alloys **A15:** 151-152
of boron, neutron radiography of **A17:** 391
carbon, in dissimilar-metal welds **A11:** 620
of carbon, iron-carbon melts **A15:** 72-73
coating, defined **A13:** 5
coating, powders used **A7:** 573
coatings **A15:** 563
coefficient **A13:** 5, 68
coefficient, concentration dependency of **A15:** 138
coefficients **A10:** 478
Diffusion (continued)
couples **A10:** 477-478, 503
current, voltammetry **A10:** 190
defined **A7:** 4 **A9:** 6 **EM2:** 1
effect in polymers **A11:** 758
effect on strength **A8:** 34
effect on tensile behavior **A8:** 35
in electrogravimetry **A10:** 198
element, in liquid **A15:** 82
Fick's first law of **A15:** 53
fields, eutectic growth **A15:** 121
force, temperature as **A15:** 74
of gases **A15:** 82-83
grain-boundary vs volume **A10:** 478
heat treatment, gaseous ferritic nitro-carburizing as **A7:** 455
of impurities, active-component fabrication **EL1:** 195
-induced grain-boundary migration, EDS/ CBED analysis of **A10:** 461-464
in interphase mass transport **A15:** 53
as leakage **A17:** 58
-limited current density, defined **A13:** 5
measurements **A10:** 243, 476-478
natural, in voltammetry **A10:** 189
of nickel into iron, radioanalytically measured **A10:** 243
organic movement by **EL1:** 680
path, in microsegregation **A15:** 137
in peritectic transformation **A15:** 126-127
in peritectic transformations **A9:** 677
in permeation **A17:** 58
phenomena, XPS analysis **A10:** 576-577
of plutonium into thorium **A10:** 249
porosity **A7:** 4
processes for alloy-coated composite powders **A7:** 174
rates, of tracer gases **A17:** 67, 69
in semiconductor development **EL1:** 961
SIMS tracer studies **A10:** 610
solid, and chemical process rates **A15:** 52
solute, in eutectic growth **A15:** 124
of solutes in precipitation reactions **A9:** 647
solutions, to heat transfer, die casting **A15:** 293-294
studies, by Mössbauer spectroscopy **A10:** 287
surface, FIM/AP study of **A10:** 583
surface property effect **A18:** 342
treatment, high-temperature, and rapid omnidirectional compaction **A7:** 545
types, bipolar junction transistor technology **EL1:** 195-196
vacancy **A13:** 68
zone, in stainless steel can production **A7:** 440
Diffusion adhesives **EM3:** 75
Diffusion aid
definition **M6:** 5
Diffusion, applications
aerospace **A6:** 387
Diffusion bonding *See also* Diffusion brazing; Diffusion welding; High-temperature solid-state welding **A6:** 149, 156-159
aluminum alloys **A6:** 157
and annealing **A6:** 157
bonding surfaces containing oxides **A6:** 157
of boron fiber MMCs **EM1:** 854
copper **A6:** 156
copper metals **M2:** 456
definition **A6:** 156
diffusion-controlled mass transport **A6:** 158-159
discontinuities from **A17:** 588-589
graphite fiber MMCs **EM1:** 869-870
with interface aids **A6:** 159
interface diffusion **A6:** 158
interface migration **A6:** 159
iron **A6:** 156
joint indistinguishable from surrounding material **A6:** 3
mechanism **A6:** 157-159
metal-matrix composites **A12:** 466
microasperity deformation **A6:** 157-158
niobium **A6:** 156
plastic flow **A6:** 158
preferred term for process **A6:** 156
procedures **A6:** 156-157
projection welding **A6:** 232-233

Diffusion bonding (continued)
sequence of metallurgical stages **A6:** 157
sheet metals **A6:** 399
SiC-titanium composite **EM1:** 862
silver **A6:** 156
surface roughness **A6:** 158
tantalum **A6:** 156
titanium alloys **A6:** 156, 157-158, 159
tungsten **A6:** 156
tungsten alloys **A6:** 581
zirconium **A6:** 156
Diffusion bonding (DB) *See also* Sintering
aluminum and aluminum alloys **A2:** 6
refractory metals and alloys **A2:** 564
titanium alloy sheet **A2:** 590-591
Diffusion bonding as a fabrication process for metal-matrix composites **A9:** 591
Diffusion bonding/superplastic forming **A14:** 844, 857, 859-860
Diffusion brazing
definition **M6:** 5
Diffusion brazing (DFB) **A6:** 343-344
advantages **A6:** 344
aluminum alloys **A6:** 343
applications **A6:** 343, 344
aerospace **A6:** 387
cobalt **A6:** 343
cobalt-base alloy **A6:** 344
critical aspects **A6:** 343
definition **A6:** 343
example, stainless steel to copper **A6:** 343
mechanically alloyed oxide dispersion-strengthened (MA ODS) alloys **A2:** 949
nickel **A6:** 343
parameters **A6:** 343
stainless steel **A6:** 343
titanium **A6:** 343
Diffusion coating
aluminum, of steel **M5:** 334-335, 339-346
temperature and time effects **M5:** 343-346
cadmium plating **M5:** 264
cementation type *See* Pack cementation processes
heat-resistant alloys **M5:** 566
oxidation protective coatings *See* Oxidation protective coating, diffusion types **M5:** 376-379
pack diffusion *See* Pack diffusion coating processes
Diffusion coatings
for cast iron **A15:** 563
Diffusion control
for gas bubble nucleation **EM1:** 757-758
Diffusion controlled (2D)
rate law expression **EM4:** 55
Diffusion controlled (3D)
rate law expression **EM4:** 55
Diffusion current **A10:** 190
Diffusion effects in aluminum alloys
etchants for examination of **A9:** 355
Diffusion measurements
advantages of AEM analysis **A10:** 476
by SIMS profiles **A10:** 610
data analysis **A10:** 476-478
diffusion coefficients **A10:** 478
diffusion couples **A10:** 477-478
radioanalytical tracer for **A10:** 243
sample preparation **A10:** 476
Diffusion pump
with split Hopkinson pressure bar **A8:** 201
Diffusion vacuum pumps
in gas mass spectrometer **A10:** 151-152
Diffusion wear
coated carbide tools **A2:** 960
Diffusion welding
applications **M6:** 682-686
for composites **M6:** 684
for iron-based alloys **M6:** 682-683
for nickel-based alloys **M6:** 683-684
for refractory metals **M6:** 684-685
Diffusion welding (continued)
for titanium and titanium alloys **M6:** 682
definition **M6:** 5
differentiation from deformation welding **M6:** 672
equipment **M6:** 673
forge welding **M6:** 675-676
heat **M6:** 673
load application **M6:** 673
metals welded **M6:** 677-678
procedures **M6:** 677
protective envelope **M6:** 673
roll welding **M6:** 676
Diffusion welding (DFW) *See also* Diffusion bonding; High-temperature solid-state welding
copper **A6:** 884-885, 886
definition **A6:** 1208
dispersion-strengthened aluminum alloys **A6:** 543, 547
dissimilar metal joining **A6:** 822
model for **A6:** 145
oxide-dispersion-strengthened materials **A6:** 1038, 1039
titanium alloys **A6:** 522
weld discontinuities **A6:** 1079-1080
Diffusion welding, procedure development and practice considerations **A6:** 883-886
advantages **A6:** 883
aluminum-base alloys **A6:** 884-885, 886
applications **A6:** 884, 885, 886
beryllium alloys **A6:** 885, 886
carbon steels **A6:** 884
cobalt-base alloys **A6:** 885
description of process **A6:** 883
dissimilar metal combinations **A6:** 885-886
ferrous-to-ferrous **A6:** 885-886
ferrous-to-nonferrous **A6:** 886
metal-ceramic joining **A6:** 886
nonferrous-to-nonferrous combinations **A6:** 886
gold **A6:** 885
high-strength steels **A6:** 884
low-alloy steels **A6:** 884
mechanism **A6:** 883
microstructure **A6:** 884
Monel **A6:** 886
nickel-base alloys **A6:** 885, 886
noble metals **A6:** 885
postweld heat treatment **A6:** 884, 885
process variants **A6:** 883-884
hot isostatic pressing (HIP) **A6:** 884
liquid-phase process **A6:** 883-884
solid-phase process **A6:** 883
superplastic forming/diffusion welding (SPF/DW) technique **A6:** 884, 885
transient liquid-phase diffusion welding **A6:** 883-884
reactive metals **A6:** 885
refractory metals **A6:** 885
stainless steels **A6:** 884, 886
tantalum alloys **A6:** 886
tensile properties **A6:** 884
titanium **A6:** 884, 885, 886
titanium alloys **A6:** 884, 885, 886
uranium alloys **A6:** 886
vanadium alloys **A6:** 886
zirconium alloys **A6:** 885, 886
Diffusion zone
defined **A9:** 6
Diffusion-alloyed steels
heterogeneity **A9:** 503
microstructures **A9:** 510-511
Diffusion-bonded powders *See also* Thermal agglomeration **A7:** 101, 102, 173
Diffusion-bonding
equipment for **A7:** 515, 517
Diffusion-induced grain-boundary migration analysis
abbreviation for **A10:** 690
in Al-4.7Cu **A10:** 461-464
by EDS/CBED **A10:** 461-464
convergent-beam electron diffraction **A10:** 462-464
EMPA dot map of **A10:** 527
x-ray microanalysis **A10:** 462
Diffusional flow **A7:** 313, 315
Diffusional flow, effect of slip lines
iron **A12:** 219
Diffusional mass transport
creep by **A8:** 310
Diffusional mixing
statistical analysis **A7:** 187
Diffusionless alloying elements *See also* Alloy selection; Alloying
iron/cerium, effect in niobium-titanium superconducting materials **A2:** 1045
Diffusionless phase transformations **A9:** 668
Diffusionless solidification
defined **A9:** 616
Diffusivity *See also* Thermal diffusivity
and average flash temperature **A18:** 41
superplastic titanium alloys **A14:** 843
and thermal conductivity **EM2:** 451-456
Diffusivity constant **A4:** 259
Digestion, as sample preparation **A10:** 165-167, 176
Digital algorithm for laminate analysis
computer program **EM1:** 274
Digital components *See also* Active digital components; Components; Digital integrated circuits (ICs); Digital systems
active-component fabrication **EL1:** 201
future trends **EL1:** 176
Digital compositional mapping
flexible processing **A10:** 528
relation of intensity to constituent **A10:** 528-529
of zinc-containing copper **A10:** 528
Digital gray-level image storages
used in scanning electron microscopy **A9:** 96
Digital image enhancement **A17:** 454-464
of color images, examples **A17:** 483-488
image capture and acquisition system **A17:** 455-456
image display **A17:** 461-463
image enhancement **A17:** 456--458
image operations **A17:** 458-460
image processing **A17:** 119, 456
information extraction **A17:** 460-461
NDE systems **A17:** 454-455
for thermal inspection **A17:** 400
Digital image processing
magnetic particles **A17:** 119
Digital images, color metallography **A9:** 138, 152-153
inkjet printout **A9:** 139
Digital imaging
and analog imaging, of iron in aluminum matrix **A10:** 448
semiautomatic analyzer for **A10:** 310
Digital instrumentation
eddy current inspection **A17:** 175
Digital integrated circuits
as gallium arsenide application **A2:** 740
Digital integrated circuits (ICs) *See also* integrated circuits (ICs)
complementary metal-oxide semiconductor (CMOS) **EL1:** 162-163
design rule for **EL1:** 161
emitter-coupled logic (ECL) **EL1:** 163-165
future trends **EL1:** 176-177
introduction **EL1:** 160-161
large-scale integration (LSI), types **EL1:** 166-168
packaging considerations **EL1:** 172-174
signal transmission **EL1:** 168-172

Digital integrated circuits (ICs) (continued)
technologies, compared **EL1:** 165-166
thermal management **EL1:** 174-176
transistor-transistor log **EL1:** 161-162
Digital logic *See also* Logic
functions, types **EL1:** 160
Digital meters
as eddy current inspection readout **A17:** 178-179
Digital oscilloscope, for torsional Kolsky bar
dynamic test results **A8:** 228
Digital position readout
coordinate measuring machine **A17:** 19
Digital radiography *See also* Radiography
color images by **A17:** 485
and computed tomography (CT) **A17:** 377
defined **A17:** 317, 383
real-time **A17:** 320
Digital signal
line, characteristics **EL1:** 169-170
processing (DSP) **EL1:** 8, 378
Digital switching systems
applications and markets **EL1:** 382-385
Digital systems *See also* High-frequency digital systems
High-frequency, design of **EL1:** 76-88
improvement technologies **EL1:** 2
Digital voltmeter
for fatigue testing machine **A8:** 368
Diglycidyl ether of bisphenol A (DGEBA) **EM1:** 66-67, 69, 399-400, 753 **EM3:** 94
for basis of epoxy resin **EM3:** 77
diacrylate properties **EM3:** 92
epoxy resin for polysulfides **EM3:** 141, 142
ester diacrylate properties **EM3:** 92
Diglycidyl ethers of bisphenol A (DGEBA) **EL1:** 811, 825-827
DIGM *See* Diffusion-induced grain-boundary migration
Dihedral angles **A9:** 604
Diisocyanates
as thermoplastic polyurethanes (TPUR) **EM2:** 204
Diisopropanol *p*-toluidine
typical formulation **EM3:** 121
Diketones, β
as extractants **A10:** 170
Dilatant
defined **A18:** 7
Dilatational Lamb waves
ultrasonic inspection **A17:** 234
Dilation during aging
cast copper alloys **A2:** 360
Dilation/voiding
SAS techniques for **A10:** 405
Dilatometer
defined **A8:** 4
and dimensional change **A7:** 310
as measure of shrinkage and densification during sintering **A7:** 310
Dillard-Kynar sensors **EM3:** 453-454
Diluent **EM3:** 11
defined **EM2:** 14
Diluents
in engine lubricant formulations **A18:** 111
for filament winding **EM1:** 505
monofunctional epoxy **EM1:** 67, 70
in polyester reaction **EM1:** 132-133
vinyl ester **EM1:** 133
Dilute copper cyanide plating process **M5:** 159, 161-162, 167
Dilute systems
EXAFS fluorescence analysis of **A10:** 418
Dilution **A6:** 500, 501
definition **A6:** 1208 **M6:** 5
Dilution factor
defined **A10:** 672
Dilution, infinite
activity coefficients **A15:** 60
Dilver-P alloy
as low-expansion alloy **A2:** 895
Dimension
in failure analysis **A11:** 747
incorrect, as casting defects **A11:** 386
-less constants, in fracture mechanics **A11:** 48-49
Dimension(s) *See also* Dimensional; Dimensional tolerance; Size
of components, process control of **EL1:** 673-674
incorrect, as casting defect **A17:** 518-519
minimum, implications **EL1:** 2
physical, by eddy current inspection **A17:** 164
physical, glass-to-metal seals **EL1:** 459
and tolerances, flexible printed boards **EL1:** 595
tube, as test variable **A17:** 175
Dimensional
accuracy, by re-pressing, aluminum and aluminum alloys **A2:** 6
change, flexible epoxy systems **EL1:** 819
changes, age-hardening, beryllium-copper alloys **A2:** 412-414
control, extrusions, aluminum and aluminum alloys **A2:** 5-6
defects, types **EL1:** 568
stability, high-performance electronic systems **EL1:** 16
Dimensional accuracy *See also* Accuracy; Allowances; Dimensional inspection; Tolerances
all-ceramic mold casting **A15:** 249
in blanking/piercing high-carbon steels **A14:** 557
of ceramic molding **A15:** 248
in coining **A14:** 180, 186
of cold extrusion **A14:** 307
of cold heading **A14:** 295-296
cut-to-length lines **A14:** 712-713
of deep drawing **A14:** 575, 589-590
defect types **A15:** 551-552
in die casting **A15:** 287-288
gray iron **A15:** 640
hot/cold box processes **A15:** 238
organic binder effect **A15:** 35
of P/M sintered parts **A7:** 480-492
of permanent mold castings **A15:** 284
of plaster molding **A15:** 242
press-brake forming **A14:** 542-543
of Replicast process **A15:** 271
rotary swaging **A14:** 140
tube bending **A14:** 668
of tube spinning **A14:** 677
Unicast process **A15:** 251
of waxes **A15:** 254
Dimensional change *See also* Dimensional control; Dimensional inspection methods; Dimensional tolerances
control by rapid omnidirectional compaction **A7:** 545
control of **A7:** 482, 545
defined **A7:** 4
during ejection, elastic springback **A7:** 480
during heat treating **A7:** 481
during P/M processing **A7:** 480
during sintering **A7:** 211, 480-481
effect of sintering temperatures **A7:** 367, 369
effect of sintering time **A7:** 370
effects of particle size distribution and mixture fluctuations **A7:** 186
in electrolytic copper powder **A7:** 115
expansion of copper powder compacts **A7:** 310
factors influencing **A7:** 291
of incoming powder, evaluation **A7:** 481-482
in sintered bronze **A7:** 377-378
of sintered metal compacts **A7:** 290-292
in sintering **A7:** 309
sintering temperature effect on **A13:** 825
in solution and reprecipitation liquid phase sintering **A7:** 320
standard test method for **A7:** 290-291
Dimensional changes
in torsional straining **A8:** 143
Dimensional characteristics
regression analysis for **A8:** 663
Dimensional control
copper alloy castings **M2:** 388-389
Dimensional flexibility
of horizontal centrifugal casting **A15:** 299
Dimensional growth
ductile iron **M1:** 46-47
gray iron **M1:** 46-47
Dimensional inspection *See also* Dimensional accuracy
of aluminum alloy forgings **A14:** 249
casting defects **A15:** 545
Dimensional inspection (continued)
computer-aided **A15:** 558-561
of titanium alloy forgings **A14:** 282
to evaluate gas turbine ceramic components **EM4:** 718
Dimensional inspection methods **A7:** 291
Dimensional measurements *See also* Laser inspection; Metrology
accuracy and conformity, in weldments **A17:** 591
by computed tomography (CT) **A17:** 364
by eddy current inspection **A17:** 164
by laser inspection **A17:** 12-16
by ultrasonic inspection **A17:** 273-274
of castings **A17:** 512, 520-521
computer-aided **A17:** 521-524
diffraction pattern technique **A17:** 13
holography **A17:** 16
interferometers **A17:** 14-15
laser triangulation sensors **A17:** 13
photodiode array imaging **A17:** 12-13
scanning laser gage **A17:** 12
sorting by **A17:** 15-16
of tubular products **A17:** 561
Dimensional measurements, periodic
for multiaxial testing **A8:** 344
Dimensional reproducibility
tolerances and allowances **A15:** 614-623
Dimensional stability **EM1:** 9, 36
cyanates **EM2:** 232
defined **EM2:** 14
of extrusion **EM2:** 386
of gray iron **A1:** 26-28 **M1:** 26-28
ionomers **EM2:** 122
maraging steels **M1:** 448, 451
of molding materials **A15:** 208
phenolics **EM2:** 242, 243
polyamide-imides (PAI) **EM2:** 128
polycarbonates (PC) **EM2:** 151
polyether sulfones (PES, PESV) **EM2:** 160, 161
polymer, as thermal property **EM2:** 59-60
polyphenylene ether blends (PPE PPO) **EM2:** 184
polyvinyl chlorides (PVC) **EM2:** 210
and process selection **EM2:** 277
RTM and SRIM, compared **EM2:** 347-348
styrene-maleic anhydrides (S/MA) **EM2:** 219
Dimensional stabilizers **EM3:** 175
Dimensional tolerances *See also* Tolerances **A7:** 291-292, 482
of copper casting alloys **A2:** 350
cost effects **EL1:** 594
leaded/leadless chip carriers **EL1:** 734
magnesium alloys **A2:** 464
malleable iron **M1:** 67
structural ceramics **A2:** 1019
Dimensioning
and tolerancing **A15:** 622-623
Dimensionless factor *See* Sommerfeld number
Dimensionless load parameter
nomenclature for lubrication regimes **A18:** 90
Dimensionless material parameter
nomenclature for lubrication regimes **A18:** 90
Dimensionless speed parameter
nomenclature for lubrication regimes **A18:** 90
Dimer **EM3:** 11
defined **A10:** 672 **EM2:** 14
parylene coatings **EL1:** 789-790
Dimerization
defined **A10:** 672
Dimerized fatty acids
epoxidized **EL1:** 818
Dimethacrylate esters
as acrylic adhesives **EL1:** 671
Dimethacrylates
as monomers for anaerobics **EM3:** 113
Dimethyl formamide
sol-gel processing **EM4:** 449
surface tension **EM3:** 181
Dimethylglyoxime
in amperometric titration **A10:** 204
complex, weighing as the, gravimetric analysis **A10:** 171
as narrow-range precipitant **A10:** 169
DIMOX directed metal oxidation process **EM4:** 232, 294

Dimple rupture
defined **A13:** 5
Dimple rupture, mode
of ductile fracture **A14:** 364
Dimple rupture(s) *See also* Dimple shape; Dimple size; Dimple(s); Ductile fracture(s); Microvoids
as aging effect, iron alloy **A12:** 458
alloy steels **A12:** 291, 334
biaxial tension effects **A12:** 31, 39
cast aluminum alloys **A12:** 413
cemented carbides **A12:** 470
corrosive environmental effects **A12:** 24-29
defined **A12:** 12-13
ductile iron **A12:** 235-237
environmental effects **A12:** 22-35
equiaxed, precipitation-hardening stainless steels **A12:** 371
exposure to low-melting metals **A12:** 29-30
hydrogen effects **A12:** 22-24
intergranular, in steel **A12:** 14
iron-aluminum alloys **A12:** 365
iron-base alloys **A12:** 365, 459-460
materials illustrated in **A12:** 217
and microvoid coalescence **A12:** 12-13
as partially oriented surface **A12:** 203
precipitation-hardening stainless steels **A12:** 370-371
shear band formation **A12:** 444-445
stereo pair, titanium **A12:** 171
and strain rate **A12:** 31-33
stress state effects **A12:** 30-31
surface contour cavities, iron **A12:** 223
surface, profile angular distribution **A12:** 203
temperature effects **A12:** 33-35
titanium alloys **A12:** 171, 443-445
tool steels **A12:** 382
transition to cleavage fracture **A12:** 33, 44
true area value, steel **A12:** 203
wrought aluminum alloys **A12:** 417, 418
Dimple shape
determined by loading **A12:** 173
effect of direction, principal stress **A12:** 30
effect of stress state **A12:** 12
high-purity copper **A12:** 399-400
Dimple size
acicular needles, effect of **A12:** 328
AISI/SAE alloy steels **A12:** 334
and asymmetrical strain **A12:** 12, 16
average, titanium alloys **A12:** 454
defined **A12:** 206
magnification effect **A12:** 425
microvoid effects **A12:** 12
and particle size **A12:** 101
range, titanium alloys **A12:** 449
temperature effects **A12:** 34, 46
titanium alloys **A12:** 449, 454
wrought aluminum alloys **A12:** 424
Dimple(s) *See also* Dimple ruptures; Dimple shape; Dimple size
alloy steels **A12:** 292, 293, 303-304, 308, 319, 324-325, 338
austenitic stainless steels **A12:** 353, 359
by nondestructive profiling **A12:** 199
cast aluminum alloys **A12:** 409-410
as cavities, in iron **A12:** 220
clusters, wrought aluminum alloys **A12:** 423
complex, precipitation-hardening stainless steels **A12:** 372
copper/copper alloys **A12:** 399-400, 402
as ductile fracture mechanism **A12:** 4
elongated **A12:** 12-16, 173-174, 304, 338, 353, 442
and fatigue striations **A12:** 177
fine, on void margins **A12:** 338
flat shear **A12:** 304
formation in copper **A12:** 173
in fractured steel, SEM fractograph **A12:** 207
and grain-boundary cavities, compared **A12:** 20
grain-boundary, in low-carbon steel **A12:** 240
Dimple(s) (continued)
height and width measurement **A12:** 207
high-purity copper **A12:** 399-400
with inclusions **A12:** 65, 67, 219
intergranular facet transition to **A12:** 306
intergranular fracture with, titanium alloys **A12:** 441
in irons **A12:** 219, 220
maraging steels **A12:** 385, 387
martensitic stainless steels **A12:** 367
on shear fractures **A12:** 12, 15-16
on tear fracture **A12:** 12, 15-16, 387, 452
oval-shaped, formation **A12:** 13
precipitation-hardening stainless steels **A12:** 371-372
pure titanium **A12:** 16
SEM for **A12:** 96
superalloys **A12:** 392
tear dimples on **A12:** 452
tension overload **A12:** 385
three-dimensional, quantitative measurement **A12:** 208
titanium/titanium alloys **A12:** 16, 441-442, 451
tool steels **A12:** 375, 379
true mean area of **A12:** 207
true mean intercept length **A12:** 207
two-dimensional, quantitative measurement **A12:** 206-207
within dimples, titanium alloys **A12:** 451
wrought aluminum alloys **A12:** 417, 423-424, 427
Dimpled rupture **A8:** 4
ductile fracture **A8:** 476
mode, stages, ductile fracture **A8:** 571
rapid overload fracture **A8:** 476
single-phase microstructures **A8:** 476
void nucleation **A8:** 479
Dimpled rupture fracture *See also* Ductile fracture; Rupture
in brittle materials **A11:** 75
defined **A11:** 3
ductile, sulfide inclusions in **A11:** 83
fracture mechanics and **A11:** 47, 47
in overstress failures **A11:** 22
and subcritical fracture mechanics (SCFM) **A11:** 47
surface, of ductile fracture **A11:** 22
Dimples
elongated **A11:** 25, 76
equiaxed, on flat-face fracture surfaces **A11:** 76
shallow, in transgranular-cleavage fracture **A11:** 79
shear, in shear-lip zone **A11:** 76
Dimpling **A14:** 4, 847
Dimpling of transmission electron microscopy specimens **A9:** 105
DIN (German) standards for steels **A1:** 157
compositions of **A1:** 175-179
cross-referenced to SAE-AISI steels **A1:** 166-174
Dings
in adhesive-bonded joints **A17:** 613
Dioctadecyldimethylammonium bromide
orientation of long-chain molecule **A10:** 119
Dioctedecyl disulfide **A18:** 532
Diode array detectors
use in Raman spectroscopy **A10:** 128-129
Diode ion plating **A18:** 840, 844- 845
Diode sputter coater
for SEM specimens **A12:** 173
Diode transistor logic (DTL) **EL1:** 160, 1141
Diode(s)
configurations, active analog components **EL1:** 145
discrete two-terminal devices as **EL1:** 429-432
failure mechanisms **EL1:** 973-974
monolithic, active analog components **EL1:** 144
removal methods **EL1:** 724-727
Schottky barrier **EL1:** 201
Diode-style electron guns **A6:** 254, 260-261
Diodes *See also* Photodiodes
gallium aluminum arsenide (GaAl) laser **A2:** 739
impact avalanche transit time (IMPATT) **A17:** 209-210
indium gallium arsenide phosphide (InGaP) laser **A2:** 739-740
laser **A2:** 739
light-emitting LEDs **A2:** 739-740
semiconductor, for microwave energy **A17:** 208
Diopside
chemical system **EM4:** 872, 873
crystal structure **EM4:** 881
DIP **EM3:** 588-589
Dip (flux bath)
relative rating of brazing process heating method **A6:** 120
Dip and look test
for solderability **EL1:** 677, 944
test standards used to evaluate solderability **A6:** 136
Dip brazing
of aluminum alloys **M6:** 1026-1027
definition **M6:** 5
modifications for aluminum alloys **M6:** 1030
of stainless steels **M6:** 1012
Dip brazing (DB) **A6:** 122-123, 336-338
aluminum **A6:** 336
aluminum alloys **A6:** 338, 939
applications **A6:** 336, 338
brazing salts **A6:** 336-337, 338
carbon steels **A6:** 336-338
carburizing and cyaniding salts **A6:** 337, 338
cast irons **A6:** 338
copper **A6:** 336
definition **A6:** 336, 1208
disadvantages **A6:** 336
ductile iron **A6:** 338
equipment maintenance **A6:** 336
ferrous alloys **A6:** 336
filler metals **A6:** 337, 338
fluxes **A6:** 337-338
fluxing agents **A6:** 337, 338
furnace construction **A6:** 336, 337
gray iron **A6:** 338
low-alloy steels **A6:** 336-338
malleable iron **A6:** 338
neutral salts **A6:** 337, 338
nickel-base alloys **A6:** 336
process details **A6:** 336
safety precautions **A6:** 338, 1191, 1202
stainless steels **A6:** 338
stainless steels in a salt bath **A6:** 911, 921-922
Dip brazing of steels in molten salt **M6:** 989-995
advantages **M6:** 989
filler metals **M6:** 991-992
fluxes **M6:** 991
furnaces **M6:** 991-992
externally heated **M6:** 991
internally heated **M6:** 991-992
pot materials **M6:** 992
joint design **M6:** 992
mechanized brazing **M6:** 994-995
preparation for brazing **M6:** 992
assemblies **M6:** 992
preheating **M6:** 992
procedures **M6:** 991-994
combination heat treating **M6:** 992-993
production examples **M6:** 992-994
use of cyanide salts **M6:** 994
safety precautions **M6:** 995
threshold limit values **M6:** 995
salts **M6:** 990-991
carburizing and cyaniding salts **M6:** 991
fluxing agents **M6:** 991
neutral salts **M6:** 990-991
Dip casting
defined **EM2:** 14

Dip coat *See also* Buckle; Investment casting; Investment precoat; Shell molding
defined **A15:** 4-5
Dip coat spall
as casting defect.......... **A11:** 385
Dip coating.......... **EM3:** 586
automotive.......... **A13:** 1015
of conformal coatings **EL1:** 764
defined **EM2:** 14
equipment, schematic **EL1:** 779, 780
of urethanes.......... **EL1:** 778-779
Dip emulsion cleaning **M5:** 34-35
Dip plating *See* Immersion plating
Dip soldering
defined **EL1:** 1141
definition.......... **M6:** 5
Dip soldering (DS) **A6:** 356
advantages.......... **A6:** 356
applications **A6:** 356
definition.......... **A6:** 356, 1208
equipment.......... **A6:** 356
fixturing **A6:** 356
personnel.......... **A6:** 356
procedure.......... **A6:** 356
safety precautions **A6:** 356
Dip test.......... **A6:** 136
Diperoxydodecanedioic acid
ATR, DRS, and PAS granular analysis of.......... **A10:** 120
Diphase cleaning
zinc alloy die castings **M5:** 677
Diphase emulsion cleaner **M5:** 33-35
Diphasic gels **EM4:** 211
Diphenyl oxide
tantalum resistance to **A13:** 727
Diphenyl oxide resins **EM3:** 11
defined **EM2:** 14
Diphenyl-methane-diisocyanate (MDI)
in polyurethanes.......... **EM2:** 257-258
Diphenylcarbazide method
UV/VIS analysis for chromium in beryllium by **A10:** 68
Diphenylmethane-4,4′-diisocyanate (MDI) **EM3:** 108, 109
Dipole
dislocation, in ferrite.......... **A10:** 469
moment, effect in solvent extraction **A10:** 164
strength of transition, molecular vibrations **A10:** 111
Dipole magnets
for niobium-titanium superconduction materials.......... **A2:** 1056
Dipole polarization
of insulators/dielectric materials **EL1:** 99- 100
Dipotassium phosphate
use in gold plating.......... **M5:** 281-282
Dipped mica capacitor
solderability defects **EL1:** 1036-1037
Dipping
acid *See* Acid dipping
bright *See* Bright Dipping
ceramic coating processes.......... **M5:** 536-538
cyanide *See* Cyanide dipping
hot *See* Hot dip
paint *See* Paint and painting, dipping process
paint coating by.......... **A7:** 460
porcelain enameling.......... **M5:** 516-517, 523, 530
porcelain enameling by.......... **A13:** 447
as silicone conformal coating application method **EL1:** 773
surface cleaning by **A13:** 381
Dipurative degassing
aluminum P/M alloys.......... **A2:** 203
Direct (Electrofax) and coated paper process
of electrostatic copying methods.......... **A7:** 582
Direct (natural) recovery processes
plating waste treatment **M5:** 315-316
Direct arc melting
plain carbon steels.......... **A15:** 706-708
Direct beam transmission ultrasonic testing **A17:** 248
Direct carbon method
TEM replication.......... **A12:** 7
Direct carbon replica technique
defined **A17:** 53
Direct carbon replicas.......... **A9:** 108
formation **A12:** 181
Direct casting methods
of carbon and low-alloy steel sheet and strip **A1:** 211
Direct chemical separation
for UV/VIS interferences.......... **A10:** 65
Direct chip interconnect by wire bonding **EL1:** 231
Direct contact method
applications **A17:** 94
bearing rollers **A17:** 116-117
of generating magnetic fields.......... **A17:** 97
head shot, magnetizing advantages limitations **A17:** 94
Direct current
differential transformer **A8:** 223
electrical potential crack monitoring system.......... **A8:** 388
electrical potential method **A8:** 389
magnetization.......... **A7:** 576
stray-current corrosion by **A13:** 87
Direct current (dc)
characteristics, WSI **EL1:** 355-356
demagnetization with.......... **A17:** 121
design requirements **EL1:** 25-26
digital system operation **EL1:** 78
effect, electrical contact materials.......... **A2:** 840
effects, magnetic particle inspection.......... **A17:** 108
injection, electric current perturbation.......... **A17:** 136
in magnetic particle inspection.......... **A17:** 91-92
magnetic properties **A2:** 777
magnetic properties nickel iron alloys **A2:** 772
probe testing **EL1:** 946
single-phase full-wave, defined.......... **A17:** 91
solid ferromagnetic conductor carrying.......... **A17:** 96
solid nonmagnetic conductor carrying **A17:** 96
vs. alternating current, magnetic particle inspection.......... **A17:** 108-110
Direct current (dc) plasma gun.......... **EM4:** 203, 205-206
Direct current arc furnace
as new technology **A15:** 367-368
Direct current cleaning *See* Cathodic electrocleaning
Direct current electrical potential method.......... **A8:** 389
for martensitic steel.......... **A8:** 390
for monitoring crack length **A8:** 417
Direct current electrode negative
definition.......... **M6:** 5
Direct current electrode negative (DCEN) **A6:** 30, 32
for air-carbon arc cutting.......... **A6:** 1176
for carbon arc welding.......... **A6:** 201
definition.......... **A6:** 1208
fluxes for welding.......... **A6:** 57
gas-metal arc welding **A6:** 183
gas-tungsten arc welding.......... **A6:** 191, 192, 193
gas-tungsten arc welding of ferritic stainless steels **A6:** 445
hardfacing.......... **A6:** 801, 804
plasma arc welding.......... **A6:** 195
plasma-MIG welding.......... **A6:** 223
shielded metal arc welding **A6:** 177
to minimize hydrogen absorption in underwater welding.......... **A6:** 1013
Direct current electrode positive
definition.......... **M6:** 5
Direct current electrode positive (DCEP) **A6:** 30, 31, 67
for air-carbon arc cutting.......... **A6:** 1172, 1175-1176
definition.......... **A6:** 1208
gas-metal arc welding **A6:** 183, 184
of ferritic stainless steels.......... **A6:** 446
gas-tungsten arc welding.......... **A6:** 191, 192
hardfacing.......... **A6:** 801
plasma arc welding.......... **A6:** 195
plasma-MIG welding.......... **A6:** 223
shielded metal arc welding **A6:** 175, 177
of ferritic stainless steels.......... **A6:** 446
to minimize hydrogen absorption in underwater welding.......... **A6:** 1013
Direct current generator-rectifier **A6:** 40
Direct current plasma (DCP) emission spectroscopy
for chemical analysis **EM4:** 553
to analyze the bulk chemical composition of starting powders.......... **EM4:** 72
Direct current plasma torches.......... **A15:** 440-443
Direct current, pulsed
electromigration effects of **A11:** 772
Direct current resistivity testing
of powder metallurgy parts **A17:** 541-542
Direct current reverse *See also* Direct current electrode positive (DCEP)
definition.......... **A6:** 1208
Direct current straight polarity *See also* Direct current electrode negative (DCEN)., definition.......... **A6:** 1208
Direct deformation
liquid impact erosion.......... **A18:** 224
Direct deposition
of powder and sponge **A7:** 71-72
Direct digitalization *See* Digital radiography
Direct dissolution
as liquid-metal corrosion **A13:** 56-57
Direct energy welding machines **M6:** 469-473
controls.......... **M6:** 470-472
contactors **M6:** 470-471
current controls **M6:** 472
heat controls.......... **M6:** 471
regulators, current and voltage.......... **M6:** 472
sequence controls.......... **M6:** 471
timers **M6:** 471
equipment.......... **M6:** 472-473
electrical system **M6:** 472
transformer.......... **M6:** 472
secondary circuit **M6:** 473
throat depth and height.......... **M6:** 473
single-phase.......... **M6:** 470
three-phase **M6:** 470
Direct extrapolation
for creep-rupture analysis **A8:** 690
Direct extrusion *See* Extrusion
Direct failures
soldering **EL1:** 943
Direct filling alloys.......... **A13:** 1348-1350
of precious metal.......... **A2:** 696
Direct forming
thermoplastic polyimides (TPI) **EM2:** 178
Direct forming single-end roving process **EM1:** 109
Direct Fourier reconstruction technique
computed tomography (CT) **A17:** 380
Direct heating refractory metal sintering furnace.......... **A7:** 627
Direct imaging *See also* Stadimetry
defined **A17:** 34
and neutron radiography **A17:** 387
Direct imaging of grain boundaries
by high resolution electron microscopy.......... **A9:** 121
Direct injection
as die casting method.......... **A15:** 286
Direct injection burner
defined **A10:** 672
Direct isostatic pressing
in aerospace applications.......... **A7:** 648-649
Direct labor
as piece cost component **EM2:** 82
Direct magnetization
in leakage field testing **A17:** 130
Direct melt process
for glass fibers.......... **EM1:** 45
Direct memory access
use in x-ray spectrometry.......... **A10:** 92
Direct metal infiltration processing
of graphite-reinforced MMCs **EM1:** 872
Direct method
to produce electron density maps.......... **A10:** 351
Direct multiple-spot welding setups **M6:** 476
Direct neutron radiography
for adhesive-bonded joints **A17:** 625-626
Direct potentiometry
ion-selective electrodes and.......... **A10:** 204
Direct pouring *See also* Pouring
as automated ladle method.......... **A15:** 498
Direct powder forming
aluminum P/M alloys.......... **A2:** 203
Direct powder precipitation
copper powders.......... **A7:** 120
Direct powder rolling
SEM analysis.......... **A7:** 235
Direct Process (General Electric Company).......... **EM3:** 49
Direct quenching **M4:** 31
Direct redrawing.......... **A14:** 584-585
Direct resistance heating
as hot pressing setup.......... **A7:** 505-507

Direct sample insertion device **A10:** 36, 690
Direct shear
as stress in fatigue fracture **A11:** 75
Direct single-spot welding setups **M6:** 475-476
Direct sintering
defined **A7:** 4
Direct stress
fatigue test specimens **A8:** 368
fixture, for rotating eccentric mass machine **A8:** 369
Direct surface replicas
defined **A17:** 53
Direct-chill (DC) semicontinuous casting
wrought copper and copper alloys **A2:** 243
Direct-chill casting
aluminum alloys **A15:** 313-314
Direct-chill semicontinuous casting of aluminum alloy ingots, center cracking
in **A9:** 634-635
Direct-cooled forging microstructures
processing of **A1:** 137, 138
Direct-current arc
defined **A10:** 25
Direct-current arc emission spectroscopy
and ICP-AES **A10:** 31
precision **A10:** 25
Direct-current motor
cam plastometer **A8:** 194
Direct-current plasma **A10:** 40, 690
Direct-current plasma atomic emission spectrometry **A10:** 21, 43
Direct-current polarograms
current-voltage curves as **A10:** 189-190
Direct-current polarography
circuit and cell arrangement **A10:** 189
Direct-drive hydraulic extrusion press **A14:** 319
Direct-drive wheels
blast cleaning **A15:** 515-516
Direct-electric drive presses **A14:** 33-34, 319
Direct-exposure method
of neutron detection **A17:** 390-391
Direct-fired batch ovens
paint curing process **M5:** 487
Direct-HIP process *See also* Hot isostatic pressing (HIP)
beryllium powder **A2:** 686
Direct-imaging ion microscopes
for SIMS analysis **A10:** 613, 614
Direct-iron blast furnace
development of **A15:** 24-25
Direct-reader spectrometer
inductively coupled plasma **A10:** 37
Direct-reduced iron **A7:** 97-98
Direct-stress fatigue testing machine *See* Axial fatigue testing machine
Directed metal oxidation **EM4:** 232-235
applications **EM4:** 232, 233-235
composite examples and applications **EM4:** 233-235
Al_2O_3 particle-filled Al_2TiO_5 composite **EM4:** 235
Nicalon fiber-reinforced Al_2O_3 .... **EM4:** 233
particle-filled AlN matrix composite **EM4:** 235
SiC fiber-reinforced Al_2O_3 matrix composites **EM4:** 233-234
SiC particle-reinforced Al_2O_3 matrix composites **EM4:** 234
ZrB_2 platelet-reinforced ZrC composites **EM4:** 234
composite processing **EM4:** 232-233
examples of ceramic-matrix systems **EM4:** 233
Direction
crack **A11:** 735-738, 744
of flux, effects **A17:** 91
of magnetization **A17:** 110-111, 129
of sensing coils, ECP **A17:** 137
and magnitude of maximum residual stress **A10:** 392
of reaction, symbol for **A10:** 691
Direction (continued)
of relative motion, surface configuration **A11:** 159
Direction of flow
in magnetic field **A17:** 90
Direction of view
of borescopes **A17:** 9
Directional properties
of steel plate **A1:** 238
Directional solidification *See also* Aluminum castings; Solidification **M2:** 150-151
castings, processing **A15:** 320-321
chills for **A2:** 348
copper alloy casting **A15:** 781-782
defects due to **A15:** 321-322
defined **A15:** 5
effect, inclusion-forming **A15:** 91
and equiaxed casting, compared **A15:** 399
of eutectic aluminum alloys **A2:** 1045
eutectic growth (cast iron) **A15:** 174-175
from the melt, methods of **A9:** 607
furnace **A15:** 321, 400-401
with high-shrinkage copper casting alloys **A2:** 346
market effects **A15:** 44
and monocrystal solidification **A15:** 319-323
nickel alloy **A15:** 817, 819, 823
particle behavior in **A15:** 142-145
planar interface **A15:** 142-144
and progressive solidification **A15:** 778
and riser location **A15:** 578-579
in succinonitrile-5.5 mole% acetone **A9:** 612
of superalloys **A2:** 429 **A15:** 418
in Unicast process **A15:** 251
used to develop crystallographic texture **A9:** 701
vertical centrifugal casting **A15:** 300
Directional solidification (DS) casting process **A1:** 995
Directionality *See also* Anisotropy
crack propagation, FSS and **A12:** 205
effect on stress-corrosion cracking **A8:** 501
HSLA steels, effect on properties **M1:** 411, 417-418
in parametric relationships, partially-oriented surfaces **A12:** 201
steel plate **M1:** 194
in tensile testing **A11:** 19
Directionally solidified eutectics *See also* Unidirectionally solidified eutectics
Directionally solidified superalloys *See also* Cast nickel-base superalloys
chemistry and DS castability **A1:** 996-997
heat treatment and mechanical properties **A1:** 997-998
Directory
of information sources **EM1:** 40-42
Dirt content
defined **A18:** 7
Dirt damage
in ball bearings **A11:** 493-494, 496
"Dirty pin" values **A8:** 60
Disadvantages of adhesive joining **EM3:** 33, 34
Disbond *See also* Debond; Delamination; Disbonding **EM3:** 11
defined **EM1:** 9 **EM2:** 14
Disbonding
interlaminar **EM1:** 234
weld **A13:** 332
Disbonding failures **EL1:** 61, 1045
Disbondment
defined **A13:** 5
Disbonds
adhesive-bonded joints **A17:** 612
by computed tomography (CT) **A17:** 361
thermal inspection **A17:** 402
Disc brake pad
from P/M friction materials **A7:** 701
Discaloy
aging **A4:** 796
aging cycle **M4:** 656
annealing **M4:** 655
broaching **A16:** 203
composition **A4:** 794 **A16:** 736 **M4:** 651-652
contour band sawing **A16:** 363
machining **A16:** 738, 741-743, 746-747, 749-758
notch sensitivity **A8:** 316
rupture time variations **A8:** 316-317
solution treating **M4:** 656
solution-treating **A4:** 796
stress relieving **M4:** 655
Discharge
in continuous stream sampling **A7:** 213
Discoloration
in bearing failures **A11:** 494
in burnup with plastic flow, roller bearings **A11:** 500-501
of fracture surface **A11:** 80
Discoloration, cast aluminum alloys
See also Colors **A12:** 408
Discoloration of aluminum alloy wrought products
effect of cooling rate on **A9:** 634
Discoloy
composition **A6:** 564 **M6:** 354
Discontinuities *See also* Cold shuts; Cracks; Defects; Flaws; Subsurface flaws; Surface flaws specific discontinuities
AISI/SAE alloy steels **A12:** 331
alloy segregation as **A11:** 121
by mechanical effects (rupture) **A11:** 383
as casting defects **A11:** 383 **A15:** 548
casting types **A17:** 512, 515
-caused brittle fracture **A11:** 85
characteristics **A17:** 348
characteristics, by color **A17:** 483-488
cold shuts **A12:** 64-65
crack initiation and **A11:** 106
deep-lying **A17:** 105
defined **A15:** 5 **A17:** 49, 103
design-related, welding **A17:** 582
detectable by eddy current inspection **A17:** 179
detectable, magnetic particle inspection **A17:** 103-105
detected by magnetic particle inspection **A7:** 575-579
detection characteristics **A17:** 104
dispersed, radiographic appearance **A17:** 349
effect on fatigue behavior **M1:** 682
effect on fatigue strength **A11:** 119-121
extremities, NDE capabilities **A17:** 677
as fracture initiation sites **A12:** 64
from welding process, types **A17:** 582
inclusions, defined **A12:** 65
internal, in shafts **A11:** 467
internal, in tire-mold casting **A11:** 358
internal, types of casting **A11:** 354-359
internal, detecting **A15:** 544-545
internal, in iron castings **A11:** 354-369
internal, magnetic particle inspection **A17:** 105
laps **A12:** 64-65
leading to fracture **A12:** 63-68
low-alloy steels **A15:** 718
metallurgical, types **A17:** 582
microwave inspection of **A17:** 212-214
multiple, defined **A17:** 663
orientation of **A17:** 105
plain carbon steels **A15:** 704
porosity **A12:** 65, 67
in pressure vessels, effects of **A11:** 646
reference, eddy current inspection **A17:** 179
revealed by liquid penetrant inspection **A17:** 86
seams **A12:** 64-65
segregation **A12:** 67
in shape, shafts **A11:** 467
subsurface **A17:** 89, 105
subsurface, effect on fatigue strength **A11:** 120

Discontinuities (continued)
subsurface, magnetic particle inspection **A17:** 89
surface, effect on fatigue strength **A11:** 119
surface, in iron castings **A11:** 352-353
surface, in shafts **A11:** 459, 467, 472
surface, iron casting failures from **A11:** 352-353
surface, magnetic particle inspection of **A17:** 89
unfavorable grain flow **A12:** 67-68
Discontinuity *See* Discontinuities **A13:** 5, 148
definition **A6:** 1208
Discontinuity, definition *See also* Weld discontinuities **M6:** 5
Discontinuous aluminum metal-matrix composites **A2:** 7, 906-907
Discontinuous ceramic fiber MMCs
See also Ceramic fibers **EM1:** 903-910
applications **EM1:** 909-910
composite fabrication **EM1:** 904-906
fibers **EM1:** 903
matrix alloys **EM1:** 904
properties **EM1:** 906-908
secondary processing **EM1:** 908-909
Discontinuous fiber composites *See also* Discontinuous fibers; Discontinuous oxide fibers; Short fiber composites **EM1:** 794-797
compressive failure **EM1:** 796-797
defined **EM1:** 27
failure analysis **EM1:** 794-797
fatigue failure **EM1:** 797
from recycled carbon fiber scrap **EM1:** 153-155
product forms **EM1:** 33
shear failure **EM1:** 797
structure of **EM1:** 794-795
tensile failure **EM1:** 795-796
Discontinuous fiber reinforced thermoplastic composites
injection molding process **EM1:** 121
Discontinuous fibers *See also* Discontinuous fiber composites; Short fibers; specific fiber, fiber forms; Whiskers
aligned, fiber-reinforced plastic with **EM1:** 153-155
aramid forms **EM1:** 115
carbon/graphite, properties **EM1:** 867
processing **EM1:** 120-121
properties **EM1:** 119-120, 903
recycled carbon fiber scrap as **EM1:** 153-156
as reinforcement **EL1:** 1119-1121
strength **EM1:** 120
vs. continuous fibers **EM1:** 105, 119
Discontinuous fibers/particles
fluidity effect **A15:** 849
in metal-matrix composites **A15:** 840
Discontinuous grain growth *See also* Grain growth **A9:** 689-690
Discontinuous graphite/aluminum metal-matrix composites **A2:** 907
Discontinuous intergranular facets **A8:** 487
Discontinuous metal-matrix composites
aluminum-base **A14:** 251
Discontinuous oxide fibers
commercially available **EM1:** 61
from chemicals/sol-gels **EM1:** 63
types/properties **EM1:** 62-63
Discontinuous phase
of composites **EM1:** 27
Discontinuous precipitation **A9:** 649
in beryllium-copper alloys **A9:** 395
Discontinuous processing
polyurethanes (PUR) **EM2:** 262-263
Discontinuous reinforced metal matrix composites (MMCs)
ceramic fiber **EM1:** 903-910
silicon carbide (SiC) whiskers **EM1:** 889-902
Discontinuous silicon carbide fibers **EM1:** 64
Discontinuous silicon carbide whiskers *See also* Silicon carbide (SiC) whiskers
properties **EM1:** 63
Discontinuous silicon carbide/aluminum metal-matrix composites **A2:** 906
Discontinuous silicon fiber MMCs **EM1:** 889-895
Discontinuous silicon nitride whiskers **EM1:** 63-64
Discontinuous sintering
defined **A7:** 4
Discontinuous yielding
defined **A8:** 4
requirements for **A9:** 684
tensile load vs. elongation **A9:** 684
Discontinuous-fiber composites
coatings for **A13:** 859, 862
Discrete chip mounting technologies *See also* Component and discrete chip mounting technologies
introduction **EL1:** 143
Discrete semiconductor
defined **EL1:** 422
Discrete semiconductor packages
description, performance, construction of **EL1:** 422-434
future directions **EL1:** 434-435
industrial consumer packages **EL1:** 422
level 1 .... **EL1:** 405
multiple-terminal devices **EL1:** 432-434
system-level package **EL1:** 434
three-terminal devices **EL1:** 422-429
TO-92, TO-3, TO 220, as specific types **EL1:** 435
two-terminal devices **EL1:** 429-432
Discrete-device testing
electrical **EL1:** 946-951
Discrimination
in sampling **A10:** 16
Discrimination ratio (DR) **EM4:** 86-87
Dished
defined **EM2:** 14
Disk *See also* Disc
forging, design curves **A7:** 411
production, P/M techniques plus forging **A7:** 522-523
Disk finishing, centrifugal *See* Centrifugal disk finishing
Disk machine *See also* Amsler wear machine
defined **A18:** 7
Disk recorder heads **A13:** 1121
corrosion failure analysis of **EL1:** 1112
Disk-forming process
simulation of **A14:** 430
Disk-pressure test
for hydrogen embrittlement **A8:** 540-541
Disk-pressure testing
for hydrogen embrittlement **A13:** 287-288
Disks
magnetizing **A17:** 94
on shafts, inspection of **A17:** 113
Disks, compressor blade
corrosion of **A13:** 999-1000
Dislocation *See also* Dislocation climb; Dislocation creep; Dislocation glide
defined **A9:** 6
in dynamic recrystallization **A8:** 173
effects, resistance-ratio test **A2:** 1096
as fracture mechanism **A12:** 12
intersections, effect on failure **A8:** 34
irradiated materials **A12:** 365
mechanics force concept in **A8:** 440
nucleation, crack tip **A12:** 30
pile-ups, microvoid coalescence at **A12:** 12
producing serrations in stress-strain curves **A8:** 35
in single-phase materials **A8:** 173
stationary, effect in crack extension **A8:** 439
structures, ordered intermetallics **A2:** 913
substructure **A12:** 52
Dislocation arrays *See also* Oriented dislocation arrays
in pure metals **A9:** 610
Dislocation Burgers vectors **A9:** 115-116 **A18:** 387
in grain boundaries **A9:** 120
Dislocation cell structure analysis
by analytical electron microscopy **A10:** 470-473
computerized misorientation determination **A10:** 471-472
limitations and conditions for orientation determination **A10:** 471
misorientation determination for small cells **A10:** 472-473
Dislocation climb
and creep at high temperatures **A8:** 309
in pure metals **A8:** 308
recovery **A8:** 310
Dislocation creep **EM4:** 295
from work hardening and thermal recovery **A8:** 309
Dislocation creep (continued)
mechanisms **A8:** 309
Dislocation densities
determination using transmission electron microscopy **A9:** 115-116
indicating degree of deformation **A9:** 685
of single crystals, etch pits used to determine **A9:** 62
in single-crystal pure metals **A9:** 607
in slip systems, contrast **A9:** 116
of specimens deformed to large strains **A9:** 693
in workhardened metals **A9:** 685
Dislocation density **A18:** 468, 469
and magnetic measurement **A17:** 131
Dislocation dipoles
contrast **A9:** 114-115
Dislocation etch pits
in copper **A9:** 127
quantitative metallography of **A9:** 126
Dislocation etch pitting reagents
for magnetic materials **A9:** 531, 534
Dislocation etching
defined **A9:** 6
Dislocation glide **A18:** 426
along crystallographic planes **A8:** 34
thermally activated **A8:** 309
viscous **A8:** 308
Dislocation lines
as linear element in quantitative metallography **A9:** 126
in surface of silicon crystal **A9:** 127
Dislocation loops
in copper single crystal deformed 10% **A9:** 689
and dislocation-precipitate reactions **A9:** 688
formation **A9:** 116
in germanium **A9:** 608
in single-crystal pure metals **A9:** 607
transmission electron microscopy **A9:** 116-117
Dislocation model
for grain boundaries **A9:** 119
Dislocation networks
formation of **A9:** 604
Dislocation pairs
contrast **A9:** 114-115
in $MnNi_3$ **A9:** 682-683
Dislocation stacking fault energy **A16:** 11
Dislocation substructure in titanium alloys
role in recrystallization studies **A9:** 461
Dislocation tangles
behind Lüders front **A9:** 685
cell structure **A9:** 685
in iron, forming cells **A9:** 685
Dislocation theory **A16:** 4, 11 **A18:** 468
Dislocation(s)
as crack initiation **A17:** 216
density, magnetic measurement of **A17:** 131
as source, acoustic emissions **A17:** 287
studies, by acoustic emission inspection **A17:** 286
Dislocation-generated antiphase boundaries **A9:** 682-683
Dislocation-precipitate reactions
types **A9:** 688
Dislocations **A13:** 5, 45-47
as a result of electric discharge machining **A9:** 27
in bright-field images, polycrystalline aluminum **A10:** 444
as carbide precipitation sites in austenitic stainless steels **A9:** 284
of cell structure, analysis of **A10:** 470-473
cell structure, by cold rolling ETP copper **A10:** 469
concentration of, in deformed metals **A9:** 684
dipoles and loops, in ferrite **A10:** 469
distribution, factors affecting **A9:** 685
edge dislocations **A9:** 682-683
effect in bcc ferrite **A11:** 84
effect of Burgers vector orientation on FIM contrast from **A10:** 588-589
effect of temperature and strain rate on **A9:** 688-691
elastic displacement field associated with **A10:** 464
in FIM samples **A10:** 587
formation **M6:** 857
formed during plastic deformation **A9:** 693

Dislocations (continued)
in germanium ... **A9:** 608
glissile, fcc material ... **A10:** 465
in gold ... **A9:** 609
in grain boundaries, transmission electron microscopy diffraction studies of ... **A9:** 119-120
imaged by x-ray topography ... **A10:** 365
imaged in aluminum alloy ... **A10:** 465
imaging by topography ... **A10:** 367-370
loops, ATEM imaged ... **A10:** 465
movement of ... **A9:** 684
moving, in $AlFe_3$ ... **A9:** 682-683
partial, stacking-fault region between ... **A10:** 589
perfect, effect in FIM images ... **A10:** 588
point defects and ... **A10:** 583
in pure metals ... **A9:** 607-608
rearrangement during recovery ... **A9:** 693
SAS techniques for ... **A10:** 405
slip deformation ... **A9:** 720
as stored energy sites in cold-worked metals ... **A9:** 692
study of alloy elements and impurities to ... **A10:** 583
in subgrain boundaries, effect on nucleated grain growth ... **A9:** 696-697
superlattice ... **A9:** 682-683
tangle of ... **A10:** 358, 467
transmission electron microscopy of ... **A9:** 113-116
types ... **A9:** 719
Disodium salt, of EDTA
use in electrogravimetry ... **A10:** 201
Disordered alpha grains in a palladium-copper alloy ... **A9:** 564
Disordered crystal structure ... **A3:** 1 • 10
Disordered materials, magnetic
exotic effects ... **A10:** 276
Disordered structure
defined ... **A9:** 6 **A10:** 672
Disordered superstructures of crystals ... **A9:** 708
Disordered systems
EXAFS analysis ... **A10:** 407
Dispensing and application equipment ... **EM3:** 693-702
classification of materials ... **EM3:** 693
dispensing system components ... **EM3:** 693-701
dispensing gun or valve ... **EM3:** 693, 699-700, 701
dispensing system automation ... **EM3:** 699-700, 701-702
future of dispensing systems ... **EM3:** 702
header system ... **EM3:** 693, 696-699
pumping system ... **EM3:** 693-696
system controls ... **EM3:** 693, 700-701
Dispensing equipment
for resin transfer molding ... **EM1:** 169
Dispersalloy
abrasion resistance ... **A18:** 669
material loss on abrasion of dental amalgams ... **A18:** 669
Dispersancy
engine oils ... **A18:** 169
Dispersant additive ... **A18:** 99-100
applications ... **A18:** 99-100
defined ... **A18:** 7
detection by infrared spectroscopy of lubricants ... **A18:** 301
effectiveness in engine oils ... **A18:** 100
in engine lubricant formulations ... **A18:** 111
formation of ... **A18:** 99-100
multifunctional nature ... **A18:** 111
in nonengine lubricant formulations ... **A18:** 111
Dispersant oil
defined ... **A18:** 7
Dispersant-viscosity improvers ... **A18:** 99, 109
Dispersants
for lubricant failure ... **A11:** 154
Dispersed shrinkage
as casting defect ... **A11:** 382
Dispersing agent
defined ... **A7:** 4
Dispersion ... **EM3:** 11
curves, leaky Lamb wave testing ... **A17:** 252
defined ... **EM2:** 14
device, x-ray spectrometers ... **A10:** 89
lineshape, and Lorentzian absorption lineshape ... **A10:** 280, 281
of microwaves ... **A17:** 204
quantifying ... **A8:** 625
Dispersion cleaning
mechanics of action ... **M5:** 23
Dispersion forces ... **EM3:** 40
Dispersion gels ... **EM4:** 449
Dispersion hardening
defined ... **A7:** 4
Dispersion, of carbide grains
in cermets ... **A2:** 991
Dispersion processing technique ... **EM4:** 449
Dispersion strengthened
abbreviation ... **A8:** 724
Dispersion strengthening *See also* Dispersioned-strengthened materials
defined ... **A7:** 4
high-energy milling ... **A7:** 69
Dispersion-casting technique (Scherer) ... **EM4:** 450
Dispersion-hardened aluminum alloys
for NbTi superconducting materials ... **A2:** 1045
Dispersion-strengthened alloys
and cermets, compared ... **A2:** 978
nickel superalloy, development ... **A2:** 429
titanium P/M products ... **A2:** 656
tungsten ... **A14:** 238
Dispersion-strengthened aluminum alloys ... **A6:** 541-547
advantages ... **A6:** 541
applications ... **A6:** 541, 542
brazing ... **A6:** 543
capacitor-discharge welding ... **A6:** 543, 544
catalytic nucleation ... **A6:** 545
characteristics of processing techniques ... **A6:** 541
composition ... **A6:** 541
description ... **A6:** 541
diffusion welding ... **A6:** 543, 547
electron-beam welding ... **A6:** 543, 544, 545, 546
filler metals ... **A6:** 543
freezing range ... **A6:** 541, 542
friction welding ... **A6:** 543, 546, 547
fusion welding ... **A6:** 542, 543-546
fusion zone ... **A6:** 542-543, 544, 545, 546
gas-tungsten arc welding ... **A6:** 543-544, 545
heat-affected zone ... **A6:** 542, 544, 545, 546
hydrogen content control ... **A6:** 543
hydrogen-induced fusion zone porosity ... **A6:** 542-543
laser-beam welding ... **A6:** 543, 544, 545, 546
metallurgy of aluminum-iron-base alloys ... **A6:** 541-542
microstructure ... **A6:** 541-542, 543-546, 547
properties ... **A6:** 541, 542
resistance welding ... **A6:** 543
solid-state welding ... **A6:** 546-547
vacuum brazing ... **A6:** 543
weld solidification behavior ... **A6:** 542
weldability ... **A6:** 542-543
Dispersion-strengthened iron-base alloys *See also* Dispersion-strengthened iron-base alloys, specific types
bar ... **A2:** 948-949
commercial alloys ... **A2:** 944-947
fabrication ... **A2:** 947-949
hot-corrosion properties ... **A2:** 947
joining of ... **A2:** 949
mechanical alloying alloy applications ... **A2:** 943
mechanical alloying process ... **A2:** 943-944
oxidation properties ... **A2:** 947
sheet ... **A2:** 949
Dispersion-strengthened iron-base alloys, specific types *See also* Dispersion-strengthened iron-base alloys
Alloy MA 754, microstructure and elevated-temperature strength ... **A2:** 944-945
Alloy MA 758, oxidation resistance properties, uses ... **A2:** 945-946
Alloy MA 760, composition and properties ... **A2:** 947
Alloy MA 760, high-temperature strength structural stability ... **A2:** 947
Alloy MA 956, properties, product forms ... **A2:** 946
Alloy MA 6000, elevated-temperature resistance ... **A2:** 946-947
strength, oxidation and sulfidation
Alloy MA 6000, microstructure and properties ... **A2:** 946-947
Dispersion-strengthened materials ... **A7:** 18, 710-727
alloys ... **A7:** 77, 522, 527-528
copper ... **A7:** 711-716, 740
copper electrodes ... **A7:** 624-626
defined ... **A7:** 4
iron-based ... **A7:** 722-727
nickel-based ... **A7:** 722-727
platinum ... **A7:** 720-722
silver ... **A7:** 716-720
superalloys ... **A7:** 722-723
Dispersion-strengthened nickel-base alloys *See also* Mechanical alloying (MA)
bars ... **A2:** 948-949
commercial alloys ... **A2:** 944-947
fabrication of MA ODS alloys ... **A2:** 947-949
hot-corrosion properties ... **A2:** 947
joining of MA ODS alloys ... **A2:** 949
mechanical alloying alloy applications ... **A2:** 943
mechanical alloying process ... **A2:** 943-944
oxidation properties ... **A2:** 947
product forms, properties ... **A2:** 946
sheet ... **A2:** 949
Dispersion-type nuclear fuel elements ... **A7:** 664-665
Dispersive EXAFS detection technique ... **A10:** 418
Dispersive infrared spectroscopy
instrumentation ... **A10:** 111
Dispersive spectrophotometers
dual-beam ... **A10:** 67-68
single-beam ... **A10:** 67
Dispersoid
defined ... **A9:** 6
Dispersoid control
aluminum alloys ... **A8:** 479
carbide distributions in steels ... **A8:** 479
Dispersoid-forming elements
and crack growth ... **A8:** 487
Displacement ... **A8:** 4
along acoustic wave train ... **A8:** 244
angle, in torsional testing ... **A8:** 146
and crack ... **A8:** 383
crack-tip, COD test for ... **A11:** 62
crosshead, as relative displacement ... **A8:** 41, 45
eddy current probe for measuring ... **A8:** 383
in-plane, by optical holographic interferometry ... **A17:** 415-416
limiting, crack arrest fracture toughness ... **A8:** 293
linear variable differential transformer for measuring ... **A8:** 383
measured by cantilever beam clip gage ... **A8:** 383
measurement at medium strain rates ... **A8:** 193
measurement by bonded strain gages ... **A8:** 193
measurement, by interferometer ... **A17:** 14
measurement, crack arrest fracture toughness ... **A8:** 292
measurement hardware, attachment ... **A8:** 384-385
measurement, in pressure bars ... **A8:** 201-202
optical holographic interferometry of ... **A17:** 405
out-of-plane, by optical holography ... **A17:** 415

SUBJECTS OF THE INDEXED VOLUMES: ASM Handbook (designated by the letter "A"): **A1:** Properties and Selection: Irons, Steels, and High-Performance Alloys (1990); **A2:** Properties and Selection: Nonferrous Alloys and Special-Purpose Materials (1990); **A3:** Alloy Phase Diagrams (1992); **A4:** Heat Treating (1991); **A6:** Welding, Brazing, and Soldering (1993); **A7:** Powder Metallurgy (1984); **A8:** Mechanical Testing (1985); **A9:** Metallography and Microstructures (1985); **A10:** Materials Characterization (1986); **A11:** Failure Analysis and Prevention (1986); **A12:** Fractography (1987); **A13:** Corrosion (1987); **A14:** Forming and Forging (1988); **A15:** Casting (1988); **A16:** Machining (1989); **A17:** Nondestructive Testing and Quality Control (1989); **A18:** Friction, Lubrication, and Wear Technology (1992). **Metals Handbook, 9th Edition** (designated by the letter "M"): **M1:** Properties and Selection: Irons and Steels (1978); **M2:** Properties and Selection: Nonferrous Alloys and Pure Metals (1979); **M3:** Properties and Selection: Stainless Steels, Tool Materials and Special-Purpose Materials (1980); **M4:** Heat Treating (1981); **M5:** Surface Cleaning, Finishing, and Coating (1982); **M6:** Welding, Brazing, and Soldering (1983). **Engineered Materials Handbook** (designated by the letters "EM"): **EM1:** Composites (1987); **EM2:** Engineering Plastics (1988); **EM3:** Adhesives and Sealants (1990); **EM4:** Ceramics and Glasses (1991); **Electronic Materials Handbook** (designated by the letters "EL"): **EL1:** Packaging (1989).

Displacement (continued)
strain energy and **A11:** 49-50
and strain, in ultrasonic testing **A8:** 242
symbols for **A8:** 726
-time and load-time curves, dynamic notched round bar testing **A8:** 281
in torsion testing **A8:** 139, 146
in torsional hydraulic actuator **A8:** 217
Displacement angle
defined **EM1:** 9 **EM2:** 14
Displacement field in imperfect crystals **A9:** 111
structure-factor contrast **A9:** 113
Displacement field of a dislocation **A9:** 113-114
Displacement interferometer
for shear wave profiles **A8:** 231
Displacement mixing **A18:** 854, 855
Displacement pendulum weighing system **A8:** 613
Displacement principle
in pycnometric theory **A7:** 265
Displacements per incident atom (dpa) **A18:** 851
Display resolution
defined **A17:** 383
Display system
SEM illuminating/imaging **A12:** 169
Disposables *See* Packaging applications
Disposal, waste *See* Waste recovery and treatment
Disposition
of arsenic, as toxic metal **A2:** 1237
of beryllium, as toxic metal **A2:** 1238
of cadmium **A2:** 1239-1240
of chromium, as toxic metal **A2:** 1242
of iron **A2:** 1252
of lead, as toxic metal **A2:** 1243
of magnesium **A2:** 1259
of mercury, as toxic metal **A2:** 1247-1248
of molybdenum **A2:** 1253
of nickel, as toxic metal **A2:** 1250
of selenium **A2:** 1254
of thallium **A2:** 1260
of tin **A2:** 1261
of titanium **A2:** 1261
Disproportionation
defined **EM2:** 14
Disregistry, lattice *See* Lattice disregistry
Dissection techniques, general
capabilities of **A10:** 380
Dissimilar material separation
bolted joints **EM1:** 716-717
bonded joints **EM1:** 717
graphite electrochemical and thermal expansion **EM1:** 716
testing **EM1:** 717-718
Dissimilar materials
drilling of **EM1:** 669-672
stress similarities when bonded **EM3:** 497-500
Dissimilar metal joining **A6:** 821, 822-828
brazing **A6:** 822
diffusion welding **A6:** 822
dilution in a joint **A6:** 821
electrodes **A6:** 824, 827
explosion welding **A6:** 822
factors influencing joint integrity **A6:** 822-824
coefficient of thermal expansion **A6:** 824, 825, 826
dilution **A6:** 822-823, 824
melting temperatures **A6:** 823-824
thermal conductivity **A6:** 824
weld metal **A6:** 822
factors responsible for cracking **A6:** 822
filler metals **A6:** 822, 823, 824-825, 827, 828
friction welding **A6:** 822
gas-metal arc welding **A6:** 824, 828
gas-tungsten arc welding **A6:** 824, 827, 828
heat-affected zone **A6:** 824, 826
hot cracking **A6:** 822, 827
service considerations **A6:** 826-827
carbon migration **A6:** 826
corrosion **A6:** 826-827
oxidation resistance **A6:** 826-827
property considerations **A6:** 826
shielded metal arc welding **A6:** 824, 827, 828
soldering **A6:** 822
specific dissimilar metal combinations **A6:** 827-828
submerged arc welding **A6:** 824

Dissimilar metal joining (continued)
ultrasonic welding **A6:** 822
welding considerations **A6:** 824-826
buttering **A6:** 825, 827
examples **A6:** 825
filler metal selection **A6:** 824-825
joint design **A6:** 825
postweld heat treatments **A6:** 825-826
preheat **A6:** 825-826
process **A6:** 824
Dissimilar metals
couples, metal/alloy compatibility **A13:** 1040, 1061
effect, brazed joints **A13:** 876
electronics industry **A13:** 1108, 1111
galvanic corrosion **A13:** 84, 440-441
joining of, electron-beam welding **A6:** 866
nickel alloys **A6:** 577-578
in pipelines, effects **A13:** 1288
Dissimilar-metal welded joints **A9:** 582
Dissimilar-metal welds
abbreviation for **A11:** 796
between austenitic and ferritic steels **A11:** 620
Dissipation factor *See also* Electrical dissipation factor **EM3:** 179, 429, 431, 432
of ceramics **EL1:** 336, 467-468
copper-clad E-glass laminates **EL1:** 535
defined **EL1:** 1141 **EM1:** 359 **EM2:** 461, 467
and dielectric constant **EM2:** 582-585
electrical **EM1:** 10
epoxy resin system composites **EM1:** 403
glass fabric reinforced epoxy resin **EM1:** 405
of glass fibers **EM1:** 46
high-temperature thermoset matrix composites **EM1:** 377
Kevlar 49 fiber/fabric reinforced epoxy resin **EM1:** 409
low-temperature thermoset matrix composites **EM1:** 398
medium-temperature thermoset matrix composites **EM1:** 385, 388
passive devices **EL1:** 997
polyether-imides (PEI) **EM2:** 158
quartz fabric reinforced epoxy resin **EM1:** 415
thermoplastic matrix composites **EM1:** 72, 367, 370
Dissipation factor, electrical *See also* Electrical displacement factor **EM1:** 10
Dissipation, total power
defined **EL1:** 5
Dissociated aluminum
surface analysis techniques **A7:** 252
Dissociated ammonia
as aluminum sintering atmosphere **A7:** 338
atmosphere for furnace brazing **M6:** 1007-1008
atmospheres **A7:** 341, 344, 361, 368, 729
brazing atmosphere source **A6:** 628
composition **A7:** 342
defined **A7:** 4
dewpoint and reduction of surface oxides **A7:** 344
effects of temperature on porosity ..... **A7:** of iron compacts, sintered in **A7:** 362
properties of P/M aluminum alloys, sintered in **A7:** 384
as sintering atmosphere **A7:** 341, 344, 361, 368, 729
stainless steels, sintered in **A7:** 252, 253
Dissociated ammonia base atmospheres
equipment **M4:** 408-409, 410
operating economics **M4:** 409
safety -precautions **M4:** 409-410
Dissociation
defined **A9:** 6
as leakage **A17:** 58
Dissociation pressure
defined **A9:** 6
Dissolution
in alloy additions **A15:** 71-74
analytical methods requiring **A10:** 4
carbon, in cast iron **A15:** 72-73
and electrometric titration **A10:** 203
of implants, as corrosion **A11:** 672
kinetics, and fluid dynamics **A15:** 73
as liquid-metal corrosion **A13:** 92
mass transfer controlled **A15:** 72-74
as melt analysis **A15:** 493

Dissolution (continued)
metal, by molten-salt corrosion **A13:** 89
metal, design for **A13:** 343
models, crack propagation **A13:** 160
with modifiers **A15:** 484
preferential **A11:** 338
rates, order of **A15:** 71
and reprecipitation, as gravimetric sample preparation **A10:** 163
sample, for classical wet analytical chemistry **A10:** 165-167
of samples, flame AAS analysis of **A10:** 56-57
slip, crack initiation by **A13:** 149
of solids, in liquid metals **A13:** 56-57
solvent, in sur-face-mount assemblies **EL1:** 666
of steel, in iron-carbon melts **A15:** 73-74
and swelling **EM2:** 771-773
thin-film chip resistors **EL1:** 1003
Dissolution etching
defined **A9:** 6
Dissolution rates in potentiostatic etching
factors affecting **A9:** 144
Dissolved gas
corrosion fatigue test specifications **A8:** 423
Dissolved gases *See also* Gases
evolution, porosity effects **A15:** 82
in water **A13:** 489
Dissolved gases, in fresh water
effect on corrosion rate **M1:** 733, 736
Dissolved hydrogen **A8:** 423
Dissolved oxygen *See also* Oxygen **A13:** 29, 221, 489, 895-898, 932
in aqueous environment synthesis **A8:** 416
control in steam with contaminants **A8:** 427
corrosion fatigue test specification **A8:** 423
effect on environmentally controlled crack growth rates **A8:** 422
pressurized water reactor specification **A8:** 423
Dissolved salts, in fresh water
effect on corrosion rate **M1:** 734-735
Dissolved salts, in soil
corrosion caused by **M1:** 725-726, 730
Dissolved salts, in water *See also* Salts **A13:** 490
Dissolved solids
effect in boiler tubes **A11:** 615-616
effect on pressure vessels **A11:** 656
in water, corrosion by **A11:** 632
Distal tips
in borescopes **A17:** 3-5
Distance
effect galvanic corrosion **A13:** 83
human vs. machine vision **A17:** 30
object-camera, machine vision **A17:** 34
Distance amplitude *See also* Ultrasonic inspection blocks **A17:** 262, 264
curves, determined, ultrasonic inspection **A17:** 265-266
Distance check
computer-aided analysis **EL1:** 133
Distance-amplitude-correction (DAC) curve method
of ultrasonic inspection forgings **A17:** 505
Distillation
efficiency measured **A10:** 243
as metal ultrapurification technique **A2:** 1094
microwave inspection **A17:** 202
separation by **A10:** 169
steam, in nitrogen determination **A10:** 172-173
Distillation column
monitoring **A13:** 202
Distillation for purifying metals **M2:** 710
Distillation retorts, historic
in Africa **A15:** 19
Distillation separation process
zirconium and hafnium **A2:** 661-662
Distilled water *See also* Water
purification **A8:** 421
titanium/titanium alloy SCC in **A13:** 689
zinc corrosion in **A13:** 760
Distorted casting
as casting defect **A11:** 386
Distortion *See also* Distortion failures **A6:** 1094-1102
allowance, pattern **A15:** 193
aluminum alloys **A6:** 727, 728
amount of **A11:** 137-138
analyses in weldments **A6:** 1095
in arc-welded aluminum alloys **A11:** 436

Distortion (continued)
austenitic stainless steels **A6:** 469
in bridge web gap **A11:** 712
buckling, in shafts **A11:** 467
buckling under compressive loading **A6:** 1102
bulging as **A11:** 139-140
by metal solidification **A15:** 615-616
casting, and gating **A15:** 589
of castings, hot isostatic pressing effects **A15:** 542
combined effects of residual stress **M6:** 886-887
control **M6:** 887-888
assembly procedures **M6:** 887-888
elastic prespring **M6:** 888
preheating **M6:** 888
presetting **M6:** 888
coppers **A6:** 757-758
core/mold, directional solidification **A15:** 321-322
corrosion of weldments **A6:** 1068
creep, in shafts **A11:** 467
defined **A8:** 4 **A11:** 3 **A15:** 5 **EM1:** 9 **EM2:** 14
design effects **A15:** 598
duplex stainless steels **A6:** 699
effect on service behavior **A6:** 1100
elastic or plastic, as failure mechanism **A11:** 75, 143
electroslag welding **A6:** 278-279
and expansion, failure of high-temperature rotary valve due to **A11:** 374-376
failures **A11:** 136-144, 489
ferritic stainless steels **A6:** 445
formation of **A6:** 1094
from mold restraint **A15:** 616-617
in gas cutting **A14:** 725
gas-tungsten arc welding **A6:** 191
in girder webs **A11:** 711-714
hardfacing **A6:** 801, 807
heat-treatable aluminum alloys **A6:** 528
hydrogen-caused **A11:** 46
hydrogen-induced cracking **A6:** 411
inelastic cyclic **A11:** 144
localized **A11:** 138
longitudinal **M6:** 880
in machining magnesium and magnesium alloy parts **A2:** 476
magnesium alloys **A6:** 774, 781
measurements **A16:** 27
mechanical cutting **A6:** 1182-1183
mold, from pouring temperatures **A15:** 283
nickel alloys **A6:** 751
out-of-plane **M6:** 886
out-of-plane, bridge components **A11:** 707, 711-714
in oxyfuel gas cutting **M6:** 904-905
oxyfuel gas welding **A6:** 287, 288, 289
permanent, in shafts **A11:** 467
in permanent mold castings **A15:** 285
porcelain enamel **A13:** 448-449
ratcheting as **A11:** 143-144
reduction by thermal treatment **M6:** 889-891
removal **M6:** 888-889
flame straightening **M6:** 888
jacking **M6:** 888-889
pressing **M6:** 888
thermal straightening **M6:** 888
and residual stress **A16:** 25
resistance welding **A6:** 834
SCC as **A11:** 567
of shafts, creep and buckling as **A11:** 467
shape, defined **A11:** 136
shape, from temperature cycling or uneven cooling **A11:** 266
shape, sheet metal **A14:** 878
sheet metals **A6:** 398, 399, 400
in shielded metal arc welds **M6:** 91-92
size, defined **A11:** 136
stainless steels **A6:** 626
steam equipment failure by **A11:** 602

Distortion (continued)
and stress ratios, from overloading **A11:** 137
thermal stresses and metal movement during welding **A6:** 1094-1095
titanium alloys **A6:** 522
in tool steels **A7:** 467
transverse shrinkage of butt welds **M6:** 870-875
types
angular change **M6:** 860-861
longitudinal shrinkage **M6:** 860-861
transverse shrinkage **M6:** 860-861
weld model **A6:** 1133, 1134, 1135
weldments **A6:** 1097-1100 **M6:** 859-860
angular distortion around the weld line **A6:** 1097, 1098, 1099
buckling **A6:** 1098-1100
longitudinal shrinkage parallel to the weld line **A6:** 1097, 1098, 1099
transverse shrinkage **A6:** 1097-1099
Distortion and safety in hardening of wrought tool steels **A1:** 777
Distortion control
in tool steels **M4:** 614-620
Distortion during heat treatment **M1:** 460, 469-470
Distortion Energy Hypothesis (DEH) **A18:** 476
Distortion failures *See also* Distortion **A11:** 136-144
analysis of **A11:** 142-143
of automotive valve spring **A11:** 138-139
defined **A11:** 136
overloading **A11:** 136-138
special types of **A11:** 143-144
specifications, failure to meet **A11:** 140-142
specifications, incorrect **A11:** 138-140
Distortion, in shadow formation
radiography **A17:** 312-313
Distortion in tool steels *See* Tool steels, defects and distortion in heat-treated parts; Tool steels, distortion **M3:** 466-469
Distortion ratio of taper sections **A9:** 450
Distributed capacitance
flexible printed boards **EL1:** 587-588
Distributed gage volume
for detecting flow localization **A8:** 589-590
to measure deformation distribution **A8:** 589-590
variation for pressed specimens **A8:** 590
Distributed impact test
defined **A18:** 7
Distributed leak
defined **A17:** 57-58
Distributed plane construction **EL1:** 625-628
Distributed power system
future trends **EL1:** 392
Distributed view
of package parasitics **EL1:** 418-419
Distribution *See also* Normal distribution; Stress distribution
(pipe)lines, high-pressure long-distance **A11:** 695
for crack-tip stresses **A11:** 49
cumulative normal **A8:** 631
form, for determining design allowables **A8:** 663
of fracture toughness, variability in **A8:** 625
introduction **A8:** 623, 629
normal **A8:** 628, 629 630 **A17:** 726-727
of- stress, and fatigue strength **A11:** 113
probability density **A17:** 675
sample means **A17:** 726-727
statistical **A8:** 628-638
t **A8:** 655
types of **A8:** 628
unknown, direct computation for **A8:** 666-667
verified by goodness-of-fit test **A8:** 637-638
Distribution coefficient **A6:** 46-47, 52, 53
Distribution contour
defined **A7:** 4
Distribution functions
acoustic emission inspection **A17:** 283

Distribution manifolds
from P/M porous parts **A7:** 700
Distribution masking
vacuum coating process **M5:** 407-408
Distribution pipelines **A13:** 1292
Distributions *See also* Statistical methods
time-to-failure, defects in **EL1:** 893
Ditallowdimethylammonium chloride, evaporated film
ATR spectrum of **A10:** 119
Ditchers, chain
of cemented carbides **A2:** 974
Ditching saws
of cemented carbide **A2:** 974
Dithiocarb
as chelator **A2:** 1236-1237
Dithizone
as solvent extractant **A10:** 170
Ditungsten carbide **EM4:** 810
Divacancies
corrosion-generated **A12:** 42-43
Divariant equilibrium
defined **A9:** 6
Divergencies
of properties **EM2:** 655-658
Divergent beam method
x-ray topography **A10:** 370-371
Dividing cone
defined **A7:** 4
Division of heat along the pin
DTHP **A18:** 280, 282-283
Divorced eutectic
in cast iron **M1:** 5-6
Divorced eutectic structure **A9:** 613
defined **A9:** 6
in the solidification structure of aluminum alloy welded joints **A9:** 579
DMPS (2,3-dimercapto-I propanesulfonic acid)
as chelator **A2:** 1236
DMT expression **A18:** 403
DMTA trace
on polysulfides **EM3:** 139
***DN* value** **A18:** 590
defined **A18:** 7
speed factor of greases **A18:** 127
DNC *See* Delayed-neutron counting
DOASIS computer program for structural analysis **EM1:** 268, 271
Dobby head
in textile looms **EM1:** 127
Doctor
bar, defined **EM1:** 9
blade, defined **EM1:** 9
Doctor bar *See* Doctor blade
Doctor bar or blade **EM3:** 11
Doctor blade *See also* Paste metering blade
defined **EM2:** 14
Doctor blade casting machine **EL1:** 462
Doctor roll **EM3:** 11
Documentation
of flexible printed wiring **EL1:** 592-593
of testing **EM1:** 300
Documentation, process
and NDE reliability **A17:** 678-679
Documents (military standards) *See also* Military standards
numerical summary **EL1:** 912
subject index **EL1:** 913
title and specification locator, military standards **EL1:** 909-911
DoD 4120.3-M
standardization documents **EM1:** 40
DoD Index of Specifications and Standards (DODISS) **EM1:** 40 **EM3:** 63-64
DoD/NASA conferences
fibrous composites **EM1:** 42
Dodder removal and yield
cleaned alfalfa seed **A7:** 591

Dodecanedioic acid, ATR
DRS, and PAS granular analysis A10: 120
DoDISS *See* Department of Defense Index of Specifications and Standards
Doehler, H.H
as inventor A15: 35
Doepp structural diagram *See* Patterson and Doepp structural diagram
"Dogbone" test bar A7: 490
Doily
defined EM1: 9 EM2: 14
Doloma
applications EM4: 46
composition EM4: 46
supply sources EM4: 46
Dolomite EM4: 44
additions, fluxes A15: 388-389
batch size EM4: 382
brick, defined A15: 5
in ceramic tiles EM4: 926
chemical composition A6: 60
composition EM4: 379
containing manganese, ESR studied A10: 264
decrepitation EM4: 379
dissolution in hydrochloric acid A10: 165
function and composition for mild steel SMAW electrode coatings A6: 60
refractory material composition EM4: 896
typical oxide compositions of raw materials EM4: 550
Dolomite brick EM4: 896
applications, refractory EM4: 900-901
Dolomite/dolomite limestone [$CaMg(CO_3)_2$#
purpose for use in glass manufacture EM4: 381
Domain EM3: 11
defined EM2: 14
SEM observed in ferromagnetic materials A10: 490
structures, evaluated by x-ray topography A10: 365
Domain boundaries
calculation of contrast in transmission electron microscopy A9: 119
Domain structures
etch pits used to determine orientation A9: 62
Domain walls
magnetic A9: 534
Domains *See* Magnetic domains
revealed by magnetic etching A9: 63-66
Dome
defined EM1: 9 EM2: 14
Dome housing, P/M Ti alloy
for Sidewinder Missile A7: 681
Dome shapes A14: 599, 636
Dome tests
hemispherical A8: 561-562
Domed
defined EM2: 14
Domestic waters, zinc corrosion in *See also* Water A13: 761
Domfer process A7: 89-92
flowchart A7: 90
functions of annealing in A7: 183, 185
P/M grade powders A7: 91
for powder production A7: 89-92, 818
for welding-grade iron powders A7: 91, 818
for welding-rod grade powders A7: 91
Donnan exclusion A10: 662, 672
Door-closer cylinder castings
fracturing A11: 363-366
Dopant EM3: 11
defined A7: 4 EM2: 15
Dopants A6: 957
atoms of, lattice location of A10: 633
low-level, GFAAS analysis of A10: 58
low-level, SIMS concentration profiles of A10: 610
metallic contaminants EM4: 117
Doped quartz glass
applications, lighting EM4: 1032, 1034, 1036
maximum operating temperature EM4: 1035
Doped semiconductors
conditions for growth A9: 612
Doped solder
definition M6: 5
Doping
defined A7: 4
Doping (continued)
effect on dynamic friction coefficient in silicon and gallium arsenide A18: 688-689
gallium arsenide (GaAs) A2: 744-745
potassium, in tungsten-rhenium thermocouples A2: 876
in tungsten powder production A7: 153
Doping method
XRPD analysis A10: 340
Doping of tungsten A9: 442
Doping, principle
alloy oxidation A13: 73
Doppler effect
defined A10: 672
in laser interferometer measurement A17: 14
microwave inspection A17: 211
velocity measurement by A17: 16-17
Doppler line broadening A10: 22
Doppler shift
defined A10: 672
DORIS
as synchrotron radiation source A10: 413
Dorn equation
for steady-state creep A8: 309
Dose-effect relationships
metal toxicities A2: 1234
Dose-response effects
metal toxicities A2: 1234
Dosimeter EM3: 11
defined A10: 672 EM2: 15
Dosimeters, pocket
for radiographic inspection A17: 301
Dot mapping *See also* Elemental mapping; Mapping; X-ray maps
for aluminum, at aluminum/brass interface of aluminum wire connections A10: 531-532
for copper, at aluminum/brass interface of aluminum wire connections A10: 531-532
effect of brightness on sensitivity of A10: 526-527
electron probe x-ray microanalysis A10: 525-527
of iron/aluminum interface in aluminum wire connections A10: 531-532
limitation of A10: 527-528
on photographic film A10: 527
for zinc, at aluminum/brass interface in aluminum wire connections A10: 531-532
of zinc, at grain boundaries A10: 527
Dot maps
from x-ray analysis A12: 168
Double action press
defined A7: 4
Double arcing
definition M6: 5
Double cantilever beam tests EM3: 386-387, 388, 389
Double carbide
tungsten and titanium A7: 158
Double carburization A7: 158
Double cone blenders and mixers A7: 4, 189
Double cone mixer
defined A7: 4
Double containment
of NAA samples A10: 236
Double crucible method EM4: 377
Double cutting
of angle sections A14: 718
Double die system A7: 327
Double diffraction A9: 109
as a result of second phase particles A9: 118
Double etching
defined A9: 6
Double fluorescence EXAFS detection technique A10: 418
Double hub forging A7: 412, 413
Double keel block *See* Keel block
Double knife edge alignment coupling
test stand A8: 312
Double layer
defined A13: 5
formation, oxide scale A13: 72
Double monochromators
stray light rejection for A10: 129
Double normalizing A4: 39
Double oxides
silicates A13: 76
Double pressing
defined A7: 4
Double resonance method
as supplemental to electron spin resonance A10: 258
Double routing complexity
WSI EL1: 354
Double salt nickel plating M5: 202-203
Double sealants
Double shear
high strain rate shear testing A8: 215
Double sintering
defined A7: 4
Double spread EM3: 11
Double submerged arc welded steel pipe
inspection of A17: 565-566
Double submerged-arc welding
steel tubular products M1: 316
Double sweep method
terne coating M5: 359
Double through transmission effect
remote-field eddy current inspection A17: 200
Double transducer system
ultrasonic testing A8: 244-245
Double upsetting and piercing A14: 87
Double-action dies
for sheet metal drawing A14: 579
Double-action mechanical press A14: 4, 495
Double-action tooling systems A7: 333-334
Double-aging
effect on embrittlement A12: 34, 47
Double-beam specimens
dimensions and tolerances A8: 516
SCC testing A8: 505, 513 A13: 255-256
Double-beam spectrophotometers
UV/VIS absorption spectroscopy A10: 67
Double-bevel groove welds
applications M6: 61
arc welding of magnesium alloys M6: 428
combinations with fillet welds M6: 66
comparison to fillet welds M6: 64-65
preparation M6: 67-68
recommended proportions M6: 69, 71-72
submerged arc welding M6: 129
Double-bevel-groove weld
definition A6: 1208
Double-block accumulating drawing machines A14: 334
Double-cantilever beam
abbreviation A8: 724
specimens, SCC testing A8: 513, 539
Double-cantilever beam geometry test A18: 404
Double-cantilever-beam tests
for fiber-resin composites EM1: 97, 98
Double-crystal spectrometers
for polycrystal rocking curve analysis A10: 371-372
Double-crystal spectrometry
as x-ray topographical A10: 371-372
Double-cup fractures
fcc metals A12: 100
Double-deflection scanning electron microscopic system A10: 493
Double-diffused junction field effect transistor EL1: 146-147
Double-drilled holes
adhesive-bonded joints A17: 612-613
Double-edge cracked specimens
geometries for A8: 251
Double-end boring machines A16: 171
and boring A16: 168
Double-end pressing
simulation of A14: 429-430
Double-end upsetting A14: 90
Double-extension dies A14: 132
Double-flare-bevel-groove weld
definition A6: 1208
Double-flare-V-groove weld
definition A6: 1208
Double-focusing mass spectrometers
in gas mass spectrometry A10: 154
Double-frame power hammer A14: 28
Double-immersion zincating process
aluminum and aluminum alloys M5: 603-604, 606

Double-inspection methods
for NDE reliability ... **A17:** 680
Double-J-groove weld
definition ... **A6:** 1208
Double-J-groove welds
applications ... **M6:** 61
recommended proportions ... **M6:** 69
Double-junction electrode ... **A8:** 417
Double-lap joints
design ... **EM3:** 474, 475
finite-element analysis ... **EM3:** 480, 481, 482
normal and shear stresses ... **EM3:** 478, 487, 491, 492
Double-layer interaction diagram ... **EM4:** 155
Double-layer theory ... **EM4:** 154-155
Double-notch shear testing
disadvantage ... **A8:** 228-229
for high shear ... **A8:** 228-229
for high strain rate shear testing ... **A8:** 187
input bar ... **A8:** 228-229
Kolsky bar apparatus, very high strain rates ... **A8:** 229
lower yield stress with strain rate ... **A8:** 229
output tube ... **A8:** 228-229
punch loading ... **A8:** 229-230
specimen ... **A8:** 228-229
statically deformed specimen ... **A8:** 229
Double-pass cylindrical mirror analyzer
for x-ray photoelectron spectroscopy ... **A10:** 571
Double-shear simple lap joint ... **EM3:** 537
Double-shear tapered-slice plate joint ... **EM3:** 536, 537
Double-shear test
aluminum alloy plate results ... **A8:** 63
for fasteners ... **A8:** 66
fixtures ... **A8:** 62, 66-67
for mill products ... **A8:** 62-63
and single-shear test, fasteners, compared ... **A8:** 66
specimen orientation ... **A8:** 62-63
Double-shot molding
defined ... **EM2:** 15
Double-sided boards
rework processes ... **EL1:** 711-712
Double-sided rigid (DSR)
defined ... **EL1:** 1141
Double-sided rigid printed wiring boards ... **EL1:** 15, 539
Double-square-groove weld
definition ... **A6:** 1208
Double-strap bonded joints ... **EM3:** 475
Double-stroke open-die headers
for cold heading ... **A14:** 292
Double-stroke solid-die headers
for cold heading ... **A14:** 292
Double-submerged arc welds
in pipe ... **A11:** 698
toe cracks in HAZ of ... **A11:** 698
Double-taper dies ... **A14:** 132
Double-U-groove butt joint, welding of titanium and titanium alloys
joint dimensions ... **A6:** 785
Double-U-groove weld
definition ... **A6:** 1208
Double-U-groove welds
arc welding of nickel alloys ... **M6:** 437, 442-443
gas metal arc welding of coppers and copper alloys ... **M6:** 416
recommended proportions ... **M6:** 69-71
submerged arc welding of nickel alloys ... **M6:** 442
Double-V-groove butt joint, welding of titanium and titanium alloys
joint dimensions ... **A6:** 785
Double-V-groove weld
definition ... **A6:** 1208
Double-V-groove welds
applications ... **M6:** 61
Double-V-groove welds (continued)
arc welding of austenitic stainless steels ... **M6:** 328
arc welding of nickel alloys ... **M6:** 337-443
electroslag welding ... **M6:** 242
gas metal arc welding of commercial coppers ... **M6:** 403
gas tungsten arc welding ... **M6:** 201
preparation ... **M6:** 67-68
recommended proportions ... **M6:** 69, 71-72
submerged arc welding ... **M6:** 129
Double-wall brazed tubing ... **A1:** 333-334 **M1:** 321
Double-wall radiographic techniques
for tube ... **A17:** 335-336
Double-welded joint
definition ... **A6:** 1208 **M6:** 5
Doubler *See also* Tabs
defined ... **EM1:** 9 **EM2:** 15
Douglas Aircraft Company
use of water-break test for preparing plastic surfaces ... **EM3:** 840
Dovetailing
definition ... **M6:** 5
Dovetails
suited to planing ... **A16:** 186
Dow Chemical Company
microdebonding indentation system developed ... **EM3:** 401
Dowel
defined ... **A15:** 5
Down hole casing
flux leakage method ... **A17:** 132
Down stroke *See* Stroke
Downdraw process ... **EM4:** 400
Downgate *See* Sprue
Downgrading ... **A1:** 1028
Downhand
definition ... **A6:** 1208
Downhole equipment
oil/gas production ... **A13:** 1237, 1248
Downslope time
definition ... **M6:** 5
Downstream sizing
in thermoplastic extrusion ... **EM2:** 382
Downtime
as reliability objective ... **EL1:** 128
Dowson and Higginson equation ... **A18:** 92
DPH *See* Diamond pyramid hardness
DPPH standard free radical
ESR spectrum of ... **A10:** 259
Draft
angle, defined ... **A14:** 4 **EM1:** 9
defined ... **A7:** 4 **A14:** 4, 49-50 **A15:** 5 **EM1:** 9 **EM2:** 15
design, effects ... **A14:** 66
forging ... **M1:** 362-363, 366-367
pattern, and casting design ... **A15:** 611-612
pattern, defined ... **A15:** 193
for precision forging ejection ... **A14:** 158
Draft angle ... **EM2:** 15, 130
Draftless forging *See* No-draft forging
Drafts, ingot
and forging defects ... **A11:** 327
Drag *See also* Ceramic molding; Cope; Flask; Mold; Patterns ... **A6:** 1162, 1163, 1164
conditions for heavy cutting ... **M6:** 910
defined ... **A15:** 5, 189, 203
definition ... **M6:** 5-6
marks, as die casting defect ... **A15:** 294
oxyfuel gas cutting ... **A14:** 721-722 **M6:** 898
Drag (thermal cutting)
definition ... **A6:** 1208
Drag angle
definition ... **M6:** 6
Dragline bucket shackle
failure of ... **A11:** 399-400
Dragline bucket tooth
failure of ... **A11:** 410
Dragout recovery process
plating waste treatment ... **M5:** 315-316
Drain casting ... **EM4:** 9, 34, 35
Drain stations
for liquid penetrant inspection ... **A17:** 78
Drain tubes
copper ... **A13:** 627
Drainage
defined ... **A13:** 5
Drained angles of repose ... **A7:** 282-283
Drape
defined ... **EM1:** 9 **EM2:** 15
of epoxy composites ... **EM1:** 73
of fiberglass fabric ... **EM1:** 110-111
prepreg ... **EM1:** 33, 139, 144
as selection criterion ... **EM1:** 77
testing of ... **EM1:** 737
of unidirectional tape prepreg ... **EM1:** 144
Drape forming
defined ... **EM2:** 15
Drape-forming machines
for stretch forming ... **A14:** 596
Draw
as casting defect ... **A11:** 384
defined ... **A15:** 5
Draw beads
deep drawing ... **A14:** 582
defined ... **A14:** 4
forces, simulated ... **A14:** 896
in press forming ... **A14:** 551-552
Draw bending
of bar ... **A14:** 661
Draw forming
defined ... **A14:** 4
Draw marks *See* Die line; Galling; Pickup; Scoring
Draw plate
defined ... **A14:** 4 **A15:** 5
Draw radius
defined ... **A14:** 4
Draw ring ... **A14:** 4, 508
Draw stock
defined ... **A14:** 4-5
Draw-down ratio
defined ... **EM2:** 15
Drawability *See also* Deep drawing; Drawing
deep, Swift cup test for ... **A8:** 562-563
defined ... **A14:** 4, 576
draw ratios ... **A14:** 576-577
earing ... **A14:** 576
factors ... **A14:** 576
of magnesium alloys ... **A14:** 828
temperature effects ... **A8:** 553
Drawbar, highway tractor-trailer
fatigue fracture ... **A11:** 127-128
Drawbead
as control, sheet metal forming ... **A8:** 547
forces ... **A8:** 567-568
simulators ... **A8:** 568
Drawbenches
for bars ... **A14:** 334-335
Drawbridging ... **A6:** 995
Drawing *See also* Deep drawing; Dies; Drawability; Dry drawing; Superconductors; Wet drawing; Wire drawing ... **A8:** 550 **A14:** 330-342
analysis ... **A14:** 918-919
approach angle ... **A14:** 330-331
of bar ... **A14:** 334-335
basic mechanics of ... **A14:** 330-331
of boxlike shells ... **A14:** 585
as bulk forming process ... **A14:** 16
commercial superconductors, manufacture of ... **A14:** 338-342
of common sizes ... **A14:** 337
of complex shapes ... **A14:** 337
of copper and copper alloys ... **A14:** 814-818
deep ... **A14:** 575-590
defined ... **A14:** 5 **EM2:** 15
dies and die materials ... **A14:** 336-337
dies, cemented carbide ... **A2:** 969

Drawing (continued)
dry-drawing continuous machines **A14:** 334
with fixed plug **A14:** 330
with floating plug **A14:** 331
force **A14:** 577, 770
forging stock, steps **A14:** 67
fundamentals **A14:** 576
heat generation during **A14:** 331
of hemispheres **A14:** 586
lubrication **A14:** 337-338, 519-520
machines **A14:** 333-334
of magnesium alloys, lubricants for **A14:** 519-520
pewter sheet **A2:** 523
preparation for **A14:** 331-333
processes, classified **A14:** 16
refractory metals and alloys **A2:** 563
of refractory metals and alloys, lubricants for **A14:** 519
of rod and wire **A14:** 333-334
shallow, Guerin process **A14:** 606
of sheet metal **A8:** 547-548
of sheet metals **A14:** 877
single-step, copper and copper alloys **A14:** 815-816
steel tubular products **M1:** 316, 317
steel wire **M1:** 260-261
test, simulative, for sheet metals **A8:** 562-563
test, Swift cup **A8:** 562-567
tests **A14:** 891-892
to size, in cold heading **A14:** 294
of tube **A14:** 335-336
workability in **A8:** 591-593
of workpieces with flanges **A14:** 585-586
of zirconium **A2:** 665
Drawing (pattern)
defined **A15:** 5
Drawing compound *See also* Die lubrication; Extreme-pressure lubricant; Lubricant(s); Lubrication **A14:** 5
defined **A18:** 7
Drawing compounds **EM3:** 41
pigmented *See* Pigmented drawing compounds
removal of, processes **M5:** 56
unpigmented *See* Unpigmented drawing compounds
Drawing dies *See* Bar drawing dies; Deep drawing dies; Tube drawing dies; Wiredrawing dies
Drawing lubricants
removal of **M5:** 71-72
Drawing of wire and rod
preferred orientation during **A10:** 359
Drawing operations, steel
phosphate coating aiding **M5:** 436-437
Drawing presses
availability **A14:** 578-579
blank size **A14:** 578
blankholding **A14:** 578
draw depth **A14:** 578
force requirements **A14:** 577-578
selection **A14:** 577-579
slide velocity **A14:** 578
Drawing quality
alloy steel sheet and strip **M1:** 164
of carbon steels **A1:** 201-202
of low-alloy steel **A1:** 208-209
low-carbon steel sheet and strip **M1:** 154, 155
Drawing quality special killed
low-carbon steel sheet and strip **M1:** 154, 155
Drawing tests **A14:** 891-892
Drawing-quality, special-killed steel
porcelain enameling of **M5:** 573
Drawn cup needle roller bearings
basic load rating **A18:** 505
Drawn fiber
defined **EM1:** 9 **EM2:** 15
Drawn polymer films
molecular orientation determined **A10:** 120
Dredges
as sampling tools **A10:** 16
Drift
defined **A9:** 6
Drill
artwork, and test information (DATI) system **EL1:** 130
bits, cemented carbide **A2:** 976
carbide metal cutting **A2:** 964-965
cost, vs hole size **EL1:** 20
Drill (continued)
tape handling **EL1:** 525-526
Drill cores
resource evaluations by neutron activation analysis **A10:** 233
Drill pipe *See* Steel tubular products **M1:** 319-320
ultrasonic inspection **A17:** 232
Drill presses *See also* Drilling
boring **A16:** 161, 170
die threading **A16:** 296, 297
die threading, dimensional control **A16:** 300
milling **A16:** 329
reaming **A16:** 240, 242
roller burnishing **A16:** 252
tapping **A16:** 255-256, 259
trepanning **A16:** 175
vertical, honing **A16:** 473, 486
Drill sizes
feeds for **A7:** 461
Drill-and-reamer combinations **A16:** 241, 245
Drilled holes
in crankshafts, fracture mechanics failure analysis **A11:** 123-125
fatigue fracture **A11:** 21
fatigue fractures from **A11:** 123
Drilling *See also* Drill presses; Drills **A16:** 212-238
adaptive control implemented **A16:** 622, 623
Al alloys **A16:** 766-767, 769, 775, 776, 778, 780-782, 782-785, 790-791
alloy steel corrosion in **A13:** 533-535
aluminum and aluminum alloys **A2:** 10
aramid composite **EM1:** 668-669
bench drilling machines **A16:** 212-213, 235, 237
by ND-YAG laser **A14:** 737
carbon steels **A16:** 673
cast irons **A16:** 648, 650, 651, 652, 654, 655, 658, 660
cemented carbide tools **A16:** 84, 85
ceramics **A16:** 102
and chip formation **A16:** 8
and chip formation analysis **A16:** 17
and chip removal **A16:** 33
CNC systems **A16:** 217-218
compared to grinding **A16:** 427
compared to reaming **A16:** 239
compared to shaped tube electrolytic machining **A16:** 554
compared to ultrasonic machining **A16:** 528
in conjunction with boring **A16:** 160, 168, 169
in conjunction with broaching **A16:** 194, 195, 204, 211
in conjunction with honing **A16:** 484
in conjunction with lapping **A16:** 494
in conjunction with milling **A16:** 304, 306
in conjunction with tapping **A16:** 256, 261
in conjunction with trepanning **A16:** 175
in conjunction with turning **A16:** 140, 153, 154
in conjunction with ultrasonic machining **A16:** 530
cost factors **A16:** 233
cross drilling **A16:** 375, 377, 386
Cu alloys **A16:** 805, 808, 811, 812, 813-816
cutting fluid flow recommendations **A16:** 127
cutting fluids **A16:** 125, 126, 127, 229-230
deep hole, and cutting fluids used **A16:** 125
deep-hole drilling machines **A16:** 215-216
defects **EL1:** 870, 1021
dial-index machines **A16:** 217
dimensional accuracy of holes **A16:** 228
of dissimilar materials **EM1:** 669-672
drill life variables **A16:** 230-232
drill point modifications **A16:** 226-228, 229
drill selections **A16:** 219
drill types **A16:** 218-225
effect, nailheading as **EL1:** 575
effect on fatigue strength **A11:** 123
equipment, locational accuracy **EL1:** 508
fixtures **A16:** 404
gang drilling machines **A16:** 214
general drill classification **A16:** 220
graphite composite **EM1:** 668-669
hafnium **A16:** 856
hardness of workpiece and cost **A16:** 232-233
heat-resistant alloys **A16:** 746-750, 751-752
horizontal, cemented carbide tools **A2:** 974
horizontal multiple-station machines **A16:** 229
indexable-insert drills **A16:** 234-237
Drilling (continued)
limits **EL1:** 508
machines and production reaming **A16:** 239-240
in machining centers **A16:** 393
as machining process **A7:** 461
Mg alloys **A16:** 821-822, 823-824
MMCs **A16:** 894, 895, 896, 900, 901
monitoring systems **A16:** 414, 415-417
multifunction machining **A16:** 366-367, 368, 374, 379-380, 384
multiple-spindle machines **A16:** 214-215
Ni alloys **A16:** 835, 837, 838, 839
numerical control (NC) **A16:** 217-218, 613-616
oil and gas, as cemented carbide application **A2:** 974-977
one-step/countersinking **EM1:** 671-672
P/M high-speed tool steels applied **A16:** 65, 67
P/M materials **A16:** 879-890
peck **EM1:** 669-671
peck drilling **A16:** 238
plated-through hole, primary **EL1:** 869-870
power consumption **A16:** 17
process capabilities **A16:** 212
radial machines **A16:** 213-214, 235
refractory metals **A16:** 860, 861, 863-865
rigid printed wiring boards **EL1:** 544, 548
rotary, cemented carbides **A2:** 974
roto-percussive, cemented carbides **A2:** 974-975
selection of drilling machines **A16:** 217
shuttle-transfer machines **A16:** 217
small-hole drilling (microdrilling) **A16:** 237-238
solid-tool **EM1:** 667-672
special drills for hard steel applications **A16:** 225-226
speed and feed **A16:** 229-232, 233, 238
stainless steels **A16:** 155, 681, 686-687, 690, 693-695, 697-699, 704
surface alterations from dull tools **A16:** 29
surface alterations produced **A16:** 23
surface finish requirements for machine tool components **A16:** 21
surface integrity effects in material removal processes **A16:** 28
and swaging, combined **A14:** 141
systems **A16:** 217-218
tape-controlled machines **A16:** 217-218
Ti alloys **A16:** 845, 846, 847, 848, 850, 851
tool adapters **A16:** 381
tool life **A16:** 218, 222, 224, 228, 233, 234
tool steel selection **A16:** 710
tool steels **A16:** 710, 711, 715-716, 718, 721, 727
tools **EM1:** 669-672
transfer drilling machines **A16:** 216-217
in transfer machines **A16:** 217, 395
trunnion-index machines **A16:** 217
types of drilling machines **A16:** 212-217
underwater **A16:** 226
upright machine spindle drives **A16:** 213
upright machines **A16:** 213-214, 235
uranium alloys **A16:** 875
vertical, cemented carbide tools **A2:** 974
vertical drill presses **A16:** 212
vertical multiple-spindle automatic chucking machines **A16:** 378
as wet chemical technique for subdividing solids **A10:** 165
zirconium **A16:** 852, 854
Zn alloys **A16:** 832, 833
Drilling fluid corrosion
oil/gas production **A13:** 1245-1246
Drills *See also* Cutting tools; Drilling
aircraft **A16:** 222
automotive series **A16:** 221
center **A16:** 221
centering **A16:** 222
chip-breaker **A16:** 222
core **A16:** 221, 229
design **A16:** 218
double-margin **A16:** 221
flank wear scars **A16:** 44
gun **A16:** 222-223, 229, 230-232
half-round **A16:** 221
high-helix **A16:** 220
left-hand **A16:** 221
life **A16:** 218, 223, 227-228, 230, 232-233, 238
low-helix **A16:** 220
oil-hole **A16:** 220-221, 229

Drills (continued)
point modification ... **A16:** 226-229
rail ... **A16:** 221
as sampling tools ... **A10:** 16
selection of ... **A16:** 219
spade ... **A16:** 223-225, 234, 235, 237
specific applications by type ... **A16:** 220-225
spotting ... **A16:** 222
step ... **A16:** 222
straight-flute ... **A16:** 221, 234
straight-shank ... **A16:** 218, 220
subland ... **A16:** 222
surface treatments ... **A16:** 219
taper-shank ... **A16:** 218, 220
TIN coatings ... **A16:** 57
twist ... **A16:** 218-224, 226-230, 234-238
types of ... **A16:** 218
underwater ... **A16:** 226
Drip applicators
for lubricant application ... **A14:** 515
Drip electron beam melting ... **A15:** 412-414
Drip feed (drop feed) lubrication
defined ... **A18:** 7
Drive axle, steel
fatigue fracture in ... **A11:** 116-117
Drive end
in torsional testing machine ... **A8:** 146
Drive gears ... **A7:** 670
Drive mechanisms *See also* Presses
capacity ... **A14:** 496
contour roll forming machines ... **A14:** 627
hydraulic presses ... **A14:** 32-33
mechanical presses ... **A14:** 31
modified, crank presses with ... **A14:** 40
screw presses ... **A14:** 33-35, 41
slider-crank, kinematics of ... **A14:** 37-38
speed change ... **A14:** 497
Drive pins, jackscrew
fatigue failure ... **A11:** 546
Drive pipe ... **M1:** 320
Drive shaft, steel splined
fatigue fracture ... **A11:** 122-123
Drive spring
for torsional impact system ... **A8:** 216-217
Drive system
in fatigue testing machine ... **A8:** 368
in load train ... **A8:** 368
tension testing machines ... **A8:** 47-48
Drive train
cam plastometer ... **A8:** 195
Drive unit
for torsional impact system ... **A8:** 222-217
Drive-gear assembly
deformation of ... **A11:** 142-143
Drive-line assembly
failed due to fatigue fracture of cap screws ... **A11:** 533, 535
Drive-pinion shaft
magnetic particle inspection of ... **A17:** 113
Driven well pipe ... **M1:** 320
Driver circuit
output impedance ... **EL1:** 26-27, 39
Driver(s)
bipolar ... **EL1:** 83
circuit ... **EL1:** 26-27, 39
circuit board line, as IC integrated ... **EL1:** 7
power dissipation, as limitation ... **EL1:** 6-7
silicon bipolar ... **EL1:** 8
Driver-pickup mode
remote-field eddy current inspection ... **A17:** 196
Drives
ultrasonic inspection ... **A17:** 232
Driving forces
and material transport sintering ... **A7:** 312-314
Driving gear, nylon
failure of ... **A11:** 764, 765
Driving severity number (DSN) ... **A18:** 578, 579
Droopers ... **A6:** 38
Drop
defined ... **A15:** 5
Drop (dropping) point ... **A18:** 136, 137
defined ... **A18:** 7
Drop (mechanical shock)
as physical testing ... **EL1:** 944
Drop bottom
cupola ... **A15:** 29
Drop etching
defined ... **A9:** 6
Drop forging
defined ... **A14:** 5
and powder forging, compared ... **A14:** 196-197
Drop hammer *See also* Air-lift hammer; Board hammer; Gravity hammer; Hammers; Steam hammer
forging, defined ... **A14:** 5
Drop hammer coining ... **A14:** 180, 655
Drop hammer forming ... **A14:** 654-658
aluminum alloys ... **A14:** 656, 803-804
blank preparation ... **A14:** 655
buckling, control of ... **A14:** 655
drop hammer coining ... **A14:** 180, 655
hammers for ... **A14:** 654
limits ... **A14:** 657-658
lubricants ... **A14:** 655
of magnesium alloys ... **A14:** 656-657, 830
multistage forming ... **A14:** 655
stainless steels ... **A14:** 774
steel processing ... **A14:** 655-656
of titanium alloys ... **A14:** 657, 847
tooling ... **A14:** 654-655
with trapped rubber ... **A14:** 608
Drop machine
development of ... **A15:** 28
Drop test
for high strain rate compression testing ... **A8:** 187
for medium-rate tests ... **A8:** 190
Drop tower compression test ... **A8:** 196-198
Drop-through
definition ... **M6:** 6
Drop-through dies
for blanking ... **A14:** 453
Drop-weight tear test ... **A8:** 724
toughness of steel ... **M1:** 701
Drop-weight tear test (DWTT) ... **A11:** 61, 796
Drop-weight test ... **A1:** 610
for steel castings ... **A1:** 367-368
use on cast steels ... **M1:** 379
Drop-weight test (DWT)
abbreviation for ... **A11:** 796
explosion-bulge test, tension test and compared ... **A11:** 59
for notch toughness ... **A11:** 57-58
Dropcut ... **A6:** 1156
Droplet erosion *See also* Erosion (erosive) wear
Droplet formation
in atomization ... **A7:** 30
Droplet sequence
flame emission spectroscopy ... **A10:** 29
Dropping mercury electrode
for polarography ... **A10:** 189
Drops
impingement ... **A11:** 165
of liquid, high-velocity impact of ... **A11:** 164
size, effect on erosion damage ... **A11:** 166
Dross ... **A6:** 1169
in aluminum alloys ... **A15:** 95
aluminum/zinc alloys, compared ... **A15:** 447
by-product, reverberatory furnace ... **A15:** 378
composition ... **A15:** 96
defined ... **A15:** 5
floating, teapot ladle for ... **A15:** 90
formation, compacted graphite irons ... **A15:** 671
formation, ductile cast iron ... **A15:** 94
formation, in copper alloys ... **A15:** 96
from solder waves ... **EL1:** 688-689, 691
particles, as contaminants ... **EL1:** 661
Dross (continued)
in permanent mold castings ... **A15:** 285
removal, in gating ... **A15:** 278, 589
Dross (nux) inclusions
as casting defect ... **A11:** 387
Dross formation
in copper casting alloys ... **A2:** 346
Dross inclusions
composition ... **A12:** 422
wrought aluminum alloys ... **A12:** 422
Drossing fluxes
aluminum alloys ... **A15:** 446
DRS *See* Diffuse reflectance spectroscopy
Drugs
GC/MS analysis of volatile compounds in ... **A10:** 639
Drum
definition ... **M6:** 6
Drum camera
to record strain in torsional impact testing ... **A8:** 217
Drum recorder
for torsional test recording ... **A8:** 147
Drum test
defined ... **A7:** 4
Drum-type blenders ... **A7:** 189
Drum-type *x-y* recorder ... **A8:** 617
Drums
rotary, for sand reclamation ... **A15:** 353-354
as sand cooling devices ... **A15:** 349-350
steel wire rope ... **A11:** 517-518
Dry ... **EM3:** 11
Dry abrasive blasting *See also* Abrasive blasting ... **M5:** 83-93
abrasives used ... **M5:** 83-88, 91
contaminant control ... **M5:** 85
metallic ... **M5:** 83-84
nonmetallic ... **M5:** 84-86
propulsion of ... **M5:** 86-88
replacement of ... **M5:** 85
selection of ... **M5:** 84-86
air blast (pressure) systems ... **M5:** 87, 90, 92-93
airless wheel systems ... **M5:** 86, 87, 90, 92-93
aluminum and aluminum alloys ... **M5:** 91, 571-572
applications ... **M5:** 91-93
continuous systems ... **M5:** 88-89, 92-93
copper and copper alloys ... **M5:** 614
cycle times ... **M5:** 90-91
equipment ... **M5:** 86-91
blasting-tumbling machines ... **M5:** 88-89
cabinet machines ... **M5:** 88-89
continuous-flow machines ... **M5:** 88-89
maintenance ... **M5:** 90
portable ... **M5:** 89
table-type machine ... **M5:** 88
heat-resistant alloys ... **M5:** 563, 565
limitations of ... **M5:** 91-93
for liquid penetrant inspection ... **A17:** 81, 82
magnesium alloys ... **M5:** 628-629
microabrasive system ... **M5:** 89-90
refractory and reactive metals ... **M5:** 652-653
soils, types of, effects ... **M5:** 92
suction systems ... **M5:** 87-88
titanium and titanium alloys ... **M5:** 652-653
tumbling system ... **M5:** 88-89, 93
wheel systems ... **M5:** 86-87, 90, 92-93
work loads, mixed, quantity, and flow of, effects of ... **M5:** 92-93
workpiece shape and size effects of ... **M5:** 92
Dry air
as converter gas ... **A15:** 426
for fracture preservation ... **A12:** 73-74
Dry and baked compression test
defined ... **A15:** 5
Dry bag isostatic pressing ... **A7:** 444-447 **EM4:** 126, 127
Dry baghouse
cupolas ... **A15:** 387

SUBJECTS OF THE INDEXED VOLUMES: ASM Handbook (designated by the letter "A"): **A1:** Properties and Selection: Irons, Steels, and High-Performance Alloys (1990); **A2:** Properties and Selection: Nonferrous Alloys and Special-Purpose Materials (1990); **A3:** Alloy Phase Diagrams (1992); **A4:** Heat Treating (1991); **A6:** Welding, Brazing, and Soldering (1993); **A7:** Powder Metallurgy (1984); **A8:** Mechanical Testing (1985); **A9:** Metallography and Microstructures (1985); **A10:** Materials Characterization (1986); **A11:** Failure Analysis and Prevention (1986); **A12:** Fractography (1987); **A13:** Corrosion (1987); **A14:** Forming and Forging (1988); **A15:** Casting (1988); **A16:** Machining (1989); **A17:** Nondestructive Testing and Quality Control (1989); **A18:** Friction, Lubrication, and Wear Technology (1992). **Metals Handbook, 9th Edition** (designated by the letter "M"): **M1:** Properties and Selection: Irons and Steels (1978); **M2:** Properties and Selection: Nonferrous Alloys and Pure Metals (1979); **M3:** Properties and Selection: Stainless Steels, Tool Materials and Special-Purpose Materials (1980); **M4:** Heat Treating (1981); **M5:** Surface Cleaning, Finishing, and Coating (1982); **M6:** Welding, Brazing, and Soldering (1983). **Engineered Materials Handbook** (designated by the letters "EM"): **EM1:** Composites (1987); **EM2:** Engineering Plastics (1988); **EM3:** Adhesives and Sealants (1990); **EM4:** Ceramics and Glasses (1991); **Electronic Materials Handbook** (designated by the letters "EL"): **EL1:** Packaging (1989).

Dry ball milling
for metallic flake pigments **A7:** 593
not recommended for mixing when SHS process involved **EM4:** 229
Dry barrel finishing
aluminum and aluminum alloys **M5:** 572
magnesium alloys **M5:** 631
Dry blend *See also* Polyvinyl chlorides (PVC)
defined **EM2:** 15
Dry blending
of uranium dioxide pellets **A7:** 665
Dry bond adhesive **EM3:** 11
Dry chlorine **A13:** 1170-1171
Dry coloring
defined **EM2:** 15
Dry corrosion *See also* Gaseous corrosion; Thermal oxidation
as atmospheric **A13:** 80
properties, ion implantation **A13:** 499
Dry cutting
of fracture surfaces **A11:** 19
of specimens **A12:** 76
Dry cutting abrasive wheels **A9:** 24
Dry cyaniding *See* Carbonitriding
Dry developers
application of **A17:** 83
stations **A17:** 79
Dry drawing *See also* Drawing
continuous machines for **A14:** 334
lubrication **A14:** 337-338
Dry etching
defined **A9:** 6
stainless steel **M5:** 560
Dry etching, polyimide
for via holes **EL1:** 328
Dry fiber
defined **EM2:** 15
Dry fiber weight
tested **EM1:** 737
Dry filament winding
of epoxy composites **EM1:** 71
Dry film
as solder mask process **EL1:** 555-556, 584
Dry fine grinding versus wet fine grinding, effect on graphite retention in cast
irons **A9:** 243
Dry fining
nickel alloys **M5:** 674
Dry friction
defined **A18:** 7
Dry friction applications **A7:** 702, 703
Dry glass bead shot peening **M5:** 144
Dry hot dip galvanized coating **M5:** 328
Dry ice
deep immersion in **A8:** 36
Dry laminate
defined **EM1:** 9 **EM2:** 15
Dry lay-up *See also* Lay up; Lay-up
defined **EM1:** 9 **EM2:** 15
laminating, of epoxy composites **EM1:** 71-73, 75
Dry lay-up method
of lamination **EL1:** 832
Dry lubricant *See* Solid lubricant
Dry nitrogen
in zoned sintering atmospheres **A7:** 347
Dry objective
defined **A9:** 6
Dry oxygen
copper/copper alloy resistance **A13:** 632-633
Dry particles *See also* Magnetic particles; Particles
magnetic **A17:** 100-101
Dry permeability
defined **A15:** 5
"Dry pin" values **A8:** 60
Dry powder developers (form A)
for liquid penetrant inspection **A17:** 76-77
Dry powder pigments
for sheet molding compounds **EM1:** 158
Dry powder technique
magnetic particle forging inspection **A17:** 500
Dry power brush cleaning **M5:** 150-153
Dry pressing **EM4:** 9, 33-34, 123, 141-146
advantages **EM4:** 141
basic compaction stages **EM4:** 142-143
bulk compression **EM4:** 142, 143
die fill **EM4:** 142
low flow and fragmentation **EM4:** 142, 143
Dry pressing (continued)
springback **EM4:** 142, 143
transitional restacking **EM4:** 142-143
compaction models **EM4:** 143-146
Cooper-Eaton model **EM4:** 143-144
empirical based models **EM4:** 143
green tensile strength **EM4:** 145-146
length-to-diameter ratio **EM4:** 145, 146
Lukasiewicz's model **EM4:** 143, 144
powder die wall coefficients of friction **EM4:** 145, 146
powder fluidity index **EM4:** 144-145, 146
quantitative models **EM4:** 143-144
semiquantitative based models **EM4:** 143
stages of compaction for agglomerated powders **EM4:** 144
upper punch hold-down pressure **EM4:** 145, 146
effect of increasing process parameters on green density and end-capping **EM4:** 146
equipment types **EM4:** 141
double-action presses **EM4:** 141, 142
floating-die presses **EM4:** 141, 142
hydraulic presses **EM4:** 141
mechanical rotary presses **EM4:** 141
punches and dies **EM4:** 141
single-action (anvil) presses **EM4:** 141, 142
state-of-the-art press **EM4:** 141
factors influencing ceramic forming process selection **EM4:** 34
mechanics of process **EM4:** 141-142
powder specifications required **EM4:** 146
binders as additives **EM4:** 146
defoamers as additives **EM4:** 146
dispersants as additives **EM4:** 146
lubricants as additives **EM4:** 146
plasticizers as additives **EM4:** 146
wetting agents as additives **EM4:** 146
properties **EM4:** 141
viscoelastic properties required **EM4:** 116
Dry radiography *See* Xeroradiography
Dry reclamation systems
for sands **A15:** 351-353
Dry rubbing test
ceramic coatings **M5:** 547
Dry running
defined **A18:** 7
Dry sand
low-stress rubber wheel abrasive wear tester **A8:** 605
Dry sand casting
defined **A15:** 5
Dry sand molding *See also* Green sand mold
and green sand molding, compared **A15:** 228
methods **A15:** 228
Dry sand molds
defined **A15:** 5
for horizontal centrifugal casting **A15:** 296
pit mold, for diesel engine **A15:** 228
vertical centrifugal casting **A15:** 301
Dry sand-rubber wheel test
defined **A18:** 7
Dry scrubbing
for sand reclamation **A15:** 227
Dry silver film
radiography **A17:** 315
Dry sliding wear *See also* Unlubricated sliding
defined **A18:** 7
Dry spike isotope dilution
for SSMS analysis **A10:** 146
Dry steam, and moisture
in steam equipment **A11:** 615
Dry strength *See also* Wet strength **EM3:** 11
defined **EM2:** 15
Dry tack **EM3:** 11
Dry time
water-base versus organic-solvent-base adhesives properties **EM3:** 86
Dry toner development
copier powders **A7:** 582-583
Dry wear **A8:** 603
Dry wear rate
sliding bearings **A18:** 515
Dry winding *See also* Wet Winding; Wet winding; Winding **EM1:** 9, 71
defined **EM2:** 15
Dry-air blast
for cleaning **A11:** 19
Dry-bag pressing method
for cermets **A2:** 982
Dry-bottom cupola
tapping **A15:** 384-385
Dry-bulb temperature *See also* Temperature(s) **EM3:** 11
defined **EM2:** 15
Dry-drawing continuous machines **A14:** 334
Dry-drawn finish **A1:** 279
Dry-film lubrication
defined **A18:** 7
Dry-film rust-preventive compounds **M5:** 459, 461-464, 466
Dry-film thickness
measurement **A13:** 417
Dry-method particles **A7:** 577-578
Dry-powder magnetic particles
defect detection **A7:** 577-578
Dry-resin coatings
fused **A15:** 565
Dry-sand investment
aluminum and aluminum alloys **A2:** 5
Dryer
for induction furnace preheating **A15:** 373
Drying *See also* Curing; Predrying; Spray drying **EM4:** 130-134
air, of rammed graphite molds **A15:** 273
air, of thick-film circuits **EL1:** 249
ceramic packaging **EL1:** 961
in chromating **A13:** 390
cluster, investment casting **A15:** 260-261
drying process **EM4:** 131-132
drying shrinkage and defects **EM4:** 132
checks **EM4:** 132
differential shrinkage **EM4:** 132
linear shrinkage **EM4:** 132
volume shrinkage **EM4:** 132
warping **EM4:** 132, 134
drying systems **EM4:** 130
conductive heating **EM4:** 130, 131
convection heating **EM4:** 130, 131
fluidized bed drying **EM4:** 130
infrared radiation **EM4:** 130
open-air drying **EM4:** 130
radiation **EM4:** 130, 131
equipment, plaster molding **A15:** 243
of gravimetric samples **A10:** 163
liquid crystal polymers (LCP) **EM2:** 181
for liquid penetrant inspection **A17:** 83
match plate patterns molds **A15:** 245
mechanisms in drying **EM4:** 130-131
evaporation **EM4:** 130-132
modes of drying **EM4:** 132-134
conduction drying **EM4:** 132, 134
controlled humidity drying **EM4:** 132-133
conventional drying **EM4:** 132, 134
freeze-drying **EM4:** 134
infrared drying **EM4:** 132
microwave drying **EM4:** 133
slurry drying **EM4:** 133-134
spray-drying **EM4:** 134
supercritical drying **EM4:** 134
vacuum-assisted drying **EM4:** 132
mold, plaster molding **A15:** 244
oil, defined **A13:** 5
polyamide-imides (PAI) **EM2:** 135
of porcelain enamels **A13:** 448
of radiographic film **A17:** 353
of solder paste **EL1:** 733
station, liquid penetrant inspection **A17:** 78-79
styrene-acrylonitriles (SAN, OSA ASA) **EM2:** 216
temperature, Antioch process **A15:** 246-247
Drying control chemical additives (DCCA) **EM4:** 449, 450
for obtaining xerogels **EM4:** 211
Drying oils
contamination of phosphate coating by **M5:** 439
Drying temperature **EM3:** 11
Drying time **EM3:** 11
Dryout, electrolyte
passive devices **EL1:** 998-999
DS *See* Directional solidification
DS MAR-M200 + Hf
aging cycle **A4:** 812

DS MAR-M247
aging cycle ... **A4:** 812
DS Nickel *See* Nickel alloys, specific types, DS Ni
DS René 80H
aging cycle ... **A4:** 812
DSC *See* Differential scanning calorimeter; Differential scanning calorimetry; Differential scanning calorimetry (DSC) analysis
DSID *See* Direct sample insertion device
DT-test *See* Dynamic tear (DT) test
DTA *See* Differential thermal analysis; Differential thermal analysis (DTA)
DTUL *See* Deflection temperature under load
DU *See* Uranium
Dual chromium *See* Microcracked chromium plating
Dual energy imaging
computed tomography (CT) ... **A17:** 378-379
defined ... **A17:** 383
Dual intensifier pump
abrasive waterjet cutting ... **A14:** 744-746
Dual refractive index method
of holographic contouring ... **A17:** 408
Dual sealing treatment
anodic coatings ... **M5:** 594-595
Dual sprocket assemblies
for farm planters ... **A7:** 673-674
Dual wavelength holographic contouring ... **A17:** 408
Dual-antenna reflection technique
microwave inspection ... **A17:** 205
Dual-beam dispersive spectrophotometers
UV/VIS absorption spectroscopy ... **A10:** 67-68
Dual-element contact-type ultrasonic search unit ... **A17:** 257-258
Dual-flow plasma cutting ... **A14:** 730
Dual-frequency eddy current inspection ... **A17:** 181
Dual-in-line packages (DIPS) ... **EL1:** 437-438
advantages ... **EL1:** 7
chip size effects ... **EL1:** 53
cooling ... **EL1:** 47
cost/performance modeling ... **EL1:** 14-15
defined ... **EL1:** 1141
failure verification/fault isolation in ... **EL1:** 1058-1061
flux contamination analysis ... **EL1:** 1109
heat flow paths ... **EL1:** 489
and high-package densities ... **EL1:** 405
package configurations ... **EL1:** 485
as package family ... **EL1:** 404
plastic postmolded, assembly sequence ... **EL1:** 487
resistor networks ... **EL1:** 178
resistores in design package ... **EL1:** 513
routability
for silicon chips ... **EL1:** 12
thermal resistance ... **EL1:** 409-410
through-hole/surface-mount assembly ... **EL1:** 437-438
vs integrated circuits (ICs) ... **EL1:** 12
Dual-mechanism cure systems ... **EL1:** 859, 863-864
Dual-phase microstructure ... **A9:** 604
Dual-phase steels ... **A1:** 148, 400, 405, 424-429 **A4:** 57, 61
annealing ... **A4:** 52
cold-forming strip ... **A1:** 417
continuous annealing ... **A4:** 62-63, 64
effects of grinding on retained austenite ... **A9:** 168-169
forming properties ... **A1:** 398, 418, 420
heat treatment of
annealing techniques and steel compositions
austenite transformation after intercritical annealing ... **A1:** 425
ferrite phase changes during intercritical annealing ... **A1:** 425-426
formation of austenite during intercritical annealing ... **A1:** 425
manganese segregation in ... **A10:** 483
Dual-phase steels (continued)
mechanical properties
ductility
tempering and strain aging ... **A1:** 427-428
work hardening and yield behavior ... **A1:** 424, 426
yield and tensile strength ... **A1:** 417, 426-427
n value determined in ... **A8:** 550
new advances in ... **A1:** 428, 429
springback in ... **A8:** 553
tensile properties ... **A8:** 555
Dual-phase steels, specific types
0.11C-1.40Mn-0.58Si-0.12Cr-0.08Mo ... **A9:** 165
0.11C-1.40Mn-0.58Si-0.12Cr-0.08Mo, heat treated and air cooled ... **A9:** 191
Dual-pitch racks
economy in manufacture ... **M3:** 853-854
Dual-property turbine wheel ... **A7:** 652
Dual-tube attachments
for flame cutting ... **A7:** 843
Dual-turret NC lathe ... **A16:** 1
Dual-wave soldering ... **EL1:** 180
Dubé scheme ... **A6:** 76
Duct assembly, medium-carbon steel
fatigue fracture ... **A11:** 118
DUCT software program
for surface model ... **A15:** 859
Ductile and brittle fractures ... **A11:** 82-101
Ductile cast iron ... **M1:** 6-7, 33-56
alloying elements ... **M1:** 34-42, 45-51, 53-55
applications ... **M1:** 34-35, 36, 47
composition ... **M1:** 33, 34-35, 36-37
compressive properties ... **M1:** 38, 41
creep data ... **M1:** 47, 48, 49
damping capacity ... **M1:** 32, 41, 45
dimensional growth ... **M1:** 46-47
electrical properties ... **M1:** 52-53, 55
elevated-temperature properties ... **M1:** 46-52
fatigue data ... **M1:** 43-44, 45, 46
ferritic grades, weldability ... **M1:** 564
fracture toughness ... **M1:** 42, 45-46
hardenability ... **M1:** 47, 49
heat treatment ... **M1:** 7, 35, 37, 42, 45
high nickel ductile iron
composition ... **M1:** 76
corrosion resistance ... **M1:** 90-91
elevated-temperature properties ... **M1:** 93-96
mechanical properties ... **M1:** 89, 92
physical properties ... **M1:** 88
impact properties ... **M1:** 39-42, 45
machinability ... **M1:** 35, 54-56
magnetic properties ... **M1:** 53-54
malleable cast iron, compared to ... **M1:** 57
matrix structure, effect of ... **M1:** 35, 45, 46, 53, 55
mechanical properties ... **M1:** 35, 36, 38-52
medium silicon ductile iron
composition ... **M1:** 76
mechanical properties ... **M1:** 92
oxidation resistance ... **M1:** 94
physical properties ... **M1:** 88
melting temperature ... **M1:** 52
metallurgical control ... **M1:** 36-38
microstructure ... **M1:** 7, 8
nodulizers ... **M1:** 37
nodulizing reactions ... **M1:** 6
notch toughness ... **M1:** 40-41, 45
oxidation resistance ... **M1:** 46
patternmakers' rules ... **M1:** 30-31, 33
pearlitic grades, weldability ... **M1:** 564
physical properties ... **M1:** 49, 52-55
section size, effect of ... **M1:** 40-41
specifications ... **M1:** 34, 35, 36
surface hardening ... **M1:** 37-38
tensile properties ... **M1:** 35, 36, 38, 41, 47, 52
testing ... **M1:** 37, 54-56
torsional properties ... **M1:** 38, 41
transition temperature ... **M1:** 40-41, 45
welding ... **M1:** 56
Ductile cast iron, grade 60-40-12, microstructure of ... **A3:** 1 • 26
Ductile cast irons
unalloyed ... **A13:** 567
Ductile cleavage mixed modes
overload failure ... **A8:** 481
Ductile crack propagation
defined ... **A8:** 4 **A11:** 3
Ductile decohesion
cyclic ... **A8:** 482-484
process, of Inconel 718 specimens ... **A8:** 484-485
Ductile, defined *See also* Dimple rupture(s); Ductile fracture(s); Ductile rupture(s); Ductile tearing; Ductility
Ductile erosion behavior
defined ... **A8:** 4 **A11:** 3
Ductile fatigue striation
in aluminum alloy ... **A8:** 484
Ductile fracture *See also* Ductility; Fracture; Fracture toughness ... **A8:** 4, 477-479 **A14:** 363-367
at cold working temperatures ... **A8:** 154, 573-574
at hot working temperatures ... **A8:** 573-574
at warm working temperatures ... **A8:** 573-574
in austenitic stainless steel ... **A8:** 154
by microvoid coalescence ... **A8:** 466
centerburst-type ... **A8:** 573-574
dead-metal zones ... **A8:** 574
defined ... **A13:** 5
fibrous tearing ... **A8:** 572
free surface ... **A8:** 573-574
hole coalescence ... **A8:** 572
and increasing temperature of deformation ... **A8:** 572
in processing map ... **A8:** 572
in repeated tension test ... **A8:** 353-354
in reversed cyclic creep test ... **A8:** 353-354
shear band tearing ... **A8:** 572
shear bands ... **A8:** 574
stages ... **A8:** 571
in static creep test ... **A8:** 353-354
stress-modified critical strain model and RKR critical stress model compared ... **A8:** 467
in tension test with range of strain rates and temperatures ... **A8:** 571
triple-point cracks and fractures ... **A8:** 574
types ... **A8:** 477-478
void growth ... **A8:** 571-572
wrought aluminum alloy ... **A2:** 41-42
Ductile fracture(s) *See also* Dimple rupture(s); Ductile rupture(s)
AISI/SAE alloy steels ... **A12:** 298
ASTM/ASME alloy steels ... **A12:** 346
by liquid lead embrittlement ... **A12:** 38
dimples in ... **A12:** 220
of fibers, metal-matrix composites ... **A12:** 468
interpretation of ... **A12:** 96-98
irons ... **A12:** 220, 223
light fractographs ... **A12:** 94, 99
low-carbon steel bolts ... **A12:** 248
maraging steels ... **A12:** 385
materials illustrated in ... **A12:** 217
metal-matrix composites ... **A12:** 467-468
micromechanism ... **A12:** 4
microscopic features ... **A12:** 97
polymer ... **A12:** 479
radial marks, SEM fractographs ... **A12:** 169
superalloys ... **A12:** 389
surface information, SEM imaging ... **A12:** 173-174
tensile, appearance ... **A12:** 173
titanium alloys ... **A12:** 443, 450, 451, 455
wrought aluminum alloys ... **A12:** 422, 425
Ductile fractures *See also* Dimpled rupture fracture
of alloy steel bolt ... **A11:** 76
appearance ... **A11:** 82-83
and brittle fractures ... **A11:** 82-101
by liquid erosion ... **A11:** 164-166
by liquid-metal embrittlement ... **A11:** 227, 231

Ductile fractures (continued)
by microvoid formation **A11:** 82
by overload fracture by **A11:** 25
by steel embrittlement **A11:** 98-101
causes of **A11:** 85-87
crack-growth rate to **A11:** 75
defined **A11:** 3
determined **A11:** 25
dimpled-rupture fracture surface of **A11:** 22
ductile-to-brittle transition **A11:** 84-85
failures, from residual stresses **A11:** 97-98
of forged steel shaft **A11:** 481-482
fractography of **A11:** 25
identification chart for **A11:** 80
improper electroplating failures from **A11:** 97
improper fabrication failures from **A11:** 87-94
improper thermal treatment failures from **A11:** 94-97
of iron T-hooks **A11:** 367-369
overload, in 63Sn-37Pb solder **A11:** 45-46
of pressure vessels **A11:** 666
propagation modes **A11:** 696
of shafts **A11:** 459, 466-467
tearing shear, fracture surface of **A11:** 697
tensile, appearance of **A11:** 75-76
topography of **A11:** 75
types of **A11:** 75-76
Ductile hole joining fracture mode
separation in **A8:** 450
Ductile intergranular fatigue **A8:** 484, 486
Ductile intergranular striations **A8:** 487
Ductile intergranular voids
particle nucleated **A8:** 487
Ductile iron *See also* Cast irons; Ductile iron, specific types **A1:** 3, 7-8, 33-55 **A9:** 245
advantages of **A1:** 33-34, 37
air-carbon arc cutting **A6:** 1172, 1176
alloys **A1:** 34
annealing **A1:** 41
application, internal combustion engine parts **A18:** 556
applications for **A1:** 34-38
arc welding **M6:** 315-316
as-cast, color etched **A9:** 157
austempered **A1:** 34, 35, 36, 37-38, 42
brake drum, brittle fracture **A11:** 370-371
brazeability **M6:** 996
brittle fracture **A12:** 227, 231-232, 235-237
chemical composition **A6:** 906
commercial pearlitic, fatigue crack propagation **A12:** 228-229
compared to gray or malleable iron **A1:** 33-34
composition limits **M6:** 309
composition of **A1:** 33, 34, 35
corrosion of **A11:** 200
cylinder head, shrinkage porosity in **A11:** 354-355
dip brazing **A6:** 338
ductile tearing **A12:** 230-236
ductile-to-brittle transition **A12:** 231, 236
etched, bull's-eye structure **A9:** 245
ferritic, high load fatigue fracture surface **A12:** 229
formation of graphite during solidification **A1:** 33
fractographs **A12:** 227-237
fracture, slow monotonic loading **A12:** 230
fracture/failure causes illustrated **A12:** 216
galling resistance with various material combinations **A18:** 596
graphite shape and distribution **A1:** 38-39
hardenability **A1:** 48-50
heat treatment **A1:** 38, 40, 41-42
high-silicon ferritic **A12:** 229
laser melting **A18:** 864, 865
laser transformation hardening **A18:** 862, 863, 864
machinability of **A1:** 52-53, 54
mechanical properties **A1:** 36, 40, 42-43
at elevated temperatures **A1:** 48, 49
composition, effect of **A1:** 40, 43-44
compressive properties **A1:** 42, 45
creep strength **A1:** 48, 49, 50
damping capacity **A1:** 45-46
fatigue strength **A1:** 39, 46-47, 48
fracture toughness **A1:** 46, 47
graphite shape, effect of **A1:** 44-45
hot-tensile properties **A1:** 48, 52
impact properties **A1:** 40, 43, 44, 45, 46

Ductile iron (continued)
notch sensitivity **A1:** 48
section size, effect of **A1:** 45
strain rate sensitivity **A1:** 47
stress-rupture properties **A1:** 48, 51, 52
tensile properties **A1:** 37, 42, 45
torsional properties **A1:** 42, 45
need for risers **A1:** 33-35
nodular, machining **A2:** 966
normalizing **A1:** 41-42
oxyacetylene welding **M6:** 604
pearlitic and ferritic, fatigue fracture surfaces **A12:** 229
physical properties
density **A1:** 50
electrical and thermal relationship **A1:** 50-51, 52
electrical resistivity **A1:** 51, 53
magnetic properties **A1:** 51-52
thermal properties **A1:** 49, 50
pipe, failure of centrifugal casting mold for **A11:** 275-276
pistons, fracture of **A11:** 360-361
postweld heat treatment practice **M6:** 314
quenching **A1:** 38, 41-42
rare earth alloy additives **A2:** 737
shrinkage allowance **A1:** 34
specifications **A1:** 34-35, 36
spur gear, brittle cleavage **A12:** 227
strength and toughness of **A1:** 33
stress relieving **A1:** 41
surface hardening **A1:** 42
T-hooks, overload failure of **A11:** 367-369
tempering **A1:** 41-42
testing and inspection **A1:** 37, 38, 39-41
thermal expansion coefficient **A6:** 907
ultimate shear stress for **A8:** 148
welding metallurgy **M6:** 308
welding of **A1:** 53-55
Ductile iron alloy production
powders used **A7:** 572
Ductile iron castings
inspection of **A17:** 532
Ductile iron, heat treating **A4:** 682-692
annealing **A4:** 685-686, 692
austempering **A4:** 682, 685, 688-689, 691, 692
austenitizing **A4:** 684-685, 686, 687, 689
characteristics **A4:** 682-684
differentiated from gray iron **A4:** 667
ductility **A4:** 684 **M4:** 545, 546
examples **A4:** 690
fatigue strength **A4:** 684, 688, 692 **M4:** 551
hardening **A4:** 684, 685, 686, 688 **M4:** 547, 548, 549, 551
hardness measurements **A4:** 669
induction hardening **A4:** 689-690, 691
microstructure **A4:** 682, 683, 685-686
nitriding **A4:** 690
normalizing **A4:** 686-687 **M4:** 546-547
proven applications for borided ferrous materials **A4:** 445
quench, comparison of severity **A4:** 687, 688-689, 690, 692 **M4:** 548
remelt hardening **A4:** 690-691
stress relieving **A4:** 689, 691-692 **M4:** 550, 551
surface hardening **A4:** 689-691 **M4:** 551
tempering **A4:** 687-688, 689, 690, 692 **M4:** 549, 550
tensile strength **A4:** 684, 685, 688, 690, 691
yield strength **A4:** 685, 690, 691
Ductile iron pipe *See also* Gray iron pipe **M1:** 98-100
aerial installations **M1:** 99, 100
applications **M1:** 98-100
casting tolerance **M1:** 98
coatings and linings **M1:** 98, 100
corrosion resistance **M1:** 98
grades **M1:** 98
joints **M1:** 98, 99, 100
manufacture **M1:** 98
mechanical properties **M1:** 98
sizes, standard **M1:** 98, 99
specifications for **M1:** 98, 100
standard laying conditions **M1:** 98, 99
testing **M1:** 98
underground installations **M1:** 99
underwater installations **M1:** 99-100

Ductile iron rolls **A14:** 353
Ductile Iron Society **A15:** 34
Ductile iron, specific types
80-60-03 induction hardened, fatigue-crack origin **A12:** 228
80-60-03 induction hardened, fatigue-test fracture **A12:** 228
AMS 5316 composition, properties and applications **M1:** 34, 35
ASME SA395 composition, properties and applications **M1:** 34, 35
ASTM A395
composition, properties and applications **M1:** 34, 35
machinability **M1:** 54-56
ASTM A476 composition, properties and applications **M1:** 34, 35
ASTM A536 composition, properties and applications **M1:** 34, 35
ASTM A536 grade 100-70-03, brittle cleavage **A12:** 227
grade 60-40-18, annealed **A9:** 247
grade 60-40-18, as-cast, etched **A9:** 247
grade 60-45-12, application of point-count grids to graphite nodules **A9:** 124
grade 60-45-12, as-cast **A9:** 247, 251-252
grade 60-45-12, surface weld **A9:** 251
grade 65-45-12, liquid nitrided and quenched **A9:** 251
grade 80-55-06, as-cast **A9:** 247, 251-252
grade 80-55-06, as-cast, effects of holding time **A9:** 252
grade 80-55-06, flame hardened **A9:** 250
grade 80-55-06, liquid nitrided **A9:** 250
MIL-I-11466B composition, properties and applications **M1:** 34, 35
MIL-I-24137 composition, properties and applications **M1:** 35
SAE J434c composition, properties and applications **M1:** 34, 35
Ductile iron tube
centrifugally cast **A9:** 161
Ductile irons *See also* Austempered ductile irons; Austenitic ductile irons; Cast iron; Cast irons **A15:** 647-666 **A16:** 648-651, 654, 658
annealing **A15:** 657-658
applications **A15:** 665
arc welding of **A15:** 527
austempering **A15:** 659-660
boring with high-speed steel and carbide tools **A16:** 655
brazing **A15:** 664-665
broaching **A16:** 656
casting and solidification **A15:** 651-652
centerless grinding **A16:** 665
composition **A15:** 656-657
composition control **A15:** 647-649
counterboring **A16:** 660
cylindrical grinding **A16:** 664
defined **A15:** 5
desulfurization **A15:** 75
development of **A15:** 35
drilling **A16:** 230, 231, 658
dross inclusions, types **A15:** 94
end milling **A16:** 663
exogenous inclusions in **A15:** 88
face milling **A16:** 662
graphite amount **A15:** 655
graphite structures **A15:** 654-655
hardening and tempering **A15:** 658-659
heat treatment **A15:** 657-660
honing **A16:** 477
horizontal centrifugal casting **A15:** 296
inclusions in **A15:** 94-95
internal grinding **A16:** 665
joining of **A15:** 664-665
machinability **A16:** 640, 643
markets for **A15:** 42
matrix microstructure effect on tool life **A16:** 650
matrix structure **A15:** 655-656
mechanical properties **A15:** 660-662
molten metal treatment **A15:** 649-651
nondestructive evaluation, castings **A15:** 663-664
normalizing **A15:** 658

Ductile irons (continued)
oxyacetylene welding **A15:** 531
physical properties **A15:** 663
planing **A16:** 657, 660
production controls, metallurgical **A15:** 652-653
property requirements **A15:** 653
raw materials **A15:** 647
reaming **A16:** 659
section size **A15:** 656
specifications **A15:** 653-654
spotfacing **A16:** 660
stress relieving **A15:** 657
surface grinding **A16:** 664
surface hardening **A15:** 659-660
tapping
turning with ceramic tools **A16:** 654
turning with single-point and box tools **A16:** 653
welding **A15:** 521, 528, 664
weldments, transverse joint properties **A15:** 528
Ductile irons, specific grades
60-40-18, boring **A16:** 655
60-40-18, broaching **A16:** 656
60-40-18, counterboring **A16:** 660
60-40-18, drilling **A16:** 658
60-40-18, end milling **A16:** 663
60-40-18, face milling **A16:** 662
60-40-18, planing **A16:** 657
60-40-18, reaming **A16:** 659
60-40-18, spotfacing **A16:** 660
60-40-18, tapping **A16:** 661
60-40-18, turning **A16:** 653, 654
65-42-12, counterboring **A16:** 660
65-45-12, boring **A16:** 655
65-45-12, broaching **A16:** 656
65-45-12, drilling **A16:** 658
65-45-12, end milling **A16:** 663
65-45-12, face milling **A16:** 662
65-45-12, planing **A16:** 657
65-45-12, reaming **A16:** 659
65-45-12, spotfacing **A16:** 660
65-45-12, tapping **A16:** 661
65-45-12, turning **A16:** 653, 654
80-55-06, boring **A16:** 655
80-55-06, broaching **A16:** 656
80-55-06, counterboring **A16:** 660
80-55-06, drilling **A16:** 658
80-55-06, end milling **A16:** 663
80-55-06, face milling **A16:** 662
80-55-06, machinability testing **A16:** 643
80-55-06, planing **A16:** 657
80-55-06, reaming **A16:** 659
80-55-06, spotfacing **A16:** 660
80-55-06, tapping **A16:** 661
80-55-06, turning **A16:** 653, 654
Ductile irons, specific types
ASTM 80-60-03, surface hardening **A4:** 689
ASTM 100-70-03
normalizing **A4:** 686
surface hardening **A4:** 689
grade 60-40-18, annealing **A4:** 686
Ductile materials
area under stress-strain curve **A8:** 23
chevron-notched specimens for **A8:** 469
effects of tension **A8:** 23
erosion and wear failure in **A11:** 155-156
static design based on yield strength **A8:** 20
tensile strength as measure of maximum load for **A8:** 20
Ductile metal powders
densification model **A7:** 299-300
milling of **A7:** 56-70
permanent lattice strain in **A7:** 61
Ductile Ni-Resist
galling resistance with various material combinations **A18:** 596
Ductile overload fracture
of extension ladder **A11:** 86-87
Ductile rupture
effect of stress-intensity factor range **A8:** 485-486
high **A8:** 571-572
Ductile rupture failure
thermal **EL1:** 56
Ductile rupture(s) *See also* Ductile fracture(s)
shear, titanium alloys **A12:** 447
titanium alloys **A12:** 443, 447
wrought aluminum alloys **A12:** 425, 428
Ductile shear
at high strain rates **A8:** 188
Ductile striation
of aluminum alloy 2024-T3 ..... **A8:** 481
fatigue cracking **A8:** 476
fatigue failure **A8:** 481-482
in Inconel 718 ..... **A8:** 482, 484
and stress-intensity range over Young's modulus **A8:** 481-482
subcritical growth under cyclic load **A8:** 476
triggering cleavage **A8:** 487
Ductile tearing
ASTM/ASME alloy steels **A12:** 345
in ductile irons **A12:** 230-236
Ductile tensile fractures
appearance **A11:** 82
flat, defined **A11:** 82
shear **A11:** 82
Ductile-brittle
behavior, and physical aging **EM2:** 756-757
transitions, as failure analysis **EM2:** 734-735
Ductile-to-brittle fracture transition **A11:** 66-71, 84-85
brittle fracture service failures **A11:** 69-71
deterioration in service **A11:** 69
effect of temperature on toughness schematic **A11:** 66
factors influencing **A11:** 68-69
in structural steels **A11:** 68-69
temperature, effect on fatigue fracture **A11:** 123
in tensile tests, ferritic steel **A11:** 66
transition temperature evaluation **A11:** 66-68
Ductile-to-brittle fracture transition of steel **M1:** 691-692
Ductile-to-brittle transition **A1:** 737-739
ductile irons **A12:** 231, 236
Ductile-to-brittle transition (DBT)
erosion/corrosion **A18:** 209
Ductile-to-brittle transition temperature
abbreviation for **A11:** 796
AISI/SAE alloy steels **A12:** 314
defined **A12:** 34
dependence on grain diameter **A11:** 68
in ductile iron **A12:** 231
effect of neutron irradiation **A12:** 127
factors influencing **A12:** 105-106
Ductile-to-brittle transition temperature (DBTT) **A6:** 101, 161
beryllium alloys **A6:** 945
ferritic stainless steels **A6:** 444, 445, 447, 452, 453, 454
molybdenum alloys **A6:** 581
niobium alloys **A6:** 581
oxide-dispersion-strengthened materials **A6:** 1039
refractory metals **A6:** 634, 941-942
refractory metals and alloys **A2:** 558
tantalum-base corrosion-resistant alloys **A6:** 599
titanium alloys **A6:** 633
tungsten **A2:** 579-580
tungsten alloys **A6:** 582
X-65 steel pipe (1.07 m) diameter **A6:** 104
Ductility *See also* Ductile fracture; Malleability; Mechanical properties, Tensile properties; Workability **EM3:** 11
aluminum and aluminum alloys **A2:** 8
aluminum and sulfur effects **A15:** 93
aluminum-lithium alloys **A12:** 440
Ductility (continued)
ASTM/ASME alloy steels, phosphorus enhancement **A12:** 349
bending tests for **A8:** 117, 125-131
bending tests, principal stress and strain directions **A8:** 127
of beryllium **A2:** 683-687
beryllium-copper alloys **A2:** 409
by hot torsion testing **A14:** 375
of cermets **A2:** 978
Charpy impact testing of **A8:** 262
of chromium carbide-based cermets **A7:** 806
in closed-die forgings **A1:** 340-341
closed-die steel forgings **M1:** 352-354, 357, 358, 374
cold finished bars **M1:** 227-232, 244
-creep data, analysis of **A8:** 693
creep, in steels **A11:** 98
and creep life, effect of multiaxial stressing **A8:** 343
for deep drawing **A14:** 575
defined **A8:** 4, 344 **A11:** 3 **A13:** 5 **EM1:** 9 **EM2:** 15
deformation heating effect **A8:** 164-165
and densification **A7:** 299-300
dependence on strain rate, cross-linked rubber **A8:** 39, 43
of dual-phase steels **A1:** 427
ductile iron **A15:** 647, 654
effect, electromagnetic forming **A14:** 646
effect, hydrogen embrittlement, nickel-base alloys **A13:** 651-652
effect of grain size **A11:** 307
effect of hold time on **A8:** 36
effect of hydrogen flaking **A11:** 316
effect of strain rate **A8:** 19
effect of temperature **A8:** 19, 34
effect of temperature, titanium alloy **A12:** 35, 48
effect of triaxial factor **A8:** 344
electrical resistance alloys **A2:** 822
elevated temperature, of hot-work die steels **A14:** 46
elevated-temperature, low-alloy and stainless steels **A11:** 265
and elevated-temperature service **A1:** 626--627, 633, 634
examining **A12:** 72, 92
and fatigue resistance **A1:** 678
and forging, effects **A7:** 414
from hot torsion tests **A8:** 165-166
galvanizing effects, steel **A13:** 438
grain growth effect **A8:** 164-165
graphite effect, in gray irons **A12:** 226
gray iron **M1:** 18
hardenable steels **M1:** 460-463, 467
hot-rolled bars **M1:** 202, 204-207, 209, 212
of heat-exchanger tubing **A11:** 628
of heat-resistant alloys **A14:** 232
of high-oxygen iron **A8:** 165-166
hydrogen damage effects **A13:** 329-330
of implant materials **A11:** 672
inclusion effect **A15:** 88
increased, in ASTM/ASME alloy steels **A12:** 349
loss, austenitic stainless steels **A13:** 171
low, by stress corrosion cracking **A7:** 254
low, of case and core, brittle fracture from **A11:** 389-391
low-alloy steels **A15:** 716
in low-carbon steels **A12:** 240
and manganese sulfide content in hot torsion tests **A8:** 166
maraging steels **M1:** 447, 448, 450
measurement, in tension testing **A8:** 26-27, 581
measures of **A8:** 21-22
and mechanical texture **A8:** 155
of metals subject to LME **A11:** 719
and multiaxial creep **A8:** 343
and multiaxial stressing **A8:** 344-345
of nickel-base alloys **A14:** 831
ordered **A2:** 913-914

Ductility (continued)
phosphorus-enhanced **A12:** 349
and plastic deformation **A7:** 298
of plated-through hole (PTH) **EL1:** 114
ratio, stress-corrosion cracking **A13:** 278
resistance to heat checking **A18:** 631
of rhenium, alloying effects **A7:** 770
of steel castings **A1:** 365
steel plate **M1:** 190-197
steel wire **M1:** 588
strain rate effect on **A8:** 38, 42
and strength **A7:** 319, 414, 691
and strength, carbon steels **A15:** 702
stress-rupture **A11:** 265
temperature dependence of **A8:** 262
tensile, effect of dissolved hydrogen on **A11:** 336-338
tensile, loss by hydrogen damage **A13:** 164
tensile, second-phase particle effects **A14:** 364
testing **A14:** 376
testing, hot compression **A8:** 583
in titanium-based alloys **A7:** 254
torsion tests for **A8:** 139
torsional, temperature and alloying effect on **A8:** 164-166
and toughness **A8:** 23 **EM2:** 554
transition temperature, defined **A11:** 59
of unalloyed uranium **A2:** 671
under tension testing **A8:** 26-27
uranium alloys **A2:** 678
variation, of polycrystalline cadmium as function of indium content **A11:** 230, 232
vs temperature, nickel-base heat-resistant alloys **A14:** 234
and workability **A8:** 571
Ductility, and volume fraction of dispersions in copper
relationship of **A9:** 125
Ductility testing **A1:** 582
Duffy clamp design
for torsional Kolsky bar **A8:** 221
Dullness
defined **EM2:** 15
Dulong and Petit values for vibrational heat capacity **EM4:** 847
Dumas method
of elemental analysis for nitrogen **A10:** 214
Dumbbell
design equations for ideal **A8:** 250
fillets and radii **A8:** 250-251
hollow, solid, for ultrasonic fatigue testing **A8:** 247, 250
Dumet
recommended glass/metal seal combinations **EM4:** 497
Dumet wire
glass/metal seals **EM4:** 1037
Dumet wire, as low-expansion
clad alloy **A2:** 894
Dummy block
defined **A14:** 5
Dummy cathode
defined **A13:** 5
Dummying
defined **A13:** 5
Duoplasmatron
defined **A10:** 672
Duplex alloys
defined **A18:** 7
horizontal centrifugal casting of **A15:** 299
nominal compositions **A15:** 724
Duplex angular-contact ball bearings **A18:** 500
Duplex anneal
cycle and microstructure **A6:** 510
Duplex annealing *See also* Annealing
wrought titanium alloys **A2:** 619
Duplex cast steels
general corrosion **A13:** 577-578
intergranular corrosion **A13:** 578-580
Duplex cells from precipitation **A9:** 648
Duplex chromium *See* Microcracked chromium plating
Duplex chromium-nickel stainless steel
principal ASTM specifications for weldable steel sheet **A6:** 399
Duplex coatings **A6:** 131
Duplex electric holders
cupolas **A15:** 384
Duplex ferritic-austenitic stainless steels **A6:** 697-699
base metals **A6:** 697-698
distortion **A6:** 699
engineering for use in the as-welded condition **A6:** 698-699
engineering for use in the postweld heat treated condition **A6:** 699
metallurgy **A6:** 697
properties **A6:** 697
Duplex grain size
defined **A9:** 6
Duplex grain structure
formation of **A9:** 603
in uranium **A9:** 481, 483
Duplex insulated thermocouple wires
color coding **A2:** 878
Duplex microstructure
defined **A9:** 6
Duplex nickel plating **M5:** 207-209
Duplex stainless steels *See also* Cast stainless steels; Stainless steels; Stainless steels, specific types; Wrought stainless steels; Wrought stainless steels, specific types
characterized **A13:** 550
compositions **A13:** 359
compositions of **A1:** 843, 847-848
corrosion of weldments **A6:** 1067
elevated-temperature properties of **A1:** 947
forgeability of **A1:** 894
formability of **A1:** 889
with high-alloy filler metals **A13:** 361
hydrogen-induced cracking **A6:** 697
intergranular corrosion **A13:** 127, 359
machinability of **A1:** 896
microstructures **A9:** 286
nondestructive testing **A6:** 1086
in pharmaceutical production facilities **A13:** 1226
pitting tests **A13:** 359-361
repair welding **A6:** 1107
resistance, localized corrosion **A13:** 563
resistance of, to stress-corrosion cracking **A1:** 727
sensitization in **A1:** 707-708
sigma phase embrittlement in **A1:** 711
stress-corrosion cracking resistance to boiling magnesium chloride **A1:** 727
stress-corrosion cracking, weldments **A13:** 361
tensile properties of **A1:** 856, 858
weldability of **A1:** 904-905
weldments **A13:** 358-361
Duplex stainless steels (DSS), wrought **A6:** 471-480
advantages **A6:** 471
alloy grades **A6:** 471-472
alloy groups, generic types **A6:** 471, 472
applicable welding processes **A6:** 479-480
applications **A6:** 472, 479
sheet metals **A6:** 398, 399
ASTM standards **A6:** 474
backing gas **A6:** 474, 477
base material properties **A6:** 471-472
brazing and soldering characteristics **A6:** 626
composition **A6:** 471, 472, 473, 474
corrosion **A6:** 626
corrosion resistance **A6:** 472
definition **A6:** 471
electrodes used **A6:** 478-479
embrittlement **A6:** 472
explosive welding **A6:** 480
Ferrite Number conversion to percentage rule **A6:** 475
filler metals **A6:** 474, 476, 479
flux-cored arc welding **A6:** 480
friction welding **A6:** 480
fusion welding **A6:** 479-480
gas-metal arc welding **A6:** 480
gas-tungsten arc welding **A6:** 476, 479, 480
heat-affected zone ... **A6:** 473, 474, 475, 476, 478, 479
liquation cracking **A6:** 478
hydrogen cracking **A6:** 474, 476, 477
interpass temperature control **A6:** 474
microstructure **A6:** 471, 472-474, 475, 476, 478
morphology **A6:** 473-474, 478
pitting **A6:** 477, 478-479
Duplex stainless steels (DSS), wrought (continued)
pitting corrosion test **A6:** 475-476
postweld heat treatment **A6:** 474, 476
precipitation of intermetallic phases **A6:** 472, 473
preheat **A6:** 474
properties **A6:** 471, 472, 478, 479
shielded metal arc welding **A6:** 476, 477, 480
solidification cracking **A6:** 474, 476, 477-478
stress-corrosion cracking **A6:** 471, 477, 479
submerged arc welding **A6:** 477, 480
"super" duplex **A6:** 471, 472
time-temperature transformation diagram **A6:** 476
weldability **A6:** 471, 474-479
welding **A6:** 474-479
Duplex tubing
inspection of **A17:** 572
Duplicate measurement
defined **A10:** 672
Duplicate sample
defined **A10:** 672
Duplicating lathes **A16:** 153
DuPont NR-150 B2 thermoplastic polyimide resin **EM1:** 79
Dupré equation **A18:** 400
Dupre's equation **EM4:** 483, 514
Durability *See also* Fatigue. of composites **EM1:** 35
assessment of **EM2:** 551-554
design requirements, damage tolerance **A17:** 666
exponential distribution for **A8:** 634
of fiber composites **EM1:** 179-180
of- E-glass **EM1:** 107
of reference standards **A17:** 677
testing, composite structures **EM1:** 340-344
tests, full-scale **EM1:** 350-351
of Western bentonite **A15:** 210
Durability assessment and life prediction for adhesive joints **EM3:** 33, 663-671
bond line corrosion **EM3:** 663, 664
component testing **EM3:** 543
corrosion-dominated durability **EM3:** 669-671
durability test techniques **EM3:** 664-666
adhesion-dominated durability **EM3:** 665-666
diffusion-dominated durability **EM3:** 664-665
wedge test **EM3:** 666-668, 669
wet peel test **EM3:** 668-669
Durability design methodologies *See also* Design; Environmental durability design methodologies
environmental **EL1:** 45-46, 65-67
mechanical **EL1:** 45-46, 55-65
thermal **EL1:** 45-55
Durability of Adhesive Bonded Structures **EM3:** 68
Durability of Structural Adhesives **EM3:** 71
Durable tools **A18:** 627
Duralumin (Al-Cu: 94-96% Al; 3-5% Cu; trace Mg)
thermal properties **A18:** 42
Duralumin alloys ..... **A3:** 1 • 25
Duralumin S(Q)
tension and torsion effective fracture strains **A8:** 168
Durand Arcoflam Clear-Line beta-quartz glass-ceramic
coefficient of thermal expansion **EM4:** 1103
composition **EM4:** 1103
Durand Arcopal (liquid-liquid opal glass)
composition **EM4:** 1101
properties **EM4:** 1101
Durand lead crystal
coefficient of thermal expansion **EM4:** 1102
composition **EM4:** 1102
softening point **EM4:** 1102
Durand Soda-Lime (soda-lime glass)
composition **EM4:** 1101
properties **EM4:** 1101
Durand Table (NaF opal glass)
composition **EM4:** 1101
properties **EM4:** 1101
Duranickel **A9:** 435-437
Duranickel 301, aged
machining **A16:** 837-843
Duranickel 301, unaged
machining **A16:** 837-843
Durapatite **EM4:** 1008

Duration *See also* Time
as parameter, acoustic emission inspection ... **A17:** 283
Durometer ... **A8:** 107-108
hardness effect on stencil printer ... **EL1:** 732
Durometer reading
defined ... **A18:** 7
Durometer readings ... **EM3:** 51
Dushman's constant ... **A6:** 44
Dust
collection, electric arc furnace ... **A15:** 360
granulation, historical use ... **A15:** 18
Dust, airborne
as contaminant ... **EL1:** 661
Dust, blast furnace
Miller numbers ... **A18:** 235
Dust clouds concentration of particles in ... **A7:** 195
duration, factors affecting ... **A7:** 196
explosion characteristics ... **A7:** 196
ignition and explosion ... **A7:** 194, 195, 198
secondary ... **A7:** 198
Dust control systems
for dry developers ... **A17:** 79
Dust, electric arc furnace (EAF)
zinc recycling from ... **A2:** 1224-1225
Dust explosion ... **A7:** 194
conditions required for ... **A7:** 195
hazards, classification and test ... **A7:** 196-198
of iron powders ... **A7:** 197
Dust pressing ... **EM4:** 9
Dust(s) *See also* Dust clouds
defined ... **A7:** 4
lofting, factors affecting ... **A7:** 196
metal ... **A7:** 495-496
Dustbulb
pigmented magnetic particles blown from ... **A7:** 578
Dusting ... **A18:** 684
defined ... **A18:** 7
Dutchman coupon rack ... **A13:** 199
Duty
defined ... **A18:** 7
Duty cycle ... **A6:** 42 **A18:** 834
definition ... **A6:** 36 **M6:** 6
Duty parameter *See* Capacity number
DVLO theory ... **EM4:** 119
Dwarf width
defined ... **EM2:** 15
Dwell
defined ... **A14:** 5 **EM1:** 9 **EM2:** 15
Dwell at liquidus ... **A6:** 354
Dwell mark ... **EM4:** 630
definition ... **EM4:** 632
Dwell pressure
isostatic compacting ... **A7:** 449
Dwell stations
for liquid penetrant inspection ... **A17:** 78
Dwell time ... **A6:** 316
in accelerated fatigue tests ... **EL1:** 741
defined ... **A7:** 4 **A12:** 59
effect on fatigue crack growth rate (*da/dN*) ... **A12:** 60
effect on hot pressed electrolytic iron powder compacts ... **A7:** 505
effect on mean stress ... **A12:** 63
effect on striation spacing ... **A12:** 60, 61
soldering ... **EL1:** 676
in spin coating ... **EL1:** 326
Dwell time, penetrant
liquid penetrant inspection ... **A17:** 82
DWTT *See* Drop-weight tear test
Dy-Fe (Phase Diagram) ... **A3:** 2 • 182
Dy-Ga (Phase Diagram) ... **A3:** 2 • 183
Dy-Ge (Phase Diagram) ... **A3:** 2 • 183
Dy-In (Phase Diagram) ... **A3:** 2 • 183
Dy-Mn (Phase Diagram) ... **A3:** 2 • 184
Dy-Ni (Phase Diagram) ... **A3:** 2 • 184
Dy-Pb (Phase Diagram) ... **A3:** 2 • 184
Dy-Pd (Phase Diagram) ... **A3:** 2 • 185
Dy-S (Phase Diagram) ... **A3:** 2 • 185
Dy-Sb (Phase Diagram) ... **A3:** 2 • 185
Dy-Sn (Phase Diagram) ... **A3:** 2 • 186
Dy-Te (Phase Diagram) ... **A3:** 2 • 186
Dy-Tl (Phase Diagram) ... **A3:** 2 • 186
Dy-Zr (Phase Diagram) ... **A3:** 2 • 187
Dye
for gage marks ... **A8:** 548
Dye and fluorescent penetrant inspections
brazed joints ... **A6:** 1119
Dye penetrant inspection ... **A6:** 97, 98, 99, 100
fitness for service evaluation ... **A6:** 376
gas-tungsten arc welding ... **A6:** 451
resistance seam welds ... **A6:** 245
shielded metal arc welding ... **A6:** 447
welding of cast CR alloys ... **A6:** 597
Dye penetrant inspection of fiber composites ... **A9:** 591
Dye penetrant leak testing
for lid seal integrity ... **EL1:** 954
Dye penetrants
visual inspection with ... **A17:** 3
Dye-penetrant enhanced x-ray radiography ... **EM1:** 770
Dye-penetrant techniques
for optical metallography specimens ... **A10:** 302
Dyes ... **A18:** 110
analytic methods for ... **A10:** 9
applications ... **A18:** 110
as colorants ... **EM2:** 501
corrosion of stainless steels in ... **M3:** 82
in engine and nonengine lubricant formulations ... **A18:** 111
functions ... **A18:** 110
Dynamic
defined ... **A11:** 3
Dynamic (flowing) solder pot ... **EL1:** 731
Dynamic analysis, numerical
in fracture mechanics ... **A8:** 445-446
Dynamic angle of repose ... **A7:** 282-283
Dynamic behavior
and viscoelasticity ... **EM2:** 414
Dynamic bend angle ... **A6:** 160
Dynamic bend tests, alternative
and Charpy V-notch impact test ... **A8:** 259
Dynamic coefficient of friction ... **A18:** 27
Dynamic compaction
aluminum P/M alloys ... **A2:** 204
Dynamic compression
apparatus ... **A8:** 218
fixture ... **A8:** 196
test, strain rate ranges for ... **A8:** 40
Dynamic compression plate
fatigue or crack initiation ... **A11:** 679, 683-686
as internal fixation device ... **A11:** 671
Dynamic corrosion tests
copper/copper alloys ... **A13:** 637-638
Dynamic critical stress-intensity
from dynamic notched round bar test ... **A8:** 275
Dynamic dielectric analysis (DDA)
of epoxies ... **EL1:** 836
Dynamic environment, body
and implants ... **A11:** 673, 676-677
Dynamic environmental stress screening ... **EL1:** 878
Dynamic fatigue, polyether sulfones (PES PESV) ... **EM2:** 161
Dynamic fatigue testing
deflection-controlled ... **EM2:** 704-705
stress-controlled ... **EM2:** 704
Dynamic fracture *See also* Dynamic fracture testing
by explosive detonation ... **A8:** 259
by impact ... **A8:** 259
damage ... **A8:** 287-288
initiation ... **A8:** 276
mechanics, as a field ... **A8:** 455
micromechanics of ... **A8:** 286-288
under rapidly applied load ... **A8:** 259
Dynamic fracture testing *See also* Dynamic fracture ... **A8:** 259-297
by Charpy impact test ... **A8:** 261-268
by crack arrest tests ... **A8:** 284-286
by crack propagation tests ... **A8:** 284
by one-point bend test ... **A8:** 271-275
by short-pulse-duration tests ... **A8:** 282-283
crack arrest fracture toughness in ... **A8:** 288-293
of ferritic materials ... **A8:** 288-293
of fracture toughness ... **A8:** 259-261
micromechanics of ... **A8:** 286-288
qualitative and quantitative ... **A8:** 259
using concept of impact response curves ... **A8:** 269-271
using dynamic notched round bar test ... **A8:** 275-282
using servohydraulic testing systems ... **A8:** 259-261
Dynamic fracture toughness
by impact response curves ... **A8:** 269
by one-point bend test ... **A8:** 271-275
data, low-alloy steels ... **A8:** 272
defined (under stress-intensity factor) ... **A13:** 12
effect of finite element analysis in ... **A8:** 282
symbol for ... **A8:** 725 **A11:** 797
Dynamic friction *See* Kinetic friction
Dynamic hardness test *See also* Rebound hardness test ... **A8:** 71
Scleroscope ... **A8:** 104-106
Dynamic hot pressing
defined ... **A7:** 4
Dynamic hydraulic torsion test
facility ... **A8:** 216
Dynamic impact fracture toughness
defined ... **A11:** 10
as function of test temperature ... **A11:** 54
Dynamic *J* fracture testing ... **A8:** 261
Dynamic leak testing
defined ... **A17:** 61
Dynamic light scattering ... **EM4:** 87
versus weighting factor ... **EM4:** 85
Dynamic load
defined ... **A18:** 7
Dynamic load capacity
bearings ... **A11:** 490
Dynamic loading ... **A16:** 21 **EL1:** 45, 62-63
slow strain rate testing ... **A8:** 496, 498-499, 519-520
Dynamic material modeling *See also* Material modeling; Process modeling ... **A14:** 370-371, 421-422
basic concepts ... **A14:** 422-423
intrinsic workability ... **A14:** 423
processing conditions, selection ... **A14:** 423-424
processing maps determined ... **A14:** 423
Dynamic mechanical analysis (DMA) ... **EM1:** 779 **EM2:** 526-527, 833-834 **EM3:** 392, 645-646
of flexible epoxies ... **EL1:** 820
to identify mesophase properties ... **EM3:** 395
Dynamic mechanical measurement ... **EM3:** 11, 318-320
defined ... **EM2:** 15
Dynamic mechanical properties
of solids ... **EM2:** 538-539
as tests ... **EM2:** 435-436
Dynamic mechanical rheometry ... **EM2:** 536-538
Dynamic mechanical spectroscopy ... **EM1:** 654-656
Dynamic modulus ... **EM1:** 9, 206 **EM3:** 11
defined ... **EM2:** 15
Dynamic modulus of elasticity ... **A8:** 249
Dynamic notched round bar testing ... **A8:** 275-282
and Charpy test, compared ... **A8:** 275, 276
data analysis ... **A8:** 281-282
for dynamic fracture ... **A8:** 275-282
fatigue apparatus for ... **A8:** 278
for high strain rate fracture toughness testing ... **A8:** 187
notch opening displacement measurement ... **A8:** 278-281
specimens ... **A8:** 277-278

Dynamic nucleation and growth
as a result of dynamic recrystallization **A9:** 690
Dynamic oscillatory shear flow
rheology **EL1:** 839-840, 844
Dynamic polarization, tests
crevice corrosion **A13:** 308-309
Dynamic random access memory (DRAM)
alpha particle induced random errors, detected **EL1:** 1056
chips, as future trend **EL1:** 390
IC trend lines **EL1:** 399
memory density **EL1:** 439
polyimides in **EL1:** 770-771
size trends **EL1:** 401
soft errors **EL1:** 805
wafer **EL1:** 393
Dynamic range
defined **A17:** 369, 383
film radiography **A17:** 299
fluorescence analysis **A10:** 76
of radiographic contrast **A17:** 299
Dynamic recovery **A8:** 172-173
at elevated temperatures **A9:** 690
effect of low temperature and high strain rate on **A9:** 688
and recrystallization in single-phase materials **A8:** 173-177
Dynamic recrystallization **A8:** 172-173
at elevated temperatures and large strains **A9:** 690
in austenitic stainless steel **A8:** 154
in processing map **A8:** 572
and recovery in single-phase materials **A8:** 173-177
Dynamic recrystallization controlled rolling (DRCR) **A1:** 117-118
Dynamic response
defined **A17:** 51
of electronic components **EL1:** 63-64
Dynamic scanning calorimetry (DSC)
on polyimides **EM3:** 156, 159
Dynamic scanning electron microscopy **A9:** 97
Dynamic seal
defined **A18:** 7-8
Dynamic shear stress/shear strain curve **A8:** 225
Dynamic sliding friction coefficient **A18:** 478
Dynamic stability
in superconductors **A2:** 1038-1039
Dynamic strain
aging, low-carbon steel **A8:** 40
rate test, Kolsky bar for **A8:** 219
role of stress in producing **A8:** 498
Dynamic tear
abbreviation **A8:** 724
toughness control tests involving **A8:** 453
Dynamic tear (DT) test
for fracture toughness **A11:** 61-62, 796
Dynamic tear properties
of compacted graphite iron **A1:** 61, 64
of ductile iron **A1:** 40, 43, 45, 46
Dynamic tension test
ringing effect **A8:** 40
strain rate ranges for **A8:** 40
Dynamic theory **A18:** 388
Dynamic torsion test
calibration and data reduction **A8:** 228
computed and measured stresses **A8:** 216-217
Dynamic toughness
determination in one-point bend test **A8:** 274-275
from strain gage records **A8:** 274-275
Dynamic viscosity *See* Viscosity **A10:** 685
Si units/symbol for **A8:** 721
Dynamic yield stress, strain rate
steel and aluminum **A8:** 41
Dynamic Young's modulus *See* Dynamic modulus of elasticity
Dynamic-condenser method, Kelvin
for adhesive-bonded joints **A17:** 611
Dynamical diffraction
in defect imaging **A10:** 367-370
Dynamical diffraction theory **A9:** 111
effect on kinematical dislocations contrast **A9:** 114
in terms of plane waves **A9:** 112
used to study grain boundaries **A9:** 120
Dynamically loaded bearing applications
beryllium-copper alloys **A2:** 418
Dynamo steels
properties of castings **M1:** 399
Dynamometer **A16:** 10
Dynamometer stator vanes
liquid erosion of **A11:** 169-170
Dysonian shapes
in ESR spectra **A10:** 261
Dysprosium *See also* Rare earth metals
conversion screens **A17:** 387, 391
as delayed-emission converter for thermal neutron radiography **EM3:** 759
properties **A2:** 1179
pure **M2:** 733
as rare earth metal **A2:** 720
TNAA detection limits **A10:** 237
Dysprosium in garnets **A9:** 538
Dystetic equilibrium *See* Eutectoid equilibrium
N,N-dimethyl + saccharin *p*-toluidine
generating free radicals for acrylic adhesives **EM3:** 120

E

2-Ethoxyethyl
uses and properties **EM3:** 126
8640 steel **A1:** 438
heat treatments for **A1:** 439
properties of **A1:** 439
processing of **A1:** 438
e *See* Electrons; Energy; Engineering strain; Modulus of elasticity
E polymers *See also* Polyaryletherketones (PAEK, PEK, PEEK, PEKK)
physical properties **EM2:** 142
E-glass *See also* Glass fibers
application **EM1:** 45, 107
cloth, effective thickness **EL1:** 599
cloth, for base materials/insulators **EL1:** 114
composition **EM1:** 45, 107
continuous-filament, in copper-clad laminates **EL1:** 534-537
defined **EM1:** 9, 29, 43 **EM2:** 15
density **EM1:** 46
durability **EM1:** 107
effect, polyester resins **EM1:** 92
effect, vinyl ester resins **EM1:** 92
epoxy matrix composite, fiber orientation effect **EM1:** 120
epoxy printed wiring boards **EL1:** 1117-1118
epoxy resin composites with/without **EM1:** 399-403
photoelastic stress pattern for **EM1:** 196
Poisson's ratio **EM1:** 46
and polyester resin composites, importance **EM1:** 43
pristine strength **EM1:** 46
properties **EM1:** 58, 175 **EM4:** 849, 850, 851, 1057
as PWB reinforcement **EL1:** 604
radiation properties **EM1:** 47
with RTM materials **EM1:** 566-567
in space and missile applications **EM1:** 817
specific heat **EM1:** 47
Young's modulus of elasticity **EM1:** 46
E11018 filler metal
hydrogen-induced cracking **A6:** 413
EAF-AOD-VAR steel-making methods **A4:** 211
EAF-VAR steel-making methods **A4:** 211
Earing **A14:** 5, 576
as changing *r* value **A8:** 550
in pewter sheet **A2:** 523
Early American foundries
Saugus Iron Works **A15:** 24
spread of **A15:** 24-27
Early manufacturing involvement (EMI)
in design **EL1:** 125
Early shakeout
as casting defect **A11:** 386
Earth elements *See* Alkaline earth elements; Rare earths
alkali, eluent suppression ion chromatography techniques for **A10:** 660
alkaline, complexometric titrations for **A10:** 164
Earthenware **EM4:** 3
colors **EM4:** 3
composition **EM4:** 5, 45
definition **EM4:** 3
Earthenware (continued)
glazing **EM4:** 3
properties **EM4:** 3, 934
properties of fired ware **EM4:** 45
subclassifications **EM4:** 3
water absorption **EM4:** 3
Earthmoving applications
crawler tractor track pins, nitrided steels for **M1:** 628, 630
proving ground tests for wear **M1:** 604, 605
wear resistant steels for **M1:** 625
EASE2 computer program for structural analysis **EM1:** 268, 271
Easy-machining steel
impact resistance and abrasion resistance properties **A18:** 759
EB melting *See* Electron beam melting and casting
EBB grade wire *See* Extra Best Best quality telephone and telegraph wire
Ebers-Moll equations
bipolar transistor analysis **EL1:** 151-152
EBIC *See* Electron beam induced current
ECAP *See* Energy compensated atom probe
Eccentric
defined **A14:** 5
gear, defined **A14:** 5
-gear drives, mechanical presses **A14:** 494
press **A14:** 5, 40
shaft drive, mechanical presses **A14:** 494
Eccentric gear
for washing machines **A7:** 623
Eccentric loading
in creep-rupture testing **A8:** 330
Eccentric-driven presses **A7:** 330, 331
Eccentricity
defined **A18:** 8
Eccentricity ratio **A18:** 90, 523- 524, 527, 528, 532
defined **A18:** 8
nomenclature for Raimondi-Boyd design chart **A18:** 91
Eccospheres **EM4:** 419
ECDM grinding *See* Electrochemical discharge grinding
E_{cell} *See* Measured cell potential
Echelle grating spectrometer **A10:** 40-41
Echo capture gating
C-scan **EM2:** 842
Echoes *See also* Pulse-echo methods; Ultrasonic inspection
in ultrasonic inspection **A17:** 240, 245-246
Eckhardt, A.G
as inventor **A15:** 34
ECL *See* Emitter-coupled logic
Economic analysis
factors in **EM2:** 293-294
Economic calculations
corrosion **A13:** 369-374
Economic process selection factors *See also* Cost(s); Economic analysis; Economics; Material costs; Materials selection; Processing
filament winding **EM2:** 373
for rotational molding **EM2:** 364-365
thermoforming **EM2:** 403
for thermoplastic injection molding ... **EM2:** 308-310
Economical casting design rules **A15:** 602-604, 611-612
Economics *See also* Corrosion economics; Cost; Cost(s); Costs
assembly costs **EM2:** 649
of cold heading **A14:** 295
of damage tolerance **EM1:** 259
estimating part cost **EM2:** 647
of explosive sheet forming **A14:** 641
final assembly costs **EM2:** 650
finishing cost **EM2:** 650
of forming welded assemblies **A14:** 642-643
forming/processing costs **EM2:** 647-648
of high-energy-rate forging **A14:** 100-101
lead frame materials **EL1:** 491-492
of materials **EM2:** 646-650
packaging cost **EM2:** 650
of precision forging **A14:** 159
prefinishing costs **EM2:** 649-650
Economizers
flue-gas corrosion in **A11:** 619
heat transfer factors **A11:** 604
steam/water-side boilers **A13:** 991-992

Economizers (continued)
tube, after penetrant testing ... **A11:** 433
tube/fin assembly removed from ... **A11:** 432
Economy in manufacture ... **M3:** 838-856
availability ... **M3:** 839, 845-846
chemical composition ... **M3:** 838, 840, 848
coatability ... **M3:** 839, 844-845
energy consumption ... **M3:** 839, 846-847
formability ... **M3:** 839, 843
heat treatment response ... **M3:** 839, 844
interactive effects ... **M3:** 839-840
machinability ... **M3:** 839, 843-844
plant standardization ... **M3:** 839
processing factors ... **M3:** 839
product form ... **M3:** 838-839, 840
quality descriptors ... **M3:** 839, 842
quantity ... **M3:** 839
raw-materials factors ... **M3:** 838-839
selection for, examples ... **M3:** 847-856
size ... **M3:** 839, 840-841
special requirements, selection for ... **M3:** 846
specifications ... **M3:** 842-843
surface finish ... **M3:** 839, 842
temper conditions ... **M3:** 839, 841-842
tolerances ... **M3:** 839
weldability ... **M3:** 839, 844
E_{const} *See* Constant cell potential
ECP *See* Electric current perturbation NDE
ECR glass
application and composition ... **EM1:** 107
ECTFE *See* Ethylene chlorotrifluoroethylene
Eczema
nickel-caused ... **A7:** 203
Eddy current
and nondestructive testing ... **A7:** 491-492
permanent magnet-coils with ... **A8:** 245
probes ... **A8:** 246, 383
Eddy current clutches
mechanical presses ... **A14:** 497
Eddy current gage
for crack arrest testing ... **A8:** 454
for fatigue crack growth testing ... **A8:** 428, 430
to monitor crack extension in corrosive environments ... **A8:** 428
Eddy current inspection *See also* NDE reliability; Remote-field eddy current inspection ... **A17:** 164-194
advantages and limitations ... **A17:** 164
of aircraft structural and engine components ... **A17:** 189-194
applications ... **A17:** 164
of arc-welded nonmagnetic ferrous tubular products ... **A17:** 566-567
automated, probability of detection (POD) curve ... **A17:** 664
of bar ... **A17:** 171-172
of billets ... **A17:** 559-560
of boilers and pressure vessels ... **A17:** 642
of bolts ... **A17:** 554
of casting surfaces ... **A17:** 512
of castings ... **A17:** 512, 525-526
codes, boiler/pressure vessels ... **A17:** 642
coil impedance ... **A17:** 166-167
of continuous butt-welded steel pipe ... **A17:** 567
of defects ... **A15:** 554-555
discontinuities detectable by ... **A17:** 179
dual-frequency ... **A17:** 181
of duplex tubing ... **A17:** 572
edge effect ... **A17:** 169
and electric current perturbation compared ... **A17:** 136
electrical conductivity ... **A17:** 167
equipment ... **A17:** 185
examples of ... **A17:** 180-182, 187-189, 190-194
for film thickness ... **A13:** 417
of finned tubing ... **A17:** 571
of forgings ... **A17:** 510-511
impedance concepts ... **A17:** 169-173
in-service, tubular products ... **A17:** 574
Eddy current inspection (continued)
and induction heating technique compared ... **A17:** 165-166
inspection coils ... **A17:** 175-177
inspection frequencies ... **A17:** 173
instruments ... **A17:** 177-179
of internal discontinuities, castings ... **A17:** 512
lift-off factor ... **A17:** 168
magnetic permeability ... **A17:** 167-168
microstructural effect ... **A17:** 51
and microwave testing, compared ... **A17:** 218
with microwaves ... **A17:** 218-219
multifrequency techniques ... **A17:** 173-175
NDE reliability models of ... **A17:** 707-709
of nonferrous tubing ... **A17:** 572-573
operating variables ... **A17:** 166-169
of pipe ... **A17:** 186, 579
of plates ... **A17:** 187-189
of powder metallurgy parts ... **A17:** 543-545
principles of operation ... **A17:** 165-166
probability of detection (POD) models in ... **A17:** 708-709
probe-flaw interaction models ... **A17:** 707-708
probes, ferromagnetic resonance ... **A17:** 220-223
process development ... **A17:** 164-165
process qualification ... **A17:** 678
resistance seam welds ... **A6:** 245
of resistance-welded steel tubing ... **A17:** 562-563
of round steel bars ... **A17:** 185-186
of seamless pipe ... **A17:** 579
of seamless steel tubular products ... **A17:** 570-571
of sheets ... **A17:** 187-189
skin effect ... **A17:** 169
of skin sections ... **A17:** 187-189
slow sweep mode ... **A17:** 190
of solid cylinders ... **A17:** 184-185
speed of inspection ... **A17:** 564
of spiral-weld steel pipe ... **A17:** 567
of steel bar and wire ... **A17:** 553-554
till factor ... **A17:** 168-169
to detect subsurface porosity ... **A6:** 1074
to detect tungsten inclusions ... **A6:** 1074
trigger pulse mode ... **A17:** 190
of tubes on solid cylinders ... **A17:** 185
of tubing ... **A17:** 169-170, 179-184, 186
and ultrasonic inspection, simultaneous ... **A17:** 272
with visual inspection ... **A17:** 3
vs. ultrasonic inspection, for primary mill products ... **A17:** 267
of weldments ... **A17:** 186, 602
Eddy current losses
ferrites ... **EM4:** 1161
Eddy current losses magnetically soft materials ... **A2:** 761
in superconductors ... **A2:** 1040
Eddy current testing *See* Eddy current inspection
Eddy currents *See also* Eddy current inspection
defined ... **A17:** 164-166
flow patterns ... **A17:** 165
for holes and small radii areas ... **A17:** 223-224
and skin effect ... **A17:** 169
Eddy sonic tests
defect detection ... **EM3:** 750, 751
Eddy-current inspection ... **M6:** 850
as failure analyses ... **A11:** 17
Eddy-current testing ... **A6:** 1081, 1082-1083, 1085, 1086, 1087, 1088
EDF tests *See* Empirical distribution function (EDF) goodness-of-fit test
Edge *See also* Absorption; K-edge; Molded edge; Shear edge
absorption, defined ... **A10:** 85
control, in roll compacting ... **A7:** 404, 406
cracking ... **A8:** 594, 596
dislocation, loop patches ... **A8:** 256
distance, in pin bearing testing ... **A8:** 4, 59
effects, in image analysis of low-carbon sheet steel ... **A10:** 316
Edge (continued)
flaws, particle ... **A7:** 59
K-shell ionization, pre- and post-structures in ... **A10:** 450
milling, cantilever beam bend test specimens ... **A8:** 132
restriction devices ... **A7:** 404, 406
shapes, characteristic EELS, for amorphous carbon and carbon in metal carbide ... **A10:** 460
stability, defined ... **A7:** 4
strength, defined ... **A7:** 4
Edge bending ... **A14:** 529
power brushing process ... **M5:** 155
Edge condition
effect on bending of steel sheet ... **M1:** 553
Edge contrast in scanning electron microscopy ... **A9:** 94
Edge defects
spring failure from ... **A11:** 559
Edge delamination test
for fiber-resin composites ... **EM1:** 97, 98
Edge detection
filters, digital image processing ... **A17:** 461
with machine vision ... **A17:** 34
remote-field eddy current inspection ... **A17:** 199
Edge dislocations ... **A13:** 45, 46
connected to antiphase boundaries ... **A9:** 682-683
in crystals ... **A9:** 719-720
diffraction contrast ... **A9:** 114
in dislocation pairs ... **A9:** 114-115
representation of ... **A9:** 682
Edge distance
defined ... **A8:** 4
Edge distance ratio
defined ... **A8:** 4 **EM1:** 9 **EM2:** 15
Edge effect ... **A18:** 206-207
in eddy current inspection ... **A17:** 169, 221
Edge enhancement
defined ... **A17:** 383
Edge grinding
of contour flanges ... **A14:** 530
Edge honing
cemented carbides ... **A18:** 796
Edge joint ... **EM3:** 11
defined ... **EM1:** 9 **EM2:** 15
definition ... **A6:** 1208
electron-beam welding ... **A6:** 260
laser-beam welding ... **A6:** 879, 880
oxyfuel gas welding ... **A6:** 286, 287
Edge joints
definition, illustration ... **M6:** 6, 60-61
electron beam welds ... **M6:** 617
gas tungsten arc welding ... **M6:** 201-202
oxyfuel gas welding ... **M6:** 589-590
Edge length of a crystal ... **A3:** 1•10
Edge lengths and interaxial angles
relationships for crystal systems ... **A9:** 706
Edge, molded *See* Molded edge
Edge preparation
carbide metal cutting tools ... **A2:** 964
definition ... **M6:** 6
Edge preparations for
arc welding of aluminum alloys ... **M6:** 375
oxyfuel gas welding ... **M6:** 589-591
Edge protection
by nickel plate ... **A11:** 24
Edge replication
as nondestructive test technique ... **EM1:** 775
Edge retention
of acrylics ... **A9:** 30
of carburized steel specimens ... **A9:** 217
in cemented carbide samples ... **A9:** 273
cleaning of specimens to be plated ... **A9:** 28
of compression-mounting epoxies ... **A9:** 29
effect of shrinkage stresses on ... **A9:** 28
in metallographic sample preparation ... **A9:** 44-45
mounting techniques ... **A9:** 31-32
of powder metallurgy specimens ... **A9:** 505

Edge retention (continued)
of tin and tin alloy specimens ... **A9:** 449
of titanium and titanium alloys ... **A9:** 458
of tungsten samples ... **A9:** 439
in wrought heat-resistant alloy specimens ... **A9:** 305
of wrought stainless steels during mounting ... **A9:** 279
Edge retention techniques ... **A16:** 28-29
in electropolishing ... **A9:** 56
for uranium and uranium alloys ... **A9:** 478
Edge rolling
wrought copper and copper alloys ... **A2:** 248
Edge, shear *See* Edgewise shear; Shear edge
Edge strain
defined ... **A8:** 4
Edge stresses
laminate ... **EM1:** 230-231
Edge weld
definition ... **A6:** 1208 **M6:** 6
Edge weld size
definition ... **A6:** 1208
Edge(s) *See also* Edge detection; Edge effect
absorption, defined ... **A17:** 309
defects, magnetic particle detection ... **A17:** 103
fluorescent indications ... **A17:** 117
human vs. machine vision ... **A17:** 30
locations, laser triangulation measurement ... **A17:** 13
offset of plate, in steel pipe ... **A17:** 565
Edge-finder probe
coordinate measuring machine ... **A17:** 25
Edge-flange weld
definition ... **A6:** 1208 **M6:** 6
electron beam welding ... **M6:** 614-615
electron-beam welding ... **A6:** 260
laser-beam welding ... **A6:** 879, 880
oxyfuel gas welding ... **A6:** 286, 287
Edge-trailing technique
defined ... **A9:** 6
Edge-trim slitters ... **A14:** 712
Edgers *See also* Fuller ... **A14:** 5, 43-44
Edges *See also* Offset parts
bending of ... **A14:** 529
blanked, characteristics ... **A14:** 447
chip ... **EL1:** 7
condition, effect, press bending ... **A14:** 523
pierced, types ... **A14:** 460-461
preparation, three-roll forming ... **A14:** 619
preparation, titanium alloy ... **A14:** 841
punch-to-die clearances for ... **A14:** 460
trimmed, press forming ... **A14:** 553
Edges, blanked *See* Blanked edges
Edgewise shear loaded subcomponents
testing ... **EM1:** 331-333
Edging
defined ... **A14:** 5
Edging rolls
in hot rolling ... **A8:** 594
EDL *See* Electroless discharge lamp
EDM *See* Electric discharge machining
EDS *See* Energy-dispersive spectrometry; Energy-dispersive x-ray Spectroscopy
EDS analysis ... **A18:** 387, 389-390, 391
EDTA *See* Ethylenediaminetetraacetic acid
EDTA titration
elements determined by ... **A10:** 173
EDXA *See* Energy-dispersive x-ray analysis
EEPROM *See* Electrically erasable programmable read-only memory
Effect
defined ... **EM2:** 599
Effective (apparent) permeability
defined for magnetic particle inspection ... **A17:** 99
Effective absorption, of x-rays
radiography ... **A17:** 310-311
Effective aperture size
computed tomography (CT) ... **A17:** 373
Effective atomic number
defined ... **A17:** 383
Effective bending stiffness matrix
laminate ... **EM1:** 224
Effective crack size ... **A8:** 4
defined ... **A11:** 3
elastic-plastic analysis for ... **A8:** 446
plane-strain estimate ... **A8:** 451
Effective crack size (continued)
resistance ... **A8:** 452
Effective draw
defined ... **A14:** 5
Effective end relief angles
shaping of tool steel die sections ... **A16:** 192
Effective flaws *See also* Flaws
interlaminar ... **EM1:** 241-244
Effective hardness number ... **A18:** 417
Effective interdiffusional distance
effect of mechanical working ... **A7:** 315
Effective leakage area
defined ... **A18:** 8
Effective length of weld
definition ... **M6:** 6
Effective load vector ... **A6:** 1133
Effective lubricant viscosity
nomenclature for Raimondi-Boyd design chart ... **A18:** 91
Effective magnetic excitation (k) ... **A6:** 43
Effective magnetization ... **A10:** 690
Effective mean potential
EXAFS analysis ... **A10:** 409
Effective medium theory ... **A6:** 143
Effective modulus, plastics
time effects ... **EM2:** 412
Effective penetration distance
x-ray diffraction ... **A18:** 464
Effective relative permittivity
defined ... **EL1:** 597-598
Effective stiffness *See also* Stiffness
of laminates ... **EM1:** 218
Effective stiffness matrix ... **A6:** 1133
Effective strain
and shear stress and strain, in torsional loading ... **A8:** 142-143
and stress, reduction of shear stress/shear strain to ... **A8:** 161
tension and torsion, for alloys ... **A8:** 168
to failure, in torsion tests ... **A8:** 165
Effective strain rate ... **A14:** 439
Effective stress
and flow stress ... **A8:** 256
and shear stress and strain in torsional loading ... **A8:** 142-143
and strain, reduction of shear stress/shear strain to ... **A8:** 161
Effective stress intensity
and fatigue striation spacing ... **A12:** 205
Effective stress-strain curves
from torsion tests using Tresca criterion ... **A8:** 163
for stainless steel in compression, tension torsion ... **A8:** 162, 164
Effective throat
definition ... **A6:** 1208 **M6:** 6
Effective true strain contour maps
aluminum alloys ... **A14:** 436
Effective volume ratio ... **A18:** 467
Effective x-ray energy
defined ... **A17:** 383
Effective yield strength *See also* Flow strength ... **A8:** 4
Effects *See* Interaction effects; Main effects
Efficiency
of combustion accelerators ... **A10:** 222
power sources ... **A6:** 36
in separations, radioanalysis measurement of ... **A10:** 243
Efficiency, x-ray detector
computed tomography (CT) ... **A17:** 369
Efflorescence ... **EM4:** 157
Effluent autoclave
chemical composition ... **A8:** 424-402
Effluent gas analysis
of residues ... **A10:** 177
Effluent polishing
plating waste treatment ... **M5:** 315-316, 318
Effluents
analytic methods applicable ... **A10:** 6, 7, 8, 11
chromatographic, IR identification of ... **A10:** 109
differential pulse polarogram in analysis of ... **A10:** 195
industrial water, x-ray spectrum ... **A10:** 95
inorganic, analytic methods ... **A10:** 7, 8
of liquid chromatographs, UV/VIS detection of species in ... **A10:** 60
Effluents (continued)
multielement fingerprinting and voltammetric analysis ... **A10:** 195
organic gas, analytic methods for ... **A10:** 11
voltammetric monitoring of metals and nonmetals in ... **A10:** 188
Eggcrate design mold
autoclave molding ... **EM1:** 578-581
Eight harness satin weave *See also* Satin (crowfoot) weave
of fiberglass fabric ... **EM1:** 111
for unidirectional/two-directional fabrics ... **EM1:** 125
Eight-harness, defined *See* Satin (crowfoot) weave; under Harness satin
Eight-harness satin
defined ... **EM2:** 15
Eikem A/S process
of gallium recovery ... **A2:** 743-744
Einstein equation
HAZ width ... **A6:** 4, 5-6
Einsteinium *See also* Transplutonium actinide metals
applications and properties ... **A2:** 1198-1201
pure ... **M2:** 733
Einzel lens
in gas mass spectrometer ... **A10:** 153, 155
Ejection *See also* Extraction
capacity ... **A7:** 324, 325
defined ... **A7:** 4
die casting ... **A15:** 294
from tooling, dimensional changes ... **A7:** 480
in hot pressing ... **A7:** 503
in lost foam casting ... **A15:** 231
mechanism, precision forging ... **A14:** 158
part, designing for ... **A7:** 329, 331
part, multiple slide forming ... **A14:** 871
of permanent molds ... **A15:** 277
pressure ... **A7:** 192, 417
stress ... **A7:** 186
stroke, defined ... **A7:** 324, 325
of workpieces, deep drawing ... **A14:** 587-588
Ejection mark
defined ... **EM2:** 15
Ejection pins
for core removal ... **A15:** 191
size and location, die casting ... **A15:** 294
Ejector
defined ... **A14:** 5 **A15:** 5
Ejector punch
defined ... **A7:** 4
Ekabor (boriding compound) ... **A4:** 441, 443
Elapsed-cycle
vs. crack-length ... **A8:** 377
Elastic analysis limit
for crack speed ... **A8:** 445
Elastic bending ... **A8:** 118-120
below yield stress ... **A14:** 881
below yield stress, as springback ... **A8:** 552
Rayleigh-Ritz method ... **A8:** 118
sign convention for bending moment ... **A8:** 119-120
simple-beam theory ... **A8:** 118
Elastic binary collision ... **A7:** 259
Elastic buckling
in creep experiments ... **A8:** 302
as distortion ... **A11:** 143
Elastic calibration
devices ... **A8:** 4, 614-615
verification method ... **A8:** 611
Elastic calibration device
defined ... **A8:** 4
Elastic collisions
as electron signals ... **A12:** 168
as energy-level diagrams ... **A10:** 127
Elastic compliance
defined ... **A18:** 8
Elastic compressive modulus *See also* compression; Elastic modulus
carbon fiber/fabric reinforced epoxy resin ... **EM1:** 411
glass fiber reinforced epoxy resin ... **EM1:** 406
graphite fiber reinforced epoxy resin ... **EM1:** 413
high-temperature thermoset matrix composites ... **EM1:** 379
initial, high-temperature thermoset matrix composites ... **EM1:** 375

Elastic compressive modulus (continued)
initial, low-temperature thermoset matrix composites ... **EM1:** 396
initial, medium-temperature thermoset matrix composites ... **EM1:** 389
medium-temperature thermoset matrix composites ... **EM1:** 387
quartz fabric reinforced epoxy resin ... **EM1:** 415
thermoplastic matrix composites ... **EM1:** 366, 368
Elastic constants ... **A10:** 382, 387, 672
amorphous materials and metallic glasses ... **A2:** 813
defined ... **A8:** 4 **A11:** 3
of steel castings ... **A1:** 374
Elastic constants, engineering of plies ... **EM1:** 237
Elastic contact and stress nomenclature ... **A18:** 487
Elastic contact dimensions and stresses, calculations of ... **A18:** 487-488
α and β parameter constants ... **A18:** 487, 488
formulas ... **A18:** 488
nomenclature ... **A18:** 487
Elastic deflection
in precision forging ... **A14:** 159
of rolls ... **A14:** 346
in spiral bevel gear forging ... **A14:** 173
in tension test machine ... **A8:** 45
tooling ... **A14:** 161
Elastic deformation ... **A7:** 58, 298 **EM3:** 11
and acoustic emissions ... **A17:** 287
average flash temperature ... **A18:** 43
defined ... **A8:** 4 **A11:** 3 **A13:** 5 **A14:** 5 **EM1:** 10 **EM2:** 15
during tensile loading ... **A9:** 684
friction during metal forming ... **A18:** 59
magnetic printing detection ... **A17:** 126
and mechanical durability ... **EL1:** 55
Elastic displacement field
associated with dislocations ... **A10:** 464
Elastic distortion
and distortion failure ... **A11:** 143-144
as failure mechanism ... **A11:** 75
temperature dependence ... **A11:** 138
Elastic effects
rheological models with ... **EL1:** 848-849
Elastic electron scatter
defined ... **A9:** 6
Elastic energy
defined ... **A8:** 4
release of ... **A11:** 50
Elastic Euler equation ... **A8:** 55
Elastic hysteresis *See* Mechanical hysteresis
Elastic limit *See also* Hugoniot elastic limit; Proportional limit; True elastic limit ... **A8:** 4, 21 **EM3:** 11
defined ... **A11:** 3 **A13:** 5 **A14:** 5 **EM1:** 10 **EM2:** 15-16
Elastic limit of chromium carbide-based cermets ... **A7:** 806
Elastic macrostrain *See* Macrostrain
Elastic modulus *See also* Elastic compressive modulus; Elastic properties; Elastic tensile modulus; Mechanical properties; Modulus of elasticity; Modulus of elasticity (E)
aluminum casting alloys ... **A2:** 150, 152-177
of beryllium ... **A2:** 683-684
cast copper alloys ... **A2:** 356-391
and density in P/M steels ... **A7:** 466
effect of filler on polymer matrices ... **A7:** 612
in failure analysis ... **EM1:** 775
of ferrous P/M materials ... **A7:** 465, 466
longitudinal, graphite laminate ... **EM1:** 184
of P/M and wrought titanium and alloys ... **A7:** 475
of S-glass ... **EM1:** 107
shear, carbon fiber/fabric reinforced epoxy resin ... **EM1:** 411
wrought aluminum and aluminum alloys ... **A2:** 62-122
Elastic modulus effect ... **EM3:** 301
Elastic modulus, plotted against complexity index ... **A9:** 130
for beryllium-aluminum alloys ... **A9:** 131
Elastic plastic model
SCC behavior ... **A13:** 278
Elastic pressure bars
in split Hopkinson pressure bar test ... **A8:** 198-199
wave propagation in ... **A8:** 199-200
Elastic properties *See also* Elasticity; Material properties; Material properties analysis; Modulus of elasticity
cast steels ... **M1:** 393
constructional steels for elevated ... **M1:** -
temperature use ... **M1:** 640-641, 647, 650
graphite-aluminum composite ... **EM1:** 188
laminate ... **EM1:** 223-226
laminate, prediction ... **EM1:** 316
of rare earth metals ... **A2:** 725
Elastic proving ring *See also* Proving ring
as elastic calibration device ... **A8:** 614-615
with precision micrometer ... **A8:** 615
Elastic recoil
preventing in tension testing ... **A8:** 48
Elastic recovery ... **A8:** 4 **EM3:** 11
affecting microhardness readings ... **A8:** 96
defined ... **EM1:** 10 **EM2:** 16
in Knoop and Vickers indentations ... **A8:** 95
Elastic recovery parameter ... **A18:** 422, 427
Elastic resilience
defined ... **A8:** 4
Elastic scattering ... **A10:** 432-433, 672
Elastic springback ... **A7:** 480
beryllium-copper alloys ... **A2:** 412
Elastic springback method
for torque ... **A8:** 327
Elastic strain *See* Elastic deformation ... **A8:** 4
effect on diffraction patterns ... **A10:** 438, 440
effect on fatigue behavior ... **M1:** 668, 670, 672
range, low-cycle fatigue ... **A8:** 364
rate, relation to strain rate ... **A8:** 41-42
specimens ... **A8:** 503-508
stress relaxation ... **A8:** 307, 323-324
Elastic strain energy
defined ... **A8:** 4
Elastic strain range
low-cycle fatigue ... **A11:** 103
Elastic strain specimens ... **A8:** 503-508
Elastic strain SSC test specimens
bent-beam ... **A13:** 248-249
C-ring ... **A13:** 249
O-ring ... **A13:** 249-250
tension ... **A13:** 250-251
tuning fork ... **A13:** 251-252
Elastic tensile modulus *See also* Elastic modulus; Tensile modulus; Tensile strength
carbon fiber/fabric reinforced epoxy resin ... **EM1:** 410
epoxy resin system composites ... **EM1:** 401
graphite fiber reinforced epoxy resin ... **EM1:** 412
high-temperature thermoset matrix composites ... **EM1:** 378
Kevlar 49 fiber/fabric reinforced epoxy resin ... **EM1:** 407
low-temperature thermoset matrix composites ... **EM1:** 395
medium-temperature thermoset matrix composites ... **EM1:** 383, 386, 388, 390
quartz fabric reinforced epoxy resin ... **EM1:** 414
thermoplastic matrix composites ... **EM1:** 365, 368, 371
Elastic theory ... **A8:** 71-73
brazing ... **A6:** 110
Elastic true strain ... **EM3:** 11
defined ... **EM2:** 16
Elastic unloading, angle and torque
shear stress ... **A8:** 183-184
Elastic wave velocity
and stress ... **A8:** 209
Elastic-plastic analysis
estimate of crack opening method ... **A8:** 440
in fracture mechanics ... **A8:** 446-447, 457-458
Rice *j*-integral method ... **A8:** 440
Elastic-plastic behavior ... **A6:** 101
in chevron-notched specimens ... **A8:** 471
in fracture toughness testing ... **A8:** 473-474
Elastic-plastic bending
equations ... **A8:** 118
factor analysis of ... **A8:** 119
strain distribution ... **A8:** 119, 121
stress distribution ... **A8:** 119-121
Elastic-plastic boundary
in indentation testing ... **A8:** 72
Elastic-plastic dynamic analysis ... **A8:** 282
Elastic-plastic finite-element method
analytical modeling ... **A14:** 425
Elastic-plastic fracture mechanics
abbreviation ... **A8:** 724
effect of *J* concept on ... **A8:** 447
fracture toughness tests ... **A8:** 455-456
Elastic-plastic fracture mechanics (EPFM) ... **A11:** 49-51
energy criterion in ... **A11:** 50
in failure analyses ... **A11:** 49-51
fracture mechanics and ... **A11:** 47
Elastic-plastic fracture toughness
crack initiation in tests ... **A8:** 390
Elastic-plastic indentation fracture mechanics
and erosion ... **A13:** 137-138
Elastic-plastic stress
as *J*-integral parameter ... **A8:** 261
Elastic-stress distribution
in pure compression ... **A11:** 461
in pure tension ... **A11:** 461
Elastically isotropic materials
small dislocation loops in ... **A9:** 117
Elasticity *See also* Anelasticity; Elastic properties, Viscoelasticity; Viscoelasticity; Young's modulus ... **A8:** 4 **EM3:** 11, 322
analysis, of fiber composites ... **EM1:** 185-188
coefficient of, defined *See* Coefficient of elasticity; Coeffient of elasticity
defined ... **A11:** 3 **A13:** 5 **A14:** 5 **EM1:** 10 **EM2:** 15
distortion of diamond pyramid indentations ... **A8:** 102
Hertz theory of ... **A8:** 72
loading conditions for evaluation ... **EM1:** 178, 186
and material selection ... **EM1:** 38-39
metals vs plastics ... **EM2:** 656
and plasticity ... **A8:** 71
solutions, in elastic bending ... **A8:** 118
temperature dependence of ... **A11:** 138
theory of ... **A11:** 103
thermoplastic polyurethanes (TPUR) ... **EM2:** 206
visco-, of polymers ... **A11:** 758
Elastohydrodynamic (EHD) viscosity-pressure coefficient ... **A18:** 83
Elastohydrodynamic lubrication
for bearings ... **A11:** 485
wear failure and ... **A11:** 150, 151
Elastohydrodynamic lubrication (EHD) *See also* Boundary lubrication; Plastohydrodynamic lubrication; Thin-film lubrication ... **A18:** 89, 92-93, 94, 477
defined ... **A18:** 8
film thickness and shape ... **A18:** 92-93
gear box applications of ferrography ... **A18:** 305
pressure distributions ... **A18:** 93
sliding traction and contact temperature ... **A18:** 93
Elastohydrodynamic lubrication theories ... **A18:** 80
concentrated contacts and film thickness of bearing steels ... **A18:** 727
Elastomer *See also* Polymer(s) ... **A13:** 5, 1164
butadiene-acrylonitrile, as sizing ... **EM1:** 124
defined ... **EM1:** 10

SUBJECTS OF THE INDEXED VOLUMES: ASM Handbook (designated by the letter "A"): **A1:** Properties and Selection: Irons, Steels, and High-Performance Alloys (1990); **A2:** Properties and Selection: Nonferrous Alloys and Special-Purpose Materials (1990); **A3:** Alloy Phase Diagrams (1992); **A4:** Heat Treating (1991); **A6:** Welding, Brazing, and Soldering (1993); **A7:** Powder Metallurgy (1984); **A8:** Mechanical Testing (1985); **A9:** Metallography and Microstructures (1985); **A10:** Materials Characterization (1986); **A11:** Failure Analysis and Prevention (1986); **A12:** Fractography (1987); **A13:** Corrosion (1987); **A14:** Forming and Forging (1988); **A15:** Casting (1988); **A16:** Machining (1989); **A17:** Nondestructive Testing and Quality Control (1989); **A18:** Friction, Lubrication, and Wear Technology (1992). **Metals Handbook, 9th Edition** (designated by the letter "M"): **M1:** Properties and Selection: Irons and Steels (1978); **M2:** Properties and Selection: Nonferrous Alloys and Pure Metals (1979); **M3:** Properties and Selection: Stainless Steels, Tool Materials and Special-Purpose Materials (1980); **M4:** Heat Treating (1981); **M5:** Surface Cleaning, Finishing, and Coating (1982); **M6:** Welding, Brazing, and Soldering (1983). **Engineered Materials Handbook** (designated by the letters "EM"): **EM1:** Composites (1987); **EM2:** Engineering Plastics (1988); **EM3:** Adhesives and Sealants (1990); **EM4:** Ceramics and Glasses (1991); **Electronic Materials Handbook** (designated by the letters "EL"): **EL1:** Packaging (1989).

Elastomer solution in methyl methacrylate
formulation **EM3:** 123
Elastomer-bonded wire-filled radial brushes **M5:** 155
Elastomer-epoxies **EM3:** 76
typical film adhesives properties **EM3:** 78
Elastomeric adhesives **EM3:** 143-150
automotive decorative trim **EM3:** 552
for automotive sound absorption **EM3:** 555-556
commercial forms **EM3:** 143
common additives and modifiers **EM3:** 150
common properties **EM3:** 143-144
cross-linking **EM3:** 143
cure mechanism **EM3:** 143
properties and characteristics **EM3:** 143-144
Elastomeric bag, and mandrel
for titanium alloys **A7:** 750
Elastomeric coatings
for prevention of erosion **A11:** 170-171
Elastomeric flexible envelopes **A7:** 300
Elastomeric materials **EL1:** 818, 824
Elastomeric mold
preshaped **A7:** 444
Elastomeric sealants
for automotive valve sealing **EM3:** 547
for differential cover sealing **EM3:** 547
semiconductor component coating **EM3:** 612
Elastomeric tooling **EM1:** 590-601
application **EM1:** 595-601
defined **EM1:** 10
details, fabricating **EM1:** 593
mandrels, design/fabrication of **EM1:** 593-594
thermal expansion molding methods **EM1:** 590-591
volumetric analysis **EM1:** 591-593
Elastomeric tooling application *See also* Composite tools; Elastomeric tooling; Tooling **EM1:** 595-601
control surface construction **EM1:** 595-596
integral structure design/fabrication **EM1:** 596-601
integral structure tooling **EM1:** 596
Elastomers **EM3:** 11, 53
chemical properties **EM3:** 52
chemistry **EM2:** 64
defined **EM2:** 16
electrical applications **EM2:** 588-589
electrical properties **EM2:** 589
entropy elasticity **EM2:** 655-656
environmental effects **EM2:** 430
no adverse effect by water-displacing corrosion inhibitors **EM3:** 641
semirigid cast, polyurethane (PUR) **EM2:** 259
thermoplastic, chemistry **EM2:** 66
thermoplastic, properties **EM2:** 451
TPUR, suppliers **EM2:** 207
Elastoplastic analysis **EM3:** 483
Elastoplastic deformation
and mechanical durability **EL1:** 55
Elastoplastics
properties of **EM2:** 451
Elbow assembly, stainless steel
weld failure in **A11:** 117
Elbow joint prosthesis
total **A11:** 670
Elbows
pipe welding **M6:** 591
ELDOR *See* Electron-electron double resonance
Electrets
with parylene coatings **EL1:** 799-800
Electric
current, SI unit/symbol for **A8:** 721
field strength, SI unit/symbol for **A8:** 721
flux density, SI unit/symbol for **A8:** 721
motors, for torsion testing **A8:** 157
Electric actuator
in closed-loop servomechanical tester **A8:** 394
Electric arc cutting
Exo-Process as **A14:** 734
Electric arc furnace (EAF) **EM4:** 44
Electric arc furnace (EAF) dust
zinc recycling from **A2:** 1224-1225
Electric arc furnaces *See also* Arc furnace
acid melting practice **A15:** 363-365
basic melting practice **A15:** 365-367
components **A15:** 357-362
cross section **A15:** 34
Electric arc furnaces (continued)
development **A15:** 32
dimensions/capacities **A15:** 358
direct current arc furnace **A15:** 367-368
laboratory testing equipment **A15:** 363
new technology **A15:** 367-368
power supply **A15:** 356-357
precision weighing **A15:** 363
scrap/alloy storage **A15:** 363
steel production by **A15:** 31
Electric arc rotating electrode process **A7:** 39
Electric arc spraying **A13:** 459-460
definition **A6:** 1208 **M6:** 6
Electric arc welding wire *See* Welding wire
Electric arc wire spraying (EAW) **A18:** 829-830, 832
Electric beam welding
as secondary operation **A7:** 456
Electric bonding
definition **A6:** 1208
Electric brazing
definition **A6:** 1208
Electric charge
density **A10:** 685
density, SI units/symbol for **A8:** 721
SI derived unit and symbol for **A10:** 685
SI unit/symbol for **A8:** 721
Electric current *See also* Alternating current (ac); Current; Direct current (dc)
density, during demagnetization **A17:** 121
density, in electric current perturbation **A17:** 136
detection sensitivities **A17:** 101
SI base unit and symbol for **A10:** 685
types, magnetic particle inspection **A17:** 108-110
Electric current perturbation NDE **A17:** 136-142
applications **A17:** 137-140
and conventional eddy-current methods compared **A17:** 136
defined **A17:** 136
flaw characterization **A17:** 140-141
principles/background **A17:** 136-137
probe configuration and orientation **A17:** 136
for small flaws **A17:** 137
system, block diagram **A17:** 137
test flaws **A17:** 137
of weldments **A17:** 602
Electric currents, and moisture
effects on roller bearing **A11:** 495, 497
Electric dipole
defined **A10:** 672
Electric dipole moment
defined **A10:** 672
Electric dipole transition
defined **A10:** 672
Electric dipole transition moment
in IR spectroscopy **A10:** 111
Electric discharge machine (EDM)
for cutting cross sections of material for TEM studies **A18:** 381
Electric discharge machining **A10:** 690
Electric discharge machining (EDM) *See also* Machining
as a sectioning method **A9:** 26-27
abbreviation for **A11:** 796
defined **A9:** 6, 26 **EM2:** 16
ductile and brittle fractures from **A11:** 91-92
effects on fatigue strength **A11:** 122
primer cup plate spalling from **A11:** 566-567
tool and die failures from **A11:** 566-567
torsional fatigue failure from **A11:** 474-475
Electric field
as field corrosion **A10:** 587
oscillating, wave theory of **A10:** 82
and stress, in field ion microscopy **A10:** 587
Electric field effect *See* Stark effect
Electric field strength **A10:** 685
copper-clad E-glass laminates **EL1:** 535
Electric fields
microwave inspection **A17:** 202-203
scanning electron microscopy study of specimens **A9:** 97
Electric flux density **A10:** 685
Electric furnace
billet casting plant **A15:** 310
defined **A15:** 5
Electric furnace steelmaking **A1:** 111
Electric furnaces, for steelmaking **M1:** 109-111, 113, 114
Electric glass *See* E-glass
Electric heating
of pouring vessels **A15:** 298, 499
Electric motor ball bearings
pitting failure of **A11:** 495, 497
Electric motor brushes **A7:** 634-636
Electric overhead traveling cranes
for open-die forging **A14:** 63
Electric overstress (OES)
as failure mechanism **EL1:** 1013-1014
in semiconductor chips **EL1:** 966
Electric potential
SI derived unit and symbol for **A10:** 685
SI unit/symbol for **A8:** 721
to monitor crack extension in corrosive environments **A8:** 428
Electric potential method apparatus **A8:** 388-389
calibration curves **A8:** 386
for crack growth in aqueous solutions **A8:** 417
crack initiation **A8:** 389-390
for crack length in acidified chloride solutions **A8:** 420
crack monitoring **A8:** 382, 386-391, 412
fatigue crack closure **A8:** 391
limitations with vacuum and gaseous fatigue tests **A8:** 412
and load **A8:** 390
measurement accuracy **A8:** 386-387
for nonconducting materials **A8:** 391
optimization parameters **A8:** 387-388
small crack behavior **A8:** 390-391
Electric power industry
fossil fuel power system corrosion **A13:** 985
Electric power industry applications *See also* Electronic industry applications; Power industry applications
of niobium-titanium superconducting materials **A2:** 1057
Electric Power Research institute **A8:** 721
run-arrest cleavage fracture experiments **A8:** 285
Electric Power Research Institute (EPRI) **EM4:** 716
data bases **A11:** 54
Electric resistance
four-arm strain gage bridge **A8:** 220
furnace, for torsion testing **A8:** 159
SI derived unit and symbol for **A10:** 685
SI unit/symbol for **A8:** 721
Electric resistance welding
for steel tubular products **A1:** 327-328 **M1:** 316
Electric resistance wire furnace-heated alloy steel die assembly **A7:** 505
Electric spark-activated hot pressing **A7:** 513
Electric welded wire fabric
for concrete reinforcement **M1:** 271
Electric-arc thermal spray coating **M5:** 365-366, 368
Electric-eye machines **A6:** 1169
Electric-motor housings, climinating cracksin ..... **A3:** 1 • 28
Electric-resistance welds
in pipe **A11:** 698
Electrical *See also* Design; Electrical design; Electrical design considerations; Electrical design methodologies; Electrical insulation; Electrical interconnection; Electrical modeling; Electrical performance testing; Electrical properties
abuse, electrostatic discharge as **EL1:** 966
applications, epoxy resin curing agents for **EL1:** 829
components, miniature, parylene applications **EL1:** 800
connectors, corrosion failure analysis of **EL1:** 1112
damage, integrated circuits **EL1:** 975-978
isolation, in optical interconnections **EL1:** 9
Electrical analog modeling **EL1:** 51
Electrical and electronic application **A7:** 624-645
of copper-based powder metals **A7:** 733
of stainless steels **A7:** 731
Electrical applications
allyls (DAP, DAIP) **EM2:** 226-227
aminos **EM2:** 230
of bulk molding compounds **EM1:** 162

Electrical applications (continued)
cast steels **M1:** 384
critical properties **EM2:** 458-459
elastomers **EM2:** 588
phenolics **EM2:** 243
polyamides (PA) **EM2:** 125
polyarylates (PAR) **EM2:** 138-139
polybenzimidazoles (PBI) **EM2:** 147
polybutylene terephthalates (PBT) **EM2:** 153
polyether sulfones (PES, PESV) **EM2:** 159
polyether-imides (PEI) **EM2:** 156
polyethylene terephthalates (PET) **EM2:** 172
polyphenylene ether blends (PPE PPO) **EM2:** 183
polyphenylene sulfides (PPS) **EM2:** 186
polysulfones (PSU) **EM2:** 200
thermoplastic fluoropolymers **EM2:** 118
thermoplastic polyimides (TPI) **EM2:** 177
for thermoplastics **EM2:** 591
thermosetting plastics **EM2:** 589
unsaturated polyesters **EM2:** 246
Electrical breakdown *See also* Electrical properties
factors influencing **EM2:** 465-467
of plastic materials **EM2:** 464-465
Electrical cables
in space boosters/satellites **A13:** 1103
Electrical ceramics
mixing operations **EM4:** 98
thermal expansion coefficient **A6:** 907
Electrical charge
defined **EL1:** 89-92
Electrical circuits
for electropolishing **A9:** 49-50
Electrical circuits, residential
EPMA failure analysis **A10:** 531
Electrical conduction *See also* Conductivity; Electrical conductivity
test methods **EM2:** 462
Electrical conductivity *See also* Conduction; Conductivity; Conductors; Electrical properties; Electrical resistivity; Superconducting materials; Superconductivity **A7:** 585 **EM3:** 33
of additives **EM2:** 474
aluminum **A2:** 3, 9
aluminum casting alloys **A2:** 153-177
of aluminum P/M parts **A7:** 742
at cryogenic temperatures, beryllium-copper alloys **A2:** 420
atomic oxygen effect in silver **A12:** 481
beryllium-copper alloys **A2:** 406
cast copper alloys **A2:** 356-391
of chromium carbide-based cermets **A7:** 806
of composite laminates **EM1:** 36
copper and copper alloys **A2:** 216, 219
of copper casting alloys **A2:** 349
of copper powders **A7:** 106, 115, 116, 712
of copper/copper alloys **A13:** 610
and corrosion rates **A11:** 199
as design consideration **EM2:** 1
of dispersion-strengthened copper **A7:** 712
in eddy current inspection **A17:** 164, 167
effect of density on **A7:** 736
electrical contact materials **A2:** 840
and electronic phenomena **EL1:** 93-96
in electrozone size analysis **A7:** 221
equivalent physical quantities **EM1:** 191
factors influencing **EM2:** 473-475
and green strength **A7:** 304
heat-treated copper casting alloys **A2:** 355
improvement with pressure **A7:** 608, 610
of lead frame alloys **EL1:** 490
lubricating effects **A7:** 192, 193
of make-break arcing contacts **A2:** 841
measurement **A7:** 485 **EM1:** 286
mechanism, fillers/reinforcements **EM2:** 469-473
polymer die attach **EL1:** 218
properties **EL1:** 89-90
silver **A2:** 699
silver contact alloys **A2:** 843
Electrical conductivity (continued)
of specialty polymers **A7:** 606
and strength, beryllium-copper alloys **A2:** 417
testing **EM2:** 585-586
wrought aluminum and aluminum alloys **A2:** 62-122
Electrical conductivity, and thermal conductivity of ductile iron
relationship between **A1:** 50-51, 52
Electrical conductor applications
steel wire for **M1:** 264-265
Electrical conductors
classification **A13:** 65
copper and copper alloys **A14:** 821
Electrical connectors **A13:** 1121-1122
beryllium-copper alloys **A2:** 416-417
Electrical contact materials *See also* Electrical contacts **A9:** 550-564
aluminum **M3:** 672
aluminum contacts **A2:** 849-850
applications **M3:** 665-672, 676-679, 686-694
availability **M3:** 690-691, 695
availability, contact alloys **A2:** 866-868
brazed assembly **A9:** 557
composite manufacturing methods **A2:** 856-858
composite material contacts **A2:** 850, 856
composite materials **M3:** 665, 672, 681, 682
contact materials, recommended **A2:** 861-866
copper contact alloys **A2:** 842-843
copper metals **M3:** 665-666
cost **M3:** 692
cost(s) **A2:** 868
damage from arcing **A9:** 563-564
defined **A2:** 840
electrolytic etching of **A9:** 551
etchants for **A9:** 551
etching procedures for **A9:** 551
failure modes of make-break contacts **A2:** 840-841
gold contact alloys **A2:** 845-846
gold metals **M3:** 668-669
life tests **A2:** 858-861
metal-graphite materials **M3:** 664, 665, 676-679, 680
microscopic examination of **A9:** 550-551
microstructures of **A9:** 551-552
molybdenum **M3:** 671-672, 673, 674, 675
molybdenum contacts **A2:** 848-849
palladium **A13:** 805
palladium metals **M3:** 669-671, 672
platinum metals **M3:** 669-671
polarized contacts, life of **A2:** 860
precious metal overlays **A2:** 848 **M3:** 671
precious metals contacts, platinum group **A2:** 846-848
preparation of specimens **A9:** 550-551
property requirements, for make-break arcing contacts **A2:** 841
selection **M3:** 662-663, 686-694
selection criteria **A2:** 840
silver contact alloys **A2:** 843-845
silver metals **M3:** 666-668
silver-cadmium oxide **M3:** 664, 665, 673, 675, 676-679, 680, 681, 684, 685, 691
sliding contacts **A2:** 841-842
tungsten **M3:** 671-672, 673, 674, 675
tungsten contacts **A2:** 848-849
Electrical contacts *See also* Electrical contact materials
contact force, factors affecting **A2:** 840
defined **A2:** 840
as electrical/magnetic applications **A7:** 630-634
failure modes **M3:** 663, 669, 682-686, 687
as infiltration products **A7:** 560
life tests **M3:** 682-686, 687
make-and-break contacts **M3:** 663-664, 676-679, 690-691
powders used **A7:** 573
precious metal overlays **M3:** 671
Electrical contacts (continued)
selection of materials **M3:** 662-663, 686-691, 692-694
sliding contacts **M3:** 664-665, 688-689
XPS analysis of surface films on **A10:** 578-579
Electrical contacts, friction and wear of **A18:** 682-684
atmospheric effects **A18:** 684
dusting **A18:** 684
brush materials **A18:** 684
"reading" or damage inspection **A18:** 684
circuit breaking **A18:** 683-684
contact resistance **A18:** 682-683
burnout **A18:** 683
insulating films effect (fritting) **A18:** 682-683
mechanical factors **A18:** 683
thermoelastic mounding **A18:** 683
tunneling **A18:** 683
development **A18:** 682
fretting wear **A18:** 243
Electrical coppers *See also* Copper; Copper alloys
applications, properties, types **A2:** 223, 230, 234
Electrical corrosion *See also* Corrosion
of aircraft accelerometers **EL1:** 1111-1112
of antenna marker beacons **EL1:** 1108
of circuit breakers **EL1:** 1108
of disk recorder heads **EL1:** 1112
of electrical connectors **EL1:** 1112
of fuses **EL1:** 1110-1111
of hybrid microcircuits **EL1:** 1115
of integrated circuits **EL1:** 1114-1115
of klystron electron tubes **EL1:** 1115-1116
of microwave detectors **EL1:** 1112-1113
of nickel/boron-plated panels **EL1:** 1113-1114
of printed wiring boards **EL1:** 1109-1110
of steering potentiometers **EL1:** 1111
of stepper motors **EL1:** 1111
of tin whiskers **EL1:** 1116
of vacuum tubes **EL1:** 1110
Electrical current
use in integrated circuit failure analysis **A11:** 767
Electrical damage
integrated circuits **EL1:** 975-978
Electrical design *See* also Design; Electrical design considerations; Electrical design methodologies
importance **EL1:** 1
of level 1 packages **EL1:** 402-403
Electrical design considerations *See also* Electrical design
device trends **EL1:** 416
flexible printed boards **EL1:** 586-588
package parasitics and effects **EL1:** 416-419
Electrical design methodologies *See also* Design
direct current (dc) design requirements **EL1:** 25-26
guideline synopsis **EL1:** 42-43
for high-performance systems **EL1:** 28-42
for low-end systems **EL1:** 26-28
performance range classifications **EL1:** 25
Electrical differential transformer extensometers **A8:** 619
Electrical discharge breakdowns
defined **EM2:** 465
Electrical discharge grinding (EDG) **A16:** 509, 565-567
applications **A16:** 566
automatic EDG machines **A16:** 565
compared to ECDG **A16:** 548, 550
diamond **A16:** 107-108
dielectric fluids **A16:** 565, 566
equipment and operation **A16:** 565
flushing **A16:** 566
process characteristics **A16:** 565-567
surface finish **A16:** 565-566
tool life **A16:** 567
Electrical discharge machines (EDM)
notches **A17:** 137-138

Electrical discharge machining *See also* Machining
notch root radius **A8:** 382
refractory metals and alloys **A2:** 561
rhenium **A2:** 561-562
Electrical discharge machining (EDM) **A16:** 19-22, 26-27, 33-34, 106, 509, 518, 557-564 **EM4:** 313, 314, 371-376
advantages **A16:** 557
application to advanced ceramics **EM4:** 375-376
changes in hardness **EM4:** 375
contamination of the surface **EM4:** 375
grain size effects **EM4:** 376
material removal rates achieved **EM4:** 376
strength **EM4:** 375-376
surface residual stresses **EM4:** 375
applications **A16:** 557, 561
carbon deposit produced **A16:** 25
cemented carbides **A18:** 796
CNC vertical EDM **A16:** 560-561, 562
compared to ECG **A16:** 542
compared to EDG **A16:** 565
compared to electron beam machining **A16:** 571
compared to electrostream and capillary drilling **A16:** 551
compared to end milling and electrochemical machining **A16:** 539
compared to laser beam machining **A16:** 574
compared to shaped tube electrolytic machining **A16:** 554
in conjunction with chemical milling **A16:** 586
diamond **A16:** 108
die casting **A16:** 560
dielectric fluids **A16:** 560, 563
drilling machines **A16:** 559, 560
effect of operating conditions on material removal rate and surface quality **EM4:** 372-374
dielectric fluid **EM4:** 373
filtration system **EM4:** 373-374
various machine parameters **EM4:** 374
electrode manufacturing processes **A16:** 559-560
electrodes
electroforming **A16:** 560
erosion process **EM4:** 371-372
fatigue strength **A16:** 31
flushing **A16:** 560, 561, 562, 563
hole drilling **A16:** 561
machine types **EM4:** 371
die-sinking machine (ram-type, plunge, or vertical erosion) ... **EM4:** 371, 372, 373
wire-cutting machines **EM4:** 371, 372, 373, 374
machines, NC implemented **A16:** 613
mechanisms involved **EM4:** 374-375
melting and evaporation **EM4:** 374
thermal spalling **EM4:** 374-375
metal oxide field effect transistor (MOS-FET) devices **A16:** 558
method of operation **A16:** 557-558
MMCs **A16:** 896
no-wear **A16:** 560
orbital abrading **A16:** 559, 561
and PCBN **A16:** 111
power supplies **A16:** 558
refractory metals **A16:** 859, 865, 868
safety of method **A16:** 564
sequential process steps **EM4:** 373
and shot peening **A16:** 35
stainless steels **A16:** 706
surface **A16:** 558-559, 563
surface alterations produced **A16:** 24, 25, 31, 34, 35
surface finish **A16:** 558, 561, 563
surface integrity **A16:** 28, 559
surface roughness arithmetic average extremes **A18:** 340
Ti alloys **A16:** 853
tooling systems **A16:** 560
traveling-wire EDM **A16:** 557, 559, 561-564
use expanding for ceramics **EM4:** 376
wire-EDM metal deposits and disadvantages **A16:** 563
wire-EDM power supplies **A16:** 563
zirconium **A16:** 852
Electrical discharge machining electrodes
powders used **A7:** 573
Electrical discharge machining preforms
powders used **A7:** 573
Electrical discharge treatment **EM3:** 42
Electrical discharge wire cutting (EDWC) **A16:** 509
Electrical discharges, static
ball bearing pitting failure from **A11:** 495, 497
Electrical dissipation factor *See also* Dissipation factor
defined **EM1:** 10 **EM2:** 16
Electrical equipment
precoated steel sheet for **M1:** 173
Electrical equivalent circuit model **A13:** 216
Electrical erosion
in sliding bearings **A11:** 486
Electrical evaluation *See also* Electrical testing; Testing
introduction **EL1:** 867-868
Electrical generator rotors
fretting wear **A18:** 243
Electrical heating **A18:** 683
Electrical industry applications
aluminum and aluminum alloys **A2:** 12-13
copper and copper alloys **A2:** 239-240
of precious metals **A2:** 693
Electrical insulation
as application **EM2:** 1
by rigid epoxies **EL1:** 810
of substrates **EL1:** 104
Electrical interconnection *See also* Interconnection(s)
direct chip interconnect **EL1:** 231-232
laser pantography **EL1:** 232-233
pressure contacts **EL1:** 232-233
tape automated bonding (TAB) **EL1:** 228-231
thermocompression bonding **EL1:** 224-225
thermosonic bonding **EL1:** 225-226
ultrasonic bonding **EL1:** 225
wire bond reliability and testing **EL1:** 226-228
as wire bonding **EL1:** 224-236
Electrical ion detection method **A10:** 144
Electrical isolation **A13:** 5, 87
Electrical laminates
circuit board materials, composition **EM1:** 73
epoxy, formulations/properties **EM1:** 76
epoxy resin **EM1:** 74-75
Electrical limit switch
fatigue testing machines **A8:** 368
Electrical modeling *See also* Modeling; Models; Simulation
of assembly/packaging **EL1:** 447-448
of VHSIC SMT devices **EL1:** 77-81
Electrical noise sources
types **A17:** 285
Electrical or conductor applications
wire for **A1:** 283
Electrical overstress failures
integrated circuits **A11:** 786-788
Electrical performance testing *See also* Electrical testing; Testing
of discrete devices **EL1:** 946-951
of monolithic microwave integrated circuits (MMICS) **EL1:** 951-952
Electrical pitting
in bearing failures **A11:** 493, 494, 496
defined **A18:** 8
Electrical plasma atomizers
atomic absorption spectrometry **A10:** 53-54
Electrical porcelain
annual sales **EM4:** 936
applications **EM4:** 935
competitive materials **EM4:** 936
development of the industry **EM4:** 936
physical properties **EM4:** 934
Electrical properties *See also* Conductivity; Electrical breakdown; Electrical conductivity; Electrical resistivity; Electrical testing; Materials and electronic phenomena; Mechanical properties; Properties; Thermal properties **EM2:** 460-480 **EM3:** 428-439
acrylonitrile-butadiene-styrenes (ABS) **EM2:** 113
actinide metals **A2:** 1189-1198
aging **EM3:** 438
discharge in gas-filled voids **EM3:** 438
electrical treeing **EM3:** 438-439
tracking **EM3:** 439
water treeing **EM3:** 439
allyls (DAP, DAIP) **EM2:** 8
Electrical properties (continued)
aluminum casting alloys **A2:** 153-177
ASTM test methods **EM2:** 334
BPA fumarate polyester **EM2:** 250
breakdown **EM3:** 437-439
electromechanical **EM3:** 438
flashover (surface) **EM3:** 438
short-term **EM3:** 438
thermal **EM3:** 438
capacitive vs transmission line environments **EL1:** 601
cast copper alloys **A2:** 356-391
cast steels **M1:** 393-394, 399
ceramics **EL1:** 335-336
commercially pure tin **A2:** 518-519
conduction **EM3:** 433-436
conduction anisotropy **EM3:** 436-437
conduction mechanisms
Hall effect **EM3:** 437
photoconduction **EM3:** 437
thermally stimulated depolarization (TSD) **EM3:** 437
conductive plastic materials **EM2:** 467-478
of conformal coatings **EL1:** 822
copper-clad E-glass laminates **EL1:** 535
cross talk **EL1:** 603-604
crystal structure effects **EL1:** 93
data sheet, typical **EM2:** 409
as design parameters **EL1:** 517-523
dielectric losses **EL1:** 603
dielectric properties **EM3:** 428-432
bridge methods **EM3:** 431
low-frequency measurements **EM3:** 429-432
radio frequency and microwave measurements **EM3:** 431-432
resonant cavities **EM3:** 432
resonant circuits **EM3:** 431
slotted transmission lines **EM3:** 432
time-domain measurements **EM3:** 430-431
time-domain reflectometry (TDR) **EM3:** 432
diffusion **EM3:** 436
ductile iron **A15:** 663 **M1:** 53, 55
elastomers/rubbers **EM2:** 589
electrical resistance alloys **A2:** 836-839
electromagnetic interference (EMI) shielding **EM2:** 476-478
of engineering plastics **EM2:** 73
epoxies **EM2:** 241
epoxy resins **EL1:** 831
fiberglass-polyester composites **EM2:** 249
insulating plastics **EM2:** 460 467
of glass fibers **EM1:** 46
gold and gold alloys **A2:** 705
of gray iron **M1:** 28, 31, 53
impedance models **EL1:** 601-603
impedance time **EL1:** 601
insulating, structural ceramics **A2:** 1019, 1021-1024
interfacial effects **EM3:** 436
of Invar **A2:** 891
lead frame materials **EL1:** 488
liquid crystal polymers (LCP) **EM2:** 181
maraging steels **M1:** 451
of materials, basic **EL1:** 89-90
measured **EM2:** 613
metals vs plastics **EM2:** 77-78
niobium alloys **A2:** 567-571
nonlinear bulk conduction **EM3:** 436-437
on supplier data sheets **EM2:** 641
palladium and palladium alloy **A2:** 715-716
para-aramid fibers **EM1:** 56
parylene coatings **EL1:** 793-794
permittivity **EL1:** 597-601
phenolics **EM2:** 245
plastics, and polymer structure **EM2:** 462-464
of plastics, characteristics **EM2:** 588-590
platinum and platinum alloys **A2:** 708-714
polyamide-imides (PAI) **EM2:** 133
polyamides (PA) **EM2:** 126
polyaryl sulfones (PAS) **EM2:** 145-146
polyarylates (PAR) **EM2:** 138, 140
of polyester resins **EM1:** 95
polyether sulfones (PES, PESV) **EM2:** 160
polyether-imides (PEI) **EM2:** 158
polyethylene terephthalates (PET) **EM2:** 173
of polyimides **EL1:** 769
polymeric substrates **EL1:** 338

Electrical properties (continued)
of polymers ... **EM2:** 62
polyphenylene sulfides (PPS) ... **EM2:** 189
polysulfones (PSU) ... **EM2:** 201
pure cobalt ... **A2:** 447
pure metals ... **A2:** 1100-1178
rare earth metals ... **A2:** 1178-1189
resistive losses (skin effect) ... **EL1:** 603
signal attenuation ... **EL1:** 603
silicone conformal coatings ... **EL1:** 822
silicones (SI) ... **EM2:** 267
silver and silver alloys ... **A2:** 699-704
static elimination testing ... **EM2:** 475-476
of steel castings ... **A1:** 374
structure sensitive and structure insensitive ... **A2:** 761-762
styrene-acrylonitriles (SAN, OSA, ASA) ... **EM2:** 215
terminology ... **EM2:** 460-461, 590-593
test methods ... **EM2:** 461-462
tests for ... **EM2:** 78, 461-462, 475-478
of thermoplastic materials ... **EM2:** 592
thermoplastic polyimides (TPI) ... **EM2:** 177
thermoplastic resins ... **EM2:** 618-619
of thermosetting molding materials ... **EM2:** 590
thermosetting resins ... **EM2:** 223
tin solders ... **A2:** 521-522
transient techniques ... **EM3:** 435-436
alternating current impedance measurements ... **EM3:** 435-436
voltage rate-of-change ... **EM3:** 435
tungsten and tungsten alloys ... **A2:** 580-581
ultrahigh molecular weight polyethylenes (UHMWPE) ... **EM2:** 169
unsaturated polyesters ... **EM2:** 247-248, 250
urethane coatings ... **EL1:** 776
versus organic-solvent-base adhesives properties ... **EM3:** 86
wrought aluminum and aluminum alloys ... **A2:** 62-122
wrought copper and copper alloys ... **A2:** 265-345
wrought magnesium alloys ... **A2:** 480-516
of zinc alloys ... **A2:** 532-542

Electrical PWB configurations
typical ... **EL1:** 597-598

Electrical resistance
of aluminum ... **A15:** 756
of cast alloy steels ... **A15:** 32
of chromate conversion coatings ... **A13:** 392
and corrosion rate ... **A11:** 198
of gray iron ... **A15:** 645
manganese cast steel ... **A15:** 32
platinum group metals ... **A2:** 846
probes ... **A13:** 199-200, 1250
relative rating of brazing process heating method ... **A6:** 120
soil corrosion relationship to ... **A13:** 762
in superconductivity ... **A2:** 1030
testing, as on-stream monitoring ... **A13:** 323

Electrical resistance alloys *See also* Electrical resistance alloys, specific types; Heating alloys; Thermostat metals ... **M3:** 640-661
atmospheres ... **A2:** 833-835
ballast resistors ... **M3:** 642
composition ... **M3:** 641
Constantan alloy ... **A2:** 825
constantans ... **M3:** 641, 643, 644
copper-manganese-nickel resistance alloys (manganins) ... **A2:** 825
copper-nickel resistance alloys ... **A2:** 823-825
electrical properties ... **M3:** 641
heating alloys ... **A2:** 827-829
iron-chromium-aluminum alloys ... **A2:** 828-829
manganins ... **M3:** 641, 642, 643-644
mechanical and physical properties ... **M3:** 641
nickel alloy ... **A2:** 433
nickel-chromium alloys ... **A2:** 825-826, 828
nickel-chromium-iron alloys ... **A2:** 828
nonmetallic materials ... **A2:** 829

Electrical resistance alloys (continued)
open resistance heaters, design of ... **A2:** 829-830
open resistance heaters, fabrication of ... **A2:** 830-831
operating temperatures, furnace ... **M3:** 647
properties ... **A2:** 823
properties of ... **A2:** 835-839
pure metals ... **A2:** 829
radio alloys, properties ... **A2:** 823
reference resistors ... **M3:** 642
requirements ... **M3:** 640
resistance alloys ... **A2:** 822-826
resistance thermometers ... **M3:** 642
resistors ... **A2:** 822-824
resistors, precision ... **M3:** 641-642
semiprecision ... **M3:** 641, 645
service life of heating elements ... **A2:** 831-833
sheathed heaters ... **A2:** 831
stability, resistors ... **M3:** 642-643
tensile strength ... **M3:** 649
thermostat metals ... **A2:** 826-827
types ... **A2:** 442

Electrical resistance alloys, specific types
22Cr-5.3Al-Fe balance, properties and applications ... **A2:** 839
35Ni-43Fe-20Cr, properties and applications ... **A2:** 838
35Ni-45Fe-20Cr
composition ... **M3:** 659
property data ... **M3:** 641, 646, 659-660
37Ni-21Cr-2Si-40Fe, properties and applications ... **A2:** 837-838
60Ni-22Fe-16Cr, properties and applications ... **A2:** 836-837
60Ni-24Fe-16Cr
composition ... **M3:** 658
property data ... **M3:** 641, 658-659
70Ni-30Cr, properties and applications ... **A2:** 836
80Ni-20Cr
composition ... **M3:** 657
life of heating elements ... **M3:** 652-654, 656
property data ... **M3:** 641, 657-658
variation of resistance with temperature ... **M3:** 648
80Ni-20Cr, properties and applications ... **A2:** 835-836
constantan
composition ... **M3:** 660
property data ... **M3:** 641, 660-661
Constantan (45Ni-55Cu), properties and applications ... **A2:** 838-839
molybdenum disilicide
composition ... **M3:** 661
property data ... **M3:** 646, 661
molybdenum disilicide ($MoSi_2$), properties and applications ... **A2:** 839

Electrical resistance gages
for elevated/low temperature tension testing ... **A8:** 35

Electrical resistance, in spot welding machine and split die assembly
See also Electrical resistivity ... **A7:** 506

Electrical resistance metals
dislocation structure in ... **A10:** 358

Electrical resistivity *See also* Electrical properties; Resistivity ... **A6:** 265
alloy cast iron ... **M1:** 88
aluminum ... **A2:** 3
of aluminum oxide-containing cermets ... **A7:** 804
carbon fibers ... **EM1:** 52
cast copper alloys ... **A2:** 357-391
of ceramics ... **EL1:** 336
changes due to recovery ... **A9:** 693
of chromium carbide-based cermets ... **A7:** 806
of copper, changes during isothermal recovery ... **A9:** 693
defined ... **A13:** 5 **EM1:** 359
of ductile iron ... **A1:** 51, 53
of high-purity nickel strip ... **A7:** 403

Electrical resistivity (continued)
magnetically soft materials ... **A2:** 761
malleable iron ... **M1:** 67
measurement ... **A2:** 1096
of metal borides and boride-based cermets ... **A7:** 811
of polyester resins ... **EM1:** 93
steels, selected grades ... **M1:** 150-151
testing ... **EM2:** 475
versus coefficient of friction for *n*-type silicon after abrasion and doping ... **A18:** 689
wrought aluminum and aluminum alloys ... **A2:** 62-122
zero, in superconductivity ... **A2:** 1030

Electrical resistivity testing
of powder metallurgy parts ... **A17:** 541-545

Electrical sensing
to analyze ceramic powder particle size ... **EM4:** 67

Electrical sheet steel ... **A9:** 196
decarburized ... **A9:** 196
internal oxidation ... **A9:** 196

Electrical steel sheet
blanking and piercing of ... **A14:** 476-482
burr height ... **A14:** 481
camber and flatness ... **A14:** 480-481
composition effects ... **A14:** 480
core plating ... **A14:** 482
lubrication ... **A14:** 481-482
materials ... **A14:** 476-477
presses and dies, for blanking/piercing ... **A14:** 477-479
stock thickness ... **A14:** 479-480

Electrical steel sheet and strip
properties and applications ... **A2:** 769

Electrical steels *See also* Magnetic materials; Magnetic materials, specific types; Silicon steels
blanking, die materials for ... **M3:** 485, 487
chemical polishing solutions ... **A9:** 533
M-36, sheet, continuously cold rolled fibering ... **A9:** 687
magnetic contrast ... **A9:** 536-537
microstructures ... **A9:** 537-538
specimen preparation ... **A9:** 531-532

Electrical surface wiring industry
amino molding compounds in ... **EM2:** 230

Electrical system pipe-type cables ... **A13:** 1292

Electrical tape
used for unmounted electropolishing specimens ... **A9:** 49

Electrical tester configuration
typical ... **EL1:** 566-567

Electrical testing *See also* Electrical evaluation; Electrical performance testing; Electrical properties
and automatic optical inspection ... **EL1:** 565-571
for defect verification ... **EL1:** 873-874
electrical properties, of plastics ... **EM2:** 588-590
electrical tests ... **EM2:** 581-588
in failure verification/fault isolation ... **EL1:** 1058-1060
high/low-temperature ... **EL1:** 1060-1061
of highly integrated systems ... **EL1:** 372-373
industry trends ... **EL1:** 565
methodology ... **EL1:** 565-566
methods ... **EL1:** 372-373, 377-378
of multilayer boards (MLBS) ... **EL1:** 566-568
reliability testing, added ... **EL1:** 567-568
system-level issues ... **EL1:** 373-377
terminology ... **EM2:** 590-593

Electrical vacuum coating ... **M5:** 394-399, 402-403, 408-409

Electrical waveform *See also* Waveform
effect, radiography ... **A17:** 304-305
effect, x-ray tubes ... **A17:** 304-305
full-wave-rectification, radiography ... **A17:** 304-305
half-wave rectification, radiography ... **A17:** 304
Villard-circuit equipment ... **A17:** 305

Electrical waveguides
of carbon fiber reinforced plastics ... **EM1:** 36

Electrical-grade glass fibers *See also* E-glass
application/composition **EM1:** 107
Electrical-grade plastics
compared **EM2:** 228
Electrical-resistance heating tape
for optical holographic interferometry **A17:** 409
Electrical-simulation software
types **EL1:** 419
Electrically conductive adhesives **EM3:** 33, 572-573
Electrically conductive inks
polymeric thick-film systems **EL1:** 346
Electrically conductive mounts
for aluminum alloys **A9:** 352
Electrically erasable programmable read-only memories (EEPROMS)
development **EL1:** 160
wafer-scale integration **EL1:** 8
Electrically heated (submerged electrodes) furnace
dip brazing **A6:** 337
Electrically induced heating
for thermal inspection **A17:** 398
Electrically nonconducting material
electrical potential methods for **A8:** 391
Electricity
conversion factors **A8:** 722 **A10:** 686
quantity of, SI derived unit and symbol for **A10:** 685
Electrification time
defined **EM2:** 592
Electro-optic ceramics and devices ... **EM4:** 1124-1130
applications **EM4:** 1128-1129
ceramics versus single crystals **EM4:** 1124
electro-optic ceramics as ferroelectrics **EM4:** 1124-1125
materials **EM4:** 1125-1128, 1129
properties **EM4:** 1124
thin films **EM4:** 1129-1130
Electro-optical interconnection conversions **EL1:** 10
Electro-osmosis **EM4:** 74
Electro-spark deposition (ESD) **A18:** 645
Electroanalytical probes
for process control **A10:** 199
Electrochemical
effects, on corrosion-fatigue **A11:** 256
etching, and circle grid analysis **A8:** 567
factors, influence in SCC **A8:** 499
machining, effect on fatigue strength **A11:** 122
magnesium separation apparatus **A15:** 82
refining, of aluminum melts **A15:** 80-81
theory, and stress-corrosion cracking **A11:** 203
Electrochemical admittance *See also* Electrochemical impedance
defined **A13:** 5
Electrochemical analysis **A10:** 181-211
capabilities, compared with classical wet analytical chemistry **A10:** 161
controlled-potential coulometry **A10:** 202
of inorganic anions and cations determined by ion chromatography **A10:** 663
voltammetry and polarography **A10:** 202
Electrochemical and corrosion effects on adhesive joints **EM3:** 628-636
effect of corrosion and adhesive joint failure **EM3:** 633
effect of dissimilar metals in contact **EM3:** 632
effect of high cathodic potentials **EM3:** 629-631
effect of impressed current **EM3:** 631-632
effect of mechanical strain **EM3:** 632-633
exposure in electrochemically inert conditions **EM3:** 628-629
increasing resistance to cathodic bond failure in adhesive joint applications **EM3:** 635-636
seawater exposure **EM3:** 629-630
theoretical models and failure mechanisms **EM3:** 633-635
Electrochemical cell **A13:** 5, 20
Electrochemical cleaning *See* specific processes by name
Electrochemical corrosion
defined **A13:** 5
testing **A13:** 212-220
zinc **M2:** 650, 652
Electrochemical corrosion testing
coating systems **A13:** 430
Electrochemical corrosion testing (continued)
of crevice corrosion **A13:** 308-309
current data, conversion to corrosion rates **A13:** 213
of environmental cracking **A13:** 217-219
fundamentals **A13:** 212
of galvanic corrosion **A13:** 215-216
limitations **A13:** 220
of localized corrosion **A13:** 216-217
polarization, conducting **A13:** 212-213
of protective coatings and films **A13:** 219-220
of SCC **A13:** 264-265
stainless steels **A13:** 564
thermodynamics of aqueous corrosion **A13:** 18-28
for uniform corrosion **A13:** 213-215, 229
Electrochemical differences, between different phases of a specimen
effects of **A9:** 47
Electrochemical discharge grinding (ECDG) **A16:** 509, 548-550
applications **A16:** 550
electrolytes **A16:** 548-549
equipment **A16:** 548
form grinding **A16:** 549
process characteristics **A16:** 548-550
profile grinding **A16:** 549
surface (plunge) grinding **A16:** 549
tool life **A16:** 549
Electrochemical equivalent
defined **A13:** 5
Electrochemical etching **A9:** 57, 60-61
defined **A9:** 6
Electrochemical grinding (ECG) **A16:** 509, 543-544
advantages and disadvantages **A16:** 546
applications **A16:** 546-547
carbon and alloy steels **A16:** 677
compared to EDG **A16:** 548, 550
compared to other machining processes **A16:** 542
current densities **A16:** 543
cylindrical grinding method **A16:** 546
electrolytes **A16:** 542, 543, 544-545
equipment **A16:** 544-546
face (plunge) grinding method **A16:** 546
feed rates **A16:** 543, 544
form grinding method **A16:** 545, 546
internal grinding method **A16:** 546
machines **A16:** 546
methods **A16:** 546
oxygen and hydrogen generation **A16:** 544
power supplies **A16:** 545-546
process characteristics **A16:** 542
surface finish **A16:** 543, 545
surface grinding method **A16:** 543, 546
surface integrity **A16:** 28
tool life **A16:** 546
wheels **A16:** 545
Electrochemical impedance *See also* Electrochemical admittance **A13:** 5, 215, 220
Electrochemical interactions between minerals and grinding media **A18:** 271
Electrochemical ion detector **A10:** 661
Electrochemical machining **EM4:** 313
refractory metals and alloys **A2:** 561
Electrochemical machining (ECM) **A16:** 14, 19, 27, 34, 509, 533-541
accuracy **A16:** 539
applications **A16:** 539-540
compared to ECG **A16:** 542
compared to shaped tube electrolytic machining **A16:** 554
computer numerical control (CNC) **A16:** 533-534
computer-aided design manufacture (CAD/CAM) **A16:** 538
effect on workpiece fatigue strength **A16:** 31
electrolytes **A16:** 533-536
equipment **A16:** 533
frontal operation **A16:** 26
insulation **A16:** 538
multiaxis ECM machines **A16:** 540
nonsludging electrolytes **A16:** 535
process control **A16:** 534
refractory metals **A16:** 868
removal rates **A16:** 534
shot peening **A16:** 35
single-axis machines **A16:** 533
sludging electrolytes **A16:** 535
Electrochemical machining (ECM) (continued)
stainless steels **A16:** 706
strong acid electrolytes **A16:** 535
surface alterations produced **A16:** 24-35
surface finish **A16:** 534, 535, 536, 537, 538, 539
Ti alloys **A16:** 846, 852
tools (cathodes) **A16:** 536-538
vertical-ram ECM machines **A16:** 541
Electrochemical machining electrodes
powders used **A7:** 573
Electrochemical metallizing *See also* Selective plating **M5:** 298
Electrochemical migration
thick-film **EL1:** 342
Electrochemical milling
rhenium **A2:** 562
Electrochemical polarization **A13:** 212-213, 271-274
high-strength steel cracking behavior from **A8:** 527, 529
for SCC testing for titanium alloys **A8:** 531
Electrochemical polishes, for surface preparation
SCC testing **A8:** 510
Electrochemical polishing
of tungsten **A9:** 45
Electrochemical potential
in acidified chloride environments **A8:** 419
and dissolved oxygen **A13:** 932
in stress-corrosion cracking **A8:** 499
used to measure anodic reactions during etching **A9:** 60
Electrochemical potentiostatic reactivation test **A13:** 564
Electrochemical principles
of potentiostatic etching **A9:** 144
Electrochemical properties
heat-treatable wrought aluminum alloy **A2:** 41
wrought aluminum, alloying effects **A2:** 44
Electrochemical protection
pulp bleach plants **A13:** 1195-1196
of zirconium/zirconium alloys **A13:** 718
Electrochemical reduction
silver powders **A7:** 148
Electrochemical series *See also* Electromotive force series
defined **A13:** 5
Electrochemical techniques
to analyze the bulk chemical composition of starting powders **EM4:** 72
Electrochemical testing **A13:** 87, 193
Electrochemical tests
for erosion **A11:** 174
stress-corrosion cracking **A8:** 532
thermodynamics and kinetics affecting SCC **A8:** 499
Electrochemical titration, electrometric titration and
compared **A10:** 202
Electrochemically accelerated mass finishing **M5:** 133
Electrochemistry
adapted for inorganic solids **A10:** 4-6
for inorganic liquids and solutions **A10:** 7
of intergranular corrosion **A13:** 123
Electrochromic device (ECD) **EM4:** 757
Electrocleaning *See* Electrolytic cleaning
Electrocoating
automotive **A13:** 1015
Electrocoating paint **M5:** 483-485, 506
Electrocorrosive wear
defined **A18:** 8
Electrode
double-junction **A8:** 417
saturated calomel **A8:** 416
silver/silver chloride **A8:** 416
Electrode burns, as flaw
defined **A17:** 562
Electrode coatings
constituents **A7:** 817
functions of **A7:** 816-817
metal powders for **A7:** 816-822
Electrode deposition **A7:** 71-72, 123
Electrode extension
definition **A6:** 1208 **M6:** 6
effect on weld attributes **A6:** 182
Electrode force
definition **A6:** 1208 **M6:** 6

Electrode holder
definition ... **A6:** 1208
Electrode holders
definitions ... **M6:** 6
Electrode holders for
flux cored arc welding ... **M6:** 98
resistance spot welding ... **M6:** 481-482
shielded metal arc welding ... **M6:** 79
Electrode indentation (resistance welding)
definition ... **A6:** 1208
Electrode lead
definition ... **A6:** 1208 **M6:** 6
Electrode melting rate (M_{rp}) ... **A6:** 28-29
Electrode potential ... **A10:** 197-198, 672
control ... **A8:** 407, 416
during corrosion fatigue tests ... **A8:** 405-407, 416-417
Luggin capillary ... **A8:** 416
measuring ... **A8:** 416-417
Electrode potentials *See also* Second-phase constituents in aluminum alloys; Solution potentials ... **M2:** 207
applied, effect on corrosion fatigue crack propagation ... **A13:** 300
in aqueous solutions ... **A13:** 17
cell potentials ... **A13:** 20-21
conventions ... **A13:** 21-22
conversion diagram ... **A13:** 22
defined ... **A13:** 5
effect, corrosion fatigue ... **A13:** 143, 298
electrode selection characteristics ... **A13:** 22-23
electromotive force series ... **A13:** 20-21
free energy ... **A13:** 19
measurement, with reference electrodes ... **A13:** 21-24
three-electrode system ... **A13:** 22
Electrode reaction *See also* Anodic reaction; Cathodic reaction
defined ... **A13:** 5
Electrode setback
definition ... **M6:** 6
Electrode skid
definition ... **M6:** 6
Electrode(s)
anodic protection ... **A13:** 464
charged interface/cation locations ... **A13:** 19
coatings ... **A7:** 816-822
composite metals for ... **A7:** 17
composition, titanium and titanium alloy castings ... **A2:** 642
consumable, in atomization process ... **A7:** 26
defined ... **A13:** 5
as infiltration products ... **A7:** 560
polarization, defined ... **A13:** 5
porous ... **A7:** 308
potentials ... **A7:** 140
powder-coated, performance of ... **A7:** 820-821
processes ... **A13:** 18-19
reaction ... **A13:** 5
reference ... **A13:** 21-24
resistance welding ... **A7:** 624-629
selection characteristics ... **A13:** 22-24
welding, classification ... **A7:** 819
welding, nickel-base ... **A2:** 443
welding, oxide dispersion-strengthened copper ... **A2:** 401
Electrodeburring ... **M5:** 308-309
Electrodeless discharge lamps
for AAS spectrometers ... **A10:** 50
Electrodeless mercury vapor lamps ... **EL1:** 864
Electrodeposited coating
to prevent edge rounding in samples ... **A9:** 45
Electrodeposited copper
chromic acid as an etchant for ... **A9:** 401
Electrodeposited copper foil
for conductors ... **EL1:** 114
Electrodeposited layers
on sleeve bearing liners ... **A9:** 567
Electrodeposited tin coatings
etching ... **A9:** 451
Electrodeposition *See also* Deposition
of amorphous materials/metallic glasses ... **A2:** 806-807
cermet ceramic coatings ... **M5:** 536
for copper foil production ... **EM3:** 591
defined ... **A13:** 5
electroplating as ... **A13:** 419
of interferences, UV/VIS absorption spectroscopy ... **A10:** 65-66
parameters, electroplated hard chromium ... **A13:** 871
porcelain enamel ... **M5:** 518
of porcelain enamels ... **A13:** 448
of precious metal overlays ... **A2:** 848
solid metal ... **A10:** 208
tin/tin alloys ... **A13:** 775
to apply interlayers for solid-state welding ... **A6:** 165, 166, 168, 169
Electrodeposition films
crystallographic texture in ... **A9:** 700
Electrodeposition primer process (E-coat) ... **EM3:** 554
Electrodes *See also* Ion-selective membrane electrodes; Potentiometric membrane electrodes
arc welding of aluminum alloys ... **A6:** 722-723
artificial graphite, resistance brazing ... **A6:** 341
boron tetrafluoride ... **A10:** 184
bromide ... **A10:** 184
calcium ... **A10:** 184
calcium/magnesium (water hardness) ... **A10:** 184
carbon steel ... **A6:** 657
submerged arc welding ... **A6:** 204, 205
carbon steel flux-cored wires ... **A6:** 188-189
carbon-graphite, resistance brazing ... **A6:** 340-341
cast iron, for shielded metal arc welding ... **A6:** 717
and cells, for electrogravimetry ... **A10:** 199-200
chloride ... **A10:** 184
chromium-copper, resistance brazing ... **A6:** 340
classifications for
flux-cored arc welding ... **A6:** 717
gas-metal arc welding ... **A6:** 719
shielded metal arc welding ... **A6:** 716, 718
shielded metal arc welding of cast irons ... **A6:** 721
coating formulations of selected SMAW electrodes ... **A6:** 59
composite flux-cored and metal cored ... **A6:** 659
copper ... **A10:** 184
copper-base, as cast iron welding consumable ... **A15:** 524
copper-based
for gas-metal arc welding ... **A6:** 719
for shielded metal arc welding ... **A6:** 718
copper-tungsten, resistance brazing ... **A6:** 340
coverings on copper-base electrodes ... **A6:** 718
cyanide ... **A10:** 185
for dc arc sources ... **A10:** 25
defective ... **EL1:** 995
defined ... **A10:** 672 **A15:** 5
definition ... **A6:** 1208
determining selectivity ... **A10:** 182-183
electric arc furnace ... **A15:** 357-359
for electrical testing ... **EM2:** 581-587
electrographite, resistance brazing ... **A6:** 341
for flux-cored arc welding of cast irons ... **A6:** 719-720
composition ... **A6:** 717
for gas-metal arc welding of cast irons ... **A6:** 718-719, 720
for gas-tungsten arc welding of cast irons ... **A6:** 720
tungsten ... **A6:** 191
tungsten alloys ... **A6:** 191
zirconium alloys ... **A6:** 191
glass membrane ... **A10:** 182
glass, pH determination by ... **A10:** 203
Electrodes (continued)
globular-to-spray transition currents ... **A6:** 182
graphite, resistance brazing ... **A6:** 339, 340
high-alloy austenitic steels, submerged arc welding ... **A6:** 204
hydrogen-induced cracking in steel weldments ... **A6:** 412, 413, 414-415
iodide ... **A10:** 185
ion-selective membrane ... **A10:** 181-183
iron-nickel, for flux-cored arc welding ... **A6:** 719
lead ... **A10:** 185
liquid membrane ... **A10:** 182
low-alloy steel flux-cored wires ... **A6:** 188-189
low-alloy steels ... **A6:** 656-657
submerged arc welding ... **A6:** 204-205, 206
mercury vs carbon bases for ... **A10:** 191
mild steel, submerged arc welding ... **A6:** 204
molybdenum, resistance brazing ... **A6:** 339, 340
nickel-base alloys, submerged arc welding ... **A6:** 205
nickel-base, as cast iron welding consumable ... **A15:** 523-524
nickel-base flux-cored wires ... **A6:** 189
nickel-based
for flux-cored arc welding ... **A6:** 719
flux-cored wires ... **A6:** 189
for gas-metal arc welding ... **A6:** 719
for shielded metal arc welding ... **A6:** 717-718
for submerged arc welding ... **A6:** 720
nickel-molybdenum ... **A6:** 743, 745, 746-747
nitrate ... **A10:** 185
perchlorate ... **A10:** 185
piezoelectric, cadmium analysis with ... **A10:** 200
platinum gauze ... **A10:** 199-200
platinum, resistance brazing ... **A6:** 340
polymer membrane ... **A10:** 182
potentiometric gas-sensing ... **A10:** 183-185
processes, reversible and irreversible ... **A10:** 191
reference ... **A10:** 185, 186, 199, 200
reference, schematic of ... **A10:** 186
refractory metal, resistance brazing ... **A6:** 340
rhenium, resistance brazing ... **A6:** 340
sample pretreatment ... **A10:** 186
for shielded metal arc welding of cast irons composition ... **A6:** 717
silver, oxidation in alkaline environments ... **A10:** 135
silver-tungsten, resistance brazing ... **A6:** 340
solid crystalline membrane ... **A10:** 182
for spark source mass spectrometry ... **A10:** 145
stainless steel
for shielded metal arc welding ... **A6:** 717
submerged arc welding ... **A6:** 204, 205
stainless steel flux-cored wires ... **A6:** 189
steel
for flux-cored arc welding ... **A6:** 719
for gas-metal arc welding ... **A6:** 719
for shielded metal arc welding ... **A6:** 717
storage requirements ... **A10:** 186
submerged arc welding composition ... **A6:** 205, 206, 207
for submerged arc welding of cast irons ... **A6:** 720, 721
sulfide ... **A10:** 185
supporting, defined ... **A10:** 683
surfaces, XPS elemental spectrum ... **A10:** 578
thermal cutting ... **A14:** 733-734
thoria-tungsten, aluminum alloys ... **A6:** 736
titanium alloy casting ... **A15:** 831-832
tool steels, submerged arc welding ... **A6:** 204
tungsten, resistance brazing ... **A6:** 339, 340
vibrating, effect in electrogravimetry ... **A10:** 200
VIM, processing routes ... **A15:** 393
zirconia-tungsten, aluminum alloys ... **A6:** 736
Electrodes AWS classification
carbon steel flux cored wires ... **M6:** 100-101
classifications for
flux cored wires ... **M6:** 99-100
gas metal arc welding ... **M6:** 162-163

Electrodes AWS classification (continued)
shielded metal arc welding **M6:** 81-82
compositions for arc welding of
heat-resistant alloys **M6:** 359
copper-based **M6:** 311
coverings on low-carbon electrodes **M6:** 81
definition **M6:** 6
high-manganese steel, composition **M6:** 120
hydrogen in flux cored wires **M6:** 102
low-alloy flux cored wires **M6:** 101-102
low-alloy steel, composition **M6:** 121
low-manganese steel, composition **M6:** 120
medium manganese steel, composition **M6:** 120
nickel-based **M6:** 311
recommended moisture on low-carbon
electrodes **M6:** 82
selection for quenched and tempered
high-strength steels **M6:** 287, 290
stainless steel composition
for flux cored arc welding **M6:** 326, 347, 349
for gas metal arc welding **M6:** 347
for shielded metal arc welding **M6:** 324, 345, 348
stainless steel flux cored wires **M6:** 101-102
strip electrode welding **M6:** 133
tungsten, arc welding of stainless steel **M6:** 334
Electrodes for
air carbon arc cutting **M6:** 919
air-carbon arc cutting **A6:** 1172, 1173-1174, 1175, 1176, 1177
arc welding
of dissimilar copper alloys **A6:** 769, 770
of magnesium alloys **A6:** 774, 778, 779
of stainless steels **A6:** 1106
arc welding of stainless steels
electrogas welding **M6:** 344-346
electroslag welding **M6:** 344-346
gas metal arc welding **M6:** 324, 331-332
shielded metal arc welding **M6:** 324-325, 345-349
submerged arc welding **M6:** 327-328
dissimilar metal joining **A6:** 824, 827
nickel alloys **A6:** 750-751
duplex stainless steels **A6:** 478-479
electrogas welding **A6:** 270 **M6:** 241
of carbon steels **A6:** 659
electroslag welding **A6:** 270, 271, 272, 273, 274, 275, 276, 277, 278 **M6:** 228-229
flash welding **M6:** 562
flux cored arc welding **M6:** 99-103
of alloy steel **M6:** 278
of carbon steel **M6:** 100-101
chemical compositions **M6:** 241
functions of the compounds **M6:** 99
of low-alloy steels **M6:** 101-102
of stainless steel **M6:** 101-102
flux-cored arc welding **A6:** 186, 188-189, 719-720
of carbon steels **A6:** 188-189
cast irons **A6:** 717, 719-720
designators by alloy types **A6:** 188
functions of the compounds **A6:** 606
heat-treatable low alloy steels **A6:** 670
of low-alloy steels **A6:** 188-189
of nickel-base alloys **A6:** 189
parameters for welding cast irons **A6:** 720
of stainless steel **A6:** 189, 705-706
usability type designators **A6:** 189
gas metal arc welding **M6:** 161-163
of alloy steels **M6:** 274-278
of aluminum alloys **M6:** 381-382
of aluminum bronzes **M6:** 420
of beryllium copper **M6:** 418
of copper nickels **M6:** 421
of coppers **M6:** 417
of nickel alloys **M6:** 439
of phosphor bronzes **M6:** 420
of silicon bronzes **M6:** 421
gas tungsten arc welding **M6:** 190-195
of aluminum alloys **M6:** 390
of aluminum bronzes **M6:** 412
of beryllium copper **M6:** 408
of copper and copper alloys **M6:** 404-405
of molybdenum **M6:** 464
of nickel alloys **M6:** 438-439
of titanium and titanium alloys **M6:** 454
Electrodes for (continued)
of tungsten **M6:** 464
gas-metal arc welding **A6:** 182, 185, 1143, 1144
of carbon steels **A6:** 655, 657
cast irons **A6:** 718-719, 720
characteristics and properties **A6:** 720
composition **A6:** 720
consumables **A6:** 719
ferritic stainless steels **A6:** 446
of nickel alloys **A6:** 743
of stainless steels **A6:** 705, 706, 707
gas-tungsten arc welding
aluminum alloys **A6:** 736
of copper alloys **A6:** 754
ferritic stainless steels **A6:** 446
nickel alloys **A6:** 742-743
of stainless steels **A6:** 705
titanium alloys **A6:** 786
hardfacing **A6:** 801, 802, 803, 804, 805, 806
molten-salt-bath dip-brazing furnaces **A6:** 337
offshore structures, undermatching
weld metal strength **A6:** 385
plasma arc welding **A6:** 196 **M6:** 217
plasma-MIG welding **A6:** 223, 224
precipitation-hardening stainless steels **A6:** 483, 484, 487, 488, 489, 492
projection welding **M6:** 509-513
railroad equipment welding **A6:** 396-397, 398
repair welding **A6:** 1105, 1106
resistance brazing **A6:** 339, 340-342 **M6:** 979-981
resistance seam welding **A6:** 238, 239, 241, 243, 244, 245 **M6:** 496-497, 530
class 1 copper **A6:** 245
class 2 copper **A6:** 245
class 3 copper **A6:** 245
class 20 copper **A6:** 245
of stainless steels **M6:** 530
resistance spot welding **A6:** 227-228 **M6:** 529-529
of stainless steels **M6:** 528
resistance welding **A6:** 833, 834, 835, 836, 837, 838, 839, 840, 843, 844, 846, 849 **M6:** 478-480
of aluminum alloys **M6:** 537-538
copper alloys **A6:** 849
of copper and copper 479-480 alloys
face shapes **M6:** 480-481
shielded metal arc welding **A6:** 61, 176-179, 1010, 1011, 1012 **M6:** 78-79, 81-85, 310-311
of alloy steels **M6:** 274-275
alternating current **M6:** 79
of aluminum bronzes **A6:** 765-766
of carbon steels **A6:** 654, 656-657
of cast irons **A6:** 716-718 **M6:** 310-311
classification **M6:** 81-82
direct current **M6:** 78
for duplex stainless steels **A6:** 477
electrode coating formulations **A6:** 59
ferritic stainless steels **A6:** 446, 447
of heat-resistant alloys **M6:** 367
heat-treatable low-alloy steels **A6:** 669, 670
high-strength low-alloy (HSLA)
structural steels **A6:** 663-664
iron-powder **M6:** 82-84
low-alloy steels for pressure vessels
and piping **A6:** 668
mild and low-alloy steels **A6:** 57
of nickel alloys **A6:** 746 **M6:** 440-441
of nickel-based heat-resistant alloys **M6:** 363
sheet metals **A6:** 398
of stainless steels **A6:** 698, 699, 700, 701, 702
of thin sections **M6:** 90
submerged arc welding **A6:** 202, 203, 204, 205, 209, 720, 721 **M6:** 119-122
of alloy steels **M6:** 278
of carbon steels **M6:** 120
of carbon-molybdenum steels **M6:** 121
of chromium-molybdenum steels **M6:** 121
of low-alloy steels **M6:** 121
of nickel alloys **A6:** 743, 748 **M6:** 121, 439, 442
weld cladding **A6:** 813, 816, 819, 821
zirconium alloys **A6:** 788
Electrodes for arc welding
of carbon steels **A6:** 641, 642, 643, 652
austenitic stainless steels **A6:** 641
low-alloy steels **A6:** 641
of dissimilar copper alloys **A6:** 769, 770
Electrodes for arc welding (continued)
high-conductivity beryllium coppers, gas-metal arc butt welding of commercial coppers and copper alloys **A6:** 760
high-strength beryllium coppers, gas-metal arc butt welding of commercial coppers and copper alloys **A6:** 760
of magnesium alloys **A6:** 774, 778, 779
for stainless steels **A6:** 1106
Electrodes for arc welding, specific types
308, for SAW of stainless steels **A6:** 703
308L
flux-cored arc welding of stainless
steels **A6:** 705
solidification cracking not promoted **A6:** 463
E7XT-1, shielded metal arc welding **A6:** 663
E7XT-4, shielded metal arc welding **A6:** 663
E7XT-5, shielded metal arc welding **A6:** 663
E7XT-6, shielded metal arc welding **A6:** 663
E7XT-7, shielded metal arc welding **A6:** 663
E7XT-8, shielded metal arc welding **A6:** 663
E7XT-11, shielded metal arc welding **A6:** 663
E7XT-G, shielded metal arc welding **A6:** 663
E70T-5, flux-cored arc welding **A6:** 651
E71T-5, flux-cored arc welding **A6:** 651
E80T1-W, flux-cored arc welding of
HSLA steels **A6:** 664
E308
SMAW of stainless steels **A6:** 699
weld cladding **A6:** 809
E308-16, for stainless steels **A6:** 696
E308L
solid-state transformations in
weldments **A6:** 83
weld cladding **A6:** 809
E308L-15, for stainless steels **A6:** 694
E308L-16, for stainless steels **A6:** 694
E308L-17, for stainless steels **A6:** 694
E308LT-3, flux-cored arc welding of
stainless steels **A6:** 705, 706
E309
shielded metal arc welding of ferritic
stainless steels **A6:** 447
shielded metal arc welding of HTLA
steels **A6:** 670
solid-state transformations in
weldments **A6:** 80, 83
weld cladding **A6:** 809
E309-16, arc welding of stainless steels **A6:** 680, 685
E309Cb, weld cladding **A6:** 809
E309L, weld cladding **A6:** 809
E309L-16, dissimilar metal joining **A6:** 825
E309Mo, weld cladding **A6:** 809
E309MoL, weld cladding **A6:** 809
E310
shielded metal arc welding of ferritic
stainless steels **A6:** 447
shielded metal arc welding of HTLA
steels **A6:** 670
solid-state transformations in
weldments **A6:** 80
weld cladding **A6:** 809
E310 ELC, shielded metal arc welding
of ferritic stainless steels **A6:** 447
E312, shielded metal arc welding of
HTLA steels **A6:** 670
E312-16
arc welding of stainless steels **A6:** 680
weld cladding **A6:** 811, 819
E316
solid-state transformations in
weldments **A6:** 83
weld cladding **A6:** 809
E316L, weld cladding **A6:** 809
E317, weld cladding **A6:** 809
E317L
gas-tungsten arc welding **A6:** 705
weld cladding **A6:** 809
E320, weld cladding **A6:** 809
E347, weld cladding **A6:** 809
E502T-X, flux-cored arc welding of
low-alloy steels **A6:** 668
E505T-X, flux-cored arc welding of
low-alloy steels **A6:** 668
E6010
characteristics and properties **M6:** 82

Electrodes for arc welding, specific types (continued)
characteristics and properties for shielded metal arc welding applications A6: 654
moisture content recommendations M6: 82-83
shielded metal arc welding A6: 59
weld metal composition and silicon deposit A6: 100
E6011
characteristics and properties M6: 82, 84-86
moisture content recommendations M6: 82-83
E6011, characteristics and properties for shielded metal arc welding applications A6: 654
E6012
characteristics and properties M6: 82, 84-85, 89
characteristics and properties for shielded metal arc welding applications A6: 654
moisture content recommendations M6: 82
shielded metal arc welding A6: 59
E6013
characteristics and properties M6: 82, 84-85
characteristics and properties for shielded metal arc welding applications A6: 654
moisture content recommendations M6: 82
repair welding of carbon steels A6: 1066
shielded metal arc welding A6: 59
underwater welding A6: 1010, 1011, 1012
E6019, characteristics and properties for shielded metal arc welding applications A6: 654
E6020
characteristics and properties M6: 82, 84
characteristics and properties for shielded metal arc welding applications A6: 654
moisture content recommendations M6: 82
shielded metal arc welding A6: 59
E6022
characteristics and properties M6: 82
moisture content recommendations M6: 82
E6027
characteristics and properties M6: 82, 85
moisture content recommendations M6: 82
E6027, characteristics and properties for shielded metal arc welding applications A6: 654
E7014
characteristics and properties M6: 82, 84-85
characteristics and properties for shielded metal arc welding applications A6: 654
moisture content recommendations M6: 82
underwater welding A6: 1013
E7015
characteristics and properties M6: 82
characteristics and properties for shielded metal arc welding applications A6: 654
moisture content recommendations M6: 82-83
shielded metal arc welding A6: 59, 651, 656, 663 M6: 265
E7016
characteristics and properties M6: 82, 84
characteristics and properties for shielded metal arc welding applications A6: 654
moisture content recommendations M6: 82-83
oxyfuel gas welding A6: 290
shielded metal arc welding A6: 651, 656, 663, 717 M6: 265
E7018
characteristics and properties M6: 82
characteristics and properties for shielded metal arc welding applications A6: 654

Electrodes for arc welding, specific types (continued)
hydrogen-induced cracking A6: 414
moisture content recommendations M6: 82-83
shielded metal arc welding A6: 651, 656, 663, 717 M6: 265
undermatching weld metal strength A6: 385
weld metal composition and silicon deposit A6: 100
E7018-Al, shielded metal arc welding of HTLA steels A6: 670
E7018M, shielded metal arc welding A6: 656
E7024
characteristics and properties M6: 82, 84-85
moisture content recommendations M6: 82
E7024, characteristics and properties for shielded metal arc welding applications A6: 654
E7027, characteristics and properties for shielded metal arc welding applications A6: 654
E7028
characteristics and properties M6: 82, 84
characteristics and properties for shielded metal arc welding applications A6: 654
moisture content recommendations M6: 82-83
shielded metal arc welding A6: 656, 663
E7048
characteristics and properties M6: 82
characteristics and properties for shielded metal arc welding applications A6: 654
moisture content recommendations M6: 82
shielded metal arc welding A6: 656, 663
E7158-K6, weld bulk composition A6: 102
E8010-G
parameters used to obtain multipass weld in X-65 steel pipe A6: 101
weld bulk composition A6: 102
E8015-XX, shielded metal arc welding A6: 663
E8016-B2, shielded metal arc welding of HTLA steels A6: 670
E8016-C1, shielded metal arc welding of HTLA steels A6: 670
E8016-XX, shielded metal arc welding A6: 663
E8018, undermatching weld metal strength A6: 385
E8018-XX, shielded metal arc welding A6: 663
E9016-B3, shielded metal arc welding of HTLA steels A6: 670
E9018, undermatching weld metal strength A6: 385
E10016-D2, shielded metal arc welding of HTLA steels A6: 670
E11016, hydrogen-induced cracking A6: 414
E11018-M
hydrogen-induced cracking in HY-80 steel A6: 412
shielded metal arc welding of HTLA steels A6: 670
E12018-M, shielded metal arc welding of HTLA steels A6: 670
EB3, submerged arc welding A6: 204-205
ECu, shielded metal arc welding of copper and copper alloys A6: 755
ECuAl-A2
characteristics and properties A6: 718
composition A6: 718
for copper alloys A6: 756
shielded metal arc welding of aluminum bronzes A6: 765-766
shielded metal arc welding of copper alloys A6: 755
shielded metal arc welding of copper-zinc alloys A6: 763

Electrodes for arc welding, specific types (continued)
weld cladding of steels A6: 820
ECuAl-B
shielded metal arc welding of aluminum bronzes A6: 765-766
shielded metal arc welding of copper and copper alloys A6: 755, 756
shielded metal arc welding of copper-zinc alloys A6: 763
ECuAl-C, shielded metal arc welding of aluminum bronzes A6: 765-766
ECuAl-D, shielded metal arc welding of aluminum bronzes A6: 765-766
ECuAl-E, shielded metal arc welding of aluminum bronzes A6: 765-766
ECuMnNiAl
characteristics and properties A6: 718
composition A6: 718
shielded metal arc welding of copper and copper alloys A6: 755, 756
ECuNi
arc welding of nickel alloys A6: 746, 749
for copper alloys A6: 756
shielded metal arc welding of copper and copper alloys A6: 755
shielded metal arc welding of copper-nickel alloys A6: 768
weld cladding of steels A6: 820
ECuNiAl
for copper alloys A6: 756
shielded metal arc welding of copper and copper alloys A6: 755
ECuSi
for copper alloys A6: 756
shielded metal arc welding of copper and copper alloys A6: 755
shielded metal arc welding of copper-zinc alloys A6: 763
weld cladding of steels A6: 820
ECuSn-A
characteristics and properties A6: 718
composition A6: 718
shielded metal arc welding of copper and copper alloys A6: 755, 756
shielded metal arc welding of copper-zinc alloys A6: 763
shielded metal arc welding of phosphor bronzes A6: 764
ECuSn-C
characteristics and properties A6: 718
composition A6: 718
shielded metal arc welding of copper and copper alloys A6: 755, 756
shielded metal arc welding of copper-zinc alloys A6: 763
shielded metal arc welding of phosphor bronzes A6: 764
EM12K, submerged arc welding A6: 204, 205
ENi-1
arc welding of cast irons A6: 717
arc welding of nickel alloys A6: 746, 749
not recommended for gas-tungsten arc welding A6: 720
weld cladding of steels A6: 820
ENi-1, arc welding of nickel alloys M6: 440-441
ENi-CI
characteristics and properties A6: 717, 718, 721
composition A6: 717
ENi-CI-A
characteristics and properties A6: 717, 721
composition A6: 717
ENiCr, joining of dissimilar metals A6: 750
ENiCrCoMo-1
arc welding of nickel alloys A6: 746, 749
shielded metal arc welding of nickel alloys A6: 746-747
ENiCrFe, joining of dissimilar metals A6: 750

Electrodes for arc welding, specific types (continued)
ENiCrFe-1
arc welding of nickel alloys **M6:** 441
composition **M6:** 359
ENiCrFe-1, arc welding of nickel alloys **A6:** 746, 749
ENiCrFe-2
arc welding of nickel alloys **A6:** 746 **M6:** 440-441
composition **M6:** 359
shielded metal arc welding of ferritic stainless steels **A6:** 447
shielded metal arc welding of HTLA steels **A6:** 670
ENiCrFe-3
arc welding of nickel alloys **A6:** 746, 749 **M6:** 440-441
composition **M6:** 359
shielded metal arc welding of HTLA steels **A6:** 670
weld cladding of steels **A6:** 820
ENiCrFe-7, arc welding of nickel alloys **A6:** 746, 749
ENiCrMo-3
arc welding of nickel alloys **A6:** 746, 749
for stainless steels **A6:** 683-685
weld cladding of steels **A6:** 820
ENiCrMo-3, arc welding of nickel alloys **M6:** 440-441
ENiCrMo-4
arc welding of nickel alloys **A6:** 746
for stainless steels **A6:** 683-685
weld cladding of steels **A6:** 820
ENiCrMo-4, arc welding of nickel alloys **M6:** 440-441
ENiCrMo-9, arc welding of nickel alloys **A6:** 746, 749 **M6:** 441
ENiCrMo-10, arc welding of nickel alloys **A6:** 746
ENiCrMo-11, arc welding of nickel alloys **A6:** 746
ENiCrMo-12, arc welding of nickel alloys **A6:** 746
ENiCu-1, nickel alloys to dissimilar metals **A6:** 751
ENiCu-4, nickel alloys to dissimilar metals **A6:** 751
ENiCu-7
arc welding of nickel alloys **A6:** 746, 749
weld cladding of steels **A6:** 820
ENiCu-A, composition **A6:** 717
ENiCu-B, composition **A6:** 717
ENiFe-CI
characteristics and properties **A6:** 717, 718, 721
composition **A6:** 717
ENiFe-CI-A
characteristics and properties **A6:** 717, 721
composition **A6:** 717
ENiFeMn-CI
characteristics and properties **A6:** 717, 718
composition **A6:** 717
ENiFeT3-CI, composition **A6:** 717
ENiMo-1
arc welding of nickel alloys **A6:** 746 **M6:** 440-441
composition **M6:** 359
ENiMo-3, composition **M6:** 359
ENiMo-7
arc welding of nickel alloys **A6:** 746
weld cladding of steels **A6:** 820
ER16-8-2, composition **A6:** 208
ER26-1, composition **A6:** 208
ER70S-2
composition **A6:** 655 **M6:** 162
gas-metal arc welding **A6:** 657
gas-tungsten arc welding **A6:** 655, 656, 658
mechanical properties **M6:** 162-163
mechanical-property requirements **A6:** 655
plasma arc welding **A6:** 658
ER70S-3
composition **A6:** 655 **M6:** 162-163
gas-metal arc welding **A6:** 657
mechanical properties **A6:** 719 **M6:** 162-163
mechanical-property requirements **A6:** 655
ER70S-4
composition **A6:** 655 **M6:** 162-163

Electrodes for arc welding, specific types (continued)
gas-metal arc welding **A6:** 657
mechanical properties **M6:** 163
mechanical-property requirements **A6:** 655
ER70S-5
composition **A6:** 655 **M6:** 162-163
gas-metal arc welding **A6:** 657
mechanical properties **M6:** 163
mechanical-property requirements **A6:** 655
ER70S-6
composition **A6:** 655 **M6:** 162-163
gas-metal arc welding **A6:** 657
mechanical properties **A6:** 719 **M6:** 163
mechanical-property requirements **A6:** 655
ER70S-7
composition **A6:** 655 **M6:** 162-163
gas-metal arc welding **A6:** 657
mechanical properties **M6:** 163
mechanical-property requirements **A6:** 655
ER70S-G
composition **A6:** 655 **M6:** 162-163
mechanical properties **M6:** 163
mechanical-property requirements **A6:** 655
ER70S-X, shielded metal arc welding **A6:** 663
ER80S-D2
composition **A6:** 655
gas-metal arc welding **A6:** 657
mechanical-property requirements **A6:** 655
ER80S-D2, classification **M6:** 276-277
ER80S-Ni1
composition **A6:** 655
mechanical-property requirements **A6:** 655
ER80S-Ni2
composition **A6:** 655
mechanical-property requirements **A6:** 655
ER80S-NI2, classification **M6:** 276
ER80S-Ni3
composition **A6:** 655
mechanical-property requirements **A6:** 655
ER80S-NI3, classification **M6:** 276
ER80S-Nil, classification **M6:** 276
ER80S-XX, shielded metal arc welding **A6:** 663
ER100S-1
composition **A6:** 655
mechanical-property requirements **A6:** 655
ER100S-2
composition **A6:** 655
mechanical-property requirements **A6:** 655
ER110S-1
composition **A6:** 655
mechanical-property requirements **A6:** 655
ER120S-1
composition **A6:** 655
mechanical-property requirements **A6:** 655
ER209, composition **A6:** 208
ER218, composition **A6:** 208
ER219, composition **A6:** 208
ER240, composition **A6:** 208
ER307, composition **A6:** 208
ER308
composition **A6:** 208
weld cladding **A6:** 809
ER308H, composition **A6:** 208
ER308L
composition **A6:** 208
weld cladding **A6:** 809
ER308Mo, composition **A6:** 208
ER308MoL, composition **A6:** 208
ER309
composition **A6:** 208
weld cladding **A6:** 809
ER309Cb, weld cladding **A6:** 809
ER309L
composition **A6:** 208
weld cladding **A6:** 809
ER309Mo, weld cladding **A6:** 809
ER309MoL, weld cladding **A6:** 809
ER310
composition **A6:** 208
weld cladding **A6:** 809
ER312
composition **A6:** 208
solid-state transformations in weldments **A6:** 80
ER316
composition **A6:** 208

Electrodes for arc welding, specific types (continued)
weld cladding **A6:** 809
ER316H, composition **A6:** 208
ER316L
composition **A6:** 208
weld cladding **A6:** 809
ER317
composition **A6:** 208
weld cladding **A6:** 809
ER317L
composition **A6:** 208
weld cladding **A6:** 809
ER318, composition **A6:** 208
ER320
composition **A6:** 208
weld cladding **A6:** 809
ER320LR, composition **A6:** 208
ER321, composition **A6:** 208
ER330, composition **A6:** 208
ER347
composition **A6:** 208
weld cladding **A6:** 809
ER349, composition **A6:** 208
ER410, composition **A6:** 208
ER410NiMo, composition **A6:** 208
ER420, composition **A6:** 208
ER430, composition **A6:** 208
ER502
composition **A6:** 208
GTAW and GMAW **A6:** 668
ER505
composition **A6:** 208
GTAW and GMAW **A6:** 668
ER630, composition **A6:** 208
ERCu
arc welding of copper and copper alloys **A6:** 755
for dissimilar copper alloy welds **A6:** 769
gas-metal arc butt welding of commercial coppers and copper alloys **A6:** 760
gas-metal arc welding of copper and copper alloys **A6:** 770
for gas-metal arc welding of coppers **A6:** 759
gas-metal arc welding of coppers to high-carbon steel **A6:** 828
gas-metal arc welding of coppers to low-alloy steels **A6:** 828
gas-metal arc welding of coppers to low-carbon steel **A6:** 828
gas-metal arc welding of coppers to medium-carbon steel **A6:** 828
gas-metal arc welding of coppers to stainless steel **A6:** 828
gas-metal arc welding of dissimilar copper alloys **A6:** 769
gas-tungsten arc welding of copper and copper alloys **A6:** 769
ERCuAl-A
for copper alloys **A6:** 756
weld cladding of steels **A6:** 820
ERCuAl-A2
arc welding of cast irons **A6:** 719
arc welding of copper and copper alloys **A6:** 755
characteristics and properties **A6:** 719, 720
composition **A6:** 720
for copper alloys **A6:** 756
for dissimilar copper alloy welds **A6:** 769
gas metal arc welding of copper nickels to low-carbon steel **A6:** 828
gas-metal arc butt welding of commercial coppers and copper alloys **A6:** 760
gas-metal arc welding of aluminum bronze to high-carbon steel **A6:** 828
gas-metal arc welding of aluminum bronze to low-alloy steel **A6:** 828
gas-metal arc welding of aluminum bronze to low-carbon steel **A6:** 828
gas-metal arc welding of aluminum bronze to medium-carbon steel **A6:** 828
gas-metal arc welding of aluminum bronze to stainless steel **A6:** 828
gas-metal arc welding of copper and copper alloys **A6:** 770

Electrodes for arc welding, specific types (continued)
gas-metal arc welding of copper nickels to high-carbon steel A6: 828
gas-metal arc welding of copper nickels to low-alloy steel A6: 828
gas-metal arc welding of copper nickels to medium-carbon steels A6: 828
gas-metal arc welding of copper nickels to stainless steels A6: 828
gas-metal arc welding of coppers A6: 762
gas-metal arc welding of coppers to high-carbon steel A6: 828
gas-metal arc welding of coppers to low-alloy steel A6: 828
gas-metal arc welding of coppers to low-carbon steel A6: 828
gas-metal arc welding of coppers to medium-carbon steel A6: 828
gas-metal arc welding of coppers to stainless steel A6: 828
gas-metal arc welding of high-zinc brasses to high-carbon steel A6: 828
gas-metal arc welding of high-zinc brasses to low-alloy steels A6: 828
gas-metal arc welding of high-zinc brasses to low-carbon steels A6: 828
gas-metal arc welding of high-zinc brasses to medium-carbon steels A6: 828
gas-metal arc welding of high-zinc brasses to stainless steel A6: 828
gas-metal arc welding of low-zinc brasses to high-carbon steel A6: 828
gas-metal arc welding of low-zinc brasses to low-alloy steel A6: 828
gas-metal arc welding of low-zinc brasses to medium-carbon steel A6: 828
gas-metal arc welding of low-zinc brasses to stainless steel A6: 828
gas-metal arc welding of silicon bronzes to high-carbon steel A6: 828
gas-metal arc welding of silicon bronzes to low-alloy steel A6: 828
gas-metal arc welding of silicon bronzes to low-carbon steel A6: 828
gas-metal arc welding of silicon bronzes to medium-carbon steels A6: 828
gas-metal arc welding of silicon bronzes to stainless steels A6: 828
gas-metal arc welding of special brasses to high-carbon steel A6: 828
gas-metal arc welding of special brasses to low-alloy steel A6: 828
gas-metal arc welding of special brasses to low-carbon steel A6: 828
gas-metal arc welding of special brasses to medium-carbon steel A6: 828
gas-metal arc welding of special brasses to stainless steel A6: 828
gas-metal arc welding of tin brasses to low-carbon steels A6: 828
gas-metal arc welding of tin brasses to medium-carbon steel A6: 828
gas-metal arc welding of tin brasses to stainless steel A6: 828
gas-tungsten arc welding of copper and copper alloys A6: 769
ERCuAl-A2, welding dissimilar copper alloys M6: 422-423
ERCuAl-A3
arc welding of copper and copper alloys A6: 755
for copper alloys A6: 756
ERCuMnNiAl, arc welding of copper and copper alloys A6: 755

Electrodes for arc welding, specific types (continued)
ERCuNi
arc welding of copper and copper alloys A6: 755
arc welding of nickel alloys A6: 745, 749
for copper alloys A6: 756
for dissimilar copper alloy welds A6: 769
gas-metal arc butt welding of commercial coppers and copper alloys A6: 760
gas-metal arc welding of copper and copper alloys A6: 770
gas-tungsten arc welding of copper and copper alloys A6: 769
submerged arc welding of nickel alloys A6: 748
weld cladding of steels A6: 820
ERCuNi, arc welding of nickel alloys M6: 439
ERCuNi-3
gas-metal arc welding of copper and copper alloys A6: 770
gas-tungsten arc welding of copper and copper alloys A6: 769
ERCuNi-7
gas-metal arc welding of copper and copper alloys A6: 770
gas-tungsten arc welding of copper and copper alloys A6: 769
ERCuNiAl
arc welding of copper and copper alloys A6: 755
for copper alloys A6: 756
ERCuSi, weld cladding of steels A6: 820
ERCuSi-A
arc welding of copper and copper alloys A6: 755
for copper alloys A6: 756
gas-metal arc butt welding of commercial coppers and copper alloys A6: 760
gas-metal arc welding for silicon bronzes A6: 766
gas-metal arc welding of copper and copper alloys A6: 770
gas-metal arc welding of coppers A6: 762
gas-tungsten arc welding of copper and copper alloys A6: 769
ERCuSn-A
arc welding of cast irons A6: 719
arc welding of copper and copper alloys A6: 755, 756
characteristics and properties A6: 719, 720
composition A6: 720
gas-metal arc butt welding of commercial coppers and copper alloys A6: 760
gas-metal arc welding of copper and copper alloys A6: 770
gas-metal arc welding of low-zinc brasses to low-carbon steels A6: 828
gas-metal arc welding of phosphor bronze to high-carbon steel A6: 828
gas-metal arc welding of phosphor bronze to low-alloy steel A6: 828
gas-metal arc welding of phosphor bronze to low-carbon steel A6: 828
gas-metal arc welding of phosphor bronze to medium-carbon steel A6: 828
gas-metal arc welding of phosphor bronze to stainless steel A6: 828
gas-tungsten arc welding of copper and copper alloys A6: 769
ERMnNiAl
arc welding of cast irons A6: 719
characteristics and properties A6: 719, 720

Electrodes for arc welding, specific types (continued)
composition A6: 720
ERNi-1
arc welding of nickel alloys A6: 743, 745, 746, 749
characteristics and properties A6: 719
composition A6: 208
submerged arc welding of nickel alloys A6: 748
weld cladding of steels A6: 820
ERNi-1, arc welding of nickel alloys M6: 439
ERNi-3
gas-metal arc welding of copper nickels to high-carbon steel A6: 828
gas-metal arc welding of copper nickels to low-alloy steel A6: 828
gas-metal arc welding of copper nickels to low-carbon steel A6: 828
gas-metal arc welding of copper nickels to medium-carbon steel A6: 828
gas-metal arc welding of copper nickels to stainless steel A6: 828
gas-metal arc welding of coppers to high-carbon steel A6: 828
gas-metal arc welding of coppers to low-alloy steel A6: 828
gas-metal arc welding of coppers to low-carbon steel A6: 828
gas-metal arc welding of coppers to medium-carbon steel A6: 828
gas-metal arc welding of coppers to stainless steel A6: 828
ERNiCr, joining of dissimilar metals A6: 750
ERNICr-3
arc welding of heat-resistant alloys M6: 359
arc welding of nickel alloys A6: 743, 745, 746, 749 M6: 439
composition A6: 208 M6: 359
submerged arc welding of nickel alloys A6: 748
weld cladding of steels A6: 820
ERNiCrCoMo-1
arc welding of nickel alloys A6: 743, 745
composition A6: 208
ERNiCrFe, joining of dissimilar metals A6: 750
ERNiCrFe-5
arc welding of heat-resistant alloys M6: 359
arc welding of nickel alloys A6: 743, 745 M6: 439
composition A6: 208 M6: 359
submerged arc welding of nickel alloys A6: 748
ERNiCrFe-6
arc welding of heat-resistant alloys M6: 359
arc welding of nickel alloys A6: 743, 745 M6: 439
composition A6: 208 M6: 359
ERNiCrFe-7
arc welding of heat-resistant alloys M6: 359
composition M6: 359
ERNiCrFe-7, arc welding of nickel alloys A6: 743, 749
ERNiCrMo-1
arc welding of nickel alloys A6: 743, 745
composition A6: 208
ERNiCrMo-2, composition A6: 208
ERNiCrMo-3
arc welding of heat-resistant alloys M6: 359
arc welding of nickel alloys A6: 749 M6: 439
composition A6: 208 M6: 359
for stainless steels A6: 683-685
submerged arc welding of nickel alloys A6: 748
weld cladding of steels A6: 820
ERNICrMo-4
arc welding of heat-resistant alloys M6: 359
arc welding of nickel alloys A6: 749 M6: 439
composition A6: 208 M6: 359

Electrodes for arc welding, specific types (continued)
for stainless steels **A6:** 683-685
weld cladding of steels **A6:** 820
ERNiCrMo-5, arc welding of nickel alloys **A6:** 745, 749
ERNiCrMo-6, arc welding of nickel alloys **A6:** 745, 749
ERNiCrMo-7
arc welding of heat-resistant alloys **M6:** 359
arc welding of nickel alloys **A6:** 745, 749
composition **A6:** 208 **M6:** 359
ERNiCrMo-8
arc welding of nickel alloys **A6:** 745, 749
composition **A6:** 208
ERNiCrMo-9
arc welding of nickel alloys **A6:** 743, 745, 749
composition **A6:** 208
ERNiCrMo-9, arc welding of nickel alloys **M6:** 439
ERNiCrMo-10
arc welding of nickel alloys **A6:** 743, 745
composition **A6:** 208
for stainless steels **A6:** 683-685
ERNiCrMo-11
arc welding of nickel alloys **A6:** 743, 745
composition **A6:** 208
ERNiCrMo-21, composition **M6:** 359
ERNiCu-7
arc welding of nickel alloys **A6:** 743, 745, 746, 749
composition **A6:** 208
submerged arc welding of nickel alloys **A6:** 748
ERNiCu-7, arc welding of nickel alloys **M6:** 439-441
ERNiFeCr-1
arc welding of nickel alloys **A6:** 743, 745
composition **A6:** 208
ERNiFeCr-1, arc welding of nickel alloys **M6:** 439
ERNiFeCr-2
arc welding of nickel alloys **A6:** 743, 749
composition **A6:** 208
ERNiMo-1
arc welding of nickel alloys **A6:** 743
composition **A6:** 208
ERNIMo-1, arc welding of nickel alloys **M6:** 439
ERNiMo-2, composition **A6:** 208
ErNiMo-3
arc welding of nickel alloys **A6:** 743
composition **A6:** 208
ERNiMo-7
arc welding of nickel alloys **A6:** 743
composition **A6:** 208
weld cladding of steels **A6:** 820
ERTi-1, composition requirements **A6:** 785
ERTi-2, composition requirements **A6:** 785
ERTi-3, composition requirements **A6:** 785
ERTi-4, composition requirements **A6:** 785
ERTi-5, composition requirements **A6:** 785
ERTi-5ELI, composition requirements **A6:** 785
ERTi-6, composition requirements **A6:** 785
ERTi-6ELI, composition requirements **A6:** 785
ERTi-7, composition requirements **A6:** 785
ERTi-9, composition requirements **A6:** 785
ERTi-9ELI, composition requirements **A6:** 785
ERTi-12, composition requirements **A6:** 785
ERTi-15, composition requirements **A6:** 785
ERXXS-G
composition **A6:** 655
mechanical-property requirements **A6:** 655
ERZr2 (R60702), chemical composition **A6:** 788
ERZr3 (R60704), chemical composition **A6:** 788
ERZr4 (R60705), chemical composition **A6:** 788
ESt, composition **A6:** 717
EWCe-2, gas-tungsten arc welding **A6:** 191
EWG, gas-tungsten arc welding **A6:** 191
EWLa-1, gas-tungsten arc welding **A6:** 191
EWP
gas-tungsten arc welding **A6:** 191
gas-tungsten arc welding of aluminum bronzes **A6:** 764-765
gas-tungsten arc welding of coppers **A6:** 761

Electrodes for arc welding, specific types (continued)
gas-tungsten arc welding of silicon bronzes **A6:** 766
EWTh-1
gas-tungsten arc welding **A6:** 191
gas-tungsten arc welding of titanium **A6:** 1107
gas-tungsten arc welding of titanium alloys **A6:** 786
EWTh-2
for copper alloys **A6:** 756, 757
gas-tungsten arc welding **A6:** 191
gas-tungsten arc welding of coppers **A6:** 761
gas-tungsten arc welding of ferritic stainless steels **A6:** 445
gas-tungsten arc welding of silicon bronzes **A6:** 767
gas-tungsten arc welding of titanium **A6:** 1107
gas-tungsten arc welding of titanium alloys **A6:** 786
gas-tungsten arc welding of zirconium alloys **A6:** 787
EWZr
gas-tungsten arc welding of aluminum bronzes **A6:** 764-765
gas-tungsten arc welding of coppers **A6:** 761
gas-tungsten arc welding of silicon bronzes **A6:** 766
EWZr-1, gas-tungsten arc welding **A6:** 191
EXIT-1, flux-cored arc welding applications **A6:** 656
EXIT-5, flux-cored arc welding applications **A6:** 656
EXIT-7, flux-cored arc welding applications **A6:** 656
EXIT-8, flux-cored arc welding applications **A6:** 656
EXIT-11, flux-cored arc welding applications **A6:** 656
EXOT-1, flux-cored arc welding applications **A6:** 656
EXOT-2, flux-cored arc welding applications **A6:** 656
EXOT-3, flux-cored arc welding applications **A6:** 656
EXOT-4, flux-cored arc welding applications **A6:** 656
EXOT-5, flux-cored arc welding applications **A6:** 656
EXOT-6, flux-cored arc welding applications **A6:** 656
EXOT-7, flux-cored arc welding applications **A6:** 656
EXOT-8, flux-cored arc welding applications **A6:** 656
EXOT-10, flux-cored arc welding applications **A6:** 656
EXOT-11, flux-cored arc welding applications **A6:** 656
EXXC-G, mechanical-property requirements **A6:** 655
F7XXX-EXXX, shielded metal arc welding **A6:** 663
F8XTX-XX, shielded metal arc welding **A6:** 663
F8XX-EXXX-XX, shielded metal arc welding **A6:** 663
Fe-55Ni, repair welding of cast iron **A6:** 1066
GMR-235, composition **M6:** 359
Hastelloy S, composition **M6:** 359
Haynes 556, composition **M6:** 359
Inconel 117, composition **M6:** 359
Inconel 601, composition **M6:** 359
Inconel 617, composition **M6:** 359
Inconel 718, composition **M6:** 359
RBCuZn-A, arc welding of copper and copper alloys **A6:** 755
RBCuZn-B, arc welding of copper and copper alloys **A6:** 755
RBCuZn-C, arc welding of copper and copper alloys **A6:** 755
René 41, composition **M6:** 359
T-1, flux-cored arc welding **A6:** 651, 657
T-2, flux-cored arc welding **A6:** 657
T-3, flux-cored arc welding **A6:** 658
T-4, flux-cored arc welding **A6:** 651, 658
T-5, flux-cored arc welding **A6:** 657
T-6, flux-cored arc welding **A6:** 658

Electrodes for arc welding, specific types (continued)
T-7, flux-cored arc welding **A6:** 651, 658
T-8, flux-cored arc welding **A6:** 651, 658
T-10, flux-cored arc welding **A6:** 658
T-11, flux-cored arc welding **A6:** 658
Waspaloy, composition **M6:** 359
Electrodes for electrogas welding, specific types
EGXXS-1, composition **A6:** 659
EGXXS-2, composition **A6:** 659
EGXXS-3, composition **A6:** 659
EGXXS-5, composition **A6:** 659
EGXXS-6, composition **A6:** 659
EGXXS-D2, composition **A6:** 659
EGXXS-G, composition **A6:** 659
EGXXT-1, composition **A6:** 659
EGXXT-2, composition **A6:** 659
EGXXT-G, composition **A6:** 659
EGXXT-Ni1 (formerly EGXXT-3), composition **A6:** 659
EGXXT-NM1 (formerly EGXXT-4), composition **A6:** 659
EGXXT-NM2 (formerly EGXXT-6), composition **A6:** 659
EGXXT-W (formerly EGXXT-5), composition **A6:** 659
Electrodes for electroslag welding **A6:** 270, 271, 272, 273, 274, 275, 276, 277, 278
Electrodes for projection welding, specific types
class 2 .... **M6:** 507-509
class 3 .... **M6:** 508-509
class 4 .... **M6:** 509
class 10 .... **M6:** 508-509
class 11 .... **M6:** 508-509
class 12 .... **M6:** 508-509
Electrodes for resistance brazing, specific types
class 2
applications **M6:** 979
production example **M6:** 988
class 13, applications **M6:** 979
class 14
applications **M6:** 979
production example **M6:** 988
Electrodes for resistance welding, specific types
class 1 .... **M6:** 479
properties **M6:** 537
resistance welding of coppers **M6:** 547-548
seam welding applications **M6:** 496
class 2 .... **M6:** 477-479
properties **M6:** 537
properties for seam welding **M6:** 496
resistance welding of coppers **M6:** 547-548
seam welding of coated steel **M6:** 502
class 3 .... **M6:** 479-480
properties **M6:** 537
properties for seam welding **M6:** 496
resistance welding of coppers **M6:** 547
class 10 .... **M6:** 480
class 11 .... **M6:** 480
resistance welding of coppers **M6:** 547
class 12 .... **M6:** 480
class 13 .... **M6:** 480
resistance welding of coppers **M6:** 547
class 14 .... **M6:** 480
resistance welding of coppers **M6:** 547
Electrodes for submerged arc welding **A6:** 202, 203, 204, 205, 209, 720, 721
high-manganese, composition **A6:** 205
low-alloy solid steel, composition **A6:** 206
low-alloy steel weld metal (both solid flux-electrode and composite flux-electrode combinations), composition **A6:** 207
low-manganese, composition **A6:** 205
medium-manganese, composition **A6:** 205
Electrodes, selection of
for welding **M1:** 562-564
Electrodes, unmelted
as forging defect **A17:** 492
Electrodialysis recovery process
plating waste treatment **M5:** 318-319
Electrodischarge machining **A14:** 247
Electrodynamic amplitude detectors **A8:** 246
Electrodynamic degassing **A7:** 181
Electrodynamic resonance systems **A8:** 240
Electroetching, of identification numbers
fatigue fracture from **A11:** 473-474

Electroformed molds ... **EM1:** 10, 584-585
defined ... **EM2:** 16
Electroformed nickel tooling ... **EM1:** 582-585
cost effectiveness ... **EM1:** 582
electroforming mold ... **EM1:** 584-585
mandrel use ... **EM1:** 583-584
Electroforming of nickel ... **M3:** 179-181
Electrogalvanized sheet steel
1006 ... **A9:** 199
specimen preparation ... **A9:** 197
Electrogalvanized steels
automotive industry ... **A13:** 1012
forming of ... **A14:** 560-561, 634
Electrogalvanizing *See also* Galvanized coatings, Zinc coatings ... **A1:** 212, 217 **A13:** 6, 766-767
adherence ... **M1:** 170-171
applications ... **M1:** 171
formability ... **M1:** 170
preparation for painting ... **M1:** 170-171
specification for ... **M1:** 171
as zinc coating ... **A2:** 528
Electrogas welding ... **M6:** 238-244
applicability ... **M6:** 239
arc length and voltage ... **M6:** 242-243
base metal thickness ... **M6:** 239
Charpy V-notch impact testing ... **M6:** 244
comparison to electroslag
impact properties ... **M6:** 244
porosity ... **M6:** 244
welding ... **M6:** 225, 244
dam, water-cooled ... **M6:** 239-240
fixed ... **M6:** 240
movable ... **M6:** 240
definition ... **M6:** 6
electrode wire guides ... **M6:** 239-240
electrode wire oscillators ... **M6:** 240
electrode wire-feed systems ... **M6:** 240
electrodes ... **M6:** 241
flux cored ... **M6:** 241
solid ... **M6:** 241
stickouts ... **M6:** 241
equipment ... **M6:** 239-241
gas boxes ... **M6:** 240-241
gas ports ... **M6:** 240-241
heat-affected zone ... **M6:** 244
inspection ... **M6:** 244
magnetic-particle ... **M6:** 244
ultrasonic ... **M6:** 244
length of joint ... **M6:** 239
metals welded ... **M6:** 239
operation setup ... **M6:** 242
current ... **M6:** 242
travel speed ... **M6:** 242
voltage ... **M6:** 242
oscillation ... **M6:** 243
postweld heat treatment ... **M6:** 244
power supplies ... **M6:** 239-240
constant-current ... **M6:** 240
constant-voltage ... **M6:** 240
preheat ... **M6:** 244
restarts and repairs ... **M6:** 243
shielding gas ... **M6:** 240-241
carbon dioxide ... **M6:** 241
weld properties ... **M6:** 244
welding of tanks ... **M6:** 243
cement slurry tanks ... **M6:** 243
storage tanks ... **M6:** 243
surge tanks ... **M6:** 243
water reservoirs ... **M6:** 243
welding parameters ... **M6:** 242
workpiece assembly ... **M6:** 241-242
back-up bars ... **M6:** 242
runoff tabs ... **M6:** 241-242
square-groove butt joints ... **M6:** 242
starting trough ... **M6:** 241
V-groove butt joints ... **M6:** 242
Electrogas welding (EGW) ... **A6:** 275-278
aluminum ... **A6:** 275, 278
Electrogas welding (EGW) (continued)
aluminum alloys ... **A6:** 738
applications ... **A6:** 270, 276-278
shipbuilding ... **A6:** 384
carbon steels ... **A6:** 276-277, 652, 653, 658-659, 660
definition ... **A6:** 270, 1208
electrodes ... **A6:** 270, 659
of carbon steels ... **A6:** 659
heat-affected zone ... **A6:** 275, 276
high-strength low-alloy structural steels ... **A6:** 664
low-alloy metals for pressure vessels and piping ... **A6:** 668
low-alloy steels ... **A6:** 276-277, 662, 664, 668
for pressure vessels and piping ... **A6:** 667
multipass ... **A6:** 275-276
not for HSLA Q & T structural steels ... **A6:** 666
power source selected ... **A6:** 37
shielding gases ... **A6:** 270, 275, 662
steel grades commonly joined ... **A6:** 656
Electrogas welding of
alloy steels ... **M6:** 239
aluminum ... **M6:** 239
carbon-manganese-silicon steels ... **M6:** 239
low-carbon steels ... **M6:** 239, 241
medium-carbon steels ... **M6:** 239
stainless steels, austenitic ... **M6:** 239, 344
stainless steels, ferritic ... **M6:** 348
stainless steels, nitrogen-strengthened austenitic ... **M6:** 345
Electrogas welds
failure origins in ... **A11:** 440
Electrogeneration
and electrometric titration ... **A10:** 202
Electrographic analysis, electrometric titration and compared ... **A10:** 202
Electrographite
electrodes, resistance brazing ... **A6:** 340, 341
oxidation resistant, resistance brazing ... **A6:** 341
Electrogravimetry ... **A10:** 197-201
accuracy and precision ... **A10:** 197
applications ... **A10:** 197, 201
capabilities, compared with voltammetry ... **A10:** 188
constant-current methods ... **A10:** 198
constant-voltage ... **A10:** 199
controlled-potential electrolysis ... **A10:** 199
and electrometric titration ... **A10:** 203
estimated analysis time ... **A10:** 197, 200
general uses ... **A10:** 197
high precision and automation ... **A10:** 199
instrumentation ... **A10:** 199-200
internal electrolysis ... **A10:** 199, 200
limitations ... **A10:** 197
microelectrogravimetry ... **A10:** 200
Nernst equation ... **A10:** 197
power supplies and circuit requirements ... **A10:** 200
principles ... **A10:** 197-198
related techniques ... **A10:** 197
samples ... **A10:** 197, 200-201
selection of method ... **A10:** 198-199
types of analysis ... **A10:** 200-201
Electrohydraulic axial fatigue machine
load train ... **A8:** 368
Electrohydraulic forming
aluminum alloys ... **A14:** 802
Electrohydraulic gravity-drop hammers ... **A14:** 25, 42
Electrohydraulic testing machine
for torsional and axial loading ... **A8:** 159-160
Electrojet thinning
as sample preparation technique for ATEM ... **A10:** 451
Electrokinetic potential
defined ... **A13:** 6
Electrokinetic properties, of ceramic powders ... **EM4:** 73, 74
electro-osmosis ... **EM4:** 74
Electrokinetic properties, of ceramic powders (continued)
electrophoresis ... **EM4:** 74
sedimentation potential ... **EM4:** 74
streaming potential ... **EM4:** 74
Electroless copper plating ... **EL1:** 545, 870
Electroless discharge lamp ... **A10:** 690
Electroless nickel
aid for diffusion bonding ... **A6:** 159
Electroless nickel coatings
resistance to cavitation erosion ... **A18:** 217
Electroless nickel plating *See also* Nickel plating ... **M1:** 102-103, 179 **M5:** 219-243
accelerators used in ... **M5:** 222, 224
acid baths ... **M5:** 220-222
adhesion ... **M5:** 225-226
agitation used in ... **M5:** 236
alkaline baths ... **M5:** 220-221
aluminum and aluminum alloys ... **M5:** 219, 221, 228, 232
amino-borane process *See* Amino-borane electroless nickel plating process
applications ... **M5:** 219, 228, 237
barrel process ... **M5:** 237
borates in ... **M5:** 223
brass ... **M5:** 238-240
buffers used in ... **M5:** 223
bulk process ... **M5:** 237
cast iron ... **M5:** 238-240
complexing agents used in ... **M5:** 221-222, 236
composition, effects of ... **M5:** 228
copper and copper alloys ... **M5:** 232-233, 621-622
corrosion resistance ... **M5:** 229-231, 237-240
for edge retention ... **A12:** 95, 100
energy ... **M5:** 221-222
equipment ... **M5:** 233-236
fatigue strength of steel affected by ... **M5:** 226, 229, 231
filtration in ... **M5:** 235-236
frictional properties ... **M5:** 227-228
hardness ... **M5:** 227-228
heat treatment, effects of ... **M5:** 224-231
heating, steam and electric ... **M5:** 234-235
hydrazine process ... **M5:** 221
hydrogen embrittlement relief ... **M5:** 237
immersion plating ... **M5:** 219
inhibitors used in ... **M5:** 222-224
lead additions, effects of ... **M5:** 223-224
limitations and advantages of ... **M5:** 219
nickel-boron coatings properties of ... **M5:** 229-231
nickel-phosphorus coatings properties ... **M5:** 223-229
orthophosphite in ... **M5:** 220, 222-223, 225
of patterns ... **A15:** 196
pH effects ... **M5:** 223-224
phosphorus content effects of ... **M5:** 223-225, 228-231
physical and mechanical properties ... **M5:** 230-231
plating rate ... **M5:** 221-224
inhibitors affecting ... **M5:** 222-223
pH affecting ... **M5:** 223-224
succinate affecting ... **M5:** 222, 224
temperature affecting ... **M5:** 221-222
pretreatment ... **M5:** 231-233
aluminum alloys ... **M5:** 232
copper alloys ... **M5:** 232-233
ferrous alloys ... **M5:** 231-232
pumps, piping, and valves ... **M5:** 234-237
racks, baskets, trays, and fixtures ... **M5:** 236-237
reaction by-products ... **M5:** 223
reducing agents used in ... **M5:** 220-222
sodium borohydride process *See* Sodium borohydride electroless nickel plating process
sodium hypophosphite process *See* Sodium hypophosphite electroless nickel plating process
solderability ... **M5:** 228
solution compositions and operating conditions ... **M5:** 219-224
solution control ... **M5:** 237

Electroless nickel plating (continued)
specifications ... **M5:** 237-240
stainless steel ... **M5:** 232
steel ... **M5:** 230-232, 238-240
stress parameters ... **M5:** 222, 224-225, 231
structure ... **M5:** 224-225
succinate baths ... **M5:** 222-224
surface activation for ... **M5:** 232-233
Taber Abraser Index values ... **M5:** 227-228
tanks ... **M5:** 233-234
temperature ... **M5:** 221-222, 224-225, 227, 231, 234-235
elevated, effects of ... **M5:** 227
heating ... **M5:** 234-235
thickness ... **M5:** 225-226, 231
uniformity of ... **M5:** 225
wear resistance ... **M1:** 634 **M5:** 225-228, 237-240
worn surface buildup with ... **M5:** 237
Electroless plating ... **A18:** 834
aluminum and aluminum alloys ... **M5:** 604-606
capabilities/limitations ... **EL1:** 510
ceramic metallization technique ... **EM4:** 534, 544
copper and copper alloys ... **M5:** 621-622
defined ... **A13:** 6
flexible printed boards ... **EL1:** 583
magnesium alloys ... **M5:** 638-639, 644, 647
nickel *See* Electroless nickel plating
for plated-through hole (PTH) ... **EL1:** 114
silver ... **M5:** 606, 621-622
thin immersion ... **EL1:** 679
Electroless plating of specimens
for edge retention ... **A9:** 32
of tin and tin alloys for edge retention ... **A9:** 449
Electrolysis *See also* Electrolytic powders; specific electrolytic powders ... **A13:** 6, 87
cell ... **A10:** 199, 207-208
cell, internal ... **A10:** 199
circuit for ... **A10:** 199
constant-current, separation and analysis of metal ions by ... **A10:** 200
controlled-potential ... **A10:** 199, 207, 208-210
copper in copper-manganese alloy ... **A10:** 201
copper powder produced by ... **A7:** 110-116, 734
of copper powders ... **A2:** 393
current, as function of time ... **A10:** 210
defined ... **A10:** 672
factors affecting in electrogravimetry ... **A10:** 197-198
Faraday's laws of ... **A10:** 203
internal ... **A10:** 199, 200
internal, copper determined by ... **A10:** 200
internal, separation of cadmium and lead by ... **A10:** 201
iron powder production by ... **A7:** 93-96
metal precipitation by ... **A7:** 54
nickel in sample containing chloride ions ... **A10:** 201
preparation for coulometric titration ... **A10:** 204
in qualitative, classical wet methods ... **A10:** 168
as second-phase test method ... **A10:** 177
in voltammetry ... **A10:** 189
Electrolysis, contaminants removed by
nickel plating baths ... **M5:** 209
Electrolyte displacement
tantalum capacitors ... **EL1:** 998
Electrolyte dryout
passive devices ... **EL1:** 998-999
Electrolytes *See also* Electrolytes for electropolishing
for aluminum alloys ... **A9:** 353
applicability for electropolishing specific metals and alloys ... **A9:** 54
for austenitic manganese steel casting specimens ... **A9:** 238
automatic jet polishing ... **A9:** 107
for beryllium-copper alloys ... **A9:** 393
classified by chemical type ... **A9:** 51-55
for cleaning of ferrous fractures ... **A12:** 75
composition ... **A9:** 105
composition, as inclusion and phase isolation technique ... **A10:** 176
composition for ECM ... **A16:** 534
conditions for electropolishing various metals and alloys ... **A9:** 52-53
conductivity for ECM ... **A16:** 534, 535-536
current density vs. applied voltage for etching and polishing ... **A9:** 61
Electrolytes (continued)
defined ... **A10:** 672 **A13:** 6, 18
effect in voltammetry ... **A10:** 189
effect on electrochemical etching ... **A9:** 60-61
electrochemical discharge grinding ... **A16:** 548-549
and electrochemical grinding ... **A16:** 542, 543, 544-545, 546
electrochemical honing ... **A16:** 488
and electrochemical machining ... **A16:** 533, 540-541
for electrolytic etching copper and copper alloys ... **A9:** 401
electrostream and capillary drilling ... **A16:** 551, 552, 553
flow, galvanic corrosion ... **A13:** 238
flow rate ... **A16:** 536, 537-538
flow rate and ECM process control ... **A16:** 534, 537
formulas ... **A9:** 52-53
groups ... **A9:** 51-55
for in situ electropolishing ... **A9:** 55
for iron-chromium-nickel heat-resistant casting alloys ... **A9:** 332
laser-enhanced etching ... **A16:** 576
layer, thickness and corrosion ... **A13:** 82
liquid, galvanic corrosion testing in ... **A13:** 237
Ni alloy electrochemical machining ... **A16:** 843
nonsludging ... **A16:** 535
for refractory metals ... **A9:** 440
resistance, corrosion rate effects ... **A13:** 234
safety precautions ... **A9:** 51-55
shaped tube electrolytic machining ... **A16:** 554, 555, 556
sludging ... **A16:** 535
strong acid ... **A16:** 535
supporting, current-voltage curve of ... **A10:** 189-190
supporting, in effluent samples ... **A10:** 195
for thinning transmission electron microscopy specimens ... **A9:** 105-106
for Ti alloys ... **A16:** 852
traces, and conductance of water ... **A10:** 203
weak, dissociated ... **A10:** 203
for wrought heat-resistant alloys ... **A9:** 307-308
for wrought stainless steels ... **A9:** 281
Electrolytes for electropolishing ... **A9:** 51-53
chemicals for ... **A9:** 67-68
copper and copper alloys ... **A9:** 400
magnetic materials ... **A9:** 533
nickel alloys ... **A9:** 435
nickel-copper alloys ... **A9:** 435
titanium and titanium alloys ... **A9:** 459
uranium and uranium alloys ... **A9:** 478
Electrolytic
cells, defined ... **A13:** 6
cleaning, defined ... **A13:** 6
corrosion test ... **A13:** 219-220
spot test ... **A13:** 421
Electrolytic action in wire sawing ... **A9:** 26
Electrolytic alkaline cleaning ... **A7:** 459
Electrolytic alkaline descaling ... **M5:** 13
Electrolytic bismuth powders
mechanical comminution ... **A7:** 56
Electrolytic brightening
acid process ... **M5:** 582
alkaline process ... **M5:** 582
aluminum and aluminum alloys ... **M5:** 580-582, 596-597, 607-608
buffing compared to ... **M5:** 581-582
chemical brightening vs ... **M5:** 581
fluoboric acid process ... **M5:** 580
process selection ... **M5:** 580-582
sealing processes ... **M5:** 580-581
sodium carbonate process ... **M5:** 580
solution compositions and operating conditions ... **M5:** 580-582
sulfuric-phosphoric-chromic acid process ... **M5:** 580-581
Electrolytic capacitors
refractory metals and alloys ... **A2:** 557, 559
Electrolytic cell ... **A9:** 49-50
Electrolytic cleaning
postforging defects from ... **A11:** 333
Electrolytic cleaning, acid ... **M5:** 60-65
molybdenum and tungsten ... **M5:** 659
tantalum and niobium ... **M5:** 663
Electrolytic cleaning, alkaline ... **M5:** 26-32
aluminum and aluminum alloys ... **M5:** 578
anodic *See* Anodic electrocleaning
Electrolytic cleaning, alkaline (continued)
brass die castings ... **M5:** 26, 28-29
cadmium plating processes ... **M5:** 263
cathodic *See* Cathodic electrocleaning
cleaning cycles ... **M5:** 28-29
cleanliness of parts ... **M5:** 30
copper and copper alloys ... **M5:** 618-620
copper displacement test ... **M5:** 30
cutting fluids removed by ... **M5:** 10
dump schedules for cleaners ... **M5:** 29-30
electropolishing processes ... **M5:** 304
equipment ... **M5:** 25, 37-38, 44-45
hafnium alloys ... **M5:** 667
heat-resistant alloys ... **M5:** 567
magnesium alloys ... **M5:** 629-630, 639, 642
mechanism of action ... **M5:** 4, 26, 33
molybdenum ... **M5:** 659
nickel alloys ... **M5:** 673
periodic reverse system ... **M5:** 27-28
pigmented drawing compounds removed by ... **M5:** 5, 7
plating process precleaning ... **M5:** 17-18
polishing and buffing compounds removed by ... **M5:** 11-12
power supply ... **M5:** 31
process ... **M5:** 22
process cycles ... **M5:** 28
safety precautions ... **M5:** 31
solution compositions and operating conditions ... **M5:** 24, 28, 578, 618-620, 629-630
solution control and testing ... **M5:** 29-30
stainless steel ... **M5:** 561
steel parts ... **M5:** 27-30
tank, construction and equipment ... **M5:** 30-31
tungsten ... **M5:** 659
unpigmented oils and greases removed by ... **M5:** 9
water-break test ... **M5:** 30
zinc alloy die castings ... **M5:** 26, 28-29
zinc alloys ... **M5:** 26, 28-29
zirconium alloys ... **M5:** 667
Electrolytic color anodizing process
aluminum and aluminum alloys ... **M5:** 595
Electrolytic conductivity
electrometric titration and ... **A10:** 203
Electrolytic copper
capacitor discharge stud welding ... **A6:** 222
Electrolytic copper, cold rolled
with recrystallization nuclei ... **A9:** 695
Electrolytic copper powders *See also* Copper; Copper alloy powders; Copper powders
addition agents ... **A7:** 112, 113
applications ... **A7:** 116
compressibility curve ... **A7:** 287
current efficiency ... **A7:** 113
dendritic structure ... **A7:** 71, 72
density of compacts ... **A7:** 310
green strength and lubrication ... **A7:** 289
production ... **A7:** 112, 113
properties ... **A7:** 114-115
Electrolytic corrosion *See also* Corrosion; Electrical corrosion
cells, caused by applied bias ... **A11:** 770
as environmental failure mechanism ... **EL1:** 493
of gold ... **A11:** 771
Electrolytic deburring ... **M5:** 308-309
Electrolytic deposition *See also* Electrolytic iron powders
iron powders produced by ... **A7:** 93-96
Electrolytic etching *See also* Anodic etching; Electrolytic etching of specific metals and alloys ... **A9:** 61
in electropolishing ... **A9:** 56
equipment set up ... **A9:** 50
for image analysis samples ... **A10:** 313
used in quantitative metallography of cemented carbides ... **A9:** 275
Electrolytic etching of specific metals and alloys
aluminum alloys, mounting for ... **A9:** 352
austenitic manganese steel casting specimens ... **A9:** 239
chromized sheet steel ... **A9:** 198
copper and copper alloys ... **A9:** 401
electrical contact materials ... **A9:** 551
heat-resistant casting alloys ... **A9:** 331-333
lead and lead alloys ... **A9:** 416
molybdenum ... **A9:** 440

Electrolytic etching of specific metals and alloys (continued)
platinum-base alloys **A9:** 551
stainless-clad sheet steel **A9:** 198
tungsten **A9:** 440
wrought stainless steels **A9:** 281-282
zinc and zinc alloys **A9:** 489
Electrolytic extractions
defined **A9:** 6
wrought heat-resistant alloys **A9:** 308-309
Electrolytic grinding *See* Electrochemical grinding (ECG)
Electrolytic iron powder
microstructure **A9:** 509
Electrolytic iron powders *See also* Electrolysis; Iron powders **A7:** 71-72
annealing **A7:** 183, 185
apparent density **A7:** 273, 297
applications and microstructure **A7:** 95, 96
commercial processes **A7:** 93-94
compacts, effect of dwell time **A7:** 505
compacts, mechanical properties of hot pressed **A7:** 511
compacts, properties **A7:** 94
for copier powders **A7:** 587
dendritic grain structure, sheet **A7:** 72
effect of particle size on green strength **A7:** 303
electrolytic deposition produced **A7:** 93-96
flow rate **A7:** 273
for food enrichment **A7:** 615
green density and green strength **A7:** 289
mechanical comminution **A7:** 56
as new food iron source **A7:** 614
pickup of oxygen, carbon, and nitrogen **A7:** 64, 65
production process **A7:** 615
properties **A7:** 94-96
Electrolytic iron single crystal
cold rolled **A9:** 694
Electrolytic iron-carbon alloy
carbon content effect on torsional ductility **A8:** 166
Electrolytic magnesium, cold rolled 50%
shear bands in **A9:** 687
Electrolytic manganese powder **A7:** 72
Electrolytic pickling
nickel alloys **M5:** 673
pitting caused by **M5:** 80
rust and scale removed by **M5:** 12
steel **M5:** 76
Electrolytic pitch-grade copper
electronic applications **A6:** 998
Electrolytic plating
capabilities/limitations **EL1:** 510-511
ceramic metallization technique **EM4:** 534, 544
as pattern coating **A15:** 196
for PTH, vias, surface wiring **EL1:** 114
quality control **EL1:** 871-872
rigid printed wiring boards **EL1:** 545
Electrolytic polishing *See also* Electrolytic brightening; Electrolytic polishing of specific metals and alloys; Electropolishing; Polishing **M5:** 303-309
abrasive blasting processes **M5:** 304, 306, 308
acid processes **M5:** 303-305, 308
agitation used in **M5:** 305
alkaline cleaning **M5:** 304
aluminum and aluminum alloys **M5:** 305-306, 308
anodizing, pretreatment for **M5:** 305
apparatus **A9:** 49-50
appearance, effect on **M5:** 303, 306
applications **A9:** 51 **M5:** 305-306, 624
attack around nonmetallic particles, voids and inhomogeneities **A9:** 51
basic laboratory setup **A9:** 61
brass and brass alloys **M5:** 305-308
by automatic jet polisher **A9:** 107
cathodes **M5:** 304

Electrolytic polishing (continued)
chromic acid solutions **M5:** 303, 305, 308
coefficients of friction affected by **M5:** 307
copper and copper alloys **M5:** 305, 308, 623-624
copper plating process **M5:** 168
corrosion resistance affected by **M5:** 306-307
current density, effects of **M5:** 304
current density vs. applied voltage for common electrolytes **A9:** 61
current-voltage curves **A9:** 105
current-voltage relationships **A9:** 48-49
deburring process (electrodeburring) **M5:** 308-309
defined **A9:** 6
determining polishing plateau **A9:** 50
effects of applied potential **A9:** 105
effects on fatigue strength **A11:** 122
electrical circuits **A9:** 49-50
electroplating, pretreatment for **M5:** 305
equipment **M5:** 304-306, 308
special **M5:** 306
fatigue strength (limit) affected by **M5:** 307-308
film formation in **M5:** 303, 307
fume ventilation **M5:** 305-306
limits of edge effects **A9:** 51
local **A9:** 55-56
magnetic etch specimens **A9:** 64
mechanical polishing compared to **M5:** 306-308
mechanism of **A9:** 48
mechanism of action **M5:** 303
of multiphase alloys **A9:** 51
nickel and nickel alloys **M5:** 305-306, 308
passivation by **M5:** 306
phosphoric acid solutions **M5:** 303, 305, 308
physical and mechanical properties affected by **M5:** 306-308
precleaning **M5:** 303-304
preparation of specimens **A9:** 49-50
preparation of specimens for local polishing **A9:** 55
procedures **A9:** 49-51 **M5:** 303-305, 624
process selection factors **M5:** 624
racks and fixtures **M5:** 304
of replication microscopy specimens **A17:** 52
safety precautions **A9:** 51-55
selection of mounting materials for specimens **A9:** 28-29, 32
silver **M5:** 306
solution compositions and operating conditions **M5:** 303-305, 309
specimens for optical metallography analysis **A10:** 301
stages in **A9:** 105
stainless steel **M5:** 305-307, 559
steel **M5:** 305-309
stress parameters **M5:** 306-308
compressive stress **M5:** 307-308
stress-relieving effect **M5:** 307-308
sulfuric acid solutions **M5:** 303-305, 308
tanks **M5:** 304
temperature **M5:** 303-304
heating and cooling **M5:** 304
test cell arrangement for **A9:** 50
time, effects of **M5:** 304
transmission electron microscopy specimens to achieve thinning **A9:** 105-107
wear, amount of, effects on **M5:** 307
work bar **M5:** 304
Electrolytic polishing fluids
effects of **A9:** 47
Electrolytic polishing of specific metals and alloys
aluminum alloys **A9:** 353
aluminum alloys, mounting for **A9:** 352
of austenitic manganese steel casting specimens **A9:** 238
beryllium **A9:** 389
beryllium-copper alloys **A9:** 393
copper and copper alloys **A9:** 400
of copper, cell voltage as a function of anode current density **A9:** 48

Electrolytic polishing of specific metals and alloys (continued)
heat-resistant casting alloys **A9:** 330
iron-cobalt and iron-nickel alloys **A9:** 533
lead and lead alloys **A9:** 416
magnesium alloys **A9:** 426
magnetic materials **A9:** 533-534
of nickel alloys **A9:** 435
of nickel-copper alloys **A9:** 435
permanent magnet alloys **A9:** 533
refractory metals **A9:** 439-440
thin-foil stainless steel specimens **A9:** 282
titanium and titanium alloys **A9:** 459
uranium and uranium alloys **A9:** 478
wrought heat-resistant alloys **A9:** 307
wrought stainless steels **A9:** 280-281
Electrolytic potentiostatic etching **A9:** 61-62
Electrolytic powder, defined **A7:** 4
production **A7:** 71-72
reduction **A7:** 148
Electrolytic protection *See* Cathodic protection
Electrolytic reagents *See* Electrolytes
Electrolytic reversibility
bipotentiometric titration and **A10:** 204
Electrolytic salt bath descaling
automated system **M5:** 98-99
nickel alloys **M5:** 672
process **M5:** 97-100
Electrolytic silver powders *See also* Silver powders **A7:** 71, 72, 148, 149
Electrolytic solution potential *See also* Electrical properties; Solution potential
aluminum casting alloys **A2:** 153-177
wrought aluminum **A2:** 62-63
wrought aluminum and aluminum alloys **A2:** 80-122
Electrolytic specimen preparation
for NDE **A17:** 52
Electrolytic stripping measurement
hard chromium plate thickness **M5:** 181
Electrolytic theory **EM3:** 300
Electrolytic tin *See* Tin alloys, specific types, commercially pure tins
Electrolytic tin coatings **M1:** 173
Electrolytic titanium sponge powder
See also Titanium powders **A7:** 167
Electrolytic tough pitch (ETP) copper *See also* Copper
fatigue life as function of stress amplitude in **A8:** 253
filler metals **A6:** 755
gas-metal arc welding **A6:** 759-760
gas-tungsten arc welding **A6:** 758-759
sheet **A6:** 146
stress-strain curve in tension **A8:** 213-214
torque/radius data **A8:** 183-184
weldability **A6:** 753
Electrolytic tough pitch copper *See also* Copper; Copper alloys; Copper alloys, specific types, C11000; Tough pitch copper; Wrought coppers and copper alloys
applications and properties **A2:** 269-272
characteristics **A2:** 230
pole figures **A9:** 702
Electrolytic tough pitch copper, anneal resistant *See* Copper alloys, specific types, C11100
Electrolytic zinc powder *See also* Zinc powders **A7:** 72
Electrolytically deposited (ED) copper **EL1:** 545, 581, 871-872
Electrolytically generated radical ions
ESR analysis **A10:** 265
Electromagnet setup, for identification of ferrite in iron-chromium-nickel heat-resistant casting alloys **A9:** 333
Electromagnetic
excitation, for axial fatigue testing **A8:** 369
gage, for shear wave profiles **A8:** 231
shakers, speed of **A8:** 43

Electromagnetic (continued)
testing machine ... **A8:** 43
Electromagnetic casting
of aluminum alloy ingots used to decrease surface defects ... **A9:** 634
aluminum alloys ... **A15:** 315
effect on dendritic structures in copper alloy ingots ... **A9:** 638
Electromagnetic circuit
magnetic field direction ... **A17:** 90
Electromagnetic enhancement *See also* Enhancement and SERS ... **A10:** 136
Electromagnetic field, molecular activity
in Raman spectroscopy ... **A10:** 127
Electromagnetic focusing device *See* Focusing device
Electromagnetic force (EMF)
inductance effects ... **EL1:** 27
Electromagnetic forming *See also* High-energy-rate forming ... **A14:** 644-653
advantages/limitations ... **A14:** 646
aluminum alloys ... **A14:** 802-803
applications ... **A14:** 646-650
of copper and copper alloys ... **A14:** 819
defined ... **A14:** 5, 644
electrical principles ... **A14:** 652
energy relations, typical ... **A14:** 652-653
equipment ... **A14:** 650-652
methods ... **A14:** 645
process description ... **A14:** 644-646
production methods ... **A14:** 646
safety ... **A14:** 650
speed ... **A14:** 644-645
workpiece design ... **A14:** 645
Electromagnetic induction
and eddy current inspection ... **A17:** 164
soft magnetic materials for ... **A2:** 761
Electromagnetic inspection *See* Electromagnetic techniques
as nondestructive testing tor fatigue ... **A11:** 134
Electromagnetic interference ... **A13:** 1108-1109
Electromagnetic interference (EMI) ... **A6:** 36
immunity of optical systems ... **EL1:** 9, 16
immunity to ... **EM4:** 409
Electromagnetic interference (EMI) protection
defined ... **EM1:** 359
Electromagnetic interference (EMI) shielding ... **EM2:** 476-478 **EM3:** 52
Electromagnetic interference shielding alternatives ... **A7:** 612
by metal-filled polymers ... **A7:** 609-610, 612
by specialty polymers ... **A7:** 606
waves ... **A7:** 609
Electromagnetic interference/radio frequency interference (EMI/RFI) ... **EM2:** 268-269
Electromagnetic lens
defined ... **A9:** 6
Electromagnetic lenses ... **A10:** 432, 672
Electromagnetic noise
sources of ... **A17:** 285
Electromagnetic radiation
atomic processes ... **A17:** 309-310
attenuation ... **A17:** 309-311
defined ... **A10:** 672
effective absorption, x-rays ... **A17:** 310-311
FMR resonant absorption of ... **A10:** 267
intensity of ... **A10:** 83
properties of ... **A10:** 83
radiographic equivalence ... **A17:** 311
and radiology ... **A17:** 295
spectrum, high-energy region ... **A10:** 83
in x-ray spectrometry ... **A10:** 83
Electromagnetic radiation emitted by electrons ... **A9:** 92
Electromagnetic sorting
and hardness testing ... **A7:** 484-485
Electromagnetic spectrum
divisions ... **A17:** 202
Electromagnetic stirring
induction furnaces ... **A15:** 369
Electromagnetic techniques
Barkhausen noise ... **A17:** 159-160
electric current perturbation as ... **A17:** 136
for forgings ... **A17:** 510-511
introduction ... **A17:** 51
magnetically induced velocity changes (MICV), for ultrasonic waves ... **A17:** 161-162
Electromagnetic techniques (continued)
nonlinear harmonics ... **A17:** 160-161
for residual stress measurement ... **A17:** 159-163
of steel bar and wire ... **A17:** 552-555
Electromagnetic testing
use in carbon control ... **M4:** 436-437
Electromagnetic theory and radiation ... **A10:** 126-127
Electromagnetic waves
as transverse ... **A17:** 203
Electromagnetic welding ... **A14:** 649 **EM2:** 724
Electromagnetic yokes *See also* Magnetic particle inspection; Yokes
defined ... **A17:** 93, 95
magnetic ... **A17:** 95
Electromagnetic-acoustic transducers
ultrasonic inspection ... **A17:** 255-256
Electromagnets
in gas mass spectrometers ... **A10:** 154
magnetic field generation by ... **A17:** 93
Electromechanical components
through-hole packages ... **EL1:** 979-981
Electromechanical fatigue systems
for axial testing ... **A8:** 369
closed-loop servomechanical ... **A8:** 394-395
comparison ... **A8:** 392
dynamic cycler ... **A8:** 395, 397
forced displacement ... **A8:** 391-392
functions ... **A8:** 391
resonance ... **A8:** 392-396
rotational bending ... **A8:** 392-393
servomechanical ... **A8:** 392
Electromechanical fatigue testing ... **A8:** 391-395
Electromechanical polishing ... **A9:** 42-43
defined ... **A9:** 6
and etch attack ... **A9:** 42-43
of molybdenum ... **A9:** 441
of tungsten ... **A9:** 441
Electrometric titration ... **A10:** 202-206
advantages ... **A10:** 202
amperometric ... **A10:** 204
applications ... **A10:** 202, 205-206
biamperometric ... **A10:** 204
bipotentiometric ... **A10:** 204
concentrations and reaction speeds in ... **A10:** 202
conductometric ... **A10:** 203
coulometric ... **A10:** 204-205
defined ... **A10:** 672
estimated analysis time ... **A10:** 202
Faraday's laws of electrolysis ... **A10:** 203
general uses ... **A10:** 202
introduction ... **A10:** 203
limitations ... **A10:** 202
oscillometric (high-frequency) ... **A10:** 203-204
potentiometric ... **A10:** 204
related techniques ... **A10:** 202
samples ... **A10:** 202, 205-206
as volumetric analysis ... **A10:** 202
Electromigration *See also* Tin Whiskers; Whiskers ... **EM3:** 581
effects at contact windows ... **A11:** 778
effects of pulsed direct current on gold film conductors ... **A11:** 771-772
as environmental failure mechanism ... **EL1:** 494
and failure kinetics, VLSI ... **EL1:** 889-890
as failure mechanism ... **EL1:** 1014-1016
failure rates, factors affecting ... **A11:** 771-772
-induced failures, effect of grain size and stripe width on ... **A11:** 772-774
in integrated circuits ... **A11:** 770-773
as semiconductor failure mechanism ... **EL1:** 963-964
with Technora fabric ... **EL1:** 535
Electromotive force
defined ... **A13:** 6
SI defined unit and symbol for ... **A10:** 685
SI unit/symbol for ... **A8:** 721
Electromotive force (emf) *See also* Thermocouple materials
defined ... **A2:** 869-871
stability ... **A2:** 881-882
and temperature relationship ... **A2:** 878
Electromotive force series *See also* Electrohemical series
and cell potentials ... **A13:** 20-21
for common metals ... **A13:** 20
defined ... **A13:** 6
metals/alloys ... **A13:** 419
Electromotive force series (continued)
for selected metals ... **A13:** 467
Electromotive series
aircraft alloys ... **EM1:** 716
Electromotive series of elements ... **A9:** 59
Electron *See* Electron beams; Electron diffraction; Electrons; Secondary electrons
defined ... **A9:** 6
energy distribution, symbol for ... **A11:** 797
mean free path, vs kinetic energy ... **A11:** 35
penetration depth, as function of voltage and sample density ... **A11:** 41
symbol for ... **A11:** 796
volt, symbol for ... **A11:** 796
Electron beam
defined ... **A9:** 6
refining, effect on inclusions ... **A11:** 340
and specimen, schematic interaction between ... **A11:** 33
voltages, spectra obtained from different ... **A11:** 41
Electron beam (EB) welding
mechanically alloyed oxide alloys ... **A2:** 949
dispersion-strengthened (MA ODS) of nickel and nickel alloys ... **A2:** 429
refractory metals and alloys ... **A2:** 560-562, 563
Electron beam analysis *See also* Electron beam induced current; Electron probe x-ray microanalysis; Microbeam analysis
local surface study by ... **A10:** 517
Electron beam brazing
stainless steels ... **M6:** 1012-1013
Electron beam convergence, effect
radiography ... **A17:** 308
Electron beam curing ... **EL1:** 786, 857
Electron beam cutting
definition ... **M6:** 6
Electron beam exposure for curing ... **EM3:** 35, 74
Electron beam gun
definition ... **M6:** 6
Electron beam gun column
definition ... **M6:** 6
Electron beam hardening treatment (EBHT) ... **A4:** 297-311
advantages ... **A4:** 297, 310
beam applications ... **A4:** 305-308, 309
beam location and duration control ... **A4:** 299-301
computer numerical control (CNC) ... **A4:** 307, 309, 310, 311
cutting edges of harvester mower blade example ... **A4:** 308
disadvantages ... **A4:** 310
electron beam hardening facilities ... **A4:** 309
electron guns ... **A4:** 308-310
Electroslag remelting (ESR) ... **A4:** 207
energy absorption ... **A4:** 297-298
energy transfer ... **A4:** 298-299, 300, 301, 303, 307, 310
grain size ... **A4:** 305, 306
hardness gradient ... **A4:** 302-303
heat conduction ... **A4:** 297-298
integration into a flexible manufacturing system ... **A4:** 310
isothermal energy transfer ... **A4:** 298, 300, 301, 302, 303, 306, 307
malleable iron ... **A4:** 695
milling machine quill example ... **A4:** 307-308
narrow tempering zone anomaly ... **A4:** 308
properties of the hardened layer ... **A4:** 301-305
shaft partially surface hardened ... **A4:** 308
surface deformation ... **A4:** 303-304
surface roughness ... **A4:** 303, 304
temperature-time cycle ... **A4:** 298-299, 300-301
versus laser beam hardening ... **A4:** 309-310
workpiece distortion ... **A4:** 304-305
Electron beam heating
surface hardening of steel parts ... **M1:** 532
Electron beam heating vacuum coating ... **M5:** 390, 404-405, 410
Electron beam induced current
abbreviation for ... **A10:** 690
arrangement using Schottky barrier technique ... **A10:** 507
as SEM special technique ... **A10:** 507
Electron beam induced current (EBIC) ... **EL1:** 372, 1094, 1100-1101

Electron beam investment casting **A15:** 417-418
Electron beam machining **EM4:** 313
Electron beam machining (EBM) **A16:** 509, 568-571
advantages and disadvantages **A16:** 571
applications **A16:** 571
backing material **A16:** 570, 571
CNC systems **A16:** 569, 571
equipment description **A16:** 568
on-the-fly drilling **A16:** 568, 570
process characteristics **A16:** 569-571
stainless steels **A16:** 705, 706
tensioning drum **A16:** 568-569
vacuum chamber **A16:** 568
Wehnelt electrodes **A16:** 568
Electron beam melting **A7:** 160-162
Electron beam melting and casting
button melting **A15:** 412
characteristics **A15:** 410-411
competing processes, compared **A15:** 410
drip melting **A15:** 412-413
equipment, drip melting **A15:** 413
heat source specifications **A15:** 411-412
melted metals, characteristics **A15:** 413-414
processes **A15:** 411
quality control **A15:** 412
Electron beam rotating disc process
Leybold-Heraeus **A7:** 167
Electron beam vaporization
ion plating process **M5:** 418
Electron beam voltage contrast
as technique in integrated circuit failure analysis **A11:** 768
Electron beam welded joints
metallography and microstructures **A9:** 581
Electron beam welding **M6:** 609-646
advantages **M6:** 610
beam oscillations **M6:** 634
in containerized hot isostatic pressing **A7:** 431
controlling heat effects **M6:** 625-626
heat-sensitive attachments and inserts **M6:** 625
crack prevention **M6:** 630, 637
definition **M6:** 6
discontinuities from **A17:** 585-587
fixturing methods **M6:** 612
hardness traverses **M6:** 637
high vacuum, welding in **M6:** 618-619
applications **M6:** 618
effect of pressure **M6:** 618-619
limitations **M6:** 619
welding conditions **M6:** 619
width of weld and heat-affected zone **M6:** 619
joining dissimilar metals **M6:** 644
indirect joining **M6:** 644
joint design **M6:** 615-618
butt joints **M6:** 615-618
butt versus corner and T-joints **M6:** 617-618
corner joints **M6:** 616-618
edge joints **M6:** 617
lap joints **M6:** 616-617
T-joints **M6:** 616-618
joint fit-up **M6:** 612
joint preparation **M6:** 612
joint tracking **M6:** 634
automatic **M6:** 634
electromechanical **M6:** 634
manual **M6:** 634
tape-controlled **M6:** 634
limitations **M6:** 611
machines **M6:** 644-646
medium vacuum, welding in **M6:** 619-620
applications **M6:** 620
comparison with high-vacuum welding **M6:** 620
comparison with nonvacuum welding **M6:** 620
Electron beam welding (continued)
penetration and weld shape **M6:** 620
multiple-tier welding **M6:** 631-632
applications **M6:** 631
difficulties **M6:** 631
nonvacuum welding **M6:** 620-624
applications **M6:** 624
operating conditions **M6:** 621-622
penetration **M6:** 622-623
procedure **M6:** 620-621
tooling **M6:** 623-624
weld shape and heat input **M6:** 623
weld shrinkage **M6:** 623
operating conditions **M6:** 613
operation principles **M6:** 609-610
operation sequence and preparation **M6:** 612-613
cleaning **M6:** 612
demagnetization **M6:** 612
preheat and postheat **M6:** 613
pumpdown **M6:** 612-613
vacuum or nonvacuum **M6:** 612
poorly accessible joints **M6:** 631
beam characteristics **M6:** 631
sidewall clearance **M6:** 631
workpiece requirements **M6:** 631
process control **M6:** 611-612
beam spot size **M6:** 611
repair welding **M6:** 635-637
application of wire-feed process **M6:** 635
future applications **M6:** 636-637
quality assurance testing **M6:** 636
wire feed and equipment **M6:** 635
safety precautions **M6:** 59, 646
scanning, use of **M6:** 632-634
accuracy of beam alignment **M6:** 632
electronic systems **M6:** 633-634
optical scanning **M6:** 633
problems **M6:** 633
procedure **M6:** 632-633
techniques **M6:** 633
without optics **M6:** 633
special joints and welds **M6:** 618
tack welding **M6:** 630
use of filler metal **M6:** 630
equipment for feeding **M6:** 630
feeding of filler wire **M6:** 630
preplacement **M6:** 630
preventing porosity **M6:** 630
prevention of cracking **M6:** 630
techniques for feeding **M6:** 630
use of pulsed beam **M6:** 634
weld geometry **M6:** 613-615
advantages of vacuum welding **M6:** 628
effects of pressure **M6:** 628
full-penetration welding **M6:** 628
melt-zone configuration **M6:** 615
part shape **M6:** 613-614
partial-penetration welding **M6:** 629
problems and flaws **M6:** 629-630
surface geometry **M6:** 614
two-pass full-penetration welding **M6:** 628
welding of thick metal **M6:** 628-630
wide welds bridging a gap **M6:** 614-615
welding of thin metal **M6:** 626-628
joining thin to thick sections **M6:** 627
partial-penetration welds **M6:** 628
welding with extreme accuracy **M6:** 632
Electron beam welding of
aluminum alloys **M6:** 641-642
heat treatable alloys **M6:** 641-642
non-heat-treatable alloys **M6:** 641
beryllium **M6:** 643-644
autogenous welds **M6:** 643-644
braze welds **M6:** 644
copper and copper alloys **M6:** 642
hardenable steel **M6:** 637-638
hardened and work-strengthened metals **M6:** 624-625
Electron beam welding of (continued)
heat-resistant alloys **M6:** 638
cobalt-based alloys **M6:** 638
iron-nickel-chromium-based alloys **M6:** 638
precipitation-hardenable nickel-based alloys **M6:** 638
solid-solution nickel-based alloys **M6:** 638
high-carbon steels **M6:** 638
high-strength alloy steels **M6:** 637-638
low-alloy steels **M6:** 637
low-carbon steel **M6:** 637
magnesium alloys **M6:** 643
medium-carbon steels **M6:** 637
refractory metals **M6:** 638-641
molybdenum **M6:** 639-640
niobium **M6:** 640-641
tantalum **M6:** 641
tungsten **M6:** 639
stainless steels **M6:** 638
titanium alloys **M6:** 643
tool steels **M6:** 638
Electron beam welds
butt **A11:** 447
failure origins **A11:** 444-447
Electron beams
50-kV **A10:** 326
artifacts, as AES limitation **A10:** 556
diffracted, intensities of **A10:** 328-329
diffracted, two-dimensional angular profile effect of differing energies **A10:** 498
energy distribution of signals generated by **A10:** 498
factors controlling shape of **A10:** 331, 332
interaction with specimens **A10:** 432-434
monochromatic, with single-crystal diffraction methods **A10:** 329-330
monochromatizing **A10:** 326
polychromatic, with single crystal diffraction methods **A10:** 329
as primary AES excitation **A10:** 550
scanning instruments **A10:** 497-500
spreading, in thin foils and bulk targets compared **A10:** 434
total, change in amplitude as function of scattering angle **A10:** 329
volume of signals produced **A10:** 498-500
in x-ray spectrometry **A10:** 82
Electron capture, and positron emission
as radioactive decay mode **A10:** 245
Electron channeling *See also* Channeling
capabilities **A10:** 365
contrast **A10:** 505
pattern, of vanadium taken in ECP mode **A10:** 504
patterns and contrast **A10:** 504-506
Electron column **A7:** 235
Electron configurations
for elements **A10:** 688
Electron density
and adhesion **A6:** 144
Electron density maps
by direct method **A10:** 350, 351
by heavy-atom method **A10:** 350
determined **A10:** 349
three-dimensional, of potassium benzyl penicillin **A10:** 350
Electron detection
as EXAFS technique **A10:** 418
Electron diffraction
analytical transmission electron microscopy **A10:** 436-440
defined **A9:** 6 **A10:** 672
and energy-dispersive spectrometry, unknown phase identification by **A10:** 455-459
EXAFS analysis as mode of **A10:** 410
fine structure effects **A10:** 438
pattern, effect of tilt **A10:** 454
patterns, indexing **A10:** 456-457

Electron diffraction (continued)
used to examine plate steels for sub-micron particles **A9:** 203
used to identify deformation twins **A9:** 688
used to investigate crystallographic texture **A9:** 701
of wrought stainless steels **A9:** 283
Electron diffraction spectroscopy
used to study x-ray emissions **A9:** 104
Electron discharge machining (EDM)
borides **EM4:** 794
Electron effect
final-state **A10:** 408
Electron energy analyzers
for x-ray photoelectron spectroscopy **A10:** 570-571
Electron energy distribution
abbreviation for **A10:** 690
Auger electron spectroscopy **A10:** 550, 551
Electron energy level diagrams *See* Energy level diagrams
Electron energy loss
transmission electron microscopy **A10:** 432, 435
Electron energy loss spectroscopy **A9:** 104 **EM1:** 285
with analytical electron microscope **A10:** 432
capabilities, and FIM/AP **A10:** 583
defined **A10:** 449, 671
light-element analysis **A10:** 459-461
limitations **A10:** 450
magnetic spectrometer and detector for **A10:** 435, 449
qualitative analysis **A10:** 450
spectrum with zero-loss, low-loss, and core-loss regions **A10:** 449-450
Electron energy loss spectroscopy (EELS) **A6:** 145 **A18:** 377, 378, 387, 390, 391 **EM3:** 237
Electron erosion
passive devices **EL1:** 999
Electron flow
defined **A13:** 6
Electron fractography
defined **A12:** 1
history **A12:** 4-8
Electron gas model **EL1:** 96-98
Electron gun **A7:** 235
defined **A9:** 6
Electron guns
for AES primary excitation **A10:** 554
analytical transmission electron microscopy **A10:** 432
conventional tungsten hairpin filament **A10:** 492
field emission guns **A10:** 432
SEM microscope **A10:** 491-492
vacuum coating process **M5:** 390, 393
Electron holes
and conductivity **EL1:** 89
Electron image
defined **A9:** 6
Electron images *See also* Images; Imaging
and x-ray area scans, microanalysis of complex structures by **A10:** 525
Electron lens
defined **A9:** 6
Electron micrograph
defined **A9:** 6
Electron microprobe
used to examine plate steels **A9:** 203
Electron microprobe images
electronic analysis **A9:** 152-153
Electron microprobes
capabilities **A10:** 102, 161
compositional analysis of welds **A6:** 100
Electron microscope *See also* Scanning electron microscope; Transmission electron microscope
defined **A9:** 6
Electron microscope column
defined **A9:** 6
Electron microscopy *See also* Scanning electron microscope; Scanning electron microscopy; Transmission electron microscope; Transmission electron microscopy **A18:** 376-391
for adhesion quality **A17:** 611
defined **A9:** 6
electron-specimen interaction **A18:** 377-378
elastic scattering **A18:** 377

Electron microscopy (continued)
inelastic scattering **A18:** 377, 387
facets in the investigation of wear **A18:** 376
for microstructural analysis **EM4:** 578
SEM and TEM instruments **A18:** 378-380
electron sources **A18:** 378
voltage selection **A18:** 378
specimen preparation **A18:** 380-382
debris specimens **A18:** 382
scanning electron microscopy **A18:** 380-381
transmission electron microscopy **A18:** 381-382
speckle method **A17:** 435
for surface replicas **A17:** 52
of tin and tin alloy coated materials **A9:** 451
to analyze ceramic powder particle sizes **EM4:** 66, 67, 69
two societies devoted to the advancement of this field **A18:** 377
used to detect omega phase in titanium alloys **A9:** 461
used to detect ordered phases in titanium and titanium alloys **A9:** 460
used to examine case hardened steels **A9:** 217
used to examine fiber composites **A9:** 592
used to identify intragranular subgrain structures **A9:** 690
of wrought stainless steels **A9:** 282-283
Electron microscopy impression *See* Impression
Electron microscopy speckle method
speckle metrology **A17:** 435
Electron multiplier analyzer
for x-ray photoelectron spectroscopy **A10:** 571
Electron multiplier phototube *See* Photomultiplier tube
Electron multipliers
in Auger spectrometer **A10:** 554
as ion detectors, gas mass spectroscopy **A10:** 154
Electron nuclear double resonance
abbreviation for **A10:** 690
as supplemental ESR technique **A10:** 258
Electron optical axis
defined **A9:** 6
Electron optical methods
analytical transmission electron microscopy **A10:** 429-489
electron probe x-ray microanalysis **A10:** 516-535
low-energy electron diffraction **A10:** 536-545
scanning electron microscopy **A10:** 490-515
Electron optical system
defined **A9:** 6
Electron optical-lens aberrations
effect on secondary electron imaging **A9:** 95
Electron optics **A13:** 1117
as advanced failure analysis techniques **EL1:** 1106
analytical transmission electron microscopy **A10:** 432
backscattered electron (BSE) imaging **EL1:** 1096-1097
columns, SEM, TEM, and AEM analysis **A10:** 432
electron beam induced current (EBIC) **EL1:** 1100-1101
of electron probe microanalyzer **A10:** 517
energy-dispersive x-ray spectroscopy (EDS) **EL1:** 1094-1095, 1097-1098
scanning electron microscope (SEM) **EL1:** 1094-1095
secondary electron imaging (SEI) **EL1:** 1095-1096
SEM examination sample **EL1:** 1099-1100
voltage contrast **EL1:** 1100-1101
wavelength dispersive x-ray spectroscopy (WDS) **EL1:** 1098-1099
Electron or x-ray spectroscopic methods
Auger electron spectroscopy **A10:** 549-567
x-ray photoelectron spectroscopy **A10:** 568-580
Electron paramagnetic resonance *See* Electron spin resonance
Electron probe
defined **A9:** 6
Electron probe microanalysis (EPMA) **A18:** 450
abbreviation for **A11:** 796
and AES failure analysis **A11:** 38-41
development of **A11:** 32-33
light-element detection using **A11:** 38
for microchemical analysis **EM4:** 25

Electron probe microanalysis (EPMA) (continued)
for microconstituents, failed extrusion press **A11:** 37
for microstructural analysis **EM4:** 578-579
for phase analysis **EM4:** 25
phosphorus content determined by **A11:** 41-42
quantitative thin-film analysis by **A11:** 41
spectra, with windowless detector **A11:** 39, 40
spectrum, intergranulated fracture in Inconel 600 .... **A11:** 40
to distinguish components of a ceramic mixture **EM4:** 96
uses **A11:** 37-38
Electron probe microanalyzer
electron optics of **A10:** 517
history **A10:** 517
schematic, with associated circuitry **A10:** 517
Electron probe x-ray micro analysis **A10:** 516-535
applications **A10:** 516, 530-535
basic microanalytical concepts **A10:** 517-518
capabilities **A10:** 102, 309, 429, 549, 610
capabilities, compared with optical metallography **A10:** 299
defined **A10:** 672
elemental mapping **A10:** 525-529
estimated analysis time **A10:** 516
flat, polished specimens for **A10:** 516
general uses **A10:** 516
homogeneity requirement of **A10:** 529
of inorganic solids **A10:** 4
introduction **A10:** 517
lateral and depth resolution **A10:** 516
limitations **A10:** 516
limits of detection **A10:** 525
measuring x-ray spectra **A10:** 518-522
nonapplicability of inhomogeneous samples for **A10:** 529
of organic solids, information from **A10:** 9
physical bases of **A10:** 518
qualitative analysis **A10:** 522-524
quantitative analysis **A10:** 524-525
related techniques **A10:** 516
samples **A10:** 516, 529-530
and secondary ion mass spectroscopy compared **A10:** 516, 517
sensitivity **A10:** 516
as spatially resolved analysis for micrometer- sized volumes **A10:** 529
specificity of spectra **A10:** 518
standards, accuracy and precision of **A10:** 524-525, 530
strategy for applying microbeam analysis **A10:** 529-530
Electron radiation *See also* Radiation properties
effects, para-aramid fibers **EM1:** 56
Electron scattering
defined **A10:** 672
Electron spectrometers
for AES analyses **A10:** 554
Electron spectroscopy
applications **EL1:** 1080-1083
Auger electron spectroscopy **EL1:** 1077-1080
imaging photoemission microscopy **EL1:** 1077
photoemission, instrumentation **EL1:** 1076-1077
for surface analysis **EL1:** 1074-1080
Electron spectroscopy for chemical analysis **A10:** 568, 689 **A13:** 1117
testing parameters **A7:** 251
Electron spectroscopy for chemical analysis (ESCA) *See also* X-ray photoelectron spectroscopy (XPS) **A11:** 35 **A18:** 445, 457 **EL1:** 1074, 1107 **EM1:** 285 **EM3:** 237
alloy identification **A18:** 305
for phase analysis **EM4:** 557
seal wear **A18:** 552
for surface analysis **EL1:** 1074
Electron spin
in ferromagnetic resonance (FMR) **A17:** 220
Electron spin resonance *See also* ESR spectrometers; Resonance methods **A10:** 253-266
acoustic **A10:** 258
anisotropies **A10:** 261-262, 265
applications **A10:** 253, 265-266
capabilities **A10:** 267
double resonance **A10:** 258
electron-electron double resonance **A10:** 258

Electron spin resonance (continued)
estimated analysis time **A10:** 253
general uses **A10:** 253
information gained using **A10:** 4, 6, 9, 10, 262
of inorganic solids **A10:** 4, 6
instrumentation **A10:** 254-257
introduction and principles **A10:** 253-254
limitations **A10:** 253
lineshapes **A10:** 261
NMR, IR, UV/VIS analysis methods and compared **A10:** 265
optical double magnetic resonance **A10:** 258
of organics **A10:** 9, 10
Planck's constant **A10:** 254
related techniques compared **A10:** 253, 257-258, 264-265
and relaxation **A10:** 257-258
samples **A10:** 253, 256, 262-266
and saturation **A10:** 257-258
sensitivity **A10:** 258-259
spectra **A10:** 259-261
supplementary experimental techniques **A10:** 257-258
typical data, summary on first transition series **A10:** 263
Electron spin resonance (ESR) **EM4:** 52
defined **EM2:** 16
in magabsorption development **A17:** 143
to analyze the surface composition of ceramic powders **EM4:** 73-74
Electron spin resonance (ESR) spectroscopy **EM3:** 11
Electron trajectory
defined **A9:** 6
Electron tubes
vs semiconductors **EL1:** 958
Electron tunneling
rate in field ionization **A10:** 585
Electron vacancy number (N_v) **A6:** 593
Electron velocity
defined **A9:** 6
Electron volt
abbreviation for **A10:** 691
Electron wavelength
defined **A9:** 6
Electron(s) *See also* Secondary electrons
in atomic structure **EL1:** 92-93
backscattered **A12:** 167-168 **EL1:** 1095
beam, in SEM imaging **A12:** 167-168
beams, radiography **A17:** 308
and conductivity **EL1:** 89
diffraction, for crystallographic description **EL1:** 93
and electrical properties **EL1:** 89
gun, in SEM imaging system **A12:** 167
mobility limits **EL1:** 95
and neutron, in radiography **A17:** 390
normal state, configurations **EL1:** 91-92
primary, as x-ray source **A17:** 305
range, for aluminum, copper, gold **EL1:** 1095
signals, SEM **A12:** 168
trajectories, tungsten and aluminum **A12:** 167
traps **EL1:** 93
Electron-beam brazing **A6:** 123-124, 922-923
advantages **A6:** 922
applications **A6:** 922
stainless steels **A6:** 923
Electron-beam coevaporation
as thin-film deposition technique **A2:** 1081
Electron-beam cutting (EBC)
definition **A6:** 1209
Electron-beam evaporation
to apply interlayers for solid-state welding **A6:** 165
Electron-beam gun
definition **A6:** 1209
Electron-beam heat treating
advantages **M4:** 518, 521
application criteria **M4:** 518-519

Electron-beam heat treating (continued)
beam control **M4:** 519-520
case depth control **M4:** 520-521
computer-control system **M4:** 520
equipment **M4:** 519, 520
quench media **M4:** 521
raster pattern **M4:** 520, 521
Electron-beam melting
of nickel-titanium shape memory effect (SME) alloys **A2:** 899
Electron-beam vacuum refining
ferritic stainless steels **A6:** 443, 444
Electron-beam welding
failure mechanisms **EL1:** 1044-1045
joint configurations **EL1:** 240
processes, applications **EL1:** 238
Electron-beam welding (EBW) **A6:** 254-261
advanced titanium-base alloys **A6:** 526
advantages **A6:** 255-256
aerospace applications **A6:** 386
aluminum **A6:** 851, 857
aluminum alloys **A6:** 739, 828
aluminum metal-matrix composites **A6:** 555, 557
aluminum-lithium alloys **A6:** 551, 552
applications **A6:** 851, 852, 854
sheet metals **A6:** 398, 400
austenitic stainless steels **A6:** 459, 462, 464
bimetal band saw blades **A6:** 1185
carbon steels **A6:** 259, 828, 860
cast irons **A6:** 720
characteristics **A6:** 254, 256
cleaning of workpiece surfaces **A6:** 257
copper **A6:** 851
"cosmetic pass" **A6:** 259-260
cryogenic service **A6:** 1017
definition **A6:** 254, 1209
demagnetization **A6:** 258
depth-to-width ratio **A6:** 255, 260
dispersion-strengthened aluminum alloys **A6:** 543, 544, 545, 546
energy conversion efficiency **A6:** 255-256
equipment **A6:** 255, 266, 258, 259, 260-261
ferritic stainless steels **A6:** 448
fixturing **A6:** 257-258
heat sources **A6:** 1144
heat-treatable aluminum alloys **A6:** 528
high or "hard" vacuum **A6:** 255, 256, 257, 260, 261
high-energy-beam welding procedures **A6:** 852
high-strength alloy steels **A6:** 258-259
iron-base alloys **A6:** 865
joint design **A6:** 260, 852-854
blind weld **A6:** 853
butt joints vs. corner and T-joints **A6:** 854
butt joints, welds in **A6:** 852-853
corner joints, welds in **A6:** 853
corner-flange weld **A6:** 853
double-square-groove weld **A6:** 853, 854
melt-through weld **A6:** 853, 854
shallow weld **A6:** 853-854
joint fit-up **A6:** 257
joint preparation **A6:** 257
limitations **A6:** 256
limitations on procedure qualifications **A6:** 1093
with low-temperature solid-state welding **A6:** 301
medium vacuum (EBW-MV) **A6:** 255, 258, 259, 260, 261
modes **A6:** 255, 260-261
molybdenum alloys **A6:** 581
nickel alloys **A6:** 587, 588
nickel-base corrosion-resistant alloys containing molybdenum **A6:** 594
nickel-chromium alloys **A6:** 587, 588
nickel-chromium-iron alloys **A6:** 587, 588
nickel-copper alloys **A6:** 587, 588
niobium alloys **A6:** 581
non-heat-treatable aluminum alloys **A6:** 538-539
nonvacuum (atmospheric) **A6:** 255, 260, 261

Electron-beam welding (EBW) (continued)
nonvacuum (EBW-NV) **A6:** 255, 256, 258, 260
process **A6:** 852, 854-855
operating conditions **A6:** 259-260
operation principles **A6:** 254-255
operation sequence **A6:** 257-260
oxide-dispersion-strengthened materials **A6:** 1038, 1039
oxidizable metals **A6:** 851
power sources **A6:** 42-44
precipitation-hardening stainless steels **A6:** 484, 487, 489, 490, 491, 492
preheat and postheat **A6:** 258-259
preparation **A6:** 257-260
pressure **A6:** 255, 260
primary components of an electron-beam welding head **A6:** 255
process control **A6:** 256
pumpdown **A6:** 258, 261
refractory metals **A6:** 851
rhenium alloys **A6:** 581, 582
safety **A6:** 261
safety precautions **A6:** 1199, 1202-1203
shallow weld **A6:** 853-854
sheet metals **A6:** 399
solidification cracking **A6:** 851
in space and low-gravity environments **A6:** 1021-1022, 1023
space welding technology **A6:** 1021
stainless steel casting alloys **A6:** 496
stainless steels **A6:** 679, 688, 698, 699, 828, 851, 853
standard procedure qualification test weldments **A6:** 1090
steel **A6:** 851, 857
stress analysis of welds **A6:** 1138
superalloys **A6:** 851
tantalum alloys **A6:** 580
temperature measurements, validation strategies **A6:** 1149
titanium alloys **A6:** 85, 512, 523, 514, 516, 517, 518, 519, 520, 521, 522, 783, 784
to solve problems in joining thin sections by oxyfuel gas welding **A6:** 288
tool steels **A6:** 258-259
tungsten alloys **A6:** 581, 582
typical floor layout **A6:** 256
vs. gas-tungsten arc welding **A6:** 254
vs. laser-beam welding (LBW) **A6:** 262
weld discontinuities **A6:** 1076-1078
weld geometry **A6:** 260
configurations for wide welds bridging a gap **A6:** 852
melt-zone configuration **A6:** 852
part configuration **A6:** 851
surface geometry **A6:** 851-852
weld production by three effects **A6:** 255
wrought martensitic stainless steels **A6:** 441
Electron-beam welding (EBW), procedure development and practice considerations **A6:** 851-873
advantages **A6:** 858-859
aluminum alloys **A6:** 855, 859, 871-872
applications **A6:** 862, 865, 866, 867, 870, 871
beam oscillations **A6:** 863
beam-broadening effect **A6:** 854
beryllium **A6:** 872-873
chill bars **A6:** 858
cobalt-base alloys **A6:** 869
controlling heat effects **A6:** 858
copper **A6:** 872
copper alloys **A6:** 855, 872
cracking **A6:** 866-867, 871-873
disadvantages **A6:** 859
dispersion-strengthening effect **A6:** 870
dissimilar metals, joining of **A6:** 866
filler metal preplacement **A6:** 860
filler metal, use of **A6:** 860-861
filler-wire feeding **A6:** 860-861

Electron-beam welding (EBW), procedure development and practice considerations (continued)
free-machining steels **A6:** 860
fusion zone **A6:** 870, 872, 873
hardenable steel **A6:** 866-867
hardened and work-strengthened metals **A6:** 867-868
heat sinks **A6:** 858
heat-affected zone **A6:** 855, 857, 858, 864, 865, 866, 867, 868, 869, 870, 871, 872
heat-resistant alloys **A6:** 869
heat-treatable alloys **A6:** 871
in high vacuum (EBW-HV) **A6:** 854-855
high-carbon steels **A6:** 867
high-nickel alloys **A6:** 865
high-strength alloy steels **A6:** 867
high-vacuum **A6:** 859
aluminum alloys **A6:** 871, 872
energy input at the weld for single-pass EBW for various depths of penetration **A6:** 857
refractory metals **A6:** 870
vs. medium vacuum EBW **A6:** 855
iron alloys **A6:** 855
iron-nickel-chromium base alloys **A6:** 869
joint design **A6:** 852-854
angular weld **A6:** 853
flush joint vs. stepped joint **A6:** 853
scarf weld **A6:** 853
shallow edge weld **A6:** 853-854
single square-groove weld **A6:** 854
slant-butt joint **A6:** 853
slant-groove weld **A6:** 853
square-groove weld **A6:** 853
T-joint, welds in **A6:** 853
joint tracking **A6:** 863
automatic **A6:** 863
electromechanical **A6:** 863
manual **A6:** 863, 864
tape-controlled **A6:** 863
keyholing **A6:** 861-862
limitations **A6:** 855, 862
low-alloy steels **A6:** 860, 867
low-carbon steel **A6:** 866
magnesium alloys **A6:** 855, 872
martensitic stainless steels **A6:** 869
medium vacuum **A6:** 855
applications **A6:** 855
disadvantages **A6:** 855
penetration **A6:** 855
titanium alloys **A6:** 872
vs. high-vacuum EBW **A6:** 855
vs. nonvacuum EBW **A6:** 855
weld shape **A6:** 855
medium-carbon steels **A6:** 867
molybdenum **A6:** 870-871
multiple-tier welding **A6:** 861-862
nickel alloys **A6:** 855
niobium **A6:** 870, 871
non-heat-treatable alloys **A6:** 871
nonvacuum EBV (workpiece out-of-vacuum) **A6:** 855-858
applications **A6:** 858
operating conditions **A6:** 857
penetration **A6:** 857, 858
pressure effect **A6:** 856
production rates **A6:** 856-857
standoff distance **A6:** 857
tooling **A6:** 857-858
vs. GTAW **A6:** 857
weld shape and heat input **A6:** 857, 858
weld shrinkage **A6:** 857
nonvacuum EBW
aluminum alloys **A6:** 871, 872
applications **A6:** 861
gun/column apparatus components **A6:** 859
vs. medium-vacuum EBW **A6:** 855
partial-penetration welding **A6:** 862
penetration, specifications for obtaining maximum **A6:** 863
poorly accessible joints **A6:** 861
beam characteristics **A6:** 861
sidewall clearance **A6:** 861
workpiece requirements **A6:** 861
precipitation-hardenable nickel-base alloys **A6:** 869

Electron-beam welding (EBW), procedure development and practice considerations (continued)
precipitation-hardenable stainless steels **A6:** 869
pressure effects **A6:** 854-855, 859
pulsed beam use **A6:** 863-864
reactive metals **A6:** 854, 855
refractory metals **A6:** 855, 869-871
as repair method **A6:** 864-866
applications **A6:** 864-865
equipment **A6:** 864
wire-feed process **A6:** 864-866
scanning use **A6:** 862-863, 864
accuracy of beam alignment **A6:** 862
electronic scanning **A6:** 863, 864
optical scanning **A6:** 863
problems **A6:** 862
procedure **A6:** 862
techniques **A6:** 862-863
without optics **A6:** 863
shielding gases **A6:** 857
solid-solution nickel-base alloys **A6:** 869
special joints and welds **A6:** 854
multiple-pass welds **A6:** 854
multiple-tier welds **A6:** 854
plug welds **A6:** 854
puddle welds **A6:** 854
tangent-tube welds **A6:** 854
three-piece welds **A6:** 854
welds using integral filler metal **A6:** 854
stainless steels **A6:** 868-869, 870
steel **A6:** 860
superalloys **A6:** 866
tack welding **A6:** 861
tantalum **A6:** 870, 871
tantalum alloys **A6:** 871
thick metal, welding of **A6:** 858-860
disadvantages **A6:** 859
full-penetration welds **A6:** 859
partial-penetration welding **A6:** 860
pressure effects **A6:** 859
problems and flaws **A6:** 860
two-pass full-penetration welding **A6:** 859-860
thin metal, welding of **A6:** 858
joining thin sections to thick sections **A6:** 858
partial-penetration welds **A6:** 858
titanium **A6:** 854
as addition to molybdenum **A6:** 870-871
alloys **A6:** 865, 872
to minimize tempering in the heat-affected zone **A6:** 868
tool steels **A6:** 867
tungsten **A6:** 870
weld size variation and HAZ **A6:** 868
welding conditions of EBW-HV **A6:** 855
welding technique when feeding filler wire **A6:** 861
welding with extreme accuracy **A6:** 862
width of weld **A6:** 855
wire-feeding equipment **A6:** 861
zinc alloys **A6:** 872
zirconium **A6:** 854
as addition to molybdenum **A6:** 870-871
alloys **A6:** 787
Electron-beam-excited electrons
used in scanning electron microscopy **A9:** 90
Electron-beam-induced current
scanning electron microscopy **A9:** 95
Electron-channeling patterns **A9:** 94
Electron-diffraction patterns
Bragg reflections excited during **A9:** 111
color enhancement **A9:** 153
Electron-dispersive analysis by x-ray (EDAX)
for preliminary identification of heavier elements **EM3:** 644
Electron-electron double resonance **A10:** 258, 690
Electron-emitter-LaB$_6$
as rare earth application **A2:** 731
Electron-hole pairs
role in scanning electron microscopy **A9:** 95
Electron-impact ionization
in gas mass spectrometer **A10:** 152-153
Electron-stimulated desorption (ESD) **A18:** 456-458 **EM3:** 237
applications **A18:** 457-458
detection limit **A18:** 456
equipment **A18:** 456-457

Electron-stimulated desorption (ESD) (continued)
fundamentals **A18:** 456
spectrum **A18:** 457
Electron-stimulated desorption-ion energy distribution (ESDIED) **A18:** 457, 458
Electron/atom (*e*/*a*) ratio **A6:** 104
Electronegativity
refined **EL1:** 93
Electronic
curve fitting, Charpy impact testing **A8:** 267
signal processing, dynamic testing machines **A8:** 40-41
Electronic alloy identification, for substrate
electroplated coatings **A13:** 421
Electronic capacitors
of tantalum powder **A7:** 160-163
Electronic ceramics **EM4:** 17-18
applications **EM4:** 17
development **EM4:** 17
dielectric constant **EM4:** 17
dielectrics/piezoelectrics **EM4:** 17
electronic packaging **EM4:** 18, 20
electrorestrictive ceramics **EM4:** 17
high-temperature superconductors **EM4:** 17
ionic conductors **EM4:** 18
optical devices **EM4:** 17
properties **EM4:** 17
pyroelectrics **EM4:** 17
semiconducting ceramics **EM4:** 17
varistors **EM4:** 18
Electronic circuit
silver diffusion and whisker formation **A9:** 101
Electronic components *See also* Components; Equipment, Microelectronic components
penetrameters for **A17:** 341
thermal inspection of **A17:** 403
thermal management **EL1:** 46-55
ultrasonic inspection **A17:** 252-254
Electronic configurations
of rare earth metals **A2:** 721-722
Electronic contacts
powders used **A7:** 573
Electronic control
and furnace chamber in closed-loop servomechanical system **A8:** 395-396
module, step-down tension testing **A8:** 324
Electronic corrosion *See also* Corrosion; Electrical corrosion
of aircraft accelerometers **EL1:** 1111-1112
of antenna marker beacon **EL1:** 1108
of circuit breakers **EL1:** 1108
of disk recorder heads **EL1:** 1112
of electrical connectors **EL1:** 1112
of fuses **EL1:** 1110-1111
of hybrid microcircuits **EL1:** 1115
of integrated circuits **EL1:** 1114-1115
of klystron electron tubes **EL1:** 1115-1116
of microwave detectors **EL1:** 1112-1113
of nickel/boron-plated panels **EL1:** 1113-1114
of printed wiring boards **EL1:** 1109-1110
of steering potentiometers **EL1:** 1111
of stepper motors **EL1:** 1111
of tin whiskers **EL1:** 1116
of vacuum tubes **EL1:** 1110
Electronic device material studies
and crystal growth **A10:** 375-376
Electronic embodiment processes
for epoxies **EL1:** 831-832
Electronic energy levels
excitation of **A10:** 86
in optical emission spectroscopy **A10:** 21-22
Electronic equipment
precoated steel sheet for **M1:** 173, 174
Electronic failure analysis *See* Failure analysis
Electronic filters, encapsulated
neutron radiography of **A17:** 394-395
Electronic heat control
definition **M6:** 6
Electronic hydrogen analysis
for corrosion rates **A11:** 199
Electronic image analysis
color metallography **A9:** 138-139, 152-153
Electronic Industries Association (EIA) **EL1:** 734
Electronic materials *See also* Electrical interconnection; Materials; Microelectronic materials
aluminum transistor base lead, fatigue failure **A12:** 483

Electronic materials (continued)
atomic structure **EL1:** 90-92
ball bond, defects **A12:** 484-485
brush/slip ring assembly, failure due to arcing **A12:** 488
chemical bonding **EL1:** 92-93
crystallography **EL1:** 93
defects, importance of **EL1:** 93
design **EL1:** 25-44
dielectric materials **EL1:** 99-100
electrical properties **EL1:** 89-90
fractographs **A12:** 481-488
fracture/failure causes illustrated **A12:** 217
insulators **EL1:** 99-100
integrated circuits, electrical discharge defect **A12:** 481
ionic and electrical conductivity **EL1:** 93-96
L-shaped flat pack leads, loading failure **A12:** 481-482
magnetic materials **EL1:** 103
microelectronic materials, physical characteristics **EL1:** 104-111
molecular electronics **EL1:** 103
ohmic contact window, defects in vacuum- deposited aluminum **A12:** 486-487
quantum mechanical band theory, solid state **EL1:** 96-103
resistive materials **EL1:** 100-101
semiconductors **EL1:** 101-103
silver solar cell interconnect, effect of atomic oxygen environment **A12:** 481
special test procedures for **EL1:** 953-955
surface phenomena **EL1:** 103-104
XRS analysis of **A10:** 82
Electronic mean free path, for inelastic scattering
as function of energy **A10:** 540
Electronic nickel, analysis of aluminum in
photometric method **A10:** 65
Electronic packages *See also* Packages; Packaging
design **EL1:** 1, 21, 25-44
Electronic packaging *See also* Electronic packages; Packages; Packaging **A6:** 618 **EM4:** 18, 20
applications **EM3:** 579-603 **EM4:** 18
board level packaging **EM3:** 587
board overcoating **EM3:** 591-592
circuit board **EM3:** 589-591
surface mounting technology (SMT) **EM3:** 589
through-hole mounting **EM3:** 588-589
design trade-offs **EL1:** 18-24
development **EM4:** 18
device packaging **EM3:** 583-587
die attach **EM3:** 583-584
encapsulation **EM3:** 585-586
hermetic versus nonhermetic packaging **EM3:** 586-587
interconnection bonding **EM3:** 584-585
package sealing **EM3:** 585
failure analysis techniques **EL1:** 957
future trends **EL1:** 390-396
goals of **EL1:** 18
hierarchy **EL1:** 397
inorganic materials: passivations and deposition methods **EM3:** 593-594
encapsulants **EM3:** 592-594
passivation properties **EM3:** 593
organic materials: adhesives, passivations, sealants, and encapsulants **EM3:** 594-602
primary technologies of **EL1:** 12
thermal stress **EL1:** 56-59
trends **EL1:** 12, 390-396
wafer processing **EM3:** 580-583
diffusion **EM3:** 580
metallization **EM3:** 581-582
passivation **EM3:** 582-583
testing and separation **EM3:** 583
wafer preparation **EM3:** 580
Electronic pencil
for SEM micrographs **A12:** 194, 207
Electronic phenomena *See also* Materials and electronic phenomena
and materials **EL1:** 89-111
Electronic polarization
of insulators/dielectric materials **EL1:** 99-100
Electronic processing and electronic devices **EM4:** 1055-1059
glass applications in electronic devices **EM4:** 1055-1059
glass applications in electronic processing **EM4:** 1055, 1056
substrate glazes **EM4:** 1061
Electronic properties
amorphous materials and metallic glasses **A2:** 815
of germanium **A2:** 734
Electronic scrap recycling
as complex ore **A2:** 1228-1229
final processing options **A2:** 1230-1231
future use trends **A2:** 1231
preliminary processing options **A2:** 1229-1230
scrap materials **A2:** 1228
Electronic signal generator
for ultrasonic inspection **A17:** 231
Electronic systems
thermal management **EL1:** 46-55
Electronic thermal control *See also* Thermal control
purpose **EL1:** 46
technologies **EL1:** 47-50
Electronic touch-trigger probes
coordinate measuring machines **A17:** 25
Electronic transducers
in leak detection **A17:** 60
Electronic transparency
of glass fiber reinforced polymers **EM1:** 36
Electronic treating
defined **EM2:** 16
Electronic vacuum coatings **M5:** 395, 398-399
Electronics
refractory metal applications in **A7:** 17, 765
Electronics applications
acrylonitrile-butadiene-styrenes (ABS) **EM2:** 111
allyls (DAP, DAIP) **EM2:** 226-227
critical properties **EM2:** 458-459
high-impact polystyrenes (PS, HIPS) **EM2:** 195
home, of homopolymer/copolymer acetals **EM2:** 101
liquid crystal polymers (LCP) **EM2:** 180
polyamides (PA) **EM2:** 125
polyarylates (PAR) **EM2:** 138-139
polybutylene terephthalates (PBT) **EM2:** 153
polyether sulfones (PES, PESV) **EM2:** 159
polyether-imides (PEI) **EM2:** 156
polyphenylene sulfides (PPS) **EM2:** 186
polysulfones (PSU) **EM2:** 200
polyurethanes (PUR) **EM2:** 259
silicones (SI) **EM2:** 266
styrene-acrylonitriles (SAN, OSA, ASA) **EM2:** 215
thermoplastic fluoropolymers **EM2:** 117
thermoplastic polyimides (TPI) **EM2:** 177
Electronics, eliminating the "purple plague" **A3:** 1•28
Electronics equipment **A13:** 1113-1126
analysis techniques **A13:** 1113-1118
chemical analysis **A13:** 1114-1117
electron optics **A13:** 1117
electronic/electric corrosion, examples **A13:** 1118-1126
Electronics industry **A13:** 1107
black boxes, moisture intrusion **A13:** 1107-1108
corrosion/prevention methods **A13:** 1107-1111
design considerations **A13:** 1111
dissimilar-metal corrosion **A13:** 1108
Electronics industry applications *See also* Electrical industry applications
copper and copper alloys **A2:** 216, 239-240
Electronics industry applications (continued)
germanium and germanium compounds **A2:** 736-737
of gold **A2:** 692
high-temperature superconductors **A2:** 1085
of precious metals **A2:** 693
pure metals **A2:** 1093
refractory metals and alloys **A2:** 558
Electronics, molecular
as future technology **EL1:** 103
Electronics soldering *See* Solderability; Soldering
Electrons
abbreviation for **A10:** 690
absorption during transmission electron microscopy **A9:** 111
attributes which affect energy distribution of backscattered electrons **A9:** 92
for Auger electron emission **A7:** 250
backscattered **A10:** 669, 689
behavior in gas mass spectrometer **A10:** 152-153
binding energy of **A10:** 569
conduction **A10:** 261, 433-434
configurations, elements **A10:** 688
deceleration, in EPMA quantitative analyses **A10:** 524
defined **A10:** 672
detected by secondary electron detector **A10:** 502
diffraction, defined **A10:** 672
effect in Auger electron spectroscopy **A7:** 250
energy loss, signal detector for **A10:** 435
energy losses studied by scanning transmission electron microscopy **A9:** 104
energy spectrum **A9:** 92
escape depth, AES analysis **A10:** 551
high-energy, and x-ray generation **A10:** 83-84
inelastic mean free path **A10:** 569-571
orbitals of, x-ray emission and **A10:** 83
photoejection of **A10:** 85
photoelectric rejection, in x-ray absorption **A10:** 84
pi, defined **A10:** 679
scattering, defined **A10:** 672
scattering volume of **A10:** 434
secondary **A10:** 86, 435, 681, 691
sources, radial distribution function analysis **A10:** 395-396
temperature of, as indication of electron kinetic energy **A10:** 24
transitions, energy-level diagrams **A10:** 569
transmitted and scattered **A10:** 434-435
unpaired, ESR analysis for **A10:** 253
velocity of, abbreviation for **A10:** 691
Electropainting
of zinc alloy castings **A15:** 796
Electrophoresis **EM4:** 74
aluminum coatings applied to steel by **M5:** 347
ceramic coatings applied by **M5:** 545-546
and electrometric titration **A10:** 203
oxidation-resistant coating applied by **M5:** 664-666
Electrophoretic painting *See* Electropainting
Electrophoretic paints **M5:** 472
Electroplated bond systems
bonded-abrasive grains **A2:** 1015
Electroplated cadmium coatings
for marine corrosion **A13:** 911-912
Electroplated chromium coatings **A13:** 636, 871-875
Electroplated coatings **A13:** 419-431 **A18:** 834-838
abrasive wear **A18:** 835
adhesive wear **A18:** 835
advantages/disadvantages **A13:** 422-424
applications **A13:** 426-427
applications, magnetic materials **A18:** 838
for carbon steels **A13:** 523
categories **A18:** 834
coefficients of friction **A18:** 835, 836, 837
for corrosion control **A11:** 195

SUBJECTS OF THE INDEXED VOLUMES: ASM Handbook (designated by the letter "A"): **A1:** Properties and Selection: Irons, Steels, and High-Performance Alloys (1990); **A2:** Properties and Selection: Nonferrous Alloys and Special-Purpose Materials (1990); **A3:** Alloy Phase Diagrams (1992); **A4:** Heat Treating (1991); **A6:** Welding, Brazing, and Soldering (1993); **A7:** Powder Metallurgy (1984); **A8:** Mechanical Testing (1985); **A9:** Metallography and Microstructures (1985); **A10:** Materials Characterization (1986); **A11:** Failure Analysis and Prevention (1986); **A12:** Fractography (1987); **A13:** Corrosion (1987); **A14:** Forming and Forging (1988); **A15:** Casting (1988); **A16:** Machining (1989); **A17:** Nondestructive Testing and Quality Control (1989); **A18:** Friction, Lubrication, and Wear Technology (1992). **Metals Handbook, 9th Edition** (designated by the letter "M"): **M1:** Properties and Selection: Irons and Steels (1978); **M2:** Properties and Selection: Nonferrous Alloys and Pure Metals (1979); **M3:** Properties and Selection: Stainless Steels, Tool Materials and Special-Purpose Materials (1980); **M4:** Heat Treating (1981); **M5:** Surface Cleaning, Finishing, and Coating (1982); **M6:** Welding, Brazing, and Soldering (1983). **Engineered Materials Handbook** (designated by the letters "EM"): **EM1:** Composites (1987); **EM2:** Engineering Plastics (1988); **EM3:** Adhesives and Sealants (1990); **EM4:** Ceramics and Glasses (1991); **Electronic Materials Handbook** (designated by the letters "EL"): **EL1:** Packaging (1989).

Electroplated coatings (continued)
corrosion prevention mechanisms **A13:** 424-426
corrosive wear **A18:** 838
defined **A13:** 419
deposition fundamentals **A18:** 834-835
deposition parameters, effects **A13:** 424
electrochemical corrosion predictions **A13:** 430
immersion plating **A13:** 430
lubrication **A18:** 835, 836, 838
of mill products **A13:** 429-430
plating design **A13:** 422
precious metal deposits **A18:** 837-838
selection **A13:** 427-429
substrates **A13:** 420-423
titanium alloys **A18:** 778, 779
tool steels **A18:** 739
Electroplated hard chromium **A13:** 871-875
applications **A13:** 875
basis metal preparation **A13:** 872
coating thickness **A13:** 871-872
corrosion resistance **A13:** 873-875
electrodeposition parameters **A13:** 871
environmental resistances **A13:** 874
postplating treatment **A13:** 872-875
Electroplated zinc coatings **A13:** 767, 911-912
Electroplating *See also* Plating **A6:** 119 **A13:** 6, 819-821, 911-912
aluminum and aluminum alloys **M5:** 600-610
applications, plated coatings **M5:** 601-602
aluminum coatings on steel **M5:** 347
anodic *See* Anodic electroplating
aqueous, oxidation-resistant coating **M5:** 664-666
atmospheric corrosion protection **M1:** 722
baths, voltammetric monitoring of compounds in **A10:** 188
bearing materials **A18:** 756
plated overlays **A18:** 756
plated silver intermediate layers **A18:** 756
as cast coating **A15:** 561-562
and chemical treatment baths, UV/VIS analyses of **A10:** 60
cleaning processes used before **M5:** 6, 16-18
as coating operation **A7:** 460
copper and copper alloys **M5:** 619-624
preparation for **M5:** 619-621
of copper statues **A15:** 22
corrosion protection **M1:** 751-754
damage to shafts **A11:** 459
defined **EM2:** 16
and drilling **A16:** 219
in electroformed nickel tooling **EM1:** 582-584
electropolishing pretreatment for **M5:** 305
flexible printed boards **EL1:** 583
hafnium alloys **M5:** 668
heat-resistant alloys **M5:** 566
hydrogen damage by **A11:** 246
improper, failures from **A11:** 97
ion plating as precursor to **M5:** 421
lead alloys **A18:** 750
magnesium alloys **M5:** 638-639, 642, 645
metal coatings, properties of **A15:** 562
metal recovery systems **M5:** 316-319
molybdenum **M5:** 660-661
music-wire spring failure from **A11:** 557
nickel, use in **M3:** 126, 179-182
niobium **M5:** 663-664
and notch toughness in wrought steels **A1:** 746
notch toughness of steels, effect on **M1:** 705
postforging defects from **A11:** 333
procedures **M5:** 603-605
refractory metals **M5:** 658-661, 663-664, 668
rinsewater recovery and recycling **M5:** 317-318
selective *See* Selective plating
solutions, controlled-potential coulometric assays of **A10:** 207
of specimens for edge retention **A9:** 32
springs, steel **M1:** 291
stainless steel **M5:** 561-562
steel
aluminum coating **M5:** 347
preparation for **M5:** 16-18
of steel bolt, hydrogen embrittlement of **A11:** 539-540
substrate considerations **M5:** 601
surface preparation for **M5:** 6, 16-18, 601-602
Electroplating (continued)
tank process, selective plating compared to **M5:** 292-293
tantalum **M5:** 663-664
tapped holes **A16:** 257-258, 259
tin alloys **A18:** 750
of tin and tin alloy specimens for edge retention **A9:** 449
titanium and titanium alloys **M5:** 658-659
tool steels **A18:** 644
tungsten **M5:** 660-661
vacuum coating in conjunction with **M5:** 394
waste disposal *See* Plating waste disposal and recovery
zinc alloys, preparation for **M5:** 677
zinc, of steel fasteners, hydrogen embrittlement in **A11:** 548
zirconium alloys **M5:** 668
Electroplating of steel
hydrogen embrittlement caused by **M1:** 687
Electroplating tests **M5:** 210-211
Electropolishing *See also* Electrolytic polishing; Polishing
current density-voltage curve for
defined **A13:** 6
for diffraction samples **A10:** 382
as FIM sample preparation **A10:** 584, 586, 598, 599
of IN 939, for FIM/AP analysis **A10:** 598
of permanent magnet alloy sample **A10:** 599
for subsurface measurement **A10:** 388
Electropolishing (ELP) **A16:** 27
and fatigue strength **A16:** 25, 31
and shot peening **A16:** 35
surface alterations produced **A16:** 24, 25
Electropolishing solutions *See* Electrolytes for electropolishing
Electropolymerization, of phenols
SERS study of **A10:** 136
Electrorefining
of copper **A7:** 120
Electroreflectance effect
microwave inspection **A17:** 217
Electroslag alloy production
powders used **A7:** 572
Electroslag casting
as innovative **A15:** 37
Electroslag melting
dissimilar metal joining **A6:** 277
effect on inclusions **A11:** 340
Electroslag remelting *See also* Consumable-electrode remelting **A6:** 277 **M1:** 111, 114
defined **A15:** 5
furnace **A15:** 402, 404
of heavy ingots **A15:** 404
ingot solidification **A15:** 403-404
nickel alloys **A15:** 820
plant design **A15:** 404
remelting under high pressure **A15:** 405
remelting under vacuum **A15:** 405
slag compositions **A15:** 402
of stainless steels **A14:** 222
steel plate **M1:** 181-182
of steels **A15:** 401-403
ultrahigh-strength steels **M1:** 422, 426, 428, 437, 439-441
vs. STAMP processing **A7:** 549
Electroslag remelting (ESR) **A1:** 930, 968-969, 970
abbreviation for **A11:** 796
of bearing steels **A11:** 490 **A18:** 731
Electroslag welding **M6:** 225-237
applicability **M6:** 226
backing bars **M6:** 227-228
circumferential welding **M6:** 226, 233
comparison to electrogas welding **M6:** 225, 244
comparison to other processes **M6:** 236
controls **M6:** 228
conventional versus consumable guide welding **M6:** 236-237
equipment cost **M6:** 236-237
joint design **M6:** 236
joint length **M6:** 236
production requirements **M6:** 236
cracking **M6:** 233-234
dams **M6:** 227-228
definition **M6:** 6
discontinuities from **A17:** 587-588
Electroslag welding (continued)
electrode wires **M6:** 228-229
metal cored wire **M6:** 229
solid wire **M6:** 228-229
fluxes **M6:** 229-230
chemical properties **M6:** 229
consumable guides **M6:** 229-230
mechanical properties **M6:** 229
physical properties **M6:** 229
starting flux **M6:** 229
guide tube holders **M6:** 227
inspection **M6:** 234-235
magnetic-particle **M6:** 235
ultrasonic **M6:** 234
joint designs **M6:** 225-226
metal transfer **M6:** 226
metallurgical considerations **M6:** 233-234
preheating and postheating **M6:** 234
weld soundness **M6:** 233-234
metals welded **M6:** 226
molds **M6:** 227-228
mounting systems **M6:** 228
oscillators **M6:** 228
porosity **M6:** 233-234
power source **M6:** 227
procedures **M6:** 225-227, 230-233
consumable guide welding **M6:** 227
conventional welding **M6:** 226-227
process conditions **M6:** 231-233
consumable guide welding **M6:** 232
control of vertical travel **M6:** 232
conventional welding **M6:** 231
deposition rate **M6:** 231
electrode wires **M6:** 232
oscillation **M6:** 231
verticality tolerances **M6:** 232-233
weld pool **M6:** 231
production examples **M6:** 235-236
consumable guide welding **M6:** 236
conventional welding **M6:** 235
replacement of a casting **M6:** 235-236
shoes **M6:** 227
stainless steels, austenitic **M6:** 344
stainless steels, ferritic **M6:** 348
stainless steels, nitrogen-strengthened austenitic **M6:** 345
weld defects **M6:** 233
weld overlaying **M6:** 807-808
weld overlays of stainless steels **M6:** 815-816
wire guides **M6:** 227
consumable guide tube **M6:** 227
nonconsumable guide tube **M6:** 227
wire-feed drive systems **M6:** 228
workpiece preparation **M6:** 230
joint edges **M6:** 230
starting cavity **M6:** 230
strongbacks **M6:** 230
Electroslag welding (ESW) **A6:** 270-279
aluminum **A6:** 278
aluminum alloys **A6:** 738
applications **A6:** 274, 275, 276-278
shipbuilding **A6:** 384
carbon steels **A6:** 273, 274, 276-277, 652, 653, 659
cast iron **A6:** 278
constitutive equations for welding current voltage and travel rate **A6:** 272, 276
consumables **A6:** 272-273
definition **A6:** 270, 1209
desulfurization **A6:** 274
distortion **A6:** 278-279
electrochemistry role **A6:** 274
electrodes **A6:** 270, 271, 272, 273, 274, 275, 276, 277, 278
energy balance in the slag phase **A6:** 272
equipment **A6:** 277
filler metals **A6:** 272, 275
fluxes **A6:** 272-273, 274, 276, 278
fusion zone **A6:** 271, 278, 279
compositional effects **A6:** 273-274
heat balance diagram **A6:** 270-271
heat input **A6:** 427
heat-affected zone **A6:** 270-271, 272, 275, 277, 279
high-productivity processes **A6:** 275
high-strength low-alloy structural steels **A6:** 664
hydrogen cracking **A6:** 278
isometric temperature distribution **A6:** 271, 272

Electroslag welding (ESW) (continued)
joint geometry effect ... **A6:** 273
low-alloy metals for pressure vessels and piping ... **A6:** 668
low-alloy steels ... **A6:** 273, 276-277, 279, 662, 664, 668
for pressure vessels and piping ... **A6:** 667
low-carbon steels ... **A6:** 270, 273, 274
Maglay process ... **A6:** 275
metallurgical and chemical reactions ... **A6:** 273-275
multipass ... **A6:** 275-276
nickel alloys ... **A6:** 740
not for HSLA Q & T structural steels ... **A6:** 666
out-of-position (nonvertical) ... **A6:** 271
postweld heat treatment ... **A6:** 277, 278, 279
power source selected ... **A6:** 37
pressure vessel manufacture ... **A6:** 379
problems ... **A6:** 278-279
process development ... **A6:** 275
process fundamentals ... **A6:** 270-272
quality control ... **A6:** 278-279
solid-state transformations ... **A6:** 275
solidification structure ... **A6:** 274-275, 276
stainless steel casting alloys ... **A6:** 496
stainless steels ... **A6:** 278
structural steels ... **A6:** 277
surfacing ... **A6:** 275
temper embrittlement ... **A6:** 278
thermal cycle ... **A6:** 270
thermochemistry role ... **A6:** 273-274
titanium ... **A6:** 278
titanium alloys ... **A6:** 783
Watanabe number ... **A6:** 278
weld cladding ... **A6:** 816, 819, 822
weld discontinuities ... **A6:** 1078
weld metal inclusions ... **A6:** 274
weld pool form factor ... **A6:** 271, 272, 273, 274-275, 278
weld pool penetration and magnetic field coupling ... **A6:** 272
welding flux conductivity ... **A6:** 271
welding parameter effect on weld metal pool shape ... **A6:** 273
X factor ... **A6:** 278
Electroslag welding electrode
definition ... **M6:** 7
Electroslag welding of
high-strength alloy steels ... **M6:** 226
low carbon steels ... **M6:** 226
low-alloy steels ... **M6:** 226
medium-carbon steels ... **M6:** 226
nickel-based alloys ... **M6:** 226
stainless steels ... **M6:** 226
steel-based alloys ... **M6:** 226
structural steels ... **M6:** 226
Electroslag welds
failure origins in ... **A11:** 439-440
Electrostatic analyzers ... **A10:** 607, 690
Electrostatic clinging
of sampling materials ... **A10:** 16
Electrostatic copier powders ... **A7:** 580-588
Electrostatic disc
powder coating ... **M5:** 485
Electrostatic discharge ... **A11:** 786-788, 796
Electrostatic discharge (ESD)
as failure mechanism ... **EL1:** 975-978, 1013-1014
in semiconductor chips ... **EL1:** 966-967
Electrostatic edge effect
xeroradiographic images ... **A17:** 315
Electrostatic fluidized bed powder coating ... **M5:** 486
Electrostatic focusing device *See* Focusing device
Electrostatic image development *See* Copier powders
Electrostatic immersion lens *See* Immersion objective
Electrostatic lens
defined ... **A9:** 7
Electrostatic nonmetallic separator (ENS) ... **A7:** 178-180
Electrostatic overstress failures
integrated circuits ... **A11:** 786-788
Electrostatic paint spraying ... **M5:** 478-480, 485-486, 494
equipment ... **M5:** 479-480
powder coating process ... **M5:** 484-486
Electrostatic porcelain enamel spraying ... **M5:** 517-518
Electrostatic powder spraying
of porcelain enamels ... **A13:** 447-448
Electrostatic precipitators ... **A7:** 73
cupolas ... **A15:** 387
Electrostatics
equivalent physical quantities ... **EM1:** 191
Electrostream drilling (ES) ... **A16:** 509, 551-553
acid electrolytes ... **A16:** 551, 552, 553
advantages ... **A16:** 551
applications ... **A16:** 551
computer control machines ... **A16:** 552
double-electrolyte system ... **A16:** 552
equipment and tooling ... **A16:** 551-553
limitations ... **A16:** 551
process capabilities ... **A16:** 551
process parameters ... **A16:** 553
stainless steels ... **A16:** 706
Electrostriction ... **EM4:** 1119
Electrothermal vaporization
for solid-sample analysis ... **A10:** 36
Electrotinning *See also* Tinning; Tinplate
defined ... **A13:** 6
Electrotransport purification
of metals ... **A2:** 1094-1095
Electrotype metal
composition ... **A2:** 549
micrograph ... **A9:** 424
Electrotyping bath
nickel plating ... **M5:** 202-204
Electrowinning
for copper powder production ... **A7:** 119-120
Electrowinning lead alloys
compositions ... **A2:** 544
Electrowinning recovery process
plating waste treatment ... **M5:** 319
Electrozone size analysis ... **A7:** 220-221
instrumentation ... **A7:** 221
vs. HIAC light obscuration analyzer ... **A7:** 223
Electrum
as gold-silver alloy ... **A15:** 18
Element test and maintenance (ETM) ... **EL1:** 376
Element testing
structural ... **EM1:** 313-329
Elemental analysis
14-MeV FNAA ... **A10:** 239
-abundance measurement by NAA ... **A10:** 233
analytical transmission electron microscopy ... **A10:** 429-489
atomic absorption spectrometry ... **A10:** 43-59
Auger electron spectroscopy ... **A10:** 549-567
by electronic image analysis ... **A9:** 153
classical wet analytical chemistry ... **A10:** 161-180
controlled-potential coulometry ... **A10:** 207-211
Dumas method ... **A10:** 214
EFG types ... **A10:** 213-215
electrochemical analysis ... **A10:** 181-211
electrogravimetry ... **A10:** 197-201
electrometric titration ... **A10:** 202-206
electron probe x-ray microanalysis ... **A10:** 516-535
electron spin resonance ... **A10:** 253-266
elemental and functional group analysis ... **A10:** 212-220
empirical formulas from ... **A10:** 213
field ion microscopy ... **A10:** 583-602
gas analysis by mass spectrometry ... **A10:** 151-157
and geometry, of particles produced by explosive detonation ... **A10:** 318-320
inductively coupled plasma atomic emission spectroscopy ... **A10:** 31-42
Elemental analysis (continued)
of inorganics ... **A10:** 4-8
ion chromatography ... **A10:** 658-667
liquid chromatography ... **A10:** 649-657
low-energy ion-scattering spectroscopy ... **A10:** 603-609
micro-, of inorganic solids, methods for ... **A10:** 4-6
of mixture composition ... **A10:** 213
molecular fluorescence spectrometry ... **A10:** 72-81
neutron activation analysis ... **A10:** 233-242
nuclear magnetic resonance ... **A10:** 277-286
optical emission spectroscopy ... **A10:** 21-30
of organic compounds by EFG ... **A10:** 213
of organics ... **A10:** 9, 10
particle-induced x-ray emission ... **A10:** 102-108
of particles produced by explosive detonation ... **A10:** 318
potentiometric membrane electrodes ... **A10:** 181-187
prompt gamma activation analysis ... **A10:** 239-240
Rutherford backscattering spectrometry ... **A10:** 628-636
scanning electron microscopy ... **A10:** 490-515
Schöniger flask method ... **A10:** 215
secondary ion mass spectroscopy ... **A10:** 610-627
spark emission spectroscopy ... **A10:** 25-26, 29
spark source mass spectrometry ... **A10:** 141-150
of surfaces, by x-ray photoelectron spectroscopy ... **A10:** 568
ultraviolet/visible absorption spectroscopy ... **A10:** 60-71
voltammetry ... **A10:** 188-196
x-ray photoelectron spectroscopy ... **A10:** 568-574
x-ray spectrometry ... **A10:** 82-101
Elemental and functional group analysis ... **A10:** 212-220
applications ... **A10:** 212, 214
Dumas method ... **A10:** 214
elemental analysis ... **A10:** 213-215
estimated analysis time ... **A10:** 212, 215
functional group analysis ... **A10:** 215-219
general uses ... **A10:** 212
of inorganic liquids and solutions, information from ... **A10:** 7
introduction ... **A10:** 212-213
Karl Fischer method for water determination ... **A10:** 219
limitations ... **A10:** 212, 214, 215, 217
of organic liquids and solutions, information from ... **A10:** 10
of organic solids, information from ... **A10:** 9
related techniques ... **A10:** 212
samples ... **A10:** 212, 213
unsaturation (alkenes) ... **A10:** 219
Elemental clusters
and surface atoms ... **A7:** 259
Elemental cobalt *See also* Cobalt; Cobalt-base alloys; Pure cobalt
linear expansion ... **A2:** 446
mining and processing ... **A2:** 446
physical properties ... **A2:** 446
uses of ... **A2:** 446
Elemental contrast
for printed board failure analysis ... **EL1:** 1039
Elemental depth profiling
Auger electron spectroscopy (AES) ... **EL1:** 1079-1080
Elemental distribution maps *See* Elemental mapping
Elemental iron powders
for food enrichment ... **A7:** 614
Elemental mapping *See also* Dot mapping; Mapping; X-ray maps
and analog mapping ... **A10:** 525-528
Auger electron microscopy (AES) ... **EL1:** 1079
Auger, gold-nickel-copper metallization system ... **A10:** 559
digital compositional ... **A10:** 528-529

Elemental mapping (continued)
of high-temperature solder **A10:** 532
SEM image and **A10:** 532-533
spark source mass spectrometry **A10:** 142
spectrometer for **A10:** 527
Elemental mercury vapor
exposure effects **A2:** 1249
Elemental sensitivity
LEISS analysis **A10:** 605-606
PIXE analysis **A10:** 104-105
Elemental surface chemistry
measuring **A7:** 258
Elemental transfer
in liquid-metal corrosion **A13:** 58-59
Elements *See also* Alloying elements
absorption **A10:** 84-85
absorption edges **A10:** 85
analysis by x-ray spectrometry **A10:** 82, 95-99
at surfaces, AES lineshapes for chemistry of **A10:** 552-553
characteristic emission as basis for x-ray spectrometry **A10:** 84
detected by surface analytical techniques **A7:** 251
detection by TNAA, optimization of **A10:** 235
heavy, depth profiles of surface impurities of **A10:** 628, 632-633
inert, implantation of **A10:** 485-486
interfering, separation in high-temperature combustion **A10:** 222
light, analysis of **A10:** 459-461, 516, 559-561
in liquid aluminum, interaction coefficients for **A15:** 59
in liquid, standard Gibbs free energies for solution **A15:** 59
location in superlattice planes, alloy IN 939 .... **A10:** 599
low atomic number, x-ray spectrometry of **A10:** 86-87
lower limit of detection **A10:** 96
mass absorption coefficients **A10:** 85
metallic and semimetallic, partitioning oxidation states in **A10:** 178
nonmetallic, partitioning oxidation states in **A10:** 178
nuclear properties **A10:** 278-279
on solid surfaces, LEISS analysis of **A10:** 603
periodic table of **A2:** 1098
periodic table of the **A10:** 688
polynuclidic, SSMS analysis **A10:** 145
removed from solution in neutral atomic form **A10:** 163
single, atomic spatial distribution by IAP analysis **A10:** 596
single, interference-free detection limits **A10:** 238
symbols for **A10:** 688
toxic, NAA analysis for **A10:** 233
toxic, SSMS analysis of natural waters for **A10:** 141
of transition series, identification by electron spin resonance **A10:** 253-266
vs depth, semiquantitative PIXE analysis for **A10:** 102
and x-rays , relationship in x-ray spectrometry **A10:** 84-85
Elements, and bonding
within mer **EM2:** 57
Elephant skin
as casting defect **A11:** 384
Elevated temperature *See also* Elevated temperature testing; Elevated-temperature compression testing; Elevated-temperature creep testing; High temperature; Temperature
acidified chloride at **A8:** 418-420
air, effect on overload fracture surfaces **A12:** 35, 49-50
apparatus for hardness testing at **A8:** 83
asymmetric rod impact test at **A8:** 205
definition **M1:** 639
design properties **A8:** 670-671
effect on dimple rupture **A12:** 34-35
effect on environments **A12:** 49
forming **A8:** 553
fracture surface, ASTM/ASME alloy steels **A12:** 349
Elevated temperature (continued)
frequency-modified fatigue equation at **A8:** 356-357
inelastic design, creep strain for **A8:** 686
and mean load, effect on corrosion fatigue **A8:** 254
Rockwell testing at **A8:** 81-83
solutions, effect on critical strain rate **A8:** 519
split Hopkinson pressure bar testing at **A8:** 202
spring steels, effect on **M1:** 284-286
springs, effect on **M1:** 296-300
strain-controlled, low-cycle fatigue tests **A8:** 346
tension testing **A8:** 34-37
titanium alloys **A12:** 442
vacuum and oxidizing gases at **A8:** 412-415
water **A8:** 421-422
Elevated temperature(s) *See also* Elevated- temperature properties; Elevated-temperature alloys; High-temperature; Temperature(s)
beryllium-nickel alloys **A2:** 426
conductivity **A2:** 1027
and ductility, of hot-work die steels **A14:** 46
effect, forgeability **A14:** 222
effect on tensile properties, magnesium alloys **A2:** 465
effects on mechanical properties, magnesium alloy **A2:** 473
embrittlement, nickel aluminides **A2:** 915
epoxy resin matrices at **EM1:** 73-74
failure, thermal analysis techniques for **EM1:** 779-780
magnesium alloy sand castings, tensile properties **A2:** 495-515
of magnesium alloys **A2:** 462
of organic matrix materials **EM1:** 33
p-aramid tensile strength at **EM1:** 55
permanent magnet materials **A2:** 801
plastic deformation resistance at **A14:** 46
polyester resins at **EM1:** 93, 95
prealloyed P/M aluminum alloy forgings **A14:** 251
resistance, mechanically alloyed oxide alloys **A2:** 943-947
dispersion-strengthened (MA ODS)
strength, mechanically alloyed oxide alloys **A2:** 943-947
dispersion-strengthened (MA ODS)
tensile properties, wrought magnesium alloys **A2:** 483
Elevated temperatures *See also* Elevated-temperature failures; High temperatures; High-temperature resins systems; Hot water; Temperature; Temperature(s); Thermal
alloying effects at **A13:** 538-539
aqueous corrosion tests, laboratory **A13:** 226
bismaleimides (BMI) applications **EM2:** 252
boiler service corrosion, carbon steels **A13:** 513-514
carburization **A11:** 406
ceramic resistance **EL1:** 335
coatings for **A13:** 461
cobalt-base alloy corrosion at **A13:** 661
creep failures **A11:** 406-407
cure systems, vinyl esters **EM2:** 275
curing **EM2:** 338
as degradation factor **EM2:** 576
and durability design **EL1:** 45
effect on heat exchangers **A11:** 640-642
effects revealed through color etching **A9:** 136
fastener performance at **A11:** 542-543
fatigue failure at **A11:** 130-133, 266
gases, corrosion testing in **A13:** 226
general oxidation under **A11:** 405
metal oxide attack **A11:** 406
organic movement at **EL1:** 680
oxidation at **A13:** 17
oxide contamination effect **EL1:** 959-960
oxide film by **EL1:** 678
petroleum refining and petrochemical operations at **A13:** 1263-1264
polyether sulfones (PES, PESV) at **EM2:** 161
precipitation embrittlement at **A11:** 407
preferential oxidation in **A11:** 405-406
resistance, of ceramics **EL1:** 335
resistance, of porcelain enamels **A13:** 450
Elevated temperatures (continued)
resistance, thermoplastic polyimides (TPI) **EM2:** 177
service, silicones (SI) **EM2:** 267
service, themal spray coatings for **A13:** 461
service, thermal degradation and **EM2:** 568-570
service, thermoset resins **EM2:** 319
silicones for **EM2:** 265
steel casting failures from **A11:** 405-408
suitability, high-performance electronic systems **EL1:** 15
sulfidation- in **A11:** 405-406
thermal fatigue at **A11:** 407-408
thermosetting resins at **EM2:** 225
water, light water reactor corrosion in **A13:** 946
zinc anode composition for **A13:** 765
Elevated-temperature alloys *See also* Elevated temperature(s)
aluminum casting alloys **A2:** 127, 130-131
mechanical properties **A2:** 147
nickel-base **A2:** 441
titanium alloys, types and characteristics **A2:** 625-626
wrought aluminum alloy **A2:** 56-58
ZGS platinum **A2:** 714
Elevated-temperature aluminum P/M alloys **A13:** 842
Elevated-temperature compression testing
aluminum oxide platens with **A8:** 196
boron nitride powder as lubricant **A8:** 201
grips **A8:** 196
Hopkinson bar **A8:** 198
in situ induction heating **A8:** 196
vacuum chamber **A8:** 196
Elevated-temperature creep testing
buttonhead specimen for **A8:** 314
Monkman-Grant relationship **A8:** 304-305
Elevated-temperature curing epoxies
in composites **A11:** 731
Elevated-temperature drawing
cold-finished bars **M1:** 245-251
Elevated-temperature failures *See also* Elevated temperatures **A11:** 263-297
analyzing techniques **A11:** 277-278
causes of premature **A11:** 281-282
in cement-mill equipment **A11:** 294
creep **A11:** 263-264
creep-fatigue interaction **A11:** 266
elevated-temperature fatigue **A11:** 266
environmentally induced **A11:** 268-277
equipment and tests for **A11:** 278-281
of gas-turbine components **A11:** 282-287
in heat-treating furnaces **A11:** 292-294
in incinerator equipment **A11:** 294
metallurgical instabilities **A11:** 266-268
in ordnance hardware **A11:** 294-296
in petroleum-refinery components **A11:** 289-292
in platinum and platinum-rhodium components **A11:** 296-297
of steam-turbine components **A11:** 287-288
stress rupture **A11:** 264-266
thermal fatigue **A11:** 266
of valves, internal-combustion engines **A11:** 288-289
Elevated-temperature properties *See also* Creep behavior; Elevated temperature(s); Elevated-temperature service; High-temperature service; Stress-rupture properties
alloy cast iron **M1:** 93-94
aluminum P/M alloys **A2:** 206-208
of austenitic stainless steels **A1:** 944-949
of carbon steels **A1:** 619, 629, 640
creep strength **A1:** 629
creep-rupture strength **A1:** 629
hardness **A1:** 640
maximum-use temperatures **A1:** 617-618
stress rupture **A1:** 619
of cast steels **A1:** 376-377
CG iron **A15:** 673
of chromium-molybdenum steels **A1:** 617-652
of compacted graphite iron **A1:** 63-66
constructional steels **M1:** 639-663
creep-fatigue interaction **A1:** 625, 632, 633, 934-935
ductile iron **A15:** 662 **M1:** 46-52
effect of composition on **A1:** 639-641
effect of heat treatment on **A1:** 638-639

- **Elevated-temperature properties (continued)**
  - effect of microstructure on ... **A1:** 638
  - of ferritic stainless steels ... **A1:** 936-937
  - of gray iron ... **A1:** 22, 26 **M1:** 26-28
  - of heat-resistant cast alloys ... **A1:** 921, 923, 927
  - of low-alloy ferritic steels ... **A1:** 620-636
    - creep behavior of $2^1/_4$Cr-1Mo steel ... **A1:** 635, 636, 647
    - creep rupture of $2^1/_4$Cr-1Mo and 9Cr-IMo steel ... **A1:** 622
    - creep strengths ... **A1:** 620, 623
    - effect of ductility on fatigue endurance ... **A1:** 633
    - elastic and shear modulus of $2^1/_4$Cr-1Mo steel ... **A1:** 628
    - fatigue properties ... **A1:** 624-626, 633, 649
    - relaxation strengths ... **A1:** 631
    - stress rupture ... **A1:** 619, 623, 629, 630, 636, 641, 642, 644
    - tensile strengths ... **A1:** 624
  - low-alloy steels ... **A15:** 720-721
  - of martensitic stainless steels ... **A1:** 939-942
  - mechanical, wrought titanium alloys ... **A2:** 624-627
  - of precipitation-hardening steels ... **A1:** 942-944
  - of steel plate ... **A1:** 238
  - tensile, heat-resistant high-alloy ... **A15:** 728-729
  - ultrahigh-strength steels ... **M1:** 431, 432, 436, 437-438, 442
  - wrought aluminum alloy ... **A2:** 59
- **Elevated-temperature service**
  - for $2^1/_4$Cr-1Mo steel ... **A1:** 618, 645-647
  - bolt steels ... **A1:** 296, 620
  - carbon and low-alloy steels for ... **A1:** 617-618
    - alloy designations and specifications ... **A1:** 618
    - maximum-use temperatures ... **A1:** 617
  - carbon steels ... **A1:** 618, 619, 629, 640
  - chromium-molybdenum steels ... **A1:** 619-620, 623
  - chromium-molybdenum-vanadium steels ... **A1:** 620-621, 624
  - corrosion ... **A1:** 629-636
    - hydrogen damage ... **A1:** 632-634, 639
    - oxidation ... **A1:** 617, 629-630, 636
    - resistance to liquid-metal ... **A1:** 634-636
    - sulfidation ... **A1:** 630-632, 636, 637
  - creep-resistant low-alloy steels ... **A1:** 619-621
  - data presentation and analysis ... **A1:** 627-629, 634, 635
  - ductility and toughness ... **A1:** 626-627
  - fatigue ... **A1:** 624-626, 632, 633
    - creep-fatigue interaction ... **A1:** 625, 632, 633
    - fatigue-crack growth ... **A1:** 625, 633
    - load frequency, effect of ... **A1:** 624
    - thermal ... **A1:** 626
  - long-term exposure ... **A1:** 623, 627
  - long-term tests ... **A1:** 622-624, 629
    - creep strength ... **A1:** 620, 623
    - relaxation tests ... **A1:** 624, 631
    - stress rupture ... **A1:** 620, 622, 623-624, 629, 630
  - mechanical properties ... **A1:** 621-622, 636-645
  - modified chromium-molybdenum steels ... **A1:** 621, 622, 625, 626
  - short-term tests ... **A1:** 622, 628
  - thermal expansion and conductivity ... **A1:** 647, 651, 652
- **Elevated-temperature testing**
  - cam plastometer ... **A8:** 196
  - feed-rod attachment ... **A8:** 384-385
  - on sintered steel, fracture-limit lines ... **A8:** 583
  - ultrasonic fatigue ... **A8:** 247
- **Elevation**
  - of nonplanar fracture surfaces ... **A12:** 211
- **Elevator cable**
  - fatigue fracture of ... **A11:** 520-521
- **Elgiloy**
  - aging cycle ... **M4:** 656
  - annealing ... **M4:** 655
  - composition ... **A6:** 929 **M4:** 651-652
  - solution treating ... **M4:** 656
  - stress relieving ... **M4:** 655
- **Elgiloy wires**
  - dental ... **A13:** 1356
- **Elinvar**
  - as low-expansion alloy ... **A2:** 889
- **Elkem electric furnace process**
  - for zinc recycling ... **A2:** 1225
- **Elkonite**
  - resistance welding ... **A6:** 833-834
- **Ellipsometers**
  - for chemical surface studies ... **A10:** 177
- **Elliptical bearing** *See* Lemon bearing (elliptical bearing)
- **Elliptical contact area** ... **A18:** 512-513
- **Elliptical cracks**
  - fracture mechanics approach to ... **A11:** 123-125
- **Elmore suspension, modified**
  - used in Bitter technique ... **A9:** 534
- **Elongated alpha**
  - defined ... **A9:** 7
- **Elongated alpha grains**
  - in titanium and titanium alloys ... **A9:** 460
- **Elongated cells**
  - formation ... **A9:** 613
- **Elongated dimples, defined** *See also* Dimple(s) ... **A12:** 173-174
- **Elongated grain**
  - defined ... **A9:** 7
- **Elongated porosity**
  - in weldments ... **A17:** 583
- **Elongated single-domain permanent magnets** ... **A7:** 643-645
- **Elongation** *See also* Ductility; Elongation at break; Elongation to failure; Engineering strain; Mechanical properties; Tensile properties; Total elongation; Ultimate elongation; Uniform elongation; Yield point elongation ... **A8:** 4 **EM3:** 11
  - aluminum alloys ... **A7:** 474
  - at break, and filler size ... **A7:** 613
  - in bend testing ... **A11:** 18
  - break, defined ... **EM2:** 16
  - and composite filler particle size ... **A7:** 612, 613
  - copper and copper alloys ... **A2:** 217-219
  - copper casting alloys ... **A2:** 348-350
  - of copper-based P/M materials ... **A7:** 470
  - in deep drawing ... **A14:** 575
  - defined ... **A11:** 4 **A14:** 5 **EM1:** 10 **EM2:** 16
  - and degree of crystallinity ... **EM1:** 101
  - of dimples ... **A11:** 25
  - ductile iron ... **A15:** 654
  - effect of gage length in uniaxial tensile testing ... **A8:** 555
  - effect of sintering temperature ... **A7:** 369
  - effect of sintering time ... **A7:** 370
  - effect of temperature on ... **A8:** 36
  - of ferrous P/M materials ... **A7:** 465, 466
  - in forming refractory metal sheet ... **A14:** 786
  - of glass fibers ... **EM1:** 262
  - heat-treated copper casting alloys ... **A2:** 355
  - inclusion effect ... **A15:** 88
  - of injection molded P/M materials ... **A7:** 471
  - -load curve, and engineering stress-strain curve ... **A8:** 20
  - local, variation of ... **A8:** 26
  - maximum, three-roll forming ... **A14:** 620
  - as measure of ductility ... **A8:** 22
  - measurements ... **A8:** 554, 655
  - of molybdenum and molybdenum alloys ... **A7:** 476
  - of natural fiber ... **EM1:** 117
  - of nickel-based, cobalt-based alloys ... **A7:** 472
  - of- steel wire strands, effect of heat treatment ... **A11:** 444
  - of P/M and wrought titanium and alloys ... **A7:** 475
  - of P/M forged low-alloy steel powders ... **A7:** 470
  - of P/M stainless steels ... **A7:** 468
  - percent, defined ... **A11:** 4 **A14:** 5
- **Elongation (continued)**
  - percent, effect of uniform elongation on ... **A8:** 27
  - percent, low-temperature thermoset matrix composites ... **EM1:** 395
  - plotted against complexity index ... **A9:** 130
  - of polyester resins ... **EM1:** 92
  - prediction from gage length ... **A8:** 26
  - and reduction in area ... **A8:** 27
  - resin matrix, and damage tolerance ... **EM1:** 262
  - of rhenium and rhenium-containing alloys ... **A7:** 477
  - and rupture life ... **A11:** 265
  - shape index used to express ... **A9:** 128
  - sintering temperature effect on ... **A13:** 825
  - of superalloys ... **A7:** 473
  - symbols and unit ... **A8:** 662
  - of tantalum wire ... **A7:** 478
  - tensile, effect of triaxiality factor ... **A8:** 344-345
  - tensile, thermoplastic matrix composites ... **EM1:** 365
  - in tension testing ... **A8:** 19
  - testing machines for ... **A8:** 47
  - thermoplastic ... **EM1:** 100-101
  - in tin and tin alloys, effect of number of grains on ... **A9:** 125
  - of titanium carbide-based cermets ... **A7:** 808
  - titanium PA P/M alloy compacts ... **A2:** 654
  - -to-fracture, effect of stress-corrosion cracking ... **A8:** 499
  - ultimate ... **EM1:** 24
  - uniform ... **A8:** 22
  - vs. tensile load, yield ... **A9:** 684
  - wrought aluminum alloy ... **A2:** 58
- **Elongation at break** ... **EM3:** 11
  - defined ... **EM1:** 10
- **Elongation, percent**
  - defined ... **A8:** 4-5
- **Elongation to failure** ... **EM1:** 32, 158
- **Eloxal anodizing process**
  - aluminum and aluminum alloys ... **M5:** 587
- **Eluent**
  - as ionic aqueous solutions for ion chromatography ... **A10:** 658
  - suppressed anion chromatography, sodium bicarbonate and sodium carbonate as ... **A10:** 659
  - suppressed cation chromatography, nitric or hydrochloric acid solution as ... **A10:** 660
  - -suppressed conductivity detection, ion chromatography ... **A10:** 659-660
- **Elutriation** ... **A7:** 4, 178, 179
- **EMA data base** ... **EM4:** 40
- **EMA transducers** *See* Electromagnetic-acoustic transducers
- **Ematal anodizing process**
  - aluminum and aluminum alloys ... **M5:** 587
- **Embeddability**
  - defined ... **A18:** 8
- **Embedded abrasive**
  - appearance in aluminum alloys ... **A9:** 353
  - defined ... **A9:** 7
  - in specimens ... **A9:** 39
  - in specimens of very soft material ... **A9:** 46-47
- **Embedded iron**
  - in stainless steel surfaces ... **A13:** 1228
- **Embedded scale**
  - in steel bar and wire ... **A17:** 549
- **Embedded technology**
  - machine vision with ... **A17:** 41
- **Embedding media** *See* Mounting materials
  - 475 °C embrittlement ... **A9:** 285
  - austenitic grades ... **A9:** 284
  - ferritic grades ... **A9:** 285
  - sigma phase ... **A9:** 284-285
- **Embossed sheet**
  - temper designations ... **A2:** 26

SUBJECTS OF THE INDEXED VOLUMES: ASM Handbook (designated by the letter "A"): **A1:** Properties and Selection: Irons, Steels, and High-Performance Alloys (1990); **A2:** Properties and Selection: Nonferrous Alloys and Special-Purpose Materials (1990); **A3:** Alloy Phase Diagrams (1992); **A4:** Heat Treating (1991); **A6:** Welding, Brazing, and Soldering (1993); **A7:** Powder Metallurgy (1984); **A8:** Mechanical Testing (1985); **A9:** Metallography and Microstructures (1985); **A10:** Materials Characterization (1986); **A11:** Failure Analysis and Prevention (1986); **A12:** Fractography (1987); **A13:** Corrosion (1987); **A14:** Forming and Forging (1988); **A15:** Casting (1988); **A16:** Machining (1989); **A17:** Nondestructive Testing and Quality Control (1989); **A18:** Friction, Lubrication, and Wear Technology (1992). **Metals Handbook, 9th Edition** (designated by the letter "M"): **M1:** Properties and Selection: Irons and Steels (1978); **M2:** Properties and Selection: Nonferrous Alloys and Pure Metals (1979); **M3:** Properties and Selection: Stainless Steels, Tool Materials and Special-Purpose Materials (1980); **M4:** Heat Treating (1981); **M5:** Surface Cleaning, Finishing, and Coating (1982); **M6:** Welding, Brazing, and Soldering (1983). **Engineered Materials Handbook** (designated by the letters "EM"): **EM1:** Composites (1987); **EM2:** Engineering Plastics (1988); **EM3:** Adhesives and Sealants (1990); **EM4:** Ceramics and Glasses (1991); **Electronic Materials Handbook** (designated by the letters "EL"): **EL1:** Packaging (1989).

[E]mbossing
aluminum alloy ... **A14:** 804
by contour roll forming ... **A14:** 634
by multiple-slide forming ... **A14:** 567
of copper and copper alloys ... **A14:** 819
defined ... **A14:** 5 **EM2:** 16
die, defined ... **A14:** 5
of stainless steels ... **A14:** 759
[E]mbrittlement *See also* Acid embrittlement; Blue brittleness; Caustic embrittlement; Corrosion; Corrosion embrittlement; Creep-rupture embrittlement; Environmental effects; Environmental embrittlement; Grain-boundary embrittlement; Hydrogen embrittlement; Liquid metal embrittlement; Neutron embrittlement; Quench-age embrittlement; Sigma-phase embrittlement; Solder embrittlement; Solid metal embrittlement; Stress-corrosion cracking; Temper embrittlement; Tempered martensite embrittlement; Thermal embrittlement ... **A14:** 785, 838, 842
400 to 500 °C ... **A11:** 99
885 °F (475 °C) ... **A12:** 136-137
adsorption-induced ... **A11:** 718
AISI 4340 steels ... **A12:** 214
aluminum alloy, by mercury ... **A12:** 30, 38
amorphous materials and metallic glasses ... **A2:** 814
in arc welds ... **A11:** 413
blue brittleness ... **A11:** 98
by aluminum ... **A11:** 234
by antimony ... **A11:** 234
by bismuth ... **A11:** 235
by cadmium ... **A11:** 235, 242
by copper ... **A11:** 235
by gallium ... **A11:** 235
by indium ... **A11:** 235
by intermetallic compounds ... **A11:** 100
by lead ... **A11:** 235-236
by lithium ... **A11:** 236
by low-melting metals ... **A12:** 29
by mercury ... **A11:** 236
by penetration of molten braze material ... **A11:** 454-455
by selenium ... **A11:** 236
by silver ... **A11:** 236
by solders and bearing metals ... **A11:** 236
by solid-metal environments ... **A11:** 239-244
by tellurium ... **A11:** 236
by thallium ... **A11:** 236
by tin ... **A11:** 236
causes ... **A12:** 123
caustic ... **A13:** 353
caustic, in pressure vessels ... **A11:** 658
chemical ... **A7:** 52, 55
couples, summary ... **A13:** 182
couples, types and summary of ... **A11:** 233-238
creep-rupture ... **A12:** 123-124
defined ... **A11:** 4 **A13:** 6
detecting by notch tensile test ... **A8:** 27
duplex stainless steels ... **A6:** 472, 626
effect of temperature ... **A12:** 29
effect of temperature, in steels ... **A11:** 239
environmental ... **A17:** 287
environmental, nickel aluminides ... **A2:** 915
environmental, nickel-base alloys ... **A13:** 647-652
environmentally assisted ... **A11:** 100-101
of ferrous metals and alloys ... **A11:** 234-238
fiber, metal-matrix composites ... **A12:** 466
from exposure to brazing temperatures ... **A6:** 622
from intergranular fractures ... **A12:** 173
grain-boundary ... **A11:** 132
grain-boundary, in refractory metals, IAP studies ... **A10:** 599
graphitization ... **A11:** 99-100 **A12:** 124
historical study ... **A12:** 2
hot dip galvanizing effect ... **A13:** 439
hydrogen ... **A11:** 100-101 **A12:** 124-126 **A13:** 163, 283-290
hydrogen, tests for ... **A8:** 537-543
intergranular corrosion ... **A12:** 126
liquid-metal ... **A8:** 486 **A11:** 225-238 **A12:** 29, 126-127 **A13:** 171-184
LME, mechanisms of ... **A11:** 226-227
LME, of nonferrous metals and alloys ... **A11:** 233-234

Embrittlement (continued)
in locomotive axles, required conditions ... **A11:** 718-719
in loose powder compaction ... **A7:** 298
low surface carbon as cause ... **A4:** 595
mechanisms, in nickel alloys ... **A10:** 561-563
neutron ... **A11:** 100
neutron irradiation ... **A12:** 127
of ordered intermetallics ... **A2:** 913-914
overheating ... **A12:** 127-129
oxygen, in iron ... **A12:** 222
oxygen, wrought titanium alloys ... **A2:** 615
in petroleum refining and petrochemical operations ... **A13:** 1265, 1274, 1277-1281
phosphorus, of brazed joints ... **A11:** 452
precipitation ... **A11:** 407
quench aging ... **A12:** 129-130
quench cracking ... **A12:** 130-132
quench-age ... **A11:** 98
radiation ... **A11:** 69
scanning Auger microscopy for ... **A7:** 254
sigma-phase ... **A11:** 99 **A12:** 132-133
SIMS analysis for materials under ... **A10:** 610
solid copper, of steel ... **A11:** 232, 234
solid-metal ... **A13:** 184-187
solid-metal, defined ... **A12:** 29-30
steel weldments ... **A6:** 420, 421, 423
of steels ... **A11:** 98-101
strain aging ... **A12:** 129-130
strain-age ... **A11:** 98
stress-corrosion cracking ... **A11:** 101 **A12:** 133-134
stress-relief ... **A11:** 98-99
surface, in polymers ... **A11:** 761
temper ... **A11:** 69, 99, 335 **A12:** 134-135
tempered martensite ... **A12:** 135-136
tempered-martensite ... **A11:** 99
thermal ... **A12:** 136
thermally induced ... **A11:** 98-100
titanium alloys ... **A12:** 448
titanium alloys, electron-beam welding ... **A6:** 872
of titanium heater tube ... **A11:** 640-642
tube ruptures by ... **A11:** 603 , 612-614
Type I ... **A13:** 184
Type II ... **A13:** 184
Embrittlement by solid-metal environments *See also* Solid metal induced embrittlement (SMIE) ... **A11:** 239-244
Embrittlement couples
types of ... **A11:** 233-238
Embrittlement of iron ... **A1:** 689-691
other impurities, effect of ... **A1:** 691
oxygen, effect of ... **A1:** 689-690
selenium and tellurium, effect of ... **A1:** 691
sulfur, effect of ... **A1:** 690-691
Embrittlement of steel *See also* specific forms by name ... **M1:** 683-688
350 °C embrittlement *See* 500 °F embrittlement
400 to 500 °C embrittlement ... **M1:** 685-686
500 °F embrittlement ... **M1:** 685
blue brittleness ... **M1:** 684
detection ... **M1:** 683-686
environmentally assisted embrittlement ... **M1:** 686-688
graphitization ... **M1:** 686
hydrogen embrittlement ... **M1:** 687
intermetallic compound embrittlement ... **M1:** 686
liquid-metal embrittlement ... **M1:** 688, 715
maraging steels ... **M1:** 446, 447, 448, 451
neutron embrittlement ... **M1:** 686-687
notch toughness, effect on ... **M1:** 699, 701-704
quench-age embrittlement ... **M1:** 684
sigma-phase embrittlement ... **M1:** 686
specific forms ... **M1:** 683
strain-age embrittlement ... **M1:** 683-684
stress-corrosion cracking ... **M1:** 687-688
temper embrittlement ... **M1:** 684-685
Embrittlement of steels *See also* Temper embrittlement in alloy steels ... **A1:** 689-736
475 °C embrittlement ... **A1:** 708
in alloy steels ... **A1:** 691-697
aluminum nitride embrittlement ... **A1:** 694-696
blue brittleness ... **A1:** 692
graphitization ... **A1:** 696-697
quench-age embrittlement ... **A1:** 692-693
strain-age embrittlement ... **A1:** 693-694
aqueous environments causing stress-corrosion cracking ... **A1:** 724

Embrittlement of steels (continued)
in austenitic stainless steels
hydrogen-stress cracking and loss of tensile ductility ... **A1:** 715
sensitization ... **A1:** 706-707
sigma phase embrittlement ... **A1:** 709-711
solution pH ... **A1:** 727
stress-corrosion cracking resistance to boiling magnesium chloride ... **A1:** 725-727
in carbon steels ... **A1:** 691-697
aluminum nitride embrittlement ... **A1:** 694-696
blue brittleness ... **A1:** 692
graphitization ... **A1:** 696-697
quench-age embrittlement ... **A1:** 692-693
strain-age embrittlement ... **A1:** 693-694
creep embrittlement ... **A1:** 626
in duplex stainless steels
sensitization ... **A1:** 707-708
sigma phase embrittlement ... **A1:** 711
stress-corrosion cracking to boiling magnesium chloride ... **A1:** 727
in ferritic stainless steels
sensitization ... **A1:** 707
sigma phase embrittlement ... **A1:** 709-711
formation of flakes in steels ... **A1:** 716-717
in heat-treated martensitic stainless steels solution pH ... **A1:** 727-728
hydrogen damage processes ... **A1:** 711
hydrogen environmental embrittlement ... **A1:** 711-712
in iron-base alloys, solid-metal embrittlement ... **A1:** 721
in iron-nitrogen and iron-carbon alloys quench-age embrittlement ... **A1:** 692-693
in leaded alloy steels, solid-metal embrittlement ... **A1:** 721-722
in line pipe steels, hydrogen-stress cracking and loss of tensile ductility ... **A1:** 716
liquid-metal embrittlement ... **A1:** 717-721
in low-carbon steels
quench-age embrittlement ... **A1:** 692
strain-age embrittlement ... **A1:** 693-694
in maraging steels
hydrogen-stress cracking and loss of tensile ductility ... **A1:** 715-716
thermal embrittlement ... **A1:** 697-698
neutron irradiation embrittlement ... **A1:** 722-723
overheating ... **A1:** 697
presence of facets ... **A1:** 697
upper shelf energy ... **A1:** 697
parameters affecting stress-corrosion cracking ... **A1:** 724
intergranular corrosion or slip dissolution ... **A1:** 724
pit geometry ... **A1:** 724
temperature ... **A1:** 724
in precipitation-hardenable stainless steels solution pH ... **A1:** 727-728
properties and conditions producing alloy susceptibility to stress-corrosion cracking ... **A1:** 723-724
static tensile stresses ... **A1:** 724
stress-corrosion cracking ... **A1:** 723-724
stress-corrosion-cracking-inducing chemical species ... **A1:** 724
quench cracking ... **A1:** 698
stress-corrosion cracking verification
chloride cracking tests ... **A1:** 725
fracture mechanics methods ... **A1:** 725
procedures ... **A1:** 724-725
slow strain rate tests ... **A1:** 725
U-bend testing ... **A1:** 725
in tool steels, hydrogen-stress cracking and loss of tensile ductility ... **A1:** 715
Embrittlement, temper
low-alloy steels ... **A15:** 721
Embrittler vapor pressures
range of ... **A11:** 243
Embrittlers
SMIE occurrences by ... **A11:** 243
Embryos, as clusters
defined ... **A15:** 103
Emery
as an abrasive in wire sawing ... **A9:** 26
defined ... **A9:** 7

Emery (continued)
as grinding abrasive before microstructural analysis **EM4:** 572
physical properties **A18:** 192
polishing with **M5:** 108
Emery paste **M5:** 117
Emery-Tate oil/pneumatic (load-measuring) system **A8:** 613-614
emf *See* Electromotive force; Thermocouple materials
EMF process *See* Electromagnetic forming
EMI *See* Electromagnetic interference
EMI shielding
by urethane hybrids **EM2:** 268-269
Emission
alpha-particle **A10:** 244-245
Auger **A10:** 550
beta-particle **A10:** 245
contrast, SEM **A10:** 502
de-excitation by **A10:** 433
defined **A10:** 85-86, 673
fluorescent yield **A10:** 86-87
from fluxes **EL1:** 644
gamma-ray, as radioactive decay mode **A10:** 245
K lines **A10:** 86
L lines **A10:** 86
lines, defined **A10:** 21, 673
M lines **A10:** 86
maxima, molecular fluorescence spectroscopy **A10:** 75
maximum depth of **A10:** 525
for photoejection of electrons in copper **A10:** 85, 87
photon **A10:** 61
positron, as radioactive decay mode **A10:** 245
profile, self-absorbed line **A10:** 22
secondary electron (low-energy) **A10:** 433-434
silver x-ray, spectrum **A10:** 90
spectra **A10:** 21, 22, 75-76, 673
studies, remote, IR for **A10:** 115
x-radiation, as basis of x-ray spectrometry **A10:** 82
x-ray, in x-ray spectrometry **A10:** 83-84
Emission control
cupolas **A15:** 386-387
Emission sources
arc
excitation mechanisms for **A10:** 24
flame **A10:** 28-29
glow discharges **A10:** 26-28
ideal **A10:** 24-25
for optical emission spectroscopy **A10:** 24-29
spark **A10:** 25-26
Emission spectrometer *See also* Spectrometers
defined **A10:** 673
Emission spectroscopy *See also* Inductively coupled plasma atomic emission spectroscopy **A13:** 1115
as advanced failure analysis technique **EL1:** 1104
defined **A10:** 673
infrared **A10:** 115
optical **A10:** 21-30
spectral interferences in **A10:** 33
for trace element analysis **A2:** 1095
Emission spectrum
defined **A10:** 673
Emission-control equipment **A13:** 1367-1370
Emissions, acoustic *See* Acoustic emission inspection; Acoustic emissions
Emissive corrosion products
acoustic emission inspection **A17:** 287
Emissive electrode
definition **M6:** 7
Emissivity
color image **A17:** 488
test surface, in thermal inspection **A17:** 396-397
Emitter diffusion
bipolar junction transistor technology **EL1:** 195
Emitter-coupled logic (ECL)
as ASIC technology **EL1:** 161
bipolar, as future trend **EL1:** 390
as circuit interface **EL1:** 160-161
and CMOS, TTL, compared **EL1:** 165-166
development **EL1:** 160
for digital ICs **EL1:** 163-165
high-speed, routing **EL1:** 76
ICs, heat flux dissipated **EL1:** 402
ICs, off-chip rinse time **EL1:** 401
parts, in design layout **EL1:** 513
Emmanuel's version of Murakami's reagent
as an etchant for heat-resistant casting alloys **A9:** 331
composition of **A9:** 331
Empennage
advanced composites for **EM1:** 34
Emphysema
from cadmium toxicity **A2:** 1240
Empirical correction software
x-ray spectrometry **A10:** 100
Empirical criterion
of fracture **A14:** 389-393
modified **A14:** 400-401
Empirical distribution function (EDF) goodness-of-fit test *See also* Chi-square goodness-of-fit test; Goodness-of-fit test **A8:** 637
Empirical formulas
determination of **A10:** 213
Empirical growth law **A9:** 697
Empirical welding factor **A18:** 540
Emulsifiable solvents
aluminum and aluminum alloys cleaned with **M5:** 576-577
cleaning process **M5:** 33
pigmented drawing compounds removed by **M5:** 4-5
Emulsification **A7:** 132
Emulsification process **A18:** 141
Emulsification time
liquid penetrant inspection **A17:** 83
Emulsifiers **A18:** 99, 106-107
application of **A17:** 82-83
applications **A18:** 107
contamination of **A17:** 85
in engine and nonengine lubricant formulations **A18:** 111
formation of **A18:** 107
for liquid penetrant inspection **A17:** 75
as lubricant additive **A14:** 514
for lubricant failure **A11:** 154
for metalworking lubricants **A18:** 141, 142, 143, 144
moieties **A18:** 106
structures **A18:** 107
Emulsifying agents **M5:** 33
Emulsifying oil
effect on bearing strength **A8:** 60
Emulsion **EM3:** 11
concentration, testing **A14:** 516-517
defined **A18:** 8 **EM2:** 16
as lubricant form **A14:** 513-514
salt-coated magnesium particles from **A7:** 132
stability, of lubricants **A14:** 516
Emulsion calibration curve
defined **A10:** 673
Emulsion cleaning **M5:** 33-39
advantages and limitations **M5:** 38-39
agitation, use of **M5:** 35
aluminum and aluminum alloys **M5:** 4-5, 35, 577
analysis of cleaners **M5:** 36
applications **M5:** 34-35, 37
brass **M5:** 35
cast iron **M5:** 35
cleaner composition and operating conditions **M5:** 33-34, 36
cleaner pH, effect of **M5:** 4-5
cleaning cycles **M5:** 35-36
concentration ranges **M5:** 34-37
Emulsion cleaning (continued)
copper and copper alloys **M5:** 618
cutting fluids removed by **M5:** 9
die castings, brass and zinc **M5:** 35
dip process **M5:** 34-35
diphase emulsion cleaners **M5:** 33-35
equipment and process control **M5:** 35-38
exposure times **M5:** 36
floating layer system **M5:** 33-34
foaming during **M5:** 10-11
immersion process **M5:** 14, 35-36
magnesium alloys **M5:** 4-5, 629-630
magnetic particle and fluorescent penetrant inspection residues removed by **M5:** 14
maintenance schedules **M5:** 37-38
materials for **EL1:** 666
mechanism of action **M5:** 4-5, 23
pigmented drawing compounds removed by **M5:** 4-7
plating process precleaning **M5:** 18
polishing and buffing compounds removed by **M5:** 10-11
rinsing process **M5:** 10-11, 36-37
rust prevention by *See also* Emulsion rust-preventive compounds **M5:** 14, 38-39
safety precautions **M5:** 21, 38-39
soak process **M5:** 34, 36, 676, 677
spray process **M5:** 33-38
stable emulsion cleaners **M5:** 33-35
steel **M5:** 35
system selection **M5:** 34-37
temperature **M5:** 33-36
unpigmented oils and greases removed by **M5:** 5, 8
unstable emulsion cleaners **M5:** 33-35
waste treatment **M5:** 39
zinc and zinc alloys **M5:** 7, 35, 676-677
Emulsion inversion
defined **A18:** 8
Emulsion polymerization *See also* Polymerization **EM3:** 11
defined **EM2:** 16
Emulsion rust-preventive compounds **M5:** 459, 461-464, 467, 469
applying, methods of **M5:** 467
duration of protection **M5:** 469
Emulsion stability **A18:** 143
Emulsions
for optical holographic interferometry **A17:** 406-407
EN 8 steel
flash welding **M6:** 557
EN 16 steel
flash welding **M6:** 557
EN 355 steel, carburized
toughness of **M1:** 536
Enamel *See also* Porcelain enameling **M5:** 495, 501, 503
defined **EL1:** 109
Enamel coatings *See also* Enamels
for cast irons **A13:** 571
Enamel film insulation
wrought copper and copper alloy products **A2:** 260
Enamel, human teeth
properties **A18:** 666
Enameled sheet steel *See* Porcelain enameled sheet steel
Enameled steel
dark-field color metallography image **A9:** 159
Enameling
zinc alloys **A2:** 530
Enameling iron
mechanical properties **M1:** 178
porcelain enameling **M1:** 177-180
porcelain enameling of **M5:** 512-513, 527
production **M1:** 178-179
sag resistance **M1:** 179-180

Enameling iron sheet ... **A9:** 201
Enameling, porcelain *See* Porcelain enameling
as cast coating ... **A15:** 563-564
Enamels *See also* Lacquers ... **EM4:** 1061-1068
ceramic decoration ... **EM4:** 1066-1068
coal tar ... **A13:** 406
compositions ... **EM4:** 1063-1065
cost factors ... **EM4:** 1068
covercoat ... **EM4:** 1066
glass enamel ... **EM4:** 1066
ground coat ... **EM4:** 1065-1066
porcelain ... **A13:** 446-452 **EM4:** 1065, 1066
product availability ... **EM4:** 1068
Enantiotropy
defined ... **A9:** 7
Encapsulants *See also* Encapsulation
composition, rigid epoxies ... **EL1:** 812-813
epoxies ... **EL1:** 283
flexible epoxy ... **EL1:** 817-831
glass ... **EL1:** 343
glob-top ... **EL1:** 803
internal stress ... **EL1:** 806-808
introduction ... **EL1:** 759-760
moisture resistance ... **EL1:** 805-806
molding compounds ... **EL1:** 803-805
overview ... **EL1:** 802-809
plastic, key properties ... **EL1:** 805-809
properties of ... **EL1:** 805-809
rigid epoxies ... **EL1:** 810
silicones ... **EL1:** 283, 773
soft error ... **EL1:** 808-809
thermal conductivity ... **EL1:** 808
thick-film ... **EL1:** 341-343
types and uses ... **EL1:** 802-805
wipe sweep ... **EL1:** 809
Encapsulate
defined ... **EL1:** 1143
Encapsulated adhesive ... **EM3:** 11
Encapsulated electronic filters
neutron radiography of ... **A17:** 394-395
Encapsulated hot isostatic pressing
applications ... **EM4:** 199-200
Encapsulated microelectronic devices
failure mechanisms ... **EL1:** 1049-1057
Encapsulated powder vacuum outgassing ... **A7:** 434-435
Encapsulation *See also* Ceramic packages; Encapsulants; Encapsulation; Encapsulation failure mechanisms; Encapsulation materials; Environment; Environmental; Plastic packages; Potting ... **EM3:** 11, 553
of bubble memory devices ... **EL1:** 819
by transfer molding ... **EL1:** 473
characteristics of ... **EL1:** 282
container design ... **A7:** 429-430
defined ... **EM1:** 10 **EM2:** 16
as environmental protection ... **EL1:** 45-46, 65-67
failures ... **EL1:** 978-979
in hot isostatic pressing ... **A7:** 428-435
of inner lead bonded chip ... **EL1:** 282-283
lead frame materials ... **EL1:** 487-488
of light emitting diodes (LEDS) ... **EL1:** 819
materials ... **A7:** 428-429
methods, of package sealing ... **EL1:** 239-243
molded-epoxy, failure mechanisms ... **EL1:** 961
outlook ... **EL1:** 449
plastic, defect ... **EL1:** 962
preliminary powder processing for ... **A7:** 435-436
of radioactive sources ... **A17:** 308
rheology ... **EL1:** 838
sheet metal ... **A7:** 425, 427, 428
tape automated bonding, plastic packages ... **EL1:** 479
undercoating, soft, properties ... **EL1:** 241
vacuum coating process ... **M5:** 397
vitreous dielectrics for ... **EL1:** 109
Encapsulation failure mechanisms
corrosion ... **EL1:** 1049-1052
interconnect ... **EL1:** 1054-1056
nuclear radiation induced device failure ... **EL1:** 1056
stress-related ... **EL1:** 1052-1054
Encapsulation materials
as environmental protection ... **EL1:** 65-66
mechanical behavior ... **EL1:** 66-67
molded plastic packages ... **EL1:** 473-475
properties ... **EL1:** 66
Encapsulation theory ... **A18:** 449
Encircling coils *See also* Coils; Probes; Sensors
eddy current inspection ... **A17:** 176
impedance ... **A17:** 169-170
for reflection/transmission methods, eddy current inspection ... **A17:** 183
remote-field eddy current inspection ... **A17:** 195-196
setup and components ... **A17:** 180
Enclosure
effect on stress state in forging ... **A8:** 588
Encoders
coordinate measuring machines ... **A17:** 24-25
Encoding, run length
machine vision process ... **A17:** 34
Encyclopedia of Polymer Science and Technology ... **EM3:** 70
End
count, defined ... **EM1:** 10
defined ... **EM1:** 10 **EM2:** 16
End brushes ... **M5:** 155
End caps
pipe welding ... **M6:** 591
End cutting-edge angle (ECEA) ... **A16:** 19, 20, 143
shaping of tool steel die sections ... **A16:** 192
End effect
in flux leakage inspection ... **A17:** 563, 564
and inspection method selection, tubular products ... **A17:** 561
ultrasonic inspection ... **A17:** 565
End effect, correction in compressive loading
extrapolation method for ... **A8:** 583
End flare
in contour roll forming ... **A14:** 633
End groups, molecular
and polymers ... **EM2:** 58
End hooks
straightening of ... **A14:** 691, 692
stresses in ... **A1:** 321
End milling *See also* Milling
Al alloys ... **A16:** 766-767, 772, 773, 784, 786, 790, 792
and boring ... **A16:** 174
carbon and alloy steels ... **A16:** 675
cemented carbides used ... **A16:** 85, 86
compared to ECM and EDM ... **A16:** 539
in conjunction with milling ... **A16:** 304, 306, 308
Cu alloys ... **A16:** 816-817
flank wear ... **A16:** 37-38, 42
flank wear scars ... **A16:** 44
grinding of end mills ... **A16:** 461
heat-resistant alloys ... **A16:** 754, 756
Mg alloys ... **A16:** 826, 827
MMCs ... **A16:** 894
P/M materials ... **A16:** 889
peripheral, cast irons ... **A16:** 651, 655, 663-664
refractory metals ... **A16:** 866, 867
stainless steels ... **A16:** 703
surface roughness arithmetic average extremes ... **A18:** 340
Ti alloys ... **A16:** 846-847, 848, 849
TiN coatings ... **A16:** 57, 58
tool steels ... **A16:** 722, 726
Zn alloys ... **A16:** 834
End mills *See* Cutting tools ... **A7:** 532
carbide metal cutting tools ... **A2:** 964-965
End plates
metal molds ... **A15:** 303
End point densification
effects of contained argon ... **A7:** 434
End product nondestructive evaluation of adhesive-bonded composite joints ... **EM3:** 777-784
acoustic emission ... **EM3:** 781
holography ... **EM3:** 782-783
radiographic techniques ... **EM3:** 781-782
shearography ... **EM3:** 782-783
state of the art technology ... **EM3:** 783
thermography ... **EM3:** 783
ultrasonic NDE ... **EM3:** 777-781
acoustic impedance ... **EM3:** 777-778
leaky Lamb waves ... **EM3:** 779-781, 782, 783
pulse-echo ... **EM3:** 778-779, 783
through-transmission ... **EM3:** 778-779, 783
ultrasonic spectroscopy ... **EM3:** 779, 781
End relief angle (ER) ... **A16:** 142
front clearance ... **A16:** 143
End relief angle (ER) (continued)
shaping of tool steel die sections ... **A16:** 192
End return
definition ... **A6:** 1209
End seal *See* Face seal
End uses *See also* Applications
casting ... **A15:** 42-43
and mold life ... **A15:** 281
End-centered
defined ... **A9:** 7
End-centered space lattice ... **A3:** 1 • 15
End-cutting reamers ... **A16:** 241, 242
End-grain attack
designing for ... **A13:** 342
End-grain exposure
in closed-die forgings ... **A1:** 342
steel forgings ... **M1:** 355-356
End-of-life data
thermal analysis ... **EM2:** 570
End-quench hardenability test ... **A1:** 452, 454, 470, 471
defined ... **A8:** 5
End-quench test ... **M1:** 457, 471-472, 474-476, 478
hardenability equivalence ... **M1:** 482-488
hardenability limits ... **M1:** 490-491
End-use requirements
defining ... **EM2:** 707
Endless-wire saw ... **A9:** 26
Endo gas *See* Endothermic gas
Endodontic failures ... **A13:** 1339
Endodontic instruments
simplified composition on microstructure ... **A18:** 666
Endogas ... **A4:** 312, 314, 318, 323
P/M induction hardening ... **A4:** 234
ENDOR *See* Electron nuclear double resonance
Endothermic atmospheres *See also* Endothermic gas ... **A7:** 4, 361
Endothermic gas
carbon potential as function of dewpoint and temperature ... **A7:** 361
composition ... **A7:** 342, 343
dewpoint ... **A7:** 343
as sintering atmosphere ... **A7:** 341-343
Endothermic reaction *See also* Exothermic reaction
defined ... **A15:** 5
Endothermic-base atmospheres
applications ... **M4:** 397
generation ... **M4:** 397-398
safety precautions ... **M4:** 398-399
Endpoints precipitation titrations ... **A10:** 164
titration, Eriochrome Black T ... **A10:** 173-174
Ends, fabric *See* Warp yarns
Ends, shaped
in multiple-slide forming ... **A14:** 569-570
Endurance ... **A8:** 5
limits, ultrasonic fatigue testing ... **A8:** 241
Endurance life
as temperature-induced stress test ... **EL1:** 498
Endurance limit *See also* Fatigue limit; Fatigue properties; Fatigue strength ... **A8:** 5 **EM3:** 11
defined ... **A11:** 4, 103 **A13:** 6
relation to residual stress and hardness in fatigue test ... **A4:** 453
Endurance ratio *See* Fatigue properties
Endurance, thermal *See* Thermal endurance
Energetics
of cluster formation ... **A15:** 103
of heterogeneous nucleation ... **A15:** 104
Energy *See also* Force requirements; Power; Power requirements; Quantity of heat; Specific energy; Work
at time of fracture ... **A11:** 49-51
available, machine size selection by ... **A14:** 85
characteristics, forming machines ... **A14:** 16
characteristics, mechanical presses ... **A14:** 38-39
characteristics, screw presses ... **A14:** 40-41
conversion factors ... **A8:** 722
converter, material system as ... **A14:** 371
criterion, as fracture condition in EPFM ... **A11:** 50
density, SI unit/symbol for ... **A8:** 721
elastic, release of ... **A11:** 50
flywheel ... **A14:** 496
folding ... **A11:** 55
forging, rapidity and intensity ... **A14:** 57

Energy (continued)
hammer blow, load-stroke curve **A14:** 42
as *J*, in toughness testing **A11:** 62-63
law of conservation of **A11:** 49
relations, electromagnetic forming **A14:** 652-653
requirements, and press selection **A14:** 491
SI unit/symbol for **A8:** 721
specific, vs strain rate, heat-resistant alloys **A14:** 233
strain, determined **A11:** 49-51
symbol for **A11:** 796
Energy (frictional work) with respect to kinematics **A18:** 478
Energy abbreviation for **A10:** 690
absorption-edge, defined **A10:** 85
analyzer, LEISS analysis **A10:** 607
basis for electromagnetic radiation **A10:** 83
density, SI derived unit and, symbol for **A10:** 685
electronic, excitation levels of **A10:** 86
impact, conversion factors **A10:** 686
kinetic, defined **A10:** 675
loss, Rutherford backscattering spectrometry **A10:** 630
nonimpact, conversion factors **A10:** 686
principal Auger electron **A10:** 551
radiant, defined **A10:** 680
SI derived unit and symbol for **A10:** 685
specific, SI derived unit and symbol for **A10:** 685
in UV/VIS absorption spectroscopy **A10:** 61
Energy absorption
of honeycomb structures **EM1:** 728
Energy analyzers
LEISS analysis **A10:** 607
Energy balance approach
evaluation of crack growth **EM3:** 382
Energy balance method
fluid flow modeling **A15:** 867-869
Energy barrier theory **A6:** 145
Energy conservation
and P/M processing **A7:** 569
Energy conversion systems
alloy steel corrosion in **A13:** 538-542
Energy delivery rate **A6:** 316
Energy dependence
of electron mean free path **A10:** 571
Energy dispersion spectrometry (EDS) process **EM4:** 375
for chemical analysis **EM4:** 550-551
to analyze electrical discharge machining process **EM4:** 375
Energy dispersive analysis of x-rays (EDAX)
for microstructural analysis **EM4:** 570
Energy dispersive spectroscopy testing **A13:** 960-962
Energy dispersive x-ray analysis
compositional analysis of welds **A6:** 100, 104
Energy dissipation
and damping **EM1:** 206, 210
reduced, by composites **EM1:** 36
Energy filtering
of backscattered electrons to improve resolution **A9:** 95
Energy heat input **A6:** 13
Energy industry applications *See also* Power industry applications
titanium and titanium alloy castings **A2:** 634, 644-645
titanium and titanium alloys **A2:** 588-589
Energy level diagrams
of an atom **A10:** 570
electron transitions **A10:** 569
electronic, copper **A10:** 85, 87
lithium **A10:** 22
of molecular light-scattering processes **A10:** 126
schematic, for an atom **A10:** 433
solvent-relaxation effects **A10:** 77
transitions of atomic spectrometries **A10:** 44
Energy level diagrams (continued)
of unpaired electron with two nuclear spins electron spin resonance **A10:** 259
Energy levels, electronic
in optical emission spectroscopy **A10:** 21-22
Energy loss **EM3:** 11
defined **EM2:** 16
electrons, signal detector for **A10:** 435
Rutherford backscattering spectrometry **A10:** 630
Energy method
matrix cracking **EM1:** 241
Energy referencing
in chemical testing **EM1:** 285
Energy release rate
crack-closure method for **EM1:** 248-250
curve, crack tip strain **EM1:** 241-242
strain, coefficients **EM1:** 242-243
strain, retention t-actor **EM1:** 244
Energy shifts, measurement
for atom chemistry **A10:** 552
Energy sources
x-ray radiography **A17:** 307-308
Energy sources, alternative
for microelectronics **EL1:** 103
Energy storage applications
magnetic, with superconducting materials **A2:** 1057
titanium and titanium alloys **A2:** 588-589
Energy, switching
defined **EL1:** 2
Energy transition temperature of steels **M1:** 691
Energy-compensated atom probe abbreviation for **A10:** 690
high-energy **A10:** 597
Energy-dispersive diffractometry used to construct pole figures **A9:** 703-706
Energy-dispersive spectrometers *See also* Spectrometers
basic components **A10:** 519
complete, diagram of **A10:** 519
dead time **A10:** 519-520
detector resolution (peak broadening) **A10:** 519
efficiency **A10:** 525
sum peaks and escape peaks **A10:** 520
Energy-dispersive spectrometry *See also* X-ray spectrometry
analysis of cartridge brass **A10:** 530
applications **A10:** 100-101
and CBED, diffusion-induced grain-boundary migration analysis by **A10:** 461-464
of cement **A10:** 99-100
continuous bremsstrahlung background, effects of **A10:** 528
dead time **A10:** 516
defined **A10:** 673
detection limits **A10:** 522
detector, cross section **A10:** 435
detector resolution **A10:** 519
dot mapping **A10:** 525-529
and electron diffraction, unknown phase identification by **A10:** 455-459
of high-temperature solder **A10:** 532
instrument selection **A10:** 522
light-element analysis **A10:** 459-461, 522
peak broadening **A10:** 519
peak identification **A10:** 522-523
peak overlap problem **A10:** 523, 530
pulse processing, long time constant in **A10:** 527
qualitative **A10:** 522-523
simultaneous multielement capabilities, effects of **A10:** 92
spectral resolution **A10:** 521-522
spectrometers for **A10:** 89-93
used to analyze x-ray line spectrum **A9:** 92
and wavelength-dispersive spectrometry, compared **A10:** 521-522
x-ray lines, 1-10 keV **A10:** 523
Energy-dispersive spectroscopy (EDS) **EM3:** 237
depth profiling **EM3:** 247
Energy-dispersive spectroscopy (EDX) analysis
abrasive wear and lubricant analysis **A18:** 308
Energy-dispersive x-ray analysis
of wrought stainless steels **A9:** 282-283
Energy-dispersive x-ray analysis (EDAX) instrumentation
for surface examination **A18:** 291
Energy-dispersive x-ray detector
use with image analysis **A10:** 318
Energy-dispersive x-ray fluorescence *See* Energy-dispersive spectrometry; Spectrometers
Energy-dispersive x-ray spectrometers
analyzer systems **A10:** 91-92
detectors **A10:** 90-91, 519
operation of **A10:** 92-93
for x-ray spectrometry **A10:** 89-93
Energy-dispersive x-ray spectrometry (EDX)
abbreviation for **A11:** 796
analysis of reheater tube rupture **A11:** 610
detectors **A11:** 32-33, 40
Energy-dispersive x-ray spectrometry (EDX) detectors
applications **A11:** 37-40
development **A11:** 32-33
x-rays generated, percentage of **A11:** 38
Energy-dispersive x-ray spectroscopy **A12:** 2, 168
Energy-dispersive x-ray spectroscopy (EDS)
description/capabilities **EL1:** 1097-1098
for intermetallic-related failure **EL1:** 1043
scanning electron microscope with **EL1:** 1094-1095
Energy-efficient operations *See* Operations, energy-efficient
Energy-loss near-edge structure (ELNES) **A18:** 390
Energy-release rate
in crack extension **A11:** 49
J as **A11:** 50
Energy-release-rate method **EM3:** 3
Energy-restricted machines **A14:** 25, 37
Engagement beads
for torsional impact system **A8:** 216
Engine
camshaft sprocket gear **A7:** 617, 619
mount support, produced by Colt-Crucible ceramic mold process **A7:** 754
oil pump driven gears **A7:** 617, 619, 672-673
rocker arm pivot **A7:** 617, 619
Engine components, aircraft
eddy current inspection **A17:** 189-194
Engine control
as commercial hybrid application **EL1:** 382
Engine lathes **A16:** 142
boring **A16:** 160, 170
and high-speed machining **A16:** 600
trepanning **A16:** 176
turning **A16:** 153, 154, 155, 157
Engine oil(s)
additives **A18:** 111, 169-170
antiwear agents used **A18:** 102
ASTM sequence IID engine test to assess corrosion-inhibiting ability **A18:** 106
defined **A18:** 8
dyes **A18:** 110
formulation **A18:** 168-169
base fluids **A18:** 168-169
physical properties of hydrofinished HVI stocks and synthetic base stocks **A18:** 169
relationship between properties and hydrocarbon structures **A18:** 168
friction modifiers used **A18:** 104
lubricant-related causes of engine malfunction **A18:** 167-168

SUBJECTS OF THE INDEXED VOLUMES: ASM Handbook (designated by the letter "A"): **A1:** Properties and Selection: Irons, Steels, and High-Performance Alloys (1990); **A2:** Properties and Selection: Nonferrous Alloys and Special-Purpose Materials (1990); **A3:** Alloy Phase Diagrams (1992); **A4:** Heat Treating (1991); **A6:** Welding, Brazing, and Soldering (1993); **A7:** Powder Metallurgy (1984); **A8:** Mechanical Testing (1985); **A9:** Metallography and Microstructures (1985); **A10:** Materials Characterization (1986); **A11:** Failure Analysis and Prevention (1986); **A12:** Fractography (1987); **A13:** Corrosion (1987); **A14:** Forming and Forging (1988); **A15:** Casting (1988); **A16:** Machining (1989); **A17:** Nondestructive Testing and Quality Control (1989); **A18:** Friction, Lubrication, and Wear Technology (1992). **Metals Handbook, 9th Edition** (designated by the letter "M"): **M1:** Properties and Selection: Irons and Steels (1978); **M2:** Properties and Selection: Nonferrous Alloys and Pure Metals (1979); **M3:** Properties and Selection: Stainless Steels, Tool Materials and Special-Purpose Metals (1980); **M4:** Heat Treating (1981); **M5:** Surface Cleaning, Finishing, and Coating (1982); **M6:** Welding, Brazing, and Soldering (1983). **Engineered Materials Handbook** (designated by the letters "EM"): **EM1:** Composites (1987); **EM2:** Engineering Plastics (1988); **EM3:** Adhesives and Sealants (1990); **EM4:** Ceramics and Glasses (1991); **Electronic Materials Handbook** (designated by the letters "EL"): **EL1:** Packaging (1989).

Engine oil(s) (continued)
lubrication classification based on end-use ... **A18:** 85, 165-167
aviation ... **A18:** 166
gasoline ... **A18:** 165
heavy-duty diesel ... **A18:** 165
marine diesel ... **A18:** 165-166
natural gas ... **A18:** 166
railroad diesel ... **A18:** 165
SAE viscosity classification ... **A18:** 85
stationary diesel ... **A18:** 165
two-stroke cycle ... **A18:** 166-167
oxidation inhibitors ... **A18:** 105
performance package ... **A18:** 169-170
performance testing ... **A18:** 170
pour-point depressants ... **A18:** 108
rust and corrosion inhibitors ... **A18:** 106
specifications ... **A18:** 162-165
automotive diesel engine service engine oil classification ... **A18:** 165
service classifications ... **A18:** 163
viscosity ... **A18:** 163, 164
winter (W) viscosity ... **A18:** 163
viscosity improvers used ... **A18:** 110
Engine structural integrity program (ENSIP) ... **A17:** 666, 668-669
Engine structural maintenance program
components ... **A17:** 666
Engineered and electronic materials
ceramics, failure analysis of ... **A11:** 744-757
continuous fiber reinforced composites, failure analysis of ... **A11:** 731-743
integrated circuits, failure analysis of ... **A11:** 766-792
polymers, failure analysis of ... **A11:** 758-765
Engineered materials
brittle fracture ... **A12:** log
effect of oxidation ... **A12:** 35
fracture modes ... **A12:** 12-22
fracture/failure causes illustrated ... **A12:** 217
types illustrated ... **A12:** 216
Engineered Materials Abstracts ... **EM3:** 72
Engineered Materials Handbook ... **EM3:** 71
Engineered products
aluminum and aluminum alloy ... **A2:** 5-7
Engineering *See also* NDE engineering; Quality control
design, tension test for ... **A8:** 19
effects, NDE reliability ... **A17:** 674-677
material, temperature and strain rate effect on ... **A8:** 19
simultaneous ... **A17:** 702
stress curve, rimmed steel ... **A8:** 554
unified life cycle ... **A17:** 702
Engineering adhesive ... **EM3:** 11
Engineering alloys
semisolid casting/forging of ... **A15:** 327
Engineering applications
low-expansion alloys ... **A2:** 896
Engineering aspects of failure and failure analysis
ductile-to-brittle fracture transition ... **A11:** 66-71
and fracture mechanics ... **A11:** 47-65
general practice in ... **A11:** 15-46
Engineering brass plating ... **M5:** 285
Engineering ceramics ... **EM4:** 16
applications ... **EM4:** 16
materials included ... **EM4:** 16
properties ... **EM4:** 16
shaping and finishing ... **EM4:** 313
Engineering change (EC) level ... **EL1:** 130
Engineering change order (ECO) wire tacking adhesives ... **EM3:** 569, 572
Engineering chromium plating *See* Hard chromium plating
Engineering contacts
recommended materials for ... **A2:** 864
Engineering design system (EDS)
collection of indexable data (CID) ... **EL1:** 129
in computer-aided design ... **EL1:** 127-132
Engineering economy
corrosion calculations ... **A13:** 369-374
Engineering index (EI)
as information source ... **EM1:** 40-41
Engineering materials
cermets as ... **A2:** 978
implantation in ... **A10:** 485
Engineering materials (continued)
stress tensors as function of depth, neutron diffraction ... **A10:** 420
Engineering plastics *See also* Design considerations; Engineering plastics families; Plastics; Polymer families; Polymers ... **EM3:** 12, 41
basic elements and analysis methods ... **EM2:** 825
blends and alloys ... **EM2:** 632-637
characteristics ... **EM2:** 68-73
compressive strength ... **EM2:** 245
defined ... **EM2:** 16, 97
design approach ... **EM2:** 74-81
effect, chemical environments ... **EM2:** 188
electrical properties ... **EM2:** 73
environmental effects ... **EM2:** 429-430
environmental factors ... **EM2:** 69-71
introduction ... **EM2:** 1
material cost ... **EM2:** 73
materials characterization ... **EM2:** 655
mechanical properties ... **EM2:** 71-73
photolytic degradation ... **EM2:** 776-782
principal constituent ... **EM2:** 1
Rockwell hardness ... **EM2:** 245
specifications and standards ... **EM2:** 89-91
testing of ... **EM2:** 515-516
thermal properties ... **EM2:** 68-69
thermal stress and physical aging ... **EM2:** 751-760
thermoset ... **EM2:** 222-225
typical, listing of ... **EM2:** 73
ultraviolet (UV) resistance ... **EM2:** 1
unfilled, coefficient of expansion ... **EM2:** 1
unit volume prices ... **EM2:** 79
Engineering plastics families *See also* Engineering plastics; Plastics; Polymer families; Polymer(s)
acrylics ... **EM2:** 103-108
acrylonitrile-butadiene-styrenes (ABS) ... **EM2:** 109-114
allyls (DAP, DAIP) ... **EM2:** 226-229
aminos ... **EM2:** 230-231
bismaleimides (BMI) ... **EM2:** 252-256
cyanates ... **EM2:** 232-239
epoxies (EP) ... **EM2:** 240-241
guide to ... **EM2:** 97-275
high-density polyethylenes (HDPE) ... **EM2:** 163-166
high-impact polystyrenes (PS, HIPS) ... **EM2:** 194-199
homopolymer and copolymer acetals (AC) ... **EM2:** 100-102
introduction ... **EM2:** 97-99, 222-225
ionomers ... **EM2:** 120-123
liquid crystal polymers (LCP) ... **EM2:** 179-182
phenolics ... **EM2:** 242-245
polyamide-imides (PAI) ... **EM2:** 128-137
polyamides (PA) ... **EM2:** 124-127
polyaryl sulfones (PAS) ... **EM2:** 145-146
polyarylates (PAR) ... **EM2:** 138-141
polyaryletherketones (PAEK, PEK, PEEK, PEKK) ... **EM2:** 142-144
polybenzimidazoles (PBI) ... **EM2:** 147-150
polybutylene terephthalates (PBT) ... **EM2:** 153-155
polycarbonates (PC) ... **EM2:** 151-152
polyether sulfones (PES, PESV) ... **EM2:** 159-162
polyether-imides (PEI) ... **EM2:** 156-158
polyethylene terephthalates (PET) ... **EM2:** 172-176
polyphenylene ether blends (PPE, PPO) ... **EM2:** 183-185
polyphenylene sulfides (PPS) ... **EM2:** 186-191
polysulfones (PSU) ... **EM2:** 200-202
polyurethanes (PUR) ... **EM2:** 257-264
polyvinyl chlorides (PVC) ... **EM2:** 209-213
reinforced polypropylenes (PP) ... **EM2:** 192-193
silicones (SI) ... **EM2:** 265-267
styrene-acrylonitriles (SAN, OSA, ASA) ... **EM2:** 214-216
styrene-maleic anhydrides (S/MA) ... **EM2:** 217-221
thermoplastic fluoropolymer ... **EM2:** 115-119
thermoplastic polyimides (TPI) ... **EM2:** 177-178
thermoplastic polyurethanes (TPUR) ... **EM2:** 203-208
thermoplastic resins, introduction ... **EM2:** 98-99
thermosetting resins, introduction ... **EM2:** 222-225
ultrahigh molecular weight polyethylenes (UHMWPE) ... **EM2:** 167-171
unsaturated polyesters ... **EM2:** 246-251
urethane hybrids ... **EM2:** 268-271
vinyl esters ... **EM2:** 272-275
Engineering properties
of steel castings ... **A1:** 376-378
Engineering properties of borides ... **EM4:** 787-801
applications ... **EM4:** 798-801
binary phase diagrams ... **EM4:** 792
chemical properties ... **EM4:** 789, 797-798, 801
classification ... **EM4:** 787-788
crystal structure ... **EM4:** 178-791
electrical properties ... **EM4:** 789, 794-796, 798, 799-800
enthalpy of formation ... **EM4:** 793, 794
erosion resistance ... **EM4:** 792
fabrication ... **EM4:** 788-789
flexural strength ... **EM4:** 791-792, 798
fracture toughness ... **EM4:** 798-799
Gibbs free energy of formation ... **EM4:** 793
hardness ... **EM4:** 789, 791, 796-798, 799
heat capacity at constant pressure ... **EM4:** 793
known borides in periodic system ... **EM4:** 788
load ... **EM4:** 796-798
magnetic properties ... **EM4:** 796-797, 800
mechanical properties ... **EM4:** 791-794, 796-799
melting point ... **EM4:** 789-791
optical properties ... **EM4:** 789, 796, 800
preparation ... **EM4:** 789-789
strength ... **EM4:** 789, 791, 798
structure ... **EM4:** 787-788, 789-791
thermal properties ... **EM4:** 789-791, 794-796
wear rate ... **EM4:** 789, 792
Young's modulus ... **EM4:** 791, 798, 799
Engineering properties of carbides ... **EM4:** 804-810
boron carbide ... **EM4:** 804-806
classes of metal carbides ... **EM4:** 804
interstitial carbides ... **EM4:** 810
mechanical properties ... **EM4:** 804
silicon carbide ... **EM4:** 806-808
tungsten carbide ... **EM4:** 808-810
Engineering properties of carbon-carbon and ceramic-matrix composites ... **EM4:** 835-843
carbon-carbon composites ... **EM4:** 835-838, 842-843
ceramic-matrix composites ... **EM4:** 838-843
Engineering properties of diamond and graphite ... **EM4:** 821-831
buckyballs ... **EM4:** 829
diamond ... **EM4:** 821-827
graphite ... **EM4:** 827-831
Engineering properties of glass-ceramics ... **EM4:** 870-877
chemical systems ... **EM4:** 870-873
compositions ... **EM4:** 870
general properties ... **EM4:** 874-877
heat-treatment cycle ... **EM4:** 870-971, 874
special applications and techniques ... **EM4:** 874, 875
Engineering properties of glass-matrix composites ... **EM4:** 858-867
composite mechanical properties ... **EM4:** 861-862
composite thermal expansion ... **EM4:** 858-859, 860
composite Young's modulus ... **EM4:** 860-861
crack-second-phase interactions ... **EM4:** 863-865
dielectric properties ... **EM4:** 859-860
effect of interfacial reaction ... **EM4:** 866-867
effect of porosity ... **EM4:** 865-866
effect of volume fraction and size of dispersed phase ... **EM4:** 866
effects of adding a second phase ... **EM4:** 862-863
hardness ... **EM4:** 867
Engineering properties of multicomponent and multiphase oxides ... **EM4:** 758-773
$BaTiO_3$ ceramics, effect of solid solution substitution on ferroelectric phase transitions ... **EM4:** 770
mullite ... **EM4:** 761-765
multiphase oxides ... **EM4:** 770-773
perovskites ... **EM4:** 766-770
silicate structures ... **EM4:** 759
silicates ... **EM4:** 758-761
spinels ... **EM4:** 765-766
Engineering properties of nitrides ... **EM4:** 812-820
additives ... **EM4:** 812
applications ... **EM4:** 812
mechanical properties ... **EM4:** 812
other nitride ceramics ... **EM4:** 819-829
properties of silicon nitride ... **EM4:** 814-819
sialon system ... **EM4:** 812
silicon nitride-based ceramics ... **EM4:** 812-814

Engineering properties of nitrides (continued)
sintering difficulties **EM4:** 812
Engineering properties of oxide glasses and other inorganic glasses **EM4:** 845-857
chemical properties **EM4:** 855-857
electrical properties **EM4:** 851-853
intermediate oxides **EM4:** 845
magnetic properties **EM4:** 854-855
mechanical properties **EM4:** 849-851
modifiers **EM4:** 845
optical properties **EM4:** 853-854
overview of glass structures **EM4:** 845-846
physical and thermal properties **EM4:** 846-849
properties of glasses **EM4:** 845
Engineering properties of single oxides **EM4:** 748-757
A_2O_3-type oxides **EM4:** 751-753
A_2O_N-type oxides **EM4:** 748
AO-type oxides **EM4:** 748-751
AO_2-type oxides **EM4:** 753-756
AO_3-type oxides **EM4:** 756-757
Engineering properties of zirconia **EM4:** 775-786
applications of zirconia-ceramics **EM4:** 784-785
fabrication of toughened ceramics **EM4:** 777-780, 783
properties of zirconia-toughened ceramics **EM4:** 780-784
transformation toughening **EM4:** 775, 776-777
Engineering specifications
and quality **A17:** 720
Engineering strain *See also* Linear strain; True strain **A8:** 5
abbreviation **A8:** 724
and area reduction **A8:** 575
at fracture, elongation as **A8:** 22
in bending **A8:** 118-119
curves, uniaxial tensile testing **A8:** 554
for engineering stress-strain curve **A8:** 20
sheet metal forming **A8:** 549
symbols for **A8:** 724
vs. true strain, creep testing **A8:** 305
Engineering stress *See also* Mean stress; Nominal stress; Normal stress; Residual stress; True stress **A8:** 4 **EM3:** 12
curves, uniaxial tensile testing **A8:** 554
defined **EM2:** 16
in tension testing **A8:** 20
and true stress **A8:** 23
Engineering stress-strain curve *See also* Stress-strain curve; True stress-true strain curve **A8:** 19
and tension testing **A8:** 20-23
Engineering tension test
mechanical behavior of materials under **A8:** 20-27
Engineering thermoplastics (ETPs) *See also* Engineering plastics; Engineering plastics families; Plastics; Polymer families; Polymer(s); Resins; Thermoplastic resins; Thermoplastics **EM2:** 98-221
advantages **EM2:** 356-357
grades of **EM2:** 98-99
introduction **EM2:** 98-99
mechanical properties **EM2:** 98
selection **EM2:** 618
systems **EM2:** 445-451
thermal and related properties **EM2:** 445-459
Engineering Thermoplastics, Properties and Applications **(Margolis)** **EM2:** 94
Engineering thermosets
thermal and related properties **EM2:** 439-444
Engineering topics
special **A2:** 1203-1204
Engineering verification
bottom-brazed flatpacks **EL1:** 992-993
Engineers
polymer science for **EM2:** 48-62
England
early metalworking in **A15:** 16, 18
Engler viscosity
defined **A18:** 8
Engraver's brass
applications and properties **A2:** 308-309
Engraving tools, electric
mechanical damage by **A11:** 342
Engulfment
of insoluble particles **A15:** 142
Enhanced corrosive attack **A18:** 178
Enhanced plastic flow theory
hydrogen damage **A13:** 165
Enhancement
effects, interelement **A10:** 97
electromagnetic, and SERS **A10:** 136
fluorescence, using organized mediums **A10:** 79
resolution, as IR method **A10:** 116-117
Enhancement mode
MOSFET **EL1:** 158
Enlargement
crack **A11:** 106
film radiography **A17:** 312
microradiography **A17:** 312
real-time radiography **A17:** 312
scattering reduction **A17:** 312
of shadow, radiography **A17:** 311-312
Enriched foodstuffs and cereals
powder used **A7:** 573
Enrichment
of impurities to grain boundaries **A13:** 156
ENS *See* Electrostatic nonmetallic separator
ENSIP *See* Engine structural integrity program
Enstatite
chain structure **EM4:** 759
chemical system, primary phase **EM4:** 872
crystal structure **EM4:** 881
maximum use temperature **EM4:** 875
Entanglements, chain
chemistry of **EM2:** 64
Entertainment electronics
commercial hybrid applications **EL1:** 385
Enthalpy ...... **A3:** 1 • 6
of fusion, of semisolid materials **A15:** 328
and heat capacity, principles of **A15:** 50
Enthalpy changes **A6:** 45
Entraining velocity
defined **A18:** 8
symbol and units **A18:** 544
Entrainment **EM4:** 414
Entrance guides
contour roll forming **A14:** 627-628
Entrapment
of liquids in secondary operations **A7:** 451
Entropy *See also* Specific entropy ...... **A3:** 1 • 7
ideal, of mixing **A15:** 101
of ideal solution **A15:** 103
SI derived unit and symbol for **A10:** 685
SI unit/symbol for **A8:** 721
Entropy changes **A6:** 45
Entropy rate ratio
determining **A14:** 440
Envelopes
measured area under rectified signal (MARSE) **A17:** 282-283
x-ray tube **A17:** 302
Environment *See also* Environmental **EM3:** 12
acid, inhibitors **A13:** 524-525
of acidified chloride solution fatigue testing **A8:** 419
-alloy systems, crack tip film formation **A13:** 146
-alloy systems, stress-corrosion cracking **A13:** 145-146, 326
aqueous, titanium/titanium alloys **A13:** 689
atmospheric **A13:** 460
atomic oxygen, in low earth orbit **A12:** 481
of borescopes, effects
as cause of stress-corrosion cracking **A8:** 495-496
caustic, nickel-base alloy SCC **A13:** 650
Environment (continued)
chemical processing plant **A13:** 1134
conventional interconnection **EL1:** 2-5
corrosion fatigue **A8:** 403
corrosive **A8:** 426-430
corrosive, nickel-base alloy behavior **A13:** 643-647
defined **A12:** 22 **A13:** 6 **EM1:** 10
distribution **EL1:** 85
distribution, high-frequency digital systems **EL1:** 85
effect, aluminum SCC **A13:** 591
effect, and fatigue **A13:** 295
effect, chemical processing corrosion **A13:** 1134
effect, corrosion fatigue **A13:** 143, 300
effect in creep experiments **A8:** 302-303
effect in hydrogen embrittlement testing **A8:** 538
effect on crack initiation **A8:** 375
effect on dimple rupture **A12:** 22-35
effect on fatigue **A12:** 35-63
effects of **A12:** 22-63
effects on fatigue strength **A8:** 374-375
elevated temperature effects **A12:** 49
end-use, corrosion in **A13:** 533-542
exposure of precracked specimens to **A8:** 517
exposure, SCC specimens **A13:** 259
as factor, coordinate measuring machines **A17:** 26-27
as factor, magnesium/magnesium alloys **A13:** 741-743
freshwater, general biological corrosion **A13:** 88
gaseous, disk-pressure test for hydrogen embrittlement in **A8:** 540-541
gaseous, effects on fatigue **A12:** 36-41
hostile, coordinate measuring machine for **A17:** 24
hot salt, SCC testing **A13:** 273-275
immersion **A13:** 460
inert, creep effects in **A12:** 59
for intergranular corrosion, austenitic stainless steels **A13:** 325
laboratory, SCC testing **A13:** 264
liquid, effects on fatigue **A12:** 36, 41-46
liquid-metal, fatigue in **A13:** 177-178
loading, effect on fatigue **A12:** 36, 53-54
marine, general biological corrosion **A13:** 88
noise, high-frequency digital systems **EL1:** 76
operating, of connectors **EL1:** 23
processing, quality control and inspection of **EM1:** 740-741
for SCC in titanium alloys **A13:** 274
for SCC of aluminum/aluminum alloys **A12:** 28
for SCC of brass **A12:** 28
for SCC of steels **A12:** 27
for SCC of titanium alloys **A12:** 28
SCC testing **A13:** 263-265
specific, stainless steel corrosion **A13:** 554-559
structural steel corrosion in various **A13:** 532
structured, for machine vision **A17:** 42
and telephone cables, interaction **A13:** 1127-1128
temperature, effects on fatigue **A12:** 36, 49-53
urban, tin-bronze preferential corrosion by **A12:** 403
vacuum, effects on fatigue **A12:** 36, 46-49
various, carbon steel corrosion prevention guide **A13:** 523
vibration, optical holographic interferometry **A17:** 413
work, and NDE reliability **A17:** 677
Environment supply system
for liquid cooling or corrosion fatigue testing **A8:** 248
Environment(s) *See also* Aging; Environmental effects; Environmental factors; Environmental resistance; Environmental stress cracking; Envi-

Environment(s) (continued)
ronmental stress crazing; Weather aging; Weather resistance; Weatherability
chemical, effect on engineering plastics **EM2:** 188
defined **EM2:** 16
effect of **EM2:** 426-428
stress cracking, high-impact polystyrenes (PS, HIPS) **EM2:** 197-198
Environmental
considerations, of hazardous cleaning materials **EL1:** 667
protect on, of plastic packages **EL1:** 470
reliability, of ceramic hybrid circuits **EL1:** 381
stability, lead frame alloys **EL1:** 491
Environmental (sustained-load)
failure **A8:** 486-488
Environmental attack
on porous parts **A7:** 697
Environmental chamber
with drop tower compression test **A8:** 197
for elevated/low temperature tension testing **A8:** 34-35
Environmental considerations unique to sealants **EM3:** 673-679
Environmental containment
methods and materials for **A8:** 411
Environmental control
corrosion protection by **A13:** 86, 379
Environmental corrosion *See also* Environment(s)
and chemical susceptibility **EM2:** 573
Environmental cracking *See also* Corrosion fatigue; Embrittlement; Environmentally induced cracking; High-temperature hydrogen attack; Hydrogen blistering; Hydrogen embrittlement; Liquid metal embrittlement; Solid metal embrittlement; Stress-corrosion cracking; Sulfide stress cracking **A13:** 6, 145-189, 217-219
Environmental cracking tests
for sour gas environments **A11:** 301-302
Environmental durability design methodologies
encapsulation materials **EL1:** 65-66
mechanical behavior, encapsulation materials **EL1:** 66-67
protection, purposes **EL1:** 65
Environmental effects *See also* Embrittlement; Environment(s); Environmental embrittlement **EM3:** 483
average stress **EM3:** 614
bare Pt-Rh thermocouples **A2:** 882
carbon fibers **EM1:** 52
crazing and fracture **EM2:** 736
designing for **EM1:** 182
effects, electrical contact materials **A2:** 840
electronic general component bonding against moisture **EM3:** 573
embrittlement, forms of **A11:** 98, 100-101
failures characterized **A11:** 78-79
full-scale damage tolerance testing **EM1:** 351
germanium and germanium compounds **A2:** 735-736
long-term test results **EM1:** 823-826
mechanical stress **EM3:** 654-655
moisture **EM3:** 654-655
moisture and molecular free volume **EM3:** 654
moisture effect in uncured adhesive film **EM3:** 559
moisture penetration, versus thickness **EM3:** 587
on automotive adhesive joints **EM3:** 613
on base-metal thermocouples **A2:** 881-882
on elasticity **EM1:** 188
on fatigue crack propagation **A8:** 403-435
on highway adhesive joints **EM3:** 613
on laminate properties **EM1:** 310
on long-term reliability **EM2:** 788-789
on molecular/physical properties **EM2:** 437
on stress-corrosion cracking **A11:** 207-212, 214
on supplier data sheets **EM2:** 642-645
on wear failure **A11:** 159-160
overview **EM3:** 613-615
para-aramid **EM1:** 56
sealants **EM3:** 678-679
seawater exposure **EM3:** 629-632, 635-636
as stress sources **A11:** 206
in stress-corrosion cracking **A8:** 499-500
synergistic interactions **EM3:** 614, 615
temperature **EM3:** 654-655
of thermoplastics **EM1:** 293

Environmental effects (continued)
and time-dependent corrosion fatigue cracking **A8:** 406
uranium and uranium alloys **A2:** 670-671
Environmental embrittlement *See also* Embrittlement; Environmental effects
acoustic emission inspection **A17:** 287
iron aluminides **A2:** 922
nickel aluminides **A2:** 915
nickel-base alloys **A13:** 647-652
Environmental exposure
during specimen preparation **EM1:** 295-296
tests and results, of mechanical properties **EM1:** 296-300, 823-826
Environmental factor
and flow rate **A7:** 280
Environmental factors *See also* Environment(s)
adjoining materials **EM2:** 71
of engineering plastics **EM2:** 69-71
long-term, properties effects **EM2:** 423-432
operating temperature **EM2:** 69-70
stress level **EM2:** 70
stress response, metals vs plastics **EM2:** 76-78
Environmental health hazard
radiation as **A10:** 247
Environmental hydrogen embrittlement **A8:** 540-542
Environmental notch tensile testing **A13:** 962
Environmental pollution
by sizings **EM1:** 123
Environmental Protection Agency **A14:** 517-518
Environmental Protection Agency (EPA) **EM1:** 36
Environmental resistance
epoxies and **EM3:** 101
styrene-acrylonitriles (SAN, OSA ASA) **EM2:** 215
urethane hybrids **EM2:** 270
Environmental stress cracking
oil/gas production **A13:** 1235-1236
of polymers **A11:** 761
Environmental stress cracking (ESC) **EM1:** 10, 97, 100 **EM3:** 12
defined **EM2:** 16
of linear medium-density polyethylene **EM2:** 361
step excitation **EM2:** 552-553
Environmental stress crazing
environmental criteria **EM2:** 798-800
material optimization **EM2:** 01
molecular mechanism **EM2:** 797-798
testing **EM2:** 801-803
Environmental stress screening (ESS) **EL1:** 875-886
applications **EL1:** 77
concepts **EL1:** 875-876
defined **EL1:** 875
design of screen **EL1:** 877-884
development **EL1:** 875, 885-886
dynamic vs static **EL1:** 884
future trends **EL1:** 886
implementation of **EL1:** 884-885
for military hardware **EL1:** 944
questionable practices **EL1:** 885
risk/results overview **EL1:** 880
scenario flow diagram **EL1:** 879
state of the art **EL1:** 885-886
thermal **EL1:** 882-884
vs sampling **EL1:** 885
Environmental test chamber
all-metal, for vacuum and gaseous fatigue tests **A8:** 411
for aqueous solutions at ambient temperatures **A8:** 415
Environmental testing *See also* Environment(s); Failure analysis; Testing
for butyl tapes **EM3:** 202
combined environments reliability testing (CERT) **EL1:** 1103
for commercial/military applications **EL1:** 493-503
as component- and board-level physical testing **EL1:** 944
failure mechanisms **EL1:** 493-494
future trends **EL1:** 503
humidity test **EL1:** 1102
natural **EM2:** 578
package-level **EL1:** 935-938
polybenzimidazoles **EM3:** 170

Environmental testing (continued)
qualification programs **EL1:** 502-503
salt fog test **EL1:** 127-128
test methods **EL1:** 494-495
test procedures **EL1:** 495-502
test standards **EL1:** 494
thermal/humidity cycling test **EL1:** 1102
Environmental variables
control, in water-recirculating systems **A13:** 487-497
effects, on aqueous corrosion **A13:** 37-44
Environmentally assisted crack growth **A18:** 403
Environmentally assisted embrittlement *See* Embrittlement **A11:** 100-101
Environmentally enhanced fatigue **A8:** 487-488
Environmentally enhanced fatigue crack propagation
acidified chloride at ambient and elevated temperatures **A8:** 418-420
aqueous solutions at ambient temperature **A8:** 415-417
corrosion fatigue **A8:** 403-410
high-temperature pure water, aerated conditions **A8:** 420-422
liquid metal environments **A8:** 425-426
steam or boiling water with contaminants **A8:** 426-430
vacuum and gaseous, at ambient temperature **A8:** 410-412
vacuum and oxidizing gases at elevated temperatures **A8:** 412-415
Environmentally important substances
ICP-AES use for **A10:** 31
MFS analysis of carcinogenic polynuclear aromatic compounds in **A10:** 72
monitoring sampling for **A10:** 12
NAA application to **A10:** 233
pollutants, GC/MS analysis of volatile compounds in **A10:** 639
radioanalysis of radioactive pollutants in **A10:** 243
sampling of **A10:** 12-18
UV/VIS trace analysis **A10:** 60
x-ray spectrometry for **A10:** 82-101
Environmentally induced cracking **A13:** 145-189
hydrogen damage **A13:** 163-171
liquid-metal embrittlement **A13:** 171-184
prediction **A13:** 145
solid metal induced embrittlement **A13:** 184-187
stress-corrosion cracking **A13:** 145-163
theory **A13:** 145
types **A13:** 79
Environmentally induced failure
elevated- temperature **A11:** 268-277
Environmentally-induced crack closure
prediction complexities **A8:** 409-410
relevant to corrosion fatigue **A8:** 408-409
Environments
aggressive and inert, effects on fatigue behavior **A11:** 253
aqueous, cracking from hydrogen charging in **A11:** 245
atmospheric, effect on stress-corrosion cracking **A11:** 208
biochemical, and body implants **A11:** 673-677
body **A11:** 673-677
carbon and low-alloy steel cracking in **A11:** 215
caustic, stress-corrosion cracking in **A11:** 217
chemical, effect on corrosion-fatigue **A11:** 255
chemical reactions of surfaces with **A11:** 160
containing hydrogen sulfide **A11:** 246
contributing to stress-corrosion cracking **A11:** 208
corrosive, and spring failures **A11:** 550
effect on corrosion-fatigue **A11:** 255
effect on crack growth and toughness **A11:** 54
effect on maraging steels **A11:** 218
effect on SCC in magnesium alloys **A11:** 223
effect on yield stress and strain hardening rate **A11:** 225
effects on pipeline failure **A11:** 701-704
effects on SCC in titanium and titanium alloys **A11:** 223-224
embrittlement forms from **A11:** 98, 100-101
field, sour gas failures in **A11:** 302-303
hydrogen failures from **A11:** 409

Environments (continued)
installation, stress-corrosion cracking in ... **A11:** 212
liquid metal ... **A8:** 425-427
liquid, polymer stress cracking in ... **A11:** 761
liquid-metal, fatigue in ... **A11:** 231-232
marine-air, effect of alloy selection and part design on SCC in ... **A11:** 309-310
polymer stress cracking in ... **A11:** 761
preservice, effects on stress-corrosion cracking ... **A11:** 211
relation of crack initiation to ... **A11:** 106
shipping and storage ... **A11:** 211-212
solid metal, embrittlement by ... **A11:** 239-244
subject to selective leaching ... **A11:** 178
testing ... **A11:** 211
wear failure from ... **A11:** 156
Enzymes
activities, UV/VIS assay of ... **A10:** 70
ESR study of ... **A10:** 264
techniques using, for determination of glucose ... **A10:** 79-80
EP *See* Epoxies; Proton energy
EP lubricant *See* Extreme-pressure lubricant
EPDM
for air conditioning heater components ... **EM3:** 57
applications ... **EM3:** 56
for extruded gaskets ... **EM3:** 58
EPFM *See* Elastic-plastic fracture mechanics
Epichlorohydrin ... **EM3:** 12, 95, 96, 97
corrosion of stainless steels in ... **M3:** 82
defined ... **EM2:** 16
epoxidization by ... **EL1:** 812
in epoxy resin manufacture ... **EM1:** 10, 66-67
for forming epoxies ... **EM3:** 94
Epicyclic straightening ... **A14:** 689
Epidemiology
PIXE analysis in ... **A10:** 102
Epidiascope
as input device for image analyzers ... **A10:** 310
Epitaxial films
rocking curve profiles ... **A10:** 375-376
Epitaxial GaAs on silicon substrate ... **EL1:** 200
Epitaxial growth ... **A6:** 45, 46, 50-51
active-component fabrication ... **EL1:** 192-193
bipolar junction transistor technology ... **EL1:** 195
evolution of crystal structure in ... **A10:** 536
Epitaxial growth, graphite
cast iron melts ... **A15:** 170
Epitaxial growth in the heat-affected zone ... **A9:** 578-579
Epitaxial layers, deposition
gallium arsenide (GaAs) ... **A2:** 745
Epitaxial mesas
by aluminum-silicon interdiffusion ... **A11:** 778
Epitaxial oxide film *See* Oxide film
Epitaxial precipitation
of silicon in Schottky barrier contacts ... **A11:** 777
Epitaxial resolidification
effect on grain size ... **A13:** 503
Epitaxially deposited films
heat tinted ... **A9:** 136
Epitaxy ... **EM3:** 12
defined ... **A9:** 7 **A10:** 673 **EM2:** 16
defined as surface phenomenon ... **EL1:** 104
molecular-beam ... **EL1:** 200
oxide structure ... **A13:** 66
Epitaxy methods
for gallium arsenide (GaAs) ... **A2:** 745
Epithermal irradiation ... **A10:** 234
Epithermal neutron activation analysis ... **A10:** 239, 689
Epithermal neutrons
defined ... **A10:** 234
with indium-resonance method ... **A17:** 392
EPK 21
flash welding ... **M6:** 558
EPK 33
flash welding ... **M6:** 558
EPMA *See* Electron probe microanalysis; Electron probe x-ray microanalysis
Epon resin/curing agent system
selection guide ... **EM1:** 134
Epoxide ... **EM3:** 12
defined ... **EM1:** 10 **EM2:** 16
homopolymerization ... **EM1:** 67-71
Epoxides
cycloaliphatic ... **EL1:** 854-866
as solder resists ... **A6:** 133
Epoxidization
of unsaturated vegetable oils ... **EL1:** 818
Epoxidized cresol novolac (ECN)
characteristics ... **EL1:** 811
Epoxidized cresol novolacs ... **EM3:** 94
Epoxidized phenol ... **EM3:** 94, 96
Epoxidized phenol novolac (EPN) ... **EM3:** 594-595
characteristics ... **EL1:** 811
Epoxidized phenol novolacs (EPN) ... **EM2:** 240
Epoxidized silicones
as ultraviolet-curable ... **EL1:** 824
Epoxies *See also* Adhesives and sealants, specific types; Epoxy; Epoxy resins; Flexible epoxies; Sealants, specific types ... **A13:** 407-408, 410, 914 **EM3:** 37, 48, 75, 94-102
absorbed water as plasticizer ... **EM3:** 623-624, 625
additives and modifiers ... **EM3:** 99-100
advantages and limitations ... **EM3:** 77
aerospace application ... **EM3:** 559
for aerospace applications ... **EM3:** 97
for aerospace honeycomb core construction ... **EM3:** 560
for aerospace honeycomb sandwich construction ... **EM3:** 560
for aerospace missile radome-bonded joint ... **EM3:** 564
for aerospace secondary honeycomb sandwich bonding ... **EM3:** 561
age-life history ... **EM3:** 736
aging effect ... **EM3:** 683, 684
for aircraft engine bonding ... **EM3:** 97
for aircraft interiors bonding ... **EM3:** 97
for aluminum skin bonding to aircraft bodies ... **EM3:** 97
aluminum-to-aluminum lap joint ... **EM3:** 327
anodized surfaces ... **EM3:** 417
applications ... **EM3:** 44, 56
armature coil sealing and bonding ... **EM3:** 612
aromatic base ... **EM3:** 95
automotive body assembly bonding ... **EM3:** 609
for automotive circuit devices ... **EM3:** 610
for automotive crash damage repair ... **EM3:** 556-557
automotive electronic tacking and sealing ... **EM3:** 610
for automotive structural bonding ... **EM3:** 554-555
for battery construction ... **EM3:** 45
for body assembly ... **EM3:** 553, 554
as body assembly sealants ... **EM3:** 608
bond line corrosion ... **EM3:** 671
bonding composites to composites ... **EM3:** 293
for bonding honeycomb to primary surfaces ... **EM3:** 563
bonds with titanium ... **EM3:** 266, 267
for building construction ... **EM3:** 97
carbon laminate composite surface contamination ... **EM3:** 846-847
casting resins ... **EM3:** 596
catalysis ... **EM3:** 95
chain propagation ... **EM3:** 95
chemical resistance properties ... **EM3:** 639
for circuit protection ... **EM3:** 592
for coating and encapsulation ... **EM3:** 580, 587
coating for carbon/graphite ... **EM3:** 288
as coatings/encapsulants ... **EL1:** 241-242, 759
cold-cured (two-part) toughened shear properties ... **EM3:** 321
commercial forms ... **EM3:** 97, 98
compared to acrylics ... **EM3:** 119, 120, 121, 124
Epoxies (continued)
competing adhesives ... **EM3:** 98, 116
for composite panel assembly or repair ... **EM3:** 46, 91, 98, 100
for concrete bonding ... **EM3:** 96, 101
as conductive adhesives ... **EM3:** 76
conformal overcoat ... **EM3:** 592
as consumer product ... **EM3:** 47
as contact material, failure mechanisms ... **EL1:** 961
copper adherends ... **EM3:** 269-270
cost factors ... **EM3:** 98
critical surface tension ... **EM3:** 180
cross-linking ... **EM3:** 361, 413
cure chemistry ... **EM3:** 95-96
cure mechanism ... **EM3:** 77, 98, 101-102, 267-268, 555, 595
design and processing parameters ... **EM3:** 100-102
DGEBA ... **EM3:** 623
moisture effect ... **EM3:** 623
as die attach adhesives ... **EM3:** 76
for die attach in device packaging ... **EM3:** 584
for die attachment and interconnection ... **EM3:** 580
domination of aerospace applications ... **EM3:** 558
durability ... **EM3:** 670
effect of fillers on moisture resistance ... **EM3:** 180
effect of formulations on properties ... **EM3:** 184
electrical contact assemblies ... **EM3:** 611
electrical insulator protection ... **EM3:** 612
electrical properties ... **EL1:** 822
for electrically conductive bonding ... **EM3:** 572, 573
for electronic bonding and encapsulation ... **EM3:** 594-596
for electronic general component bonding ... **EM3:** 573
as encapsulants, ILB chips ... **EL1:** 283
for encapsulation of electronic and electrical components ... **EM3:** 51
as engineering adhesives family ... **EM3:** 567
exhibiting polarity and hydrogen bonding ... **EM3:** 181
F16H581 fiberglass ... **EM3:** 524
as fillers in casting resins ... **EM3:** 596
film tape
mechanical properties ... **EM3:** 101
physical properties ... **EM3:** 101
for flexible printed boards ... **EL1:** 582
formulation ... **EM3:** 123
for glass and glass fibers bonding ... **EM3:** 96, 101
glass composites ... **EM3:** 282, 283
glass transition temperature ... **EM3:** 76
as glassy polymers ... **EM3:** 617
gold film deposit and bulk impedance response ... **EM3:** 436
for handle assemblies bonding ... **EM3:** 45
for hermetic packaging ... **EM3:** 587
higher temperature and fracture process ... **EM3:** 509
for highway construction ... **EM3:** 97
homopolymerization ... **EM3:** 95
hot-cured (single-part) toughened shear properties ... **EM3:** 321
for household applications ... **EM3:** 51, 97
for housing assemblies bonding ... **EM3:** 45
for lens bonding ... **EM3:** 575
for localized metal reinforcement ... **EM3:** 555
for marine applications ... **EM3:** 97
markets ... **EM3:** 97-98
mechanical properties ... **EM3:** 99, 101
medical applications ... **EM3:** 576
for metal bonding ... **EM3:** 101
microstructural analysis ... **EM3:** 412, 416
for missile assembly ... **EM3:** 97
mode I fracture ... **EM3:** 341-342, 344
mode I fracture behavior under impact loads ... **EM3:** 443
modified and fracture process ... **EM3:** 509
moisture effect ... **EM3:** 622, 736
multifunctional species ... **EM3:** 94-95, 97

Epoxies (continued)
n-aminoethylpiperazine **EM3:** 184-185
no adverse effect by water-displacing corrosion inhibitors **EM3:** 641
nylon-to-metal bonding........ **EM3:** 278
particulate composite modulus values compared **EM3:** 316
pastes, properties........ **EM3:** 98, 100, 101
phase structure **EM3:** 413
plastic energy density distribution with corner rounding........ **EM3:** 330
plasticized and fracture process **EM3:** 509
for plastics bonding **EM3:** 96
for plumbing repairs........ **EM3:** 608
polysulfide use with **EM3:** 141-142
for potting........ **EM3:** 585
prebond treatment **EM3:** 35
predicted 1992 sales **EM3:** 77
for printed board manufacture........ **EM3:** 591, 592
properties........ **EM3:** 78, 92, 101, 106
for protecting electronic components **EM3:** 59
reacting with acrylonitrile-butadiene rubber **EM3:** 148
reaction with anhydrides........ **EM3:** 99
for repair of subsonic aircraft **EM3:** 808
for repair of supersonic aircraft........ **EM3:** 808
resin chemistry........ **EM3:** 94-95
resistant to many aggressive materials **EM3:** 637
for rubber battery sealing **EM3:** 58
rubber-modified
aluminum alloys **EM3:** 329
bulk uniaxial tensile stress-strain curves **EM3:** 328
elastoplastic analysis **EM3:** 483
shear strength of fiber fragments........ **EM3:** 398-399
shear strength of mild steel joints........ **EM3:** 670
for sheet molding compound (SMC) bonding **EM3:** 46
shelf life patches **EM3:** 555
side reactions in preparation........ **EM3:** 94, 95
silane coupling agents **EM3:** 182
for sporting goods manufacturing **EM3:** 576
steel adherends **EM3:** 273
steel joints wedge tested **EM3:** 668
stress analysis........ **EM3:** 478
structural functional type **EM3:** 96-97
substrates for adhesive bonding........ **EM3:** 101
suppliers........ **EM3:** 78, 139
surface contamination of composites **EM3:** 846
for surface-mount technology bonding........ **EM3:** 570
syntactic adhesives, mechanical properties **EM3:** 101
tackifiers for **EM3:** 183
temperature effect on viscosity........ **EM3:** 701
thermal resistance........ **EM3:** 98
for thermally conductive bonding **EM3:** 571
thermally cured, for solder masking **EL1:** 599
thermoplastic additives for toughness **EM3:** 185
for thermoplastic bonding **EM3:** 50, 96
for thermosets bonding **EM3:** 50, 96
thermosetting hot melt
functional type **EM3:** 97
properties **EM3:** 101
time-temperature superposition to fracture energy **EM3:** 513
titanium adherends........ **EM3:** 268, 269
to fill surface dents in aluminum aircraft structures **EM3:** 809
as top coats........ **EM3:** 640, 641
tougheners........ **EM3:** 183, 185
undersea cable splicing **EM3:** 612
used with hot-melt adhesives **EM3:** 82
volume resistivity........ **EM3:** 45
for window assembly attachments **EM3:** 553
for wire-winding operation........ **EM3:** 574
for wood bonding **EM3:** 101

Epoxies (EP) *See also* Thermosetting resins
abrasion resistance **EM2:** 167
applications **EM2:** 240-241
characteristics........ **EM2:** 241
commercial variations **EM2:** 240
competitive materials **EM2:** 241
costs **EM2:** 240
and cyanates, co-reactivity **EM2:** 234
environmental effects **EM2:** 428
for filament winding........ **EM2:** 371
as injection-moldable........ **EM2:** 321

Epoxies (EP) (continued)
as medium-temperature resin system........ **EM2:** 441
moisture effects........ **EM2:** 766
processing/product forms **EM2:** 241
for pultrusions **EM2:** 394
as structural plastic **EM2:** 65
suppliers........ **EM2:** 241
thermoset **EM2:** 629-630

Epoxy *See also* Epoxies; Epoxy materials; Epoxy molding compounds; Epoxy packages; Epoxy resins **A7:** 606
adhesives, for surface mounting **EL1:** 670-671
attach, zone 2, package interior........ **EL1:** 1010
-based silk screen processes **EL1:** 115
copper-filled, thermal conductivity **A7:** 608, 610
defined **A13:** 6
die material for sheet metal forming........ **A18:** 628
elevated-temperature curing **A11:** 731
encapsulation, characteristics **EL1:** 810-812
glass, dielectric constant **EL1:** 506
-graphite composite fasteners for........ **A11:** 530
impurities, effects **EL1:** 962
laminates, formulations and properties **EL1:** 832
as matrix material **EL1:** 1120
molecule........ **EL1:** 670
package failure mechanisms **EL1:** 961
packages, injection molded **EL1:** 961
potting in **EL1:** 824
stripping methods **M5:** 19
transfer molding, for encapsulation........ **EL1:** 473
and urethane, acrylic, silicone, compared **EL1:** 750

Epoxy acrylates
photochemistry of **EL1:** 821-866

Epoxy cement
for plate impact testing **A8:** 234
for tubular specimen flanges, torsional Kolsky bar test **A8:** 221

Epoxy composites *See also* Epoxy resins
chemical reactions, curing **EM1:** 67-71
fabricating processes........ **EM1:** 71-73
fabrication........ **EM1:** 71-73
specific tensile strength vs. specific tensile modulus........ **EM1:** 8

Epoxy esters, and epoxy-modified alkyds
compared **A13:** 403

Epoxy fiberglass printed board substrates
properties........ **A6:** 992

Epoxy film adhesives........ **EM1:** 686

Epoxy laminates
vapor-phase soldering........ **A6:** 369

Epoxy materials *See also* Epoxies; Epoxy; Flexible epoxies; Rigid epoxies
amines, as curing agent **EL1:** 827-828
anhydrides, as curing agent........ **EL1:** 828-829
basic manufacturing processes **EL1:** 831-833
as conformal coatings........ **EL1:** 763
curing agent selection........ **EL1:** 827-831
dicyandiamide, as curing agent........ **EL1:** 830-831
epoxy resin chemistry **EL1:** 825-827
Lewis acids and bases **EL1:** 829-830
materials selection........ **EL1:** 825-827
quality assurance........ **EL1:** 831-836
rigid **EL1:** 810-816

Epoxy matrix selection
principles for........ **EM1:** 76-77

Epoxy molding compounds........ **EL1:** 803-805, 810-812
characteristics........ **EL1:** 810-812

Epoxy mounting materials........ **A9:** 30-31
for carbon and alloy steels **A9:** 166-167
for epoxy-matrix composites........ **A9:** 588
for wrought stainless steels **A9:** 279

Epoxy nitriles........ **EM3:** 44

Epoxy novolacs........ **EM1:** 67-68 **EM3:** 44, 590, 594
for printed board material systems **EM3:** 592

Epoxy paints
seawater corrosion resistance enhanced by........ **M1:** 745

Epoxy per equivalent weight (EEW)
as resin property test........ **EM1:** 736

Epoxy plastic *See also* Epoxies (EP); Plastics; Silicone plastics; Tooling resin **EM3:** 12
defined **EM1:** 10 **EM2:** 16

Epoxy primer
stripping methods **M5:** 19

Epoxy primers........ **EM3:** 640

Epoxy resin *See also* Epoxies (EP); Resin(s)
cast, as plaster molding pattern **A15:** 243
defined **EM2:** 16
as organic binder........ **A15:** 35
as pattern repair material **A15:** 194
as structural plastic, chemistry **EM2:** 65

Epoxy resins *See also* Epoxies; Epoxy; Epoxy composites; Epoxy materials; FR-4 epoxy resin; Resins; Rigid epoxies........ **EM1:** 66-77 **EM3:** 12
accelerators for........ **EM1:** 137
additives to carbon-graphite materials........ **A18:** 816
adhesives........ **EM1:** 684, 685-686
as aerospace matrix........ **EM1:** 3
for aerospace prepregs **EM1:** 139-140
for automotive electronics bonding **EM3:** 553
bismaleimide triazine, properties........ **EL1:** 534-535
BPA **EM1:** 66-67
carbon fiber reinforced **EM1:** 52, 153, 400, 410-412
in carbon fiber reinforced plastic (CFRP) removal........ **EM1:** 153
chemical and tradenames **EL1:** 826
chemistry **EL1:** 825-827
and CM-X, polysulfone, compared........ **EM1:** 103
commercially available........ **EM1:** 136
curatives for........ **EM1:** 136
curing........ **EM1:** 67-73, 134-141, 654-656
curing, agents/reactions **EM1:** 67-73
curing condition effect on glass transition temperature **EM3:** 320
curing temperatures........ **EM1:** 654
E-glass fiber reinforcement........ **EM1:** 399, 401-403
effects of modifications **EM3:** 100
and fiber-resin properties **EM1:** 399
for filament winding........ **EM1:** 135-137
formulation........ **EM1:** 134, 505
glass fabric reinforced **EM1:** 399-400, 404-405
graphite reinforced........ **EM1:** 400, 412-414
in high-strength medium-temperature thermoset matrix composites **EM1:** 399-415
high-temperature........ **EM1:** 33, 141
hot/wet in-service temperatures........ **EM1:** 33
Kevlar fiber reinforced **EM1:** 399-400, 407-409
laminate properties **EM1:** 73-76
laminating processes........ **EM1:** 71-73
for lightning strike applications **EM3:** 565
manufacture and products **EM1:** 66-67
monoepoxides **EL1:** 810
novolacs **EM1:** 67-68
photoelastic stress pattern **EM1:** 196
primer........ **EM3:** 277
properties........ **EM1:** 289, -91, 399
properties analysis and methods **EM1:** 736-737
for pultrusion **EM1:** 539
quartz fabric reinforced........ **EM1:** 399-400, 414-415
ratios of compressive to tensile yield stresses........ **EM3:** 328-329
reactions/chemical structures **EL1:** 474
and reinforcements/metals, compared **EM1:** 76
for resin transfer molding........ **EM1:** 169, 566
for RRIM technology **EM1:** 121
S-glass fiber reinforced........ **EM1:** 399-400, 405-407
sample chemical reaction........ **EM1:** 751-753
selection, principles for **EM1:** 76-77
in space and missile applications........ **EM1:** 817
suppliers........ **EM1:** 134
surface preparation **EM3:** 277
surface tension **EM3:** 181
synthesis........ **EM1:** 67
systems overview **EL1:** 825-837
tests for........ **EM1:** 289, 291
thermal stability........ **EM1:** 810
used in aerospace prepregs, types **EM1:** 140
vs phenolic novolacs, for molded plastic packages **EL1:** 474
for wet lay-up **EM1:** 134

Epoxy resins and coatings **M5:** 495, 498, 501-503, 505

Epoxy resins as mounting materials **A9:** 44, 53-54
for cemented carbides **A9:** 273
for electropolishing **A9:** 49
porcelain enameled sheet steel **A9:** 198
powder metallurgy materials........ **A9:** 504-505
for titanium and titanium alloys **A9:** 458
for tungsten **A9:** 441

Epoxy systems
DEN 438-BPA-BDMA........ **EM3:** 286

Epoxy systems (continued)
Epon 828-DTA-BDMA ... **EM3:** 286
Epon 828-NMA-BDMA ... **EM3:** 286
Epoxy thermosetting resins as mounting materials
for white irons ... **A9:** 243
Epoxy, used as filler in metallographic examination of welded joints made with
backup plates ... **A9:** 578
Epoxy-acrylates
compared with urethane-acrylates ... **EM3:** 92
for woodworking ... **EM3:** 46
Epoxy-anhydride systems
properties ... **EM3:** 95
Epoxy-aramid
for printed board material systems ... **EM3:** 592
Epoxy-aramid fiber printed board substrates
properties ... **A6:** 992
Epoxy-fiberglass
for printed board material systems ... **EM3:** 592
Epoxy-glass ... **EM3:** 601
for manufacture of printed boards ... **EM3:** 589, 590
shear stresses ... **EM3:** 402
Epoxy-Kevlar
for printed board material systems ... **EM3:** 592
Epoxy-matrix composites
applications ... **A9:** 592
mounting materials for ... **A9:** 588
Epoxy-modified alkyds ... **A13:** 403
Epoxy-novolac base phenolics ... **EM3:** 104
Epoxy-nylons ... **EM3:** 44, 75
Epoxy-phenolics ... **EM3:** 75, 76, 105
advantages and limitations ... **EM3:** 80
aircraft applications ... **EM3:** 80
bonding applications ... **EM3:** 80
properties ... **EM3:** 106
typical film adhesive properties ... **EM3:** 78
Epoxy-polyamides
moisture effect ... **EM3:** 622
Epoxy-quartz
for printed board material systems ... **EM3:** 592
Epoxy-urethanes
compared to carboxyl-terminated butadiene acrylonitrile (CTBN)-epoxy ... **EM3:** 185
electrical insulator protection ... **EM3:** 612
EPRI *See* Electric Power Research Institute
EPS *See* Expanded polystyrene patterns
Epsilon
defined ... **A9:** 7
Epsilon carbide
defined ... **A9:** 7
in lower bainite ... **A9:** 664
Epsilon martensite
in 300 series stainless steels ... **A9:** 66
in austenitic stainless steels ... **A9:** 283
Epsilon phase
in Alnico alloys ... **A9:** 539
in cobalt-base heat-resistant casting alloys ... **A9:** 334
in zinc-copper alloys ... **A9:** 489
Epsilon structure
defined ... **A9:** 7
Equality
symbols for ... **A10:** 691-692
Equality of variance
test for ... **EM1:** 305
Equalized heat treatment ... **A13:** 934
Equalizer beams
fracture of cast steel ... **A11:** 388-390
Equalizers
plating ... **EL1:** 872-873
Equalizing stretch
stainless steels ... **A14:** 777
Equalizing wedge
radiographic inspection ... **A17:** 334
Equation
abbreviation for ... **A11:** 796
Equations *See* Formulas
Equations, constitutive
for material modeling ... **A14:** 417-420
Equations, generalized
for corrosion economics ... **A13:** 372-374
Equations of state ... **A6:** 162
of ideal gas ... **A17:** 58
Equator
defined ... **EM1:** 10 **EM2:** 16
Equi-inclination contours method
for phase transformations and precipitation yields ... **A10:** 376-377
Equiaxed alpha
in isothermal titanium alloy specimen ... **A8:** 172-173
Equiaxed alpha grains
in titanium and titanium alloys ... **A9:** 460
Equiaxed castings *See* Equiaxed solidification
Equiaxed crystal growth
single-phase alloys ... **A15:** 115
Equiaxed dendrite tip
growth of ... **A15:** 135
Equiaxed dimples *See also* Dimple(s)
AISI/SAE alloy steels ... **A12:** 302, 304, 319, 339
conical ... **A12:** 14
in copper ... **A12:** 173
defined ... **A12:** 173
formation ... **A12:** 12-14
and hemispheroidal dimples ... **A12:** 173
in low-carbon iron ... **A12:** 223
maraging steels ... **A12:** 383, 387
martensitic stainless steels ... **A12:** 367
on flat-face fracture surface ... **A11:** 76
precipitation-hardening stainless steels ... **A12:** 370, 371
shape ... **A12:** 12-14
with spheroidal particles ... **A12:** 223
titanium alloys ... **A12:** 444-445
tool steels ... **A12:** 380
triaxial stress effect ... **A12:** 31, 40
wrought aluminum alloys ... **A12:** 428
Equiaxed grain structure *See also* Equiaxed grains; Grain structure
in aluminum alloy 6063 ... **A9:** 630
defined ... **A8:** 5 **A9:** 7 **A15:** 5
described ... **A15:** 133
formation of ... **A9:** 603
growth ... **A15:** 132-135, 153
growth, by big bang mechanism ... **A15:** 131
growth, by constitutional supercooling ... **A15:** 130-131
growth, vs columnar grain growth ... **A15:** 135
low-gravity growth ... **A15:** 153
steady-state analysis ... **A15:** 133-134
Equiaxed grains
as a result of dendritic growth ... **A9:** 609
crystallographic texture in ... **A9:** 700
growth, models of ... **A15:** 132-133
spherical ... **A15:** 130
in steel macrostructure ... **A9:** 623
Equiaxed nuclei
origin of ... **A15:** 130-132
Equiaxed particle shapes
defined ... **A9:** 619
Equiaxed solidification
and directional solidification/single-crystal casting ... **A15:** 399
macro-microscopic modeling ... **A15:** 887-890
particle behavior in ... **A15:** 146
Equiaxed structure
in rhenium-bearing alloys ... **A9:** 448
Equiaxed structure modeling ... **A15:** 885-890
Equicohesive temperature ... **A8:** 188
defined ... **A12:** 121
symbol for ... **A11:** 796
Equilibria, basic chemical
and analytical chemistry ... **A10:** 161-165
Equilibria chemistry
basic ... **A10:** 162-163
complexometric titrations ... **A10:** 164
Equilibria chemistry (continued)
gravimetric ... **A10:** 163
ion exchange separation ... **A10:** 164-165
oxidation-reduction reactions ... **A10:** 163-164
solvent extraction ... **A10:** 164
Equilibrium *See also* Equilibrium temperatures ... **A3:** 1 • 1
chemical, defined ... **A10:** 163
condensed phase, and activity ... **A15:** 51
conditions for ... **A15:** 50
constant, defined ... **A15:** 51
defined ... **A9:** 7
deviation from, thermal analysis ... **A15:** 183
diagram ... **A3:** 1 • 2
diagrams ... **A15:** 52-53, 57, 63-64, 136
interface, during solidification ... **A15:** 110-111
electrochemical, in aqueous solutions ... **A13:** 17
in gaseous corrosion ... **A13:** 17
liquid-solid, Vant'Hoff relation for ... **A15:** 102
local interfacial, defined ... **A15:** 101
metastable, defined ... **A15:** 101
pH, described ... **A10:** 163
phase diagram, and equilibrium partition coefficient ... **A15:** 136
in phase diagrams ... **A15:** 52, 57
reversible, metastable states as ... **A15:** 101
reversible potential, defined ... **A13:** 6
state, defined ... **A10:** 163
verification, in phase diagram determination ... **A10:** 475
Equilibrium centrifugation ... **EM3:** 12
defined ... **EM2:** 1
Equilibrium conditions
laminates ... **EM1:** 229-230
Equilibrium constant
defined ... **A15:** 51
Equilibrium decomposition into two phases ... **A9:** 655
Equilibrium diagram ... **A6:** 127
defined ... **A9:** 7
Equilibrium distribution coefficient ... **A6:** 52, 89 **A9:** 611
effect on dendritic structure in copper alloy ingots ... **A9:** 638
Equilibrium, moisture *See* Moisture equilibrium
Equilibrium partition coefficient ... **A9:** 611
Equilibrium partition ratio ... **A6:** 56
Equilibrium phase diagram ... **A6:** 127
for cubic boron nitride (CBN)/hexagonal boron nitride ... **A2:** 1009
uranium-titanium ... **A2:** 673
Equilibrium precipitates
in beryllium-copper alloys ... **A9:** 395
Equilibrium solidification
thermal analysis ... **A15:** 183
Equilibrium temperatures
cast iron, chromium, silicon, vanadium effects ... **A15:** 65
liquidus, nucleation effects ... **A15:** 101
silicon effects ... **A15:** 63
in ternary iron-base alloys ... **A15:** 65-67
Equilibrium tie lines ... **A6:** 46
Equilibrium transformation temperatures, steel
symbol for ... **A11:** 796
Equipment *See also* Automated equipment; Automation; Auxiliary equipment; Casting equipment; Electronic components; Forging equipment; Furnaces; Instruments Machines; Molding equipment; Molding machines; Pattern equipment; Portable equipment; Stationary equipment; Textile equipment; Tooling; Tools; Vacuum melting and remelting processes
acoustic emission inspection ... **A17:** 280-284, 289-290
for aluminum alloys ... **A14:** 244-245
aluminum pattern ... **A15:** 195
anodizing ... **A13:** 1314-1316
argon oxygen decarburization ... **A15:** 427
automatic pouring ... **A15:** 497-500

SUBJECTS OF THE INDEXED VOLUMES: ASM Handbook (designated by the letter "A"): **A1:** Properties and Selection: Irons, Steels, and High-Performance Alloys (1990); **A2:** Properties and Selection: Nonferrous Alloys and Special-Purpose Materials (1990); **A3:** Alloy Phase Diagrams (1992); **A4:** Heat Treating (1991); **A6:** Welding, Brazing, and Soldering (1993); **A7:** Powder Metallurgy (1984); **A8:** Mechanical Testing (1985); **A9:** Metallography and Microstructures (1985); **A10:** Materials Characterization (1986); **A11:** Failure Analysis and Prevention (1986); **A12:** Fractography (1987); **A13:** Corrosion (1987); **A14:** Forming and Forging (1988); **A15:** Casting (1988); **A16:** Machining (1989); **A17:** Nondestructive Testing and Quality Control (1989); **A18:** Friction, Lubrication, and Wear Technology (1992). **Metals Handbook, 9th Edition** (designated by the letter "M"): **M1:** Properties and Selection: Irons and Steels (1978); **M2:** Properties and Selection: Nonferrous Alloys and Pure Metals (1979); **M3:** Properties and Selection: Stainless Steels, Tool Materials and Special-Purpose Materials (1980); **M4:** Heat Treating (1981); **M5:** Surface Cleaning, Finishing, and Coating (1982); **M6:** Welding, Brazing, and Soldering (1983). **Engineered Materials Handbook** (designated by the letters "EM"): **EM1:** Composites (1987); **EM2:** Engineering Plastics (1988); **EM3:** Adhesives and Sealants (1990); **EM4:** Ceramics and Glasses (1991); **Electronic Materials Handbook** (designated by the letters "EL"): **EL1:** Packaging (1989).

Equipment (continued)
auxiliary for sheet metal forming ... **A14:** 489, 499-503
Barkhausen noise ... **A17:** 160
for beryllium forming ... **A14:** 805-806
blow molding ... **EM2:** 352-353
capabilities ... **A14:** 162
capital, blow molding, costs ... **EM2:** 29
chemical processing ... **A13:** 1137-1143
cleaning, for surface conversion ... **A13:** 381-382
for closed-die forging ... **A14:** 80-81
cold extruded copper/copper alloy parts ... **A14:** 310
cold heading ... **A14:** 291-292
communications ... **A13:** 1113-1126
compression molding, costs ... **EM2:** 297
computed tomography (CT) ... **A17:** 364-372
for computer-aided dimensional inspection ... **A15:** 559
continuous casting ... **A15:** 310-312, 314-315
continuous flow electron beam melting ... **A15:** 415-416
control, for rolling mills ... **A14:** 354-355
copper alloys, heat treating ... **M4:** 726-727
costs, thermoforming ... **EM2:** 403
cupola, refinements in ... **A15:** 384
and dies, fluid flow interaction ... **A15:** 292
for differential thermal analysis ... **A15:** 184
digital image enhancement ... **A17:** 454-455
for dimensional measurement, of castings ... **A17:** 521
for dry sand molding ... **A15:** 228
electromagnetic forming ... **A14:** 650-652
electron-beam heat treating ... **M4:** 518-521
electronics ... **A13:** 1113-1126
emission-control ... **A13:** 1367-1370
energy-efficient ... **M4:** 340-342
evaluation factors for ... **A14:** 168
for explosive forming ... **A14:** 636-638
extrusion ... **A14:** 301-302
filament winding ... **EM2:** 374-37
forging, for stainless steels ... **A14:** 227
forging, selection of ... **A14:** 36-42
forming, types/characteristics ... **A14:** 16
furnaces, titanium ... **M4:** 771-772
green sand molding ... **A15:** 341-344
handling, for open-die forging ... **A14:** 63
heat treating, for stainless steels ... **A14:** 228-229
heating, closed-die forging ... **A14:** 81
heating, for aluminum alloys ... **A14:** 247
heating, precision forging ... **A14:** 164-165
horizontal centrifugal casting ... **A15:** 296-297
hot isostatic pressing ... **A15:** 539
for hot swaging ... **A14:** 142-143
for hot upset forging ... **A14:** 83-84
hydrostatic extrusion ... **A14:** 329
impact extrusion ... **A14:** 311
in-service monitoring ... **A13:** 202
injection molding, costs ... **EM2:** 294-295
inspection, of tubes on solid cylinders ... **A17:** 185
inspection, qualification of ... **A17:** 679
for liquid penetrant inspection ... **A17:** 78-80
for magnetic particle inspection ... **A17:** 111
manual spinning ... **A14:** 599-600
martempering of steel ... **M4:** 98-100, 101, 102
melting, investment casting ... **A15:** 262
metal pattern ... **A15:** 194-195
metal-processing ... **A13:** 1311-1316
metalworking, types ... **A14:** 16
for microwave holograms ... **A17:** 225
mixing, for foamed plaster molding ... **A15:** 247
mobile units, magnetic particle inspection ... **A17:** 92-93
muller
for nickel-base alloy forming ... **A14:** 832
optical holography ... **A17:** 417-420
oxyfuel gas cutting ... **A14:** 725-728
permanent mold casting ... **A15:** 276
pickling ... **A13:** 1314-1316
for plaster molding ... **A15:** 242
plating ... **A13:** 1314-1316
pollution control ... **A15:** 384
and power sources, magnetic particle inspection ... **A17:** 92-93
prepreg, automatic tape layers (ATL) ... **EM1:** 145
pulp mill ... **A13:** 1208-1210
pultrusion ... **EM2:** 390-391

Equipment (continued)
pumping/dispensing, for resin transfer molding ... **EM1:** 169
quality control and inspection ... **EM1:** 740-741
radial forging ... **A14:** 145-149
radiographic inspection ... **A17:** 304
recording, thermal inspection ... **A17:** 399-400
resin transfer molding, costs ... **EM2:** 30
retirement-for-cause (RFC) inspection ... **A17:** 687-688
for roll forging ... **A14:** 96-97
rotational molding ... **EM2:** 365-36
salt baths ... **M4:** 293-298
scanning, ultrasonic inspection ... **A17:** 261
semisolid forming, hazards ... **A15:** 338
service life, factors influencing ... **A13:** 321
slow strain rate testing ... **A13:** 261
sports and recreational, composite applications ... **EM1:** 845-847
for stainless steel forging ... **A14:** 228-229
steam ... **A11:** 602-627
for steel bar and wire ... **A17:** 556-557
for superplastic forming ... **A14:** 860-861
tantalum, applications for ... **A13:** 726
and tests, for elevated-temperature failures ... **A11:** 278-281
thermal inspection ... **A17:** 398-400
thermoforming ... **EM2:** 301, 401-403
for titanium alloy forging ... **A14:** 273-274
tool steels, heat treating ... **M4:** 566-569, 573-574
for tube spinning ... **A14:** 676
ultrasonic inspection ... **A17:** 231, 261
ultrasonic inspection, for pressure vessels ... **A17:** 652
for ultrasonic inspection, forgings ... **A17:** 505
UV curing ... **EL1:** 787
vacuum induction degassing ... **A15:** 438-440
vacuum induction remelting and shape casting ... **A15:** 399
vacuum ladle degassing ... **A15:** 432-433
vacuum oxygen decarburization ... **A15:** 429-431, 434-435
vertical centrifugal casting ... **A15:** 307
Villard-circuit, radiography ... **A17:** 305
wave soldering ... **EL1:** 702
for wax injection, investment casting ... **A15:** 255-256

Equipment applications
aluminum and aluminum alloys ... **A2:** 13-14
copper and copper alloys ... **A2:** 239-240
refractory metals and alloys ... **A2:** 558
Equipment costs *See also* Costs; Operating costs
of flux leakage inspection ... **A17:** 564
for inspection, tubular products ... **A17:** 561-562
ultrasonic inspection ... **A17:** 565
Equipment maintenance
flame hardening ... **M4:** 496-497
Equipment maintenance programs
borescope applications ... **A17:** 9-10
Equipment testing
calibration of ... **A8:** 611-619
Equivalence, radiographic *See* Radiographic equivalence
Equivalent circuit
small signal JFETs ... **EL1:** 155-156
for WSI, power distribution system ... **EL1:** 356
Equivalent constant-amplitude stress approach
damage analysis ... **A8:** 682
Equivalent cooling rate *See* Cooling rate
Equivalent ellipse method
object orientation ... **A17:** 34
Equivalent ionic conductance
conductivity as a function of ... **A10:** 659
Equivalent radial load
defined ... **A18:** 8
Equivalent radius
of particle shape ... **A7:** 242
Equivalent series resistance (ESR)
failures ... **EL1:** 997
Equivalent stress parameters ... **A8:** 713
Equivalent weight
defined ... **A10:** 162
Equivalents/gram (eg/g) of resin ... **EM3:** 94
Er-Fe (Phase Diagram) ... **A3:** 2 • 187
Er-Ga (Phase Diagram) ... **A3:** 2 • 187
Er-Ge (Phase Diagram) ... **A3:** 2 • 188
Er-In (Phase Diagram) ... **A3:** 2 • 188

Er-Mn (Phase Diagram) ... **A3:** 2 • 188
Er-Ni (Phase Diagram) ... **A3:** 2 • 189
Er-Pd (Phase Diagram) ... **A3:** 2 • 189
Er-Pt (Phase Diagram) ... **A3:** 2 • 189
Er-Ru (Phase Diagram) ... **A3:** 2 • 190
Er-Se (Phase Diagram) ... **A3:** 2 • 190
Er-Te (Phase Diagram) ... **A3:** 2 • 190
Er-Ti (Phase Diagram) ... **A3:** 2 • 191
Er-Tl (Phase Diagram) ... **A3:** 2 • 191
Erbium *See also* Rare earth metals
properties ... **A2:** 1180
pure ... **M2:** 734
as rare earth ... **A2:** 720
as SSMS internal standard ... **A10:** 145
Erbium in garnets ... **A9:** 538
Erichsen cup test ... **A8:** 5
and hemispherical dome test, compared ... **A8:** 562
for sheet metals ... **A8:** 561
Erichsen test
defined ... **A14:** 5
Eriochrome Black T
endpoint, complexation titration ... **A10:** 173
as metallochromic indicator ... **A10:** 174
Eritwerk (EW) oxide conversion coating process ... **M5:** 598-599
ERNi-1 filler metal
addition to Alloy 301 ... **A6:** 576
ERNiCrMo-3 filler metal
fusion welding to nickel alloy 713C ... **A6:** 576
Erofeev
rate law expression ... **EM4:** 55
Erosion *See also* Abrasive wear; Ceivitation; Corrosion; Erosion-corrosion ... **A13:** 136-138 **M1:** 597, 599
at graphite/matrix interface in gray iron ... **A9:** 40
base metal, in brazed joints ... **A17:** 602
boiler and steam equipment failures by ... **A11:** 623-624
of brittle materials, wear failure from ... **A11:** 156
carbon steel boilers ... **A17:** 200-201
cavitation ... **A11:** 624
cobalt-base alloys ... **A13:** 663-664
contact, electrical contact materials ... **A2:** 840-841
core ... **A15:** 589
-corrosion, of steel castings ... **A11:** 402
cut, or wash, as casting defects ... **A11:** 381
damage ... **A11:** 164-166, 170-171
data, test-service correlations ... **A13:** 313
defined ... **A8:** 5 **A11:** 4 **A13:** 6, 136
definition ... **M6:** 7
die, in die casting ... **A15:** 289
of ductile materials, wear failures from ... **A11:** 155-156
effect, oil/gas wells ... **A13:** 479-480, 1235
effect, pollution control ... **A13:** 1367
effects of temperature and corrosion on ... **A11:** 156
electrical, in sliding bearings ... **A11:** 486
evaluation ... **A13:** 311-313
in forging ... **A11:** 341
of gray iron pump bowl ... **A11:** 373
high-temperature, breech assembly for testing ... **A11:** 282
liquid ... **A11:** 163-171
liquid-impingement ... **A11:** 623-624
mold ... **A15:** 589, 711, 821-822
mold, plain carbon steels ... **A15:** 711
pitting ... **A11:** 189
platinum group metals, as electrical contact materials ... **A2:** 846
rapid ... **A13:** 137
rates ... **A11:** 165-167
resistance ranking, for metals ... **A11:** 166
-resistant metals, use of ... **A11:** 170
steam equipment failure by ... **A11:** 602
test, for cannon tubes ... **A11:** 281
testing ... **A13:** 311-313
tube wall thinning, remote-field eddy current inspection ... **A17:** 195
volume loss over time ... **A11:** 155
and waste systems ... **A13:** 1369
as wear ... **A11:** 155-156
and wear, cobalt-base wear-resistant alloys ... **A2:** 447-448

Erosion (continued)
and wear, electrical contact material life test for **A2:** 860
Erosion (erosive wear) *See also* Cavitation erosion; Electrical pitting; Erosive wear
defined **A18:** 8
Erosion, brazing
definition **A6:** 1209
Erosion damage
characteristics of **A11:** 164-166
prevention of **A11:** 170-171
rates **A11:** 165
Erosion rate
defined **A18:** 8
Erosion rate-time curve
defined **A18:** 8
Erosion resistance, electrical
defined **EM2:** 59
Erosion resistance number (NER) **A18:** 228, 229
Erosion-corrosion *See also* Corrosion; Erosion
aircraft **A13:** 1034-1035, 1044-1045
aluminum alloys **M2:** 219
aluminum/aluminum alloys **A13:** 595-596
in brazed joints **A13:** 877-878
cast irons **A13:** 568
cavitation, material selection for **A13:** 333
in CF-8M pump impeller **A11:** 402
closed feedwater heaters **A13:** 989-990
in copper and copper alloys **A11:** 201-202
copper/copper alloys **A13:** 613
and corrosion, elevated-temperature failures **A11:** 268-271
defined **A11:** 4 **A13:** 6 **A18:** 8
in forging **A11:** 341-342
liquid **A13:** 332-333
material selection for **A13:** 332-333, 332-333
in mining/mill application **A13:** 1296
of- copper alloy heat-exchanger tubing **A11:** 634-635
of- steel castings **A11:** 402
oil/gas production **A13:** 1235
in petroleum refining and petrochemical operations **A13:** 1281-1282
stainless steels **A13:** 554
steam surface condensers **A13:** 987
steam turbines **A13:** 993-994
titanium/titanium alloys **A13:** 676, 692-693
versus liquid impingement erosion **A18:** 223
in water **A11:** 189
water-recirculating systems **A13:** 488
in wet steam flow **A13:** 964-971
Erosive wear
cemented carbides **A18:** 798-799
ceramics **A18:** 814
cobalt-base wrought alloys **A18:** 767-768
failures of **A11:** 155-156
hardfacing alloys **A18:** 762
jet engine components **A18:** 588, 592
laser-hardened cast irons **A18:** 864, 865
metal-matrix composites **A18:** 805-806, 810
pumps **A18:** 593, 597-599
seals **A18:** 549
sliding bearings **A18:** 742, 743
thermal spray coating applications **A18:** 833
titanium alloys **A18:** 781
Erosivity
defined **A18:** 8
Error *See also* Percent error **A8:** 5
estimates **A8:** 642-643
statistical, types of **A8:** 626-627
Error analysis
thermocouple extension wires **A2:** 877-878
Error detection and correction
abbreviation for **A11:** 796
Error function complement **A18:** 464
Errors
defined **A10:** 673
magnification, as distortion **A12:** 196
operator, controlling for **A10:** 12
Errors (continued)
perspective, as distortion **A12:** 196
random, in sampling **A10:** 12
in roughness parameters **A12:** 193
sampling, sources and control of **A10:** 12
ESA *See* Electrostatic analyzer
ESC *See* Environmental stress cracking; Environmental stress cracking (ESC) **EM3:** 12
ESCA *See* Electron spectroscopy for chemical analysis
Escape
probability of (electron) **A18:** 451
Escape depth
Auger electron **A10:** 551
functional dependence on kinetic energy of electrons **A10:** 551
Escape peaks **A10:** 520, 673
Esco Alloy 75 *See* Superalloys, cobalt-base, specific types, UMCo-50
Eshelby procedure **EM4:** 275
Eshelby tensor **EM4:** 735
Espy diagram **A6:** 462
ESR *See* Electron spin resonance; Electroslag remelting **EM3:** 12
ESR spectrometers *See also* Electron spin resonance
detection **A10:** 256-257
magnet **A10:** 255
microwave powered **A10:** 255-257
modulation **A10:** 255
noise elimination **A10:** 257
sample cavity **A10:** 256
scan **A10:** 255-256
typical **A10:** 254-257
variable temperatures **A10:** 257
ESS *See* Environmental stress sc,reening
Essential metals *See also* Metal(s); specific essential metals
levels of biologic activity **A2:** 1250
with potential for toxicity **A2:** 1250-1256
Essentiality
of selenium **A2:** 1254
Esso test **A8:** 259
compared with Robertson and Navy tear-test **A11:** 59-60
for notch toughness **A11:** 60
Ester **EM3:** 12
defined **EM2:** 1
Ester solutions
copper/copper alloy corrosion in **A13:** 633
Ester-cured alkaline phenolic no-bake process **A15:** 215
Esterification
microwave inspection **A17:** 202
Esters *See also* Vinyl esters
analytic methods for **A10:** 9
corrosion of stainless steels in **M3:** 82-83
determined **A10:** 218
functional group analysis of **A10:** 218
unsaturated, suppliers of **EM1:** 133
Esters, vinyl
in composites **A11:** 731
Estimate
defined **A8:** 5
Estimation **A8:** 5
of failure factors **EL1:** 900-903
parameter and percentile **A8:** 628
Estimation of population characteristics
in sampling **A10:** 13
Eta phases
in austenitic stainless steels **A9:** 284
of cemented carbides **A9:** 274-275
in nickel-base heat-resistant casting alloys **A9:** 334
in wrought heat-resistant alloys **A9:** 309-312
Eta-alumina **EM4:** 113
Etch attack
effect of load on **A9:** 42
effect of suspending liquid on **A9:** 42
Etch attack (continued)
and electromechanical polishing **A9:** 42-43
on tungsten **A9:** 45
in vibratory polishing **A9:** 42
Etch cracks
defined **A9:** 7
Etch figures
defined **A9:** 7
Etch inspection
electron-beam welding **A6:** 866
Etch pits *See also* Dislocation etch pits **A9:** 126
used to determine grain orientation **A9:** 101
Etch pitting
caused by chemically active polishing **A9:** 39-40
of rhenium and rhenium-bearing alloys **A9:** 447
for special effects **A9:** 62
to determine grain orientation in magnetic materials **A9:** 531
Etch pitting reagents
for magnetic materials **A9:** 534
Etch removal
rigid printed wiring boards **EL1:** 542, 546
Etch tanks for macroetching **A9:** 171
niobium **A9:** 440
tantalum **A9:** 440
Etch-polishing *See also* Polish-etching
carbon and alloy steels **A9:** 168-169
cast irons **A9:** 244
heat-resistant casting alloys **A9:** 331-332
lead and lead alloys **A9:** 416
metal-matrix composites **A9:** 591
powder metallurgy materials **A9:** 506
zinc and zinc alloys **A9:** 488
Etchants *See also* Etchants for specific metals and alloys
chemical grades **A9:** 67
defined **A9:** 7
disposal **A9:** 69
effect of oxidizing characteristics on electrochemical potential **A9:** 144
effects of time and temperature on life **A9:** 171-172
effects on etching practice **A9:** 63
expression of composition **A9:** 66-67
nomenclature **A9:** 57
for potentiostatic etching **A9:** 146-147
preparation and handling **A9:** 66-69
printed board coupons **EL1:** 574
safety precautions **A9:** 68-69
for specimen surface preparation **A17:** 52
for surface-mount interconnection **EL1:** 732
Etchants for specific metals and alloys *See also* Etchants
aluminum alloys, macroscopic examination **A9:** 352-354
aluminum alloys, microscopic examination **A9:** 354-357
austenitic manganese steel casting specimens **A9:** 238-239
beryllium **A9:** 390
beryllium-containing alloys **A9:** 394
carbon and alloy steels **A9:** 169-175
carbon steel casting specimens **A9:** 230
carbonitrided steels **A9:** 217
carburized steels **A9:** 217
cast irons **A9:** 244-246
cemented carbides **A9:** 274
coated sheet steels **A9:** 197-198
copper and copper alloys **A9:** 400
electrical contact materials **A9:** 550-551
fiber composites **A9:** 591
hafnium **A9:** 498
heat-resistant casting alloys **A9:** 330-332
Inconel X-750, compared **A9:** 322
iron **A9:** 170
iron-cobalt and iron-nickel alloys **A9:** 533
lead and lead alloys **A9:** 416
low-alloy steel casting samples **A9:** 230

Etchants for specific metals and alloys (continued)
magnesium alloys **A9:** 426-427
magnetic materials **A9:** 532-534
nickel alloys **A9:** 435-436
nickel copper alloys **A9:** 435-436
nitrided steels **A9:** 218
permanent magnet alloys **A9:** 533
for plate steels **A9:** 202-203
powder metallurgy materials **A9:** 508-509
refractory metals **A9:** 440
rhenium and rhenium-bearing alloys **A9:** 447
sleeve bearing materials **A9:** 565
stainless steel casting alloys **A9:** 297
stainless steels **A9:** 281-283
stainless steels welded to carbon or low alloy steels **A9:** 203
steel tubular products **A9:** 211
tin and tin alloy coatings **A9:** 451
tin and tin alloys **A9:** 450
titanium and titanium alloys **A9:** 459-460
tool steels **A9:** 257
tungsten **A9:** 441
uranium and uranium alloys **A9:** 480
for use in examining welded joints **A9:** 580
wrought heat-resistant alloys **A9:** 307-308
zinc and zinc alloys **A9:** 488
zirconium and zirconium alloys **A9:** 498

Etchback
defined **EL1:** 575, 1143
as PTH failure **EL1:** 1021
rigid printed wiring boards **EL1:** 544-545

Etched foil
as surface wiring material **EL1:** 115

Etched sections
photolighting of **A12:** 84, 87, 88

Etching *See also* Color etching; Etching of specific metals and alloys **A9:** 57-70 **EM3:** 42
as a consequence of low voltage in electrolytic polishing **A9:** 48
as a result of low applied potential in electropolishing **A9:** 105
acid *See* Acid etching
alkaline *See* Alkaline etching
with alloys susceptible to hydrogen embrittlement **A8:** 510
aluminum and aluminum alloys **M5:** 8-9, 582-586, 590, 608, 610
ammoniacal, critical parameters **EL1:** 872
amount of time needed **A9:** 63
anodic *See* Anodic acid etching; Anodic etching
argon ion **A10:** 575
by nital and picral, compared for martensite structure **A10:** 302
capabilities/limitations **EL1:** 511
characteristics of suitable specimens **A9:** 57
chemical *See* Chemical etching
chemical attack **A10:** 301
chemical, use in spark source mass spectrometry **A10:** 144
chromic sulfuric oxide acid (CSA) on aluminum 2024 .... **EM3:** 667, 668, 669
cleaning **A9:** 63
dark-field illumination **A9:** 58
defect, and open circuits **EL1:** 1018
defects **EL1:** 978, 1019
defined **A9:** 7
destructive **A9:** 60-62
of dissimilar-metal welded joints **A9:** 582
dry, stainless steel **M5:** 560
effect of electrochemical differences between phases of a specimen on **A9:** 47
effect of mechanical mount material on **A9:** 28
effect of polishing damage on **A9:** 39
effect on EPMA accuracy **A10:** 524-525
effects of etchants on etching practice **A9:** 63
electrochemical attack **A10:** 301
electrolytic, for image analysis samples **A10:** 313
in electrolytic polishing **A9:** 56
for etch pits **A12:** 96, 101
of extraction replicas **A9:** 108 **A12:** 183
FPL surface preparation peel test results **EM3:** 802, 803, 805
fractures **A12:** 96
gage marks **A8:** 548
grain boundaries **A12:** 96

Etching (continued)
hafnium alloys **M5:** 667
hard chromium plating pretreatment by **M5:** 180, 183
of integrated circuits **A11:** 769
ion sputter **A10:** 575
isotropic metals to render them optically active **A9:** 78
lead frame **EL1:** 484, 487
for macroscopic examination *See* Macroetching, magnetic **A9:** 63-66
magnesium alloys **M5:** 638-639, 645
mixing **A9:** 69
nickel alloys **M5:** 564
nomenclature **A9:** 57
nonapplicability for microbeam analysis **A10:** 530
nondestructive **A9:** 57-60
over-, alligatoring as **A12:** 351
polarized light **A9:** 58-59
and polyimides **EM3:** 161
porcelain enameling process **M5:** 514-515
powder metallurgy materials **A9:** 508-509
preferential attack on polishing scratches **A9:** 40
printed board coupons **EL1:** 573
process, quality control **EL1:** 872
rate of attack **A9:** 63
reproducibility **A9:** 63
revealing special features **A9:** 62-63
scanning electron microscopy specimens **A9:** 99
specimens for optical metallography analysis **A10:** 301
sputter, effect on AES analysis **A10:** 556
stainless steel **M5:** 560-561
steel **M5:** 16-18
suppliers of etchants **EM3:** 802, 803
surface **A10:** 575
surface, in XPS samples **A10:** 575
techniques, for image analysis samples **A10:** 313
for temper embrittlement **A12:** 134
temperatures **A9:** 63
thermal spray-coated materials **M5:** 372
time, effect on measurement of ferrite grain size **A10:** 318
to activate polytetrafluoroethylene surfaces **EM3:** 847
to detect weld defects **A9:** 581
of tools and dies **A11:** 563
vacuum cathodic **A10:** 301
wet, stainless steel **M5:** 560-561
zirconium alloys **M5:** 667

Etching alkaline cleaners *See* Alkaline cleaning, etching cleaners

Etching of specific metals and alloys *See also* Etching
aluminum alloys **A9:** 354-357
beryllium **A9:** 390
beryllium-containing alloys **A9:** 394
brass **A9:** 41
carbon steel casting specimens **A9:** 230
carbonitrided steels **A9:** 217
carburized steels **A9:** 217
cast irons **A9:** 244-245
chromized sheet steel **A9:** 198
coated sheet steels **A9:** 197-198
copper and copper alloys **A9:** 399-401
electrical contact materials **A9:** 551
of electrogalvanized sheet steel **A9:** 197
fiber composites **A9:** 591
heat-resistant casting alloys **A9:** 331-332
hot-dip aluminum coated sheet steel **A9:** 197
hot-dip galvanized sheet steel **A9:** 197
hot-dip zinc-aluminum coated sheet steel **A9:** 197
of lead and lead alloys **A9:** 415-417
low-alloy steel casting samples **A9:** 230
nickel alloys **A9:** 435
nickel-base superalloy welded joints **A9:** 580
nickel-copper alloys **A9:** 435
nitrided steels **A9:** 218
plate steels **A9:** 202-203
porcelain enameled sheet steel **A9:** 198
rhenium and rhenium-bearing alloys **A9:** 447
silicon steel transformer sheets **A9:** 62-63
sleeve bearing materials **A9:** 567

Etching of specific metals and alloys (continued)
stainless steel casting alloys **A9:** 297
stainless-clad sheet steel **A9:** 198
steel tubular products **A9:** 211
tin and tin alloy coatings **A9:** 451
tin and tin alloys **A9:** 450
titanium and titanium alloys **A9:** 459
tool steels **A9:** 256-258
in tungsten and tungsten alloys as a result of electropolishing **A9:** 440
uranium and uranium alloys **A9:** 479-480
welded joints in plate steels **A9:** 203
wrought heat-resistant alloys **A9:** 307
wrought stainless steels **A9:** 281-282
zinc and zinc alloys **A9:** 488-489

Etching reagents *See* Etchants
for beryllium-copper alloys **A2:** 405
beryllium-nickel alloys **A2:** 423

Etching techniques
double-etch method **A16:** 36
and electrochemical machining **A16:** 533
nital etch method **A16:** 36
for replication microscopy specimens **A17:** 52

Etchings for grain size visibility **A7:** 486

ETFE *See* Ethylene-tetrafluoroethylene

Ethanol
how to denature **A9:** 68
and phosphoric acid as an electrolyte for magnesium alloys **A9:** 426
substituted for methanol in etchants **A9:** 67
used in etchants **A9:** 67-68

Ethanol (anhydrous)
surface tension **EM3:** 181

Ether
description **A9:** 68

Ether linkage
molding compounds **EL1:** 804

Ethers
copper/copper alloy corrosion in **A13:** 634

Ethyl
uses and properties **EM3:** 126

Ethyl acetate ($C_4H_8O_2$)
as solvent used in ceramics processing **EM4:** 117

Ethyl alcohol **A10:** 640
for SCC of titanium alloys **A8:** 531

Ethyl alcohol (solvent)
batch weight of formulation when used in nonoxidizing sintering atmospheres **EM4:** 163

Ethyl alcohols
and SCC in titanium and titanium alloys **A11:** 224

Ethyl cellulose
as media for screening and stamping processes **EM4:** 475

Ethyl silicate
applications **EM4:** 47
as binder, investment casting **A15:** 258-259
composition **EM4:** 47
defined **A15:** 5
as silica-base bond **A15:** 212
supply sources **EM4:** 47

Ethyl trifluoracetate
XPS spectrum of carbon Is lines in **A10:** 572

Ethylene chlorotrifluoroethylene (ECTFE)
as fluoropolymer **EM2:** 115-11

Ethylene ethyl acrylate
as binder for ceramic injection molding **EM4:** 173

Ethylene glycol **EM3:** 674
as additive **EM3:** 177
description **A9:** 68
alloys **A9:** 459
electrolyte for titanium and titanium methanol and perchloric acid as an
surface tension **EM3:** 181
in water baths **A17:** 101-102

Ethylene glycol ($C_2H_6O_2$)
as solvent used in ceramics processing **EM4:** 117
as surfactant to keep powders from forming compacted layers **EM4:** 99

Ethylene glycol monobutyl ether
description **A9:** 68
surface tension **EM3:** 181

Ethylene glycol monoethyl ether *See* Cellosolve

Ethylene glycol monoethyl ether (2-ethoxy ethanol)
surface tension **EM3:** 181

Ethylene glycol monomethyl
surface tension **EM3:** 181
Ethylene glycol solutions **A13:** 635
Ethylene plastics *See also* Plastics **EM3:** 12
defined **EM2:** 16
Ethylene propylene rubber (EPR) insulation
for copper and copper alloy products **A2:** 258
Ethylene-butylene copolymer
as irradiation container material **A10:** 236
Ethylene-chlorotrifluoroethylene (E-CTFE)
surface preparation **EM3:** 279
Ethylene-chlorotrifluoroethylene (ECTFE)
in pharmaceutical production facilities **A13:** 1228
Ethylene-propylene rubber
cross-link density effect on tensile strength **EM3:** 412
Ethylene-propylene-diene monomer (EPDM) rubber
for window sealing **EM3:** 56
Ethylene-tetrafluoroethylene (ETFE)
as fluoropolymer **EM2:** 115-11
Ethylene-vinyl acetate **EM3:** 75
anodized surfaces **EM3:** 417
for body sealing an glazing materials **EM3:** 57
characteristics **EM3:** 90
polyolefin homopolymers and copolymers, typical properties **EM3:** 83
residential applications **EM3:** 675
Ethylene-vinyl acetate (EVA)
AC-400
shear modulus **EM4:** 175, 176
viscosity **EM4:** 175
with alumina, viscosity in injection molding **EM4:** 174
as binder for ceramic injection molding **EM4:** 173
Ethylene-vinyl acetate copolymers **EM3:** 80
crystallinity and composition **EM3:** 408, 411, 412
microstructure **EM3:** 407
predicted 1992 sales **EM3:** 81
properties **EM3:** 82, 412
suppliers **EM3:** 82
Ethylenediaminetetraacetic acid (EDTA)
detected by ion chromatography **A10:** 661
disodium salt of **A10:** 201
for removal of interferences, UV/VIS analysis **A10:** 65
titrations **A10:** 173
ETL
as synchrotron radiation source **A10:** 413
ETP *See* Electrolytic tough pitch (ETP) copper; Electrolytic tough pitch copper
ETPs *See* Engineering thermoplastics
Ettringite **EM4:** 12
Eu-Ga (Phase Diagram) ..... **A3:** 2 • 191
Eu-Ge (Phase Diagram) ..... **A3:** 2 • 192
Eu-In (Phase Diagram) ..... **A3:** 2 • 192
Eu-Mg (Phase Diagram) ..... **A3:** 2 • 192
Eu-Pb (Phase Diagram) ..... **A3:** 2 • 193
Eu-Pd (Phase Diagram) ..... **A3:** 2 • 193
Eu-Pt (Phase Diagram) ..... **A3:** 2 • 193
Eu-Te (Phase Diagram) ..... **A3:** 2 • 194
EUCAST software program **A15:** 888
Euctectic temperature *See also* Thermal properties
wrought aluminum and aluminum alloys **A2:** 118, 121
Euler angles
defined **A10:** 673
fiber development in **A10:** 363
relating the specimen axes with the crystal axes **A9:** 703
use in polycrystalline grain orientation **A10:** 359-361
used to describe crystallographic texture **A9:** 706
Euler equation
axial compression testing **A8:** 55
Euler plots
defined **A10:** 361
ghost peaks in **A10:** 362-363
Euler plots (continued)
ODF for copper tubing using **A10:** 361
and orientation distribution function **A10:** 360-361
Euler rotations
defining orientation **A10:** 359
Eulerian cradle
on diffractometer **A10:** 360
Eulerian frame **A6:** 1136
Europe, Western
telecommunication industry structure **EL1:** 384
European Committee for Standardization (CEN) **A2:** 16
European Polymer Journal
as information source **EM2:** 9
European standards **EM4:** 925-929, 1070
European whiteheart malleable iron
and American blackheart malleable iron **A15:** 30-31
Europium *See also* Rare earth metals
determined by controlled-potential coulometry **A10:** 209
as divalent **A2:** 720
Jones reductor for **A10:** 176
properties **A2:** 1180
pure **M2:** 734-735
as rare earth **A2:** 720
TNAA detection limits **A10:** 238
Europium in garnets **A9:** 538
Europium orthoferrite **EM4:** 53
Europium oxide
in nuclear control rods and shielding **A7:** 666
Europium oxide-urania **EM4:** 191
Eutectic **EM3:** 12
alloys ..... **A3:** 1 • 3
defined **A13:** 6 **EM2:** 1
microstructures ..... **A3:** 1 • 19-20
reaction ..... **A3:** 1 • 3, 5
soft solder ..... **A3:** 1 • 20
Eutectic alloys *See also* Eutectic structures
compositional range **A9:** 620-621
deviation from binary composition **A9:** 620-621
microstructures of **A9:** 620
schematic phase diagram **A9:** 618
solidification structures **A9:** 618-622
Eutectic arrest
defined **A9:** 7
Eutectic arrest test **A15:** 390
Eutectic attach die **EL1:** 213-215
zone 2, package interior **EL1:** 1010-1011
Eutectic bonding **EM3:** 12
Eutectic bonding as component attachment **EL1:** 349
as die attachment method **EL1:** 213
failure mechanisms **EL1:** 1045-1046
substrate metallization in **EL1:** 214
Eutectic brazing *See* Alloy brazing
Eutectic carbide *See also* Carbide particles; Cementite
appearance of, in
iron-chromium-nickel
heat-resistant casting alloys **A9:** 332
defined **A9:** 7
in gray iron **A15:** 632
Eutectic cells, in gray irons
macroetching for **A11:** 344
Eutectic die attach *See also* Die attachment methods
die back side preparation **EL1:** 213-214
gold-silicon die bond, mechanism of **EL1:** 213
manufacturing of **EL1:** 215
preforms **EL1:** 214
substrate metallization, in bonding **EL1:** 214
voids, effects **EL1:** 214-215
Eutectic die bond alloys
typical **EL1:** 1045
Eutectic fusible alloys *See also* Fusible alloys
compositions and melting temperatures **A2:** 755
Eutectic grain
defined **A9:** 620
Eutectic gray cast iron **M1:** 12-13
Eutectic growth
at very high solidification rates **A15:** 125
in cast iron **A15:** 174-180
diffusion fields **A15:** 121
irregular and regular **A15:** 121-122
simplified theory of **A15:** 124
Eutectic growth (cast iron)
coupled zone **A15:** 174
in directional solidification **A15:** 174-175
isothermal solidification **A15:** 174
microscopic particles in **A15:** 102
in multicomponent solidification **A15:** 175-180
Eutectic heating
effect on fatigue strength **A11:** 122
Eutectic melting in aluminum alloys **A9:** 358
Eutectic nucleation
of austenite-flake graphite **A15:** 170-172
austenite-spheroidal graphite **A15:** 172-173
cast iron **A15:** 169-173
Eutectic point
defined **A9:** 7
Eutectic silicon
growth **A15:** 163
Eutectic spacings as a function of growth rate
table **A9:** 619
Eutectic structures **A9:** 618-620
aluminum-silicon **A9:** 620, 622
cadmium-tin **A9:** 622
classification of **A9:** 621
colony structures **A9:** 619-620
grain structure **A9:** 620
iron-carbon **A9:** 620
phase particle structure **A9:** 619
sampling factors **A9:** 620
size ranges of, table **A9:** 619
Eutectic superalloys
effect of solidification factors on development of **A9:** 621
Eutectic temperature **A6:** 127
Eutectic tin solder
application and composition **A2:** 521
Eutectic(s)
amount, calculated **A15:** 68-69
austenite-flake graphite **A15:** 170-172
austenite-graphite **A15:** 169-170
austenite-graphite, graphite structure **A15:** 169-170
austenite-iron carbide **A15:** 173, 179-180
austenite-spheroidal graphite **A15:** 172-173
compacted/vermicular graphite **A15:** 178-179
coupled zone of **A15:** 123-124
defined **A15:** 5
and dendrites, competitive growth of **A15:** 122-124
fibrous **A15:** 119-120
flake (lamellar) graphite **A15:** 175-176
growth **A15:** 121-122, 124-125, 174-180
instabilities **A15:** 122-123
irregular ©Chinese script© type **A15:** 120
lamellar **A15:** 119-120
ledeburite (austenite iron carbide) **A15:** 180
low-gravity, composite solidification **A15:** 150-153
microstructure of **A15:** 120-121
morphology of **A15:** 119-121
nucleation, in cast iron **A15:** 169-173
on/off low-gravity models **A15:** 150-151
operating range of **A15:** 124
regular/irregular **A15:** 120
scale of **A15:** 121-122
solidification of **A15:** 119-125
spacings **A15:** 122
spheroidal graphite **A15:** 176-178
structure, and cell size, aluminum-silicon alloys **A15:** 167-168
transition, gray-to-white **A15:** 180
Eutectic-cell etching
defined **A9:** 7

Eutectics *See also* Directionally solidified eutectics; Unidirectionally solidified eutectics
in aluminum alloy castings **A9:** 358
defined **A9:** 7
effect on dendritic structures **A9:** 613
equation to predict volume fraction **A9:** 614
in tin and tin alloys **A9:** 452
types of **A1:** 3
in zinc alloys **A9:** 489-490
Eutectoid
defined **A9:** 7 **A13:** 6
of iron-copper-carbon alloy powder metallurgy materials **A9:** 510
microstructures ..... **A3:** 1 • 20-21
reaction ..... **A3:** 1 • 5
Eutectoid carbon steel **A9:** 178-179
hardness **A9:** 179
Eutectoid composition, pearlite
microstructure **A7:** 315
Eutectoid point
defined **A9:** 7
Eutectoid products of titanium alloys as a result of beta decompositions **A9:** 461
Eutectoid reactions **A9:** 658-661
Eutectoid steel
damping capacity **M1:** 32
flow softening measured in torsion **A8:** 177
mean free path between spheroidite particles **A8:** 176
Eutectoid-forming group
wrought titanium alloys **A2:** 599
Eutectoid/hypoeutectoid plain carbon steels
superplasticity **A14:** 869
Eutectoids
spacing of **A15:** 122
Evacuating ports and seals
for explosive forming **A14:** 638
Evacuation *See also* Gases
of gases, in FM process **A15:** 38
Evacuation time
for packed powders **A7:** 434
Evaluating Wood Adhesives and Adhesive Bonds: Performance Requirements, Bonding Variables, Bond Evaluation, Procedural Recommendations **EM3:** 68-69
Evaluation *See also* Corrosion testing; Fiber properties analysis; Inspection; Laboratory corrosion testing; Laminate properties; Material properties analysis; NDE methods; Nondestructive evaluation; Nondestructive evaluation methods Nondestructive evaluation techniques; Testing
of atmospheric corrosion **A13:** 207
of cavitation **A13:** 311-313
of composite materials **EM1:** 38-39
of corrosion fatigue **A13:** 291-302
of corrosion test results **A13:** 194-195
cost drivers in **EM1:** 421
of crevice corrosion **A13:** 303-310
critical, laser surface processing **A13:** 503
of erosion **A13:** 311-313
of exfoliation corrosion **A13:** 242-244
flaw **A17:** 49-50
of galvanic corrosion **A13:** 234-238
of inoculants
of intergranular corrosion **A13:** 239-241
of joints/fasteners, types **EM1:** 710
magnetic particle inspection **A17:** 103
of materials, for selection **A13:** 322
and metrology **A17:** 50
of microbiological corrosion **A13:** 314-315
of pitting corrosion **A13:** 231-233
procedure, adhesive-bonded joints **A17:** 636-637
of stress-corrosion cracking **A13:** 245-282
of surfaces, by magnetic rubber inspection **A17:** 125
test coupon **A13:** 199
titanium alloy castings **A15:** 833
of uniform corrosion **A13:** 229-230
Evaluation of brazed joints **A6:** 1117-1123
design testing, evaluation, and feedback **A6:** 1120-1121
destructive testing methods **A6:** 1120
fatigue testing under cyclic loading **A6:** 1120
impact tests **A6:** 1120
metallographic examination **A6:** 1120
peel tests **A6:** 1120
tensile and shear tests **A6:** 1120
Evaluation of brazed joints (continued)
torsion tests **A6:** 1120
nondestructive evaluation techniques **A6:** 1118-1120, 1121-1122
Evaluation of soldered joints **A6:** 1124-1128
automated inspection techniques **A6:** 1126
laser inspection **A6:** 1126
structured-light, three-dimensional vision system **A6:** 1126
x-ray laminography **A6:** 1126
destructive evaluation **A6:** 1126-1128
visual inspection **A6:** 1124-1126
bridging **A6:** 1125
dewetting **A6:** 1125
dull or rough solder surfaces **A6:** 1125
nonwetting **A6:** 1125
porosity **A6:** 1125-1126
Evans diagrams
alloying effects **A13:** 47-48
aqueous corrosion **A13:** 31-34
corrosion rate control **A13:** 1333
corrosion reaction, amorphous metals **A13:** 865
sulfide effects **A13:** 48-49
Evapograph
for microwave patterns **A17:** 208
Evaporated coatings for contrast enhancement of high-speed steel scanning electron
microscopy specimens **A9:** 99
Evaporated interference layer materials **A9:** 60
Evaporation
fading, as silicon modifier effect **A15:** 163
fields, for selected metals **A10:** 587
loss, in immersion tests **A13:** 221
microwave inspection **A17:** 215
oxide **A13:** 71
of penetrants **A17:** 85
preferential, of solute, in vacuum melting ultrapurification **A2:** 1094
as PVD process **A13:** 456-457
selection, of alloying elements **A15:** 396
of solvents, as IR sample **A10:** 112
thermal **A12:** 172-173
thermal, of amorphous materials and metallic glasses **A2:** 806
titanium alloys **A18:** 779, 780
of trace elements **A15:** 394-395
treatment, aluminum refining by **A15:** 80
Evaporation adhesives **EM3:** 75
Evaporation, and condensation
as material transport systems in sintering of compacts **A7:** 313
Evaporation deposition **A18:** 840, 841, 842-843
definition **A18:** 842
directionality **A18:** 843
evaporation coefficient **A18:** 843
gas scattering evaporation **A18:** 843
heat sources **A18:** 842-843
mean free path **A18:** 843
parameters **A18:** 841
rate of evaporation **A18:** 843
Evaporation point
defined **A9:** 7
Evaporation recovery process, plating waste treatment **M5:** 316-317
atmospheric evaporator **M5:** 316
rising film evaporator **M5:** 316
vacuum evaporators **M5:** 316-317
Evaporation vacuum coating *See* Vacuum coating, evaporation process
Evaporative cooling
of sand **A15:** 348-349
Evaporative foam casting *See also* Lost foam casting
development **A15:** 36
tolerances **A15:** 622
Evaporative pattern casting *See also* Lost foam casting
as special molding process **A15:** 37
Evaporative pattern casting (EPC) *See also* Lost foam casting
aluminum casting alloys **A2:** 140
Even tension
defined **EM1:** 10 **EM2:** 1
Event
EXAFS analysis **A10:** 410
Everdur *See* Copper alloys, specific types, C87200
properties and applications **A2:** 371
Everhart-Thornley electron detector **A12:** 93-94, 168
Evolution
of gases, porosity by **A15:** 82-87
microstructural **A15:** 883-891
Evolutionary operations (experimental) designs **EM2:** 601
Evolved gas analysis (EGA) **EM4:** 52, 53
Ewald construction, and reciprocal lattice
and LEED **A10:** 539
Ewald sphere
defined **A9:** 7
and the reciprocal space **A9:** 110-111
used in figuring the diffraction intensity **A9:** 111-112
Ewald sphere construction, for transmission electron microscopy illumination
modes **A9:** 104
E_x *See* X-ray energy
Exact damage (fatigue) model **EM1:** 247
EXAFS *See* Extended x-ray absorption fine structure
Exaggerated grain growth *See also* Grain growth **A9:** 697-698
Examination *See also* Inspection; Nondestructive evaluation (NDE); Nondestructive testing; Testing of printed board coupons; Visual examination; Visual inspection **EL1:** 574
of boilers and related equipment **A11:** 602
laboratory, of worn parts **A11:** 156-158
macroscopic **A12:** 91-93
macroscopic, for SCC **A11:** 212
macroscopic, of- shafts **A11:** 460
metallographic, pitting corrosion **A13:** 231
of metallographic sections **A11:** 24
microscopic **A11:** 173-174, 213, 629
nondestructive, of failed parts **A11:** 173
on-site, for corrosion **A11:** 173
pitting corrosion **A13:** 231-232
preliminary laboratory **A11:** 173
preliminary visual, fractures **A12:** 72-73
visual **A11:** 173, 628-629 **A12:** 91-165
visual, pitting corrosion **A13:** 231
Examples and applications *See* Applications and examples
Excess-flux deposits
control of **EL1:** 649
Exchange current
defined **A13:** 6
density, defined **A13:** 6
Exchange narrowing
as ESR line-broadening mechanism **A10:** 255
Exchange stiffness
determined **A10:** 275
as ferromagnetic resonance application **A10:** 275
FMR investigated **A10:** 267, 268
Excising
outer lead bonding **EL1:** 284
Excitation
atomic energy level diagram showing **A10:** 433
of conduction electrons leading to secondary electron (low-energy) emission **A10:** 433-434
in dc arc **A10:** 25
in eddy current inspection **A17:** 165-166
in eddy-current vs. electric current perturbation **A17:** 136-137
emission maxima, fluorescence lifetimes and **A10:** 75
index, defined **A10:** 673
mechanisms, of emission sources **A10:** 24
of phonons, as inelastic scattering process **A10:** 434
plasmon, as inelastic scattering process **A10:** 434
potential, defined **A10:** 673
primary AES mode of **A10:** 550
remote-field eddy current inspection
spark source, for elemental analysis of metal or alloy **A10:** 29
spectra, qualitative MFS analysis **A10:** 74-75
thermal, in optical emission sources **A10:** 24
volume, defined **A10:** 673
x-ray tube and secondary-target, in x-ray spectrometers **A10:** 89
Excitation error of the primary transmission electron beam **A9:** 111
Excitation, magnetostrictive
for axial fatigue testing **A8:** 369

Excitation volume
effects in SEM imaging ... **A12:** 167
Excited nuclear level
population in Mössbauer spectroscopy ... **A10:** 288
Exciter coils
remote-field eddy current inspection ... **A17:** 195-196
Exciton ... **A18:** 325
EXCO tests
for exfoliation corrosion evaluation ... **A13:** 243
Excrescence
defined ... **A18:** 8
Excretion, of mercury
as toxin ... **A2:** 1247-1248
Exfoliation *See also* Corrosion; Exfoliation corrosion
closed feedwater heaters ... **A13:** 989-990
corrosion, in thick aluminum alloy plate ... **A11:** 201
defined ... **A7:** 4 **A11:** 4 **A13:** 6
in forging ... **A11:** 338-342
and intemal oxide scale ... **A11:** 606, 610
material selection for ... **A13:** 334
ratings ... **A13:** 242-243
superheaters/reheaters ... **A13:** 992
temper effects, aluminum alloy ... **A13:** 595
tube, closed feedwater heater ... **A13:** 990
visual assessment ... **A13:** 244
Exfoliation corrosion *See also* Exfoliation
aluminum aircraft parts ... **A13:** 1022
aluminum alloys ... **M2:** 218-220
aluminum/aluminum alloys ... **A13:** 594-595
defined ... **A13:** 2918
evaluation of ... **A13:** 242-244
Exfoliation corrosion evaluation
immersion tests ... **A13:** 243
ratings, illustrated ... **A13:** 242-243
spray tests ... **A13:** 242
visual assessment ... **A13:** 244
Exfoliation corrosion, galvanic
eddy current inspection ... **A17:** 191
Exfoliation resistance *See also* Corrosion resistance
aluminum-lithium alloys ... **A2:** 187, 190-191
Exhaust booth
definition ... **M6:** 7
Eximer fluorescence
as test ... **EM2:** 426
Exo atmospheres *See* Exothermic atmospheres
Exo gas generator ... **A7:** 343
Exo-Process
electric arc cutting ... **A14:** 734
Exogenous inclusions
defined ... **A9:** 7 **A15:** 88
in ductile iron ... **A15:** 88
in steels ... **A15:** 93-94
Exogenous slag inclusion
powder forging ... **A14:** 191
Exotherm ... **EM3:** 12
defined ... **EM1:** 10 **EM2:** 17
Exothermic atmospheres ... **A7:** 4, 361
Exothermic base atmospheres
lean ... **M4:** 395, 396
operating economics ... **M4:** 396
rich exothermic atmospheres ... **M4:** 395, 396
safety considerations ... **M4:** 396-397
Exothermic brazing (EXB) ... **A6:** 123, 345-346
advantages ... **A6:** 345
aluminum alloys ... **A6:** 345
applications ... **A6:** 345
compound types ... **A6:** 345
definition ... **A6:** 345
disadvantages ... **A6:** 345
equipment ... **A6:** 345
heating vs. time characteristics ... **A6:** 346
maximum temperature vs. mass of exothermic compound ... **A6:** 346
parameters ... **A6:** 346
procedure ... **A6:** 346
refractory metals ... **A6:** 345
safety precautions ... **A6:** 345
Exothermic brazing (EXB) (continued)
stainless steels ... **A6:** 345
vs. thermite welding ... **A6:** 345
Exothermic chemical reduction
thermite as ... **A7:** 157, 194
Exothermic feeding aids ... **A15:** 586
Exothermic gas
as sintering atmosphere ... **A7:** 192, 341, 343-344
Exothermic molds
for cast Alnico alloys ... **A15:** 736
Exothermic reaction *See also* Endothermic reaction
defined ... **A15:** 5
Exothermic reactions
during mounting of carbon and alloy steels prevention of ... **A9:** 167
of epoxy resin mounting materials, effect on alloys ... **A9:** 458
hydride in titanium and titanium
Exothermic runaway
prepreg ... **EM1:** 138
Exothermic thermite reaction
tungsten carbide powder production ... **A7:** 157, 194
Exothermic-endothermic-base atmospheres
applications ... **M4:** 413
equipment ... **M4:** 412, 413
generation ... **M4:** 413
generator operation ... **M4:** 413-414
Exothermic-insulating feeding aids ... **A15:** 586
Exotic effects
as ferromagnetic resonance application ... **A10:** 276
Expandable plastic *See also* Foamed plastics; Plastics ... **EM3:** 12
defined ... **EM2:** 17
Expandable reamers ... **A16:** 241, 243-244
Expanded polystyrene molding *See* Lost foam casting
Expanded polystyrene patterns
as expendable ... **A15:** 196-197
for lost foam casting ... **A15:** 231-232
process for ... **A15:** 204-205
as Replicast process ... **A15:** 37
tooling, for Replicast process ... **A15:** 270-271
Expanded polytetrafluoroethylene (e-FTFE) ... **EL1:** 605
Expanding
aluminum alloy ... **A14:** 804
drawn workpieces ... **A14:** 586-587
nickel-base alloy tubing ... **A14:** 836
Expanding assembly
furnace brazing ... **M6:** 940
Expanding circle heat source ... **EM4:** 372
Expanding cylinder test ... **A8:** 210
Expanding ring
for high strain rate tension testing ... **A8:** 187
test ... **A8:** 210
Expansion *See* Thermal expansion ... **A7:** 4
austenite-martensite transformation with ... **A11:** 122
characteristics, iron-nickel alloys ... **A2:** 893
coefficient, defined *See* Coefficient of expansion
coefficient, of Invar ... **A2:** 889-890
of compacts on ejection ... **A7:** 309
of copper powder compacts ... **A7:** 310
differences, in dissimilar-metal welds ... **A11:** 620
and distortion, failure of high-temperature rotary valve due to ... **A11:** 374-376
during sintering ... **A7:** 311
factors affecting during homogenization ... **A7:** 315
joint, stainless steel bellows ... **A11:** 131-133
joints, bellows-type, fatigue fracture ... **A11:** 118
mold, during baking, as casting defect ... **A11:** 386
scabs, as casting defects ... **A11:** 385
Expansion, coefficient of, defined *See* Coeffecient of expansion
Expansion defects
green sand molding ... **A15:** 346
Expansion inserts ... **EM2:** 723
Expansion manufacturing technique
for honeycomb ... **EM1:** 721-722
Expansion mismatch *See* Thermal expansion
Expected life ... **EM3:** 636
Expected value
statistical ... **A8:** 624-625
Expendable molds
horizontal centrifugal casting ... **A15:** 296
processes, classified ... **A15:** 204
Expendable patterns
defined ... **A15:** 5
procedure ... **A15:** 204-205
types ... **A15:** 191-192
Experience *See* Applications
Experiment *See also* Comparative experiments; Designed experiments; Experimental design
bias in ... **A8:** 639-640
comparative, planning of ... **A8:** 623, 639-652
design, determining optimum conditions or levels for ... **A8:** 650-652
designed, analysis of ... **A8:** 623, 653-661
factorial ... **A8:** 641-643
one-at-a-time ... **A8:** 641
precision of ... **A8:** 640
proposal for ... **A8:** 639
Experimental
area ... **A8:** 640
compliance, for compact-type fatigue specimen ... **A8:** 385
design ... **A8:** 623, 640
error estimates ... **A8:** 641-643
pattern, block design as ... **A8:** 640
techniques, pin bearing testing ... **A8:** 58
Experimental alloys
designation system ... **A2:** 16
Experimental design *See also* Design; Models; NDE reliability; Quality control; Statistical analysis; Statistical methods
confounding in ... **A17:** 747, 749
controlled/uncontrolled factors ... **A17:** 693-694
factorial designs ... **A17:** 740-750, 693
factorial experiments for hit/miss data ... **A17:** 693
iterative and sequential ... **A17:** 741
multi-variable studies ... **A14:** 937-938
of NDE reliability ... **A17:** 692-694
orthogonal arrays ... **A17:** 748-750
parameter design ... **A17:** 751-752
for POD(A) function data ... **A17:** 693-694
robust, implementing ... **A17:** 750-752
selection ... **EM2:** 599
sequential ... **A17:** 741, 748
single-variable studies ... **A14:** 937
strategies ... **A17:** 742
system noise/variation ... **A17:** 742
Taguchi experiments ... **A14:** 938-939
terminology ... **EM2:** 599-600
types ... **EM2:** 600-602
of variable interactions ... **A17:** 742-743
Experimental error
internal estimates of ... **A8:** 642-643
Experimental stress analysis
as failure analysis ... **A11:** 18
Experimental techniques
quantitative fractography ... **A12:** 194-199
Experimentation
nature of ... **A8:** 639-641
Explosibility *See* Explosion(s); Explosivity; Pyrophoricity
Explosion bonding
of clad metals ... **A13:** 887
tantalum alloys ... **A6:** 580
Explosion cladding ... **M6:** 804
Explosion welding ... **M6:** 705-718
advantages ... **M6:** 707-708
applicability ... **M6:** 707-709
applications ... **M6:** 712-716
buildup and repair ... **M6:** 716
chemical process vessels ... **M6:** 712
conversion-rolled billets ... **M6:** 712-713

Explosion welding (continued)
electrical ... **M6:** 713
marine ... **M6:** 714
nonplanar specialty products ... **M6:** 715
pipeline welding ... **M6:** 715-717
transition joints ... **M6:** 713-714
tube welding and plugging ... **M6:** 715-716
tubular ... **M6:** 714-715
assembly ... **M6:** 710
definition ... **M6:** 7
explosives ... **M6:** 709
facilities ... **M6:** 710
flow sheet ... **M6:** 709
inspection ... **M6:** 710-711
nondestructive ... **M6:** 710
radiographic ... **M6:** 711
ultrasonic ... **M6:** 710
limitations ... **M6:** 708-709
metal preparation ... **M6:** 709-710
metals welded ... **M6:** 709-710
nature of bonding ... **M6:** 706-707
bond zone wave formation ... **M6:** 706
shear bands ... **M6:** 707-708
solidification defects ... **M6:** 707-708
process fundamentals ... **M6:** 705-706
cladding, parallel and angle ... **M6:** 705
jetting phenomenon ... **M6:** 705-706
product quality ... **M6:** 710-712
hardness and impact strengths ... **M6:** 712
metallography ... **M6:** 712
pressure-vessel standards ... **M6:** 711
safety ... **M6:** 59, 716-717
testing ... **M6:** 711-712
chisel testing ... **M6:** 711
ram tensile testing ... **M6:** 711
tension-shear testing ... **M6:** 711
thermal fatigue testing ... **M6:** 711
Explosion welding (EXW) ... **A6:** 160-164, 303-305
alloy steels ... **A6:** 303
aluminum ... **A6:** 303, 301
aluminum alloys ... **A6:** 739
angle bonding ... **A6:** 304
applications ... **A6:** 160, 303-304
aerospace ... **A6:** 387
bond microstructure ... **A6:** 163
bond morphology and properties ... **A6:** 162-163
bond strength determination ... **A6:** 163
bond zone morphology ... **A6:** 303
bonding fundamentals ... **A6:** 161-162
bonding parameter selection ... **A6:** 162
bonding practice ... **A6:** 160-161
carbon steels ... **A6:** 303
clad metal production ... **A6:** 160
configuration limitations ... **A6:** 303
copper ... **A6:** 303, 304
copper-nickel ... **A6:** 303
corrosion-resistant alloys ... **A6:** 303
definition ... **A6:** 303, 1209
dissimilar metal joining ... **A6:** 822
duplex stainless steels ... **A6:** 480
dynamic bend angle and material properties ... **A6:** 161
explosive parameters and shock effects ... **A6:** 161-162
explosives used ... **A6:** 305
flyer plate acceleration ... **A6:** 160
future outlook ... **A6:** 163-164
heat treatment ... **A6:** 305
impact energy ... **A6:** 160-161
interlayers ... **A6:** 303
jet formation ... **A6:** 162
Kovar ... **A6:** 304
metallurgical attributes ... **A6:** 303
metals combinations ... **A6:** 303
niobium alloys ... **A6:** 581
noise and vibration abatement ... **A6:** 304
oxide-dispersion-strengthened materials ... **A6:** 1039, 1040
"parallel gap" technique ... **A6:** 160
parallel-plate bonding ... **A6:** 304
parallel-plate bonding process sequence ... **A6:** 305
parameters ... **A6:** 304
"preset angle" technique ... **A6:** 160
process geometry ... **A6:** 304
process variations ... **A6:** 305
Explosion welding (EXW) (continued)
regulations ... **A6:** 304
safety precautions ... **A6:** 304, 1203
sequence of events in wave formation ... **A6:** 162
size limitations ... **A6:** 303
sizing considerations ... **A6:** 304-305
specifications ... **A6:** 305
stainless steel ... **A6:** 303-304
tantalum ... **A6:** 303
titanium ... **A6:** 303-304
titanium alloys ... **A6:** 522
wave formation ... **A6:** 162-163
zirconium ... **A6:** 304
Explosion welding (EXW), procedure development and process considerations ... **A6:** 896-900
commercially used metals and alloys joined ... **A6:** 896
component acceleration ... **A6:** 897-898
component collision ... **A6:** 898
critical process parameters ... **A6:** 896
explosives and explosive detonation ... **A6:** 897
fundamentals ... **A6:** 896-897
jetting and weld formation ... **A6:** 898
weld characteristics ... **A6:** 898-900
parametric limits for welding ... **A6:** 899-900
wave amplitude ... **A6:** 899
weld interface ... **A6:** 898-899
Explosion welding of
alloy steels ... **M6:** 710
aluminum ... **M6:** 710
carbon steel ... **M6:** 706-707, 710-713
copper ... **M6:** 713
copper alloys ... **M6:** 710
gold ... **M6:** 710
Hastelloy ... **M6:** 710
Incoloy 800 ... **M6:** 706-707
Inconel 718 ... **M6:** 707
magnesium ... **M6:** 710
nickel alloys ... **M6:** 710
niobium ... **M6:** 710
pipe ... **M6:** 709, 715-717
plate ... **M6:** 709
platinum ... **M6:** 710
rods ... **M6:** 709
silver ... **M6:** 710
stainless steels ... **M6:** 710-711
Stellite 6B ... **M6:** 710
tantalum ... **M6:** 710-713
titanium ... **M6:** 707, 710, 713
tube ... **M6:** 709, 715
zirconium ... **M6:** 710
Explosion(s)
hazards, plant operation ... **A7:** 197-198
metal powder, preventing ... **A7:** 197-198
output, estimating ... **A7:** 196
with sintering atmospheres ... **A7:** 348-350
suppression systems ... **A7:** 198
Explosion-bulge test ... **A11:** 58, 59
Explosions *See also* Fires; Flammability; Pyrophoricity
with aluminum-lithium alloys ... **A2:** 182-183
with sodium peroxide sinters and fusions ... **A10:** 167
and volatility, sample dissolution treatments ... **A10:** 166
Explosive
bulge test ... **A8:** 259
detonation, dynamic fracture by ... **A8:** 259
detonation, high strain rates in ... **A8:** 190
in flat plate impact test ... **A8:** 210
leaders, for torsional Kolsky bar ... **A8:** 224, 227
loading ... **A8:** 259, 276-277
Explosive actuator
SIMS determined isotopes of oxygen in ... **A10:** 625-626
Explosive bolt assemblies
neutron radiography of ... **A17:** 394
Explosive bonding *See* Explosion welding
refractory metals and alloys ... **A2:** 559
Explosive charges
position by neutron radiography ... **A17:** 394
Explosive compacting ... **A7:** 4
isostatic ... **A7:** 305
Soviet ... **A7:** 692
Explosive density ... **A6:** 161
Explosive detonation
geometric and elemental analysis of particles produced by ... **A10:** 318
Explosive devices
neutron radiography of ... **A17:** 392
Explosive forming *See also* High-energy-rate forming ... **A14:** 636-643
aluminum alloys ... **A14:** 800-802
charges for ... **A14:** 636-637, 641-642
confined systems ... **A14:** 838
defined ... **A14:** 5
dies ... **A14:** 638-639
equipment ... **A14:** 636-638
formability ... **A14:** 643
of heat-resistant alloys ... **A14:** 781-784
of heated blanks ... **A14:** 643
of plate ... **A14:** 641
safety ... **A14:** 643
of sheet metals ... **A14:** 640-641
shock-wave transmission ... **A14:** 640
transmission media ... **A14:** 639-640
of tubes ... **A14:** 641-642
unconfined systems ... **A14:** 636
of welded sheet metal preforms ... **A14:** 642-643
Explosive load factor ... **A6:** 161
Explosive mass ... **A6:** 161
Explosive pressure ... **A6:** 161
Explosive sheet forming ... **A14:** 640-641
Explosive shock synthesis
of cubic boron nitride (CBN) or diamond grit ... **A2:** 1008-1009
Explosive(s) ... **A7:** 131, 597, 600
with and without aluminum ... **A7:** 601
in high-energy-rate compacting ... **A7:** 305
metallic flake pigments for ... **A7:** 595
of typical sintering atmospheres ... **A7:** 348
Explosively loaded torsional Kolsky bar ... **A8:** 224-227
Explosives
for explosive forming ... **A14:** 638
Explosivity ... **A7:** 24, 194-198
of aluminum powders ... **A7:** 125, 127, 130
atmospheric conditions of ... **A7:** 195
chemical and physical factors n ... **A7:** 194-195
factors influencing ... **A7:** 194-196
ignition sources ... **A7:** 196-197
of magnesium powders ... **A7:** 133
ranking ... **A7:** 196-197
Exponential averaging ... **A18:** 296
Exponential distribution
cumulative distribution function ... **A8:** 629
estimated failure factors ... **EL1:** 901
failure density ... **EL1:** 896-897
mean ... **A8:** 629, 635
probability density function ... **A8:** 634-635
Exponential rate law expression ... **EM4:** 55
Exponentially weighted moving average (EWMA) charts ... **A17:** 732
Exposed underlayer
defined ... **EM2:** 17
Exposure
defined ... **A10:** 673
factors, film radiography ... **A17:** 328-330
radiographic, reciprocity law ... **A17:** 304
times, radiography ... **A17:** 317
of x-ray film ... **A17:** 324
Exposure charts *See also* Latitude charts
gamma-ray ... **A17:** 330
and reciprocity law ... **A17:** 304
x-ray radiography ... **A17:** 328-330
Exposure index
defined ... **A10:** 673
Exposure level effects
metal toxicities ... **A2:** 1234
Exposure limits
toxic metal powders ... **A7:** 202-208
Exposure meters, use of
in photomicroscopy ... **A9:** 84-85
Exposure of photomicroscopy films ... **A9:** 85
Exposure(s)
guide for Polaroid photos ... **A12:** 89
light meters and ... **A12:** 85-86
for macro Luminar lenses ... **A12:** 81
test ... **A12:** 86-87
Expoxies
analytic methods for ... **A10:** 9

Expoxies (continued)
as optical metallography
cold-mounting materials **A10:** 300
Extenal lead
embrittlement by **A13:** 181
Extend
defined **EM1:** 10 **EM2:** 17
Extendable borescopes *See also* Borescopes; Visual inspection
described **A17:** 5
Extended dies, blanking
in multiple-slide forming **A14:** 569
Extended interaction amplifier (EIA)
microwave inspection **A17:** 209
Extended solubility
of iron in aluminum **A10:** 294-295
Extended x-ray absorption fine structure **A10:** 407-419
apparatus **A10:** 411-412
applications **A10:** 407, 417-418
capabilities **A10:** 393
data analysis **A10:** 412-415
defined **A10:** 673
detection techniques **A10:** 418
estimated data analysis time **A10:** 407
estimated experimental scan time **A10:** 407
event **A10:** 410
fluorescence detection mode **A10:** 412
fluorescence enhancement technique for **A10:** 411
Fourier transform to **A10:** 412-413
general uses **A10:** 407
importance of multiple scattering **A10:** 410
introduction and fundamentals **A10:** 408-411
limitations **A10:** 407
multiple-scattering effects **A10:** 410-411
near-edge structure **A10:** 415-416
physical mechanism **A10:** 409
reflection, for subsurface study **A10:** 418
related techniques **A10:** 407
samples **A10:** 407, 417-418
scan, germanium in $GeCl_4$ molecule **A10:** 408
scan, nickel **A10:** 408
synchrotron radiation as x-ray source for **A10:** 411-412
transmission detection mode **A10:** 412
typical data analysis **A10:** 413-414
unique features of **A10:** 416-417
and x-ray diffraction, compared **A10:** 417
Extended x-ray absorption fine structure (EXAFS) **EM3:** 237
Extender *See* Lubricant
Extenders *See also* Filler; Fillers **EM3:** 12, 175-176
as additives **EM2:** 497-500
defined **EM1:** 10 **EM2:** 17
hot-melt adhesives **EM3:** 80
of reinforced polypropylenes (PP) **EM2:** 192
Extensibility **EM3:** 12
defined **EM1:** 10 **EM2:** 17
Extension **A8:** 26
acoustic horns **A8:** 245
localized and uniform **A8:** 26
Extension ladder
distortion failure in **A11:** 137
ductile overload fracture **A11:** 86-87, 91
Extension springs **A1:** 302, 320-321
Extension springs, steel
design **M1:** 303, 306, 307-308
residual stresses **M1:** 291
wire for **M1:** 288-289
Extension wires
color coding **A2:** 878
defined **A2:** 877
thermocouple **A2:** 876-878
Extensional rheometry **EM2:** 535
Extensional-bending coupling
defined **EM1:** 10 **EM2:** 17
Extensional-shear coupling **EM1:** 10, 219
defined **EM2:** 17
Extensometer **A8:** 5, 616-619
axial and diametral **A8:** 50
for creep and stress-rupture testing **A8:** 303, 313
defined **EM1:** 10 **EM2:** 17
linear variable differential transformers (LVDT) **A8:** 49, 616-617
for measuring stress and strain rates **A8:** 35, 49-50, 191
for multiaxial testing **A8:** 344
optical **A8:** 208, 210, 618
shear strain, in torsional fatigue testing **A8:** 151
for step-down tension testing **A8:** 47-51, 324
with strain gages **A8:** 549, 617
to monitor crack extension in corrosive environments **A8:** 428
Extensometers **EM3:** 12
for deformation measurement **A14:** 879
KGR 1 ..... **EM3:** 463, 464, 465, 474, 475
KGR 2 ..... **EM3:** 465, 469, 476
Extensometry
current technology **A8:** 19
systems, for creep testing **A8:** 303
in torsional fatigue testing **A8:** 151
External beam proton milliprobe
samples for **A10:** 102
External circuit
defined **A13:** 6
External corrosion gas/oil production **A13:** 1258
oxide scales, composite **A13:** 74-76
stress-corrosion prevention **A13:** 1147
tinplate **A13:** 780
treatments, stainless steels **A13:** 551-553
External gettering, as solid-state refining technique
for metals ultrapurification **A2:** 1094
External inductance
defined **EL1:** 29
External load frame
ultrasonic testing equipment **A8:** 248
External loads
modes **EM3:** 34
External pressure impregnation
infiltration **A7:** 554
External rupture
in gears **A11:** 596-597
External shrinkage
as casting defect **A11:** 382
External-to-internal
global communication **EL1:** 5
point-to-point communication **EL1:** 4
Externally pressurized seal *See also* Hydrostatic seal
defined **A18:** 8
Extinction
defined **A9:** 7
Extinction coefficient
defined **A9:** 7
Extinction distance **A10:** 367 **A18:** 388
Extinction length, diffraction parameter **A9:** 111
in structure-factor contrast **A9:** 113
Extinguishing powder composition **A7:** 198, 200
Extra Best Best quality telephone and telegraph wire **M1:** 264-265
Extra Improved Plow Steel quality rope wire **M1:** 265, 266
Extra smooth clean bright wire finish
for steel wire **A1:** 279 **M1:** 261-262
Extra-high-leaded brass
applications and properties **A2:** 310
Extra-low-carbon killed steel, for encapsulation *See also* Low carbon steels **A7:** 428
Extra-quality brass *See also* Brasses; Wrought coppers and copper alloys
applications and properties **A2:** 300-302
Extraction *See also* Ejection
of castings, vertical centrifugal casting **A15:** 306
defined **A9:** 7
efficiency measured **A10:** 243
Extraction (continued)
replicas, for sample preparation, ATEM analysis **A10:** 452
solvent, classical wet chemical analyses **A10:** 164
Extraction process *See also* Mining; Refining
beryllium **A2:** 684
Extraction replicas
formation **A12:** 182-183
of grain boundary particles, maraging steel **A12:** 183
single-stage, procedure for **A12:** 183
techniques for **A17:** 54
wrought heat-resistant alloys **A9:** 307-309
wrought stainless steels **A9:** 283-284
Extraction replication
defined **A17:** 52-53
Extraction replication techniques **A9:** 108-109
Extractive metallurgy industries *See* Mineral industry
Extractor-condenser used for removing fluids from powder metallurgy materials **A9:** 504
Extralow-carbon austenitic stainless steels **A14:** 225
Extraneous variables *See also* Nonrelevant indications
in tubular products **A17:** 561
Extrapolation
for end effects correction in compressive loading **A8:** 583
and interpolation, for creep-rupture behavior **A8:** 332-335
Extreme pressure (EP) additives **A16:** 180
grinding **A16:** 437, 438
stainless steel cutting fluids **A16:** 696
tapping **A16:** 263-264, 266, 267
thread rolling **A16:** 294
Extreme-pressure (EP) additives
antiscuff, for gears **A18:** 537, 538
gear box applications of ferrography **A18:** 305
gear lubricants **A18:** 541
gear oils **A18:** 86
in gearbox oils to arrest excessive wear **A18:** 308
grease additives **A18:** 125
high-vacuum liquid lubricants **A18:** 157
for internal combustion engine lubricants **A18:** 162
for metalworking lubricants **A18:** 141, 142, 143, 144, 145, 146, 149
for tool steel lubrication **A18:** 737, 738
Extreme-pressure (EP) lubricants
defined **A18:** 8
gear oil for bearing steels **A18:** 730, 731, 732, 733
to prevent galling in shallow forming dies **A18:** 633
Extreme-pressure additives **A14:** 512, 514-515, 696
Extreme-pressure agents *See* Antiwear and extreme-pressure (EP) agents; Extreme pressure additives
Extreme-pressure lubrication
defined **A18:** 8
Extreme-pressure oils
lubricants for rolling-element bearings **A18:** 133-134
Extrinsic conduction
defined **EL1:** 99
Extrinsic stacking faults
contrast in transmission electron microscopy **A9:** 119
and dislocation loops **A9:** 116
Extrudability
of shapes **A2:** 35
of steel **A14:** 300-301
Extruded alloys
complexity index curves used for evaluating **A9:** 130

Extruded molybdenum
elongated grain- boundary traces in **A9:** 124
Extruded shapes
characterized **A14:** 322
Extruder
defined **EM2:** 17
Extruders
capacity **EM2:** 378-379
design/operation **EM2:** 379-383
multiple-screw **EM2:** 383
thermoplastic, parameters **EM2:** 378-383
Extrusion *See also* Backward extrusion; Blown film extrusion; Chill roll extrusion; Cold extrusion; Forward extrusion; Hot extrusion; Hot forging; Hydrostatic extrusion; Impact extrusion; Sheet extrusion; Thermoplastic extrusion; Thermoplastic processes **A7:** 4 **A13:** 835, 887 **EM3:** 12 **EM4:** 33, 34, 123, 124, 151, 172
advanced ceramics **EM4:** 49
of aluminum alloys **A14:** 244
aluminum and aluminum alloys **A2:** 5-6, 34-36
aluminum, specimen locations for **A8:** 60
applications **EM4:** 166
basic methods **A14:** 316
of beryllium **A2:** 683
of beryllium powder **A7:** 759
beryllium-copper alloys **A2:** 415
billet, defined **A14:** 5
blow molding, defined **EM2:** 17
blown-film **EM2:** 383-384
as bulk forming process **A14:** 16
CAP billet stock for **A7:** 533
carbon-graphite materials **A18:** 816
central burst during **A14:** 399-400
as cermet forming technique **A7:** 800
coating **EM2:** 385-386
as coating material **A7:** 817
coextrusions **EM2:** 387
with cold heading, combined **A14:** 306
cold, nickel-base alloys **A14:** 836-837
cold, vs hot upset forging **A14:** 95
compared to tape casting **EM4:** 164
constituents of powder formulation **EM4:** 126
conventional vs hydrostatic **A14:** 327
costs **EM2:** 386
cylindrical flux by **A7:** 315
defect, in closed-die forging **A8:** 590, 594
defects **A14:** 385-386
defined **A14:** 5 **EM2:** 17
die **A7:** 442, 443
difficult **A14:** 311
disadvantages **EM4:** 166
of discontinuous fibers **EM1:** 120-121
and drawing **A8:** 591-593, 595
equipment and tooling **EM4:** 166-168
equipment compatibility with electrode coatings **A7:** 816-817
factors influencing ceramic forming process selection **EM4:** 34
filled billet process for **A7:** 518
fir tree defect **A14:** 403
flat-film/sheet **EM2:** 384-385
forging, defined **A14:** 5
forging, of a hub **A7:** 411, 412
formability, magnesium alloys **A2:** 471
forward/backward, load vs displacement curves **A14:** 37
granulated powders as feedstock **EM4:** 100
in header **A14:** 311
and heading **A14:** 297
of heat-resistant alloys **A14:** 231
high-energy-rate-forged, austenitic stainless steel **A8:** 573
of high-impact polystyrenes (PS, HIPS) **EM2:** 198
horizontal **EM4:** 170, 171
hot **A7:** 515-521 **A14:** 95, 315-326
hot, and hot upset forging **A14:** 95
hot, of cermet billets **A2:** 987
of hot upset preforms **A14:** 306
hydrostatic **A14:** 327-329
impact **A14:** 8, 311-312, 830
of interconnecting shapes, wrought aluminum alloy **A2:** 35-36
of lead and lead alloys **A2:** 547
limitations **EM4:** 166
magnesium alloy, formability **A2:** 571
Extrusion (continued)
magnesium alloy, product form selection **A2:** 463-464
mechanical consolidation **EM4:** 125, 126, 127, 128
of metal powder products **A7:** 297
of metal-polymer mixtures **A7:** 606
of mill shapes **A7:** 522
mix **EM4:** 169-170
solvent-based systems **EM4:** 169
typical plastic binder/ceramic **EM4:** 169-170
of nickel-titanium shape memory effect (SME) alloys **A2:** 899
of niobium-titanium superconducting materials **A2:** 1049-1050
pipe **EM2:** 385
pipe, defined **A14:** 5
plus isothermal forging with superalloys **A7:** 649
polyether sulfones (PES, PES **EM2:** 162
polyvinyl chlorides (PVC) **EM2:** 211-212
postextrusion processing **EM4:** 170-172
plasticating extrusion **EM4:** 171-172
solvent extrusion **EM4:** 170-171
primary testing direction, various alloys **A8:** 667
process flow **EM4:** 168-169
processes, classified **A14:** 16
products, types **EM2:** 383-386
profile **EM2:** 386
profile, of short fiber composites **EM1:** 121
quality **A14:** 301
rate vs flow stress **A14:** 317
in reaction sintering **EM4:** 292
refractory metals and alloys **A2:** 560
shear data standards for **A8:** 62
sheet, acrylonitrile-butadiene-styrenes (ABS) **EM2:** 113-114
size and shape effects **EM2:** 290
speed and temperatures **A14:** 317-319
stock, defined **A14:** 5
styrene-maleic anhydrides (S/MA) **EM2:** 220
as superplastic forming process **A14:** 857
surface finish **EM2:** 304
surface shear strain rate in torsion **A8:** 158
textured surfaces **EM2:** 305
of titanium alloys **A14:** 273
and torsion, microstructure of Udimet 700 in **A8:** 176-178
torsion testing **A14:** 373
of unalloyed uranium **A2:** 671
uranium dioxide fuel made by **A7:** 665
vertical **EM4:** 170, 171
viscoelastic properties required **EM4:** 116
von Mises effective strain rate **A8:** 158
warm, of cermet powder mixtures **A2:** 982-983
of whisker-reinforced MMCs **EM1:** 898-899
of wire **A14:** 694
workability in **A8:** 591-593
wrought aluminum alloy **A2:** 34
wrought copper and copper alloy tube shells **A2:** 249
of wrought magnesium alloy bars and shapes **A2:** 459
wrought magnesium alloy, creep properties **A2:** 485
wrought magnesium alloy, stress-strain curves **A2:** 491
wrought titanium alloys **A2:** 614
zinc and zinc alloys **A2:** 531
of zirconium tubing **A2:** 663
Extrusion coating
defined **EM2:** 17
Extrusion debonding
in squeeze casting **A15:** 326
Extrusion of wires and rods
preferred orientation during **A10:** 359
Extrusion press
casing, schematic of **A11:** 37
failed, EPMA to identify microconstituents in **A11:** 37
inclusion stringers in **A11:** 38
Extrusion process **A16:** 19
abrasive flow machining **A16:** 514, 515, 516, 519
and powder consolidation **A16:** 72
Extrusion ratio **A14:** 300, 318
Extrusion segregation
as squeeze casting defect **A15:** 325
Extrusion spinning
refractory metals and alloys **A2:** 562
Extrusion temperature
effect on notch toughness of steels **M1:** 695
Extrusion tools *See* Cold extrusion tools; Hot extrusion tools
Extrusion welding **M6:** 676-677
applications **M6:** 676-677
metals welded **M6:** 677
procedures **M6:** 677
Extrusions
aluminum **A17:** 271
aluminum alloy **M2:** 4, 5-6
aluminum alloy, ductile overload fracture **A11:** 86-87, 91
aluminum alloys **M2:** 51-58
by scanning laser gage **A17:** 12
extension-ladder side-rail, failed **A11:** 137
magnesium **M2:** 527-528, 534-535, 537
speed, cracking due to **A11:** 87, 91
ultrasonic inspection **A17:** 270-272
uranium **M3:** 776
Exudation **A7:** 4
Exudations
in aluminum alloy ingots **A9:** 633
in copper alloy ingots **A9:** 643
Eye clearance of microscope eyepieces **A9:** 73
Eye connectors
static tensile loading fracture **A11:** 390
Eye size
definition **M6:** 7
Eye terminal, corroded stainless steel
for wire rope **A11:** 521
Eyeleting
defined **A14:** 5
Eyelets
flexible printed boards **EL1:** 590
Eyepiece
defined **A10:** 673
Eyepiece graticules
for direct particle measurement **A7:** 228-229
Eyepiece of an optical microscope **A9:** 73-74
comparison of standard and high-point **A9:** 75
cross sections **A9:** 75
defined **A9:** 7
EZDA 12 *See* Zinc alloys, specific types, zinc foundry alloy ZA-12

F

475 °C embrittlement **A1:** 708
4130 steel
heat treatments for **A1:** 431
properties of **A1:** 431-432
4140 steel **A1:** 432
heat treatments for **A1:** 432
properties of **A1:** 432
4340 steel **A1:** 432-433
heat treatments for **A1:** 433
properties of **A1:** 432, 433-434
F *See* Farad; Faraday constant; Fluorescence; Force
F, as-fabricated temper
defined **A2:** 21
***F* distribution**
0.975 fractiles of **A8:** 674
***F* ratio test**
for mean and variance differences **A8:** 711
***F* statistic distributions** **A8:** 710
***F* test** **A8:** 700, 703
example of computational procedure **A8:** 672
***F*-test**
graphical display for significance of a regression **A8:** 669
graphical display for significant deviation from linearity **A8:** 670
F1, F2, F3, etc *See* Tool steels, specific types
FAA requirements, TSO-C26
aircraft brake testing standards **A18:** 586
Fabric *See also* Cloth; Nonwoven fabric; Woven fabric
of fiber reinforcements **EM2:** 506
warp face, defined **EM2:** 17
Fabric coatings
powder used **A7:** 572

Fabric construction *See also* Fabric(s); Weaves
forms, woven fabric prepregs **EM1:** 148
specification **EM1:** 148
Fabric fill face
defined **EM1:** 10 **EM2:** 17
Fabric filter dust collectors
use in dry blasting **M5:** 90
Fabric hybrid composites
impact resistance **EM1:** 149
Fabric, nonwoven *See* Nonwoven fabric
Fabric prepreg batch
defined **EM1:** 10
Fabric wear face
defined **EM1:** 10
Fabric, woven *See* Woven fabric
Fabric(s) *See also* Fabric construction; Fabric-reinforced prepreg; Mats; Multidirectional reinforced fabrics; Preforms; Rovings; Weaves; Woven fabric
aramid fiber **EM1:** 114-115
areal weight, measured **EM1:** 286
braided **EM1:** 519-527
carbon reinforced phenolic resin matrix **EM1:** 382
with combined fibers **EM1:** 126-127
construction, for woven prepregs **EM1:** 148
continuous SiC fiber MMC **EM1:** 860-861
count, fiberglass fabric **EM1:** 110
for cylindrical tube **EM1:** 569
fastener holes, technique/tools for **EM1:** 713-714
fiber finish of **EM1:** 125-126
fiber type, as parameter **EM1:** 125
fiberglass **EM1:** 110-111
graphite, forms **EM1:** 148
graphite-reinforced phenolic resin matrix **EM1:** 382
handling characteristics **EM1:** 149
hybridization **EM1:** 126-127, 149-150
knitted *See* Knitted fabrics
multidirectionally reinforced **EM1:** 119-131
nonwoven unidirectional **EM1:** 149
parameters, for textile designer **EM1:** 125
pattern, defined **EM1:** 125
preimpregnated, thermoplastic **EM1:** 547
prepreg **EM1:** 148-150, 161
pultruded **EM1:** 538
quality of **EM1:** 126
quartz **EM1:** 300, 414-415
for resin transfer molding **EM1:** 169
selvage edge, locking leno weave for **EM1:** 125-126
special, applications **EM1:** 126-127
stitched, multidirectionally reinforced **EM1:** 130
strength tests **EM1:** 732-733
styles/categories **EM1:** 126
tests for **EM1:** 291
three-dimensional **EM1:** 129-131
tubular woven **EM1:** 128
two-directional **EM1:** 125-128
unidirectional **EM1:** 125-128, 148, 169
vs fiber-epoxy tape, tensile strength **EM1:** 143-144
vs. tape, impact resistance **EM1:** 262
vs. tape prepreg, cost **EM1:** 105
yarn type, as parameter **EM1:** 125
Fabric-reinforced prepreg *See also* Fabrics; Prepregs; Tape prepregs
anisotropies **EM1:** 150
Fabricability
aluminum **A2:** 3
Fabricating *See also* Fabrication
defined **EM1:** 10
Fabrication *See also* Active component fabrication; Ceramic multilayer package fabrication; Ceramic package fabrication; Fabricability; Fabrication characteristics; In-process inspection; In-process nondestructive evaluation; Manufacture; Manufacture and assembly; Manufacturing; Manufacturing processes; Passive component fabrication; Plastic package fabrication; Primary fabrication; Processing; Rigid PWB fabrication techniques; Secondary fabrication
active-component **EL1:** 191-202
aluminum-lithium alloys **A2:** 182-184
of beryllium, formability effects **A14:** 805
boron fiber **EM1:** 58-59
caused spring failures **A11:** 555-559
of cemented carbides **A2:** 950-951
of cermets **A2:** 979-990
chip, and packaging **EL1:** 397
cobalt-base alloys **A13:** 662-663
composite fiber placement **EM1:** 33-34
of composites **EM1:** 33-34
of continuous fiber-reinforced composites **A11:** 733
copper and copper alloys **A2:** 216, 219
copper wire rod **M2:** 266-271
defined **EM1:** 10 **EM2:** 17
of dies, for HERF machines **A14:** 102
drawing, in final design package **EL1:** 523-534
effects on rolling-element bearing failure **A11:** 506-508
effects, stainless steel corrosion **A13:** 551
elastomeric mandrels **EM1:** 593-594
electronic, materials/failure mechanisms **EL1:** 1041-1048
environments, effects on stress-corrosion cracking **A11:** 211
of epoxy composites **EM1:** 71-73
errors, as failure cause **EM1:** 767
failure, dc motor armature **A11:** 421
of fiber-reinforced composites **EM1:** 179
flexible printed wiring **EL1:** 580
forging defects from **A11:** 317
fracture origins of shafts **A11:** 459
of gallium arsenide (GaAs) **A2:** 744-745
germanium and germanium compounds **A2:** 735
of glass fiber **EM1:** 108-111
of hafnium **A2:** 662-665
of impression dies **A14:** 52-53
improper, ductile and brittle failures from **A11:** 87-94
lay-up, of fiberglass hull **EM1:** 28
lead frame/lead frame strip **EL1:** 483-484
of level 1 packages **EL1:** 404-405
low-cost, and damage tolerance **EM1:** 259
material selection/evaluation for **EM1:** 38-39
mechanically alloyed oxide alloys **A2:** 947-949
dispersion-strengthened (MA ODS)
method, and costs, magnesium alloys **A2:** 467
methods for composite contact materials **A2:** 856-858
methods, reliability effects, passive components **EL1:** 180, 182
methods, thermosetting resins **EM2:** 223
MOS integrated circuit **EL1:** 147-148
multilayer interconnect **EL1:** 303-304
nickel aluminides **A2:** 918
of nickel-base alloys **A13:** 652
nondestructive inspection during **A17:** 645-646
npn transistor and MOS transistor **EL1:** 149
of passive components **EL1:** 184-188
of pressure vessels, effect on failure **A11:** 646-654
refractory metals and alloys **A2:** 558, 560-561
of rigid printed wiring boards **EL1:** 538-552
seal **EL1:** 455
semiconductor chip, introduction **EL1:** 397
semiconductor IC **EL1:** 2
and shaft failures **A11:** 472-477
stainless steels, wrought **M3:** 41-55
of steel plate **A1:** 238-239
steps, bipolar junction transistor technology **EL1:** 197
as stress source **A11:** 205
Fabrication (continued)
structural ceramics **A2:** 1020
for structural corrosion **A13:** 1303
substrate, ceramic packages **EL1:** 464-467
of superalloy metal-matrix composites **A2:** 909
tantalum **A2:** 573
tape **EL1:** 278
tape automated bonding **EL1:** 478
of ternary molybdenum chalcogenides (chevrel phases) **A2:** 1077-1079
with thermoplastic prepreg **EM1:** 103-104
of thermoplastic resin composites **EM1:** 547-552
of threaded steel fasteners **A1:** 300
titanium **M3:** 364-371
of titanium alloy turbine engine part **A14:** 275
uranium **M3:** 775-776
wafer **EL1:** 769-770, 978
of weldments, and corrosion **A13:** 344
of woven fabric prepregs **EM1:** 150
of wrought beryllium **A2:** 687
wrought copper and copper alloy products **A2:** 241-264
of wrought tool steels **A1:** 774-778
distortion and safety in hardening **A1:** 777
grindability **A1:** 775
hardenability **A1:** 775-777
machinability **A1:** 774-775
resistance to decarburization **A1:** 778
weldability **A1:** 775
wrought zinc **M2:** 636-637
of zirconium **A2:** 662-665
Fabrication characteristics *See also* Fabrication; Formability; Heat treatment; Hot working; Machinability; Weldability; Welding; Workability
actinide metals **A2:** 1189-1198
aluminum casting alloys **A2:** 153-177
beryllium-copper alloys **A2:** 411-416
cast copper alloys **A2:** 356-391
cast magnesium alloys **A2:** 482-516
electrical resistance alloys **A2:** 836-839
gold and gold alloys **A2:** 705
of make-break arcing contacts **A2:** 841
niobium alloys **A2:** 567-571
ordered intermetallics **A2:** 913-914
palladium and palladium alloys **A2:** 716-718
pewter **A2:** 522
platinum and platinum alloys **A2:** 709-714
pure metals **A2:** 1102-1178
rare earth metals **A2:** 1178-1189
silver and silver alloys **A2:** 700-704
wrought aluminum and aluminum alloys **A2:** 63-64, 66-68, 71-72, 82
wrought beryllium-nickel alloys **A2:** 424
wrought copper and copper alloys **A2:** 220-223, 265-345
wrought magnesium alloys **A2:** 480-491
of zinc alloys **A2:** 533-542
Fabrication characteristics of stainless steels **A1:** 887-905
forgeability **A1:** 889-894
machinability **A1:** 894-897
sheet formability **A1:** 888-889
weldability **A1:** 897-905
Fabrication characteristics of steels
bulk formability of carbon and low-alloy steels **A1:** 581-590
machinability of carbon and low-alloy steels **A1:** 591-602
sheet formability of carbon and low-alloy steels **A1:** 573-580
weldability of carbon and low-alloy steels **A1:** 603-613
Fabrication history
determined by optical metallography **A10:** 299
Fabrication traces
definition **EM4:** 632
Fabrication weldability tests **A1:** 611-613
Fabrications, steel *See* Steel, fabrications

Face (crystal)
defined ... **A9:** 7
Face contact pressure ... **A18:** 550
Face feed
definition ... **M6:** 7
Face milling *See also* Milling
Al alloys ... **A16:** 788-789, 791, 793
cast irons ... **A16:** 648, 651, 662
Cu alloys ... **A16:** 816
flankwear limits ... **A16:** 43
heat-resistant alloys ... **A16:** 737-738, 753, 755, 756
ledge tools used ... **A16:** 602-603
Mg alloys ... **A16:** 826, 827
P/M materials ... **A16:** 889
refractory metals ... **A16:** 859, 865, 867
surface alterations ... **A16:** 25, 30, 34
theoretical surface roughness ... **A16:** 24
theoretical surfaces produced ... **A16:** 22, 23
Ti alloys ... **A16:** 845, 846-847, 848, 850
tool steel selection ... **A16:** 710
tool steels ... **A16:** 712, 714-715, 717
Face of weld
definition ... **M6:** 7
Face opening force ... **A18:** 550
Face pressure
defined ... **A18:** 8
Face reinforcement
definition ... **A6:** 1209 **M6:** 7
Face seal
defined ... **A18:** 8
Face shield
definition ... **M6:** 7
Face tests ... **A6:** 102
Face-centered
defined ... **A9:** 7
Face-centered cubic (fcc) materials
ductile-to-brittle transition ... **A11:** 84-85
flow strength ... **A11:** 138
fractures in ... **A11:** 75
Face-centered cubic (fcc) metals
abbreviation ... **A8:** 724
adhesion measurements ... **A6:** 144
cyclic deformation in ... **A8:** 256
effect of decreasing temperature ... **A8:** 34
effect of strain rate or temperature on flow stress ... **A8:** 223-224, 226
effective stress for ... **A8:** 256
epitaxial growth direction ... **A6:** 51
fatigue limits ... **A8:** 253
textures ... **A8:** 181-182
Face-centered cubic (kc) crystal structures
abrasive wear ... **A18:** 186
adhesive wear
cavitation erosion ... **A18:** 216
cobalt-base wrought alloys ... **A18:** 768
Face-centered cubic array
density and coordination number ... **A7:** 296
Face-centered cubic crystals *See also* Face- centered cubic materials
copper, EXAFS analysis ... **A10:** 410
diffraction in ... **A10:** 360
dominant texture orientations of ... **A10:** 359
Euler angles for ... **A10:** 361
fibers in Euler angles in ... **A10:** 363
rolling textures in ... **A10:** 363-364
variance of Young's modulus with preferred orientation ... **A10:** 358
Face-centered cubic lattice
slip planes ... **A9:** 684
Face-centered cubic materials *See also* Face-centered cubic crystals
and bcc materials, Kurdjumov-Sachs orientation relationship ... **A10:** 439
formation of dislocation loops ... **A9:** 116
matrix, bright-field image and diffraction pattern ... **A10:** 11-1-440
rolling textures in ... **A10:** 363-364
Face-centered cubic metals
defined ... **A13:** 45
hydrogen embrittlement resistance ... **A13:** 330
Face-centered space lattice ... **A3:** 1 • 15
Face-centered-cubic metals
effect of low temperature on dimples in ... **A12:** 33
effects, state of stress ... **A12:** 31
embrittlement ... **A12:** 123
hydrogen embrittlement ... **A12:** 22
Face-centered-cubic metals (continued)
tensile fractures in ... **A12:** 100
Faceted dendritic morphology
as a feature of irregular eutectics ... **A9:** 620-622
Facets
acicular needle effect on ... **A12:** 328
of alloy steel fractures ... **A12:** 308, 319
aluminum-copper alloy, SEM projection ... **A12:** 195
areas measured ... **A12:** 208
of Armco iron cleavage fracture ... **A12:** 224
cleavage ... **A12:** 224, 319, 328, 380, 397, 424
cleavage, and tongues, compared ... **A12:** 424
dimpled, tool steels ... **A12:** 379
fatigue fracture ... **A12:** 14, 19
fracture appearance ... **A12:** 128
grain, in oxygen-embrittled iron ... **A12:** 222
grain-separated, high-purity copper ... **A12:** 399-400
intergranular, AISI/SAE alloy steels ... **A12:** 308
intergranular, transition to dimples ... **A12:** 306
of nickel alloy fracture ... **A12:** 397
overheating ... **A12:** 146, 147
prototype, true profile length values ... **A12:** 200
quasi-cleavage ... **A12:** 319, 338, 339
terraced, titanium alloys ... **A12:** 448
of tool steel fractures ... **A12:** 379-380
total, surface area of ... **A12:** 202
transgranular to intergranular, titanium alloys ... **A12:** 441
Facial flaws
particle ... **A7:** 59
Facilities
quality control and inspection of ... **EM1:** 740-741
Facing *See also* Mold wash
in conjunction with boring ... **A16:** 160, 164, 165, 168, 169, 174
in conjunction with drilling ... **A16:** 235
in conjunction with milling ... **A16:** 308, 322
in conjunction with turning ... **A16:** 135, 138, 140, 157
defined ... **A15:** 5
monitoring system ... **A16:** 414, 416
multifunction machining ... **A16:** 366, 369-370, 372-373, 376, 384
PCBN cutting tools ... **A16:** 115, 116
stainless steel ... **A16:** 154
turret lathe operations ... **A16:** 386
Facsimile equipment (office)
powders used ... **A7:** 573
Factor (experimental)
defined ... **EM2:** 599
Factor analysis ... **A10:** 118
Factor of safety (FS) ... **A6:** 389
Factorial designs
confounding in ... **A17:** 747, 749
for hit/miss data ... **A17:** 693
mathematical model ... **A17:** 740-742
for statistical analyses ... **A17:** 740-750
system noise/variation ... **A17:** 742
two-level fractional ... **A17:** 746-750
unreplicated ... **A17:** 746
variable interactions ... **A17:** 742-743
Factorial experimental designs ... **EM2:** 599-602
Factorial experiments ... **A8:** 641-643
analysis of ... **A8:** 633-654
estimation of main effects and interactions ... **A8:** 653-654
Factors, experimental
defined ... **A8:** 639
Factory roof
AISI/SAE alloy steels ... **A12:** 336
Fade ... **A18:** 574, 575
Fade recovery ... **A18:** 574
Fade-0-meter test method
anodic coating light fastness ... **M5:** 596
Fadeometer ... **EM3:** 12
defined ... **EM2:** 17
Fading
by evaporation, by silicon modifiers ... **A15:** 163
in grain refinement ... **A15:** 106
grain refiners for ... **A15:** 478
with modifiers ... **A15:** 484
Fahrenheit, vs. Celsius degrees
in creep/creep-rupture analyses ... **A8:** 685
Failure *See also* Adhesive failure; Catastrophic failure; Catastrophic failure. caustic cracking, low-carbon steels; Catastrophic,failure, Failure analysis; Failure analysis; Failure analysis procedures; Failures; Tensile failure/fracture ... **A8:** 526 **EM3:** 12, 37-38
Failure (continued)
aircraft, corrosion-related ... **A13:** 1022-1035, 1045-1054
amorphous materials and metallic glasses ... **A2:** 814
analysis ... **A13:** 1113-1126, 1246 **EM1:** 192-204
analysis, by acoustic emission inspection ... **A17:** 286
approximate, theories for ... **EM1:** 235
axial shear mode ... **EM1:** 198
of bolted joints ... **EM1:** 488-489
causes ... **A13:** 338 **EM1:** 767
cleavage ... **A8:** 476
cohesive; defined *See* Cohesive failure
common causes of ... **A8:** 496
compressive strength or strain to ... **A8:** 58
corrosion, prediction/consequences ... **A13:** 316-317
criteria ... **EM1:** 198-201
cumulative group mode ... **EM1:** 195
cumulative weakening ... **EM1:** 194-195
defined ... **A8:** 152, 695 **A13:** 6
delayed ... **A13:** 185-187, 329-330, 535-536, 650-652
dimpled rupture ... **A8:** 476
ductile striations ... **A8:** 476
due to overload ... **A8:** 476, 477-481
electronics/communications equipment ... **A13:** 1113-1126
environmental (sustained load) ... **A8:** 476
fatigue ... **A8:** 476 **EM1:** 201-204, 436-438, 797, 938
fatigue, surface family of ... **EM1:** 203
fiber break propagation ... **EM1:** 195
fiber mode ... **EM1:** 230-235
as function of time ... **A8:** 635
and gage length, workability ... **A8:** 156
high strain rate effect on ... **A8:** 208
in high strain rate testing ... **A8:** 187
hydrogen-induced, causes ... **A8:** 537
idealized limit cases for ... **EM1:** 235
initial, laminar ... **EM1:** 230-231
intergranular ... **A8:** 476
intraply matrix-mode ... **EM1:** 234
of laminates ... **EM1:** 230-235
loads/modes, for laminates ... **EM1:** 233
local fiber, mechanisms ... **EM1:** 198
in low-cycle torsional fatigue testing ... **A8:** 152
magnesium/magnesium alloys, causes ... **A13:** 741
of materials, high strain rate testing ... **A8:** 188
matrix mode ... **EM1:** 198, 230-235
measurement of residual stress associated with ... **A10:** 380
mechanisms ... **A8:** 714
mechanisms, in damage tolerance design ... **A17:** 702
and microstructure ... **A8:** 476
microvoid coalescence ... **A8:** 476
mode domains, potential-pH diagram ... **A13:** 153
modeling, and damage tolerance design ... **A17:** 702
modes ... **EM1:** 195, 198-200, 230-235, 240-244, 781-785
modes, of make-break contacts, electrical contact materials ... **A2:** 840-841
modes, single-shear test ... **A8:** 63-64
overload, of AISI 4340 steel threaded rod ... **A10:** 511-513
pin ... **A8:** 59
plastic strain range vs. cycles to ... **A8:** 346-347
prediction, stress based ... **A8:** 343
probability, and reliability ... **A8:** 627
quasi-cleavage ... **A8:** 476
and static strength ... **EM1:** 432-435
statistical tensile. composite geometry for ... **EM1:** 194
stress, and static-strength distribution ... **A8:** 716
structural, probabilistic fracture mechanics for predicting ... **A8:** 682
surfaces, in stress space ... **EM1:** 199
temperature effect on ... **A8:** 188
tensile fiber mode ... **EM1:** 200
tensile matrix mode ... **EM1:** 200
of thermocouples ... **A2:** 881-882
time to, in multiaxial testing ... **A8:** 344
torsion test for ... **A8:** 154
types ... **A8:** 476

Failure (continued)
weakest-link **EM1:** 194
and workability **A8:** 154
Failure analyses
for matching steel properties to requirements **A1:** 338
Failure analysis *See also* Advanced,failure analysis; Corrosion failure analysis; Corrosion failures; Electronic failure analysis methods; Failure; Failure analysis, general practice in, Failure mechanisms Failures; Failure mechanisms; Failure verification; Failure(s) **EM1:** 192-204, 765-797
and accelerated life prediction **EM2:** 788-795
accuracy of **A11:** 55
adhesive failure **EM3:** 276
radiation and vacuum effects **EM3:** 645
advanced, for corrosion failures **EL1:** 1102-1116
of aluminum wire connections **A10:** 531-532
of boilers and related equipment **A11:** 602-627
bottom-brazed flatpacks **EL1:** 992
by analysis of structure **EM2:** 824-837
by fractography **EM1:** 786-793 **EM2:** 805-81
by optical metallography **A10:** 299
by surface analysis **EM2:** 811-823
by ultrasonic nondestructive analysis **EM2:** 838-846
chemical analysis **A11:** 29-31
cohesive changed to adhesive failure **EM3:** 668
cohesive failure **EM3:** 276, 283
lap joints **EM3:** 327
radiation and vacuum effects **EM3:** 645
of continuous fiber composites **EM1:** 768-769, 781-793
of continuous fiber reinforced composites **A11:** 731-743
corrosion-fatigue **A11:** 252-262
crack-closure scheme for **EM1:** 248-250
crazing **EM2:** 734-740
of creep **EM2:** 728-730
of crystallinity **EM2:** 731-732
and damage accumulation **EM1:** 246-247
design considerations **EM2:** 1
of design vs material deficiencies **A11:** 744, 747
destructive tests **EM1:** 765-766, 774
device **EL1:** 917-918
of discontinuous fiber composites **EM1:** 794-797
engineering aspects **A11:** 15-71
of environmental stress crazing **EM2:** 796-80
failure causes **EM1:** 767
failure modes **EM1:** 195, 198-200, 230-235, 240-244, 781-785
and failure verification **EL1:** 1058
fatigue **EM1:** 244-247
of fatigue failure **EM2:** 741-75
in field **EM2:** 817-818
finite-element procedure **EM1:** 247-250
fractography in **A11:** 747
of fracture **EM2:** 734-74
and fracture mechanics **A11:** 31, 47-65
fracture surface for **A10:** 304
of fracture surfaces **A11:** 19-22
fracture types **A11:** 24-29
fracture-toughness testing and evaluation **A11:** 60-64
historical study **A12:** 2
of hydrogen damage **A11:** 245-251
interfacial failure **EM3:** 271
ceramics joined to glasses **EM3:** 298
durability assessment **EM3:** 663
introduction **EL1:** 957 **EM1:** 765-766 **EM2:** 727 **EM4:** 629
investigative operations **EM1:** 765
of laminates **EM1:** 236-251
of LME-induced fracture **A11:** 225-238
macrofractography for **A12:** 91
mechanical testing **A11:** 18-19
metallographic sections **A11:** 23-24
microanalytical techniques **A11:** 32-46
Failure analysis (continued)
of microbial degradation **EM2:** 783-787
mixed-mode **EM3:** 287
of moisture-related failure **EM2:** 761-769
nondestructive testing **A11:** 16-18
nondestructive tests **EM1:** 765, 770, 774-777
notch-toughness testing and evaluation **A11:** 57-60
objectives and stages of **A11:** 15
of organic chemical related failure **EM2:** 770-775
of photolytic degradation **EM2:** 776-782
of physical aging **EM2:** 751-760
plane stress and plane strain **A11:** 51-52
ply failure criteria, stress point **EM1:** 238
and polymer aging **EM2:** 732
of pressure vessels, procedures for **A11:** 643-644
printed board **EL1:** 1038-1040
and problem solving, microanalytical techniques **A11:** 32-46
procedures **EM1:** 770-773
procedures, iron castings **A11:** 344
of progressive ply failure, laminates **EM1:** 238-239
replication procedures for **A12:** 94-95
and report writing **A11:** 31-32
sampling and data collection **A11:** 15-16
sampling protocol for **A10:** 15-16
semiconductor, SEM for **A10:** 490
semiconductors **A10:** 490
service conditions review **A11:** 747
simulated-service testing **A11:** 31
of sliding bearings **A11:** 483-489
software **A11:** 55
solder joint **EL1:** 1034-1040
stages **EM4:** 630-631
data collection **EM4:** 630
formal report **EM4:** 631
history of the part **EM4:** 630
part inspection **EM4:** 630-631
photographs **EM4:** 630
replicas **EM4:** 630
stress approaches **EM1:** 236-237, 239-240
of stress relaxation **EM2:** 730
of stress-corrosion cracking **A11:** 203-224
submicron spatial resolution **EL1:** 1088
of surfaces **EM2:** 811, 817-822
techniques **EL1:** 954, 957
techniques for integrated circuits **A11:** 767-769
tests and data **A11:** 53-55, 57-60
thermal **EM1:** 779-780
thermal stresses **EM2:** 751-760
three-dimensional **EM1:** 239-240
types illustrated **A12:** 217
of yield **EM2:** 730-731
Failure analysis, general practice in *See also* Failure analysis; Failures **A11:** 15-46
analysis and report writing **A11:** 31-32
chemical analysis **A11:** 29-31
data collection **A11:** 15-16
failed part, preliminary examination **A11:** 16
fracture mechanics, application of **A11:** 31
fracture surfaces, macro/microscopic examination **A11:** 20-22
fracture surfaces, selection, preservation, and cleaning of **A11:** 19-20
fracture type, determination of **A11:** 24-29
mechanical testing **A11:** 18-19
metallographic sections, selection and preparation of **A11:** 23-24
microanalytical techniques **A11:** 32-46
nondestructive testing **A11:** 16-18
objectives of investigation **A11:** 15
simulated-service testing **A11:** 31
stages **A11:** 15
Failure analysis procedures *See also* Failure
fractography **EM1:** 770-771
materials characterization **EM1:** 770
nondestructive examination **EM1:** 765, 770, 774-777
Failure analysis procedures (continued)
stress analysis **EM1:** 771-773
Failure analysis specimens
mounting **A9:** 32
Failure cycles
tension-tension **A8:** 715-716
Failure density function
defined **EL1:** 896
Failure distribution function
defined **EL1:** 896
Failure estimation
two-parameter Weibull distribution for **A8:** 714-716
Failure mechanisms *See also* Failure analysis; Failure(s)
accelerated tests of **EL1:** 741
in active devices **EL1:** 1006-1017
in adhesives **EL1:** 1046-1047
bottom-brazed flatpacks **EL1:** 992-993
ceramic packages **EL1:** 961-962
in coated and encapsulated microelectronic devices **EL1:** 1049-1057
corrosion **EL1:** 1049-1052
cyclic differential of thermal expansion **EL1:** 740-742
electrolytic corrosion **EL1:** 493
electromigration **EL1:** 494, 963-964
electrostatic discharge (ESD) as **EL1:** 966-967
environmental **EL1:** 493-494
failure studies, early **EL1:** 958-959
flip-chip technology **EL1:** 1047
galvanic corrosion **EL1:** 493
interconnect **EL1:** 1054-1056
mechanical shock **EL1:** 740
multiple, as problem **EL1:** 893
overview of **EL1:** 958-968
oxide contamination effect **EL1:** 959-960
package differences **EL1:** 961-962
packaging **EL1:** 957
in passive devices **EL1:** 994-1005
plastic packages **EL1:** 962
in printed wiring boards **EL1:** 1018-1030
purple plague as **EL1:** 960
semiconductor chip **EL1:** 963-967
of silicon semiconductor devices **EL1:** 959
soft failure **EL1:** 493-494
solder joint **EL1:** 735-736
in soldering **EL1:** 1031-1040
stress-related **EL1:** 1052-1054
in structured reliability testing **EL1:** 959-961
in surface-mounted packages **EL1:** 982-993
in tape automated bonding **EL1:** 1047
temperature-sensitive **EL1:** 959-961
thermal shock **EL1:** 740-742
in through-hole packages **EL1:** 969-981
vibration (transport) **EL1:** 740-742
in welding **EL1:** 1041-1046
wire bond degradation **EL1:** 494
Failure mechanisms and related environmental factors **A11:** 75-303
brittle fractures **A11:** 82-101
corrosion **A11:** 172-202
corrosion-fatigue **A11:** 252-262
distortion **A11:** 136-144
ductile fracture(s) **A11:** 82-101
elevated-temperature **A11:** 263-297
embrittlement by solid-metal environments **A11:** 239-244
fatigue **A11:** 102-135
hydrogen-damage **A11:** 245-251
identification of **A11:** 75-81
liquid-erosion **A11:** 163-171
liquid-metal embrittlement **A11:** 225-238
in sour gas environments **A11:** 298-303
stress-corrosion cracking **A11:** 203-224
types of **A11:** 75-81
wear **A11:** 145-162
Failure mode **A8:** 476
in deformation **A8:** 473-574

SUBJECTS OF THE INDEXED VOLUMES: ASM Handbook (designated by the letter "A"): **A1:** Properties and Selection: Irons, Steels, and High-Performance Alloys (1990); **A2:** Properties and Selection: Nonferrous Alloys and Special-Purpose Materials (1990); **A3:** Alloy Phase Diagrams (1992); **A4:** Heat Treating (1991); **A6:** Welding, Brazing, and Soldering (1993); **A7:** Powder Metallurgy (1984); **A8:** Mechanical Testing (1985); **A9:** Metallography and Microstructures (1985); **A10:** Materials Characterization (1986); **A11:** Failure Analysis and Prevention (1986); **A12:** Fractography (1987); **A13:** Corrosion (1987); **A14:** Forming and Forging (1988); **A15:** Casting (1988); **A16:** Machining (1989); **A17:** Nondestructive Testing and Quality Control (1989); **A18:** Friction, Lubrication, and Wear Technology (1992). **Metals Handbook, 9th Edition** (designated by the letter "M"): **M1:** Properties and Selection: Irons and Steels (1978); **M2:** Properties and Selection: Nonferrous Alloys and Pure Metals (1979); **M3:** Properties and Selection: Stainless Steels, Tool Materials and Special-Purpose Materials (1980); **M4:** Heat Treating (1981); **M5:** Surface Cleaning, Finishing, and Coating (1982); **M6:** Welding, Brazing, and Soldering (1983). **Engineered Materials Handbook** (designated by the letters "EM"): **EM1:** Composites (1987); **EM2:** Engineering Plastics (1988); **EM3:** Adhesives and Sealants (1990); **EM4:** Ceramics and Glasses (1991); **Electronic Materials Handbook** (designated by the letters "EL"): **EL1:** Packaging (1989).

Failure mode (continued)
fatigue **A8:** 481-485
strain rate/temperature effects **EM2:** 684-687
sustained-load (environmental) **A8:** 486-488
transmission electron fractography for **A8:** 476
types and microstructure effect **A8:** 477-488
Failure model
accelerated life prediction **EM2:** 791-792
Failure modes **EM1:** 195, 198-200, 230-235, 240-244, 781-785
in-plane, types **EM1:** 781-783
laminates **EM1:** 240-244
out-of-plane, delamination as **EM1:** 781, 783-784
Failure modes, reliability and wear concepts **A18:** 493-495
dependence of failure rate on operating duration **A18:** 494-495
relationship between wear and reliability **A18:** 493
reliability characteristics **A18:** 493
reliability probability concepts **A18:** 493
statistical distributions of wear and reliability **A18:** 493-494
exponential distribution **A18:** 493
gamma distribution **A18:** 494
log normal distribution **A18:** 493
normal distribution **A18:** 493, 494
Weibull distribution **A18:** 493-494
wear and failure modes **A18:** 494
Failure rate
in accelerated testing **EL1:** 889
as "bathtub" reliability curve **EL1:** 740
defined **EL1:** 896
dependence on temperature **EL1:** 23
hybrids **EL1:** 261
Failure strength tests **EM3:** 325-334
Boeing wedge test **EM3:** 332-333
butt joints **EM3:** 330-331, 332
Charpy pendulum test **EM3:** 333
climbing drum test **EM3:** 332
creep tests **EM3:** 333
double-lap joints **EM3:** 326-328, 330
dynamic tests **EM3:** 333
EDSU/Grant method **EM3:** 328
finite-element method (FEM) **EM3:** 327-328
floating roller test **EM3:** 332
Goland and Reissner bending moment factor **EM3:** 326, 328
Izod test **EM3:** 333
joint strength predictions **EM3:** 330
lap joint test **EM3:** 330
law of complementary shears **EM3:** 326
local stress concentrations and fracture mechanics **EM3:** 329-330
napkin ring test **EM3:** 331
PABST/Hart-Smith design philosophy **EM3:** 328
peel tests **EM3:** 331-333
photoelastic stress analysis **EM3:** 327
plasticity and lap joints **EM3:** 328-329
problems associated with testing composites **EM3:** 333
recommendations to ensure accuracy **EM3:** 333-334
shear lag (Volkersen's) equations **EM3:** 325-326
single-lap joints **EM3:** 325-328, 329
T-peel test **EM3:** 332
thick-adherend test **EM3:** 330
transverse (peel) stresses **EM3:** 326
Failure studies
scanning -electron microscopy used for **A9:** 99
Failure verification *See also* Failure analysis; Failure(,V)
device history determination **EL1:** 1058
electrical testing **EL1:** 1058-1060
external visual examination, initial **EL1:** 1058
high/low-temperature electrical testing **EL1:** 1060-1061
Failure(s) *See* also Advanced failure analysis, Failure analysis; Failure mechanisms; Failure rate; Failure verification; Fatigue failure; Reliability
bulk forming **A14:** 19
catastrophic thermal, defined **EL1:** 46
defined **EL1:** 751, 1143
density distributions **EL1:** 896-897
detection, solder joint fractures **EL1:** 751
die, causes of **A14:** 55-56
Failure(s) (continued)
direct, defined **EL1:** 943
factors, estimating from test data **EL1:** 900-903
flexibilized epoxy systems **EL1:** 819
indirect, defined **EL1:** 943
intermittent, testing for **EL1:** 1060-1061
mechanical **EL1:** 45-46, 55-65
mechanical, tape automated bonding (TAB) **EL1:** 287
mechanisms, overview of **EL1:** 958-968
modes, in deformation processing **A14:** 365-366
modes, stress **EL1:** 445
on semiconductor chips **EL1:** 963-967
plated-through hole **EL1:** 1018-1030
risk categories **EL1:** 132
in rubber-die flanging **A14:** 615
in screen environments **EL1:** 875
shock and vibration **EL1:** 62-65
solder joint **EL1:** 735
statistical considerations **EL1:** 745-746
studies, early **EL1:** 958-959
Failure-free-life analysis methods **EL1:** 957
Failure-mode-and-effects analysis (FMEA)
of microcircuits **EL1:** 260-261
Failures *See also* Failure analysis; Failure mechanisms; Weld failure origins
complex **A11:** 29
defined **A11:** 4
delayed **A11:** 242-244
distortion **A11:** 136-144
engineering aspects **A11:** 15-71
inspection records of **A11:** 134
mechanisms of **A11:** 75-303
mode, shear lips as indicator of **A11:** 396
multiple-mode, boilers and steam equipment **A11:** 626
premature elevated-temperature **A11:** 281-282
of tools and dies **A11:** 563-585
types of **A11:** 75-81
Fairing
defined **EM1:** 10 **EM2:** 17
Falex tests
chromium electroplated coating applications **A18:** 835, 836
metalworking fluids **A18:** 101
Falling rate period **EM4:** 131-132
False Brinelling *See also* Brinelling; Fretting **A18:** 242, 243
defined **A8:** 5 **A11:** 4 **A18:** 8-9
in rolling-element bearings **A11:** 497-498
and true brinelling, compared in bearings **A11:** 499-500
False indications
defined **A17:** 103
False minimum rate
in creep **A8:** 330-331
False silicon peaks
in EDS spectra **A10:** 520
False switching
avoidance of **EL1:** 34
Families of x-ray lines
EPMA analysis **A10:** 522-524
Family mold
defined **EM2:** 17
Fan shafts
inspection criteria for **A11:** 107-108
reversed-bending fatigue **A11:** 476
Fansteel 80 *See* Niobium alloys, specific types; Niobium alloys, specific types, Nb-1Zr
Far east
Bronze Age in **A15:** 16-17
Far-field effects
ultrasonic beams **A17:** 239
Far-infrared radiation
defined **A10:** 673
Farad
abbreviation **A8:** 724
Faradaic currents
in pulse polarography **A10:** 193
Faraday cage
in SEM **A12:** 168
Faraday constant **EM4:** 252
96 486 C/mol, abbreviation for **A10:** 690
in electrometric titration **A10:** 203
Faraday cup
in gas mass spectrometers **A10:** 154
use in microbeam analysis **A10:** 530
Faraday effect
observation of magnetic domains **A9:** 535
Faraday's law **A13:** 6, 29, 33 **A18:** 834
controlled-potential coulometry **A10:** 210
coulometric sensors **EM4:** 1132-1133
of electrolysis **A10:** 203
Faraday's law of induction **A17:** 130
Farm equipment/machinery
feed rolls and shutoffs for **A7:** 674-675
floating cams, for planters **A7:** 673
P/M parts for **A7:** 671-676
powders used **A7:** 572
self-lubricating bearings in **A7:** 705
Farris gas dilatometer **EM2:** 657
FASOR computer program for structural analysis **EM1:** 268, 271
Fassel torches
for analytic ICP systems **A10:** 36-37
Fast axial flow (carbon dioxide) laser **A14:** 736
Fast fission nuclear reactors **A13:** 17
Fast Fourier transform **A10:** 117, 690
Fast Fourier Transform (FFT)
digital image processing **A17:** 460-461
Fast Fourier transform (FFT) algorithm **A18:** 294, 295
Fast Fourier transform (FFT) analyzer
motor-current signature analysis **A18:** 314
Fast fractures
AISI/SAE alloy steels **A12:** 295, 311, 327
high-carbon steels **A12:** 280
magnesium alloy **A12:** 456
mechanical damage in **A12:** 72
medium-carbon steels **A12:** 260-262, 264, 267
precipitation-hardening stainless steels **A12:** 371
tool steels **A12:** 376
unstable, as fatigue stage III **A12:** 175
woody **A12:** 281
Fast ion bombardment (FAB) **EM3:** 237, 241
Fast neutron activation analysis **A10:** 239, 689
to analyze bulk chemical composition of starting powders **EM4:** 72
"Fast oils" **A7:** 453
"Fast quenching" oil **A7:** 453
Fast-axial-flow (FAF) carbon dioxide (CO_2) lasers **A6:** 266, 267
Fast-passage effect *See* Electron nuclear double resonance
Fastech electronic package system
for cost determination **EL1:** 14-15
Fastener core potting **EM3:** 561
Fastener holes *See also* Fastener(s); Holes
aircraft subassemblies, eddy current inspection **A17:** 190
exit breakout damage **EM1:** 712
second-layer, crack detection **A17:** 139
techniques and tools for **EM1:** 712-715
Fastener tests **A1:** 296
Fastener wear particles
in shear testing **A8:** 68
Fastener(s) *See also* Fastener holes; Holes; Tubing
machine vision inspection **A17:** 45
protruding, eddy current inspection **A17:** 191-192
structural, schematic **A17:** 141
Fasteners *See also* Blind fastening; Fastener holes; Mechanical fasteners
for advanced composites **A11:** 530
aerospace use **EM1:** 709
alloy steel for **M1:** 256
alloy steel, hydrogen embrittlement of **A11:** 541
aluminum **EM1:** 716-717
blind **EM1:** 709-711
blind, types of failures in **A11:** 531
blind, types used in assembled components **A11:** 545-546
cadmium plating of **M5:** 261-262
carbon steel for **M1:** 254
cold finished bars for **M1:** 221
copper and copper alloys **A2:** 239
corrosion resistant, ASTM specifications **M3:** 183-185
countersunk and single-shear **EM1:** 492
economy in manufacture **M3:** 851
fabrication of **M1:** 589, 590
failure origins **A11:** 529-530
faulty repairs of **A11:** 141-142
flush head **EM1:** 707
galvanic compatibility **EM1:** 709, 717

Fasteners (continued)
head styles ... **EM1:** 706
hot rolled bars for ... **M1:** 206, 208, 209
interchangeable ... **A11:** 529
interference fit ... **EM1:** 707
ion plating of ... **M5:** 420
load share ... **EM1:** 492-493
materials for ... **EM1:** 706, 709
mechanical coating of ... **M5:** 302
mechanical, selection of ... **EM1:** 706-708
mechanical, shear load transfer through ... **EM1:** 488
performance, at elevated temperatures ... **A11:** 542-543
piercing with ... **A14:** 470
pullout, failure by ... **A11:** 548-549
special-purpose ... **A11:** 548-549
specifications ... **A11:** 529
standard galvanized ... **A13:** 443
steel *See* Steel, fasteners
strength ... **EM1:** 706
thread grinding applications ... **A16:** 278
threaded, corrosion in ... **A11:** 535-542
threaded, fatigue in ... **A11:** 531-534
Wedgelock (or Cleco-) type ... **EM1:** 711
wire for ... **A1:** 284 **M1:** 265-266
zinc-electroplated, failure of ... **A11:** 548-549
Fasteners, copper
mechanical joining ... **M2:** 440-443
Fasteners evaluation hydrogen embrittlement
evaluation hydrogen embrittlement in ... **A8:** 541-542
pin bearing testing of ... **A8:** 59
plated, hydrogen embrittlement testing Of ... **A8:** 542
shear testing of ... **A8:** 65-68
shear testing standards for ... **A8:** 62
types of ... **A8:** 68
Fasteners, threaded
elimination of ... **EL1:** 123
Fastening
aluminum ... **M2:** 202-203
mechanical ... **EM2:** 711-713
Fat
defined ... **A18:** 9
Fatigue *See also* Durability; Dynamic fatigue testing; Failure; Fatigue analysis; Fatigue crack growth; Fatigue crack growth rate; Fatigue crack growth rates; Fatigue crack initiation; Fatigue crack propagation; Fatigue crack propagation rate; Fatigue cracking; Fatigue cracks; Fatigue data analysis; Fatigue failure; Fatigue failures; Fatigue fracture; Fatigue fracture(s); Fatigue life; Fatigue limit; Fatigue loading; Fatigue properties; Fatigue resistance; Fatigue strength; Fatigue striation spacings; Fatigue striations; Fatigue test specimens; Fatigue testing; Fracture; Fracture surface(s); Fracture(s); Tensile properties; Tensile strength ... **A8:** 5, 363-365 **EM3:** 12
(*S-N*) curve, plastic ... **EM2:** 76
aluminum alloy ... **A15:** 764-765
anaerobics augmentation of performance ... **EM3:** 115
apparatus, dynamic notched round bar testing ... **A8:** 278
ASTM test methods ... **EM2:** 334
axial, in forging deformation ... **A7:** 416
behavior, in titanium-based alloys ... **A7:** 41, 44, 438
behavior of materials, compared ... **A8:** 706-712
behavior, wrought aluminum alloy ... **A2:** 42-44, 59
belted joint ... **EM1:** 440
bending machine ... **A8:** 369
brittle fracture from ... **A11:** 85
by liquid erosion ... **A11:** 164-166
carburizing surfaces ... **A8:** 373
carrier cloth effect ... **EM3:** 515-516, 518
-caused failures, measurement of associated residual stress ... **A10:** 380

Fatigue (continued)
causes of ... **A1:** 673 **M1:** 665, 673
chafing, defined ... **A11:** 2
constant lifetime diagrams ... **M1:** 666-667, 669-670
contact ... **A11:** 2, 133-134, 594
corrosion ... **A8:** 374-375, 403-410 **A11:** 2, 37, 623, 636-637
crack, defined ... **A8:** 391, 421-422, 678
crack extension, predicted ... **A13:** 298
crack growth ... **A11:** 708
crack growth rate ... **A11:** 4
crack, in clamp, fractographs of ... **A10:** 305
crack initiation ... **A8:** 363-364, 366-375 **A11:** 102-103, 106, 621
crack nucleation ... **A8:** 363
crack propagation ... **A8:** 363-365, 602-603 **A17:** 54
cracking ... **A8:** 363, 366 **A11:** 105-107
cracking, avoiding ... **A15:** 764-765
cracking, carbon steel aircraft part ... **A13:** 1024
cracks ... **A11:** 26, 85, 89, 708
cumulative damage ... **M1:** 678-679, 681
cumulative damage under varying frequencies ... **EM3:** 517
cure conditions effect ... **EM3:** 515
cycle ... **A8:** 346-347
in cycles to fracture vs. cycle period ... **A8:** 351
in cyclic torsional testing ... **A8:** 149
damage, and fatigue life prediction ... **A12:** 207
damage, cyclic differential thermal expansion ... **EL1:** 740
damage, determination of ... **A11:** 135
-damage resistance, material selection based on ... **A8:** 712
and damage tolerances ... **EM1:** 261-262
data analysis ... **A8:** 695-720 **EM1:** 441-443
data, correction factors for ... **A11:** 115
data scatter ... **M1:** 675-678
data, types ... **A13:** 292-293
decarburization of surface ... **A8:** 373
defined ... **A8:** 363, 481
defined ... **A11:** 4, 102 **A12:** 14-18, 111, 175 **A13:** 6 **EM1:** 10 **EM2:** 17
definition of ... **A1:** 673
deformation, at differing frequencies ... **A8:** 253
determined, by speckle metrology ... **A17:** 435-436
ductile iron ... **A15:** 662
ductile versus brittle behavior ... **EM3:** 51
dynamic, polyether sulfones (PES, PESV) ... **EM2:** 161
effect of frequency and wave form ... **A12:** 58-63
effect of material behavior ... **EM3:** 513-516
effect of viscoelasticity ... **EM3:** 513
effects in hip prosthesis ... **A11:** 690-693
effects of electroplating surface on ... **A8:** 373
effects of nitriding ... **A8:** 373
and elevated-temperature service ... **A1:** 624-626, 632, 633
endurance ... **EM2:** 741
endurance, aramid/carbon fiber reinforced epoxies ... **EM1:** 35
endurance limit defined ... **M1:** 666
and environment effects ... **A13:** 295
environmental effects ... **A12:** 35-63
environmental effects on ... **A1:** 624-625, 643, 645, 677, 681
environmentally enhanced ... **A8:** 487-488
erosion, and wear ... **A8:** 601
experiment synopsis ... **EM3:** 517-518
experiments ... **A8:** 695-707
failure ... **A11:** 590, 621-623 **EM1:** 201-204, 436-438, 797, 938
failure surface family ... **EM1:** 203
fatigue data, use in design ... **M1:** 675-682
fatigue life, estimating of ... **M1:** 678, 681
fatigue limit defined ... **M1:** 666
fatigue parameters, estimating of ... **M1:** 676-678
fatigue parameters of selected steels ... **M1:** 680
fatigue resistance, factors affecting ... **M1:** 665

Fatigue (continued)
fatigue resistance, surface hardening to improve ... **M1:** 527, 528, 538, 540, 541
fatigue strength defined ... **M1:** 666-667
flaking, in bearings ... **A11:** 503
fluidized-bed thermal, test for ... **A11:** 278
fracture criteria, threshold levels ... **EM3:** 516-51
fracture mechanics and ... **A11:** 47
as fracture mode ... **A12:** 35-63
fractures ... **A8:** 253, 363
frequency effect on ... **A8:** 346
fretting ... **A11:** 5, 148
in gears ... **A11:** 590-595
general aspects of ... **EM2:** 701-703
Goodman diagram for ... **EM1:** 202
helicopter rotor blade ... **EM1:** 441
high-cycle ... **A11:** 5, 106
high-cycle, AISI/SAE alloy steels ... **A12:** 296
high-impact polystyrenes HIPS) ... **EM2:** 197
high-stress, spring failure during ... **A11:** 553-555
historical studies ... **A12:** 3
initiation process, in bone plate ... **A11:** 680, 684-686
interaction with creep ... **M3:** 234-235
in iron-copper-carbon alloys ... **A14:** 201
lamellar structure and ... **A12:** 4
of lead, and lead alloys ... **A2:** 547
levels ... **EM3:** 50
life ... **EM1:** 10, 244-245, 441-443
life, defined ... **A13:** 6
life of P/M roller bearing cup ... **A7:** 620
life, of powder forged parts ... **A14:** 206
life prediction ... **A12:** 207
life, wrought aluminum alloy ... **A2:** 60
limit ... **A8:** 364 **A11:** 4, 103 **A13:** 6, 292-293, 928
limit, defined ... **EM1:** 10
limit of P/M and wrought titanium and alloys ... **A7:** 475
limits, pearlitic/martensitic malleable iron ... **A15:** 696
in liquid-metal environments ... **A11:** 231-232 **A13:** 177-178
loading ... **A8:** 374, 714, 716
low-alloy steels ... **A15:** 717
low-cycle ... **A8:** 364 **A11:** 6, 102, 103, 622 **A13:** 292
low-cycle, and fracture control philosophy ... **A17:** 666-667
low-cycle, eddy current crack inspection ... **A17:** 190
low-cycle, methods of predicting ... **M3:** 235-237
marks, on shafts ... **A11:** 461, 463
material microstructure considerations ... **EM3:** 513
materials illustrated in ... **A12:** 217
mechanical, as die failure cause ... **A14:** 47
mechanical properties affecting ... **A12:** 49
metal, as cause for shaft failure ... **A11:** 459
metal powder addition effect ... **EM3:** 515
metallurgical variables ... **M1:** 670, 672-677
microscopic fracture model ... **A8:** 477
microstructure, effect of ... **M1:** 675-677
natural rubber's resistance ... **EM3:** 145
notch blunting ... **EM3:** 515
notch effects ... **M1:** 667-668, 679-682
notch factor ... **A8:** 364, 372, 725 **A11:** 4, 103, 797
offshore oil/gas production platforms ... **A13:** 1254-1255
as operating failures, machine elements ... **A11:** 134
in P/M forging ... **A7:** 415
P/M superalloys ... **A13:** 838
para-aramid fibers ... **EM1:** 55
patches, martensitic stainless steels ... **A12:** 368-369
periodic excitation ... **EM2:** 553-554
point-stress failure by ... **EM1:** 244-246
in powder and drop forged parts ... **A14:** 197
of powder forged materials ... **A14:** 199
precracking ... **A8:** 469, 517 **A12:** 75, 236-237, 397, 479
precracking, of specimens ... **A13:** 257-258
prediction of ... **A8:** 363

Fatigue (continued)
in pressure vessels **A11:** 668-669
prevention of **A1:** 673-674
prevention of failure **M1:** 665-666
prior, effect on static creep rupture life **A8:** 353, 355
properties, CG iron **A15:** 673
properties, compared **A8:** 706-712
properties, corrosion-resistant high-alloy **A15:** 728
properties, of implant materials **A11:** 688-689
properties, P/M superalloys **A7:** 439
properties, powder forgings **A14:** 202-203
rolling-contact **A14:** 203
quasi-brittle, polycarbonate sheet **A12:** 479
ratio, defined **A11:** 4 **EM2:** 17
residual stress **M1:** 673-675, 682
resistance **A8:** 347-349, 706-713 **A11:** 31, 672
resistance, treatments for **A11:** 121
reversed-bending **A11:** 321, 462-463
rolling-contact, ball bearings **A11:** 500-505, 593-594
rotating-bending **A11:** 321, 463-464
S-N curves **M1:** 667-668
shot peening and surface rolling effect of **M1:** 674-675, 682
in situ electropolishing for tracking **A9:** 55-56
in sliding bearings **A11:** 486
as slip process **A12:** 35
solder joint **EL1:** 640, 743-744, 1031-1032
sonic **EM3:** 501, 502
spalling, in bearings **A11:** 491
specimens **A8:** 370-373, 696, 705-706
stages of **A12:** 175
stages of process **A11:** 102, 104
in STAMP-processed stainless steels **A7:** 549
static **A11:** 28
steel properties, surface conditions effect **A8:** 373
strain, basis for predicting fatigue life **M1:** 668, 670-672
strain range **A8:** 364
strategies to reduce its effect on glass and glass fibers **EM4:** 744
strength **A8:** 364, 371-374, 703-706 **A13:** 292 **EM1:** 10, 436-443
strength, aluminum alloy **A15:** 764
strength, and fatigue limit **A11:** 103
strength of ferrous P/M materials **A7:** 465, 466
and strength, plain carbon steels **A15:** 702-704
stress, basis for predicting fatigue life **M1:** 668, 671
stress-concentration factor **A8:** 364
and stress-rupture data in tension-hold-only test **A8:** 349
stress-whitening phenomenon **EM3:** 514-515
striation **A8:** 482, 484
subcase **A11:** 505
and subcritical fracture mechanics **A11:** 52-53
subsurface-initiated **A11:** 504
sudden fracture **A8:** 363
superalloys **A12:** 389
surface damage, development in stainless steel implant replica **A11:** 684, 688-689
and surface effects **A8:** 373-374, 603 **A13:** 294
surface hardening, effect of **M1:** 673-675
surface-contact, in gears **A11:** 592-593
surface-initiated, in bearing components **A11:** 501-502
surface-pitting **A11:** 134
symbols and definitions **M1:** 666-668
tension, urethane hybrids **EM2:** 269
tension-tension, Kevlar aramid fibers **EM1:** 55
test data components **A8:** 367-368
test environment effect on **A8:** 354
test specimens **A13:** 293
test, stress ratio **A8:** 363
testing **A13:** 291-293 **EM3:** 501-518
design and analysis of adhesive bonding **EM3:** 467-468, 469
FM 47 adhesive **EM3:** 502
polybenzimidazoles **EM3:** 170
testing, accelerated reliability **EL1:** 741, 747-751
testing machines **A8:** 368-370
testing, with magnetic rubber inspection **A17:** 125
tests **A11:** 281

Fatigue (continued)
then-nomechanical, test for **A11:** 278
thermal **A11:** 11, 133, 278-279, 371, 594-595, 623 **EM3:** 501
thermally induced **EL1:** 632, 640
in threaded fasteners **A11:** 531-534
titanium and titanium alloy castings **A2:** 640
of titanium and titanium alloys, effect of alpha colony size on **A9:** 460
titanium castings **A15:** 829
in titanium P/M parts **A7:** 475, 752
tooth-bending **A11:** 590-592
torsional, shafts **A11:** 464
as transgranular fracture, SEM defined **A12:** 175-176
under cyclic-stress loading, of machine parts **A11:** 371
unidirectional bending, shafts **A11:** 461-462
wear, defined **A11:** 4
wear mechanism **A8:** 602-603
wrought aluminum alloys **A12:** 418
zinc **M2:** 650-652
zone, terminating edge **A12:** 267
Fatigue (endurance) limit **EM3:** 501, 502
Fatigue acceleration transform **EL1:** 741, 744
Fatigue analysis *See also* Failure; Failure analysis; Fatigue; Fatigue failure; Fiber properties analysis; Material properties; Material properties analysis
and damage accumulation **EM1:** 146-147
failure concepts **EM1:** 244-246
full-scale tests **EM1:** 346
of laminates **EM1:** 236-251
models **EM1:** 246-217
stress approaches **EM1:** 236-237, 239-240
Fatigue characteristics
of HSLA steels **A1:** 413
Fatigue corrosion *See* Fatigue strength
Fatigue crack
growth rate, titanium P/M parts **A7:** 752
initiation sites, contaminants as **A7:** 178
Fatigue crack growth *See also* Crack growth; Crack propagation rate; Cracking; Fatigue; Fatigue crack growth rate
of ASTM A533, BI steel **A8:** 377
in bcc materials **A8:** 251
controlling random errors in analysis **A8:** 679
and damage analysis **A8:** 681-682
data analysis **A8:** 678-691
effect of vacuum **A12:** 46
fracture topography, titanium alloys **A12:** 441-443
from surface flaw **A12:** 420
in martensitic steels **A8:** 377
minimum, and threshold stress intensity **A8:** 254, 256
and plastic limit load behavior **A8:** 377
rate **A8:** 376-380, 403, 411-417, 678-682
resonant, rate of **A8:** 251
specimen **A8:** 251, 379-382, 384
testing **A8:** 427-430
threshold testing, methods **A8:** 379
with ultrasonic fatigue testing **A8:** 240
x-y recorder traces **A8:** 383-384
Fatigue crack growth rate **A8:** 5, 376-380, 403, 411-417, 678-682
abbreviation **A8:** 724
calculation **A8:** 415, 679-680
for damage analysis **A8:** 681-682
incremental polynomial method **A8:** 378
modeling **A8:** 680-681
test specimens **A8:** 678
Fatigue crack growth rate (*da/dN*)
cryogenic temperature effects **A12:** 52-53
cyclic loading effects **A12:** 53-54, 62
defined **A13:** 6
dwell time effects **A12:** 60
and fatigue striation spacing **A12:** 41
fracture mode changes with **A12:** 441
gases, effects on **A12:** 40, 52
high, microvoid coalescence formation **A12:** 119-120
measurement **A12:** 120-121
P/M superalloys **A13:** 837
second-phase particles/inclusions, effect on **A12:** 16
stress intensity factor range, effect on **A12:** 56-58

Fatigue crack growth rate (FCGR)
titanium and titanium alloy castings **A2:** 640
wrought aluminum alloy **A2:** 43, 59
wrought titanium alloys **A2:** 631
Fatigue crack growth specimen **A8:** 379-382, 384
Fatigue crack initiation *See also* Fatigue **A8:** 363-364, 366-375
applied stresses **A8:** 363-364
corrosion fatigue **A8:** 374-375
effect of test specimen size **A8:** 372-373
fatigue testing regimes **A8:** 367-368
low- and high-cycle fatigue **A8:** 367
mean stress effect **A8:** 374
SEM analysis **A12:** 169
stress concentration effect **A8:** 371-372
stress ratio **A8:** 363
surface effects and fatigue **A8:** 373-374
test specimens **A8:** 370-373
testing machines for **A8:** 368-370
Fatigue crack initiation, causes and prevention of
in threaded steel fasteners **A1:** 298, 300
Fatigue crack propagation *See also* Corrosion fatigue crack propagation; Crack propagation; Fatigue **A8:** 376-402
in acidified chloride at ambient and elevated temperatures **A8:** 418-420
in aluminum alloys **A8:** 364-365, 404
analysis **A11:** 107-110
in aqueous solutions at ambient temperature **A8:** 415-417
of austenitic stainless steel in liquid sodium **A8:** 426, 428
by hydrogen in feedwater **A8:** 422
cast aluminum alloys **A12:** 408
causes **A8:** 363
center-cracked and compact specimens **A8:** 377-378, 382
corrosion **A8:** 403-410
corrosion, P/M aluminum alloys **A12:** 440
corrosion, wrought aluminum alloys **A12:** 439
crack-length measurement techniques **A8:** 382-391
cyclic crack growth rate testing in threshold regime **A8:** 378-379
data analysis **A8:** 377-378
ductile iron **A12:** 228
effect of frequency of loading **A11:** 110
effect of inclusions, ASTM/ASME alloy steels **A12:** 346-348
electromechanical fatigue testing systems **A8:** 391-395
environmental effect on **A8:** 404
environmental effects **A8:** 377, 403-435
factors that control growth **A8:** 365
in fatigue failures **A11:** 103
fracture mechanics of **A11:** 102, 124-125
high-carbon steels **A12:** 289, 290
hydrogen-assisted **A12:** 302
in liquid metal environment **A8:** 425-426
localized failures in steel bridge components by **A11:** 707
in low-carbon steel, effect of calcium **A12:** 247
mechanism of **A12:** 15, 21
medium-carbon steels **A12:** 275
monitoring materials in elevated- temperature water **A8:** 421-422
near-threshold, hydrazine effect **A8:** 427, 429
near-threshold, load application **A8:** 428
predicted **A8:** 405
in pressurized water reactor **A8:** 402
radial, wrought aluminum alloys **A12:** 415
rate, effect of embrittling or corrosive environments **A12:** 35
rates from direct current potential measurements **A8:** 390
servohydraulic fatigue testing systems **A8:** 395-400
short crack behavior **A8:** 379-380
steam or boiling water with contaminants **A8:** 426-430
of structural alloys in liquid metal environments **A8:** 425-426
in superalloys, high-pressure hydrogen assisted **A8:** 409
test specimens **A8:** 379-382, 414-415
testing **A8:** 376-378
testing, objectives **A11:** 103

Fatigue crack propagation (continued)
in titanium alloy, hydrazine effect **A8:** 427, 429
titanium P/M and I/M alloys, compared **A2:** 652
transgranular, wrought aluminum alloys **A12:** 438-439
using fracture mechanics techniques **A8:** 376
in vacuum and gaseous environments at ambient temperature **A8:** 410-412
in vacuum and oxidizing gases at elevated temperatures **A8:** 412-415
vacuum test chamber for **A8:** 412-414
wrought titanium alloys **A2:** 624
Fatigue cracking *See also* Cracking; Fatigue
in aluminum alloy aircraft deck plate **A11:** 311-312
at defect tip **A11:** 102
at weld termination, bridge components **A11:** 707
in beam flange, bridge components **A11:** 708
brittle fracture from **A11:** 85, 89
in cold-formed parts **A11:** 307-308
crack initiation **A11:** 102-103, 106, 621
crack propagation **A11:** 106-107
effect of temper embrittlement on **A11:** 335
failures from **A11:** 105-107
forged aircraft wheel half, material defects **A11:** 323
in forging **A11:** 323
from cyclic stresses **A11:** 102
from vertical butt-weld detail **A11:** 709
front **A11:** 26
identified **A11:** 134
intergranular **A11:** 254
in Lafayette Street Bridge St. Paul, MN **A11:** 709-710
of main hoist shaft **A11:** 525
peeling-type, in compressor shaft **A11:** 471
of steel main hoist shaft **A11:** 525
steel structural member failed by **A11:** 116
thermal, in cast iron brake drum **A11:** 371
Fatigue cracks
in aircraft splice joints, eddy current inspection **A17:** 193
by liquid penetrant inspection **A17:** 86
by magnetic rubber inspection **A17:** 125
microwave inspection **A17:** 215
as planar flaws **A17:** 50
as qualification standards **A17:** 677
radiographic methods **A17:** 296
Fatigue data analysis **A8:** 695-720
at different sources or heats **A8:** 712
of composite materials **A8:** 713-718
consolidation of data **A8:** 712-713
different mean stresses or strains **A8:** 712-713
and failure mechanisms **A8:** 714
fatigue resistance at single stress or strain level **A8:** 706-712
Goodman diagram for **A8:** 713
laboratory, applications **A8:** 713
material fatigue behaviors, compared **A8:** 706-712
planning fatigue experiments **A8:** 695-697
Probit method **A8:** 702
and proof testing **A8:** 717-718
selecting stress levels **A8:** 715-716
staircase method **A8:** 703-704
statistical characterization of fatigue strength or limit **A8:** 701-706
statistical characterization of stress-life or strain-life material response **A8:** 697-701
strength degradation model **A8:** 716-717
of stress-life or strain-life curve **A8:** 696
two-point strategy **A8:** 700
with Weibull distribution **A8:** 717-718
Weibull parameters estimation **A8:** 714-716
Fatigue ductility **EM3:** 12
defined **EM2:** 17
Fatigue ductility exponent **EM3:** 12
defined **EM2:** 17
Fatigue experiments **A8:** 695-697, 700-701, 707
Fatigue failure *See also* Failure; Failure analysis; Fatigue **A8:** 5
alpha-beta interface fracture **A8:** 487
analysis **EM1:** 201-204
by mechanical failure **EM2:** 744-749
by thermal failure **EM2:** 743-744
in composite materials **A8:** 714
cyclic cleavage **A8:** 484, 487
cyclic ductile decohesion **A8:** 482-484
cyclic intergranular processes **A8:** 484
defined **EM2:** 741
delamination **EM1:** 437-438
discontinuous fiber composites **EM1:** 797
discontinuous intergranular facets **A8:** 487
ductile striations **A8:** 481-482, 487
fatigue endurance **EM2:** 741-742
fiber break/interface debonding **EM1:** 938
forked intergranular cracks **A8:** 487
layer cracking **EM1:** 436-437
mechanisms of **EM2:** 742-749
mixed fracture modes **A8:** 484-485
moisture-induced **EM2:** 765
particle nucleated ductile intergranular voids **A8:** 487
of plastics **EM2:** 702
printed boards **EL1:** 1039
of solder joints **EL1:** 740
thermal **EL1:** 56, 58-59
types of alternate microscopic fracture modes **A8:** 487
in wire bonds **EL1:** 1043
Fatigue failures **A11:** 102-135
of anchor link **A11:** 397
at elevated temperatures **A11:** 130-133
bending-, steel wire hoisting rope **A11:** 518
carbon steel counterbalance spring **A11:** 558
of carbon steel water-wall tube, at welded joint **A11:** 621-622
carbon steel wiper spring **A11:** 558-559
of carbon-molybdenum steel boiler tubes, by vibration **A11:** 621
characteristics, by macroscopy **A11:** 104-105
characteristics, by microscopy **A11:** 105
contact **A11:** 133-134
corrosion **A11:** 134
defined **A11:** 4
design, effect on strength **A11:** 115-118
of diesel-engine rocker levers **A11:** 350-352
effect of discontinuities **A11:** 119-121
effect of heat treatment **A11:** 121-122
effect of manufacturing practices **A11:** 122-130
effect of material conditions **A11:** 118-119
effect of stress concentrations **A11:** 113-115
effect of stress on **A11:** 110-115
in fasteners, carbon-graphite composite joints **A11:** 549
fatigue cracking **A11:** 105-107
fatigue damage and life, determining **A11:** 135
fatigue fracture, stages of **A11:** 104
fatigue life, prediction of **A11:** 102
fatigue-crack initiation **A11:** 102-103
fatigue-crack propagation **A11:** 103, 107-110
in forging **A11:** 321-322
from improper design **A11:** 396-397
from subsurface inclusions **A11:** 323
from weld spatter **A11:** 559
inspection schedules and techniques **A11:** 134
in integrated circuits **A11:** 774-776
laboratory, and normalized fracture band width **A11:** 762
locomotive suspension spring **A11:** 551
of machined workpiece **A16:** 21
music-wire spring **A11:** 557
phosphor bronze spring **A11:** 555-557
in shafts **A11:** 461
stage II striations typical of **A11:** 23
steel castings **A11:** 396, 397-398
Fatigue failures (continued)
steel crane shaft **A11:** 524-525
of steel elevator cable **A11:** 520-521
of steel retainer **A11:** 308-309
of steel semitrailer wheel studs **A11:** 531-532, 534
steel wire rope **A11:** 518-519
of structural bolt, from reversed-bending **A11:** 322
thermal **A11:** 133
for threaded steel fasteners **A1:** 297-299, 300, 301
of tool steel shaft **A11:** 462
Fatigue fracture *See also* Fatigue; Fractures
of aircraft fuel-tank floors **A11:** 126-127
of aircraft propeller blade **A11:** 125-126
of alloy steel lift pin, in crane **A11:** 77
of alloy steel valve springs **A11:** 551
of aluminum alloy landing-gear torque arm **A11:** 114
of angled blade plate **A11:** 683, 687
bending, of steel pump shaft **A11:** 109
in boilers and steam equipment **A11:** 621
in brazed joint, by voids **A11:** 454
by embrittlement, penetration of molten braze materials **A11:** 454-455
carbon steel pawl spring **A11:** 551-553
of cast chromium-molybdenum steel pinion **A11:** 395-396
of cast stainless steel lever **A11:** 113-114
of cast steel axle housing **A11:** 397
characterized **A11:** 77-78
of chromium-molybdenum steel integral coupling and gear **A11:** 129-130
of crankshafts **A11:** 123-125, 480
of D-6ac steel structural member **A11:** 116
defined **A11:** 75
determined **A11:** 26-27
of drive shaft **A11:** 122-123
effect of environment **M6:** 883-884
effect of residual stress **M6:** 883
of forged drive axle **A11:** 116-117
of forged steel crane-bridge wheel **A11:** 528
of forged steel rocker arm **A11:** 119-120
general features **A11:** 26-27
high-temperature **A11:** 130-133
of highway tractor-trailer steel drawbar **A11:** 127-128
identification chart for **A11:** 80
intergranular **A11:** 131-133
in knuckle pins **A11:** 128-129
mechanism, in steam equipment **A11:** 621
of pilot-valve bushing **A11:** 121
planes, single-shear and double-shear **A11:** 105
of plunger shaft, from sharp fillet **A11:** 319-320
of rolling-tool mandrel **A11:** 474-475
of spindle for helicopter blade **A11:** 125
stages in aluminum alloy **A11:** 104
steam equipment failure by **A11:** 602
of steel articulated rod **A11:** 473-474
of steel cap screws **A11:** 533, 535
of steel connecting-rod cap **A11:** 119-120
steel crankshaft **A11:** 324
of steel cross-travel shaft **A11:** 525
steel hooks **A11:** 523-524
of steel wheels for coke-oven car **A11:** 130
of stuffing box **A11:** 346-347
surface of **A11:** 21, 104, 321
surfaces, beach marks on **A11:** 104
surfaces, in drilled hole **A11:** 21
of taper pin, clutch-drive assembly **A11:** 545-546
of tool steel tube-bending-machine shaft **A11:** 462
of U-bolts **A11:** 533-535
of welded stainless steel liners, bellows-type expansion joint **A11:** 118
zones **A11:** 110
Fatigue fracture surface *See* Fatigue fracture(s); Fracture surface(s)

Fatigue fracture(s) *See also* Fatigue; Fracture surfaces; High-cycle fatigue fracture; Low- cycle fatigue fracture(s)
bending-plus-torsional, medium-carbon steels **A12:** 260
of bolt **A12:** 112, 120
cast aluminum alloys **A12:** 407-408
cobalt alloys **A12:** 398
copper **A12:** 401
in coupling pins **A12:** 113, 120
defined **A12:** 14-18
in drive shaft **A12:** 112, 120
ductile iron **A12:** 228-229
environments, types affecting **A12:** 35-63
facetlike **A12:** 120
frequency and wave form effects **A12:** 58-63
from improper heat treatment, AISI/SAE alloy steels **A12:** 309
furrow-type **A12:** 44, 54
in gaseous environments **A12:** 36-41
high-carbon steels **A12:** 277, 279-280, 283, 288
high-cycle, martensitic stainless steels **A12:** 367
hydrogen effect on surface appearance **A12:** 37, 51
illumination techniques for **A12:** 85
interpretation of **A12:** 111-121
in liquid environments **A12:** 36, 41-46
loading effect on **A12:** 36, 53-54
low-carbon steel **A12:** 243, 251
low-cycle, high-carbon steels **A12:** 283
macroscopic characteristics **A12:** 111-121
markings, interpretation **A12:** 112, 118, 119
materials illustrated in **A12:** 217
mechanisms **A12:** 4-5
medium-carbon steels **A12:** 258, 259, 260, 263, 269, 273
and monotonic fracture surfaces, compared **A12:** 229
oxygen-free high-conductivity copper **A12:** 401
profile **A12:** 15, 22
SEM fractograph of aluminum alloy **A12:** 175
sequence to **A12:** 111
stages **A12:** 14-21
and stress intensity factor range **A12:** 54-58
with striations **A12:** 16-18
superalloys **A12:** 394
surface, high-carbon steels **A12:** 279
temperature, effect on **A12:** 36, 49-53
as transgranular, with slip-plane fracture **A12:** 117
in vacuum **A12:** 36, 46-49
wrought aluminum alloys **A12:** 415-421
Fatigue fractures '
in magnesium alloys **A9:** 426
in wrought beryllium-copper alloys **A9:** 393
Fatigue life *See also* Fatigue; Normal solution **A6:** 390-391 **EM1:** 10, 244-245, 441-443 **EM3:** 12
of AISI 304 stainless steel **A8:** 348
of bearing materials **A11:** 487
data, two-parameter Weibull distribution **A8:** 633
defined **A8:** 5, 363 **A11:** 4, 102 **EL1:** 1143 **EM2:** 17
as dependent variable **A8:** 698
determination of **A11:** 135
distribution **A8:** 699
distribution for graphite-epoxy **A8:** 718
effect of strain rate, stainless steels **A12:** 59
effect of wave form **A12:** 62, 63
estimating methods **A8:** 374
flexible printed boards **EL1:** 588
as function of stress amplitude **A8:** 253
hold period in tension effect on **A8:** 352
hold-time results **A8:** 353
improper heat treatment effects **A12:** 309
influence of change in specimen surface on **A8:** 711
log normal distribution **A8:** 700-701
mean curve definition **A8:** 698-699
measured in cycles to fracture **A8:** 350
number of cycles to failure, symbol for **A8:** 725
predicting, by creep **A12:** 123
prediction of **A11:** 102
prediction techniques **A8:** 354-358, 363
reduction factor **A8:** 353
of René **A8:** 95, 352
rolling-element bearings **A18:** 134
Fatigue life (continued)
scatter about the mean curve **A8:** 699-700
and static-strength, related **A8:** 717-718
of steel parts, correction factors **A11:** 116
symbol for **A11:** 797
temperature, reversed-bending stress and **A11:** 130
test classifications **A8:** 363
truncated, by static proof test **A8:** 718
variability **A8:** 699-700
vs. hold-period time for AISI 304 stainless steel **A8:** 349
Weibull two- and three-parameter distributions **A8:** 700-701
Fatigue life for *p*% survival
defined **A8:** 5
Fatigue limit *See also* Endurance limit **A11:** 4, 103 **A13:** 6, 292-293, 928 **EM3:** 12
bcc vs. fcc **A8:** 253
data, from ultrasonic fatigue testing **A8:** 240
defined **A8:** 5, 364 **EM2:** 17
and fatigue resistance **A1:** 675
of ferritic malleable iron **A1:** 75
of gray iron **A1:** 19-22
in *S-N* curve **A8:** 364
specimen size effect on, steel in reversed bending **A8:** 372
statistical characterization of **A8:** 701-706
Fatigue limit for *p*% survival
defined **A8:** 5
Fatigue load limit **A18:** 508
Fatigue loading *See also* Dynamic fatigue testing; Fatigue; Impact loading **EM2:** 701-706
fatigue, general aspects **EM2:** 701-703
plastics vs metals **EM2:** 702
testing methods **EM2:** 703-706
Fatigue notch factor **A11:** 4, 103, 797
defined **A8:** 5, 364, 372, 725
and fatigue resistance **A1:** 675
Fatigue notch sensitivity **A11:** 4, 103, 797
defined **A8:** 5
and fatigue resistance **A1:** 675-676
in gray iron **A1:** 21-22
of steel plate **A1:** 369-370
***Fatigue of Engineering Plastics* (Hertzberg/Manson)** **EM2:** 94
Fatigue precracking
ductile irons **A12:** 236, 237
nickel alloys **A12:** 397
ultrasonic cleaning of **A12:** 75
Fatigue properties *See also* Goodman diagrams, *S-N* curves
cast steels **M1:** 383, 389, 397
closed-die forgings **M1:** 354-355, 359-361
cold finished bars **M1:** 227-233
constructional steels for elevated temperature use **M1:** 643, 657-662
of corrosion-resistant steel castings **A1:** 918, 920
ductile iron **M1:** 43-44, 45, 46
gray cast iron **M1:** 19-22
high-cycle **A2:** 624
HSLA steels, compared to hot rolled low-carbon steels **M1:** 418-419
low-cycle **A2:** 624
malleable iron **M1:** 65, 66, 70
maraging steels **M1:** 450, 451
P/M steels **M1:** 334-335, 337, 343, 346
plate **M1:** 194
springs **M1:** 291-296, 297, 303, 304, 312
of steel castings **A1:** 369-370
threaded fasteners **M1:** 275-276, 279, 282
ultrahigh-strength steels **M1:** 428, 432, 440-441
wrought titanium alloys **A2:** 623-624
Fatigue ratio **EM3:** 12
Fatigue resistance *See also* Fatigue **A8:** 347-349, 706-713
copper and copper alloys **A2:** 216
of die materials **A14:** 47
lead and lead-bearing alloys **A2:** 554
of powder forged and wrought materials **A14:** 196
Fatigue resistance function **EM3:** 33
Fatigue resistance of aluminum alloy wrought products
effect of stringers on **A9:** 635
Fatigue resistance of steels **A1:** 673-688
application of fatigue data **A1:** 673-688
comparison of fatigue testing techniques **A1:** 687
cumulative fatigue damage **A1:** 677, 686
discontinuities **A1:** 687
estimating fatigue life **A1:** 677, 684, 685-686
estimating fatigue parameters **A1:** 683-684
load data gathering **A1:** 687-688
mean stresses **A1:** 686-687
notches **A1:** 686
scatter of data **A1:** 682-683, 684
metallurgical variables of fatigue
aggressive environments **A1:** 677, 681
behavior **A1:** 678
cleanliness **A1:** 678-679, 681, 682
composition **A1:** 681
creep-fatigue interaction **A1:** 681
ductility **A1:** 678
grain size **A1:** 681
macrostructure differences **A1:** 681
microstructure **A1:** 681, 683
orientation of cyclic stress **A1:** 682, 683
residual stresses **A1:** 591, 680-681, 682
strength level **A1:** 676, 678, 679, 681
surface conditions **A1:** 677, 679
tensile residual stresses **A1:** 681
strain-based approach to fatigue **A1:** 677-678, 679
stress-based approach to fatigue **A1:** 675, 676, 677
correction factors for test data **A1:** 676-677
symbols and definitions **A1:** 674
applied stresses **A1:** 674
constant-lifetime diagram **A1:** 675
fatigue limit **A1:** 675
fatigue notch factor **A1:** 675
fatigue notch sensitivity **A1:** 675-676
fatigue strength **A1:** 675
nominal axial stresses **A1:** 674
S-N curves **A1:** 674-675
stress concentration factor **A1:** 675
stress ratio **A1:** 674
Fatigue spalling life
bearings **A18:** 258
Fatigue strength *See also* Fatigue; Mechanical properties **A8:** 5, 364, 366-374, 703-706 **A13:** 6, 295, 438-439, 595, 1264 **EM3:** 12, 502-503, 504
aluminum casting alloys **A2:** 149, 153-177
aluminum-lithium alloys **A2:** 187-188, 191-196
blended elemental titanium P/M compacts **A2:** 650-651
cast copper alloys **A2:** 356-391
in closed-die forgings **A1:** 341, 342
commercially pure tin **A2:** 518
data analysis/life prediction **EM1:** 441-443
defined **A11:** 4 **EM1:** 10 **EM2:** 17
discontinuities effect on **A11:** 119-121
of ductile iron **A1:** 39, 46-47, 48
environmental effects on **A11:** 252
and fatigue failure **EM1:** 436-438
and fatigue resistance **A1:** 675
as function of relative humidity **A11:** 252
high-impact polystyrenes (PS, HIPS) **EM2:** 197
influence of design on **A11:** 115-118
low, in bridge components **A11:** 707-708
magnesium **M2:** 531, 532
of magnesium alloys **A2:** 461-462
manufacturing practices effects **A11:** 122-130
material conditions effect on **A11:** 118-119
polyamide-imides (PAI) **EM2:** 129
prealloyed titanium P/M compacts **A2:** 652-653
and resilience, beryllium-copper alloys **A2:** 417-418
S-N relation **EM1:** 438-441
and static strength, effect of temperature **A11:** 130-133
of steel plate **A1:** 238
steel shafts **A11:** 476
of steels, bushings and **A11:** 470
stress effects **A11:** 110-115
surface finishes and roughness effects **A11:** 122
and thermal conductivity, beryllium-copper alloys **A2:** 419-420
of threaded fasteners **A1:** 298, 300
vs fatigue limit **A11:** 103
wrought aluminum and aluminum alloys **A2:** 81-122

Fatigue strength at N cycles *See also* Median fatigue strength at N cycles **A8:** 5
Fatigue strength for p% survival at N cycles
defined **A8:** 5
Fatigue strength, of machined workpiece **A16:** 25-27, 30
4340 steel after abusive grinding **A16:** 35
and abusive grinding **A16:** 24
and effect of method of machining **A16:** 31
improved by shot peening **A16:** 26-27
shot peening effect **A16:** 36
surface alterations produced **A16:** 25
thread rolling **A16:** 281
Ti-6Al-4V **A16:** 35
Fatigue striation spacings *See also* Fatigue striations
as a function of applied stress **A12:** 120
austenitic stainless steels **A12:** 353
brittle **A12:** 35
da/dN as **A12:** 16
defined **A12:** 15
determined, example case **A12:** 205
dwell time effect **A12:** 60, 61
and fatigue crack propagation rate **A12:** 41
as fatigue mechanism **A12:** 4-5
loading conditions, effect on **A12:** 15, 22
local variations **A12:** 19
martensitic stainless steels **A12:** 367
prediction of **A12:** 48
and projected images **A12:** 205
temperature, effect on **A12:** 49-50
variations, aluminum alloy **A12:** 22
wrought aluminum alloys **A12:** 431
Fatigue striations *See also* Beach marks; Crack arrest; Crack arrest marks; Fatigue striation spacings; Quasi-striations; Wallner lines **A8:** 5, 482, 484
AISI/SAE alloy steels **A12:** 331
aluminum alloy **A12:** 19, 20, 176
angle, as grain boundary locator **A12:** 430
ASTM/ASME alloy steels **A12:** 346, 348
austenitic stainless steels **A12:** 39, 52, 352-353, 359
beach marks as **A12:** 175
brittle **A12:** 119, 430, 432, 456
cast aluminum alloys **A12:** 406
copper alloys **A12:** 403
and dimples **A12:** 177
ductile **A12:** 119, 247, 346, 431-432
in ductile iron **A12:** 229
early Zapffe TEM **A12:** 6
electronic materials **A12:** 482
and fissures, compared **A12:** 294
formation, interinclusion **A12:** 247
formation, titanium alloys **A12:** 442
fracture characteristics with **A12:** 16-18
as fracture mechanism **A12:** 4-5
from sulfur-containing atmospheres **A12:** 41, 53
high-carbon steels **A12:** 290
and lamellar spacing, compared **A12:** 290
light fractographs **A12:** 94, 96-97
low-carbon iron **A12:** 220, 222
in low-carbon steel **A12:** 177, 247
maraging steels **A12:** 385, 386
martensitic stainless steels **A12:** 367
measurement **A12:** 121
in medium-density polyethylene **A12:** 480
as microscopic feature in fatigue **A12:** 118-121, 137-138
nickel alloys **A12:** 205-206, 397
in nickel, check on precision matching **A12:** 205-206
OFHC copper **A12:** 401
on joining crack fronts **A12:** 23
on plateaus, schematic **A12:** 23
roots, austenitic stainless steels **A12:** 351
shadowing technique for **A12:** 172
with slip traces **A12:** 15
superalloys **A12:** 390, 391, 392
tantalum alloys **A12:** 464
Fatigue striations (continued)
titanium alloys **A12:** 441
wrought aluminum alloys **A12:** 417-418, 426-429, 431-432, 439
Fatigue test, resistance
spot welds **M6:** 488-489
Fatigue test specimens **A8:** 370-373, 705-706
Fatigue testing *See also* Crack propagation rate; Fatigue **A8:** 5
in acidified chloride at ambient and elevated temperatures **A8:** 418-420
aluminum alloys **A15:** 764
in aqueous solutions at ambient temperatures **A8:** 415-417
axial, constant-amplitude **A8:** 149
chamber **A8:** 412-414
closed-loop servomechanical system **A8:** 396
crack initiation **A8:** 363
crack propagation **A8:** 363
cycles-to-crack-initiation **A8:** 696
cycles-to-failure **A8:** 696
data **A8:** 707, 709
electromechanical **A8:** 391-395
equipment **A8:** 696
heat-centering techniques **A8:** 712
high- and low-cycle **A8:** 696
machines **A8:** 368-370
material heat **A8:** 712
mean curve definition **A8:** 698-699
minimizing nuisance variables **A8:** 697
negative-feedback closed-loop system **A8:** 395-396, 398
overstrain data treatment **A8:** 701
regimes **A8:** 367-368
runouts **A8:** 701
sources **A8:** 712
specimens **A8:** 370-373, 705-706
staircase method **A8:** 703-704
standards and practices for **A8:** 375
ultrasonic **A8:** 240-258
in vacuum and gaseous environments **A8:** 410-412
vacuum and oxidizing gases at elevated temperatures **A8:** 412-415
Fatigue testing machines **A8:** 368-370
Fatigue wear *See also* Spalling
defined **A8:** 5 **A18:** 9
ion implantation **A18:** 856, 857, 858
stainless steels **A18:** 715
surface property effect **A18:** 342
Fatigue-crack growth
and elevated-temperature service **A1:** 625, 633
in structural steel **A1:** 663-664, 665, 666
Fatigue-stress life, data
ultrasonic **A8:** 241
Fatty acid
defined **A18:** 9
Fatty acid amides, as wax
investment casting **A15:** 254
Fatty acids
copper/copper alloy resistance **A13:** 629
for corrosion inhibitors **A13:** 481
dimerized/trimerized **EL1:** 818
Fatty acids, corrosion resistance
stainless steels **M3:** 83
Fatty compounds
in lubricants **A14:** 516
Fatty oil
defined **A18:** 9
Fatty-acid molecules
polar bonding and orientation **A11:** 152
Fault isolation *See also* Failure analysis; Failure verification
device history determination **EL1:** 1058
electrical testing **EL1:** 1058-1060
high/low-temperature electrical testing **EL1:** 1060-1061
visual examination **EL1:** 1058
Fault tolerance *See also* Dimensional tolerence; Tolerance
system level **EL1:** 377
for VLSI/VHSIC systems **EL1:** 86
wafer-scale integration **EL1:** 263, 269
Fault tree
defined **EL1:** 1143
Faulted martensite **A9:** 672-673
Fayalite
as olivine molding material **A15:** 94, 209
Faying surface *See also* Faying surface sealing **EM3:** 12
defined **EM1:** 10 **EM2:** 17
definition **A6:** 1209 **M6:** 7
Faying surface corrosion
as tension source for stress-corrosion cracking **A8:** 502
Faying surface resistance
plate **A13:** 1122
Faying surface sealing **EM1:** 719-720
of bolted joints **EM1:** 716-717
of bonded joints **EM1:** 717
inspection for full coverage **EM1:** 720
sealants for graphite-composite assemblies **EM1:** 719-720
fcc *See* Face-centered cubic (fcc) materials; Face-centered cubic lattice
Fe (Phase Diagram) .... **A3:** 2•199
Fe-C phase diagram **A4:** 3-4, 9
Fe-C system
phase diagram **A13:** 47
Fe-Ga (Phase Diagram) .... **A3:** 2•194
Fe-Gd (Phase Diagram) .... **A3:** 2•194
Fe-Ge (Phase Diagram) .... **A3:** 2•195
Fe-H (Phase Diagram) .... **A3:** 2•195
Fe-Hf (Phase Diagram) .... **A3:** 2•195
Fe-Ho (Phase Diagram) .... **A3:** 2•196
Fe-Ir (Phase Diagram) .... **A3:** 2•196
Fe-La (Phase Diagram) .... **A3:** 2•196
Fe-Lu (Phase Diagram) .... **A3:** 2•197
Fe-Mn (Phase Diagram) .... **A3:** 2•197
Fe-Mn-Ni (Phase Diagram) .... **A3:** 3•53
Fe-Mo (Phase Diagram) .... **A3:** 2•197
Fe-Mo-Nb (Phase Diagram) .... **A3:** 3•53-54
Fe-Mo-Ni (Phase Diagram) .... **A3:** 3•54-55
Fe-N (Phase Diagram) .... **A3:** 2•198
Fe-Nb (Phase Diagram) .... **A3:** 2•198
Fe-Nd (Phase Diagram) .... **A3:** 2•198
Fe-Ni (Phase Diagram) .... **A3:** 2•199
Fe-Ni-W (Phase Diagram) .... **A3:** 3•55
Fe-P (Phase Diagram) .... **A3:** 2•200
Fe-Pd (Phase Diagram) .... **A3:** 2•200
Fe-Pu (Phase Diagram) .... **A3:** 2•200
Fe-Rh (Phase Diagram) .... **A3:** 2•201
Fe-S (Phase Diagram) .... **A3:** 2•201
Fe-Sb (Phase Diagram) .... **A3:** 2•202
Fe-Sc (Phase Diagram) .... **A3:** 2•202
Fe-Se (Phase Diagram) .... **A3:** 2•202
Fe-Si (Phase Diagram) .... **A3:** 2•203
Fe-Sm (Phase Diagram) .... **A3:** 2•203
Fe-Sn (Phase Diagram) .... **A3:** 2•203
Fe-Tb (Phase Diagram) .... **A3:** 2•204
Fe-Te (Phase Diagram) .... **A3:** 2•204
Fe-Th (Phase Diagram) .... **A3:** 2•204
Fe-Ti (Phase Diagram) .... **A3:** 2•205
Fe-Tm (Phase Diagram) .... **A3:** 2•205
Fe-U (Phase Diagram) .... **A3:** 2•205
Fe-V (Phase Diagram) .... **A3:** 2•206
Fe-W (Phase Diagram) .... **A3:** 2•206
Fe-Zn (Phase Diagram) .... **A3:** 2•206
Fe-Zr (Phase Diagram) .... **A3:** 2•207
Feasibility study
in design layout **EL1:** 513
Feather
definition **EM4:** 632
Feather crystals *See* Twinned columnar growth
Feather markings
defined **A12:** 13
on chromium steel **A12:** 18

Feather markings (continued)
on cleavage fracture surface **A12:** 13
titanium alloys **A12:** 450
Feathering **EM3:** 12, 683
matrix **EM1:** 789
Feathering, matrix
in composites **A11:** 736
Feathery structures
massive transformation product **A9:** 656
Feature extraction
as properties definition process **A17:** 35
Feature size
hybrid trends **EL1:** 252-253
Feature specific measurements
made with image analyzers **A9:** 83
Feature weighing
in machine vision process **A17:** 36
Fecralloy A characteristics **A4:** 512
composition **A4:** 512
Federal and military specifications and standards **EM2:** 89-90
Federal Aviation Administration
research and development of weak bond detection techniques **EM3:** 530
Federal Motor Vehicle Safety Standards (FMVSS)
105-75, hydraulic brakes **A18:** 577
121, air brakes **A18:** 577
windshield installation **EM3:** 554
Federal Safety Standards
FMVSS 216, roof crush requirements **EM3:** 554
windshield installation **EM3:** 554
Federal specifications *See also* listings in data compilations for individual alloys
defined **EM1:** 700
Federal Specifications and Standards
Index of (FPMR 101-29.1) **EM3:** 62-63
Federal supply class (FSC) *See also* Military standards
product key words **EL1:** 914
Federal Supply Class (FSC) listing *See* DoD Index of Specifications and Standards (DoDISS)
Federal Supply Classification (FSC) system **EM3:** 62-64
adhesives, classification 8040 .... **EM3:** 63, 64
sealants, classification 8030 .... **EM3:** 63, 64
Federal Supply Classification Listing of DoD Standardization Documents **EM3:** 63-64
Feed attachments, for rotary swaging
long workpieces **A14:** 134
Feed hopper **A7:** 4
Feed mechanisms
blank **A14:** 500
blanking/piercing **A14:** 477-478
for coil stock **A14:** 499
electromagnetic forming **A14:** 646
hopper feeding **A14:** 573
mechanical **A14:** 499-500
multiple-slide machines **A14:** 567
for power spinning **A14:** 603
press **A14:** 499
rate, rotary swaging **A14:** 139
roll **A14:** 499-500
for rotary swaging **A14:** 134
for tube spinning **A14:** 677-679
Feed metal availability
design of **A15:** 582-585
Feed rate **A18:** 609
definition **M6:** 7
Feed shoe **A7:** 4
Feed systems for
capacitor-discharge stud welding **M6:** 737
stud arc welding **M6:** 732
Feedability
defined **A18:** 9
Feedback circuit
torsion testing **A8:** 158
Feedback circuits
resistance spot welding **M6:** 469
Feedback control, resistance welds
by acoustic emission inspection **A17:** 284
Feedback-controlled servohydraulic systems **A8:** 192, 426
Feeder, defined *See also* Feeding; Riser; Risering **A15:** 5
Feeder head *See* Feeder
Feeding *See also* Riser; Riser design; Risering
aids, in riser design **A15:** 586-588
for continuous flow melting **A15:** 415
copper alloys **A15:** 778-782
defined **A15:** 5
design of **A15:** 606
ductile iron **A15:** 651
metal volume **A15:** 577-578
of plain carbon steels **A15:** 711-712
in plasma arc melting/remelting **A15:** 421
of shrinkage, die casting **A15:** 291-292
system, gray iron **A15:** 640
Feeding aids
in riser design **A15:** 586-588
Feeding pellets
for injection molding compounds **EM1:** 164
Feeds
machining **A7:** 460-461
Feedthrough
in environmental test chamber **A8:** 411
Feedwater
heaters, closed, corrosion forms **A13:** 989-990
nozzles, corrosion fatigue in **A13:** 937
Feedwater heaters
corrosion of **A11:** 615
deaerating, failures in **A11:** 657-658
Feedwater tank **A8:** 423-425
Feedwater-heater tubes
ASTM specifications for **M1:** 323
Feely test *See* Esso test
FEG *See* Field emission guns
Feldspar **EM3:** 175, 176 **EM4:** 32, 44, 379
aplite **EM4:** 379
batch size **EM4:** 382
in ceramic tile **EM4:** 926, 928
chemical composition **A6:** 60
composition **EM4:** 379
flame emission sources for **A10:** 29
flux composition **EM4:** 932
function and composition for mild steel SMAW electrode coatings **A6:** 60
functions in FCAW electrodes **A6:** 188
iron content **EM4:** 379
meltability **EM4:** 379
purpose for use in glass manufacture **EM4:** 381
sintering agent for **A10:** 166
sources in U.S **EM4:** 379
in typical ceramic body compositions **EM4:** 5
typical oxide compositions of raw matefials **EM4:** 550
in U.S. sandstone deposits **EM4:** 378
Feldspar china **EM4:** 4
Feldspathic minerals **EM4:** 6
Felicity effect
acoustic emission inspection **A17:** 284, 286
Fellgett's advantage **A10:** 112, 129
Felt
defined **EM2:** 17
Felts **EM1:** 10, 115, 130
Fence
steel wire **M1:** 271
FEP *See* Fluorinated ethylene propylene; Fluorinated perfluoroethylene-propylene
Feret's diameter
for particle size measurement **A7:** 225
Fermentation
maintenance of constant conditions in **A10:** 202
Fermi energy
defined **EL1:** 97
and superconduction **A2:** 1060
Fermi level **A18:** 447
Fermi-Dirac statistical distribution **EL1:** 98
Fermium *See also* Transplutonium actinide metals
applications and properties **A2:** 1198-1201
Ferric chloride **A16:** 69-70
chemical milling etchant **A16:** 581
photochemical machining etchant **A16:** 589, 591, 593
Ferric chloride as an etchant for copper-base powder metallurgy materials **A9:** 509
nitrided steels **A9:** 218
Ferric chloride, in graphites
Raman analysis **A10:** 133
Ferric chloride tests
crevice corrosion **A13:** 304
Ferric ion corrosion **A13:** 1142
Ferric orthophosphate
as dietary iron additive **A7:** 614
Ferric salts
as impurity **A13:** 1161
Ferric sulfate **A16:** 69-70
Ferric sulfate etching
porcelain enameling process **M5:** 514-515
Ferric-nitrate pickling
magnesium alloys **M5:** 629-631, 640-641, 647
Ferrimagnetic garnets *See* Garnets
Ferrimagnetic materials
defined **A2:** 782
ESR identification of magnetic states in **A10:** 253
types **A2:** 782
Ferrimagnetic powders
magabsorption measurement **A17:** 152
Ferris-wheel disk test
and rig **A11:** 279
Ferrite *See also* Pearlite .... **A3:** 1 • 23
400 to 500 °C embrittlement role in **M1:** 685-686
500 °F embrittlement, role in **M1:** 685
abrasion resistance **M1:** 614
acicular **M6:** 39
with aligned second phase classification of in weldments **A9:** 581
blocky, distortion from **A11:** 143
in carbon and alloy steels **A9:** 177-179
in cast austenitic stainless steels **A1:** 909, 910-911
in cast irons **A13:** 566
in cast stainless steel **A15:** 724
content in submerged arc welds **M6:** 116-118
defined **A13:** 6 **A15:** 5
determination in stainless steels **M6:** 322, 344-345
in duplex alloys **A13:** 127
effect, intergranular corrosion, austenitic stainless steels **A13:** 124-125
effect on ductility of iron-chromium-nickel heat-resistant casting alloys **A9:** 333
effect on high-temperature strength of iron-alloys **A9:** 333
chromium-nickel heat-resistant casting
embrittling effect of large grains in **M1:** 701
etching to reveal, in heat-resistant casting alloys **A9:** 331-332
excess, in iron castings **A11:** 361
grain diameter, effect in low-carbon steels **A11:** 68
grain shape, effect on formability of steel sheet **M1:** 557-559
grain-boundary **M6:** 39
gray cast iron, effects in **M1:** 12, 13, 14-15, 23-25, 30
in gray iron **A15:** 632
hydrogen embrittlement, effect on susceptibility to **M1:** 687
identification of, by magnetic etching **A9:** 333-334
induction hardening of steels, effect on **M1:** 531, 532
in intergranular corrosion, casting alloys **A13:** 582
as iron allotrope **A13:** 46, 48
machinability **M1:** 571, 576
neutron embrittlement, effect on **M1:** 686
peritectic transformation to austenite **A9:** 679-680
polygonal **M6:** 39
proeutectoid **A1:** 129-130 **A11:** 359
Schoefer diagram for estimating **A15:** 725
stabilizers .... **A3:** 1 • 25 **A13:** 47
in stainless steel casting alloys **A9:** 297-298
in stainless steel, potentiostatic etching **A9:** 146
in steel revealed by color etching **A9:** 142
strengthening mechanisms in high-strength low-alloy steels **A1:** 401, 402, 403, 404, 405, 406-408
precipitation strengthening **A1:** 403
solid solution strengthening **A1:** 400
structure in microalloyed steel **A8:** 180
symbol, for temperature at which transformation upon cooling occurs **A8:** 724
temper embrittlement, role in **M1:** 685
transformation of austenite grains to **A1:** 586
transformation temperature to austenite symbol for **A11:** 796

Ferrite (continued)
transformation to austenite, symbol for temperature at **A8:** 724
wire rod, presence in decarburized layer **M1:** 254-257
Ferrite analysis **A6:** 1059
Ferrite, as carrier core
copier powders **A7:** 584
Ferrite banding
defined **A8:** 5
Ferrite carbide aggregate, classification of
in weldments **A9:** 581
Ferrite fingers
as forging flaw **A17:** 493
Ferrite grain boundaries
etching **A9:** 169-170
Ferrite grain section sizes
calculated distribution of **A9:** 133
measured distribution of **A9:** 133
Ferrite, grain size
effect of etch time on measurement **A10:** 318
Ferrite in wrought stainless steels **A9:** 281, 283
Ferrite magnet **A7:** 4
Ferrite number
definition **M6:** 120, 322
use **M6:** 344-345
Ferrite Number (FN) **A6:** 461, 463, 464, 473, 500-501, 503, 504, 685, 686, 688, 693-696, 701, 818, 819
definition **A6:** 1209
duplex stainless steels **A6:** 475
stainless steels **A6:** 677, 678, 681
Ferrite phase, changes in
during intercritical annealing **A1:** 425-426, 427
Ferrite scope **A6:** 461
Ferrite stabilizer **A6:** 100
Ferrite steels
abrasion artifacts in **A9:** 35, 38
Ferrite substrates
physical characteristics **EL1:** 106
Ferrite with nonaligned second phase [FS(NA)# **A6:** 76
Ferrite with second phase (FS) **A6:** 76
bainite [FS(B)# **A6:** 76
ferrite side plates [FS(SP)# **A6:** 76
lower bainite [FS(LB)# **A6:** 76, 77
upper bainite [FS(UB)# **A6:** 76, 77, 79
Ferrite-carbide (FC) **A6:** 76
aggregate pearlite [FC(P)# **A6:** 76
Ferrite-pearlite microstructures
processing of **A1:** 127, 130-131
Ferrite-stabilizing elements in steel
effect on pearlite growth **A9:** 661
FERRITEPREDICTOR expert system **A6:** 1059
Ferrites *See also* Iron; Iron oxide; Magnetic materials; Magnetically soft materials; Permanent magnet materials, specific types **EM4:** 18, 1161-1165
abrasive machining **EM4:** 320, 321, 322
applications **EM4:** 47, 1161, 1163-1164
electrodes **EM4:** 1164
ferrofluids **EM4:** 1164
magnetic ink **EM4:** 1164
in magnetostrictive transducers **EM4:** 1164
memory and recording **EM4:** 1165
microwave **EM4:** 1164, 1165
permanent magnets **EM4:** 1163, 1164
power transformers and inductors **EM4:** 1163
proximity sensors **EM4:** 1164
temperature sensors **EM4:** 1164
xerography powders **EM4:** 1164
combustion **EM4:** 61
commercially spray-dried granules **EM4:** 103
compositions **EM4:** 1162-1163
crystal structure influence on magnetic properties **EM4:** 1163
electrical/electronic applications **EM4:** 1106
freeze drying **EM4:** 62
functions **EM4:** 1163-1164
future technology **EM4:** 1164

Ferrites (continued)
for high-frequency applications **A2:** 776
magnetic domains revealed by the Faraday effect **A9:** 535
mechanical properties **EM4:** 316
microstructures **A9:** 538-539
physical properties **EM4:** 316
processing **EM4:** 1163
properties **EM4:** 1161-1162, 1164
soft **EM4:** 1163
specimen preparation **A9:** 533
spray roasting **EM4:** 61
synthetic **EM4:** 1161
types of **A2:** 776
ultrasonic machining **EM4:** 359
Ferrites, hard *See* Hard ferrites
Ferrites, sintering
time and temperature **M4:** 796
Ferritic alloys *See also* Casting alloys; Ferrous casting alloys
high-alloy **A15:** 723, 731-732
Ferritic bright border
fracture resistance and **A11:** 350
Ferritic cast iron
growth at high temperature **M1:** 93
machining **A2:** 966
oxidation at high temperature **M1:** 93, 94
Ferritic cast iron, nodular
salt bath nitrided **A9:** 229
Ferritic cast steels
general corrosion **A13:** 577
intergranular corrosion **A13:** 580
Ferritic chromium stainless steel
principal ASTM specifications for weldable steel sheet **A6:** 399
Ferritic chromium steel, detection of carbide and sigma phases by phase contrast
etching **A9:** 59
Ferritic ductile irons
fracture modes **A12:** 228-237
Ferritic grades
of corrosion-resistant steel castings **A1:** 913
Ferritic iron-aluminum alloys
brittleness **A12:** 365
Ferritic low-alloy steel castings
heat treatment **A9:** 231
Ferritic malleable iron *See also* Malleable cast iron; Malleable iron **A1:** 72, 75-76
alloying elements **A1:** 73, 74, 75
brazing **A1:** 76
composition **A15:** 691
corrosion resistance **A1:** 76 **A15:** 692-693
decarburized surface **A9:** 254
elevated-temperature properties of **A1:** 76
fatigue limit **A1:** 75
fine pearlite **A9:** 254
fracture toughness **A1:** 75-76, 80
graphite content **A1:** 75
heat treatment **A1:** 75
mechanical properties **A1:** 75-76 **A15:** 692
microstructure **A1:** 72
modulus of elasticity **A1:** 75
primary cementite **A9:** 254
stress rupture plot **A15:** 693
stress-rupture plot **A1:** 75
tensile properties **A1:** 73, 75
welding **A1:** 76
Ferritic microstructures
processing of **A1:** 127, 131-133
Ferritic nitrocarburizing **A4:** 264, 425-430
applications **A4:** 425, 430 **M4:** 264
diffusion zone characteristics **A4:** 425, 428 **M4:** 267, 268-269
fatigue properties **A4:** 428, 429 **M4:** 268-269
preliminary treatments **A4:** 425 **M4:** 264-265
Ferritic nitrocarburizing, gaseous **A4:** 425-430
Alnat-N process **A4:** 429
applications **A4:** 430
atmospheres, control of **A4:** 429

Ferritic nitrocarburizing, gaseous (continued)
black nitrocarburizing **A4:** 429-430
compound layer formation **A4:** 427-428
Deganit treatment **A4:** 429
furnace **A4:** 425-426 **M4:** 265-266
industrial **A4:** 428-430
limitations **M4:** 267, 268
Nitemper process **A4:** 429
Nitroc process **A4:** 429
Nitrotec process **A4:** 429-430, 431
physical metallurgy **A4:** 426-427
process **M4:** 266-267
processes **A4:** 426-430
quasiequilibrium composition of nitrocarburizing atmospheres **A4:** 426
safety precautions **A4:** 425-426 **M4:** 266
testing **M4:** 265, 266, 267-268
Ferritic rings
ductile iron **A12:** 230
Ferritic stainless steel
mill finishes **M5:** 552
porcelain enameling of **M5:** 513
Ferritic stainless steels *See also* Arc welding of stainless steels; Ferritic steels; Stainless steel(s); Stainless steels, ferritic; Steel(s); Wrought stainless steels; Wrought stainless steels, specific types **A1:** 842, 936 **A7:** 100, 185
applications, sheet metals **A6:** 400
arc-welded **A11:** 428
base metals **A6:** 683
brazing **A6:** 913
second-phase precipitation **A6:** 622
brazing and soldering characteristics **A6:** 626
in breweries **A13:** 1222
categories of **A1:** 845
characterized **A13:** 547-549
cold cracking **A6:** 677
compositions **A13:** 357 **M6:** 526
compositions of **A1:** 843, 847-848
dynamic hot hardness vs temperature (forgeability) **A14:** 226
effects, austenite and martensite **A13:** 127
elevated-temperature properties **A1:** 936-937
engineering for use in as-welded condition **A6:** 683-686
engineering for use in postweld heat-treated condition **A6:** 686
eutectic joining **EM4:** 526
forgeability of **A1:** 893
forging of **A14:** 226
formability **A14:** 759-760
formability of **A1:** 889
forming operations, suitability **A14:** 759
fracture appearance transition temperature (FATT) **A6:** 444
friction welding **M6:** 721
high frequency welding **M6:** 760
high-purity, in pharmaceutical production facilities **A13:** 1226-1227
intergranular corrosion **A13:** 125-127, 239
leaking welds **A13:** 358
machinability of **A1:** 894
machining **A2:** 967
as magnetically soft materials **A2:** 777
metallurgy **A6:** 682-683
microstructures **A9:** 284-285
notch toughness of **A1:** 859
physical properties **M6:** 527
production of **A1:** 930
repair welding **A6:** 1106
resistance, localized corrosion **A13:** 563
resistance welding **A6:** 848 **M6:** 527
sensitization **A1:** 707
sigma phase embrittlement in **A1:** 709-711
standard, hydrochloric acid corrosion **A13:** 1162
stress-corrosion cracking in **A11:** 217-218
susceptibility to hydrogen damage **A11:** 249
tensile properties of **A1:** 856, 860
thermal expansion coefficient **A6:** 907

Ferritic stainless steels (continued)
thermal properties ... **A6:** 17
ultrahigh purity/immediate purity ... **A13:** 356
weldability of ... **A1:** 900, 901-902, 903
welding to carbon steels ... **A6:** 501
welding to low-alloy steels ... **A6:** 501
weldments ... **A13:** 355-358
Ferritic stainless steels, wrought ... **A6:** 443-454
alpha phase ... **A6:** 444
alpha-prime phase ... **A6:** 444
embrittlement ... **A6:** 444
applications ... **A6:** 443, 444, 446
chemical composition ... **A6:** 443
chi phase ... **A6:** 444
classification scheme ... **A6:** 443-444
corrosion resistance ... **A6:** 450-453
distortion ... **A6:** 445
ductile-to-brittle transition temperature ... **A6:** 444, 445, 447, 452-454
electrodes ... **A6:** 445-446
electron-beam welding ... **A6:** 448
filler metals ... **A6:** 447, 448, 453
flux-cored arc welding ... **A6:** 446
friction welding ... **A6:** 448
gas-metal arc welding ... **A6:** 446
gas-tungsten arc welding ... **A6:** 445-446, 448
heat-affected zone ... **A6:** 444-450, 452, 454
hot cracking ... **A6:** 454
hydrogen embrittlement ... **A6:** 449-450, 454
interstitial element content and corrosion ... **A6:** 443, 450-452
laser-beam welding ... **A6:** 448
metallurgical characteristics ... **A6:** 444-445
plasma arc welding ... **A6:** 448
postweld heat treatments ... **A6:** 450, 451
preheating ... **A6:** 450
properties ... **A6:** 443
resistance welding ... **A6:** 448
selection of welding consumables ... **A6:** 448-449
shielded-metal arc welding ... **A6:** 446-447, 450, 451
shielding gases ... **A6:** 445-446, 449, 452, 453
sigma phase ... **A6:** 444
submerged arc welding ... **A6:** 448
weld properties ... **A6:** 449-454
weld toughness ... **A6:** 453-454
weldability ... **A6:** 445
welding procedures ... **A6:** 449-454
welding processes ... **A6:** 445-448
Ferritic steel
brittle fracture in ... **A8:** 262
creep rate ... **A8:** 331
critical strain rate for SCC in ... **A8:** 519
environments for stress-corrosion cracking ... **A8:** 527
low-alloy, stress-strain curve ... **A8:** 177
plane-strain fracture toughness ... **A8:** 479
transgranular cleavage fracture in ... **A8:** 465
upward inflection ... **A8:** 332
Ferritic steels *See also* Ferritic stainless steels; Steel(s)
acoustic emission inspection ... **A17:** 287
dissimilar-metal welds with ... **A11:** 620
ductile-to-brittle fracture transition in ... **A11:** 66
elevated-temperature properties of ... **A1:** 617-652
embrittled by zinc ... **A11:** 237-238
embrittlement by zinc ... **A13:** 184
formability of ... **A1:** 889
heat-resistant high-chromium ... **A1:** 937
boiler tubing ... **A1:** 937, 938-939
historical background ... **A1:** 937-938
turbine rotors ... **A1:** 937, 938
without vanadium ... **A1:** 931, 936-938
return bend, rupture by SCC and inclusions ... **A11:** 646
thermal expansion and conductivity of ... **A1:** 647, 651, 652
thermal expansion coefficients ... **A11:** 620
void swelling in ... **A1:** 656-657
erritic weldments
microstructure constituent classifications for the fusion zone ... **A9:** 581
erritic-austenitic (duplex) phase
stainless steels ... **A13:** 47
erritizing anneal
defined ... **A9:** 7
erro-niobium
grain refiner in steels ... **A6:** 53
Ferro-titanium
grain refiner in steels ... **A6:** 53
Ferroalloy
defined ... **A15:** 5
Ferroalloy/deoxidizer additions
in steelmaking ... **A1:** 111-112
Ferroalloys *See also* Iron alloys
chemical analysis and sampling ... **A7:** 248
sodium peroxide fusion ... **A10:** 167
Ferrochrome alloying
in nickel alloys ... **A2:** 429
Ferrochromium slags
partitioning oxidation states in ... **A10:** 178
Ferroelectric ceramics ... **EM4:** 542
Ferroelectric thin films
electrical/electronic applications ... **EM4:** 1105
Ferrofluid
used in Bitter technique ... **A9:** 534
used in magnetic etching ... **A9:** 64
Ferrograph
defined ... **A8:** 5-6 **A18:** 9
Ferrography *See also* Lubricant analysis ... **A18:** 375
corrosive wear case history ... **A18:** 310
lubricant analysis ... **A18:** 301-302
to confirm the type of wear debris ... **A18:** 310
Ferromagnetic alloys *See also* Ferrous casting alloys
magnetic particle inspection of ... **A15:** 264
Ferromagnetic antiresonance
abbreviation for ... **A10:** 690
spectrometers for ... **A10:** 271, 272
Ferromagnetic constituents of wrought stainless steels
magnetic etching ... **A9:** 282
Ferromagnetic crystals
effect on electron beams in scanning electron microscopy ... **A9:** 94
Ferromagnetic cylinders
central conductors within ... **A17:** 96
Ferromagnetic materials *See also* Nonferromagnetic materials
applied stress measurement ... **A17:** 154-155
Curie point ... **A17:** 89
defined ... **A2:** 782 **A9:** 63
eddy current inspection ... **A17:** 164
electromagnetic techniques for ... **A17:** 159-163
ESR identification of magnetic states in ... **A10:** 253
hysteresis curve ... **A17:** 99
leakage field testing ... **A17:** 129-135
magabsorption measured ... **A17:** 145, 152, 154-155
magnetic particle inspection ... **A17:** 89
magnetic printing ... **A17:** 125-126
order-disorder in ... **A10:** 284-285
properties, magnetically soft materials ... **A2:** 761-763
relaxation parameters for ... **A10:** 276
remote-field eddy current inspection ... **A17:** 195
SEM-observed grain boundaries in ... **A10:** 490
types ... **A2:** 782
Ferromagnetic nuclear resonance ... **A10:** 281
Ferromagnetic particles
in magnetic particle inspection ... **A7:** 575
Ferromagnetic parts
flaw detection by magnetic particle inspection ... **A7:** 575-579
Ferromagnetic phases
revealed by magnetic etching ... **A9:** 63
Ferromagnetic powders
magabsorption measurement ... **A17:** 152
Ferromagnetic resonance ... **A10:** 267-276
applications ... **A10:** 267, 271-276
defined ... **A10:** 673
estimated analysis time ... **A10:** 267
general uses ... **A10:** 267
homogeneity and inhomogeneity ... **A10:** 274
introduction and theory ... **A10:** 267-270
limitations ... **A10:** 267
low-temperature, probe for ... **A10:** 270
microwave spectrometers ... **A10:** 270-271
related techniques ... **A10:** 267
resonance parameters of resonance field and linewidth ... **A10:** 267-268
samples ... **A10:** 267, 268, 271-276
Ferromagnetic resonance eddy current probes
microwave inspection ... **A17:** 220-223
Ferromagnetism ... **A9:** 533-534
defined ... **A2:** 761 **A10:** 673 **A17:** 90
Ferromagnetism (continued)
onset of ... **A10:** 257
zone theory of ... **EL1:** 98
Ferromagnets ... **A10:** 253, 690
Ferromanganese *See also* Manganese
in chilled iron ... **A15:** 30
function and composition for mild steel SMAW electrode coatings ... **A6:** 60
Ferromanganese isolation
in manganese ... **A10:** 174
Ferrosilicon
in alloy cast irons ... **A1:** 90
as cast iron inoculant ... **A15:** 105
function and composition for mild steel SMAW electrode coatings ... **A6:** 60
inoculants, for gray cast iron ... **A15:** 637
use in cast iron ... **M1:** 22, 80
Ferrosilicon-based inoculant ... **A1:** 39, 90
Ferrosilicon/ferroboron, analysis for aluminum
by aluminon method ... **A10:** 68
Ferrospinels *See* Ferrites
solid-state phase transformations in welded joints ... **A9:** 580-581
solidification structures in welded joints ... **A9:** 579
Ferrous
defined ... **A15:** 5
Ferrous alloys
composition effect on corrosion ... **A13:** 537
dip brazing ... **A6:** 336
fatigue striations in ... **A12:** 176
fracture/failure causes illustrated ... **A12:** 217
grain-boundary segregation ... **A13:** 156
hydrogen damage ... **A13:** 166-169
inorganic fluxes used ... **A6:** 130
laser-beam welding ... **A6:** 263
oxyfuel gas welding ... **A6:** 281
zinc embrittlement ... **A13:** 184
Ferrous and nonferrous alloys
molten-salt dip brazing ... **A6:** 338
Ferrous casting alloys *See also* Ferrous castings
cast Alnico alloys ... **A15:** 736-739
classification of ... **A15:** 627-628
compacted graphite irons ... **A15:** 667-677
in cupolas ... **A15:** 388
ductile iron ... **A15:** 647-666
gray iron ... **A15:** 629-646
high-alloy graphitic irons ... **A15:** 698-701
high-alloy steels ... **A15:** 722-735
high-alloy white irons ... **A15:** 678-685
inclusions in ... **A15:** 91-95
low-alloy steels ... **A15:** 715-721
malleable iron ... **A15:** 686-697
plain carbon steels ... **A15:** 702-714
semisolid casting/forging of ... **A15:** 327
Western bentonite molding clays with ... **A15:** 210
Ferrous castings *See also* Ferrous casting alloys
inspection of ... **A15:** 555-556
production volume ... **A15:** 41-42
shipment tonnages ... **A15:** 41-42
Ferrous ion
for titration of oxidants ... **A10:** 205
Ferrous materials
acoustic properties ... **A17:** 235
casting, inspection of ... **A17:** 531-532
Ferrous metals
flakes in ... **A11:** 121
liquid-metal embrittlement in ... **A11:** 234
threading with circular chasers ... **A16:** 298
Ferrous metals and alloys
brazing to aluminum ... **M6:** 1030
oxyfuel gas welding
Ferrous P/M materials *See also* Iron powders ... **A7:** 682-687
alloys, annealing ... **A7:** 182-185
automotive applications ... **A7:** 617-621
class, grade, type ... **A7:** 463
density designations and ranges ... **A7:** 464
effects of steam treating on apparent hardness ... **A7:** 466
effects of steam treating on density ... **A7:** 466
infiltration systems ... **A7:** 558-559
lubrication of ... **A7:** 190-191
mechanical properties ... **A7:** 463-466
MPIF designations ... **A7:** 463
ordnance applications ... **A7:** 682
P/M structural, compositions ... **A7:** 464

Ferrous P/M materials (continued)
roll compaction **A7:** 408-409
sintering **A7:** 360-368
Ferrous phosphates
as conversion coating **A13:** 387
Ferrous powder metallurgy materials **A1:** 801-821
applications of powder forged parts **A1:** 817-818
designation of P/M materials **A1:** 804, 805
forging mode **A1:** 814
hardenability **A1:** 814
heat treatments **A1:** 814
tensile, impact, and fatigue properties **A1:** 814-815, 816
heat treatment of **A1:** 809, 810-811
infiltration **A1:** 807-808, 810
material considerations **A1:** 812
hardenability **A1:** 812-813
inclusion assessment **A1:** 813
mechanical and physical properties of **A1:** 806
composition **A1:** 809-810
porosity **A1:** 802, 805, 806, 807-809
sintered ferrous P/M materials **A1:** 805
mechanical properties **A1:** 814
compressive yield strength **A1:** 815, 817
effect of porosity on mechanical properties **A1:** 815, 817
metal injection molding (MIM)
advantages of **A1:** 819
applications **A1:** 820
factors impeding growth **A1:** 819-820
mechanical properties **A1:** 820
technology **A1:** 818-819
metal powder characteristics and
apparent density **A1:** 802
chemical composition **A1:** 803
compressibility **A1:** 802
compression flow time **A1:** 802
control **A1:** 801, 802
green strength **A1:** 802, 803
oxide content **A1:** 802
particle size distribution **A1:** 802
sampling **A1:** 801-802
sintering characteristics **A1:** 802-803
powder compacting **A1:** 802, 803
powder forging **A1:** 812
powder preparation **A1:** 803
process capabilities **A1:** 801
process considerations **A1:** 813
metal flow in powder forging **A1:** 814
powder forging **A1:** 814
preforming **A1:** 813
secondary operations **A1:** 814
sintering and reheating **A1:** 814
tool design **A1:** 814
quality assurance for P/M parts **A1:** 815-816
density **A1:** 817, 819
magnetic particle inspection **A1:** 816
metallographic analysis **A1:** 817
nondestructive testing **A1:** 817
part dimensions and surface finish **A1:** 816, 818
re-pressing **A1:** 811, 812
secondary operations **A1:** 805
sintering **A1:** 803-804
equipment **A1:** 804-805
techniques **A1:** 804
Ferrous powder metallurgy parts *See* Powder metallurgy parts, ferrous
Ferrous sulfate
pickling process
inhibitory effects of **M5:** 70, 73-74, 76-77
recovery of **M5:** 81-82
role in **M5:** 70, 73-74, 76-77, 81-82
reducing method, chromic acid wastes **M5:** 187
Ferroxplana **EM4:** 1162
Ferroxy 1 test
detection of embedded iron in nickel alloys **M5:** 672-673
Ferrules **M6:** 729
expandable **M6:** 732
Ferrules (continued)
functions during weld cycle **M6:** 732
semipermanent **M6:** 732
use in stud arc welding **M6:** 729
Ferry alloy
machining **A16:** 836-842
Fertilizer industry
pollution control **A13:** 1370
Fertilizers, corrosion resistance
stainless steels **M3:** 83
Fertilizers, liquid
corrosion caused by **M1:** 714, 715
Festooning, as take-up
SMC machines **EM1:** 160
FET *See* Field effect transistor
FFA *See* Fiberglass Fabrication Association
FFT *See* Fast Fourier transform
FIA *See* Flow injection analysis
Fiber **EM3:** 12
defined **A9:** 7
drilling **A16:** 229
Fiber alignment *See also* Alignment techniques; Fiber alignment; Fiber orientation; Fiber(s); Orientation
damping effects **EM1:** 207-208
in short fiber reinforcement **EM1:** 119-120
Fiber aspect ratios **EM3:** 397, 398
Fiber break
in composites **A8:** 714
and fatigue failure **EM1:** 438
propagation failure, analysis **EM1:** 195
Fiber breakout
from fastener hole generation **EM1:** 712-715
Fiber bridging
defined **EM2:** 17
Fiber bundle *See also* Fiber(s)
defined **EM1:** 29
dimensions, unidirectional tape prepreg **EM1:** 143
properties analysis **EM1:** 731
strength **EM1:** 194
Fiber bundling
from excessive tack **EM1:** 144
Fiber composites **A9:** 587-597
cleaning **A9:** 588
etching **A9:** 591
fabrication methods **A9:** 591
grinding **A9:** 588-589
macroexamination **A9:** 591
microexamination **A9:** 592
microstructures **A9:** 592
mounting **A9:** 587-588
on foil as a fabrication process for metal- matrix composites **A9:** 591
polishing **A9:** 589-591
sectioning **A9:** 587
specimen preparation **A9:** 587-591
Fiber composites, laminated
ultrasonic machining **A16:** 532
Fiber content **EM3:** 12
Fiber count **EM3:** 12
defined **EM2:** 18
Fiber diameter **EM3:** 12
boron fibers **EM1:** 58
carbon fibers **EM1:** 51
of conventional plastics reinforcement **EM1:** 108
defined **EM1:** 11 **EM2:** 18
in failure analysis **EM1:** 197
glass, determining **EM1:** 108
glass, nomenclature **EM1:** 109
measurement **EM1:** 286
as testing problem **EM1:** 732
Fiber direction *See also* Fiber orientation; Fiber(s), Fiber alignment; Orientation **EM3:** 12
and composite stiffness **EM1:** 179
defined **EM1:** 10 **EM2:** 18
Fiber finish *See also* Finish; Finishes; Finishing
for fabrics **EM1:** 125-126
Fiber grease
defined **A18:** 9
Fiber intermingling method
melt impregnation **EM1:** 102
Fiber length
in compounding **EM1:** 164
critical **EM1:** 119
effect, bulk molding compounds **EM1:** 162
for milled fibers
short fiber reinforced composites **EM1:** 119
Fiber metallurgy **A7:** 4
Fiber optic light guide *See also* Light guide bundle
borescope **A17:** 4
Fiber optic light sources
bifurcated **A12:** 79
for lens flare and ghost images **A12:** 84-87
for visual examination **A12:** 78
Fiber optic tubes
bifurcated **A8:** 277
Fiber optics
diffraction pattern inspection technique **A17:** 13
state-of-the-art **EM4:** 22
ultrasonic machining **EM4:** 359
in visual leak testing **A17:** 66
Fiber orientation *See also* Fiber(s); Orientation
in composites **A11:** 735-738
and direction of applied load, composites **A11:** 742
effect, composite modulus **EM1:** 120
effect, interlaminar/intralaminar fracture **EM1:** 787-790
effect on flexural fracture in graphite-epoxy lay-ups **A11:** 733
effect, specific damping capacity **EM1:** 215
effect, tensile strength **EM1:** 120
quality control of **EM1:** 730, 742-743
thermoplastic injection molding **EM2:** 311
vs. mechanical properties, tape prepregs **EM1:** 144
in woven fabric prepregs **EM1:** 149
Fiber pattern **EM3:** 12
defined **EM2:** 18
Fiber preforms *See also* Preforms
inspection **EM1:** 532
and resin injection **EM1:** 529-532
Fiber properties analysis *See also* Fiber(s); Material properties; Material properties analysis; Testing
of impregnated strands **EM1:** 731, 732
mechanical test methods **EM1:** 731-733
modulus measurement **EM1:** 731-733
of multifilament yarns **EM1:** 732-733
properties tested **EM1:** 731
of single fibers/monofilaments **EM1:** 731-732
statistical aspects **EM1:** 733-734
Fiber properties of fiber composite specimens
altering for purposes of examination **A9:** 587
Fiber pull-out and microdrop technique **EM3:** 391-394
debonding force of the microdrop **EM3:** 393-394
routes to failure **EM3:** 393
Fiber pull-out method **EM3:** 391
Fiber reinforced composite **A7:** 5
Fiber reinforcement *See also* Fiber(s); Multidirectional fiber reinforcement **EM1:** 175
effect, epoxy resins **EM1:** 75-76
effect, polyester composites **EM1:** 91, 92
and epoxy resin, compared **EM1:** 76
glass, for epoxy resin matrices **EM1:** 73
multidirectionally continuous **EM1:** 933-934
pultrusion **EM1:** 537
reinforcement direction **EM1:** 146
for resin transfer molding **EM1:** 169-171
Fiber reinforcements *See also* Fiber(s); Reinforcement(s) **EM4:** 19, 20
aramid **EM2:** 506
boron **EM2:** 505
control, process selection by **EM2:** 277

SUBJECTS OF THE INDEXED VOLUMES: **ASM Handbook** (designated by the letter "A"): **A1:** Properties and Selection: Irons, Steels, and High-Performance Alloys (1990); **A2:** Properties and Selection: Nonferrous Alloys and Special-Purpose Materials (1990); **A3:** Alloy Phase Diagrams (1992); **A4:** Heat Treating (1991); **A6:** Welding, Brazing, and Soldering (1993); **A7:** Powder Metallurgy (1984); **A8:** Mechanical Testing (1985); **A9:** Metallography and Microstructures (1985); **A10:** Materials Characterization (1986); **A11:** Failure Analysis and Prevention (1986); **A12:** Fractography (1987); **A13:** Corrosion (1987); **A14:** Forming and Forging (1988); **A15:** Casting (1988); **A16:** Machining (1989); **A17:** Nondestructive Testing and Quality Control (1989); **A18:** Friction, Lubrication, and Wear Technology (1992). **Metals Handbook, 9th Edition** (designated by the letter "M"): **M1:** Properties and Selection: Irons and Steels (1978); **M2:** Properties and Selection: Nonferrous Alloys and Pure Metals (1979); **M3:** Properties and Selection: Stainless Steels, Tool Materials and Special-Purpose Materials (1980); **M4:** Heat Treating (1981); **M5:** Surface Cleaning, Finishing, and Coating (1982); **M6:** Welding, Brazing, and Soldering (1983). **Engineered Materials Handbook** (designated by the letters "EM"): **EM1:** Composites (1987); **EM2:** Engineering Plastics (1988); **EM3:** Adhesives and Sealants (1990); **EM4:** Ceramics and Glasses (1991); **Electronic Materials Handbook** (designated by the letters "EL"): **EL1:** Packaging (1989).

Fiber reinforcements (continued)
effect on properties ... **EM2:** 281
fiberglass ... **EM2:** 504-505
graphite ... **EM2:** 505-506
low-density high-strength fibers ... **EM2:** 505-506
Fiber sheet packings ... **EM3:** 49
Fiber show ... **EM3:** 12
defined ... **EM2:** 18
Fiber sizing *See also* Sizing ... **EM1:** 122-124
chemical composition ... **EM1:** 122-123
functions/types, sizing agents ... **EM1:** 122
theory ... **EM1:** 123-124
Fiber Society (FS)
as information source ... **EM1:** 41
Fiber strength *See* Fiber(s); Strength
Fiber stress ... **EM3:** 392-394
Fiber tearing in fiber composites
coating to prevent ... **A9:** 587
mounting to prevent ... **A9:** 588
Fiber texture
in crystalline lattices ... **A9:** 701
defined ... **A9:** 7
Fiber tow *See also* Tow
characteristics, for towpreg ... **EM1:** 151
Fiber volume *See also* Fiber volume fraction
for closed-loop cure ... **EM1:** 761
Fiber volume fraction
defined ... **EM1:** 207
effect on thermal expansion, carbon-epoxy composite ... **EM1:** 190
filament winding vs. braiding ... **EM1:** 519
flexural damping variation with ... **EM1:** 208
three-dimensional braided composite ... **EM1:** 524-525
vs. Young's modulus ... **EM1:** 208
Fiber wash ... **EM3:** 13
defined ... **EM2:** 18
Fiber wet-out ... **EM1:** 132, 655
Fiber(s) *See also* Bundle; Carbon fiber; Continuous fibers; Discontinuous fibers; Fiber alignment; Fiber break; Fiber bundle; Fiber diameter; Fiber direction; Fiber length; Fiber orientation; Fiber preforms; Fiber properties analysis; Fiber reinforcement; Fiber reinforcements; Fiber volume fraction; Fiber-epoxy composites; Fiber-reinforced composites; Fiber-reinforced plastic- (FRP); Fiber-reinforced plastics (FRPs); Fiber-resin composites Fiber sizing; Filaments; Glass fibers; Glass(es); Graphite; Graphite fibers; Orientation; Reinforcements; specific fibers; Staple fibers; Strand; Textile fibers; Tow
alignment techniques for recovered ... **EM1:** 153-155
aluminum oxide ... **EM1:** 31
aramid ... **EM1:** 30
aramid, properties ... **EL1:** 615
board, as contaminants ... **EL1:** 661
boron ... **EM1:** 5, 31, 58-59
break propagation failure ... **EM1:** 195
breakage ... **EM1:** 120, 193-195, 231-232
carbon ... **EM1:** 6, 9-30, 49-53
ceramic ... **EM1:** 60-65
as chain of links ... **EM1:** 194
chemistry ... **EM2:** 64
coefficient of variation, fiber bundle strength variation with ... **EM1:** 194
for compression molding ... **EM1:** 559
constituents, software for ... **EM1:** 275
content ... **EM2:** 17, 338
content, defined ... **EM1:** 11
content, metallic composites, eddy current inspection ... **A17:** 190-191
continuous silicon carbide (SiC) ... **EM1:** 31
cost guidelines ... **EM1:** 105
count, defined ... **EM1:** 11
defined ... **EM1:** 10-11, 29 **EM2:** 17
discontinuous, in metal matrix composites ... **EM1:** 889-910
drawn, defined *See* Drawn fiber
effect, on properties ... **EM2:** 281
efficiency, and critical length ... **EM1:** 119
elastic displacement fields ... **EM1:** 187
in engineering thermoplastics ... **EM2:** 98-99
environmental effects ... **EM2:** 430-431
for filament winding ... **EM1:** 504 **EM2:** 369
fine, glass ... **EM1:** 108
in flexible fiberscopes ... **A17:** 5
forms and properties ... **EM1:** 360-362
Fiber(s) (continued)
FP, defined *See* FP fiber
glass ... **EL1:** 114 **EM1:** 12, 29, 45-48, 109
graphite ... **EM1:** 12, 49-53
high-modulus graphite ... **EM2:** 758-759
high-silica ... **EM1:** 29
introduction ... **EM1:** 29-31
lamina properties ... **EM1:** 308
local, failure mechanisms ... **EM1:** 198
machining of ... **EM1:** 667-668
mechanical properties ... **EM2:** 506
metallic, defined *See* Metallic fiber
milled, defined *See* Milled fiber
natural ... **EM1:** 117
organic ... **EM1:** 29-31, 54-57
orientation, by computed tomography (CT) ... **A17:** 363
orientation, thermoplastic injection molding ... **EM2:** 311
pattern, defined ... **EM1:** 11
placement process ... **EM1:** 33
polyphenylene sulfides (PPS) ... **EM2:** 189-190
properties ... **EM1:** 175, 731-735
properties analysis, as quality control ... **EM1:** 731-735
quartz ... **EM1:** 29
as reinforcement ... **EL1:** 1119-1121
reinforcement, types ... **EM2:** 504-506
reinforcing, comparative properties ... **EM1:** 58
reinforcing, for base/insulators ... **EL1:** 113
reinforcing, properties ... **EL1:** 615
reprocessed ... **EM1:** 153
-resin pullout, by machining/drilling ... **EM1:** 667-668
show, defined ... **EM1:** 11
silicon carbide ... **EM1:** 31, 58-59
silicon carbide, acoustic emission inspection ... **A17:** 288
single, mechanical testing of ... **EM1:** 731-733
single. strain measurement ... **EM1:** 732
size ... **EM1:** 170, 732
in SMC-R ... **EM2:** 324-325
in space and missile composite structures ... **EM1:** 817
strength, defined ... **EM1:** 193
stress-strain relationships ... **EM1:** 175
synthetic ... **EM1:** 117-118
technology, critical properties ... **EM2:** 458
Teflon-based ... **EL1:** 114
tensile failure mode ... **EM1:** 200
tensile strength, analysis ... **EM1:** 193-194
tests for ... **EM1:** 285-288
thickness, for filament winding ... **EM1:** 508
types ... **EM1:** 43, 125, 161-162, 730
volume fraction of ... **EM2:** 506
-wound composites, microwave inspection ... **A17:** 202
woven, forms ... **EM1:** 148
Fiber-epoxy composites, and metals
properties compared ... **EM1:** 178
Fiber-epoxy prepreg tape, vs. fabric
tensile strength ... **EM1:** 143-144
Fiber-matrix interface of fiber composites
examination by electron microscopy ... **A9:** 592
Fiber-reinforced aluminum
as heat sink material ... **EL1:** 130-1131
Fiber-reinforced ceramic matrix composites
aerospace applications ... **EM4:** 1004
Fiber-reinforced composite tubing ... **EM1:** 569-574
Fiber-reinforced composites *See also* Composite materials; Composites; Continuous reinforced plastic; Fiber reinforcement; fiber resin composites; Reinforcements ... **EM4:** 1
aircraft applications ... **EM1:** 801-809
aluminum metal matrix ... **A2:** 7
and cermets, compared ... **A2:** 978
defined ... **EM1:** 27
discontinuous or short fiber, defined ... **EM1:** 27
failure modes ... **EM1:** 788
marine applications ... **EM1:** 837-844
packaging ... **EL1:** 1122-1125
quasi-isotropic, properties ... **EL1:** 1123
refractory metal ... **A2:** 582-584
specific tensile strength and specific tensile modulus ... **EM1:** 28
Fiber-reinforced materials
high-temperature solid-state welding ... **A6:** 299
Fiber-reinforced plastic (FRP) ... **EM3:** 12
chemical resistance of composites ... **EM3:** 639
damping features ... **EM1:** 35
defined ... **EM1:** 11
finite-element analysis ... **EM3:** 488, 497
market ... **EM1:** 43
Fiber-reinforced plastics (FRPs) *See also* Plastics
compression molded/stamped ... **EM2:** 326-333
defined ... **EM2:** 18
molding techniques ... **EM2:** 338-343
pultrusions, properties ... **EM2:** 396
Fiber-reinforced polymers *See also* Fiber-reinforced composites; Polymer(s)
strength and stiffness ... **EM1:** 36
Fiber-reinforced sheets ... **EM3:** 41
Fiber-reinforced superalloys (FRS) ... **EM1:** 878
Fiber-reinforced thermoset molding compounds ... **EM1:** 161-163
Fiber-reinforced thermosetting composites
process molding of ... **EM1:** 500-501
Fiber-resin composites *See also* Composite materials; Composites; Fiber(s); Fiber-reinforced composites; Resins
applications ... **EM1:** 355, 356
electrical properties ... **EM1:** 359
forms and constituents ... **EM1:** 357
properties ... **EM1:** 356
test axes, definitions ... **EM1:** 357, 359
Fiberboard
milling with PCD tooling ... **A16:** 110
Fiberfrax
properties ... **A18:** 803
Fiberglass ... **EM3:** 12, 590, 592 **EM4:** 402-407
applications ... **EM1:** 27, 28 **EM4:** 379, 402-403
borate matefials used extensively ... **EM4:** 380
compound composition ... **EM4:** 566
defined ... **EM1:** 11 **EM2:** 18
definition ... **EM4:** 402
and electron beam machining ... **A16:** 571
fasteners for ... **A11:** 530
fiber forming ... **EM4:** 406-407
chopped fiber ... **EM4:** 407
textile fiber ... **EM4:** 407
wool process ... **EM4:** 406-407
fibers, in composites ... **A11:** 731, 732
-filled styrene-maleic anhydrides (S/MA) ... **EM2:** 220
furnace types ... **EM4:** 403-405, 406
direct-fired ... **EM4:** 405, 406
electric ... **EM4:** 404, 406
mixed melters ... **EM4:** 405, 406
recuperative ... **EM4:** 404-405, 406
regenerative ... **EM4:** 403-404, 406
furnaces ... **EM1:** 108
glass compositions ... **EM4:** 402-403
raw materials ... **EM4:** 403
glass preparation process ... **EM4:** 405-407
grinding by superabrasives ... **A16:** 434
historical review ... **EM4:** 402
honeycomb core material ... **EM1:** 723
iron content ... **EM4:** 378
laminates, cyanates ... **EM2:** 235
manufacturers in the U.S ... **EM4:** 406
mats, and woven roving ... **EM1:** 109
molds, rotational molding ... **EM2:** 367
paper, production ... **EM1:** 110
printed wiring board laminates properties ... **EM2:** 237
properties ... **EM4:** 566
refractive index ... **EM4:** 566
as reinforced polypropylenes (PP) ... **EM2:** 192
-reinforced pultruded products ... **EM1:** 540
as reinforcement ... **EM2:** 504-505
reinforcement, defined ... **EM1:** 11 **EM2:** 18
roving, production process ... **EM1:** 109
tapping ... **A16:** 259
weave pattern effect ... **EM1:** 150
yarns ... **EM1:** 110-111
Fiberglass Fabrication Association (FFA)
as information source ... **EM1:** 41
Fiberglass fibers
as thermocouple wire insulation ... **A2:** 882
Fiberglass in polyester composites ... **A9:** 592
Fiberglass paper ... **EM1:** 110
Fiberglass reinforced BPADCY prepolymers ... **EM2:** 232-233
Fiberglass reinforcement ... **EM3:** 12

Fiberglass storage tanks
acoustic emission inspection **A17:** 291
Fiberglass yarns
carded fibers **EM1:** 111
fabric **EM1:** 110-111
nomenclature **EM1:** 110
texturized **EM1:** 111
Fiberglass-epoxy
drilling **A16:** 227, 230
Fiberglass-epoxy composites
fabric, microstructure of **EM1:** 768
laminated, by electroformed nickel tooling **EM1:** 582-585
resin/curing agent systems **EM1:** 134
Fiberglass-epoxy printed circuit boards **EL1:** 822-824
Fiberglass-polyester resin composites
corrosion resistance **EM2:** 249
electrical properties **EM2:** 249
glass content effects **EM2:** 248
mechanical properties **EM2:** 247
Fiberglass-reinforced plastics
Epon resin/curing agent systems **EM1:** 134
history **EM1:** 27
Fiberglass-reinforced thermoset unsaturated polyesters **EM2:** 246-247
Fiberglass-resin repair procedures **EM3:** 807
Fiberglass-Teflon printed board substrates
properties **A6:** 992
Fibering *See* Directionality; Mechanical fibering; Mechanical texture
Fibering, mechanical
in forging **A11:** 319-320
Fibermax **EM1:** 63
FiberMetal
abradable seal material **A18:** 589
Fibers *See also* Fibrous fracture surface; Fibrous fractures; Tungsten fibers
α and β in rolled copper **A10:** 363
brittle-tensile fracture of **A11:** 738
buckling of **A11:** 739
continuous/discontinuous, in cast metal-matrix composites **A15:** 840
defined **A11:** 4
EFG composition analysis **A10:** 212
in Euler angles of rolled fcc materials **A10:** 363
fracture, transverse **A12:** 466
fractured, effect on fracture topography **A11:** 736
graphite **A12:** 465
lines, ODF along, as function of rolling reduction **A10:** 364
orientation of **A11:** 731, 733, 735-738, 742
pull-out **A11:** 761
reinforced composite, defined **A11:** 4
reinforcement of **A11:** 736
resin-matrix composites **A12:** 474-478
resin-matrix composites, fracture sequence **A12:** 474
in rolled fcc materials **A10:** 363
splitting, as fracture mode **A12:** 467
stresses **A11:** 4, 137
textures, in wire and rod **A10:** 359
tungsten **A12:** 466
wettability **A15:** 840-842
Fibers, aramid
laser cutting of **A14:** 742
Fiberscopes *See also* Borescopes
flexible **A17:** 5
measuring **A17:** 8
Fibertrack **EM3:** 397
Fibril rupture **EM3:** 507
Fibrillar structure
Kevlar aramid fibers **EM1:** 55
Fibrillation **EM3:** 13
defined **EM2:** 18
in polymers **A12:** 479, 480
Fibrils
as toughener **EM3:** 185
Fibrosis of lung tissue
titanium toxicity-caused **A7:** 206
Fibrous eutectic structures
as a feature of regular eutectics **A9:** 620
branching in **A9:** 620
conditions for formation of **A9:** 618
molybdenum in nickel matrix **A9:** 619
Fibrous eutectics
solidification of **A15:** 119-120
Fibrous fracture
defined **A8:** 5 **A9:** 7 **A11:** 4
Fibrous fracture surfaces *See also* Fracture surface(s)
AISI/SAE alloy steels **A12:** 311, 313-316
ductile, metal-matrix composites **A12:** 466
maraging steels **A12:** 383
precipitation-hardening stainless steels **A12:** 370
titanium alloys **A12:** 451
tool steels **A12:** 376-377
wrought aluminum alloys **A12:** 425
Fibrous fractures **A12:** 2, 313-314
Fibrous particle shapes
defined **A9:** 619
Fibrous structure **A7:** 5
defined **A8:** 5 **A9:** 7 **A11:** 4
Fibrous tearing
ductile fracture **A8:** 572
Fibrous wools
discontinuous oxide fibers as **EM1:** 62
Fick's first law **A15:** 53
Fick's Laws **A13:** 33, 68
Fickian diffusion **EM3:** 655
model **EM3:** 622
Field (bulk) testing methods
defined **EM1:** 774
Field adsorption **A10:** 588
Field compensator
in radiographic inspection **A17:** 308
Field corrosion
FIM electric field as **A10:** 587
Field crystallization failures
tantalum capacitors **EL1:** 998
Field diaphragm in optical microscope **A9:** 72
Field effect transistor
abbreviation for **A10:** 690
effect in x-ray detectors **A10:** 91
in energy-dispersive spectrometry **A10:** 519
preamplifier, EPMA analysis **A10:** 519
Field effect transistor (FET)
analysis **EL1:** 154-159
failure mechanisms **EL1:** 975
junction, double-diffused **EL1:** 146-147
junction FET, device structure **EL1:** 154-156
types, active-component fabrication technology **EL1:** 200-201
Field emission **A6:** 30, 31
Field emission guns **A10:** 432, 690
Field evaporation
in FIM/AP **A10:** 584, 585-587
in FIM/PLAP analysis of semiconductor **A10:** 601
importance **A10:** 586-587
potential energy diagram **A10:** 587
progressive, in the atom probe **A10:** 591
rate variation with temperature **A10:** 594
sequences, atom layers counted in **A10:** 590
Field experience
and corrosion testing **A13:** 194
Field flattening
digital image enhancement **A17:** 460
Field fractures
debris from **A12:** 92
Field girth welds
in pipe **A11:** 699
Field ion micrograph
stereographic projection of **A10:** 585
Field ion microscope
features **A10:** 584
improved vacuum of **A10:** 587
Field ion microscope (continued)
magnification, resolution, and image contrast **A10:** 588
modified, atom probe as **A10:** 591
working field **A10:** 587-588
Field ion microscopy **A10:** 583-602
and atom probe microanalysis **A10:** 583-602
defined **A10:** 673
disadvantage **A10:** 585
electric field and stress **A10:** 587
field evaporation **A10:** 585-587
field ionization **A10:** 585
image, absolute depth scale from **A10:** 593
image formation **A10:** 584
images of alloys and semiconductors **A10:** 589-590
images of defects in pure metals **A10:** 588-589
of IN 939 nickel-base superalloy **A10:** 598
magnification, resolution, and image contrast **A10:** 588
principles **A10:** 584
quantitative analysis of images **A10:** 590
sample preparation **A10:** 584-585
sample tip, ball model of **A10:** 585
stable operating conditions for **A10:** 587
of ternary 3:5 semiconductor **A10:** 481
working range **A10:** 587-588
Field ion microscopy (FIM) **EM3:** 237
Field ion microscopy and atom probe microanalysis **A10:** 583-602
applications **A10:** 583, 598-602
estimated analysis time **A10:** 583
general uses **A10:** 583
limitations **A10:** 583
related techniques **A10:** 583
samples **A10:** 583, 584-585, 598-602
Field ionization
critical distance **A10:** 585
defined **A10:** 584, 673
in field ion microscopy **A10:** 585
potential-energy diagram illustrating **A10:** 586
Field measurement, and magnification
effect on inclusion volume fraction **A10:** 314
Field measurements made with image analyzers **A9:** 83
Field microscopy **A9:** 83
Field modulation
use in ESR analysis **A10:** 255
Field of view of microscope eyepieces **A9:** 73
Field repairability
of cyanates **EM2:** 232
Field sampling **A10:** 16
Field scanning methods
to analyze ceramic powder particle sizes **EM4:** 66, 67
Field shapers **A14:** 650-651
Field testing
SCC evaluation **A13:** 263-264
Field-dial Hall effect device
in ESR spectrometers **A10:** 255
Field-effect transistors (FET) **A6:** 39
microwave inspection **A17:** 209
Field-emission guns
as electron source **A12:** 167
Field-emission microscopy
defined **A10:** 673
Field-of-view (FOV)
defined **A17:** 9, 383
Figure-eight cooler
for sands **A15:** 350
Filament **A7:** 5 **EM3:** 13
Filament diameter *See* Fiber diameter
Filament light sources for microscopes **A9:** 72
Filament winding *See also* Filament winding resins; Lap **EM1:** 503-518 **EM2:** 368-377 **EM4:** 224
application **EM1:** 135, 355, 356, 833-835
and braiding, compared **EM1:** 519
component winding **EM1:** 507

Filament winding (continued)
composite-to-metal joints **EM1:** 511-514
costs **EM1:** 503
defined **EM1:** 11, 33, 135, 503 **EM2:** 18
design and analysis **EM1:** 508-510
design detail factors **EM2:** 371-373
dome contours in **EM1:** 510, 514
dry **EM1:** 71
economic factors **EM2:** 373
environmental effects **EM1:** 514-517
of epoxy composites **EM1:** 71
function/property factors **EM2:** 369-371
of glass rovings **EM1:** 109
impregnation **EM1:** 505-506
low cost **EM2:** 368
mandrel preparation **EM1:** 505-507
manufacturing flow diagram **EM1:** 506
manufacturing processes **EM1:** 505-507
materials **EM1:** 504-505
molded-in color **EM2:** 306
nonstandardized test methods **EM1:** 512-514
process, advantages/disadvantages ... **EM2:** 368-369
properties effects **EM2:** 285-287
resins **EM1:** 135-138
size and shape effects **EM2:** 291, 292
standard test methods **EM1:** 510-511
surface considerations **EM2:** 373-374
surface finish **EM2:** 304
textured surfaces **EM2:** 305
of thermoplastic resin composite **EM1:** 549
tooling/equipment **EM2:** 374-377
unsaturated polyesters **EM2:** 251
wet **EM1:** 71, 135, 505
Filament winding of resin-matrix composites **A9:** 591
Filament(s) *See also* Bundle; Fiber(s); Glass filament; Monofilament; Strand; Tow; Virgin filament
defined **EM1:** 11, 29 **EM2:** 18
glass, diameter nomenclature **EM1:** 109
measurement **EM1:** 286
polyamides (PA) **EM2:** 125
tensile strength test method effect **EM1:** 286
Filament-winding resins **EM1:** 135-138
epoxy **EM1:** 135-137
future trends **EM1:** 138
processing **EM1:** 135
types **EM1:** 135-138
unsaturated polyesters **EM1:** 137-138
vinyl esters **EM1:** 137-138
Filament-wound composites
joining techniques **EM1:** 511, 514
Filamentary nickel powder **A7:** 138
Filaments
radiographic **A17:** 302
Filaments, superconductor
properties **A2:** 1052-1054
Filar eyepiece
defined **A9:** 7
electronic digital **A9:** 76
Filar factors **A18:** 414
Filar micrometer microscope eyepiece **A7:** 228, 229
Filar unit **A18:** 414
Filar units
for measuring indentations **A8:** 91
Filar-micrometer ocular **A9:** 74
File hardness
defined **A18:** 9
File hardness test **A8:** 5, 71, 107-198
File testing
of carbonitrided P/M parts **A7:** 455
Filiform corrosion
in aircraft **A13:** 107, 1028-1030
aluminum/aluminum alloys **A13:** 596-597
cell, diagrams **A13:** 106
of coated aluminum and magnesium **A13:** 107
of coated steels **A13:** 104-107
conditions for **A13:** 105
defined **A11:** 4 **A13:** 6
general appearance **A13:** 105-106
growth rates, coated metals **A13:** 107
in manned spacecraft **A13:** 1076
mechanism **A13:** 106, 107
in packaging **A13:** 107
prevention **A13:** 107-108
in space shuttle orbiter **A13:** 1066, 1076
Filigreed eutectic microstructure ...... **A3:** 1 • 20
Fill *See also* Weft
defined **EM1:** 11 **EM2:** 18
density **A7:** 5
powder **A7:** 324-325
uni-type **EM1:** 128
Fill and soak cleaning **A13:** 1139
Fill, depth **A7:** 5
and ejection stroke, rigid tool compaction **A7:** 324
Fill factor **A7:** 5
defined **A17:** 172
in eddy current inspection **A17:** 168-169
remote-field eddy current inspection **A17:** 200-201
Fill height **A7:** 5
Fill position **A7:** 5
Fill ratio **A7:** 5
Fill shoe **A7:** 5
Fill volume **A7:** 5
Fill-and-wipe
defined **EM2:** 18
Filled billet process
extrusion **A7:** 518
Filled systems
defined **EM1:** 27
Filler *See also* Additive; Adhesive modifiers; Extenders; Inert filler; Reinforcement **EM3:** 13, 49, 175-176
benefits **EM3:** 175-177
for butyls **EM3:** 202
defined **A18:** 9 **EM1:** 11
effect on adhesive coefficient of thermal expansion **EM3:** 176
effect on adhesive specific gravity **EM3:** 176
effect on structural properties of adhesives **EM3:** 180
effect, polyester resins **EM1:** 92
for epoxies **EM3:** 99
forms **EM3:** 176-177
importance in dispensing operation **EM3:** 695
influence on stress-strain behavior **EM3:** 290, 321
inorganic, for sheet molding compounds **EM1:** 141
inorganic, with polyester resin **EM1:** 92
for lip seals **A18:** 550
properties **EM3:** 175
properties analysis of **EM1:** 736-737
resin, for pultrusion **EM1:** 539
for sealants **EM3:** 674
sequence, textile loom operations **EM1:** 127-128
sheet metal compound **EM1:** 158
to reduce seal wear **A18:** 551
towpreg, for I-beam **EM1:** 152
Filler alloys
aluminum-base **A13:** 884
copper-base **A13:** 883
nickel-base **A13:** 883-884
silver-base **A13:** 880-883
systems **A13:** 884-885
Filler materials
mold **A9:** 31-32
Filler metal
brittle fracture from improper **A11:** 645-646
selection for brazed joints **A11:** 450
Filler metal materials *See also* Brazing; Soldering
applications **A7:** 840-841
concentration, matrix tensile strength ratio and **A7:** 612
metal powders for **A7:** 816-822
in polymers **A7:** 608, 611
for submerged arc process **A7:** 822
Filler metal start delay time
definition **M6:** 7
Filler metal stop delay time
definition **M6:** 7
Filler metals
aerospace materials **A6:** 386-387
for aluminum alloy brazing **A6:** 937, 938, 939
aluminum alloys, arc welding of **A6:** 722, 724-729, 730-737, 739
for aluminum metal-matrix composites **A6:** 555
aluminum-lithium alloys **A6:** 550-551
American Welding Society classifications **M6:** 275-276
beryllium alloys **A6:** 945-946
Filler metals (continued)
brazeability and solderability considerations for engineering materials **A6:** 617, 618, 619, 620-621, 624-625, 626, 627, 628-629, 630, 631, 632, 633, 634, 635-636
brazing consumables **A6:** 904-905
for brazing nickel-base alloys **A6:** 928
for brazing of ceramic materials **A6:** 949, 950, 951-952, 954, 956, 957-958
brazing of stainless steels **A6:** 911-913, 914, 915, 919, 920, 922
brazing with clad brazing materials **A6:** 347, 348
for carbon steel brazing **A6:** 906-908
for cast iron arc welding **A15:** 523-524
for cast iron brazing **A6:** 906-908, 909
cast iron, chemical compositions **A15:** 530
for cobalt-base alloys **A6:** 928
copper alloy for iron castings **M6:** 599
copper alloys **A6:** 763, 764, 765, 766, 768
for copper and copper alloy brazing.... **A6:** 931, 932, 933, 934
for coppers **A6:** 756-757, 761, 762
definition **A6:** 1209 **M6:** 7
dip brazing, 50-2, 50-3 ...... **A6:** 337, 338
dispersion-strengthened aluminum alloys **A6:** 543
dissimilar metal joining **A6:** 822, 823, 824-825, 827, 828
duplex stainless steels **A6:** 471, 474, 476, 479
effect on strength of brazed joint **A6:** 904
electrodes **A6:** 750-751
electron-beam welding **A6:** 858-859, 860-861, 868
electroslag welding **A6:** 272, 275
ferritic stainless steels **A6:** 447, 448, 453
form selection for induction brazing **M6:** 971
galvanic couples **A6:** 1065-1066
gas-metal arc welding **A6:** 194
for gas-metal arc welding of coppers **A6:** 759
gas-metal arc welding of ferritic stainless steels **A6:** 446
gas-tungsten arc welding **A6:** 193-194, 658, 805
for heat-resistant alloy brazing **A6:** 924
heat-treatable aluminum alloys **A6:** 528, 531, 533-534
heat-treatable low-alloy steels **A6:** 669-671, 672
laser-beam welding **A6:** 739, 879-880
low-alloy metals for pressure vessels and piping **A6:** 668
for low-alloy steel brazing **A6:** 929-930
low-alloy steels, welding of.... **A6:** 662-663, 665, 666, 669-671, 672, 673, 674-675
for machinery and equipment **A6:** 391-393
for magnesium alloys **A6:** 772, 774, 778, 781
modeling the addition of **A6:** 1136
for nickel alloys **A6:** 578, 750
nickel-base, compositions and uses **A2:** 444
for niobium brazing **A6:** 943
for non-heat-treatable aluminum alloys **A6:** 539
for oxide-dispersion-strengthened alloys **A6:** 928
for oxide-dispersion-strengthened materials brazing **A6:** 1040
oxyfuel gas welding **A6:** 286
plasma arc welding **A6:** 658
for precious metal brazing **A6:** 935-936
gold jewelry brazing applications **A6:** 936
for precipitation-hardening stainless steels **A6:** 483, 487, 488, 490, 491
for reactive metal brazing **A6:** 945, 946
relationship to toughness in heat-affected zone **M6:** 42
repair welding **A6:** 1105, 1106, 1107
resistance brazing **A6:** 340, 341-342
aluminum-silicon alloys **A6:** 342
copper-phosphorus alloys **A6:** 342
silver alloys **A6:** 341, 342
resistance seam welding **A6:** 244
sandwiching, induction brazing **M6:** 973-974
self-shielded flux cored wire **M6:** 813-814
shielded metal arc welding of ferritic stainless steels **A6:** 449
silver-base brazing **A2:** 560
solder alloys **A6:** 964-969
specifications **M6:** 825-826
American Welding Society **M6:** 825
federal **M6:** 826

Filler metals (continued)
other organizations ... **M6:** 826
stainless steel dissimilar welds ... **A6:** 500, 501-502
for stainless steels ... **A6:** 686, 689, 692, 693, 694, 695-696, 697, 698, 699, 703-704, 705
stainless steels for weld cladding ... **A6:** 809
steel weldments ... **A6:** 417-418, 419, 425
submerged arc welding ... **A6:** 748
for tantalum brazing ... **A6:** 943
for titanium alloys ... **A6:** 510, 783, 784
tool and die steels ... **A6:** 674-675
for tool steel brazing ... **A6:** 929-930
torch brazing ... **A6:** 328, 329
torch brazing of stainless steels ... **A6:** 914
for tungsten brazing ... **A6:** 943
ultrahigh-strength low-alloy steels ... **A6:** 673
use in electron beam welding ... **M6:** 630
weld cladding ... **A6:** 809, 816, 817, 818, 819, 820, 821
zirconium alloys ... **A6:** 788

Filler metals for
arc welding ... **M6:** 447
of aluminum alloys ... **M6:** 373-374
of austenitic stainless steels ... **M6:** 323, 327, 334-335
of ferritic stainless steels ... **M6:** 323, 346-347
of heat-resistant alloys ... **M6:** 359
of magnesium alloys ... **M6:** 427-428
of martensitic stainless steels ... **M6:** 323
of molybdenum ... **M6:** 464
of nickel alloys ... **M6:** 444-445
of nitrogen-strengthened austenitic stainless steels ... **M6:** 345
of titanium and titanium alloys ... **M6:** 447
of tungsten ... **M6:** 464
of zirconium and hafnium ... **M6:** 457
brazing of aluminum alloys ... **M6:** 1022-1023
brazing of beryllium ... **M6:** 1052
brazing of copper and copper alloys ... **M6:** 1034-1035
copper-phosphorus filler metals ... **M6:** 1034
copper-phosphorus-silver filler metals ... **M6:** 1034
copper-zinc filler metals ... **M6:** 1034
gold alloy filler metals ... **M6:** 1035
joint clearances ... **M6:** 1034
physical properties ... **M6:** 1034
silver alloy filler metals ... **M6:** 1034-1035
brazing of dissimilar refractory metals ... **M6:** 1060
brazing of heat-resistant alloys ... **M6:** 1014
powders ... **M6:** 1014
tapes and foils ... **M6:** 1014
wires ... **M6:** 1014-1015
brazing of molybdenum ... **M6:** 1057-1058
brazing of niobum ... **M6:** 1055-1056
brazing of refractory metals ... **M6:** 1056
brazing of stainless steel ... **M6:** 1001-1004
cobalt alloy filler metals ... **M6:** 1003-1004
copper filler metals ... **M6:** 1002-1004
gold alloy filler metals ... **M6:** 1003-1004
nickel alloy filler metals ... **M6:** 1003-1004
silver alloy filler metals ... **M6:** 1002-1003
brazing of steel ... **M6:** 935-938
brazing of tantalum ... **M6:** 1058
brazing of titanium ... **M6:** 1050
brazing of tungsten ... **M6:** 1059
brazing of zirconium ... **M6:** 1051-1052
dip brazing of steels ... **M6:** 991-992
flux cored arc welding ... **M6:** 311
cast irons ... **M6:** 311
furnace brazing of steels ... **M6:** 943-945
copper filler metals ... **M6:** 935-938
silver alloy filler metals ... **M6:** 943-945
gas metal arc welding ... **M6:** 311
of aluminum alloys ... **M6:** 374
of cast irons ... **M6:** 311
of copper and copper alloys ... **M6:** 417

Filler metals for (continued)
of nickel alloys ... **M6:** 440
gas tungsten arc welding ... **M6:** 200-201, 311
of aluminum alloys ... **M6:** 374, 391-392
of aluminum bronzes ... **M6:** 412
of beryllium copper ... **M6:** 408
of cast irons ... **M6:** 311
of copper and copper alloys ... **M6:** 405-406, 415
of copper nickels ... **M6:** 414
of copper-zinc alloys ... **M6:** 409
of coppers with dissimilar Metals ... **M6:** 414-415
of iron-nickel-chromium and iron-chromium-nickel heat-
of nickel alloys ... **M6:** 439
of nickel-based heat-resistant alloys ... **M6:** 358-360
of phosphor bronzes ... **M6:** 410
resistant alloys ... **M6:** 365
of silicon bronzes ... **M6:** 414
laser brazing ... **M6:** 1064
oxyacetylene braze welding ... **M6:** 596-597
plasma arc welding ... **M6:** 218
resistance brazing ... **M6:** 981-982
aluminum-silicon alloy filler metals ... **M6:** 981
copper-phosphorous alloy filler metals ... **M6:** 981
salver alloy filler metals ... **M6:** 981
shielded metal arc welding ... **M6:** 310-311
of cast irons ... **M6:** 310-311
of copper and copper alloys ... **M6:** 425
submerged arc welding ... **M6:** 311
of cast irons ... **M6:** 311
torch brazing of copper ... **M6:** 1038-1039
torch brazing of stainless steels ... **M6:** 1010
torch brazing of steels ... **M6:** 950, 961
copper-zinc filler metals ... **M6:** 961
silver alloy filler metals ... **M6:** 961
weld overlays ... **M6:** 816-817
of stainless steels ... **M6:** 813-814

Filler metals, specific types
19-9 W, composition ... **M6:** 359
40Sn-58Pb-2Sb, soldering base metals, heating methods, and flux types ... **A6:** 631
40Sn-60Pb, soldering base metals, heating methods, and flux types ... **A6:** 631
50Au-25Pd-25Ni, refractory metal brazing ... **A6:** 942
50Au-50Cu, refractory metal brazing ... **A6:** 942
50Sn-50Pb, soldering base metals, heating methods, and flux types ... **A6:** 631
52.5Cu-38Mn-9.5Ni, brazing of carbides ... **A6:** 635
52Nb-48Ni, refractory metal brazing ... **A6:** 942
53Sn-29Pb-17In-0.5Zn, soldering base metals, heating methods, and flux types ... **A6:** 631
54Ag-21.3Cu-24.7Pd, refractory metal brazing ... **A6:** 942
54Ag-42Cu-2Ni, refractory metal brazing ... **A6:** 942
60Ag-30Cu-10Sn, for beryllium alloys ... **A6:** 946
60Sn-40Pb, soldering base metals, heating methods, and flux types ... **A6:** 631
62Sn-36Pb-2Ag, soldering base metals, heating methods, and flux types ... **A6:** 631
63Sn-37Pb, soldering base metals, heating methods, and flux types ... **A6:** 631
67Ni-33Cu, refractory metal brazing ... **A6:** 942
68Ag-27Cu-4.5Ti, brazing of graphite ... **A6:** 635
68Ag-27Cu-10Sn, refractory metal brazing ... **A6:** 942
70Au-22Ni-8Pd, brazing of stainless steels ... **A6:** 916
$70SiO_2$-27MgO-$3Al_2O_3$, brazing of nitride ceramics ... **A6:** 636
70Ti-15Cu-15Ni, brazing of graphite ... **A6:** 635
71.5Ag-28.1Cu-0.75Ni, refractory metal brazing ... **A6:** 942
72Ag-28Cu, refractory metal brazing ... **A6:** 942

Filler metals, specific types (continued)
73.8Au-26.2Ni, refractory metal brazing ... **A6:** 942
75Ag-20Pb-5Mn, refractory metal brazing ... **A6:** 942
80Mo-20Ru, refractory metal brazing ... **A6:** 942
80Ni-14Cr-6Fe, refractory metal brazing ... **A6:** 942
82Ag-9Ga-9Pd, refractory metal brazing ... **A6:** 942
82Au-18Ni, refractory metal brazing ... **A6:** 942
84Ni-16Ti, refractory metal brazing ... **A6:** 942
92Au-8Pd, refractory metal brazing ... **A6:** 942
95In-5Bi, soldering base metals, heating methods, and flux types ... **A6:** 631
95Sn-5Sb, soldering base metals, heating methods, and flux types ... **A6:** 631
304L, for stainless steels ... **A6:** 705
308
for gas-tungsten arc welding ... **A6:** 705
for precipitation-hardening stainless steels ... **A6:** 488
for stainless steels ... **A6:** 686
308L
for precipitation-hardening stainless steels ... **A6:** 483, 488
for stainless steels ... **A6:** 686
309
carbide precipitation ... **A6:** 695
for precipitation-hardening stainless steels ... **A6:** 488
for stainless steels ... **A6:** 695
310, for stainless steels ... **A6:** 696
312, for stainless steels ... **A6:** 696
316L, for stainless steels ... **A6:** 705
347, for precipitation-hardening stainless steels ... **A6:** 488
617, stress-rupture strength ... **A6:** 578
A-286, composition ... **M6:** 359
Ag, brazing of titanium alloys ... **A6:** 633
Ag-4Ti, brazing of oxide ceramics ... **A6:** 636
Ag-5Al
brazing of titanium alloys ... **A6:** 633
reactive metal brazing ... **A6:** 945
Ag-5Al-0.5Mn, brazing of titanium alloys ... **A6:** 633
Ag-7.5Cu, brazing of titanium alloys ... **A6:** 633
Ag-7Cu-0.2Li, for beryllium alloys ... **A6:** 946
Ag-9Pd-9Ga
brazing of titanium alloys ... **A6:** 633
reactive metal brazing ... **A6:** 945
Ag-21.3Cu-24.7Pd, reactive metal brazing ... **A6:** 945
Ag-25Cu-15In-1Ti
brazing of nitride ceramics ... **A6:** 636
brazing of oxide ceramics ... **A6:** 636
Ag-26.7Cu-4.5Ti, reactive metal brazing ... **A6:** 945
Ag-27Cu-2Ti
brazing of nitride ceramics ... **A6:** 636
brazing of oxide ceramics ... **A6:** 636
Ag-28Cu-0.2Li, reactive metal brazing ... **A6:** 945
Ag-30Cu-10Sn, reactive metal brazing ... **A6:** 945
Ag-30Pd
brazing of refractory metals ... **A6:** 942
refractory metal brazing ... **A6:** 942
Ag-35Cu-4Ti, brazing of nitride ceramics ... **A6:** 636
Ag-Cu-In-1.25Ti, brazing of nitride ceramics ... **A6:** 636
AgCuAlTi, brazing of ceramics ... **A6:** 952
AgCuInTi, brazing of ceramics ... **A6:** 952
AgCuNiTi, brazing of ceramics ... **A6:** 952
AgCuSnTi, brazing of ceramics ... **A6:** 952
AgCuTi
brazing of ceramics ... **A6:** 952
silicon carbide brazing ... **A6:** 636
AgTi, brazing of ceramics ... **A6:** 952

Filler metals, specific types (continued)
Al
brazing of nitride ceramics **A6:** 636
brazing of oxide ceramics **A6:** 636
brazing of titanium alloys **A6:** 633
Al bronze
gas-metal arc welding of copper and copper alloys **A6:** 755
gas-tungsten arc welding of copper and copper alloys **A6:** 755
shielded metal arc welding of copper and copper alloys **A6:** 755
Al-Mn (3003), brazing of titanium alloys **A6:** 633
Al-Si (0.06-10.6), brazing of nitride ceramics **A6:** 636
Al-Si (4040), brazing of titanium alloys **A6:** 633
Au-25Ni-25Pd, for ceramic materials **A6:** 951
AuNiTi, brazing of ceramics **A6:** 952
B120VCA, brazing of niobium alloys **A6:** 634
BAg
applications **A6:** 904
base materials joined **A6:** 904
brazing of carbides **A6:** 635
brazing of cast irons and carbon steels **A6:** 906, 909, 910
brazing of cobalt-base alloys **A6:** 634
brazing of copper and copper alloys **A6:** 630, 932-934
brazing of heat-resistant alloys **A6:** 632
brazing of nickel-base alloys **A6:** 632
for brazing of precious metals **A6:** 936
brazing of stainless steels **A6:** 626, 631
for cobalt-base alloys **A6:** 929
forms **A6:** 904
for low-alloy steels **A6:** 929-930
for molten-salt dip brazing **A6:** 338
for tool steels **A6:** 929-930
BAg, 13A-8, joint clearance **A6:** 620, 621
BAg-1
brazing applications, flux classifications, and heating methods **A6:** 630
brazing of aluminum alloys **A6:** 627
brazing of cast, irons **M6:** 996
brazing of cast irons and carbon steels **A6:** 907
brazing of copper and copper alloys **A6:** 932
brazing of stainless steels **A6:** 631, 912, 914, 919, 920, 921
composition **A6:** 912, 932
composition and properties **M6:** 971
production brazing examples **M6:** 983-984
torch brazing of copper **M6:** 1038
BAg-1a
brazing applications, flux classifications, and heating methods **A6:** 630
brazing of cast irons **A6:** 626
brazing of copper and copper alloys **A6:** 932
brazing of stainless steels **A6:** 631, 912, 914, 919, 920
composition **A6:** 912, 932
composition and properties **M6:** 971
production examples **M6:** 984
resistance brazing **M6:** 981
torch brazing of copper **M6:** 1038
BAg-2
brazing applications, flux classifications, and heating methods **A6:** 630
brazing of copper and copper alloys **A6:** 932
brazing of stainless steels **A6:** 631, 912
composition **A6:** 912, 932
composition and properties **M6:** 971
torch brazing of copper **M6:** 1038
BAg-2a
brazing applications, flux classifications, and heating methods **A6:** 630
brazing of stainless steels **A6:** 631, 912
composition **A6:** 912
BAg-3
brazing applications, flux classifications, and heating methods **A6:** 630
brazing of carbides **A6:** 635
brazing of cast irons and carbon steels **A6:** 907, 908
brazing of copper and copper alloys **A6:** 932, 935

Filler metals, specific types (continued)
brazing of stainless steels **A6:** 631, 912-914, 919, 920, 922
composition **A6:** 912
BAg-3, composition and properties **M6:** 971
BAg-4
brazing applications, flux classifications, and heating methods **A6:** 630
brazing of carbides **A6:** 635
brazing of cast irons **A6:** 908
brazing of stainless steels **A6:** 631, 912
composition **A6:** 912
BAg-5
brazing applications, flux classifications, and heating methods **A6:** 630
brazing of copper and copper alloys **A6:** 932
brazing of stainless steels **A6:** 631, 912
composition **A6:** 912, 932
BAg-5, composition and properties **M6:** 971
BAg-6
brazing applications, flux classifications, and heating methods **A6:** 630
brazing of stainless steels **A6:** 631, 912
composition **A6:** 912
BAg-7
brazing applications, flux classifications, and heating methods **A6:** 630
brazing of cast irons and carbon steels **A6:** 907
brazing of stainless steels **A6:** 631, 912
composition **A6:** 912
BAg-7(a), composition and properties **M6:** 971
BAg-8
brazing applications, flux classifications, and heating methods **A6:** 630
brazing of copper and copper alloys **A6:** 932
brazing of stainless steels **A6:** 631, 912, 920
composition **A6:** 912, 932
BAg-8a
brazing applications, flux classifications, and heating methods **A6:** 630
brazing, composition **A6:** 117
brazing of copper and copper alloys **A6:** 932
brazing of stainless steels **A6:** 631, 912
composition **A6:** 912, 932
wettability indices on stainless steel base metals **A6:** 118
BAg-9
brazing applications, flux classifications, and heating methods **A6:** 630
brazing of stainless steels **A6:** 631, 912
composition **A6:** 912
BAg-10
brazing applications, flux classifications, and heating methods **A6:** 630
brazing of stainless steels **A6:** 912
composition **A6:** 912
BAg-13
brazing applications, flux classifications, and heating methods **A6:** 630
brazing of stainless steels **A6:** 631, 912, 915
composition **A6:** 912
BAg-13a
brazing applications, flux classifications, and heating methods **A6:** 630
brazing of stainless steels **A6:** 912, 920
composition **A6:** 912
BAg-18
brazing applications, flux classifications, and heating methods **A6:** 630
brazing of stainless steels **A6:** 912, 914
composition **A6:** 912
BAg-18, resistance brazing **M6:** 981
BAg-19
brazing applications, flux classifications, and heating methods **A6:** 630
brazing, composition **A6:** 117
brazing of copper and copper alloys **A6:** 932
brazing of stainless steels **A6:** 631, 912, 916, 918-919
composition **A6:** 912, 932
wettability indices on stainless steel base metals **A6:** 118
BAg-20
brazing applications, flux classifications, and heating methods **A6:** 630

Filler metals, specific types (continued)
brazing of stainless steels **A6:** 912
composition **A6:** 912
BAg-21
brazing applications, flux classifications, and heating methods **A6:** 630
brazing of stainless steels **A6:** 626, 631, 912, 913
composition **A6:** 912
BAg-22
brazing applications, flux classifications, and heating methods **A6:** 630
brazing of carbides **A6:** 635
brazing of stainless steels **A6:** 912
composition **A6:** 912
BAg-23
brazing applications, flux classifications, and heating methods **A6:** 630
brazing of carbides **A6:** 635
brazing of stainless steels **A6:** 912
composition **A6:** 912
BAg-24
brazing applications, flux classifications, and heating methods **A6:** 630
brazing of cast irons **A6:** 908
brazing of stainless steels **A6:** 631, 912
composition **A6:** 912
BAg-25
brazing applications, flux classifications, and heating methods **A6:** 630
brazing of stainless steels **A6:** 912
composition **A6:** 912
BAg-26
brazing applications, flux classifications, and heating methods **A6:** 630
brazing of stainless steels **A6:** 912
composition **A6:** 912
BAg-27
brazing applications, flux classifications, and heating methods **A6:** 630
brazing of stainless steels **A6:** 912
composition **A6:** 912
BAg-28
brazing applications, flux classifications, and heating methods **A6:** 630
brazing of stainless steels **A6:** 912
composition **A6:** 912
BAg-33, composition **A6:** 912
BAg-34, brazing of stainless steels **A6:** 912
BAlSi
applications **A6:** 904
base materials joined **A6:** 904
brazing **A6:** 627
forms **A6:** 904
joint clearance **A6:** 620, 621
BAlSi-2
brazing applications, flux classifications, and heating methods **A6:** 630
composition and solidus, liquidus, and brazing temperature ranges **A6:** 938
BAlSi-3
brazing applications, flux classifications, and heating methods **A6:** 630
brazing of aluminum alloys **A6:** 939 **M6:** 1032
composition and solidus, liquidus, and brazing temperature ranges **A6:** 938
torch, brazing of aluminum alloys **M6:** 1029
BAlSi-4
brazing applications, flux classifications, and heating methods **A6:** 630
brazing of aluminum alloys **A6:** 627, 939
composition and solidus, liquidus, and brazing temperature ranges **A6:** 938
torch brazing of aluminum alloys **M6:** 1029
use in dip brazing **M6:** 1022
BAlSi-5
brazing applications, flux classifications, and heating methods **A6:** 630
composition and solidus, liquidus, and brazing temperature ranges **A6:** 938
BAlSi-6
brazing applications, flux classifications, and heating methods **A6:** 630

Filler metals, specific types (continued)
composition and solidus, liquidus, and brazing temperature ranges A6: 938
BAlSi-7
brazing applications, flux classifications, and heating methods A6: 630
composition and solidus, liquidus, and brazing temperature ranges A6: 938
BAlSi-8
brazing applications, flux classifications, and heating methods A6: 630
composition and solidus, liquidus, and brazing temperature ranges A6: 938
BAlSi-9
brazing applications, flux classifications, and heating methods A6: 630
composition and solidus, liquidus, and brazing temperature ranges A6: 938
BAlSi-10
brazing applications, flux classifications, and heating methods A6: 630
composition and solidus, liquidus, and brazing temperature ranges A6: 938
BAlSi-11
brazing applications, flux classifications, and heating methods A6: 630
composition and solidus, liquidus, and brazing temperature ranges A6: 938
BAu
applications A6: 904
base materials joined A6: 904
brazing of copper alloys A6: 932, 933
for brazing of precious metals A6: 936
brazing of stainless steels A6: 626, 631
forms A6: 904
joint clearance A6: 620, 621
BAu-1
brazing applications, flux classifications, and heating methods A6: 630
brazing of heat-resistant alloys A6: 925
brazing of stainless steels A6: 912
composition A6: 912, 925
BAu-2
brazing applications, flux classifications, and heating methods A6: 630
brazing of heat-resistant alloys A6: 925
brazing of stainless steels A6: 912
composition A6: 912, 925
BAu-3
brazing applications, flux classifications, and heating methods A6: 630
brazing of heat-resistant alloys A6: 925
brazing of stainless steels A6: 912
composition A6: 912, 925
BAu-4
brazing applications, flux classifications, and heating methods A6: 630
brazing of copper and copper alloys A6: 932
brazing of heat-resistant alloys A6: 925
brazing of stainless steels A6: 912, 913, 915
composition A6: 912, 925, 932
BAu-5
brazing applications, flux classifications, and heating methods A6: 630
brazing of heat-resistant alloys A6: 925
brazing of stainless steels A6: 912
composition A6: 912, 925
BAu-6
brazing applications, flux classifications, and heating methods A6: 630
brazing of stainless steels A6: 912

Filler metals, specific types (continued)
composition A6: 912
BAu-7, brazing applications, flux classifications, and heating methods A6: 630
BAu-8, brazing applications, flux classifications, and heating methods A6: 630
BCo
applications A6: 904
base materials joined A6: 904
brazing of nickel-base alloys A6: 632
brazing of stainless steels A6: 626, 631
forms A6: 904
for nickel-base alloys A6: 928
BCo-1
brazing applications, flux classifications, and heating methods A6: 630
brazing of heat-resistant alloys A6: 925
brazing of stainless steels A6: 913
for cobalt-base alloys A6: 929
composition A6: 913, 925
joint clearance A6: 621
BCu
applications A6: 904
base materials joined A6: 904
brazing of carbides A6: 635
brazing of cast irons and carbon steels A6: 906, 908, 909, 910
brazing of copper and copper alloys A6: 934
brazing of heat-resistant alloys A6: 632
brazing of stainless steels A6: 626, 631
for cobalt-base alloys A6: 929
forms A6: 904
for low-alloy steels A6: 929-930
for molten-salt dip brazing A6: 338
for tool steels A6: 929-930
BCu, 13A-8
brazing of stainless steels A6: 626
joint clearance A6: 620, 621
BCu-1
brazing applications, flux classifications, and heating methods A6: 630
brazing of stainless steels A6: 912, 917
composition A6: 912
BCu-1a
brazing applications, flux classifications, and heating methods A6: 630
brazing of stainless steels A6: 912, 922-923
composition A6: 912
BCu-2
brazing of stainless steels A6: 912, 917
composition A6: 912
BCuP
applications A6: 904
base materials joined A6: 904
brazing of copper and copper alloys A6: 628, 629, 932-934
for brazing of precious metals A6: 936
forms A6: 904
joint clearance A6: 620, 621
for molten-salt dip brazing A6: 338
BCuP-1
brazing applications, flux classifications, and heating methods A6: 630
brazing of copper and copper alloys A6: 932
composition A6: 932
BCuP-2
brazing applications, flux classifications, and heating methods A6: 630
brazing of copper and copper alloys A6: 932
composition A6: 932
production brazing examples M6: 982-983
resistance brazing M6: 981
torch brazing of copper M6: 1038
BCuP-3, brazing applications, flux classifications, and heating methods A6: 630
BCuP-3, resistance brazing of copper M6: 1045

Filler metals, specific types (continued)
BCuP-4
brazing applications, flux classifications, and heating methods A6: 630
brazing of copper and copper alloys A6: 932
composition A6: 932
BCuP-4, torch brazing of copper M6: 1038
BCuP-5
brazing applications, flux classifications, and heating methods A6: 630
for brazing of copper alloys A6: 935
brazing of copper and copper alloys A6: 932-934
composition A6: 932
production brazing examples M6: 977, 982-983
resistance brazing M6: 981
torch brazing of copper M6: 1038-1039
BCuP-6, brazing applications, flux classifications, and heating methods A6: 630
BCuP-7, brazing applications, flux classifications, and heating methods A6: 630
BCuZn, joint clearance A6: 621
BMg, joint clearance A6: 620, 621
BMg-1, brazing applications, flux classifications, and heating methods A6: 630
BNi
applications A6: 904
base materials joined A6: 904
brazing of cast irons and carbon steels A6: 906
brazing of copper and copper alloys A6: 934
brazing of nickel-base alloys A6: 632
brazing of stainless steels A6: 626, 631
forms A6: 904
for low-alloy steels A6: 929-930
for molten-salt dip brazing A6: 338
for nickel-base alloys A6: 928
for tool steels A6: 929-930
BNi-1
brazing applications, flux classifications, and heating methods A6: 630
brazing of heat-resistant alloys A6: 925
brazing of stainless steels A6: 912, 917
composition A6: 912, 925
joint clearance A6: 621
BNi-1a
brazing applications, flux classifications, and heating methods A6: 630
brazing of heat-resistant alloys A6: 925
brazing of stainless steels A6: 912
composition A6: 912, 925
BNi-2
brazing applications, flux classifications, and heating methods A6: 630
brazing of heat-resistant alloys A6: 925
brazing of nickel-base alloys A6: 928
brazing of stainless steels A6: 912, 920, 921
composition A6: 912, 925
joint clearance A6: 621
BNi-3
brazing applications, flux classifications, and heating methods A6: 630
brazing of heat-resistant alloys A6: 925
brazing of stainless steels A6: 912, 917
for cobalt-base alloys A6: 928
composition A6: 912, 925
joint clearance A6: 621
BNi-4
brazing applications, flux classifications, and heating methods A6: 630
brazing of heat-resistant alloys A6: 925
brazing of stainless steels A6: 912
composition A6: 912, 925
joint clearance A6: 621
BNi-5
brazing applications, flux classifications, and heating methods A6: 630

Filler metals, specific types (continued)
brazing of heat-resistant alloys ... A6: 925
brazing of stainless steels ... A6: 912
composition ... A6: 912, 925
joint clearance ... A6: 621
BNi-6
brazing applications, flux classifications, and heating methods ... A6: 630
brazing of heat-resistant alloys ... A6: 925
brazing of stainless steels ... A6: 912
composition ... A6: 912, 925
joint clearance ... A6: 621
BNi-7
brazing applications, flux classifications, and heating methods ... A6: 630
brazing of heat-resistant alloys ... A6: 925
brazing of stainless steels ... A6: 912, 917, 921
composition ... A6: 912, 925
joint clearance ... A6: 621
BNi-8
brazing applications, flux classifications, and heating methods ... A6: 630
brazing of heat-resistant alloys ... A6: 925
brazing of stainless steels ... A6: 912
composition ... A6: 912, 925
joint clearance ... A6: 621
BNi-9
brazing of stainless steels ... A6: 912, 920
composition ... A6: 912
BPd, brazing of stainless steels ... A6: 626, 631
BPd-1, brazing applications, flux classifications, and heating methods ... A6: 630
BPt, brazing of stainless steels ... A6: 626
BVAg, joint clearance ... A6: 621
BVAu, joint clearance ... A6: 621
BVCu, joint clearance ... A6: 621
Cd-Zn, soldering base metals, heating methods, and flux types ... A6: 631
Cd-Zn-Sn, soldering base metals, heating methods, and flux types ... A6: 631
Cu
gas-metal arc welding of copper and copper alloys ... A6: 755
gas-tungsten arc welding of copper and copper alloys ... A6: 755
shielded metal arc welding of copper and copper alloys ... A6: 755
Cu-40Ag-5Ti, brazing of oxide ceramics ... A6: 636
Cu-42Ti, brazing of nitride ceramics ... A6: 636
Cu-44Ag-4Sn-4Ti, brazing of oxide ceramics ... A6: 636
Cu-71Ti, brazing of nitride ceramics ... A6: 636
Cu-Mn-Ni, brazing of stainless steels ... A6: 631
CuAlSiTi, brazing of ceramics ... A6: 952
CuNi
gas-metal arc welding of copper and copper alloys ... A6: 755
gas-tungsten arc welding of copper and copper alloys ... A6: 755
shielded metal arc welding of copper and copper alloys ... A6: 755
E17LMN
filler metals for ... A6: 693
properties ... A6: 693
E308
for arc welding of selected martensitic stainless steels ... A6: 439
dissimilar metal joining ... A6: 827
shielded metal arc welding of ferritic stainless steels ... A6: 450, 451
E308L-16, for stainless steels ... A6: 686
E309
for arc welding of selected martensitic stainless steels ... A6: 439
dissimilar metal joining ... A6: 827
E309L-16, weld cladding of stainless steels ... A6: 822
E310, for arc welding of selected martensitic stainless steels ... A6: 439
E312, for arc welding of selected martensitic stainless steels ... A6: 439
E312-16, weld cladding ... A6: 811
E316, dissimilar metal joining ... A6: 827
E347, dissimilar metal joining ... A6: 827

Filler metals, specific types (continued)
E410
for arc welding of selected martensitic stainless steels ... A6: 439
composition ... A6: 439
E410NiMo, composition ... A6: 439
E420, for arc welding of selected martensitic stainless steels ... A6: 439
EC409, gas-metal arc welding ... A6: 685
EC409Cb, gas-metal arc welding ... A6: 685
ENi-1, welding nickel alloys ... A6: 826
ENiCrFe-2
for ferritic stainless steels ... A6: 449
welding nickel alloys to carbon or low-alloy steels ... A6: 826
welding nickel alloys to stainless steels ... A6: 826
ENiCrFe-3
dissimilar metal joining ... A6: 827
welding nickel alloys to carbon or low-alloy steels ... A6: 826
welding nickel alloys to stainless steels ... A6: 826
ENiCrMo-3
for ferritic stainless steels ... A6: 449
for stainless steels ... A6: 694
welding nickel alloys to carbon or low-alloy steels ... A6: 826
welding nickel alloys to stainless steels ... A6: 826
ENiCrMo-4
welding nickel alloys to carbon or low-alloy steels ... A6: 826
welding nickel alloys to stainless steels ... A6: 826
ENiCrMo-9
welding nickel alloys to carbon or low-alloy steels ... A6: 826
welding nickel alloys to stainless steels ... A6: 826
ENiCrMo-10, for stainless steels ... A6: 694
ENiCu-7, welding nickel alloys to carbon or low-alloy steels ... A6: 826
ENiMo-7, welding nickel alloys ... A6: 826
ER AZ1O1A, composition ... M6: 428
ER AZ61A
arc welding of magnesium alloys ... A6: 772, 779
gas-tungsten arc welding of magnesium alloys ... A6: 779
ER AZ61A, composition ... M6: 428
ER AZ92A
arc welding of magnesium alloys ... A6: 772
composition ... A6: 774
repair welding of magnesium alloys ... A6: 780, 781
ER AZ92A, composition ... M6: 428
ER AZ101A
arc welding of magnesium alloys ... A6: 772, 780
composition ... A6: 774
ER EZ33A
arc welding of magnesium alloys ... A6: 772
composition ... A6: 774
ER EZ33A, composition ... M6: 428
ER16-8-2, for stainless steels ... A6: 693
ER26-1, in ferritic base metal-filler metal combination ... A6: 449
ER70S-2, gas metal arc welding ... M6: 275
ER70S-3, gas metal arc welding ... M6: 275
ER70S-4, gas metal arc welding ... M6: 275-276
ER70S-5, gas metal arc welding ... M6: 276
ER70S-6, gas metal arc welding ... M6: 276
ER70S-7, gas metal arc welding ... M6: 276
ER209, for stainless steels ... A6: 692
ER218, for stainless steels ... A6: 692
ER219, for stainless steels ... A6: 692
ER240, for stainless steels ... A6: 692
ER308
austenitic stainless steels ... A6: 462
dissimilar metal joining ... A6: 827
ferritic analysis ... A6: 1059
for stainless steels ... A6: 692, 693
ER308H, for stainless steels ... A6: 692, 693
ER308L, for stainless steels ... A6: 692, 693, 694
ER308LSi, for stainless steels ... A6: 685
ER309
austenitic stainless steels ... A6: 462
dissimilar metal joining ... A6: 825, 827

Filler metals, specific types (continued)
for stainless steels ... A6: 692, 693
ER309LSi, for stainless steels ... A6: 685, 686
ER310, for stainless steels ... A6: 692, 693
ER316
dissimilar metal joining ... A6: 827
for stainless steels ... A6: 692, 693
ER316H, for stainless steels ... A6: 692
ER316L, for stainless steels ... A6: 692, 693
ER317, for stainless steels ... A6: 692
ER317L, for stainless steels ... A6: 692
ER318, for stainless steels ... A6: 692
ER320, for stainless steels ... A6: 693
ER321, for stainless steels ... A6: 693
ER330, for stainless steels ... A6: 693
ER347
dissimilar metal joining ... A6: 827
for stainless steels ... A6: 693
ER383, for stainless steels ... A6: 693
ER385, for stainless steels ... A6: 693
ER410
for arc welding of selected martensitic stainless steels ... A6: 439
composition ... A6: 439
ER410NiMo
for arc welding of selected martensitic stainless steels ... A6: 439
composition ... A6: 439
ER420
for arc welding of selected martensitic stainless steels ... A6: 439
composition ... A6: 439
ER4043
for arc welding of aluminum metal-matrix composites ... A6: 555
gas-tungsten arc welding of dispersion-strengthened aluminum alloys ... A6: 543
ER4045
for arc welding of aluminum metal-matrix composites ... A6: 555
gas-tungsten arc welding of dispersion-strengthened aluminum alloys ... A6: 543
ER5356
for arc welding of aluminum metal-matrix composites ... A6: 555
gas-tungsten arc welding of dispersion-strengthened aluminum alloys ... A6: 543
ERCu
for copper alloys ... A6: 756, 757
gas-metal arc welding of coppers ... A6: 759, 760
gas-tungsten arc welding of coppers to high-carbon steel ... A6: 827
gas-tungsten arc welding of coppers to low-alloy steel ... A6: 827
gas-tungsten arc welding of coppers to low-carbon steel ... A6: 827
gas-tungsten arc welding of coppers to medium-carbon steel ... A6: 827
gas-tungsten arc welding of coppers to stainless steels ... A6: 827
ERCuAl-A2
gas-metal arc welding of copper-zinc alloys ... A6: 763
gas-metal arc welding of coppers ... A6: 761
gas-tungsten arc welding of aluminum bronzes ... A6: 765, 766, 827
gas-tungsten arc welding of copper nickels to high-carbon steel ... A6: 827
gas-tungsten arc welding of copper nickels to low-alloy steel ... A6: 827
gas-tungsten arc welding of copper nickels to low-carbon steel ... A6: 827
gas-tungsten arc welding of copper nickels to medium-carbon steel ... A6: 827
gas-tungsten arc welding of copper nickels to stainless steel ... A6: 827
gas-tungsten arc welding of copper-zinc alloys ... A6: 763
gas-tungsten arc welding of coppers to high-carbon steel ... A6: 827
gas-tungsten arc welding of coppers to low-alloy steel ... A6: 827
gas-tungsten arc welding of coppers to low-carbon steel ... A6: 827

Filler metals, specific types (continued)
gas-tungsten arc welding of coppers to medium-carbon steel A6: 827
gas-tungsten arc welding of coppers to stainless steel A6: 827
gas-tungsten arc welding of silicon bronzes to high-carbon steel A6: 827
gas-tungsten arc welding of silicon bronzes to low-alloy steel A6: 827
gas-tungsten arc welding of silicon bronzes to low-carbon steel A6: 827
gas-tungsten arc welding of silicon bronzes to medium-carbon steel A6: 827
gas-tungsten arc welding of silicon bronzes to stainless steel A6: 827
ERCuAl-A3
gas-metal arc welding of aluminum bronzes A6: 765
gas-tungsten arc welding of aluminum bronzes A6: 765
ERCuMnNiAl, for copper alloys A6: 756
ERCuNi
gas-metal arc welding of copper-nickel alloys A6: 768
gas-tungsten arc welding of copper-nickel alloys A6: 768
ERCuSi-A
for dissimilar copper alloy welds A6: 769
gas-metal arc welding of coppers A6: 761
gas-tungsten arc welding of copper-zinc alloys A6: 763
gas-tungsten arc welding of phosphor bronzes A6: 763
gas-tungsten arc welding of silicon bronzes A6: 767
ERCuSn-A
for dissimilar copper alloy welds A6: 769
gas-metal arc welding of copper-zinc alloys A6: 763
gas-metal arc welding of phosphor bronzes A6: 764
gas-tungsten arc welding of copper-zinc alloys A6: 763
gas-tungsten arc welding of phosphor bronzes A6: 763
gas-tungsten arc welding of phosphor bronzes to high-carbon steel A6: 827
gas-tungsten arc welding of phosphor bronzes to low-alloy steel A6: 827
gas-tungsten arc welding of phosphor bronzes to low-carbon steel A6: 827
gas-tungsten arc welding of phosphor bronzes to medium-carbon steel A6: 827
gas-tungsten arc welding of phosphor bronzes to stainless steel A6: 827
ERNi-1
welding nickel alloys to carbon or low-alloy steels A6: 826
welding nickel alloys to stainless steels A6: 826
ERNi-3
gas-tungsten arc welding of copper nickels to high-carbon steel A6: 827
gas-tungsten arc welding of copper nickels to low-alloy steels A6: 827
gas-tungsten arc welding of copper nickels to low-carbon steels A6: 827
gas-tungsten arc welding of copper nickels to medium-carbon steel A6: 827
gas-tungsten arc welding of copper nickels to stainless steel A6: 827
gas-tungsten arc welding of coppers to high-carbon steel A6: 827

Filler metals, specific types (continued)
gas-tungsten arc welding of coppers to low-alloy steels A6: 827
gas-tungsten arc welding of coppers to low-carbon steels A6: 827
gas-tungsten arc welding of coppers to medium-carbon steel A6: 827
gas-tungsten arc welding of coppers to stainless steels A6: 827
ERNiCr-3
dissimilar metal joining A6: 823, 827
resistance butt welding of nickel alloys A6: 579
welding nickel alloys to carbon or low-alloy steels A6: 826
welding nickel alloys to stainless steels A6: 826
welding of clad materials A6: 578
ERNiCrCoMo-1, carbide formation in nickel alloys A6: 577, 578
ERNiCrFe-6
for precipitation-hardening stainless steels A6: 492
welding nickel alloys to carbon or low-alloy steels A6: 826
welding nickel alloys to stainless steels A6: 826
ERNiCrMo-1
for ferritic stainless steels A6: 449
welding nickel alloys to carbon or low-alloy steels A6: 826
welding nickel alloys to stainless steels A6: 826
ERNiCrMo-3
for austenitic stainless steels A6: 467
for duplex stainless steels A6: 476
for ferritic stainless steels A6: 449
welding nickel alloys to carbon or low-alloy steels A6: 826
welding nickel alloys to stainless steels A6: 826
ERNiCrMo-4
for austenitic stainless steels A6: 467
welding nickel alloys to carbon or low-alloy steels A6: 826
welding nickel alloys to stainless steels A6: 826
ERNiCrMo-7
welding nickel alloys to carbon or low-alloy steels A6: 826
welding nickel alloys to stainless steels A6: 826
ERNiCrMo-10, for austenitic stainless steels A6: 467
ERNiMo-6, to join nickel alloys to dissimilar metals A6: 751
ERNiMo-7
welding nickel alloys to carbon or low-alloy steels A6: 826
welding nickel alloys to stainless steels A6: 826
ERZr2, filler metal for zirconium alloys A6: 788
ERZr3, filler metal for zirconium alloys A6: 788
ERZr4, filler metal for zirconium alloys A6: 788
Fe-34Si, brazing of silicon carbides A6: 636
Hastelloy W A6: 491
Haynes 188, composition M6: 359
Hf-7Mo, brazing of tantalum alloys A6: 634
Hf-19Ta-2.5Mo, brazing of tantalum alloys A6: 634
Hf-40Ta, brazing of tantalum alloys A6: 634
HS-25, composition M6: 359
Inconel 82, electroslag welding A6: 278
L-605, composition M6: 359
low-fuming brass, gas-metal arc welding of copper and copper alloys A6: 755

Filler metals, specific types (continued)
Mo-50s, refractory metal brazing A6: 942
Multimet, composition M6: 359
naval brass, gas-metal arc welding of copper and copper alloys A6: 755
NiCrFe, stainless steel dissimilar welds A6: 501
Pb-Bi, soldering base metals, heating methods, and flux types A6: 631
PbInTi, brazing of ceramics A6: 952
Phosphor bronze
gas-metal arc welding of copper and copper alloys A6: 755
gas-tungsten arc welding of copper and copper alloys A6: 755
shielded metal arc welding of copper and copper alloys A6: 755
RBCuZn
applications A6: 904
base materials joined A6: 904
brazing of carbon steels A6: 906
brazing of cast irons A6: 906, 908, 909, 910
brazing of copper and copper alloys A6: 932-934
forms A6: 904
RBCuZn, 13A-8, joint clearance A6: 621
RBCuZn-A
brazing applications, flux classifications, and heating methods A6: 630
brazing of copper and copper alloys A6: 932
composition A6: 932
composition and properties M6: 596
oxyacetylene braze welding M6: 596
RBCuZn-B
composition and properties M6: 596
oxyacetylene braze welding M6: 596
RBCuZn-C
composition and properties M6: 596-597
oxyacetylene braze welding M6: 597
RBCuZn-C, brazing applications, flux classifications, and heating methods A6: 630
RBCuZn-D
brazing applications, flux classifications, and heating methods A6: 630
brazing of carbides A6: 635
brazing of copper and copper alloys A6: 932
composition A6: 932
composition and properties M6: 596-597
oxyacetylene braze welding M6: 597
RBCuZn-E, brazing applications, flux classifications, and heating methods A6: 630
RBCuZn-F, brazing applications, flux classifications, and heating methods A6: 630
RBCuZn-G, brazing applications, flux classifications, and heating methods A6: 630
RBCuZn-H, brazing applications, flux classifications, and heating methods A6: 630
RCI, composition M6: 603
RCI-A, composition M6: 603
RCI-B, composition M6: 603
RCuSi-A, gas-tungsten arc welding of copper-zinc alloys A6: 763
Silicon bronze
gas-metal arc welding of copper and copper alloys A6: 755
gas-tungsten arc welding of copper and copper alloys A6: 755
shielded metal arc welding of copper and copper alloys A6: 755
Sn-Cd, soldering base metals, heating methods, and flux types A6: 631
Sn-Pb, soldering base metals, heating methods, and flux types A6: 631

Filler metals, specific types (continued)
Sn-Zn, soldering base metals, heating methods, and flux types **A6:** 631
SnAgTi, brazing of ceramics **A6:** 952
Ti-8.5Si
brazing of molybdenum alloys **A6:** 634
brazing of niobium alloys **A6:** 634
Ti-8.5Si, brazing of molybdenum alloys **A6:** 634
Ti-28V-4Be, brazing of niobium alloys **A6:** 634
Ti-30V, brazing of molybdenum alloys **A6:** 634
Ti-43Zr-12Ni-2Be, brazing of titanium alloys **A6:** 633
Ti-48Zr-4Be, brazing of titanium alloys **A6:** 633
Ti-65V, brazing **A6:** 943
TiCuAg, brazing of ceramics **A6:** 952
TiCuNi, brazing of ceramics **A6:** 952
V-35Nb, brazing of molybdenum alloys **A6:** 634
Zn-Al, soldering base metals, heating methods, and flux types **A6:** 631
Zr-5Be, brazing of zirconium **M6:** 1052
Filler methods, specific types
nickel titanium, silicon carbide brazing **A6:** 636
silver-copper-indium-titanium, silicon carbide brazing **A6:** 636
Filler particles in mounting materials used for epoxy matrix composites **A9:** 588
Filler sheet **EM3:** 13
Filler wire
definition **A6:** 1209
Filler(s) *See also* Additives; Extenders
as additives **EM2:** 497-500
conductive **EM2:** 469-473
effect, apparent shrinkage **EM2:** 280-281
effect, polyester resins **EM2:** 248
materials, types **EM2:** 499-500
mechanical properties **EM2:** 71-73
for reinforced polypropylenes (PP) **EM2:** 192
and reinforcements, compared **EM2:** 72
specific gravity **EM2:** 84
Fillers *See also* Filling
alumina **EM4:** 6
for coating/encapsulation **EL1:** 242-243
defined **EL1:** 1144
for flexible epoxies **EL1:** 817
for molded plastic packages **EL1:** 474-475
properties **EM4:** 7
rigid epoxy encapsulants **EL1:** 812
silica **EM4:** 6
thermal conductivity **EL1:** 813-814
wax, for investment casting **A15:** 254
Fillet **EM3:** 13
defined **A14:** 5 **A15:** 5 **EM1:** 11 **EM2:** 15
radius, defined **A15:** 5
Fillet gages **A6:** 97
Fillet scab
as casting defect **A11:** 381
Fillet sealant joints **EM3:** 550
Fillet vein
as casting defect **A11:** 381
Fillet weld
definition **A6:** 1209
electron-beam welding-NV **A6:** 260
welding of titanium and titanium alloys, joint dimensions **A6:** 785
Fillet weld break test
definition **A6:** 1209
Fillet weld leg
definition **A6:** 1209
Fillet weld size
definition **A6:** 1209
Fillet weld throat
definition **A6:** 1209
Fillet welds
allowable unit forces **M6:** 63, 67
arc welding of nickel alloys **M6:** 437
combinations with groove welds **M6:** 66-67
comparison to groove welds **M6:** 64
definition, illustration **M6:** 7, 60-61
distortion caused by angular
aluminum weldments **M6:** 878-880
change **M6:** 876-877
low-carbon steel weldments **M6:** 878-879
electron beam welding **M6:** 616-617
flux cored arc welding **M6:** 106-107, 112-113
Fillet welds (continued)
gas metal arc welding of commercial coppers **M6:** 403
gas tungsten arc welding of silicon bronzes **M6:** 413
measurement **M6:** 62-63
oxyacetylene braze welding **M6:** 597
oxyfuel gas welding **M6:** 590
recommended grooves **M6:** 70
shielded metal arc welding **M6:** 76, 85, 88-89
submerged arc welding **M6:** 129
throat size **M6:** 63, 66-67
transverse shrinkage **M6:** 875
weld placement **M6:** 63
weld size factors **M6:** 62-63
Fillets
baud streamline design **A8:** 250
in cold-formed parts **A11:** 307
in fatigue crack initiation **A8:** 371
as fracture origin, shafts **A11:** 459
gouged/fractured, adhesive-bonded joints **A17:** 612
head-to-shank, fastener, as failure origin **A11:** 529-530, 532
lack of, adhesive-bonded joints **A17:** 612, 614
noncontinuous, in brazed joints **A17:** 602
porous/frothy, adhesive-bonded joints **A17:** 612
radii, effect of- size on stress concentration **A11:** 468
radii, effect on bearings **A11:** 506-507
and radii, ultrasonic fatigue testing specimens **A8:** 250-251
radiographic inspection **A17:** 334
radius, of torsion specimen **A8:** 155-156
rolling, of shafts **A11:** 459
sharp **A11:** 85, 122, 319-320
shrinkage **A11:** 382
steel forgings **M1:** 362-363
as stress concentrator **A11:** 318, 462
vein, as casting defect **A11:** 381
Filling *See* Filler; Weft
in lost foam casting **A15:** 231
rapid **A15:** 38
Filling circuit hydraulic pumping system **A7:** 330
Filling slot ball bearings
basic load rating **A18:** 505
Filling vias *See also* Vias
in prepunched tape **EL1:** 464
Filling yarn *See also* Fiber(s); Fillers; Weft yarns; Yarn
defined **EM1:** 11 **EM2:** 18
fabric direction **EM1:** 148
in fabric pattern **EM1:** 125
in flying shuttle **EM1:** 128
Film *See also* Protective film; Thin film **EM3:** 13
amylose, fungus attack **EM2:** 785-787
in aqueous corrosion **A13:** 30
bacterial, in seawater **A13:** 900-901
biofouling organisms as **A13:** 88
cast **EM2:** 8
defined **A13:** 6 **EM2:** 18
formation and breakdown, carbon steels **A13:** 510-511
formation, and corrosion rate **A13:** 195
formation, crack tip, alloy-environments of **A13:** 146
formation, in stress-corrosion cracking **A13:** 146
formers, organic, for binders **A15:** 260
fractographic **A12:** 85, 169
growth, high-temperature corrosion **A13:** 97
high molecular weight **EM2:** 164-165
-induced cleavage model, crack propagation **A13:** 161
inhibitors **A13:** 485
iron sulfide, anaerobic **A13:** 43
microbial **A13:** 88
with modified starch additives **EM2:** 786
nylon, as polyamide application **EM2:** 125
organic **A13:** 819
paint **A13:** 528
photographic **A10:** 334, 527
plastic, for V-process **A15:** 236
polyester, thermal properties **EM2:** 449
polymeric insulation, degradation **A13:** 1100
polyphenylene sulfides (PPS) **EM2:** 189-190
protective, by alloying **A13:** 47-48
Film (continued)
protective, coatings and linings as **A13:** 400
protective, electrochemical evaluation **A13:** 219-220
rupture, in crack propagation **A13:** 160
slime **A13:** 88, 907
starch-base polyethylene **EM2:** 785
surface, stability **A13:** 17
surface, titanium/titanium alloys **A13:** 695
thermoplastic polyurethanes (TPUR) **EM2:** 205
thin, mechanisms **A13:** 67
Film (photographic)
in neutron radiography **A17:** 387, 390-391
for optical holographic interferometry **A17:** 414
Film (radiographic)
base, defined **A17:** 314
development, in radiography **A17:** 327
gradient **A17:** 299, 325-326
speed, x-ray film, radiography **A17:** 325
types, and selection **A17:** 327-328
unsharpness, defined **A17:** 300
Film adhesive
defined **EM1:** 11 **EM2:** 18
Film adhesives **EM3:** 13, 75-76
application **EM3:** 36
suppliers **EM3:** 76
Film applications
high-density polyethylenes (HDPE) ... **EM2:** 163-165
properties of **EM2:** 457-458
Film badges
for radiation monitoring **A17:** 301
Film capacitors *See also* Capacitors
failure mechanisms **EL1:** 972
"Film control
" carbon-graphite materials **A18:** 818
Film deposition *See* Interference film deposition
Film gradient
defined **A17:** 299
x-ray film, radiography **A17:** 325-326
Film radiography *See also* Radiography **A17:** 323-330
defined **A17:** 295
dynamic range **A17:** 299
enlargement effect **A17:** 312
exposure factors **A17:** 328-330
film types and selection **A17:** 327-328
gamma-ray exposure charts **A17:** 330
and real-time radiography, compared **A17:** 295
x-ray film, characteristics **A17:** 323-327
Film readers, automated
XRPD analysis **A10:** 338
Film resistance *See also* Constriction resistance; Contact resistance
defined **A18:** 9
Film resistors *See also* Resistors
construction **EL1:** 178
Film rupture
in SCC testing **A8:** 498
Film rupture model
crack propagation **A13:** 160
Film speed
effect on photomicroscopy **A9:** 85
Film strength
defined **A18:** 9
Film theory **A6:** 145
Film thickness
defined **A18:** 9
Film thickness across bearing area
nomenclature for hydrostatic bearings with orifice or capillary restrictor **A18:** 92
Film thickness measurements *See also* Films; Thick films; Thin films; Ultrathin films
Auger electron spectroscopy **A10:** 549-567
electron probe x-ray microanalysis **A10:** 516-535
optical metallography **A10:** 299-308
Rutherford backscattering spectrometry **A10:** 628-636
scanning electron microscopy **A10:** 490-515
x-ray spectrometry **A10:** 82-101
Film thickness parameter
nomenclature for lubrication regimes **A18:** 90
Film thickness-to-roughness ratio **A18:** 478
Film(s) *See also* Oxide films; Thick film; Thin filin; Thin-film materials
ceramic, defined **EL1:** 1136
conductive, as indium application **A2:** 752
cracking, semiconductor chips **EL1:** 964

Film(s) (continued)
defects **EL1:** 1000
defined **EL1:** 1144
deposition defects **EL1:** 978
growth, in thin-film materials **A2:** 1082
polyimide, as thermocouple wire insulation **A2:** 882
preparation, thin-film hybrids **EL1:** 313-316
protective, on adhesive **A17:** 614
resistance measurement **EL1:** 89
in sliding contacts **A2:** 842
thick/thin, in microcircuitry **EL1:** 89
thickness, by microwave inspection **A17:** 212
Film-forming etchants for iron-chromium- nickel heat-resistant casting alloys **A9:** 330-332
Film-forming organics
as sizing agents **EM1:** 122
Film-induced cleavage model
crack propagation **A13:** 161
Film-strength additives
for metalworking lubricants **A18:** 140-141, 142, 143, 144, 145, 146, 147
Filming inhibitors
mechanisms/application **A13:** 485-486
Films *See also* Film thickness measurement; Passive films; Thick films; Thin films; Ultrathin films
antihalation **A12:** 84
antiwear, AES characterized **A10:** 566
chemical vapor deposition, crystallographic texture in **A9:** 700
deposited, ATR spectroscopy of **A10:** 113
deposition, by shadowing **A12:** 172
drawn polymer, molecular orientation determined **A10:** 120
electrodeposition, crystallographic texture in **A9:** 700
epitaxial, rocking curve profiles **A10:** 375-376
heterogeneous surface, AES analysis of **A10:** 566
intergranular embrittlement by **A12:** 110
layer thickness, RBS analysis for **A10:** 631-632
organometallic silicate, depth profiles for **A10:** 617
organometallic silicate, on silicon substrate positive SIMS spectra for **A10:** 617
oxide, austenitic stainless steels **A12:** 352-353
passive, LEISS analysis of depth vs composition on tin-nickel substrate **A10:** 608-609
passive rupture, in SCC and corrosion fatigue **A12:** 42
photomicroscopy **A9:** 84-87
sputtering, crystallographic texture in **A9:** 700
surfaces, of electrical contacts in mercury switch, XPS analysis **A10:** 578-579
thick, analysis of **A10:** 561, 603
thin, AES analysis of **A10:** 557-561
thin, characterization of **A10:** 452-453, 536, 559, 583, 603, 631-632
thin, FIM/AP study of local composition variations **A10:** 583
thin, FIM/AP study of nucleation and growth of **A10:** 583
thin oriented, LEED determined grain size in **A10:** 536
thin, RBS compositional analysis **A10:** 631-632
thin, sample preparation for ATEM analysis **A10:** 452-453
ultrathin, LEISS analysis **A10:** 603
vinyl, identification of polymer and plasticizer materials in **A10:** 123-124
vinyl, polymer and plasticizer materials identified in **A10:** 123
Filter bed refining process
for alkali metals **A15:** 471
Filter lens definition **M6:** 7
Filter photometers
UV/VIS analysis **A10:** 67
Filter plate
definition **M6:** 7
Filter sampling
applications **A10:** 16, 94
Filter(s)
and convolutions, computed tomography (CT) **A17:** 380
edge detection, image processing **A17:** 461
effects, color images **A17:** 484
encapsulated electronic, neutron radiography of **A17:** 394-395
film radiography **A17:** 329-330
high-pass, eddy current inspection **A17:** 190
image enhancement **A17:** 459
interference, optical holography **A17:** 413
low-pass, eddy current inspection **A17:** 190
spatial, holographic **A17:** 418
Vander Lugt **A17:** 228
Filtered particle crack detection **A7:** 484
Filtered-backprojection technique
computed tomography (CT) **A17:** 380-382
Filtering
for digital image enhancement **A17:** 459
front-end, acoustic emission inspection **A17:** 286
in neutron production **A17:** 391
in servohydraulic system **A8:** 396, 399
wear mechanism **A8:** 603
Filters *See also* Filtration
for anodes, x-ray tubes **A10:** 90
colored, image analyzers **A10:** 310
composition **A10:** 94
defined **A9:** 7 **A10:** 673
detection limits **A10:** 95
foundry **A15:** 491-492
for gating systems **A15:** 596-597
ion-exchanged resin-impregnated **A10:** 94
optical microscope light source **A9:** 72
placement, gating system **A15:** 596-597
porous bronze **A2:** 401-402
for sampling **A10:** 16, 94
as sampling substrates, x-ray spectrometry **A10:** 94
selection **A15:** 491
size **A15:** 491-492
types of **A15:** 489-491, 596-597
Filters(s) **A7:** 5
of bronze P/M parts **A7:** 736-737
as porous powder applications **A7:** 17, 699
powders used **A7:** 572
Filtration *See also* Filters
of aluminum alloys **A15:** 488
benefits of **A15:** 492-493
of copper alloys **A15:** 488
defined, radiography **A17:** 315
and flow modification **A15:** 595
fundamentals **A15:** 489
in gravimetric analysis **A10:** 163
inclusion sources **A15:** 488
inherent, x-ray tubes, radiography **A17:** 306-307
of magnesium alloys **A15:** 488
melt cleanliness, determining **A15:** 493
nonferrous molten metal **A15:** 487-493
of scattered radiation **A17:** 345
screens, radiography **A17:** 298
of secondary radiation, lead screens **A17:** 315
of zinc alloys **A15:** 488-489
Filtration and washing **EM4:** 90-93
powder recovery techniques **EM4:** 90-92
centrifugation **EM4:** 90
filtration of elastic systems **EM4:** 91-92
flocculation role **EM4:** 91-92
kinetic relationship for filtration rate **EM4:** 90-91
microfiltration **EM4:** 90-92
non-elastic filtration **EM4:** 90-91
powder washing **EM4:** 92-93
Filtration processes
copper plating **M5:** 162-163, 165
FIM *See* Field ion microscopy
Fin
defined **A14:** 5 **A15:** 5 **EM2:** 18
Final annealing
defined **A9:** 7
Final breaking, of specimens *See also* Fast fractures **A12:** 77
Final cooling zone
furnaces **A7:** 352-354
Final crack size
fracture mechanics for **A11:** 56
Final current
definition **M6:** 7
Final density **A7:** 5, 463
Final design package
components **EL1:** 523-526
Final fracture
in fracture mechanics **A11:** 47
from corrosion-fatigue **A11:** 257-258
propagation of **A11:** 106-107
zone **A11:** 104-105
Final inspection *See also* Nondestructive evaluation (NDE)
defined **A17:** 49
magnetic particle inspection **A17:** 89
Final NDE
defined **A17:** 49
Final state
electronic **A10:** 408, 569
Final taper current
definition **M6:** 7
Final-polishing *See also* Polishing **A9:** 40-43
of commercially pure lead **A9:** 47
defined **A9:** 8
of electrical contact materials **A9:** 550
of hafnium **A9:** 497-498
of molybdenum **A9:** 550
of silver-cadmium oxide materials **A9:** 550
of tool steels **A9:** 257
of tungsten **A9:** 550
of very soft materials **A9:** 47
of zirconium and zirconium alloys **A9:** 497-498
FINDAP computer program
to obtain tribological Arrhenius constants **A18:** 280, 281, 283-284, 286
Fine
fibers, glass **EM1:** 108
yams, filling, in unidirectional fabric **EM1:** 148
Fine blanking presses **A14:** 502
Fine ceramics **EM4:** 39
Fine chemicals
tantalum resistance to **A13:** 728
Fine china **EM4:** 4
imports **EM4:** 935
Fine earthenware **EM4:** 3, 4
Fine features
boards with **EL1:** 561
Fine grain zone in ferrous alloy welded joints **A9:** 581
Fine hackle
definition **EM4:** 632
Fine leak hermeticity test **EL1:** 500-501
Fine leak tests **EL1:** 500-501, 930
Fine lines
boards with **EL1:** 561
Fine palladium *See also* Palladium; Platinum; Precious metals
as electrical contact materials **A2:** 847
Fine platinum *See also* Platinum; Precious metals
as electrical contact materials **A2:** 846
Fine powders
and contamination removal, seeds **A7:** 590
effect on flow rate **A7:** 297
effect on packed density **A7:** 296
modified Pechukas and Gage apparatus for flow measurement **A7:** 264-265
specific surface area measured by BET method **A7:** 262

Fine silver *See also* Pure silver; Silver alloys; Silver contact alloys; Sliver
applications **A2:** 845
as electrical contact material **A2:** 844-845
Fine steel wire **A1:** 286
designations and applications **M1:** 269
Fine stoneware **EM4:** 3, 4
Fine structure effects
atomic order **A10:** 438
electron diffraction **A10:** 438
orientation relationships **A10:** 438
satellite spots **A10:** 438
strain-induced defects **A10:** 438, 440
Fine whiteware
composition **EM4:** 5
Fine wire *See also* Wire
mechanically alloyed oxide alloys **A2:** 949
dispersion-strengthened (MA ODS)
Fine wire quality carbon steel wire rod **M1:** 254
Fine wire quality rod **A1:** 273
Fine-edge blanking *See also* Blanking **A14:** 458
applications **A14:** 475
blank design **A14:** 473
and conventional blanking, compared **A14:** 472
lubrication **A14:** 475
materials for **A14:** 474
presses **A14:** 473-474
process **A14:** 472-473
tooling setup **A14:** 472
tools **A14:** 474-475
work materials **A14:** 472-473
Fine-edge piercing *See also* Piercing
applications **A14:** 475
blank design **A14:** 473
and blanking **A14:** 472-475
and conventional blanking, compared **A14:** 472
lubrication **A14:** 475
presses **A14:** 473-474
process capabilities **A14:** 472
tools **A14:** 474-475
work materials **A14:** 472-473
Fine-grain structure
solder **EL1:** 733
Fineness, grain
AFS numbers **A15:** 209
Fines **A7:** 5 **EM3:** 13
defined **EM2:** 18
definition **M6:** 7
Fines, metal
cleaning of **A13:** 380-381
Finger buffs **M5:** 118, 126
Finger joint prosthesis
total **A11:** 670
Finger oxides
surface **A14:** 204
Fingerjoint assembly **EM3:** 13
Fingernail lacquer
powders used **A7:** 573
Fingerprint removers and neutralizers **M5:** 460-465, 467
applying, methods of **M5:** 467
Fingerprinting
half-wave potentials as **A10:** 190
multielement, and approximate quantification in effluent samples **A10:** 195
multielement, of voltammetric study in effluents **A10:** 195
for oil spill identification **A10:** 72
Finish *See also* Fiber finish; Finishes; Finishing; Glass finish; Surface finish; Surface(s) **EM3:** 13
allowance, defined **A14:** 6
control, in coining **A14:** 186
defined **A14:** 6 **EM1:** 11 **EM2:** 18
surface, types **EM2:** 303
Finish allowance
defined **A15:** 5
Finish forging
of nickel-base alloys **A14:** 262
Finish machining **A7:** 668
Finish milling **A7:** 462
Finish surface *See* Surface finish
of Brinell test workpiece **A8:** 86, 88
Finish trim
defined **A14:** 6
Finished leather
waterjet machining **A16:** 522
Finished machined parts and spare parts
rust-preventive compounds used on **M5:** 465
Finished parts *See also* Part(s)
magnetic particle inspection **A17:** 110
Finishers
defined **A14:** 6
as impression dies **A14:** 44-45
impressions, location of **A14:** 51
Finishes *See also* Finish; Sizing; Surface finish; Surface treatment; Surface(s) **A1:** 279-280
for carbon fiber **EM1:** 113
chemical, aluminum alloy **A15:** 762
chromate **A15:** 796
common gravimetric **A10:** 171
coupon **A13:** 198
for die castings **A15:** 795-797
electronics industry **A13:** 1111
for extracts **A10:** 179
green sand molds **A15:** 347
magnesium die casting **A15:** 809
solderability **EL1:** 561-564
Finishes for leaf springs **M1:** 313
Finishing *See also* Fabrication characteristics; Finish; Grinding; specific finishing methods; Surface(s); Surfaces
aluminum casting alloys **A2:** 153
aluminum-lithium alloys **A2:** 189-197
barrel for tumbling **A14:** 230
blow molding **EM2:** 358
centrifugal **A7:** 459
chemical, zinc alloys **A2:** 530
costs **EM2:** 650
dies, defined **A14:** 6
effects on fatigue strength **A11:** 122
and grinding, electrolytic copper powders **A7:** 114
nickel strip **A7:** 401-402
operations affecting surface porosity **A7:** 451
P/M parts **A7:** 295
polyether sulfones (PES, PESV) **EM2:** 162
in roll compacting **A7:** 405-406
stainless steels **M3:** 33-38
of strip **A7:** 405-406
of structural ceramics **A2:** 1021
surface, as stress source **A11:** 205
techniques **A7:** 152
temperature **A14:** 6, 81, 620
tool and die failures from **A11:** 564, 567
treatments, for shafts **A11:** 459
in tube rolling **EM1:** 574
of tungsten powders **A7:** 154
zinc alloys **A2:** 530
Finishing impression *See* Finisher
Finishing, precision metal
laser inspection **A17:** 14-15
Finishing temperature *See also* Temperature(s)
closed-die forging **A14:** 81
defined **A14:** 6
for steel **A14:** 620
FINISHR (FORTRAN)
process rough analysis tool **A14:** 411
Finite difference
codes, for spall stress **A8:** 212
method, for elastic bending **A8:** 118
Finite difference (FD) method **EM3:** 478-479
Finite element analysis
CARES computer program **EM4:** 700, 701, 702, 707
for design of the glass component shape **EM4:** 742, 744
dynamic notched round bar testing **A8:** 282
elastic-plastic **A8:** 283
with plastic zone, torsional Kolsky bar **A8:** 222-223
to analyze joint types of silicon nitride to metals in turbocharger wheels **EM4:** 724, 725
Finite element analysis (FEA)
tire tread wear **A18:** 579
Finite source theory **A6:** 8
Finite-difference (FD) method
analysis examples **EM1:** 470-476
commercial codes for **EM1:** 469-470
computer programs using **EM1:** 269
Finite-difference (FD) method (continued)
method selection **EM1:** 467-468
origins/theory development **EM1:** 463-466
solution approach **EM1:** 466-467
typical problems **EM1:** 468-469
Finite-difference analysis
computer **EL1:** 419
Finite-difference method
computer-aided **A15:** 293, 610-611
Finite-difference method (FDM) analysis **A4:** 109
Finite-element
analysis (FEA) **EL1:** 419, 442-443
modeling (FEM) **EL1:** 442-443
package design, software **EL1:** 954
Finite-element (FE) method *See* Finite-element analysis **EM1:** 247-250
analysis examples **EM1:** 470-476
commercial codes for **EM1:** 469-470
computer programs using **EM1:** 269
crack-closure scheme **EM1:** 248-250
formulation **EM1:** 247-248
LAMPS-A computer program for **EM1:** 268, 271
method selection **EM1:** 467-468
model, large-scale bolt specimen **EM1:** 337
origins/theory development **EM1:** 463-466
problems, typical **EM1:** 468-469
solution approach **EM1:** 466-467
Finite-element analysis (FEA) **EM2:** 336-337
Finite-element analysis (finite-element method) **EM3:** 477-481
butt joints **EM3:** 331
computation of maximum shear stress on end of overlap related to mean shear stress **EM3:** 665
double-lap joints **EM3:** 480, 481, 482
elastoplastic analysis **EM3:** 483
graphite-epoxy tube (small-diameter) with bonded aluminum end fitting **EM3:** 496, 498
in situ testing **EM3:** 486
for joint analysis with metallic fitting and graphite-epoxy composite tube **EM3:** 496-497, 499
lap joints **EM3:** 327-328
lap-shear coupon **EM3:** 490
mode mix for given joint geometry **EM3:** 442
modeling and design considerations **EM3:** 484-491
molding process in encapsulation **EM3:** 586
results compared to component test results **EM3:** 11-540
single-lap joints **EM3:** 480, 481
single-lap-shear specimen with and without an adhesive fillet **EM3:** 493, 494, 495, 496
stress analysis of cracked components **EM3:** 337, 341, 342
stress analysis of specimen performance **EM3:** 444
stress-strain singularities **EM3:** 484
to compute debond parameters **EM3:** 445-446, 447, 449-451
Finite-element method
analytical modeling **A14:** 425
codes, for process design **A14:** 409
computer-aided **A15:** 610-611
for die casting **A15:** 293
elastic-plastic **A14:** 425
mathematical modeling by **A14:** 159
modeling, for shape rolling **A14:** 350
problem formulation in ALPID **A14:** 425-426
process modeling/simulation **A14:** 919-924
rigid-viscoplastic **A14:** 425
software **A15:** 860
wireframe and **A15:** 858-859
Finite-element method (FEM) **A6:** 147
for flaw leakage fields **A17:** 131
Finite-element method (FEM) analysis **A4:** 109
quantitative prediction of soft interlayers **A6:** 166, 167, 168, 169, 170
Finite-element model
of torsional hydraulic actuator **A8:** 216-217
Finite-element model (FEM) **EM3:** 400, 401, 402
Finite-element modeling **A17:** 750
Finned tubing *See also* Tube(s); Tubular products
for heat exchangers **A11:** 628

Finned tubing (continued)
inspection of ... **A17:** 571-572
Finnie's equation
hard materials ... **A18:** 598
Finnie's theory
erosion rates ... **A18:** 204
Finning
as casting defect ... **A11:** 381
as rolling defect ... **A14:** 359
Fins
as forging flaws ... **A17:** 493
Fir tree cracking ... **A8:** 591, 594
Fir tree defect
in extrusion ... **A14:** 403
Fire *See also* Flame resistance; Flame retardant(s); Flammability
of magnesium powders ... **A7:** 132-133
with sintering atmospheres ... **A7:** 348-350
tests, polyester composite peformance ... **EM1:** 95
Fire point ... **A18:** 84
defined ... **A18:** 9
Fire protection
with liquid metals ... **A13:** 96-97
Fire refined copper *See* Copper alloys, specific types, C12500, C12700, C12800, C12000 and C13000
Fire refined tough pitch copper *See* Copper alloys, specific types, C12500
Fire refined tough pitch copper with silver *See* Copper alloys, specific types, C12700, C12800, C12900 and C13000
Fire refining, effect
molten metal impurities ... **A15:** 450-451, 453
Fire retardancy ... **EM3:** 590
Fire retardants *See* Combustion; Flame resistance; Flame retardants
Fire safety requirements
NFPA ... **M5:** 20-21
Fire scale
removal of ... **M5:** 76
Fire-extinguisher case
failure from overheating during spinning ... **A11:** 649
Fire-refined coppers
applications and properties ... **A2:** 275-277
Fire-refined tough pitch copper
characteristics ... **A2:** 223
Fire-side corrosion
furnace water walls ... **A13:** 995-996
steam equipment ... **A11:** 616-620
Firebox steel plate ... **A9:** 204
Firebrick *See also* Fireclay
defined ... **A15:** 5
flame emission sources for ... **A10:** 29-30
Fireclay *See also* Firebrick
aluminous, chamotte as ... **A15:** 248
as binder ... **A15:** 29
defined ... **A15:** 5
pressure calcintering for study of compaction kinetics ... **EM4:** 300
Fireclay brick
applications ... **EM4:** 899, 901, 903, 906-907
defined ... **EM4:** 255
determination of reactions during fifing ... **EM4:** 255-257
differential thermal analysis ... **EM4:** 255, 257, 258
dilatometry ... **EM4:** 255-256, 257
sintering ... **EM4:** 256-257
thermogravimetric analysis ... **EM4:** 255, 256, 258
determination of the firing curve ... **EM4:** 257-259
black coring ... **EM4:** 258
blue coring ... **EM4:** 258
Missouri superduty, properties ... **EM4:** 897, 898, 899
Firing process ... **EM4:** 242, 255-259
reactions occurring ... **EM4:** 255
Fireclay inclusions
in automobile stub axles ... **A11:** 323
Firecracker welding ... **A6:** 178
definition ... **A6:** 1209 **M6:** 7
Firecracking
of gold-nickel-copper alloys ... **A2:** 706
Fired camber
ceramic multilayer packages ... **EL1:** 468
Fired mold
defined ... **A15:** 5
Fires *See also* Explosions; Flammability; Pyrophoricity
with aluminum-lithium alloys ... **A2:** 183
Firing
and ceramic coating, Replicast process ... **A15:** 271
conditions ... **EL1:** 465
of porcelain enamels ... **A13:** 448
of rammed graphite molds ... **A15:** 273-274
Replicast process ... **A15:** 271
of substrates, ceramic packages ... **EL1:** 465-466
temperature, rammed graphite molds ... **A15:** 274
in thick-film process ... **EL1:** 332-333
Firing range
in typical ceramic body compositions ... **EM4:** 5
First aid
for liquid metals ... **A13:** 96-97
First block, second block
and finish, defined ... **A14:** 6
First fire mixes and fuzes
powders used ... **A7:** 573
First friction force ... **A6:** 888
First Law of Thermodynamics ... **A3:** 1 • 6
and crack propagation ... **EM3:** 506
First reflection switch design ... **EL1:** 42
First-degree blocking ... **EM3:** 13
First-level package
failure mechanisms ... **EL1:** 989-992
First-order Laue zone
abbreviation for ... **A10:** 690
in electron diffraction ... **A10:** 439, 442
First-order phase transition ... **A3:** 1 • 10
First-order transition ... **EM3:** 13
defined ... **EM1:** 11 **EM2:** 18
First-ply failure
laniinate ... **EM1:** 230-232
Fischer, Johann Conrad
as metallurgist ... **A15:** 31
Fish eye *See also* Window
defined ... **EM2:** 18
Fish eyes ... **A13:** 6, 164
defined ... **A8:** 5
Fish paper
defined ... **EM2:** 18
Fisher sub-sieve size ... **A7:** 123, 124
Fisher sub-sieve sizer ... **A7:** 138, 230-232, 264 **EM4:** 70
Fisher-Tropsch waxes
for investment casting ... **A15:** 253
Fisheye ... **A6:** 410, 1073
definition ... **A6:** 1209
Fisheye flaws
in weldments ... **A17:** 582
Fisheyes *See also* Cracking; Flakes
cause of ... **A11:** 92
as cleaning defect ... **EL1:** 777
defined ... **A11:** 4
definition ... **M6:** 7
from hydrogen embrittlement ... **A11:** 28, 248
Fishing
coniposite niaterial applications for ... **EM1:** 845-846
Fishing rod reels
powders used ... **A7:** 574
Fishmouthing *See* Alligatoring
Fishscale
on porcelain enameled sheet ... **M1:** 177-179
Fishtail ... **A14:** 6, 359
Fissile materials
nuclear applications ... **A7:** 664
Fission
process, neutron production ... **A10:** 421
Fission (continued)
spontaneous, as neutron source for NAA ... **A10:** 234
thermal-neutron, of uranium ... **A10:** 238
Fission, safety
in nuclear fuel pellet fabrication ... **A7:** 666
Fissure *See also* Cracking
in arc welds of nickel-based alloys ... **M6:** 363
definition ... **M6:** 7
Fissures ... **A6:** 1073
AISI/SAE alloy steels ... **A12:** 294
intergranular secondary, wrought aluminum alloys ... **A12:** 418
magnesium alloy ... **A12:** 456
OFHC copper ... **A12:** 401
titanium alloys ... **A12:** 452, 453
in weldments ... **A17:** 582
Fissuring
at grain boundaries, from hydrogen attack ... **A11:** 645
creep ... **A11:** 272
as hydrogen damage ... **A13:** 331
Fit-up, and welding
inspection techniques ... **A17:** 645-646
Fit-up stresses
as tension source for stress-corrosion cracking ... **A8:** 502
Fitness-for-service assessment of welded structures ... **A6:** 1108-1115
application of fracture-assessment procedures ... **A6:** 1110
benefits of a fitness-for-service approach ... **A6:** 1114-1115
design life ... **A6:** 1108
environmental cracking ... **A6:** 1111-1112
crack growth prevention ... **A6:** 1112
in-service monitoring ... **A6:** 1112
leak-before-break philosophy ... **A6:** 1112
remaining life prediction ... **A6:** 1112
failure modes ... **A6:** 1108
failure-assessment diagrams (FAD) ... **A6:** 1109-1110
fatigue design ... **A6:** 1110-1111, 1112
fracture ... **A6:** 1108-1110
fracture mechanics assessment procedures ... **A6:** 1109, 1111
crack-tip opening displacement (CTOD) ... **A6:** 1109
elastic-plastic fracture mechanics (EPFM) ... **A6:** 1109
J-integral ... **A6:** 1109
linear-elastic fracture mechanics (LEFM) ... **A6:** 1109
high-temperature creep ... **A6:** 1112-1114
creep-life fraction rule ... **A6:** 1113
expended-life-fraction ... **A6:** 1114
Larson-Miller parameter ... **A6:** 1113
Wedel-Neubauer method ... **A6:** 1114, 1115
life extension ... **A6:** 1108
S-N curve approach ... **A6:** 1110-1111
Fitting rust ... **A18:** 242
Fittings *See also* Joints
cast iron pipe ... **M1:** 100, 103
Five-axis robot
for leaded and leadless surface-mount joints ... **EL1:** 731
Five-point average crack length ... **A8:** 415
Five-spindle chucking machine ... **A7:** 669
Fixed base cone
as angle of repose measurement ... **A7:** 282
Fixed crack tip opening angle
in Rice *J*-integral fracture testing ... **A8:** 449
Fixed critical strain
in Rice *J*-Integral fracture testing ... **A8:** 449
Fixed fill levels ... **A7:** 323-325
Fixed height cone
as angle of repose measurement ... **A7:** 282
Fixed oil
defined ... **A18:** 9

SUBJECTS OF THE INDEXED VOLUMES: ASM Handbook (designated by the letter "A"): **A1:** Properties and Selection: Irons, Steels, and High-Performance Alloys (1990); **A2:** Properties and Selection: Nonferrous Alloys and Special-Purpose Materials (1990); **A3:** Alloy Phase Diagrams (1992); **A4:** Heat Treating (1991); **A6:** Welding, Brazing, and Soldering (1993); **A7:** Powder Metallurgy (1984); **A8:** Mechanical Testing (1985); **A9:** Metallography and Microstructures (1985); **A10:** Materials Characterization (1986); **A11:** Failure Analysis and Prevention (1986); **A12:** Fractography (1987); **A13:** Corrosion (1987); **A14:** Forming and Forging (1988); **A15:** Casting (1988); **A16:** Machining (1989); **A17:** Nondestructive Testing and Quality Control (1989); **A18:** Friction, Lubrication, and Wear Technology (1992). **Metals Handbook, 9th Edition** (designated by the letter "M"): **M1:** Properties and Selection: Irons and Steels (1978); **M2:** Properties and Selection: Nonferrous Alloys and Pure Metals (1979); **M3:** Properties and Selection: Stainless Steels, Tool Materials and Special-Purpose Materials (1980); **M4:** Heat Treating (1981); **M5:** Surface Cleaning, Finishing, and Coating (1982); **M6:** Welding, Brazing, and Soldering (1983). **Engineered Materials Handbook** (designated by the letters "EM"): **EM1:** Composites (1987); **EM2:** Engineering Plastics (1988); **EM3:** Adhesives and Sealants (1990); **EM4:** Ceramics and Glasses (1991); **Electronic Materials Handbook** (designated by the letters "EL"): **EL1:** Packaging (1989).

Fixed plug
drawing with ... **A14:** 330
Fixed probes *See also* Probes; Sensors
coordinate measuring machines ... **A17:** 25
Fixed vane
for hydraulic torsional system ... **A8:** 216
Fixed-abrasive lap
flatness compared to abrasive papers ... **A9:** 39
Fixed-bridge type
coordinate measuring machines ... **A17:** 21
Fixed-ceramic capacitors ... **EL1:** 179
Fixed-end torsion testing
axial stresses ... **A8:** 157, 180
high temperature ... **A8:** 157
vs. free-end testing ... **A8:** 181-182
Fixed-frequency continuous-wave reflection
microwave inspection ... **A17:** 206
Fixed-frequency continuous-wave transmission
microwave inspection ... **A17:** 205-206
Fixed-land bearing *See* Fixed-pad bearing
Fixed-load or fixed-displacement, crack extension force curves
defined ... **A8:** 5
Fixed-pad bearing
defined ... **A18:** 9
Fixed-plane design
computer-aided manufacturing ... **EL1:** 130
Fixed-position tester
for gear fatigue ... **A8:** 370
Fixed-table type
horizontal coordinate measuring machine ... **A17:** 23
Fixed-volume method
therilial expansion niolding ... **EM1:** 590
Fixer
defined ... **A17:** 314
radiographic, removal ... **A17:** 355-356
Fixing
of radiographic film ... **A17:** 353-356
Fixture, or set
time ... **EM3:** 13
Fixtures
aluminum and aluminum alloys ... **A2:** 14
definition ... **M6:** 7
function in torsional testing ... **A8:** 146
for holding microhardness specimens ... **A8:** 93, 96
for quenching ... **M4:** 65
Fixtures for
arc welding ... **M6:** 353-355
of aluminum alloys ... **M6:** 379-380
of heat-resistant alloys ... **M6:** 353-355
of magnesium alloys ... **M6:** 428-429
of nickel alloys ... **M6:** 437-438
brazing of heat-resisting alloys ... **M6:** 1016
dip brazing of steels ... **M6:** 992
electron beam welding ... **M6:** 612
flash welding ... **M6:** 561-564
flux cored arc welding ... **M6:** 104
furnace brazing of steels ... **M6:** 930, 940-941
gas tungsten arc welding ... **M6:** 197
of titanium ... **M6:** 454
induction brazing of steel ... **M6:** 971
laser beam welding ... **M6:** 668
manual torch brazing ... **M6:** 954-955
projection welding ... **M6:** 509-510
resistance spot welding ... **M6:** 489
shielded metal arc welding ... **M6:** 79-81
submerged arc welding ... **M6:** 134
Fixtures, graphite
for electrical integrity of a seal ... **EM4:** 539
Fixtures, heat-resistant alloys for *See* Heat-resistant alloys, fixtures
Fixturing ... **EM3:** 36-37
for electrical testing ... **EL1:** 567
radio frequency ... **EL1:** 949
and thermal energy method of deburring ... **A16:** 578
for wave soldering ... **EL1:** 590
Fizeau interferometer
application ... **A17:** 14, 16
lade potential
aqueous corrosion ... **A13:** 35
lake
defined ... **EM2:** 18
lake aluminum powders ... **A7:** 125
explosivity ... **A7:** 195
microstructure ... **A7:** 593
Flake graphite *See also* Gray iron; Quasi-flake graphite ... **A1:** 13 **A18:** 698 **M1:** 5-7, 12, 13, 14, 15, 22, 31
defined ... **A15:** 5
eutectic growth ... **A15:** 175-176
mesh-form ... **A18:** 699
microstructure ... **A18:** 698-699, 700-701
microstructure, gray cast iron ... **A15:** 120
types ... **A18:** 698
Widmanstatten ... **A18:** 699
Flake magnesium powders
explosivity ... **A7:** 195
Flake powders ... **A7:** 5
apparent density ... **A7:** 272
high-energy milling ... **A7:** 69
ignition of ... **A7:** 194
metallic pigments ... **A7:** 593-596
particle shape ... **A7:** 60, 233, 234, 593
Flaked copper
applications ... **A2:** 402
Flakes *See also* Cooling cracks; Flaking; Hydrogen embrittlement; Hydrogen, flaking; Hydrogen flaking; Thermal cracks ... **A13:** 6-7, 164, 242
AISI/SAE alloy steels ... **A12:** 310
as alloy segregation ... **A11:** 121
in alloy steel billet ... **A9:** 175
bright ... **A12:** 415
defined ... **A9:** 8 **A11:** 4
forging, as crack initiation site ... **A12:** 342
from hydrogen embrittlement ... **A11:** 28, 248
graphite ... **A11:** 360
graphite, gray iron fracture at ... **A12:** 225
high-carbon steels ... **A12:** 285
hydrogen, in forging ... **A11:** 88
hydrogen, in forgings ... **A17:** 492
internal, radiographic methods ... **A17:** 296
magnetic particle, for magnetic painting ... **A17:** 127
as notches ... **A11:** 85, 88
revealed by macroetching ... **A9:** 173
subsurface ... **A11:** 79
ultrasonic inspection of ... **A17:** 232
Flakes, formation of
in steels ... **A1:** 716-717
Flaking *See also* Flakes; Hydrogen flaking; Spalling ... **A18:** 259
in bearing failures ... **A11:** 494
in closed-die forging ... **A14:** 82
defined ... **A18:** 9
hydrogen, in tool steel ... **A11:** 574, 581
in rolling-element bearings ... **A11:** 502-503
Flame adjustment ... **M6:** 587
brazing of stainless steels ... **M6:** 1010
oxyacetylene braze welding ... **M6:** 596
Flame annealing ... **A4:** 285 **M4:** 506
defined ... **A9:** 8
Flame arrestors
P/M porous parts for ... **A7:** 700
Flame atomic absorption analysis
recommended practices ... **A7:** 249
Flame atomic absorption spectrometry
analytical sensitivities of ... **A10:** 47
flame AES and flame AFS, compared ... **A10:** 45
and GFAAS, compared ... **A10:** 58
parameters, for alloying elements in steels ... **A10:** 56
Flame atomic emission spectrometry, flame AES and flame AAS
compared ... **A10:** 45
Flame atomic fluorescence spectrometry, flame AES and flame AAS
compared ... **A10:** 45
Flame atomization ... **A10:** 48
Flame atomizers
atomic absorption spectrometry ... **A10:** 44-49
modification or salting ... **A10:** 54
technology of ... **A10:** 53
Flame classes
UL ... **EM2:** 77
Flame cutting ... **A7:** 842-845
applications ... **A7:** 844-845
definition ... **A6:** 1209
equipment ... **A7:** 842-844
of fracture surfaces ... **A11:** 19
and mechanical cutting, compared for steam equipment ... **A11:** 602
and scarfing, powders used ... **A7:** 573, 842
Flame cutting (continued)
of specimens ... **A12:** 76
Flame deposition
and drilling ... **A16:** 219
Flame emission spectroscopy ... **A10:** 28-29, 72
droplet sequence ... **A10:** 29
ionization interferences ... **A10:** 34
Flame gouges
in arc welds ... **A11:** 413
Flame hardening *See also* Gray iron, flame hardening ... **M1:** 528, 532 **M4:** 484-506
annealing ... **A4:** 285 **M4:** 506
applications ... **A4:** 268, 275-276 **M4:** 484
benefits ... **A4:** 282
burners ... **A4:** 272-274
cast irons ... **A4:** 284
combination progressive-spinning method ... **A4:** 270, 274, 280
control ... **A4:** 274-275
depth of hardness ... **A4:** 276-277
dimensional control ... **A4:** 281, 282 **M4:** 500-502
ductile iron ... **A15:** 659 **M1:** 37
equipment ... **A4:** 272-274
equipment maintenance ... **A4:** 277-278
fatigue resistance, effect on ... **M1:** 674
fuel gases ... **A4:** 270-272
gear materials ... **A18:** 261
gray cast iron ... **M1:** 29-30
gray iron ... **M4:** 539-541
hardenable steels ... **M1:** 457, 470
hardness ... **M4:** 487, 488, 494-496
malleable iron ... **A4:** 695-696 **M1:** 73
material selection ... **A4:** 283-285
medium-carbon steels ... **A12:** 265, 266
methods ... **A4:** 268-272, 274, 275, 276, 279, 280
operating procedures ... **A4:** 274-275
pattern of hardness ... **A4:** 276-277
preheating ... **A4:** 275-276 **M4:** 493-494
preventive maintenance ... **A4:** 278-279
problems ... **M4:** 499-500
problems and their causes ... **A4:** 280-281
process selection ... **A4:** 281-283 **M4:** 502-503
progressive method ... **A4:** 268-269, 271, 272, 274-276, 279
quenching ... **M4:** 498-499
quenching media ... **A4:** 280 **M4:** 499
quenching methods and equipment ... **A4:** 279-280
safety precautions ... **A4:** 279 **M4:** 497-498
scope ... **A4:** 268
spinning methods ... **A4:** 269-270, 271, 275, 276, 279
spot (stationary) method ... **A4:** 268, 269, 271, 275, 276, 279
surface conditions ... **A4:** 281 **M4:** 500, 501
tempering ... **A4:** 281 **M4:** 500
Flame hardening, burners
air-fuel gas ... **A4:** 273 **M4:** 490
construction materials for ... **A4:** 273-274 **M4:** 491-492
gas consumption ... **A4:** 272 **M4:** 487-489
high-velocity convection ... **A4:** 273 **M4:** 491
mixer-burner system ... **A4:** 273, 274 **M4:** 491
oxy-fuel gas flame heads ... **A4:** 272-273 **M4:** 489-490
radiant type ... **A4:** 273
radiant-type ... **M4:** 490-491
Flame hardening, equipment maintenance
air-gas type burners ... **A4:** 278 **M4:** 497
carbon deposit ... **A4:** 277-278 **M4:** 496
corrosion ... **A4:** 278 **M4:** 496-497
electrical components ... **A4:** 278 **M4:** 497
flame heads ... **A4:** 277-278 **M4:** 496
mechanical components ... **A4:** 278 **M4:** 497
movable holding fixtures ... **A4:** 278 **M4:** 497
piping ... **A4:** 278 **M4:** 497
preventive maintenance ... **A4:** 278-279 **M4:** 497
Flame hardening, fuel gases
depth of heating ... **A4:** 270-271 **M4:** 487
time-temperature-depth relations ... **A4:** 271-272 **M4:** 488
Flame hardening, material selection
alloy steels ... **A4:** 283, 284 **M4:** 505, 506
applications ... **A4:** 283, 284 **M4:** 503-504
carbon steels ... **A4:** 283, 284, 285 **M4:** 504-505
cast iron ... **A4:** 283, 284 **M4:** 505, 506
Flame hardening, methods
progressive ... **A4:** 268-269, 271, 272, 274, 275, 276, 279 **M4:** 485

Flame hardening, methods (continued)
progressive-spinning, combination **A4:** 270, 274, 280 **M4:** 486
spinning **A4:** 269-270, 271, 275, 276, 279 **M4:** 485-486
spot (stationary)....... **A4:** 268, 269, 271, 275, 276, 279 **M4:** 485
Flame hardening, operating procedures
coupling distance **A4:** 275 **M4:** 492
examples **A4:** 274, 275 **M4:** 493, 494, 495
flame velocity **A4:** 274-275 **M4:** 492
gas pressures **A4:** 274 **M4:** 492
hardening temperatures **A4:** 275 **M4:** 492-493
operator skill tests **A4:** 274 **M4:** 492
oxygen-to-fuel ratio **A4:** 274 **M4:** 492
Flame impingement
effect of heat fluxes on **A11:** 606
in P/M die component **A11:** 573, 579
Flame plating **A18:** 644
Flame polishing
of glass **EL1:** 104
Flame propagation rate
definition **A6:** 1209
Flame resistance *See also* Combustion; Flame retardants; Flammability; Self-extinguishing resin **EM3:** 13
defined **EM1:** 11 **EM2:** 18
polyamide-imides (PAI) **EM2:** 129
of prepregs **EM1:** 141
unsaturated polyesters **EM2:** 248-249
Flame retardant(s)
defined **EM1:** 11
polyester resins **EM1:** 96
in sheet iiiolding compounds **EM1:** 158
Flame retardants **EM2:** 503-504 **EM3:** 13
as additive, effects **EM2:** 424
defined **EM2:** 18
effect, chemical susceptibility **EM2:** 573
in engineering thermoplastics **EM2:** 98
as polymer additives **EM2:** 67
in rigid epoxies **EL1:** 813
in ultrahigh molecular weight polyethylenes (UHMWPE) **EM2:** 171
unsaturated polyesters **EM2:** 248-249
vinyl esters **EM2:** 272
Flame sources
applications **A10:** 29-30
burner selection for **A10:** 28
as emission source for optical emission spectroscopy **A10:** 28-29
Flame spraying **A7:** 5 **A13:** 7, 459
ceramic coatings **M5:** 534-536, 538-542, 546
ceramic coatings for adiabatic diesel engines **EM4:** 992
coating thickness, control of **M5:** 541-542
combustion system *See* Combustion flame spraying; Detonation gun systems
defined **EM2:** 18
definition **M6:** 7
equipment **M5:** 540-542
fuse and flame spray method **M5:** 366
gravity-feed powder system **M5:** 540-542
for metallizing by a liquid state **EM4:** 542, 543
oxygen-to-gas ratio **M5:** 367
plasma-arc *See* Plasma-arc thermal spray coating
pressure-feed powder system **M5:** 540
process steps **M5:** 539-540
in reaction sintering **EM4:** 292
rod spray system **M5:** 540-542
selective plating compared to **M5:** 300-301
surface preparation for **M5:** 539
thermal spray coating **M5:** 365-368
Flame spraying (FLSP)
cast irons **A6:** 715, 720
definition **A6:** 1209
Flame spread, and siil()ke eniission
by polyester systems **EM1:** 95
Flame straightening *See also* Straightening
defined **A14:** 6
Flame temperature
torch soldering **A6:** 135
Flame tests
as qualitative wet analyses **A10:** 168
Flame treating **EM3:** 13
defined **EM2:** 18
Flame types **M6:** 587
Flame-retarding agents
as additives **EM2:** 503-504
Flaming of surface **EM3:** 35
Flammability *See also* Explosions; Fires; Pyrophoricity; Safety precautions **EM3:** 13
aluminum-lithium alloys **A2:** 182-183
defined **EM1:** 11 **EM2:** 18
as design consideration **EM2:** 1
of fluxes **EL1:** 644
high-impact polystyrenes (PS, HIPS) **EM2:** 199
hydrogen **A7:** 345
information sources on **EM2:** 93
liquid crystal polymers (LCP) **EM2:** 181
metals vs plastic **EM2:** 77
plastic packages, additives for **EL1:** 475
of polyester resins **EM1:** 96
polyether sulfones (PES, PESV) **EM2:** 160
of polyether-imides (PEI) **EM2:** 158
polysulfones (PSU) **EM2:** 201
polyurethanes (PUR) **EM2:** 262
properties **EM2:** 78, 410, 642
rating, of laminates **EL1:** 536-537
of resins **EM1:** 135
sheet molding compounds **EM1:** 158
thermoplastic polyimides (TPI) **EM2:** 177
thermoplastic resins **EM2:** 618-619
thermosetting resins **EM2:** 223
unsaturated polyesters **EM2:** 248-250
water-base versus organic-solvent-base adhesives properties **EM3:** 86
Flammable ratings
criteria characteristics **EM1:** 358
Flammable solvent cleaners/removers
for liquid penetrant inspection **A17:** 75-76
Flammables
storage and handling **EM3:** 686
Flange
hexagonal **A8:** 221-222
for stored-torque Kolsky bar **A8:** 220
in torsional Kolsky bar test **A8:** 221-222
in upset test specimen **A8:** 580-581
Flange bearings **A18:** 741
Flange joint
laser-beam welding **A6:** 879
Flange joints
oxyfuel gas welding **M6:** 589-591
recommended grooves **M6:** 70, 72
Flange weld
definition **A6:** 1209 **M6:** 7
Flange weld size
definition **A6:** 1209
Flanged edge joints
plasma arc welding **A6:** 198
Flanged holes
piercing of **A14:** 469
Flanges
accurate spacing of **A14:** 532
curved, by rubber-die forming **A14:** 615
defined **A14:** 6
stretch, stainless steel **A14:** 773
workpieces with, drawing **A14:** 585-586
Flanges, penetrameters/identification markers
radiographic inspection **A17:** 343
Flanging
curved, bending of **A14:** 530
hole **A14:** 531, 559
limits **A14:** 530
rotary shearing **A14:** 706-707
rubber-die, failures in **A14:** 615
severe contour **A14:** 530
straight **A14:** 529
Flank wear **A18:** 610
defined **A18:** 9
Flank wear resistance
of cermets **A2:** 979
Flannel buffs **M5:** 118-119, 125
Flap polishing wheel **M5:** 109
Flap wheels
for surface preparation **A17:** 52
Flare joints
radiographic inspection **A17:** 334
Flare test
defined **A8:** 5
Flare-bevel groove weld
definition **M6:** 7
Flare-bevel-groove weld
definition **A6:** 1209
Flare-V-groove weld
definition **A6:** 1209 **M6:** 7
Flareless fittings
as tension source for stress-corrosion cracking **A8:** 502
Flares
powders used **A7:** 573
and signals, with metal fuels **A7:** 600, 602
Flaring **A14:** 6, 633, 777
Flash *See also* Closed-die forging; Flash land **A7:** 5
decorative **A15:** 17
defined **A14:** 6, 50 **A15:** 5 **EM1:** 11 **EM2:** 18
definition **A6:** 1209 **M6:** 7
design, closed-die forging **A14:** 78-79
extension, defined **A14:** 6
FIM sample rupture as **A10:** 587
from pouring temperatures **A15:** 283
gutter **A14:** 50
historic applications **A15:** 16
line, defined **A14:** 6
as metallic projection casting defect **A11:** 381
niobium forging **A14:** 237
pan, defined **A14:** 6
parting line **A15:** 192
recesses for **A14:** 89
reduction/elimination, in precision forging **A14:** 158
in sintering atmospheres **A7:** 348-349
uniform **A14:** 280
Flash, adhesive
bonded joints **A17:** 612
Flash coat
definition **A6:** 1209 **M6:** 7
Flash dewaxing, as pattern removal
investment casting **A15:** 262
Flash evaporation vacuum coating **M5:** 391-392
Flash, forging
trimming of **M1:** 366
Flash gutter **A14:** 50
Flash land **A14:** 6, 50
corrugations in **A14:** 277
impression **A14:** 50
variations, closed-die forging **A14:** 79
Flash lines
fatigue-crack origin at **A12:** 332
forging, medium-carbon steels **A12:** 258
Flash mold
defined **EM2:** 18
Flash pickling
heat-resistant alloys **M5:** 564-565
nickel and nickel alloys **M5:** 670-671
Flash plating
thin-film hybrids **EL1:** 329-330
Flash point **A18:** 84
defined **A18:** 9
environmental protection standards for lubricant disposal **A18:** 87
of organic cleaner blends **EL1:** 663
Flash radiography **A6:** 160
Flash temperature **A18:** 39-44
defined **A18:** 9
gears **A18:** 538

SUBJECTS OF THE INDEXED VOLUMES: ASM Handbook (designated by the letter "A"): **A1:** Properties and Selection: Irons, Steels, and High-Performance Alloys (1990); **A2:** Properties and Selection: Nonferrous Alloys and Special-Purpose Materials (1990); **A3:** Alloy Phase Diagrams (1992); **A4:** Heat Treating (1991); **A6:** Welding, Brazing, and Soldering (1993); **A7:** Powder Metallurgy (1984); **A8:** Mechanical Testing (1985); **A9:** Metallography and Microstructures (1985); **A10:** Materials Characterization (1986); **A11:** Failure Analysis and Prevention (1986); **A12:** Fractography (1987); **A13:** Corrosion (1987); **A14:** Forming and Forging (1988); **A15:** Casting (1988); **A16:** Machining (1989); **A17:** Nondestructive Testing and Quality Control (1989); **A18:** Friction, Lubrication, and Wear Technology (1992). **Metals Handbook, 9th Edition** (designated by the letter "M"): **M1:** Properties and Selection: Irons and Steels (1978); **M2:** Properties and Selection: Nonferrous Alloys and Pure Metals (1979); **M3:** Properties and Selection: Stainless Steels, Tool Materials and Special-Purpose Materials (1980); **M4:** Heat Treating (1981); **M5:** Surface Cleaning, Finishing, and Coating (1982); **M6:** Welding, Brazing, and Soldering (1983). **Engineered Materials Handbook** (designated by the letters "EM"): **EM1:** Composites (1987); **EM2:** Engineering Plastics (1988); **EM3:** Adhesives and Sealants (1990); **EM4:** Ceramics and Glasses (1991); **Electronic Materials Handbook** (designated by the letters "EL"): **EL1:** Packaging (1989).

Flash temperature (continued)
symbol and units **A18:** 544
Flash weld
definition **M6:** 7
Flash welding **A6:** 247-248 **M6:** 557-580
applications **A6:** 247 **M6:** 557-558
argon gas purging **M6:** 558
auxiliary equipment **A6:** 247
backups **M6:** 564
of blanks **A14:** 451
clamping dies **M6:** 561-563
horizontally sliding **M6:** 562
materials **M6:** 562
pivot **M6:** 562
shape and size **M6:** 563
vertically sliding **M6:** 562
cleaning, preweld **M6:** 564
components of machine **A6:** 247
controls **A6:** 247
definition **A6:** 247, 1209 **M6:** 7
electrode cooling **M6:** 562-563
end preparation **M6:** 564
equipment **A6:** 247-248
fixtures **M6:** 562-564
flashing cam **A6:** 248
force **M6:** 559-560
hardness of welds **M6:** 577-578
heat sources **M6:** 558-559
direct current power supplies **M6:** 559
flashing **M6:** 558-559
frequency converters **M6:** 559
three-phase power supplies **M6:** 559
wave-shaping power supplies **M6:** 559
heat-affected zone **M6:** 569-571, 576-578
machine design **M6:** 560-561
adaptive controls **M6:** 560-561
automation **M6:** 560
flashing and upsetting mechanisms **M6:** 560-561
platens **M6:** 560
transformers **M6:** 560
metals welded **M6:** 557-558
miter joints **M6:** 565-566
oxygen depletion system **A6:** 248
postweld processing **M6:** 572-573
heat treating **M6:** 573
second upset **M6:** 572
sizing, straightening, forming **M6:** 573
testing **M6:** 573
weld flash removal **M6:** 572
power sources **A6:** 248
safety precautions **M6:** 59
silicon-controlled rectifier (SCR) contactors **A6:** 247, 248
specifications **M6:** 580
state-of-the-art welding unit components **A6:** 248
strength of welds **M6:** 578-579
subsequent processing **M6:** 574-576
weld defects **M6:** 579-580
cast metal entrapment **M6:** 579-580
circumferential crevices **M6:** 579
cracking in heat-affected zone **M6:** 559
decarburization **M6:** 580
inclusions **M6:** 580
intergranular oxidation **M6:** 580
voids **M6:** 580
weld properties **M6:** 558
weld quality variables **M6:** 573-574
alloy characteristics **M6:** 573-574
alloy depletion **M6:** 574
compositional effects **M6:** 574
prior heat treatment **M6:** 574
section shape **M6:** 573
weldability ratio **M6:** 573
welding cycle **M6:** 570
welding parameters **M6:** 558
welding schedules **M6:** 576
bar **M6:** 577
plate **M6:** 577
tube **M6:** 577
welding sequence **M6:** 566-571
burnoff **M6:** 567
flashing **M6:** 568-569
preheating **M6:** 567-568
upset current **M6:** 571
upsetting **M6:** 569-571

Flash welding (continued)
workpiece extension from dies **M6:** 571
workpiece alignment **M6:** 564-565
workpiece design **M6:** 571-572
heat balance **M6:** 571-572
wrought martensitic stainless steels **A6:** 441
Flash welds
failure origins **A11:** 442-443
in pipe **A11:** 698
Flash-back **A7:** 348-349
Flash-butt welding
aluminum metal-matrix composites **A6:** 558
definition **A6:** 1209
precipitation-hardening stainless steels **A6:** 492
titanium alloys **A6:** 522
Flashback
definition **M6:** 7
in premix burners **A10:** 28
Flashback arrester
definition **M6:** 7
Flashbacks **A6:** 1201
Flashing **A6:** 842-843, 844
control **M6:** 559
definition **M6:** 558
machine design **M6:** 561-562
surfaces in flash welding **M6:** 558-559
voltages **M6:** 559
Flashing time
definition **M6:** 7
Flashless forging
with knuckle-drive mechanical press **A14:** 172-173
Flask method, Schöniger
for common elements **A10:** 215
Flask pin guides
use **A15:** 190
Flask(s) *See also* Coke bed; Cope; Drag
defined **A15:** 5-6, 341
for match plate pattern plaster mold casting **A15:** 245
molding, flaskless **A15:** 341
molds, types, green sand molding **A15:** 341
pattern, purpose **A15:** 190
in plaster molding **A15:** 243
snap *See* Snap flask
tight **A15:** 341
Unicast process **A15:** 251
Flaskless molding **A15:** 37, 341
Flat cleavage fracture
irons **A12:** 219
Flat conductor
microstripline properties **EL1:** 602
Flat deck vibratory shakeout device
green sand molding **A15:** 347
Flat die forging *See* Open-die forging
Flat dies
as CAD/CAM application **A14:** 323-324
for forging solid cylinder **A14:** 67
for four-diameter spindle **A14:** 68
for gear blank and hub **A14:** 67-68
for hot extrusion **A14:** 320, 323
for hot forging **A14:** 43
Flat frequency sensors
acoustic emission inspection **A17:** 280
Flat glass
applications **EM4:** 379, 1015
composition ranges **EM4:** 382
strength a key factor **EM4:** 741
Flat grip
for axial fatigue testing **A8:** 369
Flat part polishing machines **M5:** 122-123
Flat plate impact test **A8:** 210-211
Flat plates *See also* Plate
ECP surface flaw detection **A17:** 137-138
radiographic inspection **A17:** 332
solidification of **A15:** 606
Flat position
definition **A6:** 1209 **M6:** 7
Flat rolling *See also* Plate(s)
defined **A14:** 343
Flat sheet *See also* Sheet metals; Sheetforming
as encapsulation material **A7:** 429
rotary shearing **A14:** 705-707
shearing of **A14:** 701-707
straight-knife shearing **A14:** 701-705
Flat sheet specimen
for fatigue testing **A8:** 371-372

Flat sheet specimens
fatigue corrosion testing **A13:** 293
Flat springs **A1:** 302, 305, 306, 307
steel **M1:** 287-288, 290, 299
Flat steel wire
definition **M1:** 259
Flat tape laying **EM1:** 624-630
machines, commercial **EM1:** 626-627
machines, government sponsored **EM1:** 625-626
manual lay-up **EM1:** 624
operating parameters **EM1:** 627-628
tape placement machines **EM1:** 628
tape preparation machine **EM1:** 627
time requirements **EM1:** 629
waste strip **EM1:** 628, 630
Flat welding
arc welds of coppers **M6:** 402
flux cored arc welds **M6:** 105-107
gas metal arc welds of coppers **M6:** 417
indication by electrode classification **M6:** 84
oxyfuel gas welds **M6:** 589
shielded metal arc welds **M6:** 76, 85, 88, 441
of nickel alloys **M6:** 441
submerged arc welds **M6:** 134
Flat wire **A1:** 277
Flat-back dies
roll forging **A14:** 97
Flat-belt drive gear **A7:** 667
Flat-body two-terminal products **EL1:** 431
Flat-face tensile fractures
in aluminum alloy plate **A11:** 76
defined **A11:** 82
in ductile materials **A11:** 76, 82
equiaxed dimples on **A11:** 76
in sheet or plate **A11:** 109, 110
Flat-film extrusion
products **EM2:** 384-385
Flat-rolled products
aluminum and aluminum alloys **A2:** 5
angle-beam ultrasonic inspection **A17:** 270
mechanized inspection **A17:** 270
silicon steels, as magnetically soft materials **A2:** 766-769
straight-beam edge ultrasonic inspection **A17:** 269-270
straight-beam top ultrasonic inspection **A17:** 268-269
wrought aluminum alloy **A2:** 33
zinc and zinc alloy **A2:** 531
Flat-seam underground mining
cemented carbide tools **A2:** 975
Flat-spring failure
edge defect **A11:** 559
Flat-type Brinell indentation **A8:** 85
Flatbed *x-y* recorder **A8:** 617
Flatness **A14:** 73, 480-481
of porcelain enameling **A13:** 448
steel sheet and strip **M1:** 157, 160-161
Flatness, control of
in carbon steel sheet and strip **A1:** 205-206
Flatness of abraded surfaces **A9:** 39
effect of polishing materials on **A9:** 40
Flatpacks
bottom-brazed **EL1:** 992-993
ceramic **EL1:** 206
ceramic and metal, as package family **EL1:** 404
die attachments **EL1:** 213-217
glass-sealed **EL1:** 993
as hybrid package form **EL1:** 451
package outline **EL1:** 206
sealing of **EL1:** 237
welding processes **EL1:** 240
Flats
defined **EM2:** 18
Flatteners **A14:** 44, 713
Flattening
defined **A14:** 6
wrought titanium alloys **A2:** 619
Flattening dies **A14:** 6, 536
Flattening test
defined **A8:** 6
Flatware
copper and copper alloys **A14:** 822
Flatwise tension test **EM3:** 535, 536
polybenzimidazoles **EM3:** 171
Flavor components
IR determination of **A10:** 109

Flaw *See* Defect; Discontinuities
definition ... **A6:** 1209 **M6:** 7
Flaw characterization *See also* Cracks; Defects; Discontinuities; Flaw(s)
by electric current perturbation ... **A17:** 140-141
magnetic ... **A17:** 131
Flaw depth
effect, microwave inspection ... **A17:** 211
ultrasonic inspection ... **A17:** 240
Flaw detection ... **EM3:** 521-532
adhesive defects ... **EM3:** 521-522
categories ... **EM3:** 522
by electric current perturbation ... **A17:** 136
computed tomography (CT) ... **A17:** 364
effects of defects ... **EM3:** 522-526
and evaluation ... **A17:** 49-50
in fan-disk blade slots, by ECP ... **A17:** 138
NDE reliability models ... **A17:** 702
of steel bar and wire ... **A17:** 557
through spindle wall thickness ... **A17:** 139
ultrasonic, cold-drawn wire ... **A17:** 552
weak bond inspection ... **EM3:** 529-531
Flaw growth
in damage tolerance ... **A17:** 669
prediction, and NDE reliability ... **A17:** 663
Flaw leakage field
detection ... **A17:** 129-135
Flaw location
pressure vessels ... **A17:** 656
ultrasonic inspection ... **A17:** 240
Flaw size *See also* Size; Sizing
in damage tolerance design ... **A17:** 702
distribution ... **A17:** 663
initial, in fracture control/damage tolerance ... **A17:** 668
models of ... **A17:** 702
in NDE reliability ... **A17:** 663-664
percentage of back reflection technique ... **A17:** 263
as proportional to signal amplitude ... **A17:** 131
and sample size, in NDE reliability experiments ... **A17:** 694
and test frequency, eddy current inspection ... **A17:** 194
ultrasonic inspection ... **A17:** 240
Flaw(s) *See also* Crack detection; Crack growth; Cracks; Defects; Discontinuities Flaw characterization; Flaw detection Flaw growth; Flaw size; Flaw thickness Inclusions; Interior flaws; Nonmetallic inclusions; specific flaw types; Subsurface flaws; Surface flaws
casting, radiographic appearance ... **A17:** 348-349
casting types, ultrasonic inspection ... **A17:** 267-268
defined ... **A17:** 49, 232
in double submerged arc welded steel pipe ... **A17:** 565
forging, types ... **A17:** 491-494
as frequency shifts, microwave inspection ... **A17:** 222
liquid penetrant inspection ... **A17:** 86
location ... **A17:** 50
planar ... **A17:** 50
radiographic appearance ... **A17:** 348-351
in resistance-welded steel tubing ... **A17:** 562
reworking of ... **A17:** 86-87
in seamless steel tubular products ... **A17:** 7-568
in semiconductors, radiographic appearance ... **A17:** 350-351
shape ... **A17:** 50, 131
size ... **A17:** 50
small, electric current perturbation for ... **A17:** 137
small, impedance changes by ... **A17:** 173
in steel bar and wire ... **A17:** 549-550
thickness, ultrasonic inspection ... **A17:** 240
in tubular products ... **A17:** 561
type, in NDE reliability ... **A17:** 663
volume, and signal amplitude, magnetic characterization ... **A17:** 131
volumetric ... **A17:** 50
vs. defect, defined ... **A17:** 49
Flaws *See also* Defects; Internal defects; Material flaws
in bonded joints ... **EM1:** 486
damage tolerances of ... **EM1:** 265
effective, interlaminar ... **EM1:** 241-244
fracture-initiating ... **A11:** 748
intemal, fatigue fracture from ... **A12:** 279
interface, free-edge ... **EM1:** 241
material, in ceramics ... **A11:** 748-751
in particles ... **A7:** 59
shape parameter curves ... **A11:** 108
size, variation with crack length ... **EM1:** 255
and strength prediction ... **EM1:** 432
surface, fatigue crack growth from ... **A12:** 420
surface, fatigue fracture from ... **A12:** 279
Flax
as natural fiber ... **EM1:** 117
Flax-reinforced organic matrix composites ... **EM1:** 117
Fleet testing
dispersant effectiveness ... **A18:** 100
Flemings, M.C
semisolid metal casting/forging ... **A15:** 327
Flex roll
defined ... **A14:** 6
Flex rolling
defined ... **A14:** 6
effect on Lüders lines ... **A8:** 553
Flexed-beam impact tests ... **EM2:** 556-557
Flexed-plate impact tests ... **EM2:** 557
Flexibility ... **A8:** 6 **EM3:** 13
of epoxy composites ... **EM1:** 73
low-temperature ... **EM2:** 203
mer, and melt properties ... **EM2:** 57, 62
of metal casting ... **A15:** 39-41
of organic-coated steels ... **A14:** 565
and process selection ... **EM2:** 277
of tension testing machine ... **A8:** 50
Flexibilizer *See also* Plasticizer ... **EM3:** 13
defined ... **EM1:** 11 **EM2:** 18
as modifier. damping effects ... **EM1:** 215
Flexibilizers
types for flexible epoxies ... **EL1:** 818-819
Flexible borescopes
types ... **A17:** 5-10
Flexible envelope ... **A7:** 296
Flexible epoxies *See also* Epoxies; Epoxy; Epoxy materials; Rigid epoxies
applications, typical ... **EL1:** 818-820
available forms ... **EL1:** 817
design trade-offs ... **EL1:** 821
flexibilizers, types ... **EL1:** 818-823
materials ... **EL1:** 817-818
one- and two-component systems ... **EL1:** 817-818
premixed/frozen ... **EL1:** 818
testing ... **EL1:** 820-821
Flexible fiberscopes ... **A17:** 5
Flexible gages *See* Coordinate measuring machines (CMMs)
Flexible hinge *See* Flexure pivot bearing
Flexible inspection systems *See* Coordinate measuring machines (CMMs)
Flexible manufacturing systems
inspection of ... **A17:** 18
Flexible manufacturing systems (FMS) ... **A16:** 397-403
fixtures ... **A16:** 404, 407
Flexible molds
defined ... **EM2:** 18
defned ... **EM1:** 11
Flexible polyurethane foams ... **EM2:** 603
Flexible printed boards *See also* Boards; Flexible printed wiring (FPW); Printed wiring boards (PWBs); Rigid printed wiring boards ... **EL1:** 578-596
advantages/limitations ... **EL1:** 578-581
analysis and layout ... **EL1:** 594-596
cost effective design ... **EL1:** 592-593
cost effectiveness parameters ... **EL1:** 593-594
Flexible printed boards (continued)
design considerations ... **EL1:** 584-586
electrical design considerations ... **EL1:** 586-588
manufacturing ... **EL1:** 581-584
mechanical design considerations ... **EL1:** 588-592
rework processes ... **EL1:** 712
Flexible printed wiring (FPW) *See also* Flexible printed boards; Wiring
advantages/limitations ... **EL1:** 578-581
analysis and layout ... **EL1:** 594-596
design elements of cost-effective ... **EL1:** 592-593
design trade-offs ... **EL1:** 584-586
fabrication ... **EL1:** 580-581
protective coatings, properties ... **EL1:** 583
type comparison ... **EL1:** 580
Flexible transfer lines
as coordinate measuring machine application ... **A17:** 18
Flexible-die forming *See* Rubber-pad forming
Flexible-die forming presses ... **A14:** 503
Flexural
failure, defined ... **EL1:** 1144
fatigue, flexible printed boards ... **EL1:** 588
strength, flexible epoxies ... **EL1:** 821
Flexural characteristics
acid effects, unsaturated polyesters ... **EM2:** 247
cyanates ... **EM2:** 238
and impact strength ... **EM2:** 280
phenolics ... **EM2:** 244
polyaryl sulfones (PAS) ... **EM2:** 146
polyarylates (PAR) ... **EM2:** 140
polyaryletherketones (PAEK, PEK, PEEK, PEKK) ... **EM2:** 143
polyether imides (PEI) ... **EM2:** 157
polyether sulfones (PES, PESV) ... **EM2:** 160
polyethylene terephthalates (PET) ... **EM2:** 173
unsaturated polyesters ... **EM2:** 248
urethane hybrids ... **EM2:** 269
vinyl esters ... **EM2:** 274
Flexural creep *See also* Flexural characteristics
styrene-maleic anhydrides (S/MA) ... **EM2:** 219
Flexural damping *See also* Damping; Damping properties analysis
fiber volumee fraction effects ... **EM1:** 208
Flexural fatigue test ... **A1:** 861
Flexural fatigue tests
reliability ... **M1:** 676
Flexural modulus *See also* Flexural characteristics; Modulus ... **EM3:** 13
defined ... **EM1:** 11 **EM2:** 18
long-term exposure testing ... **EM1:** 825
polvester resins ... **EM1:** 90-92
sheet molding compounds ... **EM1:** 158
Flexural rigidity ... **A18:** 580
Flexural strength ... **EM3:** 13
defined ... **EM1:** 11 **EM2:** 18
of epoxy resin matrices ... **EM1:** 73
fabric, construction effects ... **EM1:** 149
long-term exposure testing ... **EM1:** 825
polyester resins ... **EM1:** 91, 92
retention with aging ... **EM1:** 93
sheet molding compounds ... **EM1:** 158
test ... **EM1:** 299
vs temperature, glass-polyester composites ... **EM1:** 93
Flexural test ... **EM1:** 299 **EM4:** 36
Flexure
defined ... **A8:** 6
effects in resin-matrix composites ... **A12:** 475-478
Flexure fatigue
zirconium alloys ... **A2:** 669
Flexure pivot bearing
defined ... **A18:** 9
Flexure stress
definition ... **EM4:** 632
Flexure system
for crack and lever testing machine ... **A8:** 369-370
Flight service evaluation
composite materials ... **EM1:** 823, 826-831

SUBJECTS OF THE INDEXED VOLUMES: ASM Handbook (designated by the letter "A"): **A1:** Properties and Selection: Irons, Steels, and High-Performance Alloys (1990); **A2:** Properties and Selection: Nonferrous Alloys and Special-Purpose Materials (1990); **A3:** Alloy Phase Diagrams (1992); **A4:** Heat Treating (1991); **A6:** Welding, Brazing, and Soldering (1993); **A7:** Powder Metallurgy (1984); **A8:** Mechanical Testing (1985); **A9:** Metallography and Microstructures (1985); **A10:** Materials Characterization (1986); **A11:** Failure Analysis and Prevention (1986); **A12:** Fractography (1987); **A13:** Corrosion (1987); **A14:** Forming and Forging (1988); **A15:** Casting (1988); **A16:** Machining (1989); **A17:** Nondestructive Testing and Quality Control (1989); **A18:** Friction, Lubrication, and Wear Technology (1992). **Metals Handbook, 9th Edition** (designated by the letter "M"): **M1:** Properties and Selection: Irons and Steels (1978); **M2:** Properties and Selection: Nonferrous Alloys and Pure Metals (1979); **M3:** Properties and Selection: Stainless Steels, Tool Materials and Special-Purpose Materials (1980); **M4:** Heat Treating (1981); **M5:** Surface Cleaning, Finishing, and Coating (1982); **M6:** Welding, Brazing, and Soldering (1983). **Engineered Materials Handbook** (designated by the letters "EM"): **EM1:** Composites (1987); **EM2:** Engineering Plastics (1988); **EM3:** Adhesives and Sealants (1990); **EM4:** Ceramics and Glasses (1991); **Electronic Materials Handbook** (designated by the letters "EL"): **EL1:** Packaging (1989).

Flight tube
for atom probe microanalysis **A10:** 591
Flint
natural abrasive **A16:** 434
in typical ceramic body compositions **EM4:** 5
Flint clay
refractory material composition **EM4:** 896
Flint glass
applications, glass containers **EM4:** 1082
iron content **EM4:** 378
Flip-chip assembly
failure mechanisms **EL1:** 1047
modular **EL1:** 442
on active silicon substrate **EL1:** 442
and silicon bump technology **EL1:** 440
"Flip-chip" bonding **A6:** 133
Float
defined **EM1:** 150
mechanical properties effect of **EM1:** 150
Float architectural
defect inclusion levels **EM4:** 392
Float automotive
defect inclusion levels **EM4:** 392
Float glass **EM4:** 453
defects and cost of losses **EM4:** 392
iron content **EM4:** 378
process **EM4:** 21
Float process **EM4:** 377
display glass properties **EM4:** 1048-1049
Float structure, deep-submergence vessel
inspection of **A17:** 114-115
Float-zone process **EL1:** 191-192
Floating bearing
defined **A18:** 9
Floating cams
for farm planters **A7:** 673
Floating chase
defined **EM2:** 19
Floating die **A7:** 5
defined **A14:** 6
tooling systems **A7:** 333
Floating die pressing **A7:** 5
Floating layer emulsion cleaning **M5:** 33-34
Floating plug **A14:** 6, 331
Floating soap
powder used **A7:** 573
Floating zone technique for purifying metals **M2:** 710
Floating-blade reamers **A16:** 241, 244, 246
Floating-ring bearing
defined **A18:** 9
Floating-threshold techniques
acoustic emission inspection **A17:** 285
Floating-zone refining technique
for ultrapurification **A2:** 1093-1094
Flocculants
as electrolytes for ceramic coatings **EM4:** 955
Flocculating agents
plating wastes treated with **M5:** 313-314
Flock
defined **EM2:** 19
Flock point **A18:** 87
defined **A18:** 9
Flocking **EM3:** 13
defined **EM2:** 19
Flood application
of lubricants **A14:** 515
Flood lubrication (bath lubrication)
defined **A18:** 9
Floor beam
-girder connection plates **A11:** 712
tied-arch **A11:** 714
Floor linoleum
powders used **A7:** 572
Floor materials
breweries **A13:** 1222
Floor molding
defined **A15:** 6
for dry sand molding **A15:** 228
Floor plastics
powders used **A7:** 572
Flop forging
defined **A14:** 6
Florad FC-721/FC-723 fluoropolymer coating **EL1:** 782
Flory-Rehner equation **EM3:** 413
Flotation, carbon
in iron castings **A11:** 358
Flow *See also* Creep; Flow curves; Flow lines; Flow localization; Flow rate; Flow stress; Flowability; Fluid flow; Leakage; Liquid(s); Metalflow; Plastic definition; Plastic flow; Slip; Yield **A8:** 6 **EM3:** 13
amorphous materials and metallic glasses **A2:** 813
defined **A11:** 4 **EM1:** 11 **EM2:** 19
depth, effect, crack propagation **A13:** 277
design effects **A13:** 340
detection methods **A17:** 60-61
direction, in magnetic field **A17:** 90
eddy current **A17:** 165-166
enhanced plastic **A13:** 165
of flexible epoxies **EL1:** 821
fluid **A17:** 60, 285
flux **A17:** 91
grain *See* Grain flow
in leaks, types **A17:** 58
-line, and gage penetration in specimen ends, micrograph **A8:** 590
line, defined **EM1:** 11
lines, defined **A11:** 4, 316
liquid, Reynold's numbers and **A15:** 591
marks, as casting defects **A11:** 384
marks, defined **EM1:** 11
micro-, resin, in composites **A11:** 742, 743
molecular weight effects, graphite fiber-PMR-15 composite **EM1:** 811
patterns, in forging **A11:** 319-320
plastic, tear ridges from **A12:** 224
properties, Charpy V-notch screening test for **A8:** 263
properties degradation, by hydrogen damage **A13:** 164
properties, effect of strain rate on **A8:** 38-45
resin, in cure cycle **EM1:** 753-755
strength **A11:** 138
strength, of crack growth specimen **A8:** 381
stress **A11:** 50, 102, 113
types, in gating **A15:** 591-592
of unidirectional tape prepreg **EM1:** 144
velocity, effect on erosion damage **A11:** 165-166
wet steam, erosion-corrosion **A13:** 964-971
Flow brazing
definition **M6:** 8
Flow brightening
definition **M6:** 8
Flow brightening (soldering)
definition **A6:** 1209
Flow cavitation
defined **A18:** 9
Flow cell systems
absorbance-subtraction studies of **A10:** 116
Flow chart
cleaning, of castings **A15:** 712-713
computer-aided dimensional inspection **A15:** 557
of foundry operations **A15:** 203-207
foundry, typical **A15:** 502
for green sand casting **A15:** 205
for investment casting **A15:** 206
macro-micro modeling, of solidification **A15:** 888
metal casting system **A15:** 344
for Replicast CS process **A15:** 270
for rheocasting **A15:** 207
Flow coating **A13:** 447, 1015
ceramic coatings **M5:** 538
of conformal coatings **EL1:** 764
paint *See* Paint and painting, flow coating
porcelain enamel **M5:** 516-518
of silicone conformal coatings **EL1:** 773-774
of urethanes **EL1:** 779-780
Flow continuity equation **A18:** 90
Flow curve *See also* True stress/true strain curve
for austenitic stainless steel torsion tests **A8:** 161-162
for carbon steel, torsion and tension testing **A8:** 163, 165
cyclic, grain coarsening **A8:** 175
defined **A8:** 23
isothermal high strain rate, stress-temperature plots for **A8:** 161
Flow curve (continued)
for length-to-diameter specimen ratios **A8:** 156-157
for low-carbon austenitic steel in torsion **A8:** 175
metal torsion test, dynamic recovery at hot working temperatures **A8:** 173
room temperature, in tension and torsion for copper **A8:** 162-164
shape from maximum load to fracture **A8:** 23
single-peak, grain refinement **A8:** 175
in torsion testing **A8:** 161-162
for Waspaloy **A8:** 162-163
Flow curves
generation, by hot compression testing **A14:** 439
process modeling of **A14:** 418-419
Flow diagram
environmental stress screening **EL1:** 884
Flow injection analysis **A10:** 55, 690
Flow laws
shear stress derivations for **A8:** 182-184
Flow line *See* Weld line
Flow lines **A9:** 686-687
as ALPID results **A14:** 427-428
in billet, modeling **A14:** 428
contours, micrographs **A14:** 382
defined **A8:** 6 **A9:** 8 **A14:** 6
in forged steel hook **A14:** 367
in forging **A14:** 427
revealed by macroetching **A9:** 174
Flow lines, as defect
semisolid alloys **A15:** 337
Flow lines, in wrought products
as optical metallography structural parameter **A10:** 302-303
Flow localization **A1:** 584-585
as adiabatic shear bands **A8:** 155
analyses **A8:** 169-173
by shear band formation **A8:** 589
-controlled failures, torsion sheet for **A8:** 154
controlled fracture tests **A8:** 588-589
of deformation **A8:** 573
distributed gage volume to detect **A8:** 589-590
from flow softening **A8:** 573
hot compression test for **A8:** 583
zones by sectioning and metallographic preparation **A8:** 588-589
Flow localization by shear band **A14:** 385
tests **A14:** 384-385
and workability **A14:** 364-365
Flow marks
defined **EM2:** 19
Flow meter **A7:** 5
Hall, for apparent density **A7:** 273-274
Hall, for powder flow **A7:** 278-279
Lea and Nurse permeability apparatus with **A7:** 264
Flow molding
defined **EM2:** 19
Flow plating process
selective plating **M5:** 295-297
Flow, powder *See also* Flow meter; Flow rate; Flowability **A7:** 187-189
in blending and premixing
direction under pressure **A7:** 300
effect of lubrication in compacts **A7:** 302
factor **A7:** 5
stress **A7:** 300
test **A7:** 5
Flow pressure **A18:** 43
Flow process
alkaline cleaning **M5:** 23
Flow proportional detector
wavelength- dispersive spectrometer **A10:** 520
Flow rate *See also* Flow; Pouring; Pouring time; Yield **A7:** 5, 278-281
of abrasives **A15:** 516-517
and angle of repose **A7:** 284-285
and apparent density, electrolytic iron powders **A7:** 273
in aqueous environment synthesis **A8:** 416
by empirical methods **A7:** 297
in chemical cleaning **A13:** 1142
chemical process, variable **A13:** 1136
conversion factors **A8:** 722 **A10:** 686
effect of particle size **A7:** 297
effect on cracking in elevated-temperature water

Flow rate (continued)
effects of lubricants **A7:** 190, 279, 280
of electrolytic copper powders **A7:** 114-115
fluid, aqueous corrosion **A13:** 40
Hall and Carney funnels for **A7:** 279
impact on aqueous corrosion current **A13:** 34
low alloy steel powder **A7:** 102
nonfunnel testing **A7:** 279-280
pouring **A15:** 500
testing **A7:** 278 279
variables affecting **A7:** 280-281
vs time, gating system **A15:** 597
Flow softening **A8:** 155, 172, 177, 573
Flow strength, and forgeability
effects **A14:** 76
Flow stress **A8:** 6
in 304L stainless steel **A8:** 602
aluminum alloy forgings **A14:** 241
of aluminum deformed in torsion **A8:** 178-179
at cold working temperatures **A8:** 575
at hot working temperatures **A8:** 161-162, 575
as athermal component and effective stress **A8:** 256
behavior at hot and cold working temperatures **A8:** 161-162, 575
body-centered cubic metals **A8:** 224
for closed-die forging **A14:** 75
in compression **A14:** 375-376
constrained **A8:** 577
of copper, effect of strain-rate history **A8:** 179
of copper in torsion, strain rate effect **A8:** 177-178
for cubic metals, effect of strain rate or temperature change **A8:** 223, 226
curve, compressive, with strain softening **A8:** 583
and deformation heating **A8:** 161
dependence on strain and strain rate **A8:** 44
dependence on strain rate for 1100-O aluminum **A8:** 236-237
during torsion testing **A8:** 160-163, 172
effect of strain rate **A8:** 38, 39
effect of structure **A14:** 171
effect of temperature-strain-time history on **A9:** 688-691
of forged titanium alloys **A14:** 268-270
and forging pressure **A1:** 585
of heat-resistant alloys **A14:** 232
of hexagonal close-packed metals **A8:** 223-224
of high-purity aluminum during rolling **A8:** 179
in hot compression **A8:** 582-583
for Nb-V microalloyed steel **A8:** 179
plane-strain **A8:** 577
-strain rate relationships, process modeling **A14:** 419
temperature and strain rate effect during torsion **A8:** 178
temperature and strain rate effects **A14:** 151
-temperature relations, process modeling **A14:** 419-420
torsion test for **A8:** 154
-true strain, equations **A14:** 418
-true strain rate, equations **A14:** 419
typical flow curves for **A8:** 161-162
vacuum-melted iron, continuous heating/cooling effect **A8:** 177
variations, precision forging **A14:** 161
vs extrusion rate, aluminum alloys **A14:** 317
vs strain rate, aluminum alloy forging **A14:** 242
in Waspaloy **A8:** 162-163
and workability as function of temperature **A14:** 166-170
Zener-Hollomon parameter in **A8:** 162-163
Flow through
defects **A14:** 385
defined **A14:** 6
Flow turning *See also* Shear spinning
refractory metals and alloys **A2:** 562
Flow welding
definition **M6:** 8
Flow-assisted corrosion **A18:** 223
Flow-on methods
for liquid penetrant developers **A17:** 79
Flow-through defect
in closed-die forging **A8:** 590, 594
Flow-through water-cooled copper mold
in casting machine **A15:** 311
Flowability
cellulose addition effects **A15:** 211
of composite powders **A7:** 173
control by high-energy milling **A7:** 69
defined **A15:** 6
definition **M6:** 8
effect on powder compact **A7:** 211
in milling of single particles **A7:** 59
of platinum powders **A7:** 150
Flower design
contour roll forming **A14:** 629
Flowing wells
oil/gas corrosion **A13:** 1298-1250
Flowing-gas proportional detectors
for x-ray spectrometers **A10:** 88
Fluctuating stress
in shafts **A11:** 461
Flue gas **A13:** 137, 1199-1200
corrosion **A11:** 619
thermal resistance to steam **A11:** 604
Flue gas desulfurization
alloy compositions for **A13:** 1368
dew point corrosion **A13:** 1001-1004
emission-control equipment for **A13:** 1367-1368
nickel-base alloys applications in **A13:** 654
stainless steel corrosion **A13:** 561
systems, corrosion of **A13:** 1004-1006
Flue gas desulfurization (FGD) applications **A6:** 746
Flue gas desulfurization (FGD) systems **A6:** 593
Fluid bearing *See* Hydrostatic bearing
Fluid bed cooler
for sands **A15:** 350
Fluid die(s)
can/cast **A7:** 543
carbon steel **A7:** 543
composite **A7:** 543
in forging press **A7:** 542
low-carbon steel **A7:** 543
rapid omnidirectional compaction **A7:** 542, 543
recyclable **A7:** 543
systems **A7:** 543
Fluid dynamics
and dissolution kinetics **A15:** 73
and hydrodynamic lubrication **A11:** 150-157
principles of **A17:** 58-59
Fluid erosion *See* Liquid impingement erosion
Fluid flow *See also* Flow rate; Fluid flow modeling; Fluidity; Pouring
compacted graphite irons **A15:** 671
and dies/machines, interaction of **A15:** 292
and gating, die casting **A15:** 289
and heat flow, interaction defects **A15:** 294-295
and heat/mass transfer, modeling of **A15:** 877-882
interdendritic, macrosegregation **A15:** 155
and macrostructural development **A15:** 135
modeling of **A15:** 867-876
nucleation effects **A15:** 101
and permeametry **A7:** 263
plain carbon steels **A15:** 711
principles, in gating **A15:** 590-592
of semisolid materials **A15:** 328
as source, acoustic noise **A17:** 285
system, die casting **A15:** 288-292
temperature field sensitivity **A15:** 147
turbulence by **A17:** 60
Fluid flow modeling
and actual metal flow, correlations **A15:** 876
Fluid flow modeling (continued)
energy balance methods **A15:** 867-869
examples **A15:** 874-876
of filling metal castings **A15:** 874-876
fluid domain identification **A15:** 872
with heat/mass transfer **A15:** 877-882
MAC technique **A15:** 871-872
mold filling, physical modeling **A15:** 869-870
momentum balance techniques **A15:** 870-876
velocity field, calculating **A15:** 872-874
Fluid flow phenomena during welding **A6:** 19-24
deep-penetration electron beam and laser welds **A6:** 22-23
gas-metal arc welding **A6:** 23-24
gas-tungsten arc welding **A6:** 19-22
mass transport in the arc **A6:** 19
submerged arc welding **A6:** 24
Fluid forming *See also* Fluid-cell process; Rubber-pad forming **A14:** 611-614
ASEA Quintus deep-drawing technique **A14:** 615
ASEA Quintus fluid forming press **A14:** 614-615
bulging punches **A14:** 614
presses, for deep drawing **A14:** 587
as rubber-diaphragm forming **A14:** 611
SAAB rubber-diaphragm method **A14:** 611, 614
Verson hydroform process **A14:** 612-614
Fluid friction
defined **A18:** 9
Fluid handling
components, cemented carbide **A2:** 972-973
industry applications, structural ceramics **A2:** 1019
Fluid handling applications
high-density polyethylenes (HDPE) **EM2:** 163-164
phenolics **EM2:** 243
polyether sulfones (PES, PESV) **EM2:** 159
polyphenylene ether blends (PPE, PPO) **EM2:** 183
ultrahigh molecular weight polyethylenes (UHMWPE) **EM2:** 168
unsaturated polyesters **EM2:** 246
vinyl esters **EM2:** 272-273
Fluid How rate
aqueous corrosion **A13:** 40
Fluid hydrostatic cell (FHC)
for curing laminates **EM1:** 755-757
Fluid hydrostatic pressure (FHP) **EM1:** 755-757, **EM3:** 741
Fluid iron ore reduction process **A7:** 98
Fluid jet machining **EM4:** 313, 314
Fluid lines
ASTM specifications for **M1:** 323
Fluid movement
corrosivity **A13:** 339-340
Fluid resistance **EM3:** 52
Fluid(s) *See also* Leak testing; Liquids
flow **A17:** 60, 285
as leaks **A17:** 57-58
Fluid-cell forming **A14:** 608-611
ASEA Quintus **A14:** 610-611
Demarest process **A14:** 610
Verson-Wheelon process **A14:** 609-610
Fluid-cell process *See also* Fluid forming; Fluid-cell forming; Rubber-pad forming
defined **A14:** 6
Fluid-film lubrication **A18:** 89
in bearings **A11:** 484
Fluid-flow calculations, validation strategies **A6:** 1147-1150
convection **A6:** 1147
laser-beam welding **A6:** 1148
measurement of free surface deformation **A6:** 1148-1149
"reflective topography" technique **A6:** 1148, 1149
velocity measurement **A6:** 1149
Fluid-to-solid transition plastics
as binders **A15:** 211

Fluidity *See also* Flow; Flowability; Fluid flow; Fluid flow modeling
aluminum casting alloys **A2:** 123, 145-146
of aluminum-silicon alloys **A15:** 165
of cobalt-base alloys **A15:** 811
of compacted graphite iron **A1:** 57
defined **A2:** 346 **A15:** 6, 165 **A18:** 9
gray cast iron **M1:** 11-12, 13
of gray iron **A1:** 12-13
of metal-matrix composites **A15:** 849-850
molten metal **A15:** 766-768
and pouring temperatures **A15:** 283
and solidification, relationship **A15:** 767
testing **A15:** 767-768
vs castability, copper casting alloys **A2:** 346
Fluidized bed **A7:** 5
coater, for lacquer coatings **A7:** 588
reactors **A7:** 52
Fluidized bed reduction **A7:** 5
applications **A7:** 97-98
commercial viability **A7:** 98
iron powder production **A7:** 96-98
Fluidized bed units
for sand reclamation **A15:** 354
Fluidized-bed cementation process
ceramic coating **M5:** 542, 544-545
Fluidized-bed chemical vapor deposition coating **M5:** 385
Fluidized-bed coating *See also* Coatings
defined **EM2:** 19
Fluidized-bed combustors
superheater/high-temperature air heater corrosion in **A13:** 999
Fluidized-bed equipment **A4:** 484-491
advantages of process **A4:** 490
atmospheres, control of **A4:** 486
cleaning operations **A4:** 491
defluidization **A4:** 485 **M4:** 300
external-combustion-heated furnace **A4:** 488
external-resistance-heated furnace **A4:** 488
furnace applications **A4:** 490
heat transfer **A4:** 485-486 **M4:** 301-302
heat treatment, selective **A4:** 485 **M4:** 300-301
internal-combustion gas-fired furnace **A4:** 488
internal-resistance-heated furnace **A4:** 490
principles **A4:** 484-485 **M4:** 299
safety of operations **A4:** 490-491
submerged-combustion furnace **A4:** 487-488, 489 **M4:** 304
surface treatments **A4:** 486-487
temperature, effect on minimum velocity **A4:** 484-485 **M4:** 300
tool steels, heat treating **A4:** 732-733
two-stage, internal combustion, gas-fired furnace **A4:** 489-490
velocity determination **A4:** 484 **M4:** 299-300
Fluidized-bed nitriding **A4:** 263
Fluidized-bed oxidation-resistant coating **M5:** 664-666
Fluidized-bed powder coating
conventional and electrostatic **M5:** 486
Fluidized-bed thermal fatigue tests
for gas- turbine components **A11:** 278
Fluids
body, tantalum resistance to **A13:** 728
characteristics **A13:** 479
chemical process, lead/lead alloys in **A13:** 789
corrosive, bearing damage by **A11:** 494, 496
oral **A13:** 1340-1341
produced, corrosivity **A13:** 479
systems, space shuttle orbiter **A13:** 1066-1072
used, space shuttle orbiter **A13:** 1062
working **A11:** 209
Fluids, primary/secondary
in condensation soldering **EL1:** 703
Fluoborate plating
cadmium **M5:** 256-258, 264
copper **M5:** 160-161, 166-167
lead **M5:** 273-275
nickel **M5:** 200-201
tin **M5:** 272
tin-lead **M5:** 276-278
zinc *See* Zinc fluoborate plating
Fluoborates
constituent in torch brazing flux **M6:** 962
Fluoboric acid
copper plating, use in **M5:** 166
Fluoboric acid (continued)
description **A9:** 68
as dissolution medium **A10:** 165
tin plating, use in **M5:** 271-272
tin-lead plating, use in **M5:** 276-278
Fluoboric acid electrobrightening
aluminum and aluminum alloys **M5:** 580-581
Fluorapatite
simple model system (single crystals) for human enamel in laboratory studies **A18:** 667-668
Fluorescence
abbreviation for **A10:** 690
in absorption/enhancement effects **A10:** 97
analysis, selectivity **A10:** 76
background, as problem in Raman analyses **A10:** 130
correction **A10:** 524
crystallophosphor host matrices and metals determined by inducing **A10:** 74
defined **A10:** 673 **A17:** 102-103
delayed, in molecular fluorescence spectroscopy **A10:** 73
effect in AEM-EDS microanalysis **A10:** 448
effect in microbeam analysis **A10:** 530
enhancement, using organized mediums, as MFS special technique **A10:** 79
in glow discharge emissions **A10:** 29
intensity, as measure of absorption probability, EXAFS analysis **A10:** 411
lifetimes **A10:** 75, 79
mechanism, EPMA analysis **A10:** 524
molecular, of organic compounds and atoms and inorganic atoms **A10:** 73-74
oxygen quenching of, enzymatic determination of glucose using **A10:** 79-80
quantum yields, structural effects **A10:** 74
spectrum, liquid-chromatographic fraction automobile exhaust extract **A10:** 79
in the infrared **A10:** 115
and x-ray absorption, effect in microanalysis **A10:** 448
Fluorescence correction
in EPMA analysis **A10:** 524
Fluorescence EXAFS detection technique **A10:** 418
Fluorescence lifetimes (dynamic measurements)
as MFS special technique **A10:** 79
Fluorescence spectroscopy **EM4:** 52
Fluorescent dye, added to mounting materials to detect filled defects in fiber composites **A9:** 588
Fluorescent dye penetration
soldered joints **A6:** 982
Fluorescent intensifying screens
radiography **A17:** 316-317
Fluorescent light
for optical testing illumination **EL1:** 570
Fluorescent magnetic particle testing
electron-beam welding **A6:** 866
Fluorescent magnetic-particle inspection
of feedwater heater **A11:** 658
Fluorescent metals
minimum detectable quantities for **A10:** 74
Fluorescent molecules
in molecular fluorescence spectroscopy **A10:** 73-74
Fluorescent particle inspection **A7:** 578-579
Fluorescent penetrant inspection
ceramic coatings **M5:** 546
residues, removal of **M5:** 14
to detect weld discontinuities in diffusion welding **A6:** 1080
weld compositional analysis **A6:** 100
Fluorescent penetrant inspection (FPI)
to evaluate gas turbine ceramic components **EM4:** 718
Fluorescent penetrants
selection and use **A17:** 75, 77
sensitivity levels **A17:** 75, 77
visual inspection with **A17:** 3
Fluorescent pigments
as colorants **EM2:** 501
Fluorescent screens
in real-time radiography **A17:** 319-320
Fluorescent sunlamps
for weatherometers **EM2:** 579
Fluorescent test
cleaning process efficiency **M5:** 20
Fluorescent yield **A10:** 86-87
Fluorescent-light tubes
circular **A12:** 83
Fluorescent-penetrant inspection
applications **M6:** 827
Fluoride
in chromate conversion coatings **A13:** 393-394
corrosion fatigue test specification **A8:** 423
as impurity **A13:** 1161
molten **A13:** 90
in phosphate coating baths **A13:** 384
pressurized water reactor specification **A8:** 423
salts, molten **A13:** 52-54
Fluoride (F_2)
in composition of textile products **EM4:** 403
in composition of wool products **EM4:** 403
as fining agents **EM4:** 380
Fluoride glasses
applications, optical glass products **EM4:** 1075
density **EM4:** 846
electrical properties **EM4:** 852-853
heat capacity **EM4:** 848
thermal expansion **EM4:** 847
viscosity **EM4:** 849
Fluorides
anions, separation by ion chromatography **A10:** 659
flux constituent for torch brazing **M6:** 962
flux removal after torch brazing **M6:** 963-964
gravimetric weighing as **A10:** 171
in Group VI electrolytes **A9:** 54
as impurities in uranium alloys **A9:** 477
safety precautions **M6:** 995
titration, thorium solution **A10:** 173
Fluorimetry *See* Fluorometric analysis
Fluorinated compounds
to increase electronegativity **EM3:** 181
Fluorinated epoxy resin
surface tension **EM3:** 181
Fluorinated ethylene propylene (FEP) **EM3:** 13
defined **EM2:** 19
in pharmaceutical production facilities **A13:** 1228
surface preparation **EM3:** 279
Fluorinated ethylene propylene (FEP) insulation
for copper and copper alloy products **A2:** 258
Fluorinated hydrocarbons
as cutting fluid for tool steels **A18:** 738
Fluorinated perfluoroethylene-propylene (FEP) **EM2:** 115-118
Fluorinated polyethers
lubricants for rolling-element bearings **A18:** 136
Fluorinated polymer coating
for copier powders **A7:** 588
Fluorinated polymers
environmental effects **EM2:** 428
Fluorinates
for interconnection soldering **EL1:** 180
Fluorine
as addition to carbon-graphite materials **A18:** 816
buildup in CAA oxides **EM3:** 265
determined in borosilicate glass **A10:** 179
distillation **A10:** 169
in enamel cover coats **EM3:** 304
in enamel ground coats **EM3:** 304
from surface analysis **EM3:** 294
properties **EL1:** 782-783
species weighed in gravimetry **A10:** 172
tantalum corrosion in **A13:** 731
TNAA assayed **A10:** 235-236
volumetric procedures for **A10:** 175
Fluorine (F)
component in photochromic ophthalmic and flat glass composition **EM4:** 442
properties **EM4:** 424
specific properties, imparted in CTV tubes **EM4:** 1039
in tableware compositions **EM4:** 1101
volatilization losses in melting **EM4:** 389
Fluorine, as reagent
aluminum melts **A15:** 80

Fluorine compounds
safety precautions **A6:** 1196
Fluorine, corrosion
nickel alloys **M3:** 174
Fluorine-contaminated carbon fiber-polyimide matrix composites **EM3:** 294
Fluorine-doped silica **EM4:** 211
Fluorite
chemical composition **A6:** 60
hardness **A18:** 433
slag removal poor when contained in fluxes **A6:** 61
Fluorite objective lenses **A9:** 73
Fluoroacrylate solution polymers
application methods **EL1:** 783-784
Fluoroalkylarylenesiloxanylene (FASIL)
advantages and disadvantages **EM3:** 675
performance **EM3:** 674
sealant formulation **EM3:** 677, 678
for severe environments **EM3:** 673
Fluoroaluminate glasses
optical properties **EM4:** 854
Fluoroborate glasses
optical properties **EM4:** 854
Fluorocarbon 113 cleaner **M5:** 44, 57
Fluorocarbon plastics *See also* Plastics **EM3:** 13
defined **EM2:** 19
Fluorocarbon polymers **EM3:** 594
Fluorocarbon resins **M5:** 475
Fluorocarbon sealants **EM3:** 223-226
additives and modifiers **EM3:** 225
applications **EM3:** 225-226
chemical compatibility **EM3:** 224
chemistry **EM3:** 223
commercial forms **EM3:** 223
compared to RTV type sealant **EM3:** 226
for compression packings **EM3:** 223, 225-226
cost factors **EM3:** 224, 226
for flanges **EM3:** 223, 224
for industrial gaskets **EM3:** 223, 224, 225, 226
for industrial seals **EM3:** 223
markets **EM3:** 225-226
processing parameters **EM3:** 224
properties **EM3:** 226
for pumps **EM3:** 223
for reaction vessels **EM3:** 223
resin types and applications **EM3:** 223-224
for tapes (sealant) **EM3:** 223
for tubing **EM3:** 223
for valves **EM3:** 223
Fluorocarbon telomer
thickener for grease high-vacuum application lubricants **A18:** 157
Fluorocarbon vapor detection method **EL1:** 502
Fluorocarbon(s)
by residual gas analysis **EL1:** 1066
for flexible printed boards **EL1:** 582, 583
gross leak test **EL1:** 1063
molecular structure of **EL1:** 663
physical properties **EL1:** 664
resin, as conformal coating **EL1:** 760
Fluorocarbons **EM3:** 13, 601
defined **EM2:** 19
lubricated friction testing **A18:** 48
properties **A18:** 81
surface preparation **EM3:** 291
Fluorocarbons, oil of
IR split mull **A10:** 113
Fluoroelastomer
coating for composite construction against liquid impingement erosion **A18:** 222
Fluoroelastomer sealants **EM3:** 223, 226-227
for aerospace and automotive gaskets **EM3:** 227
for aerospace and automotive packing materials **EM3:** 227
for aerospace and automotive seals **EM3:** 227

Fluoroelastomer sealants (continued)
aircraft applications with improved heat, chemical, and solvent resistance **EM3:** 226, 227
for automobile valve stem seals **EM3:** 227
chemistry **EM3:** 226-227
cost factors **EM3:** 227
for high-temperature, dynamic-type sealing **EM3:** 226
for jet engine lip seals **EM3:** 227
for O-rings **EM3:** 227
processing equipment and techniques **EM3:** 227
properties **EM3:** 226, 227
for solid rocket space booster joints **EM3:** 227
for static gasketing **EM3:** 226
suppliers **EM3:** 227
Fluoroethylene
as thermocouple wire insulation **A2:** 882
Fluoroethylene propylene
critical surface tension **EM3:** 180
Fluorogermanate glasses
electrical properties **EM4:** 852
Fluorohydrocarbon plastics *See also* Plastics **EM3:** 13
defined **EM2:** 19
Fluorometallic screens
radiography **A17:** 317
Fluorometers
single-beam **A10:** 76
Fluorometric analysis
defined **A10:** 673
Fluoroplastics *See also* Plastics **EM3:** 13
chemistry **EM2:** 66
defined **EM2:** 19
sulfuric acid corrosion **A13:** 1154
Fluoropolymer coatings *See also* Coatings
application methods **EL1:** 783-784
applications, specific **EL1:** 782
chemical/physical properties **EL1:** 782-783
conformal coatings **EL1:** 782-784
development **EL1:** 782
as encapsulants **EL1:** 759
materials, available **EL1:** 782
Fluoropolymers *See also* Thermoplastic fluoropolymers
applications **EM3:** 56
for electronic packaging applications **EM3:** 600
as engineering thermoplastics **EM2:** 447-448
in pharmaceutical production facilities **A13:** 1227-1228
products **EM3:** 601
properties **EM3:** 600-601
sealing systems for turbochargers **A18:** 567
surface preparation **EM3:** 278, 279, 290
Fluorosilicate glasses
electrical properties **EM4:** 852
Fluorosilicone sealants **EM3:** 195
chemical resistance **EM3:** 642
properties **EM3:** 227
Fluorosilicones
aerospace industry applications **EM3:** 50, 58-59
for aircraft inspection plates **EM3:** 604
component covers **EM3:** 611
for faying surfaces **EM3:** 604
for fillets **EM3:** 604
gas tank seam sealant **EM3:** 610
for rivets **EM3:** 604
room-temperature vulcanizing (RTV) **EM3:** 59
suppliers **EM3:** 59
Fluorspar
chemical composition **A6:** 60
dissolution in hydrochloric acid **A10:** 165
as flux addition **A15:** 389
function and composition for mild steel SMAW electrode coatings **A6:** 60
functions in FCAW electrodes **A6:** 188
Fluorspar (CaF_2)
as melting accelerator **EM4:** 380
purpose for use in glass manufacture **EM4:** 381

Fluorspar-aluminum, flux composition
self-shielded FCAW electrodes **A6:** 61
Fluorspar-lime, flux composition
self-shielded FCAW electrodes **A6:** 61
Fluorspar-lime-titania, flux composition
self-shielded FCAW electrodes **A6:** 61
Fluorspar-titania, flux composition
self-shielded FCAW electrodes **A6:** 61
Fluosilicaborates
constituent in torch brazing flux **M6:** 962
Fluosilicate plating
lead **M5:** 273-275
Fluosilicic acid
distillation **A10:** 169
lead corrosion by **A13:** 788
safety hazards **A9:** 69
Flurozirconate glasses
electrical properties **EM4:** 853
viscosity **EM4:** 849
Flush water
effects on stainless steel heat-exchanger tubes **A11:** 630
Flushing
inert gas **A15:** 84-85
Flushing procedures
steam generators **A13:** 944
Flute marks
in alloy steel billet **A9:** 175
revealed by macroetching **A9:** 174
Fluted bearing
defined **A18:** 9
Fluted core
defined **EM1:** 11 **EM2:** 19
Fluted mandrel
for gun barrel bore **A14:** 138-139
Flutes
defined, as fracture **A12:** 20-21
in titanium alloys **A12:** 27, 28
tool steels **A12:** 375
Fluting *See also* Fretting; Pitting
in bearing failures **A11:** 493
defined **A11:** 4 **A18:** 9
Flux *See also* Degassing; Flux density; Flux injection; Flux leakage inspection; Fluxes; Fluxing; Leak testing; Magnetic lines of flow; Mass transport **EM3:** 13 **EM4:** 6
alkali oxides **EM4:** 6
alkaline earth oxides **EM4:** 6
aqueous corrosion **A13:** 33
carbon, defined **A15:** 74
defined **A15:** 6 **A17:** 90 **EM2:** 19
definition **M6:** 8
direction, effect of **A17:** 91
entrapment, in brazed joints **A17:** 602
flow, effect on detection **A17:** 91
inclusions, effect, magnesium/magnesium alloys **A13:** 741
lines, magnetic, in leakage field testing **A17:** 129
linkage, induced current **A17:** 98
residues **A13:** 1109
Flux (dross) inclusions
as casting defect **A11:** 387
Flux activity
defined **EL1:** 643
Flux cleanability
defined **EL1:** 644
Flux cored arc welding **M6:** 96-113
applicability **M6:** 96-97
auxiliary gas shielding **M6:** 96
self-shielding **M6:** 921-922
arc spot welding **M6:** 109-110
procedures and conditions **M6:** 110, 112
thickness and position **M6:** 110
weld characteristics **M6:** 110
automatic welding **M6:** 110-112
carbon dioxide shielding gas **M6:** 103-104
cast irons **M6:** 311
characteristics **M6:** 103
containers **M6:** 103-104

Flux cored arc welding (continued)
flow rate M6: 104
purity M6: 104
of cast irons A15: 523-524
comparison to shielded metal arc welding M6: 107-108, 112-113
comparison to submerged arc welding M6: 113
corner joints M6: 113
definition M6: 8
deposition rate M6: 104
electrode holders M6: 98
cooling systems M6: 98
electrode wires, flux cored M6: 99-103
carbon steel M6: 100-101
classification of electrodes M6: 99-100
construction M6: 99-100
form and function elements M6: 103
functions of core compounds M6: 103
hydrogen content M6: 102
low-alloy M6: 101-102
manufacture M6: 99
metal transfer M6: 99
stainless steel M6: 101-102
equipment installations M6: 104
fillet welds M6: 106-107, 112-113
fluxes M6: 103
groove welding M6: 105-107
heat flow calculations M6: 31
multiple-position welding M6: 106-107
downhill and vertical-down M6: 106-107
out-of-position M6: 107
operating variables M6: 104-105
arc voltage M6: 104
current M6: 104
electrode extension M6: 105
travel speed M6: 104-105
pipe welding M6: log
fixed-position M6: log
rotation of workpieces M6: 109
power supply M6: 97-99
process fundamentals M6: 97
recommended grooves M6: 69-71
shielding gases M6: 102-104
slags M6: 103
T-joints M6: 112-113
thick sections M6: 107-109
underwater welding M6: 922
weld discontinuities M6: 830
weld overlaying M6: 807
wire-feed systems M6: 97-99
feed rolls M6: 99
maintenance M6: 99
push-type systems M6: 98-99
workpieces, holding and handling M6: 104
Flux cored arc welding (FCAW)
discontinuities from A17: 582
Flux cored arc welding of
alloy steels M6: 298-299
gray iron M6: 315, 317
hardenable carbon steels M6: 266
stainless steels, austenitic M6: 326-327
stainless steels, ferritic M6: 347
stainless steels, martensitic M6: 349
Flux cored electrode
definition M6: 8
Flux corrosivity
defined EL1: 643
Flux cover
definition M6: 8
Flux cover (metal bath dip brazing and dip soldering)
definition A6: 1209
Flux cutting
chemical A14: 728-729
Flux density *See also* Magnetic flux density
around hollow cylinder A17: 96
around solid conductors A17: 96
change, as flaw detection, electric current perturbation A17: 136
curves, during demagnetization A17: 121
and hardness A17: 134
as magnetic hysteresis A17: 131
magnetic rubber inspection applications A17: 122
with permanent-magnet yokes A17: 93
saturation, as hardness measure A17: 134
Flux, fluxes
brazing and soldering A7: 840
-coated magnesium particles A7: 132
as coating material A7: 817
density A7: 664
-powder mixtures, for brazing and soldering A7: 840
unidirectional and cylindrical in sintering A7: 315
Flux gate magnetometer
as magnetic field sensor A17: 131
Flux inclusion
overload fracture by A12: 65, 67
Flux injection
capabilities A15: 453
nonferrous molten metals A15: 452-453
process, schematic A15: 454
Flux leakage field
magnetic A17: 129-131
Flux leakage field testing *See* Flux leakage inspection
Flux leakage inspection
applications A17: 132-134
with coil A17: 130
in-service, tubular products A17: 574
of resistance-welded steel tubing A17: 563-564
of seamless steel tubular products A17: 571
speed of inspection A17: 564
Flux oxygen cutting
definition A6: 1209
Flux paste
brazing filler metals available in this form A6: 119
Flux pattern of iron filings A9: 65
Flux pinning *See also* Flux-jump stability
in niobium-titanium (copper) superconducting materials A2: 1053-1054
in superconductivity theory A2: 1034-1035
Flux stone
for foam fluxers EL1: 682
Flux tube demagging
aluminum alloys A15: 472-473
Flux(es)
activity EL1: 654, 663
air knife wipe-off EL1: 683
application methods EL1: 648
chemistry EL1: 643
components EL1: 644
contamination, printed wiring boards EL1: 1109
controls for EL1: 49
defined EL1: 1144
density controller EL1: 683
divergence, effects EL1: 888
excess, control of EL1: 649
foam fluxers EL1: 681-682
for "no-clean" applications EL1: 647-648
on board components, failure analysis EL1: 1110
physical function EL1: 643
process control EL1: 683-684
process requirements EL1: 681
properties EL1: 643-644
purpose/effect on solderability EL1: 676
reaction particles, as contaminants EL1: 661
resin fluxes EL1: 646
resin-based EL1: 645-646
resistance, of flexible epoxies EL1: 821
rotary/stationary brush fluxers EL1: 682
selection of EL1: 649-650
solder paste EL1: 653-654
and solder skip defects EL1: 647
spray fluxers EL1: 682
for surface-mount soldering EL1: 647
synthetic activated fluxes EL1: 647
technology, development EL1: 631
thermal function EL1: 643
vehicle EL1: 644
vehicle, and paste print resolution EL1: 732
water-soluble, corrosion from EL1: 1110
water-soluble organic EL1: 646-647
wave fluxers EL1: 682
wave soldering EL1: 701
Flux-core solder wire
defined EL1: 639
Flux-cored arc welding
failure origins A11: 414
Flux-cored arc welding (FCAW) A6: 186-189
advantages A6: 186
Flux-cored arc welding (FCAW) (continued)
applications A6: 186, 187
automotive A6: 393-394
base metals A6: 187-188
carbon steels A6: 643, 647, 652-654, 657, 658
cast irons A6: 716, 719-720
iron-nickel electrodes A6: 719
nickel-base electrodes A6: 719
steel electrodes A6: 719
definition A6: 1209
description A6: 186
disadvantages A6: 186
duplex stainless steels A6: 480
electrodes A6: 186, 188-189
for carbon steels A6: 656
CO_2 shielded flux compositions A6: 61
for stainless steels A6: 705-706
equipment A6: 186-187
fume-removal A6: 187
mechanized and automatic A6: 187
power supplies A6: 187
semiautomatic A6: 187
ferritic stainless steels A6: 446
flux compositions of self-shielded electrodes A6: 61
fluxes for welding A6: 58, 62
gas-shielded A6: 186
hardfacing alloys A6: 796, 799, 801, 802-803
heat sources A6: 1144
heat-treatable low-alloy steels A6: 669, 670
high-strength low-alloy quench and tempered structural steels A6: 666
high-strength low-alloy steels A6: 187
high-strength low-alloy structural steels A6: 663, 664
high-strength quenched and tempered steels A6: 188
high-temperature chromium-molybdenum steels A6: 187
low-alloy metals for pressure vessels and piping A6: 668
low-alloy steels A6: 186-188, 662, 663, 664, 666, 667, 668, 669, 670, 674, 676
matching filler metal specifications A6: 394
metallurgical discontinuities A6: 1073
mild steels A6: 187
neural network system A6: 1060
nickel-base alloys A6: 186, 187, 188
nickel-base steels A6: 187
parameters used to obtain multipass weld in X-65 steel pipe A6: 101
power source selected A6: 37
process features A6: 186
process selection guidelines for arc welding A6: 653
railroad equipment A6: 396
rating as a function of weld parameters and characteristics A6: 1104
for repair of high-carbon steels A6: 1105
safety precautions A6: 186, 187, 1192-1193
self-shielded flux-cored electrodes, compositions of A6: 61, 62
sensor systems A6: 1063, 1064
sheet metals A6: 398
shielding gases A6: 64, 67, 186-187, 189, 662
shipbuilding A6: 384
stainless steels A6: 186, 188, 680, 681, 688, 693, 698, 699, 705-706
suggested viewing filter plates A6: 1191
titanium alloys A6: 783
to prevent hydrogen-induced cold cracking A6: 436
tool and die steels A6: 674, 676
underwater welding A6: 1010, 1012
vs. gas-metal arc welding A6: 186, 187
vs. shielded metal arc welding A6: 186, 187
vs. submerged arc welding A6: 186
weld cladding A6: 816
Flux-cored electrode
definition A6: 1209
Flux-jump stability
niobium-titanium superconductors A2: 1043-1045, 1051
Flux-to-wire ratio A6: 206, 207
Fluxers
design purpose EL1: 681
for uniform flux coating EL1: 681

Fluxers (continued)
wave soldering ... EL1: 686
Fluxes *See also* Degassing; Flux; Flux injection; Fluxing; Slag ... A6: 135 EM4: 32
acidic ... A10: 167
activity tests and wiring board cleanliness guidelines ... A6: 989
agglomerated ... M6: 123
agglomerated versus fused fluxes ... M6: 811-812
alloy element, in melt additions ... A15: 72
for aluminum alloys ... A15: 445
application methods ... A6: 987
applications, by type ... A6: 130
beryllium alloys ... A6: 946
bonded ... M6: 122-123
brazeability and solderability considerations ... A6: 621-622, 625
brazing ... M2: 450-452
brazing of aluminum alloys ... A6: 937-938
brazing of cast irons ... A6: 626
brazing of copper alloys ... A6: 931, 932
brazing of low-carbon and low-alloy steels ... A6: 624
brazing of nickel-base alloys ... A6: 631
for brazing of precious metals ... A6: 936
brazing of stainless steels ... A6: 913-914
brazing types per AWS specification B2.2 ... A6: 622, 627
brazing with clad brazing materials ... A6: 347, 348
for carbon steels, submerged arc welding ... A6: 658
cleaning agents, safety precautions ... A6: 983
compatibility issues ... A6: 987
connector technology ... A6: 999
for copper alloys ... A6: 755 A15: 448-449
for copper and copper alloy brazing ... A6: 932
definition ... A6: 1209
density, heat ... A10: 685
dip brazing ... A6: 337-338
drossing ... A15: 446
effect in brazing ... A11: 451
effect on weld metallurgy ... M6: 41
electronic packaging ... A6: 618
electroslag welding ... A6: 272-273, 274, 276, 278
evaluation, soldering ... A6: 130
fluidizing slag, cupolas ... A15: 388
furnace brazing ... A6: 121
furnace soldering ... A6: 353-355
fusion of sample materials with ... A10: 93-94, 167
galvanizing ... A15: 452
halide-free ... A6: 973, 987, 988
heat ... A11: 603-605, 609
as inclusion-forming ... A15: 96
as inclusions ... A11: 451
injection ... A15: 452-453
inorganic ... A6: 129, 130
inorganic acid (IA) ... A6: 988, 990
for low-alloy steel brazing ... A6: 930
low-alloy steels, welding of ... A6: 662
low-solids ... A6: 367, 987
luminous, SI defined unit and symbol for ... A10: 685
for magnesium alloys ... A15: 447
magnetic, SI defined unit and symbol for ... A10: 685
nickel alloys ... A6: 748
no-clean (or low-residue) ... A6: 354-355, 987
organic ... A6: 136
organic acid ... A6: 129, 130, 987-988, 990, 995
for oxyacetylene welding ... A15: 530
oxyfuel gas welding ... A6: 281, 738
prefused ... M6: 122
proper use ... A15: 446-449
properties ... EM4: 7
properties, compared ... A10: 167
removal of ... M5: 57
rosin mildly activated (RMA) ... A6: 130
rosin-base ... A6: 129-130, 136, 987, 989, 994, 995
safety precautions ... A6: 1202

Fluxes (continued)
safety precautions for soldering ... M6: 1097-1100
selection guidelines for soldering ... A6: 129
soldered joints ... A6: 1124
soldering ... M2: 446-447
soldering, classification scheme ... A6: 622, 628
for soldering in electronic applications ... A6: 985, 986-988
soldering of copper alloys ... A6: 630
solid degassing ... A15: 467
specifications for soldering ... A6: 972
submerged arc welding ... A6: 202, 203
for submerged arc welding of stainless steels ... A6: 700-701, 703
surface insulation resistance (SIR) test ... A6: 987, 988
synthetic ... EM4: 44
synthetic-activated (SA) ... A6: 994
synthetically activated (resins) ... A6: 129
titanium alloys ... A6: 521
for tool steel brazing ... A6: 930
torch brazing ... A6: 328
types of ... A6: 129-130
use in process stages of brazing and soldering ... A6: 903
wave soldering ... A6: 366, 367
for weld cladding of stainless steels ... A6: 820
wettability and solderability ... A6: 128-129
Fluxes for
brazing of aluminum alloys ... M6: 1023-1025
brazing of copper and copper alloys ... M6: 1035
brazing of molybdenum ... M6: 1058
brazing of niobium ... M6: 1057
brazing of reactive metals ... M6: 1053
brazing of stainless steels ... M6: 1004
brazing of tungsten ... M6: 1059
dip brazing of steels ... M6: 991
electroslag welding ... M6: 229-230
flux cored arc welding ... M6: 103
furnace brazing of steels ... M6: 944
induction brazing of copper ... M6: 1044-1045
induction brazing of steel ... M6: 971
oxyacetylene braze welding ... M6: 597
oxyacetylene welding of cast irons ... M6: 603
oxyfuel gas welding ... M6: 583
resistance brazing ... M6: 982-983
soldering ... M6: 1081-1085
submerged arc welding ... M6: 122-127
of nickel alloys ... M6: 442
torch brazing of copper ... M6: 1039
torch brazing of stainless steels ... M6: 1011
torch brazing of steels ... M6: 950-951, 961-963
wave soldering ... M6: 1088
weld overlays of stainless steel ... M6: 811-813
Fluxes for soldering ... A6: 971-974
in electronic applications ... A6: 985, 986-988
inorganic acid ... A6: 973-974, 980, 983
organic-acid ... A6: 973, 983
rosin-base ... A6: 972-973, 976, 977, 983
Fluxes for welding ... A6: 55-62
alloy modification ... A6: 60-61
binding agents ... A6: 60-61
ferro-additions ... A6: 60
pyrochemical kinetics during welding ... A6: 59-60
recovery of elements from electrode coatings ... A6: 60
slag detachability ... A6: 61
slag formation ... A6: 61
slag viscosity ... A6: 59-60
slipping agents ... A6: 60-61
basicity index ... A6: 59
effective slag-metal reaction temperature ... A6: 55
equilibrium parameters ... A6: 55-59
activity quotient ... A6: 56, 57
delta quantity ... A6: 57, 58
equilibrium partition ratio ... A6: 56
inclusion formation ... A6: 56-57

Fluxes for welding (continued)
metal transferability during pyrochemical reactions ... A6: 55-59
nonequilibrium lever rule ... A6: 56
oxygen effect ... A6: 55-59
precipitation index ... A6: 57
shielding gas ... A6: 58-59
purposes for addition ... A6: 55
slag ... A6: 55
types of ... A6: 61-62
flux-cored arc welding ... A6: 62
for shielded metal arc welding ... A6: 62
for submerged arc welding ... A6: 62
Fluxes, specific types
1, resistance brazing ... M6: 982-983
3A
brazing of cast irons ... M6: 996
brazing of coppers ... M6: 1035
induction brazing of steel ... M6: 971
resistance brazing ... M6: 982-983
3B, brazing of coppers ... M6: 1035
4
brazing of coppers ... M6: 1035
resistance brazing ... M6: 982-983
5, brazing of coppers ... M6: 1035
alumina, submerged arc welding ... M6: 124
aluminate basic, submerged arc welding ... M6: 124
basic fluoride, submerged arc welding ... M6: 124
calcium silicate-low silica submerged arc welding ... M6: 124
calcium silicate-neutral, submerged arc welding ... M6: 124
calcium-high silica, submerged arc welding ... M6: 124
FB3-A, brazing of copper and copper alloys ... A6: 932, 933
FB3-C, brazing of copper and copper alloys ... A6: 932, 933
FB3-D, brazing of copper and copper alloys ... A6: 932, 933
FB3-E, brazing of copper and copper alloys ... A6: 933
FB3-F, brazing of copper and copper alloys ... A6: 933
FB3-G, brazing of copper and copper alloys ... A6: 933
FB3-H, brazing of copper and copper alloys ... A6: 933
FB3-I, brazing of copper and copper alloys ... A6: 932, 933
FB3-J, brazing of copper and copper alloys ... A6: 932, 933
FB3-K, brazing of copper and copper alloys ... A6: 932, 933
FB4-A, brazing of copper and copper alloys ... A6: 932, 933
manganese silicate, submerged arc welding ... M6: 124
NOCOLOK, for aluminum alloys ... A6: 939
Fluxing
of aluminum alloys ... A15: 445-447
cast iron ... M5: 353
of copper alloys ... A15: 448-451, 466, 776
defined ... A15: 445
for demagging aluminum alloys ... A15: 472-474
equipment and materials ... M5: 354
flux covers ... M5: 353
flux injection ... A15: 452-453
fused-salt, aluminum coating process ... M5: 337-339
galvanized fluxes ... A15: 452
hot dip galvanized coating process ... M5: 326-327
hot dip tin coating process ... M5: 353-354
inert gas, copper alloys ... A15: 466
of magnesium alloys ... A15: 447-448, 801
procedure ... M5: 353
solutions, composition ... M5: 353
steel ... M5: 353

SUBJECTS OF THE INDEXED VOLUMES: ASM Handbook (designated by the letter "A"): **A1:** Properties and Selection: Irons, Steels, and High-Performance Alloys (1990); **A2:** Properties and Selection: Nonferrous Alloys and Special-Purpose Materials (1990); **A3:** Alloy Phase Diagrams (1992); **A4:** Heat Treating (1991); **A6:** Welding, Brazing, and Soldering (1993); **A7:** Powder Metallurgy (1984); **A8:** Mechanical Testing (1985); **A9:** Metallography and Microstructures (1985); **A10:** Materials Characterization (1986); **A11:** Failure Analysis and Prevention (1986); **A12:** Fractography (1987); **A13:** Corrosion (1987); **A14:** Forming and Forging (1988); **A15:** Casting (1988); **A16:** Machining (1989); **A17:** Nondestructive Testing and Quality Control (1989); **A18:** Friction, Lubrication, and Wear Technology (1992). **Metals Handbook, 9th Edition** (designated by the letter "M"): **M1:** Properties and Selection: Irons and Steels (1978); **M2:** Properties and Selection: Nonferrous Alloys and Pure Metals (1979); **M3:** Properties and Selection: Stainless Steels, Tool Materials and Special-Purpose Materials (1980); **M4:** Heat Treating (1981); **M5:** Surface Cleaning, Finishing, and Coating (1982); **M6:** Welding, Brazing, and Soldering (1983). **Engineered Materials Handbook** (designated by the letters "EM"): **EM1:** Composites (1987); **EM2:** Engineering Plastics (1988); **EM3:** Adhesives and Sealants (1990); **EM4:** Ceramics and Glasses (1991); **Electronic Materials Handbook** (designated by the letters "EL"): **EL1:** Packaging (1989).

Fluxing (continued)
of zinc alloys ... **A15:** 451-452
Fluxing agents
dip brazing ... **A6:** 337, 338
Fluxing process
air knife wipe-off ... **EL1:** 683
Fluxless melting
as inclusion control ... **A15:** 96
magnesium alloys ... **A15:** 448
Fluxless vacuum brazing
aluminum alloys ... **A6:** 939
Fly ash
Miller numbers ... **A18:** 235
Fly ash systems
corrosion of ... **A13:** 1007
Fly-shuttle looms ... **EM1:** 127-128
Flyer plate ... **A6:** 160, 161 **A8:** 187, 190, 210-212, 231-235
Flyer plate dynamic bend angle ... **A6:** 162
Flyer plate hardness ... **A6:** 161
Flyer plate mass ... **A6:** 161
Flyer plate standoff distance ... **A6:** 161
Flyer plate velocity (Vp) ... **A6:** 161
Flying die
cut-to-length lines ... **A14:** 711
Flying shear
defined ... **A14:** 6
Flywheel
cam plastometer ... **A8:** 194
rotating, in torsional impact system ... **A8:** 216
Flywheel energies ... **A6:** 890
Flywheel energy
in presses ... **A14:** 496
FM *See* Ferromagnets
FM (free from Mannesmann Effect) dies ... **A14:** 61
FM method *See* Frequency modulation (FM) method
FM process
defined ... **A15:** 38
FMAR *See* Ferromagnetic antiresonance
FML (free from Mannesmann Effect with low load) dies ... **A14:** 61
FMR *See* Ferromagnetic resonance
FNAA *See* Fast neutron activation analysis
FNR *See* Ferromagnetic nuclear resonance
Foam *See* Integral skin foam; Open-cell foam; Self skinning foam; Syntactic foams
in adhesive-bonded joints ... **A17:** 614
Foam board
waterjet machinery ... **A16:** 522
Foam cleaning ... **A13:** 1139
Foam depressants
use of ... **M5:** 11
Foam filters, ceramic
as inclusion control ... **A15:** 90
Foam fluxers
through-hole soldering ... **EL1:** 681-682
Foam inhibitor ... **A18:** 99, 108
defined ... **A18:** 9
effectiveness ... **A18:** 108
in engine lubricant formulations ... **A18:** 111
in nonengine lubricant formulations ... **A18:** 111
Foam injection molding
properties effects ... **EM2:** 284
size and shape effects ... **EM2:** 290
surface finish ... **EM2:** 304
textured surfaces ... **EM2:** 305
Foam patterns
for lost foam casting ... **A15:** 231-232
Foam stop hot-melt adhesive
for refrigerator cabinet seams ... **EM3:** 45
Foam structure
$AuCu_3$ ... **A9:** 682
Foam urethane molding
size and shape effects ... **EM2:** 292
Foam vaporization
for art casting ... **A15:** 22
Foam(s) *See also* Foam injection molding; Foam urethane molding; Structural foam; Thermoplastic structural foams
filling, blow molding ... **EM2:** 358
flexable polyurethane ... **EM2:** 630
integral skin, polyurethane (PUR) ... **EM2:** 258
integral-skin rigid ... **EM2:** 630
mechanical properties ... **EM2:** 262
reinforced integral skin, PUR ... **EM2:** 261-262
rigid ... **EM2:** 630
Foam(s) (continued)
rigid integral skin, polyurethane (PUR) ... **EM2:** 259
rigid, polyurethane (PUR) ... **EM2:** 259
semirigid molded, polyurethane (PUR) ... **EM2:** 258-259
thermoplastic structural, properties ... **EM2:** 508-513
Foam-in-place
defined ... **EM1:** 11 **EM2:** 19
Foamed adhesive ... **EM3:** 13
Foamed hot-melt sealants ... **EM3:** 51
Foamed plaster molding
process of ... **A15:** 247
Foamed plastics *See also* Expandable plastic; Plastics ... **EM3:** 13
defined ... **EM1:** 11 **EM2:** 19
Foamed polystyrene, as pattern material
investment casting ... **A15:** 255
Foaming
defined ... **A18:** 9
flux ... **EL1:** 644, 648
Foaming agent ... **EM3:** 13
defined ... **EM1:** 11 **EM2:** 19
Foams
analytic methods for ... **A10:** 9
Focal lengths
and color of light ... **A9:** 75
defined ... **A9:** 8
for macrophotography ... **A9:** 87
Focal spot
defined ... **A9:** 8 **A17:** 383
definition ... **M6:** 8
Focal-plane tomography techniques ... **A17:** 379
Focus
defined ... **A9:** 8
Focused ultrasonic search units
acoustic lenses ... **A17:** 259-260
advantages and useful range ... **A17:** 260
noise ... **A17:** 260-261
Focused-beam Auger electron spectroscopy spectrum
sodium chloride in ... **A7:** 254
Focusing
automated ... **A10:** 310
of borescopes and fiberscopes ... **A17:** 8
camera ... **A12:** 79-80
diffractometer, for x-ray diffraction residual stress techniques ... **A10:** 387
effect, extended x-ray absorption fine structure ... **A10:** 411
in SEM illuminating/imaging system ... **A12:** 167
Focusing (x-rays)
defined ... **A9:** 8
Focusing cup
x-ray tubes ... **A17:** 302
Focusing device (electrons)
defined ... **A9:** 8
Focusing, optical microscope
as thickness testing ... **EL1:** 943
Foerster frequency selection method ... **A17:** 182
Foerster limit frequency methods
eddy current inspection ... **A17:** 182
Fog
on radiographic film ... **A17:** 327
Fog quenching ... **M4:** 31-32
of steel ... **M4:** 60
Fog testing *See* Salt spray (fog) testing
Fog tests
chromium plating ... **M5:** 198
Fogging ... **A10:** 143, 144
Foil ... **A8:** 618
aluminum and aluminum alloys ... **A2:** 5
aluminum, leak detection with ... **A17:** 66
beryllium ... **A7:** 759
brazing filler metals available in this form ... **A6:** 119
defined ... **A14:** 6
electrojet thinning for samples of ... **A10:** 451
finishes for ... **A1:** 848
gilding as ... **A15:** 21
Knoop minimum thickness chart for ... **A8:** 96, 101
lead and lead alloy ... **A2:** 555
mechanical properties ... **A1:** 848
platelet geometry in ... **A10:** 455
for radiographic screens ... **A17:** 316
stainless steel ... **A1:** 848
as tantalum mill product ... **A7:** 770-771
Foil (continued)
thin, AEM-EDS quantitative analysis of ... **A10:** 447
thin, electron beam spreading in ... **A10:** 434
tin ... **A2:** 525-526
-type crack growth gages ... **A8:** 246
wrought aluminum alloy ... **A2:** 33
Foil bearing ... **A18:** 530
bending-dominated continuous ... **A18:** 530, 531
bending-dominated segmented ... **A18:** 530-531
commercial varieties ... **A18:** 530
defined ... **A18:** 9
modified bending-dominated continuous ... **A18:** 531
surface coatings ... **A18:** 532
tension-dominated ... **A18:** 530
Foil butt-seam welding
of blanks ... **A14:** 450
Foil decorating
defined ... **EM2:** 19
Foil laminates, bimetal
XPS analyses ... **A11:** 44
Foil microscopy, thin *See* Thin-foil microscopy
Foils
for laminates ... **EL1:** 581-582
Fokker (Netherlands)
detection of ideal oxide configuration for adhesion on aluminum alloys ... **EM3:** 744
Fokker bond tester ... **EM3:** 522, 525, 528, 761
for adhesive-bonded joints ... **A17:** 619-620
in nondestructive testing ... **EM3:** 754-756, 766, 769-770, 772, 778
Fold
defined ... **A14:** 6
Folded chain ... **EM3:** 13
defined ... **EM2:** 19
Folding assembly
for furnace brazing ... **M6:** 940
Folding energy
in tests ... **A11:** 55
Folding over, in nonlubricated
nonisothermal hot forging ... **A8:** 582
Folds
as casting defect ... **A15:** 548
defined ... **A11:** 4
in forging ... **A11:** 327, 328, 330, 331
liquid penetrant inspection ... **A17:** 86
as notches ... **A11:** 85
in semisolid metal casting and forgings ... **A15:** 336-337
surface, as casting defects ... **A11:** 383
Follow die
defined ... **A14:** 6
Follow-up monitoring
and material selection ... **A13:** 322-323
Follower plate
defined ... **A18:** 9
FOLZ *See* First-order Laue zone
Fomblin Z25 polymer ... **A18:** 156
Food
aluminum/aluminum alloy resistance ... **A13:** 602
industry, stainless steel corrosion ... **A13:** 559-560
products, zinc/zinc alloys and coatings contact with ... **A13:** 763
tantalum resistance to ... **A13:** 728
Food additives
elemental iron ... **A7:** 614-615
Food and Drug Administration
abbreviation ... **A8:** 724
Food Chemicals Codex ... **A7:** 615
Food enrichment
iron powders for ... **A7:** 614-616
powders used ... **A7:** 572
Food equipment
metal coatings for ... **M1:** 102
Food handling and packaging equipment
precoated steel sheet for ... **M1:** 173
Food products *See also* Agricultural materials
iron powders in ... **A7:** 614
and natural products, liquid chrmoatography analysis for high molecular weight sugars ... **A10:** 649
potentiometric membrane electrode analysis of ... **A10:** 181
voltammetnc monitoring of pollutant metals and nonmetals in ... **A10:** 188

Food service applications
polycarbonates (PC) **EM2:** 151
polysulfones (PSU) **EM2:** 200
silicones (SI) **EM2:** 266
ultrahigh molecular weight polyethylenes (UHMWPE) **EM2:** 168
Foods
watedet machining **A16:** 525, 526
Footprint design pattern **EL1:** 734
Forbidden (diagram) lines
x-radiation **A10:** 86
Force *See also* Energy; Force requirements
abbreviation **A8:** 724
actuators, shape memory alloys **A2:** 900
application systems, tension testing machines **A8:** 47-48
blankholder, in drawing **A14:** 582
capacities, hot extrusion **A14:** 319
causing crack development and extension **A8:** 439-464
concept, dislocation mechanics **A8:** 440
contact, in electrical contacts **A2:** 840
conversion factors **A8:** 722 **A10:** 686
defined **EM1:** 11
drawing **A14:** 577-578
fields, atomic, use in IR normal-coordinate analysis **A10:** 110-111
measurement, tension testing machines **A8:** 48-49
moment of, SI defined unit and symbol for **A10:** 685
per unit length, conversion factors **A10:** 686
roll-separating, and torque **A14:** 345
roll-separating, estimation method **A14:** 344
SI defined unit and symbol for **A10:** 685
SI unit/symbol for **A8:** 721
stripping, in blanking **A14:** 448
symbol for **A11:** 796
units, universal testing machine **A8:** 612
vs. load, as term **A8:** 74n, 84n
Force balance
as process inodeling submodel **EM1:** 500
Force per unit length
conversion factors **A8:** 722
Force plug
defined **EM2:** 19
Force requirements *See also* Energy; Force; Power; Power requirements
in blanking **A14:** 448
drawing **A14:** 577-578
for fine-edge blanking and piercing **A14:** 474
piercing **A14:** 463
and press selection **A14:** 491
for stretch forming **A14:** 598
Force retainer plates *See also* Cavity retainer plates
Force-feed lubrication *See* Pressure lubrication
Force-restricted machines **A14:** 25, 37
Forced air cooling
ultrasonic fatigue testing **A8:** 247-248
Forced convection *See also* Convection
as aluminum melt circulation **A15:** 453-456
from flat plate, formulas **EL1:** 52
from parallel plates and ducts, formulas **EL1:** 52
infrared soldering systems **EL1:** 705
in reflow soldering **EL1:** 694
Forced cooling *See also* Cooling
selective, for stress and distortion **A15:** 616
Forced-air recirculating oven **EM3:** 37
Forced-displacement systems **A8:** 391-392
Forced-vibration systems **A8:** 391-393
Forceps, stainless steel
forging seam fracture **A11:** 329, 331
Forcing *See* Bulging
Ford anodized aluminum corrosion test **A13:** 219
Forehand welding
definition **A6:** 1209 **M6:** 8
oxyfuel gas welding **M6:** 589
technique **A6:** 183
Foreign alloy designations *See also* International Organization for Standardization (ISO) systems for **A2:** 16, 27
Foreign materials, effects
NDE reliability **A17:** 677
Foreign particles
bearing failure from **A11:** 489
Foreign specifications *See* listings in data compilations for individual alloys
Foreign structure
defined **A13:** 7
Foreline valve
for environmental test chamber **A8:** 411
Forensic studies
by x-ray spectrometry **A10:** 82
NAA analysis for **A10:** 233
PIXE analysis for **A10:** 102
Forestry tools
of cemented carbide **A2:** 974
Forge rolls **A14:** 96-97
Forge welding **M6:** 675-676
applications **M6:** 676
definition **M6:** 8, 675
metals welded **M6:** 676
procedures and sequence **M6:** 675-676
Forge welding (FOW) **A6:** 306
aluminum alloys **A6:** 306
aluminum metal-matrix composites **A6:** 558
applications **A6:** 306
automatic **A6:** 306
carbon steels **A6:** 306
cobalt-base alloys **A6:** 306
definition **A6:** 306, 1209
equipment **A6:** 306
fluxes **A6:** 306
high-alloy steels **A6:** 306
joint designs **A6:** 306
low-alloy steels **A6:** 306
manual **A6:** 306
nickel-base alloys **A6:** 306
process sequence **A6:** 306
steels **A6:** 306
titanium alloys **A6:** 306
tungsten **A6:** 306
Forge-delay time
definition **M6:** 8
Forgeability *See also* Forgeability tests; Forging; Formability **M1:** 350
of aluminum alloys **A14:** 241-242
aluminum and aluminum alloys **A2:** 8-9
carbon/alloy steels **A14:** 215-216
closed-die, of stainless steel **A14:** 223
of copper alloys **A14:** 256
defined **A14:** 6
as dynamic hot hardness vs temperature **A14:** 226
effect, isothermal/hot-die **A14:** 150-151
and flow strength, effects **A14:** 76
of heat-resistant alloys **A14:** 232
hot upsetting, of stainless steel **A14:** 223
of magnesium alloys **A2:** 1405
of metals and alloys, relative **A14:** 65
and part geometry **A14:** 77
of refractory metals **A14:** 237-238
solute content effect **A14:** 367
of stainless steel **A14:** 223-224
and strength at elevated temperatures **A14:** 222
of superalloys **A7:** 468
tests **A14:** 215, 382-384
of titanium alloys **A14:** 267-270
of tungsten alloys **A14:** 238
upset reduction vs forging pressure **A14:** 223-224
of various alloys **A14:** 75
of work metals, by high-energy-rate forging **A14:** 104
wrought aluminum alloy **A2:** 34
Forgeability of steels *See* Bulk formability of steels
Forgeability tests
notched-bar upset test **A14:** 383-384
Forgeability tests (continued)
sidepressing test **A14:** 383
truncated cone indentation test **A14:** 384
wedge-forging test **A14:** 382-384
Forged components
fatigue failures **A11:** 321
Forged parts **A1:** 455
dimensional tolerances **A7:** 482
disk, aspect ratio at fracture **A7:** 411
eddy current/nondestructive testing of **A7:** 491-492
tolerances of **A7:** 416, 482
Forged roll Scleroscope hardness number, (HFRSc or HFRSd)
defined **A8:** 6
Forged structure
defined **A9:** 8
Forged-in scale
as surface defect **A11:** 327
Forging *See also* Forgeability; Forgings, failures of; Hot forging; Solid-phase forming; Upsetting **A16:** 19
advanced titanium materials **A14:** 282-283
of alloy steels **A14:** 215-221
aluminum and aluminum alloys **A2:** 6
aluminum P/M alloys **A2:** 211-213
aluminum-lithium alloys, fatigue in **A2:** 196
anisotropy caused by **A11:** 316
and annealing, cracking after **A11:** 574, 580
automation, software tools **A14:** 410-412
of blended elemental TI-6AI-4V **A2:** 649
as bulk forming process **A14:** 16
bursts **A11:** 85, 317-318
of carbon steels **A14:** 215-221
causes of failure in **A11:** 317-342
characteristics, refractory metals **A14:** 237
chemical separation in **A11:** 314-315
closed-die **A8:** 587
cold, workability tests for **A8:** 589-591
in conjunction with broaching **A16:** 195
controlled, of carbon/alloy steels **A14:** 220-221
of copper and copper alloys **A14:** 255-258
defects **A11:** 317, 327-331, 329-333
defined **A14:** 6
described **A11:** 314
dies and presses, for SIMA process **A15:** 332
effects of billet shape and enclosure on stress state in **A8:** 588
flash, rough surface from **A11:** 472
fracture by segregation **A11:** 324-326
future trends **A14:** 20-21
hafnium **A2:** 663
hammers for **A14:** 25-29
of heat-resistant alloys **A14:** 231-236
high-definition **A14:** 243
history **A14:** 15
hot, for identification marking **A11:** 130
hot isostatic casting as replacement **A15:** 543-544
magnesium alloy, product form selection **A2:** 465-466
of magnesium alloys **A14:** 259
manufacture, tasks performed in **A14:** 409-410
mechanically alloyed oxide alloys **A2:** 948
dispersion-strengthened (MA ODS)
metal forming lubricants **A18:** 147-148
microstructure **A11:** 325-327
of molybdenum and molybdenum alloys **A14:** 237-238
new materials **A14:** 19
new processes **A14:** 16-19
of nickel-titanium shape memory effect (SME) alloys **A2:** 899
of niobium and niobium alloys **A14:** 237
of niobium-titanium ingot **A2:** 1044
of nonferrous metals **A14:** 239-287
notched-bar upset test **A8:** 588
open-die **A8:** 587
P/M high-speed tool steels **A16:** 62

Forging (continued)
post-, failures from **A11:** 331-332
preheating aluminum alloys for **A14:** 247
presses for **A14:** 25, 29-31
primary, and ingot breakdown **A14:** 222-223
primary testing direction **A8:** 667
process classifications **A14:** 15-16
process design **A14:** 409-416
process simulation **A14:** 19-20
process window for microstructure control in **A14:** 412-413
processes, classified **A14:** 16
processes, introduction to **A14:** 15-21
production goals **A14:** 407
rate, for finish forged nickel-base alloys **A14:** 262
ratio, effect on reduction **A14:** 218
recrystallization in **A14:** 231
reduction **A14:** 231
reduction rate **A14:** 231
of refractory metals **A14:** 237-238
rib-web, simulation **A14:** 428-429
of rings **A14:** 69
rolls, defined **A14:** 7
semisolid alloys, applications **A15:** 332-336, 338
sequence **A14:** 82, 413-416
severity, stainless steels **A14:** 223
shear test standards for **A8:** 62
sidepressing test **A8:** 588
spike, simulation of **A14:** 428-429
of stainless steels **A14:** 222-230 **M3:** 42-43
as superplastic forming process **A14:** 857
of tantalum and tantalum alloys **A14:** 238
tasks performed **A14:** 409-410
temperatures **A14:** 36, 231
test for flow localization controlled fracture **A8:** 588-589
and tooling for preform **A14:** 175
of tungsten and tungsten alloys **A14:** 238
of unalloyed uranium **A2:** 671
wedge test **A8:** 587
workability test for **A8:** 587-591
workpiece temperature **A14:** 231
wrought aluminum alloy **A2:** 34
of wrought magnesium alloys **A2:** 459
wrought titanium alloys **A2:** 611-614
zinc and zinc alloys **A2:** 531
zirconium **A2:** 662-663
Forging billet, defined *See also* Billets **A14:** 6
Forging brass **A14:** 255-256
applications and properties **A2:** 312
Forging cracks
by liquid penetrant inspection **A17:** 86
Forging defect
wrought aluminum alloys **A12:** 415
Forging defects *See also* Defects
in closed-die forging **A14:** 385-386
cracking in cold forging **A14:** 385
Forging dies *See also* Closed-die forging tools, Hot upset forging tools; Dies; Dies, specific types
defined **A14:** 6
Forging dies, upset *See* Hot upset forging tools
Forging envelope *See* Finish allowance
Forging equipment *See also* Equipment; Forging; Forging equipment and dies
classification and characterization **A14:** 37-42
for closed-die forging **A14:** 80-81
and dies **A14:** 23-58
hot upset forging **A14:** 83-84
precision forging **A14:** 165, 168-171
process requirements **A14:** 36
for roll forging **A14:** 96-97
selection of **A14:** 36-42
for titanium alloys **A14:** 273-274
types **A14:** 16
Forging equipment and dies **A14:** 23-58
dies and die materials for hot forging **A14:** 43-58
hammer and presses for forging **A14:** 25-35
selection of forging equipment **A14:** 36-42
Forging force **A6:** 316
Forging gap
as planar flaw **A17:** 50
Forging hammers *See* Hammers
Forging machine dies **A14:** 45
Forging machines
classified **A14:** 25
defined **A14:** 6-7
Forging machines (continued)
horizontal, die set for **A14:** 45
Forging methods
for aluminum alloys **A14:** 242-244
billets for mechanical testing **A14:** 197
selection, titanium alloy forging **A14:** 282
stainless steels **A14:** 222
for titanium alloys **A14:** 270-273
Forging mode **A7:** 413-415
effect on densification of hot forged P/M parts **A7:** 414
and stress conditions on pores **A7:** 414
Forging plane *See also* Parting plane
defined **A14:** 7
Forging presses *See also* Forging equipment; Presses
for powder forging **A14:** 194
Forging pressure *See also* Pressure
and flow stress **A1:** 585
prediction, closed-die forging **A14:** 79-80
Forging process design *See also* Design; Forging process **A14:** 409-416
hot compression testing data for **A14:** 429
interface data for **A14:** 441-442
modeling techniques used **A14:** 417-438
workability data for **A14:** 439-441
Forging processes **A14:** 59-211
closed-die forging in hammers and presses **A14:** 75-82
coining **A14:** 180-187
conditions **A14:** 423-424
data base, required **A14:** 412-413
design, data acquisition for **A14:** 439-442
design method **A14:** 413-416
development, controllable factors **A14:** 409
high-energy-rate forging **A14:** 100-107
hot upset forging **A14:** 83-95
isothermal and hot-die forging **A14:** 150-157
key results **A14:** 413-416
for magnesium alloys **A14:** 260
maps **A14:** 423
open-die forging **A14:** 61-74
parameters **A14:** 407
planning and specifications, method for **A14:** 415
powder forging **A14:** 188-211
precision forging **A14:** 158-175
radial forging **A14:** 145-149
ring rolling **A14:** 108-127
roll forging **A14:** 96-99
rotary forging **A14:** 175-179
rotary swaging of bars and tubes **A14:** 128-144
software tools **A14:** 410-412
for titanium alloys **A14:** 277-282
Forging quality
defined **A14:** 7
Forging quality plates **A1:** 236, 237
Forging steels
materials for die- casting dies **A18:** 629
Forging stock *See also* Stock
defined **A14:** 7
preparation , hot upset forging **A14:** 86
Forging temperature
effect on alpha structures in titanium and titanium alloys **A9:** 460
effect on notch toughness of steels **M1:** 695
effects of **A1:** 588-589
Forging temperatures *See also* Temperature(s)
for alloy steels **A14:** 215
for aluminum alloys **A14:** 242
for carbon steels **A14:** 215
for magnesium alloys **A14:** 259
of various alloys **A14:** 75
Forging(s) **A7:** 5, 23, 295, 410-418, 570
advantages of P/M **A7:** 417
of aluminum P/M parts **A7:** 743-744
and annealing **A7:** 456
applications **A7:** 417
automotive parts **A7:** 417
batch, frequency distribution for components of **A7:** 491
beryllium **A7:** 759
of billets **A7:** 522-529
and cold isostatic pressing **A7:** 448
commercial ferrous **A7:** 415-417
conventional, and hot isostatic pressing **A7:** 442
dimensional tolerances **A7:** 482
Forging(s) (continued)
effect of process parameters on mechanical properties **A7:** 415
effect on densification of hot forged P/M parts **A7:** 414
extrusion **A7:** 411, 412
fluid die in **A7:** 542
fracture **A7:** 410-412
history **A7:** 18
hot, flow diagram **A7:** 416
and hot isostatic pressing of superalloys **A7:** 522
of low-alloy steels **A7:** 464
modes **A7:** 413-415
plus hot extrusion, in superalloys **A7:** 523-524
preform design **A7:** 411-413
preforms **A7:** 416-417, 448
pressure and ejection pressure, at full density of part **A7:** 417
Forgings *See also* Closed-die forgings; Closed-die steel .forgings **A17:** 491-511
$2^1/_4$Cr-1Mo steel, mechanical properties **M1:** 654
alloy steel wire for **A1:** 287 **M1:** 269-270
aluminum alloy **M2:** 5-14
aluminum and aluminum alloy cleaning and finishing of **M5:** 590-591
boiler/pressure vessel, inspection **A17:** 644-645
brass *See* Brass, forgings
central burst in **A14:** 401-402
closed-die **A14:** 76-77
closed-die/upset forgings, inspection techniques **A17:** 495
copper alloy *See* Copper alloys, forgings
definition of **A1:** 337
direct-cooled forging **A1:** 137, 419
eddy current inspection **A17:** 510-511
electromagnetic inspection **A17:** 510-511
finished, by hot upset forging **A14:** 83
flow lines in **A9:** 687
forging operation flaws **A17:** 493-494
heat-resistant alloy, inspection **A17:** 496
ingot/billet processing flaws **A17:** 491-493
inspection method, selection of **A17:** 494-498
liquid penetrant inspection **A17:** 501-504
magnesium **M2:** 528-529, 537-538, 539, 540, 541
magnetic particle inspection **A17:** 112-114, 499-501
magnetizing **A17:** 94
magnetizing cable **A17:** 112
mass finishing of **M5:** 136
microalloyed forgings **A1:** 137, 358-362, 419, 588
applications and compositions of **A1:** 420
carbon contents of **A1:** 585
control of properties **A1:** 588
forging temperatures, effects of **A1:** 588
generations of **A1:** 359-360
properties of **A1:** 360-361, 588, 599
nickel alloy **A17:** 496
open-die, inspection of **A17:** 494-495
pickling of **M5:** 72- 80, 611-613
precision, computed tomography (CT) **A17:** 363
radiographic inspection **A17:** 511
shapes of **A14:** 61
shot peening of **M5:** 147
size and weight of **A14:** 61
steel *See* Steel, forgings
steel, annealing **M4:** 24-25, 26
steel forgings, inspection **A17:** 495-496
steel, normalizing **M4:** 8-11
titanium and titanium alloy, grain flow in **A9:** 460
type, and inspection method **A17:** 494-495
ultrahigh-strength steels **M1:** 422-424, 427, 429, 431, 432, 434, 438, 441
ultrasonic inspection **A17:** 268, 504-510
uranium **M3:** 776
visual inspection **A17:** 498-499
Forgings, failures of **A11:** 314-343
anisotropy caused by forging **A11:** 316-317
causes of **A11:** 317-342
defects caused by forging **A11:** 317
imperfections from ingot **A11:** 314-316
types and frequencies **A11:** 342-343
Forgings, steel
normalizing **A4:** 36, 38-39
use for metalworking rolls **M3:** 503, 506-507

Forking
definition ... **EM4:** 632
Form blocks *See also* Mandrels
for bending ... **A14:** 665
defined ... **A14:** 7
Form cutting
in green machining ... **EM4:** 183
Form die
defined ... **A14:** 7
Form factor
defined ... **A10:** 329
electroslag welding ... **A6:** 271, 272, 273, 274-275, 278
Form rolling
defined ... **A14:** 7
Form spinning
refractory metals and alloys ... **A2:** 562
Form tools ... **A16:** 380, 381
coatings and increased tool life ... **A16:** 58
Form turning
multifunction machining ... **A16:** 372, 373, 380, 381, 382, 383, 391
Form-and-spray
defined ... **EM2:** 19
Form-block method
of stretch draw forming ... **A14:** 593
Formability *See also* Forgeability; Formability testing; Formability testing, specific types; Forming; Sheet formability of steel; Workability
aluminum and aluminum alloys ... **A2:** 8
of aluminum coatings ... **A1:** 219
aluminum-lithium alloys ... **A2:** 188, 191, 196-197
of beryllium ... **A14:** 805
beryllium-copper alloys ... **A2:** 411, 416
beryllium-copper strip ... **A2:** 413
beryllium-nickel alloys ... **A2:** 427
bulk *See* Bulk formability of steels
coating effects ... **A14:** 563
of cold-rolled steels ... **A1:** 420
defined ... **A14:** 7, 19
definition of ... **A1:** 573
of dual-phase steels ... **A1:** 398, 420
effect of material properties on ... **A8:** 549-553
effect of temperature ... **A8:** 553
in explosive forming ... **A14:** 643
hardness as measure of ... **A8:** 560
hot forming ... **A14:** 620
HSLA steels ... **M1:** 409, 419
index, Fukui conical cup test for ... **A8:** 563, 565
indicators, heat-resistant alloys ... **A14:** 780
of interstitial-free steel ... **A1:** 398
lead frame alloys ... **EL1:** 490
magnesium ... **M2:** 540-541, 542
of magnesium alloys ... **A2:** 467-471
nickel- and chromium-plated steels ... **A14:** 564
optimum, material properties for ... **A8:** 549
plate ... **M1:** 194
of preprimed sheet ... **A1:** 222
problems ... **A8:** 548
rating, for cold heading ... **A14:** 291
of refractory metals ... **A14:** 785
rolling direction effect ... **A14:** 780-781
sheet *See* Sheet formability of steel ... **M1:** 545-560
aluminum coated ... **M1:** 172
bending ... **M1:** 552-555
composition, effects of ... **M1:** 553-556
forming capabilities ... **M1:** 546-547
forming limit diagrams ... **M1:** 549-553
forming requirements ... **M1:** 545-546
galvanized ... **M1:** 170-171
mechanical properties, relationship to ... **M1:** 547-549
microstructure, relationship to ... **M1:** 557-559
preferred orientation, effects of ... **M1:** 558
prepainted ... **M1:** 176
selection for ... **M1:** 559
steelmaking practices, effects of ... **M1:** 556-557
temper rolling, effects of ... **M1:** 548
terne coated ... **M1:** 174
Formability (continued)
uniaxial tensile properties, relationship to ... **M1:** 547-549
silver-base brazing filler metals ... **A2:** 702
of stainless steel ... **A1:** 888-889
of stainless steels ... **A14:** 759-761
of steel plate ... **A1:** 238
substrate, coating process impact on ... **A14:** 560
of titanium alloys ... **A14:** 838-839
vs temperature, drop hammer forming ... **A14:** 657
wrought aluminum and aluminum alloys ... **A2:** 99, 101-102
Formability index
Fukui conical cup test for ... **A8:** 563, 565
Formability test
for fracture during stretching ... **A8:** 548
intrinsic ... **A8:** 553-560
simulative ... **A8:** 553, 560-566
Formability testing *See also* Formability; Formability tests, specific types ... **A14:** 877-899
bending tests ... **A14:** 889-890
buckling tests ... **A14:** 892-893
circle grid analysis ... **A14:** 895-896
deformation measurement ... **A14:** 878-879
drawbead forces ... **A14:** 896
drawing tests ... **A14:** 891-892
formability problems ... **A14:** 878
forming limit diagram ... **A14:** 880-881, 894-895
forming types ... **A14:** 877-878
intrinsic ... **A14:** 883-889
and lubricants ... **A14:** 896-897
material properties effects ... **A14:** 879-882
shear fracture ... **A14:** 881
of sheet metals ... **A14:** 877-899
simulative tests ... **A14:** 883, 889-894
springback ... **A14:** 881-882, 893-894
strain distribution ... **A14:** 879-880
stretch-drawing tests ... **A14:** 892
stretching tests ... **A14:** 890-891
surface quality ... **A14:** 882
temperature effect on formability ... **A14:** 882-883
types of ... **A14:** 883
wrinkling ... **A14:** 881
Formability tests, specific types *See also* Formability; Formability testing; Forming
ball punch tests ... **A14:** 890-891
biaxial stretch testing ... **A14:** 887-888
Fukui conical cup test ... **A14:** 892
hardness testing ... **A14:** 889
hemispherical dome tests ... **A14:** 891
hemispherical punch method ... **A14:** 894-895
hole expansion test ... **A14:** 891
limiting dome height test ... **A14:** 891
plane-strain tensile testing ... **A14:** 887
shear testing ... **A14:** 888-889
Swift cup test ... **A14:** 891-892
Swift round-bottomed cup test ... **A14:** 892
uniaxial tensile testing ... **A14:** 883-887
wrinkling tests ... **A14:** 892
Yoshida buckling test ... **A14:** 893
Formaldehyde
emission limit for particleboard set by HUD (1984) ... **EM3:** 105
emission limit set by National Particleboard Association ... **EM3:** 105
physical properties ... **EM3:** 104
Formaldehyde (HCHO) ... **EM3:** 46
chemistry ... **EM3:** 103-104
Formaldehydes, melamine and urea
as aminos ... **EM2:** 230
Formalin ... **EM3:** 103
Formamide
as additive in sol-gel processing ... **EM4:** 446, 449
as additive used to obtain xerogels ... **EM4:** 211
surface tension ... **EM3:** 181
Forman model
crack propagation rate ... **A8:** 681
Formate solutions
copper/copper alloy SCC in ... **A13:** 634
Formation constants
as voltammetric information ... **A10:** 193
Formation energy ... **A6:** 163
Formatting
automated ... **A10:** 310
Formed rolls
for bending ... **A14:** 666
Formed-in-place gasket system
design ... **EM3:** 54
Formers *See also* Horizontal forging machines
die/workpiece temperature control ... **A14:** 171
Formetal 22 Alloy *See* Superplastic zinc; Zinc alloys, specific types, superplastic zinc alloy
Formic acid ... **A13:** 558, 646, 1141, 1157-1158
Formic acid, corrosion
stainless steels ... **M3:** 83, 84
Formica
waterjet cutting rates ... **A16:** 522
Forming *See also* Cold work; Deformation; Formability; Sheet metal forming; Solid-phase forming; Superplastic forming (SPF); Thermoforming; Vacuum forming ... **A7:** 5 **EM4:** 394-491
of aluminum alloys ... **A14:** 248, 791-804
aluminum and aluminum alloys ... **A2:** 10
aluminum sheet ... **M2:** 180
automatic container manufacturing ... **EM4:** 396-398
blow and blow ... **EM4:** 396-397
paste-mold processing ... **EM4:** 397-398
press and blow operation ... **EM4:** 397
automatic hot glass delivery ... **EM4:** 395-396
continuous glass flow ... **EM4:** 396
gob feeders ... **EM4:** 395-396
robot-like feeders ... **EM4:** 395
bars ... **A14:** 622
and bending, of tubing ... **A14:** 665, 673-674
of beryllium ... **A14:** 805-808
beryllium-copper alloys ... **A2:** 415
blanks, refractory metals ... **A14:** 788
capacity, three-roll forming ... **A14:** 619
of carbon steels, lubricants for ... **A14:** 518
casting ... **EM4:** 400-401
centrifugal casting ... **EM4:** 400, 401
gravity casting ... **EM4:** 400-401
compression ... **A14:** 594-595
in conjunction with turning ... **A16:** 140
contour roll ... **A14:** 624-635
costs ... **EM2:** 647-648
defined ... **A14:** 7 **EM2:** 19
development ... **EM4:** 394
drop hammer ... **A14:** 608, 654-658
ductile and brittle fractures from ... **A11:** 87-92
electromagnetic ... **A14:** 644-653
equipment
Hartford H-28 ... **EM4:** 397
individual section machine ... **EM4:** 396, 397
Olivotto ... **EM4:** 397, 398
ribbon machine ... **EM4:** 397-398
Turret chain paste-mold machine ... **EM4:** 398
equipment, types ... **A14:** 16
explosive ... **A14:** 636-643
extrusion ... **EM4:** 401
fluid ... **A14:** 611-614
fluid-cell ... **A14:** 608-611
fundamentals ... **EM4:** 123
future trends ... **A14:** 20-21
glass transition ... **EM4:** 394, 395
hand operations ... **EM4:** 394
cane and tubing ... **EM4:** 395
handmade containers ... **EM4:** 394
sheet ... **EM4:** 395
of heat-resistant alloys ... **A14:** 779-784
history ... **A14:** 15
hot or cold, fractures from ... **A11:** 8
HSLA steels ... **M1:** 419
inside punch ... **A7:** 337
level, multiple-slide ... **A14:** 571-572

Forming (continued)
load vs displacement curves for **A14:** 37
localized severe **A14:** 550
of low-alloy steels, lubricants for **A14:** 518-519
marbles and spheres **EM4:** 401
multiple-part **A14:** 571
multiple-slide **A14:** 519, 570-573
multistage, drop hammer **A14:** 655
new materials **A14:** 19
new processes **A14:** 16-19
of nickel-base alloys **A14:** 831-837
of P/M ferrous powders **A7:** 683
P/M materials *See* Hot forming, Re-pressing
of platinum-group metals **A14:** 520
pressed ware **EM4:** 398
applications **EM4:** 398
free pressing **EM4:** 398
ring pressing **EM4:** 398
tooling **EM4:** 398
process classification **A14:** 15-16
process simulation **A14:** 19-20
processes, introduction to **A14:** 15-21
radial draw **A14:** 595-596
radius, as press-brake forming **A14:** 536
of refractory metals **A14:** 785-788
refractory metals and alloys **A2:** 562-563
requirements, in piercing **A14:** 468-469
shapes **A14:** 622
sheet glass **EM4:** 398-399
float glass **EM4:** 399
fusion process **EM4:** 399
microsheet drawing **EM4:** 399
plate glass **EM4:** 399
redrawn sheet **EM4:** 399
sheet drawing **EM4:** 399
using the fusion process **EM4:** 399
of sheet metal **A14:** 489-503, 787
small cylinders **A14:** 621
speed **A14:** 623
of stainless steels **A14:** 519, 759-778 **M3:** 41-42
statistical analysis **A14:** 928-939
of steel strip, in multiple-slide machines **A14:** 567-574
stretch **A14:** 591-598
stretch draw **A14:** 593-594
structural ceramics **A2:** 1020
temperature and times, magnesium alloys **A2:** 472
three-roll **A14:** 616-623
of titanium alloys **A14:** 838-848 **M3:** 369
as tool steel application **A7:** 792
of tubing **A14:** 665, 673-674 **EM4:** 399-400
Danner process **EM4:** 399-400
downdraw process **EM4:** 400
updraw process **EM4:** 400
Vello process **EM4:** 400
viscosity **EM4:** 394, 395
vs machining **A14:** 778
of wire **A14:** 694-697
of wrought aluminum alloy **A2:** 41-42
wrought titanium alloys **A2:** 614-616
Forming limit analysis
copper and copper alloys **A14:** 820-823
Forming limit diagram **A8:** 577
aluminum alloys **A14:** 792
for aluminum-killed steel **A8:** 566
balanced biaxial prestrain effect of **M1:** 550-551, 554
copper and copper alloys **A14:** 822
critical wrinkling strains on **A8:** 564
defined **A8:** 6, 551 **A14:** 7
dependence on strain path **A8:** 551
determination of **A8:** 566 **A14:** 894-895 **M1:** 549-552
effect of thickness and *n* value on plane strain in **A8:** 551
for free-surface fracture **A8:** 581
including shear fracture **A8:** 551-556
limitations of use **M1:** 551-552
for low-carbon steels **A8:** 551
for maximum strain levels **A8:** 551 **A14:** 880-881
mechanical properties, correlation with **M1:** 550, 553
prediction **A14:** 911-912
thickness and strain-hardening exponent, dependence on **M1:** 553
typical **A14:** 19-20
Forming machines
and die closing speeds **A14:** 100
Forming operations, steel
phosphate coatings aiding **M5:** 436-437
Forming process
statistical analysis **A14:** 928-939
Forming properties of sheet steels **A1:** 398, 888-889
Forming rolls
contour roll forming **A14:** 628-630
Forming station
multiple-slide machines **A14:** 568-569
Forming temperature *See also* Temperature(s)
organic-coated steels **A14:** 565
Forming winder
for glass fiber **EM1:** 108
Forming-limit diagrams
aluminum alloy sheet **M2:** 182-183
Forming-limit line, and fracture-limit lines
compared **A8:** 581
Formol **EM3:** 103
Formula weight
single-crystal x-ray diffraction determined **A10:** 344
Formulas
engineering, use of **EM2:** 652-654
handbook equation examples **EM2:** 653-654
handbook equations, limitations **EM2:** 654
stress **EM2:** 652-653
Formulation
of coating cure systems **EL1:** 856-859, 861-863
Formulation verification
as quality control **EM1:** 730
Formvar
defined **A9:** 8
as thermocouple wire insulation **A2:** 882
Formvar replica
defined **A9:** 8
Forsterite
crystal structure **EM4:** 881
material to which crystallizing solder glass is applied **EM4:** 1070
material to which vitreous solder glass is applied **EM4:** 1070
as olivine molding material **A15:** 94, 209
properties **EM4:** 759
Fortafil 4R
properties **A18:** 803
FORTRAN
as process rough analysis tool **A14:** 411
Forward bowls
near-net shape consolidation **A7:** 441, 443
Forward extrusion *See also* Extrusion
of aluminum alloys **A14:** 244
hot **A14:** 315-316
load vs displacement curve **A14:** 37
of titanium alloys **A14:** 273
Forward impacting
aluminum and aluminum alloys
Forward scan potentiodynamic corrosion curves
P/M stainless steels **A13:** 829
Forward scattering **A18:** 448
Forward sigmoidal curve
fractal analysis **A12:** 213
Forward tube spinning **A14:** 676
Forward-coupled noises
defined **EL1:** 35
Fossil fuel boilers
remote-field eddy current inspection **A17:** 200-201
Fossil fuel power plants **A13:** 985-1010
alloy steel corrosion in **A13:** 538-540
ash-handling systems **A13:** 1007-1008
combined cycle plants **A13:** 986
combustion turbines **A13:** 999-1001
condenser corrosion **A13:** 986-989
deaerators and feedwater heaters **A13:** 989-990
dew point corrosion **A13:** 1001-1004
flue gas desulfurization systems **A13:** 1004-1006
gas turbines **A13:** 986
generator corrosion **A13:** 1006
hot corrosion, boilers burning municipal solid waste **A13:** 997-998
hot corrosion in coal- and oil-fired boilers **A13:** 995-996
steam power plants **A13:** 985-986
steam turbines **A13:** 993-995
Fossil fuel power plants (continued)
steam/water-side boilers **A13:** 990-993
superheaters and high-temperature air heaters **A13:** 998-999
Fossil fuels *See also* Fuels
assay for toxic elements, neutron activation analysis for **A10:** 233
Fotalite **EM4:** 440
Fotoceram **EM4:** 1057, 1058
Fotoform products ... **EM4:** 440, 871, 1055, 1056, 1057, 1058
Foucault contrast **A9:** 536
Fouling *See also* Anti-fouling; Biological corrosion
biological, water-formed **A13:** 492-494
by marine organisms **A11:** 191
defined **A13:** 7
deposits, water-formed **A13:** 492
of equipment **A13:** 1137-1138
organism **A13:** 7, 114-115
Founders
early American **A15:** 24-36
Founding, art
history of **A15:** 20-22
Foundries *See also* Automation; Early American foundries; Equipment; Foundry automation; Foundry equipment
automation of **A15:** 35-36
direct-iron blast furnace, development **A15:** 24-25
domestic, numbers of **A15:** 44-45
early American **A15:** 24-27, 31
early US, listing **A15:** 31
market trends **A15:** 44-45
mechanization of **A15:** 32-33
organizations **A15:** 34
technology development **A15:** 24-36
titanium **A15:** 826
titanium and titanium alloy castings **A2:** 636
Foundry
alloys, copper casting, as high/low-shrinkage **A2:** 346
products, aluminum **A2:** 123-151
properties, of copper alloys for sand casting **A2:** 348
type metal, as lead application **A2:** 549-550
Foundry automation
automatic pouring systems **A15:** 569-570
automatic sorting and inspection **A15:** 571-572
automatic storage and retrieval systems **A15:** 570-571
cell applications **A15:** 568-569
robotic applications **A15:** 566-568
Foundry Educational Foundation **A15:** 34
Foundry equipment *See also* Equipment
automatic pouring systems **A15:** 497-501
degassing processes (converter metallurgy) **A15:** 426-431
degassing processes (ladle metallurgy) **A15:** 432-444
foundry automation **A15:** 566-573
melting furnaces **A15:** 356-392
nonferrous molten metal processes **A15:** 445-496
and processing **A15:** 339-573
processing of castings **A15:** 502-565
sand processing **A15:** 341-355
vacuum melting and remelting **A15:** 393-425
Foundry Equipment Manufacturer's Association
See Casting Industry Supplier's Association
Foundry market **EM3:** 47
Foundry operations
flow charts of **A15:** 203-207
grain refiners in **A15:** 477-478
Foundry organizations **A15:** 34
Foundry practice
effect on gray iron fractures **A11:** 363-365
forging failure frequency from **A11:** 342
iron castings **A11:** 362
Foundry practices *See also* Foundry processing
aluminum alloy specialty castings **A15:** 755-757
for corrosion-resistant steel castings **A1:** 928
high-alloy steels **A15:** 730
low-alloy steels **A15:** 721
Foundry processing
automatic pouring systems **A15:** 497-501
degassing (converter metallurgy) **A15:** 426-431
degassing (ladle metallurgy) **A15:** 432-444
and equipment **A15:** 339-573
foundry automation **A15:** 566-573

Foundry processing (continued)
and foundry equipment ... **A15:** 339-573
of gray iron ... **A15:** 635-640
melting furnaces ... **A15:** 356-392
nonferrous molten metal ... **A15:** 445-496
physical chemistry of ... **A15:** 50-54
processing of castings ... **A15:** 502-565
sand processing ... **A15:** 341-355
vacuum melting and remelting ... **A15:** 393-425
Foundry products, aluminum *See also* Castings; Die casting; Sand Casting
Foundry returns
defined ... **A15:** 6
Foundry technology history ... **A15:** 24-36
casting alloys, advances in ... **A15:** 29-32
early American foundries ... **A15:** 24-27
equipment advances ... **A15:** 27-29
foundry mechanization ... **A15:** 32-33
twentieth century ... **A15:** 33-36
Fountain spraying ... **A7:** 73, 74
Four signal layer multilayer boards (4SL MLBs) ... **EL1:** 15
Four-arm bridge
for torsional Kolsky bar dynamic testing ... **A8:** 228
Four-arm electric resistance
strain gage bridge ... **A8:** 220
Four-ball tests
metalworking fluids ... **A18:** 101
Four-hammer radial forging machines
sizes/capacities ... **A14:** 145, 147-148
Four-harness satin
defined ... **EM2:** 19
Four-high mill *See also* Cluster mill; Two-high mill ... **A14:** 7, 351
Four-high rolling mills
for wrought copper and copper alloys ... **A2:** 244
Four-parameter element
as creep model ... **EM2:** 661-666
Four-point bend beam test, notched
for hydrogen embrittlement testing ... **A8:** 542
Four-point bend test ... **A8:** 132-136, 539-540
Four-point loaded specimens
SCC testing ... **A8:** 505
Four-post hydraulic press
components ... **A14:** 32
Four-post loading frames
cam plastometer ... **A8:** 195
Four-row cotton picker ... **A7:** 672
Four-square texture ... **A9:** 537
Four-station mechanical (radial) ring rolling mill ... **A14:** 110
Fourcault process ... **EM4:** 21, 399
display glass properties ... **EM4:** 1048-1049
Fourdrinier paper machine ... **A13:** 1186
Fourier analysis
C-scan ... **EM2:** 844-845
Fourier equation
for shape analysis ... **EM4:** 69
Fourier modulus ... **A18:** 39, 43, 44
Fourier self-deconvolution
as method of resolution enhancement ... **A10:** 116-117
Fourier transform ... **EM3:** 13
in analysis of iron and nickel ... **A10:** 416-417
data, multiple scattering effects in ... **A10:** 410
defined ... **EM2:** 19
of diffraction-peak profile ... **A10:** 386
in EXAFS data analysis ... **A10:** 412-413
to r space ... **A10:** 412-413
Fourier transform infrared (FTIR) spectroscopy ... **EM2:** 825-826
Fourier transform infrared spectroscopy ... **EM1:** 285, 730
advantages ... **A10:** 112
capabilities ... **A10:** 277
and chromatographic techniques ... **A10:** 115-116
as computerized IR ... **A10:** 109
defined ... **A10:** 673
Fourier transform infrared spectroscopy (continued)
infrared spectrum of glass ... **A10:** 122
of inorganic gases, information from ... **A10:** 8
of inorganic liquids and solutions, information from ... **A10:** 7
of inorganic solids ... **A10:** 4-6
interferometers ... **A10:** 111-112
of organic gases, information from ... **A10:** 11
of organic liquids and solutions, information from ... **A10:** 10
of organic solids, information from ... **A10:** 9
and photoacoustic spectroscopy compared ... **A10:** 115
for polymer curing reactions ... **A10:** 120
and Raman spectroscopy ... **A10:** 126
of silicon ... **A10:** 123
spectrometer, optical diagram ... **A10:** 112
Fourier transform infrared spectroscopy (FTIR) ... **A18:** 300 **EM3:** 237, 732
for analyzing organics or chemical bonding ... **EM4:** 24
depth profiling ... **EM3:** 247
for temperature-time control in polymer removal techniques ... **EM4:** 137
to analyze the surface composition of ceramic powders ... **EM4:** 73
to scan polyimides ... **EM3:** 156, 158
Fourier transform spectrometers ... **A10:** 39-40, 690
Fourier transformation
defined ... **A17:** 384
Fourier transforms ... **A18:** 352
image compression techniques ... **A18:** 347
Fourier-domain processing
in machine vision process ... **A17:** 34
Fourth reduction gear
P/M lawn equipment ... **A7:** 678
Foxall, Henry
as early founder ... **A15:** 27
FP fiber
defined ... **EM1:** 11
FR-4 (fire-resistant-4)
glass-epoxy laminate ... **A6:** 133
FR-4 epoxy resin
difunctional ... **EL1:** 534
glass boards, wave soldering ... **EL1:** 865
polyfunctional systems ... **EL1:** 534
properties ... **EL1:** 534-535
Fracjack test system
for fracture toughness testing ... **A8:** 470, 474
Fractal analysis *See also* Fractal dimensions; Fractal plot ... **EM4:** 69
of AISI 4340 steels ... **A12:** 213-214
experimental background ... **A12:** 211-212
experimental procedure ... **A12:** 212-213
fractal equation for irregular surfaces ... **A12:** 213-215
of fracture surfaces ... **A12:** 211-215
linearizaton of RSC fractal curves ... **A12:** 213
mathematical concept ... **A12:** 211
modified fractal dimensions ... **A12:** 213
profile parameters ... **A12:** 212
summary ... **A12:** 214-215
surface roughness parameters ... **A12:** 212
Fractal dimensions
applicability ... **A12:** 211, 214-215
Mandelbrot ... **A12:** 213
modified ... **A12:** 213
and roughness parameters, 4340 steel ... **A12:** 213
Fractal plot
with linearized RSCs ... **A12:** 214
RSC behavior ... **A12:** 212-213
theoretical linear, applicability ... **A12:** 211
Fractal surfaces ... **A18:** 352-353
Fractals
as descriptors of P/M systems ... **A7:** 243-245
Fraction *See also* Fracture toughness ... **A7:** 5
brittle ... **A7:** 58-59
criterion of deformation ... **A7:** 411
forging ... **A7:** 410-411
Fraction (continued)
height strain and cylinder height ... **A7:** 411
internal, in forged part ... **A7:** 412
loading to catastrophic failure ... **A7:** 58
in loose powder compaction ... **A7:** 298
as milling, process ... **A7:** 62
particle effects in milling ... **A7:** 61
resistance of ceramic grains, Ceracon process ... **A7:** 539
single ... **A7:** 59
surface ... **A7:** 251, 253, 254
and surface chemical analysis ... **A7:** 250
in upsetting ... **A7:** 411
Fraction defective, defined
in quality control ... **A17:** 735-737
Fractional crystallization
aluminum alloys ... **A2:** 4
as ultrapurification technique ... **A2:** 1093
Fractional crystallization for purifying metals ... **M2:** 709-710
Fractional damage equation
in creep-fatigue interaction ... **A8:** 355-356
Fractional distillation for purifying metals ... **M2:** 710
Fractional factorial designs *See also* Factorial designs
families of ... **A17:** 748
fractionation, consequences of ... **A17:** 747-748
orthogonal arrays ... **A17:** 748-750
sequential experimentation ... **A17:** 748
two-level ... **A17:** 746-750
Fractional factorial experiments ... **A8:** 641-644, 654-656
Fractionation
in gas mass spectroscopy ... **A10:** 152
Fractographs
bolt failure study by ... **A12:** 248
content and materials in ... **A12:** 216
dark-field illumination, light-field illumination SEM, compared ... **A12:** 92-100
fatigue cracks in clamp ... **A10:** 305
light, materials types ... **A12:** 1, 216
replica, compared ... **A12:** 95, 100
SEM, calculations of features in ... **A12:** 206-207
SEM, material types ... **A12:** 216
single SEM, in quantitative fractography ... **A12:** 194
TEM and SEM, compared ... **A12:** 185-192
TEM, material types ... **A12:** 216
Fractography *See also* Descriptive fractography; Photography ... **A8:** 6, 476 **EM1:** 786-793
advantages as SEM application ... **A12:** 8
brittle fractures ... **EM2:** 806-807
of composites ... **A11:** 740-741
of cup-and-cone fracture ... **A11:** 83
defined ... **A9:** 8 **A11:** 5 **A12:** 1 **A13:** 7 **EM1:** 766
depth-of-field effects ... **A12:** 78, 80-81, 87-88
of ductile fractures ... **A11:** 25 **EM2:** 805-806
electron ... **A12:** 1, 4-8
electron, for fatigue fractures ... **A11:** 116
for examining fast fracture in ceramics ... **EM4:** 694-695
in failure analysis ... **A11:** 747
of fiber composites ... **A9:** 592
fracture origin location by ... **A12:** 91
fracture surface features ... **EM2:** 807-810
history of ... **A12:** 1-11
of hydrogen-embrittled steels ... **A11:** 28-29
of interlaminar/intralaminar fractures ... **EM1:** 786-790
lighting for ... **A12:** 81-85
macrofractography, defined ... **A12:** 1
microfractography, defined ... **A12:** 1
modes of fracture ... **EM2:** 805-807
of- fracture surface, stage 11 striations ... **A11:** 23
of- transgranular brittle fractures ... **A11:** 25
optical, defined ... **A12:** 1
for printed board failure analysis ... **EL1:** 1039-1040
purpose of ... **A12:** 1

SUBJECTS OF THE INDEXED VOLUMES: ASM Handbook (designated by the letter "A"): **A1:** Properties and Selection: Irons, Steels, and High-Performance Alloys (1990); **A2:** Properties and Selection: Nonferrous Alloys and Special-Purpose Materials (1990); **A3:** Alloy Phase Diagrams (1992); **A4:** Heat Treating (1991); **A6:** Welding, Brazing, and Soldering (1993); **A7:** Powder Metallurgy (1984); **A8:** Mechanical Testing (1985); **A9:** Metallography and Microstructures (1985); **A10:** Materials Characterization (1986); **A11:** Failure Analysis and Prevention (1986); **A12:** Fractography (1987); **A13:** Corrosion (1987); **A14:** Forming and Forging (1988); **A15:** Casting (1988); **A16:** Machining (1989); **A17:** Nondestructive Testing and Quality Control (1989); **A18:** Friction, Lubrication, and Wear Technology (1992). **Metals Handbook, 9th Edition** (designated by the letter "M"): **M1:** Properties and Selection: Irons and Steels (1978); **M2:** Properties and Selection: Nonferrous Alloys and Pure Metals (1979); **M3:** Properties and Selection: Stainless Steels, Tool Materials and Special-Purpose Materials (1980); **M4:** Heat Treating (1981); **M5:** Surface Cleaning, Finishing, and Coating (1982); **M6:** Welding, Brazing, and Soldering (1983). **Engineered Materials Handbook** (designated by the letters "EM"): **EM1:** Composites (1987); **EM2:** Engineering Plastics (1988); **EM3:** Adhesives and Sealants (1990); **EM4:** Ceramics and Glasses (1991); **Electronic Materials Handbook** (designated by the letters "EL"): **EL1:** Packaging (1989).

Fractography (continued)
quantitative ... **A11:** 56 **A12:** 8, 193-210
as reconstruction of- fracture sequence and cause ... **A11:** 744
scanning electron microscopy used to study ... **A9:** 99
SEM ... **A11:** 22 **A12:** 8, 173-176
sources for special terminology ... **EM4:** 632
special terminology used ... **EM4:** 632-634
stereomicroscope for ... **A12:** 87-88
techniques ... **EM1:** 771
techniques, for ceramics ... **A11:** 747-749
TEM applied, history ... **A12:** 6
of translaminar fractures ... **EM1:** 790-792
Fractology *See* Fractography
Fractometer I and II tests ... **A8:** 466, 473-474
Fracture *See also* Brittle intergranular fracture; Crack; Cracking; Cracks; Ductile fracture; Failure; Fatigue; Fractography; Fracture analysis; Fracture control philosophy; Fracture mechanics; Fracture mechanics Linear elastic fracture mechanics; Fracture mechanism map; Fracture models; Fracture toughness; Fractures; Intergranular fracture; Interlaminar fracture toughness; Plane-strain fracture toughness; Stress-corrosion cracking (SCC); Tensile failure/fracture ... **EM3:** 13
as adherend defect, adhesive-bonded joints ... **A17:** 612
advanced statistical concepts in brittle materials ... **EM4:** 709-715
aluminum, particle effects ... **A8:** 478
appearance ... **A8:** 262
behavior ... **A8:** 439-441
behavior, and deformation ... **A10:** 365, 376-378
brittle ... **A8:** 476 **EM2:** 806-807
brittle intergranular ... **A17:** 287
brittle, of FIM samples ... **A10:** 587
brittleness testing ... **EM2:** 738-739
of carbon-carbon composites ... **EM1:** 913
carrier cloth effect ... **EM3:** 515-516, 518
classification ... **A8:** 477
cleavage ... **A8:** 285
compression ... **A8:** 57-58
contact surface ... **A14:** 399
crack growth model material parameters A ... **EM3:** 513
and creep life ... **A8:** 344
criteria for workability ... **A8:** 577
criteria, workability ... **A14:** 370
cure conditions effect ... **EM3:** 515
in cylindrical upset test specimens ... **A8:** 580-581
damage, micromechanics ... **A8:** 286-288
data, testing machine capacity for ... **A8:** 58
defined ... **A8:** 6 **EM1:** 11, 233 **EM2:** 19, 412
deformation heating effect on ... **A8:** 164-165
development and extension, forces causing ... **A8:** 439-464
die contact surface ... **A14:** 402-403
ductile ... **A8:** 154, 476 **EM2:** 805-806
ductile, in bulk deformation processing ... **A14:** 363-364
ductile, in overload region ... **A10:** 513
ductile versus brittle behavior ... **EM3:** 516
ductile-brittleness transition ... **EM2:** 734-735
ductility ... **A8:** 466
during analysis, pulsed-laser atom probe to overcome ... **A10:** 597
during stretching, sheet metal forming ... **A8:** 548
during swaging ... **A14:** 143
dynamic, testing of ... **A8:** 259-297
effect of material behavior ... **EM3:** 513
effect of viscoelasticity ... **EM3:** 513
and effective flaw size ... **EM1:** 242
empirical criterion of ... **A14:** 389-393
energy, measured ... **A8:** 262
environmental effects ... **EM2:** 736
environmentally-assisted ... **A8:** 499-500
experiment synopsis ... **EM3:** 517-518
failure analysis of ... **EM2:** 734-740
form and crack tortuosity ... **A8:** 480
as formability problem ... **A8:** 548
and fracture surfaces, in failure analysis ... **EM1:** 765
fracture toughness testing ... **EM2:** 739-740
grain growth effect on ... **A8:** 164-165
growth direction ... **A10:** 512

Fracture (continued)
high strain rate, testing ... **A8:** 259
hydride-induced, microstructure of ... **A8:** 487
initiation criteria ... **EM2:** 736-737
initiation, dynamic notched round bar test for ... **A8:** 275
intergranular ... **A13:** 156
intergranular, final gray area of flaw ... **A10:** 512
interlaminar ... **EM1:** 784, 786-790
internal ... **A14:** 399
intralaminar ... **EM1:** 786-790
laminar ... **EM1:** 233-235
and localized thinning, models for ... **A14:** 393
locus ... **A14:** 392
material microstructure considerations ... **EM3:** 513
mechanics ... **A8:** 439-464
mechanism ... **A8:** 154, 571-573
mechanism map ... **A14:** 363
metal powder addition effect ... **EM3:** 515
of metals ... **A8:** 154
micro-, mechanics ... **A8:** 465-468
and microstructure ... **A8:** 476-491
mode ... **A8:** 484-485, 574
models of ... **A14:** 393-396
notch blunting ... **EM3:** 515
numerical dynamic analysis ... **A8:** 445-446
origin ... **EM2:** 807-808
overload failure by catastrophic rapid ... **A8:** 476
parameters, calculating for dynamic notched round bar testing ... **A8:** 281-282
percent fibrous ... **A8:** 262
premature ... **A8:** 476, 500
premature, by SCC ... **A13:** 245
process zone, and crack front ... **A8:** 440
process zone, unloading influence in ... **A8:** 452
rapid *See also* Rapid fracture ... **A8:** 440-441
rapid overload ... **A8:** 476
regimes, deformation processing maps for ... **A8:** 154
resin shear ... **EM1:** 263
and scraping, of XPS samples ... **A10:** 575
shear, of Kovar-glass seals, XPS analysis ... **A10:** 577-578
sheet metal ... **A14:** 878
as source, acoustic emissions ... **A17:** 287
-strain data, stainless steel torsion tests ... **A8:** 167-168
strain locus ... **A14:** 393
strength, average, vs fifing atmosphere ... **A10:** 577
strength, defined ... **EM2:** 19
stress criteria ... **EM1:** 253-255
stress, defined ... **EM1:** 11 **EM2:** 19
stress intensification rates of ... **A8:** 259
stress-corrosion ... **A13:** 148
stress-whitening phenomenon ... **EM3:** 514-515
sudden, in fatigue ... **A8:** 363
surface ... **A8:** 278-279, 449
surface, hackle region ... **EM2:** 808-810
surface, mirror zone ... **EM2:** 808
surface, mist region ... **EM2:** 808
tensile, modes ... **A14:** 363
tensile stress and plastic strain for ... **A8:** 577
theories, laminates ... **EM1:** 235
torsional, interpretation of data ... **A8:** 163-169
toughness, conversion factors ... **A10:** 686
translaminar ... **EM1:** 786, 790-792
transverse, of 1075 steel railroad rail ... **A10:** 304
under high loading rates, measurement and analysis ... **A8:** 259-297
unstable ... **A8:** 496
and workability ... **A8:** 154
wrought aluminum alloy ... **A2:** 41
Fracture analysis *See also* Fracture
center cracks ... **EM1:** 57
experimental data ... **EM1:** 255-257
fracture mechanics criteria ... **EM1:** 253
of laminates ... **EM1:** 252-258
major sequential steps in ... **A11:** 49, 746
as reconstruction of fracture sequence and cause ... **A11:** 744
software for ... **A11:** 55
stress concentrations ... **EM1:** 252-253
stress fracture criteria ... **EM1:** 253-255
Fracture analysis or mechanism
Auger electron spectroscopy ... **A10:** 549-567
scanning electron microscopy ... **A10:** 549-567
x-ray topography ... **A10:** 365-379

Fracture appearance *See also* Fracture(s)
facets ... **A12:** 128
frequency, effect on ... **A12:** 58, 60
hydrogen effects on ... **A12:** 124
and impact energy vs temperature ... **A12:** 109
surface, effect of heat treatment on hydrogen- embrittled aluminum alloy ... **A12:** 33
surface, effect of microvoid nucleation ... **A12:** 12
surface, titanium alloy, heat treatment, and microstructure effects ... **A12:** 32
temperature effects on ... **A12:** 49-53
Fracture appearance transition of steel ... **M1:** 691-692
Fracture appearance transition temperature (FATT) ... **A1:** 738 **A6:** 444
effect of composition on ... **A1:** 699-700, 701, 702
Fracture control
with *J-R* measurements ... **A8:** 458
Fracture control philosophy ... **A17:** 666-673
conventional life management ... **A17:** 666-668
fracture control verification ... **A17:** 669-673
policy, and ENSIP ... **A17:** 668-669
USAF engine structural integrity program (ENSIP) ... **A17:** 666
Fracture control plan ... **A8:** 439, 450
Fracture critical tension members (FCMs) ... **A6:** 375
Fracture ductility ... **EM3:** 13
defined ... **EM2:** 19
Fracture energy ... **EM3:** 382-383, 384, 503, 509
in *J* testing ... **A11:** 62-63
measured ... **A11:** 55
and stress-corrosion cracking ... **A8:** 499
stress-whitening phenomenon and rubber content related ... **EM3:** 514
and toughness ... **A11:** 51
Fracture grain size
defined ... **A9:** 8
Fracture initiation ... **A12:** 64, 103
dynamic ... **A8:** 275, 276
Fracture interpretation *See also* Fracture(s) ... **A12:** 72
brittle fracture ... **A12:** 105-111
ductile fracture ... **A12:** 96-98
fatigue fracture ... **A12:** 111-121
high-temperature fractures ... **A12:** 121-123
tensile-test fractures ... **A12:** 98-105
Fracture lines *See* Radial marks
Fracture maps
types of ... **A12:** 33, 44
Fracture Mechanic, The
software ... **A11:** 55
Fracture mechanics *See also* Linear elastic, fracture mechanics; Linear elastic fracture mechanics ... **A8:** 439-464
accuracy of ... **A11:** 55
of aluminum alloys, fracture toughness testing of ... **A8:** 458-462
analysis, fatigue failures in crankshafts ... **A11:** 123-125
application of ... **A11:** 31
approach to corrosion fatigue ... **A8:** 403-405, 417
asymptotic continuum mechanics ... **A8:** 465
for circular hole size effect ... **EM1:** 256
composite analysis by ... **EM1:** 195-196
computations ... **A8:** 443-444
concepts and use in analyses ... **A11:** 47-53, 55-57
correlation of Charpy V-notch to ... **A1:** 753
for corrosion fatigue ... **A13:** 295
crack arrest testing ... **A8:** 453-455
crack front stresses and dynamic characterization ... **A8:** 444-445
crack initiation and growth ... **A8:** 465
crack tip stress characterization ... **A8:** 465
crack-extension force and compliance calibration ... **A8:** 441-443
of cracked bodies, and weldment testing ... **A8:** 520-521
criteria, laminates ... **EM1:** 253
for cyclic crack growth studies ... **A8:** 422-423
defined ... **A8:** 465 **A11:** 47 **A13:** 7
deformation fields ... **A8:** 465
dynamic notched round bar testing and ... **A8:** 282
effect of elastic-plastic fracture mechanics on ... **A8:** 498

Fracture mechanics (continued)
effects of scatter of the data ... **A11:** 55
elastic-plastic ... **A11:** 49-51
elastic-plastic analysis ... **A8:** 446-447, 457-458
elastic-plastic indentation ... **A13:** 137-138
estimates related to *K* ... **A8:** 443-444
and failure analysis ... **A11:** 47-65
and fatigue crack propagation ... **A8:** 376
fracture behavior ... **A8:** 440-441
fracture toughness evaluation ... **A8:** 450
fracture-toughness testing and evaluation ... **A11:** 60-64
and impact failure ... **EM2:** 90
J-R measurements ... **A8:** 456-457
K, for three-dimensional crack problems ... **A8:** 444
linear elastic ... **A11:** 47-49
linear elastic (LEFM) ... **A17:** 663-664
measurements, notched tension specimens for ... **A8:** 27
methods in crack propagation testing ... **A8:** 363
micromechanisms ... **A8:** 465
model, circular hole ... **EM1:** 253
models, of acoustic emissions and crack growth ... **A17:** 287
notch-toughness testing and evaluation ... **A11:** 57-60
numerical dynamic analysis ... **A8:** 445-446
part rejection determined by ... **A17:** 49
plane stress and plane strain ... **A11:** 51-52
plane-strain fracture toughness ... **A8:** 450-451
probabilistic ... **A8:** 682
progressive fracturing ... **A8:** 442-443
R-curve measurements ... **A8:** 451-453
requirements, for welding inspections ... **A17:** 590
Rice *J*-integral ... **A8:** 447-450
simulation and ... **A8:** 261-262
software ... **A11:** 55
specimens, SCC testing ... **A8:** 497 **A13:** 253-260
static ... **A8:** 282
of structural steels ... **A8:** 453
subcritical ... **A11:** 52-53
test piece, precracking Charpy testing as ... **A8:** 267
testing ... **A8:** 476
tests and data ... **A11:** 53-55
use in SCC testing ... **A8:** 503
use of ... **A11:** 55-57
Fracture mechanics methods ... **EM3:** 3
Fracture mechanics testing ... **A6:** 101-102
Fracture mechanics tests
low-alloy steels ... **A15:** 717
for steel castings ... **A1:** 368-369
for stress-corrosion cracking ... **A1:** 725
Fracture mechanism map ... **A14:** 363
Fracture mirror ... **EM4:** 635-636
Fracture mirrors
polycrystalline ceramics ... **A11:** 745
Fracture mode identification chart ... **A11:** 80
Fracture models *See also* Cracking; Fracture
Cockcroft ... **A14:** 394
localized thinning ... **A14:** 393
tensile stress criterion ... **A14:** 395-396
upper bound method ... **A14:** 394-395
void growth ... **A14:** 393
Fracture modes *See also* Fracture(s); specific fracture types ... **A12:** 12-22
cleavage ... **A12:** 13-14
creep damage and ... **A12:** 62
decohesive rupture ... **A12:** 18-20
dimple rupture ... **A12:** 12-13
effect of fatigue crack growth rate ... **A12:** 441
effect of low temperatures ... **A12:** 33
effect of stress state ... **A12:** 30-31
fatigue ... **A12:** 14-18
flutes ... **A12:** 20-21
miscellaneous, materials illustrated in ... **A12:** 217
mixed, materials illustrated in ... **A12:** 217
quasi-cleavage ... **A12:** 20
Fracture modes (continued)
slow, monotonic loading, ductile irons ... **A12:** 231
tearing topography surface ... **A12:** 21-22
transition ... **A12:** 33, 44
types illustrated ... **A12:** 217
unique ... **A12:** 20-22
Fracture origin *See also* Crack origin; Fracture(s) ... **A11:** 80, 87, 257, 459, 744-747 **A12:** 91
chevrons emanating from ... **A12:** 170
and fracture cause ... **A12:** 72
location, fractography for ... **A12:** 91
visual examination ... **A12:** 72
Fracture path *See also* Crack path; Fracture(s)
effect of electrochemical potential ... **A12:** 35
fractal analysis ... **A12:** 212
interdendritic, low-carbon steels ... **A12:** 249
medium-carbon steels ... **A12:** 263
and microstructure, correlated ... **A12:** 195-196
preference index ... **A12:** 201
sinusoidal ... **A12:** 109, 116
tortuosity, determined ... **A12:** 206
types, engineering alloys ... **A12:** 12
Fracture path preference index
linear ... **A12:** 201
Fracture profiles *See also* Profiles
by light microscope ... **A12:** 94, 99
examining ... **A12:** 95, 100
sections ... **A12:** 95-96
Fracture resistance
assessment of ... **A1:** 662-663
Fracture specimen(s) *See also* Specimen(s); Test specimen(s)
care and handling ... **A12:** 72
fracture-cleaning techniques ... **A12:** 73-76
nondestructive inspection ... **A12:** 77
opening secondary cracks ... **A12:** 77
preliminary visual examination ... **A12:** 72-73
preparation ... **A12:** 72-77
preservation techniques ... **A12:** 73
sectioning ... **A12:** 76-77
Fracture strain
dependence on temperature ... **A8:** 167-168
from tension and upset tests, compared ... **A8:** 581
true ... **A8:** 24
Fracture strength ... **A11:** 138, 239 **EM3:** 13
Fracture stress *See also* Rupture stress
consideration for material toughness ... **A8:** 466
defined ... **A8:** 6
and final crack size ... **A11:** 56
measured in V-notch bend bars ... **A8:** 466
of metals subject to LME ... **A11:** 719
-strain tests, for drop tower compression test ... **A8:** 197
true ... **A8:** 24
Fracture studies
nineteenth century ... **A12:** 2-3
sixteenth to eighteenth centuries ... **A12:** 1-2
twentieth century ... **A12:** 3-8
Fracture surface
air-fired, XPS survey of ... **A10:** 577
analysis in acidified chloride solutions ... **A8:** 420
of Fe-4Ni fatigue cracked ... **A8:** 486
of Fe-4Si under rapid loading ... **A8:** 479-480
Inconel 718, scanning electron micrograph ... **A8:** 482, 484
inert testing environment for ... **A8:** 37
line scan across ... **A10:** 497
low-carbon austenite, scanning electron micrograph ... **A8:** 481-483
nitrogen-fired, XPS survey ... **A10:** 577
oxygen-fired, XPS survey ... **A10:** 577
sawtooth phenomenon ... **EM3:** 511, 512
SEM analysis ... **A10:** 490, 497
showing failure origin ... **A10:** 304
for smooth copper tensile specimen in sodium nitrite ... **A8:** 490
Fracture surface (continued)
of specimens with manganese tested in hydrogen gas ... **A8:** 489
steel with manganese tested in hydrogen gas ... **A8:** 487
stress corrosion crack ... **A10:** 562-564
test environment effects on ... **A8:** 37
transgranular surfaces ... **A8:** 487
Fracture surface analysis
of an Al_2O_3-$10ZrO_2$ ceramic by instrumented stereometry ... **A9:** 96
Fracture surface analysis of optical fibers ... **EM4:** 663-668
analysis ... **EM4:** 666-668
flaw identification ... **EM4:** 666-667
fractographic analysis of subcritical crack growth ... **EM4:** 667-668
research and production aids ... **EM4:** 666-667
scanning electron microscopy ... **EM4:** 663-667
stress analysis ... **EM4:** 666
conclusions ... **EM4:** 668
experimental procedure ... **EM4:** 665-666
purpose ... **EM4:** 663
theoretical background ... **EM4:** 663-665, 667
crack branching ... **EM4:** 663-664, 667
hackle ... **EM4:** 663-664, 667
mirror ... **EM4:** 663-664, 667
mirror-to-flaw size ratio ... **EM4:** 665
mist ... **EM4:** 663-664
regions ... **EM4:** 663-664, 667
Fracture surface detail, quantification of
using semi-automatic tracing devices ... **A9:** 83
Fracture surface map *See also* Carpet plot
titanium alloy, by stereophotogrammetry ... **A12:** 198
Fracture surface markings
definition ... **EM4:** 632
Fracture surface(s) *See also* Cleaning techniques; Cleavage; Fatigue fracture(s); Fibrous fracture surfaces; Nondestructive inspection; Quantitative fractography; Sectioning; Surface(s)
arbitrary test volume enclosing ... **A12:** 199
area, importance of ... **A12:** 193
cause of failure from ... **A12:** 12
chemical etching, and ultrasonic cleaning ... **A12:** 75
cleaning, for TEM ... **A12:** 179-183
cleavage ... **A12:** 13, 17-18
computer-simulated, true profile length ... **A12:** 200
effect of microvoid coalescence ... **A12:** 12, 13
effect of photo-illumination ... **A12:** 82-89
effect of section thickness ... **A12:** 105
elevated temperature, grain-boundary cavities ... **A12:** 349
equations, basic ... **A12:** 194-196
fatigue, ductile iron ... **A12:** 228
fatigue, interpreting markings ... **A12:** 112, 118, 119
fatigue, tire tracks ... **A12:** 23
flat and shiny, AISI/SAE alloy steels ... **A12:** 304
fractal analysis of ... **A12:** 211-215
high-carbon steels ... **A12:** 279, 280, 288
history ... **A12:** 1-3
mapping, photogrammetry for ... **A12:** 197
markings, types ... **A12:** 91
mating, effect of dimples ... **A12:** 13
measurements, aluminum alloy example ... **A12:** 205-206
mechanical damage to ... **A12:** 92-93
medium-carbon steels ... **A12:** 260, 263, 273
morphologies, ASTM/ASME alloy steels ... **A12:** 345
morphologies, ductile iron ... **A12:** 235
nondestructive inspection effects ... **A12:** 77
P/M aluminum alloys ... **A12:** 440
photography of ... **A12:** 78-90
plane of polish, projection plane and correlated ... **A12:** 196
plane-strain ... **A12:** 308

Fracture surface(s) (continued)
preparation/preservation **A12:** 72-77
profile, AISI/SAE alloy steels **A12:** 327
profile, medium-carbon steels **A12:** 255
and projected images, parametric relationships **A12:** 202
prototype faceted, profile angular distribution **A12:** 203
roughness **A12:** 199-205
service fracture, AISI/SAE alloy steels **A12:** 331
Silcrome-I martensitic stainless steel **A12:** 369
steel spring wire, screw marks **A12:** 280
tool steels **A12:** 376
with triangular elements **A12:** 202
true area, importance **A12:** 211
Fracture surfaces
of aluminum alloy lug **A11:** 31
of aluminum alloys, preservation of **A9:** 351
analysis, steel castings **A11:** 396
of austenitic manganese steel casting **A9:** 238
beach marks on **A11:** 104
of beryllium-containing alloys **A9:** 393
ceramic, features of **A11:** 745
of chain link, hydrogen-assisted cracking in **A11:** 409
characteristics, for SCC **A11:** 212
with chevron marks **A11:** 21
chuck jaw tooth, white layer grain-boundary film **A11:** 576
cleavage-crack nucleation **A11:** 23
of commercially pure titanium implant, with mixed fracture morphology **A11:** 686, 689
corrosion of **A11:** 212
deposits from hydrogen damage **A11:** 250
dimpled-rupture **A11:** 22
discoloration **A11:** 80
elongated dimples on steel **A11:** 25
of failed hip prosthesis **A11:** 692, 693
of failed malleable iron rocker lever **A11:** 350
of failed shafts **A11:** 463
of failed steam-generator sample **A11:** 657
fatigue **A11:** 111, 120, 321, 398
flat-face and shear-face **A11:** 75, 76
forged steel rocker arm, fatigue fracture **A11:** 120
fractograph, with stage 11 striations **A11:** 23
for fracture stress and origin **A11:** 744
from failed steam accumulator **A11:** 648
high-cycle fatigue striations **A11:** 78
of high-strength aluminum alloy **A11:** 28
of hot-pressed ceramic **A11:** 754
of hydrogen embrittlement, fasteners **A11:** 541
hydrogen flakes on **A11:** 337
information from **A11:** 744
knuckle flange, fatigue failure **A11:** 342
light reflection of **A11:** 75
LME fractured nuts **A11:** 543
macroscopic examination of **A11:** 20
matching, determining plastic deformation **A11:** 80
microscopic examination of **A11:** 20-22
periphery, shear lip on **A11:** 83
of piping cross, intergranular cracking **A11:** 650
plastic replicas **A11:** 19
of reheat steam pipe, corrosion products on **A11:** 653
rock-candy, by stress corrosion **A11:** 28
of rotating-bending fatigue **A11:** 321
of sand-cast medium-carbon heavy-duty axle housing **A11:** 389
selection, preservation and cleaning of **A11:** 19-20
of shafts, fatigue marks on **A11:** 403
of stainless steel implant with secondary corrosion attack **A11:** 683, 688
steel connecting rod, fatigue fracture **A11:** 120
steel roller, carburized-and-hardened **A11:** 121
steel turbine disk, fatigue failure **A11:** 131
steel valve spring, torsional fatigue **A11:** 120
steel valve stem **A11:** 288
strength **A11:** 138, 239
stress **A11:** 56, 719
striations on **A11:** 75
of thermal shock failure, ceramics **A11:** 744, 752
used to study bonding of particles in powder metallurgy materials **A9:** 508

Fracture surfaces (continued)
of weld intersection, stainless steel liners **A11:** 118
zones of **A11:** 104
Fracture system
definition **EM4:** 632
Fracture test
defined **A8:** 6
Fracture testing **EM3:** 335-348
adhesive fracture property
blister specimens **EM3:** 337, 339, 340, 341
cracked-lap-shear (CLS) specimen **EM3:** 338, 339
double-cantilever beam (bending moment) **EM3:** 338-339
specimens **EM3:** 337-340
adhesive fracture property specimens
double-cantilever-beam (DCB) specimen **EM3:** 338-339, 340
end-notched-flexure (ENF) specimen **EM3:** 338-339
four-point-bend specimen **EM3:** 338-339, 340
laminated-beam specimens **EM3:** 337-339, 340, 341
mixed-mode-flexure (MMF) specimen **EM3:** 338-339
prismatic specimens **EM3:** 337, 339-340
specimen geometries **EM3:** 337
analysis of specimens **EM3:** 340-348
crack length measurements **EM3:** 342-343
direct compliance method **EM3:** 340-341
fractography **EM3:** 342, 343
hybrid compliance method **EM3:** 341-342
mode I fractographic features **EM3:** 341, 343-344
mode II fractographic features **EM3:** 343, 344-345, 348
mode III fractographic features **EM3:** 345, 346, 347, 348
ultrasonic methods **EM3:** 342
beam theory analysis **EM3:** 342
for carbon steel rod **A1:** 274
concepts
crack tip stress analysis **EM3:** 335-337
energy release rate **EM3:** 337, 339, 340, 342
stress-intensity factor **EM3:** 335-337, 338, 342
cracked-lap-shear (CLS) specimen **EM3:** 444, 445, 446, 447, 449, 450, 451
double-cantilever-beam (DCB) specimen **EM3:** 444-445, 446, 447, 462
end notched flexure (ENF) specimen **EM3:** 444, 445, 446, 447, 449, 450
mixed-mode-flexure (MMF) specimen **EM3:** 444, 445, 446, 447
wedge test **EM3:** 338, 340
Fracture topography in titanium alloys
study of **A9:** 461
Fracture toughness *See also* Dynamic fracture toughness; Fracture; Fracture toughness test; Interlaminar fracture toughness; specific fractures; Stress-intensity factor; Stress-intensity factor (K); Toughness **EM3:** 13 **EM4:** 599-605
abrasion resistant steels **M1:** 624
AISI/SAE alloy steels **A12:** 340
alloy classification **A2:** 58-59
of aluminum alloys **A7:** 474, 747 **A8:** 458-462
aluminum-lithium alloys **A2:** 184
amorphous materials and metallic glasses **A2:** 814
ASTM Committee E 24 definition **EM4:** 599
calculated for elastic plastic behavior **A8:** 473
calculating **A8:** 471
and carbon content **A8:** 481
carburized alloy steels **M1:** 536, 538
cast steels **M1:** 380, 383
cemented carbides **A2:** 956-957 **A12:** 470
of cemented carbides, effect of eta phases on **A9:** 275
of cermets, complexing cermet compositions for **A2:** 991
of closed-die forgings **A1:** 341-342
composition and microstructural effects related **A8:** 480
conversion factors **A8:** 722 **A10:** 686
copper alloys **A12:** 402
correlation with Charpy V-notch **A8:** 264
correlations of W/A with **A8:** 267

Fracture toughness (continued)
and crack length measurement **A8:** 471
crack-opening displacement (COD) test for **A11:** 62
data base **A8:** 462
defined **A8:** 6 **A11:** 5 **A13:** 7 **EM1:** 11 **EM2:** 19
of ductile iron **A1:** 46, 47 **A15:** 662 **M1:** 42, 45-46
of ductile metals and alloys **A11:** 61
and dynamic tear (DT) test **A11:** 61-62
effect of hydrogen flaking **A11:** 316
effect of neutron irradiation **A12:** 388
effect of particles, AISI 4340 steel **A8:** 479
elastic-plastic, crack initiation **A8:** 390
evaluation, in fracture mechanics **A8:** 450
examination of crack growth by electropolishing **A9:** 55-56
of ferritic malleable iron **A1:** 75-76
forged steels **M1:** 354-355
in forging **A11:** 315
and fraction of transformed microstructure in titanium alloy **A8:** 480, 482
as fracture mechanics criteria **EM1:** 253
as function of inclusion spacing **A8:** 479
of high-strength aluminum P/M alloys **A7:** 747
of high-toughness aluminum alloy plate **A8:** 461
impact response curves for **A8:** 269-271
indices, development of **A8:** 459
instrumented-impact test for **A11:** 64
interlaminar, and damage tolerance **EM1:** 262
J testing of **A11:** 62-64
linear elastic fracture mechanics **EM4:** 646
malleable irons **A15:** 696
maraging steels **M1:** 445, 448, 449-450
martensitic stainless steel **A8:** 479-480
measurements **EM4:** 599
bridge indentation method **EM4:** 603
chevron notch bend specimen **EM4:** 600, 601-602
compression precracking **EM4:** 600, 603-604
double cantilever beam **EM4:** 600, 602, 604
double torsion **EM4:** 600, 604
fractography approach **EM4:** 600, 603, 604-605
Griffith's criterion **EM4:** 599
guidelines **EM4:** 604-605
indentation crack length/ fracture **EM4:** 600, 601, 604
indentation strength **EM4:** 600, 601, 604
R-curve phenomena **EM4:** 604
single-edge notched beam **EM4:** 600, 602-603
single-edge precracked beam **EM4:** 600, 603, 604
stress-intensity factor **EM4:** 599, 600, 604
of metastable austenite, carbon content effect **A8:** 484
minimums **A8:** 459
notch toughness, relation to in steel **M1:** 690
ordered intermetallics **A2:** 913-914
of P/M and ingot metallurgy tool steels **A7:** 471, 472
of P/M and wrought titanium and alloys **A7:** 475, 752
of P/M low-alloy steel powders **A7:** 470
of pearlitic and martensitic malleable iron **A1:** 74, 80, 82
plane-strain **A8:** 450-451
plane-strain test of **A11:** 60-61
plane-strain, wrought aluminum and aluminum alloys **A2:** 74
prealloyed titanium P/M compacts **A2:** 651-652
R-curve analysis of **A11:** 64
R-curve for **A8:** 450
ratio-analysis diagram (RAD) of **A11:** 62
scatter bands for three heats stainless steel **A8:** 479-481
and SCC **A13:** 276
specimen, ultrasonic cleaning of **A12:** 74
static and dynamic, vs. temperature **A8:** 283
of steel castings **A1:** 377-378
strain rate and temperature effects on **A8:** 453
and stress fracture criteria, compared **EM1:** 256
and stress-corrosion cracking **A8:** 496-497
of structural steel **A1:** 397-398, 663, 664, 666-667, 669
sulfide **A13:** 534-535
test procedures **A8:** 459-462

Fracture toughness (continued)
testing **EM2:** 739-740
testing and evaluation **A11:** 60-64
testing of **EM1:** 264
thermoplastic polyimides (TPI) **EM2:** 178
Ti-6Al-4V, compositional and microstructural effects **A8:** 480-481
titanium and titanium alloy castings **A2:** 640
of titanium and titanium alloys, effect of alpha colony size on **A9:** 460
ultrahigh-strength steels **M1:** 426, 428, 429, 431, 437, 439, 441, 442
variability in distribution of **A8:** 625
of welded structures **A1:** 667-669, 670, 671, 672
wrought aluminum alloy **A2:** 42-44, 58-60, 74
wrought titanium alloys **A2:** 623, 631
Fracture toughness (K_{Ic}) **A6:** 101-102
Fracture toughness testing *See also* Fracture toughness
ASTM, stress intensification rate of **A8:** 259
behavior types **A8:** 473-474
chevron-notched specimens for **A8:** 469-475
compact tension specimen for **A8:** 460
data analysis in **A8:** 473-474
determining critical *J*-values and *J*-resistance curves in **A8:** 261
as dynamic fracture testing **A8:** 259-261
elastic-plastic **A8:** 455-456, 471-473
linear elastic fracture mechanics test **A8:** 470-471
mixed-mode **A8:** 460-461
plane-strain screening in **A8:** 460
rapid-load plane-strain **A8:** 260-261
specimens and test equipment **A8:** 469-470
Fracture toughness tests **A6:** 103
Fracture transition in steel **M1:** 691-692
Fracture transition temperature (T_f) **A6:** 1101
Fracture(s) *See also* Fast fractures; Fractographs; Fractography; Fracture appearance; Fracture interpretation; Fracture modes; Fracture origin; Fracture path; Fracture profiles; Fracture specimens; Fracture studies; Fracture surfaces; Fracture toughness; Fractured parts
care and handling **A12:** 72
cleaning techniques **A12:** 73-76
columnar **A12:** 2
crystalline **A12:** 2
discontinuities leading to **A12:** 63-68
effects of environment **A12:** 22-63
etching **A12:** 96
fibrous **A12:** 2
granular **A12:** 2
initiation **A12:** 64, 103
markings, types **A12:** 91
materials illustrated in **A12:** 217
and microstructure, correlating **A12:** 201
mixed modes **A12:** 176
modes of **A12:** 12-22, 176
path preference **A12:** 201
patterns, historical study **A12:** 2
photography of **A12:** 78-90
profile sections **A12:** 95-96
propagation **A12:** 91
sectioning **A12:** 76-77
sequence **A12:** 91
silky **A12:** 2
sources illustrated **A12:** 217
studies, history of **A12:** 1-8
test, as quality control application **A12:** 140-143
tests, historical **A12:** 3-4
texture, photographing **A12:** 82-85
topography, titanium alloys **A12:** 441-443
Type I through Type VII (historical) **A12:** 1
types of causes illustrated **A12:** 216
visual examination **A12:** 91-93
vitreous **A12:** 2
woody, historical **A12:** 1-3
Fracture-controlled failure *See also* Failure
torsion test for **A8:** 154
Fracture-critical components
defined **A17:** 668
Fracture-extension resistance
R-curve as **A11:** 64
Fracture-limit line **A8:** 580-583
Fracture/crack
surface property effect **A18:** 342
Fractured parts *See also* Fracture(s); Part(s)
lighting of highly reflective **A12:** 84, 88
photographic setups for **A12:** 78
photography of **A12:** 78-90
Fractures *See also* Brittle fractures; Ductile fractures; Fracture surfaces
acceleration of **A11:** 744-745
at crack tip **A11:** 47
in bearing materials **A11:** 494, 505-506
beginning, condition for **A11:** 50-51
behavior, temperature dependence of **A11:** 138
brittle **A11:** 76-77, 82-101
by mixed mechanisms **A11:** 83-84
characteristics by macroscopy **A11:** 75, 104-105
classification of **A11:** 75
cleavage, defined **A11:** 2
compression **A11:** 739
contour, differences for bend loading vs pure tensile loading **A11:** 746
contributing factors, chart for **A11:** 80
control, fracture mechanics for **A11:** 47
defined **A11:** 5
delayed **A11:** 28
determination of type **A11:** 79-80
ductile **A11:** 25, 82-101
energy **A11:** 49, 55
environmentally affected **A11:** 78-79
extension resistance, *R*-curve as **A11:** 64
fatigue **A11:** 26-27
final **A11:** 257-258
flexural, in composites **A11:** 733
hydrogen embrittlement **A11:** 28-29
impact, defined **A11:** 75
intergranular brittle **A11:** 25-26
interlaminar, in composites **A11:** 733-738
of landing-gear flat spring **A11:** 560
liquid-metal embrittlement **A11:** 27-28
LME-induced, failure analysis of **A11:** 232
mechanisms, of polymers **A11:** 758-761
mirror and hackle, ceramics **A11:** 744-747
modes **A11:** 80, 733-739
origins **A11:** 80, 87, 257, 459, 744-747
overload **A11:** 75, 367
path, transgranular **A11:** 75
photographing **A11:** 16
planes **A11:** 105
and plastic zone **A11:** 49
rock-candy **A11:** 75
shear, defined **A11:** 9
stages, from fatigue **A11:** 104
steam equipment failure by **A11:** 602
of steel crane-bridge wheel **A11:** 527-528
of steel tram-rail assembly **A11:** 526
stress-corrosion cracking **A11:** 27
stress-rupture **A11:** 75, 264-265
tensile, in composites **A11:** 734
test, defined **A11:** 5
thermal fatigue-caused **A11:** 266
transgranular brittle **A11:** 25
transition, ductile-to-brittle **A11:** 66-71
transition, transgranular-intergranular **A11:** 266
translaminar, in composites **A11:** 738-739
transverse **A11:** 79, 554
types **A11:** 24-29, 744
work, in EPFM **A11:** 50
Fractures, brittle
in steel **M1:** 689
Fracturing
as a sectioning method **A9:** 23
of cemented carbides during sectioning **A9:** 273
of tool steels for examination **A9:** 256
Fracturing, rapid *See* Rapid fracturing
Fragment ions
defined **A10:** 153
Fragmentation **A7:** 5
defined **A8:** 6 **A9:** 8
Fragmentation devices
as ordnance application **A7:** 691
Fragmented powder **A7:** 5
Frame
capacity **A14:** 495-496
defined **A8:** 6 **A14:** 7
press, types of **A14:** 492-493
stiffness, effect in SCC testing **A8:** 502-503
torsion tests for **A8:** 139
Frame integration (summing)
real-time radiography **A17:** 320
Frames
equipment, packaging/interconnecting of electronics in **EL1:** 13
interconnection levels **EL1:** 12
level, of interconnection **EL1:** 13
Frames of reference
in coordinate measuring machines **A17:** 20
Francon-Yamamoto interference contrast system **A9:** 150
Frangible bullets
powders used **A7:** 573
Frank dislocation loops
transmission electron microscopy **A9:** 116-117
Frank-Condon principle
defined **A10:** 673
Frank-Read source
in transmission topography **A10:** 370
Franke method
for free lime content **A10:** 179
Fraunhofer diffraction pattern
schematic **A7:** 217
Fraunhofer diffraction theory **EM4:** 67
Fraunhofer-Mie theories **EM4:** 66
Freckles
in wrought heat-resistant alloys **A9:** 305
Freckles, as defect
vacuum arc remelting **A15:** 407
Freckling
defined **A9:** 8
Free acid value tests
phosphate coating solutions **M5:** 442-443
Free bend **EM3:** 13
defined **A8:** 6 **EM2:** 19
Free bending **A14:** 532
Free carbon *See also* Combined carbon
in cemented carbides **A9:** 273-274
defined **A9:** 8 **A13:** 7 **A15:** 6
on a fracture surface **A9:** 276
Free corrosion potential
defined **A13:** 7
Free cyanide **M5:** 285
Free energy
change **A15:** 101, 104, 449
change in **A6:** 45, 46
composition, and temperature composition **A15:** 53
composition plot, constructed **A15:** 52
defined, in electrode potentials **A13:** 19
of formation, ideal solution **A15:** 52
Gibbs **A15:** 50-51
of metal oxide formation **A7:** 53
of particles, and size/curvature **A15:** 102
phase, thermodynamics of **A15:** 101-102
of reaction, in gases **A13:** 61-62
of solution, iron melts **A15:** 62
vs composition diagrams **A15:** 102
Free energy change
during heterogeneous nucleation **A15:** 104
oxidation reactions **A15:** 449
quantified **A15:** 101
Free energy diagram
defined **A9:** 8

Free energy of an inhomogeneous solution
expressed as an integral over volume A9: 653
Free energy of the liquid phase A6: 45
Free energy-composition diagram, of metastable and stable equilibria in
precipitation reactions A9: 650
Free energy-temperature diagrams
for gaseous corrosion A13: 17
high-temperature corrosion in gases A13: 62
oxides A13: 62-63
Free ferrite *See* Proeutectoid ferrite
defined A13: 7 A15: 6
Free growth
spherical crystal A15: 112-113
Free lime
classical wet chemical analysis in Portland cement A10: 179
Free machining
defined A13: 7
Free moisture expansion coefficient
defined EM1: 226
Free quartz EM4: 6
Free radical cure resin binder process A15: 220, 238
Free radical cure systems *See also* Chemistry; Photochemistry
dual-mechanism cure EL1: 859
formulation ingredients EL1: 856-588
formulations EL1: 858
oxygen inhibition EL1: 856
photoinitiators EL1: 854-856
Free radical polymerization EM3: 13
Free radical reactive adhesives
for magnet bonding for motors and speakers EM3: 45
Free radical type addition reactions
in solvent acrylics EM3: 208
Free radicals EM3: 91
in acrylic chemistry EM3: 120, 124
content, of fossil fuel A10: 253
defined A10: 673
electron spin resonance A10: 253-266
intermediates A10: 263
kinetic reactions of inorganic and organic A10: 253
oxidation inhibitors and metalworking lubricants A18: 141
reaction kinetics in formation and decay of A10: 266
reactions, of catalyst surfaces A10: 253
stable hydrazyl A10: 265
standard, DPPH, ESR spectrum of A10: 259
Free recovery
of shape memory alloys A2: 900
Free rolling A18: 37
defined A18: 9
Free rotation EM3: 14
defined EM2: 20
Free spread *See* Nip
Free surface fracture A14: 19-20
Free surface velocity
during spalling A8: 212
Free temper carbon graphitization
as annealing effect A15: 31
Free vibration EM3: 14
defined EM2: 20
Free wall
defined EM1: 12
Free-abrasive machining EM4: 313, 314
Free-body stress system
shafts A11: 460-461
Free-cutting brass *See also* Brasses; Wrought coppers and copper alloys
applications and properties A2: 306-307, 310-311
electrolytic etching A9: 401
recycling A2: 1214-1215
Free-cutting Muntz metal
applications and properties A2: 311
Free-cutting phosphor bronze
applications and properties A2: 325
Free-cutting yellow brass
applications and properties A2: 310-311
Free-edge delamination
in composites A8: 714
finite-element model EM1: 248-249
interlaminar EM1: 241-242
schematic EM1: 241
Free-edge delamination (continued)
structural analysis EM1: 462
Free-end torsion testing
vs. fixed-end testing A8: 181-182
Free-fall atomization A7: 26, 29
Free-flowing powder A7: 278
Free-machining beryllium-copper
as rod A2: 403
Free-machining copper *See* Copper alloys, specific types, C14500, C14700 and C18700
Free-machining copper alloys
applications and properties A2: 277-280, 291-292
defined A2: 216
Free-machining grades of austenitic stainless steels
selenium additions A9: 284
Free-machining steel
friction welding M6: 721
projection welding M6: 506, 515-516
Free-machining steel alloy production
powders used A7: 572
Free-machining steels
arc welding A6: 650-651
cemented carbide machining A2: 966
characteristics of M1: 573-578
cold extrusion of A14: 301
electron-beam welding A6: 860
powder metallurgy materials microstructures A9: 510
Free-radical polymerization *See also* Polymerization
defined EM1: 12 EM2: 19
of polyester EM1: 133
in sheet molding compounds EM1: 142
Free-running crosshead speed A8: 39
Free-surface fracture A8: 573-574
forming-limit diagram for A8: 581
Free-surface strain A8: 580
Free-surface transverse velocity A8: 235-236
Free-turning brass
applications and properties A2: 310
Freeboard requirements
secondary metallurgy A15: 433
Freedom
degrees of A8: 625-626
Freely corroding potential *See* Corrosion potential
Freeman-Carroll equation EM3: 425
Freeze (Scheil) equation A9: 614
Freeze drying
of ferrites EM4: 1163
Freeze-pump-thaw technique
for ESR studies A10: 263
Freezing *See also* Liquidus; Solidification; Solidus
coating effect A15: 281
copper alloys A15: 771, 779
equiaxed grain origin in A15: 130
and fluidity, relationship A15: 767
fronts, modeling movement of A15: 857
in magnesium powder production A7: 132
microstructure formation during A15: 119
prediction, casting geometry/topography for A15: 858
range A15: 6, 114, 771
times, compared A15: 243
Freezing curves .... A3: 1 • 16
Freezing point *See* Melting point
Freezing range *See also* Cryogenic; Low temperature
as classification, copper-base alloys A2: 346, 348
copper alloy castings A2: 348
distribution as function of A2: 348
effect on hydrogen porosity in aluminum alloy ingots A9: 633
effect on macrosegregation in copper alloy ingots A9: 639
Freezing temperature A6: 45
Freezing-point calibration
of thermocouples A2: 879-880
Freiberger decomposition
for specific sample dissolution A10: 167
French Atomic Energy Authority powder-under-vacuum process A7: 167
Frenkel defect
ionic oxides A13: 65
Frenkel defects
defined EL1: 93
Frenkel-pair defects A18: 853
Freon TF
as organic cleaning solvent A12: 74
Frequencies *See also* Multifrequency techniques; Test frequencies
acoustic microscopy methods A17: 468
bands, microwave A17: 202
bands, reflectometers A17: 214
changing, in eddy current inspection A17: 170-171
cutoff, defined, microwave inspection A17: 208
and depth of penetration, eddy current inspection A17: 169
divisions, electromagnetic spectrum A17: 202
inspection, eddy current inspection A17: 173
levels, ECP vs. eddy current inspection A17: 137
microwave equipment A17: 209
procession, FMR eddy current probes A17: 221
range, acoustic emission inspection A17: 285
remote-field eddy current inspection A17: 195
response, acoustic emission inspection A17: 280
selection, and skin effect A17: 165
skin depth vs. Foerster limit, eddy current inspection A17: 182
test, eddy current inspection A17: 174-175, 182, 194
test, for skin depth A17: 182
ultrasonic inspection A17: 232-234
Frequencies, clock *See* Clock frequencies
Frequency A6: 365
analysis for single autoclave operation A8: 425
and corrosion fatigue behavior of aluminum alloy 7079-T651 .... A8: 408
defined A10: 110, 673
defined as electrical property EL1: 598-600
effect in ultrasonic fatigue testing A8: 255-256
effect on creep-fatigue interaction A8: 346
effect on fatigue and wave form A12: 58-63
effect on fatigue behavior of lead A8: 346-347
effect on minimum creep rate A8: 347
effect on resolution in thermal- wave imaging A9: 91
and electrical breakdown EM2: 466
electrical effects EM2: 584
fatigue formation at differing A8: 253
fatigue strength at conventional and ultrasonic A8: 254
in fatigue tests A11: 112-113
group, molecular vibration as A10: 111
infrared A10: 111
of loading, effect on corrosion-fatigue A11: 255
of loading, effect on fatigue-crack propagation A11: 110
-modified fatigue equation, for creep-fatigue interaction at high temperatures A8: 356-357
natural, and damping EM1: 212-213
or Poisson's ratio, symbol for A10: 692
power reflected from microwave-resonance cavity as function of A10: 256
range, forced-displacement system A8: 392-394
separation A8: 356
SI defined unit and symbol for A10: 685
SI unit/symbol for A8: 721
of specimens, measurement of longitudinal resonance A8: 249
in ultrasonic fatigue testing A8: 101, 240-241, 249
for ultrasonic ply cutting EM1: 615-616
Frequency (x-ray)
defined A9: 8
Frequency converter A6: 41-42
Frequency discriminator
in ultrasonic hardness tester A8: 101
Frequency distribution
defined A8: 6
Frequency effects *See also* Cyclic loading
in erosion/cavitation testing A13: 313
in fretting A13: 139-140
on corrosion fatigue A13: 143, 295, 298
Frequency modulation
effect, microwave inspection A17: 202
method, ultrasonic inspection A17: 240
microwave inspection A17: 206
Frequency modulation (FM) A6: 39
Frequency plots *See* Histograms
Frequency response-small signal model
for bipolar transistors EL1: 153-154

Frequency separation
defined ... A8: 356
Frequency-modulated reflectometers
microwave inspection ... A17: 213-214
Fresh water
steel corrosion protection in ... M1: 751
Fresh water corrosion
dissolved gases, effect on ... M1: 733
dissolved salts, effect on ... M1: 734-735
natural waters ... M1: 736-738
pH, effect on ... M1: 733-738
potable water systems ... M1: 735-736
preventative measures ... M1: 738
seawater corrosion compared to ... M1: 740, 741
Freshwater
copper/copper alloys in ... A13: 621-622
galvanized coatings in ... A13: 440
general biological corrosion in ... A13: 88
stainless steel corrosion ... A13: 556
Fresnel formula
dispersion of oxide glasses ... EM4: 1079
Fresnel fringes
defined ... A9: 8
Fresnel zones
microwave holography ... A17: 224
Fretting *See also* Abrasive wear; Chafing fatigue; False brinelling; Fretting (corrosion; Fretting corrosion; Oxidative wear ... A8: 6 A18: 181, 182-183 M1: 638
in bearing failures ... A11: 493, 497-498
of cobalt-gold plated copper flats ... A13: 138
control/elimination ... A13: 613
in copper/copper alloys ... A13: 613
corrosion ... A13: 138-140
corrosion, defined ... A8: 6
damage, rollers and bearing assembly ... A11: 341
debris generation as ... A11: 148
defined ... A11: 5 A18: 9
effect of cyclic displacement on total weight loss from ... A11: 341
fatigue ... A13: 141
fatigue as mechanism of ... A11: 148
fatigue, defined ... A8: 6 A11: 5
fatigue related to ... M1: 673
and fatigue resistance ... A1: 679
in forging ... A11: 341
gold electroplating ... A18: 837
hardfacing for ... A7: 823
in implants ... A11: 672
initial adhesion as ... A11: 148
material selection for ... A13: 333
mechanical material transfer during ... A11: 688, 692
mechanism of ... A11: 148
medium-carbon steels ... A12: 262
multiple fatigue-crack initiation ... A11: 24
resistance, cast irons ... A13: 568
resistance, of material combinations ... A13: 335
of shafts ... A11: 466
in sliding bearings ... A11: 486, 487, 534-535
in stainless steel bone plate ... A11: 688, 691-692
types in ball bearing failure ... A11: 497
as wear ... A11: 148
wear, AISI/SAE alloy steels ... A12: 308
Fretting corrosion *See also* Corrosion; Fretting ... A13: 138-140
in aircraft ... A13: 1030, 1041-1043
in cast irons ... A13: 568
defined ... A11: 5 A13: 7, 138 A18: 9
factors affecting ... A13: 139-140
in forging ... A11: 341
of implants ... A11: 672, 688, 691-692
in manned spacecraft ... A13: 1082
resistance, of metal couples ... A13: 1035
in zirconium/zirconium alloys ... A13: 717
Fretting fatigue ... A18: 242
defined ... A18: 9
Fretting wear ... A18: 242-253
aluminum-silicon alloys ... A18: 791
Fretting wear (continued)
debris effect on wear ... A18: 249, 252
defined ... A18: 9, 242
environmental effects ... A18: 249-251
aqueous electrolytes ... A18: 250-251
humidity ... A18: 250
low and high temperature ... A18: 251
vacuum ... A18: 249-250
jet engine components ... A18: 588
materials ... A18: 248-249
measurement of ... A18: 242-253
axial distance measurement ... A18: 251
holographic interferometry ... A18: 251
profilometry ... A18: 251
thin-layer activation (TLA) ... A18: 251
mechanical components ... A18: 242-244
examples ... A18: 242-244
mechanism of ... A18: 251-253
parameters affecting fretting ... A18: 244-248
contact, type of ... A18: 246-247
frequency ... A18: 245-246
impact fretting ... A18: 247
normal load ... A18: 245, 249
residual stresses ... A18: 247-248
slip amplitude ... A18: 244-245, 249
surface finish ... A18: 247
vibration, type of ... A18: 247
prevention of fretting damage ... A18: 253
coatings ... A18: 253
improved design ... A18: 253
inserts ... A18: 253
lubricants ... A18: 253
surface finish ... A18: 253
stainless steels ... A18: 714-715
thermal spray coating applications ... A18: 832, 833
Fretting/wear spraying
powders used ... A7: 572
Friction
as acoustic noise source ... A17: 285
adhesion theory and AES analysis of ... A10: 566
adhesion theory of ... A11: 149
and barreling ... A8: 56
and chip formation ... A16: 10
coefficient, defined *See* Coefficient of friction
coefficient of ... EM3: 14
coefficient of, and Sommerfeld number ... A11: 485
coefficient of, defined *See* Coefficient of friction
coefficient of, symbol ... A8: 726
cold upset testing ... A8: 578-579
in compression testing ... A8: 192
consequences in upset cylinder ... A8: 573
constraint considerations in Hopkinson bar test ... A8: 200
control through lubrication ... A8: 576
as controlling workability ... A14: 368
Coulomb's law ... A8: 576
in cylindrical compression test specimen ... A8: 579-580
defined ... A18: 9-10
during deformation ... A8: 575-576
effect, in drawing ... A14: 331
effect in drop tower compression test ... A8: 197
effect, make-break arcing contacts ... A2: 841
effect of unit pressure and rubbing speed on coefficient of ... A7: 702
effect on free-surface strain ... A8: 580
effect, sliding contacts ... A2: 842
effects in cam plastometer ... A8: 195-196
factor, determining ... A14: 441
introduction to ... A18: 25-26
areas of technological interest and research ... A18: 26
friction angle ... A18: 25
friction coefficient ... A18: 25-26
static friction coefficient ... A18: 25, 26
"stick-slip" ... A18: 26
terminology and its origin ... A18: 25
in locomotive axle failures ... A11: 715
Friction (continued)
loss, notched bar impact testing ... A8: 262
and lubrication, closed-die forging ... A14: 76
plane-strain compression test for ... A14: 377-379
polyamide-imides (PAI) ... EM2: 137
polyamides (PA) ... EM2: 126
in precision forging ... A14: 159
properties, of carbon fiber reinforced polymers ... EM1: 36
reduction during axial compression testing ... A8: 56-57
in ring rolling ... A14: 108
of sliding bearing pairs, with lubricants ... A11: 486
in split Hopkinson pressure bar test ... A8: 201
and tribology ... A8: 601
in upsetting of cylinder ... A14: 365
-velocity curve, for hydrodynamic lubrication surfaces ... A8: 605
wear failures and ... A11: 148-150
wear of polymers due to ... A11: 764
and wear testing ... A8: 604
Friction and wear data, presentation ... A18: 489-492
transition diagrams ... A18: 491
determination methodology ... A18: 491
tribographs ... A18: 489-491
dependence of tribodata on interaction parameters ... A18: 491
dependence of tribodata on operational parameters ... A18: 490
dependence of tribodata on structural parameters ... A18: 490-491
friction-time master curves ... A18: 489
wear-time master curves ... A18: 489-490
tribomaps ... A18: 491-492
wear regimes ... A18: 492
Friction and wear of aircraft brakes *See* Aircraft brakes, friction and wear of
Friction and wear of aluminum-silicon alloys *See* Aluminum-silicon alloys, friction and wear of
Friction and wear of automotive and truck drive trains *See* Automotive and truck drive trains friction and wear of
Friction and wear of automotive brakes *See* Automotive brakes, friction and wear of
Friction and wear of bearing steels *See* Bearing steels, friction and wear of
Friction and wear of carbon-graphite materials *See* Carbon-graphite materials, friction and wear of
Friction and wear of cemented carbides *See* Cemented carbides, friction and wear of
Friction and wear of ceramics *See* Ceramics, friction and wear of
Friction and wear of cobalt-base alloys *See* Cobalt-base alloys, friction and wear of
Friction and wear of dies and die materials *See* Die and die materials, friction and wear of
Friction and wear of electrical contacts *See* Electrical contacts, friction and wear of
Friction and wear of hardfacing alloys *See* Hardfacing alloys, friction and wear of
Friction and wear of internal combustion engine parts *See* Internal combustion engine parts friction and wear of
Friction and wear of medical implants and prosthetic devices *See* Medical implants and prosthetic devices, friction and wear of
Friction and wear of semiconductors *See* Semiconductors, friction and wear of
Friction and wear of sliding bearing materials *See* Sliding bearing materials, friction and wear of
Friction and wear of thermoplastic composites *See* Thermoplastic composites, friction and wear of
Friction and wear of tool steels *See* Tool steels, friction and wear of
Friction angle ... A18: 27
Friction bearings, overheated
locomotive axles failure from ... A11: 715-727
Friction burn-off ... A6: 316

Friction clutches
mechanical presses A14: 497
Friction coefficient *See* Coefficient of friction
Friction damping A18: 531
Friction drive *See* Knurl drive
Friction drive press A14: 33
Friction during metal forming A18: 59-68
characteristics of friction A18: 67-68
measurement of friction A18: 65-66
flow through conical converging dies A18: 66
ring forging A18: 65-66
strip rolling A18: 66
modeling of friction A18: 59-65
hydrodynamic lubrication A18: 61-65, 68
modeling of flow through conical converging dies A18: 61-63
nomenclature for friction in metal-forming processes A18: 60
steady-state wave model A18: 59, 60, 66-68
Friction factor A18: 60, 61, 66, 67
Friction force A18: 27, 30
defined A18: 10
elastomers A18: 36
friction during metal forming A18: 67
relation to metal substrate hardness A18: 31
Friction force measurement A18: 435
Friction hill A18: 63, 65
Friction materials A7: 5, 701-703, 740
applications A7: 119, 702-703
bell and elevator furnaces for sintering A7: 357
blending A7: 701
cement copper as A7: 119, 739
compacting and sintering A7: 701-702
copper metals as A7: 119, 739
copper P/M products A2: 398-400
facing material for A7: 702
powders used A7: 573
sintered A7: 701-702, 739
Friction modifiers/antisquawk agents A18: 103-104
applications A18: 104
in engine lubricant formulations A18: 111
formation of A18: 104
for metalworking lubricants A18: 140-141, 142, 143
in nonengine lubricant formulations A18: 111
Friction oxidation *See* Fretting
Friction parts
as copper powders application A7: 105
Friction polymer
defined A18: 10
Friction reduction
by lubrication A7: 190
Friction stress A18: 60
Friction surfacing A6: 321-323
aluminum A6: 321
aluminum alloys A6: 321
applications A6: 321, 322-323
austenitic stainless steel A6: 321, 323
carbon steels A6: 321, 323
development A6: 321
equipment A6: 321-322
feed rates A6: 322
geometric arrangements A6: 323
heat-affected zone A6: 321
metal-matrix composites A6: 323
microstructure A6: 322
mild steel A6: 321, 322
monolithic structures A6: 323
nickel-base alloys A6: 321
parameters A6: 322
procedure A6: 321
steels A6: 323
Stellite alloys A6: 321, 322, 323
Friction welding M6: 719-728
applications M6: 719
continuous drive method M6: 719, 722-723
applications M6: 722-723
principles of operation M6: 722
process variables M6: 722
defined EM2: 20
definition M6: 8
discontinuities from A17: 588
inertia drive method M6: 719, 723-726
applications M6: 725-726
axial pressure M6: 724
Friction welding (continued)
effect of flywheel energy M6: 724-725
equipment M6: 723
peripheral velocity of workpiece M6: 724
principles of operation M6: 723
process variables M6: 723-724
inspection M6: 729
joint design M6: 726-728
conical joints M6: 726-727
flow of weld upset M6: 727
heat balance M6: 727
joint surface conditions M6: 726
for machine size M6: 728
tubular welds M6: 726
metals welded M6: 721-722, 724
process capabilities M6: 719-721
comparison to flash welding M6: 720
sections welded M6: 720-721
weld strength M6: 720
safety precautions M6: 59
as secondary operation A7: 456
solid-state bonding in joining non-oxide ceramics EM4: 525
Friction welding (FRW) A6: 150-154, 315-317, 321
advantages A6: 317
alumina A6: 317
aluminum A6: 317
aluminum alloys A6: 739
aluminum metal-matrix composites A6: 555, 558
applications A6: 317
aerospace A6: 387
axial pressure vs. peripheral velocities A6: 152
brittle phase formation A6: 154
continuous drive A6: 315
conventional friction welding A6: 150-151
definition A6: 150, 315, 1209
differential thermal expansion A6: 154
direct-drive A6: 315
direct-drive variables A6: 315-316
direct-drive vs. inertia drive A6: 316
direct-drive welding A6: 150-151, 154
dispersion-strengthened aluminum alloys A6: 543, 546, 547
dissimilar metal joining A6: 822
duplex stainless steels A6: 480
equipment A6: 315, 317
ferritic stainless steels A6: 448
flywheel energy A6: 152
flywheel friction welding A6: 150, 151-152
forging A6: 152
friction speed A6: 151
frictional heating A6: 315, 317
heat-affected zones A6: 152, 316, 317
inertia and direct-drive
aluminum-base alloys A6: 890
carbon steels A6: 889-890
ceramics A6: 891
cobalt-base materials A6: 891-892
copper-base materials A6: 891-892
definition A6: 888
direct-drive friction welding parameter calculations A6: 889
dissimilar metals A6: 891
inertia welding parameter calculations A6: 889
low-carbon steels, inertia welding A6: 889, 890
nickel-base materials A6: 891-892
reactive metals A6: 890-891
refractory metals A6: 890-891
stainless steels A6: 890
inertia-drive A6: 315, 316, 317
inertia-drive variables A6: 316, 317
inertia-drive welding A6: 150, 151-152, 153, 154
joint interfaces A6: 153
kinetic energy A6: 151, 152
limitations A6: 317
linear reciprocating motions A6: 150
low-melting phase formation A6: 153-154
metallurgical parameters A6: 152-154
nickel-base corrosion-resistant alloys containing molybdenum A6: 594
orbital A6: 150
oxide-dispersion-strengthened materials A6: 1038, 1039, 1040
parameters of process A6: 317
peripheral velocities A6: 152
Friction welding (FRW) (continued)
problems A6: 153-154
product evaluation A6: 316-317
qualitative factors influencing the quality A6: 150
radial A6: 150
safety precautions A6: 1203
standard procedure qualification test weldments A6: 1090
steps in process A6: 150
tantalum alloys A6: 580
technology A6: 150
thermoplastics A6: 317
titanium alloys A6: 783, 784
weld discontinuities A6: 1078-1079
weld energy A6: 151
weld quality A6: 316-317
weld upset zone A6: 152
welding parameters A6: 315-316
wrought martensitic stainless steels A6: 441
Friction welding of
alloy steels M6: 721
aluminum alloys M6: 722
carbon steels M6: 721
cast iron M6: 722
cobalt-based alloys M6: 722
copper alloys M6: 722
free-machining steels M6: 721
magnesium alloys M6: 722
molybdenum M6: 722
nickel-based alloys M6: 722
niobium M6: 722
stainless steels M6: 721-722
tantalum M6: 722
titanium and titanium alloys M6: 722
tungsten M6: 722
zirconium alloys M6: 722
Friction welding of specific materials A6: 152-154
alloy steels A6: 153
aluminum A6: 152
aluminum alloys A6: 152, 153
austenitic stainless steels A6: 152, 153, 154
brass A6: 152
bronze A6: 152
cast irons A6: 152
cemented carbides A6: 152
ceramics A6: 152, 154
cobalt A6: 154
copper A6: 152, 154
copper-nickel A6: 152
high-carbon steels A6: 153
high-speed tool steels A6: 153
iron A6: 152, 153, 154
lead alloys A6: 152
low-alloy steel A6: 152, 153
low-carbon steels A6: 153
magnesium alloys A6: 153
maraging steel A6: 152
medium-carbon steels A6: 153
nickel A6: 152, 154
stainless steel A6: 152, 153, 154
superalloys A6: 154
tantalum A6: 152
titanium A6: 152, 153
titanium alloys A6: 152, 153
tool steel A6: 152, 153
vanadium A6: 154
zirconium alloys A6: 787
Friction work A18: 27
Frictional energy A18: 438
Frictional forces A18: 476, 478
Frictional heat in sawing A9: 23
Frictional heating A18: 27, 438, 439, 513, 518
ceramics A18: 814
damage dominated by dissolution or diffusion A18: 181
electrical contacts A18: 683
mainshaft bearings in jet engines A18: 590
seals A18: 549
Frictional heating calculations A18: 39-44
correlation of experimental data with calculated values A18: 43
factors limiting the accuracy of calculations A18: 43-44
limitations of calculations A18: 44
surface layer effect A18: 44

Frictional heating calculations (continued)
transient temperature effect ... **A18:** 43-44
frictional heating nomenclature ... **A18:** 39-40
coefficient of friction ... **A18:** 39
heat partition factor ... **A18:** 39, 40, 41, 44
heat source time ... **A18:** 39
Peclet number ... **A18:** 39-40, 43, 44
real area of contact ... **A18:** 40, 41
temperature ... **A18:** 39
idealized models of sliding contact ... **A18:** 40-43
circular contact analysis with one body in motion ... **A18:** 41-43
general contact analysis ... **A18:** 40
line contact analysis with two bodies in motion ... **A18:** 40-41
Frictive track
definition ... **EM4:** 632
Friedel's law
and determination of ODF coefficient ... **A10:** 362
Friedel-Crafts mechanism
of polyimides ... **EM3:** 157
Fringe control
optical holographic interferometry ... **A17:** 412-413
Fringe pattern
produced in stacking-fault areas ... **A9:** 117
Fringes
effects of crystal thickness ... **A10:** 368
fault-causing ... **A10:** 369-370
moiré, observed by x-ray interferometry ... **A10:** 371
origin as interferences in x-ray diffraction ... **A10:** 368
Pendellösung ... **A10:** 368
Frit *See also* Glass frit
defined ... **EL1:** 109
Frit china ... **EM4:** 4
absorption ... **EM4:** 4
products ... **EM4:** 4
Frits ... **A13:** 446-447 **EM4:** 447, 953, 1069, 1071
application during dry-powder cast iron enamel ... **EM4:** 956
application during electrostatic dry-powder coatings ... **EM4:** 956
application for high-temperature torsion testing ... **A8:** 159
composition and applications for glazes of selected frits ... **EM4:** 954
electronic applications ... **EM4:** 953
from curtain and roll coating decoration method ... **EM4:** 971
in glass enamels ... **EM4:** 474
glazes ... **EM4:** 1061
to attach lightweight mirrors for space applications ... **EM4:** 1016
Frits, porcelain enamel *See* Porcelain enameling, frits
Fritted disk nebulizers ... **A10:** 36, 55
Fritting ... **A18:** 682-683
Front axle gear wear rating ... **A18:** 566
Front gages
straight-knife shearing ... **A14:** 703
Front screen
radiography ... **A17:** 315
Front-end loader
brittle fracture of support arm for ... **A11:** 69
Front-mounted mower neutral arm ... **A7:** 677
Frost and Ashby temperature-stress deformation maps ... **A18:** 426
Frost line
defined ... **EM2:** 20
Frosted area
definition ... **EM4:** 632
Frosting
defined ... **A18:** 10
Froth flotation
in nickel refining ... **A2:** 429
Froth notation
to remove RHM impurities ... **EM4:** 378
Frothing
defined ... **EM2:** 20
Frozen flexible epoxy systems ... **EL1:** 818
FRP *See* Fiber-reinforced plastic; Fiber-reinforced plastic (FRP)
FRPs *See* Fiber-reinforced plastics
Fry's reagent as an etchant
for nitrided steels ... **A9:** 218
FS *See* Fiber Society
FS-80
annealing ... **M4:** 655
composition ... **M4:** 651-652
stress relieving ... **M4:** 651-652
FS-82
annealing ... **M4:** 655
composition ... **M4:** 651-652
stress relieving ... **M4:** 655
FSS *See* Fatigue striation spacings
FSX 414
composition ... **M4:** 653
FSX-414
composition ... **A4:** 795 **A6:** 929
machining ... **A16:** 757-758
FSX-418
composition ... **A6:** 929
FSX-430
composition ... **A6:** 929
FT-IR *See* Fourier transform infrared spectroscopy
FTS *See* Fourier transform spectrometers
Fuel and oil resistance
thermoplastic polyurethanes (TPUR) ... **EM2:** 206
Fuel cells
molten carbonate ... **A13:** 1321
phosphoric acid ... **A13:** 1320
powders used ... **A7:** 573
power source corrosion ... **A13:** 1317-1323
types/corrosivity ... **A13:** 1320-1321
Fuel clad ... **A7:** 664
Fuel elements, nuclear engineering
powders used ... **A7:** 573
Fuel filters
powders used ... **A7:** 572
Fuel gases *See also* specific types
definition ... **M6:** 8
flame hardening ... **M4:** 486-489
oxyfuel gas welding ... **A6:** 281-283
oxyfuel wire spray process ... **A6:** 809
properties ... **M6:** 899-900
safety precautions ... **A6:** 1198-1199, 1200-1201
for torch brazing ... **A6:** 328
Fuel heating
as melting procedure ... **A7:** 25
Fuel injection
atomization mechanism ... **A7:** 27
Fuel injection pump tappet
material selection ... **M1:** 612-613
Fuel pellet fabrication
nuclear ... **A7:** 664-665
oxygen-to-uranium ratio ... **A7:** 665
Fuel propellants
metal powders for ... **A7:** 597-598
Fuel pump, operating levers
economy in manufacture ... **M3:** 848, 849
Fuel pump parts
powders used ... **A7:** 572
Fuel pumps
failure by vibration and abrasion ... **A11:** 465-466
Fuel rod
uranium dioxide ... **A7:** 664, 665
Fuel rods
Zircaloy-clad LWR ... **A13:** 945-948
Fuel salts
stainless steel corrosion in ... **A13:** 54
Fuel tank floors, aircraft
fatigue fracture of ... **A11:** 126-127
Fuel tanks *See* Gasoline tanks
Fuel-oil lines
ASTM specifications for ... **M1:** 323
Fuels
EFG composition analysis ... **A10:** 212
fossil, assay for toxic elements ... **A10:** 233
gases, properties ... **A14:** 723
as metals applications ... **A7:** 597-598
for nickel-base heating ... **A14:** 261
nuclear ... **A10:** 207
pyrophoricity ... **A7:** 598
selection, for steam equipment ... **A11:** 619
water-containing, corrosion in ... **A11:** 190-191
Fuels, nuclear
neutron radiography of ... **A17:** 391-392
Fugitive binder ... **A7:** 5
Fugitive vehicle process
oxidation-resistant coating ... **M5:** 664-666
Fukui conical cup test ... **A8:** 563-564, 568
Fulcher equation ... **EM4:** 849
FULDENS process ... **A1:** 781 **A16:** 60, 66
schematic and applications ... **A7:** 790
for tool steels ... **A7:** 788-789
Full custom
as ASIC product class ... **EL1:** 168
Full density *See also* Encapsulation; Hot isostatic pressing ... **A7:** 436-437
bar-shaped steel compacts ... **A7:** 505
by forging ... **A7:** 523
CAP stock plus deformation processing for ... **A7:** 533-536
containerless isostatic pressing for ... **A7:** 441
hot isostatic pressing for ... **A7:** 419
powder rolled materials ... **A7:** 401
sintering cycles for ... **A7:** 23, 375
vacuum sintering to ... **A7:** 373-374
Full die inserts ... **A14:** 47-48
Full fillet weld
definition ... **M6:** 8
Full mold
defined ... **A15:** 6
Full mold process *See* Lost foam casting
Full wave rectified current ... **A7:** 576
Full width at half maximum ... **A10:** 519, 674, 690
Full width at half maximum (FWHM)
defined ... **A17:** 384
Full-contour length ... **EM3:** 14
defined ... **EM2:** 20
Full-dip infiltration ... **A7:** 553
Full-disk buffs ... **M5:** 118, 126
Full-film lubrication *See also* Elastohydrodynamic lubrication
defined ... **A18:** 9
Full-journal bearing
defined ... **A18:** 10
Full-scale deflection (FSD) ... **A18:** 294
Full-scale tests *See also* Large-scale tests
damage tolerance ... **EM1:** 351
durability ... **EM1:** 350-351
static ... **EM1:** 347-350
static strength ... **EM1:** 432
for vehicle structural design ... **EM1:** 346
Full-slant fracture
by ductile fracture ... **A11:** 25
Full-tensor determination
plane-elastic model ... **A10:** 384
Full-volume inspection
C-scan ... **EM2:** 845
Full-wave rectified circuit
radiography ... **A17:** 304-305
Fuller
defined ... **A14:** 7
Fullering
in forging sequence ... **A14:** 43
Fullering impression *See* Fuller
Fullers ... **A14:** 43-44, 63
Fully dense
prealloyed titanium P/M compacts ... **A2:** 647
Fully dense materials *See also* Encapsulation; Full density; Hot isostatic pressing ... **A7:** 436-437
constitutive equations ... **A14:** 417

Fully dense P/M stainless steels **A13:** 833-834
Fully stabilized zirconia *See also* Zirconia; Zirconium oxide
fracture toughness **EM4:** 330
modulus of resilience **EM4:** 330
properties **EM4:** 330
Young's modulus **EM4:** 330
Fully stabilized zirconia (FSZ)
x-ray diffraction **A18:** 467
Fully supported specimens
SCC testing **A8:** 505
Fumarates *See also* Bisphenol A (BPA) fumarate resins Polyester resins
application/preparation **EM1:** 90
Fume collection
electric arc furnace **A15:** 360
Fume hood
for safety in acid dissolution treatments **A10:** 166
Fumed silica **EM3:** 49
double processing and effect on shrinkage **EM4:** 447
nonaqueous solvents for dispersion casting **EM4:** 447
Fuming
in perchloric acid **A10:** 166
Fuming nitric acid
titanium/titanium alloy resistance **A13:** 677
Fuming sulfuric acid *See* Oleum
Function
in process selection **EM2:** 279
requirements, filament winding **EM2:** 369-371
requirements, rotational molding **EM2:** 361
requirements, RTM/SRIM **EM2:** 346-349
Function control, semiconductor
SEM analysis for **A10:** 490
Function generator
for torsion testing **A8:** 158
Functional group
defined **A10:** 674
effect in ion exchange separation **A10:** 164-165
molecular, IR determination of **A10:** 109
UV/VIS identification in organic molecules **A10:** 60
Functional group analysis
acids **A10:** 215-216
alcohols **A10:** 216-217
aldehydes and ketones **A10:** 217
amines **A10:** 217-218
aromatic hydrocarbons **A10:** 218
characterization of unknowns **A10:** 215
composition of a mixture **A10:** 215
esters **A10:** 218
peroxides **A10:** 218
phenols **A10:** 219-220
purity determination **A10:** 215
types of **A10:** 215-219
Functional phase
thick-film formulations **EL1:** 249
Functional tests
of VLSI/ULSI/WSI devices **EL1:** 377-378
Functional types of adhesives **EM3:** 73
Functional variation *See* Variation (statistical)
Functionality **EM3:** 14
Functionally graded materials (FGM)
synthesized by SHS process **EM4:** 230
Functions of adhesives **EM3:** 33
Fundamental interconnection issues *See also* Interconnect(s); Interconnection
conventional interconnection environments **EL1:** 2-5
device speed, maximum usable intrinsic
future developments **EL1:** 10
interconnection-based technologies, new **EL1:** 8-10
minimum device size **EL1:** 2
physical performance issues **EL1:** 5-8
Fundamental parameter software
x-ray spectrometer calibration **A10:** 98
Fundamental Principles of Powder Metallurgy (Jones) **A7:** 18
Fundamental reflections in body-centered cubic structures **A9:** 109
Fungicides **EM3:** 674
for metalworking lubricants **A18:** 142
powder used **A7:** 572
Fungus resistance
cellophane and amylose film **EM2:** 785-787
defined **EM1:** 12
tests **EM2:** 579-580
thermoplastic polyurethanes (TPUR) **EM2:** 206
Funnels
Carney, for determining apparent density **A7:** 273-274
Scott volumeter with **A7:** 274
Furan
defined **A13:** 7
Furan (furfuryl alcohol) warm box processes
as coremaking **A15:** 238
Furan acid catalyzed no-bake binder process **A15:** 214
Furan binder system **A15:** 30, 35, 38, 238, 241
Furan hot box process
as coremaking system **A15:** 238
Furan resins *See also* Resins **EM3:** 14
defined **EM2:** 20
Furan(s)
as organic binder **A15:** 45
Furan/acid no-bake processes
as coremaking process **A15:** 238
Furan/sulfur dioxide resin binder process **A15:** 219-221, 238
Furane resins
applications **EM4:** 47
composition **EM4:** 47
supply sources **EM4:** 47
Furfural **EM3:** 103
physical properties **EM3:** 104
Furfural alcohol **EM3:** 103, 104
Furfural alcohol resins **A13:** 1154
Furfural resin *See also* Resins **EM3:** 14
defined **EM2:** 20
Furfuryl alcohol [2-(CH_4H_3O)CH_2OH#
as solvent used in ceramics processing **EM4:** 117
Furnace
-atmosphere control, tool and die failure **A11:** 573
chamber, and electronic controls in closed-loop servomechanical system **A8:** 395-396
cold-wall vacuum **A6:** 331
with constant-stress testing system **A8:** 321-322
continuous belt **A6:** 331-332
controller **A8:** 201
for creep and stress-rupture testing **A8:** 312-313
for elevated-temperature split Hopkinson pressure bar tests **A8:** 202
for high-temperature torsion testing **A8:** 158-159
mesh-belt conveyor **A6:** 331
retort (bell-type) **A6:** 331
semicontinuous **A6:** 331
tubes, elevated-temperature failures in **A11:** 290-291
vacuum **A6:** 331
Furnace (atmosphere)
relative rating of brazing process heating method **A6:** 120
Furnace (vacuum)
relative rating of brazing process heating method **A6:** 120
Furnace atmosphere, gases **A4:** 542-546
air **A4:** 543, 546, 548, 549-550, 552, 561, 562
ammonia vapor **A4:** 543, 545-548, 552, 559, 560, 561
argon **A4:** 543, 544
butane
carbon dioxide and monoxide **A4:** 543-556, 558, 562, 563
carbon dioxide plus hydrogen **A4:** 545
density **A4:** 542, 543
diffusion **A4:** 542, 565
helium **A4:** 543, 544
hydrocarbons **A4:** 544, 546, 550, 552, 555, 559, 561, 565
hydrogen **A4:** 543-545, 547-550, 552-553, 555-558, 561-562, 564, 565
inert gases **A4:** 543, 544, 546, 548, 564, 566
lithium vapor **A4:** 545-546
methane **A4:** 543, 547-552, 555, 557-558, 561, 565
methanol **A4:** 547, 548, 555, 556, 557-558, 559, 567
natural gas **A4:** 547, 549, 550, 557, 563, 567
nitrogen **A4:** 543-545, 550, 552-559, 561, 562, 564-567
Furnace atmosphere, gases (continued)
oxygen **A4:** 543-546, 548, 550, 552, 553, 561
pressure **A4:** 542, 557, 565, 566
propane **A4:** 543, 549, 550, 555, 557-546, 548, 559, 560, 562, 563
reactions **A4:** 545
specific gravity **A4:** 543
sulfur dioxide **A4:** 543
sulfurous **A4:** 546
temperature effect **A4:** 543, 558
viscosity **A4:** 542-543
water-vapor **A4:** 544-546, 548-556, 558, 561-563
Furnace atmospheres **A4:** 542-567 **M4:** 389-416
applications **A4:** 548, 550, 552, 555, 558, 559, 560, 561-562, 563
back-filling, use for **A4:** 564-566 **M4:** 414
charcoal-base **A4:** 546, 562-563 **M4:** 394, 411-412
classification **A4:** 546, 552 **M4:** 393-395
commercial nitrogen-base **A4:** 555-559, 560 **M4:** 402-408, 409
dissociated alcohols **A4:** 547
dissociated ammonia **A4:** 547, 548, 555, 556, 559-561 **M4:** 408-410
endothermic **A4:** 546, 547, 548, 550-552, 554, 555, 556, 559, 560, 567 **M4:** 397-399
exothermic **A4:** 546, 547, 548-550, 552, 555, 559, 560 **M4:** 395-397
exothermic-endothermic **A4:** 546, 560, 563-564 **M4:** 412, 413-414
flow formula **A4:** 542
furnace design, influence of **A4:** 566 **M4:** 415-416
gas-carburizing, vacuum furnace **A4:** 550, 565 **M4:** 414
generation **A4:** 549-556, 558-560, 562, 563-564, 567 **M4:** 397-398, 400, 411, 413-414
generator maintenance **A4:** 551, 554, 564 **M4:** 398, 402, 414
heating-element materials **A2:** 833-835
hydrogen **A4:** 543-544, 545, 547-550, 561-562 **M4:** 410-411
ion carburizing and ion nitriding **A4:** 565-566
ion carburizing and iron nitriding **M4:** 414-415
prepared nitrogen-base **A4:** 546, 547, 552-555 **M4:** 394, 399-402
process requirements **A4:** 562, 566-567 **M4:** 415
quantities, estimation of **A4:** 567 **M4:** 416
quenching, use for **A4:** 564-566 **M4:** 414, 415
safety precautions **A4:** 546-548, 549, 550-552, 554, 556, 560-562 **M4:** 394-395, 396-397, 398-399, 402, 409-410
steam **A4:** 546, 562 **M4:** 411
toxicity **A4:** 547
utilities ,influence of **A4:** 554, 566-567 **M4:** 416
Furnace atmospheres, gases
air **M4:** 390
ammonia vapor **M4:** 393
carbon dioxide and monoxide **M4:** 390-391
carbon dioxide plus hydrogen **M4:** 392
density **M4:** 390
diffusion **M4:** 389-390
hydrocarbons **M4:** 391
hydrogen **M4:** 391
inert gases **M4:** 391-392
nitrogen **M4:** 390
oxygen **M4:** 390
pressure **M4:** 389
reactions **M4:** 392-393
sulfurous **M4:** 393
temperature effect **M4:** 390
viscosity **M4:** 390
water vapor **M4:** 391, 392-393
Furnace atmospheres, specific types
101
applications **A4:** 547, 548 **M4:** 394
composition **A4:** 547 **M4:** 394
102
applications **A4:** 547, 548 **M4:** 394
composition **A4:** 547 **M4:** 394
201
applications **A4:** 547 **M4:** 394
composition **A4:** 547, 552 **M4:** 394
202
applications **A4:** 547 **M4:** 394
composition **A4:** 547, 552 **M4:** 394
223, composition **A4:** 552
224, composition **A4:** 552

Furnace atmospheres, specific types (continued)
301
applications **A4:** 547 **M4:** 394
composition **A4:** 547 **M4:** 394
302
applications **A4:** 547 **M4:** 394
composition **A4:** 547 **M4:** 394
402
applications **A4:** 547, 562 **M4:** 394
composition **A4:** 547 **M4:** 394
421, applications **A4:** 562
501
applications **A4:** 547 **M4:** 394
composition **A4:** 547, 563 **M4:** 394
502
applications **A4:** 547 **M4:** 394
composition **A4:** 547, 563 **M4:** 394
601
applications **A4:** 547 **M4:** 394
composition **A4:** 547, 559 **M4:** 394
621
applications **A4:** 547 **M4:** 394
composition **A4:** 547, 559 **M4:** 394
622
applications **A4:** 547 **M4:** 394
composition **A4:** 547, 559 **M4:** 394
Furnace atomizers
atomic absorption spectrometry **A10:** 49
Furnace brazing **A6:** 121, 330-332
advantages **A6:** 330
of aluminum alloys **A6:** 330, 939 **M6:** 1027-1029
applications **A6:** 331-332
atmosphere **A6:** 330
brazing **A6:** 330-331
control instrumentation **A6:** 330-331
of copper and copper alloys **A6:** 933 **M6:** 1035-1037
definition **A6:** 330, 1209 **M6:** 8
furnaces **A6:** 330-331
health and safety guidelines **A6:** 332
heat treatment **A6:** 330
limitations **A6:** 330
nickel-base alloys **A6:** 330
oxide-dispersion-strengthened materials **A6:** 1038, 1039, 1040
personnel **A6:** 330
precious metals **A6:** 936
of reactive metals **M6:** 1052
stainless steels **A6:** 911, 913, 914-918
in a vacuum atmosphere **A6:** 920-921
in an air atmosphere **A6:** 919-920
in argon **A6:** 915, 919
in dissociated ammonia **A6:** 915, 918-919
stainless steels, in dry hydrogen **A6:** 915, 916, 917
of steels *See* Furnace brazing of steels
Furnace brazing of steels **M6:** 929-949
advantages **M6:** 929-930
applicability **M6:** 929
assembly for brazing **M6:** 939-941
auxiliary fixtures **M6:** 940
crimping **M6:** 940
expanding **M6:** 940
fixture design **M6:** 941
folding or interlocking **M6:** 940
gravity locating **M6:** 939
interference or press fitting **M6:** 939
knurling **M6:** 939-940
peening **M6:** 940
riveting **M6:** 940
self-jigging **M6:** 939
spinning **M6:** 940
staking **M6:** 940
swaging **M6:** 940
tack welding **M6:** 940
thread Joining **M6:** 940
wetting of fixtures **M6:** 941
brazing furnaces **M6:** 930-933
batch-type **M6:** 931

Furnace brazing of steels (continued)
continuous-type **M6:** 932
furnace atmospheres **M6:** 931
furnace ratings **M6:** 931
retort-type **M6:** 932
vacuum-type **M6:** 932-933
brazing with copper filler metals **M6:** 935-938
brazing carburized
carburizing of assemblies **M6:** 936
components **M6:** 936-937
elimination of flux **M6:** 937
joint strength **M6:** 936
mesh belts **M6:** 937-938
selection of filler-metal form **M6:** 937
types used **M6:** 936
brazing with silver alloy filler metals **M6:** 943-945
fluxes **M6:** 944
furnaces **M6:** 944
joint clearance **M6:** 944
protective atmospheres **M6:** 944-945
types and forms **M6:** 944
joint fit and design **M6:** 941-943
change in section thickness **M6:** 943
effect on shear strength **M6:** 942
factors affecting fillet size **M6:** 943
gap principle **M6:** 941
surface condition **M6:** 943
limitations **M6:** 930
protective furnace atmospheres **M6:** 934-935
commercial nitrogen-based atmospheres **M6:** 935
endothermic-based atmosphere **M6:** 934-935
rich exothermic-based atmosphere **M6:** 934
rich prepared nitrogen-based atmosphere **M6:** 934
safety **M6:** 945-949
sequence of operations **M6:** 930
assembling and fixturing **M6:** 930
brazing **M6:** 930
cleaning **M6:** 930
cooling **M6:** 930
stopoffs **M6:** 938-939
application **M6:** 938-939
materials **M6:** 938
removal **M6:** 939
surface preparation **M6:** 938
chemical cleaning **M6:** 938
mechanical cleaning **M6:** 938
use with or replacing other processes **M6:** 945
machining and brazing **M6:** 945
substitution for forging or casting **M6:** 945
vacuum brazing **M6:** 935
venting for mesh-belt conveyor drafts **M6:** 933
effective venting **M6:** 933-934
furnaces **M6:** 529-530
poor designs **M6:** 933
Furnace butt welding
steel tubular products **M1:** 316
Furnace emission spectroscopy capabilities
molecular fluorescence spectroscopy compared **A10:** 72
Furnace heating elements
powders used **A7:** 573
Furnace parts, heat-resistant alloys for *See* Heat resistant alloys, furnace parts
Furnace safety **A4:** 657-663 **M4:** 378-385
arc suppression **A4:** 661
atmosphere furnaces **A4:** 661-662 **M4:** 384
burner operation **A4:** 658-659
burner operations **M4:** 380
electric furnaces **A4:** 659-661 **M4:** 381
fixture design **A4:** 661
flame detection **A4:** 658 **M4:** 379-380
fuel-fired, control circuits **A4:** 657-658 **M4:** 379
fuel-fired, electrical power **A4:** 657-658 **M4:** 378-379

Furnace safety (continued)
fuel-fired furnaces **A4:** 657 **M4:** 378
furnace protection **A4:** 661
ignition trials **A4:** 658 **M4:** 379
ion nitriding **A4:** 661 **M4:** 383, 384
mechanical equipment **A4:** 662-663 **M4:** 385
pilot control **A4:** 658 **M4:** 379
plasma carburizing **A4:** 661 **M4:** 383
process cooling **A4:** 662 **M4:** 384-385
purging **A4:** 658, 661-662 **M4:** 379
supervisory gas cock system **M4:** 381
supervisory gas-cock system **A4:** 659
temperature control **A4:** 659 **M4:** 380
thermocouples **A4:** 661
waste-heat recovery **A4:** 659 **M4:** 380-381
Furnace shell
electric arc furnaces **A15:** 359
Furnace shielding
powders used **A7:** 573
Furnace soldering
definition **M6:** 8
Furnace soldering (FS) **A6:** 353-355
area (linear) conduction **A6:** 353
atmospheres **A6:** 355
bare copper assembly process **A6:** 355
condensation heating **A6:** 353
definition **A6:** 353, 1209
fluxes **A6:** 353-355
infrared heating **A6:** 353
ovens used **A6:** 353
processing in inert atmosphere **A6:** 354-355
reflow profile **A6:** 353-354
reflow schedule **A6:** 354
surface-mount technology (SMT) applications **A6:** 353
vapor-phase reflow **A6:** 353, 354
Furnace(s) *See also* Heat treating; Heat treatments; Sintering; Sintering equipment; Sintering furnaces; Thermal treatments
for aluminum and aluminum alloys **A7:** 381-383
atmospheres *See also* Atmospheres; Sintering atmospheres **A7:** 453
for austenitizing **A7:** 453
brazing **A7:** 457
burn-off zone **A7:** 351
for carbonitriding **A7:** 454
continuous production **A7:** 351-359
delubrication zone **A7:** 351
design, for brass and nickel silvers **A7:** 380-381
direct heating refractory metal sintering **A7:** 627
doors, dangers near **A7:** 349
final cooling zone **A7:** 352-353
in HIP processing **A7:** 422-423
hydrogen stoking **A7:** 386
loading, in tungsten and molybdenum sintering **A7:** 391
loading, of titanium powder and compacts **A7:** 394
molybdenum resistance-type electric sintering **A7:** 690
multi-zoned, controls in **A7:** 423
preheat zone **A7:** 351
production sintering **A7:** 351-359
refractory metal composite sintering and infiltration **A7:** 628
rotation, in tungsten powder production **A7:** 153
schematic for sintering steel **A7:** 339
selection, conditions affecting **A7:** 453
for sintering cermets **A2:** 986
slandering zone **A7:** 351-352
slow cooling zone **A7:** 352
for smelting recycled lead **A2:** 1222
for tempering **A7:** 453
vacuum hot pressing and sintering **A7:** 515
zoned **A7:** 346-348

Furnaces *See also* Heat treatment; Heating; Heating equipment.... **A1:** 108-109, 110 **A14:** 249, 261
atmospheres **M6:** 931
atmospheres, analysis techniques and analyzers **A4:** 577-586
atmospheres, protective **M6:** 934-935
basic oxygen furnace **A1:** 110
batch
atmosphere sampling technique **A4:** 578, 579
carbon gradients after carburizing **A4:** 592, 593
surface carbon variability in carburized specimens **A4:** 595, 596, 597, 598, 599, 600
batch-type **M6:** 931
blast furnace **A1:** 109
carbonitriding **M4:** 181
ceramic fiber lining **A4:** 523-524, 525
cleaning of **A13:** 1139
continuous **A4:** 523
atmosphere sampling technique **A4:** 578, 579
carbon gradients **A4:** 592-594, 595, 596
shim stock analysis of carbon potential **A4:** 589
continuous, pusher-type
atmosphere sampling technique **A4:** 578, 579
efficiency **A4:** 523, 524
surface carbon variability of specimens **A4:** 596, 597, 598, 599, 600
continuous-type **M6:** 932
for copper alloys **A15:** 772-774
for directionally solidified casting **A15:** 319-321
early blast **A15:** 24
electric heating **A4:** 519, 525
electrically heated pouring **A15:** 499-500
electroslag remelting **A15:** 402, 404
energy-efficient **A4:** 519, 522-525 **M4:** 339-340
fiberglass **EM1:** 108
gas carburizing **M4:** 137-139, 172
gas fired **A4:** 519, 520, 525
gaseous nitrocarburizing **M4:** 265-266
heat treating **A13:** 1311-1314
high-temperature, use in combustion **A10:** 221-225
historic advances **A15:** 27-29
ladle, and vacuum arc degassing **A15:** 435-438
liquid carburizing **M4:** 232-239
for magnesium alloys **A15:** 800-801
modeling **A4:** 525
muffle or gas burners, for sinters and fusions **A10:** 166
for nickel alloys **A15:** 820
normalizing **A4:** 38, 39-40
pack carburizing **M4:** 224-225
pit, carbon gradients after carburizing **A4:** 592, 593
plasma cold hearth melting **A15:** 422
plasma consolidation **A15:** 421
plasma melting/casting **A15:** 420-425
radiant element, electric resistance heated pouring **A15:** 499-500
ratings **M6:** 931
retort-type **M6:** 932
salt bath **A4:** 726-727 **M4:** 293-298
simplified impulse, for IGF analysis **A10:** 228
simplified inductive, for IGF analysis **A10:** 228
for single-crystal neutron diffraction **A10:** 424
sintering **M4:** 793-794
steel, annealing **M4:** 20-21
tool steels, heat treating **A4:** 726-733 **M4:** 575-580
types for heat treating **M4:** 285-292
vacuum **A4:** 729-732 **M4:** 307-324
vacuum arc skull melting casting **A15:** 409-410
vacuum induction degassing and pouring **A15:** 396-397
vacuum induction, metallurgy of **A15:** 393-396
vacuum induction remelting and shape casting **A15:** 399
vacuum-type **M6:** 932-933
venting **M6:** 933-934
wall corrosion **A13:** 995-996
Furnaces and related equipment **EM4:** 244-254
batch furnaces **EM4:** 252
chemical safety **EM4:** 253-254
arsine **EM4:** 253-254
carbon monoxide **EM4:** 253
chemical vapor deposition **EM4:** 253-254
Furnaces and related equipment (continued)
cyanogen **EM4:** 253, 254
hydrogen atmospheres **EM4:** 254
hydrogen cyanide **EM4:** 253
phosphine **EM4:** 253-254
silane **EM4:** 253-254
titanium tetrachloride **EM4:** 253-254
combustion furnaces **EM4:** 244-244
direct heating method **EM4:** 245, 246
gaseous hydrocarbons **EM4:** 244-245
history of use of combustion **EM4:** 244
indirect heating method **EM4:** 245, 246
liquid fuels **EM4:** 244
solid fuels **EM4:** 244
continuous furnaces **EM4:** 250
electrical safety **EM4:** 253
high current **EM4:** 253
high frequency **EM4:** 253
high voltage **EM4:** 253
electrically heated furnaces **EM4:** 245-250
graphite heating elements **EM4:** 245-247, 249
non-oxide ceramic heating elements **EM4:** 245, 248-249
oxide ceramic heating elements **EM4:** 245, 247, 249
refractory metal heating elements **EM4:** 245, 247-248
furnace measurement and control **EM4:** 250, 253
proportional integral differential control **EM4:** 250, 253
furnace safety **EM4:** 253
measurement of atmosphere composition **EM4:** 252
electrochemical sensors **EM4:** 252
external techniques **EM4:** 252
gas chromatography **EM4:** 252
infrared analysis **EM4:** 252
mass spectroscopy **EM4:** 252
in situ techniques **EM4:** 252
periodic furnaces **EM4:** 250
pressure measurement **EM4:** 252
units **EM4:** 252
temperature measurement **EM4:** 250-252
as defined property **EM4:** 251
equation of state for mole of perfect gas **EM4:** 251
historical background **EM4:** 250-251
isobaric coefficient of expansion **EM4:** 251
optical perimetry **EM4:** 251-252
practical temperature measurement **EM4:** 251-252
thermocouples **EM4:** 251
types of ceramic firing and sintering furnaces **EM4:** 244
Furnaces for
brazing of steel **M6:** 930-933
brazing with silver alloy filler metal **M6:** 944
dip brazing of steels **M6:** 989-990
Furniture applications *See also* Consumer products applications
aluminum and aluminum alloys **A2:** 13
office, parts design **EM2:** 616
Furrow-type fatigue fracture
titanium alloy **A12:** 44, 54
Furrowing
in electropolishing **A9:** 50-51
Fuse and flame spray process
thermal spray coating **M5:** 366-367
Fused borax
constituent in torch brazing flux **M6:** 962
Fused ceramics
for handling and processing equipment **EM4:** 959
Fused dry-resin coatings
for cast irons **A15:** 565
Fused quartz glass
applications
electronic processing **EM4:** 1055, 1056, 1059
information displays **EM4:** 1045
properties **EM4:** 1057
Fused salt bath
cleaning of cast iron **M6:** 996-997
dip brazing of cast irons **M6:** 999-1000
Fused salt bath descaling *See* Salt bath descaling
Fused salt synthesis
of ferrites **EM4:** 1163
Fused salts *See also* Molten salts
nickel-base alloy applications in **A13:** 655
titanium/titanium alloy resistance **A13:** 681
Fused silica **EM4:** 13
applications
information displays **EM4:** 1045
military **EM4:** 1019
optical glass products **EM4:** 1074-1075, 1077, 1078
solar cell covers **EM4:** 1019
fatigue characteristics **EM4:** 1017
hot pressing and pressure densification **EM4:** 296
intrinsic strength of glass fibers **EM4:** 742
properties **EM4:** 450, 1019
silica sols to bond aggregates **EM4:** 440
sol-gel process **EM4:** 445, 447, 450
Fused silica ceramic
aerospace applications **EM4:** 1017
properties **EM4:** 1018
Fused silica dilatometer
to measure thermal expansion of glass **EM4:** 568
Fused silica fibers
importance **EM1:** 43
properties/types **EM1:** 61
thermal conductivity **EM1:** 47
Fused silica glass
applications
aerospace **EM4:** 1017, 1019, 1019
electronic processing **EM4:** 1058
compound compositions **EM4:** 566
fracture surface analysis of optical fibers **EM4:** 663-664
properties **EM4:** 566
Fused slurry process
oxidation-resistant coating **M5:** 664-666
silicide coatings **M5:** 535
Fused spray deposit
definition **M6:** 8
Fused spray deposit (thermal spraying)
definition **A6:** 1209
Fused-alumina grit blasting
ceramic coating preparation **M5:** 539
Fused-salt fluxing method
aluminum coating **M5:** 337-339
Fused-salt oxidation-resistant coating process **M5:** 654-666
Fuselage brace
from titanium-based **A7:** 754
Fuses **A13:** 1120-1121, 1132
corrosion failure analysis **EL1:** 1110-1111
Fusible alloys **M3:** 799-801
applications **M3:** 799-801
of bismuth **A2:** 755-756
casting **M3:** 800-801
eutectic
composition **M3:** 799
melting temperatures **M3:** 799
lead and lead alloy **A2:** 555
noneutectic
composition **M3:** 799-800
melting temperatures **M3:** 799
yield temperatures **M3:** 799
properties **M3:** 800-801
properties of **A2:** 756
suitability for various applications **A2:** 755
as white metal alloy **A2:** 525
Fusible finish
as surface preparation **EL1:** 679
Fusible mounting alloys
for electropolishing **A9:** 49
Fusion *See also* Lack of fusion (LOF) **EM3:** 14
with A15 superconductors **A2:** 1071-1072
bonding, of thermoplastic composite **EM1:** 552
defined **EM2:** 20
definition **A6:** 1209 **M6:** 8
fluxes for **A10:** 167
glass-forming **A10:** 94
lack, as casting defect **A15:** 548
lack, as planar flaw **A17:** 50
in lost foam casting **A15:** 231
low-temperature **A10:** 94
as sample preparation technique **A10:** 93-94
sodium peroxide **A10:** 166-167
solid sample digestion by **A10:** 166-167
thermonuclear, as ternary molybdenum chalcogenides application **A2:** 1079

Fusion (continued)
and winding **EM1:** 138
zone, arc welding, cast irons **A15:** 523
Fusion (overflow) process
display glass properties **EM4:** 1048-1049
Fusion casting
and melting/fining **EM4:** 391
Fusion, during heat treatment
as casting defect **A11:** 386
Fusion face
definition **A6:** 1209 **M6:** 8
Fusion, incomplete
in arc welds **A11:** 413
in arc-welded aluminum alloys **A11:** 435
as discontinuity **A12:** 65
in electrogas welds **A11:** 440
in electron beam welds **A11:** 445
in flash welds **A11:** 442
in slag **A11:** 440
voids caused by **A11:** 93
Fusion procedure
platinum **A7:** 16
Fusion reactors **A13:** 17
Fusion thermit welding **M6:** 694
Fusion treatment **A18:** 644
Fusion weld bead terminology **A9:** 577
Fusion welding *See also* Arc welding
of beryllium **A2:** 683
of blanks **A14:** 451
cast stainless steel alloys **A6:** 495-496, 498
copper metals **M2:** 441, 442
defined **A9:** 577
definition **A6:** 1209 **M6:** 8
dispersion-strengthened aluminum alloys **A6:** 542, 543-546
duplex stainless steels **A6:** 479-480
energy sources used **A6:** 3-6
choice of joining method **A6:** 3
energy-source intensity **A6:** 3-6
weld pool-heat source interaction times as function of heat-source intensity **A6:** 5
mechanically alloyed oxide alloys **A2:** 949
dispersion-strengthened (MA ODS)
oxide-dispersion-strengthened materials **A6:** 1039
steel tubular products **M1:** 316
Fusion zone **A6:** 99, 105
advanced titanium-base alloys **A6:** 526
aluminum alloys **A6:** 727
corrosion of weldments **A6:** 1065, 1066, 1068
definition **A6:** 1209 **M6:** 8
description **A6:** 1065
dispersion-strengthened aluminum alloys **A6:** 542-543, 544, 545, 546
electron-beam welding **A6:** 870, 872, 873
electroslag welding **A6:** 271, 278, 279
gas-metal arc welding **A6:** 183
heat-treatable aluminum alloys **A6:** 533-534, 535
high-temperature alloys **A6:** 565
machinery and equipment weldments **A6:** 391-393
martensitic stainless steels **A6:** 433, 441
nickel alloys **A6:** 588
nickel-base corrosion-resistant alloys containing molybdenum **A6:** 595-596
nickel-chromium alloys **A6:** 588
nickel-chromium-iron alloys **A6:** 588
nickel-copper alloys **A6:** 588
nonferrous high-temperature materials **A6:** 567
precipitation-hardening stainless steels **A6:** 483, 488, 490
softening, wrought aluminum alloys **A12:** 422
stainless steel casting alloys **A6:** 497, 498
tantalum alloys **A6:** 580, 581
titanium alloys **A6:** 513-514, 515, 516, 521 **A12:** 443
Fusion zone of a weldment **A9:** 577
chemical composition **A9:** 578
ferritic microstructure constituent classifications of the International Institute for Welding **A9:** 581
solid-state phase transformations in **A9:** 580-581
Fusion-containment machines
tokamaks as **A2:** 1056-1057
Fusion-zone (FZ) solidification cracking
cobalt-base corrosion-resistant alloys **A6:** 598
Future trends *See also* Applications; Trends
device trends **EL1:** 390-391
environmental stress screening **EL1:** 886
environmental testing **EL1:** 503
level 1 packages **EL1:** 405-407
in semiconductor packaging **EL1:** 449
superconductivity as **EL1:** 395
synopsis of **EL1:** 395
technology change **EL1:** 391-394
UV-curable coatings (acrylate) **EL1:** 788
Fuwa tube
for flame atomizers **A10:** 48
Fuze parts
powders used **A7:** 573
Fuzz *See also* Abrasion resistance **EM3:** 14
defined **EM1:** 12 **EM2:** 20
FWHM *See* Full width at half maximum

G

(Gas phase reaction bonding
ceramic- matrix composites **EM4:** 840
Fresh water corrosion
copper alloys **M2:** 470-471
g *See* Diffraction vector; Shear modulus
G bronze *See also* Copper casting alloys
nominal composition **A2:** 347
properties and applications **A2:** 374
"g" loading
loose powder filling **A7:** 431
G metal
recycling **A2:** 1214
G phase
in austenitic stainless steels **A9:** 284
in wrought heat-resistant alloys **A9:** 312
G-bronze
contributing to corrosion in seals **A18:** 549
seal materials **A18:** 550
g-factor
in ESR analysis **A10:** 254
as independent of temperature **A10:** 257
variation in **A10:** 262-263
G-P zones *See* Guinier-Preston (GP) zone solvus line
G-ratio
CBN grinding wheels **A16:** 462, 463, 464, 465, 466
diamond grinding wheels **A16:** 460-461, 462
G5 *See* Copper alloys, specific types, C95400
Ga-Gd (Phase Diagram) .... **A3:** 2 • 207
Ga-Ho (Phase Diagram) .... **A3:** 2 • 207
Ga-In (Phase Diagram) .... **A3:** 2 • 208
Ga-La (Phase Diagram) .... **A3:** 2 • 208
Ga-Li (Phase Diagram) .... **A3:** 2 • 208
Ga-Lu (Phase Diagram) .... **A3:** 2 • 209
Ga-Mg (Phase Diagram) .... **A3:** 2 • 209
Ga-Mn (Phase Diagram) .... **A3:** 2 • 209
Ga-Mo (Phase Diagram) .... **A3:** 2 • 210
Ga-Na (Phase Diagram) .... **A3:** 2 • 210
Ga-Nb (Phase Diagram) .... **A3:** 2 • 210
Ga-Nd (Phase Diagram) .... **A3:** 2 • 211
Ga-Ni (Phase Diagram) .... **A3:** 2 • 211
Ga-Pb (Phase Diagram) .... **A3:** 2 • 211
Ga-Pd (Phase Diagram) .... **A3:** 2 • 212
Ga-Pr (Phase Diagram) .... **A3:** 2 • 212
Ga-Pt (Phase Diagram) .... **A3:** 2 • 212
Ga-Pu (Phase Diagram) .... **A3:** 2 • 213
Ga-S (Phase Diagram) .... **A3:** 2 • 213
Ga-Sb (Phase Diagram) .... **A3:** 2 • 214
Ga-Sc (Phase Diagram) .... **A3:** 2 • 214
Ga-Se (Phase Diagram) .... **A3:** 2 • 214
Ga-Sm (Phase Diagram) .... **A3:** 2 • 215
Ga-Sn (Phase Diagram) .... **A3:** 2 • 215
Ga-Sr (Phase Diagram) .... **A3:** 2 • 215
Ga-Tb (Phase Diagram) .... **A3:** 2 • 216
Ga-Te (Phase Diagram) .... **A3:** 2 • 216
Ga-Tl (Phase Diagram) .... **A3:** 2 • 216
Ga-Tm (Phase Diagram) .... **A3:** 2 • 217
Ga-U (Phase Diagram) .... **A3:** 2 • 217
Ga-V (Phase Diagram) .... **A3:** 2 • 217
Ga-Y (Phase Diagram) .... **A3:** 2 • 218
Ga-Yb (Phase Diagram) .... **A3:** 2 • 218
Ga-Zn (Phase Diagram) .... **A3:** 2 • 218
Ga-Zr (Phase Diagram) .... **A3:** 2 • 219
Gadolinium *See also* Rare earth metals
in garnets **A9:** 538
prompt gamma activation analysis of **A10:** 240
as prompt-emission converter for thermal neutron radiography **EM3:** 759
properties **A2:** 1181
pure **M2:** 735-736
as rare earth **A2:** 702
Gadolinium oxysulfide, as scintillator
neutron radiography **A17:** 390
Gadolinium-iron garnet **A9:** 538
Gage
block, calibrated standard **A8:** 619
bridge, four-arm electric resistance strain **A8:** 220
cantilever beam clip **A8:** 383
carbon **A8:** 211
cold cathode-type **A8:** 414
defined **A14:** 7
for displacement measurements **A8:** 193
electrical resistance strain **A8:** 209
foil-type crack growth **A8:** 246
for Hugoniot elastic limit measurement **A8:** 2
incident and reflected **A8:** 220
ionization **A8:** 414
Krak **A8:** 391
manganin **A8:** 211
marks **A8:** 548-549, 579
marks, for deformation measurement **A14:** 878-879
for medium stress and strain rate measurements **A8:** 191
penetration, and flow-line contours in specimen ends **A8:** 590
piezoresistive **A8:** 211
section, in torsional testing **A8:** 145, 147, 155-157
stored-torque **A8:** 220
thermocouple vacuum **A8:** 414
transmitted **A8:** 220
types **A8:** 618
for ultrasonic testing **A8:** 245-246
Gage length **EM3:** 14
defined **EM1:** 12 **EM2:** 20
Gage length defined **A8:** 6
and failure process, workability **A8:** 156
local elongation along **A8:** 26
in stored-torque torsional Kolsky bar specimen **A8:** 222, 224
to measure flow localization **A8:** 169
-to-radius ratio **A8:** 165
for torsion specimen **A8:** 156
in uniaxial tensile testing **A8:** 555
Gage marks **A8:** 548-549, 579
Gage, scanning laser *See* Scanning laser gage
Gage section **A8:** 145, 147, 155-157
Gage(s)
bonded resistance strain **A17:** 448
defined **A17:** 18
Pirani **A17:** 68
thermal-conductivity, for leak testing **A17:** 62-63
thickness, ultrasonic inspection **A17:** 273
Gages
combination **M3:** 556-557
corrosion of **M3:** 557

Gages (continued)
diffused strain ... **EL1:** 45
hardness of ... **M3:** 555
inspection fixtures ... **M3:** 557
master ... **M3:** 557
materials for ... **M3:** 554-557
plastics, use for inspection fixtures ... **M3:** 557
precision ... **M3:** 556
production ... **M3:** 554-556
wear of ... **M3:** 554, 556, 557
Gages, high-pressure
as rare earth application ... **A2:** 731
Gaging
recessed contours, by magnifying systems ... **A17:** 11
three-dimensional, by laser ... **A17:** 12, 16
in torsional testing ... **A8:** 145, 147
visual reference ... **A17:** 10-11
Gaging, profile *See* Profile gaging
Gahnite
composition and properties ... **EM4:** 873
Galactose, in serum
indirect iodometry to determine ... **A10:** 205
Galena
as source of lead ... **A2:** 543
Galena-grinding media combination ... **A18:** 276
Galfan ... **A1:** 218
Galfan coated steel
corrosion ... **A13:** 1014
Galfan-coated steels
forming of ... **A14:** 561
Gall ... **A7:** 5
Gall-Tough *See* Stainless steels, specific types, S
Galling *See also* Abrasion; Adhesion; Adhesive wear; Die line; Pickup; Scoring; Spalling
AISI/SAE alloy steels ... **A12:** 295
in band printer wear ... **A8:** 607
of cast iron pump parts ... **A11:** 366-367
of cemented carbides ... **A7:** 779
of cobalt-base wear-resistant alloys ... **A2:** 450
cumulative material transfer as ... **A11:** 496
in deep drawing ... **A14:** 509-510
deep drawing dies ... **M3:** 496, 497, 498, 499
defined ... **A8:** 6 **A11:** 5 **A18:** 10
deformation workpiece ... **A8:** 576
as die casting defect ... **A15:** 294
electroplated soft metals ... **A18:** 838
in fasteners ... **A11:** 542
hardfacing for ... **A7:** 823
high-carbon steels ... **A12:** 285
in inner cone, roller bearing ... **A11:** 158
jet engine components ... **A18:** 591
laser hardfacing providing resistance ... **A6:** 807
n sheet metal forming ... **A8:** 548
of nickel-base alloys ... **A14:** 831
on root-attachment surface, titanium alloy compressor blade ... **A11:** 285
prevention in shallow forming dies ... **A18:** 633
pumps ... **A18:** 595
res-stance, in press forming ... **A14:** 505-507
resistance, beryllium-copper alloys ... **A2:** 418-419
resistance, of zinc alloys ... **A15:** 786
of shafts ... **A11:** 466
in shear testing ... **A8:** 68
of stainless steel castings ... **A1:** 928-929
stainless steels ... **A18:** 715-718, 723
stress, high, beryllium-copper alloys ... **A2:** 418
thermal spray coating applications ... **A18:** 831-832
of titanium alloys ... **A14:** 838
tool steels ... **A18:** 737, 739
Galling resistance
press forming dies ... **M3:** 490-493
Galling threshold load ... **A18:** 595
Gallionella
biological corrosion by ... **A13:** 117
Gallium *See also* Gallium aluminum arsenide (GaAlAs); Gallium arsenide (GaAs); Gallium arsenide phosphide; Gallium compounds; Gallium gadolinium garnet (GGG); Gallium oxide; Gallium scandium gadolinium garnet (GSGG)
alloying, wrought aluminum alloy ... **A2:** 51
as an alpha stabilizer in titanium alloys ... **A9:** 458
applications ... **A2:** 739-741
Auger electron spectroscopy map of a gallium arsenide field-effect transistor ... **EM3:** 241
Gallium (continued)
embrittlement by ... **A11:** 234, 235, 719 **A13:** 180
epithermal neutron activation analysis ... **A10:** 239
evaporation fields for ... **A10:** 587
fabrication, GaAs crystals ... **A2:** 744-745
fractional crystallization ultrapurification technique ... **A2:** 1093
gallium gadolinium garnet (GGG) ... **A2:** 740-741
and gallium m compounds ... **A2:** 739-749
integrated circuits ... **A2:** 740
as low melting embrittler ... **A12:** 29
in medical therapy, toxic effects ... **A2:** 1257
molten, tantalum resistance to ... **A13:** 733
optoelectronic devices ... **A2:** 739-740
plain carbon steel resistance to ... **A13:** 515
production ... **A2:** 747
properties and grades ... **A2:** 741
pure ... **M2:** 736-737
pure, properties ... **A2:** 1114
purification ... **A2:** 744
recovery technology ... **A2:** 741-744, 749
research and development ... **A2:** 747, 749
resources ... **A2:** 741-742, 749
secondary recovery ... **A2:** 745-746
species weighed in gravimetry ... **A10:** 172
strategic (military) factors ... **A2:** 749
thermal diffusivity from 20 to 100 °C ... **A6:** 4
TNAA detection limits ... **A10:** 237
world supply and demand ... **A2:** 746-747, 749
Gallium aluminum arsenide (GaAlAs)
applications ... **A2:** 739-740
Gallium arsenide
coefficient of friction as a function of temperature and doping ... **A18:** 688-689
cracking geometries ... **A18:** 687
FIM sample preparation of ... **A10:** 586
p-type, scratch morphology ... **A18:** 688
plastic deformation ... **A18:** 688
properties ... **EM4:** 1
in thin-film semiconductors, FIM/PLAP analysis ... **A10:** 601-602
Gallium arsenide (GaAs) *See also* Gallium; Gallium compounds
as active component fabrication technology ... **EL1:** 199-200
applications ... **A2:** 739-741
clock frequencies ... **EL1:** 76
crystal growth ... **EL1:** 200
development ... **EL1:** 160
digital active-component fabrication ... **EL1:** 201
electrical conversion efficiency ... **A2:** 740
epitaxial, on silicon substrate ... **EL1:** 200-201
fabrication ... **A2:** 744-745
high-speed electronic devices, in mixed technologies ... **EL1:** 8
ingot, wafer, device manufacturers ... **A2:** 747-748
integrated circuits ... **A2:** 740
manufacturers ... **A2:** 748
monolithic microwave integrated circuits (MMICs) ... **EL1:** 756
optoelectronic devices ... **A2:** 739-740
properties ... **A2:** 741
research and development ... **A2:** 747, 749
software for ... **EL1:** 390
technologies, high speed ... **EL1:** 8
technology, active analog components ... **EL1:** 148-150
in wafer processing ... **EM3:** 580, 581, 582, 583
Gallium arsenide (GaAs)-silicon wafer
production of ... **A2:** 747
Gallium arsenide field effect transistor (FET)
Auger electron spectroscopy scans ... **EM3:** 240, 241
Gallium arsenide phosphide (GaAsP)
applications ... **A2:** 739
Gallium compounds *See also* Gallium
applications ... **A2:** 739-741
crystal growth ... **A2:** 744
fabrication of GaAs crystals ... **A2:** 744-745
gallium gadolinium garnet (GGG) substrate ... **A2:** 740-741
gallium scandium gadolinium garnet (GSGG) ... **A2:** 741
integrated circuits ... **A2:** 740
optoelectronic devices ... **A2:** 739-740
properties and grades ... **A2:** 741
research and development ... **A2:** 747, 749
secondary recovery ... **A2:** 745-746
Gallium compounds (continued)
wafer processing and doping ... **A2:** 744-745
Gallium fluoride (GaF_2)
direct evaporation ... **A18:** 844
Gallium gadolinium garnet (GGG) bubble memory device ... **A2:** 740-741
Gallium in fusible alloys ... **M3:** 799
Gallium oxide
in binary phosphate glasses ... **A10:** 131
single-crystal garnets ... **A2:** 740
Gallium, resistance of
to liquid-metal corrosion ... **A1:** 636
Gallium scandium gadolinium garnet (GSGG)
applications ... **A2:** 741
Gallium, vapor pressure
relation to temperature ... **A4:** 495 **M4:** 310
Gallium-doped gold contact wires ... **EL1:** 958
Gallium-doped yttrium iron garnet (GaYIG)
FMR probe for ... **A17:** 221
Gallium-gadolinium-garnet (GGG) ... **EM4:** 18
GALPAT test procedures
application ... **EL1:** 375
Galvalume coated 1010 steel ... **A9:** 199
Galvanic anode
defined ... **A13:** 7
Galvanic cathodes
anodic protection of ... **A13:** 465
Galvanic cell
defined ... **A13:** 7
Galvanic cell, short-circuited
for internal electrolysis ... **A10:** 199
Galvanic coatings for pipeline ... **A13:** 1291
Galvanic corrosion *See also* Corrosion; Galvanic corrosion evaluation; Galvanic effects ... **A13:** 83-87 **M1:** 713 **M5:** 431-432
aircraft ... **A13:** 1022-1025
of alloy groupings, performance ... **A13:** 84-86
aluminum alloys ... **A6:** 729 **M2:** 209-211, 213-214
aluminum, coupled with dissimilar metals ... **M2:** 203
of aluminum/aluminum alloys ... **A13:** 587-589
of anodic members ... **A13:** 84
area/distance effects ... **A13:** 83
atmospheric ... **A13:** 237-238 **M1:** 718-719
in automotive industry ... **A13:** 1011
as beneficial ... **A13:** 324
in brazing ... **A13:** 876-877
of buried metals ... **A11:** 191
in carbon and low-alloy steels ... **A11:** 199
in carbon steel weldments ... **A13:** 363-365
cell, in aqueous solution ... **A13:** 1329
cells, caused by dissimilar metals ... **A11:** 778
in cemented carbides ... **A13:** 855-858
compatible metals ... **A11:** 185
composite-to-metal joining ... **A6:** 1041-1042
control methods ... **A13:** 186-87
of copper alloy couples, in seawater ... **A13:** 624
copper metals ... **M2:** 458-459
of copper/copper alloys ... **A13:** 612, 623-624
as corrosion mechanism ... **A13:** 234
coupons ... **A13:** 198
defined ... **A11:** 5 **A13:** 7
and design ... **A13:** 342
dissimilar metal joining ... **A6:** 826
electrochemical testing methods ... **A13:** 215-216
as environmental failure mechanism ... **EL1:** 493
evaluation of ... **A13:** 215-216, 234-238
failure of cast iron pump impeller by ... **A11:** 374
in fresh water ... **M1:** 735
galvanic series ... **A13:** 83
galvanic series in seawater ... **A11:** 185
geometry effects ... **A13:** 83
in graphite composites ... **EM1:** 706, 709, 716-718
of lead/lead alloys ... **A13:** 780
limited by tin-zinc and zinc-aluminum solders ... **A6:** 968
magnesium alloys ... **M2:** 604-609 **M5:** 628, 631, 637, 646
of magnesium/magnesium alloys ... **A13:** 747-748
in manned spacecraft ... **A13:** 1076-1079
of marine structures ... **A13:** 543
material selection to avoid ... **A13:** 324
mechanism ... **A13:** 1012
in metal-matrix composites during preparation ... **A9:** 587-588
in metallic coatings ... **A13:** 342-343
of metals ... **A11:** 185

Galvanic corrosion (continued)
metals embedded in concrete or plaster ... **A11:** 186-187
in mining/mill application ... **A13:** 1295
modes of attack ... **A13:** 84
of mounted aluminum alloy specimens ... **A9:** 352
in oil/gas production ... **A13:** 1234
on aluminum alloy spacer ... **A11:** 187-188
passivity effects ... **A11:** 185
polarization ... **A13:** 83
predicting ... **A13:** 84
protection ... **A11:** 195-197
rates, direct measurement ... **A13:** 216
ratio of cathode area to anode area ... **A11:** 186
sacrificial, of zinc ... **A13:** 755
in seawater ... **M1:** 740-741, 743, 745
with solder joining aluminum ... **A6:** 990
of soldered joints ... **A17:** 609
sources, and design considerations ... **A13:** 340
in space shuttle orbiter ... **A13:** 1060-1061, 1067-1068
of sprinkler system ... **A11:** 185-186
of stainless steels ... **A6:** 625 **A13:** 553-554 **M3:** 62-63
of steam surface condensers ... **A13:** 988
and stray-current corrosion, compared ... **A13:** 87
in threaded fasteners ... **A11:** 535
time-dependent factors, effect of ... **A11:** 187
of titanium/titanium alloys ... **A13:** 675-676, 690-692
of uranium/uranium alloys ... **A13:** 816-817
in water-recirculating systems ... **A13:** 488
of weathering steels ... **A13:** 518-519
of zirconium/zirconium alloys ... **A13:** 717
Galvanic corrosion evaluation *See also* Galvanic corrosion ... **A13:** 234-238
atmosphere testing ... **A13:** 238
component testing ... **A13:** 234
electrochemical ... **A13:** 235-236
galvanic series ... **A13:** 235
immersion tests ... **A13:** 237-238
laboratory testing ... **A13:** 234-238
modeling ... **A13:** 234-238
polarization curves ... **A13:** 235-236
specimen exposures ... **A13:** 236-238
Galvanic corrosion protection
anodic and cathodic ... **A11:** 195-197
Galvanic couple
corrosion from ... **M5:** 431-432
Galvanic couple potential *See* Mixed potential
Galvanic couples *See also* Galvanic corrosion
defined ... **A13:** 7
elimination for corrosion protection ... **M1:** 751, 752
magnesium/magnesium alloys ... **A13:** 746
seawater corrosion ... **A13:** 773
titanium/titanium alloys ... **A13:** 673
Galvanic coupling ... **A18:** 271
Galvanic currents ... **A13:** 7, 84
Galvanic effects *See also* Galvanic corrosion
in lead/lead alloys ... **A13:** 785
in magnesium/magnesium alloys ... **A13:** 741
on niobium ... **A13:** 722
on tantalum ... **A13:** 735-736
in pure tin ... **A13:** 772
telephone cables ... **A13:** 1129
uranium/uranium alloys ... **A13:** 816-817
Galvanic effects, consideration of
in selection of mechanical clamps and spacers ... **A9:** 28
Galvanic exfoliation corrosion
eddy current inspection ... **A17:** 191
Galvanic pain
oral ... **A13:** 1337
Galvanic plating *See* Immersion plating
Galvanic reduction
of palladium powders ... **A7:** 150
of silver powders ... **A7:** 148
Galvanic series *See also* Electromotive force series
aluminum alloys ... **A13:** 587
defined ... **A13:** 7, 83
in flowing seawater ... **A13:** 675
Galvanic series (continued)
galvanic corrosion evaluation ... **A13:** 235
for metals ... **A13:** 755-756
metals and alloys ... **M5:** 431
in neutral soils and water ... **A13:** 1288
in seawater ... **A13:** 83, 235, 488, 613, 718, 876, 1329
Galvanic series in seawater ... **A11:** 185 **M1:** 740
Galvanize
defined ... **A13:** 7
Galvanized bridge wire ... **A1:** 282
Galvanized coatings *See also* Electro-galvanizing, Hot dip galvanizing; Zinc and zinc alloys; Zinc coatings ... **A13:** 432, 439-441
atmospheric corrosion resistance enhanced by ... **M1:** 722
corrosion protection ... **M1:** 751-753
fresh water systems ... **M1:** 738
nails ... **M1:** 271
steel wire ... **M1:** 263-265
coating procedure ... **M1:** 263
coating weights ... **M1:** 263
tempers of wire ... **M1:** 263
tensile strength of wire ... **M1:** 263
wire rope, wire for ... **M1:** 272
zinc ... **M2:** 651-653
Galvanized iron
effect of pH in suspending liquid during polishing ... **A9:** 47
inorganic fluxes, for ... **A6:** 980
relative solderability ... **A6:** 134
relative weldability ratings, resistance spot welding ... **A6:** 834
Galvanized metal
plasma arc cutting ... **A6:** 1170
Galvanized rope wire ... **A1:** 284
Galvanized sheet duct or decking
capacitor discharge stud welding ... **A6:** 222
Galvanized sheet steel
specimen preparation ... **A9:** 197
Galvanized steel ... **A13:** 432-434
applications ... **A6:** 400
aqueous corrosion resistance ... **A13:** 433-434
atmospheric corrosion of ... **A13:** 432-433, 526-527
in automotive industry ... **A13:** 1011-1014
corrosion mechanism ... **A13:** 433
embrittlement of ... **M1:** 686
intergranular corrosion of ... **A13:** 527
phosphate coating of ... **M5:** 438, 441
polishing ... **A9:** 488
predictive equations for ... **A13:** 527
resistance seam welding ... **A6:** 244, 245
stud arc welding ... **M6:** 733
wear by corrosion and shear fracture ... **A18:** 181-182
Galvanized steels
contour roll forming ... **A14:** 634
forming applications ... **A14:** 561
mill products ... **A14:** 560
polysulfides for bonding ... **EM3:** 141, 142
press forming of ... **A14:** 560-561
sealant use for spot welding ... **EM3:** 609
toot design ... **A14:** 560
types ... **A14:** 561
Galvanized wire ... **A1:** 281
Galvanizing
definition of ... **A1:** 212
electro- ... **A13:** 766-767
Electrogalvanizing ... **A1:** 212, 217 **A2:** 528
fabrication details ... **A13:** 439
hot dip *See* Hot dip galvanized coating ... **A1:** 212, 216-217 **A13:** 436-444, 765-766
hot dip after-fabrication ... **A2:** 527-528
hot dip conventional strip ... **A2:** 527
hot-dip ... **A14:** 560
for marine corrosion ... **A13:** 910-911
mechanical ... **A2:** 528 **M5:** 300-301
Galvanizing fluxes
nonferrous molten metals ... **A15:** 452
Galvanizing vat, carbon steel
failure of ... **A11:** 273-274
Galvanneal
defined ... **A13:** 7
Galvannealed steels
forming of ... **A14:** 561
Galvanometer
defined ... **A13:** 7
Galvanostatic
defined ... **A13:** 7
polarization measurements ... **A13:** 1333
testing methods, localized corrosion ... **A13:** 217
Galvanostatic etching
electrochemical principles ... **A9:** 144
Gamma alloys *See also* Ordered intermetallics; Titanium aluminides
crystal structure and deformation ... **A2:** 927-928
future directions and applications ... **A2:** 929
material processing ... **A2:** 928
mechanical/metallurgical properties ... **A2:** 928-929
Gamma double prime ... **A1:** 951-952, 953-954
Gamma double prime phase
in wrought heat- resistant alloys ... **A9:** 309-311
Gamma iron
defined ... **A9:** 8 **A13:** 7
Gamma loss peak ... **EM3:** 14
defined ... **EM2:** 20
Gamma matrix ... **A1:** 951, 952
Gamma modulation
in SEM display systems ... **A12:** 169
Gamma phase unalloyed uranium
fabrication techniques ... **A2:** 671
Gamma phases
in Alnico alloys ... **A9:** 539
in beryllium-copper alloys ... **A9:** 395
in beryllium-nickel alloys ... **A9:** 396
of cemented carbides ... **A9:** 274
in ferrous alloys, peritectic transformation ... **A9:** 679-680
in iron-cobalt-vanadium alloys ... **A9:** 538
in nickel-base heat-resistant casting alloys ... **A9:** 334
in uranium and uranium alloys ... **A9:** 476-487
in wrought heat-resistant alloys ... **A9:** 308-311
in zinc-copper alloys ... **A9:** 489
Gamma prime ... **A1:** 951, 952-953
Gamma prime phase
electrolytic extraction and x-ray diffraction ... **A9:** 308
electropolishing ... **A9:** 307
and Ni alloys ... **A16:** 835
in nickel-base heat-resistant casting alloys ... **A9:** 334
transmission electron microscopy ... **A9:** 307
uranium alloy ... **A9:** 477-487
in wrought heat-resistant alloys ... **A9:** 309-311
Gamma prime precipitate
in Monel K-500 ... **A9:** 436
Gamma radiation *See also* Radiation
effect, polyether-imides (PEI) ... **EM2:** 157
Gamma radiation, effect
E-glass fibers ... **EM1:** 47
Gamma ray spectroscopy
defined ... **A10:** 674
Gamma ray spectrum
changes as function of time ... **A10:** 236
high-purity nickel, neutron activation analysis ... **A10:** 240
of iridium by radiochemical neutron activation analysis ... **A10:** 241
of irradiated ore ... **A10:** 235
Gamma rays
for compound polymerization ... **EL1:** 854
defined ... **A10:** 674
detector for ... **A10:** 235
emission, as radioactive decay mode ... **A10:** 245
emission, in neutron activation analysis ... **A10:** 234

Gamma rays (continued)
polarization, Mössbauer spectroscopy **A10:** 288, 295
pure cobalt as source **A2:** 446
sources, industrial radiography **A2:** 447
Gamma structure
defined **A9:** 8
Gamma transition *See* Glass transition **EM3:** 14
Gamma wrought uranium alloy phase **A9:** 477-487
Gamma-alumina **EM4:** 111, 112
Gamma-Fe_2O_3 .... **EM4:** 18
Gamma-radiation dose rate of depleted uranium **A9:** 477
Gamma-ray(s)
as computed tomography (CT) radiation source **A17:** 368
density, of powder metallurgy parts **A17:** 538-539
emissions, in neutron radiography **A17:** 391
exposure charts **A17:** 330
physical characteristics **A17:** 308
production of **A17:** 298
sources **A17:** 309
units for **A17:** 301
and x-rays, in neutron radiography **A17:** 387-388
Gandolfi camera
for XRPD analysis **A10:** 335
Gang bonding
inner lead bonding **EL1:** 278
outer lead bonding **EL1:** 286
special considerations **EL1:** 286
Gang drill presses
reaming **A16:** 247
Gantry fixture
for high-deposition submerged arc welding **A7:** 821
Gantry shape cutting system **A14:** 727
Gantry type
coordinate measuring machines **A17:** 21
Gap **A7:** 5
defined **EM1:** 12 **EM2:** 20
definition **A6:** 1209
-filling adhesive, defined **EM1:** 12
Gap glass
properties **EM4:** 1057
Gap rolls **A14:** 96-97
Gap-filling adhesion **EM3:** 14, 36
Gap-filling adhesive
defined **EM2:** 20
Gap-frame lathes **A16:** 153
Gap-frame presses **A14:** 7, 492-493
Gapasil 9
wettability indices on stainless steel base metals **A6:** 118
Gaps
measurement of **A17:** 13, 199
Garden equipment
P/M parts for **A7:** 671, 676-678
Garden tractors **A7:** 677-678
Gardner impact strength
glass length effects **EM2:** 281
Gardner impact test **EM2:** 435
Garnet **EM4:** 8, 18, 61
as abrasive **A14:** 746-748
abrasive, for lapping **A16:** 493
abrasive, for waterjet cutting of MMCs **A16:** 894
abrasive, mixed with high-pressure waterjet **A16:** 521
as an abrasive in wire sawing **A9:** 26
blasting with **M5:** 84, 94
crystal structure **EM4:** 881
defined **A2:** 740
gallium gadolinium (GGG) **A2:** 740-741
hardness **EM4:** 351
natural abrasive **A16:** 434
precipitation process **EM4:** 59
Garnets *See also* Magnetic materials
bulk composition of **A10:** 628
magnetic domains revealed by the Faraday effect **A9:** 535
specimen preparation **A9:** 533
Garrard, William
as early founder **A15:** 31
Garrett axial flow turbine engine **EM4:** 716
Garrett Ceramic Components (GCC)
fast fracture flexure strength **EM4:** 1001
stress-rupture life **EM4:** 1001
Gas *See* Gaseous corrosion; Gases
content, symbol for **A8:** 7-15
dissolved **A8:** 423
effect of evolution during casting **A9:** 643
in environmental test chamber **A8:** 411
environments **A8:** 410-412, 540-541
gun, for plate impact test **A8:** 233
oxidizing, control **A8:** 414
pressure, control **A8:** 411
Gas absorption
to analyze ceramic powder particle sizes **EM4:** 67
Gas absorption (BET) method
and pyrophoricity **A7:** 199
Gas adsorption
apparatus **A7:** 262-263
method for determining surface area **A7:** 262-263
Gas analysis
gas chromatography/mass spectrometry **A10:** 639-648
mass spectrometry **A10:** 151-157
Raman spectroscopy **A10:** 126
Gas analysis by mass spectrometry **A10:** 151-157
applications **A10:** 151, 156-157
double-focusing mass spectrometer **A10:** 154
estimated analysis time **A10:** 151
gas mass spectrometry **A10:** 151-157
of inorganics **A10:** 7, 8
ion quadrupole mass filter **A10:** 153-154
limitations **A10:** 151
of mixtures **A10:** 151
of organics **A10:** 10, 11
related techniques **A10:** 151
results of analysis **A10:** 155-156
sampling **A10:** 151-152
scanning modes **A10:** 154
spectrometer components **A10:** 151-155
Gas atomization *See also* Atomization; Gas-atomized powders; specific gas-atomized powders; Water atomization
argon entrapment **A7:** 427, 428
contamination **A7:** 180
for hardfacing powders **A7:** 142, 823
for nickel powder production **A7:** 134
for nickel-based hardfacing powders **A7:** 134, 142
of titanium powder **A7:** 167
and water atomization **A7:** 25-34
Gas bearing
defined **A18:** 10
Gas blanket
defined **EL1:** 1145
Gas boxes
electrogas welding **M6:** 240-241
Gas brazing
definition **A6:** 1210
Gas bubble nucleation
diffusion control for **EM1:** 757-759
Gas burners
for sinters and fusions **A10:** 166
Gas carbon arc welding
definition **M6:** 8
Gas carburization
case hardening by **A7:** 453-454
Gas carburized steels **A9:** 219-224
Gas carburizing **A4:** 262, 263, 312-324 **M4:** 135-175
atmospheres **A4:** 312, 313, 315-319, 320 **M4:** 143, 161
atmospheres, safe handling **A4:** 313, 318 **M4:** 140
carbon concentration gradients **A4:** 314-315, 323 **M4:** 146-154, 172-174
carbon potential, control of **A4:** 314, 316-317, 318, 319 **M4:** 143, 172-174
carbon sources
gas **A4:** 312
liquid hydrocarbon **A4:** 312
carbon sources gas **M4:** 135-136
liquid hydrocarbon **M4:** 136
carrier gas plus hydrocarbon gas **A4:** 316 **M4:** 158, 159, 160, 161-170
carrier gases **A4:** 312, 314, 316-317, 318 **M4:** 136, 137
case depth, control of **A4:** 314, 315, 319, 320, 322-323 **M4:** 154-159
cleaning of parts, prior to carburizing **A4:** 313-314 **M4:** 141
compared to pack carburizing **A4:** 325, 328
dew point **A4:** 316, 317 **M4:** 145
Gas carburizing (continued)
diffusion of carbon and nitrogen **A4:** 314-315, 318-319, 323 **M4:** 145
dimensional control **A4:** 321-322 **M4:** 168-169
direct quenching versus reheating **A4:** 320 **M4:** 163, 167-168
equilibrium compositions, calculation **A4:** 316-317 **M4:** 143-145, 146
equipment **A4:** 312-313
fixturing **A4:** 322 **M4:** 169
flow rates **A4:** 318 **M4:** 158, 161-164
in fluidized beds **A4:** 486-487, 490
furnace conditioning **A4:** 318
grain growth **A4:** 320 **M4:** 158, 172
high-temperature carburizing **A4:** 321 **M4:** 170, 171-172
homogeneous carburizing **A4:** 319 **M4:** 171
hydrocarbon gases **A4:** 316 **M4:** 156, 157, 160-161
loading methods **A4:** 318, 323 **M4:** 141
martempering **A4:** 322 **M4:** 169
mixing **A4:** 323 **M4:** 164
operating cost comparison **A4:** 357
operating cycles **A4:** 319-320 **M4:** 159, 160, 161, 162, 164-165
practices **M4:** 156, 157, 160-161
practices, case depth measurement **A4:** 322
press quenching **A4:** 320, 322 **M4:** 169-170
process parameters predicted by computer simulation **A4:** 650-654
process planning **A4:** 319-321
rate control **A4:** 318
reset rate **A4:** 318
safety precautions **A4:** 318, 323 **M4:** 139-141
selective carburizing **A4:** 320-321 **M4:** 171
selective peening **A4:** 322
sooting **A4:** 316, 317-318 **M4:** 165-167
steel composition **A4:** 320
straightening **A4:** 322
surface carbon content **A4:** 315 **M4:** 148-152, 153, 172-174
tempering **A4:** 319, 320, 321, 323 **M4:** 168, 169, 170-171
Gas carburizing, furnaces
batch, pit-type **A4:** 312-313, 314, 319 **M4:** 137-138
conditioning **A4:** 318
construction **A4:** 313 **M4:** 172
continuous **A4:** 312, 313, 314, 319 **M4:** 138-139
controllers **A4:** 318
horizontal batch **A4:** 312, 313 **M4:** 138
mesh belt, continuous **A4:** 313
pusher-type continuous **A4:** 313, 319-320 **M4:** 137, 139
roller hearth **A4:** 313
rotary hearth **A4:** 313
rotary-hearth **M4:** 138
rotary-retort batch **M4:** 138
rotary-retort continuous **A4:** 313 **M4:** 139
safety precautions, starting and stopping **A4:** 313, 323 **M4:** 140-141
sealed-quench **A4:** 313
shaker-hearth **A4:** 313 **M4:** 139
Gas carburizing, processing
atmospheres, alternatives to conventional **A4:** 314-315, 316 **M4:** 145-146
temperature, effect of **A4:** 314, 316 **M4:** 141-142
time, effect of **A4:** 314, 315, 316, 318 **M4:** 142-143
Gas cavities *See also* Cavities; Gases
cast aluminum alloys **A12:** 405-406, 412
Gas cells
IR sampling **A10:** 113
Gas chromatograph/mass spectrometer (GC/MS)
gas analysis during drying and firing **EM4:** 24, 25
Gas chromatographs
for leak detection **A17:** 64, 68
Gas chromatography *See also* Gas chromatography/mass spectrometry **A13:** 1115 **M4:** 430-431 **EM1:** 736
as advanced failure analysis technique **EL1:** 1104-1105
defined **A10:** 674
and gas analysis by mass spectrometry compared **A10:** 151
for measurement of furnace atmosphere composition **EM4:** 252
of volatile metal-organic complexes, as separation tool **A10:** 170

Gas chromatography-infrared spectroscopy **A10:** 115
Gas chromatography/mass spectrometry **A10:** 639-648
applications **A10:** 639, 647-648
capabilities **A10:** 212, 277, 649
complementary techniques **A10:** 645-647
estimated analysis time **A10:** 639
and FT-IR techniques **A10:** 115
and gas analysis by mass spectrometry compared **A10:** 151
general uses **A10:** 639
of inorganic gases **A10:** 8
of inorganic liquids and solutions **A10:** 7
interpreting mass spectrum **A10:** 641-642
introduction **A10:** 640
limitations **A10:** 639
methodology **A10:** 643-645
of organic gases **A10:** 11
of organic liquids and solutions **A10:** 10
of organic solids **A10:** 9
principles, gas chromatography **A10:** 640-641
principles, mass spectrometry **A10:** 640
pyrolysis **A10:** 647-648
related techniques **A10:** 639, 645-647
samples **A10:** 639, 644-645
for tallow-base lubricant coatings **A10:** 177
Gas classification **A7:** 5
Gas classifier **A7:** 5
Gas composition analyses
for temperature-time control in polymer removal techniques **EM4:** 137
Gas constant
defined **A10:** 674
Gas contents of aluminum alloy ingots
effects of **A9:** 634
Gas covers
cobalt-base alloys **A15:** 813
Gas cutter
definition **A6:** 1210
Gas cutting
and blanking **A14:** 458
definition **A6:** 1210
oxyfuel **A14:** 722-724
plasma arc **A14:** 730-731
for stock preparation, hot upset forging **A14:** 86
vs. planing **A16:** 186
Gas cyaniding *See* Carbonitriding
Gas cylinder
definition **A6:** 1210
Gas defects *See also* Defects; Gases
from carbon additions **A15:** 211
Gas detection *See also* Detection; Gases; Leak testing
applications, specific-gas **A17:** 64-65
devices **A17:** 61-64
ionization detectors, computed tomography (CT) **A17:** 369
Gas diffusion
as gas-liquid reaction **A15:** 82-83
Gas displacement methods
die casting **A15:** 291
Gas drilling applications
cemented carbides **A2:** 974-977
Gas elutriator
powder cleaning **A7:** 178
Gas engine exhaust purification
powder used **A7:** 573
Gas engine oils
lubricant classification **A18:** 85
Gas entrapment
collision-caused **A7:** 29-30
within particles, effect on densification and expansion during homogenization **A7:** 315
Gas environments **A8:** 410-412, 540-541
Gas explosivity **A7:** 195
Gas filters
powders used **A7:** 573
Gas fired furnace
dip brazing **A6:** 337
Gas flushing
inert **A15:** 84-85
Gas gouging
definition **A6:** 1210
Gas gun
for plate impact testing **A8:** 210, 233
for short-pulse-duration tests **A8:** 283
for split Hopkinson pressure bar facility **A8:** 201
Gas hole *See also* Gas porosity; Holes
defined **A11:** 5
Gas holes
defined **A15:** 6
zirconium alloys **A15:** 838
Gas industry
stainless steel corrosion **A13:** 560
Gas injection
pump, for demagging aluminum alloys **A15:** 474
refining, principles **A15:** 470-471
Gas ion etching **A9:** 148-150
Gas ion reaction chamber
for reactive sputtering **A9:** 148
Gas kinetic temperature **A10:** 24
Gas laser
definition **A6:** 1210
Gas lasers
suitability for welding **M6:** 651-652
Gas lines, ASTM specifications for **M1:** 323
Gasoline tanks, terne-coated steel sheet for **M1:** 173-174
Gas lubrication *See also* Pressurized gas lubrication
defined **A18:** 10
Gas mass spectrometer
analysis, summary of **A10:** 155
capability **A10:** 156
double-focusing **A10:** 154
instrument set-up and calibration **A10:** 155
introduction system **A10:** 151-152
ion detection **A10:** 154-155
ion source **A10:** 152-153
mass analyzer **A10:** 153-154
output, computerized **A10:** 155
resolution of **A10:** 155
schematic **A10:** 153
Y and Z lenses **A10:** 153
Gas mass spectrometry
defined **A10:** 674
gas chromatography and **A10:** 639-648
Gas metal arc cutting
definition **M6:** 8
Gas metal arc cutting (GMAC)
definition **A6:** 1210
Gas metal arc welding **A7:** 456 **M6:** 153-181
advantages **M6:** 154
arc spot welding **M6:** 174-175
applicability **M6:** 174
joint design **M6:** 174
operating principles **M6:** 174
automatic welding **M6:** 175-177
backing strips **M6:** 169
butt welds **M6:** 167-168
of cast irons **A15:** 523-524
causes of welding difficulty **M6:** 172-173
classification of electrodes **M6:** 162-163
comparison to gas tungsten arc welding **M6:** 397-398
comparison to other processes **M6:** 177-180
definition **M6:** 8
disadvantages **M6:** 154
electrode extension **M6:** 170
electrode position **M6:** 171
electrode size **M6:** 170-171
electrodes **M6:** 161-163
composition **M6:** 161-162
mechanical properties **M6:** 162-163
electroslag welding **M6:** 238-244
Gas metal arc welding (continued)
equipment **M6:** 156-161
motion devices **M6:** 161
water-cooling systems **M6:** 161
welding cables **M6:** 161
failure origins **A11:** 414-415
hardfacing **M6:** 785, 787
heat flow calculations **M6:** 31
improvements due to process substitution **M6:** 177-180
inadequate shielding **M6:** 172-173
intermediate and thick sections **M6:** 167
joint designs **M6:** 165-169
for dimensional control **M6:** 166
for economy **M6:** 165-166
metal transfer **M6:** 154-156
globular transfer **M6:** 154-155
pulsed current transfer **M6:** 155
rotating arc transfer **M6:** 155-156
short circuiting transfer **M6:** 155
spray transfer **M6:** 154
metals welded **M6:** 153
narrow gap welding **M6:** 173-174
advantages **M6:** 173
joint completion rates **M6:** 173-174
operating principles **M6:** 173
welding applications **M6:** 173
operation principles **M6:** 154-155
polarity **M6:** 153-154
power source variables **M6:** 156-157
inductance **M6:** 157-158
slope **M6:** 157
welding voltage (arc length) **M6:** 156-157
power sources **M6:** 156
arc initiation **M6:** 216
constant-current power supply **M6:** 156
constant-voltage power supply **M6:** 156
recommended grooves **M6:** 69-70
safety **M6:** 180-181
sections of unequal thickness **M6:** 168
shielding gas equipment **M6:** 161
accessory equipment **M6:** 161
flowmeters **M6:** 161
gas pressure regulators **M6:** 161
noses **M6:** 161
shielding gases **M6:** 163-164
argon **M6:** 163-164
argon-carbide dioxide mixtures **M6:** 164
argon-helium mixtures **M6:** 164
argon-oxygen mixtures **M6:** 164
carbon dioxide **M6:** 164
helium **M6:** 164
helium-argon-carbon dioxide mixtures **M6:** 164
nitrogen **M6:** 164
thin sections **M6:** 166-167
travel speed **M6:** 170
underwater welding **M6:** 922
weld design for improved accessibility **M6:** 168-169
weld discontinuities **M6:** 830
weld overlaying **M6:** 807
weld quality **M6:** 171-172
excessive melt-through **M6:** 172
inclusions **M6:** 172
incomplete fusion **M6:** 172
lack of penetration **M6:** 172
porosity **M6:** 172
undercutting **M6:** 172
weld metal cracks **M6:** 172
welding current **M6:** 169-170
welding guns **M6:** 158-159
machine welding **M6:** 159
semiautomatic **M6:** 158-159
welding guns, manipulation of **M6:** 159
bead type **M6:** 159
direction **M6:** 159
welding position **M6:** 153
welding torches **M6:** 216-217

Gas metal arc welding (continued)
welding variables **M6:** 169-171
wire-feed stoppages **M6:** 172
wire-feed systems **M6:** 159-161
controls **M6:** 160-161
maintenance **M6:** 161
pull-type **M6:** 160
push-type **M6:** 159-160
variable-speed **M6:** 160
workpiece holding and handling **M6:** 164-165
Gas metal arc welding (GMAW)
discontinuities from **A17:** 582
Gas metal arc welding of
alloy steels **M6:** 299-301
control of deposition **M6:** 300
electrode selection **M6:** 300
shielding gas **M6:** 300
aluminum **M6:** 153
aluminum alloys **M6:** 153, 380-390
arc characteristics **M6:** 380-381
automatic welding **M6:** 383-385
multiple-pass welding **M6:** 383
power supply and equipment **M6:** 380
repair welding **M6:** 388
shielding gases **M6:** 380
soundness of welds **M6:** 385-388
spot welding **M6:** 388-390
thicknesses **M6:** 380
weld backing **M6:** 382-383
welding schedules **M6:** 380
welding speeds **M6:** 380
aluminum bronzes **M6:** 420-421
conditions **M6:** 416, 420
electrode wires **M6:** 420
joint design **M6:** 420
preheating **M6:** 420
beryllium **M6:** 462
beryllium copper, high-conductivity **M6:** 418-419
electrode wires **M6:** 418
preheating and postweld aging **M6:** 418
properties of weldments **M6:** 418
welding conditions **M6:** 416, 418
beryllium copper, high-strength **M6:** 419-420
welding conditions **M6:** 416, 420
brasses **M6:** 420
conditions for high-zinc copper alloys **M6:** 416, 420
conditions for low-zinc brasses **M6:** 416, 420
carbon steels **M6:** 153
cast irons **M6:** 153, 310
copper and copper alloys **M6:** 153, 415-418
braze welding method **M6:** 424
electrode wires **M6:** 417
fine-wire welding **M6:** 426
joint design **M6:** 418
overlay method **M6:** 424
preheating **M6:** 418
pulsed-current welding **M6:** 426
welding conditions **M6:** 416-417
welding to dissimilar metals **M6:** 424
welding to nickel alloys **M6:** 424
welding to nonferrous metals **M6:** 424
copper nickels **M6:** 421-422
conditions **M6:** 416, 421
electrode wire **M6:** 421
joining of tubing **M6:** 421-422
joint design **M6:** 421
preheating and postheating **M6:** 421
dissimilar copper alloys **M6:** 422-424
aluminum bronze to copper-nickel **M6:** 422, 424
copper to aluminum bronze **M6:** 422
copper to copper nickels **M6:** 422
preheating **M6:** 423-424
ductile iron **M6:** 315, 317-318
gray iron **M6:** 315
hardenable carbon steels **M6:** 268-269
heat-resistant alloys **M6:** 153, 362
cobalt-based alloys **M6:** 370
iron-nickel-chromium and iron-chromium-nickel alloys **M6:** 366-367
high-strength steels **M6:** 153
low-alloy steels **M6:** 153
magnesium alloys **M6:** 153, 429-430
malleable iron **M6:** 317
nickel alloys **M6:** 439-440

Gas metal arc welding of (continued)
phosphor bronzes **M6:** 420
conditions **M6:** 417, 420
electrode wires **M6:** 420
preheating **M6:** 420
refractory metals **M6:** 153
silicon bronzes **M6:** 421
conditions **M6:** 417, 421
electrode wire **M6:** 421
joint design **M6:** 421
preheating **M6:** 421
stainless steels **M6:** 153
stainless steels, austenitic **M6:** 329-333
current **M6:** 329
full-penetration weld **M6:** 333
pulsed arc transfer **M6:** 330-331
shielding gas **M6:** 331-332
short circuiting transfer **M6:** 329-330
spray transfer **M6:** 329-331
stainless steels, ferritic **M6:** 348
stainless steels, nitrogen-strengthened austenitic **M6:** 345
titanium and titanium alloys **M6:** 153, 446, 455-456
metal transfer **M6:** 455-456
shielding **M6:** 456
Gas nitrided steels **A9:** 227-228
Gas nitriding **A4:** 263, 387-409 **M4:** 191-221
advantages **A4:** 387, 403 **M4:** 191
aluminum-containing steels **A4:** 391-392
ammonia supply **A4:** 388, 398-400 **M4:** 204-205, 208
applications **A4:** 387-388, 394-395, 396, 397, 399, 402, 409, 410 **M4:** 192, 193, 194
bright nitriding **A4:** 403 **M4:** 212-213
case depth control **A4:** 388, 390-391, 400 **M4:** 197, 198-199, 201, 202
case depth, evaluation **A4:** 388, 390-391, 392, 394, 408-409 **M4:** 219-220
case hardening process **A7:** 455
chromium-containing low-alloy steels **A4:** 392-393, 394
chromium-containing toot steels **A4:** 392-393, 395
cleaning **A4:** 388, 409
compared to plasma (ion) nitriding **A4:** 420
dimensional changes **A4:** 393-395, 400 **M4:** 199-202, 203, 204, 205
dissociation rates **A4:** 389, 390, 395, 402 **M4:** 198
distortion **A4:** 393-395, 396, 398
double-stage **A4:** 388, 389, 394, 395, 409 **M4:** 192-194, 195, 196
equipment requirements **A4:** 395-398, 399 **M4:** 204, 205, 206
exhaust gas, analysis **A4:** 399-400, 407-408 **M4:** 219
Floe process **A4:** 388
furnace cooling **A4:** 389-390 **M4:** 198
furnace purging **A4:** 388-389, 390 **M4:** 196
hardness testing **A4:** 388, 390, 391, 392-393, 395, 408 **M4:** 219
ion nitriding *See* Gas nitriding, ion process
nitridable steels **A4:** 387 **M4:** 191, 192
pack nitriding **A4:** 403 **M4:** 213
pressure nitriding **A4:** 402-403 **M4:** 211-212
prior heat treatment **A4:** 387-388, 394-395 **M4:** 191-192
problems and causes **A4:** 400-401 **M4:** 206-208
procedures **A4:** 388, 401
reasons for **A4:** 387
safety precautions **A4:** 389, 390, 399, 400 **M4:** 205-206
selective **A4:** 401, 403 **M4:** 208-209
single-stage **A4:** 388, 389, 391-392, 398, 402, 409 **M4:** 192-194, 205
stainless steels **A4:** 424
stress relieving **A4:** 394, 396, 400
surface preparation of parts **A4:** 408 **M4:** 194-196
temperatures **A4:** 387-392, 398 **M4:** 191
to reduce erosive wear in die-casting dies **A18:** 632
tool steels **A4:** 724, 752 **A18:** 739
white layer, removal from surface **A4:** 409 **M4:** 220-221
Gas nitriding furnaces **A4:** 389-390, 395-398
bell-type **A4:** 390, 396-397 **M4:** 203, 208
box **A4:** 397-398 **M4:** 203
fixtures **A4:** 398 **M4:** 203-204

Gas nitriding furnaces (continued)
temperature control **A4:** 390, 396, 398 **M4:** 203
tube retorts **A4:** 398 **M4:** 203
vertical retort **A4:** 396 **M4:** 202-203, 207
Gas nitriding, ion process
advantages **A4:** 405 **M4:** 217
applications **A4:** 406 **M4:** 217-218
arc suppression **M4:** 214
case hardening **A4:** 405 **M4:** 213, 214, 217
compared to ammonia gas nitriding **A4:** 403, 405 **M4:** 212, 216-217
control units **A4:** 403 **M4:** 214
equipment **M4:** 210, 213, 214
fixture design **M4:** 214-216
glow discharge **A4:** 407 **M4:** 213
inspection **A4:** 408 **M4:** 216
process gas **A4:** 404 **M4:** 214
processing **A4:** 403-406 **M4:** 214
quality control **A4:** 408
steel, structure and properties **A4:** 403-405 **M4:** 211, 212, 216
Gas nitriding, stainless steels **A4:** 401-402
applications **A4:** 402 **M4:** 211
austenitic, ferritic **A4:** 401 **M4:** 209-210
hardenable alloys **A4:** 401 **M4:** 210
hardness gradients **A4:** 402 **M4:** 209, 211
nitriding cycles **A4:** 402 **M4:** 211
prior condition **A4:** 401 **M4:** 210
surface preparation **A4:** 401-402 **M4:** 210-211
Gas nitriding, vacuum nitrocarburizing **A4:** 405, 406-407
cold wall furnace **A4:** 407 **M4:** 218
glow-discharge **A4:** 407 **M4:** 215, 218
hardness levels **A4:** 407 **M4:** 218
Gas nuclear filters
powders used **A7:** 573
Gas permeability
for temperature-time control in polymer removal techniques **EM4:** 137
to analyze ceramic powder particle sizes **EM4:** 67
Gas permeametry
Gas permeation
explosivity **A7:** 196
Gas phase corrosion
coal- and oil-fired boilers **A13:** 995
solid waste boilers **A13:** 997
Gas pipelines
flux leakage method **A17:** 132
HSLA steels for **A1:** 416-417
Gas pocket
defined **A15:** 6
definition **A6:** 1210
in steel pipe **A17:** 565
Gas pockets
cast aluminum alloys **A12:** 406
Gas porosity
in aluminum alloys **A9:** 358 **A15:** 79, 85-86
in arc-welded aluminum alloys **A11:** 434
in austenitic manganese steel castings causes of **A9:** 238
as casting internal discontinuity **A11:** 354
in copper/copper alloys **A15:** 86
defined **A11:** 5 **A15:** 6
detection, in copper/copper alloy castings **A17:** 534-535
as die casting defect **A15:** 294
effects in iron castings **A11:** 356-357
in electron beam welds **A11:** 445
from inclusions **A15:** 88
inspection **A15:** 557-558
overcoming of **A15:** 87
radiographic appearance **A17:** 348-349
reactions **A15:** 82
in semisolid alloys **A15:** 337
and subsurface discontinuities **A11:** 120
and weld leakage **A7:** 431
Gas ports
electrogas welding **M6:** 240-241
Gas precipitation **A7:** 52, 55
Gas Producers Association specification 2140 (grade HD5)
special-duty propane **A4:** 312
Gas production
alloy steel corrosion in **A13:** 533-538
corrosion inhibitors for **A13:** 478-484
wells **A13:** 478-480

Gas purging, hydrogen
from aluminum alloys **A15:** 459-460
Gas purity
in HIP processing **A7:** 422
Gas pycnometer **A7:** 265
Gas quenching
applications in steel **M4:** 45, 46, 58-59
equipment, gas quenching unit **M4:** 58-59
hardening tool steel **M4:** 59
recirculation of gases **M4:** 58
Gas regulator
definition **A6:** 1210
Gas runs
as casting defects **A11:** 383
Gas scattering evaporation **A18:** 843
Gas shielded arc welding
definition **A6:** 1210 **M6:** 8
sheet metals **A6:** 399
stainless steels **A6:** 1196
Gas shielded arc welding (TIG or MIG) **A18:** 644
Gas sorption
to determine open porosity **EM4:** 581-582
Gas spray pattern **A7:** 25, 27
Gas stream
heating and circulation, autoclave system **EM1:** 645-647
pressurizing., systems, autoclave system **EM1:** 647
Gas stripping **A13:** 1244
Gas system
hot isostatic pressing equipment **A7:** 420-422
Gas torch
definition **A6:** 1210
Gas torch welding **A18:** 644
Gas tungsten arc cutting
definition **M6:** 8
Gas tungsten arc weld
failure origins **A11:** 415
incomplete fusion voids in **A11:** 93
pressure vessel **A11:** 438
Gas tungsten arc welding **A7:** 456 **M6:** 182-213
advantages **M6:** 183
aluminum alloys **A15:** 763
arc systems **M6:** 184
arc length **M6:** 184
welder skill **M6:** 184
base-metal thickness **M6:** 183
of cast irons **A15:** 523-524
cobalt-base alloys **A13:** 664
comparison to other welding processes **M6:** 204-205
consumable inserts **M6:** 202
joint design **M6:** 202
typical inserts **M6:** 201-202
welding procedure **M6:** 202
in containerized hot isostatic pressing **A7:** 431
controls for welding **M6:** 188
cooling water **M6:** 196
cost **M6:** 212-213
current for welding **M6:** 184-187
alternating current **M6:** 186
direct current electrode negative **M6:** 185
direct current electrode positive **M6:** 185-186
high-frequency current **M6:** 187
oxide removal **M6:** 186
prevention for rectification **M6:** 186
pulsed-current welding **M6:** 186-187
suitability for various metals **M6:** 185
definition **M6:** 8
electrodes **M6:** 190-195
consumption of tungsten electrodes **M6:** 193
contamination of tungsten electrodes **M6:** 193-194
current-carrying capacity **M6:** 192-193
effect of tip angle **M6:** 195
electrode extension **M6:** 194
end profile **M6:** 192
finish and surface condition **M6:** 191
Gas tungsten arc welding (continued)
heating of the electrode tip **M6:** 193
size **M6:** 191-192
thoriated tungsten electrodes **M6:** 194-195
tungsten contamination of weld pool **M6:** 193
zirconiated tungsten electrodes **M6:** 195
equipment **M6:** 187-197
power supplies **M6:** 187-188
filler metals **M6:** 200-201
forms, sizes, and use **M6:** 200-201
storage and preparation **M6:** 201
filler-wire feeders **M6:** 195
cold wire system **M6:** 195
hot wire system **M6:** 195
fixtures **M6:** 197
weld backing **M6:** 197
gas flow **M6:** 199-200
gas shielding **M6:** 196
hoses **M6:** 196
mixers **M6:** 196
hardfacing **M6:** 787
heat flow calculations **M6:** 31
joint design **M6:** 201-202
butt joints **M6:** 201
corner joints **M6:** 201
edge joints **M6:** 201-202
lap joints **M6:** 201
T-joints **M6:** 202
limitations **M6:** 183
machine circumferential welding **M6:** 210
mechanized welding **M6:** 208-210
equipment **M6:** 208-209
increased productivity **M6:** 209
simultaneous high-speed welding **M6:** 209-210
metals welded **M6:** 182-183
motion devices **M6:** 196-197
oscillators **M6:** 197
rotating positioners **M6:** 197
welding heat manipulators **M6:** 196-197
process adaptability **M6:** 212
process fundamentals **M6:** 183-184
arc extinction **M6:** 184
arc initiation **M6:** 183-184
electrode and filler metal positions **M6:** 184
production examples, low-carbon steel **M6:** 204-205
recommended proportions **M6:** 69-71
safety **M6:** 58, 213
eye protection **M6:** 213
protective clothing **M6:** 213
ventilation **M6:** 213
shielding gases **M6:** 197-200
argon versus helium **M6:** 197-198
argon-helium mixtures **M6:** 198
argon-hydrogen mixtures **M6:** 198-199
nitrogen, effect of **M6:** 199
oxygen-bearing argon mixtures **M6:** 199
shielding gases, purity of **M6:** 199
cylinder gas contamination **M6:** 199
removal of air from argon **M6:** 199
spot welding **M6:** 210-212
equipment **M6:** 210
penetration **M6:** 211
rimmed versus killed steels **M6:** 211-212
use of filler-metal wire **M6:** 211
supply and control of shielding gases **M6:** 199-200
drafts and air currents **M6:** 200
maintenance of adequate shielding **M6:** 200
supply source **M6:** 200
torches for welding **M6:** 188-190
air-cooled torches **M6:** 188
centering of the electrode **M6:** 190
collets **M6:** 190
electrode sizes **M6:** 190
gas lenses **M6:** 190
gas orifice sizes in collets **M6:** 190
shape **M6:** 189-190
size **M6:** 189
Gas tungsten arc welding (continued)
types of nozzles **M6:** 189
water-cooled torches **M6:** 188-189
underwater welding **M6:** 922
water hoses **M6:** 196
weld discontinuities **M6:** 830
weld overlaying **M6:** 807
weld soundness analysis of **A10:** 478-481
welding cables **M6:** 195-196
welding dissimilar metals **M6:** 207-208
welding of carbon and alloy steels **M6:** 203-204
welding of heat-resistant alloys **M6:** 205
welding of nonferrous metals **M6:** 205-206
aluminum alloys **M6:** 205-206
beryllium **M6:** 206
copper alloys **M6:** 206
magnesium alloys **M6:** 206
nickel alloys **M6:** 206
titanium alloys **M6:** 206
zirconium alloys **M6:** 206
welding of stainless steel **M6:** 205
work-metal preparation **M6:** 202-203
aluminum alloys **M6:** 203
carbon steels **M6:** 203
low-alloy steels **M6:** 203
nickel alloys **M6:** 203
stainless steel **M6:** 203
titanium alloys **M6:** 203
workpiece shape **M6:** 183
Gas tungsten arc welding (GTAW)
of Invar **A2:** 893
mechanically alloyed oxide alloys **A2:** 949
dispersion-strengthened (MA ODS) of nickel alloys **A2:** 445
refractory metals and alloys **A2:** 563-564
Gas tungsten arc welding of
alloy steels **M6:** 182, 203, 303-304
buttering **M6:** 304-305
forgings **M6:** 303-304
weld composition **M6:** 303
aluminum alloys **M6:** 182, 203, 205-206, 390-398
alternating current welding **M6:** 392-394
automatic welding **M6:** 393-395
comparison to gas metal arc welding **M6:** 397-398
direct current electrode negative welding **M6:** 394-395
direct current electrode positive welding **M6:** 394
filler metals **M6:** 391-392
manual welding **M6:** 392-393, 395
power supply and equipment **M6:** 390
shielding gases **M6:** 390-391
thicknesses **M6:** 390
weld backing **M6:** 392
weld soundness **M6:** 395-397
aluminum bronzes **M6:** 412-413
electrodes **M6:** 412
filler metal **M6:** 412
preheating **M6:** 412
shielding gas **M6:** 412
type of current **M6:** 412
welding conditions **M6:** 412
beryllium **M6:** 206
beryllium alloys **M6:** 182
beryllium copper, high-conductivity **M6:** 408
electrodes **M6:** 408
filler metal **M6:** 408
joint design **M6:** 408
preheating **M6:** 408
type of current **M6:** 408
welding conditions **M6:** 408
beryllium copper, high-strength **M6:** 408-409
filler metal **M6:** 408
joint design **M6:** 408
preheat and postweld heat treatments **M6:** 409

SUBJECTS OF THE INDEXED VOLUMES: **ASM Handbook** (designated by the letter "A"): **A1:** Properties and Selection: Irons, Steels, and High-Performance Alloys (1990); **A2:** Properties and Selection: Nonferrous Alloys and Special-Purpose Materials (1990); **A3:** Alloy Phase Diagrams (1992); **A4:** Heat Treating (1991); **A6:** Welding, Brazing, and Soldering (1993); **A7:** Powder Metallurgy (1984); **A8:** Mechanical Testing (1985); **A9:** Metallography and Microstructures (1985); **A10:** Materials Characterization (1986); **A11:** Failure Analysis and Prevention (1986); **A12:** Fractography (1987); **A13:** Corrosion (1987); **A14:** Forming and Forging (1988); **A15:** Casting (1988); **A16:** Machining (1989); **A17:** Nondestructive Testing and Quality Control (1989); **A18:** Friction, Lubrication, and Wear Technology (1992). **Metals Handbook, 9th Edition** (designated by the letter "M"): **M1:** Properties and Selection: Irons and Steels (1978); **M2:** Properties and Selection: Nonferrous Alloys and Pure Metals (1979); **M3:** Properties and Selection: Stainless Steels, Tool Materials and Special-Purpose Materials (1980); **M4:** Heat Treating (1981); **M5:** Surface Cleaning, Finishing, and Coating (1982); **M6:** Welding, Brazing, and Soldering (1983). **Engineered Materials Handbook** (designated by the letters "EM"): **EM1:** Composites (1987); **EM2:** Engineering Plastics (1988); **EM3:** Adhesives and Sealants (1990); **EM4:** Ceramics and Glasses (1991); **Electronic Materials Handbook** (designated by the letters "EL"): **EL1:** Packaging (1989).

Gas tungsten arc welding of (continued)
welding conditions ... **M6:** 408
carbon steels ... **M6:** 182, 203
cast irons ... **M6:** 309
coated steels ... **M6:** 183
copper and copper alloys ... **M6:** 182, 206, 400, 404-415
cuprous oxide effects ... **M6:** 405
electrodes ... **M6:** 404-405
filler metal ... **M6:** 404-406
joint designs ... **M6:** 403, 406
minimizing distortion ... **M6:** 406-407
preheating ... **M6:** 406
pulsed-current welding ... **M6:** 426
shielding gases ... **M6:** 405
welding conditions ... **M6:** 403
welding technique ... **M6:** 406
copper nickels ... **M6:** 414
filler metals ... **M6:** 414
shielding gas ... **M6:** 414
welding conditions ... **M6:** 414
copper-zinc alloys ... **M6:** 409-410
filler metals ... **M6:** 409
shielding gas ... **M6:** 409
welding without filler metals ... **M6:** 410
coppers with dissimilar metals ... **M6:** 414-415
preheating ... **M6:** 415
welding with filler metal ... **M6:** 414-415
ductile iron ... **M6:** 315
gray iron ... **M6:** 315
hafnium ... **M6:** 458
hardenable carbon steels ... **M6:** 269-270
heat-resistant alloys ... **M6:** 182, 205
cobalt-based alloys ... **M6:** 368-370
iron-nickel-chromium and iron-chromium-nickel alloys ... **M6:** 365-366
nickel-based alloys ... **M6:** 358-362
lead ... **M6:** 183
low-alloy steels ... **M6:** 203
magnesium alloys ... **M6:** 182, 206, 429-432
molybdenum ... **M6:** 464
nickel alloys ... **M6:** 182, 203, 206, 438-439
nickel silver ... **M6:** 409
niobium ... **M6:** 459-460
nonferrous metals ... **M6:** 205-206
phosphor bronzes ... **M6:** 410-412
filler metal ... **M6:** 410
preheating ... **M6:** 411
shielding gas ... **M6:** 410
welding conditions ... **M6:** 410
welding without filler metal ... **M6:** 410
refractory metals ... **M6:** 182, 205
silicon bronzes ... **M6:** 413-414
filler metals ... **M6:** 414
joint design ... **M6:** 414
preheating ... **M6:** 414
welding conditions ... **M6:** 413
stainless steels ... **M6:** 182, 203, 205
stainless steels, austenitic ... **M6:** 333-342
automatic precision welding ... **M6:** 341-342
circumferential welding ... **M6:** 340
consumable inserts ... **M6:** 335-336
current ... **M6:** 333-334
filler metals ... **M6:** 334-335
heat restriction ... **M6:** 341
joint design ... **M6:** 334, 337-338
longitudinal welding ... **M6:** 340
melt-through welding ... **M6:** 342
root-pass welds ... **M6:** 336-337
shielding gas ... **M6:** 334
spot welding ... **M6:** 339
welding positions ... **M6:** 333
work metal characteristics ... **M6:** 333
stainless steels, ferritic ... **M6:** 347-348
stainless steels, martensitic ... **M6:** 349
stainless steels, nitrogen-strengthened austenitic ... **M6:** 345
structural steels high-strength ... **M6:** 203-204
tantalum ... **M6:** 460-461
titanium and titanium alloys ... **M6:** 182, 203, 206, 446, 453, 455
arc length ... **M6:** 455
electrodes ... **M6:** 454
equipment ... **M6:** 453-454
fixtures ... **M6:** 454
hot wire process ... **M6:** 455
preheating ... **M6:** 454

Gas tungsten arc welding of (continued)
shielding ... **M6:** 454
tack welding ... **M6:** 454-455
welding conditions ... **M6:** 455
tungsten ... **M6:** 464
zinc ... **M6:** 183
zirconium alloys ... **M6:** 182, 206, 458
Gas turbine engines
ceramic applications ... **EM4:** 960
as NDE reliability case study ... **A17:** 681-684
Gas turbine engines, design practices for structural ceramics *See* Design practices for structural ceramics in gas turbine engines
Gas turbines *See* Advanced gas turbines
alloy coatings for ... **A13:** 458
Gas welding
definition ... **A6:** 1210
of thermocouple thermometers ... **A2:** 871
Gas(es) *See also* Degassing; Gas porosity; Gases in metals; Porosity; specific gases
in aluminum ... **A15:** 85-86
in cast iron ... **A15:** 82-85
in computed tomography (CT) ... **A17:** 370
content, magnesium alloys ... **A15:** 802
in copper alloys ... **A15:** 86, 466
for core curing ... **A15:** 240
core venting of ... **A15:** 241
desorption, as leakage ... **A17:** 58
detectors ... **A17:** 61-65, 369
diffusion of ... **A15:** 82-83
displacement, die casting ... **A15:** 291
entrapment, in gray iron castings ... **A17:** 531
entrapment, in semisolid materials ... **A15:** 328
evacuation, in FM process ... **A15:** 38
evolution, during solidification ... **A15:** 87
flow pattern, ladle ... **A15:** 432
hydrogen, measurement of ... **A15:** 457-459
ideal, equation of state of ... **A17:** 58
leak testing of ... **A17:** 57-70
-liquid reactions, kinetics of ... **A15:** 82-83
mean free path ... **A17:** 59
mean free path lengths ... **A17:** 59
in metals ... **A15:** 82-87
and mold permeability ... **A15:** 209
producer, effect on furnaces ... **A15:** 32
purging ... **A15:** 459-460
purity, hot isostatic pressing effects ... **A15:** 542
reactive, aluminum refining with ... **A15:** 80
solubility, in copper alloys ... **A15:** 464-465
systems, at pressure, leak testing methods ... **A17:** 59
testing, in copper alloys ... **A15:** 465-466
top, analysis of ... **A15:** 384
types of flow ... **A17:** 59
use, cold box processes ... **A15:** 238
velocity measurement, laser ... **A17:** 16-17
Gas-aspirating atomization ... **A7:** 125, 127
Gas-atomized cobalt-based hardfacing powders ... **A7:** 146
Gas-atomized copper powders ... **A7:** 117
particle size measurement ... **A7:** 223, 224
properties ... **A7:** 119
Gas-atomized powders
low green strength ... **A7:** 303
microstructural characteristics ... **A7:** 38-39
Gas-atomized specialty powders
rigid tool compaction of ... **A7:** 322
Gas-atomized stainless steel powders ... **A7:** 101
Gas-atomized tool steels
mechanical properties ... **A7:** 479
nominal compositions ... **A7:** 103
Gas-cooled nuclear reactors ... **A7:** 664
Gas-filled detectors
for x-ray spectroscopy ... **A10:** 88
Gas-lift wells ... **A13:** 1248
Gas-lubricated bearings, friction and wear of ... **A18:** 522-532
advantages ... **A18:** 522
applications ... **A18:** 522, 531, 532
clearance modulus, definition ... **A18:** 523-524
compliant-surface bearings ... **A18:** 530-531
advantages ... **A18:** 530
applications ... **A18:** 531
bending-dominated continuous foil bearings ... **A18:** 531
bending-dominated segmented foil bearings ... **A18:** 530-531

Gas-lubricated bearings, friction and wear of (continued)
design analysis ... **A18:** 531
hybrid bearings ... **A18:** 531-532
modified bending-dominated continuous foil bearings ... **A18:** 531
pressurized-membrane bearings ... **A18:** 531
surface coatings ... **A18:** 532
tension-dominated foil bearings ... **A18:** 530
compressibility numbers ... **A18:** 522-523, 525, 526, 527, 528, 529
development of gas lubrication technology ... **A18:** 522
disadvantages ... **A18:** 522
eccentricity ratio, definition ... **A18:** 523-524, 527, 528, 532
friction in gas-lubricated journal bearings ... **A18:** 526
half-frequency whirl ... **A18:** 526-527
helical-grooved journal bearings ... **A18:** 528
hydrostatic gas-lubricated bearings ... **A18:** 528-529
hydrostatic journal bearings ... **A18:** 531-532
journal bearings ... **A18:** 523, 525-526
material for gas-lubricated bearings ... **A18:** 432
mean free path effect ... **A18:** 524-525
multiple-pressure sources ... **A18:** 529
pivoted-pad journal bearing ... **A18:** 527, 528
pneumatic hammer ... **A18:** 529
porous bearings ... **A18:** 529-530
Rayleigh step bearings ... **A18:** 523, 525
surface coatings ... **A18:** 532
synchronous whirl ... **A18:** 11, 526
three-sector journal bearing ... **A18:** 527-528
tilting-pad bearings ... **A18:** 523, 524
applications ... **A18:** 607
tilting-pad journal bearings ... **A18:** 527, 528
pivot circle clearance ... **A18:** 527
pivot design ... **A18:** 527
yaw stability ... **A18:** 527
Gas-metal arc welding (GMAW) ... **A6:** 124-125, 180-185, 647, 652, 653, 654, 655, 657
advantages ... **A6:** 180
all-weld-metal chemical compositions for martensitic stainless steel filler metals ... **A6:** 439
aluminum alloys ... **A6:** 180, 722-723, 724, 726, 729, 730, 731-735, 737-738, 739
aluminum bronzes ... **A6:** 754, 765
aluminum metal-matrix composites ... **A6:** 555, 556, 557
aluminum-lithium alloys ... **A6:** 551, 552
applications ... **A6:** 180, 865
automotive ... **A6:** 393, 394, 395
railroad equipment ... **A6:** 396
sheet metals ... **A6:** 398
shipbuilding ... **A6:** 384
argon in the shielding gas ... **A6:** 65
carbon dioxide in the shielding gas ... **A6:** 65
carbon steel ... **A6:** 180
cast irons ... **A6:** 716
copper-base electrodes ... **A6:** 719
nickel-base electrodes ... **A6:** 719
steel electrodes ... **A6:** 719
cobalt-base corrosion-resistant alloys ... **A6:** 598
consumables ... **A6:** 185
copper ... **A6:** 759-760, 761, 762
copper alloys ... **A6:** 180, 752, 754-755, 762, 763, 765, 768
dissimilar metals ... **A6:** 769-770
copper-nickel alloys ... **A6:** 754, 768
cross sections of an air-cooled gun ... **A6:** 183
cryogenic service ... **A6:** 1017
definition ... **A6:** 180, 1210
deposition rates ... **A6:** 180
disadvantages ... **A6:** 394
dissimilar metal joining ... **A6:** 824, 828
duplex stainless steels ... **A6:** 480
electrodes ... **A6:** 182, 185, 1143, 1144
for carbon steels ... **A6:** 655, 657
for cast irons ... **A6:** 718-719, 720
for coppers ... **A6:** 759
ferritic stainless steels ... **A6:** 446
for stainless steels ... **A6:** 705, 706, 707
for welding dissimilar copper alloys ... **A6:** 770
equipment ... **A6:** 183-185
electrode feed unit ... **A6:** 184
electrode source ... **A6:** 184

Gas-metal arc welding (GMAW) (continued)
regulated shielding gas supply **A6:** 184-185
welding control mechanism **A6:** 184
welding gun **A6:** 183-184
welding power source **A6:** 184
ferritic stainless steels **A6:** 446
filler metals **A6:** 194, 446
filler metals for coppers **A6:** 759
fluid flow phenomena **A6:** 19, 23-24
fume generation rate dependence on shielding gas composition **A6:** 68
fusion zone **A6:** 183
hardfacing alloy consumable form **A6:** 796
hardfacing alloys **A6:** 797, 798, 800, 801, 803
heat sources **A6:** 1143-1144
heat-treatable aluminum alloys **A6:** 528, 531-532
heat-treatable low-alloy steels **A6:** 669, 670, 671
limitations **A6:** 180
low-alloy metals for pressure vessels and piping **A6:** 668
low-alloy steels **A6:** 662, 663, 664, 666, 668, 669, 670, 671, 673, 674, 676
for pressure vessels and piping **A6:** 667
high-strength low-alloy steel **A6:** 180
quench and tempered structural steels **A6:** 664, 666
structural steels **A6:** 663, 664
low-carbon steels **A6:** 10, 17
magnesium alloys **A6:** 772, 776-777
matching filler metal specifications **A6:** 394
metallurgical discontinuities **A6:** 1073
nickel alloys **A6:** 180, 588, 740, 741, 742, 743-745, 746, 749, 750, 751
nickel alloys to dissimilar alloys **A6:** 751
nickel-base corrosion-resistant alloys containing molybdenum **A6:** 594
nickel-chromium alloys **A6:** 587, 588
nickel-chromium-iron alloys **A6:** 587, 588
nickel-copper alloys **A6:** 587, 588
non-heat-treatable aluminum alloys **A6:** 539
oxide-dispersion-strengthened materials **A6:** 1038, 1039
oxygen in the shielding gas **A6:** 65
parameters for type 403 martensitic stainless steel **A6:** 439
phosphor bronzes **A6:** 764
postweld heat treatment **A6:** 762
power source selected **A6:** 37, 38, 39, 40
precipitation-hardening stainless steels **A6:** 483, 487, 489
pressure vessel manufacture **A6:** 379
procedure development **A6:** 27-29
process fundamentals **A6:** 180-183
globular transfer **A6:** 180, 181, 182
metal transfer mechanisms **A6:** 180, 181-182
principles of operation **A6:** 180
process variables **A6:** 182-183
short-circuiting transfer **A6:** 180-181, 182
spray transfer **A6:** 180, 181-182
process selection guidelines for arc welding **A6:** 653
rating gas as a function of weld parameters and characteristics **A6:** 1104
for repair of high-carbon steels **A6:** 1105
for repair welding **A6:** 864
safety **A6:** 185
safety precautions **A6:** 1192-1193
schematic of process **A6:** 181
sensor systems **A6:** 1063, 1064
shielding gases **A6:** 64, 65, 66-67, 181, 183, 185, 662, 705, 706, 707
hardfacing alloys **A6:** 803
selection recommendations **A6:** 66
for silicon bronzes **A6:** 766
silicon bronzes **A6:** 754, 766
stainless steel casting alloys **A6:** 496
stainless steels **A6:** 180, 677, 680, 688, 693, 694, 697, 698, 699, 705, 706, 707
steel weldment soundness **A6:** 408, 409, 413

Gas-metal arc welding (GMAW) (continued)
suggested viewing filter plates **A6:** 1191
synergic pulsed **A6:** 41
"temper-bead" procedures **A6:** 81
titanium alloys **A6:** 783, 784, 786
to prevent hydrogen-induced cold cracking **A6:** 436
tool and die steels **A6:** 674, 676
transfer of heat and mass to the base metal **A6:** 25-29
droplet transfer modes **A6:** 27-29
droplet velocity and temperature **A6:** 27, 28
electrical and acoustic signals **A6:** 27, 28
heat transfer **A6:** 25-27
mass transfer **A6:** 25-27, 28
weld reinforcement **A6:** 27
tungsten inclusions **A6:** 1074
ultrahigh-strength low-alloy steels **A6:** 673
underwater welding **A6:** 1010
dry welding **A6:** 179
vs. EBW **A6:** 858
vs. flux-cored arc welding **A6:** 186, 187
vs. gas-tungsten arc welding **A6:** 190
vs. plasma-MIG welding **A6:** 223, 224, 225
vs. shielded metal arc welding **A6:** 180
vs. submerged arc welding **A6:** 180
zirconium alloys **A6:** 787
Gas-metal arc welding process **A18:** 653
Gas-metal arc wire electrode process **A18:** 653
Gas-metal thermochemistry
technology of **A7:** 295
Gas-metal-arc welding (GMAW)
of Invar **A2:** 893
of nickel alloys **A2:** 445
Gas-nitrided drive-gear assembly
deformation of **A11:** 142-143
Gas-phase corrosion
Raman analysis **A10:** 135
Gas-plated products
nickel tetracarbonyl **A7:** 137
Gas-pressure sintering
and HIP **EM4:** 199
pressure densification **EM4:** 299-300
conditions **EM4:** 298
parameters **EM4:** 298
pressure **EM4:** 301
temperature **EM4:** 301
to densify products of SHS process **EM4:** 320
Gas-solid reduction process
tungsten **A7:** 153, 155
Gas-sorption methods
aluminum powders **A7:** 129
Gas-tungsten arc cutting (GTAC)
definition **A6:** 1210
Gas-tungsten arc welding (GTAW) **A6:** 124-125, 190-194
12Cr-Mo-0.3V (HT9) **A6:** 435, 436
advanced titanium-base alloys **A6:** 526
advantages **A6:** 190-191
all-weld-metal chemical compositions for martensitic stainless steel filler metals **A6:** 439
alloy steels **A6:** 190
aluminum **A6:** 190, 191
aluminum alloys **A6:** 725, 729, 730, 731-735, 736, 737, 738, 739, 871, 872
castings **A6:** 192
aluminum bronzes **A6:** 754, 764-765
aluminum metal-matrix composites **A6:** 555-556
aluminum-lithium alloys **A6:** 551, 552
applications **A6:** 19, 190, 194, 394, 865
sheet metals **A6:** 398, 399
arc oscillation **A6:** 191
arc physics of **A6:** 30-35
anode **A6:** 31
arc column **A6:** 21-32, 34
arc efficiency **A6:** 31
arc length effect **A6:** 32
cathode tip shape effect **A6:** 32, 33

Gas-tungsten arc welding (GTAW) (continued)
definition of welding arc **A6:** 30
electrode regions and arc column **A6:** 30-35
electron and thermal contributions to heat transfer **A6:** 31
gas shielding **A6:** 30
relative heat transfer contributions to workpiece **A6:** 31
shielding gas composition effect **A6:** 32
austenitic stainless steels **A6:** 462-463, 464, 465, 466, 468, 1018
automatic welding **A6:** 191, 193, 194
beryllium **A6:** 192
beryllium coppers **A6:** 754
brass **A6:** 192
and brazing of stainless steels **A6:** 919
carbon steels **A6:** 190, 652, 653, 654, 655-656, 658
cast irons **A6:** 192, 716, 720
copper **A6:** 190, 756-759, 760-761, 762
copper alloys **A6:** 192, 752, 754, 755, 756, 762, 763, 764-765, 766, 767, 768
dissimilar metals **A6:** 769
copper-nickel alloys **A6:** 754, 766-768
cryogenic service **A6:** 1017, 1018
deoxidized copper **A6:** 192
depth-to-width ratio (*d/w*) **A6:** 20-21
development **A6:** 190
dispersion-strengthened aluminum alloys **A6:** 543-544, 545
dissimilar metal joining **A6:** 824, 827, 828
dissimilar metals **A6:** 190
distortion **A6:** 191
dopants **A6:** 22
duplex stainless steels **A6:** 476, 479, 480
electrodes **A6:** 191, 786
for copper alloys **A6:** 754, 756
for stainless steels **A6:** 705
electrolytic tough pitch copper **A6:** 758-759
equipment **A6:** 191
evaluation
by weld penetration tests **A6:** 608-609
using the Sigmajig test **A6:** 608
experimental observations **A6:** 20-21
ferritic stainless steels **A6:** 445-446, 447, 448
filler metals **A6:** 193-194, 658
fixtures **A6:** 786
fluid flow phenomena **A6:** 19-22
gas shielding **A6:** 30
hardfacing alloy consumable form **A6:** 796
hardfacing alloys **A6:** 798, 799, 800, 803-805, 807
heat flow in fusion welding **A6:** 8
heat sources **A6:** 1142-1143
heat-affected zone **A6:** 868
heat-resistant alloys **A6:** 192
heat-treatable aluminum alloys **A6:** 528, 531, 532, 533, 534
heat-treatable low-alloy steels **A6:** 669, 670, 671
high current effects **A6:** 22
high-carbon steel **A6:** 192
high-strength low-alloy quench and tempered structural steels **A6:** 664, 666
structural steels **A6:** 664
inert gases used **A6:** 190
interactions **A6:** 21-22
Laves phase alloys **A6:** 795
limitations **A6:** 190-191
low-alloy metals for pressure vessels and piping **A6:** 668
low-alloy steels **A6:** 662, 664, 666, 668, 669, 670, 671, 673, 674, 676
for pressure vessels and piping **A6:** 667
low-carbon steel **A6:** 192
with low-temperature solid-state welding **A6:** 301
magnesium **A6:** 190, 191
magnesium alloys **A6:** 772, 777-779, 780-781
castings **A6:** 192
manual welding **A6:** 194
mechanized welding **A6:** 191, 194

SUBJECTS OF THE INDEXED VOLUMES: ASM Handbook (designated by the letter "A"): **A1:** Properties and Selection: Irons, Steels, and High-Performance Alloys (1990); **A2:** Properties and Selection: Nonferrous Alloys and Special-Purpose Materials (1990); **A3:** Alloy Phase Diagrams (1992); **A4:** Heat Treating (1991); **A6:** Welding, Brazing, and Soldering (1993); **A7:** Powder Metallurgy (1984); **A8:** Mechanical Testing (1985); **A9:** Metallography and Microstructures (1985); **A10:** Materials Characterization (1986); **A11:** Failure Analysis and Prevention (1986); **A12:** Fractography (1987); **A13:** Corrosion (1987); **A14:** Forming and Forging (1988); **A15:** Casting (1988); **A16:** Machining (1989); **A17:** Nondestructive Testing and Quality Control (1989); **A18:** Friction, Lubrication, and Wear Technology (1992). **Metals Handbook, 9th Edition** (designated by the letter "M"): **M1:** Properties and Selection: Irons and Steels (1978); **M2:** Properties and Selection: Nonferrous Alloys and Pure Metals (1979); **M3:** Properties and Selection: Stainless Steels, Tool Materials and Special-Purpose Materials (1980); **M4:** Heat Treating (1981); **M5:** Surface Cleaning, Finishing, and Coating (1982); **M6:** Welding, Brazing, and Soldering (1983). **Engineered Materials Handbook** (designated by the letters "EM"): **EM1:** Composites (1987); **EM2:** Engineering Plastics (1988); **EM3:** Adhesives and Sealants (1990); **EM4:** Ceramics and Glasses (1991); **Electronic Materials Handbook** (designated by the letters "EL"): **EL1:** Packaging (1989).

Gas-tungsten arc welding (GTAW) (continued)
melting temperature necessary ... **A6:** 190
metallurgical discontinuities ... **A6:** 1073
molybdenum ... **A6:** 193
molybdenum alloys ... **A6:** 581
narrow groove welding ... **A6:** 194
neural network system ... **A6:** 1060
nickel alloys ... **A6:** 588, 704, 741, 742-743, 748, 749, 751
to dissimilar alloys ... **A6:** 751
nickel-base corrosion-resistant alloys containing molybdenum ... **A6:** 594
nickel-chromium alloys ... **A6:** 588
nickel-chromium-iron alloys ... **A6:** 588
nickel-copper alloys ... **A6:** 588
niobium ... **A6:** 871
niobium alloys ... **A6:** 581
numerical simulations ... **A6:** 21
oxide-dispersion-strengthened materials ... **A6:** 1038, 1039
oxygen-free copper ... **A6:** 758
personnel ... **A6:** 190
phosphor bronzes ... **A6:** 763-764
power sources ... **A6:** 36
selection ... **A6:** 36-37, 38, 39, 40
power supplies ... **A6:** 190-191
precipitation-hardening stainless steels ... **A6:** 483, 484, 487, 488, 489, 491-492
pressure vessel manufacture ... **A6:** 379
primary driving forces (four) ... **A6:** 19
process parameters ... **A6:** 191-194
process selection guidelines for arc welding ... **A6:** 653
process variations ... **A6:** 194
rating as a function of weld parameters and characteristics ... **A6:** 1104
reactive materials ... **A6:** 190
refractory metals ... **A6:** 192
for repair welding ... **A6:** 864
of high-carbon steels ... **A6:** 1105
rhenium alloys ... **A6:** 581
safety precautions ... **A6:** 1192-1193, 1200, 1202
sensor systems ... **A6:** 1063, 1064
shielding gases ... **A6:** 65, 193, 488-489, 662, 703, 704, 756, 783, 786
for coppers ... **A6:** 758
dispersion-strengthened aluminum alloys ... **A6:** 543
for hardfacing alloys ... **A6:** 803-804
selection ... **A6:** 67-68
for space and low-gravity environments ... **A6:** 1022
silicon bronzes ... **A6:** 192, 754, 766
silver ... **A6:** 192
in space and low-gravity environments ... **A6:** 1021, 1022
spot Varestraint test ... **A6:** 607-608
stainless steel casting alloys ... **A6:** 496
stainless steels ... **A6:** 190, 191, 192, 677, 679, 686, 688, 693, 697, 698, 699, 703-705, 870
strategies for controlling poor and variable penetration ... **A6:** 22
suggested viewing filter plates ... **A6:** 1191
surface-tension-drive fluid flow model ... **A6:** 19-20
tantalum ... **A6:** 190, 193
tantalum alloys ... **A6:** 580, 581
temper-bead procedure for HAZ in multipass weldments ... **A6:** 81
temperature measurement, validation strategies ... **A6:** 1149-1150
titanium ... **A6:** 190, 193
titanium alloys ... **A6:** 192, 512, 513, 514, 515, 516, 518, 520, 521, 522, 783-784, 785-786
to prevent hydrogen-induced cold cracking ... **A6:** 436
to solve problems in joining thin sections by oxyfuel gas welding ... **A6:** 288
tool and die steels ... **A6:** 674, 676
torch construction ... **A6:** 191
trace element impurities effect on weld penetration ... **A6:** 20
tungsten alloys ... **A6:** 580, 581
tungsten inclusions ... **A6:** 1074
ultrahigh-strength low-alloy steels ... **A6:** 673, 674
underwater welding ... **A6:** 1010, 1014
dry welding ... **A6:** 179
Varestraint hot crack testing ... **A6:** 603, 606-607

Gas-tungsten arc welding (GTAW) (continued)
vs. electron-beam welding (nonvacuum) ... **A6:** 857
vs. gas-metal arc welding ... **A6:** 190
weld microstructures ... **A6:** 51
weld shape variability ... **A6:** 19
weld size variation and HAZ ... **A6:** 868, 869
welding current ... **A6:** 191-193
wire feed systems ... **A6:** 191
zirconium alloys ... **A6:** 787

Gas-turbine components
elevated-temperature failures in ... **A11:** 282-287
tests for ... **A11:** 278-281

Gas/metal environments
SERS studies in ... **A10:** 136, 137

Gaseous atmosphere
impact milling in ... **A7:** 57

Gaseous contaminants
effects and removal ... **A7:** 178, 180-181
measurement ... **A7:** 181

Gaseous contamination
nickel plating baths ... **M5:** 210

Gaseous corrosion *See also* Gases
and aqueous corrosion, compared ... **A13:** 61
defined ... **A13:** 7
and oxidation ... **A13:** 17

Gaseous environments
effect on fatigue ... **A12:** 36-41
and explosivity ... **A7:** 194

Gaseous feedstocks
physical properties ... **A7:** 341

Gaseous ferritic nitrocarburizing *See* Ferritic nitrocarburizing, gaseous ... **A7:** 455

Gaseous hydrogen
cracking from ... **A11:** 247
titanium/titanium alloy corrosion by ... **A13:** 685

Gaseous nickel tetracarbonyl ... **A7:** 137

Gaseous nitrocarburizing
furnace ... **A7:** 455

Gaseous reduction ... **A7:** 5

Gaseous uranium hexafluoride
lubricant for pumping equipment bearings ... **A18:** 522

Gases *See also* Gaseous corrosion; Gases, characterization of; specific gases
absorption, titanium alloys ... **A14:** 838
acid, as gas samples ... **A10:** 152
alloy system and ... **A13:** 64
analytic methods adapted for ... **A10:** 8
at elevated temperatures, corrosion testing in ... **A13:** 226
carbon dioxide and sulfur dioxide, use in high-temperature combustion ... **A10:** 221-225
characterized ... **A10:** 1
chlorine, cast irons in ... **A13:** 570
copper/copper alloy resistance ... **A13:** 632
corrosion in ... **A13:** 17, 61-76
dissolved, in water ... **A13:** 489
effect on fatigue ... **A12:** 36-41
electroplated chromium resistances to ... **A13:** 875
filters used with ... **A10:** 94
fuel, properties of ... **A14:** 723
gold corrosion in ... **A13:** 798
helium, under pressure ... **A12:** 414
high temperature, hydrogen fluoride/hydrofluoric acid corrosion ... **A13:** 1170
high-temperature mixed ... **A13:** 1172
hydrogen, titanium embrittlement by ... **A12:** 23, 32
image, low ionization, for field ion microscopy ... **A10:** 587
inclusions in glasses, Raman analysis ... **A10:** 131
inert, for FIM operations ... **A10:** 587
inert, tantalum resistance ... **A13:** 731
infrared absorbances by, PAS analysis of ... **A10:** 115
infrared analysis of ... **A10:** 113, 115
inorganic, analytic methods for ... **A10:** 8
ionization potentials and imaging fields for selected ... **A10:** 586
kinetics of corrosion in ... **A13:** 65-70
magnesium/magnesium alloys in ... **A13:** 743
methane, along grain boundary ... **A12:** 349
molecules, three-dimensional structure of ... **A10:** 393
natural, analytic methods for ... **A10:** 11
niobium corrosion in ... **A13:** 722

Gases (continued)
organic, analytic methods for ... **A10:** 11
platinum corrosion in ... **A13:** 803
potentiometric gas-sensing electrodes for ... **A10:** 183-185
processes of, analytic methods for ... **A10:** 8, 11
properties, oxide scales ... **A13:** 70-76
RDF determination of interatomic distance distributions and coordination numbers of ... **A10:** 393
reactive, as fatigue environment ... **A12:** 35
regulators for ... **A14:** 725
removal, and chemical equilibrium ... **A10:** 163
samples of ... **A10:** 16
saturation, effect on pollution control ... **A13:** 1367
silver corrosion in ... **A13:** 798
solubility in ocean water ... **A13:** 1256
sour ... **A13:** 1257
stainless steel corrosion ... **A13:** 559
sweet ... **A13:** 1256-1257
thermal conductivity of ... **A10:** 223, 230
thermodynamics of high-temperature corrosion in ... **A13:** 61-64
titanium/titanium alloy resistance ... **A13:** 681
transmission, alloy steel corrosion ... **A13:** 536-538
types, for cutting ... **A14:** 722-724
x-ray detector for, early ... **A10:** 83
zinc/zinc alloys and coatings in ... **A13:** 763
zirconium/zirconium alloy corrosion ... **A13:** 717

Gases, characterization of *See also* Gases
extended x-ray absorption fine structure ... **A10:** 407-419
gas analysis by mass spectrometry ... **A10:** 151-157
gas chromatography/mass spectrometry ... **A10:** 639-648
infrared spectroscopy ... **A10:** 109-125
molecular fluorescence spectrometry ... **A10:** 72-81
Raman spectroscopy ... **A10:** 126-138
ultraviolet/visible absorption spectroscopy ... **A10:** 60-71

Gases, combustion
heat transfer from ... **A11:** 628

Gases, copper alloys
corrosion rate ... **M2:** 480-483

Gases, dissolved *See* Dissolved gases

Gases in metals
aluminum ... **A15:** 85-86
cast iron ... **A15:** 82-85
copper and copper alloys ... **A15:** 86
gas porosity ... **A15:** 87

Gases, liquefied
boiling points ... **M3:** 721-722

Gasification, high-temperature
of coke ... **A15:** 53

Gasket
defined ... **A18:** 10

Gasket, copper
for fatigue test chamber ... **A8:** 412

Gasketing
design ... **EM3:** 53-54

Gaskets ... **EM1:** 168, 720
of indium ... **A2:** 752

Gasless delay elements
compositions ... **A7:** 2, 604

Gasoline
copper/copper alloy resistance ... **A13:** 631, 635

Gasoline engines, design practices for structural ceramics *See* Design practices for structural ceramics in gasoline engines

Gasoline synthesis catalysts
powders used ... **A7:** 572

Gassing
defined ... **A15:** 6

Gastroenteritis
as copper toxic reactions ... **A7:** 205

Gate
arrays, defined ... **EL1:** 167-168, 1145
broken casting at ... **A11:** 386
counts per chip, growth ... **EL1:** 416
defined ... **EM2:** 20
-equivalent circuit, defined ... **EL1:** 1145
formation, MOS IC fabrication ... **EL1:** 198
-level devices, development ... **EL1:** 2
-to-gate interconnections, defined ... **EL1:** 12

Gate mark
defined ... **EM2:** 20

Gate(s) *See also* Gated pattern; Gating systems; Mold cavity; Pattern; Rigging; Risers
copper alloy casting ... **A15:** 777-778
defined ... **A15:** 6, 204
and flash grinding, Alnico alloys ... **A15:** 737
as pattern feature ... **A15:** 192
Gated desorption images
IAP analysis for ... **A10:** 596, (M-601
Gated pattern *See also* Gates; Gating system; Pattern
defined ... **A15:** 6
of early molders ... **A15:** 28
Gathering (pipe)lines
high-pressure long- distance ... **A11:** 695
Gating *See also* Gate(s); Gating systems
automated ... **A15:** 36
C-scans, pulse-echo ultrasonic inspection ... **A17:** 243
defined ... **A15:** 192
design ... **A15:** 285, 589-597
of die castings ... **A15:** 755
in directional solidified furnaces ... **A15:** 319, 321
effect on risers ... **A11:** 354
magnesium alloys ... **A15:** 805-806
methods, for inclusion control ... **A15:** 91
and mold life ... **A15:** 281
principles ... **A15:** 754
removal, economic ... **A15:** 589
size, surface finish effects ... **A15:** 285
ultrasonic inspection circuits ... **A17:** 253
Gating design
ceramic filters in ... **A15:** 594-597
fluid flow principles ... **A15:** 590-592
pressurized vs unpressurized systems ... **A15:** 593-594
runner and ingate ... **A15:** 592-593
variables ... **A15:** 589-590
vertical vs horizontal gating systems ... **A15:** 594
Gating systems *See also* Gates; Gating
air venting, die casting process ... **A15:** 291
aluminum alloys ... **A15:** 754-755
bottom gating ... **A15:** 279
copper alloys ... **A15:** 776-778
defined ... **A15:** 6
die casting ... **A15:** 288-292
directional/monocrystal solidification ... **A15:** 321
ductile iron ... **A15:** 651
feeding of shrinkage, die casting ... **A15:** 291-292
gray iron ... **A15:** 640
knife and kiss ... **A15:** 778
metal flow, computer modeling of ... **A15:** 857
metal injection, die casting ... **A15:** 288-291
misruns ... **A15:** 279
nickel alloys ... **A15:** 821-822
permanent mold casting ... **A15:** 278-279
plain carbon steels ... **A15:** 711
screens and sieves for inclusion control ... **A15:** 90
side gating ... **A15:** 279
top gating ... **A15:** 279
vertical centrifugal casting ... **A15:** 300-301
Gating techniques
for ultrasonic C-scans ... **EM2:** 840-845
Gator gard
ceramic coatings for adiabatic diesel engines ... **EM4:** 992
Gatorizing *See also* Isothermal forging ... **A1:** 974
M3: 215-216
of jet engine disks ... **A14:** 18
of superalloys ... **A7:** 468
Gauss
abbreviation ... **A8:** 724
Gauss-meter probe
in fatigue fracture analysis ... **A11:** 129
Gaussian absorption curve
ESR spectrum ... **A10:** 259
Gaussian beam diameter (dg) ... **A6:** 265
Gaussian distribution ... **A18:** 296, 297 **EL1:** 194-195
as room-temperature strength observations ... **A8:** 663
GC/MS *See* Gas chromatographylmass spectrometry
Gd-Ge (Phase Diagram) ... **A3:** 2 • 219
Gd-In (Phase Diagram) ... **A3:** 2 • 219
Gd-Mg (Phase Diagram) ... **A3:** 2 • 220
Gd-Mn (Phase Diagram) ... **A3:** 2 • 220
Gd-Ni (Phase Diagram) ... **A3:** 2 • 220
Gd-Pb (Phase Diagram) ... **A3:** 2 • 221
Gd-Pd (Phase Diagram) ... **A3:** 2 • 221
Gd-Rh (Phase Diagram) ... **A3:** 2 • 221
Gd-Sb (Phase Diagram) ... **A3:** 2 • 222
Gd-Se (Phase Diagram) ... **A3:** 2 • 222
Gd-Sn (Phase Diagram) ... **A3:** 2 • 222
Gd-Te (Phase Diagram) ... **A3:** 2 • 223
Gd-Ti (Phase Diagram) ... **A3:** 2 • 223
Gd-Tl (Phase Diagram) ... **A3:** 2 • 223
GdIG *See* Gadolinium-iron garnet
GDMS *See* Glow discharge mass spectroscopy
GE 7031
as thermocouple wire insulation ... **A2:** 882
GE diamond
erosion test results ... **A18:** 200
GE dip-form process
for copper and copper alloy wire rod ... **A2:** 255
GE RTV60 silicon rubber compound
for surface replicas ... **A17:** 53
Ge-Ho (Phase Diagram) ... **A3:** 2 • 224
Ge-In (Phase Diagram) ... **A3:** 2 • 224
Ge-K (Phase Diagram) ... **A3:** 2 • 224
Ge-La (Phase Diagram) ... **A3:** 2 • 225
Ge-Li (Phase Diagram) ... **A3:** 2 • 225
Ge-Lu (Phase Diagram) ... **A3:** 2 • 225
Ge-Mg (Phase Diagram) ... **A3:** 2 • 226
Ge-Mn (Phase Diagram) ... **A3:** 2 • 226
Ge-Mo (Phase Diagram) ... **A3:** 2 • 227
Ge-Na (Phase Diagram) ... **A3:** 2 • 227
Ge-Nb (Phase Diagram) ... **A3:** 2 • 227
Ge-Nd (Phase Diagram) ... **A3:** 2 • 228
Ge-Ni (Phase Diagram) ... **A3:** 2 • 228
Ge-P (Phase Diagram) ... **A3:** 2 • 228
Ge-Pb (Phase Diagram) ... **A3:** 2 • 229
Ge-Pd (Phase Diagram) ... **A3:** 2 • 229
Ge-Pr (Phase Diagram) ... **A3:** 2 • 229
Ge-Pt (Phase Diagram) ... **A3:** 2 • 230
Ge-S (Phase Diagram) ... **A3:** 2 • 230
Ge-Sb (Phase Diagram) ... **A3:** 2 • 230
Ge-Sc (Phase Diagram) ... **A3:** 2 • 231
Ge-Se (Phase Diagram) ... **A3:** 2 • 231
Ge-Si (Phase Diagram) ... **A3:** 2 • 231
Ge-Sm (Phase Diagram) ... **A3:** 2 • 232
Ge-Sn (Phase Diagram) ... **A3:** 2 • 232
Ge-Sr (Phase Diagram) ... **A3:** 2 • 232
Ge-Tb (Phase Diagram) ... **A3:** 2 • 233
Ge-Te (Phase Diagram) ... **A3:** 2 • 233
Ge-Ti (Phase Diagram) ... **A3:** 2 • 233
Ge-Tl (Phase Diagram) ... **A3:** 2 • 234
Ge-Tm (Phase Diagram) ... **A3:** 2 • 234
Ge-U (Phase Diagram) ... **A3:** 2 • 234
Ge-Y (Phase Diagram) ... **A3:** 2 • 235
Ge-Yb (Phase Diagram) ... **A3:** 2 • 235
Ge-Zn (Phase Diagram) ... **A3:** 2 • 235
Gear applications
cast iron, coatings for ... **M1:** 104
closed-die steel forgings ... **M1:** 370, 371, 372
machining considerations ... **M1:** 565, 579, 581, 583
nitriding for wear resistance ... **M1:** 630
phosphate coating to reduce wear ... **M1:** 632
steels for, hardenability ... **M1:** 492-496
surface-hardened steels for ... **M1:** 527, 529, 534
wear resistance vs. surface finish ... **M1:** 636
Gear blanks
economy in manufacture ... **M3:** 854
Gear bulk temperature ... **A18:** 538
Gear cutting
cast irons ... **A16:** 655
Gear drive
resistance seam welding ... **M6:** 495
Gear insert ... **A7:** 669-670
Gear manufacturing
P/M high-speed tool steel for ... **A1:** 786
Gear manufacturing (continued)
as tool steel application ... **A7:** 791
Gear oils ... **A18:** 99
additives in formulations ... **A18:** 111
antisquawk additives ... **A18:** 104
CRC L-37 and CRC L-42 axle tests ... **A18:** 101
demulsifiers ... **A18:** 107
dispersants used ... **A18:** 100
extreme-pressure agents used ... **A18:** 101
lubricant classification ... **A18:** 86
AGMA classifications ... **A18:** 86
rust and oxidation (R&O) type ... **A18:** 86
oxidation inhibitors ... **A18:** 105
pour-point depressants ... **A18:** 108
rust and corrosion inhibitors ... **A18:** 106
viscosity improvers used ... **A18:** 110
Gear pumps ... **EM3:** 693-696, 698, 719
Gear teeth
AISI/SAE alloy steels, fractured ... **A12:** 298
bending-fatigue fractures ... **A12:** 329-330
high-carbon steels, fractured ... **A12:** 277
subcase fatigue cracking ... **A12:** 322
Gear tooth
-bending impact ... **A11:** 595
chipping ... **A11:** 595
contact ... **A11:** 587-589
shear ... **A11:** 595
wear ... **A11:** 595-596
Gear(s)
magnetizing ... **A17:** 94
on shafts, inspection of ... **A17:** 113
water-base magnetic particle testing ... **A17:** 103
Gear-generating hobs
thread grinding application ... **A16:** 278
Geared safety ladle ... **A15:** 28
Gears *See also* Gear tooth; Gears,failure of ... **A7:** 667-670
bevel ... **A11:** 587
bevel, production methods ... **A14:** 124-126
blank, forging of ... **A14:** 67-68
carburizing ... **A18:** 874, 875, 876
coining dies for ... **M3:** 510
crossed-axes helical ... **A11:** 586
drive, distortion in ... **A11:** 143
electron beam weld fracture surface in ... **A11:** 447
failure modes in ... **A11:** 590
forgings, typical ... **A14:** 106-107
helical ... **A11:** 586
HERF processing of ... **A14:** 106
herringbone ... **A11:** 586
integral coupling and, fatigue fracture ... **A11:** 129-130
internal ... **A11:** 586-587
materials for ... **A11:** 590
nylon driving, failure of ... **A11:** 764, 765
oil-pump, brittle fracture of teeth ... **A11:** 344-345
and pinion, carburization failure of ... **A11:** 336
rack ... **A11:** 586-587
Rockwell hardness testing of ... **A8:** 81
spur ... **A11:** 586
stresses in ... **A11:** 589
tooth chipping failure pattern in ... **A11:** 595
and tooth contact ... **A11:** 587-589
tooth, residual porosity ... **A14:** 193
tooth section ... **A11:** 593
types of ... **A11:** 586-587
wheel, polyoxymethylene, failure of ... **A11:** 764, 765
worm-gear sets ... **A11:** 586-587
Gears and their manufacture ... **A16:** 330-355
chain sprockets ... **A16:** 334
CNC milling and hobbing machines ... **A16:** 348-350, 353
cold forming ... **A16:** 346
comparison of steels for gear cutting ... **A16:** 346
correcting for distortion in heat treating ... **A16:** 351
crossed axes helical ... **A16:** 331, 332, 333
cutter material and construction ... **A16:** 343-344

SUBJECTS OF THE INDEXED VOLUMES: ASM Handbook (designated by the letter "A"): **A1:** Properties and Selection: Irons, Steels, and High-Performance Alloys (1990); **A2:** Properties and Selection: Nonferrous Alloys and Special-Purpose Materials (1990); **A3:** Alloy Phase Diagrams (1992); **A4:** Heat Treating (1991); **A6:** Welding, Brazing, and Soldering (1993); **A7:** Powder Metallurgy (1984); **A8:** Mechanical Testing (1985); **A9:** Metallography and Microstructures (1985); **A10:** Materials Characterization (1986); **A11:** Failure Analysis and Prevention (1986); **A12:** Fractography (1987); **A13:** Corrosion (1987); **A14:** Forming and Forging (1988); **A15:** Casting (1988); **A16:** Machining (1989); **A17:** Nondestructive Testing and Quality Control (1989); **A18:** Friction, Lubrication, and Wear Technology (1992). **Metals Handbook, 9th Edition** (designated by the letter "M"): **M1:** Properties and Selection: Irons and Steels (1978); **M2:** Properties and Selection: Nonferrous Alloys and Pure Metals (1979); **M3:** Properties and Selection: Stainless Steels, Tool Materials and Special-Purpose Materials (1980); **M4:** Heat Treating (1981); **M5:** Surface Cleaning, Finishing, and Coating (1982); **M6:** Welding, Brazing, and Soldering (1983). **Engineered Materials Handbook** (designated by the letters "EM"): **EM1:** Composites (1987); **EM2:** Engineering Plastics (1988); **EM3:** Adhesives and Sealants (1990); **EM4:** Ceramics and Glasses (1991); **Electronic Materials Handbook** (designated by the letters "EL"): **EL1:** Packaging (1989).

Gears and their manufacture (continued)
cutting fluids **A16:** 343, 344-346
cyclex method process **A16:** 336, 340
double-enveloping worm **A16:** 333, 340
elliptical **A16:** 334
external **A16:** 334
face **A16:** 333, 334
face hob cutting **A16:** 335, 336, 340, 341, 343
face mill cutting **A16:** 335-336, 340, 341, 343, 344, 354
form grinding **A16:** 350, 351-352
formate cutting **A16:** 335-336, 340
G-Trac gear generator **A16:** 344
gear inspection **A16:** 355
gear shaping cutters **A16:** 193
generation grinding **A16:** 350, 352-353
grinding **A16:** 350-355
grinding fluids **A16:** 350, 354-355
grinding of bevel gears **A16:** 353-354
helical (external and internal) **A16:** 331, 333-334, 339-341, 343-344, 346, 348, 351-353
helix form cutting **A16:** 335-336, 340
herringbone (double helical) **A16:** 331
herringbone gears **A16:** 339, 341
herringbone teeth **A16:** 334
honing **A16:** 343, 350
hypoid **A16:** 332-333, 335-336, 338, 340, 350, 353-354
interlocking cutters **A16:** 337, 340, 344
internal **A16:** 332, 334, 339-340
lapping **A16:** 343
machines to cut gear teeth **A16:** 330
machining processes for other than bevel gears **A16:** 333
milling for roughing process **A16:** 335
P/M high-speed tool steels **A16:** 67, 68
planing generator process **A16:** 337-338, 340, 341
processes for bevel gears **A16:** 335-338
racks **A16:** 332, 334, 340
ratchets **A16:** 334
Revacycle process **A16:** 337, 340
rolling **A16:** 346-348
selection **A16:** 333
selection of machining process **A16:** 338-343
shaving **A16:** 341-343, 344, 350
single-enveloping worm **A16:** 333
speeds and feeds **A16:** 342, 343, 344
spiral bevel **A16:** 332-338, 340-341, 344, 349-353
spiroid **A16:** 333
spur (external and internal) **A16:** 331-334, 338-341, 344, 346, 348, 350-353
straight bevel **A16:** 332-335, 337-338, 340-341, 344, 348-354
surface finish **A16:** 340, 341-343, 350, 354-355
tangear generator method **A16:** 343
template machining **A16:** 335, 340
two-tool generator process **A16:** 337, 338, 340, 341, 350
types **A16:** 331
Waguri method **A16:** 354
worm gear sets **A16:** 331-335, 340, 352-353
Zerol bevel **A16:** 332-334, 336-338, 340, 344, 351-352, 354
Gears, failures of *See also* Gear tooth; Gears **A11:** 586-601
associated parameters **A11:** 589-590
causes of **A11:** 597-598
classification of **A11:** 590-597
final analyses and examples of **A11:** 598-601
from fatigue **A11:** 590-595
from impact **A11:** 595
from stress rupture **A11:** 596-597
from wear **A11:** 595-596
gear materials **A11:** 590
gear-tooth contact **A11:** 587-589
types of gear **A11:** 586-587
working loads **A11:** 589
Gears: friction, lubrication, and wear of **A18:** 535-545
application of gear lubricants **A18:** 542-545
gearbox example, 24-unit speed increaser **A18:** 543
pressure-fed systems **A18:** 542-543
splash lubrication systems **A18:** 542
design, factors affecting **A18:** 535
elastohydrodynamic lubrication **A18:** 538-541
Blok's contact temperature theory **A18:** 539-540
Gears: friction, lubrication, and wear of (continued)
bulk temperature **A18:** 539, 540
load-sharing factor **A18:** 539, 540
mean coefficient of friction **A18:** 540
scuffing temperature **A18:** 540-541, 543
semiwidth of Hertzian contact band **A18:** 539, 540
thermal contact coefficient **A18:** 540
gear tooth failure modes **A18:** 535-538
abrasive wear **A18:** 536, 537
adhesive wear **A18:** 536-537
carburized gears **A18:** 536, 537, 541
frosting **A18:** 536
gray staining **A18:** 536
Hertzian fatigue **A18:** 535, 536
lubrication-related **A18:** 535-538
micropitting **A18:** 536
nitrided gears **A18:** 536, 537
nonlubrication-related **A18:** 535
pitting **A18:** 536
polishing **A18:** 536, 537, 538
scuffing **A18:** 535, 536, 539, 543, 544
sulfur-phosphorus additives **A18:** 536-537, 538
three-body abrasion **A18:** 537
two-body abrasion **A18:** 537
wear **A18:** 535, 536-537
lubricant selection **A18:** 541
grease **A18:** 541
oil **A18:** 541
open-gear lubricants **A18:** 541
solid lubricants **A18:** 541
synthetic lubricants **A18:** 541, 542
nomenclature used in friction and wear of gears **A18:** 544
oil lubricant applications **A18:** 541
oil pumps in internal combustion engines **A18:** 561
rolling contact wear **A18:** 257-258
viscosity selection, gear lubricants **A18:** 541-542
Gears, induction hardened
economy in manufacture **M3:** 847
Gecim and Winer's model **A18:** 93
Geiger plate test (SAE J-671) **EM3:** 556
Geiger tube goniometer
for crystallographic structure **EL1:** 93
Gel **EM3:** 14
coat **EM1:** 12, 134, 169
defined **EM1:** 12 **EM2:** 20
point **EM1:** 12, 135
Gel coat **EM1:** 12, 134, 169 **EM3:** 14
defined **EM2:** 20
Gel permeation chromatography
of epoxies **EL1:** 833-834
Gel permeation chromatography (GPC) **EM1:** 12, 730, 736 **EM2:** 20, 518-519 **EM3:** 14
Gel point **EM1:** 12, 135 **EM3:** 14
defined **EM2:** 20
Gel time **EM2:** 20, 274 **EM3:** 14
Gel times
defined **EM1:** 132
delayed, Derakane 411-45 (vinyl ester) resin **EM1:** 133
in filament winding **EM1:** 135
polyester resins **EM1:** 133
testing of **EM1:** 737
of unidirectional tape prepregs **EM1:** 144
variation, Derakane 411-45 (vinyl ester) resin **EM1:** 137
Gel-permeation chromatography *See* Size-exclusion chromatography
Gelatin replica
defined **A9:** 8
Gelation *See also* Gel; Gel times; Working life **EM3:** 14
defined **EM1:** 12 **EM2:** 20
rheological behavior **EL1:** 850-852
time, defined **EM1:** 12
Gelation time *See also* Working life **EM3:** 14
defined **EM2:** 20
Gelling
and burn-off, Shaw process **A15:** 252
slurry, Shaw process **A15:** 250
Gelling agent *See* Thickener
Gelling, and curing
chemical **EM1:** 132
General biological corrosion *See also* Biological corrosion; Localized biological corrosion; Microbiological corrosion **A13:** 87-88
General chemical catalysts
powders used **A7:** 572
General corrosion *See also* Corrosion; specific types of corrosion; Uniform corrosion **A13:** 80-103
adjacent to leaking gaskets, testing **A13:** 962-964
of aged titanium alloys **A13:** 682
in aircraft powerplants **A13:** 1045
aluminum casting alloys **A2:** 155, 176
and atmospheric corrosion **A13:** 80-83
of austenitic cast steels **A13:** 577-578
in brazing **A13:** 876
in carbon and low-alloy steels **A11:** 199
of closed feedwater heaters **A13:** 990
of cobalt-base alloys **A13:** 658-661
of copper/copper alloys **A13:** 612
defined **A11:** 5 **A12:** 41 **A13:** 79, 80
of duplex cast steels **A13:** 577-578
in ferritic cast alloys **A13:** 577
and galvanic corrosion **A13:** 83-87
and general biological corrosion **A13:** 87-88
of heat exchangers **A11:** 630
high-temperature **A13:** 97-101
of historic Lane plate, chromium steel **A11:** 674, 680
in liquid metals **A13:** 91-97
of martensitic cast steels **A13:** 576-577
material selection to avoid/minimize **A13:** 323
and molten salt corrosion **A13:** 88-91
nickel alloys **A2:** 431-432
of offshore gas/oil production platforms **A13:** 1254
of radioactive waste containers **A13:** 971
resistance, titanium/titanium alloys **A13:** 693-694
resistance, wrought aluminum alloys **A13:** 586
resistance, wrought aluminum and aluminum alloys **A2:** 104
in stainless steels **A11:** 200 **A13:** 553
and stray-current corrosion **A13:** 87
surface, low-carbon steel **A12:** 250
of titanium/titanium alloys, specific media **A13:** 676-693
types **A13:** 79
in water-recirculating systems **A13:** 488
General design considerations *See* Design considerations
General dissection techniques
capabilities of **A10:** 380
General Dynamics Corporation
development of an adhesive bond strength classifier algorithm **EM3:** 744
General Electric tape process **A7:** 636
General information sources *See* Information sources
General oxidation
in elevated-temperature failures **A11:** 271, 405
General practice
in failure analysis **A11:** 15-46
General precipitate
defined **A9:** 8
General precipitation **A9:** 647
General rusting
in bearings **A11:** 494, 496
General spring quality wire
characteristics of **A1:** 303-304, 305-307
General yielding
fracture toughness measure with **A8:** 268
General-purpose alloys
copper casting alloys **A2:** 351-352
General-purpose resins
orthophthalic polyester as **EM2:** 246
General-purpose steel
impact resistance and abrasion resistance properties **A18:** 759
Generalized Newtonian fluid
as rheological model **EL1:** 847-848
Generator
cam plastometer **A8:** 194
Generator applications
aluminum and aluminum alloys **A2:** 13
as magnetically soft materials **A2:** 779
with niobium-titanium superconducting materials **A2:** 1057

Generator rotors
ultrasonic inspection ... **A17:** 232
Generator welding machines
gas tungsten arc welding ... **M6:** 187
Generators
blades, stainless steel ... **A13:** 560
corrosion of ... **A13:** 1006-1007
retaining rings, SCC of ... **A13:** 1006-1007
Generators, radio-frequency
for ICP systems ... **A10:** 37
Geneva pinions ... **A7:** 668
Genotoxicity
of arsenic ... **A2:** 1238
of beryllium ... **A2:** 1239
Genzel interferometer
in FT-IR spectroscopy ... **A10:** 112
Geochemical research ... **A10:** 82, 233
Geodesic
defined ... **EM1:** 12 **EM2:** 20
-isotensoid contour, defined ... **EM1:** 12
isotensoid, defined ... **EM1:** 12
ovaloid. defined ... **EM1:** 12
Geodesic isotensoid
defined ... **EM2:** 20
Geodesic ovaloid
defined ... **EM2:** 20
Geodesic-isotensoid contour
defined ... **EM2:** 20
Geographical segments
commercial thick-film hybrid market ... **EL1:** 381
Geologic samples, characterization of *See also* Geological materials
analytical transmission electron microscopy ... **A10:** 429-489
atomic absorption spectrometry ... **A10:** 43-59
Auger electron spectroscopy ... **A10:** 549-567
classical wet analytical chemistry ... **A10:** 161-180
controlled-potential coulometry ... **A10:** 207-211
electrochemical analysis ... **A10:** 181-211
electrogravimetry ... **A10:** 197-201
electrometric titration ... **A10:** 202-206
electron probe x-ray microanalysis ... **A10:** 516-535
electron spin resonance ... **A10:** 253-256
extended x-ray absorption fine structure ... **A10:** 407-419
inductively coupled plasma atomic emission spectroscopy ... **A10:** 31-42
infrared spectroscopy ... **A10:** 109-125
ion chromatography ... **A10:** 658-667
low-energy ion-scattering spectroscopy ... **A10:** 603-609
molecular fluorescence spectrometry ... **A10:** 72-81
Mössbauer spectroscopy ... **A10:** 287-295
neutron activation analysis ... **A10:** 233-242
neutron diffraction ... **A10:** 420-426
optical emission spectroscopy ... **A10:** 21-30
particle-induced x-ray emission ... **A10:** 102-108
potentiometric membrane electrodes ... **A10:** 181-187
radial distribution function analysis ... **A10:** 393-401
radioanalysis ... **A10:** 243-250
Raman spectroscopy ... **A10:** 126-138
scanning electron microscopy ... **A10:** 490-515
secondary ion mass spectroscopy ... **A10:** 610-627
single-crystal x-ray diffraction ... **A10:** 344-356
spark source mass spectrometry ... **A10:** 141-150
ultraviolet/visible absorption spectroscopy ... **A10:** 60-71
voltammetry ... **A10:** 188-196
x-ray diffraction ... **A10:** 325-332
x-ray powder diffraction ... **A10:** 333-343
x-ray spectrometry ... **A10:** 82-101
Geological analysis
of worn parts ... **A11:** 158
Geological materials *See also* Geologic samples, characterization of
brine, ion chromatography of ... **A10:** 665
crystallographic texture measurement and analysis ... **A10:** 357-364
Geological materials (continued)
ICP-AES use in ... **A10:** 31
NAA application in ... **A10:** 234
OES analysis of ... **A10:** 21
powder, PIXE analysis ... **A10:** 102
SIMS phase distribution analysis in ... **A10:** 610
Geometric
measurement, by coordinate measuring machines ... **A17:** 19
unsharpness, shadow formation radiography ... **A17:** 313
weld discontinuities, defined ... **A17:** 584
Geometric analysis of small particles by scanning electron microscopy ... **A9:** 96
Geometric dimensioning and tolerancing ... **A15:** 623
Geometric dispersion
in torsional Kolsky bar tests ... **A8:** 218
Geometric isomers
within mer ... **EM2:** 58
Geometric method
feed metal availability ... **A15:** 578, 582-584
Geometric moving average (GMA) chart ... **EM3:** 795, 796
Geometric scattering ... **A18:** 409- 410
Geometric shapes data
computer-aided analyses ... **EL1:** 133-134
Geometric uniformity
niobium-titanium superconducting materials ... **A2:** 1052-1053
Geometrical description of an interface ... **A9:** 118-119
Geometrically close-packed phases ... **M3:** 209, 223-224, 228
Geometry *See also* Specimen geometry
of 3-D orthogonal weave preform ... **EM1:** 130
of angle-interlock fabric ... **EM1:** 130
of applied field and magnetization, FMR analyses ... **A10:** 269
of Bragg-Brentano diffractometer ... **A10:** 337
broad-beam ... **A17:** 310
of chemisorbed atoms or molecules, EXAFS determined ... **A10:** 407
complex, inspection of ... **EM2:** 845
of composite, for tensile failure model ... **EM1:** 194
cone plate, and parallel, in melt rheology ... **EM2:** 535-540
crevice ... **A13:** 111-112
CT, defined ... **A17:** 384
CT scanning ... **A17:** 365-368
of deformation, plastic straining of hollow cylinder ... **A8:** 143
effect, galvanic corrosion ... **A13:** 83
effect, wireability ... **EL1:** 18
evaluation, by computer modeling ... **A13:** 234
experimental, and simple flows ... **EL1:** 841-842
fabric, braided composite ... **EM1:** 524
factors, in microbeam analysis ... **A10:** 529
flow path ... **A13:** 966
grid, resistance strain gages ... **A17:** 449
of image formation, SEM ... **A12:** 196
of lattices ... **A10:** 327-328
of leaded and leadless surface-mount joints ... **EL1:** 733-734
methods, in quantitative fractography ... **A12:** 198
of microdiffractometer ... **A10:** 338
narrow-beam ... **A17:** 310
and NDE response ... **A17:** 675
of necking ... **A8:** 26
notch, point stress/average stress criteria for ... **EM1:** 253-255
part ... **A14:** 77, 409-410 **EM2:** 707-709
of particles produced by explosive detonation ... **A10:** 318-320
platelet, in foil ... **A10:** 455
ply ... **EM1:** 458-459
of powder diffraction ... **A10:** 331
and reference standards ... **A17:** 677
representation tools ... **A14:** 410
Geometry (continued)
roll groove ... **A14:** 396
sample, and properties divergencies ... **EM2:** 655
sample, effect on X-ray diffraction residual techniques ... **A10:** 387
Seeman-Bohlin diffraction arrangement ... **A10:** 337
and shape, corrosion control ... **A13:** 339-340
of single-angle x-ray diffraction residual stress measurement ... **A10:** 384
of specimen/instrument, SEM effects ... **A12:** 168
structural, of printed wiring boards ... **EL1:** 40
substrate ... **A13:** 422
two-dimensional, of part ... **A14:** 410
of unit cells and diffraction ... **A10:** 326-327
variable wavelength, RDF analysis ... **A10:** 396
weave, multidirectionally reinforced fabrics/preforms ... **EM1:** 129-130
Geometry, and correction factors
fatigue fracture ... **A11:** 124
Geometry, coordination *See* Coordination geometry
Geothermal applications
polybenzimidazoles (PBI) ... **EM2:** 147
Gerber cutting machine
for unidirectional tape prepreg ... **EM1:** 145
Gerber's parabola
mean stress effect on fatigue strength ... **A8:** 374
Gerber's parabola and law ... **A11:** 111, 112
German Industrial Standard
abbreviation ... **A8:** 724
German silver
history of ... **A2:** 428
German silver (62% Cu; 15% Ni; 22% Zn)
Germanate glasses
composition ... **EM4:** 741
density ... **EM4:** 850
electrical properties ... **EM4:** 853
heat capacity ... **EM4:** 847
mechanical properties ... **EM4:** 850
military applications ... **EM4:** 1020
Germanates
chemical properties ... **A2:** 734
Germanes
chemical properties ... **A2:** 734
Germanides
chemical properties ... **A2:** 734
Germanium *See also* Germanium compounds
as addition to aluminum alloys ... **A4:** 843
as an alpha stabilizer in titanium and titanium alloys ... **A9:** 458
analytical and test methods ... **A2:** 736
autocorrelation functions for surface texture ... **A18:** 336, 337
carbon-coated, rain erosion applications ... **A18:** 222
chemical properties ... **A2:** 733-734
as common analyzing crystal, in x-ray spectrometry ... **A10:** 88
dislocations in ... **A9:** 608
economic aspects ... **A2:** 735-736
environmental considerations ... **A2:** 735
and germanium compounds ... **A2:** 733-738
as infrared optics material ... **A2:** 735
as internal reflection element ... **A10:** 113
intrinsic, gamma-ray detector ... **A10:** 235
laser-induced CVD for synthesis ... **A18:** 848
liquid impingement erosion protection applications ... **A18:** 222
manufacturing and processing ... **A2:** 735
ore processing ... **A2:** 735
pure ... **M2:** 737
pure, properties ... **A2:** 1115
purification ... **A2:** 735
quartz tube atomizers with ... **A10:** 49
as semiconductor material ... **A2:** 735
semiconductors, ESR studied ... **A10:** 263
sources ... **A2:** 733
specifications ... **A2:** 736

Germanium (continued)
tilt boundaries studied by high resolution electron microscopy **A9:** 121
toxicology **A2:** 736
ultrapure, by zone refining **A2:** 1093
ultrasonic machining **EM4:** 359
uses **A2:** 736-737
vapor pressure, relation to temperature **A4:** 495
Germanium (liquid), contact angles on beryllium at various test temperatures in argon and vacuum atmospheres **A6:** 116
thermal diffusivity from 20 to 100 °C **A6:** 4
Germanium carbon
protective coating against liquid impingement erosion **A18:** 222
Germanium compounds *See also* Germanium
analytical and test methods **A2:** 736
chemical properties **A2:** 733-734
economic aspects **A2:** 735-736
germanates **A2:** 734
germanes **A2:** 734
germanides **A2:** 734
and germanium **A2:** 733-738
germanium halides **A2:** 733-734
germanium oxides **A2:** 733-734
inorganic compounds **A2:** 734
manufacturing and processing **A2:** 735
ore processing **A2:** 735
organogermanium compounds **A2:** 734
sources **A2:** 733
specifications **A2:** 736
toxicology **A2:** 736
uses of **A2:** 736-737
Germanium, diffused
temperature effect **EL1:** 958
Germanium dioxide
application **A2:** 743
Germanium disulfide
chemical properties **A2:** 734
Germanium halides
chemical properties **A2:** 733
Germanium metal *See also* Germanium; Germanium compounds
chemical properties **A2:** 733
Germanium nitride
chemical properties **A2:** 734
Germanium oxide (GeO_2)
glass-forming ability **EM4:** 494
melting point **EM4:** 494
viscosity at melting point **EM4:** 494
Germanium oxides
chemical properties **A2:** 733-734
Germanium point-contact transistor **EL1:** 958
Germanium single crystals
application **A2:** 743
Germanium, vapor pressure
relation to temperature **M4:** 310
Germanium-gold alloys
applications **A2:** 743
Gettering box **A7:** 5
Gettering, external *See* External gettering
Gettering processes
monitored by x-ray topography **A10:** 376
Getters **A7:** 5
as rare earth applications **A2:** 731
TV tubes, powders used **A7:** 574
gf *See* Gram-force
GFAAS *See* Graphite furnace atomic absorption spectrometry
Ghost images
photographic **A12:** 84
Ghost peaks
in Euler plots **A10:** 362
Gibb's adsorption theory **EM4:** 70
Gibb's equation
mechanochemical effects in comminution **EM4:** 77
Gibbs energy .... **A3:** 1 • 7, 8
curves .... **A3:** 1 • 6, 7, 10
Gibbs free energy **A13:** 7, 52, 61, 63 **EM3:** 408
activity and condensed equilibrium **A15:** 51
activity coefficients **A15:** 51-52
defined **A9:** 8
equilibrium, conditions for **A15:** 50
equilibrium constant **A15:** 51
of mixing **A15:** 56
Gibbs free energy (continued)
standard, for solution, liquid aluminum **A15:** 59
standard, liquid copper **A15:** 60
thermodynamic data, tabulation of **A15:** 50-51
Gibbs free energy change **EM4:** 109
Gibbs, J. Willard .... **A3:** 1 • 2, 10
Gibbs phase rule .... **A3:** 1 • 2
Gibbs phenomenon
computed tomography (CT) **A17:** 376
Gibbs triangle
defined **A9:** 8
Gibbs-Duhem equation **A15:** 56
Gibbs-Konovalov Rule .... **A3:** 1 • 8, 10
Gibbs-Thomson coefficient **A15:** 134
Gibbs-Thomson undercooling
at solid/liquid interface **A15:** 110-111
Gibbsian adsorption **A18:** 854-855
Gibbsite **EM4:** 49, 111
decomposition **EM4:** 54
Gibs **A14:** 7, 495
GIFTS 5 computer program for structural analysis **EM1:** 268, 271
Gigajoule
abbreviation **A8:** 724
Gigapascal
abbreviation **A8:** 724
Gilding
by depletion **A15:** 19-21
weldability **A6:** 753
Gilding metal
applications and properties **A2:** 295-296
Gilding metal, 95% *See* Copper alloys, specific types, C21000
Gilding metal substitute **A7:** 17
Gill-Goldhoff method **A8:** 337
Gilsonite
as carbon mold addition **A15:** 211
Miller numbers **A18:** 235
Gilsonites **EM3:** 51
for auto body sealing and glazing materials **EM3:** 57
Ginzburg and Landau theory
of superconductivity **A2:** 1031-1034
Gircast process
as rheocasting alternative **A15:** 330
zinc alloys **A15:** 795, 797
Girder
cracked, at end of bridge cover plate **A11:** 707
cracked, at Lafayette Street Bridge St. Paul, Mn **A11:** 710
webs **A11:** 709-712
Girder sections
bend tests for **A8:** 117
Girder weldment
magnetic particle inspection of **A17:** 115
Girth pattern *See* Hoop pattern
Girth weld
center, embrittlement failure of **A11:** 438
failed, in pipeline **A11:** 422
field **A11:** 699
in steam accumulator, lack of penetration in **A11:** 648
underbead cracks in **A11:** 699
GJ *See* Gigajoule
Glacial acetic acid
austenitic manganese steel casting specimens **A9:** 238
and hydrogen peroxide as an etchant for lead and lead alloys **A9:** 415-416
and nitric acid as etchant for nickel alloys **A9:** 436
and perchloric acid (Group 11 electrolytes) **A9:** 52-54
Glancing angle
defined **A9:** 8
Glancing-angle camera
XRPD analysis **A10:** 336
Glancing-angle x-ray diffraction
capabilities compared with LEED **A10:** 536
Glass *See also* Ceramics; Glass fiber reinforcement; Glass fibers; Glassy
abrasive jet machining **A16:** 511, 512, 513
bare, defined *See* Bare glass
for brake linings **A18:** 576
broad **EM4:** 395
carbides for machining **A16:** 75
Glass (continued)
as carrier core, copier powders **A7:** 584
clean, friction coefficient data **A18:** 75
coatings, for tantalum forgings **A14:** 238
contact angle with mercury **A7:** 269
content, effect on polyester resin composites **EM2:** 248
coupling agents **EM3:** 281-283
defined **EM1:** 12, 107
defined by ASTM **EM4:** 564
diamond as abrasive for honing **A16:** 476
drilling **A16:** 230
effect of impact angle on **A11:** 155
electrode, defined **A13:** 7
for environmental test chamber **A8:** 415
epoxy bonding **EM3:** 96
erosion of steels studied **A18:** 204
finish, defined **EM1:** 12
flake, defined **EM1:** 12
flaked filler **EM3:** 177
former, defined **EM1:** 12
forms used in glass-to-metal seals **EM3:** 302
fracture toughness testing **A8:** 469
frit coatings **A14:** 237
frits, for compression testing **A8:** 195
ground by diamond wheels **A16:** 455, 456, 460
ground by superabrasives **A16:** 432, 433, 434
honing grit size selection **A16:** 478
honing stone selection **A16:** 476
hydrogen fluoride/hydrofluoric acid corrosion **A13:** 1169
incorporation in thermoplastic composites **A18:** 820
lapping **A16:** 492, 494, 499, 502
laser cutting of **A14:** 742
linings, use in breweries **A13:** 1222
liquid penetrant inspection **A17:** 71
liquidus **EM4:** 389
as lubricant during hot extrusion of tool steels **A18:** 738
as lubricant for hot forging of tool steels **A18:** 738
matrix material for ceramic-matrix composites **EM4:** 840
melting **EM4:** 386
melting and morning **EM1:** 107-108
for metalworking lubricants **A18:** 144
nitric acid corrosion **A13:** 1156
PCD tooling **A16:** 110
percent by volume **EM3:** 14
percent by volume, defined **EM1:** 12 **EM2:** 21
prebond treatment **EM3:** 35
primers **EM3:** 281-282, 283
properties **A18:** 801
in rapid solidification **A7:** 570
recommended waterjet cutting speeds **EM4:** 366
rovings, pultrusions **EM2:** 393
scales, calibrated by interferometers **A17:** 15
single-filament tensile strength **EM1:** 192
stress, defined **EM1:** 12
substrate cure rate and bond strength for cyanoacrylates **EM3:** 129
surface preparation **EM3:** 281-283
tempered, friction coefficient data **A18:** 72, 73, 75
thermal expansion rate **A7:** 611
thin fiber, friction coefficient data **A18:** 75
transition, defined **EM1:** 12
type/amount, effect on unsaturated polyesters **EM2:** 248
ultrasonic fatigue testing of **A8:** 240
ultrasonic machining **A16:** 529, 530, 531 **EM4:** 359, 360
vacuum coating of **M5:** 394, 396-397, 401-402
viewport **A8:** 411
volume resistivity and conductivity **EM3:** 45
waterjet machining **A16:** 520, 527
x-ray tube envelopes **A17:** 302
Glass (television tube)
recommended waterjet cutting speeds **EM4:** 366
Glass and carbon fiber composites
diamond for machining **A16:** 105
Glass and glass fibers, design practices *See* Design practices for glass and glass fibers
Glass balloons **EM3:** 175
Glass bead blast cleaning
aluminum and aluminum alloys **M5:** 571-572

Glass bead blast cleaning (continued)
physical properties and composition characteristics, beads **M5:** 84-85
process **M5:** 12, 19-20, 84-87, 93-94
dry **M5:** 84-87
wet **M5:** 93-94
size and roundness specifications
beads **M5:** 84-85
Glass bead peening **A16:** 34
abrasive jet machining **A16:** 512, 513
electrochemical machining **A16:** 539
to improve surface integrity **A16:** 35
Glass bead shot peening **M5:** 144-145
dry method **M5:** 144
in printer hammer guide assembly **A7:** 669
wet method **M5:** 144-145
Glass capacitors
failure mechanisms **EL1:** 998
Glass capillary tubing
used in mounting wire specimens **A9:** 31
Glass ceramics **A18:** 532
Glass cloth *See also* Scrim **EM3:** 14
defined **EM1:** 12 **EM2:** 20
Glass coatings
crystallized **M5:** 533
Glass containers **EM4:** 1082-1086
applications **EM4:** 1084-1085
compositions and colors used **EM4:** 1082
design/shape of containers **EM4:** 1082
glass types **EM4:** 1085
manufacturers identification **EM4:** 1086
quality considerations **EM4:** 1085-1086
regulatory requirements **EM4:** 1085
strength and surface treatments **EM4:** 1082-1084
test limits **EM4:** 1085
Glass die attach
in hermetic packages **EL1:** 215
Glass diodes
failure mechanism **EL1:** 973
Glass enamels **EM4:** 953
ceramic corrosion in the presence of combustion products **EM4:** 982
Glass encapsulants
thick-film pastes **EL1:** 343
Glass encapsulation **EM4:** 124
Glass fabric
reinforced epoxy resin composites **EM1:** 399-400, 404-405
reinforced phenolic resin composites **EM1:** 381-382
Glass fiber filters
for sample preparation **A10:** 94
Glass fiber reinforced plastic
nondestructive testing **A6:** 1086
Glass fiber reinforced plastics
testing problems **EM3:** 333
Glass fiber reinforced polymers *See also* Polymer(s)
electronic transparency of **EM1:** 36
Glass fiber reinforced thermoset materials **A13:** 1243-1244
Glass fiber reinforcement *See also* Glass; Glass fiber(s)
effect, polyester resins **EM2:** 248
styrene-maleic anhydrides (S/MA) **EM2:** 220-221
vinyl ester, mechanical properties **EM2:** 274
Glass fiber(s) *See also* Fiber(s); Glass; Glass fiber reinforcement
addition effects **EM2:** 70-72
defined **EM2:** 20
metallized, as reinforcements **EM2:** 473
in polypropylene **EM2:** 193
as reinforcement, styrene-maleic anhydrides (S/MA) **EM2:** 220-221
Glass fibers *See also* Fiber properties analysis; Fibers; Fibers(s); Filaments; Glass; Glass filament; Glass-phenolic matrix composites; Glass-polyester resin composites; specific glass fibers **EM1:** 45-48, 107-111 **EM4:** 1027-1030
applications **EM4:** 1015
Glass fibers (continued)
C-glass, composition/use **EM1:** 45
carded **EM1:** 111
chemical components **EM1:** 45
chopped-strand products **EM1:** 109-110
composition **EM1:** 45, 107
continuous filaments **EM4:** 1027-1029
continuous, types **EM1:** 107
critical lengths **EM1:** 120
defined **EM1:** 12
design practices *See* Design practices for glass and glass fibers
in die material for sheet metal forming **A18:** 628
E-glass, composition/use **EM1:** 45
effect, polyester resin properties **EM1:** 92
elastic properties **EM1:** 188
electrical properties **EM1:** 46-47
electronic transparency **EM1:** 36
as epoxy composite reinforcement **EM1:** 73, 75
fabrication process **EM1:** 108-111
fiberglass forming process **EM1:** 108
fiberglass mat/woven roving combinations **EM1:** 109
fiberglass paper **EM1:** 110
fiberglass roving **EM1:** 109
filament diameter nomenclature **EM1:** 109
finished forms **EM1:** 108-111
forming **EM1:** 107-108
forms **EM1:** 107-111, 360
glass wool **EM4:** 1029-1030
groups depending on fiber geometry **EM4:** 1027
health aspects **EM4:** 1030
inherent properties **EM1:** 107
introduction **EM1:** 29
longitudinal Young's modulus vs fiber volume fraction in **EM1:** 208
melting **EM1:** 107-108
milled fibers **EM1:** 110
multi-end roving process **EM1:** 109
optical properties **EM1:** 47
physical properties **EM1:** 46
properties **EM1:** 46-47, 360
as pultrusion reinforcement **EM1:** 537
radiation properties **EM1:** 47
random dispersion **EM1:** 176
reinforced epoxy resin composites **EM1:** 399-400, 405-407
as reinforcement materials, base/insulators **EL1:** 113-114
rovings, pultruded **EM1:** 537
S-glass, composition/use **EM1:** 45
in sheet molding compounds **EM1:** 157-158
single-filament tensile strength **EM1:** 192
sizing for **EM1:** 123
tensile strength vs. temperature **EM1:** 362
tension damage tolerance **EM1:** 262
textile yams **EM1:** 110-111
thermal properties **EM1:** 47
tow sizes available **EM1:** 105
types **EM1:** 29, 45
vs. aramid fibers, cost **EM1:** 105
vs. carbon fibers, cost **EM1:** 105
woven roving **EM1:** 109
Glass filament
bushing, defined **EM1:** 12
defined **EM1:** 12
Glass filament bushing
defined **EM2:** 20
Glass filters
recommended waterjet cutting speeds **EM4:** 366
Glass flake
defined **EM2:** 20
Glass former
defined **EM2:** 20
Glass frit *See also* Frit
attachment, with Alloy **EL1:** 734
devitrifying, effects **EL1:** 961-962
with precious metal powders **A7:** 149
Glass frits **EM4:** 44
Glass houseware *See* Consumer houseware
Glass industry applications *See also* Amorphous materials
of precious metals **A2:** 693
Glass insulation
waterjet machining **A16:** 522
Glass ionomers
cements (dental)
material loss on abrasion **A18:** 673
properties **A18:** 666
silver-reinforced **A18:** 673
simplified composition or microstructure **A18:** 666
Glass lamination *See* Laminated glass
Glass layers
interference films used to improve contrast **A9:** 59
Glass matrix composites **A9:** 592
with silicon-carbide fibers **A9:** 596-597
Glass melting
and forming **EM1:** 107-108
Glass membrane electrodes **A10:** 182
Glass microballoons
ion chromatography of **A10:** 665-667
Glass microspheres
as extender **EM3:** 176
Glass penetrometers **A7:** 268-269
Glass phenolic matrix composites *See also* Bulk molding compounds (BMC)
electrical applications **EM1:** 162
fabric **EM1:** 381-382
ordnance applications **EM1:** 162
prepregs, as flame-resistant **EM1:** 141
Glass powder
as incendiary **A7:** 603
Glass processing
introduction **EM4:** 377
secondary processing **EM4:** 377
Glass reinforced acrylamate composites
physical properties **EM2:** 269
Glass sealing alloy
iron-nickel-chromium **A2:** 894
mechanical properties **A2:** 892
Glass sealing sheet and strip
powders used **A7:** 574
Glass seals
heat exchangers **EM4:** 981
Glass spheres **EM3:** 175 **EM4:** 418-422
angle of repose **A7:** 282
applications **EM4:** 418, 419, 420
composition classifications **EM4:** 418
as filler material **EM2:** 499
future outlook **EM4:** 421-422
hollow glass spheres **EM4:** 419-421
controlled composition process yielding shells having uniform walls **EM4:** 419
inertial confinement fusion research programs **EM4:** 420
process for forming in combustible gas using a burner head **EM4:** 419
rotating disk process **EM4:** 420-421
porous glass beads **EM4:** 421
droplet generation method and sol-gel processing **EM4:** 421
hydrolysis of silicon and zirconium alkoxides **EM4:** 421
rotating wheel process **EM4:** 421
thixotropy **EM4:** 421
production of large spheres **EM4:** 421
solid glass spheres **EM4:** 418-419
glass beads for state-of-the-art medical treatment **EM4:** 419
marble production **EM4:** 418-419
types **EM4:** 418
Glass stress
defined **EM2:** 21
Glass substrates
applications **EL1:** 338

SUBJECTS OF THE INDEXED VOLUMES: ASM Handbook (designated by the letter "A"): **A1:** Properties and Selection: Irons, Steels, and High-Performance Alloys (1990); **A2:** Properties and Selection: Nonferrous Alloys and Special-Purpose Materials (1990); **A3:** Alloy Phase Diagrams (1992); **A4:** Heat Treating (1991); **A6:** Welding, Brazing, and Soldering (1993); **A7:** Powder Metallurgy (1984); **A8:** Mechanical Testing (1985); **A9:** Metallography and Microstructures (1985); **A10:** Materials Characterization (1986); **A11:** Failure Analysis and Prevention (1986); **A12:** Fractography (1987); **A13:** Corrosion (1987); **A14:** Forming and Forging (1988); **A15:** Casting (1988); **A16:** Machining (1989); **A17:** Nondestructive Testing and Quality Control (1989); **A18:** Friction, Lubrication, and Wear Technology (1992). **Metals Handbook, 9th Edition** (designated by the letter "M"): **M1:** Properties and Selection: Irons and Steels (1978); **M2:** Properties and Selection: Nonferrous Alloys and Pure Metals (1979); **M3:** Properties and Selection: Stainless Steels, Tool Materials and Special-Purpose Materials (1980); **M4:** Heat Treating (1981); **M5:** Surface Cleaning, Finishing, and Coating (1982); **M6:** Welding, Brazing, and Soldering (1983). **Engineered Materials Handbook** (designated by the letters "EM"): **EM1:** Composites (1987); **EM2:** Engineering Plastics (1988); **EM3:** Adhesives and Sealants (1990); **EM4:** Ceramics and Glasses (1991); **Electronic Materials Handbook** (designated by the letters "EL"): **EL1:** Packaging (1989).

Glass substrates (continued)
physical characteristics **EL1:** 106
Glass tissue
application **EM1:** 109
Glass transition **EM3:** 14
defined **EM2:** 21
Glass transition temperature *See also* Temperature(s) **EM3:** 14
and aging **EM2:** 788
and calculation of fracture energy **EM3:** 383, 384
cycle frequency effect on crack propagation rate **EM3:** 513
defined **EM2:** 21
dependence, as intrinsic **EM2:** 452
effect, moisture-related failures **EM2:** 761-764
epoxies **EM2:** 241
heterochain thermoplastic polymers **EM2:** 53
hydrocarbon thermoplastic polymers **EM2:** 50
measurement of **EM3:** 320-321
nonhydrocarbon carbon-chain thermoplastic polymers **EM2:** 51-52
polybenzimidazoles (PBI) **EM2:** 148
polyphenylquinoxalines **EM3:** 163, 164, 166
as quality assessment method **EM2:** 564-565
relaxation time related **EM3:** 422
and silicone incorporation of diphenyl groups **EM3:** 134
styrene-acrylonitriles (SAN, OSA ASA) **EM2:** 215
for temperature resistance measurement **EM2:** 559-564
thermoplastic polymers **EM2:** 54
thermoplastic polyurethanes (TPUR) **EM2:** 204
Glass transition temperature (T_g)
appropriate, thermal analysis for determining **EM1:** 779-780
change, laminate strength effects **EM1:** 226
defined **EM1:** 12
epoxy resin matrices **EM1:** 73-74
functions of **EM1:** 135
of high-temperature polymers **EM1:** 810
as matrix selection parameter **EM1:** 76
moisture absorption effects, epoxy resins **EM1:** 32
variations, causes of **EM1:** 779
wet, defined **EM1:** 32
Glass transition temperatures
defined **EL1:** 1145
dielectric **EL1:** 742
effect, epoxy resins **EL1:** 825
as laminate thermal property **EL1:** 536
resin systems **EL1:** 534-535
Glass yield
batch size **EM4:** 382
Glass(es)
defined **EL1:** 93, 109
die attach, in hermetic packages **EL1:** 215
flame polishing **EL1:** 104
insulating, SIMS analysis **EL1:** 1087-1088
phosphosilicate, and aluminum **EL1:** 965
for seals **EL1:** 455-459
silver die attach, for hermetic packages **EL1:** 215-216
wetting, on metal **EL1:** 456
Glass-base systems
for circuit protection **EM3:** 592
Glass-blowing **EM4:** 21
Glass-ceramic cookware failure analysis **EM4:** 669-673
accident reconstruction **EM4:** 672-673
accident report **EM4:** 669
breaking stress versus mirror size **EM4:** 672
fractography **EM4:** 669-671
analyses **EM4:** 669-670
discussion of analyses **EM4:** 670-671
measurement of origin mirrors **EM4:** 671
investigation of accident **EM4:** 671-672
Glass-ceramic inlays
properties **A18:** 666
Glass-ceramic ovenware
defect inclusion levels **EM4:** 392
Glass-ceramic process **EM4:** 23
Glass-ceramics **EM4:** 22-23, 439-443, 1101-1102
advantages **EM4:** 434-435
applications **EM4:** 23, 434, 1102
biomedical **EM4:** 19
dental **EM4:** 1094
as ceramic substrates **EM4:** 1110-1111
Glass-ceramics (continued)
commercial systems **EM4:** 435-438
composition **EM4:** 433, 434
crystal structure **EM4:** 30
crystallinity **EM4:** 1102
design **EM4:** 433
engineering properties *See* Engineering properties of glass ceramics
formation **EM4:** 1101-1102
high crystallinity **EM4:** 433
limitations **EM4:** 435
machinable, as bioactive material **EM4:** 1008
matrix matefial for ceramic-matrix composites **EM4:** 840
microstructure **EM4:** 433, 434
photomachinable **EM4:** 23
processing **EM4:** 433-435
properties **EM4:** 23, 30, 330, 849, 1102
Pyroceram tableware **EM4:** 1102
seals with metals **EM4:** 499-500
secondary heat treatment **EM4:** 1102
thermal properties **EM4:** 30
Glass-cleaning solutions **EM3:** 35
Glass-epoxy composites
applications **A9:** 592
etching **A9:** 591
polishing **A9:** 590
Glass-epoxy matrix composites *See also* Bulk molding compounds (BMC)
application **EM1:** 141, 163
compressive strength **EM1:** 197
fabric **EM1:** 399-400, 404-405
Glass-filled high-impact polystyrenes (PS, HIPS) **EM2:** 199
Glass-lined steels
in pharmaceutical production facilities **A13:** 1228-1229
Glass-matrix composites *See also* Engineering properties of glass-matrix composites **EM4:** 19-20
Glass-polyester resin composites
corrosion resistance **EM1:** 93, 94
dielectric strength retention **EM1:** 95
electrical properties **EM1:** 94, 95
flexural strength vs. temperature **EM1:** 93
sheet molding compound for **EM1:** 157-160
thermal stability **EM1:** 93
Glass-reinforced thermoplastic polyurethanes **EM2:** 205
Glass-sealed ceramic packages
substrates **EL1:** 203-204
Glass-to-metal seals
described **EL1:** 455-456
hybrid packages using **EL1:** 458
with low-melting temperature indium-base solders **A2:** 752
reliability and testing **EL1:** 458-459
sealing glass selection factors **EL1:** 456-458
Glass/epoxy laminate
waterjet machining **A16:** 522
Glass/metal and glass-ceramic metals seals **EM4:** 493-500
applications **EM4:** 493, 500
chemical factors in making glass/metal seals **EM4:** 496-497
reactions at glass/metal interfaces **EM4:** 496-497
wetting and spreading **EM4:** 496
classification of glass/metal seals **EM4:** 493-495
compression seal **EM4:** 494
ductile metal seal **EM4:** 494
Housekeeper seal **EM4:** 494, 495
composition and structure of glass and glass- ceramic materials **EM4:** 493
glass-ceramic/metal seals **EM4:** 499-500
glass-ceramic to metal processing **EM4:** 499-500
making of glass/metal seals **EM4:** 497-498
coefficient of thermal expansion **EM4:** 497-500
high-pressure sodium vapor lamps **EM4:** 498
metals commonly used **EM4:** 497
recommended combinations **EM4:** 497
physical factors in making glass/metal seals **EM4:** 495-496
Glass/metal seals
seal design finite-element stress- analytic techniques **EM4:** 536-538
Glass/SiO_2
methods used for synthesis **A18:** 802
Glasscutter wheels
wear of **M1:** 604, 605
Glassed steel (porcelain enamel) **M5:** 527, 529-530
Glasses *See also* Amorphous materials; Ceramics; Glasses, characterization of; Metallic glasses
analytic methods applicable **A10:** 5
applications **EM4:** 1015
As_2Te_3, K-edge EXAFS spectra of arsenic in **A10:** 411
binary phosphate, influence of cations on bonding in **A10:** 131
bond distance, coordination, and neighbors EXAFS determined **A10:** 407
borosilicate, B and F determined in **A10:** 179
bulk metallic **A2:** 819-820
C glass, effect on mechanically alloyed oxide alloys **A2:** 947
dispersion-strengthened (MA ODS)
calcium-boroaluminosilicate, SIMS depth profiles **A10:** 624
characterized **A10:** 1
defined **A2:** 804
ground, AAS analysis for silver, lead, and cadmium in **A10:** 55
high-temperature solid-state welding **A6:** 298
hydrofluoric acid as dissolution medium for **A10:** 165
hydroxyl and boron content, quantitative analysis **A10:** 121-122
inclusions that contain gases, Raman analysis **A10:** 131
K-edge EXAFS spectra of arsenic in **A10:** 411
Knight shift measurements on metallic **A10:** 284
and Kovar seals, XPS analysis **A10:** 577-578
lime glass, effect, mechanically alloyed oxide alloys **A2:** 947
dispersion-strengthened (MA ODS)
metal oxide, Raman spectroscopy **A10:** 130-131
metallic **A7:** 794
metallic, NMR Knight shift measurement on **A10:** 284
metallic, SAS applications **A10:** 405
MOLE technique for **A10:** 131
multicomponent, EDS and WDS x-ray spectra of **A10:** 521
phase separation analysis by SAXS/SANS/ SAS **A10:** 402
properties **EM4:** 1015
Raman spectroscopy for **A10:** 126, 130, 131
sales categories **EM4:** 1015
SAXS/SANS analysis of **A10:** 405
silica, RDF analysis **A10:** 397-399
silicate, bonding topologies in **A10:** 393
small-angle scattering analysis **A10:** 405
SO_2 Raman analysis **A10:** 131
spin, abbreviation for **A10:** 691
surface layer analysis in **A10:** 610, 624
typical FT-IR spectrum **A10:** 122
variation of long-range order, as function of preparation **A10:** 393
Glasses, characterization of *See also* Glasses
analytical transmission electron microscopy **A10:** 429-489
atomic absorption spectrometry **A10:** 43-59
Auger electron spectroscopy **A10:** 549-567
classical wet analytical chemistry **A10:** 161-180
controlled-potential coulometry **A10:** 207-211
electrochemical analysis **A10:** 181-211
electrogravimetry **A10:** 197-201
electrometric titration **A10:** 202-206
electron probe x-ray microanalysis **A10:** 516-535
extended x-ray absorption fine structure **A10:** 407-419
inductively coupled plasma atomic emission spectroscopy **A10:** 31-42
infrared spectroscopy **A10:** 109-125
ion chromatography **A10:** 658-667
low-energy ion-scattering spectroscopy **A10:** 603-609
neutron activation analysis **A10:** 233-242
neutron diffraction **A10:** 420-426
optical emission spectroscopy **A10:** 21-30
particle-induced x-ray emission **A10:** 102-108
potentiometric membrane electrodes **A10:** 181-187

Glasses, characterization of (continued)
radial distribution function analysis..... **A10:** 393-401
Raman spectroscopy **A10:** 126-138
Rutherford backscattering spectrometry **A10:** 628-636
scanning electron microscopy **A10:** 490-515
secondary ion mass spectroscopy **A10:** 610-627
small-angle x-ray and neutron scattering **A10:** 402-406
spark source mass spectrometry **A10:** 141-150
ultraviolet/visible absorption spectroscopy **A10:** 60-71
voltammetry **A10:** 188-196
x-ray photoelectron spectroscopy **A10:** 568-580
x-ray spectrometry **A10:** 82-101
Glasses for laboratory and process applications **EM4:** 1087-1090
glass electrodes **EM4:** 1089
glass sensors **EM4:** 1089
laboratory glassware **EM4:** 1087-1088
process systems **EM4:** 1089-1090
silica-based supports for chromatography applications **EM4:** 1088-1809
Glasses joined to ceramics
surface considerations **EM3:** 298-310
Glasses, structure and properties *See* Structure and properties of glasses
Glassiness
defined **EM2:** 20-21
Glassivation
defined **EL1:** 1145
Glassivation layers
phosphorus analyzed in **A11:** 41
Glassware
cleanliness for UV/VIS analysis **A10:** 69
Glassy carbon matrix composites
with graphite fibers **A9:** 596
polishing **A9:** 591
Glassy metals *See* Amorphous metals
Glassy plastics
ESC testing of **EM2:** 802-803
Glassy polymers *See also* Amorphous polymers; Glass
as blends **EM2:** 633
deformation **EM2:** 657
Glassy thermoplastics
stress crazing **EM2:** 797-799
Glauber salt sealing process
anodic coatings **M5:** 595
Glaze
defined **A18:** 10 **EL1:** 109
Glazed ceramics
processing defects **EM4:** 643, 644
Glazes **EM4:** 9-10, 1061-1068
application process **EM4:** 9
applications **EM4:** 10
Bristol **EM4:** 10
ceramic decoration **EM4:** 1066-1068
classification **EM4:** 1061
composition **EM4:** 45, 1063-1065
cost factors **EM4:** 1068
crystalline **EM4:** 10
FDA guidelines **EM4:** 1063-1064, 1065
firing temperatures **EM4:** 1061
fritted **EM4:** 10, 1061
heavy metal release rate **EM4:** 1062-1063, 1064-1066
lead-containing **EM4:** 1062
leadless **EM4:** 10, 1062
luster **EM4:** 10
porcelain **EM4:** 10
product availability **EM4:** 1068
properties of fired ware **EM4:** 45
raw **EM4:** 10
raw lead **EM4:** 10
refractive index of glasses **EM4:** 1064
role of specific oxides **EM4:** 1062
salt **EM4:** 10
slip **EM4:** 10
Glazes (continued)
special **EM4:** 10
zinc-containing **EM4:** 10
Glazing
brake pad material **A18:** 45
Gleeble high strain rate hot-tensile test machine **A11:** 721, 726
Gleeble system **A6:** 517
Gleeble test
low-carbon steels **A12:** 240
Gleeble test unit **A8:** 586 **A14:** 378
Gleeble unit **A1:** 581
Glide *See also* Slip
of dislocations, slip as **A8:** 34
Glide, superlattice
in ordered intermetallics **A2:** 913
Glide-plane decohesion
superalloys **A12:** 391
Glissile interfaces
shape memory alloys **A2:** 898
Glob-top encapsulants **EL1:** 803
Global (laminate) reference system **EM1:** 236
Global clocks
in WSI technology **EL1:** 9
Global communication **EL1:** 5
Global friction factor **A18:** 67
Global heat treatment
for steam generators **A13:** 942
Global illumination
MOLE/Raman analysis **A10:** 129-130
Global ply discount method
for composites **A11:** 741-742
Globular eutectic
in lead-tin alloys **A9:** 417
Globular eutectic microstructure .... **A3:** 1 • 19, 20
Globular oxide
carburizing affected by content in steels **A18:** 875
Globular particles
FeO, in iron **A12:** 223
oxide inclusions, in iron **A12:** 220
Globular sigma phase
in rhenium-bearing alloys **A9:** 448
Globular sulfides in steel **A9:** 628
Globular transfer
definition **M6:** 9
Globular transfer (arc welding)
definition **A6:** 1210
Globule test **A6:** 136, 137
for solderability **EL1:** 677, 944
standards used to evaluate stability **A6:** 136
Gloss
paint *See* Paint and painting, gloss
porcelain enamel *See* Porcelain enameling, gloss
of porcelain enamels **A13:** 451
surface, and color, testing **EM2:** 598
and toughness, high-impact polystyrenes (PS, HIPS) **EM2:** 195
Glossary *See also* Categorization; Classification; Terminology
of terms **A15:** 1-12 **EL1:** 1133-1162 **EM2:** 2-47
Glossary of terms **A4:** 948-959 **A11:** 1-11
Glossmeter, photoelectric
paint gloss testing **M5:** 491
Glow discharge cleaning
vacuum coating process **M5:** 406
Glow discharge mass spectroscopy (GDMS), for trace element measurement
ultra-high purity metals **A2:** 1095
Glow discharges **A10:** 26-29, 142
Glow-discharge nitriding *See* Ionitriding
Glucose
enzymatic determination using oxygen of fluorescence **A10:** 79
Glue **EM3:** 14
Glue line **EM3:** 14
Glue line thickness **EM3:** 14
Glue-laminated wood **EM3:** 14
Glyceregia as an etchant for
heat-resistant casting alloys **A9:** 330-331
stainless steel powder metallurgy materials **A9:** 509
wrought heat-resistant alloys **A9:** 307
Glycerin **A6:** 60
as injection molding binder **A7:** 498
Glycerol
description **A9:** 68
nitric acid and acetic acid as an etchant for tin-lead alloys **A9:** 450
and nitric acid as an etchant for magnesium alloys **A9:** 426
surface tension **EM3:** 181
Glycerol-water jets
stability of **A7:** 28, 31
Glycidyl amines **EM1:** 67, 69, 753
Glycidyl ether **EM1:** 753
Glycidyl novolac resins
as epoxy resin **EM1:** 67
Glycol **A4:** 210
incomplete removal leading to adhesion defects **EM3:** 751
Glycol phthalate
for bonding **A8:** 234
Glycolic-nitrate pickling
magnesium alloys **M5:** 630, 640-641
Glycols
as couplers **A18:** 110
Glyoxal
physical properties **EM3:** 104
GMR 235
composition **M4:** 653
GMR-235
composition **A6:** 564 **M6:** 354
electron-beam welding **A6:** 869
multiple-impulse resistance spot welding **M6:** 532
GMS *See* Gas analysis by mass spectrometry
Godbert-Greenwald furnace
explosivity testing **A7:** 197
Goiters
from cobalt toxicity **A2:** 1251
Gold *See also* Fine gold; Gold alloys; Gold alloys, specific types; Gold contact alloys, Precious metals; Precious metals; Pure gold; Pure metals **A18:** 158
(liquid), contact angles on beryllium at various test temperatures in argon and vacuum atmospheres **A6:** 116
as a reactive sputtering cathode material **A9:** 60
as alloyable coating **EL1:** 679
alloying effect **A13:** 47
applications **A2:** 704
artificial twist boundaries **A9:** 120-121
atomic interaction descriptions **A6:** 144
Auger electron microscopy map of a gallium arsenide field-effect transistor **EM3:** 241
-based filler alloy, thermal stress SCC **A13:** 879
beam leads **A11:** 775
as braze filler metal **A11:** 450
brazing, composition **A6:** 117
for brazing in ceramic/metal seals **EM4:** 504
bulk impedance response **EM3:** 436
bumps **EL1:** 277
calculated electron range **EL1:** 1095
in clad and electroplated contacts **A2:** 848
coating for dental feldspathic porcelain and ceramics **A18:** 674
coating for SEM specimens **A18:** 380
electroplate, friction coefficient data **A18:** 74
coinability of **A14:** 183
commercial fine **M2:** 679-680
as conductive thick-film material **EL1:** 249
conductor ink **EL1:** 208
contact wires **EL1:** 958
-copper alloys, EPMA analysis **A10:** 530

SUBJECTS OF THE INDEXED VOLUMES: ASM Handbook (designated by the letter "A"): **A1:** Properties and Selection: Irons, Steels, and High-Performance Alloys (1990); **A2:** Properties and Selection: Nonferrous Alloys and Special-Purpose Materials (1990); **A3:** Alloy Phase Diagrams (1992); **A4:** Heat Treating (1991); **A6:** Welding, Brazing, and Soldering (1993); **A7:** Powder Metallurgy (1984); **A8:** Mechanical Testing (1985); **A9:** Metallography and Microstructures (1985); **A10:** Materials Characterization (1986); **A11:** Failure Analysis and Prevention (1986); **A12:** Fractography (1987); **A13:** Corrosion (1987); **A14:** Forming and Forging (1988); **A15:** Casting (1988); **A16:** Machining (1989); **A17:** Nondestructive Testing and Quality Control (1989); **A18:** Friction, Lubrication, and Wear Technology (1992). **Metals Handbook, 9th Edition** (designated by the letter "M"): **M1:** Properties and Selection: Irons and Steels (1978); **M2:** Properties and Selection: Nonferrous Alloys and Pure Metals (1979); **M3:** Properties and Selection: Stainless Steels, Tool Materials and Special-Purpose Materials (1980); **M4:** Heat Treating (1981); **M5:** Surface Cleaning, Finishing, and Coating (1982); **M6:** Welding, Brazing, and Soldering (1983). **Engineered Materials Handbook** (designated by the letters "EM"): **EM1:** Composites (1987); **EM2:** Engineering Plastics (1988); **EM3:** Adhesives and Sealants (1990); **EM4:** Ceramics and Glasses (1991); **Electronic Materials Handbook** (designated by the letters "EL"): **EL1:** Packaging (1989).

Gold (continued)
corrosion applications ... **A13:** 797
corrosion in acids ... **A13:** 798
corrosion resistance ... **A13:** 796 **M2:** 669
dental solders ... **A2:** 698
determined by controlled-potential coulometry ... **A10:** 209-211
diffusion welding ... **A6:** 885
dislocations in ... **A9:** 609
electrical contacts, use in ... **M3:** 668-669
electrical resistance applications ... **M3:** 641
as electrically conductive filler ... **EM3:** 45, 178, 572, 584, 596
electrolytic corrosion of ... **A11:** 771
in electronic scrap recycling ... **A2:** 1228
electroplated layers ... **A6:** 971
electroplating ... **A18:** 837
and corrosive wear ... **A18:** 837
properties ... **A18:** 837-838
enamels ... **EM3:** 302
explosion welding ... **A6:** 896 **M6:** 710
fabrication ... **A13:** 796
field evaporation in ... **A10:** 586, 587
in filler metals used for active metal brazing ... **EM4:** 524
film conductors ... **A11:** 772
friction coefficient data ... **A18:** 71
gaseous corrosion ... **A13:** 798
glass-to-metal seals ... **EM3:** 302
gravimetric finishes ... **A10:** 171
in halogens ... **A13:** 798
hardness ... **M5:** 281
heat-affected zone fissuring in nickel-base alloys ... **A6:** 588
high-purity, SSMS analysis ... **A10:** 144
high-vacuum lubricant application ... **A18:** 153
honing stone selection ... **A16:** 476
ICP-determined in silver scrap metal ... **A10:** 41
in intermetallic compounds ... **A6:** 127
jewelry ... **M2:** 666-667
L-family x-ray lines ... **A10:** 522
lap welding ... **M6:** 673
liquid-metal embrittlement in ... **A11:** 234
in liquid-phase metallizing ... **EM3:** 306
materials for conductors ... **EM4:** 1141
in medical therapy, toxic effects ... **A2:** 1257
metal filler for polyimide-base adhesives ... **EM3:** 159
in metal powder glass frit method ... **EM3:** 305
metallization corrosion ... **A11:** 770
for metallizing ... **EM4:** 542, 544
for metallizing microwave circuitry ... **EM4:** 545
metals, for electrical contacts, properties ... **A2:** 845
-nickel-copper metallization systems, Auger elemental mapping ... **A10:** 559
nitric/hydrochloric acids as dissolution medium ... **A10:** 166
oxidation resistance ... **A13:** 796-797
particles, line scan across ... **A10:** 496
percussion welding ... **M6:** 740
photochemical machining ... **A16:** 588
photochemical machining etchant ... **A16:** 590
for plating, materials and processes selection ... **EL1:** 116
as precious metal ... **A2:** 688
in precious-metal solders ... **A6:** 968, 977
pretinning effects ... **EL1:** 731
production ... **M2:** 660
properties ... **A2:** 704-705 **A13:** 796
protective finish in electronic applications ... **A6:** 990, 991, 994-995
pure ... **M2:** 737-738
pure, properties ... **A2:** 1116
purity ... **M5:** 281
purity-hardness relationship ... **M5:** 281
recommended impurity limits of solders ... **A6:** 986
relative solderability ... **A6:** 134
as a function of flux type ... **A6:** 129
resources and consumption ... **A2:** 689
rosin flux use ... **A6:** 129
screens, radiography ... **A17:** 316
semifinished products ... **A2:** 694
as SERS metal ... **A10:** 136
single-pass sliding with amalgam material ... **A18:** 669

Gold (continued)
solderability ... **A6:** 978
in solid solution alloys ... **A6:** 127
special properties ... **A2:** 691-692 **M2:** 662-663
sponge, dealloyed ... **A13:** 130-131
stacking fault energy ... **A18:** 715
stacking-fault tetrahedra ... **A9:** 117
in stainless steel brazing filler metals ... **A6:** 911, 913, 915-916, 920
suitability for cladding combinations ... **M6:** 691
for thermal conductivity ... **EM3:** 584
thermal diffusivity from 20 to 100 °C ... **A6:** 4
thermal expansion coefficient ... **A6:** 907
thermocompression welding ... **M6:** 674
thin-film resistivity, vs film thickness ... **EL1:** 314
tin, for solder sealing ... **EM3:** 585
tin-gold eutectic ... **EM3:** 584
TNAA detection limits ... **A10:** 237
torch soldering ... **A6:** 351
trade practices ... **A2:** 690-691
TWA limits for particulates ... **A6:** 984
ultrapure, by fractional crystallization ... **A2:** 1093
ultrapure, by zone refining ... **A2:** 1094
ultrasonic welding ... **M6:** 746
usage, PWB manufacturing ... **EL1:** 510
use in thermal evaporation ... **A12:** 172
vacuum-deposited electrodes ... **EM3:** 429
in vapor-phase metallizing ... **EM3:** 306
Vickers and Knoop microindentation hardness numbers ... **A18:** 416
wettability indices on stainless steel base metals ... **A6:** 118
wetting of beryllium ... **A6:** 115
wire, and vacuum-deposited aluminum, ball bond ... **A12:** 484
for wire bonding ... **EM3:** 584-585
x-ray tubes ... **A17:** 302

Gold alloys *See also* Fine gold; Gold; Gold contact alloys
applications ... **A2:** 705-707
brazing
available product forms of filler metals ... **A6:** 119
environments that cause stress-corrosion cracking ... **A6:** 1101
joining temperatures ... **A6:** 118
corrosion of ... **A13:** 1360
dental ... **A2:** 692
for electrical contacts, properties ... **A2:** 845
PFM ... **A13:** 1355
recommended gap for braze filler metals ... **A6:** 120
tarnish ... **A13:** 1359
versus dental amalgam abrasion rates ... **A18:** 669

Gold alloys, for brazing
composition and properties ... **A7:** 839

Gold alloys, specific types *See also* Gold; Gold alloys
69Au-25Ag-6Pt ... **A9:** 562
80Au-20Cu foil ... **A6:** 146
91.7Au-8.3Ag ... **A9:** 563
Au-10Cu, as electrical contact materials ... **A2:** 846
Au-14.5Cu-8.5Pt-4.5Ag-1Zn, as electrical contact materials ... **A2:** 846
Au-25Ag-6Pt, as electrical contact materials ... **A2:** 846
Au-25Ag-9Pt-15Cu, as electrical contact materials ... **A2:** 846
Au-30Ni, cellular colonies ... **A9:** 649
Au-34Cd, martensitic structures ... **A9:** 673
fine gold (99.9 Au), as electrical contact materials ... **A2:** 845

Gold and gold alloys
aging ... **A4:** 943-944
annealing ... **A4:** 941, 942-944
colors ... **A4:** 942
compositions ... **A4:** 941, 942, 944
dental applications ... **A4:** 942, 944
electrical resistivity ... **A4:** 943, 944
hardness ... **A4:** 941, 943
mechanical properties ... **A4:** 942, 944
vapor pressure, relation to temperature ... **A4:** 495

Gold atoms in the AuCU3 Structure ... **A9:** 681

Gold brazing filler metals
properties ... **A2:** 707

Gold bronze
decorative applications ... **A7:** 595
pigments ... **A7:** 595-596

Gold bump
to gold-plate copper lead ... **EL1:** 280
to tin-plate copper lead ... **EL1:** 279-280
wafer bumping technology ... **EL1:** 277

Gold cementation
powder used ... **A7:** 573

Gold coating
on semiconductor layer ... **A9:** 97
on silver electronic circuit termination pad ... **A9:** 101

Gold coatings and films
soldering ... **A6:** 631

Gold contact alloys *See also* Fine gold; Gold alloys; Gold contact alloys; Pure gold; Pure metals
as electrical contact materials ... **A2:** 845-846
fine gold ... **A2:** 845-846
properties, applications ... **A2:** 845

Gold film conductors
electromigration effects on ... **A11:** 771-772

Gold flashing ... **M5:** 284

Gold, green *See* Gold-silver-copper alloys

Gold in dentistry ... **M2:** 684-687
cast alloys ... **M2:** 684-686, 687
applications ... **M2:** 685
composition ... **M2:** 684-685, 687
hardness ranges ... **M2:** 685, 687
mechanical properties ... **M2:** 686, 687
physical properties ... **M2:** 687
cavity filling, materials for ... **M2:** 684
gold foil, physical properties ... **M2:** 684
mat gold, physical properties ... **M2:** 684
powdered gold, physical properties ... **M2:** 684
copper addition ... **M2:** 685
lost wax process ... **M2:** 684-685
metal-ceramic technique ... **M2:** 685, 687
silver addition ... **M2:** 685
solders
applications ... **M2:** 686
composition ... **M2:** 686
wire alloys
color ... **M2:** 684, 685
composition ... **M2:** 684, 685
mechanical properties ... **M2:** 684, 686
physical properties ... **M2:** 684, 686
zinc addition ... **M2:** 685

Gold leaf
by metallic flake pigment production ... **A7:** 593

Gold plated electrical contact material ... **A9:** 562-563

Gold plating ... **M5:** 281-284 **EL1:** 638, 679
alloying metals in ... **M5:** 282-283
aluminum and aluminum alloys ... **M5:** 605
applications ... **M5:** 284
barrel ... **M5:** 282
copper and copper alloys ... **M5:** 621-623
current densities ... **M5:** 282-284, 622-623
decorative ... **M5:** 284
electroless process ... **M5:** 621-622
hot cyanide process ... **M5:** 281-283
immersion process ... **M5:** 622
magnesium alloys, stripping of ... **M5:** 647
metal distribution ... **M5:** 283
of metal-matrix composites ... **EL1:** 1127-1128
metallic impurities, effects of ... **M5:** 283
molybdenum ... **M5:** 660
niobium ... **M5:** 663-664
preparation for ... **M5:** 283
processing considerations ... **M5:** 283-284
pure gold deposits ... **M5:** 281
rigid printed wiring boards ... **EL1:** 549-550
selective ... **M5:** 283-284
solution compositions and operating
acid baths ... **M5:** 282
alloying metals in ... **M5:** 282-283
conditions ... **M5:** 281-283, 622-623, 663-664
hot cyanide ... **M5:** 281-283
immersion baths ... **M5:** 622
neutral baths ... **M5:** 282
pH ... **M5:** 282
stripping of ... **M5:** 283, 647
tantalum ... **M5:** 663-664
temperature ... **M5:** 281-282, 284
thickness ... **M5:** 283
titanium ... **M5:** 659
waste recovery ... **M5:** 284

Gold powders *See also* Gold alloys; Precious metal powders
applications **A7:** 205
exposure limits..... **A7:** 205
history..... **A7:** 14
production of **A7:** 148-150
reducing agents **A7:** 150
screen print..... **A7:** 150
thick-film..... **A7:** 149-151
toxicity **A7:** 204, 205
Gold, proof *See* Gold, commercial fine
Gold, red *See* Gold-silver-copper alloys
Gold salts
medical uses and toxicity **A2:** 1257
Gold vacuum coating..... **M5:** 388, 395, 400
Gold, vapor pressure
relation to temperature **M4:** 309, 310
Gold wafer bumping
metallization process **EL1:** 277
Gold, white *See* Gold-nickel-copper alloys
Gold wire
ball bonding **EL1:** 350
bonding..... **EL1:** 110-111
in wire bonding **EL1:** 225-226
Gold wires
scanning electron microscopy image..... **A9:** 101
Gold, yellow *See* Gold-silver-copper alloys
Gold, zone refined
impurity concentration..... **M2:** 713
Gold-aluminum wire bond system
effects of continuous voiding in..... **A11:** 776-777
Gold-base alloys
torch brazing filler metals **A6:** 328
Gold-base brazing filler metals
properties..... **A2:** 707
Gold-base soldering alloy systems
millimeter/microwave applications..... **EL1:** 755
Gold-bismuth alloys, peritectic transformations..... **A9:** 678
Bi-40Au, peritectic envelope..... **A9:** 679
Gold-cadmium alloys, martensitic structures..... **A9:** 673
as a conductive coating for scanning electron microscopy specimens..... **A9:** 97
Gold-colored brass plating..... **M5:** 285-286
Gold-nickel alloy
ion-scattering spectrometry with molybdenum contamination **A18:** 449
Gold-nickel-copper alloys **M2:** 682-683
properties..... **A2:** 706
Gold-palladium
alloy basis for dental crowns **A18:** 673
coating for SEM specimens **A18:** 380
as filler for conductive adhesives..... **EM3:** 45
Gold-palladium alloys
hydrogen embrittlement in **A11:** 45-46
Gold-palladium conductive coating..... **A9:** 98
Gold-plated stainless steel lead frame
Auger microprobe of..... **A10:** 560
Gold-platinum alloy **M2:** 683
leaching..... **A6:** 132
Gold-platinum alloy (70Au-30Pt)
properties..... **A2:** 706-707
Gold-silicon
die bond **EL1:** 213
eutectic die attach..... **EL1:** 217
phase diagram **EL1:** 213
Gold-silicon eutectic **EM3:** 584
phase diagram **EM3:** 585
Gold-silver-copper
alloy basis for dental crowns **A18:** 673
Gold-silver-copper alloys..... **M2:** 680-682
properties..... **A2:** 705-706
Gold-tin alloys
strength of..... **EL1:** 636
Gold-tin eutectic **EM3:** 584
Gold/nickel-chromium system
film preparation..... **EL1:** 314
Gold/tantalum-nitride system
film preparation..... **EL1:** 314-315
Goldhoff-Sherby parameter..... **A8:** 335
Goldstriking
copper and copper alloys **M5:** 622
Golf clubs
powders used..... **A7:** 574
Golf shoe spikes
cemented carbide **A2:** 974
Golfing
composite material applications for..... **EM1:** 846
Gomak-3 *See* Zinc alloys, specific types, AG40A
Gomak-5 *See* Zinc alloys, specific types, AC41A
Goniometer
Geiger tube..... **EL1:** 93
Goniometers
and analyzing crystals..... **A10:** 88
automatic pole-figure **A9:** 703
defined **A10:** 674
use for Raman analysis **A10:** 129
in wavelength-dispersive x-ray spectrometers..... **A10:** 89
in x-ray spectrometers..... **A10:** 87
Goodman constant-life diagram..... **A11:** 111, 112
Goodman diagram
constant-life..... **A8:** 712-713
for equivalent stress parameters **A8:** 713
for fatigue data **EM1:** 202
Goodman diagrams
springs..... **M1:** 291-295, 304
Goodness-of-fit test *See also* Chi-square tests; Empirical distribution function (EDF) goodness-of-fit tests
for normal distributions..... **A8:** 663-664
for statistical distributions..... **A8:** 637-638
statistics..... **A8:** 629
Goodness-of-fit tests..... **EM1:** 302-305
Gooseneck *See also* Hot chamber machine
defined **A15:** 6
Gooseneck punches
for press-brake forming **A14:** 536
Gorham process **EM3:** 599
parylenes formation by **EL1:** 789
Goss texture **A9:** 537
Gossler process..... **EM4:** 402
Gouge marks
alloy steels **A12:** 295
Gouges
adhesive-bonded joints..... **A17:** 613
in tubular products..... **A17:** 567
Gough model
tire wear rate..... **A18:** 580
Gouging *See also* Air carbon arc cutting
definition..... **A6:** 1210 **M6:** 9
Exo-Process for **A14:** 734
as thermal cutting **A14:** 732-734
use in oxyfuel gas repair welding **M6:** 592-593
Gouging abrasion *See also* Abrasion; Abrasive wear; High-stress abrasion; Low-stress abrasion **M1:** 599
defined **A18:** 10
simulation of **M1:** 600
Gouging amperage..... **A6:** 1175
Gouging wear ratio
of abrasion-resistant cast iron..... **A1:** 119
Gouging wear test
of abrasion-resistant cast iron..... **A1:** 120
Gouging-type corrosion
biological..... **A13:** 119
Gourdine tunnel
powder coating..... **M5:** 485-486
Gouy-Chapmann theory..... **EM4:** 119
Government Reports Announcements & Index (NTIS)..... **EM1:** 41
Government specifications *See* U.S. military and government specifications
Government-funded programs
thermoplastic..... **EM1:** 103
GPa *See* Gigapascal
GPC *See* Gel permeation chromatography
Grade **A7:** 5
of related alloys, defined **A7:** 463
Grade powders
cemented carbides..... **A2:** 951
Grade selection *See also* Material defects; Material selection
and tool and die failure **A11:** 564, 566
Graded abrasive
defined **A9:** 8
Grades
of lead..... **A2:** 543-545
Gradient elution
defined **A10:** 674
Gradient freeze technique
of GaAs crystal growth..... **A2:** 744
Gradient refractive index (GRIN) lens..... **EM4:** 460, 462-463
Gradual concentration
of corrosive substances **A11:** 211
Graduated coating
definition..... **M6:** 9
Graff-Sargent etchant used for aluminum- lithium alloys..... **A9:** 357
Graft copolymers **EM2:** 21, 58 **EM3:** 14
Grain *See also* Grain boundary; Grain boundary Grain size; Grain flow; Grain refinement; Grain refinement models; Grain size; Grain size distribution; Grain structure; Grain-boundary embrittlement; Grain-boundary separation **A7:** 5
average dislocation density..... **A10:** 358
in cast ingot, sketch **A10:** 304
defined **A9:** 8 **A11:** 5 **A13:** 7 **A15:** 6
deformation, in cold-formed parts..... **A11:** 307
direction **A8:** 61, 667-668
dropping **A12:** 126
equiaxed, in SCC fracture..... **A12:** 320
facets, dimples on..... **A12:** 399-400
facets, oxygen-embrittled iron **A12:** 222
ferrite, cleavage facets in..... **A12:** 319
flow **A11:** 5, 85, 329 **A12:** 84, 88, 434
growth, in electroslag welds **A11:** 440
growth, in weldments **A13:** 344
individual, SEM analyzed **A10:** 490
interaction stresses **A10:** 420, 424
interiors, creep through..... **A8:** 310
internal structure..... **A10:** 357
large columnar, superalloys **A12:** 391
misoriented, as defect..... **A15:** 321-322
morphology **A15:** 101, 105
morphology, topographic methods for..... **A10:** 368
multiplication, and grain structure..... **A15:** 156
orientation specifying **A10:** 359
refining, brittle fracture from..... **A11:** 85
separation, in steam-generator tubes..... **A11:** 607
shape, determination by image analysis **A10:** 309
shape, silica sand, molding effects..... **A15:** 208
structure **A8:** 501, 573-574
structure, and corrosion resistance **A13:** 193-194
structure, effect in forging failure **A11:** 339
structure, effects, aluminum SCC **A13:** 590-591
structures and dimensions, optical metallography for..... **A10:** 299
surfaces, AISI/SAE alloy steels **A12:** 340
Grain aspect ratio
mechanically alloyed oxide dispersion-strengthened (MA ODS) alloys **A2:** 944
Grain aspect ratio (GAR) **A6:** 1038, 1039
Grain boundaries *See also* Subboundaries
in Al-4.7Cu alloy **A10:** 462
anodic metal dissolution at **A13:** 46
appearance at Lüders front **A9:** 685
as Barkhausen noise source..... **A17:** 132
in beryllium, etching to reveal..... **A9:** 389-390
by channeling contrast **A10:** 505

Grain boundaries (continued)
calculated energy of, as a function of the angle of misorientation between the crystal lattices of the grains ... **A9:** 609
carbide precipitation, stainless steels ... **A13:** 551
chemistry, AES analysis ... **A10:** 549, 561-562
compositions, nickel alloys ... **A10:** 562
contrasts, FIM image of ... **A10:** 589
corrosion, defined ... **A13:** 7
crystal defects at ... **A13:** 46
defined ... **A13:** 7, 46
diffusion-induced, migration analysis by EDS/CBED ... **A10:** 461-464
digital compositional map of ... **A10:** 528
direct imaging by high resolution electron microscopy ... **A9:** 121
ditching, in intergranular corrosion ... **A13:** 582
dot map for zinc at ... **A10:** 527
effect on magnetic domain structures ... **A9:** 535
embrittlement in refractory metals, IAP studies ... **A10:** 599-600
field evaporation and ... **A10:** 587
FIM image showing gated desorption in molybdenum ... **A10:** 600
FIM/AP study of point defects in ... **A10:** 583
heat tinting to reveal ... **A9:** 136
and interfaces by fracture, AES analysis of ... **A10:** 549
as intergranular corrosion sites ... **A13:** 239
of massive transformation structures ... **A9:** 655-657
migration, diffusion-induced ... **A10:** 461-464
migration of, as grain growth mechanism ... **A9:** 697
in molybdenum, atom probe analysis ... **A10:** 600, 601
as nucleation sites ... **A9:** 694
penetration, by liquid lithium ... **A13:** 95
pinning agent, oxides as ... **A7:** 171
in polycrystalline iron ... **A9:** 609
in polycrystalline powders, changes in sintering ... **A7:** 311
in powder metallurgy materials, etching to reveal ... **A9:** 508
precipitate along ... **A10:** 307-308
precipitation and solute segregation, AES analyzed ... **A10:** 549
in pure metals ... **A9:** 610
quantitative determination by image analysis ... **A10:** 309
role in cellular precipitation ... **A9:** 647-649
segregation, analysis of ... **A10:** 481-484, 544, 549
SEM-observed ... **A10:** 490
in sensitized stainless steel ... **A13:** 930
in sintering two spheres ... **A7:** 313
transmission electron microscopy ... **A9:** 118-122
transmission electron microscopy diffraction studies ... **A9:** 119-120
in U-700 nickel-base alloy ... **A10:** 308
widening, in air-atomized copper powder ... **A7:** 117, 118
wrought heat-resistant alloys, carbide precipitation ... **A9:** 309
in wrought stainless steels, etching to reveal ... **A9:** 281
yttrium segregation at ... **A10:** 483
Grain boundary *See also* Grain; Grain size; Grain-boundary cavitation; Grain-boundary separation ... **A1:** 952
alligatoring ... **A12:** 351
as anodic in SCC process ... **A12:** 25-26
attack, preferential, in ceramic ... **A12:** 471
black voids at ... **A12:** 346
brittle fracture along ... **A11:** 100
carbide films ... **A11:** 267, 405
carbide precipitation at ... **A11:** 451
cavitation ... **A8:** 574
cementite films, embrittling effect ... **A12:** 123
cleavage fracture, iron ... **A12:** 224
copper alloy, inclusions at ... **A15:** 96
copper film ... **A12:** 157
corrosion, defined ... **A11:** 5
crack propagation along ... **A11:** 76, 77
cracking ... **A8:** 306, 501
cracking, tool steels ... **A12:** 382
creep along ... **A8:** 310
decohesion ... **A12:** 18, 23
defined ... **A9:** 8
Grain boundary (continued)
delaminations, wrought aluminum alloys ... **A12:** 418
in dynamic recovery ... **A8:** 173
effect in hydrogen embrittlement ... **A12:** 23
effect of microvoid nucleation at ... **A12:** 12
effect on cleavage fracture formation ... **A12:** 13, 17
effect on fracture ... **A8:** 35
effects, ordered intermetallics ... **A2:** 913-914
embrittlement ... **A11:** 132
embrittlement, low-carbon steel ... **A12:** 246
etching ... **A12:** 96
facets, dimples in ... **A12:** 240
fatigue cracking at ... **A11:** 102
fatigue fracture, AISI/SAE alloy steels ... **A12:** 296
ferrite, high-carbon steels ... **A12:** 278
fissuring, from hydrogen attack in low-carbon steel ... **A11:** 645
fractures, microanalysis of contaminants on ... **A11:** 41-42
mechanically alloyed oxide alloys ... **A2:** 944-945
dispersion-strengthened (MA ODS)
melting, AISI/SAE alloy steels ... **A12:** 300
melting point constituents ... **A12:** 18
methane gas along ... **A12:** 349
microvoid coalescence at ... **A12:** 12
migration, metal-matrix composites ... **A12:** 467
networks, in as-cast iron castings ... **A11:** 359
nitrides (or carbonitrides) in ... **A12:** 282
nucleation site ... **A8:** 366
paths, copper penetration following ... **A11:** 717
penetration of copper, axle failures from ... **A11:** 717
pinning agent, oxides in beryllium ... **A2:** 686
precipitation ... **A8:** 35
profile angular distributions for ... **A12:** 202
SCC fracture along ... **A12:** 320
secondary cracking ... **A12:** 222, 340, 397
separation, as casting defect ... **A15:** 548
separation, fracture mechanics of ... **A8:** 439
separation, in cobalt-chromium implant ... **A11:** 675, 681
separation, in steam tubes ... **A11:** 605
in single-phase materials ... **A8:** 173
sliding ... **A8:** 188, 306 **A12:** 19, 25, 121-123, 140-141, 364
slip lines, irons ... **A12:** 219
split ... **A11:** 667
strengthening, by embrittlement ... **A12:** 111
sulfide precipitation ... **A12:** 349
surface, iron impact fracture ... **A12:** 222
titanium alloys ... **A12:** 443 **A15:** 827
titanium and titanium alloy castings ... **A2:** 638
void formation ... **A8:** 572
zirconium ... **A2:** 663
Grain boundary diffusion ... **A7:** 5, 313
Grain boundary dislocations
reactions with lattice dislocations ... **A9:** 120
transmission electron microscopy ... **A9:** 120
Grain boundary energy ... **A18:** 400
Grain boundary etching
defined ... **A9:** 8
Grain boundary ferrite, classification of
in weldments ... **A9:** 581
Grain boundary nucleation
as a result of strain-induced boundary migration ... **A9:** 696
growth of the nucleus at the expense of polygonized subgrains ... **A9:** 696
structural detail ... **A9:** 696
Grain boundary precipitation of carbides
in austenitic stainless steels ... **A9:** 283-284
Grain boundary sliding ... **A8:** 188, 306
Grain boundary structures
thermal-wave imaging used to study ... **A9:** 91
Grain coarsening
cyclic flow curves ... **A8:** 175
defined ... **A9:** 8
Grain contrast etching
defined ... **A9:** 8
Grain density, vs. compacting pressure
beryllium powder ... **A7:** 171
Grain direction ... **A8:** 61, 667-668
Grain drills ... **A7:** 674-675
Grain dropping
defined ... **A12:** 126
Grain edges
as linear element in quantitative metallography ... **A9:** 126
Grain fineness number
defined ... **A9:** 8 **A15:** 6, 209
Grain flow
change, wrought aluminum alloys ... **A12:** 434
in closed-die forgings ... **A1:** 341, 342-343
closed-die steel forgings ... **M1:** 354, 356-357
defined ... **A9:** 8
fatigue fracture through ... **A12:** 67-68
lighting of deeply etched specimens for ... **A12:** 84, 88
in titanium and titanium alloy forgings ... **A9:** 460
unfavorable, as discontinuity ... **A12:** 67-68
Grain growth *See also* Grain size ... **EM4:** 304-310
abnormal ... **EM4:** 306-307, 310
extrinsic condition ... **EM4:** 306-307
intrinsic condition ... **EM4:** 306
attainment of high density ... **EM4:** 309-310
beryllium columnar ... **A7:** 169
by migration of high-angle boundaries ... **A9:** 696
classification of ... **A9:** 697
in cold-worked metals during annealing ... **A9:** 692, 694, 697-698
defined ... **A8:** 6-7 **A9:** 8
and densification ... **EM4:** 307, 308, 309, 310
effect of elevated temperatures and high strain on ... **A9:** 690
effect of grain-boundary free-energy on ... **A9:** 697
effect of sintering tungsten carbide/cobalt on ... **A7:** 389
effect on crystallographic texture ... **A9:** 700-701
effect on ductility ... **A8:** 164-165
effects, beryllium powder ... **A2:** 685-696
kinetic considerations ... **EM4:** 308-309
mechanisms and kinetics ... **EM4:** 304-306
driving force ... **EM4:** 304
general kinetic formulation ... **EM4:** 304
intrinsic mechanism ... **EM4:** 304
isolated second phases ... **EM4:** 305-306
kinetic laws for various controlling mechanisms ... **EM4:** 306
liquid films ... **EM4:** 305
Ostwald ripening ... **EM4:** 305, 310
parameters, particle mobility expression ... **EM4:** 306
second-phase pinning ... **EM4:** 310
solute segregation ... **EM4:** 304-305
refractory metals and alloys ... **A2:** 563
scanning electron microscopy used to study ... **A9:** 101
in shape memory effect (SME) alloys ... **A2:** 900
thermodynamic considerations ... **EM4:** 307-308
in zinc alloys ... **A9:** 489
Grain growth zone
in ferrous alloy welded joints ... **A9:** 581-581
in titanium alloy welded joints ... **A9:** 581
Grain morphology
inoculation practice for ... **A15:** 105
nucleation effects ... **A15:** 101
Grain orientation
in electrical steels, preparation of specimens ... **A9:** 531
etch pits used to determine ... **A9:** 101
and magnetic characterization ... **A17:** 131
relationship to dendritic structure ... **A9:** 638
revealed by Bitter patterns ... **A9:** 535
sensitive tint used to examine ... **A9:** 138
Grain refinement *See also* Grain; Grain refinement models; Grain structure
additions, inclusions from ... **A15:** 488
of aluminum alloys ... **A9:** 629-630 **A15:** 105, 160, 476-480, 750-751
aluminum casting alloys ... **A2:** 133-134
applications ... **A15:** 477-478
assessment, on-line ... **A15:** 166
benefits ... **A15:** 476
by alloying, wrought aluminum alloy ... **A2:** 44, 46
by flux injection ... **A15:** 453
by hexachloroethane ... **A15:** 481
by rare earth alloying ... **A15:** 481
by $TiAl_3$ constituent ... **A15:** 160
compound Al_3Ti effects ... **A15:** 105-108
of copper alloys ... **A15:** 481

Grain refinement (continued)
defined ... **A15:** 6
with dispersed precipitates, wrought aluminum alloy ... **A2:** 38
effective, and hot cracking ... **A15:** 768-769
in ferrite ... **A1:** 404, 406
of foundry alloys ... **A15:** 478-479
growth-hindering additions ... **A15:** 477
in hypoeutectic alloys ... **A15:** 167
magnesium alloy ... **A15:** 480-481, 801-802
mechanical agitation ... **A15:** 476-477
mechanism, aluminum-silicon alloys ... **A15:** 160
metastable phase effect ... **A15:** 108
models, kinetics of ... **A15:** 105-107
nitride inclusion-forming ... **A15:** 93
nucleating additions ... **A15:** 477
rapid cooling ... **A15:** 476
of wrought alloys ... **A15:** 479-480
Grain refinement models
carbide-boride ... **A15:** 106
kinetics of ... **A15:** 105-107
peritectic reaction theory ... **A15:** 106-107
Grain refiner
defined ... **A15:** 6
Grain refiners ... **A6:** 53
Grain refining inoculants
effect on aluminum alloy 1100 ... **A9:** 630-631
effect on aluminum alloy 6063 ... **A9:** 630
effect on twinned columnar growth in aluminum alloy ingots ... **A9:** 631
for use with aluminum alloys ... **A9:** 630
Grain rolls
of chilled iron ... **A14:** 353
Grain section sizes
calculated distribution of ferrite ... **A9:** 133
measured distribution of ferrite ... **A9:** 133
Grain shape *See also* Powder shape; Shape(s)
aluminum casting alloys ... **A2:** 133
beryllium ... **A2:** 683-686
Grain shape distribution in normal grain growth ... **A9:** 697
in zone-refined iron ... **A9:** 690
Grain shape, ferrite
effect on formability ... **M1:** 557-558
Grain shape in aluminum alloys
etchants for examination of ... **A9:** 355
Grain size *See also* Grain fineness number; Grain growth; Grain refinement; Powders ... **A7:** 5
in aluminum alloy ingots ... **A9:** 629-631
in aluminum alloys, etchants for ... **A9:** 355
in aluminum alloys, measurement of ... **A9:** 357
aluminum casting alloys ... **A2:** 133
aluminum P/M alloys ... **A2:** 202
austenite ... **A11:** 96, 509, 563
of austenitic manganese steel castings ... **A9:** 238
of beryllium ... **A2:** 683-686
of beryllium powders ... **A7:** 169, 756-757
beta, titanium and titanium alloy castings ... **A2:** 638
by eddy current inspection ... **A17:** 165
in carbon and alloy steels, effect on martensite ... **A9:** 178
in carbon and alloy steels, macroetching to reveal ... **A9:** 174
in carbon steel ... **A8:** 175
Charpy V-notch impact screening test for ... **A8:** 263
in closed-die forgings ... **A1:** 341
coarse, brittle fracture due to ... **A11:** 70, 96
coarse, effect in surface stress measurement ... **A10:** 387
coarsening, by laser melting ... **A13:** 503
for cold-forming operations ... **A11:** 307
comparison of measured and calculated, in aluminum-copper alloys ... **A9:** 131
in continuous cast copper alloy wirebars ... **A9:** 642
control, magnesium alloy forgings ... **A14:** 260

Grain size (continued)
copper and copper alloys, and stretch forming ... **A14:** 817-818
corrosion-etched ... **A12:** 364
defined ... **A8:** 7 **A9:** 8 **A15:** 6
determination by image analysis ... **A10:** 309, 313, 318
determination in cemented carbides ... **A9:** 274
distribution effects, superplastic alloys ... **A14:** 855
dynamically recrystallized, for copper and nickel ... **A8:** 175
effect, hot tearing ... **A15:** 160
effect in liquid-metal embrittlement ... **A11:** 229-230
effect in notched-specimen testing ... **A8:** 316-318
effect, liquid metal embrittlement ... **A13:** 175
effect, magnesium/magnesium alloys ... **A13:** 741
effect, magnetic properties ... **A2:** 763-764
effect of ductile-to-brittle transition temperature ... **A11:** 68
effect of lower yield stress ... **A11:** 68
effect of, on hardenability of steel ... **A1:** 393, 470
effect of, on notch toughness ... **A1:** 115, 744, 748-749
effect of, on precipitation strengthening ... **A1:** 402, 403
effect of, on strength of ferrite ... **A1:** 115
effect of, on strength of martensite ... **A1:** 393
effect of, on stress-corrosion cracking ... **A1:** 728
effect of, on temper embrittlement in alloy steels ... **A1:** 700-702
effect of overheating on ... **A11:** 121-122
effect of prior cold work on recrystallized ... **A10:** 308
effect of tungsten carbide on hardness ... **A7:** 775
effect on blowout failure of aluminum alloy connector tube ... **A11:** 312-313
effect on brittle fracture ... **A12:** 106-107
effect on cold-formed part failures ... **A11:** 307
effect on embrittlement ... **A12:** 29
effect on fatigue strength ... **A11:** 118-119
effect on liquid-metal embrittlement ... **A11:** 229-230, 232
effect on recrystallization kinetics in low carbon steel ... **A9:** 697
effect on stress-corrosion cracking ... **A11:** 206-208
effects, aluminum alloys ... **A15:** 751-752
effects in metal stripes ... **A11:** 773
electron probe x-ray microanalysis for ... **A10:** 516-535
epitaxial resolidification effects ... **A13:** 503
etching for visibility ... **A7:** 486
evaluation, by Shepherd fracture grain size technique ... **A11:** 563
ferrite, influence of etch time on measurement of ... **A10:** 318
fine, AISI/SAE alloy steels ... **A12:** 305
formation, and austenite cooling ... **A13:** 47
hardenability, effect on ... **M1:** 477
Hilliard's circular grid for measurement of ... **A9:** 130
historical study ... **A12:** 1, 3
increased for study of Lüders front ... **A9:** 684-685
inoculation practice ... **A15:** 105-106
large, as nonrelevant indications source ... **A17:** 106
large, rock-candy fracture as ... **A11:** 75
LEED analysis in thin oriented films ... **A10:** 536
liquid-metal embrittlement effect on ... **M1:** 688
magnesium alloys ... **A15:** 802
magnetic measurement ... **A17:** 129, 131
magnetically soft materials ... **A2:** 763
mean intercept length ... **A12:** 207
mean linear intercept measure of ... **A10:** 358
measurement ... **A15:** 476
measurements, sample preparation ... **A10:** 313
metallographic sectioning for ... **A11:** 24
neutron embrittlement, effect on resistance to ... **M1:** 686

Grain size (continued)
in nickel-base wrought heat-resistant alloys ... **A9:** 309
niobium-titanium superconducting materials ... **A2:** 1044
notch toughness of steels effect on ... **M1:** 695, 699, 701, 702
nucleation effects ... **A15:** 101
on-line computation, aluminum-silicon alloys ... **A15:** 166
and particle size, tungsten carbide ... **A7:** 153
quantitative metallography of ... **A9:** 129
in recrystallized abraded zinc ... **A9:** 37-38
reduction, in hypoeutectic alloys ... **A15:** 167
refinement, effect on hydrogen resistance ... **A13:** 170
scanning electron microscopy ... **A10:** 490-515
of shafts ... **A11:** 478
silica sand, molding effects ... **A15:** 208
in silicon irons, optimum ... **A9:** 537
in silicon irons, specimen preparation ... **A9:** 532
of silver film grown on mica ... **A10:** 543-544
steel sheet, effect on formability ... **M1:** 557-558
steel wire rod ... **M1:** 254, 255, 257
stress-life results with ... **A8:** 252
superplastic titanium alloys ... **A14:** 842-843
testing ... **A15:** 264, 751
tin, test for tinplate ... **A13:** 782
titanium alloys ... **A15:** 827
for torsion tested material ... **A8:** 155
in twisted stainless steel ... **A8:** 174-175
and workability ... **A8:** 573-574
wrought aluminum alloys ... **A12:** 437
in zinc alloys, examination ... **A9:** 490
Grain size control in beryllium ... **A9:** 390
Grain size determination in tool steels ... **A9:** 258
Grain size distribution
and AFS grain fineness number ... **A15:** 209
calculation of ... **A9:** 131-134
for ceramic molding ... **A15:** 248
molding effects ... **A15:** 208
in normal grain growth ... **A9:** 697
in zone-refined iron ... **A9:** 690
Grain size number
ASTM ... **A9:** 129-130
of austenitic manganese steel castings method to obtain ... **A9:** 238
Grain structure *See also* Equiaxed grain structure; Grain refinement; Grain size ... **A8:** 501, 573-574
aluminum alloys ... **A15:** 750
aluminum-silicon alloys ... **A15:** 159-161
cast, at hot working temperatures ... **A8:** 574
dendritic, in electrodeposition ... **A7:** 71, 72
equiaxed, growth of ... **A15:** 130-132
ferritic, annealed ... **A7:** 184
grain multiplication and ... **A15:** 156
nucleant effects ... **A15:** 105
sag-resistant, from potassium in tungsten powder production ... **A7:** 153
solder ... **EL1:** 733
zones of ... **A15:** 130
Grain structures
in aluminum alloy ingots ... **A9:** 629-631
of aluminum alloys, etching to reveal ... **A9:** 354-355
in copper alloy ingots ... **A9:** 640-642
dark-field illumination used to examine ... **A9:** 76
dihedral angles ... **A9:** 604
elongated ... **A9:** 609
equiaxed ... **A9:** 609
in eutectics ... **A9:** 620
factors controlling ... **A9:** 603
of ferrites ... **A9:** 538
in pure metals ... **A9:** 607-610
revealed by etching with polarized light ... **A9:** 59

Grain structures (continued)
three-dimensional related to two-dimensional ... **A9:** 603
types ... **A9:** 602-603
under polarized light ... **A9:** 77-78
in Waspaloy, different illumination modes compared ... **A9:** 150
Grain(s) *See also* Grain boundary; Grain growth; Grain refinement; Grain shape; Grain size
abrasive superhard ... **A2:** 1008-1015
coarsening, amorphous materials and metallic glasses ... **A2:** 811
mechanically alloyed oxide alloys ... **A2:** 944
dispersion-strengthened (MA ODS)
superabrasive ... **A2:** 1012-1015
Grain-boundary area per unit volume
image analysis determined ... **A10:** 309
Grain-boundary cavitation
ASTM/ASME alloy steels ... **A12:** 349
effect on crack propagation rate ... **A12:** 38
intergranular creep rupture by ... **A12:** 19, 26
irons ... **A12:** 219
nucleation, growth, coalescence ... **A12:** 349
Grain-boundary chemistry ... **A1:** 954
Grain-boundary ferrite (GBF) ... **A6:** 1011
Grain-boundary free-energy
effect on grain growth ... **A9:** 697
Grain-boundary grooving
observation with a hot-stage microscope ... **A9:** 82
Grain-boundary liquation
defined ... **A9:** 8
nickel-base alloys ... **A6:** 588
Grain-boundary migration
at elevated temperatures and low strain rates ... **A9:** 690
Grain-boundary precipitation ... **A13:** 154-156
Grain-boundary relief
in tin and tin alloys as a result of excess polishing ... **A9:** 449
Grain-boundary segregation ... **A13:** 156-157
Grain-boundary segregation analysis
applications ... **A10:** 483
broad segregant distributions ... **A10:** 482
by analytical electron microscopy ... **A10:** 481-484
narrow interfaces ... **A10:** 482-483
Grain-boundary separation *See also* Decohesive rupture(s); Intergranular brittle fracture
copper alloys ... **A12:** 402
final, along titanium alloys ... **A12:** 441
high-strength aluminum alloy ... **A12:** 174
phosphorus, embrittlement by ... **A12:** 29
phosphorus, tool steels ... **A12:** 375
wrought aluminum alloys ... **A12:** 418, 423
Grain-boundary sliding, at elevated temperatures and low strain rates ... **A9:** 690
in Al-1.91Mg, deformed 0.62% ... **A9:** 691
Grain-boundary sulfide precipitation
defined ... **A9:** 8
Grain-boundary traces
of a single-phase palladium solid solution ... **A9:** 126
application of test grids to measure ... **A9:** 124
in extruded molybdenum ... **A9:** 124
Grain-boundary voids
formation ... **A14:** 364
Grain-coarsened zone of HAZ (GC HAZ) ... **A6:** 80-81
Grain-corner cracks *See* Triple-point cracking; Wedge cracks
Grain-refined beryllium-copper casting alloy
properties and applications ... **A2:** 390-391
Grain-refining agents *See also* Grain refinement; Grain refiners; Master alloys
aluminum-silicon alloys ... **A15:** 160-161
Grainex process
hot isostatic pressing ... **A15:** 541, 543
Graininess
x-ray film ... **A17:** 326
Graininess in micrographs ... **A9:** 85
Graining ... **A7:** 5
Grains *See also* Subgrains
effect of thermal gradient on growth ... **A9:** 579
formation in copper alloy ingots ... **A9:** 638
Gram
abbreviation ... **A8:** 724
Gram-atom
equivalence ... **A10:** 162
Gram-equivalent weight
defined ... **A10:** 674
Gram-force
abbreviation ... **A8:** 724
Gram-force (gf)
gage calibration for microindentation hardness testers ... **A18:** 415
Gram-molecular weight
defined ... **A10:** 674
Grand average ... **EM3:** 793
Granite
high-level waste disposal in ... **A13:** 975
as lap plate material ... **EM4:** 353
PCD tooling ... **A16:** 110
ultrasonic machining ... **A16:** 532
Granular bainite ... **A9:** 665
Granular fracture *See also* Crystalline fracture; Fibrous fracture; Fracture test; Silky fracture ... **A8:** 7
defined ... **A9:** 8 **A11:** 5
Granular fractures *See also* Intergranular fracture(s); Transgranular fracture(s) Graphite
effect in malleable iron ... **A12:** 238
morphology, effect in gray irons ... **A12:** 226
nodules, ductile irons ... **A12:** 229-237
Granular materials
thieves as sampling tool for ... **A10:** 16
Granular powder ... **A7:** 5
Granular structure ... **EM3:** 14
defined ... **EM2:** 21
Granulated wire, as hybrid consolidation process
composite contact materials ... **A2:** 857
Granulating ... **A7:** 6
Granulation ... **A7:** 6
history of ... **A15:** 18-19
purpose ... **EM4:** 42
of uranium dioxide pellets ... **A7:** 665
Granulation and spray drying ... **EM4:** 100-107
bonding mechanisms ... **EM4:** 100
capillary state ... **EM4:** 100, 101
funicular state ... **EM4:** 100, 101
pendular state ... **EM4:** 100, 101
definition ... **EM4:** 100
as feedstock for other techniques ... **EM4:** 100
granulation techniques ... **EM4:** 100
agitation ... **EM4:** 100-101, 102
agitation and granule size limitations ... **EM4:** 101
agitation by mixing method ... **EM4:** 100-101
agitation by tumbling method ... **EM4:** 100
pressure ... **EM4:** 100, 101-102
pressure by extrusion ... **EM4:** 101-102
pressure by pelletizing ... **EM4:** 101, 102
pressure by roll briquetting ... **EM4:** 101
spray ... **EM4:** 100, 102
spray drying ... **EM4:** 102-107
atomization ... **EM4:** 104
binder selection ... **EM4:** 103, 106
characterization ... **EM4:** 107
droplet drying ... **EM4:** 105-106
droplet-air mixing ... **EM4:** 104-105
dryer operation ... **EM4:** 106-107
powder collection ... **EM4:** 106
powder contamination ... **EM4:** 107
slurry preparation ... **EM4:** 102-104
slurry transfer pumps ... **EM4:** 104
wall deposits ... **EM4:** 105, 106
Granules
aluminum powder grade ... **A7:** 125
Graphene *See also* Carbon fibers; Graphite; Graphite fibers
defined ... **EM1:** 49
microstructure ... **EM1:** 50
Graphical analysis
of data ... **EM2:** 603-604
Graphical method
for *da/dN* calculation ... **A8:** 680
Graphics *See* Computer aided analysis; Computer-aided design; Computer-aided manufacturing; Interactive graphics system (IGS); Software; specific graphics software; Unified shapes checking (USC)
Graphics language 1 (GL/1) ... **EL1:** 132-133
Graphics libraries
callable ... **A14:** 410
Graphite *See also* Allotropy; Carbon; Carbon fibers; Graphite fibers; Graphite flake(s); Graphite molds; Graphitic corrosion; Graphitization ... **A16:** 105, 106 **A18:** 113, 141, 698-700 **EM3:** 14 **EM4:** 45
abnormal forms, in gray iron ... **A15:** 642
additions, effect on green strength ... **A7:** 302
amount, ductile iron ... **A15:** 655
as an addition to copper-base powder metallurgy electrical contacts ... **A9:** 551
as an addition to silver-base electrical contact materials ... **A9:** 551-552
as an embedding agent ... **EM4:** 572
as antiseize additive ... **A18:** 102
applications ... **EM4:** 46, 828-829, 983
ash content ... **EM4:** 828, 829
for brake linings ... **A18:** 570
brazing *See* Brazing of carbon and graphite
brazing and soldering characteristics ... **A6:** 635
brazing of cast irons, effect on ... **M6:** 996
in bronze powder metallurgy materials etching ... **A9:** 509
brush-grade characteristics ... **A7:** 635
bulk densities ... **EM4:** 828, 829
in cast iron ... **M1:** 3-9
cast iron, effects on welding ... **M1:** 563-564
in cast iron solidification ... **A15:** 82
in cemented carbides ... **A9:** 274
ceramic/metal seals ... **EM4:** 508
chemical vapor deposition of ... **M5:** 381
clean, friction coefficient data ... **A18:** 75
coating for ... **A13:** 458
codes ... **EM4:** 827-831
coefficient of friction (powder form) ... **A18:** 825
in compacted graphite iron ... **A1:** 56, 57, 61
in composite powders ... **A7:** 174
composition ... **EM4:** 46
as compound ... **A15:** 29
content in cast irons ... **A16:** 648, 649, 654
conversion to diamond ... **A2:** 1009-1010
crab-form *See* Crab-form graphite
crystal structure ... **A16:** 453, 454
crystalline structure ... **A15:** 169
defect growth theory ... **A15:** 177
defined ... **A9:** 8 **A15:** 6 **EM1:** 12, 49 **EM2:** 21
and diamond, theoretical properties ... **EM1:** 49
in die material for sheet metal forming ... **A18:** 628
distributions, specifications ... **A15:** 631
drilling/countersinking ... **A16:** 899
in ductile iron ... **A1:** 38-39, 44-45
in ductile iron, alloying elements ... **A15:** 648
for EDG wheels ... **A16:** 565, 566
effect, mechanical properties of cast iron ... **A15:** 68-69
effect on ductile iron pistons for gun recoil mechanism ... **A11:** 360-361
effect on hardness ... **A7:** 452
for electrical conductivity ... **EM3:** 596
electrical properties ... **EM4:** 827, 828
electrode corrosion factor ... **EM4:** 388
electrode polarity in EDM ... **A16:** 558, 559, 560
engineering properties ... **EM4:** 827-831
-epoxy test structure, compression fracture ... **A11:** 742-743
and eutectic freezing ... **A16:** 650
eutectic structure ... **M1:** 5-6, 7
ferrosilicon as inoculant for ... **A15:** 105
fiber reinforced copper matrix composites for space power radiator panels ... **A2:** 922
fiber, reinforcement, aluminum metal-matrix composites ... **A2:** 7
fibers, in continuous fiber reinforced composites ... **A11:** 731
as filler for polysulfides ... **EM3:** 139
filler for seals ... **A18:** 551
as filler material ... **EM2:** 499
flake *See* Flake graphite; Quasi-flake graphite ... **A15:** 120, 631-632
flakes ... **A11:** 360
flotation ... **A15:** 698
formation ... **A11:** 99-100
formations causing cracking ... **M6:** 834
forms of *See also* specific forms by name ... **M1:** 5-6
as friction modifier ... **A18:** 104
glass-to-metal seals aid ... **EM3:** 302

Graphite (continued)
grades ... **EM4:** 827-831
in graphite/Al MMCs ... **A16:** 898-899
in graphite/steel MMCs ... **A16:** 898-899
in graphite/Ti MMCs ... **A16:** 898-899
in gray cast iron, scanning electron microscopy used to study ... **A9:** 101
in gray iron ... **A1:** 24-26 **A9:** 36, 38-40, 43
in gray iron, effects of ... **A11:** 359
as gray iron inoculant ... **A15:** 637
grease additive ... **A18:** 125
grease, effect on bearing strength ... **A8:** 60
growth, in cast iron ... **A15:** 175-179
growth, theories of ... **A15:** 177-178
hardness of gray iron, effect on ... **M1:** 19-20, 30
as heating element in HIP processing ... **A7:** 423
heating element, use in vacuum furnace ... **A4:** 500
for heating elements and hot furnace structures ... **A4:** 497, 500
for heating elements for electrically heated furnaces ... **EM4:** 245-247, 248
for heating elements used in hot isostatic pressing ... **EM4:** 195
as heating material ... **A2:** 829
high-speed tool steels used ... **A16:** 59
high-vacuum lubricant application disadvantages ... **A18:** 153
impervious ... **A13:** 1227
impressed-current anodes ... **A13:** 469
intercalated, lubricant for thermoplastic composites ... **A18:** 823, 826
intercalated, thermogravimetric analysis ... **A18:** 823
in iron castings ... **A11:** 359
kish *See* Kish graphite
lattice structure ... **A2:** 1009-1010
as layer lattice solid lubricant ... **A18:** 114-115, 117
life in furnace atmospheres ... **M3:** 655
as lubricant ... **EM4:** 121
as lubricant for hot forging of tool steels ... **A18:** 738
as lubricant in hot extrusion of tool steels ... **A18:** 738
lubricated friction testing ... **A18:** 48
machinability of gray iron effect on ... **M1:** 22
manufacturers ... **EM4:** 827
mechanical properties of gray iron effect on ... **M1:** 16, 19-20, 29
metal forming lubricant ... **A18:** 148
in metal-matrix composites ... **A18:** 803, 804, 807, 808, 810
methods used for synthesis ... **A18:** 802
mold mixture, appearance ... **A15:** 273
molded, friction coefficient data ... **A18:** 75
and molybdenum disulfide mixture ... **A14:** 238
morphology, gray irons ... **A15:** 631-632
nodules *See* Spherulitic graphite; Temper carbon
nominal compositions and properties ... **EM4:** 828
nonspheroidal, ductile iron ... **A15:** 648
operating temperatures, furnace ... **M3:** 647
outgassed, friction coefficient data ... **A18:** 75
particle size, rammed graphite molds ... **A15:** 273-274
particles, in ductile iron ... **A15:** 647
peck drilling ... **A16:** 899
as permanent mold material ... **A15:** 285
physical properties of gray iron effect on ... **M1:** 31
polymer additive ... **A18:** 154
powdered, for room-temperature compression testing ... **A8:** 195
presence in cast irons ... **M6:** 308
primary, in cast iron ... **A15:** 174
properties ... **A6:** 629 **A18:** 801, 803, 808, 817 **M3:** 646 **EM4:** 827
pyrolytic, used to redensify a carbon-carbon composite ... **A9:** 595-596
quasi-flake *See* Quasi-flake graphite

Graphite (continued)
rolling-element bearing lubricant ... **A18:** 138
scrapers used in EDG ... **A16:** 548, 549, 550
sensitive tint used to examine ... **A9:** 138
in sintered iron bearings ... **A9:** 509
size/quantity, in castings ... **A17:** 531-532
solid, as mold material ... **A15:** 285
as solid lubricant ... **A11:** 150
solid lubricant for gears ... **A18:** 541
in spheroidal cast iron ... **A15:** 120
spheroidal, solid particle effect ... **A15:** 142
spherulites *See* Spherulitic graphite
spray coating to aid oxide lubrication during hot extrusion ... **A18:** 738
stability of ... **A15:** 61
structure ... **A18:** 114 **EM4:** 827
structure, gray iron ... **M1:** 12-13
structure, in ductile iron castings ... **A17:** 532
structure, shrinkage condition ... **A11:** 359
structure, testing ... **A15:** 663-664
structure, variations classified ... **A15:** 175-176
structures, ductile iron ... **A15:** 654-655, 663-664
supply sources ... **EM4:** 46
surface effects ... **A15:** 314
thermal expansion coefficient ... **A6:** 907
thermal properties ... **A18:** 42 **EM4:** 828, 829
tongs for manual resistance brazing ... **A6:** 340
in tool steels ... **A9:** 258-259
type A ... **A11:** 364
type D ... **A11:** 360, 365
types ... **A15:** 169-170
ultrasonic machining ... **EM4:** 359
uniaxial hot pressing dies ... **EM4:** 187, 189
use in U.S. by percent ... **EM4:** 829
vermicularity, fatigue fracture from ... **A11:** 360-361
wear particles ... **A18:** 305
wear resistance of gray iron effect on ... **M1:** 24-26, 30
workpiece coatings for sheet metal forming ... **A18:** 738
for workpiece fixtures supporting hot isostatic pressing ... **EM4:** 195
Graphite coatings
nickel powders ... **A7:** 137
Graphite composite
for horizontal stabilizer stub box test ... **EM3:** 539
waterjet cutting ... **A16:** 522
Graphite composites *See also* Graphite fabric composites
drilling of ... **EM1:** 668-669
faying surface sealing in ... **EM1:** 719-720
galvanic corrosion in ... **EM1:** 706, 709, 716-718
hole generation in ... **EM1:** 712
splintering, by cutting ... **EM1:** 667
Graphite containers ... **A7:** 157
Graphite crucible, induction-heated
tungsten carbide powder ... **A7:** 157
Graphite crucibles
IGF analysis ... **A10:** 227
Graphite die
assembly, resistance heated ... **A7:** 507
in hydraulic press ... **A7:** 506
Graphite diffusion
in sintering ... **A7:** 340
Graphite fabric composites
construction effect on flexural strength ... **EM1:** 149
forms ... **EM1:** 148
hybridization ... **EM1:** 149-150
properties ... **EM1:** 149
reinforced phenolic resin, as medium temperature matrix composite ... **EM1:** 382
Graphite fiber ... **EM3:** 14
for brake linings ... **A18:** 570
Graphite fiber reinforced plastics (GFRP)
damping in ... **EM1:** 214
Graphite fiber-PMR-15 polyimide components ... **EM1:** 812-814
Graphite fiber-reinforced polyimide (GFRPI) ... **A18:** 117-118

Graphite fibers *See also* Carbon fiber; Carbon fibers; Continuous graphite fiber MMCs; Fiber(s); Fibers; Graphite; Graphite composites; Graphite-epoxy composites ... **EM1:** 49-53
and carbon fibers, compared ... **EM1:** 867
in carbon/carbon composites ... **EM1:** 916
defined ... **EM1:** 12 **EM2:** 21
elastic properties ... **EM1:** 188
forms ... **EM1:** 360-361
high-modulus, in amorphous polymers ... **EM2:** 758-759
honeycomb core material ... **EM1:** 723
importance ... **EM1:** 43
as low-density high-strength ... **EM2:** 505-506
manufacture,for continuous graphite fiber MMCs ... **EM1:** 867-868
metal-matrix composites, misaligned ... **A12:** 465
nickel-coated, as reinforcements ... **EM2:** 472
properties ... **EM1:** 361
reinforced epoxy resin composites ... **EM1:** 400, 412-414
in space and missile applications ... **EM1:** 817
in tennis rackets ... **EM1:** 31
tensile modulus vs. bulk resistivity ... **EM1:** 113
types ... **EM1:** 43
in *Voyager* aircraft ... **EM1:** 29, 30
Graphite flake(s)
in gray iron ... **A15:** 631-632
microstructure of ... **A15:** 120
sizes ... **A15:** 632
Graphite flakes
gray iron fracture at ... **A12:** 225
Graphite fluoride
as layer lattice solid lubricant ... **A18:** 116
Graphite furnace atomic absorption spectrometry
analytical sensitivities ... **A10:** 47
applications ... **A10:** 55
atomizers as air filters ... **A10:** 58
and flame AAS, compared ... **A10:** 49, 58
sample preparation ... **A10:** 55
spectrometers ... **A10:** 50-51
of trace metals in hydrogen peroxide ... **A10:** 57-58
Graphite furnace atomizers ... **A10:** 48, 49, 53
Graphite, heating element
use in vacuum furnace ... **M4:** 316
Graphite in ductile iron ... **M1:** 33, 37, 38, 40, 49, 53
Graphite in gray cast iron
effect on wear characteristics ... **M1:** 598
Graphite inclusions in
nickel alloys ... **A9:** 436-437
nickel-copper alloys ... **A9:** 436-438
Graphite iron
composition limits ... **M6:** 309
welding -metallurgy ... **M6:** 308
Graphite laminate
properties ... **EM1:** 184
Graphite molds
horizontal centrifugal casting ... **A15:** 296-297
rammed ... **A15:** 273-274
vertical centrifugal casting ... **A15:** 304
Graphite nodule count
in malleable iron ... **M1:** 59-60
Graphite nodules in cast iron
revealed under different illuminations ... **A9:** 81
Graphite nodules in ductile iron
application of point-count grids to ... **A9:** 124
Graphite paper
for heating elements used in hot isostatic pressing ... **EM4:** 195
Graphite powder and iron
microstructure ... **A7:** 315
Graphite prepreg tape ... **EM1:** 139, 714
Graphite retention, in cast irons
during specimen preparation ... **A9:** 243-244
effect of polishing cloth selection and wetness on ... **A9:** 244
effect of wet versus dry fine grinding on ... **A9:** 243

Graphite yarn
dry bundle tensile strength ... **EM1:** 361-362
Graphite-aluminum bronze composite ... **A9:** 594
Graphite-aluminum composites ... **A13:** 859-860
crenelated fiber pull-out ... **A9:** 597
elastic properties ... **EM1:** 188
fibers precoated with titanium and boron ... **A9:** 594
liquid infiltration of fiber bundles ... **A9:** 594
liquid metal infiltration (LMI) precursor production ... **EM1:** 868-869
polishing ... **A9:** 590-591
shapes ... **EM1:** 870
unidirectional ... **A9:** 594
wire ... **EM1:** 869
Graphite-carbon (70%-30%)
properties ... **A18:** 817
Graphite-containing cermets ... **A7:** 813-814
applications and properties ... **A2:** 1005
Graphite-copper composites ... **A9:** 592
Graphite-epoxy ... **EM3:** 821
AS/3501-5, 16-ply parent laminate allowables ... **EM3:** 822
AS/3501-5 tape ... **EM3:** 821
AS/3501-6, as patch material ... **EM3:** 822, 824
bonding defects from surface treatment ... **EM3:** 524
cure effect on bonding weakness ... **EM3:** 530
emphasized for commercial transport applications ... **EM3:** 835-837
laminate matrix, cracking under uniaxial tension ... **A8:** 714
residual strength data for ... **A8:** 717
specimens, weak bond inspection ... **EM3:** 530
statistical distribution of fatigue life for ... **A8:** 718
ultimate static strength data ... **A8:** 715
ultrasonic NDE ... **EM3:** 778, 779, 781
Graphite-epoxy composites *See also* Bulk molding compounds (BMC), Graphite epoxy, laminates; Graphite-epoxy tape; Graphite-epoxy tooling
96-ply, quasi-isotropic ... **A9:** 593
applications ... **A9:** 592
cloth, fastener hole techniques/tools for ... **EM1:** 713, 714
crack growth in ... **A11:** 737
with delamination crack ... **A9:** 594
edge replica of ... **EM1:** 775
effect of fiber orientation ... **A11:** 733
failed tensile surface ... **A9:** 596
failure analysis ... **EM1:** 772-773
failure surfaces in stress space ... **EM1:** 199
with glass tracer-fiber bundles ... **A9:** 593
grinding ... **A9:** 588
interlayered ... **A9:** 594
polishing ... **A9:** 590
resin toughness ... **EM1:** 262
sporting goods application ... **EM1:** 163
tapered-box structure, fractured component ... **A11:** 742
translaminar fracture in ... **A11:** 738
unidirectional ... **A9:** 593-594
Graphite-epoxy composites (Gr-Ep) ... **EM3:** 288, 289
Graphite-epoxy laminates *See also* Graphite-epoxy composites
delamination growth ... **EM1:** 784
elastic constants for ... **EM1:** 223
high-modulus, properties ... **EM1:** 223
tensile strength ... **EM1:** 233
thermal expansion coefficients ... **EM1:** 225-226
ultrasonic C-scan of ... **EM1:** 776
x-ray radiograph of ... **EM1:** 776
Graphite-epoxy tape
microstructure of ... **EM1:** 768
properties ... **EM1:** 313-320
spools of ... **EM1:** 143
unidirectional, fastener holes for ... **EM1:** 713-714
Graphite-epoxy tooling *See also* Graphite epoxy composites ... **EM1:** 586-589
manufacture, from prepregs ... **EM1:** 587-588
materials in production ... **EM1:** 589
Graphite-epoxy-aluminum/titanium substructures, fastener holes
techniques tools for ... **EM1:** 713-714
Graphite-glassy carbon ceramic composite ... **A9:** 596
Graphite-magnesium composites ... **A9:** 592
castings ... **EM1:** 871-873
liquid metal infiltration (LMI) precursor production ... **EM1:** 868-869
Graphite-polyimide (Gr-PI) composites
emphasized for higher-temperature applications ... **EM3:** 829, 832-833, 836-837
polyphenylquinoxaline bonding to graphite-polyimide composite ... **EM3:** 167
polyphenylquinoxaline bonding to titanium ... **EM3:** 167
Graphite-polyimide composites ... **A9:** 592
Graphite-PPS (polyphenylene-sulfide) composite ... **A9:** 597
Graphite-silver copper composite ... **A9:** 595, 597
Graphite-to-aluminum joints
corrosion of ... **EM1:** 716-718
Graphite/aluminum metal-matrix composites ... **A2:** 905
Graphite/epoxy laminate
waterjet cutting ... **A16:** 522
Graphite/epoxy motor case
in space boosters/satellites ... **A13:** 1102-1103
Graphite/graphite composites
for heating elements used in hot isostatic pressing ... **EM4:** 195
Graphites
activated charcoal ... **A10:** 132
determined in steel or iron ... **A10:** 178
and diamond, crystal structures of ... **A10:** 345, 355
diffraction patterns from oriented pyrolytic ... **A10:** 543
highly oriented pyrolytic ... **A10:** 132
intercalated ... **A10:** 132
powdered, effect in dc arc sources ... **A10:** 25
quantitative effect of carbon KVV lineshapes ... **A10:** 553
Raman analyses of ... **A10:** 126, 132-133
single-crystalline ... **A10:** 132
SSMS analysis ... **A10:** 144
stress-annealed pyrolytic ... **A10:** 132
structural integrity of ... **A10:** 133
vitreous carbon ... **A10:** 132
wettability of ... **A10:** 543
Graphitic carbon
defined ... **A15:** 6
Graphitic carbons
determined by selective combustion ... **A10:** 223-224
determined in steel or iron ... **A10:** 178
oxidized in resistance furnace ... **A10:** 224
surface, XPS analysis of ... **A10:** 568
Graphitic cast irons
classifications ... **A9:** 245
postweld heat treatment ... **A15:** 527
Graphitic corrosion *See also* Corrosion; Dealloying; Graphite; Graphitization; Selective leaching
of cast iron pump impeller ... **A11:** 374-375
cast irons ... **A13:** 334, 568
as dealloying of gray iron ... **A13:** 133-134
defined ... **A11:** 5 **A13:** 7
and graphitization, compared ... **A13:** 134
of gray iron pump bowl ... **A11:** 372-373
of gray iron water-main pipe ... **A11:** 372-374
in iron castings ... **A11:** 372
as selective leaching ... **A11:** 179
Graphitic materials, brazing ... **A6:** 950
chemical-vapor infiltration (CVI) ... **A6:** 950
liquid impregnation ... **A6:** 950
Graphitization *See also* Graphite; Graphitic corrosion ... **A1:** 696-697 **A13:** 7, 134 **A16:** 105 **A18:** 584 **EM3:** 14 **EM4:** 827
in ASTM A201 plate steels ... **A9:** 205
at elevated-temperature service ... **A1:** 642
at heat-affected zone ... **A11:** 614
carbon and carbon-molybdenum steels ... **M1:** 686
carbon-carbon composites ... **EM4:** 837, 838
defined ... **A9:** 8 **A15:** 6 **EM2:** 21
degree of ... **EM4:** 20
effect in carbon and low-alloy steels ... **A11:** 613
as embrittlement, interpreting ... **A12:** 124
free temper carbon ... **A15:** 31
morphology, effect on gray iron fracture ... **A12:** 226
in steam equipment ... **A11:** 613-614
of steels ... **A11:** 99-100
ternary iron-base alloys ... **A15:** 68
Graphitization (continued)
in tool steels, effect of processing procedures on ... **A9:** 258
in tool steels, test disks ... **A12:** 141, 162
Graphitization, in conversion
carbon fiber ... **EM1:** 112
Graphitizers, effect
ternary iron-base systems ... **A15:** 65
Graphitizing inoculants
alloy cast irons ... **M1:** 80
Grass staggers
as magnesium deficiency disorder ... **A2:** 1259
Graticules
microscope eyepiece ... **A9:** 73-74
Grating replicas
for SEM intemal calibration ... **A12:** 167
Gratings
diffraction ... **A10:** 23
holographic ... **A10:** 128
line, diffraction of light by ... **A10:** 345
monochromators, as wavelength sorting devices ... **A10:** 23
polychromators ... **A10:** 23, 37
in Raman spectrometer ... **A10:** 128
use to determine atom location in unit cells ... **A10:** 345
Gravel
angle of repose ... **A7:** 283
Gravelometer
paint impact resistance tests ... **M5:** 492
Graville diagram ... **A6:** 407
Gravimetric analysis *See* Gravimetry
Gravimetric finishes
common ... **A10:** 171
Gravimetric procedure
for coating weight ... **A13:** 391-392
Gravimetry
as characterization method ... **EM1:** 294
common finishes ... **A10:** 171
of compounds ... **A10:** 171
described ... **A10:** 162
goal of ... **A10:** 163
of moisture and water ... **A10:** 171
species weighed in ... **A10:** 171-172
to analyze the bulk chemical composition of starting powders ... **EM4:** 72
vs volumetric analysis ... **A10:** 171-172
weighing as the chloride ... **A10:** 171
weighing as the chromate ... **A10:** 171
weighing as the dimethylglyoxime complex ... **A10:** 171
weighing as the metal ... **A10:** 170-171
weighing as the oxide ... **A10:** 170
weighing as the phosphate or pyrophosphate ... **A10:** 171
weighing as the sulfate ... **A10:** 171
weighing as the sulfide ... **A10:** 171
Gravitational acceleration *See also* Low-gravity effects
solidification effects ... **A15:** 147-158
Gravitational force method
shot propulsion ... **M5:** 143
Gravity *See also* Counter-gravity low-pressure casting
and component removal ... **EL1:** 715-716
and dendritic spacing ... **A15:** 154-155
Gravity die casting *See also* Permanent mold casting
aluminum alloy ... **A15:** 753
aluminum casting alloys ... **A2:** 139
as permanent mold process ... **A15:** 34
zinc alloy ... **A2:** 529-530
Gravity drop hammers ... **A14:** 25, 41-42, 654
Gravity hammer
defined ... **A14:** 7
Gravity sedimentation
to analyze ceramic powder particle sizes ... **EM4:** 67
Gravity segregation
macrosegregation as ... **A15:** 139
Gravity separation
automobile scrap recycling ... **A2:** 1212
Gravity sintering ... **A7:** 661, 698
Gravity unloading
of presses ... **A14:** 500
Gravity welding ... **A6:** 178
Gravity-circulated oven ... **EM3:** 37
Gravity-feed infiltration ... **A7:** 553

Gravity-feed powder flame spraying system ... M5: 540-542
Gravity-feed welding, applications
shipbuilding ... A6: 384
Gray
abbreviation for ... A10: 691
-level thresholding ... A10: 309
-levels ... A10: 310
scale ... A10: 526, 528
as SI derived unit, symbol for ... A10: 685
Gray area
definition ... EM4: 632
Gray cast iron *See* Gray iron.... A13: 87, 567, 571 M1: 3, 5-7, 11-32
advantages of silicon-nitride-based ceramic inserts versus oxide-based ceramic inserts when machining ... EM4: 971
alloy effects ... M1: 21-22, 26-27, 28-30
applications ... M1: 11, 16-19, 22-24, 26, 30
ASTM specifications ... M1: 16-17
automatic cylinder blocks ... M1: 604
carbides ... M1: 12, 13, 14, 26
carbon content ... M1: 11, 18-20, 25-27, 29-30
castability ... M1: 11-12
casting size effects ... M1: 14-16
chill, correction of ... M1: 22
classes ... M1: 11
composition ... M1: 33
compositions, typical ... M1: 18, 19, 25-27, 30
creep ... M1: 26, 27
damping capacity ... M1: 32, 41, 45
density ... M1: 31
dimensional growth ... M1: 46-47
dimensional stability ... M1: 26-28
electrical conductivity ... M1: 53
electrical properties ... M1: 28, 31
face milling application of silicon-nitride-based ceramic inserts ... EM4: 969
fatigue ... M1: 19-22
ferritic grades, weldability ... M1: 564
flake graphite ... M1: 12, 13, 14, 22, 31
fluidity ... M1: 11-12, 13
graphite structure ... M1: 5-7
growth at high temperature chromium effect on ... M1: 92
hardenability ... M1: 29, 30
heat treatment ... M1: 13, 23, 27, 28, 29-30, 31, 32
high chromium gray iron
composition ... M1: 76
mechanical properties ... M1: 92
physical properties ... M1: 88
high nickel gray iron
composition ... M1: 76
elevated-temperature properties ... M1: 91-96
mechanical properties ... M1: 89, 92
oxidation resistance ... M1: 91-93
physical properties ... M1: 88
impact resistance ... M1: 22
inoculants ... M1: 13, 21-22, 28
machinability ... M1: 54, 55
machined with silicon nitride cutting tools ... EM4: 966
machining ... A2: 966 M1: 22-23, 27-28
magnetic properties ... M1: 31
mechanical properties ... M1: 16-23, 26-27, 28-32
medium silicon gray iron
composition ... M1: 76
mechanical properties ... M1: 92
physical properties ... M1: 88
resistance to scaling and growth ... M1: 94
microstructure ... M1: 6, 7, 12-14, 24-26, 29
Ni-Cr-Si gray iron
composition ... M1: 76
mechanical properties ... M1: 92
physical properties ... M1: 88
resistance to scaling and growth ... M1: 94
notch sensitivity ... M1: 20, 21
oxidation resistance ... M1: 46

Gray cast iron (continued)
patternmakers' rules for ... M1: 30-31, 33
pearlitic grades, weldability ... M1: 564
phosphorus, effects of ... M1: 21-22
physical properties ... M1: 30-32
pouring temperature ... M1: 11-12
pressure tightness ... M1: 21-22
prevailing sections ... M1: 15-16
residual stresses ... M1: 26-28
roughing and finishing with whisker-reinforced alumina ceramic insert cutting tools ... EM4: 972
SAE specifications ... M1: 16, 18, 19
section sensitivity ... M1: 11, 13-15
shakeout practice ... M1: 28
shrinkage allowance ... M1: 30-31
silicon, effects of ... M1: 11-13, 21-22, 28-29, 31
solidification ... M1: 12-13
specifications ... M1: 16-19
sprocket, wear of ... M1: 628
stress relief ... M1: 27, 28
stress-strain curves ... M1: 18, 20
test bars ... M1: 15-19
thermal conductivity ... M1: 31, 53
thermal expansion ... M1: 31
thermal properties ... M4: 511
volume/area ratios ... M1: 15-16
wear affected by graphite form and size ... M1: 598
wear resistance ... M1: 24-26, 30
weldability ... M1: 563-564
Gray cast iron, class 30, microstructure of ... A3: 1 • 26
Gray cast iron paper-roll driers
failures of ... A11: 653-654
Gray cast irons ... A18: 693, 748, 754
applications
internal combustion engine parts A18: 553, 556, 557
piston ring material ... A18: 557
in valve train assembly ... A18: 559
as bearing alloys ... A18: 748, 754
applications ... A18: 754
mechanical properties ... A18: 754
for brake drums and disk brake rotors ... A18: 572
for cylinder blocks of automobile internal combustion engines ... A18: 553
laser cladding ... A18: 867-868
microstructure ... A18: 754
Gray goods *See also* Greige goods
defined ... EM1: 12 EM2: 21
Gray iron *See also* Cast iron; Cast irons; Flake graphite ... A1: 3, 4-7, 12-32, 85, 100-104 A9: 245 A13: 131-134 A15: 629-646 A18: 695
abrasion damage in ... A9: 36, 38-39
acicular bainite structure ... A9: 248
air-carbon arc cutting ... A6: 1172
alloying elements ... A15: 639
alloying elements for ... A1: 6, 100-104
alloying to modify as-cast properties ... A1: 28-29
annealing ... A15: 642-643
applications ... A15: 29, 645 A18: 695, 701
internal combustion engine parts ... A18: 553
applications of ... A1: 12
arc welding of ... A15: 526-527
as-polished ... A9: 245
automotive applications of ... A1: 19, 104
bearing cap, failed ... A11: 347-348
bearing cap, hypereutectic ... A11: 349
bearings, use for *See also* Bearings, sliding ... M3: 820
cam lobe, ferritic lubricated wear ... A11: 361
carbon-silicon-oxygen equilibrium ... A15: 94
castability of ... A1: 12-13
castings, inspection of ... A15: 555-556
chemical composition ... A6: 906
classes of ... A1: 12
composition, major/minor elements A15: 629-630
composition of ... A1: 4-6, 19

Gray iron (continued)
cooling rate for ... A1: 6-7
corrosion of ... A11: 199
crankcase, failed ... A11: 362-365
creep in ... A1: 102
as cupola product ... A15: 29
cutting tool materials and cutting speed relationship ... A18: 616
cylinder blocks, cracking in ... A11: 345-346
cylinder head, microporosity cracking ... A11: 355-357
cylinder inserts, corrosion-fatigue cracking of ... A11: 371-372
defects ... A15: 640-642
defined ... A15: 6, 629
dimensional stability ... A1: 26-28
creep ... A1: 27, 102
growth ... A1: 26-27, 101
machining practice ... A1: 27-28
residual stresses ... A1: 27, 28
scaling ... A1: 27, 101-102
temperature, effect of ... A1: 26-27
dip brazing ... A6: 338
door-closer cylinder castings, fracture of ... A11: 363-366
drier head ... A11: 252-254
effect of different etchants on steadite ... A9: 246
effects of graphite in ... A11: 359
elevated-temperature properties ... A1: 26
eutectic in ... A9: 620
fatigue limit in reversed bending ... A1: 19-22
fatigue notch sensitivity ... A1: 21-22
flake graphite in ... A15: 174-176
fluidity ... A1: 12-13
foundry practice ... A15: 635-640
gage materials, wear effect on ... M3: 554, 556
gages, combination, use for ... M3: 556
gear, brittle fracture ... A11: 344-345
graphite morphology ... A15: 631-632
hardenable, application in valve train assembly ... A18: 559
heat treatment ... A1: 7, 21, 29-31 A15: 642-643
hardenability ... A1: 29, 30
localized hardening ... A1: 31
mechanical properties ... A1: 29-31
horizontal centrifugal casting ... A15: 296
hypereutectic, permanent mold casting ... A15: 275
impact resistance ... A1: 23
inclusions in ... A15: 94
inoculation ... A15: 636-639
laser melting ... A18: 863, 864, 865
laser transformation hardening ... A18: 862, 863, 864
liquid treatment of ... A1: 7
liquid-erosion resistance ... A11: 167
machinability of ... A1: 23-24
adhering sand ... A1: 23
annealing ... A1: 23-24
chill ... A1: 23
machinability rating ... A1: 23
shifted castings ... A1: 23
shrinks ... A1: 23
swells ... A1: 23
magnesium powders for nodulation in ... A7: 131
markets and volume tonnage ... A15: 41-42
matrix structure ... A15: 632-633
mechanical properties ... A15: 643-644
mechanical properties of ... A1: 19, 20
melting ... A15: 635-636
metallurgy ... A15: 629-635
microstructure ... A18: 695, 697-698, 699, 700-701
microstructure, flake graphite ... A15: 120
microstructure of ... A1: 13-14
molding ... A15: 639-640
molybdenum-clad coating and laser cladding ... A18: 867
nut, brittle fracture of ... A11: 369-370
oxyacetylene welding ... A15: 530-531
paper-drier head, surface discontinuities ... A11: 352-354

Gray iron (continued)
physical properties A1: 31-32 A15: 644-645
coefficient of thermal expansion A1: 31
damping capacity A1: 31-32
density A1: 31
electrical and magnetic properties A1: 31
thermal conductivity A1: 31
pipe, failure of centrifugal casting mold for A11: 275-276
pouring A15: 639
pouring temperatures A15: 283
pressure tightness A1: 22-23
prevailing sections A1: 16
properties A15: 643-645 A18: 695
pump bowl, graphitic corrosion A11: 372-373
retention of graphite during polishing A9: 40, 43
room-temperature structure A1: 14
scanning electron microscopy used to study graphite A9: 101
scuffing resistance A1: 24-25
chemical composition, effect of A1: 25
graphite structure, effect of A1: 24-25
matrix structure, effect of A1: 25
surface finish effects A1: 25
section sensitivity A1: 14-16 A15: 633-635
section size, effects of A1: 14, 15
volume/area ratios A1: 15-16
section thickness A18: 698
shakeout practice, effect of A1: 28
shrinkage allowance A15: 303
solidification A1: 13-14 A15: 630-631
tensile strength A18: 696
influence of CE on A1: 5
influence of composition and cooling rate on A1: 6
test bar properties A1: 16-19
compressive strength A1: 17-18, 19
elongation A1: 18
hardness A1: 18-19, 21
modulus of elasticity A1: 18, 20, 21
tensile strength A1: 18
testing precautions A1: 17
torsional shear strength A1: 18, 20
transverse strength and deflection A1: 18
typical specifications A1: 17, 18, 19
usual tests A1: 16-17, 18
thermal expansion coefficient A6: 907
ultimate shear stress for A8: 148
unalloyed, for deep-drawing dies A18: 634
for valve seats and guards for reciprocating compressors A18: 604
water-main pipe, graphitic corrosion A11: 372-374
wear A1: 24
abrasive wear A1: 24
adhesive wear A1: 24
corrosive wear A1: 24
cutting wear A1: 24
wear resistance A1: 25-26
graphite structure, effect of A1: 25-26
matrix microstructure, effect of A1: 26
welded, tensile strength A15: 527
welding metallurgy A15: 520-521
Gray iron, annealing A4: 670-671
alloy content, effect on time and temperature A4: 670, 671 M4: 530-531
ferritizing A4: 670-671 M4: 529-530
full A4: 671 M4: 530
graphitizing A4: 671 M4: 530
tensile strength, effect on A4: 670 M4: 529
Gray iron castings
inspection of A17: 51, 531
markets and volume tonnage for A15: 41-42
Gray iron, flame hardening A4: 668, 677-678
alloying elements, effects of A4: 678 M4: 540
composition A4: 678, 679 M4: 539-540
fatigue strength A4: 678 M4: 540-541
hardness A4: 678, 679 M4: 540
quenching A4: 678 M4: 541
stress relieving M4: 540
stress-relieving A4: 678
Gray iron, heat treating A4: 670-681
air cooling, effect on properties A4: 671, 672, 677 M4: 531
annealing *See* Gray iron, annealing
applications A4: 673-675, 677, 679
austempering A4: 676, 677 M4: 537
Gray iron, heat treating (continued)
austenitizing A4: 673, 675, 677 M4: 532-533, 535
composition (No. 1-32) A4: 674, 675
differentiated from ductile iron A4: 667
equipment requirements, martempering A4: 676 M4: 538
flame hardening *See* Gray iron, flame hardening
hardening A4: 672-673, 680 M4: 532-537
hardness measurement A4: 669
impact resistance, effect of tempering A4: 672-673 M4: 536-537
induction hardening A4: 678-679 M4: 541
martempering A4: 676-677 M4: 537-539
normalizing A4: 671-672 M4: 531-532, 533
properties A4: 675-676
proven applications for borided ferrous materials A4: 445
quenching A4: 673, 677 M4: 533-535
stress relieving M4: 540, 541-543
stress-relieving A4: 679-681
tempering A4: 672-673, 675, 677 M4: 535-536
tensile strength, effect of tempering A4: 679 M4: 536
Gray Iron Institute A15: 34
Gray iron pipe *See also* Ductile iron pipe M1: 97-98, 100
applications M1: 97-98
coatings and linings M1: 97-98, 100
design M1: 97
joints M1: 97, 100
manufacture M1: 98
sizes, standard M1: 98
specifications for M1: 97-98, 100
Gray iron, specific types
Class 20, type D graphite, stress-relieved A9: 247
Class 20, with porosity A9: 250
Class 30, type A graphite, as-cast A9: 247
Class 30, type D graphite, permanent mold cast A9: 247
Class 30, with slag inclusion A9: 249
Class 30B, type A graphite, austenitized and quenched A9: 248
Class 30B, type A graphite, steadite and pearlite A9: 250
Class 30B, with ferrosilicon particle A9: 249
Class 35, as-cast against a chill A9: 249
Class 40, type D graphite, as-cast A9: 247
Class 40, type D graphite, as-cast annealed A9: 247
Class 50, as-cast A9: 248
type A graphite distribution A9: 246, 249
type A, heat-resistant, as-cast A9: 255
type A, high-silicon corrosion-resistant, as- cast A9: 255
type B, abnormal structure A9: 250
type B graphite distribution A9: 246
type C graphite distribution A9: 246
type D graphite distribution A9: 246
type E graphite distribution A9: 246
Gray irons *See also* Cast irons A16: 648-652, 654, 656, 658, 664
arc welding M6: 314-315
boring A16: 169, 655
brazeability M6: 996
brittleness of A12: 123
broaching A16: 198, 203, 209, 656, 780
cemented carbide machining A16: 86, 88
centerless grinding A16: 665
cermet tools for milling A16: 97
coatings and tool life A16: 58
composition limits M6: 309
contour band sawing A16: 364
counterboring A16: 660
cylindrical grinding A16: 664
drilling A16: 223, 231, 658
electrochemical machining A16: 535
end milling A16: 325, 663
face milling A16: 662
fixtures A16: 404
fractographs A12: 225-226
fracture/failure causes illustrated A12: 216
high-speed machining A16: 604
honing A16: 477, 484, 802
hypereutectic, graphite effects A12: 226
internal grinding A16: 665
machinability A16: 640, 642
Gray irons (continued)
matrix microstructure effect on tool life A16: 650
metal removal rates A16: 652
microstructures A16: 113
milling A16: 97, 312, 313, 314, 327
oxyacetylene welding M6: 603-604
oxyfuel gas cutting M6: 112
PCBN cutting tools A16: 112, 114
pipe threading tools A16: 302
planing A16: 185-186, 658, 660
postweld heat treatment practice M6: 314
reaming A16: 243-244, 247, 248, 659
repair welding M6: 316-317
sand cast, fracture at flake A12: 225
shaping A16: 191
slab milling A16: 324
spotfacing A16: 660
surface grinding A16: 664
tapping A16: 263, 264-265, 661
thread grinding A16: 275
threading A16: 299
threading, circular chasers A16: 298
threading, tangential chasers A16: 297
tool life A16: 299
turning A16: 94, 112, 144, 146, 147, 653-654
welding metallurgy M6: 308
Gray irons, specific types
class 20, and annealing treatment A16: 651
class 20, machining A16: 275, 653-663
class 25, and chill A16: 650
class 25, stress-relieving A4: 679
class 30, annealing A4: 670
class 30, machining A16: 144-147, 275, 301, 362, 653-663
class 35, and annealing treatment A16: 651
class 35, machining A16: 144-147, 275, 362, 653-663
class 35, stress-relieving A4: 679
class 40, annealing A4: 670
class 40, machining A16: 144-147, 245, 275, 362, 644, 653-663
class 45, machining A16: 275, 323, 325
class 50, machining A16: 275, 323, 325
class 50, stress-relieving A4: 679
class 55, thread grinding A16: 275
class 60, thread grinding A16: 275
SAE J431C: grades G1800-G4000, machining A16: 144-147, 275, 323, 325
Gray levels
in SEM imaging display A12: 169
Gray solidification
ternary iron-base systems A15: 67
Gray units
radiography A17: 301
Gray value contrast, from different wavelengths of monochromatic incident
light A9: 149
Gray-level image storages
in scanning electron microscopy A9: 95-96
Gray-scale image analysis
machine vision A17: 36
Gray-scale system
image interpretation vs. algorithms A17: 36-37
in machine vision process A17: 33
Gray-to-white eutectic transition
cast iron A15: 180
Grazing angle electron incidence
effect on sample charging A10: 556
Grease A18: 99, 123-131
additives A18: 111, 124-125
applications A18: 123, 125-129
adverse conditions A18: 128
American Association of Railroads Specification 942-88 for grade 1 1/2 nonextreme-pressure grease A18: 125
bearing temperature A18: 126-127
centralized systems A18: 129
conditions, environment, and contamination A18: 127
DN value A18: 127
incompatibility A18: 129
low temperature A18: 127
paints A18: 129
pressure A18: 127
pumpability A18: 127-129
seals A18: 129

Grease (continued)
selection ... **A18:** 125, 128
speed ... **A18:** 127
viscosity as function of bearing diameter and speed ... **A18:** 127, 128
viscosity, speed, and pressure ... **A18:** 127, 128
base stocks ... **A18:** 123-124
composition ... **A18:** 123
petroleum ... **A18:** 123-124, 125
petroleum, properties ... **A18:** 124
synthetics ... **A18:** 124, 125
synthetics, temperature ranges and applications ... **A18:** 125
white mineral oil ... **A18:** 124
for bearings ... **A11:** 511
classifications ... **A18:** 125
common ... **A18:** 125
heavier ... **A18:** 125
thin ... **A18:** 125
classifications (NLGI) and consistency grades ... **A18:** 127
coloring, nickel alloys ... **M5:** 674
consistency classification ... **A18:** 136
defined ... **A18:** 10
definition of lubricating grease ... **A18:** 123
dispensing methods ... **A18:** 129-130
centralized grease systems ... **A18:** 130
hand application ... **A18:** 129
hand- or air-operated methods ... **A18:** 129-130
ferrographic applications for lubricant analysis ... **A18:** 307
frequency of relubrication ... **A18:** 130-131
handling ... **A18:** 131
lithium-base, SEM micrograph ... **A11:** 153
lubricating ... **A11:** 152
lubrication, for rolling-element bearings ... **A11:** 512
overgreasing ... **A18:** 131
removal of *See also* Vapor degreasing
aluminum and aluminum alloys ... **M5:** 576-578
babbitting process ... **M5:** 356
copper and copper alloys ... **M5:** 614, 617-620
hot dip galvanized coating process ... **M5:** 325-326, 328
hot dip tin coating process ... **M5:** 352
process types and selection ... **M5:** 5-6, 8-11
reactive and refractory alloys ... **M5:** 656-657
thermal spray coating process ... **M5:** 367
zinc alloy cleaning ... **M5:** 676-677
rust and corrosion inhibitors ... **A18:** 106
storage ... **A18:** 131
testing ... **A18:** 125
thickened, for corrosion control ... **A11:** 195
types of grease-thickener systems ... **A18:** 125-126
effect of thickener type on melting point and typical grease use ... **A18:** 126
reasons for use ... **A18:** 126
undergreasing ... **A18:** 131
use as a substitute for oils ... **A18:** 123
viscosity improvers used ... **A18:** 110
Grease life ... **A18:** 504
Grease lubrication
rolling-element bearings ... **A18:** 136-137
Greaseless buffing compounds ... **M5:** 117, 126
Greases
cleaning ... **A13:** 413-414
powders used ... **A7:** 573
Greasing
nickel alloys ... **M5:** 674
Great Bell
the (Kremlin) ... **A15:** 19
Greek alphabet ... **A10:** 692
symbols for ... **A11:** 798
Greek Ascoloy
broaching ... **A16:** 203, 204, 209
electrochemical grinding ... **A16:** 547
grinding ... **A16:** 547
milling ... **A16:** 547
wrought heat-resistant ... **A16:** 738
Greek symbols
for liquid metal processing terms ... **A15:** 49
Green *See also* Green compact; Green density; Green strength; Green strip
defined ... **A14:** 7
Green brass coloring solutions ... **M5:** 626
Green, bromcresol
acid-base indicator ... **A10:** 172
Green ceramics
diamond for machining ... **A16:** 105
Green coatings
applying ... **M5:** 377-378, 380
Green compact *See also* Compacting; Compaction; Compacts
as-pressed condition ... **A7:** 309
carbonyl nickel powder density as function of compacting pressure ... **A7:** 310
composition-depth profiles ... **A7:** 252
compression ratio for ... **A7:** 297
defined ... **A14:** 7
microstructure ... **A7:** 311, 314
pore size ... **A7:** 299
processing with composite fluid dies ... **A7:** 544
and sintered compacts, pore shapes compared ... **A7:** 311
strength of ... **A7:** 288-289, 311
transverse-rupture strength ... **A7:** 311
Green compacts *See also* Powder metallurgy parts
aluminum and aluminum alloys ... **A2:** 6
pore pressure rupture testing ... **A17:** 547
ultrasound transmission in ... **A17:** 539
Green density *See also* Density
beryllium powders ... **A7:** 172
and compacting pressure, aluminum P/M parts ... **A7:** 743
copper oxide reduction ... **A7:** 110
densification as function of ... **A7:** 310
determined ... **A7:** 310
effect of lubricated particles ... **A7:** 288
effect on sintered aluminum P/M parts ... **A7:** 385
electrolytic copper powder ... **A7:** 115
and green strength ... **A7:** 85, 289, 302
in iron powders ... **A7:** 85, 289
and pressure ... **A7:** 172, 298
shrinkage as function of ... **A7:** 310
in water-atomized iron powders ... **A7:** 85
in water-atomized tool steel powders ... **A7:** 103
Green effectiveness ... **A18:** 575
Green forming ... **A16:** 72
Green glass ... **EM4:** 1082
Green golds *See* Gold-silver-copper alloys
Green light filters
for black-and-white photography ... **A9:** 72
Green liquor
defined ... **A13:** 7
Green machining ... **EM4:** 34, 123, 124, 147, 181-185
advantages ... **EM4:** 181
applications ... **EM4:** 182
bisque firing ... **EM4:** 181
capabilities ... **EM4:** 184
surface finish ... **EM4:** 185
equipment ... **EM4:** 181-184
limitations ... **EM4:** 184-185
quality control inspection techniques ... **EM4:** 185
size requirements ... **EM4:** 184
tolerance requirements ... **EM4:** 184-185
material requirements ... **EM4:** 184
objectives ... **EM4:** 181
tooling ... **EM4:** 181-184
cutting edge specifications ... **EM4:** 183
environmental effects ... **EM4:** 183-184
feed rates ... **EM4:** 183-184
feeds ... **EM4:** 183
fixturing ... **EM4:** 183
machining processes ... **EM4:** 183
machining specifications ... **EM4:** 181-183
modulus of elasticity ... **EM4:** 183
speeds ... **EM4:** 183
Green machining (continued)
surface finish ... **EM4:** 184
Green, Nathaniel
as early founder ... **A15:** 26
Green sand *See also* Sands; Temper
defined ... **A15:** 6
investment, aluminum and aluminum alloys ... **A2:** 5
molds, copper alloy casting ... **A2:** 350
Green sand mold(s) *See also* Dry sand mold
blowing of ... **A15:** 29
clays for ... **A15:** 224-225
defined ... **A15:** 6
for horizontal centrifugal casting ... **A15:** 296
process control requirements ... **A15:** 222
raw material additions ... **A15:** 225-226
sand reclamation ... **A15:** 226-228
sand types and properties ... **A15:** 222-224
system formulation ... **A15:** 225
Green sand molding *See also* Green sand molds; Mold(s)
advantages/disadvantages ... **A15:** 804-805
computer-aided manufacture ... **A15:** 350-351
as conventional process ... **A15:** 37
and dry sand molding, compared ... **A15:** 228
equipment and processing ... **A15:** 341-351
green, defined ... **A15:** 208
impact molding as ... **A15:** 37
materials ... **A15:** 341
mold finishing ... **A15:** 347
molding media preparation ... **A15:** 344-345
molding methods ... **A15:** 341-344
molding problems ... **A15:** 345-347
operations, flow chart for ... **A15:** 205
sand uniformity ... **A15:** 35
sand/casting recovery ... **A15:** 348-350
shakeout ... **A15:** 347-348
Green strength *See also* Compressibility ... **A14:** 7, 189 **EM3:** 14, 143, 149
additive effects ... **A7:** 289
and apparent density, iron powders ... **A7:** 288
of bronze powders, lubricant effects ... **A7:** 192
and carbon effect on water-atomized iron powders ... **A7:** 85
for cemented carbides ... **A18:** 795
of compacted metal powders ... **A7:** 288-289
compacting pressure curves for ... **A7:** 302
and copper oxide reduction ... **A7:** 108, 110
defined ... **A15:** 6 **EM1:** 12 **EM2:** 21
determing ... **A7:** 211
effect of copper powder tarnishing ... **A7:** 109, 111
electrolytic copper powder ... **A7:** 115
of electrolytic iron powder ... **A7:** 94, 95
ferrous P/M materials ... **M1:** 330, 331, 345
and green density curves ... **A7:** 302
and green density in water-atomized iron powders ... **A7:** 85
measurement, standard test ... **A7:** 302
of powder compacts ... **A7:** 302-304
stainless steel powders ... **A7:** 101, 729
standard test, transverse bend ... **A7:** 288
of tantalum capacitor anodes ... **A7:** 162, 163
testing for ... **A7:** 288
theories ... **A7:** 303-304
variables affecting ... **A7:** 288-289
Green strength, adhesive
defined ... **EL1:** 673-674
Green strip
finishing ... **A7:** 405-406
nickel powder ... **A7:** 401
thickness, and work roll diameter ... **A7:** 406
Green tensile strength
testing ... **A15:** 345
Green transverse-rupture strength
water-atomized tool steel ... **A7:** 103
Greenfield sites
ferrous continuous casting ... **A15:** 309-310
Greenwood-Williamson (GW) model
surface roughness ... **A18:** 28

Greige goods *See also* Gray goods
defined **EM1:** 12 **EM2:** 21
Greinacher circuit
radiography **A17:** 305
Grid
circle analyzer **A8:** 567
jig, for round bar specimens **A8:** 279
markings, deformation measured by **A8:** 549
overlapping, dynamic notched round bar testing **A8:** 279
for strain measurement on upset cylinders **A8:** 579
Grid array, pin and pad
as package family **EL1:** 404
Grid geometries
resistance strain gages **A17:** 449
Grid mesh, development
for physical modeling **A14:** 434-437
Grid system-dividing techniques **A6:** 1095
Grid testers
as quality control **EL1:** 873
Gridless routing
as automatic trace routing **EL1:** 533
Grids, test
used in quantitative metallography **A9:** 124, 125, 130
Griffin wheel casting process
with solid graphite molds **A15:** 285
Griffith criteria of failure **EM4:** 1052
Griffith criterion **A6:** 143
Griffith criterion for fracture **A18:** 403
Griffith equation **EM4:** 780, 850, 861-862, 866
Griffith flaws **EM4:** 461
Griffith relationship
stress required for fracture **EM4:** 75
Griffith theory **EM3:** 503, 506
Griffith theory of brittle fracture **A7:** 58
Griffith's maximum tensile stress analysis for volume flaws **EM4:** 700-701
Griffith/Orowan criteria **EM4:** 709
Grignard reagents
magnesium powders application **A7:** 131
particle size **A7:** 131
Grim glow discharge emission sources **A10:** 27
Grimley-Trapnell (thin-film) theory **A13:** 67
GRIN *See* Gradient refractive index lens
Grind limit
milling **A7:** 59
Grind/polish **EM4:** 464-470
cleaning and inspection of polished components **EM4:** 469-470
advanced grinding technologies **EM4:** 470
contact measurement systems **EM4:** 470
noncontact measurement systems **EM4:** 470
shape and irregularity of the finished part **EM4:** 470
surface finish **EM4:** 470
commercial polish and precision polish processing applications **EM4:** 465
effect on silicon nitride joint bend strength **EM4:** 527
glass preparation **EM4:** 464
grinding operations **EM4:** 464
lapping (fine grinding, smoothing, or fining) **EM4:** 467-468
parameters **EM4:** 464, 465
compounds **EM4:** 468-469
pad materials **EM4:** 469
polishing **EM4:** 468-469
polishing process parameters **EM4:** 469
polishing and tooling **EM4:** 464-468
additional factors **EM4:** 467
bond type and hardness **EM4:** 466-467
concentration **EM4:** 466
diamond superabrasive grinding wheels standard marking system **EM4:** 466
setups for grinding **EM4:** 465
wheel design **EM4:** 466
processing equipment **EM4:** 464
Grindability
of ASP steels **A7:** 785
of CAP material **A7:** 535
of wrought tool steels **A1:** 775
Grinding *See also* Finishing; Grinding equipment and processes; Surfaces **A16:** 4, 19, 421-429 **EM4:** 313
abusive, and fatigue strength **A16:** 31, 35
abusive final, AISI/SAE alloy steel failure **A12:** 335
abusive or finish, effect on tool steel parts **A11:** 567, 569
abusive, residual stress **A16:** 24
adaptive control implemented **A16:** 619, 620, 621-622
after cold heading **A14:** 294
aircraft engine components, surface finish requirements **A16:** 22
Al alloys **A16:** 774, 783, 792, 798-799, 801-802
aluminum alloys **A9:** 351-352
applications, sizes of diamond/CBN grains for **A2:** 1011
austenitic manganese steel castings **A9:** 237
automatic chamfer grinder **A16:** 264
belt *See* Belt grinding
beryllium **A9:** 389
beryllium-copper alloys **A9:** 392-393
beryllium-nickel alloys **A9:** 392-393
bums, brittle fracture from **A11:** 90-91, 92
carbon and alloy steels **A9:** 168 **A16:** 676
carbon steel casting specimens **A9:** 230
carbonitrided steels **A9:** 217
carburized steels **A9:** 217
cast irons **A9:** 243 **A16:** 661-664
and cemented carbide cutting tools **A16:** 83
cemented carbides **A9:** 273 **A18:** 796
centerless **A16:** 829
centerless, cast irons **A16:** 662-663, 665
centerless, compared to honing **A16:** 486
centerless, cutting fluids used **A16:** 125, 127
centerless, in conjunction with lapping **A16:** 495
centerless, of Ti alloys **A16:** 854
ceramics **A16:** 101, 102
chrome plating removed by **M5:** 185
chromized sheet steel **A9:** 198
compared to cutting **A16:** 426-429
compared to ECG **A16:** 546, 547
compared to honing **A16:** 473
compared to lapping **A16:** 492
compounds, removal of **M5:** 14-15
conformity or equivalent diameter **A16:** 422
in conjunct-on with tapping **A16:** 264
in conjunction with boring **A16:** 162
in conjunction with broaching **A16:** 194, 195, 205, 209, 211
in conjunction with drilling **A16:** 219, 229, 238
in conjunction with EDM **A16:** 560
in conjunction with gear manufacture **A16:** 333, 339, 344, 350-355
in conjunction with lapping **A16:** 498, 502, 505
in conjunction with sawing **A16:** 358
in conjunction with thread rolling **A16:** 290
in conjunction with turning **A16:** 135, 140, 142, 153
in conjunction with ultrasonic machining **A16:** 529
coolant effects **A16:** 423-424, 426, 429
copper alloys **A16:** 818-819
of copper oxide **A7:** 107, 108
corrosive effects and abrasive wear **A18:** 189
cracks, from failure to temper **A11:** 567-568
cracks, in castings **A11:** 362
creep feed
cutting fluids used **A16:** 125
cylindrical **A16:** 421, 422, 424, 429, 829
cylindrical, cast irons **A16:** 661, 662
cylindrical, compared to microhoning **A16:** 489
cylindrical, cutting fluids used **A16:** 125
damage, AISI/SAE alloy steels **A12:** 321
damage, to shafts **A11:** 459
damage, tool steel cutter die **A11:** 567, 569
defined **A9:** 9, 35
in Domfer iron powder process **A7:** 90
ductile or brittle fractures from **A11:** 89-90
edge **A14:** 530
effect on fatigue strength **A11:** 125
effect on surfaces, Mössbauer analysis of **A10:** 287
electrical contact materials **A9:** 550
electrogalvanized sheet steel **A9:** 197
and fatigue strength **A16:** 25-26, 31, 35
Grinding (continued)
ferrites and garnets **A9:** 533
fiber composites **A9:** 588-589
fine, for samples **A10:** 16
finish, tool and die failures from **A11:** 567
and finishing, electrolytic copper powders **A7:** 114
fluids, removal of **M5:** 5, 9-10
G ratio **A16:** 107, 422-427
G-Trac generator cutters **A16:** 344
gate and flash, Alnico alloys **A15:** 737-738
gentle, and fatigue strength **A16:** 31
gentle, low stress **A16:** 31
gray cast iron **A16:** 115
in green machining **EM4:** 183
grinding fluids for thread grinding **A16:** 273
grinding wheels and workpiece parameters **A16:** 421
of hafnium **A2:** 664 **A9:** 497
hand wheel **A16:** 32
heat-resistant alloys **A16:** 757-760
heat-resistant casting alloys **A9:** 330
high-speed **A16:** 32
hobs **A16:** 344
hot-dip galvanized sheet steel **A9:** 197
hot-dip zinc-aluminum coated sheet steel **A9:** 197
impact, beryllium **A7:** 170
internal **A16:** 663, 829
internal, cast irons **A16:** 665
internal, compared to microhoning **A16:** 489
iron-cobalt and iron-nickel alloys **A9:** 532
lead and lead alloys **A9:** 415
local variations in residual stress produced by surface **A10:** 390-391
low-alloy steel casting samples **A9:** 230
low-stress procedures **A16:** 24, 26, 28, 30
machining process **A7:** 462
magnesium alloys **A9:** 425 **M5:** 628, 648-649
magnesium powders **A7:** 131, 132
martensitic cast irons **A16:** 112
mechanical *See* Mechanical grinding and finishing
Mg alloys **A16:** 827-828, 829
microhardness specimen **A8:** 93
and milling **A16:** 329
mills **A7:** 68
MMCs **A16:** 894, 896
NC implemented **A16:** 613, 616
Ni alloys **A16:** 835, 837
nickel alloys **A9:** 435 **M5:** 673-674
nickel-copper alloys **A9:** 435
of nickel-titanium shape memory effect (SME) alloys **A2:** 899
nitrided steels **A9:** 218
notch root radius **A8:** 382
P/M materials **A16:** 880, 881, 882, 889, 890, 891
permanent magnet alloys **A9:** 533
physical and mechanical properties affected by **M5:** 306, 308
plunge **A16:** 21
point, and drilling **A16:** 221
point-splitting machine and dulling **A16:** 227-228
polycrystalline diamond **A16:** 107
porcelain enamel coat repair **M5:** 524
porcelain enameled sheet steel **A9:** 198
powder metallurgy materials **A9:** 505-506
power expended **A16:** 421-429
printed board coupons **EL1:** 573
progressive, stainless steel **M5:** 555-556
Raman analysis of **A10:** 133
refractory metals **A9:** 439 **A16:** 862, 868-869
refractory metals and alloys **A2:** 560-562
relief groove, superalloys **A12:** 390
residual surface stresses from **A11:** 473-474
rhenium **A2:** 562
rhenium and rhenium-bearing alloys **A9:** 447
as sample preparation, x-ray spectrometry **A10:** 94
of sands **A15:** 32
shaper tools **A16:** 190
sizes of micron diamond powders for **A2:** 1013
sleeve bearing materials **A9:** 565
snag **A11:** 472
of soft materials, for samples **A10:** 16
specimens for optical metallography **A10:** 300-301
stainless steel **M5:** 555-559

Grinding (continued)
stainless steel casting alloys ... **A9:** 297
stainless steels ... **A16:** 705
stainless-clad sheet steel ... **A9:** 198
for statistical analysis ... **A7:** 187
steel gears ... **A16:** 354, 355
steel-bonded titanium carbide cermets ... **A2:** 997-998
step, in conjunction with drilling ... **A16:** 221, 222
superabrasives ... **A16:** 421-422
with superabrasives/ultrahard tool materials ... **A2:** 1013
surface ... **A16:** 11, 11, 28, 421, 422, 424
surface, cast irons ... **A16:** 651, 661, 664
surface, cutting fluids used ... **A16:** 125, 126
surface finish and integrity ... **A16:** 21, 28, 424-426
surface, of Mg alloys ... **A16:** 829
surface, of Ti alloys ... **A16:** 847-848, 853, 854
as surface preparation for optical microscopy ... **A9:** 34-47
surface roughness arithmetic average extremes ... **A18:** 340
system forces ... **A16:** 426-427, 428, 429
tap tooth with hardness variations ... **A8:** 98, 101
thermal spray-coated materials ... **M5:** 371
Ti alloys ... **A16:** 844, 846, 853, 854
time, effect of ... **A16:** 424, 427
tin and tin alloy coatings ... **A9:** 450
tin and tin alloys ... **A9:** 449
titanium and titanium alloys ... **A9:** 458-459
to powder, of XPS samples ... **A10:** 575
tool and die failure and ... **A11:** 564
tool life of grinding wheels ... **A16:** 421, 424, 427, 428, 441, 449, 468
tool steels ... **A9:** 257 **A16:** 722, 723, 732
traverse cylindrical ... **A16:** 28
tungsten ... **A9:** 441
ultrafine, of brittle and hard materials ... **A7:** 59
and ultrasonic cleaning ... **A7:** 462
uranium alloys ... **A16:** 874
uranium and uranium alloys ... **A9:** 478
valve spring failure from ... **A11:** 554
versus planing ... **A16:** 186
welded joints for examination ... **A9:** 578
wheel *See* Wheel grinding
wheel, bonded-abrasive grains for ... **A2:** 1013
wheel dressing ... **A16:** 422, 424, 425
wrought heat-resistant alloys ... **A9:** 305-306
wrought stainless steel ... **A9:** 279
zinc and zinc alloys ... **A9:** 488
of zirconium ... **A2:** 664 **A16:** 852, 854-855, 856
of zirconium and zirconium alloys ... **A9:** 497
Zn alloys ... **A16:** 833, 834
Grinding abrasion *See also* Abrasive wear ... **M1:** 599
high stress grindings, simulation of ... **M1:** 600
Grinding artifacts
in uranium ... **A9:** 483
Grinding balls
chilled cast iron ... **M1:** 81
martensitic Ni-Cr white cast iron ... **M1:** 81
powder-coated ... **A7:** 58
wear of ... **M1:** 601, 604, 613
Grinding, centerless
by scanning laser gage ... **A17:** 12
Grinding cracks
Barkhausen noise measurement ... **A17:** 160
by liquid penetrant inspection ... **A17:** 86
macroetching to reveal ... **A9:** 176
as planar flaws ... **A17:** 50
radiographic methods ... **A17:** 296
Grinding equipment and processes *See also* Grinding; Superabrasives ... **A16:** 430-452
abrasive wheel bonds ... **A16:** 432
abrasive wheel configurations ... **A16:** 434
abrasive wheel porosity ... **A16:** 434
abrasives ... **A16:** 431, 432
angular-approach cylindrical grinding ... **A16:** 445-446, 447
Grinding equipment and processes (continued)
bond designation and grain spacing ... **A16:** 430
camshaft grinding ... **A16:** 445
centerless grinding ... **A16:** 434, 438, 439, 440-441, 446-448, 449
coated abrasive applications ... **A16:** 436-437, 438
coated abrasive composition ... **A16:** 434-435, 439, 440
coolants ... **A16:** 434, 437, 438-439, 444
crankshaft grinding ... **A16:** 444-445
creep-feed surface grinding ... **A16:** 434, 441, 442, 443-444, 445
cutoff grinding ... **A16:** 434, 436, 440
cylindrical grinding ... **A16:** 434, 440, 444-446, 447, 449
cylindrical grinding machines ... **A16:** 448
filtering ... **A16:** 438, 439
floorstand/swing frame grinding ... **A16:** 440
flute grinding ... **A16:** 437
form grinding ... **A16:** 436, 437, 444
G ratios ... **A16:** 431, 432
gear grinding ... **A16:** 437
grinding fluid disadvantages ... **A16:** 439
grinding fluids ... **A16:** 435, 437-439
grinding wheels and disks ... **A16:** 430
internal grinding ... **A16:** 448-449, 450
internal grinding machines ... **A16:** 451
machines and processes ... **A16:** 439-452
mean particle sizes for grits used in grinding wheels ... **A16:** 431
monoset tool grinding ... **A16:** 450-452
NC camshaft grinding ... **A16:** 445
NC in continuous-dress creep-feed grinding ... **A16:** 443
organic wheel bonds ... **A16:** 433
pendulum surface grinding ... **A16:** 441
portable grinding ... **A16:** 440
precision grinding ... **A16:** 433, 437, 438, 439, 440-443
profile (contour) tool grinding ... **A16:** 450
resin bonds ... **A16:** 435, 436
roll grinding ... **A16:** 440, 444
rough grinding ... **A16:** 434, 436, 439, 440, 441
shaping in conjunction with grinding ... **A16:** 440
snagging ... **A16:** 440
standard marking systems for grinding wheels ... **A16:** 430, 431
superabrasive electroplated wheel bonds ... **A16:** 434, 448
superabrasive metal wheel bonds ... **A16:** 433-434, 448
superabrasives, types of ... **A16:** 432
surface grinding ... **A16:** 434, 440, 441, 442, 443-444, 446
tool grinding ... **A16:** 449-452
vitrified wheel bonds ... **A16:** 433, 442, 448, 450
weld grinding ... **A16:** 439, 440
wheel face grinding ... **A16:** 444
Grinding mill components *See also* Grinding balls, Liners ... **M1:** 623
Grinding wear rate
abrasive type dependence ... **A18:** 273
force, dependence on ... **A18:** 273
galvanic interaction between minerals and metal alloys, dependence on ... **A18:** 273-274
localized corrosion role ... **A18:** 274
Grinding wheels
powders used ... **A7:** 573
Grip
center-cracked tension specimen ... **A8:** 382
collet ... **A8:** 368
compact specimen ... **A8:** 382
compression testing ... **A8:** 191
for constant-load testing ... **A8:** 314
design ... **A8:** 155, 192-193
ends ... **A8:** 156-157, 370-371
for fatigue testing ... **A8:** 152, 368-369
Grip (continued)
loading capacities ... **A8:** 50
single-edge notched specimen ... **A8:** 382
types ... **A8:** 51
ultrasonic fatigue testing ... **A8:** 252
water-cooled ... **A8:** 158-159
Grip design ... **A8:** 155, 192-193
Grip ends ... **A8:** 156-157, 370-371
geometric cross sectional ... **A8:** 156-157
Gripper dies
alloy steels for ... **A14:** 86
defined ... **A14:** 7
inserts ... **A14:** 48
life of ... **A14:** 228
stroke, for machine size selection ... **A14:** 83-84
Gripping
fatigue crack growth specimens ... **A8:** 382
in hot upset forging ... **A14:** 83
systems ... **A8:** 152
techniques, tension testing ... **A8:** 50
Gripping cam, fractured
carburization effects ... **A11:** 576
Grit *See also* Blasting ... **A7:** 6
chilled iron ... **A15:** 510
defined ... **A15:** 6
high-carbon cast steel ... **A15:** 510
size, specifications ... **A15:** 514
Grit blasting *See also* Abrasive blast cleaning ... **EM3:** 259, 264, 265, 267
benefits for lubricated wear ... **M1:** 637
ceramic coating processes ... **M5:** 537, 539
defined ... **EM2:** 21
enameling ... **EM3:** 303
of maraging steels ... **A1:** 797-798 **M1:** 448
properties, characteristics and effects of grit ... **M5:** 85-87
size specifications, cast grit ... **M5:** 83-84
steel adherends ... **EM3:** 270, 271
surfaces of ceramics and glasses to be joined ... **EM3:** 300
thermal spray coating process using ... **M5:** 367-368
Grit size ... **A7:** 6
defined ... **A9:** 9
Grit-blast descaling ... **A7:** 435
Gritblasting ... **A13:** 460, 912
of titanium alloy forgings ... **A14:** 280
Grizzly bars
screening ... **A7:** 176
Groove and rotary roughening
definition ... **M6:** 9
Groove angle
definition ... **A6:** 1210 **M6:** 9
Groove face
definition ... **A6:** 1210 **M6:** 9
Groove joints
aluminum alloys, gas-shielded arc welded tensile strength ... **A6:** 729
hydrogen-induced cold cracking ... **A6:** 436
Groove radius
definition ... **A6:** 1210 **M6:** 9
Groove type
definition considerations ... **M6:** 9
Groove weld
definition ... **A6:** 1210
Groove weld size
definition ... **A6:** 1210
Groove weld throat
definition ... **A6:** 1210
Groove welds
combinations with fillet welds ... **M6:** 66-67
comparison to fillet welds ... **M6:** 64
definitions ... **M6:** 9
design considerations ... **M6:** 64-67
oxyfuel gas welding ... **M6:** 590-591
shielded metal arc welding ... **M6:** 76, 85, 89-90
Groove width ... **A18:** 433
Grooved hubs
economy in manufacture ... **M3:** 848

Grooves
arc welding of heat-resistant alloys...... **M6:** 356-357, 360, 367-368
in bearings **A11:** 485
circumferential, as corrosion attack of tube walls **A11:** 618
clearance in sheaves, wire rope **A11:** 516
definition **M6:** 9
longitudinal, in shafts **A11:** 470-471
on drums, steel wire rope **A11:** 517-518
for oxyfuel gas welding **M6:** 589-591
recommended proportions for arc welding **M6:** 69-72
sharp-edged, in bearing caps **A11:** 350
shear, in shafts **A11:** 464
weld, cracked in bridge cover plate **A11:** 710
Grooves, surface
as casting defects **A15:** 549
Grooving
carbide metal cutting tools **A2:** 965
cemented carbides **A16:** 85-86, 87
cermet tools applied **A16:** 92, 94, 95, 97
in conjunction with boring **A16:** 169
in conjunction with turning **A16:** 135, 138, 140, 143
defined **EM2:** 21
multifunction machining **A16:** 370, 372, 373, 376, 380
PCBN cutting tools **A16:** 115, 116
Grooving corrosion **A13:** 130-131, 995
Gross impact overloading
of bearings **A11:** 505-506
Gross leak tests *See also* Fine leak tests; Leak tests
hermeticity **EL1:** 500-501
for lid seal integrity **EL1:** 954
nuclear ionization detector as **EL1:** 954
package-level **EL1:** 930
Gross porosity *See also* Porosity
defined **A15:** 6
Gross sample
defined **A10:** 674
Gross yielding
as distortion **A11:** 138
Grotthuss chain reaction **EM4:** 1148
Ground bed *See also* Deep groundbed
defined **A13:** 7
Ground connection
definition **A6:** 1210
Ground copper
reduction **A7:** 107-110
Ground distribution **EL1:** 5, 27
Ground glass
blasting with **M5:** 84
Ground lead
definition **A6:** 1210
Ground noise, effects
VHSIC technology **EL1:** 76
Ground planes
design effects **EL1:** 521
Ground shot
in Domfer process **A7:** 89-91
Ground water
SSMS toxicity analysis of **A10:** 148-149
Ground-air-ground cycle in aeronautics **A1:** 687
Ground-coat porcelain enamels *See* Porcelain enameling, ground-coat enamels
Groundwater *See also* Water
saline **A13:** 621
Group
defined **A8:** 7
medians, nonparametric evaluation of **A8:** 706-707
Group frequencies
molecular vibrations as **A10:** 111
Group IV divalent oxides
direct evaporation **A18:** 844
Group theory
prediction for graphite surface analysis **A10:** 132
used to describe the symmetry of an interface **A9:** 118
Grown-junction transistors **EL1:** 958
Growth *See also* Crystal growth; Eutectic growth; Growth fundamentals
of a needle **A15:** 113
of a sphere, as free growth **A15:** 113
of beta crystals, peritectics **A15:** 125-126
Growth (continued)
by coalescence and collision, particle **A15:** 79
cast iron, defined **A15:** 6
characteristics, silicon modification **A15:** 161-162
compacted graphite irons **A15:** 673
competitive, between dendrites and eutectics **A15:** 122-124
defect growth of graphite theory **A15:** 177
dendritic, and segregation **A15:** 153-156
divorced, graphite formation from **A15:** 120
of ductile iron **A15:** 663
during delubrication, presintering, and sintering **A7:** 480
epitaxial, evolution of crystal structure in **A10:** 536
epitaxial, graphite **A15:** 170
equiaxed, models of **A15:** 132-133
eutectic, at very high solidification rates **A15:** 125
of eutectic, in cast iron **A15:** 174
eutectic silicon **A15:** 163
eutectic, simplified theory of **A15:** 124
fluctuational, and nucleation during solidification **A15:** 103
free vs constrained **A15:** 113
fundamentals of **A15:** 109-158
graphite, theories of **A15:** 177-178
kinetics, in equiaxed structure modeling **A15:** 885, 886-887
kinetics, LEED analysis **A10:** 536
kinetics, of silicon **A15:** 79
morphology, inoculation effect **A15:** 105
particle **A7:** 6
planar interface, single-phase alloys **A15:** 114-116
rate curves, austenite/graphite in cast iron **A15:** 170
rate, gray-to-white transition **A15:** 180
regular eutectic **A15:** 121-122
thermally activated, effect on isothermal phase transformations **A10:** 317
velocity, dendritic tip **A15:** 159
Growth and scaling
for compacted graphite iron **A1:** 63, 66
Growth fundamentals *See also* Crystal growth; Growth; Solidification
columnar to equiaxed transition **A15:** 130-135
insoluble particles, solid liquid interface **A15:** 142-147
low-gravity effects during solidification **A15:** 147-158
macrosegregation **A15:** 136-141
microsegregation **A15:** 136-141
and solidification, basic concepts **A15:** 109-113
solidification of eutectics **A15:** 119-125
solidification of peritectics **A15:** 125-129
solidification of single-phase alloys **A15:** 114-119
Growth in polycrystalline pure metals **A9:** 608-610
Growth law
empirical **A9:** 697
Growth rate
effect on control and size of eutectic structures **A9:** 618
Growth stresses
oxide scales **A13:** 71
Grüneisen law **EM4:** 825
GTA weld shielding gas composition
corrosion effects **A13:** 351
Guard electrode
defined **EM2:** 592
Guard-frame procedure
image analysis **A10:** 315
Guards
in torsional testing **A8:** 146
Guerin process *See also* Fluid-cell process
accessory equipment **A14:** 606
blanking **A14:** 607
defined **A14:** 7
presses **A14:** 605
procedure **A14:** 606
as rubber-pad forming **A14:** 605
shallow drawing by **A14:** 606
tools **A14:** 605-606
Guidance, of parts
by machine vision **A17:** 40-41
Guide bearing
defined **A18:** 10
Guide bundle, light *See* Light guide bundle
Guide pins
defined **A14:** 7 **EM2:** 21
mold **EM1:** 168
Guide to information sources **EM1:** 40-42
Guide to nondestructive evaluation techniques **A17:** 49-51
Guided bend
defined **A8:** 7
Guided bend test
defined **A8:** 7
Guided projectile fins **A7:** 686
Guided strippers
for piercing **A14:** 465
Guidelines
materials selection **M3:** 835-837
Guides
contour roll forming **A14:** 627-628
defined **A14:** 7
gas cutting **A14:** 726-727
Guillotine shears
for bar **A14:** 715
Guillotining
of thermoplastic composite **EM1:** 552
Guinier camera **A10:** 335-336
Guinier diffractometer
for XRPD analysis **A10:** 337
Guinier-Preston (GP) zone
beryllium-copper alloys **A2:** 404
solvus line, wrought aluminum alloy **A2:** 39
Guinier-Preston zones .... **A3:** 1 • 21 **A6:** 529, 532 **A10:** 405, 589-590
in beryllium-copper alloys **A9:** 395
defined **A9:** 9
in precipitation reactions **A9:** 650-651
Gull wing
in ceramics **A11:** 747
as lead formation **EL1:** 733-734
Gum **EM3:** 14
defined **A18:** 10
Gum tragacanth
as binder for ceramic coatings **EM4:** 955
Gun
definition **A6:** 1210
Gun angle
effect on weld attributes **A6:** 182
Gun arcing **A6:** 255
Gun drilling **A16:** 171, 173, 216, 218
Al alloys **A16:** 769, 778, 782, 784, 791
by swaging **A14:** 139
compared to gun reaming **A16:** 244-245
compared to shaped tube electrolytic machining **A16:** 554
compared to trepanning **A16:** 176, 180
Cu alloys **A16:** 814
cutting fluid flow recommendations **A16:** 127
Mg alloys **A16:** 823
multifunction machining **A16:** 397
stainless steels **A16:** 693
Gun extension
definition **M6:** 9
Gun metal *See* Copper alloys, specific types, C90500
properties and applications **A2:** 374
shrinkage allowances **A15:** 303
Gun metallizing process
oxidation-resistant coating **M5:** 665-666
Gun parts, of hardened nickel
injection molding produced **A7:** 495
Gun reamers **A16:** 239, 244-245
Gun recoil mechanism
fracture of ductile iron pistons for **A11:** 360-361
Gun technique
for metallic glasses **A2:** 805
Gunn diode oscillators
microwave inspection **A17:** 209
Gunnert drilling technique **A6:** 1095
Guns *See also* Ordnance applications **A8:** 210
bored, slid cast, historic **A15:** 26
cast steel, German **A15:** 31
casting history of **A15:** 19-20
manufacture, early American **A15:** 26
Guns for
capacitor-discharge stud welding **M6:** 736-737
electron beam welding **M6:** 609-610
gas metal arc welding **M6:** 158-159
gun/column assembly **M6:** 620-621
stud arc welding **M6:** 732

Gurney energy ... A6: 161
Gussage All Saints (England)
historical site ... A15: 16
Gusset
defined ... EM2: 21
Gutters ... A14: 7, 50
Gutterway
defined ... A18: 10
Guys
steel wire ... M1: 272
Gypsum
determining calcium content in ... A10: 173
on Mohs scale ... A8: 108
in plaster cores and molds ... A15: 242
Gypsum (calcium sulfate dihydrate)
die material for denture teeth ... A18: 675
hardness ... A18: 433
Miller numbers ... A18: 235
properties ... A18: 666
Gypsum ($CaSO_4$-$2H_2O$)
advantages ... EM4: 380
empirical formula ... EM4: 380
impurities ... EM4: 380
particle size distribution ... EM4: 380
purpose for use in glass manufacture ... EM4: 381
sources ... EM4: 380
Gypsum dies
properties ... A18: 666
simplified composition on microstructure ... A18: 666
Gyro air bearings ... A18: 532
Gyro bearings ... A18: 532
Gyro gimbal rings from titanium-based alloys ... A7: 754
Gyromagnetic ratio
symbol for ... A10: 692
Gyroscopy
beryllium ... A7: 761
Gyrotron
for microwave inspection ... A17: 209-210

H

2-Hydroxyethyl methacrylate
to increase hydrogen bonding capability in fillers ... EM3: 181
14AD *See* High aluminum defects
h *See* Planck's constant
H and D curve, defined
radiography ... A17: 324
H temper
designations for aluminum and aluminum alloys, patterned or embossed sheet ... A2: 27
strain-hardened (wrought products only) defined ... A2: 21
H-band limits ... M1: 489-490, 494, 495
H-iron process ... A7: 52, 98
H-La (Phase Diagram) ... A3: 2 • 236
H-Nb (Phase Diagram) ... A3: 2 • 236
H-Nd (Phase Diagram) ... A3: 2 • 237
H-Ni (Phase Diagram) ... A3: 2 • 237
H-number
cold form tapping ... A16: 266-267
H-Pd (Phase Diagram) ... A3: 2 • 237
H-pile system
cathodic protection system ... A13: 476
H-R-R singularity field
for crack tip characterization ... A8: 446
in *R*-curve method ... A8: 452
and Rice *J*-integral ... A8: 447-448
size of ... A8: 447
H-series ACI designations for heat-resistant casting alloys ... A9: 330
H-Sr (Phase Diagram) ... A3: 2 • 238
H-steels *See* Hardenable steels ... A1: 474, 480-481
composition ranges and limits ... M1: 127, 130-131
hardenability ... M1: 489, 494-495
H-steels (continued)
hardenability curves ... M1: 497-525
H-Ta (Phase Diagram) ... A3: 2 • 238
H-Ti (Phase Diagram) ... A3: 2 • 238
H-type cast stainless steels
corrosion behavior ... A13: 575-576
H-U (Phase Diagram) ... A3: 2 • 239
H-V (Phase Diagram) ... A3: 2 • 239
H-Zr (Phase Diagram) ... A3: 2 • 239
H1 temper, strain-hardened only
defined ... A2: 21
H2 temper
strain-hardened and partially annealed defined ... A2: 21, 25
strain-hardened and stabilized temper defined ... A2: 25
H8, H10, H11, etc *See* Tool steels, specific types
H11 die steels
at elevated temperatures ... A1: 621
H11 modified steel ... A1: 439
heat treatment for ... A1: 440-441
processing of ... A1: 439-440
properties of ... A1: 440, 441
H13 steel ... A1: 441-442
heat treatments for ... A1: 442-443
processing of ... A1: 442
properties of ... A1: 431, 442, 443-444
H112 temper
defined ... A2: 26
H_a *See* Applied magnetic field
HA alloy ... A1: 922
HA-188
composition ... M4: 651-652
HA-6510 *See* Titanium alloys, specific types, Ti-6Al-4V
HA-7146 *See* titanium alloys, specific types, Ti-7-41-4Mo
HA-8116 *See* Titanium alloys, specific types, Ti-8Al-1Mo-1V
Habit ... EM3: 14
defined ... EM2: 21
Habit plane
defined ... A9: 9
HN, stereographic projection ... A10: 455
and orientation relationships ... A10: 453-455
HAC *See* Hydrogen-assisted cracking
Hack-saw blades, development of welding technique ... A3: 1 • 26-27
Hackle *See also* River patterns
definition ... EM4: 632-633
fracture, in ceramics ... A11: 745-747
and mist, defined in ceramics ... A11: 744-745
-separation mechanisms, in composites ... A11: 737
twist ... A11: 745
velocity ... A11: 745
Hackle marks
definition ... EM4: 633
Hackle region
of fracture surface ... EM2: 808-810
Hacksaws used in sectioning ... A9: 26
Hadamard transforms ... A18: 347
Hadfield alloys ... A18: 650, 651
14Mn-1C ... A18: 651
abrasive wear of steels ... A18: 190
Hadfield manganese steel ... A18: 714, 716
abrasive/corrosive wear ... A18: 719
adhesive wear resistance ... A18: 721
applications ... A18: 702
properties ... A18: 702
Hadfield, Robert
as manganese cast steel developer ... A15: 32
Hadfield steel *See* Austenitic manganese steel
Hadfield's steel *See* Austenitic manganese steel
Hadron-electron ring anordnung (HERA)
as niobium-titanium superconducting material application ... A2: 1055
HAE, treatment
magnesium alloys ... M5: 633-634, 636-637, 639-641, 643
Hafnia insulation
for thermocouples ... A2: 883
Hafnium *See also* Hafnium alloys; Reactive metals; Zirconium; Zirconium alloys ... A9: 497-499, 501-502 A16: 844-857
in active metal process ... EM3: 305
as addition to niobium alloys ... A9: 441
as addition to tantalum alloys ... A9: 442
as addition to zirconium ... A9: 497
air-carbon arc cutting ... A6: 1176
alloying effect in titanium alloys ... A6: 508
annealing ... A2: 663
anodizing procedure for ... A9: 498-499
applications ... A2: 668
arc welding *See* Arc welding of zirconium and hafnium
atomic interaction descriptions ... A6: 144
cemented carbide coatings ... A16: 80, 81
chemical-mechanical polishing of ... A9: 497-498
cold rolling ... A2: 663
corrosion/corrosion resistance of ... A13: 707, 720
crystal bar, twins caused by cold working ... A9: 155
distillation separation process ... A2: 661-662
electron beam drip melted ... A15: 413
etchants for ... A9: 498
evaporation fields for ... A10: 587
forging ... A2: 663
grinding ... A2: 664
grinding of ... A9: 497
HfC, properties ... A16: 72
history ... A2: 661
hot rolling ... A2: 663
hot swaged ... A2: 663
interlayer material ... M6: 681
liquid-liquid separation process ... A2: 661
machining ... A2: 664
macroexamination of ... A9: 498
melting ... A2: 661
metal processing ... A2: 661-662
microexamination of ... A9: 498-499
mounting of ... A9: 497
in nickel-base superalloys ... A1: 984
organic precipitant for ... A10: 169
oxidized in steam ... A9: 137
physical properties ... A2: 665
primary fabrication ... A2: 662-664
as pyrophoric ... A7: 199
refining ... A2: 662
secondary fabrication ... A2: 664-665
sectioning of ... A9: 497
species weighed in gravimetry ... A10: 172
thermal diffusivity from 20 to 100 °C ... A6: 4
thermal expansion coefficient ... A6: 907
twins in ... A9: 502
ultrapure, by chemical vapor deposition ... A2: 1094
weld, mechanical twins ... A9: 155
welding ... A2: 665
weldment in ... A9: 502
and zirconium ... A2: 661-669
Hafnium alloys *See also* Hafnium
applications ... A2: 668
cleaning processes ... M5: 667
electroplating ... M5: 668
finishing processes ... M5: 667-668
Hafnium alloys, specific types
T-111, electron-beam welding ... A6: 871
T-222, electron-beam welding ... A6: 871
Hafnium boride (HfB_2)
binary phase diagram ... EM4: 729
properties ... EM4: 793, 796, 797-798, 799, 800
Hafnium carbide
coating for high-speed tool steels ... A16: 57
properties ... A18: 795
tap density ... A7: 277
Vickers and Knoop microindentation hardness numbers ... A18: 416

Hafnium carbide cermets ... **A7:** 810-811
applications and properties ... **A2:** 1001-1002
Hafnium carbides
in niobium alloys ... **A9:** 441
in wrought heat-resistant alloys ... **A9:** 311
Hafnium nitride
coating for high-speed tool steels ... **A16:** 57
synthesized by SHS process ... **EM4:** 229
Hafnium sponge metal *See also* Hafnium processing ... **A2:** 662-663
Hafnium-base alloys
brazing ... **A6:** 943
Hager process ... **EM4:** 402
Hainsworth, William
as early founder ... **A15:** 31
Hair
follicles, deer, high-temperature combustion analysis of sulfur content in ... **A10:** 224
single, IR spectroscopy ... **A10:** 113
Hair grease
defined ... **A18:** 10
Hairline crack *See also* Flakes; Flaking
in forging ... **A11:** 88
from hydrogen embrittlement ... **A11:** 28
in transgranular-cleavage fracture ... **A11:** 79
Hairline cracks *See* Flakes
Hairline craze *See also* Crazing ... **EM3:** 15
defined ... **EM2:** 21
Half 6-4 *See* Titanium alloys, specific types. Ti-3Al-2.5V
Half cell
defined ... **A13:** 7
Half cells ... **A10:** 164-165
Half cooling time ... **A6:** 72
Half journal bearing
defined ... **A18:** 10
Half patterns
early practice with ... **A15:** 28
Half wave direct current ... **A7:** 576
Half-and-half solder ... **A9:** 422
Half-life
defined ... **A10:** 244, 674
of radioisotopes ... **A10:** 235
in UV/VIS analysis ... **A10:** 62
Half-penny crack ... **EM4:** 641-642
Half-wave current
defined ... **A17:** 91
Half-wave potential ... **A10:** 190
Half-wave rectification
radiography ... **A17:** 304
Halide glasses
chemical properties ... **EM4:** 855-856
elastic modulus ... **EM4:** 850
electrical properties ... **EM4:** 851
optical properties ... **EM4:** 854
structural role of components ... **EM4:** 845
viscosity ... **EM4:** 849
Halide separation
from aluminum ... **A15:** 80
Halide test ... **A6:** 130
Halide torch
as gas/leak detector ... **A17:** 61-62
Halide-flux inclusion
overload fracture by ... **A12:** 65, 67
Halides ... **A10:** 181, 658
as molten salt ... **A13:** 50
organic, stainless steel corrosion ... **A13:** 558
Hall effect
defined ... **A10:** 674
in iron ... **A2:** 1121
Hall effect sensors
electric current perturbation ... **A17:** 136
for leakage field testing ... **A17:** 130-131
for magnetic field measurement ... **A17:** 132
Hall flowmeter ... **A7:** 273-274, 278-279
Hall's nozzle design ... **A7:** 127
Hall-Heroult process ... **EM4:** 50
Hall-Heroult reduction cells ... **EM4:** 788
Hall-Heroult refining process
for aluminum casting ... **A15:** 22
Hall-Petch equation ... **A12:** 106
Hall-Petch relations ... **A1:** 115
Hall-Petch relationship
and beryllium grain size ... **A2:** 683
beryllium powder ... **A7:** 171
Halloysite ... **EM4:** 5, 6
Halocarbon plastics *See also* Plastics ... **EM3:** 15
defined ... **EM2:** 21
Halogen
for flame retardance ... **EM1:** 96
Halogen detectors
for vacuum systems ... **A17:** 67
Halogen gases
effect on bare Pt-Rh thermocouples ... **A2:** 882
Halogen(s)
defined ... **A13:** 7
gases, copper/copper alloy resistance ... **A13:** 632
gold corrosion in ... **A13:** 798
ions, high-temperature ... **A13:** 648
osmium corrosion in ... **A13:** 807
platinum corrosion in ... **A13:** 803
rhodium corrosion in ... **A13:** 805
salts, stainless steel corrosion ... **A13:** 559
tantalum corrosion by ... **A13:** 731
Halogen-containing glasses ... **EM4:** 22
development ... **EM4:** 23
fluoride glasses ... **EM4:** 22
halide glasses ... **EM4:** 22
Halogen-diode testing
as gas/leak testing device ... **A17:** 62
Halogenated hydrocarbon cleaners ... **M5:** 40, 42, 44-46
Halogenated hydrocarbons, titanium/titanium alloy SCC ... **A13:** 688
resistance to ... **A13:** 648-649
Halogenated solvents
properties of ... **EL1:** 662
Halogens ... **A10:** 162, 224, 664
causing stress-corrosion cracking ... **A11:** 207
intermediate (organic) soldering fluxes ... **A6:** 628
Hamiltonian parameters
anisotropies of ... **A10:** 262
in ESR analysis ... **A10:** 256-257
Hammer and press forgings, fundamentals of ... **A1:** 346
draft ... **A1:** 346, 347, 348
fillets and radii ... **A1:** 347-348, 349
holes and cavities ... **A1:** 348
lightening holes in webs ... **A1:** 348-349
minimum web thickness ... **A1:** 348, 349
parting line ... **A1:** 346, 347, 348
ribs and bosses ... **A1:** 346-347
scale control ... **A1:** 349
Hammer and rod mills ... **A7:** 56, 69
Hammer forging *See also* Open-die forging
defined ... **A14:** 7
and radial forging, compared ... **A14:** 148
Hammer guide assembly
for a high speed printer ... **A7:** 669
Hammer milling ... **A7:** 70, 72
brittle cathode process ... **A7:** 72
Hammer scale
removal of ... **M5:** 76
Hammer welding
definition ... **A6:** 1210
Hammer(s) ... **A14:** 25-29
air, for heat-resistant alloys ... **A14:** 234
air-lift gravity-drop ... **A14:** 25
for aluminum alloys ... **A14:** 244-245
as ancillary process, ring rolling ... **A14:** 123
blow, load-stroke curve ... **A14:** 42
board-drop ... **A14:** 25
capacities of ... **A14:** 25
characterization of ... **A14:** 41-42
for closed-die forging ... **A14:** 80
closed-die forging in ... **A14:** 75-82
for coining ... **A14:** 180
counterblow hammers ... **A14:** 28
data base ... **A14:** 413
defined ... **A14:** 7
die and die materials for ... **A14:** 43-58
die forger hammers ... **A14:** 28
for drop hammer forming ... **A14:** 654
electrohydraulic gravity-drop ... **A14:** 25
electromagnetic ... **A14:** 648-649
for forging ... **A14:** 25-29
gravity-drop ... **A14:** 25
high-energy-rate forging machines ... **A14:** 29
open-die forging ... **A14:** 28-29, 61, 64
power-drop ... **A14:** 26-28
for precision forging ... **A14:** 168
for stainless steels ... **A14:** 227
steam ... **A14:** 234
Hammer(s) (continued)
for titanium alloys ... **A14:** 273
Hammer-burst fracture
tool steels ... **A12:** 378
Hammering
defined ... **A14:** 7
Hammers
hardened steel for ... **A1:** 456
Hand
defined ... **EM1:** 12 **EM2:** 21
Hand bending
vs power bending ... **A14:** 665
Hand cutting tools
hardened steel for ... **A1:** 456
Hand dipping method
porcelain enameling ... **M5:** 516-517
Hand forge *See also* Open-die forging
defined ... **A14:** 7
Hand forging *See* Open-die forging
Hand forging, primary testing direction
various alloys ... **A8:** 667
Hand hacksawing
Mg alloys ... **A16:** 827, 828
Hand lay-up *See also* Lay-up; Spray-up ... **EM2:** 338-343
defined ... **EM1:** 12 **EM2:** 21
epoxy composite ... **EM1:** 71
molded-in color ... **EM2:** 306
processes ... **EM2:** 338-341
properties effects ... **EM2:** 287
and resin transfer molding, costs compared ... **EM1:** 170
size and shape effects ... **EM2:** 291
surface finish ... **EM2:** 303
technique ... **EM1:** 132, 134
textured surfaces ... **EM2:** 305
tooling ... **EM2:** 341-343
of unidirectional tape prepreg ... **EM1:** 144-145
unsaturated polyesters ... **EM2:** 249
wet. resins for ... **EM1:** 132-134
Hand polishing ... **A9:** 35
Hand polishing jig ... **A9:** 105
Hand shield
definition ... **M6:** 9
Hand sinking
aluminum alloy forging dies ... **A14:** 247
Hand straightening
defined ... **A14:** 7
Hand tool applications
hardenable steels ... **M1:** 458, 459
Hand tool cleaning ... **A13:** 414
Hand-faired master models
quality control of ... **EM1:** 738
Hand-held air knockout tool ... **A15:** 504
Hand-held iron soldering
as surface-mount soldering ... **EL1:** 707
Hand-shanked ladles
development of ... **A15:** 33
Handbook of Adhesives ... **EM3:** 68
Handbook of Epoxy Resins ... **EM3:** 71
Handbook of Plastics and Elastomers ... **EM3:** 71
***Handbook of Plastics Flammability and Combustion Technology* (Landrock)** ... **EM2:** 93
***Handbook of Plastics Test Methods* (*2nd*) (Brown)** ... **EM2:** 93
***Handbook of Plastics Testing Technology* (Shah)** ... **EM2:** 94
Handbooks
military for adhesives ... **EM3:** 63
on plastic technology ... **EM2:** 93-95
Handling *See also* Ejection; Equipment; Transfer equipment
of aluminum/aluminum alloys ... **A13:** 602-603
of beryllium ... **A13:** 810
and board assembly, lead frame materials ... **EL1:** 488
equipment ... **A14:** 63, 163
of fractures ... **A12:** 72
of organic-coated steels ... **A14:** 564
particles from, as contaminants ... **EL1:** 661
of patterns ... **A15:** 196
of permanent magnet materials ... **A2:** 802
systems, precision forging ... **A14:** 163
Handling damage
as formability problem ... **A8:** 548

Handling life *See also* Pot life **EM3:** 15
defined **EM1:** 12 **EM2:** 21
Handling strength **EM3:** 15
Hanging mercury drop electrode **A10:** 191, 690
Hankel function
EXAFS analysis **A10:** 410
Hanning window
in EXAFS data analysis **A10:** 413, 414
Hanning window function **A18:** 295
Hard anodizing
aluminum and aluminum alloys **M5:** 586-589, 591-593
Hard blinding
screen **A7:** 177
Hard brass
recycling **A2:** 1214
Hard buffing **M5:** 115, 557
Hard chrome plate
properties, adiabatic engine use **EM4:** 990
scuffing temperatures and coefficient of friction between ring and cylinder liner **EM4:** 991
Hard chromium
defined **A13:** 7
Hard chromium plating **A18:** 644, 645, 646 **M5:** 170-187
adhesion, poor, correcting **M5:** 177
agitation **M5:** 178
aluminum and aluminum alloys **M5:** 172, 180, 184-185
anodes **M5:** 175-176, 178, 181-182, 184
distance of, thickness related to **M5:** 181-182
special (bipolar) **M5:** 182
appearance **M5:** 175, 184
applications **M5:** 170-171, 179-180, 182-183
difficult-to-plate parts **M5:** 182
baking after **M5:** 186
barrel plating **M5:** 179-180
base metal, effects of **M5:** 172
burnt deposits, correcting **M5:** 176
cast iron **M5:** 171-172
chromic acid process **M5:** 172-175, 182, 186-187
comparison of coatings for cold upsetting **A18:** 645
conductivity **M5:** 175
contamination **M5:** 173-174
copper and copper alloys **M5:** 171, 622
cost **M5:** 184
coverage, poor, correcting **M5:** 176
cracks and crack patterns **M5:** 177, 182-183
current density **M5:** 172, 174-177
current efficiency **M5:** 175
decorative chromium plating differing from **M5:** 170, 188
deposition rates **M5:** 175-176
equipment **M5:** 173, 177-180
maintenance **M5:** 179
racks and fixtures **M5:** 179-180
tanks and linings **M5:** 177-179
etching process **M5:** 180, 183
fume exhaust **M5:** 178
hardness **M5:** 171-172, 175, 182-184
Knoop **M5:** 183-184
Vickers diamond pyramid **M5:** 172, 182-183
heating and cooling **M5:** 177-178
hydrogen embrittlement caused by **M5:** 185-186
maintenance schedule **M5:** 179
mixed catalyst bath **M5:** 172
nickel and nickel-base alloys **M5:** 171-172, 184-185
nodular deposits, correcting **M5:** 176-177
part size, effects of **M5:** 172
pitted deposits, correcting **M5:** 177
plating speed **M5:** 175-176, 182
slow, correcting **M5:** 176
porosity **M5:** 183
power source **M5:** 178
problems and corrective procedures **M5:** 176-177, 185-186
Hard chromium plating (continued)
process control **M5:** 175-176
process selection factors **M5:** 171-172
quality control testing **M5:** 183
racks and fixtures **M5:** 179-180
rinsing **M5:** 178-179
safety precautions **M5:** 178, 184, 186
selective **M5:** 187
solution compositions and operating
chromic acid content **M5:** 172-176, 182
conditions **M5:** 172-177, 181-183
conversion equivalents, chromic acid and sulfate concentration adjustment **M5:** 174
sulfate content **M5:** 172-176, 181
testing methods **M5:** 173-175, 183
solution control **M5:** 172-175
steel **M5:** 171-172, 180, 184-186
stop-off media **M5:** 187
stripping methods **M5:** 184-185
sulfate plating baths **M5:** 172-176, 181
high-concentration **M5:** 172, 175-176
low-concentration **M5:** 172, 175-176
surface preparation for **M5:** 180-181, 185-186
tanks and linings **M5:** 177-179
temperature **M5:** 175-178, 182, 184
heating and cooling **M5:** 177-178
testing methods **M5:** 173-175, 183
thickness **M5:** 170-171, 181-182
anode distance related to **M5:** 181-182
measuring methods **M5:** 181
variations in **M5:** 181-182
variations in, normal **M5:** 181
throwing power **M5:** 181-182
titanium and titanium alloys **M5:** 180-181
waste disposal and recovery **M5:** 184, 186-187
wear resistance **M5:** 170, 184
measuring method **M5:** 184
zinc and zinc alloys, removal of **M5:** 185
Hard coatings **A6:** 998-999
Hard coatings on high-speed steel, contrast enhancement in scanning electron
microscopy **A9:** 98-99
Hard drawn spring wire *See also* Steels. ASTM specific types, A
characteristics **M1:** 284, 288-289
cost **M1:** 305
seams in **M1:** 290
stress relieving **M1:** 291
tensile strength ranges **M1:** 267
Hard elastic systems **A18:** 193
Hard face
defined **A18:** 10
Hard facing
cast iron **M1:** 103
for wear resistance **M1:** 622-623, 635
welding processes used in **M3:** 567
Hard facing alloys, specific types
4A, trimming tools, use for **M3:** 532
Hard facing materials **M3:** 563-567
classification **M3:** 563-564
composition **M3:** 563-564
deposition method, selection of **M3:** 566-567
product forms **M3:** 563
selection **M3:** 564-566
Hard ferrites *See also* Permanent magnet materials
as ceramic permanent magnet materials **A2:** 788-790
Hard irons
CBN for machining **A16:** 105
Hard lead (94-6) *See* Leads and lead alloys, specific types, 6% antimonial lead
Hard lead (96-4) *See* Leads and lead alloys, specific types, 4% antimonial lead
Hard magnetic materials
defined **A2:** 761
Hard materials *See also* Superabrasives; Superhard materials **A7:** 306
for arcing contacts **A2:** 841
Hard materials (continued)
milling of **A7:** 56
properties of **A2:** 1010
ultrafine grinding of **A7:** 59
"Hard metal disease"
tungsten poisoning as **A7:** 206
Hard metals *See* Cemented carbides; Hard materials **A7:** 6
mechanical comminution for **A7:** 56
production **A7:** 156
Hard nickel plating **M5:** 202-204
Hard particles
dispersion in milling **A7:** 63
Hard particles, and wear
in ball bearings **A11:** 494
Hard porcelain
physical properties **EM4:** 934
Hard powders *See also* Hard metals
vibratory and simultaneous compaction of **A7:** 306
Hard probes *See also* Fixed probes
coordinate measuring machines **A17:** 25
Hard solder *See* Gold-base brazing filler metals; Silver-base brazing filler metals
definition **A6:** 1210 **M6:** 9
Hard spots
arc strike fracture at **A11:** 97
as casting defect **A11:** 388
as welding imperfection **A11:** 92
Hard steels
cutting tool materials and cutting speed relationship **A18:** 616
Hard tin
as tin-base alloy **A2:** 525
Hard vacuum
defined **A12:** 46
Hard water
defined **A13:** 7
Hard waxes
for patterns **A15:** 197
Hard x-rays
defined **A10:** 83
Hard-drawn spring wire
characteristics of **A1:** 303-304, 305-306
Hard-face coatings
stainless steels **A18:** 716, 723
Hard-film rust-preventive compounds **M5:** 460, 465-466, 468
Hard-packed powders
sampling **A7:** 213
Hard-wired vision systems
machine vision **A17:** 44
Hardas anodizing process
aluminum and aluminum alloys **M5:** 592
Hardcoat anodize **A13:** 397
Hardenability **A7:** 451, 452
alloy steel wire rod **M1:** 257
alloying elements, effects on **M1:** 476-477
calculation of **M1:** 474
carbon, effect of **M1:** 474-476
carburized and carbonitrided steels **M1:** 534-538
case hardening **M1:** 473, 491-496
cast steels **M1:** 377, 380
cooling rates equivalent **M1:** 471, 482-488, 491, 492
critical diameter **M1:** 474, 476
curves **M1:** 471, 474-475
defined **A9:** 9 **A13:** 7
depth of hardening **M1:** 478-480
of die materials **A14:** 45
of ductile iron **A1:** 48-50 **M1:** 47, 49
ductile iron, alloying **A15:** 649, 659
effect of, on weldability **A1:** 603-604, 606
of end-quench **A7:** 452
equivalence table **M1:** 484-488
grain size, effect of **M1:** 477
of gray iron **A1:** 29, 30 **M1:** 29, 30
H-band limits **M1:** 489-491, 494, 495
H-steels **M1:** 489, 494-495

SUBJECTS OF THE INDEXED VOLUMES: ASM Handbook (designated by the letter "A"): **A1:** Properties and Selection: Irons, Steels, and High-Performance Alloys (1990); **A2:** Properties and Selection: Nonferrous Alloys and Special-Purpose Materials (1990); **A3:** Alloy Phase Diagrams (1992); **A4:** Heat Treating (1991); **A6:** Welding, Brazing, and Soldering (1993); **A7:** Powder Metallurgy (1984); **A8:** Mechanical Testing (1985); **A9:** Metallography and Microstructures (1985); **A10:** Materials Characterization (1986); **A11:** Failure Analysis and Prevention (1986); **A12:** Fractography (1987); **A13:** Corrosion (1987); **A14:** Forming and Forging (1988); **A15:** Casting (1988); **A16:** Machining (1989); **A17:** Nondestructive Testing and Quality Control (1989); **A18:** Friction, Lubrication, and Wear Technology (1992). **Metals Handbook, 9th Edition** (designated by the letter "M"): **M1:** Properties and Selection: Irons and Steels (1978); **M2:** Properties and Selection: Nonferrous Alloys and Pure Metals (1979); **M3:** Properties and Selection: Stainless Steels, Tool Materials and Special-Purpose Materials (1980); **M4:** Heat Treating (1981); **M5:** Surface Cleaning, Finishing, and Coating (1982); **M6:** Welding, Brazing, and Soldering (1983). **Engineered Materials Handbook** (designated by the letters "EM"): **EM1:** Composites (1987); **EM2:** Engineering Plastics (1988); **EM3:** Adhesives and Sealants (1990); **EM4:** Ceramics and Glasses (1991); **Electronic Materials Handbook** (designated by the letters "EL"): **EL1:** Packaging (1989).

Hardenability (continued)
hardenable steels **M1:** 456-457
heats, variations within **M1:** 477-479
J_{ec} equivalent cooling rates **M1:** 482-484, 489, 491
low-alloy steels **A15:** 715
malleable iron **M1:** 71-73
of martensite in steel, factors affecting **A9:** 178
of martensitic stainless steels **A14:** 226
P/M steels **M1:** 338, 343
in powder forging **A14:** 189, 202
of powder metallurgy steels **A9:** 503
quenching mediums **M1:** 480-496
requirements, determination of **M1:** 478-482
selection for **M1:** 481-491
spring steel **M1:** 297, 301, 303, 311-312
steel castings **M1:** 496
surface hardened steels factor in selection **M1:** 530
testing **M1:** 471-478
tubular parts **M1:** 488
variation, affected by heat composition **M1:** 477-478
white cast iron **M1:** 80
of wrought tool steels **A1:** 775-777
Hardenability curves **A1:** 485-570
Hardenability of carbon and low-alloy steels **A1:** 464-484
alloying elements **A1:** 394, 395, 468-469
boron **A1:** 469-470
calculation of hardenability **A1:** 467
carbon content, effect of **A1:** 465, 467-468, 469
determining hardenability requirements
as-quenched hardness **A1:** 471, 476
depth of hardening **A1:** 471
hardenability versus size and shape **A1:** 465, 467-468, 469, 473, 477
quenching media **A1:** 471-473
general hardenability selection charts **A1:** 473, 474-476, 478, 479, 480
estimating hardenability **A1:** 479
rectangular or hexagonal bars and plate **A1:** 478-479, 480, 481
scaled rounds **A1:** 478, 480
tubular parts **A1:** 479
grain size, effect of **A1:** 393, 470
H-steels **A1:** 47, 76, 480-481
low-hardenability steels **A1:** 466
hot-brine test **A1:** 466, 467
SAC test **A1:** 466-467, 468
steel castings **A1:** 483
steels for case hardening **A1:** 481
applications **A1:** 482
core hardness **A1:** 481-482
testing of **A1:** 464
air hardenability test **A1:** 465-466, 467
carburized hardenability test **A1:** 464-465, 466
continuous-cooling-transformation diagrams **A1:** 466
Jominy end-quench test **A1:** 464, 466
use of charts **A1:** 479-480, 482, 483
use of hardenability limits **A1:** 481
variations within heats **A1:** 459, 470, 471, 472
hot working **A1:** 470-471, 473
Hardenable carbon steels
arc welding **M6:** 247-306
classification **M6:** 261
high-carbon steels **M6:** 262
joint preparation **M6:** 264-265
low-carbon steels **M6:** 261
medium-carbon steels **M6:** 261
postweld stress relieving **M6:** 264
preheat temperatures **M6:** 263-264
processes **M6:** 265-270
weldability **M6:** 262-263
Hardenable low-expansion alloys *See also* Low-expansion alloys
properties and heat treatment **A2:** 895
Hardenable steel
electron-beam welding **A6:** 866-867
Hardenable steels *See also* Hardenability of carbon and low-alloy steels; Ultrahigh-strength steels **A1:** 451-463 **M1:** 455-470
alloying elements **M1:** 455-456, 459-460, 466-469
applications **M1:** 457-459, 470
carbon content, effect of **M1:** 455-459, 463
characteristics, general **M1:** 455-456
Hardenable steels (continued)
compositions of **A1:** 452, 453
alloy steels **A1:** 453
carbon and carbon-boron steels **A1:** 452
cracking, control of **M1:** 457, 460, 470
distortion during heat treating **A1:** 462
distortion during heat treatment **M1:** 460, 469-470
effect of alloying on quenching **A1:** 456-457
effect of carbon content on
hardenability **A1:** 454-456
high-carbon content **A1:** 455-456
low-carbon content **A1:** 454
medium-carbon content **A1:** 454-455
fabrication of parts and assemblies **M1:** 470
flame and induction hardening **A1:** 463
hardenability **M1:** 456-457
hardness vs. tempering temperature **M1:** 461, 463, 466-469
hardness-strength correlations **M1:** 455, 460-461
heat treatment **M1:** 457-461, 463, 466-470
high-carbon steels **M1:** 458-459
impact energy **M1:** 469
low-carbon steels **M1:** 457
machinability **M1:** 457, 458, 470
manganese content **M1:** 456
martensite, relation to hardness and carbon content **M1:** 457, 458
mechanical properties **M1:** 460-469
medium-carbon steels **M1:** 457-458
notch toughness **M1:** 469
quenching, alloy effects **M1:** 460
section size, effect on **M1:** 460-466
selection of alloy H-steels **A1:** 460-461
temper brittleness **M1:** 468-469
tempering **M1:** 463, 466-469
tempering of **A1:** 458-459
Hardenable steels (H-steels)
mechanical properties **A4:** 20
Hardened and work-strengthened metals
electron-beam welding **A6:** 867-868
Hardened forged steel rolls **A14:** 353-354
Hardened steel
ball indenter, for Rockwell hardness testing **A8:** 74
ground with superabrasives **A2:** 1013
machined with ultrahard tool materials **A2:** 1013
sheet, thickness for Scleroscope hardness testing **A8:** 105
Hardened steels
recommended machining specifications for rough and finish turning with HIP metal-oxide ceramic insert cutting tools **EM4:** 969
Hardened zones in carbon and alloy steels
macroetching to reveal **A9:** 175
Hardener *See also* Catalyst; Curing agent **EM3:** 15
defined **A15:** 6 **EM1:** 12 **EM2:** 21
Hardeners **EL1:** 474, 818
Hardening *See also* Heat treatment, Surface hardening; Through-hardening
bath, Unicast process **A15:** 252
by quenching, plain carbon steels **A15:** 713
copper alloys **M4:** 724-726, 736
copper and copper alloys **A2:** 236
defined **A9:** 9 **A13:** 47
depth of **A7:** 451
ductile iron **A15:** 658-659
gray cast iron **M1:** 26, 29-30
induction **A11:** 395, 478-479
maraging steels **M1:** 445-446, 448-449
neutral, growth or shrinkage during **A7:** 480
notch toughness of steels effect on **M1:** 703, 706, 709
of polyester resins **EM1:** 133
salt bath **A7:** 375
springs, steel **M1:** 287-288
stainless steels **M3:** 47
of steel-bonded cermets **A2:** 997
surface, abrasive waterjet cutting effects **A14:** 748
surface, postforging defects from **A11:** 333
of tool steels **A9:** 258-259
ultrahigh-strength steels **M1:** 423, 424, 427, 430, 431, 434-436, 439, 441
Hardening by quenching
for hot-rolled steel bars **A1:** 241
Hardening in laser processing **M6:** 793-803
laser alloying **M6:** 793-796
laser cladding **M6:** 796-798
laser melt/particle inspection **M6:** 798-802
Hardening law
defining **A8:** 343
Hardening, secondary
austenitizing effects **A12:** 341
Hardening, steel
effects of **M5:** 325
Hardening wrenches
hardened steel for **A1:** 456
Hardfacing **A6:** 789-807, 808 **A7:** 823-836 **M6:** 771-793
advantages **A6:** 797
alloy selection **M6:** 777-778
alloys **A7:** 144-146, 826
applications **A6:** 789, 790, 791, 793, 795-796, 797, 800, 802, 803 **M6:** 771-772
arc welding **A6:** 801-806 **M6:** 783-787
gas metal arc welding **M6:** 785
gas tungsten arc welding **M6:** 785
open arc welding **M6:** 784-785
plasma arc welding **M6:** 785-787
shielded metal arc welding **M6:** 783-784
submerged arc welding **M6:** 784
austenitic steels **A6:** 790, 791, 798
boride-containing nickel-base alloys **A6:** 790, 794, 795, 796
and brazing rods of nickel-based alloys **A7:** 398
build up **A6:** 789, 791, 803
by thermal spray, cobalt -base alloys **A13:** 664-665
carbide-containing alloys **A6:** 790, 792, 793, 794, 796
carbides **A6:** 792-793
as cast coating **A15:** 562-563
categories of alloys **A6:** 790
build up alloys **A6:** 790
metal-to-earth abrasion alloys **A6:** 790, 792
metal-to-metal wear alloys **A6:** 790
nonferrous hardfacing alloys **A6:** 790
tungsten carbides **A6:** 790, 791, 792, 793
chromium carbides **A6:** 792
coatings, powders used **A7:** 572
cobalt-base alloys **A6:** 789, 790, 792, 793-794, 796, 797, 799, 803 **A13:** 663-664
comparison of welding
gas metal arc welding **M6:** 787
gas tungsten arc welding **M6:** 787
oxyacetylene welding **M6:** 787
plasma arc welding **M6:** 787
processes **M6:** 781, 787
copper-base alloys **A6:** 789, 795-796
corrosion-resistant plasma spray **A7:** 797
cost advantage, calculation of **A6:** 800
defined **A13:** 7
definition **A6:** 789 **M6:** 9
distortion **A6:** 807
electrodes **A6:** 805, 806
electrodes for **A6:** 801, 802, 803, 804
flux-cored arc welding **A6:** 799, 801, 802-803
gas-metal arc welding **A6:** 800, 801, 803
gas-tungsten arc welding **A6:** 798, 799, 800, 803-805, 807
high-alloy irons **A6:** 790
high-chromium irons **A6:** 791-792, 796-797
high-silicon stainless steels **A6:** 795
iron-base alloys **A6:** 789, 790-792, 796, 799, 806
laser **A6:** 803, 806-807
Laves phase alloys **A6:** 790, 792, 793-794, 795, 796
manual powder torch welding **A6:** 800, 801
martensitic steels **A6:** 790, 791, 798
materials **A6:** 789-790 **M6:** 773-777
carbides **M6:** 776-777
cobalt-based alloys **M6:** 774
copper-based alloys **M6:** 777
iron-based alloys **M6:** 775-776
nickel-based alloys **M6:** 775
molybdenum carbides **A6:** 792
nickel-base alloys **A6:** 789, 790, 794-795, 796, 799
niobium carbides **A6:** 793
oxyacetylene welding **A6:** 804, 805 **M6:** 602

Hardfacing (continued)
oxyfuel gas welding **A6:** 800-801 **M6:** 782-783
common defects **M6:** 783
common flaws **A6:** 801
limitations **A6:** 800
operating procedures **A6:** 800-801 **M6:** 783
pearlitic steels **A6:** 790-791
plasma arc welding **A6:** 796, 798, 800, 805-806
porosity **A6:** 801
postweld heat treatments **A6:** 798
powder welding **A6:** 800, 801 **M6:** 783
principles of operation **M6:** 782
powder characteristics **A7:** 823-824
powder composition **A7:** 824-825
process selection **A6:** 796-800 **M6:** 778-782
cost **A6:** 798-800 **M6:** 781
hardfacing product forms **A6:** 798 **M6:** 780-781
metallurgical characteristics of the base metal **A6:** 798 **M6:** 779-780
physical characteristics of the workpiece **A6:** 797-798
physical characteristics of workpiece **M6:** 779
property and quality requirements **M6:** 778-779
property requirements **A6:** 796-797
quality requirements **A6:** 796-797
welder skill **A6:** 798 **M6:** 781
processes, equipment applications **A7:** 830-836
processes, powder size ranges for **A7:** 824
purpose **A6:** 789
selection of alloys **A6:** 796, 797
shielded metal arc welding **A6:** 797, 799, 800, 801-802, 803
spraying **M6:** 787-793
detonation gun 1344 spraying **M6:** 792-793
plasma spraying **M6:** 790-792
spray-and-fuse process **M6:** 788-789
submerged arc welding **A6:** 800, 801, 802
tantalum carbides **A6:** 792
techniques, specialized **A7:** 835-836
techniques, specific **A7:** 830-835
titanium carbides **A6:** 792, 793
tungsten-base carbides **A6:** 790, 791, 792, 793, 799
vanadium carbides **A6:** 792, 793
wear **M6:** 772-773
abrasive **M6:** 772-773
adhesive **M6:** 772-773
erosion **M6:** 772-773
fretting **M6:** 773
welding processes **A13:** 664
worn casting repair by **A15:** 529-530
Hardfacing, alloy improvement .... **A3:** 1 • 27
Hardfacing alloys **A7:** 144-146, 826
abrasive wear, thermal spray coatings used **A6:** 808
adhesive wear, thermal spray coatings used **A6:** 808
cavitation, thermal spray coatings used **A6:** 808
CBN for machining **A16:** 105
cobalt-based, compositions and hardness **A7:** 145, 146
erosion, thermal spray coatings used **A6:** 808
gas-metal arc welding **A6:** 800
gas-tungsten arc welding **A6:** 800
laser **A6:** 789, 800
nickel-based **A7:** 142
open arc **A6:** 800
oxyacetylene welding **A6:** 800
PCBN cutting tools **A16:** 112
plasma arc welding **A6:** 800
shielded-metal arc welding **A6:** 800
submerged arc welding **A6:** 800
surface fatigue wear, thermal spray coatings used **A6:** 808
systems, selection guidelines **A7:** 826
wear-resistant, cobalt powders in **A7:** 144
Hardfacing alloys, friction and wear **A18:** 693, 758-765
abrasive wear **A18:** 758, 759, 760-761, 763, 764, 765
applications **A18:** 758, 759, 760, 761
consumable forms **A18:** 758-765
corrosive wear **A18:** 759, 765
definition of hardfacing **A18:** 758
families (or categories) **A18:** 758
build-up alloys **A18:** 758, 759
metal-to-earth abrasion alloys **A18:** 758, 759-760, 761
metal-to-metal wear alloys **A18:** 758-759
nonferrous alloys **A18:** 758, 761-765
tungsten carbides **A18:** 758, 760-761
laser cladding components and techniques **A18:** 869
tool steels **A18:** 644-645
weld overlay material categories **A18:** 758
weld overlay processes for applying the hardfacing materials **A18:** 759
Hardfacing coatings, fusible
spray material for oxyfuel powder spray method **A18:** 830
Hardfacing materials
ground/machined with superabrasives/ultrahard tool materials **A2:** 1013
Hardness *See also* Apparent hardness; Barcol hardness; Brinell hardness number; Brinell hardness test; Hot hardness; Indentation hardness; Knoop (microindentation) hardness number; Knoop hardness; Knoop hardness test; Mechanical properties; Microhardness; Microindentation hardness number; Mohs hardness; Rockwell hardness; Rockwell hardness number; Rockwell hardness test; Scleroscope hardness test; Shore hardness; Vickers (microindentation) hardness number; Vickers hardness test **EM3:** 15, 51 **EM4:** 605
abrasive minerals **M1:** 189
of adiabatic shear bands **A12:** 32
aluminum casting alloys **A2:** 152-177
aluminum oxide-containing cermets **A7:** 804
amorphous materials and metallic glasses **A2:** 813-814
apparent, and microhardness **A7:** 489
of bearing materials **A11:** 508-509
beryllium-copper alloys **A2:** 409
Brinell **A7:** 312
by eddy current inspection **A17:** 164
of carbides and materials for hardfacing **A7:** 828
carbon content, function of **M1:** 472, 478, 480
of carbon steel
as a function of carbon content **A1:** 127
as a function of temperature **A1:** 640
of carbon steel, and swageability **A14:** 128
cast copper alloys **A2:** 356-391
cemented carbides **A2:** 955
of ceramic granules, Ceracon process **A7:** 539
CG irons **A15:** 672-673
Charpy V-notch test for **A8:** 263
chilled cast iron, effect of annealing on **M1:** 82
of chromate conversion coatings **A13:** 392-393
of chromium carbide-based cermets **A7:** 806
chromium coatings **A13:** 871
of coated carbide tools **A2:** 960
and cold extrusion **A14:** 301
cold-finished bars **M1:** 222-224, 227-234, 240
common, scales for **A7:** 489
compact, and sintering temperature **A7:** 312
and compressibility **A7:** 286
of contact metal, for electrical contact materials **A2:** 840
copper casting alloys **A2:** 348-350
and corrosion resistance **A13:** 48
corrosion-resistant cast irons **M1:** 89
corrosion-resistant high-alloy **A15:** 726-728
defined **A8:** 7, 71 **A11:** 5 **A18:** 10 **EM1:** 12
Hardness (continued)
degree of (for water), symbol for **A11:** 798
and demagnetizing **A17:** 122
determined by Knoop and Vickers testing **A8:** 90
determined by residual stress techniques **A10:** 380
determined in thin layered steels **A10:** 380, 389-390
of diamond **A2:** 1010
die, for stainless steels **A14:** 228
and die life **A14:** 99
distribution, copper casting alloys **A2:** 350
distributions and subsurface residual stress steel shaft **A10:** 389-390
ductile iron **A15:** 654
effect of abrasion damage on **A9:** 39
effect of lattice **A12:** 32
effect of recovery on **A9:** 693
effect on cratering **EL1:** 1043
and electrical conductivity, eddy current inspection **A17:** 168
equivalent hardness numbers for steel **A4:** 458, 459
evaluation **A7:** 452-453
evaluation of **M1:** 635-636
excessive, brittle fracture of chain links by **A11:** 522
of extrusion tools **A14:** 320
of failed pressure vessel **A11:** 663
fatigue properties, relation to **M1:** 665, 672, 673, 675, 676
and flux density **A17:** 134
as formability measure **A8:** 560
galvanized coating **A13:** 438
gradient, from microhardness testing **A8:** 96, 101
and grain size, tungsten carbide **A7:** 775
of gray iron **A1:** 18-19, 21
heat-resistant cast iron **M1:** 92
heat-treated copper casting alloys **A2:** 355
high, forging die fracture from **A11:** 324-325
hot **A14:** 46, 226
hot rolled bars **M1:** 201-204
indentation testing **A7:** 61, 312, 452
investigation by oscilloscope screen **A7:** 485
of iron powder compacts **A7:** 511
low, wear failure of steel bolt by **A11:** 160-161
machinability, relation to **M1:** 566, 567, 569, 571, 574, 576, 578
of magnesium alloys **A2:** 461
magnetic measurement **A17:** 129
malleable iron **M1:** 61, 63, 64, 67-70, 72, 73
maraging steels **M1:** 448, 450
matrix **A11:** 160
matrix, malleable irons **A15:** 697
maximum, by carbon content at austenitizing temperature **A7:** 451, 452
maximum, by oil quenching **A7:** 453
as measure of cold working **A7:** 61
as measure of temper **A2:** 219
measurements **EM4:** 605
guidelines **EM4:** 605
Knoop (HK) **EM4:** 605
Rockwell superficial (HR) **EM4:** 605
Vickers (HV) **EM4:** 605
mechanical, and magnetic hysteresis **A17:** 134
of metal borides and boride-based cermets **A7:** 812
minimum, for Rockwell hardness testing **A8:** 77
as resistance to deformation **A8:** 71
of molybdenum and molybdenum alloys **A7:** 476
of nickel-based, cobalt-based alloys **A7:** 472
of organic coatings **A14:** 565
of P/M and ingot metallurgy tool steels **A7:** 471, 472
of P/M forged low-alloy steel powders **A7:** 470

Hardness (continued)
palladium ... **A2:** 715
for petroleum refining and petrochemical operations ... **A13:** 1264
phase transformation ... **A12:** 32-33
of phases, revealed by differential interference contrast ... **A9:** 59
phenolics ... **EM2:** 244
plotted against complexity index ... **A9:** 130
polyamide- (PAI) ... **EM2:** 133
polyaryl sulfones (PAS) ... **EM2:** 146
ranges ... **A7:** 485
relationship to depth of artifact-containing layer ... **A9:** 37
of rhenium and rhenium-containing alloys ... **A7:** 477
Rockwell C ... **A7:** 452
of roll dies ... **A14:** 99
of rotary workpiece materials ... **A14:** 177
for RWMA group B refractory metal electrodes ... **A7:** 628
saturation flux density as measure ... **A17:** 134
springs, steel
testing ... **M1:** 283, 286
thickness of strip, relation to ... **M1:** 283, 287
yield strength, relation to ... **M1:** 301
steel plate, relation to tensile properties ... **M1:** 194, 198
steel tin mill products with temper designations ... **A4:** 53
of straight knife shears ... **A14:** 704
and strength, carbon steels ... **A15:** 702
and strength, low-alloy steels ... **A15:** 716
structural ceramics ... **A2:** 1019
and subsurface residual stress distributions ... **A10:** 389
superhard tool materials ... **M3:** 453, 455, 456, 457-458, 459, 460, 461, 463-464
surface, of bearing steels ... **A11:** 490
temperature effects ... **A14:** 55
tempering temperature and carbon content effect on ... **A14:** 198
testing ... **A11:** 18 **A14:** 889
testing, and electromagnetic sorting ... **A7:** 453, 484-485
testing, introduction ... **A8:** 71-73
testing, of castings ... **A17:** 521
testpiece, electric current effects ... **A17:** 110
tests, for coating cure ... **A13:** 418
thermoplastic polyurethanes (TPUR) ... **EM2:** 203
threaded fasteners ... **M1:** 274, 276, 278-281
of titanium carbide cermets ... **A2:** 996
of titanium carbide-based cermets ... **A7:** 808
of titanium carbide-steel cermets ... **A7:** 810
for tool steel perforator punches ... **A14:** 485
of tools and dies ... **A11:** 563
of tungsten carbide/cobalt grades ... **A7:** 775
for tungsten carbide/copper refractory metal composite materials ... **A7:** 628
of uranium alloys ... **A2:** 676
used to identify sample constituents ... **A9:** 71
values, sintered materials ... **A7:** 489
variation with carbon content and percent martensite ... **M1:** 457-458, 529-530, 561, 607-608
variations, in swaging ... **A14:** 143
variations in tap tooth during grinding ... **A8:** 98, 101
versus wear resistance
vs temperature, lead frame materials ... **EL1:** 492
wear resistance determined by ... **M1:** 603-606, 608-609
weld, preheat effects ... **A15:** 525
white cast iron ... **M1:** 77, 86-87, 89
of work metal, in piercing ... **A14:** 462
working, for coining ... **A14:** 182
wrought aluminum and aluminum alloys ... **A2:** 63-122
zinc alloys ... **A15:** 786
Hardness conversion numbers
nonaustenitic steels ... **A8:** 106
Hardness conversion table
white cast iron ... **M1:** 87
Hardness conversion tables ... **A8:** 109-113
Hardness indentation ... **A7:** 61, 312, 452
Hardness number *See also* Brinell hardness numbers
conversion, for nonaustenitic steels ... **A8:** 106
Hardness number (continued)
determining ... **A8:** 91
standard reporting procedures ... **A8:** 109-110
vs. load, microhardness testing ... **A8:** 94-96
Hardness profiles
induction hardened parts ... **M1:** 530-532
nitrided cases ... **M1:** 340
Hardness ratio
dies and die materials ... **A18:** 621
Hardness scales ... **EM3:** 189
Hardness techniques ... **A6:** 1095, 1096
Hardness test
for gray iron ... **A1:** 16-17, 18-19, 21
for hot-rolled steel bars ... **A1:** 242, 243
Hardness testing *See also* Indentation test ... **A8:** 73 **A14:** 889
Brinell ... **A8:** 84-89
of casting defects ... **A15:** 545, 553
with durometer ... **A8:** 106-107
file ... **A8:** 108
of gray iron ... **A15:** 643
as intrinsic formability test for sheet metals ... **A8:** 560
introduction ... **A8:** 71-73
of investment castings ... **A15:** 264
measuring method ... **A8:** 72
miscellaneous ... **A8:** 104-108
Mohs ... **A8:** 108
plowing ... **A8:** 107
thermal spray coatings ... **M5:** 371
types ... **A8:** 71
use in carbon control ... **M4:** 432-433
Hardness testing with optical microscope attachments ... **A9:** 91
reactive sputtered ... **A9:** 158
reactive sputtering of interference film ... **A9:** 148-149
Hardness tests
cast steel microstructure, effectiveness of correlation ... **M1:** 389, 398
gray iron ... **M1:** 16, 19, 20, 30
Hardware
applications for stainless steels ... **A7:** 731
applications of copper-based powder metals ... **A7:** 733
design methodology ... **EL1:** 78
for environmental stress screening ... **EL1:** 885
galvanized steel for ... **M1:** 167, 169
powders used ... **A7:** 572
reliability ... **EL1:** 79
system structure constraints by ... **EL1:** 127
Hardware, computer *See* Computer hardware; Computers
Hardware design methodology
high-frequency digital systems ... **EL1:** 78
Hardware green brass coloring solution ... **M5:** 626
Hardwood
carbides for machining ... **A16:** 75
Harkins-Jura method ... **EM4:** 70
Harlan process ... **A7:** 120
Harmonic bond tester
for adhesive-bonded joints ... **A17:** 621
Harmonic-oscillator approximation
IR normal-coordinate analysis ... **A10:** 110-111
Harmonics *See* Spherical harmonics
of frequency content, magabsorption ... **A17:** 148
nonlinear, for residual stress measurement ... **A17:** 160-161
Harness satin *See also* Eight-harness satin; Four-harness satin; Satin (crowfoot) weave
defined ... **EM1:** 12 **EM2:** 21
Harrison number *See also* Compressibility number
defined ... **A18:** 10
Hart-Smith A4EG/F computer analysis ... **EM1:** 338-339
Hartley transform ... **A18:** 294
Hartmann dust chamber ... **A7:** 196, 197
Hartmann lines *See* Lüders lines; Lüders Lines
Hartshorn and Ward capacitance variation method ... **EM3:** 431
Hashin's composite moduli formula
composite Young's modulus ... **EM4:** 860, 861
Hastelloy
compounds/properties in composition ... **EM4:** 499
explosion welding ... **M6:** 710
nonmetallic fusion process ... **EM3:** 306
Hastelloy alloy A
contour band sawing ... **A16:** 363
cutoff band sawing ... **A16:** 361
Hastelloy alloy B
chemical milling ... **A16:** 584
composition ... **A16:** 736, 737
contour band sawing ... **A16:** 363
cutoff band sawing ... **A16:** 361
machining ... **A16:** 738, 741-743, 746-747, 749-758
photochemical machining ... **A16:** 588
Hastelloy alloy B-2
composition ... **A16:** 736
machining ... **A16:** 738, 741-743, 746-747, 749-757
Hastelloy alloy C
chemical milling ... **A16:** 584
composition ... **A16:** 736, 737
contour band sawing ... **A16:** 363
cutoff band sawing ... **A16:** 361
machining ... **A16:** 738, 741-743, 746-747, 749-758
photochemical machining ... **A16:** 588
Hastelloy alloy C-276
composition ... **A16:** 736
machining ... **A16:** 738, 741-743, 746-747, 749-758
Hastelloy alloy G
machining of ... **A16:** 738, 741-743, 746-747, 749-758
Hastelloy alloy X
chemical milling ... **A16:** 584
composition ... **A16:** 736
electrochemical grinding ... **A16:** 547
grinding ... **A16:** 547
machining ... **A16:** 738, 741-743, 746-747, 749-757
manufacturing ratings ... **A16:** 739
milling ... **A16:** 547
production time ... **A16:** 739
shaping ... **A16:** 192
Hastelloy alloys *See also* Nickel alloys, cast, specific types; Nickel alloys, specific types; Nickel alloys, specific types, Hastelloy ... **A1:** 971
alloy C-276, cavitation erosion ... **A18:** 763
alloy X
material for jet engine components ... **A18:** 588, 591
open-faced honeycomb, abradable seal material ... **A18:** 589
corrosion resistance ... **A6:** 585
discovery and characteristics ... **A2:** 429
friction surfacing ... **A6:** 323
galling resistance with various material combinations ... **A18:** 596
Hastelloy alloys, specific types
A, cutoff band sawing with bimetal blades ... **A6:** 1184
B
cutoff band sawing with bimetal blades ... **A6:** 1184
electron-beam welding ... **A6:** 869
weld overlay for hardfacing alloys ... **A6:** 820
C ... **A6:** 593
cutoff band sawing with bimetal blades ... **A6:** 1184
weld overlay for hardfacing alloys ... **A6:** 820
C-22, in ferritic base metal-filler metal combination ... **A6:** 449
C-276
in ferritic base metal-filler metal combination ... **A6:** 449
laser weld joining, YAG ... **A6:** 88
N
composition ... **A6:** 564
electron-beam welding ... **A6:** 869
S
composition ... **A6:** 564, 573
mill annealing temperature range ... **A6:** 573
solution annealing temperature range ... **A6:** 573
W
composition ... **A6:** 573
filler metal for precipitation-hardening stainless steels ... **A6:** 491
filler metal wire, electron-beam welding ... **A6:** 868
X
composition ... **A6:** 564, 573
constitutional liquation in multicomponent systems ... **A6:** 568
electron-beam welding ... **A6:** 869
mill annealing temperature range ... **A6:** 573

Hastelloy alloys, specific types (continued)
solution annealing temperature range ... **A6:** 573
Hastelloy B
aging cycle ... **M4:** 656, 657
annealing ... **M4:** 655
composition ... **A4:** 794 **M4:** 651-652, 653
solution treating ... **M4:** 656, 657
stress relieving ... **M4:** 655
Hastelloy B-2
annealing ... **A4:** 908
composition ... **A4:** 908
stress-equalizing ... **A4:** 908
stress-relieving ... **A4:** 908
Hastelloy C
aging cycle ... **M4:** 656, 657
aging cycles ... **A4:** 813
annealing ... **M4:** 655
composition ... **M4:** 651-652, 653
solution treating ... **M4:** 656, 657
stress relieving ... **M4:** 655
Hastelloy C-276
annealing ... **A4:** 908
composition ... **A4:** 794, 908
pitting potentials on unstrained specimens ... **A8:** 418
stress-equalizing ... **A4:** 908
stress-relieving ... **A4:** 908
Hastelloy N
composition ... **A4:** 794
Hastelloy R-235
aging cycle ... **M4:** 656
annealing ... **M4:** 655
composition ... **M4:** 651-652
solution treating ... **M4:** 656
stress relieving ... **M4:** 655
Hastelloy S
composition ... **A4:** 794
mill annealing ... **A4:** 810
solution annealing ... **A4:** 810
Hastelloy, specific types
C, flash welding ... **M6:** 557
C-4, composition ... **M6:** 354
C-276, composition ... **M6:** 354
N
composition ... **M6:** 354
flash welding ... **M6:** 557
S
composition ... **M6:** 354
flash welding ... **M6:** 557
W, flash welding ... **M6:** 557
X
composition ... **M6:** 354
flash welding ... **M6:** 557
multiple-impulse resistance spot welding ... **M6:** 532
Hastelloy W
aging cycle ... **M4:** 656
annealing ... **M4:** 655
composition ... **A4:** 794 **M4:** 651-652
solution treating ... **M4:** 656
stress relieving ... **M4:** 655
Hastelloy X
age-hardening ... **A4:** 911
aging cycle ... **M4:** 656
annealing ... **A4:** 811, 908 **M4:** 655
composition ... **A4:** 794, 795, 908 **M4:** 651-652, 653
grain growth ... **A4:** 811, 812
mill annealing ... **A4:** 810
solution annealing ... **A4:** 810
solution treating ... **M4:** 656
solution-treating ... **A4:** 911
stress relieving ... **M4:** 655
stress-equalizing ... **A4:** 908
stress-relieving ... **A4:** 811, 908
Hastelloy-X
in air and vacuum, comparison of fatigue crack growth rates ... **A8:** 413
effect of mean load on stress intensity ... **A8:** 255
Hastelloy-X (continued)
fatigue crack growth ... **A8:** 255
Hat cracks
in weldments ... **A17:** 585
Hauffe-Ilschner (thin-film) theory ... **A13:** 67
Haulage rope
steel wire ... **M1:** 272
Hausmannite
electrical properties ... **EM4:** 765
Hausner ratio
and angle of repose ... **A7:** 285
Hausner shape factors ... **A7:** 239
Hawsers
steel wire ... **M1:** 272
Haynes 21
composition ... **A4:** 795
Haynes 25 ... **A1:** 967
aging, effect on properties ... **A4:** 800
composition ... **A4:** 794, 795
Haynes 88
notch sensitive in stress-rupture test ... **A8:** 315-316
Haynes 188 ... **A1:** 967
composition ... **A4:** 794
mill annealing ... **A4:** 810
solution annealing ... **A4:** 810
stress-relieving ... **A4:** 811
Haynes 214
aging ... **A4:** 810
composition ... **A4:** 794
Haynes 230
composition ... **A4:** 794
mill annealing ... **A4:** 810
solution annealing ... **A4:** 810
solution-treating ... **A4:** 810
Haynes 242
composition ... **A4:** 794
Haynes 556
annealing ... **A4:** 811
composition ... **A4:** 794
effect of temperature ... **A12:** 34, 47
grain growth ... **A4:** 811
mill annealing ... **A4:** 810
solution annealing ... **A4:** 810
Haynes alloy ... **A18:** 188
erosion test results ... **A18:** 200
for hot-forging dies ... **A18:** 626-627
material for jet engine components ... **A18:** 588, 591
Haynes alloy 6B
composition ... **A18:** 806
Haynes alloy 21
composition ... **A16:** 737
grinding ... **A16:** 759
machining ... **A16:** 738, 741-743, 746-757
Haynes alloy 25
drilling ... **A16:** 750
machining ... **A16:** 757, 758
production times ... **A16:** 739
Haynes alloy 25 (L605)
band sawing ... **A16:** 738
composition ... **A16:** 736
drilling ... **A16:** 738, 739, 750
end milling ... **A16:** 738, 739
machining ... **A16:** 738, 741-743, 746-747, 749-757
machining characteristics compared in table ... **A16:** 738, 739
reaming ... **A16:** 738, 739
straddle milling ... **A16:** 738
tapping ... **A16:** 738
threading ... **A16:** 738, 739
Haynes alloy 93
erosion test results ... **A18:** 200
Haynes alloy 188
composition ... **A16:** 736
machining ... **A16:** 738, 741-743, 746-747, 749-757
Haynes alloy 263
thread grinding ... **A16:** 275
Haynes Alloy No. 150 *See* Superalloys, cobalt-base, specific types, UMCo-50
Haynes alloys, specific types
25
brazing effect on mechanical properties ... **A6:** 929
composition ... **A6:** 564, 573, 929
filler metals ... **A6:** 928, 929
mill annealing temperature range ... **A6:** 573
solution annealing temperature range ... **A6:** 573
150, composition ... **A6:** 573, 929
188
composition ... **A6:** 564, 573, 929
mill annealing temperature range ... **A6:** 573
solution annealing temperature range ... **A6:** 573
230
composition ... **A6:** 564, 573
mill annealing temperature range ... **A6:** 573
solution annealing temperature range ... **A6:** 573
556, composition ... **A6:** 564
HR-160, composition ... **A6:** 573
Haynes, specific types
25
composition ... **M6:** 354
effect of brazing on mechanical properties ... **M6:** 1020
joint design for arc welding ... **M6:** 368
microcrack elimination ... **M6:** 369
188, composition ... **M6:** 354
556, composition ... **M6:** 354
Stellite alloys, laser cladding ... **M6:** 797
Haynes Stellite alloy No. 1
laser cladding ... **A18:** 867
Haynes Stellite alloys *See* Cobalt alloys, specific types, Haynes Stellite
HAZ *See* Heat-affected zone
HAZ CALCULATOR (software package)
Creusot-Loire systematization of CCT curves ... **A4:** 22
Hazardous materials ... **EM3:** 685-686
Hazardous metal powder operations ... **A7:** 197-198
Hazards *See* Safety
material safety data sheet (MSDC) to satisfy hazard communication standard of OSHA ... **A18:** 87
Haze ... **EM2:** 21, 594 **EM3:** 15
Hazelett cast machine
as twin band ... **A15:** 314
Hazelett process
for copper and copper alloy wire rod ... **A2:** 255
HB *See* Brinell hardness; Brinell hardness number
HB gallium arsenide crystal growth method ... **A2:** 744
HC alloy ... **A1:** 922
HCL *See* Hollow cathode lamp
hcp *See* Hexagonal close-packed; Hexagonal close-packed (hcp) materials; Hexagonal close-packed lattice
H_d *See* Demagnetizing field
HD 430
erosion test results ... **A18:** 200
HD 435
erosion test results ... **A18:** 200
HD alloy ... **A1:** 922
HDPE *See* High-density polyethylene
Head
pressure vessels, failures of ... **A11:** 644
and shell cracking, gray cast iron paper machine dryer rolls ... **A11:** 653-654
and shell cracking, paper machine dryer rolls ... **A11:** 653-654
Head and tailstock magnetizing equipment ... **A7:** 576
Head shot *See also* Direct contact method method of magnetization ... **A17:** 99, 130
Head-to-head ... **EM3:** 15
defined ... **EM2:** 21

Head-to-tail **EM3:** 15
defined **EM2:** 21
Headability *See also* Cold heading; Warm heading
defined **A14:** 291
Headboxes
corrosion of **A13:** 1186-1187
Header-slide (stock) gather
machine size selection by **A14:** 84-85
Headers *See also* Forging machines; Horizontal forging machines; Upsetters
for cold extrusion **A14:** 303
for cold heading **A14:** 292
Heading *See also* Cold heading; Upsetting; Warm heading
alloy steel wire for **A1:** 287 **M1:** 269-270
beryllium-copper alloys **A2:** 415
defined **A14:** 7
and extrusion, combined **A14:** 296-297
in header **A14:** 311
in hot upset forging **A14:** 83
severity, decreasing **A14:** 297
tools **A14:** 48, 90-91, 293-296
of wire **A14:** 694
Heading quality steel wire rod **M1:** 254, 257
Heading tools
design **A14:** 293
inserts for **A14:** 48
recessed, for upsetting with sliding dies **A14:** 90-91
Headlights
economy in manufacture **M3:** 848
Heads
polishing and buffing **M5:** 119-120, 126
Healing
of bone fractures **A11:** 673-674
Health *See also* Flammability; Safety precautions; Toxicity
and parylene coatings **EL1:** 800
and safety, with beryllium **A2:** 687
Health, environmental and occupational, and lubricants *See also* Safety **A14:** 517-518
Health hazards
radiation as **A10:** 247
Health precautions *See* Safety precautions
Health studies
NAA application in **A10:** 234
Health-care applications *See also* Medical applications
critical properties **EM2:** 457-458
Hearing safety
ultrasonic fatigue testing effect on **A8:** 240
Heart pacemaker
as hybrid medical application **EL1:** 387
Hearth
defined **A15:** 6
Hearth furnaces
reverberatory **A15:** 374-376
Heat
-affected zone, abbreviation **A8:** 724
build-up **EM1:** 13
capacity, SI unit/symbol for **A8:** 721
-centering techniques, in fatigue testing **A8:** 712
cleaned **EM1:** 13
content, conversion factors **A8:** 722 **A10:** 686
control, autoclave **EM1:** 704
conversion during curing reaction **EM3:** 323
defined **A15:** 6
degradation of polymers by **A11:** 761
effect on aluminum alloy phase structure **A9:** 359
evolution in preparing etchants **A9:** 69
extraction, of alloy additions **A15:** 71
flow, effect in medium strain rate regime **A8:** 191
flux density, SI unit/symbol for **A8:** 721
of formation, liquid/solid copper-tin alloys **A15:** 59
from mounting materials, effect on carbon and alloy steels **A9:** 166-167
of fusion, during solidification **A15:** 109
generated during curing of castable resins **A9:** 30-31
input, conversion factors **A10:** 686
quantity of, SI derived unit and symbol for **A10:** 685
Heat (continued)
release, during solidification **A15:** 109
removal, die casting **A15:** 292-294
resistance **EM1:** 13 **EM3:** 52
scanning electron microscopy study of specimens subjected to **A9:** 97
sealing **EM1:** 13
Heat aging *See also* Aging; Thermal aging
long-term, metals vs plastics **EM2:** 78
polyethylene terephthalates (PET) **EM2:** 174
Heat analysis
for classifying steels **A1:** 141
Heat and pull methods
component removal **EL1:** 717-718
Heat and shake component removal method **EL1:** 716
Heat balance
as process modeling submodel **EM1:** 500
Heat balance, overall
as cooling analysis **A15:** 293
Heat buildup *See also* Hysteresis **EM3:** 15
defined **EM2:** 21
Heat capacity ..... **A3:** 1 • 6 **A10:** 685
amorphous materials and metallic glasses **A2:** 812-813
beryllium **A2:** 683
Heat capacity, and enthalpy
principles of **A15:** 50
Heat check *See also* Checks
defined **A13:** 7
Heat check fins
as casting die defect **A15:** 295
Heat checking *See* Checking **A18:** 621-622, 623, 625
defined **A18:** 10
die wear in hot forging dies **A18:** 635, 639-640
seals **A18:** 549
wear and failure in die-casting dies **A18:** 629
Heat cleaned
defined **EM2:** 21
Heat composition, variations
effect on hardenability **M1:** 477-479
Heat conduction
in heat exchangers **A11:** 628
Heat conductivity
symbol and units **A18:** 544
Heat content *See* Enthalpy ..... **A3:** 1 • 6
Heat cure *See also* Cure; Curing
of adhesives **EL1:** 672
of conformal coatings **EL1:** 764-765
of silicone-base coatings **EL1:** 773
Heat damage
in tool steels from sectioning **A9:** 256
Heat deflection temperature **EM1:** 8, 91, 92
Heat dissipation model
TAB **EL1:** 286
Heat distortion
point **EM1:** 13
sheet molding compounds **EM1:** 158
Heat distortion temperature *See* Deflection temperature under load
Heat exchangers *See also* Heat exchangers, failures of **EL1:** 47 **EM4:** 978-985
anodic protection of **A13:** 465
application areas **EM4:** 979-980
brazeability and solderability **A6:** 618
brazing with clad brazing materials **A6:** 347, 348
causes of failures in **A11:** 629-630
ceramic materials **A6:** 948 **EM4:** 980-981
ceramic prototype systems **EM4:** 982-984
clad brazing materials, applications of **A6:** 962-963
cleaning **A13:** 1138-1139
copper and copper alloys **A14:** 822-823
corrosion of **A11:** 630-635
corrosion protection before service **A11:** 628
design of **A11:** 637-639
dip brazing **A6:** 338
economics of ceramic heat exchangers **EM4:** 984-985
eddy current inspection of **A17:** 180-181
fabricated by roll welding **A6:** 312, 314
failed, examination of **A11:** 628-629
function of tubes in **A11:** 628
furnace brazing **A6:** 915, 916-917
heat transfer considerations **EM4:** 978-979
market share **EM4:** 985
Heat exchangers (continued)
materials and applications **A11:** 628
resistance seam welding **A6:** 238-239
saltwater, hydrogen sulfide pitting in **A11:** 631
stainless steel **A13:** 560-561
stiffening in **A11:** 640
test methods **EM4:** 981-982
titanium **A2:** 588-589
torch brazing **A6:** 328
types and functions of **A11:** 628
welding practices for **A11:** 639-640
Heat exchangers, failures of *See also* Heat exchangers **A11:** 628-642
causes **A11:** 629-630
corrosion **A11:** 630-635
corrosion fatigue **A11:** 636-637
design, effects of **A11:** 637-639
elevated temperatures, effects of **A11:** 640-642
failed parts, examination of **A11:** 628-629
stress-corrosion cracking **A11:** 635-636
tubing, characteristics of **A11:** 628
welding practices, effects of **A11:** 639-640
Heat flow
conditions, single-phase alloys **A15:** 114-119
control, die casting **A15:** 292-293
control, directional solidification **A15:** 320
and fluid flow, interaction defects **A15:** 294-295
multidirectional, solidification rates **A15:** 145
paths, die casting **A15:** 293
total rate of **A18:** 43
Heat flow in fusion welding **A6:** 7-18
condition at infinity **A6:** 9
conduction equation **A6:** 8
engineering solutions and empirical correlation **A6:** 9-17
cooling rate **A6:** 10-11
general solutions **A6:** 9-10
modified temperature solution **A6:** 12
peak temperature **A6:** 11
practical application of heat flow equations **A6:** 12
solidification rate **A6:** 11-12
finite source theory **A6:** 8
gas-tungsten arc welding **A6:** 8
general approach **A6:** 8
heat-affected zone **A6:** 7, 8, 10, 11, 12, 13, 16
heat-source formulation **A6:** 8-9
isotemperature contours **A6:** 14, 15, 16, 17
literature review **A6:** 8
low-carbon steels **A6:** 10, 13, 16, 17
mathematical formulations **A6:** 8-9
parametric effects **A6:** 17
quench and tempered steels **A6:** 13
relation to welding engineering problems **A6:** 7
surface heat loss **A6:** 9
thermophysical properties of selected engineering materials **A6:** 18
thick-plate equation **A6:** 14
thin-plate equation **A6:** 13-14, 15
underwater welding **A6:** 9
welding control **A6:** 7-8
welding distortion **A6:** 8
welding heat efficiency **A6:** 9
welding thermal process **A6:** 7
metallurgical zones formed **A6:** 7
thermal states **A6:** 7
Heat flow paths
dual-in-line packages **EL1:** 489
Heat flux
at device level **EL1:** 413
effect light water reactor corrosion **A13:** 947-948
Heat flux density **A10:** 685
Heat flux distribution **A18:** 40
Heat fluxes
effects on tube-wall temperature, boilers **A11:** 605
scale thickness vs temperature **A11:** 609
uniform **A11:** 603-604
Heat forming *See* Forming; Thermoforming
Heat input
conversion factors **A8:** 722
equation **A6:** 427
Heat input per length of weld **A6:** 29
Heat insertion **EM2:** 724
Heat mark *See also* Sink-mark
defined **EM2:** 21

Heat of fusion
ductile iron ... **M1:** 52
Heat partition factors ... **A18:** 39, 40, 41, 44
Heat process scale
salt bath descaling removal of ... **M5:** 100-102
Heat reflecting paint
powder used ... **A7:** 572
Heat removal
radiography ... **A17:** 305-306
Heat removal technology *See also* Cooling
air cooling with fans ... **EL1:** 309-310
bonding to frame/cooling surface ... **EL1:** 310
methods, summary ... **EL1:** 406
water cooling ... **EL1:** 310-311
Heat resistance
cast steels ... **M1:** 400
polyphenylene ether blends (PPE PPO) ... **EM2:** 184
styrene-maleic anhydrides (S/MA) ... **EM2:** 219
thermoplastic polyimides (TPI) ... **EM2:** 177
thermoplastic polyurethanes (TPUR) ... **EM2:** 204
thermosetting resins ... **EM2:** 223
Heat sealing *See also* Impulse sealing
defined ... **EM2:** 21
Heat sealing adhesive
defined ... **EM1:** 13
Heat shield coatings
powder used ... **A7:** 572
Heat shields
powders used ... **A7:** 572
Heat sink *See also* Graphite
defined ... **EM1:** 13 **EM2:** 21
properties, of carbon-carbon composites ... **EM1:** 36
Heat sink assembly methods
cost impact ... **EM3:** 572
Heat sink, copper alloy
sliding electrical contacts ... **A2:** 842
Heat sink welding
in boiling water reactors ... **A13:** 931
Heat sinks ... **EL1:** 1129-1131
aluminum alloys ... **A6:** 726
composites for ... **EL1:** 1122
electron-beam welding ... **A6:** 869
PWB, polymer matrix composites ... **EL1:** 1117-1118
used in package cooling ... **EL1:** 414
Heat source, characterization and modeling of ... **A6:** 1141-1146
arc efficiency ... **A6:** 1142
arc welding ... **A6:** 1142-1144
electron-beam welding ... **A6:** 1144
flux-cored arc welding ... **A6:** 1144
gas-metal arc welding ... **A6:** 1143-1144
gas-tungsten arc welding ... **A6:** 1142-1143
high-energy-density welding ... **A6:** 1144-1145
laser-beam welding ... **A6:** 1142, 1144
"quasi-stationary" temperature response ... **A6:** 1141
resistance spot welding ... **A6:** 1145-1146
shielded-metal arc welding ... **A6:** 1144
simplified modeling of the heat source ... **A6:** 1141-1142
submerged arc welding ... **A6:** 1144
Heat spreaders
molded inside plastic packages ... **EL1:** 412
Heat stabilizers
as additive, effects ... **EM2:** 424-425
as additives ... **EM2:** 95
in engineering thermoplastics ... **EM2:** 98
as polymer additives ... **EM2:** 67
Heat time
definition ... **M6:** 9
Heat tinting
austenitic stainless steels ... **A9:** 282
defined ... **A9:** 9
for interference colors ... **A9:** 61
for replicas ... **A12:** 181
to produce interference films ... **A9:** 136
Heat tinting (continued)
used in quantitative metallography of cemented carbides ... **A9:** 275
wrought heat-resistant alloys ... **A9:** 307
Heat transfer
analysis, computer-aided ... **A15:** 610
by heat exchangers ... **A11:** 628
casting vs. welding solidification ... **A6:** 46
coefficient, determining ... **A14:** 441-442
coefficient, steam-side, functions of ... **A11:** 603
in design ... **EM2:** 1
design problem, die casting ... **A15:** 293-294
effects, alloy additions ... **A15:** 71-72
factors, internal scale deposits ... **A11:** 604
and flow localization ... **A8:** 170-173
fluidized-bed ... **M4:** 301-302
forced convectional, aluminum melt ... **A15:** 453-456
and mass transfer, fluid flow modeling of ... **A15:** 877-882
material, characteristics ... **A17:** 396
mechanisms, thermal inspection ... **A17:** 396
modes of ... **A7:** 341
properties, of molten metals ... **A13:** 56
in squeeze casting ... **A15:** 323
and temperature, aqueous corrosion ... **A13:** 39-40
to tooling ... **A14:** 161
tubes, erosion ... **A13:** 136
Heat transfer coefficients ... **EL1:** 24, 414
Heat transfer onto slip ring, electrical contacts
friction and wear of ... **A18:** 683
Heat transfer theory
cure cycle ... **EM1:** 746-748
Heat transport
during solidification ... **A15:** 111-113
treatment, in equiaxed grain growth ... **A15:** 133
Heat treatable steels
weldability ... **M1:** 562-563
Heat treated HSLA steels *See also* As-rolled structural steels; Microalloyed HSLA steels ... **M1:** 403, 409-410
compositions, typical ... **M1:** 404
mechanical properties ... **M1:** 409-410, 414, 417
Heat treating *See also* Annealing; Hardening; Heat treatment; Sintering; Tempering ... **EM3:** 15
aluminum alloys ... **M2:** 28-43
atmospheres, potential hazards ... **A7:** 348-350
copper metals ... **M2:** 252-260
defined ... **EM1:** 13 **EM2:** 21
dimensional change during ... **A7:** 481
furnace brazing ... **A6:** 330
heating medium used ... **A7:** 453
in hot pressing ... **A7:** 503
P/M parts ... **A7:** 295
as secondary operation ... **A7:** 451-456
and sintering, differences ... **A7:** 339
of tungsten and molybdenum sintering ... **A7:** 390-391
Heat treating furnaces, equipment for
atmospheres ... **M4:** 289
direct-fired ... **M4:** 289
electrically heated ... **M4:** 289-291
gas-fired ... **M4:** 289
heating elements ... **M4:** 289-290
maintenance ... **M4:** 292
oil-fired ... **M4:** 289
radiant-tube-heated ... **M4:** 291-292
Heat treating furnaces, types
batch ... **M4:** 285, 286
bogie hearth ... **M4:** 287, 288
car ... **M4:** 287, 288
continuous ... **M4:** 285-286
pusher ... **M4:** 286-287
walking-beam ... **M4:** 287-288
Heat treating, of tool steels
effects on microstructure ... **A9:** 258-259
Heat treatment *See also* Annealing; Fabrication characteristics; Heat treating; Heat-treatable alloys; Overheating; Post heat treating; Postweld heat treatment; Precipitation hardening; Preheating; Sintering; specific metal or process; Thermal processing
Heat treatment (continued)
50% aluminum-zinc alloy coatings ... **M5:** 349-350
of abrasion-resistant cast iron ... **A1:** 91-92 **M1:** 81-83
and acoustic emission inspection ... **A17:** 287
after overheating ... **A12:** 127
of all-ceramic mold castings ... **A15:** 249
alloy steel sheet and strip ... **M1:** 164-165
Alnico alloys ... **A15:** 738
of aluminum alloy forgings ... **A11:** 340 **A14:** 249
of aluminum alloy ingots, effect on hydrogen porosity ... **A9:** 633
of aluminum alloy rivets ... **A11:** 544-545
aluminum alloy sand and permanent mold castings ... **A2:** 137-138
for aluminum alloy sand/permanent mold castings ... **A15:** 758-759
of aluminum alloys ... **A15:** 757-762
aluminum alloys, anodizing affected by ... **M5:** 591
aluminum bronze alloys ... **A15:** 782
aluminum casting alloys ... **A2:** 136, 148, 154-177
aluminum coating process ... **M5:** 334, 339
aluminum-lithium alloys ... **A2:** 183
applications, nickel-base alloys ... **A13:** 655
of ASP steels ... **A7:** 785-787
austenitic ductile irons ... **A15:** 700-701
for austenitic manganese steel ... **A1:** 829-831
of austenitic steels ... **A11:** 393
baths, molten nitrate ... **A13:** 90-91
of bearing materials ... **A11:** 508-510
beryllium ... **A2:** 683-686
beryllium copper alloys ... **A15:** 782
beryllium-copper alloys ... **A2:** 405-408
of carbon steel parts ... **A11:** 392
carburized and carbonitrided parts process heat treating of ... **M1:** 533, 538, 539
carburized parts, selective heat treating of ... **M1:** 532
carburizing, in steel castings ... **A11:** 396
cast steels ... **M1:** 379, 381, 382, 388-389, 392, 398
of castings ... **A11:** 366
change to prevent corrosion ... **A11:** 193-194
cobalt base alloys ... **A13:** 662
for cobalt-base alloys ... **A15:** 813-814
cold heading properties improved by ... **M1:** 590-591
cold rolled sheet and strip mill heat treatment of ... **M1:** 156-157
cold-finished bars ... **M1:** 219, 232-235
of compacted graphite irons ... **A1:** 9
complexity index for assessing effects of ... **A9:** 130
condition, by eddy current inspection ... **A17:** 164
contamination, magnesium/magnesium alloys ... **A13:** 740, 741
copper alloys ... **A15:** 782
copper and copper alloys ... **A2:** 223
copper and copper alloys, formability effects ... **A14:** 809-810
cracks ... **A17:** 50, 296
cracks, martensitic stainless steels ... **A12:** 366
decarburization ... **A11:** 122
distortion in ... **A1:** 462
in drawing ... **A14:** 331-333
ductile and brittle fracture from ... **A11:** 94-97
of ductile iron ... **A1:** 8, 9, 38, 40, 41-42 **A15:** 657-660 **M1:** 35, 37, 42, 45
effect, AISI/SAE alloy steels ... **A12:** 298
effect, CG iron tensile properties ... **A15:** 673
effect, cold-heading properties ... **A14:** 293-294
effect, galvanized steel ... **A13:** 433
effect, hydrogen-embrittled aluminum alloy ... **A12:** 33
effect of, at elevated-temperature service ... **A1:** 638-639, 642

Heat treatment (continued)
effect of enthalpy and heat capacity **A15:** 50
effect of, on bearing steels **A1:** 383, 384, 385
effect on carbon steel casting microstructures **A9:** 231
effect on copper-infiltrated iron and hypoeutectoid steel compacts **A7:** 558
effect on corrosion-fatigue **A11:** 256
effect on distortion **A11:** 140
effect on fatigue strength **A11:** 119, 121-122
effect on fracture appearance, low-copper aluminum alloy **A12:** 33
effect on fracture appearance, titanium alloy **A12:** 32
effect on microstructure of austenitic manganese steel castings **A9:** 239
effects, magnesium/magnesium alloys **A13:** 741
effects on ductile-to-brittle transition temperature, structured steels **A11:** 68
effects on Waspaloy rupture time **A8:** 316
effects, stainless steel corrosion **A13:** 551
electroless nickel plate **M5:** 224-231
equalized and aged **A13:** 934
erosion rate effect **A18:** 204
eutectic melting **A11:** 122
factors influencing dimensional change **A7:** 291
failures, from surface hardening **A11:** 395
fatigue behavior related to **M1:** 672, 673-675, 679
fatigue failure at spot welds from **A11:** 310-311
faulty, and distortion **A11:** 140
of ferritic e iron **A1:** 75
of ferrous powder metallurgy materials **A1:** 809, 810-811
flame hardening **M1:** 532
flame impingement during **A11:** 573, 579
forging failures from **A11:** 332-336, 340
forging, of stainless steels **A14:** 228-229
fracture effect, medium-carbon steels **A12:** 272
fusion or melting during, as casting defect **A11:** 386
global **A13:** 942
grain size of steel, effect on **M1:** 701
of gray iron.... **A1:** 7, 21, 29-31 **A15:** 642-643 **M1:** 13, 23, 27, 28, 29-30, 31, 32
hardenable alloy steel, interrelation with machinability **M1:** 580-581
of hardenable alloys, distortion from **A11:** 140
hardenable steels **M1:** 457-461, 463, 466-470
of heat-resistant alloys **A14:** 235
high-alloy steels **A15:** 730-731
high-chromium white irons **A15:** 684
high-silicon irons **A15:** 699, 701
high-temperature, white irons **A15:** 680
history **A12:** 1-3
and hot isostatic pressing **A7:** 440
hot rolled bars and shapes **M1:** 200-201
hot upset forging **A14:** 87
of hot work tool steels **A14:** 53-55
of hot-rolled steel bars **A1:** 241
HSLA steels **M1:** 409-410
improper, AISI/SAE alloy steels **A12:** 309
induction hardening **M1:** 532
of Invar **A2:** 890-891
in investment casting **A15:** 263
ladles, processes compared **A15:** 437
lasers used for **M1:** 628-629
of low-alloy steel castings.......... **A9:** 231 **A11:** 392-393
of magnesium alloy forgings **A14:** 260
and magnetic hysteresis **A17:** 134
magnetically soft materials **A2:** 762-763
of malleable iron.......... **A1:** 10-11, 75, 76-80 **M1:** 57-63, 68-71, 73
of maraging steels **A1:** 795-797 **A11:** 218 **M1:** 445-448
of martensitic stainless steels, effects **A9:** 285
metallographic sectioning for **A11:** 24
as metallurgically influenced corrosion **A13:** 123
and mold life **A15:** 281
molybdenum **A14:** 238
of mortar tubes **A11:** 294
multiple-step **A14:** 281
nickel alloy applications in **A2:** 430
nickel alloys **A15:** 822-823
nickel-chromium white iron **M1:** 81-82
nickel-chromium white irons **A15:** 680
nickel-iron alloys **A2:** 774

Heat treatment (continued)
niobium-titanium (copper) superconducting materials **A2:** 1053-1054
nitrided parts **M1:** 540-542
nitrided/nitrogenized layer from **A12:** 282
notch toughness of steels, effect on **M1:** 699, 701, 703, 706, 708, 709
in open-die forging **A14:** 64
overheating, effect on fatigue strength **A11:** 121-122
P/M steels **M1:** 338-339, 343-346
of pearlitic and martensitic malleable iron **A1:** 76-80
for phosphate conversion coatings **A13:** 387-388
of pipe and tubing for upsetting **A14:** 92
of piping system cross failure **A11:** 649-650
of plain carbon steels **A15:** 713-714
post-HIP **A15:** 543
postcast, titanium and titanium alloy castings **A2:** 639
postforging defects from **A11:** 333
postforming, of copper and copper alloys **A14:** 810
postweld **A15:** 526-527, 530, 534-535
postweld, microstructural effects **A12:** 375
powder forging **A14:** 201
powder metallurgy high-speed tool
annealing **A1:** 783
hardening **A1:** 783
steels **A1:** 782-783
stress-relieving (before tempering) **A1:** 783
tempering **A1:** 783
prealloyed titanium P/M compacts **A2:** 653-654
precipitation, effect on SCC in aluminum alloys **A11:** 220
precipitation, wrought aluminum alloy **A2:** 40-41
of precipitation-hardenable stainless steels effects **A9:** 285
precision forging **A14:** 164-165
of pressure vessels, effects of **A11:** 648-649
principles **A13:** 48
principles of **A15:** 759-762
prior, effect on depth of induction hardening **M1:** 531-532
quench cracks **A11:** 122, 568, 570
rapid, effects in AISI/SAE alloy steels **A12:** 328
reaction, in A15 superconductor assembly **A2:** 1068-1069
refractory metals, and formability **A14:** 785-786
roughing tool cracked after **A11:** 571
schedules, factors governing **A11:** 93-94
as seawater pollutant **A13:** 899-900
and shaft failures **A11:** 475-476
silver-magnesium-nickel alloys **A2:** 702
in SIMA process **A15:** 332-333
of sintered high-speed steels **A7:** 374-375
of sintered tungsten heavy alloys **A7:** 393
solution, boiling water reactors **A13:** 931
solution, for age-hardenable alloys **A11:** 122
springs, steel **M1:** 287-288, 296, 313
of stainless steel casting alloys, effect on microstructure **A9:** 298
of stainless steel castings **A1:** 911-912
stainless steels **M3:** 46-48
corrosion resistance, effect on **M3:** 57-59
steel casting failure from **A11:** 392-396
of steel castings **A1:** 367, 370-371
of steel plate **A1:** 230-232 **M1:** 182
steel wire **M1:** 262
steel wire rod **M1:** 253, 255, 257
sterling silver **A2:** 701
of stone, prehistoric **A15:** 15
stress relief, brittle fracture and **A11:** 664
subcritical **A15:** 684
of suction shells **A13:** 1207
Heavy metals
surface hardening of steel **M1:** 527-542
threaded fasteners **M1:** 275-277
of tin and tin alloys, effect on microstructure **A9:** 452
of titanium alloy forgings **A14:** 281-282
titanium alloys **A15:** 833
of titanium alloys, effect on microstructures **A9:** 460-461
titanium and titanium alloy castings **A2:** 643-644
tool and die failure from **A11:** 564, 567-574
tool steel surface after **A11:** 573, 578

Heat treatment (continued)
tool steels **M3:** 434
of type 304 stainless steel, effect on delta ferrite **A9:** 65-66
ultrahigh-strength steels **M1:** 423-425, 427, 429-430, 431-432, 433, 434-436, 438-439, 441
unanticipated **A11:** 395
uranium **M3:** 776-777
of uranium alloys **A2:** 672-674
vacuum coating process **M5:** 409
verification **A14:** 249, 282
verification, aluminum alloy forging **A17:** 497
verification, wrought titanium alloys **A2:** 620-622
warping during **A11:** 141
wear resistance versus hardness of materials **A18:** 708
weld preheat/postweld, magnesium castings **A2:** 476
wrought aluminum alloys **A12:** 417
of wrought heat-resistant alloys **A14:** 236
wrought titanium alloys **A2:** 618-622
Heat treatment cracks
as planar flaws **A17:** 50
radiographic inspection methods **A17:** 296
Heat treatments
of alloy steels, atom probe composition profiles for **A10:** 594
explosion welding **A6:** 305
history by optical metallography **A10:** 299
Heat welding
and sealing **EM2:** 724-725
Heat-activated adhesive **EM3:** 15, 35
defined **EM1:** 12 **EM2:** 21
Heat-activated plastics
as binders **A15:** 211
Heat-affected zone
885 °F embrittlement **M6:** 527
arc welds of coppers **M6:** 404
carbide precipitation, as weld decay **A13:** 349-350
in cast iron arc welds **M6:** 309-310
caustic SCC **A13:** 354
characteristics in friction welds **M6:** 720
characteristics in laser beam welds **M6:** 656
characteristics in percussion welds **M6:** 740
chromium carbide precipitation in **M6:** 527
cold cracks in **A12:** 137, 155-156
control of toughness **M6:** 41-42
corrosion of **A13:** 652-653
crack susceptibility in precipitation-hardenable alloys **M6:** 438
cracking in
arc welds of nickel alloys **M6:** 443
electroslag welds **M6:** 233-234
flash welds **M6:** 579
nickel-based heat-resistant alloys **M6:** 355
shielded metal arc welds **M6:** 93
cracking in stainless steels
austenitic stainless steels **M6:** 321, 323
ferritic stainless steels **M6:** 346
martensitic stainless steels **M6:** 348
cracking, medium-carbon steels **A12:** 254
decohesive rupture **A12:** 18
defined **A13:** 7
definition **M6:** 9, 27-28
depth in gas cut carbon steels **M6:** 903
effect of cooling rate and alloying **M6:** 40
effect of electrode travel speed **M6:** 86
electrogas welds **M6:** 244
electron beam welds **M6:** 618-619
comparison of size **M6:** 625
effect of pressure **M6:** 618
width **M6:** 619
elimination of microcracks **M6:** 369
flash welding **M6:** 569, 571, 576-578
weld defects **M6:** 579
gas cutting effects **A14:** 724, 731
grain size in beryllium **M6:** 572-573
grain structure in thermit welds **M6:** 697
graphite formation in **A12:** 124
heat flow calculations **M6:** 31-32
influence of preheat in cast irons **M6:** 312-313
intergranular SCC, location **A13:** 929
microstructure **M6:** 35-371408
minimization by pulsed-current welding **M6:** 187
plasma arc cutting, effect on **M6:** 917
prediction of microstructure **M6:** 40

Heat-affected zone (continued)
preferential corrosion, carbon steel weldments ... **A13:** 363
preheating to reduce residual stress ... **M6:** 890
recrystallization/grain growth ... **A13:** 344
resistance spot welding ... **M6:** 486-487
shallowness in percussion welds ... **M6:** 740
sigma precipitation in ... **A13:** 350
stress-relief cracking ... **A12:** 139
striations ... **A12:** 21
structure in stud arc welding ... **M6:** 733
submerged arc welding
grain coarseness ... **M6:** 118-119
relationship of heat input ... **M6:** 118-119
toughness improvement ... **M6:** 116
variation, corrosivity ... **A13:** 344
weakness in aluminum alloys ... **M6:** 373
Heat-affected zone (HAZ) ... **A6:** 5-6
acetylene fuel gas ... **A6:** 1157
advanced titanium-base alloys ... **A6:** 526, 527
aerospace materials ... **A6:** 386
air-carbon arc cutting ... **A6:** 1176
aluminum alloys ... **A6:** 725, 726, 727, 729
aluminum-lithium alloys ... **A6:** 551, 552
arc welding of carbon steels ... **A6:** 641, 642, 647, 648, 649, 652, 658, 659, 660
austenitic stainless steels ... **A6:** 456, 464-465, 466, 467, 468
brazing ... **A6:** 110
and soldering characteristics of engineering materials ... **A6:** 623-624
capacitor discharge stud welding ... **A6:** 221
carbon and low-alloy steels ... **A6:** 405, 406, 407
cast irons ... **A6:** 709, 710, 712, 713-714, 717, 718, 720
cobalt-base corrosion-resistant alloys ... **A6:** 598
copper alloys ... **A6:** 754
coppers ... **A6:** 759
corrosion of weldments ... **A6:** 1065, 1066, 1067, 1068
cracking ... **A6:** 90-93
advanced titanium-base alloys ... **A6:** 526
machinery and equipment weldments ... **A6:** 391
nickel alloys ... **A6:** 742
cracks, by liquid penetrant inspection ... **A17:** 86
defined ... **A11:** 5
definition ... **A6:** 1210
description ... **A6:** 1065
dispersion-strengthened aluminum alloys ... **A6:** 542, 544, 545, 546
dissimilar metal joining ... **A6:** 824, 826
duplex stainless steels ... **A6:** 482, 483, 484, 485, 486, 488, 489
electrogas welding ... **A6:** 275, 276
electron-beam welding ... **A6:** 255, 258, 855, 857, 858, 864, 865, 866, 867, 868, 869, 870, 871, 872
electroslag welding ... **A6:** 270-271, 272, 275, 277, 279
ferritic stainless steels ... **A6:** 445, 446, 447, 448, 449, 450, 452, 454
friction surfacing ... **A6:** 321
friction welding ... **A6:** 316, 317
gas-tungsten arc welding ... **A6:** 868
graphitization at ... **A11:** 614
heat flow in fusion welding ... **A6:** 7, 8, 10, 11, 12, 13, 16
heat-treatable aluminum alloys ... **A6:** 530, 531, 532-534
heat-treatable low-alloy steels ... **A6:** 671, 672
high-strength low-alloy quench and tempered structural steels ... **A6:** 666
high-strength low-alloy, structural steels ... **A6:** 663, 664
high-temperature alloys ... **A6:** 563, 564-565
hot-cracking in ... **A11:** 401
hydrogen-assisted cracking in ... **A15:** 532-535
inertia welding ... **A6:** 889
intergranular carbide precipitation ... **A11:** 132
intergranular cracking ... **A11:** 132

Heat-affected zone (HAZ) (continued)
lamellar tearing ... **A6:** 1073
lamellar tearing in ... **A11:** 92
lamellar tearing, in weldments ... **A17:** 582, 585
laser welding of plastics ... **A6:** 1051
laser-beam welding ... **A6:** 262, 263
machinery and equipment weldments ... **A6:** 391
magnesium alloys ... **A6:** 780, 782
magnetic particle indications ... **A17:** 107-108
magnetic printing detection ... **A17:** 126
metallographic sections for ... **A11:** 24
microstructure evolution ... **A6:** 1136, 1137, 1138
nickel alloys ... **A6:** 587-590
nickel-base alloys ... **A6:** 577
nickel-base corrosion-resistant alloys containing molybdenum ... **A6:** 594, 595, 596, 597
nickel-chromium alloys ... **A6:** 587-590
nickel-chromium-iron alloys ... **A6:** 587-590
nickel-copper alloys ... **A6:** 587-590
niobium alloys ... **A6:** 581
non-heat-treatable aluminum alloys ... **A6:** 537, 539
nonferrous high-temperature materials ... **A6:** 566, 567, 569
oxyfuel gas cutting of base metal ... **A6:** 1159, 1160
oxyfuel gas welding ... **A6:** 290
plasma arc cutting ... **A6:** 1169
precipitation-hardening stainless steels ... **A6:** 483-484, 487, 488, 489, 490
preheating effects, cast iron ... **A15:** 525
repair welding ... **A6:** 1105, 1106, 1107
residual stresses ... **A6:** 1102
sensitization during welding of cobalt-base wrought alloys ... **A18:** 766
solid-state transformations in weldments ... **A6:** 70, 71, 72, 73, 74, 75, 79, 80-81, 82, 83, 84
stainless steel casting alloys ... **A6:** 497, 498
stainless steels ... **A6:** 679, 681, 683, 686, 689, 695, 697, 698-699
steel weldment soundness ... **A6:** 408, 409, 410, 411, 412
steel weldments ... **A6:** 416-417, 418, 419, 420, 421, 423-424, 425, 426-427
submerged arc welding ... **A6:** 202-203
tantalum alloys ... **A6:** 581
titanium alloys ... **A6:** 85, 509, 510, 513, 514, 516, 519, 521, 783, 784, 786
titanium-base corrosion-resistant alloys ... **A6:** 599
tool and die steels ... **A6:** 675
tungsten alloys ... **A6:** 582
ultrahigh-strength low-alloy steels ... **A6:** 673-674
ultrasonic welding ... **A6:** 327
ultrasonic welding, nickel-base alloys ... **A6:** 895
underwater welding ... **A6:** 1010, 1011-1012
weld characterization ... **A6:** 97, 99, 100, 101, 102, 103, 104
weld discontinuities ... **A6:** 1075, 1076, 1078
weld, pressure vessel failure from cracks in ... **A11:** 650-652
weld procedure qualification ... **A6:** 1090, 1091, 1092
wrought martensitic stainless steels ... **A6:** 433, 435, 436, 438, 440, 441
zirconium alloys ... **A6:** 787
Heat-affected zone cracking ... **A6:** 90-93
advanced titanium-base alloys ... **A6:** 526
machinery and equipment weldments ... **A6:** 391
nickel alloys ... **A6:** 742
Heat-affected zone hot cracking
evaluation by spot Varestraint test ... **A6:** 608
stainless steel casting alloys ... **A6:** 497, 498
Heat-affected zone hydrogen cracking
oxide-dispersion strengthened materials ... **A6:** 1039, 1040
Heat-affected zone liquation cracking
austenitic stainless steels ... **A6:** 464-465
duplex stainless steels ... **A6:** 478

Heat-affected zone liquation cracking (continued)
nonferrous high-temperature materials ... **A6:** 567-570
Heat-affected zone microstructure, effect of on weldability ... **A1:** 605-606, 608
Heat-affected zone of weldments
defined ... **A9:** 9
epitaxial growth in ... **A9:** 578-579
transformation behavior in ferrous alloys ... **A9:** 580-581
Heat-affected zone, validation strategies ... **A6:** 1147-1150
model validation and boundary condition ... **A6:** 1150
temperature measurement ... **A6:** 1149-1150
electron-beam welding ... **A6:** 1149
gas-tungsten arc welding ... **A6:** 1149-1150
narrow-band IR pyrometry ... **A6:** 1149-1150
optical spectral radiometric/laser reflectance (OSRLR) ... **A6:** 1150
thermocouples ... **A6:** 1149
thermography ... **A6:** 1149-1150
validation of residual calculation ... **A6:** 1150
interferometry ... **A6:** 1150
neutron diffraction method ... **A6:** 1150
strain gages ... **A6:** 1150
x-ray diffraction method ... **A6:** 1150
Heat-affected zones (HAZ)
fracture toughness improved by stress-relief heat treating ... **A4:** 33
Heat-checked die
as casting defect ... **A11:** 381
Heat-curable rubber (HCR)
as silicone form ... **EM2:** 267
Heat-cured processes *See also* Hot box processes
as coremaking systems ... **A15:** 238
Heat-deflection temperature *See also* Deflection temperature under load (DTUL); Temperature(s) ... **EM3:** 15
cyanates ... **EM2:** 238
defined ... **EM2:** 21
of engineering plastics ... **EM2:** 68
polyaryletherketones (PAEK, PEK PEEK, PEKK) ... **EM2:** 143
polybenzimidazoles (PBI) ... **EM2:** 148
polyphenylene ether blends (PPE, PPO) ... **EM2:** 184
polysulfones (PSU) ... **EM2:** 200
Heat-disposable pattern
defined ... **A15:** 6
Heat-down package
MCNC ... **EL1:** 310-311
Heat-exchanger tubes
ASTM specifications for ... **M1:** 323
copper alloy, failure of ... **A11:** 634-635
U-bend, corrosion fatigue failure of ... **A11:** 637
Heat-fail temperature ... **EM3:** 15
Heat-flow path
through boiler tubes ... **A11:** 603
Heat-loss coefficient ... **A6:** 8, 9
Heat-resistant alloy
primary testing direction ... **A8:** 667
Heat-resistant alloys *See* Wrought heat-resistant alloys ... **A14:** 231-236, 779-784 **A16:** 206, 208, 736-760
abrasive belt grinding ... **A16:** 760
advantages ... **A4:** 512-513
alloy group designations ... **A16:** 757
applications ... **M3:** 196-197, 199-200, 204, 205-206
applications, low-alloy ... **A4:** 39
austenitic stainless steels ... **M3:** 190, 204-206
boring ... **A16:** 742
brazing and soldering characteristics ... **A6:** 632-633
broach design ... **A16:** 744, 745
broach tool modification ... **A16:** 744
broaching ... **A16:** 743-746
carburizing container material ... **A4:** 327
for casings in air-fuel gas burners ... **A4:** 274
cast, arc welding of ... **A15:** 529

Heat-resistant alloys (continued)
categories ... **A16:** 736-737
centerless grinding ... **A16:** 758, 760
circular sawing ... **A16:** 756-757
climb milling ... **A16:** 752, 754
coated drills used ... **A16:** 58
cobalt-base alloys, forming practice ... **A14:** 783-784
cold reduction swaging effects ... **A14:** 129
combined creep and fatigue ... **M3:** 194
compositions ... **A16:** 736-737
contour band sawing ... **A16:** 363
cooling practice ... **A14:** 235
corrosion in ... **A11:** 200-201
counterboring ... **A16:** 751, 752
creep and stress-rupture ... **M3:** 192-194
creep damage and stress rupture ... **A11:** 290-292
cutting fluids ... **A16:** 740-741, 743-747, 750-751, 753-757
cylindrical grinding ... **A16:** 758
design properties ... **M3:** 191, 192, 198, 201
die forgings, materials for forging tools ... **M3:** 529, 530, 532, 534
dies ... **A14:** 234-235
drilling ... **A16:** 218, 746-750
electrodischarge machining ... **A16:** 752
electron-beam welding ... **A6:** 869
end milling ... **A16:** 754, 755
face milling ... **A16:** 753, 755
face milling comparisons ... **A16:** 737-738
ferritic stainless steels ... **M3:** 190-191, 194, 195-196
fixturing ... **A16:** 749-750
forging alloys ... **A14:** 232-234
forging machines ... **A14:** 234
forging methods ... **A14:** 231-232
forging of ... **A14:** 231-236
formability, alloy conditions and ... **A14:** 779-780
forming of ... **A14:** 519, 779-784
G-ratio (grinding ratio) ... **A16:** 758-760
gas-tungsten arc welding ... **A6:** 192
grinding ... **A16:** 757-760
grinding fluid identification and classification ... **A16:** 759
grinding fluids ... **A16:** 757, 758-759, 760
grinding mechanically alloyed products ... **A16:** 760
grinding wheel selection ... **A16:** 758
group designations ... **A16:** 738
gun drills ... **A16:** 747, 749
heat treatment ... **A14:** 235-236 **M3:** 194, 195, 197-198, 201-206
heating of dies ... **A14:** 235
hot forging ... **A18:** 625
IN-625, HAZ hot cracking ... **A11:** 401
ingot production ... **M3:** 189-190
internal grinding ... **A16:** 758, 760
Iron-base alloys, forming practice ... **A14:** 782
iron-base, wrought ... **M3:** 189-206
lubrication ... **A14:** 235, 519, 781
machining comparisons ... **A16:** 737-739
manufacturing ratings ... **A16:** 739
manufacturing time for parts production ... **A16:** 739
martensitic stainless steels ... **M3:** 190-191, 196-199
mechanical properties ... **M3:** 192-195, 197, 199-205
methods and tools ... **A14:** 781
milling ... **A16:** 327, 329, 754, 755
nickel-base ... **A15:** 816-819
nickel-base, forming practice ... **A14:** 782-783
oil-hole drills ... **A16:** 747, 749
oxidation resistance ... **M3:** 195-196, 197-198, 205-206
planing ... **A16:** 742-743
plasma melting/casting ... **A15:** 420
power band sawing ... **A16:** 755, 756
power hacksawing ... **A16:** 757
precipitation-hardening stainless steels ... **M3:** 190-191, 199-204
press-formed parts, materials for forming tools ... **M3:** 492, 493
product forms ... **M3:** 190-191
reaming ... **A16:** 750-751
for retort in bell-type hot-wall furnace ... **A4:** 496
rolling direction, effect on formability ... **A14:** 780-781
sawing ... **A16:** 358, 755, 756-757
for shallow forming dies ... **A18:** 623
shaping ... **A16:** 191, 192, 742-743

Heat-resistant alloys (continued)
slab milling ... **A16:** 755, 756
speed, effect on formability ... **A14:** 781
spotfacing ... **A16:** 751, 752
stock preparation ... **A14:** 235
stress-corrosion failure in ... **A11:** 309
structure control phase and working temperatures ... **A14:** 266
structure/property correlations ... **A15:** 818-819
surface finish ... **A14:** 236
surface grinding ... **A16:** 759-760
susceptibility to hydrogen damage ... **A11:** 249
tapping ... **A16:** 259, 263, 751-754
tapping machines ... **A16:** 752
temperature vs ductility ... **A14:** 234
tempering in service ... **M3:** 197-198
thread grinding ... **A16:** 273
thread milling ... **A16:** 751-754
torch brazing ... **A6:** 328
trepanning ... **A16:** 742
turning ... **A16:** 739-742
twist drills ... **A16:** 747-749
welding ... **M3:** 201, 203, 206

Heat-resistant alloys (cast)
thermal expansion coefficient ... **A6:** 907

Heat-resistant alloys, ACI, specific types *See also* Superalloys, specific types
HA alloy ... **M3:** 273
composition ... **M3:** 269, 285
machining data ... **M3:** 279
property data ... **M3:** 285-286
HB alloy, corrosion rates in air and flue gas ... **M3:** 270
HC alloy ... **M3:** 273
composition ... **M3:** 269, 286
corrosion rates in air and flue gas ... **M3:** 270
machining data ... **M3:** 279
mechanical properties ... **M3:** 270
property data ... **M3:** 286-287
use in equipment for making glass fibers ... **M3:** 280
HD alloy ... **M3:** 273
composition ... **M3:** 269, 288
corrosion rates in air and flue gas ... **M3:** 270
machining data ... **M3:** 279
mechanical properties ... **M3:** 270
property data ... **M3:** 288, 289
HE alloy ... **M3:** 273
composition ... **M3:** 269, 289
corrosion rates in air and flue gas ... **M3:** 270
machining data ... **M3:** 279
mechanical properties ... **M3:** 270
property data ... **M3:** 289, 290
HF
composition ... **A4:** 511 **M4:** 326
properties, elevated temperature ... **M4:** 326
properties, elevated-temperature ... **A4:** 511
recommended for furnace parts and fixtures ... **A4:** 513
HF alloy ... **M3:** 273
composition ... **M3:** 269, 291
corrosion rates in air and flue gas ... **M3:** 270
machining data ... **M3:** 279
mechanical properties ... **M3:** 270
property data ... **M3:** 291-293
use in equipment for making glass fibers ... **M3:** 280
HH
composition ... **M4:** 326
composition alloys ... **A4:** 511
properties, elevated temperature ... **M4:** 326
properties, elevated-temperature ... **A4:** 511
recommended for furnace parts and fixtures ... **A4:** 513
HH alloy ... **M3:** 273-274, 275
balancing of composition ... **M3:** 273-274
composition ... **M3:** 269, 293-294
corrosion rates in air and flue gas ... **M3:** 270
machining data ... **M3:** 279
mechanical properties ... **M3:** 270
property data ... **M3:** 293-295
short-time tensile properties ... **M3:** 274, 275
HI alloy ... **M3:** 274
composition ... **M3:** 269, 296
corrosion rates in air and flue gas ... **M3:** 270
machining data ... **M3:** 279
mechanical properties ... **M3:** 270

Heat-resistant alloys, ACI, specific types (continued)
property data ... **M3:** 296-297
HI, composition ... **A4:** 510
HK
composition ... **A4:** 511 **M4:** 326
properties, elevated temperature ... **M4:** 326
properties, elevated-temperature ... **A4:** 511
for radiant-tube-heated furnace equipment ... **A4:** 473
recommended for carburizing and carbonitriding furnace parts ... **A4:** 513
recommended for furnace parts and fixtures ... **A4:** 513
HK alloy ... **M3:** 274-275, 276
composition ... **M3:** 269, 297-298
corrosion rates in air and flue gas ... **M3:** 270
machining data ... **M3:** 279
mechanical properties ... **M3:** 270, 274-275, 276
property data ... **M3:** 298-300
use in equipment for making glass fibers ... **M3:** 280
HL
composition ... **A4:** 510
recommended for furnace parts and fixtures ... **A4:** 513
HL alloy ... **M3:** 275-276
composition ... **M3:** 269, 300
corrosion rates in air and flue gas ... **M3:** 270
machining data ... **M3:** 279
mechanical properties ... **M3:** 270
property data ... **M3:** 300-301
HN
composition ... **A4:** 511 **M4:** 326
properties, elevated temperature ... **M4:** 326
properties, elevated-temperature ... **A4:** 511
HN alloy ... **M3:** 276
composition ... **M3:** 269, 301
corrosion rates in air and flue gas ... **M3:** 270
machining data ... **M3:** 279
mechanical properties ... **M3:** 270
property data ... **M3:** 301-303, 304
HP alloy ... **M3:** 276
composition ... **M3:** 269, 303
corrosion rates in air and flue gas ... **M3:** 270
machining data ... **M3:** 279
mechanical properties ... **M3:** 270
property data ... **M3:** 303-305, 306, 307
HP-50WZ alloy
composition ... **M3:** 269, 305
property data ... **M3:** 305, 307
HT
for carburizing furnace trays ... **A4:** 514
composition ... **A4:** 511 **M4:** 326
cost ... **A4:** 517
liquid nitriding equipment ... **A4:** 414
muffle application ... **A4:** 517
properties, elevated temperature ... **M4:** 326
properties, elevated-temperature ... **A4:** 511
for radiant-tube-heated furnace equipment ... **A4:** 473
recommended for carburizing and carbonitriding furnace parts ... **A4:** 513
recommended for furnace parts and fixtures ... **A4:** 513
recommended for parts and fixtures for salt baths ... **A4:** 514
HT alloy ... **M3:** 276
composition ... **M3:** 269, 305
corrosion rates in air and flue gas ... **M3:** 270
machining data ... **M3:** 279
mechanical properties ... **M3:** 270
property data ... **M3:** 305-307, 308, 309
HU
composition ... **A4:** 511
cost ... **A4:** 517
properties, elevated-temperature ... **A4:** 511
recommended for carburizing and carbonitriding furnace parts ... **A4:** 513
recommended for furnace parts and fixtures ... **A4:** 513
recommended for parts and fixtures for salt baths ... **A4:** 514
HU alloy ... **M3:** 276
composition ... **M3:** 269, 310
corrosion rates in air and flue gas ... **M3:** 270
machining data ... **M3:** 279

Heat-resistant alloys, ACI, specific types (continued)
mechanical properties M3: 270
property data M3: 308, 309, 310
HV
composition M4: 326
properties, elevated temperature M4: 326
HW
cost A4: 517
recommended for furnace parts and fixtures A4: 513
HW alloy M3: 276-277
composition M3: 269, 310
corrosion rates in air and flue gas M3: 270
machining data M3: 279
mechanical properties M3: 270
property data M3: 310-312
HX
composition A4: 511 M4: 326
properties, elevated temperature M4: 326
properties, elevated-temperature A4: 511
recommended for carburizing and carbonitriding furnace parts A4: 513
recommended for furnace parts and fixtures A4: 513
HX alloy M3: 276, 277
composition M3: 269, 312
corrosion rates in air and flue gas M3: 270
machining data M3: 279
mechanical properties M3: 270
property data M3: 312, 313

Heat-resistant alloys, fixtures
applications A4: 513, 514, 515, 516 M4: 330, 331-333
cast, compared to wrought alloys A4: 511-513 M4: 327-330
cost A4: 517
life expectancy A4: 511
materials recommended A4: 511, 513 M4: 329
materials recommended for use in salt baths A4: 514 M4: 331

Heat-resistant alloys, forged
inspection of A17: 496, 503

Heat-resistant alloys, furnace parts
air-line attack A4: 517
applications A4: 513, 514, 515-517 M4: 333-336
baskets A4: 514-515
belting A4: 513, 516 M4: 328, 334-335
carbonitriding furnaces, material for A4: 515, 516 M4: 329
carburizing furnaces, materials for A4: 511, 514, 515 M4: 329
cast, compared to wrought alloys A4: 511-513 M4: 327-330
chain link A4: 513
conveyance system A4: 515, 516 M4: 333-335
cost A4: 511, 513, 517 M4: 336
electrodes A4: 514, 517 M4: 335-336
guides A4: 513
high temperature properties M4: 325, 326, 327
high-temperature properties A4: 516
materials recommended A4: 513 M4: 328, 329
materials recommended for use in salt baths A4: 514 M4: 331
muffles A4: 513, 517 M4: 328, 334, 336
nonmetallic material radiant tubes A4: 517-518
pots A4: 514, 517 M4: 335
radiant tubes A4: 513, 516-518 M4: 328, 333, 335
retorts A4: 513, 517
rolls A4: 513, 516
service life A4: 511
sprockets A4: 513
thermocouple protection tubes A4: 514
welding techniques A4: 513

Heat-resistant alloys, grids *See* Heat-resistant alloys, trays

Heat-resistant alloys, heat treating *See also* Chromium-containing alloys, specific types; Refractory metals, heat treating; Superalloys, heat treating

Heat-resistant alloys, heat treating (continued)
liquid nitriding A4: 419
stress-relieving A4: 510, 511

Heat-resistant alloys, specific types
DS MAR-M200+Hf, aging cycle A4: 812
DS MAR-M247, aging cycle A4: 812
DS Rend 80H, aging cycle A4: 812

Heat-resistant alloys, trays
applications A4: 513-514 M4: 329, 330-331
life expectancy A4: 511, 513-514 M4: 327-328, 330-331
materials recommended A4: 511, 513 M4: 328

Heat-resistant alloys, use for furnace parts
trays and fixtures M4: 325-336

Heat-resistant alloys, used for furnace parts
trays, and fixtures A4: 510-518

Heat-resistant alloys, wrought, specific types
214
characteristics A4: 512
composition A4: 512
mesh belt applications A4: 516
radiant tube applications A4: 517
recommended for carburizing and carbonitriding furnace parts A4: 513
recommended for furnace parts and fixtures A4: 513
230
characteristics A4: 512
composition A4: 512
radiant tube applications A4: 517
recommended for carburizing and carbonitriding furnace parts A4: 513
recommended for furnace parts and fixtures A4: 513
253 MA
characteristics A4: 512
composition A4: 512
recommended for furnace parts and fixtures A4: 513
304, recommended for furnace parts and fixtures A4: 513
309
composition M4: 327
properties, elevated temperature M4: 327
radiant tube applications A4: 517
recommended for furnace parts and fixtures A4: 513
recommended for parts and fixtures for salt baths A4: 514
310
composition M4: 327
properties, elevated temperature M4: 327
radiant tube applications A4: 517
recommended for furnace parts and fixtures A4: 513
314, recommended for furnace parts and fixtures A4: 513
316, recommended for furnace parts and fixtures A4: 513
330
composition M4: 327
properties, elevated temperature M4: 327
330, for radiant-tube-heated furnace equipment A4: 473
330HC
composition M4: 327
properties, elevated temperature M4: 327
333
composition M4: 327
properties, elevated temperature M4: 327
347, recommended for furnace parts and fixtures A4: 513
430, recommended for furnace parts and fixtures A4: 513
556
characteristics A4: 512
composition A4: 512

Heat-resistant alloys, wrought, specific types (continued)
recommended for carburizing and carbonitriding furnace parts A4: 513
recommended for furnace parts and fixtures A4: 513
recommended for parts and fixtures for salt baths A4: 514
600
composition M4: 327
mesh belt applications A4: 516
properties, elevated temperature M4: 327
recommended for carburizing and carbonitriding furnace parts A4: 513
recommended for furnace parts and fixtures A4: 513
recommended for parts and fixtures for salt baths A4: 514
601
composition M4: 327
mesh belt applications A4: 516
properties, elevated temperature M4: 327
radiant tube applications A4: 517
for radiant-tube-heated furnace equipment A4: 473
recommended for carburizing and carbonitriding furnace parts A4: 513
recommended for furnace parts and fixtures A4: 513
617
recommended for carburizing and carbonitriding furnace parts A4: 513
recommended for furnace parts and fixtures A4: 513
800
composition M4: 327
properties, elevated temperature M4: 327
800H, radiant tube application A4: 517
800H/800HT
recommended for carburizing and carbonitriding furnace parts A4: 513
recommended for furnace parts and fixtures A4: 513
802
composition M4: 327
properties, elevated temperature M4: 327
802, recommended for furnace parts and fixtures A4: 513
HR-120
characteristics A4: 512
composition A4: 512
recommended for carburizing and carbonitriding furnace parts A4: 513
recommended for furnace parts and fixtures A4: 513

Heat-resistant applications
nickel-base alloys A2: 430-431

Heat-resistant cast iron
creep strength M1: 92-93
growth M1: 91-93
high aluminum iron M1: 96
high chromium cast iron M1: 93-94
high nickel iron M1: 94-96
high silicon iron M1: 94
high-temperature strength M1: 93, 94
mechanical properties M1: 92
physical properties M1: 88
rupture strength M1: 93-95
scaling M1: 91-92
tensile strength M1: 92-94

Heat-resistant cast irons *See* Alloy cast irons A9: 245-246
as-cast A9: 255

Heat-resistant cast steel
corrosion of A13: 573

Heat-resistant cast steels
galling A1: 928-929
general properties A1: 909, 919, 920-921
iron-chromium-nickel A1: 922-925

Heat-resistant cast steels (continued)
iron-nickel-chromium ... **A1:** 921, 923, 925-927
magnetic properties ... **A1:** 929
manufacturing characteristics ... **A1:** 919, 928, 929
metallurgical structures ... **A1:** 921-922
properties of heat-resistant alloys ... **A1:** 920, 921, 923, 924, 925, 926, 927-928
straight chromium heat-resistant castings ... **A1:** 922
Heat-resistant casting alloys *See also* Casting alloys ... **A9:** 330-350
cobalt-base, compositions of ... **A9:** 331
cobalt-base, microstructures of ... **A9:** 334
compositions ... **A13:** 576
corrosion behavior ... **A13:** 575-576
delineating etchants for ... **A9:** 330
as H-type alloys ... **A13:** 575-576
iron-chromium-nickel, compositions of ... **A9:** 330
iron-chromium-nickel, microstructures of ... **A9:** 333-334
nickel-base, compositions of ... **A9:** 331
nickel-base, microstructures of ... **A9:** 334
specimen preparation ... **A9:** 330-332
staining etchants for ... **A9:** 330-331
Heat-resistant casting alloys, cobalt-base, specific types
98M2 Stellite, as investment cast, different magnifications compared ... **A9:** 349
Haynes 21, as-cast and aged ... **A9:** 348
Haynes 31, as-cast and aged, thin and thick sections compared ... **A9:** 348
Haynes 151, as-cast and aged ... **A9:** 348
MAR-M 302, as-cast ... **A9:** 349
MAR-M 509, as-cast and aged ... **A9:** 350
WI-52, as-cast ... **A9:** 349
Heat-resistant casting alloys, iron-chromium- nickel, specific types
9Cr- 1Mo, air cooled and tempered ... **A9:** 335
9Cr-1Mo ... **A9:** 333
9Cr-1Mo, as sand cast ... **A9:** 335
HE-14, creep tested ... **A9:** 335
HF, microstructure ... **A9:** 333
HF-25, as cast and creep tested ... **A9:** 337
HF-33, different sections of a casting compared, as cast and creep tested ... **A9:** 336-337
HF-33, fractured ... **A9:** 336-337
HF-34, lamellar structure in creep test specimen ... **A9:** 337
HH, different sections of a casting compared, as sand cast and creep tested ... **A9:** 337-338
HH, fractured ... **A9:** 337-338
HH, microstructure ... **A9:** 333
HK, microstructure ... **A9:** 333
HK-28, lamellar structures regenerated by slow cooling ... **A9:** 340
HK-35, as-cast and fractured ... **A9:** 339-340
HK-44, creep tested and fractured ... **A9:** 339
HN, as-cast and fractured ... **A9:** 340
HN, etching ... **A9:** 331-332
HN, microstructure ... **A9:** 333
HP, etching ... **A9:** 331-332
HT, etching ... **A9:** 331-332
HT, microstructure ... **A9:** 333
HT-44, as-cast and after creep testing ... **A9:** 341
HT-56, after creep testing ... **A9:** 341
HT-57, as-cast ... **A9:** 341
HU, etching ... **A9:** 331-332
HW, as-cast and after aging ... **A9:** 340
HW, etching ... **A9:** 332
HW, microstructure ... **A9:** 333
HX, etching ... **A9:** 332
Heat-resistant casting alloys, nickel-base, specific types
Alloy 713C, as-cast, different carbide structures ... **A9:** 344
Alloy 713C, different holding temperatures and times compared ... **A9:** 345
Alloy 713C, exposed to sulfidation ... **A9:** 345
Alloy 718, vacuum cast and solution annealed ... **A9:** 345
B-1900, as-cast, different magnifications compared ... **A9:** 341
B-1900, shell mold casting, as-cast ... **A9:** 342
B-1900, shell mold casting, solution annealed and aged ... **A9:** 342
Heat-resistant casting alloys, nickel-base, specific types (continued)
Hastelloy B, as-cast and annealed ... **A9:** 342
Hastelloy C, as-cast and annealed ... **A9:** 342
IN-100, as-cast, different magnifications compared ... **A9:** 342-343
IN-100, creep-rupture tested ... **A9:** 344
IN-100, different holding temperatures and times compared ... **A9:** 343-344
IN-738, after holding ... **A9:** 345
In-738, as-cast ... **A9:** 346
IN-738, solution annealed ... **A9:** 346
MAR-M 246, after high temperature exposure ... **A9:** 346-347
MAR-M 246, as-cast ... **A9:** 346
MAR-M 246, different cooling times compared ... **A9:** 346
TRW-NASA VIA, as-cast ... **A9:** 347
U-700, as-cast, and solution annealed ... **A9:** 347
U-700, cast blade after cyclic sulfidation- erosion testing ... **A9:** 347
Heat-resistant castings *See also* Heat-resistant alloys, ACI specific types; Superalloys ... **M3:** 269-284
alloy compositions ... **M3:** 269, 271
applications ... **M3:** 269, 272, 279-280, 281
ASTM specifications ... **M3:** 269
casting design ... **M3:** 280-284
corrosion rates in flue gas ... **M3:** 270
general properties ... **M3:** 269-272
machining data ... **M3:** 279
manufacture ... **M3:** 272, 277
mechanical properties ... **M3:** 270
metallurgical structure ... **M3:** 272-279
oxidation rates ... **M3:** 270
tolerances ... **M3:** 280, 283
Heat-resistant coatings
cast iron ... **M1:** 102, 105, 106
Heat-resistant ductile iron
compositions ... **M1:** 76
elevated-temperature properties ... **M1:** 94-96
mechanical properties ... **M1:** 92
physical properties ... **M1:** 88
Heat-resistant gray iron
compositions ... **M1:** 76
elevated-temperature properties ... **M1:** 91-96
mechanical properties ... **M1:** 92
physical properties ... **M1:** 88
Heat-resistant high-alloy steels
compositions ... **A15:** 724
iron-chromium ... **A15:** 724, 733
properties ... **A15:** 728-729
resistance to hot gas corrosion ... **A15:** 730
weldability ... **A15:** 730
Heat-resistant low-alloy steels
arc welding ... **M6:** 247, 292-294
ASTM classification ... **M6:** 292
filler metals ... **M6:** 294
weldability ... **M6:** 292-293
welding processes ... **M6:** 293-294
electroslag welding ... **M6:** 294
flux cored arc welding ... **M6:** 294
gas metal arc welding ... **M6:** 294
gas tungsten arc welding ... **M6:** 293-294
plasma arc welding ... **M6:** 293-294
shielded metal arc welding ... **M6:** 293
submerged arc welding ... **M6:** 294
Heat-resistant materials
metallurgy ... **A4:** 510-511
product forms ... **A4:** 510-511
Heat-resistant metals and alloys
arc welding *See* Arc welding of heat-resistant alloys
arc welding of
cobalt-based alloys ... **M6:** 367-370
iron-chromium-nickel alloys ... **M6:** 364-367
iron-nickel-chromium alloys ... **M6:** 364-367
nickel-based alloys ... **M6:** 355-364
brazing *See* Brazing of heat-resistant alloys
electron beam welding ... **M6:** 638
gas metal arc welding ... **M6:** 153
gas tungsten arc welding ... **M6:** 182, 205
shielded metal arc welding ... **M6:** 75
Heat-resistant stainless steel casting alloys, specific types *See* Stainless steel heat-resistant casting alloys, specific types
Heat-resistant steels
broaching ... **A16:** 203
Heat-resisting alloys
abrasive blasting of ... **M5:** 563-565
acid etching of ... **M5:** 564
cadmium plating of ... **M5:** 264
ceramic coating of ... **M5:** 537-538, 566
cleaning processes ... **M5:** 563-568
metallic contaminant removal ... **M5:** 563-564
problems and solutions ... **M5:** 567-568
reduced oxide and scale removal ... **M5:** 564-568
tarnish removal ... **M5:** 564
diffusion coating of ... **M5:** 566
electrolytic cleaning of ... **M5:** 567
electroplating of ... **M5:** 566
finishing processes ... **M5:** 566-568
applicable types ... **M5:** 566-567
problems and solutions ... **M5:** 567-568
hydrogen embrittlement during descaling ... **M5:** 565
oxide coatings, function of ... **M5:** 564, 566
pickling of ... **M5:** 564-568
polishing of ... **M5:** 566
salt bath, descaling of ... **M5:** 564-568
shot peening of ... **M5:** 566-567
wet tumbling of ... **M5:** 563, 565-566
wire brushing of ... **M5:** 566
Heat-resisting alloys, heat treating
aging ... **M4:** 654-655
aging, effect of cold working ... **M4:** 659-660
annealing ... **M4:** 650-654, 655
carbon pickup ... **M4:** 657
cold working ... **M4:** 659
fixturing ... **M4:** 659
furnace equipment ... **M4:** 658
grain size, effect of ... **M4:** 659
hardening rates ... **M4:** 660
intergranular oxidation ... **M4:** 657, 658
protective atmospheres ... **M4:** 657-658
quenching ... **M4:** 654, 656, 657
solution treating ... **M4:** 654, 656, 657
stress relieving ... **M4:** 650, 655
surface contamination ... **M4:** 656-657
welding ... **M4:** 660
Heat-resisting alloys, specific types *See* commercial designations
Heat-resisting aluminum coatings ... **M1:** 172
Heat-sealing adhesive ... **EM3:** 15
defined ... **EM2:** 21
Heat-shrinkable labeling ... **EM4:** 475
Heat-source diameter ... **A6:** 5
Heat-source formulation ... **A6:** 8-9
Heat-tint oxides, effect on corrosion resistance
austenitic stainless steels ... **A13:** 351
Heat-transfer efficiency ... **A6:** 25
Heat-treat cracks
martensitic stainless steels ... **A12:** 366
secondary, as intergranular ... **A12:** 366
types ... **A12:** 65
Heat-treatable age hardening alloys
aluminum-silicon ... **A15:** 159
Heat-treatable alloys *See also* Heat treatable wrought aluminum alloys; Heat treatment
aluminum bronzes, quenching/tempering temperatures ... **A2:** 356
aluminum, in niobium-titanium superconducting materials ... **A2:** 1045
electron-beam welding ... **A6:** 871
single-shear and blanking shear tests for ... **A8:** 65
steel-bonded carbides and cobalt-bonded tungsten carbide, compared ... **A2:** 966-997
temper designations ... **A2:** 15-27
Heat-treatable aluminum alloys ... **A6:** 528-535
aging ... **A6:** 529, 532-533, 534
applications ... **A6:** 528, 529-530, 535
buttering of edges ... **A6:** 535
characteristics ... **A6:** 529-530
coefficient of thermal expansion ... **A6:** 530
composition ... **A6:** 528, 529, 533
constitutional liquation ... **A6:** 75
corrosion resistance ... **A6:** 534-535
crack sensitivity vs. weld composition ... **A6:** 530, 533
cracking ... **A6:** 528
cryogenics ... **A6:** 535

Heat-treatable aluminum alloys (continued)
description of ... **A6:** 528-530
designation systems ... **A6:** 528-529
distortion ... **A6:** 528
electron-beam welding ... **A6:** 528
filler metals ... **A6:** 528, 531, 533-534
fusion zone ... **A6:** 533-534, 535
gas-metal arc welding ... **A6:** 528, 531-532
gas-tungsten arc welding ... **A6:** 528, 531, 532, 533, 534
Guinier-Preston zones ... **A6:** 529, 532
heat-affected zone ... **A6:** 530, 531, 532-533
degradation ... **A6:** 532
joint characteristics ... **A6:** 535
laser-beam welding ... **A6:** 528
liquation cracking ... **A6:** 531
metallurgy ... **A6:** 529
partial phase diagram for aluminum-copper system ... **A6:** 532
porosity during welding ... **A6:** 531-532
postweld heat treatment ... **A6:** 532-533, 534
precipitation ... **A6:** 529, 532
properties ... **A6:** 528, 529-530, 533-534, 535
resistance welding ... **A6:** 528
solidification cracking ... **A6:** 530-531, 533
stress-corrosion cracking ... **A6:** 534-535
temper designations ... **A6:** 528-529
weldability ... **A6:** 533-535
welding of ... **A6:** 530-533
Heat-treatable commercial wrought aluminum alloys
solution potentials ... **A13:** 584
Heat-treatable low-alloy (HTLA) steels ... **A6:** 668-673
applications ... **A6:** 670
classification and group description ... **A6:** 405, 406
composition ... **A6:** 668, 669, 670
cracking ... **A6:** 669, 670, 671, 672
filler metals ... **A6:** 669-671, 672
flux-cored arc welding ... **A6:** 669, 670, 669, 670
gas-metal arc welding ... **A6:** 669, 670, 671
gas-tungsten arc welding ... **A6:** 669, 670, 671
heat input ... **A6:** 672
heat-affected zones ... **A6:** 671, 672
microstructure ... **A6:** 671-672
postweld heat treatment ... **A6:** 672-673
preheating ... **A6:** 669, 671-672
shielded metal arc welding ... **A6:** 669-670
submerged arc welding ... **A6:** 669, 670-671
welding processes and practices ... **A6:** 669
Heat-treatable low-alloy steels
weldability of ... **A1:** 609
Heat-treatable wrought aluminum alloys *See* also Precipitation hardening
artificial aging ... **A2:** 40
commercial alloys ... **A2:** 40-41
effects on physical and electrochemical properties ... **A2:** 41
natural aging ... **A2:** 39-40
for precipitation strengthening ... **A2:** 39
strengthening mechanisms ... **A2:** 36, 39-41
Heat-treated aluminum alloys *See also* Aluminum alloys
identification of temper ... **A9:** 358
soluble phases ... **A9:** 358
Heat-treated steel
machined by high-speed steel cutting tools ... **A18:** 615
Heat-treated steels
submerged arc welding ... **A6:** 202
Heat-treating fixtures
carburization in ... **A11:** 272-273
components, elevated-temperature failures in ... **A11:** 292-294
Heat-treating furnace accessories
corrosion of ... **A13:** 1311-1314
Heat-treating furnaces, equipment for
atmospheres ... **A4:** 467, 469, 471-472, 473, 474
direct-fired ... **A4:** 466, 471
Heat-treating furnaces, equipment for (continued)
electrically heated ... **A4:** 466, 471-473
gas-fired ... **A4:** 471, 473
heating elements ... **A4:** 472
maintenance ... **A4:** 473-474
metallic resistance heating elements ... **A4:** 472
nonmetallic resistance heating elements ... **A4:** 473
oil-fired ... **A4:** 471, 473
radiant-tube-heated ... **A4:** 472, 473
Heat-treating furnaces, types ... **A4:** 465-474
batch ... **A4:** 465-466
bell-type ... **A4:** 466
bogie hearth ... **A4:** 465
box-type ... **A4:** 465
car ... **A4:** 465-466
car-bottom ... **A4:** 465, 466
continuous ... **A4:** 465, 466-471
continuous-belt ... **A4:** 469
conveyor-type ... **A4:** 467, 469, 498
elevator-type ... **A4:** 466
overhead monorail systems ... **A4:** 467, 471
pit-type (pot) ... **A4:** 465, 466
pusher ... **A4:** 465, 467-468, 469, 470, 498
pusher-type and rotary-hearth heat-treat systems ... **A4:** 470-471
roller-hearth ... **A4:** 469, 498
rotary-hearth ... **A4:** 467, 470
rotary-retort ... **A4:** 467, 470
shaker-hearth ... **A4:** 467, 469-470
straight-chamber ... **A4:** 467
strand-type ... **A4:** 471
walking beam ... **A4:** 467, 468-469, 498
Heat-treating industry ... **A13:** 1311-1314
Heated blanks
explosive forming of ... **A14:** 643
Heated volatile materials
brazing atmosphere source ... **A6:** 628
Heated-roll rolling
of shapes ... **A14:** 357-358
of sheet and strip ... **A14:** 356-357
Heater block
for component removal ... **EL1:** 718
Heaters *See also* Heating elements
as metal-processing equipment ... **A13:** 1316
open resistance ... **A2:** 829-831
sheathed ... **A2:** 831
Heaters, deaerating feedwater
failures in ... **A11:** 657-658
Heath, J.M
and open-hearth furnace ... **A15:** 32
Heathcote slip ... **A18:** 510
Heating *See also* Die heating; Heat treatment; Heating equipment; Heating time; Overheating; Reheating
of blocks, in ring rolling ... **A14:** 123
continuous, effect on flow stress vacuum-melted iron ... **A8:** 177
of copper alloy billets/slugs ... **A14:** 257
effect, in direct current electric potential method ... **A8:** 389
for forging magnesium alloys ... **A14:** 259
for forging, nickel-base alloys ... **A14:** 261-263
in forging of ingots ... **A14:** 222
of heat-resistant alloys ... **A14:** 235
induction ... **A14:** 152, 165
infrared ... **A14:** 152
ingot ... **A14:** 222
of magnesium alloys ... **A14:** 826-827
nickel-base alloy applications in ... **A13:** 655
straightening by ... **A14:** 681-682
system, for torsion testing ... **A8:** 158-159
Heating alloys *See also* Electrical resistance alloys ... **M3:** 646-651
applications ... **A2:** 827-828
heaters, fabrication ... **M3:** 651
heating elements, service life ... **M3:** 646, 650-653, 654, 656
iron-chromium-aluminum ... **A2:** 828-829
Heating alloys (continued)
nickel-chromium ... **A2:** 828
nickel-chromium-iron ... **A2:** 828
nonmetallic materials ... **A2:** 829
operating temperatures ... **M3:** 647
pure metals ... **A2:** 829
resistance heaters, design ... **M3:** 648-651
service environments ... **M3:** 653-657
Heating blankets and assembled vacuum bags ... **EM3:** 815-816, 817
Heating curve
vibratory milling in stainless steel chamber ... **A7:** 61, 62
Heating elements *See also* Heaters; Open resistance heaters
service life ... **A2:** 831-833
Heating elements, improving performance of ... **A3:** 1 • 27-28
Heating equipment *See also* Furnaces; Heat treatment; Heating
for aluminum alloys ... **A14:** 247
closed-die forging ... **A14:** 81
precoated steel sheet for ... **M1:** 167, 172-173, 176
for titanium alloy forgings ... **A14:** 278
Heating equipment, cast iron
coatings for ... **M1:** 104
Heating equipment for macroetching ... **A9:** 171
Heating force ... **A6:** 316
Heating gate
definition ... **M6:** 9
Heating methods
for thermal inspection ... **A17:** 398
Heating rate *See also* Cooling rate
effect on massive transformation ... **A9:** 655
Heating tape
electrical resistance ... **A17:** 409
Heating time *See also* Heat treatment; Heating; Overheating; Reheating
for aluminum alloy forgings ... **A14:** 247-248
closed-die forging ... **A14:** 81
copper alloy billets/slugs ... **A14:** 257
for stainless steel, section thickness effects ... **A14:** 228
for titanium alloys ... **A14:** 278
Heats
in fatigue testing ... **A8:** 712
Heats of formation
of inorganic oxides ... **A7:** 598
Heavy alloys ... **A7:** 6
sintering ... **A7:** 309
Heavy clay products
mixing operations ... **EM4:** 98
Heavy drafts
cold finished bars ... **M1:** 244
Heavy elements
depth profiles of impurities ... **A10:** 628, 632-633
Heavy leaded brass
applications and properties ... **A2:** 308-309
Heavy machinery parts
hardened steel for ... **A1:** 455-456
Heavy metal poisoning ... **A7:** 666
Heavy metal shadowing in replicas ... **A9:** 108
Heavy metals ... **A7:** 6
ion promoting oxidation ... **A18:** 105
radiographic methods ... **A17:** 296
Heavy phosphate coating ... **A13:** 383, 387
Heavy sections
in permanent molds ... **A15:** 277
Heavy water nuclear filters
powders used ... **A7:** 573
Heavy-atom method
to produce electron density maps ... **A10:** 350-351
Heavy-duty gas turbine engine (HDGTE) study ... **EM4:** 716
design/manufacturing trade-offs ... **EM4:** 720
Heavy-duty oil
defined ... **A18:** 10
Heavy-metal tungsten alloys
properties ... **A2:** 578-579, 581-582

Hecla process
of gallium recovery **A2:** 743-744
Heddles
in textile looms **EM1:** 127-128
Hedvall Effect **EM4:** 55
Heel block
defined **A14:** 7
Heel breaks
inner/outer tape automated bonding **EL1:** 287
Heel effect, x-ray tubes
radiography **A17:** 305
Height
of dimples, measurement in fractured steel **A12:** 207
of fracture surface roughness **A12:** 200
three-dimensional measurement **A12:** 207
Height differences
measured with polarization interferometer **A9:** 150-151
of specimens, interference techniques for measuring **A9:** 80
Height profiles
obtained from stereomicroscopy **A9:** 97
Height reduction
as near defect **EL1:** 568
HeliArc welding *See* Gas-tungsten arc welding
Helical, and modified helical
winding patterns **EM1:** 509, 514
Helical compression spring
stress-relaxation test **A8:** 328
Helical gear compaction
using anvil die closure **A7:** 324, 327
Helical gears
described **A11:** 586
gear-tooth contact in **A11:** 588
tooth **A11:** 592
Helical probes
for holes **A17:** 223
Helical springs, steel
design **M1:** 303-304, 306-308
fabrication of **M1:** 588
fatigue properties **M1:** 291-296, 297
hardenability requirements **M1:** 297, 301
relaxation **M1:** 296-300
wire for **M1:** 288-290, 301
Helical winding **EM2:** 21-22, 373
Helical-grooved journal bearings **A18:** 528
Helical-tooth cutters
for milling procedures **A7:** 462
Helicopter blade
fatigue fracture of spindle for **A11:** 126
Helium
by residual gas analysis **EL1:** 1066
characteristics in a blend **A6:** 65
as cutting fluid for tool steels **A18:** 738
electrogas welding shielding gas **A6:** 275
electron-beam welding atmosphere **A6:** 857
fine line leak test **EL1:** 1063
gas atomization with **A7:** 27
gas mass analysis of **A10:** 155
in gas pycnometry **A7:** 266
gas-metal arc welding shielding gas **A6:** 181, 185, 489
ferritic stainless steels **A6:** 446
for nickel alloys **A6:** 743, 745, 746
gas-tungsten arc welding shielding gas **A6:** 193, 488-489, 756, 1022
for dispersion-strengthened aluminum alloys **A6:** 543, 736
ferritic stainless steels **A6:** 445
in inert shielding gas mixture for GTAW **A6:** 32
ionization potential **A6:** 64
laser-beam welding shielding gas **A6:** 878
magnesium alloy shielding gas **A6:** 772, 779
hermeticity testing **EL1:** 930
as image gas, field ion microscopy **A10:** 585, 600
implantation **A10:** 485
-ionization leak detector **A17:** 64
ionization potentials and imaging fields for **A10:** 586
leakage flow rates **A7:** 433
mass spectrometers **A17:** 65, 67
mean free path **A17:** 59
mean free path value at atmospheric conditions **A18:** 525
Helium (continued)
in plasma arc powder spraying process **A18:** 830
for plasma arc spraying **A6:** 811
shielding gas
electron-beam welding, high-strength alloy steels **A6:** 867
electron-beam welding, stainless steels **A6:** 868, 869
for GMAW **A6:** 66, 67
for GMAW of cast irons **A6:** 718
for GTAW **A6:** 67
for plasma arc welding **A6:** 197
properties **A6:** 64
purity and moisture content **A6:** 65
steel weldment soundness in gas-shielded processes **A6:** 408
stud arc welding purging gas **A6:** 211
thermal conductivity **A6:** 64
valve thread connections for compressed gas cylinders **A6:** 1197
for zirconium alloys **A6:** 787-788
sigma values for ionization **A17:** 68
to propel particles in laser melt/particle injection **A18:** 868
as tracer gas, vacuum systems **A17:** 66-67
in weld relay, gas mass spectroscopy of **A10:** 156
Helium gas
exploded under pressure **A12:** 414
Helium, liquid
as coolant for superconductors **A2:** 1030, 1085
Helium pressure
disk-pressure testing **A8:** 540-541
Helium pycnometer
to measure powder volume of ceramic powders **EM4:** 71
Helium shielding gas
comparison to argon **M6:** 197-198
effect on weld metallurgy **M6:** 41
Helium shielding gases for
arc welding of
austenitic stainless steels **M6:** 331-332, 334
coppers **M6:** 402
magnesium alloys **M6:** 428
molybdenum and tungsten **M6:** 464
titanium and titanium alloys **M6:** 447-448
electron beam welding **M6:** 622
gas metal arc welding **M6:** 164
of aluminum alloys **M6:** 380
of nickel alloys **M6:** 439-440
of nickel-based heat-resistant alloys **M6:** 362
gas tungsten arc welding **M6:** 197-198
of aluminum alloys **M6:** 390
of nickel alloys **M6:** 438
of nickel-based heat-resistant alloys **M6:** 358
plasma arc welding **M6:** 217
Helium, used for ion-beam thinning of transmission electron microscopy
specimens **A9:** 107
Helium-argon
gas-metal arc welding shielding gas for aluminum alloys **A6:** 738
gas-tungsten arc welding shielding gas for aluminum alloys **A6:** 736
Helium-cadmium lasers
for optical holographic interferometry **A17:** 417
Helium-ionization leak detector **A17:** 64
Helium-neon laser beam
for analysis of integrated circuits **A11:** 768
Helium-neon lasers
continuous-wave **A10:** 128
in FT-IR spectroscopy **A10:** 112
for optical holography **A17:** 417
Helium-neon pointing lasers **M6:** 665
Helium/argon
shielding gas for laser cladding **A18:** 867
Helix traveling wave tube
microwave inspection **A17:** 209
Helmet
definition **M6:** 9
Helmholtz free energy **EM3:** 420
Helmholtz, Hermann von .... **A3:** 1 • 6
"Hem flange" adhesives **EM3:** 48
Hematite
chemical composition **A6:** 60
defined **A13:** 7
Heme synthesis, schema
and lead toxicity **A2:** 1244
Hemispheres
drawing of **A14:** 586
explosive forming of **A14:** 636
hot spinning of **A14:** 604
power spinning of **A14:** 603-604
stainless steel **A14:** 770
Hemispherical analyzers
for AES analysis **A10:** 554
Hemispherical dome tests
for lubricant evaluation **A8:** 568
and Olsen/Erichsen cup tests compared **A8:** 562
as simulative stretching test for sheet metals **A8:** 561-562
tooling for **A8:** 561
Hemispherical punch method
for determining forming limit diagrams **A8:** 566
Hemispheroidal dimples
defined **A12:** 173
Hemming
as bending **A8:** 547
defined **A14:** 7, 529
finite-element analysis **A14:** 921-922
Hemming dies
one-stroke **A14:** 537
for press bending **A14:** 527
Hemoglobin
ESR study of **A10:** 264
Hemolysis
from stibine (gaseous antimony) **A2:** 1259
Hemotologic effects
of lead **A2:** 1244
Henrian activity
defined **A15:** 51
Henry's law constant
defined **A15:** 51, 56
HEP *See* High-energy physics
Heptane
for milling cemented carbides **A16:** 72
Heptatone [$CH_3(CH_2)_4COCH_3$#
as solvent used in ceramic processing **EM4:** 117
Herbage
voltammetric monitoring of metals and nonmetals in **A10:** 188
Herbert-Gottwein technique **A18:** 440
Herbicide residues detected in plant and animal tissue **A10:** 188
Herbicides
as arsenic toxicity **A2:** 1237
Hercules A
properties **A18:** 803
Herculor
properties and applications **A2:** 371-372
Herculoy *See* Copper alloys, specific types, C87200
Hercynite inclusions in steel **A9:** 185
HERF *See* High-energy-rate forging; High-energy-rate forming
HERF machines *See* High-energy-rate forging machines
HERF processing *See* High-energy-rate forging
Hermann-Mauguin letter symbols for lattice arrangements **A9:** 706-707
Hermann-Mauguin space group
in EXAFS application **A10:** 417
Hermetic
defined **EL1:** 1145-1146
Hermetic package failures
integrated circuits **A11:** 767
Hermetic packages *See also* Hermetic; Hermeticity
costs **EL1:** 468
defined **EL1:** 453
eutectic die attach **EL1:** 213-215
glass die attach **EL1:** 215
medical and military applications **EL1:** 386-387
microelectronic **EL1:** 455
polymer die attach **EL1:** 216
silver-glass die attach **EL1:** 215-216
solder die attach **EL1:** 216-217
standards **EL1:** 213
test methods **EL1:** 494
vs nonhermetic **EL1:** 260
Hermetic sealing
for samples **A10:** 16

Hermetic seals
ceramic packages **EL1:** 961-962
chemical cleaning of **EL1:** 678
glass-to-metal **EL1:** 455-459
leaded and leadless surface-mount joints **EL1:** 734
Hermeticity *See also* Hermetic; Hermetic packages; Hermetic seals; Hermeticity testing; Reliability **EM3:** 586-587, 588
ceramic multilayer packages **EL1:** 468
considerations **EL1:** 243-244
loss, passive devices **EL1:** 997
package, and gas analysis **EL1:** 1062-1066
of plastic encapsulants **EL1:** 805
plastic packages, capacitative ratio test **EL1:** 953
rigid epoxy devices **EL1:** 815
test **EL1:** 500-502
Hermeticity, in weld relays
gas mass spectroscopy tested **A10:** 156
Hermeticity testing *See also* Hermeticity
failure characterization **EL1:** 1063
fine and gross, package-level **EL1:** 929-930
fine leak tests **EL1:** 1062-1063
gross leak tests **EL1:** 1063
helium **EL1:** 930
Hermite polynomial **A6:** 877
Herring's scaling law **EM4:** 270
Herringbone bearing
defined **A18:** 10
Herringbone gears
described **A11:** 586
Herringbone pattern *See* Chevron marks; Chevron pattern
Herringbone structure
shape memory alloys **A2:** 898
Hershey number *See also* Ocvirk number; Sommerfeld number; Stribeck curve
defined **A18:** 10
Hertz contact stress state per impact **A18:** 266, 268
Hertz elastic theory **A8:** 72
Hertz-Knudsen evaporation **A18:** 843
Hertzian cone **EM4:** 642-643
crack **EM4:** 630, 633
Hertzian cone crack
ceramics **A11:** 753
Hertzian conjunction **A18:** 79
Hertzian contact area
defined **A18:** 10
Hertzian contact pressure
defined **A18:** 10
Hertzian contact theory **A18:** 41, 263
average pressure **A18:** 40
Hertzian stress
definition **EM4:** 633
Heterocarbonyls **A7:** 135
Heterochain polymers
chemical structure **EM2:** 49-52
thermoplastic, chemical structures **EM2:** 53
Heterodyning method
microwave inspection **A17:** 211
optical holography **A17:** 417
Heteroepitaxial structures
thermal failures **EL1:** 59-60
Heteroepitaxy layers
RBS interfacial studies on **A10:** 628
Heterogeneities, anodic passivation at
in electrolytic polishing **A9:** 48
Heterogeneity
compositional, micrometer scale in single- phase materials **A10:** 516
defined **A10:** 674
detected in solids/liquids **A10:** 402
SAS techniques for **A10:** 402, 405
Heterogeneity in powder metallurgy materials **A9:** 503
Heterogeneity scatter **A8:** 329-330
Heterogeneous **EM3:** 15
defined **EM1:** 13 **EM2:** 22
Heterogeneous catalysis
FIM/AP study of **A10:** 583
Heterogeneous equilibrium
defined **A9:** 9
Heterogeneous nucleation **A6:** 45, 46, 50, 51, 53 **A9:** 647 **EM3:** 15
activation energy for **A6:** 46
constitutional supercooling driven **A15:** 130-131
defined **EM2:** 22
of equiaxed grains **A15:** 130-131
and homogeneous nucleation, compared **A15:** 105
kinetics of **A15:** 103-105
of peritectic structures **A9:** 676
Heterogeneous surface films
AES analyzed **A10:** 566
Heterojunction bipolar transistor (HBT)
research **A2:** 747
Heterojunctions
topographic imaging techniques for **A10:** 376
Heterophase boundaries *See also* Interfaces **A9:** 118
transmission electron microscopy contrast **A9:** 121-122
Heteropolytungstates
heavy-atom method to determine crystal structure of **A10:** 351
Heuristic test generation methods
system-level **EL1:** 375
Heusler alloy **A9:** 539
Heusler alloys
as trialuminide ordered intermetallic **A2:** 932-933
Heveaplus **EM3:** 145
Hex collet
for hydraulic torsional system **A8:** 216
Hexa **EM3:** 15
defined **EM1:** 13 **EM2:** 22
Hexacelsian
maximum use temperature **EM4:** 875
Hexachloroethane
degassing, for hydrogen removal **A15:** 460
for grain refinement, magnesium alloys **A15:** 481
Hexachloroplatinic acid
as allergen **A2:** 1258
Hexafluoro-antimonic acid
silicone coating effect **EL1:** 824
Hexagonal
close-packed metals **A8:** 223-224, 725
flanges, in torsional Kolsky bar test **A8:** 221-222
grip ends, for torsion specimen **A8:** 157
Hexagonal boron nitride (HBN) *See also* Cubic boron nitride (CBN) **A16:** 105, 106, 107
crystal structure **A16:** 453, 454
properties of **A2:** 1010
Hexagonal cells
formation **A9:** 613
Hexagonal close-packed
defined **A9:** 9
Hexagonal close-packed (hcp) materials
ductile-to-brittle transition **A11:** 84-85
fractures in **A11:** 75
HIC *See* Hydrogen-induced cracking
Hexagonal close-packed (hcp) metals
cavitation erosion **A18:** 216
epitaxial growth direction **A6:** 51
lower adhesive wear **A18:** 179
Hexagonal close-packed lattice
slip planes **A9:** 684
Hexagonal close-packed materials
formation of dislocation loops **A9:** 116
transmission electron microscopy contrast **A9:** 121
Hexagonal close-packed metals
defined **A13:** 45
Hexagonal closed packed array
density and coordination number **A7:** 296
Hexagonal crystal system ..... **A3:** 1 • 10, 15 **A9:** 706
Hexagonal crystals under polarized light **A9:** 77
Hexagonal lattices for crystals
defined **A9:** 9
Hexagonal ring stretching
Raman microprobe analysis **A10:** 133
Hexagonal structures
in hafnium **A9:** 497-498
in zirconium and zirconium alloys **A9:** 497-498
Hexagonal-close-packed metals
effect of low temperature on dimples in **A12:** 33
effects, state of stress **A12:** 31
hydrogen embrittlement **A12:** 22
Hexamethyidisiloxane, plasma-polymerized
NMR analysis of structure and degradation of **A10:** 285-286
Hexamethylene tetramine (HMTA, "hexa")
curing agent for novolacs **EM3:** 103
Hexamethylenetetramine *See* Hexa
as chemical etchant **A12:** 75-76
Hexane **EM3:** 1-141-42
as solvent for volatile material removal **A10:** 575
surface tension **EM3:** 181
Hexavalent chromium
cathodic electrocleaning solutions contaminated by **M5:** 34, 618-619
formation of **M5:** 191, 196
hard chromium baths contaminated by **M5:** 174, 186-187
phosphate coating solutions contaminated by **M5:** 455-456
reduction to trivalent chromium **M5:** 186-187, 302, 311-312
Hexavalent chromium plating **M5:** 188-189
Hexoloy KT SiC
properties **A18:** 548
Hexoloy SA
applications, advanced gas turbines **EM4:** 998
fast fracture flexure strength **EM4:** 1000
Hexoloy SA (alpha SiC)
properties **A18:** 548
Heyn-Bauer successive machining technique **A6:** 1095
Heywood ratios
of particle shapes **A7:** 239
Heywood shape factor
of particles **A7:** 239
Hf (Phase Diagram) ..... **A3:** 2 • 242
HF alloy **A1:** 922
Hf-Ir (Phase Diagram) ..... **A3:** 2 • 240
Hf-Mn (Phase Diagram) ..... **A3:** 2 • 240
Hf-Mo (Phase Diagram) ..... **A3:** 2 • 240
Hf-N (Phase Diagram) ..... **A3:** 2 • 241
Hf-Nb (Phase Diagram) ..... **A3:** 2 • 241
Hf-Ni (Phase Diagram) ..... **A3:** 2 • 241
Hf-Os (Phase Diagram) ..... **A3:** 2 • 242
Hf-Rh (Phase Diagram) ..... **A3:** 2 • 242
Hf-Si (Phase Diagram) ..... **A3:** 2 • 243
Hf-Ta (Phase Diagram) ..... **A3:** 2 • 243
Hf-U (Phase Diagram) ..... **A3:** 2 • 243
Hf-V (Phase Diagram) ..... **A3:** 2 • 244
Hf-W (Phase Diagram) ..... **A3:** 2 • 244
Hf-Zr (Phase Diagram) ..... **A3:** 2 • 244
HFRSc Scleroscope calibration **A8:** 104
HFRSd Scleroscope calibration **A8:** 104
hfs *See* Hyperfine structure
Hg-In (Phase Diagram) ..... **A3:** 2 • 245
Hg-K (Phase Diagram) ..... **A3:** 2 • 245
Hg-La (Phase Diagram) ..... **A3:** 2 • 245
Hg-Li (Phase Diagram) ..... **A3:** 2 • 246
Hg-Mg (Phase Diagram) ..... **A3:** 2 • 246
Hg-Na (Phase Diagram) ..... **A3:** 2 • 246
Hg-Pb (Phase Diagram) ..... **A3:** 2 • 247
Hg-Rb (Phase Diagram) ..... **A3:** 2 • 247
Hg-S (Phase Diagram) ..... **A3:** 2 • 247
Hg-Se (Phase Diagram) ..... **A3:** 2 • 248
Hg-Sn (Phase Diagram) ..... **A3:** 2 • 248

Hg-Sr (Phase Diagram) A3: 2•248
Hg-Te (Phase Diagram) A3: 2•249
Hg-Tl (Phase Diagram) A3: 2•249
Hg-Zn (Phase Diagram) A3: 2•249
HH
composition M4: 653
HH alloy A1: 922-924
HI alloy A1: 924
HI-B process for improving Goss texture A9: 537
HIAC (light obscuration) analyzer
for particle size measurement A7: 223-224
Hickory wood, waxed
friction coefficient data A18: 75
HID *See* High interstitial defects
Hide glue
wheel polishing process using M5: 208
Hierarchy *See also* Levels; Packaging
of electronic packaging EL1: 397-398
interconnection EL1: 2-4, 12-13
of plastics EM2: 68
High alkali glass fibers *See also* A-glass EM1: 107
composition
High aluminum defects
defined A9: 9
in Ti-6Al-4V alpha-beta billet appearance A9: 470
in titanium and titanium alloys A9: 459
High aspect ratio holes
rigid printed wiring boards EL1: 550-551
High brass
applications and properties A2: 306
High chrome
alloy composition and abrasion resistance A18: 189
High convection cooling system
furnace A7: 352, 354
High density contacts EL1: 440-441
High electron mobility transistor (HEMT) EL1: 188
High energy rate forming A8: 170
High explosives
for explosive forming A14: 638
High field niobium-based superconductive compounds A7: 636
High frequency, effects
WSI EL1: 362
High frequency resistance welding of
alloy steels M6: 760
aluminum and aluminum alloys M6: 760
copper and copper alloys M6: 760
low-carbon steels M6: 760
medium-carbon steels M6: 760
nickel M6: 760
titanium M6: 760
zirconium M6: 760
High frequency welding M6: 757-768
advantages and limitations M6: 760
applications M6: 760-762
finned tube M6: 762-763
pipe and tube welding M6: 760-761
spiral seam pipe and tube M6: 761-762
structural sections M6: 761
currents M6: 757
design applications M6: 758
equipment M6: 764-765
contacts M6: 764-765
impeders M6: 765
induction coils M6: 765
high frequency induction welding M6: 757, 759-760
continuous seam tube welding M6: 759-760
tube butt end joining M6: 760
high frequency resistance welding M6: 757-759
continuous seam welding M6: 759
finite length butt welding M6: 759
metallographic examination M6: 767
microstructures of welds M6: 767
operation principles M6: 757-758
procedures for welding M6: 763
joint fit-up and alignment M6: 763
postweld procedures M6: 763
safety M6: 58-59, 767-768
testing, destructive M6: 766-767
flaring test M6: 766-767
flattening test M6: 766
High frequency welding (continued)
reverse flattening test M6: 766
testing, nondestructive M6: 765-766
eddy current testing M6: 765
flux leakage examination M6: 765-766
hydrostatic testing M6: 766
ultrasonic testing M6: 765
weld quality M6: 763-764
High frequency welding of
carbon steels M6: 760
ferritic stainless steels M6: 760
High gas-jet cool (H-GJC) techniques A4: 58
High interstitial defects
defined A9: 9
in Ti-6Al-4V alpha-beta billet appearance A9: 470
in titanium and titanium alloys A9: 459
High iron briquettes process A7: 98
High Modulus Fibers and Their Composites (ASTM Committee D-30) EM1: 40
High molecular weight *See also* Molecular weight
blow molding EM2: 164
film EM2: 165
pipe EM2: 163-164
sheet EM2: 165
High nickel alloys *See also* Nickel-base alloys
metallurgical corrosion effects A13: 128-130
weldments A13: 361-362
High pH precipitation
contaminants M5: 209
High polymer EM3: 15
defined EM2: 22
High production presses A14: 502-503
High pulse current
definition M6: 9
High pulse time
definition M6: 9
High resolution
for analysis of Jominy bar A10: 508
High resolution electron microscopy A9: 121
High room-temperature strength SCC/corrosion-resistant aluminum P/M alloys A13: 839-842
High speed steels *See* Tool steels, high speed
High speed tool steel springs M1: 296, 300
High speed tool steels *See* Tool steels, high speed
High static tensile stresses
in threaded fasteners A11: 537
High strain rate
computer data acquisition systems for A8: 192
effect on yield and failure A8: 208
flow curve, isothermal, stress-temperature plots for A8: 161
fracture testing A8: 259
Hopkinson bar techniques for compression tests A8: 190
inertial constraint effect on test validity A8: 190
Kolsky bar for testing single crystals in shear at A8: 219
projectile impacts for generating A8: 190
regime, adiabatic conditions in A8: 191
rod impact (Taylor) test for A8: 190
in shock fronts A8: 190
wave propagation for generating A8: 190
High strain rate compression testing *See also* High strain rate tension testing
cam plastometer A8: 193-196
control A8: 392
conventional load frames at medium strain rates A8: 192-193
drop tower compression test A8: 196-198
Hopkinson bar test techniques A8: 198-203
measurement of stress and strain A8: 191-192
rod impact (Taylor) test A8: 203-206
stress wave propagation A8: 191
High strain rate shear testing A8: 215-239
double shear A8: 215
double-notch A8: 228-229
high-speed hydraulic torsional machines A8: 215-216
for macroscopic constitutive models A8: 215
plate impact testing A8: 230-238
punch loading A8: 215, 228-230
torsional Kolsky bar A8: 218-228
High strain rate tension testing A8: 208-214
conventional load frames A8: 208-210
High strain rate tension testing (continued)
equipment A8: 208
expanding tests A8: 210
flyer plate and short duration pulse loading A8: 210-212
inertia effects A8: 208-209
load cell ringing A8: 209
plastic wave velocity A8: 209
split-Hopkinson bar in tension A8: 212-214
strain measurement A8: 208-210
tensile test configuration A8: 208
wave propagation effects A8: 208-209
High strain rate testing A8: 187-189, 190-206, 208-214, 215-219
High strength-high temperature applications
ordered phases used for A9: 683
High temperature *See also* Elevated temperatures; Temperatures
High Temperature Materials Information Analysis (HTMIAC) EM4: 40
High temperatures *See also* Elevated temperatures
effect on austenitic stainless steels A9: 284
effect on ferritic stainless steels A9: 284-285
effect on gamma prime in wrought heat resistant alloys A9: 311
High velocity ovens
paint curing process M5: 488
High viscosity index (HVI) A18: 168, 169
High-acceleration fatigue tests
for solder attachments EL1: 741
High-alloy cast irons, heat treating A4: 697-708
advantages A4: 697
austenitic nickel-alloyed graphitic
applications A4: 698
austenitic ductile irons A4: 698-699
austenitic gray irons A4: 697-698
dimensional stabilization A4: 699
high-temperature stabilization A4: 699
irons A4: 697-699
properties A4: 697, 698
reaustenitization A4: 699
refrigeration A4: 699
solution treating A4: 699
spheroidize annealing A4: 699
stress-relieving A4: 698-699
high-alloy white cast irons A4: 700-708
applications A4: 702-703
hardness A4: 703-704
hardness of minerals and microconstituents A4: 703
heat treatments A4: 698-699, 700, 702, 706-708
high-chromium white irons A4: 702-708
microstructure A4: 704-706
Ni-Hard A4: 700, 701, 702
nickel-chromium white irons A4: 700-702
high-silicon irons A4: 699-700
advantages A4: 699
applications A4: 699, 700
corrosion-resistant A4: 700
heat treatments A4: 699-700
Nicrosilal A4: 699
nodular A4: 699
Silal A4: 699
stress-relieving A4: 700
Ni-Resist irons A4: 697, 699
High-alloy cast irons, UNS, specific types
F41000 (type 1)
composition A4: 698
ductile family not available A4: 698
mechanical properties A4: 697, 698
no stabilization heat treatments A4: 699
F41001 (type 1b)
composition A4: 698
mechanical properties A4: 697, 698
F41002 (type 2)
composition A4: 698
mechanical properties A4: 698
F41003 (type 2b)
composition A4: 698
mechanical properties A4: 697, 698
F41004 (type 3)
composition A4: 698
mechanical properties A4: 697, 698
F41005 (type 4)
composition A4: 698

High-alloy cast irons, UNS, specific types (continued)
mechanical properties **A4:** 697-698
F41006 (type 5)
composition **A4:** 698
mechanical properties **A4:** 697, 698
F41007 (type 6)
composition **A4:** 698
mechanical properties **A4:** 698
F43000 (type D-2)
composition **A4:** 698
mechanical properties **A4:** 698
refrigeration and reaustenitization **A4:** 699
F43001(type D-2b)
composition **A4:** 698
mechanical properties **A4:** 698
F43002 (type D-2c) composition **A4:** 698
F43003 (type D-3)
composition **A4:** 698
mechanical properties **A4:** 698
F43004(type D-3a) composition **A4:** 698
F43005 (type D-4)
composition **A4:** 698
mechanical properties **A4:** 698
F43006 (type D-5)
composition **A4:** 698
mechanical properties **A4:** 698
F43007 (type D-5b) composition **A4:** 698
F45000(type A)
composition **A4:** 701
designation **A4:** 701
mechanical requirements **A4:** 701
F45001(type B)
composition **A4:** 701
designation **A4:** 701
mechanical requirements **A4:** 701
F45002(type C)
composition **A4:** 701
designation **A4:** 701
mechanical requirements **A4:** 701
F45003(type D)
composition **A4:** 701
designation **A4:** 701
hardening **A4:** 702
mechanical requirements **A4:** 701
microstructure after refrigeration **A4:** 702
F45004(type A)
composition **A4:** 701
designation **A4:** 701
mechanical requirements **A4:** 701
F45005(type B)
composition **A4:** 701
designation **A4:** 701
mechanical requirements **A4:** 701
F45006(type C)
composition **A4:** 701
designation **A4:** 701
mechanical requirements **A4:** 701
F45007(type D)
composition **A4:** 701
designation **A4:** 701
mechanical requirements **A4:** 701
F45008(type E)
composition **A4:** 701
designation **A4:** 701
mechanical requirements **A4:** 701
F45009 (type A)
composition **A4:** 701
designation **A4:** 701
mechanical requirements **A4:** 701
Type D-5s, composition **A4:** 698
High-alloy ferrous metals
as hardfacing alloys **A7:** 828, 829
Osprey process for **A7:** 530
High-alloy graphitic irons *See also* Compacted graphite irons
aluminum-alloyed, effects **A15:** 698, 701
applications **A15:** 698-699, 701
austenitic ductile irons **A15:** 700-701
High-alloy graphitic irons (continued)
austenitic gray irons **A15:** 699-700
austenitic nickel-alloy irons **A15:** 699-701
corrosion resistance **A15:** 698, 701
flake graphite and nodular graphite **A15:** 698
high-silicon ductile irons **A15:** 698-699
high-silicon gray irons **A15:** 698
high-silicon irons **A15:** 698-699, 701
high-temperature service **A15:** 698-699
oxidation resistance **A15:** 698
High-alloy irons
hardfacing **A6:** 790
High-alloy stainless steels
hydrogen fluoride/hydrofluoric acid corrosion **A13:** 1168
High-alloy steel forging
localized coarse grains **A9:** 176
High-alloy steels
austenitic manganese steels **A15:** 733-735
C-type, applications **A15:** 731-732
chromium/nickel contents **A15:** 722
corrosion-resistant **A15:** 722-723
ferrite in **A15:** 724-726
ferrographic application to identify wear particles **A18:** 305
forge welding **A6:** 306 **M6:** 676
foundry practice **A15:** 730
H-type alloys, applications **A15:** 733
heat resistant, properties **A15:** 728-730
heat treatment **A15:** 730-731
heat-resistant **A15:** 723-724
mechanical properties, corrosion-resistant **A15:** 726-728
nondestructive testing **A6:** 1086
oxyfuel gas cutting **A6:** 1155 **M6:** 897
roll welding **A6:** 312 **M6:** 676
shielded metal arc welding **A6:** 176
weldability **A15:** 535-537, 730
High-alloy white irons
alloy grades **A15:** 678
applications **A15:** 681
casting design **A15:** 680
composition control **A15:** 679-680
heat treatment **A15:** 680
high-chromium white irons **A15:** 681-685
machining **A15:** 680-681
melting practice **A15:** 678-679
molds **A15:** 680
nickel-chromium alloys, special **A15:** 681
nickel-chromium white irons **A15:** 678-681
patterns **A15:** 680
pouring **A15:** 680
shakeout **A15:** 680
High-alumina
applications **EM4:** 902, 903, 906, 915
High-alumina ceramics
mixing operations **EM4:** 98
properties **A18:** 548
High-alumina fiber
as synthetic reinforcement **EM1:** 117-118
High-aluminum cast iron
composition **M1:** 76
corrosion resistance **M1:** 96
mechanical properties **M1:** 92, 94
physical properties **M1:** 88
High-aluminum irons **A1:** 103
High-angle boundaries
formation during incubation **A9:** 694
migration, during secondary recrystallization **A9:** 698
migration, effect on grain growth **A9:** 696
High-bandwidth
designs **EL1:** 85-86
systems **EL1:** 77
High-borate sealing glasses
chemical properties **EM4:** 857
High-carbon bearing steels **A1:** 381, 382
High-carbon cast steels **A1:** 372
High-carbon, high chromium
cold-work wrought tool steels **A1:** 765
High-carbon, high-chromium cold-work steels
composition limits **A18:** 735
High-carbon high-chromium cold-work tool steels
forging temperatures **A14:** 81
High-carbon, high-chromium steels
composition of tool and die steel groups **A6:** 674
High-carbon iron (etched)
decarburized **A7:** 182, 183
High-carbon low-alloy (HCLA) steels
corrosive wear rates in grinding media **A18:** 274, 275, 276
High-carbon, medium-alloy steels
composition of tool and die steel groups **A6:** 674
High-carbon spring steel
resilience of **A8:** 22, 23
High-carbon steel
base for 50% aluminum-zinc alloy coated steel wire **M5:** 350
etching of **M5:** 180
Fe-0.75C, air cooled **A9:** 190
pickling of **M5:** 68-69
plating process precleaning **M5:** 16-18
porcelain enameling of **M5:** 512-513
quenched in a hot stage **A9:** 83
rust and scale removal **M5:** 13-14
High-carbon steel parts
sintering **A7:** 343
High-carbon steel wire
carbon content **M1:** 259
mechanical tensioning, for prestressing concrete **M1:** 264, 265
High-carbon steels *See also* Carbon steel; High-carbon steels, specific types **A18:** 651, 653
air-carbon arc cutting **A6:** 1176
applications **A15:** 714
band saw blade material **A6:** 1185
for bearings **A1:** 24-25, 380-381
carbon content **A15:** 702
carbon content isolated **A10:** 177
castings **M1:** 384, 386, 393
characteristics and applications **M1:** 458-459
Cr-Mo **A18:** 651
pearlite **A18:** 651
definition of **A1:** 148
drawn wire, distortion in **A11:** 139
electron beam welding **M6:** 638
electron-beam welding **A6:** 867
flash welding decarburization **M6:** 580
forging temperatures **A14:** 81
fractographs **A12:** 277-290
fracture/failure causes illustrated **A12:** 216
friction welding **A6:** 153 **M6:** 721
gas-metal arc welding
of aluminum bronzes **A6:** 828
of copper nickels **A6:** 828
of coppers **A6:** 828
of high-zinc brasses **A6:** 828
of low-zinc brasses **A6:** 828
of phosphor bronzes **A6:** 828
of silicon bronzes **A6:** 828
of special brasses **A6:** 828
of tin brasses **A6:** 828
gas-tungsten arc welding **A6:** 192
of aluminum bronzes **A6:** 827
of copper nickels **A6:** 827
of coppers **A6:** 827
of phosphor bronzes **A6:** 827
of silicon bronzes **A6:** 827
hardfacing alloys **A6:** 798
hydrogen flaking **A12:** 125
overheating damage in **A11:** 122
oxyacetylene welding **A6:** 281
oxyfuel cutting **M6:** 904
oxyfuel gas cutting **A6:** 1159

High-carbon steels (continued)
oxyfuel gas cutting of **A14:** 724
postweld heat treatment **A6:** 649
press forming of **A14:** 556-559
repair welding **A6:** 1105
resistance seam welding **M6:** 494
resistance welding **A6:** 833, 837, 847
shell crack and detail fracture **A12:** 288
spring grades **M1:** 284, 285
stud arc welding **M6:** 733
thermal spray coating material **A18:** 832
weldability **M1:** 561, 562-563
weldability of **A1:** 609
wire rod **M1:** 253-257
High-carbon steels, specific types
AISI 10 B62, oxidation effects **A12:** 280
AISI 10 B62, steel wire, fatigue fracture surface **A12:** 280
AISI 1053, fatigue fracture surface, fracture origin **A12:** 277
AISI 1055, fatigue fracture surface **A12:** 279
AISI 1060, fatigue fracture from seam **A12:** 281
AISI 1060, fatigue fracture from surface flaw **A12:** 279, 281
AISI 1060, torsional overload fracture **A12:** 278
AISI 1070, complex fatigue fracture **A12:** 281
AISI 1070, heat-treating failure **A12:** 282
AISI 1070, inadequate removal of blanking fracture **A12:** 285
AISI 1070, torsional fatigue fracture **A12:** 282
AISI 1074, galling failure **A12:** 285
AISI 1074, hydrogen embrittlement **A12:** 284
AISI 1074, low-cycle fatigue fracture **A12:** 283
AISI 1074, mechanical or lubrication failure **A12:** 283
AISI 1074, scab failure **A12:** 284
AISI 1075, embrittlement failure by hydrogen- assisted flakes **A12:** 285
AISI 1095, fatigue fracture surface **A12:** 288
ASTM A228, hydrogen embrittlement failures **A12:** 286
ASTM A230, atypical fracture surface **A12:** 287
ASTM A230, fatigue failure from seam **A12:** 287
High-carbon wire for mechanical tensioning **A1:** 283
High-cellulose, gas shielded
chemical composition, SMAW electrode coverings **A6:** 61
High-chloride nickel plating **M5:** 200
High-chromium alloy steels
erosion resistance for pump components **A18:** 598
High-chromium alloys
resistance welding **A6:** 833
High-chromium cast iron
corrosion resistance **M1:** 90-91
heat resistance **M1:** 93-94
heat treatment, structure developed by **M1:** 86
mechanical properties **M1:** 89, 92
physical properties **M1:** 88
High-chromium cast irons
corrosion resistance **A13:** 567-568, 569
High-chromium irons **A1:** 103
advantages **A6:** 797
applications **A6:** 797
austenitic, advantages and applications of materials for surfacing, build-up, and hardfacing **A18:** 650
hardfacing **A6:** 791-792, 796-797
martensitic, advantages and applications of materials for surfacing, build-up, and hardfacing **A18:** 650
tungsten-molybdenum, advantages and applications of materials for surfacing build-up, and hardfacing **A18:** 650
High-chromium low-carbon white irons
annealing **M4:** 558
austenitizing **M4:** 558
High-chromium stainless steel
workability **A8:** 165, 575
High-chromium white iron
wear surface **A11:** 377
High-chromium white irons
applications **A15:** 684-685
austenitic **A15:** 682
classes **A15:** 681
composition and structures **A15:** 682

High-chromium white irons (continued)
heat treatment **A15:** 684
machining **A15:** 684
martensitic **A15:** 682
melting practice **A15:** 682-683
microstructures **A15:** 681-682
molds, patterns, casting design **A15:** 683
pouring practice **A15:** 683
shakeout practice **A15:** 683-684
special, corrosion resistant **A15:** 685
special, high-temperature service **A15:** 685
High-chromium/high-carbon stainless steels **A18:** 649
High-conductivity beryllium coppers
gas-metal arc butt welding **A6:** 760
High-conductivity bronze
applications and properties **A2:** 313
High-conductivity copper
degassing **A15:** 469
electronic applications **A6:** 998
High-conductivity coppers
for conductors **A2:** 251
High-copper alloys **A6:** 752
properties and applications **A2:** 224
High-cycle fatigue
defined **A11:** 5
as overstress effect **A11:** 109-110
service fracture from **A11:** 106
striations **A11:** 78
wrought titanium alloys **A2:** 624
High-cycle fatigue fracture(s) *See also* Fatigue; Fracture(s)
AISI/SAE alloy steels **A12:** 337
martensitic stainless steels **A12:** 367
precipitation-hardening stainless steels **A12:** 371, 373
titanium alloys **A12:** 452
tool steels **A12:** 381
wrought aluminum alloys **A12:** 426
High-cycle fatigue testing **A8:** 367
of D-6 AC steel **A8:** 485
load-control tests for **A8:** 696
as stress vs. cycles-to-failure **A8:** 367
torsional **A8:** 149-150
High-cycle torsional fatigue **A8:** 149-150
High-definition forgings
aluminum alloy **A14:** 243-244
High-density concrete
powder used **A7:** 573
High-density electronic packaging *See also* Packaging; Surface mount technology (SMT)
military/commercial, demand for **EL1:** 730
High-density molding
defined **A15:** 346
High-density polyethylene (HDPE) **EM3:** 15
interfacial zone shear and solid friction **A18:** 36
wear properties **A18:** 241
High-density polyethylenes (HDPE) *See also* Polyethylenes (PE); Thermoplastics resins
additives **EM2:** 166
applications **EM2:** 163-165
characteristics **EM2:** 165-166
costs **EM2:** 163
defined **EM2:** 22
grades **EM2:** 163
high molecular weight **EM2:** 163
processing **EM2:** 165-166
rotational molding **EM2:** 361
suppliers **EM2:** 166
High-density rubber
powders used **A7:** 573
High-density static random-access memory (RAM) module **EL1:** 88
High-deposition submerged arc welding **A7:** 821-822
High-efficiency sodium and potassium cyanide copper plating **M5:** 159-165, 167-169
High-electron-mobility transistor (HEMT)
research **A2:** 747
High-energy ball mill
schematic **A7:** 723
High-energy deburring **A7:** 459
High-energy milling *See also* Milling
applications **A7:** 69
homogenization of aluminum alloy **A7:** 67

High-energy physics
with A15 superconductors **A2:** 1071
as niobium-titanium superconducting material application **A2:** 1043, 1055-1056
refractory metal and alloy applications **A2:** 558
ternary molybdenum chalcogenides application **A2:** 1079
High-energy scattering **A7:** 259
High-energy wet scrubber
cupolas **A15:** 387
High-energy x-ray sources *See also* Radiographic inspection; Radiography
machine designs for **A17:** 307
machines **A17:** 388-389
R-output, high-energy sources **A17:** 307-308
High-energy-beam brazing
stainless steels **A6:** 922-923
High-energy-rate compacting **A7:** 6, 304-306
High-energy-rate forging *See also* Forging; High-energy-rate forming
application **A14:** 100, 105-107
of austenitic stainless steel extrusion **A14:** 365
blanks, preparation of **A14:** 104
defined **A14:** 7, 100
development work **A14:** 107
dies **A14:** 101-104
economics of **A14:** 100-101
lubrication **A14:** 104
machines for **A14:** 29, 101
metal flow in **A14:** 105
processing **A14:** 104-105
production examples **A14:** 102-104
High-energy-rate forging machines **A14:** 29, 101
High-energy-rate forming *See also* Electromagnetic forming; Explosive forming
defined **A14:** 7
High-energy-rate forming (HERF) machine **A18:** 636, 637, 638, 639
High-energy-rate-forged extrusion **A8:** 573
High-flash naphtha cleaners **M5:** 40, 42-43
High-frequency acoustic imaging (HAIM) **A18:** 406, 409-410, 411
applications **A18:** 410
principles **A18:** 409-410
High-frequency butt welding
of blanks **A14:** 451
High-frequency digital systems *See also* Digital systems; Very-high-speed integrated circuits (VHSICS)
analysis of PWB structures **EL1:** 81-82
controlled-impedance connector system **EL1:** 86
design guidelines and methodology **EL1:** 82-85
error-free design capability **EL1:** 85-86
factors limiting **EL1:** 79-80
hybrid wafer scale integration advantages **EL1:** 86-88
interconnection level description **EL1:** 76
modeling/simulation requirements **EL1:** 77-81
package options **EL1:** 76-77
High-frequency drive core knockout machine **A15:** 506
High-frequency furnace **A7:** 157
High-frequency furnaces **A10:** 221-222
High-frequency healing **EM3:** 15
High-frequency heating
defined **EM1:** 13 **EM2:** 22
High-frequency induction welds
failure origins **A11:** 449
High-frequency inductors
construction **EL1:** 179
High-frequency resistance welding *See also* High frequency, welding
definition **A6:** 1210 **M6:** 9
High-frequency shakeout table **A15:** 503
High-frequency testing
ultrasonic as **A8:** 240
High-frequency vibrations
fatigue-fracture surface of **A11:** 112
High-frequency welding (HFW) **A6:** 252-253
advantages **A6:** 252
aluminum **A6:** 252
applications **A6:** 252-253
brass **A6:** 252
carbon steels **A6:** 252
chromium/molybdenum alloys **A6:** 252
copper **A6:** 252

High-frequency welding (HFW) (continued)
definition **A6:** 252
disadvantages **A6:** 252
edge "V" generation **A6:** 252
equipment **A6:** 253
high-frequency induction welding (HFIW) **A6:** 252
high-frequency resistance welding (HFRW) **A6:** 252
inspection and quality control **A6:** 253
personnel **A6:** 253
proximity effect **A6:** 252
reactive metals **A6:** 252
safety **A6:** 253
skin effects **A6:** 252
stainless steels **A6:** 252, 253
titanium **A6:** 252, 253
High-gloss coatings
radiation-cure formulation **EL1:** 858
High-gloss impact polystyrenes *See also* High-impact polystyrenes (PS, HIPS)
applications and properties **EM2:** 199
High-impact polystyrene (HIPS) **EM3:** 15
High-impact polystyrenes (PS, HIPS) *See also* Polystyrenes (PS); Thermoplastic resins
alloys and blends **EM2:** 194
applications **EM2:** 194-196
characteristics **EM2:** 165-166
commercial forms **EM2:** 194
composite structures **EM2:** 195
costs and production volume **EM2:** 194
defined **EM2:** 22
design properties **EM2:** 196-198
processing **EM2:** 198
resin compound types **EM2:** 198-199
specialty, properties **EM2:** 199
suppliers **EM2:** 199
temperature effects **EM2:** 197
thermoforming **EM2:** 198-199
High-lead tin bronzes
applications **A18:** 751
composition **A18:** 750, 751
designations **A18:** 751
mechanical properties **A18:** 752
product form **A18:** 751, 752
High-leaded brass
applications and properties **A2:** 306-307
High-leaded bronze
nominal compositions **A2:** 347
High-leaded naval brass
applications and properties **A2:** 321
High-leaded tin bronze *See also* Copper casting alloys
applications and properties **A2:** 227, 278-382
corrosion ratings **A2:** 353-354
foundry properties for sand casting **A2:** 348
nominal composition **A2:** 347
High-leaded tin bronzes
applications **A15:** 784
composition/melt treatment **A15:** 772, 776
High-level design
defined **EL1:** 129
High-level simulation
defined **EL1:** 129
High-level waste
disposal **A13:** 971-980
High-low hydraulic pumping system **A7:** 330
High-manganese steels **A18:** 650
High-modulus aluminum P/M alloys
types **A2:** 209-210
High-modulus composites *See also* Composite(s)
thermoplastic suitability for **EM1:** 98-100
High-modulus graphite fibers
in amorphous polymers **EM2:** 758-759
High-nickel alloy
hardness conversion tables **A8:** 109
High-nickel alloys
abrasive wear **A18:** 767
electron-beam welding **A6:** 865
High-nickel alloys (continued)
inorganic fluxes used **A6:** 130
as magnetically soft materials **A2:** 770-771
High-nickel austenitic cast irons
corrosion resistance **A13:** 567
High-nickel cast iron
corrosion resistance **M1:** 91
growth in superheated steam **M1:** 93
heat resistance **M1:** 94-96
mechanical properties **M1:** 89, 92
physical properties **M1:** 88
stress rupture **M1:** 95
High-nickel cast irons
corrosion in **A11:** 200
hydrochloric acid resistance **A13:** 569
High-nickel irons **A1:** 103
High-nickel steels for low-temperature service **A1:** 392, 396-397, 398
High-oxygen iron
ductility from hot torsion tests **A8:** 165-166
High-palladium ceramic alloys
dental **A13:** 1361
High-palladium PFM alloys **A13:** 1355-1356
High-pass filters
eddy current inspection **A17:** 190
High-performance active devices
with bipolar technology **EL1:** 146-147
High-Performance Adhesive Bonding **EM3:** 67
High-performance alumina fiber **EM1:** 61
High-performance aluminum powders
degassing **A7:** 526
High-performance bearings
materials for **A11:** 483
High-performance caulks **EM3:** 50
High-performance composite materials *See also* Advanced composites
applications **EM1:** 206
defined **EM1:** 27
High-performance electronic systems
analytic modeling **EL1:** 16
High-performance engineering thermoplastics
types **EM2:** 98
High-performance liquid chromatography
and IR spectroscopy **A10:** 116
and mass spectrometry, with gas chromatography/mass spectroscopy **A10:** 645-646
MFS detection for **A10:** 72
of organic gases **A10:** 11
of organic liquids and solutions **A10:** 10
of organic solids **A10:** 9
High-performance liquid chromatography (HPLC) **EL1:** 833-834 **EM1:** 736, 737 **EM2:** 517-518
High-performance liquid chromatography (HPLC) techniques **EM3:** 732
chemical content analysis and quality control **EM3:** 735
High-performance MOS (HMOS)
development **EL1:** 160
High-performance steels
rolling-element bearing material **A18:** 508
High-performance systems
design considerations **EL1:** 28-42
isolated signal line **EL1:** 28-34
packaging effects **EL1:** 397
parallel signal lines **EL1:** 34-41
signal attenuation/rise time degradation **EL1:** 41-42
skin effect resistance **EL1:** 41
High-performance thermoplastic resins/resin systems **EM1:** 43, 100-101, 544
High-precision
high-strength gears **A7:** 667-668
High-pressure compaction and hot working (STAMP process) **A1:** 780
High-pressure compound buffing systems **M5:** 116
High-pressure die casting
as permanent mold process **A15:** 34
High-pressure die casting magnesium alloys
types **A2:** 456
High-pressure dies
cemented carbide **A2:** 972
High-pressure drain
nuclear power reactors **A13:** 958
High-pressure gage
as rare earth application **A2:** 731
High-pressure gas compressor
HIP processing **A7:** 422
High-pressure high-temperature (HPHT) processing
of diamond/cubic boron nitrides **A2:** 1008
High-pressure laminates *See also* Laminate(s) **EM3:** 15
defined **EM1:** 13 **EM2:** 22
High-pressure liquid chromatography (HPLC)
silica-based supports **EM4:** 1088-1089
High-pressure liquid chromatography, reverse phase
for formulation verification **EM1:** 730
High-pressure lubricant
in axial compression testing **A8:** 56
High-pressure molding **EM3:** 15
defined **A15:** 346 **EM2:** 22
devices, development of **A15:** 29
High-pressure molding defect
casting **A11:** 384
High-pressure pump **A8:** 419
High-pressure punches
cemented carbide **A2:** 972
High-pressure self-combustion sintering (HPCS) **EM4:** 199
High-pressure spot *See* Resin-starved area
High-pressure steam **A8:** 427, 429
High-pressure tanks
holographic measurement of **A17:** 16
High-pressure water blast core knock-out device **A15:** 506
High-pressure water jets
for descaling **A14:** 87
High-production peening
as straightening **A14:** 681
High-purity (HP) aluminum
fretting wear **A18:** 243, 244
High-purity alumina, applications
protection tubes and wells **A4:** 533
High-purity aluminum, alloying
superconducting materials **A2:** 1045
High-purity aluminum powders **A7:** 125
High-purity arsenic production
and recovery **A2:** 747
High-purity atomized aluminum powder **A7:** 125, 130
High-purity austenitic stainless steels
atom probe analysis **A10:** 595
High-purity copper
effects of stress and temperature **A12:** 399
fracture modes **A12:** 399-400
fracture surfaces and mechanisms **A12:** 399-400
intergranular separation **A12:** 399
High-purity deaerated water
wrought aluminum alloys **A12:** 438
High-purity iron
flat cleavage fracture **A12:** 219
as magnetically soft material **A2:** 764-765
High-purity metals
voltammetric analysis **A10:** 188
High-purity nickel
trace impurities in **A10:** 240
High-purity nickel strip *See also* Nickel powder strip
heart pacemaker battery parts **A7:** 403
High-purity niobium
preparation for superconductors **A2:** 1043-1044
High-purity oxygenated water **A8:** 420

SUBJECTS OF THE INDEXED VOLUMES: ASM Handbook (designated by the letter "A"): **A1:** Properties and Selection: Irons, Steels, and High-Performance Alloys (1990); **A2:** Properties and Selection: Nonferrous Alloys and Special-Purpose Materials (1990); **A3:** Alloy Phase Diagrams (1992); **A4:** Heat Treating (1991); **A6:** Welding, Brazing, and Soldering (1993); **A7:** Powder Metallurgy (1984); **A8:** Mechanical Testing (1985); **A9:** Metallography and Microstructures (1985); **A10:** Materials Characterization (1986); **A11:** Failure Analysis and Prevention (1986); **A12:** Fractography (1987); **A13:** Corrosion (1987); **A14:** Forming and Forging (1988); **A15:** Casting (1988); **A16:** Machining (1989); **A17:** Nondestructive Testing and Quality Control (1989); **A18:** Friction, Lubrication, and Wear Technology (1992). **Metals Handbook, 9th Edition** (designated by the letter "M"): **M1:** Properties and Selection: Irons and Steels (1978); **M2:** Properties and Selection: Nonferrous Alloys and Pure Metals (1979); **M3:** Properties and Selection: Stainless Steels, Tool Materials and Special-Purpose Materials (1980); **M4:** Heat Treating (1981); **M5:** Surface Cleaning, Finishing, and Coating (1982); **M6:** Welding, Brazing, and Soldering (1983). **Engineered Materials Handbook** (designated by the letters "EM"): **EM1:** Composites (1987); **EM2:** Engineering Plastics (1988); **EM3:** Adhesives and Sealants (1990); **EM4:** Ceramics and Glasses (1991); **Electronic Materials Handbook** (designated by the letters "EL"): **EL1:** Packaging (1989).

High-purity sodium carboxymethylcellulose
application or function optimizing powder treatment and green forming **EM4:** 49
High-purity water
aluminum/aluminum alloys in **A13:** 597
High-rate regime
in compression testing **A8:** 190
High-removal rate (HRR) machining **A16:** 607-609
cutting tools and parameters **A16:** 608-609
machine requirements **A16:** 607-608
superlathe CNC machines **A16:** 607
tool life **A16:** 608
tooling of machines **A16:** 608
High-resolution electron energy loss spectroscopy **A10:** 109, 126
High-resolution electron microscopy (HREM) **A6:** 145 **A18:** 389
High-resolution energy-compensated atom probe
microanalysis **A10:** 597
with pulsed-laser capabilities **A10:** 598
spectrum of tungsten by **A10:** 597
High-resolution infrared imaging systems
for thermal inspection **A17:** 398
High-resolution thermography **EL1:** 954-955
High-shrinkage alloys
copper casting **A2:** 346
High-silica fibers
defined **EM1:** 29
High-silica glasses
applications
electronic processing **EM4:** 1058
laboratory and process **EM4:** 1087-1088
lighting **EM4:** 1032
optical fibers as transmission media for telecommunications **EM4:** 1050-1054
laboratory glassware, composition and properties **EM4:** 1088
High-silicon alloys
cleaning solutions for substrate materials **A6:** 978
High-silicon aluminum
cutting tool material selection based on machining operation **A18:** 617
High-silicon cast iron
characteristics **M1:** 78
corrosion in **A11:** 200
corrosion resistance **M1:** 89-90
heat resistance **M1:** 93
mechanical properties **M1:** 89
physical properties **M1:** 88
High-silicon cast irons
alkali resistance **A13:** 570
corrosion rates **A13:** 567
hydrochloric acid resistance **A13:** 569
impressed-current anodes **A13:** 469, 921
nitric acid resistance **A13:** 569
High-silicon irons *See also* Ductile irons; Gray iron; High-alloy graphitic irons **A1:** 102-103
compositions **A15:** 701
for corrosion resistance **A15:** 701
ductile irons **A15:** 698-699
gray irons **A15:** 698
for high-temperature service **A15:** 698-699
High-silicon stainless steels
hardfacing **A6:** 795
High-solids paints **M5:** 472
High-speed brass plating **M5:** 286-287
High-speed compaction
effect on green strength **A7:** 302
High-speed electrical performance
and dielectric properties **EL1:** 597-610
electrical properties **EL1:** 597-604
materials properties **EL1:** 604-608
printed board manufacturability **EL1:** 608-609
recommendations **EL1:** 608
High-speed forging *See* High-energy-rate forging
High-speed hydraulic torsional machines **A8:** 215-216
High-speed machining **A16:** 597-606
aircraft engine propulsion applications **A16:** 604
airframe and defense applications **A16:** 603
alternate cutting tool geometries **A16:** 602
applications **A16:** 603
automotive applications **A16:** 604
chip formation **A16:** 597, 598-600, 603

High-speed machining (continued)
cutting fluids **A16:** 602
cutting tool materials **A16:** 601
future needs **A16:** 605
historical background **A16:** 597-598
implementation **A16:** 605
ledge tools **A16:** 602-603
machining centers **A16:** 604
nickel alloys **A16:** 837
parameters **A16:** 600-601
rotary tool machining **A16:** 603
stability principle **A16:** 599
tool life **A16:** 604
variables **A16:** 597-598
High-speed printer
metal band/metal platen wear test for **A8:** 607
High-speed printers hammer guide assembly for **A7:** 669
P/M parts for **A7:** 667
High-speed resin injection
properties effects **EM2:** 287
High-speed resin transfer molding
size and shape effects **EM2:** 292
High-speed steel
brazing temperature effect on hardness **A6:** 908
High-speed steel (HSS) *See* Toot steels, high-speed
classification and composition of hardfacing alloys **A18:** 652
compositions and properties **A16:** 51-59
cutting speed and work material relationship **A18:** 616
cutting tools **A16:** 41
for cutting toot material **A18:** 614, 615
end mills **A16:** 41
flank wear limits **A16:** 43
material yield strength **A16:** 39
P/M high-speed steels **A16:** 60-68
part material for ion implantation **A18:** 858
physical vapor deposition, critical normal force versus substrate surface roughness **A18:** 436
surface damage involving gain of material **A18:** 182
versus 60-40 brass adhesive wear **A18:** 237-238
High-speed steel cutting tools, types used
thermal spray-coated materials **M5:** 369-370
High-speed steel, rehardened
brittle fracture of **A11:** 574
High-speed steels *See also* High-speed steels, specific types; High-speed tool steels; Steels
as-sintered and wrought **A7:** 377
extrusion **A7:** 525
hot extrusion **A7:** 524-525
hot isostatic pressing plus hot rolling to mill shape **A7:** 524
mechanical properties **A7:** 376
sintering temperatures and compositions for **A7:** 375
tempering curves **A7:** 376
typical parts **A7:** 371
High-speed steels, coatings for contrast enhancement in scanning electron
microscopy **A9:** 99
High-speed steels, specific types
M2, mechanical properties **A7:** 376
M15, controlled spray deposited **A7:** 531
M35, mechanical properties **A7:** 376
T15, as-sintered microstructure **A7:** 378
T15, mechanical properties **A7:** 376
T15, microstructures **A7:** 545
High-speed tool steels *See also* High-speed steels; Steels; Tool steels; Tool steels, high-speed; Tool steels, specific types
annealing **A7:** 274
cold isostatic pressing applications **A7:** 450
cold isostatic pressing dwell pressures **A7:** 449, 450
extrusion processes for **A7:** 525
friction welding **A6:** 153
heat tinting **A9:** 136
hot extrusion **A7:** 524-525
hot isostatic pressing plus hot rolling to mill shape **A7:** 524
microstructure **A7:** 38, 39
P/M and forging/rolling techniques **A7:** 522

High-speed tool steels (continued)
powder metallurgy **A1:** 781-786
alloy development **A1:** 784-785
applications **A1:** 785-786
cutting tool properties **A1:** 784
heat treatment **A1:** 782-783
manufacturing properties **A1:** 783-784
sintered tooling **A1:** 786
reductions by cold swaging **A14:** 128
sintering **A7:** 370-376
wrought **A1:** 759-762
molybdenum **A1:** 759
tungsten **A1:** 759-762
High-stacking fault energy material **A8:** 173
High-strength alloy steels
electron beam welding **M6:** 637-638
preheating **M6:** 613
electron-beam welding **A6:** 258- 259, 867
electroslag welding **M6:** 226
shielded metal arc welding **M6:** 75
High-strength alloys
for ships and submarines **A13:** 546
High-strength alloys, conventional
fracture toughness of **A8:** 458-459
High-strength aluminum alloys *See also* Aluminum alloy powders; Aluminum powders **A7:** 745-748
applications **A7:** 748
available product forms **A7:** 746
chemical composition **A7:** 745
plane-strain fracture toughness **A8:** 451
SCC failure in **A11:** 27
High-strength aluminum P/M alloys
alloy design research **A2:** 204-210
aluminum P/M processing **A2:** 201-204
aluminum-lithium alloys **A2:** 209
aluminum-lithium-beryllium alloys **A2:** 209
ambient-temperature strength **A2:** 204-206
can vacuum degassing **A2:** 202-203
conventionally pressed and sintered alloys **A2:** 210-213
corrosion resistance **A2:** 204-206
dipurative degassing **A2:** 203
direct powder forming **A2:** 203
dynamic compaction **A2:** 204
elevated-temperature properties **A2:** 206-208
high-modulus and/or low-density alloys **A2:** 209-210
hot isostatic pressing (HIP) **A2:** 203
intermetallics **A2:** 210
introduction **A2:** 200
mechanical attrition process **A2:** 202
metal-matrix composites **A2:** 209-210
P/M technology, advantages **A2:** 200-201
part processing **A2:** 210-213
powder degassing and consolidation **A2:** 202-204
powder production **A2:** 201-202
rapid omnidirectional consolidation **A2:** 203-204
rapid solidification (RS) alloys **A2:** 204-207
strengthening features **A2:** 200, 202
stress-corrosion cracking (SCC) **A2:** 204-206
superplastic forming (SPF) **A2:** 210
vacuum degassing in reusable chamber **A2:** 203
High-strength beryllium coppers
gas-metal arc butt welding **A6:** 760
High-strength beryllium-copper casting alloys
composition **A2:** 403-404
High-strength carbon and low-alloy steels **A1:** 389
quenched and tempered low-alloy steel **A1:** 391-392
effects of alloying elements on **A1:** 392-394, 395, 396
high-nickel steels for low-temperature service **A1:** 392, 396-397, 398
mechanical properties **A1:** 389, 392, 396, 397
microalloyed quenched and tempered grades **A1:** 394-396
structural carbon steels **A1:** 389
high-strength structural carbon steels **A1:** 390-391
hot-rolled carbon-manganese structural steels **A1:** 390, 391
mild steels **A1:** 390

High-strength cast steels ... A1: 374
as low-alloy ... A15: 716
High-strength controlled-expansion alloys *See also* Low-expansion alloys
properties ... A2: 895
High-strength low-alloy
abbreviation for ... A11: 797
High-strength low-alloy (HSLA) steels
See also Microalloyed steel ... A1: 148, 151, 154, 358-362, 397-398, 399
annealing ... A4: 52
applications of ... A1: 399, 415-423
automotive ... A1: 416-417
castings ... A1: 419, 420
cold-forming strip ... A1: 418
forgings ... A1: 420
offshore structures ... A1: 417-418
oil and gas pipelines ... A1: 416
railway tank cars ... A1: 419
shipbuilding ... A1: 418
structural ... A1: 399, 401, 418, 420, 421
categories and specifications ... A1: 398-405
ASTM specifications ... A1: 399, 406, 411
SAE categories ... A1: 401
classification of ... A1: 148
acicular-ferrite ... A1: 148, 399, 404-405
as-rolled pearlitic ... A1: 404
dual-phase steels ... A1: 148, 398, 405, 424-429
hydrogen-induced cracking resistant steels ... A1: 400, 416
inclusion-shape controlled steels ... A1: 400, 405, 412-413
microalloyed ferrite-pearlite steels ... A1: 400-404, 585-588
weathering steels ... A1: 400
cold-rolled ... A1: 419-420
compositions ... A1: 401, 406, 410
in ASTM specifications ... A1: 406
of normalized European HSLA steels ... A1: 410
in SAE specifications ... A1: 401
continuous annealing ... A4: 63
control of properties ... A1: 405-411, 588
with alloying elements ... A1: 406-408
by controlled rolling ... A1: 408-409, 586-588
cooling, effect of ... A1: 402
definition of ... A1: 148
fatigue characteristics of ... A1: 413
forgings ... A1: 137, 358-362, 419, 420, 588
forming of ... A1: 402, 408, 413-414
forming properties of ... A1: 398, 418
bending radii ... A1: 413, 414
mechanical properties ... A1: 410-413
of acicular ferrite steels ... A1: 404-405
of cold-forming strip ... A1: 417
of cold-rolled sheet ... A1: 420
compared with carbon steel ... A1: 389
of control-rolled steels ... A1: 409
directionality of properties ... A1: 412-413
of dual-phase steels ... A1: 398
effect of manganese on ... A1: 402
of HSLA forgings ... A1: 360-361, 588, 599
of microalloyed ferrite-pearlite steels ... A1: 411-412
of normalized HSLA steels ... A1: 409-410
of weathering steels ... A1: 400
metallurgical effects ... A1: 359-361
first-generation microalloy steels ... A1: 359
second-generation microalloy steels ... A1: 359-360
third-generation microalloy steels ... A1: 360-361
microalloying elements ... A1: 358-359, 400-404, 419
molybdenum ... A1: 358, 403, 407
niobium ... A1: 358, 402-403, 407, 419
titanium ... A1: 231-232, 359, 403-404, 408
vanadium ... A1: 358, 401-402, 403, 408, 419
processing methods ... A1: 130-131, 398, 408-410, 586, 587, 588
selection guidelines ... A1: 415-416

High-strength low-alloy (HSLA) steels (continued)
solution-strengthened and microalloyed ... A4: 61-62
steelmaking ... A1: 405-406
welding of ... A1: 414--415, 609
yield strength ... A4: 62
High-strength low-alloy (HSLA) structural steels ... A6: 662-664
chemical compositions ... A6: 663
electrogas welding ... A6: 664
electroslag welding ... A6: 664
filler metal selection ... A6: 662-663, 664
flux-cored arc welding ... A6: 663, 664
gas-metal arc welding ... A6: 663, 664
heat-affected zones ... A6: 663, 664
postweld heat treatment ... A6: 664
properties ... A6: 662, 663
shielded metal arc welding ... A6: 663-664
specifications ... A6: 662, 664
submerged arc welding ... A6: 663, 664
welding procedures and practices ... A6: 664
High-strength low-alloy quenched and tempered (HSLA Q&T) structural steels ... A6: 662, 664-666
filler metals ... A6: 665
flux-cored arc welding ... A6: 660
gas-metal arc welding ... A6: 664, 666
gas-tungsten arc welding ... A6: 664, 666
heat input ... A6: 666
heat-affected zones ... A6: 666
postweld heat treatment ... A6: 666
preheat and interpass temperature control ... A6: 665-666
properties ... A6: 664
shielded metal arc welding ... A6: 663-664, 666
specifications ... A6: 664
weld design ... A6: 665
weld metal hydrogen ... A6: 665
welding processes ... A6: 664
High-strength low-alloy quenched and tempered structural steels, specific types
A 514
maximum welding heat input for butt joints ... A6: 666
postweld heat treatments ... A6: 666
A 517, postweld heat treatments ... A6: 666
High-strength low-alloy steel
effect on economy in manufacture ... M3: 851
hot-rolled ... A9: 185
scratch testing ... A18: 431
High-strength low-alloy steel bars ... A1: 246, 588
High-strength low-alloy steel forgings ... A1: 137, 358-362, 419, 420, 588
High-strength low-alloy steel plate ... A1: 235-236, 587
High-strength low-alloy steels *See also* As-rolled structural steels; HSLA steels; Low,-alloy steels; Steels; Steels, specific types
AMS classification ... M6: 291-292
base-metal treatment ... M6: 292
filler metals ... M6: 292
prewelding heat treatment ... M6: 292
susceptibility to cracking ... M6: 291-292
welding procedures ... M6: 292
welding processes ... M6: 291
AOD-refined, composition control ... A15: 429
applications
automotive ... A6: 395
machinery and equipment ... A6: 390
arc welding ... M6: 270-282
atmospheric corrosion ... A13: 518, 531
classification and group description ... A6: 405-406, 407
cold cracking ... A6: 73
corrosion losses, chemical plant atmospheres ... A13: 533
flux cored electrodes ... M6: 101-102
flux-cored arc welding ... A6: 187

High-strength low-alloy steels (continued)
gas-metal arc welding ... A6: 180
hydrogen-induced cracking ... A6: 73, 94
laser beam welding ... M6: 647
mechanical cutting ... A6: 1179
minimum bend radius ... A14: 524
nickel alloys welding to ... A6: 578
precipitate stability and grain boundary pinning ... A6: 73
resistance seam welding ... M6: 494
resistance spot welding ... A6: 228 M6: 478, 484-488
resistance welding ... A6: 837, 838, 840
for rolling ... A14: 355
SCC testing ... A13: 270-272
shielded metal arc welding ... A6: 746 M6: 95
steel weldment soundness ... A6: 411
stud arc welding ... M6: 733
submerged arc welding ... M6: 115-116
microstructure characteristics ... M6: 116-117
types
as-rolled pearlitic structural ... M6: 271
microalloyed ... M6: 271
type A ... M6: 271
type Q ... M6: 271
type R ... M6: 271
type W, T ... M6: 271
as weathering steels ... A13: 516
weldability ... A6: 418-419 M6: 270-271
High-strength low-alloy steels, specific types
300 M, laser-beam welding ... A6: 264
A 242
composition and carbon content ... A6: 406
composition and strength properties ... A6: 663
low-hydrogen welding processes ... A6: 664
shielded metal arc welding ... A6: 663
A 441
composition and strength properties ... A6: 663
low-hydrogen welding processes ... A6: 664
shielded metal arc welding ... A6: 663
A 514, quenching and tempering ... A6: 665, 666
A 517, quenching and tempering ... A6: 665, 660
A 543, quenching and tempering ... A6: 665, 666
A 572
composition and carbon content ... A6: 406
composition and strength properties ... A6: 663
low-hydrogen welding processes ... A6: 664
shielded metal arc welding ... A6: 663
A 588
composition and carbon content ... A6: 406
composition and strength properties ... A6: 663
low-hydrogen welding processes ... A6: 664
shielded metal arc welding ... A6: 663
A 633
composition and strength properties ... A6: 663
low-hydrogen welding processes ... A6: 664
shielded metal arc welding ... A6: 663
A 710, composition and strength properties ... A6: 663
grade 42
composition and strength properties ... A6: 663
low-hydrogen welding processes ... A6: 664
grade 50
composition and strength properties ... A6: 663
low-hydrogen welding processes ... A6: 664
grade 60
composition and strength properties ... A6: 663
low-hydrogen welding processes ... A6: 664
grade 65
composition and strength properties ... A6: 663
low-hydrogen welding processes ... A6: 664
grade A
composition and strength properties ... A6: 663
low-hydrogen welding processes ... A6: 664
grade B
composition and strength properties ... A6: 663
low-hydrogen welding processes ... A6: 664
grade C
composition and strength properties ... A6: 663

High-strength low-alloy steels, specific types (continued)
low-hydrogen welding processes **A6:** 664
grade D, composition and strength
properties **A6:** 663
grade E
composition and strength properties **A6:** 663
low-hydrogen welding processes **A6:** 664
grade F, composition and strength
properties **A6:** 663
grade G, composition and strength
properties **A6:** 663
grade H, composition and strength
properties **A6:** 66
grade J, composition and strength
properties **A6:** 663
HP9-4-20
composition **A4:** 207
heat treatments **A4:** 215, 216
HP9-4-30
composition **A4:** 207
heat treatments **A4:** 215, 216-217
mechanical properties **A4:** 216
HSLA-80
chemical composition **A6:** 425
heat analysis compositions **A6:** 406
heat input **A6:** 426, 427, 428
properties **A6:** 428
resistance to brittle fracture **A4:** 252
steel composition effect on susceptibility to cold cracking **A6:** 74
submerged arc welding **A6:** 426
weldability **A4:** 239
HSLA-100
heat analysis compositions **A6:** 406
heat-affected zone toughness **A6:** 79
resistance to brittle fracture **A4:** 252
steel composition effect on susceptibility to cold cracking **A6:** 74
weldability **A4:** 239
HY-80
chemical composition **A6:** 425
heat input **A6:** 426, 427-428
hydrogen-induced cracking **A6:** 412
military specifications **A6:** 664
properties **A6:** 427-428
quenching and tempering **A6:** 665, 666
submerged arc welding **A6:** 426
transformation plasticity **A6:** 1138
weldability **A6:** 424
HY-80, weldability **A4:** 239
HY-100 steel
electrogas welding **A6:** 277
electroslag welding **A6:** 277
military specifications **A6:** 664
quenching and tempering **A6:** 665, 666
HY-130
military specifications **A6:** 664
quenching and tempering **A6:** 665, 666
High-strength magnesium-base alloy
constitutional liquation in multicomponent systems **A6:** 568
High-strength manganese bronze *See also* Copper casting alloys; Manganese bronze
nominal composition **A2:** 347
High-strength modified copper *See* Copper alloys, specific types, C19400
applications and properties **A2:** 293-294
High-strength prealloyed P/M aluminum alloy forgings **A14:** 250-251
High-strength quenched and tempered steels
arc welding **M6:** 247, 282-292
High-strength sheet molding compounds (HMC)
properties effects **EM2:** 286-287
size and shape effects **EM2:** 291
High-strength steels **A18:** 649
coated, press forming of **A14:** 565-566
cobalt-enhanced toughness **A7:** 144
electrochemical polarization **A8:** 527
embrittlement mechanisms in **A8:** 529
gas metal arc welding **M6:** 153
low-alloy **A14:** 355, 524
low-alloy, transfer gears **A7:** 668
notch tensile test for **A8:** 27
pin bearing testing of **A8:** 59
precleaning embrittlement **A17:** 81
High-strength steels (continued)
SCC testing of **A8:** 526-527
sodium chloride stress cracking **A8:** 527
springback in **A8:** 553
sulfide stress cracking **A8:** 527
tensile properties **A8:** 555
thermal-mechanical processing for **A14:** 19
High-strength steels, applications
aerospace **A6:** 385-388
automotive **A6:** 395
diffusion welding **A6:** 884
plasma arc cutting **A6:** 1170
shielded metal arc welding **A6:** 176
ultrasonic welding **A6:** 893
High-strength steels, specific types
100S-1, chemical composition **A6:** 425
High-strength structural carbon steels **A1:** 390-391
High-strength structural steels *See* As-rolled structural steels
High-strength wrought aluminum alloys *See also* Wrought aluminum alloys
fracture toughness **A2:** 54, 59-60
High-strength yellow brasses *See also* Cast copper alloys
corrosion ratings **A2:** 353-354
properties and applications **A2:** 367-370
High-stress abrasion *See also* Low-stress abrasion
defined **A18:** 10
High-sulfate nickel plating **M5:** 202-203
High-temperature *See also* Elevated-temperature; Low-temperature; Temperature(s)
high-shear (HTHS) viscosity **A18:** 85
High-temperature adhesives **EM3:** 171, 172
advantages and limitations **EM3:** 80
aircraft and aerospace applications **EM3:** 80
chemistry **EM3:** 80
curing method **EM3:** 80
suppliers **EM3:** 80
High-temperature aerated water testing **A8:** 420-422
High-temperature air heaters
corrosion of **A13:** 998-999
High-temperature alloy steels **M1:** 284, 286
High-temperature alloys *See also* Heat-resistant alloys; Stainless steels **A6:** 563-565 **A13:** 91, 1312
applications **A6:** 563
argon oxygen decarburization **A15:** 426
avoidance of copper chills **A6:** 565
boron additions **A6:** 563
compositions and forging temperature ranges **A14:** 224
contaminant effect on weld soundness **A6:** 564-565
cutting fluids used **A16:** 125
cutting tool materials and cutting speed relationship **A18:** 616
definition **A6:** 563
depth-of-cut notching **A16:** 77
drilling **A16:** 227
effects of eutectic melting **A11:** 122
electropolishing of **M5:** 306-307
enamels **EM3:** 303
fusion zone fissuring **A6:** 565
heat-affected zone **A6:** 563, 564-565
honing **A16:** 477
isolation of nickel and cobalt in **A10:** 174
lead causing weld metal fissuring **A6:** 565
machinability test matrix **A16:** 639-640
macrofissuring **A6:** 563, 564
microfissuring **A6:** 563, 565
milling **A16:** 312, 313, 314
for nickel-base alloy cold-formed parts **A14:** 837
nickel-base, compositions **A14:** 262
nitric/hydrochloric acid as dissolution medium **A10:** 166
nonmetallic inclusions in **A11:** 316
notch sensitivity testing of **A8:** 27
oxide stability **A11:** 452
postweld heat treatment **A6:** 563-564
preweld heat treatment **A6:** 563-564
recommended machining specifications for rough ceramic insert cutting tools **EM4:** 969
and finish turning with HIP metal-oxide
refractory oxide on surface **A6:** 565
semisolid metal casting and forging of **A15:** 327
High-temperature alloys (continued)
strain age cracking **A6:** 563, 564
turning and milling recommended ceramic grade inserts for cutting tools **EM4:** 972
welding characteristics **A6:** 563-565
welding effect on performance and properties **A6:** 563
zirconium additions **A6:** 563
High-temperature applications *See also* Elevated temperature; Temperature(s) **EM1:** 810-815
condensation-type polyimides **EM1:** 810
of epoxy resins **EM1:** 141, 810
PMR polyimides **EM1:** 812-814
High-temperature bake
glass-to-metal seals **EL1:** 459
High-temperature chromium-molybdenum steels
flux-cored arc welding **A6:** 187
High-temperature cobalt alloys
contrasting by interference layers **A9:** 60
High-temperature cofired ceramic packages
burnout/firing **EL1:** 465-466
lead/pin attach **EL1:** 466
physical properties **EL1:** 468
testing **EL1:** 467
High-temperature combustion *See also* Combustion; Combustion method **A10:** 221-225
applications **A10:** 224
combustion principles **A10:** 221-222
defined **A10:** 674
detection of combustion products **A10:** 222-223
determinator systems, automatic, manual, or semiautomatic **A10:** 222
estimated analysis time **A10:** 221
general use **A10:** 221
of inorganic solids, types of information from **A10:** 4, 6
limitations **A10:** 221
of organic solids **A10:** 9
related techniques **A10:** 221
sample preparation **A10:** 223-224
samples **A10:** 221, 223
selective and total combustion **A10:** 223-224
separation of interfering elements **A10:** 222
High-temperature composites *See also* Composites; Elevated-temperature
refractory metal fiber-reinforced **A2:** 582-584
High-temperature conductivity
in rare earth cuprates **A2:** 1027
High-temperature corrosion **A13:** 97-101
in aircraft powerplants **A13:** 1038-1041
in brazed joints **A13:** 878
by alloying **A13:** 56, 59
by compound reduction, in liquid metals **A13:** 56, 59
by impurities, in liquid metals **A13:** 56-59
by interstitial reactions, in liquid metals **A13:** 56-59
carburization **A13:** 99-100
free energy of reaction **A13:** 61-62
free-energy-temperature diagrams **A13:** 62-63
in gases, thermodynamics **A13:** 61-64
hot corrosion **A13:** 100-101
hydrogen effects **A13:** 100
isothermal stability diagrams **A13:** 63-64
in liquid metals **A13:** 56-60
materials selection, molten metals **A13:** 59
in molten salts **A13:** 50-55, 89
oxidation **A13:** 98-99
in petroleum refining and petrochemical operations **A13:** 1270-1274
sulfidation **A13:** 99
types, liquid-metal **A13:** 56-59
High-temperature curing adhesives **EM3:** 36
High-temperature deaerated water testing **A8:** 422-425
High-temperature electrical testing **EL1:** 1060-1061
High-temperature epoxy resins
temperature range **EM1:** 141
High-temperature fiberglass fibers
as thermocouple wire insulation **A2:** 882
High-temperature fractures
effect of temperature **A12:** 52
interpretation of **A12:** 121-123

High-temperature gas
hydrogen fluoride/hydrofluoric acid corrosion ... **A13:** 1170
High-temperature gaseous reactions
space shuttle orbiter ... **A13:** 1092-1094
High-temperature halogen ion solutions
nickel-base alloy resistance ... **A13:** 648-649
High-temperature high-pressure fatigue test
for cannon tubes ... **A11:** 281
High-temperature hot dip galvanized coating ... **M5:** 66, 329-330
High-temperature hydrodynamic tension test
for cannon tubes ... **A11:** 281
High-temperature hydrogen attack
defined ... **A13:** 7
High-temperature intermetallics *See also* Ordered intermetallics
specific gravity vs melting point diagrams for ... **A2:** 935
High-temperature lubricants
powders used ... **A7:** 573
High-temperature materials
high-temperature solid-state welding ... **A6:** 299
machined by high-speed steel cutting tools ... **A18:** 615
High-temperature materials, specific types
Incoloy 800, effect of stress intensity factor range on fatigue crack growth rate ... **A12:** 58
Incoloy 800, effects of temperature ... **A12:** 52
Incoloy 800, ridges and striations in sulfidizing atmosphere ... **A12:** 41, 53
Incoloy 800, wedge cracking ... **A12:** 26
Inconel 600, effect of frequency and wave form on fatigue properties ... **A12:** 60
Inconel 625, wedge cracking ... **A12:** 26
Inconel 718, effects of stress intensity factor range on fatigue crack growth rate ... **A12:** 58
Inconel X-750, effect of frequency and wave form on fatigue properties ... **A12:** 59-60
Inconel X-750, effect of stress intensity factor range on fatigue crack growth rate ... **A12:** 57
Inconel X-750, effect of temperature ... **A12:** 52
Inconel X-750, effect of temperature on double- aged ... **A12:** 47
Inconel X-750, effect of vacuum ... **A12:** 48
High-temperature microscopy *See* Hot-stage microscopy
High-temperature mixed gases
chlorine corrosion ... **A13:** 1172
High-temperature nickel alloys
contrasting by interference layers ... **A9:** 60
High-temperature nickel-base superalloys *See also* Nickel-base superalloys; Superalloys
ordered intermetallic effects ... **A2:** 914
High-temperature oxidation ... **A13:** 98-99
acoustic emission inspection ... **A17:** 287
surface effects ... **A12:** 35, 72
High-temperature oxidation-resistant coating *See* Oxidation-resistant coating, high-temperature type
High-temperature oxide superconductors *See* High-temperature superconductors for wires and tapes
High-temperature powders
for hardfacing ... **A7:** 830
High-temperature properties *See* Creep behavior; Elevated-temperature properties; Stress-rupture properties
High-temperature resin systems *See also* Elevated temperatures; High-temperature service; Temperature(s)
bismaleimides (BMI) ... **EM2:** 444
polyimides (PI) ... **EM2:** 443-444
High-temperature resistance
of nickel alloys ... **A2:** 429-431
High-temperature reverse bias (HTRB)
as screening method ... **EL1:** 1060
High-temperature sands
recovery of ... **A15:** 348-349
High-temperature service *See also* Elevated-temperature properties
high-chromium white irons ... **A15:** 685
thermoplastic polymers for ... **EM2:** 54
High-temperature sintering ... **A7:** 366-368
tolerances ... **A7:** 482
High-temperature solder
elemental mapping of ... **A10:** 532
High-temperature solid-state welding ... **A6:** 297-299
advantages ... **A6:** 277, 298-299
aluminum ... **A6:** 298, 299
anisotropic materials ... **A6:** 299
applications ... **A6:** 297, 298-299
atmosphere ... **A6:** 297, 298
austenitic stainless steels ... **A6:** 298
bond-surface extension ... **A6:** 297
ceramics ... **A6:** 298, 299
disadvantages ... **A6:** 297, 298-299
elimination of the original joining surface ... **A6:** 297
equipment ... **A6:** 297, 298
fiber-reinforced materials ... **A6:** 299
glasses ... **A6:** 298
high-temperature materials ... **A6:** 299
lead ... **A6:** 298
molybdenum ... **A6:** 298
parameters ... **A6:** 297-298
primary bonding ... **A6:** 297
procedures ... **A6:** 297-298
stainless steels ... **A6:** 298, 299
tungsten ... **A6:** 298
High-temperature stage, used for sintering studies by scanning electron
microscopy ... **A9:** 100
High-temperature steam
alloy steel corrosion ... **A13:** 574
High-temperature steel
impact resistance and abrasion resistance properties ... **A18:** 759
High-temperature steels
thermal expansion coefficient ... **A6:** 907
High-temperature storage life
as temperature-induced stress test ... **EL1:** 498
High-temperature strain-controlled fatigue test
hold periods ... **A8:** 346
High-temperature strength, of iron-chromium-nickel heat-resistant casting alloys
influence of microstructure ... **A9:** 333
High-temperature structural adhesives
suppliers ... **EM3:** 508
High-temperature structural materials
ordered intermetallic ... **A2:** 913-914
High-temperature superalloys ... **A8:** 159
High-temperature superconductivity
thin-film materials ... **A2:** 1083
High-temperature superconductors ... **EM4:** 17, 1156-1158
applications ... **EM4:** 1156, 1158
composite structures ... **EM4:** 1158, 1159
critical parameters of superconductivity ... **EM4:** 1156, 1157-1158
melt-solidification processing ... **EM4:** 1157-1158
plastic extrusion ... **EM4:** 1156-1157
powder synthesis ... **EM4:** 1156, 1157
High-temperature superconductors for wires and tape *See also* Superconducting materials; Superconductivity; Superconductors
aerosol pyrolysis technique ... **A2:** 1086
anisotropy influences ... **A2:** 1087-1088
applications ... **A2:** 1085
Bi-Sr-Ca-Cu-O (BSCCO bismuth) systems processing ... **A2:** 1085
introduction ... **A2:** 1085
microstructural influences ... **A2:** 1087-1088
powder precursor preparation ... **A2:** 1086
High-temperature superconductors for wires and tape (continued)
powder-in-tube processing ... **A2:** 1086
primary oxide compounds, processing of ... **A2:** 1085-1086
primary technical challenge ... **A2:** 1085
shake-and-bake method ... **A2:** 1086
Ti-Ba-Ca-Cu-O (TBCCO thailium) system processing ... **A2:** 1085-1086
vapor deposition processing ... **A2:** 1087
weak-link influences ... **A2:** 1087-1088
Y-Ba-Cu-O (YBCO/123 compound) systems processing ... **A2:** 1085
High-temperature test
ceramic coatings ... **M5:** 546
High-temperature testing
with Kolsky bar ... **A8:** 222-223, 225
High-temperature thermoplastic resin matrices *See also* Thermoplastic resins
defined ... **EM1:** 32
for prepreg ... **EM1:** 142
types ... **EM1:** 138
High-temperature thermoset matrix composites
bismaleimide resin and fiber-resin ... **EM1:** 373-380
polyimide resin and fiber-resin ... **EM1:** 373
High-temperature torsion testing ... **A8:** 158-159
for dynamic recovery and recrystallization ... **A8:** 173
stress-strain curves for nickel ... **A8:** 174
torque-twist behavior of titanium alloy ... **A8:** 169
High-temperature water corrosion
of light water reactors ... **A13:** 946
High-temperature wear ... **A8:** 603
High-tensile hard drawn wire
characteristics ... **M1:** 289
springs, wire for ... **M1:** 284
High-tensile hard-drawn wire
characteristics of ... **A1:** 307
High-tin babbitt
recycling ... **A2:** 1219
High-titanium gas-slag shield, chemical composition
SMAW electrode coverings ... **A6:** 61
High-vacuum applications of lubricants *See* Lubricants, high-vacuum applications
High-vacuum diffusion pump
SEM ... **A12:** 171
High-velocity compacting
rigid die ... **A7:** 305
High-velocity forging *See* High-energy-rate forging
High-velocity gas jet
liquid sheet disintegration by ... **A7:** 28, 30
High-velocity oxyfuel (HVOF) powder spray process ... **A18:** 829, 830, 831, 832
High-velocity oxygen flame spraying
ceramic coatings for adiabatic diesel engines ... **EM4:** 992
High-voltage breakdown
from solder hairs ... **EL1:** 561
High-volume production
millimeter/microwave applications ... **EL1:** 757-758
High-zinc alloys *See also* Zinc; Zinc alloys
inclusion-forming in ... **A15:** 96
High-zinc brasses
corrosion in various media ... **M2:** 468-469
High-zinc brasses, gas-metal arc butt welding ... **A6:** 760
resistance spot welding ... **A6:** 850
resistance welding ... **A6:** 849-850
to high-carbon steel ... **A6:** 828
to low-alloy steel ... **A6:** 828
to low-carbon steel ... **A6:** 828
to medium-carbon steel ... **A6:** 828
to stainless steels ... **A6:** 828
weldability ... **A6:** 753
High-zinc copper alloys
gas-metal arc welding ... **A6:** 763
Higher austenitic stainless steels
sulfuric acid corrosion ... **A13:** 1151-1152

Higher-alloy metals
nitric acid corrosion ... **A13:** 1156
Higher-level integration *See also* Hybrids and higher-level integration
and hybrids ... **EL1:** 249
Higher-order Laue zone
for aluminum-copper alloy ... **A10:** 463
in electron diffraction ... **A10:** 439-440
Higher-order phase transition ... **A3:** 1 • 10
Highest contact temperatures ... **A18:** 438, 440
Highest numerical gear ratio ... **A18:** 566
Highest point of single-tooth contact (HPSTC) ... **A18:** 543
Highly accelerated stress testing (HAST) ... **EL1:** 495-497
Highly accelerated stress tests (HAST) ... **EM3:** 434
Highly alloyed steels
precoated before soldering ... **A6:** 131
Highly deformed layer *See also* Beilby layer; White layer
defined ... **A18:** 10
Highly integrated systems
test procedures ... **EL1:** 372-373
Highly oriented pyrolytic graphite ... **A10:** 132, 690
Highly oriented pyrolytic graphite (HOPG)
for atomic image used in testing scanning tunneling microscopy performance ... **A18:** 395, 396
Hightower algorithm
as automatic trace routing ... **EL1:** 531
Highway tractor-trailer steel drawbar
fatigue fracture of ... **A11:** 127-128
Highway truck equalizer beam
mechanical cracking of ... **A11:** 390
Hildebrand theory stress dependence of solubility ... **EM3:** 362
Hilger-Watts stereoscope ... **A12:** 171
Hill maximum reduced stress criterion ... **A8:** 344
Hilliard's circular test figures
for measurement of grain size ... **A9:** 130
Hillocks
and electromigration ... **EL1:** 964
Hindered contraction
as casting defect ... **A11:** 386
Hinge stress
definition ... **EM4:** 633
Hinge-like knee joint prosthesis ... **A11:** 670
HIP *See* Hot isostatic pressing
Hip and knee joints
metal powder ... **A7:** 657, 754
Hip plate for osteotomies
as internal fixation device ... **A11:** 671
Hip prostheses
effect of fatigue and stem loosening in ... **A11:** 690-693
fractures of ... **A11:** 692-693
types of ... **A11:** 670
Hip screws
austenitic stainless steels ... **A12:** 359-364
HIPS *See* High-impact polystyrene
Histogram
for IA particle analysis ... **A10:** 320
Histogram equalization
color image ... **A17:** 484
defined ... **A17:** 384
digital image enhancement ... **A17:** 456
Histogram(s)
for accessing capability ... **A17:** 523
count rate, acoustic emission inspection ... **A17:** 281-284
defined ... **A17:** 384
Histograms ... **EM3:** 786, 789, 790
of facet areas ... **A12:** 208
of profile angular distributions ... **A12:** 203
Historic Lane plate, chromium steel
general corrosion of ... **A11:** 674, 680
Historical studies
of books and artifacts, PIXE analysis for ... **A10:** 102
hkl tables, in single-crystal analysis ... **A10:** 346
milliprobe PIXE analysis ... **A10:** 107
Historical tracking ... **A14:** 928-937
control charting ... **A14:** 928-931
statistical deformation control ... **A14:** 932-937
History
of casting ... **A15:** 15-23
History (continued)
of computer simulation, foundry industry ... **A15:** 858
of foundry development ... **A15:** 24-36
of metalworking ... **A14:** 15
powder metallurgy ... **A7:** 14-20
History, of fractography ... **A12:** 1-11
before twentieth century ... **A12:** 1-3
electron fractography ... **A12:** 4-8
microfractography, development of ... **A12:** 3-4
quantitative fractography ... **A12:** 8
History plots
acoustic emission inspection ... **A17:** 283
Hit-driven systems
acoustic emission inspection ... **A17:** 282
Hit/miss data *See also* NDE reliability data analysis
analysis examples ... **A17:** 696
confidence bound calculation ... **A17:** 696
factorial experiments for ... **A17:** 693
parameter estimation ... **A17:** 695
POD(a) function from ... **A17:** 689-691
sample size requirements ... **A17:** 694
Hitachi FACOM M-780 mainframe computer ... **EL1:** 49
Hitachi silicon-carbide RAM module ... **EL1:** 49
HK *See* Knoop hardness
composition ... **M4:** 653
HK alloy ... **A1:** 925
HK-40
arc welding of tubing ... **M6:** 366
erosion test results ... **A18:** 200
HL alloy ... **A1:** 925
Hm 3000
properties ... **A18:** 803
HMDE *See* Hanging mercury drop electrode
HMU (untreated high modulus)-borosilicate
shear stress ... **EM3:** 402
HMX (explosive)
friction coefficient data ... **A18:** 75
HN alloy ... **A1:** 923, 925-926
Ho-In (Phase Diagram) ... **A3:** 2 • 250
Ho-Mn (Phase Diagram) ... **A3:** 2 • 250
Ho-Pd (Phase Diagram) ... **A3:** 2 • 250
Ho-Sb (Phase Diagram) ... **A3:** 2 • 251
Ho-Te (Phase Diagram) ... **A3:** 2 • 251
Ho-Tl (Phase Diagram) ... **A3:** 2 • 251
Hobbing
carbon and low-alloy steel gears with high-speed steel tools ... **A16:** 345
and gear manufacture ... **A16:** 330, 333-335, 343-344, 348, 351
and machining of helical gears ... **A16:** 339, 341
and machining of herringbone gears ... **A16:** 339
and machining of spur gears ... **A16:** 338-339, 341
and machining of worm gears ... **A16:** 340
Hobbing steels *See also* Tool steels
effect of composition on microstructure of ... **A9:** 258
Hobs *See* Cutting tools
high-speed tool steels ... **A16:** 57, 58, 59
Hockman cycle ... **EM3:** 189
Hoganas (Hoeganaes) process ... **A7:** 79-82
annealing ... **A7:** 182
flowchart ... **A7:** 80
process conditions ... **A7:** 79-82
Hohman A-6 wear machine ... **A18:** 11
defined ... **A18:** 10
Hoisting rope
failure of steel wire ... **A11:** 518, 519
steel wire ... **M1:** 272
Holcomb Steel Company *See* Crucible Steel Company
Hold period
in compression ... **A8:** 349-353
effect at peak strain on Rene 95 fatigue life ... **A8:** 352
effect in tension-hold-only testing on fatigue resistance of stainless steel ... **A8:** 349
effect on low-cycle fatigue resistance ... **A8:** 348
in fatigue cycle ... **A8:** 346-347
length, vs. time-to-fracture ... **A8:** 350
and strain waveform effect ... **A8:** 347-348
in tension ... **A8:** 349-353
vs. fatigue life for AISI 304 stainless steel ... **A8:** 349
Hold steps
in cure cycles ... **EM1:** 141
Hold time *See* Hold period
definition ... **M6:** 9
Hold times *See* Dwell time
Hold-down clamps
distortion in ... **A11:** 140
Hold-down plate
defined ... **A14:** 8
Hold-downs
straight-knife shearing ... **A14:** 703
Hold-time test
creep-fatigue interaction diagram ... **A8:** 356
Holding and clamping fixtures ... **A8:** 93, 96
Holding blanks
multiple-slide forming ... **A14:** 571
Holding furnace
defined ... **A15:** 6
Holding time
definition ... **M6:** 9
of samples ... **A10:** 16
Holding times *See also* Time
as inclusion-forming ... **A15:** 90
for inoculation ... **A15:** 105
Hole drilling
capabilities of ... **A10:** 380
Hole expansion test ... **A8:** 7, 562
Hole flanging ... **A14:** 8, 531, 559, 804
Hole geometry
secondary-tension test ... **A8:** 584-585
Hole machining
applications of P/M high-speed tool steel for ... **A1:** 785
of P/M tool steels ... **A7:** 789-791
Hole punching
of P/M tool steels ... **A7:** 792-793
Hole size effect ... **EM1:** 252-254, 256
Hole(s) *See also* Fastener holes; Fasteners; Tubing
as adherend defect, adhesive-bonded joints ... **A17:** 612-613
diameters, laser triangulation measurement ... **A17:** 13
magnetic rubber inspection of ... **A17:** 123
measured by CMMs ... **A17:** 18
in short parts, magnetizing ... **A17:** 94
and small radii areas, eddy currents/microwave inspection ... **A17:** 223-224
as volumetric flaw ... **A17:** 50
Holes *See also* Circular holes; Fastener holes; Gas porosity; Pierced holes; Plated-through hole (PTH)
accurate location/form, in press bending ... **A14:** 531-532
bolt, deburring drum failure from ... **A11:** 346-348
circular, fracture analysis ... **EM1:** 252-257
clearance around ... **EL1:** 19
clearance fit, fasteners for ... **EM1:** 706-707
drilled ... **A11:** 123-125, 489
elliptical, far-field force ... **EM1:** 252
evolution of ... **EL1:** 541
in fatigue crack initiation ... **A8:** 371
flanged, piercing of ... **A14:** 469
fracture surface, beach marks ... **A11:** 87
gas ... **A11:** 120
high aspect ratio ... **EL1:** 550-551
for interference fit ... **EM1:** 715
laminate theory calculations ... **EM1:** 234
lubrication, fatigue fracture at ... **A11:** 114, 346
made with pointed punch ... **A14:** 470
making, materials and processes ... **EL1:** 115
as notches ... **A11:** 85
pierced ... **A14:** 459-471
in polyamide-imides (PAI) ... **EM2:** 132
punched, brittle fracture from ... **A11:** 85, 90
quality, fastener sensitivity to ... **EM1:** 711
radius, effect on stress distribution ... **EM1:** 254
size effect ... **EM1:** 252-254, 256
size, pierced ... **A14:** 467
size, vs drill cost ... **EL1:** 20
sizes ... **EM1:** 714
spacing of ... **A14:** 468
stress concentrations around ... **EM1:** 234
as stress concentrators ... **A11:** 318
testing/evaluation ... **EL1:** 574-575
vias, formation ... **EL1:** 326-328
Holes in forgings ... **M1:** 363-364

Holidays
defined **A13:** 7
testing, of coating systems **A13:** 417
Hollandite, in ceramic waste form simulant
EPMA analysis for **A10:** 532-535
Holloman equation
applied to rapid-*n* test **A8:** 556
"Hollow bead" (defect) **A6:** 99
Hollow bolts, as ordnance application **A7:** 680
Hollow castings
by slush casting **A15:** 35
Hollow cathode lamp **A10:** 50, 690
Hollow copper electrode plasma torch **A15:** 420
Hollow cylindrical parts
magnetic particle inspection of **A17:** 111-112
Hollow injection molding **EM1:** 557
properties effects **EM2:** 284
size and shape effects **EM2:** 290
surface finish **EM2:** 304
Hollow milling
refractory metals **A16:** 862, 867
Hollow shapes
wrought aluminum alloy **A2:** 33
Hollow spheres
as extender **EM3:** 176
Hollow ware
copper and copper alloys **A14:** 822
Hollow waves
in wave soldering **EL1:** 688
Hollow-spindle lathes **A16:** 153
Holloware
forming methods **EM4:** 741-742
Holmium *See also* Rare earth metals
epithermal neutron activation analysis of **A10:** 239
properties **A2:** 1181
pure **M2:** 738-739
as rare earth **A2:** 720
Holmium in garnets **A9:** 538
Holocamera
defined **A17:** 406
Holograms
complex **A17:** 226
defined **A17:** 224, 405
three dimensional laser gaging **A17:** 12, 16
Holographic contouring *See also* Contouring
methods of **A17:** 408
multiple-index **A17:** 428
multiple-source **A17:** 425-427
multiple-wavelength **A17:** 427-428
Holographic gratings **A10:** 37, 128
Holographic inspection system **EL1:** 954
Holographic interferometric techniques **A6:** 1150
Holographic interferometry **A6:** 34, 35
for adhesive-bonded joints **A17:** 627-628
in wide sample plane-strain tensile testing **A8:** 557
Holographic plates
optical sensors with **A17:** 10
Holography *See also* Acoustical holography; Liquid-surface holography; Optical holographic interferometry; Optical holography; Scanning acoustical holography **A18:** 411-412
acoustical **A17:** 240, 438-447, 630-631
contract (purchased) **A17:** 430
defined **A17:** 405
in-house **A17:** 430
intensity **A17:** 225-226
as laser dimensional measurement **A17:** 16
microwave **A17:** 207, 224-228
millimeter wave **A17:** 226
optical **A17:** 224
Holtzmann constant
in ESR analysis **A10:** 254
HOLZ *See* Higher-order Laue zones
Home appliances *See also* Appliances
bronze P/M parts for **A7:** 736
Home Construction Projects With Adhesives and Glue **EM3:** 69
Home scrap
recycling of **A1:** 1023
Homgenizing
copper metals **M2:** 252-253
Homodyne detection system
microwave **A17:** 219
Homogeneity
defined **A10:** 674
as material assumption **EM1:** 308-309
of niobium-titanium superconductor alloys **A2:** 1044, 1052-1054
of single-phase materials, EPMA micrometer analysis **A10:** 516
techniques to determine **A7:** 315-316
Homogeneous
defined **EM1:** 13
Homogeneous flow
amorphous materials and metallic glasses **A2:** 813
Homogeneous materials
Rockwell testing of **A8:** 80, 82
Homogeneous metal powders
sintering compacts of **A7:** 309-312
Homogeneous nucleation **A6:** 45, 46 **A9:** 646 **EM3:** 15
defined **EM2:** 22
of Guinier-Preston zones in precipitation reactions **A9:** 650
and heterogeneous nucleation, compared **A15:** 105
kinetics of **A15:** 103
Homogeneous population
defined **A8:** 662
Homogeneous resistance-inductance-capacitance (RLC) line **EL1:** 357-359
Homogeneous strain
in tubular specimen **A8:** 224
Homogenization
of aluminum alloy ingots **A9:** 632
by high-energy milling Al-Fe-Ce alloy **A7:** 67
complete, austenite formation as **A7:** 314
complete, during sintering **A7:** 308
in copper alloys **A9:** 639-640
during sintering **A7:** 308, 314-315
high-alloy steels **A15:** 731
inclusion formation **A12:** 347
kinetics **A7:** 315
later stages, single peak formation **A7:** 316
and microstructure of nickel-tungsten powder compacts **A7:** 316, 317
procedures and sampling **A10:** 16
as solid-state diffusional process **A7:** 308
of stainless steel castings **A1:** 911
in steel **A9:** 625
variation of sintered density during **A7:** 314
Homogenizing
defined **A9:** 9
die-casting dies **A18:** 632
Homolite 100
crack speed and crack tip *K* **A8:** 445, 446
Homologous **EM3:** 15
defined **EM2:** 22
Homologous pairs
defined **A10:** 674
Homologous temperature **A6:** 145
defined **A8:** 34
Homolytic fragmentation-type photoinitiators **EL1:** 855
Homophase boundaries **A9:** 118
Homopolar generator
upset welding energy supply **A6:** 250
Homopolymer **EM3:** 15
surface preparation **EM3:** 291
Homopolymer and copolymer acetals (AC) *See also* Acetals; Copolymer acetals; Homopolymer acetals; Thermoplastic resins
alloyed and blended forms **EM2:** 100
applications **EM2:** 100-101
Homopolymer and copolymer acetals (AC) (continued)
characteristics **EM2:** 101-102
commercial forms **EM2:** 100
competitive materials **EM2:** 101
costs and production volume **EM2:** 100
design considerations **EM2:** 101
markets **EM2:** 100-101
processing **EM2:** 101-102
product properties **EM2:** 101
suppliers **EM2:** 102
Homopolymer(s)
chemistry of **EM2:** 63
cyanate, matrix properties **EM2:** 235
defined **EM2:** 22
polystyrenes (PS) **EM2:** 194
Homopolymerization **EM1:** 67-71, 78-80
of bismaleimides (BMI) **EM2:** 254
Honeycomb *See also* Honeycomb structures; Sandwich constructions **EM3:** 15
core characteristics **EM1:** 722-724
defined **EM1:** 13, 721 **EM2:** 22
as fastener application **A11:** 530
materials **EM1:** 722
properties **EM1:** 724
terminology **EM1:** 721
test methods **EM1:** 724
Honeycomb core repair materials **EM3:** 807
Honeycomb pattern
in fiberscope images **A17:** 6-7
Honeycomb sandwich edge strength test **EM3:** 732
Honeycomb structure **EM1:** 721-728
air flow directionality **EM1:** 728
core characteristics/types **EM1:** 722-724
corrugated manufacturing method **EM1:** 722
energy absorption **EM1:** 728
expansion manufacturing method **EM1:** 721-722
radio frequency shielding **EM1:** 728
sandwich structures **EM1:** 726-728
special processing **EM1:** 726
test methods for **EM1:** 724-726
tooling **EM1:** 728
Honeycomb structures **A16:** 893, 899-901
adhesive-bonded joints, defects **A17:** 613-615
aluminum, neutron radiography **A17:** 393
brazed panels, magnetic printing inspection **A17:** 126
microwave inspection **A17:** 202
optical holography of **A17:** 405
ultrasonic inspection **A17:** 232
Honeycombs
cuts with wire saws **A9:** 24
Honing **A16:** 19, 472-491 **EM4:** 313, 351-358
abrasive service life **A16:** 477
air gages **A16:** 480, 481
bar gages **A16:** 481, 482
cast irons **A16:** 664-665
centerless microhoning **A16:** 473
compared to lapping **A16:** 497
compared to reaming **A16:** 239
in conjunction with boring **A16:** 172
in conjunction with grinding **A16:** 432
in conjunction with lapping **A16:** 494
crosshatch angle control **A16:** 482-483
Cu alloys **A16:** 819
cubic boron nitride **A16:** 112
cutting edges for cast iron workpieces **A16:** 658
cutting fluid flow recommendations **A16:** 127
cylindrical microhoning **A16:** 473
dimensional accuracy **A16:** 484
electrochemical honing **A16:** 488
expanding gages **A16:** 481, 482
external honing **A16:** 486
flat honing **A16:** 487-488 **EM4:** 356-358
applications **EM4:** 357-358
coolant **EM4:** 357
double-side flat process **EM4:** 357

Honing (continued)
wheels used ... **EM4:** 357
gages ... **A16:** 480-482, 488
gear-tooth honing ... **A16:** 486
gears ... **A16:** 343, 351
grit and particle sizes used ... **A16:** 475, 476, 478, 490
hone forming ... **A16:** 488
honing fluids ... **A16:** 477, 482, 483-484, 485, 486
honing stones (sticks) ... **A16:** 475-478
horizontal hydraulic honing machine ... **A16:** 474, 475, 479
machine selection ... **A16:** 475
as machining process ... **A7:** 462
manual stroke honing machine ... **A16:** 474, 475
manual stroking ... **A16:** 475, 478-480, 482-486, 489
microhoning ... **A16:** 472, 488-491
MMCs ... **A16:** 896
Ni alloys ... **A16:** 835
plateau honing ... **A16:** 487
plug gages ... **A16:** 481, 482
power stroking ... **A16:** 475, 478-480, 482-486, 489
practice for internal diameters ... **A16:** 484-486
pressure ... **A16:** 483
process capabilities ... **A16:** 472
reciprocation speed ... **A16:** 482
ring gages ... **A16:** 480-481, 482
rotation speed ... **A16:** 481-482
selection of abrasive ... **A16:** 476-477, 478
special applications ... **A16:** 486
step honing ... **A16:** 486
and superabrasives ... **A16:** 454, 456
for surface integrity ... **A16:** 33
toot life of honing stones ... **A16:** 484
vertical spindle honing machine ... **A16:** 474, 475
Honing compounds
removal of ... **M5:** 14
Hooded anodes
for stem radiation ... **A17:** 305
Hoods, automotive
economy in manufacture ... **M3:** 852-853
Hook cracks
in electric-resistance weld ... **A11:** 698
in high-frequency induction welds ... **A11:** 449
Hook cracks, as flaw
defined ... **A17:** 562
Hooke's Law *See also* Elastic limit; Modulus of elasticity; Proportional limit ... **A1:** 318 **A7:** 336 **A8:** 7, 149, 198 **EM2:** 412 **EM4:** 375
defined ... **A10:** 674 **A11:** 5 **A14:** 8
in fracture mechanics ... **A11:** 50, 51
molecular vibrations and ... **A10:** 111
Hooks
C ... **A11:** 523
carbon steel, brittle fracture ... **A11:** 332-333
control and testing ... **A11:** 523
crane ... **A11:** 523
ductile iron T- , overload failure of ... **A11:** 367-369
failures of ... **A11:** 522-524
steel coil, fatigue fracture ... **A11:** 523-524
steel crane, fatigue fracture ... **A11:** 523
types and materials for ... **A11:** 522-524
Hooks, crane *See* Crane hooks
Hoop patterns
filament winding ... **EM1:** 509-510, 514 **EM2:** 373
Hoop stress ... **A6:** 374 **EM3:** 15
defined ... **EM1:** 13 **EM2:** 22
direction, in main steam line ... **A11:** 669
residual ... **A11:** 98
Hoop tension
as failure in fastened sheet ... **A11:** 531
Hoopes cell operation
aluminum alloys ... **A2:** 4
HOPG *See* Highly oriented pyrolytic graphite
Hopkinson bar
for elevated-temperature tests ... **A8:** 198
experimental procedures based on ... **A8:** 198
with precracked Charpy specimen ... **A8:** 268
pressure ... **A8:** 187
principles ... **A8:** 198-200
single pressure bar ... **A8:** 198-199
split, technique ... **A8:** 198
strain gage measurements and stress-strain behavior related ... **A8:** 198-199
surface displacement measurement ... **A8:** 198
Hopkinson bar (continued)
techniques ... **A8:** 190, 198-203
in torsion high strain rate shear testing ... **A8:** 187
for torsional loading ... **A8:** 198
Hopkinson tube
for explosively loaded torsional Kolsky bar ... **A8:** 227
Hopper dryer
defined ... **EM2:** 22
Hopper feeding
multiple-slide forming ... **A14:** 573
Horger data
for steel shafts in reversed bending ... **A8:** 372
Horizontal bending machines
for bar ... **A14:** 662
Horizontal boring mills ... **A16:** 160, 165, 167, 173, 174
Horizontal centrifugal casting
applications ... **A15:** 300
equipment ... **A15:** 296-297
materials ... **A15:** 299-300
molds ... **A15:** 296-297
as permanent mold process ... **A15:** 34
process advantages ... **A15:** 299
solidification ... **A15:** 298-299
techniques ... **A15:** 297-298
Horizontal centrifugal casting machine ... **A15:** 296
Horizontal continuous casting
of steels ... **A15:** 313
wrought copper and copper alloys ... **A2:** 243
Horizontal coordinate measuring machines
advantages/limitations ... **A17:** 21-22
fixed-table type ... **A17:** 23
horizontal-arm types ... **A17:** 22-23
moving-ram type ... **A17:** 23
moving-table type ... **A17:** 23
Horizontal drilling
cemented carbide tools ... **A2:** 974
Horizontal drilling machines
trepanning ... **A16:** 176
Horizontal fixed position
definition ... **M6:** 9
Horizontal fixed position (pipe welding)
definition ... **A6:** 1210
Horizontal flaskless molding machines
green sand molding ... **A15:** 343
Horizontal forging machines *See also* Headers; Upsetters
die set for ... **A14:** 45
heading tool and gripper inserts in ... **A14:** 48
for precision forging ... **A14:** 171
Horizontal lamination
and green strength ... **A7:** 302
Horizontal load train
on lathe bed ... **A8:** 159
Horizontal parting
devices, permanent mold casting ... **A15:** 275-276
permanent mold casting ... **A15:** 275-276
Horizontal position
definition ... **M6:** 9
Horizontal position (fillet weld)
definition ... **A6:** 1210
Horizontal position (groove weld)
definition ... **A6:** 1210
Horizontal return straight-line polishing and buffing machines ... **M5:** 121-123
Horizontal rolled position
definition ... **M6:** 9
Horizontal rolled position (pipe welding)
definition ... **A6:** 1210
Horizontal sections of a ternary diagram ... **A3:** 1 • 5
Horizontal spindle machines *See also* Machining centers ... **A16:** 1
Horizontal welding
arc welds of coppers ... **M6:** 402
flux cored arc welding ... **M6:** 106-107
gas tungsten arc welds of aluminum alloys ... **M6:** 391-392
indication by electrode classification ... **M6:** 84
oxyfuel gas welding ... **M6:** 589
fixed position for pipe ... **M6:** 591-592
rolled position for pipe ... **M6:** 591
recommended grooves ... **M6:** 71
shielded metal arc welding ... **M6:** 76, 85, 88-89
submerged arc welds ... **M6:** 134
Horn
definition ... **M6:** 9
Horn presses
for piercing ... **A14:** 463
Horn spacing
definition ... **M6:** 9
Horns, metal
microwave inspection ... **A17:** 208-209
Horseshoe thrust bearing
defined ... **A18:** 10
Hose-reinforcing wire ... **A1:** 286
Hoskins 875
as HIP furnace element ... **A7:** 422-423
Hot air knives *See also* Air knives; Hot air solder leveling (HASL)
Hot air nozzles
for reworking processes ... **EL1:** 117
Hot air solder leveling (HASL) *See also* Air knives
rigid printed wiring boards ... **EL1:** 550
solder masking ... **EL1:** 554
as solderability preservation ... **EL1:** 563
Hot ash-handling systems ... **A13:** 1007-1008
Hot bending *See also* Bending; Tube bending
dry sand filler ... **A14:** 670
temperatures and procedures, steel tubes ... **A14:** 670
thinning of outer wall ... **A14:** 670
of tube ... **A14:** 669-671
vs cold bending ... **A14:** 671
Hot blast
furnace, fuels for ... **A15:** 30
introduction of ... **A15:** 30
Hot box failures
of locomotive axles ... **A11:** 715
Hot box process
defined ... **A15:** 6
Hot box resin binder processes ... **A15:** 218
Hot brine
pipe fracture from ... **A11:** 432
Hot carriers
semiconductor chip ... **EL1:** 965
Hot cathode gun *See* Thermionic cathode gun
Hot caustic washing
as cleaning method ... **A7:** 459
Hot cell SCC
irradiated stainless steels ... **A13:** 936
Hot chamber machine *See also* Cold chamber machine; Die casting; Gooseneck; Hot chamber process
defined ... **A15:** 6
die casting ... **A15:** 35
Hot chamber process
as die casting method ... **A15:** 286
gating system ... **A15:** 290
schematic ... **A15:** 287
Hot clamp
for component removal ... **EL1:** 718
Hot components
gas-turbine ... **A11:** 283-284
Hot compression testing *See also* Compression tests ... **A8:** 581-584
for forging process design ... **A14:** 439
Hot consolidation ... **A7:** 501-521
contamination during ... **A7:** 180
hot pressing as ... **A7:** 501
processes ... **A7:** 309
Hot corrosion *See also* Gaseous corrosion; Sulfidation
aircraft powerplants ... **A13:** 1038-1040
of coal- and oil-fired boilers ... **A13:** 995-996
defined ... **A13:** 7, 100
forms ... **A13:** 100-101
of heat-resisting alloys ... **A11:** 200
of nickel-base alloys ... **A11:** 269
P/M superalloys ... **A13:** 836
superalloys ... **A12:** 391
test for resistance, by sulfidation-oxidation ... **A11:** 280
testing, equipment for ... **A11:** 280
of tungsten-reinforced composites ... **EM1:** 882-883
turbine blade failure by ... **A11:** 269, 285-287
of turbine blades and vanes ... **A13:** 1000-1001
Hot corrosion, superalloys
oxidation protective coating against ... **M5:** 376-377
Hot crack *See also* Hot cracking; Hot tear; Hot tearing
defined ... **A9:** 9 **A15:** 6-7

Hot cracking *See also* Cold cracking; Lamellar tearing; Stress-relief cracking **A1:** 608 **A6:** 53, 54
aluminum alloy **A15:** 768-769
in aluminum alloys, effect of grain refinement on **A9:** 629-631
in arc welds **A11:** 413
in arc-welded low-carbon steel **A11:** 416
as casting defect **A11:** 383 **A15:** 548
constitutional supercooling and **A6:** 48, 49
copper alloys **A6:** 754
defined **A12:** 18 **A13:** 7-8
definition **A6:** 1210
in die casting **A15:** 294
dissimilar metal joining **A6:** 822, 827
electron beam weld **A12:** 157
in electron beam welds **A11:** 446
in electroslag welds **A11:** 440
examination/interpretation **A12:** 138
ferritic stainless steels **A6:** 454
in flash welds **A11:** 443
HAZ, failures related to **A11:** 401
in iron castings **A11:** 353
nickel alloys **A6:** 749
nickel-base alloys **A6:** 588, 589-590
nickel-chromium alloys **A6:** 588, 589
nickel-chromium-iron alloys **A6:** 588, 589
phosphor bronzes **A6:** 764
stainless steel dissimilar welds **A6:** 504
in stainless steels **A6:** 677, 693, 695, 696, 699 **A12:** 123
susceptiability of leaded brasses to **A11:** 450
in weld beads **A9:** 578
in weld metal **A11:** 440
Hot cyanide bath gold plating **M5:** 281-283
alloying metals in **M5:** 282-283
barrel plating **M5:** 282
pH, plating solution **M5:** 281-282
solution composition and operating conditions **M5:** 281-283
Hot deformation
wrought titanium alloys **A2:** 612
Hot deformation, effect of
on notch toughness **A1:** 741, 742-744
Hot densification **A7:** 6
Hot die forging
superalloys **A7:** 523
Hot dip aluminum coating of steel **M5:** 333-339
batch process **M5:** 334, 337-339
continuous process **M5:** 335-337
limitations **M5:** 338-339
Hot dip coating of cast iron and steel **M5:** 351-355
abrasive blasting process **M5:** 353
applications **M5:** 351
cast irons for **M5:** 351-352
cleaning processes **M5:** 352-353
degreasing **M5:** 352
equipment and materials **M5:** 354
fluxing **M5:** 353-354
ingot tin quality **M5:** 354
pickling process **M5:** 352-353
safety precautions **M5:** 355
selective **M5:** 354-355
single-pot **M5:** 353
steels for **M5:** 351
two-pot **M5:** 353-354
wipe process **M5:** 354
Hot dip coatings **A13:** 432-445
aluminum-coated (aluminized) steel **A13:** 434-435
aluminum-zinc alloy coated steel **A13:** 435-436
for carbon steel **A13:** 522-523
cast iron **M1:** 103
for castings **A15:** 562
continuous **A13:** 432-436
corrosion protection **M1:** 752
defined **A13:** 8
galvanizing specifications **A13:** 444
hot dip galvanizing, batch process **A13:** 436-444
for marine corrosion **A13:** 910-911
Hot dip coatings (continued)
of steel **A13:** 432-445
zinc-coated (galvanized) steel **A13:** 432-434
Hot dip galvanized coating **M5:** 323-332
abrasive blasting process **M5:** 326, 329, 332
alkaline cleaning process **M5:** 325
alloying elements **M5:** 324, 330
aluminum, use as alloying element in **M5:** 324
appearance, high-temperature coatings **M5:** 329
applications **M5:** 323-324
batch process **M5:** 326-331
equipment (kettle) **M5:** 330-331
high-temperature operation **M5:** 329-330
procedures, general **M5:** 326-328
cooling **M5:** 328-329
corrosion products produced by **M5:** 331-332
corrosion protection by **M5:** 323-324, 331-332
dipping process **M5:** 327-328
dry process **M5:** 328
equipment **M5:** 330-331
fluxing **M5:** 326-327
high-temperature process **M5:** 329-330
hydrogen embrittlement caused by **M5:** 325
immersion time **M5:** 327, 329
coating thickness controlled by **M5:** 327
iron **M5:** 323-332
iron content, effect of **M5:** 324, 330
iron-zinc layer formed characteristics of **M5:** 323-329
kettle **M5:** 330-331
lead, use as alloying element in **M5:** 324
mechanical galvanizing compared to **M5:** 300-301
metallographic structure
high-temperature coatings **M5:** 329
metallurgical characteristics of coatings **M5:** 324
painting over galvanizing **M5:** 331-332
cleaning and surface preparation **M5:** 332
paint choice **M5:** 332
pickling process **M5:** 325-326, 328
post treatments **M5:** 331-332
precleaning processes **M5:** 325-326, 328
solution composition and operating conditions **M5:** 324, 328-330
steel **M5:** 323-332
silicon steels **M5:** 324, 327-330
substrate metal **M5:** 324-325
controls **M5:** 331
effects on **M5:** 325
high-temperature process **M5:** 329-330
temperature **M5:** 326-327, 329-331
thickness, coating **M5:** 324, 327, 330
immersion time controlling **M5:** 327
iron content affecting **M5:** 330
measurement of **M5:** 324
wet process **M5:** 326-328
wet storage stain inhibitors **M5:** 331
withdrawal rate **M5:** 324, 327-328
thickness determined by **M5:** 327-328
zinc quality **M5:** 324
Hot dip galvanized steel
acrylic properties **EM3:** 122
Hot dip galvanized steels
automotive industry **A13:** 1012
Hot dip galvanizing **A1:** 212, 216-217, 218 **A13:** 436-444
after-fabrication **A2:** 527-528
applications **A13:** 443-444
coating weight/thickness **A13:** 437
conventional **A2:** 572
galvanizing **A13:** 437
mechanical properties **A13:** 438
painting galvanized steel **A13:** 442-443
steel substrate **A13:** 438-439
structure **A13:** 437-438
surface preparation **A13:** 436
welding for **A13:** 439
zinc/zinc alloys and coatings **A13:** 765-766
Hot dip galvanizing, steel sheet **M1:** 169-171
ASTM specifications **M1:** 170
band-test requirements **M1:** 170, 171
designations **M1:** 168
layers, identification of **M1:** 169
mechanical properties **M1:** 170-171
silicon, effect of **M1:** 170
spangle **M1:** 170
structure of coating **M1:** 169-170
Hot dip lead alloy coating of steel *See also* Terne coatings **M5:** 358-360
Hot dip oxidation-resistant coating **M5:** 665-666
Hot dip tin coatings
steel sheet **M1:** 173
Hot dipped products
spangles formation **A8:** 548
Hot drawing *See also* Drawing
of aluminum alloys **A14:** 797
of titanium alloys **A14:** 844-845
Hot ductility testing **A6:** 91
Hot ductility tests
on carbon-manganese steels **A1:** 696
Hot dynamic degassing **A7:** 180, 181
Hot equipment
as thermal inspection application **A17:** 402
Hot etching
defined **A9:** 9
of tools and dies **A11:** 563
Hot extraction
capabilities **A10:** 226
Hot extrusion **A7:** 515-521 **A14:** 315-326
backward **A14:** 316
CAD/CAM applications **A14:** 323-326
of cermet billets **A2:** 987-988
conventional **A14:** 315-326
forward **A14:** 315-316
of high-speed steels **A7:** 524-525
and hot upset forging **A14:** 95
hydrostatic **A14:** 328
lubricated **A14:** 316
materials for **A14:** 321-322
metal flow in **A14:** 316-317
methods **A7:** 517
nonlubricated **A14:** 315-316
operating parameters **A14:** 322-323
presses for **A14:** 319-320
schematic **A14:** 315
shapes, characterized **A14:** 322
speeds and temperatures **A14:** 317-319
for steel tubular products **A1:** 328 **M1:** 316
tool steels **A18:** 738
tooling **A14:** 320-321
Hot extrusion forgings **M1:** 373-375
Hot extrusion forgings, design of **A1:** 355, 356
machining allowance **A1:** 357
mechanical properties **A1:** 356, 357
mismatch tolerances **A1:** 356-357
Hot extrusion tools
breakage **M3:** 539-540
containers **M3:** 537-538, 539
die assemblies **M3:** 538, 539
die life **M3:** 539-540
dummy blocks **M3:** 538, 539
hardness **M3:** 538, 540
liners **M3:** 538
mandrels **M3:** 538-539
materials for **M3:** 537-541
rams **M3:** 537-538, 539
Hot finishing temperature
and impact strength **A14:** 220
Hot forge rolling *See* Hot forging; Roll forging
by HERF processing **A14:** 104
dies and die materials for **A14:** 43-58
nonisothermal, deformation patterns **A14:** 375
in presses, process/machine variables **A14:** 36
tool and die materials for **A14:** 43
Hot forging *See also* Extrusion; Forging **A7:** 6
CAP billet stock for **A7:** 533
pressure densification **EM4:** 300-301

SUBJECTS OF THE INDEXED VOLUMES: ASM Handbook (designated by the letter "A"): **A1:** Properties and Selection: Irons, Steels, and High-Performance Alloys (1990); **A2:** Properties and Selection: Nonferrous Alloys and Special-Purpose Materials (1990); **A3:** Alloy Phase Diagrams (1992); **A4:** Heat Treating (1991); **A6:** Welding, Brazing, and Soldering (1993); **A7:** Powder Metallurgy (1984); **A8:** Mechanical Testing (1985); **A9:** Metallography and Microstructures (1985); **A10:** Materials Characterization (1986); **A11:** Failure Analysis and Prevention (1986); **A12:** Fractography (1987); **A13:** Corrosion (1987); **A14:** Forming and Forging (1988); **A15:** Casting (1988); **A16:** Machining (1989); **A17:** Nondestructive Testing and Quality Control (1989); **A18:** Friction, Lubrication, and Wear Technology (1992). **Metals Handbook, 9th Edition** (designated by the letter "M"): **M1:** Properties and Selection: Irons and Steels (1978); **M2:** Properties and Selection: Nonferrous Alloys and Pure Metals (1979); **M3:** Properties and Selection: Stainless Steels, Tool Materials and Special-Purpose Materials (1980); **M4:** Heat Treating (1981); **M5:** Surface Cleaning, Finishing, and Coating (1982); **M6:** Welding, Brazing, and Soldering (1983). **Engineered Materials Handbook** (designated by the letters "EM"): **EM1:** Composites (1987); **EM2:** Engineering Plastics (1988); **EM3:** Adhesives and Sealants (1990); **EM4:** Ceramics and Glasses (1991); **Electronic Materials Handbook** (designated by the letters "EL"): **EL1:** Packaging (1989).

Hot forging (continued)
refractory metals and alloys ... **A2:** 560
thermal contraction and dimensional change during ... **A7:** 480
tool steels ... **A18:** 738
Hot forging tools, closed-die *See* Closed-die forging tools, materials for
Hot forming *See also* Hot working ... **M1:** 339-346, 419
of automobile parts ... **A7:** 618, 620
and cold forming, combined ... **A14:** 620
ductile and brittle fractures from ... **A11:** 88
formability problems ... **A14:** 620
of magnesium alloys ... **A14:** 825-826
out-of-roundness ... **A14:** 622
of P/M forged low-alloy steels ... **A7:** 464
and press forming ... **A14:** 554
pressures, nickel-base alloys ... **A14:** 262
temperatures, for steel ... **A14:** 620
of titanium alloys ... **A14:** 841-842
vs cold forming, three-roll forming ... **A14:** 619-620
wrought titanium alloys ... **A2:** 615-616
Hot gas corrosion
resistance ... **A15:** 730
Hot gas soldering ... **A6:** 361-362
applications ... **A6:** 361
atmospheres ... **A6:** 361
characteristics ... **A6:** 361
definition ... **A6:** 361
parameters ... **A6:** 361-362
preheating methods for assemblies ... **A6:** 362
processing concerns ... **A6:** 362
reliability concerns ... **A6:** 362
Hot hardness *See also* Hardness ... **A18:** 614, 615, 616
of CPM alloys ... **A7:** 789 **A16:** 65
dynamic, vs temperature, as forgeability ... **A14:** 226
of hot-work tool steels ... **A14:** 46
of iron-based microcrystalline alloys ... **A7:** 796
tungsten carbide / cobalt ... **A7:** 775
Hot heading *See* Hot upset forging
bolts ... **M1:** 274, 275, 280
Hot heated manifold mold
defined ... **EM2:** 22
Hot hydrostatic extrusion ... **A14:** 328
Hot isostatic pressing
after investment casting ... **A15:** 263
of aluminum alloys ... **A15:** 541
casting distortion ... **A15:** 542
and casting mechanical properties ... **A15:** 539-541
for casting refurbishment ... **A15:** 544
as casting repair ... **A15:** 264
for casting replacement ... **A15:** 543-544
casting salvage ... **A15:** 543
of castings ... **A15:** 538-544
cemented carbides ... **M3:** 452
defined ... **A14:** 8
equipment ... **A15:** 539
gas purity ... **A15:** 542
new applications ... **A15:** 543
of nickel-base superalloys ... **A15:** 539
parameter selection ... **A15:** 541-542
post, heat treatment ... **A15:** 543
of prealloyed P/M aluminum alloys ... **A14:** 250-251
of stainless steels ... **A15:** 541
surface defects ... **A15:** 542
of titanium ... **A15:** 539-540
titanium alloys ... **A15:** 832
zirconium alloys ... **A15:** 838
Hot isostatic pressing (HIP) *See also* Cold isostatic pressing; Pressure-impregnation-carbonization (PIC) ... **A1:** 973, 993-994 **A6:** 883 **A7:** 6, 18, 295, 419-443 **A16:** 60 **EM4:** 124, 147, 194-200, 296
advanced ceramics ... **EM4:** 49
aluminum and aluminum alloys ... **A2:** 7
aluminum casting alloys ... **A2:** 141
aluminum P/M alloys ... **A2:** 203
aluminum-oxide based ceramic cutting tools ... **EM4:** 966
applications ... **A7:** 419, 437-443 **EM4:** 196, 199-200
cladless HIP ... **EM4:** 199
encapsulated hot isostatic pressing ... **EM4:** 199-200
for attainment of high density in grain growth ... **EM4:** 309
beryllium ... **A16:** 870
Hot isostatic pressing (HIP) (continued)
beryllium powder ... **A2:** 685-686 **A7:** 172
of blended elemental Ti-6Al-4V ... **A2:** 649
boron carbide ... **EM4:** 805
brazing for joining non-oxide ceramics ... **EM4:** 528
cemented carbides ... **A16:** 72 **A18:** 796
ceramics ... **A16:** 98, 101
as cermet forming technique ... **A7:** 800
of cermets ... **A2:** 986-987
cladless ... **EM4:** 194, 196-197, 199
applications ... **EM4:** 199
of composite materials ... **A7:** 442-443
of containerized powder ... **A7:** 522
containerless ... **A7:** 436
controls and instrumentation ... **A7:** 423
cycles ... **A7:** 436-438
defined ... **EM1:** 13
direct-HIP ... **EM4:** 197
duplex ... **A7:** 441
elevated temperatures ... **A7:** 419
encapsulation ... **EM4:** 194, 197-198, 199
encapsulation techniques ... **A7:** 428-435
equipment ... **A7:** 419-424 **EM4:** 194-195
convection heat losses ... **EM4:** 194-195
heating elements ... **EM4:** 195
pressure-vessel furnace construction ... **EM4:** 194-195
pressure-vessel safety guidelines ... **EM4:** 195
temperature measurement devices ... **EM4:** 195
workpiece fixtures ... **EM4:** 195
flowchart for manufacturing ... **A7:** 424-425
of forged products ... **A7:** 647-648
formation of Si_3N_4 balls and high-vacuum applications of lubricants ... **A18:** 158
fractional volume shrinkage ... **A7:** 426
gas purity ... **A7:** 422
and gas-pressure sintering ... **EM4:** 199
heat treatments and ... **A7:** 440
high-speed tool steels by ... **A7:** 144
linear dimensional shrinkage in ... **A7:** 426-427
and liquid-phase sintering ... **EM4:** 288
of MERL ... **A7:** 76, 440
minicomputer, control of ... **A7:** 424
metal-oxide composites ... **EM4:** 966
micrograin high-speed steels ... **A18:** 616
for near-net shapes ... **A7:** 437
particle-type encapsulation methods ... **EM4:** 197, 198, 199
plus conventional forging ... **A7:** 442
plus forging, superalloys ... **A7:** 522-523
plus hot rolling to mill shape, high-speed steels ... **A7:** 524-525
plus isothermal and hot die forging superalloys ... **A7:** 523
powder properties ... **A7:** 425-428
pressure densification ... **EM4:** 299, 300
advantages ... **EM4:** 242
conditions ... **EM4:** 298
parameters ... **EM4:** 298
pressure ... **EM4:** 301
temperature ... **EM4:** 301
pressure medium ... **A7:** 420-422
process parameter selection ... **A7:** 437
processing sequence ... **A7:** 424
quality control ... **EM4:** 199
and rapid omnidirectional compaction low-carbon Astroloy ... **A7:** 545
and rigid tool compaction ... **A7:** 323
of sample bars fractured and analyzed by CARES program and IEA Annex II agreement members ... **EM4:** 704-705
self-lubricating powder metallurgy composites ... **A18:** 120
silicon carbide-tetanin composite ... **EM1:** 862
silicon nitride ... **EM4:** 812, 815
silicon-nitride ceramic cutting tools ... **EM4:** 966
sinter-HIP ... **EM4:** 194, 196-197, 198, 199
structural ceramics ... **A2:** 1020
technology applied to ceramics ... **EM4:** 195-199
cladless hot isostatic pressing ... **EM4:** 194, 196-197, 199
encapsulation conforming to body configuration ... **EM4:** 197-198, 199
filling into a shaped capsule ... **EM4:** 195-196
high-precision processing ... **EM4:** 198-199
removal of adsorbed substances ... **EM4:** 198
Hot isostatic pressing (HIP) (continued)
special processing options ... **EM4:** 199
temperature selection ... **EM4:** 198
temperature guidelines ... **A7:** 437
titanium and titanium alloy castings ... **A2:** 643
of titanium powders ... **A7:** 164
of titanium-based alloys ... **A7:** 438-439
to densify ceramic coatings in molten particle deposition ... **EM4:** 207, 208
tooling used ... **A7:** 423
TTZ ... **EM4:** 512
unit ... **A7:** 419, 421
whisker-reinforced alumina cutting tools ... **EM4:** 966-967
workpiece thermocouple installation ... **A7:** 438
Hot isostatic pressing unit ... **A7:** 423
Hot isostatic pressure system ... **A7:** 420
Hot isostatic pressure vessels ... **A7:** 419-421
closure ... **A7:** 420
loading of ... **A7:** 423
Hot isostatic pressure welding
definition ... **M6:** 9
Hot isostatically pressed silicon nitride (HIPSN) ... **EM4:** 813-814
additives ... **EM4:** 814
formation ... **EM4:** 813-814
hot isostatic pressing ... **EM4:** 814
properties ... **EM4:** 815
sintering ... **EM4:** 814
strength ... **EM4:** 817
Hot melt adhesives ... **EM1:** 13, 684, 687
Hot melt process
for woven fabric prepregs ... **EM1:** 149
Hot metal desulfurization ... **A1:** 108, 109
Hot mill finishing temperature
effects of ... **A1:** 588, 590
Hot molding
SiC aluminum composite stiffeners ... **EM1:** 861
Hot oil fountains
for reworking processes ... **EL1:** 117
Hot platen soldering
as surface-mount soldering ... **EL1:** 706
Hot press bonding *See* Diffusion welding
Hot press setups ... **A7:** 504-509
Hot pressed silicon carbide (HPSC) ... **EM4:** 191, 192
key features ... **EM4:** 676
properties ... **EM4:** 330, 677
Hot pressing *See also* Hot isostatic pressing; Uniaxial hot pressing ... **A7:** 6, 501-521 **EM4:** 124, 126, 127, 186-192
advanced ceramics ... **EM4:** 49
advantages ... **EM4:** 186
applications ... **EM4:** 190-192
ceramic armor ... **EM4:** 191
ceramic bearings ... **EM4:** 191
cutting tools ... **EM4:** 191
electro-optic materials ... **EM4:** 191
ferrites ... **EM4:** 1192
heat engine components ... **EM4:** 191
microelectric ceramic packages ... **EM4:** 192
microwave absorbers ... **EM4:** 191
nuclear industry components ... **EM4:** 191
optical windows ... **EM4:** 189-190
properties ... **EM4:** 191
resistors ... **EM4:** 192
sputtering targets ... **EM4:** 191
titanates ... **EM4:** 192
tooling ... **EM4:** 191
traditionally hot-pressed ceramics ... **EM4:** 190
varistors ... **EM4:** 191
for attainment of high density in grain growth ... **EM4:** 309
boron carbide ... **EM4:** 805
ceramic materials ... **EM4:** 36
of ceramic matrix composites ... **EM4:** 224
ceramics ... **A16:** 98, 99, 100
cermets ... **A7:** 800
of clad metals ... **A13:** 887
disadvantages ... **EM4:** 186
equipment ... **A7:** 502
for gas turbine component fabrication ... **EM4:** 718
history ... **A7:** 15
in joining non-oxide ceramics ... **EM4:** 528
and liquid-phase sintering ... **EM4:** 288
pressure densification ... **EM4:** 297-299
and pressure sintering ... **A7:** 501
products ... **A7:** 509-515

Hot pressing (continued)
properties of electric spark-activated **A7:** 513
requirements **A7:** 502
silicon nitride **EM4:** 812
silicon-nitride cutting tools **EM4:** 966
of structural ceramics **A2:** 1020
to high density **A7:** 522
to solid-state join a silicon-base ceramic to itself **EM4:** 527
TTZ **EM4:** 512
tungsten alloys **A2:** 579
uniaxial **A7:** 309
vacuum **A7:** 507, 509, 517
whisker-reinforced alumina cutting tools **EM4:** 966-967
Hot pressure welding
definition **M6:** 9
Hot pressure welding (HPW)
definition **A6:** 1210
niobium alloys **A6:** 581
Hot processing
ternary molybdenum chalcogenides (chevrel phases) **A2:** 1077-1079
Hot quenching
defined **A9:** 9
Hot re-pressing *See also* Re-pressing
as powder forging **A14:** 188
thermal contraction and dimensional change during **A7:** 480
Hot roll bonding
of clad metals **A13:** 887
Hot roll bonding, clad brazing materials
brazing with **A6:** 347, 348
Hot rolled bars **M1:** 199-212
aircraft quality **M1:** 209
AMS specifications **M1:** 209
applications **M1:** 206-209, 211-212
ASTM specifications **M1:** 204, 205, 207-212
bend test requirements **M1:** 205, 206, 212
coatings for **M1:** 200
concrete-reinforcing bars **M1:** 211-212
decarburization **M1:** 200
dimensions **M1:** 199, 211
grades **M1:** 203, 205, 212
heat treatment **M1:** 200-201
HSLA steel bars **M1:** 210, 211
magnetic-particle testing **M1:** 200, 201, 209
mechanical properties **M1:** 201-209, 211, 212
merchant quality bars **M1:** 203-205
product categories **M1:** 202-212
quality descriptions **M1:** 202-212
sizes **M1:** 203-206, 212
special quality bars **M1:** 205-206
special shapes **M1:** 213
straightening **M1:** 203
structural quality **M1:** 204, 209-212
surface imperfections **M1:** 199-200
surface treatment **M1:** 200
tolerances **M1:** 199
Hot rolled low-carbon steel
fatigue properties compared with HSLA steels **M1:** 418-419
Hot rolled sheet and strip
ASTM specifications for **M1:** 154, 155
characteristics **M1:** 154
flatness **M1:** 157, 160-161
leveling **M1:** 157, 160-161
mechanical properties **M1:** 155-156, 158-159
modified low-carbon steels **M1:** 161-162
Olsen ductility **M1:** 156, 162
production of **M1:** 153-154
quality descriptors **M1:** 154-155
standard sizes **M1:** 154
strain aging **M1:** 154, 157, 162
stretcher strains **M1:** 157
surface characteristics **M1:** 157
thickness **M1:** 153, 154, 159
width range **M1:** 153, 154
Hot rolled steel
porcelain enameling of **M1:** 178 **M5:** 513-514
Hot rolled steel sheet
characteristics affecting formability **M1:** 556
mechanical properties related to formability **M1:** 547-549
minimum bend radii, selected grades **M1:** 554
Hot rolling **A1:** 115-120
CAP billet stock for **A7:** 533
controlled rolling **A1:** 115, 117-118, 130-131, 408-409
of controlled spray deposition processed shapes **A7:** 532
critical temperatures **A1:** 115-116, 126-127, 130
of direct-chill semicontinuous-cast copper and copper alloy slabs **A2:** 243
effect on notch toughness of steels **M1:** 695, 696, 699
effect on solidification structures in steel **A9:** 626-628
hafnium **A2:** 663
P/M high-speed tool steels **A16:** 62
products **A14:** 343
zirconium **A2:** 663
Hot salt environment
SCC testing of nickel alloys in **A8:** 530-531
Hot salts
stress-corrosion cracking, aircraft **A13:** 1039-1040
titanium/titanium alloy SCC **A13:** 273-274, 688-689
Hot shearing *See also* Shearing
blade materials for **A14:** 716
forgings **M1:** 367-368
for stock preparation, hot upset forging **A14:** 86
tooling set-up for **A14:** 86
Hot short .... **A3:** 1 • 19
Hot shortness **A12:** 2, 126-127
copper casting alloys **A2:** 350
cracks, in friction welds **A11:** 444
cracks, in gas metal arc weld of aluminum alloy **A11:** 435
defined **A13:** 8 **A15:** 7
effect on strength and ductility **A8:** 35
failure, of heat-exchanger shell **A11:** 640
fracture mode **A8:** 574
from sulfur **A14:** 779
sulfur effects **A15:** 29
in thermocouples **A2:** 882
Hot sizing
titanium alloys **A14:** 842
Hot solder dipping *See also* Tinning
as solderability surface preparation **EL1:** 79
Hot spinning
of hemispheres **A14:** 604
Hot spots
in integrated circuits **A11:** 767-768
Hot spraying
paint **M5:** 477-478
Hot stamping
defined **EM2:** 22
Hot start current
definition **M6:** 9
Hot static degassing **A7:** 180, 181
Hot swaging, use
materials and equipment for **A14:** 142-143
Hot tear *See also* Hot crack; Hot cracking; Hot tearing
defined **A11:** 5 **A15:** 7
fatigue fracture from **A11:** 388
as forging defect **A11:** 317
fracture at **A11:** 388, 397
in iron castings **A11:** 353
in semisolid alloys **A15:** 337
Hot tearing *See also* Hot tear; Tearing
aluminum alloy **A15:** 617
copper alloy castings, inspection of **A15:** 558
grain size effect **A15:** 160
resistance, aluminum-silicon alloys **A15:** 159
Hot tearing (continued)
in squeeze casting **A15:** 326
Hot tears
in aluminum alloy castings **A17:** 535
in austenitic manganese steel castings causes of **A9:** 238-239
by liquid penetrant inspection **A17:** 86
in copper alloy ingots **A9:** 642
Hot tensile testing
for forgeability **A14:** 215
Hot tension testing **A8:** 586-587 **A14:** 378
Hot tinning *See* Tin coatings
Hot top
defined **A15:** 7
Hot topping
copper alloys **A15:** 780-781
Hot torsion
machine, servo-controlled **A8:** 159
specimen, macrographs of twist **A8:** 174
Hot torsion tests
for ductility **A14:** 375
Hot trepanning *See also* Trepanning
in open-die forging **A14:** 63
Hot trimming *See also* Trimming
in closed-die forging **A14:** 82
copper and copper alloy forgings **A14:** 257
crankshaft fatigue fracture from **A11:** 472
defined **A14:** 8
of stainless steel forgings **A14:** 230
Hot twist testing **A1:** 583, 584
Hot uniaxial pressing
ceramics **A16:** 101
Hot upset forging *See also* Forging; Upsetting **A14:** 83-95
applicability **A14:** 83
defined **A14:** 8, 83
descaling **A14:** 87
die cooling **A14:** 87
die lubrication **A14:** 87
double-end upsetting **A14:** 90
heating **A14:** 87
and hot extrusion **A14:** 95
hot upsetting vs alternative processes **A14:** 95
inserts for **A14:** 48
machines **A14:** 83-85
metal saving techniques **A14:** 86-87
offset upsetting **A14:** 90
safety in **A14:** 95
simple upsetting **A14:** 87-88
stock, preparation of **A14:** 86
tolerances, assignment of **A14:** 93-94
tolerances, effect on cost **A14:** 94-95
tools **A14:** 85-86
upsetting and piercing **A14:** 88-89
upsetting pipe and tubing **A14:** 91-93
upsetting with sliding dies **A14:** 90-91
vs alternative processes **A14:** 95
vs cold extrusion **A14:** 95
vs cold heading **A14:** 95
Hot upset forging tools
auxiliary tools **M3:** 535-536
gripper dies **M3:** 533, 534-535
header dies **M3:** 534-535
lubrication, effect of **M3:** 535
materials for **M3:** 533-536
trimming **M3:** 535-536
wear resistance **M3:** 535, 536
Hot upset forgings **M1:** 369-373
Hot upset preforms
by extrusion **A14:** 306
Hot upset testing
of nonlubricated cylinder **A8:** 582
Hot upsetting *See also* Hot upset forging
forgeability, of stainless steel **A14:** 223
as powder forging **A14:** 188
Hot water *See also* Elevated temperatures; Temperature(s)
resistance, porcelain enamels **A13:** 451-452
zinc corrosion in **A13:** 761

SUBJECTS OF THE INDEXED VOLUMES: ASM Handbook (designated by the letter "A"): **A1:** Properties and Selection: Irons, Steels, and High-Performance Alloys (1990); **A2:** Properties and Selection: Nonferrous Alloys and Special-Purpose Materials (1990); **A3:** Alloy Phase Diagrams (1992); **A4:** Heat Treating (1991); **A6:** Welding, Brazing, and Soldering (1993); **A7:** Powder Metallurgy (1984); **A8:** Mechanical Testing (1985); **A9:** Metallography and Microstructures (1985); **A10:** Materials Characterization (1986); **A11:** Failure Analysis and Prevention (1986); **A12:** Fractography (1987); **A13:** Corrosion (1987); **A14:** Forming and Forging (1988); **A15:** Casting (1988); **A16:** Machining (1989); **A17:** Nondestructive Testing and Quality Control (1989); **A18:** Friction, Lubrication, and Wear Technology (1992). **Metals Handbook, 9th Edition** (designated by the letter "M"): **M1:** Properties and Selection: Irons and Steels (1978); **M2:** Properties and Selection: Nonferrous Alloys and Pure Metals (1979); **M3:** Properties and Selection: Stainless Steels, Tool Materials and Special-Purpose Materials (1980); **M4:** Heat Treating (1981); **M5:** Surface Cleaning, Finishing, and Coating (1982); **M6:** Welding, Brazing, and Soldering (1983). **Engineered Materials Handbook** (designated by the letters "EM"): **EM1:** Composites (1987); **EM2:** Engineering Plastics (1988); **EM3:** Adhesives and Sealants (1990); **EM4:** Ceramics and Glasses (1991); **Electronic Materials Handbook** (designated by the letters "EL"): **EL1:** Packaging (1989).

Hot water rinsing
of cold extruded part **A14:** 304
Hot wire test
quenching media **M4:** 37-38
Hot wire welding
definition **M6:** 9
titanium welding **A6:** 786
Hot work **A9:** 684
Hot work die steels
stress-corrosion cracking environments **A8:** 526
Hot work tool steels *See* Tool Steels, hot work
Hot workability ratings
for specialty steels and superalloys **A8:** 586
Hot worked structure
defined **A9:** 9
Hot working *See also* Cold working; Hot-working temperature; Warm working **EM3:** 15
beryllium-copper alloys **A2:** 421-423
of CAP material **A7:** 535
and consolidation of aluminum powders **A7:** 526-527
of controlled spray deposition material **A7:** 532
copper and copper alloys **A2:** 223
copper metals **M2:** 241
damage from sectioning **A9:** 23
defined **A9:** 9 **A13:** 8 **A14:** 8 **EM1:** 13 **EM2:** 22
discontinuous ceramic fiber MMCs **EM1:** 908-909
effect of copper **A11:** 722
effect on tool steels **A9:** 258
effects, alloy segregation **A11:** 121
forgings produced by **A11:** 314-343
hardenability affected by **M1:** 478, 480
of heat-resistant alloys **A14:** 232
of high-temperature alloys **A14:** 224
of Invar **A2:** 890
of iridium **A14:** 851
of maraging steels **A1:** 795 **M1:** 447
of palladium **A14:** 850
of platinum **A14:** 849
platinum alloys **A14:** 851
plus sintering to high density **A7:** 525
processes, beryllium-copper alloys **A2:** 415
reduction, effect on impact strength **A14:** 218
of rhodium **A14:** 850
temperature, palladium **A2:** 716
tool and die failures in **A11:** 574
of tool steels **A7:** 467
Hot working temperature **A8:** 573-574
Hot wound springs **M1:** 297, 301-303
hot rolled bars, use of **M1:** 290
steel grades for **M1:** 301
Hot-air heating
for optical holographic interferometry **A17:** 410
Hot-applied butyl sealants **EM3:** 190
for insulated glass industry **EM3:** 190
Hot-applied organisols **A13:** 400
Hot-brine test **A1:** 466, 467
Hot-cell metallograph **A9:** 83
Hot-cell microscopy **A9:** 83
Hot-corrosion properties
mechanically alloyed oxide dispersion-strengthened (MA ODS) alloys **A2:** 497
Hot-die forging **A14:** 150-157
advantages **A14:** 150-151
cost **A14:** 150, 155-157
defined **A14:** 150
die systems **A14:** 154-155
forging alloys **A14:** 152
forging design guidelines **A14:** 155
gas-fired infrared heating setup for **A14:** 152
of heat-resistant alloys **A14:** 232
induction heating system **A14:** 152
lubrication **A14:** 153-154
as new metalworking process **A14:** 17-18
process description **A14:** 151-152
process design **A14:** 153-154
process selection **A14:** 152-153
production forging **A14:** 157
vs conventional forging **A14:** 156
wrought titanium alloys **A2:** 613-614
Hot-dip aluminum-coated sheet steel
specimen preparation **A9:** 197
Hot-dip aluminum-coated sheet steel, specific types
1008, with type 1 coating **A9:** 200
1008, with type 2 coating **A9:** 200
Hot-dip galvanized sheet steel
specimen preparation **A9:** 197
Hot-dip galvanized sheet steel, specific types
1006 .... **A9:** 199
1008 .... **A9:** 199
Hot-dip galvanizing
effect on press forming **A14:** 560
tin additions **A9:** 490
Hot-dip tin coatings
etching **A9:** 451
Hot-dip tinplate
steel-tin interface **A9:** 455
Hot-dip zinc-aluminum coated sheet steel
specimen preparation **A9:** 197
Hot-filament ionization gage
for gas/leak detection **A17:** 64
Hot-gas corrosion, resistance to
in heat-resistant alloys **A1:** 925, 928
Hot-gas reflow equipment
for rework processes **EL1:** 727-728
Hot-gas welding **EM3:** 15
defined **EM2:** 22
Hot-hollow cathode (HHC) evaporation
to apply interlayers for solid-state welding **A6:** 165, 168-169, 170, 171
Hot-machining tools
cermets as **A2:** 978
Hot-melt adhesive
defined **EM2:** 22
Hot-melt adhesives **EM3:** 15, 35, 74, 75
advantages and limitations **EM3:** 75
application methods **EM3:** 47, 80
for assembly bonding **EM3:** 82
for assistance bonding **EM3:** 82
automotive applications **EM3:** 46
for automotive bonding and sealing **EM3:** 609
for automotive carpet bonding **EM3:** 46
bonding applications **EM3:** 45, 81-82
bonding composites to composites **EM3:** 293
for book binding **EM3:** 82
characteristics **EM3:** 45, 74, 80
chemical families **EM3:** 82
chemistry **EM3:** 80
as consumer product **EM3:** 47
for containers **EM3:** 82
for diaper construction **EM3:** 47
diluents **EM3:** 80
EVA copolymer crystallinity **EM3:** 411
for fiberglass insulation bonding **EM3:** 45
foamable **EM3:** 82
foundry industry applications **EM3:** 47
for freezer coils bonding **EM3:** 45
induction periods in oxygen uptake **EM3:** 620
for industrial bonding **EM3:** 82
limitations **EM3:** 81
markets **EM3:** 81-82
nonpressure sensitive **EM3:** 81
for packaging **EM3:** 45, 82
for packaging large appliances in containers **EM3:** 45
"piggy-backing" of urethanes **EM3:** 568
polymers **EM3:** 80
predicted 1992 sales **EM3:** 81
pressure sensitive **EM3:** 81
properties **EM3:** 80-81
for release-coated papers **EM3:** 82
for sheet molding compound bonding **EM3:** 46
silane coupling agents **EM3:** 182
for sound-deadening material bonding **EM3:** 45
suppliers **EM3:** 82
tackifiers used **EM3:** 183
for woodworking **EM3:** 46
Hot-melt thermoplastics
for lighting subcomponent bonding **EM3:** 552
Hot-press molding
size and shape effects **EM2:** 291
Hot-pressed silicon nitride (HPSN) **EM4:** 191, 192
abrasive machining **EM4:** 318, 321, 322, 325, 326
applications **EM4:** 813
insulation for engines **EM4:** 990
creep-feed grinding **EM4:** 344
erosion resistance **A18:** 205
formation **EM4:** 813
grinding **EM4:** 334
key features **EM4:** 676
modulus of resilience **EM4:** 316, 330
proof testing **EM4:** 596
Hot-pressed silicon nitride (HPSN) (continued)
properties **EM4:** 326, 330, 677, 815, 816-817
reciprocating surface grinding **EM4:** 344
rolling contact fatigue life **A18:** 815
scanning acoustic microscopy **A18:** 408, 409
thermal cycling behavior **EM4:** 818
thermal diffusivity **EM4:** 816
thermal shock resistance **EM4:** 818
Hot-rolled carbon-manganese structural steels **A1:** 390, 391
Hot-rolled mild steel plate
die material for sheet metal forming **A18:** 628
Hot-rolled steel
effects of steelmaking practices on formability of **A1:** 577
Hot-rolled steel bars and shapes **A1:** 240-247
allowance for surface imperfections in machining applications **A1:** 241
alloy steel bars **A1:** 245-246
aircraft quality and magnaflux quality **A1:** 246
axle shaft quality **A1:** 246
ball and roller bearing quality and bearing quality **A1:** 246
cold-shearing quality **A1:** 246
cold-working quality **A1:** 246
regular quality **A1:** 246
structural quality **A1:** 246
carbon steel bars for specific applications **A1:** 245
axle shaft quality **A1:** 245
cold-shearing quality **A1:** 245
cold-working quality **A1:** 245
structural quality **A1:** 245
decarburization **A1:** 241
dimensions and tolerances **A1:** 240
heat treatment **A1:** 241
annealing for specified microstructures **A1:** 241
hardening by quenching **A1:** 241
normalizing **A1:** 241
ordinary annealing **A1:** 241
stress relieving **A1:** 241
tempering **A1:** 241
merchant quality bars **A1:** 243
grades **A1:** 243
sizes **A1:** 243
product categories **A1:** 242-243
product requirements **A1:** 241-242
special quality bars **A1:** 244-245
special shapes **A1:** 247
structural shapes **A1:** 247
surface imperfections **A1:** 240
laps **A1:** 240
seams **A1:** 240
slivers **A1:** 240-241
surface treatment **A1:** 241
Hot-runner mold *See also* Insulated-runner
defined **EM2:** 22
Hot-salt corrosion
in stainless steels **A11:** 200
Hot-setting adhesive **EM3:** 15
defined **EM1:** 13 **EM2:** 22
Hot-stage microscopy **A9:** 82
high-carbon steel quenched in **A9:** 83
phase contrast etching **A9:** 59
for temperature-time control in polymer removal techniques **EM4:** 137
thermal etching **A9:** 62
Hot-tensile properties
of ductile iron **A1:** 48, 52
Hot-twist testing
for forgeability **A14:** 215-216
Hot-wall plasma-assisted chemical vapor deposition **EM3:** 593, 594
Hot-wire analyzers
atmospheres **M4:** 429-430
Hot-work die steels **A18:** 636-637
for hot-forging dies **A18:** 626
materials for dies and molds **A18:** 622
Hot-work tool steels
heat checking **A18:** 630
materials for die-casting dies **A18:** 629
powder metallurgy **A1:** 789-790
wrought **A1:** 762-763
chromium **A1:** 762
molybdenum **A1:** 763

Hot-work tool steels (continued)
tungsten ... A1: 762-763
Hot-workability ratings
qualitative ... A14: 381
Hot-working temperature *See also* Fabrication characteristics
wrought aluminum and aluminum alloys ... A2: 88, 91-93, 96, 98, 99-100
Hot-working temperatures
effect on plastic deformation ... A9: 688-691
Hot-wound springs ... A1: 315
Hotel china ... EM4: 4
absorption ... EM4: 4
composition ... EM4: 5
physical properties ... EM4: 934
products ... EM4: 4
Hotelware
composition ... EM4: 45
properties of fired ware ... EM4: 45
Household appliance applications *See also* Electric applications; Electronic applications
aluminum and aluminum alloys ... A2: 13
Household applications
glass fibers ... EM4: 1029
"Housekeeper seal" ... EM3: 300-301
Housing and Urban Development, U.S. Department of
particleboard formaldehyde emission standard (1984) ... EM3: 105
Housings, eliminating cracks in ... A3: 1 • 28
Hovercraft
composite structures for ... EM1: 839
How to use the Handbook ... A10: 2-11
Hoya LE-30 aluminoborosilicate glass ... EM4: 1056
HP 9-4-20
composition ... M1: 422
properties and characteristics ... M1: 440-441
HP 9-4-30
composition ... M1: 422 M4: 120
heat treatment ... M1: 441
heat treatments ... M4: 129
mechanical properties ... M1: 441-442 M4: 128, 129
processing ... M1: 441
HP alloy ... A1: 926
HP-9-4-20
composition ... M4: 120
HP-9-4-30 steel ... A1: 431, 444-445
heat treatments for ... A1: 445-446
properties of ... A1: 444, 446
HPLC *See* High-performance liquid chromatography
HPZ fiber ... EM4: 223
composition ... EM4: 225
mechanical properties, room temperature ... EM4: 225
SEM photomicrograph ... EM4: 225
H_r *See* Magnetic resonance; Rockwell hardness
HRC hardness test ... A18: 436
HRD flame reactor process
for zinc recycling ... A2: 1225
HREM *See* High resolution electron microscopy
HS 21
composition ... M4: 653
HS 31
aging cycle ... M4: 657
composition ... M4: 653
solution treating ... M4: 657
HS 36
composition ... M4: 651-652
HS-21
electrochemical grinding ... A16: 547
grinding ... A16: 547
machining ... A16: 757, 758
milling ... A16: 547
HS-25
manufacturing ratings ... A16: 739
milling ... A16: 314
production time ... A16: 739
shaping ... A16: 192
HS-25, aging
effect on properties ... M4: 659
HS-31
grinding ... A16: 759, 760
surface alterations from material removal processes ... A16: 27
HS-188
flash welding ... M6: 557
HSc *See* Scleroscope hardness number
HSd *See* Scleroscope hardness number
HSLA *See* High-strength low-alloy
HSLA plate steels
fatigue behavior ... M1: 670, 672
HSLA steel plate ... M1: 183, 188
ASTM specifications ... M1: 183, 184, 188, 189
mechanical properties ... M1: 191, 193
HSLA steels
500 °F embrittlement of ... M1: 685
alloying elements ... M1: 410-411, 417
applications ... M1: 405-406
ASTM specifications ... M1: 403, 405-407
atmospheric corrosion of ... M1: 720-723
bars, hot rolled ... M1: 210, 211
bending requirements ... M1: 406, 408, 419
brittle fracture transition ... M1: 415, 417-418
carbon steel coupled to, galvanic effect in seawater ... M1: 740, 741
characteristics ... M1: 403-406, 408-410
classifications and designations ... M1: 124, 132, 135-138
composition ranges and limits ... M1: 132, 135-138, 403-404, 407-408
corrosion in fresh water ... M1: 737-738
corrosion rate in seawater ... M1: 739, 742, 744, 745
corrosion resistance ... M1: 405-406, 409-411
directionality of properties ... M1: 411, 417-418
fatigue properties ... M1: 418-419
forming ... M1: 419
forms available commercially ... M1: 403, 405-406, 408
heat treatment ... M1: 409-410
hot forming ... M1: 419
hydrogen embrittlement of ... M1: 687
inclusion shape control ... M1: 411
manufacture ... M1: 563
mechanical properties ... M1: 403, 406, 408-410, 411-419
microalloyed, characteristics of ... M1: 409, 411
minimum bend radii ... M1: 554, 555
notch toughness ... M1: 403, 409-410, 414-415, 417-418, 695
SAE specifications ... M1: 403, 408-409
selected grades, tensile and fatigue properties ... M1: 680
selection for corrosion resistance ... M1: 751
thermomechanical treatment ... M1: 563
weldability ... M1: 563
welding ... M1: 409, 415, 419-420
yield-strength basis for classifying ... M1: 403, 409-410
HSLA steels, ASTM specific types *See* Steels, ASTM specific types
HSLA steels, SAE specific types
942X
bend radii ... M1: 408, 419, 555
composition ... M1: 132, 210, 408
mechanical properties ... M1: 408
tensile properties, bars and shapes ... M1: 211
945A
bend radii ... M1: 408, 419, 555
composition ... M1: 132, 210, 408
mechanical properties ... M1: 408
tensile properties, bars and shapes ... M1: 211
945C
bend radii ... M1: 408, 419, 555
composition ... M1: 132, 210, 408
mechanical properties ... M1: 408
HSLA steels, SAE specific types (continued)
tensile properties, bars and shapes ... M1: 211
945X
bend radii ... M1: 408, 419, 555
composition ... M1: 132, 210, 408
mechanical properties ... M1: 408
tensile properties, bars and shapes ... M1: 211
950A
bend radii ... M1: 408, 419
composition ... M1: 132, 210, 408
mechanical properties ... M1: 408
tensile properties, bars and shapes ... M1: 211
950B
bend radii ... M1: 408, 419, 555
composition ... M1: 132, 210, 408
mechanical properties ... M1: 408
tensile properties, bars and shapes ... M1: 211
950C
bend radii ... M1: 408, 419, 555
composition ... M1: 132, 210, 408
mechanical properties ... M1: 408
tensile properties, bars and shapes ... M1: 211
950D
bend radii ... M1: 408, 419, 555
composition ... M1: 132, 210, 408
mechanical properties ... M1: 408
tensile properties, bars and shapes ... M1: 211
950X
bend radii ... M1: 408, 419, 555
composition ... M1: 132, 210, 408
mechanical properties ... M1: 408
tensile properties, bars and shapes ... M1: 211
955X
bend radii ... M1: 408, 419, 555
composition ... M1: 132, 210, 408
mechanical properties ... M1: 408
tensile properties, bars and shapes ... M1: 211
960X
bend radii ... M1: 408, 419, 555
composition ... M1: 132, 210, 408
mechanical properties ... M1: 408
tensile properties, bars and shapes ... M1: 211
965X
bend radii ... M1: 409, 419, 555
composition ... M1: 132, 210, 408
mechanical properties ... M1: 408
tensile properties, bars and shapes ... M1: 211
970X
bend radii ... M1: 408, 419, 555
composition ... M1: 132, 210, 408
mechanical properties ... M1: 408
tensile properties, bars and shapes ... M1: 211
980X
bend radii ... M1: 408, 419, 555
composition ... M1: 132, 408
mechanical properties ... M1: 408
HSM copper *See* Copper alloys, specific types, C19400
applications and properties ... A2: 293-294
HT
composition ... M4: 653
HT alloy ... A1: 926-927
HT9 (12Cr-1Mo-0.3V)
applications ... A6: 433
chemical composition ... A6: 432
gas-tungsten arc welding ... A6: 435, 436
laser welding ... A6: 441
microstructure ... A6: 435
orientation and PWHT effect ... A6: 437
specific welding recommendations, filler metals ... A6: 440
tempering behavior ... A6: 440
HTO *See* Hydrogen deterioration
HTX composites
wear factors ... A18: 824
HTX-PDX-88599
composition ... A18: 821
mechanical properties ... A18: 823
wear and friction properties ... A18: 822

HU alloy ... **A1:** 926, 927
Hub *See also* Boss
defined ... **A14:** 8
forging of ... **A14:** 67-68
forgings, shapes of ... **A14:** 61
front wheel, by precision forging ... **A14:** 63
HERF processed ... **A14:** 105
rotary hot forged ... **A14:** 179
Hub extrusion ... **A7:** 412
Hub forging
central fracture ... **A7:** 412
deformation limits at fracture ... **A7:** 413
Hubbing
defined ... **A14:** 8
Huber Guinier camera
for XRPD analysis ... **A10:** 336
Huey test corrosion data
P/M extruded stainless steels ... **A13:** 835
Hugoniot elastic limit (HEL) *See also* Elastic limit
abbreviation ... **A8:** 725
in flyer plate impact test ... **A8:** 211
measuring ... **A8:** 211
Hull cell
in electropolishing ... **A9:** 51
Hull cell evaluation
as bath control ... **EL1:** 680
Hull diagram ... **A6:** 458, 462
Hull-smoothing ship cements
powder used ... **A7:** 574
Hultgren ball
as Brinell indenter ... **A8:** 84
Human body *See* Body environment; Implants; Metallic orthopedic implants, failures of
Human body burden
of chromium ... **A2:** 1242
Human body model (HBM)
defined ... **EL1:** 966
Human factors *See also* Management; Operators; Personnel; Safety
in NDE reliability ... **A17:** 679
Human oil, effect on bearing strength
aluminum alloy sheet ... **A8:** 60
Human vision, vs. machine vision
capabilities ... **A17:** 30
Humans
hair, NAA forensic studies of ... **A10:** 233
NAA analysis of toxic element retention in ... **A10:** 233
Hume-Rothery rules ... **M6:** 21
Humid air
fracture effects ... **A12:** 72
Humidity *See also* Absolute humidity; Moisture; Moisture absorption; Relative humidity; Specific humidity; Water absorption
absolute ... **EM3:** 15
in CERDIP package ... **EL1:** 962
and corrosion rate ... **A13:** 511
critical level ... **A13:** 511-512
dependence, aluminum corrosion ... **EL1:** 1050
-dependent corrosion, in accelerated testing ... **EL1:** 891
effect, atmospheric corrosion, carbon steels ... **A13:** 511-512
effect, electrical testing ... **EM2:** 584
effect, uranium/uranium alloys ... **A13:** 815
effects, electrical resistance alloys ... **A2:** 824
effects, UV conformal coating ... **EL1:** 788
-induced stress tests ... **EL1:** 495-497
ratio ... **EM3:** 15
relative ... **EM3:** 15
relative, and filiform corrosion ... **A13:** 107-108
relative, effects on low-expansion alloys ... **A2:** 892
resistance, silicone conformal coatings ... **EL1:** 822
specific ... **EM3:** 15
-temperature chambers, simulated atmosphere testing ... **A13:** 226
test, as failure analysis ... **EL1:** 1102
testing, ceramic packages ... **EL1:** 468
tests ... **A13:** 8, 1113
and thermal cycling test ... **EL1:** 1102
Humidity control
cupolas ... **A15:** 384
Humidity cycling test ... **A13:** 1113
Humidity ratio
defined ... **EM2:** 22
Humidity, relative
and explosivity ... **A7:** 195, 196
Humidity test *See also* Salt-fog test
defined ... **A13:** 8
electronic/communications equipment ... **A13:** 1113
Humidity tests
rust-preventive compounds ... **M5:** 469-470
Hump tables
cut-to-length lines ... **A14:** 711
Humpback furnaces ... **A7:** 351, 353-356
Hunter water leaching
as titanium powder process ... **A7:** 164
Hunting knives
powders used ... **A7:** 574
Hurd Shaker Process
titanium powder ... **A7:** 167
Hutchinson-Rice-Rosengren singularity field *See* H-R-R singularity field
Huygenian microscope eyepieces ... **A9:** 73
Huygens (point) source
ultrasonic-transducer crystal ... **A17:** 239
HV *See* Vickers hardness
HVOF thermal spraying
coatings for jet engine components ... **A18:** 592
HW
composition ... **M4:** 653
HW alloy ... **A1:** 926, 927
HX
composition ... **M4:** 653
HX alloy ... **A1:** 921, 926, 927
Hx11 temper
defined ... **A2:** 26
Hy Tuf
yield strength ... **A4:** 208
HY-80 ... **A6:** 406, 425, 426, 427, 428
laser-beam welding ... **A6:** 264
microstructure ... **A6:** 75
steel composition effect on susceptibility to cold cracking ... **A6:** 74
HY-130 steel ... **A6:** 664, 665, 666
laser-beam welding ... **A6:** 264
HY-180 steel (HP9-4-20)
laser-beam welding ... **A6:** 264
Hybrid *See also* Hybrid composites; Interply hybrid; Intraply hybrid; Urethane hybrids
binders, ceramic shell molds ... **A15:** 259
defined ... **EM1:** 13 **EM2:** 22
processes, permanent mold, types ... **A15:** 34
Hybrid ball bearings ... **A18:** 503
Hybrid bearings ... **A18:** 531-532
Hybrid circuits
advanced ceramic and silicon substrate ... **EL1:** 8
conformal coatings ... **EL1:** 761-762
parylene coatings ... **EL1:** 799
planar, advantages ... **EL1:** 8
size reduction of multichip system ... **EL1:** 8
Hybrid components
integral and add-on ... **EL1:** 250-259
Hybrid composites
fabric, properties ... **EM1:** 149
fastener holes, technique/tools for ... **EM1:** 715
fiber-reinforced plastic, and aluminum, weights compared ... **EM1:** 35
impact resistance ... **EM1:** 262
laminate, damping in ... **EM1:** 214
Hybrid consolidation
of shiver-base composite contact materials ... **A2:** 857
Hybrid digital-to-analog (D/A) converter
corrosion failure analysis ... **EL1:** 1115
Hybrid integrated circuitry (HIC) *See also* Hybrid; Integrated circuits (IC)
introduction ... **EL1:** 249
unique features ... **EL1:** 333-334
Hybrid laminates *See* Hybrid composites; Sandwich laminates
Hybrid microcircuits ... **A13:** 1124-1125
advantages/limitations ... **EL1:** 257-259
amplifier, corrosion failure analysis ... **EL1:** 1115
automatic optical inspection of ... **EL1:** 941
digital-to-analog (D/A) converter ... **EL1:** 1115
electronic/electrical corrosion failure analysis of ... **EL1:** 1115
package functions ... **EL1:** 451
packages, types ... **EL1:** 451-454
packaging of ... **EL1:** 259-260
Hybrid microcircuits (continued)
technology ... **EL1:** 250-262
thick-film ... **EL1:** 255-256
thin-film ... **EL1:** 257
thin-film generic ... **EL1:** 313
Hybrid microelectronics ... **EL1:** 89, 250
Hybrid packages
advanced, disadvantages ... **EL1:** 7
classification ... **EL1:** 451-454
forms of ... **EL1:** 451-452
functional classification ... **EL1:** 452-453
with glass-to-metal seals, manufacture ... **EL1:** 458
level 1 ... **EL1:** 405
materials classification ... **EL1:** 453-454
with metal-matrix composites ... **EL1:** 1128
and single-chip packages, compared ... **EL1:** 451
solder alloys for ... **EL1:** 454
thermal properties ... **EL1:** 454
thick-film technology ... **EL1:** 208
Hybrid permanent mold processes
aluminum casting alloys ... **A2:** 141-145
Hybrid technology
advanced ceramic and silicon substrates ... **EL1:** 8
advantages/limitations ... **EL1:** 257-259
design for performance ... **EL1:** 260-261
electrical requirements ... **EL1:** 261-262
interconnects, selection criteria ... **EL1:** 250-255
packaging hybrid circuits ... **EL1:** 259-260
reworkability/repairability ... **EL1:** 261
silicon, for higher bandwidth communications ... **EL1:** 8
status of ... **EL1:** 262
thick-film ... **EL1:** 206-208
thick-film, medical and military applications ... **EL1:** 389
thick-film vs thin-film ... **EL1:** 255-257
Hybrid tee
as microwave device ... **A17:** 211
Hybrid wafer-scale integration (H-WSI) *See also* Hybrid(s)
advantages ... **EL1:** 86-88
cooling path ... **EL1:** 363-364
defined ... **EL1:** 8, 354
high-density static RAM module with ... **EL1:** 88
level of ... **EL1:** 76
as new interconnection technology ... **EL1:** 8
optical clocks ... **EL1:** 9-10
optical interconnections ... **EL1:** 10
Hybrid(s) *See also* Hybrid microcircuits; Hybrid packages; Hybrid technology; Hybrid wafer-scale integration (H-WSI); Microcircuits; Thick-film hybrids; Thin-film hybrid
advantages ... **EL1:** 250
ceramic, medical and military applications ... **EL1:** 386
circuit construction ... **EL1:** 386-387
commercial applications ... **EL1:** 381-385
component attachment ... **EL1:** 386-387
consumption by application, U.S ... **EL1:** 254
conventional, test procedures ... **EL1:** 372-373
defined ... **EL1:** 250
electrical requirements ... **EL1:** 261-262
failure rates ... **EL1:** 261
and higher-level integration ... **EL1:** 249
industrial thick-film, defined ... **EL1:** 381
large-scale, instrumentation/testing ... **EL1:** 365
marketplace ... **EL1:** 253
materials technology ... **EL1:** 386
medical and military applications ... **EL1:** 387-389
microcircuit, typical ... **EL1:** 258
multilevel structures, thin-film ... **EL1:** 322-324
packaging evolution ... **EL1:** 259-260
silicon-on-silicon test procedures ... **EL1:** 372-373
technological trends ... **EL1:** 249
thick-film ... **EL1:** 258, 332-353, 386
thin-film ... **EL1:** 313-331
Hybridization
and cost ... **EM1:** 35
of woven fabrics ... **EM1:** 126-127
Hybrids ... **EM3:** 15
sealant formulation ... **EM3:** 677
Hycar additives
as tougheners ... **EM3:** 185
Hydrafilm process
hydrostatic extrusion ... **A14:** 328-329

Hydrated alumina
dentifrice abrasive ... **A18:** 665
as filler ... **EM3:** 179
Hydrated alumina filler
for sheet molding compounds ... **EM1:** 158
Hydrated forms of solid salts and acids ... **A9:** 68
Hydration envelopes
in clay-water bonding ... **A15:** 212
Hydraulic actuator, torsional
finite-element model ... **A8:** 217
Hydraulic Brinell hardness tester ... **A8:** 87
Hydraulic bronze
properties and applications ... **A2:** 365
Hydraulic bulge test ... **A8:** 558-559
Hydraulic dynamometer stator vanes
liquid erosion ... **A11:** 169-170
Hydraulic energy
in die casting ... **A15:** 286
Hydraulic equalizing system ... **M6:** 476
Hydraulic fluid ... **A18:** 99
biocides ... **A18:** 110
defined ... **A18:** 10-11
demulsifiers ... **A18:** 107
industrial hydraulic fluids ... **A18:** 99
antiwear ... **A18:** 99
fire-resistant ... **A18:** 99
rust and oxidation-inhibited (R&O) oils ... **A18:** 99
pour-point depressants ... **A18:** 108
rust-preventive function ... **M5:** 460-464
seal-swell agents ... **A18:** 110
tractor hydraulic fluids ... **A18:** 99
viscosity improvers used ... **A18:** 110
Hydraulic fluid filters
powders used ... **A7:** 572
Hydraulic forming
aluminum alloy ... **A14:** 803
of copper and copper alloys ... **A14:** 819
Hydraulic hammer *See also* Hammers
defined ... **A14:** 8
Hydraulic hot press lamination
techniques ... **EL1:** 510
Hydraulic motors ... **A8:** 157-158
Hydraulic oil cooler
crevice corrosion of tubing in ... **A11:** 632-633
Hydraulic oils
lubricant classification ... **A18:** 86
Hydraulic penetration
liquid impact erosion ... **A18:** 224
Hydraulic press ... **EM3:** 15
defined ... **EM1:** 13 **EM2:** 22
Hydraulic press brakes ... **A14:** 8, 534
Hydraulic press forging ... **A8:** 158
Hydraulic presses *See also* Mechanical presses; Presses; specific presses ... **A7:** 6, 330-332 **A14:** 31-33
for aluminum alloys ... **A14:** 245
for bar bending ... **A14:** 662
characterization ... **A14:** 37
for coining ... **A14:** 180-181
compared with mechanical presses ... **A7:** 331-332
defined ... **A14:** 8
and electric resistance wire furnace-heated alloy steel dies ... **A7:** 505
as force-restricted machines ... **A14:** 25, 37
forgings, torsion testing ... **A14:** 373
for heat-resistant alloys ... **A14:** 234
induction-heated alloy steel/graphite built into ... **A7:** 506
load vs displacement curves ... **A14:** 38
load vs time and displacement curves ... **A14:** 38
for magnesium alloys ... **A14:** 259
mandrel forging with ... **A14:** 70-71
for open-die forging ... **A14:** 61
for precision forging ... **A14:** 169
resistance-heated graphite die assembly built into ... **A7:** 507
for titanium alloys ... **A14:** 274, 276
Hydraulic presses (continued)
triple-action, for fine-edge blanking and piercing ... **A14:** 473
vs mechanical presses ... **A14:** 492
Hydraulic pressing
to make compacts of silicon ... **EM4:** 237
to make silicon oxynitride shapes by reaction-bonding ... **EM4:** 239
Hydraulic pressure
as tension source for stress-corrosion cracking ... **A8:** 502
Hydraulic rams
force rating ... **A8:** 399
for servohydraulic system ... **A8:** 399-400
in subpress assembly for medium strain rate testing ... **A8:** 193
Hydraulic rotary actuator ... **A8:** 157-158
Hydraulic servoactuator
flash welding ... **A6:** 247
Hydraulic shears ... **A14:** 8, 701-702
Hydraulic testing machine
components ... **A8:** 193, 612
fracture surface of Fe-4Si under rapid loading ... **A8:** 479-480
packless type ... **A8:** 613
torsional ... **A8:** 215-217
Hydraulic torsion
test facility ... **A8:** 216
Hydraulic valves
seizing in ... **A11:** 141
Hydraulic wheel motor manifolds ... **A7:** 675, 676
Hydraulic-mechanical press brake
defined ... **A14:** 8
Hydraulics
law of ... **A15:** 590-591
Hydrazine electroless nickel plating process ... **M5:** 221
Hydrazyl
stable free radical ... **A10:** 265
stable free radical, ESR analysis of ... **A10:** 265
Hydride decomposition ... **A7:** 52, 55
Hydride formation
embrittlement by ... **A12:** 124
Hydride formation, as hydrogen damage ... **A13:** 164, 166
Hydride generator
for ICP spectrometers ... **A10:** 36
Hydride phase
defined ... **A9:** 9
of titanium and titanium alloys, mounting of ... **A9:** 458
Hydride powder ... **A7:** 6
Hydride process ... **A7:** 6
Hydride reduction process
oxidation-resistant coating ... **M5:** 665-666
Hydride-crush-degas process
niobium powders ... **A7:** 163
Hydride-crush-dehydride process
tantalum powder ... **A7:** 161
Hydride-crush-hydride process
for refractory metals and alloys ... **A2:** 560
Hydride-dehydride process ... **A7:** 55, 162, 165
Hydride-generation systems
in analytic ICP systems ... **A10:** 36
atomic absorption spectrometry ... **A10:** 50
Hydride-induced fracture
in titanium alloy microstructure ... **A8:** 487
Hydrided TiFe
phase analysis of ... **A10:** 293-294
Hydrides deposition
neutron radiography of ... **A17:** 391
Hydrides formation
cracking from ... **A11:** 248-249
Hydrides, in zirconium alloys
preparation for examination of ... **A9:** 498
Hydriding of zirconium ... **M3:** 788-790
Hydridopolysilazane polymer
for making HPZ fiber ... **EM4:** 223
Hydroblasting *See* Water blast cleaning
Hydrobromic acid
with bromine, as sample dissolution medium ... **A10:** 166
-phosphoric acid method of analysis for copper in magnesium alloys ... **A10:** 65
Hydrocarbon cleaners
chlorinated *See* Chlorinated hydrocarbon cleaners
halogenated *See* Halogenated hydrocarbon cleaners
Hydrocarbon contamination
of microanalytical samples ... **A11:** 36
Hydrocarbon oils
as dielectric for diesinking machines ... **EM4:** 373
Hydrocarbon plastics *See also* Hydrocarbon thermoplastic polymers; Plastics ... **EM3:** 15
defined ... **EM2:** 22
Hydrocarbon polymers
high-vacuum lubricant applications ... **A18:** 156-157
Hydrocarbon processing
alloy steel corrosion in ... **A13:** 535
Hydrocarbon resins
as tackifiers ... **EM3:** 182
Hydrocarbon soils
removal of ... **M5:** 34
Hydrocarbon thermoplastic polymers ... **EM2:** 49-50
melting temperatures ... **EM2:** 50
Hydrocarbons
analytic methods for ... **A10:** 10
aromatic, determined ... **A10:** 218
as fluorescing surface impurities in Raman analyses ... **A10:** 130
long-chain, analytic methods for ... **A10:** 9
oil of, IR split mull of ... **A10:** 113
oxidation, Raman analysis ... **A10:** 133
oxidized in high-temperature combustion resistance furnaces ... **A10:** 224
polynuclear aromatic ... **A10:** 74
sigma values for ionization ... **A17:** 68
surface, determination by selective combustion ... **A10:** 223-224
Hydrochloric acid
in alcohol as an etchant for tin ... **A9:** 450
as an etchant for heat-resistant casting alloys ... **A9:** 330-331
as an etchant for wrought stainless steels ... **A9:** 281
as an etchant for zinc and zinc alloys ... **A9:** 488
cast iron resistance ... **A13:** 569
and cellosolve as an electrolyte for magnesium alloys ... **A9:** 426
cemented carbides resistance to ... **A13:** 852
as chemical cleaning solution ... **A13:** 1140
in chemical etching cleaning ... **A12:** 75-76
in color etchants ... **A9:** 142
copper/copper alloys in ... **A13:** 628-629
corrosion ... **A13:** 1160-1166
corrosion by ... **M1:** 733, 734
corrosion inhibitors used in ... **M1:** 755-756
and grinding ... **A16:** 424, 427
lead corrosion in ... **A13:** 787
nickel alloys, corrosion ... **M3:** 173
nickel-base alloy resistance ... **A13:** 644-645
residue isolation using ... **A10:** 176
sample component losses in ... **A10:** 165
for sample dissolution ... **A10:** 165
and SCC in titanium and titanium alloys ... **A11:** 223
for SCC of titanium alloys ... **A8:** 531
silver corrosion in ... **A13:** 794
solution, as eluent for suppressed cation chromatography ... **A10:** 660
stainless steel corrosion ... **A13:** 558
stainless steels, corrosion ... **M3:** 83-84
sulfuric acid, and nitric acid as an etchant for wrought heat-resistant alloys ... **A9:** 307
and sulfuric acid as an etchant for carbon and alloy steels ... **A9:** 171

Hydrochloric acid (continued)
tantalum corrosion in ... **A13:** 725-726
tantalum-tungsten corrosion in ... **A13:** 737
used in photochemical machining ... **A16:** 593
water and hydrogen peroxide as an etchant for austenitic manganese steel casting specimens ... **A9:** 238
and water as a macroetchant for plate steels ... **A9:** 203
and water as a macroetchant for tool steels ... **A9:** 256
and water as an electrolyte for platinum- base alloys ... **A9:** 551
and water as an etchant for carbon and alloy steels ... **A9:** 171, 174-176
zinc/iron corrosion in ... **A13:** 467
zirconium/zirconium alloy corrosion ... **A13:** 710
Hydrochloric acid, and ammonia reaction
as gas/leak detection ... **A17:** 61
Hydrochloric acid pickling
copper and copper alloys ... **M5:** 611-613
hot dip galvanized coating process ... **M5:** 325-326, 328
iron ... **M5:** 68-69, 74-76, 78-80, 82
nickel and nickel alloys ... **M5:** 669-670
process variables affecting scale removal ... **M5:** 74-75, 78-80
solution compositions and operating conditions ... **M5:** 68-69, 74-76, 611-613
steel ... **M5:** 68-69, 74-76, 78-80, 82
waste recovery ... **M5:** 82
Hydroclave ... **EM3:** 713
Hydrocracking
powders used ... **A7:** 574
Hydrocyanic acid
copper/copper alloy resistance ... **A13:** 629
Hydrocyanic acid, corrosion
stainless steels ... **M3:** 84
Hydrodynamic chromatography
to analyze ceramic powder particle size ... **EM4:** 68-69
Hydrodynamic intensity
reduction of ... **A11:** 170
Hydrodynamic journal bearings ... **A18:** 559-561
Hydrodynamic lubrication *See also* Elastohydrodynamic lubrication; Gas lubrication ... **A18:** 101, 516
in bearings ... **A11:** 484
and boundary lubrication ... **A11:** 484
definition ... **A18:** 11, 89
dynamic loads ... **A18:** 91
film, development in journal bearing ... **A11:** 150
friction-velocity curve ... **A8:** 605
high-speed effects ... **A18:** 91
lubricant film thickness ... **A18:** 90
reciprocating motion ... **A18:** 91
for sheet metal forming ... **A14:** 512
squeeze film action ... **A18:** 89-90, 91
thermal effects ... **A18:** 90-91
and wear ... **A11:** 150-151
wedging film action ... **A18:** 89, 90, 91
Hydrodynamic lubrication theory ... **A18:** 516, 517
Hydrodynamic machining *See* Waterjet machining
Hydrodynamic pressure
in liquid-erosion failures ... **A11:** 163-171
Hydrodynamic seal
defined ... **A18:** 11
Hydrodynamic tension test
high-temperature ... **A11:** 281
Hydrofluoric acid ... **A13:** 1166-1170
as an etchant for aluminum alloys ... **A9:** 354
as an etchant for glass-epoxy composites ... **A9:** 591
as an etchant for wrought stainless steels ... **A9:** 281
analysis of solutions in ... **A10:** 35
as chemical cleaning solution ... **A13:** 1140
copper/copper alloy corrosion ... **A13:** 629
in Group VI electrolytes ... **A9:** 54
nickel-base alloy corrosion ... **A13:** 648
nickel-base alloy resistance ... **A13:** 645-646
nitric acid and water, as an etchant for carbon and alloy steels ... **A9:** 171
nitric acid and water, as an etchant for titanium powder metallurgy materials ... **A9:** 509
Hydrofluoric acid (continued)
and nitric acid, used for polishing hafnium ... **A9:** 497
and nitric acid, used for polishing zirconium and zirconium alloys ... **A9:** 497
as nonoxidizing dissolution medium ... **A10:** 165
polyester resistance to ... **EM1:** 93-94
safety precautions ... **A9:** 69
sulfuric acid and nitric acid, as an etchant for beryllium ... **A9:** 390
tantalum corrosion in ... **A13:** 726-727
in water, as an etchant for aluminum-coated sheet steel ... **A9:** 197
Hydrofluoric acid (HF)
chemical milling etchant ... **A16:** 584, 852
Hydrofluoric acid pickling
iron and steel ... **M5:** 68, 73, 79-80
magnesium alloys ... **M5:** 630, 640-642
Hydrofoils
composite structures for ... **EM1:** 839
Hydroforming
beryllium-copper alloys ... **A2:** 411
refractory metals and alloys ... **A2:** 562
thermoplastic resin composites ... **EM1:** 548
Hydrogel
sol-gel transition role ... **EM4:** 210
Hydrogen *See also* Degassing; Embrittlement; Gases; Hydrogen damage; Hydrogen embrittlement; Hydrogen flaking; Hydrogen porosity; Hydrogen removal; Hydrogen-damage failures; Hydrogen-induced cracking ... **A7:** 54
absorption, aluminum alloys ... **A15:** 79, 456-457
absorption, at elevated temperatures, titanium embrittlement by ... **A11:** 641
addition to argon shielding gas ... **M6:** 217
alloying, wrought aluminum alloy ... **A2:** 51
analysis by SIMS ... **A10:** 610
analysis, zirconium-steel couple ... **A13:** 715
analyzed in microcircuit fabrication process ... **A10:** 156-157
-assisted cracking ... **A11:** 399, 408-410
-assisted fracture, microscopic models for ... **A8:** 466
atmosphere, AISI/SAE alloy steels in ... **A12:** 292
atmosphere for furnace brazing ... **M6:** 1004-1007
atmospheric effect helping control dusting ... **A18:** 684
atoms, by focusing effect in EXAFS ... **A10:** 411
atoms, located in organometallic or intermetallic compounds ... **A10:** 420
attack ... **A11:** 248, 290 **A13:** 164, 166, 1279-1280
attack, in ASTM/ASME alloy steels ... **A12:** 349-350
austenitic stainless steels ... **A12:** 39, 356
avoided in graphite electrically heated furnaces ... **EM4:** 246, 248
blistering ... **A11:** 5, 247-248
brazing atmosphere source ... **A6:** 628
cause of porosity ... **M6:** 839
in nickel alloy welds ... **M6:** 442-443
characteristics in a blend ... **A6:** 65
-charged specimen, cycled near K threshold ... **A8:** 487-488, 491
-charged titanium alloy, cleaved alpha grains ... **A8:** 490
charging ... **A13:** 329, 535, 536, 662
charging, cracking from ... **A11:** 245-247
codeposited with chromium in electroplating ... **A18:** 835
combustion method for elemental analysis of ... **A10:** 214
concentration profiles ... **A10:** 610
contamination in aluminum alloys ... **M6:** 385-386
contamination, in aluminum melts ... **A15:** 79
contamination, wrought titanium alloys ... **A2:** 620
content, forgings ... **A17:** 491-492
content, high-purity oxygenated water ... **A8:** 420
content, in forging ... **A11:** 315-316
content in weld metals ... **A6:** 1011-1012
contents in weld metal, and flux choice ... **A6:** 58, 59
copper/copper alloy resistance ... **A13:** 632
corrosion, bolts ... **A12:** 248
cracking ... **A8:** 539-540 **A11:** 248, 410, 413, 440
cracking induced by ... **M6:** 830-831
cracks, in welds ... **A11:** 413, 440
damage ... **A13:** 163-171
Hydrogen (continued)
damage, defined ... **A8:** 7 **A11:** 5
degassing, in vacuum melting ultrapurification ... **A2:** 1094
determined by combustion ... **A10:** 214
determined in copper ... **A10:** 231-232
diffusion in iron alloys ... **M6:** 23
diffusivity, and precipitate particles ... **A13:** 166
dissociation and recombination ... **A6:** 64
dissolved ... **A8:** 423
dissolved, effect on steel tensile ductility ... **A11:** 336-338
dissolved in TiFe crystal structure ... **A10:** 294
effect, delayed failure ... **A13:** 329-330
effect, modification ... **A15:** 485-486
effect on aluminum alloy ingots ... **A9:** 632-633
effect on antimony-doped Ni-Cr alloy steels ... **A12:** 350
effect on dimple rupture ... **A12:** 22-24
effect on fatigue fracture appearance ... **A12:** 30, 37, 51
effect, space boosters/satellite corrosion ... **A13:** 1105
effects as image gas in FIM ... **A10:** 587, 588
effects, at elevated temperatures ... **A13:** 100
effects on underwater welding ... **M6:** 922
electrode potentials ... **A7:** 140
electronic analysis ... **A11:** 199
entrapment and steel weldment soundness ... **A6:** 408
entry, damage forms ... **A13:** 329
environment embrittlement ... **A13:** 283
environmental failures ... **A11:** 409, 410
evolution during porcelain enameling ... **M1:** 179
explosive range ... **A7:** 348
failures, in steel castings ... **A11:** 409
fissuring, low-carbon steels ... **A11:** 645
flakes ... **A11:** 88
flaking ... **A12:** 125
flux cored electrode content ... **M6:** 102
gas mass analysis of ... **A10:** 155
gas-metal arc welding shielding gas for nickel alloys ... **A6:** 746
gaseous, cracking from ... **A11:** 247
gaseous, titanium embrittlement by ... **A12:** 23, 32
gaseous, titanium/titanium alloy corrosion ... **A13:** 685
high-pressure, fatigue crack propagation in purified gaseous, crack growth ... **A8:** 403
superalloys ... **A8:** 409
hydrogen-induced cracking ... **A6:** 93-95
IGF determination ... **A10:** 226, 231
as impurity in uranium alloys ... **A9:** 477
-induced blistering ... **A11:** 5, 247-248
induction of cold cracking ... **M6:** 44-46
in inorganic solids, applicable analytical methods ... **A10:** 4, 6
interaction coefficient, ternary iron-base alloys ... **A15:** 62
ion, monitoring by acid-base titration ... **A10:** 172
ionization potentials and imaging fields for ... **A10:** 586
iron sintered in, effect of hydrogen chloride ... **A7:** 319
loss, in aluminum-silicon melts ... **A15:** 164-165
mean free path ... **A17:** 59
mean free path value at atmospheric conditions ... **A18:** 525
measurement ... **A15:** 457-459
monitoring ... **A13:** 201
nickel alloy cracking in ... **A12:** 3%
overvoltage, defined ... **A13:** 8
in oxide reduction ... **A7:** 52, 53, 345-346
oxyfuel gas welding fuel gas ... **A6:** 281, 282, 283-284, 285
-oxygen equilibrium, molten copper ... **A15:** 467
pickup, in aluminum-silicon melts ... **A15:** 164-165
in plasma arc powder spraying process ... **A18:** 830
for plasma arc spraying ... **A6:** 811
porosity, in aluminum casting alloys ... **A2:** 134-135
protective atmosphere ... **M6:** 1016
reductant of metal oxides ... **M6:** 693
relative, susceptibilities ... **A13:** 288
removal, bake-out cycle ... **A13:** 330
removal, by solid-state refining ... **A2:** 1094
residual gas analysis of ... **EL1:** 1065-1066

Hydrogen (continued)
segregation, during solidification ... **A15:** 82
service, Nelson curve, for steels ... **A13:** 537
shielding gas from fluxes ... **A6:** 58
shielding gas properties ... **A6:** 64
shielding gas purity and moisture content ... **A6:** 65
sigma values for ionization ... **A17:** 68
in solid solution ... **A15:** 748
solubility in aluminum alloys ... **A6:** 722
solubility in steel ... **M1:** 687
solubility, Sievert's law ... **A13:** 329
sources, for hydrogen embrittlement ... **A13:** 284
specifications ... **A7:** 344
-stress cracks, in line pipe ... **A11:** 701
tantalum dissolution in ... **A13:** 731
testing for ... **A15:** 457-459
thermodynamic pressure ... **A15:** 84
-to-water ratio ... **A7:** 342, 344
as tracer gas ... **A17:** 68
use in gas tungsten arc welding ... **M6:** 358
use in resistance spot welding ... **M6:** 486
use of oxyfuel gas welding ... **M6:** 584
valve thread connections for compressed gas cylinders ... **A6:** 1197
in weld relay, gas mass spectrometry of ... **A10:** 156

Hydrogen abstraction-type photoinitiators ... **EL1:** 855

Hydrogen and hydrogen compounds
atmosphere for WC powder preparation ... **A16:** 71
in CVD process ... **A16:** 80
and electrochemical machining hazards ... **A16:** 536
H_2SO_4, addition to emulsions ... **A16:** 127

Hydrogen atmospheres ... **A7:** 341, 344-345, 361, 369
applications ... **M4:** 410, 411
of cemented carbides ... **A7:** 385-386
composition ... **A7:** 342
generators, in-plant ... **M4:** 411
impurities ... **M4:** 410
metal-to-metal oxide equilibria ... **A7:** 498
in ordnance applications ... **A7:** 690
safety precautions ... **M4:** 411
supply ... **M4:** 411

Hydrogen attack
and elevated-temperature service ... **A1:** 633-634, 639

Hydrogen blistering ... **A11:** 5, 247-248 **A13:** 8, 331, 1277-1278
and elevated-temperature service ... **A1:** 634

Hydrogen bonding
destruction of ... **EM2:** 773

Hydrogen brazing
definition ... **A6:** 1210 **M6:** 10

Hydrogen chloride
corrosion ... **A13:** 1160-1166
as crude oil contaminant ... **A13:** 1266-1267
dry, cast irons in ... **A13:** 570
effect on iron sintered in hydrogen ... **A7:** 319
and SCC in titanium and titanium alloys ... **A11:** 223
for SCC of titanium alloys ... **A8:** 531

Hydrogen chloride, corrosion
nickel alloys ... **M3:** 174

Hydrogen chloride gas ... **A13:** 1165-1166

Hydrogen coolers
finned tubing for ... **A11:** 628

Hydrogen cracking
duplex stainless steels ... **A6:** 474
electroslag welding ... **A6:** 278
postweld heat treatment as preventive practice ... **A6:** 1069
quenched and tempered steels ... **A6:** 384
steel weldments ... **A6:** 416, 424
submerged arc welding ... **A6:** 209
underwater welding ... **A6:** 1010, 1011-1012, 1014

Hydrogen cyanide (HCN)
produced in graphite furnaces as toxic reaction ... **EM4:** 246

Hydrogen cyanide gas in alkaline electrolytes ... **A9:** 54

Hydrogen damage *See also* Hydrogen ... **A13:** 163-171
AISI/SAE ahoy steels ... **A12:** 302
in aluminum alloys ... **A13:** 169-170
bright flakes as ... **A12:** 415
by embrittlement, in steam equipment ... **A11:** 612
in carbon and low-alloy steels ... **A11:** 126
in copper alloys ... **A13:** 170
defined ... **A11:** 5 **A13:** 8
effect of M_3C carbide in ... **A1:** 632-633
and elevated-temperature service ... **A1:** 632-634, 639
as environmentally assisted failure ... **A13:** 163
failure, pharmaceutical production ... **A13:** 1230
hydrogen environmental embrittlement ... **A1:** 711-712, 713
hydrogen-stress cracking and loss of tensile ductility ... **A1:** 712-717
in iron-base alloys ... **A13:** 166-169
material selection to avoid/minimize ... **A13:** 329-333
in nickel alloys ... **A13:** 169
in niobium/niobium alloys ... **A13:** 171
in petroleum refining and petrochemical operations ... **A13:** 1277
and rupture of boiler tubes ... **A11:** 612-613
in shafts ... **A11:** 459
spring fatigue fracture from ... **A11:** 558
steam equipment failure by ... **A11:** 602
steam/water-side boilers ... **A13:** 99! -992
in tantalum/tantalum alloys ... **A13:** 171
theories ... **A13:** 164-166
in threaded fasteners ... **A11:** 539
titanium/titanium alloys ... **A13:** 170-171, 673-674, 685-686
types ... **A13:** 163-164
types of ... **A11:** 245
in vanadium/vanadium alloys ... **A13:** 171
in zirconium alloys ... **A13:** 171

Hydrogen degradation
classification ... **A13:** 165

Hydrogen deterioration in aluminum alloys ... **A9:** 358

Hydrogen diffusion
NMR study in metals ... **A10:** 277

Hydrogen embrittlement
AISI/SAE alloy steels ... **A12:** 293, 301, 305-307
of aluminum ... **A12:** 23-24
austenitic stainless steels ... **A12:** 355
causes ... **A12:** 22-23
in decohesive rupture ... **A12:** 18, 24
effects on dimple rupture ... **A12:** 22-24
examination and interpretation ... **A12:** 124-126
high-carbon steels ... **A12:** 284, 285
low-carbon steels ... **A12:** 248
maraging steels ... **A12:** 387
precipitation-hardening stainless steels ... **A12:** 372
premature spring failures from ... **A12:** 286
as SCC mechanism, steels ... **A12:** 23-25
of stainless steels ... **A12:** 30, 355, 372
superalloys ... **A12:** 389
of titanium ... **A12:** 23, 448
tool steels ... **A12:** 381-382

Hydrogen embrittlement *See also* Hydrogen ... **M1:** 356
abbreviation ... **A8:** 725
accumulator ring fracture from ... **A11:** 337, 338
in acids ... **M1:** 687
of alloy steel fasteners ... **A11:** 541
alloying element additions ... **A8:** 487
in amorphous metals ... **A13:** 869
at high stress-intensity range, in steel ... **A8:** 409-410
in austenitic stainless steels ... **A1:** 715
brazeability of base metals ... **A6:** 622, 623

Hydrogen embrittlement (continued)
brazing and ... **A6:** 117
brittle fracture of ... **A11:** 85
cadmium plating causing ... **M5:** 199, 206-207, 268-269
cadmium-plated steel nut failed by ... **A11:** 246-247
cantilever beam test for ... **A8:** 537-538
in carbon steel pipe ... **A11:** 645
of casting alloys ... **A13:** 581
cathodic and periodic reverse electrocleaning causing ... **M5:** 34-35
as cause of premature cracking ... **A8:** 496
causes ... **A8:** 537 **M1:** 687
chromium plating causing ... **M5:** 185-186
cleaning processes causing ... **M5:** 12-13, 18-19, 27-28, 34-35
cobalt-base alloys ... **A13:** 661-662
contoured double-cantilever beam test for ... **A8:** 538-539
control ... **A13:** 289
copper ... **M2:** 239-240
copper alloys, electron-beam welding ... **A6:** 872
copper and copper alloys ... **A2:** 216
copper annealing ... **M2:** 255
and corrosion fatigue ... **A13:** 143-144
countermeasures ... **M1:** 687
cracking, in large alloy steel vessel ... **A11:** 661-663
defined ... **A8:** 7, 537 **A11:** 5 **A13:** 8, 283-284
delayed cracking ... **M1:** 687
delayed failure of bolt from ... **A11:** 539-540
detection of ... **M1:** 687
determined ... **A11:** 28-29
disk-pressure testing for ... **A8:** 540-541
dissimilar metal joining ... **A6:** 826
effect of strain rate on ... **M1:** 687
electrochemical testing ... **A13:** 218
and elevated-temperature service ... **A1:** 634
environment ... **A13:** 163, 283
etching and pickling effects ... **A8:** 510
evaluation of ... **A13:** 283-290
failure, manned spacecraft ... **A13:** 1083-1084
ferritic stainless steels ... **A6:** 449-450
and formation of flakes in steels ... **A1:** 716-717
four-point bend test for ... **A8:** 539-540
from machining, refractory metals and alloys ... **A2:** 561
from precleaning ... **A17:** 81
of gold-palladium metallization layer ... **A11:** 45-46
heat-resistant alloys, scale removal causing ... **M5:** 565
hot dip galvanized coating causing ... **M5:** 325
as hydrogen damage ... **A11:** 245-247
hydrogen sources ... **A13:** 284
inhibitors ... **A13:** 524
inhibitors to prevent ... **M1:** 755-756
internal reversible ... **A13:** 283
in line pipe steels ... **A1:** 716
localized, flakes and fisheyes as ... **A11:** 248
in maraging steels ... **A1:** 715-716 **M1:** 447, 448, 451
martensitic stainless steels ... **A6:** 626
mechanical coating and ... **M5:** 302
model, crack propagation ... **A13:** 161-162
nickel plating causing ... **M5:** 217-218
nickel plating, electroless relief of ... **M5:** 237
nickel-base alloys ... **A13:** 650-652
niobium/niobium alloys ... **A13:** 722
notch tensile test for ... **A8:** 27
parameters ... **A13:** 283
parts affected ... **A8:** 537
in petroleum refining and petrochemical operations ... **A13:** 1277
phosphate coating causing ... **M5:** 435
pickling process causing ... **M5:** 70, 81
potentiostatic slow strain rate tensile test for ... **A8:** 541
prevention ... **A13:** 289
process causes/prevention ... **A13:** 329
reduction of, by baking ... **M1:** 687
relative resistance, various alloys ... **A13:** 1104

Hydrogen embrittlement (continued)
rising step-load test for **A8:** 539-540
and SCC/corrosion fatigue cracking, compared **A13:** 291
slow strain rate tensile tests for **A8:** 499
in soils **A13:** 210
solid-state-welded interlayers **A6:** 171
springs, steel **M1:** 291
steam surface condensers **A13:** 988-989
steel **M5:** 185-186, 237, 268-269, 325
steel, welding as cause of **M1:** 562
steel weldment soundness **A6:** 410
in steels **A11:** 28-29, 100-101
and stress ratio increase **A8:** 406
stress sources **A13:** 284
and stress-corrosion cracking compared **A8:** 495, 529, 537
sulfide stress cracking **M1:** 687
susceptibility, alloy effect **A13:** 330
of tantalum/tantalum alloys **A13:** 735-736
testing **A13:** 284-288
tests for **A8:** 537-543
three-point bend test for **A8:** 539-540
threshold stress intensity parameter for **A8:** 537-542
titanium alloys **A6:** 516-517
in tool steels **A1:** 715
types **A13:** 283
wedge-opening test for **A8:** 538
wrought martensitic stainless steel **A6:** 438-439
zinc plating causing **M5:** 251, 253
of zinc-electroplated steel fastener **A11:** 548-549
Hydrogen embrittlement evaluation **A13:** 283-290
cantilever beam test **A13:** 284
contoured double-cantilever beam test **A13:** 285-286
disk-pressure testing method **A13:** 287-288
interpretation of test results **A13:** 288-289
potentiostatic slow strain rate tensile testing **A13:** 288
rising step-load test **A13:** 286-287
slow strain rate tests **A13:** 288
three-point/four-point bend tests **A13:** 286
wedge-opening load test **A13:** 284-285
Hydrogen environment embrittlement **A13:** 163, 283
Hydrogen flakes *See* Hydrogen embrittlement
Hydrogen flakes, defined
in forgings **A17:** 492
Hydrogen flaking *See also* Hydrogen
in alloy steel bar **A11:** 316, 337
in carbon tool steels **A11:** 574
defined **A12:** 125
in forging **A11:** 315-316
fractograph of **A11:** 337
in steel **A11:** 79
in tool steels **A11:** 574, 581
Hydrogen fluoride **A13:** 1166-1170, 1268-1269
Hydrogen fuel gas
in high-velocity oxyfuel powder spray process **A18:** 830
Hydrogen gas
effect on fracture surfaces tested in **A8:** 487, 489
and tramp elements **A8:** 487
Hydrogen in steel **M1:** 116
notch toughness, effect on **M1:** 694
Hydrogen ion concentration
in lubricants **A14:** 516
Hydrogen lamp
for UV/VIS analysis **A10:** 66
Hydrogen loss **A7:** 6
Hydrogen loss testing **A7:** 246-247
Hydrogen, max
chemical compositions per ASTM specification B550-92 .... **A6:** 787
Hydrogen peroxide **A16:** 29
and ammonium hydroxide as an etchant **A9:** 551
as an etchant for cemented carbides **A9:** 274
GFAAS analysis of trace tin and chromium in **A10:** 57-58
and glacial acetic acid as an etchant for lead and lead alloys **A9:** 415-416
with hydrochloric acid, as sample dissolution medium **A10:** 166
identification labelling **A9:** 67
safety hazards **A9:** 69
Hydrogen pick-up *See* Hydriding of Zirconium
Hydrogen pinholes *See* Pinholes
Hydrogen porosity
aluminum alloys **A15:** 747
sources, copper alloys **A15:** 464
Hydrogen porosity, in aluminum alloy ingots **A9:** 633
alloy 5052 .... **A9:** 632
alloy 6063 .... **A9:** 633
Hydrogen probes **A13:** 201
Hydrogen pusher furnace
for carburization of tungsten carbide powders **A7:** 157
Hydrogen reaction embrittlement **A13:** 283-284
Hydrogen reduction **A6:** 926
of atomized copper powders **A7:** 117, 118
in carbonyl vapormetallurgy processing **A7:** 92
of composite powder **A7:** 173
of copper oxide **A7:** 107, 109
of copper powders **A7:** 120
of iron powders for food enrichment **A7:** 615
of nickel-coated composite powders **A7:** 173-174
in Pyron process **A7:** 83
Hydrogen relief treatment
of steel springs **A1:** 312
Hydrogen removal
aluminum alloys **A15:** 457-462, 747-749
by gas purging **A15:** 460
by VID processing **A15:** 439
plain carbon steels **A15:** 714
porous plug degassing **A15:** 461-462
theory of **A15:** 457
vacuum induction furnace **A15:** 395
Hydrogen selenide
toxic effects **A2:** 1254
Hydrogen solubility
carbon effects **A15:** 82
in cast iron **A15:** 82
in copper **A15:** 466
in copper alloys **A15:** 86, 464-465
in magnesium alloys **A15:** 462, 465
and removal, in aluminum alloys **A15:** 85, 747-749
Hydrogen stoking furnace **A7:** 386
Hydrogen storage alloys
as rare earth application **A2:** 731
Hydrogen stress cracking
curves **A13:** 168
defined (under Hydrogen embrittlement) **A13:** 8
as hydrogen damage **A13:** 163-164
of linepipe steel **A13:** 538
in petroleum refining and petrochemical operations **A13:** 1278-1279
stainless steel splice case bolts **A13:** 1132-1133
strain rate effects, schematic **A13:** 261
of telephone cables **A13:** 1130
in telephone stainless steel clamp **A13:** 1133
Hydrogen sulfate
as carcinogenic precipitant **A10:** 169
Hydrogen sulfide
concentration, and cracking **A11:** 298
contamination, condenser tube pitting by **A11:** 631
copper/copper alloy resistance **A13:** 632
corrosion **A11:** 631
corrosion caused by **M1:** 726
corrosion in fresh water, effect on **M1:** 733
Hydrostatic testing, steel castings **M1:** 402
corrosion, in oil/gas production **A13:** 1232-1233, 1247
as crude oil contaminant **A13:** 1266
effect, copper corrosion in seawater **A13:** 906
environments containing **A11:** 246
formation in cutting fluids **A16:** 132
high-strength steel contamination with **A8:** 527
SCC, of pressure vessel welds **A11:** 426
titanium/titanium alloy SCC in **A13:** 690
wet, cracking resistance **A13:** 330
Hydrogen sulfides
effect on alloy steels **A12:** 299
Hydrogen trapping **A13:** 166-167
Hydrogen water chemistry
for boiling water reactor corrosion **A13:** 932-933
Hydrogen-assisted cold cracking
relative susceptibility of steels **A6:** 407
Hydrogen-assisted cracking
welded cast steels **A15:** 532-534
Hydrogen-assisted stress-corrosion cracking
defined (under Hydrogen embrittlement) **A13:** 8
of radioactive waste containers **A13:** 971
Hydrogen-bonded pyridine
Raman analyses **A10:** 134
Hydrogen-damage failures *See also* Hydrogen **A11:** 245-251
analysis of **A11:** 250
of cadmium-plated alloy steel bolts **A11:** 540-541
cracking from hydride formation **A11:** 248-249
cracking, from precipitation of internal hydrogen **A11:** 248
in forging **A11:** 336-338
hydrogen attack **A11:** 248
hydrogen embrittlement **A11:** 245-247
hydrogen-induced blistering **A11:** 247-248
metal susceptibility **A11:** 249-250
in nonferrous alloys **A11:** 338
in pipelines **A11:** 701
prevention of **A11:** 250-251
in steel forging **A11:** 336-337
types of **A11:** 245
Hydrogen-induced cold cracking **A6:** 436-437, 438-439
Hydrogen-induced cracking *See also* Hydrogen stress cracking; Underbead cracking **A1:** 606-607 **A6:** 379, 1111
carbon steels **A6:** 641, 642-649
defined **A13:** 8
duplex stainless steels **A6:** 697
evaluation by implant testing **A6:** 606
high-strength low-alloy steels **A6:** 73
low-alloy steels for pressure vessels and piping **A6:** 668
mild steels **A6:** 649
partially melted zone **A6:** 75
residual stresses **A6:** 1102
solid-state transformations in weldments **A6:** 79, 80
steel weldment soundness **A6:** 408, 410-415
arc energy input **A6:** 412
carbon equivalence **A6:** 412, 413
control **A6:** 411-413
cooling rates **A6:** 412, 413, 414
critical stress **A6:** 410-411, 414
distortion **A6:** 411
fatigue life **A6:** 411
filler metals **A6:** 413
measurement of hydrogen **A6:** 413
mechanism **A6:** 410-411
microcracks **A6:** 410
postweld heat treatment **A6:** 411
preheat temperature effect **A6:** 412
rebaking of electrodes **A6:** 415
residual lubricants **A6:** 413
residual moisture **A6:** 414-415
residual stresses **A6:** 410-411
stress relief annealing (SRA) **A6:** 411
temperature effect on hydrogen solubility in iron **A6:** 410, 411
variations of **A6:** 410
welding process importance **A6:** 413-415
steel weldments **A6:** 424
Hydrogen-induced cracking (HIC) *See also* Hydrogen
flakes **A11:** 79
in pipeline **A11:** 701
root, in steel weldment **A11:** 92, 93
in sour gas environments **A11:** 299
tests **A11:** 302
in weld toe, low-carbon steel **A11:** 251
Hydrogen-induced cracking resistant steels **A1:** 400, 416
Hydrogen-induced cracking technique **A6:** 1095
Hydrogen-induced delayed cracking *See* Hydrogen embrittlement
defined **A8:** 7 **A11:** 5
Hydrogen-induced failures
aircraft **A13:** 1032-1034
Hydrogen-ion activity, negative logarithm
symbol for **A8:** 725
Hydrogen-stress cracking **A1:** 342
characteristics of **A1:** 711

Hydrogen-stress cracking (continued)
and loss of tensile ductility **A1:** 712-717
of pipeline **A11:** 701
Hydrogen-to-water ratios **A7:** 343, 346
Hydrogen/argon
shielding gas for laser cladding **A18:** 867
Hydrogenated amorphous silicon **EM4:** 22
applications **EM4:** 22
Hydrogenated styrene-diene (STD)
as viscosity improvers **A18:** 109
Hydrogenation
powders used **A7:** 165, 574
Hydrogenation catalysts
powders used **A7:** 572
Hydrogenation of olefines
powder used **A7:** 574
Hydrohalides **A6:** 130
Hydrolloy
cavitation resistance **A18:** 600
Hydrolysis **EM3:** 15 **EM4:** 448
defined **A13:** 8 **EM2:** 22
of organofunctional silanes **EM1:** 123
sample, nitric acid to prevent **A10:** 166
thermoplastic resins **EM2:** 619
to oxides, precipitation by **A10:** 169
Hydrolytic resistance
polyaryl sulfones (PAS) **EM2:** 145-146
polybenzimidazoles (PBI) **EM2:** 147
polyether sulfones (PES, PESV) **EM2:** 161
Hydrolytic stability
polyphenylene ether blends (PPE PPO) **EM2:** 183-184
polysulfones (PSU) **EM2:** 200
thermoplastic polyurethanes (TPUR) **EM2:** 203-204
Hydrolytic stability test
for solder masks **EL1:** 554
Hydrolyzed ethyl silicate
manufacture of **A15:** 212
Hydromechanical press **EM3:** 15
defined **EM1:** 13 **EM2:** 22
Hydrometallugical nickel powders **A7:** 173, 299, 401, 587
Hydrometallurgical cobalt powders **A7:** 144-145, 402
Hydrometallurgical copper powders **A7:** 118-120, 734
Hydrometallurgical processing **A7:** 54, 118-120, 134, 138-142, 144-145
Hydrometallurgy
copper powders **A2:** 393-394
of nickel and nickel alloys **A2:** 429
Hydrometer
for flux specific gravity measurement **EL1:** 649
Hydroperoxide + thiourea and metal salt
generating free radicals for acrylic adhesives **EM3:** 120
Hydroperoxides **EM3:** 113
Hydrophile-lipophile balance (HLB) **A18:** 141
required HLB **A18:** 141
Hydrophilic **EM3:** 15
defined **A13:** 8 **EM1:** 715 **EM2:** 22
Hydrophilic/lipophilic (Methods B and D) post emulsifiable penetrants **A17:** 75, 77
Hydrophobic **EM3:** 15
defined **A13:** 8 **EM1:** 13
Hydroplaning
defined **A8:** 604
Hydropneumatic die cushions **A14:** 498
Hydroscopics
Hydrosol/coupling agent
polyaramid fibers **EM3:** 286
Hydrostatic bearing
defined **A18:** 11
Hydrostatic compacting **A7:** 6
Hydrostatic extrusion *See also* Extrusion **A14:** 327-329 **A18:** 59, 61
applications **A14:** 329
defined **A14:** 8
Hydrostatic extrusion (continued)
densification by **A7:** 300
equipment **A14:** 329
hot **A14:** 328
hydrafilm process **A14:** 328-329
simple **A14:** 327
simple, of brittle materials **A14:** 328
tooling **A14:** 329
Hydrostatic gas-lubricated bearings **A18:** 528-529
Hydrostatic journal bearings **A18:** 531-532
Hydrostatic leak testing
of pressure systems **A17:** 66
Hydrostatic lubrication *See also* Pressurized gas lubrication **A18:** 89, 91-92
applications **A18:** 91
in bearings **A11:** 484
defined **A18:** 11
wear and **A11:** 151
Hydrostatic measuring devices **A4:** 507
Hydrostatic modulus *See* Bulk modulus of elasticity; Bulk modulus of elasticity (*K*)
Hydrostatic mold **A7:** 6
Hydrostatic pressing **A7:** 6
rolling, and extrusion **A7:** 517
and uniform stress **A7:** 300
Hydrostatic pressure
noninfluence in material behavior **A8:** 343
in pressure-shear plate impact testing **A8:** 237
Hydrostatic seal
defined **A18:** 11
Hydrostatic testing
gas-tungsten arc welding **A6:** 451
of heat exchangers **A11:** 629
and microbiological corrosion **A13:** 314
Hydrotesting
failure of utility boiler drum during **A11:** 647
Hydrous alumina **EM3:** 175
Hydroxide alkaline cleaners **M5:** 24
Hydroxide precipitation process
plating waste treatment **M5:** 313-314
Hydroxides
as alkalies **A13:** 1179
molten, corrosion **A13:** 1200
as molten salt **A13:** 50, 91
as precipitants **A10:** 168-169
as salt precursors **EM4:** 113
solutions, copper/copper alloy SCC **A13:** 634
in torch brazing flux **M6:** 962
Hydroxyacetic-forming acid **A13:** 1141
Hydroxyapatite
coating for metal porous surfaces on prostheses **A18:** 658
in composition of human enamel (dental) **A18:** 667, 668, 669
Hydroxyapatite ceramic
approximated properties of human enamel (dental) in wear studies **A18:** 668
Hydroxyethylcellulose
removal **EM4:** 137
Hydroxyl
content in glass, quantitative analysis **A10:** 121-122
groups, content in glass **A10:** 121-122
ion, monitoring by acid-base titration **A10:** 172
Hydroxyl group **EM3:** 15
defined **EM2:** 22
Hydroxyl sulfide **EM3:** 100
Hydroxylapatite **EM4:** 208, 1009
applications, medical **EM4:** 1011, 1012
bioactive ceramic coating **EM4:** 1007, 1008
bonding mechanisms **EM4:** 1012
Hydroxypropyl cellulose
as binder for injection molding **EM4:** 173
Hydroxypropylmethyl cellulose
as binder **EM4:** 474
Hygroelastic resin matrix composites
defined **EM1:** 226
Hygroscopic **EM3:** 16
analysis, of laminates **EM1:** 226-227
Hygroscopic (continued)
defined **A13:** 8 **EM1:** 13 **EM2:** 22
Hygrothermal
behavior **EM1:** 13, 190, 460
elasticity **EM1:** 310
Hygrothermal effect **EM3:** 16
defined **EM2:** 22
HyMu 80, 800
photochemical machining etchant **A16:** 590
Hyperbaric welding
pressure effect **A6:** 58
underwater welding **A6:** 1010, 1011, 1014
Hyperbolic-weight constant-stress
test apparatus **A8:** 318-319
Hypereutectic alloy
defined **A9:** 9
Hypereutectic alloys ...... **A3:** 1 • 3
aluminum-silicon, refinement of **A15:** 753
modification of **A15:** 482
modifier additions **A15:** 484
Hypereutectic gray iron
structure of bearing cap cast from **A11:** 349
Hypereutectic silicon refinement
by flux injection **A15:** 453
Hypereutectoid alloy
defined **A9:** 9
Hypereutectoid alloys ...... **A3:** 1 • 21
Hypereutectoid carbon steels *See also* Carbon steels **A9:** 178
Hypereutectoid compositions
Bagaryatski orientation relationship **A9:** 658
decomposition **A9:** 659
Hypereutectoid plain carbon steels
super-plasticity **A14:** 869
Hyperfine interaction constants
effect of low temperature on **A10:** 257
Hyperfine splitting
in analysis of transition group metals **A10:** 260
and sensitivity reduction **A10:** 258
Hyperglycemia
from cobalt toxicity **A2:** 1251
Hyperrine structure
abbreviation for **A10:** 690
determination of intensity ratios **A10:** 261
ESR spectrum **A10:** 259
patterns **A10:** 260
unresolved, as ESR line-broadening mechanism **A10:** 255
Hypersonic combustion flame spray (HCFS) guns **EM4:** 204-205, 206, 207
Jet-Kote method **EM4:** 204
Hypersonic flame spraying **EM4:** 203, 204-205, 206, 207
Jet-Kote method **EM4:** 204
Hypertension
from cadmium exposure **A2:** 1241
Hypervelocity oxyfuel powder spray process *See* High-velocity oxyfuel powder spray process
Hypochlorites **A13:** 727, 1179-1180, 1194
Hypodermic needles
burrs detected on **A17:** 13
Hypoeutectic alloys ...... **A3:** 1 • 3
aluminum-silicon, structural effects **A15:** 167-168
flux injection **A15:** 453
modification of **A15:** 453, 482
modifier additions **A15:** 484
Hypoeutectic aluminum-silicon alloys
modification of **A2:** 134
Hypoeutectic cast iron
bearing structure **A11:** 349
Hypoeutectic iron
solidification in **A1:** 13-14
Hypoeutectoid alloys ...... **A3:** 1 • 21
Hypoeutectoid steel compacts
heat treating **A7:** 558
Hypoeutectoid/eutectoid plain carbon steels
as superplastic **A14:** 869
Hypoid
bevel gears, described **A11:** 587

Hypoid (continued)
gears, gear-tooth contact in **A11:** 588
pinion, extreme wear in **A11:** 597
Hypoid gear lubricant (hypoid oil)
defined **A18:** 11
Hypophosphite reduced electroless nickel plating *See* Sodium hypophosphite electroless nickel plating process
Hypophosphorous acid
as reducing agent **A10:** 169
Hypotheses testing **A8:** 626-627
Hypothesis, testing
and systematic samples **A10:** 12-13
Hypothetical standard state
thermal properties for **A15:** 56-57
Hysteresis *See also* Damping; Demagnetization; Heat buildup; Hysteresis loops; Magnetic hysteresis ... **EM3:** 16
applications, permanent magnet materials **A2:** 793-794
defined **A8:** 7 **A17:** 100, 384 **EM1:** 13 **EM2:** 22
effect, defined **A2:** 782
as effect, orientation stability of domain **A17:** 145
energy loss, magnetically soft materials **A2:** 761
in fracture toughness testing **A8:** 474
loop, permanent magnet materials **A2:** 782, 784
losses, in superconductors **A2:** 782, 784
magnetic **A2:** 782, 784 **A17:** 99-100
magnetic induction distortion by **A17:** 161
of magnetoresistance of nickel **A17:** 144
measurement **A2:** 782
properties, and magnetic measurement **A17:** 129
in torsional testing **A8:** 147
Hysteresis index **A18:** 423
Hysteresis loop **A7:** 6 **EM3:** 16
compression-going part **A8:** 356
defined **EM2:** 22-23
partitioned strain-range components **A8:** 357-358
tension-going part **A8:** 356
torsional **A8:** 150-151
Hysteresis loops
in magabsorption theory **A17:** 145
for magnetic materials, Barkhausen noise **A17:** 159
measurement **A17:** 132-133
shape effects **A17:** 100
stress effects on **A17:** 161
superimposed **A17:** 146
under magnetomotive forces **A17:** 147
Hysteretic viscoelastic effects
in polymers **A11:** 758
Hytrel **EM4:** 665, 666
n-Hexane (C_6H_{14})
as solvent used in ceramic processing **EM4:** 117
The Handbook of Adhesive Raw Materials **EM3:** 68
The Handbook of Pressure-Sensitive Adhesive Technology **EM3:** 70
The Handbook of Surface Preparation **EM3:** 70

I

acoustic-emission **A11:** 18
eddy-current **A11:** 17
liquid-penetrant **A11:** 17
magnetic-particle **A11:** 16-17
nondestructive testing as **A11:** 134
practices, of pressure vessels **A11:** 654
procedures, effect on heat-exchanger failure **A11:** 629-630
schedules and techniques, for fatigue failures **A11:** 134
ultrasonic **A11:** 17
visual, schedules and techniques **A11:** 134
In cake filtration
inclusion removal **A15:** 489
fn-Pr (Phase Diagram) .... **A3:** 2 • 256
I *See* Intensity
I beams
wrought aluminum alloy **A2:** 34
I-beam
towpreg filler in **EM1:** 152
I-lead
defined **EL1:** 1146
I-pores *See also* Internal particle porosity
$I/I_{corundum}$ method
XRPD analysis **A10:** 340
I/M *See* Ingot metallurgy
I/O *See* Input/output
I^2R heating
thermostat metals **A2:** 826
IA *See* Image analysis
IAA *See* International Aerospace Abstracts
IACS *See* International Annealed Copper Standard
IAP *See* Imaging atom probe
IBM thermal conduction modules **EL1:** 48
I_c *See* Coherent atomic scattering intensity; Ion chromatography
Ice
friction coefficient data **A18:** 75
Icicles **EL1:** 642, 684, 688
ICP *See* Inductively coupled plasma; Inductively coupled plasma atomic emission spectroscopy
ICP sample introduction
to analytical system **A10:** 34-36
ICP torch
and gas supplies **A10:** 34, 36-37
ICP-AES *See* Inductively coupled plasma atomic emission spectroscopy
ICP-MS *See* Inductively coupled plasma mass spectroscopy
ICPMS *See* Inductively coupled plasma mass spectroscopy
Ideal crack
defined **A8:** 7
model, progressive fracturing **A8:** 440
Ideal diameter *See* Critical diameter
Ideal entropy of mixing
solidification thermodynamics **A15:** 101
Ideal fracture energy **A6:** 143
Ideal gas
equation of state **A17:** 58-59
Ideal linear elastic behavior
in fracture toughness testing **A8:** 474
Ideal solution
described **A15:** 101-102
entropy of **A15:** 103
free energy of formation **A15:** 52
Ideal water content
silica-base bonds **A15:** 212
Ideal-crack-tip stress field *See also* Modes
Ideality
and activity coefficients **A15:** 51-52
Identification *See also* Identification marks
of carbonitrides **A11:** 39
chart, of fracture modes **A11:** 80
in corrosion analysis **A11:** 174
of fatigue fracture, in boilers and steam equipment **A11:** 621
of iron oxide inclusions **A11:** 38-39
of microconstituents, in failed extrusion press **A11:** 37
of presses **A14:** 489
systems, stainless steels **A13:** 547
of test specimens **A13:** 205
of types of failures **A11:** 75-81
Identification etching
defined **A9:** 9
Identification markers
placement of **A17:** 341-343
radiographic inspection **A17:** 338-341
Identification marks *See also* Identification stamps
cracks in shaft from **A11:** 472-473
effect on fatigue strength **A11:** 130
as fatigue fracture source **A11:** 130
on threaded fasteners **A11:** 529, 531
as postforging failure process **A11:** 331
Identification stamps, steel
cracking from **A11:** 472-473
Identification systems
forgings **M1:** 351
Identification tags
for material traceability **EM1:** 741
Identity operation
defined in crystal symmetry **A10:** 346
Idiomorphic crystal
defined **A9:** 9
Idiomorphic particles .... **A3:** 1 • 20
Idiopathic hemochromotosis
due to iron toxicity **A2:** 1252
Idler arms
farm machinery **A7:** 674
Idler pivots
garden equipment **A7:** 677
IGBT devices **A6:** 43
IGF *See* Inert gas fusion
Igniters
compositions **A7:** 604
with metal fuels **A7:** 600, 604
Ignition
autogenous and pyrophoric **A7:** 194, 198-200
factors affecting **A7:** 196
pellet **A7:** 604
secondary sources **A7:** 197
sources for explosions **A7:** 195-197
temperatures, for magnesium powders **A7:** 133
Ignition loss *See also* Loss on ignition **EM3:** 16
defined **EM1:** 13 **EM2:** 23
Ignition-resistant high-impact polystyrenes (PS HIPS) **EM2:** 199
Ignitron tubes
resistance spot welding **M6:** 470-471
Ignitrons **A6:** 42
I_i *See* Incoherent atomic scattering intensity
Iionization gage
in vacuum pumping system **A8:** 414
IIW standard reference blocks
ultrasonic inspection **A17:** 264
Illite **EM4:** 6
as green sand molding clay **A15:** 224
Illium alloys *See* Nickel alloys, cast, specific types; Stainless steels, cast, specific types
Illium H *See* Superalloys, cobalt-base, specific types, UMCo-50
Illuminance
SI derived unit and symbol for **A10:** 685
SI unit/symbol for **A8:** 721
Illumination *See also* Bright-field images; Dark-field images
ambient, optical holographic interferometry **A17:** 413
in borescopes and fiberscopes **A17:** 8-9
bright-field **A10:** 310, 689
dark-field **A10:** 690
effect in fracture surface, medium-carbon steels **A12:** 257
global and punctual, in MOLE/Raman analyses **A10:** 129-130
machine vision, schematic **A17:** 32
for optical testing **EL1:** 570
phase-contrast **A12:** 438
photographic, effects on fracture surfaces **A12:** 82-89
in SEM imaging **A12:** 167-168
shift, photo effects **A12:** 86
stroboscopic, of rotating parts **A17:** 13
ultraviolet **A12:** 84-85
vision machine **A17:** 31
Illumination modes
optical etching **A9:** 57-59
for René 95, compared **A9:** 323
for Waspaloy, compared **A9:** 150
for wrought stainless steels compared **A9:** 290-291, 294
Illumination system
of a transmission electron microscope **A9:** 103-104
of an optical microscope **A9:** 72
Illumination techniques
compared **A9:** 78-82
for porcelain enameled sheet steel **A9:** 199
Illuminators
Nicholas **A12:** 81
stereomicroscope **A12:** 78
Ilmenite
chemical composition **A6:** 60
as silica sand impurity **A15:** 208
ILZRO 16 zinc alloy die castings **A2:** 529
properties **A2:** 538
Image **A18:** 346
defined **A9:** 9
Image analysis *See also* Image(s); Imaging; Imaging analysis; Metallographic identification; Surface topography and image analysis (area) **A10:** 309-322
applications **A10:** 309, 316-320
by material contrast **A9:** 94

Image analysis (continued)
computed tomography (CT) **A17:** 378
data analysis **A10:** 313
defined **A10:** 674
electronic **A9:** 138-139, 152-153
estimated analysis time **A10:** 309
general uses **A10:** 309
image analyzers **A10:** 310-313
of inorganic solids **A10:** 4-6
introduction **A10:** 309-310
limitations **A10:** 309
in machine vision process **A17:** 34-35
for microstructural analysis **EM4:** 26, 578
NMR, ESR, and UV/VIS analysis compared **A10:** 265
of organic solids **A10:** 9
for particle sizing **A7:** 225-230
possible errors **A10:** 313-316
procedure, block diagram **EL1:** 366
in radiographic interpretation **A17:** 348
related techniques **A10:** 309
samples ..... **A10:** 309 , 313, 316-320
thermal **EL1:** 368-369
types of image analyzers **A10:** 309
visual **EL1:** 366

Image analysis system
connected to scanning electron microscope **A9:** 138

Image analyzer
in quantitative metallography **A9:** 83, 85
used for scanning electron microscopy **A9:** 96

Image analyzers
components of **A10:** 310-313
data analysis **A10:** 313
detection and measurement **A10:** 311-313
input devices **A10:** 310
scanners **A10:** 310-311

Image artifacts
aliasing and Gibbs phenomenon **A17:** 375-376
defined **A17:** 375

Image contrast *See also* Etching **A17:** 373-375
defined **A10:** 674
different films compared **A9:** 86-87
different illuminations compared **A9:** 78-82
different paper grades compared **A9:** 86-87
field ion microscope **A10:** 588
linear attenuation coefficient values **A17:** 374
in micrograph print making **A9:** 85
objective contrast **A17:** 374
in scanning electron microscopy **A10:** 500-504
techniques in optical microscopy **A9:** 76-82

Image conversion *See also* Image conversion media; Images
incoherent-to-coherent, microwave holography **A17:** 227-228
intensifying and filtration screens **A17:** 298
radiography **A17:** 298
real-time imaging media **A17:** 298
recording media **A17:** 298

Image conversion media *See also* Image(s)
fluorescent intensifying screens **A17:** 316-317
fluorometallic screens **A17:** 317
lead oxide screens **A17:** 316
lead screens **A17:** 315-316
metal screens **A17:** 316
radiographic **A17:** 314-317
recording media **A17:** 314-315

Image display *See also* Image(s)
concept **A17:** 455
CRT displays **A17:** 461-462
printers **A17:** 462-463
pseudo three-dimensional images **A17:** 463
videotape/videodisk **A17:** 463

Image distortions **A9:** 77

Image enhancement *See also* Digital image enhancement **A17:** 456-458
charge-coupled device (CCD) **A17:** 10
digital **A17:** 454-464
as failure analysis technique **EL1:** 1073

Image enhancement (continued)
filtering **A17:** 459
geometric processes **A17:** 458
histogram equalization, color **A17:** 484
image combination **A17:** 458-459
and image processing **A17:** 454

Image gas atom
potential energy of outer electron **A10:** 586

Image gases, low ionization
for field ion microscopy **A10:** 587

Image isocons
in optical sensors **A17:** 10

Image modification
in scanning electron microscopy **A9:** 95

Image orthicons
in optical sensors **A17:** 10

Image preprocessing
image analysis **A10:** 310

Image processing *See also* Digital image enhancement **A17:** 456
C-scan **EM2:** 843-845
capabilities, computers **A17:** 455
of color images **A17:** 483-488
computed tomography (CT) **A17:** 378
in digital image enhancement **A17:** 454
frame integration (summing) **A17:** 320
information extracted **A17:** 460-461
NDE functions **A17:** 454
real-time radiography **A17:** 320-323

Image processing arrays **EL1:** 8

Image quality *See also* Image(s)
computed tomography (CT) **A17:** 372-377
deficient, causes/corrections **A17:** 346, 355
defined, radiographic inspection **A17:** 338
detail perceptibility, of images **A17:** 300
DIN standards **A17:** 341
radiographic contrast **A17:** 298-299
radiographic definition **A17:** 299-300
and radiographic sensitivity **A17:** 298-300
in scanning electron microscopy **A9:** 95

Image reconstruction algorithm
computed tomography (CT) **A17:** 359-360, 380

Image resolution **A7:** 580

Image restoration
in machine vision process **A17:** 33

Image rotation
defined **A9:** 9

Image shearing eyepieces **A7:** 229

Image unsharpness *See* Unsharpness

Image(s) *See also* Digital image enhancement; Image analysis; Image contrast; Image conversion; Image conversion media Image display; Image enhancement; Image processing; Image quality; Imaging Imaging system
acoustic microscopy **A17:** 469
artifacts **A17:** 375-376
capture and acquisition, digital image enhancement **A17:** 455-456
color, by various NDE methods **A17:** 483-488
composite, with digital image enhancement **A17:** 457
computed tomography (CT) **A17:** 360
contrast **A17:** 373-375
detail perceptibility, radiography **A17:** 300
digitization devices **A17:** 456
display **A17:** 455, 461-463
filtering **A17:** 379, 459-461
formation, geometry of **A12:** 196
formation, machine vision **A17:** 31-33
intensifiers **A17:** 318-319
machine vision, interpretation **A17:** 35-37
as NDE response, and NDE reliability **A17:** 674
organization, human vs. machine vision **A17:** 30
preprocessing, machine vision **A17:** 33-34
projected, quantitative fractography **A12:** 194-196
restoration, machine vision **A17:** 33
shapes, human vs. machine vision **A17:** 30

Image(s) (continued)
statistics and measurement, digital imaging processing **A17:** 460
stereo, fractographic **A12:** 87-88
subtraction, in real-time radiography **A17:** 320
thermal inspection, interpretation **A17:** 400
three-dimensional, holograms as **A17:** 405
whole, parallel processing of **A17:** 44
width, photomacrographic **A12:** 80

Image-processing equipment
ultrasonic inspection **A17:** 253-254

Image-quality indicators *See* Penetrameters

Image-verification procedure **A10:** 438, 440

Images
contrast **A10:** 500-504, 588, 674
FIM, formation of **A10:** 584
FIM, quantitative analysis **A10:** 590-591
from scanning electron microscopy **A9:** 90
typical field ion micrograph (tungsten) **A10:** 585

Imaging *See also* Stereo imaging
backscattered electron (BSE) imaging **EL1:** 1094, 1096-1098
capabilities/limitations **EL1:** 508-510
of crystalline structure of integrated circuit Auger electron spectroscopy **A10:** 555
defects, x-ray topography **A10:** 367-368
detectors **A10:** 144, 493-494
digital micro **A10:** 447, 448
dual energy **A17:** 378-379
flicker-free **A10:** 525
image-verification procedure **A10:** 438, 440
infrared **A8:** 247, 248
ion, as SIMS application **EL1:** 1085
in neutron radiography **A17:** 387-388
partial angle **A17:** 379
photodiode array **A17:** 12-13
photoimaging **EL1:** 509-510
positive, for automatic optical inspection **EL1:** 942
primary **EL1:** 548
process, rigid printed wiring boards **EL1:** 545
real time, as neutron detection method **A17:** 391
rigid printed wiring boards **EL1:** 542
screen printing **EL1:** 508-509
second phase, by scanning electron microscopy **A10:** 490
secondary electron, AES detector for **A10:** 554
SEM, clarity **A12:** 168
stereo, SEM display systems **A12:** 171
stereoscopic, in quantitative fractography **A12:** 196-197
surface features, by scanning electron microscopy **A10:** 490
system, molecular optical laser examiner **A10:** 129-130
system, SEM **A12:** 167-168
TEM and STEM modes, relationship between **A10:** 442
in the analytical electron microscope **A10:** 440-446
in the STEM mode **A10:** 442
in the TEM mode **A10:** 440-442
thermal wave, powder metallurgy parts **A17:** 541
thermal-wave **A12:** 169
of topographic or microstructural features **A10:** 299
ultrasonic, of powder metallurgy parts **A17:** 540

Imaging analysis *See also* Image analysis; Imaging
acoustic microscopes **EL1:** 1069-1071
infrared microscopes **EL1:** 1068-1069
infrared thermography **EL1:** 1071
liquid crystals **EL1:** 1071-1072
optical microscopy **EL1:** 1067-1068
photography **EL1:** 1072-1073
special techniques **EL1:** 1067-1073

Imaging atom probe
advantage for single elements **A10:** 596

Imaging atom probe (continued)
and atom probe analysis, as complementary **A10:** 596-597
of interfacial segregation in molybdenum **A10:** 599-601
schematic.......... **A10:** 595
Imaging methods
for use with cemented carbides.......... **A9:** 275
Imaging modes
in transmission electron microscopy **A9:** 103-104
Imaging systems
high-resolution infrared **A17:** 398
pressure vessel.......... **A17:** 654
real-time radiography.......... **A17:** 322
Imaging technique
secondary ion mass spectroscopy as **A7:** 257
IMC *See* Intermetallic-matrix composites
IMI 834
solution treating **A4:** 917, 918-919
Imidazoles **EM3:** 95
Imide monomers
formation **EM1:** 78
Imidization *See also* Curling
and cure, compared **EL1:** 772
kinetics **EL1:** 772
Immersed arcing
refractory wear **A15:** 437
Immersed electrodes furnace
dip brazing **A6:** 337
Immersion *See also* Immersion coatings; Immersion plating; Immersion tests; Total immersion tests
for chromate conversion coatings **A13:** 390
cleaning, for urethane coatings.......... **EL1:** 777
cooling **EL1:** 2, 49, 55
corrosion, cemented carbides.......... **A13:** 857
depths, wave soldering **EL1:** 688
environments, thermal spray coatings for **A13:** 460
plates, as surface preparation **EL1:** 679
plating **A13:** 8, 430
soft solder corrosion by **A13:** 774
solder plating.......... **EL1:** 564
Immersion cleaning
acid *See* Acid cleaning, immersion process
alkaline *See* Alkaline cleaning, immersion process
emulsion *See* Emulsion cleaning, immersion process
Immersion coatings
for marine corrosion **A13:** 916-918
organic.......... **A13:** 916-918
tin **A13:** 776
Immersion cooling.......... **EL1:** 2, 49, 55
Immersion etching
defined **A9:** 9
using nitric acid solution **A16:** 36
Immersion lens *See* Immersion objective
Immersion objective
defined **A9:** 9
Immersion oil
used in magnetic etching.......... **A9:** 64
Immersion paint-stripping method **M5:** 18-19
Immersion phosphate coating systems.......... **M5:** 434-437, 440-443, 445-448, 450-453
equipment.......... **M5:** 445-448, 450-451
immersion time.......... **M5:** 441, 448, 456
safety precautions **M5:** 453-454
Immersion plating.......... **A13:** 8, 430
aluminum and aluminum alloys.......... **M5:** 601-607
copper and copper alloys **M5:** 601-606
double immersion process.......... **M5:** 603-604, 606
magnesium alloys **M5:** 638-639, 644-646
nickel **M5:** 219
Immersion pulse-echo ultrasonic testing, to detect cracks
electron-beam welding.......... **A6:** 1077
Immersion techniques *See also* Immersion ultrasonic inspection
basic, ultrasonic inspection.......... **A17:** 258
for pressure systems.......... **A17:** 60
transmission ultrasonic inspection.......... **A17:** 248
ultrasonic, with solvents **A17:** 81, 82
Immersion tests *See also* Total immersion tests **A13:** 114, 207, 220-224, 236-238, 265-266
Immersion time
alloying and bouyancy effects **A15:** 72
Immersion tin coating **A13:** 776
Immersion ultrasonic inspection *See also* Immersion techniques; Immersion-type ultrasonic search units; Ultrasonic inspection
of adhesive-bonded joints.......... **A17:** 617-619
of forged aluminum alloy.......... **A17:** 510
of nonferrous tubing.......... **A17:** 574
Immersion-type ultrasonic search units *See also* Search units
basic immersion.......... **A17:** 258
water-column designs **A17:** 258-259
wheel-type **A17:** 259
Immiscible.......... **EM3:** 16
defined **EM2:** 23
Immiscible blend
defined **EM2:** 632
Immiscible elements
alloyed by high- energy milling.......... **A7:** 70
Immiscible liquids
rare earth metals.......... **A2:** 726-727
Immunity
defined **A13:** 8
Immunity, of iron
in water and dilute aqueous solutions.......... **A11:** 198
Immunoassays
MFS use of fluorescent labels for.......... **A10:** 72
Impact
bar test, strain rate ranges for.......... **A8:** 40
behavior, effect of hydrogen flaking **A11:** 316
distortion from.......... **A11:** 138
dynamic fracture by.......... **A8:** 259
effect in ceramics.......... **A11:** 753-755
effect of.......... **A7:** 60
energy, defined.......... **A8:** 7 **A11:** 5
faces, optical alignment for plate testing **A8:** 235
as failure mode in gears **A11:** 590, 595
force as proportional to mass of milling medium **A7:** 60
fracture **A11:** 75, 76, 753
fracture toughness.......... **A11:** 54
Hertzian cone crack from **A11:** 753
improvers, effect, chemical susceptibility.......... **EM2:** 572
load, defined **A8:** 7, 259 **A11:** 6
modifiers, as additives **EM2:** 497
properties, correlated with toughness.......... **A11:** 54-55
properties, defined **EM2:** 434
resistance, low, brittle fracture from.......... **A11:** 90-91
response curves **A8:** 269-272
rod, in symmetric rod testing **A8:** 203-204
wear **A8:** 603
Impact attrition mill.......... **A7:** 757
Impact avalanche transit time (IMPATT) diode
microwave inspection.......... **A17:** 209
Impact breaker bar
rapid wear from retained austenite.......... **A11:** 367-368
Impact bruise
definition.......... **EM4:** 633
Impact compaction **A7:** 56-59
Impact cutoff machines.......... **A14:** 714, 718-719
Impact damage *See also* Damage; Damage tolerance
adhesive-bonded joints.......... **A17:** 615
cross section **EM1:** 260
delamination as **EM1:** 259-260
and design configuration.......... **EM1:** 263-264
material effects.......... **EM1:** 262-264
residual strength, resin effects **EM1:** 262
of specimens.......... **EM1:** 296
tolerance requirements.......... **EM1:** 265
Impact energy *See also* Charpy, test; Impact response curves; Impact testing; Izod testing **A8:** 7
of ferrous P/M materials.......... **A7:** 465, 466
Impact extrusion *See also* Extrusion
defined **A14:** 8
equipment and tooling.......... **A14:** 311
of magnesium alloys.......... **A14:** 311-312, 830
pressures **A14:** 312
procedure.......... **A14:** 311-312
thermal expansion.......... **A14:** 312
tolerances **A14:** 312
Impact extrusions, magnesium alloy, product form
forming, aluminum and aluminum alloys **A2:** 6
grinding, beryllium.......... **A2:** 684
selection.......... **A2:** 464-465
Impact fatigue.......... **A18:** 242
Impact fracture
Charpy, shear dimples in shear-lip zone of **A11:** 76
defined **A11:** 75
Impact fracture toughness
types compared **A11:** 54
Impact fractures
AISI/SAE alloy steels **A12:** 314, 319, 324, 328
iron.......... **A12:** 222
tool steels **A12:** 377, 378
Impact grinding system
beryllium powders.......... **A7:** 170, 171
Impact line
defined **A14:** 8
Impact loading **EM2:** 679-700
design and analysis techniques **EM2:** 691-700
effects, polymer deformation **EM2:** 680-681
metal vs plastic **EM2:** 75-76
response, material considerations **EM2:** 680-691
thin plastic components **EM2:** 691-700
Impact loading techniques **A8:** 259
Impact molding
development of.......... **A15:** 37
Impact parts, aluminum alloy
cold extrusion of.......... **A14:** 309-310
Impact polystyrene *See* High-impact polystyrenes (PS, HIPS)
Impact processes
as acoustic noise source **A17:** 285
Impact properties *See also* Notch toughness.......... **A1:** 61-62, 63, 64
carburized steels.......... **M1:** 534-536
cast steels **M1:** 378-381, 382, 389, 390, 392-393, 398, 399
CG iron.......... **A15:** 673
cold finished bars.......... **M1:** 224, 227-229, 231, 235, 246-249
corrosion-resistant cast irons.......... **M1:** 89
density effects **A14:** 202-203
of ductile iron **A1:** 40, 43, 44, 45, 46 **A15:** 661-662 **M1:** 39-42, 45
gray cast iron **M1:** 22
heat-resistant cast irons.......... **M1:** 92
hot finishing temperature effects **A14:** 220
hot-working reduction effects.......... **A14:** 218
malleable cast irons.......... **M1:** 65, 70-71
P/M steels **M1:** 334-335, 337-339, 342-343
powder forgings **A14:** 202-203
steel plate.......... **M1:** 194
ultrahigh-strength steels **M1:** 422, 423, 426-431, 433-436, 438, 442
Impact resistance *See also* Fiber properties analysis; Impact damage; Impact strength; Material properties analysis
of aluminum oxide-containing cermets **A7:** 804
of aramid fibers **EM1:** 36
of chromium carbide-based cermets.......... **A7:** 806
and composite design.......... **EM1:** 33
effects of iron oxide **A7:** 464
fabric vs. tape.......... **EM1:** 262
of gray iron.......... **A1:** 23
hot dip galvanizing.......... **A13:** 438
hybrid fabric composites.......... **EM1:** 149
of steel castings.......... **A1:** 365, 367-369
of titanium carbide-based cermets.......... **A7:** 808
of titanium carbide-steel cermets **A7:** 810
Impact response curves
concept of.......... **A8:** 259, 269-271
Impact sintering.......... **A7:** 6
Impact strength *See also* Impact loading; Impact tests; Impact toughness; Izod impact strength; Mechanical properties.......... **EM3:** 16
aluminum casting alloys.......... **A2:** 152
ASTM test methods **EM2:** 334
cast copper alloys.......... **A2:** 357-391
of cemented carbides for wear applications.......... **A7:** 778
commercially pure tin **A2:** 518
compared **EM2:** 168
as damage tolerance property **EM1:** 99
defined **EM1:** 13 **EM2:** 23
flexible epoxies **EL1:** 821
and flexural modulus **EM2:** 280
as fracture toughness, testing **EM2:** 739
Gardner **EM2:** 281
glass addition effect.......... **EM2:** 72
high-impact polystyrenes (PS, HIPS).......... **EM2:** 197

Impact strength (continued)
Izod, sheet molding compounds ... **EM1:** 158
in P/M forging ... **A7:** 415
polyester resins ... **EM1:** 91, 92
polyether sulfones (PES, PESV) ... **EM2:** 161
polyvinyl chlorides (PVC) ... **EM2:** 210
structural foams ... **EM2:** 509
styrene-maleic anhydrides (S/MA) ... **EM2:** 219
thermoplastics ... **EM1:** 292-293
of tungsten-reinforced composites ... **EM1:** 883-884
Impact test *See also* Charpy impact test; Izod impact test; Reverse impact test ... **EM3:** 16, 447-448
defined ... **EM1:** 13
Impact testing *See also* Charpy V-notch impact test; Izod testing ... **M1:** 689-691
defined ... **A8:** 7
pressure-shear plate ... **A8:** 230-238
stress-intensity ... **A8:** 453
torsional ... **A8:** 216-218
used for fracturing ... **A9:** 23
Impact tests *See also* Charpy impact test; Izod impact test; Reverse impact test; Tup impact test
defined ... **EM2:** 23
and tough-brittle transition ... **EM2:** 554-556
types ... **EM2:** 687-688
Impact toughness *See also* Impact strength; Mechanical properties
pure titanium ... **A2:** 596
wrought aluminum and aluminum alloys ... **A2:** 109-110
Impact toughness testing ... **A6:** 101
Impact value ... **EM3:** 16
Impact velocity
defined ... **A18:** 11
Impact wear ... **A18:** 263-270
defined ... **A18:** 11, 263
experimental background ... **A18:** 263
jet engine components ... **A18:** 588, 591
linear impact wear ... **A18:** 264-265
machine contacts ... **A18:** 265-266
measurable wear ... **A18:** 266
zero-wear limit ... **A18:** 265-266, 268
model for compound impact ... **A18:** 264
plotting a wear curve ... **A18:** 268-270
solution methods for measurable wear ... **A18:** 266-268
computational procedures ... **A18:** 267-268
examples ... **A18:** 266-267
Impact-modified acrylics ... **EM2:** 103, 105, 107
Impacting balls ... **A7:** 59, 60
Impaction ratio *See* Collection efficiency
Impacts
aluminum alloy ... **M2:** 8-10
Impedance *See also* Controlled impedance; Impedance models ... **EM3:** 428, 431, 435
acoustic ... **A17:** 234-235, 238, 476
acoustic, for various materials ... **A17:** 476
and admittance matrices ... **EL1:** 35-36
alternating current, measurement ... **A13:** 200-201
apparent ... **EL1:** 37
bridge, eddy current inspection ... **A17:** 176-178
calculated values ... **EL1:** 419
changes, by small flaws ... **A17:** 173
characteristic ... **EL1:** 29-30, 35-36, 588
coil, in eddy current inspection ... **A17:** 166-167
components ... **A17:** 166-167
concepts, eddy current inspection ... **A17:** 169-173
conductor ... **EL1:** 83
controlled, connector system ... **EL1:** 86
controlled, for high-bandwidth digital systems ... **EL1:** 76
defined ... **EL1:** 1146
diagrams, eddy current inspection ... **A17:** 170-172
driver circuit output ... **EL1:** 26
electrochemical, spectroscopy ... **A13:** 220
electrochemical test methods ... **A13:** 215, 220
Impedance (continued)
formulas for ... **EL1:** 29
line, VHSIC interconnects ... **EL1:** 388
matched-load ... **EL1:** 11, 111, 1- 40
-measuring mode, remote-field eddy current inspection ... **A17:** 196
normalization ... **A17:** 170
output, driver circuit effect ... **EL1:** 39
-plane diagram, eddy current inspection ... **A17:** 166
selection ... **EL1:** 83
of signal line ... **EL1:** 29-30
of solid cylindrical bar ... **A17:** 171-172
surface, modulation ... **A17:** 217
test, for anodized aluminum ... **A13:** 220
of tube ... **A17:** 169-170
ultrasonic beams ... **A17:** 238
ultrasonic inspection ... **A17:** 234-235
Impedance, and high-temperature testing
Kolsky bar ... **A8:** 222-223
Impedance models ... **EL1:** 601-603
asymmetric stripline properties ... **EL1:** 602-603
coated microstrip ... **EL1:** 602
microstripline properties ... **EL1:** 602
stripline properties ... **EL1:** 602
Impeller
austenitic cast iron, shrinkage porosity in ... **A11:** 355-356
cast iron pump, graphitic corrosion of ... **A11:** 374-375
centrifugal compressors ... **A18:** 606, 607, 608
water pump, cavitation damage ... **A11:** 167-168
Impeller, steel
radiographic inspection ... **A17:** 333-334
Imperfection
defined ... **A9:** 9
Imperfections, weld *See also* Weld defects ... **A11:** 92-93
Impervious graphite ... **A13:** 1154, 1227
Impingement *See also* Impingement corrosion
attack ... **A11:** 6, 634
copper/copper alloys ... **A13:** 613
corrosion, defined ... **A13:** 8
defined ... **A18:** 11
drop ... **A11:** 164-165
flame, in D2 P/M die component ... **A11:** 573, 579
microjet ... **A11:** 165
water drop ... **A13:** 142, 339, 519
Impingement attack
versus impingement erosion ... **A18:** 223
Impingement corrosion *See also* Corrosion; Impingement; Liquid impingement erosion; Liquid-erosion, failures
of copper alloy heat-exchanger tubing ... **A11:** 634-635
in malleable iron elbow, leakage and failure at bend ... **A11:** 189
in water ... **A11:** 189
Impingement erosion
defined ... **A18:** 11
hardfacing for ... **A7:** 823
Impingement grain structure
defined ... **A9:** 602-603
Impingement umbrella
defined ... **A18:** 11
Implant alloys
dental ... **A13:** 1357-1359
Implant alloys (dental)
of precious metals ... **A2:** 696
Implant fixation
porous coatings for ... **A7:** 659-661
Implantation *See also* Ion implantation
as effect of primary ion bombardment ... **A10:** 611
helium, in bcc iron alloy ... **A10:** 485
of inert elements ... **A10:** 475
of inert gas elements ... **A10:** 485
surface modification by ... **A13:** 498-500
Implantation defects ... **EL1:** 978
Implants *See also* Metallic orthopedic implants, failures of
bending of ... **A11:** 672
breakage or disintegration ... **A11:** 672
combined dynamic and biochemical attack on ... **A11:** 676-677
complications related to ... **A11:** 672
deficiencies, failures related to ... **A11:** 680-681
estimated cyclic loading on ... **A11:** 673
failed historic Lane plate ... **A11:** 674
failures of ... **A11:** 670-694
for intemal fixation ... **A11:** 671
interaction with body environment ... **A11:** 673
and ionic composition of chlorine ... **A11:** 672, 673
loaded in appropriate cyclic fatigue range ... **A11:** 684, 689
loosening of ... **A11:** 672
materials for ... **A11:** 672-673, 684-689
metallic orthopedic ... **A11:** 670-694
orthopedic ... **A7:** 657-659
orthopedic, degradation of ... **A11:** 689-692
Implants, orthopedic
austenitic stainless steels ... **A12:** 359-364
Impregnate
defined ... **EM1:** 13 **EM2:** 23
Impregnated fabric *See also* Prepreg
defined ... **EM1:** 13 **EM2:** 23
Impregnation *See also* Wetting ... **A7:** 6
centrifugal pressure ... **A7:** 554
defined ... **A15:** 7
external pressure ... **A7:** 554
in filament winding processes ... **EM1:** 505-506
procedure, die casting ... **A15:** 295
in pultrusion ... **EM1:** 534
of resin-coated glass cloth ... **EL1:** 114
shear-thinning method ... **EM1:** 102
sintered bronze bearings ... **A2:** 395
of thermoplastic resins ... **EM1:** 101-103
Impregnation, resin
in pultrusion ... **EM2:** 390
Impressed current
cathodic protection system ... **A13:** 467-469, 476
defined ... **A13:** 8
test, SCC in aluminum alloys ... **A13:** 265
Impressed electrical current
SCC tests with ... **A8:** 532
Impressed-current anodes ... **A13:** 469, 921-922
Impression
defined ... **A9:** 9 **A14:** 8
Impression dies
benders ... **A14:** 44
blockers ... **A14:** 44
edgers ... **A14:** 43-44
fabrication of ... **A14:** 52-53
finishers ... **A14:** 44-45
flash-and-gutter, for HERF processing ... **A14:** 101-104
flatteners ... **A14:** 44
fullers ... **A14:** 43
rollers ... **A14:** 44
splitters ... **A14:** 44
Impression replica
define ... **A9:** 9
Impression size
Brinell test ... **A8:** 84
Impression-die forging *See* Closed-die forging; Impression dies
"Improved Gas Shielding," patent (No 5 081,334) ... **A6:** 446
Improved plow steel quality rope wire ... **M1:** 265, 266
Improvement, process
defined ... **A17:** 740
Impulse
definition ... **M6:** 10
Impulse (resistance welding)
definition ... **A6:** 1210
"Impulse response" ... **A18:** 600

SUBJECTS OF THE INDEXED VOLUMES: ASM Handbook (designated by the letter "A"): **A1:** Properties and Selection: Irons, Steels, and High-Performance Alloys (1990); **A2:** Properties and Selection: Nonferrous Alloys and Special-Purpose Materials (1990); **A3:** Alloy Phase Diagrams (1992); **A4:** Heat Treating (1991); **A6:** Welding, Brazing, and Soldering (1993); **A7:** Powder Metallurgy (1984); **A8:** Mechanical Testing (1985); **A9:** Metallography and Microstructures (1985); **A10:** Materials Characterization (1986); **A11:** Failure Analysis and Prevention (1986); **A12:** Fractography (1987); **A13:** Corrosion (1987); **A14:** Forming and Forging (1988); **A15:** Casting (1988); **A16:** Machining (1989); **A17:** Nondestructive Testing and Quality Control (1989); **A18:** Friction, Lubrication, and Wear Technology (1992). **Metals Handbook, 9th Edition** (designated by the letter "M"): **M1:** Properties and Selection: Irons and Steels (1978); **M2:** Properties and Selection: Nonferrous Alloys and Pure Metals (1979); **M3:** Properties and Selection: Stainless Steels, Tool Materials and Special-Purpose Materials (1980); **M4:** Heat Treating (1981); **M5:** Surface Cleaning, Finishing, and Coating (1982); **M6:** Welding, Brazing, and Soldering (1983). **Engineered Materials Handbook** (designated by the letters "EM"): **EM1:** Composites (1987); **EM2:** Engineering Plastics (1988); **EM3:** Adhesives and Sealants (1990); **EM4:** Ceramics and Glasses (1991); **Electronic Materials Handbook** (designated by the letters "EL"): **EL1:** Packaging (1989).

Impulse sealing *See also* Heat sealing
defined ... **EM2:** 23
Impurities *See also* Inclusions ... **A6:** 146
cast metal sensitivity to ... **A15:** 74
in compound form ... **A7:** 246
as crack initiation sites ... **A17:** 216
defined ... **A9:** 9
depth profiles of heavy element ... **A10:** 632-633
determined in nickel ... **A10:** 240
diffusion, active-component fabrication ... **EL1:** 194-195
effect, chemical processing ... **A13:** 1136
effect, intergranular embrittlement, nickel ... **A13:** 164
effect, kinetics of gaseous corrosion ... **A13:** 68-69
effect on boundary migration ... **A9:** 696-697
effect on metal and metal-to-ceramic adhesion ... **A6:** 144
effect, zirconium alloys ... **A13:** 709
enrichment of ... **A13:** 156
as function of position ... **EL1:** 146
grain-boundary segregation, and intergranular corrosion ... **A13:** 239
heavy-metal, effect in magnesium/magnesium alloys ... **A13:** 740
in heterogeneous nucleation ... **A15:** 103
in HIP processing ... **A7:** 425-428
in hydrogen chloride/hydrochloric acid ... **A13:** 1161-1162
leaded red brass, fire refining effects ... **A15:** 451
liquid chromatography compound analysis for ... **A10:** 649
liquid-metal corrosion by ... **A13:** 56-59, 92
LPCVD thin film analysis for ... **A10:** 624
metal, in synthetic diamond ... **A10:** 417
metallic and nonmetallic, chemical analysis ... **A7:** 246-249
molten copper, fire refining effect ... **A15:** 450
in molten salts ... **A13:** 50
oxygen effect in flux ... **A15:** 451
powder behavior and physical properties ... **A7:** 246
RBS analysis of surface ... **A10:** 628
removal, by continuous flow melting ... **A15:** 414-415
in silica sand ... **A15:** 208
solder, composition effects ... **EL1:** 637-639
in UO_2, determined ... **A10:** 149-150
volatile, aluminum melts ... **A15:** 80
Impurities aluminum and aluminum alloys ... **A2:** 3, 16, 44
concentrations, of purified metals ... **A2:** 1096
concentrations, titanium and chromium ... **A2:** 1097
effect in magnetically soft materials ... **A2:** 762-763
effects, commercially pure titanium ... **A2:** 594
effects, wrought aluminum alloy ... **A2:** 44
interstitial, reduction of ... **A2:** 1094-1095
limit, niobium-titanium superconducting materials ... **A2:** 1044
limits, exceeding, cast copper alloys ... **A2:** 365
removal ... **A2:** 1093-1095
specific elements, wrought aluminum alloy ... **A2:** 46-57
in tin solders ... **A2:** 520-521
tolerance, casting vs wrought copper alloy ... **A2:** 346
unalloyed uranium ... **A2:** 672
Impurities, molecular
and polymers ... **EM2:** 58
Impurity atoms
and point defects ... **A13:** 46
Impurity effect
on fatigue crack growth rate ... **A8:** 411
Impurity elements *See also* Residual elements
temper embrittlement, role in ... **M1:** 684-685
Impurity removal *See* Composition control; Impurities
in situ **induction heating**
for elevated- temperature compression testing ... **A8:** 196
In situ **measurement technique**
for fiber-matrix interphase characterization ... **EM3:** 391
In situ **studies**
SEM imaging ... **A12:** 169
In vivo **measurement**
by neutron activation analysis ... **A10:** 233
IN-100
composition ... **A16:** 737
for hot-forging dies ... **A18:** 625, 626
machining ... **A16:** 738, 741-743, 746-758
surface alterations from material removal processes ... **A16:** 27
IN-102
composition ... **A16:** 736
machining ... **A16:** 738, 741-743, 746-747, 749-758
thread grinding ... **A16:** 275
IN-713 C
composition ... **M6:** 354
IN-713C
composition ... **A6:** 564
electron-beam welding ... **A6:** 869
repair welding ... **A6:** 564
strain age cracking ... **A6:** 564
IN-738 *See* Nickel-base superalloys, specific types
grinding ... **A16:** 760
machining ... **A16:** 738, 741-743, 746-758
IN-738X
composition ... **A16:** 737
IN-792
composition ... **A16:** 737
machining ... **A16:** 738, 741-743, 746-757
In-bed superheaters/air heaters ... **A13:** 999
In-K (Phase Diagram) ... **A3:** 2 • 252
In-La (Phase Diagram) ... **A3:** 2 • 252
In-ladle alloying
mechanized ladles ... **A15:** 498
In-Li (Phase Diagram) ... **A3:** 2 • 252
In-line drawing and straightening machine
for bars ... **A14:** 333
In-line horizontal process equipment
condensation (vapor phase) soldering ... **EL1:** 703-704
In-line process monitors
for materials analysis ... **EL1:** 917
In-Lu (Phase Diagram) ... **A3:** 2 • 253
In-Mg (Phase Diagram) ... **A3:** 2 • 253
in-Mn (Phase Diagram) ... **A3:** 2 • 253
In-mold coating
SMC parts ... **EM2:** 306
In-motion radiography *See also* Radiography
motion unsharpness ... **A17:** 338
view selection ... **A17:** 337-338
In-Na (Phase Diagram) ... **A3:** 2 • 254
In-Nb (Phase Diagram) ... **A3:** 2 • 254
In-Nd (Phase Diagram) ... **A3:** 2 • 254
In-Ni (Phase Diagram) ... **A3:** 2 • 255
In-P (Phase Diagram) ... **A3:** 2 • 255
In-Pb (Phase Diagram) ... **A3:** 2 • 255
In-Pd (Phase Diagram) ... **A3:** 2 • 256
In-phase component
microwave inspection ... **A17:** 205
In-phase signal
nuclear magnetic resonance ... **A17:** 144
In-plane deformation
measured by speckle metrology ... **A17:** 432-434
In-plane determination
forming limit diagrams ... **A8:** 566
In-plane displacements
by optical holography ... **A17:** 415-416
In-plane failure mode
types ... **EM1:** 781-783
In-plane shear
fractures ... **EM1:** 789-790
ply ... **EM1:** 238
In-plane shear fractures
in composites ... **A11:** 736-738
In-Plant Powder Metallurgy Association ... **A7:** 19
In-process corrosion
of beryllium ... **A13:** 810
In-process inspection *See also* In-process nondestructive evaluation; In-service inspection; In-service nondestructive evaluation
defined ... **A17:** 49
machine vision as gaging tool ... **A17:** 29
magnetic particle inspection ... **A17:** 89
parts, by coordinate measuring machines ... **A17:** 18
surface flaws, by laser ... **A17:** 17
In-process nondestructive evaluation *See also* In-process inspection; In-service inspection; In-service nondestructive evaluation
defined ... **A17:** 49
In-process testing *See also* Testing
function ... **EL1:** 131
life cycle ... **EL1:** 139-140
In-Pt (Phase Diagram) ... **A3:** 2 • 256
In-Pu (Phase Diagram) ... **A3:** 2 • 257
In-Rb (Phase Diagram) ... **A3:** 2 • 257
In-S (Phase Diagram) ... **A3:** 2 • 257
In-Sb (Phase Diagram) ... **A3:** 2 • 258
In-Sc (Phase Diagram) ... **A3:** 2 • 258
In-Se (Phase Diagram) ... **A3:** 2 • 259
In-service defects
adhesive-bonded joints ... **A17:** 16
In-service failures
system degradation during ... **EL1:** 9
In-service inspection *See also* In-process inspection
acoustic emission inspection, of pressure vessels ... **A17:** 656
of boilers ... **A17:** 641
defined ... **A17:** 49
of pressure vessels, quantitative evaluation ... **A17:** 653-654
radiographic, of pressure vessels ... **A17:** 649
of tubular products ... **A17:** 574-581
of weldments ... **A17:** 600-601
In-service monitoring ... **A13:** 197-203
analysis of process streams ... **A13:** 201
of corrosion, strategies ... **A13:** 202-203
equipment ... **A13:** 202
interpretation and reporting ... **A13:** 203
selecting method of ... **A13:** 197-201
sentry holes ... **A13:** 201
side-stream (bypass) loops ... **A13:** 201
In-service nondestructive evaluation *See also* In-process inspection
defined ... **A17:** 49
In-Si (Phase Diagram) ... **A3:** 2 • 259
In-Sm (Phase Diagram) ... **A3:** 2 • 260
In-Sn (Phase Diagram) ... **A3:** 2 • 260
In-Sr (Phase Diagram) ... **A3:** 2 • 260
In-Tb (Phase Diagram) ... **A3:** 2 • 261
In-Te (Phase Diagram) ... **A3:** 2 • 261
In-Th (Phase Diagram) ... **A3:** 2 • 261
In-the-mold treatment
ductile iron ... **A15:** 650
In-Ti (Phase Diagram) ... **A3:** 2 • 262
In-Tl (Phase Diagram) ... **A3:** 2 • 262
In-Tm (Phase Diagram) ... **A3:** 2 • 262
In-tolerance parts
coordinate measuring machines for ... **A17:** 20
In-V (Phase Diagram) ... **A3:** 2 • 263
In-Y (Phase Diagram) ... **A3:** 2 • 263
In-Yb (Phase Diagram) ... **A3:** 2 • 263
In-Zn (Phase Diagram) ... **A3:** 2 • 264
IN909
coefficient of expansion and jet engine applications ... **A18:** 589
Inadequate joint penetration
definition ... **M6:** 10
Incendiary bombs
powders used ... **A7:** 573
Incendiary materials
comparison ... **A7:** 681
with metal fuels ... **A7:** 600, 603
Inceram ... **EM4:** 1096
Incident bar
in split Hopkinson pressure bar ... **A8:** 198-199, 201
in torsional Kolsky bar dynamic test ... **A8:** 228
Incident electrons
Monte Carlo projections ... **A12:** 167
Incident energy
in microwave inspection ... **A17:** 203
Incident gage
for bar testing ... **A8:** 220-221, 277
Incident illumination
in a transmission electron microscope ... **A9:** 103
Incident light
for discontinuities ... **A12:** 63
meter, effects ... **A12:** 85
Incident light microscopy
Köhler illumination principle ... **A9:** 58-59
Incident light, monochromatic
gray value contrast at different wavelengths ... **A9:** 149

Incident wave
in split Hopkinson pressure bar **A8:** 200-201
Incident-light microscopes
light paths **A9:** 71, 81
optical path **A9:** 80
relationship between resolution and of light **A9:** 78
numerical aperture for four wavelengths
Incinerator equipment
elevated-temperature failure in **A11:** 294
Incinerator wall tube corrosion **A13:** 997-998
Incipient melting *See* Burning
in austenitic manganese steel castings **A9:** 239
casting failure from **A11:** 407
revealed by differential interference contrast **A9:** 152
Incipient melting temperature *See also* Thermal properties
cast copper alloys **A2:** 360, 361, 377, 379
wrought aluminum and aluminum alloys **A2:** 74, 81
Inclination
milling **A16:** 318, 329
Inclined position
definition **A6:** 1210 **M6:** 10
Inclined position (with restriction ring)
definition **A6:** 1210
Inclined vibrating screens **A7:** 177
Inclined-surface mounting plug
used to mount tin and tin alloy coated materials **A9:** 450-451
Included angle
definition **A6:** 1210
Included angle of groove welds **M6:** 65-67
Inclusion *See also* Voids **EM3:** 16
before and after twisting **A8:** 156
defined **EM1:** 13 **EM2:** 23
definition **EM4:** 633
level effect on torsional fracture strain **A8:** 155
parallel to torsion axis, round bar **A8:** 155
spacing, fracture toughness as function of **A8:** 479
Inclusion count
defined **A9:** 9
Inclusion particles
as nucleation sites **A9:** 694
Inclusion shape control **A15:** 91, 709
Inclusion shape controlled steels **A1:** 400, 405, 412-413
Inclusion-forming reactions *See also* Defects; Inclusions **A15:** 88-97
in aluminum alloys **A15:** 95
in cast irons **A15:** 94-95
control of **A15:** 90-91
in copper alloys **A15:** 96
in ferrous alloys **A15:** 91-95
inclusion types **A15:** 88-89
in magnesium alloys **A15:** 96
in nonferrous alloys **A15:** 95-96
physical chemistry **A15:** 89-90
in steels **A15:** 91-94
Inclusions *See also* Defects; Flaws; Gas defects; Impurities; Inclusion-forming reactions; Nonmetallic inclusions; Oxide inclusions; Oxides; Slag; Slag inclusions **A9:** 59
AISI/SAE alloy steels **A12:** 333
in aluminum alloy **A12:** 19
in aluminum alloys **A15:** 95-96, 488, 749
aluminum casting alloys **A2:** 146
analytical transmission electron microscopy **A10:** 429-489
in arc-welded aluminum alloys **A11:** 435-436
in arc-welded low-carbon steel **A11:** 416
Auger electron spectroscopy **A10:** 549-567
brittle fracture by **A11:** 85, 89
by liquid penetrant inspection **A17:** 86
in carbon steel castings, examination for **A9:** 230
in cast irons **A15:** 94-96

Inclusions (continued)
as casting defect **A17:** 519-520
as casting defects **A11:** 384, 387-388
as casting internal discontinuity **A11:** 354
casting, types **A15:** 552-553
characterized by image analyzers **A9:** 83
cluster, spring failure from **A11:** 554
complex, EPMA detected **A11:** 39
complex, in steels **A15:** 92-93
computed tomography (CT) **A17:** 361
control of **A15:** 90-91
in copper alloys **A15:** 96
in copper and copper alloys, examination **A9:** 401
copper, effect in low-carbon steel **A12:** 249
defined **A9:** 9 **A10:** 176 **A11:** 6 **A13:** 8 **A15:** 7, 88
diborides in aluminum alloys **A15:** 95
in dimples **A12:** 65, 67, 174
as discontinuities, defined **A12:** 65
and drawability **A14:** 575
dross **A12:** 422
dross or flux, as casting defect **A11:** 387
in duct-le cast iron **A15:** 94-95
eddy current inspection of **A17:** 164
effect in bearing element failure **A11:** 504
effect in SCC **A12:** 26
effect on aluminum alloy fluidity **A15:** 767
effect on cold-formed part failures **A11:** 307
effect on fatigue crack propagation **A12:** 346, 347
effect on striation **A12:** 16
in electrogas welds **A11:** 440
in electron beam welds **A11:** 445
electron probe x-ray microanalysis **A10:** 516-535
exogenous **A15:** 88
exogenous slag **A14:** 191
in ferrous alloys **A15:** 91-95
fireclay **A11:** 323
in flash welds **A11:** 442
flotation times, calculated **A15:** 79
as fracture origin, AISI/SAE alloy steels **A12:** 303
from pipe **A12:** 427
globular oxide, in iron **A12:** 220
in gray cast iron **A15:** 94
in gray iron castings **A17:** 531
high-carbon steels **A12:** 277
identification with polarized light **A9:** 76
illuminated and photographed **A12:** 86
image analysis of **A10:** 314-316
indigenous **A15:** 88-89
influence on fatigue resistance **M1:** 672-674, 682
and insoluble particles, in liquid/solid interface **A15:** 142
intentional **A15:** 88
intermetallic **A15:** 95-96
intermetallic, wrought aluminum alloys **A12:** 423
as internal defects, in FIM samples **A10:** 587
in iron castings **A11:** 357-358
in iron, dimples from **A12:** 219
isolation of residues **A10:** 176
lead, effect on steel embrittlement **A11:** 239, 242
in low-alloy steel castings, examination for **A9:** 230
in low-carbon steels, effect of calcium **A12:** 247
macroetching to reveal **A9:** 173
in magnesium alloys **A15:** 96
magnetic field testing detection **A17:** 129
manganese oxide/sulfide, in iron **A12:** 221
manganese sulfide **A11:** 322
melting, remelting, and refining processes affecting **A11:** 340
metallographic sectioning of **A11:** 24
microvoid coalescence at **A12:** 12
microwave inspection **A17:** 202, 212
multiphase spinels **A11:** 322
in nonferrous alloys **A15:** 95-96
nonmetallic **A11:** 316, 322-323, 477-478, 504 **A17:** 89, 492, 535

Inclusions (continued)
nonmetallic, in ceramic castings **A15:** 248
nonmetallic, in cold extrusions **A14:** 301
nonmetallic, in permanent mold castings **A15:** 285
nonmetallic, powder forged parts **A14:** 204
optical metallography **A10:** 299-308
oxide, analysis in steel alloys **A10:** 162
oxide, as squeeze casting defect **A15:** 325
oxide or flux, in brazing **A11:** 451
oxide-sulfide, in shafts **A11:** 462
oxygen **M6:** 838-839
in plain carbon steels **A15:** 709
plastic **A14:** 358
in plate steels, examination for **A9:** 203
in powder forging **A14:** 190-191, 204
radiographic appearance **A17:** 349
radiographic methods **A17:** 296
ratings, by image analysis **A10:** 309
refinement of residues **A10:** 176-177
refractory **A14:** 358
removal, in gating **A15:** 589
in resistance welds **A11:** 440
in return bend, rupture by **A11:** 646
ribbonlike, AISI/SAE alloy steels **A12:** 326
as rolling defects **A14:** 358
sand, as casting defect **A11:** 387
scanning electron microscopy **A10:** 490-515
second-phase, wrought aluminum alloys **A12:** 423
selective attack on **A11:** 182-183
separation techniques **A15:** 90-91
service fracture, steel forging **A12:** 65, 66
shape control, HSLA steels **M1:** 411
shape, control of **A15:** 91, 709
shapes, common **A15:** 93
size and distribution **A12:** 333
size, effect in ferrous melts **A15:** 79
slag **A11:** 440 **M6:** 837-838, 843-844
slag, in steel pipe **A17:** 565
slag, radiographic appearance **A17:** 350
sorting by dot maps **A12:** 168
as source, acoustic emissions **A17:** 287
sources of **A15:** 488
spall from **A11:** 89
spheroidal oxide, in iron **A12:** 220
and spring failures **A11:** 554
in steel bar and wire **A17:** 549
in steel, isolating **A10:** 176
steel plate, effect on machinability **M1:** 197
stringers **A12:** 140-141, 161
stringers, EPMA identified **A11:** 38
subsurface **A11:** 120-121, 323
sulfide **A11:** 83, 477-478, 723
sulfide, in white iron **A12:** 239
surface, SEM analysis **A10:** 490
testing **A10:** 176-177
testing methods **A15:** 493
titanium, in maraging steels **A12:** 383
tungsten **M6:** 839
tungsten, in weldments **A17:** 582, 584
types **A15:** 88-89
types, powder forging **A14:** 191
ultrasonic cleaning effects **A12:** 75
ultrasonic inspection of **A17:** 232
in uranium alloys **A9:** 477, 479, 481, 484
as volumetric flaw **A17:** 50
wrought aluminum alloys **A12:** 418
in wrought stainless steels, etching to examine **A9:** 281
Inclusions in
arc welds of magnesium alloys **M6:** 435
arc welds of nickel-based alloys **M6:** 363
brazements of stainless steel **M6:** 1001
flash welds **M6:** 580
gas metal arc welds **M6:** 172
Inco 625
friction surfacing **A6:** 323

Inco alloy 020
composition ... **A16:** 836
machining ... **A16:** 836-840, 842, 843
Inco alloy 330
composition ... **A16:** 836
machining ... **A16:** 836-840, 842, 843
INCO alloy 600, applications
heat exchangers ... **EM4:** 984
Inco alloy 718
EDM ... **A16:** 868
Inco alloy C-276
composition ... **A16:** 836
machining ... **A16:** 836-840, 842, 843
Inco alloy G-3
composition ... **A16:** 836
machining ... **A16:** 836-840, 842, 843
Inco alloy HX
composition ... **A16:** 836
machining ... **A16:** 42, 843
Inco alloy MS 250
composition ... **A16:** 836
machining ... **A16:** 40, 842, 843
Inco Ltd
nickel powders ... **A7:** 137-138
Incoherent atomic scattering intensity
abbreviation for ... **A10:** 690
Incoherent interface
between matrix and precipitate ... **A9:** 648
defined ... **A9:** 604, 647
Incoherent scattering
defined ... **A9:** 9
Incoherent-to-coherent image converters
microwave holography ... **A17:** 227-228
Incology, specific types
800
composition ... **M6:** 354
explosion welding ... **M6:** 706-707
800H, composition ... **M6:** 354
801
composition ... **M6:** 354
flash welding ... **M6:** 557
802, composition ... **M6:** 354
901, composition ... **M6:** 354
903, composition ... **M6:** 354
DS, flash welding ... **M6:** 557
MA 956, brazing ... **M6:** 1020
Incoloy
recommended for parts and salt bath fixtures ... **A4:** 514
Incoloy 800
composition ... **A4:** 512 **A6:** 564
erosion test results ... **A18:** 200
for radiant-tube-heated furnace equipment ... **A4:** 473
welding to carbon, low-alloy or stainless steels ... **A6:** 826
Incoloy 800 H
composition ... **A6:** 564
Incoloy 800 HT
composition ... **A6:** 564
Incoloy 800H
erosion test results ... **A18:** 200
Incoloy 801
composition ... **A6:** 564
Incoloy 802
composition ... **A4:** 512 **A6:** 564
Incoloy 901
aging ... **A4:** 804
aging cycle ... **M4:** 656
aging precipitates ... **A4:** 796
aging treatments, effect on properties ... **M4:** 662, 663
annealing ... **M4:** 655
composition ... **A4:** 794 **A6:** 564 **M4:** 651-652
creep-rupture properties ... **A4:** 805
double-aging ... **A4:** 804
electron-beam welding ... **A6:** 865
fatigue strength ... **A4:** 801
grain size effect on fatigue properties ... **A4:** 802
for hot-forging dies ... **A18:** 626
mechanical properties after third aging treatment ... **A4:** 804
nickel content and alloy classification ... **A4:** 800
solution heat treatment ... **A4:** 801
solution treating ... **M4:** 656
stabilization effect on mechanical properties ... **A4:** 804
Incoloy 901 (continued)
stress relieving ... **M4:** 655
tensile properties ... **A4:** 801, 802, 803
Incoloy 903
aging ... **A4:** 796
composition ... **A4:** 794 **A6:** 564
nickel content and alloy classification ... **A4:** 800
oxidation ... **A4:** 798
sleeve material for brazed joint in turbocharger application ... **EM4:** 724
solution-treating ... **A4:** 796
Incoloy 909
brazing ... **A6:** 949, 956-957
composition ... **A6:** 564
Incoloy 925
composition ... **A6:** 564
Incoloy alloy 800
composition ... **A16:** 736, 836
machining ... **A16:** 738, 741-743, 746-747, 749-758, 837-840, 842, 843
Incoloy alloy 800 H
machining ... **A16:** 757, 758
Incoloy alloy 800 HT
composition ... **A16:** 836
machining ... **A16:** 836-840, 842, 843
Incoloy alloy 801
composition ... **A16:** 736
machining ... **A16:** 738, 741-743, 746-747, 749-758
Incoloy alloy 802
composition ... **A16:** 736, 836
machining ... **A16:** 738, 741-743, 746-747, 749-758, 837-840, 842, 843
Incoloy alloy 804
machining ... **A16:** 757, 758
Incoloy alloy 825
composition ... **A16:** 836
machining ... **A16:** 757, 758, 837-840, 842, 843
Incoloy alloy 901
composition ... **A16:** 736
machining ... **A16:** 203, 204, 275, 738, 741-743, 746-747, 749-757
Incoloy alloy 903
chemical milling ... **A16:** 583
composition ... **A16:** 836
electrochemical machining ... **A16:** 843
machining ... **A16:** 738, 741-743, 746-747, 749-757, 837-840, 842, 843
Incoloy alloy 907
composition ... **A16:** 836
electrochemical machining ... **A16:** 843
machining ... **A16:** 837-840, 842, 843
Incoloy alloy 909
composition ... **A16:** 836
electrochemical machining ... **A16:** 843
machining ... **A16:** 837-840, 842, 843
Incoloy alloy 925
composition ... **A16:** 836
machining ... **A16:** 837-840, 842, 843
Incoloy alloy DS
composition ... **A16:** 836
machining ... **A16:** 842, 843
Incoloy alloy MA 956
composition ... **A16:** 836
machining ... **A16:** 837-840, 842, 843
Incoloy alloys *See* High-temperature materials, specific types; Nickel alloys, specific types, Incoloy; Nickel-base superalloys, specific types
Incoloy MA 956 ... **A6:** 928
brazing ... **A6:** 632
characteristics ... **A4:** 512
composition ... **A4:** 512
Incomplete block experimental plans ... **A8:** 644-649
Incomplete block experiments ... **A8:** 646-649, 657-661
Incomplete blocks ... **A8:** 640
Incomplete casting
as casting defect ... **A11:** 385-386 **A17:** 512, 517
Incomplete casting, as defect
types ... **A15:** 550-551
Incomplete fusion *See also* Fusion; Fusion, incomplete; Lack of fusion (LOF)
definition ... **M6:** 10
as discontinuity ... **A12:** 65
radiographic appearance ... **A17:** 350
in steel pipe ... **A17:** 565
Incomplete penetration *See also* Lack of penetration (LOP); Penetration
radiographic methods ... **A17:** 296
root, radiographic appearance ... **A17:** 350
in steel pipe ... **A17:** 565
Inconel
back reflection intensity ... **A17:** 238
cast
ceramic-bonded fluoride coatings for lubricants ... **A18:** 118
graphite as solid lubricant ... **A18:** 115
cutoff band sawing with bimetal blades ... **A6:** 1184
environments that cause stress-corrosion cracking ... **A6:** 1101
extension of tool life via ion implantation, examples ... **A18:** 643
nonmetallic fusion process ... **EM3:** 306
oxyacetylene welding ... **A6:** 281
relative solderability as a function of flux type ... **A6:** 129
seal materials ... **A18:** 550
seawater exposure effect on adhesives ... **EM3:** 632
tooling for pressed ware ... **EM4:** 398
ultrasonic welding ... **A6:** 326
for valve springs for reciprocating compressors ... **A18:** 604
weld overlay for hardfacing alloys ... **A6:** 820
Inconel 52
filler metal for oxide-dispersion-strengthened materials ... **A6:** 1039
Inconel 601
annealing ... **A4:** 908
composition ... **A4:** 512, 908
stress-equalizing ... **A4:** 908
Inconel 82
filler metal for stainless steel casting alloys ... **A6:** 496
Inconel 92
filler metal for stainless steel casting alloys ... **A6:** 496
Inconel 100
aging cycle ... **A4:** 812
composition ... **A4:** 794, 795 **A6:** 573 **M4:** 653
Inconel 100 gatorized
composition ... **A6:** 573
Inconel 102
composition ... **A4:** 794 **A6:** 573
Inconel 112
in ferritic base metal-filler metal combination ... **A6:** 1149
Inconel 182
filler metal for stainless steel casting alloys ... **A6:** 496
Inconel 600
annealing ... **A4:** 908 **M4:** 655
applications, protection tubes and wells ... **A4:** 533
bright annealing ... **A4:** 910
composition ... **A4:** 512, 908 **A6:** 564, 573 **M4:** 651-652
constitutional liquation in multicomponent systems ... **A6:** 568
electron-beam welding ... **A6:** 869
erosion test results ... **A18:** 200
fixture for solution treating uranium bars ... **A4:** 937
as fixtures in carburizing furnaces ... **A4:** 907
joined to silicon carbide ... **A6:** 636
laser welding ... **A6:** 441
thermal diffusivity from 20 to 100 °C ... **A6:** 4
materials for parts and fixtures in nitriding furnaces ... **A4:** 398
noncyanide liquid nitriding furnace liners ... **A4:** 414
probe material for quenching cooling curve analysis ... **A4:** 68, 69, 88, 91
probe material used to find quenching temperatures ... **A4:** 12, 13
radiant tube applications ... **A4:** 518
springs, strip for ... **M1:** 286
springs, wire for ... **M1:** 285
stress relieving ... **M4:** 655
stress-equalizing ... **A4:** 908
stress-relieving ... **A4:** 908

Inconel 600 (continued)
trace element impurity effect on GTA weld penetration ... **A6:** 20
welding to carbon, low-alloy or stainless steels ... **A6:** 826
Inconel 601
composition ... **A6:** 564, 573
Inconel 617
annealing ... **A4:** 908
characteristics ... **A4:** 512
composition ... **A4:** 512, 794, 908 **A6:** 564, 573 **M4:** 651-652
filler metal for oxide-dispersion-strengthened materials ... **A6:** 1038
filler metal for stainless steel casting alloys ... **A6:** 496
mill annealing ... **A4:** 810
mill annealing temperature range ... **A6:** 573
solution annealing ... **A4:** 810
solution annealing temperature range ... **A6:** 573
stress-equalizing ... **A4:** 811, 908
Inconel 625
aging ... **A4:** 796, 810
annealing ... **A4:** 908
composition ... **A4:** 794, 908 **A6:** 564, 573 **M4:** 651-652
electron-beam welding ... **A6:** 869
laser melt/particle injection ... **A18:** 869, 870, 871
mill annealing ... **A4:** 810
solution annealing ... **A4:** 810
solution-treating ... **A4:** 796
stress-relieving ... **A4:** 908
weld overlay for hardfacing alloys ... **A6:** 820
Inconel 626, applications
aerospace ... **A6:** 387
Inconel 671
erosion test results ... **A18:** 200
Inconel 700
aging cycle ... **M4:** 656
annealing ... **M4:** 655
ceramic cutting tool cost-effectiveness ... **EM4:** 967
composition ... **M4:** 651-652
electron-beam welding ... **A6:** 867
solution treating ... **M4:** 656
stress relieving ... **M4:** 655
Inconel 702
composition ... **A4:** 794 **A6:** 564, 573 **M4:** 651-652
Inconel 706
aging ... **A4:** 796
composition ... **A4:** 794 **A6:** 564, 573 **M4:** 651-652
for hot-forging dies ... **A18:** 626
niobium content and alloy classification ... **A4:** 800
solution-treating ... **A4:** 796, 803
Inconel 713
aging cycle ... **A4:** 812 **M4:** 657
composition ... **M4:** 653
solution treating ... **M4:** 657
Inconel 713C
composition ... **A4:** 795
Inconel 713C, for press forging heat-resistant alloys
nickel-base alloys ... **A18:** 625
Inconel 713LC
composition ... **A4:** 795
for hot-forging dies ... **A18:** 625, 626
Inconel 718
age hardening ... **A6:** 564
aging ... **A4:** 796, 804, 805, 911
aging cycle ... **A4:** 812 **M4:** 656
aging cycles ... **A6:** 574
aging temperature effect on mechanical properties ... **A4:** 806
annealing ... **A4:** 908 **M4:** 655
applications ... **A4:** 804 **A6:** 928
composition ... **A4:** 794, 795, 908 **A6:** 564, 573 **M4:** 651-652
compounds/properties in composition ... **EM4:** 499
direct aging ... **A4:** 804
Inconel 718 (continued)
electron-beam welding ... **A6:** 869
friction welding ... **A6:** 153
glass-ceramic/metal seals ... **EM4:** 499, 500
glass-metal seals ... **EM4:** 875
heat treatments ... **A6:** 928
hermetic glass-ceramic/metal seating ... **EM4:** 479
with HIP, aging cycle ... **A4:** 812
for hot-forging dies ... **A18:** 626
laser-beam welding ... **A6:** 263
material for jet engine components ... **A18:** 588, 591
mechanical properties ... **A4:** 803, 804, 805, 807 **EM4:** 316
mechanical properties, function of processing ... **A4:** 806
nickel flashing treatment ... **A6:** 926
with niobium carbides, different illuminations compared ... **A9:** 82
niobium content and alloy classification ... **A4:** 800
physical properties ... **EM4:** 316
precipitation strengthening and grain size ... **A4:** 799
solution heat treatment ... **A4:** 796, 803, 804, 911
solution treating ... **M4:** 656
solution treatment ... **A6:** 574
springs, strip for ... **M1:** 286
springs, wire for ... **M1:** 285
strain age cracking resistance ... **A6:** 84
stress relieving ... **M4:** 655
stress-equalizing ... **A4:** 908
stress-relieving ... **A4:** 908
temperature measurement, validation strategies ... **A6:** 1149
tensile properties ... **A4:** 806, 807
thermomechanical properties ... **A4:** 798
trace element impurity effect on GTA weld penetration ... **A6:** 20
transformation diagram ... **A4:** 803
Inconel 721
composition ... **A4:** 794
Inconel 722
composition ... **A4:** 794 **A6:** 564, 573
Inconel 725
aging ... **A4:** 796
composition ... **A4:** 794
solution-treating ... **A4:** 796
Inconel 738
aging cycle ... **A4:** 812
composition ... **A4:** 795
Inconel 751
composition ... **A4:** 794 **A6:** 573
Inconel 792
aging cycle ... **A4:** 812
composition ... **A4:** 795
Inconel 800
composition ... **M4:** 651-652
Inconel 901
aging ... **A4:** 796
precipitation strengthening and grain growth ... **A4:** 799
thermomechanical processing ... **A4:** 798
Inconel 907
aging ... **A4:** 796
cold working effect on aging ... **A4:** 800
composition ... **A4:** 794
oxidation ... **A4:** 798
solution-treating ... **A4:** 796
Inconel 909
aging ... **A4:** 796
cold working effect on aging ... **A4:** 800
composition ... **A4:** 794
nickel content and alloy classification ... **A4:** 800
oxidation ... **A4:** 798
precipitation strengthening and grain size ... **A4:** 800
solution-treating ... **A4:** 796
Inconel 925
aging ... **A4:** 796
Inconel 925 (continued)
composition ... **A4:** 794
solution-treating ... **A4:** 796
Inconel 939
aging cycle ... **A4:** 812
Inconel alloy
abrasive waterjet machining ... **A16:** 527
broaching ... **A16:** 209
electrochemical grinding ... **A16:** 542
grinding ... **A16:** 437
honing stone selection ... **A16:** 476
photochemical machining ... **A16:** 588
photochemical machining etchant ... **A16:** 590
sawing ... **A16:** 361, 363
Inconel alloy 100
drilling ... **A16:** 237
electron beam drilling ... **A16:** 570
Inconel alloy 600
chemical milling ... **A16:** 584
composition ... **A16:** 736, 836
machining ... **A16:** 738, 741-743, 746-747, 749-758, 837-840, 842, 843
Inconel alloy 601
composition ... **A16:** 736, 836
machining ... **A16:** 738, 741-743, 746-747, 749-757, 837-840, 842, 843
Inconel alloy 617
composition ... **A16:** 736, 836
electrochemical machining ... **A16:** 843
machining ... **A16:** 738, 741-743, 746-747, 749-757, 837-840, 842, 843
Inconel alloy 625
chemical milling ... **A16:** 584
composition ... **A16:** 736, 836
electrochemical machining ... **A16:** 843
machining ... **A16:** 738, 741-743, 746-747, 749-757, 837-840, 842, 843
Inconel alloy 690
composition ... **A16:** 836
machining ... **A16:** 837-840, 842, 843
Inconel alloy 700
composition ... **A16:** 736
machining ... **A16:** 738, 741-743, 746-747, 749-758
thread grinding ... **A16:** 275
Inconel alloy 702
thread grinding ... **A16:** 275
Inconel alloy 706
composition ... **A16:** 736, 836
electrochemical machining ... **A16:** 843
machining ... **A16:** 738, 741-743, 746-747, 749-757, 837-840, 842, 843
Inconel alloy 713C
composition ... **A16:** 737
machining ... **A16:** 741-743, 746-757
Inconel alloy 718
chemical milling ... **A16:** 584
composition ... **A16:** 37, 836
effect of EDM and grinding on fatigue strength ... **A16:** 35
electrochemical grinding ... **A16:** 547
electrochemical machining ... **A16:** 540, 541, 843
electrochemical machining and EDM ... **A16:** 25
fatigue strength and method of machining ... **A16:** 31
fatigue strength and shot peening ... **A16:** 36
feed rates for electrochemical machining ... **A16:** 534
grinding ... **A16:** 462, 464, 547, 760, 843
high removal rate machining ... **A16:** 608
high-speed machining and chip formation ... **A16:** 598, 599
honing with CBN ... **A16:** 479
laser beam drilling ... **A16:** 32
LBM-produced heat-affected zone ... **A16:** 24
machinability ... **A16:** 640, 737
machinability and notching ... **A16:** 642, 646
machining ... **A16:** 738, 741-743, 746-747, 749-758
milling ... **A16:** 547, 842
planing ... **A16:** 839

Inconel alloy 718 (continued)
sawing **A16:** 360
spade and gun drilling **A16:** 839
surface alterations from material removal processes **A16:** 26, 27, 34
surface characteristics **A16:** 31
tapping **A16:** 840
thread grinding **A16:** 275
threading **A16:** 840
turning **A16:** 837, 838
Inconel alloy 721
thread grinding **A16:** 275
Inconel alloy 722
thread grinding **A16:** 275
Inconel alloy 751
composition **A16:** 736, 836
electrochemical machining **A16:** 843
machining **A16:** 738, 741-743, 746-747, 749-757, 837-840, 842, 843
Inconel alloy 901
broaching **A16:** 209
Inconel alloy MA 754
composition **A16:** 738, 836
electrochemical machining **A16:** 843
machinability **A16:** 737
machining **A16:** 738, 741-743, 746-747, 749-758, 837-840, 842, 843
Inconel alloy MA 956
composition **A16:** 738
drilling **A16:** 746, 747, 749
machinability **A16:** 737
turning **A16:** 741
Inconel alloy MA 6000
composition **A16:** 738
grinding **A16:** 760
machinability **A16:** 737
machining **A16:** 737, 738, 741-743, 746-757
Inconel alloy X
broaching **A16:** 209
electrochemical grinding **A16:** 547
grinding **A16:** 547
milling **A16:** 547
Inconel alloy X-750
composition **A16:** 736, 836
electrochemical machining **A16:** 843
machining **A16:** 738, 741-743, 746-747, 749-758, 837-840, 842, 843
Inconel alloys *See also* High-temperature materials, specific types; Nickel alloys, specific types; Nickel-base superalloys, specific types
arc welding **M6:** 354
for basket used in liquid carburizing **A4:** 338
for casings of high-velocity convection burners **A4:** 274
development and characteristics **A2:** 429
pot material to hold noncyanide carburizing process **A4:** 331
for pots for liquid pressure nitriding **A4:** 415
reaction with graphite hearths in vacuum heat treating **A4:** 503
resistance spot welding **M6:** 480
work load support material in vacuum heat treating **A4:** 502-503
Inconel MA 754
brazing **A6:** 632, 928
composition **A6:** 577
gas-tungsten arc weld **A6:** 928
properties **A6:** 578
Inconel MA 6000
brazing **A6:** 632, 928
composition **A6:** 557
properties **A6:** 578
Inconel nickel-base alloys *See* Nickel-base alloys, specific types
Inconel, specific types
600
composition **M6:** 354
flash welding **M6:** 557
gas metal arc welding **M6:** 362
601, composition **M6:** 354
617, composition **M6:** 354
625
composition **M6:** 354
flash welding **M6:** 557
700, flash welding **M6:** 557
702, composition **M6:** 354
Inconel, specific types (continued)
706
composition **M6:** 354
flash welding **M6:** 557
718
brazing **M6:** 1020
composition **M6:** 354-355
explosion welding **M6:** 707
flash welding **M6:** 557
722, composition **M6:** 354
MA 754, brazing **M6:** 1020
MA 956, brazing **M6:** 1020
W, flash welding **M6:** 558
X-750
composition **M6:** 354
flash welding **M6:** 557
flash welding schedule **M6:** 577
Inconel X
ultrasonic welding **A6:** 326, 895
Inconel X-750
age hardening **A4:** 796, 803, 911
aging cycle **M4:** 656
aging cycles **A6:** 574
aging, effect on properties **M4:** 659
annealing **A4:** 908 **M4:** 655
applications **A4:** 911
ceramic-bonded fluoride coatings for lubricants **A18:** 118
composition **A4:** 794, 908 **A6:** 564, 573 **M4:** 651-652
cutoff band sawing with bimetal blades **A6:** 1184
electron-beam welding **A6:** 869
graphite as solid lubricant **A18:** 115
material for jet engine components **A18:** 588
nickel content and alloy classification **A4:** 800
solution treating **M4:** 656
solution treatment **A6:** 574
solution-treating **A4:** 796, 803, 911
springs, strip for **M1:** 286
springs, wire for **M1:** 285
stabilization treating **A4:** 803
stress relieving **M4:** 655
stress-relieving **A4:** 908
thermal treatments for precipitation hardening **A4:** 805
Inconels *See* Nickel alloys, cast, specific types; Nickel alloys, specific types
Incongruent phase change .... **A3:** 1 • 4
Inconol alloys *See* Nickel alloys, specific types, Inconel
Increment
defined **A10:** 674
Incremental optical encoder
in electrohydraulic testing machine **A8:** 160
Incremental permeability *See also* Permeability
defined and magnetically measured **A17:** 134
Incremental strain rate testing **A8:** 219, 223-226
Incremental testing
on explosively loaded torsional Kolsky bar **A8:** 225
polynomial, crack propagation rate **A8:** 378, 415, 518, 678-679
strain rate, Kolsky bar for **A8:** 219, 223-224
Incubation
formation of recrystallization nuclei during **A9:** 694, 698
period, defined **A13:** 8
prediction, from test data **A13:** 316
of stress-corrosion cracking **A13:** 245
Incubation and nucleation
stress-corrosion cracking **A8:** 496
Incubation period *See also* Cavitation erosion; Impingement erosion
in alloy addition **A15:** 71-74
defined **A18:** 11
Incubation resistance number (NOR) **A18:** 228
Incubation time
crack initiation **A11:** 242, 244
Incuro 60
wettability indices on stainless steel base metals **A6:** 118
Incuro 60, brazing
composition **A6:** 117
Incusil 10
brazing, composition **A6:** 117
Incusil 10 (continued)
wettability indices on stainless steel base metals **A6:** 118
Incusil 15
brazing, composition **A6:** 117
wettability indices on stainless steel base metals **A6:** 118
Indentation *See also* Impression; Indentation testing; Indenter
barrel-shaped, with diamond pyramid indenter **A8:** 100, 102
in bearing failures **A11:** 499
Brinell **A8:** 85-86
definition **M6:** 10
with equal diameters, different areas **A8:** 102
Knoop and Vickers, compared **A8:** 90
measurement **A8:** 85, 91
perfect **A8:** 100, 102
pincushion **A8:** 100, 102
spacing .... **A8:** 80 , 85, 88, 94, 105
testing **A8:** 71-73, 99, 102
true brinelling, spalling by **A11:** 500
types, in rolling-element bearings **A11:** 499
Indentation creep **A18:** 421
Indentation depth **A18:** 421, 422
on-load elasto-plastic **A18:** 422
Indentation, effects
in ring rolling **A14:** 119
Indentation fracture technique **A18:** 421
Indentation hardness *See also* Brinell hardness test; Hardness; Knoop (microindentation) hardness number; Microindentation hardness number Nanohardness test; Rockwell hardness number; Vickers (microindentation) hardness test **A7:** 61 **A18:** 33, 434 **EM3:** 16
defined **A8:** 7 **A18:** 11 **EM2:** 23
Indentation hardness testing **A7:** 312
measurement errors **A7:** 452
as quality control tool **A7:** 452-453
Indentation size
affecting microhardness readings **A8:** 96
vs. load, hardness testing **A8:** 94
Indentation size effect (ISE) exponent **A18:** 424
Indentation testing **A8:** 71-73, 99, 102
Indentation welding *See* Lap welding
Indenter
ball **A8:** 72
defined **A18:** 11
diamond **A8:** 74-75
elastic theory of blunt **A8:** 72
geometry **A8:** 81
Knoop **A8:** 90
methodology **A8:** 79-80
for Rockwell hardness testing **A8:** 74, 75
selection **A8:** 84, 90-91
shape, in microhardness testing **A8:** 96
types of blunt **A8:** 71
verification **A8:** 88
Vickers **A8:** 90-91
Independent compounders *See* Suppliers
Independent moving press platens **A7:** 323
Independent variables
in fatigue testing **A8:** 698
Independently loaded mixed-mode specimen (ILMMS) **EM3:** 509-510, 511, 512
opening load versus in-plane shear load **EM3:** 517
Index of Aerospace Materials Specifications **EM3:** 72
Index of plasticity **A18:** 424, 425, 426
Index of refraction *See* Refractive index
Index of surface roughness
defined **A12:** 201
Indexes
for hand lay-up **EM1:** 144
Indexing
of cold extrusion process **A14:** 303
of electron diffraction patterns **A10:** 456-457
mechanism, multiple-slide forming **A14:** 570
Indexing rotary automatic polishing and buffing machines **M5:** 120-123, 125, 127
Indialite **EM4:** 759
Indication(s) *See also* Nonrelevant indications; Relevant indications
defined **A17:** 103
false, defined **A17:** 103
magnification of **A17:** 125

Indicator tapes
as gas detection devices **A17:** 61
Indicator tissues *See also* Biologic indicators
for metal toxicity **A2:** 1234
Indicators
acid-base **A10:** 172
Indices *See* Miller indices
Indigenous inclusions *See* Deoxidation products
defined **A15:** 88-89
sulfide **A15:** 89
Indirect (backward) extrusion *See* Backward extrusion; Extrusion
Indirect compliance
for crack growth in aqueous solutions **A8:** 417
Indirect determination
of bromine **A10:** 70
Indirect failures
soldering **EL1:** 943
Indirect furnace heating
as hot pressing setup **A7:** 504-505
Indirect ion chromatography **A10:** 661
Indirect labor
as piece cost component **EM2:** 82
Indirect precipitation **A7:** 54
Indirect resistance heating
as hot pressing setup **A7:** 505-507
Indirect sintering **A7:** 6
Indirect-fired batch ovens
paint airing process **M5:** 488
Indirect-fired convection continuous oven
paint airing process **M5:** 487-488
Indium **A13:** 179-181, 186, 515
as a reactive sputtering cathode material **A9:** 60
as addition to aluminum alloys **A4:** 843
as addition to aluminum-silicon alloys **A18:** 790
as addition to tin-lead solders **A6:** 968
compatibility in bearing materials **A18:** 743
in Cu-Ni-In, thermal spray coating material **A18:** 832
as delayed-emission converter for thermal neutron radiography **EM3:** 759
determined by controlled-potential coulometry **A10:** 209
effect on crack propagation in steel **A11:** 244
in electroplated coatings **A18:** 838
electroplating of bearing materials **A18:** 756
embrittled by mercury **A11:** 234
embrittlement by **A11:** 235
epithermal neutron activation analysis **A10:** 239
evaporation fields for **A10:** 587
friction coefficient data **A18:** 71
in fusible alloys **M3:** 799
high-vacuum lubricant application **A18:** 153
and indium alloys **EL1:** 636-637
lap welding **M6:** 673
in lead-base alloys **A18:** 750
lead-indium **EM3:** 584
as low-melting embrittler **A12:** 29
-mercury solutions, LME of iron-aluminum alloys by **A11:** 225
price per pound **A6:** 964
pure **M2:** 739-740
safety standards for soldering **M6:** 109
segregation and solid friction **A18:** 28
as solid lubricant **A18:** 31
species weighed in gravimetry **A10:** 172
thermal diffusivity from 20 to 100 °C **A6:** 4
TNAA detection limits **A10:** 237
as tramp element **A8:** 476
TWA limits for particulates **A6:** 984
used to measure degree of surface cleanliness **EM3:** 322
vapor pressure, relation to temperature **A4:** 495
Indium alloying, wrought aluminum alloy **A2:** 51-52
applications **A2:** 750-753, 1259
and bismuth **A2:** 750-757
Indium alloying, wrought aluminum alloy (continued)
conductive films **A2:** 752
corrosion resistance **A2:** 752
glass-to-metal seals **A2:** 752
large temperature differentials **A2:** 752
low-melting temperature indium-base solders **A2:** 751-752
occurrence **A2:** 750
pricing history **A2:** 751
production of **A2:** 750-751
properties **A2:** 751
pure, properties **A2:** 1117
recovery methods **A2:** 750
semiconductors **A2:** 752-753
silver-palladium compatibility **A2:** 752
thermal fatigue resistance **A2:** 752
Indium chloride
biologic effects **A2:** 1259
Indium conversion screens
for neutron radiography **A17:** 387, 391
Indium gallium arsenide phosphide (InGaAsP)
laser diodes **A2:** 739-740
photodiodes **A2:** 740
Indium oxide
conductive films **A2:** 752
hydrated, biologic effects **A2:** 1259
Indium, resistance of
to liquid-metal corrosion **A1:** 636
Indium, vapor pressure
relation to temperature **M4:** 310
Indium-resonance technique *See also* Neutron radiography
applied **A17:** 391-392
for elements, nuclear fuels **A17:** 391-392
Indium-tellurium alloys
martensitic structures **A9:** 673
Indium-tin alloy (50-50) .... **A3:** 1 • 19
Indium-tin-oxide
conductive film of **A2:** 752
Individual chuck-spindle-drive rotary automatic polishing and buffing machines **M5:** 121
Individuals
defined **A10:** 674
Indoor atmospheres **A13:** 746-747, 763
Induced current method
applications **A17:** 94, 98-99
for bearing rings **A17:** 117
direct versus alternating current **A17:** 98
of generating magnetic fields **A17:** 97-99
magnetizing, advantages/limitations **A17:** 94
Induced eddy currents
remote-field eddy current inspection **A17:** 195
Induced pressure system **A7:** 125
Inductance
defined **EL1:** 1147
external/internal, defined **EL1:** 29
formulas for **EL1:** 29
power supply/ground distribution network **EL1:** 27-28
SI derived unit and symbol for **A10:** 685
SI unit/symbol for **A8:** 721
Inductance, changes
in magabsorption **A17:** 149
Induction
relative rating of brazing process heating method **A6:** 120
Induction and tungsten mesh heating **A7:** 389
Induction bonding
defined **EM2:** 23
Induction brazing
definition **M6:** 1
Induction brazing (IB) **A6:** 121-122, 333-335, 947
advantages **A6:** 333
applications **A6:** 333, 334-335
brass **A6:** 333, 335
coil and joint configurations **A6:** 333, 334
copper **A6:** 333
Induction brazing (IB) (continued)
copper and copper alloys **A6:** 934-935
definition **A6:** 333, 1210
equipment **A6:** 333-334
fixturing **A6:** 333-334
heat content (mass basis) vs. temperature **A6:** 333, 334
human factors, engineering ergonomics **A6:** 335
joint design **A6:** 334-335
limitations **A6:** 333
precious metals **A6:** 936
stainless steel **A6:** 335, 911, 921, 922
steel **A6:** 333, 334
Induction brazing of
carbon steels **M6:** 966
copper and copper alloys **M6:** 966-967, 1042-1045
low-alloy steels **M6:** 966
reactive metals **M6:** 1052
stainless steels **M6:** 966, 1011-1012
Induction brazing of steels **M6:** 965-975
assembly **M6:** 971
brazing of dissimilar metals **M6:** 973-97
comparison to oxyacetylene welding **M6:** 97
comparison to shielded metal arc welding **M6:** 974-975
filler metals **M6:** 970-971
alloy selection **M6:** 970-971
form selection **M6:** 971
preforms **M6:** 971
filler-metal sandwich **M6:** 973-97
fixturing **M6:** 971
material selection **M6:** 971
fluxes **M6:** 971
frequency range **M6:** 965
hardening simultaneously **M6:** 972-973
heat treating combination **M6:** 96
inductors **M6:** 967-968
coil turn spacing **M6:** 97
cooling **M6:** 97
coupling **M6:** 97
design **M6:** 968-97
heating patterns **M6:** 967-968
magnetic fields **M6:** 967-968
tubing **M6:** 97
work coils **M6:** 96
matching impedance **M6:** 96
metals brazed **M6:** 965
operation principles **M6:** 966-967
brazing of steel **M6:** 966
brazing of steel to copper **M6:** 966
power supply **M6:** 967
frequency selection **M6:** 967
motor-generator units **M6:** 967
solid-state power supplies **M6:** 967
vacuum-tube units **M6:** 967
process capabilities **M6:** 965-966
quantity limitations **M6:** 966
shape limitations **M6:** 965-966
size limitations **M6:** 965
tight joint brazing **M6:** 972
interference fit **M6:** 972
two-process brazing **M6:** 975
Induction bridge system
eddy current inspection **A17:** 178
Induction coil
in electrohydraulic testing machine **A8:** 160
Induction coil method
in electric current perturbation **A17:** 136
Induction curing **EM3:** 553-554
Induction furnaces *See also* Furnaces; Induction heating **A15:** 368-374
defined **A15:** 7
electric, development **A15:** 32
electromagnetic stirring **A15:** 369
high-frequency **A10:** 221-222
lining material **A15:** 372-373
melting operations **A15:** 373-374
power supplies **A15:** 369-371

SUBJECTS OF THE INDEXED VOLUMES: ASM Handbook (designated by the letter "A"): **A1:** Properties and Selection: Irons, Steels, and High-Performance Alloys (1990); **A2:** Properties and Selection: Nonferrous Alloys and Special-Purpose Materials (1990); **A3:** Alloy Phase Diagrams (1992); **A4:** Heat Treating (1991); **A6:** Welding, Brazing, and Soldering (1993); **A7:** Powder Metallurgy (1984); **A8:** Mechanical Testing (1985); **A9:** Metallography and Microstructures (1985); **A10:** Materials Characterization (1986); **A11:** Failure Analysis and Prevention (1986); **A12:** Fractography (1987); **A13:** Corrosion (1987); **A14:** Forming and Forging (1988); **A15:** Casting (1988); **A16:** Machining (1989); **A17:** Nondestructive Testing and Quality Control (1989); **A18:** Friction, Lubrication, and Wear Technology (1992). **Metals Handbook, 9th Edition** (designated by the letter "M"): **M1:** Properties and Selection: Irons and Steels (1978); **M2:** Properties and Selection: Nonferrous Alloys and Pure Metals (1979); **M3:** Properties and Selection: Stainless Steels, Tool Materials and Special-Purpose Materials (1980); **M4:** Heat Treating (1981); **M5:** Surface Cleaning, Finishing, and Coating (1982); **M6:** Welding, Brazing, and Soldering (1983). **Engineered Materials Handbook** (designated by the letters "EM"): **EM1:** Composites (1987); **EM2:** Engineering Plastics (1988); **EM3:** Adhesives and Sealants (1990); **EM4:** Ceramics and Glasses (1991); **Electronic Materials Handbook** (designated by the letters "EL"): **EL1:** Packaging (1989).

Induction furnaces (continued)
types ... **A15:** 368-369
water cooling systems ... **A15:** 371-372
Induction hardening ... **M1:** 528-532
applications, relation to cost ... **M4:** 475-476
case depth ... **M4:** 470-471, 472
cast iron *See also* Cast iron, induction heating ... **M4:** 476-480
cracking ... **M4:** 473
defined ... **A9:** 9
distortion ... **M4:** 473
ductile iron ... **A15:** 659 **M1:** 37
effect of seams during ... **A11:** 478-479
equipment selection ... **M4:** 480
fatigue resistance ... **M1:** 674-675
fatigue strength, improved by ... **M4:** 452
frequency selection ... **M4:** 454-455, 456
grain coarsening ... **M4:** 469, 470
gray cast iron ... **M1:** 29-30
hardenable steels ... **M1:** 457, 470
hardening temperatures ... **M4:** 469-470
heating duration ... **M4:** 457-459, 559
high-frequency resistance ... **M4:** 476
machining ... **M4:** 463-464
magnetic fields ... **M4:** 451
malleable cast iron ... **M1:** 73
operating practices ... **M4:** 458, 471-473
power selection ... **M4:** 456-457
quenching ... **M4:** 463, 464-465
quenching oils ... **M4:** 464, 465
residual stress ... **M4:** 473
safety precautions ... **M4:** 467
steel selection for ... **M4:** 467-468, 469
surface hardness, control ... **M4:** 469-470
through hardening ... **M4:** 452
transverse cracking during ... **A11:** 395
wear resistance ... **M4:** 451-452
Induction hardening and tempering ... **M4:** 451-483
Induction hardening equipment
capacitors ... **M4:** 462
coil coolants ... **M4:** 461
coil design ... **M4:** 459
inductor coils, design ... **M4:** 457-459
line-frequency units ... **M4:** 452
maintenance ... **M4:** 467
motor-generator units ... **M4:** 452-454, 461
quenching systems ... **M4:** 464-465
scanning devices ... **M4:** 463
solid state units ... **M4:** 453-454, 455, 461
static frequency converters ... **M4:** 452
transformers ... **M4:** 461-462
transmission cables ... **M4:** 460-461
vacuum tube units ... **M4:** 454, 455, 462
work-handling ... **M4:** 463
Induction hardening, improper
effect in alloy steel ... **A12:** 3m
Induction hardening, process control
cooling of equipment ... **M4:** 466-467
delay time ... **M4:** 466
hardening cycle ... **M4:** 465-466
heating time ... **M4:** 466
power density ... **M4:** 466
quenching cycle ... **M4:** 466
Induction hardening, steels ... **A4:** 184-191
austenitizing temperatures ... **A4:** 184-186
case depth ... **A4:** 188, 189
Curie point ... **A4:** 187, 188
electrical properties ... **A4:** 186-187
energy requirements ... **A4:** 188, 189
frequency selection ... **A4:** 188-189
heating parameters ... **A4:** 187-188
induction heating temperatures for metalworking processes ... **A4:** 188
induction tempering ... **A4:** 186, 191, 194
magnetic properties ... **A4:** 186-187
operating conditions for through hardening ... **A4:** 190, 192
power density and heating time ... **A4:** 189-191
power ratings for surface hardening ... **A4:** 190
residual stresses ... **A4:** 185-186
temperatures required ... **A4:** 188
time-temperature relations ... **A4:** 184-186
Induction hardening, through hardening
density ... **M4:** 456, 457, 473-474
heating rate ... **M4:** 474, 475
methods ... **M4:** 473
Induction heat treating of steel *See* Steel, induction heat treating
Induction heaters
for steam generator corrosion ... **A13:** 942
Induction heating *See also* Induction furnace
of alloy steels ... **A7:** 506
of alloys ... **A8:** 159
for copper alloy powders ... **A7:** 121
defined ... **A9:** 9 **A15:** 7
for high-temperature torsion testing ... **A8:** 158-159
in situ, for elevated-temperature compression testing ... **A8:** 196
infrared controller for ... **A8:** 414
of polymers ... **A7:** 607
for precision forging ... **A14:** 165
radio frequency generators ... **A8:** 414
system, hot-die/isothermal forging ... **A14:** 152
temperature control ... **A8:** 36, 414
in vacuum and oxidizing environments ... **A8:** 414
Induction heating stress improvement
for boiling water reactors ... **A13:** 931
Induction heating technique, and eddy current inspection
compared ... **A17:** 165-166
Induction melting *See also* Induction heating
of gray iron ... **A15:** 636
plain carbon steels ... **A15:** 708
vacuum, in irons ... **A12:** 219
Induction precurable adhesives
automotive applications ... **EM3:** 553-554
Induction radiant heating
tungsten and molybdenum sintering ... **A7:** 392
Induction sintering ... **A7:** 6
Induction soldering ... **A6:** 363-365
advantages ... **A6:** 363
applications ... **A6:** 365
coil configurations ... **A6:** 364
coupling efficiency ... **A6:** 363
Curie temperature effects ... **A6:** 363
definition ... **A6:** 1210 **M6:** 10
iron ... **A6:** 363
limitations ... **A6:** 363
mechanical property relationships ... **A6:** 364
nickel ... **A6:** 363
penetration depth of heating ... **A6:** 365
personnel ... **A6:** 365
preplaced solder, 5DD-2-5DD-3 ... **A6:** 364-365
resistivity effects ... **A6:** 363
safety concerns ... **A6:** 365
safety precautions ... **A6:** 1191
setup parameters ... **A6:** 365
skin depth ... **A6:** 365
skin effect ... **A6:** 363
workpiece geometry ... **A6:** 363-364
Induction tempering
advantages ... **M4:** 477, 480
application ... **M4:** 480
cycles ... **A4:** 186
electric-furnace tempering, compared to ... **M4:** 482-483
power density ... **M4:** 477, 480-481
radiation pyrometer ... **M4:** 481
results ... **M4:** 482
schematic diagrams ... **M4:** 479, 481
tempering cycles ... **M4:** 481, 482
tempering parameter (T.P.) ... **A4:** 186, 194
versus induction hardening ... **A4:** 186
voltage regulator ... **M4:** 481-482
Induction welding *See* Electromagnetic welding
definition ... **M6:** 10
high frequency welds ... **M6:** 757, 759-760
Induction welding (IW)
definition ... **A6:** 1210
Induction welds, high-frequency ... **A11:** 449
Induction-hardened steel shaft
subsurface residual stress and hardness distributions in ... **A10:** 389-390
Induction-hardening
gear materials ... **A18:** 261
Induction-heated graphite crucible
tungsten carbide powders ... **A7:** 157
Inductive coil sensors *See also* Coils; Probes; Sensor(s)
for leakage field testing ... **A17:** 130
Inductively coupled plasma *See also* Inductively coupled plasma atomic emission spectroscopy
electric and magnetic fields ... **A10:** 32
inhomogeneous, nomenclature of zones ... **A10:** 32
polychromator or direct-reader spectrometer for ... **A10:** 37
Inductively coupled plasma (ICP)
as trace element analysis ... **A2:** 1095
Inductively coupled plasma (ICP) emission spectroscopy
radio frequency, for chemical analysis ... **EM4:** 553, 554, 555
to analyze the bulk chemical composition of starting powders ... **EM4:** 72
for trace element analysis ... **EM4:** 24
Inductively coupled plasma atomic emission spectroscopy ... **A10:** 31-42
analytical characteristics ... **A10:** 33-34
applications ... **A10:** 31, 41
and atomic absorption spectrometry compared ... **A10:** 31, 43
atomic theory ... **A10:** 33
calibration curves ... **A10:** 33, 34
capabilities ... **A10:** 141, 233, 333
capabilities, compared with molecular fluorescence spectroscopy ... **A10:** 72
capabilities, compared with X-ray spectrometry ... **A10:** 82
defined ... **A10:** 674
detection electronics and interface ... **A10:** 39
detection limits ... **A10:** 33
and direct-current arc emission spectrography ... **A10:** 31
direct-current plasma ... **A10:** 40
direct-reading spectrometer ... **A10:** 37-38
and emission spectroscopy ... **A10:** 21
estimated analysis time ... **A10:** 31
of inorganic liquids and solutions ... **A10:** 7
of inorganic solids ... **A10:** 4-6
interference effects ... **A10:** 33-34
introduction ... **A10:** 31-32
limitations ... **A10:** 31
nebulizer ... **A10:** 34-36
neutron activation analysis and compared ... **A10:** 233
new developments ... **A10:** 39-40
plasma ... **A10:** 32
polychromator ... **A10:** 34, 37-38
precision and accuracy ... **A10:** 33
principles of operation ... **A10:** 32
procedure ... **A10:** 34
related techniques ... **A10:** 31
samples ... **A10:** 31, 34-36
scanning monochromator ... **A10:** 38
system components ... **A10:** 34-39
system computer for ... **A10:** 39
zones of plasma ... **A10:** 32
Inductively coupled plasma emission (ICPE)
spectrometric metals analysis ... **A18:** 300
Inductively coupled plasma mass spectroscopy
capabilities ... **A10:** 233
instrumentation for ... **A10:** 40
neutron activation analysis and, compared ... **A10:** 233
as new development ... **A10:** 39-40
Inductively coupled plasma mass spectroscopy (ICPMS), for trace element measurement
ultra-high purity metals ... **A2:** 1095
Inductor(s)
chip, types ... **EL1:** 179
fabrication ... **EL1:** 187-188
failure mechanisms ... **EL1:** 1003-1005
implementation at microwave frequency ... **EL1:** 178
materials selection ... **EL1:** 182-183
in passive components ... **EL1:** 179
removal methods ... **EL1:** 724
thin-film ... **EL1:** 320-321
through-hole packages, failure mechanisms ... **EL1:** 979-980
types and construction ... **EL1:** 179
wire-wound chip, structure ... **EL1:** 179
Inductors
basic designs for brazing ... **M6:** 968-969
for brazing of steel ... **M6:** 967
cooling ... **M6:** 970

Inductors (continued)
coupling and coil turn spacing **M6:** 970
tubing **M6:** 970
work coils **M6:** 969
Industrial acoustic microscopy techniques
compared **A17:** 470
Industrial and material handling applications
homopolymer/copolymer acetals **EM2:** 100
liquid crystal polymers (LCP) **EM2:** 180
of part design **EM2:** 616
phenolics **EM2:** 242
polyamides (PA) **EM2:** 125
polybutylene terephthalates (PBT) **EM2:** 153
polyether-imides (PEI) **EM2:** 156
polyethylene terephthalates (PET) **EM2:** 172
polyphenylene ether blends (PPE PPO) **EM2:** 183
silicones (SI) **EM2:** 266-267
styrene-maleic anhydrides (S/MA) **EM2:** 218
thermoplastic polyurethanes (TPUR) **EM2:** 205
ultrahigh molecular weight polyethylenes (UHMWPE) **EM2:** 167-168
unsaturated polyesters **EM2:** 246
urethane hybrids **EM2:** 268
Industrial applications *See also* Applications
of bronze P/M parts **A7:** 736
bulk molding compounds **EM1:** 163
commercial hybrids **EL1:** 381-385
copper machinery and equipment **A2:** 239
of copper-based powder metals **A7:** 733
of hybrid **EL1:** 254-255
of precious metals **A2:** 693-694
for stainless steels **A7:** 731
thermocouple selection for **A2:** 885
thick-film hybrids **EL1:** 381
zirconium alloys **A2:** 668-669
Industrial applications for adhesives **EM3:** 567-578
adhesives technologies **EM3:** 567-568
categories **EM3:** 567
construction **EM3:** 577-578
cost considerations **EM3:** 568
electrical **EM3:** 573-575
electronics **EM3:** 568-573
general component bonding **EM3:** 573
major properties **EM3:** 568
medical **EM3:** 575-576
performance considerations **EM3:** 568
sporting goods manufacturing **EM3:** 576
systematic approach to adhesive/application selection **EM3:** 568
Industrial Applications of Adhesive Bonding **EM3:** 70
Industrial argyria
as silver toxicity **A2:** 1260
Industrial atmosphere
effect on stress-corrosion cracking **A8:** 499
Industrial atmospheres
alloy steel corrosion in **A13:** 531
aluminum weathering data **A13:** 607
austenitic stainless steel atmospheric corrosion in **A13:** 554
contaminants in **A13:** 81
corrosion in **A11:** 193
defined **A13:** 8
galvanized coatings in **A13:** 440
magnesium/magnesium alloys in **A13:** 743
SCC of aluminum alloy in **A13:** 266
simulated service testing **A13:** 204
steel corrosion in **A13:** 1304
telephone cables in **A13:** 1127
zinc/zinc alloys and coatings in **A13:** 757
Industrial computed tomography *See also* Computed tomography (CT) **A17:** 358-386
applications **A17:** 362-363
and backscatter/Compton imaging **A17:** 362
capabilities/disadvantages **A17:** 360-361
of castings **A17:** 528-529
Industrial computed tomography (continued)
equipment **A17:** 364-372
glossary **A17:** 383-385
historical background **A17:** 358-359
image quality **A17:** 372-377
microstructure effect **A17:** 51
and nuclear magnetic resonance (NMR) **A17:** 362
and nuclear tracer imaging **A17:** 362
of powder metallurgy parts **A17:** 538
principles of **A17:** 359-360
radiation sources **A17:** 368-369
and radiography, compared **A17:** 295, 361-362
reconstruction techniques **A17:** 379-382
special features **A17:** 377-379
system design **A17:** 364-372
Industrial diamond *See* Diamond; Polycrystalline diamond; Synthetic diamond
Industrial Fasteners Institute (IFI) **A1:** 289
Industrial hydraulic fluids
additives in formulation
Industrial lasers **A14:** 735-742
Industrial lubricants
antiwear agents used **A18:** 102
Industrial machinery
codes governing **M6:** 825
Industrial market
sealants **EM3:** 58
Industrial materials
quantitative elemental analysis by classical wet chemistry **A10:** 162-179
raw, sampling of **A10:** 12-18
waste products, sampling of **A10:** 12-18
Industrial metal-graphite brushes **A7:** 634
Industrial oils
antisquawk additives **A18:** 104
gear, demulsifiers **A18:** 107
Industrial or standard-quality wire **A1:** 282
Industrial pollution **A7:** 208
Industrial quality carbon steel wire rod **M1:** 254
Industrial quality low-carbon steel wire **M1:** 264
Industrial quality rod **A1:** 273
Industrial robots
for press loading, blanks **A14:** 50
for press unloading **A14:** 500-501
as transfer equipment **A14:** 501
Industrial solder alloys
lead and lead alloy **A2:** 553
Industrial toxicity *See* Occupational metal toxicity
Industrial waste waters
cast iron resistance **A13:** 570
Industries
affected by microbiological corrosion **A13:** 118
Industry standards **EM3:** 61-64
issued by professional and trade organizations **EM3:** 61
Industry structure
automotive electronics **EL1:** 382
consumer electronics **EL1:** 385
telecommunications **EL1:** 383-385
Inelastic buckling
in axial compression testing **A8:** 55-56
Inelastic collisons
as electron signals **A12:** 168
Inelastic cyclic buckling
as distortion **A11:** 144
Inelastic electron scatter *See* Incoherent scatter
Inelastic mean free path **A18:** 453
electron **A10:** 569-571
Inelastic scattering
analytical transmission electron microscopy **A10:** 433-434
bremsstrahlung **A10:** 433
defined **A10:** 674
excitation of conduction electrons and emission **A10:** 433-434
secondary electron (low-energy)
inner-shell ionization **A10:** 433
Inelastic strain
range **A8:** 347
rate **A8:** 357
Inelastically scattered electrons
Kikuchi lines **A9:** 109-110
Inert (vacuum) gas infusion
as trace element analysis **A2:** 1095
Inert anode
defined **A13:** 8
Inert atmosphere **EM3:** 16
Inert atmosphere chamber thermal spray coating **M5:** 364-365
Inert carriers
and liquid metal embrittlement **A13:** 177
and liquid-metal embrittlement **A11:** 230-231
Inert elements
implantation of **A10:** 475
Inert filler *See also* Filler; Filler(s) **EM3:** 16
defined **EM1:** 13 **EM2:** 23
Inert gas
defined **A15:** 7
definition **A6:** 1210 **M6:** 10
effect, Cosworth process **A15:** 38
Inert gas atomized powders **A13:** 833
Inert gas blanketing
for liquid metal fires **A13:** 96
Inert gas elements
implanted **A10:** 485
Inert gas flushing
of aluminum melts **A15:** 80
for hydrogen removal from aluminum and copper **A15:** 87
nitrogen and hydrogen removal by **A15:** 84-85
Inert gas fluxing
copper alloys **A15:** 466-467
for hydrogen pickup/loss, aluminum-silicon melts **A15:** 165
Inert gas fusion **A10:** 226-232
defined **A10:** 674
detection of fusion gases **A10:** 229-230
determination of gases **A10:** 229-231
estimated analysis time **A10:** 226
general use **A10:** 226
of inorganic solids **A10:** 4, 6
introduction **A10:** 226-227
limitations **A10:** 226
operation, principles of **A10:** 227-228
related techniques **A10:** 226
samples **A10:** 226, 231-232
selective fusion **A10:** 231
separation of fusion gases **A10:** 228-229
Inert gas ion sputtering
LEISS analysis **A10:** 603
Inert gas metal arc welding
definition **A6:** 1210
Inert gas tungsten arc welding
definition **A6:** 1210
Inert gas(es)
content, explosion characteristics **A7:** 196
as explosion extinguisher **A7:** 200
in explosion prevention **A7:** 197
melting **A7:** 25
Inert gas-purged polychromator **A10:** 37
Inert gases
laser-beam welding of aluminum alloys **A6:** 739
shielding gas for laser cladding **A18:** 867
Inert lubricants
as corrosion control **A11:** 194
Inert plasma spraying (IPS)
molten particle deposition **EM4:** 205, 206, 207
Inert-gas atomization process
beryllium powders **A2:** 685
Inertia
effects in strain rate testing **A8:** 40-41, 208-209
Inertia friction welding
titanium alloys **A6:** 522
Inertia spike **A18:** 47

Inertia welding
bending fatigue fracture after ... **A11:** 469
heat-affected zone ... **A6:** 889
Inertial
constraint, effects on high strain rate test validity ... **A8:** 190
loading, and Charpy V-notch impact test ... **A8:** 259, 267
Infant mortality *See also* Bathtub reliability curve; Reliability
reliability life cycle ... **EL1:** 244, 897
Infection of open wounds
prevention ... **A7:** 573
Infiltrant ... **A7:** 6, 552
Infiltrated P/M steels
composition ... **M1:** 333
infiltration process ... **M1:** 337-339
mechanical properties ... **M1:** 334-335, 344
Infiltrated powder metallurgy materials
artifacts ... **A9:** 504
Infiltrated steel powders ... **A7:** 564
composition ... **A7:** 464
microstructural analysis ... **A7:** 487-488
Infiltration ... **A7:** 6, 551-566
capillary methods ... **A7:** 552
carbon, carbide infiltration test ... **A7:** 556
as cermet forming technique ... **A7:** 800-801
of cermets ... **A2:** 989-990
effect on transverse-rupture ... **A7:** 564
localized ... **A7:** 564
mechanical parts ... **A7:** 564-565
mechanism ... **A7:** 551-553
of porous P/M parts ... **A7:** 105
powders used ... **A7:** 573
process for multifilamentary super-conducting wire ... **A7:** 638
process for multifilamentary tape ... **A7:** 637
products ... **A7:** 560-565
systems ... **A7:** 554-560
techniques ... **A7:** 17-18, 553-554
vacuum ... **A7:** 554
Infiltration, liquid metal *See* Liquid metal infiltration
Infinite dilution, liquid metals
activity coefficients at ... **A15:** 60
Infinitely thick/thin samples
x-ray spectrometry ... **A10:** 93
Inflection point
defined ... **A9:** 10
Influent water
monitoring in acidified chloride solutions ... **A8:** 419
Information *See also* Data sheets; Information sources
design ... **EM2:** 410
engineering, types needed ... **EM2:** 411
materials, data base on ... **EM2:** 95
property ... **EM2:** 405
qualitative and quantitative, evaluating ... **EM2:** 405
triangular flow of ... **EM2:** 1
Information display ... **EM4:** 1045-1049
applications of glass ... **EM4:** 1045
display glass properties ... **EM4:** 1047-1049
dot-matrix printers and ink-jet printers ... **EM4:** 1047
flat panel display substrate requirements ... **EM4:** 1046-1047
flat panel display technologies ... **EM4:** 1045-1046
Information Handling Services
VSMF Data Control Services ... **EM3:** 64
Information sources *See also* Data sheets
books, handbooks, monographs ... **EM2:** 93-95
data base, materials information ... **EM2:** 95
government information center ... **EM2:** 95
guide to ... **EM1:** 40-42
journals, trade magazines, periodicals ... **EM2:** 92-93
manufacturer literature ... **EM2:** 92
recommended reading ... **EM2:** 95
short courses, seminars, conferences ... **EM2:** 95
Information system (SEM imaging)
electron signals ... **A12:** 168
in situ studies ... **A12:** 169
specimen/instrument geometry effects ... **A12:** 168
thermal-wave imaging ... **A12:** 169
x-ray signals ... **A12:** 168
Infrared
controller, for induction heating ... **A8:** 414
imaging, ultrasonic testing ... **A8:** 247-248
defined ... **EM1:** 13
relative rating of brazing process heating method ... **A6:** 120
Infrared (IR) ... **EM3:** 16
curing, of polymers ... **EL1:** 786
defined ... **EM2:** 23
emission, defined ... **EL1:** 1147
energy, for surface-mount reflow ... **EL1:** 694
heating, for solder reflow ... **EL1:** 117
reflow soldering ... **EL1:** 286
soldering ... **EL1:** 180, 704-705
spectra, parylene coatings ... **EL1:** 795
Infrared (IR) inspection
of solder joints ... **EL1:** 737, 942
Infrared (IR) microscopes
for optical imaging ... **EL1:** 1068-1069
Infrared (IR) preheaters
quartz ... **EL1:** 685-686
Infrared (IR) solder joint inspection ... **EL1:** 737, 942
Infrared (IR) soldering ... **EL1:** 180, 704-705
Infrared (IR) spectroscopy ... **A18:** 460-461 **EM1:** 736, 737 **EM2:** 23, 521-522, 826
as advanced failure analysis ... **EL1:** 1103-1104
of epoxies ... **EL1:** 834-835
lubricant analysis ... **A18:** 300-301, 311
Infrared (quartz) brazing ... **A6:** 123, 124
Infrared (thermal transfer) imaging
soldered joints ... **A6:** 981
Infrared absorption
as detector for C and S in high-temperature combustion ... **A10:** 221-222
sulfur determination in high-temperature combustion by ... **A10:** 221-225
Infrared absorption (IR) spectroscopy
applications ... **EM4:** 561-562
for phase analysis ... **EM4:** 561
Infrared absorption spectrophotometry
residue analysis by ... **A10:** 177
Infrared analysis
for measurement of furnace atmosphere composition ... **EM4:** 252
Infrared analyzers
atmospheres ... **M4:** 424-426
Infrared Astronomy Satellite
beryllium P/M parts ... **A7:** 760
Infrared brazing
definition ... **M6:** 10
Infrared brazing (IRB)
definition ... **A6:** 1210
Infrared detection
of carbon and sulfur, high-temperature combustion ... **A10:** 223
inert gas fusion ... **A10:** 230
Infrared detectors
used in thermal-wave imaging ... **A9:** 90
Infrared diode lasers
applications ... **A10:** 112
Infrared domes ... **EM4:** 18
compositions ... **EM4:** 18
fabrication processes ... **EM4:** 18
properties ... **EM4:** 18
Infrared emission spectroscopy ... **A10:** 115
Infrared heating system
for hot-die forging ... **A14:** 152
Infrared linear dichroism spectroscopy
for molecular orientation in drawn polymer films ... **A10:** 120
Infrared micosampling ... **A10:** 116
Infrared microscopes ... **A10:** 116
Infrared optics
as beryllium application ... **A2:** 684
germanium and germanium compounds ... **A2:** 735, 737
Infrared photosensors
for automatic pouring ... **A15:** 500
Infrared radiation
defined ... **A10:** 674
definition ... **M6:** 10
gold applications ... **A2:** 692
use in integrated circlet failure analysis ... **A11:** 767
Infrared radiation inspection *See also* Infrared thermography
of adhesive-bonded joints ... **A17:** 628-630
gas analyzers, as gas/leak detection devices ... **A17:** 63
imaging equipment, for thermal inspection ... **A17:** 398-399
infrared radiation, evapograph-recorded ... **A17:** 208
radiometer testing, of adhesive-bonded joints ... **A17:** 628-629
Infrared reflection-absorption spectroscopy ... **A10:** 114, 119
Infrared reflectivity
of beryllium ... **A2:** 684
Infrared signature ... **A6:** 360
Infrared soldering
definition ... **M6:** 10
Infrared soldering (IRS) *See also* Furnace soldering ... **A6:** 353-355
definition ... **A6:** 1210
Infrared spectra ... **A10:** 110, 116, 674
Infrared spectrometers
defined ... **A10:** 674
Infrared spectroscopy ... **A10:** 109-125
absorbance ... **A10:** 110, 117
applications ... **A10:** 109, 118-124
attenuated total reflectance spectroscopy ... **A10:** 113-114
basic principles ... **A10:** 110-111
Beer's law ... **A10:** 117
capabilities ... **A10:** 212, 333, 649
chromatographic techniques ... **A10:** 115-116
computerized, Fourier transform infrared spectroscopy as ... **A10:** 109
curve filling ... **A10:** 117-118
defined ... **A10:** 674
degrees of freedom ... **A10:** 110
depth profiling ... **A10:** 113, 115
diffuse reflectance spectroscopy ... **A10:** 114
dipole moment ... **A10:** 111
dispersive ... **A10:** 111
emission ... **A10:** 115
estimated analysis time ... **A10:** 109
Fourier-transform infrared spectroscopy ... **A10:** 109-110, 111-112
and gas analysis by mass spectroscopy compared ... **A10:** 151
general uses ... **A10:** 109
for glasses, compared with Raman ... **A10:** 130-131
infrared reflection-absorption spectroscopy ... **A10:** 114
of inorganic gases ... **A10:** 8
of inorganic liquids and solutions ... **A10:** 7
instrumentation ... **A10:** 110-112
introduction ... **A10:** 109-110
limitations ... **A10:** 109
microsampling for ... **A10:** 116
molecular vibrations ... **A10:** 111
of organic gases ... **A10:** 11
of organic liquids and solutions ... **A10:** 10
of organic solids ... **A10:** 9
for orientation of DTDMAC on nonmetallic surfaces ... **A10:** 119
photoacoustic spectroscopy ... **A10:** 115
polarization modulation ... **A10:** 114-115
qualitative analysis ... **A10:** 116-117
quantitative analysis ... **A10:** 117-118
and Raman, compared for polymer analyses ... **A10:** 131
and Raman spectroscopy ... **A10:** 126-127
reflectance methods ... **A10:** 113-115
reflection-absorption spectroscopy ... **A10:** 114
related techniques ... **A10:** 109
samples ... **A10:** 109, 112-116
sampling and sample preparation ... **A10:** 112-116
Snell's law ... **A10:** 113
specular reflectance ... **A10:** 115
use of Fourier transform spectrometers in ... **A10:** 39
as vibrational surface probe ... **A10:** 136
Infrared spectroscopy (IR) ... **EM4:** 53
for analyzing organics and chemical bonding ... **EM4:** 24
for chemical analysis of polymer fibers ... **EM4:** 223
to analyze the surface composition of ceramic powders ... **EM4:** 73

Infrared spectroscopy or spectrometry **EM3:** 16
Infrared spectrum
defined **A10:** 110, 674
Infrared thermography
of castings **A17:** 512
for optical imaging **EL1:** 1071
Infrared welding
safety precautions **A6:** 1191
Infrared-transparent pressing powder
sample pellets of **A10:** 113
Ingate *See also* Gate
design, in gating systems **A15:** 589-590, 592-593
Inglis equation **EM4:** 850
Ingot **A7:** 6
alloying element and impurity specifications **A2:** 16
aluminum-lithium alloys **A2:** 182
designation system, aluminum and aluminum alloy **A2:** 15-16
niobium-titanium, forging and inspection **A2:** 1044
titanium, production **A2:** 594-595
unalloyed and alloyed aluminum compositions **A2:** 22-25
uranium, as derbies **A2:** 670
Ingot casting **M1:** 114
Ingot defects
copper alloy ingots **A9:** 642-645
Ingot metallurgy (I/M)
vs. powder metallurgy (P/M) **A7:** 745, 747
Ingot metallurgy alloy rod **A7:** 468
Ingot metallurgy alloys, specific types
2014, rotating beam fatigue strength **A7:** 469
7075, rotating beam fatigue strength **A7:** 469
Ingot metallurgy tool steel alloys, specific types *See also* Tool steels; Tool steels, specific types
D2, mechanical properties and compositions **A7:** 471
M2, mechanical properties and compositions **A7:** 471
M4, mechanical properties and compositions **A7:** 471
M42, mechanical properties and compositions **A7:** 471
T15, mechanical properties and compositions **A7:** 471
Ingot pattern, revealed by macroetching **A9:** 173
in low-carbon steel billet **A9:** 174
Ingot processing
wafer preparation **EL1:** 192
Ingot size
effect on center cracking of aluminum alloys **A9:** 635
effect on macrosegregation of aluminum alloy ingots **A9:** 633
Ingot steels
early fractographs **A12:** 5
Ingot(s)
defects **A15:** 404, 407
defined **A15:** 7
electron beam **A15:** 412
freezing, nucleation effects **A15:** 101
heavy, electroslag remelting of **A15:** 404
molds **A15:** 43-44
solidification, by electroslag remelting **A15:** 403-404
VAR, structure **A15:** 406
VIM, processing routes **A15:** 393
Ingots *See also* Billets
breakdown, and primary forging **A14:** 222-223
breakdown, for stainless steel forging **A14:** 222-223
breakdown, of tantalum **A14:** 238
chemical segregation in **A17:** 491
defined **A14:** 8
electroslag-remelted (ESR), forging burst in **A11:** 317-318
forging defects and failures from **A11:** 314-316
Ingots (continued)
heating, for forging **A14:** 222
macrosegregation **A14:** 64
for open-die forging **A14:** 64-65
pipe defect **A17:** 491
pipes, shrinkage from **A11:** 315
piping in, schematic **A11:** 315
processing flaws **A17:** 492-493
shapes forged from **A14:** 61
titanium **M3:** 362-364
transverse distribution of solute in **A11:** 324-325
Inherent filtration, x-ray tubes
radiography **A17:** 306-307
Inherent flaw size **EM1:** 253, 255
Inhibited acid cleaners **M5:** 60
Inhibited admiralty metal **A13:** 614, 626
Inhibited alkaline cleaners **M5:** 7, 9-10
Inhibited alloys
corrosion resistance **A13:** 611
Inhibited aluminum brass
for heat exchangers/condensers **A13:** 626
Inhibited nitric acid, red fuming
SCC resistance in testing **A8:** 522-523
Inhibited pickling solutions **M5:** 70, 74-77
Inhibited recrystallization structure
formation of **A9:** 603
Inhibited sulfuric acid
in cathodic cleaning **A12:** 75
Inhibitive primers **A13:** 913
Inhibitor *See also* Catalyst **EM3:** 16
defined **EM1:** 13 **EM2:** 23
Inhibitor materials
for solid propellants **A7:** 598
Inhibitors *See also* Anodic inhibitor
in acid environments **A13:** 524-525
acid environments, use in **M1:** 756
in acid pickling **A13:** 524
anionic molecular structures **A13:** 480
anodic **A11:** 197 **A13:** 494-495
application **A13:** 485-486
application methods **A13:** 480-482
automotive applications **A13:** 525 **M1:** 757
for bacteria-induced corrosion **A13:** 482-483
for breweries **A13:** 1222-1223
for carbon steels **A13:** 524-525
cathodic **A11:** 197 **A13:** 495-497
cationic molecular structures **A13:** 479
composition **A13:** 485
copper **A13:** 497
corrosion, as barrier protection **A13:** 378
for corrosion control **A11:** 197-198
corrosion protection of steel **M1:** 751, 754-757
for crude oil refineries **A13:** 485-486
defined **A13:** 8, 1140
effect, corrosion kinetics **A13:** 379
effect of velocity **A11:** 198
effectiveness **A13:** 1143
fatty acids sources for **A13:** 481
formulations **A13:** 478
fresh-water systems **M1:** 733-736, 738
for gas wells **A13:** 1249
for general corrosion resistance **A13:** 693
hydrogen embrittlement **A13:** 524
hydrogen embrittlement prevention of **M1:** 755-756
laboratory testing **A13:** 483
mechanisms **A13:** 485
monitoring results of **A13:** 197, 483
neutralizing **A13:** 485
for oil and gas production **A13:** 478-484, 1240-1243
oxidation, for lubricant failures **A11:** 154
packaging applications **A13:** 525 **M1:** 757
process water, use in **M1:** 749-750
quality control **A13:** 483
seawater corrosion, effect on **M1:** 745
selection of **A11:** 198
steam systems, use in **M1:** 748
steel pickling process, use in **M1:** 755-756
Inhibitors (continued)
treating programs, computerization **A13:** 483
types **A13:** 478-480, 1141
volatile **A13:** 525 **M1:** 756-757
in waterfloods **A13:** 482
for zinc/zinc alloys and coatings **A13:** 761
Inhomogeneities
which affect electrochemical potential **A9:** 60-61
Inhomogeneity
experimental plans for **A8:** 643
as ferromagnetic resonance application **A10:** 274-275
magnetic, determined **A10:** 274-275
of particle surfaces **A7:** 250
in rolling **A8:** 593-595
in surfaces, AES analysis for **A10:** 549
verifying surface **A7:** 260
Inhomogeneous flow
amorphous materials and metallic glasses **A2:** 813
Inhomogeneous solutions
free energy expressed as an integral over volume **A9:** 653
Initial (instantaneous) stress
defined **EM1:** 13
Initial adhesion
as fretting **A11:** 148
Initial current
definition **M6:** 10
Initial failure
laminar **EM1:** 230-231
Initial intensity
abbreviation for **A10:** 690
Initial modulus *See also* Modulus; Young's modulus **EM3:** 16
defined **EM1:** 13 **EM2:** 23
Initial permeability
in magnetic particle inspection **A17:** 99
Initial pitting
defined **A18:** 11
Initial propagation
crack and fatigue **A11:** 106
Initial recovery
defined **A8:** 7
Initial solidification point **A18:** 695
Initial staircase testing **A8:** 706
Initial strain **EM3:** 16
defined **A8:** 7 **EM1:** 13 **EM2:** 23
Initial stress **EM3:** 16
defined **A8:** 7 **EM2:** 23
Initial tangent modulus *See also* Modulus of elasticity **A8:** 7
Initialization commands **EL1:** 5
Initiation *See also* Crack initiation; Fatigue fracture initiation
crack **EM1:** 201
fatigue cracks **A12:** 112
of fracture **A12:** 103
phase, crevice corrosion **A13:** 303
of stable crack growth, defined **A8:** 7
in stress-corrosion cracking **A8:** 496 **A13:** 245-246
Initiator **EM3:** 16
defined **EM1:** 13 **EM2:** 23
Injecting contacts
development **EL1:** 958
Injection *See also* Die casting
carbon dioxide **A13:** 1253
defined **A15:** 7
direct, as die casting method **A15:** 286
of fine coke, cupolas **A15:** 384
high-velocity, in die casting **A15:** 288-292
metal, gating system, die casting **A15:** 288-291
oxygen enrichment and, cupolas **A15:** 384
of plastic patterns, investment casting **A15:** 256
of wax patterns, investment casting **A15:** 254-256
Injection blow molding
defined **EM2:** 23
properties effects **EM2:** 285

Injection compression molding
properties effects EM2: 283-284
size and shape effects EM2: 289-290
Injection laser diodes
as gallium compound application A2: 739
Injection molded P/M materials
as-sintered mechanical properties A7: 471
mechanical properties A7: 466-467, 471
Injection molded P/M materials, specific types
17-4PH, as-sintered mechanical properties A7: 471
316L, as-sintered mechanical properties A7: 471
IN-100, as-sintered mechanical properties A7: 471
Iron-nickel, as-sintered mechanical properties A7: 471
Injection molding *See also* Pressure casting; Reaction injection molding (RIM); Reciprocating-screw injection molding; Reinforced reaction injection molding (RRIM); Screw plasticating injection molding; Structural reaction injection molding (SRIM); Thermoplastic injection molding; Thermoplastic structural foams; Thermoset injection molding; Vacuum injection molding A7: 6, 495-500 EM1: 555-558 EM4: 33, 34, 123, 124, 173-179, 188-189
acrylics EM2: 106-107
acrylonitrile-butadiene-styrenes (ABS) EM2: 112-113
advanced ceramics EM4: 49
applications EM4: 173
binder formulation EM4: 173-174
water-soluble binder EM4: 173
binder removal EM4: 178-179
constituents of powder formulation EM4: 126
pyrolysis EM4: 179
solvent extraction of binder EM4: 178-179
supercritical extraction EM4: 179
thermal degradation EM4: 178
of bulk molding compounds EM1: 161
for coating/encapsulation EL1: 240-241
compression EM2: 83-284
costs, compared EM1: 170
defects in dewaxed parts EM4: 179
blistering EM4: 179
cracks EM4: 179
crazing EM4: 179
delamination EM4: 179
knit lines EM4: 179
pinholes EM4: 179
skin formation EM4: 179
slumping EM4: 179
speckles EM4: 179
defined A15: 7 EM1: 13 EM2: 23
definition EM4: 173
of discontinuous fibers EM1: 120-121
economic factors EM2: 294-296
epoxy packages EL1: 961
equipment EM1: 121
factors influencing ceramic forming process selection EM4: 34
foam EM2: 284, 290
for gas turbine component fabrication EM4: 718, 720
granulated powders as feedstock EM4: 100
helicopter pilot helmet EM1: 121
high-performance ceramics, binders EM4: 121
with HIP EM4: 198
hollow EM2: 284, 290
ionomers EM2: 121-123
liquid crystal polymers (LCP) EM2: 181-182
materials and properties A7: 498-499
mechanical consideration EM4: 125, 126, 127
of metal-polymer mixtures A7: 606
mix preparation EM4: 176
equipment EM4: 176
milling EM4: 176
mixing operations EM4: 176
molded-in color EM2: 305-306
molding equipment EM4: 176-177
plunger-type machine EM4: 176, 177
screw-type machine EM4: 176, 177
molding machines for ceramics EM4: 177-178
organics removal EM4: 135, 136
P/M injection molding (MIM) process A2: 984-985
Injection molding (continued)
polyaryletherketones (PAEK, PEK PEEK, PEKK) EM2: 144
polyether sulfones (PES, PESV) EM2: 161-162
polysulfones (PSU) EM2: 201-202
polyvinyl chlorides (PVC) EM2: 210-211
porosity of parts A7: 499
powder characteristics EM4: 173
dispersion of particles EM4: 173
particle packing EM4: 173
powder loading EM4: 173
powder selection and production A7: 495-496
process EM4: 173
processing A7: 495-498
properties effects EM2: 282-284, 287, 288-289, 291-292
in reaction sintering EM4: 292
reciprocating screw injection molding machine EM1: 555-557
rheological behavior EM4: 174
Bingham plastic EM4: 174
dilatant EM4: 174, 176
measurements EM4: 174
pseudoplastic EM4: 174, 176
St. Venant EM4: 174
rheology of mix formulations EM4: 174-176
dynamic shear modulus EM4: 175-176
temperature EM4: 176
viscosity EM4: 174-175, 176
and rigid tool compaction A7: 323
size and shape effects
styrene-maleic anhydrides (S/MA) EM2: 220
surface finish EM2: 303
technology A7: 18, 23
textured surfaces EM2: 305
thermoplastic EM1: 555-557 EM2: 302
thermoplastic polyimides (TPI) EM2: 178
thermoset EM1: 558 EM2: 302
to make compacts of silicon EM4: 237
tooling application, beryllium-copper alloys A2: 419
variations A7: 498
viscoelastic properties required EM4: 116
Injection molding compounds
chopped glass for EM1: 110
discontinuous fiber reinforced fibers for EM1: 33
feeding EM1: 164-165
flowing EM1: 166-167
injecting EM1: 166
path of EM1: 165
screws EM1: 165-166
transporting EM1: 165-166
Injection molding machine
polyamide-imides (PAI) EM2: 133-135
Injection ports
of molds EM1: 168
Injection pressures
and process selection EM2: 277-278
Ink
for gage marks A8: 548
Ink jet printing EM4: 475
Ink marking
for component identification A11: 130
Inkjet printout of digitized image A9: 139
Inks *See also* Thick films
conductive EL1: 207
dielectric EL1: 208, 346
electrically conductive EL1: 346
noble-metal conductor EL1: 207-208
nonnoble metal thick-film conductor EL1: 208
resistive EL1: 208
thick-film EL1: 207
Inks for magnetic paper and tape
powder used A7: 573
Inlays, dental
as investment cast A15: 35
Inlet film thickness A18: 94
nomenclature for lubrication regimes A18: 90
Inmold process
for ductile iron A15: 650
Inner diameter (ID) sawing
semiconductor friction and wear A18: 685-686
Inner lead bonding (ILB)
plastic packages EL1: 478-479
process EL1: 230-231
Inner lead bonding (ILB) (continued)
in tape automated bonding EL1: 228, 278-279, 478-479
Inner noise *See also* Noise
defined A17: 723
Inner raceway A18: 499, 500, 501
Inner-shell ionization
as inelastic scattering process A10: 433
Inoculant *See also* Inoculation
defined A15: 7
gray iron A15: 637-638
Inoculants A6: 53
cast iron M1: 80
gray cast iron M1: 13, 21-22, 28
Inoculants, effect of
on alloy cast irons A1: 90
Inoculation *See also* Grain refinement; Inoculant
addition methods A15: 638-639
of cast iron A15: 170
defined A15: 7
of ductile iron A15: 650-651
of gray iron .... A15: 35, 636, --639
of high-alloy graphitic irons A15: 698
late, vs ladle inoculation A15: 639
mold, of gray iron A15: 638-639
practice, kinetics of A15: 105
Inorganic
defined A13: 8
Inorganic acids, as corrosive
copper casting alloys A2: 352
Inorganic acids, corrosion
zirconium M3: 784, 785-786, 787
Inorganic atoms
MFS analysis of A10: 74
Inorganic binders
in spray drying A7: 74
Inorganic chemical-setting ceramic linings A13: 453-455
Inorganic coatings
as corrosion control A11: 194
magnesium/magnesium alloys A13: 749-742
for structural corrosion A13: 1304-1305
zinc-rich A13: 411-412, 769
Inorganic colloidal magnesium aluminum silicate
application or function optimizing powder treatment and green forming EM4: 49
Inorganic compounds
gas analysis of A10: 151
with germanium A2: 734
Inorganic Crystal Structure Data Base A10: 355
Inorganic elements
SSMS analysis of A10: 141
Inorganic fillers
as wire forming lubricants A14: 696
Inorganic gases
analytic methods for A10: 8
Inorganic glasses
compared to naturally occurring glasses EM4: 1
Inorganic insulation
of electrical steel sheet A14: 482
Inorganic interlaminar insulation
magnetic cores A2: 780
Inorganic liquids
analytic methods for A10: 7
Inorganic materials *See also* Inorganic materials, characterization of; Inorganic solid materials
for identification and structure determination in A10: 109
properties EL1: 470
single-crystal, Raman analysis of A10: 129
Inorganic materials, characterization of *See also* Inorganic materials; inorganic solid materials
analytical transmission electron microscopy A10: 429-489
atomic absorption spectrometry A10: 43-59
Auger electron spectroscopy A10: 549-567
classical wet analytical chemistry A10: 161-180
controlled-potential coulometry A10: 207-211
crystallographic texture measurement and analysis A10: 357-364
electrochemical analysis A10: 181-211
electrogravimetry A10: 197-201
electrometric titration A10: 202-206
electron probe x-ray microanalysis A10: 516-535
electron spin resonance A10: 253-266

Inorganic materials, characterization of (continued)
extended x-ray absorption fine structure **A10:** 407-419
ferromagnetic resonance **A10:** 267-276
field ion microscopy **A10:** 583-602
inductively coupled plasma atomic emission spectroscopy **A10:** 31-42
infrared spectroscopy **A10:** 109-125
ion chromatography **A10:** 658-667
low-energy electron diffraction **A10:** 536-545
low-energy ion-scattering spectroscopy **A10:** 603-609
molecular fluorescence spectrometry **A10:** 72-81
Mössbauer spectroscopy **A10:** 287-295
neutron activation analysis **A10:** 233-242
neutron diffraction **A10:** 420-426
nuclear magnetic resonance **A10:** 277-286
optical emission spectroscopy **A10:** 21-30
optical metallography **A10:** 299-308
particle-induced x-ray emission **A10:** 102-108
potentiometric membrane electrodes **A10:** 181-187
radial distribution function analysis **A10:** 393-401
Raman spectroscopy **A10:** 126-138
Rutherford backscattering spectrometry **A10:** 628-636
scanning electron microscopy **A10:** 490-515
secondary ion mass spectroscopy **A10:** 610-627
single-crystal x-ray diffraction **A10:** 344-356
small-angle x-ray and neutron scattering **A10:** 402-406
spark source mass spectrometry **A10:** 141-150
ultraviolet/visible absorption spectroscopy **A10:** 60-71
voltammetry **A10:** 188-196
x-ray diffraction **A10:** 325-332
x-ray diffraction residual stress techniques **A10:** 380-392
x-ray photoelectron spectroscopy **A10:** 568-580
x-ray powder diffraction **A10:** 333-343
x-ray spectrometry **A10:** 82-101
x-ray topography **A10:** 365-379
Inorganic mixtures
liquid chromatography for **A10:** 649
Inorganic molecules
MFS analysis of **A10:** 74
Inorganic oxides
heat of formation **A7:** 598
Inorganic pigments **EM3:** 16
defined **EM1:** 13-14 **EM2:** 23
Inorganic polymers
as thermoplastic **EM2:** 66
Inorganic solid materials *See also* Inorganic materials; Inorganic materials, characterization of
analytic methods for **A10:** 4-6
chemical reagents, composites, and catalysts analytic methods for **A10:** 6
glasses and ceramics, analytic methods for **A10:** 5
metals, alloys, and semiconductors, analytic methods for **A10:** 4
minerals, ores, and slags, analytic methods for **A10:** 6
pigments, compounds, and effluents, analytic methods for **A10:** 6
Inorganic solutions
analytic methods for **A10:** 7
Inorganic zinc coatings **M5:** 475, 501-505
Inorganic zinc-rich paint
defined **A13:** 8
Inorganic-inorganic composites
laser cutting of **EM1:** 679-680
Inorganic-organic composites
laser cutting of **EM1:** 679
Inorganics
formulation **EM3:** 123
as grease thickeners **A18:** 126
Input bar
double-notch shear testing **A8:** 228-229
Input connections
low thermal electromotive force cadmium **A8:** 389
Input devices
image analyzers **A10:** 310
Input-output signal pin count *See* Pin count
Inputs/outputs (I/Os)
basic structures **EL1:** 161-166
count, reduced **EL1:** 311
functions **EL1:** 128
interfaces, types **EL1:** 160
and mounting technologies **EL1:** 143
pad connection **EL1:** 7
of surface-mount components **EL1:** 730
Insecticidal spraying
atomization mechanism **A7:** 27
Insert **EM3:** 16
defined **EM1:** 14
Insert collar
cracked gray iron **A11:** 372
Insert rack
coupon testing **A13:** 199
Insert(s)
cast-in, magnesium alloy parts **A2:** 466
indexable carbide **A2:** 963
magnesium alloy parts **A2:** 466-467
negative-rake **A2:** 966
press-fit and shrink-fit, magnesium alloy parts **A2:** 466-467
screwed-in, magnesium alloy parts **A2:** 467
Inserts *See* Die inserts
blow molding **EM2:** 358
defined **A15:** 7 **EM2:** 23
for die casting **A15:** 287
for friction reduction **A8:** 197
mold, permanent mold casting **A15:** 280
molded-in, polyamide-imides (PAI) **EM2:** 131
types **EM2:** 722-724
Inserts, load bearing
in magnesium parts **M2:** 538-540, 541, 542
Inside diameter
abbreviation **A8:** 725
Inside mold line (IML) surfaces
resin transfer molding for **EM1:** 168
Inside-outside contrast
in dislocation loops **A9:** 116
in dislocation pairs **A9:** 114
Insoluble particles *See also* Particles
interfacial, during solidification **A15:** 142-147
Insoluble phases in aluminum alloys **A9:** 358
Inspection *See also* Chemical analysis; Die inspection; Evaluation; Examination; Formability testing; Investigation procedures; Nondestructive evaluation; Nondestructive evaluation (NDE); Nondestructive inspection (NDI); Nondestructive testing; Testing; Ultrasonic nondestructive analysis; Visual examination; Visual inspection
Alnico alloys **A15:** 737-739
of aluminum alloy castings **A15:** 556-557
of aluminum alloy forgings **A14:** 249
analysis, thermoplastic injection molding **EM2:** 313
of casting defects **A15:** 544-561
of composite tool fabrication **EM1:** 739
of copper alloy castings **A15:** 557-558
cost drivers in **EM1:** 421
design requirements for **EM1:** 183
destructive **A13:** 417
devices, computer-aided manufacturing **EL1:** 131
dimensional, computer-aided **A15:** 558-561
dissimilar metal joining **A6:** 825
of ductile iron castings **A15:** 556
eddy current **A13:** 417 **A15:** 554-555
of equipment **A13:** 322
for erosion-corrosion, wet steam flow **A13:** 968-969
and failure analysis techniques **EL1:** 954
Inspection (continued)
of ferrous castings **A15:** 555-556
of fiber preforms/resin injection **EM1:** 532
fluorescent-penetrant inspection **M6:** 827
of gray irons **A15:** 555-556
hot techniques, for continuous casting **A15:** 315
of investment castings **A15:** 264
lay-up, types **EM1:** 739
of layers, ceramic packages **EL1:** 464
liquid fluorescent penetrant **A15:** 264
liquid penetrant **A15:** 264, 553
liquid-penetrant inspection **M6:** 826-827
magnetic particle **A14:** 204 **A15:** 264, 553-554
of magnetic particles **A7:** 575-579
magnetic-particle inspection **M6:** 827
of malleable irons **A15:** 556
monocrystal casting **A15:** 322-323
of niobium-titanium ingot **A2:** 1044
offshore gas/oil production platforms **A13:** 1254
organic coatings and linings **A13:** 416-418
of P/M materials **A7:** 480-492
of P/M products **A7:** 295
preliminary, cast Alnico alloys **A15:** 737
qualification tests for overlays **M6:** 817-818
radiographic **A15:** 554-555, 557
radiographic inspection **M6:** 827
rigid printed wiring boards **EL1:** 542-543, 547
sand, in media preparation **A15:** 345
of second-generation patterns **EM1:** 738-739
and sorting, automatic **A15:** 571
techniques **EM1:** 743-744
of thermal spray coatings **A13:** 462
of titanium alloy forgings **A14:** 282
of tooling master models **EM1:** 738
ultrasonic **A15:** 555, 557
ultrasonic inspection **M6:** 877
x-ray radiography **A15:** 264
x-ray, real time system, schematic **A15:** 554
Inspection coils *See also* Coils; Encircling coils; Probes; Sensors
eddy current inspection **A17:** 175-177
Inspection equipment *See also* Equipment
Inspection equipment and techniques *See also* Equipment; Nondestructive evaluation methods; Nondestructive evaluation techniques **A17:** 1-17
coordinate measuring machines **A17:** 18-28
laser inspection **A17:** 12-17
machine vision **A17:** 29-45
qualification of **A17:** 679
robotic inspection systems **A17:** 29-45
visual inspection **A17:** 3-11
Inspection fixtures *See* Gages
Inspection for
capacitor-discharge stud welds **M6:** 738
electrogas welds **M6:** 244
electroslag welds **M6:** 234-235
explosion welding **M6:** 710-711
laser beam welds **M6:** 668-669
repair welds of electron beam welding **M6:** 636
resistance welds of aluminum alloys **M6:** 543
soldering **M6:** 1089-1091
stud arc welds **M6:** 734
torch brazing of steels **M6:** 951-952
underwater welds **M6:** 924
weld overlays **M6:** 817-818
Inspection frequencies *See* Frequencies
Inspection log
adhesive-bonded joints **A17:** 633
Inspection materials
control of **A17:** 678
Inspection, nondestructive
of cast steels **M1:** 401-402
Inspection of welded joints **A6:** 1081-1088
acoustic emission (AE) testing **A6:** 1080, 1084
advanced methods **A6:** 1088
image processing software **A6:** 1088
neural network software **A6:** 1088

Inspection of welded joints (continued)
probe scanners for creating ultrasonic images **A6:** 1088
applications **A6:** 1081, 1084
eddy-current testing **A6:** 1081, 1082-1083, 1085, 1087, 1088
evaluation of test results **A6:** 1086-1087
leak testing **A6:** 1081, 1084
magnetic-particle testing (MPT) **A6:** 1081, 1082, 1083, 1085, 1086, 1087, 1088
nondestructive evaluation performance **A6:** 1085-1086
nondestructive evaluation method selection **A6:** 1086, 1087-1088
nondestructive evaluation techniques **A6:** 1081-1084
penetrant testing **A6:** 1081, 1082, 1084, 1085, 1086, 1087, 1088
principal test methods **A6:** 1081
radiographic testing **A6:** 1081, 1083, 1085, 1086, 1087, 1088
double-wall, double-image technique **A6:** 1083
image quality indicator (IQI) **A6:** 1086
test procedures **A6:** 1084-1085
test-operator training **A6:** 1085
ultrasonic testing **A6:** 1081, 1083-1084, 1085-1086, 1087, 1088
visual inspection **A6:** 1081-1082, 1085
Inspection records
for macroetched specimens **A9:** 172-173
Inspection scheduling
damage analyses for **A8:** 682
Inspection standards *See* Specifications; Standards
Inspection techniques *See* Inspection Equipment and Techniques
Inspection testing for acceptance **EM3:** 684
Inspectors *See also* Human factory; Management; Personnel; Safety
and NDE reliability **A17:** 677
Instability
anisotropic plate **EM1:** 446
in composite structures **EM1:** 445-449
in dendritic/eutectic growth **A15:** 122-123
from shear deformation **EM1:** 447
orthotropic plate **EM1:** 445-446
plate, from postbuckling behavior **EM1:** 447-449
shell panel **EM1:** 448-449
single-phase, dendritic and eutectic **A15:** 122
in tension **A8:** 25
two-phase **A15:** 122-123
types, planar solid/liquid eutectic interface **A15:** 123
unsymmetric laminate **EM1:** 446-447
Installation
of acoustic emission sensors **A17:** 281
of fasteners **EM1:** 710, 711
Installation environments
stress-corrosion cracking in **A11:** 212
Installation, wet *See* Wet installation
Instant photographic processes used in photomicroscopy **A9:** 84-86
Instantaneous erosion rate
defined **A18:** 11
Instantaneous recovery *See* Initial recovery **EM3:** 16
Instantaneous strain *See* Initial strain **EM3:** 16
Instantaneous stress *See* Initial (instantaneous) stress; Initial stress
Institute for Interconnecting and Packaging Electronic Circuits (IPC) **EL1:** 734
Institute of Electrical and Electronics Engineers (IEEE) Holm Conference on Electrical Contacts **A18:** 682
Institute of Scrap Recycling Industries (ISRI)
scrap specifications of **A1:** 1026
Instron 1331 servohydraulic testing machine
for fracture testing **EM3:** 510
Instrument applications
aluminum and aluminum alloys **A2:** 14
of precious metals **A2:** 693
Instrument grades
beryllium .... **A2:** 68©87 **A9:** 390
Instrument response time
defined **A10:** 674
Instrumentation *See also* Testing
applications, for hybrids **EL1:** 254
for Charpy V-notch impact test **A8:** 259
Instrumentation (continued)
data, use in failure analysis **A11:** 747
for deformation recording **A8:** 146
electrical test methods **EL1:** 372-378
geometry effects, in SEM **A12:** 168
for macroscopic fracture analysis **A11:** 256-257
physical test methods **EL1:** 365-372
of rolling mills **A14:** 354
SEM **A12:** 166-171
for split Hopkinson pressure bar testing **A8:** 202
and testing **EL1:** 365-380
for torsional testing **A8:** 146-147
of x-ray spectrometry **A10:** 87-93
as XRPD source of error **A10:** 341
Instrumented impact testing
Charpy, for dynamic fracture testing **A8:** 264-267, 276
defined **A8:** 7
recording equipment **A8:** 197
Instrumented-impact test
for fracture toughness **A11:** 64
Instruments *See also* Equipment; Machines
eddy current inspection **A17:** 177-179
types, eddy current **A17:** 176-178
Insulated-gate bipolar transistors (IG BTs) **A6:** 39
Insulated-gate field effect transistor (IGFET)
defined **EL1:** 1147
Insulating ceramics **EM4:** 17
Insulating function **EM3:** 33, 36
Insulating glass compositions
SIMS analysis of **EL1:** 1087-1088
Insulating pads *See also* Chill
defined **A15:** 7
Insulating plastics *See also* Conductive plastic materials; Electrical properties; Plastics
electrical breakdown **EM2:** 464-467
electrical values, significance **EM2:** 467
polymer structure and electrical properties **EM2:** 462-464
terminology **EM2:** 460-461
test methods **EM2:** 461-462
Insulating sleeves *See also* Chill
defined **A15:** 7
Insulation *See also* Core plating; Electrical properties; Insulation resistance
applications, polyurethanes (PUR) **EM2:** 259
ceramic, thermocouple wire **A2:** 883
copper wire and cable **M2:** 274
corrosion in **A13:** 340-343
corrosivity of improper use **A13:** 341
discontinuous oxide fibers for **EM1:** 62-63
of electrical steel sheet **A14:** 482
improved, of composites **EM1:** 36
inorganic **A14:** 482
materials, pharmaceutical production **A13:** 1231
organic **A14:** 482
as plastic characteristic **EM2:** 460
and protection, of thermocouples **A2:** 882-884
resistance **EM1:** 14
thermal, corrosion beneath **A13:** 1144-1147, 1230-1231
for titanium alloy forging **A14:** 279
in tungsten-rhenium thermocouples **A2:** 876
for wrought copper and copper alloy products **A2:** 258-260
Insulation board
waterjet machinery **A16:** 522
Insulation resistance **EM3:** 16
allyls (DAP, DAIP) **EM2:** 227
ceramic multilayer packages **EL1:** 468
defined **EM2:** 23
package-level testing **EL1:** 938
testing **EM2:** 585-586
of urethane coatings **EL1:** 776
Insulation, thermal
corrosion under **A11:** 184
Insulative adhesives
applications **EM3:** 45
Insulator **EM3:** 16
defined **EM1:** 14 **EM2:** 23
Insulator(s)
as dielectrics, defined **EL1:** 90
materials **EL1:** 113-114, 1041
polarization **EL1:** 99-100
processes selection **EL1:** 113-114
Insulator(s) (continued)
quantum mechanical band theory (solid state) **EL1:** 99-100
Insulin production
powder used **A7:** 573
Integral actuator shaft
in hydraulic torsional system **A8:** 215-216
Integral color anodizing
aluminum and aluminum alloys **M5:** 595-596, 609-610
Integral composite structure
defined **EM1:** 14
Integral coupling and gear
fatigue fracture in **A11:** 129-130
Integral heat exchangers
as thermal control **EL1:** 47
Integral hybrid components
advantages/limitations **EL1:** 258-259
Integral molar free energy
vs partial molar free energy **A15:** 55
Integral of
symbol for **A10:** 692
Integral skin foam **EM2:** 23, 258 **EM3:** 16
defined **EM1:** 14
Integral structure **EM3:** 16
defined **EM2:** 23
Integral thermal properties
liquid aluminum-magnesium alloys **A15:** 57
liquid aluminum-silicon alloys **A15:** 57
liquid copper-aluminum alloys **A15:** 58
liquid copper-nickel alloys **A15:** 58
liquid copper-tin alloys **A15:** 59
liquid copper-zinc alloys **A15:** 59
Integral-finned tube
in heat exchangers **A11:** 628, 635
Integrally bladed rotor (IBR)
abrasive flow machining **A16:** 519
Integrally heated
defined **EM1:** 14
Integrated beam stops
as optic structure **EL1:** 10
Integrated blade inspection system
as NDE reliability case study **A17:** 686-687
Integrated circuit chip(s) *See also* Chip(s); Integrated circuits (ICs)
defined **EL1:** 1147
engineering design system for **EL1:** 127
mounting technologies **EL1:** 143
top, cooling **EL1:** 307-308
Integrated circuit defects
electronic materials **A12:** 481
Integrated circuits *See also* Integrated circuits, failure analysis of **A13:** 1124
beam leads **A11:** 775-776
chip, silicon rubber encapsulated **A11:** 775
cross section, LPCVD tungsten layers **A10:** 513
dead-on-arrival (DOA) failures **A11:** 766
device failures, types of **A11:** 766
device-operating failures (DOF) **A11:** 766
failure rates **A11:** 766
infant mortality and steady-state life failures **A11:** 766
memory, SEM analysis of **A11:** 768
oxygen in wafers for **A10:** 122-123
plating peeling, XPS analysis of **A11:** 43-45
secondary electron micrograph **A10:** 555
SEM analysis of **A10:** 513-514
Integrated circuits (ICs) *See also* Advanced CMOS logic; Complementary MOS; Digital integrated circuits; Electrically erasable programmable read-only memories; Emitter-coupled logic; FR-4 glass-epoxy boards; Gallium arsenide (GaAs); High-performance MOS; High-speed MOS; Hybrid integrated circuitry; Integrated circuit chips(s); Integrated semiconductor packages; Large-scale integration; Medium-scale integration; Metal-oxide semiconductor (MOS); Microprocessing units; Microprocessor peripherals; Monolithic microwave integrated circuits; N-channel MOS; P-channel MOS; Programmable read-only memories; Random access memory; Read-only memories; Small-scale integration; Transistor-transistor logic **EM3:** 579, 583, 587
acoustic microscopy **A17:** 474-480
analog, application **A2:** 740
by SEM with EDS attachment **EL1:** 1095

Integrated circuits (ICs) (continued)
chip top, cooling **EL1:** 307-308
chips, mounting technologies **EL1:** 143
complexity **EL1:** 399
defined **EL1:** 1147
design rule for **EL1:** 161
digital **A2:** 740
and discrete semiconductor, compared **EL1:** 422
drivers integrated on **EL1:** 7
electrical damage **EL1:** 975-978
electronic/electrical corrosion failure analysis of **EL1:** 1115-1116
encapsulation failures **EL1:** 978-979
fabrication, metal-oxide semiconductor **EL1:** 196-198
failure locations and mechanisms **EL1:** 975-979
hierarchy **EL1:** 398
interconnection with **EL1:** 386
interfacing with **EL1:** 386
lines-and-spaces reduction **EL1:** 253
logic forms, evolution of **EL1:** 160-161
material properties **EL1:** 207
microwave **A2:** 740
monolithic microwave (MMIC) **A2:** 740, 747
monolithic wafer-scale **EL1:** 8-9
MOS, fabrication **EL1:** 147-148
npn bipolar junction transistor technology **EL1:** 144-145
packaged, physical hierarchy **EL1:** 2
packages, chip-level, characteristics **EL1:** 404
packages, performance ranking **EL1:** 404
packaging, automatic optical inspection of **EL1:** 941
power as parameter of **EL1:** 401
semiconductor chip fracture **EL1:** 978-979
semiconductor, decreasing size **EL1:** 297
single, monolithic **EL1:** 2
substitution, thick-film hybrids as **EL1:** 386
system level, trends **EL1:** 177
technologies, market share trends **EL1:** 399
temperature-sensitive failure mechanisms **EL1:** 959-961
testing/instrumentation **EL1:** 365
thermal acceleration factor **EL1:** 46
thin-film chromium resistors, corrosion failure analysis **EL1:** 1114-1115
thin-film resistor in photodetector, corrosion failure analysis **EL1:** 1115
through-hole packages **EL1:** 975-979
trends **EL1:** 399-401
vs dual-in-line packages (DIPS) **EL1:** 12
wafer fabrication defects **EL1:** 978
wire bonding
wire bonding and tape automated bonding (TAB) **EL1:** 977-978
Integrated circuits, failure analysis of **A11:** 766-792
electrostatic/electrical overstress failures **A11:** 786-788
metallic interface failures **A11:** 776-778
metallization failures **A11:** 769-776
plastic-package failures **A11:** 788-789
silicon oxide failures **A11:** 778-781
silicon oxide interface failures **A11:** 781-782
silicon p-n junction failures **A11:** 782-786
techniques of **A11:** 767-769
Integrated intensity ratios **A18:** 467
Integrated laminating center (ILC)
broadgoods prepreg dispenser **EM1:** 63
ply handier **EM1:** 63
translaminar stitcher **EM1:** 636-638
Integrated planar lenses
defined **EL1:** 10
Integrated semiconductor packages *See also* Integrated circuits (ICs)
assembly/packaging **EL1:** 438-449
cost reduction **EL1:** 448-449
differentiation of **EL1:** 438-441
outlook for **EL1:** 449

Integrated semiconductor packages (continued)
quality **EL1:** 441-448
semiconductor packaging road map **EL1:** 436-437
through-hole and surface-mount assembly **EL1:** 437-438
Integrating spheres
as radiation collector in mid-infrared region **A10:** 114
Integration *See also* Hybrid(s); Integrated circuitry; Integrated circuits (ICs); Macrointegration; Multilayer boards (MLBs); Printed wiring boards (PWBs); Thick-film circuits; Thin-film circuits; Ultralarge-scale integration; Very-large-scale integration (VLSI)
higher-level, and hybrids **EL1:** 249
knowledge-based **A14:** 413-416
types **EL1:** 160
Intelligent automation for joining technology **A6:** 1057- 1064
artificial intelligence (AI) techniques **A6:** 1057
components included **A6:** 1057
development of systems **A6:** 1057
interface between off-line planners and real-time control systems **A6:** 1061
computer files **A6:** 1061
trajectory **A6:** 1061
weld job **A6:** 1061
weld set **A6:** 1061
off-line planning system **A6:** 1057-1061
computer-aided design (CAD) system **A6:** 1057-1058
sequential operation of **A6:** 1057
real-time control intelligent system **A6:** 1061-1064
hardware controller **A6:** 1062
job editor **A6:** 1061-1062
job interpreter **A6:** 1061
standards **A6:** 1061
welding cell hardware **A6:** 1062-1064
welding cell sensors **A6:** 1062-1064
WELDEXCELL system **A6:** 1057-1064
backward chaining **A6:** 1059
blackboard **A6:** 1060
data base systems **A6:** 1058-1059
expert systems **A6:** 1059
FERRITEPREDICTOR expert system **A6:** 1059
forward chaining **A6:** 1059
frames **A6:** 1059
HEATFLOW neural network system **A6:** 1060
knowledge sources **A6:** 1058
neural network system (NNS) **A6:** 1059-1060, 1063
parts designer **A6:** 1057-1058
path planner **A6:** 1057, 1058
production rules **A6:** 1059
system integrator **A6:** 1057, 1060-1061
WELDBEAD neural network system **A6:** 1060
WELDHEAT expert system **A6:** 1059, 1060
welding schedule developer **A6:** 1057, 1058-1060
WELDPROSPEC expert system **A6:** 1059
WELDSELECTOR expert system **A6:** 1059
Intelligent vision *See* Machine vision
Intense Pulsed Neutron Source
Argonne National Laboratory **A10:** 424
Intensification *See also* Image intensifiers
lead screens **A17:** 315-316
Intensifying screens
radiography **A17:** 298
Intensiostatic *See* Galvanostatic
Intensity
abbreviation for **A10:** 690
absorption/enhancement effects on **A10:** 97
calculating crystal structure from **A10:** 349
coherent atomic scattering, abbreviation for **A10:** 690
defined **A17:** 384
of diffracted beams **A10:** 328-329
diffracted, resolving of **A10:** 424

Intensity (continued)
diffraction, in single-crystal x-ray diffraction **A10:** 348-349
diffractional, kinematic, and dynamic effects in **A10:** 366-367
of electromagnetic radiation **A10:** 83
hydrodynamic, reduction of **A11:** 170
incoherent atomic scattering, abbreviation for **A10:** 690
infrared **A10:** 111, 117
initial, abbreviation for **A10:** 690
in IR spectra **A10:** 110
luminous, SI base unit and symbol for **A10:** 685
measured by diffractometers **A10:** 351
measurement, as XRPD source of error **A10:** 341
observed, in single-crystal analysis **A10:** 346
profiles, rocking curves as **A10:** 372
radiant, defined **A10:** 62, 680, 685
ratio, defined **A10:** 674
ratio methods, for thin-film sample preparation **A10:** 95
relative Auger, quantitative AES analysis based on **A10:** 553
relative, x-ray spectral lines **A10:** 98
scattered **A10:** 604
theoretical vs thickness, single-element x-ray spectrometry **A10:** 100
threshold stress **A11:** 204-205
total diffracted, abbreviation for **A10:** 690
ultrasound **A17:** 248
vs concentration nickel, nickel ores **A10:** 99
Intensity (x-rays)
defined **A9:** 10
Intensity holography
microwave **A17:** 225-226
Intensity image **A18:** 346, 347- 348
Intensity of scattering
defined **A9:** 10
Intensive mixer
green sand preparation **A15:** 344, 346
Inter failure
in composites **A12:** 466, 478
Interaction
damage rule **A8:** 358
experiments **A8:** 642, 653-654
third-order or higher, as estimate of experimental error **A8:** 642
Interaction coefficients
aluminum and copper alloys **A15:** 55-56, 59-60
for elements in liquid aluminum **A15:** 59
for elements in liquid copper alloys **A15:** 60
Interaction effects
as variables **A17:** 742-750
Interaction/additive (experimental)
defined **EM2:** 600
Interactions
parallel signal lines **EL1:** 34-35
Interactive effects of alloying elements
on notch toughness **A1:** 742, 743
Interactive graphics system (IGS) **EL1:** 127-129
Interatomic bond distances
determined **A10:** 344, 409
Interatomic bond lengths
and physical properties **A10:** 355
Interatomic bond rupture rate
in stress-corrosion cracking **A13:** 147
Interatomic bonding forces **A7:** 304
Interaxial angle of a crystal .... **A3:** 1 • 10
Interaxial angles and edge lengths
relationships for crystal systems **A9:** 706
Intercalation
defined **A10:** 345
in graphites **A10:** 132
Intercept method
defined **A9:** 10
used for measuring aluminum alloy grain size **A9:** 357-358
Intercommunicating porosity **A7:** 6

Interconnect(s)
defects ... EL1: 1002
defects, passive devices ... EL1: 1004-1005
defects, wire-wound resistors ... EL1: 1002
failure mechanism ... EL1: 1054-1056
length, threshold ... EL1: 403
-related failure mechanisms ... EL1: 974
resistance, self-inductance, load capacitance ... EL1: 417
spacing ... EL1: 47
Interconnected pore volume ... A7: 6
Interconnected porosity ... A7: 6
Interconnecting extrusions
applications ... A2: 36
Interconnecting shapes
wrought aluminum alloy ... A2: 35-36
Interconnection *See also* Conventional interconnections; Electrical interconnection; Fundamental interconnection uses; Interconnect(s)
-based technologies, new ... EL1: 8-10
chip, schemes ... EL1: 231-233
chip-to-chip, future trends ... EL1: 177
choice, cost/performance effects ... EL1: 14
compared ... EL1: 299
conductive polymer ... EL1: 15
conventional, environments of ... EL1: 2-5
defined ... EL1: 12, 1147
density, effects ... EL1: 7
electrical ... EL1: 224-236
electrical performance ... EL1: 18
environments, conventional ... EL1: 2-5
flexible printed boards ... EL1: 588
flexible printed wiring ... EL1: 579
from IC packages to frames ... EL1: 12-17
hierarchy ... EL1: 398
hybrid, technology selection criteria ... EL1: 250-255
of ICs, thick-film ceramic hybrids for ... EL1: 386
issues, fundamental ... EL1: 2-11
in leaded and leadless surface-mount joints ... EL1: 732
as level 4 definition ... EL1: 76
levels, high-frequency digital systems ... EL1: 76
levels, illustrated ... EL1: 13
lines, dynamic switching energy ... EL1: 2
lines, length ... EL1: 25
materials and processes selection ... EL1: 116-117
methods, passive components ... EL1: 179-180
modeling ... EL1: 12-17
and multichip packaging ... EL1: 2
optical ... EL1: 9-10
of packaged WSI-based planar/three-dimensional circuit modules ... EL1: 10
physical, hierarchy ... EL1: 2-4
physical performance issues ... EL1: 5-8
point-to-point ... EL1: 2-4
propagation time ... EL1: 20-21
reliability ... EL1: 261
reliability, CTE effect ... EL1: 730
reliability effects, passive components ... EL1: 180-182
scheme, determinants ... EL1: 18-20
in semiconductor packaging, defined ... EL1: 397
separation ... EL1: 1020-1021
switching energy ... EL1: 2
system options ... EL1: 984-986
system requirements ... EL1: 12-17
technology, future developments ... EL1: 10
thermal expansion mismatch ... EL1: 611
thick-film multilayer ... EL1: 347
through-hole, evolution ... EL1: 507
wire-bonded ... EL1: 226-228, 472
wireability ... EL1: 18-20
Interconnection hierarchy ... EL1: 12-13
defined ... EL1: 12
in high frequency digital systems ... EL1: 76
levels of ... EL1: 12-13
packaging ... EL1: 15-16
trends and technology drivers ... EL1: 12
Interconnection network simulator ... EL1: 14
Interconnection parasitic effects *See* Analog parasitic effects
Interconnection system modeling
future of ... EL1: 15-17
interconnection hierarchy ... EL1: 12-13
for packaging requirements ... EL1: 15-16
of performance ... EL1: 13-15
Interconnection system modeling (continued)
trends ... EL1: 12
Interconnection system requirements
future challenges ... EL1: 15-17
Interconnection-based technologies
new ... EL1: 8-10
Intercritically reheated grain-coarsened (ICGC) zone ... A6: 81
Intercrystalline
defined ... A9: 10
Intercrystalline corrosion *See also* Intergranular corrosion; Intergranular Corrosion
on cerclage wire of sensitized steel ... A11: 676, 681
Intercrystalline cracking *See* Intergranular cracking
defined ... A13: 8
Intercrystalline cracks
defined ... A9: 10
Intercrystalline oxidation
of Invar ... A2: 890
Interdendritic
defined ... A9: 10
Interdendritic cavities, in aluminum alloy ingots, result of high hydrogen
content ... A9: 632-633
Interdendritic corrosion *See also* Corrosion ... A13: 8, 590
defined ... A11: 6
Interdendritic fluid flow
macrosegregation ... A15: 155
Interdendritic networks, of second phase constituents, in aluminum alloy
ingots ... A9: 631-632
Interdendritic porosity
defined ... A9: 10
Interdendritic preferential oxidation
from chemistry gradients ... A11: 406
Interdendritic shrinkage
tin bronzes ... A2: 348
Interdiffusion
alloying, in HIP encapsulation ... A7: 429
at aluminum-silicon contacts, integrated circuits ... A11: 777-778
distance ... A7: 315
metal-metal ... A11: 776-777
Interelement effects *See also* Inductively coupled plasma atomic emission spectroscopy
absorption of x-rays as cause ... A10: 97
in emission spectroscopy ... A10: 33-34
in x-ray spectrometry ... A10: 87
Interface *See also* Liquid/solid interface ... A7: 6 EM3: 16
cellular/dendritic, particle behavior at ... A15: 144-145
corrosion, in brazed joints ... A11: 451
defined ... A11: 6 EM2: 23
dendritic solid/liquid, shape of ... A15: 155-156
growth rate at ... A15: 115
melting, electronic materials ... A12: 488
perturbed shape ... A15: 116
planar, directional solidification ... A15: 142-144
shape, effect on particle pushing/entrapment ... A15: 144
solid/liquid, eutectic ... A15: 122
traps, silicon oxide interface failures ... A11: 782
twin-matrix, iron ... A12: 224
velocity, single-phase alloys ... A15: 114-119
Interface activity ... A7: 6
Interface debonding
in composites ... A8: 714
Interface diffusion
in massive transformations ... A9: 655
Interface shear strength ... A18: 34
Interface stability model ... A6: 52
Interface thermal resistances
in package thermal design ... EL1: 412-413
Interface(s)
with coordinate measuring machines ... A17: 20
defects, in adhesive-bonded joints ... A17: 610
with machine vision system ... A17: 37
reflection, of ultrasonic waves ... A17: 231
Interfaces *See also* Heterophase boundaries
between design/tooling/manufacturing personnel ... EM1: 428-431
cementite/ferrite, atom probe composition profile across, in pearlitic steel ... A10: 593
Interfaces (continued)
defined ... EM1: 14
delamination at ... EM1: 24
in discontinuous ceramic MMCs ... EM1: 90
FIM/AP study of segregation of alloy elements and impurities to ... A10: 583
flaws at ... EM1: 241
future of ... EL1: 177
idealized constructions ... A9: 119
manufacturing ... EM1: 429, 430
mixed ... EL1: 160
narrow, AEM analysis ... A10: 482-483
resin, damage tolerance testing ... EM1: 26
segregation, atom probe composition profile for ... A10: 593
SERS analysis of ... A10: 136
and superlattice studies ... A10: 634-635
thermoplastic matrix processing ... EM1: 545
thick-film hybrids ... EL1: 386
transmission electron microscopy ... A9: 118-122
types ... EL1: 160
Interfacial bonding
carbon fibers ... EM1: 5
Interfacial connection methods ... EL1: 590
Interfacial corrosion
in brazed joints ... A13: 877
susceptibility, brazing alloy/stainless steels ... A13: 878
Interfacial debonding
by cumulative group mode failure ... EM1: 195
detection of ... EM3: 392
and fatigue failure ... EM1: 438
Interfacial energy ... A18: 399-400
in precipitation hardening reactions ... A9: 646
Interfacial polarization
of insulators/dielectric materials ... EL1: 99-100
Interfacial specific free energy ... A18: 435
Interfacial studies
on heteroepitaxy layers ... A10: 628
segregation in molybdenum ... A10: 599-601
superlattices ... A10: 634
Interference
defined ... A9: 10
Interference bushings ... A8: 502
Interference colors
enhancement using polarized light ... A9: 78
etching techniques ... A9: 61
Interference, destructive
x-ray spectrometers ... A10: 88
Interference effects in color metallography ... A9: 136
Interference fasteners ... A8: 502
Interference film deposition *See also* Anodizing; Color etching; Heat tinting; Potentiostatic etching; Reactive sputtering; Vacuum deposition ... A9: 135-138
Interference films
optical constants of cathode materials ... A9: 149
use in potentiostatic etching ... A9: 143-144
Interference filter
defined ... A9: 10
Interference fit fasteners ... EM1: 707-708, 715
Interference fits ... EM3: 16
defined ... EM1: 14 EM2: 23
Interference fringes
produced in differential interference contrast microscopy ... A9: 151
Interference grounding
beryllium-copper alloys ... A2: 417
Interference layers, coated on anisotropic materials to improve grain contrast ... A9: 59-60
refractive indices ... A9: 60
Interference microscope ... A9: 80
Interference microscopes ... A17: 3, 17
Interference microscopy
for microstructural analysis ... EM4: 578
Interference of waves
defined ... A10: 675
Interference techniques of optical microscopy ... A9: 80
Interference vapor-deposited films
used for cemented carbides ... A9: 275
Interference-contrast illumination ... A9: 79
Interference-current effects ... A13: 1128-1129, 1288-1289

Interferences *See also* Spectral interferences
in collection of x-ray lines in 2.30-keV spectral vicinity **A10:** 522
effect of complexation reactions **A10:** 164
effects in ICP **A10:** 33-34, 40
electrodeposition of **A10:** 65-66
in EXAFS analysis **A10:** 408
filters for **A10:** 67
fringes as **A10:** 368
intentional addition of **A10:** 66
ionization **A10:** 29, 33, 34
separation by complexation for **A10:** 65
simultaneous UV/VIS analysis for **A10:** 65
spectral, wavelength-dispersive spectrometry **A10:** 521-522
of waves, defined **A10:** 675

Interferograms
F-F-IR **A10:** 112
as resolution enhancement **A10:** 117
time-average **A17:** 416

Interferograms, laser
of low gravity effects **A15:** 148

Interferometer
for extensometer calibration **A8:** 619
for Hugoniot elastic limit measurement **A8:** 211
normal velocity **A8:** 233
records, melted 4340 steel **A8:** 234-235
transverse displacement **A8:** 210, 233

Interferometer(s)
applications **A17:** 15
laser inspection by **A17:** 14-15
schematic **A17:** 14

Interferometers *See also* Optical microscopes
changing optical path length **A10:** 112
defined **A10:** 675
FT-IR **A10:** 111-112
Genzel **A10:** 112
Michelson **A10:** 39, 112
types **A9:** 80

Interferometry *See* Interferometer(s); Multiple-exposure interferometry; Optical holographic interferometry; Real-time interferometry; Time-average interferometry **A6:** 1150

Interfragmentary corrosion
aluminum alloy **A13:** 590

Intergranular *See also* Intercrystalline; Transgranular **A7:** 6
attack, on STAMP-produced products **A7:** 549
defined **A13:** 8

Intergranular attack *See also* Exfoliation
austenitic stainless steels **A6:** 465-466
in forging **A11:** 338

Intergranular beta
defined **A9:** 10

Intergranular brittle fracture
determined **A11:** 25-26

Intergranular carbides
cracking through **A11:** 407
precipitation, from overheating **A11:** 132

Intergranular cavities
iron **A12:** 219

Intergranular corrosion *See also* Corrosion; Interdendritic corrosion; Intergranular corrosion evaluation
in aircraft **A13:** 1028, 1045
aluminum alloys **M2:** 212
of aluminum/aluminum alloys **A13:** 130, 589-590
of austenitic cast alloys **A13:** 578-580
of austenitic stainless steels **A1:** 912, 945 **A11:** 180 **A12:** 364 **A13:** 124, 325
in austenitic stainless steels, result of welding **A9:** 283
of brazed joints **A13:** 879
by microbiological deposit **A13:** 315
of cast irons **A13:** 568
cast stainless steel neck fitting failure by **A11:** 404

Intergranular corrosion (continued)
as casting defect **A11:** 383
in CN-7M casting **A11:** 405
of copper alloy C26000 ..... **A11:** 182
of copper/copper alloys **A13:** 613-614
corrosion-resistant casting alloys **A13:** 578-580
crack initiation by **A13:** 149
crack propagation **A17:** 54
critical crevice temperature, cast/wrought alloys **A13:** 581
defined **A11:** 6, 180 **A13:** 8
of duplex cast alloys **A13:** 578-580
in duplex stainless steel weldments **A13:** 359
eddy current inspection **A17:** 191
effect on stress-corrosion cracking **A1:** 724, 725
electrochemistry **A13:** 123
as embrittlement, interpretation and examination **A12:** 126
evaluation of **A13:** 239-241
ferrite/austenite grain-boundary ditching **A13:** 582
of ferritic alloys **A1:** 916-917
in forging **A11:** 338
from liquid lithium **A13:** 93, 95
from welding **A13:** 49
of galvanized steel **A13:** 527
intergranular attack as **A13:** 942
intergranular penetration as **A13:** 942-944
of light metals and alloys **A11:** 182
in manned spacecraft **A13:** 1079-1080
material selection to avoid **A13:** 324-325
mechanisms **A13:** 123
metallographic sections of **A11:** 24
in mining/mill applications **A13:** 1295
nickel alloys **A2:** 432 **A11:** 181-182
in recrystallized wrought aluminum alloy **A13:** 590
in space shuttle orbiter **A13:** 1068
of stainless steels **A13:** 123, 239-240, 554, 562
in steam generators **A13:** 942
of steel castings **A11:** 403-404
and stress-corrosion cracking **A13:** 123, 148, 942
susceptibility **A13:** 123, 239-240
testing for **A13:** 124, 239-240, 562
of titanium **A11:** 182
n wrought heat-resisting alloys **A11:** 277
and uniform corrosion resistance **A13:** 48
weight loss, stainless steel **A13:** 123
in zinc alloys **A9:** 489-490
zinc/zinc alloys and coatings **A13:** 765

Intergranular corrosion evaluation *See also* Intergranular corrosion
of aluminum alloys **A13:** 240-241
of copper alloys **A13:** 241
of magnesium alloys **A13:** 241
of nickel-base alloys **A13:** 239-240
purpose **A13:** 239
simulated-service and accelerated tests **A13:** 239
of stainless steels **A13:** 239-240
testing media **A13:** 240
of zinc die casting alloys **A13:** 241

Intergranular corrosion resistance in stainless steel casting alloys **A9:** 297

Intergranular cracking *See also* Transgranular cracking
brittle fracture by **A11:** 82
defined **A11:** 6
and deformation creep **A11:** 29
from electroplated coating **A11:** 337
matrix **A8:** 487
micro-fracture mechanics morphology **A8:** 465
microscopic models for **A8:** 466
in steam tubes **A11:** 605
stress-corrosion **A13:** 148
velocities **A13:** 160

Intergranular cracks
in copper-bearing lead **A9:** 418
in wrought beryllium-copper alloys **A9:** 393

Intergranular creep fractures
austenitic stainless steels **A12:** 364
by grain-boundary cavitation **A12:** 19, 26
by triple-point cracking **A12:** 19, 25
by wedge cracking **A12:** 26

Intergranular decohesion fractures **A12:** 24, 31
aluminum alloys **A12:** 35
austenitic stainless steel **A12:** 27, 34
bands, fatigue fracture in hydrogen **A12:** 37, 51
of brass **A12:** 28, 36
by SCC **A12:** 24, 27-28, 36
effect of corrosive or embrittling environment **A12:** 35
embrittling effect, low-melting metals **A12:** 29
and hydrogen embrittlement, steel **A12:** 31
of steels **A12:** 27

Intergranular dimple rupture
effect of microvoid nucleation **A12:** 12
in steel **A12:** 14

Intergranular embrittlement
by films or segregation **A12:** 110

Intergranular facets
discontinuous **A8:** 487

Intergranular failure **A8:** 476

Intergranular fracture *See also* Cleavage; Transgranular fracture
acoustic emission inspection **A17:** 287
brittle, of ordered intermetallics **A2:** 913-914
brittle, surfaces of **A11:** 77
by fatigue cracking **A11:** 131-133
by overheating **A11:** 436
by SCC, in stainless steel bolts **A11:** 536-537
in cadmium-plated steel **A11:** 29
causes of **A11:** 42
defined **A11:** 6 **A13:** 8
definition **EM4:** 633
and embrittlement **A11:** 82
EPMA spectrum, failed Inconel 600 bellows **A11:** 39-41
microscopic examination **A11:** 75
and Ni_3X alloys **A2:** 931-932
nickel **A13:** 157-158
nickel aluminides **A2:** 914-915
in piping system cross, from improper heat treatment **A11:** 649-650
process **A8:** 486-487
stress-corrosion cracking **A13:** 156
stress-intensity factor range effect **A8:** 485-486
in temper-embrittled steel **A11:** 22

Intergranular fracture, in austenitic manganese steel castings
causes of **A9:** 238

Intergranular fracture process **A8:** 486-487

Intergranular fracture(s) *See also* Cleavage fracture(s); Creep fractures; Intergranular corrosion; Intergranular decohesion fractures; Intergranular dimple rupture
AISI/SAE alloy steels **A12:** 293, 299-300, 305, 307-308, 335
ASTM/ASME alloy steels **A12:** 349-350
austenitic stainless steels **A12:** 352-353, 355
austenitizing effect **A12:** 339
in bcc metals **A12:** 123
of bolts **A12:** 299
brittle **A12:** 30, 38, 109, 174-175
by grain-boundary cavitation **A12:** 349
by grain-boundary sliding **A12:** 121-123, 140-141
by liquid cadmium embrittlement **A12:** 30, 39
cleavage, high-carbon steels **A12:** 290
copper alloys **A12:** 402
decohesive **A12:** 24
in engineering alloys **A12:** 12
fight fractographs **A12:** 94, 98
from creep rupture **A12:** 18-19, 25
high-carbon steels **A12:** 284
high-purity copper **A12:** 400
hydrogen-assisted, low-carbon steel **A12:** 248
iridium and iridium ahoy **A12:** 462-463
irons **A12:** 219

Intergranular fracture(s) (continued)
low-carbon steels **A12:** 240
medium-carbon steels **A12:** 271
nickel alloys **A12:** 397
oxygen-embrittled Armco iron **A12:** 222
from quench cracking **A12:** 131, 148-149
rock-candy appearance **A12:** 110-111
salt corrosion assisted, superalloys **A12:** 389
SCC, in precipitation-hardening stainless steels **A12:** 373
separation, iron alloy **A12:** 459
sintered tungsten **A12:** 462
superalloys **A12:** 391, 393
titanium alloys **A12:** 441
tool steels **A12:** 375
and transcrystalline cleavage, iron **A12:** 222
types **A12:** 110, 121-123
wrought aluminum alloys **A12:** 431, 434, 439
Intergranular fractures, effect of
on stress-corrosion cracking **A1:** 724
Intergranular microvoid coalescence
defined **A12:** 128
Intergranular networks
in iron castings **A11:** 359
Intergranular oxidation
in flash welds **M6:** 580
Intergranular penetration
definition **A6:** 1210 **M6:** 10
Intergranular secondary cracking
OFHC copper **A12:** 401
Intergranular segregation
and tramp elements **A8:** 476
Intergranular separation
alloy steels **A12:** 341
high-purity copper **A12:** 399
iron alloy **A12:** 459
Intergranular stress-corrosion cracking
aircraft part **A13:** 1031
in austenitic stainless steels **A13:** 124
in boiling water reactors **A13:** 928-933
in collet retainer tube **A13:** 933
crack propagation **A13:** 155-157
critical potentials **A13:** 151-152
defined **A11:** 6 **A13:** 8
in HAZ **A13:** 929-930
in jet pump beams **A13:** 933-935
keyway **A13:** 952-953
quantitative model, boiling water reactors **A13:** 930
stainless steel **A13:** 152
steam generators **A13:** 942
stress dependence of **A13:** 930
vs. temperature **A13:** 154
Intergranular stress-corrosion cracking (IGSCC) **A6:** 379
austenitic stainless steels **A6:** 466
Intergranular-transgranular fracture transition
in elevated-temperature failures **A11:** 266
Interior flaw(s) *See also* Defect(s); Discontinuities; Flaws
NDE detection methods **A17:** 50
Interior probes
remote-field eddy current inspection **A17:** 195-201
Interiors, unconsolidated
tool and die failures from **A11:** 574-575
Interlaboratory studies
recommended practices for chemical analysis **A7:** 249
Interlamella spacing
high-carbon steels **A12:** 290
Interlamellar spacing *See also* True spacing
in pearlite **A9:** 660
Interlaminar **EM3:** 16
defined **EM1:** 14 **EM2:** 23
Interlaminar cracking
free-edge delamination **EM1:** 241-242
shear lag effect **EM1:** 243-244
transverse cracks **EM1:** 242-244
Interlaminar fracture
crack directions/fiber orientation **EM1:** 787-790
fractography of **EM1:** 786-790
Interlaminar fracture toughness *See also* Fracture toughness; Toughness
Interlaminar fracture toughness (G_{Ic}) *See also* Fracture toughness; Toughness
of carbon fiber composites **EM1:** 98
Interlaminar fracture toughness (G_{Ic}) (continued)
and damage tolerance **EM1:** 262
specimens **EM1:** 343-345
of thermoplastic composites **EM1:** 100
vs. resin modulus, thermoplastics **EM1:** 99
vs. strain, thermoplastics **EM1:** 100
Interlaminar fractures *See also* Delaminations
in composites **A11:** 733-738
crack directions and fiber orientation **A11:** 735-738
effect of fiber reinforcement **A11:** 736
in-plane shear **A11:** 736-738
tensile, schematic without stress concentration **A11:** 737
tension **A11:** 735-736
Interlaminar insulation
magnetic cores **A2:** 780
Interlaminar normal stress
around hole **EM1:** 234
Interlaminar shear **EM3:** 16
defined **EM1:** 14 **EM2:** 23
fracture, fractography of **EM1:** 789-790
Interlaminar shear strength
by matrix resin **EM1:** 31
glass fabric reinforced epoxy resin **EM1:** 404
graphite fiber reinforced epoxy resin **EM1:** 413
short fiber based CFRP **EM1:** 156
test **EM1:** 299
Interlaminar shear stress
around hole **EM1:** 234
defined **EM1:** 229
thickness distribution **EM1:** 240
Interlaminar stresses
analysis **EM1:** 229 230
and moment equilibrium **EM1:** 229
normal, distribution **EM1:** 239-240
as out-of-plane failure cause **EM1:** 783-784
Interlaminar tensile stress
and delamination under fatigue loading **A8:** 714
Interlayer sizing **EM1:** 122-124
Interlayers
material selection **M6:** 680-681
materials commonly used **M6:** 681
reasons for use **M6:** 680
Interleaf layers **EM3:** 489
Interligament cracking
in failed secondary superheater outlet header from boiler **A11:** 667
Interlocking
of blanks **A14:** 450
in composite compression fracture **A11:** 739
from translaminar fracture **EM1:** 792
Interlocking joints
wrought aluminum alloy **A2:** 36
Interlocks **A18:** 589-590
Intermediate annealing **A14:** 620-621, 779
defined **A9:** 10
Intermediate dielectrics
effect of phosphorus on **A11:** 771
Intermediate electrode
defined **A13:** 8
Intermediate flux
definition **M6:** 10
Intermediate lead-tin babbit alloys
compositions **A2:** 524
Intermediate phase
defined **A9:** 10
Intermediate phases ..... **A3:** 1 • 4
Intermediate temperature setting adhesive
defined **EM1:** 14 **EM2:** 23
Intermediate-temper wire **A1:** 850
Intermediate-temperature-setting adhesive **EM3:** 16
Intermetallic
aluminum P/M alloys **A2:** 210
phases, wrought aluminum alloy **A2:** 37
Intermetallic compound
defined **A9:** 10
Intermetallic compound, cleaved
crack growth change **A8:** 482
Intermetallic compound embrittlement **M1:** 686
Intermetallic compound layers
in soldered joints **A17:** 608
Intermetallic compounds ..... **A3:** 1 • 4 **A6:** 127, 128, 129, 134, 163
for A15 superconductors **A2:** 1060-1062
Intermetallic compounds (continued)
and alloys, compared **A2:** 910
between tin and tin alloy coatings and the substrate, examination of **A9:** 451
braze filler metals and base materials forming **A11:** 452
diffusion welding **A6:** 886
embrittlement by **A11:** 100
explosion welding **A6:** 163, 896
inclusions in uranium alloys **A9:** 477, 479
low-solubility zirconium **A2:** 666
neutron diffraction analysis **A10:** 420
permanent magnets **A9:** 539
in soldering operations **EL1:** 634-636
TiFe, hydrided, phase analysis by Mssbauer spectroscopy **A10:** 293
titanium-base **A2:** 590
Intermetallic inclusions *See also* Inclusions
in aluminum alloys **A15:** 11, 195-96
copper alloys **A15:** 96
from compound precipitation **A15:** 488
particles as **A15:** 95
property effects, aluminum-silicon alloys **A15:** 167
wrought aluminum alloys **A12:** 423
Intermetallic layer
of electrolytic tinplate **A9:** 456
formed beneath tin-lead coating **A9:** 456
Intermetallic phases **A13:** 551, 642-643
in aluminum alloys **A9:** 358-360
aluminum-silicon alloys **A15:** 166
defined **A9:** 10
in tin-indium alloys **A9:** 452
in wrought stainless steels **A9:** 284-285
Intermetallic powders *See also* Metallic elements; Metallic glass powders **A7:** 514
Intermetallic-matrix composites
development of **A2:** 909-911
Intermetallic-phase precipitation
in elevated- temperature failures **A11:** 267
Intermetallic-related failures
thermocompression bonding **EL1:** 1042-1043
Intermetallics
application in future jet engine components **A18:** 592
compatibility with steel **A18:** 743
effect on crack growth and toughness **A11:** 54
heat-affected-zone cracks **A6:** 92
phase contrast imaging **A18:** 389
Intermittent furnace
porcelain enameling **M5:** 520
Intermittent immersion tests **A13:** 222-223
Intermittent life tests **EL1:** 494, 499
Intermittent service motors and generators
as magnetically soft material application **A2:** 779
Intermittent weld
definition **M6:** 10
Intermolecular forces
in polymers **EM2:** 64
Internal bursts *See also* Bursts
effect on cold-formed part failure **A11:** 307
effect on fatigue strength **A11:** 121
radiographic methods **A17:** 296
Internal can lacquer
types **A13:** 780
Internal cavities *See also* Cavities
for copper castings, cost/design considerations **A2:** 355
Internal chill *See* Inverse chill
Internal coils
eddy current inspection **A17:** 183
Internal combustion engine applications
of structural ceramics **A2:** 1019
Internal combustion engine lubricants **A18:** 162-170
additives **A18:** 162, 169-170
applications **A18:** 162
formulation **A18:** 168-169
base fluids **A18:** 168-169
physical properties of hydrofinished HVI stocks and synthetic base stocks **A18:** 169

Internal combustion engine lubricants (continued)
relationship between wear properties and hydrocarbon structures ... **A18:** 168
lubricant classification based on end-use ... **A18:** 165-167
aviation engine oils ... **A18:** 166
gasoline engine oils ... **A18:** 165
heavy-duty diesel engine oils ... **A18:** 165
marine diesel engine oils ... **A18:** 165-166
natural gas engine oils ... **A18:** 166
railroad diesel engine oils ... **A18:** 165
stationary diesel engine oils ... **A18:** 165
two-stroke cycle engine oils ... **A18:** 165, 166-167
lubricant-related causes of engine malfunction ... **A18:** 167-168
corrosion ... **A18:** 168
deposit formation ... **A18:** 167
mechanism of deposit formation ... **A18:** 167
oil consumption ... **A18:** 167-168
oil thickening ... **A18:** 167
ring sticking ... **A18:** 167, 168
wear ... **A18:** 167, 168
performance package ... **A18:** 169-170
performance testing ... **A18:** 170
specifications ... **A18:** 162-165
service classifications ... **A18:** 163
viscosity ... **A18:** 163, 164
winter (W) viscosity ... **A18:** 163
types ... **A18:** 162
boundary lubrication ... **A18:** 162
elastohydrodynamic lubrication ... **A18:** 162
fluid-film lubrication ... **A18:** 162
hydrodynamic lubrication ... **A18:** 162
mixed-film lubrication ... **A18:** 162
types of internal combustion engines ... **A18:** 162
viscosity, speed, and equipment load ... **A18:** 162, 163
world lubricant market and consumption percentages ... **A18:** 162
Internal combustion engine parts, friction and wear of ... **A18:** 553-561
abrasive wear ... **A18:** 555, 558, 559
adhesive wear ... **A18:** 555, 559
applications ... **A18:** 553
corrosive wear ... **A18:** 555, 556, 558
crankshaft bearings ... **A18:** 559-561
engine types and design considerations ... **A18:** 553-554
engine types ... **A18:** 553
general features ... **A18:** 553-554
principles of operation ... **A18:** 553
valve train designs ... **A18:** 560
friction and wear of engine components ... **A18:** 554
contribution of major components to engine friction ... **A18:** 554
emission and fuel economy improvements ... **A18:** 554
engine wear ... **A18:** 555
lubrication regimes for friction components ... **A18:** 554-555
future outlook ... **A18:** 561
in-line automotive engine, six-cylinder, cross sectional view ... **A18:** 554
lubrication ... **A18:** 555, 556, 557, 558-559, 561
lubrication regimes ... **A18:** 555
oil pumps ... **A18:** 561
oxidational wear ... **A18:** 556
pistons and piston ring assembly ... **A18:** 555-558
cylinder liner materials ... **A18:** 556-557
factors affecting piston ring wear ... **A18:** 557-558
piston ring-cylinder liner scuffing ... **A18:** 557
piston rings ... **A18:** 555-556
pistons ... **A18:** 55, 556
typical cylinder bore materials ... **A18:** 557
pitting ... **A18:** 555, 559
schematic of a four-stroke engine cycle ... **A18:** 553
scuffing ... **A18:** 555, 556, 557, 559

Internal combustion engine parts, friction and wear of (continued)
valve train assembly ... **A18:** 558-559
valve train designs ... **A18:** 560
valve train friction ... **A18:** 558-559
valve train wear ... **A18:** 559
Internal combustion engines
abrasive wear in ... **M1:** 602, 604
elevated-temperature valve failure ... **A11:** 288-289
in pipe ... **A11:** 704
Internal conversion
as radioactive decay mode ... **A10:** 245
Internal cracks *See also* Cracks; Discontinuities; Flaws
flaw shape parameter curve for ... **A11:** 108
forging ... **A11:** 317 **A17:** 494
radiographic methods ... **A17:** 296
sizes at various stress levels ... **A11:** 108
Internal defects *See also* Defects; Discontinuities; Flaws
in ceramics ... **A11:** 748-751
resin-matrix composite failure from ... **A12:** 474
ultrasonic inspection ... **A17:** 530
Internal delamination *See also* Delamination
by machining and drilling ... **EM1:** 667-668
Internal discontinuities
casting ... **A15:** 544-545
in castings ... **A17:** 512
in iron castings ... **A11:** 354-359
magnetic particle inspection ... **A17:** 105
in shafts ... **A11:** 459, 467, 477
Internal electrolysis ... **A10:** 199-201
cell for ... **A10:** 199
for copper determination ... **A10:** 200
separation of cadmium and lead by ... **A10:** 201
Internal energy ... **A3:** 1 • 5
Internal fixation devices
analysis of failed ... **A11:** 680
defined ... **A11:** 670
degrees of stability in ... **A11:** 673
design of ... **A11:** 677-680
implants as ... **A11:** 671
typical orthopedic examples ... **A11:** 671
unstable, adapting or bridging functions ... **A11:** 673
Internal fluorescence peak
in EDS spectra ... **A10:** 520
Internal force spike ... **A18:** 47, 48
Internal fracture
prediction ... **A14:** 399
Internal gears
described ... **A11:** 586-587
Internal global communication
conventional interconnection ... **EL1:** 5
Internal grinding *See* Grinding
Internal inductance
defined ... **EL1:** 29
Internal lead
embrittlement by ... **A13:** 181-182
Internal lockseams
forming of ... **A14:** 573
Internal open circuits
passive devices ... **EL1:** 997
Internal oxidation *See also* Manufacturing methods ... **A7:** 6
conventional ... **A7:** 717
defined ... **A9:** 10 **A11:** 6 **A13:** 8
for silver-base composites with dispersed oxides ... **A2:** 857
Internal particle porosity (I-pores)
and compaction ... **A7:** 299
and low compressibility ... **A7:** 298
in water atomized iron powders ... **A7:** 85
Internal point-to-point communication ... **EL1:** 4
Internal rate of return economic analysis ... **A13:** 370
Internal reflection elements
in ATR spectroscopy ... **A10:** 113-114

Internal reflection elements (continued)
micro KRS-5 (thallium bromide/thallium iodide) ... **A10:** 113
and sample interface, angle of incidence ... **A10:** 114
top view, micro KRS-5 ... **A10:** 113
Internal reversible hydrogen embrittlement ... **A13:** 283
Internal rupture
in gears ... **A11:** 596
Internal scatter, shadow formation
radiography ... **A17:** 313
Internal shapes, tube
by swaging ... **A14:** 138-139
Internal shrinkage
by radiographic methods ... **A17:** 296
as casting defect ... **A11:** 382
defined ... **A15:** 7
Internal shrinkage cracks
defined ... **EM2:** 23
Internal standard
defined ... **A10:** 675
Internal standard line
defined ... **A10:** 675
Internal standard method
XRPD analysis ... **A10:** 340
Internal stress *See* Residual stress
by plastic encapsulants ... **EL1:** 806-808
Internal stresses *See also* Stress(es)
in aluminum alloy ingots, effect of homogenization on ... **A9:** 632
classified ... **EM2:** 751-752
Internal surfaces
cleaning of ... **A15:** 520
Internal sweating
as casting defect ... **A11:** 387
Internal tension stresses
in aluminum alloy ingots, effect on center cracking ... **A9:** 634-635
Internal upsets
hot upset forging ... **A14:** 92
Internal-hydrogen failures
of steel castings ... **A11:** 408-409
International Aerospace Abstracts (IAA)
as information source ... **EM1:** 41
International Annealed Copper Standard ... **A8:** 725
International classification
of casting defects ... **A11:** 381-388 **A15:** 545-553
International Committee of Foundry Technical Associations (Zurich, Switzerland) ... **A15:** 34
International Conferences on Jet Cutting Technology ... **A18:** 222
International designations and specifications ... **A1:** 156-159, 166-174, 174-194
British (BS) standards ... **A1:** 158
compositions of BS alloy steels ... **A1:** 185-186
compositions of BS carbon steels ... **A1:** 182-184
cross-referenced to SAE-AISI steels ... **A1:** 166-174
French (AFNOR) standards ... **A1:** 158-159
composition of AFNOR alloy steels ... **A1:** 189-190
composition of AFNOR carbon steels ... **A1:** 186-188
cross-referenced to SAE-AISI steels ... **A1:** 166-174
German (DIN) standards ... **A1:** 157
compositions of DIN alloy steels ... **A1:** 178-179
compositions of DIN carbon steels ... **A1:** 175-178
cross-referenced to SAE-AISI steels ... **A1:** 166-174
Italian (UNI) standards ... **A1:** 159
compositions of UNI alloy steels ... **A1:** 192-193
compositions of UNI carbon steels ... **A1:** 190-191
cross-referenced to SAE-AISI steels ... **A1:** 166-174
Japanese (JIS) standards ... **A1:** 157-158
compositions of JIS alloy steels ... **A1:** 181
compositions of JIS carbon steels ... **A1:** 180
cross-referenced to SAE-AISI steels ... **A1:** 166-174
Swedish (SS_{14}) standards ... **A1:** 159
compositions of SS_{14} alloy steels ... **A1:** 194

International designations and specifications (continued)
compositions of SS_{14} carbon steels ... **A1:** 193
cross-referenced to SAE-AISI steels ... **A1:** 166-174
International Electrotechnical Commission (IEC) ... **EM2:** 461
International Institute for Welding
microstructure constituent classifications for ferritic weldments ... **A9:** 481
International Institute of Welding
instrumented Charpy impact test, standardizing by ... **A8:** 266
International Journal of Powder Metallurgy and Powder Technology ... **A7:** 19
International Nickel Company
ductile iron of ... **A15:** 35
International Organization for Standardization ... **A8:** 625
International Organization for Standardization (ISO) ... **EM2:** 91, 461-462
basic load rating equations for radial ball bearings ... **A18:** 505
cross referencing system ... **A2:** 16, 17-25
friction test standards ... **A18:** 53
instrument calibration specimens for measuring surface texture (5436) ... **A18:** 341
lubricant viscosity grade (3448-1975 E) ... **A18:** 134
specifications for threaded fasteners ... **A1:** 289
viscosity grades for specifying miscellaneous industrial oils ... **A18:** 99
viscosity grades of lubricants ... **A18:** 85
International Organization of Standardization (ISO)
R513 classification of carbides ... **A16:** 75
***International Plastics Handbook* (Saechtling)** ... **EM2:** 93
International Practical Temperature Scale (IPTS 68, amended 1975) ... **A2:** 879
International Research Group on Wear of Engineering Materials (IRG-OECD)
transition diagram determination methodology ... **A18:** 491
International Society for Hybrid Microelectronics (IC) ... **EL1:** 734
International Standards (IS) ... **EM3:** 62
International Standards Organization ... **EM3:** 62
flow rate test ... **A7:** 279-280
global sealant standards as goal ... **EM3:** 188
sampling procedures ... **A7:** 212
International System of grade designation
specifications for ductile iron ... **A1:** 35, 36
International Temperature Scale (ITS 27) ... **A2:** 878
International Temperature Scale of 1990 (ITS-90) ... **A2:** 879
Interparticle friction
and flow rate ... **A7:** 280
Interparticle oxides
powder forged parts ... **A14:** 204
Interparticle porosity
and compaction ... **A7:** 299
Interpass temperature
cast iron arc welding ... **A15:** 525-526
definition ... **A6:** 1210 **M6:** 10
Interpenetrating polymer networks (IPNs) ... **EM3:** 161, 602
Interphase ... **EM3:** 16, 395, 397
composite materials ... **EM3:** 391
defined ... **EM1:** 14 **EM2:** 23
fiber-matrix characterization methods ... **EM3:** 391-392
Interphase boundary cracking
due to creep deformation ... **A8:** 306
Interphase interfaces
FIM/AP study of point defects in ... **A10:** 583
Interphase interfaces between matrix and precipitate ... **A9:** 648
Interphase mass transport
defined ... **A15:** 52-54
Interplanar distance
defined ... **A9:** 10
Interplanar spacing d_{hkl} *See* d-spacings
Interply hybrid *See also* Hybrid
defined ... **EM1:** 14 **EM2:** 23
Interpolation
to determine creep-rupture behavior ... **A8:** 332-335
Interposer contact/contact module
detail of ... **EL1:** 394
Interpretation
of cooling curves (thermal analysis) ... **A15:** 182-185
of fractures ... **A12:** 96-123
magnetic particle inspection ... **A17:** 103
of monitoring and test results ... **A13:** 203, 316-317
and review, magnetic rubber inspection ... **A17:** 123
Interpulse time
definition ... **M6:** 10
Interrupted aging
defined ... **A9:** 10
Interrupted pour
as casting defect ... **A11:** 383
Interrupted quenching
defined ... **A9:** 10
Intersection of phase-field boundaries ... **A3:** 1 • 8, 10
Intersection scarp
definition ... **EM4:** 633
Interstitial atoms
and point defects ... **A13:** 46
Interstitial diffusion
solid-state ... **A13:** 68
Interstitial elements, effects of
on notch toughness ... **A1:** 742
Interstitial elements, notch toughness of steel
effect on ... **M1:** 694, 697
Interstitial fluid
and dental alloys ... **A13:** 1340
ionic composition of ... **A11:** 672
Interstitial impurity
reduction of ... **A2:** 1094-1095
Interstitial ion
illustrated ... **A13:** 65
Interstitial nitrogen
determined in steels ... **A10:** 178
Interstitial pneumonitis ... **A7:** 203
Interstitial reactions
liquid-metal corrosion by ... **A13:** 56-59
Interstitial solid solution ... **A3:** 1 • 15, 16
Interstitial solid solutions
defined ... **A9:** 10
Interstitial solid-solution strengthening
ion implantation strengthening mechanisms ... **A18:** 855, 856, 858
postimplantation low-temperature aging treatment ... **A18:** 856
Interstitial-free (IF) steels ... **A4:** 61
Interstitial-free steel ... **A1:** 112-113, 131-132, 405, 578
cold-rolled strip ... **A1:** 417
composition of ... **A1:** 417
deep-drawing properties of ... **A1:** 398
effects of steelmaking on formability of ... **A1:** 578
mechanical properties ... **M1:** 178
porcelain enameling of ... **M1:** 178 **M5:** 512-513
production ... **M1:** 179
production of ... **A1:** 112-113, 131-132
tensile and yield strengths of ... **A1:** 417
tensi
Interstitials *See* Point defects; Vacancies
Intersystem crossing
defined ... **A10:** 675
Interval
estimate, defined ... **A8:** 7
and probability ... **A8:** 624
statistical ... **A8:** 626
tests, and stress amplitude ... **A8:** 374
Interval erosion rate
defined ... **A18:** 11
Intracellular fluid
ionic composition of ... **A11:** 672
Intracrystalline
defined ... **A9:** 10
Intracrystalline cracking *See* Transcrystalline cracking; Transgranular cracking
Intragranular precipitation
of carbides, in austenitic stainless steels ... **A9:** 284
of Laves phase, in austenitic stainless steels ... **A9:** 284
Intragranular subgrain structure
as a result of dynamic recovery ... **A9:** 690
Intralaminar ... **EM3:** 16
defined ... **EM2:** 23
Intralaminar fracture
crack directions/fiber orientations ... **EM1:** 787-790
Intralaminar fracture (continued)
fractography of ... **EM1:** 786-790
Intralaminar fractures
in composites ... **A11:** 733-738
crack directions and fiber orientation ... **A11:** 735-738
in-plane shear fractures ... **A11:** 736-738
tension ... **A11:** 735-736
Intramedullary rods
as implant nails ... **A11:** 671
Intramedullary tibia nail
as internal fixation device ... **A11:** 671, 680
Intraply hybrid *See also* Hybrid
defined ... **EM1:** 14 **EM2:** 24
Intrinsic (electrical) breakdown
defined ... **EM2:** 464
Intrinsic conduction
defined ... **EL1:** 99
Intrinsic device speed
and intrinsic device delay ... **EL1:** 2
Intrinsic dielectric breakdown
integrated circuits ... **A11:** 779
Intrinsic dielectric strength
defined ... **EM2:** 460
Intrinsic formability tests *See also* Workability ... **A8:** 553-560 **A14:** 883-889
dynamic material modeling ... **A14:** 423
Intrinsic induction
in permanent magnet materials ... **A2:** 782, 784
Intrinsic properties
thermal ... **EM2:** 451
Intrinsic stacking faults
contrast in transmission electron microscopy ... **A9:** 119
and dislocation loops ... **A9:** 116
Intrinsic viscosity *See also* Viscosity ... **EM3:** 16
defined ... **EM2:** 24
Introduction
casting advantages ... **A15:** 37-45
casting applications ... **A15:** 37-45
foundry technology, development ... **A15:** 24-36
and historical development ... **A15:** 13-45
history of casting ... **A15:** 15-23
market size ... **A15:** 37-45
Introduction system
gas mass spectrometer ... **A10:** 151-152
Introduction system complex ... **A10:** 152
Introduction to composites
general information sources ... **EM1:** 40-42
general use considerations ... **EM1:** 35-37
glossary of terms ... **EM1:** 3-26
introduction ... **EM1:** 27-34
selection and evaluation ... **EM1:** 38-39
Introfaction ... **EM3:** 16
defined ... **EM2:** 24
Introfier ... **EM3:** 16
Intrusion alarms
as germanium application ... **A2:** 743
Intumescence
defined ... **A13:** 8
Intumescents
applications ... **EM3:** 56
Invar *See also* Low-expansion alloys
applications ... **M3:** 798
composition effects on expansion coefficient ... **A2:** 889-890
copper-clad, as heat sink ... **EL1:** 1129-1131
corrosion resistance ... **A2:** 891
defined ... **EL1:** 1147
effects of processing ... **A2:** 890-891
electrical properties ... **A2:** 891
expansion coefficients ... **M3:** 793, 794
heat treatment ... **M3:** 794, 795
as low-expansion alloy ... **A2:** 889-893
machinability ... **A2:** 891-892
magnetic properties ... **A2:** 891
physical/mechanical properties ... **A2:** 891-892
processing ... **M3:** 794-795
properties ... **A2:** 889
thermoelastic coefficient ... **A2:** 891
welding ... **A2:** 892-893
Invar: 36% Ni
thermal properties ... **A18:** 42
Invariant
equilibrium ... **A3:** 1 • 2
point ... **A3:** 1 • 2
reactions ... **A3:** 1 • 5

Invariant melting temperature of the solvent element (T_m) **A6:** 89
Invariant phase field determination
by Auger electron spectroscopy **A10:** 474
Inventory control systems **EM3:** 685
Inverse bainite **A9:** 665
Inverse chill *See also* Chill; Chilled iron
defined **A15:** 7
ductile iron fracture-e from **A12:** 227
in iron castings **A11:** 362
Inverse gas chromatographic studies **EM3:** 289
Inverse logarithmic reaction rates
high-temperature corrosion, gases **A13:** 66-67
Inverse pole figures
generation of **A9:** 703
Inverse segregation
in Bronze Age **A15:** 16
in copper alloy ingots **A9:** 643-644
defined **A9:** 10 **A15:** 7
tin sweat from **A15:** 16
Inverse skin-doubler coupon **EM3:** 471-472, 475, 476
Inverse-square law, and shadow intensity
radiography **A17:** 313
Inversion center
in single-cells **A10:** 347
Inverted bench microscope **A9:** 72
Inverted chill *See* Inverse chill
Inverted dies
for blanking **A14:** 453
Inverted incident-light microscope **A9:** 72
light path **A9:** 71
with television monitor attached **A9:** 84
Inverted metallograph **A9:** 74
Inverted metallographic microscope
used in identification of ferrite in heat-resistant casting alloys **A9:** 333
Inverted microscope
defined **A9:** 10
magnetic etching setup **A9:** 64-65
Inverted optics
for liquid x-ray spectrometry **A10:** 95
in wavelength-dispersive x-ray spectrometers **A10:** 88
Inverted research-quality optical microscopes **A9:** 73
Inverted "T" fillet weld test **A6:** 725
Inverted-chip reflow (flip-chip) technology **EM3:** 584, 585
Inverters **A6:** 39, 40, 44
Inverters, series/parallel
for induction furnaces **A15:** 370-371
Investigation procedures
of pipeline failures **A11:** 695-697
Investing *See also* Investment casting
defined **A15:** 7
lost foam pattern **A15:** 233
Investment
defined **A15:** 7
Investment (lost wax) casting *See also* Investment casting; Precision casting
of Alnico alloys **A15:** 737
applications **A15:** 266
basic process steps **A15:** 204-206
by electron beam melting **A15:** 417-418
ceramic cores, manufacture **A15:** 261
and ceramic molding, compared **A15:** 248
ceramic shell molds, manufacture **A15:** 257-261
colloidal silica bonds in **A15:** 212
design considerations **A15:** 264-266
development of **A15:** 35
inspection and testing **A15:** 264
magnesium alloy **A15:** 799, 806-807
market trends **A15:** 44
melting and casting **A15:** 262-263
of metal-matrix composites **A15:** 847-848
mold firing and burnout **A15:** 262
of nickel-base superalloys **A15:** 207
pattern and cluster assembly **A15:** 257
Investment (lost wax) casting (continued)
pattern materials **A15:** 253-255
pattern removal **A15:** 261-262
patternmaking **A15:** 195, 255-257
postcasting operations **A15:** 263-264
as precision molding **A15:** 37
process, steps in **A15:** 253
Replicast CS process, as special process **A15:** 267
and Replicast CS process, compared **A15:** 270
Shellvest system, as special process **A15:** 266-267
titanium alloys **A15:** 825-826
titanium, with vacuum arc skull melting **A15:** 409-410
tolerances **A15:** 621-622
Investment cast cobalt-chromium- molybdenum alloy
for orthopedic implants **A7:** 658
Investment cast superalloys *See* Polycrystalline cast superalloys
Investment cast turbine blades
neutron radiography of **A17:** 388, 393-394
Investment casting *See also* Aluminum alloys; Castings; Dip coat; Expendable pattern; Foundry products; Investing; Investment (lost wax) casting **A15:** 253-261 **M2:** 147
aluminum casting alloys **A2:** 140-141
defined **A15:** 7
magnesium alloy, cost-quantity relationships **A2:** 463
tungsten-reinforced composites **EM1:** 885
Investment Casting Institute **A15:** 34
Investment casting techniques **A7:** 428
Investment precoat *See also* Dip coat
defined **A15:** 7
Investment shell
defined **A15:** 7
molds, ceramic, manufacture of **A15:** 257-261
Inward diffusion coatings
superalloys **M5:** 378
I_0 *See* Initial intensity
Iodide process for purifying metals **M2:** 711
Iodide process, of chemical vapor deposition
for metal ultrapurification **A2:** 1094
Iodides
determined by precipitation titration **A10:** 164
as electrode **A10:** 185
as fining agents **EM4:** 380
solvent extractant for **A10:** 170
Iodimetric titration
indirect **A10:** 174
Iodimetry
as class of redox titration **A10:** 174
Iodine
and absolute methanol, to isolate inclusions in steels **A10:** 176
determined in water by coulometric titration **A10:** 205
etchant for laser-enhanced etching **A16:** 576
and methanol, second-phase test method **A10:** 177
species weighed in gravimetry **A10:** 172
tantalum resistance to **A13:** 731
TNAA detection limits **A10:** 237
Iodine bromide, in graphites
Raman analysis **A10:** 133
Iodine chloride, in graphites
Raman analysis **A10:** 133
Iodoantimonite method
analysis for antimony n copper alloys by **A10:** 68
Iodometry
indirect **A10:** 205
Ion beam
assisted deposition **A13:** 498
LEISS analysis **A10:** 606-607
milling, transmission electron microscopy **A10:** 451-452
mixing **A13:** 498
Ion beam (continued)
mixing, as sputtering artifact **A10:** 556
in spark source mass spectrometer **A10:** 142
sputtering **A13:** 498
sputtering, Auger electron emission by **A10:** 550
Ion beam deposition **A18:** 842
Ion beam etching **A9:** 62
Ion bombardment coating *See* Sputtering
Ion bombardment etching
of fiber composites **A9:** 591
Ion bombardment, of transmission electron microscopy specimens to achieve
dimpling **A9:** 105
Ion carburizing *See* Plasma (ion) carburizing
Ion chamber detectors
computed tomography (CT) **A17:** 370
Ion channeling
for damage depth profiles **A10:** 632
Ion chromatogram
typical total **A10:** 644
Ion chromatographs
and AAS instruments **A10:** 55
major components **A10:** 658-659
Ion chromatography **A10:** 658-667
applications **A10:** 658, 665-667
calibration curves **A10:** 666
capabilities **A10:** 181, 649, 663
defined **A10:** 675
estimated analysis time **A10:** 658
exclusion **A10:** 662
general uses **A10:** 658
indirect **A10:** 661
of inorganic liquids and solutions, information from **A10:** 7
of inorganic solids, types of information from **A10:** 4-6
instrumentation **A10:** 665
introduction **A10:** 658-659
limitations **A10:** 658
modes of detection **A10:** 659-662
modes of separation **A10:** 662-663
of organic solids, information from **A10:** 9
related techniques **A10:** 658
reversed-phase **A10:** 663
sample preparation and standardization **A10:** 663-665
samples **A10:** 658, 663-667
with spectrophotometric detection **A10:** 661
standard, separation mode **A10:** 662
to analyze the bulk chemical composition of starting powders **EM4:** 72
Ion chromatography exclusion
separation in **A10:** 662
Ion cluster beam technique
for gallium arsenide (GaAs) **A2:** 745
Ion conduction
and ternary molybdenum chalcogenides **A2:** 1077
Ion detection
in gas mass spectrometers **A10:** 153, 154-155
methods **A10:** 143-144
methods, electrical and photometric **A10:** 143-144
Ion energy
effect on sputtering yield **A9:** 107
Ion etching **A9:** 61-62
defined **A9:** 10
used in quantitative metallography of cemented carbides **A9:** 75
Ion exchange
defined **A10:** 675
principles of **A10:** 658-659
separation **A10:** 66, 164-165, 170, 249
Ion exchange recovery process, plating waste treatment **M5:** 317-318
reciprocal flow process **M5:** 317
resins, characteristics **M5:** 317
Ion exchange resins *See also* Resins **EM3:** 16
defined **EM2:** 24

Ion gun
for sputtering and compositional depth profiling **A7:** 255
Ion guns
aligned, by AES analysis **A10:** 550
for sputter removal of atoms, AES analysis **A10:** 554
used in thinning transmission electron microscopy specimens **A9:** 107
Ion imaging
as SIMS application **EL1:** 1085
Ion implantation *See also* Implantation **A13:** 498-500 **A18:** 850-858 **M5:** 422-426 **EL1:** 197-198, 201
advantages **A18:** 850
AEM determination of microstructures in **A10:** 484-487
aluminum-silicon alloys **A18:** 791
applications **A18:** 857-858 **M5:** 424-426
cutting **A18:** 858
end-use **A18:** 858
metalforming **A18:** 858
capabilities **M5:** 422-423
comparison of coatings for cold upsetting **A18:** 645
corrosion resistance **M5:** 425-426
damage, defect depth distribution for **A10:** 628
definition **A18:** 850
dose regimes for effective hardening **A18:** 858
equipment **A18:** 856-857 **M5:** 422-423
acceleration tube **A18:** 857
analyzing magnet **A18:** 857
components in schematics **A18:** 857
ion extractor **A18:** 856-857
system differences **A18:** 857
target chamber **A18:** 857
of gallium arsenide (GaAs) **A2:** 745
heavy, depth distribution by Rutherford backscattering spectrometry **A10:** 628
limitations **A18:** 850
limitations of **M5:** 424
microstructural properties of implanted regions **A18:** 852-855
crystalline-to-amorphous transformation **A18:** 853-854
displacement mixing **A18:** 854, 855
Gibbsian adsorption **A18:** 854-855
modeling of microstructural changes **A18:** 854-855
near-surface region defects **A18:** 852-853
preferential sputtering **A18:** 854, 855, 857
radiation-enhanced diffusion **A18:** 854
radiation-induced segregation **A18:** 854, 855
secondary phase formation **A18:** 855
microstructure effect on tribological properties **A18:** 855
abrasive wear **A18:** 855-856, 857, 858
adhesive wear **A18:** 855-856, 857, 858
corrosive wear **A18:** 856, 857, 858
fatigue wear **A18:** 856, 857, 858
penetration process **M5:** 422-424
channeling **M5:** 423-424
depth of distribution **M5:** 423-424
diffusion **M5:** 424
lattice damage **M5:** 423
range **M5:** 423
plasma-source ion implantation **A18:** 850
processes **A18:** 850-852
basic principles **A18:** 850-851
defects affecting tribological performance **A18:** 850
energy deposition profiles **A18:** 851
energy transfer **A18:** 850-851
range **A18:** 851, 852
sputtering **A18:** 851-852, 857
profile, phosphorus, in silicon **A10:** 623-624
Raman analysis **A10:** 133
resistance to cavitation erosion **A18:** 217
SIMS phosphorus depth profiles **A10:** 624
stainless steels **A18:** 716
titanium alloys **A18:** 778-780
tool steels **A18:** 642-643, 645
use of transmission electron microscopy for surface studies **A18:** 381
wear resistance, tests of **M5:** 425
Ion implantations **EM4:** 23
ceramic coatings for adiabatic diesel engines **EM4:** 992
Ion implanters **A13:** 499
Ion incidence angle
variation of sputtering yield with **A9:** 107
Ion lasers
for optical holographic interferometry **A17:** 417
Ion mass
effect on sputtering yield **A9:** 107
Ion mass spectroscopy and ion-scattering spectroscopy **A7:** 260
Ion microprobes **A10:** 161, 614
Ion microscopes **A10:** 614, 615
Ion milling
as FIM sample preparation **A10:** 584-585
Ion neutralization
defined **A10:** 675
Ion neutralization spectroscopy (INS) **EM3:** 237
Ion nitrided steels **A9:** 229
Ion nitriding *See* Plasma (ion) nitriding
Ion or plasma nitriding
jet engine components **A18:** 591
tool steels **A18:** 641, 739
Ion plating **M5:** 417-421
aluminum coatings applied by **M5:** 420-421
applications **M5:** 420-421
beneficial effects **M5:** 419-420
corrosion protection **M5:** 420-421
electroplating, precursor to **M5:** 421
equipment **M5:** 417-418
evaporation sources and
electron beam vaporization **M5:** 418
impact ionization, sputtering targets **M5:** 418
radiofrequency induction ionization **M5:** 418
reactive ionization **M5:** 418
resistance heating ionization **M5:** 418
techniques **M5:** 418
fasteners **M5:** 420
film properties, control of **M5:** 417
ion-gun system **M5:** 419
plasma density, enhanced **M5:** 419-420
process **M5:** 417-419
variables, control of **M5:** 418-419
stainless steel **M5:** 420-421
steel **M5:** 420-421
temperature, substrate, control of **M5:** 419
throwing power **M5:** 417, 420
titanium **M5:** 420-421
titanium nitride and titanium carbide coatings **M5:** 421
of uranium/uranium alloys **A13:** 821
vacuum coating process **M5:** 387
as vapor deposition coating **A13:** 457
Ion plating process **A18:** 840, 844-845
Ion quadrupole mass filter
in gas mass spectrometers **A10:** 153-154
Ion scattering
spectra **A10:** 604-606, 675
surface structure study by channeling/ blocking **A10:** 633
Ion scattering spectrometry (ISS) **EL1:** 1090-1092, 1107
Ion scattering spectroscopy *See also* Low- energy, ion-scattering spectroscopy **A7:** 259-260 **A13:** 1118
capabilities **A10:** 549
defined **A10:** 675
low-energy **A10:** 603-609
P/M applications **A7:** 260
with secondary ion mass spectroscopy **A7:** 260
Ion scattering spectroscopy (ISS)
for surface analysis **EM4:** 25
Ion sputtering **A10:** 565, 575 **A18:** 449, 453-454, 455
Ion(s) *See also* Anion; Cation
activity, defined **A13:** 1
adsorption, at electrode **A13:** 18
chloride, as cause, corrosion in concrete **A13:** 513
concentration, and corrosion rate **A13:** 229
and conductivity **EL1:** 89
defined **A13:** 8
diffusion, through scale **A13:** 70
exchange, defined **A13:** 8
interstitial **A13:** 65
major/minor, in seawater **A13:** 893-898
mobile, contamination by **EL1:** 45
nobility, and molten salt corrosion **A13:** 50
reduction potentials **A13:** 589
release, surgical implants **A13:** 1329
Ion(s) (continued)
zinc, as cathodic inhibitors **A13:** 495
Ion-beam machining **EM4:** 313
Ion-beam thinning
of transmission electron microscopy specimens **A9:** 107-108
Ion-carburized steels **A9:** 225-226
Ion-exchange **EM4:** 460-463
applications **EM4:** 461-463
chemical strengthening **EM4:** 461-462
dicing **EM4:** 462
eyeglass **EM4:** 460, 462
ophthalmics **EM4:** 462-463
optics **EM4:** 462-463
stuffing **EM4:** 461-462
channel waveguides **EM4:** 462-463
interdiffusion **EM4:** 460, 462
kinetics **EM4:** 460
potassium-for-sodium exchange accomplishing strengthening **EM4:** 463
practical aspects **EM4:** 461
ochre as carrier **EM4:** 461
vapor phase methods **EM4:** 461
process to strengthen glass containers **EM4:** 1084
refractive index increased by ions **EM4:** 463
self-diffusion **EM4:** 460, 462
self-diffusion coefficients **EM4:** 460, 461
thermodynamics **EM4:** 460-461
equilibrium constant **EM4:** 460-461
selectivity coefficient **EM4:** 460-461
Ion-exchange chromatography
defined **A10:** 675
as separation technique **A10:** 168, 170, 653, 658-659
Ion-exchange column
simulated pressurized water reactor water system **A8:** 423-424
Ion-exchange resins
core, materials for **A10:** 662
defined **A10:** 675
function in ion-exchange chromatography **A10:** 658-659
as sampling substrates, x-ray spectrometry **A10:** 94
Ion-induced Auger electron spectroscopy (IAES) **EM3:** 237
Ion-induced threshold drift
silicon oxide interface failures **A11:** 782
Ion-pair chromatography **A10:** 653-654, 675
Ion-plating using sputtering
to apply interlayers for solid-state welding **A6:** 165
Ion-scattering spectrometry (ISS) **A18:** 448-450
applications **A18:** 449-450, 451
equipment **A18:** 449
fundamentals **A18:** 448-449
neutralization of ions **A18:** 449
resolution **A18:** 449
sensitivity **A18:** 449
spectrum **A18:** 449, 450, 451
Ion-scattering spectroscopy (ISS) **EM3:** 237
Ion-selective electrode *See also* Ion-selective membrane electrodes
analysis of inorganic solids **A10:** 6
and direct and titrimetric potentiometry **A10:** 204
Ion-selective membrane electrodes *See also* Ion selective electrodes
determining selectivity **A10:** 182-183
membrane potential **A10:** 182
types of **A10:** 182
Ionic bond
defined **A10:** 675
definition **M6:** 10
Ionic bonding
chemical **EL1:** 92
Ionic charge
defined **A10:** 675
Ionic composition, of blood plasma
interstitial fluid, and intracellular fluid **A11:** 672
Ionic conductivity
in aqueous corrosion **A13:** 17
and electronic phenomena **EL1:** 93-96
Ionic conductors **EM4:** 18
applications **EM4:** 18
Ionic contamination
as PTH failure mechanism **EL1:** 1026-1027

Ionic contamination (continued)
silicon oxide interface failures by ... **A11:** 781
Ionic crystals
ESR studied ... **A10:** 263
Ionic displacement
in qualitative classical wet analysis ... **A10:** 168
as second-phase test method ... **A10:** 177
vs dilute acid, as digestion method ... **A10:** 176
Ionic film
formed by electrolytes ... **A9:** 50
Ionic materials
in electronic manufacturing ... **EL1:** 660
Ionic oxides
defect structure ... **A13:** 65
Ionic polarization
of insulators/dielectric materials ... **EL1:** 99-100
Ionic residues
cleaning of ... **EL1:** 660-661
Ionitriding ... **M1:** 542, 627
Ionization
defined ... **A17:** 384 **EM2:** 592
electron-impact, in gas mass spectrometer ... **A10:** 152-153
gage testing, as gas/leak detection device ... **A17:** 64
gages ... **A17:** 64, 67-68
inner-shell, as inelastic scattering process ... **A10:** 433
interferences ... **A10:** 29, 33, 34
limit, optical emission spectroscopy ... **A10:** 22
post-, EEL imaging ... **A10:** 450
potential, defined ... **A10:** 153
pre-, EEL imaging ... **A10:** 450
self-, of water ... **A10:** 203
suppressants, atomic absorption spectrometry ... **A10:** 48
suppressed in flame emission sources ... **A10:** 30
Ionization potential
shielding gases ... **A6:** 64
shielding gases for GTAW ... **A6:** 67
Ionizing effects
radiation units ... **A17:** 300
Ionizing radiation
defined ... **EL1:** 854
Ionograph
for residue measurement ... **EL1:** 667
Ionomer
polyaramid composites ... **EM3:** 284
surface preparation ... **EM3:** 279
Surlyn resins ... **EM3:** 279
Zn^{2+} ... **EM3:** 283
Ionomer resins ... **EM3:** 16
Ionomers *See also* Thermoplastic resins
applications ... **EM2:** 120
characteristics ... **EM2:** 120-123
competitive materials ... **EM2:** 120
costs ... **EM2:** 120
defined ... **EM2:** 24
design considerations ... **EM2:** 121-122
mechanical properties ... **EM2:** 120-121
melt viscosity vs shear rate ... **EM2:** 123
processing ... **EM2:** 122-123
production volume ... **EM2:** 120
suppliers ... **EM2:** 123
thermal properties ... **EM2:** 123
Ions
beam, LEISS analysis ... **A10:** 606-607
bombardment, effects ... **A10:** 611
bombardment, surface oxide enhancement ... **A7:** 259
as catalysts in decomposition of electrolytes ... **A9:** 51
chloride, determining nickel in samples containing ... **A10:** 201
common effect, in gravimetric analysis ... **A10:** 163
commonly used in secondary ion mass spectroscopy ... **A7:** 258
daughter, GC/MS scans of ... **A10:** 646
defined ... **A10:** 675
Ions (continued)
detection methods, spark source mass spectrometry ... **A10:** 143-144
detectors, electron multiplier as ... **A10:** 153-155
dichromate, UV/VIS analysis of chromium in ... **A10:** 70
effect of Chelons on ... **A10:** 164
exchange principles ... **A10:** 658-659
exchange separation, classical wet chemical analysis ... **A10:** 164-165
formation ... **A10:** 142, 640
fragment, in gas mass spectrometer ... **A10:** 153
hydrogen or hydroxyl, monitoring by acid-base titration ... **A10:** 172
inorganic, determined by ion chromatography ... **A10:** 663
interactions, determined by single-crystal x-ray diffraction ... **A10:** 344
metal, separation and determination of ... **A10:** 197, 200-201
monitoring ... **A10:** 153-154, 644
negatively charged, or anions ... **A10:** 659
numbers for atom probe microanalysis ... **A10:** 594
optical aberrations, in atom probe analysis ... **A10:** 595
parent, GC/MS scans ... **A10:** 646
positively charged, or cations ... **A10:** 659
probe, defined ... **A10:** 679
quantitative determination by electrogravimetry ... **A10:** 197
radical ... **A10:** 263, 265
rate of flow to electrode, effects in electrogravimetry ... **A10:** 198
removal, by electrogravimetry ... **A10:** 197, 200
sample, ion chromatography ... **A10:** 658, 663-665
secondary, defined ... **A10:** 681
separation, in constant current electrogravimetry ... **A10:** 198
solvated molecular, UV/VIS analyzed ... **A10:** 61
as sources, mass spectrometer ... **A10:** 142-143, 152-153
species of, defined ... **A10:** 675
transition-element, identification of valence states of ... **A10:** 253-266
transition-metal, ESR analysis of ... **A10:** 253, 254
Ions, specific
in stress-corrosion cracking ... **A11:** 207-208
Iosipescu shear test ... **EM3:** 392, 401, 402, 403
method and specimen ... **EM1:** 299-300
IPTS 68 *See* International Practical Temperature Scale
IR *See* Infrared; Infrared spectroscopy; Infrared spectroscopy
IR drop
defined ... **A13:** 24
IR ferrography
lubricant indicators and range of sensitivities ... **A18:** 301
Ir-La (Phase Diagram) ... **A3:** 2 • 264
Ir-Mo (Phase Diagram) ... **A3:** 2 • 264
Ir-Nb (Phase Diagram) ... **A3:** 2 • 265
Ir-Ni (Phase Diagram) ... **A3:** 2 • 265
Ir-Pd (Phase Diagram) ... **A3:** 2 • 265
Ir-Pt (Phase Diagram) ... **A3:** 2 • 266
Ir-Rh (Phase Diagram) ... **A3:** 2 • 266
Ir-Ru (Phase Diagram) ... **A3:** 2 • 266
Ir-Ta (Phase Diagram) ... **A3:** 2 • 267
Ir-Th (Phase Diagram) ... **A3:** 2 • 267
Ir-Ti (Phase Diagram) ... **A3:** 2 • 267
Ir-U (Phase Diagram) ... **A3:** 2 • 268
ir-v (Phase Diagram) ... **A3:** 2 • 268
Ir-W (Phase Diagram) ... **A3:** 2 • 268
Ir-Zr (Phase Diagram) ... **A3:** 2 • 269
IRE *See* Internal reflection elements
IRECA stainless steel alloys
cavitation erosion ... **A18:** 217, 218
IRG transition diagram ... **A18:** 491
defined ... **A18:** 11
Iridescence
defined ... **EM2:** 24
Iridium *See also* Precious metals
in acids ... **A13:** 805
annealing ... **A4:** 945, 946-947 **M4:** 760
anomaly, at Cretaceous-Tertiary boundary ... **A10:** 240-241
atomic interaction descriptions ... **A6:** 144
concentration found as function of depth in strata ... **A10:** 241
corrosion applications ... **A13:** 804-805
corrosion resistance ... **A13:** 804 **M2:** 669
destructive TNAA for ... **A10:** 239, 241
determined by controlled-potential coulometry ... **A10:** 209
electrical circuits for electropolishing ... **A9:** 49
environmental corrosion ... **A13:** 806
fabrication ... **A13:** 802-804
field evaporation in ... **A10:** 586, 587
fractured sheet, secondary cracks ... **A12:** 462
as gamma-ray source ... **A17:** 308
gravimetric finishes ... **A10:** 171
in halogens ... **A13:** 806
hardness ... **A4:** 947
intergranular fracture ... **A12:** 463
mechanical properties ... **A4:** 944
in medical therapy, toxic effects ... **A2:** 1258
in metal powder-glass frit method ... **EM3:** 305
oxidation resistance ... **A13:** 804
point defects observed by field ion microscopy ... **A10:** 588
polycrystalline, brittle intergranular fracture ... **A12:** 462
as precious metal ... **A2:** 688
properties ... **A13:** 804
pure ... **M2:** 740
pure, properties ... **A2:** 1117
as pyrophoric ... **A7:** 199
resources and consumption ... **A2:** 689
special properties ... **A2:** 694
thermal diffusivity from 20 to 100 °C ... **A6:** 4
thermal expansion coefficient ... **A6:** 907
toxicity ... **A7:** 207
vapor pressure, relation to temperature ... **A4:** 495 **M4:** 310
working of ... **A14:** 851
Iridium alloys
laser beam welding ... **M6:** 663
Iridium coating
molybdenum and tungsten ... **M5:** 660-661
Iridium-rhodium thermocouples
insulation ... **A2:** 883
properties and applications ... **A2:** 874
Iron *See also* ATOMET iron powders; ATOMET iron powders, specific types; Atomized iron powders; Cast iron; Cast irons; Ductile iron; Electrolytic iron powders; Ferrous casting alloys; Galvanized iron; Gray irons; Iron alloy powders; Iron alloy powders, specific types; Iron alloys; Iron alloys, specific types; Iron powder metallurgy materials, specific types; Iron powders; Iron powders, specific types; Iron-base alloys; Iron-carbon alloys; Iron-carbon melts; Iron-carbon-silicon alloys; Magnetic materials; Pure iron; Reduced iron powders; Sponge iron; Sponge iron powders; Ternary iron-base alloys; White irons; Wrought iron
as a beta stabilizer in titanium alloys ... **A9:** 458
as a reactive sputtering cathode material ... **A9:** 60
abrasive blasting of ... **M5:** 91
abrasive wear ... **A18:** 189
absorptivity ... **A6:** 265
acid cleaning of ... **M5:** 59-67
as addition to aluminum alloys ... **A4:** 843
addition to solid-solution nickel alloys ... **A6:** 575
additives ... **A7:** 614
age-hardened, FIM/AP study of precipitates in ... **A10:** 583

SUBJECTS OF THE INDEXED VOLUMES: ASM Handbook (designated by the letter "A"): **A1:** Properties and Selection: Irons, Steels, and High-Performance Alloys (1990); **A2:** Properties and Selection: Nonferrous Alloys and Special-Purpose Materials (1990); **A3:** Alloy Phase Diagrams (1992); **A4:** Heat Treating (1991); **A6:** Welding, Brazing, and Soldering (1993); **A7:** Powder Metallurgy (1984); **A8:** Mechanical Testing (1985); **A9:** Metallography and Microstructures (1985); **A10:** Materials Characterization (1986); **A11:** Failure Analysis and Prevention (1986); **A12:** Fractography (1987); **A13:** Corrosion (1987); **A14:** Forming and Forging (1988); **A15:** Casting (1988); **A16:** Machining (1989); **A17:** Nondestructive Testing and Quality Control (1989); **A18:** Friction, Lubrication, and Wear Technology (1992). **Metals Handbook, 9th Edition** (designated by the letter "M"): **M1:** Properties and Selection: Irons and Steels (1978); **M2:** Properties and Selection: Nonferrous Alloys and Pure Metals (1979); **M3:** Properties and Selection: Stainless Steels, Tool Materials and Special-Purpose Materials (1980); **M4:** Heat Treating (1981); **M5:** Surface Cleaning, Finishing, and Coating (1982); **M6:** Welding, Brazing, and Soldering (1983). **Engineered Materials Handbook** (designated by the letters "EM"): **EM1:** Composites (1987); **EM2:** Engineering Plastics (1988); **EM3:** Adhesives and Sealants (1990); **EM4:** Ceramics and Glasses (1991); **Electronic Materials Handbook** (designated by the letters "EL"): **EL1:** Packaging (1989).

Iron (continued)
alloyed with Ni ... **A16:** 835
as alloying addition to zirconium ... **A9:** 497
alloying, aluminum casting alloys ... **A2:** 132
alloying effect in titanium alloys ... **A6:** 508-509
alloying effect on nickel-base alloys ... **A6:** 589
alloying effects on copper alloys ... **M6:** 402
alloying, in microalloyed uranium ... **A2:** 677
alloying, in wrought titanium alloys ... **A2:** 599
alloying, nickel-base alloys ... **A13:** 641
alloying, ordered intermetallics ... **A2:** 913
alloying, wrought aluminum alloy ... **A2:** 52
alloying, wrought copper and copper alloys ... **A2:** 242
alloys, radial rim cracking ... **A11:** 285
as aluminum alloying element ... **A2:** 16
in aluminum alloys ... **A15:** 745
in aluminum, extended solubility of ... **A10:** 294-295
in aluminum matrix, x-ray maps of ... **A10:** 448
amorphous, cut with a wire saw ... **A9:** 25
analysis in copper-beryllium alloys, by thiocyanate method ... **A10:** 68
analysis, in lead alloys, by phenanthroline method ... **A10:** 66
aqueous corrosion ... **A13:** 512, 515
art casting of ... **A15:** 24
atmospheric corrosion ... **A13:** 82
atomic interaction descriptions ... **A6:** 144
atomic interactions and adhesion ... **A6:** 144
Auger chemical map for ... **A10:** 557
in austenitic stainless steels, formation of intermetallic phases ... **A9:** 284
-based alloys, relative hydrogen susceptibility ... **A8:** 542
bearing material systems ... **A18:** 746
  applications ... **A18:** 746
  bearing performance characteristics ... **A18:** 746
  load capacity rating ... **A18:** 746
binary, relative potency factors ... **A6:** 89
binder for WC ... **A16:** 72
biologic effects and toxicity ... **A2:** 1252
biological corrosion ... **A13:** 116-117
boriding ... **A4:** 441
cathodic protection by zinc ... **A13:** 467
cavitation erosion ... **A18:** 216, 217
chilled ,honing stone selection ... **A16:** 476
chilled, recommended machining specifications for rough and finish turning with HIP metal oxide ceramic insert cutting tools ... **EM4:** 969
chromium plating baths contaminated by ... **M5:** 175
cleaning solutions for substrate materials ... **A6:** 978
in composition of aluminum- silicon alloys ... **A18:** 785-786, 787, 788
compressed powdered ... **A2:** 765
concentration, pickling solutions effects of ... **M5:** 69-70
constant-current electrolysis ... **A10:** 200
contaminant of optical fibers ... **EM4:** 413-414
contamination ... **M6:** 321
contamination, in P/M stainless steels ... **A13:** 826-827
content in nickel-base and cobalt-base high-temperature alloys ... **A6:** 573
controlled amount present in glasses used for lighting ... **EM4:** 1034
in copper alloys ... **A6:** 753
in copper, analysis of- phases of ... **A10:** 294
corrosion, factors ... **A13:** 37-38
  corrosion in acid solution ... **A13:** 29
corrosion fatigue test specification ... **A8:** 423
corrosion, in dissolved oxygen ... **A13:** 29
corrosion of ... **M5:** 430
corrosion rate, and relative humidity ... **A13:** 908
corrosion rates, in seawater ... **A13:** 893
  oil/gas production monitoring ... **A13:** 1250
counts, oil/gas production monitoring
critical relative humidity ... **A13:** 82
crystals, liquid-lithium corrosion ... **A13:** 95
cutting tool materials and cutting speed relationship ... **A18:** 616
cyclic stress-strain curve for ... **A8:** 256
deformed 5%, twin band in ... **A9:** 690
deformed 14%, dislocation structure ... **A9:** 690
desulfurization ... **A15:** 75
determined by controlled-potential coulometry ... **A10:** 209
dietary, and lead toxicity ... **A2:** 1246
diffusion bonding ... **A6:** 156
diffusion welding ... **M6:** 677
dislocation tangles at 9% and 20% strain ... **A9:** 693
E-pH diagram ... **A13:** 22
effect, corrosion resistance, sintered austenitic stainless steels ... **A13:** 831
effect of acid concentration on corrosion rate of ... **A11:** 175
effect on Cu alloy machinability ... **A16:** 808
effect on maraging steels ... **A4:** 222
effects, electrolytic tough pitch copper ... **A2:** 269
effects, in cartridge brass ... **A2:** 301
electrical resistance applications ... **M3:** 641
electrochemical grinding ... **A16:** 543
electrochemical machining removal rates ... **A16:** 534
electroslag welding, reactions ... **A6:** 274
embedded ... **A13:** 1228
embedded, nickel alloys removal of ... **M5:** 672-673
emission spectrum, spectral complexity of ... **A10:** 22
enamels ... **EM3:** 302
energy-dispersive x-ray diffraction pattern ... **A9:** 703
erosion resistance ... **A18:** 201
as essential metal ... **A2:** 1250, 1252
estimated volume of signals produced by 20-keV electron beam ... **A10:** 500
etchants for ... **A9:** 170
evaporation fields for ... **A10:** 587
in Fe-Mo-C, thermal spray coating material ... **A18:** 832
ferromagnetism ... **A9:** 533
friction coefficient data ... **A18:** 71, 72
friction welding ... **A6:** 153, 152, 154
galvanic corrosion ... **A13:** 84
in gas-metal eutectic method ... **EM3:** 305
glass/metal seals ... **EM4:** 1037
gold plating baths contaminated by ... **M5:** 283
grain boundaries in ... **A9:** 609
gray, graphitic corrosion ... **A13:** 133-134
in hardfacing alloys ... **A18:** 759-760, 763, 765
in heat-resistant alloys ... **A4:** 510, 512
heating element, use in vacuum furnace ... **A4:** 500 **M4:** 316
hot dip galvanized coating content in, effects of ... **M5:** 324, 330
hot dip galvanizing of ... **M5:** 323-332
hydrochloric acid corrosion ... **A13:** 467
ICP-determined in plant tissues ... **A10:** 41
impurity found in gypsum ... **EM4:** 380
impurity in solders ... **M6:** 1072
as inclusion-forming, copper alloys ... **A15:** 96
induction soldering ... **A6:** 363
as inoculant ... **A15:** 105
iosotope composition and intensity ... **A10:** 146
K-absorption edge of ... **A10:** 416
for laser alloying ... **A18:** 866
lath martensite ... **A9:** 671
liftoff effect ... **A17:** 223
lubricant indicators and range of sensitivities ... **A18:** 301
in magnesium alloys with manganese ... **A9:** 427
martensitic structures ... **A9:** 668-672
melt, carbon activity ... **A15:** 62
melt, free energies of solution ... **A15:** 62
microstructure of annealed material ... **M6:** 22, 24
in moist chlorine ... **A13:** 1172
molten ... **A11:** 275
molten, and silica, slag from ... **A15:** 208
Monte Carlo electron trajectory simulation for EPMA effects in ... **A10:** 518
neutron and x-ray scattering, and absorption' compared ... **A10:** 421
and nickel, activities of ... **A15:** 51
nickel diffusion into, measurement of ... **A10:** 243
nickel plating bath contamination by ... **M5:** 208-210
nonmetallic elements determined in ... **A10:** 178
oriented dislocation arrays in thin foil ... **A9:** 128
oxide scale on, effect of abrading and polishing ... **A9:** 47
oxygen effect on coefficient of friction ... **A18:** 32
peritectic structures ... **A9:** 679-680
phases ... **A13:** 48
phosphate coating solutions concentration and removal ... **M5:** 442-443
photometric analysis methods ... **A10:** 64
physical properties related to thermal stresses ... **A4:** 605
pickling of ... **M5:** 68-82
polishing pure powder materials ... **A9:** 507
powder metallurgy materials, etching ... **A9:** 508-509
powder metallurgy materials microstructure ... **A9:** 509-511
precoating ... **A6:** 131
pressurized water reactor specification ... **A8:** 423
properties ... **A6:** 629
pure ... **M2:** 741-760
  composition ... **A18:** 806
  thermal properties ... **A18:** 42
pure, dimensional change on sintering ... **A7:** 480
pure, magnetization curve ... **A17:** 168
pure, properties ... **A2:** 1118-1127
pure, x-ray tubes ... **A17:** 302
qualitative tests to identify ... **A10:** 168
quantitative determination of carbon and sulfur ... **A10:** 223-224
recommended impurity limits of solders ... **A6:** 986
redox titration ... **A10:** 175
Rockwell hardness scale for ... **A8:** 76
rolled, shear bands ... **A9:** 686
room temperature bend strength of silicon nitride metal joints ... **EM4:** 526
rot, defined ... **A13:** 8
rusting, humidity and atmospheric pollution effects ... **A13:** 511
saltwater corrosion ... **A13:** 623, 893
secondary phase formation ... **A18:** 854
segregation and solid friction ... **A18:** 28
single crystal, cold rolled ... **A9:** 694
sintered, for brake linings ... **A18:** 570
in sintered metal powder process ... **EM3:** 304
sodium hydroxide corrosion ... **A13:** 1174
softening, early ... **A15:** 26
as solder impurity ... **EL1:** 638
solderable and protective finishes for substrate materials ... **A6:** 979
soldering ... **M6:** 1075
species weighed in gravimetry ... **A10:** 172
spectrometric metals analysis ... **A18:** 300
sponge ... **A7:** 14, 287
and steel, anaerobic corrosion ... **A13:** 43
structure, principles ... **A13:** 46-47
substitute for cobalt composition in hardfacing alloys ... **A18:** 762
temperature effect, corrosion rate ... **A13:** 910
thermal diffusivity from 20 to 100 °C ... **A6:** 4
in thermite, AAS analysis for ... **A10:** 56
as tin solder impurity ... **A2:** 520
TNAA of ... **A10:** 236, 238
toxicity ... **A2:** 1252 **A7:** 203-204
transgranular cleavage ... **A11:** 25
triggering anaerobic curing mechanism ... **EM3:** 113, 118
TWA limits for particulates ... **A6:** 984
ultrasonic welding ... **M6:** 746
use in flux cored electrodes ... **M6:** 103
used to microalloy beryllium ... **A9:** 390
vacuum heat-treating support fixture material ... **A4:** 503
vacuum nitrocarburizing ... **A4:** 407
vapor degreasing of ... **M5:** 45, 54
vapor pressure, relation to temperature ... **A4:** 495 **M4:** 309, 310
Vickers and Knoop microindentation hardness numbers ... **A18:** 416
volumetric procedures for ... **A10:** 175
in water, potential-pH diagram ... **A13:** 153
and WC ... **A16:** 71
weld overlay material ... **M6:** 806-807
wire, 98% reduction, cell structure ... **A9:** 688
wire, curly grain structure in ... **A9:** 688
workability ... **A8:** 165, 575
wrought 0.5% C, thermal properties ... **A18:** 42
wrought, ocean corrosion rate ... **A13:** 898
x-radiation, sulfur determination from ... **A10:** 94

Iron (continued)
in zinc alloys **A9:** 489 **A15:** 788
zinc plating baths contaminated by............. **M5:** 253
zone-melted, changes in residual strain hardening during isothermal recovery **A9:** 693
zone-refined, cold rolled and annealed **A9:** 698
zone-refined, grain shape distribution during isothermal grain growth **A9:** 698
zone-refined, grain size distribution during isothermal grain growth **A9:** 698
zone-refined, normal grain growth during isothermal anneals **A9:** 697
Iron (II) sulfate
decomposition temperatures......................... **EM4:** 55
Iron (III) nitrate
decomposition temperatures......................... **EM4:** 55
Iron (III) oxalate
decomposition temperatures......................... **EM4:** 55
Iron (III) sulfate
decomposition temperatures......................... **EM4:** 55
Iron (unpurified)
permeability **EM4:** 1162
resistivity.. **EM4:** 1162
saturation flux density **EM4:** 1162
Iron alloys *See also* Iron; Iron alloys, specific types
analysis for copper in, neocuproine method.. **A10:** 65
bcc, EXAFS spectra above K-edges............. **A10:** 416
bcc, helium implantation analysis............... **A10:** 485
diffraction techniques, elastic constants, and bulk values for **A10:** 382
diffusion of hydrogen...................................... **M6:** 45
diffusion welding **M6:** 682-683
electron beam welding **M6:** 638
eutectic joining **EM4:** 526
false silicon peak in EDS spectra of............ **A10:** 520
general welding characteristics...................... **A6:** 563
glow discharges for.. **A10:** 28
hardfacing **A6:** 789, 790-792, 796, 799, 806
hardfacing material **M6:** 773, 775-776
austenitic steels.. **M6:** 776
high-alloy irons **M6:** 776
martensitic steels.. **M6:** 776
pearlitic steels **M6:** 775-776
hydrogen peroxide dissolution medium for... **A10:** 166
laser beam welding... **M6:** 647
laser melt/particle inspection **M6:** 801-802
not readily diffusion bondable **A6:** 156
oxide-dispersion-strengthened materials... **A6:** 1039
oxyfuel gas cutting.. **M6:** 897
phase diagram **M6:** 22-24
resistance brazing.. **A6:** 341
seal design techniques.................................. **EM4:** 534
sodium peroxide fusion **A10:** 167
solderable and protective finishes for substrate materials.................................. **A6:** 979
spraying for hardfacing **M6:** 789
steel, AAS analysis for trace metals.............. **A10:** 55
sulfuric acid as sample dissolution medium **A10:** 165
trace element impurity effect on GTA weld penetration...................................... **A6:** 20
Iron alloys, specific types *See also* Iron; Iron alloys
17-4PH, diffraction techniques, elastic constants, and bulk values for **A10:** 382
17-14 CuMo, composition **A6:** 564
316, diffraction techniques, elastic constants and bulk values for **A10:** 382
410, diffraction techniques, elastic constants and bulk values for **A10:** 382
4340, diffraction techniques, elastic constants and bulk values for **A10:** 382
4340, local variations in residual stress from surface grinding................... **A10:** 390-391
6260, diffraction techniques, elastic constants and bulk values for **A10:** 382

Iron alloys, specific types (continued)
9310, diffraction techniques, elastic constants and bulk values for **A10:** 382
52100, diffraction techniques, elastic constants, and bulk values for **A10:** 382
A286, solidification structures in welded joints **A9:** 580
Armco iron friction welded to carbon steel **A9:** 156
Fe-0.2C-12Cr-1Mo
Charpy V-notch data for weld joints of EB welded alloy.............................. **A6:** 440
microhardness traverse test results **A6:** 441
Fe-0.8C, TEM micrograph............................. **A10:** 509
Fe-3Si, kinetics of secondary recrystallization during isothermal annealing **A9:** 698
Fe-3Si, single crystal, cold rolled.................. **A9:** 692
Fe-3Si, single crystal, cold rolled and annealed **A9:** 694-695
Fe-3Si, single crystal, dislocation substructure changes.................................. **A9:** 693
Fe-9Ni
Charpy V-notch absorbed energy vs. strength **A6:** 1018
cryogenic service.. **A6:** 1018
yield strength vs. temperature **A6:** 1016
Fe-10Ni-8Co, embrittlement testing in **A8:** 538-539
Fe-15Mo-4C-1B, refractory metal brazing, filler metal.. **A6:** 942
Fe-20Cr-25Ni-4.5Mo (GMAW), gas-metal arc welding........................... **A6:** 1018
Fe-20Ni and Fe-25Ni, diffusion measurements in... **A10:** 477
Fe-21Cr-6Ni-9Mn
Charpy V-notch absorbed energy vs. strength **A6:** 1018
composition.. **A6:** 1017
cryogenic service............................... **A6:** 1017-1018
Fe-21Cr-6Ni-9Mn-0.3N
composition.. **A6:** 1017
yield strength vs. fracture toughness....... **A6:** 1017
yield strength vs. temperature **A6:** 1016
Fe-22Mn
composition.. **A6:** 1017
cryogenic service.. **A6:** 1017
yield strength vs. fracture toughness....... **A6:** 1017
Fe-24.8Zn, cellular precipitation **A9:** 649
Fe-25Be (at%), spinodally decomposed......... **A9:** 654
Fe-30Ni-6Ti, lamellar precipitate **A9:** 129
Fe-35Ni-16Cr, precipitated particles in **A9:** 126
Fe-50Si, addition to weld pool **A6:** 60
Fe-60Mn-30Si, addition to weld pool **A6:** 60
Fe-80Mn, addition to weld pool **A6:** 60
Fe-B alloy, hardfacing...................................... **A6:** 807
Fe-Cr-Co permanent-magnet, FIM/AP images **A10:** 600
Fe-Cr-Ni-B alloys, laser hardfacing................ **A6:** 806
Haynes 556, AEM-EDS two-phase microanalysis of..................................... **A10:** 447
Incoloy 800, diffraction techniques, elastic constants, and bulk values for **A10:** 382
Inconel 600 U-bend, residual stress **A10:** 390
Inconel 718, minimum and maximum principal residual stress profiles......... **A10:** 392
Inconel 718, x-ray elastic constant determination for.................................. **A10:** 388
Invar, diffraction techniques, elastic constants and bulk values for **A10:** 382
JBK75, solidification structures in welded joints **A9:** 580
Kovar, elemental mapping **A10:** 532
M50, diffraction techniques, elastic constants and bulk values for **A10:** 382
M50 high-speed tool, diffraction-peak breadth at half height **A10:** 387

Iron alloys, specific types (continued)
René 95, diffraction-peak breadth at half height, as function of percent cold work **A10:** 387
Iron aluminides *See also* Ordered intermetallics
alloying effects **A2:** 923-924
corrosion resistance.................................. **A2:** 924-925
ductility **A2:** 922
environmental embrittlement......................... **A2:** 922
mechanical behavior **A2:** 920-922
phase stability **A2:** 920-925
potential applications **A2:** 925
slip behavior.. **A2:** 922
structural use, potential **A2:** 920
weldability **A2:** 924-925
Iron blast furnace
early **A15:** 24
Iron brake drum
brittle fracture of **A11:** 370
Iron Bridge (England).................................... **A15:** 18
Iron carbide *See* Cementite
Iron carbides
in as-cast iron castings **A11:** 359
Iron carbonyl powders
apparent density... **A7:** 297
effect of particle size on apparent density **A7:** 273
magabsorption measurement....................... **A17:** 144
Iron, cast *See* Cast iron
Iron castings *See also* Ductile cast iron; Gray cast iron
cleaning of **M1:** 101
coatings for... **M1:** 101-106
conversion coatings for **M1:** 104
diffusion coatings for..................................... **M1:** 104
dry-resin coatings for **M1:** 106
electroplating of.. **M1:** 101-103
flame spraying of **M1:** 103-104
hard facing of.. **M1:** 103
hot dip coatings for **M1:** 103
organic coatings for **M1:** 105-106
porcelain enameling of........................... **M1:** 104-105
Iron castings, failures of **A11:** 344-379
analysis procedures.. **A11:** 344
design, faulty... **A11:** 344-352
foundry practice and processing, effects of **A11:** 362-367
internal discontinuities **A11:** 354-358
material selection, faulty **A11:** 344-352
microstructure, effect of **A11:** 359-362
service conditions related to **A11:** 367-378
surface discontinuities **A11:** 352-353
Iron chip combustion accelerators................ **A10:** 222
Iron chloride
pickling solutions affected by **M5:** 74-76
Iron copper powders
composition **A7:** 464
as diffusion-bonded **A7:** 173
effect of particle size and shape **A7:** 188
effect of stabilizer on iron-iron contact formation.. **A7:** 187
liquid-phase sintering...................................... **A7:** 319
sintering **A7:** 365-366
Iron, cupola *See* Cupola iron
Iron, embrittlement of **A1:** 689-691
by oxygen **A1:** 689-690
by selenium **A1:** 691
by sulfur.. **A1:** 690-691
by tellurium... **A1:** 691
Iron, enameling *See* Enameling iron
Iron ferrites **A9:** 538
Iron flake powders
width change with vibratory ball milling time **A7:** 60
Iron, high-purity
magnetic applications..... **M3:** 591, 598, 599, 603, 607
Iron intermetallic compounds
in uranium alloys **A9:** 477

Iron, iron alloys
sintering, time and temperature **M4:** 796
Iron municipal castings
market trends **A15:** 44
Iron ore
US history of **A15:** 24-27
Iron ore (or concentrate)
Miller numbers **A18:** 235
Iron ores
acid dissolution mediums **A10:** 165
Iron oxide *See also* Oxides
angle of repose **A7:** 183
in ceramic tiles **EM4:** 926
as colorants **EM4:** 380
content, acid/basic slags **A15:** 357
effect on color of ceramics **EM4:** 5
Fe_2O_3 .... **EM4:** 3
in composition of textile products **EM4:** 403
in composition of wool products **EM4:** 403
purpose for use in glass manufacture **EM4:** 381
for temperature sensors **EM4:** 17
FeO
in composition of textile products **EM4:** 403
in composition of wool products **EM4:** 403
fluorescent magnetic particle inspection **A7:** 579
freeze drying **EM4:** 62
from steel-steam interaction **A11:** 603
function and composition for mild steel SMAW electrode coatings **A6:** 60
as heat-exchanger tube desposit **A11:** 633
and impact resistance **A7:** 464
inclusions, identification of **A11:** 38-39
magnetic, thermal conductivity **A11:** 604
metal-to-metal equilibria **A7:** 340
as pigment **EM3:** 179
reduction *See* Iron oxide reduction
as refractory, core coatings **A15:** 240
removal of **M5:** 76
rouge (FeO, Fe_2O_3, Fe_3O_4), purpose for use in glass manufacture **EM4:** 381
solubility, for cleaning **A13:** 381
specific properties imparted in CTV tubes **EM4:** 1040, 1042
to improve thermal conductivity **EM3:** 178
toxicity **A7:** 203
in vacuum-degassed steels **A11:** 328
Iron oxide (Fe_3O_4)
and oxidational wear **A18:** 287, 288
presence with O-ring corrosion **A18:** 549
Iron oxide, effects
magabsorption **A17:** 143
Iron oxide films
composition determined **A10:** 135
Iron oxide reduction **A7:** 52-53
ancient **A7:** 14
powder production by **A7:** 79-82
Iron oxide scale
on porcelain enameled steel sheet **M1:** 179
Iron oxide with manganese oxide **A9:** 184
Iron pentacarbonyl **A7:** 92, 135
properties **A7:** 92
Iron phase diagrams, discussion of
iron-chromium .... **A3:** 1 • 26
iron-chromium-nickel .... **A3:** 1 • 27
iron-manganese .... **A3:** 1 • 27
iron-manganese-carbon .... **A3:** 1 • 27
iron-nickel .... **A3:** 1 • 26
Iron phosphate coating
characteristics of **M5:** 434-435
crystal structure **M5:** 450, 455
equipment **M5:** 445, 448-449
immersion systems **M5:** 435, 440
room-temperature precleaning processes **M5:** 15
solution composition and operating conditions **M5:** 435, 442-444, 448-449
spray system **M5:** 434-435, 441, 448-449
weight, coating **M5:** 435-436, 448, 456
Iron phosphate coatings **A1:** 222
steel corrosion protection by **M1:** 754
steel sheet **M1:** 174
Iron phosphate prepaint treatment **M5:** 476-477, 490
Iron phosphating **A13:** 383, 385-386
Iron plantations
as early foundries **A15:** 26
Iron plating
nickel-iron alloy plating **M5:** 206-207
solution compositions and operating conditions **M5:** 663-664
tantalum and niobium **M5:** 663-664
Iron plating of specimens
for edge retention **A9:** 32
Iron powder *See* P/M steels
compressibility **A14:** 190
contamination **A14:** 190-191, 204
function and composition for mild steel SMAW electrode coatings **A6:** 60
for metal powder cutting **A14:** 727
Iron powder alloys *See also* Iron; Iron powder alloys, specific types; Iron powders; Iron powders, specific types
chemical analysis and sampling **A7:** 248
compound of friction materials **A7:** 701
dimensional changes on sintering **A7:** 480-481
P/M replacement **A7:** 670
sponge, green density and green strength **A7:** 289
Iron powder alloys, specific types *See also* ATOMET iron powders; Atomized iron powders; Cast iron; Electrolytic iron powders; Iron; Iron powder alloys; Iron powders; Iron powders, specific types; reduced iron powders; Sponge iron powders
A-Met 1000, flow rate through Hall and Carney funnels **A7:** 279
Ancor MH-100, increasing three lubricants **A7:** 191
Ancor MH-100, lubricant effect **A7:** 190
Ancor MH-100, lubricant mixing time and compacting pressure, effect on stripping pressure **A7:** 191
F-0000, effects of steam treating **A7:** 466
F-0000, mechanical properties **A7:** 465
F-0005, mechanical properties **A7:** 465
F-0008, effects of steam treating **A7:** 466
F-0008, mechanical properties **A7:** 465
FC-0200, mechanical properties **A7:** 464
FC-0205, mechanical properties **A7:** 465
FC-0208, mechanical properties **A7:** 465
FC-0505, mechanical properties **A7:** 465
FC-0508, mechanical properties **A7:** 465
FC-0700, effects of steam treating **A7:** 466
FC-0708, effects of steam treating **A7:** 466
FC-0808, mechanical properties **A7:** 465
FC-1000, mechanical properties **A7:** 465
FN-0200, mechanical properties **A7:** 465
FN-0205, mechanical properties **A7:** 465
FN-0208, mechanical properties **A7:** 465
FN-0400, mechanical properties **A7:** 465
FN-0405, mechanical properties **A7:** 465
FN-0408, mechanical properties **A7:** 466
FN-0700, mechanical properties **A7:** 466
FN-0705, mechanical properties **A7:** 466
FN-0708, mechanical properties **A7:** 466
FX-1005, mechanical properties **A7:** 466
FX-1008, mechanical properties **A7:** 466
FX-2000, mechanical properties **A7:** 466
FX-2005, mechanical properties **A7:** 466
FX-2008, mechanical properties **A7:** 466
Hoeganaes MH-100, pressure and green density for **A7:** 298
MH-100, flow rate (Hall and Carney funnels) **A7:** 279
MP-35HD, flow rate (Hall and Carney funnels) **A7:** 279
RZ strip, rolling **A7:** 401
Iron Powder Core Council **A7:** 19
Iron powder metallurgy materials, specific types
Ancormet 100, pressed and sintered **A9:** 517
Ancormet 101, carbon-reduced iron ore **A9:** 513
Ancorsteel 1000, fracture surface **A9:** 515
Ancorsteel 1000, pressed and sintered **A9:** 517
Ancorsteel 1000, pressed, unsintered **A9:** 514
Ancorsteel 1000, water-atomized and annealed **A9:** 513, 528
Ancorsteel 1000B, water-atomized and double-annealed **A9:** 513
Atomet 28, porosity **A9:** 513
Atomet 28, powder **A9:** 528
Iron powder metallurgy materials, specific types (continued)
Atomet 28, pressed and sintered **A9:** 515-517
Atomet 28, with different amounts of copper and carbon added **A9:** 518-519
copper infiltrated **A9:** 519
Fe-1.5 graphite **A9:** 527
MH-100, carbon-reduced iron ore **A9:** 513
MH-100, carbon-reduced sponge iron powder **A9:** 528
MH-100, pressed and sintered **A9:** 516
MP35, pressed and sintered **A9:** 516
MP35HD, porosity **A9:** 513
Pyron 100, hydrogen reduced **A9:** 513
Pyron 100, pressed and sintered **A9:** 516
Pyron 100, with different amounts of copper added **A9:** 518
Pyron D63, hydrogen reduced **A9:** 513
Pyron D63, hydrogen reduced mill scale **A9:** 528
Pyron D63, pressed and sintered **A9:** 516
SCM A283, electrolytic powder **A9:** 514
sulfur added **A9:** 521
water-atomized powder **A9:** 513
water-atomized, with MnS added, pressed and sintered **A9:** 521
Iron powder strip
roll compacted **A7:** 408
Iron powder(s) *See also* ATOMET iron powders; ATOMET iron powders, specific types; Atomized iron powders; Cast iron; Electrolytic iron powders; Iron powder alloys; Iron powder alloys, specific types; Iron powders, specific types; Reduced iron powders; Sponge iron powders
acid insoluble chemical analysis **A7:** 242
annealing **A7:** 182-184
apparent and tap densities **A7:** 189, 297
applications **A7:** 17, 76-77, 79, 203, 617-621, 667-670
atomized, compressibility curve **A7:** 287
in automotive applications **A7:** 17, 617-621
brittle cathode process **A7:** 72
business machine applications **A7:** 667-670
by carbonyl vapormetallurgy processing **A7:** 92-93
by Domfer process **A7:** 89-92
by electrolysis **A7:** 93-96
by fluidized bed reduction **A7:** 96-98
by Hoganas process **A7:** 79-82
by Pyron process **A7:** 82-83
by QMP process **A7:** 86-89
by water atomization of low carbon iron **A7:** 83-86
as carrier core, copier powders **A7:** 584
coercive force of cores from **A7:** 640
cold isostatic pressing dwell pressures **A7:** 449
compacted by supersonic vibrations **A7:** 306
compacting grade, properties **A7:** 94
compacts, effects of lubricant on density and strength of **A7:** 289
composition **A7:** 464
composition-depth profile **A7:** 256
contact angle with mercury **A7:** 269
decomposition of carbonyls **A7:** 54
density and hardness **A7:** 511
dependence of green strength on apparent density in **A7:** 288
dimensional change **A7:** 292
dust explosion **A7:** 197
effect of hot pressing temperature and pressure on density of **A7:** 504
effect of lubrication on flow **A7:** 191
electrical and thermal conductivity **A7:** 742
electrolytic **A7:** 71, 72, 273
explosive isostatic compaction of **A7:** 305
for flame cutting **A7:** 842
fluorescent magnetic particle inspection of **A7:** 579
for food enrichment **A7:** 614-616
functional parts, appliances **A7:** 623
green strength and green density in **A7:** 289
high-ductility, sintering **A7:** 340-341
as incendiary **A7:** 603
internal particle porosity **A7:** 85
joint fill **A7:** 821-822
low-carbon **A7:** 83-86

Iron powder(s) (continued)
in magnetic separation of seeds **A7:** 589
magnetization of **A7:** 640
markets **A7:** 571
maximum permeability of cores from **A7:** 640
metal-to-metal oxide equilibria **A7:** 340
milling in water **A7:** 63-64
moderate explosivity class **A7:** 196-197
particles, by hydrogen reduction **A7:** 247
Poisson's ratio vs. theoretical density **A7:** 414
position in P/M industry **A7:** 23
pressure and green density **A7:** 298
pressure requirements for hot pressing **A7:** 505
pressure-density relationships **A7:** 298, 299
production **A7:** 23, 79-99
production capacities **A7:** 23
properties for metal powder cutting **A7:** 843
pyrophoricity **A7:** 199
reduced **A7:** 273, 289
regulation of **A7:** 615
relative biological values **A7:** 615
relative density **A7:** 305
shipments **A7:** 23, 24, 570
sintered in hydrogen, effect of hydrogen chloride **A7:** 319
sintering **A7:** 361-362
soft magnetic components from **A7:** 640
sponge, compressibility curve **A7:** 287
spray drying applications **A7:** 76-77
in starter motor case **A7:** 620
structural, compositions **A7:** 464
for submerged arc process **A7:** 822
sulfides or manganese sulfide additives for machinability **A7:** 486, 487
tap density **A7:** 189, 277, 297
technology **A7:** 17-18
testing iron content **A7:** 247-248
toxicity, exposure limits **A7:** 203-204
toxicity, reactions and disease symptoms **A7:** 203-204
undersintered, atomized and diffusion alloyed **A7:** 486
for welding **A7:** 817-818
Iron powder(s), specific types *See also* Iron powder alloys; Iron powder alloys, specific types
45P, dimensional change on sintering **A7:** 480
Ancorsteel 1000, Charpy impact response **A7:** 415
Ancorsteel 1000 powder, mercury porosimetry determinations **A7:** 268
ATOMET 28, dimensional change on sintering **A7:** 481
F-0000, dimensional change on sintering **A7:** 480
F-0008, dimensional change on sintering **A7:** 480
FC-0208, dimensional change on sintering **A7:** 481
FN-0208, as-sintered characteristics **A7:** 485
FN-0208, dimensional change on sintering **A7:** 481
M P 61 LA welding rod grade **A7:** 91
MH100, dimensional change on sintering **A7:** 481
MP 32, particles **A7:** 92
MP 32, properties **A7:** 91
MP 35HD, properties **A7:** 91
MP 36S, properties **A7:** 91
MP 39, properties **A7:** 91
MP 52, properties **A7:** 91
MP 55, properties **A7:** 91
MP 62 welding rod grade **A7:** 91
MP 64 welding rod grade **A7:** 91
MP 65 welding rod grade **A7:** 91
SCM A-210, electrolytic, compacting properties **A7:** 94
Iron rotating bands **A7:** 683-684
mandrel expansion testing **A7:** 684
Iron scrap, recycling of **A1:** 1023
factors influencing scrap demand **A1:** 1024
purchased scrap supply **A1:** 1024-1026
scrap use by industry **A1:** 1023-1024
Iron sheet
with enameled coating **A9:** 201
inverse pole figure **A9:** 704
Iron sheet, electrodeposited
dendritic structure **A7:** 72
Iron silicoborides **A4:** 440
Iron soldering
definition **M6:** 10
Iron soldering (INS) **A6:** 349-350
applications **A6:** 349
constant-voltage soldering irons **A6:** 349
damage **A6:** 349-350
definition **A6:** 1210
equipment **A6:** 349
materials for soldering iron tips **A6:** 349
soldering iron selection **A6:** 349-350
soldering iron tip and joint thermal profile **A6:** 350
temperature-controlled soldering irons **A6:** 349
variable-temperature soldering irons **A6:** 349
Iron solution test
for tinplate **A13:** 781
Iron sulfate
use in color etching **A9:** 142
Iron sulfate, pickling inhibited by *See* Ferrous sulfate, pickling inhibited by
Iron sulfide
particles **A12:** 219
Iron sulfides
mixed with manganese sulfides **A9:** 184
on a steel corrosion test coupon **A9:** 162
Iron toxicity
disposition **A2:** 1252
Iron ‡m+ chromium
chemical compositions per ASTM specification B550-92 .... **A6:** 787
Iron-adhesion measurements **A6:** 144
Iron-alloy microstructures
iron-0.8% carbon .... **A3:** 1 • 21
iron-24.8% zinc .... **A3:** 1 • 22
Iron-aluminum
welding (bonding) **A6:** 145
Iron-aluminum alloys
environmental effect on yield stress and strain hardening rate in **A11:** 225
experimental **A12:** 365
for flame cutting and lancing **A7:** 843
magnetic applications **M3:** 598, 599, 601-602
as magnetically soft materials **A2:** 769-770
stabilizer effect on degree of blending **A7:** 188
surfactant effect **A7:** 188
Iron-aluminum interfacial layer, aluminum coating
characteristics and effects of **M5:** 333-336, 338, 345-347
Iron-base alloys *See also* Binary iron-base alloys; Cast irons; Ferrous casting alloys; Iron; Iron-carbon alloys; Iron-carbon melts; Iron-carbon-silicon alloys; Iron-copper-carbon alloys; Superalloys, iron-base; Ternary iron-base alloys
aluminum coating of **M5:** 341
binary, thermodynamics of **A15:** 61-62
bismuth effect, gas solubility **A15:** 83
cavitation erosion **A18:** 768, 769
composition **M4:** 797
corrosion of **A11:** 632
corrosion rate in water **A13:** 207
corrosion rates, molten salts **A13:** 51
corrosion resistance **A2:** 778
crystal structure effect, on corrosion **A13:** 126
effect, oxyfuel gas cutting **A14:** 722
electron-beam welding **A6:** 855, 865
electronic applications **A6:** 990-991, 998
embrittlement by copper during welding **A11:** 721
Iron-base alloys (continued)
explosive forming **A14:** 782
forging temperatures and forgeability **A14:** 232
forming practice **A14:** 782
fretting wear **A18:** 248, 250
heat-resistant, forging of **A14:** 104, 232-233
HERF forgeability **A14:** 104
hydrogen damage prevention **A13:** 329
lubrication of tool steels **A18:** 738
machining, ultrahard materials for **A2:** 1010
melting heat for **A15:** 376
peritectic transformations **A15:** 129
pouring temperatures **A15:** 283
reductions by cold swaging **A14:** 128
sintering temperatures **M4:** 797
solid-metal embrittlement of **A1:** 721
stripping of **M5:** 218
sulfur effect, nitrogen solubility **A15:** 83
superplasticity in **A14:** 868-872
susceptibility to embrittlement **A1:** 689
tellurium effect, nitrogen solubility **A15:** 83
temperature-time curve **A15:** 185
thermal spray coating recommended **A18:** 832
thermodynamic properties **A15:** 61-70
Unicast process for **A15:** 251
Iron-base alloys, heat treating *See also* Heat-resisting alloys, heat treating
aging **M4:** 661-662
annealing **M4:** 655, 660
castings **M4:** 660-661
procedure modification **M4:** 661-664
solution treating **M4:** 661-662
stress relieving **M4:** 655, 660
temperatures **M4:** 655, 656, 657, 660-665
Iron-base alloys, specific types
Fe-B, laser cladding **A18:** 867
Fe-Cr-Mn-C
laser cladding **A18:** 867
laser melting **A18:** 865
Iron-base hardfacing alloys
laser cladding materials **A18:** 866-867
Iron-base heat-resistant alloys
machining **A16:** 738, 741-743, 746-747, 749-758
Iron-base magnet alloys
FIM/AP analysis of **A10:** 597-599
spinodal decomposition of **A10:** 598
Iron-base shape memory alloys
future prospects **A2:** 901
Iron-base superalloys
AEM analysis of γ-austenite matrix **A10:** 453, 455
alloying elements, effect of **A1:** 951
ATEM image, spot diffraction pattern, and for **A10:** 438, 440
indexed schematic diffraction pattern
compositions of **A1:** 965
diffraction pattern and bright-field and centered dark-field images of **A10:** 442
dispersion-strengthened alloys
compositions **A1:** 973
properties **A1:** 976
fracture/failure causes illustrated **A12:** 217
with high nickel content **A1:** 959
microstructure **A1:** 959, 961-962
physical properties **A1:** 963-964
stress-rupture properties **A1:** 962
tensile properties **A1:** 958-959, 960-961
two-phase microanalysis in **A10:** 447
Iron-based
dispersion-strengthened materials **A7:** 722-727
friction materials, nominal compositions **A7:** 702, 703
hardfacing alloys **A7:** 828-829
microcrystalline alloys **A7:** 795-797
P/M parts, secondary operations performed on **A7:** 451-462
powder alloys, for automotive applications **A7:** 617-621

SUBJECTS OF THE INDEXED VOLUMES: ASM Handbook (designated by the letter "A"): **A1:** Properties and Selection: Irons, Steels, and High-Performance Alloys (1990); **A2:** Properties and Selection: Nonferrous Alloys and Special-Purpose Materials (1990); **A3:** Alloy Phase Diagrams (1992); **A4:** Heat Treating (1991); **A6:** Welding, Brazing, and Soldering (1993); **A7:** Powder Metallurgy (1984); **A8:** Mechanical Testing (1985); **A9:** Metallography and Microstructures (1985); **A10:** Materials Characterization (1986); **A11:** Failure Analysis and Prevention (1986); **A12:** Fractography (1987); **A13:** Corrosion (1987); **A14:** Forming and Forging (1988); **A15:** Casting (1988); **A16:** Machining (1989); **A17:** Nondestructive Testing and Quality Control (1989); **A18:** Friction, Lubrication, and Wear Technology (1992). **Metals Handbook, 9th Edition** (designated by the letter "M"): **M1:** Properties and Selection: Irons and Steels (1978); **M2:** Properties and Selection: Nonferrous Alloys and Pure Metals (1979); **M3:** Properties and Selection: Stainless Steels, Tool Materials and Special-Purpose Materials (1980); **M4:** Heat Treating (1981); **M5:** Surface Cleaning, Finishing, and Coating (1982); **M6:** Welding, Brazing, and Soldering (1983). **Engineered Materials Handbook** (designated by the letters "EM"): **EM1:** Composites (1987); **EM2:** Engineering Plastics (1988); **EM3:** Adhesives and Sealants (1990); **EM4:** Ceramics and Glasses (1991); **Electronic Materials Handbook** (designated by the letters "EL"): **EL1:** Packaging (1989).

Iron-based (continued)
self-lubricating sintered bearings
chemical composition **A7:** 706
tool steels, HIP temperatures and process times **A7:** 437
Iron-based heat-resistant alloys *See* specific types and Heat-resistant metals and alloys
Iron-bronze, bearing material systems **A18:** 746
applications **A18:** 746
bearing performance characteristics **A18:** 746
load capacity rating **A18:** 746
Iron-bronze sintered bearings **A7:** 706, 708
Iron-bronze-graphite
bearing material systems **A18:** 746-747
Iron-carbon
bearing material systems **A18:** 746-747
phase diagram .... **A3:** 1 • 25
system .... **A3:** 1 • 23-25
transformation temperatures .... **A3:** 1 • 24, 25
Iron-carbon (Fe-C) alloys
microstructure affected by carbon content **A18:** 874
Iron-carbon alloy
1.4% C, different heat treatments compared **A9:** 194
Iron-carbon alloys
cooling curve, interpretation **A15:** 182
diagram, with letter notations **A15:** 67-68
effects of combined carbon **A7:** 363
heat treatability **A7:** 79
liquid, structure of **A15:** 168-169
-lubricant systems **A7:** 186
miter gears for postage meters **A7:** 668
P/M drive gear **A7:** 667
phase diagram **A15:** 629
quench-aging in **A1:** 692-693
solubility of hydrogen and nitrogen **A15:** 82
thermodynamics of **A15:** 61-62
Iron-carbon alloys, image analysis plots
effects of detected area fraction **A10:** 309
Iron-carbon eutectic **A9:** 620
Iron-carbon melts
carbon addition to **A15:** 72-73
carbon concentration profiles **A15:** 73
steel dissolution in **A15:** 73-74
Iron-carbon phase diagram **A1:** 126-127
Iron-carbon-manganese alloys
thermodynamics of **A15:** 65
Iron-carbon-phosphorus alloys
thermodynamics of **A15:** 65
Iron-carbon-silicon alloys
composition effects **A15:** 172
coupled growth lines, constructed **A15:** 171
gray irons as **A15:** 629
phase diagram, simplified **A15:** 630
thermodynamics of **A15:** 64-65
Iron-carbon-sulfur alloys
nitrogen solubility of **A15:** 84
Iron-cast
oxyacetylene welding **A6:** 281
Iron-cementite
phase diagram .... **A3:** 1 • 25
system .... **A3:** 1 • 23-25
transformation temperatures .... **A3:** 1 • 24, 25
Iron-chromium alloys *See also* Ferritic stainless steels **M3:** 269, 272-273
chromium effect on corrosion **A13:** 124
corrosion resistance **A13:** 47-48
crystal structure effects on corrosion **A13:** 126
electropolishing of **M5:** 307
glass-to-metal seals **EM3:** 302
as heat-resistant high alloy **A15:** 724, 733
oxide scale formation **A13:** 98
polarization curve **A9:** 146
properties of **A1:** 920
sigma-phase embrittlement **A12:** 132
Iron-chromium phase diagram **A6:** 447
Iron-chromium powders
in automotive industry **A7:** 620-621
by STAMP process **A7:** 548
Iron-chromium-aluminum alloy
solidification in **A12:** 140, 160
Iron-chromium-aluminum alloys
creep rupture testing of **A8:** 302
electrical resistance, properties **A2:** 823
as heating alloys **A2:** 828-829
in sheathed heaters **A2:** 831
Iron-chromium-aluminum-yttrium coating
superalloys **M5:** 376
Iron-chromium-carbon alloys
boriding **A4:** 441
carburizing **A4:** 368-369
Iron-chromium-carbon equilibrium diagram **A6:** 433, 436
Iron-chromium-cobalt alloys
composition profiles for **A10:** 600
as permanent magnet materials **A2:** 790
Iron-chromium-nickel alloys *See* Nickel-iron-chromium alloys **M3:** 269, 273-276
arc welding **M6:** 364-367
composition of **A1:** 912
effect of nickel additions on SCC resistance **A8:** 530
as heat-resistant high alloy **A15:** 724, 733
properties of **A1:** 920
sigma phase **A9:** 136
Iron-chromium-nickel heat-resistant casting alloys
compositions of **A9:** 330
etchants for **A9:** 330-332
identification of ferrite **A9:** 333-334
identification of secondary phases **A9:** 332-333
microstructures **A9:** 333-334
Iron-chromium-nickel heat-resistant casting alloys, specific types
9Cr-1Mo **A9:** 333
9Cr-1Mo, air cooled and tempered **A9:** 335
9Cr-1Mo, as sand cast **A9:** 335
HE-14, creep tested **A9:** 335
HF, microstructure **A9:** 333
HF-25, as cast and creep tested **A9:** 337
HF-33, different sections of a casting compared, as cast and creep tested **A9:** 336-337
HF-33, fractured **A9:** 336-337
HF-34, lamellar structure in creep test specimen **A9:** 337
HH, different sections of a casting compared, as sand cast and creep tested **A9:** 337-338
HH, fractured **A9:** 337-338
HH, microstructure **A9:** 333
HK, microstructure **A9:** 333
HK-28, lamellar structures regenerated by slow cooling **A9:** 340
HK-35, as-cast and fractured **A9:** 339-340
HK-44, creep tested and fractured **A9:** 339
HN, as-cast and fractured **A9:** 340
HN, etching **A9:** 331-332
HN, microstructure **A9:** 333
HP, etching **A9:** 331-332
HT, etching **A9:** 331-332
HT, microstructure **A9:** 333
HT-44, as-cast and after creep testing **A9:** 341
HT-56, after creep testing **A9:** 341
HT-57, as-cast **A9:** 341
HU, etching **A9:** 331-332
HW, as-cast and after aging **A9:** 340
HW, etching **A9:** 332
HW, microstructure **A9:** 333
HX, etching **A9:** 332
Iron-chromium-nickel heat-resistant castings **A1:** 922-925
Iron-chromium-nickel system .... **A3:** 1 • 25
Iron-cobalt alloys *See* Magnetic materials
magnetic applications **M3:** 603, 605, 608
magnetic properties **A2:** 775
as magnetically soft materials **A2:** 774-776
strip, properties **A2:** 777
Iron-cobalt alloys, specific types
Alloy 2V-49Co-49Fe, as magnetically soft materials **A2:** 774-775
Alloy 27Co-0.6Cr-Fe, as magnetically soft materials **A2:** 775
Iron-cobalt-chromium
alloys, as special, low-expansion alloys **A2:** 895
Iron-copper
bearing material systems **A18:** 746-747
Iron-copper alloys, powder metallurgy materials
microstructures **A9:** 510
Iron-copper-carbon alloys *See also* Iron-base alloys
cooling rates **A14:** 200
mechanical property/fatigue data **A14:** 200-201
sulfur/carbon effects **A14:** 200
Iron-copper-carbon alloys, powder metallurgy materials
microstructures **A9:** 510
Iron-copper-carbon-lubricant systems **A7:** 186
Iron-copper-graphite powders
sintering **A7:** 366
Iron-deficiency anemia **A7:** 614
Iron-graphite mixtures
microstructures **A9:** 509-510
Iron-graphite powders
dimensional change on sintering **A7:** 480
microstructure **A7:** 315
sintering of **A7:** 362-365
Iron-iron carbide-silicon system **M1:** 3-4
Iron-iron carbide-silicon ternary phase diagram **A13:** 566
Iron-iron contact formation
effect of stabilizer **A7:** 187
Iron-lithium mixtures
decomposition temperatures **EM4:** 55
Iron-magnesium-silicon alloy, cerium bearing
digitized image **A9:** 139, 162
Iron-nickel alloy vacuum-coated films **M5:** 395
Iron-nickel alloy wires, transmission pinhole photographs **A9:** 702
microstructures **A9:** 309
Iron-nickel alloys *See* Magnetic materials
expansion characteristics **A2:** 893
magnetic applications **M3:** 602-603, 604, 608
thermal expansion **A2:** 890
Iron-nickel alloys, Fe-14-Ni
boriding **A4:** 441
Iron-nickel alloys, specific types
Fe 3.5Ni
Poisson's ratio **M3:** 748
Young's modulus **M3:** 743
Fe-5Ni
Poisson's ratio **M3:** 748
Young's modulus **M3:** 743
Fe-9Ni
Poisson's ratio **M3:** 748
Young's modulus **M3:** 743
Iron-nickel Dumet alloys
glass-to-metal seals **EM3:** 302
Iron-nickel powders
composition **A7:** 464
prealloyed, powder pole piece from **A7:** 641
Iron-nickel-carbon alloys
atomized, particle boundaries **A7:** 486
microstructure **A7:** 488
Iron-nickel-chromium alloys **M3:** 269, 276-277
applications and properties **A2:** 439-441
arc welding **M6:** 364-367
composition of **A1:** 912-913
Fe-35Ni-15Cr
heat-resistant alloy applications **A4:** 515
salt pot composition **A4:** 335, 337
Fe-35Ni-18Cr-44Fe, ribbon material in heat-treating furnaces **A4:** 472
Fe-35Ni-20Cr, heat-resistant alloy applications **A4:** 515
as heat-resistant high alloy **A15:** 724, 733
as low-expansion alloys **A2:** 894
properties of **A1:** 920-921
Iron-nickel-chromium base alloys
electron-beam welding **A6:** 869
Iron-nickel-chromium glass seal alloys
types, compositions, and thermal expansions for **A2:** 894
Iron-nickel-chromium ternary diagram **A6:** 458-459, 460, 461
Iron-nickel-chromium-titanium alloys
as hardenable low-expansion alloys **A2:** 895
thermoelastic coefficients **A2:** 895
Iron-nickel-cobalt alloys
as low-expansion alloys **A2:** 894-895
Iron-nickel-cobalt ASTM F 15 alloy
composite speedbrake **EM3:** 563
glass-to-metal seals **EM3:** 301
Iron-nickel-copper-molybdenum powders
diffusion-bonded **A7:** 173
Iron-nitrogen alloys
quench-aging in **A1:** 692-693
Iron-oxygen phase diagram **A15:** 89
Iron-phosphorus alloys
microstructural analysis **A7:** 488

Iron-phosphorus alloys (continued)
powder metallurgy materials
microstructures **A9:** 510
powder metallurgy materials, pressed and sintered **A9:** 520
Iron-silicon alloys
thermodynamics of **A15:** 62
Iron-silicon soft magnetic alloys
microstructural analysis **A7:** 488
Iron-silicon transformer sheet
effect of texture on anisotropic properties **A9:** 700
Iron-to-iron oxide equilibria
sintering **A7:** 340
Iron-vanadium carbide alloys
insoluble particle effects **A15:** 142
Iron-wrought
oxyacetylene welding **A6:** 281
Iron-zinc intermetallic compound
brittle fracture from **A11:** 100
Ironing *See also* Tube drawing
aluminum alloys **A14:** 797
in deep drawing **A14:** 508
defined **A14:** 8
multipass, tube drawing **A14:** 335-336
multiple-die, for beverage cans **A14:** 335
for reducing shells **A14:** 586
Ironmaking **A1:** 107, 109
Irons
bending fracture, oxide inclusions **A12:** 220
blowholes **A12:** 221
bright-field, dark-field, and SEM images compared **A12:** 92
cast, granular brittle fracture **A12:** 103
cleavage fractures, twist boundary **A12:** 17
coinability of **A14:** 183
ductile, figure numbers for **A12:** 216
embrittlement sources **A12:** 123
enamels **EM3:** 303
fractographs **A12:** 219-224
fracture types, historical **A12:** 1
fracture/failure causes illustrated **A12:** 216
grain-boundary cavitation **A12:** 219
gray, figure numbers for **A12:** 216
high-purity, flat cleavage fracture **A12:** 219
intergranular fracture and transcrystalline cleavage **A12:** 222
light fractography images, compared **A12:** 93-94
low-carbon, dimpled ductile rupture **A12:** 220
low-carbon, high oxygen, tensile-test fracture **A12:** 223
low-carbon, oxide inclusion **A12:** 222
malleable, figure numbers for **A12:** 216
slip lines **A12:** 219
transgranular cleavage fracture **A12:** 460
white, figure numbers for **A12:** 216
woody fractures **A12:** 1-3
wrought, impact fracture **A12:** 224
Irons, specific types
Armco, cleavage fracture **A12:** 224
Armco, oxygen-embrittled **A12:** 222
Armco, shear step **A12:** 224
Armco, slip steps **A12:** 224
Armco, slip-band cracks **A12:** 224
Armco, tear-ridges **A12:** 224
Armco, tilt boundary, cleavage steps, river patterns **A12:** 18
Fe-0.3C-0.6Mn-5.0Mo, quasi-cleavage fracture **A12:** 26
Fe-0.3Ni, fracture surface **A12:** 457
Fe-0.6Mn-5.0Mo, quasi-cleavage fracture **A12:** 26
Fe-3.9Ni, cleavage fracture **A12:** 457
Fe-4Al, fracture mode transition **A12:** 462
Fe-8Ni-2Mn-0.1Ti, facet areas measured **A12:** 208
Fe-Cr-Ta, cleavage and quasi-cleavage **A12:** 460
Fe-Cr-Ta, tension overload fracture **A12:** 460
Irradiance **A10:** 675, 685
comparative distribution **EM2:** 579
factors affecting **EM2:** 578
SI unit/symbol for **A8:** 721
Irradiated channel fracture
austenitic stainless steels **A12:** 365
Irradiated materials
ESR analysis **A10:** 254, 263
x-ray diffraction residual stress techniques **A10:** 385
Irradiation *See also* Radiation properties **EM3:** 17
of atoms, decay rate **A10:** 235
container, as contamination source **A10:** 235
defined **EM1:** 14 **EM2:** 24
epithermal vs thermal neutron **A10:** 234
gamma-ray spectrum of ores after **A10:** 235
of sample, NAA analysis as **A10:** 234
TNAA detection limits for rock and soil after **A10:** 236-238
Irradiation-assisted SCC, in nuclear power industry **A13:** 935-936
IRRAS *See* Infrared reflection-absorption spectroscopy
Irregular eutectics **A9:** 621-622
and regular eutectics **A15:** 120
Irregular holes *See also* Holes
adhesive-bonded joints **A17:** 612-613
Irregular powder **A7:** 6
Irregular-shaped particles **A7:** 233, 234
Irregularity *See also* Defects; Discontinuities; Flaw(s)
defined **A17:** 49
Irreversible
defined **EM1:** 14
Irreversible process .... **A3:** 1 • 7
Irrigation applications
homopolymer/copolymer acetals **EM2:** 101
Irrigation pipe and tools
aluminum and aluminum alloys **A2:** 14
Irrigation pipe, polymer
failure of **A11:** 762-763
Irwin's crack tip plastic deformation zone concept **EM3:** 363
Isaichev orientation relationship
in upper bainite **A9:** 664
ISE *See* Ion-selective electrode
Isentropic secant bulk modulus **A18:** 82
ISO *See* International Organization for Standardization **EM2:** 91
ISO 2738
powder metallurgy materials **A9:** 504
ISO Classification R513
of carbides **A16:** 75
ISO equivalents of Aluminum Association international alloy designations **A2:** 26
ISO property classes
threaded fasteners **M1:** 274, 275, 277
Iso resins *See also* Polyester resins
clear casting mechanical properties **EM1:** 91
electrical properties **EM1:** 94, 95
in fiberglass-polyester resin composites **EM1:** 91
glass content effects **EM1:** 91
mechanical properties **EM1:** 90
preparation/properties **EM1:** 90
ISO Technical Committee 61 (ISO TC 61) **EM3:** 62
Iso-*G*-specimens **EM3:** 342
Isobar
defined **A10:** 675
Isobutyraldehyde
physical properties **EM3:** 104
Isochronous graphs **EM3:** 317
Isocons
image **A17:** 10
Isocorrosion diagram
austenitic/duplex cast steels **A13:** 579
defined **A13:** 8
ferritic cast steels **A13:** 577-578
Isocorrosion diagram (continued)
fully austenitic cast steels **A13:** 580
high-silicon cast irons **A13:** 569
for Nickel 200/Nickel 201 in alkalies **A13:** 651
nickel-base alloys in hydrochloric acid **A13:** 646
nickel-base alloys in nitric acid **A13:** 646
for titanium alloys in hydrochloric acid **A13:** 680
for zirconium in hydrochloric acid **A13:** 709
for zirconium in phosphoric acid **A13:** 717
for zirconium in sulfuric acid **A13:** 708
Isocratic elution
defined **A10:** 675
Isocyanate plastics *See also* Plastics; Polyurethane; Urethane plastics **EM3:** 17
defined **EM1:** 14 **EM2:** 24
in polyurethanes (PUR) **EM2:** 257
Isocyanates **EM3:** 46
applications **EM3:** 45
characteristics **EM3:** 45
as health hazard **EM3:** 203
for packaging **EM3:** 45
use in urethane sealant manufacture **EM3:** 203
Isocyanates, as urethane coating **A13:** 409
Isoelectric point **EM4:** 74
of oxide ceramics **EM4:** 154
Isoimides **EM3:** 157
Isokinetic sampling
devices for **A10:** 16
Isolated porosity
in arc welds **A11:** 413
Isolated signal line
design considerations **EL1:** 28-34
Isolation
dielectric **EL1:** 199
diffusion **EL1:** 195
junction **EL1:** 199
Isomer **EM3:** 17
Isomer shift
in Mössbauer spectroscopy **A10:** 288-290, 292
Isomerism, mer
and melt properties **EM2:** 62
Isomerization
during PMR-15 polymerization **EM1:** 83
Isomers *See also* Stereoisomer
cis/trans, chemistry of **EM2:** 64
defined **EM2:** 24
determined by GC-IR **A10:** 115
geometric, polyisoprene **EM2:** 58
geometric, within mer **EM2:** 58
identification and quantification by NMR 277
shift, in Mössbauer spectroscopy **A10:** 288-290
Isometric
defined **A9:** 10
Isometric graphs **EM3:** 317
Isomorphic blends
defined **EM2:** 632
Isomorphous
defined **A9:** 10
Isomorphous group
titanium alloys **A2:** 599
Isomorphous series in aluminum alloys **A9:** 359
Isomorphous system
defined **A9:** 10
Isopentyl ether
as solvent used in ceramics processing **EM4:** 117
Isophorone diisocyanate **EM3:** 203
Isophthalic polyester resin *See also* Unsaturated polyesters
properties **EM2:** 246
Isophthalic resins *See* Iso resins
Isopleths of a ternary phase diagram .... **A3:** 1 • 5
Isopropanol (anhydrous)
surface tension **EM3:** 181
Isopropyl acetate
surface tension **EM3:** 181
Isopropyl alcohol
as solvent used in ceramics processing **EM4:** 117

Isopropyl alcohol (IPA)
as PWB cleaning agent **EL1:** 777
Isostatic bonding *See* Diffusion welding
Isostatic compacting *See also* Cold isostatic pressing; Hot isostatic pressing; Isostatic pressing
and lubrication **A7:** 190
of porous parts **A7:** 698
and rigid tool compaction **A7:** 323
Isostatic compaction
of NbTi superconducting materials **A2:** 895
Isostatic mold **A7:** 6
Isostatic multiaxial compaction
of cermets **A2:** 979
Isostatic pressing *See also* Hot isostatic pressing; Isostatic compacting; isostatic pressing **EM4:** 9, 34
advanced ceramics **EM4:** 49
of beryllium powders **A7:** 758
carbon-graphite materials **A18:** 816
and Ceracon process **A7:** 537
and compaction in rigid dies, densities compared **A7:** 298-299
defined **A7:** 6 **A14:** 8 **EM1:** 14 **EM2:** 24
explosive **A7:** 317
in flexible envelopes **A7:** 297
pressure and green density **A7:** 298
pressure-density relationships with **A7:** 298
in reaction sintering **EM4:** 292
reduced friction in **A7:** 300
with tapping and vibration **A7:** 297
to make compacts of silicon **EM4:** 237
to make silicon oxynitride shapes by reaction- bonding **EM4:** 239
in triaxial compression **A7:** 304
viscoelastic properties required **EM4:** 116
Isostress lines
for rupture data extrapolation **A8:** 333
Isotactic stereoisomerism **EM3:** 17
defined **EM2:** 24
Isotensoid *See* Geodesic isotensoid; Geodesic-isotensoid contour
Isotensoid, geodesic *See* Geodesic isotensoid
Isotherm
creep-rupture test data **A8:** 332
of porous material **EM4:** 582
Isothermal aging
of polybenzimidazoles **EM3:** 170, 171
Isothermal annealing
defined **A9:** 10
Isothermal annealing time
thickness of the peritectic layer as a function of **A9:** 677
Isothermal behavior
metals **A8:** 45
Isothermal compressibility
mercury **A7:** 268, 269
Isothermal contour lines .... **A3:** 1 • 5
Isothermal conversion
in steel dissolution **A15:** 74
Isothermal flow curve
stress-temperature plots for **A8:** 161
Isothermal forging *See also* Forging; Gatorizing **A1:** 971 **A8:** 158, 170 **A14:** 150-157
advantages **A14:** 150-151
cost **A14:** 150, 155-157
defined **A14:** 8, 150
die systems **A14:** 154-155
forging alloys **A14:** 152
forging design guidelines **A14:** 155
gatorizing **A14:** 18
of heat-resistant alloys **A14:** 232
induction heating system **A14:** 152
lubrication **A14:** 153-154
of near-net shapes **A7:** 522
as new metalworking process **A14:** 17-18
of nickel-base alloys **A14:** 265-266
process **A14:** 151-152
process design **A14:** 153-154
process selection **A14:** 152-153
production forgings **A14:** 157
torsion testing **A14:** 373
wrought titanium alloys **A2:** 613-614
Isothermal formation of bainite in steel **A9:** 662
Isothermal grain growth
grain shape distribution in zone-refined iron **A9:** 698
Isothermal grain growth (continued)
grain size distribution in zone-refined iron **A9:** 698
in pure metals **A9:** 697
in single-phase alloys **A9:** 697
Isothermal holding, and rheocasting
compared **A15:** 330
Isothermal oxidative stability
prediction **EM2:** 566
Isothermal plane-strain sidepressing **A8:** 172-173
Isothermal processing
CAP billet stock for **A7:** 533
Isothermal ratcheting
defined **A11:** 143-144
Isothermal recovery of mechanical and physical properties **A9:** 693
Isothermal recrystallization curves **A9:** 694
Isothermal rolling
of near-net shapes **A7:** 522
Isothermal secant bulk modulus **A18:** 82
Isothermal sections of a ternary diagram .... **A3:** 1 • 5
Isothermal solidification
eutectic growth (cast iron) **A15:** 174
Isothermal stability diagrams
alloy system and one gas **A13:** 64
for gaseous corrosion **A13:** 17
high-temperature corrosion, in gases **A13:** 63-64
one metal and two gases **A13:** 63-64
predominance area diagrams **A13:** 64
Isothermal technique (IT)
hot press densification **EM4:** 189
Isothermal test
conditions, heated subassembly for **A8:** 582
departure, in high strain rate compression testing **A8:** 191
Isothermal transformation
defined **A9:** 10
Isothermal transformation (IT) diagram
defined **A9:** 10
Isothermal uniaxial compression
rate of flow localization **A8:** 172
Isotherms
adsorption, Raman analysis as probe for **A10:** 134
phase transformation **A10:** 317
Isotone
defined **A10:** 675
Isotope abundances
in mass spectra **A10:** 641-642
Isotope dilution analysis
by ICP-MS **A10:** 40
capabilities **A10:** 243
neutron activation analysis and compared **A10:** 233
Isotope exchange technique
as nitrogen dissociation measure **A15:** 83
Isotope factor
detained **A10:** 236
Isotopes
defined **A10:** 675
dilution **A10:** 40, 145, 233, 243
dry spike dilution, for spark source mass spectrometry **A10:** 146
mass numbers of most stable per element **A10:** 688
multielement, for SSMS analysis **A10:** 145-146
natural composition, iron, chromium, nickel and manganese **A10:** 146
naturally occurring, percent abundance and atomic mass **A10:** 643
oxygen, determined in explosive actuator **A10:** 625-626
radioactive **A10:** 243, 244
ratio measurement **A10:** 40, 233
as tracers or "spikes" **A10:** 145
Isotopes, radioactive
as gamma-ray source **A17:** 308
Isotopic analysis *See also* Mass analysis
of inorganic gases, analytic methods for **A10:** 8
of inorganic liquids and solutions, analytic methods for **A10:** 7
of inorganic solids, analytical methods for **A10:** 4-6
of organic solids, analytic methods for **A10:** 9
Isotopic analysis (continued)
of organic solids and liquids, analytic methods for **A10:** 10
Isotropic **EM3:** 17
defined **A9:** 10 **EM1:** 14 **EM2:** 24
Isotropic alloys
Alnico **A15:** 738
Isotropic elasticity theory **A18:** 469
Isotropic metals *See also* Optically isotropic metals
anodic oxidation to improve contrast under polarized light **A9:** 59
examination with crossed-polarized light **A9:** 78
Isotropic pitch-based precursor fibers
properties **EM1:** 52
Isotropic thermal motions
and crystal structure determination **A10:** 352, 353
Isotropically decomposed microstructure
computer simulation **A9:** 653
Isotropy
defined **A8:** 7 **A9:** 10
as material behavior **A8:** 343
I_t *See* Total diffracted intensity
IT-WOL modified compact specimens
SCC testing **A8:** 512
Item
defined **A8:** 7
Iterative experimentation
defined **A17:** 741
Iterative least squares procedure
for runouts
Iterative reconstruction
computed tomography (CT) **A17:** 359, 382, 384
Iterative technique **A8:** 715
ITS-90 *See* International Temperature Scale of 1990
Izod impact strength *See also* Impact strength
high-performance engineering thermoplastics **EM2:** 98
polyarylates (PAR) **EM2:** 139
polycarbonates (PC) **EM2:** 151
polyvinyl chlorides (PVC) **EM2:** 211
Izod impact test *See also* Impact strength; Impact test; Impact tests **EM2:** 24, 434 **EM3:** 17
defined **EM1:** 14
of polyester resins **EM1:** 91
Izod impact testing
hot forged disks **A7:** 410
Izod test *See also* Charpy test; Charpy V-notch impact test; Impact properties **M1:** 689
defined **A11:** 6
for notch toughness **A11:** 57-60
Izod testing *See also* Charpy V-notch impact test; Impact testing
defined **A8:** 7
specimen, for impact cantilever bend **A8:** 262
In roller leveling
to control flatness in carbon steel sheet **A1:** 206
In situ
discontinuous copper-matrix composites **A2:** 909
In situ film growth
thin-film materials **A2:** 1082
In situ microstructural analysis
by surface replication **A17:** 52-56
In temper rolling
to control flatness in carbon steel sheet **A1:** 205-206
The International Adhesion Conference 1984 Proceedings **EM3:** 69
The International Journal of Adhesion and Adhesives **EM3:** 65

J

***J* analysis**
for plastic-elastic analysis **A8:** 446-447
***J* testing**
for fracture toughness **A11:** 62-64
J-1650
composition **A4:** 795
J-bend configuration
ceramic packages **EL1:** 205
J-groove joints
radiographic inspection **A17:** 334

J-groove weld
definition **A6:** 1210
J-groove welds
arc welding of nickel alloys **M6:** 706
definition, illustration **M6:** 60-61
double, applications of **M6:** 61
gas tungsten arc welding of heat-resistant alloys **M6:** 356-358, 360
oxyfuel gas welding **M6:** 590-591
preparation **M6:** 67-68
single, applications of **M6:** 61
***J*-integral**
for crack growth **A8:** 377
defined **A8:** 7-8 **A11:** 6 **A12:** 15
fracture mechanics of **A11:** 51
relation to *K* **A8:** 440
for toughness evaluation **A8:** 457
J-lead
component removal **EL1:** 726
defined **EL1:** 1148
as lead formation **EL1:** 734
packages, small-outline **EL1:** 117
***J-R* curve** **A8:** 456-457
***J-R* measurements** **A8:** 456-457
***J*-resistance curve, and *J*-values**
for dynamic toughness testing **A8:** 261
J1570
composition **A16:** 736
grinding **A16:** 759, 760
machining **A16:** 738, 741- 746-747, 749-758
Jablonsky diagram, molecular absorption and de-excitation processes
molecular fluorescence spectroscopy **A10:** 73
Jaccarino-Peter effect
ternary molybdenum chalcogenides **A2:** 1077
Jacket
defined **EL1:** 1148
Jacketing **A18:** 738
for wrought copper and copper alloy products **A2:** 258-260
Jacketing, wire/cable
polyamide **EM2:** 125
Jacking oil **A18:** 518, 519
Jackscrew
defined **EL1:** 1148
Jackscrew drive pins
SCC of **A11:** 546-548
Jacquet's electropolishing solution
preparation of silicon irons **A9:** 532
Jacquinot's advantage
in FT-IR spectroscopy **A10:** 112
Jade
lapping process applied **A16:** 492
Jahn-Teller distortion
insertion electrodes for lithium batteries of spinels **EM4:** 766
Jammers, solid-state phase-array
with gallium arsenide MMICs **A2:** 740
Jamming
gear tooth fracture by **A12:** 277
Jander
rate law expression **EM4:** 55
Japan
casting in **A15:** 21, 312
major programs in ceramic gas turbine development **EM4:** 716-717
telecommunication industry structure **EL1:** 384
value engineering in **EL1:** 121
Japanese Automotive Standards Organization
performance testing of engine oils **A18:** 170
Japanese Industrial Standard **A8:** 725
Japanese-type explosives
aluminum powder containing **A7:** 601
Jarring molding machine
development of **A15:** 28
Jasper stoneware **EM4:** 3, 4
Jaw crusher
defined **A7:** 6
Jaw crusher test
of abrasion-resistant cast iron **A1:** 97
Jaws, low-alloy steel
brittle fracture **A11:** 389-391
JCL-4036
composition **A18:** 821
wear and friction properties **A18:** 822
JCL-4063
composition **A18:** 821
wear and friction properties **A18:** 822
JCPDS card, x-ray powder diffraction data
mineral quartz **A10:** 341
JCPDS Powder Diffraction File
use in electron diffraction/EDS analysis **A10:** 455-459
J_{ec} equivalent cooling rates **M1:** 482-484, 489, 491
Jeffries' method
defined **A9:** 10
J_{eh} equivalent hardness **M1:** 481, 490
Jelly roll method
of superconductor assembly **A2:** 1067
Jelly-roll superconductor manufacture method
modified **A14:** 341-342
Jemez mountains
gamma-ray spectrum of ores from **A10:** 235
Jenks, Joseph
as early founder **A15:** 24, 33
Jenks, Jr. Joseph
as early founder **A15:** 25
Jernkontoret fracture tests (Arpi) **A12:** 3
Jernkontoret system **A18:** 875
JESD-22
as environmental testing standard **EL1:** 494
JESD-26
as environmental testing standard **EL1:** 494
Jet engine
bearing failure by misalignment **A11:** 507-508
military, LME failure of steel nuts in **A11:** 543
turbine blade, creep deformation and cracking in **A11:** 29
turbine blade, high-temperature fatigue fracture **A11:** 131
turbine disk, fracture surface **A11:** 131
Jet engine components **A7:** 18, 646-652
infiltration processed **A7:** 562-563
powders used **A7:** 572
titanium and titanium alloy **A2:** 587-588
Jet engine components, wear of **A18:** 588-592
abrasive wear **A18:** 588, 590
adhesive wear **A18:** 588, 590
blade midspan stiffeners and tip shrouds **A18:** 589-590
combustor and nozzle assemblies **A18:** 591
compressor airfoil erosion **A18:** 591-592
discussion and summary **A18:** 592
dovetails **A18:** 591
erosive wear **A18:** 588, 592
fretting wear **A18:** 588
gas path seals **A18:** 588, 589
abradable seal materials **A18:** 589
abrasive coatings for clearance control **A18:** 589
impact wear **A18:** 588, 591
lubrication **A18:** 588
mainshaft bearings **A18:** 590-591
major engine subsystems **A18:** 588
oxidational wear **A18:** 588, 591
particle erosion **A18:** 588
rolling contact fatigue **A18:** 588, 590
sliding wear **A18:** 588, 591
Jet engine parts
ultrasonic inspection **A17:** 232
Jet engines
components, by hot-die/isothermal forging **A14:** 150
disks, gatorized **A14:** 18
Jet Kote surfacing system
for hardfacing **A7:** 835
Jet molding
defined **EM2:** 24
Jet plating **A18:** 835
Jet polisher
used to prepare transmission electron microscopy specimens **A9:** 107
Jet pulverizer
defined **A7:** 6
Jet pump beams
SCC of alloy X-750 .... **A13:** 933-935
Jet size
effect on erosion damage **A11:** 166
Jet, water *See* Waterjet
Jet-Kote (JK) method
ceramic coatings for adiabatic diesel engines **EM4:** 992
molten particle deposition **EM4:** 204
Jet-producing apparatus, Hall's
for atomized aluminum powders **A7:** 127
Jethete M152
self-shielding blade alloy to protect against liquid impingement erosion **A18:** 222
Jethete steel
flash welding **M6:** 557
Jetting **A6:** 898, 899 **EM2:** 24, 181
critical angle for **A6:** 162
definition **M6:** 705-706
in explosion welding **M6:** 705-706
in injection molding **EM1:** 166
Jewel bearing
defined **A18:** 11
Jewelry
gold cast **A15:** 18
golds **A2:** 690
with platinum settings **A2:** 695
precious metal **A2:** 695
white metal **A2:** 525
Jewelry applications
of stainless steels **A7:** 731
Jewelry bronze *See also* Bronzes-, Wrought coppers and copper alloys
applications and properties **A2:** 297-298
Jewelry striking die
effect of excessive carburization **A11:** 571-572
Jewett nail
austenitic stainless steels **A12:** 361, 363-364
cobalt alloy **A12:** 398
Jewett nail plate with three-flanged nail
as internal fixation device **A11:** 671
JFL-4036
composition **A18:** 821
wear and friction properties **A18:** 822
Jib crane
historic **A15:** 33
Jig **EM3:** 17, 36-37
defined **EM2:** 24
grid, for round bar specimens **A8:** 279
sheet compression **A8:** 56
Jig for hand polishing **A9:** 105
Jiggering **EM4:** 8, 34
viscoelastic properties required **EM4:** 116
Jigging system
automobile scrap recycling **A2:** 1213
Jigs
aluminum and aluminum alloys **A2:** 14
Jigs for
furnace brazing of steels **M6:** 939
gas tungsten arc welding **M6:** 197
shielded metal arc welding **M6:** 79-81
JIS *See* Japanese Industrial Standard
JIS (Japanese) standards for steels **A1:** 157-158
compositions of **A1:** 180-181
cross-referenced to SAE-AISI steels **A1:** 166-174
JIS S45C
friction welding **A6:** 441
JK flip-flop
defined **EL1:** 1148
JKR theory of pull-off **A18:** 403

Jobber's reamer **A16:** 242
Jobbing casting plants
sandslingers in **A15:** 28
Joggling
in rotary shearing **A14:** 706-707
of titanium alloys **A14:** 847
Johns Hopkins University Applied Physics Laboratory **EM1:** 40
Johnson noise *See* Thermal noise
Joining *See also* Fabrication characteristics; Joints; Weldability; Welding
aluminum **M2:** 191-203
aluminum and aluminum alloys **A2:** 9
aluminum casting alloys **A2:** 157-177
applications **EM4:** 477
of beryllium **A2:** 683
beryllium-nickel alloys **A2:** 424-425
by electromagnetic forming **A14:** 646
copper metals **M2:** 440-457
of copper-based powder metals **A7:** 733
copper/copper alloys, ease of **A13:** 610
and dimensional change **A7:** 292
of ductile iron castings **A15:** 664-665
effect of composition on **A11:** 315
of electrical contact materials **A2:** 841
electrical resistance alloys **A2:** 822, 824
failure mechanisms **EL1:** 1041-1048
of galvanized structural members **A13:** 441-442
in granulation **A15:** 18-19
LME by copper during **A11:** 721
of magnesium alloys **A2:** 471-475
materials **EL1:** 1041
mechanical means **A7:** 456
of mechanically alloyed oxide alloys **A2:** 949
dispersion-strengthened (MA ODS)
of open resistance heaters **A2:** 831
of P/M parts **A7:** 295
palladium **A2:** 716
refractory metals and alloys **A2:** 563-564
as secondary operation **A7:** 451, 456-458
stainless steel **M3:** 48-50
with tin solders **A2:** 520-521
titanium alloys **M3:** 368, 369-370
of whisker-reinforced MMCs **EM1:** 889-900
of wood laminate patterns **A15:** 194
wrought titanium alloys **A2:** 616-618
Joining metallurgy
control of toughness in the heat-affected zone **M6:** 41-42
filler metal **M6:** 42
improving toughness **M6:** 42
microstructure **M6:** 42
welding process **M6:** 42
general metallurgy **M6:** 21-26
age hardening **M6:** 23, 25
cold working **M6:** 23
eutectic binary alloy systems **M6:** 21-22
eutectoid reaction **M6:** 22
grain growth **M6:** 23
heat-affected zone **M6:** 27-28
Hume-Rothery rules **M6:** 21
partially melted zone **M6:** 27
peritectic reaction **M6:** 22
phases in metals **M6:** 21-22
recrystallization **M6:** 23
single-phase alloy systems **M6:** 21
single-phase metals **M6:** 21
tempering precipitation **M6:** 23, 25
unaffected base metal **M6:** 27-28
unmixed zone **M6:** 26-28
weld interface **M6:** 27-28
heat flow calculations **M6:** 31-34
cooling rates in fusion zone **M6:** 32-34
heat input **M6:** 31-32
thermal cycle of heat-affected zone **M6:** 34
microstructure of weld and heat-affected zone **M6:** 35-37
grain size **M6:** 35-36
influence of solidification structure **M6:** 36-37
multiple-pass welds **M6:** 36
prediction of microstructures **M6:** 39-40
weld metal composition **M6:** 39-40
postweld heat treatment **M6:** 43-44
aging **M6:** 44
normalizing **M6:** 43-44
quenching and tempering **M6:** 44
Joining metallurgy (continued)
solution treating **M6:** 44
preheating **M6:** 42-43
calculation of temperatures **M6:** 43
reduction of distortion and residual stress **M6:** 43
stress relieving **M6:** 43
procedures for welding **M6:** 40-41
fluxes **M6:** 41
shielding gases **M6:** 40-41
solidification of welds **M6:** 28-31
cells, dendrites, and microsegregation **M6:** 29-31
epitaxial growth **M6:** 28
rate of solidification **M6:** 31
solute banding **M6:** 31-32
weld pool shape **M6:** 28-29
weld, definition of **M6:** 26-28
composite zone **M6:** 26-27
continuous cooling transformation diagrams **M6:** 25-26, 39-40
time-temperature-transformation diagrams **M6:** 25
welding defects **M6:** 44-48
chevron cracking **M6:** 48
ductility-dip cracking **M6:** 48
ferrite vein cracking **M6:** 46
hot cracking in weld metal and heat-affected zone **M6:** 47
hydrogen-induced cold cracking **M6:** 44-46
intergranular corrosion **M6:** 48
lamellar tearing **M6:** 46
porosity **M6:** 44
reheat cracking **M6:** 46-47
welding effects on
cooling rates **M6:** 38-39
distance from weld interface **M6:** 40
ferrite formation **M6:** 39
microstructure **M6:** 37-40
stainless steel welds **M6:** 40
weld-metal composition **M6:** 39-40
Joining non-oxide ceramics **EM4:** 523-530
applications **EM4:** 523, 529-530
factors to consider for applications **EM4:** 523
interfacial reactions **EM4:** 523
joining methods **EM4:** 523-526
active metal brazing **EM4:** 523-524, 529
eutectic joining **EM4:** 526
solid-state bonding **EM4:** 525, 528
SQ brazing **EM4:** 525
joining specific materials **EM4:** 527-529
microwave heating **EM4:** 528
non-oxide ceramics **EM4:** 528-529
silicon-base ceramic to itself **EM4:** 527-528
joint properties **EM4:** 526-527
interfacial strength **EM4:** 526, 527
roughness **EM4:** 527
thermal stress relaxation **EM4:** 526-527
Joining of
aluminum metal-matrix composites **A6:** 554-558
ceramics **A6:** 617, 618, 619
composites to metals **A6:** 1041-1047
dissimilar metals **A6:** 821, 822-828
nickel alloys to dissimilar metals **A6:** 749-751
nitride ceramics **A6:** 636
organic-matrix composites **A6:** 1026-1036
oxide-dispersion-strengthened materials **A6:** 1037-1040
plastics **A6:** 1048-1055
Joining of Advanced Composites **EM3:** 67
Joining of Composite Materials **EM3:** 67
Joining oxide ceramics **EM4:** 511-520
applications **EM4:** 511
brazing with filler metals **EM4:** 516-519
direct brazing **EM4:** 517-519
indirect brazing **EM4:** 517
brazing with glasses **EM4:** 519-520
applications **EM4:** 520
fracture toughness measurements **EM4:** 520
ceramic materials **EM4:** 511
high-alumina ceramics **EM4:** 512
SiC_w-reinforced ceramic-matrix composites **EM4:** 512
transformation-toughened zirconias **EM4:** 512
diffusion welding **EM4:** 514-516
applied pressure **EM4:** 515
bonding temperature **EM4:** 515-516
Joining oxide ceramics (continued)
ceramic/ceramic joints **EM4:** 514, 516
ceramic/metal joints **EM4:** 514-515
surface roughness **EM4:** 516
mechanical properties of joints and their measurement **EM4:** 512-514
critical stress intensity factor **EM4:** 512-514
fracture mechanics testing **EM4:** 514
oxide ceramics' properties **EM4:** 512
Joining process **EM3:** 36-37
Joining processes
applications, automotive **A6:** 393
as principle of material selection **A6:** 373
sheet metals **A6:** 398-400
properties **A6:** 399-400
specifications **A6:** 399, 400
surface finishing **A6:** 400
Joining processes, selection of **M6:** 50-59
base-metal properties **M6:** 52-55
chemical composition **M6:** 52
effect of fabrication **M6:** 52, 55
mechanical properties **M6:** 52
physical properties **M6:** 52
end use applications **M6:** 56-57
aircraft and aerospace construction **M6:** 56-57
automotive and railroad **M6:** 57
piping, pressure vessel, boiler, and storage tank construction **M6:** 56
shipbuilding construction **M6:** 56
structural welding **M6:** 56
equipment costs **M6:** 56
fusion and nonfusion processes **M6:** 50-51
industrial usage **M6:** 56
joint design and preparation **M6:** 51-52
location of work place **M6:** 55
quality requirements **M6:** 57
review of processes **M6:** 50-51, 55
safety **M6:** 57-59
clothing **M6:** 57-58
electrical installations **M6:** 58
eyes and face **M6:** 57
fire protection **M6:** 58
gas cylinders **M6:** 024858
personnel protection **M6:** 57-58
respiratory **M6:** 58
for specific processes **M6:** 58-59
toxic materials **M6:** 58
training **M6:** 58
weld joint properties **M6:** 57
welder skill **M6:** 55-56
Joining Technologies for the 1990s-Welding, Brazing, and Soldering, Mechanical, Explosive, Solid-State, Adhesive **EM3:** 69
Joint **EM3:** 17
butt **EM3:** 17
definition **A6:** 1210 **M6:** 10
edge **EM3:** 17
lap **EM3:** 17
scarf **EM3:** 17
starved **EM3:** 17
Joint Army-Navy-NASA-Air Force (JAN-NAF) Interagency Propulsion Committee **EM1:** 40
Joint brazing procedure
definition **M6:** 10
Joint buildup sequence
definition **M6:** 10
joint clearance
definition **A6:** 1210 **M6:** 10
Joint Committee on Powder Diffraction Standards **EM4:** 25
Joint deformities
from molybdenum toxicity **A7:** 204
Joint design **EM3:** 42-43, 54
aerospace applications **A6:** 385, 386-387
aluminum bronzes, gas-metal arc welding **A6:** 765
applications, railroad equipment **A6:** 395
brazeability and solderability considerations **A6:** 621, 622
brazing of aluminum alloys **A6:** 938
brazing of cast irons and carbon steels **A6:** 907
brazing of copper and copper alloys **A6:** 932
carbide tool brazing **A6:** 635
combination of groove and fillet welds **M6:** 66-67
composite-to-metal joining **A6:** 1041, 1042-1044, 1045-1046

Joint design (continued)
copper-nickel alloys, gas-metal arc welding **A6:** 768
definition **M6:** 10
design considerations for weld joints **M6:** 61-62
accessibility **M6:** 62
weld metal **M6:** 61-62
dissimilar metal joining **A6:** 825
edge preparation **M6:** 67-68
back gouging **M6:** 68
backing bars **M6:** 68
methods of cutting **M6:** 67-68
root faces **M6:** 64, 67-68
spacer bars **M6:** 68
fillet welds **M6:** 61
gas-metal arc welding of coppers **A6:** 759
gas-tungsten arc welding **A6:** 762
of coppers **A6:** 757, 761
groove-weld preparations **M6:** 64, 66
bevel angles **M6:** 65
double versus single **M6:** 65-66
included angle **M6:** 64-65
joint preparation after assembly **M6:** 66
root opening **M6:** 64-65
joining processes **M6:** 51-55
laser-beam welding **A6:** 879-880
magnesium alloys **A6:** 772-774
nickel alloys **A6:** 740-741, 743, 744, 745, 748
nomenclature **M6:** 60
reactive metals, brazing of **A6:** 946
reducing distortion **M6:** 887-889
reducing residual stress **M6:** 887-889
refractory metals **A6:** 946
brazing **A6:** 634
resistance spot welding **A6:** 834-835
self-jigging soldering joint configurations **A6:** 982
silicon bronzes, gas-tungsten arc welding **A6:** 766, 767
soldering **A6:** 974-977
in electronic applications **A6:** 991-999
structural solder joints **A6:** 980
submerged arc welding of nickel alloys **A6:** 748
types of joints **M6:** 60-61
butt **M6:** 60-61
corner **M6:** 60-61
edge **M6:** 60-61
lap **M6:** 60-61
T **M6:** 60-61
types of welds **M6:** 60-61
bevel groove **M6:** 60-61
fillet **M6:** 60-61
J-groove **M6:** 60-61
square-groove welds **M6:** 60-61
U-groove **M6:** 60-61
V-groove **M6:** 60-61
Joint design for
arc welding of
aluminum alloys **M6:** 374-375, 379
beryllium **M6:** 462
coppers **M6:** 402
magnesium alloys **M6:** 428-429
nickel alloys **M6:** 437
stainless steels, austenitic **M6:** 328, 334, 337-339
dip brazing of steels **M6:** 992
electron beam welding **M6:** 615-618
friction welding **M6:** 726-728
furnace brazing of steels **M6:** 941-943
gas metal arc welding **M6:** 165-169
of aluminum bronzes **M6:** 420
of copper nickels **M6:** 421
of coppers **M6:** 418
of heat-resistant alloys **M6:** 356-357
of nickel alloys **M6:** 437-440
of nickel-based heat-resistant alloys **M6:** 362
of silicon bronzes **M6:** 421
gas tungsten arc welding **M6:** 201-202
of beryllium copper **M6:** 408
Joint design for (continued)
of coppers **M6:** 403, 406
of heat-resistant alloys **M6:** 356-357
of nickel alloys **M6:** 437-439
of nickel-based heat-resistant alloys **M6:** 360
of silicon bronzes **M6:** 414
oxyacetylene pressure welding **M6:** 595
oxyfuel gas welding **M6:** 589-591
plasma arc welding **M6:** 218
shielded metal arc welding **M6:** 356-357
of heat-resistant alloys **M6:** 356-357
of nickel alloys **M6:** 437, 441-442
submerged arc welding **M6:** 133-134
of nickel alloys **M6:** 437
torch brazing of copper **M6:** 1041
torch brazing of steels **M6:** 950
Joint distortion
from adhesives **EM1:** 687
Joint efficiency
definition **A6:** 1210 **M6:** 10
Joint failure criterion **EM3:** 483-484
Joint fill, iron powder
arc welding with **A7:** 821-822
Joint flash
as casting defect **A11:** 381
Joint geometry
definition **M6:** 10
Joint Industry Conference press classification system **A14:** 489
Joint penetration
in arc welds **A11:** 413
definition **A6:** 1210 **M6:** 10
in electrogas welds **A11:** 440
in electron beam welds **A11:** 445
in laser beam welds **A11:** 447
in resistance welds **A11:** 441
in slag **A11:** 440
Joint preparation **M6:** 60-72, 257-259, 264-265
Joint prostheses
elbow **A11:** 670
sliding knee **A11:** 670
total finger **A11:** 670
total hip, fractures of **A11:** 692-693
total shoulder **A11:** 670
types of **A11:** 670-671
Joint root
definition **A6:** 1210
Joint test action group (JTAG) boundary scan network **EL1:** 376
Joint type
definition **A6:** 1210
joint(s) *See also* Adhesive joint; Adhesive-bonded joints; Butt joint; Edge joint; Joining; Lap joint; Scarf joints; Solder joints; Soldered joints; Soldering; specific joints; Starved joints
adhesive, defined *See* Adhesive joint
adhesive, design of **EM1:** 683
adhesive, glue-line thickness inspection **A17:** 188-189
adhesively bonded **EM1:** 480-481, 486-487
adhesively-bonded, thermal stress **EL1:** 57-58
annular snap **EM2:** 719-720
arc-welded, radiographic inspection **A17:** 334-335
blind fastening **EM1:** 709-711
bolted, material separation **EM1:** 716-717
bonded, dissimilar material separation **EM1:** 717
bonded, surface preparation **EM1:** 681
bonded, ultrasonic inspection **A17:** 272-273
brazed, flaws in **A17:** 602-603
bulbous, fatigue life of **EL1:** 642
butt, defined *See* Butt joint
cantilever snap **EM2:** 714-718
clastic-plastic adhesive model **EM1:** 484
clearance, soldered joints **A17:** 608
for composite structures **EM1:** 479-480
composite-to-metal, filament-wound **EM1:** 511-514
data, composite materials and **EM1:** 313-319
defects, and solder impurities **EL1:** 642
joint(s) (continued)
design, niobium and tantalum alloys **A2:** 563-564
distortion **EM1:** 687
edge, defined *See* Edge joint
elements, testing of **EM1:** 325-326
elimination by composites **EM1:** 35
fastener hole considerations **EM1:** 712-715
flawed/damaged **EM1:** 486-487
graphite-to-aluminum, corrosion of **EM1:** 716-718
graphite-to-graphite **EM1:** 717
inspection, solder **EL1:** 735-739, 942
integrity, of brazed joints **A17:** 603
of interconnecting shapes, wrought aluminum alloy **A2:** 35-36
lap *See* Lap joint
lap, stresses in **EL1:** 57
large thermal mass, desoldering **EL1:** 721
lid-base, for electron beam/laser welding **EL1:** 240
in magnesium alloys, shear strength **A2:** 478
magnetic particle inspection **A17:** 106-110
mechanical fastener selection for **EM1:** 706-708
multirow bolted composite **EM1:** 491-492
nonuniformity of load transfer through **EM1:** 481-484
overlapping, in woven fabric prepregs **EM1:** 150
practical considerations **EM1:** 492-494
properties, of solder **EL1:** 640-642
quality **EL1:** 632, 735
ring-and-plug, properties **EL1:** 640-641
scarf *See* Scarf joint
shear load transfer **EM1:** 480-481, 488
single-hole bolted composite **EM1:** 488-490
single-lap adhesively bonded **EM1:** 485
solder, inspection of **EL1:** 942
solder, thermal failures **EL1:** 60
soldered **A17:** 605-609
splice, eddy current inspection **A17:** 193
spot-welded, Lamb wave inspection **A17:** 250-251
stepped-lap adhesively bonded **EM1:** 485-486
surface-mount, leaded and leadless **EL1:** 730-734
surface-mount, properties **EL1:** 641-642
surface-mount solder **EL1:** 117
thermally induced fatigue/creep failures **EL1:** 632
torsion snap **EM2:** 718-719
tubesheet rolled, eddy current inspection **A17:** 180-181
types, radiographic inspection **A17:** 334
visual inspection **A17:** 3
welded, neutron radiography of **A17:** 393
welded, ultrasonic inspection **A17:** 272
in wrought copper and copper alloy tube and pipe **A2:** 249
Y *See* Knuckle area
Joint-aging time **EM3:** 17
Joint-conditioning time **EM3:** 17
Joints *See also* Fittings
aluminum brazed **A13:** 1081
axially loaded, EMF made **A14:** 647-648
bell-and-spigot **A11:** 424
bellows expansion, fatigue cracking **A11:** 131-133
bolted, crevice corrosion in **A11:** 184
bolted, springlike effect of loading **A11:** 530-531, 533
brazed, corrosion of **A13:** 876-886
brazed, failures of **A11:** 450-455
by electromagnetic forming **A14:** 646
cast iron pipe **M1:** 97, 98, 99, 100
comer, lamellar tearing in **A11:** 92
design of **A11:** 116-117, 442, 450, 530-531, 639
design, to minimize corrosion **A13:** 879
electronic industry
expansion, bellows-type, fatigue fracture in **A11:** 118
flash-welded, failed **A11:** 442
mechanical, in heat exchangers **A11:** 639
riveted **A11:** 184, 544-545
solder-alloy, microstructure **A13:** 1357

Joints (continued)
in space shuttle orbiter **A13:** 1063-1065
spacing, weathering steels **A13:** 521
surface condition of **A11:** 450-451, 639
systems, total, dissimilar materials in **A11:** 670
torque, EMF made **A14:** 648
transverse properties, ductile iron weldments **A15:** 528
welded **A11:** 117, 621
welded, gray iron **A15:** 527
Joints, biological
artificial **A11:** 670-671
knee **A11:** 670
replacement of **A11:** 671
Joints, brazed and soldered
mechanical properties **A7:** 841
Joints, integrity of
determined by optical metallography **A10:** 299
Jolleying **EM4:** 8
Jolt
power, source of **A15:** 28
rollovers, squeeze with **A15:** 29
-type molding machines **A15:** 341-342
Jolt ramming
defined **A15:** 7
Jolt squeeze
molding machines, green sand molding **A15:** 342
for small molds **A15:** 28
Jolt-squeezer machine
defined **A15:** 7
Jominy bar
high-resolution analysis for **A10:** 508
Jominy end-quench test **A1:** 452, 464, 466
Jominy equivalent hardenability
use in selection of steel for carburized parts **M1:** 537
Jominy hardenability *See also* Hardenability
powder forgings **A14:** 197, 202
Jominy hardenability bands
carburizing steels **A18:** 875
Jominy hardenability specimen
heat flux data after quenching **A4:** 73, 74
Jominy test *See* End-quench hardenability, test; End-quench test
Jones reductor
for redox volumetric methods **A10:** 175-176
Josephson effects
in superconductors **A2:** 1040-1041, 1088
Joule **A10:** 685, 691
abbreviation **A8:** 725
Joule, James .... **A3:** 1 • 6
Joule-Thomson effect
defined **A10:** 675
Joule-Thomson expansion
defined **A10:** 675
Journal
defined **A18:** 11
Journal bearing **A18:** 523, 525- 526, 741
defined **A18:** 11
wedging film action of hydrodynamic lubrication **A18:** 89
Journal bearings
barrel-shape and hourglass-shape, failure in **A11:** 489
brass, locomotive axle failures with **A11:** 715
bronze backed **A11:** 715
hydrodynamic lubrication film development in **A11:** 150
Laudig iron-backed **A11:** 716
ultrasonic inspection of **A11:** 717
Journal boxes
early P/M technology for **A7:** 16
Journal of Applied Polymer Science **EM3:** 68
Journal of Applied Polymer Science (JAPS)
as information source **EM2:** 93
Journals
axle shaft, reversed-bending fatigue fracture **A11:** 321
crankpin, fatigue fracture from metal spraying **A11:** 481
as information source **EM2:** 92-93
main-bearing, failures in **A11:** 358-359, 477-478
P/M professional **A7:** 19
shafts, grinding bums **A11:** 89
-to-head failure, cast iron paper-roll dryer **A11:** 655
Joystick controller
coordinate measuring machine **A17:** 22
Judder *See also* Spragging
defined **A18:** 11
Jumbo tube trailers
acoustic emission inspection **A17:** 291
Jump frequency
in homogeneous nucleation **A15:** 103
Jumper wire
defined **EL1:** 1148
"Jumping the gap" **A6:** 37
Jumps, Barkhausen *See also* Barkhausen noise
defined **A17:** 159
Jumps, crack *See* Crack jumps
Junction field effect transistors (JFETs) *See also* Field effect transistor
defined **EL1:** 1148
device structure **EL1:** 154-156
devices **EL1:** 147
fabrication **EL1:** 196
I-V characteristics **EL1:** 155
isolation, techniques **EL1:** 199
small-signal parameters and equivalent circuit **EL1:** 156-159
Junction growth theory **A18:** 33
Junction measuring
in thermocouple thermometer **A2:** 870-871
Junction(s)
capacitor structures, active analog components **EL1:** 144
coatings, flexible epoxy **EL1:** 819-820
defined **EL1:** 1148
temperature, thermal conductivity impact **EL1:** 814
transistor, thermal stress effects **EL1:** 56-57
Junction-to-ambient thermal resistance **EL1:** 410-411
Junctions
leakage, integrated circuits **A11:** 784
silicon *p-n*, failures of **A11:** 782-786
Just In Time (JIT) manufacturing
furnace systems used **A4:** 470
Just-freezing solid phase
in ultrapurification by zone refining **A2:** 1093
Just-in-time (JIT) production
fixturing **A16:** 404, 410
Just-in-time production systems **EM3:** 797
Jute
defined **EM2:** 24
as natural fiber **EM1:** 117
The Journal of Adhesion **EM3:** 66
The Journal of Adhesion Science and Technology **EM3:** 66

K

K *See* Bulk modulus of elasticity; Stress-intensity factor
***K* calibration** *See* Stress-intensity calibration
***K* factor**
defined **EM1:** 14
K lines
fluorescent yield vs atomic number for **A10:** 87
intensity in titanium **A10:** 97
relative emission intensities of **A10:** 86, 87
K lines, Laplanche
for structural diagrams **A15:** 69-70
K radiation
defined **A9:** 10
K series
defined **A9:** 10
K shell
defined **A12:** 168
K-42-B
notch effects in stress rupture **M3:** 230
K-absorption
edge, pure nickel **A10:** 408
of nickel and iron **A10:** 416
K-band equipment
microwave inspection **A17:** 214
***K*-decreasing method**
for near-threshold fatigue crack growth rates **A8:** 379-380
vs. *K*-increasing, SCC tests **A8:** 517-518
K-edge
absorption spectrum, krypton gas **A10:** 410
K-edge (continued)
aluminum, in light-clement analysis by AEM methods **A10:** 461
EXAFS of nickel metal **A10:** 408
EXAFS spectra **A10:** 411
fine structure **A10:** 409
iron and nickel, EXAFS spectra above the **A10:** 416
of magnesium-, iron-, chromium-containing compounds, EXAFS studies **A10:** 408
normalized nickel **A10:** 417
XANES, vanadium **A10:** 415
***k*-factor**
defined **A10:** 675
K-family x-ray lines
EPMA analysis **A10:** 522
***K*-increasing, vs. *K*-decreasing**
SCC tests **A8:** 517-518
K-matrix method
as IR multicomponent analysis **A10:** 117, 118
K-Monel alloy *See also* Nickel alloys, specific types
discovery and characteristics **A2:** 429
K-Na (Phase Diagram) .... **A3:** 2 • 269
k-out-of-n reliability block diagram **EL1:** 899-900
K-Pb (Phase Diagram) .... **A3:** 2 • 269
K-photoelectron, germanium
backscattering of **A10:** 408
K-radiation
defined **A10:** 675
K-Rb (Phase Diagram) .... **A3:** 2 • 270
K-S (Phase Diagram) .... **A3:** 2 • 270
K-Sb (Phase Diagram) .... **A3:** 2 • 270
K-Se (Phase Diagram) .... **A3:** 2 • 271
K-series
defined **A10:** 675
K-shell
defined **A10:** 675
K-Sn (Phase Diagram) .... **A3:** 2 • 271
K-Te (Phase Diagram) .... **A3:** 2 • 271
K-Tl (Phase Diagram) .... **A3:** 2 • 272
K1A magnesium-aluminum casting alloy
for high damping capacity **A2:** 456
Kα aluminum or magnesium x-ray lines
characteristic **A10:** 570
Kα doublet **A10:** 385-386, 570
Kα radiation
pure **A10:** 326
Kahn tear-test *See* Navy tear-test
Kaiser effect
acoustic emission inspection **A17:** 284, 286
Kalcolor anodizing process
aluminum and aluminum alloys **M5:** 592
Kalling's reagent
as an etchant for wrought heat-resistant alloys **A9:** 307
Kanofsky-Srinivasan 90% confidence band values **EM4:** 701, 702, 704, 705, 706, 707
Kanthal A-1
as HIP furnace element **A7:** 422-423
Kaolin **A6:** 60 **EM3:** 175, 176 **EM4:** 5-6, 32
ceramic fiber from **EM1:** 60
in ceramic tiles **EM4:** 926, 928
composition **EM4:** 5-6
dickite **EM4:** 5, 6
filler for urethane sealants **EM3:** 205
formula **EM4:** 5
halloysite **EM4:** 5, 6
illite **EM4:** 6
kaolinite **EM4:** 5-7
crystal structure **EM4:** 882
sheet structure **EM4:** 759
margarite **EM4:** 6
microstructure **EM4:** 5-6
Miller numbers **A18:** 235
muscovite mica **EM4:** 6
nacrite **EM4:** 5, 6
paragonite mica **EM4:** 6
Kaolin (china clay)
as filler material **EM2:** 499-500
Kaolinite **EM4:** 5, 6, 7
formula **EM4:** 6
as molding clay **A15:** 210-211, 224
Kaolinite, recrystallized
abrasive in commercial prophylactic paste **A18:** 666
Kaowool discontinuous fibers **EM1:** 63

Kapton Skybond 701 condensation polyimide resin **EM1:** 79
Kara-Kumi braiding **EM1:** 519
Karat levels
gold alloys **A2:** 705-706
Karl Fischer
method for water determination **A10:** 219
reagent for surface oxides **A10:** 177, 204
titration **A10:** 177
Karsten, Karl
as metallurgist **A15:** 29
Kawasaki basic oxygen process (K-BOP) operation **A1:** 111
Kayem 12 *See* Zinc alloys, specific types, zinc foundry alloy ZA-12
Keel block
defined **A8:** 8 **A15:** 7
Keel splice
former **A7:** 439
titanium **A7:** 682
Kel-F
material for surface force apparatus **A18:** 402
Keller's equation
for wear rate **A13:** 968-969
Keller's etchant
used for aluminum-lithium alloys **A9:** 357
Keller's reagent
as an etchant for aluminum powder metallurgy materials **A9:** 509
Kellogg diagrams
defined **A13:** 63
Kelly, William
as inventor **A15:** 31-32
Kelvin
abbreviation **A8:** 725
Kelvin dynamic-condenser method
adhesive-bonded ©joints **A17:** 611
Kelvin equation
pore size distribution given by desorption **EM4:** 213
relation of capillary radius to vapor pressure **EM4:** 71
to calculate pore size and pore size distributions in open porosity **EM4:** 581-582
Kelvin, Lord .... **A3:** 1 • 7
Keramid 601/353
constituent material properties **EM1:** 86
Kerf **A18:** 686
in cutting **A14:** 731, 751-752
defined **EM1:** 14 **EM2:** 24
definition **A6:** 1211 **M6:** 10
Kerimids *See also* Bismaleimides (BMI)
types **EM2:** 252-253
Kernel
defined **A17:** 384
Kernel multiplication
C-scan **EM2:** 844
Kerner relation
glass-particulate zirconia composites thermal expansion **EM4:** 858
Kerosene
as a coolant/lubricant for diamond wheels **A9:** 25
as dielectric for diesinking machines **EM4:** 373
in liquid penetrant inspection **A17:** 73
Kerosene-and-whiting test
as penetrant inspection **A17:** 73
Kerosine oil
addition to titanium alloys as lubricant **A18:** 778, 780
Kerr effect
observation of magnetic domains **A9:** 535-536
Keto/enol functionality **EM3:** 289
Ketones *See* Polyaryletherketones (PAEK, PEK, PEEK, PEKK)
copper/copper alloy corrosion in **A13:** 634
as organic cleaning solvent **A12:** 74
Ketones, and aldehydes
determined **A10:** 217
Kevlar **EM3:** 17
defined **EM1:** 14 **EM2:** 24
fasteners for **A11:** 530
fiber properties **EL1:** 615
fibers, in composites **A11:** 731, 732
laminates **EL1:** 618
microwave inspection **A17:** 214
PCD tooling **A16:** 110
as PWB reinforcement **EL1:** 605
as reinforcement **EL1:** 535
Kevlar aramid fibers *See also* Aramid fibers; Aramid fibers, specific, types; Para-aramid fibers
composition **EM1:** 54
defined **EM1:** 14
as honeycomb core material **EM1:** 724
properties **EM1:** 54-56
reinforced epoxy resin composites **EM1:** 399-400, 407-409
Kevlar cut
with wire saws **A9:** 25
Key
defined **EL1:** 1148
-parameter trends, in thermal design of packages **EL1:** 409
Key curve
as load-displacement curve **A8:** 261
Key results
of process design and analysis **A14:** 413-416
Keyhole
definition **M6:** 10
Keyhole Charpy tests **A11:** 57-58
Keyhole formation by an electron or laser beam
theory of **A6:** 22-23
Keyhole specimen *See also* Charpy, V-notch impact test **A8:** 8
Keyhole welding, plasma arc
welds **M6:** 218-220
Keying **A7:** 7
defined **EL1:** 1148
improved, and green strength **A7:** 304
slot, defined **EL1:** 1148
Keyway
defined **EL1:** 1148
fatigue fracture origin at **A12:** 260
fretting wear, AISI/SAE alloy steels **A12:** 308
in tapered shaft, peeling fracture at **A12:** 253
Keyway intergranular stress-corrosion cracking **A13:** 952-953
Keyways
effect in gears and gear trains **A11:** 589
in fatigue crack initiation **A8:** 371
as fracture origin, shafts **A11:** 459
as notches **A11:** 85
shaft cracking from **A11:** 471-472
sharp, fatigue failure from **A11:** 321
of steel pinion shaft **A11:** 524
as stress raisers **A11:** 115
torsional stresses in shaft **A11:** 115
KI SHELL
knowledge-based integration shell **A14:** 412
Kick-out relays
as eddy current inspection readout **A17:** 178
Kickers
for unloading presses **A14:** 500
Kidney
cadmium toxicity **A2:** 1240
Kidney filter
waterjet machining **A16:** 522
Kikuchi diffraction pattern **A18:** 386
Kikuchi diffraction patterns
analytical transmission electron microscopy **A10:** 437-438
defined **A10:** 675
deformed OFHC copper, dislocation cell structure analysis **A10:** 471
Kikuchi lines *See* Kikuchi diffraction patterns **A9:** 109-110
defined **A9:** 10
Killed low-carbon steel
roll welding **A6:** 312
Killed steel **A1:** 142, 226, 227 **M1:** 112, 124
carbon steel wire rod **M1:** 254, 255, 257
cold heading **M1:** 590
corrosion in seawater **M1:** 741
notch toughness of **M1:** 694, 695, 698, 706, 708
plate **M1:** 181, 194
porcelain enameling of **M5:** 512-514
Killing
to control nitrogen and produce sound weld metal **A6:** 69
Kiln
coremaking with **A15:** 32
defined **A15:** 7 **EL1:** 1148
Kiln operations
pollution control **A13:** 1370
Kilobar
abbreviation **A8:** 725
Kilocalorie
abbreviation **A8:** 725
Kiloelectron volt
defined **A17:** 384
Kilogram-force
abbreviated **A8:** 725
Kilohertz
abbreviated **A8:** 725
Kilonewton
abbreviated **A8:** 725
Kilopascal
abbreviated **A8:** 725
Kimax glass molds **A7:** 533
Kimble glass, TM-9
glass/metal seals **EM4:** 536
Kinematic SEM analysis
of deformation **A12:** 166
Kinematic theory **A18:** 388
Kinematic viscosity *See* Viscosity
SI derived unit and symbol for **A10:** 685
SI unit/symbol for **A8:** 721
Kinematic wear marks
defined **A18:** 11
Kinematical contrast integral **A9:** 112
evaluated for different dislocations **A9:** 114
Kinematical diffraction
in defect imaging **A10:** 367-370
Kinematical diffraction amplitude **A9:** 111-112
Kinematical diffraction theory **A9:** 111-112
Kinematics
collision, in Rutherford backscattering spectrometry **A10:** 629
of diffraction **A10:** 366
Kinematics, slider-crank mechanism
mechanical presses **A14:** 37-38
Kinetic blurring
radiography **A17:** 320
Kinetic compensation effect **EM4:** 56
Kinetic energy
AES and XPS measurement as function of **A10:** 550
amorphous materials and metallic glasses **A2:** 811-812
of Auger and photoemitted electrons **A10:** 569-570
of bombarding electrons, effect in x-ray emission **A10:** 84
converted to radiation **A10:** 83
decomposition, effect in shape memory alloys **A2:** 900
defined **A10:** 675
and electron escape depth **EL1:** 1077
of electrons, electron temperature for **A10:** 24
in forging machines **A14:** 25
of heavy particles, gas kinetic temperature for **A10:** 24
measurement, as basis of x-ray photoelectron spectroscopy **A10:** 85
neutrons **A17:** 387
as pass energy **A10:** 571

Kinetic energy (continued)
penetrators, uranium alloys as **A2:** 673
Raman spectroscopy analysis **A10:** 129
range, Auger electron **A10:** 551
in scattering **A17:** 390
stem radiation **A17:** 305
of styrene polymerization reaction **A10:** 132
Kinetic energy (ejected electron) **A18:** 445, 446, 447, 452, 454
Kinetic energy penetrators **A7:** 688-691
Kinetic energy vs electron mean free path
in metals **A11:** 34
Kinetic friction
defined **A18:** 11
Kinetic friction coefficients for selected materials **A18:** 27, 46, 47, 48, 70-75
defined (as kinetic coefficient of friction) **A18:** 11
factors affecting relative contributions **A18:** 70
and lubrication **A18:** 58
Kinetic stability **A18:** 143
Kinetic temperature, gas
and kinetic energy of heavy particles **A10:** 24
Kinetics **A8:** 497, 499
absorbance-subtraction studies of **A10:** 116
of alloy additions **A15:** 71-74
of aqueous corrosion **A13:** 17, 29-36
of atmospheric corrosion **A13:** 512
of cellular decomposition of martensite in uranium alloy **A10:** 316-318
of charge transfer reactions **A13:** 32
chemical, and rheology **EL1:** 838
chemical, principles of **A15:** 52-54
chemical reaction, potentiometric membrane electrodes as detectors for **A10:** 181
corrosion **A12:** 41 **A13:** 18-19
of corrosion in gases **A13:** 65
of corrosion process **A11:** 769
corrosion protection by **A13:** 377
crack-growth, steels **A13:** 279
crystal, by x-ray topography **A10:** 376
of crystals, and material transformations **A10:** 376
dispersive EXAFS analyses **A10:** 418
dissolution, boundary layer thickness effect **A15:** 73
failure, for VLSI mechanisms **EL1:** 889-893
first-order, ESP study of free radicals **A10:** 266
of gas-liquid reactions **A15:** 82-83
GC-IR study of **A10:** 115
growth **A15:** 79, 885-887
of growth, LEED analysis **A10:** 536, 544
of heterogeneous nucleation **A15:** 103-105
of homogeneous nucleation **A15:** 103
of imidization **EL1:** 772
of inclusion-forming reactions **A15:** 89-90
of inert gas flushing **A15:** 87
inhibitor effects, aqueous solutions **A13:** 379
laws of corrosion **A13:** 17
linear oxidation **A13:** 66
liquid/solid interface
logarithmic and inverse logarithmic oxidation **A13:** 67
mass transfer limited **A15:** 53
NMR analysis **A10:** 277
nucleation **A15:** 101-108
of overlayer growth at submonolayer level and above, LEED analysis **A10:** 544
of oxidation **A13:** 17, 67
parabolic **A13:** 67
parabolic oxidation **A13:** 67
of particle behavior, interfacial **A15:** 142-143
of particle separation **A15:** 79
of peritectic systems **A15:** 125-129
of radical production and decay **A10:** 265-266
rate laws, types **A13:** 17
reaction, analyses **A10:** 109, 628
reaction, of desulfurization **A15:** 75
repassivation **A12:** 42
resistance, of liquids to crystallization **A15:** 103
of solidification **A15:** 52-53
stress intensity effects **A13:** 246
of stress-corrosion cracking **A13:** 153-155, 246
swelling **EM2:** 771
of texture development **A10:** 424
voltammetric analysis **A10:** 188
Kinetics, chemical
in composite curing **EM1:** 748-751
King roll *See* Radial roll
Kingery equation
thermal expansion coefficient of a two-phase composite **EM4:** 858
Kingsbury bearing *See* Tilting-pad bearing
Kink bands **A9:** 688
defined **A9:** 10, 685
in zinc single crystal **A9:** 689
Kink sites
dissolution **A13:** 46
Kink-band formation
as failure mode **EM1:** 197
in p-aramid fibers **EM1:** 55, 56
Kinking (interfacial crack propagation) **A6:** 146
Kirchhofrs equation **EM4:** 868
Kirkendall porosity **A7:** 302
dispersion-strengthened aluminum alloys **A6:** 547
formation and coarsening in nickel blend compacts **A7:** 314
Kirkendall void formation
refractory metals **A6:** 943
Kirkendall voids
effect on intermetallic formation **A11:** 776
Kirkendall voids/void clusters **EL1:** 680, 1012-1013, 1148
Kirksite zinc alloy
gravity casting **A2:** 530
Kish **A18:** 698
Kish graphite *See* Carbon flotation **M1:** 4, 5, 12-13
Kish tracks
as casting defect **A11:** 388
Kiss and knife gates
copper alloy casting **A15:** 778
Kiss roll coating *See also* Coatings; Film(s)
defined **EM2:** 24
Kissing weld
laser-beam welding **A6:** 879
Kit method
of multifilamentary conductor assembly **A2:** 1047
Kits *See* Preweighed, packaged kits; Proportioned kits
Kjeldahl determination
applications **A10:** 214
estimated analysis time **A10:** 215
in ion-selective electrode methods **A10:** 186
limitations **A10:** 215
for nitrogen **A10:** 172-173, 214-215
use with bromine and methyl acetate dissolution **A10:** 177
Kjeldahl method
to analyze the bulk chemical composition of starting powders **EM4:** 72
$KL_{2,3}$ transition
Auger electron emission by **A10:** 550
K_{Ic} *See* Fracture toughness
Klemm's reagent
as an etchant for silicon-iron alloys **A9:** 531
Klemm's reagents for color etching **A9:** 142
KLM markers
for qualitative x-ray spectrometric analysis **A10:** 95-96
Kloeckner metallurgy scrap (KMS) process **A1:** 111
K_{Iscc}
maraging steels **M1:** 451
ultrahigh-strength steels **M1:** 426, 428, 431, 436, 441
Klystron **A10:** 255, 675
Klystron electron tubes **A13:** 1125-1126
corrosion failure analysis **EL1:** 1115-1116
Klystrons
for microwave generation **A17:** 208-209
KM-CAL (cooling system of Kawasaki Steel Corporation) **A4:** 58
Knapp nickel-phosphorus alloy plating bath **M5:** 204
Kneading
in milling **A7:** 63
Knee
definition **M6:** 10
Knee joint prosthesis
hinge-like **A11:** 670
sliding **A11:** 670
Knife and kiss gating systems
copper alloys **A15:** 778
Knife coating *See also* Coatings
defined **EM2:** 24
Knife-edge
attachment, crack-opening displacement transducer **A8:** 384
for constant-stress compression testing **A8:** 320-322
support **A8:** 320-331
Knife-edge contacts, machined
as attachment **A8:** 384-385
Knife-line attack **A6:** 466, 1066 **A13:** 8, 124
Knight Shift measurements **A10:** 284
in A15 superconductors **A2:** 1060
Knit line *See* Weld line
Knitlines
in injection molding **EM1:** 167
Knitted fabrics
defined **EM1:** 14
Knives *See also* Shear blades
for cutting fabrics **M1:** 625
slitting **A14:** 708-709
straight shear **A14:** 703-705
Knives, shear
service condition failures **A11:** 575, 582-583
Knob
Robertson **A11:** 60
Knobbly structure
in alloy steel fracture surface **A12:** 33, 43
aluminum alloy fracture surface **A12:** 33
cast aluminum alloys **A12:** 409
formation **A12:** 33
Knock
defined **A18:** 11
Knockout *See also* Core knockout; Shakeout **A7:** 7
defined **A14:** 8 **A15:** 7
mark, defined **A14:** 8
pin, defined **A14:** 8
as postcasting operation, investment casting **A15:** 263
punch **A7:** 7
Knockout devices
for molds **EM2:** 615
Knockout machine, defined *See* Core knockout machine
Knoop (microindentation) hardness number
defined **A18:** 11
Knoop hardness *See also* Hardness **EM3:** 17
abbreviation **A8:** 725
abbreviation for **A11:** 797
defined **EM2:** 24
number (HK), defined **A11:** 6
test, defined **A11:** 6
Knoop hardness (HK)
P/M materials **A16:** 882
polycrystalline diamond **A16:** 108
Knoop hardness indentation **A7:** 61
Knoop hardness number (HK)
Brinell conversions **A8:** 111
and case depth **A8:** 96, 101
defined **A8:** 8, 90
equivalent Rockwell B numbers **A8:** 109-110
for indentations with same test load **A8:** 91-95
Vickers conversions **A8:** 112-113
vs. load **A8:** 94-96
Knoop hardness scale
abrasives **A16:** 431, 432
Knoop hardness testing *See also* Brinell hardness testing; Knoop microhardness testing; Rockwell hardness testing; Scleroscope hardness testing; Vickers hardness testing **A8:** 8
as static indentation test **A8:** 71
Knoop indentation
compared with Vickers **A8:** 90
elastic recovery **A8:** 95
filar units for measuring **A8:** 91
in microconstituents of tool steel **A8:** 97, 101
surface preparation **A8:** 93
Knoop indentation hardness **A18:** 433
Knoop indenter *See also* Vickers indenter
defined .... **A8:** go
for glass testing **A8:** 90
hardness number and load, related **A8:** 95-96
indentations **A8:** 91

Knoop indenter mark, used to measure rate of material removal while polishing powder
metallurgy materials **A9:** 507
Knoop microhardness testing *See also* Vickers microhardness testing **A8:** 90-98
applications **A8:** 94-98
determining hardness number **A8:** 90-91
indentations **A8:** 90-91, 94
indenter selection **A8:** 90-91
machines for **A8:** 91-93
number vs. load **A8:** 94-96
surface preparation **A8:** 93
Knoop microindentation equation **A18:** 416
Knoop microindentation tests **A18:** 415, 416, 417, 418
proportionality constants as a function of hardness, force, and diagonal length **A18:** 416
Knoop minimum thickness chart
for sheet or foil **A8:** 96, 101
Knot-type twisted wire radial brushes **M5:** 151
Knotek-Feibelman (KF) model
electron- stimulated desorption (ESD) **A18:** 456, 457
Knowledge-based expert systems
for complex components **A15:** 860
Knowledge-based expert systems, for rough analysis
process design **A14:** 409-412
Knowledge-based integration
objectives of **A14:** 413
Knuckle area
defined **EM1:** 14 **EM2:** 24
Knuckle flange, steel steering
fracture surface **A11:** 342
Knuckle lever drives
mechanical presses **A14:** 494
Knuckle pins
fatigue fracture in **A11:** 128-129
rod failure in bore wall between **A11:** 477
Knuckle-drive presses
for coining **A14:** 180
flashless forging with **A14:** 172-173
tension forging, schematic **A14:** 174
Knuckle-joint mechanical press **A14:** 40
Knudsen flow **A7:** 264
Knudsen number **A18:** 524
for gas flow types **A17:** 59
Knudsen's cosine law **A18:** 843
Knurl drive
resistance seam welding **M6:** 495
Knurled disk
in spring-material test apparatus **A8:** 134-135
Knurling
in conjunction with turning **A16:** 135
Cu alloys **A16:** 816
furnace brazing of steels **M6:** 939-940
MMCs **A16:** 896
multifunction machining **A16:** 372, 373, 375, 376, 386, 387
Kohler illumination principle **A9:** 58
Kohler's rule
for copper magnetoresistance in niobium-titanium superconducting materials **A2:** 1045
Koirtyohann/Pickett continuum-source background correction system **A10:** 51, 52
Kolbe's process
alkylsalicylic acid formation **A18:** 100
Kolene process **A4:** 668
Kolmogorov-Smirnov (K-S) goodness-of-fit tests **EM4:** 701, 702, 704, 705, 707
Kolmogorov-Smirnov critical values
for normal distribution **EL1:** 902
Kolsky bar *See also* Split Hopkinson pressure bar; Torsional Kolsky bar
explosively loaded torsional **A8:** 224-225
Konovalov, Dmitry ..... **A3:** 1 • 10
Korloy 2570 *See* Zinc alloys, specific types, zinc foundry alloy ZA-12
Korloy 2684 *See* Superplastic zinc; Zinc alloys, specific types, superplastic zinc alloy
Kossel pattern
as CBEDP **A10:** 439, 442
Kossel-Mollenstaedt (K-M) diffraction patterns **A10:** 439, 441, 690
Kovar *See also* Iron-base alloys; Low-expansion alloys
ceramic/metal seals **EM4:** 502, 506, 507
chemical integrity of seals **EM4:** 540, 541
defined **EL1:** 1148
electronic applications **A6:** 998
explosion welding **A6:** 304
glass-to-metal seals **EL1:** 455, 958
glass/metal seals **EM4:** 494, 495, 497, 498
lead, EDS spectra **EL1:** 1098
as lead frame material **EL1:** 731
leaded microelectronic devices, failure analysis **EL1:** 1036-1037
as low-expansion alloy **A2:** 889
as metal-matrix material **EL1:** 1126-1127
and nickel alloy 42, applications **A2:** 443-444
nominal composition and applications **A2:** 895
oxidation **EL1:** 457 **EM4:** 497
properties **A6:** 992
properties of **EM4:** 503
recommended glass/metal seal combinations **EM4:** 497
seal design techniques **EM4:** 534, 535, 538
solderability **A6:** 978
ultrasonic welding **A6:** 895
Kovar alloy
in ceramic packages **EM3:** 585
glass-to-metal seals **EM3:** 302
surface atomic values **A10:** 579
Kovar lead material
chloride ion pitting of **A11:** 770
Kovar, photochemical
machining etchant **A16:** 588, 590
Kovar-glass seals
shear fraction studies of **A10:** 577-578
Kozeny-Carman equation
for permeametry **A7:** 264
Kraft digesters **A13:** 1208-1210
Kraft paper
as honeycomb core material **EM1:** 723
Kraft process
defined **A13:** 8
Kraft pulping liquors
corrosion by **A13:** 1208-1214
Krak-Gages
for corrosion fatigue testing **A8:** 428
for crack length **A8:** 391
Kroll magnesium-reduced titanium powder **A7:** 164
Kroll process **A7:** 7, 55
for commercially pure titanium **A2:** 1044
Kroll's reagent
as an etchant for titanium and titanium alloys **A9:** 459
Kronecker Delta
use in ODF analysis **A10:** 362
KRS-5
as internal reflection element **A10:** 113
Krumbein shape factors **A7:** 239
Krupp Works (Germany) **A15:** 31, 32, 34
Krypton
continuous-wave gas lasers **A10:** 128
gas, EXAFS analysis **A10:** 409, 410
gas, K-edge absorption spectrum **A10:** 410
ionization potentials and imaging fields for **A10:** 586
Krypton ion lasers
for optical holographic interferometry **A17:** 417
Krypton, used for ion-beam thinning of transmission electron microscopy
specimens **A9:** 107
K_u *See* Uniaxial anisotropy
Kulenkampff expression
x-ray emission **A10:** 84
Kunzl's law
energy/chemical shifts in EXAFS analysis following **A10:** 416
Kurchatov
as synchrotron radiation source **A10:** 413
Kurdjumov-Sachs orientation relationship **A4:** 221 **A10:** 438-439
in ferrous martensite **A9:** 670
Kurtosis **A18:** 348
Kurtosis coefficient **A18:** 296, 297
Kyanite
as aluminum silicate molding sand **A15:** 209
island structure **EM4:** 758
refractory material composition **EM4:** 896

L

cast steels **M1:** 54, 55, 56, 400
cold finished bars **M1:** 216, 235
ductile iron **M1:** 35, 54-56
gray iron **M1:** 22-23, 54, 55
hardenable steels **M1:** 457, 458, 470
malleable iron **M1:** 55, 63, 66, 67, 71
measures of **M1:** 565-567
ratings alloy steel bars, cold finished **M1:** 239
carbon steel bars, cold finished **M1:** 236-238
selected steels **M1:** 568-570, 576, 582
scatter in ratings **M1:** 567-568
steel plate **M1:** 194, 197
l *See* Length
L & N Fyrestan, applications
protection tubes and wells **A4:** 533
L lines **A10:** 86, 87
L shell
defined **A10:** 676 **A12:** 168
L'vov platform **A10:** 49, 55
L-605
aging cycle **M4:** 656
annealing **M4:** 655
chemical milling **A16:** 584
composition **A4:** 794, 795 **A6:** 573, 929 **M4:** 651-652
electrochemical machining removal rates **A16:** 534
mill annealing **A4:** 810
mill annealing temperature range **A6:** 573
solution annealing **A4:** 810
solution annealing temperature range **A6:** 573
solution treating **M4:** 656
stress relieving **M4:** 655
L-band frequencies
microwave reflectometer **A17:** 214
L-direction
defined **EM1:** 15
L-edges, of cesium to neodymium
EXAFS determined **A10:** 408
L-family x-ray lines
EPMA analysis **A10:** 522
L-radiation
defined **A10:** 675
L-sections
as basic casting shape **A15:** 599, 604-610
solidification sequence **A15:** 604-606
L-series
defined **A10:** 676
L-shape hooks
materials for **A11:** 522-523
L/d *See* Length-to-diameter ratio
L/D ratio
defined **A18:** 12
L1, L2, L3, etc *See* Tool steels, specific types
$L1_2$-ordered trialuminide alloys
properties **A2:** 929-930
$L_{2,3}$ transition
Auger electron emission by **A10:** 550

SUBJECTS OF THE INDEXED VOLUMES: ASM Handbook (designated by the letter "A"): **A1:** Properties and Selection: Irons, Steels, and High-Performance Alloys (1990); **A2:** Properties and Selection: Nonferrous Alloys and Special-Purpose Materials (1990); **A3:** Alloy Phase Diagrams (1992); **A4:** Heat Treating (1991); **A6:** Welding, Brazing, and Soldering (1993); **A7:** Powder Metallurgy (1984); **A8:** Mechanical Testing (1985); **A9:** Metallography and Microstructures (1985); **A10:** Materials Characterization (1986); **A11:** Failure Analysis and Prevention (1986); **A12:** Fractography (1987); **A13:** Corrosion (1987); **A14:** Forming and Forging (1988); **A15:** Casting (1988); **A16:** Machining (1989); **A17:** Nondestructive Testing and Quality Control (1989); **A18:** Friction, Lubrication, and Wear Technology (1992). **Metals Handbook, 9th Edition** (designated by the letter "M"): **M1:** Properties and Selection: Irons and Steels (1978); **M2:** Properties and Selection: Nonferrous Alloys and Pure Metals (1979); **M3:** Properties and Selection: Stainless Steels, Tool Materials and Special-Purpose Metals (1980); **M4:** Heat Treating (1981); **M5:** Surface Cleaning, Finishing, and Coating (1982); **M6:** Welding, Brazing, and Soldering (1983). **Engineered Materials Handbook** (designated by the letters "EM"): **EM1:** Composites (1987); **EM2:** Engineering Plastics (1988); **EM3:** Adhesives and Sealants (1990); **EM4:** Ceramics and Glasses (1991); **Electronic Materials Handbook** (designated by the letters "EL"): **EL1:** Packaging (1989).

$L2_1$ Heusler alloys
as ordered intermetallics **A2:** 932-933
L_{10} life *See* Rating life
L605
flash welding **M6:** 557
La-Mg (Phase Diagram) .... **A3:** 2 • 272
La-Mn (Phase Diagram) .... **A3:** 2 • 272
La-Ni (Phase Diagram) .... **A3:** 2 • 273
La-Pb (Phase Diagram) .... **A3:** 2 • 273
La-S (Phase Diagram) .... **A3:** 2 • 273
La-Sb (Phase Diagram) .... **A3:** 2 • 274
La-Sc (Phase Diagram) .... **A3:** 2 • 274
La-Se (Phase Diagram) .... **A3:** 2 • 274
La-Sn (Phase Diagram) .... **A3:** 2 • 275
La-Tl (Phase Diagram) .... **A3:** 2 • 275
La-Zn (Phase Diagram) .... **A3:** 2 • 275
Labelers
automated **EM1:** 620, 622
Labeling
and ply cutting **EM1:** 619-623
of powder metallurgy material specimens **A9:** 504-505
Labor
blow molding, costs **EM2:** 299
costs **EM2:** 82, 278
fringes, as burden (overhead) **EM2:** 86
requirements, compression molding **EM2:** 298
requirements, injection molding **EM2:** 296
Laboratory and process applications
glassware **EM4:** 1087-1090
Laboratory animals
NAA analysis for retention of toxic elements in **A10:** 233
Laboratory characterization techniques
introduction **A18:** 333
Laboratory corrosion testing *See also* Corrosion testing; Evaluation **A13:** 212-228
acceptability criteria development with **A13:** 316
aqueous, at elevated temperatures/pressures **A13:** 226-227
of corrosion inhibitors **A13:** 483
electrochemical methods **A13:** 212-220
of galvanic corrosion **A13:** 234-238
in gases at elevated temperatures **A13:** 226
immersion tests **A13:** 220-224, 303
liquid metals **A13:** 227
salt spray **A13:** 224-226
of SCC .... **A13:** 2(A
simulated atmosphere **A13:** 226
of steam turbine SCC **A13:** 955
uniform corrosion **A13:** 229-230
weldments **A13:** 346
Laboratory inspection
of tubular products **A17:** 561
Laboratory radiography
of boilers/pressure vessels **A17:** 648-649
Laboratory sample
defined **A10:** 676
Laboratory testing equipment
for weight **A15:** 363
Laboratory testing methods for solid friction **A18:** 45-58
friction databases **A18:** 57-58
friction measurements **A18:** 56-57
friction models **A18:** 46
friction nomenclature **A18:** 47-48
kinetic coefficient of friction **A18:** 47, 48
lubricated friction **A18:** 47, 48
static friction **A18:** 47-48, 57
stick-slip behavior **A18:** 47, 48, 56, 57
friction testing techniques **A18:** 46-47
capstan test **A18:** 46
inclined plane test **A18:** 46
historical development of techniques **A18:** 45-46
performing a valid test **A18:** 53-55
material documentation **A18:** 54-55
surface condition **A18:** 55
system modeling **A18:** 53-54
reporting system losses **A18:** 57
standard friction tests **A18:** 48-53
ASTM standards and specifications **A18:** 48-53
non-ASTM standards **A18:** 53
test parameters **A18:** 55-56
Laboratory tests
cracking, for hydrogen damage **A11:** 250
fatigue **A11:** 102
preliminary, of failed parts **A11:** 173
Laboratory tests (continued)
for wear **A11:** 161-162
Labyrinth seal wear
bearing failure from **A11:** 511-512
Lachance-Trail equation
calibration curves for x-ray spectrometry **A10:** 98
Lacing
machine for **EM1:** 130-131
operation **EM1:** 131
Lack of bonding
as planar flaw **A17:** 50
Lack of fill
in brazed joints **A17:** 602
Lack of fusion (LOF) **A6:** 1073, 1075, 1078, 1079
as planar flaw **A17:** 50
in weldments **A17:** 582, 584
Lack of penetration (LOP) **A6:** 1073, 1075, 1076, 1077
in weldments **A17:** 582, 584
Lack of resin fill-out
defined **EM2:** 24
Lacquer
defined **A18:** 12
Lacquer and lacquering
air-drying types **M5:** 626-627
aluminum **M5:** 572, 583
compatability, multilayer coatings **M5:** 500
copper and copper alloys **M5:** 626-627
corrosion protection **M5:** 474-475
selective cadmium plating using **M5:** 268
stop-off medium, chrome plating **M5:** 187
thermosetting types **M5:** 626-627
types **M5:** 495-496, 500-502
vacuum coatings **M5:** 400
zinc plating **M5:** 255
Lacquer coating
copier powders **A7:** 588
Lacquer sealing process
anodic coatings **M5:** 595
Lacquer(s) *See also* Enamels
-coated steel, filiform corrosion **A13:** 104
coatings, oil/gas pipes **A13:** 1259
on tinplate **A13:** 779-780
Lagging, corrosivity of **A13:** 341
Lacquering
zinc alloys **A2:** 530
Lacquers
corrosion protection with **M1:** 752-755
defined **EM2:** 24
as mounting materials for electropolishing **A9:** 49
powders used **A7:** 572
Lacquers, clear
for zinc alloy castings **A15:** 796
Lacquers, clear acrylic
as fracture preservatives **A12:** 73
Lactic acid, corrosion
stainless steels **M3:** 84
Ladder
extension, ductile overload fracture **A11:** 86-87, 91
extension, overloading to distortion failure **A11:** 137
Ladder diagrams
in atom probe microscopy **A10:** 594
of nickel-base superalloy IN 939 .... **A10:** 598, 599
Ladder polymer **EM3:** 17
defined **EM2:** 24
Ladle brick
defined **A15:** 7
Ladle coating
defined **A15:** 7
Ladle furnace and vacuum arc degassing
equipment/processing **A15:** 436-438
furnace design **A15:** 436-438
movable vacuum vessel **A15:** 436
stationary vessel **A15:** 436
Ladle hooks
materials for **A11:** 522
Ladle metallurgy
direct current arc furnace **A15:** 368
Ladle preheating
defined **A15:** 7
Ladle refining **A1:** 480
Ladle steelmaking **A1:** 112-113
Ladle treatments **A1:** 930
Ladle(s) *See also* Pouring
automated dip and pour **A15:** 498
defined, and types
desulfurization systems **A15:** 75-77
development of **A15:** 27-28
direct bottom pour **A15:** 498
electric arc furnace **A15:** 362-363
gas flow patterns in **A15:** 432
inoculation **A15:** 638-639
insulating, types **A15:** 363
metallurgy, degassing procedures **A15:** 432-444
for nickel alloy pouring **A15:** 820-821
one-man, development **A15:** 27
pouring, magnesium alloy **A15:** 800-801
pouring, types of **A15:** 497
robotic **A15:** 498
safety foundry **A15:** 28
shanked, development **A15:** 27
sizes, plasma heating and degassing **A15:** 444
skimming, as inclusion control **A15:** 90
slag, skimming, for inclusion control **A15:** 90
tilting of **A15:** 498
transfer, and pouring temperatures **A15:** 283
treatments, compared **A15:** 437
vacuum oxygen decarburization **A15:** 429-431
wheeled, 19th century **A15:** 28
Ladles
in steelmaking **M1:** 111
Lafayette Street Bridge, over Mississippi River at St. Paul
fatigue cracks in **A11:** 709
Lagrangian diagram
for flat plate impact test **A8:** 211
of stress waves in flyer plate impact test **A8:** 211
for tensile loading apparatus **A8:** 212-213
Lagrangian reference frame **A6:** 1135
LaGuerre polynomial **A6:** 877
LAM programmable calculator program
for composite materials analysis **EM1:** 277-279
Lamb waves
inspection, of spot-welded joint **A17:** 250-251
leaky, ultrasonic inspection **A17:** 251-252
testing, ultrasonic inspection **A17:** 234, 250-251
Lamb's wool paint rollers **M5:** 503
Lamb-Mössbauer factor *See* Recoil-free fraction
Lambda ratio **A18:** 30, 260, 517, 518
Lamella **EM3:** 17
defined **EM2:** 24
Lamellae **EM3:** 17
defined **EM2:** 24
Lamellae, in unidirectionally solidified eutectics as oriented surface in
quantitative metallography **A9:** 128
Lamellar alpha structures
in titanium and titanium alloys **A9:** 460
Lamellar constituents
in iron-chromium-nickel heat-resistant casting alloys **A9:** 332-334
Lamellar corrosion *See* Exfoliation corrosion
Lamellar cracking **A1:** 607-608, 609
Lamellar distances **A9:** 129
Lamellar eutectic *See also* Eutectic(s)
graphite growth, in multidirectional solidification **A15:** 175-176
solidification of **A15:** 119-120
spacing of **A15:** 120
Lamellar eutectic microstructure .... **A3:** 1 • 19, 20
Lamellar eutectic structures
as a feature of regular eutectics **A9:** 621
aluminum and Mg_2Al_3 .... **A9:** 619
branching in **A9:** 620
conditions for formation of **A9:** 618
$CuAl_2$-Al **A9:** 620
in Ni_3Al-Ni_3Nb **A9:** 619
niobium-carbide in nickel matrix **A9:** 619
Lamellar faults
in eutectic microstructures **A9:** 620
Lamellar particle shapes
defined **A9:** 621
Lamellar phases
in lead alloys **A9:** 417
Lamellar sigma phase
in rhenium-bearing alloys **A9:** 448
Lamellar spacings **A9:** 129
lamellar structure **A13:** 28, 47, 830
ductile iron **A12:** 229

lamellar structure (continued)
as fatigue mechanism **A12:** 4
fine, irons **A12:** 221
fractured pearlite, and striations, compared **A12:** 119
pearlite, high-carbon steels **A12:** 290
Lamellar tear
defined **A9:** 10
definition **A6:** 1211
Lamellar tearing *See also* Cold cracking; Hot cracking; Stress-relief cracking **A6:** 95-96, 1073, 1076 **A13:** 8-9, 521 **M6:** 248-249, 252-254, 832
at weld root **A11:** 665
carbon steels **A6:** 651, 652
conditions for **A6:** 95
deoxidation practice **M6:** 252-253
effect of joint restraint **M6:** 254
reduction of component restraint **M6:** 254-255
reduction of joint restraint **M6:** 254-255
through-thickness strains **M6:** 253-254
effect of preheating **M6:** 253
effect of welding procedure **M6:** 253
examination/interpretation **A12:** 138-139
formation **A12:** 158
low-carbon steel plates **A11:** 416
matrix properties **M6:** 253
hydrogen embrittlement **M6:** 253
notch toughness **M6:** 253
strength **M6:** 253
offshore structural steels **A6:** 384-385
orientation of members **M6:** 252
relationship to through-thickness reduction in area **M6:** 252
section thickness **M6:** 252
surface condition **M6:** 252-253
thermal contraction strain **M6:** 252
as weld imperfection **A11:** 92
in weldments **A17:** 582, 585
Lamellar thickness **EM3:** 17
defined **EM2:** 24
Lamina
allowables, material assumptions **EM1:** 308-310
defined **EM1:** 14 **EM2:** 24
strength **EM1:** 432-433
Laminae
defined **EM1:** 14, 218 **EM2:** 24
Laminar composite materials
microwave inspection **A17:** 202
Laminar composites *See also* Composite(s); Laminate(s)
defined **EM1:** 27
Laminar defects
radiography of **A11:** 17
Laminar flow
defined **EM2:** 25
in leaks **A17:** 58
nu values **EL1:** 52
Laminar fluid flow **A6:** 162
Laminate **A8:** 714
Laminate coordinates
defined **EM2:** 25
Laminate Material Properties (LAM) programmable calculator program
for composite materials analysis **EM1:** 277-279
Laminate model
accelerated life prediction **EM2:** 792-793
Laminate orientation
defined **EM2:** 25
Laminate ply **EM3:** 17
defined **EM2:** 25
Laminate property analysis *See also* Laminate(s)
of failure **EM1:** 230-235
hygroscopic analysis **EM1:** 226-227
lamination theory **EM1:** 220-222
properties **EM1:** 222-226
of strength **EM1:** 230-235
stress analysis **EM1:** 227-230
stress-strain relations **EM1:** 218-220
thermal analysis **EM1:** 226-227
Laminate ranking
computer program **EM1:** 274
notation **EM1:** 450-451
parameter sensitivity **EM1:** 455-457
vs. optimization **EM1:** 451-455
Laminate sizing
by laminate ranking method **EM1:** 450-457
Laminate(s) *See also* Angle-ply laminate; Anisotropic laminate; Anisotropy of laminate; Balanced laminate; Bidirectional laminate; Cross laminate; Cross-ply laminate; Dry laminate; Laminate property analysis; Laminate ranking; Layer Symmetrical laminate; Low-pressure laminate; Parallel laminate; Quasi-isotropic laminate; Reinforced plastics; Unidirectional laminate; Unsymmetric laminate
additive, rigid printed wiring boards **EL1:** 548
analysis, applications **EM1:** 460-462
analysis, software for **EM1:** 276-277
angle-ply, defined *See* Angle-ply laminate
anisotropic, defined *See* Anisotropic laminate
aramid fiber **EL1:** 615-618
available, for PWBs **EL1:** 607-608
behavior, prediction **EM1:** 310-311
carbon fiber, wet/dry properties **EM1:** 76
carbon-epoxy, ultimate tensile strength **EM1:** 147
composition, flexible printed boards **EL1:** 581
computer analysis programs **EM1:** 269, 274, 275-281
conditions, printed board coupons **EL1:** 576
construction **EM1:** 221, 229
coordinate systems **EM1:** 14, 219
copper-clad E-glass **EL1:** 534-537
damage tolerance, testing **EM1:** 264
defects **EL1:** 1022
defined **EL1:** 1147 **EM1:** 14 **EM2:** 25
design tests **EM1:** 310-311
dethylaminopropylamine-cured (DEAPA-cured) strength **EM1:** 74
dry, defined *See* Dry laminate
elastic constants **EM1:** 223
epoxy, formulations/properties **EL1:** 832
epoxy, properties **EM1:** 71-76
failure analysis **EM1:** 236-251
failure load/modes **EM1:** 233
failure modes **EM1:** 240-244
failure surfaces **EM1:** 232
fatigue analysis **EM1:** 236-251
fatigue failure in **EM1:** 246
fiberglass, and cyanates **EM2:** 235
fiberglass printed wiring board, properties **EM2:** 237
fracture analysis **EM1:** 252-258
geometric variables **EM1:** 236
as heat sinks **EL1:** 1129-1131
high-pressure, defined *See* High-pressure laminates
hybrid, damping in **EM1:** 214
load-strain response **EM1:** 231
low-pressure, defined *See* Low-pressure laminates
material properties, for design **EM1:** 183
notation **EM1:** 222
orientation ,defined **EM1:** 14
parallel, defined *See* Parallel laminate
ply, defined **EM1:** 14
process behavior, model **EM1:** 745
progressive ply failures **EM1:** 238-239
properties **EL1:** 608 **EM1:** 222-226, 287, 308-314
properties analysis of **EM1:** 218-235
quartz, advantages/disadvantages **EL1:** 619
quasi-isotropic, defined *See* Quasi-isotropic laminate
of recycled carbon fiber **EM1:** 154-155
reference system **EM1:** 236
resin effects **EM1:** 287
sandwich, damping analysis **EM1:** 214
simple loading of **EM1:** 208-209
size **EM1:** 236, 450-457
SMT, properties **EL1:** 616
Laminate(s) (continued)
special, in structural analysis **EM1:** 460
stacking arrangements **EM1:** 209
strength analysis **EM1:** 236-251
stress/moment resultants **EM1:** 221
symmetric, properties of **EM1:** 222-224
tensile strength measurement **EM1:** 286-287
unsymmetric **EM1:** 224, 446-447
woven **EM1:** 125-128
Laminated coatings
aluminum and aluminum alloys **M5:** 609-610
coated carbide tools **A2:** 959
Laminated composite analysis
computer software for **EM1:** 269, 275-281
Laminated construction
of continuous fiber reinforced composites **A11:** 732
Laminated glass **EM4:** 423-426
applications **EM4:** 423
cladding and body glasses **EM4:** 424-425
compositions of glasses **EM4:** 424
properties of glasses **EM4:** 424
thermal expansion curves **EM4:** 425
viscosity curves **EM4:** 425
composite stresses **EM4:** 425-426
delayed breakage **EM4:** 426
dicing **EM4:** 425
surface compression magnitude **EM4:** 425
glass requirements **EM4:** 423-424
chemical durability **EM4:** 424
composition compatibility **EM4:** 424
thermal expansion **EM4:** 424
viscosity **EM4:** 424
weathering characteristics **EM4:** 424
process of lamination **EM4:** 423, 424
hot-lamination **EM4:** 423
recommended waterjet cutting speeds **EM4:** 366
stress profiles **EM4:** 423
Laminated metal sandwiches
as heat sinks **EL1:** 1129-1131
Laminated-ceramic packages
thermal performance **EL1:** 409-410
Laminates **EM3:** 17, 504-505 **EM4:** 1101
advantages **EM4:** 1101
Corelle **EM4:** 1101
formation **EM4:** 1101
graphite-epoxy **EM3:** 821
optical holographic interferometry of **A17:** 423
properties **EM4:** 1101
resin-matrix composites **A12:** 478
strength **EM4:** 1101
Laminating cells
generations of **EM1:** 639-640
Laminating process
epoxy composites **EM1:** 71-73
Laminating resins
amino **EM2:** 628
Lamination **A7:** 7 **EM3:** 17 **EM4:** 377
cells, generation development **EM1:** 639-640
defect by insufficient green strength **A7:** 302
defined **A9:** 10 **A11:** 6
effect on *x-y* shrinkage **EL1:** 465
electrical processes, wet/dry lay-up methods **EL1:** 832
flash (bead) removal **EL1:** 544
hydraulic vs autoclave systems **EL1:** 510
mass, rigid printed wiring boards **EL1:** 551
multilayer **EL1:** 543-544
multilayer ceramic capacitors **EM4:** 1116
pipeline failure from **A11:** 697
ply, automated **EM1:** 639-641
sequential process of **EL1:** 133-134
in sheet and plate, brittle fracture by **A11:** 85
specification **EL1:** 523
in substrate fabrication **EL1:** 464-465
technology, capabilities/limitations **EL1:** 510
thermoplastic polyimides (TPI) **EM2:** 178
two-step process **EL1:** 464

Lamination flash (bead) removal
rigid printed wiring boards ... **EL1:** 544
Lamination iron
heat treatment ... **A2:** 762
Lamination theory ... **EM1:** 218, 220-222, 234, 458-460
Laminations *See also* Blanking; Electrical steel sheet; Piercing
blanked and pierced ... **A14:** 477
by liquid penetrant inspection ... **A17:** 71, 86
fabrication ... **A2:** 780
heat treatments ... **A2:** 762-763
interlaminar insulation ... **A2:** 780
matched, by Guerin process ... **A14:** 606-607
as planar flaw ... **A17:** 50
radiographic methods ... **A17:** 296
stacking ... **A14:** 478
in steel bar and wire ... **A17:** 549
strips, iron-cobalt alloy, properties ... **A2:** 777
tubular products ... **A17:** 567
ultrasonic inspection of ... **A17:** 232
Laminographic radiography
as x-ray solder joint inspection ... **EL1:** 738
Laminography techniques
development of ... **A17:** 379
LAMMA *See* Laser microprobe mass analysis
Lamp bases
as slush cast ... **A15:** 35
Lamp filaments ... **A7:** 16, 153, 631
Lamp IR systems
infrared soldering ... **EL1:** 704-705
Lampblack ... **A7:** 7
Lamps
sodium ... **A2:** 752
Lamps, pulsed xenon/ultraviolet
for curing ... **EL1:** 864
LAMPS-A computer program for structural analysis ... **EM1:** 268, 271
LamRank computer program ... **EM1:** 451-454
Lancing
for blanks ... **A14:** 446
iron-aluminum blends for ... **A7:** 843
powders used ... **A7:** 572
in press brake ... **A14:** 538
in stretch drawing ... **A14:** 593
Land
defined ... **EM2:** 25
definition ... **A6:** 1211
Land impression
flash ... **A14:** 50
Land patterns
dimensions, control of ... **EL1:** 672-673
outer lead bonding ... **EL1:** 283-285
Landing gear
deflection yoke, SCC failure of ... **A11:** 23
flat spring, fracture of ... **A11:** 560-561
torque arm, fatigue fracture of ... **A11:** 114
Lane plate (implant)
failed historic ... **A11:** 674, 680
Lang topography
near-surface analysis ... **A10:** 377
Pendellosüng fringes obtained by ... **A10:** 368
for polygonization of elastic strain relaxation ... **A10:** 370, 377
Langelier saturation index
defined ... **A13:** 9
Langer equation
strain-life ... **A8:** 698
Langmuir Torch atomizer
atomic absorption spectrometry ... **A10:** 54
Langmuir-Blodgett deposition ... **A18:** 401
Langmuir-Blodgett deposition technique ... **A10:** 119-120
Langmuir-type isotherms ... **EM4:** 71
Lanthanide carbonyls ... **A7:** 135
Lanthanides
on periodic table ... **A10:** 688
species weighed in gravimetry ... **A10:** 172
use in flame atomizers ... **A10:** 48
weighed as the fluoride ... **A10:** 171
Lanthanum *See also* Rare earth metals
additions to flame AAS samples ... **A10:** 48
adhesion and solid friction ... **A18:** 33
classification in tungsten alloy electrodes for GTAW ... **A6:** 191
in compacted graphite iron ... **A1:** 56
in composition of cobalt-base wrought alloys ... **A18:** 766
Lanthanum (continued)
in ferrite ... **A1:** 408
fluoride separation ... **A10:** 169
in heat-resistant alloys ... **A4:** 512
internal flaw analysis ... **EM4:** 666
melting point ... **A2:** 720
properties ... **A2:** 1182
pure ... **M2:** 760-761
as rare earth ... **A2:** 720
TNAA detection limits ... **A10:** 237
vapor pressure, relation to temperature ... **A4:** 495
Lanthanum chromite ($LaCr_2O_4$)
for oxide ceramic heating elements of electrically heated furnaces ... **EM4:** 249
Lanthanum hexaboride
as electron source ... **A12:** 167
Lanthanum sesquioxide-yttrium oxide (La_2O_3-Y_2O_3) solid-state sintering ... **EM4:** 277, 278
Lanthanum, vapor pressure
relation to temperature ... **M4:** 310
Lanxide process ... **EM4:** 35
ceramic-matrix composites ... **EM4:** 840, 842
Lap *See also* Filament winding
defined ... **A9:** 10 **EM1:** 14 **EM2:** 25
Lap defect
in closed-die forging ... **A8:** 590, 592
Lap joint
analysis ... **EM1:** 338-341
brazing ... **A6:** 120
defined ... **EM1:** 14 **EM2:** 25
definition ... **A6:** 1211
electron-beam welding ... **A6:** 260
laser-beam welding ... **A6:** 879, 880
oxyfuel gas welding ... **A6:** 286, 287
soldering ... **A6:** 130
Lap joint test ... **A18:** 404
Lap joints ... **EM3:** 17, 34
arc welding of
aluminum alloys ... **M6:** 374-375, 379
nickel alloys ... **M6:** 437
stainless steels, austenitic ... **M6:** 334
brazing of
aluminum alloys ... **M6:** 1026
beryllium ... **M6:** 1053
niobium ... **M6:** 1056
definition, illustration ... **M6:** 10, 60-61
dip brazing of steels ... **M6:** 992
electron beam welds ... **M6:** 616-617
furnace brazing of steels ... **M6:** 943
gas metal arc welding of commercial coppers ... **M6:** 403
gas tungsten arc welding ... **M6:** 201, 361
of aluminum alloys ... **M6:** 396
of heat-resistant alloys ... **M6:** 356-357, 360
of silicon bronzes ... **M6:** 413
oxyfuel gas welding ... **M6:** 589-590
radiographic inspection ... **A17:** 334
resistance brazing ... **M6:** 983
seam welding ... **M6:** 501
strain ... **EM3:** 546
stresses in ... **EL1:** 57
tolerances for laser beam welds ... **M6:** 663-664
ultrasonic welding ... **M6:** 747-751
Lap shear failure tests
for strength ... **EM1:** 710
Lap welding ... **M6:** 673
indentors ... **M6:** 673
metals welded ... **M6:** 673
Lap welds
in pipe ... **A11:** 698
Lap-joint shear test
for sheet metal ... **A8:** 67
Lap-seam welding
of blanks ... **A14:** 450
Lap-shear coupon test ... **EM3:** 471, 472-473, 476
contaminated coupon ... **EM3:** 524, 526
Lap-shear strength
steel-propylene joints ... **EM3:** 62
Lap-shear testing ... **EM3:** 773-775, 804, 809
LaPlace equation ... **A7:** 312
in computer modeling ... **A13:** 234
Laplace pressure
negative ... **A18:** 400
Laplace transforms
of viscoelastic matrix properties ... **EM1:** 191
Laplace's equation
electrical potential ... **A8:** 386
Laplanche diagram
structural ... **A15:** 68-70
Lapping ... **A16:** 492-505 **EM4:** 313, 351-358
abrasive cutting mechanism ... **EM4:** 351-356
charged plate abrasives ... **EM4:** 352, 354
cylindrical lapping between flat laps ... **EM4:** 356
double-sided flat lapping ... **EM4:** 355-356
lap plate materials ... **EM4:** 352
lapping processes and equipment ... **EM4:** 353-356
lapping vehicle fluids ... **EM4:** 353
processing equipment ... **EM4:** 354, 355, 356
rolling abrasives ... **EM4:** 351-352
single-side flat lapping ... **EM4:** 353-355
sliding abrasives ... **EM4:** 352
slurry ... **EM4:** 351, 352, 355
anvil and flyer plates before ... **A8:** 235
balls for ball bearings ... **A16:** 504-505
barrel finishing ... **A16:** 503
bearing races ... **A16:** 501, 502
cast irons ... **A16:** 664-665
cemented carbides ... **A18:** 796
centerless lapping with bonded abrasives ... **A16:** 496-497
centerless roll lapping ... **A16:** 496-497
compared to honing ... **A16:** 486, 491
coolants ... **A16:** 495, 496, 497, 502, 505
crankshafts ... **A16:** 497
defined ... **A18:** 12
of diamond tools for machining Zn alloys ... **A16:** 832
end surfaces
fixture, modified diamond-stop ... **A8:** 234-235
flat lapping using manual methods ... **A16:** 498, 503
flat lapping using mechanical methods ... **A16:** 498-500, 503, 504
flat surfaces ... **A16:** 498-502
gages ... **A16:** 495-496
gears ... **A16:** 343, 351, 505
glazing ... **A16:** 503
individual-piece lapping ... **A16:** 492
inner cylindrical surfaces ... **A16:** 497-498
load capacities ... **A16:** 501
machine lapping between plates ... **A16:** 494, 497
as machining process ... **A7:** 462
matched-piece lapping ... **A16:** 492
MMCs ... **A16:** 896
Ni alloys ... **A16:** 835
outer cylindrical surfaces ... **A16:** 494-497
outer surfaces of piston rings ... **A16:** 497
P/M materials ... **A16:** 889, 891
problems in flat and end lapping ... **A16:** 502-503
process capabilities ... **A16:** 492
ring lapping ... **A16:** 494
select on of vehicle ... **A16:** 493-494
selection of abrasive ... **A16:** 492-493
size tolerance and parallelism ... **A16:** 500
spherical surfaces ... **A16:** 503-504
springlike parts ... **A16:** 505
and superabrasives ... **A16:** 456
synthetic superabrasives ... **A2:** 1012
to accelerate wear-in ... **A16:** 505
workpiece size and shape ... **A16:** 500
Lapping compounds
removal of ... **M5:** 14
Laps
in billets, magnetic particle inspection ... **A17:** 115
brittle fractures from ... **A11:** 85
in closed-die forging ... **A14:** 385
in connecting engine rod ... **A11:** 328-329
as crack origin, steel ... **A12:** 64, 65
defined ... **A11:** 6
as discontinuities, defined ... **A12:** 64-65
eddy current inspection of ... **A17:** 164
effect on cold-formed part failure ... **A11:** 307
as forging defect ... **A11:** 317, 327, 328
as forging process flaws ... **A17:** 493
forging/rolling, as planar flaw ... **A17:** 50
in hot-rolled steel bars ... **A1:** 240
in integrated circuits ... **A11:** 769
liquid penetrant inspection ... **A17:** 71, 86
low-carbon steel fracture from ... **A12:** 252
magnetic particle inspection ... **A17:** 107
micrograph of ... **A11:** 317
preventing, in hot upset forging ... **A14:** 88
radiographic methods ... **A17:** 296

Laps (continued)
rolling, fracture from ... **A12:** 64
in semisolid metal castings and forgings ... **A15:** 336-337
in sledge-hammer head ... **A11:** 574, 580
in steel bar and wire ... **A17:** 550
tool and die failure from ... **A11:** 574
tool steels ... **A12:** 378
tubular products ... **A17:** 567
Laps in rolled steel
revealed by macroetching ... **A9:** 176
LaQue Center for Corrosion Technology
history ... **A2:** 429
LARC TPI
temperature capabilities ... **EM1:** 78
LARC-160 polyimide resin ... **EM1:** 662
Large area back contacts
development ... **EL1:** 958
Large aspect ratio elements ... **EM3:** 491
Large components *See* Large parts; Parts
Large parts *See* Part(s)
acoustic emission inspection ... **A17:** 278
demagnetizing ... **A17:** 121
magnetic particle inspection of ... **A17:** 89
Large scale integration (LSI) *See also* Integration
packaging, development ... **EL1:** 961
types and trade-offs ... **EL1:** 166-168
Large-end bearing *See* Big-end bearing
Large-scale integrated (LSI) circuits ... **A18:** 685-689
Large-scale integration (LSI) devices
abbreviation for ... **A11:** 797
chip failures in ... **A11:** 766
failure analysis in ... **A11:** 767
latch-up detection in ... **A11:** 768
Large-scale tests *See also* Full Scale tests ... **EM1:** 336-338
Larmor frequency
defined ... **A10:** 676
Larmor period
defined ... **A10:** 676
Larson-Miller Parameter *See* Time temperature parameters
and elevated-temperature service ... **A1:** 627-628
rupture stress as a function of ... **A6:** 440
scale thickness and oxide penetration as functions of ... **A11:** 605, 797
Larson-Miller parameter (LMP) ... **A8:** 303-304, 340
extrapolation abilities ... **A8:** 335
method of creating master for Inconel 718 ... **A8:** 333, 335
time-temperature ... **A8:** 333-334
Larson-Miller plot
2$^1/_4$ Cr-1Mo steel ... **M1:** 646, 656
Laser
definition ... **A6:** 1211
interferograms, of low gravity effects ... **A15:** 148
pouring control ... **A15:** 500-501
surface treatment, of eutectic alloys ... **A15:** 125
to groove ceramic coatings to better resist thermal shock ... **EM4:** 207
Laser ablation
of high-temperature superconductors ... **A2:** 1087
as thin-film deposition technique ... **A2:** 1082
Laser ablation deposition ... **A18:** 844
Laser acoustic microscopes
scanning ... **EL1:** 370
Laser alloying ... **M6:** 793-796
alloying procedures ... **M6:** 794-795
experimental application ... **M6:** 793-794
liquid flow ... **M6:** 795-796
microstructures ... **M6:** 795
Laser beam
helium-neon ... **A11:** 768
scanner, optical schematic ... **A11:** 768-769
welds, failure origins ... **A11:** 447-449
Laser beam cutting
definition ... **M6:** 10
safety precautions ... **M6:** 59
Laser beam cutting, defined *See also* Laser cutting; Lasers ... **A14:** 8
Laser beam machining (LBM) ... **A16:** 19, 34, 509, 572-576 **EM4:** 313, 314, 367-370
carbon dioxide lasers ... **A16:** 572-573, 574, 575, 576
cutting ... **A16:** 574-575
debris removal ... **EM4:** 369-370
drilling compared to electron beam machining ... **A16:** 571
drilling of carbon and alloy steels ... **A16:** 677
drilling of Inconel ... **A16:** 32
effect on ceramic adhesion properties ... **EM4:** 370
electrolytes ... **A16:** 576
equipment ... **A16:** 572 **EM4:** 367-368
cost ... **EM4:** 367
focusing lens specifications ... **EM4:** 368
Gaussian mode waves ... **EM4:** 367-368
power requirements ... **EM4:** 367
wavelength specifications ... **EM4:** 367
etchant solutions for laser-enhanced etching ... **A16:** 576
excimer lasers ... **A16:** 572
fiber optic cables ... **A16:** 576
future developments ... **A16:** 576
gas jets ... **A16:** 573-574, 575-576
gas-assisted LBM ... **A16:** 573
in green machining ... **EM4:** 183
laser lens protection ... **EM4:** 369-370
laser welding ... **A16:** 572, 575
laser-enhanced etching ... **A16:** 576
lens selection ... **A16:** 573, 574, 575
MMCs ... **A16:** 893, 894, 895, 896
neodymium-doped YAG lasers ... **A16:** 572, 573, 574, 575, 576
neodymium-glass lasers ... **A16:** 572
PCBN ... **A16:** 111
percussion drilling ... **A16:** 573-574
process capabilities of lasers ... **A16:** 572
processes ... **EM4:** 368-369
contour machining ... **EM4:** 369
drilling ... **EM4:** 369, 370
precision machining ... **EM4:** 369
scribing ... **EM4:** 368-369, 370
through-hole metallization ... **EM4:** 369
robotic devices ... **A16:** 576
shrinkage tolerance problem ... **EM4:** 367
stainless steels ... **A16:** 706
surface alterations produced ... **A16:** 24
surface integrity effects ... **A16:** 28
surface treatment ... **A16:** 575-576
Ti alloys ... **A16:** 846, 852, 855
trepanning ... **A16:** 573, 574
versus ultrasonic machining ... **EM4:** 361
Laser beam sorting system
dimensional ... **A17:** 15-16
Laser beam weld
solidification cracking ... **A12:** 345
wrought aluminum alloys ... **A12:** 422
Laser beam welding ... **M6:** 647-671
advantages ... **M6:** 648
applications ... **M6:** 647-649
continuous wave welding ... **M6:** 657-658
definition ... **M6:** 10
design for welding ... **M6:** 663-664
butt joints ... **M6:** 663-664
kissing welds ... **M6:** 664
lap joints ... **M6:** 663-664
sheet ... **M6:** 663-664
wires ... **M6:** 664-665
differences of processes ... **M6:** 656
comparison of test results ... **M6:** 658
gas shielding ... **M6:** 659
general welding considerations ... **M6:** 655-656
in-process inspection ... **M6:** 668-669
acoustic emission ... **M6:** 668
video ... **M6:** 668
ionization potentials ... **M6:** 659
keyhole welding ... **M6:** 658
laser material interactions ... **M6:** 654-655
Laser beam welding (continued)
laser safety ... **M6:** 669-670
chemical hazards ... **M6:** 670
electrical hazards ... **M6:** 669
eye hazards ... **M6:** 669-670
skin exposures ... **M6:** 670
training, medical examinations, and documentation ... **M6:** 670
lasers suitable for welding ... **M6:** 651-653
carbon dioxide lasers ... **M6:** 651-652
gas lasers ... **M6:** 651-652
neodymium: yttrium aluminum garnet lasers ... **M6:** 651
solid-state lasers ... **M6:** 651
metals welded ... **M6:** 647
operating costs ... **M6:** 669
optics design and beam transport ... **M6:** 665-667
atmospheric effects ... **M6:** 666
beam transport ... **M6:** 665
focal point travel ... **M6:** 667
focusing optic selection ... **M6:** 666-667
helium-neon pointing lasers ... **M6:** 665
optics material ... **M6:** 665-666
optics mounting ... **M6:** 666-667
power measurement ... **M6:** 665
plasma effects ... **M6:** 659-661
fundamental concepts ... **M6:** 659
suppression techniques ... **M6:** 660-661
welding performance ... **M6:** 659-660
process fundamentals ... **M6:** 649-651
explanation of lasers ... **M6:** 649-650
laser cavities ... **M6:** 650
laser characteristics ... **M6:** 650-651
product improvement ... **M6:** 656-657
pulsed welding ... **M6:** 656-657
reduced cost ... **M6:** 657
safety precautions ... **M6:** 59, 669-670
seam welding ... **M6:** 657
selection of lasers ... **M6:** 652-654
cost ... **M6:** 654
expandability ... **M6:** 653-654
penetration ... **M6:** 652-653
productivity ... **M6:** 653
weld quality ... **M6:** 653
spot welding ... **M6:** 657
threshold effects ... **M6:** 655-656
relationship of thickness and welding speed ... **M6:** 655-656
weld properties ... **M6:** 661-663
aluminum and its alloys ... **M6:** 661
iridium alloys ... **M6:** 663
steels ... **M6:** 662
titanium and its alloys ... **M6:** 662-663
work and beam handling devices ... **M6:** 667-668
positioning ... **M6:** 668
speed requirements ... **M6:** 668
vibration or oscillation ... **M6:** 668
Laser beam welding of
aluminum and aluminum alloys ... **M6:** 647, 661-662
carbon steels ... **M6:** 647
copper ... **M6:** 647
copper alloys ... **M6:** 647
high-strength low-alloy steels ... **M6:** 647
iridium alloys ... **M6:** 663
iron-based alloys ... **M6:** 647
lead ... **M6:** 647
low-alloy steels ... **M6:** 647
nickel alloys ... **M6:** 647
precious metals and alloys ... **M6:** 647
refractory metals ... **M6:** 647
stainless steels ... **M6:** 647, 662
penetration in welds ... **M6:** 654, 656
welding rates ... **M6:** 658
titanium and titanium alloys ... **M6:** 647, 662-663
Laser beam xerographic printers ... **A7:** 580
Laser boriding
titanium alloys ... **A18:** 781

Laser brazing A6: 123-124 M6: 1064-1066
comparison to conventional brazing M6: 1065
limitations M6: 1065-1066
procedure
filler metal M6: 1064
protective atmosphere.............................. M6: 1065
Laser cladding *See* Laser hardfacing M6: 796-798
alumina M6: 798
aluminum-silicon alloys A18: 791
dense matrix.. M6: 798, 800
experimental results.. M6: 796
Haynes Stellite alloys M6: 797
silicon.. M6: 797
Tribaloy alloys M6: 796-797
Laser confocal scanning microscope (LCSM) A18: 357
diagram A18: 359
hardware configurations.............................. A18: 358
Laser cutting *See also* Lasers A14: 735-742 EM1: 676-680
applications A14: 741-742
with CO_2 laser EM1: 676
competing cutting methods.......................... A14: 737
of composites EM1: 678-680
cutting principles A14: 737-740
defined A14: 8, 735
focusing laser beams.................................... EM1: 676
gas-assisted cutting EM1: 676-677
laser types A14: 735-737
with neodymium-YAG laser EM1: 676
as new metalworking process........................ A14: 18
process variables A14: 738-740
system equipment A14: 740-741
systems EM1: 677-678
Laser diodes
with gallium compounds................................ A2: 739
Laser Doppler velocimetry
to analyze ceramic powder particle sizes EM4: 67
Laser Doppler velocity gage A17: 16-17
Laser excitation
optical testing EL1: 570
Laser hardening
aluminum-silicon alloys A18: 791
resistance to cavitation erosion.................... A18: 217
Laser hardfacing A6: 789, 799-800, 802, 806-807
applications A6: 803, 806-807
cracking A6: 807
materials A6: 806
processing A6: 806-807
shielding gases A6: 807
substrates A6: 806
Laser inspection *See also* Laser(s) A6: 1126 A17: 12-17
beam sorting system A17: 15-16
dimensional measurements A17: 12-16
of solder joints EL1: 942
of soldered joints A17: 606
of surfaces A17: 17
velocity measurements A17: 16-17
Laser interferometers A17: 14-15
for strain measurement A8: 193
Laser ionization mass spectroscopy (LIMS), for trace element measurement
ultra-high purity metals................................ A2: 1095
Laser melt/particle inspection M6: 798-802
alloying materials M6: 800
aluminum M6: 802
iron-based alloys M6: 801-802
mechanical characterization M6: 802-803
microstructures M6: 800-801
process description M6: 799-800
Laser metrology systems
seal wear A18: 552
Laser microprobe mass analysis.... A10: 142, 516, 583
Laser microprobe mass analysis (LAMMA)
abbreviation for A11: 797
components A11: 36
development of A11: 35-36
uses of A11: 37
Laser microprobe mass spectrometry (LMMS) EL1: 1088-1090
Laser nitriding
titanium alloys A18: 781
Laser pantography
as chip interconnect EL1: 232-233
Laser photoacoustic shockwave test A18: 434
Laser soldering A6: 359-360
advantages A6: 359
applications A6: 359
blind laser soldering A6: 359
key attributes A6: 359
carbon dioxide laser.. A6: 359
disadvantages... A6: 359
fine pitch technology (FPT) A6: 359
heat-sinking efficiency A6: 359
helium-neon (He-Ne) laser A6: 359
infrared (IR) detector A6: 359, 360
infrared signature A6: 360
intelligent laser soldering A6: 359-360
key attributes A6: 360
lasers utilized A6: 359
Nd:YAG laser A6: 359
surface-mount EL1: 706-707
tape automated bonding (TAB) A6: 359
Laser speckle patterns
speckle metrology A17: 432
Laser surface processing A13: 501-503 A18: 861-871
laser alloying A18: 865-866
applications A18: 866
examples of laser-alloyed surface microstructures A18: 866
laser gas alloying A18: 866
processing A18: 865-866
wear behavior of laser-alloyed surfaces A18: 866
laser characteristics A18: 861
laser cladding A18: 861, 866-868
applications A18: 868
examples of layer microstructures A18: 867
processing A18: 866-867
wear behavior of layers A18: 867-868
laser melt/particle injection A18: 868-871
applications A18: 871
microstructures of layers A18: 869
processing A18: 868-869
wear behavior of layers A18: 869-871
laser melting A18: 863-865
applications A18: 865
examples of laser-melted surface microstructures A18: 864
processing A18: 863-864
wear behavior of laser-melted surfaces A18: 864-865
laser surface modification techniques A18: 861
process variables and methods of measuring A18: 861
laser transformation hardening A18: 861-863
applications A18: 863
examples of laser-hardened surface microstructures A18: 862
processing A18: 861-862
wear behavior of laser-hardened surfaces A18: 862-863
Laser surface transformation hardening A4: 265, 286-295
advantages A4: 287
fundamentals....................... A4: 286-287 M4: 507-508
heat flow A4: 287, 288-289 M4: 509, 510-511
laser hardening, metallurgy of A4: 287-288 M4: 508-509
metalworking lasers........... A4: 290-291 M4: 512-513
optical systems.......... A4: 288, 290, 291-293, 294, 295 M4: 512, 513-514
process conditions M4: 511-512
processing parameters A4: 290
surface hardening, specific parts.... A4: 293-295 M4: 514-517
cast iron camshaft lobes M4: 515, 516-517
cylinder with conical top (4140 steel) M4: 515-516
large gear (1045 steel) M4: 516, 517
steel plates (1045 steel)......................... M4: 514-515
Laser Technology, Inc.
exclusive licensing of electronic shearography EM3: 762
Laser triangulation sensors A17: 13-14
Laser welded joints
metallography and microstructures............... A9: 581
Laser welding
joint configurations EL1: 240
processes, applications EL1: 238
systems, failure mechanisms............. EL1: 1044-1045
Laser welding (continued)
tube/pipe rolling A14: 631
Laser(s) *See also* Laser inspection
bench micrometer, self-contained A17: 13
coherence length limitations A17: 411-412
dimensional measurement applications........ A17: 12
effect, on speckle metrology......................... A17: 432
as holographic components................... A17: 417-420
inspection A17: 12-17
interferometric micrometer, for length measurement A17: 15
light, in optical holography......................... A17: 224
neodymium-doped yttrium aluminum garnet (Nd:YAG) A17: 410, 418
probes, in coordinate measuring machines A17: 25
semiconductor, effect on optical interconnections EL1: 10
sources, optical holographic interferometry A17: 417-418
speckle patterns A17: 432
spot pulsing EL1: 368
triangulation sensors A17: 13-14
types, optical holographic interferometry A17: 407
as UV source, for radiation curing EL1: 864
Laser-based light scattering techniques EM4: 77
Laser-beam brazing
stainless steels A6: 922
Laser-beam cutting (LBC)
definition A6: 1211
safety precautions A6: 1203
Laser-beam welding
failure mechanisms EL1: 1044-1045
Laser-beam welding (LBW)........ A6: 262-268, 874-880
absorptivity A6: 878
advanced titanium-based alloys A6: 526
advantages A6: 262, 395, 874
aluminum alloys A6: 739
aluminum and aluminum alloys A6: 263
aluminum metal-matrix composites A6: 555, 556-557
aluminum-lithium alloys A6: 551
applications A6: 262, 874
automotive A6: 393, 395
sheet metals A6: 398, 400
austenitic stainless steels................. A6: 459, 463, 464
cast irons A6: 720
conduction-mode welding A6: 264
consumables, use of.................................. A6: 879-880
deep-penetration-mode welding A6: 264-265, 266
definition A6: 262, 874, 1211
dispersion-strengthened aluminum alloys A6: 543, 544, 545, 546
efficiency A6: 262
energy consumption A6: 262
ferritic stainless steels A6: 448
filler metals A6: 739, 879-880
fluid-flow calculation, model of A6: 1148
fundamentals A6: 264-266
hardfacing alloy consumable form A6: 796
health and safety A6: 267-268
heat sources A6: 1142, 1144
heat-affected zone A6: 262, 263
heat-treatable aluminum alloys A6: 528
interaction time A6: 875
joint fit-up A6: 879
joint preparation A6: 879
joint weld design A6: 879-880
laser safety officer designation A6: 267
laser welding parameters........................... A6: 265-266
laser-beam diameter A6: 875-876
laser-beam power A6: 875
laser-beam spatial distribution A6: 876-878
limitations A6: 262
limitations on procedure qualification A6: 1093
microwelding with pulsed lasers A6: 263, 266, 267
modes, indexing of A6: 876-877
nickel-base corrosion-resistant alloys containing molybdenum A6: 594
niobium alloys A6: 581
non-heat-treatable aluminum alloys A6: 538
oxide-dispersion-strengthened materials A6: 1039
parameters for selected pulsed and continuous wave lasers A6: 263

Laser-beam welding (LBW) (continued)
peak penetration ... **A6:** 262
penetration welding ... **A6:** 263
polarized beam application developed by Fraunhofer Institute ... **A6:** 263
power density ... **A6:** 875
precipitation-hardening stainless steels ... **A6:** 489
procedure development ... **A6:** 874-879
process applications ... **A6:** 262-264
process selection ... **A6:** 874
processing equipment ... **A6:** 266-267
safety precautions ... **A6:** 1203
shielding gases ... **A6:** 878
in space and low-gravity environments ... **A6:** 1021, 1022
special welding practices ... **A6:** 879-880
stainless steel casting alloys ... **A6:** 496
stainless steels ... **A6:** 262-263, 679, 688, 697, 698, 699
standard procedure qualification test weldments ... **A6:** 1090
steels ... **A6:** 263
stress analysis of welds ... **A6:** 1138
titanium and titanium alloys ... **A6:** 85, 263-264, 512, 513, 514, 516, 517, 518, 520, 521, 783, 784
training, medical examinations, and documentation ... **A6:** 268
traverse speed ... **A6:** 878
ventilation of metal fumes, limit values ... **A6:** 268
vs. arc welding ... **A6:** 262
vs. electron-beam welding (EBW) ... **A6:** 262
vs. oxyacetylene welding (OAW) ... **A6:** 262
vs. plasma arc welding (PAW) ... **A6:** 262
wrought martensitic stainless steels ... **A6:** 440-441
zirconium alloys ... **A6:** 787
Laser-enhanced plating ... **A18:** 835
Laser-induced chemical vapor deposition (LCVD) ... **A18:** 846, 848
photolytic ... **A18:** 848
pyrolytic ... **A18:** 848
Laser-induced fluorescence spectroscopy ... **A10:** 80-81
Laser-induced resonance ionization mass spectroscopy ... **A10:** 141, 142
Laser-jet plating ... **A18:** 835
Laser-trimmable resistors
lamination process for ... **EL1:** 464
Laser/vision inspection systems
robotic ... **EM2:** 845
Lasers *See also* Laser beam cutting; Laser cutting; Laser welding
ablation for solid-sample analysis ... **A10:** 36
broadband tunable dye ... **A10:** 128
cavity operation ... **M6:** 650
characteristics ... **M6:** 650-651
continuous-wave gas ... **A10:** 128
in corrosion analysis ... **A10:** 135
cost ... **M6:** 654
defined ... **A14:** 735
definition ... **M6:** 647
description of operation ... **M6:** 649-650
expandability ... **M6:** 653-654
exposure to preactivate Raman analysis samples ... **A10:** 130
helium-neon, in FT-IR spectroscopy ... **A10:** 112
interactions of materials ... **M6:** 654-655
irradiation, of 202 steel ... **A10:** 623
as light source in vibrational spectroscopy ... **A10:** 128
low-powered, for Raman analysis of graphites ... **A10:** 132
in molecular fluorescence spectroscopy ... **A10:** 76, 80-81
neodymium-doped, yttrium-aluminum-garnet (Nd:YAG), in spark source mass spectrometry ... **A10:** 142, 597
nitrogen gas, for atom probe analysis ... **A10:** 597
operating costs of welding ... **M6:** 669
Lasers (continued)
optics design and beam transport ... **M6:** 665-667
parts required ... **M6:** 649
penetration ... **M6:** 653-654
productivity ... **M6:** 653
properties affecting interaction ... **M6:** 654-655
radiation, use in Raman spectroscopy ... **A10:** 128
in Raman analyses of polymers ... **A10:** 131
Raman molecular microprobe ... **A10:** 129
safety in welding ... **M6:** 669-670
stability of ... **A10:** 112
to provide heat for soldering ... **A6:** 112
treatment of stainless steel, SIMS surface analysis of ... **A10:** 622-623
tunable infrared ... **A10:** 112
tunable radiation from ... **A10:** 142
types ... **A14:** 735-737
types suitable for welding ... **M6:** 651-652
weld quality ... **M6:** 653
welding by ... **A10:** 156
Lasers, quantum-well
research ... **A2:** 747
Lashing wire ... **A1:** 852
Lasing crystals ... **EM4:** 18
Lass *See* Seams
Lasser anodizing process
aluminum and aluminum alloys ... **M5:** 592
Last pass heat sink welding
boiling water reactors ... **A13:** 932
Latch-up
charge transfer during ... **A11:** 786
in CMOS, silicon *p-n* junction failures ... **A11:** 785-786
as failure mechanism ... **EL1:** 977, 1017
in large-scale integrated CMOS devices ... **A11:** 768
Latent curing agent *See also* Curing agent ... **EM3:** 17
defined ... **EM2:** 25
Latent defects
defined ... **EL1:** 867
Latent heat method
solidification modeling ... **A15:** 887-888
Latent heat of fusion *See also* Thermal properties ... **A6:** 45
aluminum casting alloys ... **A2:** 153, 165, 168, 173
Lateral chromatic aberration ... **A9:** 75
effect of focal length on magnification ... **A9:** 77
Lateral connection plate, of bridges
cracking of ... **A11:** 707
Lateral crack
definition ... **EM4:** 633
Lateral extrusion
defined ... **A14:** 8
Lateral outflow jetting
liquid impact erosion ... **A18:** 224
Lateral resolution
atom probe analysis ... **A10:** 595-596
Latex ... **EM3:** 17
acrylic sealants ... **EM3:** 188-191
additives ... **EM3:** 210-211
applications ... **EM3:** 145, 211-212
artificial ... **EM3:** 86
and backup material use ... **EM3:** 212
for bonding decorative wall panels ... **EM3:** 577
chemistry ... **EM3:** 210-211
cost factors ... **EM3:** 211-212
cure mechanism ... **EM3:** 212
design ... **EM3:** 53
fillers ... **EM3:** 211
forms ... **EM3:** 211-212
formulation ... **EM3:** 210-211
for metal building construction ... **EM3:** 57
methods of application ... **EM3:** 211, 212, 213
performance ... **EM3:** 674
pigment-to-binder ratio ... **EM3:** 211
and primer use ... **EM3:** 212
properties ... **EM3:** 210-214
sealant applications ... **EM3:** 177
Latex (continued)
sealant characteristics (wet seals) ... **EM3:** 57
shelf life ... **EM3:** 213
suppliers ... **EM3:** 211
surface preparation ... **EM3:** 212
testing methods ... **EM3:** 212-213
total joint movement ... **EM3:** 211
vinyl acrylic, advantages and disadvantages ... **EM3:** 675
Latex caulks
applications ... **EM3:** 211
for exterior seals in construction ... **EM3:** 56
Lath martensite ... **A9:** 671
basic arrangements ... **A9:** 706
defined ... **A9:** 10-11, 706
Hermann-Mauguin symbols ... **A9:** 707
ordered and disordered ... **A9:** 708
parameters for simple metallic crystals ... **A9:** 716-718
Pearson symbols ... **A9:** 707
in steel ... **A9:** 178
Lath martensite [M(L)# ... **A6:** 76
Lathe
bed ... **A8:** 159
grip, for fatigue testing ... **A8:** 368
tool test, CPM alloys ... **A7:** 789
turning, for comminution of solid magnesium ... **A7:** 131
Lathe tools *See also* Lathes ... **A16:** 19-20
flank wear ... **A16:** 37-38
high-speed tool steels ... **A16:** 58
and shaping ... **A16:** 190
TiN coating ... **A16:** 57
Lathe turning
physical and mechanical properties affected by ... **M5:** 306, 308
Lathes *See also* Lathe tools ... **A16:** 1
automatic ... **A16:** 136, 137, 140-141, 367-393
design of ... **A16:** 135
die threading ... **A16:** 296, 297
for drilling ... **A16:** 212, 229, 235
high-precision, rigid high-power ... **A16:** 98
honing ... **A16:** 473, 486
manual ... **A16:** 136, 137
for manual spinning ... **A14:** 599
for milling ... **A16:** 329
NC, drilling ... **A16:** 235
roller burnishing ... **A16:** 252
semiautomatic ... **A16:** 136-137
surface finish requirements for machine tool components ... **A16:** 21
thread rolling ... **A16:** 284, 285, 286
for turning ... **A16:** 135, 428
Latin square experimental plan ... **A8:** 640, 643, 647-650
for two sources of inhomogeneity ... **A8:** 643
Latitude
charts, x-ray film ... **A17:** 329
of radiographic contrast ... **A17:** 299
Lattice
dislocations, defined ... **A13:** 45
plane ... **A13:** 45
Lattice constants ... **A3:** 1 • 10
defined ... **A9:** 10
Lattice defects
bright-dark image oscillations ... **A9:** 112
diffraction in imperfect crystals ... **A9:** 111
imaging ... **A9:** 110
Lattice diffusion ... **A7:** 7
sintering-caused ... **A7:** 314
Lattice dislocations
reactions with grain boundary dislocations ... **A9:** 120
Lattice disregistry
inoculation effects ... **A15:** 105
Lattice distortion
effect on nucleation sites ... **A9:** 694
Lattice fringes, production of
by transmission electron microscopy ... **A9:** 103-104

Lattice hardening **A12:** 32
Lattice parameter
defined **A9:** 10
Lattice parameters .... **A3:** 1 • 10
Lattice pattern
defined **EM1:** 14 **EM2:** 25
Lattice points .... **A3:** 1 • 15
Lattice strain
permanent **A7:** 61
Lattice structure
of graphite **A2:** 1009-1010
ordered, intermetallics **A2:** 913-914
Lattice-image contrast transmission electron microscopy **A9:** 103
beam diagram **A9:** 104
Lattice-parameter method
XRPD analysis **A10:** 339
Lattices *See also* Crystal lattices; Superlattices **EM3:** 17
as block design **A8:** 640
cubic, singular and vicinal surfaces of **A10:** 537
defects imaged by x-ray topography **A10:** 365
determining geometries of **A10:** 327-328
diffraction in **A10:** 327
distortion in **A10:** 365, 368
image of zinc oxide by combined beams **A10:** 446
imaged by phase contrast **A10:** 445
imaging, transmission electron microscopy **A10:** 445-446
location in **A10:** 628, 633-634
MFS study of constituents **A10:** 72
modulations, in photoacoustic spectroscopy **A10:** 115
-parameter and lattice-type determinations, by XRPD analysis **A10:** 333
reciprocal, and Ewald construction **A10:** 539
space, in crystal systems **A10:** 347
spacing **A10:** 384, 690
strain measurement by dechanneling in **A10:** 634-635
strain measurement in **A10:** 633-635
vibrations **A10:** 126, 130
Laudig iron-backed journal bearings **A11:** 716
Laue camera
for XRPD analysis **A10:** 334-335
Laue case
reflection topography **A10:** 366
synchrotron radiation with **A10:** 374
transmission patterns of aluminum **A10:** 376
Laue equations
defined **A9:** 11
Laue method
defined **A9:** 11
Laue x-ray technique used to investigate crystallographic texture **A9:** 701-702
Laue zones
first-order (FOLZ) **A10:** 439, 442
higher-order (HOLZ) **A10:** 439, 442
zero-order (ZOLZ) **A10:** 439
Launder
defined **A15:** 7
Lauter tubs
in breweries **A13:** 1223-1224
Laves phase
in austenitic stainless steels **A9:** 284
in cobalt-base wrought alloys **A18:** 766
influence on wear resistance **M1:** 613
in Tribaloy alloy (T-800) **A2:** 449
in wrought heat-resistant alloys **A9:** 309, 312
Laves phase alloys
for hardfacing **A6:** 790, 792, 793-794, 795, 796 **A7:** 825
microstructure **A7:** 833
Laves phase formation
in superalloys **A8:** 479
Laves-type alloy compositions **A18:** 762-763
compositions **A18:** 762
microstructure **A18:** 762-763, 764
properties **A18:** 762, 763
Law of Conservation of Energy .... **A3:** 1 • 6
and stress analyses **A11:** 49
Law of mass action **A6:** 58
Law of normal tension **EM4:** 635, 636
Lawn equipment, P/M parts **A7:** 671, 676-678
powders use **A7:** 572
self-lubricating bearings in **A7:** 705
Lawn equipment, P/M parts (continued)
tractors **A7:** 677-678
Laws
of continuity **A15:** 591
of hydraulics **A15:** 590-591
Newton's first **A15:** 591
of thermodynamics **A15:** 50
Lay *See also* Twist
defined **EM2:** 25
Lay-up *See also* D lay-up; Dry lay-up; Hand lay-up; Manual lay-up; Wet lay-up
automated, for thermoplastics **EM1:** 103-104
automatic, of unidirectional tape prepregs **EM1:** 145
complex **EM1:** 642-643
defined **EM1:** 14-15 **EM2:** 25
dry, defined *See* Dry lay-up
equipment, for unidirectional tape prepregs **EM1:** 145
fiberglass fabrication by **EM1:** 28
hand/machine, defined **EM1:** 33
hand/manual **EM1:** 33, 144, 602-604
inspection, types **EM1:** 739
mechanically assisted **EM1:** 605-607
preparation for cure **EM1:** 642-644
procedures, quality control in **EM1:** 740-744
reinforcing material, quality control in **EM1:** 740-744
simple **EM1:** 642
unidirectional vs. quasi-isotropic **EM1:** 146
vacuum bag **EM1:** 703
wet, resins for **EM1:** 132-134
wet/dry, for epoxy composites **EM1:** 71-73
Lay-up fabrication technique
for resin-matrix composites **A9:** 591
Lay-up method, wet and dry
of lamination **EL1:** 832
Lay-ups
graphite-epoxy **A11:** 733-734
Layer *See also* Laminate(s)
analysis, software for **EM1:** 275-276
cracking **EM1:** 436
cracking, computer program for **EM1:** 277
definition **M6:** 10
properties, computing **EM1:** 276
Layer bearing *See also* Bimetal bearing; Trimetal bearing
defined **A18:** 12
Layer formation
during oxidation **A8:** 603
Layer growth
in A15 compounds **A2:** 1063
Layer injection molding **EM1:** 557
Layer pantography techniques **EM3:** 594
Layer personalization
ceramic multilayer packages **EL1:** 463-464
Layer-lattice material
defined **A18:** 12
Layer-type dezincification
as selective leaching **A11:** 633
Layered mechanical coatings **M5:** 301
Layered structures
RBS analysis for **A10:** 628
Layers
in physical interconnection hierarchies **EL1:** 2-5
Layout
design process **EL1:** 513-515
flexible printed boards **EL1:** 594-596
master, requirements **EL1:** 516
multilayer printed board, requirements **EL1:** 516
panel, rigid printed wiring boards **EL1:** 540-541
steps, for digital, ECL, through-hole board **EL1:** 515-516
LC *See* Liquid chromatography
LCP *See* Liquid crystal polymer
LDH *See* Limiting dome height
LDPE *See* Low-density polyethylene
LDR *See* Limiting draw ratio
Le Châtelier, Henri .... **A3:** 1 • 7
Le Chatelier's principle **A6:** 80
Lea and Nurse permeability apparatus
manometer and flowmeter **A7:** 264
Leachates **A10:** 7, 658
Leached substances
as corrosive **A11:** 210
Leached-glass fibers **EM1:** 61
Leaching *See* Selective leaching **A6:** 132
batch curves, Sherritt process **A7:** 140
of composite powders **A7:** 173
of copper oxide for copper powders **A7:** 119
of copper powders **A7:** 119
in hydrometallurgical nickel powder production **A7:** 134, 138-139
in hydrometallurgical processing of cobalt and cobalt alloy powders **A7:** 145
Lead *See also* Lead alloys; Lead powders; Lead recycling; Lead toxicity; Lead-base alloys; specific leaded alloys **A13:** 784-792
as a reactive sputtering cathode material **A9:** 60
abrasion of **A9:** 41
-acid batteries **A10:** 135
addition to aluminum-base bearing alloys **A18:** 752
addition to aluminum-silicon alloys **A18:** 790
additive for copper alloys **M6:** 400
additive improving machinability of steels **A16:** 125
air-acetylene flame atomizer for **A10:** 48
alloy impressed-current anodes **A13:** 469
alloying, aluminum casting alloys **A2:** 132
alloying, copper and copper alloys **A2:** 236
alloying, copper casting alloys **A2:** 346
alloying effect on nickel-base alloys **A6:** 590
alloying for machinability in nickel silver powders **A7:** 122
alloying, wrought aluminum alloy **A2:** 52
alloying, wrought copper and copper alloys **A2:** 242
alloys, corrosion-fatigue limits **A11:** 253
alloys, for soft metal bearings **A11:** 483
in alloys, oxyfuel gas cutting **A6:** 1165
in aluminum alloys **A15:** 746
aluminum-silicon-lead alloys, mixed bearing microstructure **A18:** 744
ammunition **A2:** 554
anodes **A2:** 555
applications **A2:** 548-555
atmospheric corrosion **A13:** 82, 787
attachment to acidified chloride solution specimen **A8:** 419
attachments of resistor networks, hydrogen embrittlement **A11:** 45-46
battery grids **A2:** 548-549
in blood, GFAAS analysis **A10:** 55
blood levels, national estimates **A2:** 1243
in building **A15:** 20-21
cable sheathing **A2:** 550-551
and cadmium separation, by internal electrolysis **A10:** 201
in carbon-graphite materials **A18:** 816
in cast iron **A1:** 5, 8
casting temperatures **A2:** 547-548
chemical, in industrial/domestic waters **A13:** 785
codeposited with silver for electroplated bearings **A18:** 838
commercially pure, corrosive wear **A18:** 744
compatibility in bearing materials **A18:** 743
compositions and grades **A2:** 543-545
compounds, solubility of **A13:** 786
compounds, toxicity **A15:** 96
conformability and embeddability **A18:** 743
contamination **M6:** 321
content additions to P/M materials **A16:** 885
content in solders **M6:** 1069-1071
content in stainless steels **A16:** 682-683, 684
content, manganese bronzes **A2:** 348
in copper alloys **A6:** 753
in copper-base alloys **A18:** 750
corrosion, forms **A13:** 784
corrosion rate, chemical environments **A13:** 781
corrosion resistance **A2:** 547
corrosion resistance of **A1:** 221
creep-fatigue experiments **A8:** 346
density **A2:** 545
deposition of **A10:** 201
determined by 14-MeV FNAA **A10:** 239
determined by controlled-potential coulometry **A10:** 209
determined in coal fly ash **A10:** 147
determined in plant tissues **A10:** 41
EDTA titration **A10:** 173-174

Lead (continued)
effect, electrolytic tough pitch copper **A2:** 270
effect of, on machinability of carbon
steels **A1:** 599-600
effect of, on steel **A1:** 145
effect on brasses **M6:** 1033
effect on Cu alloy machinability **A16:** 805
effect on fracture morphology, steel **A12:** 30, 38
effect on macrosegregation in copper
alloys **A9:** 639
effects, cartridge brass **A2:** 300
as electrode **A10:** 185
electrodeposited coatings **A13:** 427
electroless nickel plating use in **M5:** 223-224
electroplated coatings for bearings **A18:** 838
electroplating of bearing materials **A18:** 756
electropolishing with alkali hydroxides **A9:** 54
embrittlement **A13:** 179
embrittlement by **A11:** 235-236 **A13:** 181-182
embrittlement in nickel alloys **M6:** 437
environments that cause
stress-corrosion cracking **A6:** 1101
extrus'on characteristics **A2:** 547
extrusion of **A14:** 318, 321
fatigue properties **A2:** 547
ferrographic analysis **A18:** 306
final-polishing **A9:** 47
in flake graphite composition **A18:** 699
foil **A2:** 555
fractured ingots, history **A12:** 1
and free-cutting grades of carbon or
low-alloy steels **A16:** 149
in free-machining metals **A16:** 389
free-machining steel additive **A16:** 672, 673, 674, 675, 676, 677, 679
frequency effect on fatigue behavior of **A8:** 346-347
friction coefficient data **A18:** 71
fusible alloys **A2:** 555
galvanic corrosion of **A13:** 85
galvanic corrosion with magnesium **M2:** 607
gas tungsten arc welding **M6:** 183
gaseous hydride, for ICP sample
introduction **A10:** 36
gold plating baths contaminated by **M5:** 283
gravimetric finishes **A10:** 171
heat-affected zone fissuring in
nickel-base alloys **A6:** 588
high-temperature solid-state welding **A6:** 298
high-vacuum lubricant application **A18:** 153, 154
history **A2:** 543
hot dip galvanized coating, use as
alloying element in **M5:** 324
inclusion in steels, embrittlement
effect **A11:** 239, 242
influence on fracture morphology of
steel **A11:** 226
in iron-based alloys, AAS analysis for **A10:** 55, 56
lap welding **M6:** 673
laser beam welding **M6:** 647
LEISS segregated to surface of tin-lead
solder **A10:** 607-608
liquid, application **A13:** 92
liquid, brittle fracture induced by **A11:** 225
liquid, embrittling effects **A11:** 28
liquid, tantalum resistance to **A13:** 733
as low-melting embrittler **A12:** 29
lubricant indicators and range of
sensitivities **A18:** 301
machinability additive **A16:** 685, 689
macroscopic examination **A9:** 416-417
as major toxic metal with multiple
effects **A2:** 1242-1247
malleability, softness, lubricity **A2:** 545
melting point **M5:** 275
microscopic examination **A9:** 416-417
microstructural wear **A11:** 161
microstructures **A9:** 417-418
as minor element, ductile iron **A15:** 648

Lead (continued)
molten **A11:** 274
molten, applications **A13:** 56
mounting materials for **A9:** 415
nickel alloy surface, removal from **M5:** 672
nickel plating bath contamination by **M5:** 208-210
as oxide-forming, copper alloys **A15:** 96
oxyacetylene welding **A6:** 281
oxyfuel gas welding **A6:** 281, 282, 285
photometric analysis methods **A10:** 64
pipe **A2:** 552-553
plain carbon steel resistance to **A13:** 515
pouring temperature/rate of cooling **A2:** 545
preparation for metallographic
examination **A9:** 415-416
price per pound **A6:** 964
processing **A2:** 543
produced by laser alloying **A18:** 866
products **A2:** 548-555
properties of **A2:** 545-548
protective film **A13:** 82
pure **M2:** 761-762
pure, properties **A2:** 1129
pyrophoricity **A7:** 196-197, 199
qualitative tests to identify **A10:** 168
quartz tube atomizers with **A10:** 49
recovery from brass **A10:** 200
recycling **A2:** 1221-1223
refining of **A15:** 474-476
relative solderability **A6:** 134
relative solderability as a function of
flux type **A6:** 129
relative weldability ratings, resistance
spot welding **A6:** 834
release into glass housewares limited **EM4:** 1100
removal, by slags **A15:** 452
resistance of, to liquid-metal corrosion **A1:** 636
resistance to chemicals **A13:** 780
safety regulations affecting soldering **A6:** 984
safety standards for soldering **M6:** 1098
sheet **A2:** 551-552
sheet, foil, and wire, stop-off media
chrome plating **M5:** 187
shielded metal arc welding **A6:** 179 **M6:** 75
shrinkage allowance **A15:** 303
in sleeve bearing liners **A9:** 567
soil corrosion of **A13:** 789
solder characteristics **M6:** 1070-1072
soldering **A6:** 631
solders **A2:** 553
as solid lubricant inclusion for stainless steels **A18:** 716
solution potential **M2:** 207
sound control materials **A2:** 556
species weighed in gravimetry **A10:** 172
spectrometric metals analysis **A18:** 300
in steel chips, GFAAS analysis **A10:** 55
strength **A2:** 545
structures **A2:** 555-556
sulfate ion separation **A10:** 169
sulfuric acid as dissolution medium **A10:** 165
sulfuric acid corrosion of **A13:** 1153
surface corrosion, Raman analysis **A10:** 135
terne coatings **A2:** 554-555
tests for assessing soldering exposure **M6:** 1099-1100
thermal diffusivity from 20 to 100 °C **A6:** 4
thermal expansion **A2:** 545, 547
thermal expansion coefficient **A6:** 907
thermal properties **A18:** 42
to collimate or shield sources for storage and shipment **A18:** 325
toxicity **A6:** 1195, 1196
toxicity and exposure limits **A7:** 207-208
as trace element, cupolas **A15:** 388
TWA limits for particulates **A6:** 984
type metals **A2:** 549-550
ultrapure, by zone-refining technique **A2:** 1094
in underground ducts **A13:** 787-789

Lead (continued)
vapor pressure **A6:** 621
Vickers and Knoop microindentation
hardness numbers **A18:** 416
volatilization losses in melting **EM4:** 389
volumetric procedures for **A10:** 175
in water **A13:** 784-787
in zinc alloys **A9:** 489 **A15:** 788
in zinc/zinc alloys and coatings **A13:** 759
Lead (Pb)
in enamel cover coats **EM3:** 304
penetrated by neutrons in thermal
neutron radiography **EM3:** 759
tin-lead alloys, solder sealing **EM3:** 585
tin-lead solders **EM3:** 584
Lead alloys *See also* Lead
ammunition **A2:** 554
analysis for iron by phenanthroline
method **A10:** 66
anodes **A2:** 555
anodes, chrome plating **M5:** 175, 182
applications **A2:** 548-555
battery grids **A2:** 548-549
bearings, use for *See also* Bearings,
sliding **M3:** 814-815
cable sheathing **A2:** 550-551
chromium plating anodes, use as **M5:** 191
coatings *See* Hot dip lead alloy coating
compositions and grades **A2:** 543-545
creep characteristics **A2:** 551
foil **A2:** 555
Freiberger decomposition in **A10:** 167
friction welding **A6:** 152
fusible alloys **A2:** 555
hot extrusion, billet temperatures for **M3:** 537
inoculants for **A15:** 105
lead-base bearing alloys (babbitt
metals) **A2:** 553-554
macroscopic examination **A9:** 416
melting heat for **A15:** 376
melting of **A15:** 474-476
microscopic examination **A9:** 416-417
microstructures **A9:** 417-424
mounting materials **A9:** 415
pipe **A2:** 552-553
plumbum series **A2:** 555-556
preparation for metallographic
examination **A9:** 415-416
processing **A2:** 543
products **A2:** 548-555
properties **A2:** 545-548
room-temperature tensile properties **A2:** 550
sample dissolution medium **A10:** 166
sheet **A2:** 551-552
soldering **A6:** 631
solders **A2:** 553
sound control materials **A2:** 556
structures **A2:** 555-556
terne coatings **A2:** 554-555
thermal expansion coefficient **A6:** 907
type metals **A2:** 549-550
Lead alloys, specific types *See also* Sleeve bearing materials, specific types
40Sn-60Pb **A6:** 351
63Sn-37Pb solder **A6:** 113, 351
phase diagram **A6:** 128
Pb-0.26Sb, cellular solidification
structure **A9:** 613
SAE alloy 13 .... **A9:** 419
SAE alloy 14 .... **A9:** 419
Lead and lead alloys
age-hardening **M4:** 740-741, 742
applications **M2:** 495-499
battery grids **M4:** 742
cold storage **M4:** 743
corrosion resistance **M2:** 495, 511-522
acids **M2:** 515-522
atmospheric **M2:** 512-513
chemicals **M2:** 515-522

Lead and lead alloys (continued)
differential aeration **M2:** 514
galvanic corrosion **M2:** 513-514, 515-516
soil **M2:** 514-515
underground ducts **M2:** 513-514
water **M2:** 511-512
dispersion hardening **M4:** 742
fabrication **M4:** 742-743
grades of lead **M2:** 494
hardness stability **M4:** 741, 742
hardness testing **M4:** 741-742
heat treating **M4:** 740-743
lead-base alloys *See also* Leads and lead alloys, specific types **M2:** 494
lead-base solders **M2:** 497, 505-506
pig leads **M2:** 494
plumtum series **M2:** 498-499
products **M2:** 495-499
properties of lead *See also* Lead, pure and Leads and lead alloys, specific types **M2:** 494-495
quenching **M4:** 741
refining of lead **M2:** 493-494
service temperatures **M4:** 743
solid-solution hardening **M4:** 740
solution treating **M4:** 740-741
sound-control materials **M2:** 498, 499
sources of lead **M2:** 493
Lead and lead alloys, heat treating **A4:** 925-927
age-hardening **A4:** 925, 926, 927
alloys susceptible to f-ire cracking **A4:** 884
battery grids **A4:** 925, 926-927
cold storage **A4:** 925, 926, 927
dispersion hardening **A4:** 927
fabrication **A4:** 927
hardness stability **A4:** 926
hardness testing **A4:** 925-926
for molten metal baths used in tempering of steel **A4:** 128-129, 133
for pots and workpieces for austenitizing tool steels **A4:** 721
quenching **A4:** 925, 926
service temperatures **A4:** 927
solid-solution hardening **A4:** 925, 927
solution-treating **A4:** 925, 926
vapor pressure of lead **A4:** 493
vapor pressure of lead, relation to temperature **A4:** 495
Lead azide ($Pb(N_3)_2$) (explosive)
friction coefficient data **A18:** 75
Lead babbitt **A18:** 748, 749-750
bearing material microstructure **A18:** 743, 744
in bimetal bearing material systems **A18:** 747
casting processes **A18:** 754-755
in trimetal bearing material systems **A18:** 748
Lead bend fatigue
lead frame alloys **EL1:** 491
Lead blocks as mounting materials for tungsten **A9:** 441
Lead borate glasses
electrical hardness **EM4:** 852
hardness **EM4:** 851
Lead borate, properties
non-CRT applications **EM4:** 1048-1049
Lead borosilicate glasses
metallizing by thick-film adhesion **EM4:** 544
properties **EM4:** 1057
non-CRT applications **EM4:** 1048-1049
Lead burning
definition **A6:** 1211 **M6:** 10
Lead cable sheathing
as lead and lead alloy application **A2:** 550-551
Lead carbonate **A6:** 965
Lead coatings *See also* Terne coatings
cast irons **M1:** 102, 103
corrosion protection **M1:** 752-754
Lead compounds
applications **A2:** 548
Lead coplanarity
in leaded and leadless surface-mount joints **EL1:** 731
Lead counts **EL1:** 203, 211
Lead crystal
composition and properties **EM4:** 742, 1102
defect inclusion levels **EM4:** 392
properties **EM4:** 742
Lead dioxide
deposited by potentiostatic etching **A9:** 146
Lead ferrites **A9:** 539
Lead fluoborate plating **M5:** 273-275
equipment **M5:** 275
maintenance and control **M5:** 275
solution composition and operating conditions **M5:** 273-275
Lead fluorosilicate glasses
electrical properties **EM4:** 853
Lead fluosilicate plating **M5:** 274-275
equipment **M5:** 275
maintenance and control **M5:** 274
solution composition and operating conditions **M5:** 274-275
Lead foil
characteristics and applications **A2:** 555
Lead formations and forming **EL1:** 487, 731
lead frame materials **EL1:** 488
molded plastic packages **EL1:** 475
out lead bonding **EL1:** 284-285
types **EL1:** 733-734
Lead frame materials
alloys **EL1:** 203-204, 490-491
lead frame fabrication **EL1:** 484
package assembly **EL1:** 484-485
package function **EL1:** 485-486
requirements **EL1:** 486-489
trends and economics **EL1:** 491-492
Lead frame strip
manufacture of **EL1:** 483-484
Lead frame(s) *See also* Lead; Lead frame materials
alloys **EL1:** 203-204, 490-491
fabrication **EL1:** 484
material **EL1:** 210-211, 483-492
molded plastic packages **EL1:** 471
nuclear radiation induced failure **EL1:** 1056
positioning **EL1:** 483
strip, manufacture of **EL1:** 483-484
Lead frits and glazes
typical oxide compositions **EM4:** 550
Lead germanate
for barium titanate capacitors **A2:** 743
Lead germanate glasses
optical properties **EM4:** 854
Lead glasses
applications, lighting **EM4:** 1032, 1034, 1037
composition **EM4:** 1102
when used in lamps **EM4:** 1033
for drinkware **EM4:** 1102
properties **EM4:** 1033
Lead in copper **M2:** 242-243
Lead in fusible alloys **M3:** 799
Lead in steel **M1:** 115, 576-578, 580-581, 583
Lead integrity **EL1:** 459, 938
Lead lanthanum zirconate titanate (PLZT) **EM4:** 191
applications **EM4:** 48
key product properties **EM4:** 48
pressure densification
pressure **EM4:** 301
technique **EM4:** 301
temperature **EM4:** 301
raw materials **EM4:** 48
Lead magnesium niobate (PMN) **EM4:** 16
for actuators and transducers **EM4:** 17
applications **EM4:** 48
electrical/electronic applications of formulations **EM4:** 1105
electromechanical parameters in ceramics **EM4:** 1120
electrostriction **EM4:** 1119
electrostrictor for piezoelectric actuator **EM4:** 1121
key product properties **EM4:** 48
properties **EM4:** 1, 1119
raw materials **EM4:** 48
Lead magnesium titanium niobate (PMTN) **EM4:** 16
for actuators and transducers **EM4:** 17
Lead metaniobate
as transducer element **A17:** 255
Lead monoxide
in lead-containing glazes **EM4:** 1062
role in glazes **EM4:** 1062
rolling-element bearing lubricant **A18:** 138
Lead monoxide (continued)
specific properties imparted in CTV tubes **EM4:** 1039
Lead mount packages
two-terminal **EL1:** 429-430
Lead oxide
in binary phosphate glasses **A10:** 131
Lead oxide screens, as image conversion medium
radiography **A17:** 316
Lead oxides
applications **EM4:** 380
purpose for use in glass manufacture **EM4:** 381
versus lead silicates **EM4:** 380
Lead pipe and traps
applications **A2:** 552-553
Lead pitch, fine
as trend **EL1:** 438
Lead plating **M5:** 273-275
anodes, purity of **M5:** 275
applications **M5:** 275
barrel process **M5:** 275
corrosion plating **M5:** 275
equipment **M5:** 275
fluoborate, fluosilicate, and sulfamate processes **M5:** 273-275
maintenance **M5:** 274-275
process control **M5:** 274-275
solution compositions and operating conditions **M5:** 273-275
steel, stripping of **M5:** 275
stripping from steel **M5:** 275
Lead poisoning **A7:** 207-208
Lead pollution
industrial sources **A7:** 208
Lead powders *See also* Lead; Leaded brass powders, Lead-platinum alloys
chemical analysis and sampling **A7:** 248
explosivity **A7:** 196-197
as pyrophoric **A7:** 199
toxicity and exposure limits **A7:** 207-208
Lead preparation
leaded and leadless surface-mount joints **EL1:** 731
Lead, pure
for radiographic screens **A17:** 316
Lead recycling battery-recycling chain **A2:** 1221-1222
government regulations **A2:** 1223
lead scrap, sources **A2:** 1221
lead-smelting process **A2:** 1222
new processes **A2:** 1223
recycled lead, production **A2:** 1221
refining **A2:** 1222
specifications **A2:** 1222-1223
Lead screens
filtration of secondary radiation **A17:** 315
intensification **A17:** 315-316
precautions **A17:** 316
for scattered radiation **A17:** 344
Lead separation
glass capacitors **EL1:** 998
Lead sheet
as lead and lead alloy application **A2:** 551-552
Lead shot **A7:** 282, 296
Lead silicate glasses
electrical properties **EM4:** 851
glass-to-metal seals **EM3:** 302
Lead silicates ($2PbO{\cdot}SiO_2$, $PbO{\cdot}SiO_7$, $4PbO{\cdot}SiO_2$)
applications **EM4:** 380
purpose for use in glass manufacture **EM4:** 381
versus lead oxides **EM4:** 380
Lead sulfamate plating **M5:** 274-275
maintenance and control **M5:** 274
solution composition and operating conditions **M5:** 274-275
Lead sulfide interference film
formation **A9:** 141-142
Lead tarnish
removal of **A9:** 416
used to differentiate phases **A9:** 416
Lead telluride (PbTe)
hot pressing applications **EM4:** 192
Lead termination
common methods **EL1:** 713
Lead titanate
as refractory filler **EM4:** 1072

Lead titanate zirconate
dielectric constants ... **EM4:** 773
elastic constants ... **EM4:** 773
piezoelectric constants ... **EM4:** 773
Lead toxicity biologic indicators ... **A2:** 1246
carcinogenesis ... **A2:** 1245-1246
chelatable lead ... **A2:** 1246
disposition ... **A2:** 1243
hematologic effects ... **A2:** 1244
heme metabolism ... **A2:** 1246
interaction with other minerals ... **A2:** 1246
neurologic effects ... **A2:** 1243-1244
organolead compounds ... **A2:** 1246
renal effects ... **A2:** 1244-1245
sources ... **A2:** 1243
in teeth ... **A2:** 1246
toxicity ... **A2:** 1243-1247
treatment ... **A2:** 1246-1247
Lead, vapor pressure
relation to temperature ... **M4:** 309, 310
Lead wire
attached to strain gages ... **A8:** 202
hyperbolic-weight constant-stress tests ... **A8:** 319
Lead wires ... **A7:** 715
defined ... **EL1:** 1148
oxide dispersion-strengthened copper ... **A2:** 401
thermocouple ... **A2:** 876-878
Lead zinc borate
elastic modulus ... **EM4:** 850
properties, non-CRT applications ... **EM4:** 1048-1049
Lead zirconate titanate
as transducer element ... **A17:** 255
Lead zirconate titanate (PZT)
for actuators and transducers ... **EM4:** 17
applications ... **EM4:** 48, 1119
characteristics ... **EM4:** 1121
compounds ... **EM4:** 1105
electrical/electronic applications ... **EM4:** 1105
electromechanical parameters ... **EM4:** 1119, 1120
gas pressure sintering for pressure densification ... **EM4:** 299
hot pressing applications ... **EM4:** 192
key product properties ... **EM4:** 48
materials for electro-optic ceramics and devices ... **EM4:** 1125
metallization by electroless deposition of nickel ... **EM4:** 544
modified formulations ... **EM4:** 1119
permittivity ... **EM4:** 1120
piezoelectrics ... **EM4:** 47, 770
planar coupling factor ... **EM4:** 1120
properties ... **EM4:** 1
raw materials ... **EM4:** 48
strain curves ... **EM4:** 1119, 1120
for surface acoustic wave devices ... **EM4:** 1119, 1121
thin-film capacitors ... **EM4:** 1117
Lead zirconium titanate (PZT)
for piezoelectric positioners for scanning tunneling microscopy ... **A18:** 394
Lead(s)
attach, ceramic packages ... **EL1:** 466
bonding, plastic packages ... **EL1:** 478-479
broken ... **EL1:** 1007
clip on ... **EL1:** 453
defined ... **EL1:** 1148
effects, solders ... **EL1:** 638
finish, in lead frame assembly ... **EL1:** 488
form factors ... **EL1:** 991
hybrid packages without ... **EL1:** 452
solderability testing ... **EL1:** 954
strength, first-level package ... **EL1:** 991-992
unclinching, by desoldering ... **EL1:** 722
Lead-acid storage batteries
as lead application ... **A2:** 548-549
Lead-antimony alloys
compositions ... **A2:** 545-546
Lead-antimony microstructures ... **A9:** 417, 421
Lead-antimony-tin microstructures ... **A9:** 417, 424
Lead-arsenic alloys
compositions ... **A2:** 544
Lead-barium alloys
compositions ... **A2:** 544
Lead-base alloys *See also* Lead ... **A13:** 784-792
atmospheric corrosion of ... **A13:** 787
bearing ... **A13:** 774
as bearing alloys ... **A18:** 748, 749-750
applications ... **A18:** 750
composition ... **A18:** 749, 750
corrosion resistance from additions ... **A18:** 750
designations ... **A18:** 749, 750
electroplated overlays ... **A18:** 750
mechanical properties ... **A18:** 749
microstructure ... **A18:** 749
in bimetal bearing material systems ... **A18:** 747
corrosive wear ... **A18:** 744
heat and temperature effects on strength retention ... **A18:** 745
resistance to chemicals ... **A13:** 780
soil corrosion of ... **A13:** 786, 789
tin-lead solder ... **A13:** 780-781
in underground ducts ... **A13:** 787-789
in water ... **A13:** 784-787
Lead-base babbitt ... **A9:** 419
Lead-base babbitts as bearing alloys ... **A2:** 523-524
characteristics and compositions ... **A2:** 553-554
Lead-base bearing alloys characteristics and compositions ... **A2:** 553-554
tin additives ... **A2:** 523-524
Lead-base porcelain enamels
composition of ... **M5:** 510-511
Lead-bismuth alloys, liquid
application ... **A13:** 92
Lead-cadmium alloys
composition ... **A2:** 544
Lead-calcium alloys
for battery corrosion ... **A13:** 1317-1318
Lead-calcium alloys for batteries and casting ... **A2:** 545
compositions ... **A2:** 544
Lead-calcium microstructures ... **A9:** 417-418, 420
Lead-calcium-tin phase precipitate
in lead- calcium alloys ... **A9:** 417
Lead-containing silicate glass
applications ... **EM4:** 1070
eight-point analysis of profile ... **EM4:** 1070-1071
joined using a vitreous solder glass ... **EM4:** 1070
Lead-copper alloys
electroless nickel plating of ... **M5:** 233
Lead-copper microstructures ... **A9:** 417-418
Lead-filter screens *See* Lead screens
Lead-fluoroborate glasses
electrical properties ... **EM4:** 852
Lead-free brass
capacitor discharge stud welding ... **A6:** 222
Lead-free rolled copper
capacitor discharge stud welding ... **A6:** 222
Lead-halosilicate glasses
electrical properties ... **EM4:** 852, 853
Lead-indium ... **EM3:** 584
Lead-induced inclusion bodies
renal tubular lining cell ... **A2:** 1245
Lead-lithium alloys, liquid
application ... **A13:** 92
Lead-platinum alloys ... **A7:** 15
Lead-sheathed telephone cables ... **A13:** 1127
Lead-silver alloys
compositions ... **A2:** 544
Lead-smelting processes
for recycled lead scrap ... **A2:** 1222
Lead-tin alloy *See* Terne coating; Tin-lead plating
Lead-tin alloy-coated steels
automotive industry ... **A13:** 1014
Lead-tin alloys
as overlays for aluminum- silicon alloys ... **A18:** 791
Lead-tin alloys, eutectic
electropolished SEM section ... **A10:** 510
Lead-tin coatings *See* Terne coatings
Lead-tin microstructures ... **A9:** 417, 421-424
Lead-tin plating *See* Tin-lead plating
Lead-zirconium-titanate (PZT) sonic converters ... **A8:** 244
Leaded alloy steels
solid-metal embrittlement of ... **A1:** 719, 721-722
Leaded brass powders
dimensional change ... **A7:** 292
horn ring-adjusting nut for industrial paint sprayer ... **A7:** 738
rack, stereo three-dimensional microscope ... **A7:** 738
reticle ... **A7:** 738-739
Leaded bronzes *See* Bearing bronzes
Leaded chip carriers
ceramic, package configuration ... **EL1:** 485
future trends ... **EL1:** 407
as package family ... **EL1:** 404
thermal expansion mismatch problem ... **EL1:** 611
Leaded commercial bronze *See also* Bronzes; Wrought coppers and copper alloys
applications and properties ... **A2:** 305-309
Leaded commercial nickel-bearing bronze
applications and properties ... **A2:** 305-306
Leaded copper *See* Copper alloys, specific types, C18700
Leaded copper alloys *See also* Wrought coppers and copper alloys
applications and properties ... **A2:** 228, 291-292
Leaded copper-zinc alloys
forgeability ... **A14:** 256
Leaded high-strength yellow brasses *See also* Cast copper alloys
corrosion rating ... **A2:** 353-354
properties and applications ... **A2:** 368-369
Leaded manganese bronze *See also* Copper casting alloys; Manganese bronze
nominal composition ... **A2:** 347
properties and applications ... **A2:** 225
Leaded Muntz metal
applications and properties ... **A2:** 311
Leaded naval brass *See also* Copper casting alloys; Naval brass
applications and properties ... **A2:** 320-321
brazing ... **A6:** 629-630
nominal composition ... **A2:** 347
Leaded nickel brass *See also* Cast copper alloys
corrosion ratings ... **A2:** 353-354
properties and applications ... **A2:** 388
Leaded nickel bronze
corrosion ratings ... **A2:** 353-354
properties and applications ... **A2:** 388-389
Leaded nickel-silver *See also* Copper casting alloys
nominal compositions ... **A2:** 347
Leaded nickel-tin bronze *See also* Cast copper alloys
nominal composition ... **A2:** 347
properties and applications ... **A2:** 378
Leaded package technology
surface mount packaging ... **EL1:** 982-983
Leaded phosphor bronze *See also* Copper casting alloys
nominal composition ... **A2:** 347
Leaded quad flatpack ... **EL1:** 404
Leaded red brasses *See also* Copper casting alloys; Leaded semired brasses; Red brasses
composition/melt treatment ... **A15:** 772, 776
corrosion ratings ... **A2:** 353-354
fire refining effects ... **A15:** 451
foundry properties, for sand casting ... **A2:** 348
nominal composition ... **A2:** 347
properties and applications ... **A2:** 225, 364
Leaded red bronzes
as general-purpose copper casting alloys ... **A2:** 351
Leaded resistor networks
defined ... **EL1:** 178

Leaded semired brasses *See also* Copper casting alloys
composition/melt treatment **A15:** 772, 776
corrosion ratings **A2:** 353-354
foundry properties for sand casting **A2:** 348
as general-purpose copper casting alloy **A2:** 351
nominal composition **A2:** 347
properties and applications **A2:** 225, 365-366
Leaded silicon brass *See also* Copper casting alloys; Silicon brass
nominal composition **A2:** 347
Leaded surface-mount joints
assembly equipment **EL1:** 732-733
board solderability **EL1:** 731
component preparation **EL1:** 731
curing and reflow **EL1:** 733
interconnection **EL1:** 732
material and process control **EL1:** 731-733
SMT design guidelines **EL1:** 733-734
technology trends **EL1:** 730
tinning **EL1:** 731
Leaded tin bronze *See also* Copper casting alloys; Tin bronzes
corrosion ratings **A2:** 353-354
foundry properties for sand casting **A2:** 348
as general-purpose copper casting alloy **A2:** 352
nominal composition **A2:** 347
properties and applications **A2:** 226, 376-382
Leaded tin bronze, as part of bimetal bearing *See also* Sleeve bearing materials, specific types **A9:** 567
Leaded tin bronzes
composition/melt treatment **A15:** 772, 776
Leaded yellow brasses *See also* Cast copper alloys
corrosion ratings **A2:** 353-354
foundry properties, for sand casting **A2:** 348
as general-purpose copper casting alloy **A2:** 352
nominal composition **A2:** 347
properties and applications **A2:** 225, 366-370
Leaded-type fixed resistors *See* Resistors
Leader
explosive **A8:** 224, 227
Leadframe
failure of integrated circuit **A11:** 43-45
Leading **A18:** 574
Leadless chip carriers (LCC)
ceramic, heat sinks for **EL1:** 1129-1131
ceramic, introduction **EL1:** 506
defined **EL1:** 1148
die attachments **EL1:** 213-217
interconnection system options **EL1:** 984-986
joints for **EL1:** 730-734
leaded **EL1:** 985
in multilayer ceramic packages **EL1:** 205
as package family **EL1:** 404
package outline **EL1:** 206
as package without leads **EL1:** 452
solderjoints **EL1:** 987
technology trends **EL1:** 730
testing **EL1:** 988-989
thermal expansion control **EL1:** 983-984
Leadless packaging technology
failure mechanism **EL1:** 983-989
Leadless resistor networks
defined **EL1:** 178
Leadless surface-mount joints
assembly equipment **EL1:** 732-733
board solderability **EL1:** 731
curing and reflow **EL1:** 733
interconnection **EL1:** 732
material and process control **EL1:** 731-733
SMT design guidelines **EL1:** 733-734
technology trends **EL1:** 730
tinning **EL1:** 731
Leads and lead alloys, specific types
1% antimonial lead **M2:** 506
4% antimonial lead **M2:** 506-507
5-95 solder **M2:** 505
6% antimonial lead **M2:** 507
8% antimonial lead **M2:** 508
9% antimonial lead **M2:** 508
20-80 solder **M2:** 505
50-50 solder **M2:** 506
acid-copper lead **M2:** 494
Leads and lead alloys, specific types (continued)
arsenical lead **M2:** 494, 502-503
calcium lead, Pb-0.07Ca **M2:** 503
calcium lead, Pb-0.09Ca-0.3Sn **M2:** 503
calcium lead, Pb-0.09Ca-0.5Sn **M2:** 503-504
calcium lead, Pb-0.09Ca-1.0Sn **M2:** 504
calcium lead, Pb-0.065Ca-0.7Sn **M2:** 504
calcium lead, Pb 0.065Ca-1.3Sn **M2:** 504-505
chemical lead **M2:** 494, 501-502
common lead **M2:** 494
copper-bearing lead *See* Leads and lead alloys, specific types, acid-copper lead
corroding lead **M2:** 494, 500, 501
lead-base babbitt (alloy 7) **M2:** 508-509
lead-base babbitt (alloy 8) **M2:** 509
lead-base babbitt (alloy 13)
lead-base babbitt (alloy 15) **M2:** 510
silver-lead solder **M2:** 505
Leaf nodes
in interconnection hierarchy **EL1:** 5
Leaf spring
fretting wear **A18:** 243
Leaf springs **A1:** 321-322
types of **A1:** 323-324
Leaf springs, steel
coatings and finishes for **M1:** 313
energy storage **M1:** 308-311
mechanical prestressing **M1:** 312-313
mechanical properties **M1:** 312
steel grades for **M1:** 311-312
types **M1:** 309-311
working stress **M1:** 288
Leafing powders **A7:** 594
Leak rate
defined **A17:** 57
tested in weld relays **A10:** 156
Leak size
back pressuring effects **A17:** 65
effect, leak detection method **A17:** 68
effect, system response **A17:** 69
Leak testing *See also* Environmental testing **A6:** 1081, 1084 **A17:** 57-70
acoustic **A17:** 59-60
adhesive-bonded joints **A17:** 630
brazed joints **A6:** 1119, 1122
by acoustic emission inspection **A17:** 284
by quantity loss **A17:** 65-66
with calibrated leaks **A17:** 70
for casting defects **A15:** 555
of castings **A17:** 531
choosing optimum system of **A17:** 68
common errors in **A17:** 70
containerized hot isostatic pressing **A7:** 431-434
of containerless hot isostatic pressing **A7:** 436
conversion factors **A17:** 57
decision tree guide for **A17:** 69
defined **A17:** 57
dynamic, defined **A17:** 61
fine and gross **EL1:** 500-502
flow in leaks, types **A17:** 58
fluid dynamics, principles of **A17:** 58-59
with gas detectors **A17:** 61-65
gases, at pressure **A17:** 59
as hermeticity testing **EL1:** 1062-1063
liquids, at pressure **A17:** 59
objectives **A17:** 57-58
package-level **EL1:** 929-930
for packages **EL1:** 501
of pressure systems **A17:** 59-66
for seal integrity **EL1:** 953-954
sensitivity ranges **A17:** 59
static, defined **A17:** 61
system response **A17:** 68-70
terminology of **A17:** 57
in titanium sintering **A7:** 394
typical setup for **A7:** 433, 434
of vacuum systems **A17:** 66-68
weld-assembled hot isostatic pressing container in **A7:** 433, 434
of weldments **A17:** 602
Leak valve
in environmental test chamber **A8:** 411
Leak(s) *See also* Flaws; Leak size; Leak testing; Leakage; Leakage fields
calibrated, leak rate measurement by **A17:** 70
defined **A17:** 57
detection and evaluation **A17:** 50
Leak(s) (continued)
distributed, defined **A17:** 57-58
environment, importance **A17:** 70
location, effect on leak testing method **A17:** 68
minimum detectable, defined **A17:** 57
procedures for testing **A17:** 57-70
rate **A17:** 57
real and virtual, compared **A17:** 57-58
tightness requirements **A17:** 70
types of **A17:** 57-58
Leakage *See also* Flow; Leak testing; Leakage fields; Leaks **A18:** 295
defined **A17:** 57
flow rate in containerized hot isostatic pressing **A7:** 431-434
magnetic particle detection of **A7:** 577
measurement of **A17:** 57
monitoring of **A17:** 57
rate of **A17:** 57
source detection, semiconductor devices **EL1:** 1089-1090
surface electrical, from ionic residues **EL1:** 660
Leakage fields *See also* Leak testing; Leakage
data analysis **A17:** 131
defined **A17:** 90
illustrated **A17:** 90
magnetic contrast in materials with **A9:** 536
magnetic flux, in magnetic field testing **A17:** 129-131
in magnetic particle inspection **A17:** 89
testing, principles **A17:** 129-131
theoretical models **A17:** 131
Leaky Lamb wave (LLW) testing
transmission ultrasonic inspection **A17:** 251-252
Lean exothermic gas atmospheres
for brazing furnaces **A7:** 457
composition **A7:** 342
Learning curve (LC)
for cost projection **EM1:** 424-425, 427
Least count **EM3:** 17
defined **A8:** 8 **EM2:** 25
Least reading *See* Least count
Least squares
parameter, confidence limits on **A8:** 700
regression analysis **A8:** 698-699
response curve, for Probit fatigue data **A8:** 703
technique **A8:** 676, 688
Least squares data analysis method **EM2:** 602
Least squares fit
curve fitting **A10:** 118
in EXAFS data analysis **A10:** 414
in surface stress measurement **A10:** 386
use in determining crystal structure **A10:** 351, 353
for x-ray spectrometry **A10:** 97-98
Leather
friction coefficient data **A18:** 75
Leather (artificial)
electron beam machining **A16:** 570
Leatherhard liquid content **EM4:** 132
Leaves
aluminum alloys **A12:** 433
Ledeburite .... **A3:** 1 • 24 **A6:** 75 **A18:** 697
defined **A9:** 11 **A13:** 9
from laser melting **A18:** 864
Ledeburite, growth
as eutectic **A15:** 180
Ledge tool
and Ti alloys **A16:** 844
Ledges, diverging
and fracture origin **A11:** 80
LEDs *See* Light emitting diodes
Lee algorithm
in computer-aided design **EL1:** 529-531
Lee-Kuhn workability test **A7:** 410, 411
LEED *See* Low-energy electron diffraction
LEFM *See* Linear-elastic fracture mechanics
Leforte aqua regia
as dissolution medium for sulfide minerals **A10:** 166
Leg of a fillet weld
definition **M6:** 11
Legends
as applied to solder mask **EL1:** 559
Legging **EM3:** 17
Lehigh bend test
for notch toughness **A11:** 59

Lehigh restraint test ... **A1:** 612
Leinfelder technique ... **A18:** 671
Leis equivalent stress parameter
Goodman diagram plot for ... **A8:** 713
LEISS *See* Low-energy ion-scattering spectroscopy
Leisure applications *See* Recreation applications
Leisure equipment *See* Sports and recreational equipment
Leisure products
of bronze P/M parts ... **A7:** 736
LEL *See* Lower explosive limit
Lemon (statistical) method
modified ... **EM1:** 302, 305
Lemon bearing (elliptical bearing)
defined ... **A18:** 12
Lengendre addition theorem ... **A10:** 362
Length *See also* Crack length ... **A10:** 685, 686, 690
conversion factors ... **A8:** 722
intercept ... **A12:** 194
of linear feature ... **A12:** 195
mean intercept ... **A12:** 195
mean, of discrete linear features ... **A12:** 195
mean perimeter, of closed figures ... **A12:** 195
perimeter ... **A12:** 194, 195
projected, fractal analysis ... **A12:** 212
ratios, for partially oriented surfaces ... **A12:** 201
SI unit/symbol for ... **A8:** 721
stereological relationships ... **A12:** 196
true, and true area, parametric relationships ... **A12:** 204
true, defined ... **A12:** 199-200
true, fractal analysis ... **A12:** 212
true profile, defined ... **A12:** 199-200
true, values for dimpled and prototyped faceted 4340 steel ... **A12:** 203
true, values for various materials ... **A12:** 200
Length, critical *See* Critical length
Length, fiber *See* Fiber length
Length measurement
by interferometer ... **A17:** 14-15
by laser interferometric micrometer ... **A17:** 15
scales, calibration ... **A17:** 15
in unidirectional flow process, by lasers ... **A17:** 16-17
Length of recess
nomenclature for hydrostatic bearings with orifice or capillary restrictor ... **A18:** 92
Length regions
on-chip interconnection lines ... **EL1:** 6
Length-to-diameter (*L*/*d*) ratio
for cam plastometer specimen ... **A8:** 195
effect on specimen temperature ... **A8:** 156
Leno weave
defined ... **EM2:** 25
locking ... **EM1:** 125-127
Lens aperture
selection ... **A12:** 80-81
Lens arrays
spherical microintegrated ... **EM4:** 440
Lens housing
Maverick missile ... **A7:** 682
Lenses
coated ... **A12:** 84
convergent magnetic, in SEM imaging ... **A12:** 167
defined ... **A10:** 676
and deflector system, ECAP ... **A10:** 597
Einzel, in gas mass spectrometer ... **A10:** 153
electromagnetic, analytical transmission electron microscopy ... **A10:** 432
flare problem ... **A12:** 84-87
holographic ... **A17:** 419
interchangeable, in optical holography ... **A17:** 414
macro luminar ... **A12:** 81
Macro-Nikkor ... **A12:** 81
photographic ... **A12:** 79
Ray diagram ... **A10:** 492
selection of apertures ... **A12:** 80-81
SEM microscopes ... **A10:** 492
stigmators ... **A12:** 167
Lensing action
scanning electron microscopy ... **A10:** 492
Lepidolite ($LiF{\cdot}KF{\cdot}Al_2O_3{\cdot}3SiO_2$) ... **EM4:** 1039
as melting accelerator ... **EM4:** 380
purpose for use in glass manufacture ... **EM4:** 381
Lerch and Bogue method
of free lime content ... **A10:** 179
Let-go ... **EM3:** 17
defined ... **EM2:** 25
Lettering
by multiple-slide forming ... **A14:** 567
Letterpress inks ... **A7:** 595
Leucite ... **EM4:** 7
Level
defined in comparative experiments ... **A8:** 641
determining experimental optimum ... **A8:** 650-652
Level 1 package(s) *See also* Hierarchy; Level(s); Packages; Packaging
current ... **EL1:** 403-405
design ... **EL1:** 401-403
discrete semiconductor/hybrid ... **EL1:** 405
fabrication technologies ... **EL1:** 404-405
package families ... **EL1:** 403-404
trends ... **EL1:** 405-407
Level control
of fluxes ... **EL1:** 648
Level winding *See* Circumferential winding
Level wound
definition ... **M6:** 11
Level(s)
of assembly, for environmental stress screening ... **EL1:** 884
of electronic packaging, defined ... **EL1:** 397
Leveler lines
defined ... **A14:** 8
Levelers
for strip ... **A14:** 713
Leveling ... **A18:** 347
defined ... **A14:** 8
wrought copper and copper alloys ... **A2:** 247-248
Leveling factors
quality design ... **A17:** 722
Lever arm
creep testing machine ... **A8:** 313
in step-down tension testing ... **A8:** 324
Lever, cast stainless steel
fatigue failure in ... **A11:** 114
Lever rule ... **A3:** 1 • 17, 18-19 **A6:** 47, 81, 819
defined ... **A9:** 11
for eutectic, iron-carbon alloys ... **A15:** 68-69
volume fraction, peritectics ... **A15:** 125
Levigation
defined ... **A9:** 11
Levitation
by superconducting magnets ... **A2:** 1027
Lewis acid sites, pyridine adsorption at
Raman studies ... **A10:** 134
Lewis acid-catalyzed epoxide homopolymerization ... **EM1:** 66-71
Lewis acids ... **EM3:** 95
Lewis acids and bases ... **EL1:** 829-830, 860
Lexan
tools for shaped tube electrolytic machining ... **A16:** 555
Leybold-Heraeus electron-beam rotating disc process
titanium powder production ... **A7:** 167
LF/VD-VAD *See* Ladle furnace and vacuum arc degassing
Li-Mg (Phase Diagram) ... **A3:** 2 • 276
LI-Na (Phase Diagram) ... **A3:** 2 • 276
Li-Pb (Phase Diagram) ... **A3:** 2 • 276
Li-Pd (Phase Diagram) ... **A3:** 2 • 277
Li-S (Phase Diagram) ... **A3:** 2 • 277
Li-Se (Phase Diagram) ... **A3:** 2 • 277
Li-Si (Phase Diagram) ... **A3:** 2 • 278
Li-Sn (Phase Diagram) ... **A3:** 2 • 278
LI-Sr (Phase Diagram) ... **A3:** 2 • 278
Li-Te (Phase Diagram) ... **A3:** 2 • 279
Li-Tl (Phase Diagram) ... **A3:** 2 • 279
Li-Zn (Phase Diagram) ... **A3:** 2 • 279
Libby soda-lime
coefficient of thermal expansion ... **EM4:** 1102
composition ... **EM4:** 1102
softening point ... **EM4:** 1102
Liberty Bell
history of ... **A15:** 27
Licensing
for radiation protection ... **A17:** 301
Lid seals ... **EL1:** 953-954, 1058
Liechti-COD techniques ... **EM3:** 452-453
Life *See also* Bathtub curve; Life, cycle; Reliability
acceleration, silicon transistors ... **EL1:** 960
assessment, from extrapolation of temperature at service stress ... **A8:** 338
in creep/creep-rupture analyses ... **A8:** 685
-cycle optimization, design for ... **EL1:** 127-141
estimates, calculating ... **A8:** 682
fraction rule, for determining remaining service life ... **A8:** 337-338
median/calculated median ... **EL1:** 963
normal, in bathtub reliability curve ... **EL1:** 244
-prediction methods ... **A8:** 346
PTH size effects ... **EL1:** 988
time, exponential distribution applied to ... **A8:** 634
vs current density ... **EL1:** 963-964
vs. shear strain of hot rolled and normalized steel ... **A8:** 151
vs. torsional stress in high-cycle regime wrought aluminum alloy ... **A8:** 150
Life (remaining)
carbide composition as indicator ... **A17:** 55
determination by replication ... **A17:** 52
Life cycle *See also* Life cycle testing; Life-cycle optimization
curve ... **EL1:** 897-899
optimization ... **EL1:** 127-141
phases ... **EL1:** 897
reliability ... **EL1:** 897-899
testing for ... **EL1:** 135-140
Life cycle testing
accelerated thermal cycle testing ... **EL1:** 136-139
in-process testing ... **EL1:** 139-140
zero-risk analysis ... **EL1:** 135-136
zero-risk stress testing ... **EL1:** 136
Life management
fracture control philosophy and ... **A17:** 666, 672
Life prediction
accelerated ... **EM2:** 788-795
Life tests *See also* Service life
ASTM, for product control ... **A2:** 831
in circuit breakers ... **A2:** 859
examples, electrical contact materials ... **A2:** 858-861
on peened spindles ... **A11:** 126
using a movable-coil relay ... **A2:** 858-859
using ASTM microcontact tester ... **A2:** 858
Life-cycle optimization *See also* Design
computer-aided analysis ... **EL1:** 132-135
computer-aided design (CAD) ... **EL1:** 127-129
computer-aided manufacturing (CAM) ... **EL1:** 127, 129-132
design for ... **EL1:** 127-141
testing for life cycle ... **EL1:** 135-140
Life-limit criterion ... **A18:** 177
Lifeline
determined ... **EM2:** 570
Lifetime
composite, evaluation ... **EM1:** 203
defined ... **EM1:** 201
prediction, under fatigue ... **EM1:** 244
Lifetimes, electronic excited-state
MFS determined ... **A10:** 72
Lift beam furnace ... **A7:** 7
Lift pin
reversed bending in ... **A11:** 77
Lift rod ... **A7:** 7

Lift-off circle
defined ... **A17:** 219
Lifters
for unloading presses ... **A14:** 500-501
Lifting booms
materials for ... **A11:** 515
Lifting equipment, failures of ... **A11:** 514-528
chains ... **A11:** 515, 521-522
cranes and related members ... **A11:** 515, 525-528
failure mechanisms and origins ... **A11:** 514
hooks ... **A11:** 515, 522-524
investigation of ... **A11:** 514-515
materials for equipment ... **A11:** 515
shafts ... **A11:** 515, 524-525
steel wire rope ... **A11:** 515-521
Lifting fork arm
microstructural fracture ... **A11:** 325-326
Lifting rod ... **A7:** 7
Lifting-sling
brittle fracture ... **A11:** 527
Liftoff
circle, defined ... **A17:** 219
curves, impedance-plane diagram ... **A17:** 168
defined ... **A17:** 137
effect, aluminum and iron ... **A17:** 223
as error, gap measurement ... **A17:** 200
factor, in eddy current inspection ... **A17:** 168
probe, ECP sensitivity to ... **A17:** 137
variation, as test variable ... **A17:** 175
Liftout
defined ... **A14:** 8
Ligament-shaped particles ... **A7:** 32
Ligand
defined ... **A13:** 9
Ligands ... **A10:** 70, 676
defined ... **A2:** 1235
preferred for removal of toxic metals ... **A2:** 1236
Light *See also* Illumination; Lighting
defined ... **A10:** 676
degradation of polymers by ... **A11:** 761
diffuse/reflected, for automatic optical inspection ... **EL1:** 942
-element sensitivity, Auger emission and ... **A10:** 550
fractography ... **A12:** 93-96
incident, for discontinuities ... **A12:** 63
meters, fractographic ... **A12:** 85-86
path, through Czerny-Turner monochromator ... **A10:** 23
sources, fractographic ... **A12:** 81-82
transmitted, defined ... **EL1:** 1067
in UV/VIS analysis ... **A10:** 61
Light blockage *See also* Light obscuration
Light brown coloring solutions
copper and brass ... **M5:** 626
Light copper
as recycling scrap ... **A2:** 1215
Light elements
EPMA detected ... **A11:** 38
identification of ... **A10:** 459-461
metallographic identification ... **A10:** 558-561
Light emitting diodes (LEDS)
encapsulation ... **EL1:** 819
failure mechanisms ... **EL1:** 973-974
junction coating, specification ... **EL1:** 820
Light filters *See also* Color filter
for use with optical microscopes ... **A9:** 72
Light fraction
defined ... **A18:** 12
Light fractography
deep-field microscopy ... **A12:** 96
defined ... **A12:** 93
etching fractures ... **A12:** %
fracture profile sections ... **A12:** 95-96
replicas for light microscopy ... **A12:** 94-95
taper sections ... **A12:** 96
Light guide bundle
borescope ... **A17:** 3-4
Light intensity
distribution, for object orientation ... **A17:** 34
variations, for object position ... **A17:** 35
Light intensity distribution method
of object orientation ... **A17:** 34
Light interferometry, spectral
laser, and monochromatic to determine temperatures in seal wear ... **A18:** 552
Light metals
galvanic corrosion ... **A13:** 84
radiographic methods ... **A17:** 296
Light microscope photograph
nichrome thin-film resistor ... **EL1:** 1099
Light microscopy ... **A18:** 370-375
analytical procedures ... **A18:** 370-371
dark-field fractograph, and SEM image compared ... **A12:** 92
embrittlement phenomena ... **A12:** 123-137
equipment ... **A18:** 370
historical study ... **A12:** 4
interpretation of fractures ... **A12:** 96-123
light fractography ... **A12:** 93-96
metallographic sections ... **A18:** 374-375
quality control applications ... **A12:** 140-143
replicas for ... **A12:** 94-95
specimen preparation ... **A18:** 371-373
surface replication ... **A18:** 373-374
acetate cement films ... **A18:** 374
examples ... **A18:** 374
film replicas ... **A18:** 373-374
taper sectioning ... **A18:** 373
and visual examination ... **A12:** 91-165
wear debris ... **A18:** 375
weld cracking ... **A12:** 137-140
Light microscopy, and scanning electron
failure mode with ... **A8:** 476
Light obscuration
particle size analysis by ... **A7:** 221-225
Light paths
in a Nomarski differential interference microscope ... **A9:** 151
in incident-light microscopes ... **A9:** 71, 81
Light pens
used to make stereological measurements ... **A9:** 83
Light photomacrography
scanning ... **A12:** 81
Light power contacts
recommended materials for ... **A2:** 864
Light reflection
of fracture surface ... **A11:** 75
Light scattering
for assessing particle size and particle size distribution ... **A7:** 124, 216-218
energy-level diagram of ... **A10:** 127
fundamentals ... **A10:** 126-130
theory and instrumentation ... **A7:** 217-218
to measure particle size ... **EM4:** 66
Light sources
for color photography ... **A9:** 139-140
Light spot analyzers
adjustable ... **A7:** 229
Light stabilizers
as additives ... **EM2:** 495
Light, structured *See* Structured light
Light turbidimetry ... **A7:** 219
Light water reactors
fuel rods, corrosion of ... **A13:** 945-948
Zircaloy-clad ... **A13:** 945-948
Light-element analysis
by EDS/UTW-EDS/EELS ... **A10:** 459-461
combined EDS/EELS analysis ... **A10:** 460-461
with EDS and WDS systems, compared ... **A10:** 522
for precipitate identification in stainless steel ... **A10:** 459-461
results ... **A10:** 461
sensitivity, Auger emission and ... **A10:** 550
Light-emitting diodes (LDS)
application, characteristics ... **A2:** 739
defined ... **A2:** 739
and liquid crystal displays (LCDs) ... **A2:** 740
Light-field illumination *See* Bright-field illumination
Light-field Illumination fractographs
dark-field and SEM, compared ... **A12:** 92-100
material types in ... **A12:** 216
texture in ... **A12:** 83
Light-section microscope ... **A9:** 82
Light-section microscopy ... **A9:** 80-82
Lightening holes
in forgings ... **M1:** 364
Lighter flints
as rare earth metal application ... **A2:** 729
Lighting ... **EM4:** 1032-1037
application of glass in lamps ... **EM4:** 1034-1037
Lighting (continued)
glass manufacture ... **EM4:** 1032-1033, 1034
glass/metal seals ... **EM4:** 1037
glasses used ... **EM4:** 1032
for macrophotography ... **A9:** 86-87
properties of glasses ... **EM4:** 1033-1034
Lighting applications
aluminum and aluminum alloys ... **A2:** 13
of polyarylates (PAR) ... **EM2:** 139
Lighting fixtures
precoated steel sheet for ... **M1:** 172, 173, 176
Lighting, photographic
basic, illustrated ... **A12:** 82
direct and oblique ... **A12:** 78
for etched sections ... **A12:** 84
for highly reflective parts ... **A12:** 84
parallel ... **A12:** 83
ring ... **A12:** 83
techniques ... **A12:** 82-85
tent ... **A12:** 88
with ultraviolet illumination ... **A12:** 84-85
Lightly coated electrode
definition ... **M6:** 11
Lightning strike protection
for mechanical fasteners ... **EM1:** 708
Lightweight metals
P/M ... **A7:** 741-764
Lignite ... **EM4:** 7
Miller numbers ... **A18:** 235
Lignosulfonates ... **EM4:** 44, 45
applications ... **EM4:** 47
composition ... **EM4:** 47
supply sources ... **EM4:** 47
LIM *See* Liquid injection molding
Lime
sulfate pickle liquor treated with ... **M5:** 81
Lime ($CaCO_3$)
(basic), flux composition, CO_2 shielded FCAW electrodes ... **A6:** 61
chemical composition ... **A6:** 60
functions in FCAW electrodes ... **A6:** 188
Lime bright annealing ... **A1:** 280
steel wire ... **M1:** 262
Lime buffing compound ... **M5:** 117
Lime coatings
hot rolled bars ... **M1:** 200
steel wire ... **M1:** 262, 266
steel wire rod ... **M1:** 253
Lime feldspar ($CaAl_2Si_2O_8$) ... **EM4:** 6
Lime, free
analysis in Portland cement ... **A10:** 179
Lime glass, effect
mechanically alloyed oxide dispersion-strengthened (MA ODS) alloys ... **A2:** 947
Lime kiln/kiln operations
pollution control ... **A13:** 1370
Lime neutralization
plating wastes ... **M5:** 313
Lime-titania (basic or metal), flux composition
CO_2 shielded FCAW electrodes ... **A6:** 61
Limestone
acid insoluble fraction testing ... **EM4:** 379
angle of repose ... **A7:** 283
aragonite ... **EM4:** 379
batch size ... **EM4:** 382
burned lime ... **EM4:** 379
calcite ... **EM4:** 379
chemical analysis ... **EM4:** 379
chemical composition ... **A6:** 60
composition ... **EM4:** 379
decrepitation ... **EM4:** 379
deposits in Caribbean and U.S ... **EM4:** 379
dissolution in hydrochloric acid ... **A10:** 165
dolomite quicklime ... **EM4:** 379
flame emission sources for ... **A10:** 30
in Hoeganaes process ... **A7:** 79-82
iron content ... **EM4:** 379
lime ... **EM4:** 379
Miller numbers ... **A18:** 235
mining techniques ... **EM4:** 379
particle size ... **EM4:** 379
quantitative spectrographic analysis ... **EM4:** 379
quicklime ... **EM4:** 379
refractory heavy metal (RH) deposits ... **EM4:** 379
Limestone additions
fluxes ... **A15:** 388

Limestone/calcite ($CaCO_3$) purpose for use in glass manufacture **EM4:** 381
Limit
endurance, defined **A11:** 4
fatigue, defined **A11:** 4
load, and limit stress **A11:** 50
stress, and limit load **A11:** 50
Limit analysis, of distortion
low-carbon steels **A11:** 136
Limit analysis of plasticity
theorems **EM1:** 198
Limit, endurance *See* Fatigue limit
Limitations **EM3:** 33
Limited dispense output volume **EM3:** 718-719
Limited frequency response
in Charpy impact testing **A8:** 267
Limited solid solution
defined **A9:** 11
Limited-coordination specification
defined **EM2:** 25
Limited-coordination specification (or standard) **EM3:** 17
Limited-range mass spectrometry techniques **A7:** 434
Limiting blank diameter (LBD)
in drawing tests **A8:** 563
Limiting current density
defined **A13:** 9
Limiting dome height (LDH)
defined **A8:** 8
test, as stretching test for sheet metals **A8:** 562
Limiting dome height (LDH) test **A1:** 576
Limiting draw ratio (LDR)
correlation with material properties **A8:** 565
drawability expressed as **A8:** 562
equation **A8:** 563
Limiting drawing ratio
in deep drawing **A14:** 575
defined **A14:** 8
Limiting shear modulus **A18:** 93
nomenclature for lubrication regimes **A18:** 90
Limiting shear stress **A18:** 93
nomenclature for lubrication regimes **A18:** 90
Limiting stability rate **A6:** 52
Limiting static friction **A18:** 47-48, 57
defined **A18:** 12
Limonite
Miller numbers **A18:** 235
LIMS *See* Laser ionization mass spectroscopy
Lincoln ventmeter test **A18:** 128
Line
intensity **A10:** 328
pair, defined **A10:** 676
position **A10:** 328
profile **A10:** 331-332
scan, double-deflection system **A10:** 493-495
Line (in x-ray diffraction patterns)
defined **A9:** 11
Line blackening *See also* Blackening; Intensity
in spark source mass spectrometry **A10:** 142
Line broadening *See also* Broadening
analysis, and microbeam method **A10:** 374
analysis, factors controlling **A10:** 331, 332
collisional, emission spectroscopy **A10:** 22
Doppler, in emission spectroscopy **A10:** 22
mechanisms, electron spin resonance **A10:** 255
stark, emission spectroscopy **A10:** 22
in x-ray diffraction residual stress techniques **A10:** 386-387
Line compounds ..... **A3:** 1•4, 18
Line contact brushing **M5:** 151, 153
Line defects
in crystals **A9:** 719
Line delay, intrinsic
defined **EL1:** 2
Line etching **A9:** 62-63
electrical steels **A9:** 531
Line gratings *See* Gratings
Line heating
for thermal inspection **A17:** 398
Line indices
defined **A9:** 11
Line of contact, between mating parts
interference at **A11:** 116
Line pair
defined **A10:** 676
Line, parting *See* Parting line
Line pipe *See also* Pipe; Pipelines, failures of; Steel tubular products **M1:** 315, 319
Charpy V-notch impact tests **A11:** 705-706
corrosion **A11:** 703
defect lengths in **A11:** 704
electric-resistance welded **A11:** 699
hydrogen-stress cracks in **A11:** 701
SCC fracture surface **A11:** 702
Line pipe plate
API X60 ..... **A9:** 209
Line pipe steels
hydrogen embrittlement in **A1:** 716
Line pipe steels, specific types
API X60 steel, mechanical properties **A4:** 239
API X70 steel, mechanical properties **A4:** 239
Line projection method
as visual inspection **A17:** 8
Line spall
in steels **A12:** 113-115, 129-134
Line stretch **A18:** 47
Line-shafting **A18:** 595
Lineage structure **A9:** 604
defined **A9:** 11
Linear arrays, used in quantitative metallography **A9:** 124
Lineal roughness parameter
R_L **A12:** 199-200
Linear
defined **EM1:** 15
Linear absorption coefficient
symbol for **A10:** 692
Linear absorption coefficients **A18:** 463, 466
Linear actuator
in electrohydraulic testing machine **A8:** 160
Linear attenuation coefficient
defined **A17:** 384
Linear ball bearings
with drop tower compression system **A8:** 196
Linear coefficient of thermal expansion *See also* Coefficient of thermal expansion; Thermal expansion; Thermal properties
aluminum casting alloys **A2:** 153
wrought aluminum and aluminum alloys **A2:** 62-122
Linear creep-fatigue interaction
diagram **A8:** 355-356
Linear cumulative damage rule **A8:** 337-338, 355, 374
Linear dispersion
defined **A10:** 676
Linear elastic fracture behavior
and bend tests **A8:** 117
Linear elastic fracture mechanics *See also* Fracture mechanics; Stress-intensity factor
defined **A8:** 8, 376 **A13:** 9
equation for crack propagation and stress intensity **A8:** 678
for mechanical driving force of cracks **A8:** 497
test, for fracture toughness **A8:** 470-471
Linear elastic fracture mechanics (LEFM) **A11:** 47-49 **EM3:** 503
crack-tip stresses by **A11:** 47
defined **A11:** 6
delamination analysis **EM1:** 784
design philosophy **A17:** 664
in failure analyses **A11:** 47-49
and fracture mechanics **A11:** 47
inspection requirements **A17:** 663
of laminates **EM1:** 252-253
Linear elastic fracture mechanics theory
used by CARES design methodology **EM4:** 700, 703
Linear elastic fracture toughness
by precracked Charpy test **A8:** 268
Linear expansion *See also* Coefficient of thermal expansion **EM3:** 17
defined **EM1:** 15 **EM2:** 25
Linear ferrites
as magnetically soft materials **A2:** 776
Linear fracture mechanics (LFM)
invalidity in cases **EM3:** 382
Linear kinetic rate law **A13:** 17
Linear low-density polyethylene (LLDPE) *See also* Polyethylene (PE)
in rotational molding **EM2:** 361
Linear medium-density polyethylene (LMDPE) *See also* Polyethylene (PE)
in rotational molding **EM2:** 361
Linear model dummy variable approach **A8:** 712
Linear oxidation
kinetics and reaction rates **A13:** 66
Linear plastic model
SCC behavior **A13:** 278
Linear polarization **A13:** 33, 1250-1251
Linear polarization resistance
and corrosion rates **A11:** 199
Linear polyesters
surface preparation **EM3:** 278
Linear polyethylene
crack-growth mechanisms **A12:** 479
Linear polymers
formation **EM1:** 751
Linear porosity
in arc welds **A11:** 413
in weldments **A17:** 583
Linear ratio
equality of volume fraction to **A9:** 125
Linear regression *See also* Least squares fit; Regression analysis **A8:** 669
Linear regression analysis **A6:** 90
Linear shrinkage **A7:** 7
of thermoplastic resins **A9:** 30
of thermosetting resins **A9:** 29
Linear strain *See also* Engineering strain; True strain **A8:** 8
tensile or compressive **EM3:** 17
Linear strain, compressive/tensile
defined **EM2:** 25
Linear stress
beam distribution **A8:** 119-120
Linear sweep voltammetry
vs DME polarography **A10:** 191
Linear theory of thermoelasticity **EL1:** 55
Linear thermal expansion, coefficients
oxides and metals **A13:** 71
Linear tolerances *See also* Allowances; Tolerances
die casting components **A15:** 619-620
investment castings **A15:** 622
Linear traces
in the plane of polish **A9:** 126
Linear variable differential transformer (LVDT) **A18:** 336-337 **EM3:** 463, 464, 465
for creep testing **A8:** 303
for displacement measurement **A8:** 383
with drop tower compression test **A8:** 197
extensometers **A8:** 35, 616-617
frequency response limitations **A8:** 193
for medium strain rate testing **A8:** 193
for medium stress and strain rates **A8:** 191-192
strain transducer **A8:** 313
to monitor crack extension in corrosive environments **A8:** 428
Linear wear **A18:** 239
abrasive sliding process **A18:** 264
Linear-damage law **A11:** 112
Linearity
calibration, ultrasonic inspection **A17:** 266

Linearity (continued)
detector, defined ... **A17:** 384
as material assumption ... **EM1:** 309
of thermoplastics ... **EM1:** 100
Linearization
of reversed sigmoidal curves ... **A12:** 212-215
Lined pipe
sulfuric acid corrosion ... **A13:** 1153
Linepipe steels
arc welding ... **M6:** 278-280
longitudinal seam, submerged arc welding ... **M6:** 280-282
Liners
defined ... **A14:** 8
Liners, bellows
fatigue fracture in ... **A11:** 118
Liners, for griding mills
material selection for wear resistance ... **M1:** 621, 623-624
Lines
-and-spaces reductions, ICs/PWBs ... **EL1:** 253
energy reduced ... **EL1:** 2
interconnection ... **EL1:** 2
lengths shortened ... **EL1:** 2
number per channel, interconnection ... **EL1:** 19
Lines per channel (LPC) ... **EL1:** 20
Lineshapes
absorption and dispersion, in nuclear magnetic resonance ... **A10:** 280
Auger spectra, and bonding changes ... **A10:** 552
changes, measurement ... **A10:** 552
Lorentzian and dispersion ... **A10:** 281
Linewidth
as FMR resonant phenomenon ... **A10:** 267, 268
Lining cure effect ... **A18:** 571
Lining(s)
defined ... **A15:** 7
ladle, for plain carbon steels ... **A15:** 710
materials, induction furnaces ... **A15:** 372-373
refractory, cupolas ... **A15:** 386
refractory, VIM crucibles ... **A15:** 394
Linings *See also* Coatings
cast iron pipe ... **M1:** 97-98, 100
chemical-setting ceramic ... **A13:** 453-455
dual ... **A13:** 453-454
for hydrogen damage ... **A11:** 251
monolithic and membrane dual ... **A13:** 453-454
organic ... **A13:** 399-418
phenolic ... **A13:** 408
sulfuric acid corrosion ... **A13:** 1153-1154
welded metal, corrosion of ... **A13:** 652
Linishing
defined ... **A18:** 12
Link segments
interconnection ... **EL1:** 3
Links, chain *See* Chain links
Linnik-type interferometer ... **A9:** 80
Linoleum
versus tile whiteware ... **EM4:** 929
Linotype machine
development of ... **A15:** 35
Linotype metal
as lead and lead alloy application ... **A2:** 523
micrograph ... **A9:** 424
Linseed oil
use in painting ... **M5:** 498
Linseed oil, copper/copper alloy resistance ... **A13:** 631
Linters
defined ... **EM2:** 25
Linze-Donovitz (LD) method ... **A1:** 111
Lip-pour ladle *See also* Ladle(s)
defined ... **A15:** 7
Lipophilic *See also* Hydrophilic; Hydrophobic
defined ... **A13:** 9
Lipophilic/hydrophilic (methods B and D) postemulsifiable penetrants ... **A17:** 75, 77
Lipowitz alloy
LME failures in ... **A11:** 719
Liquation ... **A3:** 1 • 19
defined ... **A9:** 11 **A15:** 7
definition ... **A6:** 1211 **M6:** 11
Liquation cracking ... **A6:** 1065
aluminum alloys ... **A6:** 83
austenitic stainless steels ... **A6:** 456, 464-465
evaluation by spot Varestraint test ... **A6:** 608
heat-treatable aluminum alloys ... **A6:** 531
Liquation cracking (continued)
nonferrous high-temperature materials ... **A6:** 567-569
Liquefaction systems
nickel alloy applications ... **A2:** 430
Liquefied petroleum gas (LPG)
fuel gas for oxyfuel gas cutting ... **A6:** 1161, 1162
Liquid *See also* Liquid erosion; Liquid erosion ,failures; Liquid impingement erosion; Liquid-metal embrittlement
heat transfer to ... **A11:** 628
level, effect on crevice corrosion ... **A11:** 184
metals, and solid-metal interactions ... **A11:** 718
metals, effect of temperature on embrittlement ... **A11:** 239
metals, role in crack propagation ... **A11:** 227, 232
Liquid alloys
NMR analysis of electronic structure of ... **A10:** 284
Liquid aluminum *See also* Aluminum; Aluminum alloys
standard Gibbs free energies for solution in ... **A15:** 59
tantalum reaction with ... **A13:** 733
thermodynamic properties ... **A15:** 55-60
Liquid aluminum-magnesium alloys
thermal properties ... **A15:** 57
Liquid aluminum-silicon alloys
integral thermal properties ... **A15:** 57
thermodynamic properties ... **A15:** 57
Liquid ammonia
for ammonia dissociator ... **A7:** 344
Liquid argon
for hot isostatic pressing ... **A7:** 421, 422
Liquid bismuth
tantalum reaction with ... **A13:** 733
Liquid buffing compounds ... **M5:** 116-118
characteristics and advantages of ... **M5:** 117-118
low-pressure systems ... **M5:** 116
Liquid butadiene-acrylonitrile copolymers ... **EM3:** 100
Liquid cadmium
low-alloy steel embrittlement by ... **A12:** 30, 39
Liquid calcium
ultrapurification by external gettering ... **A2:** 1094
Liquid carbonitriding ... **A4:** 330-331 **M4:** 227, 229, 230, 231, 232, 245
Liquid carbonyl compounds
safety hazards ... **A7:** 138
Liquid carburizing ... **A4:** 329-347
applications ... **A4:** 340-341, 344, 345 **M4:** 245-246, 247
bath composition ... **A4:** 329-331, 339 **M4:** 239
carbon bath ... **A4:** 330, 331 **M4:** 227
carbon gradients ... **A4:** 331, 334 **M4:** 230
case depth, control ... **A4:** 331, 332, 334, 335, 340, 341 **M4:** 232, 239, 240, 241-242
combined with brazing ... **A4:** 345-346 **M4:** 248
compared to pack carburizing ... **A4:** 325
cyanide wastes, disposal ... **A4:** 347 **M4:** 248-249
cyanide-containing baths ... **A4:** 329-330, 343, 345 **M4:** 227
cyaniding ... **A4:** 330-331 **M4:** 227, 229, 230, 231, 232, 245
cyaniding time and temperature ... **A4:** 334-335 **M4:** 232
daily maintenance routines ... **A4:** 339-340
dimensional changes ... **A4:** 340-341, 343 **M4:** 234, 240-244
Durofer process ... **A4:** 333, 334
equipment maintenance ... **A4:** 335-338 **M4:** 239-240
examples ... **A4:** 340-341
graphite cover ... **A4:** 333, 339 **M4:** 239
hardness gradients ... **A4:** 334 **M4:** 230-232
high-temperature baths, cyanide type ... **M4:** 228
high-temperature, low-temperature combination baths ... **A4:** 330 **M4:** 228-229
low-temperature baths, cyanide type ... **A4:** 329-330
low-temperature baths, cyanide-type ... **M4:** 227-228
low-toxicity regenerable salt bath process ... **A4:** 332-334
noncyanide ... **A4:** 331-334, 344, 345, 346, 347 **M4:** 229-230
noncyanide salts, disposal ... **A4:** 344, 347 **M4:** 249
partial immersion ... **A4:** 345, 346
process control ... **A4:** 338-340
quenching cyanided parts ... **A4:** 334 **M4:** 245
Liquid carburizing (continued)
quenching media ... **A4:** 341-344 **M4:** 244
safety precautions, cyanide salts ... **A4:** 344, 346-347
safety precautions, cyanide slats ... **M4:** 248
salt baths, externally heated ... **A4:** 335-336, 338 **M4:** 239
salt baths, internally heated ... **A4:** 336-337, 338 **M4:** 239
salt removal (washing) ... **A4:** 344
selective carburizing ... **A4:** 344 **M4:** 247-248
stop-offs ... **A4:** 344-345
Liquid carburizing, furnaces
design ... **A4:** 335-336
electrical-resistance ... **A4:** 335 **M4:** 233
gas or oil ... **M4:** 232-233
immersed-electrode ... **A4:** 336-337, 346 **M4:** 235
lines, automatic ... **A4:** 338 **M4:** 236-239
operating factors ... **A4:** 335 **M4:** 234-235
salt pots ... **A4:** 335, 336, 337, 340 **M4:** 233-234
shutdown and restarting ... **A4:** 340
submerged-electrode ... **A4:** 337 **M4:** 235
temperature ... **A4:** 335 **M4:** 234
Liquid carriers
wet abrasive blasting ... **M5:** 94-95
Liquid chromatographs
and AAS instruments ... **A10:** 55
components ... **A10:** 650
UV/VIS detection of species in ... **A10:** 60
Liquid chromatography ... **A10:** 649-657 **A13:** 1115
as advanced failure analysis technique ... **EL1:** 1104-1105
applications ... **A10:** 649, 655-657
chromatographs, liquid ... **A10:** 55, 650-651
defined ... **A10:** 676
estimated analysis time ... **A10:** 649
fundamental concepts ... **A10:** 651
general uses ... **A10:** 649
introduction ... **A10:** 649-650
limitations ... **A10:** 649
mobile phase programming ... **A10:** 654
modes of ... **A10:** 651-654
preparative ... **A10:** 654
qualitative and quantitative analysis by ... **A10:** 654
related techniques ... **A10:** 649
samples ... **A10:** 649
as separation tool ... **A10:** 170
Liquid cleaners
for surfaces ... **A13:** 382
Liquid cooling *See also* Cooling; Heat removal
environment supply system for ... **A8:** 248
multichip structures ... **EL1:** 310
ultrasonic fatigue testing ... **A8:** 247-248
vs air cooling ... **EL1:** 50
Liquid copper alloys
interaction coefficients for elements in ... **A15:** 60
thermodynamic properties ... **A15:** 55-60
Liquid copper penetration
in HAZ of weld in stainless steels ... **A11:** 431
Liquid copper-aluminum alloys
thermodynamic properties ... **A15:** 58
Liquid copper-nickel alloys
integral thermal properties ... **A15:** 58
thermodynamic properties ... **A15:** 58
Liquid copper-tin alloys
heats of formation ... **A15:** 59
thermodynamic properties ... **A15:** 58
Liquid copper-zinc alloys
integral thermal properties ... **A15:** 59
thermodynamic properties ... **A15:** 59
Liquid corrosion
of copper casting alloys ... **A2:** 352
Liquid corrosion service
C-type cast alloys for ... **A13:** 576-582
Liquid crystal
for optical imaging ... **EL1:** 1071-1072
polymers, as reinforcement ... **EL1:** 605
resins, for printed wiring boards ... **EL1:** 607
Liquid crystal, cholesteric
use in integrated circuit failure analysis ... **A11:** 767-768
Liquid crystal displays (LCDs), and light-emitting diodes (LEDs)
compared ... **A2:** 740
Liquid crystal polymer ... **EM1:** 15, 54
Liquid crystal polymer (LCP) ... **EM3:** 17

Liquid crystal polymers (LCP) *See also* Thermoplastic resins
applications ... **EM2:** 180-181
blends and alloys ... **EM2:** 180
chemistry ... **EM2:** 66, 179
commercial forms ... **EM2:** 179-180
competitive materials ... **EM2:** 180-181
compound types ... **EM2:** 181-182
costs and production volume ... **EM2:** 180
defined ... **EM2:** 25
design considerations ... **EM2:** 181
liquid crystalline state ... **EM2:** 179
processing ... **EM2:** 181
properties ... **EM2:** 181
supplies ... **EM2:** 182
Liquid crystals *See also* Microwave liquid crystal display (MLCD)
for adhesive-bonded joints ... **A17:** 629-630
cholesteric, as temperature sensors ... **A17:** 399
display, microwave ... **A17:** 208
Liquid cutting fluids
machining ... **A7:** 461
Liquid dewaxing, as pattern removal
investment casting ... **A15:** 262
Liquid diffusion coefficient of the solute ... **A6:** 52
Liquid disintegration ... **A7:** 7
Liquid dye penetrant inspection *See also* Liquid penetrant inspection
of powder metallurgy parts ... **A17:** 546-547
Liquid dynamic compaction
aluminum P/M alloys ... **A2:** 204
Liquid dynamics
effect in liquid erosion ... **A11:** 163
Liquid entrapment
in secondary operations ... **A7:** 451
Liquid epoxies
for flexible epoxy systems ... **EL1:** 817
Liquid erosion
defined ... **A11:** 163
failures ... **A11:** 163-171
rate, variation with exposure time ... **A11:** 165
shield, effects ... **A11:** 170-171
Liquid erosion-corrosion
material selection for ... **A13:** 332-333
Liquid feedstocks
physical properties ... **A7:** 342
Liquid film-forming systems
curing of ... **EL1:** 854
Liquid filters
powders used ... **A7:** 573
Liquid fire method
of wet washing organic substances ... **A10:** 166
Liquid flow induced segregation
mushy region ... **A15:** 39-140
Liquid flow technique
and permeametry ... **A7:** 263
Liquid fluorescent penetrant inspection
of investment castings ... **A15:** 264
Liquid gaskets ... **EM3:** 53
Liquid helium coolant
for superconductors ... **A2:** 1030, 1085
Liquid hydrogen ... **A7:** 344
Liquid impact erosion *See* Erosion (erosive wear)
Liquid impingement erosion *See also* Erosion (erosive wear) ... **A18:** 221-230
conjoint chemical action ... **A18:** 221
definition ... **A18:** 221
factors affecting erosion severity ... **A18:** 227-229
dependence on droplet size ... **A18:** 227-228
dependence on impact angle ... **A18:** 227
dependence on liquid properties ... **A18:** 228
dimensionless parameters for describing erosion ... **A18:** 227
erosion resistance ... **A18:** 228
velocity dependence/threshold considerations ... **A18:** 227
liquid impact erosion mechanisms ... **A18:** 223-225
corrosion interactions ... **A18:** 225
Liquid impingement erosion (continued)
liquid/solid interaction-impact pressures ... **A18:** 223-224
material response-development of damage ... **A18:** 224-225
means for combatting erosion ... **A18:** 230
material selection ... **A18:** 230
materials used for protection ... **A18:** 230
modification of impingement conditions ... **A18:** 230
protective shielding ... **A18:** 230
occurrences in practice ... **A18:** 221-222
aircraft rain erosion ... **A18:** 222
coating applications ... **A18:** 222
moisture erosion ... **A18:** 221-222
rain erosion ... **A18:** 221, 222
steam turbine blade erosion ... **A18:** 221-222
relationship to other erosion processes ... **A18:** 222-223
cavitation erosion ... **A18:** 222, 225-226
continuous jet impingement ... **A18:** 222
erosion-corrosion ... **A18:** 223
impingement attack ... **A18:** 223
solid particle erosion ... **A18:** 222-223
test methods for erosion studies ... **A18:** 229-230
time dependence of erosion rate ... **A18:** 225-227
implications for testing and prediction ... **A18:** 226-227
qualitative description ... **A18:** 225-226
reasons for time dependence ... **A18:** 226
Liquid impregnation *See also* Impregnation
of carbon/carbon composites ... **EM1:** 918
Liquid inclusion sampler
as melt analysis technique ... **A15:** 493
Liquid Infiltrant
system compatibility ... **A7:** 552
Liquid injection molding (LIM)
defined ... **EM2:** 25
Liquid iron-carbon alloys
structure of ... **A15:** 168-169
Liquid junction potential ... **A13:** 23-24
Liquid lead, embrittling effect
alloy steel ... **A12:** 30, 38
Liquid lubricants ... **A18:** 81-88
functions ... **A18:** 81
health, safety, and environment ... **A18:** 87
toxicity ... **A18:** 87
lubricant classification ... **A18:** 85-87
API engine service classifications ... **A18:** 86
automotive manual transmissions and axles, API system of lubricant service ...
circulation oils ... **A18:** 87
compressor lubricants ... **A18:** 86-87
designations ... **A18:** 86
energy conserving classification ... **A18:** 85
engine oils ... **A18:** 85
engine tests for API classification ... **A18:** 86
gear oils ... **A18:** 86
hydraulic oils ... **A18:** 86
misting oils ... **A18:** 87
transmission and torque-converter fluids ... **A18:** 85
turbine oils ... **A18:** 86
viscosity grades ... **A18:** 85
methods of lubricant application ... **A18:** 87-88
drop oilers ... **A18:** 87
manual application ... **A18:** 87
mist lubrication ... **A18:** 88
oil carrier ... **A18:** 88
pressurized feed ... **A18:** 88
splash lubrication ... **A18:** 88
properties ... **A18:** 81-85
acidity ... **A18:** 84
alkalinity ... **A18:** 84
ash ... **A18:** 84
color ... **A18:** 82
corrosivity ... **A18:** 84
density ... **A18:** 82
detergency ... **A18:** 84-85
Liquid lubricants (continued)
gravity ... **A18:** 82
oxidative stability ... **A18:** 84
of specific lubricants ... **A18:** 82
stability ... **A18:** 84
thermal stability ... **A18:** 84
viscosity ... **A18:** 82-84
volatility ... **A18:** 84
Liquid magnesium
comminution of ... **A7:** 132
Liquid membrane electrodes ... **A10:** 182
Liquid mercury
decohesive fracture from ... **A12:** 18, 25
Liquid metal *See also* Metallic glasses
by-pass sample station for purity ... **A8:** 426
controlling purity ... **A8:** 425-426
embrittlement, sustained-load failure ... **A8:** 486
environmental chamber on circulating loop system ... **A8:** 427
environments, fatigue crack propagation behavior ... **A8:** 425-426
history ... **A2:** 804
plugging indicator for ... **A8:** 426
synthesis and processing methods ... **A2:** 805-809
Liquid metal assisted cracking *See* Liquid metal embrittlement
Liquid metal baths
kinetic paths for melting ... **A15:** 71
Liquid metal embrittlement *See also* Embrittlement; Liquid metal corrosion; Liquid metals ... **A13:** 171-184
of aluminum ... **A13:** 178
of austenitic stainless steels, by zinc ... **A13:** 184
in brazed joints ... **A13:** 878-879
by aluminum ... **A13:** 179-180
by antimony ... **A13:** 180
by bismuth ... **A13:** 180
by cadmium ... **A13:** 180
by copper ... **A13:** 180
by gallium ... **A13:** 180
by indium ... **A13:** 180-181
by lead, internal/external ... **A13:** 181-182
by lithium ... **A13:** 182
by mercury ... **A13:** 182
by silver ... **A13:** 182
by solders and bearing metals ... **A13:** 183
by tellurium ... **A13:** 183
by tin ... **A13:** 183-184
of copper ... **A13:** 178-179
defined ... **A13:** 9
effect, selenium ... **A13:** 182
effect, sodium ... **A13:** 182-183
effect, thallium ... **A13:** 183
as environmentally induced cracking ... **A13:** 145, 171-184
environments, fatigue in ... **A13:** 177-178
of ferritic steels, by zinc ... **A13:** 184
of ferrous metals and alloys ... **A13:** 179-184
grain size ... **A13:** 175
inert carriers and ... **A13:** 177
liquid role, crack propagation ... **A13:** 174-175
material selection for ... **A13:** 334-335
mechanisms ... **A13:** 172-174
metallurgical, mechanical, physical factors ... **A13:** 175-177
of nonferrous metals and alloys ... **A13:** 178-179
occurrence ... **A13:** 175
strain rate, effects ... **A13:** 175-177
stress effects ... **A13:** 177
susceptibility ... **A13:** 178
of tantalum ... **A13:** 179
temperature effects ... **A13:** 175-177
of titanium ... **A13:** 179
titanium/titanium alloy SCC ... **A13:** 689
of zinc ... **A13:** 178
Liquid metal environments ... **A8:** 425-427
Liquid metal forging *See* Squeeze casting
Liquid metal infiltration (LMI)
defined ... **EM1:** 15

Liquid metal infiltration (LMI) (continued)
of graphite fiber MMCs **EM1:** 868-869
tungsten-reinforced composites **EM1:** 885
Liquid metal infiltration, and diffusion bonding as a fabrication process for metal-matrix
composites **A9:** 591
Liquid metal processing
aluminum-base alloys, thermodynamic properties **A15:** 55-60
composition control **A15:** 71-81
copper-base alloys, thermodynamic properties **A15:** 55-60
gases in metals **A15:** 82-87
inclusion-forming reactions **A15:** 88-97
introduction **A15:** 49
iron-base alloys, thermodynamic properties **A15:** 61-70
physical chemistry, principles of **A15:** 50-54
principles of **A15:** 49
Liquid metal tests
sealed environmental chamber for **A8:** 425
Liquid metals *See also* Liquid metal embrittlement; Liquid metal processing; Liquid(s); Liquid-metal corrosion
activity coefficients at infinite dilution in **A15:** 60
atomization **A7:** 34
containment, materials selection **A13:** 56, 59
corrosion by **M1:** 714-715
corrosion in **A13:** 17, 91-97
embrittlement, titanium **M3:** 416
fretting wear **A18:** 248
heat transfer properties **A13:** 56
high-temperature corrosion in **A13:** 56-60
laboratory corrosion tests in **A13:** 227
niobium resistance in **A13:** 722
NMR analysis of electronic structure of **A10:** 284
production, by vacuum induction melting **A15:** 393-399
reaction with oxygen **A13:** 94-95
role, crack propagation **A13:** 174-175
selective dissolution **A13:** 134
spills and accidents, recovery **A13:** 96
stainless steel corrosion **A13:** 559
tantalum/tantalum alloy resistance to **A13:** 731-735
titanium/titanium alloy resistance **A13:** 681
zirconium, corrosion in **M3:** 790
Liquid nitrided 1010 steel **A9:** 229
Liquid nitriding *See also* Liquid nitrocarburizing **A4:** 410-419
advantages **A4:** 410
aerated bath **M4:** 252-254
aerated cyanide-cyanate bath **A4:** 411-413, 414, 415, 419
applications **A4:** 410
case depth **A4:** 413, 419 **M4:** 254
case hardening **A4:** 413 **M4:** 254, 256
cyanide waste neutralization **A4:** 414
equipment **A4:** 414-415 **M4:** 257
equipment maintenance **M4:** 258
equipment maintenance schedules **A4:** 415
examples **A4:** 410
liquid pressure nitriding **A4:** 411, 414, 415 **M4:** 251-252, 253
noncyanide **A4:** 412, 414 **M4:** 252, 253, 254, 255
operating procedures **A4:** 413-414 **M4:** 251, 254-257
safety precautions **A4:** 414, 415 **M4:** 257
salt baths **A4:** 413, 415, 419 **M4:** 255-257
steel composition, effect of **A4:** 412-413 **M4:** 253, 254, 255
systems **A4:** 410-411 **M4:** 250-251
tool steels **A4:** 410-411, 724, 752
uses **M4:** 250, 251
Liquid nitrocarburizing **A4:** 415-418
compound layer **A4:** 415-416, 417, 418-419 **M4:** 259, 260, 263
Falex tests **A4:** 418
high cyanide baths **M4:** 258-259
high cyanide with sulfur **A4:** 415-416, 417 **M4:** 258-259
high-cyanide baths **A4:** 415, 416
low-cyanide baths **A4:** 416-418
noncyanide bath **M4:** 263
noncyanide baths **A4:** 418-419
Liquid nitrocarburizing (continued)
nontoxic salt bath treatment **A4:** 416-418 **M4:** 254, 259-260
salt baths, liquid **A4:** 416-419 **M4:** 261-262
wear testing **A4:** 416, 417, 418 **M4:** 260-261, 262
Liquid nitrogen *See* Nitrogen
Liquid nitrogen coolant
for superconductors **A2:** 1030
Liquid organic compounds
ESR study of **A10:** 263
Liquid particle behavior
at interface **A15:** 142-147
Liquid penetrant inspection *See also* Magnetic particle inspection; NDE reliability Penetrants; Penetrant inspection **A17:** 71-88
of aluminum alloy castings **A17:** 534
of arc-welded nonmagnetic ferrous tubular products **A17:** 567
of boilers and pressure vessels **A17:** 642
of brazed assemblies **A17:** 604
of casting defects **A15:** 553, 556-557
of castings **A17:** 512, 524-525
codes, boilers/pressure vessels **A17:** 642
costs/sensitivity **A17:** 77
defined **A17:** 71
developers **A17:** 76-77
electron-beam welding **A6:** 866
emulsifiers **A17:** 75
equipment requirements **A17:** 78-80
of finned tubing **A17:** 571
of forgings **A17:** 501-504
of heat-resistant forged alloys **A17:** 503-504
leak detection with **A17:** 66
materials maintenance **A17:** 85
materials used **A17:** 74-77
niobium-titanium superconducting materials **A2:** 1044
of nonferrous tubing **A17:** 574
of open-die forgings **A17:** 494-495
oxyfuel gas welding **A6:** 289
penetrant method selection **A17:** 77-78
penetrant methods **A17:** 73-74
penetrants used **A17:** 77-75
personnel, training and certification **A17:** 85-86
physical principles **A17:** 71-73
postcleaning **A17:** 84
of powder metallurgy parts **A17:** 546-547
precleaning **A17:** 80-82
process description **A17:** 74
process evolution **A17:** 73
process qualification **A17:** 678
processing parameters **A17:** 82-84
quality assurance, of inspection materials **A17:** 84-85
of resistance-welded steel tubing **A17:** 565
of seamless steel tubular products **A17:** 571
solvent cleaner/removers **A17:** 75-76
specifications and standards **A17:** 87-88
of steel bar and wire **A17:** 550-551
of steel forgings **A17:** 495-496
to detect weld and HAZ discontinuities in nickel alloys **A6:** 577
to detect weld cracks **A6:** 1076
to detect weld metal and base metal cracks **A6:** 1075
for weld characterization **A6:** 97, 98, 99, 100, 102
of weldments **A17:** 592
workpiece, inspection and evaluation **A17:** 86-87
Liquid penetrant testing, to detect weld discontinuities
electroslag welding **A6:** 1078
Liquid permeametry **A7:** 263
Liquid phase methods
composite processing **A2:** 583
Liquid phase sintering **A7:** 7, 309, 316, 319-321
anisotropic **A7:** 320
common systems involving **A7:** 319
of copper powders **A7:** 735
dimensional change **A7:** 320
isotropic **A7:** 320
phase accommodation **A7:** 320
phase diagram **A7:** 319
sintering time versus density with **A7:** 320
and solid phase sintering **A7:** 308
solution and reprecipitation method **A7:** 320
swelling in **A7:** 320
transient **A7:** 319
Liquid phase sintering (continued)
of tungsten heavy alloys **A7:** 392-393
Liquid photoimageable process
for solder masks **EL1:** 556-557
Liquid polishing and buffing compounds
removal of **M5:** 5, 10-11
Liquid polymeric binders
for sand **A15:** 211
Liquid polymers
metal powders dispersed in **A7:** 607
Liquid polysulfides **EM3:** 100
applications **EM3:** 676
formulations **EM3:** 676
hardener effect on impact resistance **EM3:** 184, 185
as sealants **EM3:** 676
Liquid potassium
tantalum resistance to **A13:** 735
Liquid processing of steel **A1:** 107, 108
blast furnace stove use **A1:** 107-108
cokemaking **A1:** 107
future technology for **A1:** 114
Liquid pycnometer **A7:** 265
Liquid resin *See also* Resins **EM3:** 17
defined **EM2:** 25
Liquid resin-to-bisphenol-A ratios **EM3:** 94
Liquid sheet
disintegration by high- velocity gas jet **A7:** 28, 30
Liquid shim
defined **EM1:** 15
Liquid shrinkage
defined **A11:** 6
Liquid shrinkage, defined *See* Casting shrinkage
Liquid sodium **A13:** 90, 735
Liquid sodium-cooled fast breeder reactors **A7:** 666
Liquid sodium-potassium nuclear filters
powders used **A7:** 573
Liquid treatment
of compacted graphite irons **A1:** 9
of ductile iron **A1:** 8, 9
of gray iron **A1:** 7
of malleable iron **A1:** 10
Liquid(s) *See also* Flow; Fluid flow; Liquid metals
alloy addition as **A15:** 71-72
corrosive **A12:** 35, 77
environments, effect on fatigue **A12:** 36, 41-46
-gas reactions, kinetics of **A15:** 82-83
intrusion, thermal inspection of **A17:** 402
kinetic resistance to crystallization **A15:** 103
leak testing of **A17:** 57-70
-liquid contraction, in solidification **A15:** 598
oil suspending **A17:** 101
as penetrants, effects **A12:** 77
phase, movement of **A15:** 138
-solid contraction, in solidification **A15:** 599
-solid equilibrium, Vant'Hoff relation **A15:** 102
-solid interface, insoluble particles at **A15:** 142-147
state, solidification of **A15:** 109-110
suspending, for magnetic particle inspection **A17:** 100-102
systems, at pressure, leak testing methods **A17:** 59
thermal properties of **A15:** 55-60
velocity measurement, laser **A17:** 16-17
water-suspending **A17:** 101-102
Liquid-cooled integral heat exchanger
as thermal control **EL1:** 47
Liquid-encapsulated Czochralski (LEC) **EM3:** 580
Liquid-encapsulated Czochralski (LEC) method
of GaAs crystal growth **A2:** 744
Liquid-encapsulated Czochralski (LEC) puller **EM3:** 581
Liquid-erosion failures **A11:** 163-171
analysis of **A11:** 167-170
cavitation **A11:** 163-164
corrosion, effect of **A11:** 167
damage resistance, metals **A11:** 166-167
erosion damage **A11:** 164-166, 170-171
liquid-impingement erosion **A11:** 164
Liquid-gas reactions
kinetics of **A15:** 82-83
Liquid-immersion corrosion
of threaded fasteners **A11:** 535
Liquid-impingement erosion
in boilers and steam equipment **A11:** 623-624

Liquid-impingement erosion (continued)
components and structures affected **A11:** 163
damage processes **A11:** 164
as liquid erosion **A11:** 164
Stellite 6B shield against **A11:** 170-171
Liquid-liquid chromatography **A10:** 652, 676
Liquid-liquid contraction
in solidification **A15:** 598
Liquid-liquid opal glass
composition **EM4:** 1101
properties **EM4:** 1101
Liquid-liquid separation process
zirconium and hafnium **A2:** 661
Liquid-lithium corrosion resistance **A13:** 92-93, 733
Liquid-metal corrosion *See also* Liquid metal embrittlement; Liquid metals; Specific liquid metals **A13:** 91-97
and aqueous/molten-salt corrosion, compared **A13:** 17
by alloying **A13:** 56, 59
by compound reaction **A13:** 56, 59
by dissolution **A13:** 56-57
by impurities **A13:** 56-59
by interstitial reactions **A13:** 56-59
of carbon steels **A13:** 514-515
forms **A13:** 92-94
manned spacecraft **A13:** 1094-1097
in material susceptibility **A13:** 59
resistance to **A1:** 634-636
safety considerations **A13:** 94-97
Liquid-metal embritflement
defined **A12:** 29-30
examination and interpretation **A12:** 126-127
four forms **A12:** 126, 143
Liquid-metal embrittlement **A1:** 635, 717-721 **M1:** 688, 715
Liquid-metal embrittlement (LME) **A11:** 225-238
in aluminum alloy plates, by mercury **A11:** 79
brittle fracture by **A11:** 85
classic and diffusional, compared **A11:** 232, 234
cobalt-base corrosion-resistant alloys **A6:** 598
crack growth and propagation rates **A11:** 719, 726
defined **A11:** 6
delayed failure in **A11:** 243
determined **A11:** 27-28
effects of grain size on **A11:** 229-230, 232
environment, fatigue in **A11:** 231-232
in ferrous metals and alloys **A11:** 234-238
forms of **A11:** 717-718
-induced fracture, failure analysis of **A11:** 232
inert carriers and **A11:** 230-231
initiation time vs temperature **A11:** 244
intergranular **A11:** 234
liquid effect in crack propagation **A11:** 227
in locomotive axles **A11:** 717-723
mechanical factors **A11:** 229-231
in mechanical fasteners **A11:** 543-544
mechanisms **A11:** 226-227, 724-725
metallurgical factors **A11:** 229-231
in nonferrous metals and alloys **A11:** 233-234
occurrence of **A11:** 227-229
physical factors **A11:** 229-231
and solid metal induced embrittlement **A11:** 240
specificity of **A11:** 718
steel failures caused by **A11:** 101, 719-723
and stress-corrosion cracking, compared **A11:** 718
susceptibility to **A11:** 27-28, 233
Liquid-nitrogen cooling
for optical holographic interferometry **A17:** 410
Liquid-partition chromatography *See* Liquid-liquid chromatography
Liquid-penetrant crack detection **A7:** 484
Liquid-penetrant inspection **M6:** 847-848
applications **M6:** 826-827
codes, standards, and specifications **M6:** 827
as failure analysis **A11:** 17
Liquid-penetrant inspection (continued)
as nondestructive test for fatigue **A11:** 134
Liquid-phase epitaxy (LPE)
for gallium arsenide (GaAs) **A2:** 745
to make ferrite films **EM4:** 1163
to make garnet films **EM4:** 1163
Liquid-phase mass transfer
gas porosity by **A15:** 82
kinetics **A15:** 83
Liquid-phase sintering
of cermets **A2:** 986
Liquid-phase sintering (LPS) *See also* Sintering fundamentals **EM4:** 285-289
applications **EM4:** 285
disadvantages **EM4:** 285
driving force for sintering and densification mechanisms **EM4:** 285-288
densification rate **EM4:** 286
pore removal **EM4:** 286, 287-288
rearrangement **EM4:** 285-286, 288
solution-precipitation **EM4:** 286-286
future outlook **EM4:** 289
general requirements **EM4:** 285
grain growth **EM4:** 288
liquid film migration **EM4:** 288
particle shape **EM4:** 272-273
phase diagram use **EM4:** 288
reactive **EM4:** 288-289
real powder compacts **EM4:** 288
sintering aids **EM4:** 288
stages **EM4:** 285, 287
transient **EM4:** 288-289
Liquid-resin conformal coatings **EL1:** 762-764
Liquid-solid chromatography **A10:** 651-652, 676
of epoxies **EL1:** 834
Liquid-solid chromatography (LSC) **EM2:** 519-520
Liquid-solid contraction
in solidification **A15:** 599
Liquid-solid interface
conditions for a flat **A9:** 612
Liquid-state rheology
for epoxies **EL1:** 835
Liquid-surface acoustical holography *See also* Acoustical holography
acoustical system **A17:** 438-439
for adhesive-bonded joints **A17:** 631
commercial equipment **A17:** 441
object size and shape **A17:** 439
optical system **A17:** 439
and scanning acoustical holography compared **A17:** 443-444
sensitivity and resolution **A17:** 439-440
Liquid-vapor metal coolants
corrosion effects **A13:** 93
Liquid/solid interface *See also* Interface kinetics of
Liquidized-bed equipment
atmosphere control **M4:** 302
carburizing **M4:** 303
cleaning operations **M4:** 306
surface treatments **M4:** 302
Liquidized-bed equipment, furnaces
applications **M4:** 305-306
direct-resistance-heated **M4:** 304
external-combustion-heated **M4:** 303, 304
external-resistance-heated **M4:** 303, 304
internal-combustion gas-fired, two-stage **M4:** 304-305, 306
internal-resistance-heated **M4:** 303
safety, operational **M4:** 306
Liquids *See also* Liquids, characterization of; Solutions
immiscible, permeability **A10:** 164
inorganic, and solutions, analytic methods for **A10:** 7
nonvolatile, as IR samples **A10:** 112
organic, analytic methods for **A10:** 10
organic volatile, analytic methods for **A10:** 10
RDF coordination numbers for **A10:** 393
Liquids (continued)
as samples **A10:** 95, 664
SAS analysis of **A10:** 402
surface tensions **EM3:** 181
with suspended solids, sample preparation **A10:** 95
volatile, analytical methods for **A10:** 7
Liquids, characterization of *See also* Liquids
atomic absorption spectrometry **A10:** 43-59
classical wet analytical chemistry **A10:** 161-180
controlled-potential coulometry **A10:** 207-211
electrochemical analysis **A10:** 181-211
electrogravimetry **A10:** 197-201
electrometric titration **A10:** 202-206
electron spin resonance **A10:** 253-266
elemental and functional group analysis **A10:** 212-220
extended x-ray absorption fine structure **A10:** 407-419
gas analysis by mass spectrometry **A10:** 151-157
gas chromatography/mass spectrometry **A10:** 639-648
inductively coupled plasma atomic emission spectroscopy **A10:** 31-42
infrared spectroscopy **A10:** 109-125
ion chromatography **A10:** 658-667
liquid chromatography **A10:** 649-657
molecular fluorescence spectrometry **A10:** 72-81
neutron activation analysis **A10:** 233-242
neutron diffraction **A10:** 420-426
nuclear magnetic resonance **A10:** 277-286
optical emission spectroscopy **A10:** 21-30
potentiometric membrane electrodes **A10:** 181-187
radial distribution function analysis **A10:** 393-401
Raman spectroscopy **A10:** 126-138
ultraviolet/visible absorption spectroscopy **A10:** 60-71
voltammetry **A10:** 188-196
x-ray spectrometry **A10:** 82-101
Liquidus *See also* Freezing range; Melting range; Solidus .... **A3:** 1 • 2 **A6:** 127
defined **A9:** 11 **A15:** 7-8
definition **M6:** 11
gray cast iron **M1:** 11-12
lines, single-phase alloy **A15:** 114
slope, in phase diagrams **A15:** 102
surfaces, calculated **A15:** 64-65
Liquidus composition (TL) **A6:** 89
Liquidus dwell **A6:** 354
Liquidus line **A6:** 46-47, 48
Liquidus temperature *See also* Thermal properties **EM4:** 424
aluminum casting alloys **A2:** 153-177
aluminum-copper alloy, determined **A15:** 185
cast copper alloys **A2:** 356-391
defined **A9:** 611
determined by cooling curves/thermal analysis **A15:** 184-185
effects of constitutional supercooling on **A9:** 611-612
glass fibers **EM1:** 107
wrought aluminum and aluminum alloys **A2:** 62-122
Liquidus temperatures
cast irons **A18:** 695, 698, 701
Liquidus troughs *See also* Monovariant liquidus troughs
Liquified petroleum gas (LPG) **A18:** 85
Liquor finish
steel wire **M1:** 262
Lissajous figures
eddy current inspection **A17:** 174-175
magabsorption signals in **A17:** 144
in stress corrosion measurement, microwave inspection **A17:** 218
List price, part
defined **EM2:** 83

SUBJECTS OF THE INDEXED VOLUMES: ASM Handbook (designated by the letter "A"): **A1:** Properties and Selection: Irons, Steels, and High-Performance Alloys (1990); **A2:** Properties and Selection: Nonferrous Alloys and Special-Purpose Materials (1990); **A3:** Alloy Phase Diagrams (1992); **A4:** Heat Treating (1991); **A6:** Welding, Brazing, and Soldering (1993); **A7:** Powder Metallurgy (1984); **A8:** Mechanical Testing (1985); **A9:** Metallography and Microstructures (1985); **A10:** Materials Characterization (1986); **A11:** Failure Analysis and Prevention (1986); **A12:** Fractography (1987); **A13:** Corrosion (1987); **A14:** Forming and Forging (1988); **A15:** Casting (1988); **A16:** Machining (1989); **A17:** Nondestructive Testing and Quality Control (1989); **A18:** Friction, Lubrication, and Wear Technology (1992). **Metals Handbook, 9th Edition** (designated by the letter "M"): **M1:** Properties and Selection: Irons and Steels (1978); **M2:** Properties and Selection: Nonferrous Alloys and Pure Metals (1979); **M3:** Properties and Selection: Stainless Steels, Tool Materials and Special-Purpose Materials (1980); **M4:** Heat Treating (1981); **M5:** Surface Cleaning, Finishing, and Coating (1982); **M6:** Welding, Brazing, and Soldering (1983). **Engineered Materials Handbook** (designated by the letters "EM"): **EM1:** Composites (1987); **EM2:** Engineering Plastics (1988); **EM3:** Adhesives and Sealants (1990); **EM4:** Ceramics and Glasses (1991); **Electronic Materials Handbook** (designated by the letters "EL"): **EL1:** Packaging (1989).

Lister, Thomas
as Liberty Bell caster ... **A15:** 27
Literature
powder metallurgy ... **A7:** 18-19
Litharge ... **A18:** 572
Litharge (PbO)
component in photochromic ophthalmic and flat glass composition ... **EM4:** 442
in drinkware compositions ... **EM4:** 1102
in glaze composition for tableware ... **EM4:** 1102
purpose for use in glass manufacture ... **EM4:** 381
Lithia
as melting accelerator ... **EM4:** 380
Lithia minerals
for use in glass manufacture ... **EM4:** 381
Lithia-alumina-calcia-silica (LACS) glass ceramics, applications
dental ... **EM4:** 1094
Lithium *See also* Aluminum-lithium alloys
as addition to aluminum alloys ... **A4:** 843, 844, 845
alloying, wrought aluminum alloy ... **A2:** 52
in aluminum-base alloys being diffusion welded ... **A6:** 885
aluminum-lithium alloys, corrosion resistance ... **EM3:** 671
aluminum-lithium alloys, weldability ... **A6:** 726
-base grease, SEM micrograph ... **A11:** 153
cations, in glasses, Raman analysis ... **A10:** 131
corrosion fatigue test specification ... **A8:** 423
deoxidation, of copper alloys ... **A15:** 469-470
deoxidizing, copper and copper alloys ... **A2:** 236
-doped germanium gamma-ray detector ... **A10:** 235
-drifted silicon detector ... **A10:** 89, 90
embrittlement by ... **A11:** 236 **A13:** 182
energy-level diagram, emission lines ... **A10:** 22
flame emission sources for ... **A10:** 30
glass-to-metal seals ... **EM3:** 302
hazards ... **A13:** 96
as inclusion-forming ... **A15:** 96
liquid ... **A13:** 92
as low-melting embrittler ... **A12:** 29
in medical therapy, toxic effects ... **A2:** 1257-1258
molten, applications ... **A13:** 56
molten, Gibbs free energies of formation ... **A13:** 52
molten, tantalum resistance to ... **A13:** 733
plain carbon steel resistance to ... **A13:** 515
pure ... **M2:** 762-763
pure, properties ... **A2:** 1131
species weighed in gravimetry ... **A10:** 172
ultrapure, by distillation ... **A2:** 1094
and uranium mononitride fuel ... **A13:** 733
in vacuum, as SERS metal ... **A10:** 136
vapor pressure, relation to temperature ... **A4:** 495
vapor pressures ... **A6:** 621
wrought alloys containing ... **A13:** 585
Lithium 12-hydroxystearate ... **A18:** 129
Lithium 12-hydroxystearate soap ... **A18:** 136, 137
Lithium aluminosilicate
aerospace applications ... **EM4:** 1016
electrical properties ... **EM4:** 852
in formation of glass-ceramic ovenware ... **EM4:** 1103
Lithium aluminum silicate, porous
fracture features ... **A11:** 745
Lithium ambient-temperature batteries ... **A13:** 1318-1319
Lithium borate
glass-forming fusions ... **A10:** 94
Lithium carbonate
for acidic samples in flux ... **A10:** 94
as binding agent for samples, x-ray spectrometry ... **A10:** 94
decomposition temperatures ... **EM4:** 55
as fluxes ... **A10:** 167
as melting accelerator ... **EM4:** 380
to control dusting ... **A18:** 684
Lithium carbonate, as treatment for depression
toxic effects ... **A2:** 1257
Lithium complex soap ... **A18:** 126, 129
Lithium disilicate ($Li_2O{\cdot}2SiO_2$)
primary phase glass-ceramics based on ... **EM4:** 870, 871
thermal expansion coefficient ... **EM4:** 499
Lithium ferrites ... **A9:** 538
Lithium fluoride
for base samples in flux ... **A10:** 94
as common analyzing crystal, in x-ray spectrometry ... **A10:** 88
impregnation effects on typical carbon- graphite base material ... **A18:** 817
rolling-element bearing lubricant ... **A18:** 138
Lithium metaborate
as flux ... **A10:** 167
Lithium metasilicate
information-display applications ... **EM4:** 1049
photo-nucleated phase ... **EM4:** 440, 441
Lithium nephrotoxicity
chronic ... **A2:** 1257
Lithium niobate ($LiNbO_3$) ... **EM4:** 18, 52, 56
characteristics ... **EM4:** 1121
freeze drying ... **EM4:** 62
synthesized by SHS process ... **EM4:** 227
to modify optical signals ... **EM4:** 17
used for surface acoustic wave devices ... **EM4:** 1119
Lithium nitrate, apparent threshold stress values
low-carbon steel ... **A8:** 526
Lithium nitride ... **A16:** 105
Lithium oxide (Li_2O)
component in photochromic ophthalmic and flat glass composition ... **EM4:** 442
in composition of glass-ceramics ... **EM4:** 499
in glaze compositions for tableware ... **EM4:** 1102
in ovenware compositions ... **EM4:** 1103
in tableware compositions ... **EM4:** 1101
Lithium powders
-containing copper powders ... **A7:** 118
flammability in dust clouds ... **A7:** 195
as pyrophoric ... **A7:** 199
Lithium, resistance of
to liquid-metal corrosion ... **A1:** 635
Lithium silicate glasses
density ... **EM4:** 846
electronic processing ... **EM4:** 1058
optical properties ... **EM4:** 854
Lithium soap ... **A18:** 126, 129
grease lubrication for rolling-element bearings ... **A18:** 503-504
Lithium stearate
for brass and nickel-silver P/M parts ... **A7:** 191
burn-off ... **A7:** 351, 352
lubricant ... **A7:** 190-193
Lithium sulfate
as transducer element ... **A17:** 255
Lithium tantalate ($LiTaO_3$)
applications ... **EM4:** 17
characteristics ... **EM4:** 1121
to modify optical signals ... **EM4:** 17
used for surface acoustic wave devices ... **EM4:** 1119
Lithium tetraborate
as flux ... **A10:** 93, 167
glass-forming fusions with ... **A10:** 94
Lithium, vapor pressure
relation to temperature ... **M4:** 310
Lithium-alumina-silicate glasses
nonlinear refractive index ... **EM4:** 1080
Lithium-aluminum alloys
heat treating ... **A4:** 843
Lithium-aluminum-silicate ... **EM4:** 676
applications
aerospace ... **EM4:** 1004
flow separator housing in advanced gas turbines ... **EM4:** 998
key features ... **EM4:** 676
properties ... **EM4:** 677
thermal expansion ... **EM4:** 686
Lithium-doped germanium Ge(Li) detectors ... **A10:** 235, 241
Lithium-doped silicon detectors ... **A10:** 83, 89-91, 389, 519
Lithium-fluoride-beryllium fluoride salts
corrosion ... **A13:** 52-54
Lithium-gaiiosilicate glasses
electrical properties ... **EM4:** 852
Lithium-oxide-based glass
x-ray transparency of ... **EM1:** 107
Lithium-silicate based glass-ceramics
glass- ceramic/metal sealing applications ... **EM4:** 479
Lithium/iodine batteries ... **A13:** 1319
Lithium/oxyhalide cells ... **A13:** 1319
lithium/sulfur dioxide batteries ... **A13:** 1319
Lithium/vanadium pentoxide cells ... **A13:** 1319
Lithographing (transfer coating) ... **A7:** 460
Lithography ... **EL1:** 193, 978
Little-end bearing *See also* Big-end bearing
defined ... **A18:** 12
Live time
in x-ray spectrometry ... **A10:** 92
LLDPE *See* Linear low-densitiy polyethylene
LLW testing *See* Leaky Lamb wave testing
LME *See* Liquid metal embrittlement; Liquid-metal embrittlement
LMMS *See* Laser microprobe mass spectrometry
LMP *See* Larson-Miller parameter
Load *See also* Applied load; Loading; Unloading ... **EM3:** 17
accuracy ... **A8:** 612
application, ambient temperatures ... **A8:** 417
application and measurement, vacuum and gaseous fatigue tests ... **A8:** 411
application, cam plastometer ... **A8:** 194
application, Rockwell hardness testing ... **A8:** 80
application, stress-relaxation spring test ... **A8:** 328
application, universal testing machine ... **A8:** 134, 612
applied, and distortion ... **A11:** 136
applied, effect in composites ... **A11:** 733
applied, in near-threshold fatigue crack propagation testing ... **A8:** 428
axial, rolling-contact fatigue from ... **A11:** 503-504
of bench-mounted tester ... **A8:** 91
buckling ... **A11:** 137
-carrying capacity, bearings ... **A11:** 485
changes during creep-rupture testing ... **A8:** 337-339
characteristics, mechanical presses ... **A14:** 38-39
characteristics, metalworking machines ... **A14:** 16
characteristics, screw presses ... **A14:** 40-41
choice in microhardness testing ... **A8:** 96
in cold extrusion ... **A14:** 302
Considére's construction for point of maximum ... **A8:** 25
cycles, for one leg, during walking ... **A11:** 672
defined ... **A8:** 8 **A18:** 12
-deflection, for spring-tempered phosphor bronze strip ... **A8:** 136
deflection temperature under *See* Deflection temperature under load
deformation under, defined *See* Deformation under load
displacement curve, strain energy under ... **A11:** 49-50
effect in vibratory polishing ... **A9:** 42, 44
and electrical potential vs. load point displacement ... **A8:** 390
-elongation curve, and engineering stress-strain curve ... **A8:** 20
-elongation diagram, for low-carbon steel ... **A8:** 22
-elongation measurements, engineering stress-strain curve from ... **A8:** 20
end, in torsional testing machine ... **A8:** 146
in fiber composites, study of ... **A9:** 592
fiber-to-fiber ... **A11:** 731
force, low, constant-stress testing system for ... **A8:** 321-322
frame, rising step-load test ... **A8:** 541
frequency, and stress-intensity range, effect on corrosion fatigue ... **A8:** 404-406
as function of crack mouth opening displacement ... **A8:** 451
in high-pressure steam environments ... **A8:** 428
interactions, in damage analyses ... **A8:** 682
limit, and distortion ... **A11:** 136-138
magnesium alloy forgings ... **A14:** 260
maximum to failure, in fracture toughness testing ... **A8:** 470
measurement ... **A8:** 193, 554
measuring systems ... **A8:** 193, 612-614
minor or major, in Rockwell hardness testing ... **A8:** 74-75

Load (continued)
nomenclature for hydrostatic bearings with orifice or capillary restrictor **A18:** 92
operating, of gears and gear drives **A11:** 589
point displacement vs. load and electrical potential **A8:** 390
range, forced-vibration system **A8:** 8, 392
ranges of **A8:** 47
rapidly applied, dynamic fracture by **A8:** 259
ratio and overloading, effect on corrosion- fatigue **A11:** 255
requirements, drop tower compression test **A8:** 197
ringing effect **A8:** 40, 44, 193, 209
rolling, for titanium **A14:** 357
secondary, and pipeline failure **A11:** 704
selection, Brinell test **A8:** 84
selection, Rockwell hardness testing **A8:** 74
states, in composites **A11:** 735
step size, during precracking **A8:** 382
stress corrosion at **A11:** 53
in three- and four-point bend tests **A8:** 134
-time and displacement-time curves **A8:** 281
-time response, medium-strength steel **A8:** 266
-torsional moment, in cyclic torsional testing **A8:** 149
transducer, in hydraulic torsional system **A8:** 215-216
transfer, fastener system for **A11:** 529
uniaxial *See* Uniaxial load
upset reduction effects **A14:** 232
in vacuum and gaseous fatigue tests **A8:** 410
verification **A8:** 88, 611
very light, microhardness testing **A8:** 96
vs displacement curve, screw press **A14:** 41
vs displacement curves, various **A14:** 38-39
vs ram displacement, nonlubricated extrusion **A14:** 316
vs. deflection data, three- and four-point bending **A8:** 135-136
vs. elongation tests, forces for **A8:** 47
vs. elongation, yield **A9:** 684
vs. force, as terms **A8:** 74n, 84n
vs. hardness number, microhardness testing **A8:** 94-96
vs. indentation size, hardness testing **A8:** 94
vs. mouth opening displacement, for ideal plastic behavior **A8:** 471
Load applicator **A8:** 134-135
Load cell
amplifier **A8:** 48
cam plastometer **A8:** 195-196
capacity **A8:** 400
compression test fixture **A8:** 198
cross-sectional view **A8:** 49
and digital load indicator **A8:** 615
with drop tower compression system **A8:** 196
as fatigue testing machine sensor **A8:** 368
features for servohydraulic systems **A8:** 400
follow-the-load method of calibration **A8:** 616
force vs. time, for ringing phenomenon **A8:** 44
high vibrational frequency type **A8:** 193
load determination method **A8:** 193
as load-measuring system **A8:** 613
for measuring medium rate stress and strain **A8:** 191
natural frequency **A8:** 193
quartz load washer **A8:** 193
ringing **A8:** 40-41, 44, 191, 193, 209
selection for servohydraulic systems **A8:** 400
strain-gage type **A8:** 48
system, for universal testing machine calibration **A8:** 614
torsional, and rotary actuator **A8:** 151
for vacuum fatigue test chamber **A8:** 414-415
for verification-, Brinell test **A8:** 88

Load control
data, linear model dummy variable approach for **A8:** 712
in resonant system **A8:** 393
test, for high-cycle fatigue testing **A8:** 696
Load cycles, for one leg
during walking **A11:** 672
Load data **EM3:** 36
Load displacement
computerized data-acquisition system **A8:** 385
curve, dynamic notched round bar testing **A8:** 281
curve, for commercial-purity aluminum **A8:** 229, 231
nonlinear record, rapid-load fracture test **A8:** 260
record, crack arrest fracture toughness **A8:** 292-293
Load factor **A18:** 101, 104
Load frame
alignment in fatigue evaluations **A8:** 400
conventional **A8:** 192-193, 208-210
in fatigue testing machine **A8:** 368
four-post with cam plastometer **A8:** 195
in load train **A8:** 368
in servohydraulic systems **A8:** 400
Load frames, conventional *See* Conventional load frames
Load frequency **A8:** 404-406
Load maintainer piston
for crank and lever testing machine **A8:** 369-370
Load measuring systems **A8:** 193, 612-614
Load paths, redundant
in NDE reliability **A17:** 702
Load per unit projected bearing area
nomenclature for Raimondi-Boyd design chart **A18:** 91
Load range, forced displacement system **A8:** 8, 392
Load ratio **A18:** 94
Load ratio (*R*) *See also* Stress ratio
in cyclic ductile cohesion **A8:** 484
defined **A8:** 8
Load sharing factor **A18:** 539, 540
symbol and units **A18:** 544
Load train
for creep test stand **A8:** 312
design **A8:** 159-160
in fatigue testing machine **A8:** 368
horizontal, on lathe bed **A8:** 500
test specimen **A8:** 368
vertical, in modified test machines **A8:** 159
Load transfer
nonuniform **EM1:** 491-484
shear **EM1:** 280-481, 488
Load weighing system **A8:** 48-49
Load(s) *See also* Fatigue loading; Impact loading
cyclic, metals vs plastics **EM2:** 76
impact, metals vs plastics **EM2:** 75-76
long-term, metals vs plastics **EM2:** 75
short-term, metals vs plastics **EM2:** 75
Load-bearing additives
for metalworking lubricants **A18:** 140-141, 142, 143
Load-bearing area **A18:** 432, 433
Load-bearing structures
engineering formulas **EM2:** 652-654
Load-carrying capacities
self-lubricating bearing materials **A7:** 709
Load-carrying capacity (of a lubricant) **A18:** 12
Load-deflection curve
defined **EM1:** 15
Load-extension curves for steel sheet **M1:** 548
Load-strain recorder system **A8:** 617
Loading *See also* Cyclic loading; Frequency effects; Reversed torsional loading; Shock loading; Torsional loading
of acoustic emission inspection specimens **A17:** 286

Loading (continued)
axial **A11:** 102, 109
conditions **A12:** 4-5, 12-15, 72
conditions, fractures from **A11:** 75
control, acoustic emission inspection **A17:** 286
cyclic, environment and temperature effects **A8:** 377
cyclic tension-tension fatigue **A8:** 717-718
cyclic-stress, machine part fatigue under **A11:** 371
defined **A7:** 7 **A18:** 12
direction **A8:** 63-64, 302
displacement control, in elastic-plastic fracture toughness tests **A8:** 455
dynamic, slow strain rate testing **A13:** 260-263
dynamic, slow strain rate testing of SCC **A8:** 496, 498-499
dynamic, vs. quasi-static loading **A8:** 259
effect of mixed-mode, corrosion-fatigue **A11:** 255
effect on crack propagation and striation **A12:** 15
effect on fatigue **A12:** 15, 36, 53-54
end **A12:** 332
end, of explosively loaded torsional Kolsky bar **A8:** 224, 227
examining **A12:** 72, 92
failure, electronic materials **A12:** 481-482
as fatigue mechanism **A12:** 4-5
fatigue parameters based on **A12:** 54
fixtures, for fatigue testing machines **A8:** 368
of flat components **A11:** 109
fracture, modes **A12:** 14
frequency, effect on fatigue-crack propagation **A11:** 110
high-impact, brittle fracture from **A11:** 344-345
imbalance, effect in medium-carbon steels **A12:** 273
impact, notch toughness under **A11:** 57-60
inertial, Charpy impact testing **A8:** 267
mean stress, effect on fatigue cracking **A12:** 15
mechanical, of soldered joints **A17:** 608
method, effects **A13:** 247
methods, precracked SCC specimens **A8:** 714
modes of **A11:** 47
modified wedge-opening **A11:** 64
monotonic tensile, of polymers **A11:** 759
parameters, effect on corrosion-fatigue **A11:** 254-255
parameters, fatigue corrosion testing **A13:** 292
of precracked SCC specimens **A13:** 259-263
pressure, mean **A8:** 71
proof test followed by fatigue **A8:** 717-718
proportional **A8:** 344, 447
punch **A8:** 229-230
pure tensile vs bend, fracture contour and **A11:** 746
rapid, fracture surface of Fe-4Si under rapid **A8:** 479-480
rate, effect on toughness and crack growth **A11:** 54
rates, dynamic notched round bar testing **A8:** 276
rates, high, and stress intensity **A8:** 259-260
repeated, acoustic emission inspection **A17:** 286
scanning electron microscopy study of specimens **A9:** 97
service, as cause for stress-corrosion cracking **A8:** 496
service, estimates from crack extension behavior **A8:** 439
sinusoidal, *S-N* curve **A8:** 364
sinusoidal, *S-N* curves for **A11:** 103
slow, in four-point bending **A12:** 350
slow, notch toughness under **A11:** 57-60
springlike, of bolted joints **A11:** 530-531, 533
static, of precracked (fracture mechanics) SCC specimens **A13:** 253-260
static, of smooth SCC specimens **A13:** 246-253

Loading (continued)
static, SCC testing ... **A13:** 246-253
static tensile ... **A11:** 390
static tension ... **A8:** 717-718
stress intensity factor range, effect on ... **A12:** 54
subcritical, effect on drop tower compression test specimen ... **A8:** 197
synchronous device for stressing specimens ... **A8:** 507
system, for liquid metals ... **A8:** 426
torsional ... **A11:** 109
torsional, AISI/SAE alloy steels ... **A12:** 330
train, crack arrest fracture toughness testing ... **A8:** 289
type, correction factors ... **A11:** 116
type, effect on fatigue-crack propagation ... **A11:** 108-109
uniform and nonuniform, crack effects ... **A12:** 175-176
visual examination, effect on ... **A12:** 72
Loading bar
for quick-release clamp ... **A8:** 221
Loading conditions
as fatigue mechanism ... **A12:** 4-5
mechanical damage from ... **A12:** 72
types of ... **A12:** 12-15
Loading cycles ... **EM3:** 637
Loading, cyclic *See* Cyclic loading
Loading frame, two-post
cam plastometer ... **A8:** 194
Loading rate ... **A18:** 436
Loading sheet
defined ... **A7:** 7
Loading weight
defined ... **A7:** 7
Loadings
basic, with effective elastic properties ... **EM1:** 186
differing, for carbon-epoxy composites ... **EM1:** 201
for full-scale static test ... **EM1:** 349
mechanical, laminate stresses from ... **EM1:** 227-228
off-axis, defined ... **EM1:** 209
for structural-element testing ... **EM1:** 320-329
for subcomponent testing ... **EM1:** 329-340
thermally induced, laminates ... **EM1:** 224-225
Loam
defined ... **A15:** 8
as mixed molding material ... **A15:** 32
molding, described ... **A15:** 228-229
molds ... **A15:** 31, 34
Lobe curve
resistance spot welding ... **M6:** 478, 484, 486
Lobed bearing
defined ... **A18:** 12
Local action
defined ... **A13:** 9
Local adaptive-thresholding technique
magnetic particles ... **A17:** 118-119
Local annealing
for gas cutting ... **A14:** 724-725
Local arrest
in cleavage fracturing ... **A8:** 453
Local asperity contact pressure
nomenclature for lubrication regimes ... **A18:** 90
Local asperity contact shear stress
nomenclature for lubrication regimes ... **A18:** 90
Local asperity contact temperature rise
nomenclature for lubrication regimes ... **A18:** 90
Local brittle zones (LBZ) ... **A6:** 81
Local cells
defined ... **A13:** 9
Local elongation
variation with position along gage length ... **A8:** 26
Local friction factor
friction during metal forming ... **A18:** 67
Local interfacial equilibrium
defined ... **A15:** 101
Local necking strain
true ... **A8:** 24-25
Local physical performance
of interconnections ... **EL1:** 5-6
Local preheating
definition ... **M6:** 11
for gas cutting ... **A14:** 724
Local principal tension ... **EM4:** 636
Local seizure
and wear measurement ... **A18:** 367
Local solidification *See also* Solidification
and insoluble particles ... **A15:** 142
times, low-gravity ... **A15:** 154
Local stress concentrations
design details for ... **A13:** 343
Local stress relief heat treatment
definition ... **M6:** 11
Local structure
determination by EXAFS analysis ... **A10:** 407
Local thermal equilibrium model (ion calculation) ... **A7:** 258
Local thermodynamic equilibrium
abbreviation for ... **A10:** 690
emission source in ... **A10:** 24
Local thermodynamic equilibrium (LTE) ... **A6:** 32
welding arcs ... **A6:** 32
Localized alloying/melting
as failure mechanism ... **EL1:** 1013-1014
Localized biological corrosion *See also* Biological corrosion; General biological corrosion; Microbiological corrosion ... **A13:** 114-120
of aluminum ... **A13:** 118-119
of austenitic stainless steel ... **A13:** 117
of copper alloys ... **A13:** 119
industries ... **A13:** 115-116
of iron and steel ... **A13:** 116-117
organisms involved ... **A13:** 115-116
of stainless steel ... **A13:** 117-118
tuberculation ... **A13:** 119-120
Localized corrosion ... **A13:** 104-122
of austenitic stainless steels ... **A13:** 563
biological ... **A13:** 114-120
in chemical cleaning ... **A13:** 1142
of cobalt-base alloys ... **A13:** 661
of corrosion-resistant casting alloys ... **A13:** 580-581
and crevice corrosion ... **A13:** 108-113
defined ... **A13:** 9, 79, 104
and design ... **A13:** 339
electrochemical testing methods ... **A13:** 216-217
of ferritic/duplex stainless steels ... **A13:** 563
and filiform corrosion ... **A13:** 104-108
material selection to avoid/minimize ... **A13:** 323-324
of niobium/niobium alloys ... **A13:** 722-723
and pitting corrosion ... **A13:** 113-114
potentiostatic and galvanostatic testing methods ... **A13:** 217
and premature cracking ... **A8:** 496
of radioactive waste containers ... **A13:** 971
scratch-repassivation testing method ... **A13:** 217
of stainless steels ... **A13:** 562
testing ... **A13:** 195
types ... **A13:** 79
and uniform corrosion ... **A13:** 48
water-recirculating system ... **A13:** 488
Localized corrosion (pitting)
of cobalt-base corrosion-resistant alloys ... **A2:** 453
Localized deformation *See also* Deformation; Flow localization; Necking
and instability in tension ... **A8:** 25
Localized distortion
temperature dependence of ... **A11:** 138
Localized extension
in ductility measurement ... **A8:** 26
Localized heating *See also* Heat treatment; Heating
for straightening ... **A14:** 681-682
Localized pitting
copper/copper alloys ... **A13:** 612
Localized severe forming ... **A14:** 550-551
Localized strains
in workability theory ... **A14:** 389-390
Localized thinning model
of fracture ... **A14:** 393
Locating boss
defined ... **A15:** 8
Locating points
as pattern feature ... **A15:** 192
Locating ring
defined ... **EM2:** 25
Location
codes/standards/requirements for ... **A17:** 49
crack, by microwave inspection ... **A17:** 203
determination and evaluation ... **A17:** 50
discontinuity, magnetic effects ... **A17:** 100
of flaws ... **A17:** 50, 86
Location (continued)
leak ... **A17:** 68
Location plots
acoustic emission inspection ... **A17:** 283-284
Locational accuracy
technological capabilities ... **EL1:** 508
Lock
defined ... **A14:** 8
Lock components
powders used ... **A7:** 573
Lock forming quality galvanized steel sheet
formability ranking ... **M1:** 547
Lock seam joint
soldering ... **A6:** 130
Lock-in amplifier
Auger apparatus ... **A7:** 251
Lock-seam dies
for press-brake forming ... **A14:** 537
Lockalloy ... **A6:** 945
Lockalloy sheet and plate manufacturing ... **A7:** 761
Lockbolts
pull-type, materials and composite applications ... **A11:** 530
stump-type, materials and composite applications ... **A11:** 530
Locked dies ... **M1:** 361
defined ... **A14:** 8
Locked-up stress
definition ... **A6:** 1211
Locking collar, steel
failure from fibering or banding ... **A11:** 320
Locking leno weave
unidirectional/two-directional fabrics ... **EM1:** 125-127
"Locking the metal out" ... **A8:** 548
Locks
and counterlocks ... **A14:** 48-49
for mismatch ... **A14:** 49
Lockseaming *See also* Can seaming
internal, forming of ... **A14:** 573
of metal strip ... **A14:** 572-573
Locomotive *See also* Locomotive axles, failures of
axles, failed, structure and microstructure ... **A11:** 721-722
diesel engine, corrosion fatigue cracking ... **A11:** 371-372
spring, fatigue fracture ... **A11:** 551-552
suspension spring, failure of ... **A11:** 551
Locomotive axles, failures of ... **A11:** 715-727
axle studies, results of ... **A11:** 723-724
background ... **A11:** 715-723
conclusions ... **A11:** 725-726
results of analyses ... **A11:** 725-726
simulation of LME mechanism ... **A11:** 724-725
Lodex alloys *See* Permanent magnet materials, specific types
log *See* Common logarithm (base 10)
Log normal distribution ... **A8:** 628, 630-631, 700-701
Log ratio scanning
analysis by ... **A10:** 144
Log rupture time ... **A8:** 188, 332
Log secondary creep rate
vs. log stress isotherm ... **A8:** 332
Log stress
vs. log rupture time ... **A8:** 188, 332
vs. log secondary creep rate isotherm ... **A8:** 332
vs. rupture life for aluminum alloy ... **A8:** 332
Log-log rupture
upward inflection at long times ... **A8:** 332-333
Log-log scale ... **A8:** 697-698
Logan Manufacturing Company, Phoenixville PA, sand blast ... **A15:** 33
Logarithm base, natural
symbol for ... **A8:** 724
Logarithmic
kinetic rate law ... **A13:** 17
oxidation, of scales ... **A13:** 72
reaction rates, gaseous corrosion ... **A13:** 66-67
Logarithmic creep ... **A8:** 308, 309
Logarithmic decrement factor ... **A6:** 1053
Logic *See also* Digital logic
advanced Schottky/FAST ... **EL1:** 82
chip selection ... **EL1:** 128
circuits, future trends ... **EL1:** 177
complementary metal-oxide semiconductor (CMOS) ... **EL1:** 76

Logic (continued)
conventional, minimum device size for **EL1:** 2
design **EL1:** 129
families **EL1:** 601
forms of **EL1:** 160-161
gates **EL1:** 6
n-channel MOS (NMOS) **EL1:** 76
network circuitry **EL1:** 2
nondeterministic, smaller devices for **EL1:** 2
optical **EL1:** 10
structure, design of **EL1:** 129
transistor-transistor (TTL) **EL1:** 76
Logic-wiring design
computer-aided manufacturing **EL1:** 130
Lognormal distribution
estimated failure factors **EL1:** 902
failure density **EL1:** 897
scaling, failure plot **EL1:** 889
Lomakin effect **A18:** 594, 595
London dispersion forces **EM3:** 17
defined **EM2:** 25
Long bar machine **A7:** 39-41
Long bar plasma rotating electrode process machine **A7:** 39-41
Long parts
straightening of **A14:** 680-689
Long period **EM3:** 17
defined **EM2:** 25
Long tapers *See* Tapers
Long terne coated steels
automotive industry **A13:** 1014
Long terne coatings **M5:** 358-359
Long terne sheet *See also* Lead; Lead alloys
characteristics and applications **A2:** 554-555
Long terne steel sheet *See* Terne coatings, steel sheet
Long transverse *See* Transverse
Long transverse testing direction **A8:** 672
Long wave radiation
cathodoluminescence **A9:** 90
Long-chain branching **EM3:** 17
defined **EM2:** 215
Long-chain hydrocarbons
analytic methods for **A10:** 9
Long-chain polymers
chemical structure **EM2:** 52
Long-line current
defined **A13:** 9
Long-period ordering .... **A3:** 1 • 11
Long-period superlattices **A9:** 710, 719
Long-range order
analyses for **A10:** 277, 393
Long-taper dies **A14:** 132
Long-term environmental factors
properties effects **EM2:** 423-432
Long-term etching
defined **A9:** 11
Long-term exposure
and elevated-temperature service **A1:** 623, 627
Long-term heat aging
metals vs plastics **EM2:** 78
Long-term loads
metals vs plastics **EM2:** 75
Long-term temperature resistance
of engineering plastics **EM2:** 68
Longitudinal chromatic aberration **A9:** 75
in an uncorrected lens **A9:** 77
Longitudinal crack
definition **A6:** 1211
Longitudinal cracks
in weldments **A17:** 585
Longitudinal direction
defined **A8:** 8 **A9:** 11 **A11:** 6
Longitudinal drilling
in conjunction with turning **A16:** 135
Longitudinal flaws
in steel bar and wire **A17:** 550
Longitudinal grooves
shafts **A11:** 470-471
Longitudinal Kerr effect **A9:** 535-536
Longitudinal magnetization
for leakage field testing **A17:** 130
in magnetic fields **A17:** 91
Longitudinal profiles
distorted steel shotgun barrel **A11:** 139
Longitudinal properties
carbon steels **A14:** 219
Longitudinal resistance seam welding
definition **M6:** 11
Longitudinal sequence
definition **M6:** 11
Longitudinal shear
and damping **EM1:** 207
Longitudinal shrinkage
bending distortion **M6:** 877-879
aluminum welds **M6:** 878-879
low-carbon steel plate **M6:** 878
butt welds **M6:** 875-876
fillet welds **M6:** 876
Longitudinal stretch forming machines **A14:** 596
Longitudinal tensile test
of titanium embrittlement **A11:** 642
Longitudinal wave, at impact
x-t diagram **A8:** 232
Longitudinal waves
ultrasonic **A17:** 233, 505
Longitudinal welds *See also* Weld(s); Welding
eddy current inspection **A17:** 186
Longitudinal welds, pipe
failure of **A11:** 698-699, 704
Longos
defined **EM1:** 15 **EM2:** 25
Looms
fly-shuttle **EM1:** 127, 128
rapier **EM1:** 127, 128
Loop classifier **A7:** 7
Loop tenacity
defined **EM1:** 15 **EM2:** 25
Looping pit
cut-to-length lines **A14:** 711
Loops, hysteresis
defined **A17:** 100
Loose abrasive grains
for polishing **A2:** 1012-1013
synthetic lapping abrasives **A2:** 1012
Loose fillers
for bending **A14:** 666
Loose metal
as formability problem **A8:** 548
sheet **A14:** 878
Loose patterns
described **A15:** 189-190
Loose pieces
in die casting **A15:** 287
Loose powder *See also* Loose powder filling; Loose powder sintering
compaction, stages **A7:** 297-298
containerization, powder processing **A7:** 435-436
effect of tapping on density **A7:** 297
feeding into cold die **A7:** 502-503
porosity **A7:** 555
sampling **A7:** 213
Loose powder filling
in encapsulation containers for hot isostatic pressing **A7:** 431, 433
practices **A7:** 431, 433
with predensified compacts **A7:** 431, 433
Loose powder sintering **A7:** 7, 296, 308
Loose tolerance
defined **A14:** 307
Loose-powder method
hot extrusion of powder mixtures **A2:** 988
Loosely packed powders *See* Loose powder
Lorentz force
effect on magnetic contrast **A9:** 95
gas-tungsten arc welding **A6:** 22
submerged arc welding **A6:** 24
Lorentz microscopy
for imaging magnetic domain boundaries **A10:** 446
study of magnetic domains **A9:** 536
Lorentz polarization, and absorption
surface stress measurement correction for **A10:** 385-386
Lorentz reciprocity theorem **A17:** 218
Lorentzian absorption curve
ESR spectrum **A10:** 259
Lorentzian absorption lineshape
and dispersion lineshape **A10:** 281
Los Alamos National Laboratory
split Hopkinson pressure bar test facility **A8:** 200-201
Loss angle *See* Phase tingle
Loss factor *See also* Tan delta **EM3:** 428, 429
defined **EM1:** 15 **EM2:** 26
of glass fibers **EM1:** 46
microwave inspection **A17:** 204
Loss function concept
of quality **A17:** 720-721
Loss index *See also* Loss factor
of glass fibers **EM1:** 46
Loss modulus *See also* Complex modulus; Modulus **EM1:** 15, 761 **EM3:** 17, 318-319, 320
defined **EM2:** 26
Loss, of back reflection
ultrasonic inspection **A17:** 246
Loss on ignition *See also* also Sizing content; Sizing content
defined **EM1:** 15 **EM2:** 26
Loss on reduction (LOR)
oxygen plus water **A7:** 155
Loss tangent *See* Dissipation factor; Electrical dissipation factor; Tan delta **EM3:** 319, 32
microwave inspection **A17:** 204
Lossy signal transmission line
analytical solution **EL1:** 41-42
Lost foam casting *See also* Evaporative foam casting; Expendable pattern **A15:** 230-234
advantages **A15:** 234
coating types **A15:** 232
defined **A15:** 8
foam pattern **A15:** 231
investing pattern **A15:** 233
pattern molding **A15:** 231-232
pouring **A15:** 233-234
process technique **A15:** 230-231
processing parameters **A15:** 231-234
sand system **A15:** 233-234
Lost pattern process *See* Lost foam casting
Lost wax process *See also* Investment (lost wax) casting; Precision casting
defined **A15:** 8
historical use **A15:** 16, 19-20, 22
and neutron radiography **A17:** 393
Lost-foam pattern casting *See also* Evaporative pattern casting (EPC)
aluminum casting alloys **A2:** 5, 140
Lost-wax investment molding
titanium and titanium alloy castings **A2:** 635-636
Lot *See also* Batch **EM3:** 17
averages, use in normal distribution computation **A8:** 664-666
-centered regression analysis for creep-rupture data **A8:** 691-693
defined **A7:** 7 **A8:** 8 **A10:** 676 **EM1:** 15 **EM2:** 26
size, economics of **EM1:** 420-421
-to-log variation, in creep/creep-rupture analyses **A8:** 684
Lot qualification radio frequency testing **EL1:** 946-949
Lot sample
defined as gross sample **A10:** 674

Low brass *See also* Brasses; Wrought coppers and copper alloys
applications and properties **A2:** 299-300
resistance spot welding **A6:** 850
weldability **A6:** 753
Low brass, 80%, microstructure of ...... **A3:** 1 • 22
Low carbon sheet steel
effects of insufficient grinding **A9:** 169
electroless nickel plated, different mounts compared **A9:** 167
Low carbon steel billet
ingot pattern **A9:** 174
Low carbon steels *See also* Magnetic materials; Plate steels
calcium aluminate inclusions in **A9:** 628
calcium sulfide inclusions in **A9:** 628
cold rolled, annealed, with recrystallized grains **A9:** 695
deep drawn aerosol can, Lüders lines in **A9:** 688
effect of penultimate grain size on recrystallization kinetics **A9:** 697
line etching **A9:** 62
Lüders front **A9:** 685
powder metallurgy materials, etching **A9:** 508-509
precipitation of cementite **A9:** 179
turning and milling recommended ceramic grade inserts for cutting tools **EM4:** 972
Low carbon steels, specific types
0.05% C, Fe_3C carbide at ferrite grain boundaries **A9:** 179
0.06% C, cold rolled and annealed, carbide particles **A9:** 180
0.10% C, cold rolled 90%, annealed **A9:** 181
1% Si, titanium bearing, delineating ferrite grain boundaries **A9:** 170
1020, laser welded to 70600 copper-nickel **A9:** 408
chromium-molybdenum-vanadium thermally etched grain boundaries **A9:** 83
Low earth orbit
atomic oxygen environment of **A12:** 481
Low energy electron diffraction (LEED) **EM3:** 23
Low expansion alloys
applications **M3:** 792, 798
composition, effect on expansivity **M3:** 792, 793
expansion coefficients **M3:** 792, 793, 794, 795, 798
magnetic properties **M3:** 794
mechanical properties **M3:** 794, 795, 797
Low frequency cycle
definition **M6:** 11
Low gravity effects *See also* Gravitational acceleration
convection, and solute redistribution **A15:** 148-149
convection, in liquid **A15:** 147-148
dendrite spacing **A15:** 150
during solidification **A15:** 147-158
eutectic alloys **A15:** 150-153
experimental systems **A15:** 147
morphological stability **A15:** 153-156
for on-eutectic interphase spacing **A15:** 149
temperature gradient **A15:** 147-148
thermal convention **A15:** 147-148
Low molecular weight fragments **EM3:** 41
Low pass filter
to reduce load cell ringing **A8:** 193
Low pulse current
definition **M6:** 11
Low pulse time
definition **M6:** 11
Low temperature
design properties **A8:** 670-671
tension testing **A8:** 34-37
Low temperature properties
magnesium **M2:** 531-532, 533, 534
Low temperature tension testing **A8:** 34-37
Low temperatures
alloys for structural applications **M3:** 721-772
Low yttria-doped high-purity silicon nitride
processed by glass-encapsulated HIP processing **EM4:** 199
Low-acceleration fatigue tests
for solder attachments **EL1:** 741
Low-alloy cast steels **A1:** 372-374, 375
Low-alloy chromium-molybdenum powders
tensile properties of steel railroad wheel rings **A7:** 548
Low-alloy metals for pressure vessels and piping
flux-cored arc welding **A6:** 668
gas-metal arc welding **A6:** 668
gas-tungsten arc welding **A6:** 668
Low-alloy nickel
applications and characteristics **A2:** 435, 437, 441
Low-alloy special-purpose tool steel *See* Tool steels, low alloy
Low-alloy special-purpose tool steels **A1:** 767
composition limits **A18:** 736
forging temperatures **A14:** 81
Low-alloy steel *See also* Alloy steel **A1:** 201, 207, 208-211
air-hardening of **A1:** 644-645, 646
alloying elements in **A1:** 144-147
castings and **A1:** 363-364
classification of **A1:** 149
composition of **A1:** 152-153
creep-resistance **A1:** 619-621
definition of **A1:** 149
direct castings methods **A1:** 211
electropolishing of **M5:** 305, 308
for elevated-temperature service **A1:** 618
ferritic, stress-strain curve **A8:** 177
forgings *See* High-strength, low-alloy steel forgings
hardenable low-alloy steels **A1:** 453
high-strength, rolling during torsion test **A8:** 179
International designations and specifications
British (BS) steel compositions ... **A1:** 158, 166-174, 185
for **A1:** 156-159, 166-194
French (AFNOR) steel compositions **A1:** 158, 166-174, 188
German (DIN) steel compositions **A1:** 157, 166-174, 178-179
Italian (UNI) steel compositions **A1:** 159, 166-174, 192
Japanese (JIS) steel compositions **A1:** 157, 166-174, 181
Swedish (SS) steel compositions **A1:** 159, 166-174, 194
measured times-to-fracture for **A8:** 272
mechanical properties **A1:** 209-211, 396
mill heat treatment **A1:** 209
phosphate coating of **M5:** 437
physical properties of **A1:** 195-200
plate *See* Steel plate
plating, preparation for **M5:** 16-18
porcelain enameling of **M5:** 513
production of **A1:** 930
production of sheet and strip **A1:** 208
quality descriptors **A1:** 208-209
quenched and tempered **A1:** 391-397
SAE-AISI designations **A1:** 152-153
sheet and strip **A1:** 207-209
shot peening of **M5:** 141-142, 145
stress-corrosion cracking environments **A8:** 526
tensile properties **A8:** 555
workability **A8:** 165, 575
Low-alloy steel castings *See also* Austenitic manganese steel castings; Carbon steel castings **A9:** 230-231, 235-236
abrasives for **A9:** 230
aluminum deoxidized, normalized, cooled and tempered **A9:** 235
aluminum deoxidized, quenched and tempered **A9:** 235
compositions of **A9:** 230
etchants **A9:** 230
etching **A9:** 230
grinding **A9:** 230
heat treatment **A9:** 231
microstructures **A9:** 231
mounting **A9:** 230
polishing **A9:** 230
sectioning **A9:** 230
specimen, annealed **A9:** 236
specimen, normalized, quenched and tempered **A9:** 235
Low-alloy steels *See also* Alloy steels; ASP steels; Carbon steels; High-strength low-alloy steels; Low-alloy steels, specific types; Plate steels;
Low-alloy steels (continued)
Steels; Steels, specific types; Tool steels; Tool steels, specific types
air-carbon arc cutting **A6:** 1175
alloying elements **A15:** 702, 715-716
applications, sheet metals **A6:** 399
atmospheric corrosion resistance **A13:** 82
austenitic stainless-clad, welding of **A6:** 501-502
bainitic microstructure **A11:** 393
boriding **A4:** 438
brazing and soldering characteristics **A6:** 624
brazing properties **M6:** 966
brazing temperature effect on hardness **A6:** 908
C, P, and S determined in **A10:** 29
case hardening of **M1:** 491-496
cast, weldability **A15:** 532-534
casting **A13:** 573-575
castings **M1:** 377, 379, 386, 388-389, 390, 392-393, 394-399
castings, failures **A11:** 392-393
cladding of austenitic stainless steel to **A6:** 502-504
composition **A7:** 102
compositions **A14:** 54
compressibility and properties **A7:** 102
corrosion of **A11:** 199
corrosion properties **A15:** 720
covering for welding electrodes **A6:** 176, 177, 178
defined **A15:** 715
degassing procedures **A15:** 428
dendritic structure **A9:** 623
deoxidation **A15:** 721
diffusion welding **A6:** 884
diffusion-bonded grades **A7:** 102
dip brazing **A6:** 336-338
discontinuity effects **A15:** 718
dissimilar metal joining **A6:** 821, 824, 825
drop hammer forming of **A14:** 655-656
electrodes
for flux-cored arc welding **A6:** 188
submerged arc welding **A6:** 204-205, 206
electrodes for submerged arc welding
electrogas welding **A6:** 276-277
electron beam welding **M6:** 662
electron-beam welding **A6:** 860, 867
electroslag welding ... **A6:** 273, 276-277, 279 **M6:** 226
elevated temperature properties **M1:** 639-663
elevated-temperature ductility **A11:** 265
elevated-temperature properties **A15:** 720-721
eutectic joining **EM4:** 526
ferritic, alloying effects on SCC behavior **A13:** 270
ferrographic application to identify wear particles **A18:** 305, 306
ferrous, composition effect on corrosion **A13:** 537-538
filler metals for torch brazing **M6:** 953
flash welding **M6:** 558
flux cored electrodes **M6:** 101-102
flux-cored arc welding **A6:** 186, 187, 188
designator **A6:** 189
forge welding **A6:** 306 **M6:** 676
forged, mechanical properties **A7:** 464, 470
foundry practices **A15:** 721
friction welding **A6:** 152, 153
furnace brazing *See* Furnace brazing of steels
fusion welding to stainless steels **A6:** 826, 827
gas metal arc welding **M6:** 153
gas tungsten arc welding **M6:** 203
gas-metal arc welding
of aluminum bronzes **A6:** 828
of copper nickels **A6:** 828
of coppers **A6:** 828
of high-zinc brasses **A6:** 828
of low-zinc brasses **A6:** 828
of phosphor bronzes **A6:** 828
of silicon bronzes **A6:** 828
of special brasses **A6:** 828
of tin brasses **A6:** 828
gas-tungsten arc welding
of aluminum bronzes **A6:** 827
of copper nickels **A6:** 827
of coppers and copper-base alloys **A6:** 827
of phosphor bronzes **A6:** 827
of silicon bronzes **A6:** 827
hardenability **A15:** 715

Low-alloy steels (continued)
hardfacing alloys for **A6:** 791, 798
heat treating of **A14:** 53-55
HERF forgeability **A14:** 104
high-strength, horizontal centrifugal casting of **A15:** 299
high-temperature erosion testing for **A11:** 282
highway-truck equalizer beam of **A11:** 390
for hot forging **A14:** 43
hydrogen fluoride/hydrofluoric acid corrosion **A13:** 1166-1167
hydrogen-induced cracking **A6:** 94
ingot tub, thermal fatigue in **A11:** 408
,jaws, brittle fracture of **A11:** 389-391
laser beam welding **M6:** 647
liquid nitriding **A4:** 419
machinery and equipment weldments **A6:** 391, 393
macrostructure of **A9:** 623
in marine atmosphere **A13:** 541
material for die forging tools **M3:** 529, 530, 534
materials for die-casting dies **A18:** 629
materials for dies and molds **A18:** 622, 625
melting **A15:** 721
microstructure of **A9:** 623-624
molten nitrate corrosion resistance **A13:** 90-91
molybdenum-vanadium, creep curves **A11:** 263
notch toughness **M1:** 689-709
nut, LME service failure of **A11:** 228
oxide stability **A11:** 452
oxyacetylene welding **A6:** 281, 284
oxyfuel gas cutting **M6:** 897
oxyfuel gas welding **A6:** 286
in petroleum refining and petrochemical operations **A13:** 1262-1263
plasma (ion) nitriding **A4:** 423
plasma and shielding gas compositions **A6:** 197
plasma arc welding **M6:** 214
plate, chevron patterns in **A11:** 76, 77
powder, contamination of **A14:** 191
powder metallurgy materials **A9:** 503
prealloyed grades **A7:** 102
principal ASTM specifications for weldable sheet steels **A6:** 399
processed by STAMP process **A7:** 548
production **A7:** 100-103
projection welding **A6:** 233 **M6:** 506
recommended guidelines for selecting PAW shielding gases **A6:** 67
Replicast process for **A15:** 272
resistance seam welding **A6:** 241, 245
resistance soldering **A6:** 357
resistance spot welding **M6:** 477, 486
resistance welding **A6:** 837, 840, 841
roll welding **A6:** 312 **M6:** 676
SCC testing **A13:** 270
seawater, corrosion in **M1:** 739-746
section size and mass effects **A15:** 717-718
segregation during dendritic growth **A9:** 625-626
service temperature of die materials in forging **A18:** 625
shielded metal arc welding **A6:** 57, 61, 176 **M6:** 75
soil corrosion **M1:** 725-731
soldering **A6:** 624
in sour gas environments **A11:** 300
stress-corrosion cracking in **A11:** 214-215
stress-rupture crack in welds of **A11:** 427
structure and property correlations **A15:** 716-717
submerged arc welding **A6:** 204-205
sulfide stress-corrosion cracking **A13:** 532
superplasticity **A14:** 871
susceptibility to hydrogen damage **A11:** 126, 249
temperature measurement, validation strategies **A6:** 1149
tensile fracture from shrinkage porosity in **A11:** 389-390
test coupon vs casting properties **A15:** 719

Low-alloy steels (continued)
thermal stress relief for SCC in **A13:** 328
thermoreactive deposition/diffusion process **A4:** 448, 452
tool steels **A14:** 81
torch brazing *See* Torch brazing of steels
transfer gear **A7:** 668
ultrasonic welding **A6:** 893
water-atomized **A7:** 101, 102, 302
water-quenched **A11:** 393
weld-metal toughness **M6:** 42
weldability **A6:** 420, 424 **A15:** 719-720 **M1:** 563
welding to austenitic stainless steels **A6:** 500-501
welding to ferritic stainless steels **A6:** 501
welding to martensitic stainless steels **A6:** 501
weldment properties **A6:** 417, 424
yield strength vs. tempering temperature **A13:** 954
zinc stearate effect on green strength **A7:** 302

Low-alloy steels for pressure vessels and piping **A6:** 666-668
electrogas welding **A6:** 667, 668
electroslag welding **A6:** 667, 668
filler metal selection **A6:** 668
flux-cored arc welding **A6:** 667
gas-metal arc welding **A6:** 667
gas-tungsten arc welding **A6:** 667
heat resistance **A6:** 667
hydrogen-induced cracking **A6:** 668
postweld heat treatments **A6:** 668, 669
preheating **A6:** 669
properties **A6:** 667
shielded metal arc welding **A6:** 667, 668
specifications **A6:** 668
submerged arc welding **A6:** 667, 668
welding procedures and practices **A6:** 667-668

Low-alloy steels, specific types *See also* ASP steels; Carbon steels; Low-alloy steels; Low-carbon steel powders; Steels; Steels, specific types; Structural steels, specific types; Tool steels; Tool steels, specific types
0.15%C, artifact structures from overheating **A9:** 166
1520, mechanical properties **A7:** 470
4120, mechanical properties **A7:** 470
4130, mechanical properties **A7:** 470
4150, STAMP process **A7:** 548
4600, effect of admixed lubricant on green strength **A7:** 303
4600 series, powder metallurgy materials microstructures **A9:** 510
4620, workability test **A7:** 411
4630 modified, mechanical properties **A7:** 470
4640, mechanical properties **A7:** 470
A 203, applications, welding of pressure vessels and piping **A6:** 667, 669
A 204, applications, welding of pressure vessels and piping **A6:** 667, 669
A 302, applications, welding of pressure vessels and piping **A6:** 669
A 333
applications, welding of pressure vessels and piping **A6:** 669
SMAW **A6:** 668
A 335, applications, welding of pressure vessels and piping **A6:** 669
A 353
applications, welding of pressure vessels and piping **A6:** 669
SMAW **A6:** 668
A 369, applications, welding of pressure vessels and piping **A6:** 667
A 387, applications, welding of pressure vessels and piping **A6:** 667, 669
A 420, applications, welding of pressure vessels and piping **A6:** 667
A 508, applications, welding of pressure vessels and piping **A6:** 667

Low-alloy steels, specific types (continued)
A 522, applications, welding of pressure vessels and piping **A6:** 667
A 533, applications, welding of pressure vessels and piping **A6:** 667
A 541, applications, welding of pressure vessels and piping **A6:** 667
A 672, applications, welding of pressure vessels and piping **A6:** 669
Ancorsteel 2000, composition **A7:** 102
Ancorsteel 2000, properties **A7:** 102
Ancorsteel 4600V, composition **A7:** 102
Ancorsteel 4600V, prealloyed powder water-atomized and annealed **A9:** 514
Ancorsteel 4600V, properties **A7:** 102
ASTM A352, grade LC3, water quenched and tempered **A9:** 235
ASTM A487, class 2, normalized by austenitizing **A9:** 235
ASTM A487, normalized by austenitizing **A9:** 235
CA6NM, heat treatment of **A9:** 231
class 2, applications, welding of pressure vessels and piping **A6:** 667
class 3, applications, welding of pressure vessels and piping **A6:** 667
class 4, applications, welding of pressure vessels and piping **A6:** 667
class 5, applications, welding of pressure vessels and piping **A6:** 667
class 6, applications, welding of pressure vessels and piping **A6:** 667
class 7, applications, welding of pressure vessels and piping **A6:** 667
class 8, applications, welding of pressure vessels and piping **A6:** 667
Distaloy 4600 A, pressed and sintered **A9:** 514-515
Fe-0.37Mn-1.80Ni-0.63Mo, pressed and sintered **A9:** 527
Fe-0.42Ni-0.62Mo-0.2C, prealloyed, pressed and sintered, carbon content varied **A9:** 519-520
Fe-1.8Ni-0.5Mo-0.4C, pressed and sintered **A9:** 527
Fe-1.75Ni-0.5Mo-1.5Cu 0.5C, diffusion alloyed, pressed and sintered **A9:** 522-523
Fe-1.85Ni-0.6Mo-0.5C, prealloyed powder **A9:** 520
Fe-1.85Ni-0.60Mo-0.2C, prealloyed, pressed and sintered, carbon content varied **A9:** 520
Fe-2.0Cu-0.8C, blisters formed from sintering **A9:** 526, 530
Fe-2.0Cu-0.8C, pressed and sintered **A9:** 527
Fe-2.0Cu-0.8C, pressed and sintered, steam blackened **A9:** 528
Fe-2.0Ni-0.5Mo-0.2C, powder-forged gear **A9:** 526
Fe-2MCM, mechanical properties **A7:** 470
Fe-2Ni-0.3C, injection molded and sintered **A9:** 526
Fe-2Ni-0.5Mo-0.5C, pressed, sintered forged **A9:** 526
Fe-2NI-0.8C, pressed and sintered **A9:** 527
Fe-2Ni-0.8C, pressed and sintered tensile bar **A9:** 527
grade 2, applications, welding of pressure vessels and piping **A6:** 667
grade 3, applications, welding of pressure vessels and piping **A6:** 669
grade 5, applications, welding of pressure vessels and piping **A6:** 667, 669
grade 7, applications, welding of pressure vessels and piping **A6:** 669
grade 8, applications, welding of pressure vessels and piping **A6:** 669
grade 9, applications, welding of pressure vessels and piping **A6:** 667

Low-alloy steels, specific types (continued)
grade 11, applications, welding of pressure vessels and piping A6: 667, 669
grade 12, applications, welding of pressure vessels and piping A6: 667, 669
grade 21, applications, welding of pressure vessels and piping A6: 667
grade 22, applications, welding of pressure vessels and piping A6: 667, 669
grade FP 1, applications, welding of pressure vessels and piping A6: 667
grade FP 2, applications, welding of pressure vessels and piping A6: 667
grade FP 5, applications, welding of pressure vessels and piping A6: 667
grade FP 7, applications, welding of pressure vessels and piping A6: 667
grade FP 9, applications, welding of pressure vessels and piping A6: 667
grade FP 11, applications, welding of pressure vessels and piping A6: 667
grade FP 12, applications, welding of pressure vessels and piping A6: 667
grade FP 21, applications, welding of pressure vessels and piping A6: 667
grade FP 22, applications, welding of pressure vessels and piping A6: 667
grade WPL 3, applications, welding of pressure vessels and piping A6: 667
grade WPL 8, applications, welding of pressure vessels and piping A6: 667
grade WPL 9, applications, welding of pressure vessels and piping A6: 667
Iron-copper, mechanical properties A7: 470
Iron-copper-manganese-nickel- molybdenum powders, mechanical properties A7: 470
Iron-manganese-molybdenum-nickel mechanical properties A7: 470
Iron-nickel, mechanical properties A7: 470
Iron-nickel-molybdenum-manganese-chromium, mechanical properties A7: 470
type A, applications, welding of pressure vessels and piping A6: 667, 669
type B, applications, welding of pressure vessels and piping A6: 667, 669
type C, applications, welding of pressure vessels and piping A6: 667
type D, applications, welding of pressure vessels and piping A6: 667, 669
type E, applications, welding of pressure vessels and piping A6: 667, 669
type F, applications, welding of pressure vessels and piping A6: 667, 669
type H 75, applications, welding of pressure vessels and piping A6: 669
type H 80, applications, welding of pressure vessels and piping A6: 669
type I, applications, welding of pressure vessels and piping A6: 667
type II, applications, welding of pressure vessels and piping A6: 667
type P1, applications, welding of pressure vessels and piping A6: 669
type P5, applications, welding of pressure vessels and piping A6: 669
type P11, applications, welding of pressure vessels and piping A6: 669
type P12, applications, welding of pressure vessels and piping A6: 669
type P22, applications, welding of pressure vessels and piping A6: 669
Low-alloy steels, welding of A6: 662-676
electrogas welding A6: 662, 664, 668
electroslag welding A6: 662, 664, 668
factors determining procedures and practices A6: 662
filler metals A6: 662-663, 665, 668, 669-671, 672, 673, 674-675
flux-cored arc welding A6: 662, 663, 664, 666, 668, 669, 670, 674, 676
fluxes A6: 662
gas-metal arc welding A6: 662, 663, 664, 666, 668, 669, 670, 671, 673, 674, 676, 828
gas-tungsten arc welding A6: 662, 664, 666, 668, 669, 670, 671, 673, 674, 676, 828

Low-alloy steels, welding of (continued)
heat-treatable low alloy steels A6: 662, 668-673
applications A6: 670
composition A6: 668, 669, 670
cracking A6: 669, 670, 671, 672
filler metals A6: 669-671
heat input A6: 672
microstructure A6: 671-672
postweld heat treatment A6: 672-673
preheating A6: 669, 671-672
welding processes and practices A6: 669
high-strength low-alloy (HSLA) structural steels A6: 662- 664
chemical compositions A6: 663
electrogas welding A6: 664
electroslag welding A6: 664
filler metal selection A6: 662-663, 664
flux-cored arc welding A6: 663, 664
gas-metal arc welding A6: 663, 664
heat-affected zone A6: 663
properties A6: 662, 663
shielded metal arc welding A6: 663-664
specifications A6: 662, 664
submerged arc welding A6: 663, 664
welding procedures and practices A6: 664
high-strength low-alloy quenched and tempered (HSLA Q&T) structural steels A6: 662, 664-666
filler metals A6: 665
gas-metal arc welding A6: 664
gas-tungsten arc welding A6: 664
heat input A6: 666
postweld heat treatment A6: 666
preheat and interpass temperature control A6: 665-666
properties A6: 664
specifications A6: 664
weld design A6: 665
weld metal hydrogen A6: 665
welding processes A6: 664
medium-carbon heat-treatable (quenched and tempered) low-alloy (HTLA)
for pressure vessels and piping A6: 662, 666-668
filler metal selection A6: 668
heat resistance A6: 667
postweld heat treatment A6: 668, 669
preheating A6: 669
properties A6: 667
welding procedures and practices A6: 668
welding processes and practices A6: 668
shielded metal arc welding A6: 662, 663-664, 666, 668, 669-670, 674, 676
steels A6: 668-673
submerged arc welding A6: 662, 663, 664, 668, 669, 670-671, 674, 676
tool and die steels A6: 662, 674-676
composition A6: 674, 675
cracking A6: 675
description of steels A6: 674
filler metals A6: 674-675
postweld heat treatment A6: 675-676
preheating A6: 675, 676
repair practices A6: 675, 676
welding applications A6: 674
welding procedures and practices A6: 675-676
welding processes A6: 674
ultrahigh-strength low-alloy steels A6: 662, 673-674
filler metals A6: 673
gas-tungsten arc welding A6: 673
microstructure A6: 673-674
plasma arc welding A6: 673
postweld heat treatment A6: 674
preheating A6: 673-674
properties A6: 673
specifications A6: 673
welding consumables A6: 662
welding procedures and practices A6: 664
welding processes A6: 673
welding processes A6: 662
Low-alloy tool steel
brazing temperature effect on hardness A6: 908
Low-alloy tool steels
for hot-forging dies A18: 622-623, 624

Low-beryllium copper *See* Copper alloys, specific types, C17500
applications and properties A2: 288-289
Low-carbon austenite
scanning electron micrograph of fracture surface A8: 481-483
transmission electron fractograph of surface A8: 481-483
Low-carbon bainite A1: 404-405
Low-carbon cast steels A1: 364, 371-372
Low-carbon copper-bearing age-hardening steels, specific types
A710/A710M, heat analysis compositions A6: 406
A736/A736M, heat analysis compositions A6: 400
Low-carbon copper-flashed steel
capacitor discharge stud welding A6: 222
Low-carbon hardenable steels
characteristics and applications M1: 457
Low-carbon iron
ductile rupture A12: 220
oxide inclusion A12: 222
Low-carbon iron powders
annealing A7: 182
properties A7: 85-86
water atomization A7: 83-86
Low-carbon lamination steel
as magnetically soft materials A2: 765-766
Low-carbon mold steels
composition limits A18: 736
forging temperatures A14: 81
Low-carbon nickel steel
porosity of injection molded A7: 499
Low-carbon quenched and tempered steels A1: 149
Low-carbon sheet iron powder strip roll-compacted A7: 408
thin-gage roll-compacted properties A7: 408-409
Low-carbon steel
aluminum coating of M5: 333-334
austenitic, flow curve A8: 175
base for 50% aluminum-zinc coated steel sheet and wire M5: 349-350
boiler water embrittlement detector testing A8: 526
bulk-processed and racked parts plating of, preparation for M5: 16-17
cavitation resistance A18: 600
cleavage fracture in A8: 466
colombium-alloyed, corrosion resistance M5: 334
combined effects of strain rate and temperature in A8: 38, 40
creep-fatigue interaction diagram A8: 356-357
effect of temperature on strength and ductility A8: 36
electropolishing of M5: 305, 308
forming limit diagram A8: 551
galling in shallow forming dies A18: 633
galvanic corrosion with magnesium M2: 607
hot dip tin coating of M5: 351
hot rolled, *r* value A8: 550
microhardness traverses A8: 229-230
Modul-*r* measure of modulus of elasticity in A8: 557
phosphate coating of M5: 437
pickling of M5: 69, 72-80
plating process, preparation for M5: 16-18
porcelain enameling of M5: 512-514
potentiodynamic polarization curves A8: 532
rust and scale removal M5: 14
SCC testing of A8: 526
shear zone A8: 229-230
spall stress data A8: 211-212
springback in A8: 552
stress-strain curve for A8: 21-22
testing mediums A8: 526
threshold stress intensity A8: 256
titanium-alloyed, corrosion resistance M5: 334
true yield stress at various strains A8: 38, 29
Low-carbon steel containers
corrosion resistance of M1: 714-715
Low-carbon steel powders
arc welding electrodes, function and composition A7: 817
fluid dies A7: 543

Low-carbon steel powders (continued)
parts, as encapsulation material **A7:** 429
Low-carbon steel powders, specific types
1010, for encapsulation **A7:** 428
1018, for encapsulation **A7:** 428
1020, for encapsulation **A7:** 428
Low-carbon steel wire *See also* Baling wire; Manufacturers' wire
carbon content **M1:** 259
flat wire **M1:** 259
for general use **A1:** 282
tensile strength **M1:** 262-263
w-cycle fatigue **M1:** 668, 670, 682
Low-carbon steels *See also* Carbon steel; Carbon steels; Low-carbon steels, specific types; Mild steels; Steel(s); Steels; Steels, specific types
acoustic emission inspection, of welds **A17:** 599-600
anion/temperature effects, hydrogen absorption **A13:** 330
applications **A15:** 714
applications, railroad equipment **A6:** 396
arc welding with nickel alloys **M6:** 443
arc-welded, failure in **A11:** 415-422
base metal, clad brazing materials **A6:** 961
for bearings **A1:** 24-25, 480-481
bfittle fracture **A12:** 249
blanking of **A14:** 445-458
boiler tubes, rupture from overheating **A11:** 607-608
brazeability of **A11:** 450
brazing and soldering characteristics **A6:** 624
brazing temperature effect on hardness **A6:** 908
capacitor discharge stud welding **M6:** 738
carbon content **A15:** 702
case-hardening of **M1:** 491-496
castings **M1:** 382-384, 386-388
Charpy V-notch impact energy, effect of specimen orientation on **A11:** 68
clad-metal corrosion **A13:** 889
classification and group description **A6:** 405, 406, 407
cleaning **A12:** 76
cleavage crack path **A12:** 117
cleavage, fractograph **A12:** 174
coextrusion welding **A6:** 311
cold heading of **A14:** 291
cold-finished bars **M1:** 215-251
composition, effect on blanking and piercing **A14:** 480
compressed powdered iron **A2:** 765
copper contamination **A12:** 249
corrosion fatigue fracture surface **A12:** 250
corrosion in river water **M1:** 737-738
cross wire projection welding **M6:** 518
deep drawing lubricants **A14:** 583
deep drawing of **A14:** 584
definition of **A1:** 147
distortion **A6:** 1098
ductile-to-brittle transition **A11:** 84, 85
effect of clearance on piercing and stripping force **A14:** 462
effect of ferrite grain diameter in **A11:** 68
for electrical steel sheet **A14:** 476
electrogas welding **M6:** 239, 241
electron beam welding **M6:** 637
electron-beam welding **A6:** 866
electroslag welding **A6:** 270, 273, 274 **M6:** 226
embrittlement by intermetallic compounds **A11:** 100
extrusion welding **M6:** 677
fatigue fracture of cold-formed part **A11:** 308-309
filler metals for torch brazing **M6:** 953
forge welding **M6:** 676
as forged **A14:** 218
forging pressures **A14:** 217
fractographs **A12:** 240-252
fracture/failure causes illustrated **A12:** 216

Low-carbon steels (continued)
friction welding **A6:** 153, 890 **M6:** 721
furnace brazing *See* Furnace brazing of steels
gas porosity in electron beam welds of **A11:** 445
gas-dryer piping, corrosion product on **A11:** 631
gas-metal arc welding **A6:** 10, 17
of aluminum bronzes **A6:** 828
of copper nickel **A6:** 828
of coppers **A6:** 828
of high-zinc brasses **A6:** 828
of low-zinc brasses **A6:** 828
of phosphor bronzes **A6:** 828
of silicon bronzes **A6:** 828
of special brasses **A6:** 828
of tin brasses **A6:** 828
gas-tungsten arc welding **A6:** 192
of aluminum bronzes **A6:** 827
of copper nickels **A6:** 827
gas-tungsten arc welding of coppers and copper-base alloys **A6:** 827
of phosphor bronzes **A6:** 827
of silicon bronzes **A6:** 827
globular-to-spray transition currents for electrodes **A6:** 182
heat flow in fusion welding **A6:** 10, 13, 16, 17
high frequency resistance welding **M6:** 760
hydrogen absorption **A13:** 329
influence of detection setting on detection area fraction **A10:** 312
intergranular fracture **A12:** 240
limit analysis for **A11:** 136-137
machinability **M1:** 573
magnetic applications **M3:** 598-599, 609, 611
as magnetically soft materials **A2:** 765-766
marine pitting **A13:** 906
mechanical cutting **A6:** 1178, 1179, 1180
microstructure, under hydrogen attack **A11:** 290
multiple-slide forming **A14:** 571
nipple, fissuring at grain boundaries from hydrogen attack **A11:** 645
nondestructive testing **A6:** 1086
notch toughness **M1:** 691, 695, 696, 700, 703, 704
oxyacetylene braze welding **M6:** 598
oxyacetylene welding **A6:** 281, 419
oxyfuel cutting **M6:** 903-904
oxyfuel gas cutting **A6:** 1159
oxyfuel gas cutting of **A14:** 724
oxyfuel gas welding **A6:** 285, 286, 287, 288, 289, 800
percussion welding **M6:** 740
piercing of **A14:** 459-471
pipe, bending of **A14:** 667
plasma arc cutting **M6:** 916
press bending **A14:** 523-532, 667
press forming of **A14:** 545-555
press-formed parts, materials for forming tools **M3:** 492, 493
pressure vessel, failure by caustic embrittlement by potassium hydroxide **A11:** 658-660
principal ASTM specifications for weldable sheet steels **A6:** 399
production examples of gas tungsten arc welding **M6:** 204-205
projection welding **M6:** 506-507, 509, 513-514, 518, 520
properties **A6:** 992
quench-age embrittlement **A1:** 692
quench-age embrittlement of **M1:** 684
repair welding **A6:** 1105
resistance brazing **M6:** 976
resistance seam welding **A6:** 239, 240, 241, 243, 244, 245 **M6:** 494, 497-498, 502
resistance spot welding **M6:** 477, 479-480, 486
resistance welding **A6:** 834, 835, 836, 837, 838, 840, 841, 842, 843, 847
roll welding **A6:** 312, 313 **M6:** 676
for rolling **A14:** 355
seawater corrosion of **M1:** 739-745

Low-carbon steels (continued)
shear deformation and shear lips **A12:** 244
sheet and strip **M1:** 153-162
ASTM specifications **M1:** 154, 155
characteristics **M1:** 154
flatness **M1:** 157, 160-161
formability **M1:** 545-560
leveling **M1:** 157, 160-161
mechanical properties **M1:** 155-156, 158-161
microstructure, effect on formability **M1:** 557-558
mill heat treatment **M1:** 156-157
minimum bend radii **M1:** 554
nonstandard grades **M1:** 161-162
Olsen ductility **M1:** 156, 161, 162
production of **M1:** 153-154
quality descriptors **M1:** 154-155
selection for formed parts **M1:** 559
standard sizes **M1:** 154
steelmaking practice, effect on formability **M1:** 556-557
strain aging **M1:** 154, 157, 162
stretcher strains **M1:** 157
surface characteristics **M1:** 157
thickness **M1:** 153, 154, 159, 161
width range **M1:** 153, 154
sheet, edge effects and etching influence image analysis **A10:** 316, 318
sheet metals **A6:** 399
shielded metal arc welding **A6:** 10, 176 **M6:** 75
solderable and protective finishes for substrate materials **A6:** 979
soldering **A6:** 624
spheroidized cementite particles pinning a recrystallization front **A10:** 471
strain-age embrittlement of **M1:** 683-684
stress-relieving treatments **A6:** 1101
stress-strain behavior, and limit analysis **A11:** 136
structural, SCC failure in **A11:** 27
for structures **A13:** 1299
stud arc welding **A6:** 215, 216 **M6:** 733
stud material **M6:** 731, 735
submerged arc welding **A6:** 10 **M6:** 115
temperamm effect on fracture modes **A12:** 33, 45
tension fractures **A12:** 240
threaded fasteners **M1:** 273-277
to avoid intergranular corrosion **A13:** 325
transgranular brittle fracture **A11:** 25
ultra-, effect of temperature on fracture mode **A12:** 34, 46
weld model **A6:** 1132
weld overlay materials **M6:** 816-817
weldability **A6:** 419
weldability of **A1:** 608 **M1:** 562, 563
wire rod **M1:** 254-257
tensile strength **M1:** 256, 257
Low-carbon steels enamel application **EM3:** 301
enamels **EM3:** 303
surface preparation **EM3:** 271
Low-carbon steels, specific types
AISI 15 B22, delayed fracture **A12:** 248
AISI 1019 shaft, fatigue fracture **A12:** 243
AISI 1020, centerline cracks **A12:** 244
AISI 1020 shaft, brittle fracture **A12:** 243
AISI 1025, bfittle intergranular fracture **A12:** 245
AISI C-1080, stress-corrosion fracture **A12:** 27, 35
ASME SA178, intemal corrosion fatigue cracking **A12:** 245
ASTM A178, grain-boundary embfitlement failure **A12:** 246
ASTM A516-70, calcium effects on inclusions and fatigue crack propagation **A12:** 247
ASTM A517-70, fatigue crack propagation **A12:** 247
SAE 1010 tie rod, bending impact fracture **A12:** 242

Low-carbon steels, specific types (continued)
SAE 1010 tie rod, in-service fatigue fracture **A12:** 241
Low-coefficient-of-expansion alloys
precipitation hardenable nickel alloys **A6:** 576
Low-cost processes
types **EL1:** 448-449
Low-CTE metal planes
constraining **EL1:** 625-628
Low-cycle fatigue *See also* Fatigue; Fatigue cracks
in boilers and steam equipment **A11:** 622
crack initiation testing **A8:** 367
cracking in **A11:** 102, 284
data **A13:** 292
data results in tension-hold-only test **A8:** 351
defined **A11:** 6
eddy current crack inspection **A17:** 190
fatigue-crack initiation **A11:** 103
ferris-wheel testing rig for **A11:** 279
in fracture control **A17:** 666-667, 672-673
lead, effect of frequency of cycling **A8:** 347
and pressure vessels **A8:** 347
strain range **A8:** 364
strain range vs cycles-to-failure for **A11:** 103
tests **A8:** 364
thermal failure, stainless steel tee fitting **A11:** 622
transition fatigue life **A8:** 712
Low-cycle fatigue (LCF)
wrought titanium alloys **A2:** 624
Low-cycle fatigue curve
for 347 stainless steel **A8:** 367
Low-cycle fatigue fracture(s) *See also* Fatigue fracture(s)
AISI/SAE alloy steels **A12:** 308
bending, austenitic stainless steels **A12:** 362
high-carbon steels **A12:** 283
maraging steels **A12:** 386
metal-matrix composites **A12:** 469
precipitation-hardening stainless steets **A12:** 371
titanium alloys **A12:** 452
tool steels **A12:** 377, 380
wrought aluminum alloys **A12:** 426
Low-cycle fatigue resistance
and inelastic strain range **A8:** 347
Low-cycle fatigue testing **A8:** 367, 696
Bauschinger effect **A8:** 367
Coffin-Manson relationship **A8:** 367
elevated-temperature, strain-controlled **A8:** 346
hold periods **A8:** 347-348
stress-amplitude vs. time-to-fracture **A8:** 350
time-to-fracture in tension-hold- only **A8:** 351-352
Low-cycle fatigue tests **A1:** 626
Low-cycle torsional fatigue *See also* High-cycle torsional fatigue **A8:** 150-152
Low-density aluminum P/M alloys
types **A2:** 209-210
Low-density flexible RIM systems **EM2:** 260
Low-density high-stiffness P/M aluminum alloys **A13:** 841-842
Low-density polyethylene (LDPE) *See also* Polyethylene (PE)
in rotational molding **EM2:** 361
Low-end systems
design considerations **EL1:** 26-28
Low-energy electron diffraction **A10:** 536-545 **EM1:** 285
acronym **A10:** 689
applications **A10:** 536, 543-544
capabilities, FIM/AP and **A10:** 583
defined **A10:** 676
diffraction measurements **A10:** 539-541
estimated analysis time **A10:** 536
general uses **A10:** 536
introduction **A10:** 537
limitations **A10:** 536, 542-543
and multiple-scattering effects in EXAFS analysis **A10:** 410
principles, diffractions from surfaces **A10:** 538-539
related techniques **A10:** 536
sample preparation **A10:** 541-542
samples **A10:** 536, 541-542
surface crystallography vocabulary **A10:** 537-538
surface-sensitive electron diffraction limitations of **A10:** 542-543
use with Auger electron spectroscopy **A10:** 554
Low-energy electron diffraction (LEED) **A6:** 145 **A18:** 450
Low-energy electron diffraction spot profile analysis (SPA-LEED) **A6:** 145
Low-energy ion-scattering spectrometry (LEISS) **A18:** 449
Low-energy ion-scattering spectroscopy **A10:** 603-609
applications **A10:** 603, 607-609
Auger electron spectroscopy and **A10:** 554
basic elements of system **A10:** 607
capabilities **A10:** 568
capabilities, compared with Rutherford backscattering spectrometry **A10:** 628
effect of improved mass resolution **A10:** 605
electrostatic analyzers for **A10:** 607
estimated analysis time **A10:** 603
general uses **A10:** 603
of inorganic solids, information from **A10:** 4-6
instrumentation **A10:** 606-607
introduction **A10:** 603-604
limitations **A10:** 603
of organic solids, information from **A10:** 9
quantitative analysis **A10:** 605-606
related techniques **A10:** 603
samples **A10:** 603, 607-609
scattering principles **A10:** 604
spectra **A10:** 604-606
standards and correction factors **A10:** 606
Low-expansion alloys
42% Ni-irons **A2:** 893-894
43% to 47% Ni-iron alloys **A2:** 894
applications, engineering **A2:** 896
Dumet wire **A2:** 894
engineering applications **A2:** 896
hardenable low-expansion alloys **A2:** 895
with high-expansion alloys, applications **A2:** 889
high-strength, controlled-expansion alloys **A2:** 895
Invar **A2:** 889-893
iron-cobalt-chromium **A2:** 895
iron-nickel alloys **A2:** 889-894
iron-nickel-chromium alloys **A2:** 894
iron-nickel-cobalt alloys **A2:** 894-895
Kovar **A2:** 895
nickel alloy **A2:** 433
nickel-iron **A2:** 443-444
special alloys **A2:** 895
Super-Invar **A2:** 894-895
tradenames for **A2:** 896
types **A2:** 889
Low-expansion nickel alloys
thermal expansion coefficient **A6:** 907
Low-expansion substrates
characteristics **EL1:** 611-613
Low-friction bearing materials
composite powders **A7:** 175
Low-hardenability steels **A1:** 466
Low-head direct-chill casting, of aluminum alloy ingots used to decrease surface defects **A9:** 634
Low-hydrogen
high-strength electrodes **A7:** 819
Low-hydrogen iron powder, chemical composition
SMAW electrode coverings **A6:** 61
Low-hydrogen welding rods
for hydrogen embrittlement **A11:** 251
Low-lead tin bronzes
applications **A18:** 750, 751
casting processes **A18:** 754, 755
composition **A18:** 751
designations **A18:** 751
mechanical properties **A18:** 750, 752
product form **A18:** 751, 752
Low-leaded brass *See also* Brasses; Wrought coppers and copper alloys
applications and properties **A2:** 306, 307
Low-level design
defined **EL1:** 129
Low-melting alloys
for thin-wall tube swaging **A14:** 137
Low-melting metals
effect on dimple rupture **A12:** 29-30
Low-melting oxides
types **A13:** 73
Low-melting temperature solders
indium-base **A2:** 751-752
Low-modulus rayon/isotropic pitch precursor fibers **EM1:** 52
Low-nickel alloys
as magnetically soft materials **A2:** 771-772
Low-pass filters
eddy current inspection **A17:** 190
Low-performance thermoplastic resins **EM1:** 169
Low-pressure chamber thermal spray coating **M5:** 364-365
Low-pressure chemical vapor deposition (LPCVD)
hot wall vacuum furnaces used **A4:** 495
Low-pressure dewaxing
for cermets **A2:** 989
Low-pressure die casting
magnesium alloy **A15:** 807
as permanent mold process **A15:** 34, 276
Low-pressure laminates *See also* Laminate(s) **EM3:** 17
defined **EM1:** 15 **EM2:** 26
Low-pressure liquid compound buffing systems **M5:** 116
Low-pressure molding **EM3:** 17
defined **EM2:** 26
Low-pressure sputter chambers
atomic absorption spectrometry **A10:** 54
Low-pressure synthesis
of superhard coatings **A2:** 1009
Low-pressure turbine rotors
corrosion design **A13:** 953
Low-pressure water atomization
for electrical contact composites **A2:** 857
Low-pressure/high-pressure compound buffing systems **M5:** 116
Low-profile resins *See also* Resins
defined **EM2:** 26
Low-shear agitated-type blenders **A7:** 189
Low-shrink resins *See* Low-profile resins
Low-silicon bronze
applications and properties **A2:** 334
Low-solubility intermetallic compound
of zirconium **A2:** 666
Low-stacking fault energy material
dynamic recrystallization for **A8:** 173
Low-strength iron alloy **A8:** 487-488, 491
Low-stress abrasion *See also* High-stress abrasion
defined **A18:** 12
Low-stress sliding
cracks in copper crystal from **A8:** 603
Low-temperature alpha alloy *See* Titanium alloys, specific types, Ti-5Al-2.5Sn
Low-temperature aluminum-tin alloy *See* Titanium alloys, specific types, Ti-5A-2.5Sn
Low-temperature cofired ceramic packages
burnout/firing **EL1:** 466
lead/pin attach **EL1:** 466
physical properties **EL1:** 468
testing **EL1:** 467
Low-temperature corrosion **A13:** 1266-1270
steam equipment **A11:** 618-619
Low-temperature electrical testing **EL1:** 1060-1061
Low-temperature fusion
sample preparation **A10:** 94
Low-temperature impact energy
of steel plate **A1:** 238
Low-temperature niobium-base superconductors
history **A2:** 1027-1028
Low-temperature properties *See also* Cryogenic properties; Temperature(s)
of magnesium alloys **A2:** 462
in superconductivity **A2:** 1030
of thin-film materials **A2:** 1082-1083
ultrahigh-strength steels **M1:** 426-428, 431, 435
wrought aluminum alloy **A2:** 59-60
wrought titanium alloys **A2:** 628-631
Low-temperature properties of structural steel **A1:** 662-672
advances in steel technology **A1:** 665-666, 668, 669
assessment of fracture resistance **A1:** 662-663
design and failure criteria **A1:** 662
fatigue crack growth in structural steel **A1:** 663-664, 665, 666
fracture toughness characteristics **A1:** 664, 666-667, 669

Low-temperature properties of structural steel (continued)
fracture toughness of welded structures A1: 667-669, 670, 671, 672
fracture toughness requirements for A1: 663, 664
structural steel specifications A1: 664-665, 667, 668
Low-temperature resins systems
allyls EM2: 440
aminos EM2: 439
polyurethanes (PUR) EM2: 439
thermoplastic polyurethanes (TPUR) EM2: 203
thermoset polyesters EM2: 440
Low-temperature sensitization (LTS) A6: 466
Low-temperature separation method
auto scrap recycling A2: 1212
Low-temperature service
high-nickel steels for A1: 392, 396-397, 398
Low-temperature soldering EL1: 686, 694-695
Low-temperature solid-state welding A6: 300-302
advantages A6: 300
aluminum A6: 300, 301
aluminum-silicon alloys A6: 300
applications A6: 300
beryllium A6: 300, 301
copper A6: 300
definition A6: 300
disadvantages A6: 300
dissimilar materials A6: 300, 301, 302
with electron-beam welding A6: 301
final machining of welded parts A6: 302
with gas-tungsten arc welding A6: 301
interlayer fabrication method A6: 301
interlayer materials A6: 300
joint geometry A6: 300
maraging steels A6: 300, 301
silver A6: 300
stainless steels A6: 300, 301
surface preparation A6: 300-301
welding methods A6: 301-302
isostatic pressure A6: 301-302
uniaxial compression A6: 301
Low-temperature storage
of samples A10: 16
Low-temperature superconducting materials
thin-film A2: 1082-1083
Low-temperature techniques
in molecular fluorescence spectroscopy A10: 78-79
Low-temperature tensile properties *See also* Tensile properties
casting, magnesium alloys A2: 464
ductile iron A15: 662
sheet/plate magnesium alloys A2: 463
Low-temperature thermoset matrix composites EM1: 392-398
fabric type effects/weaves EM1: 393
properties EM1: 392
thermoset polyester resins for EM1: 392-398
Low-temperature toughness
of steel castings A1: 376, 377-378
Low-tin aluminum-base alloys
composition A2: 524
Low-viscosity sealers M5: 369
Low-voltage accelerators
as neutron sources A17: 388
Low-voltage connectors
corrosion failure analysis EL1: 1112
Low-voltage failures EL1: 999
Low-zinc alloys
applications A13: 614
Low-zinc brasses
corrosion in various media M2: 468-469
gas-metal arc butt welding A6: 760, 763
to high-carbon steel A6: 828
to low-alloy steel A6: 828
to low-carbon steel A6: 828
to medium-carbon steel A6: 828
Low-zinc brasses (continued)
to stainless steel A6: 828
resistance welding A6: 849-850
weldability A6: 753
Low-zinc silicon brass *See also* Cast copper alloys
properties and applications A2: 372
Low-zinc zinc phosphate A13: 386
Lower bainite A9: 664-665
defined A9: 179
Lower bound method
analytical modeling A14: 425
Lower control limit (LCL)
in control chart method A17: 726
Lower explosive limit (LEL)
aluminum powder A7: 130
Lower limit of detection
of an element A10: 96
Lower punch
defined A7: 7 A14: 8
Lower ram
defined A7: 7
Lower yield point
defined A8: 21-22
Lower-die materials
for press forming A14: 506-507
Lowest point of single-tooth contact (LPSTC) A18: 543
LPCVD thin films *See also* Films; Thin films
quantitative impurity analysis in A10: 624
LSC *See* Liquid-solid chromatography
LTE *See* Local thermodynamic equilibrium
Lu-Pb (Phase Diagram) .... A3: 2 • 280
Lu-Tl (Phase Diagram) .... A3: 2 • 280
Lubaloy *See* Copper alloys, specific types, C41100
applications and properties A2: 314-315
Lubber's process EM4: 398
Lubricant *See* Lubricant failures; Lubricants; Lubrication
with cam plastometer A8: 195
commercial metalworking glass, for specimen protection A8: 159
for compression testing A8: 195
defined A8: 8 EM1: 15
effects on bearing strength of aluminum alloys A8: 60
evaluation, sheet metal forming A8: 568
in forming operations A8: 567-568
for gripping specimens A8: 382
high-pressure, in axial compression testing A8: 56
in sheet metal forming A8: 547, 567-568
Lubricant analysis A18: 299-311
case histories A18: 307-310
abrasive wear A18: 308
cast iron wear in a diesel engine A18: 308-309
corrosive wear A18: 309-310
gearbox wear A18: 307-308
nonferrous metal wear in a marine engine A18: 310
water in the oil in a reduction gearbox A18: 308
failure analysis programs A18: 310-311
causes of machine failure A18: 310
ferrography, applications of A18: 305-307
aircraft gas turbine engines A18: 306-307
compressors A18: 307
diesel engines A18: 306
gasoline engines A18: 307
gear boxes A18: 305-306
grease A18: 307
hydraulic systems A18: 307
oil/wear particle analysis methods A18: 299-302
applications and purposes A18: 299
ferrography A18: 301-302, 310
infrared (IR) spectroscopy A18: 300-301, 311
magnetic plug/chip detection (MCD) A18: 301
particle counting A18: 301
Lubricant analysis (continued)
physical inspection A18: 299-300
physical testing A18: 300
sampling of service lubricants A18: 299
spectrometric metals analysis A18: 300
preventive maintenance programs A18: 310
guidelines for establishing a condition monitoring program A18: 310
types of wear particles A18: 302-305
alloy identification: heat treatment of slides A18: 305
black oxide particles A18: 303, 304
corrosive wear particles A18: 304
cutting wear particles A18: 302, 303, 308
dark metallo-oxide particles A18: 303, 304
fatigue spalls and chunks A18: 302, 303
fibers A18: 305
friction polymer A18: 304
glass fibers A18: 304, 305
inorganic crystalline minerals A18: 304
laminar particles A18: 302-303
organic crystalline materials A18: 304
other particles A18: 304, 305
red iron oxide (Fe_2O_3) Particles A18: 304
red oxide sliding wear particles A18: 303-304
rubbing wear particles A18: 302, 303
sliding wear particles A18: 302, 303
spheres A18: 303
Lubricant coatings
measured A10: 177
Lubricant compatibility *See* Compatibility (lubricant)
Lubricant failures
from contamination A11: 153
from transition temperature A11: 154
from viscosity A11: 153-154
mechanical design and A11: 154
in pressurized lubricating systems A11: 153
prevention of A11: 154
Lubricant film thickness (*h*) A18: 146
Lubricant flow
nomenclature for hydrostatic bearings with orifice or capillary restrictor A18: 92
Lubricant forms *See also* Lubricant(s); Lubrication
emulsions A14: 513-514
pastes, suspensions, coatings A14: 514
solutions A14: 513
Lubricant parameter A18: 83
Lubricant roll
for wrought copper and copper alloys A2: 245
Lubricant(s) *See also* Liquid lubricants; Lubricant forms; Lubricated bearings; Lubricating; Lubrication; Solid lubricants; specific lubricants
additives A14: 514-515
for aluminum alloy forgings A14: 248
for aluminum alloy forming A14: 793
application of A14: 161, 248, 502, 515
applicators of A14: 502
bearing steels A18: 727-728, 730, 731, 732, 733
for bearings A7: 709
for bending A14: 664
burnoff A7: 191
carbon-graphite materials A18: 818
carriers (bases) A14: 514
cemented carbides A18: 796
for centrifugal compressors A18: 607, 608
characteristics, control of A14: 515-517
chemistry of A14: 514
coatings, effect on explosivity A7: 194
for coining A14: 181
compacting A7: 352
for contour roll forming A14: 625
control, procedures A14: 516-517
for copper and copper alloy forging A14: 257
of copper-based powder metals A7: 733
for deep drawing A14: 509, 583
of deep-drawing dies A18: 635
defined A7: 7 A14: 8 A18: 12

SUBJECTS OF THE INDEXED VOLUMES: ASM Handbook (designated by the letter "A"): **A1:** Properties and Selection: Irons, Steels, and High-Performance Alloys (1990); **A2:** Properties and Selection: Nonferrous Alloys and Special-Purpose Materials (1990); **A3:** Alloy Phase Diagrams (1992); **A4:** Heat Treating (1991); **A6:** Welding, Brazing, and Soldering (1993); **A7:** Powder Metallurgy (1984); **A8:** Mechanical Testing (1985); **A9:** Metallography and Microstructures (1985); **A10:** Materials Characterization (1986); **A11:** Failure Analysis and Prevention (1986); **A12:** Fractography (1987); **A13:** Corrosion (1987); **A14:** Forming and Forging (1988); **A15:** Casting (1988); **A16:** Machining (1989); **A17:** Nondestructive Testing and Quality Control (1989); **A18:** Friction, Lubrication, and Wear Technology (1992). **Metals Handbook, 9th Edition** (designated by the letter "M"): **M1:** Properties and Selection: Irons and Steels (1978); **M2:** Properties and Selection: Nonferrous Alloys and Pure Metals (1979); **M3:** Properties and Selection: Stainless Steels, Tool Materials and Special-Purpose Materials (1980); **M4:** Heat Treating (1981); **M5:** Surface Cleaning, Finishing, and Coating (1982); **M6:** Welding, Brazing, and Soldering (1983). **Engineered Materials Handbook** (designated by the letters "EM"): **EM1:** Composites (1987); **EM2:** Engineering Plastics (1988); **EM3:** Adhesives and Sealants (1990); **EM4:** Ceramics and Glasses (1991); **Electronic Materials Handbook** (designated by the letters "EL"): **EL1:** Packaging (1989).

Lubricant(s) (continued)
diagnostic guidelines for detecting and rectifying service lubricant deterioration **A18:** 300
die, for titanium alloy forging **A14:** 279
for drop hammer forming **A14:** 655-657
dry helical lobe rotary compressors and wear **A18:** 605, 606, 608
effect in stainless steel powders **A7:** 729
effect of P/M parts after sintering **A7:** 191
effect on friction **A18:** 45-46
effect on iron premixes **A7:** 190
effect on sintered nickel silvers and brasses **A7:** 379
effect on surface condition in tests for solid friction **A18:** 55
effectiveness of **A14:** 515
forms of **A14:** 513-514
gears **A18:** 537
gray cast irons for bearing materials **A18:** 754
for heat-resistant alloy forging **A14:** 235
high-temperature, to modify steel brakes **A18:** 583
hot forging **A14:** 220
hot-forging dies and wear resistance **A18:** 637
hydrocarbon, removing in sintering **A7:** 339-340
hydrodynamic **A18:** 516
included in microscopic analysis **A18:** 376
increasing types with iron powders **A7:** 191
internal combustion engine parts **A18:** 555, 556, 557, 558-559, 561
introduction to **A18:** 79-80
application methods **A18:** 79
basic geometries for surfaces **A18:** 79
jet engine components **A18:** 588
loss in sintering **A7:** 309
Lubricant additives and their functions **A18:** 98-111
additives **A18:** 99-111
antiwear and extreme-pressure (EP) agents **A18:** 99, 101-103, 111
base fluid **A18:** 98
biocides **A18:** 110, 111
couplers **A18:** 110, 111
demulsifiers **A18:** 99, 106-107
detergents **A18:** 99, 100-101, 102, 111
dispersants **A18:** 99-100, 111
dyes **A18:** 110, 111
emulsifiers **A18:** 99, 106-107, 111
engine lubricants
foam inhibitors **A18:** 99, 108, 111
formulations **A18:** 111
friction modifiers/antisquawk agents **A18:** 103-104, 111
functions of lubricants **A18:** 98
gear oils **A18:** 99
greases **A18:** 99
hydraulic fluids **A18:** 99
introduction of a new additive **A18:** 111
lubricant formulation **A18:** 111
metalworking fluids **A18:** 99
miscellaneous industrial oils **A18:** 99
multifunctional nature **A18:** 111
nonengine lubricants **A18:** 98, 111
oxidation inhibitors **A18:** 99, 101, 104-105, 106, 111
performance package **A18:** 98
performance specifications **A18:** 98
pour-point depressants **A18:** 99, 107-108, 111
power steering fluid **A18:** 98
rust and corrosion inhibitors **A18:** 99, 105-106, 111
seal-swell agents **A18:** 110, 111
shock absorber fluids **A18:** 98-99
subcategories **A18:** 98
transmission fluid **A18:** 98
viscosity improvers **A18:** 99, 108-110, 111
for magnesium alloys **A14:** 827
mandrel **A14:** 140
for manual spinning **A14:** 600
melting point **A7:** 190
microbiology of **A14:** 517
in milling environment **A7:** 63
molybdenum disulfide and colloidal graphite **A14:** 238
for nickel-base alloy forming **A14:** 831-832

Lubricant(s) (continued)
nitrided surfaces and wear resistance **A18:** 879-880
nonmetallic bearing materials **A18:** 754, 755
nonuse in encapsulation hot isostatic pressing **A7:** 425
for power spinning **A14:** 604
for precision forging **A14:** 161
premixed with metal powders **A7:** 190
for press forming organic-coated steels **A14:** 564
pumps **A18:** 595, 597
quantity, copper P/M parts **A7:** 193
recommended inspection intervals for selected engines, drive systems, and power generating units **A18:** 299
for refractory metal forming **A14:** 788
removal **A7:** 91, 339-340, 386
removal, in cemented carbide sintering **A7:** 386
residue, defined **A14:** 8
rolling contact wear **A18:** 257, 259, 260, 261
selection **A7:** 190, 709
selection and use, sheet metal forming **A14:** 512-520
semiconductors **A18:** 685, 686, 687
for shallow forming dies **A18:** 633
silver reinforcement of glass ionomers **A18:** 673
sliding bearing materials **A18:** 744, 747, 750
solid, effect on compressibility **A7:** 286
solid, types and uses **A14:** 515
for stainless steel forming **A14:** 760-761
for steel, cold extruded **A14:** 304
surface texture **A18:** 336, 342-343
cutting **A18:** 738-739
hot extrusion **A18:** 738
hot forging **A18:** 738
sheet metal forming **A18:** 737-738
thermoplastic composites **A18:** 820, 822-826
thin-film, compatibility **A18:** 743
titanium alloys **A18:** 778, 780, 781-783
to improve PV capability and reduce wear rate of metallic bearings **A18:** 515, 516
to lessen fretting wear of aluminum wire wound on steel support rope **A18:** 243
to prevent fretting damage **A18:** 253
to reduce wear of die-casting dies **A18:** 629
tool steels **A18:** 736, 737-739
valve train friction analysis characteristics **A18:** 559
and wear measurement **A18:** 365, 367, 369
testing of **A14:** 896-897
for three-roll forming machines **A14:** 619
for titanium alloy forming **A14:** 840
toxicity **A14:** 517
for tube spinning **A14:** 678
tungsten oxide as **A14:** 238
Verson hydroform process **A14:** 612
for wire forming **A14:** 696-697
Lubricant-speed parameters **A18:** 83
Lubricants
as additives **EM2:** 496-497
bearing **A11:** 485
breakdown temperature, effect on bearing steels **A11:** 490
chemical analysis **A13:** 960
copper bearing **A13:** 964
effect, chemical susceptibility **EM2:** 572
effect of particles in **A11:** 503-504
failures of **A11:** 152-154
for fasteners **A11:** 542
inert, as corrosion control **A11:** 194-195
loss, bearing failure from **A11:** 511-512
lubricating grease **A11:** 152
lubricating oils **A11:** 152
molybdenum disulfide **A13:** 959-964
nuclear, effect on corrosion **A13:** 959-964
petroleum-base, for corrosion control **A11:** 194
properties of **A11:** 151-152
-related failures, nuclear reactors **A13:** 959-960
silicone fluids **EM2:** 266
solid **A11:** 152
solid, analyses **A13:** 961-962
thin-film, roller bearing microspalling from **A11:** 502
viscosities for ball and roller bearings **A11:** 511
and wear failure **A11:** 151-152

Lubricants for rolling-element bearings **A18:** 132-138
functions **A18:** 132
grease lubrication **A18:** 136-137
advantages **A18:** 136
base fluid viscosity **A18:** 136
compatibility **A18:** 137
composition **A18:** 136
consistency **A18:** 136
corrosion prevention behavior **A18:** 137
load-carrying ability **A18:** 137
penetration grades **A18:** 136
regreasing procedures **A18:** 137
relubrication intervals **A18:** 136-137
speed limits **A18:** 136
temperature range **A18:** 136, 137
liquid lubricants **A18:** 132-134
extreme-pressure oils **A18:** 133-134
fluid lubrication for rolling bearings **A18:** 132-133
methods (containment systems) **A18:** 132-133
mineral oils **A18:** 133-134, 135, 136, 137
rust and oxidation (R&O) inhibited oils **A18:** 133-134
synthetic hydrocarbon fluids **A18:** 132, 134
viscosity **A18:** 134, 135
polymeric lubricants **A18:** 137
markets **A18:** 137
solid lubricants **A18:** 137-138
types and properties of nonpetroleum oils **A18:** 134-136
dibasic acid esters **A18:** 135-136
fluorinated polyethers **A18:** 135, 136
phosphate esters **A18:** 134
polyglycols **A18:** 134, 135
silicone fluids **A18:** 135-136
Lubricants, high-vacuum applications **A18:** 150-159
grease **A18:** 157
channeling **A18:** 157
slumping **A18:** 157
ideal tribological situations and considerations **A18:** 151-152
acceptable creep or migration **A18:** 151
long life **A18:** 151
low friction and wear **A18:** 151-152
low vapor pressures **A18:** 151
no wear/no significant deformation **A18:** 151-152
suitable electrical conductivity **A18:** 151, 152
temperature insensitivity **A18:** 151, 152
liquid lubricants **A18:** 152, 154-157
additives **A18:** 157
advantages **A18:** 154
application methods **A18:** 155
disadvantages **A18:** 154-155
hydrocarbon polymers **A18:** 155-156
mineral oils **A18:** 155
perfluoropolyalkylether (PFPE) **A18:** 156, 157, 158, 159
poly-α-olefin **A18:** 155, 156-157
polyimide polymers **A18:** 159
polyolester **A18:** 155, 156
properties **A18:** 154-155
silahydrocarbons **A18:** 156, 157
silicone oils **A18:** 155, 157
scope of the problem **A18:** 150-151
boundary contact **A18:** 150
electrical conductivity **A18:** 151
performance range **A18:** 150
volatility problem **A18:** 150
solid lubricants **A18:** 152-154
advantages **A18:** 152
application methods **A18:** 152-153
disadvantages **A18:** 152, 153
fluorides **A18:** 152
lamellar solids **A18:** 152, 153-154
nonpolymer-based composites **A18:** 154
oxides **A18:** 152, 154
polymers **A18:** 152, 154
soft metals **A18:** 152, 153
sulfides **A18:** 152
versus liquid **A18:** 152
space environment applications **A18:** 158, 159
surface modification with and without lubrication **A18:** 157-158
ceramic and hard coat contact **A18:** 158

Lubricants, high-vacuum applications (continued)
ion implantation **A18:** 158
terrestrial ultrahigh vacuum (UHV) environments (chambers) **A18:** 159
types of vacuum lubricants **A18:** 152-158
Lubricated bearings *See also* Bearings; Lubricant(s); Lubrication; Self-lubricating bearings **A7:** 704-709
Lubricated extrusion
CAD/CAM application for **A14:** 325-326
Lubricated friction **A18:** 47, 48
Lubricated rollers
wedging film action of hydrodynamic lubrication **A18:** 89
Lubricated wear *See also* Wear **M1:** 604-606
roller chain components, variation with lubricant type **M1:** 629-631
shafting for journal bearings **M1:** 606
types of lubrication **M1:** 604-605
Lubricating
defined **A7:** 7
Lubrication *See also* Lubricant forms; Lubricant(s); Lubricating; specific lubricants
as additive, ultrahigh molecular weight polyethylenes (UHMWPE) **EM2:** 171
and adhesive wear **A8:** 602
AES analysis of **A10:** 565
of aluminum alloy forgings **A14:** 248
for bar bending **A14:** 664
bearings, methods of **A11:** 484
in beryllium forming **A14:** 806
bi-lubricant systems **A7:** 192
in blanking **A14:** 456
boundary **A11:** 151, 484 **A14:** 512
for cam plastometer specimen **A8:** 195
coatings, permanent mold castings **A15:** 282
of cold extruded copper/copper alloy parts **A14:** 311
in cold heading **A14:** 294
control, precision forging **A14:** 161
in deep drawing **A14:** 508, 583
defined **A18:** 12
die, zinc alloy casting **A15:** 790
of dies **A14:** 87, 229, 508
during strain rate compression testing **A8:** 192
effect in pin bearing testing of aluminum/magnesium alloys **A8:** 60
effect on density and stress distribution, in rigid dies **A7:** 302
effect on green strength **A7:** 288, 289, 302-303
effect on Hall flow rate **A7:** 279, 280
effect on surfaces **A11:** 154, 484
elastohydrodynamic **A11:** 150, 151
of electrical steel sheet processing **A14:** 481-482
electroplated coatings **A18:** 835, 836, 838
environmental and occupational health concerns **A14:** 517-518
extreme-pressure **A14:** 512
film thickness, and surface roughness **A11:** 152
in fine-edge blanking and piercing **A14:** 475
fluid-film, for bearings **A11:** 484
for forging titanium alloys **A14:** 279
and friction, closed-die forging **A14:** 76
for friction control **A8:** 576
for heat-resistant alloy forming **A14:** 781-782
hole, fatigue fracture at **A11:** 114, 346
for hot swaging **A14:** 143
in hot upset forging **A14:** 87
hot-die/isothermal forging **A14:** 153-154
hydrodynamic **A11:** 150-151, 484
hydrodynamic, friction-velocity curve for surfaces capable of **A8:** 605
hydrostatic **A11:** 151, 484
inadequate, bearing adhesive wear by **A11:** 496, 498
loss, in locomotive axle failures **A11:** 715
for magnesium alloys **A14:** 260
mechanisms **A14:** 512-513

Lubrication (continued)
modes of **A11:** 150-151
for nickel-base alloys **A14:** 261
oil and grease, for rolling-element bearings **A11:** 512
in open-die forging **A14:** 64
overheating and spalling from improper **A11:** 130
in powder compacts, effects on compressibility and powder flow **A7:** 302
for powder stability **A7:** 182
in precious metal powders **A7:** 149
of press bending **A14:** 528-529
in press forming **A14:** 504-505, 545
-property improvers **A11:** 154
regimes, source for **A8:** 568
of rolling-element bearings **A11:** 509-512
rotary swaging **A14:** 139-140
of secondary pressing operations **A7:** 338
self-, of engineering plastics **EM2:** 1
in sheet formability testing **A8:** 547, 567-568
solid, bearing designs for **A11:** 153
solid-film **A14:** 512-513
in split Hopkinson pressure bar test **A8:** 201
squeeze casting **A15:** 324
squeeze-film, for bearings **A11:** 485
for stainless steel dies **A14:** 229
surface chemical analysis and **A7:** 250
thermal spray coating applications **A18:** 832
thick-film (hydrodynamic) **A14:** 512
thin-film **A11:** 510
thin-film (quasi-hydrodynamic) **A14:** 512
and tribology **A8:** 601
for tube bending **A14:** 672-673
in tube mandrel swaging **A14:** 139
and wear **A8:** 604
and wear failure **A11:** 150-154
Lubrication breakdown
high-carbon steels **A12:** 283
Lubrication engineer
hardness defined by **A8:** 71
Lubrication oils
rust-preventive uses **M5:** 460-464
Lubrication regimes *See also* Boundary lubrication; Elastohydrodynamic lubrication Full-film lubrication; Hydrodynamic lubrication; Quasi-hydrodynamic lubrication **A18:** 89-96
boundary lubrication **A18:** 89, 94, 96
defined **A18:** 12
for internal combustion engine parts **A18:** 555
modes of asperity lubrication **A18:** 94-96
thick-film lubrication **A18:** 89-94
elastohydrodynamic lubrication **A18:** 89, 92-93
hydrodynamic lubrication **A18:** 89-91
hydrostatic lubrication **A18:** 89, 91-92
plastohydrodynamic lubrication **A18:** 89, 93-94
thin-film lubrication **A18:** 89, 94-96
Lubrication viscosity
nomenclature for lubrication regimes **A18:** 90
Lubricious (lubricous)
defined **A18:** 12
Lubricity
defined **A18:** 12
lead and lead alloys **A2:** 545
Lubricomp O-BG
composition **A18:** 821
material of choice for wear and friction applications **A18:** 826
mechanical properties **A18:** 823
wear and friction properties **A18:** 822
wear factor **A18:** 824
Lubriquip Houdaille grease test **A18:** 128
Lubronze *See* Copper alloys, specific types, C42200
applications and properties **A2:** 315-316
Lucalox (translucent Al_2O_3) fabrication using grain growth suppression **EM4:** 305, 309
Lucas-Tooth and Pyne model
calibration of x-ray spectrometry **A10:** 98

Lucite
honing stone selection **A16:** 476
Lucite as a mounting material for
powder metallurgy materials **A9:** 504
titanium and titanium alloys **A9:** 458
Lüders bands *See also* Deformation bands; Lüders lines
in annealed steel sheet **A9:** 685
appearance **A9:** 687
defined **A8:** 8, 21-22 **A9:** 11
from strain aging **A12:** 129, 148
as nonhomogeneous deformation **A8:** 44
in rimmed 1008 steel **A8:** 22
in sheet metal forming **A8:** 548, 553
Lüders front
in low-carbon steel, ultrafine-grained **A9:** 685
optical microscopy used to monitor **A9:** 684
thin-foil transmission electron microscopy **A9:** 685
Lüders lines *See also* Deformation bands; Plastic deformation; Stretcher strains **A1:** 574 **A9:** 686-687
appearance of **A9:** 687
in deep drawn low-carbon steel aerosol can **A9:** 688
defined **A9:** 11 **A11:** 6 **A14:** 8
detected by magnetic printing **A17:** 126
in ductile fracture **A11:** 25
as formability problem **A8:** 548
Lüders strain **M1:** 683-684
Machinability *See also* Machining
Ludox
as one-component sol **EM4:** 209
Ludwik equation **A8:** 24
Luerkens equations **EM4:** 69
Luggin capillary
for measuring electrode potential **A8:** 416
reference electrodes **A13:** 24
Luggin probe
defined **A13:** 9
Lugs
materials and fracture of **A11:** 31, 515
Luminar lenses
and optimum aperture concept **A12:** 81
Luminous flux
SI unit/symbol for **A8:** 721
Luminous intensity
SI unit/symbol for **A8:** 721
Luminous point patterns
and hardness ranges **A7:** 485
Lumped mass-linear spring model **EM3:** 448
Lumped resistance capacitance (RC) analysis
of circuit approximation **EL1:** 30-31
as performance analysis **EL1:** 25
Lumped view of package parasitics **EL1:** 418-419
Lunar surface
composition analysis by neutron activation analysis **A10:** 234
destructive TNAA analysis **A10:** 239
Lundin aluminum dip coating process **M5:** 335-336
Lungs
beryllium toxicity to **A2:** 1238
Lusterloy *See* Copper alloys, specific types, C61500
Lustrous carbon defects
Replicast process for **A15:** 270
Lustrous carbon films
as casting defect **A11:** 388
Lutetium *See also* Rare earth metals
melting point **A2:** 720
properties **A2:** 1182
pure **M2:** 763
as rare earth **A2:** 720
TNAA detection limits **A10:** 237, 238
LVDT *See* Linear variable-differential transformer
Lyotropic liquid crystal **EM3:** 18
defined **EM2:** 26

M

2-Methoxy-1 -methylethyl
uses and properties **EM3:** 126
2-Methoxyethyl
uses and properties **EM3:** 126
M *See* Bending moment; Magnetization; Magnification; Molal solution; Molar solution
M lines
emission **A10:** 86
M metal
recycling **A2:** 1214
M shell
defined **A10:** 676 **A12:** 168
***m* value** *See also* Strain rate sensitivity,
methods of determining, sheet metals **A8:** 557
role in strain distribution, sheet metal forming **A8:** 550
of sheet metals **A8:** 555-556
M-1 Abrams tank
P/M equipment **A7:** 688
M-252
composition **A4:** 794, 795 **A6:** 564, 573 **A16:** 736, 737 **M6:** 354
electrochemical machining **A16:** 534, 539
machining **A16:** 203, 275, 738, 741-743, 746-759
M-308
broaching **A16:** 204
***m*-cresol**
physical properties **EM3:** 104
M-glass
defined **EM1:** 15 **EM2:** 26
M-M-0011 alloy *See* Superalloys, nickel- base, specific types, MAR-M 247
***m*-phenylenediamine (MDA)**
in epoxy curing **EM1:** 70
m-value *See* Strain-rate sensitivity
M1, M2, M3, etc *See* Tool steels, specific types
M2 concept **A6:** 875
MA 754
applications **A6:** 1039
arc welding **A6:** 1038
composition **A6:** 577, 1037
filler metals for **A6:** 1039
furnace brazing **A6:** 1038
furnace-brazed weld properties **A6:** 1039
gas-tungsten arc weld **A6:** 928
postweld annealing **A6:** 1039
properties **A6:** 928
welding consumables **A6:** 1038
MA 758
applications **A6:** 1039
composition **A6:** 1037
weld transverse properties **A6:** 1038
welding consumables **A6:** 1038
MA 956
composition **A6:** 577, 1037
electron-beam welding **A6:** 1039
explosion welding **A6:** 1040
filler metals for **A6:** 1039
furnace-brazed weld properties **A6:** 1039
gas-tungsten arc welding **A6:** 1038, 1039
grain structure **A6:** 1039
laser-beam welding **A6:** 1039
properties **A6:** 578
weld transverse properties **A6:** 1038
welding consumables **A6:** 1038
MA 6000
brazing **A6:** 632, 928
composition **A6:** 577, 1037
postweld annealing **A6:** 1039
properties **A6:** 578
welding processes **A6:** 1039
MA ODS alloys *See* Dispersion-strengthened iron-base alloys; Dispersion-strengthened nickel-base alloys; Dispersion-strengthened nickel-base alloys, specific types; Mechanical alloying (MA)
MAC software program **A15:** 867, 871-872
MacCoull-Walther equation **A18:** 83
Macerate
defined **EM1:** 15 **EM2:** 26
Machinability *See also* Machining **A1:** 68-69, 240, 591-602 **A18:** 617
aluminum alloys **A15:** 766 **M2:** 187-190
Machinability (continued)
aluminum and aluminum alloys **A2:** 7-8
aluminum casting alloys **A2:** 150, 153-177
of aluminum P/M parts **A7:** 742
aluminum-silicon alloys **A15:** 159, 167
of austenitic manganese steel **A1:** 838-839
austenitic manganese steels **A15:** 734-735
of carbon steels **A1:** 595-597
of carbon steels with other additives **A1:** 599
calcium **A1:** 599
nitrogen **A1:** 599
phosphorus **A1:** 599
selenium and tellurium **A1:** 599
of carbon/alloy steels **A14:** 218-219
of carburizing steels **A1:** 600
cast copper and copper alloys **A2:** 224-228, 348-351, 356-391
of cold-drawn steel **A1:** 601
of compact infiltrated copper alloy **A7:** 565
of compacted graphite iron **A1:** 68-69
of copper **A12:** 401
copper casting alloys **M2:** 388-389
of ductile iron **A1:** 52-53, 54 **A15:** 665
of gray iron **A1:** 23-24 **A15:** 644
heat-resistant casting alloys **M3:** 279
high-silicon irons **A15:** 701
of Invar **A2:** 891-892
of leaded carbon and resulfurized steels **A1:** 599-600
magnesium alloys **M2:** 549, 551
of magnesium and magnesium alloys **A2:** 475-476
of maraging steels **A1:** 795
measures of **A1:** 591-592
cutting speed **A1:** 591-592
machinability testing for screw machines **A1:** 593
power consumption **A1:** 591, 592
quality of surface finish **A1:** 592-593
tool life **A1:** 591-592
microstructure and **A1:** 595, 596
resulfurized carbon steels **A1:** 597-599
control and effect of sulfide morphology **A1:** 598
economic **A1:** 598-599
manganese content **A1:** 597
scatter in machinability ratings **A1:** 593
of stainless steels **A1:** 894-897
of steel castings **A1:** 378
of steel plate **A1:** 238
sulfides and manganese sulfides for **A7:** 486
of through-hardening alloy steels **A1:** 600-601
tungsten **M3:** 330-332
wrought aluminum alloys **A2:** 30-32, 104, 111
of wrought tool steels **A1:** 774-775
ZGS platinum **A2:** 714
Machinability ratings
of gray iron **A1:** 23
of steels **A1:** 593-595
Machinability test methods **A16:** 639-647
application/grade selection **A16:** 645-647
cutting conditions **A16:** 642-643
cutting tool material test methods **A16:** 639-642
facing test method **A16:** 643
grade selection and insert and chip groove selection **A16:** 646
grade selection and machining economics **A16:** 647
grade selection and operation determination **A16:** 645-646
machinability ratings **A16:** 643-645
machine test matrix **A16:** 643
nodular iron impact test method **A16:** 643
operation matrix test method **A16:** 642
property matrix impact test **A16:** 641
property matrix test method **A16:** 640-642
property matrix turning test **A16:** 641-642
Taylor's tool life tests **A16:** 644
workpiece matrix test method **A16:** 639-640
Machinability testing **A18:** 617
for screw machines **A1:** 593
Machinable ceramic
recommended waterjet cutting speeds **EM4:** 366
Machine
capacity, testing of **A8:** 58
fatigue testing, classified **A8:** 368-370
high-speed hydraulic torsional **A8:** 215-216
Machine (continued)
tension testing, characteristics **A8:** 208
Machine brazing
definition **M6:** 11
Machine deflection
in tube spinning **A14:** 678
Machine Design
as information source **EM2:** 92
Machine drive gears **A7:** 667
Machine finish allowance
pattern **A15:** 193
Machine forging *See* Hot upset forging
Machine guns
accelerators for **A7:** 685-686
Machine inspection
of tubes on solid cylinders **A17:** 185
Machine knives *See* Shearing and slitting tools
Machine oxygen cutting
definition **M6:** 11
Machine parts
net shape **A7:** 295
Machine pins
failure in **A11:** 545
Machine rates
parts per hour **EM2:** 85
Machine screws
economy in manufacture **M3:** 851-852
Machine shot capacity
defined **EM2:** 26
Machine tool components
surface finish requirements **A16:** 21
Machine tool industry applications
of structural ceramics **A2:** 1019
Machine tools
interferometer error characterization **A17:** 15
Machine vision *See also* Machine vision process **A17:** 29-45
applications **A17:** 29, 37-41
future outlook **A17:** 41-45
limitations **A17:** 42
process **A17:** 30-37
schematic **A17:** 31
system, inspection of **A17:** 118-119
systems market, by industry **A17:** 29
vs. human vision, capabilities **A17:** 30
Machine vision process
image analysis **A17:** 34-35
image formation **A17:** 31-33
image interpretation **A17:** 35-37
image preprocessing **A17:** 33-34
interfacing **A17:** 37
Machine welding
definition **A6:** 1211 **M6:** 11
Machined aluminum molds
rotational molding **EM2:** 367
Machined bars **A1:** 249-250
Machined parts
as XRS samples **A10:** 95
Machined patterns
described **A15:** 195
Machined surfaces
examination by light section microscopy **A9:** 80
Machined tooling, pattern
investment casting **A15:** 256
Machinery *See* Equipment
rust-preventive compounds used on **M5:** 460-465
Machinery Adhesives for Locking, Retaining and Sealing **EM3:** 71
Machinery and equipment **A6:** 389-393
Machinery applications
aluminum and aluminum alloys **A2:** 13-14
copper and copper alloys **A2:** 239
Machinery materials
ultrasonic inspection **A17:** 232
Machinery steel
electrochemical grinding **A16:** 547
grinding **A16:** 547
milling **A16:** 547
Machines *See* Equipment; specific equipment types
Machining *See also* Assembly and assembly forms; Electric discharge machining (EDM); Machinability; Machining applications; specific machining techniques **M1:** 565-585 **EM1:** 665-682
abrasive water-jet cutting **EM1:** 673-675
abrasives for shaping ceramics **EM4:** 313

Machining (continued)
for adhesive bonding surface preparation ... **EM1:** 681-682
allowance ... **A14:** 125-126, 158
allowances, centrifugal casting molds ... **A15:** 302
alloy steels ... **A2:** 967 **M1:** 568, 579-583
Alnico alloys ... **A2:** 785
aluminum and aluminum alloys ... **A2:** 7-8, 966
aluminum, chemical milling vs mechanical milling ... **M2:** 14, 16
aluminum part, relative cost ... **M2:** 13, 15
applications of cemented carbides ... **A7:** 777
austenitic manganese steel ... **M3:** 586-588
austenitic stainless steel ... **A2:** 966
beryllium-copper alloys ... **A2:** 415-416
brittle fracture from ... **A11:** 85, 89-92
with CAM ... **A14:** 908-909
carbon steels ... **M1:** 572-580
of castings ... **A11:** 362
centers, as coordinate measuring machine application ... **A17:** 18
cold drawn steel ... **M1:** 581-582, 584
and cold heading, compared ... **A14:** 291
copper alloys *See* Copper alloys, machined parts
of copper-based powder metals ... **A7:** 733
damage, in ceramics ... **A11:** 750-752
defects, by computed tomography (CT) ... **A17:** 361
of densified ceramics ... **EM4:** 124
determination of maximum residual stress produced by ... **A10:** 392
discontinuous ceramic fiber MMCs ... **EM1:** 909
and drilling, solid-tool ... **EM1:** 667-672
ductile fractures from ... **A11:** 89-92
ductile nodular iron ... **A2:** 966
effect on fatigue strength ... **A11:** 122
electrochemical ... **A11:** 122
ferritic stainless steels ... **A2:** 967
finished machined parts, rust- preventive compounds for ... **M5:** 465
forgings, allowances for ... **M1:** 368-369, 370, 375
free-abrasive ... **EM4:** 313, 314
free-machining steels ... **A2:** 966 **M1:** 573-581
gray cast iron ... **A2:** 966
gray iron ... **M1:** 27-28
of hafnium ... **A2:** 664
hardness, effect on ... **M1:** 566, 567, 569, 571, 574, 576, 578
heading/extrusion as ... **A14:** 296-297
high-chromium white irons ... **A15:** 684
hole ... **A7:** 789-791
of hole, as fatigue fracture origin ... **A11:** 474
improper, of shafts ... **A11:** 459, 472
influence, tool and die failure ... **A11:** 564, 566-567
introduction ... **EM1:** 665
with isothermal and hot-die forging ... **A14:** 150
laser cutting ... **EM1:** 676-680
machining allowances, case hardened parts ... **M1:** 627
machining costs ... **M1:** 565, 582-585
maraging steels ... **M1:** 447
marks, in shafts ... **A11:** 459
martensitic stainless steels ... **A2:** 967
metals, abrasive waterjet cutting for ... **A14:** 752-754
microstructure, effect of ... **M1:** 570-574, 580-581
nickel-base alloys ... **A2:** 966
nickel-chromium white irons ... **A15:** 680-681
non-abrasive methods ... **EM4:** 313
nonmetallics, abrasive waterjet cutting for ... **A14:** 754-755
numerical control ... **A15:** 198-199
of- silica-silica composites ... **EM1:** 936
P/M parts ... **A7:** 295
parameters, cemented carbides ... **A2:** 968
of patterns, investment casting ... **A15:** 256
of permanent mold cavity ... **A15:** 280
plain carbon steels ... **A2:** 967
postcasting, investment casting ... **A15:** 263-264

Machining (continued)
postforging defects from ... **A11:** 333
pre-sintering, of structural ceramics ... **A2:** 1020
of precracked SCC specimens ... **A13:** 257
precracked SCC testing specimens ... **A8:** 516
preplate, hard chromium plating ... **M5:** 180
process characteristics ... **M1:** 565-571
in quasi-static torsional testing ... **A8:** 145
refractory metals and alloys ... **A2:** 560-562
as secondary operation ... **A7:** 451, 461-462
solid-tool ... **EM1:** 667-672
stainless steels ... **M3:** 44-46
of stainless steels, vs forming ... **A14:** 778
steel-bonded titanium carbide cermets ... **A2:** 997-998
with superabrasives/ultrahard tool materials ... **A2:** 1013
thermal spray-coated
feed and speed ranges ... **M5:** 369-370
materials ... **M5:** 369-370
of thermoplastic composite ... **EM1:** 552
titanium alloys ... **A2:** 966
tolerances, and dimensional change ... **A7:** 480
and turning, of P/M drive gear ... **A7:** 667
ultrahigh-strength steels ... **M1:** 422, 424, 427, 429, 431, 432, 434, 438, 441
uranium ... **M3:** 777-778
vs coining ... **A14:** 185-186
water-jet cutting ... **EM1:** 673-675
of whisker-reinforced MMCs ... **EM1:** 899
of zirconium ... **A2:** 664
zirconium alloys ... **A15:** 838
Machining allowance *See also* Allowances
in hot forgings ... **A14:** 158
in ring rolling ... **A14:** 125-126
Machining allowances for wrought tool steels ... **A1:** 778-779
Machining applications
cemented carbides ... **A2:** 965-968
cemented carbides for ... **A2:** 951-968
parameters, cemented carbides ... **A2:** 968
workpiece materials ... **A2:** 965-967
Machining centers ... **A16:** 2, 393-394
fixturing ... **A16:** 404, 410
Machining damage
definition ... **EM4:** 633
Machining marks
AISI/SAE alloy steels ... **A12:** 296
coarse, low-carbon steel ... **A12:** 251
leading to fatigue fracture ... **A12:** 251
Machining technology
capabilities/ limitations ... **EL1:** 508
Machinist
hardness defined by ... **A8:** 71
Mack EO-K[2 specification
engine oil requirement by an OEM ... **A18:** 165
Mack T-6
multicylinder engine tests ... **A18:** 101
Mack Truck GO-H
gear oil specifications ... **A18:** 86
Mackintosh-Hemphill Company
as early foundry ... **A15:** 28
Macor
composition ... **EM4:** 873
properties ... **EM4:** 873, 876
thermal properties ... **EM4:** 876
Macor type
dielectric properties ... **EM4:** 877
elastic constant ... **EM4:** 875
maximum use temperature ... **EM4:** 875
Macro
defined ... **EM1:** 15 **EM2:** 26
Macro-Nikkor lenses
aperture selection ... **A12:** 80-81
Macro-SIMS instrument
for qualitative analysis ... **A10:** 613
Macroalloying
of ordered intermetallics ... **A2:** 913

Macroanalysis *See also* Bulk analysis
of inorganic gases, analytic methods for ... **A10:** 8
of inorganic liquids and solutions, analytic methods for ... **A10:** 7
of inorganic solids, analytic methods for ... **A10:** 4-6
macrograph, as-cast aluminum ingot ... **A10:** 302
optical metallography ... **A10:** 301-303
of organic solids and liquids, analytic methods for ... **A10:** 9, 10
Macrocrack *See also* Cracking; Microcrack ... **A8:** 57
Macrocrack propagation
in fatigue ... **A12:** 117-118
Macrodefects
titanium alloy forgings ... **A17:** 498
Macroelectronics
development ... **EL1:** 89
Macroetch test
definition ... **A6:** 1211
Macroetch testing ... **A1:** 275
for carbon steel rod ... **A1:** 274
Macroetchants *See also* Etchants
aluminum alloys ... **A9:** 352
copper and copper alloys ... **A9:** 399
for stainless steels ... **A9:** 279-281
for titanium and titanium alloys ... **A9:** 460
for tool steels ... **A9:** 256
wrought heat-resistant alloys ... **A9:** 306
Macroetched disks
macrophotography of ... **A9:** 86-87
Macroetching *See also* Chemical etching ... **A9:** 62
aluminum alloys ... **A9:** 354
austenitic manganese steel casting specimens ... **A9:** 238
beryllium-copper alloys ... **A9:** 393-394
beryllium-nickel alloys ... **A9:** 393-394
carbon and alloy steels ... **A9:** 170-177
of cast iron ... **A11:** 344
defined ... **A9:** 11
equipment ... **A9:** 171
forgings ... **M1:** 354
inspection records ... **A9:** 172-173
iron-nickel alloys ... **A9:** 532
magnifications ... **A9:** 62
nickel alloys ... **A9:** 435-436
nickel-copper alloys ... **A9:** 435-436
reveal as-cast solidification structures in steel ... **A9:** 624
steel wire rod ... **M1:** 255, 257
of tool steels ... **A9:** 256
uranium and uranium alloys ... **A9:** 479
wrought heat-resistant alloys ... **A9:** 305
wrought stainless steels ... **A9:** 279
Macroexamination
aluminum alloys ... **A9:** 353-354
cemented carbides ... **A9:** 273
rhenium and rhenium-bearing alloys ... **A9:** 447
stereomicroscope for ... **A9:** 88
tool steels ... **A9:** 256
uranium and uranium alloys ... **A9:** 479
visual ... **A12:** 72-73
Macrofabrication
performance/cost effects ... **EL1:** 10
Macrofouling films
in seawater ... **A13:** 901-902
Macrofractography ... **A12:** 1, 3, 91
defined *See* Fractography,
of fractured shaft ... **A11:** 123
Macrograph
defined ... **A9:** 11
Macrohardness test *See also* Microindentation hardness number ... **A8:** 73
defined ... **A18:** 12
Macrohardness testing
thermal spray coatings ... **M5:** 371

Macroinclusions *See also* Defects; Inclusion-forming reactions; Inclusions
defined **A15:** 88
Macrointegration *See also* integration
defined **EL1:** 2
with optical interconnections **EL1:** 9
Macroorganisms
general biological corrosion by **A13:** 88
Macrophotography of specimens **A9:** 86-87
Macropore
defined **A7:** 7
Macroporosity
microwave inspection **A17:** 202
Macroprobe analysis **A7:** 258
Macroscopes
illustrated **A12:** 79
mounted **A12:** 79
techniques **A12:** 91-93
Macroscopic
defined **A9:** 11 **A11:** 6 **A13:** 9
Macroscopic constitutive models **A8:** 215
Macroscopic crescents *See* Beach marks
Macroscopic examination
of failure types **A11:** 75
of fracture surfaces **A11:** 20
of fractures **A11:** 104-105, 256-257
of shafts **A11:** 460
for stress-corrosion cracking **A11:** 212-213
Macroscopic failure
types **A8:** 476
Macroscopic metallurgical features of deformed structures **A9:** 686-687
Macroscopic solids
solidification of **A15:** 101-102
Macroscopic stress
defined **A10:** 676
Macroscopy **EM3:** 18
defined **EM2:** 26
of metallographic sections **A10:** 303-304
Macrosegregation
in copper alloy ingots **A9:** 639, 643-644
defined **A9:** 614, 625
during solidification **A15:** 136-141
evaluation **A15:** 138
gravity segregation **A15:** 139
ingot **A14:** 64
inhomogeneous solid distribution/channel segregation **A15:** 140-141
interdendritic fluid flow **A15:** 155
liquid flow induced segregation, mushy region **A15:** 139-140
low gravity effect **A15:** 157
and microsegregation, during solidification **A15:** 136-141
plane front solidification **A15:** 138-139
solid phase movement/segregation **A15:** 141
Macroshear stress
in medium-carbon steets **A12:** 253
Macroshrinkage *See also* Shrinkage
defined **A11:** 6 **A15:** 8
Macroshrinkage, in austenitic manganese steel castings
causes of **A9:** 238
Macroslip *See also* Microslip
defined **A18:** 12
Macrostrain **EM3:** 18
defined **A8:** 8 **EM2:** 26
Macrostresses
measurement by x-ray diffraction residual stress techniques **A10:** 380
states analyzed by neutron diffraction **A10:** 424
in x-ray diffraction residual stress techniques **A10:** 381
Macrostructural characterization of a sectioned weld **A6:** 97, 102-103, 104
Macrostructural examination *See* Macroexamination
Macrostructure
defined **A9:** 11 **A11:** 6 **A13:** 9
examples **A9:** 604-605
fracture **A11:** 75
of steel **A9:** 623
Macrostructure, casting
parameters affecting **A15:** 130
Macrostructure differences
and fatigue resistance **A1:** 681
Macroviewer
as input device on image analyzers **A10:** 310
Magabsorption NDE **A17:** 143-158
applications **A17:** 152-158
of bar **A17:** 155
bridge detector **A17:** 149-150
circuit, illustrated **A17:** 148
concept illustrated **A17:** 145
defined **A17:** 145
detection **A17:** 148-152
detection methods **A17:** 148-149
marginal oscillator detector **A17:** 150-152
measurements **A17:** 146-148, 152-158
probe specimen geometry **A17:** 155
signals **A17:** 148, 152
theory **A17:** 145-148
of wire **A17:** 154-155
Magamp control **A6:** 38, 41
Magic squares
in fractional factorial designs **A17:** 748
Magic-angle spinning nuclear magnetic resonance
capabilities **A10:** 407
Maglay process **A6:** 275
Magnaflux quality **A1:** 254
Magnaflux-quality alloy steel bars **M1:** 209
Magnatite and ethylenedia minetetra-acetic acid
decomposition temperatures **EM4:** 55
Magne gage **A6:** 461
Magnesia **A16:** 98 **EM4:** 6, 45
applications **EM4:** 46
refractory **EM4:** 901, 903, 906
as ceramic substrate **EL1:** 337
composition **EM4:** 46
gas phase reactions **EM4:** 62
properties **A18:** 814
refractory material composition **EM4:** 896
refractory physical properties **EM4:** 897, 898, 899
sintering aid for nearly inert crystalline ceramics **EM4:** 1008
supply sources **EM4:** 46
Magnesia, as lining material
induction furnaces **A15:** 372
Magnesia brick **EM4:** 895-896
Magnesia ceramics
chemical etching **EM4:** 575
Magnesia-doped silicon nitride **EM4:** 197
Magnesia-PSZ **A6:** 956, 957
Magnesia-silicon carbide
hot pressing **EM4:** 191
Magnesia-stabilized zirconia **EM4:** 1138
Magnesite **EM4:** 13
Magnesite, as refractory
core coatings **A15:** 240
Magnesium *See also* Cast magnesium alloys; Cast magnesium alloys, specific types; Demagging; Magnesium alloys; Magnesium alloys, specific types; Magnesium alloys, specific, types; Magnesium alloys, specific types; Magnesium powder(s); Magnesium recycling; Wrought magnesium alloys; Wrought magnesium alloys, specific types **A13:** 740-754 **M2:** 525-552
as addition to aluminum alloys **A4:** 843, 844, 845, 846, 847
addition to aluminum-base alloys **A18:** 752
alloying addition to heat-treatable aluminum alloys **A6:** 528, 529, 530, 531, 534
alloying, aluminum casting alloys **A2:** 132
alloying and contaminant effects **A13:** 741
alloying effect on nickel-base alloys **A6:** 590
alloying in aluminum alloys **M6:** 373
alloying, wrought aluminum alloy **A2:** 52-53
alloys, analysis for copper in, hydrobromic acid/phosphoric acid method **A10:** 65
as aluminum alloying element **A2:** 16
in aluminum alloys **A15:** 746
in aluminum, effect on torsional ductility **A8:** 167
in aluminum powder metallurgy alloys **A9:** 511
as an addition to zinc alloys **A9:** 490
applications **A2:** 455, 1259 **M2:** 525
brazing of aluminum alloys effect on **M6:** 1022
in cast iron **A1:** 5
as casting material **A15:** 35
chemical resistance **M5:** 4
chrome plating, hard, removal of **M5:** 185
coated, types **A13:** 107
in compacted graphite iron **A1:** 56
Magnesium (continued)
in composition of aluminum-silicon alloys **A18:** 786, 787, 789
copper plating of **M5:** 160-161, 163, 168-169
corrosion rate, and relative humidity **A13:** 908
degassing of **A15:** 462-464
deoxidizing, copper and copper alloys **A2:** 236
design and weight reduction **A2:** 476-479
as desulfurization reagent **A15:** 75
determined in plant tissues **A10:** 41
dietary sources **A2:** 1259
in ductile iron **A15:** 648, 650
EDTA titration **A10:** 173
effect on aluminum alloy soldering **A6:** 628
effect on ductile iron welds **M6:** 604
effects, in wrought/cast aluminum alloys **A13:** 586
electrochemical activity of **A13:** 740
electrodes, nylon liners **A6:** 184
electrolytic, cold rolled 50%, shear bands in **A9:** 687
etching by polarized light **A9:** 59
explosion welding **A6:** 896 **M6:** 710
filiform corrosion of **A13:** 107
forgings, corner/fillet radii **A2:** 468-469
friction coefficient data **A18:** 71
galvanic corrosion of **A13:** 84
gas-tungsten arc welding **A6:** 190, 191
hydrogen fluoride/hydrofluoric acid corrosion **A13:** 1169
hydrogen solubility **A15:** 465
as inclusion-forming, aluminum alloys **A15:** 95
inserts **A2:** 466-467
ion removal from **A10:** 200
ions, exchanged in water softeners **A10:** 658-659
in iron-base alloys, flame AAS analysis **A10:** 56
joined to aluminum alloys **A6:** 739
lubricant indicators and range of sensitivities **A18:** 301
machinability **A2:** 475-476
and magnesium alloys **A15:** 798-810
and magnesium alloys, selection application **A2:** 455-479
in malleable iron **A1:** 10
mechanical properties **A2:** 460-462
metal-matrix composites **A2:** 460
as minor toxic metal, biologic effects **A2:** 1259
molten, to produce titanium **A13:** 56
in nodular cast iron, inclusions from **A15:** 90
in nodular graphite composition **A18:** 699
oxyfuel gas welding **A6:** 281, 282
photoejection of electrons in **A10:** 85
photometric analysis methods **A10:** 64
pin bearing testing **A8:** 59
plasma arc cutting **M6:** 916
precipitated as ternary phosphate **A10:** 173
production **M2:** 525, 526
in production of spheroidal graphite **A1:** 7
pure **M2:** 763-768
pure, properties **A2:** 1132
pure, thermal properties **A18:** 42
radiographic absorption equivalence **A17:** 311
radiographic film selection **A17:** 328
rare earth alloy additives **A2:** 728
recycling **A2:** 1216-1218
selection of product form **A2:** 462-466
reductant of metal oxides **M6:** 692, 694
relative solderability as a function of flux type **A6:** 129
sacrificial anodes **A13:** 468-469
separation, in aluminum melts **A15:** 80
shielding gas purity **A6:** 65
shipments, structural and nonstructural applications **A2:** 455
shrinkage allowance **A15:** 303
slip planes **A9:** 62
solderability **A6:** 971, 978
soldering **A6:** 631
solution potential **M2:** 207
species weighed in gravimetry **A10:** 172
spectrometric metals analysis **A18:** 300
standard designations **M2:** 525-526, 527, 528
stress corrosion, microwave inspection **A17:** 215
suitability for testing, various substances **A13:** 744
sum peaks in EDS spectrum of **A10:** 520

Magnesium (continued)
superplasticity of A8: 553
systems, beryllium in A2: 426
tantalum resistance to A13: 733
thermal diffusivity from 20 to 100 °C A6: 4
TNAA detection limits A10: 237
treatment, of ductile iron A15: 649-650
treatment, of high-alloy graphitic irons A15: 698
ultrapure, by distillation and zone refinement A2: 1094
used to make detergents A18: 100
vapor degreasing of M5: 45-46, 55
vapor pressure A4: 493 A6: 621
vapor pressure, relation to temperature A4: 495
Vickers and Knoop microindentation hardness numbers A18: 416
volumetric procedures for A10: 175
weakened or distorted by welding EM3: 33
weighed as the phosphate A10: 171
in zinc alloys A15: 788
in zinc/zinc alloys and coatings A13: 759

Magnesium alloys *See also* Cast magnesium alloys; Demagging; Magnesium; Magnesium alloys, specific types; Magnesium casting alloys A6: 772-782 A9: 425-434 A13: 740-754 A14: 259-260, 825-830 A16: 30 M2: 525-552
abrasion resistance, coatings M5: 638
abrasive blasting of M5: 628-629
in acids/alkalies A13: 742
additions, in ductile iron A15: 649-650
adhesive bonding M2: 547-549, 550
adhesive bonding of A2: 474
aerospace applications A13: 753
air-carbon arc cutting A6: 1176
alkaline cleaning of M5: 630-631, 633, 635, 637-639, 645, 647
alloy designations A15: 798-799
anodic coating of M5: 628-629, 632-638, 640-641, 643-644, 647
problems and corrections M5: 636-638, 642-643
process control of M5: 635-636, 640-641
repair of M5: 635
surface preparation M5: 632
applications A2: 455-462 A15: 799
arc welding *See* Arc welding of magnesium alloys
atmospheric corrosion M5: 628, 637, 646
atmospheric effects A13: 742, 747
automotive applications A13: 753
auxiliary pouring equipment A15: 800-801
back reflection intensity A17: 238
band sawing A16: 827, 828
bar and tube, die materials for drawing M3: 525
bars and shapes A2: 459
bearing strength A2: 460-461 M2: 528, 530
bending M2: 552
blanking, die materials for M3: 485, 486
brazing to aluminum M6: 1031
broaching A16: 203, 206, 823
cadmium plating, stripping of M5: 647
as case in plaster molds A15: 243
cast, cleaning and finishing M5: 630, 632, 637, 646, 648
cast magnesium alloys, properties of A2: 491-516
casting alloys A2: 456-459 M2: 526-527
castings A2: 462-463 M2: 533-534, 536
gas-tungsten arc welding A6: 192
centerless grinding A16: 829
chemical analysis and sampling A7: 248
chemical composition A13: 740-741
chemical conversion coatings A13: 751 M5: 628-629, 632-638, 640-641, 643-644, 647
chemical milling A16: 579, 582-583
chemical polishing procedures A9: 426
chemical treatment M5: 628-629, 632-638, 640-641, 643-644
problems and corrections M5: 636-638, 642-643

Magnesium alloys (continued)
process control of M5: 635-636, 640-641
Chemical Treatment No. 9 .... M5: 632, 636-637, 640-641, 643
Chemical Treatment No. 17 .... M5: 632-633, 636-638, 640-641, 643
chip formation A16: 820, 821-822, 824, 825, 828
chrome plating, copper-nickel-chromium and decorative chromium systems M5: 646-647
stripping of M5: 646
chrome plating, hard, removal M5: 185
circular sawing A16: 827, 828
classes of A9: 427
cleaning of A13: 750
cleaning processes M5: 628-631, 648-649
chemical M5: 629-631
mechanical M5: 628-630, 648-649
cold forming of A14: 825
compositions of A9: 427
compressive strength A2: 460 M2: 528, 529-530, 531
content in MMCs and machinability A16: 897
contour band sawing A16: 363
copper striking of M5: 639, 645-647
stripping of M5: 646
corrosion fatigue fracture A12: 456
corrosion fatigue in A13: 745
corrosion resistance
in-process corrosion M5: 636
and protection M5: 628-629, 631-634, 636-638, 646
corrosion tests A13: 745
counterboring A16: 823, 825
Cr-22 treatment M5: 634-635, 640-641, 644
cracking A6: 780, 781, 782
critical strain rate for SCC in A8: 519
cutting fluids A16: 820-829
cutting fluids (machining coolants) A2: 475
cylindrical grinding A16: 829
damping capacity A2: 462
deep drawing M2: 543, 545
deep drawing, formability A2: 468-469
deep drawing of A14: 827-828
defects, permanent mold castings A15: 285
design and weight reduction A2: 476-479
designations A2: 455-456
die casting A15: 286, 807-810
die castings, cleaning and finishing M5: 628, 632, 646, 648
zinc undercoating for M5: 646
die cutting speeds A16: 301
die forgings, materials for forging tools M3: 529, 530
die heating A14: 259
die threading A16: 825-826
distortion A6: 774, 781
distortion (machining) A2: 476
drilling A16: 220, 222, 225, 229, 230, 823-824
drop hammer forming A14: 656-657, 830
dust collection system, polishing and grinding dust M5: 648-649
effect of lubrication during pin bearing testing A8: 60
electrochemical grinding A16: 543
electrochemical machining removal rates A16: 534
electrodes A6: 774, 778, 779
electrolytes for A9: 426
electrolytic polishing procedures A9: 426
electron beam welding M6: 643
electron-beam welding A6: 855, 872
electronic/computer applications A13: 753
elevated temperature properties M2: 532-533, 534, 535
elevated temperatures A2: 462
embrittled by zinc A11: 234
emulsion cleaning of M5: 4-5, 629-630
end milling A16: 826, 827
environmental factors A13: 741-743

Magnesium alloys (continued)
environments that cause stress-corrosion cracking A6: 1101
etchants for A9: 426-427
etching of M5: 638-639, 645
eutectics A9: 427-428
extrusions A2: 459, 463-464 M2: 527-528, 534-535, 537, 546
face milling A16: 826, 827
fatigue strength A2: 461-462 M2: 531, 532
fatigue strength as function of relative humidity A11: 252
ferrous chloride treatment method A16: 829-830
filler metals A6: 772, 774, 778, 781
finishing processes M5: 628-629, 631-649
chemical and anodic M5: 628-629, 632-638, 640-641, 643-644
mechanical M5: 629, 631-632, 648-649
organic coating M5: 629, 647-648
plating M5: 629, 638-639, 642, 644-647
fire extinguishing powders A16: 830
fire hazard A16: 820, 822, 828-830
flash welding M6: 558
fluxing of A15: 447-448
forging of A14: 259-260
forgings A2: 459, 465-466 M2: 528-529, 537-538, 539, 540, 541
forgings, flaws and inspection methods A17: 497
formability A2: 467-471 M2: 540-541, 542, 546
forming, lubricants for A14: 519-520
forming of A14: 825-830
fracture surface characteristics A9: 426
in fresh water A13: 742
friction welding A6: 153 M6: 722
galvanic corrosion M5: 628, 631, 637, 646
galvanic corrosion of A13: 747-748
galvanic couples, in salt/marine atmospheres A13: 746
gas metal arc welding M6: 153
gas tungsten arc welding M6: 182, 206
gas-metal arc welding A6: 772, 776, 777
shielding gases A6: 66
gas-tungsten arc welding A6: 772, 777-779, 780-781
in gases A13: 743
general biological corrosion of A13: 87
gold plating, stripping of M5: 647
grain refinement of A15: 480-482
grinding A16: 827-828, 829
grinding of A9: 425 M5: 628, 648-649
gun drilling A16: 823, 824
HAE treatment M5: 633-637, 636-637, 639-641, 643
hand hacksawing A16: 827, 828
handling of chips and fines A16: 829-830
hard-anodizing treatments for A13: 752
hardness M2: 528, 530-531
hardness and wear resistance A2: 461
heat treatment A14: 260
heat treatment following growth of FPL oxide EM3: 260
heat-affected zone A6: 780, 782
heating for forging A14: 259
HERF forgeability A14: 104
with high zinc levels, casting A2: 456-457
high-pressure die casting alloys A2: 456
honing stone selection A16: 476
hot extrusion, billet temperatures for M3: 537
hot extrusion of A14: 321-322
hot extrusion, tool materials for dies M3: 538
hot forming of A14: 825-826
hydrogen formation A16: 822, 829, 830
impact extrusion of A14: 311-312, 830
impact extrusions A2: 464-465 M2: 535, 537
inclusions in A15: 96, 488
inoculants for A15: 105
inserts A2: 466-467 M2: 538-540, 541, 542
intergranular corrosion evaluation A13: 241

SUBJECTS OF THE INDEXED VOLUMES: ASM Handbook (designated by the letter "A"): **A1:** Properties and Selection: Irons, Steels, and High-Performance Alloys (1990); **A2:** Properties and Selection: Nonferrous Alloys and Special-Purpose Materials (1990); **A3:** Alloy Phase Diagrams (1992); **A4:** Heat Treating (1991); **A6:** Welding, Brazing, and Soldering (1993); **A7:** Powder Metallurgy (1984); **A8:** Mechanical Testing (1985); **A9:** Metallography and Microstructures (1985); **A10:** Materials Characterization (1986); **A11:** Failure Analysis and Prevention (1986); **A12:** Fractography (1987); **A13:** Corrosion (1987); **A14:** Forming and Forging (1988); **A15:** Casting (1988); **A16:** Machining (1989); **A17:** Nondestructive Testing and Quality Control (1989); **A18:** Friction, Lubrication, and Wear Technology (1992). **Metals Handbook, 9th Edition** (designated by the letter "M"): **M1:** Properties and Selection: Irons and Steels (1978); **M2:** Properties and Selection: Nonferrous Alloys and Pure Metals (1979); **M3:** Properties and Selection: Stainless Steels, Tool Materials and Special-Purpose Materials (1980); **M4:** Heat Treating (1981); **M5:** Surface Cleaning, Finishing, and Coating (1982); **M6:** Welding, Brazing, and Soldering (1983). **Engineered Materials Handbook** (designated by the letters "EM"): **EM1:** Composites (1987); **EM2:** Engineering Plastics (1988); **EM3:** Adhesives and Sealants (1990); **EM4:** Ceramics and Glasses (1991); **Electronic Materials Handbook** (designated by the letters "EL"): **EL1:** Packaging (1989).

Magnesium alloys (continued)
internal grinding ... **A16:** 829
investment casting ... **A15:** 806-807
joining ... **M2:** 546-549
joining of ... **A2:** 471-475
joint design ... **A6:** 772-774
lapping ... **A16:** 499
low-temperature properties ... **A2:** 462 **M2:** 531-532, 533, 534
lubrication ... **A14:** 260, 827
lubrication of tool steels ... **A18:** 738
machinability ... **A2:** 475-476 **M2:** 549, 551
machines and dies ... **A14:** 259
macroetching of ... **A9:** 426
manual spinning of ... **A14:** 828-829
in marine atmospheres ... **A13:** 745-746
markets for ... **A15:** 42
mass (barrel) finishing of ... **M5:** 631
materials for die-casting dies ... **A18:** 629
mechanical polishing procedures ... **A9:** 425
mechanical properties ... **A2:** 457, 460-462 **M2:** 529-533
melting ... **A15:** 801-803
melting furnaces ... **A15:** 800-801
metal-matrix composites ... **A2:** 460
metallurgical factors ... **A13:** 740-741
Mg-matrix composites ... **A16:** 821
microscopic examination of ... **A9:** 426-427
military specifications finishes ... **M5:** 646, 648
milling ... **A16:** 312, 313, 326, 327, 826-827, 828
mold coatings for ... **A15:** 282
mounting materials for ... **A9:** 425
multiple-operation machining ... **A16:** 826
nickel undercoating
of ... **M5:** 638-639, 645-647
stripping of ... **M5:** 646-647
nominal compositions ... **A2:** 457 **A15:** 799
in organic compounds ... **A13:** 742-743
organic finishing of ... **M5:** 629, 647-648
surface preparation ... **M5:** 647-648
oxide films ... **A9:** 427
painting of ... **M5:** 629, 647-648
paint selection ... **M5:** 647-648
primers for ... **M5:** 629, 647-648
stripping of paint ... **M5:** 648
performance, saltwater exposures ... **A13:** 745
peripheral milling ... **A16:** 826-827
permanent mold casting ... **A15:** 799, 807
photochemical machining etchant ... **A16:** 588, 590
pickling of ... **M5:** 629-633, 635-642, 645, 647-648
pickle rate ... **M5:** 635-636
solution, magnesium content ... **M5:** 635-636
planing ... **A16:** 823
plate buckling ... **M2:** 550-552
plating of ... **M5:** 629, 638-639, 642, 644-647
electroless ... **M5:** 638-639, 642, 644-647
electrolytic ... **M5:** 638-639, 645
preparation for ... **M5:** 638
stripping of plate ... **M5:** 646-647
thickness, total ... **M5:** 646
uses of plated alloys ... **M5:** 646
polishing and buffing ... **M5:** 628, 631-632
postweld heat treatments ... **A6:** 777, 781, 782
pouring temperatures ... **A15:** 283
power band sawing ... **A16:** 827, 828
power hacksawing ... **A16:** 827, 828
power spinning ... **A14:** 829
preheating ... **A6:** 776, 777, 780
press-brake forming of ... **A14:** 827
press-formed parts, materials for forming tools ... **M3:** 492, 493
primary testing direction ... **A8:** 667
process parameters ... **A15:** 801-803
properties of ... **A2:** 480-516
protection of assemblies ... **A13:** 748-749
protection systems, proven ... **A13:** 753-754
protective coating systems ... **A13:** 749-753
quality control inspection coatings ... **M5:** 637-638, 646
radiographic inspection ... **A6:** 781-782
in real/simulated environments ... **A13:** 743-747
reaming ... **A16:** 822, 823, 824-825
recommended gap for braze filler metals ... **A6:** 120
recommended hot extrusion tool steels and hardnesses ... **A18:** 627
relative power for machining ... **A16:** 820

Magnesium alloys (continued)
removal, from permanent molds ... **A15:** 284
repair welding of castings ... **A6:** 779-782
resistance seam welding ... **A6:** 245 **M6:** 494
resistance spot welding ... **M6:** 479, 480
resistance welding ... **A6:** 833, 834, 840, 841
riveting ... **M2:** 549
riveting of ... **A2:** 474-475
Rockwell scale for ... **A8:** 76
rubber-pad forming of ... **A14:** 829-830
safe practice (machining) ... **A2:** 475-476
safety ... **A14:** 826 **A16:** 828
safety precautions, cleaning and finishing processes ... **M5:** 648-649
in salt solutions ... **A13:** 742
sand and permanent mold casting alloys ... **A2:** 456-459
sand casting ... **A15:** 798-799, 803-806
satin finishing of ... **M5:** 632
sawing ... **A16:** 364-365, 827, 828
SCC testing ... **A13:** 273, 745
SCC testing of ... **A8:** 529-530
seam welds ... **A2:** 473
sectioning procedures ... **A9:** 425
selection and application ... **A2:** 455-479
selection of product form ... **A2:** 462-466
semisolid forging/casting of ... **A15:** 327
shaping ... **A16:** 823
shear strength ... **A2:** 461 **M2:** 528-530
shear stresses and HP ... **A16:** 15
sheet and plate ... **A2:** 459-460, 467-468 **M2:** 529, 541-543, 544
shielding gases ... **A6:** 772, 778
shot peening of ... **M5:** 145-146
shrinkage allowances ... **A15:** 303
silverplating, stripping of ... **M5:** 647
slotting ... **A16:** 827, 828
in soils ... **A13:** 743
soldering ... **A6:** 631
solvent cleaning of ... **M5:** 629
spot welds ... **A2:** 473 **M2:** 547, 548
squeeze casting of ... **A15:** 323
stress relieving ... **A2:** 472-473 **M2:** 547
stress-corrosion cracking in ... **A11:** 223
stretch forming ... **A2:** 469-471 **M2:** 543, 545-546
stretch forming of ... **A14:** 830
stripping, plated deposits, paints and coatings ... **M5:** 646-647
surface activation processes ... **M5:** 642, 644, 646
surface finish ... **A16:** 820-827
surface grinding ... **A16:** 829
surface preparation ... **A6:** 774-776
systems, and nominal compositions ... **A13:** 741
tantalum resistance to ... **A13:** 733
tapping ... **A16:** 825-826
thermal expansion coefficient ... **A6:** 907 **A16:** 820
thermal properties of Mg-Al (electrolytic) ... **A18:** 42
tool life ... **A16:** 820, 821, 822, 823, 826
tools ... **A16:** 821, 822, 823, 825, 826
turning ... **A16:** 94, 135, 823
Unicast process for ... **A15:** 251
vapor degreasing of ... **M5:** 629
vibratory finishing of ... **M5:** 631
warpage as result of cold working ... **A16:** 820
weight reduction ... **M2:** 551-552
weldability ... **A6:** 772
welded joints ... **A9:** 433-434
welding ... **A2:** 471-472 **M2:** 546-547, 548
weldments, cost of ... **A2:** 473-474
wire brushing of ... **M5:** 629, 649
wire, die materials for drawing ... **M3:** 522
for wire-drawing dies ... **A14:** 336
workability ... **A8:** 165, 575
workpieces, heating of ... **A14:** 826-827
wrought alloys ... **A2:** 459-460, 480-491
wrought, cleaning and finishing ... **M5:** 630, 646
zinc plating of ... **M5:** 638-639, 642, 644-647
solution composition and operating conditions ... **M5:** 644-645, 647
stripping of ... **M5:** 646
ZW3, tension and torsion effective fracture strains ... **A8:** 168

Magnesium alloys, corrosion resistance ... **M2:** 596-609
acids ... **M2:** 600

Magnesium alloys, corrosion resistance (continued)
alkalis ... **M2:** 600
at elevated temperature ... **M2:** 601-602
atmospheric ... **M2:** 597-598, 599, 600
cold working, effects of ... **M2:** 597
composition, effects of ... **M2:** 596-597
corrosion protection ... **M2:** 602, 603, 607, 609
corrosion testing ... **M2:** 598, 603-604, 607
fresh water ... **M2:** 598
galvanic corrosion ... **M2:** 604-609
gases ... **M2:** 601
heat treatment, effects of ... **M2:** 597, 598
organic compounds ... **M2:** 600-601
salt solutions ... **M2:** 599-600, 601
soils ... **M2:** 601
temperature, effects of ... **M2:** 602

Magnesium alloys, heat treating ... **A4:** 899-906
aging ... **A4:** 899, 900, 901, 904 **M4:** 746, 747
annealing ... **A4:** 899, 900 **M4:** 745
dimensional stability ... **A4:** 900, 904 **M4:** 749, 751, 752
distortion control ... **A4:** 900, 904 **M4:** 749
fires, prevention and control ... **A4:** 903, 906 **M4:** 751-753
furnace loading ... **A4:** 903 **M4:** 748
furnace(s) ... **A4:** 902-903, 906
furnaces ... **M4:** 748
hardness testing ... **A4:** 905 **M4:** 750
mechanical properties ... **A4:** 899, 900, 901, 905 **M4:** 744-745
microscopic examination ... **A4:** 905 **M4:** 750
problems, prevention of ... **A4:** 903, 904-905 **M4:** 749, 750
protective atmospheres ... **A4:** 901-902, 905, 906 **M4:** 746-747
quenching media ... **A4:** 903 **M4:** 748-749
reheat treating ... **A4:** 900 **M4:** 746
section size ... **A4:** 900
solution treating ... **M4:** 744, 746, 747
solution-treating ... **A4:** 899, 900, 901, 903, 904, 905
stress-relieving, castings ... **A4:** 899, 900, 905 **M4:** 745
stress-relieving, wrought ... **A4:** 899, 900 **M4:** 745, 746
temper designations ... **A4:** 899, 900 **M4:** 744, 745
temper determination ... **A4:** 899, 904-905, 906 **M4:** 750-751
temperature control ... **A4:** 903 **M4:** 748
tensile tests ... **A4:** 905 **M4:** 750
time and temperature ... **A4:** 900, 901 **M4:** 746, 747, 748, 749, 751
weld-repaired castings ... **A4:** 905

Magnesium alloys, specific types *See also* Magnesium, Magnesium alloys
$2MgO2 \bullet 2Al_2O_3 \bullet 5SiO_2$, properties ... **A6:** 949
9980 A, composition ... **A6:** 773
9980 B, composition ... **A6:** 773
9990 A, composition ... **A6:** 773
9990 B, composition ... **A6:** 773
9995 A, composition ... **A6:** 773
9998 A, composition ... **A6:** 773
A3A, composition ... **A6:** 773
alloy PE, for special-quality sheet ... **A2:** 459
AM60A ... **A9:** 430 **M2:** 569
composition ... **M2:** 526, 528
mechanical properties ... **M2:** 526, 528
AM60A, as die casting, composition ... **A15:** 286
AM60A, composition ... **A6:** 773
AM60B, high-purity die cast alloy for ductility ... **A2:** 456
AM80A, composition ... **A6:** 773
AM90A, composition ... **A6:** 773
AM100A ... **A9:** 431 **M2:** 570, 571
aging ... **A4:** 901 **M4:** 747
atmospheric corrosion ... **M2:** 599
composition ... **A4:** 899 **A6:** 773 **M2:** 526, 528
heat treatment ... **A4:** 899 **M4:** 745
mechanical properties ... **M2:** 526, 528
postweld treatments to obtain 80-95% stress relief ... **A6:** 782
preheat temperatures and postweld heat treatments ... **A6:** 777
relative arc weldability ... **A6:** 772
selection guide of filler alloys for welding ... **A6:** 775
solution treating ... **M4:** 747

Magnesium alloys, specific types (continued)
solution-treating ... A4: 901
AM100A, sand and permanent mold casting alloy ... A2: 456
AM100B, composition ... A6: 773
AS21, for creep strength ... A2: 456
AS41A ... A9: 430 M2: 571
composition ... M2: 526, 528
mechanical properties ... M2: 526, 528
AS41A, as die casting, composition ... A15: 286
AS41A, composition ... A6: 773
AS41A, for creep strength ... A2: 456
AZ 31
selection guide of filler alloys for welding ... A6: 775
thermal diffusivity from 20 to 100 °C ... A6: 4
AZ10 A
composition ... A6: 773
relative arc weldability ... A6: 772
selection guide of filler alloys for welding ... A6: 775
AZ10A ... M2: 553
composition ... M2: 528
mechanical properties ... M2: 528
AZ10A, for wrought extruded bars/shapes ... A2: 459
AZ10A-F, postweld treatments to obtain 80-95% stress relief ... A6: 782
AZ21A, composition ... A6: 773
AZ21X1 ... M2: 554
applications ... M2: 528
composition ... M2: 528
mechanical properties ... M2: 528
AZ21X1, for battery applications ... A2: 459
AZ31 B ... A9: 427-429, 433-434
AZ31, corrosion rates vs. exposure and iron content ... A13: 747
AZ31, galvanic attack ... A13: 747
AZ31A, composition ... A6: 773
AZ31B ... A16: 820 M2: 554-555
annealing ... M4: 745
annealing temperatures ... A4: 900
atmospheric corrosion ... M2: 599, 600, 604, 606, 607
composition ... A6: 773, 774 M2: 528, 529
corrosion rate in 3% NaCl ... M2: 598
electrode potential in 3% NaCl ... M2: 598
galvanic corrosion ... M2: 605-607
gas-tungsten arc welding ... A6: 779, 780
marine corrosion ... M2: 600, 603-604, 606, 607
mechanical properties ... M2: 528, 529
relative arc weldability ... A6: 772
selection guide of filler alloys for welding ... A6: 775
stress-relieving treatments ... A4: 900 M4: 746
AZ31B, cracking ... A13: 1050
AZ31B, for forgings ... A2: 459
AZ31B, for hammer forgings ... A2: 459
AZ31B, for sheet and plate ... A2: 459
AZ31B, for wrought bars/shapes ... A2: 459
AZ31B, galvanic corrosion ... A13: 748
AZ31B-F, -O, stress relief ... A16: 821
AZ31B-F, fatigue strength as function of relative humidity ... A11: 252
AZ31B-F, postweld treatments to obtain 80-95% stress relief ... A6: 782
AZ31B-F, stress-relieving treatments ... A4: 900 M4: 746
AZ31B-H24
gas-tungsten arc welding ... A6: 778, 779
postweld treatments to obtain 80-95% stress relief ... A6: 782
AZ31B-H24, stress relief ... A16: 821
AZ31B-O, postweld treatments to obtain 80-95% stress relief ... A6: 782
AZ31C ... M2: 554-555
composition ... A6: 773, 774 M2: 528
mechanical properties ... M2: 528
selection guide of filler alloys for welding ... A6: 775
AZ31C, annealing temperatures ... A4: 900 M4: 745
AZ31C, for wrought bars/shapes ... A2: 459
AZ31C, relative arc weldability ... A6: 772
AZ61A ... A9: 429 M2: 555-556
annealing ... M4: 745
annealing temperatures ... A4: 900
atmospheric corrosion ... M2: 599
composition ... A6: 773, 774 M2: 528
corrosion in salt solutions ... M2: 601
corrosion potential in 3% NaCl ... M2: 598
electrode potential in 3% NaCl ... M2: 598
galvanic corrosion ... M2: 607
mechanical properties ... M2: 528
relative arc weldability ... A6: 772
selection guide of filler alloys for welding ... A6: 775
stress-relieving treatments ... A4: 900 M4: 746
AZ61A, for forgings ... A2: 459
AZ61A, for wrought extruded bars/shapes ... A2: 459
AZ61A, galvanic corrosion ... A13: 748
AZ61A-F, postweld treatments to obtain 80-95% stress relief ... A6: 782
AZ61A-F, stress relief ... A16: 821
AZ61A-F, stress-relieving treatments ... A4: 900 M4: 746
AZ63A ... A9: 431-432 M2: 571-572
aging ... A4: 901, 904 M4: 747, 751
applications ... A4: 904-905
composition ... A4: 899 A6: 773, 774 M2: 528
corrosion potential in 3% NaCl ... M2: 598
dimensional stability ... M4: 752
electrode potential in 3% NaCl ... M2: 598
heat treatment ... A4: 899 M4: 745
marine corrosion ... M2: 603-604
mechanical properties ... M2: 528
postweld heat treatment ... M4: 753
postweld heat treatments ... A4: 905
postweld treatments to obtain 80-95% stress relief ... A6: 782
preheat temperatures and postweld heat treatments ... A6: 777
relative arc weldability ... A6: 772
selection guide of filler alloys for welding ... A6: 775
solution treating ... M4: 747
solution-treating ... A4: 900, 901
yield strength ... A4: 902
AZ63A, sand and permanent mold casting alloy ... A2: 456
AZ80A ... A9: 429-430 A16: 820 M2: 556, 557
annealing temperatures ... A4: 900
atmospheric corrosion ... M2: 599
composition ... A4: 899 A6: 773, 774 M2: 528, 529
forgeability ... M2: 528, 529
heat treatment ... A4: 899
mechanical properties ... M2: 528, 529
relative arc weldability ... A6: 772
selection guide of filler alloys for welding ... A6: 775
AZ80A, annealing ... M4: 745
AZ80A, for forgings ... A2: 459
AZ80A, for wrought extruded bars/shapes ... A2: 459
AZ80A, forgeability ... A2: 459
AZ80A-F, postweld treatments to obtain 80-95% stress relief ... A6: 782
AZ80A-F, stress-relieving temperatures ... M4: 746
AZ80A-F, stress-relieving treatments ... A4: 900
AZ80A-T5, postweld treatments to obtain 80-95% stress relief ... A6: 782
AZ80A-T5, stress relief ... A16: 821
AZ80A-T5, stress-relieving treatments ... A4: 900 M4: 746
AZ81 ... A9: 427
AZ81A ... M2: 573, 574
aging ... A4: 901 M4: 747
composition ... A4: 899 A6: 773, 774 M2: 526, 528
heat treatment ... A4: 899 M4: 745
postweld heat treatment ... M4: 753
postweld heat treatments ... A4: 905
postweld treatments to obtain 80-95% stress relief ... A6: 782
preheat temperatures and postweld heat treatments ... A6: 777
relative arc weldability ... A6: 772
selection guide of filler alloys for welding ... A6: 775
solution treating ... M4: 747
solution-treating ... A4: 901, 905
AZ81A, sand and permanent mold casting alloy ... A2: 456
AZ90A, composition ... A6: 773, 774
AZ91, corrosion rates vs. exposure and iron content ... A13: 474
AZ91, die cast automotive application ... A15: 807
AZ91, die casting applications ... A15: 808
AZ91 die casting, nickel, iron, and copper content effects ... A13: 741, 745
AZ91, high purity, application of ... A15: 798
AZ91 sand cast, heavy-metal contamination ... A13: 741, 743
AZ91, thermal diffusivity from 20 to 100 °C ... A6: 4
AZ91A ... A9: 431-433 M2: 574, 575
composition ... M2: 526, 528
mechanical properties ... M2: 526, 528
AZ91A, composition ... A6: 773, 774
AZ91A, die casting ... A15: 799
AZ91A-T6, tensile strength ... A2: 460
AZ91B ... M2: 574, 575
AZ91B (AZ91D) ... A16: 820
AZ91B, as die casting, composition ... A15: 286, 799
AZ91B, as produced from scrap/secondary metal ... A2: 456
AZ91B, composition ... A6: 773, 774
AZ91C ... A9: 434 A16: 820 M2: 574, 575, 576
aging ... A4: 901, 904 M4: 747
composition ... A4: 899 A6: 773, 774 M2: 526, 528
dimensional stability ... M4: 752
gas-tungsten arc welding ... A6: 780
heat treatment ... A4: 899 M4: 745
marine corrosion ... M2: 603, 604
mechanical properties ... M2: 526, 528
postweld heat treatment ... M4: 753
postweld heat treatments ... A4: 905
postweld treatments to obtain 80-95% stress relief ... A6: 782
preheat temperatures and postweld heat treatments ... A6: 777
relative arc weldability ... A6: 772
selection guide of filler alloys for welding ... A6: 775
solution treating ... M4: 747
solution-treating ... A4: 901, 905
temper determination ... A4: 906
AZ91C, permanent mold casting of ... A15: 275
AZ91C, sand and permanent mold casting alloy ... A2: 456
AZ91C-T4, gas-tungsten arc welding ... A6: 780-781
AZ91C-T6 aircraft-generator gearbox, corrosion fatigue fracture ... A11: 260
AZ91C-T6, gun drilling ... A16: 824
AZ91D, galvanic corrosion ... A13: 746
AZ91D, most commonly used magnesium die casting alloy ... A2: 456
AZ91E ... A16: 820
AZ91E, sand and permanent mold casting alloy ... A2: 456
AZ92A ... A9: 431, 433-434 A16: 820 M2: 577-578
aging ... A4: 901, 904 M4: 747, 751
atmospheric corrosion ... M2: 599
composition ... A4: 899 A6: 773, 774 M2: 526, 528

Magnesium alloys, specific types (continued)
corrosion potential in 3% NaCl M2: 598
dimensional stability M4: 752
electrode potential in 3% NaCl M2: 598
heat treatment A4: 899 M4: 745
mechanical properties M2: 526, 528
postweld heat treatment M4: 753
postweld heat treatments A4: 905
postweld treatments to obtain 80-95% stress relief A6: 782
preheat temperatures and postweld heat treatments A6: 777
relative arc weldability A6: 772
selection guide of filler alloys for welding A6: 775
solution treating M4: 747
solution-treating A4: 900-901, 905
temper determination A4: 906
yield strength A4: 902
AZ92A, roller burnishing A16: 253
AZ92A, sand and permanent mold casting alloy A2: 456
AZ92A-T6, repair welding A6: 781, 782
AZ101A, composition A6: 773, 774
AZ125A, composition A6: 773, 774
AZCOML
postweld treatments to obtain 80-95% stress relief A6: 782
relative arc weldability A6: 772
EK30A
composition A6: 773, 774
preheat temperatures and postweld heat treatments A6: 777
relative arc weldability A6: 772
EK41A
composition A6: 773, 774
preheat temperatures and postweld heat treatments A6: 777
relative arc weldability A6: 772
selection guide of filler alloys for welding A6: 775
EQ21
postweld treatments to obtain 80-95% stress relief A6: 782
preheat temperatures and postweld heat treatments A6: 777
relative arc weldability A6: 772
EQ21A
aging A4: 901
heat treatment A4: 899
postweld heat treatments A4: 905
solution-treating A4: 901
EQ21A, magnesium-silver casting alloy A2: 458-459
EZ33, postweld treatments to obtain 80-95% stress relief A6: 782
EZ33A A9: 432 M2: 578, 579, 580, 581
aging A4: 901 M4: 747
composition A4: 899 A6: 773, 774 M2: 527, 528
contraction at elevated temperatures A4: 905
gas-tungsten arc welding A6: 782
heat treatment A4: 899 M4: 745
mechanical properties M2: 527, 528
postweld heat treatment M4: 753
postweld heat treatments A4: 905
preheat temperatures and postweld heat treatments A6: 777
relative arc weldability A6: 772
repair welding A6: 782
selection guide of filler alloys for welding A6: 775
solution treating M4: 747
solution-treating A4: 901
EZ33A, magnesium-rare-earth-zirconium A2: 458
EZ33A, moderate-temperature use A15: 799
EZ33A, sand casting pressure tightness A15: 799
EZ33A-T5, contraction at elevated temperatures M4: 751
HK31A A9: 429, 432
aging A4: 901 M4: 747
annealing temperatures A4: 900 M4: 745
composition A4: 899 A6: 773, 774
contraction at elevated temperatures A4: 905
heat treatment A4: 899 M4: 745
postweld heat treatments A4: 905

Magnesium alloys, specific types (continued)
preheat temperatures and postweld heat treatments A6: 777
reheat treating A4: 900
relative arc weldability A6: 772
selection guide of filler alloys for welding A6: 775
solution treating M4: 747
solution-treating A4: 900, 901
stress-relieving treatments A4: 900
HK31A, cast M2: 557, 558, 559, 560, 561
HK31A, for sheet and plate A2: 459
HK31A, forgeability A2: 459
HK31A, magnesium-thorium-zirconium casting A2: 458
HK31A, wrought M2: 557, 558, 559, 560, 561
composition M2: 528, 529
corrosion potential in 3% NaCl M2: 598
electrode potential in 3% NaCl M2: 598
mechanical properties M2: 528, 529
HK31A-H24, postweld treatments to obtain 80-95% stress relief A6: 782
HK31A-T6, contraction at elevated temperatures M4: 751
HM21A A9: 428, 430 M2: 561-562, 563, 564, 565
annealing temperatures A4: 900
composition A4: 899 A6: 773, 774 M2: 528, 529
corrosion potential in 3% NaCl M2: 598
electrode potential in 3% NaCl M2: 598
forgeability M2: 528, 529
heat treatment A4: 899
mechanical properties A4: 899 M2: 528, 529
relative arc weldability A6: 772
selection guide of filler alloys for welding A6: 775
stress-relieving treatments A4: 900
HM21A, annealing M4: 745
HM21A, for forging A2: 459
HM21A, for sheet and plate A2: 459
HM21A, forgeability A2: 459
HM21A-T5, stress-relieving treatments M4: 746
HM21A-T8, postweld treatments to obtain 80-95% stress relief A6: 782
HM21A-T8, stress-relieving treatments M4: 746
HM21A-T81, postweld treatments to obtain 80 to 95% stress relief A6: 782
HM21A-T81, stress relieving treatments M4: 746
HM31A A9: 430 M2: 565, 566
annealing M4: 745
annealing temperatures A4: 900
composition A4: 899 A6: 773, 774 M2: 528
heat treatment A4: 899
mechanical properties M2: 528
postweld heat treatment M4: 747
relative arc weldability A6: 772
selection guide for filler alloys for welding A6: 775
stress-relieving treatments A4: 900 M4: 746
HM31A-T5, postweld treatments to obtain 80-95% stress relief A6: 782
HM31A-T5, stress-relieving treatments M4: 746
HZ32, capabilities A15: 799
HZ32A A9: 432 M2: 584, 585-586
aging A4: 901 M4: 747
composition A4: 899 A6: 773, 774 M2: 527, 528
contraction at elevated temperatures A4: 905
heat treatment A4: 899 M4: 745
mechanical properties M2: 527, 528
postweld heat treatment M4: 753
postweld heat treatments A4: 905
preheat temperatures and postweld heat treatments A6: 777
relative arc weldability A6: 772
selection guide of filler alloys for welding A6: 775
solution treating M4: 747
solution-treating A4: 901
HZ32A, magnesium-thorium-zirconium casting A2: 458
HZ32A-T5, contraction at elevated temperatures M4: 751
K1A A9: 430 M2: 587
composition A6: 773, 774 M2: 526, 528
mechanical properties M2: 526, 528
preheat temperatures and postweld heat treatments A6: 777

Magnesium alloys, specific types (continued)
relative arc weldability A6: 772
selection guide of filler alloys for welding A6: 775
K1A, casting alloy, for high damping capacity A2: 456
LA141A A9: 428
composition A6: 773, 774
selection guide of filler alloys for welding A6: 775
LS141A, composition A6: 773, 774
LZ145A, composition A6: 773, 774
M1A M2: 567, 568
atmospheric corrosion M2: 599
composition A6: 773, 774 M2: 528
corrosion potential in 3% NaCl M2: 598
electrode potential in 3% NaCl M2: 598
mechanical properties M2: 528
selection guide of filler alloys for welding A6: 775
M1A, contour band sawing A16: 363
M1A, for hammer forgings A2: 459
M1A, for wrought extruded bars/shapes A2: 459
M1B, composition A6: 773, 774
M1C, composition A6: 773, 774
Mg-0.5Zr creep by atom diffusion A9: 691
Mg-32A1 eutectic alloy, transverse section through parallel rods A9: 128
MG1, selection guide of filler alloys for welding A6: 775
PE M2: 555
composition M2: 528, 529
mechanical properties M2: 528, 529
QE-22A, corrosion fatigue A13: 1037
QE22, postweld treatments to obtain 80-95% stress relief A6: 782
QE22A A9: 432-433 M2: 587-588, 589
aging A4: 901 M4: 747
composition A4: 899 A6: 773, 774 M2: 527, 528
heat treatment A4: 899 M4: 745
mechanical properties A4: 902, 903 M2: 527, 528
postweld heat treatment M4: 753
postweld heat treatments A4: 905 A6: 782
preheat temperatures and postweld heat treatments A6: 777
quenching media A4: 903
relative arc weldability A6: 772
selection guide of filler alloys for welding A6: 775
solution treating M4: 747
solution-treating A4: 901
QE22A, engine gearbox A15: 805
QE22A, magnesium-silver casting alloy A2: 458-459
QE22A, mechanical properties A15: 799
QE22A-T6
quenching medium, effect on tensile properties M4: 748
tensile properties M4: 748, 749
QH21A M2: 589, 590, 591
aging A4: 901 M4: 747
composition A4: 899 A6: 773, 774 M2: 527, 528
heat treatment A4: 899 M4: 745
mechanical properties M2: 527, 528
postweld heat treatment M4: 753
postweld heat treatments A4: 905
quenching media A4: 903
relative arc weldability A6: 772
solution treating M4: 747
solution-treating A4: 901
RZ 5
thermal diffusivity from 20 to 100 °C A6: 4
TA54A, composition A6: 773, 774
WE43
postweld treatments to obtain 80-95% stress relief A6: 782
preheat temperatures and postweld heat treatments A6: 777
relative arc weldability A6: 772
WE43A
aging A4: 901
heat treatment A4: 899
postweld heat treatments A4: 905

Magnesium alloys, specific types (continued)
solution-treating ... **A4:** 901
WE54
postweld treatments to obtain 80-95% stress relief ... **A6:** 782
preheat temperatures and postweld heat treatments ... **A6:** 777
relative arc weldability ... **A6:** 772
WE54A
aging ... **A4:** 901, 903
heat treatment ... **A4:** 899
postweld heat treatments ... **A4:** 905
solution-treating ... **A4:** 901
WE54A, capabilities ... **A15:** 799
ZC63
postweld treatments to obtain 80-95% stress relief ... **A6:** 782
preheat temperatures and postweld heat treatments ... **A6:** 777
relative arc weldability ... **A6:** 772
ZC63, magnesium-aluminum alloy ... **A2:** 457
ZC63A
aging ... **A4:** 901
heat treatment ... **A4:** 899
postweld heat treatments ... **A4:** 905
solution-treating ... **A4:** 901
ZC71, capabilities ... **A2:** 459
ZC71A
aging ... **A4:** 901
composition ... **A4:** 899
heat treatment ... **A4:** 899
solution-treating ... **A4:** 901
stress-relieving ... **A4:** 900
ZE10A ... **A9:** 428
composition ... **A6:** 773, 774
relative arc weldability ... **A6:** 772
selection guide of filler alloys for welding ... **A6:** 775
ZE10A-0, postweld treatments to obtain 80-95% stress relief ... **A6:** 782
ZE10A-H24, postweld treatments to obtain 80-95% stress relief ... **A6:** 782
ZE41, postweld treatments to obtain 80-95% stress relief ... **A6:** 782
ZE41A ... **M2:** 591-592
aging ... **A4:** 901 **M4:** 747
composition ... **A4:** 899 **A6:** 773, 774 **M2:** 526, 527, 528
heat treatment ... **A4:** 899
mechanical properties ... **M2:** 526, 527, 528
postweld heat treatment ... **M4:** 753
postweld heat treatments ... **A4:** 905
preheat temperatures and postweld heat treatments ... **A6:** 777
relative arc weldability ... **A6:** 772
selection guide of filler alloys for welding ... **A6:** 775
solution treating ... **M4:** 747
solution-treating ... **A4:** 901
temper determination ... **A4:** 906
ZE41A, applications ... **A15:** 799
ZE41A, gearbox housing ... **A15:** 805
ZE41A, magnesium-aluminum casting alloy ... **A2:** 457
ZE41A-T5, main transmission housing ... **A15:** 805
ZE63A ... **M2:** 592-593
aging ... **A4:** 901 **M4:** 747
composition ... **M2:** 526, 528
heat treatment ... **A4:** 899
mechanical properties ... **M2:** 526, 528
solution treating ... **M4:** 747
solution-treating ... **A4:** 901
ZE63A, composition ... **A6:** 773, 774
ZE63A, high-strength grade casting alloy ... **A2:** 457
ZH62A ... **A9:** 432 **M2:** 593
aging ... **A4:** 901 **M4:** 747
composition ... **A4:** 899 **A6:** 773, 774 **M2:** 526, 528
heat treatment ... **A4:** 899

Magnesium alloys, specific types (continued)
mechanical properties ... **M2:** 526, 528
postweld heat treatment ... **M4:** 753
postweld heat treatments ... **A4:** 905
preheat temperatures and postweld heat treatments ... **A6:** 777
relative arc weldability ... **A6:** 772
selection guide of filler alloys for welding ... **A6:** 775
solution treating ... **M4:** 747
solution-treating ... **A4:** 901
ZH62A, capabilities ... **A15:** 799
ZH62A, high zinc level casting alloy ... **A2:** 456-457
ZK21A ... **A9:** 429 **M2:** 568
composition ... **A6:** 773, 774 **M2:** 528
mechanical properties ... **M2:** 528
relative arc weldability ... **A6:** 772
ZK21A-F, stress relief ... **A16:** 821
ZK40A ... **M2:** 568
composition ... **M2:** 528
mechanical properties ... **M2:** 528
ZK40A, composition ... **A6:** 773, 774
ZK51A ... **A9:** 432 **M2:** 594
aging ... **A4:** 901 **M4:** 747
composition ... **A4:** 899 **A6:** 773, 774 **M2:** 526, 528
heat treatment ... **A4:** 899
mechanical properties ... **M2:** 526, 528
postweld heat treatment ... **M4:** 753
postweld heat treatments ... **A4:** 905
preheat temperatures and postweld heat treatments ... **A6:** 777
relative arc weldability ... **A6:** 772
selection guide of filler alloys for welding ... **A6:** 775
solution treating ... **M4:** 747
solution-treating ... **A4:** 901
ZK51A, high zinc level casting alloy ... **A2:** 456-457
ZK51A, mechanical properties ... **A15:** 799
ZK60-T5, peripheral milling ... **A16:** 826
ZK60A ... **A9:** 430 **M2:** 568-569
annealing temperatures ... **A4:** 900
composition ... **A4:** 899 **A6:** 773, 774 **M2:** 528, 529
corrosion potential in 3% NaCl ... **M2:** 598
electrode potential in 3% NaCl ... **M2:** 598
forgeability ... **M2:** 528, 529
heat treatment ... **A4:** 899
mechanical properties ... **M2:** 528, 529
relative arc weldability ... **A6:** 772
selection guide of filler alloys for welding ... **A6:** 775
stress-relieving ... **A4:** 900
ZK60A, annealing ... **M4:** 745
ZK60A, applications ... **A2:** 459
ZK60A, for forgings ... **A2:** 459
ZK60A, forgeability ... **A2:** 459
ZK60A-F, stress relief ... **A16:** 821
ZK60A-F, stress-relieving treatments ... **A4:** 900 **M4:** 746
ZK60A-T5, stress relief ... **A16:** 821
ZK60A-T5, stress-relieving treatments ... **M4:** 746
ZK60A-T6, stress relief ... **A16:** 821
ZK60B
composition ... **A6:** 773, 774
ZK61A ... **A9:** 433 **M2:** 595
aging ... **A4:** 901 **M4:** 747
composition ... **A4:** 899 **A6:** 773, 774 **M2:** 526, 528
heat treatment ... **A4:** 899
mechanical properties ... **M2:** 526, 528
preheat temperatures and postweld heat treatments ... **A6:** 777
relative arc weldability ... **A6:** 772
selection guide of filler alloys for welding ... **A6:** 775
solution treating ... **M4:** 747
solution-treating ... **A4:** 901
ZK61A, high zinc level casting alloy ... **A2:** 456-457
ZK61A, mechanical properties ... **A15:** 799
ZK63A, high zinc level casting alloy ... **A2:** 456-457

Magnesium alloys, specific types (continued)
ZM21A
composition ... **M2:** 528
mechanical properties ... **M2:** 528
ZM21A, for wrought extruded bars/shapes ... **A2:** 459
ZW 1, thermal diffusivity from 20 to 100 °C ... **A6:** 4
ZX21A, selection guide of filler alloys for welding ... **A6:** 775
Magnesium alumina spinel ... **EM4:** 61
Magnesium aluminate ($MgAl_2O_4$)
complex arc-rib mirror and fracture surface ... **EM4:** 640
coprecipitation process ... **EM4:** 59
Magnesium aluminosilicate
as S-glass composition ... **EM1:** 45
Magnesium aluminum oxide
reactive hot pressing for pressure densification ... **EM4:** 300
Magnesium aluminum silicate (MAS)
applications, regenerator cores for advanced gas turbines ... **EM4:** 1001
ceramic regenerator disks lacking strength ... **EM4:** 721
chemical durability ... **EM4:** 877
dielectric properties ... **EM4:** 877
hardness ... **EM4:** 876
maximum use temperatures ... **EM4:** 875
Magnesium aluminum silicates, hydrated
as fillers ... **EM3:** 178
Magnesium anodes
for cathodic protection ... **M1:** 745
Magnesium carbonate
for abrasive jet machining ... **A16:** 512
Magnesium carbonate ($MgCO_3$)
decomposition ... **EM4:** 109, 112
kinetics ... **EM4:** 109-110
stress effects present in uniaxial pressing of a powder compact ... **EM4:** 145
Magnesium casting alloys
high-pressure die casting ... **A2:** 456
sand and permanent mold casting ... **A2:** 456-459
Magnesium chloride
stress-corrosion cracking resistance to boiling ... **A1:** 725-727
stress-corrosion cracking verification procedures ... **A1:** 725
Magnesium chloride solution
SCC testing in ... **A13:** 272-273
Magnesium deficiency
biologic effects ... **A2:** 1259
Magnesium deoxidation
of copper alloys ... **A15:** 470
Magnesium ferrite ... **EM4:** 59
Magnesium ferrites ... **A9:** 538
Magnesium fluoride
direct evaporation ... **A18:** 844
Magnesium fluoride (MgF_2) ... **EM4:** 191
application ... **EM4:** 1
crack patterns and fracture origins ... **EM4:** 641
Magnesium germanate
as phosphor ... **A2:** 743
Magnesium granules
mechanically comminuted ... **A7:** 131
Magnesium hydrate
Miller numbers ... **A18:** 235
Magnesium hydroxide ($Mg(OH)_2$)
pressure densification ... **EM4:** 300
Magnesium hydroxides
as sheet molding compound thickeners ... **EM1:** 158
Magnesium in cast iron ... **M1:** 6, 8, 37
Magnesium nitrate, as sample modifier
GFAAS analysis ... **A10:** 55
Magnesium oxide
in 98.58Ag-0.22MgO-0.2Ni ... **A9:** 559
as a constituent of silver-base electrical contact materials ... **A9:** 551-552

Magnesium oxide (continued)
in binary phosphate glasses ... **A10:** 131
cermets, applications and properties ... **A2:** 993
insulation, for thermocouples ... **A2:** 883
tool materials for cast iron machining ... **A16:** 656
toxicity of ... **A2:** 1259
vacuum heat-treating support fixture material ... **A4:** 503
Magnesium oxide (MgO) *See also* Engineering properties of single oxides ... **EM4:** 57
added to extend glass working range ... **EM4:** 379
additive to aid pressure densification ... **EM4:** 298
calcination ... **EM4:** 110-112
composition ... **EM4:** 14
in composition of glass-ceramics ... **EM4:** 499
in composition of textile products ... **EM4:** 403
in composition of wool products ... **EM4:** 403
corrosion resistance of refractories ... **EM4:** 391
as dopant in ceramic/ceramic joints ... **EM4:** 516
in drinkware compositions ... **EM4:** 1102
electrical integrity of seals ... **EM4:** 539
erosion of ceramics ... **A18:** 205
erosion of steels ... **A18:** 204
in glaze composition for tableware ... **EM4:** 1102
grain-growth inhibitor ... **EM4:** 188
hot pressed for pressure densification ... **EM4:** 296
impurity found in gypsum ... **EM4:** 380
ion sputtering effect ... **EM3:** 245
in ovenware compositions ... **EM4:** 1103
pressure densification
pressure ... **EM4:** 301
technique ... **EM4:** 301
temperature ... **EM4:** 301
properties ... **A18:** 801 **EM4:** 14, 24
sintering aid ... **EM4:** 188
in tableware compositions ... **EM4:** 1101
to improve thermal conductivity ... **EM3:** 178
Vickers and Knoop microindentation hardness numbers ... **A18:** 416
Magnesium oxide abrasives
for hand polishing ... **A9:** 35
reclamation of ... **A9:** 353
Magnesium oxide, as refractory
core coatings ... **A15:** 240
Magnesium oxide-containing cermets ... **A7:** 803
Magnesium oxide-enriched surface films ... **A7:** 254
Magnesium oxides
as sheet molding compound thickeners ... **EM1:** 158
Magnesium particles
atomized ... **A7:** 132
Magnesium powder(s)
applications ... **A7:** 131
chemical analysis and sampling ... **A7:** 248
chemical requirements ... **A7:** 602
-coated grinding balls ... **A7:** 58
explosive reaction with moisture ... **A7:** 194-195, 199
flammability in dust clouds ... **A7:** 195
high explosivity class ... **A7:** 196
as incendiary ... **A7:** 603
intoxication ... **A7:** 204
lathe turnings ... **A7:** 131
and moisture, pyrophoricity ... **A7:** 194-195, 199
production ... **A7:** 131-133
as pyrophoric ... **A7:** 199
pyrotechnic requirements ... **A7:** 601
as reactive material, properties ... **A7:** 597
safety hazards ... **A7:** 132-133
solid, comminution of ... **A7:** 131
toxicity, diseases, exposure limits ... **A7:** 204
types and forms ... **A7:** 602
Magnesium recycling
melting practices ... **A2:** 1218
scrap sources ... **A2:** 1216-1217
secondary magnesium, properties ... **A2:** 1218
technology ... **A2:** 1217-1218
Magnesium spinel ... **EM4:** 61
Magnesium sulfamate
use in rhodium plating ... **M5:** 290-291
Magnesium, vapor pressure
relation to temperature ... **M4:** 309, 310
Magnesium zirconate
ceramic coatings for dies ... **A18:** 643, 644
Magnesium-37% tin alloy, microstructure of ... **A3:** 1 • 19
Magnesium-aluminosilicate glass-ceramic
aerospace applications ... **EM4:** 1018
Magnesium-aluminum alloys
effect of zinc on ... **A9:** 428
voids in ... **A9:** 427
zincating process ... **M5:** 603-604
Magnesium-aluminum casting alloys
specific types ... **A2:** 456
Magnesium-aluminum-manganese alloys ... **A9:** 427
Magnesium-aluminum-silicate type
elastic constant ... **EM4:** 875
Magnesium-aluminum-zinc alloys, brazing
available product forms of filler metals ... **A6:** 119
brazing, joining temperatures ... **A6:** 118
Magnesium-aluminum-zinc alloys, casting
fatigue properties ... **A2:** 461
Magnesium-base metal matrix composites ... **A13:** 861
Magnesium-bodied atmospheric deep-sea diving suit ... **A13:** 753
Magnesium-ceramic particle composites ... **A2:** 460
Magnesium-containing alloys ... **A1:** 39
Magnesium-didymium alloys ... **A9:** 428
Magnesium-lithium-aluminum alloys ... **A9:** 427
Magnesium-manganese alloying
wrought aluminum alloy ... **A2:** 52
Magnesium-matrix composites
development and production ... **A2:** 907-908
Magnesium-matrix composites with carbon fibers
polishing ... **A9:** 590
Magnesium-mischmetal alloys ... **A9:** 428
Magnesium-partially stabilized zirconia (Mg-PSZ)
brazing with glasses ... **EM4:** 520
cyclic fatigue crack-growth rates ... **EM4:** 696
direct brazing for joining oxide ceramics ... **EM4:** 517-519
Magnesium-rare earth alloys
etchants for ... **A9:** 427
Magnesium-rare earth metal-zirconium alloys ... **A9:** 427-428
Magnesium-rare-earth zirconium alloys
casting ... **A2:** 458
Magnesium-rich phases in aluminum alloys
appearance of ... **A9:** 353
Magnesium-sialon type
dielectric properties ... **EM4:** 877
Magnesium-silicide alloying
wrought aluminum alloy ... **A2:** 52-53
Magnesium-silver casting alloys ... **A2:** 458-459
Magnesium-thorium alloy
etchants for ... **A9:** 427
Magnesium-thorium-zirconium alloys ... **A9:** 427-428
casting ... **A2:** 458
Magnesium-zinc alloys
workability ... **A8:** 575
Magnesium-zinc-zirconium alloys ... **A9:** 427-428
Magnesium-zirconium alloys ... **A9:** 427-428
sand mold pouring ... **A15:** 802
Magnesium/calcium oxide, as refractory
core coatings ... **A15:** 240
Magnet alloys
as commercial permanent magnet materials ... **A2:** 785
Magnet coils
for amplitude detection in ultrasonic testing ... **A8:** 245-246
Magnet steels *See* Permanent magnet materials, specific types
as commercial permanent magnet materials ... **A2:** 785
Magnet yoke assembly
test part magnetization ... **A7:** 575
Magnetic
anisotropy, determined ... **A10:** 272-273
defined ... **A13:** 9
disordered materials, exotic effects in ... **A10:** 276
inhomogeneities, determined ... **A10:** 274
samples, ESR analysis ... **A10:** 264
separation ... **A10:** 177
states, identification ... **A10:** 253, 267
-structure analysis, Mössbauer spectroscopy ... **A10:** 287
Magnetic accelerators
in flat plate impact test ... **A8:** 210
Magnetic alignment
defined ... **A9:** 11
Magnetic alloys
sintering atmospheres ... **A7:** 341, 345
Magnetic anisotropy
determined ... **A10:** 272-273
Magnetic anisotropy forces
effect on preferred alignment of magnetic moments ... **A9:** 533
Magnetic applications ... **A7:** 624-645
Magnetic bearing
for centrifugal compressors ... **A18:** 607
defined ... **A18:** 12
Magnetic bridge comparator ... **A14:** 205
Magnetic bridge comparator testing
of powder metallurgy parts ... **A17:** 545
Magnetic bridge sorting ... **A7:** 490, 491
Magnetic brush
development, copper powders ... **A7:** 582-583
equipment, copper powders ... **A7:** 582-583
systems ... **A7:** 586
Magnetic bubble material
backscattering spectrum ... **A10:** 631
Magnetic bubble technology ... **EM4:** 1164
applications ... **EM4:** 18
composition control ... **EM4:** 18
crystal structure and orientation ... **EM4:** 18
fabrication processes ... **EM4:** 18
mixing operations ... **EM4:** 98
properties ... **EM4:** 18
Magnetic ceramics *See also* Ferrites
applications ... **EM4:** 18
composition control ... **EM4:** 18
crystal structure and orientation ... **EM4:** 18
fabrication processes ... **EM4:** 18
mixing operations ... **EM4:** 98
properties ... **EM4:** 98
Magnetic classifications of materials ... **A9:** 63
Magnetic clutch fluids
powders used ... **A7:** 573
Magnetic coil application
beryllium-copper alloys ... **A2:** 420
Magnetic confinement
for thermonuclear fusion ... **A2:** 1056-1057
Magnetic contrast ... **A10:** 506, 676
scanning electron microscopy ... **A9:** 94, 536-537
Magnetic copier powders ... **A7:** 585
Magnetic core
defined ... **A7:** 7
Magnetic core materials
for high frequencies ... **A7:** 643-645
Magnetic cores
core selection and ease of fabrication ... **A2:** 780
design and fabrication ... **A2:** 780
Interlaminar insulation ... **A2:** 780
Magnetic domain
defined ... **A2:** 761
Magnetic domain motion
studied by x-ray topography with synchrotron radiation ... **A10:** 365
Magnetic domain patterns
revealed by magnetic etching ... **A9:** 63-66
Magnetic domain structures
in ferrites ... **A9:** 539
in ferrites, specimen preparation ... **A9:** 532
methods of study ... **A9:** 534-537
Magnetic domains
defined ... **A17:** 159
interpretation, field and tension effects on resistance ... **A17:** 145
in magabsorption theory ... **A17:** 145
orientation stability, and hysteresis ... **A17:** 145
physical theory ... **A17:** 131
revealed by magnetic contrast ... **A9:** 94
Magnetic drum system
seed separation ... **A7:** 589
Magnetic electron microscopes ... **A12:** 6
Magnetic energy
defined ... **A2:** 784
Magnetic energy storage
superconducting ... **A2:** 1057
Magnetic etching ... **A9:** 63-66
of iron-chromium-nickel heat-resistant casting alloys to identify ferrite ... **A9:** 333-334
wrought stainless steels ... **A9:** 282

Magnetic field
applications, ternary molybdenum chalcogenides (chevrel phases) **A2:** 1079-1080
effect in ESR **A10:** 254
effects, niobium-titanium superconductors **A2:** 1043, 1045-1046, 1054-1057
mass analyzers, in gas mass spectrometers **A10:** 153-154
oscillating, wave theory of **A10:** 83
strength, in ESR analysis **A10:** 253, 685
structural changes as function of **A10:** 420
in superconducting materials **A2:** 1030, 1033-1034
Magnetic field disturbance (MFD) inspection
of steel reinforcements **A17:** 133
Magnetic field, pulsed
in high-energy-rate compacting **A7:** 306
Magnetic field strength
SI unit/symbol for **A8:** 721
Magnetic field testing *See also* Magnetic fields; Magnetic measurement **A17:** 129-135
applications **A17:** 132-134
magnetic characterization of materials **A17:** 131-132
magnetic leakage field testing principles **A17:** 129-131
Magnetic field(s) *See also* Magnetic field testing; Magnetic leakage field testing
with Barkhausen noise jumps **A17:** 160
circular magnetization **A17:** 90-91
direction of **A17:** 90
effect on magnetoresistance **A17:** 144
effect on resistivity, ferromagnetic materials **A17:** 144
electromagnetic yokes **A17:** 95
generation, magnetic field testing **A17:** 129-131
intensity, defined **A17:** 145
interaction, microwave inspection **A17:** 202
lines, remote-field eddy current inspection **A17:** 195-196
magnetized bar **A17:** 90
magnetized ring **A17:** 90
methods of generating **A17:** 93-99
strength, and magnetic characterization **A17:** 131
Magnetic fields
in electromagnetic forming **A14:** 644
scanning electron microscopy examination of specimens **A9:** 97
Magnetic flux
SI unit/symbol for **A8:** 721
Magnetic flux density *See also* Flux density
in magabsorption measurement **A17:** 145
SI unit/symbol for **A8:** 721
Magnetic hyperfine interactions
transitions and relative line intensities **A10:** 293
Magnetic hysteresis *See also* Demagnetization; Hysteresis; Hysteresis loops
curve, ferromagnetic material **A17:** 99
in magnetic particle inspection **A17:** 99-100
measurable characteristics of **A17:** 131
Magnetic hysteresis curve
for soft magnetic materials **A7:** 639
Magnetic induction **A7:** 640
in permanent magnet materials **A2:** 782, 784
Magnetic inhomogeneities **A10:** 274-275
Magnetic interaction
in Mössbauer effect **A10:** 293
Magnetic iron oxide
thermal conductivity of **A11:** 604
Magnetic iron oxide (Fe_2O_3, Fe_3O_4)
categorized as ferrites **EM4:** 1161
Magnetic leakage field testing *See also* Magnetic field testing
analysis of data **A17:** 131
defect leakage fields, origin **A17:** 129
experimental techniques **A17:** 129-131
Magnetic lens
defined **A9:** 11
Magnetic lines of flow *See also* Flux
defined **A17:** 90
Magnetic lines of force
schematics **A17:** 90
Magnetic mass spectrometers
SIMS **A11:** 35
Magnetic materials *See* Magnetically soft materials; Permanent magnet materials **A9:** 531-549
electrolytes for electropolishing **A9:** 533
etchants **A9:** 532-534
in electroplated coatings **A18:** 838
elemental cobalt alloying **A2:** 446
FIM/AP study of **A10:** 583
influence of spinodal decomposition **A9:** 654
microexamination **A9:** 533-537
microstructures of magnetically soft materials **A9:** 537-538
microstructures of permanent magnets **A9:** 538-539
permeability of **A17:** 99
quantum mechanical band theory **EL1:** 103
rare earth alloy additives **A2:** 729-731
specimen preparation **A9:** 531-533
Magnetic materials, specific types
1% Si iron electrical sheet steel, cold reduced 70% **A9:** 542
2.5% Si flat-rolled electrical sheet **A9:** 542
2.5% Si flat-rolled electrical sheet, different heat treatments compared **A9:** 543
2V-Permendur, iron-nickel substitutes for **A9:** 538
3% Si flat-rolled electrical strip, 70% reduction **A9:** 543-544
3% Si flat-rolled, oriented electrical hot band, as hot rolled **A9:** 544
3% Si flat-rolled, oriented electrical strip, different reductions and heat treatments **A9:** 544
3% Si steel, cubic etch pits identifying different orientations **A9:** 544
3% Si steel, different heat treatments compared **A9:** 544
3% Si steel, thermal faceting and pitting **A9:** 545
3-79 Moly Permalloy cold-rolled strip **A9:** 545
3-79 Moly Permalloy, magnetic test-ring specimen **A9:** 545
3.25% Si cold-rolled electrical strip **A9:** 544
3.25% Si flat-rolled, oriented electrical strip **A9:** 544
4-79 Moly Permalloy, textured structure in **A9:** 700
330FR, ferritic stainless steel, solenoid-quality **A9:** 545
Alnico 5, cast with directional grain **A9:** 547
Alnico 5, cast with random grains **A9:** 547
Alnico 5, casting, solution annealed, cooled in a magnetic field **A9:** 547
Alnico 5, microstructure **A9:** 539
Alnico 5, pressed from powder **A9:** 547
Alnico 5E, microstructure **A9:** 539
Alnico 6B, microstructure **A9:** 539
Alnico 8, microstructure **A9:** 539
Alnico 9, cast with directional grain **A9:** 547
barium ferrite, anisotropic, pressed and sintered **A9:** 549
Chromindur 11, deep drawn, solution annealed **A9:** 547
Chromindur 11, telephone receiver magnet **A9:** 546
$Co_{3.45}Fe_{0.25}Cu_{1.35}SM$, diffractometer traces **A9:** 702
Cunife, cold rolled 80% reduction, aged **A9:** 549
electrical iron, decarburization annealed **A9:** 542
Magnetic materials, specific types (continued)
Fe-1.9V, annealed **A9:** 546
Fe-27Co, cold-rolled strip, annealed **A9:** 546
Fe-28.5Cr-10.6Co, isotropic spinodal structure **A9:** 653
Fe-50Ni, cold-rolled strip **A9:** 545
$Fe_{80}B_{18.3}P_{1.7}$ amorphous metal ribbon, as-cast **A9:** 546
garnet, magnetic domains **A9:** 546
nickel ferrite, pressed and sintered **A9:** 546
Platinax 11, microstructure **A9:** 539
Remalloy **A9:** 538
Remendur 27, wire **A9:** 546
samarium-cobalt, different illumination modes compared **A9:** 548-549
silicon core iron, bar **A9:** 543
silicon iron, nonoriented **A9:** 542
Vicalloy II, effects of cold working **A9:** 539
Magnetic measurement *See also* Magnetic field testing
experimental techniques **A17:** 132
of metallurgical/magnetic properties **A17:** 131-132
nondestructive characterization **A17:** 134
Magnetic measuring method
cadmium plate thickness **M5:** 267
Magnetic methods
capabilities of **A10:** 380
Magnetic molding **A15:** 234-235
Magnetic moments
and ferromagnetism **A9:** 533
Magnetic orientations
permanent magnet materials **A2:** 791
Magnetic painting *See also* Magnetic particle inspection
advantages **A17:** 127
applications **A17:** 128
defined **A17:** 126
performance **A17:** 127-128
Magnetic particle and eddy current testing of steel springs **A1:** 309
Magnetic particle crack detection **A7:** 484
Magnetic particle inspection *See also* Liquid penetrant inspection; Magnetic painting Magnetic printing; Magnetic rubber inspection; NDE reliability **A6:** 98, 100 **A7:** 575-579 **A17:** 89-128
advantages/limitations **A17:** 89-90
applications, specific **A17:** 116-120
automated equipment **A17:** 116-120
of billets **A17:** 115-116, 558-559
of boilers and pressure vessels **A17:** 642
of casting defects **A15:** 553-554
of casting surfaces **A17:** 512
of castings **A17:** 112-114, 512
codes, boilers/pressure vessels **A17:** 642
demagnetization after **A17:** 120-122
detectable discontinuities **A17:** 103-105
dry powder technique, for forgings **A17:** 500
equipment **A7:** 575-576
of finned tubing **A17:** 571
fitness for service evaluation **A6:** 376
of forgings **A17:** 112-114, 499-501
of hollow cylindrical parts **A17:** 111-112
hot rolled bars and shapes **M1:** 200, 201
of investment castings **A15:** 264
magnetic fields, description of **A17:** 90-91
magnetic fields, generating **A17:** 93-99
magnetic hysteresis **A17:** 99-100
magnetic painting **A17:** 126-128
magnetic particles/suspending liquids **A17:** 100-102
magnetic printing **A17:** 125-126
magnetic rubber inspection **A17:** 122-125
magnetizing current **A17:** 91-92
nomenclature of **A17:** 103
nonrelevant indications **A17:** 105-108
of open-die forgings **A17:** 494-495
oxyfuel gas welding **A6:** 289
permeability of magnetic materials **A17:** 99

Magnetic particle inspection (continued)
of powder forged parts **A14:** 204
powder used **A7:** 573
power sources **A17:** 92-93
procedures for **A17:** 108-111
process qualification **A17:** 678
proprietary methods **A17:** 122-128
of resistance-welded steel tubing **A17:** 565
of seamless pipe **A17:** 579
of seamless steel tubular products **A17:** 571
of steel bar and wire **A17:** 550
of steel forgings **A17:** 495
steel wire rod **M1:** 257
suspending liquids for **A7:** 578
to detect subsurface slag inclusions **A6:** 1074
to detect weld cracks **A6:** 1076
to detect weld discontinuities, electroslag welding **A6:** 1078
to detect weld metal and base metal cracks **A6:** 1075
ultraviolet light **A17:** 102-103
vs. ultrasonic inspection, for primary mill products **A17:** 267
of welded chain links **A17:** 116
of weldments **A17:** 114-115, 591-592
wet technique, for forgings **A17:** 500
Magnetic particle inspection residues
removal of **M5:** 53-54
Magnetic particle(s) **A7:** 577-578
applications **A7:** 577-578
buildup as defect detection **A7:** 577
separation in encapsulation hot isostatic pressing **A7:** 431
Magnetic particles *See also* Particle(s); Powder(s)
in a colloidal suspension for magnetic etching **A9:** 63-64
application **A17:** 89
digital image processing **A17:** 119
dry **A17:** 100-101
magnetic properties **A17:** 100
in oil suspending liquid **A17:** 101
shape, effects of **A17:** 100
size, effects of **A17:** 100
types **A17:** 100-101
visibility and contrast **A17:** 100
for visual inspection **A17:** 3
in water suspending liquid **A17:** 101-102
wet **A17:** 101
wet bath, strength of **A17:** 102
Magnetic permeability
cast copper alloys **A2:** 356-391
in eddy current inspection **A17:** 164, 167-168
effect of carburization on **A11:** 272
in microwave inspection **A17:** 203
Magnetic perturbation methods *See* Magnetic field testing
Magnetic phase diagrams
ferromagnetic resonance **A10:** 268
Magnetic polarization *See* intrinsic induction; Permanent magnet materials
Magnetic printing *See also* Magnetic particle inspection
applications **A17:** 125-126
brazed honeycomb panels **A17:** 126
for elastic/plastic deformation detection **A17:** 126
procedure **A17:** 125
Magnetic properties
actinide metals **A2:** 1189-1198
alloying additions, effect **A2:** 762
amorphous materials and metallic glasses **A2:** 815-816
anisotropic Alnico alloys **A15:** 739
annealed carbon steel **M1:** 150
of austenitic manganese steel **A1:** 837
cartridge brass **A2:** 302
cast copper alloys **A2:** 356-391
cast steels **M1:** 399
cemented carbides **A2:** 957-958
change, as source, nonrelevant indications **A17:** 105
cobalt and rare-earth permanent magnet materials **A2:** 788
of ductile iron **A1:** 51-52 **A15:** 663 **M1:** 53-54
effect of crystallographic texture on **A9:** 700-701
effect of impurities **A2:** 762
gilding metal **A2:** 295
Magnetic properties (continued)
of gray iron **A1:** 31 **M1:** 31
of high-alloy castings **A1:** 929
HSM copper **A2:** 294
of Invar **A2:** 891
of iron-cobalt alloys **A2:** 775-776
magnetic characterization **A17:** 131-132
malleable iron **M1:** 31
maraging steels **M1:** 451
nickel plating **M3:** 182
nominal, permanent magnet materials **A2:** 792
pure cobalt **A2:** 447
pure metals **A2:** 1100-1178
rare earth metals **A2:** 1178-1189
red brass **A2:** 299
of steel castings **A1:** 374
and stress, in electromagnetic techniques **A17:** 159
and superconductivity **A2:** 1035-1036
transplutonium actinide metals **A2:** 1200
white cast irons **M1:** 31
wrought aluminum and aluminum alloys **A2:** 84, 87
Magnetic pulse forming *See* Electromagnetic forming
Magnetic refrigerants
as rare earth application **A2:** 730-731
Magnetic resonance *See also* Electron spin resonance; Resonance methods
defined **A10:** 676
field **A10:** 690
linewidth, symbol for **A10:** 692
principles of **A10:** 254
Magnetic resonance imaging (MRI)
application **A2:** 1027
with niobium-titanium superconducting materials **A2:** 1054-1055
Magnetic resonance sensors
for leakage field testing **A17:** 131
Magnetic rubber inspection *See also* Magnetic particle inspection **A17:** 122-125
advantages/limitations **A17:** 122
for areas of limited visual accessibility **A17:** 123
of castings **A17:** 525
of difficult-to-inspect shapes or sizes **A17:** 124-125
for fatigue test monitoring **A17:** 125
indications, magnification of **A17:** 125
on coated surfaces **A17:** 123-124
procedure **A17:** 122-123
surface evaluation by **A17:** 125
Magnetic saturation
nickel-iron alloys **A2:** 770
Magnetic seal
defined **A18:** 12
Magnetic sector mass spectrometers **A7:** 258
Magnetic sensor housing applications
beryllium-copper alloys **A2:** 418-419
Magnetic separation **A1:** 1026
with niobium-titanium superconducting materials **A2:** 1057
of seeds **A7:** 589-592
Magnetic separator
for sand **A15:** 32
schematic **A7:** 589
Magnetic shielding
defined **A9:** 11
of magnetically soft materials **A2:** 761
Magnetic shields
powders used **A7:** 572
Magnetic spectrometer, and detector
EELS analysis **A10:** 435
Magnetic steels
sintering **A7:** 340
Magnetic structure
determined by neutron diffraction **A10:** 420
refined by Rietveld method **A10:** 423
Magnetic susceptibility
cast copper alloys **A2:** 359
low, beryllium-copper alloys **A2:** 418-419
Magnetic switching
use in spark source mass spectrometry **A10:** 144
Magnetic tape recorders
as eddy current inspection readout **A17:** 179
Magnetic temperature compensation
alloys for **A2:** 773-774
Magnetic test
quenching media **M4:** 61-62
Magnetic testing **A1:** 1029-1030
Magnetic testing methods
magnetically soft materials **A2:** 763-778
Magnetic thickness gages
paint dry film thickness tests **M5:** 491
Magnetic writing
as nonrelevant indication source **A17:** 106
Magnetic-fixed probes
for film measurement **A13:** 417
Magnetic-force percussion welding **M6:** 740, 745
applications **M6:** 745
arc starters **M6:** 745
arc time **M6:** 745
weld areas **M6:** 745
welding force **M6:** 745
Magnetic-particle inspection **A1:** 274-275 **M6:** 848
advantages and limitations **A11:** 16-17
applications **M6:** 827
codes, standards, and specifications **M6:** 827
effect on fatigue fracture **A11:** 129
effect on fatigue strength **A11:** 128
of electrogas welds **M6:** 244
of electroslag welds **M6:** 235
fluorescent **A11:** 658
as nondestructive fatigue test **A11:** 134
of rocket-motor case **A11:** 96
of slag inclusions **A11:** 322
surface effects **A12:** 77
Magnetic-particle testing (MPT) **A6:** 1081, 1082, 1083, 1085, 1086, 1087, 1088
resistance seam welds **A6:** 245
Magnetically induced velocity changes (MIVC)
for residual stress measurement **A17:** 161-162
for ultrasonic waves **A17:** 161-162
Magnetically soft materials *See also* Soft magnetic alloys **A14:** 476-477
alloy classifications **A2:** 763-778
alloy selection, for power generation applications **A2:** 778-780
alloying additions **A2:** 762
corrosion resistance **A2:** 778
defined **A2:** 761
demagnetization resistance **A2:** 784
ferromagnetic properties **A2:** 761-763
grain size, maximizing **A2:** 763
heat treatment effects **A2:** 762-763
high-purity iron **A2:** 764-765
impurity effects **A2:** 762-763
iron-aluminum alloys **A2:** 769-770
iron-cobalt alloys **A2:** 774-776
low-carbon steels **A2:** 765-766
magnetic cores, design and fabrication **A2:** 780
magnetic testing methods **A2:** 763-778
motors and generators **A2:** 779
nickel alloy **A2:** 433
nickel-iron alloys **A2:** 770-774
and permanent magnet materials, compared **A2:** 784
residual stress, minimizing **A2:** 763
silicon iron bar and heavy strip **A2:** 769
silicon steels (flat-rolled products) **A2:** 766-769
stainless steels **A2:** 776-778 **M3:** 605, 607, 609, 610
stress effects **M3:** 609
temperature stability **M3:** 607, 608
transformers **A2:** 779-780
Magnetism
conversion factors **A8:** 722 **A10:** 686
effect in fatigue fracture **A11:** 129-130
in electrozone size analysis **A7:** 221
fundamental concepts **M3:** 615-618
fundamentals of **A2:** 782, 784
magabsorption measurement **A17:** 156-157
and permeability **A17:** 96
residual, defined **A17:** 100
Magnetite
for base metal conductors **EM4:** 1142
chemical composition **A6:** 60
defined **A9:** 11
electrical properties **EM4:** 765
flame emission sources for **A10:** 29-30
green hydrated or black anhydrous **A11:** 632
Miller numbers **A18:** 235
poor properties limiting its applications **EM4:** 1161

Magnetite (continued)
produced by spinel ferrite compositions ... **EM4:** 1162
as silica sand impurity ... **A15:** 208
specific properties imparted in CTV tubes ... **EM4:** 1040
Magnetite powder ... **A9:** 534
Hoeganaes process ... **A7:** 81
Magnetization *See also* Demagnetization; Magnetic properties; Magnetism; Magnets; Permanent magnet materials ... **A7:** 575-577
abbreviation for ... **A10:** 690
alternating and direct current ... **A7:** 576
circular ... **A17:** 90-91
curve, hysteresis ... **A17:** 99
curves ... **A2:** 787-788
curves, pure iron and nickel ... **A17:** 168
determined ... **A10:** 271-272
direct, in leakage field testing ... **A17:** 130
direction of ... **A17:** 110-111, 129
domain theory ... **A17:** 131-132
effective ... **A10:** 275, 690
equipment ... **A7:** 575-577
as ferromagnetic resonance application ... **A10:** 271-272
FMR measurement of ... **A10:** 267
head-shot ... **A17:** 99
indirect, in leakage field testing ... **A17:** 130
irreversible changes, permanent magnet materials ... **A2:** 795-797
longitudinal ... **A7:** 576 **A17:** 91, 130
in magnetic painting ... **A17:** 127
for magnetic rubber inspection ... **A17:** 123
methods ... **A17:** 91, 94, 110, 130
multidirectional ... **A17:** 93
net nuclear ... **A10:** 280
optimum ... **A7:** 575
optimum, leakage field testing ... **A17:** 130
prior to use, permanent magnet materials ... **A2:** 802
reversible changes, permanent magnet materials ... **A2:** 797-800
saturation ... **A17:** 131-132
saturation, as function of concentration ... **A10:** 272
temperature dependence ... **A10:** 273
temperature effects ... **A2:** 798-801
total, defined ... **A2:** 761
yoke ... **A17:** 130
Magnetized bar
defined ... **A17:** 90
Magnetized ring
defined ... **A17:** 90
Magnetizing current
alternating current ... **A17:** 92
direct current ... **A17:** 91-92
in magnetic particle inspection ... **A17:** 91-92
Magnetizing force
excessive ... **A17:** 105
Magneto-optical Faraday effect
observation of magnetic domains ... **A9:** 535
Magneto-optical Kerr effect
observation of magnetic domains ... **A9:** 535-536
Magnetoabsorption
as term ... **A17:** 144
Magnetocrystalline aniosotropy
FMR study of ... **A10:** 267
Magnetoelastic effect
defined ... **A17:** 159
Magnetographic sensors
for leakage field testing ... **A17:** 131
Magnetohydrodynamic (MHD) power generation
with niobium-titanium superconduction materials ... **A2:** 1057
Magnetohydrodynamic casting
as semisolid metalworking ... **A15:** 331
and SIMA process, compared ... **A15:** 332
Magnetohydrodynamic lubrication
defined ... **A18:** 12
Magnetometer
flux gate ... **A17:** 131
Magnetometers
defined ... **A10:** 676
vibrating sample, abbreviation for ... **A10:** 691
Magnetomotive force
electron-beam welding ... **A6:** 42
Magnetomotive forces
hysteresis loops under ... **A17:** 147
Magneton
defined ... **A10:** 676
Magnetooptical materials
as rare earth application ... **A2:** 731
Magnetoplumbites ... **EM4:** 18
Magnetoresistance
of copper in niobium-titanium superconducting materials ... **A2:** 1045
defined ... **A17:** 144
Magnetoresistive sensors
for leakage field testing ... **A17:** 131
Magnetostatic modes
FMR eddy current probes ... **A17:** 220
Magnetostatics
equivalent physical quantities ... **EM1:** 191
Magnetostriction
constants, effects in magabsorption measurement ... **A17:** 154
defined ... **A2:** 761 **A10:** 676
transducers, ultrasonic inspection ... **A17:** 256
Magnetostrictive alloy
SAM analyzed ... **A10:** 510
Magnetostrictive cavitation test device
defined ... **A18:** 12
Magnetostrictive methods
capabilities of ... **A10:** 380
Magnetron deposition ... **A18:** 841, 842
Magnetron sputtering systems ... **M5:** 413-415
Magnets *See also* Permanent magnet materials ... **A7:** 624-644
Chromidur ductile permanent, FIM/AP analysis of ... **A10:** 598-599
commercial designations and suppliers ... **A2:** 783
commercial, with A15 superconductors ... **A2:** 1070
directional solidification used to develop properties ... **A9:** 701
in gas mass spectrometers ... **A10:** 153-154
mass analyzer ... **A10:** 153-154
permanent ... **A7:** 17
rare earth applications ... **A2:** 729-731
saturation ... **A10:** 255
superconducting ... **A2:** 1027-1029, 1054-1057
technology ... **A2:** 1028
in Zeeman-corrected spectrometer ... **A10:** 52
Magnification
abbreviation for ... **A10:** 690
of an optical microscope ... **A9:** 74-76
at the film plane for photomicroscopy ... **A9:** 84
borescopes ... **A17:** 9
camera, detenning ... **A12:** 80
defined ... **A9:** 11 **A10:** 676 **A11:** 6
defined, stereoscopic methods ... **A12:** 196
effect, image analysis ... **A10:** 313-314
effect on inclusion volume fraction, image analysis ... **A10:** 314
error ... **A12:** 196
field ion microscope ... **A10:** 588
and focusing, photomacrographic ... **A12:** 79-80
of indications, by magnetic rubber inspection ... **A17:** 125
in macrophotography ... **A9:** 87
miniborescopes ... **A17:** 4
projective, with microfocus x-ray sources radiography ... **A17:** 300
rigid borescopes ... **A17:** 4
of scanning electron microscopes ... **A9:** 89
in SEM illuminating/imaging system ... **A12:** 167
symbol for ... **A10:** 692
Magnification system of a transmission electron microscope ... **A9:** 103
Magnifiers
in visual leak testing ... **A17:** 66
Magnitude of strain
as x-ray diffraction analysis ... **A10:** 325
Mahogany
as pattern material ... **A15:** 194
Mail handling equipment
P/M parts for ... **A7:** 667
Main bearing
defined ... **A18:** 12
Main effects
as variables ... **A17:** 745-750
Main hoist shaft
fatigue cracking of ... **A11:** 525
Main landing-gear deflection yoke
SCC failure of ... **A11:** 23
Main metal zinc alloy
gravity castings ... **A2:** 530
Main roll *See* Radial roll
Main steam line
cracks oriented to hoop stress direction in ... **A11:** 669
failure by thermal fatigue ... **A11:** 668-669
of power plant, cross section through weld failure ... **A11:** 668
of power-generating station, failure of ... **A11:** 667-668
Main-bearing journals, crankshaft
micrographs ... **A11:** 358-359
Mainframe computers *See also* Computers
multichip assemblies in ... **EL1:** 298
thermal conductivity ... **EL1:** 308
Maintainability *See also* Maintenance; Repair
as material selection parameter ... **EM1:** 39
Maintenance *See also* Maintainability; Repair
of Brinell hardness testers ... **A8:** 88
chemicals, hydrogen embrittlement testing for ... **A8:** 541
costs, of composites ... **EM1:** 259
costs, structural, probabilistic fracture mechanics for predicting ... **A8:** 682
Maintenance, improper
and steam equipment failure ... **A11:** 603
Major component analysis
analytical transmission electron microscopy ... **A10:** 429-489
atomic absorption spectrometry ... **A10:** 43-59
Auger electron spectroscopy ... **A10:** 549-567
classical wet analytical chemistry ... **A10:** 161-180
controlled-potential coulometry ... **A10:** 202, 207-211
electrochemical analysis ... **A10:** 181-211
electrogravimetry ... **A10:** 197-201
electrometric titration ... **A10:** 202-206
electron probe x-ray microanalysis ... **A10:** 516-535
electron spin resonance ... **A10:** 253-266
elemental and functional group analysis ... **A10:** 212-220
field ion microscopy ... **A10:** 583-602
gas analysis by mass spectroscopy ... **A10:** 151-157
gas chromatography/mass spectrometry ... **A10:** 639-648
inductively coupled plasma atomic emission spectroscopy ... **A10:** 31-42
infrared spectroscopy ... **A10:** 109-125
of inorganics ... **A10:** 4-6, 7, 8
ion chromatography ... **A10:** 658-667
liquid chromatography ... **A10:** 649-657
low-energy ion-scattering spectroscopy ... **A10:** 603-609
molecular fluorescence spectrometry ... **A10:** 72-81
Mössbauer spectroscopy ... **A10:** 287-295
neutron diffraction ... **A10:** 420-426
optical emission spectroscopy ... **A10:** 21-30
of organics ... **A10:** 9, 10
particle-induced x-ray emission ... **A10:** 102-108
potentiometric membrane electrodes ... **A10:** 181-187

Major component analysis (continued)
Raman spectroscopy ... **A10:** 126-138
Rutherford backscattering spectrometry ... **A10:** 628-636
scanning electron microscopy ... **A10:** 490-515
secondary ion mass spectroscopy ... **A10:** 610-627
spark source mass spectrometry ... **A10:** 141-150
ultraviolet/visible absorption spectroscopy ... **A10:** 60-71
voltammetry ... **A10:** 188-196
x-ray diffraction ... **A10:** 325-332
x-ray photoelectron spectroscopy ... **A10:** 568-580
x-ray powder diffraction ... **A10:** 333-343
x-ray spectrometry ... **A10:** 82-101
Major mismatches, as term *See also* Coefficient of thermal expansion; CTE-matched materials ... **EL1:** 958
Make-break contacts *See also* Electrical contact materials
arcing, property requirements for ... **A2:** 841
failure modes of ... **A2:** 840-841
power circuits, recommended materials ... **A2:** 863
properties of composites for ... **A2:** 851-853
Maleimide-terminated thermosetting polymers ... **EM2:** 631
Malignant neoplastic lesions
as metal powder toxic reaction ... **A7:** 201
Malleability *See also* Ductility ... **A8:** 8
of copper ... **A15:** 26-27
defined ... **A11:** 6
of electrical contact materials ... **A2:** 841
lead and lead alloys ... **A2:** 545
Malleable cast iron ... **M1:** 3, 9
annealing ... **M1:** 57-63
applications ... **M1:** 57, 63, 70, 71, 73
brazing ... **M1:** 66-67, 71
bull's-eye structure ... **M1:** 60, 61
composition ... **M1:** 58
corrosion resistance ... **M1:** 66
damping capacity ... **M1:** 32
density ... **M1:** 67
dimensional tolerances ... **M1:** 67
ductile iron, compared to ... **M1:** 57
elevated-temperature properties ... **M1:** 65-66, 70-72
fatigue resistance ... **M1:** 65, 66, 70
general characteristics ... **M1:** 57-58, 64, 67-68
hardenability ... **M1:** 71-73
heat treatment ... **M1:** 57-63, 68-71, 73
impact properties ... **M1:** 65, 70-71
machinability ... **M1:** 55, 63, 66, 67, 71
magnetic properties ... **M1:** 31
manganese content, effects of ... **M1:** 58, 64, 73
mechanical properties ... **M1:** 64-73
microstructure ... **M1:** 9, 58-63, 66
mottle, control of ... **M1:** 58
nodule count, control of ... **M1:** 59-60
oxidation resistance ... **M1:** 46
patternmakers' rules for ... **M1:** 33
physical properties ... **M1:** 67
production method ... **M1:** 9
shrinkage allowance ... **M1:** 67
specifications ... **M1:** 63-64, 68
sprocket, pearlitic, wear of ... **M1:** 628
surface hardening ... **M1:** 73
uses ... **M1:** 9
wear resistance ... **M1:** 71
welding ... **M1:** 66-67, 71
white iron, conversion from ... **M1:** 57-59
Malleable cast irons
unalloyed ... **A13:** 567
Malleable iron *See also* Cast iron; Cast irons; Ferritic malleable iron; Pearlitic malleable iron; White iron ... **A1:** 9-11, 71-84 **A9:** 245
air-carbon arc cutting ... **A6:** 1172, 1176
alloying elements ... **A1:** 10, 71-72
American blackheart, development ... **A15:** 30-31
annealing of ... **A1:** 71, 72-73
application, internal combustion engine parts ... **A18:** 553, 556, 557
application, piston ring materials ... **A18:** 557
applications ... **A1:** 73, 74, 83, 84 **A15:** 690-691
arc welding ... **M6:** 315-316
arc welding of ... **A15:** 528
blackheart malleable iron ... **A1:** 74
brazeability ... **M6:** 996
castings, inspection of ... **A15:** 556
Malleable iron (continued)
castings, markets for ... **A15:** 42, 44
castings, properties ... **A15:** 693
chemical composition ... **A6:** 906
compared to ductile iron ... **A1:** 71
composition limits ... **M6:** 309
composition of ... **A1:** 5, 9-10, 71-72
control of mottle ... **A1:** 71
control of nodule count ... **A1:** 73-74
cooling rate of ... **A1:** 10
damping capacity ... **A1:** 82, 84
defined ... **A15:** 8
development of ... **A15:** 30-31
dip brazing ... **A6:** 338
ferritic ... **A15:** 691-693
ferritic malleable iron ... **A1:** 75-76
alloying elements ... **A1:** 75
corrosion resistance ... **A1:** 76
fatigue limit ... **A1:** 75
fracture toughness ... **A1:** 75-76, 80
graphite content ... **A1:** 75
heat treatment ... **A1:** 75
microstructure ... **A1:** 72
modulus of elasticity ... **A1:** 75
stress-rupture plot ... **A1:** 75
tensile properties ... **A1:** 73, 75
welding and brazing of ... **A1:** 76
gear materials, surface treatment and minimum surface hardness ... **A18:** 261
grades of ... **A1:** 73, 74
heat treatment of ... **A1:** 10-11, 75, 76-80
liquid treatment of ... **A1:** 10
martensitic ... **A15:** 693-697
melting practices ... **A1:** 72 **A15:** 686-687
metallurgical factors ... **A1:** 71
microstructure ... **A1:** 72, 76, 77 **A15:** 687-690 **A18:** 695, 700-701
oxyacetylene braze welding ... **M6:** 596-598
oxyacetylene welding ... **M6:** 604-605
oxyacetylene welding of ... **A15:** 531
pearlitic ... **A15:** 693-697
pearlitic-martensitic malleable irons ... **A1:** 76-84
Charpy V-notch impact energy ... **A1:** 81
compressive strength ... **A1:** 82
fatigue properties ... **A1:** 81
fracture toughness ... **A1:** 74, 80, 82
hardness ... **A1:** 80-82
heat treatment ... **A1:** 76-80
mechanical properties at elevated temperatures ... **A1:** 80, 82
microstructure ... **A1:** 76, 77
modulus of elasticity ... **A1:** 82
shear strength ... **A1:** 82
stress-rupture plot ... **A1:** 81
tempering times ... **A1:** 80
tensile properties ... **A1:** 73, 78, 79, 82
torsional strength ... **A1:** 11-80, 82
unnotched fatigue limits ... **A1:** 81, 83
wear resistance ... **A1:** 83, 84
welding and brazing of ... **A1:** 83-84
properties ... **A18:** 695
properties of ... **A1:** 73, 74-83
solidification of ... **A1:** 72
thermal expansion coefficient ... **A6:** 907
types of ... **A1:** 74-75
welding metallurgy ... **A15:** 520-521 **M6:** 308
whiteheart malleable iron ... **A1:** 74
Malleable iron castings
inspection of ... **A17:** 531-532
Malleable iron, heat treating
annealing ... **M4:** 552
bainitic ... **M4:** 554
hardening ... **M4:** 552-553, 554
hardness, ferritic ... **M4:** 555
hardness, pearlitic ... **M4:** 555
martempering ... **M4:** 553
tempering ... **M4:** 553, 554, 555
Malleable Iron Research Institute ... **A15:** 34
Malleable iron rocker lever
failed ... **A11:** 350-352
Malleable irons *See also* Cast irons
boring ... **A16:** 167, 655
broaching ... **A16:** 206, 656
centerless grinding ... **A16:** 665
corrosion of ... **A11:** 200, 350-352
counterboring ... **A16:** 660
cutting fluid effect on tool life ... **A16:** 652
Malleable irons (continued)
cylindrical grinding ... **A16:** 664
drilling ... **A16:** 229, 230, 658
dry machining ... **A16:** 392
end milling ... **A16:** 663
face milling ... **A16:** 662
fractographs ... **A12:** 238-239
fracture/failure causes illustrated ... **A12:** 216
honing ... **A16:** 477
internal grinding ... **A16:** 665
milling ... **A16:** 97, 327
planing ... **A16:** 657, 660
reaming ... **A16:** 248, 659
spotfacing ... **A16:** 660
surface grinding ... **A16:** 664
tapping ... **A16:** 263, 265, 661
turning ... **A16:** 94, 653, 654
turret and engine lathes for machining ... **A16:** 383
Malleable irons, heat treating ... **A4:** 693-696
annealing ... **A4:** 693, 694, 695
bainitic ... **A4:** 695
examples ... **A4:** 695-696
first-stage graphitization (FSG) ... **A4:** 693
hardening ... **A4:** 693, 694-696
hardness, ferritic ... **A4:** 694-695, 696
hardness, pearlitic ... **A4:** 694-696
manufacture with charcoal-based atmospheres ... **A4:** 562, 563
martempering ... **A4:** 695
second-stage graphitization (SSG) ... **A4:** 693
tempering ... **A4:** 693, 694-695, 696
Malleable irons, specific grades
32510, machining ... **A16:** 362, 649, 650, 653-656, 658-663
35018, machining ... **A16:** 362, 653-663
40010, machining ... **A16:** 653-659, 661, 662, 664
45006, machining ... **A16:** 653-659, 661, 662, 664
45008, machining ... **A16:** 653-659, 661, 662, 664
48004, matrix microstructure effect on tool life ... **A16:** 650
50005, machining ... **A16:** 2, 664
53004, contour band sawing ... **A16:** 362
60003, contour band sawing ... **A16:** 362
60003, matrix microstructure effect on tool life ... **A16:** 650
ASTM 80002, matrix microstructure effect on tool life ... **A16:** 650
M3210, machining ... **A16:** 362, 653-663
M4504, machining ... **A16:** 653-659, 661, 662, 664
M5003, machining ... **A16:** 653-659, 661, 662, 664
Malleable irons, specific types
32510 grade
composition ... **A4:** 696
hardness ... **A4:** 696
45007 grade
composition ... **A4:** 696
hardness ... **A4:** 696
45010 grade
composition ... **A4:** 696
hardness ... **A4:** 695-696
60003 grade
composition ... **A4:** 696
hardness ... **A4:** 696
80002 grade
composition ... **A4:** 696
hardness ... **A4:** 695, 696
Malleable irons, speciric types
ASTM A47 grade 32510, fracture sequence ... **A12:** 238
ASTM A220 grade 50005, microcracking ... **A12:** 239
Malleable/nodular iron, application
piston ring materials ... **A18:** 557
Malleableizing
defined ... **A9:** 11
Malleablizing *See also* White iron
defined ... **A15:** 8
Management *See also* Human factors; Personnel; Safety
functions, in quality control ... **A17:** 719-720
NDE reliability system ... **A17:** 674-675
role, NDE reliability ... **A17:** 680
Mandelbrot fractal dimension ... **A12:** 213
Mandelic acid
as narrow-range precipitant ... **A10:** 169
Mandrel ... **A7:** 7
defined ... **EM1:** 15

Mandrel (continued)
and elastomeric bag, for titanium alloys **A7:** 750
elastomeric, design/fabrication **EM1:** 593-594
in filament winding **EM1:** 135
preparation, filament winding **EM1:** 505-507
removal, filament winding **EM1:** 507
stress-relaxation bend test **A8:** 326
for tube rolling **EM1:** 573
-type wipe bending device **A8:** 125
use, in electroformed nickel tooling ... **EM1:** 582-584

Mandrel forging *See also* Forging **A14:** 70-71
of aluminum alloy **A14:** 244
defined **A14:** 8
shapes of **A14:** 61
wrought aluminum alloy **A2:** 34

Mandrel swaging
and drilling, combined **A14:** 141
of tubes **A14:** 137-138

Mandrel test
hard chromium plating bath **M5:** 174-175, 183

Mandrels *See* Tube drawing dies **EM2:** 26, 374-375
ball **A14:** 667
for bending **A14:** 666
carburization cracking **A11:** 576, 583
defined **A14:** 8
effect on machine capacity **A14:** 137-138
fluted, for gun barrel bore **A14:** 138-139
full-length, for tube swaging **A14:** 137
lubricants for **A14:** 140
for manual spinning **A14:** 599-600
materials for **A14:** 668
moving, drawing with **A14:** 331, 335-336
plug and formed **A14:** 667
for power spinning of cones **A14:** 602
radial forging over **A14:** 146
tolerances, swaging bar/tube **A14:** 140
tube bending with **A14:** 666-668
tube bending without **A14:** 668
for tube spinning **A14:** 676-677
for tube swaging **A14:** 137, 138
tube swaging with **A14:** 137-139
tube swaging without **A14:** 134-137
types using blind fasteners **A11:** 546

Manganese *See also* Ferromanganese; Manganese cast steels; Manganese steels
as a beta stabilizer in titanium alloys **A9:** 458
as a carbide former in steel **A9:** 178
in active fluxes for submerged arc welding **A6:** 204
added to reduce solubility product in austenite **A4:** 245
addition affecting stainless steel machinability **A16:** 689, 690
as addition to aluminum alloys **A4:** 843
as addition to austenitic manganese steel castings **A9:** 239
addition to ductile cast iron **A6:** 709
addition to low-alloy steels for pressure vessels and piping **A6:** 667
as addition to nickel-iron alloys for electrodes **A6:** 717-718, 719
additions **A8:** 487, 489
in alloy cast irons **A1:** 87
alloying addition to heat-treatable aluminum alloys **A6:** 528
alloying, aluminum casting alloys **A2:** 132
alloying effect in titanium alloys **A6:** 508
alloying effect on nickel-base alloys **A6:** 589, 590
alloying effects on copper alloys **M6:** 402
alloying effects, stainless steels **A13:** 550
alloying in aluminum alloys **M6:** 373
alloying, wrought aluminum alloy **A2:** 53-54
in aluminum alloys **A15:** 746
in aluminum-silicon alloys **A18:** 786, 788
as an addition to low-carbon electrical steels **A9:** 537
as an addition to nickel-iron alloys **A9:** 538

Manganese (continued)
as an addition to permanent magnet alloys **A9:** 538-539
as an austenite-stabilizing element ',n wrought stainless steels **A9:** 283
as an austenite-stabilizing element in steel **A9:** 177-178
at elevated-temperature service **A1:** 640
as austenite stabilizer **A13:** 47
in austenitic manganese steel **A1:** 822-824, 825
in austenitic stainless steels **A6:** 457, 458, 461, 462, 463
availability of **A1:** 1021
biologic effects and toxicity **A2:** 1252-1253
in cast iron **A1:** 5, 28
cause of temper embrittlement **A4:** 135
in chilled iron **A15:** 30
in compacted graphite iron **A1:** 59
complexation titration for **A10:** 174
in composition, effect on ductile iron **A4:** 686, 689
in composition, effect on flame hardening **A4:** 277
in composition, effect on gray irons **A4:** 670, 671, 672, 673, 678
content effect on solidification cracking **A6:** 90
content in carbon steels and alloy steels **A16:** 670, 672
content in heat-treatable low-alloy (HTLA) steels **A6:** 670
content in HSLA Q&T steels **A6:** 665
content in stainless steels **M6:** 320, 322
content in tool and die steels **A6:** 674
content in ultrahigh-strength low-alloy steels **A6:** 673
content loss in electrodes after rebaking **A6:** 415
content of weld deposits **A6:** 675
in copper alloys **A6:** 753
-copper alloys, isolation of manganese in **A10:** 174
in cupolas **A15:** 388, 390
deoxidizing, copper and copper alloys **A2:** 236
determined by controlled-potential coulometry **A10:** 209
determined in stainless steel **A10:** 146
distribution in pearlite **A9:** 661
in ductile iron **A1:** 43 **A15:** 648, 649
effect of, on hardenability **A1:** 393, 394, 395
effect of, on machinability of carbon steels **A1:** 598
effect of, on notch toughness **A1:** 740
effect on base metal color matching in aluminum alloys **A6:** 730
effect on borided steels **A4:** 441
effect on Curie point **A4:** 187
effect on hardness of tempered martensite **A4:** 124, 128-129
effect on maraging steels **A4:** 222, 224
effect on oxidation resistance in cobalt-base heat-resistant casting alloys **A9:** 334
effect on sigma formation in ferritic stainless steels **A9:** 285
effect, shape memory effect (SME) alloys **A2:** 900
effect, ternary iron-base alloys **A15:** 65
electroslag welding, reactions **A6:** 273, 274, 278
in enamel cover coats **EM3:** 304
enameling ground coat **EM3:** 304
erosion resistance **A18:** 228
as essential metal **A2:** 1250, 1252-1253
evaporation fields for **A10:** 587
in ferrite **A1:** 402, 406
flux composition effect **A6:** 57, 58
formation of intermetallic phases in austenitic stainless steels **A9:** 284
functions in FCAW electrodes **A6:** 188

Manganese (continued)
gamma spectrum radionuclide **A18:** 326
in hardfacing alloys **A18:** 759, 760
in heat-resistant alloys **A4:** 510
in high-alloy white irons **A15:** 680
high-Mn steels, drilling **A16:** 220, 221, 229
ICP-determined in plant tissues **A10:** 41
induction hardening cracking tendency **A4:** 202
inorganic fluxes for **A6:** 980
interference with copper **A10:** 201
in iron-base alloys, flame AAS analysis for **A10:** 56
isolation of ferromanganese in **A10:** 174
isotope composition and intensity **A10:** 146
in laser cladding material **A18:** 867
in limestone **EM4:** 379
loss effect on welding parameters **A6:** 68
in low-carbon steel forgings double normalized **A4:** 39
in magnesium alloys **A9:** 427
in malleable iron **A1:** 10
malleable iron content composition limits **A4:** 694
microalloying of **A14:** 220
in moly-manganese paste process **EM3:** 304
in nickel-chromium white irons **A15:** 680
oxygen cutting, effect on **M6:** 898
in P/M alloys **A1:** 810
photometric analysis methods **A10:** 64
pickup in submerged arc welding **M6:** 116
in powder metallurgy steels **A9:** 503
pure **M2:** 768
pure, properties **A2:** 1135
recovery from selected electrode covering **A6:** 60
redox titration **A10:** 175
relationship to hot cracking **M6:** 833
relationship to toughness **M6:** 42
roll welding **A6:** 314
segregation, in dual-phase steels **A10:** 483
species weighed in gravimetry **A10:** 172
spraying for hardfacing **M6:** 789
in stainless steels **A18:** 712, 714
in steel **A1:** 144, 576-577
in steel weldments **A6:** 416-417, 418, 419
in structural steels **A1:** 407
submerged arc welding **A6:** 206 **M6:** 115
effect on cracking **M6:** 128
promotion of acicular ferrite **M6:** 117
transfer due to flux content **M6:** 124-125
suitability for cladding combinations **M6:** 691
sulfur scavenging by **A15:** 18
TNAA detection limits **A10:** 237
to form simple and complex carbides **A16:** 667
to improve hardenability in carburized steels **A4:** 366
in tool steels **A18:** 734, 735-736
in toot steels **A16:** 52, 53
toxicity **A6:** 1195, 1196
use in flux cored electrodes **M6:** 103
vapor pressure **A4:** 493, 494
vapor pressure, relation to temperature **A4:** 495
volumetric procedures for **A10:** 175
wear resistance of austenitic steels related to content **A18:** 708
weighed as the phosphate **A10:** 171
weld-metal content, underwater welding **A6:** 1010-1011
in wrought stainless steels **A1:** 872
in zinc alloys **A15:** 788
in zirconium alloys, by periodate method **A10:** 69

Manganese acetate
effect on removal rate of PMMA from tape-cast films **EM4:** 137

Manganese alloys
weld overlay material **M6:** 807

Manganese brass
resistance spot welding ... **A6:** 850
Manganese bronze *See also* Copper casting alloys; High-strength manganese bronze; Leaded manganese bronze; Manganese toxicity
applications and properties ... **A2:** 225
foundry properties for sand casting ... **A2:** 348
freezing range ... **A2:** 348
as high-shrinkage foundry alloy ... **A2:** 346
high-strength, applications ... **A2:** 355
melt treatment ... **A15:** 774
nominal composition ... **A2:** 347
properties and applications ... **A2:** 367-370
recycling ... **A2:** 1214
shrinkage allowances ... **A15:** 303
Manganese bronze A
weldability ... **A6:** 753
Manganese bronzes
shielded metal arc welding ... **A6:** 755
Manganese cast steels ... **A15:** 32, 303, 535, 716
Manganese conversion coating
for shafts ... **A11:** 482
Manganese diffusion coating
for wear resistance ... **M1:** 635
Manganese dioxide
as colorant ... **EM4:** 380
deposited by potentiostatic etching ... **A9:** 146
sulfur dioxide removal in high-temperature combustion by ... **A10:** 222
Manganese dioxide (MnO_2)
XPS spectrum ... **A18:** 447
Manganese dioxide/pyrolusite (MnO_2)
purpose for use in glass manufacture ... **EM4:** 381
Manganese ferrite ... **EM4:** 59
Manganese in cast iron ... **M1:** 78
depth of chill ... **M1:** 77
ductile iron ... **M1:** 38, 41, 54
gray iron ... **M1:** 28
malleable iron ... **M1:** 58, 64, 73
Manganese in stainless steels ... **M3:** 57
Manganese in steel
castings ... **M1:** 384, 386-391, 394, 399
constructional steels for elevated temperature use ... **M1:** 647
distribution in steel plate ... **M1:** 189, 195-196
hardenability affected by ... **M1:** 477
hardenable steels ... **M1:** 456
hydrogen solubility, effect on ... **M1:** 687
machinability of resulfurized steels ... **M1:** 573
modified low-carbon steels ... **M1:** 161-162
notch toughness, improvement of ... **M1:** 194, 692 694, 697
sheet, effect on formability ... **M1:** 553-554
temper embrittlement, role in ... **M1:** 684
Manganese oxide
chemical analysis ... **A7:** 256
function and composition for mild steel SMAW electrode coatings ... **A6:** 60
inclusions in iron ... **A12:** 221
with iron oxide causing internal reflection ... **A9:** 184
with manganese sulfide tails ... **A9:** 184
metal-to-metal oxide equilibria ... **A7:** 340
Manganese oxysulfide
effect in low-carbon steel ... **A12:** 249
Manganese phosphate coating ... **M5:** 435-439, 441-444, 448, 450-451, 455-456
characteristics of ... **M5:** 435
crystal structure ... **M5:** 450-451, 455
equipment ... **M5:** 445, 448, 451, 453
immersion systems ... **M5:** 435, 438, 440
iron concentration and removal ... **M5:** 443
solution composition and operating conditions ... **M5:** 435, 443-445, 448, 451, 453
wear-resistance applications ... **M5:** 436-437
weight, coating ... **M5:** 435-437, 448, 451
Manganese phosphate coatings
corrosion protection by ... **M1:** 754
steel sheet ... **M1:** 174
Manganese phosphates
as conversion coating ... **A13:** 387
Manganese pneumonitis
from manganese toxicity ... **A2:** 1253
Manganese powders
brittle cathode process ... **A7:** 72
composition-depth profiles ... **A7:** 256
containing low alloy steel powder ... **A7:** 101
Manganese powders (continued)
metal-to-metal oxide equilibria ... **A7:** 340
pyrotechnic requirements ... **A7:** 601
requirements ... **A7:** 602
tap density ... **A7:** 277
Manganese selenides
in austenitic stainless steels ... **A9:** 284
in electrical steels ... **A9:** 537
in type 303 stainless steel ... **A9:** 291
Manganese silicate
flux viscosity, and weld surface pocking ... **A6:** 60
fluxes used for SAW applications ... **A6:** 62
Manganese stainless steels
inclusion content ... **A6:** 1018
Manganese steels *See also* Manganese; Manganese cast steels
abrasive wear ... **A18:** 190, 705
applications ... **A18:** 759
austenitic, properties ... **A18:** 759
fretting wear (1.5%) ... **A18:** 250
hardfacing alloys for ... **A6:** 790, 791
hardfacing material for mining and mineral industry applications (14 wt%) ... **A18:** 653
impact resistance and abrasion resistance properties ... **A18:** 759
shrinkage allowance ... **A15:** 303
wear resistance relation to toughness ... **A18:** 707
wear-resistant, weldability ... **A15:** 535
Manganese steels, specific types
16MnCr5 steel
carburized ... **A18:** 864
case-hardened ... **A18:** 864
nitrided ... **A18:** 864
20MnCr5, nominal compositions ... **A18:** 725
Manganese suicide
inclusions ... **A12:** 221, 263
Manganese sulfide
in austenitic manganese steel castings ... **A9:** 239
content in P/M materials ... **A16:** 884, 885, 886, 888, 889
in electrical steels ... **A9:** 537
in free-machining powder metallurgy steels ... **A9:** 510
in free-machining stainless steels ... **A16:** 685, 686
free-machining steel additive ... **A16:** 672, 673, 674, 675, 676, 677
as inclusion ... **A11:** 322
in steel ... **A9:** 625-626, 628
in type 303 stainless, steel ... **A9:** 291
in wrought stainless steels, sulfur printing to reveal ... **A9:** 279
Manganese sulfide, content
and ductility in hot torsion tests ... **A8:** 166
Manganese sulfide inclusions, in tool steels
effects of ... **A9:** 258
Manganese sulfide stringers
in 1213 steel ... **A9:** 627
Manganese sulfide tails
in manganese oxide ... **A9:** 184
Manganese sulfides
as inclusions ... **A15:** 92, 633
Manganese toxicity
chronic manganese poisoning ... **A2:** 1253
manganese pneumonitis ... **A2:** 1253
Manganese, vapor pressure
relation to temperature ... **M4:** 310
Manganese-aluminum bronze *See also* Cast copper alloys
properties and applications ... **A2:** 386
Manganese-aluminum compound in magnesium alloys ... **A9:** 427
Manganese-bismuth films
study by Kerr effect ... **A9:** 535
Manganese-bronze (high tensile)
filler metals ... **A6:** 756
relative solderability as a function of flux type ... **A6:** 129
Manganese-chromium-molybdenum-boron alloys
SCC resistance ... **A13:** 535
Manganese-copper alloys
martensitic structures ... **A9:** 673
Manganese-gold alloys
martensitic structures ... **A9:** 673
Manganese-modified zinc phosphate ... **A13:** 386
Manganese-molybdenum cast steels ... **A1:** 373
as low-alloy ... **A15:** 716
Manganese-molybdenum steel, flux-cored arc welding
designator ... **A6:** 189
Manganese-nickel-aluminum bronzes
shielded metal arc welding ... **A6:** 755
Manga-nese-nickel-chromium-molybdenum cast steels ... **A1:** 373
as low-alloy ... **A15:** 716
Manganese-nickel-chromium-molybdenum- niobium alloys
SCC resistance ... **A13:** 535
Manganese-zinc (MnZn) ferrite ... **EM4:** 18
applications ... **EM4:** 199
frequency ranges ... **EM4:** 1162
permeability ... **EM4:** 1162
precipitation process ... **EM4:** 59, 60
processing ... **EM4:** 1163
resistivity ... **EM4:** 1162
saturation flux density ... **EM4:** 1162
solid-state sintering ... **EM4:** 277-278
variation with temperature ... **EM4:** 1163, 1164
Manganese-zinc ferrites ... **A9:** 538
Manganin gages
for Hugoniot elastic limit measurement ... **A8:** 211
Manganins *See also* Electrical resistance alloys
properties and applications ... **A2:** 823, 825
Manganism
as chronic manganese poisoning ... **A2:** 1253
Manhattan-type routing patterns ... **EL1:** 76
Manifold
definition ... **M6:** 11
Manipulators ... **A14:** 8, 63
Manned spacecraft
atomic oxygen in low earth orbit effects ... **A13:** 1099-1100
case histories ... **A13:** 1075-1100
crevice corrosion ... **A13:** 1080-1082
filiform corrosion ... **A13:** 1076
fretting corrosion ... **A13:** 1082
galvanic corrosion ... **A13:** 1076-1079
high-temperature gaseous corrosion ... **A13:** 1092-1094
hydrogen embrittlement ... **A13:** 1087-1092
intergranular corrosion ... **A13:** 1079-1080
liquid-metal cracking ... **A13:** 1094-1097
mechanical systems of ... **A13:** 1072-1074
oxygen ignition in ... **A13:** 1092
pitting attack ... **A13:** 1075-1076
precipitation, corrosion products ... **A13:** 1097-1099
and space shuttle orbiter ... **A13:** 1058-1075
stress-corrosion cracking ... **A13:** 1082-1087
Mannesmann Effect
in open-die forging ... **A14:** 61
Mannesmann process
defined ... **A14:** 8
Mannich bases ... **EM3:** 95
Mannich products
dispersants ... **A18:** 99, 100
Manometer
Lea and Nurse permeability apparatus with ... **A7:** 264
Manometers
as leak detectors ... **A17:** 67
Manson-Haferd parameter *See* Time temperature parameters
extrapolation abilities ... **A8:** 335
time-temperature ... **A8:** 333-334
Manson-Succop parameter *See* Time temperature parameters
extrapolation abilities ... **A8:** 335
for time-temperature ... **A8:** 333-334
Manual arc welding ... **A18:** 644
Manual bending
machines, for bar ... **A14:** 661-662
of wire ... **A14:** 695-696
Manual brazing
definition ... **M6:** 11
Manual inspection
of printed circuits ... **EL1:** 127
Manual lay-up *See also* Hand lay-up ... **EM1:** 602-604
compacting ... **EM1:** 604
flat tape ... **EM1:** 624
mold release ... **EM1:** 602-603

Manual lay-up (continued)
orientation accuracy **EM1:** 603-604
ply count **EM1:** 603
ply flipping **EM1:** 602-603
resin removal **EM1:** 604
tape/fabric prepregs, compared **EM1:** 602
Manual metal arc (MMA) welding **A6:** 82
Manual oxygen cutting
definition **M6:** 11
Manual peening
as straightening **A14:** 681
Manual polishing *See* Hand polishing
Manual powder torch welding **A6:** 800, 801
for hardfacing **A7:** 834-835
Manual radiographic Rim processing
steps **A17:** 351-353
Manual soldering
flexible printed boards **EL1:** 590
Manual spinning *See also* Spinning **A14:** 559-600
applicability **A14:** 599
equipment **A14:** 599-600
of magnesium alloys **A14:** 828-829
practice **A14:** 600
stainless steels **A14:** 771-772
tools **A14:** 600
Manual straightening **A14:** 680-681
Manual torch brazing **A6:** 121
Manual transmission synchronizer gear and keys **A7:** 617, 619
Manual turret lathes **A16:** 370-371
Manual ultrasonic inspection
of boilers/pressure vessels **A17:** 649
Manual weighing and mixing **EM3:** 687-688
disadvantages **EM3:** 687-688
Manual welding
definition **A6:** 1211 **M6:** 11
Manufacturability *See also* Design for manufacturability (DFM); Manufacture and assembly; Manufacturing
design materials, and **EL1:** 1
of level 1 packages **EL1:** 403
of printed boards **EL1:** 608-609
Manufacture *See also* Assembly and manufacture; Die making; Fabrication
and assembly, design for **EL1:** 119-126
of blanks **A14:** 120
commercial, of glass-to-metal seal hybrid packages **EL1:** 458
die, for hot-die/isothermal forging **A14:** 154
forging, tasks of **A14:** 409-410
of superplastic metals **A14:** 867-868
Manufacture and assembly *See also* Manufacturability; Manufacturing
design for **EL1:** 119-126
design for manufacturability **EL1:** 120-125
early manufacturing involvement (EMI) **EL1:** 125
future directions **EL1:** 125-126
historical background **EL1:** 119-120
value engineering **EL1:** 120-121
Manufactured components and assemblies
boilers and related equipment, failures of **A11:** 602-627
bridge components, failures of **A11:** 707-714
gears, failures of **A11:** 586-601
heat exchangers, failures of **A11:** 628-642
lifting equipment, failures of **A11:** 514-528
locomotive axles, failures of **A11:** 715-727
mechanical fasteners, failures of **A11:** 529-549
metallic orthopedic implants, failures of **A11:** 670-694
pipelines, failures of **A11:** 695-706
pressure vessels, failures of **A11:** 643-669
rolling-element bearings, failures of **A11:** 490-513
shafts, failures of **A11:** 459-482
sliding bearings, failures of **A11:** 483-489
springs, failures of **A11:** 550-562
tools and dies, failures of **A11:** 563-585
Manufactured unit **EM3:** 18
Manufacturers, gallium arsenide ingot
wafer, devices **A2:** 748
Manufacturers, literature
as information source **EM2:** 92
Manufacturers' wire
annealed low-carbon steel **M1:** 264
Manufacturing *See also* Fabrication; In-process inspection; Manufacturability; Manufacture and assembly; Production
additive, rigid printed wiring boards **EL1:** 549
bottom-brazed flatpacks **EL1:** 993
capabilities, and material selection **EM1:** 38-39
competition, by quality design and control **A17:** 719
costs, of surface-mount components **EL1:** 730
defect and device failure analysis **EL1:** 917-918
design requirements for **EM1:** 182
early involvement (EMI) **EL1:** 125
effect, wiring design **EL1:** 581
effects, microvoids as **EL1:** 83
effects, of PWB structures **EL1:** 81-82
epoxy materials **EL1:** 831-836
of eutectic die attach **EL1:** 215
factors, high-frequency digital systems **EL1:** 82
first-level package **EL1:** 991
flexible printed boards **EL1:** 581-584
goals **EL1:** 632
of lead frame strip **EL1:** 483-484
limitations, high-frequency digital systems **EL1:** 80
management, functions, quality control **A17:** 719-720
perspective, of quality control **A17:** 719-720
phase, environmental stress screening **EL1:** 876-877
of polymer die attach **EL1:** 220
printed board **EL1:** 505, 539-540, 869-874
printed circuit **EL1:** 540-548
process control, acoustic emission inspection **A17:** 289-290
quality control in **EL1:** 869-874
-related failures, silicon nodules **EL1:** 1016-1017
test, system-level **EL1:** 373-374
time-frame, for new products **EL1:** 390
variables, conformal coatings **EL1:** 762
Manufacturing cells, as application
coordinate measuring machine **A17:** 18
Manufacturing economy *See* Economy in manufacture
Manufacturing Handbook and Buyers' Guide **EM3:** 66
as information source **EM2:** 92
Manufacturing practices
cleaning **A11:** 126-127, 616, 630
drilling **A11:** 123
grinding **A11:** 89-92, 125, 472-474, 567-569
identification marking **A11:** 130, 331, 472-473, 529, 531
machining **A11:** 85, 89-92, 122, 362, 459, 472, 750-752
magnetic-particle inspection **A11:** 16-17, 96, 128-129, 134, 322, 658
plating **A11:** 24, 43-45, 126, 308, 450, 459
secondary, and failure of heat exchangers **A11:** 629
straightening **A11:** 88, 125-126
surface compression **A11:** 125-126
welding **A11:** 127, 411-449
Manufacturing practices, effects of, on notch toughness **A1:** 742
cast steels **A1:** 746-747
wrought steels **A1:** 741, 742-746
Manufacturing process selection *See also* Manufacturing processes; Processing; Secondary manufacturing processes **EM2:** 277-404
design detail factors **EM2:** 288-292
economic factors **EM2:** 293-301
function/properties factors **EM2:** 279-287
introduction **EM2:** 277-278
Manufacturing process selection (continued)
shape factors **EM2:** 288-292
size factors **EM2:** 288-292
surface requirement factors **EM2:** 302-307
Manufacturing processes *See also* Fabrication; Manufacturing process selection; Processing; Production; Secondary manufacturing processes **EM1:** 497-663
aerospace **EM1:** 575-663
autoclave cure systems **EM1:** 645-648
autoclave molding tooling for **EM1:** 578-581
automated integrated system for **EM1:** 636-638
automated ply lamination **EM1:** 639-641
blow molding **EM2:** 352-359
boron fiber **EM1:** 58-59
braiding **EM1:** 519-518
carbon fiber **EM1:** 49-50
for composite structures, cost drivers in **EM1:** 419-427
compression molding **EM1:** 559-563 **EM2:** 302-303
compression molding and stamping **EM2:** 324-337
computer-controlled ply cutting/labeling **EM1:** 619-623
computerized autoclave cure control **EM1:** 649-653
consumer product **EM1:** 554-574
contoured tape laying **EM1:** 631-635
curing BMI resins **EM1:** 657-661
curing polyimide resins **EM1:** 662-663
discontinuous fibers **EM1:** 120-121
effect on design process **EM1:** 428-431
elastomeric tooling **EM1:** 590-601
electroformed nickel tooling **EM1:** 582-585
for epoxy resins **EM1:** 66-67, 654-656
errors, as failure cause **EM1:** 767
fiber preforms/resin injection **EM1:** 529-532
filament winding **EM1:** 503-518 **EM2:** 368-377
flat tape laying **EM1:** 624-630
of glass fibers **EM1:** 45
hand lay-up **EM2:** 338-343
injection molding **EM1:** 555-558
introduction **EM1:** 497
manual lay-up **EM1:** 602-604
material control in **EM1:** 741-742
mechanically assisted lay-up **EM1:** 605-607
personnel, interfaces with design/tooling personnel **EM1:** 428-431
preparation for cure **EM1:** 642-644
prepreg molding **EM2:** 338-343
prepreg tow **EM1:** 151
process modeling and optimization **EM1:** 499-502
pultrusion **EM1:** 533-543 **EM2:** 289-398
resin transfer molding **EM1:** 564-568 **EM2:** 349-351
rotational molding **EM2:** 360-367
sheet molding compound **EM1:** 141-142
spray-up **EM2:** 338-343
structural reaction injection molding **EM2:** 344-351
thermoforming **EM2:** 303, 399-403
thermoplastic extrusion **EM2:** 303, 378-388
thermoplastic injection molding **EM2:** 302, 308-318
thermoplastic matrix processing **EM1:** 544-553
thermoset injection molding **EM2:** 302, 319-323
thermosetting pultrusion **EM2:** 303
tooling effects **EM1:** 430
tooling, for autoclave molding **EM1:** 578-581
tube rolling **EM1:** 569-574
ultrasonic ply cutting **EM1:** 615-618
unidirectional tape prepregs **EM1:** 143
wound tube **EM1:** 135
Manufacturing records
shafts **A11:** 460
Manufacturing systems, flexible *See* Flexible manufacturing systems

Manufacturing use of adhesives
introduction to facility and equipment requirements **EM3:** 681-682
Manufacturing-to-cost (MTC) process **EM1:** 419, 423
Many-beam theory **A18:** 388
Mapp gas
for oxyfuel gas cutting **A14:** 724
Mapping *See also* Dot mapping; Elemental mapping; X-ray maps
analog **A10:** 525-528
chemical state, Auger electron spectroscopy **A10:** 555-556
compositional, electron probe x-ray microanalysis **A10:** 516, 525-529
elemental, Auger electron spectroscopy (AES) **EL1:** 1079
physically large systems **EL1:** 2
two-dimensional topographic **A10:** 372
Mapping applied to fracture studies **A9:** 99
Maps
phosphorus **A12:** 349
photogrammetry for **A12:** 197
sulfur **A12:** 349
x-ray **A12:** 167, 473
MAR-M 200
composition **M4:** 653
MAR-M 247
composition **M4:** 653
MAR-M 509
composition **M4:** 653
MAR-M-509
material for jet engine components **A18:** 588
MAR-M200
composition **A4:** 795 **A16:** 737
machining **A16:** 738, 741-743, 746-758
MAR-M246
composition **A4:** 795 **A16:** 737
machining **A16:** 738, 741-743, 746-757
MAR-M246+Hf
aging cycle **A4:** 812
MAR-M247
aging cycle **A4:** 812
composition **A4:** 795
MAR-M302
composition **A4:** 795 **A6:** 929 **A16:** 737
machining **A16:** 738, 741-743, 746-758
MAR-M322
composition **A4:** 795 **A6:** 929 **A16:** 737
machining **A16:** 738, 741-743, 746-757
MAR-M432
machining **A16:** 757, 758
MAR-M509
composition **A4:** 795 **A6:** 929 **A16:** 737
machining **A16:** 738, 741-743, 746-758
MAR-M905
machining **A16:** 757, 758
MAR-M918
composition **A4:** 795 **A6:** 573, 929
Maraging steel
18% Ni (300 CVM), gas nitrided, different I etchants compared **A9:** 228
for case hardening, composition of **A9:** 219
constitutional liquation **A6:** 75
etchants for **A9:** 218
friction welding **A6:** 152, 153
low-temperature solid-state welding **A6:** 300, 301
for pressure bar construction **A8:** 200
SCC environments **A8:** 526
void sheet formation **A8:** 479
Maraging steels *See also* Maraging steels, specific types; Steel(s) **A1:** 793-800 **M1:** 445-452
age hardening **A1:** 793-794 **A4:** 220, 221, 222, 223, 224, 225, 226-227 **M1:** 445-446, 448, 449 **M4:** 130-131, 132
applications **A1:** 800 **M1:** 451
bainitic formation **A4:** 219
cleaning after heat treatment **A4:** 228 **M4:** 132
cobalt-free, grain-boundary precipitates **A12:** 183
cobalt-free high-titanium, thermal embrittled **A12:** 136, 154
commercial alloys **A1:** 795
commercial alloys, composition **M1:** 447
composition **A4:** 219, 220
corrosion resistance **M1:** 450-451
Maraging steels (continued)
dimensional stability **M1:** 448, 451
effects of section thickness on fracture toughness **A11:** 54
electrical resistivity **A4:** 223-224
embrittlement **M1:** 446, 447, 448, 451
fractographs **A12:** 383-387
fracture toughness **M1:** 445, 448, 449, 450
fracture/failure causes illustrated **A12:** 217
heat treating **A4:** 219-228 **M4:** 130-132
heat treatment **M1:** 445-448
for hot-forging dies **A18:** 625
hydrogen-stress cracking and loss of tensile ductility **A1:** 715-716
K_{Iscc} values **M1:** 451
liquid-erosion resistance **A11:** 167
machining **M1:** 447
martensite aging **A4:** 222-224
martensite formation **A4:** 219, 220-222
mechanical properties **A1:** 799 **A4:** 220, 226 **M1:** 448, 449-451
molten salt corrosion **A13:** 53
nitriding of **M1:** 448, 541
notch toughness **M1:** 450, 451, 697, 701
overaging **A4:** 220, 222, 223-224, 227, 228 **M4:** 131
overaging, effects of **M1:** 446, 448
phase transformations **A4:** 220 **M4:** 130, 131
physical metallurgy **A1:** 793-795 **M1:** 445
physical properties **A1:** 800 **M1:** 451
powder metallurgy products **M1:** 449
processing **M1:** 447-449
cold working **A1:** 795
heat treating **A1:** 795-797
hot working **A1:** 795
machining **A1:** 795
melting **A1:** 795
powder metallurgy products **A1:** 798-799
surface treatment **A1:** 797-798
welding **A1:** 798
resistance to corrosion and stress corrosion **A1:** 799
service temperature of die materials in forging **A18:** 625
shallow dimples **A12:** 14
solution treatment **A4:** 224-225, 226 **M4:** 131-132
stress-corrosion cracking **M1:** 450-451
stress-corrosion cracking in **A11:** 218
surface treatment **M1:** 448, 450-451
then-nal embrittlement **A12:** 136, 154
thermal cycling **A4:** 225-226
thermal embrittlement **A4:** 225
thermal embrittlement of **A1:** 697-698
tool applications **M3:** 446-447, 513
tooling use **A4:** 765-766
welding **M1:** 445, 446, 449, 563
yield strengths **A4:** 219, 221, 224, 227
Maraging steels, 18Ni
cryogenic service **A6:** 1017
yield strength vs. fracture toughness **A6:** 1017
Maraging steels, specific types
18% Ni grade 300, fibrous fracture **A12:** 383
18% Ni grade 300, fracture toughness **A12:** 385
18% Ni grade 300, low-cycle fatigue fracture **A12:** 386
18% Ni grade 300, slow-bend fracture **A12:** 387
18% Ni grade 300, tensile-test fracture **A12:** 384
18Ni (200)
composition **M4:** 130
heat treatment **M4:** 132
mechanical properties **M4:** 132
18Ni (250)
composition **M4:** 130
heat treatment **M4:** 132
mechanical properties **M4:** 132
18Ni (300)
composition **M4:** 130
heat treatment **M4:** 132
mechanical properties **M4:** 132
18Ni (350)
composition **M4:** 130
heat treatment **M4:** 132
mechanical properties **M4:** 132
18Ni (cast)
composition **M4:** 130
heat treatment **M4:** 132
mechanical properties **M4:** 132
18Ni, heat treatment **A4:** 219
Maraging steels, specific types (continued)
18Ni(200)
composition **A4:** 220
hardness **A4:** 766
heat treatment **A4:** 220, 221, 224, 226, 766
mechanical properties **A4:** 220, 226
molybdenum-bearing precipitate **A4:** 223
yield strengths **A4:** 219
18Ni(250)
cold work effect on fasteners **A4:** 228
composition **A4:** 220 **M1:** 447
density **M1:** 145
electrical resistivity **M1:** 150-151
hardness **A4:** 766
heat treatment **A4:** 220, 221, 224, 226, 227, 766
mechanical properties **A4:** 220, 226, 227
microstructure **A4:** 221, 222
molybdenum-bearing precipitate **A4:** 223
seizure resistance **M1:** 611
solution annealing **A4:** 225, 226
tensile properties .... **A4:** -125
thermal conductivity **M1:** 148
thermal expansion **M1:** 146-147
yield strengths **A4:** 219
18Ni(300)
composition **A4:** 219, 220
hardness **A4:** 766
heat treatment **A4:** 220, 221, 224, 226, 766
mechanical properties **A4:** 220, 226
molybdenum-bearing precipitate **A4:** 223
solution annealing **A4:** 225, 226
yield strength **A4:** 219, 223
18Ni(300) composition **M1:** 447
18Ni(350)
composition **A4:** 219, 220
hardness **A4:** 766
heat treatment **A4:** 220, 221, 226, 766
mechanical properties **A4:** 226
molybdenum-bearing precipitate **A4:** 223
short-range ordering **A4:** 222
18Ni(cast)
composition **A4:** 220
mechanical properties **A4:** 226
Marandet-Sanz
three-step Charpy/fracture toughness correlation **A8:** 265
Marangoni convection (surface-tension-driven thermocapillary flow) **A6:** 19, 264
Marble
chemical composition **A6:** 60
drilling **A16:** 230
Marble grinding balls **A7:** 58
Marble melt process
for glass fibers **EM1:** 45
Marble's reagent as an etchant for
beryllium-containing alloys **A9:** 394
heat-resistant casting alloys **A9:** 330
nitrided steels **A9:** 218
permanent magnet alloys **A9:** 533
wrought heat-resistant alloys **A9:** 307
MARC computer program for structural analysis **EM1:** 268, 271
MARC software program
for heat analysis problems **A15:** 861
Marcel Dekker, Inc.
as information source **EM2:** 93
Marciniak biaxial stretching test **A8:** 558
Marciniak in-plane sheet torsion test
as shear test **A8:** 559-560
Marform process **A14:** 9, 607-608
defined **A14:** 9
presses **A14:** 607
procedure **A14:** 608
tools **A14:** 607-608
Margarite **EM4:** 6
Marginal fracture **A18:** 669
Marginal oscillator magabsorption detector **A17:** 150-152
Marine
applications for stainless steels **A7:** 731
atmospheres, corrosion in **A11:** 193, 209
environment, titanium test panels for **A7:** 681
equipment, with self-lubricating bearings **A7:** 705
organisms, corrosion and fouling by **A11:** 191

Marine applications *See also* Boating; Saltwater corrosion resistance **EM1:** 837-844
aluminum and aluminum alloys **A2:** 11
cobalt-base wear-resistant alloys **A2:** 451
composite masts **EM1:** 841
hovercraft **EM1:** 839
HSLA steels **M1:** 406
hydrofoils **EM1:** 839
laminated sailcloths **EM1:** 841-842
mine warfare vessels **EM1:** 837
navigational aids **EM1:** 839
offshore engineering **EM1:** 839
passenger ferries **EM1:** 839
pleasure boats/luxury yachts **EM1:** 841
powerboats **EM1:** 839-840
racing yachts **EM1:** 840-841
ship hulls **EM1:** 837-838
sonar domes **EM1:** 838
submarine structures **EM1:** 838
submersibles **EM1:** 838-839
titanium and titanium alloys **A2:** 588
unsaturated polyesters **EM2:** 246
Marine atmospheres *See also* Atmospheres; Atmospheric corrosion; Marine corrosion; Seawater **A13:** 902-906
alloy content and **A13:** 906
alloy steel corrosion in **A13:** 542-544
carbon dioxide in **A13:** 904
chloride airborne contamination **A13:** 903
corrosion data, various metals/alloys **A13:** 915-916
corrosion rates, copper/copper alloys **A13:** 616
crevice corrosion by **A13:** 112
galvanic couples, magnesium/magnesium alloys **A13:** 746
galvanized coatings **A13:** 440
location, corrosion effects **A13:** 904-905
low-alloy steel corrosion in **A13:** 541
moisture of **A13:** 902-903
nickel-chromium/copper-nickel-chromium coatings **A13:** 430
orientation (to earth's surface), effects **A13:** 905
rusting of various coatings in **A13:** 778
SCC of aluminum alloys in **A13:** 265
shallow crevice corrosion, stainless steel
stainless steel corrosion in **A13:** 303, 555
sulfur dioxide as airborne contaminant **A13:** 903-904
sunlight effects **A13:** 905
telephone cables in **A13:** 1127
temperature effects **A13:** 905
time effects **A13:** 906
wind effects **A13:** 905-906
zinc/zinc alloys and coatings in **A13:** 757
Marine corrosion *See also* Atmospheres; Marine atmospheres; Seawater **A13:** 893-926
by seawater **A13:** 893-902
cathodic protection **A13:** 919-924
in marine atmospheres **A13:** 902-906
metallic coatings for **A13:** 906-912
organic coatings for **A13:** 912-918
Marine environment
alternate immersion test for Al SCC-susceptibility in **A8:** 523
stress-corrosion cracking in **A8:** 499
Marine environments *See also* Environment
classification **A13:** 1255
corrosivity **A13:** 510-511
crevice corrosion in stainless steels **A13:** 303
general biological corrosion in **A13:** 88
types **A13:** 893
Marine equipment, cast iron
coatings for **M1:** 104, 105
Marine structures
tidal zones/immersion depth **A13:** 542
Marion-Cohen technique
plane-stress elastic model **A10:** 384
Marker-and-Cell software programs **A15:** 867, 871-872
Market(s)
for hybrids **EL1:** 381, 384
for medical and military applications **EL1:** 389
and technology trends, thick-film hybrids **EL1:** 386-389
worldwide, telecommunications **EL1:** 384-385
Markets *See* Applications **EM3:** 44-47, 76-77
aircraft/aerospace industry **EM3:** 44
appliance **EM3:** 46
auto sealant **EM3:** 57
automotive **EM3:** 45-46
casting, development **A15:** 34
construction **EM3:** 46-47, 56-58
consumer products **EM3:** 47, 56-58
for copper and copper-based powders **A7:** 572
for iron powders **A7:** 571
pressure-sensitive adhesives (PSA) **EM3:** 47
for resin binder processes **A15:** 215
sealants **EM3:** 56
woodworking **EM3:** 46
Marking
and electrochemical machining **A16:** 533
as vitreous dielectric application **EL1:** 109
Marongoni effect **A6:** 468
Marongoni surface tension gradient induced convection forces **A6:** 46
MARSE, as parameter
acoustic emission inspection **A17:** 282-283
Marsh, A.L
as metallurgist **A15:** 32
Marshall's reagent as an etchant for
carbon and alloy steels **A9:** 169-170
electrical steels **A9:** 531
Martempering **A1:** 455, 457 **M1:** 431, 460
defined **A9:** 11
effect on distortion **A11:** 141
gray iron **M4:** 537-539
Martempering of steel *See* Steel, martempering
Martensite *See also* Strain-induced martensite **A1:** 127, 133-134
acicular needles of **A12:** 328
arc welds of cast irons **M6:** 313
as-quenched hardness affected by **M1:** 472, 478-480
at tip of failed shear blade **A11:** 575, 583
in austenitic manganese steel castings **A9:** 239
in austenitic stainless steels **A9:** 283
in carbon and alloy steels, etching to reveal **A9:** 170
in carbon and alloy steels, microstructure **A9:** 178
carbon content, influence on as-quenched hardness **M1:** 457, 458, 529, 530, 561, 607-608
case structure of carburized or carbonitrided parts **M1:** 533, 534, 538
in cast iron **M1:** 6, 7, 9, 563-564
in cast irons **A13:** 566
defined **A9:** 11 **A13:** 9
effect, ferritic stainless steels **A13:** 127
effect, intergranular corrosion, austenitic stainless steels **A13:** 124-125
effect of carbon content **M6:** 247-248
fatigue resistance, effect on **M1:** 675, 676
ferrous **A9:** 668-672
formation **A13:** 47 **M6:** 247-248
formation during welding **M1:** 561-563
formation, symbols for **A11:** 797
fretting wear **A18:** 248
grain size effect on strength of **A1:** 393
hardness of **A1:** 394
hydrogen embrittlement of **M1:** 687
neutron embrittlement susceptibility to **M1:** 686
nonferrous **A9:** 672-674
notch toughness, effect on **M1:** 693, 701-702
percentage as function of hardness **A7:** 452
polarized light used to examine **A9:** 79
in precipitation-hardenable stainless steels **A9:** 285
Martensite (continued)
present with massive transformation structures **A9:** 655-656
relation to hardenability **M1:** 493
in stainless steels, resulting from plastic deformation **A9:** 66
structure, forcing, as hardening **A13:** 47
tempered, in ductile iron, crack growth **A12:** 228
tempering **M1:** 564
tempering of **A1:** 134-136, 137
in titanium and titanium alloys **A9:** 460-461
transformation in steel **M4:** 33, 34
transformed **A12:** 26, 32, 42
untempered white, metallographic sectioning **A11:** 24
in uranium alloys **A9:** 476-477, 485-486
wear resistance affected by service temperature **M1:** 608
wear resistance affected by tempering temperature **M1:** 613-614
wear resistance compared with pearlite **M1:** 611-612
zones, weld failures from **A11:** 426
Martensite (M) **A6:** 76
Martensite completion point (M_f) **A6:** 437, 438
Martensite decomposition in uranium alloys **A10:** 316
Martensite malleable irons *See* Malleable cast irons
Martensite range
defined **A9:** 11
Martensite start and finish temperatures in steel **A9:** 178
Martensitic
defined **A9:** 11
Martensitic alloy irons
advantages **A6:** 797
applications **A6:** 797
Martensitic alloy irons, specific types
Cr-Mo, advantages and applications of materials for surfacing, build-up, and hardfacing **A18:** 650
Cr-W, advantages and applications of materials for surfacing, build-up, and hardfacing **A18:** 650
Martensitic alloy steel
classification and composition of hardfacing alloys **A18:** 652
Martensitic alloys
high-alloy **A15:** 722, 731-732
malleable iron, mechanical properties **A15:** 693-696
Martensitic cast steels
corrosion fatigue **A13:** 581
general corrosion **A13:** 576-577
intergranular corrosion **A13:** 580
Martensitic ductile iron, grinding media material
mining industry **A18:** 654
Martensitic finish temperature **A9:** 669
Martensitic grades
of corrosion-resistant steel castings **A1:** 913
Martensitic iron
classification and composition of hardfacing alloys **A18:** 652
Martensitic malleable irons *See* Pearlitic-martensitic malleable iron
Martensitic precipitation-hardenable stainless steels *See also* Wrought stainless steels; Wrought stainless steels, specific types **A9:** 285
Martensitic stainless steel *See also* Cast stainless steels; Wrought stainless steels **A1:** 841-842
compositions of **A1:** 843, 847-848
elevated-temperature properties **A1:** 939-942
forgeability **A1:** 892, 893
machinability of **A1:** 894
notch toughness of **A1:** 859
tensile properties of **A1:** 858, 862-863
weldability of **A1:** 902

Martensitic stainless steels *See also* Stainless steel(s); Stainless steels; Stainless steels, martensitic; Stainless steels, wear of; Steel(s); Wrought stainless steels; Wrought stainless steels, specific types **A6:** 678-682 **A14:** 226, 759
annealing **A7:** 185
arc welding *See* Arc welding of stainless steels
arc-welded **A11:** 427
base metals **A6:** 679
brazing **A6:** 913
brazing and soldering characteristics **A6:** 626
characterized **A13:** 550
composition **M6:** 52
electron beam welding **M6:** 638
electron-beam welding **A6:** 869
engineering for use after postweld heat treatment **A6:** 680- 682
engineering for use in the as-welded condition **A6:** 679-680
fractographs **A12:** 366-369
fracture/failure causes illustrated **A12:** 217
machining **A2:** 967
as magnetically soft materials **A2:** 777-778
metallurgy **A6:** 678-679
microstructures **A9:** 285
mill finishes **M5:** 552
physical properties **M6:** 527
properties **A7:** 100
repair welding **A6:** 1106
resistance welding **A6:** 848 **M6:** 527
in sour gas environments **A11:** 300
stress-corrosion cracking in **A11:** 217
stud arc welding **M6:** 733
susceptibility to hydrogen damage **A11:** 249
thermal expansion coefficient **A6:** 907
thermal properties **A6:** 17
welding to carbon steels **A6:** 501
welding to low-alloy steels **A6:** 501
Martensitic stainless steels, specific types
AISI 410, fracture surfaces **A12:** 366
AISI 431, high-cycle fatigue fracture **A12:** 367
AISI 501, mating segments, fatigue fracture **A12:** 368-369
AISI 4340, light fractographs **A12:** 83
Silcrome-1, fatigue fracture surfaces **A12:** 369
Martensitic stainless steels, wrought *See* Wrought martensitic stainless steel selection
Martensitic start temperature **A9:** 668-669
as a function of carbon content in steel **A9:** 670
Martensitic start temperature (Ms) **A6:** 437, 438
Martensitic steel
cooling in ultrasonic testing **A8:** 247
fatigue crack growth data **A8:** 376-377
SCC environments **A8:** 526
Martensitic steels
advantages **A6:** 797
applications **A6:** 797
hardfacing **A6:** 790, 791, 798
as hardfacing alloys **A7:** 828, 829
Martensitic steels, advantages and applications of materials for surfacing
build-up, and hardfacing **A18:** 650
Martensitic steels, specific types
medium-carbon Cr-Mo **A18:** 651
Martensitic structures
in titanium alloy welded joints **A9:** 581
Martensitic surface layer in abraded steel **A9:** 38
Martensitic transformation **A12:** 26, 32, 42
copper alloys **M2:** 259
in shape memory alloys **A2:** 897
Martensitic transformations
as a result of plastic deformation **A9:** 686
scanning electron microscopy used to study **A9:** 97
Martensitic white cast iron *See* White cast iron, martensitic
Martin hard coat anodizing process
aluminum and aluminum alloys **M5:** 592
Martin's diameter
particle size measurement **A7:** 225
Maryland
early American foundries **A15:** 25
Mash resistance seam welding
definition **M6:** 11
Mash-seam welding
of blanks **A14:** 450-451
Mask
definition **M6:** 11
Mask (thermal spraying)
definition **A6:** 1211
Mask set
defined **EL1:** 1149
Masking *See also* Solder mask; Stopping-off
cadmium plating **M5:** 267-268
components, for conformal coating **EL1:** 764
mass finishing processes **M5:** 136
for UV-curable coatings **EL1:** 787
vacuum coating process **M5:** 407-408
Masks
for scattered radiation **A17:** 344
Masonry walls
metallic anchors and ties in **A13:** 1302, 1306-1310
Mass
conversion factors **A8:** 722 **A10:** 686
effect on hardness **A4:** 36, 39
measurements, GC/MS analysis **A10:** 642-643
minimum detectable, abbreviation for **A10:** 690
numbers, isotopes **A10:** 688
per unit area, conversion factors **A8:** 722
per unit length, conversion factors **A8:** 722
per unit time, conversion factors **A8:** 722
per unit volume, conversion factors **A8:** 722
resolution, LEISS analysis **A10:** 605
SI base unit and symbol for **A10:** 685
SI unit/symbol for **A8:** 721
vs x-ray production, PIXE analysis **A10:** 104
Mass absorption
coefficients **A10:** 85, 99, 676
vs wavelength, absorption edges as discontinuities in **A10:** 85
vs x-ray energy, copper **A10:** 85, 87
in x-ray spectrometry **A10:** 84
Mass absorption coefficient
of beryllium **A2:** 683-684
Mass accumulation
to analyze ceramic powder particle sizes **EM4:** 67
Mass analysis
analytic methods **A10:** 4-5, 7, 8, 9, 10
gas analysis by **A10:** 151-157
Mass analyzer
in gas mass spectrometer **A10:** 153-154
Mass attenuation coefficient
defined **A17:** 387
Mass bonding
as inner lead process **EL1:** 278-281
outer lead bonding **EL1:** 286
Mass characteristics *See also* Density
actinide metals **A2:** 1189-1198
aluminum casting alloys **A2:** 153-177
cast copper alloys **A2:** 356-391
cast magnesium alloys **A2:** 492-516
electrical resistance alloys **A2:** 836-839
gold and gold alloys **A2:** 704-707
niobium alloys **A2:** 567-571
palladium-silver alloys **A2:** 716
platinum and platinum alloys **A2:** 708
pure metals **A2:** 1099-1178
rare earth metals **A2:** 1178-1189
silver and silver alloys **A2:** 699-704
wrought aluminum and aluminum alloys **A2:** 62-122
wrought copper and copper alloys **A2:** 265-345
wrought magnesium alloys **A2:** 480-491
of zinc alloys **A2:** 532-542
Mass concentration (in a slurry)
defined **A18:** 12
Mass density
SI derived unit and symbol for **A10:** 685
Mass discrimination
corrected by Einzel lens system **A10:** 155
in gas mass spectrometer **A10:** 153
Mass, effect of
cast steels **M1:** 384-386, 392-393, 398
iron castings *See* Section sensitivity
Mass filter, quadrupole
in gas mass spectrometer **A10:** 153
Mass finishing **M5:** 128-137
abrasives, used in **M5:** 614
aluminum and aluminum alloys **M5:** 129-130, 134
applications **M5:** 131, 133-135
automated **M5:** 129, 136
Mass finishing (continued)
barrels **M5:** 128-129
action **M5:** 128-129
types used, size and shape **M5:** 129
burnishing process **M5:** 134
castings **M5:** 135
centrifugal disk finishing *See* Centrifugal disk finishing
cleaning process **M5:** 136
cleanliness standards **M5:** 136
compounds used in **M5:** 134-135
addition to equipment, methods of **M5:** 134-135
copper and copper alloys **M5:** 614-616
deburring process **M5:** 128, 134, 614-616
dry barrel operations *See* Dry barrel finishing
electrochemically accelerated **M5:** 133
equipment **M5:** 128-134, 136
auxiliary **M5:** 136
maintenance of **M5:** 136
selection of **M5:** 133-134
fixtures used in **M5:** 136
forgings **M5:** 135
limitations and advantages of **M5:** 128, 134
magnesium alloys **M5:** 631
masking processes **M5:** 136
media for **M5:** 128, 134-135
types used **M5:** 135
problems in **M5:** 131, 136
processes **M5:** 128-133, 136
automated **M5:** 129, 136
control **M5:** 136
principles of operation **M5:** 128-129
types *See also* specific types by name **M5:** 128, 133
safety precautions **M5:** 136
shine rolling process **M5:** 134
spindle finishing *See* Spindle finishing
stainless steel **M5:** 554-555
surface finishes **M5:** 615
titanium and titanium alloys **M5:** 656, 659
vibratory finishing *See* Vibratory finishing
waste disposal **M5:** 136-137
wet barrel operations *See* Wet barrel finishing
Mass lamination *See also* Laminates; Lamination
defined **EL1:** 1149
rigid printed wiring boards **EL1:** 551
Mass loss **A13:** 229, 323
brake linings **A18:** 571
Mass measurements
gas analysis by mass spectrometry **A10:** 151-157
gas chromatography/mass spectrometry **A10:** 639-648
low-energy ion-scattering spectroscopy **A10:** 603-609
Rutherford backscattering spectrometry **A10:** 628-636
secondary ion mass spectroscopy **A10:** 610-627
spark source mass spectrometry **A10:** 141-150
Mass media finishing, aircraft engine components
surface finish requirements **A16:** 22
Mass memory
use in x-ray spectrometry **A10:** 92
Mass peak
defined for gas mass spectrometer **A10:** 155
Mass production machinery
for magnetic particle inspection **A17:** 111
Mass resolution *See also* Resolution
LEISS analysis **A10:** 605
Mass resolution detector **A7:** 258
Mass scale
atom probe, calibration of **A10:** 592
RBS energy scale as translated into **A10:** 629
Mass spectra
atom probe microanalysis interpretation **A10:** 591-593
atom probe, of IN 939 ..... **A10:** 599
Mass spectrometers
as attachments **A17:** 67
defined **A10:** 151
helium **A17:** 65
SIMS **A11:** 34-36
spark source **A10:** 142
testing, for gas/leak detection **A17:** 63
time-of-flight **A10:** 142
as vacuum leak testing method **A17:** 67
Mass spectrometry **EM1:** 736
capabilities **A10:** 212, 226, 649

Mass spectrometry (continued)
capabilities, compared with IR spectroscopy **A10:** 109
defined **A10:** 676
gas analysis by **A10:** 151-157
and high-performance liquid chromatography use with GC/MS **A10:** 645-646
-isotope dilution, capabilities **A10:** 243
laser-induced resonance ionization **A10:** 142
and mass spectrometry, and gas chromatography **A10:** 646-647
for molecular structure **A10:** 116
spark source **A10:** 141-150
for temperature-time control in polymer removal techniques **EM4:** 137
to analyze the bulk chemical composition of starting powders **EM4:** 72
to measure hydrogen sources in welds **A6:** 413
Mass spectrometry (MS) **EM3:** 18
defined **EM2:** 26
Mass spectrometry/mass spectrometry
block diagram of spectrometer system for **A10:** 646
nonvolatile compound analysis **A10:** 639
Mass spectroscopy *See* Mass spectrometry **A13:** 1115-1116
as advanced failure analysis technique **EL1:** 1105-1107
for measurement of furnace atmosphere composition **EM4:** 252
for trace element analysis **A2:** 1095
of ultra-high purity metals, measurement techniques **A2:** 1095-1096
Mass spectrum
atom probe **A10:** 591, 592
collision-activated dissociation **A10:** 647
complex, atom probe microanalysis **A10:** 591-592
defined **A10:** 676
ethyl alcohol **A10:** 640
in gas analysis **A10:** 151, 155
naphthalene **A10:** 643
pentachlorobiphenyl **A10:** 642
scanning modes, gas mass spectrometers **A10:** 154
tabulated output, for GC/MS analysis of extracted shale oil **A10:** 643
Mass transfer
controlled dissolution **A15:** 73-74
deposits, from liquid lithium corrosion **A13:** 93
and dissolution in liquid metals **A13:** 56-57
effect, alloy additions **A15:** 72
and heat transfer, fluid flow modeling of **A15:** 877-882
in liquid-metal corrosion **A13:** 17
liquid-phase **A15:** 83
profile **A13:** 57
temperature-gradient **A13:** 51
Mass transfer limited dissolution process
alloy additions **A15:** 72-74
Mass transfer limited kinetics
boundary layer model for **A15:** 53
Mass transport *See also* Flux
control, aqueous corrosion **A13:** 33-35
during solidification **A15:** 111-113
interphase, defined **A15:** 52-54
as parameter, stress-corrosion cracking **A13:** 147
Mass transport mechanisms
injection molded materials **A7:** 466
Mass-absorption coefficient
neutron radiography **A17:** 309, 390
Massive forming *See* Bulk forming
Massive projections
as casting defects **A11:** 381
Massive transformation structures **A9:** 655-657
allotropy **A9:** 655
congruent points **A9:** 655
feathery grains **A9:** 656
two-phase fields **A9:** 655-656
Massmann graphite furnace atomizers **A10:** 49
Master alloy powder **A7:** 7
Master alloy processing
kinetics of **A15:** 107-108
thermal analysis techniques **A15:** 108
Master alloy(s) *See also* Hardeners; Master alloy processing
defined **A15:** 8
for grain refinement **A15:** 160-161, 477
processing **A15:** 107-108
as silicon modifiers, forms of **A15:** 164
strontium-base, in aluminum-silicon alloys **A15:** 164
Master block
defined **A14:** 9
Master curve
compact parameter **A8:** 333, 335
Larson-Miller method of creating **A8:** 333, 335
Master drawing *See also* Artwork; Design
defined **EL1:** 1149
in final design package **EL1:** 524-525
flexible printed boards **EL1:** 593
printed wiring boards **EL1:** 516
Master pattern
defined **A15:** 8
match plate pattern plaster mold castings **A15:** 245
Mat *See* Mats
defined **EM2:** 26
Mat frits and glazes
typical oxide compositions **EM4:** 550
Match
defined **A14:** 9
Match plate
defined **A15:** 8
development **A15:** 28
pattern machines, green sand molding **A15:** 343
patterns, characteristics **A15:** 190, 240
Match plate pattern plaster mold casting **A15:** 242, 245-246
Match/search methods
for unknown phases or particles **A10:** 455-459
Matched approach technique, as interconnection option **EL1:** 985
Matched CTE materials *See* Coefficient of thermal expansion (CTE); CTE-matched materials
Matched draft *See* Blend draft
Matched edges
defined **A14:** 9
Matched lines *See* Matched edges
Matched metal die molding
defined **EM2:** 26
Matched metal molding
defined **EM1:** 15
Matched weld lip
container **A7:** 431
Matched-die molding
with discontinuous fiber **EM1:** 121
Matched-mold thermoforming **EM2:** 400
Matching
fracture-surface **A11:** 80
Matching draft
defined **A14:** 9
Material (ply) reference system **EM1:** 236
Material anisotropy *See* Anisotropy
Material balance, as process modeling
submodel **EM1:** 500
Material chemistry
of crack propagation **A13:** 155-158
Material conditions
alloying **A11:** 119
effect on fatigue strength **A11:** 118
grain size **A11:** 118-119
second phases **A11:** 119
solid-solution strengthening **A11:** 119
Material contrast
scanning electron microscopy **A9:** 93-94
Material control
epoxy materials **EL1:** 833-836
in-process **EM1:** 741-742
Material control (continued)
in leaded and leadless surface-mount joints **EL1:** 731-733
maintaining, traceability **EM1:** 741-742
storage conditions **EM1:** 741
Material converter
information for **EM2:** 1
Material cost *See* also Cost(s)
of composites **EM1:** 35
Material costs *See also* Cost; Cost(s); Economic process selection factors; Economics
components **EM2:** 83-84
defined **EM2:** 82
of engineering thermoplastics **EM2:** 99
parts, costing breakdown **EM2:** 83-85
and process selection **EM2:** 278
thermoforming **EM2:** 403
thermoplastic injection molding **EM2:** 309
Material data sheets *See* Data sheets
Material defects
failures, cold-formed parts **A11:** 307
fatigue cracking at **A11:** 323
forging failures from **A11:** 317
spring failures caused by **A11:** 551-553
in steam equipment **A11:** 603
Material deficiency
in failure analysis **A11:** 744, 747
Material displacement straightening
high-production peening **A14:** 681
manual peening **A14:** 681
manual straightening **A14:** 680-681
stretch straightening **A14:** 681
Material flaws *See also* Defects; Internal defects; Material defects
in ceramics **A11:** 748-751
fracture-initiating **A11:** 748
thermal shock failure from **A11:** 753
Material flow *See* Metalflow
Material forms *See also* Product forms
of composites **EM1:** 33
of composites and metals **EM1:** 35
Material handling conveyor
wave soldering **EL1:** 702
Material inspection test
bending ductility test for **A8:** 117
Material modeling *See also* Dynamic material modeling; Modeling; Process modeling
basic concepts (yield criteria) **A14:** 417-418
constitutive equations for **A14:** 417-420
flow curves **A14:** 418-419
flow stress-strain rate relationships **A14:** 419
flow stress-temperature relations **A14:** 419-420
temperature/strain rate, combined effects **A14:** 420
Material modeling environment
as knowledge-based expert system **A14:** 411
Material price *See* Material Costs
Material processing *See also* Processing
titanium aluminides **A2:** 926-928
Material properties *See also* Fiber properties analysis; Material properties analysis; Material(s)
analysis **EM1:** 185
computational procedures **EM1:** 302
correlation with simulative tests **A8:** 565
damage tolerances of **EM1:** 262-264
description **EM1:** 355-359
as design category **A11:** 115
determining strain distribution **A8:** 550
of die attachments **EL1:** 213
effect in uniaxial tensile testing **A8:** 555-556
effect on formability **A8:** 549-553
effect on wear **A11:** 158-159
IC packages **EL1:** 207
information sources on **EM1:** 40-42
long-term environmental effects **EM1:** 823-826
as material selection parameter **EM1:** 38-39
numerical computation models **EM1:** 187
probabilistic nature **EM1:** 309-310
selection matrix guidelines for **EM1:** 38

SUBJECTS OF THE INDEXED VOLUMES: ASM Handbook (designated by the letter "A"): **A1:** Properties and Selection: Irons, Steels, and High-Performance Alloys (1990); **A2:** Properties and Selection: Nonferrous Alloys and Special-Purpose Materials (1990); **A3:** Alloy Phase Diagrams (1992); **A4:** Heat Treating (1991); **A6:** Welding, Brazing, and Soldering (1993); **A7:** Powder Metallurgy (1984); **A8:** Mechanical Testing (1985); **A9:** Metallography and Microstructures (1985); **A10:** Materials Characterization (1986); **A11:** Failure Analysis and Prevention (1986); **A12:** Fractography (1987); **A13:** Corrosion (1987); **A14:** Forming and Forging (1988); **A15:** Casting (1988); **A16:** Machining (1989); **A17:** Nondestructive Testing and Quality Control (1989); **A18:** Friction, Lubrication, and Wear Technology (1992). **Metals Handbook, 9th Edition** (designated by the letter "M"): **M1:** Properties and Selection: Irons and Steels (1978); **M2:** Properties and Selection: Nonferrous Alloys and Pure Metals (1979); **M3:** Properties and Selection: Stainless Steels, Tool Materials and Special-Purpose Materials (1980); **M4:** Heat Treating (1981); **M5:** Surface Cleaning, Finishing, and Coating (1982); **M6:** Welding, Brazing, and Soldering (1983). **Engineered Materials Handbook** (designated by the letters "EM"): **EM1:** Composites (1987); **EM2:** Engineering Plastics (1988); **EM3:** Adhesives and Sealants (1990); **EM4:** Ceramics and Glasses (1991); **Electronic Materials Handbook** (designated by the letters "EL"): **EL1:** Packaging (1989).

Material properties (continued)
and shear fracture ... **A8:** 551
software for ... **EM1:** 275
and springback or shape distortion ... **A8:** 552-553
static metallic, design allowables for ... **A8:** 662-677
tape automated bonding assembly ... **EL1:** 289
and wrinkling ... **A8:** 551, 564
Material properties analysis *See also* Fiber properties analysis; Laminate properties analysis
failure ... **EM1:** 192-204
long-term testing ... **EM1:** 823-826
physical, of fiber composites ... **EM1:** 185-192
strength ... **EM1:** 192-204
Material removal rate (MRR) ... **A16:** 17
electrical discharge machining ... **EM4:** 373, 374, 376
electrostream drilling ... **A16:** 552
erosion process of electrical discharge machining ... **EM4:** 372
high removal rate machining ... **A16:** 607
laser-beam machining ... **A16:** 572, 574, 575
and polishing wear ... **A18:** 191, 194, 195, 197
superabrasive grinding ... **A16:** 460, 463
Material requirements for service conditions ... **A6:** 373-400
aerospace ... **A6:** 385-388
advanced materials ... **A6:** 388
damage tolerance ... **A6:** 387-388
filler metals ... **A6:** 386-387
heat-affected zone ... **A6:** 386
joinability ... **A6:** 386-387
material properties of importance ... **A6:** 386, 387
material selection criteria ... **A6:** 385-386, 387
postweld heat treatments ... **A6:** 386
specifications ... **A6:** 385-386, 387-388
automobiles ... **A6:** 393-395
matching filler specifications ... **A6:** 394
safety standards ... **A6:** 393
bridges and buildings ... **A6:** 375-377
codes for steel structures ... **A6:** 375
environment ... **A6:** 375-376
fitness for service ... **A6:** 376-377
material selection ... **A6:** 375
crack-opening displacement (COP) test ... **A6:** 376-377
machinery and equipment ... **A6:** 389-393
applications ... **A6:** 389
design considerations ... **A6:** 389-391
material properties of weldments ... **A6:** 391-393
stress category classification ... **A6:** 392
stress ranges allowable ... **A6:** 391
stresses allowable in weld metal ... **A6:** 390
weldability classes of steels used ... **A6:** 390
postweld heat treatments ... **A6:** 375
pressure vessels and piping ... **A6:** 377-381
codes and specifications ... **A6:** 377
cracking behavior ... **A6:** 379
environment ... **A6:** 377-379
fabricability ... **A6:** 379-381
material properties ... **A6:** 378
mechanical properties of weldments ... **A6:** 381
oxidation in steam service ... **A6:** 379
postweld heat treatment ... **A6:** 381
steels for high-temperature service ... **A6:** 378-379
steels for low-temperature service ... **A6:** 377-378
steels for ordinary-temperature service ... **A6:** 377, 378
principle attributes of a material ... **A6:** 374
principles of material selection ... **A6:** 373
properties relating to selection for service of ASTM steels approved in AWS D1.1-92 ... **A6:** 376
railroad equipment ... **A6:** 395-398
freight cars ... **A6:** 395-396
locomotives ... **A6:** 397
material selection ... **A6:** 395
specifications ... **A6:** 395
T-rail ... **A6:** 397
track ... **A6:** 397
service conditions ... **A6:** 374-375
sheet metals ... **A6:** 398-400
coated metals ... **A6:** 400
distortion ... **A6:** 398-400
fabrication codes ... **A6:** 400
joining processes ... **A6:** 398-399
Material requirements for service conditions (continued)
materials ... **A6:** 399-400
shipbuilding and offshore structures ... **A6:** 381-385
ABS requirements for higher-strength hull structural steel ... **A6:** 383
ABS requirements for ordinary-strength hull structural steel ... **A6:** 382
container ships ... **A6:** 381-382
liquefied natural gas/liquefied petroleum gas ships ... **A6:** 382-383
materials for cryogenic applications ... **A6:** 383
offshore structures ... **A6:** 383-384
ships ... **A6:** 381-383
tankers ... **A6:** 381-382, 383
weld considerations ... **A6:** 384-385
welding process selection ... **A6:** 384
weld thermal cycle effect ... **A6:** 375
Material safety data sheet (MSDS)
for lubricants to satisfy hazard communication standard ... **A18:** 87
Material Safety Data Sheets ... **A6:** 68
Material selection *See also* Dissimilar materials; Material(s)
of adhesives ... **EM1:** 683-688
alloy, and steel casting failure ... **A11:** 391
alloy, effect on SCC in aluminum alloys ... **A11:** 219
alloy, failure from ... **A11:** 391
of alloy, for buried metals ... **A11:** 192
of compatible metals, for galvanic corrosion ... **A11:** 185
and composite material requirements ... **EM1:** 38
cost guidelines ... **EM1:** 105
design guidelines, for RTM ... **EM1:** 168
Epon resin/curing agent guide, fiberglass-reinforced plastics ... **EM1:** 134
of epoxy matrices ... **EM1:** 76-77
for fan shafts, improper ... **A11:** 476-477
fiber-matrix, for continuous-fiber MMCs ... **EM1:** 879-880
flame hardening ... **M4:** 503-505
in forging ... **A11:** 320-321
for gears ... **A11:** 590
for implants ... **A11:** 672-673, 680
improper, effect on microstructure ... **A11:** 359
of inhibitors ... **A11:** 198
for iron castings ... **A11:** 344-352, 344
and laminate ranking ... **EM1:** 456-457
matrix guidelines ... **EM1:** 38
mechanical fasteners ... **EM1:** 706-708
parameters ... **EM1:** 38-39
poor, and steel casting failure ... **A11:** 391
poppet-valve stem failure from improper ... **A11:** 320-321
for resin transfer molding ... **EM1:** 168-171
in shafts ... **A11:** 459
for sour gas environments ... **A11:** 300
and steel casting failure ... **A11:** 391
to prevent hydrogen damage ... **A11:** 250-251
Material substitution
as design philosophy ... **EM1:** 313
Material transformations ... **A10:** 376, 566
Material transport systems
during sintering ... **A7:** 312-314
Material utilization
improved by HIP ... **A7:** 442, 443
and STAMP process ... **A7:** 550
Material waste
of composites ... **EM1:** 37
Material(s) *See also* Dielectric materials; Electronic materials; Engineering plastics; Engineering plastics families; Lead frame materials; Material control; Material costs; Material properties; Material selection; Materials analysis; Materials and electronic phenomena; Materials selection; Matrix materials; Microelectronic materials; Polymer families; Polymers; specific materials
alternative, for PWBs ... **EL1:** 604-608
behavior, in large-strain range ... **EM2:** 681-684
behavior, rheology characteristics ... **EM2:** 612-614
characterization, of composites ... **A11:** 739-740
for coating/encapsulation ... **EL1:** 241-243
composition control, epoxies ... **EL1:** 833
Material(s) (continued)
condensation soldering ... **EL1:** 703
conditions ... **A11:** 118-119
costs ... **EM2:** 648-649
defects ... **A11:** 307, 317, 323, 551-553, 603, 744, 747
and design ... **EM2:** 612-617
and design, introduction ... **EL1:** 1
dielectric, properties ... **EL1:** 604-608
economics of ... **EM2:** 646-650
effects, in PWB structures ... **EL1:** 81-82
and electronic data ... **EL1:** 89-111
flaws ... **A11:** 748-751, 753
for flexible epoxy systems ... **EL1:** 817-818
as hybrid package classification ... **EL1:** 453-454
information, data base ... **EM2:** 95
packaging, thermal properties ... **EL1:** 454
in plastic packages ... **EL1:** 210-212
properties ... **A11:** 115, 158-159
for silicone conformal coatings ... **EL1:** 773
and solderability ... **EL1:** 676-677
specification, improper, of heat exchangers ... **A11:** 635-636
strength, and corrosion-fatigue ... **A11:** 259
for thick-film hybrid technology ... **EL1:** 386
toughness property, defined ... **A11:** 60
transfer, cumulative, in rolling-element bearings ... **A11:** 496
wave soldering ... **EL1:** 701
Materials *See also* Die materials; Materials selection; Metals; Nonmetallic materials; Selection; Tool materials; Work metal
characteristics, of NDE methods ... **A17:** 51
characterization, by magnetic field testing ... **A17:** 131-132
development, corrosion testing for ... **A13:** 193
fully dense, constitutive equations ... **A14:** 417
functional requirements of ... **A13:** 338
permeability, in magnetic particle inspection ... **A17:** 99
studies, acoustic emission inspection in ... **A17:** 289
Materials analysis
analytical methods ... **EL1:** 918-926
in microelectronic manufacturing ... **EL1:** 917-918
Materials and electronic phenomena ... **EL1:** 89-111
atomic structure ... **EL1:** 90-92
chemical bonding ... **EL1:** 92-93
crystallography ... **EL1:** 96
electrical properties of materials ... **EL1:** 89-90
insulators and dielectric materials ... **EL1:** 99-100
ionic and electrical conductivity ... **EL1:** 93 96
magnetic materials ... **EL1:** 103
microelectronic materials, physical characteristics ... **EL1:** 104-111
molecular electronics ... **EL1:** 103
quantum mechanical band theory, solid state ... **EL1:** 96-103
resistive materials ... **EL1:** 100-101
semiconductors ... **EL1:** 101-103
surface phenomena ... **EL1:** 103-104
Materials and Processes Technical Information System (MAPTIS) ... **EM4:** 692
Materials Business File ... **EM3:** 72
Materials characterization
defined ... **A10:** 1
Handbook articles, organization of ... **A10:** 1
introduction ... **A10:** 1
qualitative chemical tests ... **A10:** 168
sampling for ... **A10:** 12-18
tandem methods of ... **A10:** 1
Materials dust explosion hazard testing ... **A7:** 198
Materials Engineering ... **EM3:** 66
concepts ... **M3:** 825-834
total life cycle ... **M3:** 826, 829-830
Materials for friction and wear applications
introduction ... **A18:** 693-694
adaptation of tools ... **A18:** 693
material state ... **A18:** 693-694
selection guidelines ... **A18:** 693
Materials Research Society (MRS)
as information source ... **EM1:** 41
Materials science
microbeam analysis strategy for ... **A10:** 529-530
NAA application in ... **A10:** 234
PIXE studies in semiconductor industry ... **A10:** 107
Materials Science newsletter (NTIS) ... **EM1:** 41

Materials selection *See also* Alloy selection; Materials; Raw materials; Selection ... **A13:** 321-337
adjoining material effects ... **EM2:** 71
aluminum casting alloys ... **A2:** 126-131
base material/insulator ... **EL1:** 113-114
basic concepts ... **M3:** 825-834, 835-837
blanking and piercing dies ... **M3:** 484-488
blends and alloys, engineering plastics ... **EM2:** 632-637
for brazed joints ... **A13:** 879
of cast irons ... **A13:** 571
closed-die forging tools ... **M3:** 526-532
coating system ... **A13:** 412
coining dies ... **M3:** 508-511
cold extrusion tools ... **M3:** 514-520
cold heading tools ... **M3:** 512-513
computer data banks for ... **EM2:** 1
conductors ... **EL1:** 114
core system ... **A15:** 237-241
core/mold, permanent mold casting ... **A15:** 280
corrosion testing for ... **A13:** 193
of corrosive media, corrosion testing ... **A13:** 194
cost-effective ... **A13:** 335
and costs ... **EM2:** 1
cutting tools ... **M3:** 470-477
deep drawing dies ... **M3:** 494-499
and design ... **A13:** 321, 338-339, 343
for dew point corrosion ... **A13:** 1003
die casting dies ... **M3:** 542-543
dies for drawing wire, bar, and tubing ... **M3:** 521-525
economics ... **A13:** 335
economics, of materials ... **EM2:** 646-650
economy in manufacture ... **M3:** 838-856
electrical conductivity ... **EM2:** 475
electrical contacts ... **M3:** 662-695
in electronics industry
epoxy materials ... **EL1:** 825-827
filament winding ... **EM2:** 371-372
for flue gas desulfurization ... **A13:** 1367-1368
gages ... **M3:** 554-557
for galvanic corrosion ... **A13:** 86, 324
for gas wells ... **A13:** 1249
hole making and circuitization ... **EL1:** 115
hot extrusion tools ... **M3:** 537-541
hot upset forging tools ... **M3:** 533-536
for hydrogen damage ... **A13:** 329-333
initial material/processing choice ... **EM2:** 79-80
introduction ... **EM2:** 611
for liquid-metal containment ... **A13:** 59
literature ... **A13:** 321
material and design ... **EM2:** 612-617
metalworking rolls ... **M3:** 502-507
for mining and mill application ... **A13:** 1294
molds for plastics and rubber ... **M3:** 546-550
for molten salt corrosion ... **A13:** 91
for oil/gas production ... **A13:** 1235-1236
for operating conditions ... **A13:** 321
organic coatings ... **A13:** 529
for passive components ... **EL1:** 182-184
permanent magnet materials ... **A2:** 792-802
petroleum refining and petrochemical operations ... **A13:** 1262
for pipeline corrosion ... **A13:** 1289-1292
of plastics ... **EM2:** 68, 475
plating for PTH, vias, surface wiring ... **EL1:** 113-115
powder-compacting tools ... **M3:** 544-545
precious metal plating ... **EL1:** 117
press forming dies ... **M3:** 489-493
process ... **A13:** 321-323
process flow, as model for ... **EL1:** 112-113
and process selection ... **EM2:** 611
for radiation control ... **A13:** 951
refractory metals and alloys ... **A2:** 557-560
resins, for injection molding ... **EM2:** 322-323
shear spinning tools ... **M3:** 500-501
shearing and slitting tools ... **M3:** 478-483

Materials selection (continued)
for shipped second-level package ... **EL1:** 113-117
for simulated service testing ... **A13:** 204-205
sliding bearings ... **M3:** 802-822
solder mask or protective coat ... **EL1:** 115-116
soldering/interconnection ... **EL1:** 116-117
for stress-corrosion cracking ... **A13:** 325-329
structural applications at subzero temperatures ... **M3:** 721-772
structural ceramics ... **A2:** 1019-1020
structural components, use of tool materials for ... **M3:** 558-559
for suction rolls ... **A13:** 1208
supplier data sheets, interpreting ... **EM2:** 638-645
for thermocouples ... **A2:** 885 **M3:** 696-720
thermoplastic injection molding ... **EM2:** 308
thermoplastic resins ... **EM2:** 618-625
thermoset resins ... **EM2:** 55, 626-631
for thermosets ... **EM2:** 55
thread-rolling dies ... **M3:** 551-553
to avoid erosion-corrosion ... **A13:** 332-333
to avoid intergranular corrosion ... **A13:** 324-325
to avoid/minimize corrosion ... **A13:** 323-335
for UV coatings ... **EL1:** 785-786
wax, for investment casting ... **A15:** 254-255
wear applications ... **M3:** 563-594
Materials Technology Institute tests
crevice corrosion ... **A13:** 304
Materials Week ... **EM3:** 71
Mathar-Soete drilling technique ... **A6:** 1095
Mathematical modeling *See also* Computer applications; Modeling; Process modeling; Simulation
with machine vision ... **A17:** 37
of open-die forging ... **A14:** 65
precision forging ... **A14:** 159
Mathematical models *See also* Models; NDE reliability models
derivation ... **A17:** 751
fitting ... **A17:** 751
model building ... **A17:** 741
parameter design ... **A17:** 750-752
for predicting NDE reliability ... **A17:** 702-715
for quality control ... **A17:** 740-742
variable effects ... **A17:** 741-742
Mathematics
of reliability ... **EL1:** 895-896
Matheson standard gas sample ... **A10:** 155
Mating fracture surface(s)
AISI/SAE alloy steels ... **A12:** 294, 318, 323
automotive bott ... **A12:** 274
cast aluminum alloys ... **A12:** 408, 411-412
crack origins ... **A12:** 13, 323
drive shaft ... **A12:** 258
effect of dimples ... **A12:** 13
elongated manganese sulfide inclusions ... **A12:** 26
fatigue, low-carbon steel ... **A12:** 251
fracture origin ... **A12:** 13, 323
martensitic staintess steels ... **A12:** 368
matching dimples on ... **A12:** 12
medium-carbon steels ... **A12:** 255, 258, 268, 274
overload ... **A12:** 411-412
sectioned ... **A12:** 268
sudden overload failure ... **A12:** 294
torsional overload fracture, high-carbon steets ... **A12:** 278
wrought aluminum alloys ... **A12:** 416
Mating-die method
stretch draw forming ... **A14:** 593
Matrices *See also* Composite(s), Composite materials; Constituent materials; Matrix; Resin systems; Resins; specific matrices
for aerospace ... **EM1:** 32-33
bismaleimide resins ... **EM1:** 32
for commercial applications ... **EM1:** 31-32
constituents, software for ... **EM1:** 275
cracking, energy method ... **EM1:** 241
defined ... **EM1:** 89
elastic displacement fields ... **EM1:** 187

Matrices (continued)
elasticity, measured ... **EM1:** 188
epoxy resins ... **EM1:** 32, 74-77
feathering in ... **EM1:** 789
forms ... **EM1:** 175-176
formulations ... **EM1:** 290-291
interlaminar, fracture toughness testing ... **EM1:** 264
introduction ... **EM1:** 31-33
lamina properties ... **EM1:** 308
mode strength of ... **EM1:** 198
polyester resins ... **EM1:** 31-32
polyimide resins ... **EM1:** 32
of prepreg forms ... **EM1:** 291-292
for pultrusion product ... **EM1:** 538-540
resin properties tests ... **EM1:** 289-294
tensile failure mode ... **EM1:** 100
thermal analysis techniques for ... **EM1:** 779-780
thermoplastic resins ... **EM1:** 32-33, 292-294
thermoset, properties and tests for ... **EM1:** 32, 289-290
type, effect, bulk molding compounds ... **EM1:** 162
vinyl ester resins ... **EM1:** 31-32
as weak link ... **EM1:** 31
Matrimid 5282 bismaleimide ... **EM1:** 81-82, 86
Matrix *See also* Matrices; Matrix effects ... **EM3:** 18
absorption, as XRPD source of error ... **A10:** 341
complex, nontrace components in ... **A10:** 109
crazing, in polymers ... **A11:** 759, 761
creep, alloy steels ... **A12:** 349
defined ... **A9:** 11 **A10:** 676 **A11:** 6 **A13:** 46 **A17:** 384 **EM1:** 15 **EM2:** 26
definition ... **M6:** 11
feathering, in composites ... **A11:** 736
guidelines, for material selection ... **EM1:** 38
methods, for quantitative IR analysis ... **A10:** 117
reduced lamina stiffness ... **EM1:** 219
resin tests ... **EM1:** 289-294
sample ... **A10:** 83, 162, 168-170
separation, effect of crack tip plastic zone ... **A12:** 16
Matrix alloys *See also* Composites; Metal-matrix composites
aluminum-base discontinuous ... **A14:** 251
Matrix composites *See* Composites; Matrix materials; Metal-matrix composites
Matrix coupon encapsulation method ... **EM3:** 391
Matrix cracking
as failure mechanism ... **A8:** 696
Matrix densification ... **A7:** 314
Matrix digestion test method ... **EM1:** 765-767, 774
Matrix effects *See also* Interferences
in AFS ... **A10:** 46
in emission spectroscopy ... **A10:** 33
in titanium ... **A10:** 97
in x-ray spectrometry ... **A10:** 97
Matrix isolation
defined ... **A10:** 676
Matrix materials *See also* Composites; Metal-matrix composites; Metal-matrix Composites; Subcomposites
composition, in cermets ... **A2:** 991
for niobium-titanium superconducting materials ... **A2:** 1044-1046
for packaging applications ... **EL1:** 1120
and reinforcement ... **EL1:** 1119-1121
Matrix metal ... **A7:** 7
Matrix method of calculating diffraction contrast for translation interfaces ... **A9:** 118-119
Matrix phase, identification of
in eutectic microstructures ... **A9:** 620
Matrix properties
cyanate homopolymer ... **EM2:** 235
thermosetting resins ... **EM2:** 235
Matrix properties of fiber composite specimens, altering for purposes of
examination ... **A9:** 587
Matrix reinforcements *See* Reinforcements

SUBJECTS OF THE INDEXED VOLUMES: ASM Handbook (designated by the letter "A"): **A1:** Properties and Selection: Irons, Steels, and High-Performance Alloys (1990); **A2:** Properties and Selection: Nonferrous Alloys and Special-Purpose Materials (1990); **A3:** Alloy Phase Diagrams (1992); **A4:** Heat Treating (1991); **A6:** Welding, Brazing, and Soldering (1993); **A7:** Powder Metallurgy (1984); **A8:** Mechanical Testing (1985); **A9:** Metallography and Microstructures (1985); **A10:** Materials Characterization (1986); **A11:** Failure Analysis and Prevention (1986); **A12:** Fractography (1987); **A13:** Corrosion (1987); **A14:** Forming and Forging (1988); **A15:** Casting (1988); **A16:** Machining (1989); **A17:** Nondestructive Testing and Quality Control (1989); **A18:** Friction, Lubrication, and Wear Technology (1992). **Metals Handbook, 9th Edition** (designated by the letter "M"): **M1:** Properties and Selection: Irons and Steels (1978); **M2:** Properties and Selection: Nonferrous Alloys and Pure Metals (1979); **M3:** Properties and Selection: Stainless Steels, Tool Materials and Special-Purpose Materials (1980); **M4:** Heat Treating (1981); **M5:** Surface Cleaning, Finishing, and Coating (1982); **M6:** Welding, Brazing, and Soldering (1983). **Engineered Materials Handbook** (designated by the letters "EM"): **EM1:** Composites (1987); **EM2:** Engineering Plastics (1988); **EM3:** Adhesives and Sealants (1990); **EM4:** Ceramics and Glasses (1991); **Electronic Materials Handbook** (designated by the letters "EL"): **EL1:** Packaging (1989).

Matrix resin tests **EM1:** 289-294
Matrix resins *See* Resins
for pultrusion **EM2:** 394
Matrix structure
austenitic, ductile iron **A15:** 649
ductile iron **A15:** 655-656
in gray iron **A1:** 25 **A15:** 632-633
Matrix structure, ductile iron **M1:** 35, 45, 46, 53, 55
Matrix-plus-dispersed-phase structure **A9:** 604
Mats
chopped-strand, for resin transfer molding **EM1:** 169
continuous-strand, for resin transfer molding **EM1:** 169
defined **EM1:** 15
E-glass, pultruded **EM1:** 537
fiberglass, and woven roving **EM1:** 109
needled, defined *See* Needled mat
random-fiber, pultruded **EM1:** 537
short-fiber **EM1:** 121
tests for **EM1:** 291-292
Matte, and specular surface finish
aluminum alloy **A15:** 762-763
Matte surfaces, exposed
painting over **M5:** 332
Mattsson's pH 7.2 solution
for SCC testing of copper alloys **A8:** 525
Maturation room
sheet molding compounds **EM1:** 160
Mature grain structure
defined **A9:** 603
Maturing temperature **EM3:** 18
Maurer, E
as metallurgist **A15:** 32
Maurer structural diagrams
for cast iron production **A15:** 69-70
MAX
as synchrotron radiation source **A10:** 413
MAX-D index
use for unknown phase/ particle identification **A10:** 456
Maximum
abbreviation for **A10:** 690
Maximum available engine torque **A18:** 566
Maximum available skid torque **A18:** 566
Maximum compressive stress
in fatigue testing **A11:** 02
Maximum contact temperature **A18:** 40
Maximum crack length (MCL) parameter **A6:** 89, 90
Maximum deflection
in differing systems **A8:** 392
Maximum elongation **EM3:** 18
defined **EM2:** 26
Maximum energy content
permanent magnetic materials **A2:** 763, 782
Maximum erosion rate
defined **A18:** 12
Maximum flash temperature **A18:** 40, 41
Maximum flash temperature of test gears
symbol and units **A18:** 544
Maximum Hertzian contact pressure **A18:** 41
design chart **A18:** 91
Maximum likelihood estimation method
with Weibull distribution **A8:** 715
Maximum likelihood least squares procedure
for runouts **A8:** 701
Maximum load
Considére's construction for point of **A8:** 25
defined **A8:** 8
nomenclature for Raimondi- Boyd design chart **A18:** 91
Maximum normal strain
as creep rupture criterion **A8:** 344
Maximum normed residual (MNR) statistic **EM1:** 303
Maximum permeability
in magnetic particle inspection **A17:** 99
Maximum permissible dose
radiation **A17:** 301
Maximum permissible exposure (MPE) level
potential eye exposure during laser-beam welding **A6:** 267
Maximum pore size **A7:** 7
Maximum principal stress **EM3:** 491-492, 493, 494, 495, 496, 497
Maximum rate period
defined **A18:** 12
Maximum reduced stress
as creep rupture criterion **A8:** 344
Maximum relative shear **EM3:** 827, 828
Maximum shear stress
as creep rupture criterion **A8:** 343-344
Maximum strain levels
as creep rupture criterion **A8:** 344
forming limit diagrams for **A8:** 551
Maximum strength *See* Ultimate strength
Maximum stress
direction effect on dimple shape **A12:** 13
symbol for **A11:** 797
Maximum stress criterion
ply failure **EM1:** 238
Maximum stress level
and composite material age **A8:** 716
defined **A8:** 8
Maximum stress method **EM3:** 481
Maximum stress-intensity factor, K_{min}
defined **A8:** 8
Maximum tensile stress
in fatigue testing **A11:** 102
Maximum upper use temperature
composite **EM1:** 43
Maximum usable intrinsic device speed
defined **EL1:** 2
Maxwell element
as creep/viscoelasticity model..... **EM2:** 414, 659-660
Maxwell equation **EM3:** 429
Maxwell's equations **EL1:** 601
Mayer, Jacob
as discoverer **A15:** 31
Mayer, Julius von .... **A3:** 1 • 6
Mazak 3, zinc alloy
properties **A2:** 532-533
Mazak-3 *See* Zinc alloys, specific types, AG40A
Mazak-5 *See* Zinc alloys, specific types, AC41A
MCA *See* Multichannel analyzers
McClintock model
of void coalescence **A14:** 393
McConomy curves
and elevated-temperature service **A1:** 630, 631
McCormick reaper
of crucible steel **A15:** 31
McDonnell Douglas mast-mounted helicopter site
beryllium P/M parts **A7:** 760
McDonnell Douglas/Air Force Primary Adhesively Bonded Structures Technology (PABST) program **EM3:** 804
McGraw-Hill Seminary Center (New York) **EM2:** 95
McKee total elbow joint prosthesis **A11:** 670
McQuaid-Ehn grain size
defined **A9:** 11
McQuaid-Ehn test
to determine austenitic grain size **A1:** 241
MCR *See* Minimum creep rate
MDA BMI *See* Bis(4-maleimidodiphenyl) methane
MDI *See* Diphenyl-methane-diisocyanate
ME *See* Mössbauer effect
Mealing
defined **EL1:** 1149
Mean **EM3:** 786
binomial distribution **A8:** 647
curve **A8:** 665-667
exponential distribution **A8:** 635
as first moment of population **A8:** 628
normal distribution **A8:** 630-631
Poisson distribution **A8:** 637
statistical **A8:** 624-625
of statistical distributions **A8:** 629
Weibull distribution **A8:** 633
Mean coefficient of friction
symbol and units **A18:** 544
Mean contact temperature **A18:** 438, 440
Mean current of a train of rectangular pulses **A6:** 40
Mean curvature
convex figures **A12:** 195
Mean depth of penetration **A11:** 165, 797
Mean deviation
computed for normal distribution **A8:** 664
Mean effective pressure (MEP) **A18:** 554
Mean fatigue curve
confidence limits on **A8:** 700
Mean fatigue curve (continued)
scatter **A8:** 699-700
Mean fatigue life
vs. percent failure for different stresses **A8:** 699
Mean fatigue strength
staircase method for **A8:** 704
Mean free distance
stereological relationships **A12:** 195
Mean free distance between particles **A9:** 129
Mean free path
for gases and pressures **A17:** 59
Mean intercept length of grains **A9:** 129-130
Mean linear intercept measure
grain size **A10:** 358
Mean load
capability, portable corrosion-fatigue machine with **A8:** 243
and high temperatures, effect on corrosion fatigue **A8:** 254
pressure **A8:** 71
ultrasonic test facility with external load frame for **A8:** 248
zero, tensile strength accumulation in at **A8:** 353, 355
load-controlled continuous-cycling tests
Mean matrix intercept distance **A9:** 130
Mean particle size
in HIP **A7:** 425
Mean particle spacing **A9:** 129-130
Mean peak spacing **A18:** 334, 335
Mean rubbing interfacial temperature **A18:** 572
Mean square roughness **A18:** 335
Mean square strain **A18:** 468
Mean, statistical
in quality control **A17:** 727
Mean strain
effect on fatigue resistance of materials **A8:** 712-713
Mean strength
measure of **A8:** 626
Mean stress *See also* Engineering stress; Nominal stress; Normal stress; Residual stress; True stress
defined **A8:** 8
effect of dwell time **A12:** 63
effect on corrosion-fatigue **A11:** 255
effect on fatigue resistance of materials **A8:** 712-713
effect on fatigue strength **A8:** 374
effect on stress amplitude **A8:** 374 **A11:** 112
fatigue fracture from **A11:** 131
fatigue tests at **A11:** 110-111
Gerber's parabola and **A8:** 374
as loading condition, effect on fatigue cracking **A12:** 15
modified Goodman law **A8:** 374
and ramp rates **A12:** 63
Soderberg's law **A8:** 374
symbol for **A11:** 797
vs. strain range, René 95 .... **A8:** 353-354
Mean surface extrusion distance **A18:** 465
Mean tangent diameter
convex figures **A12:** 195
Mean time before failure (MTBF) predictions
microcircuitry **EL1:** 260
Mean time to failure (MTTF) **A18:** 493-494
Mean time-to-failure
abbreviation for **A11:** 797
Mean-free-path damping
EXAFS analysis **A10:** 410
Mean-time-to-failure (MTTF)
defined **EL1:** 896
Measling
as PTH failure mechanism **EL1:** 1025, 1149
Measurability parameter
defined **A8:** 387
Measure
angular, symbol for **A10:** 691
of central tendency **A8:** 624-625
SI standardized **A10:** 685
units of **A10:** 691
of variability **A8:** 625
Measured area under the rectified signal envelope (MARSE) **A17:** 282-283
Measured cell potential
abbreviation for **A10:** 690

Measured quantities
relationship to calculated quantities **A9:** 124
Measurement *See also* Corrosion testing; Evaluation
alternating current impedance **A13:** 200-201
case depth **M4:** 275-281
of corrosion potentials **A13:** 200
of corrosion test results **A13:** 194-195
depth **A8:** 85
for dimensional inspection **A15:** 559-560
experimental, or longitudinal resonance frequency of specimens **A8:** 249
of hydrogen, in aluminum alloys **A15:** 748
of indentation, Brinell test **A8:** 85
laser level, in pouring systems **A15:** 570
and load application in vacuum and gaseous fatigue tests **A8:** 411
of load at medium strain rates **A8:** 193
polarization resistance **A13:** 200, 214-215
quantitative, of dimples **A12:** 206-207
of soil corrosion **A13:** 209-210
statistical vs individual **A12:** 207-208
of strain at medium rates **A8:** 193
and temperature control in creep furnaces **A8:** 312-313
unit, in fractal analysis **A12:** 211-215
in wear test **A8:** 604, 606
Measurement circuit electronics
electrical testing **EL1:** 567
Measurement, dimensional *See* Dimensional measurements
Measurement of surface forces and adhesion **A18:** 399-405
atomic force microscopy **A18:** 401, 402
basic concepts **A18:** 399-400
interfacial energy **A18:** 399-400
surface energy **A18:** 399-400
surface energy and surface forces **A18:** 399
surface forces **A18:** 399
work of adhesion **A18:** 400, 403, 404, 405
measuring adhesion **A18:** 402-404
adhesion between curved surfaces **A18:** 403
fracture experiments **A18:** 403
fundamental adhesion measurements **A18:** 403
history dependence and sample preparation **A18:** 403
modes of separation **A18:** 404
practical adhesion measurements **A18:** 403-404
rate-dependent effects **A18:** 403
measuring surface forces **A18:** 400-402
Derjaguin approximation **A18:** 400, 401
environments **A18:** 402
instrumental requirements **A18:** 400
other measurements (adsorption, friction, refractive index, viscosity) **A18:** 402
preparation of surfaces and fluids **A18:** 402
pull-off force **A18:** 401, 403
substrate materials **A18:** 401
techniques **A18:** 400-401
state of the art **A18:** 404-405
Measurement uncertainty (σ) **A18:** 296
Measurements
on a planar section used for particle-size distribution calculations **A9:** 131
for quantitative metallography **A9:** 123-125
Measurements of film thickness *See* Film thickness measurements
Measurements of thickness *See* Thickness measurements
Measures
metric and conversion data **A8:** 721-723
Measuring junction
thermocouple thermometer **A2:** 870-871
Mechanical (peen) plating **A7:** 459
powders used **A7:** 572
Mechanical abrasion **EM3:** 42
sanding **EM3:** 42
Mechanical abrasion (continued)
shotblasting **EM3:** 42
Mechanical abuse
iron castings failure from **A11:** 375
Mechanical activation
defined **A18:** 12
Mechanical adhesion **EM3:** 18
defined **EM1:** 15
Mechanical agitation
for grain refinement **A15:** 476-477
Mechanical alignment
defined **A9:** 11
Mechanical alloying *See also* Oxide dispersion-strengthened alloys **A7:** 7 **M3:** 215
alloy applications **A2:** 943
aluminum P/M alloys **A2:** 202
commercial alloys **A2:** 944-947
equipment **A7:** 723
and fabrication **A2:** 947-949
in high-energy milling **A7:** 69
mechanically alloyed oxide alloys **A2:** 943-949
dispersion-strengthened (MA ODS) oxidation and hot-corrosion properties **A2:** 947
oxide-dispersion-strengthened superalloy products by **A7:** 527, 528
process **A2:** 943-944 **A7:** 723-725
Mechanical amorphization **EM4:** 23
Mechanical applications
polyphenylene sulfides (PPS) **EM2:** 186
Mechanical attrition
for beryllium powders **A7:** 170
for copier powders **A7:** 587
Mechanical attrition process
mechanical alloying **A2:** 202
reaction milling **A2:** 202
sinter-aluminum-pulver (SAP) technology **A2:** 202
Mechanical behavior
of materials under tension **A8:** 20-27
in torsion testing **A8:** 155
Mechanical belt grinding
stainless steel **M5:** 556
Mechanical bond
definition **M6:** 11
Mechanical bond (thermal spraying)
definition **A6:** 1211
Mechanical bonding process
babbitting **M5:** 356
Mechanical capping **M1:** 112
Mechanical cleaning *See also* Cleaning; specific processes by name
methods, liquid penetrant inspection **A17:** 81
types **A13:** 1143
Mechanical coating **A7:** 459 **M5:** 300-302
alloy coatings **M5:** 300-301
applications **M5:** 301-302
cadmium and cadmium combination coatings **M5:** 300-302
chromate coatings used with **M5:** 301-302
combination coatings **M5:** 300-302
copper flash **M5:** 302
corrosion resistance **M5:** 300-302
equipment **M5:** 302
fasteners **M5:** 302
galvanizing process **M5:** 300-301
hydrogen embrittlement and **M5:** 302
layered coatings **M5:** 301
peen plating **M5:** 300
powder metallurgy parts **M5:** 301-302
procedures **M5:** 300-302
sandwich coatings **M5:** 301
steel **M5:** 300, 302
tin and tin combination coatings **M5:** 300-302
waste treatment **M5:** 302
zinc and zinc combination coatings **M5:** 300-302
Mechanical comminution
of hard metals and oxide powders **A7:** 56
Mechanical comminution (continued)
of silver powders **A7:** 148
Mechanical conditions *See also* Material conditions
of shafts **A11:** 459-460
of stress-corrosion cracking **A11:** 214
Mechanical consolidation **EM4:** 125-129
consolidation methods and use of additives **EM4:** 126
dry consolidation methods **EM4:** 126-127
particle compact **EM4:** 126
powder formulation constituents **EM4:** 126
powder packing **EM4:** 125
starting powder characteristics **EM4:** 125-126
test methods **EM4:** 128-129
compaction rate diagram **EM4:** 128-129
compaction response diagram **EM4:** 128
wet consolidation methods **EM4:** 127-128
Mechanical cracks
defined **A11:** 7
in iron castings **A11:** 353
Mechanical crimp terminations
flexible printed boards **EL1:** 590-591
Mechanical cutting for welding preparation **A6:** 1178-1186
accessory equipment for straight-knife shearing **A6:** 1179- 1180
back gages **A6:** 1179-1180
front gages **A6:** 1180
hold-downs **A6:** 1179, 1180-1181
squaring arms **A6:** 1180
accuracy in straight-knife shearing **A6:** 1180
band saw blade selection **A6:** 1185-1186
bimetal blades **A6:** 1185
carbon steel blades **A6:** 1185, 1186
pitch **A6:** 1185-1186
saw width **A6:** 1185
tooth form **A6:** 1186
band saws **A6:** 1184-1186
blade composition **A6:** 1184
blade physical properties **A6:** 1184
blade selection **A6:** 1185-1186
parameter adjustment after visual examination of chips **A6:** 1185
types of machines **A6:** 1184-1185
blade design and production practice **A6:** 1182-1183
combination machines **A6:** 1181, 1182, 1183
cutting rate selection **A6:** 1186
distortion **A6:** 1182-1183
fixturing **A6:** 1182
guillotine machines **A6:** 1181-1182, 1183
iron workers (heavy-duty shears) **A6:** 1181-1183
knife clearance **A6:** 1180
knife rake **A6:** 1180
nibblers **A6:** 1183-1184
punching and shearing machines **A6:** 1182
ram speed in straight-knife shearing **A6:** 1180
safety **A6:** 1180-1181
shears **A6:** 1178-1179
applicability **A6:** 1178
capacity **A6:** 1179
guillotine **A6:** 1178, 1179
hydraulic **A6:** 1179
machines for straight-knife shearing **A6:** 1178
mechanical **A6:** 1178-1179
resquaring **A6:** 1178, 1179
squaring **A6:** 1178, 1179, 1180-1181
speed selection **A6:** 1186
in space and low-gravity environments **A6:** 1023
Mechanical damage
AISI/SAE alloy steels **A12:** 305, 337
cast aluminum alloys **A12:** 406, 407, 413
common types **A12:** 72
defined **A12:** 72
during shear deformation, tool steels **A12:** 380
high-carbon steels **A12:** 283
scab, high-carbon steels **A12:** 284
wrought aluminum alloys **A12:** 426

Mechanical defects
variable resistors ... **EL1:** 1002
Mechanical deformation *See also* Deformation
effect on texturing ... **A10:** 358
Mechanical degradation
polymer chains ... **EM2:** 424
Mechanical design *See also* Design
categories of ... **A11:** 115
costs of ... **EM2:** 83
and lubricant failure ... **A11:** 154
and tool and die failure ... **A11:** 564-566
Mechanical design considerations
flexible printed boards ... **EL1:** 588-592
Mechanical disintegration ... **A7:** 7
Mechanical durability ... **EL1:** 45, 55-65
Mechanical failure
defined ... **EL1:** 56
as fatigue failure mechanism ... **EM2:** 744-749
and package reliability ... **EL1:** 45
Mechanical fasteners *See also* Blind fastening; Fastener holes; Fasteners; Joint(s)
clamp-up ... **EM1:** 706-707
corrosion compatibility of ... **EM1:** 706
head configuration ... **EM1:** 706
interference fit ... **EM1:** 707-708
lightning strike protection ... **EM1:** 708
selection of ... **EM1:** 706-708
shear load transfer through ... **EM1:** 488
strength of ... **EM1:** 706
tipping or cocking of ... **EM1:** 706
Mechanical fasteners, failures of ... **A11:** 529-549
blind fasteners ... **A11:** 545
by liquid-metal embrittlement ... **A11:** 543-544
causes of ... **A11:** 530-531
corrosion in threaded fasteners ... **A11:** 535-541
corrosion protection ... **A11:** 541-542
fastener performance at elevated temperatures ... **A11:** 542-543
fastener types ... **A11:** 529
fatigue in threaded fasteners ... **A11:** 531-534
from fretting ... **A11:** 534-535
origins of ... **A11:** 529-530
pin fasteners ... **A11:** 545
rivets ... **A11:** 544-545
semipermanent pins ... **A11:** 545-548
special-purpose fasteners ... **A11:** 548-549
Mechanical fastening
types ... **EM2:** 711-713
Mechanical fastening function ... **EM3:** 33
conventional method ... **EM3:** 33
spot welding ... **EM3:** 33
Mechanical feeds ... **A7:** 460-461
Mechanical fibering ... **A9:** 686
in forging ... **A11:** 319
in M-36 electrical steel sheet ... **A9:** 687
Mechanical filter
for explosively loaded torsional Kolsky bar ... **A8:** 224
Mechanical finishing *See* specific processes by name
Mechanical forging presses *See also* Mechanical presses; Presses ... **A14:** 29-31
accuracy of ... **A14:** 39-40
characterization ... **A14:** 37-40
crank presses with modified drives ... **A14:** 40
kinematics, slider-crank mechanism ... **A14:** 37-38
knuckle-joint ... **A14:** 40
load and energy characteristics ... **A14:** 38-39
as stroke-restricted machines ... **A14:** 25, 37
time-dependent characteristics ... **A14:** 39
total deflection in ... **A14:** 39
Mechanical galvanizing *See also* Mechanical coating ... **M5:** 300-301
as zinc coating ... **A2:** 528
Mechanical grinding and finishing
molybdenum ... **M5:** 659-660
niobium ... **M5:** 662-663
tantalum ... **M5:** 662-663
tungsten ... **M5:** 659-660
Mechanical hands
for unloading presses ... **A14:** 500
Mechanical hysteresis ... **EM3:** 18
defined ... **A8:** 8 **EM2:** 26
Mechanical impedence, and high-temperature testing
Kolsky bar ... **A8:** 222-223
Mechanical implant failures ... **A11:** 681-687
Mechanical integrity
of system package ... **EL1:** 21
Mechanical interlocking ... **A7:** 303
as mechanism of green strength ... **A7:** 303
as milling process ... **A7:** 62
Mechanical joints
of heat exchangers ... **A11:** 639
Mechanical keying ... **EM3:** 416
Mechanical metallurgy
defined ... **A8:** 8
Mechanical mounting devices ... **A9:** 28-29
Mechanical mounting of thin-sheet specimens ... **A9:** 31
Mechanical multilevel press ... **A7:** 670
Mechanical packaging *See also* Packaging
for midrange computer ... **EL1:** 22
trade-off factors ... **EL1:** 21
Mechanical parts
critical properties ... **EM2:** 458
Mechanical plating *See also* Mechanical coating ... **A13:** 9, 767 **M5:** 300-301
of steel springs ... **A1:** 312 **M1:** 291
Mechanical polishing *See also* Polishing ... **A9:** 33-47 **A16:** 34
of aluminum alloys ... **A9:** 352-353
damage from ... **A9:** 39-40
defined ... **A9:** 11
electropolishing compared to ... **M5:** 306-308
lead and lead alloys ... **A9:** 416
of magnesium alloys ... **A9:** 425
Raman analysis of ... **A10:** 133
of replication microscopy specimens ... **A17:** 52
uranium and uranium alloys ... **A9:** 478
Mechanical press brakes ... **A14:** 9, 533-534
Mechanical press forging ... **A8:** 158
Mechanical presses *See also* Hydrautic presses; Mechanical forging presses; Presses; ,specific presses ... **A7:** 7, 329-332
for aluminum alloys ... **A14:** 245
for bar bending ... **A14:** 662
for blanking ... **A14:** 451, 456
clutches and brakes ... **A14:** 496-497
for coining ... **A14:** 180
for copper and copper alloys ... **A14:** 256
defined ... **A14:** 9
drives for ... **A14:** 489-495
forging ... **A14:** 29-31
forging, torsion testing ... **A14:** 373
gear drives ... **A14:** 489-490
for heat-resistant alloys ... **A14:** 234
for magnesium alloys ... **A14:** 259
nongeared drive ... **A14:** 489-490
for precision forging ... **A14:** 170
shut height adjustment ... **A14:** 497
slide actuation in ... **A14:** 493-495
for titanium alloys ... **A14:** 273-274
vs hydraulic presses ... **A14:** 492
Mechanical prestressing
of leaf springs ... **A1:** 325-326 **M1:** 312-313
Mechanical proof testing
crack detection by ... **A7:** 484
Mechanical properties *See also* A-basis; B-basis; Electrical properties; Hardness; listings of specific properties in data compilations for individual metals and alloys; Material properties; Materials; Mechanical variables; Physical properties; Properties; S-basis; specific property by name; Tensile properties; Thermal properties; Thermomechanical analysis (TMA); Thermomechanical properties; Typical basis ... **A6:** 100-102 **EM2:** 433-438 **EM3:** 18
abrasion resistant cast irons ... **M1:** 619-620
abrasion resistant steels ... **M1:** 617-618
of abrasion-resistant cast irons ... **A1:** 94
acrylonitrile-butadiene-styrenes (ABS) ... **EM2:** 111-113
actinide metals ... **A2:** 1189-1198
after tube spinning ... **A14:** 678
aging effects ... **EM2:** 756
alloy steel sheet and strip ... **M1:** 165
of alloy steels ... **A14:** 165
aluminum ... **A2:** 3
aluminum alloy ... **A2:** 49-51
aluminum alloys ... **A15:** 765-767 **M2:** 55, 58, 59-62
of aluminum alloys, effect of melting on ... **A9:** 358
Mechanical properties (continued)
aluminum casting alloys ... **A2:** 49, 152-177 **M2:** 148-151
aluminum coated steel sheet ... **M1:** 171, 172
aluminum-lithium alloys ... **A14:** 250
amorphous materials and metallic glasses ... **A2:** 813-815
of AOD-refined steels ... **A15:** 428
ASTM test methods ... **EM2:** 334
at elevated temperatures, ductile iron ... **A15:** 662
austenitic manganese steels ... **A15:** 734
beryllium, instrument grades ... **A2:** 687
beryllium-copper alloys ... **A2:** 409-411
bismaleimides (BMI) ... **EM2:** 256
of carbon steel casting specimens ... **A9:** 230-231
of carbon steels ... **A1:** 202, 205, 206 **A14:** 164
cast copper and copper alloys ... **A2:** 224-228, 348, 350, 356-391
cast magnesium alloys ... **A2:** 491-516
cast steels ... **M1:** 378-399
cast structural effects ... **A15:** 167-168
and casting design ... **A15:** 765
ceramics ... **EL1:** 335-336
characterizing, technical meetings for ... **A17:** 51
closed-die steel forgings ... **M1:** 354-359, 374, 375
cobalt-base corrosion-resistant alloys ... **A2:** 454
cobalt-base high-temperature alloys ... **A2:** 452
cobalt-base wear-resistant alloys ... **A2:** 451
coefficients of variation ... **M1:** 676
cold finished bars ... **M1:** 221-234, 239-251
commercially pure tin ... **A2:** 518-519
commercially pure titanium ... **A2:** 594
of compacted graphite iron ... **A1:** 57-66
compressive strength
gray cast irons ... **M1:** 17, 19
malleable cast irons ... **M1:** 70
constructional steels for elevated temperature use ... **M1:** 639-663
copper alloy castings ... **M2:** 385-389, 387
copper and copper alloys ... **A2:** 217-219
copper-clad E-glass laminates ... **EL1:** 535-536
correlation to microstructure in beryllium- aluminum alloys ... **A9:** 130
corrosion-resistant alloys ... **A15:** 726-728
of corrosion-resistant cast irons ... **A1:** 99
of corrosion-resistant steel castings ... **A1:** 909, 914, 917-920
data sheet, typical ... **EM2:** 408
of deep drawn sheet product ... **A14:** 575
defined ... **A8:** 8 **A11:** 7 **EM1:** 15 **EM2:** 26
density effects ... **A14:** 189
as design parameters ... **EL1:** 517-523
of ductile iron ... **A1:** 40, 42-43, 48, 49 **A15:** 660 **M1:** 35, 36, 38-52
dynamic ... **EM2:** 435-436, 538
effect, compression molding/stamping ... **EM2:** 333
effect of crystallographic texture on ... **A9:** 700-701
effect of recovery changes ... **A9:** 693
effects, electrical contact materials ... **A2:** 840
effects, thermoplastic injection molding ... **EM2:** 311
electrical resistance alloys ... **A2:** 836-839
of engineering plastics ... **EM2:** 71-73
of engineering thermoplastics ... **EM2:** 98
and environmental exposure tests ... **EM1:** 295-301
epoxy resins ... **EL1:** 831
estimation of ... **A17:** 51
factors affecting, in high-temperature service ... **A1:** 636-644
of ferritic malleable iron ... **A1:** 75-76 **A15:** 692
ferrous P/M materials ... **M1:** 329, 332-346
fiberglass-polyester composites ... **EM2:** 247
of fillers ... **EM2:** 71-73
and flow localization ... **A8:** 169
flux effects ... **A15:** 448
forged titanium alloy part ... **A14:** 272
and formability ... **A1:** 573-575
and gating, die casting ... **A15:** 289
gold and gold alloys ... **A2:** 704-707
of gray iron ... **A1:** 29-31 **A15:** 643-644 **M1:** 16-23, 26-27, 28-32
hafnium products ... **A2:** 666
hardenable steels ... **M1:** 460-463, 466-469
of heat-resistant cast irons ... **A1:** 99
high-impact polystyrenes (PS, HIPS) ... **EM2:** 196
hot dip galvanized steel sheet ... **M1:** 171, 172
hot isostatic pressing, effects ... **A15:** 539-541

Mechanical properties (continued)
hot rolled steel bars and shapes **M1:** 201-209, 210, 211
hot-rolled bar **A14:** 172
HSLA steels **M1:** 406, 408, 411-419
inclusion effects **A15:** 88
Invar and glass sealing alloy **A2:** 892
ionomers **EM2:** 120-121
of iron-copper-carbon alloys **A14:** 200, 201
lead frame materials **EL1:** 489, 731
of leaf springs **A1:** 325
liquid crystal polymers (I CP) **EM2:** 181
long terne steel sheet **M1:** 171, 174
long-term **EM2:** 612-613
long-term elevated-temperature tests **A1:** 620, 622-624, 629, 630
of low-alloy cast steels **A1:** 374
of low-alloy steel sheet/strip **A1:** 209-210
low-carbon steel sheet and strip **M1:** 155-156, 158-161
low-expansion alloys **M3:** 794, 795
magnesium **M2:** 529-533
magnesium alloys **A2:** 457, 460-462
and magnetization, physical theory **A17:** 131-132
of make-break arcing contacts **A2:** 841
malleable irons **M1:** 64-73
maraging steels **M1:** 448, 449-451
martensitic malleable iron **A15:** 693-697
mean free distance between particles used to study **A9:** 129
measurement **EM2:** 433-435
measurement of used to investigate crystallographic texture **A9:** 701
mechanically alloyed oxide alloys **A2:** 946-947
dispersion-strengthened (MA ODS)
metal phase **A17:** 289
moisture effects **EM2:** 766-769
molecular/physical properties of **EM2:** 436-437
molybdenum and molybdenum alloys **A2:** 575-577
nickel aluminides **A2:** 917-918
nickel-titanium shape memory effect (SME) alloys **A2:** 899
niobium alloys **A2:** 567-571
of nonferrous alloys, compared **A15:** 787
on supplier data sheets **EM2:** 639-641
palladium and palladium alloys **A2:** 715
parylene coatings **EL1:** 793
of pearlitic and martensitic malleable iron **A1:** 74, 81
pearlitic malleable iron **A15:** 693-696
permanent magnet materials **A2:** 793
pewter sheet **A2:** 523
phenolics **EM2:** 245
plate, steel **M1:** 188-198
of plate steels **A9:** 203
platinum and platinum alloys **A2:** 707-714
polyamide-imides (PAI) **EM2:** 129, 133
polyamides (PA) **EM2:** 126
polyaryl sulfones (PAS) **EM2:** 145
polyarylates (PAR) **EM2:** 138-139
polyetheretherketone (PEEK) **EM2:** 144
polyethylene terephthalates (PET) **EM2:** 173
polyimides, thin-film hybrids **EL1:** 325-326
of polymer blends **EM2:** 634-636
polymeric substrates **EL1:** 338
of polymers **EM2:** 60-61
polysulfones (PSU) **EM2:** 201
porosity effects, powder forgings **A14:** 203
of powder forgings **A14:** 198-203
of powder metallurgy (P/M) superalloys **A1:** 974, 975-976
of precision forgings **A14:** 158
of pultrusions **EM2:** 395
pure cobalt **A2:** 447
of quenched and tempered alloy steels **A1:** 389, 392, 396, 397
range, engineering materials **EM2:** 433

Mechanical properties (continued)
of rare earth metals **A2:** 725, 1178-1189
refractory metal fiber-reinforced composites **A2:** 583
refractory metals, forming effects **A14:** 785-786
of reinforced RIM (PUR) materials **EM2:** 263
of reinforcement fibers **EM2:** 506
of reinforcements **EL1:** 1119, 1120 **EM2:** 71-73
rhenium **A2:** 582
as selection criterion, electrical contact materials **A2:** 840
semisolid forged aluminum alloys **A15:** 333
shape memory alloys **A2:** 898
short-term **EM2:** 612
short-term elevated-temperature tests **A1:** 622, 627, 628
silver and silver alloys **A2:** 699-704
of solder **EL1:** 639-640
specialty HIPS **EM2:** 199
springs, steel **M1:** 283-288, 291-296, 297, 300, 301, 302, 303, 304, 312
of squeeze cast products **A15:** 323
stainless steels **M3:** 6, 17, 18-28
statistical analyses of **EM1:** 302-307
of steel castings **A1:** 367, 375, 376
of steel plate **A1:** 237, 238
structural beryllium grades **A2:** 686
structural ceramics **A2:** 1019, 1021-1024
of structural RIM **EM2:** 263
styrene-acrylonitriles (SAN, OSA ASA) **EM2:** 215
tantalum **A2:** 573
terms and symbols **A8:** 662
test methods **EM1:** 286-287, 296-300, 731-733
thermoplastic polyimides (TPI) **EM2:** 177
thermoplastic resins **EM2:** 618
tin solders **A2:** 521-522
titanium alloy castings **A2:** 639
titanium aluminides **A2:** 926-929
titanium casting alloy **A2:** 639-640
titanium castings **M3:** 408-410
of titanium P/M products **A2:** 647-654
transplutonium actinide metals **A2:** 1199
of tubular products **A17:** 561
ultrahigh molecular weight polyethylenes (UHMWPE) **EM2:** 169
ultrahigh-strength steels **M1:** 421-442
unsaturated polyesters **EM2:** 247
uranium **M3:** 774-775
uranium alloys, quenched **A2:** 674
uranium, unalloyed **A2:** 671-672
vinyl esters **EM2:** 273-274
vs carbon content, iron-carbon alloys **A14:** 200
vs temperature **EM2:** 752
welded joints, ductile iron **A15:** 665
white metal **A2:** 525
wrought aluminum and aluminum alloys **A2:** 57, 62-122
wrought copper an copper alloys **A2:** 265-345
wrought magnesium alloys **A2:** 480-491
of wrought stainless steels **A1:** 930, 931, 932-935
of wrought steels **A14:** 196
wrought titanium alloys **A2:** 621-622
of zinc alloys **A2:** 532-542

Mechanical properties, loss
from pitting corrosion **A13:** 232

Mechanical properties, powder
and ultrasonic velocity **A7:** 484

Mechanical pulping equipment
corrosion of **A13:** 1214-1218

Mechanical randomization
for estimating differences between key variables **A8:** 696

Mechanical satin finishing
aluminum and aluminum alloys **M5:** 574-575, 577

Mechanical scrubbing
copper and copper alloys **M5:** 617
sand reclamation by **A15:** 227, 352

Mechanical seal *See* Face seal

Mechanical shears
for plate and flat sheet **A14:** 701

Mechanical shock testing
as failure verification **EL1:** 1061
package-level **EL1:** 937-938
as physical testing **EL1:** 944

Mechanical sieve shaker **A7:** 215

Mechanical skimmers
induction furnaces **A15:** 374

Mechanical spring and electrical switch applications
beryllium-copper alloys **A2:** 417-418

Mechanical spring wire **M1:** 266, 267
for general use **A1:** 284-285
for special applications **A1:** 285

Mechanical stability (of a grease)
defined **A18:** 12

Mechanical stage
defined **A9:** 11

Mechanical steel tubing **A1:** 334-336

Mechanical stirring
used to achieve random orientation in castings **A9:** 701

Mechanical strength
of ceramics **EL1:** 336
of substrates **EL1:** 104

Mechanical strength, particle size distribution and mixture fluctuations
effects on **A7:** 186

Mechanical stress
and chip failure, plastic packages **EL1:** 480
-related failures, and accelerated testing **EL1:** 891-892

Mechanical stress relieving **A6:** 1100

Mechanical stress response
metals vs plastics **EM2:** 74-76

Mechanical stresses
effect on SCC of T-bolt **A11:** 538
in heat exchangers **A11:** 636-637

Mechanical stressing
for optical holographic interferometry **A17:** 410

Mechanical tensioning
high-carbon steel wire for **M1:** 264, 265

Mechanical testing **EM1:** 286-287, 296-300, 731-733 **EM2:** 544-558
background **EM2:** 544-546
defined **A8:** 8
as failure analysis **A11:** 18-19
of heat-exchanger failed parts **A11:** 629
metric and conversion data for **A8:** 721-723
modulus, assessment of **EM2:** 548-551
of shafts **A11:** 460
standard, for strain rate regime compression testing **A8:** 190
strength, assessment of **EM2:** 551-554
and tensile testing **A7:** 489-491
tensile tests, information from **EM2:** 546-548
tensile tests, limitations of **A11:** 18-19
of tool steels **A11:** 564
toughness, assessment of **EM2:** 554-557
using test bars **A7:** 489-491

Mechanical texture **A8:** 155

Mechanical toughness
of polyimides **EL1:** 768

Mechanical transducers
as mercury displacement measure **A7:** 268

Mechanical treatment, reduction of residual stresses **M6:** 889-89
proofstressing **M6:** 884, 89
vibratory stress relief **M6:** 89

Mechanical tubing *See* Steel tubular products **M1:** 323-326

Mechanical twin
defined **A9:** 11 **A11:** 7

Mechanical twin bands in Fe-3Si with indentations **A9:** 690

Mechanical twinning *See also* Microtwins; Twinning
in body-centered cubic materials **A8:** 34-35
effect in liquid erosion **A11:** 165

Mechanical twinning (continued)
effect of high strain rate on **A9:** 688
effect of low temperature on **A9:** 688
in rhenium and rhenium-bearing alloys **A9:** 447
in titanium and titanium alloys as a result of sectioning **A9:** 458
Mechanical twins
in 18Cr-8Ni stainless steel **A9:** 686
as a result of shearing **A9:** 23
in magnesium alloys **A9:** 425, 428
in metals with noncubic crystal structure **A9:** 37-38
as nucleation sites **A9:** 694
in zinc **A9:** 37-38
Mechanical upsetter
defined **A14:** 9
Mechanical variables *See also* Materials; Mechanical properties
in eddy current inspection **A17:** 563
ultrasonic inspection **A17:** 565
Mechanical wear
defined **A18:** 12
Mechanical winders
for glass fibers **EM1:** 45
Mechanical working
copper and copper alloys **A2:** 219, 223
defined **A14:** 9
effect on effective interdiffusional distance **A7:** 315
effect on homogenization kinetics **A7:** 315
segregation from **A11:** 315
Mechanical-chemical polishing process **A9:** 39-40
Mechanical/environmental combined testing **EM3:** 389-390
Mechanically assisted degradation **A13:** 136-144
cavitation erosion **A13:** 142
corrosion fatigue **A13:** 142-144
defined **A13:** 136
erosion **A13:** 136-138
fretting corrosion **A13:** 138-140
fretting fatigue **A13:** 141
types **A13:** 79
water drop impingement **A13:** 142
Mechanically assisted lay-up **EM1:** 605-607
complex-shape seating devices **EM1:** 606-607
locating partial plies **EM1:** 606
ply cutting **EM1:** 606
ply sorting/stacking **EM1:** 606
tape-laying machines **EM1:** 605-606
Mechanically capped steel **A1:** 143
Mechanically filed powders
brazing and soldering **A7:** 837
Mechanically foamed plastic *See also* Plastics **EM3:** 18
defined **EM2:** 26
Mechanically induced heating
for thermal inspection **A17:** 398
Mechanics
micro-fracture **A8:** 465-468
Medallions
production of **A14:** 184
Medals and medallions
of copper-based powder metals **A7:** 733
Media *See* Image conversion media; Real-time imaging media; Recording media
Median
statistical **A8:** 625
Median crack
definition **EM4:** 633
Median fatigue life
defined **A8:** 8
Median fatigue strength
at *N* cycles, defined **A8:** 8
identifying values **A8:** 701
value, trends, with increasing sample size **A8:** 706
Median fatigue-limit estimate
small-sample procedures for **A8:** 706
Median life, rolling-element bearing
symbol for **A11:** 797
Medical analysis
elemental content in toxicology epidemiology, PIXE analysis **A10:** 102
and materials characterization **A10:** 1
voltammetric monitoring of metals and nonmetals in **A10:** 188
Medical applications
blow molding **EM2:** 359
circuit applications **EL1:** 386
circuit construction **EL1:** 386-387
component attachment **EL1:** 387-388
liquid crystal polymers (LCP) **EM2:** 180
market size/outlook **EL1:** 389
materials technology **EL1:** 388-389
polycarbonates (PC) **EM2:** 151
polyether sulfones (PES, PESV) **EM2:** 159
polyether-imides (PEI) **EM2:** 156
polysulfones (PSU) **EM2:** 200
of radiography **A17:** 314
reliability and performance criteria **EL1:** 386
styrene-acrylonitriles (SAN, OSA ASA) **EM2:** 215
thermoplastic fluoropolymers **EM2:** 117
trends **EL1:** 386-389
ultrahigh molecular weight polyethylenes (UHMWPE) **EM2:** 168
Medical implants and prosthetic devices, friction and wear of **A18:** 656-662
alternative materials **A18:** 662
future prospects **A18:** 662
historical background **A18:** 656-658
alternative metals **A18:** 657
Austin Moore femoral implant **A18:** 657
ceramics **A18:** 657-658
compositions of implant materials **A18:** 658
current status **A18:** 658
early excision arthroplasty **A18:** 656
femoral head replacements **A18:** 656-657
interposition arthroplasty **A18:** 656
Judet prosthesis **A18:** 657
metal-on-metal implants **A18:** 657
metal-on-polymer implants **A18:** 657
physical and mechanical properties of materials **A18:** 659
replacement arthroplasty **A18:** 657-658
total hip replacements **A18:** 657
implant material properties **A18:** 658-662
ceramics **A18:** 658, 662
metals **A18:** 658, 661, 662
ultrahigh molecular weight polyethylene **A18:** 658, 662
ion-implanted Ti-6Al-4V in contact with UHMWPE **A18:** 779
wear lives of implants **A18:** 662
Medical industry applications *See also* Medical therapy
nickel alloys **A2:** 430
shape memory alloys **A2:** 901
structural ceramics **A2:** 1019
titanium and titanium alloy surgical implants **A2:** 589
titanium and titanium alloys **A2:** 589
of titanium P/M products **A2:** 647
Medical P/M applications **A7:** 657-663
for stainless steels **A7:** 731
Medical products **EM4:** 1007-1013
bioactive glasses **EM4:** 1009-1011
bioceramics present uses **EM4:** 1011
calcium phosphate ceramics **EM4:** 1011-1012
carbon-base implant materials **EM4:** 1012-1013
glass fibers **EM4:** 1029
glass-ceramics **EM4:** 1009-1011
nearly inert crystalline ceramics **EM4:** 1008-1009
porous ceramics **EM4:** 1009
resorbable calcium phosphates **EM4:** 1012
tissue attachment mechanisms **EM4:** 1007-1008
Medical therapy *See also* Medical industry applications
with aluminum, toxic effects **A2:** 1256
with bismuth, toxic effects **A2:** 1256-1257
with gallium, toxic effects **A2:** 1257
with gold, toxic effects **A2:** 1257
with lithium, toxic effects **A2:** 1257-1258
metals with toxicity related to **A2:** 1256-1258
with platinum, toxic effects **A2:** 1258
with vanadium, toxic effects **A2:** 1262
Medicine
ceramic applications **EM4:** 960
Medium bronze
properties and applications **A2:** 382
Medium strain rate compression testing **A8:** 192-196
Medium strain rate regime
drop test **A8:** 190
Medium strain rate regime (continued)
heat flow effect **A8:** 191
Medium tolerance
defined **A14:** 307
Medium-alloy air-hardening steels **M1:** 422, 434-439
Medium-alloy air-hardening tool steels *See* Tool steels, medium-alloy air-hardening
Medium-alloy steels **A14:** 81, 871
ferrographic application to identify wear particles **A18:** 305
Medium-carbon cast steels **A1:** 364, 372, 373
Medium-carbon low-alloy steels **A1:** 430-431 **M1:** 421-434
composition of tool and die steel groups **A6:** 674
electron-beam welding **A6:** 867
Medium-carbon steel
friction welding **A6:** 153
gas-metal arc welding
of aluminum bronzes **A6:** 828
of copper nickels **A6:** 828
of coppers **A6:** 828
of high-zinc brasses **A6:** 828
of low-zinc brasses **A6:** 828
of phosphor bronzes **A6:** 828
of silicon bronzes **A6:** 828
of special brasses **A6:** 828
of tin brasses **A6:** 828
gas-tungsten arc welding
of aluminum bronzes **A6:** 827
of copper nickels **A6:** 827
of coppers and copper-base alloys **A6:** 827
of phosphor bronzes **A6:** 827
of silicon bronzes **A6:** 827
hardfacing alloys **A6:** 798
HAZ microstructure formed by assembly-weld deposit **A11:** 423
hot dip tin coating of **M5:** 351
oxyacetylene welding **A6:** 281
oxyfuel gas cutting **A6:** 1159
postweld heat treatment **A6:** 648-649
repair welding **A6:** 1105
welded bellows liners, fatigue fracture in **A11:** 118
Medium-carbon steel, forged
ultrasonic inspection **A17:** 506-507
Medium-carbon steels *See also* Medium-carbon steels, specific types
applications **A15:** 714
automotive bolt, fatigue failure **A12:** 274
carbon content **A15:** 702
castings **M1:** 377, 384-386, 392, 393, 401
characteristics and applications **M1:** 457-458
cold finished bars **M1:** 215-251
definition of **A1:** 148
electrogas welding **M6:** 23
electron beam welding **M6:** 63
electroslag welding **M6:** 22
fatigue fracture **A12:** 258
forge welding **M6:** 67
fractographs **A12:** 253-276
fracture/faiture causes illustrated **A12:** 216
friction welding **M6:** 721
high frequency resistance welding **M6:** 76
I-beam, fatigue fracture surface **A12:** 263-264
inclusion effects **A15:** 92-93
machinability **M1:** 573
notch toughness **M1:** 695, 705
oxyfuel cutting **M6:** 90
oxyfuel gas cutting **A14:** 724
percussion welding **M6:** 74
plate, shear bands **A12:** 42
resistance spot welding **M6:** 477, 48
single-overload torsional fracture **A12:** 275
spring grades **M1:** 285
steel axle housing, fatigue fracture **A12:** 276
for structures **A13:** 1299
submerged arc welding **M6:** 115-11
threaded fasteners **M1:** 273-277
weld, HAZ cracking **A12:** 254
weldability **M1:** 561, 562, 563
weldability of **A1:** 608-609
Medium-carbon steels, specific types
AISI 1030 tapered shaft, torsional fatigue or "peeling" fracture **A12:** 253

Medium-carbon steels, specific types (continued)
AISI 1033, effect of temperature on fracture A12: 254
AISI 1035, brittle fracture A12: 258
AISI 1035, cup-and-cone tensile fracture A12: 253
AISI 1038 modified, fatigue fracture A12: 259
AISI 1039 shaft, fatigue fracture surface A12: 262
AISI 1040, fatigue fracture surface A12: 260
AISI 1041, fatigue fracture surface A12: 260, 261
AISI 1041, fracture by reverse stressing A12: 262
AISI 1041, fretting in keyed spindle A12: 262
AISI 1041, torsional fatigue fracture surface A12: 261
AISI 1045 crane gear, effects of flame hardening A12: 265
AISI 1045, fracture surfaces A12: 266-268
AISI 1046, fatigue fracture surface A12: 274
AISI 1046, reversed bending fatigue A12: 269
AISI 1050, bending overload fracture A12: 272
AISI 1050, fatigue fracture surface A12: 273
AISI 1050, fatigue fractures A12: 269, 273
AISI 1050, rotating bending failure A12: 273
AISI 1144, fatigue fracture surface A12: 263
ASTM A515 grade 70, crack mating surfaces A12: 255
ASTM A515 grade 70, fractured shell A12: 256-257
SAE 1050 modified, brittle fracture A12: 270-271
Medium-carbon structural steel
stress-strain curve for A8: 23
Medium-carbon ultrahigh-strength steels *See also* Ultrahigh-strength steels
classification of A1: 149
compositions of A1: 157
Medium-density P/M stainless steels
mechanical properties A13: 825
Medium-high-carbon steel wire
carbon content M1: 259
Medium-lead tin bronzes
applications A18: 750, 751
casting processes A18: 754, 755
composition A18: 751
designations A18: 751
mechanical properties A18: 752
product form A18: 751, 752
Medium-leaded brass *See also* Brasses; Wrought coppers and copper alloys
applications and properties A2: 307-308, 309
Medium-leaded naval brass
applications and properties A2: 320-321
Medium-low-carbon steel wire
carbon content M1: 259
Medium-modulus RTV silicones
for gasketing EM3: 54
Medium-pressure mercury vapor lamps
for radiation curing EL1: 864
Medium-rate compression testing A8: 190
Medium-scale integration (MSI) *See also* Integration development EL1: 160
Medium-silicon cast irons
compositions M1: 76
mechanical properties M1: 92
oxidation resistance M1: 94
physical properties M1: 88
Medium-temperature resin systems
epoxy EM2: 441
phenolics EM2: 441-442
silicones (SI) EM2: 442-443
Medium-temperature thermoset matrix composites EM1: 381-391
carbon fabric reinforced phenolic resin EM1: 382
glass fabric reinforced phenolic resin EM1: 381-382
graphite fiber reinforced phenolic resin EM1: 382
phenolic resin EM1: 381-382
Medium-viscosity sodium carboxymethylcellulose added to Veegum
application or function optimizing powder treatment and green forming EM4: 49
Mediums
for sample dissolution A10: 165, 166, 168
Meehanite metal
lapping A16: 492
milling with PCBN tools A16: 112
M_{eff} *See* Effective magnetization
Mega electron volt ions A7: 259
Megaelectron volt (MeV)
defined A17: 384
Megapact vibratory ball mill A7: 66
Meinhard nebulizer
for analytic ICP systems A10: 35
Meissner/Meissner-Ochsenfeld effect A2: 1030
MEKP *See* Methyl ethyl ketone peroxide
Melamine EM4: 933, 935
Melamine as a mounting material A9: 29
Melamine formaldehyde EM3: 104
Melamine plastics *See also* Plastics EM3: 18
defined EM2: 26
Melamine-formaldehyde resins EM2: 230, 321
Melilite
crystal structure EM4: 881
Melon
regenerator in liquid nitrocarburizing baths A4: 418
Melt EM3: 18
acid/base behavior A13: 89
characteristics, thermoplastic polyurethanes (TPUR) EM2: 207
defined EM1: 15 EM2: 26
fluoride A13: 90
processes, direct and marble, for glass fibers EM1: 45
properties ,of polymers EM2: 62
thermal gradients A13: 89
values, thermoplastic resins EM1: 294
Melt blending
of polymer-polymer mixtures EM2: 489
Melt extraction A7: 48, 49
techniques in titanium powder production A7: 167
Melt flow rate
for molecular weight EM2: 534
Melt impregnation
of thermoplastics EM1: 101
Melt index EM3: 18
defined EM2: 26
Melt infiltration
ceramic-matrix composites EM4: 840, 842
Melt infiltration process
for whisker-reinforced MMCs EM1: 898
Melt lubrication (phase-change lubrication)
defined A18: 12
Melt penetration
liquid-phase sintering A7: 319
Melt processing
of high-temperature superconductors A2: 1088
polyaryl sulfones (PAS) EM2: 146
Melt purification A15: 74-81
aluminum melts A15: 79-81
ferrous metals A15: 75-79
Melt refining
of aluminum alloys A15: 470-471
of copper alloys A15: 449-450
Melt rheology
cone/plate/parallel geometries in EM2: 535-540
Melt shop
steel castings A15: 310
Melt spinning
aluminum P/M alloys A2: 202
of metallic glasses A2: 806
Melt spinning technology A7: 48
Melt strength EM3: 18
defined EM2: 26
Melt treatments
copper alloys A15: 774-782
magnesium alloys A15: 802
Melt viscosity
cyanate resins EM2: 238
ionomers EM2: 122-123
polyvinyl chlorides (PVC) EM2: 210
Melt(s)
aluminum, purification of A15: 79-81
cleanliness, determining A15: 493
ferrous, purification of A15: 75-79
purification of A15: 71, 74-81
quality, reverberatory furnaces A15: 379
temperature, alloy addition A15: 72
vacuum arc remelting A15: 407
VIM, cleanliness of A15: 396
volume, in squeeze casting A15: 324
Melt-quench growth (MQG) technique
high-temperature superconductors A2: 1088
Melt-through A6: 1073
definition A6: 1211 M6: 11, 83
elimination by low peak temperatures M6: 59
in gas metal arc welds M6: 17
in weldments A17: 582
Melt-through welding
gas tungsten arc welding M6: 34
Melt/particle injection, laser *See* Laser processing techniques, laser melt/particle injection
Meltable solids
as IR samples A10: 112-113
Meltback time
definition M6: 11
Meltdown period
acid steelmaking A15: 364
Melting *See also* Heat treatment; Melting furnaces; Remelting
as a result of electric discharge machining A9: 27
acid, practice for A15: 363-365
of alloys, and typical gas/metal spray pattern A7: 25, 27
aluminum alloys A15: 746-747
in aluminum alloys, effect on mechanical properties and quench cracking A9: 358
aluminum and aluminum alloys A2: 9
austenitic ductile irons A15: 700
beryllium-copper alloys A2: 409, 421-423
beryllium-nickel casting alloys A2: 425
and casting, investment casting A15: 262-263
compacted graphite irons A15: 668
of consumable electrode under vacuum A15: 406
continuous flow vs drip method A15: 415
of copper A7: 116
copper alloys A15: 772-774
corrosion-resistant high-silicon irons A15: 701
crucible furnace A15: 383
during heat treatment, as casting defect A11: 386
energy requirements, reverberatory furnace A15: 376
eutectic A11: 122
fluxless, as inclusion control A15: 96
granulated powders as feedstock EM4: 100
of gray iron A15: 635-636
hafnium A2: 662
heats, various alloys A15: 376
high-chromium white irons A15: 682-683
high-silicon irons A15: 698-699, 701
incipient, casting failure from A11: 407
of lead alloys
in liquid metal baths, kinetic paths A15: 71
localized surface, tool steel A11: 573, 579
low-alloy steels A15: 721
magnesium alloys A15: 800-801
of malleable iron A1: 72 A15: 686-687
of maraging steels A1: 795
of metals, history A15: 15-23

Melting (continued)
mode, VIM process ... **A15:** 398
nickel alloys ... **A2:** 429
nickel-chromium white irons ... **A15:** 678-679
of niobium-titanium composite ... **A2:** 1044
operations, induction furnaces ... **A15:** 373-374
plasma cold hearth ... **A15:** 424
rate, and particle size distribution ... **A7:** 41, 43
and refining, Domfer iron powder process ... **A7:** 89-90
in steel plate production ... **A1:** 228
of superalloys ... **A1:** 968, 970-971, 986-988
times, estimated, for alloy additions ... **A15:** 71-74
of tin powders ... **A7:** 123
titanium and titanium alloy castings ... **A2:** 642
titanium and titanium alloys ... **A2:** 590
for titanium ingot production ... **A2:** 595-596
wrought copper and copper alloys ... **A2:** 242
zirconium ... **A2:** 662
of zirconium alloys ... **A15:** 837
Melting bath agitation
vacuum induction furnace ... **A15:** 397
Melting curves ... **A3:** 1 • 2, 16
Melting curves, rods
iron-carbon baths ... **A15:** 73
Melting furnaces
crucible furnaces ... **A15:** 374, 381-383
cupolas ... **A15:** 383-392
electric arc furnaces ... **A15:** 356-368
induction furnaces ... **A15:** 368-374
investment casting ... **A15:** 262
reverberatory furnaces ... **A15:** 374-381
Melting heats
basic steelmaking ... **A15:** 366-367
and raw materials, acid steelmaking ... **A15:** 364
Melting, localized
as failure mechanism ... **EL1:** 1013-1014
Melting point ... **EM3:** 18
adhesion ... **A6:** 144
aluminum casting alloys ... **A2:** 123
beryllium ... **A2:** 683-684
defined ... **A9:** 11 **A15:** 8 **EM2:** 26
of electrical contact materials ... **A2:** 840
of epoxy resin ... **EM1:** 736
high-temperature intermetallics ... **A2:** 935
of metal borides and boride-based cermets ... **A7:** 812
of rare earths ... **A2:** 720
in reduction reactions ... **A7:** 53
and solidification temperature, nucleation effects ... **A15:** 101
Melting point isotherm
in pure metals ... **A9:** 610
Melting points, crystallographic transformation and thermodynamic values ... **EM4:** 883-890
data listings ... **EM4:** 888-890
data sources ... **EM4:** 887-888
free energy of formation ... **EM4:** 886-887, 889-890
thermodynamic properties ... **EM4:** 883-887-889-890
types of invariant point ... **EM4:** 883, 884
Melting pot
polished-and-etched section of failed ... **A11:** 38
steel, failure of ... **A11:** 38-39
x-ray maps of failed ... **A11:** 39
Melting practice
effect on notch toughness ... **M1:** 706, 708
Melting pressure
defined ... **A9:** 11
Melting range *See also* Liqidus; Solidus
defined ... **A15:** 8
definition ... **M6:** 11
Melting rate
definition ... **M6:** 11
Melting stock
electrolytic iron powder as ... **A7:** 93
Melting temperature *See also* Fabrication characteristics; Melting point
aluminum casting alloys ... **A2:** 155-177
ductile iron ... **M1:** 52
incipient, cast copper alloys ... **A2:** 360
low, of indium- and bismuth-base alloys ... **A2:** 750
maraging steels ... **M1:** 447, 450, 451
of rare earth metals ... **A2:** 723
symbol for ... **A11:** 798
symbols for ... **A8:** 726
Melting temperatures
heterochain thermoplastic polymers ... **EM2:** 53
hydrocarbon thermoplastic polymers ... **EM2:** 50
nickel-base alloys ... **A14:** 265
nonhydrocarbon carbon-chain thermoplastic polymers ... **EM2:** 51-52
thermoplastic polymers ... **EM2:** 54
Melting times
calculated ... **A15:** 72
effect, induction-stirred melts ... **A15:** 74
Melting/fining ... **EM4:** 386-393
competing processes ... **EM4:** 388-389
devitrification ... **EM4:** 388, 389
electrode corrosion ... **EM4:** 388
reboil ... **EM4:** 388
refractory corrosion ... **EM4:** 388
volatilization ... **EM4:** 388-389, 393
emerging processes ... **EM4:** 393
furnaces for specific applications ... **EM4:** 391-392
container glass melters ... **EM4:** 391-392
fiberglass melters ... **EM4:** 392
float glass melters ... **EM4:** 392
melting rates ... **EM4:** 391
specialty glass melters ... **EM4:** 392
fundamentals ... **EM4:** 386-387
batch consolidation methods ... **EM4:** 386
batch melting ... **EM4:** 386, 387, 390
convection in melters ... **EM4:** 387-388
fining ... **EM4:** 386-387
heat transfer to batch materials ... **EM4:** 386
homogenizing ... **EM4:** 387
melting accelerators ... **EM4:** 386
furnace parameters ... **EM4:** 392
environmental impact ... **EM4:** 392-393
inclusion level ... **EM4:** 392
melting (glass preparation) cost ... **EM4:** 392
melting rate ... **EM4:** 392
melting defects ... **EM4:** 389, 392
bubbles ... **EM4:** 389, 392
chemical inhomogeneities ... **EM4:** 389
solid inclusions ... **EM4:** 389, 392
melting furnaces ... **EM4:** 389-391
all-electric cold-top melters ... **EM4:** 390-391
conditioning and delivery systems ... **EM4:** 391
control systems ... **EM4:** 391
electronically boosted ... **EM4:** 390
energy saving methods ... **EM4:** 390
fuel-fired tank furnaces ... **EM4:** 389
pot furnaces ... **EM4:** 389
purposes ... **EM4:** 386
refractories ... **EM4:** 391
Membrane
defined ... **A9:** 11
Membrane filters
for sample preparation ... **A10:** 94
Membrane potential
ion-selective electrode ... **A10:** 182
Membrane tests ... **EM3:** 373-378
blister test ... **EM3:** 373-374
adhered layers under nonzero in-plane stress ... **EM3:** 374-375
adhered layers under zero in-plane stress ... **EM3:** 373-374
fabrication of samples ... **EM3:** 373
Griffith energy balance ... **EM3:** 373, 376, 379
constrained blister test (CBT) ... **EM3:** 375, 376
cutout tests ... **EM3:** 377-378
indentation techniques ... **EM3:** 377, 378
inverted blister test ... **EM3:** 377
island blister test ... **EM3:** 375, 377
Peninsula blister test ... **EM3:** 377
Membrane-type ion chromatography suppressor ... **A10:** 660
Memory
cells, failure mechanisms ... **EL1:** 966
circuits, future trends ... **EL1:** 176-177
density, DRAM, SIP ... **EL1:** 439
mass ... **A10:** 92
standard types ... **EL1:** 160
Memory circuits
SEM analysis of ... **A11:** 768
Memory effects
furnace atomizers ... **A10:** 49
ultrasonic and fritted disk nebulizers ... **A10:** 36
Memory, plastic *See* Plastic memory
Mendelev number ... **A6:** 143
Menhaden fish oil (dispersant)
batch weight of formulation when used in nonoxidizing sintering atmospheres ... **EM4:** 163
Meniscograph solderability testing
for package leads ... **EL1:** 954
Meniscus
reading level of ... **A10:** 172
Meniscus rise test
for solderability ... **EL1:** 677
test standards used to evaluate solderability ... **A6:** 136
Menkes' disease (Menkes' "kinky-hair syndrome")
from copper toxicity ... **A2:** 1251-1252
Menstruum method ... **A7:** 7
of tungsten carbide powder production ... **A7:** 157
of tungsten/titanium carbide powder production ... **A7:** 158
Mensuration
as characterization method ... **EM1:** 294
Menzel-Gomer-Redhead (MGR) model
electron-stimulated desorption ... **A18:** 456
Mer *See also* Polymer(s) ... **EM2:** 26, 57-58 **EM3:** 18
defined ... **EM1:** 15
Merchant hybrid marketplace ... **EL1:** 253
Merchant quality
bars ... **M1:** 203-205
steel grades, compositions ... **M1:** 126
Merchant quality hot-rolled carbon steel bars ... **A1:** 243
grades of ... **A1:** 243
sizes of ... **A1:** 1-243
Merchant quality steels
compositions of ... **A1:** 150
Merchant wire ... **A1:** 282 **M1:** 264
Mercuric iodide detectors
capabilities ... **A10:** 95
Mercuric mercury
toxicity of ... **A2:** 1248
Mercuric oxide, on mercury film electrode
Raman spectroscopy ... **A10:** 136
Mercurous bromide and chloride, on mercury film electrode
Raman analysis ... **A10:** 136
Mercurous compounds
toxicity of ... **A2:** 1248
Mercury *See also* Mercury toxicity
alloying, aluminum casting alloys ... **A2:** 132
alloying, wrought aluminum alloy ... **A2:** 54
in alloys, oxyfuel gas cutting ... **A6:** 1165
in aluminum alloys ... **A15:** 746
amalgamation with aluminum ... **A13:** 589
anodic attack of ... **A10:** 204
cathodes, in electrogravimetry ... **A10:** 199-200
as cause of failure by cracking ... **A8:** 522
cavitating ... **A11:** 165
compressibility of ... **A7:** 268-269
contact angle with P/M materials ... **A7:** 269
-damage, LME in aluminum alloy plate ... **A11:** 79
in dental amalgam ... **A18:** 669
determined by controlled-potential coulometry ... **A10:** 209
effect in dropping mercury electrode ... **A10:** 189
as electrode in voltammetry ... **A10:** 189, 191
embrittlement by ... **A13:** 182
as embrittlement source ... **A11:** 234, 236
as embrittler, and material selection ... **A13:** 335
as embrittler of tantalum and titanium alloys ... **A11:** 234
in furnace atomizers ... **A10:** 49
-indium, cadmium embrittlement by ... **A11:** 230, 232
-indium solutions, LME of iron-aluminum alloys by ... **A11:** 225
lamp sources, spectral output of ... **A10:** 76
liquid, applications ... **A13:** 92
liquid, as LME embrittler ... **A11:** 226
liquid, decohesive fracture from ... **A12:** 18, 25
as low-melting embrittler ... **A12:** 29
as major toxic metal with multiple effects ... **A2:** 1247-1250
nitrate solution immersion of copper alloys for SCC testing ... **A8:** 525
plain carbon steel resistance to ... **A13:** 515
as platinum alloy ... **A7:** 15

Mercury (continued)
poisoning, in gilding ... **A15:** 21
porosity measures ... **A7:** 266-270
pure ... **M2:** 769-770
pure, properties ... **A2:** 1138
quartz-tube atomizers for ... **A10:** 49
solutions, copper/copper alloy SCC in ... **A13:** 634
species weighed in gravimetry ... **A10:** 172
sulfuric acid as dissolution medium ... **A10:** 165
tantalum resistance to ... **A13:** 733-735
TNAA detection limits ... **A10:** 238
toxicity ... **A6:** 1195 **A7:** 204-205
vapor, as embrittler ... **A12:** 30, 38, 424
vapor detection, atomic absorption spectrometry for ... **A10:** 43
volume displacements, measurement ... **A7:** 267-268
volumetric procedures for ... **A10:** 175
weighed as the chloride ... **A10:** 171
weighed as the sulfide ... **A10:** 171
Mercury cadmium telluride ... **EM3:** 594
Mercury cathodes ... **A10:** 170, 199-200
Mercury chloride
photochemical machining etchant ... **A16:** 591
Mercury intrusion
to analyze ceramic powder particle sizes ... **EM4:** 67
Mercury porosimetry ... **A7:** 267-270
applications ... **A7:** 269-270
limitations ... **EM4:** 581
as measure of pore size and distribution ... **A7:** 262, 266-271
reliability ... **A7:** 268-269
for temperature-time control in polymer removal techniques ... **EM4:** 137
to determine bulk density ... **EM4:** 582
to determine open porosity in porous materials ... **EM4:** 580-581, 582
Mercury pump
Toepler pump as ... **A10:** 152
Mercury, resistance of
to liquid-metal corrosion ... **A1:** 635
Mercury salt solutions
copper/copper alloy SCC in ... **A13:** 634
Mercury switch
analysis of surface films on electrical contacts in ... **A10:** 578-579
Mercury toxicity
alkyl mercury ... **A2:** 1249
biologic indicators ... **A2:** 1249
cellular metabolism ... **A2:** 1248
disposition ... **A2:** 1247
mercuric mercury ... **A2:** 1248
mercurous compounds ... **A2:** 1248
mercury vapor ... **A2:** 1248
metabolic transformation an excretion ... **A2:** 1247-1248
metallic mercury ... **A2:** 1249
organic mercury ... **A2:** 1248-1249
toxicology ... **A2:** 1248-1249
treatment ... **A2:** 1249-1250
Mercury vapor
toxicity of ... **A2:** 1248-1249
Mercury vapor lamps
for radiation curing ... **EL1:** 864
Mercury-sensitized radiation stimulation ... **EM3:** 594
Mercury-vapor light sources for microscopes ... **A9:** 72
Mergenthaler, Ottmar
as inventor ... **A15:** 35
Mesh ... **A7:** 7
screening ... **A7:** 176
Mesh generation, automated
die casting ... **A15:** 293
Mesh number ... **A7:** 7
Mesh size ... **A7:** 7
Mesh-belt conveyor furnace ... **A7:** 7, 351-355
Mesh-connected array
future of ... **EL1:** 8-9
Mesnager-Sachs boring-out technique ... **A6:** 1095
Mesophase ... **EM3:** 395
defined ... **EM1:** 15 **EM2:** 26
Mesophase pitch-based precursor fibers
carbon fiber properties ... **EM1:** 51
and polyacrylonitrile (PAN), compared ... **EM1:** 50
processing sequence ... **EM1:** 50
Metabolic transformation, of mercury
as toxin ... **A2:** 1247
Metabolites
GC/MS analysis of ... **A10:** 639
Metadynamic recrystallization ... **A9:** 691
Metakaolin ... **EM4:** 7
Metal *See also* specific types by name
chemical vapor deposition of ... **M5:** 381
cleaning processes, general description ... **M5:** 5
contamination by *See* Metallic impurities
corrosion protection *See* Corrosion protection
galvanic series ... **M5:** 431
recovery from plating wastes systems for ... **M5:** 315-319
Metal alloys
composition determined by ICP-AES analysis ... **A10:** 31
microstructural changes studied by x-ray topography ... **A10:** 366
Metal and metal-to-ceramic adhesion ... **A6:** 143-147
adhesion energy ... **A6:** 143
adhesion measurement ... **A6:** 444
adhesion, theory of ... **A6:** 143-144
grain boundary energy ... **A6:** 143
interface formation ... **A6:** 145-146
interfacial analysis ... **A6:** 144-145
interfacial energy ... **A6:** 143
properties affecting adhesion ... **A6:** 144
strength of interfaces ... **A6:** 146-147
welding (bonding) ... **A6:** 145
Metal arc cutting
definition ... **M6:** 11
Metal arc cutting (MAC)
definition ... **A6:** 1211
Metal band/metal platen
wear test for high-speed printer ... **A8:** 607
Metal bond systems
bonded-abrasive grains ... **A2:** 1015
Metal borides and boride-base cermets
properties ... **A2:** 1003
Metal borides, and boride-based cermets
properties ... **A7:** 812
Metal cans *See also* Metal packages; Metal-body devices
die attachments ... **EL1:** 213-217
as package ... **EL1:** 958
Metal carbide powders production ... **A7:** 156-158
tap densities ... **A7:** 277
Metal carbonyl powders ... **A7:** 135
formation and decomposition ... **A7:** 135-136
stability ... **A7:** 137
Metal casting *See also* Casting; Molding and casting processes
advantages ... **A15:** 37-45
applications ... **A15:** 37-45
computer applications in ... **A15:** 855-891
functional advantages ... **A15:** 39-41
market size ... **A15:** 37-45
market trends/end uses ... **A15:** 41-45
process developments ... **A15:** 38
versatility ... **A15:** 37-39
worldwide production ... **A15:** 42
Metal casting industry
automation of ... **A15:** 33-36
Metal charge
cupolas ... **A15:** 388
Metal cladding *See* Roll welding
Metal coatings *See also* Coatings; Electroplating
cobalt ... **A7:** 174
copper ... **A7:** 174
corrosion protection ... **M1:** 751-754
Metal coatings (continued)
electroplated ... **A15:** 562
nickel ... **A7:** 174
Metal composition factor (MCF) ... **A6:** 421
Metal compound vacuum coating ... **M5:** 399, 401
Metal conductivity, in electrozone size analysis *See also* Electrical conductivity ... **A7:** 221
Metal containers
corrosion ... **A13:** 88
Metal cored electrode
definition ... **A6:** 1211 **M6:** 11
Metal cores
constraining, thermal expansion properties ... **EL1:** 619-625
substrates ... **EL1:** 337-338
Metal cutting and grinding fluids
antifoaming additives ... **A16:** 124
antimicrobial agents ... **A16:** 124
antimisting additives ... **A16:** 124
application methods of flooding and misting ... **A16:** 126
biocides ... **A16:** 124
biological effects ... **A16:** 131-132
chemistry of ... **A16:** 87-88
control and test methods ... **A16:** 126-128
corrosion inhibitors ... **A16:** 124
cutting fluids flow recommendations ... **A16:** 127
detergents of long-chain alcohols ... **A16:** 124
disposal of ... **A16:** 131
dyes ... **A16:** 124
emulsions ... **A16:** 122, 123
emulsions of soaps ... **A16:** 123
extreme-pressure (EP) additives ... **A16:** 123-124, 125
fluid cleaning ... **A16:** 129-131
health practices ... **A16:** 131-132
microbes present ... **A16:** 132
odor masks ... **A16:** 124
recycling ... **A16:** 129-131
selection of ... **A16:** 124-125
solutions of cutting oils ... **A16:** 122-123
solutions of synthetic fluid lubricants ... **A16:** 123
solutions, water-base ... **A16:** 123
storage and distribution ... **A16:** 128-129
system cleaning ... **A16:** 129
Metal cutting, by refractory metals *See also* Flame cutting ... **A7:** 765
Metal dissolution ... **A13:** 29, 89, 343
Metal dust(s)
cloud explosion ... **A7:** 194
in injection molding ... **A7:** 495-496
production ... **A7:** 496
and sintered densities ... **A7:** 496
Metal dusting ... **A13:** 9, 380, 1312-1313
Metal dusting, stainless steel
in elevated temperatures ... **A11:** 272
Metal electrode
definition ... **A6:** 1211 **M6:** 11
Metal electrode face bonding (MELF) ceramic capacitors ... **EL1:** 178, 187
Metal electrode face bonding (MELF) chip resistors ... **EL1:** 178, 184
Metal electrode potentials ... **A7:** 140
Metal electrodeposition ... **A7:** 71-72
Metal encapsulation ... **A7:** 428-435
Metal fines
surface cleaning for ... **A13:** 380-381
Metal finishing, precision *See* Precision metal finishing
Metal flow *See also* Flow; Flowability; Fluid flow; Gating systems
ALPID simulations of ... **A14:** 427
in closed-die forging ... **A14:** 78
die casting ... **A15:** 288-292
in drawing ... **A14:** 576
during spike forging ... **A14:** 428
during swaging ... **A14:** 128-129
finite-element modeling ... **A14:** 352

Metal flow (continued)
in forgings ... **M1:** 353, 354, 361
in HERF processing ... **A14:** 105
in high-energy-rate forging ... **A14:** 102
in hot extrusion ... **A14:** 316-317
localization, workability effects ... **A14:** 364-365
parting line effects ... **A14:** 48
in powder forging ... **A14:** 194-197
in precision forging ... **A14:** 159
rates ... **A15:** 311
restraint, in deep drawing ... **A14:** 581-582
for simple parts ... **A14:** 50
steel wire fabrication ... **M1:** 589, 590, 591
two-dimensional, strain computation for ... **A14:** 433-434
and workability ... **A14:** 369-370
Metal flow patterns
used to identify plastic deformation modes ... **A9:** 686
Metal fume fever ... **A6:** 1196
as zinc toxicity ... **A2:** 1255
Metal halides
as additive to metalworking lubricants ... **A18:** 141
Metal horns
microwave inspection ... **A17:** 208-209
Metal hub flap polishing wheel ... **M5:** 109
Metal in-line treatment degassing system ... **A15:** 461, 463-464, 470
Metal injection *See also* Gating; Injection
chamber ... **A15:** 288-289
and gating design ... **A15:** 289-291
overflow ... **A15:** 289
sprues and runners ... **A15:** 289
Metal injection molding (MIM) technology ... **A1:** 818-819
advantages of ... **A1:** 819
applications ... **A1:** 820
factors impeding growth of ... **A1:** 819-820
mechanical properties of ... **A1:** 820
Metal ion deposition
galvanic corrosion ... **A13:** 84
Metal ions
determination of ... **A10:** 197, 200-201
Metal joints
nondestructive evaluation of adhesive bonds ... **EM3:** 743-776
Metal magnetism
in electrozone size analysis ... **A7:** 221
Metal matrix composites *See also* Composite materials
acoustic emission inspection ... **A17:** 287-288
aluminum-base ... **A13:** 859-861
coatings ... **A13:** 861-862
copper-base ... **A13:** 861
design, for corrosion prevention ... **A13:** 862-863
fiber-reinforced, cross section ... **A13:** 859
magnesium-base ... **A13:** 861
structural characteristics ... **A13:** 859
titanium, ultrasonic inspection ... **A17:** 250
Metal matrix composites (MMCs) ... **EM1:** 849-910
alumina fiber reinforced ... **EM1:** 31, 118
boron fibers in ... **EM1:** 31, 117, 851-857
continuous aluminum oxide fiber ... **EM1:** 874-877
continuous boron fiber ... **EM1:** 851-857
continuous graphite fiber ... **EM1:** 867-873
continuous silicon carbide fiber ... **EM1:** 858-866
continuous tungsten fiber ... **EM1:** 878-888
discontinuous ceramic fiber ... **EM1:** 903-910
discontinuous silicon fiber ... **EM1:** 889-895
elastic properties ... **EM1:** 187
introduction ... **EM1:** 849
whisker-reinforced ... **EM1:** 896-902
Metal Matrix Composites Information Analysis Center (MMCIAC) ... **EM1:** 41
Metal mold reaction
as casting defect ... **A11:** 384
Metal molds
end plate dimensions ... **A15:** 303
in permanent mold casting ... **A15:** 275
pretreatment ... **A15:** 303-304
vertical centrifugal casting ... **A15:** 301-302
Metal movement *See* Shrinkage
Metal oxides *See* Oxides ... **A18:** 144
for accelerating adhesive cure ... **EM3:** 179
as additive to metalworking lubricants ... **A18:** 141
particles, separation in high-temperature combustion ... **A10:** 222
Metal oxides (continued)
polymer additives ... **A18:** 154
Raman analysis of ... **A10:** 130-135
surface films, XPS determined oxidation states in ... **A10:** 568
Metal oxides (MO)
hydrogen-reduced ... **A7:** 340
Metal packages *See also* Metal cans; Metal-body devices ... **EM3:** 585, 588
defined ... **EL1:** 453-454
isolation, in testing/reliability ... **EL1:** 459
sealing methods ... **EL1:** 237-238
Metal particle segregation
polymers ... **A7:** 606
Metal patterns
equipment and processes for ... **A15:** 194-195
materials for ... **A15:** 197
Metal penetration *See also* Burn-in; Burn-on; Burned-on sand
as casting defect ... **A11:** 385
in cores ... **A15:** 240
defined ... **A11:** 7 **A15:** 8
vs inclusion effects ... **A15:** 90
Metal pipe
sampling trainload for percentage of alloying element ... **A10:** 15
Metal powder cutting ... **A7:** 842-845
apparatus ... **A7:** 843-844
applications ... **A7:** 844-845
definition ... **M6:** 11, 913-91
Metal powder cutting (POC) definition ... **A6:** 1211
Metal Powder Industries Federation, (MPIF) ... **A7:** 19
parts classification ... **A7:** 332-333
sampling procedures ... **A7:** 212
standardized ferrous materials ... **A7:** 463
Metal Powder Industries Federation Test Method 37
for case depth ... **A9:** 508
Metal Powder Producers Association ... **A7:** 19
Metal powder production ... **A7:** 23-24
basic processes ... **A7:** 24
individual powders ... **A7:** 24
Metal powder shipments
North American (1981) ... **A7:** 25
Metal powder slip casting
schematic ... **A2:** 984
Metal powder(s) ... **A7:** 7, 571
annealing ... **A7:** 182-185
apparent density ... **A7:** 272-275
applications, major ... **A7:** 572-574
auxiliary, in tungsten carbide powder production ... **A7:** 157
behavior under pressure ... **A7:** 297-304
blending ... **A7:** 186-189
bulk chemical analysis ... **A7:** 246-249
bulk properties ... **A7:** 211
characterization and testing ... **A7:** 211
chemical analysis and sampling ... **A7:** 248
cleaning ... **A7:** 178-181
compacted, green strength ... **A7:** 288-289
compressibility ... **A7:** 286-287
consolidation ... **A7:** 295-307
electrodeposition of ... **A7:** 71-72
explosions, preventing ... **A7:** 197-198
explosivity ... **A7:** 194-200
for filler materials ... **A7:** 816-822
flow rate ... **A7:** 278-281
for hardfacing ... **A7:** 823-826
high-energy compacting ... **A7:** 305
history ... **A7:** 14-20
homogeneous, sintering compacts ... **A7:** 309-312
hot pressing fully dense compacts ... **A7:** 501-521
mechanical fundamentals consolidation ... **A7:** 296-307
mixing ... **A7:** 186-189
optical sensing zone methods for ... **A7:** 223-225
packing ... **A7:** 296-297
physical and chemical properties ... **A7:** 211
premixing ... **A7:** 186-189
pyrophoricity ... **A7:** 194-200
roll compacting ... **A7:** 401-409
sampling ... **A7:** 212-213
spray drying of ... **A7:** 73-78
surface chemical analysis ... **A7:** 250-261
tap density ... **A7:** 276-277
toxicity ... **A7:** 201-208
Metal powder(s) (continued)
unconsolidated, rigid tool compaction ... **A7:** 322-328
Metal powders *See also* Powder forging; Powder metallurgy materials; Powders
for accelerating adhesive cure ... **EM3:** 179
as additive to metalworking lubricants ... **A18:** 141
cutting ... **A14:** 728
hot extrusion of ... **A14:** 322
mounting ... **A9:** 31
refractory metals and alloys ... **A2:** 560-565
sintering ... **M4:** 793-797
Metal preforms
in Ceracon process ... **A7:** 537-541
Metal processing *See also* Processing
mills, nickel alloy applications ... **A2:** 430
nickel alloy applications ... **A2:** 430
zirconium and hafnium ... **A2:** 661-662
Metal, purity characteristics
resistance-ratio test ... **M2:** 711-712, 713
trace-element analysis ... **M2:** 711
Metal recovery
in hydrometallurgical processing of cobalt and cobalt alloy powders ... **A7:** 145
in hydrometallurgical processing of nickel powder production ... **A7:** 134, 140-141
in Sherritt nickel powder production ... **A7:** 140-141
Metal removal rates
and adaptive control ... **A16:** 618, 619, 621, 622, 624
Al alloys ... **A16:** 764, 791
chemical milling ... **A16:** 581, 583, 585, 586
electrical discharge grinding ... **A16:** 566
electrical discharge machining ... **A16:** 557, 558, 560, 561
electrochemical discharge grinding ... **A16:** 548, 549, 550
electrochemical grinding ... **A16:** 542-543, 545, 546
grinding ... **A16:** 421, 422, 423, 424, 425, 426, 427, 428, 429
high removal rate machining ... **A16:** 607, 608
high-speed machining ... **A16:** 603, 604
photochemical machining ... **A16:** 589, 590, 592
Metal resistors
failure mechanisms ... **EL1:** 971, 999-1001
Metal screen filters ... **A15:** 490
Metal separation
from sands ... **A15:** 350
Metal shadowing *See also* Shadowing ... **A12:** 7
defined ... **A9:** 12
Metal smearing
in shafts ... **A11:** 463
Metal spray
for marine corrosion ... **A13:** 907-909
Metal spraying
as controlled spray deposition ... **A7:** 531
crankshaft, fatigue fracture from ... **A11:** 480
pattern, during gas atomization ... **A7:** 25, 27
of porous powders ... **A7:** 698
of shafts ... **A11:** 459
Metal spraying (metallizing)
of patterns ... **A15:** 196
Metal structure
and shrinkage ... **A7:** 310
Metal systems ... **A7:** 570
Metal transfer *See* Automatic pouring systems; Pouring
in electrical contact materials ... **A2:** 840
Metal vacuum coatings ... **M5:** 399-400
Metal vessel testing
acoustic emission inspection ... **A17:** 291-292
Metal whiskers *See also* Tin whiskers
in electronics industry ... **A13:** 1110
Metal(s) *See also* Base metals; Core(s); Metal body devices; Metal cans; Metal matrix composites (MMCs); Metal packages; Metal-matrix composites; Metallization; Metallurgy; Plating; Pure metals; specific metals and alloys; Toxic metals; toxicity; Ultrapure metals
acoustic microscopy ... **A17:** 472-473
aircraft alloys, thermal coefficient of expansion for ... **EM1:** 716
alternative, and costs ... **EM2:** 85
base, cold cracks in ... **A11:** 440
base-, cracking in laser beam welds ... **A11:** 449
buried, corrosion of ... **A11:** 191-192
claddings, flexible printed boards ... **EL1:** 581
coinability of ... **A14:** 183

Metal(s) (continued)
cold-extruded **A14:** 300
compatible, for galvanic corrosion **A11:** 185
and composites, compared **EM1:** 35, 216, 259-260
and composites, damage tolerances compared **EM1:** 259-260
and composites, damping properties compared **EM1:** 216
corrosion, electrical measurement of **EL1:** 953
damage resistance of **A11:** 166-167
deposition, thin-film hybrids **EL1:** 329
dissimilar, eddy current inspection of **A17:** 164
early melting of **A15:** 15
effect of temperature on toughness 'in **A11:** 66
elasticity of **EM2:** 656
electron mean free path vs kinetic energy in **A11:** 34
embedded in concrete or plaster, galvanic corrosion of **A11:** 186-187
engineering, stress-strain curve **EM2:** 74
and epoxy resin, compared **EM1:** 76
erosion-resistant **A11:** 170
essential, with potential for toxicity **A2:** 1250-1256
fatigue loading **EM2:** 702
ferrous, LME in **A11:** 234-238
fiber reinforcement of **EM1:** 59
and fiber-epoxy composites, properties compared **EM1:** 178
fracture modes of **A11:** 80
gases in **A15:** 82-87
for glass-to-metal seals **EL1:** 455
ground/machined with superabrasives and ultrahard tool materials **A2:** 1013
light, intergranular corrosion of **A11:** 182
light/heavy, radiographic methods **A17:** 296
liquid **A2:** 804-809
matrix, for whisker-reinforced MMCs **EM1:** 896-897
minor toxic **A2:** 1258-1262
-mold interface, inclusion control at **A15:** 90
molten, in elevated-temperature failures **A11:** 273-276
monolithic, properties **EL1:** 1120
nonferrous, liquid-metal embrittlement of **A11:** 233
parts, holographic inspection **A17:** 423-424
percent, effect on solder paste print resolution **EL1:** 732
planes, low-CTE, constraining **EL1:** 625-628
porcelain-enameled, as substrate material **EL1:** 106
powder, additives **EM2:** 170
removal, from PWBs **EL1:** 511
for RTM tooling **EM1:** 168-169
-saving techniques, hot upset forging **A14:** 86-87
solid, embrittlement by **A11:** 239-244
specific damping capacities **EM1:** 216
structure, effect on stress-corrosion cracking **A11:** 206-207
susceptibility to hydrogen damage **A11:** 249
susceptibility to stress-corrosion cracking **A11:** 206-207
thermal conductivity **EM1:** 924
thermal-sprayed, as coating **A15:** 563
and thermosetting resins, compared **EM2:** 222
toxicity of **A2:** 1233-1269
with toxicity related to medical therapy **A2:** 1256-1258
usage in PWB manufacturing **EL1:** 510
use, history of **A15:** 15-23
vs composites, compared **EM2:** 371-373
vs plastic, by competitive pairs **EM2:** 87
vs plastic, in design **EM2:** 74-78
vs plastics, costs **EM2:** 83
vs polyamide-imides (PAI) **EM2:** 128
white, for bearings **A11:** 483
wires, as reinforcement **EL1:** 1119-1121
work hardening of **A14:** 299-300

Metal-bearing ores
photometric methods for analysis **A10:** 64
Metal-body devices *See also* Metal cans; Metal packages
power packages **EL1:** 425-427
small-signal **EL1:** 422-423
Metal-bonded abrasive wheels **A7:** 797
Metal-ceramic composites *See also* Cermets
future and problems **A2:** 1024
Metal-coated steel wire **M1:** 263
Metal-excess semiconductors
types **A13:** 65
Metal-feed location
effect on aluminum alloy 1100 .... **A9:** 631
effect on grain size in aluminum alloy ingots **A9:** 631
Metal-filled polymer composites **A7:** 606-613
as electromagnetic interference shields **A7:** 609
Metal-film deposition
development of crystallographic texture during **A9:** 700-701
Metal-gas interfaces **A10:** 136
Metal-graphite brushes **A7:** 634-636
Metal-graphite electrical contact materials
properties **A2:** 842
Metal-head pressure system
atomized aluminum powders **A7:** 125
Metal-induced embrittlement
solid-state-welded interlayers **A6:** 171
Metal-induced embrittlement of steels **A1:** 717-722
of liquid metal **A1:** 717-721
of solid metal **A1:** 719, 721-722
Metal-inert gas (MIG) welding *See* Gas-metal arc welding
Metal-matrix composites *See also* Cast metal-matrix composites; Composites; Matrix alloys
abrasive waterjet cutting of **A14:** 752-754
aluminum-base **A14:** 251
applications, aerospace **A6:** 388
cast, development **A15:** 36
casting techniques **A15:** 842-848
ceramic, low-gravity effects **A15:** 152-153
defined **A15:** 840
etching **A9:** 591
fabrication methods **A9:** 591
fractographs **A12:** 465-469
fracture/failure causes illustrated **A12:** 217
friction surfacing **A6:** 323
friction welding **A6:** 153, 154
galvanic corrosion during preparation **A9:** 588
grinding **A9:** 588-589
insoluble particle effects **A15:** 142
market effects **A15:** 44
microstructure **A9:** 592
mounting materials for **A9:** 588
polishing **A9:** 589-591
preparation, semisolid alloys **A15:** 328
production techniques **A15:** 840
semisolid metal casting and forging **A15:** 327, 336
SiC whisker-reinforced aluminum **A14:** 20
specific strength/modulus **A15:** 840
stress rupture **A12:** 468
titanium **A14:** 283
with zinc alloy matrices **A15:** 797
Metal-matrix composites (MMCs) *See also* Composites; Laminates; Subcomposites **A16:** 893-901 **EM4:** 47
for A15 superconductors **A2:** 1064-1065
abrasive waterjet cutting **A16:** 893-894, 896, 897
Al-B composites **A16:** 895-896
Al-matrix composites **A16:** 896-898
Al-SiC composites **A16:** 895-896, 897
aluminum P/M alloys **A2:** 209-210
aluminum-matrix composites **A2:** 7, 126, 904-907
application, in packages **EL1:** 1126-1128
boring **A16:** 896
and cermets, compared **A2:** 978

Metal-matrix composites (MMCs) (continued)
chemical milling **A16:** 896, 897
circular sawing **A16:** 894
climb milling **A16:** 900
composite trimming parameters **A16:** 894
continuous fiber aluminum MMC **A2:** 904-906
continuous graphite/copper MMCs **A2:** 909
continuous tungsten fiber reinforced copper MMC **A2:** 908-909
contour band sawing **A16:** 895
coolants **A16:** 894, 895, 896, 897, 898, 899
copper-matrix composites **A2:** 908-909
defined **A2:** 903
die threading **A16:** 896
discontinuous aluminum MMC **A2:** 906-907
dissimilar-material laminates **A16:** 898-899
drilling **A16:** 894, 895, 896, 897, 898-899
electrical discharge machining **A16:** 896
end milling **A16:** 894
Fiber FP Al MMC **A16:** 897, 898
grinding **A16:** 894, 896
honing **A16:** 896
intermetallic-matrix composites **A2:** 909-911
knurling **A16:** 896
lapping **A16:** 896
laser cutting **A16:** 893, 894, 895, 896
machining guidelines **A16:** 893-898
magnesium-matrix composites **A2:** 907-908
matrix materials **A16:** 895
Mg-matrix composites **A16:** 896-898
microstructure **A16:** 893-894
milling **A16:** 894, 895, 896, 900, 901
oxide-reinforced composites **A16:** 896-898
peck drilling **A16:** 894, 898, 899
power band sawing **A16:** 894, 895, 897, 900
power hacksawing **A16:** 895
processing methods **A2:** 903-904
properties **EL1:** 1126
property prediction **A2:** 903
reaming **A16:** 896, 899
recast material **A16:** 895
reinforcements for **A2:** 903
sawing **A16:** 897
in situ discontinuous copper MMC **A2:** 909
superalloy-matrix composites **A2:** 909
surface finish **A16:** 896, 897, 899, 900
surface grinding **A16:** 896
tapping **A16:** 896, 897
thread rolling **A16:** 896
threading **A16:** 896
Ti-SiC MMCs **A16:** 896
titanium-matrix composites **A2:** 590
tool life **A16:** 893, 894, 895, 896, 897, 898
turning **A16:** 894, 896, 897-898
waterjet machining **A16:** 894
wire EDM **A16:** 895-896, 897
of wrought magnesium alloys **A2:** 460
Metal-matrix composites, friction and wear of **A18:** 693, 801-810
applications **A18:** 801
in future jet engine components **A18:** 592
composition **A18:** 801
friction coefficient **A18:** 803, 804, 805, 806, 807, 808-809, 810
from aluminum-silicon alloys **A18:** 789-791
mechanical properties **A18:** 801-803
in steel brakes **A18:** 583
synthesis techniques **A18:** 801
tribological behavior of fiber-reinforced
abrasive wear conditions **A18:** 808
adhesive wear conditions **A18:** 807-808
fiber type effect **A18:** 808-809
metal-matrix composites **A18:** 807-809
orientation effect **A18:** 808-809
tribological behavior of metal-matrix particulate composites **A18:** 803-807
contacting conditions effect **A18:** 807
friction and abrasive wear **A18:** 804-805, 806, 807

Metal-matrix composites, friction and wear of (continued)
friction and erosive wear **A18:** 805-806
friction and sliding wear **A18:** 803-804, 806, 807
particle size effect **A18:** 806
testing parameters effect..................... **A18:** 806-807
wear mechanisms **A18:** 809-810
abrasive wear with hard particles **A18:** 809
sliding wear with hard particles.............. **A18:** 809
sliding wear with soft particles......... **A18:** 809-810
Metal-matrix composites, specific types
carbon (graphite)-magnesium, tensile fracture **A12:** 465
Fe-24Cr-4Al-1Y with W-1ThO fibers, low-cycle fatigue fracture.................... **A12:** 469
Ni-15Cr-25W-2Al-2Ti with tungsten fibers ductile fracture.......................... **A12:** 468
NS-55 tungsten fibers-AISI 1010 carbon steel tensile fracture **A12:** 466
NS-55 tungsten-Al 6061 matrix, tensile fracture **A12:** 466
tungsten fibers with silver matrix, ductile and transverse cleavage fractures.. **A12:** 469
Metal-matrix diamond blades used in abrasive cutting **A9:** 24-25
Metal-matrix high-temperature superconductor cermets
applications and properties............................ **A2:** 967
Metal-metal interdiffusion
as integrated circuit failure mechanism............ **A11:** 776-777
Metal-metal systems *See also* Metallic glasses
processing techniques...................................... **A2:** 806
Metal-metalloid systems *See also* Metallic glasses
processing methods **A2:** 806
Metal-organic chemical vapor deposition (MOCVD)
for GaAs crystal growth **EL1:** 200
Metal-organic molecular beam epitaxy (MOMBE)
research **A2:** 747
Metal-oxide semiconductor (MOS)
capacitor structures.. **EL1:** 156
defined **EL1:** 1150
development.. **EL1:** 160
-device-focused technologies, integrated **EL1:** 2
integrated circuit fabrication......................... **EL1:** 198
transistors **EL1:** 147, 157-158
Metal-oxide semiconductor (MOS) devices *See also* Integrated circuits; Integrated circuits, failure analysis of; Semiconductors; Silicon semiconductor devices
chip failures in.. **A11:** 766
distribution of malfunctions for **A11:** 767
oxide failures in.. **A11:** 766
surface inversion in.. **A11:** 766
transistor, in saturation **A11:** 782
Metal-oxide semiconductor field-effect transistor (MOSFET) **A6:** 43
defined **EL1:** 1150
depletion mode.. **EL1:** 8-159
enhancement mode... **EL1:** 158
gate breakdown **EL1:** 159
limit, network logic circuitry effect **EL1:** 2
n-channel, for GaAs digital circuits............ **EL1:** 201
in saturation mode.. **EL1:** 966
small-signal equivalent circuit..................... **EL1:** 159
twin-on voltage instability............................ **EL1:** 159
Metal-oxide semiconductor field-effect transistors (MOSFETs).......................... **EM3:** 583
Metal-oxygen systems
combustion temperature................................. **A7:** 597
Metal-processing equipment............... **A13:** 1311-1316
carburization in.................................... **A13:** 1311-1313
and industry, niobium applications............ **A13:** 723
molten-metal corrosion of........................... **A13:** 1314
molten-salt corrosion of **A13:** 1313-1314
oxidation **A13:** 1311
plating, anodizing, and pickling equipment **A13:** 1314-1316
sulfidation **A13:** 1313
Metal-reinforced ceramic composites **EM1:** 927-929
Metal-semiconductor contact
example of **A11:** 778
Metal-sheathed thermocouples
assemblies for.. **A2:** 884-885
Metal-spray babbitting...................................... **M5:** 357
Metal-spray ceramic coating **M5:** 539
Metal-to-earth abrasion alloys........ **A18:** 758, 759-760
applications **A18:** 760
compositions **A18:** 759, 760, 761
microstructure **A18:** 760
primary function **A18:** 758
properties... **A18:** 760
Metal-to-metal
adhesive joint, eddy current inspection............ **A17:** 188-189
defects, adhesive-bonded joints................... **A17:** 612
voids, in adhesive-bonded joints................. **A17:** 610
Metal-to-Metal Adhesive Bonding **EM3:** 68
Metal-to-metal wear alloys..................... **A18:** 758, 759
abrasive wear **A18:** 759
applications **A18:** 759
compositions **A18:** 759
properties... **A18:** 759
Metal/ceramic brazing assembly interface
molybdenum particles on **A10:** 457
Metal/metal
sliding... **A8:** 602
wear **A8:** 604
Metalforming applications
cemented carbides..................................... **A2:** 968-971
Metallic abrasives, types
dry blasting **M5:** 83-84
Metallic binder phase
cermets **A2:** 979
superhard boron/silicon metalloids **A2:** 1008
Metallic bond
definition... **A6:** 1211 **M6:** 11
Metallic bonding
chemical **EL1:** 92-93
Metallic bonding forces **A7:** 303-304
apparent density, shearing, and heating in farming **A7:** 303-304
Metallic coated steels............................... **A13:** 526-527
Metallic coatings *See also* Chemical conversion coatings; Coatings; Conversion coatings; Elastomeric coating; Electroplating coating; Protective coatings; specific coatings; specific types by name; Surface coatings
for cast irons............................ **A13:** 570 **M1:** 101-104
cladding **A11:** 195
electroplated **A11:** 195
electroplating.. **A13:** 911-912
for galvanic corrosion...................................... **A13:** 84
hot-dip... **A13:** 910-911
for marine corrosion **A13:** 906-912
metal spray **A13:** 907-909
of nickel-base alloys.. **A14:** 832
for pipeline **A13:** 1291
sprayed metal.. **A11:** 195
steel sheet.. **M1:** 167-174
Metallic coatings, effect of
on formability **A1:** 579-580
Metallic composite materials *See* Composite materials; Metal matrix composites
Metallic elements
in sintering tungsten and molybdenum **A7:** 390
Metallic engineering alloys
fracture toughness testing **A8:** 469
Metallic fiber *See also* Fiber(s)........................ **EM3:** 18
defined **EM1:** 15 **EM2:** 26
Metallic flake pigments **A7:** 593-596
production **A7:** 593-594
Metallic glass
cut with a wire saw **A9:** 25
Metallic glass powders..................................... **A7:** 795
as amorphous powder metal **A7:** 794
microcrystalline alloys from.................... **A7:** 794-795
Metallic glasses *See also* Amorphous metals **EM4:** 22
amorphous superconductors................... **A2:** 816-817
applications **A2:** 818-820 **EM4:** 22
brazing materials **A2:** 819
bulk metallic glasses.................................. **A2:** 819-820
chemical properties................................... **A2:** 817-818
coatings **A2:** 819
controlled crystalline microstructures **A2:** 820
crystallization **A2:** 809-811
defined **A10:** 676 **A13:** 9
deformation mechanisms................................ **A2:** 813
diffraction experiments **A2:** 809-811
Metallic glasses (continued)
electrical conductivity.................................. **EM4:** 566
electrical transport properties **A2:** 815
electrodeposition **A2:** 806-807
failure, fracture toughness, embrittlement **A2:** 814
future developments... **A2:** 820
glass transition and crystallization **A2:** 812
heat capacity-two level systems **A2:** 812-813
historical introduction/background **A2:** 804-805
Knight shift measurements on.................... **A10:** 284
magnetic properties **A2:** 815-816
mechanical properties.............. **A2:** 813-815 **EM4:** 22
preparation by nonconventional techniques **EM4:** 22
rapid quenching from the melt **A2:** 805-806
reinforcing fibers **A2:** 817
SAS applications **A10:** 405
short-range ordering... **A2:** 810
soft magnetic materials **A2:** 818-819
solid-state amorphitization...................... **A2:** 807-809
structural models.. **A2:** 809-811
structure dependence on synthesis/thermal history................................ **A2:** 811-812
synthesis and processing methods......... **A2:** 805-809
technology **A2:** 818-820
thermal transport... **A2:** 813
thermodynamic properties **A2:** 812-814
vapor quenching.. **A2:** 806-807
yield strength, hardness, elastic constants.. **A2:** 813-814
Metallic implant materials
as corrosion resistant and biocompatible **A11:** 672-673
Metallic implants................................. **A13:** 1324-1335
background.. **A13:** 1324-1329
biocompatibility of **A13:** 1328-1329
corrosion forms.................................... **A13:** 1330-1332
corrosion significance **A13:** 1328-1329
corrosion testing **A13:** 1332-1333
electrochemistry.................................... **A13:** 1329-1330
metals/alloys.. **A13:** 1325-1328
Metallic impurities
chromium plating baths................................. **M5:** 189
nickel plating baths.................................. **M5:** 207-210
plating waste disposal removal procedures **M5:** 313-314
tin-lead plating baths, removal of................ **M5:** 278
Metallic inclusions *See also* Inclusions
as casting defect... **A11:** 387
Metallic interface failures
in integrated circuits.............................. **A11:** 776-778
interdiffusion at aluminum-silicon contacts.. **A11:** 777-778
metal-metal interdiffusion **A11:** 776-777
Metallic letterpress inks
powders used.. **A7:** 574
Metallic magnesium treatment
of ductile iron **A15:** 649
Metallic material, static
design allowables for.............................. **A8:** 662-677
Metallic materials
ordered metallic compounds as **A2:** 913
Metallic mercury, as biologic indicator
mercury toxicity.. **A2:** 1249
Metallic microcontacts
green strength and electrical conductivity **A7:** 304
Metallic nickel
partitioning oxidation states in.................... **A10:** 178
Metallic offset inks
powders used.. **A7:** 574
Metallic orthopedic implants, failures of **A11:** 670-694
analysis of.. **A11:** 680
complications related to **A11:** 672
degradation of implants......................... **A11:** 689-692
fatigue properties, implant materials **A11:** 688-689
internal fixation............................ **A11:** 671, 677-680
and interaction with body environment **A11:** 673-677
materials of.. **A11:** 672-673
prosthetic devices..................................... **A11:** 670-671
related to implant deficiencies.............. **A11:** 680-681
related to mechanical or biomechanical conditions...................................... **A11:** 681-687

Metallic orthopedic implants, failures of (continued)
total hip joint prostheses ... **A11:** 692-693
Metallic paint
shielding alternatives ... **A7:** 612
Metallic particles
fuel pump shaft wear from ... **A11:** 465
Metallic platings
uranium/uranium alloys ... **A13:** 819-821
Metallic projections
as casting defects ... **A11:** 381 **A15:** 546 **A17:** 512-513
Metallic radius
of rare earth elements ... **A2:** 722
Metallic resistors
types and construction ... **EL1:** 178
Metallic rotogravure inks
powders used ... **A7:** 574
Metallic salts
from decomposed electrolytes ... **A9:** 51
Metallic wear *See also* Adhesive wear; Severe wear
defined ... **A18:** 12-13
Metallic whisker *See also* Whiskers
defined ... **EM2:** 26
Metallic wires
properties/toxicity ... **EM1:** 118
Metalliding
for wear resistance ... **M1:** 635
MetalLife treatment
tool steels ... **A18:** 643
Metallizability
of polyamide-imides (PAI) ... **EM2:** 129
of substrates ... **EL1:** 105
Metallization *See also* Plating
aluminum ... **EL1:** 195, 303, 965
of aluminum nitride and silicon carbide ... **EL1:** 306-307
chip, stresses in ... **EL1:** 445
chromium, thin-film hybrids ... **EL1:** 326
computer modeling ... **EL1:** 442-444
in hybrid wafer scale integration ... **EL1:** 88
integrity, testing ... **EL1:** 953
notching and voiding in ... **EL1:** 892
thick-film, ceramic packages ... **EL1:** 1-463
thin-film ... **EL1:** 299
time-temperature dependence ... **EL1:** 1043
and transfer pattern, thin-film hybrids ... **EL1:** 329-330
Metallization failures
by corrosion ... **A11:** 769-770
integrated circuits ... **A11:** 769-776
Metallized ceramic
electron diffraction/EDS analysis for unknown phases in ... **A10:** 457-458
Metallized glass fibers
as reinforcements ... **EM2:** 473
Metallized-paper capacitors ... **EL1:** 179
Metallizing ... **EM4:** 542-545
adhesion ... **EM4:** 543-545
chemical ... **EM4:** 545
compound ... **EM4:** 544-545
definition of good adhesion ... **EM4:** 543
mechanical ... **EM4:** 544, 545
properties ... **EM4:** 543-544
defined ... **A13:** 9 **EM2:** 26
definition ... **A6:** 1211
electrical conductivity property ... **EM4:** 542
electrical connection ... **EM4:** 742
mechanical connection ... **EM4:** 542
methods ... **EM4:** 542-543
atomistic deposition process ... **EM4:** 542-543
bulk metallization ... **EM4:** 542, 543
coating formation steps ... **EM4:** 542
electron beam evaporation ... **EM4:** 543
flash evaporation ... **EM4:** 543
particulate depositions ... **EM4:** 542, 543
vacuum evaporation ... **EM4:** 543
Metallizing (metal spraying)
of patterns ... **A15:** 196
Metallo-organic chemical vapor deposition ... **A10:** 601, 602, 690
Metallo-organic chemical vapor deposition (MOCVD)
gallium arsenide (GaAs) ... **A2:** 745
of high-temperature superconductors ... **A2:** 1087
Metallochromic indicators
common ... **A10:** 174
Metallograph *See also* Optical microscope ... **A9:** 72
defined ... **A9:** 11-12
hot cell ... **A9:** 83
inverted ... **A9:** 74
research-quality, with projection screen ... **A9:** 74
used in macrophotography ... **A9:** 86
Metallographic analysis
of powder forged parts ... **A14:** 204
Metallographic analysis and sectioning
examination and analysis of ... **A11:** 24
fracture mode identification chart for ... **A11:** 80
selection and preparation of ... **A11:** 23-24
for stress-corrosion cracking ... **A11:** 213-214
of weldments ... **A11:** 412
of worn parts ... **A11:** 159
Metallographic evaluation
printed board coupon ... **EL1:** 572-577
Metallographic finish
for microhardness specimen ... **A8:** 93
Metallographic identification
of cellular decomposition of martensite in uranium alloy ... **A10:** 316
of light elements ... **A10:** 549, 559-561
optical metallography ... **A10:** 299-308
Metallographic reagents *See* Etchants
Metallographic sectioning methods
for profiles ... **A12:** 198-199
Metallographic sections
macroscopy of ... **A10:** 303-304
Metallographic test methods ... **M5:** 371-372, 596
Metallography ... **A10:** 517, 676 **A18:** 371, 372
crack detection by ... **A7:** 484
development of ... **A15:** 27
and fracture mechanics testing ... **A8:** 476
for melt cleanliness ... **A15:** 493
and microstructure ... **A7:** 485-489 **A8:** 476
of rare earth metals ... **A2:** 725
to determine shear band formation in titanium alloy ... **A8:** 589
Metallography, use in phase-diagram determination ... **A3:** 1 • 18
Metalloids
implantation of ... **A10:** 485
Metalloproteins
EXAFS structural analysis of ... **A10:** 407
Metallothionein
role in cadmium toxicity ... **A2:** 1240-1241
Metallurgical attach, zone 2
package integrity ... **EL1:** 1010-1011
Metallurgical bond
definition ... **A6:** 1211
Metallurgical burn
defined ... **A18:** 13
Metallurgical chemistry
development of ... **A15:** 27
Metallurgical compatibility *See* Compatibility (metallurgical)
Metallurgical condition
testing ... **A13:** 193-194
Metallurgical control
in production of ductile iron ... **A1:** 38
Metallurgical design
corrosion protection by ... **A13:** 379
Metallurgical details
magnetic printing detection ... **A17:** 126
Metallurgical discontinuities
welding ... **A17:** 582
Metallurgical factors affecting weldability of steels ... **A1:** 603-606, 607
chemical composition effect ... **A1:** 606
hardenability and weldability ... **A1:** 603-604
heat-affected zone microstructure ... **A1:** 605-606
preweld and postweld heat treatments ... **A1:** 606
weld metal microstructure ... **A1:** 604-605
Metallurgical hydrogen
refining ... **A7:** 344
Metallurgical instabilities
from elevated- temperature failures ... **A11:** 266-268
Metallurgical microscopy
defined ... **EL1:** 1067
Metallurgical parameters
effect on corrosion fatigue ... **A11:** 256
in shaft failures ... **A11:** 477-480
Metallurgical properties
magnetic characterization of ... **A17:** 131-132
Metallurgical stability
electrical resistance alloys ... **A2:** 824
Metallurgical structures
of heat-resistant cast steels ... **A1:** 921-922
Metallurgical susceptibility
to stress-corrosion cracking ... **A8:** 495
Metallurgical variables of fatigue behavior ... **A1:** 678
Metallurgically influenced corrosion ... **A13:** 123-135
of aluminum alloys ... **A13:** 130
aqueous corrosion ... **A13:** 45-49
dealloying corrosion ... **A13:** 131-134
grooving, in carbon steel ... **A13:** 130-131
of high-nickel alloys ... **A13:** 128-130
intergranular corrosion, mechanisms ... **A13:** 123
of stainless steels ... **A13:** 124-127
Metallurgy *See also* Secondary metallurgy
of cast iron ... **A1:** 3-4
of cermet system conductors ... **EL1:** 340
connector ... **EL1:** 22
of gray iron ... **A15:** 629-635
of monocrystal casting ... **A15:** 322
unconventional ... **A7:** 570
of vacuum induction furnace ... **A15:** 393-396
of zirconium and zirconium alloys ... **A2:** 665-667
Metallurgy, joining *See* Joining metallurgy
Metals *See also* Metals and alloys, characterization of; Molten metals
alkali, oxygen reactions ... **A13:** 94
and alloys, compatible dissimilar-metal couples ... **A13:** 340-343, 1040, 1061
in alloys, electrogravimetric determination ... **A10:** 197
amorphous ... **A13:** 864-870
amphoteric, stray-current corrosion ... **A13:** 87
analytic methods for ... **A10:** 4
carbon in, determined by high-temperature combustion ... **A10:** 221-225
composition, testing ... **A13:** 193-194
controlled-potential coulometry for ... **A10:** 202
corrosion analysis of ... **A10:** 134-135
crystallographic texture measurement and analysis ... **A10:** 357-364
deformation, by oxidation ... **A13:** 72
deposition of small amounts ... **A10:** 198
detecting adsorbed monolayers on ... **A10:** 114
determined fluorimetrically by complexation ... **A10:** 74
dissimilar ... **A13:** 84, 340-343, 1040, 1061
dissolution ... **A13:** 29, 89, 343
dust ... **A13:** 9, 380, 1312-1313
effect of oxygen on positive secondary ion yields in ... **A10:** 612
electromotive force series ... **A13:** 20
electronegative, polishing wear without abrasives ... **A18:** 197
erosion of ... **A18:** 201-204
embedding of erodent fragments ... **A18:** 203
mechanisms ... **A18:** 201-203
micromachining ... **A18:** 202, 203
particle flux ... **A18:** 204

Metals (continued)
particle hardness **A18:** 203-204
particle shape **A18:** 203
particle size **A18:** 203
platelet mechanism **A18:** 199-203
rate **A18:** 199-203
resistance **A18:** 201, 202
sequential observation technique **A18:** 202
temperature **A18:** 204
FIM/AP study of point defects in **A10:** 583
finished, AAS analysis of **A10:** 43
fluorescent, MFS analysis **A10:** 72
form, testing **A13:** 193-194
friction coefficient data **A18:** 75
functional requirements **A13:** 338
galvanic series of **A13:** 755-756
-gas environments, SERS for **A10:** 137
high-purity, voltammetric analysis **A10:** 188
hydrogen susceptibility **A13:** 288
implantation of **A10:** 485
instability, in aqueous solutions **A13:** 377
laser processing of **A13:** 501-504
light **A13:** 84
linear thermal expansion, coefficients **A13:** 71
liquid, corrosion in **A13:** 17
liquid, electronic structure of **A10:** 284
low-melting, embrittling by **A12:** 29-30
material parameters that should be documented to ensure repeatability when testing tribosystems **A18:** 55
and metal surfaces **A13:** 45-46
metal wear generated by plastics **A18:** 240
microgram amounts, electrogravimetric measurement **A10:** 197
multiphase, image analysis of **A10:** 309
no adverse effect by water-displacing corrosion inhibitors **EM3:** 641
non-, Raman analysis of surface species of **A10:** 134
nonferrous, coatings on **A13:** 776
on periodic table **A10:** 688
and oxygen, principal reactions **A13:** 61
particle size effect of erosion rate **A18:** 200
partly converted, fractures **A12:** 1-3
penetration, from pitting **A13:** 232
perchloric acid as dissolution medium **A10:** 166
photometric methods for analysis **A10:** 64
porcelain fused to **A13:** 1353
precious, SSMS analysis in geological ores **A10:** 141
prompt gamma activation analysis of **A10:** 240
pure, structures and thermal properties **A13:** 62-63
relative erosion factors **A18:** 201
residual, removal from corrosion specimens **A13:** 96
samples, treatment in mineral acids **A10:** 165
sampling of **A10:** 12-18
SAS applications **A10:** 405
SIMS analysis of surface layers **A10:** 610
SIMS phase distribution analysis in **A10:** 610
single-phase, image analysis of **A10:** 309
sliding and adhesive wear **A18:** 237-239, 241
small-angle scattering analysis **A10:** 405
in solutions, concentration range of **A10:** 188
spark source excitation for elemental analysis of **A10:** 29
stacking faults **A13:** 45
standard conditions for sliding **A18:** 236
standard emf series **A13:** 1329
substrates, and filiform corrosion **A13:** 108
sulfur in, determined by high-temperature combustion **A10:** 221-225
surface preparation **EM3:** 259-275
surfaces, FIM/AP study of **A10:** 583
surfaces, IR study of monolayers adsorbed on **A10:** 118-120
as surgical implants **A13:** 1325
types illustrated **A12:** 216
use of ICP-AES for trace impurities in **A10:** 31
used in electronic systems **A13:** 1108
vapor-deposited, corrosion-resistant **A13:** 458
wear generated by plastics **A18:** 239-240
weighing as the, gravimetric analysis **A10:** 170-171

Metals (continued)
x-ray diffraction residual stress techniques for crystalline **A10:** 381
Metals and alloys, characterization of *See also* Alloys; Alloys, specific types; Metals
analytical transmission electron microscopy **A10:** 429-489
atomic absorption spectrometry **A10:** 43-59
Auger electron spectroscopy **A10:** 549-567
classical wet analytical chemistry **A10:** 161-180
controlled-potential coulometry **A10:** 207-211
crystallographic texture measurement and analysis **A10:** 357-364
electrochemical analysis **A10:** 181-211
electrogravimetry **A10:** 197-201
electrometric titration **A10:** 202-206
electron probe x-ray microanalysis **A10:** 516-535
electron spin resonance **A10:** 253-266
extended x-ray absorption fine structure **A10:** 407-419
ferromagnetic resonance **A10:** 267-276
field ion microscopy **A10:** 583-602
inductively coupled plasma atomic emission spectroscopy **A10:** 31-42
inert gas fusion **A10:** 226-232
low-energy electron diffraction **A10:** 536-545
low-energy ion-scattering spectroscopy **A10:** 603-609
Mössbauer spectroscopy **A10:** 287-295
neutron activation analysis **A10:** 233-242
neutron diffraction **A10:** 420-426
nuclear magnetic resonance **A10:** 277-286
optical emission spectroscopy **A10:** 21-30
particle-induced x-ray emission **A10:** 102-108
potentiometric membrane electrodes **A10:** 181-187
radial distribution function analysis **A10:** 393-401
radioanalysis **A10:** 243-250
Rutherford backscattering spectrometry **A10:** 628-636
scanning electron microscopy **A10:** 490-515
secondary ion mass spectroscopy **A10:** 610-627
single-crystal x-ray diffraction **A10:** 344-356
small-angle x-ray and neutron scattering **A10:** 402-406
spark source mass spectrometry **A10:** 141-150
ultraviolet/visible absorption spectroscopy **A10:** 60-71
voltammetry **A10:** 188-196
x-ray diffraction **A10:** 325-332
x-ray diffraction residual stress techniques **A10:** 380-392
x-ray photoelectron spectroscopy **A10:** 568-580
x-ray powder diffraction **A10:** 333-343
x-ray spectrometry **A10:** 82-101
x-ray topography **A10:** 365-379
Metals Data File **A10:** 355-356
Metals Properties Council **A8:** 725
Metals, purification methods
chemical vapor deposition **M2:** 711
degassing **M2:** 710
distillation **M2:** 710
floating zone technique **M2:** 710
fractional crystallization **M2:** 709-710
iodide process **M2:** 711
sublimation **M2:** 710
vacuum melting **M2:** 710
zone refining **M2:** 710
Metals recovered from solution
powder used **A7:** 573
metalworking *See also* Forging; Forming
defined **A14:** 15
equipment, types **A14:** 16
future trends **A14:** 20-21
glass lubricant, for specimen protection **A8:** 159
history of **A14:** 15
new materials **A14:** 19
new processes **A14:** 16-19
process simulation **A14:** 19-20
processes **A8:** 575-576
processes, classification of **A14:** 15-16
torsional rotation rates in **A8:** 157-158
Metalworking fluids **A18:** 99
additives in formulation **A18:** 111
biocides **A18:** 110
emulsifiers **A18:** 107

Metalworking fluids (continued)
extreme-pressure agents used **A18:** 101
Falex tests **A18:** 101
four-ball tests **A18:** 101
friction modifiers **A18:** 104
rust and corrosion inhibitors **A18:** 106
Timken tests **A18:** 101
Metalworking lubricants **A18:** 139-149
additives **A18:** 140-142
antifoams **A18:** 141-142, 144, 147
antimicrobial agents **A18:** 142, 143-144
antioxidants **A18:** 141, 143
antiwear additives **A18:** 141
biocides **A18:** 142, 143, 144
boundary additives **A18:** 140-141, 142, 143
corrosion inhibitors **A18:** 141, 143, 144, 147
defoamers **A18:** 141-142, 144, 147
emulsifiers **A18:** 141, 142, 143, 144
extreme-pressure (EP) additives **A18:** 141, 142, 143, 144, 145, 146, 149
film-strength additives **A18:** 140-141, 142, 143, 144, 145, 146, 147
friction modifiers **A18:** 140-141, 143, 144, 145, 146, 147
fungicides **A18:** 142
load-bearing additives **A18:** 140-141, 142, 143
oiliness additives **A18:** 140-141, 142, 143
oxidation inhibitors **A18:** 141, 143, 147
passive extreme-pressure (PEP) additives **A18:** 141
surfactants **A18:** 141, 142, 143, 144
suspended solids **A18:** 141
common functions **A18:** 139
formulations **A18:** 139-140
mineral oils **A18:** 139
petroleum oils **A18:** 139, 140, 142
synthetic fluids **A18:** 139-140
metal forming lubricants **A18:** 146-148
control **A18:** 148
friction **A18:** 146-147
handling **A18:** 148
lubricant films **A18:** 146
lubrication regimes **A18:** 146
reversing mill lubricants **A18:** 147
selection **A18:** 147-148
metal removal lubricants **A18:** 144-146
built-up edge **A18:** 145
disposal **A18:** 145-146
lubricant application **A18:** 145
lubricant maintenance **A18:** 145
reclamation **A18:** 145-146
selection **A18:** 145
metal removal operations **A18:** 139
metals subjected to either removal or forming processes **A18:** 139
quality **A18:** 148-149
control charts **A18:** 148
histogram **A18:** 148-149
Pareto diagram **A18:** 148, 149
scatter diagram **A18:** 148, 149
statistical methods **A18:** 148, 149
total quality management **A18:** 149
requirements, special **A18:** 139
types **A18:** 142-144
emulsions **A18:** 142, 143-144, 145, 146, 147
micellar solutions **A18:** 142, 144, 145, 146, 147
microemulsions **A18:** 142, 144, 145, 146, 147
solid-lubricant suspensions **A18:** 144
staining oils **A18:** 142-143
straight oils **A18:** 142-143, 144, 145, 147
true solutions **A18:** 144
viscosity **A18:** 140, 147
factors in estimating the optimum viscosity **A18:** 140
kinematic viscosity measurement **A18:** 140
Metalworking rolls **M3:** 502-507
arrangement in mill housings **M3:** 503, 504
cast iron rolls **M3:** 504-506
cast steel rolls **M3:** 506
cemented tungsten carbide rolls **M3:** 507
design **M3:** 504
forged rolls, miscellaneous **M3:** 507
forged steel rolls **M3:** 503, 506
misuse **M3:** 504
parts **M3:** 502, 503, 507
sleeve rolls **M3:** 507
strength **M3:** 503-504

Metalworking rolls (continued)
types of rolls ... **M3:** 502-503
wear resistance ... **M3:** 503
Metastability, thermodynamic
defined ... **A15:** 52
Metastable
defined ... **A9:** 12
equilibrium ... **A3:** 1 • 1, 4
phases ... **A3:** 1 • 1
Metastable alloys
effects of plastic deformation on ... **A9:** 686
Metastable and stable solvus curves
phase diagram ... **A9:** 650
Metastable austenite
carbon content effect on fracture toughness ... **A8:** 484
Metastable beta
defined ... **A9:** 12
Metastable equilibrium
defined ... **A15:** 101
Metastable oxides, formation
by gases ... **A13:** 62
Metastable phase diagrams
used with rapidly solidified alloys ... **A9:** 615
Metastable phases
as a result of partial massive transformation ... **A9:** 656-657
in beryllium-copper alloys ... **A9:** 395
in beryllium-nickel alloys ... **A9:** 396
in uranium alloys ... **A9:** 476-477
Metastable precipitates in beryllium-copper alloys ... **A9:** 395
Metastable precipitation
of beta ... **A15:** 128-129
Metastable solid phase
eutectoid reactions ... **A9:** 658-661
Metastable solutions
spinodal decomposition ... **A9:** 652-653
Metastable solvus curves
influence on age hardening ... **A9:** 651
Metatectic reaction ... **A3:** 1 • 5
Metatorbernite, and turquoise
ESR analysis ... **A10:** 265
Meter ... **A10:** 685, 691
Meter-movement relays
recommended microcontact materials ... **A2:** 866
Metering and mixing equipment ... **EM3:** 687-692
material transfer pumps ... **EM3:** 691
meter/mix machines ... **EM3:** 690-692
mixing systems ... **EM3:** 692
Metering devices
from P/M porous parts ... **A7:** 700
Meters
analog, eddy current inspection ... **A17:** 178
digital, eddy current inspection ... **A17:** 178-179
Meters, light
for photography ... **A12:** 85-86
Metglas cut
with a wire saw ... **A9:** 25
Methacryl phosphate
formulation ... **EM3:** 123
Methacrylatebutadiene
surface preparation ... **EM3:** 291
Methacrylated urethanes
used as modifiers ... **EM3:** 121
Methacrylates ... **EM3:** 91
for bonding together galvanized steel sheets ... **EM3:** 577
cure mechanism ... **EM3:** 567
for electronic general component bonding ... **EM3:** 573
heat-accelerated competing with anaerobics ... **EM3:** 116
for lens bonding ... **EM3:** 575
for loud speaker assembly ... **EM3:** 575
as monomers for anaerobics ... **EM3:** 113
for motor magnet bonding ... **EM3:** 574, 575
for photovoltaic cell construction ... **EM3:** 578
properties compared ... **EM3:** 92
Methacrylates (continued)
for sporting goods manufacturing ... **EM3:** 577
for thermally conductive bonding ... **EM3:** 571, 572
UV-curing ... **EM3:** 568
Methacrylates/silicones
for solar panel construction ... **EM3:** 578
Methacrylic acid
formulation ... **EM3:** 122
to increase electronegativity or hydrogen bonding ... **EM3:** 181
typical formulation ... **EM3:** 121
Methacryloxyethyl phosphate
formulation ... **EM3:** 122
Methane
analyzed in microcircuit fabrication process ... **A10:** 157
in CVD coating process ... **A16:** 80
explosive range ... **A7:** 348
formation avoided in graphite electrically heated fumaces ... **EM4:** 246, 248
Methane (CH_4)
fuel gas for torch brazing ... **A6:** 328
Methane chemical group
and polymer naming ... **EM2:** 56
Methane gas bubbles
cavitation ... **A12:** 37, 51
rapid coalescence of ... **A12:** 349
Methanol
absolute, to isolate inclusions from steel ... **A10:** 176
ethylene glycol and perchloric acid as an alloys ... **A9:** 459
electrolyte for titanium and titanium
explosive range ... **A7:** 348
and iodine, second-phase test methods ... **A10:** 177
and nitric acid (Group VIII electrolytes) ... **A9:** 52-55
substitutes for, in etchants ... **A9:** 67
titanium/titanium alloy SCC ... **A13:** 687
used in etchants ... **A9:** 67-68
vapor, physical properties as atmosphere ... **A7:** 341
Methocel
as binder in ceramics extrusion process ... **EM4:** 169
Method A
water-washable penetrants ... **A17:** 75, 77-78
Method B and D
lipophilic/hydrophilic postemulsifiable penetrants ... **A17:** 75, 77-78
Method C
solvent-removable penetrants ... **A17:** 75, 78
Method D
hydrophilic emulsifiers ... **A17:** 75
Methodology
of design ... **EL1:** 82-87
Methods based on sputtering or scattering phenomena
atom probe microanalysis ... **A10:** 583, 591-602
field ion microscopy ... **A10:** 583-591, 598-602
low-energy ion-scattering spectroscopy ... **A10:** 603-609
Rutherford backscattering spectrometry ... **A10:** 628-636
secondary ion mass spectroscopy ... **A10:** 610-627
Methyl
uses and properties ... **EM3:** 126
Methyl alcohol (CH_3OH)
as solvent used in ceramics processing ... **EM4:** 117
Methyl alcohols
and SCC in titanium and titanium alloys ... **A11:** 224
for SCC of titanium alloys ... **A8:** 531
Methyl cellulose
as injection molding binder ... **A7:** 498
Methyl chemical group
and polymer names ... **EM2:** 57
Methyl ethyl ketone
epoxy resin removal by ... **EM1:** 153
Methyl ethyl ketone (continued)
surface tension ... **EM3:** 181
Methyl ethyl ketone (C_4H_8O) solvent
batch weight of formulation when used in nonoxidizing sintering atmospheres ... **EM4:** 163
used in ceramics processing ... **EM4:** 117
Methyl ethyl ketone peroxide (MEKP) ... **EM1:** 133
as vinyl ester cure ... **EM2:** 274
Methyl formate
for core curing ... **A15:** 240
Methyl mercury
dose-response relationship for ... **A2:** 1249
Methyl metbacrylate as a mounting material ... **A9:** 29-30
for copper and copper alloys ... **A9:** 399
Methyl methacrylate ... **EM3:** 18
as acrylic plastic ... **EM2:** 103
defined ... **EM2:** 26
formulation ... **EM3:** 122, 123
Methyl methacrylate (MMA) ... **EM1:** 94
Methyl orange
as acid-base indicator ... **A10:** 172
Methyl red
as acid-base indicator ... **A10:** 172
Methyl thymol blue
as metallochromic indicator ... **A10:** 174
Methylacetylene
chemical bonding ... **M6:** 90
Methylacetylene-propadiene (MPS)
valve thread connections for compressed gas cylinders ... **A6:** 1197
Methylacetylene-propadiene-stabilized (MPS or MAPP) gas
cost ... **A6:** 1158
deep water cutting ... **A6:** 1158-1159
fuel gas for oxyfuel gas cutting ... **A6:** 1157, 1158-1159, 1160, 1162
heat content ... **A6:** 1158
properties ... **A6:** 1158, 1160
Methylacetylene-propadiene-stabilized gas
chemical bonding ... **M6:** 90
properties as fuel gas ... **M6:** 89
use in cutting ... **M6:** 901-90
use in underwater cutting ... **M6:** 922-92
Methylcellulose
as binder ... **EM4:** 474
as binder for injection molding ... **EM4:** 173
as binder in ceramics extrusion process ... **EM4:** 169, 171
removal ... **EM4:** 137
Methylene
stretching vibrations ... **A10:** 111
Methylene blue index ... **EM4:** 117
to measure surface area in traditional ceramics ... **EM4:** 5
Methylene chloride
photochemical machining stripper ... **A16:** 590
Methylene chloride solvent cleaners ... **M5:** 40-41, 44-49, 57
cold solvent cleaning process use in ... **M5:** 40-41
flash point ... **M5:** 40
vapor degreasing, use in ... **M5:** 44-49, 57
Methylene chloride test
for chemical resistance ... **EL1:** 536-537
Methylene dianiline (DNA)
bismaleimide system ... **EM3:** 154-155
Methylisobutyl ketone
solvent extraction with ... **A10:** 169
Methylsilicone fluids
applications/properties ... **EM2:** 266
Methyltrichlorosilane (MTS)
chemical vapor deposition ... **EM4:** 217
Metric and conversion data ... **A8:** 721-723
Metric conversion guide ... **A1:** 1035-1037 **A2:** 1270-1272 **A10:** 685-687 **A11:** 793-795 **A13:** 1371-1374 **A14:** 941-943 **A15:** 893-895 **A17:** 755-757 **EL1:** 1163-1165 **EM1:** 945-947 **EM2:** 847-849 **EM3:** 849-851

Metric conversions
guide and references ... **A12:** 489-491
Metric fasteners
property class designations of ... **A1:** 289-290, 291
Metric-conversion guide ... **A18:** 884
Metrology *See also* Dimensional measurement and evaluation ... **A17:** 50
image analysis use ... **A10:** 316
speckle ... **A17:** 432-437
Meyer index ... **A18:** 424
Meyer index of austenitic manganese and stainless steel ... **A1:** 832
MFD *See* Magnetic field disturbance (MFD) inspection
MFS *See* Molecular fluorescence spectroscopy
Mg-Mn (Phase Diagram) ... **A3:** 2 • 280
Mg-Ni (Phase Diagram) ... **A3:** 2 • 281
Mg-Pb (Phase Diagram) ... **A3:** 2 • 281
Mg-Sb (Phase Diagram) ... **A3:** 2 • 281
Mg-Sc (Phase Diagram) ... **A3:** 2 • 282
Mg-Si (Phase Diagram) ... **A3:** 2 • 282
Mg-Sm (Phase Diagram) ... **A3:** 2 • 282
Mg-Sn (Phase Diagram) ... **A3:** 2 • 283
Mg-Sr (Phase Diagram) ... **A3:** 2 • 283
Mg-Th (Phase Diagram) ... **A3:** 2 • 283
Mg-Ti (Phase Diagram) ... **A3:** 2 • 284
Mg-Y (Phase Diagram) ... **A3:** 2 • 284
Mg-Yb (Phase Diagram) ... **A3:** 2 • 284
Mg-Zn (Phase Diagram) ... **A3:** 2 • 285
Mg-Zr (Phase Diagram) ... **A3:** 2 • 285
MgO partially stabilized zirconia (PSZ)
heat-treatable ceramic ... **A18:** 814
MHOST computer program for structural analysis ... **EM1:** 268, 271
Mica ... **A6:** 60 **A18:** 144 **EM4:** 6
as additive to metalworking lubricants ... **A18:** 141
chemical composition ... **A6:** 60
cleaved, friction coefficient data ... **A18:** 75
contaminated, friction coefficient data ... **A18:** 75
crystal structure ... **EM4:** 882
effect on rheological behavior of casting slips ... **EM4:** 5
as filler ... **EM3:** 175, 177, 179
as filler material ... **EM2:** 192, 500
function and composition for mild steel SMAW electrode coatings ... **A6:** 60
grain size of silver film grown on ... **A10:** 543-544
properties ... **A18:** 801
surface force measurement material ... **A18:** 401
Mica, as refractory
core coatings ... **A15:** 240
Mica capacitors
defined ... **EL1:** 179
Mica/talc
methods used for synthesis ... **A18:** 802
Micelle
defined ... **A10:** 676
Michael addition bismaleimides (BMIs) ... **EM1:** 79, 80, 82
Michells pockets ... **A18:** 551
Michelson interferometer ... **A10:** 39, 112
Michigan Solidification Simulator ... **A15:** 861
Micro ... **EM3:** 18
defined ... **EM1:** 15 **EM2:** 26-27
Micro-fracture mechanics ... **A8:** 465-468
Micro-lap joints
plasma arc welding ... **A6:** 198
Microabrasive dry blasting system ... **M5:** 89-90
Microabsorptions
as XRPD source of error ... **A10:** 341
Microalloyed carbon-manganese steels
weldability ... **A6:** 420
Microalloyed forging steels ... **A14:** 219-221
Microalloyed HSLA steels *See also* As-rolled structural steels, Heat treated HSLA steels ... **M1:** 403, 409
compositions, typical ... **M1:** 404
inclusion shape control ... **M1:** 411
mechanical properties ... **M1:** 408, 412
Microalloyed steel *See also* High strength low-alloy (HSLA) steels
ASTM specifications of ... **A1:** 399, 406, 411
compositions ... **A1:** 406
mechanical properties ... **A1:** 411
brittle fracture of ... **A1:** 412
comparison of plate and bar products ... **A1:** 588
control of properties ... **A1:** 405-410
Microalloyed steel (continued)
controlled rolling of ... **A1:** 115, 117-118, 131, 408-409, 586-588
deformation-temperature-time sequence during torsion testing ... **A8:** 179
directionality of properties ... **A1:** 412-413
elements ... **A1:** 358-359
ferrite structure in ... **A8:** 180
metallurgical effects ... **A1:** 359-362
microalloyed bars ... **A1:** 246, 419-420
alterative strengthening mechanisms in ... **A1:** 587-588
applications and compositions ... **A1:** 420
high-strength low-alloy bar products ... **A1:** 588
processing of ... **A1:** 587-588
microalloyed castings, applications and compositions ... **A1:** 420
microalloyed forgings ... **A1:** 137, 358-362, 419, 588
applications ... **A1:** 420
carbon contents of ... **A1:** 585
control of properties ... **A1:** 588
forging temperature, effects of ... **A1:** 588
generations of ... **A1:** 359-360
properties of ... **A1:** 360-361, 588, 599
microalloyed plates
compositions ... **A1:** 406, 410
mechanical properties ... **A1:** 409, 410, 411
processing of ... **A1:** 586-587
microalloyed quenched and tempered grades ... **A1:** 394-396
microalloying elements ... **A1:** 358-359, 400-404, 419
effects on properties ... **A1:** 358
molybdenum ... **A1:** 358, 403, 407
niobium ... **A1:** 358, 402-403, 407, 419
titanium ... **A1:** 231-232, 359, 403-404, 408
vanadium ... **A1:** 358, 401-402, 403, 408, 419
notch toughness
of acicular ferrite ... **A1:** 404-405
compared with carbon steel ... **A1:** 389
effect of reheating on ... **A1:** 414
of microalloyed ferrite-pearlite ... **A1:** 412
of microalloyed forgings ... **A1:** 360, 361
processing of ... **A1:** 130-131, 408-410, 586, 587, 588
coiling temperature and nitrogen content, effect of ... **A1:** 412
control of properties ... **A1:** 408, 588
forging temperature, effect of ... **A1:** 588-589
hot mill finishing temperature, effect of ... **A1:** 412, 588, 590
roll forces and roll torques, compared ... **A8:** 180
strengthening mechanisms of ... **A1:** 405, 586-587
grain refinement ... **A1:** 405
precipitation strengthening ... **A1:** 403, 586-587
Microalloyed steel bars ... **A1:** 246, 419-420
applications ... **A1:** 420
processing of ... **A1:** 587-588
alternative strengthening mechanisms in ... **A1:** 587-588
high-strength low-alloy bar products ... **A1:** 588
Microalloyed steels
precipitate stability and grain boundary pinning ... **A6:** 73
Microalloying
cold heading steels ... **A14:** 221
elements, effects ... **A14:** 219-220
of ordered intermetallics ... **A2:** 913-914
uranium ... **A2:** 677
Microanalysis *See also* Microanalysis, methods for
atomic probe ... **A10:** 583, 591-602
basic concepts, EPMA analysis ... **A10:** 517-518
by x-ray photoelectron spectroscopy ... **A7:** 257
chemical analysis and microscopy in ... **A10:** 517
chemical, at atomic level, FIM/AP for ... **A10:** 583
of crystal structure, inorganic solids analytical methods for ... **A10:** 4-6
of defects, inorganic solids, analytical methods for ... **A10:** 4-6
defined ... **A10:** 676
elemental, of inorganic solids, analytical methods for ... **A10:** 4-6
for inhomogeneous compositional structures ... **A10:** 529
of inorganic solids, analytical methods for ... **A10:** 4-6
of morphology, inorganic solids, analytical methods for ... **A10:** 4-6
Microanalysis (continued)
of phase distribution, inorganic solids analytical methods for ... **A10:** 4-6
of phase identification, inorganic solids analytical methods for ... **A10:** 4-6
spatial resolution in ... **A10:** 525
x-ray, in analytical electron microscopy ... **A10:** 446-449
x-ray, of diffraction-induced grain-boundary migration ... **A10:** 462
Microanalysis, methods for *See also* Microanalysis
analytical transmission electron microscopy ... **A10:** 429-489
Auger electron spectroscopy ... **A10:** 549-567
crystallographic texture measurement and analysis ... **A10:** 357-364
electron probe x-ray microanalysis ... **A10:** 516-535
field ion microscopy ... **A10:** 583-602
image analysis ... **A10:** 309-322
low-energy electron diffraction ... **A10:** 536-545
optical emission spectroscopy ... **A10:** 21-30
optical metallography ... **A10:** 299-308
particle-induced x-ray emission ... **A10:** 102-108
radial distribution function analysis ... **A10:** 393-401
Raman spectroscopy ... **A10:** 126-138
scanning electron microscopy ... **A10:** 490-515
x-ray powder diffraction ... **A10:** 333-343
x-ray topography ... **A10:** 365-379
Microanalytical techniques
compared ... **A11:** 35, 36-37
examples ... **A11:** 37-46
for failure analysis and problem solving ... **A11:** 32-46
history ... **A11:** 32-35
samples and limitations ... **A11:** 36
types ... **A11:** 32-35
Microballoons, glass
ion chromatography analysis of ... **A10:** 665-667
Microband segments
nucleation in ... **A9:** 694
Microbands
defined ... **A9:** 12, 685
development of ... **A9:** 685
Microbeam analysis *See also* Electron probe x-ray microanalysis
applicability ... **A10:** 529
applying, electron probe x-ray microanalysis ... **A10:** 529-530
compositional gradients ... **A10:** 530
and line broadening analysis ... **A10:** 374
multiphase samples ... **A10:** 529-530
sample preparation ... **A10:** 529-530
sampling strategy ... **A10:** 529-530
Microbes
corrosion by ... **A13:** 43, 314
Microbial corrosion *See also* Biological corrosion; Local biological corrosion; Microbiological corrosion ... **A13:** 88, 115, 118, 120
Microbial degradation
biodegradable mechanisms ... **EM2:** 783-784
biodegradation, defined ... **EM2:** 784
biodegradation, measured ... **EM2:** 784
biodeterioration, defined ... **EM2:** 784
biodeterioration, measured ... **EM2:** 784-785
experimental example ... **EM2:** 785-787
of polymers ... **EM2:** 424
Microbial films ... **A13:** 88
Microbiocides
in cooling water systems ... **A13:** 493
Microbiological attack
as degradation factor ... **EM2:** 576
Microbiological corrosion *See also* Biological corrosion; Localized biological corrosion; Microbes; Microbial corrosion
aircraft ... **A13:** 1031-1032
austenitic stainless steel weldments ... **A13:** 353-355
chemical analysis ... **A13:** 314
corrosion testing ... **A13:** 314-315
defined ... **A13:** 314
paper machine ... **A13:** 1190
risk analysis ... **A13:** 315
Microbiologically induced corrosion
of austenitic stainless steel weldments ... **A13:** 353-355
Microbiologically influenced corrosion (MIC) ... **A6:** 1068
austenitic stainless steels ... **A6:** 467

Microbiology
of lubricants **A14:** 517
Microbuckling **EM1:** 196, 792
Microcast-X process **A1:** 992
hot isostatic pressing **A15:** 541, 543
Microcharpy, unnotched
of aluminum oxide-containing cermets **A7:** 804
Microcircuit process gas
analyzed **A10:** 156-157
Microcircuitry
defined **EL1:** 89, 250
as miniaturization **EL1:** 89
package evaluation requirements **EL1:** 458-459
Microcircuits
hybrid **EL1:** 250-262, 451-454
hybrid packaging **EL1:** 259-260, 451-454
Microcline ($K_2O{\cdot}Al_2O_3{\cdot}6SiO_2$) purpose for use in glass manufacture **EM4:** 381
Microcompact
compaction **A7:** 58
particle size , surface area, and surface forces **A7:** 58
Microcomputers *See also* Automation; Computer applications; Computer-aided design/Computer-aided manufacture
-based data systems, for Raman spectrometers **A10:** 128
effect on x-ray spectrometry **A10:** 83
materials analysis software for **EM1:** 269, 274-281
use, molding machines **A15:** 350-351
Microcomputers, for corrosion data analysis *See also* Computers; Personal computers **A13:** 317
Microconstituents
of D2 tool steel, Knoop indentation **A8:** 97, 101
EPMA identified **A11:** 37
in low-carbon steel **A1:** 579
measuring hardness **A8:** 97
Microcontacts *See also* Electrical contact materials
recommended contact materials **A2:** 864, 866
Microcontamination
in plated-through holes **EL1:** 1028-1029
Microcrack
to macrocrack coalescence **A8:** 57
Microcracked chromium plating **M5:** 189-191, 193, 197
advantages of **M5:** 189
crack pattern **M5:** 189
solution compositions and operating conditions **M5:** 189-191
Microcracking *See also* Microfracture **A18:** 184, 185-186 **EM3:** 18
at slip bands **A12:** 360
austenitic stainless steels **A12:** 356
brittle, polycarbonate sheet **A12:** 479
defined **EM1:** 15 **EM2:** 27
deformation zone friction and polymers **A18:** 36
ductile irons **A12:** 235
from brittle matrix fracture **EM1:** 787, 789
initiation, ductile iron **A12:** 233, 234
laminate **EM1:** 230
malleable iron **A12:** 239
of matrices **EM1:** 31
rolling contact wear **A18:** 259-260
semiconductors **A18:** 685-689
through sulfide inclusions, malleable iron **A12:** 239
tool steels **A18:** 737
Microcracks *See also* Crack(s)
in composites, acoustic emission inspection **A17:** 288
in continuous fiber reinforced composites **A11:** 736
defined **A9:** 12 **A11:** 7
in electroplated hard chromium deposits **A13:** 871
formation, as source, acoustic emissions **A17:** 287
Microcracks (continued)
hydrogen-induced cracking **A6:** 410
magnetic particle detection **A17:** 102
in nylon driving gear **A11:** 764
in polymers **A11:** 759
in polyoxymethylene **A11:** 761
upon quenching **A11:** 324
Microcrazing
in all-ceramic mold casting **A15:** 249
in linear polyethylene **A12:** 479
Microcrystalline alloys
corrosion resistance **A7:** 797
from metallic glasses **A7:** 794-795
heat-treatable bulk **A7:** 794
iron-based **A7:** 795-797
microstructures **A7:** 795
nickel-molybdenum-iron-boron hot hardness **A7:** 796
for tool applications **A7:** 796
Microcrystalline cellulose and sodium carboxymethylcellulose
application or function optimizing powder treatment and green forming **EM4:** 49
Microcrystalline model
of amorphous materials and metallic glasses **A2:** 809, 810
Microcrystalline wax emulsions
as binders for spray drying before dry pressing **EM4:** 146
Microcrystalline waxes
for patterns **A15:** 197, 253-255
Microcrystalline waxes (with modifiers)
as binder for ceramic injection molding **EM4:** 173
Microdebonding **EM3:** 401
Microdebonding/microindentation technique **EM3:** 399-401, 403
Microdensitometer
spark source mass spectrometry **A10:** 143
for topographic intensity profiles **A10:** 372
Microdiffraction **A6:** 145
defined **A10:** 438-439
Microdiffraction modes **A18:** 386, 387
Microdiffractometer **A10:** 338
Microduplex alloys
formation of **A9:** 604
Microduplex stainless steels
superplasticity **A14:** 871
Microdynamometer **EM3:** 18
defined **EM2:** 27
Microelectric packages
leak testing **EL1:** 953-954
Microelectrogravimetry
conduction of **A10:** 200
Microelectronic Center/North Carolina (MCNC) heat-down package **EL1:** 310-311
Microelectronic components *See also* Electronic components
acoustic microscopy of **A17:** 473-481
custom, machine vision process **A17:** 45
integrated circuits (ICs) **A17:** 474-480
Microelectronic devices
failure mechanisms **EL1:** 1049-1057
hermetic, with glass-to-metal seals **EL1:** 455
in plastic encapsulants **EL1:** 805
thermal-wave imaging of **A12:** 169
Microelectronic materials *See also* Electrical materials; Electronic materials; Material(s); specific microelectronic materials
adhesives for **EL1:** 110
bonding wires (chip/wire assembly) **EL1:** 110
conductive materials **EL1:** 106-107
gold and aluminum wire selection **EL1:** 110-111
physical characteristics **EL1:** 104-111
polymeric-film dielectrics **EL1:** 108-109
resistive materials **EL1:** 107-108
substrates **EL1:** 104-106
thick-film dielectrics **EL1:** 108-109
Microelectronic materials (continued)
thin-film dielectrics **EL1:** 108-109
Microelectronics *See also* Materials and electronic phenomena; Microelectronic materials
advanced, quality control/assurance **EL1:** 867-868
AES in-depth surface analyses for **A10:** 549
defined **EL1:** 89, 250, 1150
IC thick- and thin-film hybrid **EL1:** 89
industry **EL1:** 458, 509
periodic chart **EL1:** 91, 94
thermal stress failures **EL1:** 55-62
Microelectronics industry **EL1:** 458, 509
Microelectronics tests **EM3:** 378-380
direct pull-off test **EM3:** 380
electromagnetic tensile test **EM3:** 380
island blister test **EM3:** 379-380
peel tests **EM3:** 378-379
Microelemental analysis **A13:** 1116-1117
as advanced failure analysis technique **EL1:** 1106
Microenthalpy method
solidification modeling **A15:** 887-888
Microetchants *See* Etchants
Microetching
carbon and alloy steels **A9:** 169-170
of cast irons to improve as-polished surfaces **A9:** 244
defined **A9:** 12
magnifications **A9:** 62
of tool steels **A9:** 257-258
Microetching solutions
chemicals for **A9:** 67-68
Microfilming
of radiographs **A17:** 356
Microfissure *See* Microcrack; Microcracks
Microfissure crevice corrosion, and pitting
compared **A13:** 349
Microfissures
in shielded metal arc welds **M6:** 9
type of weld discontinuity **M6:** 83
in weld overlays **M6:** 81
Microfocus radiography
to evaluate gas turbine ceramic components **EM4:** 718
Microfocus x-ray(s)
projective magnification with radiography **A17:** 300
tubes, radiography **A17:** 302-303
Microforging **A7:** 62
particle effects in milling **A7:** 61
Microfractography **A12:** 1, 3-4
defined *See* Fractography
interpretation **A11:** 22
Microfracture *See also* Microcracking **A18:** 184, 185-186
defined **A18:** 13
Microfracture topography *See* Surface roughness
Microgap welding processes
failure mechanisms **EL1:** 1044
Micrograin high-speed steels
for cutting tool materials **A18:** 615-616
Micrograph **A10:** 302, 676
Micrographs *See also* Photomicroscopy
defined **A9:** 12
Microhardness *See also* Hardness; Microhardness test
and apparent hardness **A7:** 489
defined **A8:** 8-9
fixtures **A8:** 93, 96
measurement **A7:** 61
of nickel powders, effect of milling time **A7:** 61
testing **A7:** 453-455
traverses, in low-speed test **A8:** 230
Microhardness number
defined **A18:** 13
Microhardness test
for analyzing metal failure **A8:** 97-98
applications **A8:** 90, 96-98, 102

Microhardness test (continued)
bench-mounted testers **A8:** 91-92
for carburized and nitrided cases **A8:** 96
of cast iron **A8:** 97
with decarburized workpieces **A8:** 83
defined **A8:** 9, 90
depth **A8:** 102
diagonal or diameter **A8:** 102
fixtures **A8:** 93, 96
force applied **A8:** 73
hardness gradient **A8:** 96, 101
indenters **A8:** 73, 102
Knoop and Vickers **A8:** 90-98
load **A8:** 94, 96, 102
method of measurement **A8:** 102
for microconstituent hardness **A8:** 97
for nonferrous metals **A8:** 89
optical equipment **A8:** 92-93
surface hardening operations **A8:** 96
surface preparation **A8:** 93, 102
techniques compared **A8:** 102
testers **A8:** 91-93
ultrasonic **A8:** 98-103
Microhardness testing **A8:** 90-103
by indenter attachments on optical microscopes **A9:** 82
as mechanical test **A11:** 18
of rocket-motor case fracture **A11:** 96
thermal spray coatings **M5:** 371
Microhardness tests **A6:** 104, 105
Microhardness traverse test **A6:** 103, 105
wrought martensitic stainless steels **A6:** 441
Microhardness traverses
in copper **A8:** 230
in high-speed test **A8:** 230
in low-carbon steel specimen **A8:** 229-230
Microinclusions *See also* Defects; Inclusion-forming reactions; Inclusions
deep shell crack from **A12:** 288
defined **A15:** 88
Microindentation hardness number *See also* Knoop (microindentation) hardness number; Nanohardness test; Vickers (microindentation) hardness number
defined **A18:** 13
Microindentation hardness testing **A18:** 414-418
common uses **A18:** 414
correlation of microindentation hardness numbers with wear **A18:** 418
microindentation hardness numbers of materials **A18:** 416
microindentation testing of coatings **A18:** 416-417
multilayer coating systems **A18:** 417
single-layer coated surfaces **A18:** 416-417
principles of microindentation testing **A18:** 414-415
factors affecting accuracy and reliability **A18:** 414
procedure, steps in **A18:** 414
relation of force and pressure to microindentation hardness units **A18:** 415
standard reference materials for microindentation hardness **A18:** 415
Vickers microindentation hardness test **A18:** 415-416, 417
proportionality constants as function of hardness, force, and diagonal length **A18:** 416
Vickers indenter versus Knoop indenter **A18:** 415-416
wear measurement using microindentations **A18:** 417-418
Microindentation test **EM3:** 399-401
Microinhomogeneity
from internal particle boundaries **A7:** 29
Microjet impingement
effect in ductile materials **A11:** 165
Microlaminations and cracks **A7:** 486
Microlithography
as interferometer application **A17:** 14
Micromachining
erosion mechanism **A18:** 202, 203
metal-matrix composites **A18:** 809, 810
Micromechanics
of dynamic fracture **A8:** 286-288
Micromechanisms
of fracture **A12:** 179
Micromerography analysis **A7:** 129, 218-219
Micromesh **A7:** 7
sieves **A7:** 124, 216
sizing **A7:** 7
vacuum siever **A7:** 216
Micrometer **A7:** 7
laser bench **A17:** 13
laser interferometric **A17:** 15
Micrometer screw **A8:** 619
Micrometer screws
thread grinding application **A16:** 278
Micrometer spindles **A16:** 110-111
Micrometer test method **M5:** 596
Micromodification
photographic **EL1:** 732
Micron **A7:** 7
Micron bar
on micrographs **A12:** 167
Microorganisms, biological corrosion by **A13:** 88, 118, 314
Micropen **EM4:** 1141
Microperforation
as hydrogen damage **A13:** 164
Microphase separation
SAS techniques for **A10:** 405
Microphases **A6:** 77
Microphone pickup, for amplitude detection
ultrasonic testing **A8:** 246
Microplasma welding **A6:** 755
Microplastic deformation
ductile irons **A12:** 236
Micropolishing *See also* Polishing
as FIM sample preparation **A10:** 584
Micropore **A7:** 7
Microporosity *See also* Porosity; Voids
in brittle cleavage fracture, ductile iron **A12:** 227
cemented carbides **A12:** 470
defined **A9:** 12 **A11:** 7
dendritic **A15:** 119
in directional solidification **A15:** 321
effects on cracking in gray iron **A11:** 355-357
from solidification shrinkage **A15:** 109
Micropotentiometers
recommended microcontact materials **A2:** 866
Microprobe
analysis, NBS standards for **A10:** 530
Auger, of gold-plated stainless steel lead frame **A10:** 560
electron, capabilities **A10:** 161
ion, capabilities **A10:** 161
laser Raman molecular **A10:** 129
proton, and PIXE analysis **A10:** 107
Microprobe analysis **A7:** 258
Microprobe examinations
copper and copper alloy specimens for **A9:** 400
Microprobe microanalyzer
used in structural analysis of lead alloys **A9:** 417
Microprocess analysis code (MIPAC) **EM1:** 268, 271
Microprocessing units (MPUS) **EL1:** 160, 177
Microprocessor control
iron powder production **A7:** 23
Microprocessor peripherals
as standard circuits **EL1:** 160
Microprocessor-controlled vacuum coating system **M5:** 410
Micropulverizer **A7:** 7
Microradiography *See also* Radiography
enlargement effect **A17:** 312
to reveal as-cast solidification structures in steel **A9:** 624
Microsampling
infrared **A10:** 116
Microscope
affecting microhardness readings **A8:** 96
for measuring indentation size **A8:** 91
traveling low-power, optical crack measuring **A8:** 382
verification, Brinell test **A8:** 88
Microscopes *See also* Macroscopes; Microscopy; Optical microscope; Scanning electron microscope(s); Scanning electron microscopes; specific microscopy methods; Stereomicroscopes; Transmission electron microscopes
for acoustic microscopy **A17:** 465
analytical electron **A10:** 430-432
defined **A9:** 12 **A10:** 676
detectors and image formation in **A10:** 492
direct-imaging ion **A10:** 613
field ion **A10:** 584
infrared **A10:** 116
interference, for surface **A17:** 17
ion **A10:** 614
laser interference, with interferometers **A17:** 16
optical **A11:** 20
for plastic replicas **A17:** 53
scanning electron **A11:** 22
SEM **A10:** 491-494
toolmakers' **A17:** 10-11
transmission electron **A11:** 20-22
upright, with image analyzers **A10:** 310
in visual leak detection **A17:** 66
Microscopic
defined **A9:** 12 **A11:** 7 **A13:** 9
Microscopic analysis
of metals **A15:** 27
Microscopic examination
of failed parts **A11:** 173-174, 629
of failure types **A11:** 75
of fracture surfaces **A11:** 20-22
of fractures **A11:** 105, 258-259
magnesium alloys **M4:** 751
ofshafts **A11:** 460
use in carbon control **M4:** 433
of wom parts **A11:** 158-159
Microscopic fracture modes **A8:** 476-477
Microscopic metallurgical features
of deformed structures **A9:** 686-688
Microscopic method
paint dry film thickness tests **M5:** 492
Microscopic model
for predicting cleavage fracture **A8:** 466
Microscopic solids
solidification of **A15:** 102-103
Microscopic stress *See also* Microstresses
defined **A10:** 676
Microscopy *See also* Scanning electron microscopy (SEM); Transmission electron microscopy (TEM)
acoustic **EL1:** 369-371
and acoustic emission inspection **A17:** 286
acoustical, and ultrasonic inspection **A17:** 240
deep-field **A12:** 96
defined **A10:** 676
for degree of homogenization **A7:** 316
development of **A15:** 27
light **A12:** 91-165
optical **A7:** 227
for particle size analysis **A7:** 225-230
residue analysis by **A10:** 177
scanning electron **A7:** 228
transmission electron **A7:** 227-228
Microscopy/image analysis
versus weighting factor **EM4:** 85
Microsecond
symbol for **A10:** 692
Microsectioning
printed board coupons **EL1:** 572-577
Microsegregation *See also* Segregation **A15:** 136-138
in aluminum alloy ingots **A9:** 631-632
in cast structures **A9:** 611-617
defined **A9:** 12, 625 **A15:** 8
dendrite coarsening **A15:** 138
dendrite morphology/diffusion path **A15:** 137
during solidification **A15:** 136-141
equilibrium phase diagram/partition coefficient **A15:** 136
liquid phase, movement **A15:** 138
and macrosegregation, during solidification **A15:** 136-141
nickel-base alloys **A6:** 588
phase transformation **A15:** 137-138
in rapid solidification processing **A15:** 138
solidification mode/structure **A15:** 137
solute redistribution, nonequilibrium solidification **A15:** 136-137
temperature/concentration dependency of diffusion coefficient **A15:** 138

Microsegregation (continued)
third solute element, effect **A15:** 138
undercooling **A15:** 138
Microshrinkage *See also* Shrinkage
computed tomography (CT) **A17:** 361
defined **A11:** 7 **A15:** 8
radiographic appearance **A17:** 348
of tin bronzes **A2:** 348
Microshrinkage cavity **A7:** 7
Microshrinkage pores
by liquid penetrant inspection **A17:** 86
Microsiemen
as common ion chromatography unit **A10:** 659
Microslip *See also* Macroslip; Slip
defined **A18:** 13
Microspalling *See also* Peeling; Spalling
in roller bearing **A11:** 502
Microspectroscopy **A6:** 145
Microsphorite
Miller numbers **A18:** 235
Microstrain **EM3:** 18
defined **A8:** 9 **EM2:** 31
Microstrain as a result of electric discharge machining **A9:** 27
Microstresses
associated with cold working **A10:** 386
determined **A10:** 380, 386-387
states analyzed **A10:** 424
in x-ray diffraction residual stress techniques **A10:** 380, 381
Microstrip, coated
in impedance models **EL1:** 602
Microstrip lines
properties, impedance models **EL1:** 602
terminations **EL1:** 520
Microstructural AEM analysis of ion- implanted alloys
implantation of two species: ternary alloys **A10:** 486
inert elements, implantation of **A10:** 485-486
metalloids, implantation of **A10:** 485
metals, implantation of **A10:** 485
precipitation **A10:** 485-486
radiation-induced changes **A10:** 484
sample preparation **A10:** 484
thermal treatments **A10:** 486-487
Microstructural analysis **A6:** 97, 103, 104 **EM3:** 406-418 **EM4:** 570-579
in atomic detail **A10:** 583
cross-link density **EM3:** 418
cross-linked adhesives **EM3:** 412-413
crystalline morphology **EM3:** 409-412
crystallinity **EM3:** 407-409
EPMA compositional analysis of phases at **A10:** 516
field ion microscopy and atom probe **A10:** 583
fracture energy **EM3:** 406
fracture surface specimens **EM4:** 577
replication **EM4:** 577
of ion-implanted alloys **A10:** 484
microstructure **EM3:** 406-407
phase structure **EM3:** 413-414
phase structure influence **EM3:** 418
preparation of thin sections for transmitted illumination **EM4:** 576-577
revealing of microstructure **EM4:** 574-576
cathodic vacuum etching **EM4:** 575
chemical etching **EM4:** 575, 577
differential interference contrast microscopy **EM4:** 576
electrolytic etching **EM4:** 575
thermal etching **EM4:** 575
vapor deposited interference films **EM4:** 575-576
sample preparation **EM4:** 570-574
coarse and fine polishing **EM4:** 573-574
impregnation and mounting **EM4:** 570-572
mechanical grinding **EM4:** 572-573, 574
sampling **EM4:** 570
Microstructural analysis (continued)
sectioning **EM4:** 570
standard methods for incident light examination **EM4:** 574
substrate **EM3:** 415-417
substrate influence **EM3:** 417-418
techniques **EM4:** 577-579
electron microscopy **EM4:** 578
electron probe microanalysis **EM4:** 578-579
image analysis **EM4:** 578
interference microscopy **EM4:** 578
polarizing microscopy **EM4:** 578
reflected light microscopy **EM4:** 578
scanning electron microscopy **EM4:** 578
undercooling **EM3:** 408-409
use of stereological relationships **A10:** 309
voids **EM3:** 414--415
x-ray topography **A10:** 366
Microstructural evolution
modeling of **A15:** 883-891
Microstructural evolution modeling
of columnar structures **A15:** 884-885
of equiaxed structures **A15:** 885-890
macroscopic **A15:** 883
Microstructural inhomogeneity
effect on annealing stages **A9:** 692
Microstructural morphology
scanning electron microscopy used to study **A9:** 101
Microstructural overload failure
cleavage **A8:** 479-481
ductile fracture **A8:** 477-479
Microstructure **EM3:** 18, 320-321
of abrasion-resistant cast iron **A1:** 92-94, 95
acoustic emission inspection of **A17:** 286-289
AISI/SAE alloy steels, austenitization effects **A12:** 292
alloy steels **M1:** 455-456, 459-460
Alnico alloy **A12:** 461
alterations, in bearing materials **A11:** 505
aluminum alloy weldments, variation effects **A13:** 344
aluminum coatings **A13:** 434
of aluminum P/M alloys **A2:** 200
aluminum P/M parts **A7:** 385
analysis, by replication **A17:** 54-55
analysis, for problem solving **A7:** 485
annealed ductile iron **A15:** 658
as-cast and cast + HIP, Ti alloy castings **A2:** 638-639
as-sintered P/M tool steels **A7:** 377
austempered ductile iron, strength **A12:** 232
austempered iron **A15:** 660
austenitic stainless steel
high-energy-rate- forged extrusion **A14:** 365
of bainite **A1:** 128, 129
-based classification, corrosion-resistant high-alloy steels **A15:** 722-723
beryllium-copper alloys **A2:** 286-290, 404-405
boron fibers **EM1:** 59
break-up in two-phase alloys **A8:** 177
brittle cleavage fracture, ductile iron **A12:** 227
brittle fracture of roll-assembly sleeve by **A11:** 327
of brittle fractured high-speed tool **A11:** 574, 579
brittleness from **A11:** 325-326
bronze shell **A12:** 403
carbon fiber **EM1:** 50-52
of carbon steel, and swageability **A14:** 128
of carburizing bearing steels **A1:** 381-382, 383
cartridge brass **A2:** 302
cast copper alloys **A2:** 384
cast iron breaker bar, premature wear **A11:** 368
cast irons, and weldability **A15:** 522
of cast metal-matrix composites **A15:** 850-851
of cast stainless steels **A1:** 909, 910 **A13:** 574-575
of cemented carbides **A7:** 780-783 **A13:** 847
Microstructure (continued)
cemented carbides for machining applications **A2:** 951-953
cermet **A7:** 801-802
of cermets **A2:** 990-992
changes, sintering time and temperature **A7:** 311
characteristics of weld and heat-affected zone **M6:** 35-3
characterization **A17:** 50-51
Charpy V-notch as screening test for **A8:** 263
coated carbide tools **A2:** 962
coating **A13:** 526
cobalt-base alloys **A15:** 813
of cobalt-base wear-resistant alloys **A2:** 449
cold finished bars **M1:** 233
commercial bronze **A2:** 297
commercially pure titanium **A2:** 594
of compacted blend copper particles **A7:** 314
compacted graphite irons **A15:** 667
of composite materials **EM1:** 768
and compositional effects on toughness TI-6Al-4V **A8:** 480-481
continuous emissions from **A17:** 287
control, process window for **A14:** 412-413
controlled crystalline, amorphous materials and metallic glasses **A2:** 820
conventionally cast vs semisolid cast zinc alloys **A15:** 796-797
of copper powder compact **A7:** 311
in corrosion analysis **A11:** 173-174
and corrosion properties, uranium/uranium alloys **A13:** 816
of corrosion-resistant steel castings **A1:** 909, 913-915
of crack propagation **A13:** 155-158
of creep void **A17:** 55
crystallographic texture measurement and analysis for **A10:** 358
defined **A9:** 12 **A11:** 7 **A13:** 9 **EM1:** 15 **EM2:** 27
development, during deformation processing **A8:** 173-178
development in multiphase alloys **A8:** 177-178
direct-cooled forging **A1:** 137, 138
discontinuous graphite-aluminum MMC **EM1:** 868
ductile iron **A15:** 655
during creep **A8:** 305-306
effect, carbon fiber modulus **EM1:** 50
effect in fatigue **A12:** 14
effect, mechanical properties, titanium and titanium alloy castings **A2:** 639
effect of, at elevated-temperature service **A1:** 638, 641
effect of high temperatures **A12:** 121
effect of, on notch toughness **A1:** 744, 747-749, 750
effect of, on weldability **A1:** 604-606, 607, 608
effect on corrosion-fatigue **A11:** 256
effect on damage resistance **A11:** 166
effect on metal properties **A11:** 325
effect on stress-corrosion cracking **A8:** 501
effect on tensile ductility **A12:** 101
effects, aluminum/aluminum alloys **A13:** 585-587
effects, laser surface processing **A13:** 503
effects of sintering tungsten and molybdenum **A7:** 391
effects of welding **M6:** 37-4
effects on ductile-to-brittle transition temperature, structured steels **A11:** 68-69
effects on wear failure **A11:** 160-161
eutectic, interpretation of **A15:** 120-121
of failed outlet header, linked voids and split grain boundaries in **A11:** 667
of failed reheater tube **A11:** 611
of failed water pipe **A11:** 374
fatigue behavior, effect on **M1:** 675-677
of ferrite-pearlite **A1:** 127, 130-131
ferritic **A1:** 127, 131-133

Microstructure (continued)
ferritic malleable iron **A15:** 687
fibrous eutectic **A15:** 119
of forged titanium **A14:** 271
forging failures from **A11:** 325-327
forgings, by eddy current/electromagnetic inspection **A17:** 511
and formability, correlation between **A1:** 578-579
formation, during freezing **A15:** 119
fractal analysis **A12:** 212-215
and fracture **A8:** 476-491 **A11:** 75
fracture appearance, hydrogen-embrittled titanium alloy **A12:** 32
of fracture materials **A8:** 466-467
and fracture path, correlated **A12:** 195-196, 201
gage length to measure flow localization **A8:** 169
galvanized coatings **A13:** 432
gas-atomized cobalt-based hardfacing powder **A7:** 146
and gating, die casting **A15:** 289
graphitized, of plain carbon steel **A11:** 613
gray cast iron, flake graphite **A15:** 120
of gray iron **A1:** 13-14 **M1:** 12-14, 24-26, 29
of gray iron product **A11:** 354, 363-365
of green compacts **A7:** 311
hardenable carbon steels **M1:** 455-459
high-alloy graphitic iron **A15:** 698-700
of high-carbon bearing steels **A1:** 381, 382, 383
high-carbon steel, torsional overload fracture **A12:** 278
high-chromium white irons **A15:** 681-682
high-purity copper **A12:** 399-400
high-temperature superconductors **A2:** 1088
hydrogen-assisted cracking, welded cast steel **A15:** 534-536
hypereutectic gray iron **A12:** 226
impact on macroscopic fracture behavior **A8:** 439
importance for superalloys in high-pressure hydrogen gas **A8:** 408
inadequate heat treatment effects, cast aluminum alloys **A12:** 410
and infiltration **A7:** 552
influence of, on temper embrittlement in alloy steels **A1:** 702
influencing corrosion fatigue crack propagation **A8:** 405, 408
of initial condition **A14:** 418
intergranular, AISI/SAE alloy steels **A12:** 300
of ion-implanted alloys, AEM determined **A10:** 484-487
of iron castings **A11:** 359
lamellar eutectic **A15:** 119
laser alloyed casings **M6:** 79
laser clad Haynes stellite **M6:** 79
lead-base bearing alloys **A2:** 553
and leakage field fluctuation **A17:** 129
low-carbon steel after rolling **A14:** 355
low-strength, effect on gray iron bearing cap **A11:** 347-348
and machinability of steels **A1:** 595, 596
and magnetic characterization **A17:** 131
and magnetic hysteresis **A17:** 134
and magnetic properties, FIM/AP analysis of relationship **A10:** 599
of malleable iron **A1:** 72, 76, 77 **A15:** 687-690
malleable irons **M1:** 58-63, 66
of martensite **A1:** 127, 133-134
tempering of **A1:** 134-136, 137
martensitic malleable iron **A15:** 689
and mechanical/physical properties **A17:** 51
mechanically alloyed oxide alloys **A2:** 944-945, 947
dispersion-strengthened (MA ODS)
medium-carbon steels, shell fracture **A12:** 256-257
of metallic second phase, sintered polycrystalline diamond **A2:** 1011
metallographic investigation **A8:** 476
and metallography **A7:** 485-489
modification, Ti and Ti alloy castings **A2:** 639
molten salt corrosion **A13:** 89
nickel-chromium white cast iron **A15:** 681
nondendritic, in rheocasting **A15:** 327
normalized iron **A15:** 658
of overheated carbon steel boiler tubes **A11:** 608
of overheated tool steel **A11:** 574, 581

Microstructure (continued)
of overheating ruptures, steam equipment **A11:** 606-607
P/M stainless steels **A13:** 827
palladium-silver alloys **A2:** 716
of pearlite **A1:** 127, 128-129
pearlitic malleable iron **A15:** 688
in planar sections **A12:** 198
platinum **A2:** 709
platinum alloys **A2:** 709-714
platinum-iridium alloys **A2:** 710
platinum-palladium alloys **A2:** 709
of powder forged parts **A14:** 204
of precipitates, with chemical analysis **A17:** 55
proeutectoid ferrite and cementite **A1:** 127, 129-130
properties associated with **A1:** 126
pultruded MMC tubing **EM1:** 871
of pump-impeller **A11:** 375
quenched and tempered **A1:** 136-137, 138
of quenched uranium alloys **A2:** 674
relation to machinability of steel **M1:** 570-574, 580-581
relationship to toughness in heat-affected zone **M6:** 41-4
result of laser melt/particle inspection **M6:** 800-801
of SCC-failed stainless steel bolts **A11:** 537
second-phase constituents, wrought aluminum alloy **A2:** 37
semisolid aluminum alloy, MHD cast/forged, compared **A15:** 331
of shafts **A11:** 478
single-phase alloys **A15:** 114-119
single-phase, dimpled rupture **A8:** 476
sintered **A7:** 375-377
of sintered PCBN **A2:** 1012
of sintered polycrystalline diamond **A2:** 1012
solid/liquid interface, graphite spheroid growth **A15:** 173
solidification, nucleation kinetics of **A15:** 101
of spalled hole from improper EDM **A11:** 566
spiking defect **A11:** 351
spinoidal, coarsening in copper alloys **A12:** 402
of squeeze castings **A15:** 326
stainless steel, sigma phase in **A17:** 55
steel sheet, formability related to **M1:** 557-559
stereological parameters to determine influence of processing on **A10:** 316
stirring effects **A15:** 329
strain impact, SIMA aluminum alloy **A15:** 330
study in multiphase alloys **A8:** 177-178
styrene-maleic anhydrides (S/MA) **EM2:** 217
of surface crack **A17:** 54
surface hardened steel **M1:** 531, 533-534, 538, 540
surface, LEED analysis of **A10:** 536
surface, of carburized steel **A11:** 326-327
and surface topography **A12:** 393
tempered martensite **A15:** 660
thermit rail welds **M6:** 69
of Ti-6Al-4V **A2:** 637-639
tin-base bearing alloys **A2:** 524-525
titanium alloy base plate **A12:** 443
of titanium carbide cermets **A2:** 991
titanium carbonitride cermet **A2:** 999-1000
titanium P/M compact **A2:** 652
tool steel alloy **A7:** 103, 104
of tool steel ring forging **A11:** 570
transverse, high-carbon steels **A12:** 278
Udimet 700, in torsion and extrusion **A8:** 176-178
ultrasonic inspection **A17:** 274-275
upper-bainite, stereo-pair photographs **A12:** 89
wear resistance influenced by **M1:** 608, 611-615
weld metal, AEM interpretation **A10:** 478-481
weld-metal, from inadequate pre- and postweld heat treatment **A12:** 375
of worn cast iron pump parts **A11:** 365-367
wrought aluminum alloys, fusion zone softening and pores **A12:** 422
of wrought cobalt-base superalloys **A1:** 965-967
of wrought iron-base superalloys **A1:** 951, 959, 961-962
of wrought nickel-base superalloys **A1:** 951-956
wrought titanium alloys **A2:** 605-608
zone formation **A15:** 118

Microtearing *See also* Tearing
in ferritic ductile iron **A12:** 234

Microtomes, for specimen preparation
ATEM analysis **A10:** 452

Microtongues *See also* Tongues
in ductile irons **A12:** 232

Microtopography *See also* Topography
Fizeau interferometer for **A17:** 14
of surfaces, laser-detected **A17:** 17

Microtrac **EM4:** 67

Microtrac (light obscuration) analyzer
for particle size measurement **A7:** 223-224

Microtwinning **A6:** 163

Microtwins
iron alloy **A12:** 457

Microvoid
in aluminum fracture **A8:** 478
coalescence **A8:** 465-466, 476

Microvoid coalescence *See also* Dimple rupture(s); Dimple(s); Voids
AISI/SAE alloy steels **A12:** 293
by ductile rupture, titanium alloys **A12:** 443
crack growth by **A11:** 227, 230
as creep damage **A11:** 290
ductile fracture by **A11:** 82
ductile iron **A12:** 230-231
as ductile mechanism **A12:** 96
effect on dimple size **A12:** 12
as failure mechanism **A12:** 12
in fatigue fracture **A12:** 119-120
fracture by **A11:** 25
growth, effect on fracture surface **A12:** 13
high-purity copper **A12:** 399-400
intergranular **A12:** 14, 128
intergranular dimple rupture, steel **A12:** 14
iron-base alloy fracture by **A12:** 220, 460
linking by creep, austenitic stainless steels **A12:** 364
in low-carbon steel bolts **A12:** 248
maraging steels **A12:** 383
martensitic stainless steels **A12:** 369
as micromechanism of ductile fracture **A12:** 4
schematic under Modes I, II **A12:** 16
in stainless steel **A11:** 84, 86
strain-controlled, ASTM/ASME alloy steels **A12:** 350
superalloys **A12:** 388, 393
titanium alloys **A12:** 441, 445
tool steel fracture by **A12:** 381

Microvoid nucleation
effect on fracture surface appearance **A12:** 12

Microvoids *See also* Bubbles; Defects; Microvoid coalescence; Voids
defined **EL1:** 82
effect in PWBs **EL1:** 82
in tungsten-rhenium thermocouples **A2:** 876

Microwave
absorption intensity, in ESR analysis **A10:** 253
curing, of polymers **EL1:** 786
defined **EL1:** 1158
detector, corrosion failure analysis **EL1:** 1112-1113
integrated circuits **EL1:** 1150
losses, FMR determination of **A10:** 267
radiation, defined **A10:** 676-677
spectrometers, in FMAR measurements in transmission **A10:** 272
spectroscopy, capabilities **A10:** 253

Microwave applications
alloy characteristics **EL1:** 755
alloys for gold-base systems **EL1:** 755
commercial vs military **EL1:** 754-755
device attachment **EL1:** 755-756
high-volume production **EL1:** 757-758
passive components fabrication **EL1:** 188-189
resins and reinforcements **EL1:** 534-537
secondary/tertiary attachment alloys **EL1:** 756-757

Microwave brazing **A6:** 124

Microwave detector **A13:** 1122-1123

Microwave drying
applications **EM4:** 133

Microwave excitation **A18:** 840

Microwave ferrites
electrical/electronic applications **EM4:** 1106
as magnetically soft materials **A2:** 776
properties **EM4:** 1106

Microwave heating
effect on non-oxide ceramic joints **EM4:** 528

Microwave holography *See also* Holography; Microwave inspection; Optical holography
examples **A17:** 226-227
incoherent-to-coherent image converters **A17:** 227-228
instrumentation **A17:** 207-211
and millimeter wave/optical holographies compared **A17:** 226
optical **A17:** 224
practice **A17:** 224-228
process **A17:** 224-227
zone plates **A17:** 224
Microwave inspection *See also* Microwaves **A17:** 202-230
applications **A17:** 202-203
chemical composition, dielectric materials **A17:** 215
discontinuities detected **A17:** 212-214
eddy currents for holes and small radius areas **A17:** 223-224
ferromagnetic resonance eddy current probes **A17:** 220-223
instrumentation **A17:** 207-211
material anisotropy measurement **A17:** 215
microwave eddy current testing **A17:** 218-219
microwave holography **A17:** 207, 224-228
moisture analysis with microwaves **A17:** 215
physical principles **A17:** 203-205
plastic material properties **A17:** 51
special techniques **A17:** 205-207
stress-corrosion measurement **A17:** 215-218
surface cracks in metals, detection **A17:** 214-215
thickness gaging **A17:** 211-212
and ultrasonic inspection, compared **A17:** 202
and x-ray radiographic inspection compared **A17:** 202
Microwave integrated circuits *See also* Monolithic microwave integrated circuits (MMICs)
of gallium compounds **A2:** 740
Microwave liquid crystal display (MLCD)
capabilities **A17:** 208
Microwave packages *See also* Monolithic- microwave integrated circuits (MMIC)
defined **EL1:** 452
polymer matrix composites in **EL1:** 1117-1118
Microwave radiation
defined **A10:** 676-677
Microwave surface impedance
microwave inspection **A17:** 205
Microwave thermography
instrumentation **A17:** 208
Microwave(s) *See also* Microwave inspection
absorption and dispersion **A17:** 204
crack detection system **A17:** 219
defined **A17:** 202
detection instruments **A17:** 207-211
detection, of stress corrosion **A17:** 205
eddy current testing **A17:** 218-219
frequency bands **A17:** 202
material anisotropy measurement by **A17:** 215
in microwave holography **A17:** 224-227
moisture analysis using **A17:** 215
physical principles **A17:** 203-205
reflection and refraction laws **A17:** 204
scattering of **A17:** 204-205
source, modulation **A17:** 217
stress-corrosion measurement **A17:** 215-218
supercomponent concept **A17:** 210
surface cracks detection **A17:** 214-215
Microwave-resonant cavity
modes **A10:** 256
power reflected, as function of frequency **A10:** 256
Microwelding **A6:** 193
Microwelding with pulsed lasers ... **A6:** 263, 266, 267
Microyield strength *See also* Yield strength
beryllium **A2:** 684
MICV *See* Magnetically induced velocity changes (MICV)
Middle-infrared radiation
defined **A10:** 677
Middle-roll offset straightening **A14:** 692
Middlings
in powder cleaning **A7:** 178
Midvale Company (Philadelphia)
steel castings **A15:** 31
Mie scattering process **EM4:** 853
spectral transmittance of colloidally colored glasses **EM4:** 1081
Mie scattering theory **EM4:** 67
Miedema formula **A6:** 144
Miedema technique **A6:** 145
MIG welding *See* Gas metal arc welding
definition **A6:** 1211
Migration *See also* Electrochemical migration; Solid-state migration; Transference
diffusion-induced grain-boundary, EDS/CBED analysis of **A10:** 461-464
electrical, in voltammetry **A10:** 189
as leakage **A17:** 58
tests, ceramic packages **EL1:** 468
in thick-film pastes **EL1:** 342
Mil
defined **EM1:** 15
MIL specifications *See also* listings in data compilations for individual alloys
MIL-ST **EL1:** 6-883
for environmental testing **EL1:** 494
glass-to-metal seals **EL1:** 458
MIL-STD ..... **EL1:** -454
and federal supply class (FSC) **EL1:** 914-915
individual requirements **EL1:** 915
requirements, summary **EL1:** 908
total requirements **EL1:** 915-916
versatility/utilization of **EL1:** 913
MIL-STD-38510
glass-to-metal seals **EL1:** 458
Mild Plow Steel quality rope wire **M1:** 265, 266
Mild steel *See* Low-carbon steel
abrasion resistance **M3:** 582, 583
abrasive wear **A18:** 189
corrosion in sulfuric acid **M3:** 88, 89
erosion test results **A18:** 200
laser cladding **A18:** 867, 868
laser melting **A18:** 865
as lubricant during hot extrusion of tool steels **A18:** 738
press forming dies, use for **M3:** 489, 492
solution potential **M2:** 207
wear rates for test plates in drag conveyor bottoms **A18:** 720
Mild steels *See also* Carbon steels; Low-carbon steels **A1:** 390
covering for welding electrodes **A6:** 176
flux-cored arc welding **A6:** 187
friction surfacing **A6:** 321, 322
hardfacing **A6:** 807
hydrogen-induced cracking **A6:** 646
induction soldering, physical properties **A6:** 364
plasma and shielding gas compositions **A6:** 197
plasma-MIG welding **A6:** 224, 225
projection welding **A6:** 233
relative solderability **A6:** 134
resistance seam welding **A6:** 243
shielded metal arc welding **A6:** 57, 61
submerged arc welding **A6:** 204
turning and milling recommended ceramic grade inserts for cutting tools **EM4:** 972
weldability **A6:** 419-420
Mild wear *See also* Normal wear; Severe wear **A8:** 603
defined **A18:** 13
Mildewcides **EM3:** 674
Military *See* Ordnance applications: U.S. Military specifications
Military and Federal Specifications and Standards **EM3:** 72
Military applications *See also* Ordnance applications
advanced composites for **EM1:** 804-806
aluminum-lithium alloys **A2:** 182
circuit applications **EL1:** 386
circuit construction **EL1:** 386-387
component attachment **EL1:** 387-388
design criteria **EL1:** 516-517
environmental testing for **EL1:** 493-503
and federal specifications and standards **EM2:** 89-90
future trends **EL1:** 392-393
of germanium and germanium compounds **A2:** 743
of hybrids **EL1:** 254
liquid crystal polymers (LCP) **EM2:** 180
market size/outlook **EL1:** 389
materials technology **EL1:** 388-389
mine warfare vessels **EM1:** 837
multidirectionally reinforced ceramics **EM1:** 933-940
polybenzimidazoles (PBI) **EM2:** 147
polybutylene terephthalates (PBT) **EM2:** 153
qualification programs **EL1:** 502-503
reliability and performance criteria **EL1:** 386
solder joints, millimeter/microwave **EL1:** 754-758
of titanium P/M products **A2:** 647
trends **EL1:** 386-389
Military electronic systems
as defined by levels **EL1:** 76
Military Handbook 17 ..... **EM1:** 40
Military Handbook MIL-HDBK-337, Adhesive Bonded Aerospace Structure Repair **EM3:** 67
Military Handbook MIL-HDBK-691B, Adhesive Bonding **EM3:** 67
Military handbooks for adhesives **EM3:** 63
Military projectiles
P/M history **A7:** 17
Military Specification MIL-STD-1942MR
bend strength tests of ceramics **EM4:** 710
Military specifications
aircraft brake testing standards (MIL-W-5013) **A18:** 586
austenitic manganese steel **M3:** 585
automotive diesel engine services engine oil classification **A18:** 165
defined **EM1:** 700
engine oils **A18:** 162-163, 164, 165
gear oils (MIL-L-2105C; MIL-L-2105D) **A18:** 86
lubricant selection for aviation engine oils (MIL-L-7808G; MIL-L-23699C) **A18:** 166
MIL-J-24445, ram tensile procedure, explosion welds **A6:** 305
MIL-S-24645 ..... **A6:** 406
Military Standard 1944 ..... **EM1:** 40
Military standardization documents **EL1:** 906-911
Military Standardization Handbook, The **A8:** 61
Military standards
standardization documents **EL1:** 906-911
system, utilizing **EL1:** 911-916
Military vehicles
P/M parts for **A7:** 687-688
Milk/milk products
pure tin resistance **A13:** 772
Mill **A7:** 7
defined **A14:** 9
edge, defined **A14:** 9
finish, defined **A14:** 9
primary function **A7:** 59
Mill anneal
cycle and microstructure **A6:** 510
Mill annealing
solid-solution-strengthened alloys **A6:** 572, 573, 574
wrought titanium alloys **A2:** 619

Mill defect
stress-rupture cracking in ... **A11:** 603
Mill finishes, stainless steel ... **M5:** 551-552
grade limitations ... **M5:** 552
preservation of ... **M5:** 552
Mill heat treatment
of cold-rolled steel products ... **A1:** 202-204
of low-alloy steel ... **A1:** 209
Mill inspection
of tubular products ... **A17:** 561
Mill processes *See also* Aluminum mill and engineered wrought products
aluminum ... **A2:** 29-61
Mill product
defined ... **A14:** 9
Mill products ... **A13:** 193-194, 429-430 **M1:** 111
aluminum ... **A2:** 29 **M2:** 44-62
commercial wrought aluminum, types ... **A2:** 33-34
commercially pure titanium specifications ... **A2:** 594
defined ... **A2:** 33
forging and rolling ... **A7:** 522-529
mechanically alloyed oxide alloys ... **A2:** 947-949
dispersion-strengthened (MA ODS)
refractory metals and alloys ... **A2:** 557-559
rust-preventive compounds used on ... **M5:** 465-466
shear testing of ... **A8:** 62-65
tungsten ... **A2:** 577
wrought copper, copper base ... **A2:** 238
Mill scale ... **A13:** 9, 84, 524
atmospheric corrosion resistance affected by ... **M1:** 722
defined ... **A14:** 9
as nonrelevant indication source ... **A17:** 105
removal rates in uninhibited acids ... **M1:** 755
seawater corrosion, effect on ... **M1:** 739
soil corrosion related to ... **M1:** 729, 731
Mill scale, corrosion from ... **M5:** 431, 476
removal of ... **M5:** 476
Mill scale powder ... **A7:** 7
Mill scales
in Pyron process ... **A7:** 82
Mill shapes, of superalloys
manufacturing ... **A7:** 527-528
Milled fiber *See also* Fiber(s)
defined ... **EM1:** 15 **EM2:** 27
Milled glass fibers ... **EM1:** 110
as filler for polysulfides ... **EM3:** 139
Milled zircon opacifier ... **EM4:** 50
Miller index
of crystallographic direction ... **A10:** 359
single-crystal analysis ... **A10:** 346
unit meshes and two-dimensional ... **A10:** 537
Miller indices ... **A18:** 465-466
defined ... **A9:** 12
for designating crystal planes ... **A9:** 708
Miller numbers *See also* Slurry abrasion response number ... **A18:** 234-235
abrasivity of a fluid ... **A18:** 597
defined ... **A18:** 13
Miller-Bravais indices
defined ... **A9:** 12
for designating planes in hexagonal crystals ... **A9:** 708-710
Millett, Eli
as early founder ... **A15:** 32
Millimeter applications
alloys for gold-base systems ... **EL1:** 755
commercial vs military ... **EL1:** 754-755
device attachment ... **EL1:** 755-756
high-volume production ... **EL1:** 757-758
secondary/tertiary attachment alloys ... **EL1:** 756-757
Millimeter wave holography, and optical holography
compared ... **A17:** 226
Milling *See also* End milling; Face milling Milling machines, specific types ... **A7:** 7
abusive, and resulting fatigue strength ... **A16:** 31
adaptive control implemented ... **A16:** 618-624
aircraft engine components, surface finish requirements ... **A16:** 22
Al alloys ... **A16:** 766-769, 772-773, 784-785, 791-800
applications of P/M high-speed tool steel for ... **A1:** 785

Milling (continued)
and arithmetic roughness average ... **A16:** 14
automatic feed mechanisms ... **A16:** 309
Be alloys ... **A16:** 870
of brittle single particles ... **A7:** 60
carbon and alloy steels ... **A16:** 675-676
cast irons ... **A16:** 648, 656, 661
CBN tooling ... **A16:** 112
cemented carbides used ... **A16:** 75, 76, 79
ceramic tooling ... **A16:** 102
cermet tool parameters ... **A16:** 96
cermet tooling ... **A16:** 92, 97
chemical activity ... **A7:** 64
chemical, and resulting fatigue strength ... **A16:** 31
chemical, titanium and titanium alloy castings ... **A2:** 643
chip formation ... **A16:** 8
chip formation analysis ... **A16:** 17
chip removal operations for surface integrity ... **A16:** 33
climb (down-thread) ... **A16:** 268, 269, 309, 319, 321, 327
compared to band sawing ... **A16:** 356, 363, 364
compared to broaching ... **A16:** 194, 196, 209
compared to drilling ... **A16:** 234
compared to ECG ... **A16:** 546, 547
compared to EDM ... **A16:** 564
compared to grinding ... **A16:** 426, 427
compared to shaping and slotting ... **A16:** 187, 192, 193
compared to thread grinding ... **A16:** 270
compared to ultrasonic machining ... **A16:** 528
compared with alternative processes ... **A16:** 329
conditions, for precious metal powders ... **A7:** 149
in conjunction with broaching ... **A16:** 209
in conjunction with drilling ... **A16:** 216, 218
in conjunction with EDM ... **A16:** 560
in conjunction with superabrasive grinding ... **A16:** 460
in conjunction with tapping ... **A16:** 263
in conjunction with turning ... **A16:** 135, 140, 142, 153
in conjunction with ultrasonic machining ... **A16:** 530, 531
contour, Al honeycomb ... **A16:** 604
conventional (up-thread) ... **A16:** 268, 269
Cu alloys ... **A16:** 805, 808
cutter design effect on efficiency ... **A16:** 318-319
cutting fluid flow recommendations ... **A16:** 127
cutting fluids ... **A16:** 125, 327-328
CVD-coated tools ... **A16:** 87
diamond (PCD) tooling ... **A16:** 110
of ductile single spherical particles ... **A7:** 60
end ... **A16:** 311, 314, 320-329
end, for gear manufacture ... **A16:** 333, 339
energy relationships ... **A7:** 61-62
environment ... **A7:** 63-65
equipment ... **A7:** 65-70
face ... **A16:** 20, 311-313, 315-322, 327-329
fixtures ... **A16:** 405
fluid ... **A7:** 7
flycut milling test ... **A16:** 640-641
forces in ... **A7:** 56
gang ... **A16:** 309, 319, 320, 321
gentle, and fatigue strength ... **A16:** 31
geometrical relation of cutter to work ... **A16:** 317-318
hafnium ... **A16:** 856
heat-resistant alloys ... **A16:** 752
heating curves ... **A7:** 61, 62
helical gears ... **A16:** 339, 341
herringbone gears ... **A16:** 339
high-energy ... **A7:** 23, 67, 69-70
high-speed ... **A16:** 329
internal gears ... **A16:** 339
liquid ... **A7:** 7
in machining centers ... **A16:** 308-309, 393
as machining process ... **A7:** 67
for magnesium powder production ... **A7:** 131
material for milling cutters ... **A16:** 313-317
and maximum peak-to-valley roughness height ... **A16:** 26
mechanical, electric automatic controls ... **A16:** 309
mechanical, electric-hydraulic controls ... **A16:** 310

Milling (continued)
mechanical, hydraulic automatic controls ... **A16:** 310
mechanism ... **A7:** 62-63
medium ... **A7:** 60-61
Mg alloys ... **A16:** 820, 821-822, 826-827, 828
MMCs ... **A16:** 894, 895, 896, 900, 901
multifunction machining ... **A16:** 366-367, 374, 377, 381, 386
NC implemented ... **A16:** 613, 614, 616, 617
Ni alloys ... **A16:** 837, 840-841
nonreactive ... **A7:** 64-65
notch root radius ... **A8:** 382
numerical control (NC) ... **A16:** 305, 306, 310
objectives ... **A7:** 56
optimization of machine setup ... **A16:** 309
P/M high-speed tool steels ... **A16:** 65, 66, 67
P/M materials ... **A16:** 880, 883, 889
of P/M tool steels ... **A7:** 789
parameters, and powder characteristics ... **A7:** 60-65
particle effects in microforging ... **A7:** 61
particle effects of fracturing ... **A7:** 61
particles effects of welding ... **A7:** 61
PCBN cutting tools ... **A16:** 112, 116
peripheral ... **A16:** 311, 319-322, 327-329
peripheral, Al alloys ... **A16:** 786, 793
peripheral, Cu alloys ... **A16:** 815, 816
peripheral, for gear manufacture ... **A16:** 333
peripheral, Mg alloys ... **A16:** 826-827
peripheral, refractory metals ... **A16:** 865, 866, 867
peripheral, Ti alloys ... **A16:** 846, 848, 849
peripheral, tool steels ... **A16:** 721
physical and mechanical properties affected by ... **M5:** 308
powder characteristics ... **A7:** 65
power consumption ... **A16:** 17-18
power requirements ... **A16:** 319
principles ... **A7:** 56-60
process selection ... **A7:** 56
processes ... **A7:** 56, 62
PVD-coated tools ... **A16:** 83
in QMP iron powder process ... **A7:** 87
reactive ... **A7:** 64-65
refractory metals ... **A16:** 860, 861, 867
sample, for chemical surface studies ... **A10:** 177
setup rigidity ... **A16:** 320
slab (peripheral) ... **A16:** 320, 324, 327, 328
slab (peripheral), carbon and alloy steels ... **A16:** 675
slab (peripheral), heat-resistant alloys ... **A16:** 755, 756
slab (peripheral), Hf ... **A16:** 856
speed, feed, and depth of cut ... **A16:** 323-327, 329
spur gears ... **A16:** 338, 339, 341
stainless steels ... **A16:** 692, 695, 699, 701, 703-704
straddle ... **A16:** 308, 309, 320
straddle, of tool steels ... **A16:** 715, 717
surface alterations produced ... **A16:** 23
surface finish and integrity ... **A16:** 21, 28, 327
Ti alloys ... **A16:** 845, 846-847
time ... **A7:** 59-60
tool life ... **A16:** 314, 315, 318, 321, 322, 326-329
tool steels ... **A16:** 715, 718-719, 721, 723, 726, 727
tooling material choice ... **A16:** 317
uranium alloys ... **A16:** 875
and vertical multiple-spindle automatic chucking machines ... **A16:** 379
vs blanking ... **A14:** 458
as wet chemical technique for subdividing solids ... **A10:** 165
worm gears ... **A16:** 340
wrought copper and copper alloys ... **A2:** 243-244
zirconium ... **A16:** 852, 853-854, 855
Zn alloys ... **A16:** 834
Milling, chemical
aluminum alloy ... **A15:** 763
Milling, copper alloys
lead frame strip ... **EL1:** 483
Milling cutters *See* Cutting tools
Milling machines *See also* Milling ... **A16:** 1, 303-309, 326, 329
bed-type ... **A16:** 305, 308, 321
chucking ... **A16:** 303
die threading ... **A16:** 297
and drilling ... **A16:** 212
gantry-type ... **A16:** 306-307, 308

Milling machines (continued)
gear manufacture **A16:** 330, 333-334, 335, 348
knee-and-column **A16:** 304-305, 306
lapping **A16:** 503
moving-bridge **A16:** 306-307, 308
multiple-spindle bar **A16:** 303
nonreversing tapping attachments **A16:** 255
planer-type **A16:** 306-307
planetary **A16:** 308, 309
profilers **A16:** 308
rise-and-fall **A16:** 305
rotary millers **A16:** 308
single-spindle bar machines **A16:** 303
special-purpose **A16:** 307-309
tape-controlled **A16:** 304
Milliprobe, for historical studies
PIXE analysis **A10:** 107
MIM process *See* P/M injection molding process
Mindlin analysis
partial slip **A18:** 244
Mine waters **A13:** 1293-1294
Minelbite
composition **EM4:** 873
manufacturer **EM4:** 873
properties **EM4:** 873
Miner's law **A11:** 112
Miner's rule **A8:** 374 **EM3:** 517
for lifetime evaluation **EM1:** 203
Mineral abrasives
use in dry blasting **M5:** 84
Mineral acid cleaning *See also* Acid cleaning **M5:** 59-65
barrel process **M5:** 60-65
cleaner composition and operating conditions **M5:** 59-60, 61-62
corrosivity factors **M5:** 65
electrolytic process **M5:** 60-65
equipment and process control **M5:** 61-65
immersion process **M5:** 60-65
iron and steel **M5:** 59-65
maintenance schedules **M5:** 65
process types *See also* specific processes by name **M5:** 60-62
spray process **M5:** 60-65
wiping process **M5:** 60-65
Mineral acids
cemented carbide corrosion in **A13:** 852
corrosion, casting alloys **A13:** 575
defined **A13:** 1140
sample dissolution in **A10:** 165
stainless steel corrosion in **A13:** 557-558
Mineral beneficiation
of uranium **A2:** 670
Mineral industry **A13:** 1293-1298
cyclic loading machinery **A13:** 1297
materials selection for **A13:** 1294
pump and pumping systems **A13:** 1294-1296
reactor vessels **A13:** 1297
roof bolts **A13:** 1294
tanks **A13:** 1296-1297
wire rope **A13:** 1294
Mineral oil **A16:** 122-123
approximate temperature exposure limits **A18:** 84
for bearing steels **A18:** 732
critical temperature **A18:** 96
defined **A18:** 13
high-vacuum lubricant applications **A18:** 155
lubricant for tool steels **A18:** 738
lubricants for rolling-element bearings **A18:** 133-134, 135, 136, 137
as metalworking lubricants **A18:** 139, 140, 142, 144, 146, 147
properties **A18:** 81
removal of **M5:** 10, 439, 577
thermal expansion **A18:** 83
for tool steel lubrication **A18:** 737, 738
vapor degreasing solvents percentage in **M5:** 47-48
Mineral oils
for machining **A7:** 461
Mineral processing **EM4:** 961-965
applications **EM4:** 963-965
materials **EM4:** 961-962
property comparison of various ceramic materials and abrasive-resistant steel **EM4:** 962
smelting and spray drying **A7:** 76
wear of ceramics **EM4:** 962-963
Mineral processing industry applications
structural ceramics **A2:** 1019
Mineral quartz
XRPD analysis **A10:** 341
Mineral seal oil
to dilute cutting oils **A16:** 186
Mineral spirit cleaners **M5:** 40-42, 44-46, 57
Mineral spirits
as cleaning solvents **EL1:** 663
Mineral spirits as a coolant/lubricant for diamond wheels **A9:** 25
Mineral-filled plastics
as mounting materials **A9:** 45
Minerals *See also* Minerals, characterization of
analytic methods applicable **A10:** 6
Crystallinity characterized by electron spin resonance **A10:** 253
with defects, ESR studied **A10:** 264
effects of composition on mass absorption **A10:** 97
EXAFS analysis of **A10:** 407
sample dissolution mediums **A10:** 166
sampling of **A10:** 12-18
scale-forming **A13:** 1137
Minerals, characterization of *See also* Minerals
analytical transmission electron microscopy **A10:** 429-489
atomic absorption spectrometry **A10:** 43-59
Auger electron spectroscopy **A10:** 549-567
classical wet analytical chemistry **A10:** 161-180
controlled-potential coulometry **A10:** 207-211
electrochemical analysis **A10:** 181-211
electrogravimetry **A10:** 197-201
electrometric titration **A10:** 202-206
electron probe x-ray microanalysis **A10:** 516-535
electron spin resonance **A10:** 253-266
extended x-ray absorption fine structure **A10:** 407-419
inductively coupled plasma atomic emission spectroscopy **A10:** 31-42
infrared spectroscopy **A10:** 109-125
ion chromatography **A10:** 658-667
low-energy ion-scattering spectroscopy **A10:** 603-609
molecular fluorescence spectrometry **A10:** 72-81
Mössbauer spectroscopy **A10:** 287-295
neutron activation analysis **A10:** 233-242
neutron diffraction **A10:** 420-426
optical emission spectroscopy **A10:** 21-30
particle-induced x-ray emission **A10:** 102-108
potentiometric membrane electrodes **A10:** 181-187
radial distribution function analysis **A10:** 393-401
radioanalysis **A10:** 243-250
Raman spectroscopy **A10:** 126-138
scanning electron microscopy **A10:** 490-515
secondary ion mass spectroscopy **A10:** 610-627
single-crystal x-ray diffraction **A10:** 344-356
spark source mass spectrometry **A10:** 141-150
ultraviolet/visible absorption spectroscopy **A10:** 60-71
voltammetry **A10:** 188-196
x-ray diffraction **A10:** 325-332
x-ray powder diffraction **A10:** 333-343
x-ray spectrometry **A10:** 82-101
Mines
acoustic emission inspection **A17:** 290
Mini-tuning fork specimens **A8:** 508
Miniature angle-beam standard reference blocks
ultrasonic **A17:** 267
Miniature L-plate as internal fixation device **A11:** 671
Miniaturization **EL1:** 89, 631
Miniboat *See* L'vov platform
Miniborescopes *See also* Borescopes; Visual inspection **A17:** 4
Minicomputers
for image analysis **A10:** 310
Minielectronics
development **EL1:** 89
Minifocus x-ray tubes
radiography **A17:** 302
Minimum bend radius
for cold forming aluminum alloys **A8:** 129-130
defined **A8:** 125
subjectivity of **A8:** 128
Minimum commitment method
for creep-rupture data extrapolation **A8:** 334-335
Minimum contact length
symbol and units **A18:** 544
Minimum creep rate **A8:** 689-691, 725
Minimum design value
for low- and elevated-temperature properties **A8:** 670-671
for static metallic materials **A8:** 662-677
Minimum detectable leak
defined **A17:** 57
rate, defined **A17:** 57
Minimum dynamic bend angle **A6:** 161
Minimum fatigue cycle
and proof loading relationship **A8:** 582
Minimum film thickness **A18:** 90, 92, 539
nomenclature for lubrication regimes **A18:** 90
nomenclature for Raimondi-Boyd design chart **A18:** 91
symbol and units **A18:** 544
Minimum friction force
nomenclature for Raimondi-Boyd design chart **A18:** 91
Minimum load
defined **A8:** 9
Minimum mechanical property value
as design value **A8:** 662
Minimum required speed **A18:** 65
Minimum specific film thickness
symbol and units **A18:** 544
Minimum stress **A8:** 9
in fatigue testing **A11:** 102
symbol for **A11:** 797
Minimum stress-intensity factor
K_{min} **A8:** 9
Minimum wafer thickness **A18:** 686
Minimum wall thickness **A6:** 374
Minimum work-metal hardness
for Rockwell hardness testers **A8:** 77
Minimum-draft forgings **A14:** 258
Mining
beryllium **A2:** 684
of elemental cobalt **A2:** 446
Mining and mineral industries, friction and wearing **A18:** 649-654
ball and rod grinding media **A18:** 654
damage, types of **A18:** 649-651
abrasive wear **A18:** 649-650, 651, 652
adhesive wear **A18:** 649, 651
corrosive wear **A18:** 649, 650-651
testing for types of wear damage **A18:** 651
methods to improve wear resistance **A18:** 651-654
design use **A18:** 653-654
ferrous materials for grinding mill liners **A18:** 651
hardfacing deposition **A18:** 652, 653
material selection **A18:** 651-652
wear plate use **A18:** 652-653
Mining applications
cemented carbides **A2:** 974-977

Mining applications (continued)
vinyl esters in **EM2:** 272
Mining explosives
aluminum powder containing **A7:** 601
Mining industry *See* Mineral industry
Minitorches
for the ICP **A10:** 37
Minor component analysis
analytical transmission electron microscopy **A10:** 429-489
atomic absorption spectrometry **A10:** 43-59
Auger electron spectroscopy **A10:** 549-567
classical wet analytical chemistry **A10:** 161-180
controlled-potential coulometry **A10:** 207-211
electrochemical analysis **A10:** 181-211
electrogravimetry **A10:** 197-201
electrometric titration **A10:** 202-206
electron probe x-ray microanalysis **A10:** 516-535
electron spin resonance **A10:** 253-266
elemental and functional group analysis **A10:** 212-220
field ion microscopy **A10:** 583-602
gas analysis by mass spectrometry **A10:** 151-157
gas chromatography/mass spectrometry **A10:** 639-648
inductively coupled plasma atomic emission spectroscopy **A10:** 31-42
infrared spectroscopy **A10:** 109-125
of inorganic gases, analytical methods for **A10:** 8
of inorganic liquids and solutions, analytical methods for **A10:** 7
of inorganic solids, analytical methods for **A10:** 4-6
ion chromatography **A10:** 658-667
liquid chromatography **A10:** 649-657
low-energy ion-scattering spectroscopy **A10:** 603-609
molecular fluorescence spectrometry **A10:** 72-81
neutron activation analysis **A10:** 233-242
neutron diffraction **A10:** 420-426
optical emission spectroscopy **A10:** 21-30
of organic solids, analytical methods for **A10:** 9
of organic solids and liquids, analytic methods for **A10:** 10
particle-induced x-ray emission **A10:** 102-108
potentiometric membrane electrodes **A10:** 181-187
Raman spectroscopy **A10:** 126-138
Rutherford backscattering spectrometry **A10:** 628-636
scanning electron microscopy **A10:** 490-515
secondary ion mass spectroscopy **A10:** 610-627
spark source mass spectrometry **A10:** 141-150
ultraviolet/visible absorption spectroscopy **A10:** 60-71
voltammetry **A10:** 188-196
x-ray diffraction **A10:** 325-332
x-ray photoelectron spectroscopy **A10:** 568-580
x-ray powder diffraction **A10:** 333-343
x-ray spectrometry **A10:** 82-101
Minor elements
in aluminum melts **A15:** 79
Minor toxic metals
types and effects **A2:** 1258-1262
MINT degassing system **A15:** 461, 463-464
Minus mesh **A7:** 8
Minus sieve **A7:** 8
MIPAC computer program for structural analysis **EM1:** 268, 271
Mirau interferometer
in laser interference microscope **A17:** 16
Mirror analyzer
cylindrical **A7:** 251
Mirror finish, buffing to
nickel alloys **M5:** 674-675
Mirror, fracture
in ceramics **A11:** 745-747
Mirror illuminator
defined **A9:** 12
Mirror plane
defined in crystal symmetry **A10:** 346
Mirror region
definition **EM4:** 633
Mirror sheaths
as rigid borescopes **A17:** 5
Mirror silvering
powder used **A7:** 574
Mirror zone
fracture surface **EM2:** 808
Mirrors
for optical holography **A17:** 418-419
photographic effects **A12:** 83
rear surface **A8:** 234
sloping, for Hugoniot elastic limit measurement **A8:** 211
in visual leak testing **A17:** 66
Misaligned parts
CMM measurement of **A17:** 19
Misalignment
assembly, brittle fracture of gray iron nut **A11:** 369-370
of bearing and shaft, overheating failure by **A11:** 507-508
of roller-element bearings **A11:** 489, 507
shaft/crankshaft fracture from **A11:** 475
as tension source for stress-corrosion cracking **A8:** 502
Miscellaneous industrial oils **A18:** 99
Misch metal in HSLA steel
effect on toughness **M1:** 418
Mischmetal **M2:** 770-771
Mischmetal, as rare earth metal
properties **A2:** 1183
Mischromes
causes and correction **M5:** 190, 195
Miscibility .... **A3:** 1 • 2
between oil and refrigerant **A18:** 87
polymer, and plasticizers **EM2:** 496
of polymer blends **EM2:** 487 488, 632
SAS techniques for **A10:** 405
Miscibility gap
defined **A9:** 12
in solid state and spinodal lines **A9:** 652
Miscible blends
defined **EM2:** 632
Miscible solids .... **A3:** 1 • 2
Mises criterion **A18:** 476
Mises stress contours
at beam lead **EL1:** 288, 290
Mismatch **A14:** 9, 49
Mismatch tolerance
steel forgings **M1:** 364-365, 366, 375
Misorientation determination
in dislocation cell structure analysis **A10:** 471-472
Misoriented grains
from directional solidification **A15:** 321-322
Misregistration *See also* Registration
defined **EL1:** 1150
in printed board coupons **EL1:** 576
as PTH failure **EL1:** 1021-1022
Misrun
as casting defect **A11:** 385
defined **A11:** 7
in iron castings **A11:** 353
Misrun(s)
defined **A15:** 8
permanent mold casting **A15:** 279
of permanent mold castings **A15:** 285
and pouring temperature **A15:** 283
Misruns
radiographic appearance **A17:** 349
Missile filters
powders used **A7:** 573
Missile lathes **A16:** 153
Missile nose cones **A7:** 680
Missile systems
composite applications **EM1:** 819-822
types **EM1:** 816-817
Missile wing
titanium P/M **A7:** 681
Mist
in salt spray (fog) testing **A13:** 224-226
Mist, defined
for glass and ceramics **A11:** 745
Mist hackle
definition **EM4:** 633
Mist lubrication
defined **A18:** 13
Mist region
fracture surface **EM2:** 808
Mist/velocity hackle **EM4:** 636
MITAS-II software program
for heat transfer problems **A15:** 861
Mitchell bearing *See* Tilting-pad bearing
Miter gears **A7:** 668-669
Miter joints
flash welding **M6:** 565-56
Mitsubishi high thermal conduction module **EL1:** 48
Mitsubishi process **EM4:** 903
Mix (noun) **A7:** 8
Mix (verb) **A7:** 8
Mix flow, and bulk density
lubricant effects **A7:** 190
Mixed anion effect **EM4:** 852
Mixed aromatic amines
formulation **EM3:** 122
Mixed catalyst hard chromium plating **M5:** 172
Mixed crystal
tungsten carbide; titanium carbide **A7:** 158
Mixed fracture modes **A8:** 484-486
SEM **A12:** 176
Mixed grain size *See* Duplex grain size
Mixed lubrication *See* Quasi-hydrodynamic lubrication
Mixed mechanisms
fracture by **A11:** 83-84
Mixed nepheline **EM4:** 6
Mixed phthalates (plasticizer)
batch weight of formulation when used in oxidizing sintering atmospheres **EM4:** 163
Mixed potential *See also* Galvanic couple potential
defined **A13:** 9
theory, of aqueous corrosion **A13:** 17
Mixed sol monoliths **EM4:** 447
Mixed technology (MT)
boards **EL1:** 681
boards, adhesives for **EL1:** 670-674
boards, wave fluxers for **EL1:** 682
combined processes for **EL1:** 694-695
and through-hole soldering **EL1:** 681
Mixed-acid etchants
used for aluminum alloys **A9:** 354
Mixed-alkali effect **EM4:** 852
Mixed-film lubrication **A18:** 518
Mixed-mode
fractures, fractographs of **A11:** 84
loading, effect on corrosion fatigue **A11:** 255
Mixed-mode crack propagation **EM3:** 509, 510, 512
Mixed-mode cyclic loading **EM3:** 509
and crack propagation **EM3:** 512-513
Mixed-mode fracture toughness testing **A8:** 460-461
Mixed-mode screening tests
fracture toughness testing **A8:** 461
Mixed-oxide coatings
as barrier protection **A13:** 379
Mixed-stress criteria **A8:** 344
Mixers
and blenders **A7:** 189
Mixes, ramming
types of **A15:** 373
Mixing *See also* Mulling **A7:** 8, 24, 186-189 **EM4:** 95-98
analysis techniques used **EM4:** 96
and blending, quality **A7:** 186-187
of bulk molding compounds **EM1:** 161
by intensive mixer **A15:** 344, 346
demixing **EM4:** 95, 96, 97-98
demixing due to size difference **EM4:** 97-98
dispersive mixing **EM4:** 176
effect of baffles in cylindrical mixer **A7:** 189
equipment **EM4:** 98, 99
equipment, for foamed plaster molding **A15:** 247
explosive hazards **A7:** 197
Gibbs free energy of **A15:** 56
in injection molding **A7:** 496-497 **EM4:** 176
equipment **EM4:** 176
milling **EM4:** 176
mixing operations **EM4:** 176
mixing index **EM4:** 96-97
mixing practice **EM4:** 98-99
mixer selection **EM4:** 98, 99
mixing operations **EM4:** 98-99
mixture behavior **EM4:** 97, 98
convective mixing **EM4:** 97, 98

Mixing (continued)
diffusive mixing **EM4:** 97, 98
rheological behavior **EM4:** 97, 98
shear mixing **EM4:** 97, 98
near random **A7:** 187
nonuniformity assessment **EM4:** 95-97
direct approach **EM4:** 95, 96
indirect approach **EM4:** 95, 96-97
scale of size **EM4:** 95-96
objectives **EM4:** 95
ordering **EM4:** 95, 98
random homogeneous mixture variance **EM4:** 96
of samples **A17:** 730
of sands **A15:** 32, 238-239
slurry, match plate pattern plaster mold casting **A15:** 245
in stress-corrosion cracking **A13:** 147
techniques, SMC resin pastes **EM1:** 159
in transport, injection molding **EM1:** 165
tumbling **EM4:** 99
variables, green sand preparation **A15:** 344-345
Mixing and dispensing equipment
suppliers **EM3:** 604
Mixing chamber
definition **A6:** 1211 **M6:** 11
manual torch brazing **M6:** 952-95
oxyfuel gas welding **M6:** 585-58
Mixing equipment **EM3:** 604, 687-692
Mixing of etchants **A9:** 69
Mixture **A7:** 8
A1A pyrotechnic **A7:** 604
explosion characteristics **A7:** 196
rule of, and green strength **A7:** 302
Mixtures .... **A3:** 1 • 8
acid, nonoxidizing, sample dissolution by **A10:** 165
composition determined **A10:** 213
determination of molecular components in **A10:** 109
interpreting spectra of **A10:** 116
liquid chromatography isolation of pure compounds from **A10:** 649
matrix methods to analyze complex **A10:** 117
organic and inorganic, gas analysis of **A10:** 151
oxidizing acids, sample dissolution by **A10:** 166
separation and component analysis by liquid chromatography **A10:** 649
of volatile compounds, GC/MS analysis of **A10:** 639
Miyauchi shear test
for sheet metals **A8:** 559-560
MMA-1942 *See* Titanium alloys, specific types Ti-Pd alloys
MMA-5137 *See* Titanium alloys, specific types, Ti-5Al-2.5Sn
MMC *See* Metal matrix composites
MMCIAC *See* Metal Matrix Composites Information Analysis Center
MMCs *See* Metal matrix composites; Metal-matrix composites
Mn (Phase Diagram) .... **A3:** 2 • 287
Mn-Mo (Phase Diagram) .... **A3:** 2 • 285
Mn-N (Phase Diagram) .... **A3:** 2 • 286
Mn-Nd (Phase Diagram) .... **A3:** 2 • 286
Mn-Ni (Phase Diagram) .... **A3:** 2 • 286
Mn-P (Phase Diagram) .... **A3:** 2 • 287
Mn-Pd (Phase Diagram) .... **A3:** 2 • 287
Mn-Pr (Phase Diagram) .... **A3:** 2 • 288
Mn-Pu (Phase Diagram) .... **A3:** 2 • 288
Mn-Sb (Phase Diagram) .... **A3:** 2 • 288
Mn-Si (Phase Diagram) .... **A3:** 2 • 289
Mn-Sm (Phase Diagram) .... **A3:** 2 • 289
Mn-Sn (Phase Diagram) .... **A3:** 2 • 290
Mn-Ti (Phase Diagram) .... **A3:** 2 • 290
Mn-U (Phase Diagram) .... **A3:** 2 • 290
Mn-V (Phase Diagram) .... **A3:** 2 • 290
Mn-Y (Phase Diagram) .... **A3:** 2 • 291
Mn-Zn (Phase Diagram) .... **A3:** 2 • 291
Mn-Zr (Phase Diagram) .... **A3:** 2 • 291
Mo-N (Phase Diagram) .... **A3:** 2 • 292
Mo-Nb (Phase Diagram) .... **A3:** 2 • 292
Mo-Nb-Ti (Phase Diagram) .... **A3:** 3 • 56
Mo-Ni (Phase Diagram) .... **A3:** 2 • 292
Mo-Ni-Ti (Phase Diagram) .... **A3:** 3 • 56
Mo-Ni-W (Phase Diagram) .... **A3:** 3 • 56
Mo-O (Phase Diagram) .... **A3:** 2 • 293
Mo-Os (Phase Diagram) .... **A3:** 2 • 293
Mo-P (Phase Diagram) .... **A3:** 2 • 293
Mo-Pd (Phase Diagram) .... **A3:** 2 • 294
Mo-Pt (Phase Diagram) .... **A3:** 2 • 294
Mo-Pu (Phase Diagram) .... **A3:** 2 • 294
MO-RE 5 *See* Superalloys, cobalt-base, specific types, UMCo-50
Mo-Rh (Phase Diagram) .... **A3:** 2 • 295
Mo-Ru (Phase Diagram) .... **A3:** 2 • 295
Mo-S (Phase Diagram) .... **A3:** 2 • 295
Mo-Si (Phase Diagram) .... **A3:** 2 • 296
Mo-Ta (Phase Diagram) .... **A3:** 2 • 296
Mo-Ti (Phase Diagram) .... **A3:** 2 • 296
Mo-Ti-W (Phase Diagram) .... **A3:** 3 • 57
Mo-U (Phase Diagram) .... **A3:** 2 • 297
Mo-V (Phase Diagram) .... **A3:** 2 • 297
Mo-W (Phase Diagram) .... **A3:** 2 • 297
Mo-Zr (Phase Diagram) .... **A3:** 2 • 298
Mobile carbon
determined in iron and steels **A10:** 178
Mobile communications
as telecommunication hybrid application **EL1:** 383
Mobile homes
aluminum and aluminum alloys **A2:** 11
Mobile ions
contaminants, and durability design **EL1:** 45
surface inversion, accelerated testing **EL1:** 892-893
Mobile nitrogen
determined in steels **A10:** 178
Mobile phase
defined **A10:** 677
Mobile phase programming
functional group analysis **A10:** 654
Mobile ridge concept **A18:** 66
Mobile units
magnetic particle inspection **A17:** 92-93
Mobility
and angle of repose **A7:** 284-285
Mobility, carrier
defined **EL1:** 90
Mock leno weave
defined **EM2:** 27
MOCVD *See* Metallo-organic chemical vapor deposition
Modacrylics
as engineering thermoplastic **EM2:** 448
Mode *See also* Ideal-crack-tip stress field
defined **A8:** 9 **A11:** 7
of failure, determined by torsional testing **A8:** 145
forced-displacement system **A8:** 392
forced-vibration system **A8:** 392
in-plane shear fractures, in composites **A11:** 736-738
of loading, and fracture mechanics **A11:** 47
mixed, loading and fractures **A11:** 84, 255
resonance system **A8:** 392
rotational bending system **A8:** 392
servomechanical system **A8:** 392
statistical **A8:** 625
tension fractures, in composites **A11:** 735-736
Mode 1
crack deformation **A8:** 441-442
Mode 2
crack deformation **A8:** 442
Mode 3
crack deformation **A8:** 443
Mode coherence coefficient (MCC) **A6:** 876
Mode I loading condition
tear effect on dimple shape **A12:** 12-14
Mode II loading condition
shear effect on dimple shape **A12:** 12-14
Mode III loading conditions
shear effect on dimple shape **A12:** 12-14
Model building
as data analysis **EM2:** 602
Model C Carbon Calibration **A8:** 105
Model C Scleroscope **A8:** 104
Model D Scleroscope **A8:** 104, 105
Model simulation
in computer-aided design **EL1:** 129
Model(s) *See also* Analytic modeling; Modeling; Simulation
analytical, of solder shear fatigue **EL1:** 743-746
frequency response-small signal, for bipolar transistors **EL1:** 153-154
heat dissipation, TAB **EL1:** 286
high-level, as engineer defined **EL1:** 129
impedance **EL1:** 601-603
of materials and processes selection **EL1:** 112-113
of rheological behavior **EL1:** 847-852
thermal **EL1:** 174-175
of thermal analysis **EL1:** 14
Modeling *See also* Analytic modeling; Computer applications; Electrical modeling; Fluid flow modeling; Interconnection system modeling; Material modeling; Mathematical modeling; Model(s); Modeling materials; Models; Physical modeling; Process modeling; Simulation; Software tools
advanced forging process **A14:** 409-416
advantages **A15:** 857
analytical **A14:** 425
closed-loop cure **EM1:** 762
of combined fluid flow and heat/mass transfer **A15:** 877-882
computer and physical **A13:** 234
in computer-aided design **EL1:** 129
computerized, for assembly packaging **EL1:** 442-444
cure **EM1:** 500-501, 704-705, 758-759
deformation, of open-die forging **A14:** 65-67
dynamic material, for workability **A14:** 370-371
electrical **EL1:** 447-448
of equiaxed solidification **A15:** 887-890
of equiaxed structures **A15:** 885-890
finite-element analysis (FEA/FEM) **EL1:** 442-443
finite-element, for shape rolling **A14:** 350
of fluid flow **A15:** 867-876
of fracture **A14:** 393-396
high-frequency digital systems **EL1:** 77-81
interconnection system **EL1:** 12-17
macroscopic, of solidification **A15:** 883
of manufacturing processes **EM1:** 499-502
mathematical, precision forging **A14:** 159
methods, for preform design **A14:** 51
microscopic, of equiaxed structures **A15:** 885-887
of microstructural evolution **A15:** 883-891
physical **A14:** 159-160, 431-437
physical, of mold filling **A15:** 869-870
plastic stress minimization **EL1:** 444
predictive **EM2:** 527
of solidification **A15:** 36
of solidification heat transfer **A15:** 858-866
solidification, software for **A15:** 198
stress, as process control tool **EL1:** 446
stress, in chip metallization passivation **EL1:** 443-444
surface **A15:** 859
techniques, forging process design **A14:** 417-438
thermal **EL1:** 446-447
of thermal durability **EL1:** 50-52
time-to-failure **EL1:** 887-888
Modeling materials
aluminum **A14:** 432-433
for physical simulation **A14:** 431-432
plasticine **A14:** 432

Modeling materials (continued)
strain-rate sensitive **A14:** 432
Models *See also* Computer applications; Mathematical models; Modeling; NDE reliability; NDE reliability models
beam, for radiographic inspection **A17:** 710
beam, of ultrasonic inspection **A17:** 705
detector, for radiographic inspection **A17:** 710
of equiaxed growth **A15:** 132 133
future impact of **A17:** 711-713
grain refinement **A15:** 105-107
mathematical, machine vision analyses **A17:** 37
measurement, of ultrasonic inspection **A17:** 704-705
for predicting NDE reliability **A17:** 702-715
probe-flaw, eddy current inspection **A17:** 707-708
sample interaction, for radiographic inspection **A17:** 710
Modern pewter *See* Tin alloys, specific types, pewter
Modern Plastics **EM3:** 66
trade magazine **EM2:** 92, 95
Modern Plastics Encyclopedia **EM3:** 66
as information source **EM2:** 92
Modes
common **EL1:** 37-39
decoupled, propagation **EL1:** 36
differences **EL1:** 37
of fracture **A12:** 12-71
of mechanical failure, TAB assembly **EL1:** 287
vs mechanism, in active devices **EL1:** 1006
Modification
of aluminum-silicon alloys **A15:** 482-483
degree, on-line assessment **A15:** 166
effects, aluminum-silicon alloys **A15:** 752-753
hydrogen and degassing interactions **A15:** 485-486
of hypoeutectic aluminum-silicon alloys **A2:** 134
microstructure, titanium alloys **A15:** 828
and refinement **A15:** 753
theory of **A15:** 482
of titanium alloy microstructure **A15:** 828
Modifications
approaches **EM3:** 34
by additives **EM2:** 493-507
by polymer-polymer mixtures **EM2:** 487-492
to polymers **EM2:** 66-67
Modified acrylics **EM3:** 18, 75
advantages and limitations **EM3:** 77, 78
chemistry **EM3:** 78
curing method **EM3:** 78
predicted 1992 sales **EM3:** 77
properties **EM3:** 92
suppliers **EM3:** 78
Modified beryllium cupro-nickel alloy 72C
properties and applications **A2:** 391
Modified chrome pickle
magnesium alloys **M5:** 632, 636, 640-641
Modified compact specimens
SCC testing **A8:** 512-513 **A13:** 254-255
Modified creep
in elevated-temperature failures **A11:** 264
Modified difference method
for crack propagation rate **A8:** 680
Modified double-beam specimens
dimensions for various plate thicknesses **A8:** 504
SCC testing **A8:** 505
Modified E-glass *See* ECR glass
Modified empirical criterion
for ductile fracture **A14:** 401
Modified epoxies **EM3:** 104
for flexible printed boards **EL1:** 582
Modified G bronze
nominal composition **A2:** 347
Modified Goodman diagram method **A6:** 390
Modified Goodman's law
mean stress effect on fatigue strength **A8:** 374
and tensile strength **A8:** 374
Modified jelly roll process
for superconductor assembly **A2:** 1066
Modified Lemon (statistical) method **EM1:** 302, 305
Modified low-carbon steel sheet and strip **A1:** 206, 208
Modified silicone sealants **EM3:** 191
Modified solid-solution copper alloys **A2:** 234-235
Modified Villard circuit
radiography **A17:** 305
Modified WOL compact specimens
SCC testing **A8:** 512
Modifier additions
of aluminum-silicon alloys **A15:** 162-165
antimony **A15:** 484
calcium **A15:** 484
chemical, for aluminum-silicon alloys **A15:** 751-752
chemical, of core coatings **A15:** 240
sodium **A15:** 484
strontium **A15:** 484
for wax patterns **A15:** 197
Modifiers **EM3:** 18, 73
properties analysis of **EM1:** 736-737
types **EM1:** 737
Modmor I
properties **A18:** 803
Modul-*r* test
for sheet metals **A8:** 557
Modular rectangular polishing and buffing machines **M5:** 122
Modular tray set
for multiple small parts **A7:** 423
Modulation *See also* Frequency modulation
in ESR spectrometer **A10:** 255
microwave inspection **A17:** 217
polarization **A10:** 114-115
Modulation transfer function (MTF)
defined **A17:** 298, 373, 383
Modules
capacity/specifications **EL1:** 128
engineering design system for **EL1:** 127
environmental stress screening **EL1:** 878
as interconnections, defined **EL1:** 76
test **EL1:** 140
as three-terminal devices **EL1:** 429
Modulus *See also* Dynamic modulus; Flexural modulus; Initial modulus; Modulus of elasticity; Offset modulus; Secant modulus; Tangent modulus
assessment of **EM2:** 548-551
complex, sinusoidal excitation test methods **EM2:** 551
initial **EM3:** 18
offset **EM3:** 18
of resilience, defined **EM1:** 15
of rigidity, defined **EM1:** 15
of rupture, in bending, defined **EM1:** 15
of rupture, in torsion, defined **EM1:** 15
secant **EM3:** 18
sustained stress exposure testing **EM1:** 825
tangent **EM3:** 18
vs temperature, polyvinyl chlorides (PVC) **EM2:** 210
Modulus, bulk *See* Bulk modulus
Modulus, complex shear *See* Complex shear modulus
Modulus, complex Young's *See* Complex Young's modulus
Modulus, compressive *See* Compressive modulus
Modulus, dynamic *See* Dynamic modulus
Modulus, flexural *See* Flexural modulus
Modulus, initial *See* Initial modulus
Modulus, loss *See* Loss modulus
Modulus of elasticity *See also* Chord modulus; Elastic; Elastic modulus; Elastic properties; Elasticity; Offset modulus; Secant modulus; Superplasticity; Tangent modulus; Young's modulus **A1:** 58, 61, 62, 63 **EM3:** 18, 51
abbreviation for **A10:** 690
aluminum and aluminum alloys **A2:** 9
aluminum casting alloys **A2:** 150
of aluminum oxide-containing cermets **A7:** 804
chromium carbide based
anisotropic **A10:** 358
carbon and alloy steels, selected grades **M1:** 680
cermets **A7:** 806
in deep drawing **A14:** 575
defined **A10:** 677 **A14:** 9 **EM1:** 15 **EM2:** 434
of ferritic malleable iron **A1:** 75
of glass fibers **EM1:** 46
of gray iron **A1:** 18, 20, 21 **M1:** 18-19, 27
malleable iron **M1:** 65, 70
Modulus of elasticity (continued)
maraging steels **M1:** 451
mechanically alloyed oxide alloys **A2:** 945
dispersion-strengthened (MA ODS)
of metal borides and boride-based cermets **A7:** 812
of P/M and I/M alloys **A7:** 747
P/M steels **M1:** 334-335, 339, 340
of pearlitic malleable iron **A1:** 82
in rhenium and rhenium-containing alloys **A7:** 477
spring steel **M1:** 284-286
in stress measurement **A10:** 382
for threaded steel fasteners **A1:** 295-296
of titanium carbide-based cermets **A7:** 808
of titanium carbide-steel cermets **A7:** 810
variance with preferred orientation **A10:** 358
Modulus of elasticity, (E) *See also* Chord modulus; Secant modulus; Tangent modulus; Young's modulus
in bending **A8:** 133-136
in cantilever beam bend test **A8:** 132
defined **A8:** 9, 22 **A11:** 7
design values for **A8:** 662
effect in fastener performance at elevated temperatures **A11:** 542
and engineering stress-strain curve **A8:** 22
in four-point bend tests **A8:** 132, 134
and hardness conversions **A8:** 109
in shear, torsion tests for **A8:** 139
of sheet metals **A8:** 555
in split Hopkinson bar test **A8:** 202
in springback tests **A8:** 565
step-loading curve **A8:** 315
symbol for **A11:** 796
testing machines for **A8:** 47
in three-point bend test **A8:** 132, 134
in uniaxial tensile testing **A8:** 555
values at different temperatures **A8:** 22
variability in **A8:** 623
Modulus of elasticity of gears **A18:** 539
symbol and units **A18:** 544
Modulus of elasticity of pinions **A18:** 539
Mohs hardness, defined **A18:** 13
Mohs hardness number **A18:** 433
Mohs hardness scale **A18:** 433
Molecular beam epitaxy (MBE) system **A18:** 159
Molecular seal, defined **A18:** 13
Molybdenite **A18:** 113
Molybdenum
as addition to cemented carbides **A18:** 800
addition to cylinder liner materials for strength **A18:** 556
bond coatings **A18:** 831
in cobalt-base alloys **A18:** 766
codeposited with nickel in electroplating **A18:** 836
dichalcogenides **A18:** 113
erosion resistance **A18:** 201
erosion test results **A18:** 200
friction coefficient data **A18:** 71
in hardfacing alloys **A18:** 759-760, 762
for laser alloying **A18:** 866
in laser cladding material **A18:** 867
lubricant indicators and range of sensitivities **A18:** 301
nitride-forming element **A18:** 878
oxidational wear **A18:** 287
paint for laser-alloyed stainless steel **A18:** 866
plasma-spray coating for piston rings **A18:** 556
relative erosion factor **A18:** 200
secondary phase formation **A18:** 854
spectrometric metals analysis **A18:** 300
spray material for oxyfuel wire spray process **A18:** 829
in stainless steels **A18:** 710, 712
thermal properties **A18:** 42
in thermal spray coating materials **A18:** 832
thermal spray coating recommended **A18:** 832
in tool steels **A18:** 734, 735-736
Vickers and Knoop microindentation hardness numbers **A18:** 416
wear resistance of die material **A18:** 635-636

Modulus of elasticity of pinions (continued)
x-ray characterization of surface wear results for various microstructures ... **A18:** 469
Molybdenum carbide (Mo_2C), properties ... **A18:** 795
Molybdenum disulfide (MoS_2)
as addition to carbon-graphite materials ... **A18:** 816
as additive to metalworking lubricants ... **A18:** 141
as antiseize additive ... **A18:** 102
coating for titanium alloys ... **A18:** 781, 782-783
effect of testing parameters ... **A18:** 806
ferrographic application to identify wear particles ... **A18:** 305, 307
grease additive ... **A18:** 125
high-vacuum lubricant applications ... **A18:** 153-154, 159
as layer lattice solid lubricant ... **A18:** 115-116
lubricated friction testing ... **A18:** 48
lubrication for thermal spray coating ... **A18:** 591
in metal-matrix composites, friction coefficient ... **A18:** 804
methods used for synthesis ... **A18:** 802
oxidation ... **A18:** 591
properties ... **A18:** 801
rolling-element bearing lubricant ... **A18:** 138
sliding wear mechanisms in metal-matrix composites ... **A18:** 809-810
solid lubricant for gears ... **A18:** 541
to control dusting ... **A18:** 684
wear particles ... **A18:** 305
Molybdenum high-speed steels, composition limits ... **A18:** 735
Molybdenum hot-work steels
composition limits ... **A18:** 735
for hot-forging dies ... **A18:** 623
resistance to thermal fatigue ... **A18:** 640
service temperature of die materials in forging ... **A18:** 625
Molybdenum selenide ($MoSe_2$), methods used for synthesis ... **A18:** 802
Molybdenum suicide, coatings for gas-lubricated bearings ... **A18:** 532
Molybdenum-chromium-nickel, plasma-spray coating for pistons ... **A18:** 556
Molydisulfide
filler for lip seals ... **A18:** 550
filler for seals ... **A18:** 551
Monel
contributing to corrosion in seals ... **A18:** 549
seal materials ... **A18:** 550
spray material for oxyfuel wire spray process ... **A18:** 829
symbol and units ... **A18:** 544
Modulus of resilience *See also* Elastic energy; Resilience; Strain energy ... **EM3:** 18
defined ... **A8:** 9, 22
for steel grades ... **A8:** 22
Modulus of resilience (MOR) ... **EM4:** 329
Modulus of resistance
defined ... **EM2:** 27
Modulus of rigidity *See also* Shear modulus; Torsional modulus ... **EM3:** 18
defined ... **EM2:** 27
maraging steels ... **M1:** 451
springs, steel ... **M1:** 284-286, 296-297, 300, 307
Modulus of rigidity, (*G*) *See also* Shear modulus
from shear testing ... **A8:** 65
torsion-shear test for ... **A8:** 64-65
Modulus of rupture ... **A8:** 9
defined ... **A11:** 7 **EM2:** 27
gray iron test bars ... **M1:** 17-18
malleable iron ... **M1:** 70
Modulus of rupture (in bending) ... **EM3:** 18
Modulus of rupture (in torsion) ... **EM3:** 18
Modulus of rupture (MOR) bending bar
incorporated into the CARES computer program ... **EM4:** 701, 704, 705-706
Modulus of toughness *See also* Toughness
defined ... **A8:** 9, 22-23
Modulus, offset *See* Offset modulus
Modulus, secant *See* Secant modulus
Modulus, shear *See* Shear modulus
Modulus, tangent *See* Tangent modulus
Modulus-directed tensile tests ... **EM2:** 546-547
Mohair paint rollers ... **M5:** 503
Mohr titration
for chloride ... **A10:** 173
defined ... **A10:** 164
Mohr's circle
interlaminar shear/tension stresses ... **EM1:** 336
tensile stress for applied shear ... **A11:** 734
Mohr's circle for stress ... **A10:** 381
Mohr-Coulomb failure law ... **EM4:** 145
Mohs hardness *See also* Hardness ... **EM3:** 18
defined ... **EM1:** 16 **EM2:** 27
Mohs hardness test
as scratch test ... **A8:** 71
Mohs scale ... **A8:** 9, 71, 107-108
Moiety
defined ... **A10:** 677
multielement, classical wet analysis of ... **A10:** 162
Moil ... **EM4:** 397
Moil point
surface roughness from heat treatment ... **A11:** 573, 578
Moiré diffraction patterns ... **A9:** 110
of second-phase precipitates ... **A9:** 117
Móire fringes ... **A10:** 371, 677
in dynamic notched round bar testing ... **A8:** 277, 279-280
mechanical analog of ... **A8:** 280
Moire interferometry ... **A6:** 1150 **EM3:** 450-451, 452
Moiré pattern
defined ... **A10:** 677
Moist air
as corrosive environment ... **A12:** 24
Moisture *See also* Marine atmospheres; Moisture absorption; Moisture content; Moisture curing; Moisture resistance; Moisture swelling analysis; Moisture-related failure; Rainwater; Steam; Vapor; Water; Water absorption; Water vapor
absorber, forms ... **EL1:** 958
absorption, encapsulant ... **EL1:** 470
conductivity, laminate ... **EM1:** 226
as contaminant in microcircuit process gases ... **A10:** 156-157
contamination, of welds ... **A13:** 344
content ... **EL1:** 82, 1106-1107
content in spray drying ... **A7:** 74
content, sand ... **A15:** 238
corrosion effect in steam equipment ... **A11:** 615
corrosion of gas-dryer piping ... **A11:** 631
cycling, atmospheric ... **EM1:** 296
as degradation factor ... **EM2:** 576
diffusion, equivalent physical quantities ... **EM1:** 191
diffusion, in UDCs ... **EM1:** 191-192
and dirt, bearing damage by ... **A11:** 494, 496
and durability design ... **EL1:** 45
effect, carbon fibers ... **EM1:** 52
effect, glass fiber ... **EM1:** 46
effect, laminate strength ... **EM1:** 227
effect, marine atmospheres ... **A13:** 902-903
effect, nuclear reactor erosion-corrosion ... **A13:** 965
effect, soil corrosion, carbon steels ... **A13:** 512
effects, core-oil sand mixes ... **A15:** 219
effects, epoxy resin matrices ... **EM1:** 32, 76, 736, 750
effects, full-scale static test ... **EM1:** 348
effects, injection molding compounds ... **EM1:** 164
electrical breakdown from ... **EM2:** 465
equilibrium ... **EM1:** 16
Moisture (continued)
equilibrium/relative humidity, para-aramid fibers ... **EM1:** 56
expansion, laminate ... **EM1:** 226
Fick's law of ... **EM1:** 227
galvanic corrosion from ... **A13:** 86
and gas porosity ... **A15:** 82
and glass transition temperature ... **EM2:** 761-764
and hermeticity ... **EL1:** 243
-induced plastic-package failures, integrated circuits ... **A11:** 788-789
internal concentration, defined ... **EM1:** 190
internal, sources ... **EL1:** 1064-1065
intrusion, electronic black boxes ... **A13:** 1107-1108
laminate stresses from ... **EM1:** 228-229
monitor, in-situ on-chip ... **EL1:** 953
package, content analysis ... **A13:** 1117
and passivation ... **EL1:** 243
in plastic encapsulation ... **EL1:** 962
producing flammable gases or vapors ... **A7:** 194
and pyrophoricity ... **A7:** 194, 199
reaction with liquid metals ... **A13:** 94-95
removal in high-temperature combustion ... **A10:** 222
and sampling ... **A10:** 16
in semiconductor development ... **EL1:** 958-959
special test procedures ... **EL1:** 953
stress-corrosion cracking in environment of ... **A8:** 409, 499
tests, in magnetic separation of seeds ... **A7:** 591
trapped, as surface blow defect ... **A15:** 337
wicking ... **A13:** 519, 521
Moisture absorption *See also* Absorption; Moisture; Water absorption ... **EM3:** 18
acrylics ... **EM2:** 106
of aramid fibers ... **EM1:** 190
cyanates ... **EM2:** 234
defined ... **EM1:** 16 **EM2:** 27
effect on glass transition temperature. epoxy resins ... **EM1:** 32
long-term exposure testing ... **EM1:** 823
polyamide-imides (PAI) ... **EM2:** 130
polyamides (PA) ... **EM2:** 126
polyether sulfones (PES, PESV) ... **EM2:** 161
Moisture analysis
with microwaves ... **A17:** 215
Moisture content ... **EL1:** 82, 1106-1107 **EM3:** 18
defined ... **EM1:** 16 **EM2:** 27
of mixed resin systems, tested ... **EM1:** 737
test, epoxy resins ... **EM1:** 736
Moisture curing
of room-temperature vulcanizing (RTV) silicones ... **EL1:** 823
Moisture effects on adhesive joints ... **EM3:** 622-627
for curing ... **EM3:** 35
effect on organic adhesives ... **EM3:** 36
improvement of joint durability ... **EM3:** 625-627
application of primer (coupling agents) ... **EM3:** 625-627
hydration inhibition or retardation ... **EM3:** 625
increasing barrier to water diffusion ... **EM3:** 625
mechanism of strength loss ... **EM3:** 623-625
displacement of adhesive by water ... **EM3:** 623-624
hydration of oxide layers ... **EM3:** 624-625
migration of water to adhesive joints ... **EM3:** 622-623
critical water concentration ... **EM3:** 623
liquid water versus vapor-sorption isotherm ... **EM3:** 623
water diffusion in polymers ... **EM3:** 622-623
strength degradation and failure mode ... **EM3:** 623
Moisture equilibrium ... **EM3:** 18
defined ... **EM2:** 27
Moisture regain ... **EM3:** 18
Moisture resistance
conformal coatings ... **EL1:** 776
of ECR-glass and S-glass ... **EM1:** 107
of flexible epoxies

Moisture resistance (continued)
glass-to-metal seals **EL1:** 459
of natural fibers **EM1:** 117
package-level testing **EL1:** 936-937
of plastic encapsulants **EL1:** 805-806
silicone base coatings **EL1:** 773
of silicone conformal coatings **EL1:** 822-823
sizing effects **EM1:** 123
tests **EL1:** 494-497
Moisture separator drain
nuclear reactors **A13:** 958
Moisture swelling analysis
of fiber composites **EM1:** 188-190
Moisture vapor transmission
defined **EM1:** 16
Moisture vapor transmission (MVT) **EM3:** 18
defined **EM2:** 27
Moisture vapor transmission rate (MVTR) **EM3:** 50
Moisture-cured urethanes **A13:** 410
Moisture-related failure *See also* Moisture; Moisture absorption; Water absorption
in composites **EM2:** 765-766
creep and stress relaxation effects **EM2:** 763-765
effect on mechanical properties **EM2:** 766-769
glass transition temperature effects **EM2:** 761-764
moisture-induced damage mechanisms **EM2:** 762-766
moisture-induced fatigue failure **EM2:** 765
in thermoplastics **EM2:** 767-768
in thermosets **EM2:** 766-767
Molal solution
defined **A13:** 9
Molality
defined **A10:** 677
Molar absorptivity
in UV/VIS analysis **A10:** 62-63
Molar energy
SI unit/symbol for **A8:** 721
Molar entropy
SI unit/symbol for **A8:** 721
Molar extinction coefficients **A10:** 47
Molar heat capacity
SI unit/symbol for **A8:** 721
Molar solution
defined **A13:** 9
Molarity
defined **A10:** 162, 677
and titrant standardization **A10:** 172
Mold *See also* Automatic mold; Cored mold; Deep-draw mold; Sprayed metal molds **A7:** 8
cracked or broken, as casting defect **A11:** 381
defect, high pressure **A11:** 384
defined **EM1:** 16
design/construction, for resin transfer molding **EM1:** 168-169
electroformed, defined *See* Electroformed molds
flexible, defined *See* Flexible molds
for hand wet lay-up technique **EM1:** 132
moisture effects **EM1:** 164
pipe, overheating failure in **A11:** 275-276
repair, poor, as casting defect **A11:** 385
Mold assembly
Antioch process **A15:** 247
match plate pattern plaster casting **A15:** 245
plaster molding **A15:** 244
plastic molding **A15:** 244
and tolerances **A15:** 618
Mold blowoff, as finishing
green sand molding **A15:** 347
Mold casting *See* Permanent mold casting; Plaster mold casting
Mold cavities *See also* Gate; Riser; Runner; Sprue
coring methods for **A15:** 279
machining of **A15:** 280
surface, finish effects **A15:** 284-285
Mold cavity
resin transfer molding **EM1:** 168
Mold cleaners **EM1:** 168-169
Mold closing, as finishing
green sand molding **A15:** 347
Mold coating *See also* Coatings
defined **A15:** 8
Mold creep
as casting defect **A11:** 387
Mold design *See also* Design; Design considerations
directional solidification **A15:** 321
effect, dimensional accuracy **A15:** 284
effect, mold life **A15:** 281
finish effects, permanent molds **A15:** 285
and hot cracking **A15:** 768
permanent mold casting **A15:** 277-278
vertical centrifugal casting **A15:** 300-304
Mold dressing *See* Mold coating
equipment, plaster molding **A15:** 243
plaster molding **A15:** 244
Mold drop
as casting defect **A11:** 381
Mold element cutoff
as casting defect **A11:** 381
Mold erosion
nickel alloys **A15:** 821-822
plain carbon steels **A15:** 711
Mold expansion, during baking
as casting defect **A11:** 386
Mold facing *See* Mold coating
Mold filler materials **A9:** 31-32
Mold filling
heat loss, modeling of **A15:** 879-880
physical modeling of **A15:** 869-870
rapid **A15:** 589
rheology **EL1:** 838-842
techniques, FM process **A15:** 38
Mold fluxes *See also* Fluxes
copper alloys **A15:** 450-451
Mold hardness, uniform
development of **A15:** 29
Mold inserts
permanent mold castings **A15:** 280
Mold jacket
defined **A15:** 8
Mold materials
aggregate **A15:** 208-211
Antioch process **A15:** 246
compacted graphite irons **A15:** 671
as inclusion-forming **A15:** 90
and mold life **A15:** 281
permanent molds, recommended **A15:** 280
selection, permanent mold casting **A15:** 280
solid graphite as **A15:** 285
Mold materials for castable resins **A9:** 30
Mold metallurgy **A1:** 114
Mold pouring techniques *See also* Pouring
vs mold filling techniques **A15:** 38
Mold properties
effect on pure metal solidification structures **A9:** 608-610
Mold release agents *See also* Release agents **EM1:** 16, 158 **EM3:** 41
as additives **EL1:** 475
Mold releases **EM1:** 168-169, 602-603
Mold restraint
as casting distortion source **A15:** 616-617
Mold shift
defined **A15:** 8
Mold shrinkage *See also* Mold(s); Shrinkage
defined **EM1:** 16 **EM2:** 27
glass addition effect **EM2:** 72
rammed graphite molds **A15:** 274
statistical analysis of **EM2:** 604-605
Mold spraying
robotic **A15:** 568
Mold stabilization *See also* Burn-off
in all-ceramic mold casting **A15:** 249
defined **A15:** 250
Unicast process **A15:** 252
Mold steels *See* Tool steels, mold steels **A1:** 767-768
Mold surface
defined **EM1:** 16
Mold temperature *See also* Pouring temperature; Temperature(s)
control of **A15:** 283
horizontal centrifugal casting **A15:** 297
permanent mold casting **A15:** 282-283
surface finish effects, permanent mold castings **A15:** 285
Mold wall casting deficiencies **A17:** 531
Mold walls
inspection of **A15:** 555-556
movement, as defect **A15:** 29
thickness, and riser location **A15:** 580-581
Mold warpage *See* Warpage
Mold wash *See also* Facing
defined **A15:** 8
horizontal centrifugal casting **A15:** 296
for inclusions **A15:** 90
permanent molds **A15:** 304
Mold(s) *See also* Coatings; Mold assembly; Mold cavities; Mold design; Mold drying mold types
bivalve, historic use **A15:** 16
blow molding **EM2:** 353-356
carbon, vertical centrifugal casting **A15:** 304
colloidal silica **A15:** 212
complexity, and design **A15:** 611-613
composition, match plate patterns **A15:** 245
construction, thermoplastic injection molding **EM2:** 313-317
copper, horizontal centrifugal casting **A15:** 296
defined **A15:** 8 **EM2:** 27
development of **A15:** 28
dilation, and feed metal volume **A15:** 577
distortion, directional solidification **A15:** 321-322
erosion **A15:** 589, 711, 821-822
ethyl silicate **A15:** 212
fill simulation **EM2:** 312
finishes, types **A15:** 347
firing and burnout, investment casting **A15:** 262
flow-through water-cooled copper **A15:** 311
for gray iron **A15:** 639-640
high-alloy white irons **A15:** 680
high-chromium white irons **A15:** 683
high-silicon irons **A15:** 699
horizontal centrifugal casting **A15:** 296-297
influences, in sand casting **A15:** 618
level measurement **A15:** 501
life, permanent mold castings **A15:** 280-281
loam **A15:** 31
metal **A15:** 275, 301-303
-metal interface, inclusion control at **A15:** 90
monocrystal casting **A15:** 322-323
operation, effect, dimensional accuracy **A15:** 284
produced by shell process **A15:** 217-218
properties, water content variation **A15:** 212-213
rammed graphite **A15:** 273-274
removal, permanent mold casting **A15:** 283-284
restraint, distortion from **A15:** 616-617
rotational molding **EM2:** 366-367
silicate bonded **A15:** 229
sodium silicate **A15:** 213
spalling, from thermal expansion **A15:** 208
speed curves, vertical centrifugal casting **A15:** 305-306
steel **A15:** 296, 366
stone, Bronze Age **A15:** 15-16
surface, defined **EM2:** 27
surfaces **EM2:** 614
transportation, green sand molding **A15:** 347
type, foundry processes classified by **A15:** 204
types **EM2:** 614
various, freezing times, compared **A15:** 243
venting **A15:** 29, 568
for zirconium alloys **A15:** 836-837
Mold-wall movement
as casting defect **A11:** 386
Moldability
of phenolics **EM2:** 243
Molded composite materials
properties **EM1:** 161
Molded edge
defined **EM1:** 16 **EM2:** 27
Molded net
defined **EM1:** 16 **EM2:** 27
Molded plastic gear **A7:** 669-670
Molded plastic packages
assembly of **EL1:** 471-475
silicon chip, thermomechanical considerations **EL1:** 415-416
thermal performance of **EL1:** 409-410
without cavities **EL1:** 452
Molded printed wiring boards
defined **EL1:** 505
Molded-epoxy encapsulations **EL1:** 961
Molded-in inserts **EM2:** 722-723
Molding *See also* Autoclave molding; Compression molding; Contact molding; Injection molding; Matched metal molding; Mold shrinkage; Mold(s); Molding compounds; Molding conditions; Postmolding; Premolding; Pressure bag

Molding (continued)
molding; Processing; Thermal expansion molding; Transfer molding; Vacuum bag molding **A7:** 8
19th century development **A15:** 32
Alnico alloys **A15:** 736
application **EM1:** 355, 356
austenitic ductile irons **A15:** 700
compression **EM2:** 331-333
cost, thermoplastic injection molding **EM2:** 309
cycle, basic operations **EL1:** 803
defined **EM1:** 16 **EM2:** 27
as electronic embedment, epoxies **EL1:** 832
of gray iron **A15:** 639-640
high-density, defined **A15:** 346
high-pressure, defined **A15:** 346
high-silicon irons **A15:** 701
in injection molding **A7:** 497
innovative, types **A15:** 37
limitations **EM2:** 614-615
magnetic **A15:** 234-235
methods, titanium alloys **A15:** 825-826
methods, titanium and titanium alloy castings **A2:** 635-636
mixed materials from **A15:** 32
pattern, in lost foam casting **A15:** 231
primary process costs **EM2:** 84
printed board coupons **EL1:** 573
problems **A15:** 345-347
process, powder injection, for cermets **A2:** 984-985
processes, common **EM2:** 85
pulp, defined *See* under Pulp
recent developments **A15:** 37-38
secondary processing costs **EM2:** 84-86
special processes, types **A15:** 37
springback at **A7:** 480
thermal expansion methods **EM1:** 590-591
vacuum **A15:** 235-236

Molding aggregates *See also* Aggregate molding materials; Clays; Plastic materials; Sands
bonds formed in **A15:** 212-213

Molding compound *See also* Molding powder
defined **EM1:** 16
with epoxy resins **EM1:** 400
resins **EM1:** 164
thermoplastic, for injection **EM1:** 165
thermoset, for injection **EM1:** 165

Molding compounds *See also* Reinforced molding compound
allyls (DAP, DAIP), properties **EM2:** 228
amino **EM2:** 230-231
amino resin **EM2:** 628
defined **EM2:** 27
as encapsulant **EL1:** 803-805
epoxy **EL1:** 813-815
extra high strength, properties effects **EM2:** 286
phenolic **EM2:** 242-245, 627
in plastic packages **EL1:** 211-212
thermosetting, electrical properties **EM2:** 590

Molding conditions
acrylics **EM2:** 106-107
acrylonitrile-butadiene-styrenes (ABS) **EM2:** 113-114
high-impact polystyrenes (PS, HIPS) **EM2:** 198
liquid crystal polymers (LCP) **EM2:** 182

Molding cycle
defined **EM1:** 16 **EM2:** 27

Molding equipment *See also* Equipment
effect, green sand system **A15:** 226
high-pressure, first **A15:** 29
sand, development of **A15:** 35

Molding machines *See also* Equipment
air operated, development **A15:** 29
computer-aided **A15:** 350-351
cope and drag **A15:** 343
defined **A15:** 8
development of **A15:** 28
first-generation **A15:** 341-342

Molding machines (continued)
horizontal flaskless **A15:** 343
jolt squeeze **A15:** 342
jolt-type **A15:** 341-342
match plate pattern **A15:** 343
pressure wave method **A15:** 343
problems **A15:** 345-347
rap-jolt **A15:** 342-343
sand slinger **A15:** 342
second-generation **A15:** 342-344
vertically parted **A15:** 344

Molding media preparation *See also* Sand preparation
green sand casting **A15:** 344-345

Molding powder
defined **EM1:** 16

Molding pressure
defined **EM1:** 16 **EM2:** 27

Molding processes *See also* Casting processes
aggregate molding materials **A15:** 208-211
centrifugal casting **A15:** 296-307
ceramic molding **A15:** 248-252
classified **A15:** 203-207
continuous casting **A15:** 308-316
coremaking **A15:** 238-241
development of **A15:** 35
die casting **A15:** 286-295
expendable, types **A15:** 204
flow charts for **A15:** 203-207
investment casting **A15:** 253-269
molding aggregates, bonds formed in **A15:** 212-213
new and emerging processes **A15:** 317-338
permanent mold casting **A15:** 275-285
permanent, types classified **A15:** 204
plaster molding **A15:** 242-247
rammed graphite molds **A15:** 273-274
Replicast process **A15:** 270-272
resin binder processes **A15:** 214-221
sand molding **A15:** 222-237

Molding release agent *See also* Release agent
defined **EM2:** 27

Molding sands *See also* Alumina; Naturally bonded molding sand; Sand(s)
adhering, blast cleaning of **A15:** 506
defined **A15:** 8
magnesium **A15:** 804

Moldless casting
nonferrous **A15:** 315

Molds
use in solid sample preparation **A10:** 93

Molds for
electroslag welding **M6:** 227-22
repair thermit welding **M6:** 699-701

Molds for mounting fiber composites **A9:** 588

Mole *See* Molecular optical laser examiner **A10:** 677, 685, 690
defined **A13:** 9
equivalence of **A10:** 162

Molecular
mass, defined **EM2:** 27
orientation, process effects on **EM2:** 281-282
properties, temperature and molecular structure **EM2:** 37
structure, thermoplastic resins **EM2:** 619-621
weight, defined **EM2:** 27

Molecular absorption
Jablonsky diagram indicating **A10:** 73
as requirement for fluorescence **A10:** 73-74

Molecular analyses
of bulk inorganic solids, analytical methods for **A10:** 6
of inorganic gases, analytical methods for **A10:** 8
of inorganic liquids and solutions, analytical methods for **A10:** 7
IR quantitative determination, in mixtures **A10:** 109

Molecular analyses (continued)
of organic solids, analytical methods for **A10:** 9
of organic solids and liquids, analytical methods for **A10:** 10
Raman spectroscopy **A10:** 126-138

Molecular bands
in emission spectroscopy **A10:** 23
in flame emissions **A10:** 29

Molecular beam epitaxy (MBE)
for gallium arsenide (GaAs) **A2:** 745

Molecular clusters
and surface atoms **A7:** 259

Molecular conformation, and stereochemistry
IR determination of **A10:** 109

Molecular electronics *See also* Microelectronics
development goal of **EL1:** 89
as future technology **EL1:** 103

Molecular emission
minimizing **A10:** 23
in optical emission spectroscopy **A10:** 22-23

Molecular flow
defined **A10:** 152

Molecular fluorescence spectroscopy **A10:** 72-81
applications **A10:** 72, 79-81
capabilities, compared with UV/VIS analysis **A10:** 60
defined **A10:** 677
detection limits and dynamic range **A10:** 76
estimated analysis time **A10:** 72
excitation and emission spectra, qualitative analysis **A10:** 74-75
fluorescent molecules **A10:** 73-74
general uses **A10:** 72
indirect, of nonfluorescing atoms molecules **A10:** 76
of inorganic and organic materials **A10:** 4-5, 7-10
instrumentation **A10:** 76-77
introduction and theory **A10:** 73
lasers **A10:** 76, 80-81
limitations **A10:** 72, 76
monochromators **A10:** 76-77
possible errors **A10:** 77-78
practical considerations **A10:** 77-78
quantitative analysis **A10:** 75
radiation sources **A10:** 76
related techniques **A10:** 72, 78-79
samples **A10:** 72, 76, 79-81
special techniques **A10:** 78-79

Molecular free volume **EM3:** 654

Molecular How
in leaks **A17:** 58

Molecular hydrogen lamp
for continuous- source background correction **A10:** 51

Molecular information
from surface analytical techniques **A7:** 251

Molecular light-scattering processes
energy- level diagram for **A10:** 127

Molecular mass **EM3:** 18

Molecular models
from rubber network theory **EL1:** 849

Molecular nitrogen **A7:** 345

Molecular optical laser examiner
instrument layout for **A10:** 130
for Raman spectroscopy **A10:** 129-130
use in glasses **A10:** 131

Molecular orientation
in drawn polymer films **A10:** 120
IR determination of **A10:** 109

Molecular reorientation
as NMR kinetic process **A10:** 277

Molecular scattering
by Raman spectroscopy **A10:** 126-130

Molecular species
as adsorbed on surfaces **A10:** 109

Molecular spectroscopy
methods of **EM2:** 825-828

Molecular spectrum
defined ... **A10:** 677
Molecular structure
balls and massless springs model ... **A10:** 110, Ill
compound, IR spectroscopy for ... **A10:** 109
defined ... **A10:** 677
determined by single-crystal analysis ... **A10:** 345
mass spectrometry for ... **A10:** 116
nuclear magnetic resonance ... **A10:** 116
in UV/VIS analysis ... **A10:** 61
Molecular vibrations
of diatomic molecule ... **A10:** 111
group frequencies as ... **A10:** 111
in infrared spectroscopy ... **A10:** 111
normal-coordinate analysis ... **A10:** 110-111
with Raman spectroscopy ... **A10:** 126-138
Molecular weight ... **EM3:** 18
average, defined ... **EM2:** 5
defined ... **A10:** 677 **EM1:** 16
and ESC resistance ... **EM2:** 800
high, high-density polyethylenes (HDPE) ... **EM2:** 163
polyether sulfones (PES, PESV) ... **EM2:** 161
structural analysis ... **EM2:** 828-830
for viscoelastic analysis ... **EM2:** 533-535
Molecular weight, number-average
in polymers ... **A11:** 758
Molecular-beam epitaxy (MBE)
for GaAs crystal growth ... **EL1:** 200
Molecular-cage effect
XANES analysis ... **A10:** 415-416
Molecules
acrylic ... **EL1:** 671
adsorbed, on smooth metal surfaces, SERS analysis of ... **A10:** 137
adsorbed, orientation on single crystals ... **A10:** 407
aggregate, in complexometric titrations ... **A10:** 164
aromatic, in ESR analysis ... **A10:** 259
bonding of ... **EM2:** 63
chemisorbed, EXAFS geometry of ... **A10:** 407
defined ... **A10:** 677
diatomic, molecular vibrations ... **A10:** 111
electronic structure, UV/VIS analysis ... **A10:** 60
and emission spectroscopy ... **A10:** 22-23
epoxy ... **EL1:** 670
functional groups in ... **A10:** 109
gas, three-dimensional structure of ... **A10:** 393
inorganic, MFS analysis ... **A10:** 74
interactions between, determined by single- crystal x-ray diffraction ... **A10:** 344
odd electron, ESR analysis of ... **A10:** 254
organic and inorganic fluorescent ... **A10:** 72-74
organic, UV/VIS analysis of functional groups in ... **A10:** 60
with overlapping fluorescence spectra ... **A10:** 75
polarization of ... **A10:** 127
polymer, SAS applications ... **A10:** 405
primary/secondary solvent ... **A13:** 19
structure within ... **EM2:** 58
surfactant, in water, structural changes in ... **A10:** 118
symmetry of ... **A10:** 348
Molten
iron, in elevated-temperature failures ... **A11:** 275
lead, in elevated-temperature failures ... **A11:** 274
metals, in elevated-temperature failures ... **A11:** 273-276
salts, in elevated-temperature failures ... **A11:** 276-277
-solids failures, in engine valves ... **A11:** 289
zinc, in elevated-temperature failures ... **A11:** 273
Molten carbonate fuel cell ... **A13:** 1321
Molten chemical-bath dip brazing
definition ... **M6:** 11
Molten fluorides ... **A13:** 90
Molten glass
melting/forming of ... **EM1:** 108
Molten metal
fluidity ... **A15:** 766-768
Molten metal flame spraying
definition ... **M6:** 11
Molten metal pumps
aluminum melts ... **A15:** 454-456
systems ... **A15:** 486-487
Molten metal-bath dip brazing
definition ... **M6:** 11
Molten metals
cast irons in ... **A13:** 570
corrosion, metal-processing equipment ... **A13:** 1314
effects on tantalum ... **A13:** 736
high-temperature corrosion in ... **A13:** 56-60
salts, types ... **A13:** 90-91
zirconium/zirconium alloy resistance ... **A13:** 717
Molten particle deposition ... **EM4:** 202-208
applications ... **EM4:** 207-208
color deposits ... **EM4:** 208
composite parts ... **EM4:** 208
corrosion protection ... **EM4:** 208
electrical property special surfaces ... **EM4:** 208
free-standing ceramic bodies ... **EM4:** 208
medical implants ... **EM4:** 208
protection against antifretting and wear ... **EM4:** 208
self-lubricating coatings ... **EM4:** 208
thermal barrier coatings ... **EM4:** 207-208
approach ... **EM4:** 124
coating generation ... **EM4:** 206-207
coating formation ... **EM4:** 206, 207
coating heat treatments ... **EM4:** 207
cryogenic cooling ... **EM4:** 207
effects of particulate flattening ... **EM4:** 206-207
grooving ... **EM4:** 207
material-dependent effect ... **EM4:** 206
deposition methods ... **EM4:** 202-206
transfer of heat and momentum ... **EM4:** 202-203
typical spraying processes for ceramics ... **EM4:** 203-206
factors determining the feasibility of using process in an application ... **EM4:** 202
future trends ... **EM4:** 208
lamellar structure of a coating ... **EM4:** 203
properties of ceramic and cermet coatings ... **EM4:** 203
Molten salt *See* Dip brazing of steels in molten salt
Molten salt bath cleaning
for liquid penetrant inspection ... **A17:** 81, 82
Molten salt bath descaling *See* Salt bath descaling
Molten salt baths
for precoat cleaning ... **A15:** 561
Molten salt corrosion *See also* Molten salts ... **A13:** 17, 88-91
in coal- and oil-fired boilers ... **A13:** 995-996
fluorides ... **A13:** 52-54
high-temperature ... **A13:** 50-55
kinetics ... **A13:** 50-51
and liquid-metal corrosion, compared ... **A13:** 17
literature ... **A13:** 54
mechanisms ... **A13:** 17, 89
metal-processing equipment ... **A13:** 1313-1314
molten salt types ... **A13:** 90-91
nitrates/nitrites ... **A13:** 51-52
prevention ... **A13:** 91
purification ... **A13:** 51
rates, iron-base alloys ... **A13:** 51
solid waste boilers ... **A13:** 997
test methods ... **A13:** 51
thermodynamics ... **A13:** 50-51
in titanium/titanium alloys ... **A13:** 689
in zirconium/zirconium alloys ... **A13:** 717
Molten salt descaling *See also* Descaling; Scaling
of titanium alloy forgings ... **A14:** 280
Molten salt electrolysis ... **A4:** 668
Molten salt paint stripping method ... **M5:** 18-19
Molten salt quenching
of steel ... **M4:** 49, 60
Molten salts *See also* Molten salt corrosion
analysis ... **A10:** 131
as corrosive environment ... **A12:** 24
purification ... **A13:** 51
sampler, natural circulation loop ... **A13:** 50
selective dissolution ... **A13:** 134
types ... **A13:** 90-91
Molten weld pool
definition ... **A6:** 1211 **M6:** 11
Moly Permalloy
photochemical machining etchant ... **A16:** 590
Moly-manganese (Mo-Mn) process
brazing process for joining alumina ceramics ... **EM4:** 517
ceramic/metal seals ... **EM4:** 503-504, 506
chemical integrity of seals ... **EM4:** 540
Moly-manganese paste
to provide a metallizing layer for a ceramic ... **EM3:** 299
Molybdate films
deposited by color etching ... **A9:** 141
Molybdate oranges
as pigment ... **EM3:** 179
Molybdates
as anodic inhibitors ... **A13:** 494
Molybdena catalysts
analyses ... **A10:** 134
Molybdenite ... **A7:** 154-155
Molybdenum *See also* Molybdenum alloys; Molybdenum alloys, specific types; Molybdenum powders; Pure molybdenum; Refractory metals; Refractory metals and alloys; Refractory metals and alloys, specific types
as a beta stabilizer in titanium alloys ... **A9:** 458
as a carbide-forming element in steel ... **A9:** 661
in a nickel matrix ... **A9:** 619
as addition to austenitic stainless steels ... **A6:** 689
addition to improve weldability of tungsten ... **A6:** 870
addition to low-alloy steels for pressure vessels and piping ... **A6:** 667
addition to restrict graphitization ... **M6:** 834
addition to solid-solution nickel alloys ... **A6:** 575
additions to martensitic stainless steels ... **M6:** 348
in age hardening in maraging steels ... **A1:** 794
in alloy cast irons ... **A1:** 89, 90
alloy, tension-overload fracture ... **A12:** 464
alloying effect in titanium alloys ... **A6:** 508, 512
alloying effect on electron beam welding ... **M6:** 639-640
alloying effects, stainless steels ... **A13:** 550
alloying, in cast irons ... **A13:** 567
alloying, in microalloyed uranium ... **A2:** 677
alloying, magnetically soft materials ... **A2:** 762
alloying, nickel-base alloys ... **A13:** 641
alloying, uranium/uranium alloys ... **A13:** 814
alloying, wrought aluminum alloy ... **A2:** 54
alloys, with oxides, workability ... **A8:** 165
as an addition to austenitic manganese steel castings ... **A9:** 239
as an addition to cobalt-base heat-resistant casting alloys ... **A9:** 334
as an addition to iron-chromium-nickel heat-resistant casting alloys ... **A9:** 332-333
as an addition to nickel-base heat-resistant casting alloys ... **A9:** 334
as an addition to nickel-iron alloys ... **A9:** 538
as an addition to niobium alloys ... **A9:** 441
as an addition to silver-base switchgear materials ... **A9:** 552
as an addition to stainless steel casting alloys ... **A9:** 298
as an addition to tantalum alloys ... **A9:** 442
as an alloying addition to wrought heat- resistant alloys ... **A9:** 310-312
as an alloying addition to wrought stainless steels ... **A9:** 283-285
applications ... **A2:** 557-559, 574 **M3:** 320-321
applications, sheet metals ... **A6:** 400
arc welding *See* Arc welding of molybdenum and tungsten
at elevated-temperature service ... **A1:** 640
atomic interaction descriptions ... **A6:** 144
in austenitic manganese steel ... **A1:** 824-825
in austenitic stainless steels ... **A6:** 457, 458, 461, 465, 467
bcc, fatigue limits ... **A8:** 253
bend transition temperature ... **M6:** 463
binary alloys ... **A9:** 442
biologic effects and toxicity ... **A2:** 1253-1254
as bond coats for thermal spray coatings ... **A6:** 813
brazed joint, thermal stress SCC ... **A13:** 879
brazing ... **A2:** 564 **M6:** 1057-1058
filler metals and their properties ... **M6:** 1057-1058
fluxes and atmospheres ... **M6:** 1058
precleaning and surface preparation ... **M6:** 1057-1058
processes and equipment ... **M6:** 1058
brazing and soldering characteristics ... **A6:** 634
in cast iron ... **A1:** 6, 28-29

Molybdenum (continued)
characteristics and weldability **M6:** 462-463
chromium plating of **M5:** 660-661
cleaning processes **M5:** 659-660
in cobalt-base alloys **A1:** 985
commercially pure, pressed from powder and sintered **A9:** 445
commercially pure sheet **A9:** 445
in compacted graphite iron **A1:** 57, 59
components in cold-wall vacuum furnace **A6:** 331
in composition, effect on ductile iron **A4:** 686, 688, 689
in composition, effect on gray irons **A4:** 672, 673, 676, 678
composition similar to ceramics and glasses **EM3:** 306
in compounds providing flame retardance and smoke suppression **EM3:** 179
consumption **A2:** 557
content, and crevice corrosion, nickel-base alloys **A13:** 305
content in heat-treatable low-alloy (HTLA) steels **A6:** 670
content in HSLA Q&T steels **A6:** 665
content in nickel-base and cobalt-base high-temperature alloys **A6:** 573
content in stainless steels **M6:** 320
content in tool and die steels **A6:** 674
content in ultrahigh-strength low-alloy steels **A6:** 673
content of weld deposits **A6:** 675
corrosion resistance **M3:** 321
creep rupture testing of **A8:** 302
crystal, x-ray fractography for fracture surface of **A10:** 377
determined by controlled-potential coulometry **A10:** 209
determined in molybdenum-tungsten alloys **A10:** 207
in diffusion-alloyed steel powder metallurgy materials **A9:** 510
in ductile iron **A1:** 45
in duplex stainless steels **A6:** 471-472, 478
early TEM **A12:** 6
effect, amorphous metals **A13:** 868
effect of, on corrosion resistance **A1:** 912
effect of, on hardenability **A1:** 395, 413, 468
effect of, on notch toughness **A1:** 741
effect on borided steels **A4:** 441
effect on Curie point **A4:** 187
effect on sigma formation in ferritic stainless steels **A9:** 285
effect, Stellite alloys **A13:** 658
as electrical contact materials **A2:** 848-849
electrical contacts, use in **M3:** 671-672, 673, 674, 675
electrical discharge machining **A2:** 561
electrical resistance applications **M3:** 641, 646, 647, 655
electrolytic etching **A9:** 440
electromechanical polishing **A9:** 441
electron beam drip melted **A15:** 413
electron beam welding **M6:** 639-640
electron-beam welding **A6:** 870-871
electroplating of **M5:** 660-661
electroslag welding, reactions **A6:** 274
elongated grain-boundary traces in extruded **A9:** 124
embrittlement sources **A12:** 123
epithermal neutron activation analysis **A10:** 239
as essential metal **A2:** 1250, 1253-1254
evaporation fields for **A10:** 587
fabrication **M3:** 314, 321
in ferrite **A1:** 408
as ferrite stabilizer **A13:** 47
finishing processes **M5:** 659-662
forging of **A14:** 237-238

Molybdenum (continued)
forming **A2:** 562
friction welding **M6:** 722
functions in FCAW electrodes **A6:** 188
furnace elements **A7:** 423
in gas-metal eutectic method **EM3:** 305
gas-tungsten arc welding **A6:** 193
glass-to-metal seals **EL1:** 455 **EM3:** 302
gold plating of **M5:** 660
gravimetric finishes **A10:** 171
in gray iron **A1:** 22
as gray iron alloying element **A15:** 639
heat treating **A14:** 238
in heat-resistant alloys **A4:** 512, 515
heating element, use in vacuum furnace **A4:** 500 **M4:** 316
for heating elements and hot furnace structures **A4:** 497, 500, 501, 502
high-temperature solid-state welding **A6:** 298
hot swaging of **A14:** 142
hydrogen fluoride/hydrofluoric acid corrosion **A13:** 1169
implanted in aluminum **A10:** 484
influence of, on phosphorus-induced temper embrittlement **A1:** 699-700
influence of, on sigma phase embrittlement **A1:** 710
interface with molybdenum carbide **A9:** 121-122
interfacial segregation studies in **A10:** 599-601
iridium plating of **M5:** 660
in iron-base alloys, flame AAS analysis of **A10:** 56
Jones reductor for **A10:** 176
laser surface alloying of **A13:** 504
liquid-metal embrittlement of **A11:** 234
machining **A2:** 560
in maraging steels **A4:** 220, 222, 223, 224
mechanical properties **A2:** 575-577 **M3:** 321
in microalloy steel **A1:** 358
microstructures **A9:** 442
in moly-manganese paste process **EM3:** 304
for multiple radiation shields of cold-wall vacuum furnaces **A4:** 498
neutron and x-ray scattering, and absorption compared **A10:** 421
in nickel-base corrosion-resistant alloys containing molybdenum **A6:** 595
in nickel-base superalloys **A1:** 984
in nickel-chromium white irons **A15:** 680
oxidation-resistant coatings high-temperature **M5:** 661-662
oxygen cutting, effect on **M6:** 898
in P/M alloys **A1:** 810
particules, on metal/ceramic brazing assembly interface **A10:** 457
percussion welding **M6:** 740
photometric analysis methods **A10:** 64
physical properties **A6:** 941
pitting effect, amorphous metals **A13:** 868
plasma-MIG welding **A6:** 224
polish-etching **A9:** 441
precautions **M3:** 321
in precipitation-hardening steels **M6:** 350
production **A2:** 574-575
properties **A6:** 629
pure **M2:** 771-776
pure, properties **A2:** 1140
qualitative tests to identify **A10:** 168
reaction of hearth with nickel-bearing alloys in vacuum heat treating **A4:** 503
reactions with gases and carbon **M6:** 1055
recovery from selected electrode coverings **A6:** 60
recrystallization temperatures **M6:** 1055
as refractory metal, commercial uses **A7:** 17
resistance-type electric sintering furnace **A7:** 690
rhodium plating of **M5:** 660
rolling **A2:** 562

Molybdenum (continued)
sheath, in ternary molybdenum chalcogenides (chevrel phases) **A2:** 1077
sheet forming **A14:** 787-788
sintering, time and temperature **M4:** 796
skeletons, composites with **A2:** 855
with sodium samples, epithermal neutron activation analysis of **A10:** 239
solderability **A6:** 978
solubility of rhenium in during etching **A9:** 447
species weighed in gravimetry **A10:** 172
spraying for hardfacing **M6:** 789
sputter-induced reduction of oxides **EM3:** 245
in steel **A1:** 146, 577
strengths of ultrasonic welds **M6:** 752
submerged arc welding promotion of acicular ferrite **M6:** 117
temperatures, mill processing **M3:** 317
thermal diffusivity from 20 to 100 °C **A6:** 4
tilt boundaries studied by high resolution electron microscopy **A9:** 121
TNAA detection limits **A10:** 237
to improve hardenability in carburized steels **A4:** 366, 367
to promote hardness **A4:** 124, 127, 128-129, 130
tongs for manual resistance brazing **A6:** 340
as trace element, cupolas **A15:** 388
transition temperatures **M6:** 1055
tubing, production techniques **A2:** 562-563
ultrapure, by chemical vapor deposition **A2:** 1094
ultrapure, by zone-refining technique **A2:** 1094
ultrasonic welding **A6:** 327
unalloyed sintered compact **A9:** 562
use in flux cored electrodes **M6:** 103
used as heating elements in vacuum furnaces **A4:** 499-500
vacuum heat-treating support fixture material **A4:** 503
vacuum-arc-cast high oxygen **A12:** 6
vapor pressure, relation to temperature **A4:** 495 **M4:** 309, 310
vibratory polishing of **A9:** 550
volumetric procedures for **A10:** 175
welds, ductility **A2:** 564
in wrought stainless steels **A1:** 872
Molybdenum alloy 362 *See* Molybdenum alloys, specific types, Mo-0.5Ti
Molybdenum alloy 363 *See* Molybdenum alloys, specific types, TZM
Molybdenum alloy 364 *See* Molybdenum alloys, specific types, TZM
Molybdenum alloy TZM **A7:** 768-769
Molybdenum alloys *See also* Molybdenum; Molybdenum alloys, specific types **A9:** 442
annealing **A4:** 816-817, 818, 819 **A6:** 581
applications **A2:** 557-559, 574
arc cast, ultrasonic welding **A6:** 326
bar and tube, die materials for drawing **M3:** 525
ceramic coating of **M5:** 542-545
cleaning **A4:** 817
cleaning processes **M5:** 659-660
corrosion resistance **A2:** 575
density and melting temperature **M5:** 380
ductile-to-brittle transition temperature **A6:** 581
electron-beam welding **A6:** 581
electroplating of **M5:** 660-661
filler metals for **M3:** 320
FIM sample preparation of **A10:** 586
finishing processes **M5:** 659-662
forging characteristics **A14:** 237
forging of **A14:** 237-238
furnaces **A4:** 816-817
gas-tungsten arc welding **A6:** 581

Molybdenum alloys (continued)
high-temperature behavior in various gas atmospheres **A4:** 816-817, 818
interstitial impurities effect **A6:** 581
mechanical properties **A7:** 476-477, 768-769
Ni_3Mo effect on 18Ni maraging steels **A4:** 223, 224
nickel-nickel molybdenum for control thermocouples in vacuum heat treating **A4:** 506
nitriding **A4:** 816, 818
oxidation protective coating of **M5:** 380
oxidation-resistant coatings high-temperature **M5:** 661-662
principal ASTM specifications for weldable nonferrous sheet metals **A6:** 400
production **A2:** 574
sintered, ultrasonic welding **A6:** 326, 327
stress-relieving **A4:** 817
tensile strength **A6:** 580
tool steels **A14:** 81
ultrasonic welding **A6:** 894
welding conditions effect **A6:** 581
wire, die materials for drawing **M3:** 522
Molybdenum alloys, specific types *See also* Molybdenum; Molybdenum alloys
70Mo-30W, cold worked and annealed **A9:** 446
composition **M3:** 342
Doped Mo, annealing **A4:** 816
Mo, annealing **A4:** 816
Mo-0.5Ti, cold rolled and annealed **A9:** 445
Mo-0.5Ti composition **M3:** 316, 341
property data **M3:** 341-342
temperatures, mill processing **M3:** 317
Mo-0.5Ti, mechanical properties **A2:** 575-576
Mo-0.5Ti, physical properties **A6:** 941
Mo-0.5Ti-0.02C, mechanical properties **A2:** 575-576
Mo-0.5Ti-0.008Zr, physical properties **A6:** 941
Mo-12.50S, twinning **A9:** 127
Mo-30W, annealing **A4:** 816
Mo-35Re, twin-trace length as a function of angle **A9:** 128
Mo-35Re, twinned structure in single crystal **A9:** 127
Mo-36W, composition **M3:** 316
Mo-50Re, machining **A16:** 858-864, 866-869
Mo-50Re, physical properties **A6:** 941
Mo-MHC, annealing **A4:** 816
Mo-Re, face milling **A16:** 863
Mo-TZM annealing **A4:** 816, 817, 819
recrystallization **A4:** 817, 819
MZC (Mo-1Ti-0.3Zr), mechanical properties **A2:** 576
Ni-Cr-Mo, distortion in heat treatment **A4:** 614
property data **M3:** 342
TZC
TZC, machining **A16:** 858-864, 866-869
TZM **A9:** 446
TZM (Mo-0.5TI-O.lZr), properties **A2:** 576-577
TZM composition **M3:** 316, 343
property data **M3:** 342, 343
temperatures, mill processing **M3:** 317
TZM, machining **A16:** 858-864, 866-869
TZM, rhodium plating of **M5:** 660
Molybdenum and molybdenum alloys
abrasive cutoff sawing **A16:** 868
boring **A16:** 859
in commercial CPM tool steel compositions **A16:** 63
content affecting tool steel grindability **A16:** 732
content in stainless steels **A16:** 682-684
counterboring **A16:** 860
drilling **A16:** 860
electrochemical grinding **A16:** 543
electrochemical machining **A16:** 534, 535
electron beam machining **A16:** 570
end milling-slotting **A16:** 866
face milling **A16:** 863
in high-speed tool steels **A16:** 52, 59
internal grinding **A16:** 869
in low-alloy steels **A16:** 150
milling **A16:** 312, 313, 864
oil hole or pressurized-coolant drilling **A16:** 861
in P/M high-speed tool steels **A16:** 61
peripheral end milling **A16:** 866
photochemical machining **A16:** 588, 590, 592-593
Molybdenum and molybdenum alloys (continued)
power band sawing **A16:** 867
power hacksawing **A16:** 867
reaming **A16:** 862
spade drilling **A16:** 861
spotfacing **A16:** 860
surface grinding **A16:** 868
tapping **A16:** 862
thermal expansion coefficient **A6:** 907
to form simple and complex carbides **A16:** 667
trepanning **A16:** 860
turning **A16:** 858
Molybdenum and niobium fine powders **A7:** 437
Molybdenum boride cermets
application and properties **A2:** 1004
Molybdenum boride-based cermets **A7:** 812-813
Molybdenum carbide
in cermets **A16:** 91-97
interface with molybdenum **A9:** 121-122
Molybdenum carbides
quantitative effect of carbon KVV spectra in **A10:** 553
Molybdenum coating
cadmium plate **M5:** 269
Molybdenum dioxide
deposited by potentiostatic etching **A9:** 146
Molybdenum disilicide *See also* Electrical resistance alloys **M3:** 646, 647, 655
+ 10% ceramic additives, properties and applications **A2:** 839
as heating material **A2:** 829
maximum operating temperatures **EM4:** 250
for non-oxide ceramic heating elements of electrically heated furnaces **EM4:** 248-249, 250
as ordered intermetallic **A2:** 934-935
properties and applications **A2:** 839
for rod elements in furnaces **A4:** 472
thermal expansion coefficient **A6:** 907
Molybdenum disulfide **A7:** 154-155
as a solid lubricant **A16:** 123
as lubricant **A14:** 238, 481-482
lubricant for room-temperature compression testing **A8:** 195, 201
Molybdenum disulfide lubricant
nuclear reactor **A13:** 959-964
Molybdenum electrode corrosion factor **EM4:** 388
for base metal conductors **EM4:** 1142
for cofiring the metallization with the alumina **EM4:** 544
glass-metal seals **EM4:** 875, 1037
for heating elements for electrically heated furnaces **EM4:** 247, 248, 249
vapor pressure **EM4:** 249
for heating elements used in hot isostatic pressing **EM4:** 195
non-oxide ceramic joining **EM4:** 480
recommended glass/metal seal combinations **EM4:** 497
room-temperature bond strength of silicon nitride metal joints **EM4:** 1526
in solid-state bonding for joining non-oxide ceramics **EM4:** 529
solid-state bonding in joining non-oxide ceramics **EM4:** 525
Molybdenum high-speed tool steels **A1:** 759, 764
forging temperatures **A14:** 81
Molybdenum hot-work tool steels **A1:** 763
Molybdenum in cast iron **M1:** 77, 79-80, 83-86, 95-96
ductile iron **M1:** 40, 47-49, 51
gray iron **M1:** 21, 26, 28, 29, 30
Molybdenum in stainless steel **M3:** 57, 58
Molybdenum in steel **M1:** 115, 411, 417
castings **M1:** 388, 394, 395
constructional steel for elevated temperature use **M1:** 647
graphitization affected by **M1:** 686
hardenability affected by **M1:** 477, 478
maraging steels **M1:** 445, 446, 447
nitriding, effect on **M1:** 540-541
notch toughness, effect on **M1:** 694
P/M materials **M1:** 337
seawater corrosion affected by **M1:** 741-742, 745
soil corrosion affected by **M1:** 729-730
steel sheet, effect on formability **M1:** 554
Molybdenum in steel (continued)
temper embrittlement, suppression of **M1:** 684-685
Molybdenum oxide
metal-to-metal oxide equilibria **A7:** 340
Molybdenum oxide alloys
workability **A8:** 575
Molybdenum oxide catalysts
Raman analysis **A10:** 133
Molybdenum oxide reduction **A7:** 52-53
Molybdenum powders *See also* Molybdenum; Molybdenum alloys
applications **A7:** 156
chemical analysis and sampling **A7:** 248
compositions **A7:** 155
density with high-energy compacting **A7:** 305
explosivity **A7:** 196-197
exposure limits **A7:** 204
finishing **A7:** 155-156
flowchart, production process **A7:** 156
as high-temperature material **A7:** 767-769
mechanical properties **A7:** 476
melting point and density **A7:** 152
metal-to-metal oxide equilibria **A7:** 340
microstructure **A7:** 156
plasma spray powder **A7:** 77
powder and compact purity **A7:** 389-390
processing sequence flowchart **A7:** 767
production **A7:** 152-156
properties **A7:** 155, 768
pure, mechanical properties **A7:** 476
reduction sequence **A7:** 155
relative density **A7:** 390
shipment tonnage **A7:** 24
sintering **A7:** 389-392
spray dried **A7:** 77
tap density **A7:** 277
as tooling material **A7:** 423
toxicity, diseases, exposure limits **A7:** 204
Molybdenum silicide ($MoSi_2$)
adiabatic temperatures **EM4:** 229
applications **EM4:** 230
synthesized by SHS process **EM4:** 229
Molybdenum steels
composition of tool and die steel groups **A6:** 674
Molybdenum toxicity
biologic effects **A2:** 1253-1254
Molybdenum trioxide
two stage reduction **A7:** 52
Molybdenum tubing extrusion
cast Co alloy tools used **A16:** 70
Molybdenum, zone refined
impurity concentration **M2:** 713
Molybdenum-base electrical contact materials
microstructures of **A9:** 552
Molybdenum-base powder metallurgy material
60Mo-40Ag **A9:** 562
Molybdenum-cobalt high-speed steel
by STAMP process **A7:** 548
Molybdenum-copper powders
infiltration of **A7:** 555
Molybdenum-niobium alloys **A13:** 534-538
Molybdenum-rhenium
annealing practices for microexamination **A9:** 447
metallographic techniques for **A9:** 447-448
Molybdenum-rhenium alloy
mechanical properties **A7:** 477
Molybdenum-rhenium alloys, polycrystalline
bend contours in **A10:** 445
Molybdenum-silver composites, properties
for electrical make-break contacts **A2:** 851
Molybdenum-silver powders **A7:** 555, 561
Molybdic acid
description **A9:** 68
use in color etching **A9:** 142
Moment, and stress
structural analysis **EM1:** 460
Moment of force
SI unit/symbol for **A8:** 721
Moment of inertia
abbreviation **A8:** 724
Moment scale
for cantilever beam bend test **A8:** 132-133
Moment-curvature relationship
in elastic bending **A8:** 118

Moment-stress relationships
in elastic bending **A8:** 118
Momentum
effects in gating **A15:** 591-592
Momentum balance
as process modeling submodel **EM1:** 500
Momentum balance techniques
fluid flow modeling **A15:** 867, 870-876
Mond-Langer process
nickel separation **A7:** 135, 137-138
Monel *See* Nickel alloys, cast, specific types, M-35
alloy 400, expansion joint, cleaning **A12:** 76
die cutting speeds **A16:** 301
diffusion welding **A6:** 886
drilling **A16:** 229
electrochemical grinding **A16:** 542
environments that cause stress-corrosion cracking **A6:** 1101
failed in liquid mercury **A12:** 25
as fastener material **A11:** 530
galvanic corrosion with magnesium **M2:** 607
honing stone selection **A16:** 476
inorganic fluxes for **A6:** 980
liquid-erosion resistance **A11:** 167
oxyacetylene welding **A6:** 281
photochemical machining **A16:** 588, 590
relative solderability as a function of flux type **A6:** 129
relative weldability ratings, resistance spot welding **A6:** 834
solderability **A6:** 978
tension and torsion effective fracture strains **A8:** 168
thread rolling **A16:** 282
threading **A16:** 299
tool for electrochemical machining **A16:** 533
tooling for ultrasonic machining **EM4:** 359
weld overlay for hardfacing alloys **A6:** 820
Monel 400 .... **A9:** 435-437
composition **A16:** 836
machining **A16:** 837-840, 842, 843
sawing **A16:** 361, 363, 842
springs, strip for **M1:** 286
springs, wire for **M1:** 284
Monel 400 tube
eddy current inspection **A17:** 182-183
Monel 401
composition **A16:** 836
machining **A16:** 837-840, 842, 843
Monel 450
composition **A16:** 836
machining **A16:** 837-840, 842, 843
Monel 501
contour band sawing **A16:** 363
cutoff band sawing **A16:** 361
Monel alloys *See also* Nickel alloys, specific types; Nickel alloys, specific types, Monel 400
arc welding to stainless steel **M6:** 444
conditions for gas metal arc welding **M6:** 444
conditions for manual arc welding **M6:** 437
explosion welding **M6:** 713
brazing **A6:** 627-628
brazing to aluminum **M6:** 1031
development and characteristics **A2:** 429
projection welding **M6:** 506
resistance seam welding **A6:** 245
resistance spot welding **M6:** 480
Monel alloys, specific types
400
applications **A6:** 586
characteristics **A6:** 586
composition **A6:** 586
cutoff band sawing with bimetal blades **A6:** 1184
friction welding **A6:** 153
thermal diffusivity from 20 to 100°C **A6:** 4
501, cutoff band sawing with bimetal blades **A6:** 1184
Monel alloys, specific types (continued)
K-500, cutoff band sawing with bimetal blades **A6:** 1184
R-405, cutoff band sawing with bimetal blades **A6:** 1184
Monel K-500 .... **A9:** 435-438
contour band sawing **A16:** 363
cutoff band sawing **A16:** 361
springs, strip for **M1:** 286
springs, wire for **M1:** 284
Monel K-500, unaged
composition **A16:** 836
machining **A16:** 837-840, 842, 843
Monel K500, aged
composition **A16:** 836
machining **A16:** 837-840, 842, 843
Monel R-405 .... **A9:** 435-437
composition **A16:** 836
machining **A16:** 837-840, 842, 843
sawing **A16:** 361, 363
Money and time
engineering economy **A13:** 369
Monitor screen
machine vision system **A17:** 41
Monitoring *See also* Corrosion testing; Evaluation; In-service monitoring; Simulated service testing
acoustic emission, of pressure vessels **A17:** 654-656
acoustic emission, of weldments **A17:** 598-602
of corrosion inhibitors **A13:** 197, 483, 486
corrosion, ultrasonic inspection **A17:** 275-276
crack, by ultrasonic inspection **A17:** 273
during welding **A17:** 289, 601
of equipment **A13:** 322
fatigue test, by magnetic rubber inspection **A17:** 125
follow-up **A13:** 322-323
gas/oil production **A13:** 1246, 1250-1251
in-service **A13:** 197-203
on-stream **A13:** 322-323
of phosphate bath **A13:** 384
postweld **A17:** 599
radiation **A17:** 301-302
resistance welds, acoustic emission inspection **A17:** 284
ultrasonic inspection variables **A17:** 231
Monitoring systems *See* Tool condition monitoring systems
Monkman-Grant relationship **A6:** 168
advantages **A8:** 336
for elevated-temperature tensile creep-rupture testing **A8:** 304-305
extrapolation abilities **A8:** 335
for rupture life **A8:** 335-337
Monoaluminide oxidation protective coating
superalloys **M5:** 376
Monoaluminide-chromium- aluminum-yttrium coatings
superalloys **M5:** 376-379
Monochromatic
defined **A9:** 12 **A10:** 677
Monochromatic beams
single-crystal diffraction methods **A10:** 329-330
Monochromatic incident light
gray value contrast at different wavelengths **A9:** 149
Monochromatic objective
defined **A9:** 12
Monochromatic reflection topography **A10:** 366
Monochromator **EM3:** 19
defined **EM2:** 27
Monochromators
for atomic absorption spectrometry and related techniques **A10:** 44, 50-52
crystal, for neutron diffraction **A10:** 422
Czerny-Turner design **A10:** 23, 38
defined **A10:** 677
dispersive infrared **A10:** 111
and flame emission sources **A10:** 30
Monochromators (continued)
grating, as wavelength sorting device **A10:** 23
ICP-AES analysis **A10:** 34
for molecular fluorescence spectroscopy **A10:** 76-77
with molecular optical laser examiners **A10:** 129-130
scanning **A10:** 38, 128
stray light rejection for single, double triple **A10:** 129
Monoclinic
defined **A9:** 12
Monoclinic crystal system .... **A3:** 1 • 10, 15 **A9:** 706
Monoclinic structure in uranium alloys **A9:** 477
Monoclinic unit cells **A10:** 346-348
Monocrystal casting *See also* Single-crystal casting
metallurgy of **A15:** 322-323
processes **A15:** 323
Monocrystal solidification
an directional solidification **A15:** 319-323
Monocrystalline silicon (MOSi)
grinding **EM4:** 335
Monoculture
wafer-scale integration **EL1:** 354
Monoethanolamine, corrosion
stainless steels **M3:** 84
Monofilament
defined **EM2:** 27-28
Monofilamentary conductors
assembly techniques **A2:** 1046-1047
Monofilamentary wire *See also* Multifilamentary wire; Niobium-titanium superconductors; Wire
niobium-titanium **A2:** 1043, 1046-1047
Monofilaments **EM1:** 16, 731-732
Monofunctional epoxy diluents **EM1:** 67, 70
Monolayer
defined **EM1:** 16 **EM2:** 28
Monolayers
adsorbed on metal surfaces, examination of **A10:** 118-120
Langmuir-Blodgett **A10:** 120
sub-, effect of adsorbed oxygen on metals **A10:** 552
top, AES chemical analysis **A10:** 551
Monolithic
capacitors, active analog components **EL1:** 144
diodes, active analog components **EL1:** 144
-like WSI approaches, summary **EL1:** 271-272
and membrane dual linings **A13:** 453-455
metals, properties **EL1:** 1120
resistors, active analog components **EL1:** 144
wafer-scale integrated circuits **EL1:** 8
Monolithic and fibrous refractories
applications **EM4:** 910, 912-917
fibrous refractories **EM4:** 911-912, 913
monolithic refractory materials **EM4:** 910-911, 913, 915
Monolithic forged hot isostatic pressure vessels **A7:** 420
Monolithic microwave integrated circuits (MMICS) *See also* Integrated circuits (ICs)
active analog components **EL1:** 149-150
in active-component fabrication **EL1:** 201
characteristics, applications **A2:** 740
circuit elements, active analog components **EL1:** 150
defined **EL1:** 946
device attachment **EL1:** 755-756
electrical performance testing **EL1:** 946, 951-952
elements of **EL1:** 150
phased-array radar systems **A2:** 740
radio frequency testing **EL1:** 952
research and development **A2:** 747
Monoliths
drying problem **EM4:** 210, 211, 212, 213
Monomer **EM3:** 19
defined **A10:** 677 **A13:** 9
multifunctional, for solder masking **EL1:** 555
parylene coatings **EL1:** 790-791

Monomer(s)
in acrylonitrile-butadiene-styrenes **EM2:** 109
defined **EM2:** 28
Monomeric compounds
in free radical cure formulations **EL1:** 857
Monomeric dicyanates *See also* Cyanates
chemical structure **EM2:** 232
Monomers
conversion **EM1:** 132
defined **EM1:** 16
effect, polyester resin thermal stability **EM1:** 93
of PMR- 15, chemical structure **EM1:** 141
structure, PMR-15 polyimide **EM1:** 811
Monoperoxydodecanedioic acid, ATR
DRS, and PAS granular analysis of **A10:** 120
Monosized titania (TiO_2)
sintering **EM4:** 264, 265
Monosubstituted benzene derivatives
substituent effects on fluorescence quantum yields for **A10:** 74
Monotectic alloys
low gravity effects **A15:** 152
Monotectic reaction .... **A3:** 1 • 5
Monotectoid reaction .... **A3:** 1 • 5
Monotonic fracture
and fatigue fracture, compared **A12:** 229
slow, pearlitic ductile irons **A12:** 230
Monotonic yield strength
for crack growth specimen **A8:** 380-381
Monotron hardness test
defined **A8:** 9
Monotropism
defined **A9:** 12
Monotype metal
as lead and lead alloy application **A2:** 549
micrograph **A9:** 424
Monovariant equilibrium .... **A3:** 1 • 2
Monovariant liquidus troughs, on a nickel- aluminum-molybdenum isothermal section **A9:** 621
MONPAC acoustic emission tests **A17:** 292
Montan wax
for investment casting **A15:** 254
Monte Carlo electron trajectory simulation
of EPMA effects **A10:** 518
Monte Carlo projection
electron trajectories **A12:** 167
Monte Carlo techniques
for 20-keV electrons **A10:** 499
defined **A10:** 677
Montmorillonite
as bonding clay **A15:** 210, 224
Montmorillonites **EM4:** 5, 6
formula **EM4:** 5
microstructure **EM4:** 6
in solvents **EM4:** 117
"Moonlight" project
Japan **EM4:** 717
Moore fatigue-test data
steel **A11:** 115-116
Moore hip endoprosthesis
classic **A11:** 670
Moore pins, broken adjustable
from cobalt- chromium alloy **A11:** 675, 681
Moquette
versus tile whiteware **EM4:** 929
"Morning sickness"
in carbon aircraft brakes **A18:** 585
Morphic templates
for particle shapes **A7:** 241-242
Morpholine derivatives
as biocides **A18:** 110
Morphological analysis
of particle shape **A7:** 241-243
Morphology **EM3:** 19
carbide **A11:** 575
classification **EM2:** 489-490
crack, in hydrogen damage **A11:** 250
crack, titanium alloys **A11:** 240, 241
of crystal structure, organic solids, analyses for **A10:** 9
defined **EM1:** 16 **EM2:** 28
of dendrite and diffusion path microsegregation **A15:** 137
eutectic silicon, aluminum-silicon alloys **A15:** 163
of eutectics **A15:** 119-121
Morphology (continued)
fracture, of nuclear steam-generator wall **A11:** 657
grain, topographic methods for **A10:** 368
graphite, in gray iron **A15:** 631-632
high-impact polystyrenes (PS, HIPS) **EM2:** 194
image analysis for **A10:** 309-322
of inorganic solids, analytical methods for **A10:** 4-6
molecular structure of organic solids, analyses for **A10:** 9
optical microscope and SEM imaging for **A10:** 521
of organic solids, analytical methods for **A10:** 9
phase distribution of organic solids, analyses for **A10:** 9
planar to equiaxed solidification shift **A15:** 153
stability, of solid/liquid interface, in low gravity **A15:** 153-156
and structure, of polymers **A11:** 758
Morris, Colonel Lewis
as early founder **A15:** 25
Morrison-Miller effect **A18:** 233
Mortar shell bodies
as ordnance application **A7:** 684-685
Mortar tubes
alloy compositions for **A11:** 294
ductility of **A11:** 295
heat treatments for **A11:** 294
transverse yield strength **A11:** 295
Mortars **EM4:** 895, 911
bonding in structural caly products ... **EM4:** 947-948
performance characteristics and properties **EM4:** 921 923
portland cement-lime **EM4:** 947
MOS *See* Silicon metal-oxide semiconductors
MOS transistors *See also* Transistor(s) **EL1:** 149, 157-158
Mosaic crystal
defined **A9:** 12
Mosaic crystal structure
defined **A10:** 351
Mosaic structure
defined **A9:** 12
Mosely's law
representation of **A10:** 433
MOSFET *See* Metal-oxide semiconductor field-effect transistor
Mössbauer effect *See also* Mössbauer spectroscopy
abbreviation for **A10:** 690
capabilities **A10:** 277
defined **A10:** 677
as method **A10:** 287-295
sources, principal methods used for producing **A10:** 291-292
Mössbauer spectroscopy *See also* Mössbauer effect **EM4:** 52, 53
absorption spectra **A10:** 294
applications **A10:** 287, 293-295
capabilities **A10:** 253, 267, 277
capabilities, compared with classical wet analytical chemistry **A10:** 161
defined **A10:** 677
estimated analysis time **A10:** 287
experimental arrangement **A10:** 293
general uses **A10:** 287
introduction and principles **A10:** 287-293
limitations **A10:** 287
for partitioning oxidation states, iron **A10:** 178
related techniques **A10:** 287
selection rules **A10:** 288
sources, principal methods for producing **A10:** 291-292
spectra components **A10:** 293
transitions, properties of **A10:** 289-290
Mössbauer spectrum
defined **A10:** 677
Mössbauer transitions
some properties of **A10:** 289-290
Most vulnerable component (MVC) **A6:** 354
Most vulnerable component (MVC) maximum peak temperature **A6:** 354
Mother boards *See also* Backplanes
defined **EL1:** 1150
materials and processes selection **EL1:** 117
Motifs **A18:** 346, 347
Motion
human vs. machine vision **A17:** 30
relative, object position defined by **A17:** 34-35
-sensing, with machine vision **A17:** 44
unsharpness **A17:** 338
whole-body, effect in optical holography **A17:** 412-413
Motion brazing
aluminum alloys **M6:** 1030
Motor pole pieces
powders used **A7:** 573
Motor shafts
economy in manufacture **M3:** 854
Motor springs **A1:** 302
steel **M1:** 287
Motor Vehicles Manufacturers Association
performance testing of engine oils **A18:** 170
Motor-current signature analysis (MCSA) **A18:** 313-318
application to motor-operated valves **A18:** 314-318
actuation of wedge-gate valves under differential pressures **A18:** 317-318
degraded worm and worm-gear lubrication **A18:** 316
gate-valve unseating **A18:** 317
gear-tooth wear **A18:** 316
stem taper effect on packing friction **A18:** 316
stem-nut wear **A18:** 314-316
equipment **A18:** 313-314
signal acquisition and recording **A18:** 313-314
signal analysis **A18:** 313-314
signal processing **A18:** 313-314
future trends **A18:** 318
method **A18:** 314
operating principles **A18:** 313
Motor-generator (M-G) set **A6:** 39
Motor-generator units
induction brazing of steel **M6:** 967
shielded metal arc welding **M6:** 77-78
submerged arc welding **M6:** 131
Motor-operated valves (MOVs)
motor- current signature analysis **A18:** 313-318
Motorboat-engine connecting rod
fractured **A11:** 328, 330
Motorcycle-transmission shaft
high-cycle fatigue in **A11:** 106
Motors **A8:** 157-158
aluminum and aluminum alloys **A2:** 13
as magnetically soft material application **A2:** 779
with niobium-titanium superconducting materials **A2:** 1057
Mottled cast iron
defined **A15:** 8
Mottled iron **A18:** 695, 700, 701 **M1:** 3, 12, 14, 22
Mottling *See* Radiation diffraction
Mount Joy Forge *See* Valley Forge Company
Mounting
angle **A10:** 576
of bearings, abuse in **A11:** 508
defects, through-hole packages **EL1:** 979
design, effect on bearing failure **A11:** 506
frame, in forced-vibration systems **A8:** 391
microhardness specimen **A8:** 93
optical metallography sample, in Bakelite **A10:** 300
powders, for XPS analysis **A10:** 575
printed board coupons **EL1:** 572-573
specimens for optical metallography **A10:** 300
Mounting and mounting materials, specific metals and alloys *See also* Mounting of specimens
aluminum alloys **A9:** 352
austenitic manganese steel castings **A9:** 237-238
beryllium-copper alloys **A9:** 392
beryllium-nickel alloys **A9:** 393
carbon and alloy steels **A9:** 166
carbon steel casting specimens **A9:** 230
carbonitrided steels **A9:** 217
carburized steels **A9:** 217
cast irons **A9:** 243
cemented carbides **A9:** 273
electrical contact materials **A9:** 550
ferrites and garnets **A9:** 533
fiber composites **A9:** 587-588
hafnium **A9:** 497
iron-cobalt and iron-nickel alloys **A9:** 532

Mounting and mounting materials, specific metals and alloys (continued)
lead and lead alloys ... **A9:** 415
low-alloy steel casting samples ... **A9:** 230
magnesium alloys ... **A9:** 425
nitrided steels ... **A9:** 218
porcelain enameled sheet steel ... **A9:** 198
refractory metals ... **A9:** 439
sleeve bearing materials ... **A9:** 565
tin and tin alloys ... **A9:** 449
titanium and titanium alloys ... **A9:** 458
tool steels ... **A9:** 256-257
tungsten ... **A9:** 441
uranium and uranium alloys ... **A9:** 478
wrought heat-resistant alloys ... **A9:** 305
wrought stainless steels ... **A9:** 279
zinc and zinc alloys ... **A9:** 488
zirconium and zirconium alloys ... **A9:** 497
Mounting artifact
defined ... **A9:** 12
Mounting brackets, economy in manufacture ... **M3:** 850
alloying elements ... **M3:** 598, 599
amorphous materials ... **M3:** 603-604
applications ... **M3:** 601, 605-606, 607, 609-611, 612, 613
ferrites ... **M3:** 605-606
grain size and orientation ... **M3:** 597-598
heat treatment ... **M3:** 609, 610
impurities ... **M3:** 597
iron, high-purity ... **M3:** 597, 598, 599, 603, 607
iron-aluminum alloys ... **M3:** 598, 599, 601-602
iron-cobalt alloys ... **M3:** 603, 605, 608
iron-nickel alloys ... **M3:** 602-603, 605, 606, 607, 608, 613
low-carbon steels ... **M3:** 598-599, 609, 611
magnetic temperature compensation ... **M3:** 607-608
nickel irons ... **M3:** 603, 605, 608, 609, 610, 611, 612, 613
P/M iron ... **M3:** 608-609, 610, 613
selection for specific applications ... **M3:** 609-611, 612, 613
silicon steels ... **M3:** 599-601, 603-605, 608-613
Mounting clamp ... **A9:** 167-168
Mounting devices
mechanical ... **A9:** 28-29
Mounting materials *See also* Mounting and mounting materials, specific metals and alloys
conductive ... **A9:** 32
heat generated during curing ... **A9:** 31
for powder metallurgy materials ... **A9:** 504-505
for use with perchloric acid ... **A9:** 53-54
for use with sulfuric acid ... **A9:** 54
for welded joints ... **A9:** 577, 581
Mounting molds for fiber composites ... **A9:** 588
Mounting of specimens *See also* Mounting and mounting materials, specific metals and alloys ... **A9:** 28-32
with castable resins ... **A9:** 30
cleaning ... **A9:** 28
defined ... **A9:** 12
for edge retention of sample ... **A9:** 44-45
for electrolytic polishing ... **A9:** 49, 53-54
labelling ... **A9:** 32
materials ... **A9:** 28
mechanical devices ... **A9:** 28-29
metal powders ... **A9:** 31
powder metallurgy materials ... **A9:** 504-505
reasons for ... **A9:** 28
of scanning electron microscopy specimens ... **A9:** 97
selection of ... **A9:** 28-30
sheet-metal specimens ... **A9:** 198
storage ... **A9:** 32
thin sheet specimens ... **A9:** 31
tubes ... **A9:** 31
vacuum impregantion ... **A9:** 31
wire ... **A9:** 31
Mounting systems
electroslag welding ... **M6:** 228
Mounting technology *See also* Soldering technology
component and discrete chip, introduction ... **EL1:** 143
introduction ... **EL1:** 631-632
Mouth opening displacement, vs. load
as ideal elastic-plastic behavior ... **A8:** 471
Movable magnetic cores ... **A6:** 38
Movable-coil relay
life test using ... **A2:** 858-859
Movement capability
measurement of ... **EM3:** 189
Movement, sudden
acoustic emissions from ... **A17:** 278
Moving bridge type
coordinate measuring machines ... **A17:** 21
Moving dislocations
in AlFe3 ... **A9:** 682-683
Moving mandrel
tube/cup drawing with ... **A14:** 335-336
Moving parts, cable
optical sensor monitoring ... **A17:** 10
Moving-insert straightening ... **A14:** 688
Moving-ram type
horizontal coordinate measuring machine ... **A17:** 23
Moving-range control charts ... **A17:** 732-734
Moving-table type
horizontal coordinate measuring machine ... **A17:** 23
Mower
front-mount ... **A7:** 677
Mower hub
P/M ... **A7:** 677
MOX-1 explosives
aluminum powder containing ... **A7:** 601
MP 35 (conventional iron powder)
composition ... **A16:** 884
drilling time required and drill failure ... **A16:** 885
sintering results ... **A16:** 886
MP 36
composition ... **A16:** 885
MP 36S (P/M material)
composition ... **A16:** 884
in free-machining steels ... **A16:** 884
inclusions ... **A16:** 885
percentage of inclusions by type ... **A16:** 885
properties, sintering performance and chemical composition of transverse rupture bars ... **A16:** 884
MP 37 (P/M material)
composition ... **A16:** 884
drilling time required and drill failure ... **A16:** 885
in free-machining steels ... **A16:** 884
inclusions ... **A16:** 885
percentage of inclusions by type ... **A16:** 885
prealloyed MnS iron powder ... **A16:** 879, 880
properties, sintering performance and chemical composition of transverse rupture bars ... **A16:** 884
sintering results ... **A16:** 886
MP-35N
composition ... **M6:** 354
MP-159
composition ... **M6:** 354
MP35-N
composition ... **A4:** 794 **A6:** 564, 573, 929
MP35-N Multiphase
composition ... **A6:** 598
MP159
composition ... **A4:** 794 **A6:** 564, 573
MPC *See* Metals Properties Council
MPIF *See* Metal Powder Industries Federation
MPIF (Metal Powder Industries Federation) designations for ferrous P/M materials ... **A1:** 805
MPIF designation/classification *See also* Metal Powder Industries Federation
class I (simple) parts, basic geometries ... **A7:** 322
class I parts, typical ... **A7:** 332
class II parts, typical ... **A7:** 332
class III parts, typical ... **A7:** 332
class IV (complex) parts, basic geometries ... **A7:** 322
class IV arts, typical ... **A7:** 333
standard molded "dogbone" test bar ... **A7:** 489, 490
MPIF designations
P/M materials ... **M1:** 332-335
MRS *See* Materials Research Society
MS *See* Mass spectrometry
MSA-Whitby centrifuge
powder measurement ... **A7:** 129
MSC NASTRAN
finite-element analysis code ... **EM3:** 479, 480
Mu phase in wrought heat-resistant alloys ... **A9:** 309, 312
Mucilage ... **EM3:** 19
Mud cracks
austenitic stainless steels ... **A12:** 361
Mud, drilling
Miller numbers ... **A18:** 235
Muffle furnaces ... **A7:** 8
for sinters and fusions ... **A10:** 166
Muffle-type continuous production furnace ... **A7:** 351, 353
Mull preparation
for IR samples ... **A10:** 113
Müller total hip prosthesis ... **A11:** 670
Mullers *See also* Mulling; Sand preparation
additions, to core-oil sands ... **A15:** 219
batch-type, green sand preparation ... **A15:** 344-345
continuous, green sand preparation ... **A15:** 344-345
processes with ... **A15:** 238
revolving ... **A15:** 32
sand cooling by ... **A15:** 349
Mulling *See also* Mixing; Mullers; Sand preparation
of angular sands ... **A15:** 208
defined ... **A15:** 8
of green sands ... **A15:** 341
of sands ... **A15:** 32
Mullite *See also* Aluminum silicate; Mullite-cordierite composites ... **EM4:** 19, 49
applications ... **EM4:** 764-765
refractory ... **EM4:** 906-907
with beta-spodumene ... **EM4:** 870
for brake linings ... **A18:** 570
ceramic corrosion in the presence of combustion products ... **EM4:** 982
ceramic substrates ... **EM4:** 1107
ceramic/metal seals ... **EM4:** 502
chemical system ... **EM4:** 780
chemically stable intermediate phase at atmospheric pressure ... **EM4:** 761
coating formation in molten particle deposition ... **EM4:** 206
composition ... **EM4:** 761
from aluminum silicate ... **A15:** 209-210
glass-free prepared by sol-gel method ... **EM4:** 763
heat treatments ... **A18:** 814
island structure ... **EM4:** 758
matrix material for ceramic-matrix composites ... **EM4:** 810
mechanical properties ... **EM4:** 762
melting behavior of solid solution and various surrounding phase fields ... **EM4:** 763
as mold refractory, investment casting ... **A15:** 258
monolithic and Si-C reinforced ... **EM4:** 762-763
mullite-glass composite, effect of interfacial reaction ... **EM4:** 866-867
processing ... **EM4:** 762-764
properties ... **EM4:** 14, 30, 46, 503, 512, 759, 761, 762-764, 765
refractory compositions ... **EM4:** 896
as refractory, core coatings ... **A15:** 240

Mullite (continued)
result of pyrolphyllite phase transformations **EM4:** 761
sintering **EM4:** 762
specific types **EM4:** 873
structure **EM4:** 761, 762, 764
substrate properties **EM4:** 1108
as thermostructural ceramic **EM4:** 1003
ultrafine powder from CVD **EM4:** 763-764
variation in solid-solution range **EM4:** 761-762
versus sillimanite **EM4:** 762
Mullite inclusions in steel **A9:** 185
Mullite-cordierite composites
composite Young's modulus **EM4:** 860, 861
dielectric constant **EM4:** 859, 861
fracture toughness **EM4:** 865
thermal expansion **EM4:** 859, 860
Mullite/zirconia
applications **EM4:** 771
flexural strength **EM4:** 770-771
production of **EM4:** 770-771
properties **EM4:** 770
reaction-sintered property data **EM4:** 773
Multi-cation materials **EM4:** 60, 61
Multi-end roving process **EM1:** 109
Multi-wall forged hot isostatic pressure vessels **A7:** 420
Multiaxial creep theories
experimental methods and ductility **A8:** 343-344
Multiaxial fatigue testing machines **A8:** 370
Multiaxial stressing **A8:** 343-344
Multichannel acoustic emission inspection **A17:** 283
Multichannel analyzers
defined **A10:** 677
effect in EPMA analysis **A10:** 519
Multichannel detectors
Fellgett's advantage and **A10:** 129
Raman spectrometer with **A10:** 128
Multichip assemblies *See also* Multichip modules (MCMs); Multichip packaging; Multichip technology
in mainframe computers **EL1:** 298
Multichip hybrids
advantages **EL1:** 270-271
Multichip modules (MCMS)
advantages/limitations **EL1:** 258
defined **EL1:** 1150
future trends **EL1:** 392
as hybrid **EL1:** 250
hybrid technology **EL1:** 252
materials **EL1:** 301-307
structure **EL1:** 300-301
tape automated bonding (TAB) technology **EL1:** 277
thermal control **EL1:** 47-49
thermal parameters **EL1:** 48
thermal via effect **EL1:** 53-54
thin-film **EL1:** 391
Multichip packaging
system level **EL1:** 2
with tape automated bonding (TAB) **EL1:** 275
Multichip technology
capabilities/limitations **EL1:** 510
cofired ceramic, Cooling in **EL1:** 307-308
connections to PWBs and system **EL1:** 311
cooling **EL1:** 307-309
heat removal **EL1:** 309-311
large-scale system, communication environment **EL1:** 2-5
materials **EL1:** 301-307
module structures, new **EL1:** 300-301
modules with ceramic substrates, hybrid packages for **EL1:** 451-454
new **EL1:** 299-300
planar **EL1:** 307
types, compared **EL1:** 297-298
Multicircuit winding
defined **EM1:** 900 **EM2:** 28
Multicolored workpieces
for physical modeling **A14:** 437
Multicomponent glasses, sol gel processing **EM4:** 451
applications **EM4:** 451
formulations **EM4:** 451
Multicomponent iron-carbon systems
thermodynamics of **A15:** 68-69
Multicomponent materials
advantages **EM3:** 687
Multicylinder engine tests, NTC-400
Mack T **A18:** 101
Multidetector translate-rotate systems
computed tomography (CT) **A17:** 366
Multidimensional shape characterization **A7:** 240-241
Multidirectional fiber reinforcement
fabrication methods **EM1:** 934
fiber selection **EM1:** 934-936
formulation/fabrication parameters **EM1:** 935
properties **EM1:** 937
Multidirectional magnetizing
magnetic particle inspection **A17:** 93
Multidirectional solidification *See also* Solidification
austenite-iron carbide eutectic **A15:** 179-180
compacted/vermicular graphite eutectic **A15:** 178-179
defect growth of graphite theory **A15:** 177
eutectic growth (cast iron) **A15:** 175-180
flake (lamellar) graphite eutectic **A15:** 175-176
particle behavior in **A15:** 145-146
spheroidal graphite eutectic **A15:** 176-178
surface adsorption theory **A15:** 178
Multidirectional tape prepregs **EM1:** 146-147
applications **EM1:** 147
product forms **EM1:** 146
properties **EM1:** 146-147
strength/weight vs. material form compared **EM1:** 146
vs. unidirectional tape prepregs **EM1:** 146
Multidirectionally reinforced carbon/graphite matrix composites **EM1:** 915-919
Multidirectionally reinforced ceramics **EM1:** 933-940
applications **EM1:** 939-940
composite formulation/processing **EM1:** 934-937
fiber reinforcement **EM1:** 933-934
properties **EM1:** 937-939
Multidirectionally reinforced fabrics
reinforcement materials **EM1:** 129
testing of **EM1:** 131
three-dimensional weaving machines **EM1:** 130-131
weave geometry **EM1:** 129-130
Multidirectionally reinforced preforms
reinforcement materials **EM1:** 129
testing of **EM1:** 131
three-dimensional weaving machines **EM1:** 130-131
weave geometry **EM1:** 129-130
Multielement analysis
energy-dispersive x-ray spectrometry **A10:** 82-101
simultaneous, ICP-AES as **A10:** 32, 33
Multielement isotopes
for spark source mass spectrometry **A10:** 145
Multifilament superconducting alloy
color etched **A9:** 157
Multifilament yarn
defined **EM2:** 28
Multifilament yarns *See also* Yarns
defined **EM1:** 16
testing **EM1:** 732-733
yarn/strand test methods **EM1:** 732
Multifilamentary
composite wire, niobium-titanium **A2:** 1043, 1047-1049
conductors, assembly techniques **A2:** 1047-1049
NbTi superconducting composite fabrication **A2:** 1043-1052
Multifilamentary tapes
production **A7:** 637
Multifilamentary wire
production **A7:** 637-638
Multiform process **EM4:** 1056
Multifrequency techniques
eddy current inspection **A17:** 173-175, 200
Multifunctional mercaptoesters **EM3:** 626
Multifunctional parts
design of **EL1:** 124
Multigrade oil
defined **A18:** 13
Multijet attachments for flame cutting **A7:** 843
Multilayer boards (MLBS)
design, and via density **EL1:** 613
design requirements **EL1:** 516
Multilayer boards (MLBS) (continued)
fabrication , development **EL1:** 510
printed, defined **EL1:** 1150
resins and reinforcements **EL1:** 534-535
rework processes **EL1:** 712
sequential design **EL1:** 621
Multilayer ceramic capacitors
fabrication **EL1:** 185-187
Multilayer ceramic capacitors applications **A7:** 151
Multilayer ceramic packages
die attachments **EL1:** 213-217
and substrates **EL1:** 204-206
types **EL1:** 213
Multilayer ceramics, cofired
as substrated material **EL1:** 106
Multilayer chip inductors **EL1:** 179, 187
Multilayer dielectrics
thick-film **EL1:** 341-342
Multilayer hole-to-internal-feature registration
rigid printed wiring boards **EL1:** 552
Multilayer inner layer processes
baking **EL1:** 541
buried via inner layers **EL1:** 543
developing **EL1:** 542
etch and resist removal **EL1:** 542
imaging **EL1:** 542
photoresist, application **EL1:** 542
printing **EL1:** 542
registration holes, creating **EL1:** 542
standard **EL1:** 539
surface treatment **EL1:** 543
testing/inspection/repair **EL1:** 542-543
Multilayer interconnect technology **EL1:** 249, 347
Multilayer lamination
rigid printed wiring boards **EL1:** 543-544
Multilayer printed wiring board
development **EL1:** 631
Multilayer(s)
devitrifying dielectrics for **EL1:** 109
rigid printed wiring boards **EL1:** 539
stack-up, rigid printed wiring boards **EL1:** 543
Multilevel hybrid structures
thin-film **EL1:** 322-324
Multimaterial structures
thermal failures **EL1:** 59
Multimet
composition **A4:** 794
Multimet (N-155)
composition **A6:** 564 **M6:** 354
Multimet alloy
metal dusting of **A13:** 1313
Multioriented prepregs *See* Multidirectional tape prepregs
Multipair telephone cables **A13:** 1127
Multiparameter techniques
eddy current inspection **A17:** 175
Multipass weldments **A1:** 603, 605
Multiphase alloys
developing electropolishing procedures for **A9:** 50-51
microstructure development in **A8:** 177-178
torsion test to simulate die chilling effects on workability on **A8:** 180
Multiphase ceramics
SIMS phase distribution analysis **A10:** 610
Multiphase materials
image analysis of **A10:** 309
neutron diffraction analysis **A10:** 420
XRPD-determined weight fraction of crystalline phases in **A10:** 333
Multiphase microstructures **A9:** 604
Multiphase oxides
alumina/zirconia **EM4:** 770
dielectric properties **EM4:** 771-772
effect of isovalent substitutions on transition temperature **EM4:** 771
equations of state for poled piezoelectric ceramics **EM4:** 772
flexural strength as a function of temperature **EM4:** 772
mullite/zirconia **EM4:** 770-773
pseudobinary, subsolidus $PbZrO_3$-$PbTiO_3$ diagram **EM4:** 772
resistivity coefficient effect **EM4:** 771
Multiphase spinals
as inclusions **A11:** 322

Multiple *See also* Blank; Cutoff; Multiple dies
defined ... **A14:** 9
Multiple coils
eddy current inspection ... **A17:** 176
Multiple cracking *See also* Crack(s); Cracking
from corrosion fatigue ... **A11:** 79
initiation, ratchet marks from ... **A11:** 77
Multiple die pressing ... **A7:** 8
Multiple dies ... **A14:** 51
for blanking ... **A14:** 455-456
for closed-die forging ... **A14:** 44
for drawing ... **A14:** 579-580
Multiple discontinuities
defined ... **A17:** 663
Multiple discrete free-fall jet atomization nozzles ... **A7:** 26-27
Multiple etching
defined ... **A9:** 12
Multiple fiber fracturing phenomenon ... **EM3:** 394-396
Multiple heats ... **A8:** 130
Multiple internal reflectance (MIR) spectroscopy ... **A18:** 461
Multiple lead heating and pulling method
component removal ... **EL1:** 717-718
Multiple population means ... **A8:** 711-712
Multiple punch presses ... **A7:** 8
Multiple ring-liner hot isostatic pressure vessels ... **A7:** 420
Multiple scattering
effects, in EXAFS analysis ... **A10:** 415
event, defined ... **A10:** 677
Multiple target sputtering
as thin-film deposition technique ... **A2:** 1082
Multiple test piece method
for shear stress ... **A8:** 182-184
Multiple-alkylated cyclopentane (MAC), high-vacuum lubricant applications ... **A18:** 155, 156
molecular structures ... **A18:** 156
properties ... **A18:** 155
Multiple-beam interferometer *See also* Interferometers
principle of ... **A17:** 16
Multiple-cavity die *See* Combination die
Multiple-coat paint systems
for filiform corrosion ... **A13:** 108
Multiple-component
alloys, platinum group, as electrical contact materials ... **A2:** 848
composites, as electrical contact materials ... **A2:** 856
Multiple-crevice assembly testing ... **A13:** 305-308
Multiple-diameter reamers ... **A16:** 241, 245
Multiple-electrode resistance welding machines ... **M6:** 475
Multiple-expansion volume introduction system
gas mass spectrometer ... **A10:** 152
Multiple-exposure interferometry *See also* Optical holography
as optical holographic interferometry ... **A17:** 408
Multiple-girder diaphragm connection plates ... **A11:** 712-714
Multiple-hearth furnace
for sand reclamation ... **A15:** 227-228
Multiple-impulse weld timer
definition ... **M6:** 11
Multiple-impulse welding
definition ... **M6:** 11
Multiple-index holographic contouring ... **A17:** 428
Multiple-layer adhesive ... **EM3:** 19
Multiple-mandrel ring rolling mills ... **A14:** 111, 114
Multiple-mode failure *See also* Mixed mechanisms; Mixed mode; Mode
in boilers and steam equipment ... **A11:** 626
of pressure vessels ... **A11:** 650
Multiple-motion adjustable stop presses ... **A7:** 335
Multiple-motion die set presses ... **A7:** 335
Multiple-operation machining ... **A16:** 366-403
Al alloys ... **A16:** 761, 764, 783, 793, 797
automatic guided vehicle (AGV) material-handling devices ... **A16:** 401, 402
automatic lathes ... **A16:** 366, 367, 368, 384
automatic shape turners ... **A16:** 368
automatic tracers ... **A16:** 11-11-11, 368
automatic turret indexing ... **A16:** 371-372, 376
automatic turret lathes ... **A16:** 371-372, 373, 377, 381, 387
choice of production techniques ... **A16:** 401-402
CNC machine tools ... **A16:** 393, 397
CNC Swiss-type automatic bar machines ... **A16:** 375-376, 378
Cu alloys ... **A16:** 815
cutting fluids ... **A16:** 384, 386, 392, 393
dimensional control ... **A16:** 390-392
direct numerical control (DNC) ... **A16:** 403
engine lathes ... **A16:** 367, 368, 370, 380, 383, 384
flexible manufacturing systems ... **A16:** 366-367, 397-403
form tools ... **A16:** 380-381
machine classifications ... **A16:** 368
machining centers ... **A16:** 366, 368, 389, 390, 391, 393-394, 398, 399, 400
manpower requirements for system ... **A16:** 402-403
manual turret lathes ... **A16:** 369-371, 373, 387
material handling system ... **A16:** 397
Mg alloys ... **A16:** 826
multifunctional systems ... **A16:** 366
NC lathes ... **A16:** 368
NC machines ... **A16:** 394, 398, 399, 402, 403
Ni alloys ... **A16:** 840, 841
process capabilities ... **A16:** 369
production techniques ... **A16:** 400
ram-type turret lathes ... **A16:** 370
saddle-type turret lathes ... **A16:** 370
safety and protection ... **A16:** 392-393
screw machines ... **A16:** 367, 369, 373, 375
selection of equipment and procedure ... **A16:** 383-388
single-spindle automatic bar and chucking machines ... **A16:** 371-374
single-spindle automatic lathes ... **A16:** 367-369
Swiss-type automatic bar machines ... **A16:** 374-378
tool adapters and mountings ... **A16:** 381-382
tool life ... **A16:** 369, 381, 386, 388-389
tool material and design ... **A16:** 388-389
tools, standard cutting ... **A16:** 379-380
transfer machines (transfer lines) ... **A16:** 366, 393, 394, 395-397, 398
vertical multiple-spindle automatic chucking machines ... **A16:** 378-379, 382, 384
work-in-process inventory ... **A16:** 397-398
workpiece supports ... **A16:** 382-383
Multiple-part dies
applications ... **A14:** 51
Multiple-part forming
multiple-slide ... **A14:** 571
Multiple-pass weld
epitaxial growth in ... **A9:** 579
Multiple-plate systems
copper plating ... **M5:** 161, 169
Multiple-pressure sources ... **A18:** 529
Multiple-ram presses ... **A14:** 34-35
Multiple-roll rotary straightening ... **A14:** 691-693
entry and delivery tables ... **A14:** 692-693
middle-roll offset ... **A14:** 692
roll angle ... **A14:** 692
tube deflection ... **A14:** 692
Multiple-screw extruders ... **EM2:** 28, 383
Multiple-slide forming
applicability ... **A14:** 567
assembly operations ... **A14:** 573
blanking ... **A14:** 569-570
defined ... **A14:** 567
forming ... **A14:** 570-573
of high-carbon steels ... **A14:** 559
machines ... **A14:** 502, 567-569, 573-574
Multiple-slide forming (continued)
operations by ... **A14:** 567
rotary machines for ... **A14:** 573-574
stainless steels ... **A14:** 766-767
of stainless steels, lubricants for ... **A14:** 519
of wire ... **A14:** 696
Multiple-slide machines ... **A14:** 9, 502, 567-569, 573-574
cutoff unit ... **A14:** 568
forming station ... **A14:** 568-569
for high production ... **A14:** 502
for press forming ... **A14:** 545
press station ... **A14:** 567-568
stock straighteners ... **A14:** 567
stock-feed mechanism ... **A14:** 567
for wire ... **A14:** 696
Multiple-slide press
defined ... **A14:** 9
Multiple-slide rotary forming machines ... **A14:** 573-574
Multiple-source evaporation vacuum coating ... **M5:** 391, 393
Multiple-source holographic contouring ... **A17:** 425-427
Multiple-spindle automatic bar and chucking machines ... **A16:** 376-378
Multiple-station air knockout tool fixture ... **A15:** 505
Multiple-station tooling
cold extrusion ... **A14:** 303
Multiple-step heat treatments
of titanium alloy forgings ... **A14:** 281
Multiple-terminal devices *See also* Three-terminal devices; Two-terminal devices
performance ... **EL1:** 423
power modules ... **EL1:** 432-434
small-signal/opto modules ... **EL1:** 432
Multiple-wavelength holographic contouring ... **A17:** 427-428
Multiple-zone resistance wound tube furnaces ... **A8:** 36
Multiplet splitting
in XPS analysis ... **A10:** 572
Multiplex advantage
in FT-IR spectroscopy ... **A10:** 112
in Raman spectroscopy ... **A10:** 129
Multiplexed system
eddy current inspection ... **A17:** 174
Multiplier phototube *See* Photomultiplier tube
Multiport nozzle
definition ... **M6:** 12
Multiport nozzle (plasma arc welding and cutting)
definition ... **A6:** 1211
Multipurpose shear machines ... **A14:** 715-716
Multiroll rotary straighteners ... **A14:** 686-687
Multislide forming
lubricants for ... **A14:** 519
Multistage cleaning and chemical treatment ... **EM3:** 34
Multistage drop hammer forming ... **A14:** 655
Multistation automatic swaging transfer machine ... **A14:** 136
Multivariate analysis methods ... **A8:** 653
Municipal solid waste
boiler hot corrosion with ... **A13:** 997-998
Muntz metal *See also* Copper alloys, specific types, C28000; Wrought coppers and copper alloys ... **A6:** 752
antimonial leaded *See* Copper alloys, specific types, C36700
antimonial leaded, applications and properties ... **A2:** 311
applications and properties ... **A2:** 304-305
arsenical, applications and properties ... **A2:** 311
arsenical leaded *See* Copper alloys, specific types, C36600
characteristics of ... **M5:** 285
free-cutting *See* Copper alloys, specific types, C37000 ... **A2:** 311

Muntz metal (continued)
inhibited leaded *See* Copper alloys, specific types, C36600, C36700 and C36800
leaded *See* Copper alloys, specific types, C36500, C36600, C36700 and C36800
phosphorized leaded *See* Copper alloys, specific types, C36800
phosphorized leaded, applications and properties **A2:** 311
resistance spot welding **A6:** 850
uninhibited leaded *See* Copper alloys, specific types, C36500
uninhibited leaded, applications and properties **A2:** 311
weldability **A6:** 753
Muntz metal (60-40 lead-free brass)
for flame head for oxy-fuel gas flame heating **A4:** 274
Muntz metal, 60%, microstructure of .... **A3:** 1 • 22
Munz's straight-through crack assumption **EM4:** 602
Murakami's reagent
etching with **A11:** 404
Murakami's reagent as etchant for
cemented carbides, classification of reaction rates **A9:** 274
copper-tungsten mixtures **A9:** 551
electrical contact materials **A9:** 551
heat-resistant casting alloys **A9:** 331-332
molybdenum **A9:** 440-441
nitrided steels **A9:** 218
rhenium and rhenium-bearing alloys **A9:** 447
tungsten **A9:** 440-441
wrought stainless steels **A9:** 281;282
Murexide
as metallochromic indicator **A10:** 174
Muriatic acid
description **A9:** 68
Muscovite
chemical composition **A6:** 60
Muscovite mica **EM4:** 6
Mush buffing **M5:** 116, 119
Mushet, Robert
as alloy developer **A15:** 32
Mushrooming **A6:** 42
Music spring steel wire **A1:** 285
Music wire *See also* Steels, ASTM specific types, A
carbon steel wire rod for **M1:** 255
characteristics **M1:** 266, 289
characteristics of **A1:** 306
cost **M1:** 305
modulus of rigidity **M1:** 300
seams in **M1:** 290
size of **A1:** 277
stress relieving **M1:** 291
tensile strength ranges **M1:** 268
Music wire gage (MWG) **A1:** 277
Music Wire Gage system **M1:** 259
Music-wire spring
fatigue fracture of **A11:** 557
Mussels
as biofouling organism **A13:** 88
Mutagenicity
of platinum complexes **A2:** 1258
Mutual solubility
of copper and silver **A11:** 452
high-nickel base materials **A11:** 452
MVT *See* Moisture vapor transmission
MY-10B
properties **A18:** 549
MY-10K
properties **A18:** 549
Mycor
seals to metals **EM4:** 499
Mykroy
seals to metals **EM4:** 499
Mylar **EM4:** 161, 162
transmission by **A10:** 100
N-Methyl pyrrolidone (NMP)
to solvate polyimides **EM3:** 152

N

3% Ni cast iron
galling resistance with various material combinations **A18:** 596
Copper-nickel-chromium plating *See* Copper-nickel-chromium plating
N *See* Normal solution; Refractive index
N shell
defined **A10:** 677
***n* value** *See also* Strain-hardening coefficient; Work-hardening coefficient
determining, uniaxial tensile testing **A8:** 556-557
effect on plane strain **A8:** 551
effect on wrinkling **A8:** 551
role in strain distribution, sheet metal forming **A8:** 550
of sheet metals **A8:** 555-557
N(E) *See* Electron energy distribution
N-4 alloy *See* Copper alloys, specific types, C15000
N-100
as synchrotron radiation source **A10:** 413
N-155
aging **A4:** 796
aging cycle **M4:** 656
annealing **M4:** 655
arc welding **M6:** 364-365
composition **A4:** 794 **A16:** 736 **M4:** 651-652 **M6:** 354
electron-beam welding **A6:** 869
flash welding **M6:** 557
machining **A16:** 738, 741-743, 746-747, 749-758
mill annealing **A4:** 810
solution annealing **A4:** 810
solution treating **M4:** 656
solution-treating **A4:** 796
stress relieving **M4:** 655
N-benzoyl-N-phenylhydroxylamine (BPA)
as precipitant **A10:** 169
n-butanol
surface tension **EM3:** 181
n-butyl acetate
surface tension **EM3:** 181
N-butyl alcohol
description **A9:** 68
n-butyraldehyde
physical properties **EM3:** 104
n-channel metal-oxide semiconductors (NMOS) **EL1:** 76, 160
N-methyl-pyrrolidine (NMP)
to solvate polyimides **EM3:** 152
N-N-dimethyl + saccharin *p*-toluidine
generating free radicals for acrylic adhesives **EM3:** 120
N-Nb (Phase Diagram) .... **A3:** 2 • 298
N-Ni (Phase Diagram) .... **A3:** 2 • 298
N-phenyl carbazole
absorption and fluorescence emission spectra **A10:** 75
N-phenylbenzo-hydroxamic acid
as solvent extractant **A10:** 170
n-propanol
surface tension **EM3:** 181
N-Ta (Phase Diagram) .... **A3:** 2 • 299
N-Th (Phase Diagram) .... **A3:** 2 • 299
N-Ti (Phase Diagram) .... **A3:** 2 • 299
N-type oxides
alloy oxidation **A13:** 73
impurities effect **A13:** 69
semiconductor, defect structure **A13:** 66
n-type transistors
development **EL1:** 958
N-U (Phase Diagram) .... **A3:** 2 • 300
N-Zr (Phase Diagram) .... **A3:** 2 • 300
Na (Phase Diagram) .... **A3:** 2 • 300
Na-Pb (Phase Diagram) .... **A3:** 2 • 301
Na-Rb (Phase Diagram) .... **A3:** 2 • 301
Na-S (Phase Diagram) .... **A3:** 2 • 301
Na-Sb (Phase Diagram) .... **A3:** 2 • 302
Na-Se (Phase Diagram) .... **A3:** 2 • 302
Na-Sn (Phase Diagram) .... **A3:** 2 • 302
Na-Sr (Phase Diagram) .... **A3:** 2 • 303
Na-Te (Phase Diagram) .... **A3:** 2 • 303
Na-Tl (Phase Diagram) .... **A3:** 2 • 303
NA22H
for radiant-tube-heated furnace equipment **A4:** 473
NAA *See* Neutron activation analysis
Nabarro-Herring diffusional creep **EM4:** 296
NACE standards
for testing waters/materials in water **A13:** 208
Nacrite **EM4:** 5, 6
Nadimide-terminated thermosetting polymers **EM2:** 631
Nailheading **EL1:** 575, 1151
Nails
steel wire **M1:** 271
NAND gate
defined **EL1:** 1151
Nanocoulomb
abbreviation for **A10:** 691
Nanogram
abbreviation for **A10:** 691
Nanohardness test
defined **A18:** 13
Nanoindentation **A18:** 419-427
correlation (in practice) between continuous depth recording (CDR) and micro/nanoindentation **A18:** 419
definition **A18:** 419
future trends **A18:** 427
information, type obtained **A18:** 419
instruments **A18:** 419-421
basic requirements **A18:** 420-421
commercial specifications **A18:** 420
options **A18:** 420, 421
other physical measurements **A18:** 421
test procedures **A18:** 421-427
averaging of multiple tests **A18:** 423-424
deformation measurement choice **A18:** 421-422
flow behavior **A18:** 426-427
flow measurement choice **A18:** 421-422
hardness and modulus **A18:** 424-426
slow-loading test **A18:** 422-423, 426
Nanometer
abbreviation for **A10:** 691
Nanosecond
abbreviation for **A10:** 691
Naphtha
as organic cleaning solvent **A12:** 74
Naphtha cleaners **M5:** 40, 42-43, 617
Naphtha, contaminated
copper alloy corrosion **A13:** 635
Naphtha desulfurization reactor
cracking from hydrogen damage **A11:** 664
Naphthalene **A10:** 78, 643
Naphthalene, as pattern material
investment casting **A15:** 255
Naphthalene-1.5-diisocyanate (NDI)
in polyurethanes **EM2:** 257
Naphthenic acids **A13:** 271
Napkin ring test **A18:** 404 **EM3:** 321, 322, 361, 441, 443-444
Napped cloth
defined **A9:** 12
Narrow-beam geometry **A17:** 309-311
Narrowing exchange
as ESR line-broadening mechanism **A10:** 255
NASA
integrated advanced materials data base **EM4:** 692
NASA Co- W-Re
composition **A16:** 737
machining **A16:** 738, 741-743, 746-757
NASA Co-W-Re
composition **A4:** 795 **A6:** 929
NASA Lewis Research Center
directed Small Engine Components Technology Studies **EM4:** 716
NASA/CARES computer program *See* Probabilistic design of ceramic components NASA/CARES computer program
Nascent surface
defined **A18:** 13
Nasmythe, James
as early founder **A15:** 28, 33
NASTRAN computer program for structural analysis **EM1:** 268, 271
NASTRAN finite-element analysis code **EM3:** 479, 486
National Advisory Committee for Aeronautics **A8:** 725
National Aeronautics and Space Administration
structural adhesives development **EM3:** 513
National Aeronautics and Space Administration (NASA)
advanced composite standard tests **EM1:** 553
compression specimen, for damage tolerance testing **EM1:** 264

National Aeronautics and Space Administration (NASA) (continued)
gallium research A2: 747
as information source EM1: 41, 42
National Aerospace and Defense Contractor's Accreditation Program (NAD-CAP) EM3: 62
National Association of Corrosion Engineers A8: 725
National Association of Pattern Manufacturers A15: 193
National Bureau of Standards A8: 90, 725
SRM, microprobe analysis A10: 530
Standard Reference Material (NBS SRM) A10: 147
testing machine calibrator from A8: 615
National Bureau of Standards Voluntary Engineering Standards (data base) EM4: 40
National Casting Council A15: 34
National Electric Code (NEC)
main disconnect switch provision for installed welding machines A6: 36
National Electrical Manufacturers Association (NEMA) EM1: 75
arc welding power source classes (3) A6: 36
National Fire Protection Association
codes for metal powder plant operations A7: 198
National Institute of Ceramic Engineers (NICE)
as information source EM1: 41
National Institute of Standards and Technology (NIST)
computerized data base aiding x-ray photoelectron spectroscopy (XPS) EM3: 237
National Lubricating Grease Institute (NLGI)
consistency grades of greases A18: 99
grease classification grades defined in specification IID A18: 125
grease consistency classes A18: 136
service classifications of greases A18: 99
National Marine Manufacturers Association (NMMA)
oil certification A18: 85
performance testing of engine oils A18: 170
National Material Advisory Board Committee
National Research Council EM1: 101
National Materials Advisory Board A8: 187
National Particleboard Association
formaldehyde emission limit EM3: 105
National SAMPE Technical Conference/SAMPE Symposium and Exhibition EM2: 95
National Standards Association (NAS) EM1: 701
National standards laboratory EM3: 19
defined EM2: 28
National Technical information Service (NTIS) EM1: 41 EM4: 40
Natural (direct) recovery processes
plating waste treatment M5: 315-316
Natural abundance
of isotopes A10: 643
Natural aging *See also* Aging; Artificial aging
aluminum-lithium alloys A2: 184
defined A9: 12 A13: 9
wrought aluminum alloy A2: 39-40
Natural circulation loop
and salt sampler A13: 50
Natural clay earthenware EM4: 3, 4
Natural convection *See also* Connection
from vertical plane, formulas EL1: 52
Natural draft
defined A14: 9
Natural fibers
forms EM1: 175
history/application EM1: 117
and synthetic, compared EM1: 117
Natural frequency
and damping EM1: 212-213
Natural frequency (continued)
of load cells A8: 193
Natural gas
analytic method for A10: 11
chemical bonding M6: 900
fuel gas for oxyfuel gas cutting A6: 1157-1158, 1161, 1162
combustion ratio A6: 1157-1158
cost A6: 1158
heat content A6: 1158
parameters for machine oxynatural gas drop cutting of shapes from low-carbon steels A6: 1158, 1159
parameters for machine oxynatural gas shape cutting of carbon steels A6: 1158, 1159
recommended parameters for machine OFC-A of carbon steels A6: 1158
fuel gas for torch brazing M6: 950
for oxyfuel gas cutting A14: 723
oxyfuel gas welding fuel gas A6: 281, 283, 285
physical properties as atmosphere A7: 341
properties as fuel gas M6: 899
use for cutting M6: 901, 902, 903
use in oxyfuel gas welding M6: 584
Natural logarithm (base *e*)
abbreviation for A10: 690
symbol for A11: 797
Natural logarithm base
symbol for A8: 724
Natural resources recovery
with cemented carbides A2: 974-977
Natural rubber (NR) EM3: 75, 594
for art material cement EM3: 145
for automotive interior trim bonding EM3: 145
for belting EM3: 145
characteristics EM3: 90
for commercial tape foundation EM3: 145
exposure in electrochemically inert conditions EM3: 629
for footwear EM3: 145
for hoses EM3: 145
for latex compounds in paper EM3: 145
for leather shoe production EM3: 145
for off-the-road tires EM3: 145
for retread and tire patch compounds EM3: 145
substrate cure rate and bond strength for cyanoacrylates EM3: 129
for surgical plaster foundation EM3: 145
for textile binding EM3: 145
Natural rubber latex
properties EM3: 88
Natural sands *See also* Sands; Synthetic sands
vs synthetic sands A15: 208
Natural stone
diamond for machining A16: 105
Natural stoneware EM4: 3, 4
versus tile whiteware EM4: 929
Natural strain *See* True strain
Natural uranium *See also* Uranium; Uranium alloys
fissionable isotope U-235 and U-238 ..... A2: 670
processing of A2: 670
Natural waters *See also* Water
aluminum-zinc alloy coating in A13: 435-436
aluminum/aluminum alloys in A13: 597
analysis of A10: 41
chloride-containing, crevice corrosion A13: 112
corrosion A13: 17
corrosion of steel in M1: 736-738
for corrosion testing A13: 208
lead corrosion in A13: 785
zinc/zinc alloys and coatings in A13: 762
Naturalizers
as aqueous cleaners EL1: 664
Naturally bonded molding sand *See also* Molding sands; Sand(s)
defined A15: 8
early practice with A15: 28
Naturally occurring substances
as system favorable to ESR analysis A10: 263
Naval Air Systems Command EM3: 62
Naval aircraft
failure of aluminum catapult-hook attachment fitting for A11: 88, 91
Naval brass A6: 752
applications and properties A2: 319-322
brazeability M6: 1033-103
brazing A6: 629-630
projection welding M6: 50
weldability A6: 753
Naval grade aluminum
as honeycomb core material EM1: 723
Naval Publications and Forms Center (NPFC)
source of standards EM3: 63, 64
Naval Research Laboratory
drop-weight tear test (DWTT) A11: 61
Navier-Stokes equations
defined A18: 13
fluid flow A15: 867
Navigational aids
composite materials for EM1: 839
Navy G-bronze
properties and applications A2: 352, 376
Navy M bronze *See also* Leaded tin bronzes
properties and applications A2: 375-376
Navy M copper casting alloy
nominal composition A2: 347
Navy tear test A8: 259
Navy tear-test
compared with Robertson and Esso (Feely) A11: 59-60
for notch toughness A11: 60
Navy Tombasil *See* Copper alloys, specific types, C87200
properties and applications A2: 371-372
Nb tube process
for A15 superconductor assembly A2: 1066-1067
Nb-10Hf
physical properties A6: 941
Nb-28Ta-10W
physical properties A6: 941
Nb-Ni (Phase Diagram) ..... A3: 2 • 304
Nb-Os (Phase Diagram) ..... A3: 2 • 304
Nb-Pd (Phase Diagram) ..... A3: 2 • 304
Nb-Pt (Phase Diagram) ..... A3: 2 • 305
Nb-Rh (Phase Diagram) ..... A3: 2 • 305
Nb-Ru (Phase Diagram) ..... A3: 2 • 305
Nb-Si (Phase Diagram) ..... A3: 2 • 306
Nb-Ta (Phase Diagram) ..... A3: 2 • 306
Nb-Th (Phase Diagram) ..... A3: 2 • 306
Nb-Ti (Phase Diagram) ..... A3: 2 • 307
Nb-Ti-W (Phase Diagram) ..... A3: 3 • 57
Nb-U (Phase Diagram) ..... A3: 2 • 307
Nb-V (Phase Diagram) ..... A3: 2 • 307
Nb-W (Phase Diagram) ..... A3: 2 • 308
Nb-Zr (Phase Diagram) ..... A3: 2 • 308
NBD100
mechanical properties A18: 774
NBD200
rolling contact fatigue properties A18: 260-261
NBS *See* National Bureau of Standards
NBS-traceable plots
of AE sensor sensitivity A17: 280
NC machining *See* Numerical control machining
ND *See* Normal direction (of a sheet)
ND PROP computer program for structural analysis EM1: 268, 271
Nd-Fe-B permanent magnets
as rare earth application A2: 730
Nd-Ni (Phase Diagram) ..... A3: 2 • 308
Nd-Pt (Phase Diagram) ..... A3: 2 • 309
Nd-Rh (Phase Diagram) ..... A3: 2 • 309
Nd-Sb (Phase Diagram) ..... A3: 2 • 309
Nd-Si (Phase Diagram) ..... A3: 2 • 310
Nd-Sn (Phase Diagram) ..... A3: 2 • 310
Nd-Te (Phase Diagram) ..... A3: 2 • 310
Nd-Ti (Phase Diagram) ..... A3: 2 • 311

Nd-Tl (Phase Diagram) .... **A3:** 2 • 311
Nd-YAG industrial laser **A14:** 736-737
Nd-Zn (Phase Diagram) .... **A3:** 2 • 311
NDE *See* Nondestructive evaluation
NDE detection methods *See also* Detection; NDE reliability; Nondestructive evaluation Nondestructive evaluation methods Nondestructive evaluation techniques Nondestructive inspection of specific products; Quantitative nondestructive evaluation; specific nondestructive evaluation methods
for planar flaws **A17:** 50
for surface/interior flaws **A17:** 50
for volumetric flaws **A17:** 50
NDE digital image enhancement systems **A17:** 454-455
NDE engineering
inspection personnel **A17:** 677
procedure selection/development **A17:** 675
reference standards **A17:** 676-677
signal/noise relationships **A17:** 676
system/process performance characteristics **A17:** 675
NDE reliability *See also* Fracture control; NDE reliability applications; NDE reliability data analysis; NDE reliability models Nondestructive evaluation methods Nondestructive evaluation techniques Process control; Statistical methods
applications, to systems **A17:** 674-688
conditional probability in **A17:** 675
data analysis **A17:** 689-701
demonstration program design, and POD function **A17:** 664
flaw size criteria **A17:** 664
fracture control philosophy **A17:** 666-673
history **A17:** 663-664
introduction **A17:** 663-665
NDE response **A17:** 674
performance, by POD curves **A17:** 679
possible outcomes **A17:** 675
prediction models **A17:** 702-715
procedure selection/development **A17:** 675
system management and schedule **A17:** 674-675
system/process performance characteristics **A17:** 675
NDE reliability applications *See also* NDE reliability
airframes, as case study **A17:** 680-681
applications (case studies) **A17:** 680-688
engineering effects **A17:** 674
gas turbine engines, as case study **A17:** 681-684
integrated blade inspection system **A17:** 686-687
NDE process control **A17:** 677-680
NDE response **A17:** 674
NDE system management and schedule **A17:** 674-675
prior art **A17:** 674
retirement-for-cause (RFC) inspection equipment **A17:** 687-688
space shuttle program, as case study **A17:** 685-686
special inspection systems **A17:** 686
structural assessment testing applications **A17:** 686
trend identification **A17:** 731-732
NDE reliability data analysis *See also* NDE reliability **A17:** 689-701
of hit/miss data **A17:** 695-696
maximum likelihood analysis **A17:** 694-695
NDE process, statistical nature **A17:** 689-692
NDE reliability experiments, design of **A17:** 692-694
of signal response **A17:** 697-700
NDE reliability models *See also* NDE reliability
of eddy current inspection **A17:** 707-709
essential characteristics **A17:** 703
examples **A17:** 712-713
and fracture control **A17:** 702
future impact of **A17:** 711-713
optimization, with artificial intelligence **A17:** 713
predictive **A17:** 702-715
of radiographic inspection **A17:** 709-711
standardization of **A17:** 713
of ultrasonic inspection **A17:** 703-707
NDI *See* Nondestructive inspection
NDT *See* Nondestructive testing
NDT-210 bond tester
for adhesive-bonded joints **A17:** 620
NDTT *See* Nil ductility transition temperature; Nil-ductility transition temperature
NDZ *See* Copper alloys, specific types, C99400
NDZ-S *See* Cast copper alloys, specific types (C99500)
Near East
Bronze Age in **A15:** 15-16
Near net shape(s) *See also* Net shape
by cold isostatic pressing **A7:** 448
forging and rolling **A7:** 522-529, 649
forging, weight savings **A7:** 649
HIP processing for **A7:** 424, 437
isothermal forging and rolling **A7:** 522
production by rapid omnidirectional compaction **A7:** 545
for titanium alloy complex shapes **A7:** 41, 44, 546
of titanium parts **A7:** 450, 546
Near-alpha alloys
titanium **A14:** 839
Near-alpha titanium alloys **A9:** 458
Near-beta titanium alloys **A9:** 458
beta flecks in **A9:** 459-460
Near-edge structure
determined **A10:** 407, 415-416
Near-end cross talk *See also* Cross talk
defined **EL1:** 36
Near-field effects
ultrasonic beams **A17:** 239
Near-infrared
abbreviation for **A10:** 690
radiation, defined **A10:** 677
regions of spectrum, FT-IR analysis **A10:** 114
Near-net
resin content prepregs, epoxy **EM1:** 73
shape manufacturing, three-dimensional braiding **EM1:** 519
Near-net shape
by isothermal/hot-die forging **A14:** 150-157
by precision forging **A14:** 158
die temperature, and yield strength **A14:** 153
forging, defined **A14:** 158
forging design parameter **A14:** 154
titanium alloy precision forging **A14:** 285
Near-net shape casting
by rheocasting **A15:** 327
by squeeze casting **A15:** 323
market trends **A15:** 44
organic binder effect **A15:** 35
parts, metal-matrix composites for **A15:** 338
Near-net shape processes *See* Hot-die forging; Isothermal forging; Near-net shape; Net shape
Near-net shape technologies
beryllium powder **A2:** 685-686
titanium and titanium alloy castings **A2:** 634
Near-perfect crystal
diffraction in **A10:** 365-367
Near-surface
analysis **A10:** 126, 133-137
defects, single crystals **A10:** 633
stresses, effect of stress gradient correction of measurement of **A10:** 388
Near-threshold fatigue crack growth
pH effect in stainless steel and titanium alloy **A8:** 427, 430
rate, titanium alloy, oxygen content effect **A8:** 429, 430
Near-threshold fatigue crack propagation
hydrazine effect on **A8:** 427, 429
Near-threshold regime
crack closure relevancy **A8:** 408
Nearest layer model
Raman vibrational behavior of intercalated graphites **A10:** 133
Nearest neighbors, carbon
liquid iron-carbon alloys **A15:** 168
Nearest-neighbor atoms **A9:** 681-682
Nearly dense P/M stainless steels
mechanical properties **A7:** 471
Neat form
thermoplastic resins **EM2:** 98
Neat oil
defined **A18:** 13
Neat resin *See also* Matrices; Neat form; Resins **EM3:** 19
defined **EM1:** 16 **EM2:** 28
Neat resin (continued)
properties, and composite performance **EM1:** 98-99
systems, formulations **EM1:** 290-291
Nebulizers
for AAS and related techniques **A10:** 44-50
for analytical ICP systems **A10:** 34-36
argon gas flow as **A10:** 34
Babington **A10:** 36
concentric **A10:** 34, 35
conductive solids **A10:** 36, 690
corrosion-resistant **A10:** 35
crossflow **A10:** 34, 35
defined **A10:** 677
effects in flame spectroscopy **A10:** 29
efficiency of, defined **A10:** 35
fritted disk **A10:** 36, 55
for ICP-MS **A10:** 40
pneumatic **A10:** 34-36, 47
ultrasonic **A10:** 36, 55
NEC SX liquid-cooling module **EL1:** 48-49
Neck *See also* Localized deformation; Necking
defined **A7:** 8
development, sheet metal forming **A8:** 548
stress distribution at **A8:** 25-26
Neck formation
defined **A7:** 8
Neck liner, failed
SCC in **A11:** 403
Neckdown
as feature imperfection **EL1:** 732
Necking *See also* Localized deformation; Neck **EM3:** 19
aluminum alloys **A14:** 804
average true stress with **A8:** 25
in compression tests **A8:** 577
defined **A8:** 9 **A11:** 7 **A14:** 9 **EM1:** 16 **EM2:** 28
diagnosis by circle grid analysis **A8:** 566
and ductility measurement **A8:** 26
due to creep **A8:** 302
from creep, nickel-base alloy **A11:** 283
from steady-state creep **A8:** 308
geometry of **A8:** 26
in locomotive axle failures **A11:** 716-717
onset, and instability in tension **A8:** 25
in polymers **A11:** 760
for reducing diameter **A14:** 586
in sheet metal forming **A8:** 548
of split-Hopkinson bar test specimen **A8:** 214
strain, true local **A8:** 24-25
in tensile-test fractures **A12:** 98-105
in tension tests **A8:** 578
as thinning, in sheet metals **A8:** 551
triaxial stress in tensile specimen by **A8:** 25
of wrought aluminum alloys **A2:** 41
Necklace pattern
for optimum superalloy performance **A7:** 523
Needle bearing
defined **A18:** 13
Needle blow
defined **EM2:** 28
Needle crystals
growth of **A15:** 112, 113
paraboloids of revolution **A15:** 113
Needle roller bearings **A18:** 511
f_v factors for lubrication method **A18:** 511
Needle roller with machined rings bearings
basic load rating **A18:** 505
Needle tissue biopsy
GFAAS detection of metals in **A10:** 55
Needle-roller bearing
described **A11:** 490
drawn-cup, gross overload failure **A11:** 505
Needle/cone penetration test
butyl tapes **EM3:** 202
Needled
felt, multidirectionally reinforced **EM1:** 130
mat, defined **EM1:** 16
Needled mat
defined **EM2:** 28
Needles
defined **A7:** 8
diffraction pattern technique measurement **A17:** 13
Neel point *See* Neel temperature
Neel temperature
defined **A10:** 677

Negative absorption
fluorescent enhancement as ... **A10:** 98
Negative carbon extraction replication
defined ... **A17:** 54
Negative creep
in austenitic stainless steels ... **A8:** 331
in Nimonic 80A ... **A8:** 331
Negative distortion
defined ... **A9:** 12
Negative eyepiece
defined ... **A9:** 12
Negative glow
in glow discharges ... **A10:** 27
Negative ion charge
symbol for ... **A10:** 692
Negative logarithm of hydrogen-ion activity *See* pH
symbol ... **A11:** 797
Negative metal-oxide semiconductor (NMOS)
active component fabrication ... **EL1:** 196-198
process sequence ... **EL1:** 148
Negative phase contrast ... **A9:** 59
Negative replica
defined ... **A9:** 12
Negative tensile strength
in fatigue-crack initiation ... **A11:** 102-103
Negative tensile stress
in fatigue testing ... **A11:** 102
Negative-feedback closed-loop system
for fatigue testing ... **A8:** 395-396, 398
Neilson, James B
as inventor ... **A15:** 30
Nelson diagram ... **A6:** 378, 380
Nematic crystals
for optical imaging ... **EL1:** 1072
Neocuproine method
analysis for copper in iron and steel alloys by ... **A10:** 65
Neodymium *See also* Rare earth metals
internal flaw analysis ... **EM4:** 666
and praseodymium, separation of ... **A10:** 249-250
properties ... **A2:** 1184
pure ... **M2:** 776
as rare earth ... **A2:** 720
TNAA detection limits ... **A10:** 238
-YAG lasers ... **A10:** 142, 597, 601
Neodymium oxide
specific properties imparted in CTV tubes ... **EM4:** 1040
Neodymium: yttrium aluminum garnet lasers
comparison of weld results ... **M6:** 65
laser beam transport ... **M6:** 66
welding applications ... **M6:** 651
Neodymium-doped glass pulsed lasers, parameters
laser beam welding applications ... **A6:** 263
Neodymium-doped yttrium aluminum garnet (Nd:YAG) ... **A6:** 262, 263, 266
Neodymium-doped yttrium aluminum garnet (Nd:YAG) lasers
for optical holographic interferometry ... **A17:** 410, 418
Neodymium-doped yttrium aluminum garnet (Nd:YAG) pulsed lasers, parameters
laser-beam welding applications ... **A6:** 263
Neodymium-doped yttrium aluminum garnet continuous wave lasers, parameters
laser-beam welding applications ... **A6:** 263
Neodymium-doped yttrium aluminum garnet face-pumped laser ... **A6:** 266
Neodymium-doped yttrium aluminum garnet solid-state (Nd:YAG) lasers ... **A6:** 266
Neodymium-doped yttrium-aluminium-garnet (Nd:YAG) laser ... **A10:** 142, 597, 601
Neodymium-glass lasers *See* Laser beam machining
Neodymium-iron-boron alloys *See* Magnetic materials
as permanent magnet materials ... **A2:** 790-792
Neodymium-yttrium aluminum garnet (Nd-YAG) lasers ... **EL1:** 368-369, 737
Neon
adsorbed, effects of ... **A10:** 588
as image gas, for FIM analysis of semiconductor ... **A10:** 601, 602
as imaging gas, FIM/AP analysis of permanent magnet alloy ... **A10:** 599
ionization potentials and imaging fields for ... **A10:** 586
mean free path ... **A17:** 59
mean free path value at atmospheric conditions ... **A18:** 525
Neon, used for ion-beam thinning of transmission electron microscopy
specimens ... **A9:** 107
Neoparies ... **EM4:** 873, 876
Neopentyl polyester
low vapor pressures ... **A18:** 151
properties ... **A18:** 81
Neoprene
as bag material ... **A7:** 447
and flotation tanks for cleaning cutting fluids ... **A16:** 129
Neoprene phenolics ... **EM3:** 104, 105
formulations ... **EM3:** 107
properties ... **EM3:** 106
suppliers ... **EM3:** 104
Neoprene rubber
for highway construction joints ... **EM3:** 57
Neoprene rubber, maskant material
chemical milling of Al alloys ... **A16:** 803
Neoprene systems
phenolic resins as tackifiers ... **EM3:** 182
Neoprenes *See also* Chloroprenes ... **EM3:** 86
applications ... **EM3:** 44, 56
characteristics ... **EM3:** 53, 90
exhibiting polarity and hydrogen bonding ... **EM3:** 181
for extruded gaskets ... **EM3:** 58
sealant characteristics (wet seals) ... **EM3:** 57
silane coupling agents ... **EM3:** 182
substrate cure rate and bond strength, for cyanoacrylates ... **EM3:** 129
for truck trailer joints ... **EM3:** 58
for window sealing ... **EM3:** 56
Nephaline
chemical systems ... **EM4:** 870
Nepheline
as extender ... **EM3:** 175, 176
Nepheline ceramic ... **EM4:** 1101, 1102
composition ... **EM4:** 1101
properties ... **EM4:** 1101
Nepheline syenite ... **EM4:** 6, 32, 44
composition ... **EM4:** 379
flux composition ... **EM4:** 932
iron content ... **EM4:** 379
mining techniques ... **EM4:** 379
purpose for use in glass manufacture ... **EM4:** 381
source ... **EM4:** 379
Nephrotoxicity
of platinum ... **A2:** 1258
Neppiras formula, ultrasonic testing ... **A8:** 243, 250
NEPSAP computer program for structural analysis ... **EM1:** 268, 272
Neptunium
determined by controlled-potential coulometry ... **A10:** 209
pure ... **M2:** 777, 832-833
Neptunium, as actinide metal
properties ... **A2:** 1190
Nernst equation ... **A13:** 9, 30 **EM4:** 252
open-circuit cell potential ... **A15:** 80-81
Nernst equations ... **A10:** 197, 208
Nernst thickness
defined ... **A13:** 9
Nernst, Walter ... **A3:** 1 • 7
Nernst-Einstein relationship ... **EM3:** 436
Nernst-Planck equations ... **EM4:** 461
Nessler tubes
in UV/VIS absorption analysis ... **A10:** 66
Nested cloth *See* Nesting
Nested vias
concept of ... **EL1:** 304
Nesting
of blanks ... **A14:** 450
defined ... **EM1:** 16 **EM2:** 28
Net
defined ... **A14:** 158
Net active anodic potential ranges ... **A9:** 144-145
Net cathodic potential ranges ... **A9:** 144
Net list sorting
as automatic trace routing ... **EL1:** 533
Net, molded *See* Molded net
Net positive suction head (NPSH) ... **A18:** 5
defined ... **A18:** 13
pumps ... **A18:** 597, 599
Net positive suction head available (NPSHA) ... **A18:** 599, 600
Net positive suction head required (NPSHR) ... **A18:** 599, 600
Net present value analysis ... **EM2:** 335
Net scale growth rate ... **A18:** 208
Net section stress
effect of SCC in ... **A8:** 502
Net shape *See also* Near-net shape(s) ... **A7:** 8
by precision forging ... **A14:** 158
disk with internal cavities ... **A7:** 427
forging, defined ... **A14:** 158
machine parts ... **A7:** 295
parts, by isothermal/hot-die forging ... **A14:** 150-157
Net shape casting
by squeeze casting ... **A15:** 323
organic binder effects ... **A15:** 35
technologies, titanium alloy ... **A15:** 824
Net shape investment casting
of nickel and nickel alloys ... **A2:** 429
Net shape technologies ... **EM1:** 35
Net shape technology
beryllium powder ... **A2:** 685
titanium and titanium alloy castings ... **A2:** 634
for titanium P/M products ... **A2:** 647
Net-to-net risks
life cycle testing ... **EL1:** 136
Net-to-node data set
accelerated thermal cycle data generation ... **EL1:** 136-138
Netting analysis
defined ... **EM1:** 16 **EM2:** 28
for filament winding ... **EM1:** 508
of laminates ... **EM1:** 229
Network etching
defined ... **A9:** 12
Network logic circuitry ... **EL1:** 2
Network polymers
chemical structure ... **EM2:** 52
Network resistors ... **EL1:** 178, 184
Network structure ... **A7:** 8
defined ... **A9:** 12
Networked polymers
formation ... **EM1:** 751-752
Networks
continuous second-phase, in forging ... **A11:** 327
embrittling intergranular, in iron castings ... **A11:** 359
grain-boundary ... **A11:** 359
intergranular, in casting ... **A11:** 359
of projections, as casting defects ... **A11:** 381
Neumann bands
defined ... **A9:** 12
as nucleation sites ... **A9:** 694
Neumann-Kopp rule ... **EM4:** 883
Neural networks
future trends ... **EL1:** 393-394
Neurologic effects
of lead ... **A2:** 1243-1244
Neutral bath gold plating ... **M5:** 282
Neutral filter
defined ... **A9:** 12 **A10:** 677

Neutral flame
definition ... **A6:** 1211
Neutral hardening
growth or shrinkage during ... **A7:** 480
Neutral loss scans
gas chromatography/mass spectrometry ... **A10:** 646-647
Neutral oil
defined ... **A18:** 13
Neutral point ... **A18:** 147
Neutral radius ... **A18:** 60, 65
Neutral salts ... **A6:** 337, 338
Neutral-density light filters ... **A9:** 72
Neutral-flame
definition ... **M6:** 1
Neutralization
enameling ... **EM3:** 303
Neutralization process, plating waste disposal ... **M5:** 312-313
equipment ... **M5:** 314
Neutralizers
as inhibitors ... **A13:** 485
Neutron absorber
defined ... **A10:** 677
Neutron absorption
defined ... **A10:** 677
Neutron activation analysis ... **A10:** 233-242
14-MeV fast neutron activation analysis as ... **A10:** 239
applications ... **A10:** 233, 240-241
basic principles and introduction ... **A10:** 234
capabilities, compared with PIXE ... **A10:** 102
defined ... **A10:** 677
epithermal neutron activation analysis as ... **A10:** 239
estimated analysis time ... **A10:** 233
of inorganic liquids and solutions, information from ... **A10:** 7
of inorganic solids, information from ... **A10:** 4-6
limitations ... **A10:** 233
neutron sources ... **A10:** 234
nondestructive thermal neutron activation analysis as ... **A10:** 234-238
of organic liquids and solutions, information from ... **A10:** 10
of organic solids, information from ... **A10:** 9
prompt gamma activation analysis as ... **A10:** 239-240
radiochemical (destructive) TNAA as ... **A10:** 238-239
related techniques ... **A10:** 233
samples ... **A10:** 233-236, 240-241
to analyze the bulk chemical composition of starting powders ... **EM4:** 72-73
for trace elements ... **A2:** 1095
as trace-element assay ... **A10:** 233
uranium assay by delayed-neutron counting ... **A10:** 238-239
Neutron bombardment, effect
E-glass ... **EM1:** 47
Neutron capture *See also* Neutron absorption
prompt γ-ray activation analysis ... **A10:** 239-240
prompt γ-rays ... **A10:** 234
slow or thermal ... **A10:** 234
Neutron cross section
defined ... **A10:** 678
Neutron detector
defined ... **A10:** 678
Neutron diffraction ... **A10:** 420-426
applications ... **A10:** 420, 425-426
capabilities ... **A10:** 333, 365, 407
defined ... **A10:** 678
detection for ... **A10:** 422
general uses ... **A10:** 420
introduction ... **A10:** 421
limitations ... **A10:** 420
monochromators ... **A10:** 413
neutron powder diffraction ... **A10:** 422-423
neutron production ... **A10:** 421-422
pole figure ... **A10:** 423-424
related techniques ... **A10:** 420
residual stress ... **A10:** 424
Rietveld method ... **A10:** 423
samples ... **A10:** 420, 422-423
single-crystal ... **A10:** 424-425
single-crystal capabilities ... **A10:** 344
texture measurements ... **A10:** 357, 423-424
Neutron diffraction (continued)
two modes of ... **A10:** 422
used to investigate crystallographic texture ... **A9:** 701
Neutron economy
effect of beryllium in nuclear reactors ... **A7:** 664
Neutron embrittlement *See also* Embrittlement; Radiation ... **M1:** 686-687
defined ... **A11:** 7 **A13:** 9
in steels ... **A11:** 100
Neutron flux
defined ... **A10:** 678
Neutron irradiation ... **A1:** 653-661
damage to steels ... **A1:** 653
effect on fracture mode/toughness superalloys ... **A12:** 388
as embrittlement, examination and interpretation ... **A12:** 127
irradiation damage processes ... **A1:** 653
displacement damage ... **A1:** 653-654
transmutation helium ... **A1:** 654-655
mechanical properties
elevated-temperature tensile behavior ... **A1:** 657-658
fatigue ... **A1:** 660
irradiation creep ... **A1:** 660
irradiation embrittlement ... **A1:** 658-660, 722-723
irradiation-assisted SCC ... **A1:** 660-661
low-temperature tensile behavior ... **A1:** 657
thermal creep ... **A1:** 660
void swelling ... **A1:** 655-657
Neutron irradiation embrittlement ... **A1:** 658-660, 722-723
Neutron powder diffraction *See also* Neutron diffraction
data, Rietveld refinement of ... **A10:** 344
instrumentation ... **A10:** 422
Rietveld method ... **A10:** 423
samples ... **A10:** 422-423
Neutron powder diffractometers, schematic
with multidetector bank ... **A10:** 422
Neutron radiation
effect on Charpy impact properties ... **A11:** 69
Neutron radiation, effects
cast copper alloys ... **A2:** 361
Neutron radiography *See also* Neutron sources; Radiography ... **A17:** 287-295 **EM1:** 775
of adhesive-bonded joints ... **A17:** 624-626
applications and examples ... **A17:** 391-395
attenuation of neutron beams ... **A17:** 390
and conventional radiography, compared ... **A17:** 387
defined ... **A11:** 17
microstructure effect ... **A17:** 51
neutron detection methods ... **A17:** 390-391
neutron sources ... **A17:** 388-390
principles of ... **A17:** 387-388
vs. conventional radiography ... **A17:** 387-388
Neutron scattering, and x-ray scattering
compared ... **A10:** 421
Neutron sources *See also* Neutron(s)
accelerators ... **A17:** 388-389
nuclear reactors ... **A17:** 388
radioactive ... **A17:** 389-390
for radiography, of adhesive-bonded joints ... **A17:** 625
Neutron spectroscopy
defined ... **A10:** 678
Neutron spectrum
defined ... **A10:** 678
Neutron topography
capabilities ... **A10:** 365
Neutron(s) *See also* Neutron sources; Radiation; Radiography
detection methods ... **A17:** 390-391
energy ranges, and neutron radiography ... **A17:** 388
production ... **A17:** 387-389
as radiation ... **A17:** 295
Neutron-absorbing materials
properties ... **A2:** 1002
Neutrons *See also* Neutron diffraction
14-MeV, for FNAA ... **A10:** 239
capture, slow or thermal, in NAA ... **A10:** 234
cross section, defined ... **A10:** 678
defined ... **A10:** 677
delayed ... **A10:** 238
Neutrons (continued)
detector for, defined ... **A10:** 678
effect of low absorption ... **A10:** 423
epithermal ... **A10:** 234
high-energy, use for assay measurement ... **A10:** 234
-irradiated ores, y-ray spectrum of ... **A10:** 235
production for neutron diffraction ... **A10:** 421-422
reactions, in NAA ... **A10:** 234
sources for NAA ... **A10:** 234
sources, radial distribution function analysis ... **A10:** 395
spallation sources ... **A10:** 421
New Commercial Polymers **(Elias/Vohwinkel)** ... **EM2:** 94
New England
early American foundries in ... **A15:** 24-25
New England Foundrymen's Association ... **A15:** 34
New Glass Forum
Japan ... **EM4:** 692
New True Dentalloy
material loss on abrasion of dental amalgams ... **A18:** 669
Newt
defined ... **A18:** 13
Newton ... **A10:** 685, 691
Newton (N)
ISO unit for force ... **A18:** 415
Newton's first law
momentum ... **A15:** 591
Newton's law of cooling ... **A6:** 70
Newton's rings ... **A9:** 150
Newton's second law ... **A8:** 40
Newton-Raphson method
iterative technique for shape behavior ... **A8:** 715
Newtonian fluid ... **EM3:** 19
defined ... **A18:** 13
Newtonian viscosity, defined *See* under Viscosity
Newtonian viscous shear ... **A18:** 526
Nextel ceramic fiber ... **EM1:** 61- 2
Nextel mullite fiber ... **EM4:** 223
Nextel ultrafibers ... **EM1:** 63
NGLI number
defined ... **A18:** 13
NGR *See* Nuclear gamma-ray resonance
Ni-16Cr-16Mo-4W, gas-metal arc welding
Charpy V-notch absorbed energy vs. strength ... **A6:** 1018
Ni-20Cr-2.5Cb-0.5Ti
Charpy V-notch absorbed energy vs. strength ... **A6:** 1018
Ni-20Cr-2.5Cb-0.5Ti, flux-cored arc welding
Charpy V-notch absorbed energy vs. strength ... **A6:** 1018
Ni-Hard
applications, erosion resistance for pump components ... **A18:** 598
ball and rod grinding media, mining industry ... **A18:** 654
Ni-Hard 2C ... **A16:** 112
Ni-Hard cast iron
machinability testing ... **A16:** 640
milling with PCBN tools ... **A16:** 112
turning with PCBN tools ... **A16:** 112
wear resistance vs. retained austenite ... **M1:** 614
Ni-Hard cast irons ... **A1:** 89
Ni-O (Phase Diagram) ... **A3:** 2 • 312
Ni-Os (Phase Diagram) ... **A3:** 2 • 312
Ni-P (Phase Diagram) ... **A3:** 2 • 313
Ni-Pb (Phase Diagram) ... **A3:** 2 • 313
Ni-Pd (Phase Diagram) ... **A3:** 2 • 314
Ni-Pr (Phase Diagram) ... **A3:** 2 • 314
Ni-Pt (Phase Diagram) ... **A3:** 2 • 314
Ni-Pu (Phase Diagram) ... **A3:** 2 • 315
Ni-Re (Phase Diagram) ... **A3:** 2 • 315
Ni-Resist ... **A1:** 103
Ni-Resist (types 1,2)
galling resistance with various material combinations ... **A18:** 596
Ni-Rh (Phase Diagram) ... **A3:** 2 • 316
Ni-Ru (Phase Diagram) ... **A3:** 2 • 316
Ni-S (Phase Diagram) ... **A3:** 2 • 316
Ni-Sb (Phase Diagram) ... **A3:** 2 • 317
Ni-Sc (Phase Diagram) ... **A3:** 2 • 317
Ni-Se (Phase Diagram) ... **A3:** 2 • 317
Ni-Si (Phase Diagram) ... **A3:** 2 • 318

Ni-Sm (Phase Diagram) .... **A3:** 2 • 318
Ni-Sn (Phase Diagram) .... **A3:** 2 • 318
Ni-span-c 902, aged
machining **A16:** 836-840, 842, 843
Ni-span-c 902, unaged
machining **A16:** 836-840, 842, 843
Ni-Ta (Phase Diagram) .... **A3:** 2 • 319
Ni-Te (Phase Diagram) .... **A3:** 2 • 319
Ni-Ti (Phase Diagram) .... **A3:** 2 • 319
Ni-U (Phase Diagram) .... **A3:** 2 • 320
Ni-V (Phase Diagram) .... **A3:** 2 • 320
Ni-Vee bronze A
galling resistance with various material combinations **A18:** 596
Ni-Vee bronze B
galling resistance with various material combinations **A18:** 596
Ni-Vee bronze D
galling resistance with various material combinations **A18:** 596
Ni-W (Phase Diagram) .... **A3:** 2 • 320
Ni-Y (Phase Diagram) .... **A3:** 2 • 321
Ni-Yb (Phase Diagram) .... **A3:** 2 • 321
Ni-Zn (Phase Diagram) .... **A3:** 2 • 321
Ni-Zr (Phase Diagram) .... **A3:** 2 • 322
Ni_3Al
atomic interaction descriptions **A6:** 144
cohesion affected by impurities **A6:** 144
Ni_3Al aluminides *See also* Nickel aluminides; Ordered intermetallics
alloying effects **A2:** 914-915
anomalous dependence, yield strength temperature **A2:** 915-916
environmental embrittlement, elevated temperatures **A2:** 915
fabrication **A2:** 918
intergranular fracture **A2:** 914-915
mechanical properties **A2:** 917-918
processing **A2:** 918
solid-solution hardening **A2:** 916-917
structural applications **A2:** 918
yield strength **A2:** 915-916
Ni_3Si
cohesion affected by impurities **A6:** 144
Ni_3X alloys *See also* Ordered intermetallics; Trialuminides
and intergranular fracture **A2:** 931-932
NiAl
atomic interaction descriptions **A6:** 144
atomic interactions and adhesion **A6:** 144
NiAl aluminides *See also* Ordered intermetallics
alloy stoichiometry **A2:** 919-920
ductility **A2:** 919
future of **A2:** 920
grain size effect **A2:** 919-920
for high-temperature applications **A2:** 918
structure and property relationships **A2:** 919
Nib **A7:** 8
Nib-starter machines
percussion welding **M6:** 74
Nibbling **A14:** 762, 841
as wet chemical technique for subdividing solids **A10:** 165
Nicalocoat (Nippon Carbide) **EM4:** 225
Nicalon **EM1:** 60, 63-64
applications **EM4:** 983
properties **A18:** 803
Nicalon (SiC) fibers **EM3:** 402
microindentation test **EM3:** 400
Nicalon fiber **EM4:** 223, 232, 233
SEM photomicrograph **EM4:** 225
NICE *See* National Institute of Ceramic Engineers
Nicholas illuminator
fractographic **A12:** 81
Nichrome
photochemical machining **A16:** 588
relative solderability as a function of flux type **A6:** 129
Nichrome (continued)
relative weldability ratings, resistance spot welding **A6:** 834
resistance seam welding **A6:** 245
Nichrome 80
composition **A6:** 573
Nichrome heating elements, improving life of .... **A3:** 1 • 27-28
Nichrome system
oxidation- resistant coating **M5:** 665-666
Nichrome thin-film resistor **EL1:** 1099, 1100
"Nick-break tests" **A6:** 102
Nickel *See also* Cast nickel; Nickel alloy powders; Nickel alloys; Nickel alloys, specific types; Nickel powder strip; Nickel powders; Nickel silver powders; Nickel steel powders; Nickel tetracarbonyl powders; Nickel- based hardfacing alloys; Nickel-base alloys; Nickel-base alloys, specific types; Nickel-base superalloys; Nickel-based superalloys; Nickel-carbonyl powders; Nickel-iron powders **M4:** 754-759
as a beta stabilizer in titanium alloys **A9:** 458
as a reactive sputtering cathode material **A9:** 60
in active metal process **EM3:** 305
as addition to aluminum alloys **A4:** 843
addition to aluminum-base alloys **A18:** 752
as addition to brazing filler metals **A6:** 904, 905
addition to cylinder liner materials for strength **A18:** 556
addition to low-alloy steels for pressure vessels and piping **A6:** 667
additions to martensitic stainless steels **M6:** 348
adhesion and solid friction **A18:** 32
as adhesion layer for polyimides **EM3:** 158
adhesion measurement of fcc metals **A6:** 144
adhesion measurements **A6:** 144
adhesion to copper **A6:** 144
adhesion to silver **A6:** 144
age hardening **A4:** 907, 911-912 **M4:** 758-759
air-carbon arc cutting **A6:** 1172, 1176
in alloy cast irons **A1:** 89
alloying, aluminum casting alloys **A2:** 132
alloying effect in titanium alloys **A6:** 508
alloying effect on copper alloys **M6:** 400
alloying, effects, SCC **A13:** 273
alloying, effects, stainless steel **A13:** 550
alloying, in cast irons **A13:** 567
alloying, wrought copper and copper alloys **A2:** 242
in aluminum alloys **A15:** 746
as an addition to austenitic manganese steel castings **A9:** 239
as an austenite-stabilizing element in steel **A9:** 177-178
as an austenite-stabilizing element in wrought stainless steels **A9:** 283
annealing **A4:** 907-911 **M4:** 754-757
applications, protection tubes and wells **A4:** 533
applications, sheet metals **A6:** 400
atmospheric, by pollution **A7:** 203
atmospheric corrosion of **A13:** 82, 515
atomic interaction descriptions **A6:** 144
as austenite stabilizer **A13:** 47
in austenitic manganese steel **A1:** 825
in austenitic stainless steels **A6:** 457, 458, 459, 461, 465, 467
base metal solderability **EL1:** 677
batch annealing **A4:** 908, 909
bismuth in, GFAAS analysis **A10:** 57
brazing to aluminum **M6:** 1031
bright annealing **A4:** 909-910
buffing *See* Nickel, polishing and buffing of
carbide strengthening **A2:** 429-430
in cast iron **A1:** 6, 28
in cemented carbides **A18:** 795
ceramic/metal seals **EM4:** 506, 507
characteristics **A13:** 641
Nickel (continued)
chlorine corrosion of **A13:** 1171
chrome plating, hard removal of **M5:** 184-185
cleaning processes **M5:** 669-673
coating for seals **A18:** 551
for coating in ceramic/metal seals **EM4:** 504
and cobalt, corrosion in aqueous alkaline media **A10:** 135
in cobalt-base alloys **A1:** 985 **A18:** 766, 769, 770
coextrusion welding **A6:** 311
cohesion affected by impurities **A6:** 144
as colorant **EM4:** 380
commercial forms **A2:** 433-435
commercially pure *See* Nickel alloys, specific types, Nickel 200 and Nickel 201
commercially pure and low-alloy **A2:** 435, 437, 441
commercially pure, intergranular corrosion **A13:** 325
in compacted graphite iron **A1:** 59
in composition, effect on ductile iron **A4:** 686, 688, 689
in composition, effect on gray irons **A4:** 671, 672, 673, 676, 678, 681
in composition, factor affecting overheating of tool steels **A4:** 603
compositions **A2:** 436
compositions of specific types **M6:** 436
concentration vs intensity plot for **A10:** 99
conductor inks **EL1:** 208
contamination source for niobium electron-beam welding **A6:** 871
content in heat-treatable low-alloy (HTLA) steels **A6:** 670
content in HSLA Q&T steels **A6:** 665
content in nickel-base and cobalt-base high-temperature alloys **A6:** 573
content in stainless steels **M6:** 320
content in tool and die steels **A6:** 674
content in ultrahigh-strength low-alloy steels **A6:** 673
content of weld deposits **A6:** 675
continuous annealing **A4:** 909
controlled-potential electrogravimetry of **A10:** 200
in copper alloys **A6:** 752
in copper alloys, by dimethylglyoxime method **A10:** 66
in copper infiltrated powder metallurgy steels **A9:** 510
corrosion in **A11:** 202
corrosive wear testing of stainless steels in mines of **A18:** 717
crack deflection in nickel in glass **EM4:** 863
critical relative humidity **A13:** 82
deposition cycle, porcelain enameling **M5:** 514-515
determined by controlled-potential coulometry **A10:** 209
determined in samples containing chloride ions **A10:** 201
determined in stainless steel **A10:** 146
diffusion brazing **A6:** 343
diffusion into iron, measurement of **A10:** 243
in diffusion-alloyed steel powder metallurgy materials **A9:** 510
direct-current and differential pulse polarograms of **A10:** 194
dislocation network **A9:** 115
dominant texture orientations **A10:** 359
in ductile iron **A1:** 43-44 **A15:** 648, 649
in duplex stainless steels **A6:** 471, 472-473, 476
dynamically recrystallized grain size
Zener-Hollomon parameter **A8:** 175
effect of, on hardenability **A1:** 393, 395, 468
effect of, on notch toughness **A1:** 741
effect on activity coefficient in gas carburizing **A4:** 315
effect on crack formation **M6:** 833

Nickel (continued)
effect on Curie point A4: 187
effect on grain size during thermomechanical processing A4: 246-247, 248
effect on hardness of tempered martensite A4: 124, 128-129
effect on oxygen cutting M6: 898
effect on pearlite in steel A9: 661
effect on SCC of copper A11: 221
effect on sigma formation in ferritic stainless steels A9: 285
effects, electrical contact materials A2: 843
effects, in cartridge brass A2: 301
electrical resistance applications M3: 641, 645
as electrically conductive filler EM3: 178
electrodeburring of M5: 308
electrodeposited coatings A13: 426
electroforming M3: 179-181
electroless
abrasive wear A18: 837
applications A18: 837
coefficients of friction A18: 837
corrosive wear A18: 837
in electroplated coatings A18: 835, 837
magnetic electroplated coating A18: 838
properties A18: 837
reducing agents A18: 837
wear rates A18: 835, 836
electroless, coating for titanium alloys A4: 921
electroless deposition for metallization of lead zirconate titanate EM4: 544
electroless deposits, specific types
EN A18: 835, 836, 837
EN 400 A18: 835, 836, 837
EN 600 A18: 835, 836, 837
electrolytic plating in roll bonding of bearing materials A18: 756
electron-beam welding, filler metal for copper alloys A6: 860
electronic, analysis for aluminum in A10: 65
electronic applications A6: 990
electroplated
applications A18: 836
corrosive wear A18: 837
in electroplated coatings A18: 835, 836-837
properties A18: 836
wear rates A18: 836
electroplated on surface for sintered metal powder process EM3: 305
electroplated onto sleeve bearing liners A9: 567
electroplating of bearing materials A18: 756
embrittled by cadmium in cesium A11: 234
embrittlement A13: 179
in enameling ground coat EM3: 303, 304
environments that cause stress-corrosion cracking A6: 1101
epithermal neutron activation analysis A10: 239
erosion mechanisms A18: 202, 203
erosion resistance A18: 228
erosion-enhanced corrosion A18: 208
eutectic joining EM4: 526
evaporation fields for A10: 587
EXAFS scan using synchrotron radiation A10: 408
explosion welding A6: 162
extrusion welding M6: 677
in ferrite A1: 408
ferromagnetism A9: 533
in filler metals for active metal brazing EM4: 523, 524
as filler to gain electrical conductivity EM3: 572
finishing processes M5: 673-675
foil, solid-state welding A6: 169
formation of intermetallic phases in austenitic stainless steels A9: 284
forming, lubricants for A14: 520
fracture mechanism map for A14: 363
friction coefficient data A18: 71, 74
friction welding A6: 152, 154
functions in FCAW electrodes A6: 188
furnaces A4: 909-910
in gas-metal eutectic method EM3: 305
gas-tungsten arc welding shielding gas selection A6: 67
glass-to-metal seals EM3: 302

Nickel (continued)
glass/metal seals EM4: 1037
as gold alloy A2: 690
gold plating, uses in M5: 282-283
gravimetric finishes A10: 171
in gray iron A1: 22
as gray iron alloying element A15: 639
hardfacing A6: 807
in hardfacing alloys A18: 759, 763-764
in heat-resistant alloys A4: 510, 511, 512, 514, 515
in heat/corrosion-resistant casting alloys A13: 574, 577
high frequency resistance welding M6: 76
in high-alloy white irons A15: 670
high-purity, -y-ray spectrum A10: 240
high-purity, effect of relative humidity on fretting A13: 140
historical development A2: 428-429
hydride-generation AAS system for A10: 50
hydrochloric acid corrosion of A13: 1162
hydrometallurgy A2: 429
hysteresis of magnetoresistance A17: 144
ICP-determined in silver scrap metal A10: 41
-implanted aluminum A10: 485
impurities determined in A10: 240
impurity in solders M6: 1072
inclusion in white irons A18: 697
induction heating energy requirements for metalworking A4: 189
induction heating temperatures for metalworking processes A4: 188
induction soldering A6: 363
intensity vs concentration, x-ray spectrometry A10: 99
in intermetallic compounds A6: 127, 128
and iron, activities of A15: 51
in iron-base alloys, flame AAS analysis A10: 56
iron-nickel-cobalt ASTM F 15 alloy EM3: 301
isolation in high-temperature alloys A10: 174
isotope composition and intensity A10: 146
in jet engine mainshaft bearings A18: 590
K-edge A10: 414, 416
lap welding M6: 673
for laser alloying A18: 866
laser cladding A18: 867
in laser cladding material A18: 867
in limestone EM4: 379
lubricant indicators and range of sensitivities A18: 301
as major toxic metal with multiple effects A2: 1250
in maraging steels A4: 220, 221, 222, 224
material for conductors EM4: 1142
metal filler for polyimide-base adhesives EM3: 159
in metal powder-glass frit method EM3: 305
metallic, partitioning oxidation states in A10: 178
Miller numbers A18: 235
mining and refining M3: 125-126
near-surface region defects A18: 852-853, 854
neutron and x-ray scattering, and absorption compared A10: 421
Nickel 200, properties EM4: 503
and nickel alloys A15: 815-823
-nickel pair, phase and envelope function for A10: 414
normalized, K-edge EXAFS plotted A10: 417
in nickel-chromium white irons A4: 700-702 A15: 679
ores M3: 125-126
organic precipitant for A10: 169
origins and applications A7: 202-203
oxyacetylene welding A6: 281
in P/M alloys A1: 809-810
in pharmaceutical production facilities A13: 1227
phosphate coating solution contaminated by M5: 455-456
-phosphorus film, extent of coverage on platinum substrate A10: 608
photometric analysis methods A10: 64
physical metallurgy A2: 429-430
plasma and shielding gas compositions A6: 197
plating for tool steels A18: 739
plating, industrial applications M3: 179-182

Nickel (continued)
plating, uses EL1: 679
polishing and buffing of M5: 108, 112-114, 674-675
polycrystalline, cavitation in A11: 165
Pourbaix diagram A13: 28
precipitation hardening A2: 430
precoating A6: 131
process control factors in annealing A4: 909-911
products A7: 395-397
promotion of acicular ferrite M6: 117-118
properties A6: 629, 992 A18: 795
protective atmospheres A4: 909, 910
pure M2: 777, 833
pure (99.9%) A18: 42, 73, 74
pure, magnetization curve A17: 168
pure, properties A2: 1143
pure, x-ray tubes A17: 302
pyrometallurgy A2: 429
as pyrophoric A7: 199
qualitative tests to identify A10: 168
recommended impurity limits of solders A6: 986
recovery from selected electrode coverings A6: 60
reference specimens for optical reading of microindentation testing A18: 414
refining A2: 428-429
relative hydrogen susceptibility A8: 542
relative solderability as a function of flux type A6: 129
relative weldability ratings, resistance spot welding A6: 834
resistance seam welding A6: 241 M6: 49
resistance soldering A6: 357
roll welding A6: 312, 314 M6: 67
room temperature bend strength of silicon nitride metal joints EM4: 526
safety exposure limits A7: 203
as sample modifier, GFAAS analysis A10: 55
secondary phase formation A18: 854
shielded metal arc welding A6: 176
shrinkage allowance A15: 303
sintering, time and temperature M4: 796
sodium hydroxide corrosion of A13: 1176
as solder impurity EL1: 638
solderability A6: 978
solderable and protective finishes for substrate materials A6: 979
soldering A6: 631 M6: 1075
solid-solution hardening A2: 429
solid-state sintering EM4: 276
solution potential M2: 207
specific properties imparted in CTV tubes EM4: 1040, 1042
spectrometric metals analysis A18: 300
spraying for hardfacing M6: 789
stacking-fault energy A18: 715
in stainless steel brazing filler metals A6: 911, 913, 915, 917, 920
in stainless steels A18: 710, 712, 713, 716, 721 M3: 57
static lithium corrosion of A13: 94
in steel weldments A6: 417
in steels A1: 146, 577
strengths of ultrasonic welds M6: 752
stress equalizing M4: 755, 756
stress relieving M4: 755, 756, 758
stress-corrosion cracking A13: 157
stress-corrosion cracking in A11: 223
stress-equalizing A4: 907, 908, 911
stress-relieving A4: 907, 908, 911
stress-strain curves A8: 174
submerged arc welding A6: 206
submerged arc welding electrodes M6: 121
substitute for cobalt composition in hardfacing alloys A18: 762
suitability for cladding combinations M6: 691
superplasticity of A8: 553
support of SERS in vacuum A10: 136
thermal conductivity value A6: 587
thermal diffusivity from 20 to 100 °C A6: 4
thermal expansion coefficient A6: 907
in thermal spray coating materials A18: 832
thoria-doped EM4: 60
as tin solder impurity A2: 520
tin-plated A9: 456

Nickel (continued)
titration, with cyanide **A10:** 174
to improve hardenability in carburized steels **A4:** 366, 369
toxicity **A6:** 1195, 1196 **A7:** 202-203
as trace element, cupolas **A15:** 388
trace-element concentrations **A10:** 194, 240
TWA limits for particulates **A6:** 984
ultrasonic welding **A6:** 327 **M6:** 74
usage, PWB manufacturing **EL1:** 510
use in flux cored electrodes **M6:** 103
uses **M3:** 126-127
vacuum heat-treating support fixture material **A4:** 503
vapometallurgy **A2:** 429
vapor pressure, relation to temperature **A4:** 495 **M4:** 309, 310
in vapor-phase metallizing **EM3:** 306
Vickers and Knoop microindentation hardness numbers **A18:** 416
volumetric procedures for **A10:** 175
weighed as the dimethylglyoxime complex **A10:** 171
weld overlay for hardfacing alloys **A6:** 820
welding to copper and copper alloys **A6:** 769
weldments **A13:** 361-362
in wrought stainless steels **A1:** 871-872
x-rays, effects of absorption/enhancement on **A10:** 97
in zinc alloys **A15:** 788
in zinc/zinc alloys and coatings **A13:** 759
Nickel 200
composition **A16:** 836
gas metal arc welding **M6:** 439
machining **A16:** 837-840, 842-843
Nickel 201
composition **A16:** 836
machining **A16:** 837-840, 842-843
Nickel 205
composition **A16:** 836
machining **A16:** 837-840, 842-443
Nickel 212
composition **A16:** 836
machining **A16:** 837-840, 842-843
Nickel 222
composition **A16:** 836
machining **A16:** 837-840, 842-843
Nickel 270
composition **A16:** 836
machining **A16:** 837-840, 842-843
Nickel alloy eutectic **A9:** 619
Nickel alloy powders **A7:** 134, 142
Auger depth profile **A7:** 255
for brazing, composition and properties **A7:** 838
chemical analysis and sampling **A7:** 248
effect of milling time on microhardness of **A7:** 61
HIP temperatures and process times **A7:** 437
mechanical properties **A7:** 468, 472
nonconventional sintering **A7:** 398
porous **A7:** 699
refinement and blending **A7:** 63, 64
sintering **A7:** 395-398
Nickel alloys *See also* Cast nickel; Electrical resistance alloys; Electrical resistance alloys, specific types; Heat-resistant alloys, heat treating; Nickel; Nickel alloys, specific types; Nickel toxicity; Nickel-base heat- resistant casting alloys; Nickel-base superalloys; Nickel-base superalloys, specific types; Superalloys; Superalloys, heat treating; Wrought heat-resistant alloys
acid etching process **M5:** 564
age hardening **A4:** 911-912 **M4:** 758-759
alloy and market developments **A2:** 429
aluminum coating of **M5:** 340-343
annealing **A4:** 907-911 **M4:** 754-757
Nickel alloys (continued)
applications **A2:** 430-433 **A6:** 383, 576, 578 **A15:** 815, 823
sheet metals **A6:** 400
arc welding *See* Arc welding of nickel alloys
arc welding to cast irons **M6:** 307
axial tests on **A8:** 352
bar and tube, die materials for drawing **M3:** 525
batch annealing **A4:** 908, 909
binary, relative potency factors **A6:** 89
bobbing process **M5:** 674-675
brazing
available product forms of filler metals **A6:** 119
joining temperatures **A6:** 118
brazing to aluminum **M6:** 1031
bright annealed, cleaning of **M5:** 670-671
bright annealing **A4:** 909-910 **M4:** 756-757
bright dipping **M5:** 670-671
brushing **M5:** 674-675
cadmium plating of **M5:** 264
carbide strengthening **A2:** 429-430
cast, stage I fatigue fracture appearance **A12:** 19
castings, compositions **A15:** 815-817
characteristics **A2:** 430-433
chemical pitting **A2:** 432
chromium plating of **M5:** 172-173
classified **A15:** 815
cleaning processes **M5:** 669-673
copper flash prevention and removal **M5:** 673
embedded iron removal **M5:** 672-673
lead and zinc removal **M5:** 672
cleaning solutions for substrate materials **A6:** 978
color etching **A9:** 141-142
coloring process **M5:** 674-675
commercial forms **A2:** 433-445
composition **M3:** 748
compositions **A2:** 436 **M6:** 354, 43
containment autoclaves **A8:** 420
continuous annealing **A4:** 909
controlled-expansion alloys **A2:** 443-444
conversion tables **A8:** 109
corrosion **A6:** 579
corrosion fatigue **A2:** 432
corrosion resistance **A2:** 431-433 **M3:** 171-174
current densities for electropolishing **A9:** 435
dead-soft annealing **A4:** 909 **M4:** 757
dissimilar metal joining **A6:** 749-751
distortion **A6:** 751
electrical resistance alloys **A2:** 433
electrical resistance, properties **A2:** 823
electrodes **A6:** 743, 745, 746-747, 748
electrolytes for electropolishing **A9:** 435
electron-beam welding **A6:** 855
electropolishing of **M5:** 305-306, 308
electroslag remelting **A15:** 403
electroslag welding **A6:** 740
etchants for microscopic examination of **A9:** 435
explosion welding **A6:** 896 **M6:** 71
fasteners, use in **M3:** 184
fatigue at subzero temperatures **M3:** 735-736, 753
fatigue striations, precision matching **A12:** 205-206
filler metals **A2:** 444 **A6:** 578, 743, 750
finishing processes **M5:** 673-675
fluxes **A6:** 748
foundry practice **A15:** 820-823
fractographs **A12:** 396-397
fracture toughness **M3:** 735, 752
fracture/failure causes illustrated **A12:** 217
fretting wear **A18:** 248, 250
furnaces **A4:** 909-910
fusion weldability test written into purchase specifications **A6:** 577
gas tungsten arc welding **M6:** 182, 203, 20
Nickel alloys (continued)
gas-metal arc welding **A6:** 180, 740, 741, 742, 743-745, 746, 749, 750, 751
shielding gases **A6:** 66-67
gas-tungsten arc welding shielding gas selection **A6:** 67
general corrosion **A2:** 431-432
general welding characteristics **A6:** 563
gold plating of **M5:** 283
gold with molybdenum contamination, ion- scattering spectrometry **A18:** 449
grinding process **M5:** 673-674
groups **M5:** 670
hardening techniques **A4:** 911-912 **M4:** 757, 759
hardfacing **A6:** 807
heat treatments **A4:** 215-217
heat-affected zone cracking **A6:** 577
heat-resistant alloy applications **A4:** 517
heat-resistant applications **A2:** 430-431
heat-resistant, compositions **A15:** 816
high-strength low-alloy (HSLA) steels, welding to **A6:** 578
historical development **A2:** 428-429
hot cracking **A6:** 749
hot extrusion, billet temperature for **M3:** 537
hydrogen embrittlement **A12:** 124
hydrometallurgy **A2:** 429
induction heating for **A8:** 159
intergranular corrosion **A2:** 432
iron-nickel-chromium alloys **A2:** 442-443
joint design **A6:** 743, 744, 745, 748
laser beam welding **M6:** 64
laser-beam welding **A6:** 263
liquid-penetrant inspection **A6:** 577
low-expansion alloys **A2:** 433
machining **A2:** 966
machining, ultrahard materials for **A2:** 1010
material for jet engine components **A18:** 588
melting practice **A15:** 820
metal treatment **A15:** 820
minimizing weld defects **A6:** 749
mutual dissolution and erosion **A6:** 621
nickel alloys, welding of dissimilar types **A6:** 577-578
nickel-chromium alloys **A2:** 436, 441-442
nickel-chromium-iron series **A2:** 436, 441-442
nickel-copper alloys **A2:** 435-436
nickel-iron low-expansion alloys **A2:** 443-444
nominal compositions of **A9:** 436
nonmetallic inclusions in **A9:** 436-437
oxidational wear **A18:** 287
oxyfuel gas cutting **A6:** 1155
oxyfuel gas welding **A6:** 281 **M6:** 58
passive potential range **A9:** 145
percussion welding **M6:** 74
physical metallurgy **A2:** 429-430
pickling **M5:** 669-673
formulas **M5:** 669-670
specialized operations **M5:** 672-673
surface conditions, effect of **M5:** 670-672
plasma and shielding gas compositions **A6:** 197
plasma arc welding **A6:** 197, 740, 745-746
plasma-MIG welding **A6:** 224
polishing and buffing **M5:** 674-675
precipitation hardening **A2:** 430
precoated before soldering **A6:** 131
preparation of metallographic specimens **A9:** 435-438
principal ASTM specifications for weldable nonferrous sheet metals **A6:** 400
process control, annealing **A4:** 909-911
cold work, prior, effect of **A4:** 910
contamination, protection from **A4:** 911
cooling rate, effect of **A4:** 910
embrittlement **A4:** 910-911
fluctuating atmospheres, effect of **A4:** 909, 910-911
fuels **A4:** 909

Nickel alloys (continued)
furnace-temperature **A4:** 910
grain size control **A4:** 910
process-control, annealing **M4:** 757-758
cold work, prior, effect of **M4:** 758
contamination, protection from **M4:** 758
cooling rate, effect of **M4:** 758
fluctuating atmospheres, effect of **M4:** 758
fuels **M4:** 757-758
furnace-temperature **M4:** 758
grain size control **M4:** 758
proprietary, corrosion-resistant **A15:** 816
protective atmospheres **A4:** 909, 910
pyrometallurgy **A2:** 429
reaction with molybdenum hearths in vacuum heat treating **A4:** 503
recommended guidelines for selecting PAW shielding gases **A6:** 67
recommended shielding gas selection for gas-metal arc welding **A6:** 66
refining **A2:** 428-429
repair welding **A6:** 579
resistance brazing **M6:** 97
resistance seam welding **A6:** 241, 245 **M6:** 49
resistance soldering **A6:** 357
resistance spot welding **M6:** 47
resistance welding **A6:** 841
salt bath descaling **M5:** 672
SCC testing in water and aqueous solutions **A8:** 530-531
shape memory alloys **A2:** 433
shielded metal arc welding **A6:** 176, 740, 742, 744, 745, 746-747, 749, 751 **M6:** 75
shielding gas **A6:** 743-745
soft magnetic alloys **A2:** 433, 553
solderable and protective finishes for substrate materials **A6:** 979
soldering **A6:** 631
solid-solution hardening **A2:** 429
solution treating **M4:** 758-759
solution-treating **A4:** 907, 911
special purpose **A2:** 430
specialty annealing **A4:** 909
spindle finishing **M5:** 673-674
springs, cleaning of **M5:** 673
squeeze casting of **A15:** 323
stainless steels, welding to **A6:** 578
steel, welding to **A6:** 578
stress equalizing **M4:** 755, 756, 758
stress relieving **M4:** 755, 756, 758
stress-corrosion cracking **A6:** 749
stress-corrosion cracking (SCC) **A2:** 432-433
stress-equalizing ..... **A4:** 907 , 908, 911
stress-relieving **A4:** 907, 908, 911
stress-rupture strength **A6:** 578
structure and property correlations **A15:** 817-819
submerged arc welding **A6:** 740, 743, 748, 749
surface conditions, pickling affected by **M5:** 670-672
oxidized and scaled surfaces **M5:** 671-672
reduced oxide surfaces **M5:** 671
tarnish **M5:** 670-671
tarnish removal **M5:** 670-671
tensile properties at subzero temperatures **M3:** 734-735, 737, 740, 742, 749-750, 751
thermal expansion coefficient **A6:** 907
torch annealing **A4:** 909 **M4:** 757
ultrahigh-strength steel compositions **A4:** 207
vapometallurgy **A2:** 429
weld overlay material **M6:** 806, 81
welding alloys **A2:** 444-445
welding, cleaning for **M5:** 673
welding current **A6:** 743, 747, 748
welding electrodes **A6:** 176
welding techniques **A6:** 743, 745, 747
welding to copper and copper alloys **A6:** 769
wire, die materials for drawing **M3:** 522
Nickel alloys, cast, specific types *See also* Heat-resistant alloys, specific types; Nickel alloys, specific types; Stainless steels, specific types; Superalloys, specific types
CA-100, mechanical properties **M3:** 177
Chlorimet 2, composition *See also* N-12M **M3:** 176
Chlorimet 3, composition *See also* CW-12M **M3:** 176
Nickel alloys, cast, specific types (continued)
CW-12M
composition **M3:** 161, 176
corrosion resistance **M3:** 164-165
mechanical properties **M3:** 177
property data **M3:** 161-165
CY-40
composition **M3:** 163, 176
mechanical properties **M3:** 177
property data **M3:** 163
stress-rupture properties **M3:** 178
tensile properties, elevated temperature **M3:** 178
CZ-100
composition **M3:** 164, 176
property data **M3:** 164-165
H Monel
composition **M3:** 166
property data **M3:** 166
Hastelloy B *See also* N-12M
composition **M3:** 172, 176
corrosion rate in sulfuric acid ... **M3:** 171, 172, 173
Hastelloy C, composition *See also* CW-12M **M3:** 176
Hastelloy D
composition **M3:** 172, 176
corrosion rate in sulfuric acid **M3:** 171, 172
Illium 98
composition **M3:** 168, 176
corrosion resistance **M3:** 168
property data **M3:** 168
Illium B
composition **M3:** 168, 176
corrosion resistance **M3:** 169
property data **M3:** 169
Illium G
composition **M3:** 169, 176
corrosion resistance **M3:** 169
property data **M3:** 169-170
Inconel *See* CY-40
M-35
composition **M3:** 165, 176
mechanical properties **M3:** 177
property data **M3:** 165-166
N-12M
composition **M3:** 167, 176
corrosion resistance **M3:** 164-165
mechanical properties **M3:** 177
property data **M3:** 164-165, 167-168
QQ-N-288, grades A thru E
composition **M3:** 176
mechanical properties **M3:** 177
S Monel composition **M3:** 166
property data **M3:** 166, 167
Nickel alloys, forgings
inspection methods **A17:** 496
Nickel alloys, high temperature *See* High temperature nickel alloys
Nickel alloys, selection of **A6:** 586-592
alloying of elements effect **A6:** 588-590
applications **A6:** 586
compositions of selected popular alloys **A6:** 586
consumption and applications by Western nations in 1990 ..... **A6:** 586
cost as factor determining applications **A6:** 586
electron-beam welding **A6:** 587, 588
fabrication **A6:** 587
fusion zone **A6:** 588
gas-metal arc welding **A6:** 587, 588
gas-tungsten arc welding **A6:** 588
heat-affected zone **A6:** 587-590
heat-affected zone fissuring **A6:** 588
history and development of **A6:** 586
postweld heat treatment **A6:** 590
sigma formation **A6:** 587
special metallurgical welding considerations **A6:** 590-592
thermal conductivity **A6:** 587
unmixed zone **A6:** 588, 589
welding characteristics **A6:** 587
welding metallurgy **A6:** 587-590
Nickel alloys, specific types *See also* Heat-resistant alloys, specific types; Nickel alloys, cast, specific types; specific Hastelloy alloys; specific Inconel alloys; Stainless steels, specific types; Superalloys, nickel-base, specific types; Superalloys, specific types
Nickel alloys, specific types (continued)
9Ni steel
cryogenic service **A6:** 1017
10Ni-8Co-1Mo steel, crack growth **A8:** 678-679
13Cr-4Ni-0.05C, hydrogen-induced cold cracking resistance **A6:** 438
20, shielded metal arc welding **A6:** 746
20Cb3, applications **A2:** 442
20Cb3, composition **A6:** 741
20Mo-4, applications **A2:** 442
25-6Mo, composition **A6:** 741
25Cr-20Ni, hot cracking **A6:** 497
28, composition **A6:** 741
35Ni-15Cr
for cast element material in heat-treating fumaces **A4:** 472
heat-resistant alloy applications **A4:** 515, 516, 517
35Ni-18Cr
35Ni-18Cr-44Fe, ribbon material in heat-treating furnaces **A4:** 472
35Ni-19Cr, recommended for parts and fixtures for salt baths **A4:** 514
35Ni-20Cr, heat-resistant alloy applications **A4:** 515
68Ni-20Cr, for element strip material in heat-treating furnaces **A4:** 472
80Ni-20Cr
austempering **A4:** 159
for element strip material in heat-treating fumaces **A4:** 472
heat-resistant alloy applications **A4:** 516
recommended for furnace parts and fixtures **A4:** 513
recommended for parts and fixtures for salt baths **A4:** 514
35Ni-45Fe-20Cr, applications **A2:** 442
36, composition **A6:** 741
42, composition **A6:** 741
45Cr-55Ni, resistance butt welding **A6:** 578
48, composition **A6:** 741
52Ni-48Fe, plated **A9:** 562-564
60Ni-24Fe-16Cr, applications **A2:** 442
70Ni30Cr, laser melting and wear behavior **A18:** 866
75Cu-25Ni, roll welding **A6:** 314
80Ni-20Cr-1.5Si, applications **A2:** 442
200
annealing temperatures **A6:** 590
applications **A6:** 749
composition **A6:** 586, 741
electrical resistivity **A6:** 587
gas-metal arc welding **A6:** 746
heat treatment procedures recommended **A6:** 590
plasma arc welding **A6:** 746
properties **A6:** 587
shielded metal arc welding **A6:** 746, 747
sulfur embrittlement **A6:** 590
welding products for dissimilar-metal joints **A6:** 591
201
annealing temperatures **A6:** 590
applications **A6:** 586, 749
carbon content maximum % **A6:** 589
composition **A6:** 586, 741
electrical resistivity **A6:** 587
heat treatment procedures recommended **A6:** 590
properties **A6:** 587
roll welding **A6:** 313
welding products for dissimilar-metal joints **A6:** 591
201, bend-test fracture surface **A12:** 396
201, explosively bonded to tantalum **A9:** 445
205, composition **A6:** 741
233, composition **A6:** 741
270, composition **A6:** 741
300, composition **A6:** 741
301
composition **A6:** 741
postweld strain-age cracking **A6:** 576
strengthened by heat treatment **A6:** 575
330
applications **A6:** 749

Nickel alloys, specific types (continued)
composition A6: 576
330 HC, composition A6: 576
400
applications A6: 749
composition A6: 741
microstructures A6: 589
plasma arc welding A6: 746
shielded metal arc welding A6: 750
401, composition A6: 741
404, composition A6: 741
450, applications A6: 749
600
applications A6: 749
composition A6: 576, 741
electroslag welding A6: 278
shielded metal arc welding A6: 746
601
applications A6: 749
composition A6: 576, 741
617
applications A6: 749
composition A6: 576, 741
622
applications A6: 749
composition A6: 741
625
applications A6: 749
composition A6: 576, 741
filler metal A6: 467
filler metal for stainless steel casting alloys A6: 496
plasma-MIG welding A6: 224
shielded metal arc welding A6: 746
686
applications A6: 749
composition A6: 741
690
applications A6: 749
composition A6: 576, 741
shielded metal arc welding A6: 746
702, composition A6: 741
706, composition A6: 577, 741
713 C
friction welding A6: 578
postweld strain-age cracking A6: 576
718
applications A6: 749
composition A6: 577, 741
friction welding A6: 153
postweld strain-age cracking A6: 576
strengthened by heat treatment A6: 575
725, composition A6: 741
751, composition A6: 577, 741
800 alloy
applications A6: 749
composition A6: 576, 741
800 H, composition A6: 576, 741
800 HT alloy
applications A6: 749
composition A6: 576, 741
801, composition A6: 741
802
composition A6: 741
825
applications A6: 749
composition A6: 741
shielded metal arc welding A6: 746
902, composition A6: 741
903, composition A6: 577, 741
904, composition A6: 577
907, composition A6: 577, 741
908, composition A6: 741
909
composition A6: 577, 741
postweld strain-age cracking A6: 576
strengthened by heat treatment A6: 575
925, composition A6: 577, 741

Nickel alloys, specific types (continued)
alloy 042 (Dumet), and Kovar, applications A2: 443-444
alloy 042 (Dumet), as low-expansion A2: 443
alloy 052, as low-expansion A2: 443
alloy 400, characteristics A2: 435
alloy 426, as low-expansion A2: 443
alloy 600, applications and characteristics A2: 436
alloy 600, for nuclear power applications A2: 442
alloy 601, composition and characteristics A2: 441
alloy 625, alloying and applications A2: 442
alloy 690, alloying and applications A2: 442
alloy 718, alloying and applications A2: 441
alloy 800, applications A2: 442
alloy 800H, applications A2: 442
alloy 800HT, applications A2: 442
alloy 801, applications A2: 442
alloy 825, alloying and characteristics A2: 442
alloy 902, alloying and characteristics A2: 443
alloy 903, alloying and characteristics A2: 443
alloy 907, alloying and characteristics A2: 443
alloy 909, alloying and characteristics A2: 443
alloy 925, alloying and characteristics A2: 442
alloy C-22, alloying and applications A2: 442
alloy C-276, alloying and applications A2: 442
alloy G-3/G-30, alloying and application A2: 442
alloy K-500, characteristics A2: 435
alloy R-405, characteristics A2: 435
alloy X, applications A2: 441
alloy X750, alloying and applications A2: 441
B2, composition A6: 741
C-22, filler metals A6: 467
C-276 alloy
applications A6: 749
composition A6: 741
filler metal for stainless steel casting alloys A6: 496
filler metals A6: 467
C-276, slow strain rate testing of A8: 530
commercially pure nickel (N02200), welding to carbon, low-alloy or stainless steels A6: 826
CZ-100, mechanical properties A15: 817
DS, composition A6: 576
DS Ni
composition M3: 151
property data M3: 151-152
Duranickel 301 .... A9: 435-437
composition M3: 132
property data M3: 132-133
Duranickel, fracture surface A12: 397
G alloy
applications A6: 749
shielded metal arc welding A6: 746
G-3 alloy
applications A6: 749
composition A6: 741
shielded metal arc welding A6: 746
G-30, composition A6: 741
Hastelloy B
annealing M4: 756
composition M3: 748 M4: 755
stress relieving M4: 756
tensile properties at subzero temperatures M3: 749
Hastelloy B-2
composition M3: 152
property data M3: 153
Hastelloy B-2, welding to carbon, low-alloy, or stainless steels A6: 826
Hastelloy C
annealing M4: 756
composition M3: 748 M4: 755
stress relieving M4: 756

Nickel alloys, specific types (continued)
tensile properties at subzero temperatures M3: 749
Hastelloy C-4
composition M3: 153
property data M3: 154
Hastelloy C-4, welding to carbon, low-alloy or stainless steels A6: 826
Hastelloy C-276
composition M3: 154, 172
corrosion resistance M3: 172
property data M3: 154-155
Hastelloy C-276, welding to carbon, low-alloy or stainless steels A6: 826
Hastelloy G
composition M3: 155, 172
corrosion resistance M3: 172
property data M3: 156
Hastelloy G, welding to carbon, low-alloy or stainless steels A6: 826
Hastelloy G-3
composition M3: 156
property data M3: 156
Hastelloy N
composition M3: 157
property data M3: 157
Hastelloy S
composition M3: 157
property data M3: 157-158
Hastelloy series, discovery and characteristics A2: 429
Hastelloy W, property data M3: 159
Hastelloy X
age-hardening M4: 759
annealing M4: 759
composition M3: 159 M4: 755
oxidation resistance M3: 259
property data M3: 159-160
solution treating M4: 759
Hastelloy X, cleaning and finishing processes M5: 567-568
HK40, stress-rupture strength A6: 578
HX, composition A6: 576
Incoloy 800
composition A6: 564 M3: 147, 172
corrosion resistance M3: 148, 172
property data M3: 147-148
welding to carbon, low-alloy or stainless steels A6: 826
Incoloy 801
composition M3: 148
property data M3: 148-149
Incoloy 825
composition M3: 149, 172
corrosion resistance M3: 149, 150, 152, 172
property data M3: 149-150, 151, 152
Incoloy 825, welding to carbon, low-alloy or stainless steels A6: 826
Incoloy, cleaning and finishing processes M5: 670-672, 674
Incoloy series, discovery and characteristics A2: 429
Inconel
cutoff band sawing with bimetal blades A6: 1184
environments that cause stress-corrosion cracking A6: 1101
oxyacetylene welding A6: 281
relative solderability as a function of flux type A6: 129
ultrasonic welding A6: 326
Inconel 600
aluminum coating, tensile strength affected by M5: 342
annealing M4: 756
cleaning and finishing process M5: 568
composition.... A6: 654, 573 M3: 141, 172, 749 M4: 755

Nickel alloys, specific types (continued)
constitutional liquation in multicomponent systems ... A6: 568
corrosion resistance ... M3: 141-143, 171, 173
electron-beam welding ... A6: 869
fasteners ... M3: 184
joined to silicon carbide ... A6: 636
laser welding ... A6: 4, 264, 441
Poisson's ratio ... M3: 742
property data ... M3: 143-145
stress relieving ... M4: 756
tensile properties at subzero temperatures ... M3: 749
trace element impurity effect on GTA weld penetration ... A6: 20
welding to carbon, low-alloy or stainless steels ... A6: 826
Young's modulus ... M3: 740
Inconel 601
annealing ... M4: 756
composition ... M4: 755
Inconel 617
annealing ... M4: 756
composition ... M4: 755
Inconel 625
annealing ... M4: 756
composition ... M3: 143, 172 M4: 755
corrosion resistance ... M3: 143, 144, 172
property data ... M3: 143-144
Inconel 671
composition ... M3: 144
corrosion resistance ... M3: 144
property data ... M3: 144-145
Inconel 690
composition ... M3: 145
corrosion resistance ... M3: 146
property data ... M3: 145-146
Inconel 706
composition ... M3: 748
fatigue-crack-growth rate ... M3: 753
fracture toughness ... M3: 752
tensile properties at subzero temperatures ... M3: 723, 749, 752
Inconel 713, aluminum coating creep affected by ... M5: 342
Inconel 718
age hardening ... A6: 564
age-hardening ... M4: 759
aging cycles ... A6: 574
annealing ... M4: 756
applications ... A6: 928
composition ... A6: 564, 573 M3: 748 M4: 755
electron-beam welding ... A6: 869
fatigue-crack-growth rate ... M3: 753
fracture toughness ... M3: 752
friction welding ... A6: 153
heat treatments ... A6: 928
laser-beam welding ... A6: 263
nickel flashing treatment ... A6: 926
solution treating ... M4: 759
solution treatment ... A6: 574
strain-age cracking resistance ... A6: 84
temperature measurement, validation strategies ... A6: 1149
tensile properties at subzero temperatures ... M3: 750, 751
trace element impurity effect on GTA weld penetration ... A6: 20
Young's modulus ... M3: 737
Inconel, cleaning and finishing processes ... M5: 112-114, 563, 568, 670-673
Inconel series, discovery an characteristics ... A2: 429
Inconel X, ultrasonic welding ... A6: 326, 895
Inconel X-750
age-hardening ... M4: 759
annealing ... M4: 756
composition ... M3: 748 M4: 755
fatigue-crack-growth rate ... M3: 753
fracture toughness, weldments ... M3: 752
Poisson's ratio ... M3: 742
solution treating ... M4: 759
tensile properties at subzero temperatures ... M3: 750, 751
Young's modulus ... M3: 740
Inconel X-750 cleaning and finishing processes ... M5: 563, 568, 673

Nickel alloys, specific types (continued)
Invar 36
composition ... M3: 748
tensile properties at subzero temperatures ... M3: 750
Invar, thermal expansion of ... A2: 443-444
J-1500, ultrasonic welding ... A6: 326
K Monel, fatigue life ... M3: 753
K-500 alloy
applications ... A6: 749
composition ... A6: 577, 741
strengthened by heat treatment ... A6: 575
K-Monel, discovery and characteristics ... A2: 429
K-Monel, ultrasonic welding ... A6: 326
MA 754
applications ... A6: 1039
arc welding ... A6: 1038
composition ... A6: 577, 1037
filler metals for ... A6: 1039
furnace brazing ... A6: 1038
furnace-brazed weld properties ... A6: 1039
gas-tungsten arc weld ... A6: 928
postweld annealing ... A6: 1039
properties ... A6: 928
welding consumables ... A6: 1038
MA 758
applications ... A6: 1039
composition ... A6: 1037
weld transverse properties ... A6: 1038
welding consumables ... A6: 1038
MA 956
composition ... A6: 577, 1037
electron-beam welding ... A6: 1039
explosion welding ... A6: 1040
filler metals for ... A6: 1039
furnace-brazed weld properties ... A6: 1039
gas-tungsten arc welding ... A6: 1038, 1039
grain structure ... A6: 1039
laser-beam welding ... A6: 1039
properties ... A6: 578
weld transverse properties ... A6: 1038
welding consumables ... A6: 1038
MA 957
composition ... A6: 577
properties ... A6: 578
MA 6000
brazing ... A6: 632, 928
composition ... A6: 577, 1037
postweld annealing ... A6: 1039
properties ... A6: 578
welding processes ... A6: 1039
MAR M-200, directionally solidified ... A15: 319
MAR-M200, effect of temperature on strength and ductility ... A8: 36
Monel 400 ... A9: 435-437
annealing ... A4: 908 M4: 756
bright annealing ... A4: 909
composition ... A4: 908 M3: 133, 172 M4: 755
corrosion resistance ... M3: 136, 171
fasteners ... M3: 184
mechanical properties ... A4: 907, 911
property data ... M3: 134-136
stress equalizing ... M4: 756
stress relieving ... M4: 756
stress-equalizing ... A4: 907, 908, 911
stress-relieving ... A4: 908
Monel 400, cleaning and finishing of ... M5: 670, 673
Monel 400, welding to carbon, low-alloy or stainless steels ... A6: 826
Monel 405, fasteners ... M3: 184
Monel 502
composition ... M3: 140
property data ... M3: 140, 141
Monel 502, welding to carbon, low-alloy or stainless steels ... A6: 826
Monel, cleaning and finishing processes ... M5: 108, 670-671, 673-674
Monel, discovery and characteristics ... A2: 429
Monel, electropolishing of ... M5: 305
Monel K-500 ... A9: 435-438
age hardening ... A4: 910, 911
age-hardening ... M4: 759
annealing ... A4: 900, 910 M4: 756
composition ... A4: 908 M3: 137, 172, 748 M4: 755
corrosion resistance ... M3: 171
fasteners ... M3: 184

Nickel alloys, specific types (continued)
property data ... M3: 137-139
for screw-in tips for flame heads for oxy-fuel gas flame heating ... A4: 274
solution treating ... M4: 759
solution-treating ... A4: 911
stress-equalizing ... A4: 908
stress-relieving ... A4: 908
tensile properties at subzero temperatures ... M3: 749, 751
Monel K-500, welding to carbon, low-alloy or stainless steels ... A6: 826
Monel R-405 ... A9: 435-437
annealing ... A4: 908 M4: 756
composition ... A4: 908 M3: 137 M4: 755
property data ... M3: 137
stress-equalizing ... A4: 908
stress-relieving ... A4: 908
N06625, filler metal ... A6: 449
NASAIR 100, creep curves for ... A8: 305
NASAIR 100, steady-state creep rate ... A8: 306
NASAIR 100, time-to-rupture as function of steady-state creep ... A8: 306
Ni-20Cr, trace element evaporation ... A15: 395
Ni-20Pd-10Si, brazing ... A6: 945
Ni-25Cu (at %), dendritic solidification structure ... A9: 613
Ni-30Cu, nickel-copper phase diagram after rapid cooling ... A4: 832
Ni-Al bronze, galling resistance with various material combinations ... A18: 596
Ni-Cr-Al/bentonite, abradable seal material ... A18: 589
Ni-Cr-Al/nickel-graphite, abradable seal material ... A18: 589
Ni-Cr-Mo, distortion in heat treatment ... A4: 614
Ni-P-SiC, coatings ... A18: 836
Ni-ThO_2, coevaporation or cosputtering using multiple sources ... A18: 843
Ni_3Al-Ni_3Nb ... A9: 619
Ni_3Mo, effect on 18Ni maraging steels ... A4: 223, 224
Nickel 200 ... A9: 435-436
composition ... M3: 128, 172
corrosion resistance ... M3: 130, 171
property data ... M3: 128-130
tensile properties at subzero temperatures ... M3: 749
Nickel 200, cleaning and finishing of ... M5: 670-671, 673
Nickel 200, plastically deformed ... A9: 161
Nickel 200, slip bands and cracks ... A9: 159
Nickel 201
composition ... M3: 130, 172
corrosion resistance ... M3: 171
property data ... M3: 130-131
Nickel 270 ... A9: 435-437
composition ... M3: 131
property data ... M3: 131-132
tensile properties at subzero temperatures ... M3: 749
Nimonic 80A, preferential grain boundary precipitation ... A9: 648
Nimonic, discovery and characteristics ... A2: 429
Nimonic series, applications ... A2: 441
Permanickel 300 ... A9: 435-437
R-405 alloy, applications ... A6: 749
R-405, composition ... A6: 741
RA330, composition ... A6: 564
RA333
composition ... A6: 741
mill annealing temperature range ... A6: 573
solution annealing temperature range ... A6: 573
Udimet 700, electropolishing of ... M5: 306
UNS 08800
dissimilar metals, welding of ... A6: 578
resistance butt welding ... A6: 578
Waspaloy, applications ... A2: 441
X-750 alloy
applications ... A6: 749
composition ... A6: 577, 741
postweld strain-age cracking ... A6: 577
Nickel alloys, welding ... A6: 740-751
cast nickel alloys ... A6: 742-748
cleaning of workpieces ... A6: 740
electrodes ... A6: 742-743, 745

Nickel alloys, welding (continued)
filler metals ... **A6:** 743, 745
gas-tungsten arc welding ... **A6:** 740, 741, 742-743, 748, 749, 751
heat treatment ... **A6:** 740, 742
joint design ... **A6:** 740-741, 743
precipitation-hardenable alloys ... **A6:** 742
shielding gases ... **A6:** 742
welding characteristics ... **A6:** 587
welding fixtures ... **A6:** 741-742
welding metallurgy ... **A6:** 587-590
Nickel aluminide alloys, friction and wear of ordered intermetallics ... **A18:** 772-777
composition ... **A18:** 772
future outlook ... **A18:** 777
mechanical properties ... **A18:** 772-773, 774
physical properties ... **A18:** 772, 773
pin-on-disk friction and wear data ... **A18:** 775, 776
spray material for oxyfuel wire spray process ... **A18:** 829
structure ... **A18:** 772
Nickel aluminide alloys, specific types
IC-15
abrasive wear ... **A18:** 774
sliding wear ... **A18:** 774
IC-50
abrasive wear ... **A18:** 774
as-formed cavitation erosion rate ... **A18:** 774
cold worked, cavitation erosion rate ... **A18:** 774
composition ... **A18:** 772
mechanical properties ... **A18:** 772, 773, 774
pin-on-disk friction and wear data ... **A18:** 776
plasma weld overlay cavitation erosion rate
reciprocating ball-on-flat friction and wear data ... **A18:** 776
solid-particle erosion data ... **A18:** 773, 774
unlubricated reciprocating cylinder-on-flat friction and wear data ... **A18:** 776
IC-74M
composition ... **A18:** 772
pin-on-disk friction and wear data ... **A18:** 776
sliding wear ... **A18:** 775, 776
IC-218
abrasive wear ... **A18:** 774
cavitation erosion rate ... **A18:** 774
composition ... **A18:** 772
mechanical properties ... **A18:** 774
pin-on-disk friction and wear data ... **A18:** 776
sliding wear ... **A18:** 774
solid particle erosion ... **A18:** 773, 774
IC-218 LZr
composition ... **A18:** 772
mechanical properties ... **A18:** 773, 774
pin-on-disk friction and wear data ... **A18:** 776
sliding wear ... **A18:** 775, 776
IC-221
abrasive wear ... **A18:** 774
cavitation erosion rate ... **A18:** 774
composition ... **A18:** 772
mechanical properties ... **A18:** 774
pin-on-disk friction and wear data ... **A18:** 776
IC-357, composition ... **A18:** 772
IC-396M
composition ... **A18:** 772
mechanical properties ... **A18:** 774
reciprocating ball-on-flat friction and wear data ... **A18:** 776
Type 304H stainless steel, solid-particle erosion data ... **A18:** 773, 774
unlubricated reciprocating cylinder-on-flat friction and wear data ... **A18:** 776
Nickel aluminides *See also* Ordered intermetallics
NI_3AL aluminides ... **A2:** 914-918
NiAl aluminides ... **A2:** 918-919
Nickel aluminum coatings
bond coatings ... **A18:** 831
Nickel and nickel alloys *See also* Heat-resistant alloys ... **A16:** 835-843
abrasive flow machining ... **A16:** 517
age hardening ... **A16:** 835-836
band sawing ... **A16:** 842
binder for WC ... **A16:** 72
broaching ... **A16:** 837
CBN for precision grinding ... **A16:** 454, 455, 456
cemented carbides for machining ... **A16:** 87
chemical machining ... **A16:** 843
chemical milling ... **A16:** 583
chip breakers ... **A16:** 837
compositions ... **A16:** 836
content in cast Co alloys ... **A16:** 69
content in cermet tools ... **A16:** 95
content in low-alloy steels ... **A16:** 150
content in stainless steels ... **A16:** 682-684, 688, 689
cutoff band sawing ... **A16:** 360
cutting fluids ... **A16:** 125, 159, 836-843
drilling ... **A16:** 230, 835, 837, 838, 839, 840
electrical discharge machining ... **A16:** 558, 560
electrochemical discharge grinding ... **A16:** 548
electrochemical grinding ... **A16:** 543, 545
electrochemical machining ... **A16:** 534, 535, 843
electron beam machining ... **A16:** 570, 843
grinding ... **A16:** 760, 843
gun drilling ... **A16:** 838, 839
heat-resistant compositions ... **A16:** 737, 744
high-speed machining ... **A16:** 598, 602
high-speed tool steels used ... **A16:** 58, 59
hone forming ... **A16:** 488
honing ... **A16:** 477, 843
laser beam machining ... **A16:** 576, 843
machinability ... **A16:** 835
microdrilling ... **A16:** 238
milling ... **A16:** 112, 312-314, 755, 837, 840-841, 842
multiple-operation machining ... **A16:** 840
PCBN tools used ... **A16:** 114
photochemical machining ... **A16:** 588, 590, 591, 593, 843
planing ... **A16:** 837, 839
plating of drills ... **A16:** 219
power hacksawing ... **A16:** 757
reaming ... **A16:** 751, 839, 840
sawing ... **A16:** 358, 360, 841-843
shaping ... **A16:** 192, 837
solution annealing ... **A16:** 835
spade and gun drilling ... **A16:** 838, 839
surface alterations ... **A16:** 27
tapping ... **A16:** 263, 752, 753, 839-840
threading ... **A16:** 839-840
to harden and strengthen steel ... **A16:** 667
and trepanning ... **A16:** 180
turbine engine, electrostream and capillary drilling ... **A16:** 551
turning ... **A16:** 740, 741-742, 837, 838, 839, 840, 841
Nickel bronzes
melt treatment ... **A15:** 775
Nickel buffing compound ... **M5:** 117
Nickel carbonate additions
nickel plating ... **M5:** 201
Nickel carbonyl
chemical vapor deposition process ... **M5:** 382
Nickel carbonyl poisoning ... **A2:** 1250
Nickel carbonyl powders
apparent density ... **A7:** 273, 297
compact density ... **A7:** 310
decomposition ... **A7:** 134
effect of particle size on apparent density ... **A7:** 273
formation ... **A7:** 136
poisoning, chelating agents for ... **A7:** 203
production ... **A7:** 134-138
Nickel carbonyl vapor decomposition process
nickel plating ... **M5:** 219
Nickel cast steels ... **A1:** 373-374
as low-alloy ... **A15:** 716
Nickel chloride plating ... **M5:** 199-200, 204, 207-208, 212, 216-217
Nickel coating
chemical vapor deposition of ... **M5:** 382
molybdenum ... **M5:** 661
niobium ... **M5:** 663-664
tungsten ... **M5:** 661
undercoatings, rhodium plating process ... **M5:** 290
Nickel coatings
bonded-abrasive grains ... **A2:** 1015
Nickel, commercially pure *See* Nickel alloys, specific types, Nickel 200
Nickel content of iron-chromium-nickel heat- resistant casting alloys, effect on
sulfidation attack ... **A9:** 333
Nickel deposit molds
rotational molding ... **EM2:** 367
Nickel dip *See* Electroless nickel
Nickel, dispersion strengthened *See* Nickel alloys, specific types, DS Ni
Nickel equivalence ... **A6:** 457, 459-460, 462, 463, 464, 483, 503
Nickel equivalent ... **A6:** 817-818, 819, 825
Nickel ferrites ... **A9:** 538
precipitation process ... **EM4:** 59
sintering behavior ... **EM4:** 58
Nickel fibers in a silver matrix ... **A9:** 100
Nickel flake powders ... **A7:** 596
Nickel flash
enameling ... **EM3:** 303
Nickel flashing ... **M6:** 1016
Nickel fluoborate plating ... **M5:** 200-201
Nickel gear bronze
properties and applications ... **A2:** 375
Nickel in cast irons *See also* High-nickel cast irons; Ni-Hard cast iron ... **M1:** 77, 79, 91, 96
ductile iron ... **M1:** 39-41, 53, 54
gray iron ... **M1:** 21, 28, 29
Nickel in copper ... **M2:** 242
Nickel in steel ... **M1:** 115, 411, 417
atmospheric corrosion affected by ... **M1:** 717, 721-722
carburized, notch toughness data ... **M1:** 534-536
castings, effect on ... **M1:** 388, 394-395
hardenability affected by ... **M1:** 477
hydrogen solubility, effect on ... **M1:** 687
maraging steels ... **M1:** 445-447
notch toughness, effect on ... **M1:** 693, 695
P/M materials ... **M1:** 333, 337, 342-343
seawater corrosion, effect on ... **M1:** 741-742, 744
Steel sheet, effect on formability ... **M1:** 534
temper embrittlement, role in ... **M1:** 684
Nickel irons
magnetic applications ... **M3:** 603, 605, 608, 609, 610, 611, 612-613
Nickel, low-carbon *See* Nickel alloys, specific types, Nickel 201
Nickel metal coatings
for diamond adhesives ... **EM4:** 333
Nickel monoxide (NiO) *See also* Engineering properties of single oxides
pressure densification
pressure ... **EM4:** 301
technique ... **EM4:** 301
temperature ... **EM4:** 301
for temperature sensors ... **EM4:** 17
Nickel oxide
angle of repose ... **A7:** 283
Nickel painted polycarbonate
static decay rate ... **A7:** 612
Nickel pellets and powder products ... **A7:** 137
Nickel plate
relative solderability ... **A6:** 134
Nickel plating *See also* Electroless nickel plating ... **M5:** 199-243
acid buffering ... **M5:** 209
agitation used in ... **M5:** 214
all-chloride bath ... **M5:** 202-203, 217
all-sulfate bath ... **M5:** 201-202

Nickel plating (continued)
aluminum and aluminum alloys **M5:** 180, 203, 206, 216, 218, 604-605
annealing process **M5:** 217-218
anodes **M5:** 201, 206-207, 211-212, 217
conforming **M5:** 201, 217
insoluble **M5:** 211
antipitting agents used in **M5:** 200-203, 206, 209
applications **M5:** 199-204, 207, 215-216
automatic process **M5:** 214 216
maintenance schedules **M5:** 214-215
baking treatments **M5:** 217-218
barrel process **M5:** 202-203, 205-206, 213-215, 237
applications **M5:** 215
conditions **M5:** 202-203, 205-206
equipment **M5:** 213
maintenance schedules **M5:** 214
solution composition and operating
base metal, effects of **M5:** 215-216
black nickel process **M5:** 199, 204-205, 623
boric acid used in **M5:** 199-200, 204-205, 207, 209
brass **M5:** 206, 215-216
bright plating process **M5:** 199, 204-206
brighteners used in **M5:** 204-206
carriers used in **M5:** 204-205
cast irons **M1:** 102-103
chloride-sulfate bath **M5:** 201-202
chromium plating used in **M5:** 205, 207
cold (double salt) bath **M5:** 202-203
composite process (cadmium and nickel) **M5:** 264
contamination **M5:** 205, 207, 211
removal of **M5:** 208-211
copper and copper alloys **M5:** 199-200, 215-216, 218, 617, 621-623
copper-nickel and copper-nickel-chromium systems
magnesium alloys **M5:** 646
corrosion protection **M1:** 753, 754
corrosion resistance **M5:** 199-201, 204, 207-208
current density **M5:** 200-215
alloy composition affected by **M5:** 204, 206
selection factors **M5:** 212
decomposition process, nickel carbonyl vapor **M5:** 219
decorative **M5:** 199, 205, 218
decorative chromium plating process **M5:** 193-194
duplex system **M5:** 207-209
efficiency **M5:** 212
electroless *See* Electroless nickel plating
electrotyping bath **M5:** 202-203
equipment **M5:** 213-215
maintenance **M5:** 214-215
filtration systems **M5:** 213-214
fluoborate bath **M5:** 200-201
gaseous contamination effects **M5:** 210
general-purpose baths **M5:** 199-201
hafnium alloys **M5:** 668
hard nickel bath **M5:** 202-204
hardness **M5:** 200, 203, 206, 210-213, 217-218
heat-resisting alloys **M5:** 566-568
stripping method **M5:** 567-568
high-chloride bath **M5:** 200
high-sulfate bath **M5:** 202-203
hydrogen embrittlement caused by **M5:** 217-218
immersion process **M5:** 219
iron-nickel plating **M5:** 206-207
limitations **M5:** 215-216
magnesium alloys **M5:** 638-639, 645-647
stripping of **M5:** 646-647
maintenance schedules **M5:** 214-215
maraging steels **M1:** 448
metal contamination effects **M5:** 207-210
nickel carbonate used in **M5:** 201
nickel chloride baths **M5:** 199-200, 204, 207-208, 212, 216-217
nickel concentration **M5:** 200, 206, 210
nickel salt recovery **M5:** 318
nickel sulfate baths **M5:** 200-204, 207-208
nickel-iron alloy process **M5:** 206-207
nickel-phosphorus alloy bath **M5:** 202, 204
organic contaminant effects **M5:** 208-210
pH **M5:** 200-212, 214
power-generating equipment **M5:** 213
pumping equipment **M5:** 213
purification procedures **M5:** 208-209

Nickel plating (continued)
radius affecting **M5:** 216-217
rinsewater recovery **M5:** 318
selection of method **M5:** 215-216
solution compositions and operating conditions **M5:** 199-218, 663-664, 668
general-purpose baths **M5:** 199-201
special-purpose baths **M5:** 199, 201-204
tests **M5:** 210-211
solution control **M5:** 209-211, 214-215
maintenance schedules **M5:** 214-215
special-purpose baths **M5:** 199, 201-204
stainless steel **M5:** 204, 216
steel **M5:** 199-205, 216-217
still tank process **M5:** 212-215
applications **M5:** 215
equipment **M5:** 212-213
maintenance schedules **M5:** 215
strength **M5:** 200-201, 203-204
stress parameters **M5:** 200-202, 204-205, 211
stripping of **M5:** 218, 567-568, 647
sulfamate bath **M5:** 200-201, 216
surface activation of **M5:** 198, 216
tantalum **M5:** 663-664
temperature **M5:** 200-213, 216, 218
heating and cooling systems **M5:** 213
structural change varying with **M5:** 218
variation in, affects of **M5:** 211-212
thickness **M5:** 199-200, 207, 212, 216-217
corrosion resistance, recommended thicknesses **M5:** 199-200
variations in, effects of **M5:** 216-217
threaded fasteners **M1:** 279
tungsten **M5:** 660
water used in **M5:** 209
Watts bath **M5:** 199, 208-209, 212, 217
purification procedure **M5:** 209
weight **M5:** 212
wetting agents used in **M5:** 200, 206, 209
workpiece shape affecting **M5:** 216-217
zinc alloys **M5:** 199-200, 203-204, 210, 215-216, 218
contamination effects **M5:** 210
zirconium alloys **M5:** 668
Nickel plating for preservation of the white layer in nitrided steels **A9:** 218
Nickel plating of specimens for edge retention **A9:** 32
carbonitrided and carburized steels **A9:** 217
heat-resistant casting alloys **A9:** 330
nitrided steels **A9:** 218
stainless steel specimens **A9:** 279
in titanium and titanium alloys **A9:** 458
tungsten **A9:** 439
uranium **A9:** 478
wrought heat-resistant alloys **A9:** 305
Nickel powder metallurgy
Soviet **A7:** 693
Nickel powder strip *See also* High-purity nickel strip
applications **A7:** 403-404
effect of work roll diameter **A7:** 406
finished **A7:** 401-402
properties **A7:** 401-402
pure porous **A7:** 406
roll compacted **A7:** 401-402
Nickel powders *See also* Nickel alloy powders; Nickel powder strip; Nickel silver powders; Nickel steel powders; Nickel tetracarbonyl powders; Nickel-based hardfacing alloys; Nickel-based superalloys; Nickel-carbonyl powders; Nickel-iron powders
annealing **A7:** 182
apparent density **A7:** 273, 297
applications **A7:** 138, 141-142
carbon monoxide, surface **A7:** 258
carbonyl decomposition **A7:** 54
as carrier core, copier powders **A7:** 585
chemical analysis and sampling **A7:** 248
coating **A7:** 174
in commercial operation **A7:** 141
compacts, density distribution **A7:** 301, 314, 318
compacts, variation of sintered density **A7:** 314, 318
content of sintered compact densities tungsten powder **A7:** 318
cylindrical compact, density distribution **A7:** 301

Nickel powders (continued)
density with high-energy compacting **A7:** 305
filamentary **A7:** 138
filler for polymers **A7:** 606
general purpose **A7:** 138
general purpose, spikey **A7:** 138
high-density, semi-smooth **A7:** 138
hydrometallurgical processing **A7:** 118, 141
inoculant **A7:** 173
Kirkendall porosity **A7:** 314
liquid phase particles, in polymers **A7:** 607
nonconventional sintering **A7:** 398
porosity **A7:** 314
precipitation, effect of particle size on apparent density **A7:** 273
pressure and green density **A7:** 298
produced by atomization **A7:** 134, 142
produced by carbonyl decomposition **A7:** 138
produced by carbonyl vapormetallurgy processing **A7:** 134-138
produced by hydrometallurgical processing **A7:** 134, 138-142
produced by precipitation **A7:** 54, 297
produced by Sherritt process **A7:** 141
production **A7:** 134-143
products, sintered dense **A7:** 395-397
products, sintered porous **A7:** 395
properties and applications **A7:** 138
pyrophoricity **A7:** 199
recovery from sulfide concentrates **A7:** 138
reduction process **A7:** 173-174
refining, nickel tetracarbonyl for **A7:** 137
relative density **A7:** 305
rolling of strip **A7:** 406, 407
safety exposure limits **A7:** 203
shipment tonnage **A7:** 24
sintering **A7:** 395-398
sintering atmospheres **A7:** 397-398
tap density **A7:** 277
toxic reactions and disease symptoms **A7:** 203
toxicity **A7:** 202-203
vacuum atomized, spherical **A7:** 44, 45
Nickel reduction process **A7:** 173-174
Nickel salt
reduction under pressure (Sherritt Gordon process) **A7:** 134, 138-142
Nickel sheet and strip
powder used **A7:** 574
Nickel silver *See also* Copper and copper alloys
brazeability **M6:** 1034
chromic acid as an etchant for **A9:** 401
coining of **A14:** 184
composition and properties **M6:** 401
compositions of various types **M6:** 546
electrolytic etching **A9:** 401
electropolishing of **M5:** 306
gas tungsten arc welding **M6:** 409
physical properties **M6:** 546
pickling of **M5:** 612
powder metallurgy materials, etching **A9:** 509
resistance spot welding **M6:** 479
resistance welding *See* Resistance welding of copper and copper alloys-vs
Nickel silver powders
contacts, applications **A7:** 560
as copper alloy powder **A7:** 121, 122
dimensional change **A7:** 292
mechanical properties **A7:** 738, 740
microstructural analysis **A7:** 488-489
P/M parts **A7:** 737-739
P/M parts, effect of lithium stearate lubrication **A7:** 191
as pressed and sintered prealloyed powders **A7:** 464
properties **A7:** 122
sintering of **A7:** 378-381
Nickel silvers *See also* Copper-zinc-nickel alloys **A6:** 752
55-18 *See* Copper alloys, specific types, C77000
65-10 *See* Copper alloys, specific types, C74500
65-12 *See* Copper alloys, specific types, C75700
65-15 *See* Copper alloys, specific types, C75400
65-18 *See* Copper alloys, specific types, C75200
applications **A2:** 228, 342-345
brazing **A6:** 630
copper-base structural parts with **A2:** 396
corrosion in various media **M2:** 468-469

Nickel silvers (continued)
corrosion resistance ... **A13:** 611
foundry properties for sand casting ... **A2:** 348
gas-metal arc butt welding ... **A6:** 760
oxyacetylene welding ... **A6:** 281
properties ... **A2:** 228, 342-345
relative weldability ratings, resistance spot welding ... **A6:** 834
resistance welding ... **A6:** 850
solderability ... **A6:** 978
thermal conductivity ... **A6:** 754
thermal expansion coefficient ... **A6:** 907
weldability ... **A6:** 753
Nickel silvers, specific types
12% *See* Copper alloys, specific types, C97300
20% *See* Copper alloys, specific types, C97600
25% *See* Copper alloys, specific types, C97800
Nickel, specific types
213, galling resistance with various material combinations ... **A18:** 596
270, erosion-enhanced corrosion ... **A18:** 208
305, galling resistance with various material combinations ... **A18:** 596
Nickel 200
annealing ... **A4:** 908 **M4:** 756
bright annealing ... **A4:** 909
composition ... **A4:** 908 **M4:** 755
stress equalizing ... **M4:** 756
stress relieving ... **M4:** 756
stress-equalizing ... **A4:** 908
stress-relieving ... **A4:** 908
Nickel 201
annealing ... **A4:** 908 **M4:** 756
composition ... **M4:** 755
compositions ... **A4:** 908
stress equalizing ... **M4:** 756
stress relieving ... **M4:** 756
stress-equalizing ... **A4:** 908
stress-relieving ... **A4:** 908
Nickel steel
flux-cored arc welding designator ... **A6:** 189
thermal properties ... **A18:** 42
Nickel steels
powder metallurgy materials, etching ... **A9:** 508
powder metallurgy materials, heterogeneity in ... **A9:** 503
powder metallurgy materials microstructures ... **A9:** 510
powder metallurgy materials, pressed and part ... **A9:** 530
sintered, different densities from the same
powder metallurgy materials, pressed and sintered, copper and carbon varied ... **A9:** 521-522
powder metallurgy materials undersintered ... **A9:** 529
surface relief from formation of upper bainite ... **A9:** 662
Nickel striking
copper plating process ... **M5:** 160
stainless steel ... **M5:** 232
Nickel strip *See* Nickel powder strip
Nickel sulfamate plating ... **M5:** 200-201, 216
Nickel sulfate plating ... **M5:** 200-204, 207-208
Nickel tetracarbonyl powders ... **A7:** 134-137
formation as exothermic ... **A7:** 137
properties ... **A7:** 136
for refining nickel ... **A7:** 137
stability ... **A7:** 137
system pressure and temperature effect in formation ... **A7:** 136
vapor, toxicity of ... **A7:** 136-137
Nickel, thoria dispersed *See* Nickel alloys, specific types, DS Ni
Nickel toxicity
disposition and toxicology ... **A2:** 1250
Nickel, zone refined
impurity concentration ... **M2:** 713
Nickel-20% chromium-1% aluminum alloy, microstructure of ... **A3:** 1•22
Nickel-alloy castings, corrosion resistant
applications ... **M3:** 175-178
castability ... **M3:** 175
classification ... **M3:** 175
nickel-chromium-iron ... **M3:** 177
nickel-chromium-molybdenum ... **M3:** 178
nickel-copper ... **M3:** 176-177
nickel-molybdenum ... **M3:** 178
proprietary alloys ... **M3:** 176, 178, 180, 182
Nickel-aluminum allloys
as bond coats for thermal spray coatings ... **A6:** 813
shielded metal arc welding ... **A6:** 755
Nickel-aluminum alloys
301, galling resistance with various material combinations ... **A18:** 596
abradable seal material ... **A18:** 589
Nickel-aluminum bronze *See also* Copper casting alloys
applications and properties ... **A2:** 331-332
nominal composition ... **A2:** 347
Nickel-aluminum composite powder ... **A7:** 173, 174
for hardfacing ... **A7:** 829-830
thermal spray ... **A7:** 173, 175
Nickel-aluminum oxidation protective coating
superalloys ... **M5:** 376
Nickel-aluminum-molybdenum ... **A9:** 621
Nickel-aluminum-molybdenum-tantalum alloys
time-to-rupture ... **A8:** 304
Nickel-antimony system phase diagrams
peritectic transformation ... **A9:** 677
Nickel-base alloys *See also* High-nickel alloys; Nickel; Nickel-base alloys, specific types; Superalloys ... **A13:** 641-657 **A14:** 261-266, 831-837
AAS furnace atomizer for trace analysis in ... **A10:** 55
in acid media, corrosion resistance ... **A13:** 643-647
advantages ... **A6:** 797
AEM determined orientation relationships in ... **A10:** 454
age-hardened, FIM/AP study of precipitates in ... **A10:** 583
alloying elements, effects ... **A13:** 641-642
amorphous, NMR Knight shifts ... **A10:** 284
applications ... **A6:** 797 **A13:** 653-656
arc-welded, failures in ... **A11:** 433-434
bar forming ... **A14:** 836
bending of ... **A14:** 834-836
blanking of ... **A14:** 832-833
brazing ... **A6:** 927-928
brazing and soldering characteristics ... **A6:** 631-632
bulk composition of ... **A10:** 562
cavitation erosion ... **A18:** 768, 769
characteristics ... **A13:** 641
chlorine corrosion ... **A13:** 1171
coextrusion welding ... **A6:** 311
cold extrusion of ... **A14:** 836-837
cold heading of ... **A14:** 836-837
cold-formed parts, for high-temperature service ... **A14:** 837
compositions ... **A14:** 831
cooling after forging ... **A14:** 263
corrosion in ... **A11:** 200-201, 202
corrosion of ... **A13:** 641-657
corrosion performance ... **A13:** 653-656
corrosion resistance ... **A6:** 585
corrosion resistance in alkalies ... **A13:** 647
corrosion resistance, molten salts ... **A13:** 90
in corrosive environments ... **A13:** 643-647
cracking due to creep ... **A11:** 284
crevice corrosion ... **A13:** 305
critical melting and precipitation temperatures ... **A14:** 265
cutting tool material selection based on machining operation ... **A18:** 617
Nickel-base alloys (continued)
cutting tool materials and cutting speed relationship ... **A18:** 616
deep drawing of ... **A14:** 833
dendritic solidification structure in ... **A10:** 306
die materials ... **A14:** 261
diffraction techniques, elastic constants, and bulk values for ... **A10:** 382
diffusion welding ... **A6:** 885, 886
dip brazing ... **A6:** 336
electrochemical interaction, and corrosion ... **A13:** 643
electrodes for flux-cored arc welding ... **A6:** 189
electronic applications ... **A6:** 990
electroslag welding ... **A6:** 278
environmental embrittlement ... **A13:** 647-652
expanding of ... **A14:** 836
explosive forming ... **A14:** 783
fabrication and weldability ... **A13:** 652-653
FIM sample preparation of ... **A10:** 586
finish forging of ... **A14:** 262
flux-cored arc welding ... **A6:** 186, 187, 188
forge welding ... **A6:** 306
forging of ... **A14:** 261-266
forging temperature ranges ... **A14:** 263
forming, lubricants for ... **A14:** 520
forming practice ... **A14:** 782-783
fretting wear ... **A18:** 251
friction surfacing ... **A6:** 321
furnace brazing ... **A6:** 330
fusion welding to steels ... **A6:** 827-828
galling of ... **A14:** 831
galvanic corrosion of ... **A13:** 85
-boundary precipitation ... **A13:** 155
grain ... **A13:** 154
grain-boundary compositions ... **A10:** 562
grain-boundary segregation ... **A13:** 156
hardfacing ... **A6:** 789, 790, 794-795, 796, 799
heat-resistant, forging of ... **A14:** 233-234
heating for forging ... **A14:** 261-263
high, metallurgical effects on corrosion ... **A13:** 128-130
high temperature, compositions ... **A14:** 262
high-performance, hydrogen fluoride/hydrofluoric acid corrosion ... **A13:** 1168
hot corrosion of ... **A13:** 102
hot-die/isothermal forging of ... **A14:** 150-157, 265-266
hot-forming pressures ... **A14:** 262
hydrochloric acid corrosion of ... **A13:** 1162
hydrogen damage ... **A13:** 169
hydrogen flaking in ... **A11:** 337
hydrogen fluoride/hydrofluoric acid corrosion ... **A13:** 1168
hydrogen mitigation, underwater welding ... **A6:** 1011-1012
hydrogen peroxide dissolution medium ... **A10:** 166
intergranular corrosion evaluation of ... **A13:** 239-240
intergranular corrosion of ... **A11:** 181-182
intermetallic phases ... **A13:** 642-643
isothermal section, AEM determined ... **A10:** 475
jet-engine turbine blade, high-temperature fatigue fracture ... **A11:** 131
LEISS depth profile ... **A10:** 609
liquation cracking ... **A6:** 75
lubricants for ... **A14:** 261, 831-832
lubrication of tool steels ... **A18:** 737, 738
material for jet engine components ... **A18:** 591
mechanical properties of ... **A1:** 984
in moist chlorine ... **A13:** 1173
molten salt corrosion ... **A13:** 89
mutual solubility of ... **A11:** 452
nominal chemical compositions ... **A13:** 644
not readily diffusion bondable ... **A6:** 156
ordered, effect of antiphase boundaries on FIM image ... **A10:** 589, 590

Nickel-base alloys (continued)
ordered Ni-Mo, effects of antiphase boundaries on FIM image **A10:** 589-590
oxide-dispersion-strengthened materials **A6:** 1039
petroleum refining and petrochemical operations **A13:** 1263
in pharmaceutical production facilities **A13:** 1227
physical properties of **A1:** 983
piercing of **A14:** 832-833
pipe bending **A14:** 834-835
plate, as edge protection **A11:** 24
plate bending **A14:** 835-836
projection welding **A6:** 233
pulse-fraction curves for atom probe analysis **A10:** 595
rod forming **A14:** 836
roll welding **A6:** 312
SCC resistance **A13:** 328-329
shearing of **A14:** 832-833
sheet bending **A14:** 835-836
slurry erosion **A18:** 768, 770
sodium hydroxide corrosion **A13:** 1176
specialty, intergranular corrosion **A13:** 325
spinning of **A14:** 833-834
straightening of **A14:** 837
strain hardening of **A14:** 831
stress-corrosion cracking in **A11:** 223
stress-rupture strengths for **A1:** 985
stretching and necking due to creep **A11:** 283
strip bending **A14:** 835-836
sulfuric acid corrosion **A13:** 1152
tempers for **A14:** 831
thermal fatigue cracks **A11:** 284
thermal spray coating recommended **A18:** 832
for thermal spray coatings **A13:** 461
thermal-mechanical processing **A14:** 265
tools and equipment for forming **A14:** 832
trace analysis for aluminum in **A10:** 66
trace element impurity effect on GTA weld penetration **A6:** 20
tube bending **A14:** 834-835
ultrasonic welding **A6:** 895
weld cladding **A6:** 822
weld metal, SCC susceptibility **A13:** 328
weldments **A13:** 361-362
for wire-drawing dies **A14:** 336
Nickel-base alloys, heat treating *See also* Heat-resisting alloys, heat treating
aging **M4:** 668
annealing **M4:** 655, 664-665
atmospheres **M4:** 665
solution treating **M4:** 666
stress relieving **M4:** 655, 664-665
temperatures **M4:** 655, 656, 657, 664-669
weldments **M4:** 668-669
Nickel-base alloys, special metallurgical welding considerations **A6:** 575-579
creep-resistant secondary carbide-strengthened alloys welded with nickel alloys **A6:** 577
heat-affected zone **A6:** 577
overaging **A6:** 575
precipitation-hardenable nickel alloys **A6:** 575-577
low-coefficient-of-expansion alloys **A6:** 576
mechanically alloyed products **A6:** 576-577
postweld strain-age cracking (PWSAC) **A6:** 576
solid-solution nickel alloys **A6:** 575
special welded product conditions **A6:** 577-579
Nickel-base alloys, specific types *See also* Nickel; Nickel-base alloys; Superalloys, specific types
20Cb-3, intergranular corrosion **A13:** 325
80% Ni; 20% Cr, thermal properties **A18:** 42
90% Ni; 10% Cr, thermal properties **A18:** 42
200, alkali resistance **A13:** 647
200 commercially pure, intergranular corrosion **A13:** 325
200, hydrogen fluoride/hydrofluoric acid corrosion **A13:** 1168
200, oxidizing ion effect, corrosion in boiling acetic acid **A13:** 648
200, SCC susceptibility **A13:** 328
201, alkali resistance **A13:** 647
201, SCC susceptibility **A13:** 328
600, crevice corrosion **A13:** 111

Nickel-base alloys, specific types (continued)
600, flux-cored arc welding **A6:** 188
600, grain-boundary segregation measurements **A13:** 157
600, SCC susceptibility **A13:** 328
600 tubing, primary-side SCC **A13:** 941
625, crevice corrosion **A13:** 111
625, flux-cored arc welding **A6:** 188
690, crevice corrosion **A13:** 111
800, high-temperature corrosion **A13:** 99-101
800, SCC susceptibility **A13:** 328
825, crevice corrosion **A13:** 109
A-286, recommended upsetting pressures for flash welding **A6:** 843
alloy 200, forging practice **A14:** 263
alloy 301, forging practice **A14:** 263
alloy 400, forging practice **A14:** 263-264
alloy 600, forging practice **A14:** 264
alloy 625, forging practice **A14:** 264
alloy 706, forging practice **A14:** 264
alloy 718, forged and machined **A14:** 264
alloy 718, forging practice **A14:** 264
alloy 722, forging practice **A14:** 265
alloy 751, forging practice **A14:** 265
alloy 800, forging practice **A14:** 264
alloy 825, forging practice **A14:** 264
alloy 901, thermal mechanical processing **A14:** 265
alloy 903, forging practice **A14:** 265
alloy 907, forging practice **A14:** 265
alloy 909, forging practice **A14:** 265
alloy 925, forging practice **A14:** 264-265
alloy K-500, forging practice **A14:** 264
alloy X-750, forging practice **A14:** 264
B-1900, stress-rupture stress vs rupture life **A11:** 267
Hastelloy B, weldment corrosion **A13:** 362
Hastelloy B-2, weldment corrosion **A13:** 362
Hastelloy C, weldment corrosion **A13:** 362
Hastelloy C-22, AEM-determined orientation relationships **A10:** 454
Hastelloy C-22, corrosion resistance **A13:** 324
Hastelloy C-22, weld soundness in **A10:** 478-481
Hastelloy C-276, cyclic potentiodynamic polarization curves **A13:** 218
Hastelloy C-276, slow strain rate SCC testing **A13:** 274
Hastelloy C-276, weld metal corrosion **A13:** 324
Hastelloy C-276, weldment corrosion **A13:** 362
Hastelloy G, intergranular corrosion **A13:** 325
Hastelloy G-3, intergranular corrosion resistance **A13:** 325
Hastelloy N, molten salt corrosion resistance **A13:** 53-54
Incoloy 800, SCC susceptibility **A13:** 328-329
Incoloy 825, SCC resistance **A13:** 328
Incoloy 901, diffraction techniques, elastic constants, and bulk values for **A10:** 382
Inconel 600, chromium sensitization in **A10:** 483
Inconel 600, diffraction techniques, elastic constants, and bulk values for **A10:** 382
Inconel 600, EPMA spectrum, intergranular fracture **A11:** 40
Inconel 600, hydrogen fluoride/hydrofluoric acid corrosion **A13:** 1168
Inconel 600, SCC of safe-end on reactor nozzle **A11:** 660-661
Inconel 600, SCC susceptibility **A13:** 328
Inconel 600, steam line bellows failure **A11:** 38-41
Inconel 600 tubing, residual stress and cold work distribution **A10:** 390
Inconel 600 tubing, residual stress and percent cold work distributions **A10:** 390
Inconel 601, catastrophic sulfidation **A13:** 1313
Inconel 625, chemical composition as function of etching time **A10:** 559
Inconel 625, corrosion rate **A10:** 558
Inconel 625, intergranular corrosion **A13:** 325
Inconel 625, SCC resistance **A13:** 328
inconel 690, SCC susceptibility **A13:** 328
Inconel 718, diffraction techniques, elastic constants, and bulk values for **A10:** 382
Inconel 718, x-ray elastic constant determination for **A10:** 388

Nickel-base alloys, specific types (continued)
Inconel, liquid-erosion resistance **A11:** 167
Inconel, SCC failures in **A11:** 27
Inconel X-750, diffraction technique, elastic constants, and bulk values for **A10:** 382
Inconel X-750 springs, SCC failure **A11:** 559-560
MAR-M 246, effect of temperature increase on microstructure of **A11:** 282
Monel 400, hydrogen fluoride/hydrofluoric acid corrosion **A13:** 1168
Monel 400, oxidizing ions effect, corrosion in boiling acetic acid **A13:** 648
Monel 400, oxygen effect on hydrofluoric acid corrosion **A13:** 645, 647
Monel 400, SCC susceptibility **A13:** 328
Monel, effect of aeration/deaeration **A13:** 221
Monel, effect of temperature on corrosion **A13:** 220
Monel K500, hydrogen fluoride/hydrofluoric acid corrosion **A13:** 1168
Monel, pitting corrosion **A13:** 120
Ni-25Cr alloy, abrasive wear **A18:** 189-190
Ni-Cr, advantages and applications of materials for surfacing, build-up, and hardfacing **A18:** 650
Ni-Cr-B
advantages and applications of materials for surfacing, build-up, and hardening **A18:** 650
classification and composition of hardfacing alloys **A18:** 652
Ni-Cr-Mo, advantages and applications of materials for surfacing, build-up, and hardfacing **A18:** 650
Ni-Cr-Mo-W
advantages and applications of materials for surfacing, build-up, and hardening **A18:** 650
classification and composition of hardfacing alloys **A18:** 652
Ni-Ti-Co, orthodontic wires, properties **A18:** 666
Nimonic PE 16, liquid sodium corrosion **A13:** 91
René 95, thermal-mechanical processing **A14:** 265
René 96, diffraction techniques, elastic constants, and bulk values for **A10:** 382
U-700, continuous grain-boundary precipitate in **A10:** 308
U-700, stress-rupture stress vs rupture life **A11:** 267
Waspaloy, flow curves **A14:** 374
Waspaloy, hot-die/isothermal forging of **A14:** 152
Waspaloy, pseudo binary phase diagrams **A14:** 236
Waspaloy, thermal mechanical processing **A14:** 265
X-750, jet pump beams, SCC of **A13:** 933-935
Nickel-base corrosion-resistant alloys containing molybdenum, selection of **A6:** 593-597
alloy B family **A6:** 593-594
alloy C family **A6:** 593
alloy G family **A6:** 594
annealing **A6:** 596
argon-oxygen decarburization (AOD) process **A6:** 593
cast CR alloy welding **A6:** 597
chemical compositions of most widely used materials in family **A6:** 593
clad plate, joining of **A6:** 597
dissimilar metals welding **A6:** 596-597
dye-penetrant inspection **A6:** 597
electron-beam welding **A6:** 594
friction welding **A6:** 594
fusion zone **A6:** 595-596
gas hole formation **A6:** 595-596
gas-metal arc welding **A6:** 594
gas-tungsten arc welding **A6:** 594
heat-affected zone **A6:** 594, 595, 596, 597
joining to ferrous alloys **A6:** 597
laser-beam welding **A6:** 594
oxyacetylene welding not recommended **A6:** 594
plasma arc welding **A6:** 594

Nickel-base corrosion-resistant alloys containing molybdenum, selection of (continued)
postweld heat treatment **A6:** 596
repair welding of wrought alloy parts **A6:** 597
resistance spot welding **A6:** 594
shielded metal arc welding **A6:** 594
shot peening **A6:** 596
stringer bead welding **A6:** 594
submerged arc welding not recommended **A6:** 594
welding characteristics **A6:** 594
welding metallurgy **A6:** 594-596
Nickel-base corrosion-resistant alloys containing molybdenum, specific types
B-2
adverse effects of grain boundary precipitation in HAZ **A6:** 595
annealing **A6:** 596
composition **A6:** 593
development **A6:** 593
fusion zone attack **A6:** 595
metallographic cross sections of corroded surfaces **A6:** 596
microstructure **A6:** 593-594
nitrogen solubility **A6:** 596
physical properties **A6:** 593-594
shielded metal arc welding **A6:** 594
solution annealing temperatures **A6:** 596
C-22
composition **A6:** 593
development **A6:** 593
hot cracking **A6:** 596
nitrogen solubility **A6:** 596
physical properties **A6:** 594
solution annealing temperatures **A6:** 596
to refurbish corroded welds in alloy C-276 parts **A6:** 595
C-276
composition **A6:** 593
corrosion attack caused by grain boundary precipitation **A6:** 595
corrosion resistance **A6:** 593
development **A6:** 593
hot cracking **A6:** 596
refurbished in corroded welds with C-22 .... **A6:** 595
solution annealing temperatures **A6:** 596
C4
composition **A6:** 593
development **A6:** 593
hot cracking **A6:** 596
solution annealing temperatures **A6:** 596
G-3
composition **A6:** 593
development **A6:** 594
refurbished in corroded welds with alloy G-30 .... **A6:** 595
solution annealing temperatures **A6:** 596
G-30
composition **A6:** 593
development **A6:** 594
physical properties **A6:** 594
solution annealing temperatures **A6:** 596
to refurbish welds in alloy G-3 parts **A6:** 595
N
composition **A6:** 593
solution annealing temperatures **A6:** 596
Nickel-base electrodes
as cast iron welding consumable **A15:** 523-524
Nickel-base filler alloys
brazing corrosion resistance **A13:** 883-884
Nickel-base hardfacing alloy, improving .... **A3:** 1•27
Nickel-base hardfacing alloys
laser cladding materials **A18:** 866-867
Nickel-base heat-resistant alloys
HERF forgeability **A14:** 104
machining **A16:** 738, 741-743, 746-747, 749-757
Nickel-base heat-resistant casting alloys
compositions of **A9:** 331
microstructures **A9:** 334
Nickel-base heat-resistant casting alloys, specific types
Alloy 713C, as-cast, different carbide structures **A9:** 344
Alloy 713C, different holding temperatures and times compared **A9:** 345
Alloy 718, vacuum cast and solution annealed **A9:** 345
Alloy 7130, exposed to sulfidation **A9:** 345
B-1900, as-cast, different magnifications compared **A9:** 341
B-1900, shell mold casting, as-cast **A9:** 342
B-1900, shell mold casting, solution annealed and aged **A9:** 342
Hastelloy B, as-cast and annealed **A9:** 342
Hastelloy C, as-cast and annealed **A9:** 342
IN-100, as-cast, different magnifications compared **A9:** 342-343
IN-100, creep-rupture tested **A9:** 344
IN-100, different holding temperatures and times compared **A9:** 343-344
IN-738, after holding **A9:** 345
IN-738, as-cast **A9:** 346
IN-738, solution annealed **A9:** 346
MAR-M 246, after high temperature exposure **A9:** 346-347
MAR-M 246, as-cast **A9:** 346
MAR-M 246, different cooling times compared **A9:** 346
TRW-NASA VIA, as-cast **A9:** 347
U-700, as-cast, and solution annealed **A9:** 347
U-700, cast blade after cyclic sulfidation- erosion testing **A9:** 347
Nickel-base nodulizers **A1:** 39
Nickel-base steels
flux-cored arc welding **A6:** 187
relative solderability **A6:** 134
Nickel-base superalloy composites with molybdenum wires
polishing **A9:** 591
Nickel-base superalloys *See also* Nickel alloys; Specific types; Superalloys; Wrought heat-resistant alloys
applications **A6:** 83
aerospace **A6:** 386, 387
cast nickel-base superalloys **A1:** 981-985, 986-994
constitutional liquation **A6:** 75
directionally solidified superalloys **A1:** 995, 996-998
effects of reactive atmospheres **A12:** 40
elemental cobalt alloying **A2:** 446
embrittlement in **A12:** 123
etching of welded joints **A9:** 580
extreme-temperature solid lubricants **A18:** 118
fracture/failure causes illustrated **A12:** 217
heat treatments **A6:** 566-567
high-resolution energy-compensated atom probe analysis of **A10:** 597
hot isostatic pressing, effects **A15:** 539
investment cast, turbine blades of **A15:** 207
microstructures **A9:** 309
ordered intermetallic effects **A2:** 914
phase chemistry and stability in **A10:** 598
postweld heat treatment **A6:** 83, 92, 93, 566-567
powder metallurgy (P/M) superalloys **A1:** 972-976
pulse-fraction curves **A10:** 594, 595
reheat cracking **A6:** 92, 93
for rolling **A14:** 356
single-crystal superalloys **A1:** 995-996, 998-1006
solid-state transformations in weldments **A6:** 83-84
solidification structures in welded joints **A9:** 580
strain-age cracking **A6:** 83
thermal expansion coefficient **A6:** 907
Nickel-base superalloys (continued)
VIM melt protocol for **A15:** 394
wrought nickel-base superalloys **A1:** 950-959, 968-972
Nickel-base superalloys, arc-welded
failures in **A11:** 433-434
Nickel-base superalloys, specific types
718
liquation cracking **A6:** 568-569, 570
postweld heat treatments **A6:** 570
strain-age cracking **A6:** 566, 568-569, 570
731, strain-age cracking **A6:** 566
IN 939, atom probe mass spectra of γ matrix and primary γ precipitates **A10:** 598
IN 939, γ phase analysis **A10:** 598
IN 939, chemical analysis of ultrafine secondary precipitates **A10:** 598
IN 939, composition of **A10:** 598
IN 939, FIM and TEM images of **A10:** 598
IN 939, FIM/AP analysis of **A10:** 598
IN 939, four-stage heat treatment of **A10:** 598
IN 939, ladder diagrams **A10:** 599
Astroloy, effect of stress intensity factor range on fatigue crack growth rate **A12:** 57-58
Astroloy, effects of vacuum on fatigue **A12:** 48
Astroloy, fatigue fracture in air/vacuum **A12:** 48, 54
CMSX-2, single-crystal, hydrogen embrittlement **A12:** 389
illuminations compared **A9:** 81
IN-100 nickel-base, dendritic stress-rupture fracture **A12:** 391
IN-718, effect of frequency on fracture appearance **A12:** 58, 60
IN-738, effect of frequency and wave form on fatigue properties **A12:** 63
IN-738, fatigue and creep fractures **A12:** 389
Incoloy 800, effect of stress intensity factor range on fatigue crack growth rate **A12:** 58
Incoloy 800, effects of temperature **A12:** 52
Incoloy 800, ridges and striations in sulfidizing atmosphere **A12:** 41, 53
Incoloy 800, wedge cracking **A12:** 26
Incoloy X750, fatigue fracture mechanisms **A12:** 392
Inconel 600, corrosion fatigue **A12:** 45-46
Inconel 600, effect of frequency and wave form-n on fatigue properties **A12:** 60
Inconel 625, reactively sputtered **A9:** 158
inconel 625, wedge cracking **A12:** 26
Inconel 718 (UNS N07718), evaluation **A12:** 393
Inconel 718, effects of stress intensity factor range on fatigue crack growth rate **A12:** 58
Inconel 718, heat treated, with niobium carbides, different illuminations compared **A9:** 82
Inconel X-750, effect of frequency and wave form on fatigue properties **A12:** 59-60
Inconel X-750, effect of stress intensity factor range on fatigue crack growth rate **A12:** 57
Inconel X-750, effect of temperature **A12:** 52
Inconel X-750, effect of temperature on double- aged **A12:** 47
inconel X-750, effect of vacuum **A12:** 48
Inconel X-750, fatigue fracture appearance **A12:** 58
Inconel X-750, fatigue fractures in air/vacuum **A12:** 48, 55
Inconel X-750, light/SEM fractographs compared **A12:** 94, 96
Nimocast PK24, applications **A6:** 319
Nimonic 90, IAP study of **A10:** 596
RA 333, precipitated particles in **A9:** 126
Udimet 720, creep fracture **A12:** 395

Nickel-base superalloys, specific types (continued)
Udimet 720, fatigue fracture ... **A12:** 395
Udimet 720, multiple transgranular fatigue origins ... **A12:** 395
Waspaloy, solution annealed and aged different illumination modes compared ... **A9:** 150
Nickel-base/boride-type alloys ... **A18:** 762, 763-765
compositions ... **A18:** 762, 763
corrosion resistance ... **A18:** 765
galling ... **A18:** 764-765
microstructure ... **A18:** 763, 764
phases formed ... **A18:** 764
properties ... **A18:** 764
spray-and-fuse process ... **A18:** 765
Nickel-based alloys
brazing ... **M6:** 1018-102
atmospheres ... **M6:** 101
cleanliness ... **M6:** 1018-101
Inconel 718 ... **M6:** 102
precipitation-hardenable alloys ... **M6:** 101
stresses ... **M6:** 1019-102
thermal cycles ... **M6:** 102
cross-wire projection welding ... **M6:** 518
diffusion welding ... **M6:** 677
electroslag welding ... **M6:** 22
extrusion welding ... **M6:** 677
forge welding ... **M6:** 67
friction welding ... **M6:** 722
hardfacing materials ... **M6:** 775
plasma arc welding ... **M6:** 214
roll welding ... **M6:** 676
Nickel-based, cobalt-based alloys, specific types
Haynes 208, mechanical properties and compositions ... **A7:** 472
Haynes 711, mechanical properties and compositions ... **A7:** 472
Haynes N-6, mechanical properties and compositions ... **A7:** 472
Star J Metal, mechanical properties and compositions ... **A7:** 472
Stellite 3, mechanical properties and compositions ... **A7:** 472
Stellite 6, mechanical properties and compositions ... **A7:** 472
Stellite 19, mechanical properties and compositions ... **A7:** 472
Stellite 31, mechanical properties and compositions ... **A7:** 472
Stellite 98 M2, mechanical properties and compositions ... **A7:** 472
Stellite 190, mechanical properties and compositions ... **A7:** 472
Nickel-based dispersion-strengthened materials ... **A7:** 722-727
Nickel-based hardfacing alloys ... **A7:** 142, 825-827
corrosion rates ... **A7:** 828
melting ranges ... **A7:** 142
wear data ... **A7:** 828
Nickel-based heat-resistant alloys *See* specific types and Heat-resistant metals and alloys
Nickel-based superalloys *See also* Nickel-based superalloys, specific types; Superalloys; superalloys
compositions of experimental dispersion- strengthened ... **A7:** 527
consolidation by hot isostatic pressing ... **A7:** 439-441
contaminants ... **A7:** 178
creep-rupture testing of ... **A8:** 302
dendrite arm spacing and particle size in atomization ... **A7:** 33, 48
elevated temperature tension testing in air ... **A8:** 36
high-cycle corrosion behavior ... **A8:** 254, 255
microstructure in rapid solidification ... **A7:** 47, 48
specific types
Nickel-based superalloys, specific types *See also* Nickel-based superalloys; Superalloys; Superalloys, specific types
AF115, HIP for ... **A7:** 440
Astroloy, mechanical properties ... **A7:** 441
IN-100, and MERL 76 ... **A7:** 440
IN-100, boundaries by carbide precipitates ... **A7:** 428
IN-100, mechanical properties ... **A7:** 441
MAR M002, Osprey-atomized ... **A7:** 531
Nickel-based superalloys, specific types (continued)
MERL 76 ... **A7:** 440
Nimonic alloy AP1, mechanical properties of hot isostatically pressed plus conventionally forged ... **A7:** 442
René 95, mechanical properties ... **A7:** 441
René 95, tensile and creep properties of hot pressed ... **A7:** 442
Nickel-bonded titanium carbide cermets
applications and properties ... **A2:** 995
Nickel-boron
electroless coatings ... **A18:** 837
Nickel-boron coatings
electroless ... **M5:** 229-231
Nickel-carbon system
stability of diamond/graphite ... **A2:** 1009
Nickel-chrome alloys
for heat treating furnace equipment ... **A4:** 468, 471, 472
Nickel-chromium (Ni-Cr)
alloys, adhesion of porcelain to metal substrates ... **EM4:** 1093
redox reactions in metal alloy-glass sealing ... **EM4:** 489, 490
resistive thin film production ... **EM4:** 543
Nickel-chromium alloy *See* Nickel alloys, cast, specific types, CY-40
Nickel-chromium alloy vacuum coatings ... **M5:** 402-403
Nickel-chromium alloys
alloy basis for dental crowns ... **A18:** 673
applications and properties ... **A2:** 438
characteristics and types ... **A2:** 436, 441-442
cleaning of ... **M5:** 270-273
creep rupture testing of ... **A8:** 302
dissimilar metal joining ... **A6:** 825-826
fretting wear ... **A18:** 251
galling resistance with various material combinations ... **A18:** 596
as heating alloys ... **A2:** 828
joining ... **A2:** 831
laser cladding components and techniques ... **A18:** 869
Nimonic series, discovery and characteristics ... **A2:** 429
recommended gap for braze filler metals ... **A6:** 120
in sheathed heaters ... **A2:** 831
spray material for oxyfuel wire spray process ... **A18:** 829
Nickel-chromium alloys dental ... **A13:** 1351
PFM ... **A13:** 1356
weldment corrosion ... **A13:** 361-362
Nickel-chromium alloys, selection of
alloying of elements effect ... **A6:** 588-590
applications in severely corrosive environments with extreme temperatures ... **A6:** 586-587
compositions of selected popular alloys ... **A6:** 586
consumption and applications by western nations in 1990 ... **A6:** 586
electron-beam welding ... **A6:** 587, 588
fabrication ... **A6:** 587
fusion zone ... **A6:** 588
gas-metal arc welding ... **A6:** 587, 588
gas-tungsten arc welding ... **A6:** 588
heat-affected zone ... **A6:** 587-590
postweld heat treatment ... **A6:** 590
special metallurgical welding considerations ... **A6:** 590-592
thermal conductivity ... **A6:** 587
unmixed zone ... **A6:** 588, 589
welding characteristics ... **A6:** 587
welding metallurgy ... **A6:** 587-590
Nickel-chromium alloys, specific types
600
annealing temperatures ... **A6:** 590
composition ... **A6:** 586
constitutional liquation and carbide precipitation ... **A6:** 588
heat treatment procedures recommended ... **A6:** 590
hot cracking ... **A6:** 587-588
intergranular carbides and stress-corrosion cracking ... **A6:** 587
Nickel-chromium alloys, specific types (continued)
iron as ferroalloy to keep costs down ... **A6:** 589
manganese segregation ... **A6:** 588
properties ... **A6:** 587
welding products for dissimilar-metal joints ... **A6:** 591
601
annealing temperatures ... **A6:** 590
composition ... **A6:** 586
properties ... **A6:** 587
welding products for dissimilar-metal joints ... **A6:** 591
690
annealing temperatures ... **A6:** 590
composition ... **A6:** 586
intergranular carbides and stress-corrosion cracking ... **A6:** 587
properties ... **A6:** 587
welding products for dissimilar-metal joints ... **A6:** 591
Nickel-chromium cast irons
abrasion resistance ... **M1:** 87-88
compositions ... **M1:** 76, 82
heat treatment ... **M1:** 81-82
mechanical properties ... **M1:** 86
physical properties ... **M1:** 88
Nickel-chromium coatings ... **A13:** 427-429
uniformity ... **A7:** 174
Nickel-chromium sprayed undercoatings
ceramic coating process ... **M5:** 539
Nickel-chromium superalloys
ceramic-bonded fluoride coatings for lubricants ... **A18:** 118
Nickel-chromium white irons
abrasion-resistant
applications ... **A15:** 681
composition ... **M4:** 558
composition control ... **A15:** 679-680
heat treatment ... **A15:** 680
machining ... **A15:** 680-681
melting practice ... **A15:** 678-679
molds, patterns, and casting design ... **A15:** 680
pouring ... **A15:** 680
shakeout ... **A15:** 680
special ... **A15:** 681
stress relieving ... **M4:** 558
Nickel-chromium-aluminum resistance alloys *See also* Electrical resistance alloys
properties and applications ... **A2:** 823, 825, 835-839
Nickel-chromium-aluminum sprayed undercoatings
ceramic coating process ... **M5:** 539
Nickel-chromium-aluminum-yttrium coating
superalloys ... **M5:** 376
Nickel-chromium-aluminum/Bentonite
as alloy-coated composite powder ... **A7:** 174
Nickel-chromium-boron
diffusion brazing filler metal ... **A6:** 344
Nickel-chromium-iron alloys
applications ... **A15:** 823
applications and properties ... **A2:** 438, 441
chemical analysis and sampling ... **A7:** 248-249
composition ... **A15:** 816
heat treatment ... **A15:** 822-823
as heating alloys ... **A2:** 828
joining ... **A2:** 831
in sheathed heaters ... **A2:** 831
structure and property correlations ... **A15:** 818
temperature versus milling time curves ... **A7:** 62
welding ... **A15:** 822
Nickel-chromium-iron alloys, selection of
alloying of elements effect ... **A6:** 588-590
applications, oxidizing and carburizing ... **A6:** 587
compositions of selected popular alloys ... **A6:** 586
consumption and applications by western nations in 1990 ... **A6:** 586
electron-beam welding ... **A6:** 587, 588
fabrication ... **A6:** 587
fusion zone ... **A6:** 588
gas-metal arc welding ... **A6:** 587, 588
gas-tungsten arc welding ... **A6:** 588
heat-affected zone ... **A6:** 587-590
postweld heat treatment ... **A6:** 590

Nickel-chromium-iron alloys, selection of (continued)
special metallurgical welding considerations **A6:** 590-592
thermal conductivity **A6:** 587
unmixed zone **A6:** 588, 589
welding characteristics **A6:** 587
welding metallurgy **A6:** 587-590
Nickel-chromium-iron heating elements, improving life of .... **A3:** 1 • 27-28
Nickel-chromium-molybdenum
spraying for hardfacing **M6:** 789
Nickel-chromium-molybdenum alloys
applications **A15:** 823
cast steels, as low-alloy **A15:** 716
composition **A15:** 516
heat treatment **A15:** 823
structure and property correlations **A15:** 818
welding **A15:** 822
weldment corrosion **A13:** 361-362
Nickel-chromium-molybdenum cast steels **A1:** 374
Nickel-chromium-molybdenum steel
electropolishing of **M5:** 308
Nickel-chromium-molybdenum-vanadium steel rotor
effect of stress intensity range range factor on fatigue crack growth rate **A12:** 56-57
Nickel-chromium-phosphorus alloys
400
annealing temperatures **A6:** 590
composition **A6:** 586
heat treatment procedures recommended **A6:** 590
properties **A6:** 587
thermal conductivity **A6:** 587
welding products for dissimilar-metal joints **A6:** 591
brazing
available product forms of filler metals **A6:** 119
joining temperatures **A6:** 118
Nickel-chromium-silicon cast irons **M1:** 94
composition **M1:** 76
mechanical properties **M1:** 92
physical properties **M1:** 88
Nickel-chromium-silicon-boron-carbon alloy
water-atomized **A7:** 32
Nickel-chromium/chromium carbide
as thermal spray powder **A7:** 174
Nickel-chromium/diatomite alloy-coated composite powder **A7:** 174
Nickel-coated composite powders **A7:** 173-174
Nickel-coated graphite fibers
as conductive reinforcements **EM2:** 472
Nickel-coated graphite powders **A7:** 137
Nickel-cobalt sealing process
anodic coatings **M5:** 595
Nickel-cobalt-chromium-aluminum- yttrium coating
superalloys **M5:** 376
Nickel-copper
relative solderability **A6:** 134
Nickel-copper alloy *See* H Monel; M-35; Nickel alloys, cast, specific types; S Monel
Nickel-copper alloys
applications **A15:** 823
applications and properties **A2:** 437
applications, characteristics, types **A2:** 435-436, 441
chemical analysis and sampling **A7:** 248
cleaning of **M5:** 670-674
etching of **A9:** 435
galling resistance with various material combinations **A18:** 596
low-melting metal embrittlement **A12:** 29
macroetching of **A9:** 435-436
Monel, discovery and characteristics **A2:** 429
Nickel-copper alloys (continued)
nominal compositions **A9:** 436
nonmetallic inclusions in **A9:** 436-438
projection welding **M6:** 503-506
transformation of **A15:** 125-129
x-ray diffraction peaks for compacts **A7:** 316
Nickel-copper alloys, selection of **A6:** 586-592
alloying of elements effect **A6:** 588-590
applications, seawater **A6:** 586
carbon content % and hot cracking **A6:** 589
compositions of selected popular alloys **A6:** 586
consumption and applications by western nations in 1990 .... **A6:** 586
electron-beam welding **A6:** 587, 588
fabrication **A6:** 587
fusion zone **A6:** 588
gas-metal arc welding **A6:** 587, 588
gas-tungsten arc welding **A6:** 588
heat-affected zone **A6:** 587-590
microstructures **A6:** 589
postweld heat treatment **A6:** 590
special metallurgical welding considerations **A6:** 590-592
thermal conductivity **A6:** 587
unmixed zone **A6:** 588, 589
welding characteristics **A6:** 587
welding metallurgy **A6:** 587-590
Nickel-copper alloys, specific types
Monel 400 .... **A9:** 435-437
Monel K-500 .... **A9:** 435-438
Monel R-405 .... **A9:** 435-437
Nickel-graphite (75/25, 80/20, and 85/15)
abradable seal material **A18:** 589
Nickel-graphite composite powders **A7:** 137, 174, 175
thermal-sprayed abradable seals **A7:** 175
Nickel-iron
glass-to-metal seals **EM3:** 302
relative solderability **A6:** 134
relative solderability as a function of flux type **A6:** 129
solderability **A6:** 978
Nickel-iron alloy plating **M5:** 206-207
Nickel-iron alloys *See also* Magnetic materials
corrosion resistance **A2:** 778
heat treatments **A2:** 774
low-expansion **A2:** 443-444
magnetic properties **A2:** 772-774
as magnetically soft materials **A2:** 770-774
microstructure **A9:** 538
permeability **A2:** 775
physical properties **A2:** 774
producers **A2:** 775
types **A2:** 770-772
Nickel-iron base metal solderability **EL1:** 676
Nickel-iron films
Faraday effect used to study **A9:** 535
Kerr effect used to study **A9:** 535-536
Nickel-iron powders
Dual Inline Package integrated circuit **A7:** 404
lead frame **A7:** 404
powder strip, controlled expansion properties **A7:** 403
powder strip, resistor cap and band termination application **A7:** 403
resistor end caps **A7:** 404
Nickel-iron-chromium alloys
cleaning of **M5:** 670-673
Nickel-iron-cobalt powder alloys
F-15 electronic part **A7:** 404
Nickel-manganese bronze
properties and applications **A2:** 370-371
Nickel-manganese cast steels **A1:** 374
Nickel-matrix composites with silicon-carbide fibers
polishing **A9:** 591
Nickel-modified zinc phosphate **A13:** 386
Nickel-molybdenum alloys *See also* Hastelloys; Nickel alloys, specific types
development and characteristics **A2:** 429
weldment corrosion **A13:** 361
Nickel-molybdenum-chromium alloys
comparison to nonferrous alloys using vibratory cavitation test **A18:** 763
galling **A18:** 763
Nickel-molybdenum-vanadium alloys
hydrogen flaking **A11:** 337
Nickel-nickel molybdenum
for control thermocouples in vacuum heat treating **A4:** 506
Nickel-phosphorus alloy plating **M5:** 202, 204
Nickel-phosphorus alloys
brazing
available product forms of filler metals **A6:** 119
joining temperatures **A6:** 118
recommended gap for braze filler metals **A6:** 120
Nickel-phosphorus coatings
electroless **M5:** 223-229
Nickel-phosphorus electrodeposited coatings **A13:** 426
Nickel-phosphorus film
LEISS determination of coverage on platinum substrate **A10:** 608
Nickel-plated steel
resistance spot welding **M6:** 479
Nickel-plated steels
press forming of **A14:** 563-564
Nickel-silicon alloys
radiation-induced segregation **A18:** 855
Nickel-silver system
wetting and spreading **EM4:** 485
Nickel-steel alloys
discovery of **A2:** 428
Nickel-steel powders
composition **A7:** 464
effects of density on mechanical properties **A7:** 467
microstructural analysis **A7:** 488
Nickel-sulfur phase diagram .... **A3:** 1 • 27
Nickel-tin bronze *See also* Copper casting alloys
applications and properties **A2:** 227
nominal composition **A2:** 347
Nickel-tin-plated steels
for space shuttle orbiter **A13:** 1091-1092
Nickel-titanium
orthodontic wires **A18:** 666, 675-676
resistant to cavitation erosion as a coating **A18:** 217
Nickel-titanium alloys
as shape memory effect (SME) alloys **A2:** 899
Nickel-titanium silicide (G phase)
in austenitic stainless steels **A9:** 284
Nickel-to-chromium ratio of heat-resistant casting alloys, effect on ferrite
formation **A9:** 333
Nickel-tungsten powder compacts
microstructures and homogenization **A7:** 316
Nickel-vanadium cast steels **A1:** 374
Nickel-zinc ferrites
applications **EM4:** 199
properties **EM4:** 1162
Nickel/boron-plated panels **A13:** 1123-1124
corrosion failure analysis of **EL1:** 1113-1114
Nicks
in shafts **A11:** 459
Nicol prism
defined **A9:** 12 **A10:** 678
Nicoro 80
brazing, composition **A6:** 117
wettability indices on stainless steel base metals **A6:** 118
Nicrobraz 125
diffusion brazing filler metal **A6:** 343

Nicrosil/Nisil thermocouple *See* Thermocouples, materials, nonstandard
Nicrosilal *See* Gray cast iron, Ni-Cr-Si
gray iron **A1:** 103
Nicuman 23
brazing, composition **A6:** 117
wettability indices on stainless steel base metals **A6:** 118
Nicusil 3
brazing, composition **A6:** 117
wettability indices on stainless steel base metals **A6:** 118
Nicusil 8
brazing, composition **A6:** 117
wettability indices on stainless steel base metals **A6:** 118
Nielson panel test
cleaning process efficiency **M5:** 20
Nier-type ion source
in gas mass spectrometer **A10:** 153
NIFDI computer program for structural analysis **EM1:** 268, 272
NiHard I
alloy composition and abrasion resistance **A18:** 189
NiHard IV
alloy composition and abrasion resistance **A18:** 189
Nikasil treatment **A18:** 791
NIKE process analysis software **A14:** 412
Nil strength temperature (NST) **A6:** 611
Nil temperature (NDT) **A6:** 611
Nil-ductility casting
gray irons **A12:** 226
Nil-ductility transition temperature *See also* Transition temperature
and Charpy V-notch specimens for nuclear pressure vessel testing **A8:** 263-264
in crack arrest tests **A8:** 285
crack toughness temperature range as above **A8:** 453
HSLA steels **M1:** 415, 417-418
neutron embrittlement, relation to **M1:** 686
precracked Charpy *W/A* values estimated by **A8:** 267-268
toughness control tests involving **A8:** 453
Nil-ductility transition temperature (NDTT) **A1:** 367-368, 412
bulging and fracture above **A11:** 59
by DWT test **A11:** 57-58
Nilo alloys
band sawing **A16:** 842
broaching **A16:** 837
composition **A16:** 836
drilling **A16:** 839
grinding **A16:** 843
milling **A16:** 842
multiple-operation machining **A16:** 840
planing **A16:** 839
spade and gun drilling **A16:** 839
tapping **A16:** 840
turning **A16:** 837, 838
Nilo K
flash welding **M6:** 557
Nimocast 80
for hot-forging dies **A18:** 626
Nimocast 90
for hot-forging dies **A18:** 626
Nimonic 75
composition **A4:** 794 **A6:** 573 **M4:** 651-652
Nimonic 80A
aging **A4:** 796, 805
aging cycle **M4:** 656
aging cycles **A6:** 574
annealing **M4:** 655
composition **A4:** 794 **A6:** 573 **M4:** 651-652
diffusion brazing **A6:** 343
hardness **A4:** 808
non-oxide ceramic joining **EM4:** 480
rupture properties **A4:** 808
solution treating **M4:** 656
solution treatment **A6:** 574
thermal diffusivity from 20 to 100 °C **A6:** 4
solution-treating **A4:** 793, 796
stress relieving **M4:** 655
Nimonic 86
composition **A4:** 794 **A6:** 573
Nimonic 90
aging **A4:** 796, 805
aging cycle **M4:** 656
aging cycles **A6:** 574
annealing **A4:** 809 **M4:** 655
cold-working effect on recrystallization and grain growth **A4:** 808
composition **A4:** 794 **A6:** 573 **M4:** 651-652
grain growth **A4:** 809
solution treating **M4:** 656
solution treatment **A6:** 574
solution-treating **A4:** 793, 796
stress relieving **M4:** 655
wear resistance **A18:** 636
Nimonic 91
composition **A6:** 573
Nimonic 95
composition **A4:** 794
Nimonic 100
composition **A4:** 794
Nimonic 105
composition **A4:** 794 **A6:** 573
Nimonic 115
composition **A4:** 794 **A6:** 573
Nimonic alloy
hardfacing **A6:** 807
Nimonic alloys
band sawing **A16:** 842
broaching **A16:** 837
composition **A16:** 736, 836
drilling **A16:** 839
electrochemical machining **A16:** 843
grinding **A16:** 843
machining **A16:** 738, 741-743, 746-747, 749-758
milling **A16:** 842
multiple-operation machining **A16:** 840
photochemical machining etchant **A16:** 590
planing **A16:** 839
spade and gun drilling **A16:** 839
tapping **A16:** 840
thread grinding **A16:** 275
turning **A16:** 837, 838
Nimonic, specific types
75, flash welding **M6:** 557
80A, flash welding **M6:** 557
90, flash welding **M6:** 557
105, flash welding **M6:** 557
C263, flash welding **M6:** 558
C475, flash welding **M6:** 558
PE 7, flash welding **M6:** 557
PE 11, flash welding **M6:** 558
PE 13, flash welding **M6:** 558
PE 16, flash welding **M6:** 558
Nimonics
laser cladding components and techniques **A18:** 869
Niobium *See also* Niobium alloys; Niobium alloys alloying, nicket-base alloys; Niobium alloys, specific types; Niobium-titanium superconductors; Refractory metals; Refractory metals and alloys; Refractory metals and alloys, specific types
as a beta stabilizer in titanium alloys **A9:** 458
as a preferential carbide former in stainless steel casting alloys **A9:** 298
addition to ferritic stainless steels **A6:** 444, 445
addition to low-alloy steels for pressure vessels and piping **A6:** 667
addition to precipitation-hardenable nickel alloys **A6:** 575, 576
addition to solid-solution nickel alloys **A6:** 575
alloy compositions **M6:** 459
as alloying addition to zirconium **A9:** 497
alloying effect in titanium alloys **A6:** 508
alloying, nickel-base alloys **A13:** 641
alloying, uranium/uranium alloys **A13:** 814-815
alloying, wrought aluminum alloy **A2:** 54
alloying, wrought titanium alloys **A2:** 599
alloys, SEM/AES failure analysis **A11:** 41-42
alloys, workability **A8:** 165, 575
as an addition to cobalt-base heat-resistant casting alloys **A9:** 334
as an addition to wrought heat-resistant alloys **A9:** 312
as an alloying addition to wrought stainless steels **A9:** 283-285
applications **A2:** 557-559 **A13:** 723-724 **M3:** 322
Niobium (continued)
applications, sheet metals **A6:** 400
arc welding *See also* Arc welding of niobium **M6:** 459-460
at elevated-temperature service **A1:** 640-641
atomic interaction descriptions **A6:** 144
in austenitic stainless steels **A6:** 457, 458, 461, 463
-base superconductors, history **A2:** 1027-1029
brazed joints, shear test data **A13:** 885
brazing **M6:** 1055-1057
filler metals and their properties **M6:** 1055-1056
fluxes and atmospheres **M6:** 1057
precleaning and surface preparation **M6:** 1056-1057
brazing and soldering characteristics **A6:** 634
ceramic/metal joints **EM4:** 516
chemical compositions per ASTM specification B550-92 .... **A6:** 787
chemical properties **A2:** 566-567 **M3:** 322
cleaning **A2:** 563
coatings **A2:** 566 **M3:** 323, 324
in cobalt-base alloys **A1:** 985
coextrusion welding **A6:** 311
consumption **A2:** 557
content effect on alloy solidification cracking **A6:** 89, 90
content in stainless steels **M6:** 320
corrosion, aqueous media **A13:** 723
corrosion, in specific media **A13:** 722
corrosion resistance **M3:** 322
corrosion resistance mechanism **A13:** 722
creep rupture testing of **A8:** 302
diffusion bonding **A2:** 564 **A6:** 156
diffusion welding **M6:** 677
effect of hydrogen content on ductility **A11:** 338
effect of, on hardenability **A1:** 413, 419
effect of, on notch toughness **A1:** 741-742
effect on anodic dissolution of phases in wrought heat-resistant alloys **A9:** 308
effect on critical temperatures in austenite **A4:** 246, 247, 248, 250
effects of thermoreactive deposition/diffusion process **A4:** 451
electrical discharge machining **A2:** 561
in electrodes, weld metal hydrogen vs. oxygen content **A6:** 59
electron beam welding **M6:** 640-641
electron-beam welding **A6:** 870, 871
electroslag welding, reactions **A6:** 274
embrittlement, and formability **A14:** 785
evaporation fields for **A10:** 587
explosion welding **A6:** 896
extrusion welding **M6:** 677
fabrication **M3:** 323
in ferrite **A1:** 403, 408
ferrite formation **M6:** 346
as ferrite stabilizer **A13:** 47
in ferritic stainless steels **A6:** 451, 454
filler metals for **M3:** 320, 323
filler metals for brazing of **A6:** 943
forging of **A14:** 237
friction coefficient data **A18:** 71
friction welding **M6:** 722
galvanic effects **A13:** 722
and gamma double prime in wrought heat- resistant alloys **A9:** 309
gas-tungsten arc welding **A6:** 871
gravimetric finishes **A10:** 171
in hardfacing alloys **A18:** 759
for heating elements for electrically heated furnaces **EM4:** 247
high, purity, preparation for superconductors **A2:** 1043-1044
high-purity sheet, cold worked and annealed **A9:** 442
in high-strength low-alloy steel **A1:** 358, 402-403
hydrogen damage **A13:** 171
hydrogen damage in **A11:** 338
as implant material **A11:** 673
in Inconel 718 .... **A1:** 1018
intergranular lithium corrosion attack **A13:** 93, 95
interlayer material **M6:** 681
intermetallic compounds, for A15 superconductors **A2:** 1060
localized corrosion **A13:** 722
in maraging steels **A4:** 220, 222, 224
mechanical and physical properties **A2:** 567-571

Niobium (continued)
mechanical properties **M3:** 323
microalloying of **A14:** 220
microstructures **A9:** 441
monocrystals, stress-strain curves for **A8:** 38, 39
and niobium alloys **A2:** 565-571
ores, fusion flux for **A10:** 167
ores, hydrofluoric acid as dissolution medium **A10:** 165
physical properties **A6:** 941
polish-etching **A9:** 440-441
polishing **A9:** 440
powder production **A2:** 565-566
precipitate stability and grain boundary pinning **A6:** 73
in precipitation-hardening steels **M6:** 350
production **A2:** 565-566
properties **A6:** 629
pure, properties **A2:** 1144
purity **M3:** 321, 322
reactions with gases and carbon **M6:** 1055
recovery from selected electrode coverings **A6:** 60
recrystallization temperatures **M6:** 1055
reduction process **A2:** 565
refractory metal brazing filler metal **A6:** 942
resistance spot welding **M6:** 478
roll welding **A6:** 314
room-temperature bend strength of silicon nitride metal joints **EM4:** 526
sheath, in ternary molybdenum chalcogenides (chevrel phases) **A2:** 1079
sheet, forming **A2:** 562 **A14:** 787
in simulated scrubber solutions **A13:** 724
single crystals, strain rate in **A8:** 39
solid-state bonding in joining non-oxide ceramics **EM4:** 525, 526
in stainless steels **A18:** 710, 712, 713
in steel **A1:** 146, 577
in steel weldments **A6:** 417, 418, 420
strain-age cracking resistance in precipitation strengthened alloys **A6:** 573
strengthening of **A9:** 441
submerged arc welding **M6:** 118, 129
  effect on cracking **M6:** 129
  effect on microstructure toughness **M6:** 118-119
suitability for cladding combinations **M6:** 691
in sulfuric acid **A13:** 723
tantalum/niobium as dopant for tungsten carbide **A18:** 795
tantalum/titanium/niobium carbides in cemented carbides **A18:** 795
temperatures, mill processing **M3:** 317
thermal diffusivity from 20 to 100 °C **A6:** 4
thermal expansion coefficient **A6:** 907
to promote hardness **A4:** 124
tooling recommendations for machining **M3:** 323
trace amounts affecting induction hardening **A4:** 185
transition temperatures **M6:** 1055
tube process, for A15 superconductor assembly **A2:** 1066-1067
tubing, production techniques **A2:** 562-563
ultrapure, by chemical vapor deposition **A2:** 1094
ultrapure, by zone-refining technique **A2:** 1094
vapor pressure, relation to temperature **A4:** 495
welds, ductility **A2:** 564
wet chlorine resistance **A13:** 1173
x-ray characterization of surface wear results for various microstructures **A18:** 469

Niobium (Columbium)
pure **M2:** 777-779

Niobium alloys *See also* Niobium; Niobium alloys, specific types **A13:** 171, 722-724 **A14:** 237-342
annealing **A4:** 817-819
anodizing **A9:** 142
applications **A2:** 557-560
chemical properties **A2:** 566-567
cleaning **A4:** 819
cleaning processes **M5:** 662-663
coatings **A2:** 566
consumption **A2:** 557
corrosion products on intergranular fracture surface **A12:** 37
density and melting temperature **M5:** 380
ductile-to-brittle transition temperature **A6:** 581
electron-beam welding **A6:** 581
electroplating of **M5:** 663-664
explosion welding **A6:** 581
finishing processes **M5:** 663-666
furnaces **A4:** 817-819
gas-tungsten arc welding **A6:** 581
heat-affected zone **A6:** 581
hot pressure welding **A6:** 581
joint design **A2:** 563-564
laser-beam welding **A6:** 581
machining **A2:** 560
mechanical and physical properties **A2:** 567-571
microstructure effect **A6:** 581
oxidation protective coating of **M5:** 375, 379-380
oxidation-resistant coating of **M5:** 664-666
plasma arc welding **A6:** 581
powder production **A2:** 565-566
production **A2:** 565-566
resistance welding **A6:** 581
roll welding **A6:** 312
thermal expansion coefficient **A6:** 907
welding atmosphere effect **A6:** 581
welding conditions effect **A6:** 581

Niobium alloys, specific types
80Nb-10W-10Hf-0.1Y, mechanical/physical properties **A2:** 569-570
89Nb-10Hf-1Ti, mechanical/physical properties **A2:** 568-569
B-66
  annealing treatment, postweld **M3:** 319
  composition **M3:** 316, 338
  property data **M3:** 339, 340
  temperatures, mill processing **M3:** 317
  welding conditions **M3:** 319
C-103
  composition **M3:** 316, 335
  property data **M3:** 335-336
  temperatures, mill processing **M3:** 317
C-129Y
  annealing treatment, postweld **M3:** 319
  composition **M3:** 316, 336
  property data **M3:** 336-337
  temperatures, mill processing **M3:** 317
  welding conditions **M3:** 319
C102, abrasive cutoff sawing **A16:** 868
C103, annealing **A4:** 816 **M4:** 788
C103, machining **A16:** 858-864, 866-869
C129Y, annealing **A4:** 816 **M4:** 788
C129Y, machining **A16:** 858-864, 866-869
Cb-132M
  composition **M3:** 339
  property data **M3:** 339-340
Cb-752
  annealing treatment, postweld **M3:** 319
  composition **M3:** 316, 337
  property data **M3:** 337-338
  temperatures, mill processing **M3:** 317
  welding conditions **M3:** 319
Cb-752, machining **A16:** 858-864, 866-869
D-43, annealing treatment, postweld **M3:** 319
F580, annealing **M4:** 788
FS-85
  annealing treatment, postweld **M3:** 319

Niobium alloys, specific types (continued)
  composition **M3:** 316, 340
  property data **M3:** 340, 341, 342
  temperatures, mill processing **M3:** 317
  welding conditions **M3:** 319
FS-85, machining **A16:** 69
FS-291, machining **A16:** 858-864, 866-869
FS85, annealing **A4:** 816 **M4:** 788
Nb, annealing **A4:** 816
Nb-1Zr
  composition **M3:** 316, 334
  property data **M3:** 334
  temperatures, mill processing **M3:** 317
Nb-1Zr (FS80, WC1Zr, KBI-1), annealing **A4:** 816
Nb-1Zr, brazing **A6:** 943
Nb-1Zr, mechanical/physical properties **A2:** 567-568
Nb-10Hf, physical properties **A6:** 941
Nb-10W-2.5Zr, mechanical/physical properties **A2:** 570-571
Nb-28Ta-10W, physical properties **A6:** 941
Nb-28Ta-10W-1Zr, mechanical/physical properties **A2:** 571
Nb-Zr, machining **A16:** 858-864, 866-869
Nb752, annealing **A4:** 816 **M4:** 788
niobium, commercial high-purity **M3:** 321-322
  property data **M3:** 333
SCb-191
  annealing treatment, postweld **M3:** 319
  composition **M3:** 316, 340
  property data **M3:** 340-341, 342, 343
  temperatures, mill processing **M3:** 317
SNb291, annealing **A4:** 816 **M4:** 788
WC-3015, machining **A16:** 858-864, 866-869
WC291, annealing **A4:** 816

Niobium alloys, specific types I
C103, electron beam welded **A9:** 155
C103, plate, cold worked and annealed **A9:** 442
C103, plate, heat tinted **A9:** 161
FS-80, tube, vacuum annealed **A9:** 442
FS-85, sheet, extruded, warm rolled, and annealed **A9:** 442
Nb-30TI-20W, sheet **A9:** 442
Nb-46.5Ti, rod, effect of annealing on **A9:** 443

Niobium and niobium alloys
alloyed with uranium for corrosion resistance **A16:** 874
in cast Co alloys **A16:** 69
content in stainless steels **A16:** 682-683, 684, 689
electrochemical grinding **A16:** 543
electrochemical machining removal rates **A16:** 534
milling **A16:** 312, 313
photochemical machining **A16:** 588, 590
principal ASTM specifications for weldable nonferrous sheet metals **A6:** 400

Niobium, and titanium welds
neutron radiography of **A17:** 393

Niobium boride
superconducting transition temperature **EM4:** 796

Niobium borides
in wrought heat-resistant alloys **A9:** 312

Niobium carbide
in cemented carbides **A18:** 795
coating applied to die-casting dies **A18:** 632
friction coefficient data for coating **A18:** 74
properties **A18:** 795

Niobium carbide cermets **A7:** 811
applications and properties **A2:** 1001-1002

Niobium carbides
in austenitic stainless steels **A9:** 284
in cermets **A16:** 92
in complex carbides **A16:** 71-74, 80, 82, 83
in Inconel 718, different illuminations compared **A9:** 82
in wrought heat-resistant alloys **A9:** 311

Niobium carbonitride in plate steels
examination for **A9:** 203
Niobium in steel **M1:** 115, 183, 188-189, 411
400 to 500 °C embrittlements effect on **M1:** 686
constructional steels for elevated temperature use, effect on **M1:** 647
notch toughness, effect on **M1:** 694, 696, 699, 700
steel sheet, effect on formability **M1:** 555
Niobium, niobium alloys, heat treating
annealing **M4:** 788
cleaning **M4:** 788
furnaces **M4:** 788-789
Niobium nitrides
in austentic stainless steels **A9:** 284
in plate steels, examination for **A9:** 203
in wrought heat-resistant alloys **A9:** 312
Niobium pentoxide
recovery of **A2:** 1043-1044
Niobium powder
reduction and production **A2:** 565-566
Niobium powders **A7:** 18
alloys **A7:** 160-162, 771-772
-based, high field superconductive compounds **A7:** 636
chemical analysis and sampling **A7:** 248
composition **A7:** 162
forward bowls, by HIP **A7:** 441, 443
as high-temperature material **A7:** 771-772
particle shape **A7:** 162
physical properties **A7:** 162-163
production **A7:** 160, 162-163
rocket applications **A7:** 18
separation from tantalum **A7:** 160
and tantalum production flowchart **A7:** 161
Niobium selenide ($NbSe_2$)
high-vacuum lubricant application **A18:** 153-154
Niobium, vapor pressure
relation to temperature **M4:** 310
Niobium, zone refined
impurity concentration **M2:** 713
Niobium-carbide rods in nickel matrix **A9:** 619
Niobium-hafnium-titanium powders **A7:** 162
Niobium-molybdenum microalloyed steels **A1:** 403
Niobium-stabilized stainless steels
etching **A9:** 282
Niobium-tin (Nb_3Sn)
hot pressing applications **EM4:** 192
Niobium-tin multifilamentary composite wire **A14:** 341-342
Niobium-tin P/M wire
superconducting **A7:** 638
Niobium-tin superconductors
manufacture of **A14:** 340-341
Niobium-titanium alloys
homogeneity **A2:** 1044, 1052-1053
for superconductors **A2:** 1043-1044
Niobium-titanium alloys, specific types
Nb-46.5Ti, for superconductors **A2:** 1043
Nb-50Ti, for superconductors **A2:** 1043
Niobium-titanium superconductors *See also* Superconducting materials
alloy selection/preparation **A2:** 1043-1044
assembly techniques **A2:** 1046-1049
cabling **A2:** 1051-1052
extrusion of **A2:** 1049-1050
filament properties **A2:** 1052-1054
in high-energy physics **A2:** 1055-1056
isostatic compaction **A2:** 1049
for magnetic confinement for thermonuclear fusion **A2:** 1056-1057
for magnetic energy storage **A2:** 1057
for magnetic resonance imaging (MRI) **A2:** 1054-1055
manufacture of **A14:** 338-340
matrix materials **A2:** 1044-1046, 1064
niobium selection/preparation, for superconductors **A2:** 1043-1044
for power applications **A2:** 1057
sizing, final **A2:** 1051
stabilizing of **A2:** 1051
superconductor composites, processing of **A2:** 1046-1052
titanium selection/preparation, for superconductors **A2:** 1043-1044
twisting **A2:** 1051
welding of **A2:** 1049
Niobium-titanium superconductors (continued)
wire drawing **A2:** 1050
Niobium-zirconium alloys, vs niobium-titanium
as superconductors **A2:** 1043
Nioro
brazing, composition **A6:** 117
wettability indices on stainless steel base metals **A6:** 118
Nioroni
brazing, composition **A6:** 117
wettability indices on stainless steel base metals **A6:** 118
Nip (crush)
defined **A18:** 13
Nipples
pipe for **A1:** 331
steel pipe for **M1:** 318-319
Nippon Telephone and Telegraph (NTT) cooling channels **EL1:** 310-311
NIR *See* Near-infrared
NISI computer program for structural analysis **EM1:** 268, 272
Nital
as etchant **A10:** 316, 318
and picral, etchants compared **A10:** 302
substitutes for methanol in **A9:** 67
Nital as an etchant for
austenitic manganese steel casting specimens **A9:** 239
babbitted bearings **A9:** 451
carbon and alloy steels **A9:** 169-175
carbon steel casting specimens **A9:** 230
carbonitrided steels **A9:** 217
carburized steels **A9:** 217
cast irons **A9:** 244
chromized sheet steel **A9:** 198
electrical steels **A9:** 531
ferrous powder metallurgy materials **A9:** 508-509
low-alloy steel casting samples **A9:** 230
nitrided steels **A9:** 218
permanent magnet alloys **A9:** 533
plate steels **A9:** 202-203
stainless-clad sheet steel **A9:** 198
steel tubular products **A9:** 211
steel-backed aluminum-tin bearings **A9:** 451
tin-antimony alloys **A9:** 450
tool steels **A9:** 256-257
welded joints in plate steels **A9:** 203
Nitinol alloys
milling **A16:** 313
Nitralloy
honing stone selection **A16:** 476
Nitralloy 135
drilling **A16:** 237
Nitralloy 135M **M1:** 540, 542
Nitralloy 135M, for gears
resistant to scuffing **A18:** 538
Nitralloy, die-casting dies
use in **M3:** 542, 543
Nitralloy EZ
composition and heat treatment **M1:** 542
Nitralloy G
composition and heat treatment **M1:** 542
wear of roller chain pins **M1:** 631
Nitralloy N
composition and heat treatment **M1:** 542
wear compared with carburized 4620 steel **M1:** 630
wear of roller chain pins **M1:** 628, 630-631
Nitralloy, nitrided
hardness profile **M1:** 633
Nitralloy series
135, 135M, gas nitriding **A4:** 387, 394
135 type G
applications **A4:** 397
gas nitriding **A4:** 397
plasma (ion) nitriding **A4:** 423
Nitrate solutions
cracking in **A11:** 214-215
Nitrate solutions, apparent threshold stress values
low-carbon steels **A8:** 526
Nitrate(s)
molten **A13:** 90-91
and nitrites, molten salt corrosion **A13:** 51-52
solutions, copper/copper alloy SCC in **A13:** 634-635
Nitrates
anions, separation by ion chromatography **A10:** 659
calcination **EM4:** 111
as electrodes **A10:** 185
as salt precursors **EM4:** 113
Nitration
aramid fibers **EM3:** 285
Nitric acid *See also* Nital; Nitric acid corrosion
acetic acid and glycerol as an etchant for tin-lead alloys **A9:** 450
and acetic acid as an etchant for gold-base alloys **A9:** 551
and acetic acid as an etchant for palladium welded to nickel silver **A9:** 551
and acetic acid as an etchant for stainless steels welded to carbon or low alloy steels **A9:** 203
in acetone as an etchant for hot-dip galvanized sheet steels **A9:** 197
and ammonium molybdate as an etchant for lead and lead alloys **A9:** 416
as an etchant for beryllium-containing alloys **A9:** 393-394
as an etchant for silicon steel transformer sheets **A9:** 62-63
as an etchant for wrought stainless steels **A9:** 281
cast iron resistance **A13:** 569
as chemical cleaning solution **A13:** 1140
corrosion inhibitors used in **M1:** 756
as etchant for nickel-copper alloys **A9:** 436
etching of damaged areas for surface integrity **A16:** 28
in ethyl alcohol as an etchant for hot-dip galvanized sheet steels **A9:** 197
fuming **A13:** 677
and glacial acetic acid as etchant for nickel alloys **A9:** 436
and glycol as an etchant for magnesium alloys **A9:** 426
hydrochloric acid and glycerol as an etchant for plated precious metals **A9:** 551
hydrochloric acid, and sulfuric acid as an alloys **A9:** 307
etchant for wrought heat-resistant
with hydrochloric acid, as sample dissolution medium **A10:** 166
and hydrofluoric acid in methanol as an coatings **A9:** 197
etchant for hot-dip zinc-aluminum
and hydrofluoric acid, used for polishing hafnium **A9:** 497
and hydrofluoric acid, used for polishing zirconium and zirconium alloys **A9:** 497
and hydrofluoric acid with water as an etchant for carbon and alloy steels **A9:** 171
and hydrogen absorption in chemical milling **A16:** 584, 585, 586
as immersion test solution **A13:** 221
inhibited red fuming **A13:** 264
lead/lead alloys in **A13:** 788
and methanol (Group VIII electrolytes) **A9:** 54-55
and methanol as an electrolyte beryllium- copper alloys **A9:** 393
and methanol as an electrolyte for aluminum alloys **A9:** 353
in methyl alcohol as an etchant for hot-dip galvanized sheet steels **A9:** 197
nickel alloys, corrosion **M3:** 173
nickel-base alloy resistance **A13:** 645
as oxidizing sample dissolution medium **A10:** 166
plants, pollution control **A13:** 1369-1370
red fuming **A13:** 264, 687
red fuming, for SCC testing of copper alloys **A8:** 525
red fuming, for SCC testing of titanium alloys **A8:** 531
residue isolation using **A10:** 176
safety hazards **A9:** 69
as sample modifier, GFAAS analysis **A10:** 55
solubility of lead nitrate in **A13:** 786

Nitric acid (continued)
solution, as eluent for suppressed chromatography **A10:** 660
stainless steel corrosion **A13:** 557
stainless steels, corrosion resistance **M3:** 84-87
sulfuric acid and hydrofluoric acid as an etchant for beryllium **A9:** 390
tantalum resistance to **A13:** 726
tantalum-tungsten alloys in **A13:** 737
as titanium cleaning agent **A12:** 75
titanium/titanium alloy resistance **A13:** 677
volatility **A10:** 166
zirconium/zirconium alloy resistance **A13:** 710-715
Nitric acid corrosion **A13:** 1154-1156
Nitric acid pickling
iron and steel **M5:** 68, 72-73, 79-80
magnesium alloys **M5:** 629-630, 640-641
stainless steel **M5:** 553
Nitric oxide
SERS analysis of **A10:** 136
Nitric-hydrofluoric acid pickling
iron and steel **M5:** 73
stainless steel **M5:** 553-554
Nitric-sulfuric acid pickling
magnesium alloys **M5:** 629-630, 640-641
Nitric/hydrofluoric acid
chemical milling etchant **A16:** 873
photochemical machining etchant **A16:** 592, 593
Nitride ceramics
brazing and soldering characteristics **A6:** 636
hot isostatic pressing **EM4:** 197
Nitride fibers **EM1:** 60-61, 63-64
Nitride glasses
development **EM4:** 23
Nitride inclusions in
nickel alloys **A9:** 436-437
nickel copper alloys **A9:** 436-438
Nitride strengthening
in creep tests **A8:** 331-332
Nitride-base cermets
applications and properties **A2:** 1004-1005
defined **A2:** 979
Nitride-based cermets **A7:** 813
Nitride-bonded silicon carbide
applications **EM4:** 963, 964, 984
Nitride-carbide inclusion types
defined **A9:** 12
Nitride-forming elements (NFE)
thermoreactive deposition/diffusion process **A4:** 449
Nitrided cases
microhardness testing **A8:** 96
Nitrided parts
failures of **A11:** 573
steel cases, LAMMA microanalysis of **A11:** 42-43
Nitrided stainless steel
SCC in **A13:** 933
Nitrided steels
case and core microstructures **M1:** 540
case hardness **M1:** 540
characteristics of **M1:** 528
compositions and heat treating temperatures of **M1:** 542
corrosion resistance **M1:** 540
etchants for **A9:** 218
etching **A9:** 218
grinding **A9:** 218
microstructures **A9:** 218
mounting **A9:** 218
plating **A9:** 218
polishing **A9:** 218
preservation of the white layer during specimen preparation **A9:** 218
sectioning **A9:** 218
specimen preparation **A9:** 217-218
Nitrides *See also* Ceramics; Cubic boron nitride; Silicon nitride
Al nitride, additive to Si_3N_4
Nitrides (continued)
Be nitride **A16:** 100
chemical vapor deposition of **M5:** 381
complex carbonitride cermets **A16:** 91, 92, 94
effect on high-temperature strength of iron-alloys **A9:** 333
chromium-nickel heat-resistant casting
as electrical conductor **A13:** 65
formation of, in iron-chromium-nickel heat- resistant casting alloys **A9:** 333-334
in grain boundaries, high-carbon steels **A12:** 282
as inclusions **A10:** 176
inclusions formed in fluxes **A6:** 56
Li nitride **A16:** 105
Mg nitride **A16:** 100
plasma-assisted physical vapor deposition **A18:** 848
in plate steels, examination for **A9:** 203
as refractory cermet **A7:** 813
SiAlON **A16:** 101
in silicon-iron electrical steels **A9:** 537
solution, as corrosive environment **A12:** 24
in structural ceramics **A2:** 1019, 1021-1024
Ti carbonitride cermets **A16:** 90, 93, 94, 97, 98
Ti nitride cermets **A16:** 91, 95, 98, 103
in wrought heat-resistant alloys **A9:** 309, 312
in wrought heat-resistant alloys, anodic dissolution to extract **A9:** 308
in wrought heat-resistant alloys, electrolytic extraction and x-ray diffraction **A9:** 308
in zirconium alloys, preparation for examination of **A9:** 498
Zr carbonitride **A16:** 98
Nitriding **A18:** 878-882 **M1:** 540-541, 627
advantages **A18:** 878
alloy steels, hardness profiles **M1:** 632-633
bath **A18:** 881
characteristics of nitrided surfaces **A18:** 878-879
coatings for taps, Ti alloys **A16:** 847
and cold form tapping **A16:** 266
comparison of coatings for cold upsetting **A18:** 645
defined **A9:** 12 **A13:** 9
and drilling **A16:** 219
for fatigue resistance **A11:** 121
fatigue resistance, effect on **M1:** 673-674
fatigue surface **A8:** 373
in fluidized beds **A4:** 490
galling resistance with various material combinations **A18:** 596
gas **A18:** 881
gear materials **A18:** 261
growth or shrinkage during **A7:** 480
H11 tool steel **M1:** 435-436
H13 tool steel **M1:** 439
influence on wear behavior **A18:** 880
ionitriding **M1:** 542, 627
lubrication and wear resistance **A18:** 879-880
maraging steels **M1:** 448
of master gages **M3:** 557
methods **M1:** 540-541
nitriding potential effect on decarburizing **A18:** 878
as nonmetallic in steels **A11:** 316
notch toughness of stubs effect on **M1:** 704-705
optimization of wear by process technology **A18:** 880
optimum case depth for fatigue resistance **M1:** 541
plasma **A18:** 881
postforging defects from **A11:** 333
of powder metallurgy materials **A9:** 503
shallow forming dies **A18:** 633
stainless steels **A18:** 715, 716, 723
steels for **M1:** 540-541, 627
surface finish affected by **A18:** 880
surface topography affected by **A18:** 880
Nitriding (continued)
as surface treatment in wrought tool steels **A1:** 779
of taps **A16:** 259
temperatures for **M1:** 540, 627
thermal spraying limitations **A18:** 831
titanium alloys **A18:** 780-781, 866
to reduce erosive wear in die-casting dies **A18:** 632
to reduce thermal and mechanical fatigue **A18:** 640
to reduce wear of die-casting dies **A18:** 629, 630
to surface harden wear-resistant toot steels for sheet metal forming **A18:** 628
of tool and high-speed steels **A7:** 374
tool steels **A18:** 641-642, 645, 739
use of transmission electron microscopy for surface studies **A18:** 381
for valve train assembly components **A18:** 559
variable influence on wear resistance of parts **A18:** 879-880
compound layer **A18:** 879-880
diffusion layer **A18:** 880
wear resistance improved by **M1:** 540, 627-631
wear resistance of materials **A18:** 879
abrasive wear **A18:** 879, 880, 881, 882
adhesive wear **A18:** 879, 880, 881, 882
corrosive wear **A18:** 882
surface fatigue **A18:** 879, 880, 881-882
tribo-oxidation **A18:** 879, 880, 881
welding factor **A18:** 541
white layer **M1:** 540
Nitriding steels
thermal expansion coefficient **A6:** 907
Nitriding surface
chemical studies of **A10:** 177
Nitrile bag materials **A7:** 447
Nitrile mastics
for body sealing and glazing materials **EM3:** 57
Nitrile phenolics **EM3:** 44, 75, 76, 104, 105
advantages and limitations **EM3:** 79
compared to nylon-epoxies **EM3:** 78
formulations **EM3:** 107
properties **EM3:** 106, 107
suppliers **EM3:** 104-105
typical film adhesive properties **EM3:** 78
Nitrile resins (NRs)
as engineering thermoplastics **EM2:** 448
Nitrile rubber *See also* Acrylonitrile-butadiene rubber **EM3:** 76, 78, 86, 89-90
characteristics **EM3:** 90
corrosion against mild steel in seawater **A18:** 549
for platen seals **EM3:** 694
properties **EM3:** 144
substrate cure rate and bond strength for cyanoacrylates **EM3:** 129
Nitrile-butadiene rubber (NBR) *See* Acrylonitrile-butadiene rubber
Nitriles **EM3:** 51
applications **EM3:** 44
for auto body sealing **EM3:** 57
silane coupling agents **EM3:** 182
Nitriloacetic acid
detected by ion chromatography **A10:** 661
Nitrilotris methylene phosphoric acid (NTMP) **EM3:** 250-251, 625
Nitrite anions
separation by ion chromatography **A10:** 659
Nitrite solution
effect on critical strain rate **A8:** 519
Nitrites
as anodic inhibitors **A13:** 494
as corrosion inhibitors **A18:** 277
and ECG **A16:** 545
and nitrates, molten salt corrosion **A13:** 51-52
in phosphate coatings **A13:** 384
solutions, copper/copper alloy corrosion **A13:** 635

Nitrocarburizing *See also* Austenitic nitrocarburizing; Carbonitriding; Ferritic nitrocarburizing; Plasma nitrocarburizing **A7:** 455 **A18:** 878, 879, 880 **M1:** 541-542
advantages **A18:** 878
carbon potential effect on decarburizing **A18:** 878
comparison of coatings for cold upsetting **A18:** 645
defined **A9:** 12 **A13:** 9
influence on wear behavior **A18:** 880
tool steels **A18:** 645
wear resistance of materials **A18:** 879
Nitrocellulose *See* Cellulose nitrate
as incendiary **A7:** 603
silane coupling agents **EM3:** 182
Nitrocellulose coating
flaws in **A13:** 108
Nitrocellulose resins and coatings **M5:** 473-474
Nitrogen *See also* Atmospheres; Atomization; Gas-atomized powders; Nitrogen gas; Nitrogen-based atmospheres; specific gas-atomized powders
absorption, by tin-containing P/M stainless steels **A7:** 254
as addition to austenitic stainless steels **A6:** 689
addition to strengthen nickel equivalent **A6:** 100
addition to strengthen stainless steels **A6:** 100
alloying effects, stainless steels **A13:** 550-551
as an alpha stabilizer in titanium alloys **A9:** 458
as an austenite-stabilizing element in steel **A9:** 177-178
as an austenite-stabilizing element in wrought stainless steels **A9:** 283
analyzed in microcircuit fabrication process **A10:** 156-157
as austenite stabilizer **A13:** 47
in austenitic stainless steels **A6:** 457, 458, 461, 462, 465, 467, 468
backing for gas tungsten arc welds **M6:** 197
base for furnace atmospheres **M6:** 934-935
in CAP process **A7:** 533
cause of porosity in nickel alloy welds **M6:** 442-443
characteristics in a blend **A6:** 65
chemistry at surfaces, AES analysis of **A10:** 553
combustion method for elemental analysis of **A10:** 214
in composition of stainless steels **A18:** 710, 712
composition, wt% (maximum) liquation cracking **A6:** 568
contamination **M6:** 321
content affecting work-hardening rate in stainless steels **A16:** 689
content in stainless steels **A16:** 682-683, 684, 690, 691 **M6:** 322
corrosion resistance effect, sintered austenitic stainless steels **A13:** 831
as cutting fluid for tool steels **A18:** 738
degassing, in vacuum melting ultrapurification **A2:** 1094
determined by combustion **A10:** 214
determined by potentiometric membrane electrodes **A10:** 181
in duplex stainless steels **A6:** 471-472, 473
effect of, on machinability of carbon steels **A1:** 599
effect of, on notch toughness **A1:** 741
effect of, on steel composition and formability **A1:** 577
effect on ferrite formation in heat-resistant casting alloys **A9:** 333
effect on shielding gas mixture **M6:** 199
effect, P/M stainless steels **A13:** 829-830
effects in case hardening **A7:** 454
electroslag welding, reactions **A6:** 278
enrichment, of surfaces **A11:** 42
entrapment and steel weldment soundness **A6:** 408
exclusion in dc arc sources **A10:** 25
in ferrite **A1:** 406, 408
ferritic stainless steel content **M6:** 346
in ferritic stainless steels **A9:** 284-285
-fired fracture surface, XPS survey of **A10:** 577
Nitrogen (continued)
gas form for milling WC **A16:** 72
gas mass analysis of **A10:** 155
groups, causing stress-corrosion cracking **A11:** 207
in heat-resistant alloys **A4:** 512
in high-purity metals, biamperometric analysis for **A10:** 205
in high-speed tool steel melting operation **A16:** 52, 53
impurity in diamond **A16:** 454
as impurity in uranium alloys **A9:** 477
as impurity, magnetic effects **A2:** 762
incident-ion energy **A18:** 851, 852, 854
inert gas fusion system for detecting **A10:** 226, 228-231
in inorganic solids, applicable analytical methods **A10:** 4, 6
interstitial contamination **M6:** 463
ion implantation of alloys **A18:** 779, 783
Kjeldahl determination of **A10:** 172, 214
for laser alloying **A18:** 866
liquid, as coolant, for superconductors **A2:** 1030
liquid, for optical holographic interferometry **A17:** 410
lubricant indicators and range of sensitivities **A18:** 301
mean free path **A17:** 59
microalloying of **A14:** 220
mobile or interstitial, determined in steels **A10:** 178
in organic substances, ISE analysis **A10:** 186
penetration from porosity in carbonitriding **A7:** 454, 455
photometric analysis methods **A10:** 64
pickup in milling of electrolytic iron powders **A7:** 64, 65
in plasma arc powder spraying process **A18:** 830
for plasma arc spraying **A6:** 811
in precipitation-hardening steels **M6:** 350
prompt gamma activation analysis of **A10:** 240
removal **A7:** 180-181
removal, by solid state refining **A2:** 1094
resistance spot welding of steels and content effect **A6:** 228
shielding gas for arc welding of low-alloy steels **A6:** 662
shielding gas for plasma arc cutting **A6:** 1167, 1168, 1170
shielding gas properties **A6:** 64
shielding gas purity and moisture content **A6:** 65
in silicon irons **A9:** 537
-sintered aluminum P/M alloys **A7:** 742
solubility in austenitic stainless steels **A13:** 827
solubility in tantalum **A13:** 730
in stainless steels **A1:** 930
in steel weldments **A6:** 418
in superheaters **A13:** 1201
to harden and strengthen steel **A16:** 667, 674, 675, 676
use in resistance spot welding **M6:** 486
valve thread connections for compressed gas cylinders **A6:** 1197
volumetric procedures for **A10:** 175
in weld relay, gas mass spectroscopy of **A10:** 156
in wrought heat-resistant alloys **A9:** 311
in wrought stainless steels **A1:** 872
Nitrogen blanket **EM3:** 19
Nitrogen ceramics
decomposition control **EM4:** 194
Nitrogen compounds
as crude oil contaminant **A13:** 1267
Nitrogen content
by residual gas analysis **EL1:** 1065
Nitrogen dioxide
for accelerated SCC testing of copper alloys **A8:** 525
SERS analysis of **A10:** 136
Nitrogen gas *See also* Nitrogen; Nitrogen-based atmospheres
-atomized aluminum powder **A7:** 130
-atomized stainless steel powders **A7:** 101-103
drying **A7:** 76
physical properties **A7:** 341
Nitrogen in steel **M1:** 116, 410, 417
carbonitriding **M1:** 533, 536, 539-540
hardenability affected by **M1:** 477
modified low-carbon steels **M1:** 162
nitriding **M1:** 540-542
notch toughness, effect on **M1:** 693, 695, 697
steel sheet, effect on formability **M1:** 555
Nitrogen, liquid
iron cooling by **A12:** 219
Nitrogen, max
chemical composition per ASTM specification B550-92 .... **A6:** 787
Nitrogen monoxide
tantalum resistance to **A13:** 731
Nitrogen oxides
fume generation from arc welding **A6:** 68
Nitrogen peroxide
SERS analysis of **A10:** 136
Nitrogen shielding gas
gas metal arc welding **M6:** 164-851
Nitrogen tetroxide
and SCC in titanium and titanium alloys **A11:** 224
for SCC of titanium alloys **A8:** 531
titanium/titanium alloy SCC **A13:** 687
Nitrogen-base atmospheres *See* Prepared nitrogen-base atmospheres
Nitrogen-based atmospheres *See also* Atmospheres; Atomization; Gas atomized powders; Nitrogen; Nitrogen gas **A7:** 341, 345-346, 361
for aluminum sintering **A7:** 383
carbon-control agents **A7:** 346
compositions, conventional and synthetic **A7:** 342
oxidants **A7:** 346
oxide-reducing agents **A7:** 345-346
Nitrogen-doped silica **EM4:** 211
Nitrogen-strengthened austenitic stainless steels **A1:** 892-893 **A14:** 225-226
Nitrogenized steels *See also* Nitrogen in steel **A1:** 208 **M1:** 162
Nitroguanidine explosive **A6:** 161
Nitronic 30
abrasive wear **A18:** 719
composition **A18:** 711
corrosive wear **A18:** 719
Nitronic 50
cavitation resistance **A18:** 600
galling threshold load **A18:** 595
Nitronic 60 *See* Stainless steels, specific types, S21800
Nitrous oxide
tantalum resistance to **A13:** 731
Nitrous oxide-acetylene flames **A10:** 29, 48
NMOS *See* n-channel metal-oxide semiconductors
NMR *See* Nuclear magnetic resonance
NMR spectroscopy **EM2:** 826-827
N,N,N′,N′-tetraglycidyl-4,4′-diamino diphenyl methane (MY 720) **EM3:** 94-95, 97
No cleaning
as cleaning option **EL1:** 666
No nickel/no pickle system
porcelain enameling process **M5:** 515
"No-clean" applications
fluxes for **EL1:** 647-648
No-crack temperature
cast irons **A6:** 710, 711
No-draft forging
defined **A14:** 9
No-hold-test
plastic-strain fatigue resistance effect on stainless steel **A8:** 348
saturation effect **A8:** 348-349
No-standards analysis
in quantitative thin-film EPMA analysis **A11:** 41
No. 1 yellow brass
properties and applications **A2:** 366
No. 3 die zinc die casting alloy
properties **A2:** 532
Nobility, increased
by alloying **A13:** 47
Noble
defined **A13:** 9
metals **A13:** 9, 793-807
potential, defined **A13:** 9

Noble metal coating
molybdenum ... **M5:** 661
niobium ... **M5:** 664-666
tantalum ... **M5:** 664-666
tungsten ... **M5:** 661
vacuum process ... **M5:** 388, 395, 400
Noble metal conductor inks ... **EL1:** 207-208
Noble metals *See* Precious metals ... **A13:** 9, 793-807
alloying, as selective oxidation ... **A13:** 73
anodic behavior of ... **A13:** 807
clad systems ... **A13:** 888-889
compatibility ... **A13:** 342
contact, titanium/titanium alloys ... **A13:** 696
as dental alloys ... **A13:** 1350
diffusion welding ... **A6:** 885
galvanic corrosion ... **A13:** 86
gold ... **A13:** 796-800
hydrochloric acid corrosion ... **A13:** 1164
iridium ... **A13:** 802-806
materials for conductors of thick film circuits ... **EM4:** 1141
osmium ... **A13:** 806-807
palladium ... **A13:** 799-801, 804-805
PFM alloys ... **A13:** 1355
platinum ... **A13:** 797-803
polishing wear without abrasives ... **A18:** 197
properties ... **A13:** 794
rhodium ... **A13:** 801-802, 805
ruthenium ... **A13:** 805-806
silver ... **A13:** 793-798
systems, dealloying ... **A13:** 133
Node
defined ... **EM1:** 16
Node, leaf/root
in physical hierarchy ... **EL1:** 5
Node separation
adhesive-bed joints ... **A17:** 613
Nodes in cellular structures ... **A9:** 613
Nodular cast iron *See* Ductile cast iron
Nodular eutectic microstructure ... **A3:** 1•20
Nodular ferritic cast iron
salt bath nitrided ... **A9:** 229
Nodular graphite
defined ... **A9:** 12
Nodular graphite, in cast iron *See also* Temper carbon ... **M1:** 6-7, 9
Nodular iron *See* Ductile iron
application, piston ring material ... **A18:** 557
gear materials, surface treatment and minimum surface hardness ... **A18:** 261
as lap plate material ... **EM4:** 353
laser melting ... **A18:** 864
metallographic sections ... **A18:** 375
parameters for machining with HIP metal- oxide composite-grade ceramic insert cutting tools ... **EM4:** 968
turning and milling recommended ceramic- grade inserts for cutting tools ... **EM4:** 972
for valve seats and guards for reciprocating compressors ... **A18:** 604
Nodular iron (cast)
thermal expansion coefficient ... **A6:** 907
Nodular irons *See also* Ductile cast iron; Ductile irons
cermet tools for milling ... **A16:** 97
contour band sawing ... **A16:** 362
milling ... **A16:** 97, 327
reaming ... **A16:** 248
turning ... **A16:** 94
Nodular or spheroidal graphite iron ... **A18:** 695, 698
applications ... **A18:** 695, 700, 701
microstructure ... **A18:** 699-701
properties ... **A18:** 695
Nodular pearlite
defined ... **A9:** 12
Nodular powders ... **A7:** 8, 233, 234
Nodule count, control of
malleable cast iron ... **M1:** 59-60
Nodule count, control of (continued)
in malleable iron ... **A1:** 73-74
Nodulizing reaction
compacted graphite cast iron ... **M1:** 8
ductile cast iron ... **M1:** 6-7
Noduluar cast iron (NCI)
advantages of silicon-nitride-based ceramic inserts versus oxide-based ceramic inserts when machining ... **EM4:** 971
indirect brazing of PSZ for joining oxide ceramics ... **EM4:** 517
recommended machining specifications for rough and finish turning with HIP metal oxide ceramic insert cutting tools ... **EM4:** 969
rough and finish machining recommended starting conditions with silicon-nitride based ceramic insert cutting tool ... **EM4:** 971
rough and finish with whisker-reinforced alumina ceramic insert cutting tools ... **EM4:** 972
Noise *See also* Barkhausen noise; Signal-to-noise ratio
budget analysis ... **EL1:** 82
and contrast sensitivity ... **A17:** 374
coupling, schematic ... **EL1:** 418
current, defined ... **EL1:** 1151
defined ... **A10:** 678 **A17:** 384
in direct current electrical potential method ... **A8:** 389
elimination, ESR spectrometers ... **A10:** 257
environment ... **EL1:** 76
factors affecting ... **A17:** 374-375
factors, quality design ... **A17:** 722
and flaw responses, discrimination ... **A17:** 676
focused ultrasonic search units ... **A17:** 260-261
forward-coupled ... **EL1:** 35
inner, in quality design ... **A17:** 723
in inspection materials ... **A17:** 678
measurement ... **A17:** 375
in NDE reliability ... **A17:** 675
outer, in quality design ... **A17:** 722
quantum or photon, defined ... **A17:** 384
saturated backward coupled ... **EL1:** 35
signals, acoustic emission inspection ... **A17:** 284
structural, defined ... **A17:** 384
system, experimental study ... **A17:** 742
thermal, defined ... **A10:** 683
thermal, thermal inspection ... **A17:** 400
variation, in quality design ... **A17:** 723
Noise cross talk *See* Cross talk
Noise, electrical
platinum group metals ... **A2:** 846
Noise level
in bearing failures ... **A11:** 494
Noise margin analysis
for high-frequency design methodology ... **EL1:** 78-79
Noise suppression
in rotary swaging ... **A14:** 144
Noise voltage (equivalent input)
defined ... **EL1:** 1151
Noise-equivalent power
defined ... **EL1:** 1151
NOL ring
defined ... **EM1:** 16 **EM2:** 28
Nomarski contrast illumination ... **A18:** 371, 372, 374
Nomarski differential
interference microscopy ... **EM4:** 578
Nomarski interference contrast system ... **A9:** 150-152
attachment for a Reichert microscope ... **A9:** 150
Nomenclature *See also* Categorization; Classification; Definitions; Notation; Terminology
as applied to solder masks ... **EL1:** 559
applying, rigid printed wiring boards ... **EL1:** 547
chemical an trade names, epoxy resins ... **EL1:** 826
fiberglass yams ... **EM1:** 110
glass filament diameter ... **EM1:** 109
of magnetic particle inspection ... **A17:** 103
Nomenclature (continued)
of polymers ... **EM2:** 53-56
of thermal-mechanical effects ... **EL1:** 746
Nomenclatures
cast and wrought aluminum alloys ... **A2:** 4-5
Nomex *See* Aramid fibers, specific types
Nominal area (of contact) *See also* Apparent area of contact; Area of contact
defined ... **A18:** 13
Nominal axial stresses ... **A1:** 674
and fatigue resistance ... **A1:** 674
Nominal compositions *See also* Composition
aluminum casting alloys ... **A2:** 152-177
of copper casting alloys ... **A2:** 347
wrought aluminum and aluminum alloys ... **A2:** 62-122
wrought copper and copper alloys ... **A2:** 265-345
Nominal engineering stress ... **A8:** 726
Nominal normal stress on the contact path ... **A18:** 682
Nominal plastic zone size ... **A8:** 446
Nominal rate of strain *See also* Specific crosshead rate
in machine stiffness effects ... **A8:** 41
Nominal strain *See* Strain
Nominal strength *See* Ultimate strength
Nominal stress *See also* Engineering stress; Mean stress; Normal stress; Residual stress; True stress ... **EM3:** 19
carbon-manganese steel ... **A13:** 263
defined ... **A8:** 9 **EM1:** 16 **EM2:** 28
effects on fatigue fractures ... **A11:** 110-111
and high-cycle fatigue ... **A11:** 109-110
in shafts ... **A11:** 461
Nominal value ... **EM3:** 19
defined ... **EM1:** 16 **EM2:** 28
Nomographs
for coupled lines ... **EL1:** 39-41
to predict minimum oil film thickness for steadily loaded bearings ... **A18:** 517
to predict power loss in steadily loaded bearings ... **A18:** 519
Nomographs, for scale-up
explosion output values ... **A7:** 197
Non-abrasive machining methods ... **EM4:** 313, 314
Non-ambient-temperature scanning electron microscopy ... **A9:** 97
Non-dimensional voltage
as function of potential lead position ... **A8:** 388
Non-heat-treatable alloys
electron-beam welding ... **A6:** 871
single-shear and blanking shear tests for ... **A8:** 65
Non-heat-treatable aluminum alloys ... **A6:** 537-540
alloy classification ... **A6:** 537
applications ... **A6:** 537, 540
corrosion ... **A6:** 537, 540
electron-beam welding ... **A6:** 538-539
filler alloy selection ... **A6:** 537-539
filler metals ... **A6:** 539
gas-metal arc welding ... **A6:** 539
heat-affected zone ... **A6:** 537, 539
hot cracking ... **A6:** 538-539
laser-beam welding ... **A6:** 538
porosity ... **A6:** 539
properties ... **A6:** 537, 539-540
temper designations ... **A6:** 537
weld properties ... **A6:** 539-540
Non-heat-treatable stainless steels
electrochemical polarization ... **A8:** 529
SCC testing of ... **A8:** 527-529
testing in boiling magnesium chloride solution ... **A8:** 528-529
testing in polythionic acids ... **A8:** 528
wick test for ... **A8:** 529
Non-heat-treatable wrought aluminum alloys
strength improvement ... **A2:** 36-39
Non-Newtonian viscosity
defined ... **A18:** 14

Non-work hardening material
stress distribution in torsion testing of **A8:** 140
Nonaqueous solvent-suspendible developers (form D)
for liquid penetrant inspection **A17:** 77
Nonary system or diagram .... **A3:** 1 • 2
Nonasbestos organic (NAO)
for organic brake linings............... **A18:** 569, 570, 572
Nonaustenitic steel
hardness conversion numbers for................. **A8:** 106
Nonaustenitic steels
fracture transition... **A11:** 66
Nonaveraging extensometer............................ **A8:** 616
Nonchromated deoxidizers
use of **M5:** 8
Nonclassical creep **A8:** 331-332
Nonconductive resins
as mounting materials for
electropolishing.. **A9:** 49
Nonconformal surfaces
defined **A18:** 13
Noncontact bearing *See also* Gas lubrication; Magnetic bearing
defined **A18:** 13
Noncontact method *See also* Laser inspection
probe, aircraft subassemblies, eddy
current inspection................................. **A17:** 190
Noncontact methods
to measure strain... **A8:** 193
Noncontact trigger probes
coordinate measuring machines.................... **A17:** 25
Noncontinuous fillet
in brazed joints **A17:** 602
Noncorrosive flux
definition.. **M6:** 12
Noncubic phases in aluminum alloys **A9:** 359-360
Noncyanide plating
brass **M5:** 285
cadmium **M5:** 256-257
silver **M5:** 279-230
zinc *See* Zinc alkaline noncyanide plating
Noncylindrical bending **A8:** 119, 121
Nondestructive analysis
by voltammetry **A10:** 188
neutron activation analysis.................... **A10:** 233-242
surface residual stress measurement,
for quality control................................ **A10:** 380
thermal neutron activation **A10:** 234-238
thin/thick samples, PIXE as......................... **A10:** 102
ultrasonic **EM2:** 838-846
uranium assay by DNC as **A10:** 238
x-ray topography.................................... **A10:** 365-379
Nondestructive etching **A9:** 57-60
Nondestructive evaluation *See also* Formability testing, Inspection; Testing; Workability tests
of aluminum alloy forgings.......................... **A14:** 249
of powder forged parts.......................... **A14:** 204-205
of titanium alloy forgings............................ **A14:** 282
Nondestructive evaluation (NDE) *See also* Nondestructive analysis; Nondestructive evaluation methods Nondestructive evaluation techniques; Nondestructive inspection (NDI); Nondestructive inspection methods; Nondestructive testing; Nondestructive testing (NDT); Ultrasonic analysis............. **A18:** 406, 412 **EM1:** 16 **EM3:** 19 **EM4:** 32, 547-548
acoustic emission technique **EM3:** 781
composite joint end products.............. **EM3:** 777-784
acoustic emission **EM3:** 781
holography **EM3:** 782-783
radiographic techniques **EM3:** 781-782
shearography **EM3:** 782-783
state of the art technology........................ **EM3:** 783
thermography **EM3:** 783
ultrasonic **EM3:** 777-781
of composites **A11:** 739
defined **A17:** 49 **EM2:** 28
definition.. **EM3:** 743
fabrication flaws during DOE-ATTAP
program.. **EM4:** 998
of failed parts.. **A11:** 173
of gray iron paper-dryer head...................... **A11:** 352
of heat exchangers **A11:** 629 **EM4:** 981
holography **EM3:** 782-783
leaky Lamb wave (LLW) technique.................. **EM3:** 779-781, 782, 783
Nondestructive evaluation (NDE) (continued)
metal joints adhesively bonded........... **EM3:** 743-746
adherend defects **EM3:** 746-747
adhesive flash **EM3:** 746
blown core... **EM3:** 748, 749
burned adhesive.. **EM3:** 746
condensed core.. **EM3:** 747
corrosion **EM3:** 750, 751, 760
crushed core... **EM3:** 747
dents, dings, and wrinkles **EM3:** 746
disbonds **EM3:** 746, 748-750
double-drilled or irregular holes **EM3:** 746
foam intrusion **EM3:** 750
foreign objects.. **EM3:** 750
fracture (cracks)... **EM3:** 746
fractured or gouged fillets **EM3:** 746
frequency of rejectable flaws in
adhesive bonded assemblies.......... **EM3:** 745
generic flaw types and
flaw-producing mechanisms.......... **EM3:** 745
honeycomb core defects **EM3:** 748, 749, 750-751
honeycomb sandwich defects........... **EM3:** 747-750
honeycomb structure.................. **EM3:** 772-773, 774
impact damage.................................. **EM3:** 750-751
interface defects................................ **EM3:** 743-744
lack of fillets.. **EM3:** 746
metal-to-metal defects....................... **EM3:** 748-750
metal-to-metal joints......................... **EM3:** 772, 774
metal-to-metal voids.................. **EM3:** 743, 745-746
missing fillets... **EM3:** 750
node separation................................. **EM3:** 748, 749
poor fabrication................................. **EM3:** 750, 751
porosity.. **EM3:** 746, 773
porous or frothy fillets............................. **EM3:** 746
pretreatment flaws.................................... **EM3:** 743
protective film left on adhesive **EM3:** 748
repair defects **EM3:** 750
scratches and gouges **EM3:** 746
short core.. **EM3:** 750
skin-to-core voids at edge of chemically milled steps or doublers........ **EM3:** 750
ultrasonic inspection **EM3:** 748
unbonds **EM3:** 743, 746
voids in foam adhesive joints.......... **EM3:** 748, 750
water in core cells..................................... **EM3:** 747
plasma arc welding... **A6:** 198
radiographic techniques....................... **EM3:** 781-782
shearography.. **EM3:** 782-783
of solder joints **EL1:** 735-738
state of the art technology.......................... **EM3:** 783
Sundstrand Power Systems advanced
gas turbine component
development.. **EM4:** 999
thermography **EM3:** 783
turbocharger turbine wheel proof
testing **EM4:** 726
ultrasonic pulse-echo.................... **EM3:** 778-779, 783
ultrasonic spectroscopy....................... **EM3:** 779, 781
ultrasonic through-transmission......... **EM3:** 778-779, 783
Nondestructive evaluation (NDE) techniques
to evaluate gas turbine ceramic
components... **EM4:** 718
Nondestructive evaluation (NDE) testing and inspection **EM4:** 617-626
acoustic emission................................. **EM4:** 625-626
acoustic resonances **EM4:** 626
infrared inspection **EM4:** 626
limitations.. **EM4:** 617
liquid penetrants **EM4:** 625
microwaves.. **EM4:** 626
nuclear magnetic resonance **EM4:** 626
radiography **EM4:** 617-621
computed tomography **EM4:** 617, 619-620
microfocus radiography................... **EM4:** 618, 619
neutron radiography **EM4:** 618, 620-621
projection radiography **EM4:** 618-619
x-ray computed tomography........... **EM4:** 619-620
x-ray microradiography.................... **EM4:** 618-619
ultrasonics... **EM4:** 621-625
bulk porosity measurements **EM4:** 621
coupling techniques.......................... **EM4:** 623-624
defect detection by acoustic
microscopy **EM4:** 624-625
standard defect specimens **EM4:** 624
Nondestructive evaluation methods *See also* NDE detection methods; NDE reliability Nondestructive evaluation; Nondestructive evaluation techniques; Nondestructive inspection of specific products Quantitative nondestructive evaluation
acoustic emission inspection **A17:** 278-294
acoustic microscopy **A17:** 465-482
acoustical holography............................ **A17:** 438-447
codes, for boilers and pressure vessels **A17:** 641-644
color, usage of... **A17:** 483-488
digital image enhancement **A17:** 454-464
eddy current inspection **A17:** 164-194
electric current perturbation.................. **A17:** 136-142
electromagnetic, for residual stress
measurement **A17:** 159-163
industrial computed tomography **A17:** 358-386
leak testing **A17:** 57-70
liquid penetrant inspection **A17:** 71-88
magabsorption **A17:** 143-158
magnetic field testing **A17:** 129-135
magnetic particle inspection..................... **A17:** 89-128
materials, control of **A17:** 678
microwave inspection............................ **A17:** 202-230
neutron radiography............................... **A17:** 387-395
optical holography **A17:** 405-431
possible outcomes **A17:** 675
for powder metallurgy parts.................. **A17:** 537-547
qualification of **A17:** 678-679
radiographic inspection.......................... **A17:** 295-357
reliability of **A17:** 663-715
remote-field eddy current inspection **A17:** 195-201
replication microscopy **A17:** 52-56
specification requirements **A17:** 663
speckle metrology **A17:** 432-437
statistical nature of................................ **A17:** 689-692
thermal inspection.................................. **A17:** 396-404
Nondestructive evaluation techniques *See also* NDE detection methods; NDE reliability Nondestructive evaluation; Nondestructive evaluation methods; Nondestructive inspection of specific products Quantitative nondestructive evaluation
flaw detection and evaluation **A17:** 49-50
guide to **A17:** 49-51
leak testing **A17:** 50, 57-70
liquid penetrant inspection **A17:** 71-88
material characteristics, important................ **A17:** 51
mechanical and physical properties
estimated **A17:** 51
metrology and evaluation............................... **A17:** 50
for planar flaws **A17:** 50
in product cycle, adhesive-bonded
joints.. **A17:** 632-636
reasons for **A17:** 49
replication microscopy **A17:** 52-56
for residual stresses **A17:** 51
selection of method.. **A17:** 49
technical meetings about **A17:** 51
for volumetric flaws **A17:** 50
Nondestructive examination *See* Nondestructive evaluation (NDE) **M6:** 847-853
accuracy **M6:** 851
inspection methods **M6:** 847-850
acoustic emission **M6:** 849-850
eddy-current **M6:** 850
liquid-penetrant..................................... **M6:** 847-848
magnetic-particle....................................... **M6:** 848
radiographic.. **M6:** 848-849
ultrasonic.. **M6:** 849
visual.. **M6:** 847
reliability... **M6:** 851
selection of technique **M6:** 850-851
characteristics of discontinuity **M6:** 850
constraints **M6:** 850-851
fracture mechanics requirements **M6:** 850
sensitivity... **M6:** 851
Nondestructive examination (NDEx) *See* Nondestructive evaluation (NDE)
Nondestructive inspection *See also* Inspection; Nondestructive evaluation (NDE); Nondestructive evaluation methods Nondestructive evaluation techniques NDE reliability; Quantitative nondestructive evaluation; specific nondestructive evaluation methods; Visual examination inspection
brazed joints.. **A6:** 1118-1119

Nondestructive inspection (continued)
effects ... **A12:** 77
pitting corrosion ... **A13:** 231
profile generation ... **A12:** 199
soldered joints ... **A6:** 981-982
of specific products ... **A17:** 489-659
of steel castings ... **A1:** 378-379
Nondestructive inspection (NDI) *See also* Inspection; Nondestructive evaluation; Nondestructive testing ... **EM3:** 526-529
acoustic emission (AE) ... **EM1:** 777
of cured composite laminate ... **EM1:** 532
defined ... **EM1:** 16
definition ... **EM3:** 743
detail (micro) methods, defined ... **EM1:** 774
field (bulk) methods, defined ... **EM1:** 774
grading of adhesive defects ... **EM3:** 526
infrared thermography ... **EM3:** 526
methods ... **EM1:** 770
neutron radiography ... **EM3:** 524, 525, 526
pulse-echo mode of operation ... **EM3:** 527
quality control, introduction and overview ... **EM3:** 727-728
radiographic methods ... **EM3:** 528
radiography ... **EM1:** 775-776
resonance methods ... **EM3:** 527-528
selection of ... **EM3:** 521, 522
Shurtronics harmonic bond tester flat-panel tests ... **EM3:** 527, 528
Shurtronics harmonic bond tester reference-panel tests ... **EM3:** 527, 528
stress waves (ultrasonic) ... **EM3:** 530
surface/edge replication ... **EM1:** 775
tap hammer inspection method ... **EM3:** 527-528
techniques ... **EM1:** 774-777
thermographic methods ... **EM3:** 528
thermography ... **EM1:** 777
ultrasonic techniques ... **EM1:** 776-777
ultrasonic through-transmission inspection ... **EM3:** 523, 524, 526-527, 528, 529
vibrothermography ... **EM1:** 777 **EM3:** 528
x-ray computer tomography (CT) ... **EM3:** 528
x-ray radiography ... **EM3:** 526, 528
Nondestructive inspection (NDI) testing
ceramic gas turbine engine components ... **EM4:** 717
Nondestructive inspection of specific products *See also* NDE reliability
adhesive-bonded joints ... **A17:** 610-640
billets ... **A17:** 549-560
boilers ... **A17:** 641-659
brazed assemblies ... **A17:** 582-609
castings ... **A17:** 512-535
forgings ... **A17:** 491-511
powder metallurgy parts ... **A17:** 536-548
pressure vessels ... **A17:** 641-659
soldered joints ... **A17:** 582-609
steel bar ... **A17:** 549-560
tubular products ... **A17:** 561-581
weldments ... **A17:** 582-609
wire ... **A17:** 549-560
Nondestructive repair
and rework processes ... **EL1:** 710-711
Nondestructive test (NDT) methods
adhesion measurements ... **A6:** 144
Nondestructive testing
not applicable to upset welded joints ... **A6:** 250
Nondestructive testing (NDT) *See also* Nondestructive evaluation; Nondestructive evaluation (NDE); Nondestructive inspection (NDI); Nondestructive inspection methods ... **EM3:** 19, 37
210 sonic bond tester ... **EM3:** 766
acoustic emission techniques ... **EM3:** 760
acoustical holography ... **EM3:** 765-767
adhesive bond strength classifier algorithm ... **EM3:** 744
Advanced Bond Evaluator (ABE) ... **EM3:** 756-757
AGA Thermovision ... **EM3:** 763
Nondestructive testing (NDT) (continued)
applications and limitations to bonded joints ... **EM3:** 751-767
Bondascope 2100 ... **EM3:** 757-758
Bondscan Thermography Inspection System ... **EM3:** 763-765
burned-adhesive reference standard ... **EM3:** 770
calibration of bond testers ... **EM3:** 770
contact potential ... **EM3:** 744
contamination tester ... **EM3:** 744
cores ... **EM3:** 37
correlated with destructive test results ... **EM3:** 772
correlation of results for built-in defects in honeycomb structures ... **EM3:** 766
correlation of results for built-in defects in laminate panels ... **EM3:** 766
defined ... **EM1:** 16
definition ... **EM3:** 743
definition of inspection grade numbers versus void sizes ... **EM3:** 767
detailed written test procedure ... **EM3:** 768
entrapped moisture detection ... **EM3:** 745
establishing quality control standards in product cycle ... **EM3:** 767-771
evaluation and correlation of results ... **EM3:** 771-775
in failure analysis ... **A11:** 16-18
for fatigue failure ... **A11:** 134
Fokker bond tester ... **EM3:** 522, 525, 528, 754-756, 766, 769-770, 772, 778
Fokker bond tester readings correlated to bond strength ... **EM3:** 772-774, 775
harmonic bond tester ... **EM3:** 766, 769, 770
holographic interferometry ... **EM3:** 760-761, 762
honeycomb reference standards ... **EM3:** 770-771
infrared or thermal inspection ... **EM3:** 762-763
infrared radiometer testing ... **EM3:** 762-763
inspection without standards ... **EM3:** 771
inspecton log, daily ... **EM3:** 768
leak (hot-water) test ... **EM3:** 765
liquid-surface acoustical holography ... **EM3:** 765
marked by impact damage ... **EM3:** 751
metal-to-metal reference standards ... **EM3:** 769-770, 771
for moisture and corrosion damage ... **EM3:** 751
NDT-210 bond tester ... **EM3:** 755-756, 769, 770, 772
Novascope ... **EM3:** 758
proof loading ... **EM3:** 37
reference test standards ... **EM3:** 768-769
rejection or nonconformance record ... **EM3:** 768
scanning acoustical holography ... **EM3:** 765-766
schedules and techniques ... **A11:** 134
selection of test method ... **EM3:** 767
shearography ... **EM3:** 761-765
Shurtronics Mark I harmonic bond tester ... **EM3:** 756-757, 760
Sondicator ... **EM3:** 756, 757, 759, 766
Sondicator bond tester ... **EM3:** 770
sonic testing ... **EM3:** 37
specification (test method) ... **EM3:** 768
substitute standards ... **EM3:** 771
tap test ... **EM3:** 760, 766
tapping ... **EM3:** 37
thermal neutron radiography ... **EM3:** 759, 766, 769, 772, 773
ultrasonic bond testers ... **EM3:** 753-759, 766
ultrasonic inspection
contact pulse echo ... **EM3:** 752, 753, 766
contact through transmission ... **EM3:** 752, 766
immersion C-Scan method ... **EM3:** 752-754, 761-762, 766, 770, 772-773
pulse echo ... **EM3:** 748, 769, 772
ringing technique ... **EM3:** 752
ultrasonic inspection and bond test method sensitivity ... **EM3:** 758-759
ultrasonic inspection techniques ... **EM3:** 37, 766
limitations ... **EM3:** 753
wave interference effects ... **EM3:** 772
use of liquid crystals (cholesteric) ... **EM3:** 762, 763
Nondestructive testing (NDT) (continued)
use of thermochromic or thermoluminescent coatings ... **EM3:** 762, 763, 764
variable-quality reference standards ... **EM3:** 770, 771
visual inspection methods ... **EM3:** 37, 751-752
voids, porosity and unbond standards ... **EM3:** 769
of weldments ... **A11:** 412
x-ray radiography ... **EM3:** 759, 766, 770
Nondestructive testing for
explosion welds ... **M6:** 710
high frequency welds ... **M6:** 765-766
solder joints ... **M6:** logo
Nondestructive testing methods
eddy current, powder forged components ... **A7:** 491-492
examination ... **A7:** 491
magnetic particle inspection ... **A7:** 575-579
Nondestructive thermal neutron activation analysis
as common NAA analysis ... **A10:** 234-238
Nondeterministic logic ... **EL1:** 2
Nondeterministic surface ... **A18:** 346, 348
Nondezincification alloy *See also* Copper alloy castings; Copper alloys, specific types, C99400
properties and applications ... **A2:** 389
Nondrying oils
phosphate coatings supplemented with ... **M5:** 453
Nonequilibrium constituents in aluminum alloy ingots
effect of cooling rate on ... **A9:** 634
Nonequilibrium lever rule ... **A6:** 56
Nonequilibrium phases
in the fusion zone of welded joints ... **A9:** 580-581
Nonetching alkaline cleaners *See* Alkaline cleaning, nonetching cleaners
Noneutectic fusible alloys *See also* Fusible alloys
properties ... **A2:** 756
Nonferromagnetic materials *See also* Ferromagnetic materials
eddy current inspection ... **A17:** 164
electric current perturbation inspection of ... **A17:** 136
magabsorption measurement of stress in ... **A17:** 155-156
nickel-plated ,magabsorption measurement ... **A17:** 145
remote-field eddy current inspection ... **A17:** 195
tubesheet rolled joints, eddy current inspection ... **A17:** 180-181
Nonferrous alloys
coatings on ... **A13:** 776
diffraction techniques, elastic constants, and bulk values for ... **A10:** 382
forging of ... **A14:** 239-287
fracture/failure causes ... **A12:** 217
glow discharges for ... **A10:** 28
hydrogen-damage failure(s) ... **A11:** 338
laser cutting ... **A14:** 742
liquid-metal embrittlement ... **A11:** 233-234 **A13:** 178-179
occurrence of SMIE in ... **A11:** 243
recycling of ... **A2:** 1205-1232
for rolling ... **A14:** 355-356
solid-metal embrittlement ... **A13:** 185
in sour gas environments ... **A11:** 300
submerged arc welding ... **A6:** 203
Nonferrous alloys, AMS specific types *See also* Steels, AMS specific types
4544, springs, strip for ... **M1:** 286
5525, springs, strip for ... **M1:** 286
5540, springs, strip for ... **M1:** 286
5542, springs, strip for ... **M1:** 286
5596, springs, strip for ... **M1:** 286
5597, springs, strip for ... **M1:** 286
5698, springs, wire for ... **M1:** 285
5699, springs, wire for ... **M1:** 285
7233, springs, wire for ... **M1:** 284

SUBJECTS OF THE INDEXED VOLUMES: ASM Handbook (designated by the letter "A"): **A1:** Properties and Selection: Irons, Steels, and High-Performance Alloys (1990); **A2:** Properties and Selection: Nonferrous Alloys and Special-Purpose Materials (1990); **A3:** Alloy Phase Diagrams (1992); **A4:** Heat Treating (1991); **A6:** Welding, Brazing, and Soldering (1993); **A7:** Powder Metallurgy (1984); **A8:** Mechanical Testing (1985); **A9:** Metallography and Microstructures (1985); **A10:** Materials Characterization (1986); **A11:** Failure Analysis and Prevention (1986); **A12:** Fractography (1987); **A13:** Corrosion (1987); **A14:** Forming and Forging (1988); **A15:** Casting (1988); **A16:** Machining (1989); **A17:** Nondestructive Testing and Quality Control (1989); **A18:** Friction, Lubrication, and Wear Technology (1992). **Metals Handbook, 9th Edition** (designated by the letter "M"): **M1:** Properties and Selection: Irons and Steels (1978); **M2:** Properties and Selection: Nonferrous Alloys and Pure Metals (1979); **M3:** Properties and Selection: Stainless Steels, Tool Materials and Special-Purpose Materials (1980); **M4:** Heat Treating (1981); **M5:** Surface Cleaning, Finishing, and Coating (1982); **M6:** Welding, Brazing, and Soldering (1983). **Engineered Materials Handbook** (designated by the letters "EM"): **EM1:** Composites (1987); **EM2:** Engineering Plastics (1988); **EM3:** Adhesives and Sealants (1990); **EM4:** Ceramics and Glasses (1991); **Electronic Materials Handbook** (designated by the letters "EL"): **EL1:** Packaging (1989).

Nonferrous alloys, ASTM specific types *See also* Steels, ASTM specific types
B103, springs, strip for ... **M1:** 286
B159, springs ... **M1:** 284
B168, springs ... **M1:** 286
B194, springs ... **M1:** 286
B197, springs ... **M1:** 284
Nonferrous alloys, heat treating ... **A4:** 823-839
annealing of cold-worked metals ... **A4:** 826-830
cold-working effect on properties and microstructure ... **A4:** 827, 830
dislocations ... **A4:** 826-827
grain growth ... **A4:** 827-828, 829, 830
hot working ... **A4:** 827, 830
recovery ... **A4:** 827-828, 829
recrystallization ... **A4:** 827-830
diffusion in metals and alloys
activation energy ... **A4:** 825
diffusion constants ... **A4:** 825
diffusion in alloys (chemical diffusion) ... **A4:** 824
diffusion in pure metals (self-diffusion) ... **A4:** 823, 824
Fick's laws of diffusion ... **A4:** 824-825, 826, 831, 833
grain-boundary diffusion ... **A4:** 826
interstitial diffusion ... **A4:** 826
intrinsic diffusion coefficients ... **A4:** 826
Kirkendall effect ... **A4:** 826
temperature dependence of the rate of diffusion ... **A4:** 825-826
vacancies ... **A4:** 823-824
homogenization of castings ... **A4:** 830-832
chemical homogenization annealing ... **A4:** 831-832
coring ... **A4:** 831, 832
dendrite formation ... **A4:** 831, 832
precipitation hardening heat
precipitation control through heat treatment ... **A4:** 834
precipitation hardening ... **A4:** 834-836
precipitation process ... **A4:** 833
solution heat treatments ... **A4:** 833
treatments ... **A4:** 832-836
two-phase structure development ... **A4:** 836-839
Nonferrous alloys, wear applications *See* Wear-resistant alloys, nonferrous
Nonferrous applications
automotive industry ... **A7:** 617-621
Nonferrous corrosion-resistant materials, selection of
introduction ... **A6:** 585
Nonferrous hardfacing alloys ... **A18:** 758, 761-765
applications ... **A18:** 758, 759, 762
bronze-type ... **A18:** 761, 765
cavitation erosion ... **A18:** 762, 763
classifications ... **A18:** 758
cobalt-base/carbide-type alloys ... **A18:** 761-762, 764
corrosion resistance ... **A18:** 762
galling ... **A18:** 763, 765
high-silicon stainless steel alternate material ... **A18:** 762
iron and nickel substitutes for cobalt compositions ... **A18:** 762
Laves-type alloy compositions ... **A18:** 762-763
nickel-base/boride-type alloys ... **A18:** 763-765
properties ... **A18:** 758-762, 763
spray-and-fuse process ... **A18:** 765
Nonferrous high-temperature materials
constitutional liquation ... **A6:** 567-568, 569
fusion zone ... **A6:** 567
heat-affected zone ... **A6:** 566, 567, 569
liquation cracking ... **A6:** 567-569
Nonferrous high-temperature materials, postweld heat treatment of ... **A6:** 572-574
aging treatments ... **A6:** 572-574
annealing ... **A6:** 572
categories ... **A6:** 572
mill annealing ... **A6:** 572, 573, 574
nominal composition of selected nickel-base and cobalt-base high-temperature alloys ... **A6:** 573
precipitation-strengthened alloys ... **A6:** 572-573
solid-solution-strengthened alloys ... **A6:** 572, 574
solution annealing ... **A6:** 572-574
solution treating and quenching ... **A6:** 572, 574
Nonferrous high-temperature materials, postweld heat treatment of (continued)
strain-age cracking, guidelines for avoidance of ... **A6:** 573-574
stress relieving ... **A6:** 572-574
types of postweld heat treatment ... **A6:** 572
Nonferrous high-temperature materials, welding metallurgy of ... **A6:** 566-571
age hardening ... **A6:** 566, 567
grain size ... **A6:** 567
heat treatments, and liquation cracking ... **A6:** 568-570
heat-affected zone liquation cracking ... **A6:** 567
impurities ... **A6:** 568-570
liquating precipitate amount ... **A6:** 567-568
parameters affecting liquation cracking ... **A6:** 567-570
precipitate types ... **A6:** 568
reducing susceptibility to liquation cracking ... **A6:** 570
strain-age cracking ... **A6:** 566-567
Nonferrous materials
acoustic properties ... **A17:** 235
lubrication of ... **A7:** 191-192
Nonferrous metal parts
secondary operations ... **A7:** 451
Nonferrous metals
air-carbon arc cutting ... **A6:** 1172, 1173-1174
Brinell test application ... **A8:** 89
with crystalline aggregates, Rockwell testing of ... **A8:** 80
glass-to-metal seals ... **EM3:** 301
hardened steel ball indenters for ... **A8:** 74
liquid-metal embrittlement of ... **A11:** 233
microhardness testing for ... **A8:** 89
Nonferrous metals and alloys
brazing to aluminum ... **M6:** 1031
friction welding ... **M6:** 722
gas tungsten arc welding ... **M6:** 205-207
oxyfuel gas cutting ... **M6:** 897
spraying for hardfacing ... **M6:** 789
Nonferrous tubing
inspection of ... **A17:** 572-574
Nonferrous-based infiltration systems ... **A7:** 559-560
Nonfill
defined ... **A14:** 9
Nonflammable solvent cleaner/removers
for liquid penetrant inspection ... **A17:** 75-76
Nonhalogenated solvents
properties of ... **EL1:** 665
Nonhard-drying materials
phosphate coatings supplemented with ... **M5:** 453, 456
Nonheat-treatable
aluminum alloys ... **A13:** 584, 602
commercial wrought aluminum alloys, solution potentials ... **A13:** 584
stainless steels, SCC testing ... **A13:** 272-273
Nonheavy-metal systems
as cathodic inhibitors ... **A13:** 496
Nonhomogeneous deformation ... **A8:** 44
Nonhomogeneous materials
Rockwell hardness testing of ... **A8:** 77, 82
Nonhomogeneous strain
in tubular specimen ... **A8:** 222, 224
Nonhydrogen charged specimen
cycled near stress-intensity factor range threshold ... **A8:** 487-488, 491
Nonhydroscopic
defined ... **EM2:** 28
Nonhygroscopic ... **EM3:** 19
defined ... **EM1:** 16
Nonionic detergents
for surface cleaning ... **A13:** 380
Nonionic surfactants
use in alkaline cleaners ... **M5:** 24
Nonionic/nonpolar contaminants
cleaning of ... **EL1:** 658-659
Nonionic/polar residues
cleaning of ... **EL1:** 659-660
Nonisothermal sidepressed specimen
transverse metallographic sections with equiaxed alpha structure ... **A8:** 589
Nonisothermal upset test ... **A1:** 584
for flow localization ... **A8:** 588
for titanium alloys with equiaxed alpha starting microstructure ... **A8:** 589
Nonleafing
paste, as aluminum pigment ... **A7:** 594
powder, as aluminum pigment ... **A7:** 594
Nonlinear amplification used for image modification in scanning electron
microscopy ... **A9:** 95
Nonlinear elastic fracture mechanics (NLEFM) ... **A11:** 47, 797
Nonlinear harmonics
capabilities and limitations ... **A17:** 161
instrumentation ... **A17:** 161
residual stress measurement by ... **A17:** 160-161
stress dependence ... **A17:** 161
Nonlinear strain functions
rheological behavior ... **EL1:** 848-849
Nonlinear stress analysis
laminate ... **EM1:** 230
Nonlinear stress functions
rheological behavior ... **EL1:** 849
Nonlinearity, tolerable
in rapid-load fracture testing ... **A8:** 260
Nonlubricated specimen
deformation patterns in ... **A8:** 582
Nonlubricated wear
metal adhesion and cold welding as ... **A11:** 154-155
Nonmetallic abrasives, types
wet and dry blasting ... **M5:** 84-86, 93-84
Nonmetallic bearing materials ... **A18:** 748, 754
advantages ... **A18:** 754
applications ... **A18:** 754
disadvantages ... **A18:** 754
mechanical properties ... **A18:** 754
Nonmetallic elements
partitioning oxidation states in ... **A10:** 178
Nonmetallic inclusion testing
steel wire rod ... **M1:** 257
Nonmetallic inclusions *See also* Inclusions; Inclusions
in aluminum alloy castings ... **A17:** 535
in carbon and alloy steels ... **A9:** 169-170, 173, 179
in cold extrusions ... **A14:** 301
defined ... **A9:** 12
in electropolishing ... **A9:** 51
excessive segregation of ... **A11:** 477-478
forging failures from ... **A11:** 317, 322-323
identification by polarized light etching ... **A9:** 59
in ingots ... **A17:** 492
in nickel alloys ... **A9:** 436-437
in nickel-copper alloys ... **A9:** 436-438
powder forged parts ... **A14:** 204
rod fatigue cracking from ... **A11:** 477
in steel ... **A9:** 625-626
in steel, effects of hot rolling on ... **A9:** 628
subsurface, rolling-contact fatigue from ... **A11:** 504
in wrought iron ... **A9:** 39, 42
Nonmetallic materials
coatings, for carbon steels ... **A13:** 524
conductors, galvanic corrosion ... **A13:** 84
contact with aluminum/aluminum alloys ... **A13:** 600-602
functional requirements ... **A13:** 338
in moist chlorine ... **A13:** 1172
rust-preventive compounds attacking ... **M5:** 466
to avoid localized corrosion ... **A13:** 324
Nonmetallic materials, hobs
use in cutting ... **M3:** 477
Nonmetallic materials in metal powders
effect on compressibility ... **A7:** 286
Nonmetallic stringers
formation ... **A12:** 65
Nonmetallic-filled brushes ... **M5:** 152, 155-157
Nonmetallic-inclusion testing
for carbon steel rod ... **A1:** 274
Nonmetals
analysis of surface species ... **A10:** 134
concentration range of voltammetric analysis ... **A10:** 188
on periodic table ... **A10:** 688
Raman analysis of surface species on ... **A10:** 134
Nonoriented silicon steels ... **A14:** 476, 480
grades ... **A2:** 766-769
Nonoxide fibers
continuous silicon carbide ... **EM1:** 63-64
discontinuous silicon carbide whiskers ... **EM1:** 64
silicon nitride whiskers ... **EM1:** 64

Nonoxidizing acids
sample dissolution by ... **A10:** 165-166
Nonoxidizing salts *See* Salts
Nonparametric evaluation ... **A8:** 653
of group medians ... **A8:** 706-707
of percent survival values ... **A8:** 707
of phosphor bronze strip percent survival values ... **A8:** 707, 709
Nonparametric methods
for mechanical properties ... **EM1:** 304
Nonphotoimagable polyimides
thin-film hybrids ... **EL1:** 326-327
Nonplanar solidification ... **A6:** 46, 48, 49, 50, 52, 53
Nonpolar materials
in electronic manufacturing ... **EL1:** 659
Nonpolar/nonionic contaminants
cleaning of ... **EL1:** 658-659
Nonreactive contaminants ... **A7:** 178
Nonreactive milling ... **A7:** 64-65
Nonrefractory metals
effects of lowered BIV in ... **A10:** 588
Nonrelevant indications *See also* Extraneous variables; Relevant indications
defined ... **A17:** 103
liquid penetrant inspection ... **A17:** 86
of magnetic particle welding inspection ... **A17:** 592
sources ... **A17:** 105-106
versus relevant indications ... **A17:** 106-108
Nonresonant force and vibration technique
defined ... **EM2:** 28
Nonresonant forced and vibration technique ... **EM3:** 19
Nonrigid plastic *See also* Plastics ... **EM3:** 19
defined ... **EM2:** 28
NONSAP computer program for structural analysis ... **EM1:** 268, 272
Nonsilicate glasses ... **EM4:** 22
hardness ... **EM4:** 851
Nonsilicated alkaline cleaners ... **M5:** 577
Nonsoap grease
defined ... **A18:** 14
Nonsoluble/particulate contaminants
cleaning of ... **EL1:** 661
Nonspherical powders
copier ... **A7:** 587
Nonstandard thermocouples *See also* Thermocouple materials
types and properties ... **A2:** 874-877
Nonsteady-state diffusion
solid ... **A13:** 68
Nonstructural adhesives ... **EM3:** 44
Nonsulfate sulfur
removal in hydrometallurgical nickel powder production ... **A7:** 140
Nonsynchronous initiation
definition ... **M6:** 12
Nontarnishing ammonia solution
brass alloy SCC environment tested in ... **A10:** 563-564
Nonthermionic emission ... **A6:** 30, 31
Nontraditional machining processes ... **A16:** 2, 4, 509-593
classification of ... **A16:** 509
Nontraditional/densification processes ... **EM4:** 124
Nontransferred arc
definition ... **M6:** 12
Nontransferred arc (plasma arc welding and cutting, and plasma spraying)
definition ... **A6:** 1211
Nontransparent samples
surfaces characterized ... **A10:** 70-71
Nonuniform loading
effects on cracking ... **A12:** 175-176
Nonuniform strain
effect in axial compression testing ... **A8:** 55-58
Nonuniform stress
effects in axial compression testing ... **A8:** 55-58
Nonuniformity of variance ... **A8:** 699
Nonvacuum electron beam welds ... **A11:** 444-447
Nonvolatile liquids
as IR samples ... **A10:** 112
Nonwetting
as solderability mechanism ... **EL1:** 676, 1032-1034
Nonwoven fabric *See also* Fabric(s) ... **EM1:** 16, 149
defined ... **EM2:** 28
Nonwoven fabric (paper) properties
aramid fibers ... **EL1:** 615-616
Nonwoven fabric (tape) properties
aramid fibers ... **EL1:** 616
Nopco Wax
burn off ... **A7:** 191
Noranda process ... **EM4:** 903
Norbide
erosion test results ... **A18:** 200
NOREM alloys
applications ... **A18:** 723
Normal
defined ... **A9:** 12
Normal absorption of high-energy electrons ... **A9:** 111
Normal curve
area under ... **A8:** 675
Normal direction
abbreviation for ... **A11:** 797
defined ... **A9:** 12 **A11:** 7
Normal direction (of a sheet)
abbreviation for ... **A10:** 690
Normal distribution *See also* Distribution ... **A8:** 628-630 **EM3:** 790, 791, 795, 796, 797
cumulative distribution function ... **A8:** 629-630
cumulative, function ... **EL1:** 898
direct computation for ... **A8:** 664-666
estimated failure factors ... **EL1:** 901-902
failure density ... **EL1:** 897
Kolmogorov-Smirnov critical values ... **EL1:** 902
as model, fatigue crack growth analysis ... **A8:** 679
modeling of physical measurements ... **A8:** 629
parameter estimates ... **A8:** 629-631
percentile and percentile estimates ... **A8:** 629-631
probability density function ... **A8:** 629
variance ... **A8:** 629-631
Normal distribution, standard
in quality control ... **A17:** 738
Normal engineering stress ... **A8:** 726
Normal force *See* Load
Normal grain growth *See also* Grain growth ... **A9:** 697
grain shape distribution ... **A9:** 697
grain size distribution ... **A9:** 697
Normal life
in reliability curve ... **EL1:** 244
Normal load *See* Load
Normal operating load
symbol and units ... **A18:** 544
Normal phase chromatograms
liquid chromatography ... **A10:** 652
Normal probability plot
for two-point strategy data ... **A8:** 704-705
Normal pulse polarography
as improved voltammetry ... **A10:** 193
Normal relative radius of curvature ... **A18:** 539
symbol and units ... **A18:** 544
Normal seams, in billets
magnetic particle inspection ... **A17:** 115
Normal segregation
defined ... **A9:** 13
Normal solution *See also* Fatigue life
abbreviation for ... **A10:** 690
symbol for ... **A8:** 725
Normal solutions
defined ... **A13:** 9
Normal stress *See also* Engineering stress; Mean stress; Nominal stress; Principal stress; Residual stress; True stress ... **A8:** 9, 726 **EM3:** 19
defined ... **A13:** 9 **EM1:** 16 **EM2:** 28
Normal stresses
cutting tools ... **A18:** 610-611
Normal unit load
symbol and units ... **A18:** 544
Normal velocity interferometer
in plate impact testing ... **A8:** 233
record of remelted 4340 steel ... **A8:** 234-235
Normal wear
defined ... **A18:** 14
Normal-coordinate analysis
diatomic molecule ... **A10:** 111
infrared spectroscopy ... **A10:** 110-111
results ... **A10:** 111
Normal-operating-stress regime (NOSR) ... **EM3:** 652, 654, 655
Normal-phase chromatography ... **A10:** 652
defined ... **A10:** 678
Normality
in analytical chemistry ... **A10:** 162
defined ... **A10:** 678
and titrant standardization ... **A10:** 172
Normalization
impedance ... **A17:** 170
Normalized compliance ... **A8:** 385
Normalized erosion resistance
classified ... **A11:** 167
defined ... **A18:** 14
Normalized pressure ... **A18:** 281
Normalized velocity ... **A18:** 281
Normalized wear rate ... **A18:** 281
Normalizing *See also* Heat treatment
alloy steel sheet and strip ... **M1:** 165
cast steels, effect on mechanical properties ... **M1:** 384-388
cold finished bars ... **M1:** 234
cold rolled low carbon steel sheet and strip ... **M1:** 157
of cold-rolled steel products ... **A1:** 204
defined ... **A9:** 12-13 **A13:** 9
for ductile iron ... **A1:** 41 **M1:** 37
effect on toughness ... **A1:** 390
gray iron ... **M4:** 531-532, 533
of high-strength structural carbon steels ... **A1:** 390, 391
of low-alloy steel sheet/strip ... **A1:** 209
notch toughness ... **M1:** 696, 701, 703, 706-709
of steel plate ... **A1:** 231 **M1:** 182
of tool and die steels ... **A14:** 53
ultrahigh-strength steels ... **M1:** 423-424, 427, 429, 433, 435, 438, 441
Normalizing of steel *See* Normalizing; Normalizing, etc; Steel castings; Steel, normalizing
Noroc 33
erosion test results ... **A18:** 200
Norton NC-132
fast fracture flexure strength ... **EM4:** 1000-1001
stress-rupture life ... **EM4:** 1001
Norton NC-435 (RBSC)
properties ... **EM4:** 240
Norton's silicon-infiltrated silicon carbide ... **EM4:** 980
Norton/TRW NT-154
fast fracture strength ... **EM4:** 1101
stress-rupture life ... **EM4:** 1001
Nose radius (NR) ... **A16:** 143, 151, 152
boring ... **A16:** 167
of shaper tools ... **A16:** 1
shaping of tool steel die sections ... **A16:** 192
and turning ... **A16:** 158, 159
Nose-splitting
in cast aluminum-magnesium alloy ... **A8:** 595-596
Nosing
for reducing shells ... **A14:** 586
of tubing ... **A14:** 673
Notation *See also* Nomenclature; Terminology
computer, for laminate analysis ... **EM1:** 276
laminate ... **EM1:** 222, 450-451
x-ray and spectroscopic ... **A10:** 569

Notch *See also* Stress concentration
acuity, defined **A11:** 7
brittle fracture from **A11:** 85
brittleness, defined **A8:** 9 **A11:** 7
configuration, notched-specimen testing **A8:** 316
depth **A8:** 9, 221
depth, defined **A11:** 7
design stress **A8:** 316
ductility, defined **A8:** 9
effect on rupture life **A8:** 316-318
effect on stress concentration and notch-rupture strength ratio **A8:** 317
effect on superalloy rupture life **A8:** 316-318
effect on Waspaloy rupture time **A8:** 334
effects in stress-rupture tests **A8:** 333
fatigue specimen, nitrided **A8:** 373
intentional **A11:** 85-87
preparation of crack growth specimen **A8:** 382
rupture strength, defined **A11:** 7
and rupture strength ratio of notched and unnotched specimens **A8:** 316-317
sensitivity, defined **A11:** 7
sharpness severity, and plane-strain fracture toughness **A8:** 450
strength, defined **A11:** 7
as stress concentrator **A11:** 318
as stress raiser **A8:** 364
stress-concentration factor **A8:** 316-317
Notch effect
on Inconel 751 rupture life **A8:** 316-318
Notch factor **EM3:** 19
defined **EM1:** 17 **EM2:** 28
Notch fatigue *See also* Fatigue; Fatigue crack
life of hot isostatically pressed materials **A7:** 439
strength, and axial stress comparison **A7:** 468
Notch opening displacement
as function of time **A8:** 281
measurement in dynamic notched round bar testing **A8:** 278-281
Notch radius
with notch-sensitivity for steels in bending or axial fatigue loading **A8:** 373
notched-specimen testing **A8:** 315
Notch root
radius, machining **A8:** 382
thickness, cantilever beam test **A8:** 537
Notch sensitivity **A1:** 48 **EM3:** 19
A-286 superalloy **A8:** 316
in creep rupture **A8:** 315
defined **A8:** 27 **EM1:** 17 **EM2:** 28
Discaloy **A8:** 316
factor, in test specimen **A8:** 372
Haynes 88 superalloy **A8:** 315-316
stress-rupture test specimen **A8:** 315
and tensile strength **A8:** 372
of titanium alloys **A14:** 838
variation with notch radius for steels, bending or axial fatigue loading **A8:** 373
Notch strength
defined **A8:** 9
ratio, abbreviation **A8:** 725
Notch strengthening
stress-rupture tests **A8:** 315
Notch tensile strength *See* Notch strength
Notch tensile test
mechanical behavior under **A8:** 27
Notch tensile testing **A13:** 962-963
Notch toughness *See also* Impact properties **M1:** 689-709
4140 steel, vs. tempering temperature **M1:** 469
aging, effect of **M1:** 701, 704
anisotropy, effect of **M1:** 695, 696, 700
carburization, effect of **M1:** 704-706
carburized steels **M1:** 534, 536
castings **M1:** 705-709
changes, Charpy V-notch as screening test for **A8:** 263
Charpy and Izod tests of **A11:** 57
closed-die steel forgings **M1:** 358, 359
coatings, effect of **M1:** 705
composition, effect of **M1:** 692-699
constructional steels for elevated temperature use **M1:** 640, 642, 651
correlations and comparisons of **A11:** 60
Notch toughness (continued)
correlations with other mechanical properties **M1:** 703-704
decarburization, effect of **M1:** 705, 707
deoxidation, effect of **M1:** 694-695, 698
drop-weight test for **A11:** 57-58
ductile iron **M1:** 40-41, 45
ductile-to-brittle fracture transition **M1:** 691-692, 699-700, 704
embrittlement, effect of **M1:** 699, 701-703
Esso (Feely) test for **A11:** 60
explosion-bulge test for **A11:** 58-59
ferrous P/M materials **M1:** 343
finishing conditions, effect of **M1:** 699, 700
fracture toughness, relation to **M1:** 690
grain size, effects of **M1:** 695, 699, 701, 703
heat treatments, effect of **M1:** 696, 701, 703, 706, 708, 709
HSLA steels **M1:** 403, 409-410, 414-415, 417-418
Lehigh bend test for **A11:** 59
manufacturing process, effects of **M1:** 694-703
maraging steels **M1:** 450, 451
measurement of **A8:** 262
melting practice, effects of **M1:** 706, 708
microstructure, effect of **M1:** 697-698, 706, 709
Navy tear-test for **A11:** 60
nitriding, effect of **M1:** 704-705
notched slow-bend test for **A11:** 59
Robertson test for **A11:** 59-60
rolling conditions, effect of **M1:** 695, 696
Schnadt specimen test for **A11:** 57
steel plate **M1:** 194
submicroscope structures effect of **M1:** 701-703
surface condition, effects of **M1:** 704-705
testing **M1:** 689-691
testing and evaluation **A11:** 57-60
thickness effects **M1:** 696-697, 700, 701
ultrahigh-strength steels **M1:** 426, 431, 432
Notch toughness of steels **A1:** 737-754
of acicular ferrite **A1:** 404-405
comparison of a low-carbon steel and an HSLA steel **A1:** 389, 404-405
composition, effect of **A1:** 668, 739-742
correlations of, with other mechanical properties **A1:** 753
ductile-to-brittle transition **A1:** 737-739
manganese, effect of **A1:** 390
manufacturing practices, effect of **A1:** 742
normalizing **A1:** 390
wrought steels **A1:** 741, 742-746
of microalloyed forgings **A1:** 360, 361
of microalloyed structural steels **A1:** 412
microstructure, effects of **A1:** 747
grain size **A1:** 744, 748-749
microstructural constituents **A1:** 747-748
reheating on **A1:** 414
submicroscopic structure **A1:** 749, 750
variability of Charpy test results **A1:** 749-753
Notch-insensitive material
defined **EM1:** 235
Notch-rupture strength
grain size effect **A8:** 316-317
ratio **A8:** 316-317
Notch-sensitive material
defined **EM1:** 235
Notch-strength ratio (NSR)
for notch brittleness **A8:** 27
Notched bar
fatigue tests using **A8:** 254
Notched bolt
in quick-release clamp **A8:** 221
Notched impact specimen
cleavage fracture in **A11:** 22
Notched slow-bend test
for notch toughness **A11:** 59
Notched specimen *See also* Impact test; Specimens **EM3:** 19
defined **EM1:** 17 **EM2:** 28
Notched specimen testing **A8:** 315-318
bar, ultrasonic fatigue **A8:** 250-251
for corrosive environmental effects **A8:** 427
fatigue **A8:** 371-372
Notched-bar impact testing
Charpy V-notch test for **A8:** 262
Notched-bar rupture test
constructional steels for elevated temperature use **M1:** 640, 642
Notched-bar upset test **A1:** 584
for determining workability in forging **A8:** 588
for forgeability **A14:** 215, 383-384
rating system for rupture in notched areas **A8:** 588-589
specimen preparation method **A8:** 588
Notched-pin
material selection for **A8:** 221
Notched-specimen testing
grain size effect **A8:** 316-318
notch configuration **A8:** 316
notch radius **A8:** 315
notch sensitivity **A8:** 315
notch-rupture life factors **A8:** 315
root radius **A8:** 315
superalloys **A8:** 315
uses **A8:** 315
Notches
electrical discharge machined (EDM) **A17:** 137-138
fatigue failure **M1:** 679, 681-682
fatigue notch factor **M1:** 667-668
fatigue notch sensitivity **M1:** 667
gray cast iron **M1:** 20, 21
influence on fatigue resistance **M1:** 667-670, 679, 681, 682
liquid-metal embrittlement, effect on **M1:** 688
Notches, avoidance
in copper casting alloys **A2:** 346
Notching
for blanks **A14:** 446
in press brake **A14:** 538
Novaculite
blasting with **M5:** 84, 93-94
NovaScope bond tester
adhesive-bonded joints **A17:** 623
Novolac resins
defined **EM1:** 17
epoxy **EM1:** 68
phenolic, properties/tests for **EM1:** 289-290
Novolac-resols **EM3:** 105
Novolacs **EM3:** 19, 105, 595
chemistry **EM3:** 103
commercial forms **EM3:** 104
curing mechanism **EM3:** 104
epoxidized cresol **EL1:** 811
epoxidized phenol **EL1:** 811
oil-modified **EM3:** 105
phenolic and cresol, as epoxy resins **EL1:** 826-827
powdered **EM3:** 105
rubber-modified **EM3:** 105
for shell molding **EM3:** 105
Novolacs, defined **EM2:** 28
epoxidized phenol **EM2:** 240
phenolic, production **EM2:** 242
and vinyl esters **EM2:** 272
Novoston *See* Cast copper alloys, specific types (C95700); Copper alloys, specific types, C95700
Nozzle
defined **EM2:** 28
definition **A6:** 1211
definition-ion **M6:** 12
Nozzle, reactor
SCC of safe-end on **A11:** 660-661
Nozzle(s) *See also* Atomization
atomizer **A7:** 74-76
ceramic contamination from **A7:** 178
design **A7:** 26-28
effect of decreasing length-to-diameter ratio **A7:** 28, 31
freezing of **A7:** 32
technology, atomizing **A7:** 125-127
Nozzles *See* Spray nozzles
cemented carbide **A2:** 973
design, for lasers **A14:** 739
extrusion, by contour forging **A14:** 71
feedwater, corrosion fatigue **A13:** 937
overstressing **A13:** 1229-1230
particle stream erosion, waterjet cutting **A14:** 746-747
torch, by swaging **A14:** 136
***np* chart** **EM3:** 796, 797
Np-Pu (Phase Diagram) ..... **A3:** 2 • 322
Np-U (Phase Diagram) ..... **A3:** 2 • 322
npn bipolar junction transistor **EL1:** 144-145, 147-149, 191, 958

npn transistor
fabrication ... EL1: 149
npn type transistors
development ... EL1: 958
NQR *See* Nuclear quadrupole resonance
NSLS
as synchrotron radiation source ... A10: 413
NSR *See* Notch-strength ratio
NTC-400
multicylinder engine tests ... A18: 101
NTIS *See* National Technical Information Service
Nu-iron process ... A7: 97
Nuclear applications *See also* Power industry applications ... A7: 664-666, 761
of beryllium powder ... A7: 169, 761
nickel alloys ... A2: 430
refractory metals and alloys ... A2: 558
titanium and titanium alloy castings ... A2: 634
and x-ray applications ... A7: 761
zirconium and zirconium alloys ... A2: 667-668
Nuclear cermet fuel *See also* Nuclear reactor fuel cermets, specific types ... A7: 8
Nuclear charge *See* Atomic number
Nuclear control reflectors
powders used ... A7: 573
Nuclear control rods ... A7: 573, 666
Nuclear cross section (σ)
defined ... A10: 678
Nuclear fuel cycle
nickel-base alloy applications ... A13: 656
Nuclear fuel(s)
elements ... A7: 664-666
pellets ... A7: 664-666
powders packed for ... A7: 297
Nuclear fuels *See also* Boilers; Pressure vessels
assays of ... A10: 207
element details, neutron radiography for ... A17: 391-392
element size, neutron radiography
element size, neutron radiography for ... A17: 391
neutron radiographic inspection ... A17: 391-392
toxicity of ... A2: 1261
Nuclear gamma-ray resonance
acronym ... A10: 689
Nuclear ionization detector (NID)
for lid seal leaks ... EL1: 954
Nuclear lubricants
corrosion effects ... A13: 959-964
Nuclear magnetic resonance ... A10: 277-286
applications ... A10: 277, 282-286
basic equation of ... A10: 279
Bloch equations (T_1 and T_2) ... A10: 280
capabilities ... A10: 287, 649
capabilities, compared with infrared spectroscopy ... A10: 109
defined ... A10: 678
ESR, IR, and UV, compared with ... A10: 265
estimated analysis time ... A10: 277
experimental arrangement ... A10: 281-282
ferromagnetic nuclear resonance ... A10: 281
general uses ... A10: 277
inorganic applications ... A10: 7, 282-283
introduction and principles ... A10: 277-281
Knight shift measurements, metallic glass ... A10: 284
magic-angle spinning, capabilities ... A10: 407
for molecular structure ... A10: 116
nuclear quadrupole resonance ... A10: 281
organic applications ... A10: 9, 10, 285-286
polyimide resin analysis ... A10: 285
pulse-echo method ... A10: 281
related techniques ... A10: 277
samples ... A10: 277, 282-286
sensitivity ... A10: 281
Nuclear magnetic resonance (NMR)
See also NMR spectroscopy ... EM1: 736 EM3: 19
applications ... A2: 1027
defined ... EM2: 28
Nuclear magnetic resonance (NMR) (continued)
and industrial computed tomography ... A17: 362
in magabsorption development ... A17: 143
as niobium-titanium superconducting material application ... A2: 1057
as ternary molybdenum chalcogenide application ... A2: 1080
Nuclear magnetic resonance (NMR) spectroscopy ... EM4: 460
for chemical analysis of polymer fibers ... EM4: 223
to analyze the phase composition of ceramic powders ... EM4: 73
to estimate pore volume to surface area ratio ... EM4: 71, 72
Nuclear materials
radioactive elements determined in ... A10: 243
Nuclear power
brazeability and solderability applications ... A6: 618
Nuclear power applications
boiler/pressure vessels, inspection codes/methods ... A17: 641-644
components, creep defects ... A17: 54
eddy current inspection ... A17: 173, 182
fuel elements, ultrasonic inspection ... A17: 232
of neutron radiography ... A17: 387, 391-395
nuclear waste, acoustic emission inspection ... A17: 281
remote-field eddy current inspection ... A17: 199-200
waste containers, ultrasonic inspection ... A17: 275-276
Nuclear power industry ... A13: 927-984
boiling water reactors, corrosion in ... A13: 927-937
cobalt-base alloy applications ... A13: 667
containment materials, corrosion of ... A13: 971-980
erosion-corrosion in wet steam flow ... A13: 964-971
niobium applications ... A13: 723
nuclear lubricants, effect on corrosion ... A13: 959-964
radiation fields, corrosion influences on ... A13: 948-952
radioactive-waste isolation ... A13: 971-980
stainless steel corrosion ... A13: 561
steam generator failure/degradation ... A13: 937-945
steam turbine materials, SCC in ... A13: 952-959
structures, corrosion of ... A13: 1299
Zircaloy-clad LWR fuel rods ... A13: 945-948
zirconium applications ... A13: 718-719
Nuclear power plant(s)
industry, P/M developments ... A7: 18
refractory metals for ... A7: 765
technology ... A7: 18
Nuclear power reactor(s)
material fabrication for ... A7: 664
pressurized water, and boiling water ... A7: 664
shielding ... A7: 666
Nuclear power systems
alloy steel corrosion in ... A13: 540-541
Nuclear pressure vessel
impact properties of submerged arc weld ... A11: 69
Nuclear pressure vessels
Charpy V-notch test use with ... A8: 263-264
crack arrest toughness for ... A8: 284
design code ... A8: 263-264
toughness, requirements for steels ... A8: 263
Nuclear properties
actinide metals ... A2: 1189-1198
cast copper alloys ... A2: 358, 390
of elements ... A10: 278-279
pure metals ... A2: 1100-1178
rare earth metals ... A2: 1178-1189
transplutonium actinide metals ... A2: 1198
Nuclear quadrupole resonance ... A10: 281, 689
Nuclear radiation induced device failure
microelectronic devices ... EL1: 1056
Nuclear reactor fuel cermets, specific types *See also* Nuclear cermet fuel
aluminum, properties ... A7: 805
beryllium, properties ... A7: 805
chromium, properties ... A7: 805
iron, properties ... A7: 805
magnesium, properties ... A7: 805
molybdenum, properties ... A7: 805
nickel, properties ... A7: 805
niobium, properties ... A7: 805
stainless steel, type 304, properties ... A7: 805
uranium dioxide, properties ... A7: 805
zirconium, properties ... A7: 805
Nuclear reactors ... A10: 233, 459
codes governing ... M6: 823-824
fretting wear in the advanced gas-cooled reactor (AGR) ... A18: 250
impact fretting wear ... A18: 247
neutron embrittlement of materials in ... M1: 686-687
as neutron sources ... A17: 388
Nuclear Regulatory Commission ... A8: 285, 725
Nuclear Regulatory Commission (NRC)
data bases ... A11: 54
Nuclear steam systems
corrosion in ... A11: 616, 656-657
Nuclear structure
defined ... A10: 678
Nuclear tracing imaging, and computed tomography
compared ... A17: 362
Nuclear waste
acoustic emission inspection ... A17: 281
Nuclear waste ceramic forms
EPMA study of ... A10: 532-534
Nuclear waste disposal
alloy steel corrosion in ... A13: 541-542
Nuclear-grade stainless steels ... A13: 931
Nuclear-polarization techniques
and double resonance ... A10: 258
Nucleated resins
in injection molding ... EM1: 167
Nucleating agent ... EM3: 19
Nucleating agents ... EM2: 29, 502-503
added to pure metal melt to promote fine grain structure ... A9: 608
Nucleation *See* Row nucleation; Secondary nucleation ... A13: 67, 245
in age-hardening materials, FIM/AP study of ... A10: 583
at grain boundaries ... A8: 366
at twin boundaries ... A8: 366
control and powder production ... A7: 54
crack ... A8: 366
crack, AISI/SAE alloy steels ... A12: 307
crack, as fatigue stage ... A11: 102, 104, 106
defined ... A9: 13
dislocation, crack tip ... A12: 30
fatigue crack, austenitic stainless steel implants ... A12: 360
grain-boundary cavity ... A12: 20
LME and SMIE, compared ... A11: 240
of martensitic structures ... A9: 668
of massive transformation structures ... A9: 656-657
in metals, by SAXS/SANS/SAS ... A10: 402
microvoid, effects on fracture surface ... A12: 12
of peritectic structures ... A9: 676
in polycrystalline pure metals ... A9: 608-610
in precipitation hardening ... A9: 646-647
rate of ... A9: 649
SAS techniques for ... A10: 405
scanning electron microscopy used to study ... A9: 101
in stress-corrosion cracking ... A8: 496
theories, and isothermal phase transformations ... A10: 317
in titanium and titanium alloys ... A9: 460

Nucleation and growth
compared to spinodal decomposition for formation of two-phase mixtures **A9:** 652
of lower bainite **A9:** 664-665
of pearlite **A9:** 658-661
of upper bainite **A9:** 663
Nucleation sites
in cold-worked metals **A9:** 694-696
of dislocation cells **A9:** 685
for solid-state transformations in welded joints **A9:** 579
Nucleation theory
and amorphous materials and metallic glasses **A2:** 811
Nuclei
properties of **A10:** 277-278
Nucleic acids
ESR studies of **A10:** 264
Nucleus
defined **A9:** 13
Nuclide
defined **A10:** 678
Nugget
definition **A6:** 1211 **M6:** 12
Nugget area **A6:** 10-11
Nugget size
definition **M6:** 12
Nugget size (resistance welding)
definition **A6:** 1211
Nuisance variables
blocking **A8:** 697
effecting fatigue behavior **A8:** 697
Null hypotheses
defined **A8:** 626
and probability **A8:** 624, 626-627
Number 3 Die Casting Alloy *See* Zinc alloys, specific types, AG40A
Number 5 Die Casting Alloy *See* Zinc alloys, specific types, AC41A
Number of stress cycles endured
symbol for **A8:** 725
Numbered alloys
300M
composition **M4:** 120
heat treatment **M4:** 122
mechanical properties **M4:** 122, 123
Numeric control (NC) system
computer-aided manufacturing **EL1:** 131
Numerical aperture
defined **A10:** 678
Numerical aperture of objective lenses **A9:** 72-73
defined **A9:** 13
effect on depth of field **A9:** 76
relationship for four wavelengths of light **A9:** 78
relationship to depth of field and wavelength of light **A9:** 78
and resolution of an optical microscope **A9:** 75-76
Numerical aspects of modeling welds **A6:** 1131-1139
contact conductance **A6:** 1133-1134
convection **A6:** 1133-1134
energy equation and heat transfer **A6:** 1132-1136
finite-difference methods (FDM) **A6:** 1132, 1133, 1134
finite-element methods (FEM) **A6:** 1132, 1133, 1134, 1135, 1137-1138, 1139
fluid flow in the weld pool **A6:** 1138-1139
geometry of weld models **A6:** 1131-1132
material vs. spatial reference frames **A6:** 1135-1136
microstructure evolution **A6:** 1136
modeling of welds **A6:** 1131
modeling the addition of filler metal **A6:** 1136
modeling the heat source in a weld **A6:** 1134-1135
prescribed-temperature heat source **A6:** 1135
radiation boundary conditions **A6:** 1133-1134
stress analysis near the weld pool **A6:** 1138
stress analysis of welds in thin-walled structures **A6:** 1138
thermal stress analysis of welds **A6:** 1136-1137
transformation plasticity **A6:** 1138
transient vs. steady state **A6:** 1136
Numerical control (NC) **A16:** 613-617
adaptive systems **A16:** 617
advantages **A16:** 614-615
automatically programmed tool (APT) language **A16:** 615-616
axis of motion **A16:** 614, 615, 616, 617
basic length units (BLU) **A16:** 614, 616
computer numerical control (CNC) **A16:** 613, 614, 615, 616, 617
continuous path or contouring systems **A16:** 616-617
direct numerical control (DNC) **A16:** 613
evolution of **A16:** 613-614
feedback devices **A16:** 614
fundamentals **A16:** 614
industrial robot systems and NC **A16:** 616
Local Area Network **A16:** 613
machines **A16:** 614, 615, 616, 617
machining centers **A16:** 613-614
part programmer required **A16:** 614
point-to point (PTP) programming **A16:** 17
programming **A16:** 16
software, CNC systems **A16:** 613
system structure **A16:** 616
Numerical design and analysis **EM3:** 477-500
closed-form methods **EM3:** 477-478, 479, 481
design considerations in FE modeling **EM3:** 91
failure criteria **EM3:** 481-484
finite-element results of common joint geometries **EM3:** 491-493
graphite-epoxy tube (small-diameter) with bonded aluminum end fitting **EM3:** 495-496
metallic fitting and graphite-epoxy composite tube joint analysis **EM3:** 97
numerical methods **EM3:** 478-481
single-lap-shear specimen with/without adhesive fillet **EM3:** 493-495
Numerical dynamic analysis
in fracture analysis **A8:** 445-446
Numerical methods
of thermal durability modeling **EL1:** 51
Numerical process modeling **EM1:** 500
Nut formers **A14:** 292, 296
Nut steels **A1:** 291, 292-294, 295
Nutcracker shears *See* Alligator shears
Nuts *See also* Threaded fasteners and specific types by name
cadmium-plated steel, LME failure **A11:** 543
carbon steel for **M1:** 254-255
cast cobalt-chromium-molybdenum alloy, effect of differing conditions on **A11:** 675, 681
composition **M1:** 277
die-cast zinc alloy, SCC of **A11:** 538-539
fabrication **M1:** 277
failure origins in **A11:** 529-530
gray iron, brittle fracture of **A11:** 369-370
heat treatment **M1:** 277
mechanical fastening of **EM2:** 711
mechanical properties **M1:** 274, 278
proof testing **M1:** 278
selection of steel for **M1:** 265-266, 274, 276-277
strength grades and property classes **M1:** 273-277
wheel **A11:** 532
Nutshell flour **EM3:** 175
Nylon *See also* Nylon 6/6; Nylon alloys; Nylon plastics; Polyamides (PA) **A7:** 606 **EM1:** 17, 35 **EM3:** 19
applications **EM2:** 125-126
defined **EM2:** 29
die material for sheet metal forming **A18:** 628
driving gear, failure of **A11:** 764, 765
ESC testing **EM2:** 802-803
film, as PA application **EM2:** 125
glass fiber reinforced, fracture mechanisms **A11:** 759
IPN polymers **EM3:** 602
moisture effects **EM2:** 768
primers **EM3:** 278
reinforced **EM3:** 601
in rotational molding **EM2:** 361
as semicrystalline polymer **A11:** 758
solvent cements **EM3:** 567
stress crazing **EM2:** 798-801
substrate cure rate and bond strength for cyanoacrylates **EM3:** 129
Nylon (continued)
surface preparation **EM3:** 278
thermal properties **A18:** 42
vs polyamide-imides (PAI) **EM2:** 128
Nylon 6 (cast) and extruded
friction coefficient data **A18:** 73
Nylon 6/6 *See also* Caprolactam; Nylon; Polyamides (PA)
chemistry **EM2:** 66
friction coefficient data **A18:** 73
properties **EM2:** 124
as structural plastic **EM2:** 66
wear factor **A18:** 824
Nylon 6/6 (+ PTFE)
friction coefficient data **A18:** 73
Nylon alloys
properties **EM2:** 125
Nylon brushes
in remote-field eddy current inspection **A17:** 201
Nylon fibers
friction coefficient data **A18:** 75
Nylon plastics *See also* Nylon; Plastics **EM3:** 19
defined **EM1:** 17 **EM2:** 29
Nylon polishing wheels **M5:** 109
Nylon-epoxies **EM3:** 76
advantages and limitations **EM3:** 77
applications **EM3:** 78
typical film adhesive properties **EM3:** 78
Nylons, SiC-filled
PCD tooling **A16:** 110
Nyquist frequency **A18:** 294
Nyquist sampling frequency **A17:** 384
The National Institute of Ceramic Engineers (NICE) **EM4:** 40
The National Institute of Standards and Technology (NIST) **EM4:** 40

O

100% solids systems **EM3:** 35
n-Octyl alcohol ($C_8H_{17}OH$)
as solvent used in ceramics processing **EM4:** 117
O, annealed temper
defined **A2:** 21
***o*-cresol**
physical properties **EM3:** 104
O-factor, single crystals
FMR eddy current probes **A17:** 220
O-Pb (Phase Diagram) ..... **A3:** 2 • 323
O-Pr (Phase Diagram) ..... **A3:** 2 • 323
O-Pu (Phase Diagram) ..... **A3:** 2 • 323
O-ring
for fatigue test chamber **A8:** 412
grips, with elevated-temperature compression testing **A8:** 196
rubber assemblies, neutron radiography of **A17:** 388
seals, in explosive bolt assemblies, neutron radiography of **A17:** 394
specimens, SCC testing **A8:** 506
O-ring specimens
SCC testing **A13:** 249-250
O-rings
in graphite-composite assemblies **EM1:** 720
O-Sn (Phase Diagram) ..... **A3:** 2 • 324
O-Ti (Phase Diagram) ..... **A3:** 2 • 324
O-V (Phase Diagram) ..... **A3:** 2 • 325
O-W (Phase Diagram) ..... **A3:** 2 • 325
O-Y (Phase Diagram) ..... **A3:** 2 • 326
O-Zr (Phase Diagram) ..... **A3:** 2 • 326
O1, O2, O4, etc *See* Tool steels, specific types
Oak Ridge Molten Salt Reactor Experiment **A13:** 51-52
Oak Ridge National Laboratory (ORNL)
ceramic materials and component fabrication methods **EM4:** 716, 720, 997
run-arrest experiments **A8:** 285
thermal shock fracture experiments **A8:** 285-286
Oak Ridge Thermal Ellipsoid Program (ORTEP)
to illustrate crystal structure **A10:** 352, 354
Object(s) *See also* Part(s); Samples; Specimens
-camera distance determination, machine vision **A17:** 34
dynamic behavior, strain sensing for **A17:** 51
location and recognition algorithms **A17:** 37

Object(s) (continued)
orientation, machine vision ... **A17:** 34
position defined by relative motion ... **A17:** 34-35
shape, NDE method by ... **A17:** 51
size, NDE methods for ... **A17:** 51
Objective
defined ... **A9:** 13 **A10:** 678
Objective aperture
defined ... **A9:** 13
Objective lenses of microscopes ... **A9:** 72-73
in a transmission electron microscope ... **A9:** 103-104
cross sections ... **A9:** 75
defects ... **A9:** 75
degree of correction, effect on resolution ... **A9:** 75
relationship between resolution and numerical aperture ... **A9:** 78
Objectives
corrosion testing ... **A13:** 193
experimental ... **A8:** 639-640
Oblique cutting ... **A16:** 7-8, 10
Oblique evaporation shadowing
defined ... **A9:** 13
Oblique illumination
effects on fatigue fracture ... **A12:** 85
for medium-carbon steels ... **A12:** 275
photographic ... **A12:** 83
replica fractograph ... **A12:** 100
and vertical lighting, compared ... **A12:** 87
Oblique illumination in optical microscopy ... **A9:** 76
defined ... **A9:** 13
Oblique photography
by view camera ... **A12:** 78
Oblique section *See* Taper section
Oblong weave screens ... **A7:** 176
Observation *See also* Examination; Visual examination
direct, of thin-foil specimens ... **A12:** 179
Observation, visual
of physical crack size ... **A8:** 452
Observed significance level (OSL)
calculated ... **EM1:** 303
Observed value
defined ... **A8:** 59
Obsidian ... **EM4:** 1
Obsolete scrap ... **A1:** 1023
Occluded argon
effects in argon-atomized superalloys ... **A7:** 254
Occupational metal toxicity *See also* Toxic metals; Toxicity of metals
acute selenium poisoning ... **A2:** 1254
from essential metals ... **A2:** 1250-1256
hydrogen selenide ... **A2:** 1254
inhalation of iron oxide fumes ... **A2:** 1252
lithium hydride ... **A2:** 1257
of magnesium ... **A2:** 1259
of manganese ... **A2:** 1253
selenium ... **A2:** 1254-1255
of silver ... **A2:** 1260
thallium ... **A2:** 1261
tin ... **A2:** 1261
titanium ... **A2:** 1261
vanadium ... **A2:** 1262
of zinc ... **A2:** 1255-1256
Occupational radiation exposure ... **A13:** 949
Occupational Safety and Health Administration ... **A14:** 518
Occupational Safety and Health Administration (OSHA) ... **A7:** 202 **EM1:** 36
time-weighted average (TWA) concentrations for cyanoacrylates ... **EM3:** 131
Occurrence
of bismuth ... **A2:** 753
gallium ... **A2:** 741
of indium ... **A2:** 750
predicted vs. actual frequencies ... **A8:** 637
rhenium ... **A2:** 581
Ocean water *See also* Pacific Ocean; Seawater; Water
uranium/uranium alloys in ... **A13:** 815
Ochre
as carrier of cations in ion-exchange ... **EM4:** 461
OCL-4036
mechanical properties ... **A18:** 823
Octadecenoic armide ... **A18:** 532
Octanary system or diagram ... **A3:** 1 • 2
Ordered crystal structure ... **A3:** 1 • 10
Octanol ($C_8H_{17}OH$)
as solvent used in ceramics processing ... **EM4:** 117
Octonoic acid [$CH_3(CH_2)_6CO_2H$#
as solvent used in ceramics processing ... **EM4:** 117
Ocular *See* Eyepiece
Ocvirk number
defined ... **A18:** 14
OD *See* Outside diameter
ODF *See* Orientation distribution function
ODMR *See* Optical double magnetic resonance
Odor detection
of gases ... **A17:** 61
ODS copper *See* Oxide dispersion-strengthened copper
ODS superalloys *See* Oxide dispersion strengthened alloys
OES *See* Optical emission spectroscopy
Off analysis
as cause of casting failure ... **A11:** 391
Off time
definition ... **M6:** 12
Off-axis loading ... **EM1:** 209
Off-chip rinse time
ECL/CMOS ICs ... **EL1:** 401
Off-line quality control ... **A17:** 719, 725
Off-stream pressure vessel periods
protection during ... **A11:** 661
Offhand belt polishing and wheel buffing
copper and copper alloys ... **M5:** 616
Office buildings
corrosion in ... **A13:** 1299
Office equipment applications of copper- based powder metals ... **A7:** 733
for stainless steels ... **A7:** 731
Office-chair roller, rubber
failure of ... **A11:** 763-764
Offset *See also* Offset yield strength; Yield strength
arms, for torsional fatigue testing ... **A8:** 151
defined ... **A8:** 9 **A14:** 9
method, for bending yield strength determination ... **A8:** 134
Offset central conductor
application ... **A17:** 96-97
Offset dies
for press-brake forming ... **A14:** 536
Offset modulus ... **EM3:** 19
defined ... **EM1:** 17 **EM2:** 29
Offset of plate edges
steel pipe ... **A17:** 565
Offset parts *See also* Edge bending
press forming ... **A14:** 549
Offset platemaking ... **A7:** 580
Offset upsetting ... **A14:** 90
Offset yield strength ... **EM3:** 19
in bending ... **A8:** 132, 134
defined ... **A8:** 9, 21 **A14:** 9 **EM1:** 17 **EM2:** 29
temperature effect on ... **A8:** 36
Offshore applications
high-strength low-alloy steels for ... **A1:** 417-418
Offshore engineering
composite structures for ... **EM1:** 839
Offshore gas/oil production platforms
corrosion of ... **A13:** 920, 922-924, 1239-1240, 1254-1255
Offshore oil rigs
fretting wear ... **A18:** 250
Offshore structures ... **A6:** 381-385
OFHC *See* Oxygen-free high conductivity
OFHC copper
shear stress-strain curve for ... **A8:** 216-217
OFL-4036
mechanical properties ... **A18:** 823
Ohm
defined ... **EL1:** 89
Ohm, as SI derived unit
symbol for ... **A10:** 685
Ohmic contact window
vacuum-deposited aluminum defects ... **A12:** 486, 487
Ohmic contacts
development ... **EL1:** 958
Ohmic resistance energy loss ... **A10:** 199
Oil
bath ovens, temperature control in ... **A8:** 36
canning, as formability problem ... **A8:** 548
cold rolling, as ductility bending test lubrication ... **A8:** 127
contamination by, vapor degreasing solvents ... **M5:** 47-48
defined ... **A18:** 14
"fast quenching" ... **A7:** 453
flow, in hydraulic or rotary actuator motors ... **A8:** 157-158
human, effect on bearing strength, aluminum alloy sheet ... **A8:** 60
phosphate coating process *See* Phosphate coating process, oils in
pump, wear ... **A8:** 607
removal of
phosphate coating process ... **M5:** 443, 446
process types and selection ... **M5:** 5, 8-9
SAE 10 ... **A8:** 60
SAE 40 ... **A8:** 60
in servo-hydraulic test frames ... **A8:** 192
viscous, with room-temperature compression testing ... **A8:** 195
Oil and gas industry applications
cobalt-base wear-resistant alloys ... **A2:** 451
Oil bath lubrication ... **A18:** 132-133
Oil canning *See* Canning
carbon steel effects ... **A13:** 520, 522
Oil coatings
corrosion protection ... **M1:** 752, 757
hot rolled bars ... **M1:** 200
Oil content ... **A7:** 8
of sintered bronze bearings ... **A7:** 705
Oil coolers
finned tubing for ... **A11:** 628
hydraulic, crevice corrosion of tubing in ... **A11:** 632-633
Oil country tubular goods *See also* Steel tubular products ... **A1:** 329, 330, 332-333 **A9:** 210-211 **M1:** 315, 319-320
Oil cup
defined ... **A18:** 14
Oil drilling
powders used ... **A7:** 574
Oil drilling applications
cemented carbides ... **A2:** 974-977
Oil film whirl (whip)
sliding bearings ... **A18:** 520
Oil films
in sliding contacts ... **A2:** 840-841
Oil fired furnace
dip brazing ... **A6:** 337
Oil flow quantity ... **A18:** 543
Oil flow rate
defined ... **A7:** 8
symbol and units ... **A18:** 544
Oil fog lubrication *See* Mist lubrication
Oil, fresh
as fracture preservative ... **A12:** 73
Oil groove
defined ... **A18:** 14
Oil hole or pressurized-coolant drilling
refractory metals ... **A16:** 861, 863, 865
Oil holes
effect in gears ... **A11:** 589
as fracture origin, shafts ... **A11:** 459

Oil impregnation **A7:** 8, 463
Oil jet lubrication **A18:** 133
Oil mist lubrication **A18:** 133
Oil pans, automotive
forming strain analysis **M1:** 545-546
Oil permeability
defined **A7:** 8
Oil pipelines
high-strength low-alloy steels for **A1:** 416-417
Oil pocket
defined **A18:** 14
Oil production *See also* Oil refineries; Petroleum production operations; Petroleum refining and petrochemical operations
alloy steel corrosion in **A13:** 533-538
corrosion inhibitors for **A13:** 478-484
stainless steel corrosion in **A13:** 560
wells **A13:** 478-480
Oil quenching *See* Quenching
for continuous carburizers **M4:** 59, 60, 63
control of oils **M4:** 53
cooling characteristics **M4:** 38, 39, 40, 43, 44, 45-47
emulsions **M4:** 40, 45
equipment, maintenance **M4:** 63, 66-68
in gaseous nitrocarburizing **A7:** 455
martempering **M4:** 44-45
oil flow **M4:** 47, 48, 49, 60
quench loads **M4:** 49
safety precautions **M4:** 68
selection of oil **M4:** 42, 53-54
conventional oils **M4:** 44
fast quenching oils **M4:** 44
surface conditions of quenched work **M4:** 53
temperature **M4:** 40, 47-49
tool steel die crack during **A11:** 565
water contamination **M4:** 41, 49-53
Oil refineries
carbon steel weldment corrosion **A13:** 365-366
Oil refinery applications
cast steels **M1:** 388
Oil ring lubrication
defined **A18:** 14
Oil rust-preventive compounds **M5:** 459-470
applying, methods of **M5:** 466-467
duration of protection **M5:** 469
safety precautions **M5:** 470
Oil shale
GC/MS analysis of volatile compounds in **A10:** 639
Oil spills
MFS identification of **A10:** 72
Oil starvation
defined **A18:** 14
Oil suspending liquid
for magnetic particles **A17:** 101
Oil well tubing
magnetic particle inspection of **A17:** 111-112
Oil whirl
defined **A18:** 14
Oil(s)
cracked, copper alloy corrosion in **A13:** 636
solvent cleaning **A13:** 413-414
steam cleaning **A13:** 414
surface, cleaning of **A13:** 380
Oil-based sealants **EM3:** 188, 189, 191
characteristics of wet seals **EM3:** 57
Oil-cutting chemicals **EM3:** 52
Oil-filled self-lubricating parts
powders used **A7:** 574
Oil-hardening cold work tool steels *See* Tool steels, oil-hardening cold work
Oil-hardening cold-work steels
composition limits **A18:** 735
Oil-hardening cold-work tool steels **A1:** 765
forging temperatures **A14:** 81
Oil-hole reamers **A16:** 245
Oil-immersion lenses **A9:** 73
cross section **A9:** 75
Oil-impregnated bearings
density **A7:** 463
Oil-in-water microemulsions **A18:** 144
Oil-pump gears
sand-cast **A11:** 344
Oil-tempered wire *See* also, Steels, ASTM Specific types, A **A1:** 280-281
alloy steel wire **M1:** 269, 270
Oil-tempered wire (continued)
characteristics **M1:** 289
characteristics of **A1:** 306
cost **M1:** 305
seams in **M1:** 290
spring wire, tensile strength **M1:** 267, 269
steel wire **M1:** 262
stress relieving **M1:** 291
Oil/wear particle analysis methods *See also* Lubricant analysis **A18:** 299-302
applications and purposes **A18:** 299
ferrography **A18:** 301-302
infrared (IR) spectroscopy **A18:** 300-301
magnetic plug/chip detection (MCD) **A18:** 301
particle counting **A18:** 301
physical inspection **A18:** 299-300
physical testing **A18:** 300
sampling of service lubricants **A18:** 299
spectrometric metals analysis **A18:** 300
Oiliness *See* Lubricity
Oiliness additives
for metalworking lubricants **A18:** 140-141, 142, 143
Oilite bearings
in split Hopkinson pressure bar brackets **A8:** 201
Oilless bearing
defined **A7:** 8
Oils
additives, as lubricant failure preventive **A11:** 154
-ash corrosion, steam equipment **A11:** 618
for bearings **A11:** 511
detection limits of minor elements in **A10:** 101
in lubricants **A14:** 514
lubricating **A11:** 152
lubrication, for rolling-element bearings **A11:** 512
sulfur determination in **A10:** 101
vegetable, olefinic unsaturation **A10:** 205
wear metal analysis, optical emission spectroscopy **A10:** 21
Oils, soldering or tinning
wave soldering **EL1:** 689
Oklahoma sandstone
particle size distribution effect on technique **EM4:** 378
Old English finish copper coloring solution **M5:** 626
Olefin
defined **EM1:** 17 **EM2:** 29
Olefin copolymers (OCP)
hydrocarbon moiety of dispersants **A18:** 99
as viscosity improvers **A18:** 109, 110
Olefin plastics *See also* Plastics **EM3:** 19
bismaleimide reaction with **EM2:** 255
defined **EM2:** 29
Olefin polymer
properties **A18:** 81
Olefin-modified styrene-acrylonitriles (OSA) *See* Olefin plastics; Styrene-acrylonitriles (SAN, OSA, ASA)
Olefinic unsaturation in vegetable oils
electrometric titration determined **A10:** 205
Olefln **EM3:** 19
Oleic acid
copper/copper alloy resistance **A13:** 629
film, effect on green strength and electrical conductivity in copper powders **A7:** 304
processing aid affecting viscosity in injection molding **EM4:** 174
as surfactant for material flow **A7:** 188
Oleoresinous coatings **M5:** 474, 498-499, 501, 505
Oleoresins
applications **EM3:** 56
Oleorosin
for gum rosin **EL1:** 645
Oleum
tantalum corrosion in **A13:** 725, 730
Olfaction
odor/gas detection by **A17:** 61
Oligomer **EM2:** 29, 64
Oligomers **EM3:** 19, 41
formation **EM1:** 78
for solder masking **EL1:** 555
Olive green chromate conversion coating
cadmium plate **M5:** 264
Olivine
applications **EM4:** 46
composition **EM4:** 46
crystal structure **EM4:** 881
island structures **EM4:** 758
supply sources **EM4:** 46
Olivine sand
abrasive mixed with high-pressure waterjet **A16:** 521
Olsen cup test **A8:** 562, 565 **A14:** 9, 624, 792
Olsen ductility
low-carbon steel sheet and strip **M1:** 156, 161, 162
Olsen ductility test
defined **A8:** 9 **A14:** 9
Olsen-Erichsen cup test **A1:** 576
Olyphant washer test
for flexible epoxies **EL1:** 820
OM *See* Optical metallography
OMC VCA *See* Titanium alloys, specific types, Ti-13V-11Cr-3Al
Omega meter
for contaminant extraction **EL1:** 667
Omega phase
defined **A9:** 13
in titanium alloys **A9:** 461
in titanium-iron binary **A9:** 474
Omega structures
wrought titanium alloys **A2:** 606-607
On cooling testing **A8:** 586-587
On heating tests **A8:** 586-587
On-aircraft eddy current inspection
examples of **A17:** 191-194
On-chip
clock signals **EL1:** 7
connections, level **EL1:** 76
data interconnections, advantages **EL1:** 6
integration **EL1:** 365
interconnection lines **EL1:** 6
issues, physical performance **EL1:** 6-7
metallurgy **EL1:** 480
moisture monitor **EL1:** 953
On-chip built-in self-test **EL1:** 374-376
On-line activities
quality control **A17:** 725
On-line cleaning **A13:** 1143
On-line image processing in scanning electron microscopy **A9:** 96
On-line nondestructive testing **EM3:** 555
On-site examination
of boilers and related equipment **A11:** 602
of failed parts **A11:** 173
On-site photomacrography
setups for **A12:** 78
On-stream corrosion monitoring **A13:** 322-323
Once-through steam generators **A13:** 937-938
One-at-a-time experiments **A8:** 641
"One-body wear" **A18:** 263
One-component adhesive **EM3:** 19
One-component thermoset injection molding **EM1:** 558
One-dimensional analysis
of defects **A10:** 465-466
One-half fractional factorial design
defined **A17:** 747
"One-knob control" **A6:** 40-41
One-out-of N parallel reliability block diagram **EL1:** 899
One-point bend test **A8:** 271-276
for high strain rate fracture toughness testing **A8:** 187
response curve for **A8:** 20
One-shot molding *See also* Prepolymer molding
defined **EM2:** 29
One-side welding, applications
shipbuilding **A6:** 384
One-sided alternative hypotheses **A8:** 626
One-sided tolerance limits **A8:** 664-665, 700
One-stage welding **A6:** 316
One-step drilling method **EM1:** 671-672
One-step replicas
TEM **A12:** 7
One-step temper embrittlement *See* Tempered martensite embrittlement

One-stroke hemming dies
for press-brake forming ... **A14:** 537
One-way shape memory
shape memory alloys ... **A2:** 897
ONIA process ... **A7:** 98
Onion seeds
magnetically cleaned ... **A7:** 589
Onnes' technology of superconducting magnets ... **A2:** 1027-1028
Onset of recirculation ... **A18:** 594
Opacifiers ... **EM3:** 304, 308, 310
Opacity, and density
radiography ... **A17:** 324
Opals (opaque) dinnerware
advantages and disadvantages ... **EM4:** 1101
categories ... **EM4:** 1101
durability ... **EM4:** 1101
formation ... **EM4:** 1101
strength ... **EM4:** 1101
tempering ... **EM4:** 1101
Opaque frits and glasses
typical oxide compositions ... **EM4:** 550
Opaque-base film
radiography ... **A17:** 314
Opaque-stop microscope
basic components ... **A9:** 58
Open arc welding
hardfacing ... **M6:** 784
hardfacing alloys ... **A6:** 800
Open assembly time ... **EM3:** 19
Open brickwork furnaces ... **A7:** 353
Open circuit
defect causing ... **EL1:** 1018
as failure sites ... **EL1:** 127
as hard defect ... **EL1:** 568
Open crucible system
for liquid metal environment testing ... **A8:** 425
Open dies
for cold heading ... **A14:** 293
defined ... **A14:** 9
flat ... **A14:** 43
swage ... **A14:** 43
V-dies ... **A14:** 43
Open metallization
of integrated circuits ... **A11:** 768
Open nozzle blasting equipment ... **A13:** 414
Open pit mining
cemented carbide tools ... **A2:** 975
Open pore
defined ... **A7:** 8
Open porosity
defined ... **A7:** 8
Open resistance heaters
design of ... **A2:** 829-830
fabrication of ... **A2:** 830-831
Open shrinkage
as casting defect ... **A11:** 382
Open square grids
use in quantitative metallography ... **A9:** 124-125
Open time
water-base versus organic-solvent-base adhesives properties ... **EM3:** 86
Open tolerance
defined ... **A14:** 307
Open-back gap-frame presses
for piercing ... **A14:** 463
Open-cell cellular plastic *See also* Plastics ... **EM3:** 19
defined ... **EM2:** 29
Open-cell foam
defined ... **EM1:** 17
Open-circuit potential *See also* Corrosion potential
defined ... **A13:** 9
Open-circuit voltage
definition ... **M6:** 12
Open-circuit voltage (OCV) ... **A6:** 37-38
Open-cycle testing
stress oxidation from ... **A11:** 132
Open-die extrusion
schematic ... **A18:** 59, 61
Open-die forging *See also* Forging; Hand forge ... **A8:** 587 **A14:** 61-74
allowances and tolerances ... **A14:** 71-73
of aluminum alloys ... **A14:** 243
application ... **A14:** 61
auxiliary tools ... **A14:** 61-62
contour forging ... **A14:** 71
defined ... **A14:** 9
deformation modeling of ... **A14:** 65-67
die sets for ... **A14:** 44
dies for ... **A14:** 61
forgeability and ... **A14:** 65
hammers ... **A14:** 28-29, 61
hammers and presses ... **A14:** 61
handling equipment ... **A14:** 63
of heat-resistant alloys ... **A14:** 231
ingots for ... **A14:** 64-65
production and practice ... **A14:** 63-64, 67-71
ring rolling by ... **A14:** 126
of ring, with saddle ... **A14:** 127
of stainless steels ... **A14:** 222
of titanium alloys ... **A14:** 272
wrought aluminum alloy ... **A2:** 34
Open-die forging hammers ... **A14:** 28-29
Open-die forgings
inspection techniques ... **A17:** 494-495
Open-die headers
for cold heading ... **A14:** 292
Open-end wrench
preform shape and consolidated part for ... **A7:** 540
Open-hearth furnace ... **M1:** 109-110, 113
Open-hearth iron
chemical analysis and sampling ... **A7:** 249
Open-hole-compression test
for damage tolerance ... **EM1:** 97
Open-hole-tension test
for damage tolerance ... **EM1:** 97
Open-loop system
for fatigue test control ... **A8:** 368
Open-tube chemical vapor deposition
chromium ... **M5:** 383
Opened crack
austenitic stainless steels ... **A12:** 356
fracture surface, AISI/SAE alloy steels ... **A12:** 335
precipitation-hardening stainless steels ... **A12:** 374
primary ... **A12:** 77
secondary ... **A12:** 77
Opening
of secondary cracks ... **A11:** 19-20
Opening mode of deformation *See* Stress- intensity factor
Opening, quick
in crack arrest testing ... **A8:** 454
Operating conditions *See* Service conditions
Operating costs *See also* Costs; Equipment costs
of eddy current inspection ... **A17:** 563
of flux leakage inspection
ultrasonic inspection ... **A17:** 565
Operating environment
of connectors ... **EL1:** 23
Operating pitch diameter of pinion
symbol and units ... **A18:** 544
Operating pitch line velocity ... **A18:** 542
symbol and units ... **A18:** 544
Operating procedures
flame hardening ... **M4:** 492-493
Operating room air filters
powders used ... **A7:** 573
Operating systems for process design ... **A14:** 409
Operating temperature
titanium carbide- steel cermets ... **A7:** 810
Operational definitions
Shewart control chart model ... **A17:** 734
Operations, energy efficient ... **A4:** 519-525
combustion control ... **A4:** 519-521
Operations, energy efficient (continued)
combustion control by pulse techniques ... **A4:** 521
flue gas analysis ... **A4:** 520
furnace design and operation ... **A4:** 522-525
recuperators ... **A4:** 521-522, 523, 524
regenerative burners ... **A4:** 521-522, 524
waste-heat recovery ... **A4:** 519, 521-522
Operations, energy-efficient ... **M4:** 337-342
accounting practice ... **M4:** 338
checklist, conservation ... **M4:** 339
cycle temperatures ... **M4:** 339
energy requirements, evaluation ... **M4:** 337-338
equipment modification ... **M4:** 341
equipment use ... **M4:** 340-341
fixtures ... **M4:** 339
furnace utilization ... **M4:** 339-340
heat treating practices, modification ... **M4:** 339-340
strategies ... **M4:** 338-339
Operator fatigue
minimized by image analysis ... **A10:** 309
Ophthalmic and optical glasses ... **EM4:** 1074-1081
absorption and color ... **EM4:** 1080-1081
dispersion ... **EM4:** 1079
materials that compete with glass ... **EM4:** 1076-1077
nonlinear refractive index ... **EM4:** 1079-1080
ophthalmic glass products ... **EM4:** 1074
optical glass products ... **EM4:** 1074-1076
properties ... **EM4:** 1077, 1079
refractive index ... **EM4:** 1077-1079
Opposed electrode seam welding
of flatpack ... **EL1:** 237, 240
Opposite-wall detection, tubing
by remote-field eddy current inspection ... **A17:** 195
Optical aberrations, ion
in atom probe analysis ... **A10:** 595
Optical aids
in visual leak testing ... **A17:** 66
Optical and x-ray spectroscopy
atomic absorption spectrometry ... **A10:** 43-59
inductively coupled plasma atomic emission spectroscopy ... **A10:** 31-42
infrared spectroscopy ... **A10:** 109-125
molecular fluorescence spectrometry ... **A10:** 72-81
optical emission spectroscopy ... **A10:** 21-30
particle-induced x-ray emission ... **A10:** 102-108
Raman spectroscopy ... **A10:** 126-138
ultraviolet/visible absorption spectroscopy ... **A10:** 60-71
x-ray spectrometry ... **A10:** 82-101
Optical anisotropy
and color metallography ... **A9:** 138
Optical applications
critical properties of ... **EM2:** 458
Optical axis
defined ... **A10:** 678
Optical cells
selection for UV/VIS absorption analysis ... **A10:** 68
Optical ceramics ... **EM4:** 18
applications ... **EM4:** 18, 20
infrared domes ... **EM4:** 18
lasing crystals ... **EM4:** 18
mirrors ... **EM4:** 18
optical properties ... **EM4:** 18
phosphors ... **EM4:** 18
Optical clocks
and clock skew ... **EL1:** 7, 9
Optical color metallography ... **A9:** 138
Optical comparators
defined ... **A17:** 10-11
schematic ... **A17:** 10
Optical computing
in machine vision process ... **A17:** 44
Optical constants of cathode materials used in interference film metallography ... **A9:** 149

Optical crack growth
measuring systems **A8:** 246, 382-383
Optical density
in ion detection **A10:** 143
Optical devices **EM4:** 17
Optical diagram
FT-IR spectrometer **A10:** 112
Optical diffraction, and single-crystal x-ray diffraction
compared **A10:** 345-346
Optical double magnetic resonance
as ESR supplemental technique **A10:** 258, 689
Optical emission spectroscopy **A10:** 21-30
applications **A10:** 21, 29-30
and atomic absorption spectroscopy **A10:** 21
Boltzman equation **A10:** 24
capabilities, compared with classical wet analytical chemistry **A10:** 161
capabilities, compared with UV/VIS absorption spectroscopy **A10:** 60, 226, 253
for compositional analysis of welds **A6:** 100
defined **A10:** 678
and direct-current plasma emission spectroscopy **A10:** 21
electronic structure **A10:** 21-22
electronic transitions **A10:** 22
emission sources **A10:** 24-26
estimated time analysis **A10:** 21
excitation spectra **A10:** 22
general principles **A10:** 21-23
and ICP-AES **A10:** 21
of inorganic solids **A10:** 4-6
introduction **A10:** 21
monochromator **A10:** 23-24
optical systems **A10:** 23-24
polychromator **A10:** 23-24
related techniques **A10:** 21
samples **A10:** 21
of stainless alloy **A10:** 178-179
and x-ray fluorescence **A10:** 21
Optical emission spectroscopy (OES)
for constituent analyses **EM4:** 24
to analyze the bulk chemical composition of starting powders **EM4:** 72
Optical encoder transducer
for torsion testing **A8:** 158
Optical equipment
for microhardness testers **A8:** 92-93
Optical etching **A9:** 57-59
defined **A9:** 13
illumination modes **A9:** 58
Optical extensometer **A8:** 618
for deformation measurement **A8:** 548
for strain measurement **A8:** 35, 193
Optical fiber(s) *See also* Fibers
ribbon cables **EL1:** 9
Optical fibers **EM4:** 409-417
basic physics **EM4:** 409-412
absorption **EM4:** 410-412
critical angle **EM4:** 409
evanescent field **EM4:** 411
fiber index profile constant **EM4:** 410, 416
fiber index profiles **EM4:** 411, 416
mode field diameter **EM4:** 412
modes **EM4:** 410
numerical aperture (NA) **EM4:** 409
optical loss behavior **EM4:** 411
scattering **EM4:** 410
dispersion **EM4:** 412-413
bandwidth **EM4:** 413
chromatic **EM4:** 412-413
graded index profile fiber **EM4:** 412
intermodal **EM4:** 412
intramodal **EM4:** 412-413
material **EM4:** 412-413
waveguide **EM4:** 413
fiber manufacture **EM4:** 414-417
dopant selection **EM4:** 415, 416
fiber draw **EM4:** 416-417
modified chemical vapor deposition laydown and consolidation **EM4:** 414, 415, 416
outside vapor deposition laydown and consolidation **EM4:** 414-415
quality control measurements **EM4:** 417
vapor axial deposition **EM4:** 415-416
fracture surface analysis **EM4:** 663-668
Optical fibers (continued)
as germanium application **A2:** 743
high-silica glass **EM4:** 377
optical fiber glass materials **EM4:** 413-414
Optical fractography
defined **A12:** 1
Optical gages *See* Scanning laser gage
Optical glass process
powder used **A7:** 574
Optical glasses
development **EM4:** 21
diamond abrasives for grinding **EM4:** 333, 334
melting/fining **EM4:** 392
recommended waterjet cutting speeds **EM4:** 366
Optical holographic interferometry *See also* Optical holography
of composite materials **A17:** 429
continuous-wave techniques **A17:** 410
dual refractive index method **A17:** 408
holographic contouring **A17:** 408
inspection techniques **A17:** 407-408
multiple-exposure **A17:** 408
pulsed-laser techniques **A17:** 410-411
real-time **A17:** 408
stressing methods for **A17:** 408-410
time-average **A17:** 408
uses **A17:** 405-406
Optical holography *See also* Optical holographic interferometry
applications **A17:** 421
of composite materials **A17:** 429
continuous-wave techniques **A17:** 410
contract (purchase) holography **A17:** 430
of debonds, in sandwich structures **A17:** 421-429
defined **A17:** 224
equipment **A17:** 417-420
holographic components **A17:** 417-420
holographic reconstruction **A17:** 407
holographic recording **A17:** 406-407
in-house systems **A17:** 430
inspection procedures **A17:** 410-411
interferometric inspection techniques **A17:** 407-408
and millimeter wave holography compared **A17:** 226
optical holographic interferometry uses **A17:** 405-406
portable systems **A17:** 420-421
pulsed-laser techniques **A17:** 410-411
readout methods **A17:** 413-414
results, interpretation of **A17:** 414-417
selection of holographic systems **A17:** 420-421, 429-430
stressing for interferometry, methods **A17:** 408-410
systems, types of **A17:** 420-421, 429-430
test variables, effects of **A17:** 411-413
Optical inspection *See* Automatic optical inspection, Visual inspection
of solder joints **EL1:** 735
Optical interconnections
defined **EL1:** 1-11, 111, 11, 9-10
packaging requirements, modeling **EL1:** 16
Optical interferometry
interfacial debonding of joints **EM3:** 453
Optical interferometry technique **A18:** 396, 397, 402
Optical laser examiner *See* Molecular optical laser examiner (MOLE)
Optical logic
advantages **EL1:** 10
Optical mask
for echelle spectrometer **A10:** 41
Optical measuring
advantages in dynamic notched round bar testing **A8:** 281
device, dynamic notched round bar testing **A8:** 279
Optical metallography *See also* Metallographic identification **A10:** 299-308
applications **A10:** 299
capabilities **A10:** 287, 365, 429
defects observable using **A10:** 307
estimated analysis time **A10:** 299
general uses **A10:** 299
image analysis for morphology **A10:** 309
Optical metallography (continued)
of inorganic solids, types of information from **A10:** 4-6
introduction **A10:** 299-300
limitations **A10:** 299
macroanalysis **A10:** 301-303
microanalysis **A10:** 304-308
of organic solids, information from **A10:** 9
related techniques **A10:** 299
samples **A10:** 299, 300-301
and SEM/TEM, compared **A10:** 299
specimen preparation **A10:** 300-301
structure-property relationships established by **A10:** 299-300
Optical methods *See also* Electron optical methods; Optical and x-ray spectroscopy
capabilities **A10:** 102
Optical micrograph
of continuous-fiber composites **EM1:** 769
titanium and titanium alloys **A8:** 476-477
Optical microscopes **A9:** 71-88
comparison microscopes **A9:** 83-84
component parts **A9:** 71-75
condenser **A9:** 72
depth of field **A9:** 76 **A10:** 497
determining magnification **A9:** 74
illumination system **A9:** 72
with image analyzers **A10:** 310
lens defects **A9:** 75
light filters **A9:** 72
light paths **A9:** 71, 81
light section **A9:** 82
objective lens **A9:** 72-73
for plastic replicas **A17:** 53
portable **A9:** 84
research-quality **A9:** 73
resolution **A9:** 75-76
resolution limits **A10:** 495-496
with SEM imaging, for morphology studies **A10:** 521
stages **A9:** 82-83
with television monitor **A9:** 84
types of **A9:** 71-72
Optical microscopy *See also* Electron optical methods; Optical and x-ray spectroscopy **A6:** 143 **A7:** 227 **A18:** 435 **EM1:** 771
acceptable degree of edge rounding of samples **A9:** 44
advantages and types **EL1:** 1067-1068
capabilities **A10:** 490
clean room **A9:** 83
for descriptive fractography **EM4:** 640-641
and electronic image analysis **A9:** 152
of fiber composites **A9:** 587, 592
field **A9:** 83
focusing, as plating thickness inspection **EL1:** 943
of fracture surfaces **A11:** 20
for gage width estimates **A8:** 229
hot-cell **A9:** 83
identification of ferrite in heat-resistant casting alloys by magnetic etching **A9:** 333
image contrast techniques **A9:** 76-82
interference techniques **A9:** 80
for intermetallic compounds **EL1:** 1043
magnetic etching **A9:** 64-66
methods **EL1:** 366-368
for microstructural analysis **EM4:** 25-26
photographs for failure analysis **EM4:** 630
preparation of surfaces for **A9:** 33-47
SAM, SEM, and, compared **A10:** 509-510
to analyze ceramic powder particle sizes **EM4:** 66, 67, 69
to determine surface roughness in ceramic powder characterization **EM4:** 27
to obtain images of spot samples of mixtures **EM4:** 96
ultralong depth-of-field **EL1:** 954
used to identify intragranular subgrain structures **A9:** 690
used to monitor deformation at the Lüders front **A9:** 684
weld characterization **A6:** 104
Optical microscopy specimens
effect of orientation on reflectance **A9:** 77

Optical path in an incident-light research microscope **A9:** 80
Optical plastics
physical properties **EM2:** 596
spectrophotometric transmission **EM2:** 595
Optical probing
as testing **EL1:** 372
Optical properties *See also* Refractive index **EM2:** 481-486
acrylic **EM2:** 105
actinide metals **A2:** 1189-1198
cleaning **EM2:** 485-486
coatings **EM2:** 484-485
commercially pure tin **A2:** 518-519
electrolytic tough pitch copper **A2:** 271
of germanium **A2:** 734
gilding metal **A2:** 295
of glass fibers **EM1:** 47
manufacturing and production tolerances **EM2:** 482-483
measured **EM2:** 614
optical plastic materials **EM2:** 483-484
palladium and palladium alloys **A2:** 716-718
para-aramid fibers **EM1:** 56
parylene coatings **EL1:** 795
platinum and platinum alloys **A2:** 708
polyarylates (PAR) **EM2:** 140
of polymers **EM2:** 62
pure metals **A2:** 1100-1178
rare earth metals **A2:** 1179-1189
thermal considerations **EM2:** 481-482
wrought aluminum and aluminum alloys **A2:** 64-65
Optical pyrometry
for practical temperature measurement **EM4:** 251
Optical sections **A18:** 357
Optical sensing zone
for particle sizing **A7:** 221-225
Optical sensors
image sensors for **A17:** 10
Optical spectroscopy **A1:** 1030
Optical stream scanning
to analyze ceramic powder particle sizes **EM4:** 67
Optical systems *See also* Electron optical methods; Optical and x-ray spectroscopy
collection optics **A10:** 24
emission spectroscopy **A10:** 23-24
wavelength sorters **A10:** 23-24
Optical testing
ad hoc testing **EM2:** 598
birefringence **EM2:** 596-597
and characterization **EM2:** 594-598
refractive index **EM2:** 595-596
surface gloss and color **EM2:** 598
surface irregularity and contamination **EM2:** 597-598
transmission and haze **EM2:** 594
yellowness **EM2:** 594-595
Optical tracing machines **A6:** 1169
Optical vacuum coatings **M5:** 395-396, 399-403
Optical-lens aberrations
effect on secondary electron imaging **A9:** 95
Optically anisotropic materials, examination of
with crossed-polarized light **A9:** 72
using polarized light **A9:** 76-78
Optically anisotropic metals and phases
polarized light used to etch **A9:** 58-59
Optically isotropic metals
activation of surfaces to respond to polarized light **A9:** 138
Optics
beam-condensing **A10:** 113
collection **A10:** 24, 128
electron, of electron probe microanalyzer **A10:** 432, 517
inverted, in wavelength-dispersive x-ray spectrometers **A10:** 88
Optics, infrared
as germanium/germanium compounds application **A2:** 735, 737
Optimal die angle **A18:** 66
Optimization
computer program **EM1:** 452-454
and process modeling **EM1:** 499-502
programs, for joints **EM1:** 479
property, in cure **EM1:** 657-658
vs. laminate ranking **EM1:** 451-455
Optimization (experimental) designs **EM2:** 601
Optimization parameters
for test piece geometries **A8:** 387-388
Optimizations, cost
as engineering function **EM2:** 87
Optimum absorbance
in UV/VIS analysis **A10:** 68
Optimum-aperture concept **A12:** 80-81
Opto coupler
schematic **EL1:** 433
Opto devices
two and multiple terminal **EL1:** 432
Optoelectronic circuit packages **EL1:** 8, 453
Optoelectronic devices
gallium aluminum arsenide laser diodes **A2:** 739
of gallium compounds **A2:** 739-740
indium gallium arsenide phosphide laser diodes **A2:** 739-740
light-emitting diodes (LEDS) **A2:** 739-740
microanalysis of **A10:** 601-602
photodiodes **A2:** 740
solar cells **A2:** 740
Optoelectronic materials
plasma-assisted physical vapor deposition **A18:** 848
Oral corrosion
in fluid environments, dental alloys **A13:** 1348
processes **A13:** 1344-1346
Oral fluids **A13:** 1340-1341
Oralloy *See* Uranium
Orange, methyl
as acid-base indicator **A10:** 172
Orange peel
as a result of plastic deformation **A9:** 686-687
as casting defect **A11:** 384
in cold-formed parts **A11:** 308
cracking **A8:** 57
defined **A8:** 9 **A11:** 7 **EM1:** 17 **EM2:** 29
as formability problem **A8:** 548
on surface of austenitic manganese steel casting specimens **A9:** 238
in sheet metal forming **A8:** 553
Orbital forging *See also* Forging; Rotary forging
as new metalworking process **A14:** 17
wrought aluminum alloy **A2:** 34
Orbital scanning
borescopes **A17:** 4-5
Orbiter
space shuttle **A13:** 1058-1075
Orbitest machine
for tubes and solid cylinders **A17:** 185
Orboresonant cleaning and finishing **M5:** 134
Orchard heater
brittle fracture of **A11:** 100
Order
long- and short-range, EXAFS studied **A10:** 407, 408
long-range, intermetallic compounds, NMR analysis **A10:** 277
long-range, RDF, in amorphous materials **A10:** 393
-order transitions **A10:** 544
Order (in x-ray reflection)
defined **A9:** 13
Order-disorder
analysis, NMR, in ferromagnetic alloys **A10:** 284
in ferromagnetic alloys **A10:** 284-285
transitions, LEED analyzed **A10:** 544
Order-disorder transformation
defined **A9:** 13
Ordered alloys *See also* Ordered intermetallics
deformation **A2:** 913-914
effect of antiphase boundaries on FIM images **A10:** 589
ladder diagrams of **A10:** 594
Ordered beta structure *See also* Beta structure
in palladium-copper alloy **A9:** 564
Ordered crystal structures **A9:** 681-683
prototype structures **A9:** 681
superlattice dislocations **A9:** 682-683
Ordered crystals
diffraction patterns **A9:** 109
Ordered domains **A9:** 681-683
$AuCu_3$ structures **A9:** 681-682
BiF_3 .... **A9:** 682-683
CsCl **A9:** 682
Ordered intermetallics
ductility **A2:** 919-920, 922
high-temperature, specific gravity vs melting point diagrams **A2:** 935
introduction **A2:** 913-914
iron aluminides **A2:** 920-925
nickel aluminides **A2:** 914-920
silicides **A2:** 929, 933-935
structure **A2:** 929-930
summary **A2:** 935
titanium aluminides **A2:** 655, 925-929
titanium aluminides, and titanium alloys, compared **A2:** 655
trialuminides **A2:** 929-935
Ordered lattices
ordered intermetallics **A2:** 913-914
Ordered particle pattern **A7:** 186
Ordered structure .... **A3:** 1 • 10 **A10:** 407, 678
defined **A9:** 13
Ordered superstructure of a crystal **A9:** 708
Ordered systems
EXAFS analysis of **A10:** 407
Ordering
composition profile for **A10:** 593
ferri-, ferro-, antiferro-, or complex magnetic by neutron diffraction **A10:** 420
long- and short-range, by neutron diffraction **A10:** 420
sublattice, in intermetallic compounds **A10:** 283-284
Ordnance applications *See also* Military applications **A7:** 679-695
acoustic emission inspection **A17:** 290
adhesive-bonded joints **A17:** 633
bulk molding compounds **EM1:** 162
of copper-based powder metals **A7:** 733
neutron radiography **A17:** 391
P/M ferrous materials **A7:** 682-687
Ordnance hardware
elevated-temperature failure in **A11:** 294-296
Ordnance springs **M1:** 304
Ores *See also* Mining
analytic methods applicable **A10:** 6
direct AAS analysis of **A10:** 43
effects of composition on mass absorption **A10:** 97
germanium **A2:** 733, 735
metallic, cemented carbide mining tools **A2:** 975
mineral, potentiometric membrane electrode analysis **A10:** 181
powder or briquet sample preparation, x-ray spectroscopy **A10:** 93
samples, crushed **A10:** 165
samples, resource evaluations by NAA **A10:** 233
settling, and sampling **A10:** 14
sodium peroxide fusion for **A10:** 167
treatment in mineral acids **A10:** 165
Orford Tops and Bottoms process
for nickel refining **A2:** 428-429

Organ pipes
tin alloys for ... **A2:** 525
Organic *See also* Inorganic; Organic acid corrosion; Organic acid(s); Organic coatings ... **EM3:** 19
defined ... **A13:** 9 **EM1:** 17 **EM2:** 29
materials, porcelain enamels, resistance ... **A13:** 449-450
media, zirconium/zirconium alloy resistance ... **A13:** 717
Organic acid (OA) fluxes
furnace soldering ... **A6:** 353
Organic acid cleaning *See also* Acid cleaning ... **M5:** 65-67
advantages of, vs mineral acid cleaning ... **M5:** 65
applications ... **M5:** 65-66
corrosivity factors ... **M5:** 65
safety considerations ... **M5:** 65
Organic acid corrosion
of alloy steels ... **A13:** 544
cast iron resistance ... **A13:** 570
characteristics ... **A13:** 1157
formic acid ... **A13:** 1157-1158
nickel-base alloy resistance ... **A13:** 646
Organic acid(s)
and compounds, stainless steel corrosion ... **A13:** 558-559
defined ... **A13:** 9, 1140
Organic acids
conductometric titration of ... **A10:** 203
in Group VI electrolytes ... **A9:** 54
in isopropanolic medium, determined by electrometric titration ... **A10:** 205
Organic acids, as corrosive
copper casting alloys ... **A2:** 352
Organic acids, nickel alloys, corrosion
See also specific type of acid by name ... **M3:** 173
Organic adhesives
as component attachment technology ... **EL1:** 348-349
as die attachment ... **EL1:** 213
Organic binders
for slurry in spray drying ... **A7:** 74
Organic carbon ... **A8:** 423
Organic chemical related failure *See also* Chemical analysis; Chemical properties; Environmental effects; Failure analysis
additives, leaching of ... **EM2:** 774
chemical interactions ... **EM2:** 770
dissolution and swelling ... **EM2:** 771-773
hydrogen bonding ... **EM2:** 773
physical interactions ... **EM2:** 770-774
of plastics ... **EM2:** 775
solvent recrystallization ... **EM2:** 773-774
surface energy effects ... **EM2:** 773
swelling kinetics ... **EM2:** 771
Organic chemicals, zirconium
corrosion in ... **M3:** 785-786
Organic chlorides
in petroleum refining and petrochemical operations ... **A13:** 1268
Organic coated steels ... **A13:** 528-529, 1014-1015
Organic coating *See also* Paint and painting
composition and characteristics of ... **M5:** 497-500
copper and copper alloys ... **M5:** 621, 626-627
corrosion resistance ... **M5:** 474
magnesium alloys ... **M5:** 629, 647-648
surface preparation ... **M5:** 647-648
salt bath descaling removal of ... **M5:** 102-103
zinc ... **M5:** 502, 505
Organic coatings *See* Coatings ... **A13:** 399-418, 912-918
alkyd resins ... **A13:** 400-403
application ... **A13:** 415-416
auto-oxidative cross-linked resins ... **A13:** 400
as barrier protection ... **A13:** 378
for carbon steels ... **A13:** 524
for cast irons ... **A13:** 571 **M1:** 105-106
for copper/copper alloys ... **A13:** 636
corrosion protection ... **M1:** 731, 738, 745, 751, 754
cross-linked thermosetting coatings ... **A13:** 406-410
debonded, cathodic protection ... **A13:** 468
immersion ... **A13:** 916-918
industry, and legislation ... **A13:** 399-400
and linings ... **A13:** 399-418
Organic coatings (continued)
for magnesium/magnesium alloys ... **A13:** 752
for marine corrosion ... **A13:** 912-918
materials ... **A13:** 400-410
as preservation ... **EL1:** 563
quality assurance ... **A13:** 416-418
selection ... **A13:** 412, 529
in solderable systems ... **EL1:** 680
for structural protection ... **A13:** 1303-1304
surface preparation for ... **A13:** 412-415, 912-913
thermoplastic resins ... **A13:** 403-406
topside coating systems ... **A13:** 913-914
zinc-rich ... **A13:** 410-412
Organic complexing agents
determined fluorimetrically ... **A10:** 74
Organic composite coated steels
automotive industry ... **A13:** 1014-1015
Organic composite coatings ... **A1:** 223
Organic composites *See also* Composites; Organic materials; Organic materials, characterization of
analytic methods for ... **A10:** 9
Organic compounds *See also* Compounds; Organic materials; Organic materials, characterization of
containing nitro groups, assayed by controlled-potential coulometry ... **A10:** 207
determination of empirical formula of ... **A10:** 212
direct method for determining crystal structure of ... **A10:** 351
EFG analysis for ... **A10:** 212
elemental analysis ... **A10:** 181
gas analysis of ... **A10:** 151
gold corrosion in ... **A13:** 800
hyperfine splitting in ESR analysis ... **A10:** 260
identified ... **A10:** 213
magnesium/magnesium alloys in ... **A13:** 742-743
MFS analysis of ... **A10:** 73-74
NMR analysis of ... **A10:** 277
platinum corrosion in ... **A13:** 803
silver corrosion resistance in ... **A13:** 797
titanium/titanium alloy resistance ... **A13:** 680-681
UV/VIS analysis of ... **A10:** 60
zinc/zinc alloys and coatings in ... **A13:** 763
Organic compounds, copper alloys
corrosion rate ... **M2:** 479-480
Organic compounds, neutral
as copper casting application ... **A2:** 352
Organic contaminants
removal from XPS samples ... **A10:** 575
Organic contamination
as PTH failure mechanism ... **EL1:** 1027-1030
surface, SIMS analysis ... **EL1:** 1086-1087
Organic contamination, effects
precious metal electrical contacts ... **A2:** 846
Organic emulsions
as cleaning solutions ... **A13:** 1141
Organic fibers ... **EM1:** 54-57
compressive properties ... **EM1:** 55
creep and fatigue ... **EM1:** 55
electrical/optical properties ... **EM1:** 56
environmental behavior ... **EM1:** 56
material properties ... **EM1:** 54-56
para-aramid ... **EM1:** 54
tensile modulus ... **EM1:** 54
tensile properties, hot/wet conditions ... **EM1:** 55
tensile strength ... **EM1:** 54-55
thermal properties ... **EM1:** 56
toughness ... **EM1:** 55
Organic films *See also* Film
for uranium/uranium alloys ... **A13:** 819
Organic fluids
SCC testing in ... **A13:** 275
stress-corrosion cracking in ... **A8:** 531
Organic gases *See also* Gases; Organic materials, characterization of
analytic methods for ... **A10:** 11
Organic halides
stainless steel corrosion ... **A13:** 558
Organic insulation
of electrical steel sheet ... **A14:** 482
Organic interlaminar insulation
magnetic cores ... **A2:** 780
Organic lead
toxicity ... **A7:** 208
Organic linings ... **A13:** 399-418
alkyd resins ... **A13:** 400-403
application ... **A13:** 415-416
Organic linings (continued)
auto-oxidative cross-linked resins ... **A13:** 400
and coatings ... **A13:** 399-418
cross-linked thermosetting coatings ... **A13:** 406-410
materials ... **A13:** 400
quality assurance ... **A13:** 416-418
selection ... **A13:** 412
surface preparation for ... **A13:** 412-415
thermoplastic resins ... **A13:** 403-406
zinc-rich ... **A13:** 410-412
Organic liquids *See also* Liquids; Liquids, characterization of; Organic materials, characterization of
analytic methods for ... **A10:** 10
in Group VI electrolytes ... **A9:** 54
pure tin resistance ... **A13:** 772
Organic materials *See also* Organic materials, characterization of
in aircraft friction materials ... **A18:** 582
carbon and sulfur in ... **A10:** 221-225
characterized ... **A10:** 1
codeposited, plated coating effects ... **EL1:** 679-680
deer hair follicles, determination of sulfur in ... **A10:** 224
determining inorganic materials in ... **A10:** 167
fusion techniques for ... **A10:** 167
in hybrid microelectronics ... **EL1:** 103
identified, and structure determined in ... **A10:** 109
liquid fire method of wet washing ... **A10:** 166
microanalytical elemental analysis ... **A10:** 186
properties ... **EL1:** 470
single-crystal, Raman analysis of ... **A10:** 129
Organic materials, characterization of
analytical transmission electron microscopy ... **A10:** 429-489
Auger electron spectroscopy ... **A10:** 549-567
classical wet analytical chemistry ... **A10:** 161-180
controlled-potential coulometry ... **A10:** 207-211
electrochemical analysis ... **A10:** 181-211
electrogravimetry ... **A10:** 197-201
electrometric titration ... **A10:** 202-206
electron spin resonance ... **A10:** 253-266
elemental and functional group analysis ... **A10:** 212-220
extended x-ray absorption fine structure ... **A10:** 407-419
gas analysis by mass spectrometry ... **A10:** 151-157
gas chromatography/mass spectrometry ... **A10:** 639-648
infrared spectroscopy ... **A10:** 109-125
liquid chromatography ... **A10:** 649-659
low-energy ion-scattering spectroscopy ... **A10:** 603-609
molecular fluorescence spectrometry ... **A10:** 72-81
neutron activation analysis ... **A10:** 233-242
neutron diffraction ... **A10:** 420-426
nuclear magnetic resonance ... **A10:** 277-286
particle-induced x-ray emission ... **A10:** 102-108
potentiometric membrane electrodes ... **A10:** 181-187
radial distribution function analysis ... **A10:** 393-401
Raman spectroscopy ... **A10:** 126-138
secondary ion mass spectroscopy ... **A10:** 610-627
single-crystal x-ray diffraction ... **A10:** 344-356
ultraviolet/visible absorption spectroscopy ... **A10:** 60-71
voltammetry ... **A10:** 188-196
x-ray diffraction ... **A10:** 325-332
x-ray photoelectron spectroscopy ... **A10:** 568-580
x-ray powder diffraction ... **A10:** 333-343
x-ray spectrometry ... **A10:** 82-101
Organic matrix composites
with boron filaments ... **EM1:** 117
Organic mercury
toxicity of ... **A2:** 1248-1249
Organic mixtures *See also* Mixtures; Organic materials, characterization of
liquid chromatography of ... **A10:** 649
Organic outgassing products
electronics industry ... **A13:** 1109-1110
Organic phosphating process ... **A13:** 386
Organic pigments
as colorants ... **EM2:** 501
Organic polymers *See also* Polymer(s)
thermal degradation mechanisms ... **EM2:** 423
Organic polysulfides, recrystallized
single- crystal analysis of ... **A10:** 353

Organic single crystals *See also* Crystals; Organic materials; Organic materials, characterization of; Single crystals
ESR studied **A10:** 263
Organic solutions *See also* Organic materials; Organic materials, characterization of; Solutions
analytic methods for **A10:** 10
as corrosive environment **A12:** 24
Organic solvent cleaners **EL1:** 662-663
Organic solvents *See also* Organic materials; Organic materials, characterization of; Solvents
chlorinated, carcinogenic **A12:** 74
as cleaning solutions **A13:** 1141
flame atomic absorption spectrometry for **A10:** 47
for fracture cleaning **A12:** 74
in ICP sample introduction **A10:** 35
Organic solvents, resistance to
porcelain enamel **M5:** 529
Organic substrates
aramid materials **EL1:** 532
Organic thin films
in multichip technology **EL1:** 299
Organic vehicle
cermet systems **EL1:** 341
Organic zinc-rich coatings **A13:** 411, 768-769
Organic zinc-rich paint
defined **A13:** 9
Organic-coated steels
press forming of **A14:** 564-565
Organic-fiber brush
for fracture cleaning **A12:** 74
Organic-matrix composites
application in future jet engine components **A18:** 592
composite-to-metal joining **A6:** 1041-1047
PABST bonded metal fuselage program **A6:** 1026-1027, 1033
Organic-matrix composites, joining of **A6:** 1026-1036
absorbed moisture, effects of **A6:** 1027-1028
adhesion failures **A6:** 1029
adhesive bonds, examples of good and bad **A6:** 1028-1029
alternatives to peel-ply surfaces **A6:** 1030
bonding process control for thermoset adhesives **A6:** 1032- 1034
partial-vacuum cures **A6:** 1033-1034
prebond moisture effect on bonded joint strength **A6:** 1032-1033
temperature variation effect on bond strength **A6:** 1032
defective bonds
difficulties in detecting **A6:** 1028
sources of **A6:** 1026-1027
grit blasting **A6:** 1030
inspection methods **A6:** 1029
Nomex honeycomb cores **A6:** 1028, 1033
peel-ply surfaces **A6:** 1029-1030, 1031
problems encountered **A6:** 1026-1036
proper processing, importance of **A6:** 1026
properly processed bonds **A6:** 1028-1029
Redux bonding **A6:** 1026
surface preparation for thermoset composites **A6:** 1029-1030, 1031
thermoplastic composite panel bonding **A6:** 1034-1036
bonding with thermoset adhesives
fusion bonding of thermoplastic composites **A6:** 1035
fusion bonding using the dual resin approach **A6:** 1035- 1036
water-break test **A6:** 1030-1032
thermoplastic composites, bonding of **A6:** 1028
Organic-organic composites
laser cutting of **EM1:** 678-679
Organic-solvent-base adhesives **EM3:** 74
Organics removal **EM4:** 135-138
organics degradation and removal principles **EM4:** 136
oxidative degradation **EM4:** 136
theoretical aspects **EM4:** 136
thermal degradation **EM4:** 136
polymer removal techniques **EM4:** 137-138
capillary action (wicking) **EM4:** 138
pressure-temperature control **EM4:** 138
solvent extraction **EM4:** 138
supercritical extraction **EM4:** 138
temperature-time control **EM4:** 137, 138
weight-loss control **EM4:** 137-138
removal of binders from powder compacts and tape-cast films **EM4:** 136-137
powder compacts **EM4:** 136-137
tape-cast films **EM4:** 137
removal process overview **EM4:** 135-136
debinding techniques **EM4:** 135
thermal degradation **EM4:** 135-136
Organisms *See also* Anti-fouling; Biofouling
aerobic, effects **A13:** 43
biological corrosion by **A13:** 41-43, 87-88, 115-116
macro, effects **A13:** 88
micro, effects **A13:** 88, 118
Organofunctional silanes
as sizing **EM1:** 123
Organogermanium compounds
chemical properties **A2:** 734
Organolead compounds
toxicity of **A2:** 1246
Organometallics **EM4:** 62
analytic methods for **A10:** 9
silicate film, positive SIMS spectra for **A10:** 617
Organosilane coupling agents **EM3:** 42, 590
to improve wet strength durability **EM3:** 625-626
Organosol **EM3:** 19
defined **EM2:** 29
Organosulfur compounds
collision-activated dissociation mass spectra **A10:** 647
Organotitanates **EM3:** 626
Orientation *See also* Anisotropy; Fiber orientation; Preferred orientation; Texture, crystallographic **EM3:** 19
accuracy, in manual lay-up **EM1:** 603-604
circumferential, eddy current inspection **A17:** 186
codes/standards/requirements for **A17:** 49
crystallographic, in sheet metal forming **A8:** 553
defined **A9:** 13 **EM1:** 17 **EM2:** 29
design for ease of **EL1:** 121-125
discontinuity, detection effect **A17:** 105
discontinuity, magnetic particle inspection **A17:** 105
ductility effects, beryllium powder **A2:** 684
effect **A10:** 119
effect, radiographic inspection **A17:** 295
effects, from processing **EM2:** 754-755
extrusion **EM2:** 387
fiber, and discontinuous fiber strength **EM1:** 119-120
fiber, effect on tensile strength **EM1:** 120
fiber, thermoplastic injection molding **EM2:** 311
of fibers, effect in composites **A11:** 733
flow-induced, effects **EM2:** 751
grain, and magnetic measurement **A17:** 131
and hardness, of diamond **A2:** 1010
human vs. machine vision **A17:** 30
laminar **EM1:** 218-222
magnetic domain, in magabsorption **A17:** 145
molecular, determined in drawn polymer films **A10:** 120
molecular, IR determination of **A10:** 109
molecular, process effects on **EM2:** 281-282
multi, reinforcement effect **EM1:** 146
number, in sublamination **EM1:** 456
object, in machine vision process **A17:** 34
Orientation (continued)
options, pultrusions **EM2:** 393-395
ply-angle **EM1:** 456
precracked SCC testing specimens **A8:** 516
preferred **A11:** 6, 316
preferred, in plastic torsion **A8:** 143
rigid printed wiring boards **EL1:** 541
specimen, effect on Charpy V-notch impact energy **A11:** 68
stresses **EM2:** 751-752
vertical of chips
Orientation distribution function
along fiber lines as function of rolling reduction **A10:** 363-364
analysis, series method of **A10:** 361-363
coefficient, determining **A10:** 362
for copper tubing, Euler plots method **A10:** 361
defined **A10:** 360
and Euler plots **A10:** 360-361
series representations **A10:** 361-362
used to describe crystallographic texture **A9:** 704
Orientation effects
SEM image recording **A12:** 169
Orientation of domain structures, etch pits used to determine **A9:** 62
for bainitic structures **A9:** 663-664
effect on massive transformation structures **A9:** 655-657
of ferrous martensite **A9:** 669-671
Orientation, preferred
of beryllium powders **A7:** 756
Orientation relationships
crystallographic, SEM evaluated **A10:** 490
determined **A10:** 453
in dislocation cell structure analysis **A10:** 470-473
as fine structure effect **A10:** 438
and habit plane **A10:** 453-455
Kurdjumov-Sachs, for fcc/bcc materials **A10:** 438-439
and misorientation determination **A10:** 471-472
as x-ray diffraction analysis **A10:** 325
Oriented dislocation arrays
in copper **A9:** 128
equations for quantitative metallography of special microstructures **A9:** 126-129
in iron **A9:** 28
use of circular and parallel linear test grids for quantitative metallography of **A9:** 124
Oriented materials **EM3:** 19
defined **EM1:** 17 **EM2:** 29
Oriented silicon steels **A14:** 476-477, 480
magnetically soft materials **A2:** 767-769
Oriented strand board
construction method **EM3:** 105
Oriented surfaces
structural information by EXAFS **A10:** 407
Orifice gas
definition **M6:** 12
for plasma arc welding **M6:** 217
Orifice gas (plasma arc welding and cutting)
definition **A6:** 1211
Orifice restrictor area
hydrostatic gas- lubricated bearings **A18:** 528
Orifice throat length
definition **M6:** 12
Origin *See also* Beach marks; Crack(s); Fracture(s)
fracture **A11:** 257
fracture, in ceramics **A11:** 744-747
fracture, in pipelines **A11:** 695-696
region, elliptical cracks in **A11:** 124
Original crack size
defined **A8:** 9 **A11:** 7
Original equipment manufacturer (OEM)
engine oil requirements **A18:** 162-163, 165
lubricant performance establishment **A18:** 166
performance requirements for lubricants **A18:** 98-99

Oriskany quartzite **EM4:** 378
Ormocers
alkoxide-derived gels **EM4:** 210-211
Orotron
for microwave inspection **A17:** 209-210
Orowan method **A8:** 194
Orr-Sherby-Dorn parameter *See* Time temperature parameters **A8:** 333-334
Orsat analyzers
atmospheres **M4:** 429
Orthicons
image **A17:** 10
Ortho resins *See also* Polyester resins
at elevated temperatures **EM1:** 93
clear casting mechanical properties **EM1:** 91
in fiberglass-polyester resin composites **EM1:** 91
flame retarded **EM1:** 96
glass content effects **EM1:** 91
preparation/properties **EM1:** 90
types **EM1:** 43
Orthochromatic filter
defined **A9:** 13
Orthochromatic photographic films **A9:** 85-86
different films compared **A9:** 86-87
Orthoclase **EM4:** 6
crystal structures **EM4:** 882
hardness **A18:** 433
Orthoclase (feldspar)
on Mohs scale **A8:** 108
Orthodontic appliances
gold **M2:** 684-687
Orthodontic biomechanics **A13:** 1356
Orthodontic wire
simplified composition or microstructure **A18:** 666
Orthodontic wires, wrought
of precious metals **A2:** 696
Orthoferrites **EM4:** 59
Kerr effect used to study magnetic domain structures **A9:** 535
precipitation process **EM4:** 59
Orthogonal arrays
in fractional factorial designs **A17:** 748-750
Orthogonal machining **A16:** 7-17
chip ratio **A16:** 8
chip velocity **A16:** 8
forces **A16:** 13-16, 17
rake angle **A16:** 8-9, 11
shear angle **A16:** 8-9, 11
shear strain **A16:** 9
Orthogonal normal stresses *See* Principal stresses
Orthogonal weave **EM1:** 130
Orthopedic devices
fretting wear **A18:** 244, 250
Orthopedic external fixation system
powder used **A7:** 573
Orthopedic implants *See also* Implants; Metallic orthopedic inplants, failures of **A13:** 665-667
alloys for **A7:** 658
austenitic stainless steels **A12:** 359-364
metallic **A11:** 670-694
porous **A7:** 657-659
Orthopedic internal fixation devices
types of **A11:** 671
Orthopedic surgery
corrective **A11:** 671
Orthopedic wire
for biomechanical implant **A11:** 671
Orthophosphate
affect on electroless nickel plating **M5:** 220, 222-223, 225
Orthophosphate formation
during copper plating **M5:** 165
Orthophosphoric acid
as electrolyte for copper **A9:** 48-49
in electrolytes **A9:** 54
for oxide coating removal **A12:** 75
and water as an electrolyte for uranium alloys **A9:** 478
Orthophthalic polyester resins *See also* Unsaturated polyesters
preparation **EM2:** 246
Orthophthalic resins *See* Ortho resins
Orthorhombic
defined **A9:** 13
Orthorhombic crystal system .... **A3:** 1 • 10, 15 **A9:** 706
Orthorhombic forms, in uranium **A9:** 476-477, 479
Orthorhombic unit cells **A10:** 346-348
Orthotropic **EM3:** 20
defined **EM1:** 17 **EM2:** 29
plate, instability of **EM1:** 445-446
Orthotropy
as material assumption **EM1:** 308-309
Os-Pt (Phase Diagram) .... **A3:** 2 • 326
Os-Pu (Phase Diagram) .... **A3:** 2 • 327
Os-Re (Phase Diagram) .... **A3:** 2 • 327
Os-Rh (Phase Diagram) .... **A3:** 2 • 327
Os-Ru (Phase Diagram) .... **A3:** 2 • 328
Os-Si (Phase Diagram) .... **A3:** 2 • 328
Os-Ti (Phase Diagram) .... **A3:** 2 • 328
Os-U (Phase Diagram) .... **A3:** 2 • 329
Os-V (Phase Diagram) .... **A3:** 2 • 329
Os-W (Phase Diagram) .... **A3:** 2 • 329
Os-Zr (Phase Diagram) .... **A3:** 2 • 330
OSA *See* Styrene-acrylonitriles
Oscillating circuits
as demagnetization **A17:** 121
Oscillating electric field
wave theory of **A10:** 83
Oscillating magnetic field
wave theory of **A10:** 83
Oscillating sample **A7:** 226
Oscillating wire saw **A9:** 26
Oscillation detector
in ultrasonic hardness tester **A8:** 101
Oscillators
electroslag welding **M6:** 228
high frequency welding **M6:** 764
marginal magabsorption **A17:** 150-152
parameters for weld overlaying **M6:** 811
ultrasonic inspection **A17:** 231
Oscillatory displacement
in ultrasonic testing **A8:** 242
Oscillometric (high-frequency) titration
as electrometric **A10:** 203-204
Oscilloscope
for fatigue testing machines **A8:** 368
for incremental strain rate test **A8:** 226
for instrumented impact testing **A8:** 197
interconnection-network electrical performance simulator **EL1:** 14
pulse outputs, dynamic notched round bar testing **A8:** 277
records from Kolsky bar tests **A8:** 220-221, 225, 229
for split Hopkinson bar test instrumentation **A8:** 202
tilt pin, for plate impact test **A8:** 233
Oscilloscope screen
hardness investigating by **A7:** 485
Oscilloscopes
controls **A17:** 254
display, eddy current inspection **A17:** 189
for microwave thickness gaging **A17:** 212
ultrasonic inspection **A17:** 253
Osmium *See also* Precious metals **A13:** 806-807
distillation **A10:** 169
electrical circuits for electropolishing **A9:** 49
gravimetric finishes **A10:** 171
lamp filaments **A7:** 16
in medical therapy, toxic effects **A2:** 1258
as precious metal **A2:** 688
pure **M2:** 779-780
pure, properties **A2:** 1145
resources and consumption **A2:** 689-690
special properties **A2:** 692, 694
thermal expansion coefficient **A6:** 907
toxicity **A7:** 207
Osmium tetroxide
toxic effects **A2:** 1258
Osmium, vapor pressure
relation to temperature **A4:** 495 **M4:** 310
Osprey process (spray deposition) **A7:** 530-531
Osprey processing
aluminum-silicon alloys **A18:** 791
Osprey spray-forming technique
aluminum P/M alloys **A2:** 204
Osteotomies
stabilizing of **A11:** 671
Ostwald ripening **A1:** 637 **A6:** 74 **EM4:** 266, 288
creep testing **A8:** 306
Ostwald ripening (continued)
in peritectic nucleation **A9:** 676
in precipitation reactions **A9:** 647
Ounce metal *See also* Cast copper alloys, specific types; Copper alloys, specific types, C83600
properties and applications **A2:** 364
Out time
defined **EM1:** 17 **EM2:** 29
prepreg, defined **EM1:** 139
Out-of roundness *See also* Ovality
forming of **A14:** 622
Out-of-plane
deformation, speckle metrology measurement **A17:** 434
displacements, by optical holography **A17:** 415
Out-of-plane (delamination) failure mode **EM1:** 781, 783-784
Out-of-plane distortion
bridge components **A11:** 707, 711-714
Out-of-roundness (eccentricity) **A16:** 171
electrical discharge grinding **A16:** 565
honing **A16:** 484, 485
and lapping **A16:** 494, 495, 496, 497
multifunction machining **A16:** 390
P/M high-speed tool steels **A16:** 62-63
thread rolling **A16:** 293
Outboard motors
powders used **A7:** 574
Outdoor atmospheres *See* Air; Atmospheres; Atmospheric corrosion
Outdoor cooking grill
forming strain analysis **M1:** 545
Outdoor furniture
failure of steel fasteners in **A11:** 548-549
precoated steel sheet for **M1:** 167, 172
Outer lead bonding (OLB)
process **EL1:** 231
sequence **EL1:** 286
in TAB assembly process **EL1:** 283-286
in tape automated bonding **EL1:** 228
tape automated bonding, plastic packages **EL1:** 479
Outer noise *See also* Noise
defined **A17:** 722-723
Outer raceway **A18:** 499, 500, 501
Outer-Heimholz Plane (OHP) **A13:** 19
Outgassing *See also* Degassing **EM3:** 20
adhesive **EL1:** 674
defined **EM2:** 29
in encapsulation **A7:** 434-435
in encapsulation hot isostatic pressing **A7:** 431
as preheating defect **EL1:** 687
Outgassing products
organic **A13:** 1109-1110
Outlet header , in secondary superheater
interligament cracking in **A11:** 667
Outlet piping
elevated-temperature failures in **A11:** 291
Outlier
defined **A10:** 678
Outliers
in statistical analyses **EM1:** 303
Outline drawing
in final design package **EL1:** 523
Outlines, package *See* Packages; specific package types
Outokumpu process
for copper and copper alloy wire rod **A2:** 255
Output bar
with split Hopkinson pressure bar **A8:** 199, 201
Output tube
double-notch shear testing **A8:** 228-229
Outside diameter
abbreviation for **A11:** 797
Outside mold line (OML) surfaces
resin transfer molding for **EM1:** 168
Outward diffusion coatings
superalloys **M5:** 378
Oval-shaped dimples
formation **A12:** 13
Ovality *See also* Out-of-roundness
in four-piece dies **A14:** 133
nominal values for computing **A14:** 133
in rotary straighteners **A14:** 693
in two-piece dies **A14:** 132-133
Ovaloid
defined **EM1:** 17

Ovaloid, geodesic *See* Geodesic ovaloid
Oven curing adhesives
for automotive applications **EM3:** 553
Oven dry **EM3:** 20
defined **EM1:** 17 **EM2:** 29
Oven soldering
definition **A6:** 1211
Over-and-under polishing and buffing machines **M5:** 121
Over-tumbling
effects in rotary tumbling and vibratory processing **A7:** 458
Overaging *See also* Aging **A6:** 575
of aluminum alloys, etchants for examination for **A9:** 355
defined **A9:** 13 **A13:** 9
and elevated-temperature failures **A11:** 267
kinetics of **A10:** 317
Overaging treatments
beryllium-copper alloys **A2:** 407
Overall critical current density
as function of transverse magnetic field **A7:** 638
Overaustenitizing
effects on tool steel parts **A11:** 570
tool and die **A11:** 569-571
Overbasing
defined **A18:** 14
Overblending **A7:** 188
Overcorrection, of signal
AAS spectrometers **A10:** 51
Overfill
defined **A7:** 8
Overfills
as forging flaw **A17:** 493
Overgrind
dependence of diametral expansion on **A14:** 142
Overhang-type roll forging machine **A14:** 96
Overhaul effect
environmental stress screening **EL1:** 877
Overhead crane
magnetic particle inspection **A17:** 115
Overhead door spring steel wire
tensile strength ranges **M1:** 268
Overhead position
definition **A6:** 1211 **M6:** 12
Overhead welding
arc welding of coppers **M6:** 402
indication by electrode classification **M6:** 84
oxyfuel gas welding **M6:** 589
shielded metal arc welding **M6:** 76, 85, 441
of nickel alloys **M6:** 441
Overheating *See also* Burning; Heat treatment; Heating; Heating time
aluminum alloy air bottle failure by **A11:** 436
of aluminum alloys, etchants for examination for **A9:** 355
boiler ruptures by **A11:** 603-614
in boilers and steam equipment, causes **A11:** 608-609
brittle fracture of carbon steel hook from **A11:** 332-333
of copper forgings **A11:** 332, 334
defined **A9:** 13 **A13:** 9
as die failure cause **A14:** 56
during hot working, tool and die failures **A11:** 574
during spinning, fire-extinguisher case failure from **A11:** 649
effect on fatigue strength **A11:** 121-122
as embrittlement, examination interpretation **A12:** 127-129
facets, steel alloy **A12:** 146
failures, locomotive axles **A11:** 716
in forging **A17:** 493
forging failures from **A11:** 332
fracture, alloy steel **A12:** 144
fracture, vanadium-niobium plate steel **A12:** 145
Overheating (continued)
of friction bearings, effect on locomotive axles **A11:** 715
from improper lubrication **A11:** 130
from misalignment, bearing failure from **A11:** 507-508
from wear, bearing failure from **A11:** 511-512
historical study **A12:** 2
inicrostructural characteristics **A11:** 581
localized, rupture of reheater tube from **A11:** 610
in low-carbon steel **A12:** 246
organic-coated steels **A14:** 565
of pressure vessels, effect of **A11:** 649
rapid, in steam equipment **A11:** 610
rapid, thin-lip ruptures as **A11:** 606
rapid, tube-wall thinning by **A11:** 606
rupture of low-carbon steel boiler tubes from **A11:** 607
steam generator failures from **A11:** 603
of steel, macroetching to reveal **A9:** 176
of steel pipe mold **A11:** 275-276
steel wire rope failure by **A11:** 519-520
thin-lip ruptures from **A11:** 606
of titanium alloys **A14:** 838
of titanium tubing, embrittlement by **A11:** 642
transverse fracture **A12:** 142, 163
tube ruptures by **A11:** 603
Overheating, of steels **A1:** 697
presence of facets **A1:** 697
upper-shelf energy **A1:** 697
Overlap **A6:** 1073, 1075
definition **A6:** 1211 **M6:** 12
detection, with machine vision **A17:** 43-44
as discontinuity, weldments **A17:** 582
shielded metal arc welds **M6:** 93-94
weld discontinuity **M6:** 837
Overlaps
as rolling defect **A14:** 359
Overlay coatings
stripping of **M5:** 19
Overlay oxidation protective coating *See* Oxidation protective coating, overlay type
Overlay sheet
defined **EM1:** 17 **EM2:** 29
Overlayer growth
LEED kinetic analysis **A10:** 544
Overlaying
cast irons **A6:** 720-721
definition **A6:** 1211
Overlays, precious metal
for electrical contact materials **A2:** 848
Overload *See also* Overload failures; Overload fractures; Overloading; Overstress
thermal contraction, in ductile iron brake drum **A11:** 370-371
thermal-stress, in castings **A11:** 370
Overload failure
of quench-cracked AISI 4340 steel threaded rod **A10:** 511-513
Overload failures *See also* Overload fractures
by shear **A11:** 398-399
ductile fracture of T-hook **A11:** 367-369
needle-roller bearing **A11:** 505
service ductile fracture as **A11:** 85
in solder joints **EL1:** 1031-1032
of steel castings **A11:** 396-399
of tooth adapter **A11:** 398
Overload fracture **A8:** 477-481 **A12:** 12
AISI/SAE alloy steels **A12:** 294, 298-299
at flux inclusion **A12:** 65, 67
cast aluminum alloys **A12:** 409, 411-413
dimple rupture, effect of elevated temperature **A12:** 35, 49-50
ductile **A12:** 101, 299, 443
high-carbon steels **A12:** 278
macrograph **A12:** 101
medium-carbon steels **A12:** 258
sudden **A12:** 294, 413
Overload fracture (continued)
titanium alloys **A12:** 443
torsional **A12:** 258, 278
Overload fractures *See also* Overload failures
in alloy steel bolt **A11:** 76
aluminum alloy extrusions **A11:** 86-87, 91
brittle, of gray iron nut **A11:** 369-370
defined **A11:** 75
ductile **A11:** 25, 85-87, 91, 137
ductile, in 63Sn-Pb solder **A11:** 45-46
ductile, in extension ladder **A11:** 86-87, 137
failed in torsion **A11:** 399
in iron castings **A11:** 367
single-, behavior in shafts **A11:** 460-461
Overloading
as die failure cause **A14:** 55-56
as distortion failure **A11:** 136-138
effect of load ratio and, corrosion-fatigue **A11:** 255
final fracture by **A11:** 104
gross impact, of bearings **A11:** 505-506
of rollers, spalling and surface deterioration **A11:** 501
of side rails, in extension ladders **A11:** 137
spall formation by **A12:** 114
of steel wire rope **A11:** 514
Overmix (verb)
defined **A7:** 8
Overpickling
effects of **M5:** 80
Overpotential
defined **A13:** 30
in electrogravimetry **A10:** 198
Overpressed powder metallurgy materials
microstructures **A9:** 512
Overpressure sintering
for cermets **A2:** 989
Overpressurization
fracture by **A12:** 345
Overscanning, field of mixed phases
errors observed by **A10:** 529
Oversinter (verb)
defined **A7:** 8
Oversize powder
defined **A7:** 8
Overstrain
effect on fatigue behavior **M1:** 678-679, 681
Overstrain, periodic
treatment in fatigue testing **A8:** 701
Overstress *See also* Overload
defined **A11:** 109
effects on fatigue-crack propagation **A11:** 109-110
failures, dimpled fracture in **A11:** 22
failures, electrostatic/electrical **A11:** 786-788
n fatigue testing **A8:** 367
Overstress, electrical
as failure mechanism **EL1:** 1013-1014
Overstressing
of nozzles **A13:** 1229-1230
Overtempered martensite (OTM) **A16:** 25-27, 36
Overtempering
alloy steels **A12:** 301
Overvoltage
defined **A13:** 9
effects in EDS **A10:** 523
Oxacetylene **A6:** 121
carbide-containing nickel-base alloys, poor weldability **A6:** 795
Oxal anodizing process
aluminum and aluminum alloys **M5:** 587
Oxalate coprecipitation **EM4:** 58, 59
Oxalates
decomposition of **EM4:** 56
as salt precursors **EM4:** 113
Oxalic acid
as an electrolytic reagent for wrought stainless steels **A9:** 281
used in attack-polishing of beryllium **A9:** 389

Oxalic acid (continued)
used to electrolytically etch heat resistant casting alloys **A9:** 331
Oxalic acid, stainless steels
corrosion resistance **M3:** 86, 87
Oxalic anodizing process
aluminum and aluminum alloys **M5:** 592
Oxazolidines
for chemical curing of urethane sealants **EM3:** 207
Oxidants
in nitrogen-based atmospheres **A7:** 346
Oxidation *See also* Gaseous corrosion; High-temperature corrosion; Oxidation corrosion; Oxidation resistance; Oxidation-reduction; Oxide scales; Oxides; Photooxidation; Redox; Reduction **A6:** 374 **EM3:** 20
in active-component fabrication **EL1:** 195
AISI/SAE alloy steels **A12:** 305
alloy cast irons **M1:** 93-94
alloy: doping principle **A13:** 73
in Alnico alloys **A9:** 539
in aqueous corrosion **A13:** 29
in arc welds **A11:** 413
of ASTM F-15 (Kovar) alloy **EL1:** 457
of atomized powders **A7:** 36-38
and average crack propagation rate **A13:** 155
biochemical **A13:** 897
breakaway **A13:** 72
in carbon and alloy steels, etching **A9:** 170
of carbon fibers **EM1:** 52
of carbon-carbon composites **EM1:** 913-914, 920-921
cast iron **M1:** 93
of cast irons **A1:** 101
catastrophic, of scales **A13:** 72-73
chemical reaction products identified **A10:** 549
chromium effects on **A13:** 97, 578
of chromium-molybdenum steels **A1:** 617, 629-630, 636
of clean fracture surface, for replicas **A12:** 181
of copper films, using oxygen **A10:** 609
of copper powder **A7:** 106-109
-corrosion, in ceramics **A11:** 755-757
defined **A11:** 7 **A13:** 10, 17, 61 **EM1:** 17 **EM2:** 29
deformation by **A13:** 72
as degradation factor **EM2:** 576
as effect of high temperature **A12:** 35
effect on dimple rupture **A12:** 35
effect on FMR in single-crystal iron whisker **A10:** 274
effect on slip reversal **A12:** 15
effects in creep testing **A8:** 301
of electrical contact materials **A2:** 841
electrical resistance alloys **A2:** 824
embrittlement, high-temperature **A13:** 1093
fracture, cleaning of **A12:** 73-76
free radical induced **EM2:** 777-779
in fretting **A13:** 138
from dewetting **EL1:** 676
general, in elevated-temperature failures **A11:** 271
and green strength **A7:** 288
heat-resistant cast alloys **A13:** 576
in heat-resistant casting alloys, effect of chromium on **A9:** 333-334
high-carbon steels **A12:** 280
high-temperature **A12:** 72 **A13:** 17, 61, 98-99, 1311
high-temperature, acoustic emission inspection **A17:** 287
inhibitors, for lubricant failure **A11:** 154
initial, gaseous corrosion **A13:** 67
intercrystalline, of Invar **A2:** 890
intergranular, AISI/SAE alloy steels **A12:** 300
internal **A13:** 73
internal and conventional **A7:** 717
internal, of silver-base contact composites **A2:** 857
internal, SIMS analysis of second-phase distribution **A10:** 610
iron, partitioning states in **A10:** 178
isothermal stability diagrams **A13:** 64
kinetics **A13:** 17, 67
layer formation **A8:** 603
layer formation during **A8:** 603
LEISS analysis of **A10:** 603, 609
linear, reaction rates **A13:** 66

Oxidation (continued)
logarithmic, of scales **A13:** 72
in low-carbon steel **A12:** 246
in magnesium alloys **A9:** 427
mechanically alloyed oxide alloys **A2:** 947
dispersion-strengthened (MA ODS) mechanism **A13:** 17, 65
microwave inspection **A17:** 202
molecular structure and orientation determined in **A10:** 109
in molten-salt corrosion **A13:** 89
MOS IC fabrication **EL1:** 198
of niobium **A14:** 237
P/M superalloys **A13:** 835-836
parabolic, kinetics **A13:** 67
in petroleum refining and petrochemical operations **A13:** 1273-1274
of polyester resins **EM1:** 93
preferential **A11:** 405-406 **A13:** 17
pure tin **A13:** 770-771
rapid **A13:** 1040
rates **A13:** 74, 97
and reduction during electrochemical etching **A9:** 60-61
-reduction reactions, classical wet chemistry **A10:** 163-164
resistance, and fatigue **A11:** 131
resistance of nonoxide fibers **EM1:** 64
resistant carbon-carbon composites **EM1:** 920-921
role in electrolytic etching of heat-resistant casting alloys **A9:** 331
selection **A13:** 17
selective **A13:** 73-74
SEM fractographs of spiking defect without **A11:** 351
SERS studies of **A10:** 136
of silver **A13:** 794
of silver electrodes in alkaline environments **A10:** 135
SIMS tracer studies **A10:** 610
of single-crystal superalloys **A1:** 1004, 1006
of sintered iron/steel parts, with superheated steam **A13:** 823
of sliding contacts **A2:** 842
species, analysis of **A10:** 201
of species soluble in solution **A10:** 208
stainless steel **M1:** 93-94
stainless steel pan failure from **A11:** 274
of stainless steels **A13:** 554, 559
states, detection of **A7:** 256
states, of elements **A10:** 688
states of metal atoms in metal oxide surface films, XPS for **A10:** 568
states, partitioning **A10:** 178
static, fracture surface of ceramic exposed to **A11:** 754
steam treatment **A16:** 266
steam treatment of taps **A16:** 259, 260, 261
stress, from open-cycle testing **A11:** 132
-sulfidation test, for hot corrosion resistance **A11:** 280
sulfuric acid formation by **A13:** 1197
surface, FIM/AP study of **A10:** 583
of tantalum **A14:** 238
of tantalum, protection against **A13:** 730-731
test, cyclic **A13:** 1312
thermal **A13:** 499, 694, 696
and thermal fatigue cracking, cast ductile iron rotor **A11:** 374-376
titanium alloys **A12:** 442
of tungsten-reinforced composites **EM1:** 882-883
in unalloyed uranium **A2:** 672
uranium/uranium alloys **A13:** 813-815
in voltammetry **A10:** 189
Wagner theory of **A13:** 69-70
and wear **A8:** 601, 603
of wrought cobalt-base superalloys **A1:** 965-966, 968
of wrought nickel-base superalloys **A1:** 957, 959, 965
alloying for surface stability **A1:** 956-957
protection against oxidation **A1:** 957, 959
and x-ray spectrometry **A10:** 82
Oxidation corrosion
pitting, in ceramics **A11:** 755
test, for gas-turbine components **A11:** 279-280

Oxidation diffusion processes **A16:** 41
Oxidation grain size
defined **A9:** 13
Oxidation inhibitors **A18:** 99, 101, 104-105, 106
applications **A18:** 105
decomposition **A18:** 104-105
in engine lubricant formulations **A18:** 111
for metalworking lubricants **A18:** 141, 143, 147
in nonengine lubricant formulations **A18:** 111
oxidation mechanism **A18:** 104
Oxidation protective coatings **M5:** 375-380
aluminum **M5:** 333-335, 337, 344-345
applications **M5:** 375-380
applying, methods of **M5:** 377-380
ceramic **M5:** 535-537, 543, 546
cobalt-aluminum type **M5:** 376
cobalt-chromium-aluminum- yttrium type **M5:** 377-378
corrosion resistance **M5:** 375-376, 379-380
diffusion type **M5:** 376-380
aluminide **M5:** 376-380
applying, methods of **M5:** 377-378, 380
costs **M5:** 379
inward **M5:** 378
outward **M5:** 378
refractory metals **M5:** 380
service characteristics **M5:** 376-377, 379-380
silicide **M5:** 380
superalloys **M5:** 377-379
ductility **M5:** 376-378
green coating, applying **M5:** 377-378, 380
iron-chromium-aluminum-yttrium type superalloys **M5:** 376
monoaluminide-chromium-aluminum-yttrium types, superalloys **M5:** 376-379
nickel-chromium-aluminum-yttrium type, superalloys **M5:** 376
nickel-cobalt-chromium-aluminum-yttrium type, superalloys **M5:** 376
overlay type **M5:** 376-379
applying, methods of **M5:** 379-380
composition and microstructure **M5:** 376-377
costs **M5:** 379
service characteristics **M5:** 379
superalloys **M5:** 376-379
pack cementation process **M5:** 377-378, 380
physical vapor deposition method overlay types **M5:** 379
plasma-spray method **M5:** 379
refractory metals **M5:** 375-376, 379-380
applying, methods of **M5:** 380
requirements for **M5:** 379
service characteristics **M5:** 380
substrate types **M5:** 380
slurry processes **M5:** 379-380
sputtering process **M5:** 379
superalloys **M5:** 375-378
applying, methods of **M5:** 377-379
requirements for **M5:** 376
service characteristics **M5:** 376-377
thermal mechanical fatigue behavior of **M5:** 376-378
thermal spray process **M5:** 362, 379
thickness **M5:** 376
types of **M5:** 376-380
Oxidation resistance *See also* Oxidation **EL1:** 773, 822
of aluminum alloying of metal coatings **A7:** 74
of aluminum oxide-containing cermets **A7:** 804
cast steel **M1:** 46
cemented carbides **A13:** 855
cobalt-base high-temperature alloys **A2:** 452
of copper alloys for electrical contact materials **A2:** 843
cyclic, P/M superalloys **A13:** 839
dissimilar metal joining **A6:** 826
ductile iron **M1:** 46
from composite powders **A7:** 175
gold **A13:** 796-797
gray iron **M1:** 46
heating elements, electrical resistance alloys **A2:** 831
malleable iron **M1:** 46
mechanically alloyed oxide alloys **A2:** 943-947

Oxidation resistance (continued)
dispersion-strengthened (MA ODS)
nickel alloys ... **A2:** 442
osmium ... **A13:** 807
P/M stainless steels ... **A13:** 832
palladium ... **A13:** 800
platinum ... **A13:** 798-799
rhodium ... **A13:** 802
sintered stainless steels ... **A13:** 832
ZGS platinum ... **A2:** 714
Oxidation stability
brazing and ... **A6:** 117
Oxidation-reduction (redox) *See also* Classical wet analytical chemistry; Oxidation; Redox; Reduction
of potential contaminant ions, in gravimetric samples ... **A10:** 163
reactions, as classical wet chemical analysis ... **A10:** 163-164
Oxidation-resistant coating
applying, methods of ... **M5:** 664-666
high-temperature types ... **M5:** 661-662, 664-665
molybdenum ... **M5:** 661-662
niobium ... **M5:** 664-665
refractory metals ... **M5:** 661-662, 664-665
tantalum ... **M5:** 664-665
tungsten ... **M5:** 661-662
Oxidational wear ... **A18:** 280-288
application to practical tribosystems ... **A18:** 287-288
calculation of heat flow ... **A18:** 282-283
definition ... **A18:** 280-281
FINDAP computer program ... **A18:** 280, 281, 283-284, 286
future trends ... **A18:** 288
jet engine components ... **A18:** 588, 591
mechanism of mild oxidational wear ... **A18:** 281-282
mechanisms of wear ... **A18:** 281
mild oxidational wear theory ... **A18:** 282
OXYWEAR computer program ... **A18:** 280, 283, 285-286, 288
piston rings ... **A18:** 556
theory and experimental results comparison ... **A18:** 283-287
tool steels ... **A18:** 739
Oxidative stability ... **A18:** 84
Oxidative wear *See also* Abrasive wear; Fretting; Fretting corrosion; Oxidational wear
defined ... **A8:** 10 **A11:** 7 **A18:** 14
progression ... **A8:** 603-604
Oxide
chemical vapor deposition of ... **M5:** 381
formation of
electropolishing processes ... **M5:** 307
hot dip galvanized coating process ... **M5:** 327-328
protection against *See* Oxidation-protective coating; Oxidation-resistant coating
removal of
aluminum and aluminum alloys ... **M5:** 8, 12-13
chromium plating ... **M5:** 185
copper and copper alloys ... **M5:** 611-614, 619-620
heat-resisting alloys ... **M5:** 564
hot dip galvanized coating process ... **M5:** 326-327
nickel alloys ... **M5:** 671
processes ... **M5:** 65, 97-103
reactive and refractory alloys ... **M5:** 650-654, 659, 662-667
stainless steel ... **M5:** 553-554
Oxide additives
hot pressed silicon nitride ... **A7:** 516
Oxide breakdown
semiconductor chips ... **EL1:** 965
Oxide ceramic coatings ... **M5:** 534-536, 546
hardness ... **M5:** 546-547
melting points ... **M5:** 534-535
Oxide ceramics
brazing and soldering characteristics ... **A6:** 636
as tooling material ... **A7:** 423
Oxide cermets ... **A7:** 798, 802-804
aluminum oxide cermets ... **A2:** 992-993
beryllium oxide cermets ... **A2:** 993
defined ... **A2:** 979
iron and cordierite cermets ... **A2:** 995
magnesium oxide cermets ... **A2:** 993
metal-matrix high-temperature superconductor cermets ... **A2:** 995
silicon oxide cermets ... **A2:** 992
thorium oxide cermets ... **A2:** 993
uranium oxide cermets ... **A2:** 993-994
zirconium oxide cermets ... **A2:** 993
Oxide coating
spring wire ... **M1:** 262
Oxide coatings
heat-resisting alloys ... **M5:** 564, 566
molybdenum ... **M5:** 662
niobium ... **M5:** 664-666
removal of ... **A12:** 75
tantalum ... **M5:** 664-666
tungsten ... **M5:** 662
Oxide colors
as colorants ... **EM2:** 501
Oxide compounds
for high-temperature superconductors ... **A2:** 1085-1086
Oxide conversion coating ... **M5:** 598-599
Oxide dispersion strengthened superalloys ... **A13:** 834-838
Oxide dispersion-strengthened (ODS) alloys
brazing ... **A6:** 632, 928
fusion welding ... **A6:** 632
Oxide dispersion-strengthened alloys
alloy classes ... **M6:** 1020
definition ... **M6:** 1020
filler metals ... **M6:** 1020
Oxide dispersion-strengthened alloys (MA ODS alloys)
aluminum and aluminum alloys ... **A2:** 7
bar ... **A2:** 948-949
commercial alloys ... **A2:** 944-947
copper alloys ... **A2:** 400-401
fabrication ... **A2:** 946-949
hot-corrosion properties ... **A2:** 947
joining of ... **A2:** 949
mechanical alloying alloy applications ... **A2:** 943
mechanical alloying process ... **A2:** 943-944
oxidation properties ... **A2:** 947
rare earth alloy additives ... **A2:** 729
sheet ... **A2:** 949
Oxide dispersion-strengthened alloys (MA ODS alloys), specific types
alloy MA 754, microstructure elevated-temperature strength ... **A2:** 944-945
alloy MA 758, oxidation resistance properties ... **A2:** 945-946
alloy MA 760, high-temperature strength structural stability, oxidation resistance ... **A2:** 947
alloy MA 6000, elevated-temperature resistance ... **A2:** 946-947
strength, oxidation resistance, sulfidation
Oxide dispersion-strengthened copper alloys
manufacture ... **A2:** 400
properties ... **A2:** 400
uses ... **A2:** 401
Oxide dispersion-strengthened materials
dimple size ... **A12:** 12
Oxide fibers *See* Aluminum oxide fibers
Oxide film *See also* Barrier film; Film; Oxides
aluminum ... **A13:** 583
aluminum casting alloys ... **A2:** 145-146
formation, zirconium ... **A13:** 718
thermodynamic stability, by Pourbaix diagram ... **A13:** 583
in welds ... **A13:** 344
Oxide film replica
defined ... **A9:** 13
Oxide films *See also* Surface oxide films
deposited by color etching ... **A9:** 141
dispersion in milling ... **A7:** 63
heat tinting ... **A9:** 61
High-temperature ... **EL1:** 678
in magnesium alloys ... **A9:** 427
on aluminum alloys used to reveal microstructural features ... **A9:** 351
on uranium and uranium alloys to increase polarized light contrast ... **A9:** 480
as solder defects ... **EL1:** 642
thickness, on pure aluminum ... **A7:** 248
thin, effect on green strength ... **A7:** 302
in zirconium alloys, enhancement by anodizing ... **A9:** 498
Oxide fluxing
mechanically alloyed oxide dispersion-strengthened (MA ODS) alloys ... **A2:** 947
Oxide glasses
engineering properties *See* Engineering properties of oxide glasses and other inorganic glasses
for joining non-oxide ceramics ... **EM4:** 528
structures ... **EM4:** 846
Oxide inclusions *See also* Oxide(s) ... **A6:** 1073 **M6:** 838
in brazing ... **A11:** 451
as casting defect ... **A11:** 388
flash welds ... **M6:** 580
flux deposits in submerged arc welds ... **M6:** 117
in weldments ... **A17:** 582
Oxide inclusions in
aluminum alloy 1100 ... **A9:** 635
aluminum alloy 2024 ... **A9:** 635
Nickel 200 ... **A9:** 436
Oxide layers
explosion characteristics ... **A7:** 196
Oxide layers from reactive sputtering
absorption coefficients ... **A9:** 60
refractive indices ... **A9:** 60
Oxide powders ... **A7:** 171, 194, 303
for brazing and soldering ... **A7:** 837-840
mechanical comminution for ... **A7:** 56
Oxide reduction *See also* Solid-state reduction ... **A7:** 52-53
by advanced atomization or thermomechanical processing ... **A7:** 256
cobalt and cobalt alloy powders by ... **A7:** 145-146
of copper powders ... **A7:** 734
equilibria for reactions ... **A7:** 52, 53
and internal porosity ... **A7:** 299
and sintering ... **A7:** 309
standard free energy in formation ... **A7:** 53
suggested accuracies ... **A7:** 53
temperatures ... **A7:** 52
Oxide reduction process
oxidation-resistant coating ... **M5:** 664-665
Oxide removal
arc welding of nickel alloys ... **M6:** 437
by direct current electrode positive ... **M6:** 186
resistance welding of aluminum alloys ... **M6:** 539
titanium and titanium alloys ... **M6:** 449
Oxide scale
as casting defect ... **A11:** 385
on high-purity iron, effects of abrading and polishing ... **A9:** 47
source of uranium contamination ... **A9:** 477
Oxide scales *See also* Gaseous corrosion; Oxides, Oxidation ... **A13:** 70-76
alloy oxidation: doping principle ... **A13:** 73, 75-76
catastrophic oxidation ... **A13:** 72-73
in gaseous corrosion ... **A13:** 17
high-temperature ... **A13:** 97-101
oxide evaporation ... **A13:** 71
paralinear oxidation ... **A13:** 70-71
protective, characteristics ... **A13:** 97
relative thickness ... **A13:** 70
selective oxidation ... **A13:** 73-74
spalling ... **A13:** 98

Oxide scales (continued)
stress relief ... **A13:** 71-72
structure, variation ... **A13:** 97
Oxide skins
as casting defect ... **A11:** 388
Oxide strengthening
in creep tests ... **A8:** 331-332
Oxide stringers
in aluminum alloy wrought products ... **A9:** 635
as linear element in quantitative metallography ... **A9:** 126
Oxide superconductors
hot isostatic pressing ... **EM4:** 197
Oxide texture ... **A13:** 66
Oxide(s)
attack, in steel castings at elevated temperatures ... **A11:** 406
breakdown, in MOS and CMOS ... **A11:** 778-779
effect in forging ... **A11:** 316
effect in hydrogen damage ... **A12:** 125-126
effect on fatigue crack growth rate ... **A12:** 41
failures, in MOS devices ... **A11:** 766
-filled intergranular cracks, steam line ... **A11:** 669
film, austenitic stainless steels ... **A12:** 352-353
films, in carbon and low-alloy steels ... **A11:** 199
flaws, in integrated circuits ... **A12:** 481
formation on heat-exchanger tubes ... **A11:** 629
as inclusions ... **A12:** 65, 220
layer, on cast ductile iron rotor ... **A11:** 376
martensitic stainless steels ... **A12:** 366
penetration, grain boundary, thick-lip ruptures by ... **A11:** 605
penetration, in steam-cooled tubes ... **A11:** 604-607
residual, in brazing ... **A11:** 452
scaling and perforation ... **A11:** 191
in seams, fasteners ... **A11:** 530
spheroidal ... **A12:** 220, 300
stability, of brazed joints ... **A11:** 452-453
in structural ceramics ... **A2:** 1019, 1021-1024
surface, effects, metal-matrix composites ... **A12:** 466
temperature, effect on ... **A12:** 35
trapped, in friction welds ... **A11:** 444
volatile, with oil-ash corrosion ... **A11:** 618
wrought aluminum alloys ... **A12:** 435
Oxide-based cermets ... **A7:** 798, 802-804
Oxide-dispersed additives
in tungsten and molybdenum sintering ... **A7:** 391
Oxide-dispersion techniques ... **A7:** 717-719
Oxide-dispersion-strengthened alloys
spray drying applications ... **A7:** 77
Oxide-dispersion-strengthened materials ... **A6:** 1037-1040
applications ... **A6:** 1037
composition ... **A6:** 1037
design strategy ... **A6:** 1037-1039
diffusion welding ... **A6:** 1038, 1039
ductile-to-brittle transition temperatures ... **A6:** 1039
electron-beam welding ... **A6:** 1038, 1039
explosion welding ... **A6:** 1039, 1040
friction welding ... **A6:** 1038, 1039, 1040
furnace brazing ... **A6:** 1038, 1039, 1040
fusion welding ... **A6:** 1039
gas-metal arc welding ... **A6:** 1039
gas-tungsten arc welding ... **A6:** 1038, 1039
grain structure ... **A6:** 1038-1039
heat-affected zone hydrogen cracking ... **A6:** 1039, 1040
iron-base alloys ... **A6:** 1039
laser-beam welding ... **A6:** 1039
mechanical alloying (MA) ... **A6:** 1037
nickel-base alloys ... **A6:** 1039
porosity ... **A6:** 1037
postweld heat treatments ... **A6:** 1038, 1039
resistance welding ... **A6:** 1038, 1039-1040
shielded metal arc welding ... **A6:** 1039
weld transverse properties ... **A6:** 1038
welding consumables ... **A6:** 1038
welding processes ... **A6:** 1039-1040
Oxide-dispersion-strengthened superalloys
consolidation by hot isostatic pressing ... **A7:** 440
products, mechanical alloying of ... **A7:** 527, 528
wrought ... **A7:** 527-528
Oxide-induced crack closure ... **A8:** 409-410
Oxide-reduced powders
green strengths ... **A7:** 303
Oxide-reduced surfaces *See* Reduced- oxide surfaces
Oxide-reducing agents
in nitrogen-based atmospheres ... **A7:** 345-346
Oxide-sulfide inclusions
in shafts ... **A11:** 462
Oxide-type inclusions
defined ... **A9:** 13
Oxides *See also* Aluminum oxide; Beryllium oxide; Ceramics; Cerium oxide; Chromium oxide Iron oxide; Dross; Magnesium oxide; Oxide scales; Scales; Silicon oxide; Yttrium oxide; Zirconium oxide
in a tungsten-nickel powder metallurgy material ... **A9:** 562
amphoteric ... **A13:** 66
anodic, as barrier protection ... **A13:** 377-378
applications ... **EM4:** 203
carbon fiber reaction with ... **EM1:** 52
chromium, scale ... **A13:** 97
competing ... **A13:** 74-76
contamination, effect, structured reliability testing ... **EL1:** 959-960
cracking ... **A13:** 72
defined ... **A13:** 61
deposition ... **A10:** 201
diffusion data in ... **A13:** 68
double ... **A13:** 76
as electrical conductor ... **A13:** 65
evaporation ... **A13:** 71
as fillers ... **EM3:** 33, 179
finger, powder forged part surfaces ... **A14:** 204
formation, as interfering elements in high temperature combustion ... **A10:** 222
Gibbs energy of formation ... **A13:** 63
as grain-boundary pinning agent ... **A7:** 171
growth, and substrate preparation ... **EL1:** 197
growth, diffusion types ... **A13:** 71
heat-tint, effects on corrosion resistance of austenitic stainless steels ... **A13:** 351-353
hydrated, precipitation for ... **A10:** 169
inclusion, spotty particle ... **A14:** 191
as inclusions ... **A10:** 176
inclusions formed in fluxes ... **A6:** 56
interparticle ... **A14:** 204
ionic, defect structure ... **A13:** 65
linear thermal expansion, coefficients ... **A13:** 71
low-melting ... **A13:** 73
maximum service temperature ... **EM4:** 203
metastable ... **A13:** 62
as molten salts ... **A13:** 50
n-type ... **A13:** 66
p-type ... **A13:** 65-66
plasma spray material ... **EM4:** 203
plasma-assisted physical vapor deposition ... **A18:** 848
plastic flow ... **A13:** 72
properties ... **EM4:** 203
protective, and explosivity ... **A7:** 194
protective film ... **A13:** 517-518
pulsed laser atom probe analysis for ... **A10:** 597
removal, by flux ... **EL1:** 676
sample dissolution in hydrochloric acid ... **A10:** 165
semiconductor, defect structure ... **A13:** 65-66
solid, defect structure ... **A13:** 61, 65
structures and thermal properties ... **A13:** 64
surface, cleaning of ... **A13:** 381
surface layers, SIMS analysis ... **A10:** 610
surface, measuring ... **A10:** 177
textures ... **A13:** 66
to improve thermal conductivity ... **EM3:** 178
uniform and nodular ... **A13:** 947
vanadium K-edge XANES spectra in ... **A10:** 415
weighing as the, gravimetric analysis ... **A10:** 170
white nodules ... **A13:** 947
in wrought heat-resistant alloys ... **A9:** 312
in zirconium alloys, preparation for examination of ... **A9:** 498
Oxidic interference films
produced by reactive sputtering ... **A9:** 148-149
Oxidizable metals
electron-beam welding ... **A6:** 851
Oxidizable organic compounds in electrolytes ... **A9:** 51
Oxidized polymer ... **EM3:** 41
Oxidized surface (on steel)
defined ... **A13:** 10
Oxidized-silicon substrates
physical characteristics ... **EL1:** 106
Oxidizing
agents ... **A13:** 10, 677, 1140
potential ... **A13:** 39
power ... **A13:** 37-39
salts, copper/copper alloy corrosion ... **A13:** 630
Oxidizing acids
sample dissolution by ... **A10:** 166
Oxidizing agents
effect on staining of heat-resistant casting alloys ... **A9:** 330-331
in electrolytes ... **A9:** 51
mixing with reducing agents ... **A9:** 69
role in electrochemical etching ... **A9:** 60
Oxidizing atmospheres
heating-element materials ... **A2:** 833-834
Oxidizing flame
definition ... **A6:** 1211 **M6:** 12
Oxidizing gases ... **A8:** 412-415
Oxidizing salt bath descaling
nickel alloys ... **M5:** 672
process ... **M5:** 97-98, 653-654
Oxidizing salts *See* Salts
Oxirane ring (epoxide group)
defined ... **EL1:** 810
Oxonium ion ... **EM3:** 100
Oxy-acid etching system
porcelain enameling process ... **M5:** 514-515
Oxy-natural gas ... **A6:** 121
Oxy/methylacetylene-propadiene- stabilized gas ... **M6:** 901-903
Oxyacetylene braze welding of
cast irons ... **M6:** 599-600
cleaning by salt bath ... **M6:** 600
copper alloy filler metals ... **M6:** 599
low-carbon steels ... **M6:** 598
malleable iron ... **M6:** 596-598
steel ... **M6:** 598-599
versus arc tack welding ... **M6:** 598
Oxyacetylene braze welding of steel and cast irons ... **M6:** 596-600
advantages ... **M6:** 596
applicability ... **M6:** 596
base metals ... **M6:** 596
filler metals ... **M6:** 596-597
joint properties ... **M6:** 597
RBCuZn-A ... **M6:** 596
RBCuZn-B ... **M6:** 596
RBCuZn-C ... **M6:** 596-597
RBCuZn-D ... **M6:** 596-597
flame adjustment ... **M6:** 596
fluxes ... **M6:** 597
application ... **M6:** 597
joint preparation ... **M6:** 597-598
fillet welds ... **M6:** 597
plug welds ... **M6:** 597
slot welds ... **M6:** 597
limitations ... **M6:** 596
postheating ... **M6:** 598
preheating ... **M6:** 598
repair of iron castings ... **M6:** 599-600
surface preparation ... **M6:** 598
use of a salt bath ... **M6:** 598
Oxyacetylene cutting
definition ... **M6:** 12
Oxyacetylene gas ... **M6:** 901, 902
Oxyacetylene pressure welding ... **M6:** 595
advantages ... **M6:** 595
closed-gap technique ... **M6:** 595
heating setup ... **M6:** 595
joint design ... **M6:** 595
alloy steels ... **M6:** 595
carbon steels ... **M6:** 595
limitations ... **M6:** 595
operation sequence ... **M6:** 595
Oxyacetylene process
cobalt-base alloys ... **A13:** 664
Oxyacetylene torch brazing
of copper ... **M6:** 1039
Oxyacetylene welding
of cast irons ... **M6:** 601-605
definition ... **M6:** 12
hardfacing ... **M6:** 787

Oxyacetylene welding (continued)
induction brazing as replacement **M6:** 974
Oxyacetylene welding (OAW)
carbide-containing nickel-base alloys, poor weldability **A6:** 795
definition **A6:** 1211
hardfacing **A6:** 804, 805
hardfacing alloys **A6:** 800
Laves phase alloys, poor weldability **A6:** 795
low-carbon steels **A6:** 419
not recommended for nickel-base corrosion-resistant alloys containing molybdenum **A6:** 594
vs. laser-beam welding **A6:** 262
Oxyacetylene welding process **A6:** 5, 6
Oxydol anodizing process
aluminum and aluminum alloys **M5:** 586-587
Oxyfluoride gels **EM4:** 210, 211
Oxyfluoride glasses
optical properties **EM4:** 854
Oxyfuel detonation process *See* High-velocity oxyfuel powder spray process
Oxyfuel gas cutting
acetylene **M6:** 899-901
combustion reaction **M6:** 900
cost **M6:** 901
heat content **M6:** 901
applications **A14:** 721
chemical flux cutting **M6:** 914
chemistry of cutting **M6:** 897-899
alloying of iron **M6:** 898
chemical reactions **M6:** 899
drag **M6:** 898
oxygen consumption **M6:** 898
oxygen purity **M6:** 898
preheating **M6:** 898-899
close-tolerance cutting **M6:** 913
comparison to oxyfuel gas gouging **M6:** 914
comparison to plasma arc cutting **M6:** 918
cutting, factors affecting **A14:** 721-722
cutting speeds **M6:** 916
definition **M6:** 12
effect on base metal **A14:** 724-725 **M6:** 903-906
annealing **M6:** 904, 912
deformation **M6:** 905
distortion **M6:** 904-905
heat-affected zone **M6:** 903, 905
local preheating **M6:** 904-905
equipment **A14:** 725-728 **M6:** 906-908
cutting tips **M6:** 906-907
cutting torches **M6:** 906
gas regulators **M6:** 906
guidance **M6:** 907
hose **M6:** 906
portable cutting **M6:** 907
stationary cutting **M6:** 907
tape control **M6:** 908
tracers, manual or magnetic **M6:** 907-908
fuel gases, properties of **M6:** 899-900
combustion ratio **M6:** 900
cost analysis **M6:** 899
coupling distance **M6:** 900
flame temperature **M6:** 899-900
heat distribution in the flame **M6:** 900
heat of combustion **M6:** 900
heat transfer **M6:** 900
gas combustion **A14:** 722-724
heavy cutting **M6:** 909-911
drag **M6:** 910-911
gas flow requirements **M6:** 905, 910
preheating **M6:** 908-910
starting **M6:** 910-911
light cutting **M6:** 909
medium cutting **M6:** 909-910
cutting speed **M6:** 909
kerf angle **M6:** 910
kerf compensation **M6:** 909
machine accuracy **M6:** 909-910
plate movement **M6:** 910
Oxyfuel gas cutting (continued)
preheat **M6:** 909
surface finish **M6:** 908-909
tip design **M6:** 909
metal powder cutting **M6:** 913-914
methylacetylene-propadiene-stabilized gas **M6:** 899-903
natural gas **M6:** 899-903
nesting of shapes **A14:** 728 **M6:** 913
operation **M6:** 896-897
operation principles **A14:** 720-721
preparation of weld edges **M6:** 911-912
preheating for bevel cutting **M6:** 911-912
simultaneous trimming and beveling **M6:** 912
torch settings for bevels **M6:** 910
process capabilities **A14:** 721 **M6:** 897
applications **M6:** 897
comparison to other operations **M6:** 897
thickness limits **M6:** 897
propane **M6:** 899-900, 902-903
propylene **M6:** 903
quality of cut **M6:** 897
safety **M6:** 914-915
cylinders **M6:** 914-915
protective clothing **M6:** 914
working environment **M6:** 915
stack cutting **M6:** 911
starting the cut **M6:** 908-909
underwater cutting **M6:** 914, 921-925
Oxyfuel gas cutting (OFC) **A6:** 1155-1165
advantages **A6:** 1156
alloy steels **A6:** 1159
applications **A6:** 1155
shipbuilding **A6:** 384
carbon steels **A6:** 1159-1160
close-tolerance cutting **A6:** 1164
control of distortion **A6:** 1160-1161
cutting of bars and structural shapes **A6:** 1164
cutting tips **A6:** 1161-1162
cutting torches **A6:** 1161, 1162, 1163
definition **A6:** 1211
deformation **A6:** 1160
description **A6:** 1155
distortion **A6:** 1160
drag **A6:** 1156
effect on base metal **A6:** 1159-1161
equipment **A6:** 1161-1162
equipment selection factors **A6:** 1162
fuel gas properties **A6:** 1156-1157
gas regulators **A6:** 1161
heavy cutting **A6:** 1163-1164
high-carbon steels **A6:** 1159
high-low regulators **A6:** 1161
hose **A6:** 1161
kerf angle **A6:** 1163
kerf compensation **A6:** 1162-1163
light cutting **A6:** 1162
limitations **A6:** 1156
local preheating **A6:** 1159, 1161
low-carbon steel **A6:** 1159
machine torch piercing **A6:** 1162
medium cutting **A6:** 1162-1163
medium-carbon steels **A6:** 1159
oxygen consumption **A6:** 1156
preheating **A6:** 1156
principles of operation **A6:** 1155
process capabilities **A6:** 1155-1156
safety **A6:** 1164-1165
cylinders **A6:** 1164
protective clothing **A6:** 1164
working environment **A6:** 1165
safety precautions **A6:** 1200-1201
stainless steels **A6:** 1196
starting the cut **A6:** 1162
thickness limits **A6:** 1156
Oxyfuel gas cutting (steel)
suggested viewing filter plates **A6:** 1191
Oxyfuel gas cutting of
alloy steels **M6:** 904
Oxyfuel gas cutting of (continued)
bars and structural shapes **M6:** 913
cast iron **M6:** 897, 912-913
gray iron **M6:** 912-913
high-carbon steels **M6:** 904-905
low-carbon steels **M6:** 903-905
medium-carbon steels **M6:** 904
stainless steels **M6:** 897, 903, 912, 914
Oxyfuel gas powder
for austenitic stainless steel **A7:** 844
Oxyfuel gas spraying
definition **A6:** 1211
Oxyfuel gas welding **M6:** 583-594
advantages **M6:** 583
bridging gaps in poor fit-ups **M6:** 593
capabilities **M6:** 583
combustion of natural gas and propane **M6:** 588
definition **M6:** 12
edge preparation **M6:** 589-591
equipment **M6:** 583-586
gas storage **M6:** 584
goggles **M6:** 586
hoses **M6:** 585
mixing chambers **M6:** 585-586
torch inlet valves **M6:** 586
welding tips **M6:** 586
welding torches **M6:** 585-586
flame adjustment **M6:** 587
positive-pressure outfit **M6:** 587
fluxes **M6:** 583
fuel gases **M6:** 583-584
gas pressure selection **M6:** 586
gases **M6:** 583-584
acetylene **M6:** 584
hydrogen **M6:** 584
natural gas **M6:** 584
oxygen **M6:** 584
propane **M6:** 584
proprietary gases **M6:** 584
hardfacing deposition in mining and mineral industries **A18:** 653
joint design **M6:** 589-591
corner-edge joints **M6:** 589-590
fillet welds **M6:** 589-590
groove welds **M6:** 590-591
lap joints **M6:** 589-590
plate **M6:** 590
sheet **M6:** 589
short-flanged-edge butt joints **M6:** 589
square-groove butt joints **M6:** 599
T-joints **M6:** 589-590
limitations **M6:** 583
metals welded **M6:** 583
oxyacetylene combustion **M6:** 587-588
acetylene flame **M6:** 587
carburizing flame **M6:** 587
neutral flame **M6:** 588
reducing flame **M6:** 587-588
separated flame **M6:** 587-588
shape of flame cone **M6:** 588
oxyhydrogen combustion **M6:** 588
oxidizing flame **M6:** 587-588
pipe welding **M6:** 591-592
applications **M6:** 69
fittings **M6:** 591
horizontal-fixed position **M6:** 591-592
horizontal-rolled position **M6:** 591
tack welds **M6:** 591
vertical position **M6:** 591-592
postheating **M6:** 593
preheating **M6:** 593
recommended grooves **M6:** 592-594
repairs and alterations **M6:** 592-593
backing **M6:** 592
gouging **M6:** 592-593
safety **M6:** 58, 593-594
sheet **M6:** 592

Oxyfuel gas welding (continued)
techniques M6: 589
backhand welding M6: 589
flat and horizontal positions M6: 589
forehand welding M6: 589
vertical and overhead positions M6: 589
tip-orifice sizes M6: 586
tube, thin-walled M6: 592
use in hardfacing M6: 782-783
welding rods M6: 588-589
class RG45 .... M6: 588-589
class RG60 .... M6: 588-589
class RG65 .... M6: 588-589
specification M6: 588
weld-metal strengthening M6: 588
Oxyfuel gas welding (OFW) A6: 281-290
advantages A6: 281
aluminum alloys A6: 738-739
applications A6: 288-289
railroad equipment A6: 398
capabilities A6: 281
cast irons A6: 720
definition A6: 281, 1211
equipment A6: 282-283
flame adjustment A6: 283-284
fluxes A6: 281
fuel gases A6: 281-283
gas pressure selection A6: 283
gas-tungsten arc welding A6: 758
gases A6: 281-282, 285
hardfacing alloys A6: 796, 797, 798, 799
limitations A6: 281
low-carbon steel A6: 800
mixing chambers A6: 283
personnel training A6: 281
safety precautions A6: 282, 283, 290, 1190, 1193, 1200-1201
tip-orifice size selection A6: 283
Oxyfuel gas welding (steel)
suggested viewing filter plates A6: 1191
Oxyfuel gas welding and cutting
safety precautions A6: 1191
Oxyfuel gas welding of
aluminum alloys M6: 583
carbon M6: 583
cast iron M6: 583
copper alloys M6: 583
ferrous metals alloys M6: 583
nickel alloys M6: 583
nonferrous metals M6: 583
steels M6: 583-594
zinc alloys M6: 583
Oxyfuel powder (OFP) spray — method A18: 829, 830, 832
spray materials A18: 830
Oxyfuel welding (OFW) A6: 124, 125
alterations of workpieces A6: 289
alternative metal-joining processes for thin sheet A6: 288
aluminum alloys A6: 281, 282, 285
applications A6: 281
austenitic stainless steels A6: 284
backhand welding A6: 286, 287, 288
bridging gaps in poor fit-ups A6: 290
carbon backing A6: 289
carbon steel A6: 281, 286
cast irons A6: 281, 714-715
combustion of natural gas and propane A6: 285
copper A6: 285
copper alloys A6: 281, 285, 756
distortion A6: 287, 288, 289
edge preparation A6: 286-289
ferrous alloys A6: 281
filler metals A6: 281, 286
forehand welding A6: 286, 287, 288
heat-affected zone A6: 290
joint design A6: 286-289
lead A6: 281, 282, 285
low-alloy steels A6: 286
low-carbon steel A6: 285, 286, 287, 288, 289
magnesium A6: 281, 282
magnetic-particle inspection A6: 289
melt-through weld A6: 286, 288
nickel alloys A6: 281
oxyacetylene combustion A6: 284-285
oxyacetylene welding (OAW) torch A6: 281
Oxyfuel welding (OFW) (continued)
oxyfuel gouging A6: 289
oxyhydrogen combustion A6: 285
postweld heat treatment A6: 289-290
precious metals A6: 281
preheating A6: 289-290
proprietary gases A6: 281, 282, 283
repair of iron castings A6: 1105
repair welding A6: 289
stainless steel A6: 285
stainless steels A6: 281
steel A6: 281, 282, 284, 288
two-pass technique A6: 287-288
welding rods A6: 285-286
welding techniques A6: 286
zinc alloys A6: 281
Oxyfuel wire (OFW) spray process A18: 829, 830, 832
spray materials A18: 829
Oxyfuel/oxyacetylene (OFW/OAW) welding
hardfacing alloy consumable form A6: 796
Oxygas cutting
definition A6: 1211
Oxygen *See also* Air; Atmospheres; Oxygen content; Oxygen corrosion
as a reaction gas in reactive sputtering A9: 148-149
absorption, at elevated temperatures, titanium embrittlement by A11: 641
addition, cure system EL1: 857
alloying effects on carbon alloys M6: 400, 402
as an alpha stabilizer in titanium and titanium alloys A9: 458
analysis by inert gas fusion, for aluminum- killed steel A10: 231
analysis, zirconium-steel couple A13: 715
analyzed in microcircuit fabrication process A10: 156-157
atmospheric corrosion, role in M1: 718
atomic, in low earth orbit A12: 481 A13: 1099-1100
Auger chemical map for A10: 557
in austenitic stainless steels A6: 468
basic oxygen process A1: 110, 111, 112
bombardment, SIMS spectra for A10: 615-616
by residual gas analysis (RGA) EL1: 1065
cause of porosity in nickel alloy welds M6: 442-443
-cell attack, as selective leaching A11: 628
characteristics in a blend A6: 65
and chemical concentration cells, biological corrosion A13: 42-43
chemistry at surfaces, AES analysis of A10: 553
combustion mechanisms of metals in A7: 597
as common reactant A13: 61
composition, wt% (maximum) liquation cracking A6: 568
composition-depth profile A7: 256
consumption, oxyfuel gas cutting A14: 721
consumption, with moisture/carbon dioxide generation EL1: 1066
contamination A8: 408 M6: 321
contamination, weld embrittlement from A11: 438
content affecting machinability of carbon and alloy steels A16: 672
content effect on near-threshold fatigue crack growth rates A8: 427, 430
content, high-purity water A8: 420
content in P/M materials A16: 887, 888
contents in weld metal and flux choice A6: 58, 59
in copper alloys A6: 753
corrosion of gas-dryer piping A11: 631
cutting, types A14: 720-729
degassing, in vacuum melting purification A2: 1094
determined by 14-MeV FNAA A10: 239
diffusion, as oxide growth A13: 65
dissociation and recombination A6: 64
dissolved A8: 416, 423, 427 A13: 29, 221, 489, 895-898, 932
dissolved, as corrosive, copper casting alloys A2: 352
effect, atmospheric contaminants A13: 81
effect, copper alloys in seawater A13: 624
effect, corrosion resistance, sintered austenitic stainless steels A13: 831
Oxygen (continued)
effect, liquid-metal corrosion A13: 58
effect, nickel-base alloy corrosion hydrofluoric acid A13: 645, 647
effect, nuclear reactor erosion-corrosion A13: 965
effect of, on steel composition and formability A1: 577
effect of temperature A12: 35
effect on argon shielding gas M6: 163-164
effect on basicity index M6: 41
effect on eutectic joining EM4: 526
effect on fluxes when introduced A6: 55-59
effect on positive secondary ion yields A10: 612
effect, P/M stainless steels A13: 830
effect, pulp bleach plants A13: 1194
effect, soil corrosion, carbon steels A13: 512-513
effect, space booster/satellite corrosion A13: 1105
effect, titanium and titanium alloy castings A2: 639
effect, zinc corrosion in water A13: 760
in electrical steels A9: 537
electroslag welding, reactions A6: 273, 274
-embrittled iron A12: 222
embrittlement of iron by A1: 689-690
as embrittler of fcc metals A12: 123
enrichment A11: 574
exclusion, oil/gas production A13: 1245
-fired fracture surface, XPS survey A10: 577
flooding in secondary ion mass spectroscopy A7: 258
in flue gas, effects A13: 1200
freshwater corrosion, effect on M1: 733, 738
fugacity during active metal brazing EM4: 524-525
as gas assist, laser cutting A14: 739
gas mass analysis of A10: 155
gas-metal arc welding shielding gas A6: 185
groups, causing stress-corrosion cracking A11: 207
in heat-resistant alloys A4: 512
high-energy neutron irradiation of A10: 234
in high-velocity oxyfuel powder spray process A18: 830
ignition, manned spacecraft A13: 1092
as impurity in iron powder A7: 615
as impurity in uranium alloys A9: 477
as impurity, magnetic effects A2: 762
inert gas fusion systems for detecting A10: 226, 228-229, 231
influence on fracture toughness of Ti-6Al-4V A8: 480-481
influence on friction coefficient of clean iron surfaces A18: 32
inhibition, free radical cure systems EL1: 856
as inhibitor for anaerobics EM3: 114
injection A10: 221
in inorganic solids, applicable analytical methods A10: 4, 6
interstitial contamination M6: 463
isotopes, determination in explosive actuator A10: 625-626
isotopes in explosive actuator, SIMS determined A10: 625-626
-leak detector A17: 64
mean free path A17: 59
in molten salts A13: 50
oxidation of copper films using A10: 609
oxyfuel gas cutting *See* Oxyfuel gas cutting
oxyfuel gas welding fuel gas A6: 281-282, 283, 284, 285, 287-288, 290
penetration, intergranular A11: 132
pickup in milling of electrolytic iron powders A7: 64, 65
plus water, loss on reduction (LOR) A7: 155
principal reactions with metals A13: 61
purity, effect on cutting A14: 722
quenching of fluorescence A10: 79-80
reaction with liquid metals A13: 94-95
reactivity/oxidation potential A6: 64
reduction, annealing as A7: 182
removal A7: 180-181
removal, by solid state refining A2: 1094
removal from water A13: 482
role in polymerization and degradation reactions EM3: 620
scavenging A11: 615 A13: 482

Oxygen (continued)
sensors **EM3:** 609
shielding gas for plasma arc cutting **A6:** 1167
shielding gas properties **A6:** 64
shielding gas purity and moisture content **A6:** 65
in silicon wafers, quantitative analysis of **A10:** 122-123
soil corrosion, effect on **M1:** 725, 731, 739-740, 743
solubility, in seawater **A13:** 900
solubility in tantalum **A13:** 728-731
solubility in water **M1:** 733, 734
in steel, effect on nonmetallic inclusions **A9:** 179
submerged arc welding
flux potential **M6:** 116-117
influence of flux on weld-metal content **M6:** 125
surface treatment for drills **A16:** 219
in titanium, determined **A10:** 231
tolerance, tantalum weldment corrosion **A13:** 346-347
ultrahigh-purity **A10:** 224
uranium oxidation rate in **A13:** 813
use in high-temperature combustion **A10:** 221-222, 224
use in oxyfuel gas welding **M6:** 584
use in resistance spot welding **M6:** 486
valve thread connections for compressed gas cylinders **A6:** 1197
weld-metal content, underwater welding **A6:** 1010-1011
in zirconium, role **A2:** 667
Oxygen acetylene powder method **A18:** 644
Oxygen acetylene rod method **A18:** 644
Oxygen arc cutting **A14:** 734
definition **M6:** 12, 919-920
Oxygen arc cutting (AOC)
definition **A6:** 1211
Oxygen cell *See* Differential aeration cell
Oxygen concentration cell *See* Differential aeration cell
Oxygen content
effect of particle size **A7:** 37
effect on densification and expansion during homogenization **A7:** 315
effect on dynamic properties of forged parts **A7:** 415
effect on sintering **A7:** 372
and explosivity **A7:** 195, 196
in superalloy powders **A7:** 434
and surface area of atomized aluminum powder **A7:** 130
tested by hydrogen loss testing **A7:** 246-247
in uranium dioxide fuel **A7:** 665
of water-atomized metal powders **A7:** 37
Oxygen content of
of copper alloy wirebar, controlling **A9:** 642
copper alloys, effects on microstructure **A9:** 405-406
Oxygen corrosion *See also* Oxygen
control, oil/gas production **A13:** 1246
dry, copper/copper alloy resistance **A13:** 632-633
from waterfloods, oil/gas wells **A13:** 482
oil/gas production **A13:** 1232
Oxygen cutter
definition **A6:** 1211 **M6:** 12
Oxygen cutting **A14:** 720-729
definition **M6:** 12
Oxygen cutting operator
definition **A6:** 1211 **M6:** 12
Oxygen gouging
definition **A6:** 1212 **M6:** 12
Oxygen grooving
definition **A6:** 1212
Oxygen in steel **M1:** 116
notch toughness, effect on **M1:** 694
steel sheet, effect on formability **M1:** 556
Oxygen index
polyether-imides (PEI) **EM2:** 157
Oxygen lance
definition **A6:** 1212 **M6:** 12
Oxygen lance cutting
definition **M6:** 12
Oxygen lance cutting (LOC)
definition **A6:** 1212
Oxygen lancing
definition **A6:** 1212
Oxygen, max
chemical compositions per ASTM specification B550-92 .... **A6:** 787
Oxygen plasma dry ashing
for sample dissolution **A10:** 167
Oxygen probes
atmospheres **M4:** 428-429
for monitoring endo gas in sintering atmospheres **A7:** 343
Oxygen quenching of fluorescence **A10:** 79-80
Oxygen sensors **EM4:** 1131-1138
applications **EM4:** 1131
challenges for future developments **EM4:** 1138
characteristics of sensor response ... **EM4:** 1135-1136
materials considerations and processing techniques **EM4:** 1136-1138
operational and environmental effects on sensor performance **EM4:** 1136
principle of operation **EM4:** 1131
semiconductor (conductimetric) sensors **EM4:** 1134-1135, 1136
solid electrolyte sensors **EM4:** 1131, 1134, 1135-1136, 1137, 1138
Oxygen-containing coppers
brazing **A6:** 931
Oxygen-flask combustion *See* Schöniger flask method
Oxygen-free copper *See* Copper alloys, specific types, C10100 and C10200; Copper alloys, specific types, C10200 .... **A6:** 752
dispersion-strengthened **A7:** 713
effect of impurities on conductivity **A7:** 106
electronic applications **A6:** 998
gas-metal arc welding **A6:** 759-760
thermal conductivity **A6:** 754
weldability **A6:** 753
Oxygen-free coppers *See also* Copper; Copper alloys; Wrought coppers and copper alloys
applications and properties **A2:** 265-268
characteristics **A2:** 230
Oxygen-free electronic copper *See* Copper alloys, specific types, C10100
applications and properties **A2:** 265
Oxygen-free extra-low-phosphorus copper *See* Copper alloys, specific types, C10300
applications and properties **A2:** 265
Oxygen-free halogen glasses
electrical properties **EM4:** 853
Oxygen-free high conductivity (OFHC) copper
as braze material for ceramic/metal seal **EM4:** 538
Oxygen-free high-conductivity copper
fatigue test fracture surfaces **A12:** 401
Oxygen-free low-phosphorus copper *See* Copper alloys, specific types, C10800
applications and properties **A2:** 268-269
Oxygen-free silver copper *See* Copper alloys, specific types, C10400, C10500 and C10700
applications and properties **A2:** 267-268
Oxygen-plasma chemical etching of carbon- carbon composites **A9:** 591
Oxygenation
of pure water **A8:** 421
Oxyhalide glasses
electrical properties **EM4:** 851, 852
structural role of components **EM4:** 845
structures **EM4:** 846
Oxyhydrogen cutting
definition **M6:** 12
Oxyhydrogen gas
torch brazing **M6:** 1039
Oxyhydrogen welding
definition **M6:** 12
Oxyhydrogen welding (OHW)
definition **A6:** 1212
Oxyhydroxide boehmite **EM3:** 262
Oxynatural gas **M6:** 901-903
torch brazing of copper **M6:** 1039
Oxynatural gas cutting
definition **M6:** 12
Oxynitride gels **EM4:** 210, 211
Oxynitride glasses **EM4:** 22
Oxyplex formation
in radiation curing **EL1:** 857
Oxypropane cutting
definition **M6:** 12
Oxypropane gas **M6:** 902-904
torch brazing of steel **M6:** 1039
Oxypropylene gas cutting **M6:** 903-905
Oxysulfides
machinability of carbon and alloy steels **A16:** 672, 675
OXYWEAR computer program **A18:** 280, 283, 285-286, 288
Ozone
defined **A13:** 10
fume generation from arc welding **A6:** 68
Ozone resistance
silicone-base coatings **EL1:** 773, 822
thermoplastic polyurethanes (TPUR) **EM2:** 206

P

***Pulvermetallurgie und Sinterwerkstoffe* (Kieffer and Hotop)** **A7:** 18-19
©Pendellosung© fringes **A18:** 386
P *See* Phosphorescence; Polarization
P grades, pressure pipe
composition **M1:** 324
tensile properties **M1:** 325
P polarization *See* Polarization
P-658RC (carbon-graphite)
properties **A18:** 549, 551
P-692
properties **A18:** 549
P-1700 thermoplastic **EM1:** 99
p-aramid fibers *See also* Aramid fibers; Aramid fibers, specific types; Para-aramid fibers
elevated-temperature tensile properties **EM1:** 55
polyethylene **EM1:** 54, 56
p-channel MOSFET
compatible **EL1:** 147
enhancement mode **EL1:** 158
P-channel transistors (PMOS)
development **EL1:** 160
p-charts *See also* Control charts; Quality control
for fraction defective **A17:** 735-737
***p*-cresol**
physical properties **EM3:** 104
P-D alloy *See* Titanium alloys, specific types, Ti-Pd alloys
P-F test, Shepherd
for tool steels **A12:** 141, 162
***p-n* junctions** *See also* Silicon *p-n* junction failures
in integrated circuits **A11:** 768
***p*-nonylphenol**
physical properties **EM3:** 104
P-Pd (Phase Diagram) .... **A3:** 2•330
P-Pr (Phase Diagram) .... **A3:** 2•330
P-Q^2 diagram
for metal volume flow **A15:** 290
P-Ru (Phase Diagram) .... **A3:** 2•331
P-Sn (Phase Diagram) .... **A3:** 2•331
P-T diagram
defined **A9:** 15
P-T-X diagram
defined **A9:** 15

p-tert butylphenol
physical properties **EM3:** 104
p-tert octylphenol
physical properties **EM3:** 104
P-Ti (Phase Diagram) .... **A3:** 2 • 331
P-type oxides
alloy oxidation **A13:** 73
defect structure **A13:** 65-66
impurities effect **A13:** 69
p-type transistors
development **EL1:** 958
P-X diagram
defined **A9:** 15
P-X projection
defined **A9:** 15
P-Zn (Phase Diagram) .... **A3:** 2 • 332
P/M
defined **A7:** 8
P/M aluminum alloys *See also* Aluminilm alloys; Aluminum; P/M aluminum alloys, specific types **A13:** 838-842
experimental **A12:** 440
fractographs **A12:** 440
fracture/failure causes illustrated **A12:** 217
high-performance classes **A13:** 839-842
P/M aluminum alloys, specific types
Al-4.2Mg-2.1Li, corrosion-fatigue cracking **A12:** 440
Al-4.2Mg-2.1Li, fracture along powder particle boundaries **A12:** 440
P/M forging *See* Forging **M1:** 339-346
P/M friction materials *See also* Copper P/M products; Powder metallurgy
defined **A2:** 398-400
P/M injection molding (MIM) process
for cermets **A2:** 984-985
P/M materials *See* Powder metallurgy alloys; Powder metallurgy materials
P/M molybdenum alloy
fatigue fracture probability curves **A8:** 253
threshold stress intensity **A8:** 256
P/M parts *See* Part(s); Powder metallurgy parts
P/M processing
superalloys **M3:** 214-216
P/M stainless steels *See* Sintered (porous) P/M stainless steels
P/M steels
alloying elements **M1:** 333, 337-339
ASTM designations **M1:** 330, 332, 333
blending **M1:** 331
carbon content **M1:** 333, 336-338, 340, 342
carbonitriding **M1:** 342, 343
compacting **M1:** 329, 331
composition **M1:** 333, 334-339
compressibility **M1:** 327-330
copper content **M1:** 333, 337, 340
density designations **M1:** 333
fatigue data **M1:** 334-335, 337, 342, 346
green strength **M1:** 330, 131, 345
heat treatment **M1:** 339-343, 343-346
hot forming **M1:** 339-346
infiltration **M1:** 339, 342, 343
maraging steels **M1:** 449
mechanical properties **M1:** 329, 332-346
modulus of elasticity **M1:** 334-335, 339, 340
MPIF designations **M1:** 330, 332, 333
nickel content **M1:** 333, 337, 342, 343
oxide content in powder, test for **M1:** 330
particle size **M1:** 328, 330
phosphorus content **M1:** 338
physical properties **M1:** 329, 333-337, 340, 343-346
porosity **M1:** 334-337, 343, 344
powder characteristics **M1:** 327-331
process capabilities **M1:** 327
re-pressing **M1:** 339, 343, 345
SAE designations **M1:** 332, 333
secondary operations **M1:** 331-332
sintering **M1:** 329, 330-331
steam treating
sulfur content **M1:** 338
tempering temperature, effect of **M1:** 344
testing **M1:** 327-331
P/M superalloys *See* Superalloys **A13:** 834-838
aerospace applications **A13:** 836
compositions **A13:** 837
cyclic oxidation resistance **A13:** 839
P/M Technology Newsletter **A7:** 19
P/M tool steel alloys *See* Tool steel alloys, specific types
P/M tool steels *See* Tool steels; Tool steels, specific types
P1, P2, P3, etc *See* Tool steels, specific types
PA *See* Polyaiiides; Polyamides
PA P/M technology *See* Prealloyed titanium P/M compacts/products
PAC 78 computer program for structural analysis **EM1:** 268, 272
Pacemaker, heart
as hybrid medical application **EL1:** 387
Pacific Ocean **A13:** 895, 897-898, 901
Pack (can)
for rolling titanium and nickel-base alloys **A14:** 356
Pack carburization
for case hardening **A7:** 453
Pack carburized steels **A9:** 224
Pack carburizing **A4:** 262, 263 **A18:** 873
advantages **A4:** 325 **M4:** 222
applications **A4:** 326 **M4:** 224
carbon potential **A4:** 325-326, 327 **M4:** 223
case depth **A4:** 325, 326, 328 **M4:** 224
compound selection **A4:** 325 **M4:** 223
compounds, carburizing **A4:** 325 **M4:** 222-223
containers **A4:** 325, 326, 327, 328 **M4:** 225
disadvantages **A4:** 325 **M4:** 222
distortion **A4:** 325, 326, 327 **M4:** 224
furnaces **A4:** 326-327 **M4:** 224-225
packing **A4:** 327-328 **M4:** 225-226
process **A4:** 325, 327 **M4:** 222-223
selective carburizing **A4:** 328 **M4:** 223
steel, effect of composition **A4:** 326 **M4:** 223
temperature **A4:** 325, 326 **M4:** 223
time, relation to case depth **A4:** 325, 326 **M4:** 223, 224
work-load densities **A4:** 327
Pack cementation
aluminum coating **M5:** 340
ceramic coating **M5:** 542-544
chromium coating **M5:** 383
equipment **M5:** 543-544
high-pressure, oxidation-resistant coating **M5:** 664-666
oxidation protective coatings **M5:** 377-378, 380
oxidation-resistant coatings **M5:** 664-666
process steps **M5:** 543
surface preparation for **M5:** 542-543
Pack diffusion
aluminum coating, steel **M5:** 340-341
colorizing **M5:** 340
oxidation protective coatings **M5:** 377, 380
reaction agents and products **M5:** 340
Pack mounting of aluminum alloys **A9:** 352
Package
defined **EM1:** 17
forming glass fiber **EM1:** 108
Package design
connectors **EL1:** 23
electronic **EL1:** 25-26
finite-element, software **EL1:** 954
Package failures
hermetic and nonhermetic **A11:** 767
integrated circuits **A11:** 766
plastic, integrated circuit chip in **A11:** 767
Package families
level 1 packages **EL1:** 403-404
Package forms
cavity packages **EL1:** 452
flatpacks **EL1:** 451
packages without leads **EL1:** 452
plug-in **EL1:** 451-452
Package moisture content analysis **A13:** 1117
as advanced failure analysis technique **EL1:** 1106-1107
Package parasitics **EL1:** 952
effects of **EL1:** 18
lumped vs distributed view of **EL1:** 418-419
Package reliability *See also* Reliability
environmental factors **EL1:** 46, 65-67
mechanical factors **EL1:** 45-46, 55-65
thermal factors **EL1:** 45-55
Package sealing
by brazing **EL1:** 237-239
by polymer coating/encapsulation **EL1:** 239-243
Package sealing (continued)
by soldering **EL1:** 237-239
by welding **EL1:** 237-238
and decapsulation **EL1:** 243
methods **EL1:** 237-243
and passivation coatings **EL1:** 237-248
Package(s)
application of composites in **EL1:** 1126-1128
capacitance **EL1:** 7
cavity **EL1:** 448, 452
cavity, solutions for **EL1:** 448
ceramic **EL1:** 203-206, 454
chip-level, elimination of **EL1:** 407
construction elements **EL1:** 471
cracking, and stress **EL1:** 480
crystal, metal can as **EL1:** 979
defined **EL1:** 1152
development **EL1:** 446
differences, and failure mechanisms **EL1:** 961-962
discrete semiconductor **EL1:** 422-435
electrical effect **EL1:** 76
final design **EL1:** 523-526
flatpacks **EL1:** 451
hermetic **EL1:** 213-217, 453
hermeticity, and gas analysis **EL1:** 1062-1066
hybrid, forms of **EL1:** 451-452
integrated circuit, material properties **EL1:** 207
introduction **EL1:** 397
key-parameter trends **EL1:** 409
level 1, design of **EL1:** 401-403
as limiting high-performance systems **EL1:** 39
metal **EL1:** 237-238, 453-454
with metal-matrix composites **EL1:** 1126-1128
microwave **EL1:** 453
options, high-frequency digital systems **EL1:** 76-77
optoelectronic circuits **EL1:** 453
performance ranking **EL1:** 404
physical size, reduced **EL1:** 25
plastic **EL1:** 209-212, 245-246, 454, 471
plastic dual-in-line (PDIP) **EL1:** 210
plastic leaded chip carrier (PLCC) **EL1:** 209
plastic, materials in **EL1:** 210-212
plastic pin-grid array (PGA) **EL1:** 210
plastic quad flatpack **EL1:** 209-210
plastic-body **EL1:** 427-428
plastic-encapsulated **EL1:** 217-221
plug-in **EL1:** 451-452
power hybrid **EL1:** 452-453
radio-frequency (RF) **EL1:** 428-429
sealing methods **EL1:** 237-243
second-layer, materials and processes selection **EL1:** 113-117
side-brazed, outline **EL1:** 205
single-in-line (SIP) **EL1:** 210
small-outline **EL1:** 209
small-signal **EL1:** 422-424, 452
substrates and **EL1:** 203-212
thermal expansion mismatch problem **EL1:** 611
thermal resistance **EL1:** 409, 412-413
thick-film hybrid technology **EL1:** 206-208
visual examination of **EL1:** 1058
without leads, as hybrid package form **EL1:** 452
Package-level physical test methods
analytical considerations **EL1:** 928
data interpretation **EL1:** 928
die attach tests **EL1:** 931-934
environmental testing **EL1:** 935-938
equipment and techniques **EL1:** 927-928
fine and gross hermeticity tests **EL1:** 929-930
particle impact noise detection (PIND) **EL1:** 930-931
and residual gas analysis **EL1:** 927
wire bond strength tests **EL1:** 934-935
Packaging *See also* Electronic packaging; Mechanical packaging; Packages **EM3:** 45
in active component fabrication **EL1:** 199
alternatives, second-level packages **EL1:** 116
aluminum and aluminum alloys **A2:** 10
aluminum foil, vapor barrier penetration **A13:** 108
applications, steel wire **M1:** 264
architecture, design/materials of **EL1:** 1
architecture/system, design and materials **EL1:** 1
beryllium **A13:** 810

Packaging (continued)
bipolar junction transistor technology **EL1:** 195-196
chip-level, overview of **EL1:** 398-407
costs of **EM2:** 650
customized **EL1:** 440-441
defects, through-hole packages **EL1:** 979
defined **EL1:** 398
dense VLSI **EL1:** 269-270
of digital ICs, future trends **EL1:** 177
digital integrated circuits **EL1:** 172-174
elimination of **EL1:** 449
failure mechanisms **EL1:** 957
fields of **EL1:** 398
filiform corrosion **A13:** 107
future trends **EL1:** 45, 390-396
goals and purposes **EL1:** 18, 449
hierarchical levels, defined **EL1:** 12-13, 397
of hybrid microcircuits **EL1:** 259-260
inhibitor applications **A13:** 525
inhibitors used in **M1:** 757
innovative solutions **EL1:** 448
integrated circuit, automatic optical inspection **EL1:** 941
integrated circuit chip **A11:** 775
interconnect (P/1) structure, thermal expansion **EL1:** 611
levels of **EL1:** 12-13, 397
materials **EL1:** 1041
materials, integrated circuit, analyses of **A11:** 783
mechanical, factors **EL1:** 21
military, future trends **EL1:** 392-393
molybdenum powders **A7:** 156
polymide coatings in **EL1:** 770-771
for quality **EL1:** 441-448
-related failure mechanisms **EL1:** 974
reliability **EL1:** 127
requirements, modeling **EL1:** 15-16
selection **EL1:** 128
semiconductor, defined **EL1:** 397
of silicon chips, as interconnection level **EL1:** 13
system-level, multichip **EL1:** 2
thermal expansion mismatch problem **EL1:** 611
thermal stress in **EL1:** 56-59
three-dimensional **EL1:** 441
trends **EL1:** 45, 390-396
of tungsten powders **A7:** 154
wafer-scale integration **EL1:** 270, 272-273
Packaging **(formerly *Modern Packaging*)** **EM3:** 66
Packaging and container applications
wire for **A1:** 282
Packaging applications *See* Electronic packaging applications
high-impact polystyrenes (PS, HIPS) **EM2:** 194-195
of parts design **EM2:** 616-617
polyether-imides (PEI) **EM2:** 156
properties of **EM2:** 457
styrene-acrylonitriles (SAN, OSA, ASA) **EM2:** 215
Packaging density
analysis **EL1:** 13-14
by through-hole vs surface mount technologies **EL1:** 730
defined **EL1:** 1152
factors **EL1:** 438-440
polymer matrix composites **EL1:** 1117-1118
wiring capacity **EL1:** 13-14
Packaging materials *See also* Materials; Materials analysis; Materials and electronic phenomena; Materials selection **EL1:** 104
composites **EL1:** 1122-1125
CTEs for **EL1:** 58
dielectric constants of **EL1:** 21
Packed density *See also* Density **A7:** 8
of ceramic powders **A7:** 297
increasing **A7:** 296
Packed density (continued)
types of **A7:** 297
Packing *See also* Packaging; Packed density **A7:** 8, 296-297
in HIP processing **A7:** 426
into metal cans **A7:** 517
Packing density **A7:** 8, 426
in HIP processing **A7:** 426
Packing material *See also* Packaging; Packed density; Packing **A7:** 8
Packing out **EM3:** 718-719
Packless-type hydraulic testing machine **A8:** 613
Packout rust formation
weathering steels **A13:** 519
Pad
defined **A14:** 9
Pad lubrication
defined **A18:** 14
Paddle mixer **A7:** 8
Pads **A18:** 569
capacitance, chip edge effects **EL1:** 7
design of **EL1:** 520
printed board coupons **EL1:** 576
PAEK *See* Polyaryletherketone; Polyaryletherketones
PAFEC computer program for structural analysis **EM1:** 268, 272
Pagers
as telecommunication hybrid application **EL1:** 383
PAI *See* Polyamide-imide; Polyamide-imide resins
Pai-Thong (white copper)
as early nickel alloy **A2:** 429
Paint
14 KRS-5 ATR spectrum **A10:** 121
adherence, effect of surface carbons on **A10:** 224
automobile, NAA forensic studies of **A10:** 233
dissolved in methylene chloride solvent, ATR analysis **A10:** 121
for gage marks **A8:** 548
potassium, calcium, and titanium determination in **A10:** 98
strippers, hydrogen embrittlement testing for **A8:** 541
temperature-sensitive, for ultrasonic testing **A8:** 247
Paint adhesion
on a scribed surface test **A13:** 220
Paint and painting **M5:** 471-508
abrasion resistance **M5:** 490-491
abrasive blasting process, wire brushing compared with **M5:** 476
acrylic resins and coatings **M5:** 473-474, 498, 500, 504
adhesion **M5:** 490, 492
advantages and limitations of, various systems **M5:** 471-472, 483-484
air-oxidizing coatings **M5:** 500-501
alkyd resins and coatings **M5:** 473-474, 495, 498-499, 505
modified **M5:** 475, 498-499
aluminum and aluminum alloys **M5:** 17, 19, 457, 606-607, 609
pretreatment for **M5:** 17, 457
shipping of **M5:** 19
application viscosity and efficiency **M5:** 493-494
applying, methods of *See also* specific methods by name **M5:** 474, 477-486, 492, 502
limitations **M5:** 474
autophoretic paints **M5:** 472
baking *See* Paint and painting, curing methods
beading **M5:** 493
blistering **M5:** 489, 491, 493, 495
brittleness **M5:** 489, 493
brushing and rolling processes **M5:** 489, 493
equipment **M5:** 503
checking **M5:** 493
chemical resistance **M5:** 474-475
Paint and painting (continued)
chemically reactive paints **M5:** 500-501
chlorinated rubber resins and coatings **M5:** 473-475, 495, 498, 500-502, 505
coal tar resins and coatings **M5:** 495, 500-502, 504-505
color defects **M5:** 489, 493
color testing **M5:** 490-491, 499
composition and characteristics **M5:** 497-500
corrosion protection **M5:** 474-476, 491, 500-501, 504
cost factors **M5:** 474, 494, 504
coverage, calculating **M5:** 494
cracking and cratering **M5:** 493
curing methods, paint films **M5:** 479, 486-489, 505-506
batch ovens, direct- and indirect-fired **M5:** 487
coatings classified by **M5:** 500-501
continuous ovens, convection and radiant **M5:** 487-488
convection baking time and temperature **M5:** 479, 488
defects attributable to improper baking and corrections **M5:** 488-489
equipment other than ovens **M5:** 488-489
heat recovery **M5:** 488
high velocity ovens **M5:** 488
prebake solvent evaporation **M5:** 488
curtain coating process **M5:** 482-483, 494
equipment **M5:** 482-483
defects, paint films **M5:** 488-489, 492-493
dipping process **M5:** 480, 494
agitation used in **M5:** 480
equipment **M5:** 481-482, 503
dry film tests **M5:** 490-491
durability, film **M5:** 476-477
electrocoating process **M5:** 483-484, 494
cathodic systems **M5:** 474, 484
equipment **M5:** 484-485
resins used **M5:** 483
electrophoretic paints **M5:** 472
elongation properties, film **M5:** 491
enamel *See also* Porcelain enameling **M5:** 495, 501, 503
environmental precautions **M5:** 472-473, 491, 494
epoxy resins and coatings **M5:** 498-499, 501-505
equipment
batch ovens, direct- and indirect-fired **M5:** 487-488
brushing and rolling **M5:** 503
continuous ovens, convection and radiant **M5:** 487-488
curing ovens and other equipment **M5:** 487-489
curtain coating **M5:** 473, 482-483
dipping **M5:** 480-487
electrocoating **M5:** 484-485
flow coating **M5:** 481-482
powder coating **M5:** 484-486
roller coating **M5:** 482
spraying **M5:** 479-480
exterior-exposure tests **M5:** 491
flow coating process **M5:** 481-482, 494
equipment **M5:** 481-482
fluorocarbon resins **M5:** 473
functions, basic **M5:** 473-475
gloss defects **M5:** 489, 493
gloss testing **M5:** 490-491
glossary of terms **M5:** 495-497
hardness, film **M5:** 490, 492
high-solids paints **M5:** 472
hot dip galvanized coatings painting over **M5:** 331-332
impact resistance **M5:** 492
lacquer *See* Lacquer and lacquering
linseed oil **M5:** 498
magnesium alloys **M5:** 629, 647-648
maintenance program **M5:** 507
marine atmospheres **M5:** 474, 492

SUBJECTS OF THE INDEXED VOLUMES: ASM Handbook (designated by the letter "A"): **A1:** Properties and Selection: Irons, Steels, and High-Performance Alloys (1990); **A2:** Properties and Selection: Nonferrous Alloys and Special-Purpose Materials (1990); **A3:** Alloy Phase Diagrams (1992); **A4:** Heat Treating (1991); **A6:** Welding, Brazing, and Soldering (1993); **A7:** Powder Metallurgy (1984); **A8:** Mechanical Testing (1985); **A9:** Metallography and Microstructures (1985); **A10:** Materials Characterization (1986); **A11:** Failure Analysis and Prevention (1986); **A12:** Fractography (1987); **A13:** Corrosion (1987); **A14:** Forming and Forging (1988); **A15:** Casting (1988); **A16:** Machining (1989); **A17:** Nondestructive Testing and Quality Control (1989); **A18:** Friction, Lubrication, and Wear Technology (1992). **Metals Handbook, 9th Edition** (designated by the letter "M"): **M1:** Properties and Selection: Irons and Steels (1978); **M2:** Properties and Selection: Nonferrous Alloys and Pure Metals (1979); **M3:** Properties and Selection: Stainless Steels, Tool Materials and Special-Purpose Materials (1980); **M4:** Heat Treating (1981); **M5:** Surface Cleaning, Finishing, and Coating (1982); **M6:** Welding, Brazing, and Soldering (1983). **Engineered Materials Handbook** (designated by the letters "EM"): **EM1:** Composites (1987); **EM2:** Engineering Plastics (1988); **EM3:** Adhesives and Sealants (1990); **EM4:** Ceramics and Glasses (1991); **Electronic Materials Handbook** (designated by the letters "EL"): **EL1:** Packaging (1989).

Paint and painting (continued)
matte surfaces, exposed ... **M5:** 332
mechanical resistance ... **M5:** 474-475
multiple layer application, compatability parameters ... **M5:** 500-501
nitrocellulose resins and coatings ... **M5:** 473-503
oleoresinous coatings ... **M5:** 498-499, 501, 503, 505
phenolic resins and coatings ... **M5:** 473-474, 478, 498-499, 501
phosphate coating bases for ... **M5:** 434-436, 442, 448-450, 454-456, 476-478
pigments, contamination by, vapor degreasing solvents ... **M5:** 48
pinholing ... **M5:** 489, 493
polyester resins and coatings ... **M5:** 478, 483, 496, 502
polyurethane resins ... **M5:** 473, 498-500, 503
powder coating process ... **M5:** 485-486, 494
electrostatic spray application ... **M5:** 485
equipment ... **M5:** 485-486
fluidized bed methods, conventional and electrostatic ... **M5:** 485-486
powder paint ... **M5:** 472
prepaint treatments ... **M5:** 476, 478
primer *See* Primer
quality control ... **M5:** 489-492, 505-507
radiation cure coatings ... **M5:** 486
regulations governing ... **M5:** 472, 494
resins ... **M5:** 472-473, 483, 495-500, 503
roller coating (machine) process ... **M5:** 482, 494
equipment ... **M5:** 482
rust, effects of ... **M5:** 473, 499, 503
pictorial representation of rust classification ... **M5:** 499, 503
safety precautions ... **M5:** 472, 494
sagging ... **M5:** 493
sampling and testing, coating material, preapplication ... **M5:** 505
service environment effects of ... **M5:** 472-473
silicone alkyd coatings ... **M5:** 474, 500-501
silicone resins and coatings ... **M5:** 473-475
spraying process ... **M5:** 477-480, 485-486, 494
airless ... **M5:** 478-494
electrostatic ... **M5:** 478-480, 485-486, 494
equipment ... **M5:** 478-479
hot ... **M5:** 477
spreading rate ... **M5:** 494
steel ... **M5:** 474-477, 481, 499, 504-505
stripping of ... **M5:** 19
storage, materials ... **M5:** 505-506
stripping methods ... **M5:** 18-19, 648
abrasive ... **M5:** 18-19, 648
brush on, wipe on, or squirt on ... **M5:** 18-19
immersion ... **M5:** 18-19
spray ... **M5:** 18-19
structural design, workpiece effects of ... **M5:** 476, 504-505
substrate and surface conditions effects of ... **M5:** 473
surface preparation ... **M5:** 5, 17, 332, 476, 478, 503-505
inspection procedures ... **M5:** 499, 503
methods, summary of ... **M5:** 505-506
smoothness parameters ... **M5:** 476
system selection ... **M5:** 472-475, 504
examples of ... **M5:** 474-475
temperature ... **M5:** 482, 488-490
terminology, glossary of ... **M5:** 495-497
tests ... **M5:** 489-492, 505-507
coating materials, preapplication ... **M5:** 505
coatings, monitoring and evaluating ... **M5:** 489-492
thickness, coat ... **M5:** 483, 489-491, 493
dry film, testing ... **M5:** 490-491
wet film, testing ... **M5:** 492
tumble coating process ... **M5:** 500
two-component coatings ... **M5:** 472, 500-502
types of paint used ... **M5:** 471-472, 500-503
urethane resins and coatings ... **M5:** 474, 498, 502-505
varnish ... **M5:** 497-499, 503
vinyl resins and coatings ... **M5:** 474-475, 497-498, 500-502, 505
viscosity ... **M5:** 482, 489-490, 493-494
volume solids content, coating ... **M5:** 494
water-borne paints ... **M5:** 471-472, 500-501
weathering tests, artificial ... **M5:** 491

Paint and painting (continued)
wet film tests ... **M5:** 490, 492
wire brushing process, abrasive blasting compared with ... **M5:** 476
wrinkling ... **M5:** 489, 493
zinc coatings
inorganic ... **M5:** 475, 501-505
organic ... **M5:** 502, 505
Paint base
aluminum coating, suitability ... **M1:** 173
galvanized steel for ... **M1:** 169, 170
long terne sheet ... **M1:** 174
phosphate coating for ... **M1:** 174-175
preprimed steel sheet ... **M1:** 175
Paint industry applications
quality control design ... **A17:** 733-734
Paint mitts
for paint application ... **A13:** 415
Paint systems ... **EM3:** 640-641
coating system selection ... **EM3:** 640-641
failure ... **EM3:** 640
formulation
function ... **EM3:** 640
moisture effect ... **EM3:** 640-641
top coats ... **EM3:** 41
types ... **EM3:** 640
Paint(s)
acrylic ... **A13:** 400, 404-406
adhesion ... **A13:** 220, 389, 394, 442-443
for automotive industry ... **A13:** 1015-1017
cleaning for ... **A13:** 414-415
as coating ... **A7:** 459-460
conductive copper powder ... **A7:** 105
copper and gold bronze pigments for ... **A7:** 595
dipping ... **A7:** 460
exposure tests ... **A13:** 522
films, corrosion protection ... **A13:** 528
multiple-coat, for filiform corrosion ... **A13:** 108
oil, Portland cement in ... **A13:** 443
pre-, processing ... **A13:** 528
as protective film ... **A13:** 400
spraying, atomization mechanism ... **A7:** 27
water-base acrylic ... **A13:** 405-406
weakening, phosphate coatings ... **A13:** 385
zinc chromate ... **A13:** 769
zinc dust-zinc oxide ... **A13:** 443
zinc-bearing ... **A13:** 768-769
Paint-marking systems
Orbitest machine ... **A17:** 185
Paintability
phosphate coatings ... **A13:** 385
Paintbrush transducer contact-type ultrasonic search units ... **A17:** 258
Painted weathering steels ... **A13:** 519-520
Painting
design effects ... **A13:** 340
galvanized steel ... **A13:** 442
of shafts ... **A11:** 459
spray, coating application ... **A13:** 415-416
zinc alloys ... **A2:** 530
Painting, magnetic *See* Magnetic painting
Paints
atmospheric corrosion resistance enhanced by ... **M1:** 722
corrosion protection ... **M1:** 751, 754, 755
definition and classification of types ... **M1:** 105-106
seawater corrosion resistance enhanced by ... **M1:** 745
Pair distribution functions
separation of RDF into ... **A10:** 397-398
Pair production, as attenuation process
radiography ... **A17:** 309
Palavital ... **EM4:** 1008
Palco
brazing, composition ... **A6:** 117
wettability indices on stainless steel base metals ... **A6:** 118
Palcusil 5
brazing, composition ... **A6:** 117
wettability indices on stainless steel base metals ... **A6:** 118
Palcusil 10
brazing, composition ... **A6:** 117
wettability indices on stainless steel base metals ... **A6:** 118
Palcusil 15
brazing, composition ... **A6:** 117

Palcusil 15 (continued)
wettability indices on stainless steel base metals ... **A6:** 118
Palcusil 20
brazing, composition ... **A6:** 117
wettability indices on stainless steel base metals ... **A6:** 118
Palcusil 25
brazing, composition ... **A6:** 117
wettability indices on stainless steel base metals ... **A6:** 118
Palette(s)
for color images ... **A17:** 485
control, digital image enhancement ... **A17:** 457-458
Palisades
broken graphite fibers as ... **A12:** 465
Palladium *See also* Palladium alloys; Palladium powders; Precious metals; Pure metals ... **A13:** 799-801
as a conductive coating for scanning electron microscopy specimens ... **A9:** 97
as a reactive sputtering cathode material ... **A9:** 60
Ag-4Pd brazed interlayers, solid-state welding ... **A6:** 166
alloying effect in titanium alloys ... **A6:** 508
alloying effects ... **A13:** 47, 799-800
annealing ... **M4:** 760, 761, 762
applications ... **A2:** 714-715
as braze filler metal ... **A11:** 450
in clad and electroplated contacts ... **A2:** 848
commercially pure ... **M2:** 699-701
as conductive thick-film material ... **EL1:** 249
corrosion application ... **A13:** 800-801
corrosion resistance ... **A13:** 799 **M2:** 669
determined by controlled-potential coulometry ... **A10:** 209
electrical circuits for electropolishing ... **A9:** 49
electrical contact materials ... **A13:** 805
for electrical contacts ... **A2:** 847 **A9:** 563-564
electroplating properties ... **A18:** 838
etchants for ... **A9:** 551
evaporation fields for ... **A10:** 587
fabrication ... **A13:** 799
gravimetric finishes ... **A10:** 171
in liquid-phase metallizing ... **EM3:** 306
low-melting temperature indium-base solder compatibility ... **A2:** 752
materials for conductors ... **EM4:** 1141
in medical therapy, toxic effects ... **A2:** 1258
in metal powder-glass frit method ... **EM3:** 305
for metallizing ... **EM4:** 542, 544
organic precipitant for ... **A10:** 169
oxidation resistance ... **A13:** 800
photometric analysis methods ... **A10:** 64
for plating, materials and processes selection ... **EL1:** 116
as precious metal ... **A2:** 688
properties ... **A2:** 692, 694, 714-718, 847, 1146 **A13:** 799
providing absorption for architectural glass ... **EM4:** 450
pure ... **M2:** 780-781
pure, properties ... **A2:** 1146
as pyrophoric ... **A7:** 199
relative solderability as a function of flux type ... **A6:** 129
resources and consumption ... **A2:** 689-690
semifinished products ... **A2:** 694
single-phase solid solution, recrystallized grains in ... **A9:** 126
solderability ... **A6:** 978
soldering ... **A6:** 631
special properties ... **A2:** 692, 694
substituted for gold in electroplating ... **A18:** 837
support of SERS in vacuum ... **A10:** 136
thermal diffusivity from 20 to 100 °C ... **A6:** 4
thermal expansion coefficient ... **A6:** 907
toxicity ... **A7:** 207
ultra pure, by fractional crystallization ... **A2:** 1093
ultrasonic welding ... **M6:** 746
vapor pressure, relation to temperature ... **M4:** 310
in vapor-phase metallizing ... **EM3:** 306
weighed as the dimethylglyoxime complex ... **A10:** 171
wire ... **A9:** 564

Palladium (continued)
wire, temperature effects on tensile strength **A13:** 804
working of **A14:** 850
Palladium alloys *See also* Palladium; Precious metals; Titanium alloys, specific types, Ti-Pd alloys
applications **A2:** 716-718
brazing
available product form of filler metals **A6:** 119
brazing, joining temperatures **A6:** 118
electrical contacts, use in **M3:** 669-671, 672
etchants for **A9:** 551
properties **A2:** 714-718
recommended gap for braze filler metals **A6:** 120
Palladium alloys, specific types *See also* Palladium; Palladium alloys
35Pd-10Pt-10Au-30Ag-14Cu-1Zn **A9:** 564
40Ag-30Pd-30Au, properties **A2:** 717-718
50Pd-50Ag **A9:** 564
60Pd-40Cu **A9:** 564
60Pd-40Cu, properties **A2:** 717
95.5Pd-4.5Ru, properties **A2:** 718
95Pd-5Ru **A9:** 564
Pd-2.1Be
in argon and vacuum atmospheres **A6:** 116
contact angles on beryllium at various test temperatures
wetting of beryllium **A6:** 115
Palladium and palladium alloys
annealing **A4:** 944, 945, 946
applications **A4:** 946
hardness **A4:** 944, 945, 946
mechanical properties **A4:** 946
tensile strength **A4:** 944, 945
vapor pressure of palladium, relation to temperature **A4:** 495
Palladium chloride
toxic effects **A2:** 1258
Palladium foil hydrogen probe **A13:** 201
Palladium powders *See also* Palladium; Precious metal powders
applications **A7:** 150
effect as activator in tungsten sintering **A7:** 318
history **A7:** 15
partlcles, small-sized **A7:** 150
production of **A7:** 148, 150-151
reducing agents **A7:** 150
specific surface area **A7:** 151
Palladium-barrier
detector, for gas/leak measurement **A17:** 64
gages, for vacuum leak testing **A17:** 68
Palladium-copper alloy **M2:** 702-703
Palladium-gold
coating for TEM specimens **A18:** 382
Palladium-gold conductor inks **EL1:** 208
Palladium-nickel
electroplating properties **A18:** 838
substituted for gold in electroplating **A18:** 837
Palladium-nickel alloys
for plating, materials and processes select-on **EL1:** 116
usage, PWB manufacturing **EL1:** 510
Palladium-ruthenium alloy **M2:** 704-705
Palladium-ruthenium alloys
as electrical contact materials **A2:** 847
Palladium-shadowed plastic-carbon replica **A12:** 185
Palladium-silicon (Pd-Si) **EM4:** 22
Palladium-silver
substituted for gold in electroplating **A18:** 837
Palladium-silver alloys **M2:** 701-702
PFM **A13:** 1355
properties **A2:** 716-717
Palladium-silver conductor inks **EL1:** 208
Palladium-silver powders
thick-film **A7:** 151
Palladium-silver-copper alloys **M2:** 703
properties **A2:** 717
Palladium-silver-copper alloys, brazing
available product forms of filler metals **A6:** 119
joining temperatures **A6:** 118
Palladium-silver-gold alloys **M2:** 703-704
properties **A2:** 717-718
Palmansil 5, wettability
indices on stainless steel base metals **A6:** 118
Palmgren-Miner rule
for lifetime evaluation **EM1:** 203
Palmqvist crack model equation **EM4:** 601
Palni
brazing, composition **A6:** 117
wettability indices on stainless steel base metals **A6:** 118
Palnicusil
brazing, composition **A6:** 117
wettability indices on stainless steel base metals **A6:** 118
Palniro 1
brazing, composition **A6:** 117
wettability indices on stainless steel base metals **A6:** 118
Palniro 4
brazing, composition **A6:** 117
wettability indices on stainless steel base metals **A6:** 118
Palniro 7
brazing, composition **A6:** 117
wettability indices on stainless steel base metals **A6:** 118
Palsil 10
brazing, composition **A6:** 117
wettability indices on stainless steel base metals **A6:** 118
PAN *See* Polyacrylonitrile
as metallochromic indicator **A10:** 174
Pancake coil
eddy current inspection **A17:** 172
Pancake forging **A14:** 9, 61, 73
Pancake grain structure
defined **A9:** 13
Panchromatic photographic films **A9:** 85-86
Panel
defined **EL1:** 1152
layout, rigid printed wiring boards **EL1:** 540-541
and natural convection IR systems, infrared soldering **EL1:** 705
plating, vs pattern plating, rigid printed wiring boards **EL1:** 540
Panel cracking
and aluminum nitride embrittlement **A1:** 695
Panel shear testing
polybenzimidazoles **EM3:** 170
Panoramic radiographic technique
for curved plate **A17:** 332
Pantograph
optical comparators with **A17:** 11
Pantographs **A6:** 1169
Paper
acidity-basicity measured in **A10:** 172
aramid fiber (Nomex) **EM1:** 115
neutron radiography of **A17:** 392
radiographic **A17:** 314
water-soluble, for leak detection **A17:** 66
Paper and printing industry applications *See also* Pulp and paper industry applications
aluminum and aluminum alloys **A2:** 14
structural ceramics **A2:** 1019
Paper and pulp industry **A13:** 1186-1220
Paper, blanking
die materials for **M3:** 485, 487
Paper chromatography
as qualitative separation technique **A10:** 168
Paper clip wire **A1:** 286 **M1:** 269
Paper coatings
powders used **A7:** 574
Paper, copier
friction coefficient data **A18:** 75
Paper copies
plain and coated **A7:** 584
Paper, fiberglass *See* Fiberglass paper
Paper, Film and Foil Converter **EM3:** 66
Paper industry applications
silicones (SI) **EM2:** 266
vinyl esters **EM2:** 272
Paper machine corrosion **A13:** 1186-1190
machine components **A13:** 1186-1188
mechanisms **A13:** 1189-1190
white water **A13:** 1188-1189
Paper machine dryer rolls
shell and head cracking **A11:** 653-654
Paper radiography *See also* Radiographic inspection; Radiography
defined **A17:** 295
Paper stuffing
for coiled metals **A14:** 710
Paper-and-oil insulation
for wrought copper and copper alloy products **A2:** 260
Paper-base film *See* Radiographic paper
Paper-dryer head
surface discontinuities in **A11:** 352-354
Paperback strain gages **A8:** 201
Papermaking
constituents of powder formulation **EM4:** 126
mechanical consolidation **EM4:** 125, 127-128
Papers, background
for photomacrography **A12:** 78
PAR *See* Polyarylates
as metallochromic indicator **A10:** 174
Para-aramid
for brake linings **A18:** 569, 570, 576
Para-aramid fibers *See also* Aramid fibers; p-aramid fibers **EM1:** 54
chemical structure **EM1:** 54
importance **EM1:** 43
polymer chain orientation **EM1:** 54
stress-strain behavior **EM1:** 55
thermal properties **EM1:** 56
Parabens
as antimicrobial agents **A10:** 655
liquid chromatography analysis in baby lotion **A10:** 655-656
Parabolic kinetics **A13:** 17, 67
Parabolic markings
on fracture surface **EM2:** 810
Parabolic oxidation **A18:** 208
kinetics **A13:** 67
Parabolic peak location method
x-ray diffraction residual stress techniques **A10:** 386
Parabolic rate constant **A18:** 208
Parabolic reflector **EM3:** 20
Paraffin
burn-off **A7:** 351, 352
-impregnated sintered iron driving bands **A7:** 17
Paraffin waxes (with modifiers)
as binder for ceramic injection molding **EM4:** 173
Paralinear oxidation **A13:** 70-71
Parallax
determination, by stereo imaging **A12:** 197
excessive, in SEM imaging **A12:** 167
Parallel
with option for signal vias **EL1:** 112
with power reinforcement **EL1:** 113
process flow **EL1:** 136
Parallel duplex microstructure; conditions for **A9:** 618
Parallel fiber reinforced ring *See* NOL ring
Parallel gap explosive bonding technique **A6:** 161, 162
Parallel laminate *See also* Laminate(s) **EM3:** 20
defined **EM1:** 17 **EM2:** 29

Parallel lighting
photomacrographic ... **A12:** 83
Parallel linear test grids
used in quantitative metallography ... **A9:** 124-126
Parallel mixing rule ... **EM4:** 859
Parallel moiré fringes ... **A9:** 110
Parallel plate geometries
in melt rheology ... **EM2:** 535-540
Parallel processing, of whole image
machine vision process ... **A17:** 44
Parallel rods in unidirectionally solidified eutectics
in Mg-32A1, transverse section ... **A9:** 128
quantitative metallography of ... **A9:** 127
Parallel seam welding
of flatpack ... **EL1:** 240
Parallel signal lines
capacitive loading ... **EL1:** 37-39
design considerations ... **EL1:** 34-41
simulation procedure ... **EL1:** 35
Parallel termination
defined ... **EL1:** 522
for reflection ... **EL1:** 170-171
Parallel welding
definition ... **A6:** 1212
Parallel-axes photography method ... **A12:** 88
Parallel-plate plasma-assisted chemical vapor deposition ... **EM3:** 593
Parallel-rail straightening ... **A14:** 688-689
Parallel-roll straightening ... **A14:** 684-686, 691
Parallelism
and press accuracy ... **A14:** 495
Parallelism gage
for adhesive bonded assemblies ... **A7:** 457
Paramagnetic centers and properties
ESR characterized ... **A10:** 253, 263
Paramagnetic materials
defined ... **A9:** 63
Paramagnetic resonance *See* Electron spin resonance
Paramagnetism ... **A10:** 257, 678
Parameter
defined ... **A8:** 10
estimating unknown ... **A8:** 628-637
experimental, defined ... **A8:** 639
plot, for stress-rupture data ... **A8:** 315
Parameter (in crystals) *See* Lattice parameter
Parameter design
analytical approaches ... **A17:** 750-751
experimental approaches ... **A17:** 751-752
stage, quality design ... **A17:** 722
Parameter estimates
binomial distribution ... **A8:** 636
exponential distribution ... **A8:** 635
normal distribution ... **A8:** 629-631
Poisson distribution ... **A8:** 637
of statistical distributions ... **A8:** 628, 629
Weibull distribution ... **A8:** 633
Parameter estimation
NDE reliability data analysis ... **A17:** 695-696
Parameters
defining background, RDF analysis ... **A10:** 396
short distance, RDF analysis ... **A10:** 351
standard definitions/symbols ... **A13:** 369
Parameters, profile and surface roughness
fractal analysis ... **A12:** 212
Parametric analysis
of creep-rupture ... **A8:** 690
of differences between two groups of fatigue data ... **A8:** 707-708
Parametric relationships
fracture surface and projected images ... **A12:** 202
profile and surface roughness, equation ... **A12:** 212
true area/length, plotted ... **A12:** 204
Parasitic corrosion
aluminum/air batteries ... **A13:** 1319-1320
Parasitic devices
in integrated circuits ... **A11:** 785
Parasitic peaks *See* Escape peaks
Parasitics, package *See* Analog parasitic,effects; Package parasitics
Paraxylyiene
as conformal coating ... **EL1:** 761
Parent ion scans
gas chromatography/mass spectrometry ... **A10:** 646
Parent metal
definition ... **A6:** 1212
Parent population
defined ... **A10:** 12
Parent-daughter relationship
radioactive isotope decay ... **A10:** 244
Parfocal eyepiece
defined ... **A9:** 13
Parfocal lens systems ... **A9:** 73
Pargonite mica ... **EM4:** 6
Paris equation ... **A1:** 663 **A11:** 52, 53
Paris law ... **A6:** 1111
Paris power law ... **A8:** 378, 680-681
Parison ... **EM4:** 396, 397
defined ... **EM2:** 29
Parison programming
blow molding ... **EM2:** 353
Parison swell
defined ... **EM2:** 29
Parking structures
corrosion in ... **A13:** 1299, 1309
Parr oxygen bombs
use in ion chromatography ... **A10:** 664
Part design *See also* Design
applications ... **EM2:** 617
approach ... **EM2:** 78-81
detailed ... **EM2:** 79-80
initial material/processing choice ... **EM2:** 79-80
prototyping ... **EM2:** 80-81
Part ejection *See also* Ejection; Part(s) ... **A7:** 325
multiple-slide forming ... **A14:** 571
part geometries, Ceracon process ... **A7:** 540-541
pressures ... **A7:** 190
Part geometry ... **A14:** 77, 409-410
design guidelines for ... **EM2:** 707-709
ionomers ... **EM2:** 121
nominal wall thickness ... **EM2:** 709
and process selection ... **EM2:** 278
processing and tolerances ... **EM2:** 707-708
Part handling *See also* Parts
real-time radiography ... **A17:** 322-323
Part program
for coordinate measuring machines ... **A17:** 20
Part salvage
hard chromium plating for ... **M5:** 171
Part shape
in cold isostatic pressing ... **A7:** 448
in hot isostatic pressing ... **A7:** 424-425
measured ... **A8:** 549
Part size
in cold isostatic pressing ... **A7:** 448
in hot isostatic pressing ... **A7:** 424
Part tolerances
units of analysis ... **A17:** 18
Part(s) *See also* Component; Components; Computer-aided design (CAD); Cylinder(s); Design; Fractured parts; P/M parts, specific applications; Part design; Part geometry; Powder metallurgy parts; Tubular products
of acrylonitrile-butadiene-styrenes (ABS) ... **EM2:** 114
alignment, by interferometer ... **A17:** 14
aluminum alloy, cold extrusion of ... **A14:** 307-310
aluminum alloy, precision-forged ... **A14:** 252
for appliances ... **A7:** 622-623
automotive, closed-die forging ... **A14:** 82
in automotive industry ... **A7:** 617-621
bimetallic, processing ... **A7:** 544-545
blow molded, surface finish ... **EM2:** 304
blow molding of ... **EM2:** 357
for business machines ... **A7:** 667-670
by radial forging ... **A14:** 146
classification ... **A7:** 332
cold formed nickel-base alloy, for high-temperature service ... **A14:** 837
cold-extruded steel, ultrasonic inspection ... **A17:** 271
with complex shapes, cold extruded ... **A14:** 310
composite control surfaces ... **EM1:** 595-601
composite, cost analyses ... **EM1:** 422-423
composite prototype ... **EM1:** 36-37
cone-shaped, press forming ... **A14:** 549
consolidation ... **EM1:** 36 **EM2:** 85
coordinate measuring machine inspection ... **A17:** 20
cost estimating ... **EM2:** 709-710
costing breakdown ... **EM2:** 83-85
cuplike, cold extruded ... **A14:** 305
for cure, in autoclave ... **EM1:** 702
Part(s) (continued)
cured, matrix content fluctuation ... **EM1:** 308
cylindrical ... **A14:** 529, 572
deep cuplike, cold extruded ... **A14:** 309-310
density, with injection molding ... **A7:** 495
design detail, as process selection factor ... **EM2:** 288-292
design of ... **EM2:** 615-616
dimension, and die cavity ... **A14:** 159
dimensions and surface finish, powder forgings ... **A14:** 204
dimensions, in precision forging ... **A14:** 159
dish-shaped, press forming ... **A14:** 549
drawing and ironing ... **A14:** 509
effect of lithium stearate lubricant ... **A7:** 191
effect of lubricants after sintering ... **A7:** 191
elastic mating ... **EM2:** 720-721
estimating costs of ... **EM2:** 647
fabrication, brasses for ... **A7:** 121
fabrication, costs ... **EM2:** 82
failure, design considerations ... **EM2:** 1
fracture sources illustrated ... **A12:** 217
fractured, photography of ... **A12:** 78-90
geometry ... **A14:** 77, 409-410 **EM2:** 707-709
hardness determination ... **A7:** 452
HERF processed ... **A14:** 103
with high diameter-to-thickness ratios, rotary forging for ... **A14:** 176
identification, machine vision ... **A17:** 40
impact ... **A14:** 309-310
inspection and quality control ... **A7:** 295
iron powder production capacities ... **A7:** 23
irregular, flux leakage inspection ... **A17:** 133
large, blow molded HDPE ... **EM2:** 164
large, cup-shape ... **A14:** 550
large, defined ... **EM2:** 277
large, extrusion of ... **A14:** 307
large, materials for press forming ... **A14:** 507
large, rotational molding for ... **EM2:** 360
large, RTM/SRIM for ... **EM2:** 344-351
large, stretch forming machines for ... **A14:** 596
large, structural-foam injection of ... **EM2:** 508
long, straightening of ... **A14:** 680-689
-making system, designed ... **A7:** 335-337
mechanical, critical properties ... **EM2:** 458
minute, by coining ... **A14:** 184
misaligned *See* Misaligned parts
multiple, forming of ... **A14:** 571
new, economic factors ... **EM2:** 293
photolighting of highly reflective ... **A12:** 84, 88
-piece cost analysis ... **EM2:** 358-359
planes of metal flow for ... **A14:** 50
porous ... **A7:** 105, 451, 696-700, 731
powder forged, quality assurance for ... **A14:** 203-205
precision and precision sintered ... **A7:** 9
production ... **A7:** 571
recessed, press forming ... **A14:** 549
rectangular stainless steel, drawing ... **A14:** 770
reproducibility, of unidirectional tape prepreg ... **EM1:** 145
resin transfer molded ... **EM1:** 168
rudder composite, production ... **EM1:** 596-601
secondary operations performed on ... **A7:** 451-462
shallow cuplike, cold extruded ... **A14:** 309
shallow, Guerin process drawing ... **A14:** 607
shape ... **A7:** 424-425, 448
shape, as process selection factor ... **EM2:** 288-292
simplification, by composites ... **EM1:** 35
size ... **A7:** 424, 448
size, as process selection factor ... **EM2:** 288-292
size, in press forming ... **A14:** 504
sizes, for thermoforming ... **EM2:** 399
small complex stainless steel ... **A14:** 765
small, magnetic particle inspection methods ... **A17:** 117
small, materials for press forming ... **A14:** 506
stationary/moving, laser inspection of ... **A17:** 12-16
symmetric, rotating-die machines for ... **A14:** 178
testing, by magnetic particle inspection ... **A7:** 575-579
thermoplastic injection molding of ... **EM2:** 313-315
of thermoplastic structural foams ... **EM2:** 508-510
tube-spun ... **A14:** 678
tubular, backward/forward extruded ... **A14:** 305
typical, forging machines for ... **A14:** 101
Y-shaped, inspection of ... **A17:** 113-114

Part-through cracks
fracture mechanics of **A11:** 51
Partial angle imaging
computed tomography (CT) **A17:** 379
Partial annealing *See also* Annealing
defined **A13:** 10
effect, corrosion resistance **A13:** 49
Partial denture alloys
of precious metals **A2:** 696
Partial dentures
removable **A13:** 1338
Partial discharge
defined **EM2:** 592
Partial dislocations
Burgers vector **A9:** 685
Partial hydrodynamic lubrication *See* Quasihydrodynamic lubrication
Partial immersion tests
laboratory **A13:** 222
Partial joint penetration
definition **A6:** 1212 **M6:** 12
Partial journal) bearing
defined **A18:** 14
Partial pressure
effect on stress-corrosion cracking **A8:** 499
Partial pressure analyzers
as gas chromatographs **A17:** 64
as vacuum testing method **A17:** 68
Partial reflection, defined
ultrasonic inspection **A17:** 231
Partial volume artifacts
computed tomography (CT) **A17:** 376
defined **A17:** 384
Partial-width indentation test **A1:** 583 **A8:** 584-585
Partially deterministic surface **A18:** 346
Partially melted zone **A6:** 750
Partially oriented surfaces
dimpled fracture as **A12:** 203
parametric methods for **A12:** 204
Partially stabilized toughened aluminas
chemical integrity of seals **EM4:** 540
Partially stabilized zirconia
application, internal combustion engine parts **A18:** 558
cubic debris particles, x-ray diffraction **A18:** 469
dependence of wear on velocity and temperature **A18:** 490
erosion resistance **A18:** 205
heat treatments **A18:** 814
mechanical and physical properties **A18:** 813
x-ray diffraction **A18:** 464, 465-466, 467
Partially stabilized zirconia (2 wt% Y_2O_3) properties **EM4:** 503
Partially stabilized zirconia (PSZ) **A6:** 951, 952, 956, 957
Partially stabilized zirconia (ZrO_2) (PSZ) **EM4:** 19, 676, 756
applications **EM4:** 977
engine insulation **EM4:** 990
brazing with glasses **EM4:** 519
ceramic fracture and mechanical integrity **EM4:** 532
chemical integrity of seals **EM4:** 540
coefficient of thermal expansion **EM4:** 685-686
direct brazing for joining oxide ceramics **EM4:** 518
fabrication **EM4:** 777-779
fracture toughness **EM4:** 330, 586
grinding **EM4:** 335
indirect brazing for joining oxide ceramics **EM4:** 517
joining oxide ceramics **EM4:** 512
key features **EM4:** 676
mineral processing **EM4:** 961
properties **EM4:** 330, 512, 677
property comparison, mineral processing **EM4:** 962
specialty zirconia refractories **EM4:** 907
strength changing with temperature **EM4:** 682
Partially stabilized zirconia (ZrO_2) (PSZ) (continued)
thermal properties when used as engine wall insulator lining **EM4:** 992
wear resistance **EM4:** 974
Partially stabilized zirconia ceramics **EM4:** 548
Partially transformed zone
in ferrous alloy welded joints **A9:** 581
in titanium alloy welded joints **A9:** 581
Particle
collection system, with drop tower compression test **A8:** 197
effect on fracture toughness, AISI 4340 steel **A8:** 476
effects in aluminum fracture **A8:** 478
Particle absorption
as XRPD source of error **A10:** 341
Particle accelerator
defined **A10:** 678
Particle analyzer
electronic flow diagram **A7:** 218
equipment **A7:** 218
Particle boundaries in powder metallurgy materials **A9:** 503
Particle density *See also* Density; Packed density
explosion characteristics **A7:** 196
nickel-coated composite powder **A7:** 174
Particle formation
during atomization **A7:** 31
stages, schematic **A7:** 237
Particle fracture model
fatigue failure **A12:** 207
Particle growth
effect in gravimetric analysis **A10:** 163
mechanisms in tungsten powder production **A7:** 153
in precipitation from solution **A7:** 54
Particle hardness **A7:** 8
Particle impact noise detection (PIND) **EL1:** 930-931, 954
Particle inspection
fluorescent **A7:** 578-579
magnetic **A7:** 575-578
Particle pore size *See also* Pore size; Porosity
and green strength **A7:** 303
internal, effect on compressibility **A7:** 269, 270
Particle porosity *See also* Porosity
internal, in water atomized iron powders **A7:** 85
and tap density **A7:** 276
Particle shape *See also* Shape **A7:** 8
analysis **A7:** 233-245
and apparent density **A7:** 272
in atomization process **A7:** 30-32
in atomized powders **A7:** 25
beryllium powder **A7:** 170
ceramic, in Ceracon process **A7:** 539
change, effect of milling time in titanium- based alloys **A7:** 60
collisions effect **A7:** 29-30, 34
common **A7:** 234
control by high-energy milling **A7:** 69
conventional factors **A7:** 236-237, 239
of copier powders **A7:** 585
effect in blending and premixing **A7:** 188
effect of additions on **A7:** 31-32
effect of collisions **A7:** 32
effect on compressibility **A7:** 286
effect on explosivity **A7:** 194-196
effect on powder compact **A7:** 211
electrolytic copper powder **A7:** 115
explosion characteristics **A7:** 196
and flow rate **A7:** 280
of galvanic silver **A7:** 148
in HIP **A7:** 425
irregular, determining density **A7:** 296-297
magnesium powders **A7:** 132
in magnetic separation of seeds **A7:** 590
and manufacturing methods **A7:** 233
Particle shape (continued)
in milling of single particles **A7:** 59
niobium **A7:** 162
and principles of milling **A7:** 58
scanning electron microscopy of **A7:** 234-236
scanning electron microscopy used to determine **A9:** 99-100
stainless steel powder **A7:** 100-102, 728
stereological characterization of **A7:** 237-241
and tap density **A7:** 276
terms **A7:** 236, 239
tin powders **A7:** 124
tungsten carbide/cobalt system, microstructure after liquid-phase sintering **A7:** 320
water-atomized tool steel powder **A7:** 102, 104
Particle size *See also* Particle size distribution; Particle(s); Particles **A7:** 8, 23, 214-232
abrasive, effect on wear **M1:** 601-602
analysis **A7:** 8
apparent density **A7:** 272
in atomization process **A7:** 30-33
average, in atomized powders **A7:** 25
average, permeametry measure **A7:** 262-265
of ball milled QMP iron **A7:** 86
bimodal distribution **A7:** 41, 43
by morphological analysis **A7:** 241
and chemical analysis **A7:** 246
classification **A7:** 8
collisions effect **A7:** 29-30, 34
control, by high-energy milling **A7:** 70
control, in rotary furnaces **A7:** 153
control, in tungsten powder production **A7:** 153-154
of copier powders **A7:** 585
copper, in reduction of oxide **A2:** 393
copper powders **A2:** 392
data **A7:** 229-230
determining characteristics **A7:** 239
dimensions **A7:** 225-226
dispersed, effect on homogenization kinetics **A7:** 315
distribution *See* Particle size distribution
effect in blending and premixing **A7:** 188
effect in gravimetric analysis **A10:** 163
effect in oxygen content of water-atomized copper **A7:** 37
effect of copper powder addition agents **A7:** 112, 113
effect on bioavailability of iron powders for food enrichment **A7:** 615
effect on densification and expansion during homogenization **A7:** 315
effect on green strength **A7:** 303
effect on homogenization kinetics **A7:** 315
effect on powder compact **A7:** 211
effects, and sample preparation for x-ray spectrometry **A10:** 93
electrozone size analysis **A7:** 220-221
and explosibility of aluminum powders **A7:** 130
filler, and elongation **A7:** 605, 613
and flow rate **A7:** 280
geometric and elemental analysis **A10:** 318-320
and grain size of tungsten carbide **A7:** 153
and hazardous ignition temperatures **A7:** 133
image analysis for **A7:** 225-230
light scattering techniques **A7:** 216-218
low alloy steel powder **A7:** 102
and lubrication in iron premixes **A7:** 190
in magnetic separation of seeds **A7:** 590
mean **A7:** 230, 425
measurement **A7:** 214, 216-218, 263-265
mechanically alloyed oxide alloys **A2:** 943-944
dispersion-strengthened (MA ODS)
microcompact **A7:** 58
microscopy and image analysis for **A7:** 225-230
in milling of single particles **A7:** 59

SUBJECTS OF THE INDEXED VOLUMES: ASM Handbook (designated by the letter "A"): **A1:** Properties and Selection: Irons, Steels, and High-Performance Alloys (1990); **A2:** Properties and Selection: Nonferrous Alloys and Special-Purpose Materials (1990); **A3:** Alloy Phase Diagrams (1992); **A4:** Heat Treating (1991); **A6:** Welding, Brazing, and Soldering (1993); **A7:** Powder Metallurgy (1984); **A8:** Mechanical Testing (1985); **A9:** Metallography and Microstructures (1985); **A10:** Materials Characterization (1986); **A11:** Failure Analysis and Prevention (1986); **A12:** Fractography (1987); **A13:** Corrosion (1987); **A14:** Forming and Forging (1988); **A15:** Casting (1988); **A16:** Machining (1989); **A17:** Nondestructive Testing and Quality Control (1989); **A18:** Friction, Lubrication, and Wear Technology (1992). **Metals Handbook, 9th Edition** (designated by the letter "M"): **M1:** Properties and Selection: Irons and Steels (1978); **M2:** Properties and Selection: Nonferrous Alloys and Pure Metals (1979); **M3:** Properties and Selection: Stainless Steels, Tool Materials and Special-Purpose Materials (1980); **M4:** Heat Treating (1981); **M5:** Surface Cleaning, Finishing, and Coating (1982); **M6:** Welding, Brazing, and Soldering (1983). **Engineered Materials Handbook** (designated by the letters "EM"): **EM1:** Composites (1987); **EM2:** Engineering Plastics (1988); **EM3:** Adhesives and Sealants (1990); **EM4:** Ceramics and Glasses (1991); **Electronic Materials Handbook** (designated by the letters "EL"): **EL1:** Packaging (1989).

Particle size (continued)
nickel-coated composite powder **A7:** 174
optical sensing zone for **A7:** 221-225
P/M materials **M1:** 328, 330
and packed density **A7:** 296
PIXE analysis of atmospheric aerosols by **A10:** 102
precious metal powder **A2:** 694
and pyrophoricity **A7:** 198-199
range **A7:** 8
reduction as milling objective **A7:** 56
screening determined **A7:** 176-177
sedimentation techniques **A7:** 218-220
shrinkage of compacts as function of **A7:** 309
of silver powders **A7:** 147
and sintered density of tungsten-nickel compacts **A7:** 318
in thermal decomposition **A7:** 55
in tungsten and molybdenum sintering **A7:** 389
in tungsten powders, effects of reduction temperature and powder depth **A7:** 153
typical, PIXE analysis **A10:** 103
Particle size analysis
lubricant indicators and range of sensitivities **A18:** 301
Particle size distribution *See also* Particle size; Particles **A7:** 8, 214-232 **A9:** 131-134
as analysis, in second-phase testing **A10:** 177
and apparent density **A7:** 273
and atomization **A7:** 30-33, 75-76
atomized aluminum powder grades **A7:** 129
of atomized powders **A7:** 25
bimodal **A7:** 41, 43
chemical analyses of **A7:** 246
collisions effect **A7:** 29-30, 34
control by screening **A7:** 176-177
control in copper alloy powders **A7:** 121
copper powders **A7:** 188
cumulative plot **A7:** 40, 42
curves, iron powder lots **A7:** 590
during sintering of tungsten and molybdenum **A7:** 391
effect in blending and premixing **A7:** 186
effect on apparent density **A7:** 297
effect on compressibility **A7:** 286
effect on powder compact **A7:** 211
effects on explosivity **A7:** 194, 196
electrolytic copper powder **A7:** 114
electrozone size analysis **A7:** 220-221
as function of melting rate **A7:** 41, 43
as function of rotation rate **A7:** 40, 42
of green compacts **A7:** 311
in HIP **A7:** 425-428
image analysis **A10:** 309-322
image analysis for **A7:** 225-230
light scattering techniques for **A7:** 216-218
measurement **A7:** 214, 216-218
methods of measuring **A7:** 124
microscopy and image analysis for **A7:** 225-230
in milling of single particles **A7:** 59
narrow, in milling process **A7:** 63, 64, 148
optical metallography **A10:** 299-308
optical sensing zone for **A7:** 221-225
in powders, image analysis of **A10:** 309
primary, in milling **A7:** 56
and pyrophoricity **A7:** 198-199
as residue analysis **A10:** 177
in rotating electrode process **A7:** 40-41
scanning electron microscopy **A10:** 490-515
scanning electron microscopy used to study **A9:** 99-100
sedimentation techniques **A7:** 218-220
and tap density **A7:** 276
tin powders **A7:** 124
for tungsten powders and cemented tungsten powders **A7:** 154
Particle size distribution (PSD)
dimensional characterization of powder role **EM4:** 41
Particle sizing **EM4:** 83-87
application of measurement and SPC to ceramics **EM4:** 87
general characteristics of instruments **EM4:** 83-85
analytical principles versus weighting factor **EM4:** 85

Particle sizing (continued)
Coulter counter **EM4:** 85
J concept for means and distribution **EM4:** 83-85
size range **EM4:** 83-84
weighting factor **EM4:** 83-85
sampling **EM4:** 83
statistical uses of particle-size measurements **EM4:** 85-87
computations **EM4:** 85-87
designed experiment response variable **EM4:** 85, 87
requirements of a good measurement **EM4:** 87
sampling inspection by variables **EM4:** 86
specification testing/acceptance sampling **EM4:** 85, 86-87
statistical process control **EM4:** 85-86, 87
total testing cost per batch **EM4:** 86-87
Particle(s) *See also* Magnetic particles; Powder(s) **A7:** 8
adherence, from magnetizing force **A17:** 105
agglomeration *See also* Agglomeration **A7:** 54
analyzers **A7:** 218
application, in magnetic particle inspection **A7:** 577
beryllium, size and shape **A2:** 864-865
board, as contaminants **EL1:** 661
bonding, in sintering **A7:** 340
bonding, lubricant effects **A7:** 192
boundaries, in undersintered condition **A7:** 486
classification **A7:** 59
collisions, reduction **A7:** 254
colored, magnetic **A17:** 101
contamination identification, LMMS analysis **EL1:** 1089
dimensionality of **A7:** 233-245
dross, as contaminants **EL1:** 661
dry, magnetic particle inspection **A17:** 100-101
elastic and plastic deformation **A7:** 58
extraction replicas of **A17:** 54
fineness **A7:** 59
flaw types in **A7:** 59
flux reaction, as contaminants **EL1:** 661
fracture equation **A7:** 59
from handling, as contaminants **EL1:** 661
hard, dispersion in milling **A7:** 63
inhomogeneity at surface **A7:** 250
iron carbonyl, marginal-oscillator signals **A17:** 153
in laminar and turbulent flow **A17:** 58
large, tendency to fracture **A7:** 64
ligament-shaped **A7:** 32
loose, as failure mechanism **EL1:** 1011
magnetic, inspection of **A7:** 575-579
mechanical interlocking of **A7:** 303
in mechanically alloyed oxide alloys **A2:** 943-944
dispersion-strengthened (MA ODS)
morphology **A7:** 8
patterns in powder mixtures **A7:** 186-187
patterns, yielding nonrelevant indications **A17:** 105-108
powder properties tests on **A7:** 211
precious metal powder **A2:** 694
projection shape **A7:** 240
qualitative SEM examination of **A7:** 234-236
as reinforcement **EL1:** 1119-1121
as reinforcements, aluminum metal-matrix composites **A2:** 7
roughness, effect on apparent density **A7:** 297
screening **A7:** 176-177
second-phase, replicas **A17:** 54
separator, superfine **A7:** 179
shape, copper powders **A2:** 392
shape, magnetic particle inspection **A17:** 100
single, fracture mechanics **A7:** 59
single, milling of **A7:** 59-60
size, magnetic particle inspection **A17:** 100
small, tendency to weld **A7:** 64
spacing **A7:** 8
structure, effect on powder compact **A7:** 211
studies by surface analytical techniques **A7:** 251
surface analysis techniques **A7:** 250-261
surface composition and explosivity **A7:** 195
surface contour **A7:** 233-245

Particle(s) (continued)
suspension, for magnetic particle inspection **A7:** 578
unbonded, and cracks, detected by metallography **A7:** 484
wet, magnetic particle inspection **A17:** 101
Particle-induced x-ray emission **A10:** 102-108
applications **A10:** 102, 106-107
Binary Encounter Model **A10:** 104
calibration **A10:** 105
capabilities, compared with Rutherford backscattering spectrometry **A10:** 628
characteristic x-rays **A10:** 103
data reduction **A10:** 105
defined **A10:** 678
detection limits **A10:** 104
estimated analysis time **A10:** 102
general uses **A10:** 102
introduction and principles **A10:** 103-105
limitations **A10:** 102
and milliprobe and historical studies **A10:** 107
and neutron activation analysis, compared **A10:** 233
and other elemental analyses, compared **A10:** 106
Plane Wave Born Approximation **A10:** 104
and proton microprobe **A10:** 107
quality assurance protocols **A10:** 105
RACE code **A10:** 105
related techniques **A10:** 102
samples **A10:** 102
sensitivity **A10:** 104-105
stopping distance **A10:** 103
typical set-up for **A10:** 103
x-ray cross section **A10:** 104
and x-ray fluorescence, compared **A10:** 105-106
Particles *See also* Particle size; Particle size distribution
absorption of **A10:** 341
accelerator, defined **A10:** 678
atmospheric, analysis of **A10:** 106
boundaries, P/M aluminum alloys **A12:** 440
charged, detectors for **A10:** 245-246
cleavage fracture from **A12:** 457
combined geometric and elemental analysis **A10:** 318-320
contaminant, in bearing lubricants **A11:** 486
density and spacing, effect on tensile fracture **A12:** 100-101
in discontinuous oxide battings **EM1:** 62
effect on fatigue striation **A12:** 16
of elemental categories, data and statistical summaries **A10:** 320-322
energy-dispersive spectrum from **A10:** 319
foreign, and bearing wear **A11:** 489
gold, scan line across **A10:** 496
growth of **A10:** 163
hard, and ball bearing wear **A11:** 494
heavy, kinetic energy of **A10:** 24
located, image analysis for **A10:** 319
magnetic, inspection of **A11:** 16-17
metallic, fuel pump shaft wear from **A11:** 465
opaque, AISI/SAE alloy steels **A12:** 308
PIXE analysis **A10:** 102-108
produced by explosive detonation, image analysis of **A10:** 318-320
scanning electron micrograph **A10:** 319
second-phase **A12:** 16, 219, 423
second-phase, brittle fracture and **A11:** 26
shape, geometric and elemental analysis of **A10:** 318-320
size **A12:** 101, 207
small, examination by ATEM **A10:** 452
small, single-stage extraction replicas for ATEM analysis **A10:** 452
soap, in lithium-base grease **A11:** 153
spheroidal, iron **A12:** 223
sulfide, effect on ridge formation **A12:** 41, 53
sulfur, in asphalt **A12:** 473
surface defects, iron **A12:** 220
suspended, PIXE analysis **A10:** 102
unknown, identification by electron diffraction/EDS **A10:** 455-459
Particles, quantitative metallography
measurements of **A9:** 123-125
relationships **A9:** 129-130
size distribution **A9:** 131-134

Particulate composite **EM1:** 17, 27
Particulate erosion
jet engine components **A18:** 588
pumps **A18:** 597-599
Particulate reinforcements **EM4:** 19
Particulates
airborne, in coordinate measuring machines **A17:** 1031
machines **A17:** 27
behavior, and toxicity **A7:** 210
characteristics and toxicity **A7:** 201
effects and removal **A7:** 178-180
PIXE phase analysis of **A10:** 102
radiation, and radiology **A17:** 295
as reinforcements **EL1:** 1119-1121
removal **A10:** 664
toxic reactions **A7:** 201
Parting *See* Alloying; Dealloying; Selective leaching
for blanks **A14:** 445
defined **A11:** 8
Parting agent *See* Mold release agent; Mold release agents; Release agents
Parting dies
multiple-slide forming **A14:** 570
Parting line
defined **A14:** 9, 48 **EM1:** 17 **EM2:** 29
forgings **M1:** 361-363
metal flow effects **A14:** 48
Parting plane *See also* Forging plane
defined **A14:** 9
Partitioned strain range
components of hysteresis loop **A8:** 357-358
life relationships, hysteresis loops for **A8:** 357
vs. cycles to failure in creep-fatigue range **A8:** 357
Partitioning **EL1:** 128-129
in gas chromatography/mass spectrometry **A10:** 641
oxidation states **A10:** 178
in solvent extraction **A10:** 164
Partitionless solidification
defined **A9:** 616
Partmaking system
designing **A7:** 335-337
Parts
ECL, in layout **EL1:** 513
fewer, and manufacturability **EL1:** 123-124
marking, visual examination of **EL1:** 1058
misapplication, through-hole packages **EL1:** 970
mismarked, through-hole packages **EL1:** 970
multifunctional **EL1:** 124
numbers, in design for manufacturability (DFM) **EL1:** 122
self-locating **EL1:** 124
separable component, defined **EL1:** 1156
standard, as design for manufacturability rule **EL1:** 122-123
Parts consolidation
and costs **EM2:** 85
Parts list
in final design package **EL1:** 523
Parts per billion
defined **A13:** 10
Parts per million
defined **A13:** 10
Parylene coatings
applications **EL1:** 797-800
barrier properties **EL1:** 794-795
as conformal coatings **EL1:** 763
crystallinity **EL1:** 796
defined **EL1:** 789
dimer **EL1:** 789-790
electrical properties **EL1:** 793-794
as encapsulant **EL1:** 242
health and safety issues **EL1:** 800
introduction **EL1:** 759
mechanical properties **EL1:** 793
monomer **EL1:** 790-791
polymer properties **EL1:** 792-797
polymerization mechanism **EL1:** 791-792
Parylene coatings (continued)
repair **EL1:** 798
solvent resistance **EL1:** 796-797
surface energy **EL1:** 795-796
thermal properties/endurance **EL1:** 794
ultraviolet and infrared spectra/optical properties **EL1:** 795
Parylenes **EM3:** 599-600
for circuit protection **EM3:** 592
for coating and encapsulation **EM3:** 580
conformal over coat **EM3:** 592
properties **EM3:** 599-600
PAS *See* Photoacoustic spectroscopy; Polyaryl sulfone; Polyaryl sulfones; Polyarylsulfone
Pascal **A10:** 685, 691
Paschen-Runge polychromators
for ICP **A10:** 37
Pass energy
as kinetic energy **A10:** 571
PASS test **A13:** 220
Passenger ferries
composite structures for **EM1:** 839
Passes
defined **A14:** 9
number, contour roll forming **A14:** 628-629
number, for out-of-roundness **A14:** 622
rotary straightening **A14:** 690
Passivating film in electrolytic polishing **A9:** 48
Passivating solutions
for cleaning **A13:** 1141
Passivation *See also* Activation
amorphous metals **A13:** 868
aqueous corrosion **A13:** 35-36
base metal **EL1:** 675
of chip **EL1:** 244
chromium alloying effects **A13:** 48
coatings, cleaning of **A13:** 381
computer modeling **EL1:** 443-444
copper and copper alloys **M5:** 624
for corrosion in stainless steel **A11:** 195
cracks **EL1:** 1054
defined **A13:** 10 **EL1:** 1152
and dewetting **EL1:** 642
electropolishing accomplishing **M5:** 306
and hermeticity, considerations **EL1:** 243-244
of iron in water and dilute aqueous solutions **A11:** 198
phosphate coating process affected by **M5:** 438
potential, aqueous corrosion **A13:** 35
safety precautions **M5:** 560
solutions **A13:** 552, 1141
stainless steel **M5:** 306, 431-432, 558-560
corrosion and **M5:** 558-560
techniques, and stainless steel corrosion **A13:** 552
of tin **A13:** 772
Passivation, surface
rare earth metals **A2:** 735
Passivation treatment
to increase corrosion resistance of stainless steel weldments **A6:** 1069
Passivator
defined **A13:** 10
Passivators
as lubricant additives **A14:** 515
Passive
defined **A13:** 10
Passive anodic potential ranges **A9:** 144
Passive clearance control
blade tips of jet engines **A18:** 589
Passive component fabrication *See also* Passive devices
device construction, generic **EL1:** 178-179
fabrication methods **EL1:** 184-188
interconnection methods **EL1:** 179-180
material selection **EL1:** 182-184
passive microwave components **EL1:** 188-189
reliability factors **EL1:** 180-182
Passive corrosion
defined **A12:** 41-42
Passive devices *See also* Passive component fabrication
capacitor failure mechanisms **EL1:** 994-999
capacitors **EL1:** 994
carbon composition resistors **EL1:** 1002-1003
defined **EL1:** 109
failure mechanisms **EL1:** 994-1005
inductor failure mechanisms **EL1:** 1004-1005
inductors **EL1:** 1003-1004
radio frequency **EL1:** 1005
resistor failure mechanisms **EL1:** 999-1003
resistors **EL1:** 999
thin-film chip resistors **EL1:** 1003
Passive element
defined **EL1:** 1152
Passive extreme pressure (PEP) additives
for metalworking lubricants **A18:** 141
Passive film rupture
in SCC and corrosion fatigue **A12:** 42
Passive films *See also* Films; Thick films; Thin films
composition vs depth, on tin-nickel substrate **A10:** 608-609
study of **A10:** 557-558
Passive microwave components
fabrication of **EL1:** 188-189
Passive substrates *See also* Substrates
Imitation **EL1:** 8
Passive-active cell
defined **A13:** 10
Passivity
defined **A13:** 10
effects in galvanic corrosion **A11:** 185
Paste
dispersion pigments, sheet molding compounds **EM1:** 158
doctor blades, SMC machines **EM1:** 159
metering, SMC machines **EM1:** 159-160
and resin, for sheet molding compounds **EM1:** 159-160
Paste (nonfluxing)
brazing filler metals available in this form **A6:** 119
Paste brazing filler metal
definition **M6:** 13
Paste compound **A7:** 8
for brazing and soldering **A7:** 840
Paste feeders
automatic torch brazing **M6:** 959
Paste soldering filler metal
definition **M6:** 13
Pastes *See also* Solder pastes **EM3:** 20, 47
as aluminum alloy application **A2:** 14
defined **EL1:** 1152
as lubricant form **A14:** 514
stiff, forming structural ceramics from **A2:** 1020
Pasteurizers
for breweries **A13:** 1223
Pasty range of the alloy **A6:** 964, 965
PATCHES-III computer program for structural analysis **EM1:** 268, 272
Patenting **A4:** 55, 162
steel wire **M1:** 262
steel wire rod **M1:** 253
Patents
acrylics **EM3:** 122, 124
Path length
in UV/VIS absorption spectroscopy **A10:** 62
Path length factor **A18:** 464
Patina
defined **A13:** 10
Patina copper coloring solutions **M5:** 625
PATRAN
finite-element analysis code **EM3:** 479, 480
PATRAN-G computer program for structural analysis **EM1:** 268, 272
Pattern *See* Diffraction pattern

Pattern error
as casting defect A11: 386
Pattern mounting error
as casting defect A11: 386
Pattern plating EL1: 115, 540
Pattern-fit algorithm
as automatic trace routing EL1: 531-532
Pattern/follower wheel-shaped cutting machines A6: 1169
Patterned sheet
temper designations A2: 26
Patterning, and selective plating
thin-film hybrids EL1: 329
Patternmaker's shrinkage
cast copper alloys A2: 356-391
Patternmakers' rules M1: 30-31, 33
Patterns
aluminum and aluminum alloys A2: 14
Pauli exclusion principle EL1: 90
Pauling electronegativity A6: 144
Pawl spring
fatigue fracture A11: 551-553
Pawls, cast iron
coatings for M1: 104
Pawls, wrought steel
economy in manufacture M3: 848, 849
Pb-Pd (Phase Diagram) .... A3: 2 • 332
Pb-Pr (Phase Diagram) .... A3: 2 • 332
Pb-Pt (Phase Diagram) .... A3: 2 • 333
Pb-Pu (Phase Diagram) .... A3: 2 • 333
Pb-Rb (Phase Diagram) .... A3: 2 • 333
Pb-Rh (Phase Diagram) .... A3: 2 • 334
Pb-S (Phase Diagram) .... A3: 2 • 334
Pb-Sb (Phase Diagram) .... A3: 2 • 334
Pb-Sb-Sn (Phase Diagram) .... A3: 3 • 57-58
Pb-Se (Phase Diagram) .... A3: 2 • 335
Pb-Sn (Phase Diagram) .... A3: 2 • 335
Pb-Sn-Zn (Phase Diagram) .... A3: 3 • 58
Pb-Sr (Phase Diagram) .... A3: 2 • 335
Pb-Te (Phase Diagram) .... A3: 2 • 336
Pb-Tl (Phase Diagram) .... A3: 2 • 336
Pb-Y (Phase Diagram) .... A3: 2 • 336
Pb-Yb (Phase Diagram) .... A3: 2 • 337
Pb-Zn (Phase Diagram) .... A3: 2 • 337
PBI *See* Polybensimidazole; Polybenzimidazole; Polybenzimidazoles
PBT *See* Polybutylene terephthalate
PBZT
for printed board material systems EM3: 592
PC *See* Polycarbonate; Polycarbonates
PCBN *See* Cubic boron nitride, polycrystalline; Polycrystalline cubic boron nitride (PCBN)
PCD *See* Diamond, polycrystalline
Pd-Pt (Phase Diagram) .... A3: 2 • 337
Pd-Pu (Phase Diagram) .... A3: 2 • 338
Pd-Rh (Phase Diagram) .... A3: 2 • 338
Pd-Ru (Phase Diagram) .... A3: 2 • 338
Pd-S (Phase Diagram) .... A3: 2 • 339
Pd-Sb (Phase Diagram) .... A3: 2 • 339
Pd-Se (Phase Diagram) .... A3: 2 • 339
Pd-Si (Phase Diagram) .... A3: 2 • 340
Pd-Sm (Phase Diagram) .... A3: 2 • 340
Pd-Sn (Phase Diagram) .... A3: 2 • 340
Pd-Te (Phase Diagram) .... A3: 2 • 341
Pd-Ti (Phase Diagram) .... A3: 2 • 341
Pd-Tl (Phase Diagram) .... A3: 2 • 342
Pd-U (Phase Diagram) .... A3: 2 • 342
Pd-V (Phase Diagram) .... A3: 2 • 342
Pd-W (Phase Diagram) .... A3: 2 • 343
Pd-Y (Phase Diagram) .... A3: 2 • 343
Pd-Yb (Phase Diagram) .... A3: 2 • 343
Pd-Zn (Phase Diagram) .... A3: 2 • 344
PDA *See* Photodiode arrays
PDIP *See* Plastic dual-in-line package (PDIP)
Peak averaging A18: 296
Peak broadening
in EPMA measurement A10: 519
as XRPD source of error A10: 341
Peak contact temperature A18: 438, 440
Peak liquidus temperature A6: 354
Peak overlap
defined A10: 678
major component analysis with A10: 530
spectral, AES A10: 556
Peak penetration
laser-beam welding A6: 262
Peak switching technique
for ion current signals A10: 144
Peak temperature-cooling time (PTCT) diagram A6: 71, 72-73
Peak-age treatment
beryllium-copper alloys A2: 407
Peaks
artifact, as AEM-EDS microanalytic limitation A10: 448
Auger electron A10: 551
breadth and position, micro- and macrostresses from A10: 387
broadening A10: 341, 519
diffraction, in surface stress measurement A10: 385
EPMA, gold-copper alloys A10: 530
escape or parasitic A10: 520
false silicon or internal fluorescence A10: 520
identification, EDS A10: 522-523
internal fluorescence, EDS spectra A10: 520
sum, defined A10: 92, 520
switching A10: 144
well-resolved, and high concentrations A10: 530
width, liquid chromatography A10: 651
in x-ray spectrometer detectors A10: 92
Pearlite .... A1: 127, 128-129 A3: 1 • 21 A13: 10, 47, 566
500 °F embrittlement, susceptibility to M1: 685
in atomized iron A7: 487
in austenitic manganese steel castings A9: 239
in carbon and alloy steels A9: 170, 178-179
in cast iron M1: 4, 6-9
in cast irons, magnifications to resolve A9: 245
decomposition by spheroidization and graphitization A11: 613
defined A9: 13
effect on cast iron M6: 1000
growth A10: 508
hydrogen embrittlement, effect on M1: 687
machinability influenced by M1: 571-573
microstructure, eutectoid composition A7: 315
microstructures A9: 658-661
notch toughness, effect on M1: 695, 699, 702, 704, 706
SEM resolution of A10: 494
steel as oriented surface in quantitative metallography A9: 128
surface hardened steels M1: 531, 533
temperature of austenite transformation symbol A11: 796
wear resistance compared with martensite M1: 611-612
Pearlite, spheroidized
high-carbon steel microstructure A12: 278
Pearlite-reduced steels A1: 148
Pearlite/ferrite ratio
in compacted graphite iron A1: 57, 59
Pearlitic ductile irons
fracture modes A12: 228-230, 235-236
Pearlitic gray iron, application
internal combustion engine parts A18: 556
Pearlitic gray irons
machining A2: 966
Pearlitic malleable iron
annealed and oil quenched A9: 253
annealed and tempered A9: 253
austenitized, air cooled A9: 252
centrifugally cast and annealed A9: 254
effects of increasing tempering temperature A9: 253
flame hardened A9: 253
with gray iron rim A9: 254
oil quenched and tempered A9: 253
Pearlitic malleable iron, heat treating *See* Malleable iron, heat treating
Pearlitic malleable iron, specific type
Grade 45008, annealed and tempered A9: 252
Pearlitic malleable irons *See* Malleable cast irons
Pearlitic steel
decarburization effects A11: 77-78
fatigue behavior of M1: 675, 677
transition temperature, effect of alloying elements on M1: 417
Pearlitic steels
abrasion artifacts in A9: 35-36, 38
acoustic emission inspection A17: 287
advantages A6: 797
applications A6: 797
Pearlitic steels (continued)
hardfacing A6: 790-791
as hardfacing alloys A7: 828
Pearlitic steels, specific types
low-alloy steels, advantages and applications of materials for surfacing, build-up and hardfacing A18: 650
mild steels, advantages and applications of materials for surfacing, build-up and hardfacing A18: 650
Pearlitic structure
defined A9: 13
Pearlitic-martensitic malleable iron A1: 76-84
brazing A1: 83-84
Charpy V-notch impact energy A1: 81
compressive strength A1: 82
fracture toughness A1: 74, 80, 82
heat treatment A1: 76-80
mechanical properties A1: 74, 78, 79, 80-83
shear strength A1: 82
tensile properties A1: 82
torsional strength A1: 82
unnotched fatigue limits A1: 81, 83
microstructure A1: 76, 77
modulus of elasticity A1: 82
rehardened-and-tempered malleable iron A1: 79
selective surface hardening A1: 84
stress-rupture plot A1: 81
tempering times A1: 80
wear resistance A1: 83-84
welding A1: 83-84
Pearson IV distribution function A18: 851
Pearson symbols .... A3: 1 • 15
conversion to Strukturbericht symbols A9: 707
for identifying space lattices A9: 706-707
Pearson VII distribution functions
in surface stress measurement A10: 386
Pearson, William B .... A3: 1 • 15
Pebbles *See* Orange peel
Pechukas and Gage apparatus
modified A7: 265
Peck drilling
for dissimilar materials EM1: 669-671
MMCs A16: 894, 898, 899
Peclet number A18: 39-40, 43, 44
Pedestal bearing
defined A18: 14
PEEK *See* Polyaryletherketones; Polyether etherketone; Polyetherketoneketone
Peel bond
defined EL1: 1152
Peel, or stripping
strength EM3: 20
Peel ply EM1: 17, 642, 682 EM3: 20
defined EM2: 29
Peel strength EM2: 29, 237
defined EL1: 1152 EM1: 17
Peel stress
epoxy design and EM3: 101
testing for latex EM3: 213
Peel test A18: 404
for adhesion failures by thin-film contaminants A11: 43
definition M6: 13
for resistance spot welds M6: 487
Peel test procedures
ceramic/metal seals EM4: 505, 506, 507
Peel tests *See also* Microelectronics tests EM3: 383-384, 385, 386, 809
raw materials quality control EM3: 732
using surface preparation procedures EM3: 805
Peeling *See also* Microspalling A18: 259
in roller bearing A11: 502
in rotary swaging A14: 143-144
-type cracks, in shafts A11: 471
Peeling fracture
medium-carbon steel A12: 253
Peeling, plating
as PTH failure mechanism EL1: 1025-1026
Peen plating *See also* Mechanical coating A7: 459 M5: 300
Peening *See also* Shot peening; Shot peening, Stress peening
abrasive jet machining A16: 512, 513
in bearingizing A16: 254
cast iron welds M6: 313

Peening (continued)
definition ... **M6:** 13
embrittlement in nickel alloys ... **M6:** 437
furnace brazing of steels ... **M6:** 940
high-production ... **A14:** 681
manual ... **A14:** 681
reduction of residual stress ... **M6:** 891-892
butt welds ... **M6:** 891-892
stainless steels ... **M6:** 891
Peening, shot *See* Shot peening
Peening wear
defined ... **A18:** 14
Peenscan measurement method
shot peen coverage ... **M5:** 139
Peg topple test ... **A18:** 404
Pegmatite
in ceramic tiles ... **EM4:** 926
PEI *See* Polyether-imide; Polyetherimides
PEKK *See* Polyaryletherketones; Polyetherketoneketone
Pellet ... **A7:** 8
Pellets
KBr, as IR samples ... **A10:** 113
as samples, x-ray spectrometry ... **A10:** 93
Pellets, feeding
for injection molding compounds ... **EM1:** 164-165
PEM *See* Photoelastic modulators
Pen points
powders used ... **A7:** 574
Pen recorder
with drop tower compression test ... **A8:** 197
Pencil glide ... **A9:** 684
Pencil-type electropolishing chamber ... **A9:** 55
Pendant drop process
titanium powders ... **A7:** 167
Pendellösung fringes ... **A10:** 368, 370
Pendellosung oscillations
effect on diffraction contrast images ... **A9:** 111
Pendulum load-measuring system ... **A8:** 132-133, 613
Penetrameters
for electronic components ... **A17:** 341
placement of ... **A17:** 341-343
plaque-type ... **A17:** 339-340
radiographic inspection ... **A17:** 338-341
step wedge ... **A17:** 341
for weldments ... **A17:** 592-593
wire-type ... **A17:** 340-341
Penetrant inspection *See also* Liquid penetrant inspection; Penetrant(s)
materials used ... **A17:** 74-77
Penetrant inspection testing
brazed joints ... **A6:** 1119
Penetrant testing ... **A6:** 1081, 1082, 1084, 1085, 1086, 1087, 1088
cracking from ... **A11:** 433
Penetrant(s) *See also* Liquid penetrant inspection
application of ... **A17:** 82
bleedback ... **A17:** 74
characteristics, physical/chemical ... **A17:** 75, 77
classification ... **A17:** 77
dye and fluorescent, as visual inspection ... **A17:** 3
excess, removal of ... **A17:** 74
liquid, leak detection with ... **A17:** 65
maintenance of ... **A17:** 85
method A, water-washable ... **A17:** 75, 77
method B and D, lipophilic/hydrophilic postemulsifiable ... **A17:** 75, 77-78
method C, solvent-removable ... **A17:** 75, 77-78
methods, liquid ... **A17:** 73-74
quality assurance of ... **A17:** 84-85
selection and use ... **A17:** 75
type I, fluorescent ... **A17:** 75, 77-78
type II visible ... **A17:** 75, 77
types ... **A17:** 74-75, 77-78
Penetration *See also* Depth of penetration; Incomplete penetration; Lack of penetration (LOP) ... **A13:** 33, 232, 545, 942; **EM3:** 20
degree of ... **A18:** 185
depth, electron ... **A11:** 41
depth of ... **A18:** 185
incomplete ... **A17:** 50, 86, 296, 350
of liquid penetrants ... **A17:** 74
metal, as casting defect ... **A11:** 385
of molten braze material, fatigue fracture by ... **A11:** 454-455
radiation, in neutron radiography ... **A17:** 387
time *See* Dwell time
of ultrasonic inspection ... **A17:** 231
Penetration (of a grease)
defined ... **A18:** 14
Penetration, depth of
in ATR spectroscopy ... **A10:** 113
Penetration distance ... **A18:** 463, 467
x-ray diffraction ... **A18:** 464, 470
Penetration grades (greases) ... **A18:** 136
Penetration hardness number
defined ... **A18:** 14
Penetration losses
in superconductors ... **A2:** 1039-1040
Penetration method
paint dry film thickness tests ... **M5:** 491
Penetration, solvent
surface-mount assemblies ... **EL1:** 666
Penetration welding
laser-beam welding ... **A6:** 263
Penetrator techniques
to prevent can folding ... **A7:** 518
Penetrometer *See also* Penetration (of a grease)
defined ... **A18:** 14
Penetrometers
assembly ... **A7:** 268
glass ... **A7:** 267, 269
for mercury volume displacement measurement ... **A7:** 267
Penicillamine
as chelator ... **A2:** 1236
Penning gage
for gas/leak detection ... **A17:** 64
Penny bronze
applications and properties ... **A2:** 313
Pentachlorobiphenyl
mass spectrum for ... **A10:** 642
Pentaerithritol tetranitrate (PETN)
friction coefficient data ... **A18:** 75
Pentaerythritol
as common crystal analyzing crystal ... **A10:** 88
Penumbra, defined
for radiographic definition ... **A17:** 313
PEP
as synchrotron radiation source ... **A10:** 413
Pepper blister *See* Blister
Pepperhoff interference film technique
for examination of tool steels ... **A9:** 258
Peptone
use in tin-lead plating baths ... **M5:** 277-278
Peracid epoxidation
of olefins ... **EM1:** 66-68
Peracid epoxides *See* Cycloaliphatic epoxides
Percent by volume glass *See* Glass, percent by volume
Percent cold work
diffraction-peak breadth at half height as function of ... **A10:** 387
gradient, and maximum residual stress ... **A10:** 380
and residual stress caused by stress-relieving heat treatment and forming, measured ... **A10:** 380
and residual stress distribution ... **A10:** 390
surface or subsurface, by x-ray diffraction residual stress techniques ... **A10:** 380, 390
Percent elongation *See also* Elongation; Total elongation; Uniform elongation
effect of exposure time and temperature on ... **A8:** 37
effect of uniform elongation on ... **A8:** 27
thermoset matrix composites ... **EM1:** 395
vs. strain rate sensitivity ... **A8:** 42
Percent error
defined ... **A8:** 10
Percent failure
vs. life for different stresses ... **A8:** 699
Percent of large particles (PLP) ... **A18:** 302
Percent oil ... **A18:** 143
Percent reduction
of drawability ... **A8:** 562
Percent replication
test program ... **A8:** 696-697
Percent survival (curve)
in stress-corrosion cracking ... **A13:** 276
Percent survival values
nonparametric evaluation of ... **A8:** 707
Percent theoretical density ... **A7:** 8
Percentage defective (*p*) chart ... **EM3:** 795-797
Percentage of back reflection technique
ultrasonic inspection ... **A17:** 263
Percentile
estimates ... **A8:** 629-635
exponential distribution ... **A8:** 635
normal distribution ... **A8:** 630-631
population ... **A8:** 629
of statistical distributions ... **A8:** 629
Weibull distribution ... **A8:** 633
Percentile estimates ... **A8:** 628-635
Weibull distribution ... **A8:** 633-634
Percentiles of the studentized range
g ... **A8:** 658
Percentiles of the *t* distribution ... **A8:** 655
Perchlorate
as electrode ... **A10:** 185
Perchloric acid
and acetic anhydride electrolyte ... **A9:** 51
and alcohol (Group I electrolytes) ... **A9:** 52-54
as electrolyte ... **A9:** 51-54
as electrolyte for aluminum ... **A9:** 48, 353
and glacial acetic acid (Group II electrolytes) ... **A9:** 53-54
and glacial acetic acid as an etchant for lead and lead alloys ... **A9:** 416
methanol and ethylene glycol as an electrolyte for titanium and titanium alloys ... **A9:** 459
mounting materials for use with ... **A9:** 49, 53-54
as oxidant and sample dissolution medium ... **A10:** 166
residue isolation using ... **A10:** 176
Perchloroethylene solvent cleaners ... **M5:** 40, 44-48, 617
cold solvent cleaning process, use in ... **M5:** 40
flash point ... **M5:** 40
magnesium alloy cleaning ... **M5:** 629
vapor degreasing, use in ... **M5:** 10, 44-45, 47-48
Percussion
definition ... **EM4:** 633
Percussion weld
definition ... **M6:** 13
Percussion welding *See also* Attachment methods; Joining; Welding ... **M6:** 739-745
applications ... **M6:** 739-740
workpiece condition ... **M6:** 740
workpieces, design and size ... **M6:** 739-740
arc starting ... **M6:** 741
alternating current ... **M6:** 741
direct current ... **M6:** 741
starter nib ... **M6:** 741
arc time ... **M6:** 740-741
as attachment method, sliding contacts ... **A2:** 842
capacitor-discharge welding ... **M6:** 739-745
control ... **M6:** 742
displacement, current, and voltage ... **M6:** 742-743

Percussion welding (continued)
high-voltage machines M6: 744-745
low-voltage machines M6: 743-744
preparation of workpieces M6: 742
sequence of steps M6: 742
voltage M6: 741-742
combinations of work metals M6: 740
comparison to stud welding M6: 739
current for welding M6: 741
polarity M6: 741
definition M6: 13
force for welding M6: 741
damping M6: 741
impact velocity M6: 741
peak loading M6: 1382741
heat-affected zone M6: 740
magnetic-force welding M6: 740, 745
applications M6: 745
arc starters M6: 745
arc time M6: 745
force for welding M6: 745
weld areas M6: 745
metals welded M6: 740
power supplies M6: 740
high-voltage capacitors M6: 740
low-voltage capacitors M6: 740
resistance welding transformers M6: 740
safety M6: 59, 745
welding energy M6: 741
Percussion welding (PEW)
definition A6: 1212
Percussion welding of
aluminum alloys M6: 399, 740
copper M6: 745
copper alloys M6: 740
copper-tungsten M6: 740
gold M6: 740
low-carbon steels M6: 740
medium-carbon steels M6: 740
molybdenum M6: 740
nickel alloys M6: 740
silver M6: 740
silver-cadmium oxide M6: 740, 745
silver-tungsten M6: 740, 745
stainless steels M6: 740
tantalum M6: 740
thermocouple alloys M6: 740
Perester dibasic acid + metal ion
generating free radicals for acrylic adhesives EM3: 120
Perfectly oriented surface
true area and length A12: 204
Perfilming *See also* Films
Perfluorinated polyalkylether oils
creep or migration A18: 151
Perfluoro alkoxy alkane (PFA)
as fluoropolymer EM2: 116
Perfluoroalkoxy (PFA)
surface preparation EM3: 279
Perfluoroalkoxytetrafluoroethylene (PFA)
in pharmaceutical production facilities A13: 1227-1228
Perfluoroalkyl ethers EM3: 677
Perfluoropolyalkylether (PFPE)
factors influencing fluid degradation A18: 156
high-vacuum lubricants A18: 156, 157, 158
molecular structures A18: 156
properties A18: 155, 156
Perforating *See also* Piercing; Punching
defined A14: 9
Perforator bushings
materials for A14: 485
Perforator punches
material for A14: 485
Performance *See also* Electrical performance testing; NDE reliability; Physical performance; Reliability; specific inspection methods; Testing
analytic modeling EL1: 14-15
characteristics, NDE system/process A17: 675
chip EL1: 439
connector EL1: 23
and design, testing EL1: 954-955
digital systems, improvement technologies EL1: 2
discrete semiconductor packages EL1: 422
electrical, testing EL1: 946-952
electrical/package EL1: 402
Performance (continued)
and end use EL1: 1
factors, product/process, in quality design A17: 722
fiber properties and EM1: 43
logic function EL1: 2
as material selection parameter EM1: 38-39
measure , signal-to-noise ratio A17: 750
mechanical, rigid epoxies EL1: 810
microcircuit, design for EL1: 260-261
multiple/three/two-terminal packages EL1: 423
NDE system, validation of A17: 675
physical, issues of EL1: 5-8
and prior art, NDE reliability A17: 674
probability of detection (POD) curves for A17: 679
properties, rigid epoxies EL1: 813-815
quantification A17: 674
range classifications, for electrical design EL1: 25
ranking, of IC packages EL1: 404
relative operating characteristic (ROC) curves for A17: 679-680
requirements, high I/O controlled-impedance connector EL1: 87
requirements, structural assessment A17: 686
semiconductor chips, introduction EL1: 397
speed as parameter of EL1: 400
system, modeling of EL1: 13-15
system, with fixed, minimum device size EL1: 2
and tension testing A8: 19
thermal EL1: 409-411
thermomechanical EL1: 414-416, 814-815
of UV-curable coatings EL1: 787-788
variations A17: 674
Performance package A18: 111
Performance, Use of phase diagrams to improve .... A3: 1 • 27-28
Periclase
topotactic relationship with brucite ill
Perimeter
length, as basic figure quantity A12: 194-195
mean, of closed figures A12: 195
Periodate method
analysis for manganese in zirconium alloys by A10: 69
Periodic chart
microelectronic EL1: 91, 94
Periodic marks
by SCC of brass A12: 28
Periodic overstrain
of SAE 1045 hot rolled bar A8: 701
scatter in data A8: 701
Periodic table
for analytic sensitivities of AAS A10: 46
of the elements A10: 688
Periodic table of the elements A2: 1098
Periodic-reversal copper plating M5: 159-160, 162, 164-165, 168
cycle efficiency M5: 164-165
Periodic-reverse electrocleaning
materials and processes M5: 27-28, 34-35
Periodicals
as information source EM2: 92-93
Peripheral milling *See* Willing
Peripherals applications
for hybrids EL1: 254
Perishable tools A18: 627
Peristaltic pumps
use with concentric nebulizers A10: 35
Peritectic
defined A9: 13
Peritectic equilibrium
defined A9: 13
Peritectic reaction .... A3: 1 • 5
during solidification A15: 125-126
theory, as grain refinement model A15: 106-107
vs peritectic transformation A15: 125
Peritectic reactions A9: 676-677
effects on dendritic structures A9: 613
formation of aluminum alloy phases by A9: 359
in tin-antimony alloys, effect on microstructure A9: 452
Peritectic structures A9: 675-680
phase diagrams A9: 675
Peritectic temperature A9: 675
Peritectic transformation
during solidification A15: 126-127
in multicomponent systems A15: 129
vs peritectic reaction A15: 125
Peritectic transformations A9: 677
Peritectoid phase equilibrium
defined A9: 13, 675
Peritectoid reactions and transformations A9: 678
Peritectold reaction .... A3: 1 • 5
Permalloy
boriding A4: 441, 445
microstructural effects EM4: 1161
Permalloy (Ni-Fe), application
advanced magnetic storage A18: 838
Permalloy, on ceramic substrate
x-ray spectrometry A10: 100-101
Permanence EM3: 20
defined EM1: 17 EM2: 29
Permanent dipole bond
definition M6: 13
Permanent dipole interaction EM3: 40
Permanent distortion
in shafts A11: 467
Permanent lattice strain A7: 61
Permanent magnet alloys *See* Platinum-cobalt permanent magnet alloy
Permanent magnet alloys, specific types *See also* Magnetic materials, specific types
$Co_{3.45}Fe_{0.25}Cu_{1.35}SM$ diffractometer traces A9: 702
Permanent magnet materials M3: 615-639
aging *See* Stability
alloy usage, optimum A2: 793-802
Alnico alloys A2: 785-787
applications A2: 792-802 M3: 629-632
changes, irreversible and reversible A2: 795-799
classified by application-relevant properties A2: 796
cobalt and rare-earth alloys A2: 787-788
commercial designations and suppliers A2: 783
commercial materials A2: 784-787
Cunife, commercial A2: 785
Curie temperature M3: 624
demagnetization curves M3: 620, 621-623, 635, 638
design considerations A2: 799-802
designations M3: 616-619
economic considerations A2: 792-793
fundamentals of magnetism A2: 782, 784
hard ferrite (ceramic) materials A2: 788-790
hysteresis applications A2: 793-794
introduction A2: 782
iron-chromium-cobalt alloys A2: 790
magnet alloys A2: 785
magnet steels A2: 785
magnetic energy M3: 615, 617-618, 620, 621
magnetic hysteresis M3: 616, 620, 630-632
magnetic properties M3: 625
and magnetically soft materials compared A2: 784
magnetization losses *See* Stability
maximum energy content A2: 782
mechanical properties M3: 626
neodymium-iron-boron A2: 730
neodymium-iron-boron alloys A2: 790
nominal composition M3: 624
physical properties M3: 626
platinum-cobalt A2: 713
platinum-cobalt alloys A2: 787
samarium-cobalt A2: 729-730
selection A2: 792-802 M3: 629-632
stability M3: 632-639
stabilization and stability A2: 794-795
stress effects M3: 638-639
temperature effects *See* Stability
Permanent magnet materials, specific types
Alcomax alloys, loss in magnetism M3: 637, 638
Alnico alloys M3: 626-627
applications M3: 630-632
compositions M3: 624
Curie temperature M3: 624
demagnetization curves M3: 621-622
hysteresis loss M3: 630
loss of magnetism M3: 633-637
magnetic properties M3: 625, 629

Permanent magnet materials, specific types (continued)
mechanical properties M3: 626
physical properties M3: 626
ceramics *See* Ferrites
Co-RE alloys *See* Cobalt/rare earth alloys
cobalt/rare earth alloys M3: 628-629
Co-RE alloys *See* Cobalt/rare earth
Curie temperature M3: 624
demagnetization curves M3: 623
loss magnetism M3: 635, 636
magnetic properties M3: 625
mechanical properties M3: 626
physical properties M3: 626
Cunico M3: 621, 623
composition M3: 624
Curie temperature M3: 624
demagnetization curves M3: 622
magnetic properties M3: 625
mechanical properties M3: 626
physical properties M3: 626
Cunife M3: 621, 623
applications M3: 623
composition M3: 624
Curie temperature M3: 624
demagnetization curves M3: 622
loss of magnetism M3: 634
magnetic properties M3: 625
mechanical properties M3: 626
physical properties M3: 626
stress, effect on magnetization curves M3: 638, 639
ferrites M3: 629
applications M3: 631-632
Curie temperature M3: 624
demagnetization curves M3: 622, 635
loss of magnetism M3: 633-636
magnetic properties M3: 625
mechanical properties M3: 626
physical properties M3: 626
Lodex alloys M3: 627-628
compositions M3: 624
Curie temperature M3: 624
demagnetization curves M3: 622-623
loss of magnetism M3: 634
magnetic properties M3: 625
mechanical properties M3: 626
physical properties M3: 626
magnet steels M3: 619-620
applications M3: 631-632
compositions M3: 624
Curie temperatures M3: 624
demagnetization curves M3: 621
hysteresis loss M3: 630
loss of magnetism M3: 633, 634, 637
magnetic properties M3: 625
mechanical properties M3: 626
physical properties M3: 626
P-6 alloy
composition M3: 624
hysteresis loss M3: 630
magnetic properties M3: 625
mechanical properties M3: 626
physical properties M3: 626
platinum-cobalt M3: 628
composition M3: 624
Curie temperature M3: 624
demagnetization curve M3: 623
loss of magnetism M3: 635, 636
magnetic properties M3: 625
mechanical properties M3: 626
physical properties M3: 626
Remalloy M3: 620-621
applications M3: 632
composition M3: 624
Curie temperature M3: 624
demagnetization curve M3: 622
magnetic properties M3: 625, 629
mechanical properties M3: 626

Permanent magnet materials, specific types (continued)
physical properties M3: 626
semihard alloys M3: 628
Vicalloy M3: 621, 623, 626
composition M3: 624
Curie temperature M3: 624
demagnetization curves M3: 622
hysteresis loss M3: 630
loss of magnetism M3: 633, 634
magnetic properties M3: 625
mechanical properties M3: 626
physical properties M3: 626
stress, effect on magnetization curves M3: 638, 369
Permanent magnet(s) *See also* Magnet; Magnetic; Permanent magnetic materials
defined A2: 782
hard magnetic materials as A2: 761
magnetic field generation by A17: 93
yokes, applications A17: 93
Permanent magnets *See* Magnetic materials A7: 8
cobalt powders in A7: 144
demagnetization curve for A7: 639
powders used A7: 573
yoke assembly A7: 575
Permanent magnets, alloy development of ..... A3: 1 • 26
Permanent mold casting *See also* Castings, Foundry products; Permanent mold processes; Permanent mold(s) A15: 275-285
alloying element and impurity specifications A2: 16
aluminum alloy, heat treatments for A15: 758-759
aluminum alloys M2: 144, 145, 147
of aluminum alloys, weights A15: 275
aluminum and aluminum alloys A2: 5
aluminum casting alloys A2: 139
aluminum-silicon, characteristics of A15: 159
casting design A15: 284
casting methods A15: 275-277
casting removal, from molds A15: 283-284
centrifugal casting method A15: 276-277
constant-level pouring method A15: 276
continuous casting method A15: 277
of copper alloys A2: 346 M2: 384
core materials, selection A15: 280
cores A15: 279-280
costs A15: 285
defects A15: 285
dimensional accuracy A15: 284
gating systems A15: 278-279
horizontal parting/tilt casting methods A15: 275-276
hybrid processes A2: 141-145
low-pressure die method A15: 276
of magnesium alloys A15: 275, 799, 807
magnesium alloys, specific types A2: 456-459
market trends A15: 44
mold coatings A15: 281-282
mold design A15: 277-278
mold life A15: 280-281
mold materials, selection A15: 280
mold temperature A15: 282-283
pouring temperature A15: 283
processes A15: 34-35, 205-206
semiautomatic A15: 275
semisolid-metal processing A2: 142
solid graphite, machined A15: 285
squeeze casting A2: 141-142
squeeze casting method A15: 277
surface finish A15: 284-285
tolerances A15: 620-621
turntables A15: 276
vacuum casting method A15: 276
vs sand casting A15: 285
zinc alloys A15: 797
Permanent mold casting machines A15: 276

Permanent mold castings, aluminum and aluminum alloys
anodizing M5: 590
Permanent mold processes *See also* Permanent mold casting; Permanent mold(s)
centrifugal casting A15: 34
defined A15: 34
development A15: 34-35
procedure A15: 205-206
slush casting A15: 34-35
types A15: 34, 204
Permanent mold(s) *See also* Permanent mold casting; Permanent mold processes
aluminum, and semisolid forging compared A15: 333-334
defined A15: 9
horizontal centrifugally cast A15: 296-297
processes, development of A15: 34-35
as reusable reverse patterns A15: 192
vertical centrifugal casting A15: 301-304
wash for A15: 304
Permanent patterns
in ceramic molding A15: 248
processes A15: 204
Permanent radio magnets
powders used A7: 574
Permanent set EM3: 20
defined A8: 10 A14: 9 EM1: 17 EM2: 29
Permanent television magnets
powders used A7: 574
Permanent viscosity loss A18: 84, 110
Permanent-magnet setups
for identification of ferrite in heat-resistant casting alloys A9: 334
Permanganate titration
for chromium and vanadium A10: 176
Permanickel 300 ..... A9: 435-437
Permeability *See also* Breathing A7: 8 EM3: 20
constant, with changing temperature nickel-iron alloys A2: 773
defined A15: 9 A17: 96 EM1: 17 EM2: 29
dry, defined *See* Dry permeability
effect, remote-field eddy current inspection A17: 197
effective (apparent), defined A17: 99
high, of magnetically soft materials A2: 761
of immiscible liquids A10: 164
incremental, measured A17: 134
initial, defined A17: 99
in magabsorption theory A17: 145
magnetic, in eddy current inspection A17: 167
of magnetic materials A17: 99
magnetic printing A17: 125
maximum, defined A17: 99
and microwave inspection A17: 202
mold, and porosity A15: 209
nickel-iron alloys A2: 711-772, 775
of plaster molds A15: 242
of polymers EM2: 61-62
of porous parts A7: 696
radio frequency, magabsorption theory A17: 148
and resistivity, compared A17: 145
reversible A17: 145-147
SI derived unit and symbol for A10: 685
SI unit/symbol for A8: 721
Permeability coefficients EM3: 623
Permeameter
Blaine or air A7: 264
Permeametry
apparatus and limitations A7: 264-265
as measure of specific surface area and average particle size A7: 262-265
Permeation EM4: 136
as flow in leaks A17: 58
Permissible variation *See also* Tolerance
defined A8: 10
Permittivity *See* Tan delta
defined EL1: 597-601
and microwave inspection A17: 202, 205

SUBJECTS OF THE INDEXED VOLUMES: ASM Handbook (designated by the letter "A"): **A1:** Properties and Selection: Irons, Steels, and High-Performance Alloys (1990); **A2:** Properties and Selection: Nonferrous Alloys and Special-Purpose Materials (1990); **A3:** Alloy Phase Diagrams (1992); **A4:** Heat Treating (1991); **A6:** Welding, Brazing, and Soldering (1993); **A7:** Powder Metallurgy (1984); **A8:** Mechanical Testing (1985); **A9:** Metallography and Microstructures (1985); **A10:** Materials Characterization (1986); **A11:** Failure Analysis and Prevention (1986); **A12:** Fractography (1987); **A13:** Corrosion (1987); **A14:** Forming and Forging (1988); **A15:** Casting (1988); **A16:** Machining (1989); **A17:** Nondestructive Testing and Quality Control (1989); **A18:** Friction, Lubrication, and Wear Technology (1992). **Metals Handbook, 9th Edition** (designated by the letter "M"): **M1:** Properties and Selection: Irons and Steels (1978); **M2:** Properties and Selection: Nonferrous Alloys and Pure Metals (1979); **M3:** Properties and Selection: Stainless Steels, Tool Materials and Special-Purpose Materials (1980); **M4:** Heat Treating (1981); **M5:** Surface Cleaning, Finishing, and Coating (1982); **M6:** Welding, Brazing, and Soldering (1983). **Engineered Materials Handbook** (designated by the letters "EM"): **EM1:** Composites (1987); **EM2:** Engineering Plastics (1988); **EM3:** Adhesives and Sealants (1990); **EM4:** Ceramics and Glasses (1991); **Electronic Materials Handbook** (designated by the letters "EL"): **EL1:** Packaging (1989).

Permittivity (continued)
SI derived unit and symbol for **A10:** 685
SI unit/symbol for **A8:** 721
Perovskite catalysts
freeze drying **EM4:** 62
Perovskite, in ceramic waste form simulant
EPMA analysis for **A10:** 532-535
Perovskites **EM4:** 61
applications **EM4:** 768
dielectric properties **EM4:** 768-769
examples including unit cell parameters **EM4:** 769
ferroelectric properties **EM4:** 768
piezoelectric properties **EM4:** 769-770
positive temperature coefficient of resistivity **EM4:** 769
properties **EM4:** 766
structure **EM4:** 766-768
structure-property relationship **EM4:** 768
tolerance factor **EM4:** 767
Peroxides **A13:** 677, 1194
addition to acrylics **EM3:** 120, 121-122
determined **A10:** 218
functional group analysis of **A10:** 218
for preparation of polysulfides **EM3:** 50
Peroxides, as initiator
polyester resins **EM1:** 133
Peroxy compounds **EM3:** 20
defined **EM2:** 30
Perpendicular section
defined **A9:** 13
Persistent slip bands
defined **A12:** 117
Personal computer based laminate analysis program **EM1:** 274
Personal computers *See also* Automated; Automation; Computers
for electrical testing **EL1:** 567
Personal computers, for corrosion data analysis *See also* Computers; Microcomputers **A13:** 317
Personal products
of copper-based powder metals **A7:** 733
Personnel *See also* Management; Operators; Safety
for boilers/pressure vessel fabrication **A17:** 641
inspection, and NDE reliability **A17:** 677
liquid penetrant inspection, training and certification **A17:** 85-86
radiograph interpretation **A17:** 347
radiographic, safety of **A17:** 297
training, with coordinate measuring machines **A17:** 28
ultrasonic inspection **A17:** 232
Perspective
distortion, effects of stereo imaging **A12:** 171
effect in SEM imaging **A12:** 169, 171
error, as distortion **A12:** 196
Persulfate hydroxide and cyanide as an etchant for beryllium-containing alloys **A9:** 394
Perturbation *See also* Stress(es)
in planar interface growth **A15:** 114-116
of stresses, at broken fiber end **EM1:** 193
PES *See* Polyether sulfones
Pesticides
as arsenic toxicity **A2:** 1237
GC/MS analysis **A10:** 639
liquid chromatography analysis of thermally unstable **A10:** 649
residues, detected in plant and animal tissues **A10:** 188
PESV *See* Polyether sulfones
PET *See* Polyethylene terephthalate
Petalite
specialty refractory **EM4:** 908
PETN (explosive)
friction coefficient data **A18:** 75
PETRA
as synchrotron radiation source **A10:** 413
Petro-forge presses **A7:** 305
Petrochemical applications
polybenzimidazoles (PBI) **EM2:** 147
thermoplastic fluoropolymers **EM2:** 117
Petrochemical industry
nickel-base alloy applications **A13:** 655-656
Petrochemical industry applications *See also* Oil industry
nickel alloys **A2:** 430
titanium and titanium alloys **A2:** 588
Petroff equation
defined **A18:** 14
Petrography
defined **A9:** 13
residue analysis by **A10:** 177
Petrolatum
packing for seals **A18:** 551
Petrolatum rust-preventive compounds **M5:** 459, 461-469
applying, methods of **M5:** 466
duration of protection **M5:** 469
film thickness, factors influencing **M5:** 467-468
flow characteristics determination of **M5:** 469-470
Petroleum *See also* Petroleum products
analytic methods for **A10:** 10
derivatives, analytic methods for **A10:** 10
oil, GC/MS analysis of volatile compounds in **A10:** 639
voltammetric monitoring of metals and nonmetals in **A10:** 188
Petroleum cleaners
aliphatic **M5:** 40-41
Petroleum fuels reforming
powder used **A7:** 574
Petroleum industry
horizontal centrifugal casting in **A15:** 300
Petroleum industry applications
computed tomography (CT) **A17:** 362-363
for flux leakage method **A17:** 132
tubular products **A17:** 577-578
Petroleum jelly
effect on bearing strength in aluminum alloy sheet **A8:** 60
Petroleum lubricating oil
surface tension **EM3:** 181
Petroleum oil *See* Mineral oil
Petroleum production operations **A13:** 1232-1261
cast steels, corrosion in **A13:** 575
corrosion causes **A13:** 1232-1235
corrosion control methods **A13:** 1235-1245
corrosion-resistant alloys **A13:** 1236
industry standards **A13:** 1259-1260
nonmetallic materials **A13:** 1243-1244
primary production **A13:** 1247-1251
problems/protective measures **A13:** 1245-1259
secondary recovery **A13:** 1251-1253
Petroleum products *See also* Petroleum
acidity-basicity measured in **A10:** 172
analysis of **A10:** 100-101
sulfur determination by XRS **A10:** 82
Petroleum refining and petrochemical operations **A13:** 1262-1287
alloy steel corrosion in **A13:** 535
codes and standard specifications **A13:** 1263
corrosion **A13:** 1266-1274
corrosion control **A13:** 1282-1284
erosion-corrosion **A13:** 1281-1282
high-temperature corrosion **A13:** 1270-1274
low-temperature corrosion **A13:** 1266-1270
materials selection **A13:** 1262
principal materials **A13:** 1262-1266
SCC and embrittlement **A13:** 1274-1281
Petroleum-base compounds
solvent- cutback **A12:** 73
Petroleum-base lubricants
for corrosion control **A11:** 194
Petroleum-based oils
phosphate coatings supplemented with **M5:** 453
Petroleum-based rust-preventive compounds *See* Solvent-cutback petroleum-based rust-preventative compounds
Petroleum-refinery components
elevated- temperature failures in **A11:** 289-292
Petrov equation
friction coefficient **A18:** 45, 46
Pettifor structure map **A6:** 143
Pewter *See also* Tin; Tin alloys; Tin alloys, specific types, pewter; Tin and tin alloys, specific types; Tin-antimony-copper alloys; White metal **A13:** 774 **M2:** 614
heat treating **M4:** 776
Pewter (continued)
properties **A2:** 522-523
recycling **A2:** 1219
PFA *See* Perfluoro alkoxy alkane
PFM alloy systems **A13:** 1353-1356
PGAA *See* Prompt gamma activation analysis
pH *See also* Acidity; Bases **EM3:** 20
ADV-pH test, for reclaimed sand **A15:** 355
of aqueous environment synthesis solution **A8:** 416
of body fluids, shifts in **A11:** 673
cemented carbide corrosion rate as function of **A13:** 850
change, effect in analyte extraction **A10:** 164
changes, for corrosion control **A11:** 198
of chromate conversion coatings **A13:** 394
constant, maintained by electrometric titration **A10:** 202
control, in boiler tubes **A11:** 615
control, in EDTA titration **A10:** 173
corrosion fatigue test specification **A8:** 423
and corrosion rates, in boiler tubes **A11:** 612-613
corrosive effects in seawater **A13:** 896-898
defined **A13:** 10 **EM1:** 17 **EM2:** 30
determination by glass electrode **A10:** 203
effect, boiler corrosion **A13:** 517
effect, corrosion rate of steel in water **A13:** 991
effect, erosion-corrosion, nuclear reactors **A13:** 965
effect, ethyl silicate slurries **A15:** 212
effect, in aqueous corrosion **A13:** 37-39, 489-490, 512, 896-898, 1304
effect in chelometric titration **A10:** 164
effect in inorganic precipitation **A10:** 169
effect, in stress-corrosion cracking **A13:** 147
effect in sulfide-stress cracking **A11:** 298
effect, in water **A13:** 489-490
effect, iron corrosion rate in aerated soft water **A13:** 1301
effect of overpotential in electrogravimetry **A10:** 198
effect on fatigue crack growth in steam with contaminants **A8:** 427
effect on fatigue crack growth rate **A8:** 416
effect on near-threshold fatigue crack growth alloy **A8:** 427, 430
rates in stainless steel and titanium
effect on reduction potential **A7:** 54
effect on time-to-facture by stress-corrosion cracking **A11:** 220
effect, pollution control **A13:** 1367
effect, pulp bleach plants **A13:** 1193
effect, water-soluble flux properties **EL1:** 647
effect, zinc corrosion **A13:** 526, 1304
electrode, for high-purity water tests **A8:** 422
and equilibrium, in classical wet analysis **A10:** 163
of fluxes **EL1:** 644
high-purity oxygenated water **A8:** 420
influence in acidified chloride environments **A8:** 419
low, and stress-corrosion cracking **A8:** 499-500
Mattsson's 7.2 solution, for copper alloys **A8:** 525
meters, vs acid-base indicators **A10:** 173
negative logarithm of hydrogen-ion activity as **A10:** 690
neutral conditions, aqueous corrosion **A13:** 38
pressurized water reactor specification **A8:** 423
of sands **A15:** 208
solution, effect on stress-corrosion cracking **A8:** 499
vs. boric acid concentration for pressurized water reactor **A8:** 423
of water vapor, effect on copper alloy tubing **A11:** 634-635
PH 13-8 Mo *See* Stainless steels, specific types, S13800
PH 15-7Mo *See* Stainless steels
PH 17-7 *See* Stainless steels, specific types, S17700
pH effect on potentiostatic etching **A9:** 144-147
of polishing fluids effects of **A9:** 47
Phantom
defined **A17:** 384
Phantom-emitter transistor structure
schematic **A11:** 788

Pharmaceutical final filters
powders used **A7:** 573
Pharmaceutical industry **A13:** 1226-1231
bismuth applications **A2:** 1256
construction materials **A13:** 1226-1228
products, aluminum/aluminum alloys resistance to **A13:** 602
products, tantalum resistance to **A13:** 728
stainless steel corrosion **A13:** 560
Pharmaceutical mixtures
liquid chromatography of **A10:** 649
voltammetric analysis of metals in **A10:** 188
Pharmaceuticals, stainless steels
corrosion resistance **M3:** 91-92
Phase **EM3:** 20
defined **A9:** 13 **EM2:** 30
Phase accommodation, in liquid-phase sintering
tungsten-nickel-iron alloys **A7:** 320
Phase analysis **EM4:** 557-563
by Mössbauer spectroscopy **A10:** 287-295
of hydrided TiFe, Mössbauer effect **A10:** 293-294
material applications **EM4:** 561
nuclear magnetic resonance **A10:** 277-286
phase identification by diffraction **EM4:** 557-560
electron diffraction **EM4:** 560
x-ray diffraction **EM4:** 558-560
purposes **EM4:** 557
quantitative analysis **EM4:** 562
differential scanning calorimetry **EM4:** 562
requirements **EM4:** 557
spectroscopic methods **EM4:** 560-561
infrared absorption spectroscopy **EM4:** 561-562
Raman spectroscopy **EM4:** 561
thermal analysis **EM4:** 561-562, 563
differential thermal analysis **EM4:** 561-562
differential thermogravimetric analysis **EM4:** 561, 563
measurement capabilities **EM4:** 562
thermogravimetric analysis **EM4:** 561, 562, 563
Phase angle **EM3:** 20
defined **EM2:** 30
Phase angle firing
as power control for Mo furnaces **A7:** 423
Phase boundaries
of a two-phase field, application of volume- fraction measurements to establish **A9:** 125
determined **A10:** 474
effect on lateral resolution, atom probe analysis **A10:** 595
heat tinting to reveal **A9:** 136
stress-corrosion cracking along **A8:** 501
Phase boundary structures
thermal-wave imaging used to study **A9:** 91
Phase change **EM3:** 20
defined **EM2:** 30
Phase changes
as a result of electric discharge machining **A9:** 27
Phase compositions
quantitative analysis of peak shapes for **A7:** 316
Phase contrast
analytical electron microscopy **A10:** 445, 446
defined **A10:** 678
microscopy, Lorentz microscopy as **A10:** 446
and potentiostatic etching **A9:** 144
Phase contrast etching **A9:** 59
Phase contrast illumination **A9:** 79
defined **A9:** 13
Phase contrast imaging **A18:** 389
Phase contrast microscopy
principles of **A9:** 59
Phase contrast transmission electron microscopy *See* Lattice-image contrast transmission electron microscopy
Phase diagram **EM1:** 750
gold-silicon **EL1:** 213
Phase diagram determination
equilibrium verification **A10:** 475-476
experimental procedure **A10:** 474
phase boundaries **A10:** 474-475
traditional and modem probe-forming transmission electron method **A10:** 473-474
Phase diagrams **A6:** 127
A15 superconducting materials **A2:** 1062
beryllium-copper alloys **A2:** 404
binary Fe-C system **A15:** 61
binary iron-chromium equilibrium **A6:** 678, 681
binary isomorphous **A6:** 46, 47
chemical principles **A15:** 52
construction errors ..... **A3:** 1 • 9, 10
and cooling curve, relationship **A15:** 182
defined **A9:** 13 **A15:** 9
description ..... **A3:** 1 • 2
determination ..... **A3:** 1 • 17-18
determination of **A10:** 473-476
as determined by thermal analysis **A15:** 182-185
effect on dendritic structures **A9:** 613
eutectic, coupled zones **A15:** 123
eutectic, schematic **A15:** 121
Fe-C-P, liquidus surfaces calculated **A15:** 64
features ..... **A3:** 1 • 7-10
iron-carbon **A15:** 629
iron-carbon binary **A4:** 43, 45, 48, 49
iron-chromium-nickel pseudo-binary **A6:** 688
iron-oxygen **A15:** 89
and laws of thermodynamics **A15:** 50
lead-tin **A6:** 128
lines and labels ..... **A3:** 1 • 8
magnetic **A10:** 268
micro- and macrosegregation in **A15:** 136
partition coefficients in **A15:** 102, 136
peritectic reaction **A15:** 125
reading of ..... **A3:** 1 • 18-22
schematic of binary eutectic **A9:** 618
single-phase region, liquidus/solidus lines **A15:** 114
ternary iron-chromium-nickel **A6:** 686
tin-lead **A2:** 552
used in electropolishing **A9:** 49
XRPD determined **A10:** 333
Phase diagrams defined **A13:** 46-47
for Fe-C system **A13:** 47
iron-iron carbide-silicon ternary **A13:** 566
Phase diagrams, pseudo binary
for Waspaloy **A14:** 236
Phase differences
effect of polarized light on image **A9:** 78
Phase distribution
of inorganic solids, methods for analysis **A10:** 4-6
of organic solids, methods for analysis **A10:** 9
SIMS analysis **A10:** 610
Phase extraction used to determine shapes of eutectic structures **A9:** 620
Phase field
description ..... **A3:** 1 • 2
rule ..... **A3:** 1 • 7
Phase identification
aided by polarized light **A9:** 79
by anodizing **A9:** 142
by polarized light etching **A9:** 59
of inorganic solids, applicable analytical methods **A10:** 4-6
material contrast in scanning electron microscopy **A9:** 94
second-phase testing, classical wet chemistry **A10:** 176-177
surface, LEED analysis **A10:** 536
transmission electron microscopy **A9:** 307-308
transmission electron microscopy diffraction patterns **A9:** 109-110
unknown, by electron diffraction/EDS analysis **A10:** 455-459
in wrought heat-resistant alloys **A9:** 307-309
of wrought stainless steels **A9:** 281-282
Phase interfaces
between matrix and precipitate **A9:** 648
types **A9:** 604
Phase or compound identification
analytical transmission electron microscopy **A10:** 429-489
electron probe x-ray microanalysis **A10:** 516-535
elemental and functional group analysis **A10:** 212-220
field ion microscopy **A10:** 583-602
gas analysis by mass spectrometry **A10:** 151-157
gas chromatography/mass spectrometry **A10:** 639-648
infrared spectroscopy **A10:** 109-125
liquid chromatography **A10:** 649-659
molecular fluorescence spectrometry **A10:** 72-81
Mössbauer spectroscopy **A10:** 287-295
neutron diffraction **A10:** 420-426
nuclear magnetic resonance **A10:** 277-286
optical metallography **A10:** 299-308
Raman spectroscopy **A10:** 126-138
single-crystal x-ray diffraction **A10:** 344-356
small-angle x-ray and neutron scattering **A10:** 402-406
ultraviolet/visible absorption spectroscopy **A10:** 60-71
x-ray diffraction **A10:** 325-332
x-ray powder diffraction **A10:** 333-343
Phase particle shapes **A9:** 619
determination of **A9:** 620
Phase particle structure in eutectics **A9:** 619
Phase particles
arrangement in eutectic colony structures **A9:** 619-620
Phase problem
in single-crystal x-ray diffraction **A10:** 349-351
Phase relief
in tin and tin alloys as a result of excess polishing **A9:** 449
Phase rule
defined **A9:** 13
description ..... **A3:** 1 • 2
violations ..... **A3:** 1 • 9, 10
Phase separated glasses **EM4:** 433
applications **EM4:** 433
microstructure **EM4:** 433
opal glasses **EM4:** 433
properties **EM4:** 433
Phase separation **EM3:** 20
in titanium alloys as a result of beta decomposition **A9:** 461
Phase shift
defined **EL1:** 1152
Phase splitting
in titanium alloys as a result of beta decomposition **A9:** 461
Phase transformation
and hardness **A12:** 32-33
Phase transformation, by welding
corrosion effects **A13:** 49
Phase transformations *See also* Solid-state phase transformations
as a result of heat tinting **A9:** 136
effect of temperature on **A10:** 318
effect on as-cast solidification structures in steel **A9:** 624
hot-stage microscopy used to study **A9:** 82
ion implantation strengthening mechanisms **A18:** 855, 856, 857
pressure- or temperature-induced, XRPD analysis **A10:** 333
revealed by differential interference contrast **A9:** 59
solid-state, XRPD analysis **A10:** 333
studied by FIM/AP **A10:** 583
studied by x-ray topography **A10:** 365, 376
Phase transition temperature *See also* Temperatures
stainless steels, and magabsorption measurement **A17:** 152

Phase transitions
crystallographic, variable-temperature ESR studies of A10: 257
observing by neutron diffraction A10: 420
Phase(s) *See also* Solid phases
equilibria, and inclusion-forming inclusions A15: 89
in equilibrium system A15: 50-54
intermetallic A15: 166-167
stability, phase diagrams of A15: 57, 62
transformation, in microsegregation A15: 137-138
Phase-change lubrication *See* Melt lubrication
Phase-dependent voltage contrast
use in integrated circuit failure analysis A11: 768
Phase-discrimination technique
eddy current inspection A17: 172-173
Phase-field-boundary
curvatures ...... A3: 1 • 9, 10
extensions ...... A3: 1 • 3, 4
intersections ...... A3: 1 • 8, 10
Phase-fraction lines ...... A3: 1 • 17, 19
Phase-rotator control EM3: 758
Phase-sensitive detector
microwave inspection A17: 205, 208
Phase-stepping methods
optical holography A17: 416
Phase/grain size and distribution
image analysis A10: 309-322
optical metallography A10: 299-308
scanning electron microscopy A10: 490-515
Phased-array radar systems
with monolithic microwave integrated circuits (MMICS) A2: 740
Phases *See also* Phase or compound identification; Phase transformations; Phase transitions ...... A3: 1 • 1
amorphous, TEM bright-field images of ceramic containing A10: 445
amount of, as x-ray diffraction analysis A10: 325
changes detected in A10: 277, 282-283
changes, using single-crystal x-ray diffraction for A10: 354
chemistry of A10: 445, 446, 678
compositional analysis, EPMA A10: 516
crystalline, XRPD analysis A10: 333
defined, for inclusion and second-phase testing A10: 176
differences in scattering from different electrons within an atom A10: 328
precipitate, SEM analysis A10: 490
problem in single-crystal x-ray diffraction A10: 349-351
in rapidly solidified alloys A9: 615-617
separation of A10: 402, 405
size and shape determined by contrasting interference layers A9: 60
stability of A10: 598
structure, characterization by optical metallography A10: 299
unknown, identification of A10: 455
wrought aluminum alloy A2: 36-37
Phases in aluminum alloys
designations A9: 356-359
formation of A9: 359
identification of A9: 355-360
possible phases of various systems A9: 359
substitution of elements A9: 359
PHBV-biodegradable plastic
development of EM2: 786
Phenacite
crystal structure EM4: 881
island structure EM4: 758
Phenanthroline method
for iron in lead alloys A10: 66
Phenoformaidehyde resins
formation of A10: 132
Phenol A13: 1270
Phenol formaldehyde EM3: 104
formulations EM3: 107
for particle board production EM3: 106
properties EM3: 106
Phenol formaldehyde novolac (PN)
characteristics EL1: 811
Phenol formaldehyde resins
applications EM4: 47
Phenol formaldehyde resins (continued)
composition EM4: 47
supply sources EM4: 47
Phenol formaldehyde/resorcinol formaldehyde EM3: 104
Phenol red
as acid-base indicator A10: 172
Phenol, stainless steel
corrosion resistance M3: 87
Phenol-aralkyl bonds
bonded-abrasive grains A2: 1014
Phenol-formaldehyde as mounting material for electropolishing A9: 49
Phenol-formaldehyde novolac EM3: 595
Phenol-formaldehyde resins
chemical resistance properties EM3: 639
formulations EM3: 107
sulfuric acid corrosion A13: 1154
Phenolic acid catalyzed no-bake binder process A15: 214-215, 238
Phenolic compounds
lubricant analysis A18: 300
Phenolic epoxy EM3: 106
Phenolic ester cold box resin binder process A15: 220-221
Phenolic fiber
abrasive blasting of M5: 91
Phenolic hot box processes
as coremaking system A15: 238
Phenolic linings A13: 408
Phenolic novolacs EM3: 594, 595
for coating/encapsulation EL1: 242
as epoxy resin EL1: 826-827
vs epoxy resin, for molded plastic packages EL1: 474
Phenolic resin *See also* Phenolics; Resins
defined EM2: 30
Phenolic resin binders EM3: 47
for foundry sand patterns EM3: 47
Phenolic resins EM3: 20
additives to carbon-graphite materials A18: 816, 817
adhesives EM1: 684
application EM1: 32
defined EM1: 17
die materials for sheet metal forming A18: 628
fabric reinforced, cage material for rolling- element bearings A18: 503
fiber-reinforced composites, properties EM1: 381
flame resistance of EM1: 141
formation A13: 408, 1154
impregnation effects on typical carbon- graphite base material A18: 817
impregnation effects on typical graphite-base material A18: 817
as medium-temperature thermoset EM1: 381-391
as modifiers EM3: 181
properties EM1: 290, 292, 381
sample chemical reactions EM1: 751-753
as tackifiers EM3: 182
test methods for EM1: 292
tests for EM1: 290
as thermosetting EM1: 32
types EM1: 289
Phenolic resins and coatings M5: 473-475, 496, 498-499, 501
Phenolic resins as mounting materials A9: 44
for cast irons A9: 243
Phenolic silicones
suppliers EM3: 105
Phenolic spheres
as extender EM3: 176
Phenolic urethane cold box process A15: 219, 238
Phenolic urethane no-bake binder system A15: 216-217
Phenolic/carbon microballoons
abradable seal material A18: 589
Phenolic/epoxy-novolacs EM3: 104
Phenolics *See also* Resins; Thermosetting resins EM3: 75, 103-107
additives and modifiers EM3: 107
advantages and limitations EM3: 79
for aerospace honeycomb core construction EM3: 560
for aerospace industry applications EM3: 105
aircraft applications EM3: 79, 559
for aircraft skins EM3: 105
Phenolics (continued)
applications EM2: 242-243
for auto transmission blades EM3: 105
for automobile brakeshoes EM3: 79
based on modified phenols and/or aldehydes EM3: 104
for bonding aluminum oxide and silicon carbide EM3: 105
for bonding coated abrasives EM3: 105
for bonding paper to wood, plastics, and metals EM3: 105
for bonding plywood EM3: 79
for brake blocks EM3: 105
for brake linings EM3: 105, 639
by-products from cure EM3: 74
characteristics EM2: 243-245
chemical resistance properties EM3: 639
chemistry EM3: 79, 103-104
chipboard construction EM3: 105
for cloth bonding EM3: 105
for clutch disks EM3: 79
for clutch facings EM3: 105, 639
for coated and bonded abrasives EM3: 105
commercial forms EM3: 104
compared to epoxies EM3: 98
composites EM3: 105
consumption (1989) EM3: 105
cost factors EM3: 106
costs (1989) EM3: 105
costs and production volume EM2: 242
cross-linking EM3: 413
cure rate reduction EM3: 105
curing methods EM3: 79, 103
for disk pads EM3: 105
environmental effects EM2: 428
formaldehyde EM3: 103
formulations EM3: 107
for foundry and shell moldings EM3: 105
for friction materials EM3: 105
for glass-phenolic laminate EM3: 107
as glassy polymers EM3: 617
hardboard construction EM3: 105
for hot pressing weather-resistant plywood EM3: 107
industrial applications EM3: 567
as injection-moldable EM2: 321
for insulation materials EM3: 105
for laminating EM3: 103, 105
for leather bonding EM3: 105
markets EM3: 105-106
as medium-temperature resin system EM2: 441-442
for metal bonding EM3: 79
microstructural analysis EM3: 412
modified EM3: 106
moisture effect
molding compounds EM2: 627
novolacs EM3: 103
as organic binders A15: 35
oriented strand board (OSB) construction EM3: 105
for particle board production EM3: 106
particleboard construction EM3: 105
physical properties EM3: 104
for plastic bonding EM3: 105
for plywood construction EM3: 105
prebond treatment EM3: 35
predicted 1992 sales EM3: 77
primers EM3: 101
processing EM2: 244-245
product forms EM2: 245
properties EM3: 106-107
reinforced, properties EM2: 245
resistant to many aggressive materials EM3: 637
resols EM3: 103
rigidity EM3: 106
for rocket motor nozzles EM3: 105-106
for rubbers bonding EM3: 105
shelf life EM3: 79
for shell molding EM3: 105
silane coupling agents EM3: 182
stainless steel bonding EM3: 106
steel joint showing adhesion-dominated durability EM3: 666, 667
as structural plastic EM2: 65
as structural plastic, chemistry EM2: 65
for structural wood bonding EM3: 105

Phenolics (continued)
substrate cure rate and bond strength for cyanoacrylates **EM3:** 129
suppliers **EM2:** 245 **EM3:** 80, 104-105
surface preparation **EM3:** 277
tackifiers for **EM3:** 183
thermal resistance **EM3:** 98
for tire cord adhesion **EM3:** 105
tougheners **EM3:** 183
vinyl **EM3:** 106
wafer board construction **EM3:** 105
wood adhesives (plywood) **EM3:** 103
for wood, fibrous and granulated **EM3:** 105
for wood product bonding **EM3:** 106
for wood veneer plywood production **EM3:** 107
Phenolphthalein
as acid-base indicator **A10:** 172
Phenols **EM3:** 103, 104
determined **A10:** 218-219
electropolymerization of, SERS study of **A10:** 136
in isopropanolic medium, determined by electrometric titration **A10:** 205
Phenomenological theory
of anisotropy **A8:** 143
Phenoxy resins *See also* Resins **EM3:** 20, 181
defined **EM2:** 30
Phenyl glycidyl ether (PGE)
as epoxy diluent **EM1:** 67, 70
Phenylgroups
in polymers, Raman analysis **A10:** 131
in silicone, Raman analysis **A10:** 132
Phenylhydrazine
as precipitant **A10:** 169
Phenylsilane resins *See also* Resins **EM3:** 20
defined **EM2:** 30
Phenylthiohydantoic acid
as precipitant **A10:** 169
Philbrook, B.F
investment casting by **A15:** 35
Philips PW-1410 sequential x-ray spectrometer
spectrum from **A10:** 88
Phillips vacuum gages
for gas/leak detection **A17:** 64
Phonons, excitation
as inelastic scattering process **A10:** 434
Phosphate
coatings for valve train assembly components **A18:** 559
Miller numbers **A18:** 235
Phosphate alkaline cleaners **M5:** 23-24, 28, 35
Phosphate bonded molds
as inorganic binder system **A15:** 229-230
Phosphate coating **M5:** 434-456
accelerators used in **M5:** 435, 442
acidity, baths **M5:** 435, 442-443, 455
alkaline cleaning process **M5:** 438-439, 450-451, 453-454
alloys coated by **M5:** 437-438
aluminum and aluminum alloys **M5:** 438, 597-599
applications **M1:** 174 **M5:** 434-442, 444, 447-450, 453-454
applying methods of *See also* Phosphate coating, immersion; Spray **M5:** 440-441
baskets **M5:** 446, 448
cast iron **M5:** 436-438, 441, 448
cast irons **M1:** 104
ceramic coatings *See* Phosphate-bonded ceramic coatings
chromate concentration **M5:** 443
chromic acid used in **M5:** 439, 442-443, 448, 454
chromium, hexavalent contamination, limits **M5:** 455-456
for cold extruded parts **A14:** 304
contaminants, limits and treatment of **M5:** 454-456
conveying equipment **M5:** 446-447, 449

Phosphate coating (continued)
corrosion protection **M1:** 754
corrosion resistance and protection **M5:** 435-436, 441-442, 453, 455-456
crystal size, control of **M5:** 449-451, 455
cycle times **M5:** 448-450, 453-454
drawing and forming operations aided by **M5:** 436-437
drums **M5:** 445-446
equipment **M5:** 444-449, 453-454
use of **M5:** 447-449, 453-454
etching by **M5:** 8
galvanized steel **M1:** 169
hydrogen embrittlement caused by **M5:** 435
hypoid gear wear, effect on **M1:** 632
immersion systems *See* Immersion phosphate coating systems
iron concentration and removal **M5:** 442-443
iron phosphate process *See* Iron phosphate coating
limitations, shape and size imposing **M5:** 452-453
low-temperature coatings **M5:** 456
maintenance **M5:** 443-444
solution **M5:** 443-444
tank **M5:** 444
manganese phosphate process *See* Manganese phosphate coating
nickel contamination, limits **M5:** 455-456
oils in **M5:** 435, 439, 442-443, 448, 453, 455-456
contamination by **M5:** 439, 456
immersion tanks **M5:** 448, 453, 455
supplemental coatings **M5:** 435-436, 442, 448, 453, 455-456
operating control schedules **M5:** 444
painting base applications **M5:** 434-436, 442, 449-450, 454-456
painting over **M1:** 174-175
phosphating time **M5:** 437, 441, 456
phosphoric acid cleaners compared with **M5:** 436
pickling process **M5:** 438-439, 450, 453-454
posttreatment processes **M5:** 442-443
precleaning processes **M5:** 438-439, 447, 450-451
prepaint treatment **M5:** 476-478
process, chemical control of **M5:** 442-443
process steps **M5:** 438-442
production of **M1:** 174-175
quality control inspection methods **M5:** 451-452
racks **M5:** 447
repair, coatings **M5:** 452
rinsing processes **M5:** 439-443, 445, 447-448, 450-451, 453-454
chromic acid **M5:** 442-443, 448
immersion **M5:** 439-440
postcleaning **M5:** 439-442
postphosphating **M5:** 441-442
spray **M5:** 440
tanks **M5:** 445, 448
roller wear in torque converter affected by **M1:** 631
room-temperature precleaning procedures **M5:** 15
safety precautions **M5:** 453-454
solution break-in **M5:** 444-445
solution compositions and operating conditions **M5:** 435, 441-445, 448-450, 453-454
spray system *See* Spray phosphate coating system
stainless steel **M5:** 437-438
steel **M5:** 434-439, 441-443, 449
steel wire **M1:** 262, 266
steel wire rod **M1:** 253
surface preparation for **M5:** 5, 15, 17
tanks and accessories **M5:** 444-445, 448, 455
temperature **M5:** 438, 441, 456
low-temperature coatings **M5:** 456
tests, chemical solution control **M5:** 442-443
thickness, coating **M5:** 434, 436, 438, 441, 448-451
types, characteristics of **M5:** 434-435
voids, coating **M5:** 451-452

Phosphate coating (continued)
waste recovery and disposal **M5:** 448, 454-456
wax coating, supplementary **M5:** 435
wear resistance **M5:** 436-437
wear resistance improved by **M1:** 631-632, 634
weight, coating **M5:** 434-437, 441-442, 448-449, 451, 454
control of **M5:** 449, 454
determination of **M5:** 451
immersion time, function of **M5:** 441
work-supporting equipment **M5:** 446-448
zinc contamination, limits **M5:** 455-456
zinc phosphate process *See* Zinc phosphate coating
Phosphate coatings **A1:** 222
application methods **A13:** 387
automotive **A13:** 1015
as barrier protection, aqueous solutions **A13:** 378-379
bath, testing **A13:** 384-385
for carbon steel **A13:** 523-524
characteristics **A13:** 386
equipment **A13:** 387-388
formation **A13:** 383
heavy phosphates **A13:** 387
iron phosphating **A13:** 385-386
processing sequence **A13:** 383
types **A13:** 383
weight of **A13:** 385
zinc phosphating **A13:** 386-387
Phosphate conversion coatings **A13:** 383-388
Phosphate esters
lubricants for rolling-element bearings **A18:** 134, 135
properties **A18:** 81
Phosphate frits
melting/fining **EM4:** 392
Phosphate glasses
composition **EM4:** 741
elastic modulus **EM4:** 850
seal design techniques with chemical integrity **EM4:** 539
Phosphate porcelain enamels
composition of **M5:** 510-511
Phosphate rhodium plating solutions **M5:** 290
Phosphate treatment
cadmium plating **M5:** 264
Phosphate-bonded ceramic coatings **M5:** 535-537
densities and maximum service temperatures **M5:** 537-538
Phosphate-fluoride (PF) etch
of polyphenylquinoxalines **EM3:** 166, 167
Phosphate-free electrocleaners **M5:** 28, 35
Phosphated steels
welding factor **A18:** 541
Phosphates
anions, separation by ion chromatography **A10:** 659
as anodic inhibitor **A13:** 494-495
compounds, in conversion coatings **A13:** 383
as fining agents **EM4:** 380
general theory **A13:** 383-384
rock, nitric acid as dissolution medium for **A10:** 166
weighing as the, gravimetry analysis **A10:** 171
Phosphating **A18:** 62
defined **A13:** 10
Phosphazenes
advantages and disadvantages **EM3:** 675
chemistry **EM2:** 66
Phospher bronzes
corrosion in various media **M2:** 468-469
Phosphide embrittlement
in brazing **A11:** 452
Phosphide sweat
as casting defect **A11:** 387
Phosphonates
as cathodic inhibitors **A13:** 495
Phosphonic esters
dispersants **A18:** 99, 100

Phosphonitrile-fluoroelastomers (PNF) **EM3:** 677, 678
Phosphonitrilic fluoroelastomers (PNF)
for severe environments **EM3:** 673
Phosphor bronze *See* Copper alloys, specific types, C51000
photochemical machining **A16:** 588, 590
strip rolled from static cast ingot **A9:** 644
thread rolling **A16:** 282
Phosphor bronze A
springs, strip for **M1:** 286
springs, wire for **M1:** 284
Phosphor bronze spring
fatigue fracture **A11:** 555-557
Phosphor bronze strip
nonparametric evaluation of percent survival values for **A8:** 707, 709
spring-tempered, load deflection plot **A8:** 136
spring-tempered, modulus of elasticity and proof strength in bending **A8:** 136
Phosphor bronzes *See also* Bearing bronzes; Copper-tin alloys **A6:** 752
51000, roll welding **A6:** 313
applications **A15:** 784
applications and properties **A2:** 321-325
for bearings, wear-resistant applications **A2:** 352, 354
brazeability **M6:** 1034
brazing **A6:** 630, 931, 934
cladding material for brazing **A6:** 347
composition and properties **M6:** 401
corrosion resistance **A13:** 611
as electrical contact materials **A2:** 843
gas metal arc welding **M6:** 420
gas tungsten arc welding **M6:** 410-412
gas-metal arc butt welding **A6:** 760
gas-metal arc welding **A6:** 764
to high-carbon steel **A6:** 828
to low-alloy steel **A6:** 828
to low-carbon steel **A6:** 828
to medium-carbon steel **A6:** 828
to stainless steel **A6:** 828
gas-tungsten arc welding **A6:** 763-764
to high-carbon steel **A6:** 827
to low-alloy steel **A6:** 827
to low-carbon steel **A6:** 827
to medium-carbon steel **A6:** 827
to stainless steel **A6:** 827
hot cracking **A6:** 764
recycling **A2:** 1214
relative weldability rating, resistance spot welding **A6:** 834
resistance spot welding **A6:** 850
shielded metal arc welding **A6:** 754, 755, 763, 764 **M6:** 425
shrinkage allowance **A15:** 303
weldability **A6:** 753
Phosphor gear bronze
properties and applications **A2:** 374-375
Phosphor silicon bronzes
thermal expansion coefficient **A6:** 907
Phosphorescence **A10:** 679, 690
Phosphoric acid
applications **EM4:** 47
cast iron resistance **A13:** 569-570
as chemical cleaning solution **A13:** 1140
composition **EM4:** 47
copper/copper alloys in **A13:** 627-628
derivatives, as solvent extractant **A10:** 169-170
as electrolyte **A9:** 51
and ethanol as an electrolyte for magnesium alloys **A9:** 426
as ferrous cleaning agent **A12:** 75
hydrogen peroxide and methanol as an electrolyte for beryllium-copper alloys **A9:** 393
nickel alloys, corrosion **M3:** 173
in organic solvent (Group III electrolytes) **A9:** 52-54
pure, nickel-base alloy corrosion **A13:** 645, 647
as sample dissolution medium **A10:** 165
as sample modifier, GFAAS analysis **A10:** 55
stainless steel corrosion **A13:** 558
stainless steels, corrosion **M3:** 86, 87
supply sources **EM4:** 47
tantalum corrosion in **A13:** 725
in water (Group III electrolytes) **A9:** 52-54
Phosphoric acid (continued)
for wrought heat-resistant alloys **A9:** 308
zirconium/zirconium alloy resistance **A13:** 715-716
Phosphoric acid and water
for rust removal **A9:** 172
Phosphoric acid anodization (PAA) **EM3:** 42, 52, 249-250, 251, 558
aluminum **EM3:** 41, 625, 845
surface preparation, processing quality control **EM3:** 733, 738
Phosphoric acid bonds
characteristics **A15:** 213
Phosphoric acid chemical brightening baths **M5:** 580
Phosphoric acid cleaners **M5:** 8, 10, 436
phosphate coatings compared with **M5:** 436
Phosphoric acid cleaning process **M5:** 59-65
Phosphoric acid electropolishing solutions **M5:** 303, 305, 308
Phosphoric acid fuel cells **A13:** 1320-1321
Phosphoric acid pickling
magnesium alloys **M5:** 630-631, 640-641
Phosphoric anodizing process
aluminum and aluminum alloys **M5:** 592
Phosphoric-nitric acid chemical brightening baths **M5:** 579-580
Phosphoric-sulfuric acid chemical brightening baths **M5:** 580
Phosphorized admiralty metal
applications and properties **A2:** 318-319
Phosphorized leaded Muntz metal
applications and properties **A2:** 311
Phosphorized naval brass
applications and properties **A2:** 319-320
Phosphorous compounds
as flame retardants **EM2:** 504
Phosphors **EM4:** 18
material compositions used **EM4:** 18
thermally quenched **A17:** 399, 604-605
Phosphorus *See also* Dephosphorization
as addition to aluminum-silicon alloys **A18:** 788
as addition to brazing filler metals **A6:** 904
as addition to carbon-graphite materials **A18:** 816
in alloy cast irons **A1:** 87-88
alloying, aluminum casting alloys **A2:** 132
alloying effect on copper alloys **M6:** 402
alloying effect on nickel-base alloys **A6:** 590
alloying, magnetic property effect **A2:** 762
alloying, wrought copper and copper alloys **A2:** 242
in aluminum alloys **A15:** 746
aluminum-phosphorus-oxygen SBD **EM3:** 250, 251
as an addition to low-carbon electrical steels **A9:** 537
analyzed in glassivation layers **A11:** 41
at elevated-temperature service **A1:** 640
in austenitic manganese steel **A1:** 822, 824
in austenitic stainless steels **A6:** 457, 458, 463
back-titration determination **A10:** 173
in cast iron **A1:** 5
in cast iron composition **A18:** 695
cast iron content, effect on machinability **A16:** 649, 652, 654
cause of hot cracks **A6:** 409
cause of temper embrittlement **A4:** 124, 135
in composition, effect on ductile iron **A4:** 686
in composition, effect on gray irons **A4:** 671
in composition, factor affecting overheating of tool steels **A4:** 602
composition, wt% (maximum), liquation cracking **A6:** 568
in compounds providing flame retardance **EM3:** 179
concentrations, cupolas **A15:** 390
content, atmospheric corrosion effects **A13:** 514
content in high-strength low-alloy quenched and tempered steels **A6:** 665
content in stainless steels **A16:** 682-683, 685, 688 **M6:** 320, 322
in copper alloys **A6:** 753
in copper alloys, inclusion-forming **A15:** 90
copper-phosphorus alloys, resistance brazing filler metals **A6:** 342
Phosphorus (continued)
cracking sensitivity in stainless steel casting alloys **A6:** 497
deoxidation **A15:** 468-469
-deoxidized coppers **A13:** 615, 627
deoxidizing, copper and copper alloys **A2:** 236
determined by 14-MeV FNAA **A10:** 239
detrimental effects in thermit welds **M6:** 693-694
detrimental to welding of alloy systems **A6:** 89
in ductile iron **A15:** 649
for ductility enhancement, ASTM/ASME alloy steels **A12:** 349
in duplex stainless steels **A6:** 472
effect, cartridge brass **A2:** 301
effect, gas dissociation **A15:** 83
effect in copper alloys **A11:** 221, 635
effect in intermediate dielectrics on integrated circuit reliability **A11:** 771
effect of, on machinability of carbon steels **A1:** 599
effect of, on notch toughness **A1:** 740
effect on austenitic manganese steel castings **A9:** 239
effect on crack formation **M6:** 833
effect on hardness of tempered martensite **A4:** 124, 128-129
effect on macrosegregation in copper alloys **A9:** 639
effect, ternary iron-base systems **A15:** 65
electroless nickel plating, content effects of **M5:** 223-225, 228-231
electroslag welding reactions **A6:** 273, 274, 278
embrittlement, of brazed joints **A11:** 452
in embrittlement of iron **A1:** 691
as embrittler **A12:** 29
in ferrite **A1:** 401, 407-408
in ferritic stainless steels **A6:** 454
in free-machining metals **A16:** 389
glow discharge sources for **A10:** 29
glow discharge to determine, in low-alloy steels and cast iron **A10:** 29
grain boundary adhesion **A6:** 144
in gray iron **A1:** 22
and grinding **A16:** 437, 438
heat-affected zone fissuring in nickel-base alloys **A6:** 588, 589
high-energy neutron irradiation of **A10:** 234
in high-speed tool steels **A16:** 52
ICP-determined in natural waters **A10:** 41
as impurity **EL1:** 638, 965
impurity in solders **M6:** 1072
as inoculant **A15:** 105
ion chromatography analysis in organic solids **A10:** 664
ion-implantation profile in silicon, SIMS analysis **A10:** 623-624
lubricant indicators and range of sensitivities **A18:** 301
in magnetic electroplated coatings **A18:** 838
maps **A12:** 349
as minor element, gray iron **A15:** 630
as modifier, aluminum-silicon alloys **A15:** 752
nickel-phosphorus electroless coatings properties of **M5:** 223-229
oxygen cutting, effect on **M6:** 898
in P/M alloys **A1:** 810
photometric analysis methods **A10:** 64
radiochemical, destructive TNAA of **A10:** 238
removal, from melt **A15:** 366
resistance spot welding of steels and content effect **A6:** 228
segregation, tool steels **A12:** 375
as silicon modifier **A15:** 79, 752
species weighed in gravimetry **A10:** 172
specifications, cast copper alloys **A2:** 378
spectrometric metals analysis **A18:** 300
in steel **A1:** 144, 577
in steel weldments **A6:** 418, 420
submerged arc welding **M6:** 128
effect on cracking **M6:** 128
surface segregation during heating **A10:** 564-565
tantalum phosphide formation in **A13:** 728
as tin solder impurity **A2:** 520-521
to harden and strengthen steel **A16:** 667, 672, 674, 675, 676

Phosphorus (continued)
to increase intergranular fracture in carburized steels A4: 368, 369
as tramp element A8: 476
use in resistance spot welding M6: 486
volumetric procedures for A10: 175
Phosphorus chlorides
tantalum resistance to A13: 727
Phosphorus copper *See* Copper alloys, specific types, C12200
Phosphorus deoxidation
of copper alloys A15: 468-469
Phosphorus embrittlement
brazing and A6: 117
Phosphorus in cast iron
alloy effects M1: 77, 78
ductile iron M1: 39-41
gray iron M1: 21-22
Phosphorus in steel M1: 115-116, 411, 417
500 °F embrittlement, role in M1: 685
atmospheric corrosion affected by M1: 721-722
castings, effect on M1: 399
machinability influenced by M1: 575-576
modified low-carbon steels M1: 162
neutron embrittlement, effect on susceptibility to M1: 686
notch toughness, effect on M1: 693
P/M materials M1: 338
seawater corrosion, effect on M1: 744
steel sheet, effect on formability M1: 554
temper embrittlement, role in M1: 684
Phosphorus pentoxide (P_2O_5)
in composition of glass-ceramics EM4: 499
in tableware compositions EM4: 1101
Phosphorus printing A9: 177
Phosphorus segregation
in carbon and alloy steels, revealed by macroetching A9: 176-177
color etching A9: 142
Phosphorus-deoxidized, tellurium-bearing copper *See* Copper alloys, specific types, C14500
Phosphorus-deoxidized tellurium-bearing coppers
properties A2: 277-278
Phosphorus-doped silicon substrate
high- resolution SIMS spectra A10: 623
Phosphosilicate glass
aluminum reaction with EL1: 965
Phosphosilicate glass (PSG) EM3: 582
Phosphosilicate glass (PSG), applications
electronic processing EM4: 1056
Photoacoustic spectroscopy EM4: 52
applications A10: 115
depth profiling a granular sample using A10: 120
and F-F-IR, compared A10: 115
Photochemical machining (PCM) A16: 509, 587-593
advantages and disadvantages A16: 591-592
applications A16: 587, 592
Be alloys A16: 872-873
compared to chemical milling A16: 579, 581
design considerations A16: 590
etchability ratings of metals and alloys A16: 588
etchant composition effect A16: 591
etchants A16: 587, 588, 589-590, 591, 592, 593
etching A16: 589-590, 591, 592, 593
etching machines A16: 589-590
masking with photoresists A16: 588-589
metal defect effects on process A16: 588
photoresists A16: 587, 588-589, 590, 593
preparation of masters A16: 588
printed circuit etching application A16: 593
process description A16: 587-590
stripping and inspection A16: 590
tolerances of metals A16: 591
Photochemical sensitivity
engineering plastics EM2: 575
Photochemical systems
ESR analysis of A10: 256
Photochemistry *See also* Chemistry
of cycloaliphatic epoxides/epoxy acrylates EL1: 854-866
defined EL1: 854
of polymers EM2: 777-780
Photochromic ophthalmic crown glasses EM4: 1078
Photoconductors
ESR studied A10: 263
Photocopying applications *See also* Copier powders A7: 89
Photocross-linking
polyimides EM3: 157
Photocuring
defined EL1: 854
Photodegradation
control of EM2: 780
Photodetector resistors
corrosion failure analysis EL1: 1115
Photodetectors
for echelle spectrometer A10: 41
laser inspection with A17: 12
Photodiode
array imaging, as laser inspection A17: 12-13
defined A17: 384
in plate impact testing A8: 234
Photodiode arrays
in ICP spectrometers A10: 38, 690
Photodiodes *See also* Diodes
gallium aluminum arsenide (GaAlAs) A2: 740
Photoejection of electrons
by x-radiation A10: 84-85
Photoelastic coating method
of stress analysis A17: 51, 450–453
Photoelastic coating-drilling technique A6: 1095
Photoelastic coatings
for residual stress study A11: 134
Photoelastic fringe patterns
color image A17: 488
Photoelastic modulators A10: 114, 115, 690
Photoelastic stress
E-glass fibers EM1: 196
Photoelastic technique of isochromatic fringes A8: 269
Photoelasticity
in gears A11: 589-591
Photoelectric absorption A18: 324, 325
in EXAFS analysis A10: 409
Photoelectric effect
absorption and A10: 97
as basis of electron spectroscopy for chemical analysis A18: 445-446
defined A10: 679
and radiationless transitions A10: 568
in x-ray photoelectron spectroscopy (XPS) A11: 35
in x-ray spectrometry A10: 84
Photoelectric effect, atomic attenuation
radiography A17: 309-310
Photoelectric electron multiplier *See* Photomultiplier tube
Photoelectric glossmeter
paint gloss testing M5: 491
Photoelectric penetration distance
effect in defect imaging A10: 367
Photoelectron emission process
binding energy in A7: 255
Photoelectron spectroscopy
adhesion interfacial analysis A6: 144-145
Photoelectrons
and Auger electrons A10: 550
K-, backscattering of A10: 408
Photoemiss on
defined EL1: 1074-1075
Photoemission
principles and nomenclature A10: 569
Photoemission electron microscopy (PEEM) A6: 145
Photoetching
refractory metals and alloys A2: 561
Photoextinction
to analyze ceramic powder particle sizes EM4: 67
Photoflash
bombs A7: 131
compositions with metal fuels A7: 600, 604
Photogrammetry *See also* Stereophoto- grammetry methods, quantitative fractography A12: 197-198
stereo-, contour map and profiles by A12: 197
Photographic film
as detectors A10: 326
dot maps recorded on A10: 527
Photographic films *See also* Films
in neutron radiography A17: 387, 390-391
Photographic films for photomicroscopy A9: 84-86
different films compared A9: 86-87
Photographic filters
color filter nomograph A9: 140
Photographic micromodification EL1: 732
Photographic papers
effect on print contrast A9: 85-87
grades A9: 85-87
Photographic procedures
photomicroscopy A9: 84-86
Photographic reduction dimension
defined EL1: 1152
Photographs
materials illustrated A12: 216
Photography *See also* Fractography; Optical microscopy; Photomicroscopy
35-mm single-lens-reflex cameras A12: 78-79
of aluminum alloy macrospecimens A9: 354
auxiliary equipment A12: 89
of color etched specimens A9: 142
for corrosion examination A11: 173
depth of field A9: 87
depth-of-field effects A12: 78, 80-81, 87-88
exposures, test A12: 86-87
as failure analysis technique EL1: 1072-1073
film A12: 85
filters A9: 72
focusing A12: 79-80
of fractured parts/surfaces A12: 78-90
of fractures A11: 16
high-speed, and ring displacement A8: 210
high-speed, for strain measurement single pressure bar test A8: 199
lens aperture selection A12: 80-81
lens conversion tables A12: 81
lenses A12: 79
light meters A12: 85-86
light sources A12: 81-82
lighting techniques A12: 82-83
macrophotography of specimens A9: 86-87
magnification, determining A12: 80
microscope systems A12: 79
of phase transformations A9: 82
records, as analysis data A11: 15-16
resolution A9: 87
scanning light photomacrography A12: 81
setups for fractured parts A12: 78
stereo images A12: 87-88
test exposures A12: 86-87
view cameras A12: 78-79
visual examination, preliminary A12: 78
Photography applications
of silver A2: 691
Photography flash bulbs
powders used A7: 574
Photoimageable solder masks
vs screened solder masks EL1: 116
Photoimaging *See also* Imaging
capabilities/limitations EL1: 509-510
liquid, for solder masking EL1: 556-558
Photoinitiators EM3: 90, 91, 124
added to anaerobics EM3: 114

SUBJECTS OF THE INDEXED VOLUMES: ASM Handbook (designated by the letter "A"): **A1:** Properties and Selection: Irons, Steels, and High-Performance Alloys (1990); **A2:** Properties and Selection: Nonferrous Alloys and Special-Purpose Materials (1990); **A3:** Alloy Phase Diagrams (1992); **A4:** Heat Treating (1991); **A6:** Welding, Brazing, and Soldering (1993); **A7:** Powder Metallurgy (1984); **A8:** Mechanical Testing (1985); **A9:** Metallography and Microstructures (1985); **A10:** Materials Characterization (1986); **A11:** Failure Analysis and Prevention (1986); **A12:** Fractography (1987); **A13:** Corrosion (1987); **A14:** Forming and Forging (1988); **A15:** Casting (1988); **A16:** Machining (1989); **A17:** Nondestructive Testing and Quality Control (1989); **A18:** Friction, Lubrication, and Wear Technology (1992). **Metals Handbook, 9th Edition** (designated by the letter "M"): **M1:** Properties and Selection: Irons and Steels (1978); **M2:** Properties and Selection: Nonferrous Alloys and Pure Metals (1979); **M3:** Properties and Selection: Stainless Steels, Tool Materials and Special-Purpose Materials (1980); **M4:** Heat Treating (1981); **M5:** Surface Cleaning, Finishing, and Coating (1982); **M6:** Welding, Brazing, and Soldering (1983). **Engineered Materials Handbook** (designated by the letters "EM"): **EM1:** Composites (1987); **EM2:** Engineering Plastics (1988); **EM3:** Adhesives and Sealants (1990); **EM4:** Ceramics and Glasses (1991); **Electronic Materials Handbook** (designated by the letters "EL"): **EL1:** Packaging (1989).

Photoinitiators (continued)
free radical cure systems **EL1:** 854-856
Photoionization
x-ray detectors **A10:** 90
Photolithography **A18:** 835 **EL1:** 193-194, 197, 1152
to form the mask before the ion-exchange process begins **EM4:** 463
Photoluminescence
defined **A10:** 679
Photolysis
in free radical cure systems **EL1:** 855-856
Photolytic degradation
defined **EM2:** 776
polymer photochemistry **EM2:** 777-780
protection of plastics **EM2:** 780-782
sunlight **EM2:** 776-777
ultraviolet light **EM2:** 776-777
Photolytic laser-induced chemical vapor deposition **A18:** 848
Photolytic systems
ESR analysis for **A10:** 256
Photomacrographs
preparation of **A12:** 78-90
Photomacrography *See also* Fractography; Photography; Photomacrographs
auxiliary equipment **A12:** 89
central optical path **A12:** 79
as optical imaging **EL1:** 1072
scanning light **A12:** 81
view camera systems **A12:** 79
Photometers
defined **A10:** 679
filter **A10:** 67
Photometric methods
for analysis of metals and metal-bearing ores **A10:** 64
ion detection technique **A10:** 143-144
Photometric methods of chemical analysis
recommended practices **A7:** 249
Photomicrograph
defined **A9:** 13
Photomicrographs
SEM **A12:** 194
Photomicroscopy **A9:** 84-87
Photomontage
austenitic stainless steels **A12:** 354
Photomultiplier tube
for AAS and related techniques **A10:** 44
abbreviation for **A10:** 691
defined **A10:** 679
for molecular fluorescence spectroscopy **A10:** 77
use in ICP polychromator **A10:** 38
for UV/VIS analysis **A10:** 66
in x-ray spectrometers **A10:** 89
Photomultiplier tube (PMT) **A17:** 371, 384
Photon
defined **A17:** 384
Photon detectors **A10:** 246, 326
Photon energies
electromagnetic spectrum **A17:** 202
high, scattering at **A17:** 345
as radiation source **A17:** 298
Photon excitation **A18:** 840
Photon factory
as synchrotron radiation source **A10:** 413
Photon scanning tunneling microscopy (PSTM) **A6:** 145
Photoncorrelation spectroscopy (PCS)
to analyze ceramic powder particle size **EM4:** 68
Photonics
as primary electronic packaging technology **EL1:** 12
Photons *See also* X-ray photon emission
absorbed or scattered **A10:** 84-85
absorption of **A10:** 61
colliding, in Raman spectroscopy **A10:** 127
defined **A10:** 679
detectors for **A10:** 246
EDS measurement of **A10:** 519-520
effect in EDS detectors **A10:** 90
emission **A10:** 61
and fluorescent yield **A10:** 86
infrared light as **A10:** 109
numbers for dot mapping **A10:** 527
scattering of **A10:** 84
Photooxidation
and chemical susceptibility **EM2:** 573
as radiation degradation **EM2:** 424
Photoplate blackening and ion exposure for SSMS
relationship between **A10:** 143
Photopolymer
defined **EL1:** 1152
Photopolymerization
of acrylics **A13:** 406
Photoprinting process
quality control in **EL1:** 870-871
Photoresist
for grating application **A8:** 234
Photoresist, application
rigid printed wiring boards **EL1:** 542
Photoresists
chemical milling of Al alloys **A16:** 803
Photosedimentation
to analyze ceramic powder particle size **EM4:** 67-68
Photosensitive glasses **EM4:** 439-443
composition **EM4:** 440
nonbridging oxygens **EM4:** 439, 440
photochromic glasses **EM4:** 441-443
applications **EM4:** 442
bistable **EM4:** 442
borosilicates as bases **EM4:** 441
chemical strengthening **EM4:** 442
darkening **EM4:** 441, 442, 443
extrusion **EM4:** 442
fading **EM4:** 441, 442, 443
ion-exchange to increase impact strength **EM4:** 462
optical bleaching **EM4:** 442
physical strengthening **EM4:** 442
polarizing **EM4:** 442
processing **EM4:** 441
property alteration **EM4:** 442
redraw **EM4:** 442
silver halide particles role **EM4:** 441-443
transmittance versus wavelength **EM4:** 443
variations **EM4:** 442-443
photosensitive mechanisms **EM4:** 439
processing **EM4:** 440-441
products **EM4:** 440
types
developed photosensitive **EM4:** 439-440
direct photosensitive **EM4:** 439
photosensitive nucleated **EM4:** 439, 440
Photosensitive liquid solder masks
flexible printed boards **EL1:** 584
Photosensitive polyimides **EL1:** 327-328, 767
Photosensitivity
of polyimides **EM3:** 157, 160
Photosynthesis
and biochemical oxidation **A13:** 897
ESR studied **A10:** 264
Photovoltaic materials
plasma-assisted physical vapor deposition **A18:** 848
Phthalate esters **EM3:** 20
defined **EM2:** 30
Phthalic acid salts
as ion chromatography eluents **A10:** 660
Phthalocyanine
as pigment **EM3:** 179
Physical adsorption *See* Physisorption
Physical aging *See also* Aging **EM2:** 751, 755-758
defined **EM2:** 751-752
in polymers **A11:** 758
quenching stresses, defined **EM2:** 751
and thermal stresses **EM2:** 751-760
Physical analysis
of thermoplastic resins **EM2:** 533-540
Physical blowing agent **EM2:** 30, 503 **EM3:** 20
Physical catalyst **EM3:** 20
defined **EM1:** 17
Physical chemistry *See also* Chemistry; Thermodynamics
chemical kinetics **A15:** 52-54
chemical thermodynamics **A15:** 50-52
enthalpy and heat capacity **A15:** 50
Gibbs free energy **A15:** 50-51
of inclusion-forming reactions **A15:** 89-90
interphase mass transport **A15:** 53-54
of melt purification **A15:** 74-81
nucleation **A15:** 52-53
Physical chemistry (continued)
phase diagrams **A15:** 52
principles of **A15:** 50-54
Physical crack size *See also* Crack; Crack length; Crack size
defined **A8:** 10 **A11:** 8
visual observation of **A8:** 452
Physical design *See also* Technology rules
of level 1 packages **EL1:** 401-402
subsystem, in computer-aided design **EL1:** 129
Physical etching **A9:** 57, 61-62
Physical interconnection hierarchy
large scale systems **EL1:** 2-4
Physical metallurgy *See also* Physical properties
aluminum-lithium alloys **A2:** 179-180
Physical modeling *See also* Modeling
approaches **A14:** 431
experimental procedure, example **A14:** 435-436
grid mesh development **A14:** 434-435
model materials **A14:** 431-433
of open-die forging **A14:** 66-67
precision forging **A14:** 159
segmented die design **A14:** 435
strain for two-dimensional flow computation **A14:** 433-434
three-dimensional grid for strain calculation **A14:** 436-437
tool setup, for plane-strain wedge testing **A14:** 435
Physical objective aperture
defined **A9:** 13-14
Physical partitioning
in computer-aided design **EL1:** 128-129
Physical performance *See also* Performance; Physical properties; Physical test methods; Reliability; Testing
issues **EL1:** 5-8
local **EL1:** 5-6
selected chip-to-chip issues **EL1:** 7-8
selected on-chip issues **EL1:** 6-7
Physical principles
of liquid penetrant inspection **A17:** 71-73
Physical properties *See also* Fiber properties analysis; Laminate properties analysis; Material properties; Material properties analysis; Mechanical properties; specific type of alloy
of abrasion-resistant cast iron **A1:** 96
of acrylics **EM2:** 104
and aging **EM2:** 756
allyls (DAP, DAIP) **EM2:** 228
aluminum **A2:** 3
aluminum alloys **A15:** 764-765 **M2:** 53-54, 58
aluminum and aluminum alloys **A2:** 45-46
aluminum casting alloys **A2:** 145-147
ASTM test methods **EM2:** 334
beryllium-copper alloys **A2:** 407-409
by eddy current inspection **A17:** 164
of carbon and alloy steels **A1:** 195-199 **M1:** 145-151
of cast stainless steels **A1:** 927
of cast steels **A1:** 374, 376 **M1:** 393-400
of cast superalloys **A1:** 983
ceramic multilayer packages **EL1:** 467-468
characterizing, technical meetings for **A17:** 51
cobalt-base corrosion-resistant alloys **A2:** 454
cobalt-base high-temperature alloys **A2:** 452
cobalt-base wear-resistant alloys **A2:** 451
commercial pure tin **A2:** 518-519
of compacted graphite iron **A1:** 66-69
conformal coatings **EL1:** 762
of corrosion-resistant cast iron **A1:** 96
data sheet, typical **EM2:** 407
of deep drawn sheet product **A14:** 575
defined **A8:** 10 **A11:** 8
design values for **A8:** 662
of ductile iron **A1:** 49, 50 **A15:** 661, 663 **M1:** 49, 52-55
effect of crystallographic texture on **A9:** 700-701
effect of recovery changes **A9:** 693
estimation of **A17:** 51
ferrous P/M materials **M1:** 333-339, 340, 345
of fiber composites **EM1:** 185-192, 730
fluoropolymer coatings **EL1:** 782-783
of gallium arsenide (GaAs) **A2:** 741
of germanium **A2:** 734
of gray iron **A1:** 31-32 **A15:** 644-645 **M1:** 30-32
of heat-resistant cast iron **A1:** 96

Physical properties (continued)
heat-treatable wrought aluminum alloy ... **A2:** 41
liquid crystal polymers (LCP) ... **EM2:** 181
loss, from heat aging ... **EM2:** 78
malleable iron ... **M1:** 67
maraging steels ... **M1:** 451
measured ... **EM2:** 613-614
measurement of, used to investigate crystallographic texture ... **A9:** 701
mechanically alloyed oxide alloys ... **A2:** 943-947
dispersion-strengthened (MA ODS)
microelectronic materials ... **EL1:** 104-111
of nickel alloys ... **A2:** 441
nickel-iron alloys ... **A2:** 774
nickel-titanium shape memory effect (SME) alloys ... **A2:** 899
niobium alloys ... **A2:** 567-571
on supplier data sheets ... **EM2:** 641-642
of optical plastics ... **EM2:** 596
permanent magnet materials ... **A2:** 793
pewter ... **A2:** 522
phenolics ... **EM2:** 245
polyaryl sulfones (PAS) ... **EM2:** 145
polyarylates (PAR) ... **EM2:** 138-139
polyaryletherketones (PAEK, PEK, PEEK, PEKK) ... **EM2:** 142
polyethylene terephthalates (PET) ... **EM2:** 173
polysulfones (PSU) ... **EM2:** 200
of pultrusions ... **EM2:** 395-396
reinforced polypropylenes (PP) ... **EM2:** 192-193
rhenium ... **A2:** 582
silicones (SI) ... **EM2:** 266
solder ... **EL1:** 639
in solidification ... **A15:** 109-110
of steel castings ... **A1:** 374, 376
tantalum ... **A2:** 573
temperature and molecular structure ... **EM2:** 436-437
tool steels ... **M4:** 563, 573, 574
tubular products ... **A17:** 561
uranium alloys, quenched ... **A2:** 674
urethane coatings ... **EL1:** 776
urethane hybrids ... **EM2:** 269
wrought aluminum, alloying effects ... **A2:** 44-57
of wrought stainless steels ... **A1:** 871
of wrought superalloys ... **A1:** 963-964
of wrought tool steels ... **A1:** 774, 775
Physical properties, use in phase-diagram determination ... **A3:** 1 • 18
Physical quality *See* Structural quality
Physical quality steel sheet
formability of ... **M1:** 547
Physical scale modeling ... **A13:** 234
Physical strength
of flexible epoxies ... **EL1:** 821
Physical test methods *See* Board-level physical test methods; Component-level physical test methods; Package-level physical test methods; Testing; Wafer-level physical test methods
automatic optical inspection ... **EL1:** 941-942
environmental testing ... **EL1:** 944
package level ... **EL1:** 927-940
for plating thickness ... **EL1:** 943
solder joint inspection ... **EL1:** 942
for solderability ... **EL1:** 943-944
standards ... **EL1:** 941
wafer-level ... **EL1:** 917-926
Physical testing
defined ... **A8:** 10
Physical tests *See also* Physcial properties; Testing
of reinforcement fibers ... **EM1:** 285-286
Physical vapor deposition *See also* PVD and CVD coatings ... **A18:** 840-846
abbreviation for ... **A11:** 797
activated reactive evaporation ... **A18:** 840, 844, 845
alloy deposition ... **A18:** 843
ARE (BARE) process ... **A18:** 845, 848, 849
Physical vapor deposition (continued)
cemented carbides ... **A18:** 796
coating method for valve train assembly components ... **A18:** 559
coatings for jet engine components ... **A18:** 592
decomposition of compounds ... **A18:** 843-846
defined ... **A13:** 10
diode ion plating ... **A18:** 840, 844-845
direct evaporation ... **A18:** 844
hybrid PVD processes ... **A18:** 844-846
ion plating ... **A18:** 844-845
laser ablation deposition ... **A18:** 844
molybdenum disulfide application to surfaces ... **A18:** 114
parameters ... **A18:** 841
plasma-assisted reactive evaporation ... **A18:** 840, 843-844
processes ... **A18:** 840-843
PVD-TiX, comparison of coatings for cold upsetting ... **A18:** 645
reactive ion plating ... **A18:** 840, 844, 845-846
reactive PVD processes ... **A18:** 844
reactive sputtering (RS) ... **A18:** 840, 844, 848
single-element species deposition ... **A18:** 843
techniques for deposition of metals, alloys, and compounds ... **A18:** 843-846
titanium alloys ... **A18:** 778-780
titanium nitride coatings, critical normal force versus substrate surface roughness ... **A18:** 436
tool steels ... **A18:** 645, 739
unbalanced magnetron sputter deposition ... **A18:** 849
Physical vapor deposition (PVD) ... **A16:** 51 **EM4:** 124
application of coatings to cemented carbides ... **A16:** 71
ceramic coatings for adiabatic diesel engines ... **EM4:** 992
coated carbide tools ... **A2:** 961-962
and drilling ... **A16:** 219
high-speed steel tool coatings ... **A16:** 83, 87
Physical vapor deposition method
oxidation protective coating ... **M5:** 379
Physical wear mechanism
abrasive ... **A8:** 602
fatigue ... **A8:** 602-603
interaction ... **A8:** 603
oxidation ... **A8:** 603
Physical work rile (PWF)
in design process ... **EL1:** 129
Physically clean surfaces ... **A9:** 28
Physically deposited interference layers ... **A9:** 59-60
Physics, atmospheric
PIXE studies in ... **A10:** 106
Physiochemical methods
of powder production ... **A7:** 52-55
Physisorption *See also* Chemisorption
as contamination in gravimetric samples ... **A10:** 163
defined ... **A10:** 679 **A13:** 10
of pyridine, Raman analysis ... **A10:** 134
Pi *See* Polyimide; Polyimides
bonding, defined ... **A10:** 679
electron, defined ... **A10:** 679
PIA *See* Plastics Institute of America
PIC *See* Pressure-impregnation-carbonization
Pick and place operations
robotic ... **A15:** 568
Pick count *See also* Count
defined ... **EM1:** 17, 286 **EM2:** 30
measured ... **EM1:** 286
Pick-up roll ... **EM3:** 20
Picker bar components
for cotton pickers ... **A7:** 673
Pickle
defined ... **A13:** 10
Pickle-lag test
for tinplate ... **A13:** 781
Pickling *See also* Acid pickling ... **M5:** 68-82 **EM3:** 42
acid cleaning compared to ... **M5:** 59-60
acid concentration, effects of ... **M5:** 69-70, 73-76, 79-80
agitation used in ... **M5:** 73
alkaline precleaning process ... **M5:** 70-71, 81
analysis of solution ... **M5:** 69, 70-73
anodic-cathodic system ... **M5:** 76
atmospheric corrosion resistance affected by ... **M1:** 722
automatic systems ... **M5:** 613
batch process ... **M5:** 68-69, 71
bath life ... **M5:** 669-670
blistering in, effects ... **M5:** 80-81
castings ... **M5:** 326
caustic-permanganate process ... **M5:** 73
continuous process ... **M5:** 68-69, 71-72
copper and bronze contamination ... **M5:** 80
of copper and copper alloy forgings ... **A14:** 258
copper and copper alloys ... **M5:** 611-615, 619-620
cycles, stainless steels ... **M5:** 73-74
defects ... **M5:** 80-81
descaling time, effects on ... **M5:** 72-76, 78-79
electrolytic *See* Electrolytic pickling
equipment and process control ... **M5:** 72-73, 76-77, 80
equipment, corrosion of ... **A13:** 1314-1316
ferrous sulfate, role in ... **M5:** 70, 73-74, 76-77, 81-82
inhibitory effects of ... **M5:** 70, 73-74, 76-77
recovery of ... **M5:** 81-82
flash *See* Flash pickling
forgings ... **M5:** 72, 80
hafnium alloys ... **M5:** 667
heat-resistant alloys ... **M5:** 564-568
heating methods ... **M5:** 79-80
hot dip galvanized coating process ... **M5:** 325-326, 328
hot dip tin coating process using ... **M5:** 352-353
hydrochloric acid process *See* Hydrochloric acid pickling
hydrofluoric acid process *See* Hydrofluoric acid pickling
hydrogen damage by ... **A11:** 246
and hydrogen embrittlement ... **A8:** 510
hydrogen embrittlement caused by ... **M1:** 687 **M5:** 70, 81
inhibited solutions ... **M5:** 70, 74-77
iron ... **M5:** 68-82
iron concentration, solution effects of ... **M5:** 69-70
for liquid penetrant inspection ... **A17:** 81
magnesium alloys ... **M5:** 629-633, 635-638, 635-636
pickle rate
solution magnesium content ... **M5:** 635-636
maraging steels ... **M1:** 448
molybdenum ... **M5:** 659
nickel and nickel alloys ... **M5:** 669-673
formulas ... **M5:** 669-670
specialized operations ... **M5:** 672-673
niobium ... **M5:** 663
nitric acid process *See* Nitric acid pickling
nitric-hydrofluoric acid process *See* Nitric-hydrofluoric acid pickling
overpickling, effects of ... **M5:** 80
of palladium ... **A14:** 850
phosphate coating process ... **M5:** 438-439, 449, 453-454
phosphoric acid process ... **M5:** 630-631, 640-641
pickling time, effects on ... **M5:** 72-76
pitting in, effects of ... **M5:** 80
plating process precleaning ... **M5:** 16-18
of platinum ... **A14:** 850
porcelain enameling process ... **M5:** 514-515
precleaning procedures ... **M5:** 70-72
for precoat cleaning ... **A15:** 561
preheating process ... **M5:** 73, 75
process *See also* specific processes by name ... **M5:** 68-69
process variables, effects of ... **M5:** 73-78
refractory metals ... **M5:** 654-655, 659, 663

Pickling (continued)
rinsing process ... **M5:** 72-73, 79
rust and scale removed by ... **M5:** 12-13, 15
safety precautions ... **M5:** 21, 81, 613-614, 669
salt bath descaling process
concentration and temperature of acids ... **M5:** 97-98
using ... **M5:** 97-99, 102
salt bath pretreatment in ... **M5:** 672
sand removal by ... **M5:** 69, 72
scale removal by ... **M5:** 68-69, 72-80
scalebreaking used in ... **M5:** 74, 76, 78
solution compositions and operating conditions ... **M5:** 68-82, 325-326, 564-567, 611-613, 615, 635-636, 669-670
solution life ... **M5:** 73
stainless steel ... **M5:** 72-73, 76, 553-554, 561
steel *See* Steel, pickling of
storage tanks for acid ... **M5:** 79-80
strip speed effects ... **M5:** 74-75, 78-79
sulfate liquor, killing of ... **M5:** 81
sulfuric acid process *See* Sulfuric acid pickling
surface conditions, effect of ... **M5:** 670-672
surfactants used in ... **M5:** 69, 71
tanks, construction ... **M5:** 76-79
tantalum and niobium ... **M5:** 663
temperature ... **M5:** 72-76, 79-80
control of ... **M5:** 79-80
effects of ... **M5:** 72-76, 79-80
titanium alloys ... **A6:** 785
titanium and titanium alloys ... **M5:** 654-655
tungsten ... **M5:** 659
uninhibited solutions, use of ... **M5:** 70
waste recovery and treatment ... **M5:** 81-82
water rolling process ... **M5:** 615
weight loss tests ... **M5:** 70, 73-77
weld areas, selective attack during ... **M5:** 565
zirconium alloys ... **M5:** 667
Picks *See* Filling yarns
Pickup *See also* Die line; Galling; Scoring
defined ... **A14:** 10
Pickup joint
magnetic particle inspection ... **A17:** 109
Pickup weld
nonrelevant indications from ... **A17:** 106-107
Picometer
abbreviation for ... **A10:** 691
Picral
labelling of ... **A9:** 67
Picral and hydrochloric acid as an etchant for
carbonitrided steels ... **A9:** 217
carburized steels ... **A9:** 217
Picral and nital
etchants compared ... **A10:** 302
Picral and nital as an etchant for
carbon and alloy steels ... **A9:** 169
nitrided steels ... **A9:** 218
Picral as an etchant for
austenitic manganese steel casting specimens ... **A9:** 239
babbitted bearings ... **A9:** 451
carbon and alloy steels ... **A9:** 169-170
carbon steel casting specimens ... **A9:** 230
carbonitrided steels ... **A9:** 217
carburized steels ... **A9:** 217
chromized sheet steel ... **A9:** 198
electrical steels ... **A9:** 531
ferrous powder metallurgy materials ... **A9:** 508-509
low-alloy steel casting samples ... **A9:** 230
nitrided steels ... **A9:** 218
plate steels ... **A9:** 202
stainless-clad sheet steel ... **A9:** 198
steel tubular products ... **A9:** 211
tin and tin alloy coatings ... **A9:** 451
tool steels ... **A9:** 257
Picral with zephiran chloride as an etchant for carbon and alloy steels ... **A9:** 169
Picric acid *See also* Picral
description ... **A9:** 68
in water as an etchant for hot-dip galvanized sheet steels ... **A9:** 197
Picture element *See* Pixel; Pixels
Picture points
IA scanners ... **A10:** 310
Picture-frame test ... **EM1:** 324
Piece part stress screening
factors in ... **EL1:** 877-878
Piece-part cost analysis
blow molding ... **EM2:** 358-359
Pieced buffs ... **M5:** 118, 126-127
Pierced fabric method
ceramic-ceramic composites ... **EM1:** 934
Pierced holes *See also* Hole flanging; Holes
at angle to surface ... **A14:** 469
characteristics ... **A14:** 459
size of ... **A14:** 467
spacing of ... **A14:** 468
wall, quality of ... **A14:** 459
Piercing *See also* Blanking; Fine-edge piercing; Holes; Pierced hole; Piercing dies; Punching
accuracy of ... **A14:** 466-467
aluminum alloy ... **A14:** 519, 793-794
auxiliary equipment ... **A14:** 477-478
and blanking, compared ... **A14:** 459
and blanking, with compound dies ... **A14:** 456-457
for blanks ... **A14:** 446
of carbon/low-alloy steels, lubricants for ... **A14:** 518
clearance and tool size ... **A14:** 461-463
compound dies, use ... **A14:** 465-466
of copper and copper alloys ... **A14:** 811-812
defined ... **A14:** 10, 459
die clearance, selection ... **A14:** 459-460
edges ... **A14:** 460-461
of electrical steel sheet ... **A14:** 476-482
with fastener ... **A14:** 470
and fine-edge blanking ... **A14:** 472-475
flanged holes ... **A14:** 469
force requirements ... **A14:** 463
forgings ... **M1:** 368
forming requirements, effect ... **A14:** 468-469
in header ... **A14:** 311
of heat-resistant alloys, lubricants for ... **A14:** 519
high-carbon steel ... **A14:** 556-558
hole spacing ... **A14:** 468
hole wall ... **A14:** 459
holes ... **A14:** 459, 468-469
of low-carbon steel ... **A14:** 459-471
of nickel-base alloys ... **A14:** 520, 832
of organic-coated steels ... **A14:** 565
pierced holes, characteristics ... **A14:** 459
with pointed punch ... **A14:** 470
in press brake ... **A14:** 538
presses ... **A14:** 463-464
progressive dies, use ... **A14:** 466
rotary ... **M2:** 262-263
safety ... **A14:** 471
and shaving ... **A14:** 470
single-operation dies, use ... **A14:** 465
special techniques ... **A14:** 469-470
of stainless steels ... **A14:** 519, 759, 761-762
of thick stock ... **A14:** 467
of thin stock ... **A14:** 467-468
tool dulling, effect ... **A14:** 461
tool materials for ... **A14:** 539
tools ... **A14:** 464-465
transfer dies ... **A14:** 466
tube, and slotting ... **A14:** 470
and upsetting ... **A14:** 88-89
vs alternative methods ... **A14:** 470-471
work metal hardness, effect ... **A14:** 462
Piercing dies *See also* Blanking and piercing dies; Piercing
applications ... **A14:** 485-486
material selection for ... **A14:** 483-486
tool materials ... **A14:** 484-485
Piezoelectric
plastic converters ... **A8:** 244
sonic converters ... **A8:** 244
transducers ... **A8:** 240, 243-245
Piezoelectric ceramics ... **EM4:** 1119-1123
applications ... **EM4:** 199, 1119-1123
piezoelectric actuators ... **EM4:** 1119-1123
piezoelectric vibrators ... **EM4:** 1119, 1121, 1122
surface acoustic wave filters ... **EM4:** 1119, 1121
transformers ... **EM4:** 1119, 1121
piezoelectric and electrostrictive materials ... **EM4:** 1119
polymers ... **EM4:** 1119
Piezoelectric dynamometer ... **A16:** 678
Piezoelectric effect, in deposition
electrogravimetry ... **A10:** 198
Piezoelectric sensors
for acoustic emission inspection ... **A17:** 280-281
Piezoelectric transducers
ultrasonic inspection ... **A17:** 254-255
Piezoelectricity
defined ... **A17:** 254
Piezoelectrics
applications ... **EM4:** 1105-1106
Piezoelectro polymers *See also* Polymers
defined ... **EM2:** 30
Piezoresistive gages
for Hugoniot elastic limit measurement ... **A8:** 208
Pig iron ... **M1:** 109
anthracite/bituminous ... **A15:** 30
for early foundries ... **A15:** 25
grades for ductile iron ... **A15:** 647
in iron plantations ... **A15:** 26
as metal charge ... **A15:** 388
refining, and converter development ... **A15:** 31
Pig lead *See also* Lead
composition ... **A2:** 543-545
Pigmented drawing compounds
removal of ... **M5:** 4-8, 56
Pigmented oils and greases
for nickel-base alloy forming ... **A14:** 831-832
Pigments ... **EM3:** 41, 179
aluminum flake ... **A7:** 594-595
copper and gold bronze ... **A7:** 595-596
effect, chemical susceptibility ... **EM2:** 572-573
inorganic ... **EM1:** 13-14
liquid resin containing ... **EM1:** 133-134
metallic flake ... **A7:** 593-596
oxide colorants ... **EM4:** 45
for sheet metal compounds ... **EM1:** 158
for sheet molding, compounds ... **EM1:** 141
sphere-based ... **EM4:** 45
spinel-based ... **EM4:** 45
for ultrahigh molecular weight polyethylenes (UHMWPE) ... **EM2:** 171
x-ray detectable, in tracer yarns ... **EM1:** 149
zircon-based ... **EM4:** 45
"Pigtail
" copper ... **A6:** 895
Pileups
effect on failure ... **A8:** 34
Pilger tube-reducing process *See* Tube reducing
Piling
pipe ... **M1:** 318
seawater corrosion of ... **M1:** 744
soil corrosion of ... **M1:** 730-731
zones of seawater corrosion ... **M1:** 740, 742-744
Piling pipe ... **A1:** 331
Pilkington's float glass process ... **EM4:** 21
three varieties of solar-cell cover glasses ... **EM4:** 1019
Pilling-Bedworth volume ratios ... **A13:** 65, 97
Pillow block bearings
in constant-stress testing system ... **A8:** 320
Pillow-block bearing
misalignment of ... **A11:** 475
Pilot arc
definition ... **M6:** 13
Pilot arc (plasma arc welding)
definition ... **A6:** 1212
Pilot-size vibratory mill ... **A7:** 67
Pilot-valve bushing, steel
fatigue fracture of ... **A11:** 121
Piloted dies ... **A14:** 132
Pimple
defined ... **EM2:** 30
Pin
and bolt, bearing strength ... **EM1:** 314-316
defined ... **A8:** 10
deformation, avoiding ... **A8:** 59
"dirty", values for ... **A8:** 60
distortion, with titanium or high-strength steel ... **A8:** 59
"dry", values for ... **A8:** 60
failure, with titanium or high-strength steel ... **A8:** 59
holes, defined ... **EM1:** 17
loading, fatigue test specimens ... **A8:** 371
Pin bearing testing ... **A8:** 59-61
cleaning procedures ... **A8:** 60
diameter to thickness ratio (D/t) ... **A8:** 59
preferred measurement method ... **A8:** 60
test specimens ... **A8:** 59-61
Pin Brinell hardness tester ... **A8:** 88

Pin count
pin density ... **EL1:** 691, 1153
proliferation, trends ... **EL1:** 405
vs complexity ... **EL1:** 400-401
Pin expansion test
defined ... **A8:** 10
Pin fasteners
defined ... **A11:** 529
failures in ... **A11:** 545-548
semipermanent ... **A11:** 545-548
Pin grid array (GP) ... **EM3:** 588-589
Pin hooks
materials for ... **A11:** 522
Pin test
of abrasion-resistant cast iron ... **A1:** 97
Pin transfer
as adhesive application ... **EL1:** 671
in surface-mount soldering ... **EL1:** 700
Pin(s)
attach, ceramic packages ... **EL1:** 466
contact, defined ... **EL1:** 1152-1153
defined ... **EL1:** 1152
density ... **EL1:** 691, 1153
electrical effects ... **EL1:** 76
signal, and system complexity ... **EL1:** 87
visual examination ... **EL1:** 1058
Pin-grid array (PGA) *See also* Ceramic pin-grid array, Plastic pin-grid array
defined ... **EL1:** 1153
elements of ... **EL1:** 475
future trends ... **EL1:** 407
multichip packaging ... **EL1:** 311
outline ... **EL1:** 205
as package family ... **EL1:** 404
plastic (PPGAS) ... **EL1:** 210
plastic, assembly of ... **EL1:** 475-476
Pin-in-hole plastic packaging
fabrication ... **EL1:** 470-482
Pin-on-disk bench tests ... **A18:** 117
Pin-on-disk machine
defined ... **A18:** 14
Pin-on-disk tests ... **A18:** 116, 119, 120
abrasion of dental amalgams ... **A18:** 669
adhesive wear of ceramics ... **A18:** 241
molybdenum disulfide in high-vacuum lubricant applications ... **A18:** 153
x-ray characterization of surface wear ... **A18:** 470
Pin-socket interconnection ... **EM3:** 588
Pin-to-hole connection process
materials and processes selection ... **EL1:** 117
Pinch trimming
of blanks ... **A14:** 446
Pinch-off
defined ... **EM2:** 30
Pinch-type three-roll forming machines ... **A14:** 616-617
Pincushion image distortion *See also* Positive distortion ... **A9:** 77
Pincushion indentation
with pyramid indenter ... **A8:** 100, 102
Pine
as pattern material ... **A15:** 194
Pinhole cameras
for XRPD analysis ... **A10:** 334-335
Pinhole eyepiece
defined ... **A9:** 14
Pinhole photographs of iron-nickel alloy wires ... **A9:** 702
Pinhole photography used to determine crystal orientation ... **A9:** 702
schematic ... **A9:** 701
Pinhole porosity *See also* Defects; Porosity
defined ... **A15:** 9
in gray iron ... **A15:** 641-642
Pinhole system
defined ... **A9:** 14
Pinhole testing
of coating systems ... **A13:** 417
Pinholes ... **EL1:** 687, 1153
as casting defects ... **A11:** 382
defined ... **A9:** 14 **A11:** 8 **A17:** 562 **EM2:** 30
surface, as casting defects ... **A11:** 382
Pinion
countershaft, fatigue fracture ... **A11:** 396
failures, in forging ... **A11:** 336
steel, fatigue fracture of cast chromium- molybdenum ... **A11:** 395
tooth, profile ... **A11:** 596
wear ... **A11:** 597
Pinion speed
symbol and units ... **A18:** 544
Pinning effect ... **A6:** 73-74
Pins *See also* Pin fasteners
cast, fatigue failure in ... **A11:** 398
fasteners ... **A11:** 545-548
and gripping cam, fractured, carburization effects ... **A11:** 573, 576
knuckle, fatigue fracture in ... **A11:** 128-129
lift, reversed bending in ... **A11:** 77
orthopedic ... **A11:** 671
piston, grinding bums ... **A11:** 89
pivot, fatigue fracture ... **A11:** 308
positive-locking ... **A11:** 548
push-pull ... **A11:** 548
quick-release ... **A11:** 548
radial-locking ... **A11:** 548
steel jackscrew drive, SCC of ... **A11:** 546-548
taper, service failure ... **A11:** 545-546
Pins, cold forged
economy in manufacture ... **M3:** 854
Piobert effect
defined ... **A8:** 21
Piobert lines *See* Lüders lines
Pipe *See also* Cast iron pipe, Steel pipe; Fluid handling applications; Line pipe; Pipeline corrosion; Pipelines; Pipelines, failures of; Piping; Piping systems; Tubular products
$2^1/_4$ Cr-1Mo steel, mechanical properties ... **M1:** 654
aluminum and aluminum alloys ... **A2:** 5
applications, ductile iron ... **M1:** 36
bending, nickel-base alloys ... **A14:** 834-835
body, defects in ... **A11:** 697, 699-701
body, preservice test failures in ... **A11:** 697-698
boiling water reactor ... **A13:** 928-933
by liquid penetrant inspection ... **A17:** 86
carbon steel, aqueous corrosion of ... **A13:** 512
carbon steel, fracture in cooling tower ... **A11:** 639-640
carbon steel, weld inspection ... **A17:** 112
cast iron
coatings for ... **M1:** 103, 104
pipe laying conditions for ... **M1:** 98, 99
cast steel, comparison with wrought steel ... **M1:** 400
chip-conveyor, fracture from poor fit-up ... **A11:** 426-427
cleaning of ... **A13:** 1139
codes governing ... **M6:** 824
compositions ... **A13:** 1289
corrugated steel, natural water corrosion ... **A13:** 433
defined ... **A15:** 9
double-submerged arc welds in ... **A11:** 698
drill, ultrasonic inspection ... **A17:** 232
as encapsulation material ... **A7:** 429
explosion welding ... **M6:** 709, 715-717
flash welding ... **M6:** 558
flux cored arc welding ... **M6:** 109
in forging ... **A11:** 327
forming dies, for press-brake forming ... **A14:** 537
from rolling ... **A14:** 358
galvanized steel for ... **M1:** 167, 170
gray iron, graphitic corrosion ... **A13:** 131-132
gray iron water-main, graphitic corrosion ... **A11:** 372-374
Pipe (continued)
high frequency resistance welding ... **M6:** 759, 760-762
high molecular weight ... **EM2:** 163-164
hydrogen embrittlement in carbon steel ... **A11:** 645
ingot, from forging ... **A11:** 315
joining processes ... **M6:** 56
lead, applications ... **A2:** 552-553
-line, cathodic protection system ... **A13:** 472
magnetic paint inspected ... **A17:** 128
marine, cathodic protection ... **A13:** 922
mechanically alloyed oxide alloys ... **A2:** 949
dispersion-strengthened (MA ODS)
mold, overheating failure of ... **A11:** 275-276
multipoint cutting tools used ... **A16:** 59
oxyfuel gas welding ... **M6:** 591-592
pile structure, zinc sacrificial anode protection system ... **A13:** 472-476
plasma arc welding ... **M6:** 221
polyethylene, fracture surface of ... **A11:** 761
polymer irrigation, failure of ... **A11:** 762-763
portable irrigation, aluminum and aluminum alloys ... **A2:** 14
pressure, ultrasonic inspection ... **A17:** 232
radiographic methods ... **A17:** 296
residual shrinkage ... **A11:** 553
rolling, contour roll forming ... **A14:** 630-632
seamless ... **A17:** 272, 579
secondary, effect on cold-formed part failures ... **A11:** 307
service failures of ... **A11:** 699-704
soil corrosion of ... **M1:** 727, 729-730
soldering ... **M6:** 1093-1095
spring fatigue fracture from ... **A11:** 551
standard ... **M1:** 315, 318
steel, galvanic corrosion ... **A13:** 86
steel, ultrasonic inspection ... **A17:** 271-272
submerged arc-welded ... **A17:** 578-579
upsetting of ... **A14:** 91-93
vinyl ester ... **EM2:** 272
welded, eddy current weld inspection ... **A17:** 186
weldments, failure of ... **A11:** 704
wrought aluminum alloy ... **A2:** 33 **A12:** 427
wrought copper and copper alloys ... **A2:** 248-250
Pipe (defect)
ingot ... **A17:** 491
in steel bar and wire ... **A17:** 549
ultrasonic inspection of ... **A17:** 232, 272
Pipe extrusion
products ... **EM2:** 385
Pipe in ingots
revealed by macroetching ... **A9:** 174
Pipe joint
soldering ... **A6:** 130
Pipe joint compounds
powders used ... **A7:** 572
Pipe, metal
sampling trainload of ... **A10:** 15
Pipe reamers ... **A16:** 241, 245
Pipe sizes and specifications
for steel tubular products ... **A1:** 328, 329-331, 333
Pipe, steel
pickling of ... **M5:** 69
Pipe steels *See also* Steel pipe, specific types
API compositions ... **A9:** 211
ASTM compositions ... **A9:** 211
Pipe steels, specific types
A53
composition ... **A6:** 642
mechanical properties ... **A6:** 642
recommended preheat and interpass temperatures ... **A6:** 644
A106
composition ... **A6:** 642
mechanical properties ... **A6:** 642

Pipe steels, specific types (continued)
recommended preheat and interpass temperatures **A6:** 644
A381
composition **A6:** 642
mechanical properties **A6:** 642
recommended preheat and interpass temperatures **A6:** 644
X-65
multipass SMAW/flux-cored arc weld characterized **A6:** 102-104
parameters used to obtain a multipass weld in 1.07 m diameter pipe **A6:** 101-102
Pipe taps
coatings and increased tool life **A16:** 58
Pipe threading
speed **A16:** 302
Pipeline corrosion *See also* Pipe **A13:** 1288-1292
bacteriological **A13:** 1288
causes **A13:** 1288-1289
control/prevention **A13:** 1289-1292
gas/oil **A13:** 1255
mining **A13:** 1294-1296
of specific pipelines **A13:** 1292
steel, new and old **A13:** 1288
Pipeline girth welds
inspection of **A17:** 579-581
Pipeline steel
stress-corrosion behavior **A8:** 521
Pipelines *See also* Pipe; Pipeline corrosion; Pipelines, failures of
acoustic emission inspection **A17:** 290
bends, buckles, or wrinkles in **A11:** 704
corrosion failures **A11:** 703-704
effect of environment on **A11:** 701-704
failed girth weld in **A11:** 422
fatigue cracks **A11:** 704
fracture, arrest of **A11:** 704-706
fracture origin location **A11:** 695-696
gas, flux leakage method **A17:** 132
high-pressure long-distance, characteristics of **A11:** 695
holographic measurement of **A17:** 16
hydrogen-stress cracking **A11:** 701
internal combustion in **A11:** 704
longitudinal weld defects in **A11:** 704
magnetic paint inspected **A17:** 128
sabotage of **A11:** 704
secondary loads on **A11:** 704
stainless steel, grain-boundary embrittlement **A11:** 132
stress-corrosion cracking in **A11:** 701-703
types of **A11:** 695
weldments , as failure cause **A11:** 704
Pipelines, failures of *See also* Pipe; Pipelines **A11:** 695-706
causes of **A11:** 697-704
defects causing **A11:** 697
high-pressure long-distance pipelines characteristics of **A11:** 695
investigation procedures of **A11:** 695-697
pipeline fractures, arrest of **A11:** 704-706
Piper, Walter
as early founder **A15:** 28
Pipet
defined **A10:** 679
Pipette gravity sedimentation
to analyze ceramic powder particle sizes **EM4:** 67
Pipework materials
for aqueous solutions at ambient temperatures **A8:** 415-416
Piping *See also* Pipe; Pipelines **A6:** 377-381
in bearing failures **A11:** 494
boiler-feed, corrosion **A11:** 615
gas-dryer, corrosion product on **A11:** 631
high-pressure, hydrogen damage in **A11:** 612
in ingots, schematic **A11:** 315
line, reheat steam, failure at power-generating station **A11:** 652-653
pressure vessels, failures of **A11:** 644
primary and secondary, forging **A11:** 315
stainless steel, SCC failure at welds **A11:** 216
system cross failure, intergranular cracking **A11:** 649-650
system, Incoloy 800, failure of **A11:** 291-292
Piping systems
acoustic emission inspection **A17:** 298
Pirani gages
for thermal conductivity of gases **A17:** 62-63
for vacuum leak testing **A17:** 68
Piston
load maintainer **A8:** 369-370
rod, length in servohydraulic systems **A8:** 399-400
Piston alloys
aluminum casting, properties **A2:** 127, 130-131
mechanical properties **A2:** 147
Piston and anvil technique
for metallic glasses **A2:** 805
Piston extrusion **EM4:** 9
Piston pumps **EM3:** 693-696, 698, 700
Piston rings
relaxation of **M1:** 296, 299
Piston rods and shafts
failures of **A11:** 459
Piston speed
average **A18:** 603
Piston-pin bearings *See* Little-end bearing
Pit
defined **EM1:** 17 **EM2:** 30
definition **EM4:** 633
Pit and fissure sealants
properties **A18:** 666
simplified composition on microstructure **A18:** 666
Pit geometry, effect of
on stress-corrosion cracking **A1:** 724
Pit molding **A15:** 9, 228
Pitch **EM3:** 20
and adjustable solid dies **A16:** 302
for carbon aircraft brakes **A18:** 584, 585
defined **A17:** 18 **EM1:** 17-18 **EM2:** 30
precursors, for carbon fiber **EM1:** 112
of thread chasers **A16:** 300, 301
Pitch line speed
gears **A18:** 542
Pitch-base carbon fiber reinforced copper
as metal-matrix packaging composite **EL1:** 1127
Pitch-base carbon fibers *See also* Carbon fibers
ultrahigh-module, properties **EL1:** 1122
Pitch-catch testing
ultrasonic inspection **A17:** 249
Pitchblende (uranium)
toxicity of **A2:** 941-942
Pitches
for surface-mount package options **EL1:** 77, 730
Pitchline damage, in gears
forms of **A11:** 150
Pitman arm press drive
in mechanical presses **A14:** 31
Pits *See also* Erosion; Erosion-corrosion; Pitting; Pitting corrosion; Preferential pitting
depth measurement **A13:** 232
initiation **A13:** 43, 49
as localized corrosion site **A13:** 112-113
standard rating chart **A13:** 231
types **A13:** 231
Pitsch-Petch orientation relationship in pearlite **A9:** 658
Pitting *See also* Corrosion; Corrosion pitting; Etch pitting; Pits; Pitting corrosion; Preferential pitting **A18:** 257-258, 259
AISI/SAE alloy steels **A12:** 329, 332
aluminum alloys **M2:** 204-206
of aluminum coatings **A11:** 542
in amorphous metals **A13:** 867-868
austenitic stainless steels **A6:** 467 **A12:** 358
in boiler service corrosion, carbon steels **A13:** 514
in brazed joints **A13:** 877
by arcing **A12:** 488
by differential aeration **A11:** 632
by molten salts **A13:** 50
in carbon and low-alloy steels **A11:** 199
carbon steel **A17:** 200-201
cast irons **A13:** 568
causes, detection and stages of **A11:** 176-177
ceramic **A12:** 471
chloride, paper machine corrosion **A13:** 1189-1190
closed feedwater heaters **A13:** 990
of cobalt-base alloys **A2:** 453
Pitting (continued)
of condenser tube, saltwater heat exchanger **A11:** 631
copper alloys **A12:** 403
copper metals **M2:** 459-461
copper/copper alloys **A13:** 612-613
corrosion **A12:** 41, 43, 243, 358
corrosion of weldments **A6:** 1067, 1068
corrosion, radiographic methods **A17:** 296
crack initiation at **A13:** 43, 149
critical temperature **A13:** 114
cross section, heat-exchanger tube **A11:** 630
cross section, steel forging **A11:** 341
defined **A8:** 10 **A11:** 8 **A13:** 10, 231
definition **A6:** 1067
effect in surface stress measurement **A10:** 387
electrical, in bearing failures **A11:** 493-495, 497
electrochemical testing methods **A13:** 216
in electropolishing **A9:** 49-51
failure, stainless steel storage tank **A11:** 177
in fatigue fracture **A11:** 127
fatigue, resistance in bearing steels **A11:** 490
in forging **A11:** 340-341
formation, at sulfide inclusions **A13:** 48-49
fresh water corrosion **M1:** 734-737
in gears **A11:** 592-593 **A18:** 257-258
in heat-exchanger tubes, from chloride ions **A11:** 630
internal combustion engine parts **A18:** 555, 559
in iron castings **A11:** 353
iron, unalloyed titanium **A13:** 673
kinetics of **A8:** 497
in lead-antimony-tin alloys during etching **A9:** 417
as liquid-erosion damage **A11:** 167-170
localized, copper/copper alloys **A13:** 612
marine **A13:** 906
material selection to avoid/minimize **A13:** 323
in mechanical fasteners **A11:** 542
in medium-carbon steels **A12:** 259
metal penetration **A13:** 232
and microfissure crevice corrosion, compared **A13:** 349
on nylon driving gear **A11:** 764, 765
oxidation-corrosion, in ceramics **A11:** 755
oxygen, in superheater tubes **A11:** 615
in polycrystalline nickel **A11:** 165
potential **A8:** 418, 532
potential, defined **A13:** 583
potential testing, titanium/titanium alloys **A13:** 672-673
precipitation-hardening stainless steels **A12:** 373
rolling-element bearings **A18:** 258
in SCC **A12:** 25, 27
seawater corrosion **M1:** 739, 742, 744-745
small-scale, in cavitating mercury **A11:** 165
in stainless steels **A11:** 200 **A13:** 554, 559
of STAMP processed products **A7:** 549
standard charts **A13:** 232
steam equipment failure by **A11:** 602
steam generators **A13:** 944
steam surface condensers **A13:** 987-988
steam turbines **A13:** 993
steam/water-side boilers **A13:** 992
in steel bar and wire **A17:** 549
surface, as contact fatigue **A11:** 133-134
surface, in shafts **A11:** 467
surface, schematic of **A11:** 341
temperature effects **A13:** 114
tests, duplex stainless steel weldments **A13:** 359-361
in titanium **A11:** 202
tube, remote-field eddy current inspection **A17:** 195
tubular products **A17:** 567
in ultrasonic fatigue testing **A8:** 254
underalloyed weld metal **A13:** 349
and uniform corrosion resistance **A13:** 48
of various metals **A11:** 177-178
Pitting as a result of etching **A9:** 450-451, 478
electropolishing **A9:** 105
of rhenium and rhenium-bearing alloys **A9:** 447
of electropolishing **A9:** 440
in tungsten and tungsten alloys as a result
Pitting contact
by electromigration **EL1:** 964

Pitting corrosion *See also* Corrosion; Corrosion pitting; Crevice corrosion; Pits; Pitting; Pitting resistance **M5:** 432-433
aircraft powerplants **A13:** 1043-1044
of aluminum/aluminum alloys **A13:** 118-119, 583-584
austenitic stainless steel weldments **A13:** 348
autocatalytic **A13:** 112, 113
biological **A13:** 120
of carbon steel superheater tube **A11:** 614-615
causes, detection, and stages **A11:** 176-177
in chloride environments, nickel-base alloys **A13:** 644-645
copper alloys **A12:** 403
in corrosion fatigue **A13:** 614
cyclic potentiodynamic polarization measurement **A13:** 231
defined **A12:** 41
evaluation of **A13:** 231-233
examination **A13:** 231-232
from stainless steel wire brush cleaning **A13:** 350, 352
low-carbon steel **A12:** 243
in manned spacecraft **A13:** 1075-1076
martensitic stainless steels **A12:** 43
mechanisms/theories **A13:** 113-114, 1012
of metals **A11:** 176-177
in mining/mill applications **A13:** 1295
occurrence and testing **A13:** 114
on stainless steel bone screw **A11:** 687, 691
pickling process causing **M5:** 80
preventive agents *See* Antipitting agents
schematic **A13:** 1331
in soil, measurement **A13:** 209-210
space shuttle aluminum alloy **A13:** 1065
space shuttle orbiter **A13:** 1066
stainless steel **A13:** 113
testing **A13:** 216, 231-233, 359-361, 559, 672-673
thiosulfate **A13:** 349, 352
water-recirculating systems **A13:** 488-489
zirconium/zirconium alloys **A13:** 717
Pitting factor
defined **A13:** 10
Pitting potential **A6:** 1067
Pitting resistance **A13:** 114
P/M stainless steels **A13:** 834
test method **A13:** 231
Pitting resistance equivalent (PRE) **A6:** 471, 472, 474, 479, 697
Pittsburgh forming process **EM4:** 21
Pittsburgh Iron and Steel Foundries Company *See* Mackintosh-Hemphill Company
Pittsburgh Reduction Company
aluminum casting by **A15:** 35
Pittsburgh Steel Casting Company
steel castings by **A15:** 31
Pivot bearing
defined **A18:** 14
Pivot pins
fatigue fracture of **A11:** 308
Pivot shaft
ceramic mold shapemaking of **A7:** 429
Pivot shears *See* Alligator shears
Pivoted-pad bearing *See* Tilting-pad bearing
PIXE *See* Particle-induced x-ray emission
Pixel
defined **A17:** 18
Pixel size
for automatic optical inspection **EL1:** 942
effect on nonfractal behavior **A12:** 211
Pixels
density, in elemental mapping of high-temperature solder **A10:** 532
in dot mapping **A10:** 526-527
image analysis scanners **A10:** 310
picture elements as **A10:** 690
Placement
accuracy **EL1:** 732
of decoupling capacitors **EL1:** 28
Placement (continued)
in surface-mount technology **EL1:** 730
and transporting, outer lead bonding **EL1:** 285
Plackett Burman experimental design **EM2:** 601
Plain and coated paper copiers **A7:** 584
Plain bearing
defined **A18:** 14-15
Plain carbon steel
dynamic yield stress vs. strain rate for **A8:** 41
torsional ductility **A8:** 166
Plain carbon steels *See also* Carbon steels; Plate steels **A14:** 128, 869-871 **A15:** 702-714
applications **A15:** 714
blue brittleness in **A11:** 98
boiler tube, corrosion-fatigue cracks in **A11:** 79
chemical composition **A6:** 906
cleaning operations **A15:** 712-713
desulfurization **A15:** 709-710
effect of disturbed metal on metallographic appearance of **A10:** 301
foundry practice, specifics **A15:** 710-712
fracture map **A12:** 44
graphitized microstructure **A11:** 613
heat treatment **A15:** 713-714
liquid-metal corrosion resistance **A13:** 515
machining **A2:** 967
macrostructure of **A9:** 623
melting practice **A15:** 705-710
microstructural analysis **A7:** 487
microstructure of **A9:** 623-624
molten nitrate corrosion **A13:** 90-91
segregation during dendritic growth **A9:** 625-626
specimen size effect on fatigue limit **A13:** 293
structure and property correlations **A15:** 702-705
sulfur dioxide corrosion **A13:** 81
temperature effects on erosion **A13:** 313
thermal stress relief for SCC prevention **A13:** 328
Plain journal bearing
defined **A18:** 15
Plain thrust bearing
defined **A18:** 15
Plain weave *See also* Fabric(s); Weaves
defined **EM1:** 18 **EM2:** 30
for fiberglass fabric **EM1:** 111
open-selvage, carbon fabric **EM1:** 128
unidirectional/two-directional fabrics **EM1:** 125
for woven fabric prepregs **EM1:** 148
yarn interlacing **EM1:** 125
Planar **EM3:** 20
defined **EL1:** 1153 **EM1:** 18 **EM2:** 30
Planar anisotropy *See also* Anisotropy; Plastic strain ratio **A1:** 575
in deep drawing **A14:** 576, 584
defined **A8:** 10
and *r* value **A8:** 550
of sheet metals **A8:** 555
steel sheet **M1:** 549
Planar assembly
materials and processes selection **EL1:** 116-117
Planar chip tape automated bonding (TAB)
defined **EL1:** 275
Planar circuit areas
for interconnections **EL1:** 6
Planar defects, effect on transmitted wave amplitudes in transmission electron microscopy **A9:** 119
Planar diode glow discharge deposition **A18:** 841
Planar electrode sputtering systems **M5:** 413-414
Planar extension
rheology **EL1:** 839
Planar flaws *See also* Flaw(s)
defined **A17:** 50
NDE detection methods **A17:** 50
Planar flow casting technique
amorphous materials and metallic glasses **A2:** 806
Planar growth of alloys **A9:** 612
Planar helix winding
defined **EM1:** 18 **EM2:** 30
Planar hybrid circuits
advantages **EL1:** 8
Planar image reformation
computed tomography (CT) **A17:** 378
Planar interface *See also* Interface
growth, single-phase alloys **A15:** 114-116
particle behavior at **A15:** 142-144
Planar lead tape automated bonding (TAB)
defined **EL1:** 275
Planar magnetron sputtering technique **A18:** 841, 842
Planar multichip technology **EL1:** 307
Planar reformation
defined **A17:** 384
Planar sections
measurements used for particle-size distribution curves **A9:** 131-133
for profile generation **A12:** 198
of spheres, limits for **A9:** 133
Planar shape **A7:** 240
Planar slip
defined **A9:** 687
Planar solidification **A6:** 49, 50, 52, 53
Planar winding *See also* Polar winding
defined **EM1:** 18 **EM2:** 30
Planar-magnetron sputter deposition
to fabricate solid-state welded silver interlayers **A6:** 165, 166, 168, 169, 170
Planar-mounted components
desoldering of **EL1:** 722
Planck constant **A18:** 441
Planck, Max ..... **A3:** 1•7
Planck's constant
abbreviation for **A10:** 690
defined **A10:** 679
in electromagnetic radiation **A10:** 83
in ESR analysis **A10:** 254
Planck's law **A18:** 441
Plane
defined **A17:** 18
Plane (crystal)
defined **A9:** 14
Plane angle
SI supplementary unit and symbol for **A10:** 685
SI unit/symbol for **A8:** 721
Plane bending
fatigue test specimens **A8:** 368
Plane cleavage
defined **A11:** 2
Plane front solidification *See also* Solidification growth during
and heat transfer **A15:** 112-113
macrosegregation mechanism **A15:** 138-139
Plane glass illuminator
defined **A9:** 14
Plane grating
defined **A10:** 679
Plane of focus *See* Focusing
Plane of working
defined **A9:** 14
Plane strain
brittle flat-face fracture by **A11:** 75
condition for **A11:** 51
defined **A8:** 10 **A10:** 679 **A11:** 8 **A13:** 10
during deformation **A8:** 576
effect of thickness and *n* value on **A8:** 551
fracture toughness, defined (under Stress-intensity factor) **A13:** 12
mode, fatigue fracture **A11:** 105
and plane stress, compared **A11:** 51-52
region **A8:** 547, 549
screening tests, for fracture toughness **A8:** 460
sidepressing, isothermal, titanium alloy specimen **A8:** 172
stretching, as sheet metal forming **A8:** 547
test, for fracture toughness **A11:** 60-61

Plane stress *See also* Crack tip opening displacement
condition for **A11:** 51
defined **A8:** 10 **A10:** 679 **A11:** 8 **A13:** 10
during deformation **A8:** 576
fracture toughness, defined (under Stress-intensity factor) **A13:** 12
mode, fatigue fracture **A11:** 105
and plane strain, compared **A11:** 51-52
shear-face fracture by **A11:** 75
Plane stress modulus **A18:** 33
Plane waves
dynamical diffraction theory **A9:** 112
Plane-matching model
for grain boundaries **A9:** 119
Plane-polarized light
and observation of magnetic domains **A9:** 535-536
in optical microscopy **A9:** 72
Plane-strain bending
stress **A8:** 120-121, 123
Plane-strain compression test **A1:** 582-583 **A8:** 160-164, 584 **A14:** 377-379
Plane-strain crack arrest toughness **A8:** 725
Plane-strain deformation **A1:** 121
Plane-strain flow stress **A8:** 577
Plane-strain forging
flow localization rate **A8:** 172
Plane-strain fracture surface
slip bands **A8:** 481-482
Plane-strain fracture toughness *See also* Fracture toughness; Mechanical properties
AISI 4130 steel **A8:** 478
defined **A8:** 10
ferritic steels **A8:** 479
in fracture mechanics **A8:** 450-451
testing **EM2:** 739-740
wrought aluminum and aluminum alloys **A2:** 74, 109, 113
Plane-strain fracture toughness (K_{Ic})
defined **A11:** 8, 10
in fracture mechanics **A11:** 51
for high strength and low toughness **A11:** 60
standard specimens for determining **A11:** 61
stress-intensity rate, effect on **A11:** 54
variation with tensile strength **A11:** 54
Plane-strain fracture toughness test
plain carbon steels **A15:** 702
Plane-strain fracture toughness testing
for high strain rate testing **A8:** 187
martensitic stainless steel **A8:** 479-480
symbol for **A8:** 725
test procedures **A8:** 459-460
Plane-strain fractures
AISI/SAE alloy steels **A12:** 308
tension-overload, maraging steels **A12:** 385
wrought aluminum alloys **A12:** 423
Plane-strain plastic zone **A8:** 471
Plane-strain rapid-load fracture toughness testing **A8:** 260-261
Plane-strain stretching
of sheet metals **A14:** 877
Plane-strain tensile testing **A8:** 553, 557-558 **A14:** 887
Plane-strain wedge testing
physical modeling **A14:** 435
Plane-stress elastic model
full-tensor determination **A10:** 384
Marion-Cohen technique **A10:** 384
$\sin^2 \phi$ technique **A10:** 384
single-angle technique **A10:** 383-384
two-angle technique **A10:** 384
of x-ray diffraction stress measurement **A10:** 382-384
Plane-stress fracture toughness *See also* *R*-curve; Stress-intensity factor
defined **A8:** 10
symbol for **A8:** 725
test specimen for **A8:** 462
Plane-stress fracture toughness (K_c) *See also* Stress-intensity factor
defined **A11:** 8, 10
Plane-stress plastic zone **A8:** 471
Plane-stress screening tests
fracture toughness testing **A8:** 462
Planers *See also* Planing
adjustable rail mills **A16:** 182
clamping hardware **A16:** 182
Planers (continued)
duplex tables **A16:** 183
hydraulic **A16:** 181
magnetic chucking **A16:** 182
mechanical-drive **A16:** 181
setup plates **A16:** 182
tool capacity **A16:** 182
workpiece capacity **A16:** 182
Planetary gear assemblies **A7:** 675
Planetary rolling mills **A14:** 351-352
Planimetric method *See* Jeffries' method
Planing *See also* Planers **A16:** 181-186
Al alloys **A16:** 773, 778
carbon and alloy steels **A16:** 676
cast irons **A16:** 657, 659-660
cemented carbides used **A16:** 75
compared to shaping and slotting **A16:** 187
in conjunction with milling **A16:** 329
contouring **A16:** 186
Cu alloys **A16:** 810-811, 813
and cutting fluids **A16:** 186
and fixturing **A16:** 405
and gear manufacture **A16:** 343
hafnium **A16:** 856
heat-resistant alloys **A16:** 742-743
machining, process **A7:** 461
Mg alloys **A16:** 821-822, 823
Ni alloys **A16:** 837, 839
process capabilities **A16:** 181
semifinish and finish **A16:** 184
and shaping **A7:** 461
speed, feed, and depth of cut **A16:** 185-186
tandem (gang) planing **A16:** 181, 182-183
tool design **A16:** 183-185
tools, high-speed tool steels used **A16:** 58
triple planing **A16:** 185
vs. band sawing **A16:** 186
vs. broaching **A16:** 186
vs. gas cutting **A16:** 186
vs. grinding **A16:** 186
vs. milling **A16:** 186
vs. sawing **A16:** 186
workpiece setup **A16:** 183
zinc **A16:** 855
Planned grouping
in comparative experiments **A8:** 640-645
effect on experimental bias **A8:** 640
in randomized and block experimental designs **A8:** 643
Planned interval (immersion) testing **A13:** 223-224
Planning
of comparative experiments **A8:** 639-652
corrosion tests **A13:** 193-196
statistical, and analysis **A13:** 316-317
Plano lens
definition **M6:** 13
Plano objective lenses **A9:** 73
cross sections **A9:** 75
PLANS computer program for structural analysis **EM1:** 268, 272
Plant operation hazards **A7:** 197-198
Plant tissues
AAS analysis of trace metals **A10:** 55
analysis of **A10:** 41
ICP-sequential monochromator analysis of **A10:** 41
powdered, PIXE analysis **A10:** 102
voltammetric detection of herbicides/pesticides in **A10:** 188
Plantations, iron *See* Iron plantations
Planters (farm)
P/M parts for **A7:** 673
Plaque
and corrosion products **A13:** 1347
Plaque-type penetrameters
radiographic inspection **A17:** 339-340
Plasma
definition **M6:** 13
effects in laser beam welding **M6:** 659-661
suppression in laser beam welding **M6:** 660-661
Plasma (ion) carburizing **A4:** 262-263, 352-362
advantages **A4:** 352, 356-357
applications, industrial **A4:** 352, 359-362
carbon mass flow **A4:** 361
carbon profiles **A4:** 355, 356, 358, 359
carbon source **A4:** 361
case depth **A4:** 357
Plasma (ion) carburizing (continued)
characteristics **A4:** 352-359
coverage and wrap-around effect **A4:** 354
description **A4:** 352
diffusion characteristics **A4:** 355-356
dissociation of methane to carbon **A4:** 354-355
down-hole carburizing **A4:** 361-362
efficiency in utilization of gas **A4:** 357-359
equipment requirements **A4:** 359
glow-discharge plasma properties **A4:** 352-353
glow-discharge plasma range and limitations **A4:** 353-354
hollow-cathode effect **A4:** 361, 362
hydrocarbon utilization efficiency **A4:** 358, 359
loading requirements and limitations **A4:** 359
minimum power density **A4:** 361
operating cost comparison **A4:** 357
Paschen curves **A4:** 353, 354
process parameters **A4:** 359-360
production equipment **A4:** 359
properties of parts **A4:** 362
sooting **A4:** 358, 361
time-temperature cycle **A4:** 360
voltage levels **A4:** 354
Plasma (ion) nitriding **A4:** 263, 420-424
advantages **A4:** 420, 424
applications **A4:** 424
atmosphere and pressure control **A4:** 423
auxiliary heating **A4:** 423
case depth **A4:** 424
case structures **A4:** 420-421
compound layers **A4:** 420-421, 423-424
cooling **A4:** 422
for dimensional control **A4:** 424
disadvantages **A4:** 424
equipment **A4:** 422-423
fatigue strength **A4:** 424
fixturing **A4:** 423
formation **A4:** 420-421
glow (discharge) process **A4:** 422
hardness profiles **A4:** 423-424
hot wall vacuum furnaces used **A4:** 495
power supply and control **A4:** 423
process description **A4:** 421-422
suitability of materials **A4:** 423
tool steels **A4:** 724, 754
Plasma arc
cupola **A15:** 36, 392
remelting **A15:** 424
Plasma arc (PA) powder spray **A18:** 829, 830, 831, 832
Plasma arc cutting
applications **A14:** 731-732
bevel cutting **M6:** 917
comparison to oxyfuel gas cutting
definition **M6:** 13
heat-affected zone **M6:** 917
operating principles and parameters **A14:** 729-731
operation **M6:** 914-916
cutting speeds **M6:** 915-916
gas, selection of **M6:** 912, 916
power supply **M6:** 915-916
quality of cut **M6:** 917
technique **M6:** 917
torches **M6:** 914-916
water injection **M6:** 915-917
work metals **M6:** 916-917
pierce capacity **A14:** 731
safety **M6:** 58, 917-918
types **A14:** 730
Plasma arc cutting (PAC) **A6:** 1166-1171
aluminum **A6:** 1167, 1169, 1170, 1171
applications **A6:** 1170
shipbuilding **A6:** 384
brass **A6:** 1170
carbon steels **A6:** 1168, 1169-1170
characteristics of a plasma arc cut **A6:** 1169, 1170
copper alloys **A6:** 752, 754, 1169, 1170
definition **A6:** 1166, 1212
equipment **A6:** 1166-1167
controls **A6:** 1167
coolant system for torch **A6:** 1166-1167
leads **A6:** 1166
manifold assembly **A6:** 1167
optional equipment **A6:** 1167
power supply **A6:** 1167

Plasma arc cutting (PAC) (continued)
torches ... **A6:** 1166, 1168
galvanized metal ... **A6:** 1170
gases ... **A6:** 1167
heat-affected zone ... **A6:** 1169
high-strength steels ... **A6:** 1170
operating sequence ... **A6:** 1167-1168
process considerations ... **A6:** 1168-1170
cut quality ... **A6:** 1169-1170
gas shielded ... **A6:** 1168, 1170
process capabilities ... **A6:** 1168-1169
process mechanization ... **A6:** 1169
process variations ... **A6:** 1168
water-injection ... **A6:** 1168, 1170
water-shielded ... **A6:** 1168, 1170
process description ... **A6:** 1166
safety ... **A6:** 1170-1171
compressed gas cylinders ... **A6:** 1171
electric shock ... **A6:** 1171
explosions ... **A6:** 1171
fire ... **A6:** 1171
fumes ... **A6:** 1170
gases ... **A6:** 1170
noise ... **A6:** 1170
radiant energy ... **A6:** 1170
safety precautions ... **A6:** 1201
stainless steels ... **A6:** 1167, 1169, 1170
suggested viewing filter plates ... **A6:** 1191
Plasma arc cutting of
aluminum ... **M6:** 914, 916-917
aluminum alloys ... **M6:** 916
carbon steels ... **M6:** 914, 916
copper ... **M6:** 916
low-carbon steels ... **M6:** 916
magnesium ... **M6:** 916
stainless steels ... **M6:** 914, 916-917
titanium ... **M6:** 916
Plasma arc machining
compared with laser beam machining ... **A16:** 572
stainless steels ... **A16:** 705, 706
Plasma arc melting
and refining, nickel and nickel alloys ... **A2:** 429
for shape memory effect (SME) alloys ... **A2:** 899
Plasma arc welding ... **A7:** 465 **M6:** 214-224
accessory equipment ... **M6:** 218
advantages ... **M6:** 215
applicability ... **M6:** 214-215
circumferential pipe welding ... **M6:** 221
comparison of processes ... **M6:** 221-224
copper and copper alloys ... **M6:** 214, 444
definition ... **M6:** 13
discontinuities from ... **A17:** 587
electrodes ... **M6:** 217
failure origins ... **A11:** 415
filler metals ... **M6:** 218
hot-wire systems ... **M6:** 218
hardfacing ... **M6:** 785-787
joint design ... **M6:** 218-219
butt joints in thin metal ... **M6:** 218
edge-flange welds ... **M6:** 218
machined-groove joints ... **M6:** 218
square-groove butt joints ... **M6:** 218
keyhole welding ... **M6:** 218-220
backing requirements ... **M6:** 220
starting the weld ... **M6:** 220
terminating the weld ... **M6:** 220
limitations ... **M6:** 215
manufacture of stainless steel tubing ... **M6:** 220-221
metals welded ... **M6:** 214
multiple-pass welding ... **M6:** 221
orifice and shielding gases ... **M6:** 217-218
power sources ... **M6:** 216
process fundamentals ... **M6:** 215
arc modes ... **M6:** 215
current ... **M6:** 215
distance from orifice ... **M6:** 215
heat-energy concentration ... **M6:** 215

Plasma arc welding (continued)
plasma generation ... **M6:** 215
stainless steels, austenitic ... **M6:** 342-344
circumferential pipe welding ... **M6:** 344
tube welding ... **M6:** 342-343
vessel welding ... **M6:** 344
stainless steels, ferritic ... **M6:** 348
stainless steels, nitrogen-strengthened austenitic ... **M6:** 345
transferred, for hardfacing ... **A7:** 833-834
underwater welding ... **M6:** 922
weld overlaying ... **M6:** 807-808
weld overlays of stainless steel ... **M6:** 814-815
welding positions ... **M6:** 214
welding stainless steel foil ... **M6:** 221
welding torches ... **M6:** 216-217
arc-constricting nozzles ... **M6:** 217
machine torches ... **M6:** 216
orifice diameter ... **M6:** 217
torch cooling systems ... **M6:** 217
torch position ... **M6:** 216
work-metal thickness ... **M6:** 214-215
Plasma arc welding (PAW) *See also* Plasma transferred arc welding ... **A6:** 124-125, 195-199
advantages ... **A6:** 195-196
alloy steels ... **A6:** 197
aluminum ... **A6:** 197
aluminum alloys ... **A6:** 195, 197, 199, 735, 736-737
aluminum bronzes ... **A6:** 754
aluminum-lithium alloys ... **A6:** 551, 552
applications ... **A6:** 195, 197
butt joints ... **A6:** 197, 198
carbon steels ... **A6:** 197, 652, 653, 654, 658
components ... **A6:** 198
copper alloys ... **A6:** 197, 756
copper-nickel alloys ... **A6:** 754
current and operating modes ... **A6:** 195
definition ... **A6:** 195, 1212
disadvantages ... **A6:** 195-196
electrodes ... **A6:** 196
equipment ... **A6:** 196-197
ferritic stainless steels ... **A6:** 448
filler metals ... **A6:** 658
flanged edge joints ... **A6:** 198
hardfacing alloys ... **A6:** 796, 798, 800, 805-806
health and safety precautions ... **A6:** 199
inspection ... **A6:** 198
joints ... **A6:** 198
keyhole mode ... **A6:** 195, 197, 198, 199
maximum current with selected electrode diameter, vertex angle, and nozzle bore diameter ... **A6:** 197
melt-in mode ... **A6:** 195, 198
micro-lap joints ... **A6:** 198
microplasma mode ... **A6:** 195, 198, 199
nickel alloys ... **A6:** 197, 740, 745-746
nickel-base corrosion-resistant alloys containing molybdenum ... **A6:** 594
niobium alloys ... **A6:** 581
noise level and welding safety ... **A6:** 1192
personnel requirements ... **A6:** 198-199
plasma (orifice) and shielding gases ... **A6:** 196-197
power source ... **A6:** 37, 196
precipitation-hardening stainless steels ... **A6:** 490
principles of operation ... **A6:** 195
process operating procedure ... **A6:** 198
process selection guidelines for arc welding ... **A6:** 653
for repair welding ... **A6:** 1103, 1107
safety precautions ... **A6:** 1192-1193, 1196
shielding gases ... **A6:** 65, 662
silicon bronzes ... **A6:** 754
single-V butt joints ... **A6:** 198
square butt joints ... **A6:** 198
stainless steel casting alloys ... **A6:** 496
stainless steels ... **A6:** 197, 199, 698, 699
suggested viewing filter plates ... **A6:** 1191
of tantalum ... **A6:** 197

Plasma arc welding (PAW) (continued)
titanium alloys ... **A6:** 197, 198, 512, 513, 514, 516, 519, 520, 521, 522, 783, 784, 786
to solve problems in joining thin sections by oxyfuel gas welding ... **A6:** 288
tolerance to variation in current and gas flow rate ... **A6:** 195
tool and die steels ... **A6:** 674, 676
troubleshooting ... **A6:** 198
ultrahigh-strength low-alloy steels ... **A6:** 673
variable polarity plasma arc (VPPA) welding ... **A6:** 195, 197, 199
vs. laser-beam welding ... **A6:** 262
vs. plasma-MIG welding ... **A6:** 224
weld cladding ... **A6:** 816, 818, 819
weld discontinuities ... **A6:** 1078
weld quality control ... **A6:** 198
welding torches ... **A6:** 196
of zirconium ... **A6:** 197
of zirconium alloys ... **A6:** 198, 787
Plasma arc welding of
alloy steels ... **M6:** 304-306
applications ... **M6:** 305-306
effect of arc length ... **M6:** 305
immediate current application ... **M6:** 305-306
joint preparation ... **M6:** 305
keyhole plasma welding ... **M6:** 306
pulsed plasma arc welding ... **M6:** 306
shielding gas ... **M6:** 305
aluminum alloys ... **M6:** 214
carbon steels ... **M6:** 214
cobalt-based alloys ... **M6:** 214
copper nickel ... **M6:** 916
low-alloy steels ... **M6:** 214
nickel alloys ... **M6:** 440
nickel-based alloys ... **M6:** 214
stainless steels ... **M6:** 214, 220, 221
titanium and titanium alloys ... **M6:** 214, 446, 456
Plasma cold crucible casting ... **A15:** 424-425
Plasma cold hearth melting ... **A15:** 424
Plasma etching
as photolithographic process, active-component fabrication ... **EL1:** 194
thin-film hybrids ... **EL1:** 327
Plasma gas welding
shielding gas selection ... **A6:** 67, 68
Plasma heating and degassing *See also* Plasma melting and casting
equipment/processing ... **A15:** 440-444
ladle furnace, three-phase ac unit ... **A15:** 440
as ladle metallurgy ... **A15:** 440-444
ladle sizes ... **A15:** 444
nickel alloys ... **A15:** 820
plain carbon steels ... **A15:** 710
plasma torches ... **A15:** 440
Plasma ion deposition ... **A13:** 498
Plasma ladle reheater ... **A15:** 442
Plasma melting and casting *See also* Plasma heating and degassing
atmosphere control ... **A15:** 420-423
furnace equipment ... **A15:** 420
melting and remelting ... **A15:** 423-425
plasma torch ... **A15:** 419-420
processes ... **A15:** 423-425
Plasma metallization
and process plating ... **EL1:** 511
Plasma metallizing
definition ... **A6:** 1212
Plasma nitrocarburizing ... **A4:** 431-435
advantages ... **A4:** 432
applications ... **A4:** 433, 434
equipment ... **A4:** 432, 434
in fluidized beds ... **A4:** 490
history of process ... **A4:** 431-432
masking arrangement ... **A4:** 435
physical metallurgy ... **A4:** 432
powder metallurgy (P/M) components ... **A4:** 434-435

Plasma rotating electrode process
beryllium powder ... A2: 685
Plasma rotating electrode process (PREP) ... A1: 973
advantages and applications ... A7: 42
in hot isostatic pressing ... A7: 438
particles produced ... A7: 167
of titanium and titanium alloy powder production ... A7: 165, 167, 469
Plasma smear test procedure
for laminates ... EL1: 536-537
Plasma source ion implantation (PSII) ... A18: 642
Plasma spheroidized magnetite
in copier powders ... A7: 587
Plasma spray coating ... A7: 8
Plasma spray coatings
molybdenum for piston rings ... A18: 556
resistance to cavitation erosion ... A18: 217
titanium alloys ... A18: 778, 780
Plasma sprayed coatings
for wear resistance ... M1: 635
Plasma spraying ... A7: 8 A13: 10, 460 EM4: 32
ceramic coatings for adiabatic diesel engines ... EM4: 992
as cermet forming technique ... A7: 800
comparison with polymer-derived coatings ... EM4: 225
definition ... M6: 13
of fibers as a fabrication process for metal-matrix composites ... A9: 591
for hardfacing ... A7: 797, 832-833
Plasma spraying (PS)
surface preparation of metals ... EM3: 265, 266
Plasma spraying (PSP)
cast irons ... A6: 720
definition ... A6: 1212
Plasma torch
design ... A15: 419-420
direct current ... A15: 440-443
heating and degassing ... A15: 440
high-power, steel melting/ladle heating ... A15: 442
hollow copper electrode design ... A15: 420
ICP, structure of ... A10: 32
plasma generation ... A15: 419
tungsten tip design ... A15: 419
use in ICP-AES ... A10: 31
Plasma transfer arc spraying
ceramic coatings for adiabatic diesel engines ... EM4: 992
Plasma transferred arc (PTA) welding ... A6: 805
advantages ... A6: 805
disadvantages ... A6: 805-806
hardfacing alloy consumable form ... A6: 796
hardfacing alloys ... A6: 799
Laves phase alloys ... A6: 795
Plasma transferred arc hardfacing process ... A6: 802
Plasma transferred arc process
for hardfacing ... A7: 833-834
Plasma treatment ... EM3: 42, 847
Plasma-arc thermal spray coating
ceramic coatings ... M5: 535, 541-542, 546-547
oxide-resistant coating ... M5: 665-666
selective plating compared to ... M5: 292-293
transferred arc process ... M5: 363-364
Plasma-assisted chemical vapor deposition (PACVD) ... A18: 840, 846, 847-848
advantages over CCVD ... A18: 848
compounds and synthesized deposition rates ... A18: 848
definition ... A18: 847
ion bombardment energy ... A18: 848
limitations ... A18: 848
neutral radicals ... A18: 847
substrate temperature ... A18: 848
Plasma-assisted physical vapor deposition (PAPVD) ... A18: 840
compounds and synthesized deposition rates ... A18: 848
Plasma-assisted reactive evaporation ... A18: 840, 843-844
Plasma-emission spectrophotometry
for chemic analysis ... EM4: 553, 554
Plasma-enhanced chemical vapor deposition (PE-CVD)
electronic processing of glasses ... EM4: 1056
Plasma-fired cupolas ... A15: 392
Plasma-MIG welding ... A6: 223-225
advantages ... A6: 223
aluminum ... A6: 224
aluminum alloys ... A6: 223, 224
applications ... A6: 223-225
definition ... A6: 223
deposition rates ... A6: 223, 224-225
disadvantages ... A6: 223
electrodes ... A6: 223, 224
equipment ... A6: 223-224
metal transfer ... A6: 223
mild steels ... A6: 224, 225
molybdenum ... A6: 224
nickel alloys ... A6: 224
personnel ... A6: 225
power sources ... A6: 223-224
principles of operation ... A6: 223
procedure ... A6: 224
inspection ... A6: 224
process operating procedure ... A6: 224
troubleshooting ... A6: 224
weld quality control ... A6: 224
safety ... A6: 225
shielding gases ... A6: 224
spray transfer ... A6: 223
stainless steels ... A6: 224, 225
steel ... A6: 224
tungsten ... A6: 224
vs. gas-metal arc welding ... A6: 223, 224, 225
vs. plasma arc welding ... A6: 224
Plasma-polymerized hexamethyidisiloxane
structure and degradation of ... A10: 285-286
Plasma-spray method
oxidation protective coating ... M5: 379
Plasma-sprayed coating
on lead wires in acidified chloride solutions ... A8: 420
Plasma-sprayed layers
contrasting by interference layers ... A9: 60
Plasma-transferred arc process
cobalt-base alloys ... A13: 664
Plasmajet ... A10: 40, 679
Plasmas
defined ... A10: 679
mixed gas ... A10: 37
Plasmon
defined ... A10: 679
excitation, as inelastic scattering process ... A10: 434
loss, Auger electron ... A10: 551
loss, peak structures, alumina and aluminum ... A10: 552
Plasmon effects
cathodoluminscence used to detect ... A9: 91
Plasmon peaks ... A18: 390
PLASTEC Adhesives ... EM3: 71
Plaster
galvanic corrosion of metals embedded in ... A11: 186-187
Plaster casting *See also* Castings, Foundry products
aluminum alloys ... M2: 146
Plaster castings
aluminum and aluminum alloys ... A2: 5
copper alloys for ... A2: 348
Plaster mold casting
copper alloys ... M2: 384-385
Plaster molding ... A15: 242-247
Antioch process ... A15: 246-247
applications ... A15: 242
calcium sulfate, characteristics ... A15: 242-243
conventional, sequence of operations ... A15: 243-245
defined ... A15: 9
flasks ... A15: 243
foamed plaster molding process ... A15: 247
match plate patterns ... A15: 245-246
metals cast in ... A15: 243
mold drying equipment ... A15: 243
patterns and coreboxes ... A15: 243
plaster mold compositions ... A15: 242-243
tolerances ... A15: 622
Plasters
in Neolithic period ... A15: 15
Plastic *See also* Alkyd plastic; Allyl plastic; Crystalline plastic; Foamed plastic; Isocyanate plastics; Polymer(s); Silicone plastics
defined ... EM1: 18
for environmental test chamber for aqueous solutions at ambient temperatures ... A8: 415
flammability characteristics ... EM1: 358, 359
memory
reinforced, defined ... EM1: 20
response, in low-cycle torsional fatigue ... A8: 150
vacuum coating of ... M5: 394, 400-401
Plastic (powder coat) finishing
of zinc alloy ... A2: 530
Plastic abrasives
use in dry blasting ... M5: 85
Plastic bending equations ... A8: 118
Plastic binders
for sand ... A15: 211
Plastic bond
brazing filler metals available in this form ... A6: 119
Plastic buckling
by overloading ... A11: 137
Plastic clay
applications ... EM4: 47
composition ... EM4: 47
supply sources ... EM4: 47
Plastic cloth screen surfaces ... A7: 176
Plastic coatings
for fracture surfaces ... A12: 73
for zinc alloy castings ... A15: 796
Plastic composites
environmental effects ... EM2: 428-429
Plastic deformation *See also* Deformation; Deformation, Ductility, Closed-die steel forgings; Microplastic deformation; Workability Yielding ... A16: 4, 7-10, 23-24, 30 A18: 176, 181, 183 EM3: 20
of aluminum mill and engineered products ... A2: 29
and average flash temperature ... A18: 43
in compaction ... A7: 58, 298
continuous (acoustic) emission ... A17: 287
in creep curves ... A8: 308
in cup-and-cone fracture ... A11: 82-83
defined ... A8: 10 A9: 14 A11: 8 A13: 10 A14: 10 EM1: 18 EM2: 30
deformation modes ... A9: 686-688
development of ... A9: 693
development of crystallographic texture during ... A9: 700-701
as die failure cause ... A14: 47
in ductile irons ... A12: 227-237
of ductile/brittle fractures ... A12: 173
effect, hydrogen absorption, low-carbon steels ... A13: 329
effect of composition on ... A9: 685
effect of crystal structure on ... A9: 684
effect on fatigue cracks, wrought aluminum alloys ... A12: 420
effect on fringe patterns ... A10: 368
effect on linear elastic fracture mechanics ... A11: 47
effect on martensite formation in austenitic stainless steels ... A9: 283
elevated temperatures ... A9: 688-691
energy stored during cold working ... A9: 692
extension ladder collapse by ... A11: 137
fracture during ... A7: 410
full densification of powder compact by ... A7: 502
gross, under tension ... A8: 20
and hardness testing ... A8: 71
high-purity copper ... A12: 399-400
impact ... A12: 336
in liquid erosion ... A11: 165
in loose powder compaction ... A7: 58, 298
low temperature and high strain rate ... A9: 688
in magnesium alloys ... A9: 427
and magnetic hysteresis ... A17: 134
magnetic printing detection ... A17: 126
martensite in stainless steel formed by ... A9: 66
mechanical energy in ... A7: 61
in medium-carbon steels ... A12: 258
microstructural features ... A9: 685
in pin bearing testing ... A8: 60

Plastic deformation (continued)
point defects created by ... **A9:** 116
residual stresses from ... **A13:** 255-256
resistance, of die materials ... **A14:** 46
shear bands, titanium alloys ... **A12:** 445
slip ... **A9:** 684-685
specimens ... **A8:** 510, 513
specimens, SCC testing ... **A13:** 253
stress effects ... **A15:** 616
surface property effect ... **A18:** 342
theory ... **A18:** 281
and thermal stress ... **EL1:** 56
warping from ... **A11:** 141
Plastic deformation structures in
hafnium ... **A9:** 499
zirconium and zirconium alloys ... **A9:** 499
Plastic deformation zone ... **EM3:** 508-509, 511, 514
Plastic denture teeth
properties ... **A18:** 666
simplified composition on microstructure ... **A18:** 666
Plastic distortion *See also* Creep; Distortion
as failure mechanism ... **A11:** 75
Plastic dual-in-line packages (PDIP)
defined/outline ... **EL1:** 210
die attachments ... **EL1:** 217-221
through-hole/surface-mount assembly ... **EL1:** 437-438
Plastic electrical tape used for unmounted electropolishing specimens ... **A9:** 49
Plastic encapsulants
key properties ... **EL1:** 805-809
Plastic encapsulation
defined ... **EL1:** 1153
Plastic film
in vacuum molding ... **A15:** 236
Plastic flow *See also* Flow; Plastic deformation; Yield; Yielding ... **EM3:** 20
AISI/SAE alloy steels ... **A12:** 329
and burnup, tapered roller bearing ... **A11:** 500-501
by slip process ... **A13:** 46
defined ... **A14:** 10 **EM1:** 18 **EM2:** 30
enhanced ... **A13:** 165
high-purity copper ... **A12:** 400
of loose powders in densification ... **A7:** 298
microscopic, tear ridges from ... **A12:** 224
of oxide ... **A13:** 72
in pure compression ... **A8:** 576-577
in pure tension ... **A8:** 576-577
rolling-element bearings, failure by ... **A11:** 499-500
Tresca criterion for ... **A8:** 576
and workability ... **A14:** 369
wrought aluminum alloys ... **A12:** 420
Plastic foam *See* Cellular plastic ... **EM3:** 20
for foam vaporization ... **A15:** 22
Plastic forming *See* Extrusion; Injection molding
Plastic hinge formation
during ring rolling ... **A14:** 113
Plastic hysteresis index ... **A18:** 422, 426
Plastic impression
contrast enhancement by coating ... **A9:** 98
Plastic inclusions
in rolling ... **A14:** 358
Plastic instability
in compression testing ... **A8:** 583-584
defined ... **A8:** 10
in high strain rate testing ... **A8:** 188
Plastic instability in compression
test of ... **A14:** 376-377
Plastic leaded chip carrier (PLCC) *See also* Chip carriers (CC)
component removal ... **EL1:** 726
defined/package outline ... **EL1:** 209
die attachments ... **EL1:** 217-221
as surface-mount package option ... **EL1:** 7
thermal expansion mismatch problem ... **EL1:** 611
thermal resistance ... **EL1:** 409-41
Plastic limit load behavior
and crack growth ... **A8:** 377
Plastic macrodeformation
determined ... **A11:** 80
Plastic materials
acoustic emission inspection ... **A17:** 291
chemical composition ... **A17:** 215
components, neutron radiography of ... **A17:** 391
liquid penetrant inspection ... **A17:** 71
microwave inspection ... **A17:** 202
Plastic memory ... **EM3:** 21
defined ... **EM2:** 30
Plastic microstrain
in SCC testing ... **A8:** 498
Plastic package fabrication
assembly methods ... **EL1:** 471-47
molded plastic packages ... **EL1:** 471-47
plastic pin-grid arrays ... **EL1:** 475-47
reliability issues ... **EL1:** 479-48
requirements of ... **EL1:** 470-471
tape automated bonding (TAB) ... **EL1:** 476-47
Plastic packages *See also* Encapsulation; Plastic package fabrication
defined ... **EL1:** 451
environmental tests ... **EL1:** 494-49
failure mechanisms ... **EL1:** 96
plastic leaded chip carrier (PLCC) ... **EL1:** 20
plastic quad flatpack (PQFP) ... **EL1:** 209-21
reliability ... **EL1:** 245-24
semiconductor, structure ... **EL1:** 241
small-outline packages (SOPS) ... **EL1:** 20
Plastic parts
abrasive blasting of ... **M5:** 91
Plastic patterns
equipment ... **A15:** 195
injection, investment casting ... **A15:** 256
investment casting ... **A15:** 255
Plastic pin-grid arrays *See also* Pin-grid arrays (PGA)
assembly methods ... **EL1:** 475-47
chip attach ... **EL1:** 47
defined/outline ... **EL1:** 21
reliability ... **EL1:** 48
sealing ... **EL1:** 47
substrate ... **EL1:** 47
Plastic pressing ... **EM4:** 34
Plastic quad flatpack package (PQFP)
die attachments ... **EL1:** 217-221
and substrates ... **EL1:** 209-21
thermal resistance ... **EL1:** 409-41
Plastic replica technique
schematic ... **A17:** 53
Plastic replicas
defined ... **A9:** 14
of fracture surfaces ... **A11:** 19
used in local electropolishing ... **A9:** 55
Plastic sealants
for electroplating ... **A7:** 460
Plastic sheet, stop-off medium
chrome plating ... **M5:** 187
Plastic strain ... **A13:** 252-253
defined ... **A8:** 10
deformation for, hollow cylinder ... **A8:** 143
described ... **A11:** 50
effect on plastic deformation structures ... **A9:** 686
fatigue resistance ... **A8:** 348
fatigue resistance affected by ... **M1:** 665, 668, 670-672
range, low-cycle fatigue ... **A11:** 103
ratio, steel sheet ... **M1:** 548, 549, 552
specimens, SCC testing ... **A8:** 508-509
in stress relaxation ... **A8:** 307, 323-324
topographic methods for ... **A10:** 368
Plastic strain, determined
by speckle metrology ... **A17:** 435
Plastic strain range
of AISI 304 stainless steel ... **A8:** 348
low-cycle fatigue ... **A8:** 364
Plastic strain range (continued)
vs. cycles to failure ... **A8:** 346-347
Plastic strain rate
relation to strain rate ... **A8:** 41-42
in ultrasonic testing ... **A8:** 256
Plastic strain ratio *See also* Anisotropy factor; Planar anisotropy; r value
defined ... **A8:** 10
of sheet metals ... **A8:** 550, 555-557
Plastic stress minimization
computer modeling of ... **EL1:** 44
Plastic strip zone
in elastic-plastic analysis ... **A8:** 446
Plastic surface replicas ... **A17:** 53
Plastic tooling
die material for sheet metal forming ... **A18:** 628
Plastic torsion
anisotropy in ... **A8:** 143
Plastic true strain ... **EM3:** 21
defined ... **EM2:** 30
Plastic wave propagation
of aluminum and alpha-titanium ... **A8:** 231
in Hopkinson bar test ... **A8:** 200
test limitations ... **A8:** 231
Plastic wave velocity
and stress ... **A8:** 209
Plastic welding, applications
automobiles ... **A6:** 393
Plastic work ... **A18:** 422, 423, 427
heat conversion in ... **A8:** 45
Plastic work hardening
in torsion testing ... **A8:** 140
Plastic zone
plane strain/plane stress ... **A8:** 471
in polymers ... **A11:** 762
size at crack tip, determined ... **A11:** 49
and subcritical fracture mechanics (SCFM) ... **A11:** 52
in tubular specimen, finite-element analysis ... **A8:** 222-223
Plastic(s) *See also* Engineering plastics; Engineering plastics families; Epoxies (EP); Plasticizers; Polymer families; Polymer(s); Resins; specific plastics ... **A7:** 606-613
aerospace material specifications ... **EM2:** 91
as aggregate of properties ... **EM2:** 405
aluminum flake for ... **A7:** 595
ASTM standard test methods ... **EM2:** 90
ASTM standards ... **EM2:** 90
categorization ... **EM2:** 68
chemical compatibility ... **EM2:** 1
chemistry ... **EM2:** 64
composites ... **EM2:** 428-429
conductive ... **EM2:** 589-590
creep modulus ... **EM2:** 75
creep rupture strength ... **EM2:** 75
defined ... **EM2:** 30
effect on powder metallurgy ... **A7:** 463
electrical properties ... **EM2:** 588-590
electrical properties tests ... **EM2:** 78
electrical-grade, compared ... **EM2:** 228
fatigue (*S-N*) curve ... **EM2:** 76
fatigue failure of ... **EM2:** 702-703
fatigue loading ... **EM2:** 702
general-purpose ... **EM2:** 64-66
glassy, ESC testing of ... **EM2:** 802-803
gold bronze and copper pigments in ... **A7:** 595
hierarchy of ... **EM2:** 68
as insulating ... **EM2:** 460
optical ... **EM2:** 483-484, 596
PHBV-biodegradable ... **EM2:** 786
plastics, powder metal-filled ... **A7:** 606-613
sealants, for electroplating ... **A7:** 460
semiconductive ... **EM2:** 589-590
strength, design guidelines ... **EM2:** 709
stress-strain curve ... **EM2:** 74
structural, chemistry ... **EM2:** 65-66
thermal expansion rate ... **A7:** 611
time-dependent behavior ... **EM2:** 405

Plastic(s) (continued)
UL standards for ... **EM2:** 91
viscoelastic behavior ... **EM2:** 63, 412, 659
vs metals, by competitive pairs ... **EM2:** 87
vs metals, costs ... **EM2:** 83
vs polymers, as terms ... **EM2:** 1
Plastic-body devices
three-terminal ... **EL1:** 427-428
Plastic-bonded sheet
brazing filler metals available in this form ... **A6:** 119
Plastic-carbon replicas ... **A9:** 108
Plastic-clad space-frame concept ... **EM3:** 554
Plastic-encapsulated
devices, types ... **EL1:** 217
diodes, failure mechanisms ... **EL1:** 973
Plastic-filled metals lubricants
powders used ... **A7:** 573
Plastic-filled self-lubricating parts
powders used ... **A7:** 574
Plastic-package failures
integrated circuits ... **A11:** 788-789
moisture-induced ... **A11:** 788-789
stress-induced ... **A11:** 789
Plastic-starch blends
biodisintegration/biodegradation of ... **EM2:** 786-787
Plastic-strain ratio ... **A1:** 575 **A14:** 10, 575
Plastic-zone adjustment
defined ... **A8:** 10
Plastically deformed surface layer as a result of abrasion ... **A9:** 37-40
Plasticine
as physical modeling material ... **A14:** 432
Plasticity *See also* Inelastic strain ... **A14:** 10, 911 **EM3:** 20
and crack propagation rate ... **A8:** 678
defined ... **A8:** 110 **A9:** 14 **A13:** 10
and elasticity ... **A8:** 71-72
equations, for ring geometries ... **A8:** 585
and green strength ... **A7:** 302
of kaolinite ... **A15:** 210
temperature dependence of ... **A11:** 138
theorems of limit analysis of ... **EM1:** 198
theory ... **A8:** 71-72, 559
Plasticity adjustment factor
in elastic-plastic analysis ... **A8:** 446
for plane stress fracture testing ... **A8:** 449
Plasticity factor, *P*
as fracture toughness adjustment ... **A8:** 472-474
Plasticity theory
brazing and ... **A6:** 110
Plasticization, polymer
as solubility ... **EM2:** 61
Plasticized metal dust feedstocks
in injection molding ... **A7:** 498
Plasticizer *See also* Flexibilizer
defined ... **EM1:** 18
for thermoplastics ... **EM1:** 103
water as ... **EM1:** 76, 141
Plasticizers *See also* Flexibilizer ... **A7:** 8 **EM3:** 20, 41, 49, 150
additive for sealants ... **EM3:** 673
for butyls ... **EM3:** 199, 202
defined ... **EL1:** 818 **EM2:** 30
effect, chemical susceptibility ... **EM2:** 572
GC/MS analysis of ... **A10:** 639
infrared spectrum ... **A10:** 124
and polymer materials, identified in vinyl film ... **A10:** 123-124
and polymer miscibility ... **EM2:** 496
for slurry in spray drying ... **A7:** 74, 75
and solubility ... **EM2:** 61
types, for flexible epoxies ... **EL1:** 818
for urethane sealants ... **EM3:** 205
Plasticizers, resinous
for wax patterns ... **A15:** 197
Plastics ... **A6:** 1048-1055 **EM3:** 20
abrasive jet machining ... **A16:** 511
acidity-basicity measured in ... **A10:** 172
acrylic, drilling ... **A16:** 227, 229, 230
adhesive bonding ... **A6:** 1048
analytic methods for ... **A10:** 9
applications ... **A6:** 1050, 1051, 1052, 1053, 1054
blanking, die materials for ... **M3:** 485, 487
carbon fiber reinforced, with PCD tooling ... **A16:** 110

Plastics (continued)
categories of thermoplastics and composites ... **A6:** 1048
coatings for tools before hot-tool welding ... **A6:** 1049
contact-angle testing, surface treatment effects ... **EM3:** 277
cutting fluids used ... **A16:** 125
cutting tool material selection based on machining operation ... **A18:** 617
deep drawing dies, use for ... **M3:** 499
dielectric welding ... **A6:** 1054
as drawing tool material ... **A14:** 511
drilling ... **A16:** 237
EFG composition analysis ... **A10:** 212
electrofusion welding ... **A6:** 1053-1054
electromagnetic forming with ... **A14:** 649-650
electromagnetic welding ... **A6:** 1053-1054
electron beam machining ... **A16:** 570
engineering ... **EM3:** 21
epoxy bonding ... **EM3:** 96
etching ... **EM3:** 277
evaluation of welds ... **A6:** 1054-1055
light microscopy ... **A6:** 1055
Moire interferometric method ... **A6:** 1055
scanning electron microscopy ... **A6:** 1055
x-ray techniques ... **A6:** 1055
extrusion welding ... **A6:** 1050
focused infrared welding ... **A6:** 1050-1051
friction and wear data tested against polycarbonates ... **A18:** 58
friction welding ... **A6:** 1051-1053
fusion welding of thermoplastics only ... **A6:** 1048, 1049
fusion-welding techniques ... **A6:** 1049-1051
GC/MS analysis of volatile compounds in ... **A10:** 639
ground by diamond wheels ... **A16:** 455, 460
guide shoe material for honing ... **A16:** 478
hobs, use in cutting ... **M3:** 477
honing ... **A16:** 472
hot-gas welding ... **A6:** 1050
hot-tool welding ... **A6:** 1049-1050
hydrogen fluoride/hydrofluoric acid corrosion ... **A13:** 1169
implant welding ... **A6:** 1053-1054
induction welding ... **A6:** 1054
inspection fixtures, use for ... **M3:** 557
lap-shear strength, surface preparation effects ... **EM3:** 276
lapping ... **A16:** 499
laser welding ... **A6:** 1051
lightweight fiber-reinforced, waterjet machining ... **A16:** 525
low surface energy ... **EM3:** 42
material for jet engine components ... **A18:** 588
material parameters that should be documented to ensure repeatability when testing tribosystems ... **A18:** 55
mechanical fastening ... **A6:** 1048
microwave welding ... **A6:** 1054
molded, drilling ... **A16:** 219, 221, 229, 230
PCD tooling ... **A16:** 110
plasma treatment ... **EM3:** 277
press forming dies, use for ... **M3:** 489, 490, 492, 493
primers ... **EM3:** 277
resistance welding ... **A6:** 1053-1054
Rockwell hardness testing of ... **A8:** 76
scrapers used in ECDG ... **A16:** 548, 549, 550
spin welding ... **A6:** 1051, 1052
surface parameters ... **EM3:** 41
surface preparation ... **EM3:** 276-280
thermal energy method of deburring ... **A16:** 578
thermoplastics, weldability ... **A6:** 1053
ultrasonic fatigue testing of ... **A8:** 240
ultrasonic welding ... **A6:** 1051, 1052-1053
use in breweries ... **A13:** 1222
for valve plates for reciprocating compressors ... **A18:** 604
vibration welding ... **A6:** 1051, 1052
Plastics and rubber
mold materials for ... **M3:** 546-550
Plastics Compounding ... **EM3:** 66
as periodical ... **EM2:** 93
Plastics Compounding Redbook ... **EM3:** 66
as information source ... **EM2:** 93

Plastics Design Forum ... **EM3:** 66
as trade magazine ... **EM2:** 93
Plastics Engineering *See* Engineering plastics ... **EM3:** 66
as information source ... **EM2:** 93
***Plastics Engineering Handbook* (SPI)** ... **EM2:** 94
Plastics Focus: An Interpretive News Report ... **EM2:** 93
Plastics for mounting *See also* Resins ... **A9:** 29
ceramic-filled ... **A9:** 45
mineral-filled ... **A9:** 45
Plastics in Building Construction
as information source ... **EM2:** 93
Plastics Institute of America (PIA)
as information source ... **EM1:** 41
Plastics Institute of America, Inc. (Hoboken NJ) ... **EM2:** 95
Plastics Packaging ... **EM3:** 66
***Plastics Process Engineering* (Throne)** ... **EM2:** 94
***Plastics Products Design Handbook* (Miller)** ... **EM2:** 94
Plastics Technical Evaluation Center (PLASTEC) ... **EM2:** 95
Plastics Technology ... **EM3:** 66
trade magazine ... **EM2:** 92
***Plastics Technology Handbook* (Chanda/Roy)** ... **EM2:** 94
Plastics World ... **EM3:** 66
as information source ... **EM2:** 92
PlasticTrends
as information source ... **EM2:** 93
Plastigel ... **EM3:** 21
defined ... **EM2:** 30
Plastisol coating
steel sheet ... **M1:** 176
Plastisols ... **EM3:** 21
for auto body sealing and glazing materials ... **EM3:** 57
for automobile interior seam sealing ... **EM3:** 720
automotive applications ... **EM3:** 46
cross-linking ... **EM3:** 46
cure properties ... **EM3:** 51
defined ... **EM2:** 30
dual-mechanism radiation cure formulation ... **EL1:** 85
suppliers ... **EM3:** 58, 59
to form gaskets for auto air filters ... **EM3:** 46
Plastohydrodynamic lubrication (PHL) ... **A18:** 89, 93-94
defined ... **A18:** 15, 94
lubricant film thickness ... **A18:** 94
mixed-film ... **A18:** 94
Plastometer ... **EM3:** 21
defined ... **EM2:** 30
Plate ... **M1:** 181-198
$2^1/_4$ Cr-1Mo steel, mechanical properties ... **M1:** 654
alloy steel for ... **M1:** 183
aluminum alloy, fracture toughness ... **A8:** 461
aluminum alloy, pin bearing testing of ... **A8:** 61
aluminum alloy, specimen location for ... **A8:** 60
aluminum and aluminum alloys ... **A2:** 5
aluminum-lithium alloys, fatigue in ... **A2:** 195-196
applications ... **M1:** 181
ASTM specifications ... **M1:** 183-184
austenitic stainless steel, hardness conversion tables ... **A8:** 109
bending, of nickel-base alloys ... **A14:** 835-836
bending strength tests for ... **A8:** 117, 132
bending test specimens ... **A8:** 125
beryllium-copper alloys ... **A2:** 403, 411
boiler/pressure vessel, inspection ... **A17:** 644-645
by multiple-slide forming ... **A14:** 567
carbon content, distribution ... **M1:** 189, 195-196
carbon steel for ... **M1:** 183
compression test fixture ... **A8:** 198
defined ... **A14:** 343
definition ... **M1:** 181
directionality ... **M1:** 194
eddy current inspection ... **A17:** 187
and etch, for surface wiring ... **EL1:** 11
explosion welding ... **M6:** 709
explosive forming of ... **A14:** 641
fabrication of ... **M1:** 194, 197-198
fatigue properties ... **M1:** 194
fatigue testing of ... **A13:** 293
flash welding ... **M6:** 558, 577

Plate (continued)
flat, ECP detection of surface flaws **A17:** 137-138
formability, of magnesium alloys **A2:** 467-468
magnesium alloy **A14:** 825-826
mechanically alloyed oxide
alloys **A2:** 948-949
dispersion-strengthened (MA ODS)
notch toughness, thickness, effects on................ **M1:** 696-697, 700-701
oxyfuel gas welding... **M6:** 590
penetrameters/identification markers,
radiographic inspection **A17:** 343
platemaking practices..................................... **M1:** 182
primary testing direction, various
alloys **A8:** 667
quality descriptors **M1:** 182-183
radiographic methods **A17:** 296
reflowed solder, as preservation **EL1:** 56
rolled, straight-beam top ultrasonic
inspection **A17:** 268-269
rolling, mechanics of...................................... **A14:** 346
rotary shearing.. **A14:** 705-707
shearing of **A14:** 701-707
specimen **A8:** 314, 371-372
steelmaking practices for **M1:** 181-182
straight-knife shearing of........................ **A14:** 701-705
stress analysis of rolling................................ **A14:** 347
thickness, effect on mechanical
properties **M1:** 188-189, 194, 196
ultrahigh-strength steel for............................ **M1:** 188
wrought aluminum alloy **A2:** 33, 60
wrought beryllium-copper alloys.................. **A2:** 409
of wrought magnesium alloys **A2:** 459-460
wrought titanium alloys **A2:** 610-611
Plate buckling
magnesium **M2:** 550-552
of magnesium and magnesium alloy
parts **A2:** 477-479
Plate castings, shapes for **A15:** 599
flat, solidification of....................................... **A15:** 606
Plate dies *See* Steel-rule dies
Plate geometries
in melt rheology **EM2:** 535-540
Plate glass
erosion of steels **A18:** 204
Plate glass process **EM4:** 21
Plate impact experiments................................. **A8:** 287
Plate impact facility
for pressure-shear impact testing........... **A8:** 233-234
Plate impact testing
anvil preparation... **A8:** 234
Carpenter Hampden steel properties **A8:** 234
Carpenter Stentor steel properties **A8:** 234
copper tilt pins **A8:** 234
diamond paste **A8:** 234
diffraction grating **A8:** 233
flyer and anvil properties **A8:** 234
gas gun for **A8:** 233
for high strain rate shear testing **A8:** 215
modified diamond stop lapping
fixture... **A8:** 234-235
pressure-shear **A8:** 230-238
shear flow stress **A8:** 236
shear strain rate **A8:** 236
tilt pin oscilloscope record **A8:** 233
Plate martensite
defined **A9:** 14
ferrous **A9:** 671-672
nonferrous **A9:** 672
in steel **A9:** 178
Plate materials
aluminum alloy, flat-face tensile frac-
ture in **A11:** 76
aluminum alloy, LME by mercury in............ **A11:** 79
cadmium-plated steel, arc striking at
hard spot in **A11:** 97
fatigue-fracture surface marks **A11:** 111
flat-face fracture....................................... **A11:** 109-110
heat exchangers, application **A11:** 628
Plate materials (continued)
punched hole brittle fracture in...................... **A11:** 90
Plate mill
effect on ferrite structure in microal-
loyed steel **A8:** 180
Plate rolling
mechanics of.. **A14:** 346
spread **A14:** 346
stress/roll-separating force, prediction **A14:** 346
Plate steel
hydrogen flaking **A12:** 141
Plate steels *See also* Alloy steels; Carbon steels; Low-alloy steels; Low-carbon steels; Plain carbon steels
ASTM compositions.. **A9:** 202
classification **A9:** 203
etchants for **A9:** 202-203
etching **A9:** 202-203
examination for carbides................................ **A9:** 203
examination for inclusions **A9:** 203
examination for nitrides................................. **A9:** 203
grinding.. **A9:** 202
macroexamination **A9:** 203
mechanical properties.................................... **A9:** 203
microstructures **A9:** 203
mounting.. **A9:** 202
polishing **A9:** 202
sectioning **A9:** 202
specimen preparation **A9:** 202-203
welded joints, examination of................ **A9:** 202-203
Plate steels, specific types
API X60, for line pipe, control-rolled **A9:** 209
ASTM A36, as-rolled **A9:** 204
ASTM A201, Grade A, graphitization
after five years' service......................... **A9:** 204
ASTM A201, Grade B, crack in a weld **A9:** 204
ASTM A285, Grade C, blistering............ **A9:** 204-205
ASTM A285, Grade C, hot rolled.................. **A9:** 204
ASTM A285, Grade C, weld metal
cracks **A9:** 204
ASTM A387, Grade D, normalized and
compared **A9:** 205
tempered, optical and TEM micrographs
ASTM A515, Grade 70..................................... **A9:** 205
ASTM A516, Grade 70..................................... **A9:** 205
ASTM A517, Grade B, austenitized
quenched and tempered, optical and
TEM micrographs compared **A9:** 206
ASTM A517, Grade J, welded joint **A9:** 206
ASTM A517, Grade M, quenched and
tempered **A9:** 206
ASTM A533, Grade B, different speci-
mens from same plate **A9:** 206-207
ASTM A533, Grade B, optical and
TEM micrographs compared................ **A9:** 206
ASTM A537, Grade A, normalized,
optical and TEM micrographs
compared **A9:** 207
ASTM A537, Grade B, quenched and
compared **A9:** 207
tempered, optical and TEM micrographs
ASTM A542, Class 2, quenched and
compared **A9:** 207-208
tempered, optical and TEM micrographs
ASTM A553, Grade A, quenched and
compared **A9:** 208
tempered, optical and TEM micrographs
ASTM A562, normalized and cooled in air
compared **A9:** 208
optical and TEM micrographs
ASTM A572, Grade 55, as hot rolled............ **A9:** 208
ASTM A572, Grade 65..................................... **A9:** 208
ASTM A633, Grade C....................................... **A9:** 209
ASTM A710, Grade A, Class 3 .... **A9:** 209
ASTM A737, Grade B **A9:** 209
ASTM A808, as rolled **A9:** 209
Plate theory **EM3:** 385, 512
Plate waves *See also* Lamb waves
ultrasonic inspection...................................... **A17:** 234
Plate(s) **A7:** 8
beryllium.. **A7:** 759
Plate-bending theory
for laminates.. **EM1:** 220
Plate-out
of additives **EM2:** 494
Plateability
of compact infiltrated copper alloy................ **A7:** 565
Plateau honing process................................. **A18:** 336
Plateau velocity
for stress-corrosion cracking **A8:** 497
Plateaus
ASTM/ASME alloy steels............................. **A12:** 347
in ductile irons **A12:** 231
fatigue striations on **A12:** 23
multiple, crack propagation on **A12:** 16
Plated coatings
corrosion prevention mechanisms **A13:** 424-426
Plated finishes
for zinc alloy castings............................ **A15:** 796-797
Plated solder
as preservation **EL1:** 562-56
Plated steel
brazing to aluminum...................................... **M6:** 1030
Plated through-holes
abbreviation for **A11:** 797
Plated-through hole (PTH) *See also* Through-substrate plated-through holes (TSPTH)
aramid fiber reliability **EL1:** 61
components, removal of................................... **EL1:** 72
drilling defects **EL1:** 87
failures **EL1:** 1018-103
flexible printed boards **EL1:** 589-59
knee, thinning at... **EL1:** 67
leadless packaging **EL1:** 98
low-CTE metal planes **EL1:** 62
materials and processes selection............ **EL1:** 113-11
metal core construction **EL1:** 622-62
poor filling, from fluxes **EL1:** 68
probability of survival..................................... **EL1:** 98
quartz fabrics **EL1:** 61
reliability considerations....... **EL1:** 617, 619, 622-623, 627, 699-700
size, life effect **EL1:** 98
soldered, illustrated **EL1:** 11
soldering, methods of.............................. **EL1:** 681-68
technologies, solder joint inspection............. **EL1:** 73
types, multilayer structure **EL1:** 55
vs surface-mount device designs **EL1:** 55
Plated-through hole drilling
primary.. **EL1:** 869-87
Plated-through hole structure test
defined **EL1:** 115
Platelet alpha structure
defined **A9:** 14
Platelet thickness
and spacing in transformed
microstructure **A8:** 480
Platelets.. **A7:** 8
cast aluminum alloys..................................... **A12:** 409
microstructure, titanium alloys..................... **A12:** 442
Platelets, inclined through foil
AEM analysis **A10:** 453, 455
Platen
defined **A14:** 10
definition.. **M6:** 13
Platen force
definition.. **M6:** 13
Platen press
hydraulically or pneumatically
actuated **EM3:** 37
Platen spacing
definition.. **M6:** 13
Platens
aluminum oxide, with ele-
vated-temperature compression
testing **A8:** 196
cam plastometer **A8:** 195-196

Platens (continued)
compression test fixture **A8:** 198
defined **EM1:** 18 **EM2:** 30
with drop tower compression system **A8:** 196
in subpress assembly for medium strain rate testing with conventional load frames **A8:** 192-193

Plates
anisotropic, instability of **EM1:** 446
damping analysis of **EM1:** 209-213
damping data **EM1:** 212
with holes, stresses **EM1:** 234
laminated, damping analysis **EM1:** 210-213
orthotropic, instability **EM1:** 445-446
postbuckling behavior **EM1:** 447-449
structural analysis **EM1:** 461
thin, theory of **EM1:** 220

Plates, caul *See* Caul plates

Platform-type package **EL1:** 237, 45

Platinel thermocouple *See* Thermocouples, materials, nonstandard

Platinel thermocouples
types/properties/ applications **A2:** 875-876

Plating *See also* Coatings; Electroplated deposits; Metallization; Plated coatings; Plated-through hole (PTH); specific coatings
adhesion, as PTH failure mechanism **EL1:** 102
alloyable coatings **EL1:** 67
barrel *See* Barrel plating
barrier platings **EL1:** 67
of beryllium **A2:** 683
capabilities/limitations **EL1:** 510-51
chemical *See* Chemical plating
chromium **A13:** 871-875
chromium, effect in AISI/SAE alloy steel fracture **A12:** 297
codeposited organics in **EL1:** 679-68
in cold-formed parts **A11:** 308
connector **EL1:** 2
copper, multichip structures **EL1:** 303-30
cracks, as PTH failure **EL1:** 102
defects, types **EL1:** 102
defined **EL1:** 115
effect on fatigue strength **A11:** 126
electro- *See* Electroplating
electroless *See* Electroless plating **EL1:** 510, 545, 870
electroless copper **EL1:** 545, 870
electroless nickel, for edge retention **A12:** 95, 100
electrolytic **EL1:** 510-511, 545-546, 871-872
electrolytic copper **EL1:** 545-546, 871-872
equalizers **EL1:** 872-873
equipment, corrosion of **A13:** 1314-1316
first-level packages **EL1:** 989-991
flexible printed boards **EL1:** 583
folds, printed board coupons **EL1:** 576
gold **EL1:** 549-550
hydrogen embrittlement by **A12:** 22, 30
hydrogen entry **A13:** 330
hydrogen-charging, precipitation-hardening stainless steels **A12:** 372
immersion *See* Immersion plating **A13:** 430
immersion solder, as preservation **EL1:** 564
ion **A13:** 457, 821
lead frame **EL1:** 484, 487
materials and processes selection **EL1:** 113-116
materials, hydrogen damage susceptibility **A11:** 126
mechanical *See* Mechanical plating **A13:** 767
metallic, uranium/uranium alloys **A13:** 819-821
of mill products **A13:** 430
nickel, for edge protection **A11:** 24
nodules, printed board coupons **EL1:** 575
of P/M parts **A7:** 460
pattern vs panel, rigid printed wiring boards **EL1:** 540
peeling in integrated circuits **A11:** 43-45
plasma process **EL1:** 511
precious metals, materials and processes selection **EL1:** 116
problems, first-level package **EL1:** 991
selective *See* Selective plating
of shafts **A11:** 459
of springs **A1:** 311-312
steel, hydrogen blistering **A13:** 332
strikes **EL1:** 679
for surface preparation **EL1:** 679

Plating (continued)
as surface treatment in tool steels **A1:** 779
thickness of **EL1:** 942-943
of thin-film hybrids **EL1:** 313, 329-330
tin-lead, rigid printed wiring boards **EL1:** 546
to CIC layers **EL1:** 627-62
voids, as PTH failure mechanism **EL1:** 1022-1023
zinc **A13:** 767
zinc alloys **A2:** 530

Plating, arsenic
historic **A15:** 16

Plating baths
acid -basicity measured **A10:** 172
contamination **EL1:** 679
hydrogen embrittlement testing for **A8:** 541
potentiometric membrane electrode analysis of **A10:** 181
solution analysis, by ion chromatography **A10:** 658
wet chemical analysis of **A10:** 165

Plating, core
of electrical steel sheet **A14:** 482

Plating cracks
as planar flaws **A17:** 50

Plating efficiency **A18:** 834

Plating for edge retention *See also* Nickel plating **A9:** 32
cleaning of specimens **A9:** 28
nitrided steels **A9:** 218
of tool steels **A9:** 256

Plating for preservation of the white layer in nitrided steels **A9:** 218

Plating of aluminum alloys containing copper as a result of using magnesium oxide
abrasives **A9:** 353

Plating slivers
as PTH failure mechanism **EL1:** 1024-1025

Plating, stainless steel *See also* Electroplating **M3:** 55

Plating waste disposal and recovery
See also Waste recovery and treatment **M5:** 310-319
chromium reduction process **M5:** 311-312
clarification process **M5:** 313-314
complex wastewater treatment systems **M5:** 314-315
conventional wastewater treatment systems **M5:** 311-314
cyanide oxidation process **M5:** 312
direct (natural) recovery **M5:** 315-316
dragout recovery **M5:** 315-316
dragout, reduction of **M5:** 310-311
effluent polishing **M5:** 318
electrodialysis recovery **M5:** 318-319
electrowinning recovery **M5:** 319
evaporation rates **M5:** 316
evaporation recovery **M5:** 316-317
ion exchange recovery **M5:** 317-318
neutralization process **M5:** 312-313
oxidation-reduction potential measurements **M5:** 312
recovery systems **M5:** 315-319
regulations governing **M5:** 310-311
sludge de-listing **M5:** 310
reverse osmosis recovery **M5:** 318
rinsewater flows, minimizing **M5:** 311
rinsewater recovery and recycling **M5:** 318
sludge dewatering **M5:** 313-314
treatment load, minimizing **M5:** 310-311
vapor recompression **M5:** 316

Platings *See also* Coatings; Electroplating; Metalic coatings
cast irons **M1:** 101-103
corrosion protection **M1:** 752-754
springs, steel **M1:** 291

Platinized titania
Auger electron spectroscopy application **A18:** 449-450, 451
ion-scattering spectrometry application **A18:** 449, 450

Platinosis
as platinum toxic reaction **A7:** 207

Platinum *See also* Platinum alloys; Platinum-group metals; Precious metals **A13:** 797-799
as a conductive coating for scanning electron microscopy specimens **A9:** 97

Platinum (continued)
as a reactive sputtering cathode material **A9:** 60
alloying **A13:** 47, 798
alloys, FIM sample preparation of **A10:** 586
alloys, relative hydrogen susceptibility **A8:** 542
annealing **M4:** 760, 761, 762
anodes showing electrochemical and corrosion effects **EM3:** 629-631
antitumor applications **A2:** 1258
applications **A2:** 707
atomic interaction descriptions **A6:** 144
as braze filler metal **A11:** 450
brazing with glasses **EM4:** 520
catalyst for silicones **EM3:** 598
catalyst for silicones PSAs **EM3:** 135, 136, 137
catalyst for vulcanization of silicones **EM3:** 217
chemical properties **A2:** 846
in clad and electroplated contacts **A2:** 848
-clad niobium, corrosion control **A13:** 888-889
for coating surfaces before transmission electron microscopy **EM3:** 242
commercially pure **M2:** 688-690
components, elevated-temperature failure in **A11:** 296-297
corrosion in acids **A13:** 801
corrosion in gases **A13:** 803
corrosion in halogens **A13:** 803
corrosion in organic compounds **A13:** 803
corrosion in salts **A13:** 802
corrosion resistance **A13:** 798 **M2:** 668-669
corrosion weight loss **A13:** 804
as crucible material for glass melting **EM4:** 21
determined in silver scrap metal **A10:** 41
electrical circuits for electropolishing **A9:** 49
as electrical contact materials **A2:** 846-848
electrical contacts, use in **M3:** 669-671
electrical resistance applications **M3:** 641, 646, 647, 655
electrodes for resistance brazing **A6:** 340
electrodes, in biamperometric titration **A10:** 204
evaporation fields for **A10:** 587
explosion welding **A6:** 896 **M6:** 710
fabrication **A13:** 797
friction coefficient data **A18:** 71
glass-to-metal seals **EM3:** 302
gravimetric finishes **A10:** 171
for heating elements used in HIP **EM4:** 195
joining **EM4:** 487
lead, for acidified chloride solutions **A8:** 419
in liquid-phase metallizing **EM3:** 306
material for conductors **EM4:** 1141
material for electrode in commercial oxygen sensors **EM4:** 1137
material to which crystallizing solder glass seal is applied **EM4:** 1070
in medical therapy, toxic effects **A2:** 1258
in metal powder-glass frit method **EM3:** 305
for metallizing **EM4:** 542, 544
nitric/hydrochloric acid dissolution medium **A10:** 166
oxidation resistance **A13:** 798-799
photochemical machining etchant **A16:** 590
polycrystalline on a titania substrate, ion- scattering spectrometry application **A18:** 449
as precious metal **A2:** 688
primary bond metal with alumina **EM3:** 300
properties **A13:** 797-798
pure **M2:** 781-783
pure, properties **A2:** 1147
recommended glass/metal seal combinations **EM4:** 497
relative solderability **A6:** 134
relative solderability as a function of flux type **A6:** 129
resources and consumption **A2:** 689
semifinished products **A2:** 694
soldering **A6:** 631
special properties **A2:** 692
substrate, extent of coverage of nickel-phosphorus film on **A10:** 608
suitability for cladding combinations **M6:** 691
thermal diffusivity from 20 to 100 °C **A6:** 4
thermal expansion coefficient **A6:** 907
TWA limits for particulates **A6:** 984
ultrapure, by fractional crystallization **A2:** 1093

Platinum (continued)
ultrasonic welding **M6:** 746
in vacuum, as SERS metal **A10:** 136
vapor pressure, relation to temperature **M4:** 309, 310
in vapor-phase metallizing **EM3:** 306
wire, temperature effect on tensile strength **A13:** 800
working of **A14:** 520, 849-850
Platinum alloy vacuum coating **M5:** 388, 390-391, 394
Platinum alloys *See also* Platinum-group metals (PGM)
applications **A2:** 709-714
as electrical contact materials **A2:** 846-848
electrolytic etching of **A9:** 551
magnetic **A9:** 539
working of **A14:** 520, 851
Platinum alloys, specific types
75Pt-25Ir **A9:** 563
89Pt-11Ru **A9:** 563
Platinum and platinum alloys
age hardening **A4:** 946
annealing **A4:** 944, 945-946
hardness **A4:** 943, 944
mechanical properties **A4:** 944, 945, 946
tensile strength **A4:** 944
vapor pressure of platinum, relation to temperature **A4:** 495
Platinum black powders **A7:** 150
Platinum coating molybdenum **M5:** 661
Platinum complexes
mutagenic and carcinogenic effects **A2:** 1258
Platinum dispersion-strengthened alloys
applications **A7:** 722
Platinum group elements and alloys
annealing **A4:** 944-947
applications **A4:** 944-945
Platinum group metals
electrical circuits for electropolishing **A9:** 49
toxicity **A7:** 207
Platinum oxide layer sputtered onto Sn-18Ag-15Cu **A9:** 61
Platinum plating
niobium **M5:** 663-664
solution compositions and operating conditions **M5:** 663-664
tantalum **M5:** 663-664
titanium **M5:** 658-659
Platinum powders *See also* Precious metal powders
ancient P/M practices **A7:** 14
black production and characteristics **A7:** 150
chemically precipitated **A7:** 150
compaction by Wollaston process **A7:** 15-16
dispersion-strengthened **A7:** 720-722
fusion procedure **A7:** 16
history **A7:** 14-16
production **A7:** 148, 150
as pyrophoric **A7:** 199
reducing agents **A7:** 150
thick-film **A7:** 151
toxicity **A7:** 207
Platinum salts
as allergens **A2:** 1258
Platinum x-ray tubes **A17:** 302
Platinum-carbon alloys
for thermal evaporation **A12:** 172-173
Platinum-clad niobium
as anode material **A13:** 889
Platinum-cobalt alloys *See* Magnetic materials
Platinum-cobalt permanent magnet alloy **M2:** 697-698
Platinum-cobalt permanent magnet alloys **A2:** 713, 787
Platinum-gold conductor inks **EL1:** 208
Platinum-gold powders
thick-film **A7:** 151
Platinum-group metals
iridium **M2:** 664
Platinum-group metals (continued)
jewelry **M2:** 666-667
osmium **M2:** 665
palladium **M2:** 663-664
platinum **M2:** 663
production **M2:** 660-661
rhodium **M2:** 664
ruthenium **M2:** 664-665
special properties **M2:** 660-661
Platinum-group metals (PGM) *See also* Platinum; Platinum alloys
in electronic scrap recycling **A2:** 1228
in medical therapy, toxic effects **A2:** 1258
resources and consumption **A2:** 689-690
special properties **A2:** 692, 694
trade practices **A2:** 691
Platinum-group metals (PGM), specific types
79Pt-15Rh-6Ru, properties **A2:** 711-712
Pd-9.5Pt-9.0Au-32.4Ag, as electrical contact materials **A2:** 848
Pd-26Ag-2Ni, as electrical contact materials **A2:** 848
Pd-30Ag-l4Cu-10Au-10Pt-1Zn, as electrical contact materials **A2:** 848
Pd-38Ag-16Cu-1Pt-lZn ,as electrical contact materials **A2:** 848
Pd-40Ag, as electrical contact materials **A2:** 847-848
Pd-40Cu, as electrical contact materials **A2:** 847
platinum 67, as thermocouple reference standard **A2:** 870
Pt-18.4Pd-8.2Ru, as electrical contact materials **A2:** 847
Platinum-iridium alloys *See also* Platinum; Platinum alloys **M2:** 691-693
as electrical contact materials **A2:** 846-847
properties **A2:** 709-710
for tip materials for scanning tunneling microscopy **A18:** 395
Platinum-molybdenum thermocouples
properties/applications **A2:** 874-875
Platinum-nickel alloys **M2:** 695-696
properties **A2:** 713
Platinum-palladium alloys **M2:** 690-691
properties **A2:** 709
Platinum-palladium-gold powders
for multilayer ceramic capacitors **A7:** 151
Platinum-rhenium alloys
for tip materials for scanning tunneling microscopy **A18:** 395
Platinum-rhodium
for noble metal thermocouples used in vacuum heat treating **A4:** 506
Platinum-rhodium alloys **M2:** 693-694
elevated-temperature failure **A11:** 296-297
properties **A2:** 710-711
Platinum-rhodium dispersion-strengthened materials **A7:** 721
Platinum-rhodium thermocouples
bare, effect of environment **A2:** 882
ceramic insulation **A2:** 883
Platinum-rhodium-ruthenium alloy **M2:** 695
Platinum-ruthenium alloys **M2:** 694-695
as electrical contact materials **A2:** 847
properties **A2:** 711
Platinum-silver conductor inks **EL1:** 337
Platinum-tungsten alloys **M2:** 696-697
properties **A2:** 712-713
PLC units *See* Programmable logic control (PLC) units
PLCC *See* Plastic leaded chip barrier
Plenum **A6:** 1166
definition **M6:** 13
of uranium dioxide fuel rod **A7:** 665
Plenum chamber (plasma arc welding and cutting, and plasma spraying)
definition **A6:** 1212
Plied yarn *See also* Plies; Ply; Yarn; Yarns
defined **EM1:** 18 **EM2:** 30
Plies *See also* Fiber(s) Ply; Prepreg; Yarn
numbers, for laminate ranking **EM1:** 455-456
orientation **EM1:** 218
partial, composite tooling **EM1:** 581
partial, locating **EM1:** 606
UDC, for laminates **EM1:** 218
Plots
carpet **A12:** 172
contour **A12:** 172
fractal **A12:** 211-214
Plotting
defined **EL1:** 1153
Plow Steel quality rope wire **M1:** 265, 266
Plowing *See also* Scratching
defined **A8:** 10 **A11:** 8
test **A8:** 107
Plowing (ploughing) **A18:** 34, 35, 184-185, 186
component of friction **A18:** 432
defined **A18:** 15
stress **A18:** 432, 433
term (F_p) **A18:** 33
Plug and formed mandrels, for tube **A14:** 137, 667
Plug forming
defined **EM2:** 30
Plug gages
scanning laser gages for **A17:** 12
Plug joint
laser-beam welding **A6:** 879
Plug scores
tubular products **A17:** 568
Plug weld
definition **A6:** 1212
Plug welds **M6:** 13
cracking in **A11:** 655-656
definition **M6:** 13
electron beam welding **M6:** 618
oxyacetylene braze welding **M6:** 597
Plug-in packages **EL1:** 451-452
Plug-type
dealloying, copper/copper alloys **A13:** 614
zincification **A13:** 128, 129, 132
Plug-type dezincification
as selective leaching **A11:** 633
Plug-type die inserts **A14:** 47
Plugging indicator
for liquid metals purity **A8:** 426
Plugs
defined **A14:** 10
fixed, drawing with **A14:** 330
floating, drawing with **A14:** 331
Plumber's wiping solder
micrograph **A9:** 422
Plumbicon tubes
dynamic range, radiography **A17:** 318
television, as optical image sensors **A17:** 10
Plumbing
copper and copper alloys **A2:** 239-240
Plumbing applications *See also* Construction applications; Fluid handling applications
homopolymer/copolymer acetals **EM2:** 101
of part design **EM2:** 616
Plumbing goods brass
properties and applications **A2:** 365
Plumbism **A7:** 297-298
Plumbum coatings
applications **A2:** 555-556
Plumbum series **M2:** 498-499
of lead and lead alloy structures **A2:** 555-556
Plunge quenching
of uranium alloys **A2:** 673
Plunger *See also* Die casting; Force plug; Port
defined **A15:** 9
Plunger shaft
fatigue fracture from sharp fillet **A11:** 319-320
Plus mesh **A7:** 9
Plus sieve **A7:** 9
Plutonium
-beryllium reactions, as neutron source for NAA **A10:** 234

Plutonium (continued)
determined by controlled-potential coulometry ... **A10:** 209
diffusion into thorium ... **A10:** 249
pure ... **M2:** 783-785, 832-833
as pyrophoric ... **A7:** 199
thermal diffusivity from 20 to 100 °C ... **A6:** 4
Plutonium alloys
tantalum corrosion in ... **A13:** 735
Plutonium, as actinide metal
properties ... **A2:** 1192
Plutonium oxide powders
packed density ... **A7:** 297
Plutonium-gallium alloys
as embrittlement source ... **A11:** 234
Ply *See also* Fiber(s); Plies; Preply; Prepreg; Yarn ... **EM3:** 21
count, in manual lay-up ... **EM1:** 603
cutting, mechanically assisted ... **EM1:** 606
defined ... **EM1:** 18 **EM2:** 30
drop-off ... **EM1:** 322, 435
elastic constants ... **EM1:** 237
failure criteria ... **EM1:** 138
forming station, mechanized ... **EM1:** 605
geometry ... **EM1:** 458-459
lamination, automated ... **EM1:** 639-641
peel, in cure preparation ... **EM1:** 642
progressive failures ... **EM1:** 238-239
properties ... **EM1:** 313-314, 316-318
reference system ... **EM1:** 236
sorting/stacking, mechanically assisted ... **EM1:** 606
strength proper-ties ... **EM1:** 237-238
stress at a point theories ... **EM1:** 238
stress-strain law ... **EM1:** 459
stresses, at first-ply failure ... **EM1:** 232
thickness, cure ... **EM1:** 761
Ply buckling
resin-matrix composites ... **A12:** 478
Ply cutters ... **EM1:** 619, 622
Ply cutting
automated system specifications ... **EM1:** 622-623
CAD/CAM integrated manufacturing center for ... **EM1:** 621-622
computer-controlled ... **EM1:** 619-623
mechanically assisted ... **EM1:** 606
ultrasonic ... **EM1:** 615-618
Ply die cutting ... **EM1:** 608-614
automation of ... **EM1:** 613-614
die-cutting system ... **EM1:** 608-613
Ply flipping
in manual lay-up ... **EM1:** 602-603
Ply, laminate *See* Laminate ply
Ply pattern
cutting systems, automated ... **EM1:** 619-620
labeling systems, automated ... **EM1:** 620
Ply, peel *See* Peel ply
Ply termination tests ... **EM3:** 822, 823
Plying
of glass textile yams ... **EM1:** 110
Plywood ... **EM3:** 21
as laminate ... **EM1:** 218
PM 1000
composition ... **A6:** 1037
PM 2000
composition ... **A6:** 1037
PM 3030
composition ... **A6:** 1037
PM rule *See* Palmgren-Miner rule
PMMA *See* Polymethyl methacrylate
PMR acetylene end-capped Thermid AL-600 resin ... **EM1:** 84
PMR polyimides *See also* Polyimide ... **EM1:** 82-83, 85, 89 **EM3:** 21
applications ... **EM1:** 812-814
autoclave cure cycle ... **EM1:** 662
chemical structure ... **EM1:** 141
constituent properties ... **EM1:** 89
cure cycle, with Advanced Cure Monitor (ACM) ... **EM1:** 761
cure cyle ... **EM1:** 141
defined ... **EM1:** 18 **EM2:** 30
development ... **EM1:** 810-811
isomerization during polymerization ... **EM1:** 83
monomer structure ... **EM1:** 811
press-molding cure cycle ... **EM1:** 663
properties ... **EM1:** 290, 810-812
PMR polyimides (continued)
reverse Diels-Aider, reaction scheme ... **EM1:** 82
temperature capabilities ... **EM1:** 78
tests for ... **EM1:** 290
PMS technology *See also* Ternary molybdenum chalcogenides (chevrel phases)
fabrication ... **A2:** 1077-1079
superconducting properties, wire filaments ... **A2:** 1079
PMT *See* Photomultiplier tube
Pneumatic die cushions ... **A14:** 498
Pneumatic hammer ... **A18:** 529
Pneumatic nebulizers ... **A10:** 34-36, 47
Pneumatic powder dispenser
for metal powder cutting ... **A7:** 843
Pneumatic press ... **A7:** 9
Pneumatic scrubbing
for sand reclamation ... **A15:** 227, 352
Pneumatic shears
for plate and flat sheet ... **A14:** 702
Pneumatic-action grips ... **A8:** 51
Pneumatic-hydraulic system ... **A8:** 586
Pneumoconiosis ... **A7:** 202
pnp transistors ... **EL1:** 146-147, 958
Pochhammer-Chree oscillations
in elastic pressure bars ... **A8:** 199-200
Pocket
defined ... **A18:** 15
Pocket dosimeters
for radiation monitoring ... **A17:** 301
Pocket pressure
hydrostatic lubrication ... **A18:** 91
Pocket pressure in hydrostatic bearing
nomenclature for lubrication regimes ... **A18:** 90
Pocket-thrust bearing
defined ... **A18:** 15
Pockmarks
tool steels ... **A12:** 381
POD *See* Probability of detection
POD(a) function *See also* Probability of detection (POD) confidence bounds ... **A17:** 695
defined ... **A17:** 689-690
experimental design for ... **A17:** 693-694
Point analysis
beryllium-copper alloy ... **A10:** 559
cold-rolled steel ... **A10:** 556-558
of solids ... **A7:** 257
x-ray scanning electron microscopy ... **A9:** 92-93
Point charge
for explosive forming ... **A14:** 636
Point contacts
in sliding contact wear tests ... **A8:** 605
Point defect agglomerates ... **A9:** 116-117
Point defects *See also* Interstitials; Vacancies ... **A13:** 46
annealing out during recovery ... **A9:** 693
crystalline ... **EL1:** 93
in crystals ... **A9:** 719
effect, resistance-ratio test ... **A2:** 1096
FIM images in pure metals ... **A10:** 588-589
formation of ... **A9:** 116
internal grain structure ... **A10:** 358
observed using FIM ... **A10:** 588
in radiation damage, FIM/AP study of ... **A10:** 583
as stored energy sites in cold-worked metals ... **A9:** 692
transmission electron microscopy ... **A9:** 116-117
Point estimate
defined ... **A8:** 10
Point heating
for thermal inspection ... **A17:** 398
Point location
as stereoscopic method ... **A12:** 196-198
Point modification
on drills ... **A16:** 226-228
Point plots
acoustic emission inspection ... **A17:** 283-284
Point ratio
equality of volume fraction to ... **A9:** 125
Point spread function (PSF)
in computed tomography (CT) ... **A17:** 372-373
defined ... **A17:** 384
Point stress
analysis, software for ... **EM1:** 275-281
criterion, for fracture ... **EM1:** 254, 255
failure, by fatigue ... **EM1:** 244-246
theory, for laminates ... **EM1:** 235
Point stress (continued)
theory, ply ... **EM1:** 238
Point symmetries of crystal structure ... **A9:** 708
Point-count grids ... **A9:** 123-125
Point-count method of quantitative metallography ... **A9:** 123-125
Point-counting method, metallo- graphic evaluation
thermal spray coatings ... **M5:** 371-372
Point-to-point interconnection ... **EL1:** 2-4
Pointing
in drawing ... **A14:** 332-333
Poise
defined ... **A18:** 15
Poiseuille's law
kinetics of infiltration as controlled by viscous flow through pores ... **EM4:** 239-240
Poisoning *See also* Toxicity
by sintering atmospheres ... **A7:** 348-350
by toxic sintering atmospheres ... **A7:** 348-349
of grain refiner ... **A15:** 108
heavy metal, nuclear fuel pellets ... **A7:** 666
Poisseuille
defined ... **A18:** 15
Poisson distribution, percentile ... **A8:** 629
cumulative distribution function ... **A8:** 629
cumulative probability function ... **A8:** 637
mean ... **A8:** 629, 637
parameter estimates ... **A8:** 629, 637
percentile estimates ... **A8:** 629
probability density function ... **A8:** 629
probability mass function ... **A8:** 636-637
variance ... **A8:** 629, 637
Poisson distribution, use
quality control ... **A17:** 736
Poisson expansion
in axial compression testing ... **A8:** 56
Poisson's ratio *See also* Elastic properties; Mechanical properties; Tensile Poisson's ratio ... **A11:** 8, 51 **EM3:** 21 **EM4:** 424
aluminum casting alloys ... **A2:** 150, 152-177
cast copper alloys ... **A2:** 357-391
of chromium carbide-based cermets ... **A7:** 806
defined ... **A8:** 10 **A10:** 679 **A14:** 10 **EM1:** 18, 186 **EM2:** 30-31
defined, directions f()r ... **EM1:** 358-359
and density in P/M steels ... **A7:** 466
and density of low-alloy ferrous parts ... **A7:** 463
design values for ... **A8:** 662
ductile iron ... **M1:** 38
effect in torsional testing ... **A8:** 218
for elastic bending ... **A8:** 118
and extensional-shear ratio, laminates ... **EM1:** 223
in flexure of aluminum oxide-containing cermets ... **A7:** 804
and forging modes ... **A7:** 414
of glass fibers ... **EM1:** 46
maraging steels ... **M1:** 451
or frequency, symbol for ... **A10:** 692
P/M steels ... **M1:** 339
in stress measurement ... **A10:** 382
symbol for ... **A8:** 726
tensile, medium-temperature thermoset matrix composites ... **EM1:** 386
in torsion testing ... **A8:** 139
unbalanced, in composites ... **EM1:** 260
wrought aluminum and aluminum alloys ... **A2:** 62-122
Poisson's ratio of gear
symbol and units ... **A18:** 544
Poisson's ratio of pinion
symbol and units ... **A18:** 544
Poker chips
powders used ... **A7:** 574
Polar additives
grinding ... **A16:** 437
Polar backscattering
as angle-beam ultrasonic inspection ... **A17:** 248
Polar coordinate vision *See also* Machine vision
application ... **A17:** 40
defined ... **A17:** 3
Polar crystals
LEISS identification of faces ... **A10:** 603
Polar Kerr effect ... **A9:** 535
Polar lubricants
for wire forming ... **A14:** 696
Polar moment of inertia ... **A8:** 725

Polar weave
3-D, geometry of **EM1:** 129
for multidirectionally reinforced fabrics/preforms **EM1:** 129-130
process, summary **EM1:** 131
Polar winding *See also* Planar winding
defined **EM1:** 18
filament winding **EM2:** 31, 373
pattern **EM1:** 13, 508, 514
Polar/nonionic materials
in electronic manufacturing **EL1:** 660
Polar/nonionic residues
cleaning of **EL1:** 659-660
Polarimeter **EM3:** 21
defined **EM2:** 31
Polariscope **EM3:** 21
defined **EM2:** 31
Polarity
definition **A6:** 1212
Polarizability
defined **EL1:** 99
Polarizability, molecular
effect in Raman spectroscopy **A10:** 127
Polarization **EL1:** 99-100, 1153 **EM3:** 634
of a molecule, abbreviation for **A10:** 690
admittance, defined **A13:** 10
analyzer, in Raman spectrometer **A10:** 128
anodic **A13:** 123
γ-ray, Mössbauer spectroscopy **A10:** 288
characteristics, crevice corrosion of stainless steels **A13:** 304
concentration **A13:** 4, 34
curve **A13:** 10, 152
defined **A13:** 10
diagrams, coupled potential and galvanic current from **A13:** 236
electrochemical **A13:** 271-274
in galvanic corrosion **A13:** 83
galvanostatic **A13:** 1333
measurements **A13:** 84
method, electrochemical corrosion testing **A13:** 214
microwave **A17:** 20
modes, dynamic diffraction **A10:** 367
modulation **A10:** 114-115
molecular **A10:** 127
P/S, in interferometer measurement **A17:** 1
potentiostatic, curve for inconel **A10:** 558, 625
prevention by stirring in electrogravimetry **A10:** 200
resistance measurement **A13:** 200, 214
resistance technique **A13:** 209
scrambler, in Raman spectrometer **A10:** 128
tests, crevice corrosion **A13:** 308-309
in voltammetry **A10:** 189
Polarization curves
of 18-8 austenitic stainless steel **A9:** 144-145
of an alloy in a deaerated-acid environment **A9:** 144
anodic, iron-chromium alloys **A13:** 48
aqueous corrosion **A13:** 34-35
defined **A13:** 10
galvanic corrosion evaluation by **A13:** 235-236
of iron-chromium alloys **A9:** 146
reference electrodes **A13:** 23
Polarization interferometer **A9:** 150-151
Polarization modulation
as infrared spectroscopic method **A10:** 114-115
Polarization potential **A18:** 274
Polarization resistance
aqueous corrosion **A13:** 33
defined **A13:** 10
measurement **A13:** 214-215
methods **A13:** 214
modeling of **A13:** 234
polarization curves **A13:** 216
soil corrosion **A13:** 209
Polarization resistance techniques **A18:** 274, 275
Polarized ceramics
as tranducer elements **A17:** 25
Polarized contacts
life of **A2:** 860-861
Polarized light
and color etching **A9:** 136
optical color metallography **A9:** 138
in optical microscopy **A9:** 76-79
in potentiostatic etching **A9:** 144
used in microscopic examination of lead and lead alloys **A9:** 416
used to examine porcelain enameled sheet steel **A9:** 499
Polarized light etching **A9:** 58-59
Polarized light illumination
defined **A9:** 14
used to identify deformation twins **A9:** 688
Polarized light microscopy *See also* Differential-interference contrast technique
of fiber composites **A9:** 592
principles of **A9:** 58
uranium and uranium alloys **A9:** 479-480
used for hafnium **A9:** 497-499
used for zirconium and zirconium alloys **A9:** 497-499
used to examine cuprous oxide inclusions in copper and copper alloys **A9:** 400
Polarized ultrasonic wave technique **A6:** 1095, 1096
Polarizer
defined **A9:** 14
Polarizing element
defined **A10:** 679
Polarizing filters **A9:** 72
for optical images **EL1:** 1067-1068
Polarizing microscopy
for microstructural analysis **EM4:** 578
Polarograms, dc and differential pulse
of nickel in cobalt nitrate **A10:** 194
Polarographic wave
defined **A10:** 190
Polarography
capabilities **A10:** 207
current-sampled **A10:** 193
defined **A10:** 189, 679
differential pulse **A10:** 193
electrometric titration and, compared **A10:** 202
potentiostat for **A10:** 199
pulse **A10:** 193
Polaroid films
fractographic **A12:** 85, 169
Polaroid filters **A9:** 76
Polaroid Instant 35-mm slide camera system **A12:** 78-79
Polaroid instant process photographic films **A9:** 84-86
Polaroid photographs
exposure guide **A12:** 89
Pole densities
determination of distribution **A9:** 703
Pole figure
defined **A9:** 14
Pole figures
(111) from copper tubing **A10:** 363
defined **A10:** 679
determining, in crystallographic measurement and analysis **A10:** 360-361
expected orientations **A10:** 360
from inside wall, copper tubing **A10:** 363
from midwall, copper tubing **A10:** 362
measured (111) for Cu-3Zn **A10:** 360
neutron diffraction **A10:** 423
thick-sample reflection measurement **A10:** 360
Pole pieces
with induced current method **A17:** 9
Pole pieces, magnetic
powders used **A7:** 573
Pole vaulting
composite material applications for **EM1:** 846-847
Pole-figure construction by point plotting **A9:** 703
Pole-figure goniometers **A9:** 703
Pole-figure techniques used to determine crystallographic texture **A9:** 702-706
Polepiece
defined **EM2:** 30
Poling **EM4:** 17
Polish attack of tin and tin alloy coatings **A9:** 451
Polish-etching *See also* Etch-polishing
molybdenum **A9:** 441
niobium **A9:** 440-441
tantalum **A9:** 440-441
tungsten **A9:** 441
Polished steel dies
for cold upset testing **A8:** 579
Polished surface
defined **A9:** 14, 35
Polishing *See also* Buffing; Finishing; Polishing of specific metals and alloys **A7:** 593 **A16:** 19, 23 **M5:** 107-117, 119-127 **EM4:** 313, 351-358
abrasive flow machining **A16:** 518, 519
abrasives used **M5:** 108
aluminum *See* Aluminum, polishing and buffing of
applications **M5:** 107-108, 120-121
automatic equipment **A10:** 313
automatic systems **M5:** 115, 120-122
limitations **M5:** 115
rotary machines, types used **M5:** 120-122
belt *See* Belt polishing
brass *See* Brass, polishing and Sub
bronze **M5:** 112-114
cast iron **M5:** 112-114
chemical *See* Chemical brightening; Chemical polishing
chemical, as FIM sample preparation **A10:** 584
chemical/mechanical **EM4:** 358
chemically active **A9:** 39-40
compounds
contamination by **M5:** 439
removal of **M5:** 576-577
copper *See* Copper, polishing and buffing of
damage **A10:** 301
defined **A9:** 14, 35 **A18:** 15
of dissimilar-metal welded joints **A9:** 582
edge retention of sample during **A9:** 44-45
effect of cloth selection and wetness on graphite retention in cast irons **A9:** 244
effects of, on oxide scale on iron **A9:** 47
of electrogalvanized sheet steel **A9:** 197
electrolytic *See* Electrolytic brightening; Electrolytic polishing
electrolytic, of replication microscopy specimens **A17:** 5
equipment **M5:** 107, 119-125
fiber composites **A9:** 589-591
final-polishing **A9:** 40-43
flat **EM4:** 358
flat part machines, types used **M5:** 122-123
glossary of terms **M5:** 125-127
hand procedures **A9:** 35
heads **M5:** 119-120
heat-resistant alloys **M5:** 566
for inclusions in wrought iron **A9:** 42
limitations **M5:** 115
lubricants used **M5:** 115
magnesium alloys **M5:** 628, 631-632
of magnetic etching specimens **A9:** 64
materials used **M5:** 107-108
mechanical **EM4:** 358
mechanical, electropolishing compared to **M5:** 306-08
mechanical, of replication microscopy specimens **A17:** 5

Polishing (continued)
mechanical procedures ... **A9:** 35-47
mechanical, Raman analysis ... **A10:** 133
microbeam analysis samples ... **A10:** 529-530
of microhardness specimen ... **A8:** 93
nickel *See* Nickel, polishing and buffing of
of polarized light specimens ... **A9:** 58
powder metallurgy materials ... **A9:** 506-507
precision flat ... **EM4:** 358
printed board coupons ... **EL1:** 573
problems ... **M5:** 124-125
refractory metals ... **M5:** 655-656
rhenium ... **A2:** 562
rotary automatic systems ... **M5:** 120-122
rough and fine, optical metallography specimen preparation ... **A10:** 301
safety precautions ... **M5:** 648-649
of samples, x-ray spectrometry ... **A10:** 93
scanning electron microscopy specimens ... **A9:** 97
semiautomatic systems ... **M5:** 115, 120
sizes of micron diamond powders for ... **A2:** 1013
skid polishing ... **A9:** 42
stainless steel *See* Stainless steel, polishing and buffing of
steel *See* Steel, polishing and buffing of
straight-line machines, types used ... **M5:** 121-123
superabrasive grains for ... **A2:** 1012-1013
for surface oxide layers ... **A9:** 46
terminology, glossary of ... **M5:** 125-127
thermal spray-coated materials ... **M5:** 370-371
titanium and titanium alloys ... **M5:** 655-656
for very hard materials ... **A9:** 45-46
for very soft materials ... **A9:** 45-47
vibratory polishing ... **A9:** 42
of welded joints for examination ... **A9:** 578
wheel *See* Wheel polishing
work-holding mechanisms ... **M5:** 119-120
zinc *See* Zinc, polishing and buffing of
zinc alloys ... **A2:** 530
of zinc castings ... **A15:** 796
Polishing abrasives
effect on flatness ... **A9:** 40
Polishing artifacts
in carbon and alloy steels, prevention of ... **A9:** 169
defined ... **A9:** 14
scratch traces ... **A9:** 40
Polishing chamber
pencil-type for local electropolishing ... **A9:** 55
Polishing cloth
effect on flatness ... **A9:** 40
effect on graphite retention in cast irons ... **A9:** 244
effect on retention of graphite in gray iron ... **A9:** 40
effect on sample edge rounding ... **A9:** 45
Polishing cloth wetness
effect on graphite retention in cast irons ... **A9:** 244
Polishing damage ... **A9:** 39-40
effect on etching ... **A9:** 39-40
on brass ... **A9:** 41
scratch traces ... **A9:** 40
Polishing defects
in cast irons ... **A9:** 244
Polishing film
electrolytic ... **A9:** 48, 50
Polishing jig for hand polishing ... **A9:** 105
Polishing of specific metals and alloys
aluminum alloys ... **A9:** 351-353
aluminum-silicon alloys ... **A9:** 42
austenitic manganese steel castings ... **A9:** 237-238
beryllium ... **A9:** 389
beryllium-copper alloys ... **A9:** 392-393
beryllium-nickel alloys ... **A9:** 392-393
carbon and alloy steels ... **A9:** 168-169
carbon steel casting specimens ... **A9:** 230
carbonitrided steels ... **A9:** 217
carburized steels ... **A9:** 217
cast irons ... **A9:** 243-244
cemented carbides ... **A9:** 273
chromized sheet steel ... **A9:** 198
copper and copper alloys ... **A9:** 400
of electrical contact materials ... **A9:** 550
ferrites and garnets ... **A9:** 533
galvanized steel ... **A9:** 488
Polishing of specific metals and alloys (continued)
hot-dip aluminum-coated sheet steel ... **A9:** 197
hot-dip galvanized sheet steel ... **A9:** 197
hot-dip zinc-aluminum coated sheet steel ... **A9:** 197
iron-cobalt and iron-nickel alloys ... **A9:** 532-533
lead and lead alloys ... **A9:** 415-416
low-alloy steel casting samples ... **A9:** 230
nickel alloys ... **A9:** 435
nickel-copper alloys ... **A9:** 435
niobium ... **A9:** 440
permanent magnet alloys ... **A9:** 533
porcelain enameled sheet steel ... **A9:** 198
refractory metal composites ... **A9:** 550
refractory metals ... **A9:** 439-440
rhenium and rhenium-bearing alloys ... **A9:** 447
sleeve bearing materials ... **A9:** 565-567
stainless steel casting alloys ... **A9:** 297
stainless-clad sheet steel ... **A9:** 198
tantalum ... **A9:** 440
tin and tin alloy coatings ... **A9:** 450-451
tin and tin alloys ... **A9:** 449
tin plate ... **A9:** 198
titanium and titanium alloys ... **A9:** 459
tool steels ... **A9:** 257
tungsten composites ... **A9:** 550
uranium and uranium alloys ... **A9:** 478
wrought heat-resistant alloys ... **A9:** 306-307
zinc and zinc alloys ... **A9:** 488
Polishing rate
defined ... **A9:** 14
Polishing wear ... **A18:** 191-197
with abrasives ... **A18:** 191-197
chip machining mechanism ... **A18:** 194
delamination mechanism ... **A18:** 195-196
erosion mechanisms ... **A18:** 196
fracture toughness ... **A18:** 192-193, 196
indentation hardness ... **A18:** 192
melting point ... **A18:** 193
multiple-pass mechanisms ... **A18:** 195-196
parameters required to generate specularly reflecting topographies ... **A18:** 193-194
property requirements ... **A18:** 191-193
summary ... **A18:** 196
wear rates ... **A18:** 196-197
defined ... **A18:** 15, 191
without abrasives ... **A18:** 197
chemical-mechanical mechanisms ... **A18:** 197
surface flow ... **A18:** 197
Pollutants
atmospheric, effects on carbon steels ... **A13:** 511-512
in seawater, effects ... **A13:** 898-900
Polluted cooling waters
copper/copper alloys in ... **A13:** 625
Pollution *See also* Effluents ... **A7:** 203, 206
environmental, GFAAS analysis of ... **A10:** 58
shielding gas and fume generation ... **A6:** 68
studies, by NAA ... **A10:** 233
Pollution control *See also* Waste recovery and treatment
cleaning process selection influenced by ... **M5:** 20
plating wastes *See* Plating waste disposal and recovery
vinyl esters for ... **EM2:** 272
Pollution control equipment
cupolas ... **A15:** 384
Poly (4-methylpentene) (PMP)
as engineering thermoplastic ... **EM2:** 446
Poly para-phenyleneterephthalamide (PPD-T) ... **EM1:** 30
structure/orientation ... **EM1:** 54
Poly(amido amines) ... **EM3:** 95
Poly(arylene ether imidazole)
chemistry ... **EM3:** 172
properties ... **EM3:** 171
synthesis ... **EM3:** 173
Poly(butyl acrylate) adhesives
applications ... **EM3:** 203
automotive decorative trim ... **EM3:** 552
Poly(methylmethacrylate)
as amorphous polymer ... **A11:** 758
Poly(*p*-xylene) ... **EM3:** 594, 599-600
Poly(p-xylene) polymers (PPXs)
Gorham process formation ... **EL1:** 789
Poly(propylene glycol)diglycidyl ether ... **EM3:** 100
Poly(vinylchloride) (PVC)
as brittle polymer ... **A11:** 761
reactor, ductile fracture of stub-shaft assembly in ... **A11:** 481-482
water filter housing, failed, fracture surface ... **A11:** 762-763
Poly-1H
1H-pentadecafluoroctyl methacrylate (PFOM) ... **EL1:** 782-784
Poly-α-olefin (PAO) oils
critical temperatures ... **A18:** 96
high-vacuum lubricant applications ... **A18:** 155, 156-157
molecular structures ... **A18:** 156
properties ... **A18:** 155
low vapor pressure ... **A18:** 151
Poly-l-methylstyrene ... **EM3:** 619
Polyacetal
substrate cure rate and bond strength for cyanoacrylates ... **EM3:** 129
Polyacetals
as engineering plastic ... **EM2:** 429
Polyacetylene
conduction anisotropy ... **EM3:** 436
Polyacrylamide
percent conversion to poly(N-dimethylaminomethylacrylamide), Raman analysis for ... **A10:** 132
Polyacrylamides
as flocculants ... **EM4:** 92
Polyacrylate
defined ... **EM2:** 31
Polyacrylates ... **EM3:** 21, 76, 82-83, 86
cross-linking ... **EM3:** 85
as dispersants ... **A18:** 111
as foam inhibitors ... **A18:** 108
hydrocarbon moiety of dispersants ... **A18:** 99
as pour-point depressants ... **A18:** 111
for printed board material systems ... **EM3:** 592
properties ... **EM3:** 85
removal ... **EM4:** 137
as viscosity improvers ... **A18:** 109, 110, 111
Polyacrylics
as flocculants ... **EM4:** 92
Polyacrylonitrile (PAN) *See also* Carbon fiber ... **EM3:** 21
-based fiber, schematic three-dimensional ... **EM1:** 51
for carbon aircraft brakes ... **A18:** 584, 585
as carbon fiber precursor ... **EM1:** 112
defined ... **EM1:** 18 **EM2:** 31
as graphite fiber precursor ... **EM1:** 43
and mesophase pitch-based precursor fibers compared ... **EM1:** 50
precursor fibers, carbon fiber properties with ... **EM1:** 49
processing sequence ... **EM1:** 50
properties ... **EM1:** 867
Polyalkylene glycol
properties ... **A18:** 81
Polyalphaolefin (PAO) fluids ... **A18:** 134
Polyalphaolefins ... **A16:** 123
Polyamic acids ... **EM1:** 78
chemistry of ... **EL1:** 767
Polyamide
defined ... **EM1:** 18
Polyamide (PA)
nylon ... **A18:** 6
cage material for rolling-element bearings ... **A18:** 503
friction coefficient data ... **A18:** 73
nylon 6/6 + 15% polytetrafluoroethylene (PTFE), friction coefficient data ... **A18:** 73
nylon 6/6 polytetrafluoroethylene (PTFE)/glass, friction coefficient data ... **A18:** 73
wear properties ... **A18:** 241
Polyamide copolymer
typical properties ... **EM3:** 83
Polyamide hardeners
effect on toughness ... **EM3:** 185
Polyamide hot melts
applications ... **EM3:** 44
Polyamide plastic *See* Nylon; Nylon plastics; Polyamides (PA) ... **EM3:** 21

Polyamide-imide (PAI) **EM3:** 601
surface contamination **EM3:** 847
Polyamide-imide (PAI) resins **EM3:** 21
Polyamide-imide coatings **EL1:** 759
Polyamide-imides (PAI) *See also* Thermoplastic resins **EM2:** 128-137
applications **EM2:** 128
characteristics **EM2:** 128-130
competitive materials **EM2:** 128
costs **EM2:** 128, 130
defined **EM2:** 31
design considerations **EM2:** 130-132
mechanical properties **EM2:** 133
processing **EM2:** 130, 132-137
resin compounds **EM2:** 137
safety and handling **EM2:** 137
Polyamideimide resins (PAI)
defined **EM1:** 18
hot/wet in-service temperatures **EM1:** 33
matrix processing **EM1:** 544
as thermoplastic aerospace matrix **EM1:** 33
Polyamides (PA) *See also* Nylon; Thermoplastic resins **EM2:** 124-127 **EM3:** 21, 75, 80
advantages and limitations **EM3:** 82
alloys **EM2:** 124-125
applications **EM2:** 125-126
automotive applications **EM3:** 46
bonding applications **EM3:** 82
characteristics **EM2:** 126-127
commercial forms **EM2:** 124
competitive materials **EM2:** 126
costs **EM2:** 125
critical surface tension **EM3:** 180
for curing epoxies **EM3:** 95
defined **EM2:** 31
design considerations **EM2:** 127
electrical properties **EM2:** 126
as engineering plastic **EM2:** 429
mechanical properties **EM2:** 126-127
melting point **EM3:** 618
predicted 1992 sales **EM3:** 81
processing **EM2:** 127
production volume **EM2:** 125
properties **EM3:** 82, 83
resin types **EM2:** 127
suppliers **EM2:** 127
surface preparation **EM3:** 291
thermal properties **EM2:** 448
thermosetting forms **EM2:** 124
Polyamines **EM3:** 98
as flocculants **EM4:** 92
Polyaminoamides **EM3:** 100
Polyaramid
gas plasma treatment **EM3:** 285-286
Kevlar **EM3:** 283, 284
surface preparation **EM3:** 283-286
tensile properties **EM3:** 285
Polyaramid fibers
and epoxy resins **EM1:** 75
Polyaryl sulfone (PAS) **EM3:** 21
Polyaryl sulfone resins
Astrel **EM3:** 279
surface preparation **EM3:** 279
Polyaryl sulfones (PAS) *See also* Thermoplastic resins
characteristics **EM2:** 145-146
defined **EM2:** 31
properties **EM2:** 145
Polyarylate resins
Ardel **EM3:** 279
surface preparation **EM3:** 279
Polyarylates **EM3:** 21, 601
Polyarylates (PAR) *See also* Thermoplastic resins
applications **EM2:** 138-139
characteristics **EM2:** 139-141
chemistry **EM2:** 138
commercial amorphous, types **EM2:** 138
defined **EM2:** 31
electrical properties **EM2:** 138
Polyarylates (PAR) (continued)
mechanical properties **EM2:** 138-139
physical properties **EM2:** 139
processing **EM2:** 140-141
suppliers **EM2:** 141
thermal properties **EM2:** 138-139
Polyarylates (PAR), specific types *See also* Polyarylates (PAR)
Ardel D-100, tensile creep **EM2:** 139
Xydar SRT-300, mechanical properties **EM2:** 139
Polyaryletherketone (PAEK) **EM3:** 21
Polyaryletherketones (PAEK, PEK, PEEK, PEKK) *See also* Thermoplastic resins **EM2:** 142-144
applications **EM2:** 142
characteristics **EM2:** 142-144
chemical resistance **EM2:** 144
defined **EM2:** 31
processing **EM2:** 142-143
product forms **EM2:** 143
Polyarylsulfone (PAS)
defined **EM1:** 18
Polybenzimidazole (PBI) **EM1:** 18, 30 **EM3:** 21
Polybenzimidazoles (PBI) *See also* Thermoplastic resins **EM2:** 147-150
applications **EM2:** 147-148
characteristics **EM2:** 148-150
chemical resistance **EM2:** 150
commercial forms **EM2:** 147
competitive materials **EM2:** 148
costs **EM2:** 147
defined **EM2:** 31
design considerations **EM2:** 150
heat-deflection temperature **EM2:** 148
processing **EM2:** 150
production volume **EM2:** 147
properties **EM2:** 149
suppliers **EM2:** 150
tensile strength **EM2:** 149
thermosets **EM2:** 147
Polybenzimidazoles (PBIs) **EM3:** 75, 76, 161, 165, 169-174
for advanced aircraft
advantages and limitations **EM3:** 80
for automotive industries **EM3:** 171
for battery separators **EM3:** 169
chemistry **EM3:** 169
compared to epoxies **EM3:** 98
cost factors **EM3:** 169
for domestic households **EM3:** 171
for electronics **EM3:** 171
for fabricated structures **EM3:** 169
for fire-resistant aircraft seats **EM3:** 169
for flight suits **EM3:** 169
for foams **EM3:** 169
forms **EM3:** 169
for heat-protective apparel **EM3:** 169
markets **EM3:** 169
for membranes **EM3:** 169
for missiles **EM3:** 171-172
for oil recovery systems **EM3:** 171
for orbital vehicles **EM3:** 172
for printed circuit boards **EM3:** 169
processing parameters **EM3:** 172-173
product design considerations **EM3:** 171-172
properties **EM3:** 171
for satellites **EM3:** 171-172
for space vehicles **EM3:** 171
suppliers **EM3:** 169
tensile shear strength **EM3:** 170
Polybutadiene resins
for resin transfer molding **EM1:** 169
Polybutenes
appliance market applications **EM3:** 59
characteristics **EM3:** 53
properties **EM3:** 82, 677
for recreational vehicle sealing **EM3:** 58
suppliers **EM3:** 59, 82
Polybutylene terephthalate (PBT) *See* under Thermoplastic polyesters **EM3:** 21, 601
inter-facial shear strength of embedded fibers **EM3:** 394
Polybutylene terephthalates (PBT) *See also* Thermoplastic resins
alloys, blends, compounds **EM2:** 153
applications **EM2:** 153-154
characteristics **EM2:** 154-155
commercial forms **EM2:** 153
competitive materials **EM2:** 154
costs and production volume **EM2:** 153
defined **EM2:** 31
processing **EM2:** 155
resin compound types **EM2:** 155
suppliers **EM2:** 155
Polybutylenes **EM3:** 21
Polybutylenes (PB)
defined **EM2:** 31
as engineering thermoplastic **EM2:** 446
Polycaprolactam *See* Caprolactam
Polycarbonate **A7:** 606, 612
crazing in **A11:** 759
sheet, effect of thickness **A11:** 762
as substrate **EL1:** 339
waterjet machining **A16:** 522
yielding and necking in **A11:** 760
Polycarbonate (PC)
deformation zone friction **A18:** 36
friction and wear data **A18:** 58
Polycarbonate resin
defined **EM1:** 18
Polycarbonate sheet
quasi-brittle fatigue crack propagation **A12:** 479
Polycarbonates (PC) *See also* Thermoplastic resins **EM3:** 21, 576
applications **EM2:** 151
characteristics **EM2:** 151-152
chemistry **EM2:** 65
chemistry/variations **EM2:** 151
competitive materials **EM2:** 151
contact-angle testing **EM3:** 277
costs and production volumes **EM2:** 151
critical surface tensions **EM3:** 180
defined **EM2:** 31
as engineering plastic **EM2:** 430
as engineering thermoplastic **EM2:** 449
interfacial shear strength of embedded fibers **EM3:** 394
moisture effects **EM2:** 768
polyurethane shear strength when bonded to steel **EM3:** 663, 664
resin compound types **EM2:** 152
rotational molding **EM2:** 361
solvent cements **EM3:** 567
as structural plastic **EM2:** 65
and styrene-maleic anhydrides (S/MA) alloy **EM2:** 221
substrate cure rate and bond strength for cyanoacrylates **EM3:** 129
suppliers **EM2:** 152
surface preparation **EM3:** 279, 291
Polycarbosilane
cross-linking **EM4:** 223
as joining agent material for non-oxide ceramics **EM4:** 528
Polycarboxylate **EM4:** 1092-1093
Polychloroprene
properties **EM3:** 144
Polychloroprene rubber (CR)
additives and modifiers **EM3:** 149-150
for auto trim bonding **EM3:** 149
for belt lamination **EM3:** 149
chemistry **EM3:** 149
commercial forms **EM3:** 149
for contact bonding **EM3:** 149
cross-linking **EM3:** 150
cure mechanism **EM3:** 150
for furniture assembly **EM3:** 149

Polychloroprene rubber (CR) (continued)
for hose and belting production **EM3:** 149
for leather shoe sole bonding **EM3:** 149
markets **EM3:** 149
properties **EM3:** 149, 150
for truck and trailer roof and floor bonding **EM3:** 149
Polychlorotrifluoroethylene (CTFE)
as fluoropolymer **EM2:** 66, 115-116
Polychlorotrifluoroethylene (PCTFE) **EM3:** 223
Polychromatic
artifacts, defined **A17:** 38
Corning Inc **EM4:** 440, 442
defined **A17:** 38
x-ray spectra, defined **A17:** 38
Polychromatic radiation
beams, single-crystal diffraction methods **A10:** 329
interelement effects **A10:** 97
Polychromatic reflection topography **A10:** 366
Polychromators
computerized **A10:** 24
defined **A10:** 679
ICP-AES **A10:** 34, 37-38
Paschen-Runge **A10:** 37
as wavelength sorting device **A10:** 23-24
Polychrome body finishes
powder used **A7:** 572
Polycondensation *See* Condensation polymerization **EM3:** 21
Polycrystal rocking curve analysis **A10:** 371-373
Polycrystal scattering topography
basic principle **A10:** 374
Soller slit arrangements for **A10:** 375
Polycrystalline
defined **A9:** 14 **A11:** 8
Polycrystalline aggregate, preferential alignment of the crystalline lattice *See* Texture, crystallographic
Polycrystalline alumina reinforced aluminum alloys **EM1:** 890-891
Polycrystalline aluminum
strain rate sensitivity **A8:** 237-238
ultrasonic fatigue testing **A8:** 252
Polycrystalline cast superalloys *See also* Cast cobalt-base superalloys; Cast nickel-base superalloys **A1:** 981-994
age hardening **A1:** 981
application of **A1:** 981
composition and density **A1:** 982
cobalt-base **A1:** 983
nickel-base **A1:** 982
control of casting microstructure **A1:** 990-992
carbides **A1:** 990-991
dendrites **A1:** 990
eutectic segregation **A1:** 991
grain size **A1:** 992
porosity **A1:** 991-992
design of **A1:** 983-986
cobalt-base **A1:** 985-986
nickel-base **A1:** 983-985
heat treatment **A1:** 993
hot isostatic pressing **A1:** 993-994
effect on fatigue strength **A1:** 992, 994
investment casting of **A1:** 989-990
mechanical properties of **A1:** 984
fatigue properties **A1:** 991, 992, 994
stress-rupture properties **A1:** 985, 986, 987, 991, 993
tensile properties **A1:** 984
physical properties of **A1:** 983
quality considerations **A1:** 988-989
tramp element content **A1:** 989
stress-rupture strengths for selected **A1:** 985
vacuum induction melting of **A1:** 986-988
Polycrystalline ceramics
fracture of **A11:** 26
stress-corrosion failure **EM4:** 659
Polycrystalline copper
nucleation of recrystallized grains **A9:** 694-695
Polycrystalline cubic boron nitride (PCBN)
synthesis of **A2:** 1009
tool applications **A2:** 1016
tool blanks **A2:** 1016-1017
Polycrystalline diamond
cutting tools **A2:** 1016
synthesis of **A2:** 1009
Polycrystalline diamond (continued)
tool applications **A2:** 1016
tool blanks **A2:** 1015-1016
Polycrystalline diamond (PCD) **EM4:** 822
cutting speed and work material relationship **A18:** 616
cutting toot material **A18:** 614, 617
Polycrystalline diffraction methods **A10:** 331-332
Polycrystalline glass ceramics
mirror radius relation to failure stress **EM4:** 656-657
Polycrystalline materials
described **A10:** 358
double-crystal diffractometry for **A10:** 372, 373
engineering, residual stress and texture determined by neutron diffraction **A10:** 420
EXAFS analysis **A10:** 410
measurement and analysis **A10:** 357-364
preferred orientation in **A10:** 358
RDF-determined interatomic distance distributions and coordination numbers **A10:** 393
temperature-dependent properties of **A11:** 138
texture as measure of average grain orientation in **A10:** 358
x-ray topography of aggregates of **A10:** 365
Polycrystalline metal
crack nucleation **A8:** 366
fracture surface **A8:** 479-480
in torsional Kolsky bar experiment **A8:** 222
Polycrystalline metals
cleavage fractures **A12:** 252
grain-boundary sliding n **A9:** 690
Polycrystalline nickel
cavitation in **A11:** 165
Polycrystalline pure metals
solidification structures of **A9:** 608-610
Polycrystalline solids
defined **EL1:** 93
Polycrystalline specimens, deformed to large strains
dislocation densities of **A9:** 693
microband regions **A9:** 694
microstructure of **A9:** 693
Polycrystalline specimens, deformed to small strains
recrystallization by strain-induced boundary migration **A9:** 696
Polycrystals
Berg-Baffett reflection topographic method for **A10:** 369
in torsional Kolsky bar experiments **A8:** 221-222
Polycyanosiloxanes **EM3:** 678
Polycythemia
as cobalt toxic reaction **A7:** 204
from cobalt toxicity **A2:** 1251
Polydimethylsiloxane *See* Silicones (SI)
Polydimethylsiloxane (PDMS)
as basis for silicone PSAs **EM3:** 134
Polydimethylsiloxane-base silicones
solvent resistance/repairability **EL1:** 822
Polydisperse particle size systems **EM4:** 153
Polyelectrolytes **EM4:** 155
Polyester **A13:** 10, 410
die material for sheet metal forming **A18:** 628
waterjet machining **A16:** 522
Polyester adhesives
for flexible printed boards **EL1:** 582
Polyester amide **EM3:** 109
Polyester cloth
for overwraps for plumbing repairs **EM3:** 608
Polyester coating
steel sheet **M1:** 176
Polyester composites *See also* Glass fiber polyester resin composites; Polyester resins **EM1:** 91
Polyester fibers
manufactured with germanium dioxide **A2:** 734
Polyester film
thermal properties **EM2:** 449
Polyester film adhesives
for flexible printed wiring boards (PWBS) **EM3:** 45
Polyester films
flexible printed boards **EL1:** 583
Polyester hot-melt adhesives
cooling rate and crystallinity **EM3:** 412
Polyester matrix composites **A9:** 592
Polyester mounting materials **A9:** 30-31
used for tin and tin alloys **A9:** 449
Polyester plastics *See also* Plastics; Polyesters; Unsaturated polyesters **EM3:** 21
defined **EM2:** 31
Polyester polybutylene terephthalate (PBT)
friction coefficient data **A18:** 73
Polyester polyols
as fillers **EM3:** 181
Polyester resin
coating for aramid fibers **EM3:** 285
Polyester resins **EM1:** 90-96
additives to carbon-graphite materials **A18:** 816
at elevated temperatures **EM1:** 93
bisphenol A (BPA) fumarates **EM1:** 90
catalyst-promoter-inhibitor systems **EM1:** 133
chemical resistance **EM1:** 93-94
chemistry **EM1:** 90
chlorendics **EM1:** 90
for commercial application **EM1:** 31-32
components of **EM1:** 132
cost and thermal stability **EM1:** 43
defined **EM1:** 90, 90
and E-glass composites, importance **EM1:** 43
electrical properties **EM1:** 95
and fiber-resin composites **EM1:** 363-372
for filament winding **EM1:** 137-138
flame-retardant **EM1:** 96
flexural modulus, acid ratio effect **EM1:** 90
formulations **EM1:** 133, 141
gel times **EM1:** 142
glass fiber reinforced, sheet molding compound as **EM1:** 157-160
iso resins **EM1:** 90
for low-temperature thermoset matrices **EM1:** 392-398
mechanical properties **EM1:** 90-93
monomer effects, thermal stability **EM1:** 93
natural fiber reinforcement **EM1:** 117
ortho resins **EM1:** 90
oxidative stability **EM1:** 93
properties **EM1:** 175, 290, 292, 392-398
for pultrusion **EM1:** 538
reinforcing fibers **EM1:** 91-93
for resin transfer molding **EM1:** 169, 566
in RRIM technology **EM1:** 121
sample chemical reactions **EM1:** 751-753
for sheet molding compounds **EM1:** 141-142
sulfuric acid corrosion **A13:** 1154
tests for **EM1:** 290, 292
thermal stability **EM1:** 93
tonnage used **EM1:** 90
types **EM1:** 138, 290
ultraviolet (UV) resistance **EM1:** 94-95
unsaturated **EM1:** 90, 137-138, 538
vinyl ester **EM1:** 90
for wet lay-up **EM1:** 132-133
Polyester resins and coatings **M5:** 473, 496, 498, 502
Polyester styrene, thermosetting
impregnating with **M5:** 621
Polyester-urethane no-bake resins **A15:** 216
Polyesters *See also* Polyester plastics; Thermoplastic resins; Thermosetting resins; Unsaturated polyesters; Water-extended polyester **EM3:** 75, 80, 594, 601
advantages and limitations **EM3:** 82
aromatic thermoplastic, chemistry **EM2:** 65-66
as basis for polyols **EM3:** 109-110
bonding applications **EM3:** 82
for coating/encapsulation **EL1:** 242
contact-angle testing **EM3:** 277
electrical contact assemblies **EM3:** 611
as engineering thermoplastic **EM2:** 440, 448-449
environmental effects **EM2:** 428
industrial applications **EM3:** 567
IPN polymers **EM3:** 602
moisture effects **EM2:** 766-768
as polymeric substrate **EL1:** 338-339
predicted 1992 sales **EM3:** 81
price compared to urethane sealants **EM3:** 203
properties compared **EM3:** 92
silane coupling agents **EM3:** 182
surface preparation **EM3:** 277, 279

Polyesters (continued)
thermoplastic **EM2:** 42, 429 **EM3:** 21
thermosetting **EM2:** 42-43, 628-629 **EM3:** 21
undersea cable splicing **EM3:** 612
unsaturated **EM2:** 65, 246-251 **EM3:** 21
cure mechanism **EM3:** 323
microstructural analysis **EM3:** 412
for porosity sealing for engine blocks **EM3:** 57
for protecting electronic components **EM3:** 59
Polyesters, thermoplastic *See* Thermoplastic polyesters
Polyesters, thermosetting *See* Thermosetting polyesters
Polyether etherketone (PEEK) in compression molding **EM1:** 562
defined **EM1:** 18
fiber wet out/interface of **EM1:** 99
hot/wet in-service temperatures **EM1:** 33
interlaminar fracture toughness **EM1:** 100
matrix processing **EM1:** 544
melt viscosity **EM1:** 102
properties **EM1:** 101
as thermoplastic aerospace matrix **EM1:** 32-33
Polyether resin
coating for aramid fibers **EM3:** 285
Polyether silicones (silane-modified RTV) **EM3:** 228-233
additives **EM3:** 228
adhesion characteristics **EM3:** 231
aerospace applications **EM3:** 230
automotive applications **EM3:** 230
chemistry **EM3:** 228
compared to urethanes **EM3:** 231
compounding **EM3:** 228-229
construction applications **EM3:** 229
cost factors **EM3:** 233
cross-linking **EM3:** 228
cure properties **EM3:** 229-230, 231
durability **EM3:** 231
fillers used **EM3:** 228
formulations **EM3:** 228-229
limitations **EM3:** 232
processing parameters **EM3:** 231-232
product types **EM3:** 229-231
properties **EM3:** 228, 229-231
shelf life **EM3:** 232
strength properties **EM3:** 230-231
suppliers **EM3:** 232-233
weathering **EM3:** 231, 232
windshield glass bonding **EM3:** 229, 231
Polyether sulfone
as substrate **EL1:** 339
Polyether sulfones (PES, PESV) *See also* Thermoplastic resins
applications **EM2:** 159-160
characteristics **EM2:** 160
competitive materials **EM2:** 159-160
compound types **EM2:** 162
costs and production volume **EM2:** 159
defined **EM2:** 31
designing with **EM2:** 160-162
dynamic fatigue **EM2:** 161
impact strength **EM2:** 161
processing **EM2:** 161-162
suppliers **EM2:** 162
Polyether sulfones (PESV) **EM3:** 21, 576, 601
surface preparation **EM3:** 280
Polyether-amide
properties **EM3:** 82
Polyether-imides (PEI) *See also* Thermoplastic resins
applications **EM2:** 156
characteristics **EM2:** 156-157
commercial forms **EM2:** 156
defined **EM2:** 31
electrical properties **EM2:** 158
processing **EM2:** 158
thermal properties **EM2:** 158
vs polyamide-imides (PAI) **EM2:** 128
Polyether-modified silicones
as sealants **EM3:** 188
Polyether-polyurethane silicone sealant formulation **EM3:** 229
Polyetherimides (PEI) **EM3:** 21, 576, 601
surface preparation **EM3:** 280
Polyetheretherketone (PEEK)
+ 15% polytetrafluoroethylene (PTFE), friction coefficient data **A18:** 73
carbon laminate composite surface contamination **EM3:** 847
chemistry **EM2:** 66
coefficient of friction **A18:** 824, 825, 826
compositions **A18:** 821
defined *See* Polyaryletherketones (PAEK)
as engineering plastic **EM2:** 430, 449
as engineering thermoplastic
and fracture process **EM3:** 509
friction coefficient data **A18:** 73
interfacial shear strength of embedded fibers **EM3:** 394
key properties **EM2:** 144
lubricant effect on wear factors **A18:** 823, 824
mechanical properties **A18:** 823 **EM2:** 144
piston ring materials, reciprocating compressors **A18:** 603
polytetrafluoroethylene (PTFE)/glass, friction coefficient data **A18:** 73
for printed board material systems **EM3:** 592
processing **EM2:** 144
surface preparation **EM3:** 279
for valve plates for reciprocating compressors **A18:** 604
vs polyamide-imides (PAI) **EM2:** 128
wear and friction data **A18:** 821, 822-823
wear factors of various composites **A18:** 823-824
Polyetherimide (PEI)
+ 15% polytetrafluoroethylene (PTFE), friction coefficient data **A18:** 73
friction coefficient data **A18:** 73
liquid impingement erosion protection applications **A18:** 222
polytetrafluoroethylene (PTFE)/glass, friction coefficient data **A18:** 73
Polyetherimide resins (PEI)
as aerospace thermoplastic matrix **EM1:** 33
Polyetherketone **EM3:** 601
coefficient of friction **A18:** 826
liquid impingement erosion protection applications **A18:** 222
wear factors **A18:** 824
Polyetherketoneketone (PEKK) *See also* Polyaryletherketones (PAEK)
chemistry **EM2:** 66
defined *See* Polyaryletherketones (PAEK)
Polyethers
as basis for polyols **EM3:** 109-110
as urethane coating **A13:** 410
Polyethersulfone
liquid impingement erosion protection applications **A18:** 222
Polyethylene **A7:** 606 **A13:** 329, 1154
as container for electrolytes **A9:** 51
copolymer pipe, slow crack growth failure **A12:** 479
crystallization, Raman analysis of **A10:** 132
ductile-to-brittle failure mechanisms **A11:** 761
as electropolishing mounting material **A9:** 49
high-density, for irradiation containers **A10:** 236
linear, crack growth mechanisms **A12:** 479
mechanical properties **EM4:** 316
medium-density, fatigue striations **A12:** 480
medium-density, tearing and fibrillation **A12:** 480
as mounting material **A9:** 53-54
nickel composites resistivity **A7:** 609
oxidative degradation **EM4:** 136
physical properties **EM4:** 316
Polyethylene (continued)
pipe, fracture band width as function of crack length **A11:** 762
pipe, fracture surface **A11:** 761
as semicrystalline polymer **A11:** 758
ultrahigh molecular weight **A7:** 607
Polyethylene (PE) *See also* High-density polyethylenes (HDPE); Plastics **EM3:** 21, 80, 594, 601
adhesive bonding **EM3:** 290, 291, 292
anodized surfaces **EM3:** 417
bonded to AS4 carbon fibers by epoxy resin **EM3:** 393, 395-396, 402
coatings for copper and steel **EM3:** 412
copper adherends **EM3:** 269
critical surface tensions **EM3:** 180
crystalline morphology **EM3:** 410
crystallinity **EM3:** 411
defined **EM2:** 31
friction coefficient data **A18:** 73
as general-purpose polymer **EM2:** 65
high-density **EM2:** 163-166
contact-angle testing **EM3:** 277
effect of zinc-modified primer **EM3:** 283
lap-shear strength **EM3:** 276, 280
hot melt, microstructural analysis **EM3:** 416
influence of transcrystalline layer in adhesion **EM3:** 414
microstructural analysis
microstructure **EM3:** 407
polyaramid composites **EM3:** 284
as polymer modification **EM2:** 65
predicted 1992 sales **EM3:** 81
properties **EM3:** 82, 412
in rotational molding **EM2:** 361
stress crazing **EM2:** 797-800
as structural plastic **EM2:** 65
suppliers **EM3:** 82
surface preparation **EM3:** 43, 279, 291
tape, as bond breaker **EM3:** 549
thermally stimulated depolarization (TSD) **EM3:** 437
as thermoplastic system, grades **EM2:** 446
for tool steel lubrication **A18:** 737-738
ultrahigh molecular weight (UHMWPE) **EM2:** 167-171
vs polyvinyl chlorides (PVC), usage **EM2:** 209
water treeing as problem **EM3:** 439
Polyethylene coatings
microwave thickness gaging **A17:** 21
Polyethylene encasement
for cast iron pipe **M1:** 100
Polyethylene fibers *See also* Para-aramid fibers
high performance **EM1:** 31
properties **EM1:** 54, 56
specific strength/modulus **EM1:** 57
Polyethylene glycol
application or function optimizing powder treatment and green forming **EM4:** 49
batch weight of formulation when used in nonoxidizing sintering atmospheres **EM4:** 163
as binder **EM4:** 474
binder used in silicon powder **EM4:** 237
partial evaporation resulting in hard granules **EM4:** 107
removal from silicon nitride compacts **EM4:** 136
Polyethylene insulation
wrought copper and copper alloy products **A2:** 258
Polyethylene, low-molecular weight
as binder for ceramic injection molding **EM4:** 173
Polyethylene oxide nonyl phenol **EM3:** 210
Polyethylene terephthalate (PET) **EM3:** 21, 601
critical surface tensions **EM3:** 180
crystallinity **EM3:** 409
and germanium dioxide **A2:** 743

Polyethylene terephthalate (PET) (continued)
surface preparation **EM3:** 291
Polyethylene terephthalate (PET), defined *See* Thermoplastic polyesters
Polyethylene terephthalates (PET) *See also* Thermoplastic resins
applications **EM2:** 172
characteristics **EM2:** 172-176
commercial forms **EM2:** 172
competitive materials **EM2:** 172
costs and production volume **EM2:** 172
defined **EM2:** 31
engineering properties **EM2:** 172
processing **EM2:** 175
product design **EM2:** 172-175
product types **EM2:** 176
suppliers **EM2:** 176
Polyethylene waxes
for investment casting **A15:** 253, 254
Polyethylene-jacketed telephone cables **A13:** 1127
Polyethyleneoxide
as flocculant **EM4:** 92
Polyethyleneterephthalate
+ 15% polytetrafluoroethylene (PTFE), friction coefficient data **A18:** 73
controlled indent as useful general cleaning procedure **A18:** 424
friction coefficient data **A18:** 73
polytetrafluoroethylene (PTFE)/glass, friction coefficient data **A18:** 73
Polyfilm **EM3:** 44
Polyfluorotrichloroethylene
surface preparation **EM3:** 279
Polyfunctional epoxy resin systems **EL1:** 534-535
Polyglycol materials
in assembly processing **EL1:** 659
Polyglycols
lubricants for rolling-element bearings **A18:** 134, 135
Polygonal ferrite, classification of
in weldments **A9:** 581
Polygonization
Lang topography for **A10:** 377
tungsten fibers **A12:** 467
Polygonization during recovery **A9:** 693-694
Polygonized structure in grains **A9:** 604
Polygranular fused silica
strength **EM4:** 755
thermal properties **EM4:** 754
Polyimide
fracture surface **A12:** 480
Polyimide (PI) *See also* Thermoplastics; Thermosets **EM3:** 21, 44, 75, 76, 151-162
acetylene-terminated **EM3:** 155-156, 157, 159
additives **EM3:** 161
advantages and limitations **EM3:** 80
aerospace industry applications **EM3:** 559
aerospace structural parts adhesive **EM3:** 151
for aircraft structural parts and repair **EM3:** 160
for alpha-particle memory chip protection **EM3:** 161
aminosilane adhesion promoter **EM3:** 160
aromatic **EM3:** 151
bismaleimides **EM3:** 44
for bonding composites **EM3:** 160
bonds with titanium **EM3:** 266
by-products from cure **EM3:** 74
chemical cross-linking reaction **EM3:** 159
chemical resistance properties **EM3:** 639
chemistry **EM3:** 151
for circuit protection **EM3:** 592
for coating and encapsulation **EM3:** 580
compared to epoxies **EM3:** 98
compared to polyphenylquinoxalines **EM3:** 165
competition and future trends **EM3:** 161
composites bonding to composites **EM3:** 293
as conductive adhesives **EM3:** 76
conformal overcoat **EM3:** 592
copolymerization reaction **EM3:** 155
cost factors **EM3:** 160
critical surface tensions **EM3:** 180
cure process **EM3:** 597
development **EM3:** 154, 158, 159, 160, 161
as die attach adhesives **EM3:** 159, 161, 580
for die attach in device packaging **EM3:** 584
for dielectric interlayer adhesion **EM3:** 161
electrical properties **EM3:** 151
Polyimide (PI) (continued)
electrical/electronics applications **EM3:** 44
electronic packaging applications **EM3:** 594, 596-597
for electronics/microelectronics end-uses **EM3:** 151
fastest sales growth **EM3:** 77
flexible **EM3:** 677
advantages and disadvantages **EM3:** 675
properties **EM3:** 678
for severe environments **EM3:** 673
for flexible printed wiring boards (PWBs) **EM3:** 45, 161
forms **EM3:** 158-160
for hybrid circuit encapsulants **EM3:** 159, 161
for integrated circuit dielectric films **EM3:** 161
isoimides **EM3:** 155-156
for lamination **EM3:** 161
LARC TPI, titanium adherend **EM3:** 266-267
markets **EM3:** 160-161
mechanical properties **EM3:** 151, 161
for metal-to-metal bonding **EM3:** 160
for microelectronics industry protective coating **EM3:** 160
for multichip modules **EM3:** 160
photosensitivity **EM3:** 597, 598
poly(imide siloxane) **EM3:** 159
polyamic acid form **EM3:** 159
for polymer thick film **EM3:** 161
precursors **EM3:** 161
predicted 1992 sales **EM3:** 77
for printed board material systems **EM3:** 592
processing parameters **EM3:** 161
properties **EM3:** 106, 598
ready-for-use electronic-grade solutions **EM3:** 160
rubber-toughened systems **EM3:** 155
service temperature **EM3:** 621
shelf life **EM3:** 161
for substrate attach adhesive **EM3:** 161
suppliers **EM3:** 158-160, 161
surface contamination **EM3:** 845-846
surface preparation **EM3:** 277, 291
synthesis **EM3:** 151
for tape automated bonding (TAB) **EM3:** 161
in tape backings for silicone applications **EM3:** 134
Thermid EL acetylene polyimide oligomer formulations **EM3:** 159
Thermid IP-6001, FTIR study **EM3:** 158
thermoplastic **EM3:** 22, 158, 185
titanium adherends **EM3:** 268, 270
toughened by comonomers **EM3:** 155
toughened by reactive diluents **EM3:** 155
ultrapure, in microelectronics industry **EM3:** 160
UV-curable **EM3:** 161
Polyimide adhesives
for flexible printed boards **EL1:** 582
Polyimide coatings
applications **EL1:** 326, 767
chemistry **EL1:** 767
curing of **EL1:** 771-772
effectiveness **EL1:** 1056
flexible printed boards **EL1:** 583
insulation, for wrought copper and copper alloy products **A2:** 258
introduction **EL1:** 759
packaging **EL1:** 770-771
photosensitivity **EL1:** 767
processing **EL1:** 769-772
properties **EL1:** 767-769
for thermocouple wire **A2:** 882
thickness **EL1:** 761
thin-film hybrids **EL1:** 326
wafer fabrication **EL1:** 769-770
Polyimide dry etch process
for via holes **EL1:** 328
Polyimide fiber, synthetic
as thermocouple wire insulator **A2:** 882
Polyimide fiberglass printed board substrates
properties **A6:** 992
Polyimide film
as thermocouple wire insulation **A2:** 882
Polyimide quartz printed board substrates
properties **A6:** 992
Polyimide resin
NMR analysis of curing mechanism of **A10:** 285
Polyimide resins *See also* PMR polyimides **EM1:** 78-89
addition-type, chemistry **EM1:** 78-79
adhesives **EM1:** 684
as aerospace matrix **EM1:** 32
for aerospace prepregs **EM1:** 141
Avimid K-III, matrix processing **EM1:** 544
bismaleimides **EM1:** 78-79
chemistry **EM1:** 78-80
condensation-type, chemistry **EM1:** 78
constituent material properties **EM1:** 80-83
curing **EM1:** 662-663
defined **EM1:** 18
and fiber-resin composites **EM1:** 373-380
high temperature resistant, types/applications **EM1:** 810-815
as high-temperature thermoset **EM1:** 373-380
hot/wet in-service temperatures **EM1:** 33
mechanical/physical properties **EM1:** 78
oven imidization before cure **EM1:** 662-663
postcure **EM1:** 663
preparation for autoclave cure **EM1:** 662
press molding **EM1:** 663
for printed wiring boards **EM3:** 569
properties **EM1:** 290, 292, 373-380
for resin transfer molding **EM1:** 169
sample chemical reactions **EM1:** 751-753
sizing for **EM1:** 123
in space and missile applications **EM1:** 817
temperature capabilities **EM1:** 78
tests for **EM1:** 290, 292
thermoplastic **EM1:** 78
types **EM1:** 79, 290
Polyimide sulfone
effect of high temperatures **EM3:** 513
thermoplastic **EM3:** 355-357, 358-359
mechanical properties **EM3:** 360
overlap edge stress concentration for single-lap joints **EM3:** 360-361
Polyimide-aramid
for printed board material systems **EM3:** 592
Polyimide-aramid fiber printed board substrates
properties **A6:** 992
Polyimide-fiberglass
for printed board material systems **EM3:** 592
Polyimide-glass **EM3:** 601
dielectric constant **EL1:** 506
as dielectric medium **EL1:** 83
for manufacture of printed boards **EM3:** 589, 590
PWB structures **EL1:** 82
Polyimide-Kevlar
for printed board material systems **EM3:** 592
Polyimide-quartz
for printed board material systems **EM3:** 592
Polyimide-silicones **EM3:** 601
Polyimides *See also* Coatings; Polymide coatings
chemistry **EL1:** 324-326, 767
as epoxy replacement for PTH reliability **EL1:** 114
high-vacuum lubricant applications **A18:** 159
laminate capabilities **EL1:** 114
for multichip structures **EL1:** 302
nonphotolmageable **EL1:** 326-327
photosensitive **EL1:** 327-328, 767
properties, thin-film hybrids **EL1:** 324-326
resin properties **EL1:** 534, 606
as solder resists **A6:** 133
as substrates **EL1:** 339
Polyimides (PI) *See also* Thermoplastic polyimides (TPI); Thermoplastic resins
defined **EM2:** 31
as engineering plastic **EM2:** 430
as high-temperature resin system **EM2:** 443-444
monomeric constituents **EM2:** 560
thermoset **EM2:** 630
vs polyamide-imides (PAI) **EM2:** 128
Polyisobutylenes (PIB) **EM3:** 146, 190
additive applications **EM3:** 198-199
as barrier of silicone sealant **EM3:** 196
as binders **EM3:** 199
chemistry **EM3:** 198-199
cross-linking **EM3:** 198, 199
as fillers **EM3:** 199
glass transition temperature **EM3:** 198
for insulated glass construction **EM3:** 58
polymer properties **EM3:** 198-199

Polyisocyanates
in polyurethanes...........................EM2: 257
Polyisoprene
geometric isomers ofEM2: 58
Polymer
constant rate of extension testing
machines for.............................A8: 47
hypothetical failure envelope forA8: 43
metal wear.................................A8: 603
strain rate effect on strengthA8: 38, 42
strain rate sensitivityA8: 38, 42
tensile behavior.............................A8: 38, 43
Polymer anisotropy
microwave inspection........................A17: 21
Polymer blend system
factor analysis and curve fitting
applied..................................A10: 118
Polymer blends See also Alloys; Blends; Copolymers;
Polymer(s)
chemistryEM2: 66-67
Polymer chemistry
modifications to polymersEM2: 66-67
overviewEM2: 63-67
polymer categorization........................EM2: 63-64
polymer familiesEM2: 64-66
Polymer coating See also Coatings
cleaning for.................................EL1: 239-240
molding methodsEL1: 240-243
for package sealingEL1: 239-243
properties...................................EL1: 239
Polymer Communications
as information sourceEM2: 93
Polymer die attach See also Die attachment methods
bond integrityEL1: 217-218
electrical conductivity.......................EL1: 218
for hermetic packages........................EL1: 216
for plastic-encapsulated devices............EL1: 217-220
reliability..................................EL1: 218-220
thermal conductivity.........................EL1: 218
Polymer Engineering and Science
as information sourceEM2: 93
Polymer families See also Engineering plastic-s;
Engineering plastics families; Polymer(s)
acetal resinsEM2: 65
acrylic polymersEM2: 65
amino resinsEM2: 65
aromatic polyarylates (PARS)EM2: 66
aromatic polysulfones (PSU)EM2: 66
aromatic thermoplastic polyestersEM2: 65-66
epoxy resinsEM2: 65
fluoroplasticsEM2: 66
general-purpose plasticsEM2: 64-65
liquid crystal polymersEM2: 66
nylonEM2: 66
phenolic resinsEM2: 65
plastics, general-purposeEM2: 64-65
polycarbonate (PC)...........................EM2: 65
polyetherketoneEM2: 66
polyethyleneEM2: 65
polyphenylene ether (PPE)EM2: 65
polyphenylene oxide (PPO)EM2: 65
polyphenylene sulfide (PPS)EM2: 66
polypropyleneEM2: 65
polystyreneEM2: 64
polyurethanes................................EM2: 66
polyvinyl chlorideEM2: 64-65
structural plasticsEM2: 65-66
styrene-maleic anhydride (SMA)
copolymers...............................EM2: 66
unsaturated polyestersEM2: 65
Polymer films See also Film(s); Films
conductive, as interconnectionEL1: 16
dielectrics, physical characteristics........EL1: 108-109
drawn, infrared dichroism spectros-
copy to determine molecular orientation inA10: 120
stretched, infrared determination of
molecular orientation inA10: 109

Polymer glasses
SAS applications............................A10: 405
Polymer infiltration....................EM4: 35
Polymer: International Journal for the Science and
Technology of Polymers
as information sourceEM2: 93
Polymer matrix................................EM3: 22
definedEM2: 32
Polymer matrix composites
as packaging material...................EL1: 1117-1118
Polymer membrane electrodesA10: 182
Polymer monomer syrup
formulation................................EM3: 123
Polymer names
chemicalEM2: 53
commercialEM2: 53-56
customaryEM2: 53
systematicEM2: 53
Polymer Process Engineering
as information sourceEM2: 93
Polymer pyrolysis............................EM4: 35
for advanced ceramics.......................EM4: 47
for ceramic-matrix composites............EM4: 840, 842
Polymer quenchants...........................A1: 455
Polymer science
chemical structureEM2: 48-52
polymer namesEM2: 52-56
polymers, properties ofEM2: 59-62
polymers, types ofEM2: 48
structure and propertiesEM2: 57-59
Polymer thick film (PTF)
for flexible printed boardEL1: 581-582
Polymer(s) See also Alloys; Amorphous polymers;
Blends; Branched polymer; Carbon fiber reinforced polymers; Copolymer; Crystalline
polymers; E polymers; Elastomers; Fiber reinforced polymers; Glass fiber reinforced
polymers; Homopolymers; Liquid crystal polymer; Mer; Piezoelectro polymers; Plastic; Polymer matrix; Polymer science; Prepolymer;
Terpolymer; Thermoplastic polymers; Unsaturated polymers
acetylene-terminated thermosetting..........EM2: 631
acrylic, chemistry...........................EM2: 65
additives....................................EM2: 7
aging, and failure analysis..................EM2: 732
amorphous, high-modulus graphite
fibers in...............................EM2: 758-759
analysis methodsEM2: 825
analysis, scheme for........................EM2: 835-836
with aromatic rings.........................EM2: 52
basic structuresEM1: 751
bond energies................................EM2: 57
branched, defined See Branched polymer
carbon-chainEM2: 49
categorization...............................EM2: 63-64
chain entanglementsEM2: 64
chemical groups in naming....................EM2: 56-57
chemical properties..........................EM2: 61-62
chemical resistance..........................EM2: 62
chemical structureEM2: 48-52
chemical tests for..........................EM1: 285
chemistry of.................................EM2: 63-67
classificationEM2: 489
coefficients of linear thermal
expansionEM2: 456
composition, effect on properties........EM2: 566-567
conductivity.................................EM2: 62
definedEM1: 18 EM2: 31-32
deformationEM2: 680-681
electrical propertiesEM2: 62
entanglement, requiredEM1: 101
familiesEM2: 64-66
fatigue failure of...........................EM2: 741-750
as filled systemsEM1: 27
film-forming, as sizing agentsEM1: 122
formation, platinum, as electrical contact materialsA2: 846
general behavior ofEM2: 734

Polymer(s) (continued)
heterochainEM2: 49-52
high temperature resistant, polyimides
asEM1: 810-815
high-performance, thermal
characteristicsEM2: 559
hydrocarbonEM2: 49
inorganicEM2: 66
intermolecular forcesEM2: 64
key, summaryEM2: 561
liquid crystalline, para-aramid fibers
asEM1: 54
long-chain vs network........................EM2: 52
maleimide-terminated thermosettingEM2: 631
mechanical properties........................EM2: 60-61
in milligram quantities, analysisEM2: 836-837
miscibilityEM2: 496
modifications toEM2: 66-67
moisture absorptionEM1: 189
nadimide-terminated thermosettingEM2: 631
namesEM2: 52-56
optical properties...........................EM2: 62
organic, oxidation effect, electrical contact materialsA2: 841
permeabilityEM2: 61-62
photochemistry ofEM2: 777-780
powderedEM1: 102
properties, and structureEM2: 57-59
properties ofEM2: 57-62, 758
reference sourcesEM2: 405
resins, types...............................EM1: 43
rigid-rod fiber-formingEM1: 29-31
science, for engineersEM2: 48-62
solubilityEM2: 61
stiffnessEM2: 61
strengthEM2: 61
structureEM2: 57-59, 462-464, 571-572
structure, and electrical properties EM2: 462-464
structure, properties influencingEM2: 57-59
thermal properties...........................EM2: 59-60
thermoplastics, chemistry ofEM2: 63, 66
thermosets, chemistry ofEM2: 63
toughnessEM2: 61
typesEM2: 48
vs plastics, as termsEM2: 1
water absorptionEM2: 761
Polymer-base adhesives
Epon 1001/V115 films.....................EM3: 362, 363
moisture ingression on joint
performance.............................EM3: 362
Polymer-base bearing material
SP-21M Vespel properties....................A18: 549
Polymer-curing reactions
ATR monitoringA10: 120-121
Polymer-derived ceramics....................EM4: 223-226
ceramics from organosilicon polymers EM4:
223-226
ceramic matrix composites....................EM4: 224
coatingsEM4: 224-225
developmentEM4: 223
fibersEM4: 223-224
nonfugitive bindersEM4: 225-226
Polymer-on-metal (POM) construction
thermal expansion propertiesEL1: 619-622
Polymer-Plastics Technology and Engineering
as information sourceEM2: 93
Polymer-polymer mixtures
combination technologyEM2: 487-489
commercialEM2: 491-492
polymer selection/property
modificationEM2: 489-490
preparation ofEM2: 489
properties modification byEM2: 487-492
Polymeric binders, liquid
for sandA15: 211
Polymeric coatings
for oil/gas pipesA13: 1259
Polymeric composites
for RTM tooling..............................EM1: 168-169

SUBJECTS OF THE INDEXED VOLUMES: ASM Handbook (designated by the letter "A"): A1: Properties and Selection: Irons, Steels, and High-Performance Alloys (1990); A2: Properties and Selection: Nonferrous Alloys and Special-Purpose Materials (1990); A3: Alloy Phase Diagrams (1992); A4: Heat Treating (1991); A6: Welding, Brazing, and Soldering (1993); A7: Powder Metallurgy (1984); A8: Mechanical Testing (1985); A9: Metallography and Microstructures (1985); A10: Materials Characterization (1986); A11: Failure Analysis and Prevention (1986); A12: Fractography (1987); A13: Corrosion (1987); A14: Forming and Forging (1988); A15: Casting (1988); A16: Machining (1989); A17: Nondestructive Testing and Quality Control (1989); A18: Friction, Lubrication, and Wear Technology (1992). Metals Handbook, 9th Edition (designated by the letter "M"): M1: Properties and Selection: Irons and Steels (1978); M2: Properties and Selection: Nonferrous Alloys and Pure Metals (1979); M3: Properties and Selection: Stainless Steels, Tool Materials and Special-Purpose Materials (1980); M4: Heat Treating (1981); M5: Surface Cleaning, Finishing, and Coating (1982); M6: Welding, Brazing, and Soldering (1983). Engineered Materials Handbook (designated by the letters "EM"): EM1: Composites (1987); EM2: Engineering Plastics (1988); EM3: Adhesives and Sealants (1990); EM4: Ceramics and Glasses (1991); Electronic Materials Handbook (designated by the letters "EL"): EL1: Packaging (1989).

Polymeric impressed-current anodes **A13:** 469
Polymeric insulation
for wrought copper and copper alloy products **A2:** 258, 260
Polymeric lubricants, rolling-element bearings **A18:** 137
markets **A18:** 137
Polymeric paste systems
dielectric inks **EL1:** 346
electrically conductive inks **EL1:** 346
inks **EL1:** 345-346
resistive inks **EL1:** 346-347
solvents and polymers **EL1:** 346
Polymeric polysulfides
in concrete **A12:** 472
Polymeric substrates
materials **EL1:** 338-339
Polymeric thick-film systems
links **EL1:** 345-346
solvents and polymers **EL1:** 346
Polymerization *See also* Addition polymerization; Condensation polymerization; Copolymer polymerization; Copolymerization; Free-radical polymerization **EM3:** 22, 619-620, 729, 730
acrylate, free radical cure system **EL1:** 857
anionic **EM3:** 35
of aramid fibers **EM1:** 30
of bismaleimides **EM1:** 79
of butadiene and styrene, Raman analysis **A10:** 132
chemical composition **A17:** 21
cycloaliphatic epoxide, with Lewis acids **EL1:** 860
defined **EM1:** 18 **EM2:** 32
degree of, defined *See* Degree of polymerization
entropies **EM3:** 620
heats **EM3:** 620
induced by adhesives **EM3:** 35
mechanism, parylene coatings **EL1:** 791-792
microwave inspection **A17:** 20
of phenols, SERS studies of **A10:** 136
PMR-15, isomerization during **EM1:** 83
radiation **EL1:** 854
Raman analysis **A10:** 131
ring-opening **EL1:** 805
types **EM1:** 752
of unsaturated compounds **EL1:** 854-864
vapor-phase **EL1:** 759
vinyl esters **EM2:** 275
x-ray topography and synchrotron radiation studies **A10:** 365
Polymers *See also* Polymers, failure analysis of **A7:** 606-613 **A18:** 693 **EM3:** 22
abrasion resistance **A18:** 490
additives to carbon-graphite materials **A18:** 816
amorphous, mechanical and electrical relaxation times **EM3:** 40
analysis of **A10:** 9, 131-132, 639, 647-648
blends **A10:** 118, 405
block, SAS applications **A10:** 405
as coatings to improve glass strength **EM4:** 743-744
common, relative resistance to deterioration **A11:** 761
compaction, and metal filler **A7:** 607
compounding techniques **A7:** 606-607
crack propagation in **A11:** 762
crazing in **A11:** 759-760
crystallinity **EM3:** 409
curing agent's role in adhesive bonding **EM3:** 290
curing, ATR monitoring of **A10:** 115, 120-121
curing methods **EL1:** 786
damage dominated by tearing **A18:** 180
defined **A13:** 10
for dielectrics **EM3:** 378
dilational deformation types **A11:** 759
drawn films, molecular orientation determined **A10:** 120
effect of filler on properties **A7:** 606, 611-613
EFG determination of unsaturation in **A10:** 212
electrically conductive **A7:** 607-608
ESR analysis **A10:** 263
fractographs **A12:** 479-480
fracture/failure causes illustrated **A12:** 217
fretting wear **A18:** 248
grease additives **A18:** 125
Polymers (continued)
identification in vinyl film **A10:** 123-124
impregnation-pyrolysis technique, to prepare ceramic-matrix composites **EM4:** 224
impressed-current anodes **A13:** 922
infrared spectrum **A10:** 123
inserts for automotive transfer cases **A18:** 566-567
for interlevel insulators **EM3:** 378
irreversible deformation types **A11:** 759
linear polyethylene, crack growth mechanisms **A12:** 479
liquid chromatography monitoring of stability during aging **A10:** 649
liquid crystal, as reinforcement **EL1:** 605
liquid, metal powders dispersed in **A7:** 607
low molecular weight **EM3:** 41
network, cross-linked, coatings as **EL1:** 854
nondilational deformation types **A11:** 759
organic **EM3:** 40
for photoresists **EM3:** 378
and plasticizers identified in vinyl film **A10:** 123
polycarbonate sheet, quasi-brittle fatigue crack propagation **A12:** 479
polyimide, failure origin **A12:** 480
properties, parylene coatings **EL1:** 792-797
PVC and PVA, copolymer formation **A13:** 404
pyrolysis GC/MS analysis of **A10:** 647-648
for radiation masks **EM3:** 378
Raman analysis **A10:** 131-132
resins, use in ion exchange separation **A10:** 164
resistivity **A7:** 607, 608
SAXS/SANS analysis of **A10:** 405
seal materials **A18:** 550
silicone **EM3:** 40
sliding and adhesive wear **A18:** 237, 239-240, 241
break-in period **A18:** 240
linear wear **A18:** 239
PV limit **A18:** 239
rubber **A18:** 240
sliding severity **A18:** 240
wear properties of selected polymers **A18:** 241
sliding bearings and their applications **A18:** 516
solid-film, for plastic-encapsulated devices **EL1:** 220-221
solubility parameters **A11:** 761
standard surface conditions for sliding **A18:** 236
structure and morphology **A11:** 758
substrates ,thick film circuits **EL1:** 249
surface parameters **EM3:** 41
thermally conductive **A7:** 608-609
thick-film systems **EL1:** 346
to fabricate integrated circuits **EM3:** 378
viscoelasticity of **A11:** 758
wear due to friction **A11:** 764
work material for ion implantation **A18:** 858
Polymers, failure analysis of **A11:** 758-765
brittlelike fracture **A11:** 761
case studies **A11:** 762-765
crack propagation in polymers **A11:** 762
deformation and fracture mechanisms **A11:** 758-761
environmental stress cracking **A11:** 761
polymer structure and morphology **A11:** 758
viscoelasticity of polymers **A11:** 758
***Polymers for Engineering Applications* (Seymour)** **EM2:** 94
Polymers/Ceramics/Composites Alert **EM3:** 72
Polymethacrylates **EM3:** 86 **EM4:** 1093
Polymethacrylates (PMA)
infrared spectroscopy data **A18:** 301
pour-point depressants, detection by infrared spectroscopy of lubricants **A18:** 301
as viscosity improvers **A18:** 109, 110
Polymethyl methacrylate *See also* Acrylic plastic **A7:** 606
defined **EM1:** 18
Polymethyl methacrylate (PMMA) *See also* Acrylic plastic; Plastics **EM3:** 617
abrasion resistance **EM2:** 167
commercial, as engineering thermoplastic **EM2:** 447-448
critical surface tensions **EM3:** 180
defined **EM2:** 32
degradation **EM3:** 620
Polymethyl methacrylate (PMMA) (continued)
dependence of relaxation modulus on time **EM3:** 618
Lucite **EM3:** 280
medical applications **EM4:** 1009
Plexiglas **EM3:** 280
surface preparation **EM3:** 280, 291
thermal degradation **EM4:** 136
used as a binder in tape casting **EM4:** 137
Polymethyl methacrylate syrup
typical formulation **EM3:** 121
Polymethylmethacrylate (PMMA)
deformation zone friction **A18:** 36
friction coefficient data
liquid impingement erosion protection applications **A18:** 222
Polymethylpentene
surface preparation **EM3:** 291
Polymorph
defined **A10:** 679
Polymorphic transformations
by milling **A7:** 56
Polymorphism .... **A3:** 1 • 1 **EM3:** 22
defined **A9:** 14 **A10:** 679 **EM2:** 32
in uranium and uranium alloys **A9:** 476
Polynary uranium alloys **A13:** 817-818
Polynomial regressions
use in x-ray spectrometry calibration **A10:** 97
Polynuclidic elements
SSMS analysis **A10:** 145
Polyol-isocyanate resin binder system **A15:** 217, 238
Polyolefins **EM3:** 22, 75, 147
adhesive **EM3:** 290
catalytic effect by steel and copper **EM3:** 418
cooling rate and crystallinity **EM3:** 412
defined **EM2:** 32
environmental effects **EM2:** 426-427
highly crystalline prebond treatments **EM3:** 35
moisture effects
as substrate for silicones **EM3:** 136
surface preparation **EM3:** 278, 291
thermal degradation **EM2:** 423
Polyolerins
thermal degradation **EM4:** 136
Polyolester (POE)
high-vacuum lubricant applications **A18:** 155, 156
molecular structures **A18:** 156
properties **A18:** 155
Polyols **EM3:** 22, 108, 109
defined **EM2:** 32
epoxidized **EL1:** 818
in polyurethanes (PUR) **EM2:** 257
steps in manufacture of **EM3:** 203
as thermoplastic polyurethanes (TPUR) **EM2:** 204
as urethane coatings **A13:** 409
Polyoxymethylene (POM) *See also* Acetal (AC) resins
+ 15% polytetrafluoroethylene (PTFE), friction coefficient data **A18:** 73
+ polytetrafluoroethylene composite, wear properties **A18:** 241
defined **EM2:** 32
friction coefficient data **A18:** 73
moisture effects **EM2:** 768
polytetrafluoroethylene (PTFE)/glass, friction coefficient data **A18:** 73
sliding bearings, thick bonded to porous bronze with steel backing **A18:** 516
wear properties **A18:** 241
Polyoxymethylene gear wheel
failure of **A11:** 764, 765
Polyoxymethylenes (POM) **EM3:** 22
Polyperfluoroalkylether (PFPE)
addition to titanium alloys as lubricant **A18:** 778, 779
Polyphase alloys
defined **A13:** 46
Polyphenyl ether
properties **A18:** 81
Polyphenylene ether **EM3:** 576
Polyphenylene ether (PPE) *See also* Polyphenylene ether blends (PPE, PPO)
as blend **EM2:** 183-185
chemistry **EM2:** 65

Polyphenylene ether blends (PPE, PPO) *See also* Polyphenylene ether (PPE); Polyphenylene oxide (PPO); Thermoplastic resins
alloys and blends **EM2:** 183
applications **EM2:** 183-184
characteristics **EM2:** 184-185
competitive materials **EM2:** 184
costs and production volume **EM2:** 183
design guidelines **EM2:** 185
processing **EM2:** 184-185
as structural plastics **EM2:** 65
suppliers **EM2:** 185
Polyphenylene oxide (PPO) *See also* Polyphenylene ether blends (PPE, PPO)
as blend **EM2:** 183-185
chemistry **EM2:** 65-66
defined **EM2:** 32
as engineering plastic **EM2:** 430
Polyphenylene oxides (PPO) **EM3:** 22
Noryl polyphenylene oxide **EM3:** 279
surface preparation **EM3:** 279, 291
Polyphenylene sulfide (PPS)
+ 15% polytetrafluoroethylene (PTFE), friction coefficient data **A18:** 73
polytetrafluoroethylene (PTFE)/glass, friction coefficient data **A18:** 73
wear factor **A18:** 824
Polyphenylene sulfide (PPS) resins
in compression molding **EM1:** 562
defined **EM1:** 18
matrix processing **EM1:** 544
as thermosetting aerospace matrix **EM1:** 32-33
in woven sheeting **EM1:** 32, 33
Polyphenylene sulfides
for coating/ encapsulation **EL1:** 242
Polyphenylene sulfides (PPS) *See also* Thermoplastic resins **EM3:** 22, 601
applications **EM2:** 186
characteristics **EM2:** 186-190
chemical resistance **EM2:** 186
chemistry **EM2:** 66
competitive materials **EM2:** 186
composites **EM2:** 190
compound types **EM2:** 185
costs and production volume **EM2:** 186
critical surface tensions **EM3:** 180
crystallinity **EM2:** 189
curing **EM2:** 187
defined **EM2:** 32
electrical properties **EM2:** 189
as engineering plastic **EM2:** 430
film and fiber **EM2:** 180-190
interfacial shear strength of embedded fibers **EM3:** 394
nominal properties **EM2:** 188
processing **EM2:** 189
properties **EM2:** 187-189
suppliers **EM2:** 190
surface preparation **EM3:** 280, 291
thermal stability **EM2:** 186
vs polyamide-imides (PAI) **EM2:** 128
Polyphenylquinoxalines **EM3:** 163-168
for aerospace protective coating **EM3:** 164
chemistry **EM3:** 163
compared to polyimides **EM3:** 165
cost factors **EM3:** 164-165
cured properties **EM3:** 165-167
elevated-temperature plasticity **EM3:** 163
forms **EM3:** 163-164
functional types **EM3:** 163-164
general preparation **EM3:** 163
for high-temperature wire insulation **EM3:** 164
markets **EM3:** 164-165
non-crosslinked thermoplastic nature **EM3:** 163
pot life **EM3:** 167
processing parameters **EM3:** 167
properties **EM3:** 163, 165
shelf life **EM3:** 167
storage conditions **EM3:** 167
Polyphenylquinoxalines (continued)
suppliers **EM3:** 164
thermally crosslinkable **EM3:** 163, 164
thermooxidative stability temperature **EM3:** 166
thermoplasticity a limiting factor **EM3:** 163
for titanium bonding with advanced composites **EM3:** 163
uncured properties **EM3:** 165
Polyphosphate/HEDP
cathodic inhibitor **A13:** 496
Polyphosphates **A13:** 495
detected by ion chromatography **A10:** 661
Polyphosphonates
detected by ion chromatography **A10:** 661
Polyphthalate carbonate **EM3:** 576
Polypropylene **EM4:** 940
as binder for ceramic injection molding **EM4:** 173
as container for electrolytes **A9:** 51
crack propagation **A11:** 762
crazes in **A11:** 759
defined
effect on strain rate **A8:** 38, 42
isostatic, infrared linear dichroism spectroscopy of **A10:** 120
plastic zone **A11:** 762
transmission by **A10:** 100
waterjet machining **A16:** 522
Polypropylene fluxers **EL1:** 681
Polypropylene-cement system
fiber pull-out experiment **EM3:** 392
Polypropylenes (PP) *See also* Reinforced polypropylenes **EM3:** 22
adhesive bonding **EM3:** 290
for appliance seal and gasket applications **EM3:** 51
for auto battery casings **EM3:** 58
chemistry **EM2:** 65 **EM3:** 50
contact-angle testing **EM3:** 277
crystallinity **EM3:** 411
cure properties **EM3:** 51
defined **EM2:** 32
as engineering thermoplastic **EM2:** 446
as general-purpose polymer **EM2:** 65
lap-shear strength **EM3:** 276
mechanical keying **EM3:** 416
medical applications **EM3:** 576
methods of application **EM3:** 59
microstructure **EM3:** 407
physical properties **EM2:** 192
predicted 1992 sales **EM3:** 81
properties **EM3:** 50
for refrigerator and freezer cabinet sealing **EM3:** 59
reinforced **EM2:** 192-193 **EM3:** 22
stereoisomers of **EM2:** 58
structures of **EM2:** 65
suppliers **EM3:** 59
surface energies correlation **EM3:** 294
surface preparation **EM3:** 42, 280, 291
for water deflector sealing **EM3:** 57
Polyquinoxalines **EM3:** 161, 163
Polysaccharides
removal **EM4:** 137
Polysilastyrene
used in silicon-base ceramics **EM4:** 223
Polysilazane
application or function optimizing powder treatment and green forming **EM4:** 49
infiltrant to fill voids in RBSN material **EM4:** 296
used in silicon-base ceramics **EM4:** 223
Polysilicic acid sols **EM4:** 446
Polysilicon **EM3:** 593
Polysilicon-silicide (polycide) structures
integrated circuits **A11:** 773-774
Polysiloxane network film **EM3:** 626
Polystyrene *See also* Expanded polystyrene patterns
as amorphous polymer **A11:** 758
Polystyrene (continued)
and compressive yield **A11:** 759
crack propagation **A11:** 762
crazing in **A11:** 761
film, crazing in thin **A11:** 758
film, deformation behavior **A11:** 759
foamed, investment casting **A15:** 255
friction coefficient data **A18:** 73
as organic binder **A15:** 35
pattern assembly, investment casting **A15:** 257
pattern injection, investment casting **A15:** 256
as solder resist **A6:** 133
Polystyrene (PS)
as binder for ceramic injection molding **EM4:** 173
thermal degradation **EM4:** 136
Polystyrene blend
analysis of **A10:** 118
Polystyrene microspheres
measurement **A7:** 223-224
Polystyrene phase associating resins **EM3:** 183
Polystyrene, with benzene
for replicas **A12:** 180
Polystyrenes (PP) *See also* High-impact polystyrenes (PS, HIPS)
effect, trichlorofluoromethane **EM2:** 195
as engineering thermoplastic **EM2:** 446
as general-purpose polymer **EM2:** 64
high-impact **EM2:** 194-199
Polystyrenes (PS) **EM3:** 22, 75, 594
aramid composites **EM3:** 286
critical surface tensions **EM3:** 180
high-impact **EM3:** 22
microstructure **EM3:** 407
rubber-toughened **EM3:** 413
solvent cements **EM3:** 567
surface preparation **EM3:** 279-280
volume resistivity and conductivity **EM3:** 45
Polysulfide
adhesives **EM1:** 687
defined **EM1:** 18 **EM2:** 32
faying surface sealants **EM1:** 719
Polysulfide corrosion
oil/gas production **A13:** 1232-1233
Polysulfide sealants
one-part **EM3:** 190, 191
properties **EM3:** 191
two-part manually mixed **EM3:** 190, 191
two-part mechanically meter-mixed **EM3:** 190, 191
Polysulfide-base primers **EM3:** 640
Polysulfides **EM3:** 22, 44, 138-142, 193-197, 815
3-phenyl-1, 1-dimethyl urea **EM3:** 142
additives **EM3:** 139-140
additives and modifiers **EM3:** 196
advantages and disadvantages **EM3:** 675
for aerodynamic smoothing compounds **EM3:** 194
for aerospace industry **EM3:** 193, 194, 196
for aircraft assembly sealing **EM3:** 808
for aircraft inspection plates **EM3:** 604
for aluminum bonding **EM3:** 138
application parameters **EM3:** 196
applications **EM3:** 50, 56
automotive market applications **EM3:** 608
as body assembly sealants **EM3:** 609
for carbon bonding **EM3:** 138
for ceramics bonding **EM3:** 138
characteristics **EM3:** 53
chemical properties **EM3:** 52
chemical resistance **EM3:** 641
chemistry **EM3:** 50, 138, 193-194
for civil engineering **EM3:** 194
commercial forms **EM3:** 194, 195
competing with urethane sealants **EM3:** 205
as concrete adhesives **EM3:** 138
for concrete and mortar patch repair **EM3:** 138
for concrete crack repair **EM3:** 138
for construction industry **EM3:** 193-194, 196

Polysulfides (continued)
cost factors **EM3:** 139, 195, 196
cross-linking **EM3:** 138, 193
curatives for LP-epoxy reaction **EM3:** 142
cure properties **EM3:** 51
as curing agents for epoxy resins **EM3:** 184-185
for curtain wall construction **EM3:** 549
cycloaliphatic amine **EM3:** 142
degradation **EM3:** 679
DGEBA epoxy resin **EM3:** 142
Dicyandiamide **EM3:** 142
for electrically conductive sealants **EM3:** 194
epoxidized **EL1:** 818
epoxy resin **EM3:** 142
epoxy-terminated **EM3:** 140-141
for faying surface **EM3:** 604
for fiber-reinforced plastic (FRP) bonding, polysulfides **EM3:** 141, 142
fillers **EM3:** 139, 141
for fillets **EM3:** 604
formulations **EM3:** 676
for fuel tank sealants **EM3:** 194
for fuel tank structures (airframe) bonding **EM3:** 195
functional types **EM3:** 193-194
for galvanized steel bonding **EM3:** 141
for galvanized steel G-60 bonding **EM3:** 142
gas tank seam sealant **EM3:** 610
for glass bonding **EM3:** 138, 141, 142
glass transition temperature **EM3:** 139
for glass unit insulation **EM3:** 194-195, 196
for grouting compounds **EM3:** 138
for high-performance military aircraft **EM3:** 195
for insulated glass construction **EM3:** 46, 58
for integral fuel tank sealing **EM3:** 58
for lead-free automotive body solder **EM3:** 138
liquid epoxy concrete adhesive use **EM3:** 138, 140
LP-epoxy ratios **EM3:** 140, 141, 142
markets **EM3:** 194-195
mercaptan-terminated **EM3:** 138, 141
modifiers **EM3:** 139-140
mortar white silica (HDS-100) **EM3:** 139
for packaging **EM3:** 58-59
performance **EM3:** 674
physical properties of liquids **EM3:** 140
for plastics bonding **EM3:** 138
for plywood bonding **EM3:** 141, 142
pot life **EM3:** 139, 141, 142, 194, 196
for pressure sealants **EM3:** 194
processing parameters **EM3:** 196-197
properties **EM3:** 50, 139, 196, 677
for protecting electronic components **EM3:** 59
for quick-repair sealants **EM3:** 194
as reactive diluent **EM3:** 138
for rivets **EM3:** 604
sealant characteristics **EM3:** 57, 188
secondary seal for polyisobutylene **EM3:** 190
shelf life **EM3:** 141, 195, 196
silane Coupling agents **EM3:** 182
solvent use **EM3:** 140
for steel bonding **EM3:** 138
suppliers **EM3:** 138-139, 193, 195
surface preparation **EM3:** 291
for tack coat for concrete **EM3:** 138
tertiary amine **EM3:** 142
thermal properties **EM3:** 52
for truck trailer joints **EM3:** 58
for windshield sealing **EM3:** 194, 608
for wood bonding **EM3:** 138
Polysulfone resins
surface preparation **EM3:** 279
Polysulfone thermoplastic resin
and CM-X, epoxy, compared **EM1:** 103
defined **EM1:** 18
and fiber-resin composite **EM1:** 363-372
properties **EM1:** 364
Polysulfones (PSU) *See also* Thermoplastic resins **EM3:** 22, 576, 601
applications **EM2:** 200
characteristics **EM2:** 200-202
coatings for carbon/graphite **EM3:** 289
commercial forms **EM2:** 200
critical surface tensions **EM3:** 180
defined **EM2:** 32
design considerations **EM2:** 201
as engineering plastic **EM2:** 430

Polysulfones (PSU) (continued)
flammability **EM2:** 201
impact-modified properties **EM2:** 497
processing **EM2:** 201-202
properties **EM2:** 200-201
solvent cements **EM3:** 567
surface contamination **EM3:** 847
vs polyamide-imides (PAI) **EM2:** 128
Polyterephthalate **EM3:** 22
Polytetrafluoroethylene
as incendiary **A7:** 603
in pharmaceutical production facilities **A13:** 1227
Polytetrafluoroethylene (PTFE) **EM3:** 22, 223, 601
abrasion resistance **EM2:** 167
as additive to metalworking lubricants **A18:** 141
additives and modifiers **EM3:** 225
as bearing material **A18:** 754, 755
bronze-filled **A18:** 253
chemical compatibility **EM3:** 224
chemical compatibility of seals **A18:** 550
chemistry **EM2:** 66 **EM3:** 223
coating for gears, lubrication **A18:** 541
codeposited in electroplated coatings **A18:** 834-835
coefficient of friction **A18:** 824, 825
commercial forms **EM3:** 223
-compatable chemicals and solvents **EM2:** 115
for compression packings **EM3:** 225-226
cost factors **EM3:** 224, 226
critical surface tensions **EM3:** 180
damage caused by chip formation **A18:** 182
damage dominated by tearing in a piston pump **A18:** 180
defined **EM2:** 32
effect on wear factors **A18:** 824
filler for lip seals **A18:** 546, 550-551
filler for seals **A18:** 551
films **A18:** 117
flex-circuit usage **EM3:** 591
friction coefficient, tribotest example **A18:** 483, 484, 485
friction wear of metals **A18:** 239
GORE-TEX **EM3:** 225
Gylon **EM3:** 225
high-vacuum lubricant applications **A18:** 154
incorporation in thermoplastic composites **A18:** 820, 822-823
interfacial zone shear **A18:** 36
liquid impingement erosion protection applications **A18:** 222
markets **EM3:** 225
methods used for synthesis **A18:** 802
molding techniques **EM2:** 115
for printed board material systems **EM3:** 592
processing parameters **EM3:** 224
properties **A18:** 801
resin types and applications **EM3:** 223-224
sliding bearings **A18:** 516
sliding wear mechanisms in metal-matrix composites **A18:** 809-810
surface contamination **EM3:** 847
surface preparation **EM3:** 279, 290
tape **EM3:** 51, 53
in tape backings for silicone applications **EM3:** 134
thermal degradation **EM4:** 136
thermogravimetric analysis **A18:** 823
as thermoplastic fluoropolymer **EM2:** 115-119
for tool steel lubrication **A18:** 737-738
use in metal-on-polymer total hip replacements **A18:** 657, 661
vs polyamide-imides (PAI) **EM2:** 128
wafers used during leaky Lamb wave technique **EM3:** 780
Polytetrafluoroethylene (PTFE) insulation
for wrought copper products **A2:** 258
Polytetrafluoroethylene fluorocarbon
prebond treatment **EM3:** 35
Polytetrafluoroethylene vessels
for sample dissolution treatment **A10:** 165, 166
Polytetrafluoroethylene-encapsulated silicone rubber
for platen seals **EM3:** 694
Polytetranuoroethylene (PTFE) *See also* Teflon
as laminating resin, properties **EL1:** 534-535
structural repeat unit **EL1:** 605

Polythioethers, terminal mercapto groups in
Raman analysis **A10:** 132
Polythionic acid stress-corrosion cracking
austenitic stainless steels **A12:** 354
Polythionic acids
SCC testing in **A13:** 273
SCC testing of stainless steels in **A8:** 528-529
Polytitanates **EM4:** 60
Polytitanocarbosilane
for making Tyranno fiber **EM4:** 223
Polyurea **A18:** 126, 129 **EM3:** 203
Polyurethane *See also* Isocyanate plastics; Urethane plastics **A7:** 447, 606
adhesives **EM1:** 684
coating for composite constructions against liquid impingement erosion **A18:** 222
damage dominated by tearing **A18:** 180
defined **EM1:** 18
friction coefficient data **A18:** 73
medical applications **EM4:** 1009
as organic binder **A15:** 35
pattern block materials, use **A15:** 194
in RRIM technology **EM1:** 121
as thermocouple wire insulation **A2:** 882
Polyurethane resins **M5:** 473, 496, 498, 502
Polyurethanes *See also* Urethane coatings; Urethanes
and acrylic, silicone, epoxy coatings, compared **EL1:** 775
as coating/encapsulant **EL1:** 759
as conformal coating **EL1:** 761, 763
electrical properties **EL1:** 822
potting in **EL1:** 824
stripping methods **M5:** 19
Polyurethanes (PUR) *See also* Isocyanate plastics; Thermoplastic polyurethanes (TPUR); Thermoplastic urethanes; Thermosetting resins
applications **EM2:** 258-260
characteristics **EM2:** 260-264
chemical structure **EM2:** 204
chemistry **EM2:** 66
commercial forms **EM2:** 257-258
competitive materials **EM2:** 259-260
costs and production volume **EM2:** 258
defined **EM2:** 32
environmental effects **EM2:** 428
flexible, foams **EM2:** 603
as injection-moldable **EM2:** 322
as low-temperature resin system **EM2:** 439
processing **EM2:** 262-264
properties **EM2:** 260-263
resin compound types **EM2:** 264
solid **EM2:** 630
as structural plastic **EM2:** 66
suppliers **EM2:** 164
thermoplastic (TPUR) **EM2:** 203-208, 258
thermoset **EM2:** 630
Polyvinyl acetals **EM3:** 22
bonding applications **EM3:** 82
defined **EM2:** 32
properties **EM3:** 82
suppliers **EM3:** 82
Polyvinyl acetate **A7:** 606
Polyvinyl acetate (PVA)
binder used in silicon powder **EM4:** 237
Polyvinyl acetate (PVAC) *See also* Latex **EM3:** 22, 75, 80, 86, 210-214
applications **EM3:** 56
characteristics **EM3:** 90
defined **EM2:** 32
for plumbing seals **EM3:** 608
properties **EM3:** 677
residential applications **EM3:** 675
typical properties **EM3:** 83
Polyvinyl acetate emulsion adhesive **EM3:** 22
for composite panel construction **EM3:** 46
Polyvinyl acetate-base caulks **EM3:** 188, 190
Polyvinyl alcohol
as sample binding agent **A10:** 94
Polyvinyl alcohol (PVA)
application or function optimizing powder treatment and green forming **EM4:** 49
as binder **EM4:** 474
as binder for ceramic coatings **EM4:** 955
as binder for spray drying before dry pressing **EM4:** 146

Polyvinyl alcohol (PVA) (continued)
as binder in slurry preparation for spray drying ... **EM4:** 103, 107
diazo-sensitized, for screen making ... **EM4:** 472
removal ... **EM4:** 137
Polyvinyl alcohol (PVAL) ... **EM3:** 22
coatings for carbon/graphite ... **EM3:** 289
defined ... **EM2:** 32-33
Polyvinyl and acrylic resin emulsions
applications ... **EM3:** 45
characteristics ... **EM3:** 45
industrial applications ... **EM3:** 567
for packaging ... **EM3:** 45
Polyvinyl butyral (binder)
batch weight of formulation when used in oxidizing sintering atmospheres ... **EM4:** 163
Polyvinyl butyral (PVB) *See also* Polyvinyl acetals ... **EM3:** 22
defined ... **EM2:** 33
in laminated glass ... **EM4:** 453
thermal degradation ... **EM4:** 136
used as a binder in tape casting ... **EM4:** 137
Polyvinyl butyral in pine oil
as media for screening and stamping processes ... **EM4:** 475
Polyvinyl butyral-phenolics ... **EM3:** 75
advantages and limitations ... **EM3:** 79-80
bonding applications ... **EM3:** 79-80
Polyvinyl carbazole ... **EM3:** 23
defined ... **EM2:** 33
Polyvinyl chlordic acetate
defined ... **EM2:** 33
Polyvinyl chloride ... **A7:** 447, 606, 609, 610
sulfuric acid corrosion ... **A13:** 1153
waterjet machining ... **A16:** 522
Polyvinyl chloride (PVC) ... **EM3:** 23, 594
applications ... **EM3:** 56
contact-angle testing ... **EM3:** 277
critical surface tensions ... **EM3:** 180
medical applications ... **EM3:** 576
solvent cements ... **EM3:** 567
substrate cure rate and bond strength for cyanoacrylates ... **EM3:** 129
surface preparation ... **EM3:** 291
verifilm material ... **EM3:** 736, 737
Polyvinyl chloride acetate ... **EM3:** 23
Polyvinyl chloride coatings
steel fence wire ... **M1:** 271
steel sheet ... **M1:** 176
Polyvinyl chloride fluxers ... **EL1:** 681
Polyvinyl chloride plastisol ... **EM3:** 59
shear strength of mild steel joints ... **EM3:** 670
Polyvinyl chlorides (PVC) *See also* Dry blend; Thermoplastic resins
abrasion resistance ... **EM2:** 167
alloys and blends ... **EM2:** 209
applications ... **EM2:** 209
characteristics ... **EM2:** 209-212
commercial forms ... **EM2:** 209
competitive materials ... **EM2:** 209
costs and production volume ... **EM2:** 209
custom compounding ... **EM2:** 210
defined ... **EM2:** 33
dimensional stability ... **EM2:** 210
as engineering thermoplastic ... **EM2:** 447
extrusion ... **EM2:** 211-212
as general-purpose polymer ... **EM2:** 64-65
impact-modified properties ... **EM2:** 497
injection molding ... **EM2:** 210-211
plasticizer effects ... **EM2:** 496
processing ... **EM2:** 210-212
resin compound types ... **EM2:** 212
in rotational molding ... **EM2:** 361
suppliers ... **EM2:** 212-213
thermal and related properties ... **EM2:** 447
thermal degradation ... **EM2:** 423-424
vinyl degradation ... **EM2:** 212
vs polyethylene (PE), usage ... **EM2:** 209
Polyvinyl fluoride
chemistry ... **EM2:** 66
surface preparation ... **EM3:** 291
Polyvinyl formal
defined ... **EM2:** 33
Polyvinyl formal (PVF) ... **EM3:** 23
Polyvinyl formal as a mounting material ... **A9:** 29-30
for metal-matrix composites ... **A9:** 588
Polyvinyl formal, with ethylene dichloride
for replicas ... **A12:** 180
Polyvinyl formal-phenolics ... **EM3:** 75
advantages and limitations ... **EM3:** 79-80
Polyvinyl methyl ether
used as modifiers ... **EM3:** 121
Polyvinyl-chloride
as a mounting material ... **A9:** 29-30, 44, 49, 53-54
Polyvinylalkylethers ... **EM3:** 6, 82-83
Polyvinylchloride
determined in vinyl film ... **A10:** 123-124
Polyvinylchloride (PVC)
as insulation, copper and copper alloy products ... **A2:** 258, 260
as thermocouple wire insulation ... **A2:** 882
for tool steel lubrication ... **A18:** 738
Polyvinylidene chloride (PVDC) ... **EM3:** 23
defined ... **EM2:** 33
Polyvinylidene fluoride (PVDF) ... **EM3:** 23
chemistry ... **EM2:** 66
defined ... **EM2:** 33
as fluoropolymer ... **EM2:** 115-119
Kynar ... **EM3:** 279
Kynar piezoelectric film ... **EM3:** 454
surface preparation ... **EM3:** 279
Polyxylyienes *See also* Parylene coatings
for coating/encapsulation ... **EL1:** 242
POM *See* Polyoxymethylenes
Pontachrome Black TA *See* Eriochrome Black T
Pooling technique
S-N fatigue relations ... **EM1:** 441
Pop-in precracking methods ... **A8:** 517
Popoff
on porcelain enamel steel sheet ... **M1:** 179
Poppet valves
check-, redesign of ... **A11:** 70-71
stems, fracture in ... **A11:** 320-321
thermal fatigue failure ... **A11:** 289
Poppet valves, steel
aluminum coating process ... **M5:** 335, 339-341
Population *See also* Sample
defined ... **A8:** to **A10:** 679
for determining design allowables ... **A8:** 662-663
effect of measurement process on ... **A8:** 624
of excited nuclear level, Mössbauer spectroscopy ... **A10:** 288
first/second movements of ... **A8:** 628
mean ... **A8:** 628
means , differences, with different standard deviations ... **A8:** 711
means, differences, with similar standard deviations ... **A8:** 709-711
means, multiple, differences ... **A8:** 711-712
percentiles ... **A8:** 629
skewed or asymmetric ... **A8:** 629
statistical ... **A8:** 624
symbol for ... **A8:** 629
target and parent, in random sampling ... **A10:** 12
variance ... **A8:** 628
Population mean ... **A18:** 481
Population parameters ... **A18:** 481
Population standard deviation ... **A18:** 481, 482
Porcelain ... **EM4:** 4
absorption ... **EM4:** 4
carbides for machining ... **A16:** 75
composite restorative material (dental), combined with ... **A18:** 670
composition ... **EM4:** 4, 5, 45
electrical, composition ... **EM4:** 5
fracture surface ... **EM4:** 644
Porcelain (continued)
fused-to-metal, for crowns (dental) ... **A18:** 673
glazes ... **EM4:** 1061
glazing ... **EM4:** 4
hard, composition ... **EM4:** 5
honing stone selection ... **A16:** 476
process ... **EM4:** 4
products ... **EM4:** 4
properties ... **EM4:** 4
properties (PFM) ... **A18:** 666
properties of fired ware ... **EM4:** 45
simplified composition or microstructure (PFM) ... **A18:** 666
steatite, composition ... **EM4:** 5
tender, composition ... **EM4:** 5
versus acrylics for denture teeth ... **A18:** 673-674
Porcelain denture teeth
properties ... **A18:** 666
simplified composition or microstructure ... **A18:** 666
Porcelain enamel ... **EM4:** 937-942
alkaline clean-rinse-neutralize method ... **EM4:** 937
appplications ... **EM4:** 937, 939
electronic ... **EM4:** 939
home laundry equipment ... **EM4:** 940-942
household ... **EM4:** 939-940
"clean only" preparation system ... **EM4:** 937, 938
coating materials ... **EM4:** 937
dry applications method ... **EM4:** 1065
electrostatic dry powder process ... **EM4:** 937
future outlook ... **EM4:** 942
heavy metal release performance rating ... **EM4:** 1065
history and development ... **EM4:** 937
low-carbon steel compositions for substrates ... **EM4:** 938
metal substrates ... **EM4:** 937-938
powder application ... **EM4:** 940
product categories ... **EM4:** 939-940
service properties ... **EM4:** 938-939
service temperature limits ... **EM4:** 939
Porcelain enamel on iron sheet ... **A9:** 201
Porcelain enameled sheet steel
illumination of ... **A9:** 199
single coating on extra-low carbon steel ... **A9:** 201
specimen preparation ... **A9:** 198
Porcelain enameling ... **M1:** 177-180 **M5:** 509-531
abrasion resistance ... **M5:** 526-527, 529
abrasive blasting process ... **M5:** 515
acid resistance ... **M5:** 509-511, 520-521, 525-526, 528-529
adherence ... **M5:** 526-527
aging stability, slips ... **M5:** 524
alkaline cleaning process ... **M5:** 514-516
alkaline resistance ... **M5:** 510-511, 525-528
aluminum *See* Aluminum, porcelain enameling of
appearance, indoor exposure uses ... **M5:** 525-526
auxiliary coating procedures ... **M5:** 519
brushing method ... **M5:** 519
carbon boiling ... **M1:** 177-179
as cast coating ... **A15:** 563-564
cast iron *See* Cast iron, porcelain enameling of
cast irons ... **M1:** 104-105
chemical resistance ... **M5:** 525-529
chipping resistance ... **M5:** 527, 529
color ... **M5:** 509, 522-523, 527-529
matching and control ... **M5:** 522-523
specifications for ... **M5:** 527
types ... **M5:** 509
consistency ... **M5:** 523
continuity ... **M5:** 527, 529-530
corrosion protection provided by ... **M1:** 754
corrosion resistance ... **M5:** 525-526
cover-coat enamels ... **M5:** 509-513, 515-517, 520-521, 523-524, 530
in-process repair of ... **M5:** 524
types ... **M5:** 509-510
design parameters ... **M5:** 524-525

Porcelain enameling (continued)
dipping method **M5:** 516-517, 523, 530
distortion, enameled parts **M5:** 523-524
drain time, measuring **M5:** 523
dry processes **M5:** 510-511, 517-521, 523, 530
drying process **M5:** 518-519
electrical properties, enamels **M5:** 526
electrodeposition method **M5:** 518
electrostatic spray processes **M5:** 517-518
enamel types, mixed-oxide compositions **M5:** 509-511
enamelability of steel **M5:** 512-513
etching process **M5:** 514-515
expansion patterns, metal and enamel **M5:** 525
firing time **M5:** 521-522, 530
flatness, enameled parts **M5:** 523-525
flow coating method **M5:** 516-518, 530
formability considerations, sheet steel **M5:** 512
frits **M5:** 509- 12, 523-524, 530
mixed-oxide compositions **M5:** 509-511
particle size **M5:** 523-524
preparation of **M5:** 510-512
weight **M5:** 523
furnaces **M5:** 519-521
batch **M5:** 519-521
continuous **M5:** 520
forced convection heating **M5:** 520-521
intermittent **M5:** 520
gas evolution during **M1:** 179
gloss, specifications for **M5:** 527-528
glossary of terms **M5:** 530-531
grinding and blending, frits **M5:** 510-511
grinding process, surface preparation **M5:** 524
ground-coat enamels **M5:** 509-512, 519-521, 523-524, 530
in-process repair of **M5:** 524
types **M5:** 509-510
hardness, enamel **M5:** 525-526
heat treatment **M1:** 182
hot water resistance **M5:** 530
HSLA steel for **M1:** 183, 188
imperfections **M1:** 182
low-carbon steel sheet and strip surface condition for **M1:** 157
manganese content distribution ... **M1:** 189, 195, 196
mechanical properties **M1:** 188-198
mechanical properties enamels **M5:** 525-527
metals suitable for, characteristics and selection factors **M5:** 512-514
methods **M5:** 516-519
mill additions, frits **M5:** 511-512
mixed oxide composition enamels **M5:** 509-511
nickel deposition cycle **M5:** 514-515
no nickel/no pickle system **M5:** 515
organic solvent resistance **M5:** 529
particle size, frits **M5:** 523-524
pickling process **M5:** 514-515
plumbing fixture standards **M5:** 527
process variables and control **M5:** 521-524
production of steel sheet for **M1:** 178-179
properties of enamels **M5:** 525-527
reinforcing process **M5:** 519
repairs, in-process **M5:** 524
rigidity, enameled parts **M5:** 510, 524
sag, characteristics **M5:** 512-513, 523-525
slips **M5:** 511, 516-519, 523-524, 531
aging stability **M5:** 524
consistency **M5:** 523
specific gravity **M5:** 523
spalling resistance **M5:** 527
specific gravity, slips **M5:** 523
specifications and standards **M5:** 526-530
spraying method **M5:** 517-518, 520-521, 523
stainless steel **M5:** 512-513
steel *See* Steel, porcelain enameling of
steel selection, factors, in **M1:** 177, 179-180
strength considerations, steel **M5:** 513-514, 526
stress patterns, metal and enamel **M5:** 525-526
surface defects, effects of **M5:** 512
surface imperfections **M1:** 177, 179
surface preparation for **M5:** 514-516, 527
tests related to **M5:** 527
suspension stability **M5:** 523
temperature, service **M5:** 509-510, 520-522, 525-526
high, resistance to **M5:** 509-510, 525-526
terminology, glossary of **M5:** 530-531

Porcelain enameling (continued)
testing methods **M5:** 523, 526-530
thermal shock resistance **M5:** 509-511, 525, 527, 529
thickness, coating **M5:** 519, 521, 527
torsion resistance **M5:** 525-527, 529
types of steel sheet for **M1:** 177-178
typical applications **M1:** 177-179
water resistance **M5:** 510-511, 530
weather resistance **M5:** 509-511, 525, 526-529
weight, deposited frit **M5:** 523
weldability characteristics substrate metal **M5:** 514
wet processes ... **M5:** 510-512, 517, 519, 521, 523, 531
workpiece size, maximum **M5:** 524-525
Porcelain enamels **A13:** 446-452
applications **A13:** 446, 449
coating evaluation **A13:** 450-452
corrosion resistance **A13:** 449-450
enameling process **A13:** 447-448
fused to dental alloys **A13:** 1352-1356
process variables **A13:** 448-449
service temperatures, maximum **A13:** 450
surface preparation for **A13:** 447
test methods, specifications, standards **A13:** 450
types **A13:** 446-447
wet-process, workability of **A13:** 449
Porcelain fused to metal (dental) alloys
of precious metals **A2:** 696
Porcelain overlay dental restorations
powder used **A7:** 573
Porcelain-enameled metal
as substrate material **EL1:** 106, 249, 337
Porcelainized metal interconnect boards
thermal expansion **EL1:** 615
Pore
defined **A11:** 8
definition **EM4:** 633
Pore pressure rupture testing
of green compacts **A17:** 54
Pore size *See also* Particle pore size **A7:** 9
control in roll compacting **A7:** 401
and green strength **A7:** 303
measured by mercury porosimetry **A7:** 262, 266-270
and pressure **A7:** 299
Pore size distribution **A7:** 9
as function of pressure **A7:** 299
measured by mercury -270 porosimetry **A7:** 262, 266
range **A7:** 9
Pore size, fine/coarse sands
compared **A15:** 223
Pore volume of cemented carbides **A9:** 275
Pore(s) **A7:** 9
area **A7:** 9
blind, measuring **A7:** 264-265
channels **A7:** 9
constrictive, in mercury porosimetry **A7:** 269
forging mode and stress conditions on **A7:** 414
formation **A7:** 9
forming material **A7:** 9
interconnected or isolated **A7:** 486, 487
rounding **A7:** 486, 488
size *See* Pore size
size distribution *See* Pore size distribution
structure **A7:** 9, 312
volume of **A7:** 265, 464
wall **A7:** 9
Pores
radiographic methods for **A17:** 29
Porosimeter **A7:** 9, 266-270
Porosimetry **EM4:** 71
Porosity *See also* Density; Gas porosity; Hydrogen porosity; Microporosity; Pinhole porosity; Shrinkage cavities; Shrinkage porosity; Surface porosity; Voids **A6:** 1073, 1079 **A7:** 9 **M6:** 839-840 **EM3:** 23 **EM4:** 580
in 60Ag-40Ni **A9:** 558
in 85Ag-15Cd0 .... **A9:** 557
in 90W-10Ag **A9:** 561
adhesive-bonded joints **A17:** 61
in aluminum alloys **A9:** 358
in aluminum-silicon alloys **A15:** 164-165
as-cast titanium castings **A15:** 828
blowholes **M6:** 839
brittle fracture from **A11:** 85

Porosity (continued)
by Fisher sub-sieve sizer **A7:** 230-232
by thermal neutron radiography **A17:** 39
in carbon and alloy steels, revealed by macroetching **A9:** 173
carbon and nitrogen penetration and **A7:** 454
cast aluminum alloys **A12:** 67, 405-406
in castings **A11:** 24
causes **M6:** 839-840
cemented carbides **A2:** 958 **A7:** 779 **A16:** 79
closed **EM4:** 580, 582
closed, powder processing and **A7:** 435
cluster **A11:** 413
of composites **EM1:** 35-36
computed tomography (CT) **A17:** 36
constitutive equations **A14:** 417-418
content, as quality control variable **EM1:** 730
in copper alloy ingots **A9:** 643
in copper alloys **A15:** 86, 464
defined ... **A9:** 14 **A11:** 8 **A15:** 9 **EM1:** 18-19 **EM2:** 33
definition **A6:** 1212 **M6:** 13 **EM4:** 580
determination in cemented carbides **A9:** 274
as discontinuities, defined **A12:** 65, 67
ductile iron cleavage fracture from **A12:** 227
effect, hot isostatic pressing **A15:** 540-541
effect, insoluble particles **A15:** 142
effect of cooling rate **M6:** 44
effect of pressure on compacts **A7:** 269
effect on bond testing **A17:** 63
effect on fatigue behavior **M1:** 682
effect on secondary operations **A7:** 451
effect, polar backscattering **A17:** 24
in electrogas welds **A11:** 440
in electron beam welds **A11:** 445
of electroslag **A11:** 440
evaluated by magnetic bridge sorting **A7:** 491
in fatigue fractures **A11:** 128
fine, as casting defect **A15:** 547
in flash welds **A11:** 442
flux effects **A15:** 448
in friction welds **A11:** 444
from inclusions **A15:** 88
gas, and subsurface discontinuities **A11:** 120
gas, causes **A15:** 87
gas hole, zirconium alloys **A15:** 838
gas, in arc-welded aluminum alloys **A11:** 434
gas, in copper/copper alloy castings **A17:** 534-53
gas, in iron castings **A11:** 356-357
gas, overcoming **A15:** 86
and green strength **A7:** 303
gross, defined **A15:** 6
and hardness of P/M materials **A7:** 262
herringbone **M6:** 839
hot isostatic pressing for **A15:** 263-264
as hydrogen damage **A12:** 124
hydrogen, in aluminum **A15:** 747
hydrogen, in aluminum casting alloys **A2:** 134-135
hydrogen, in copper alloys **A15:** 464
I-pores **A7:** 299
inaccessible **A7:** 262
induced, oxide formation **A15:** 91
interdendritic, wrought aluminum alloys **A12:** 431
internal, effect on compressibility **A7:** 286
internal, low compressibility **A7:** 298
isolated **A11:** 413
Kirkendall **A7:** 314
of laser beam welds **A11:** 447
linear **A11:** 413
liquid penetrant inspection **A17:** 71, 8
methods to determine **A7:** 262-271
microwave inspection **A17:** 21
mold, grain size distribution effect **A15:** 208-209
and mold permeability **A15:** 209
in molybdenum sintered compact **A9:** 562
open **EM4:** 580-582
oxide scale **A13:** 72
P/M steels **M1:** 333-335, 337, 343, 344
and paint adhesion **A7:** 459
permanent mold casting **A15:** 279, 285
of plated coatings **A13:** 425-426
in polymers, control of **A7:** 606
and pouring temperature **A15:** 283
of powder forgings, mechanical effects **A14:** 203
in powder metallurgy materials **A9:** 503

Porosity (continued)
in powder metallurgy materials, effect of polishing on ... **A9:** 506
in powder rolling ... **A7:** 401
preserving, in machining ... **A7:** 461
prevention ... **M6:** 44
in electron beam welds ... **M6:** 630
radiographic appearance ... **A17:** 35
random ... **A12:** 66, 67
relationship to hydrogen in aluminum alloys ... **A9:** 633
residual interparticle during sintering ... **A7:** 314
residual, powder forgings ... **A14:** 193
in resistance welds ... **A11:** 440-441
in rhenium and rhenium-bearing alloys ... **A9:** 447
in rolling ... **A14:** 358
and shear strength ... **A7:** 697
shrinkage ... **A12:** 405-406
shrinkage, aluminum casting alloys ... **A2:** 136
shrinkage, effect on fatigue strength ... **A11:** 120
shrinkage, in iron castings ... **A11:** 354
solidification, wrought aluminum alloys ... **A12:** 431
sources ... **M6:** 44
as squeeze casting defect ... **A15:** 325
in steel bar and wire ... **A17:** 54
in steel pipe ... **A17:** 56
stress conditions on ... **A14:** 190
surface, blast cleaning healing ... **A15:** 506
susceptible metals ... **M6:** 44
and tap density ... **A7:** 276
thermally induced ... **A7:** 181
in tin bronzes ... **A2:** 348
of tin coatings ... **A13:** 78
titanium and titanium alloy castings ... **A2:** 638-639
in titanium and titanium alloys, associated with high interstitial defects ... **A9:** 459
titanium BE compact ... **A2:** 649
in titanium powder ... **A7:** 164, 165
in tool steels ... **A7:** 427-428
total, and I-pores, V-pores ... **A7:** 299
types in arc welds ... **A11:** 413
of ultrahigh molecular weight polyethylenes (UHMWPE) ... **EM2:** 170
V-pores ... **A7:** 299
void volume ... **EM4:** 580
as volumetric flaw ... **A17:** 5
weld, corrosivity ... **A13:** 344
in weldments ... **A9:** 578, 581 **A17:** 582-58
wormholes ... **M6:** 839
wrought aluminum alloys ... **A12:** 422, 425
in zinc and zinc alloys, determining ... **A9:** 490
Porosity comparison chart ... **A9:** 274
Porosity in
arc welds of
coppers ... **M6:** 402-403
magnesium alloys ... **M6:** 435
nickel alloys ... **M6:** 442-443
nickel-based heat-resistant alloys ... **M6:** 363
stainless steels ... **M6:** 324
electrogas welds ... **M6:** 244
electroslag welds ... **M6:** 233-234
gas metal arc welds ... **M6:** 172
of aluminum alloys ... **M6:** 386
oxyacetylene welds of cast irons ... **M6:** 601
oxyfuel gas welds for hardfacing ... **M6:** 783
resistance welds of aluminum alloys ... **M6:** 543-544
shielded metal arc welds ... **M6:** 92
wormhole ... **M6:** 92-93
solder joints ... **M6:** 1090
solid-state welds ... **M6:** 677
submerged arc welds ... **M6:** 127-128
Porosity sealing ... **EM3:** 54
low viscosity methacrylate resins ... **EM3:** 54
styrene-base unsaturated polyesters ... **EM3:** 54
water solutions of sodium silicates ... **EM3:** 54
Porous
applications ... **A7:** 449
bearings ... **A7:** 9, 17, 704, 706
coatings ... **A7:** 659-663
electrodes ... **A7:** 308
metal filters ... **A7:** 17
nickel products ... **A7:** 395-397
orthopedic implants ... **A7:** 657-659
particles ... **A7:** 233, 234
parts ... **A7:** 105, 451, 696-700, 731
sinter cake, hammer milling of ... **A7:** 70
solids, penetration by mercury ... **A7:** 266
Porous and reconstructed glasses ... **EM4:** 427-431
applications ... **EM4:** 429, 431
composition of leachable alkali-borosilicate glasses ... **EM4:** 427, 428
graded seals ... **EM4:** 431
leaching ... **EM4:** 427-428
phase separation ... **EM4:** 427
processing ... **EM4:** 427
properties of porous glasses ... **EM4:** 428-430
adsorption of water vapor ... **EM4:** 429
cleaning ... **EM4:** 429
commercially available compositions ... **EM4:** 428
controlled-pore glass ... **EM4:** 429-430
enlarging pores ... **EM4:** 428, 429
removal of OH^- groups ... **EM4:** 429
reconstructed glass products
colored reconstructed glasses ... **EM4:** 431
transmittance properties ... **EM4:** 430-431
Porous bearing ... **A18:** 529-530
defined ... **A18:** 15
Porous bronze filters
fabrication ... **A2:** 402
powders ... **A2:** 401
properties and applications ... **A2:** 402
Porous chromium plating ... **M5:** 183
Porous filters
mixing operations ... **EM4:** 98
Porous materials
scanning electron microscopy used to study ... **A9:** 99-100
Porous molds
defined ... **EM2:** 33
Porous plug degassing
for hydrogen removal ... **A15:** 461-462
Porous region
definition ... **EM4:** 633
Porous seam
definition ... **EM4:** 633
Porous specimens
electroless plating ... **A9:** 32
mounting ... **A9:** 31
Port *See also* Cold chamber machine; Plunger defined
defined ... **A15:** 9
Portability
of composites ... **EM1:** 37
Portable abrasive blasting equipment ... **M5:** 89
Portable Brinell hardness tester ... **A8:** 87-88
Portable cutting machines ... **M6:** 907
Portable equipment *See also* Equipment
eddy current inspection ... **A17:** 18
for liquid penetrant inspection ... **A17:** 7
for magnetic particle inspection ... **A17:** 92, 11
Portable magnetic particle inspection equipment ... **A7:** 575
Portable power tools
with self-lubricating bearings ... **A7:** 705
Portable resistance welding machines ... **M6:** 474-475
Portable Rockwell hardness tester ... **A8:** 78, 79
Portable ultrasonic fatigue testing ... **A8:** 242
Portascan ... **EM3:** 758
Porter bars
for open-die forging ... **A14:** 63
Portland cements and concrete ... **EM4:** 918-924
admixtures ... **EM4:** 920-921
Portland cements and concrete (continued)
applications ... **EM4:** 923-924
and benefits from characteristics ... **EM4:** 922
blended cement ingredients other than portland-cement clinker ... **EM4:** 919
cement paste simulated structure in concrete ... **EM4:** 924
characteristics in which improvements benefit specific applications ... **EM4:** 922
characteristics of cements ... **EM4:** 918, 920
chemistry and physics of cements ... **EM4:** 918-920
composition ... **EM4:** 919
compressive and tensile strength ranges of moist-cured concretes ... **EM4:** 922
concrete microstructure (reflected light micrograph) ... **EM4:** 923
fineness ranges ... **EM4:** 919
new emerging materials ... **EM4:** 923-924
portland cement clinkers microstructure ... **EM4:** 920
portland cement paste microstructure ... **EM4:** 921
properties and characteristics ... **EM4:** 921-923
standards ... **EM4:** 924
tests ... **EM4:** 924
U.S. consumption of cement and other construction materials (1950-1987) ... **EM4:** 923
uses ... **EM4:** 918, 919
Portland cement ... **EM4:** 10-12
air-entraining ... **EM4:** 12
analysis of free lime ... **A10:** 179
applications ... **EM4:** 12
ASTM specifications ... **EM4:** 12
blast-furnace slag ... **EM4:** 12
cement chemistry ... **EM4:** 11
color ... **EM4:** 11, 12
composition ... **EM4:** 11
definition ... **EM4:** 12
development ... **EM4:** 10
expanding cement ... **EM4:** 12
flame emission sources for ... **A10:** 30
hydration ... **EM4:** 11-12
manufacturing process ... **EM4:** 10-11
masonry cement ... **EM4:** 12
microstructure ... **EM4:** 11
in oil paints ... **A13:** 443
oil well cement ... **EM4:** 12
Portlandite ... **EM4:** 12
Ports, evacuating
explosive forming ... **A14:** 638-639
Poschenrieder analyzer
in ECAP analysis ... **A10:** 597
Position measurement
by ultrasonic inspection ... **A17:** 27
Position-sensitive detector
abbreviation for ... **A10:** 691
effect in double-crystal spectrometry ... **A10:** 372-374, 377
neutron diffraction ... **A10:** 422
Positioned weld
definition ... **M6:** 13
Positioners
shielded metal arc welding ... **M6:** 79-81
Positioning
of explosive charges, by neutron radiography ... **A17:** 39
of markers and penetrameters, radiographic inspection ... **A17:** 341-34
of ultrasonic beam search units ... **A17:** 26
Positive carbon extraction replication
method ... **A17:** 5
steps for ... **A17:** 5
Positive clutches
mechanical presses ... **A14:** 496-497
Positive distortion
defined ... **A9:** 14
Positive eyepiece
defined ... **A9:** 14 **A10:** 679

Positive imaging
for automatic optical inspection **EL1:** 942
Positive ion charge
symbol for **A10:** 692
Positive mold *See also* Molds
defined **EM2:** 33
Positive phase contrast **A9:** 59
Positive replica
defined **A9:** 14
Positive secondary ion yields
effect of oxygen on **A10:** 612
Positive-contact bushing
defined **A18:** 15
Positive-contact seal *See also* Face seal
defined **A18:** 15
Positive-locking pins
failure of **A11:** 548
Positron emission, and electron capture
as radioactive decay mode **A10:** 245
Post heat treating *See also* Heat treatment
wrought titanium alloys **A2:** 620
Post, steel
fatigue fracture of **A11:** 116
Post-compacting processing
in rapid omnidirectional compaction **A7:** 544
Post-fatigue test analysis
vacuum and gaseous **A8:** 412
Post-general yield fracture toughness
from precracked Charpy test **A8:** 267
Post-Moire interferometry **EM3:** 451-452
Postage meter
miter gears **A7:** 668
Postbuckling
of plates **EM1:** 447-449
Postcasting operations die casting **A15:** 295
investment casting **A15:** 263-264
Postcleaning *See also* Cleaning; Precleaning
in liquid penetrant inspection **A17:** 8
Postcompaction *See also* Compaction
treatments, prealloyed titanium P/M compacts **A2:** 653-654
Postcure *See also* Cure **EM3:** 23
defined **EM1:** 19 **EM2:** 33
polyimide resins **EM1:** 663
Postcured water-base inorganic silicates **A13:** 411
Postcut method
of contour roll forming **A14:** 624-625
Postdynamic recrystallization **A9:** 690-691
Postemulsifiable fluorescent penetrant method
o liquid penetrant inspection **A17:** 7
Postemulsifiable liquid penetrant inspection
of steel forgings **A17:** 501-50
Postfatigue testing **EM3:** 822
Postfilling buoyant convection
modeling of **A15:** 880-881
Postfired printing
ceramic multilayer packages **EL1:** 466
Postflow time
definition **M6:** 13
Postforging *See also* Forging
in hot-die/isothermal forging **A14:** 154
Postforging processes
failures from **A11:** 331-332
Postforming **EM3:** 23
defined **EM1:** 19 **EM2:** 33
Posthandling
thermoplastic polyurethanes (TPUR) **EM2:** 207
Postheat current
definition **M6:** 13
Postheat time
definition **M6:** 13
Postheating
definition **A6:** 1212 **M6:** 13
Postiamination baking
rigid printed wiring boards **EL1:** 544
Postionization structure
typical K-shell edge **A10:** 450
Postmolding *See also* Molding
operations **EM2:** 85
Postplating treatment
electroplated hard chromium **A13:** 872-875
Postsintering *See also* Heat treatment; Sintering
cemented carbides **A2:** 951
Postsolder properties
of adhesives **EL1:** 674
Posttensioning anchorage
structural corrosion **A13:** 1308, 1310
Postwash stations
liquid penetrant inspection **A17:** 7
Postweld cleaning
corrosion **A13:** 350
Postweld heat treatment *See also* Heat treatment; specific processes
advanced titanium-base alloys **A6:** 526, 527
aerospace materials **A6:** 386
aluminum alloys **A6:** 83, 726-727, 728 **A15:** 763
aluminum-lithium alloys **A6:** 551
arc welding of carbon steels **A6:** 641, 645-647, 648, 649
austenitic stainless steels **A6:** 466, 467, 469
carbon content in wrought martensitic stainless steels **A6:** 438
cast irons **A6:** 713-714 **A15:** 526
cast steels **A15:** 534-535
cobalt-base corrosion-resistant alloys **A6:** 599
coppers **A6:** 761
definition **A6:** 1212 **M6:** 13
diffusion welding **A6:** 884, 885
dissimilar metal joining **A6:** 825, 826
duplex stainless steels **A6:** 474, 476
electroslag welding **A6:** 277-279
ferritic stainless steels **A6:** 450, 451
friction welding and **A6:** 152, 153
gas-metal arc welding of coppers **A6:** 762
for graphitic cast irons **A15:** 527
heat-affected zone cracks **A6:** 92, 93
heat-affected zone in multipass weldments **A6:** 81
heat-treatable aluminum alloys **A6:** 532-533, 534
heat-treatable low-alloy steels **A6:** 672-673
high-strength low-alloy quench and tempered structural steels **A6:** 666
high-strength low-alloy structural steels **A6:** 664
high-temperature alloys **A6:** 563-564
low-alloy steels for pressure vessels and piping **A6:** 668-669
machinery and equipment **A6:** 393
magnesium alloys **A6:** 777, 781, 782
martensitic stainless steels **A6:** 433, 435, 437, 438, 439, 440, 441
material requirements for service conditions **A6:** 375
nickel alloys **A6:** 740, 742
nickel-base alloys **A6:** 590
nickel-base corrosion-resistant alloys containing molybdenum **A6:** 596
nickel-base superalloys **A6:** 83, 566-567
oxide-dispersion-strengthened materials **A6:** 1038, 1039
oxyacetylene welding, castings **A15:** 530
oxyfuel gas welding **A6:** 289
precipitation-hardening stainless steels **A6:** 483-484, 487, 488, 489, 490, 492, 493
pressure vessels **A6:** 381
residual stresses **A6:** 1102
resistance seam welding **A6:** 239
roll welding **A6:** 312, 314
stainless steel casting alloys **A6:** 497, 498
stainless steel welded to carbon steel **A6:** 501
stainless steels **A6:** 677, 680-682, 686, 688, 695, 697
steel weldments **A6:** 416, 419, 420, 421, 422, 423, 424, 426-427
submerged arc welding **A6:** 426
titanium alloys **A6:** 85-86, 508, 509, 510, 512, 514, 515, 517, 518-519, 786
titanium-base corrosion-resistant alloys **A6:** 599
to reduce corrosion susceptibility in weldments **A6:** 1069
to reduce stress-corrosion cracking **A6:** 1068
tool and die steels **A6:** 675-676
treatment
ultrahigh-strength low-alloy steels **A6:** 674
zirconium alloys **A6:** 788 **A15:** 838
Postweld heat treatment cracking
examination **A12:** 139-140
Postweld heat-treat cracking *See* Stress relief, embrittlement
Postweld interval
definition **M6:** 13
Postweld strain-age cracking (PWSAC)
precipitation hardenable nickel alloys **A6:** 576
Postwelding operations
ductile and brittle fractures from **A11:** 94
Pot **EM3:** 23
defined **A15:** 9 **EM2:** 33
Pot annealing
steel wire **M1:** 262
Pot life *See also* Handling life; Working life **EM3:** 23
defined **EM1:** 19 **EM2:** 33
in filament winding **EM1:** 135
resin transfer molding **EM1:** 169
Potable water systems
corrosion in **M1:** 735-736
Potash
Miller numbers **A18:** 235
Potash (K_2O)
component in photochromic ophthalmic and flat glass composition **EM4:** 442
in composition of glass-ceramics **EM4:** 499
in composition of textile products **EM4:** 403
in composition of wool products **EM4:** 403
in drinkware compositions **EM4:** 1102
in glaze composition for tableware **EM4:** 1102
in ovenware compositions **EM4:** 1103
properties **EM4:** 424
purpose for use in glass manufacture **EM4:** 381
in tableware compositions **EM4:** 1101
Potash feldspar ($KAlSi_3O_8$) **EM4:** 6, 7
Potash mining tools
cemented carbide **A2:** 976
Potash-SiO_2 self-diffusion coefficients of alkali ions **EM4:** 461
Potash-soda-lime-zinc silicate glass
Corning glass code 8361 derived **EM4:** 463
Potassium *See also* Liquid potassium **A13:** 92, 94-95, 735
acid-base titration **A10:** 173
additions to flame AAS samples **A10:** 48
cations, in glasses, Raman analysis **A10:** 131
determination in paint, absorption and enhancement effects **A10:** 98
in enamel cover coats **EM3:** 304
in enameling ground coat **EM3:** 304
flame emission sources for **A10:** 30
as flux **A10:** 167
lubricant indicators and range of sensitivities **A18:** 301
organic precipitant for **A10:** 169
pure **M2:** 786-787
pure properties **A2:** 1148
pyrophoricity **A7:** 194, 199
Raman vibrational behavior **A10:** 133
species weighed in gravimetry **A10:** 172
spectrometric metals analysis **A18:** 300
TNAA detection limits **A10:** 237
use in flux cored electrodes **M6:** 103
in vacuum, as SERS metal **A10:** 136
vapor pressure **A6:** 621
volatilization losses in melting **EM4:** 389
Potassium benzyl penicillin
three-dimensional electron density map of **A10:** 350
Potassium borate
ECG **A16:** 545
Potassium bromide
pellets, as IR samples **A10:** 113
pellets, as Raman samples **A10:** 131
sample, for Raman spectroscopy **A10:** 129
Potassium carbonate
carburizing role **A4:** 325
ECG **A16:** 545
use in gold plating **M5:** 281-282
used to neutralize etching acids **A9:** 172
Potassium carbonate (K_2CO_3)
purpose for use in glass manufacture **EM4:** 381
Potassium chloride
ECM electrolyte **A16:** 533, 535, 536
Potassium chloride zinc plating system **M5:** 251-252
Potassium cyanide and ammonium persulfate as an etchant for
palladium and palladium alloys **A9:** 551
Potassium cyanide copper plating **M5:** 159-165, 167-169
Potassium cyanide gold plating **M5:** 281-282

Potassium cyanide in an aqueous solution as an electrolyte for
gold plate **A9:** 551
silver plate **A9:** 551
Potassium dichromate
as an etchant for copper-base powder metallurgy materials **A9:** 509
Potassium dichromate ($K_2Cr_2O_7$)
purpose for use in glass manufacture **EM4:** 381
Potassium ferricyanide
as an etchant for wrought stainless steels **A9:** 281
Potassium fluorrichterite
in glass-ceramics **EM4:** 1102
Potassium germanate glasses
density **EM4:** 846
Potassium hydrogen difluoride
as fusion flux **A10:** 167
Potassium hydroxide **A13:** 1178
as an electrolyte for tungsten **A9:** 440
as an etchant for wrought stainless steels **A9:** 281-282
embrittlement by **A11:** 658-660
etchant for laser-enhanced etching **A16:** 576
safety hazards **A9:** 69
used to electrolytically etch heat-resistant casting alloys **A9:** 331
Potassium hydroxide fusion
crucibles for **A10:** 167
Potassium lead silicate glass
properties **EM4:** 1057
Potassium nitrate
EDG electrolyte **A16:** 548
electrolyte for Ni alloy ECM **A16:** 843
impact treatment bath for photochromic glasses **EM4:** 462
ion-exchange in melts **EM4:** 461
specific properties imparted in CTV tubes **EM4:** 1040-1041
Potassium nitrate, apparent threshold stress values
low-carbon steel **A8:** 526
Potassium perchlorate
as incendiary **A7:** 603
Potassium phosphate
ECG electrolyte **A16:** 545
Potassium pyrosulfate
as acidic flux **A10:** 167
for low-temperature fusions **A10:** 94
Potassium salts
ECG electrolyte **A16:** 545
Potassium silicate
binding agent **A6:** 61
chemical composition **A6:** 60
function and composition for mild steel SMAW electrode coatings **A6:** 60
Potassium silicate gel network **EM4:** 446
Potassium silicate gels **EM4:** 450
Potassium stannate tin plating process **M5:** 271
Potassium titanate
function and composition for mild steel SMAW electrode coatings **A6:** 60
Potassium, vapor pressure
relation to temperature **A4:** 495 **M4:** 310
Potassium-aluminosilicate glass
redox reactions in metal alloy-glass sealing **EM4:** 489
Potassium-baria-phosphate glass
nonlinear refractive index **EM4:** 1080
Potassium-richterite glass-ceramic
composition **EM4:** 1101
properties **EM4:** 1101
Potassium-rubidium lead silicate glass
properties **EM4:** 1057
Potassium-silicate glasses
density **EM4:** 846
Potassium-sodium lead silicate glass
properties **EM4:** 1057
Potency factor **A6:** 809
Potential
between working and reference electrodes **A10:** 661
half-wave, as polarographic wave parameter **A10:** 190
membrane, ion-selective membrane electrode **A10:** 182
standard, in controlled-potential electrolysis **A10:** 208
use in voltammetry **A10:** 189
vs time, potentiometric membrane electrodes **A10:** 186
Potential buffers
use in electrogravimetry **A10:** 200
Potential difference
SI derived unit and symbol for **A10:** 685
SI unit/symbol for **A8:** 721
Potential drop technique *See* Electrical potential method
Potential energy diagrams **A10:** 586
Potential in electropolishing
effects **A9:** 105
Potential lead position
non-dimensional voltage as function **A8:** 388
Potential measurement probes
and current input, test specimen geometries **A8:** 386
spot-welded **A8:** 389
Potential ranges identified on a polarization curve **A9:** 144-145
Potential requirements
in design process **EL1:** 128
Potential versus pH (Pourbaix) diagrams *See* Pourbaix (potential-pH) diagram
Potential(s) *See also* Active potential; Chemical potential; Corrosion potential; Critical pitting potential; Decomposition potential; Electrochemical potential; Electrode potential; Electrokinetic potential; Equilibrium (reversible) potential; Free corrosion potential; Noble potential; Open circuit potential; Protective potential; Redox potential; standard electrode potential
control, anodic protection **A13:** 464
coupled, galvanic corrosion evaluation **A13:** 237
-current relationship, aqueous corrosion **A13:** 30-31
decay **A13:** 24
defined **A13:** 10
effect on crack growth rate **A13:** 155
electrochemical **A13:** 932
electrode **A13:** 19-21, 298
in flowing seawater **A13:** 557
in galvanic series **A13:** 235
measurement **A13:** 21-24, 84, 920
pitting, defined **A13:** 583
reduction, magnesium/magnesium alloys **A13:** 740
repassivation, titanium/titanium alloys **A13:** 684-685
stable and reproducible **A13:** 22-23
Potential-pH diagram *See* Pourbaix (potential pH) diagram
copper-ammonia- water system **A7:** 54
Potentiodynamic (potentiokinetic)
defined **A13:** 10
Potentiodynamic polarization
curves, carbon-manganese steel, SCC testing **A13:** 264
curves, zirconium in hydrochloric acid **A13:** 709
galvanic corrosion **A13:** 235
methods, cyclic **A13:** 217
tests, crevice corrosion **A13:** 308
Potentiodynamic polarization curves
low-carbon steel **A8:** 532
Potentiometer
defined **A9:** 14
Potentiometric gas-sensing electrodes **A10:** 183-185
Potentiometric membrane electrodes *See also* Classical, electrochemical and radiochemical analysis; Potentiometry **A10:** 181-187
additional techniques used with **A10:** 183
advantages **A10:** 181
analysis methods **A10:** 183
applications **A10:** 181, 186
calibration curves **A10:** 183
capabilities, compared with voltammetry **A10:** 188
defined **A10:** 679
estimated analysis time **A10:** 181
experimental arrangement for **A10:** 186
general uses **A10:** 181
introduction **A10:** 181
ion-selective membrane electrodes **A10:** 181-183
limitations **A10:** 181
possible errors **A10:** 185-186
potentiometric gas-sensing electrodes **A10:** 183-185
related techniques **A10:** 181
samples **A10:** 181, 186
subtraction techniques used with **A10:** 183
titration methods **A10:** 183
Potentiometric titration **A10:** 183, 203, 204
to analyze the bulk chemical composition of starting powders **EM4:** 72
Potentiometry *See also* Potentiometric membrane electrodes
capabilities, compared with voltammetry **A10:** 188
Potentiostat
for acidified chloride solutions **A8:** 419-420
automatic, for controlled-potential analysis **A10:** 200
defined **A9:** 14 **A13:** 10
galvanostat, power **A10:** 199
used to measure current-voltage curve in electropolishing **A9:** 105
Potentiostatic
defined **A13:** 10
testing methods **A13:** 217, 288
Potentiostatic control
SCC tests with **A8:** 532
Potentiostatic etchants **A9:** 146-147
Potentiostatic etching **A9:** 61, 137, 143, 147
defined **A9:** 14
principles of **A9:** 62
wrought stainless steels **A9:** 282
Potentiostatic polarization
measurement **A13:** 1333
tests, crevice corrosion **A13:** 308-309
Pottery
definition **EM4:** 3
description **EM4:** 3
Potting *See also* Encapsulation **EM3:** 23, 553, 585
bubble memory device **EL1:** 820
of circuits **EL1:** 824
compounds, as encapsulants **EL1:** 802-803
defined **EM2:** 33
as electronic embedment, epoxies **EL1:** 832
materials, rigid epoxies **EL1:** 810
printed board coupons **EL1:** 572-573
Potting fill levels
neutron radiography of **A17:** 394-39
Poultice corrosion *See also* Deposit corrosion
automotive industry
defined **A11:** 8 **A13:** 10
mechanism **A13:** 1012
Pound-force
abbreviated **A8:** 725
Pound/pound force
abbreviation for **A11:** 797
Pour, interrupted
as casting defect **A11:** 383
Pour point **A18:** 83, 86
defined **A18:** 15
engine oils **A18:** 169
gear lubricants **A18:** 542

Pour-point depressant A18: 99, 107-108
applications A18: 108
characteristics A18: 107
defined A18: 15
in engine lubricant formulations A18: 111
multifunctional nature A18: 111
in nonengine lubricant formulations A18: 111
performance range A18: 108
structural features A18: 108
tests of performance A18: 108
Pour-point depressants
for lubricant failure A11: 154
Pourbaix (potential-pH) diagrams A13: 24-28
for aluminum with oxide film A13: 583
application, crack propagation A13: 152-155
aqueous corrosion A13: 35
for cobalt A13: 27
computation and construction A13: 25
for copper A13: 28
copper-zinc alloy A13: 133-134
and corrosion on lead surfaces A10: 135
defined A10: 679 A13: 10
for iron A13: 22, 37-38, 153, 1301
for iron-water system A13: 153, 1301
for molten salts A13: 50-51
for nickel A13: 28
for nickel-water system A13: 378
for niobium in water A13: 724
practical use A13: 27-28
for titanium A13: 1330
for titanium-water system A13: 670
for uranium A13: 816
for water A13: 27
Pourbaix diagrams
conditions for corrosion, passivation, and immunity of iron A11: 198
for pH and applied potential changes A11: 198
Poured short
as casting defect A11: 385
Pouring *See also* Automatic pouring systems; Ladle(s); Metal transfer; Pouring temperature; Pouring time; Risering
in all-ceramic mold casting A15: 249
aluminum alloys A15: 754
in Antioch process A15: 247
basin A15: 9, 91
cobalt-base alloys A15: 813-814
computer modeling of A15: 857
constant-level, permanent mold method A15: 276
control parameters A15: 500
copper alloys A15: 776-778
of copper casting alloys A2: 346
crucible furnace A15: 383
defined A15: 9
devices, development of A15: 27-28
direct pouring A15: 498
in foamed plaster molding process A15: 247
and gating A15: 589
high-alloy white irons A15: 680
high-chromium white irons A15: 683
high-silicon irons A15: 699
in horizontal centrifugal casting A15: 297
ladles, crucible furnaces A15: 382-383
lost foam casting A15: 233-234
of magnesium alloys A15: 800-801, 803
in magnetic rubber inspection A17: 122-12
mechanized ladle pouring A15: 498
of molten gray iron A15: 639
nickel alloys A15: 820-821
of plain carbon steels A15: 710
in plaster molding A15: 244-245
profile A15: 501
rate and duration A15: 500
in Replicast process A15: 271
slurry, plaster molding A15: 244-245
of solid graphite molds A15: 285
spout design/positioning, centrifugal casting A15: 305
titanium and titanium alloy castings A2: 642
in vertical centrifugal casting A15: 306
Pouring basin
defined A15: 9
Pouring gates
shrinkage porosity and A11: 354
Pouring temperature
aluminum castings A15: 238
Pouring temperature (continued)
ductile iron A15: 651
effect, mold life A15: 281
effect, mold temperature A15: 282
effect on grain size of austenitic manganese steel castings A9: 238
gray cast iron M1: 11-12
gray iron A15: 639
inclusion-forming effect A15: 94
of lead and lead alloys A2: 545
lead-base bearing alloys A2: 553
permanent mold casting A15: 283
Pouring time *See also* Flow rate; Fluidity
nickel alloys A15: 821
plain carbon steels A15: 710-711
and rate A15: 500
Powder *See also* Powder cutting; Powder forging
alloys, heat-resistant, forging of A14: 234
brazing filler metals available in this form A6: 119
compacts, preferred orientations in A10: 358
consolidation, double-end pressing simulation of A14: 429-430
diffraction, geometry and detection methods A10: 331
forged parts, tolerances A14: 204
metallurgy, abbreviation for A10: 691
pattern A10: 262, 265
purity A14: 189
samples, ESR analysis of A10: 262
x-ray diffraction of A10: 333-343
Powder bed
cross section, reduction furnace A7: 155
fluid flow through, permeametry as A7: 263
maximum density in compaction A7: 57, 58
Powder cleaners
for surfaces A13: 382
Powder cleaning A7: 178-179
gaseous contaminant removal A7: 180-181
particulate contaminant removal A7: 178-180
Powder cloud development
copier powders A7: 583
Powder coating
in copier powders A7: 588
of electrodes A7: 820-821
Powder coating paint *See* Paint and painting, powder coating process
Powder coatings
applications A13: 400
Powder compacting
dies, cemented carbide A2: 971
punches, cemented carbide A2: 971
Powder compaction *See also* Compacting; Compaction A7: 297, 304-307, 401-409
Powder compaction processes EM4: 123-124
Powder compacts *See also* Compacting; Compaction; Compacts A7: 288-289, 298-304, 453
Powder compression molding
size and shape effects EM2: 291
Powder consolidation *See also* Consolidation; Powder consolidation methods
direct powder forming A2: 203
dynamic compaction A2: 204
hot isostatic pressing (HIP) A2: 203
rapid omnidirectional consolidation A2: 203-204
Powder consolidation methods *See also* Consolidation
beryllium A2: 685
vacuum hot pressing and hot isostatic pressing, compared A2: 685
Powder cutting
definition A6: 1212
metal A14: 728
Powder cutting torches A7: 843-845
Powder degassing
can vacuum degassing A2: 202-203
dipurative degassing A2: 203
vacuum degassing in reusable chamber A2: 203
Powder density *See also* Density A7: 188, 296
Powder developers, dry
for liquid penetrant inspection A17: 76-7
Powder Diffraction File (PDF) A10: 327
Powder feeder
definition M6: 13
Powder fires A7: 132-133
Powder flame guns
thermal spray coating M5: 366-367
Powder flame spray process A13: 459
Powder flame spraying
definition M6: 13
Powder flow *See* Flow
Powder flowability *See* Flowability
Powder forged parts
applications of A14: 205-207
Powder forging *See also* Forging; Powder; Powder metallurgy A14: 188-211
alloy development A14: 189
applications A14: 205-207
and competitive processes, compared A14: 196
defined A14: 10
and ferrous powder metallurgy materials A1: 812
heat treatment A14: 201-202
hot re-pressing as A14: 188
hot upsetting as A14: 188
inclusion assessment A14: 190
iron powder contamination A14: 190-191, 204
materials A14: 188-191
mechanical properties A14: 198-203
metal flow in A14: 194
mode A14: 201
as new metalworking process A14: 17
powder characteristics A14: 189
as precision forging A14: 158
preforming A14: 191-192
presses for A14: 194
problems and causes A14: 192
process A14: 188, 191-198
quality assurance, parts A14: 203-205
requirements A14: 189
secondary operations A14: 197
sintering and reheating A14: 192-194
tool design A14: 197
Powder injection
plain carbon steel A15: 709-710
Powder metal joint fill
submerged arc welding M6: 133
Powder metal-filled plastics A7: 606-613
Powder metallurgy *See also* Powder; Powder forging
alloys, prealloyed A14: 250-251
defined A13: 10, 823
for discontinuous fiber MMCs EM1: 897-898, 904-905
electrical contacts, use in M3: 690-691
magnetic applications M3: 608-609, 610, 613
manufacture of cemented carbides by M3: 451-452
as metalworking A14: 15
methods, solid-solution alloys by A14: 238
steels, for coining A14: 182-183
titanium alloy A15: 824
titanium alloys M3: 370-371
tool steels M3: 441, 442, 443, 444-445
Powder metallurgy (P/M) A1: 801 A7: 9, 14
aluminum alloys A6: 724
applications A7: 246-249, 569-574
commercial developments A7: 16-17
compacting presses A7: 329-338
consolidation techniques A7: 719-720
contact materials A7: 624-645
ecological considerations A7: 569
high-temperature materials A7: 765-772
history A7: 14-20
industry A7: 17-18, 23
literature A7: 18-19
in making tool steels A1: 757
P/M materials A7: 463-479, 482
parts *See also* Part(s) A7: 9
presses A7: 329-338
production sintering practices A7: 360-400
products A7: 257, 295, 451-462, 569-574
systems and applications A7: 246-249, 569-574
techniques and powder properties A7: 211
trade association A7: 19
Powder metallurgy (P/M) alloys
dispersion-strengthened A2: 943-949
Stellite alloys, application A2: 449
titanium and titanium alloy A2: 590
Powder metallurgy (P/M) cobalt-base alloys A1: 977-980
compositions A1: 978
grinding of A1: 979-980

Powder metallurgy (P/M) cobalt-base alloys (continued)
machining of **A1:** 978, 980
mechanical properties **A1:** 977, 978
physical properties **A1:** 977, 978
Powder metallurgy (P/M) parts
aluminum and aluminum alloys **A2:** 6-7
copper-base **A2:** 396-399
Powder metallurgy (P/M) processes
Alloy 2024, blended with aluminum-silicon alloys **A18:** 791
aluminum-silicon alloys **A18:** 785, 791
bearing materials **A18:** 755
mainshaft bearings of jet engines **A18:** 590
for micrograin high-speed steels **A18:** 615-616
self-lubricating, composites **A18:** 120
steel
applications, internal combustion engine parts **A18:** 553
wear resistance and cost effectiveness **A18:** 706
Powder metallurgy (P/M) processing *See also* Aluminum P/M processing
of A15 superconductors **A2:** 1067
advantages **A2:** 840
of aluminum alloy parts **A2:** 210-213
atomization **A2:** 201
can vacuum degassing **A2:** 202-203
cemented-carbide products **A2:** 980
cermets **A2:** 979-980
dynamic compaction **A2:** 204
electrical contact composite materials **A2:** 856-857
elemental P/M, titanium **A2:** 647-651
high-strength aluminum alloys **A2:** 200-215
high-temperature superconductors **A2:** 1086
liquid dynamic compaction **A2:** 204
mechanical alloying process **A2:** 202
mechanical attrition process **A2:** 202
mechanically alloyed oxide alloys **A2:** 943-944
dispersion-strengthened (MA ODS)
melt-spinning techniques **A2:** 202
neodymium-iron-boron permanent magnet materials **A2:** 790-791
of nickel superalloys, development **A2:** 429
Osprey process **A2:** 204
powder degassing and consolidation **A2:** 202-204
powder production **A2:** 201-202
prealloyed P/M, titanium **A2:** 647, 651-653
reaction milling **A2:** 202
sinter-aluminum-pulver (SAP) technology **A2:** 202
splat cooling **A2:** 201-202
tin and tin alloy powders, applications **A2:** 519-520
titanium and titanium alloy castings **A2:** 634
vacuum plasma structural deposition **A2:** 204
Powder metallurgy (P/M) superalloys *See also* Powder metallurgy (P/M) cobalt-base alloys
mechanical properties of **A1:** 974-976
fatigue properties **A1:** 974, 975
stress-rupture properties **A1:** 974
tensile properties **A1:** 974, 975
oxide dispersion strengthened alloys **A1:** 972, 973, 974-975, 976
compositions of **A1:** 973
physical properties **A1:** 976
stress-rupture properties **A1:** 976
tensile properties **A1:** 976
powder consolidation **A1:** 973
powder production **A1:** 970, 972-973
thermomechanical working **A1:** 973-974
Powder metallurgy (PM)
thermoplastic polyimides (TPI) **EM2:** 178
Powder metallurgy alloys *See also* Powder metallurgy tool steels **A16:** 879-892
additives **A16:** 879, 885
boring **A16:** 881, 882, 889
brass, composition **A16:** 881

Powder metallurgy alloys (continued)
bronze, composition **A16:** 881
chip formation **A16:** 883, 885
classification system **A16:** 880
comparison of machinability parameters **A16:** 890
composition **A16:** 884-885
compositions **A16:** 880
Cu-base material compositions **A16:** 881
Cu-base structural materials **A16:** 881
cutting fluids (coolants) **A16:** 881, 886, 889-890
cutting speed **A16:** 879, 885, 886, 889
deburring **A16:** 880, 881
density effect **A16:** 883, 886, 890
design **A16:** 880
drilling **A16:** 879-890
electrostream and capillary drilling **A16:** 551
end milling **A16:** 889
face milling **A16:** 889
ferrous structural materials **A16:** 880
forging **A16:** 879
grinding **A16:** 880, 881, 882, 889, 890, 891
grinding ratio **A16:** 882
hardness and density values **A16:** 882, 888, 889, 890
honing **A16:** 889, 891
lapping **A16:** 889, 891
machinability **A16:** 879, 881-889
machinability factors **A16:** 881-889
machinability indexes **A16:** 887
machining guidelines **A16:** 889-891
machining variables **A16:** 885-886
material guidelines **A16:** 880
microstructure **A16:** 9, 883, 884
milling **A16:** 880, 883, 889
Ni-Ag, composition **A16:** 881
optimizing part machining **A16:** 879, 881, 883
porosity factor **A16:** 881-882, 883, 890
presintering **A16:** 879
process effects **A16:** 883-884
properties vs. machinability **A16:** 886-889
reaming **A16:** 889, 890, 891
roller burnishing **A16:** 889, 890, 891
shaped tube electrolytic machining **A16:** 554
sintering results **A16:** 886, 887, 888, 889
stainless steel compositions **A16:** 880
strength correlation **A16:** 890
surface finish **A16:** 879, 881, 883, 884, 885, 890, 891
tapping **A16:** 880, 881, 889, 891
temperature factor **A16:** 882, 883, 884
tool life **A16:** 879, 881, 882, 883, 884, 885, 886, 887, 890
turning **A16:** 880, 881, 882, 884, 886, 889
Powder metallurgy Cu-base alloys
composition **A16:** 880
hardness and density **A16:** 882
Powder Metallurgy Equipment Association
Powder metallurgy Fe-base alloys
composition **A16:** 880
hardness and density **A16:** 882
Powder Metallurgy Industries Association **A7:** 19
Powder metallurgy materials **A9:** 503-530
bonding of particles, examination of **A9:** 508
development of **A13:** 823
etching **A9:** 508-509
fully dense P/M stainless steels **A13:** 833-834
grinding **A9:** 505-506
macroexamination **A9:** 507-508
magnetic **A9:** 539
microexamination **A9:** 508-509
microstructures **A9:** 509-512
mounting **A9:** 504-505
P/M aluminum alloys **A13:** 838-842
P/M superalloys **A13:** 834-838
polishing **A9:** 506-507
preparation **A9:** 503-507

Powder metallurgy materials (continued)
preparation of scanning electron microscopy specimens **A9:** 97
rhenium powder **A9:** 447
scanning electron microscopy **A9:** 508
scanning electron microscopy used to study **A9:** 99-100
sectioning **A9:** 503-504
sintered (porous) P/M stainless steels **A13:** 824-832
sintered iron-base P/M parts **A13:** 823-824
sleeve bearing liners **A9:** 567
use of magnetic field to orient magnetic powder particles **A9:** 701
wax impregnation **A9:** 504
Powder metallurgy P/M tungsten
commercial grade **A2:** 577
Powder metallurgy parts **A17:** 536-54
acoustic methods **A17:** 53
aluminum alloy **M2:** 10-13
Brinell test application **A8:** 89
computed tomography (CT) of **A17:** 3
defect types **A17:** 536-53
electrical resistivity testing **A17:** 541-54
liquid penetrant inspection **A17:** 7
mechanical coating of **M5:** 301-302
microhardness testing **A8:** 97
nondestructive tests **A17:** 537-54
porosity, effects and correlation of **M5:** 620-621
pressure testing **A17:** 545-54
radiographic techniques **A17:** 537-53
Rockwell hardness testing of **A8:** 83
surface preparation for plating **M5:** 620-621
testing, current status **A17:** 53
visual inspection **A17:** 545-54
Powder Metallurgy Parts Association **A7:** 19
Powder metallurgy parts, ferrous
carbonitrided parts, hardness **M4:** 799
carbonitrided parts, tempering **M4:** 799-800
carbonitriding **M4:** 799
equipment **M4:** 799
hardness **M4:** 798-799, 800
hardness evaluation **M4:** 798-799
heat treating **M4:** 798-800
heating media **M4:** 798
process selection **M4:** 798
quenching **M4:** 798
surface hardening **M4:** 799
techniques **M4:** 799
tempering **M4:** 799
Powder metallurgy products *See* P/M steels **A1:** 798-799
sintering atmospheres **M4:** 794-796
Powder metallurgy steels, heat treating **A4:** 229-236
alloy content **A4:** 230, 232
carbonitriding **A4:** 230, 231, 232, 233-234
carburizing **A4:** 233
case depth **A4:** 230
dispersion strengthening of lead alloys **A4:** 927
furnace atmospheres **A4:** 548, 558, 561-562, 563
hardenability **A4:** 229-230
hardenability, effect of alloy content **A4:** 230
hardening cycles **A4:** 234-235
high-temperature sintering **A4:** 232
induction hardening **A4:** 234
liquid nitriding **A4:** 419
material effect on heat-treated properties **A4:** 230-232
material properties **A4:** 229-230, 236
mechanical properties **A4:** 233, 236
neutral hardening **A4:** 232-233
nitrocarburizing **A4:** 234-235
parts guidelines **A4:** 235-236
plasma (ion) nitriding **A4:** 423, 424
plasma nitrocarburizing **A4:** 434-435
porosity, effect on case depth **A4:** 230
porosity, effect on material properties **A4:** 229-230

Powder metallurgy steels, heat treating (continued)
prealloyed powders ... **A4:** 231-232
processing, effect on heat-treated properties ... **A4:** 230-232
quenching media ... **A4:** 233
resistance index (R) ... **A4:** 230
sintering atmosphere hazards ... **A4:** 547-548
steam treating ... **A4:** 235, 236
tempering ... **A4:** 234
tool steel production ... **A4:** 765, 766
Powder metallurgy steels, specific types
F-0000 carbon steel, nitrocarburizing ... **A4:** 235
FC-0205-HT
composition ... **A4:** 234
mechanical properties ... **A4:** 234
nitrocarburizing ... **A4:** 235
FC-0208-HT
composition ... **A4:** 234
mechanical properties ... **A4:** 234
FL-4205-HT
composition ... **A4:** 234
mechanical properties ... **A4:** 231, 234
FL-4605-HT
composition ... **A4:** 234
mechanical properties ... **A4:** 231, 234
FN-0205-HT
composition ... **A4:** 234
mechanical properties ... **A4:** 234
SINT-D35, plasma nitrocarburizing ... **A4:** 434
Powder metallurgy techniques as a fabrication method for continuous-filament metal-
matrix composites ... **A9:** 591
Powder metallurgy tool steels ... **A1:** 780-792 **A16:** 60-68, 733-735
advantages ... **A16:** 733
advantages over conventional tool steels ... **A1:** 780
Anti-Segregation Process (ASP) ... **A16:** 733
applications of high-speed tool steels ... **A1:** 785-786
broaching ... **A1:** 785-786
gear manufacturing ... **A1:** 786
hole machining ... **A1:** 785
milling ... **A1:** 785
carbon effect on machinability ... **A16:** 734, 735
chip formation ... **A16:** 735
classification ... **A1:** 781-792
cold-work steels ... **A1:** 786-789
high-speed steels ... **A1:** 781-786
hot-work steels ... **A1:** 789-790
cold-work tool steels ... **A1:** 786-78
commercial ASP tool steel compositions ... **A16:** 733
composition ... **A1:** 781
Crucible Particle Metallurgy (CPM) ... **A16:** 733
grinding ... **A16:** 734-735
grinding ratio ... **A16:** 735
heat treatment ... **A16:** 734
heat treatment of H13, effect on size change of ... **A1:** 7
heat treatment of high-speed tool
annealing ... **A1:** 78
austenitizing temperature of ASP 23 ... **A1:** 78
hardening ... **A1:** 78
steels ... **A1:** 782-78
stress relieving (before hardening) ... **A1:** 783
tempering ... **A1:** 783
high-speed, annealing ... **A16:** 61
high-speed, application of the FULDENS process ... **A16:** 5
high-speed, applications ... **A16:** 8
high-speed, ASP steels heat treatment ... **A16:** 61-62
high-speed, broaching ... **A16:** 66-67, 68
high-speed, comparison of cutting edge wear ... **A16:** 61
high-speed, CPM process used ... **A16:** 62-64
high-speed, drilling ... **A16:** 65, 67
high-speed, gear manufacturing ... **A16:** 67, 68
high-speed, grindability index ... **A16:** 62, 63
high-speed, hardening ... **A16:** 61-62, 63
high-speed, hole machining ... **A16:** 65
high-speed, milling ... **A16:** 65, 66
high-speed, reaming ... **A16:** 65, 67
high-speed, stress relieving ... **A16:** 61, 62
high-speed, tapping ... **A16:** 65, 67
high-speed, tempering ... **A16:** 62, 63
high-speed tool steels ... **A1:** 781-78
alloy development ... **A1:** 784-78
Powder metallurgy tool steels (continued)
applications ... **A1:** 785-78
cutting tool properties ... **A1:** 78
heat treatment ... **A1:** 782-783
manufacturing properties ... **A1:** 783-784
sintered tooling ... **A1:** 786
hot-work tool steels ... **A1:** 789-790
machinability, tool life of CPM alloys ... **A1:** 786
machining conditions ... **A16:** 733
machining operations ... **A16:** 734
mechanical properties of CPM alloys
bend fracture strength ... **A1:** 786
Charpy C-notch toughness ... **A1:** 786, 790
hot hardness ... **A1:** 785
temper resistance ... **A1:** 785
wear resistance ... **A1:** 790
mechanical properties of H13
Charpy V-notch impact strength ... **A1:** 791
hardness ... **A1:** 790
tensile strength ... **A1:** 791
thermal fatigue resistance ... **A1:** 792
sulfur effect on machinability ... **A16:** 734
surface finish ... **A16:** 735
vanadium effect on machinability ... **A16:** 734
Powder metals
amorphous ... **A7:** 794-797
Powder method
defined ... **A9:** 14
Powder method/slurry infiltration
ceramic-matrix composites ... **EM4:** 840
Powder mixtures ... **A7:** 186-189, 291, 302
Powder molding
defined ... **EM2:** 33
size and shape effects ... **EM2:** 291
Powder packing *See also* Packing
tap density and ... **A7:** 276
Powder paints ... **M5:** 472
Powder polymers ... **A7:** 606-613
Powder preparation
of cermets ... **A2:** 979-980
and mechanical properties, titanium P/M compacts ... **A2:** 654
Powder pressing
granulated powders as feedstock ... **EM4:** 100
Powder production ... **A7:** 9, 23-24
atomization ... **A2:** 201
chemical methods ... **A7:** 52-55
electrolytic ... **A7:** 71-72
melt-spinning techniques ... **A2:** 202
splat cooling ... **A2:** 201-202
Powder properties ... **A7:** 73-74, 211, 302
Powder rill ... **A7:** 9, 323-325
control ... **A7:** 323, 325
density and ... **A7:** 324
effect on quality of mixing ... **A7:** 189
fixed levels ... **A7:** 323, 325
multilevel parts ... **A7:** 325
multiple lower punches ... **A7:** 325
single-level parts ... **A7:** 325
Powder rolling *See also* Powder preparation; Roll compacting ... **A7:** 9, 401-409
of cermets ... **A2:** 983-984
cobalt strip, properties ... **A7:** 402
direct, SEM analysis ... **A7:** 235
mill ... **A7:** 401, 408
process ... **A7:** 406, 407
sleeve bearing fabrication by ... **A7:** 407-408
Powder shape *See also* Grain shape
beryllium ... **A2:** 683-686
Powder slip casting
schematic ... **A2:** 984
Powder welding
hardfacing ... **A6:** 800, 801
Powder(s) *See also* Coarse powders; Fine powders; Green compacts; Hard powders; Magnetic particles; Metal powders; Powder metallurgy (P/M); Powder preparation; Powder production; Prealloyed powders; Soft materials ... **A7:** 9
as aluminum alloy application ... **A2:** 14
applications ... **A7:** 569-574
blends for flame cutting ... **A7:** 842
cermet mixtures, warm extrusion of ... **A2:** 982-983
characteristics and milling parameters ... **A7:** 60-65
classification ... **A7:** 59
cleanliness ... **A7:** 178-181
consolidation, cemented carbides ... **A2:** 951
Powder(s) (continued)
consolidation techniques and production ... **A7:** 23
containers *See also* Containers ... **A7:** 542
copper, production of ... **A2:** 392-394
cupronickel ... **A2:** 402
degassing ... **A7:** 180-181
designation ... **A7:** 9
developer ... **A17:** 7
direct deposition of ... **A7:** 71-72
dispensers, for flame cutting ... **A7:** 843
encapsulation *See* Encapsulation; Hot isostatic pressing
feeding, in roll compacting ... **A7:** 403, 405, 406
fires ... **A7:** 133
free-flowing, defined ... **A7:** 278
grade, production, cemented carbides ... **A2:** 951
handling, in rotating disk atomization ... **A7:** 47
for implant production ... **A7:** 658
incoming, dimensional changes ... **A7:** 481-482
lancing ... **A7:** 843, 845
lubricant ... **A7:** 9
magabsorption measurements of ... **A17:** 152-15
mechanically alloyed oxide alloys ... **A2:** 943
dispersion-strengthened (MA ODS)
mixtures and blending ... **A7:** 186-189, 291, 302
P/M ferrous ... **A7:** 682-683
particle morphology, changes in ... **A7:** 62
particle sizes ... **A7:** 23
polyethylene, microwave inspection ... **A17:** 21
precious metal ... **A2:** 694-695
precursor preparation, high-temperature superconductors ... **A2:** 1086
processing ... **A7:** 435, 436
producers, industries supplied by ... **A7:** 23
production ... **A7:** 9, 23-24, 52-55, 71-72
properties, and spray drying ... **A7:** 73-74
purity ... **A7:** 23
refinement ... **A7:** 63, 64
refractory metal and alloys ... **A2:** 557-565
scarfing ... **A7:** 844-845
self-heating and igniting ... **A7:** 194
shape, atomized aluminum ... **A7:** 129
shipments ... **A7:** 24
size, atomized aluminum ... **A7:** 129
size classification ... **EL1:** 562
size distribution, in rotating disk atomization ... **A7:** 47
soft, terminal density ... **A7:** 299
solder ... **EL1:** 651-653
stability, lubrication and surface treatment for ... **A7:** 182
structural ceramic forming with ... **A2:** 1020
superalloy, mechanically alloyed oxide alloys ... **A2:** 943-944
dispersion-strengthened (MA ODS)
surface ... **A7:** 250-261
systems ... **A7:** 569-574
technology ... **A7:** 9
tin and tin alloy ... **A2:** 519-520
titanium P/M ... **A2:** 647-648
treatments ... **A7:** 24
ultrafine, by high-energy milling ... **A7:** 70
unconsolidated, rigid tool compaction of ... **A7:** 322-328
-under-vacuum process (Ti powder) ... **A7:** 167
Powder-compacting tools
materials for ... **M3:** 544-545
Powder-graded seal ... **EM4:** 527
Powder-in-tube processing
high-temperature superconductors ... **A2:** 1086-1087
Powder-metallurgy parts
porosity during welding ... **A6:** 531, 532
Powdered biological materials
analysis of ... **A10:** 106-107
Powdered graphite
for room-temperature compression testing ... **A8:** 195
Powdered metals
cutting tool materials and cutting speed relationship ... **A18:** 616
Powdered polymers
fabrication ... **EM1:** 102
Powders
for fully dense P/M stainless steels ... **A13:** 833

Powders (continued)
image analysis of particle size distributions in ... A10: 309
mounting, for XPS analysis ... A10: 575
production and composition, P/M parts ... A13: 825
pure, Raman analysis ... A10: 129
for sample dissolution ... A10: 165
samples, fluxes for fusing of ... A10: 167
as samples for x-ray spectrometry ... A10: 93-94
as samples, Raman analyses ... A10: 130
SIMS analysis of surface layers ... A10: 610
XPS samples ground to ... A10: 575
Powdrex process ... A1: 780-781
Power *See also* Energy; Force requirements; Power requirements
conversion factors ... A10: 686
density, conversion factors ... A10: 686
dissipation, by dynamic material modeling ... A14: 37
as distributed from root ... EL1: 5
distribution ... EL1: 168-169, 354-357
efficiency, by dynamic material modeling ... A14: 371
for electric arc furnaces ... A15: 356-357
as IC parameter ... EL1: 401
for induction furnaces ... A15: 369-371
radiant flux, SI derived unit and symbol for ... A10: 685
requirements, for ECL and MLB ... EL1: 521
for reverberatory furnaces ... A15: 376
supplies, electrogravimetry ... A10: 200
three-terminal devices ... EL1: 424-428
Power band sawing
Al alloys ... A16: 795, 797
Cu alloys ... A16: 818
heat-resistant alloys ... A16: 755
Mg alloys ... A16: 827
MMCs ... A16: 894, 895, 897, 900
refractory metals ... A16: 867
stainless steels ... A16: 705
Ti alloys ... A16: 846, 851
tool steels ... A16: 714, 723
zirconium ... A16: 855
Power bending *See also* Bending
and hand bending ... A14: 665
of wire ... A14: 695-696
Power boiler steam generators
remote-field eddy current inspection ... A17: 20
Power brushing ... M5: 150-156
applications ... M5: 150-156
brush types ... M5: 150-156
deburring process ... M5: 151-156
dry brush cleaning ... M5: 150-153
edge blending ... M5: 155
finishing machine selection ... M5: 156
flat surfaces ... M5: 156
performance characteristics and problems ... M5: 150-151
round surfaces ... M5: 156
safety precautions ... M5: 150
small brush techniques ... M5: 154
speeds ... M5: 150-156
wet brush cleaning ... M5: 152-153
Power circuits
recommended contact materials ... A2: 861-862
Power consumption ... A1: 592
steel machining ... M1: 566-567
Power cores
fabrication steps ... EL1: 133
Power cycling
effect on thermocompression ball bond ... A12: 484
Power cycling, and temperature cycling
compared ... EL1: 961
Power density
conversion factors ... A8: 722 A10: 686
as problem ... EL1: 177
Power dissemination
total ... EL1: 5
Power dissipation
calculating ... EL1: 175-176
driver, as limitation ... EL1: 6-7
optical interconnections ... EL1: 10
static, avoidance ... EL1: 6
vs time, chip ... EL1: 46
Power distribution ... EL1: 168-169, 354-357
Power equipment
ultrasonic inspection ... A17: 23
Power factor
defined ... EM1: 19, 359 EM2: 33, 461, 467, 592
Power forging hammers
for open-die forging ... A14: 61
Power generating applications *See* Power industry applications
Power generation applications
polybenzimidazoles (PBI) ... EM2: 147
vinyl esters ... EM2: 272
Power hacksawing ... A16: 29
Al alloys ... A16: 796, 800, 801
Cu alloys ... A16: 818
heat-resistant alloys ... A16: 757
Mg alloys ... A16: 827
MMCs ... A16: 895
refractory metals ... A16: 867
stainless steels ... A16: 704, 705
Ti alloys ... A16: 846, 851, 854
tool steels ... A16: 724
zirconium ... A16: 855
Power hacksaws used in sectioning ... A9: 23
Power hybrid packages
defined ... EL1: 452
Power inductors
failure mechanisms ... EL1: 1004-1005
Power industry
stainless steel corrosion ... A13: 560-561
Power industry applications *See also* Energy industry applications; Nuclear applications
of A15 superconductors ... A2: 1070-1071
cobalt-base wear-resistant alloys ... A2: 451
high-temperature superconductors ... A2: 1085
magnetically soft materials ... A2: 778-780
nickel alloys ... A2: 430
of niobium-titanium superconducting materials ... A2: 1057
structural ceramics ... A2: 1019
of titanium and titanium alloys ... A2: 588-589
transmission, with A15 superconductors ... A2: 1071
utilities, copper and copper alloys ... A2: 239
Power laminates
fabrication steps ... EL1: 133
Power law
creep ... A8: 304, 310
Paris ... A8: 680-681
rate law expression ... EM4: 55
Power law model of Ostwald and de Waele ... EM3: 322
Power loss ... EM3: 23
defined ... EM2: 33
Power modules
multiple terminal ... EL1: 432-434
Power packages
three-terminal discrete semiconductor ... EL1: 424-428
Power plant *See also* Nuclear power applications
components, creep defects ... A17: 54
Power plants
coal-fired ... A13: 985
combined cycle ... A13: 986
failure analysis procedures for ... A11: 602
fossil fuel ... A13: 985-1010
gas turbines ... A13: 986
main steam line failure in ... A11: 667-669
steam ... A13: 985-986
Power presses *See also* Hydraulic presses; Mechanical presses; Presses
for sheet metal forming ... A14: 489-491
Power pulsing
as failure mechanism ... EL1: 1012
Power requirements *See also* Force requirements; Power
contour roll forming ... A14: 625
forming, stainless steels ... A14: 760
laser cutting ... A14: 738-739
shearing, of bar ... A14: 714
straight-knife shearing ... A14: 702-703
three-roll forming ... A14: 620
Power socket wiring ... EL1: 132
Power sources ... A6: 36-44
arc welding ... A6: 36-41
disconnect switch provision ... A6: 36
fuse and conductor size recommendations ... A6: 36
multiple operator (MO) power sources ... A6: 41
nameplate specifications ... A6: 36
power source selection ... A6: 36-37
pulsed power supplies ... A6: 40-41
ratings and standards ... A6: 36
source characteristics ... A6: 37-40
electron-beam welding ... A6: 42-44
resistance welding power sources ... A6: 40, 41-42
Power spectral density (PSD) ... A18: 335, 336
antiwear agents used ... A18: 101
dispersants used ... A18: 100
friction modifiers ... A18: 104
viscosity improvers used ... A18: 110
Power spinning ... A14: 600-604
cone spinning mechanics ... A14: 601
of cones, tools for ... A14: 602-603
of hemispheres ... A14: 603-604
lubricants and coolants ... A14: 604
machines ... A14: 601-602
of magnesium alloys ... A14: 829
speeds and feeds ... A14: 603
of titanium alloys ... A14: 845
of tube, mechanical properties ... A14: 678
Power springs
steel ... M1: 287, 288
Power steering metering pumps
farm tractors ... A7: 671-672
Power steering pressure plate ... A7: 617, 619
Power supplies for
capacitor discharge stud welding ... M6: 737
electrogas welding ... M6: 239-240
electroslag welding ... M6: 228
flash welding ... M6: 559
flux cored arc welding ... M6: 97-99
gas metal arc welding ... M6: 156-158, 429
of aluminum alloys ... M6: 380
of magnesium alloys ... M6: 429
gas tungsten arc welding ... M6: 187-188
generators ... M6: 187
three-phase rectifiers ... M6: 187-188
transformers ... M6: 188
gas tungsten arc welding, hot wire ... M6: 455
gas tungsten arc welding of
aluminum alloys ... M6: 390
magnesium alloys ... M6: 429-430
titanium and titanium alloys ... M6: 453-454
induction brazing of copper ... M6: 1043
induction brazing of steel ... M6: 967
percussion welding ... M6: 740
plasma arc cutting ... M6: 917, 919
plasma arc welding ... M6: 216
resistance seam welding ... M6: 495
shielded metal arc welding ... M6: 76-78
stud arc welding ... M6: 732
submerged arc welding ... M6: 130-132
Power supply
electronic ultrasonic inspection ... A17: 252
inductances ... EL1: 27-28
for ultrasonic testing ... A8: 243
Power system
distributed ... EL1: 392
future trends ... EL1: 393

Power, thermoelectric
defined ... **A2:** 870
Power three-lead package
construction ... **EL1:** 425
Power tool cleaning
surfaces ... **A13:** 414
Power train control
as automotive hybrid application ... **EL1:** 382
Power-delay product
and logic function performance ... **EL1:** 2
Power-driven hammer *See also* Hammer; Power forging hammers; Power-drop hammers
defined ... **A14:** 10
Power-drop hammers ... **A14:** 26-28, 41-42
capacities ... **A14:** 25
for forming ... **A14:** 654
operation ... **A14:** 42
Power-law dislocation creep ... **EM4:** 296
Power-temperature cycling
as stress test ... **EL1:** 499
Power-train efficiency ... **A18:** 566
Powerboats
composite structures for ... **EM1:** 839
Powered carrier hanger blast cleaning machine ... **A15:** 508
Powered rotary benders
for tube ... **A14:** 668
Powerplants, aircraft
corrosion of ... **A13:** 1037-1045
Poynting vector field
in RFEC probe operation ... **A17:** 196, 198
Pozzolans ... **EM4:** 919, 920
PP *See* Polypropylenes; Reinforced polypropylenes
PPE *See* Polyphenylene ether; Polyphenylene ether blends
PPO *See* Polyphenylene ether blends; Polyphenylene oxide; Polyphenylene oxides
PPS *See* Polyphenylene sulfide; Polyphenylene sulfides
PQR data base ... **A6:** 1059
Pr-Sb (Phase Diagram) ... **A3:** 2•344
Pr-Se (Phase Diagram) ... **A3:** 2•344
Pr-Si (Phase Diagram) ... **A3:** 2•345
Pr-Sn (Phase Diagram) ... **A3:** 2•345
Pr-Te (Phase Diagram) ... **A3:** 2•345
Pr-Tl (Phase Diagram) ... **A3:** 2•346
Pr-Zn (Phase Diagram) ... **A3:** 2•346
Practical drying rate ... **EM4:** 131
Prandtl number (Pr) ... **A6:** 264, 1133
Praseodymium *See also* Rare earth metals
in compacted graphite iron ... **A1:** 56
in ferrite ... **A1:** 408
properties ... **A2:** 1185
pure ... **M2:** 787-788
pure, properties ... **A2:** 1149
as rare earth ... **A2:** 720
Praseodymium and neodymium
separation by radioanalysis ... **A10:** 249-250
Pratt & Whitney F-100 engine
connector link arm for ... **A7:** 750
Pratt and Leachworth Company
early steel casting by ... **A15:** 31
Pratt and Whitney Gatorizing process ... **A1:** 974
Pre-edge absorption
oxide series ... **A10:** 415
Pre-sinter machining
of structural ceramics ... **A2:** 1020
Preaging ... **A1:** 259
Prealloyed (PA) powders
for sintering ... **EM4:** 268
Prealloyed P/M alloys
forging of ... **A14:** 250-251
Prealloyed powders ... **A7:** 9
aluminum ... **A7:** 509
atomized bronze ... **A2:** 392-393
of brass and nickel silver ... **A2:** 392
copper-tin ... **A7:** 736
copper-zinc, mechanical properties of hot pressed ... **A7:** 510
homogenization of ... **A7:** 308
low-alloy steel ... **A7:** 101, 102
microstructural analysis ... **A7:** 488
shapemaking ... **A7:** 750-752
shrinkage during sintering ... **A7:** 480, 481
titanium ... **A7:** 165-167, 749, 754-755
Prealloyed titanium P/M compacts
mechanical properties ... **A2:** 651-653
Prealloyed titanium P/M products
types and processes ... **A2:** 655
Preamplifiers
acoustic emission inspection ... **A17:** 280-281
and amplifiers, in x-ray spectrometer detectors ... **A10:** 91
field effect transistor, for EPMA ... **A10:** 519
pulsed optical ... **A10:** 91
Prearc period
defined ... **A10:** 679
Prebond treatments ... **EM3:** 23, 34-35
Precarburizing ... **A4:** 327
Precautions *See also* Toxicity
with aluminum-lithium alloys ... **A2:** 182-184
cast copper alloys ... **A2:** 357-391
Precession camera
diffraction pattern photographs by ... **A10:** 345-346
Precession method
net nuclear magnetization ... **A10:** 280
single-crystal diffraction ... **A10:** 330
Precious metal plating
materials/process selection ... **EL1:** 116
Precious metal powders *See also* Gold powders; Palladium powders; Platinum powders; Silver powders ... **A7:** 147-151
manufacturing ... **A7:** 149
production of ... **A7:** 147-151
properties ... **A7:** 148-149
thermal expansion rate ... **A7:** 611
Precious metals *See also* Electrical contact materials; Gold; Gold alloys; Iridium; Noble metals; Osmium; Palladium; Palladium alloys; Platinum; Platinum alloys; Pure metals; Ruthenium; Silver; Silver alloys ... **M2:** 659-667
annealing ... **A4:** 939-947 **M4:** 760, 761, 762
application methods for banding decorating ... **EM4:** 472
brazes used in ceramic/metal joining ... **EM4:** 479
coatings ... **M2:** 666, 667
coatings, types and uses ... **A2:** 695
commercial forms ... **M2:** 665-667
commercial forms and uses ... **A2:** 694-695
defined ... **A13:** 10
dental alloys ... **A2:** 695-698
differential-interference contrast technique of examination ... **A9:** 552
electronic applications ... **A6:** 991, 994-995
in electronic scrap recycling ... **A2:** 1228
in electroplated coatings ... **A18:** 837-838
gold ... **M2:** 660
gold and gold alloys ... **A2:** 704-707
hydrogen fluoride/hydrofluoric acid corrosion ... **A13:** 1169
impressed-current anodes ... **A13:** 469, 921
industrial applications ... **A2:** 693-694
industrial uses ... **M2:** 664-666
jewelry ... **M2:** 666-667
microstructures of ... **A9:** 552
nonfusion joining processes for sheet metals ... **A6:** 399
overlays, electrical contact materials ... **A2:** 848
oxyfuel gas welding ... **A6:** 281
palladium and palladium alloys ... **A2:** 714-718
platinum and platinum alloys ... **A2:** 707-714
platinum group, as electrical contact materials ... **A2:** 846-848
platinum-group ... **M2:** 660-661
powders, types and uses ... **A2:** 694-695
precoating ... **A6:** 131
properties ... **A2:** 688, 699-719
resistance brazing ... **A6:** 340
resources and consumption ... **A2:** 689-690
silver ... **M2:** 659-660
silver and silver alloys ... **A2:** 699-704
special properties ... **A2:** 691-694 **M2:** 662-665
SSMS analysis in geological ores ... **A10:** 141
surface treatments, titanium/titanium alloys ... **A13:** 693-694
trade practices ... **A2:** 690-691 **M2:** 661-662
ultrasonic welding ... **A6:** 895
uses of ... **A2:** 688-698
Precious metals and alloys
laser beam welding ... **M6:** 647
soldering ... **M6:** 1075
Precious metals, corrosion resistance
gold ... **M2:** 669-670
iridium ... **M2:** 669
Precious metals, corrosion resistance (continued)
palladium ... **M2:** 669
platinum ... **M2:** 668-669
rhodium ... **M2:** 669
silver ... **M2:** 670
Precious metals, industrial applications
ceramics ... **M2:** 664-665
chemical ... **M2:** 664-666
coatings ... **M2:** 666, 667
containers ... **M2:** 666
crucible ... **M2:** 665-666
electrical/electronic ... **M2:** 664
electrochemical ... **M2:** 665
glass ... **M2:** 664-665
instruments ... **M2:** 664
powder ... **M2:** 666
reflectors ... **M2:** 666
safety devices ... **M2:** 666
Precipitate contrast
transmission electron microscopy ... **A9:** 117-118
Precipitate(s)
analysis, by replication ... **A17:** 54-55
debonding, as acoustic emission source ... **A17:** 287
as source, Barkhausen noise ... **A17:** 132
Precipitate-hardened phase
stainless steels ... **A13:** 47
Precipitated carbide *See* Carbide precipitates
Precipitated particles *See also* Boundary precipitates
in grain boundaries of Fe-35Ni-16Cr ... **A9:** 126
in grain boundaries of RA 333 ... **A9:** 126
in grain boundaries of Waspaloy ... **A9:** 126
as linear element in quantitative metallography ... **A9:** 126
Precipitates
as alloying problem ... **A13:** 48
defined ... **A13:** 46
discrete, along grain boundaries in nickel-base alloy ... **A10:** 307
FIM/AP study of nucleation, growth, and coarsening of ... **A10:** 583
in gravimetric analysis ... **A10:** 163
identification in stainless steels, by light- element analysis ... **A10:** 459-461
in intergranular corrosion ... **A13:** 239
lost, in gravimetric analysis ... **A10:** 163
measurement of persistence depth, as FIM application ... **A10:** 590
optical micrograph, in stainless steel tube ... **A10:** 459
particles, growth in gravimetric analysis ... **A10:** 163
phases, SEM analysis ... **A10:** 490
pure, quantitative removal for weighing gravimetry as ... **A10:** 163
secondary, from welding ... **A13:** 49
transmission electron microscopy ... **A9:** 117-118
ultrafine secondary, atom probe analysis of IN 939 ... **A10:** 598
weld metal, preferential attack ... **A13:** 347-348
Precipitates, dispersed
wrought aluminum alloy ... **A2:** 38
Precipitating inhibitors ... **A13:** 495
Precipitation ... **A7:** 9
ammonium hydroxide ... **A10:** 168
atom probe composition profile of ... **A10:** 593
brazeability ... **A6:** 622, 625, 626
by cupferron ... **A10:** 169
by hydrolysis to oxides ... **A10:** 169
carbide, HAZ ... **A13:** 349-350
of chemical compound, as chemical equilibrium ... **A10:** 163
co-, in gravimetric analysis ... **A10:** 163
coherent, effect on diffraction pattern ... **A10:** 438, 440
of corrosion products, manned spacecraft ... **A13:** 1097-1099
defined ... **A9:** 14
effect in SCC ... **A12:** 26
efficiency measured ... **A10:** 243
embrittlement, steel castings, elevated temperatures ... **A11:** 407
from solution, powder production by ... **A7:** 52, 54
grain-boundary, AES analysis ... **A10:** 549
grain-boundary, in nickel alloys ... **A13:** 154-155
hardening, and infiltration ... **A7:** 558
heat treatment, defined ... **A13:** 10

Precipitation (continued)
heat treatment, effect on SCC in aluminum alloys ... **A11:** 220
of implanted alloys ... **A10:** 485-486
indirect ... **A7:** 54
inorganic ... **A10:** 169
of intermetallic compounds, inclusions from ... **A15:** 488
of intermetallic phases, stainless steels ... **A13:** 551
intermetallic-phase ... **A11:** 267
of internal hydrogen, cracking from ... **A11:** 248
methods for hydrometallurgical processing of copper powder ... **A7:** 118-119
nonmetallic inclusions by ... **A11:** 316
particles, and hydrogen diffusivity ... **A13:** 166
primary metastable, of beta ... **A15:** 128
processes, interaction of ... **A11:** 267-268
of R_2O_3 group, by ammonium hydroxide ... **A10:** 168
separations ... **A10:** 168
SIMS analysis of second-phase distribution due to ... **A10:** 610
sodium hydroxide ... **A10:** 168
successive ... **A7:** 54
sulfide ion ... **A10:** 169
techniques ... **A10:** 168-170
in tin-antimony alloys resulting from surface strain energy ... **A9:** 450
in titanium and titanium alloys ... **A9:** 460-461
titrations ... **A10:** 173
in weldments ... **A13:** 344
within dimples, x-ray analysis ... **A12:** 174
Precipitation etching ... **A9:** 61
defined ... **A9:** 14
and line etching ... **A9:** 62
of silicon steel transformer sheets ... **A9:** 62-63
Precipitation hardenable nickel-base wrought heat-resistant alloys ... **A9:** 309
Precipitation hardened magnetic alloys ... **A9:** 538-539
Precipitation hardening *See also* Aging; Heat-treatable; Precipitation strengthening ... **A3:** 1 • 22
aluminum alloys ... **M2:** 29-30, 34, 38-39, 40-42, 43
of beryllium copper alloys ... **A14:** 809
beryllium-copper alloys ... **A2:** 403 **A9:** 395
of beryllium-nickel alloys ... **A9:** 395
copper alloys ... **M2:** 256
of copper and copper alloys ... **A14:** 810
defined ... **A9:** 14 **A13:** 10
of nickel and nickel alloys ... **A2:** 429-430
of nickel-base heat-resistant casting alloys effect on strength ... **A9:** 334
refractory metals and alloys ... **A2:** 563
stainless steels, as magnetically soft materials ... **A2:** 777-778
stainless steels, characterized ... **A13:** 550
wrought aluminum alloys ... **A13:** 586
Precipitation hardening alloys
high-alloy steel ... **A15:** 723, 731-733
Precipitation heat treating
aluminum alloys ... **A15:** 761
Precipitation heat treatment *See also* Artificial aging
defined ... **A9:** 14
wrought aluminum alloy ... **A2:** 40
Precipitation index ... **A6:** 57
Precipitation processes
plating waste treatment ... **M5:** 313-315
Precipitation reactions ... **A9:** 646-651
free energy-composition diagram of metastable and stable equilibria ... **A9:** 650
microstructural features ... **A9:** 651
phase diagram configurations ... **A9:** 646
sequence ... **A9:** 649-651
Precipitation strengthening *See also* Precipitation hardening
in copper ... **A1:** 411
dependence on precipitate size ... **A1:** 402, 403
effect of cooling on ... **A1:** 119, 402
Precipitation strengthening (continued)
in elevated-temperature service ... **A1:** 637
in ferrite ... **A1:** 402, 404, 406
wrought aluminum alloy ... **A2:** 39
Precipitation temperatures
nickel-base alloys ... **A14:** 265
Precipitation temperatures of aluminum alloys ... **A9:** 351
Precipitation titrations
of industrial materials ... **A10:** 173
Volhard, of silver ... **A10:** 173
Precipitation-hardenable stainless steel
SCC causing environments ... **A8:** 526
workability ... **A8:** 165, 575
Precipitation-hardenable stainless steels *See also* Wrought stainless steels; Wrought stainless steels, specific types
microstructures ... **A9:** 285
pickling of ... **M5:** 72
Precipitation-hardening (PH) stainless steels, wrought ... **A6:** 482-493
aging ... **A6:** 482, 483, 484, 487, 488, 492
austenitic PH steels, welding of ... **A6:** 482, 485, 487, 490-493
definition ... **A6:** 482
electrodes ... **A6:** 483, 484, 487, 488, 489, 492
electron-beam welding ... **A6:** 484, 487, 489, 490, 491, 492
filler metals ... **A6:** 483, 487, 488, 490, 491
fusion zone ... **A6:** 483, 488, 490
gas-metal arc welding ... **A6:** 483, 487, 489
gas-tungsten arc welding ... **A6:** 483, 484, 487, 488, 489, 491-492
heat-affected zone ... **A6:** 483-484, 487, 488, 489, 490
laser-beam welding ... **A6:** 489
martensitic PH steels, welding of ... **A6:** 482-488
microstructure ... **A6:** 482-483, 488
postweld heat treatment ... **A6:** 483-484, 487, 488, 489, 490, 492, 493
resistance welding ... **A6:** 489, 490, 491, 492
semiaustenitic PH steels, welding ... **A6:** 482, 485, 486-487, 488-490
shielded metal arc welding ... **A6:** 483, 489, 490
solidification cracking ... **A6:** 484, 487, 490
spot and seam welding ... **A6:** 489
submerged arc welding ... **A6:** 489
types ... **A6:** 482
weldability ... **A6:** 483-488
Precipitation-hardening alloys
nickel-base alloys ... **A6:** 869, 927
wear resistance versus hardness of materials ... **A18:** 708
Precipitation-hardening stainless steel *See also* Stainless steel(s); Steel(s)
arc-welded ... **A11:** 428
stress-corrosion cracking in ... **A11:** 217-218
susceptibility to hydrogen damage ... **A11:** 249
Precipitation-hardening stainless steels *See also* Stainless steels, precipitation hardening; Wrought stainless steels ... **A6:** 695-697 **A14:** 226-227 **M3:** 25-28, 30-31
arc welding *See* Arc welding of stainless steels
brazing and soldering characteristics ... **A6:** 626
compositions ... **M6:** 526
compositions of ... **A1:** 843, 847-848
electron beam welding ... **M6:** 638
electron-beam welding ... **A6:** 869
elevated-temperature properties ... **A1:** 942-944
flash butt welding ... **A6:** 492
forgeability of ... **A1:** 893-894
fractographs ... **A12:** 370-374
fracture toughness properties ... **A1:** 865
fracture/failure causes illustrated ... **A12:** 217
friction welding ... **M6:** 721
furnace brazing ... **A6:** 918-919
machinability of ... **A1:** 896-897
physical properties ... **M6:** 527
plasma arc welding ... **A6:** 490
repair welding ... **A6:** 1106-1107
Precipitation-hardening stainless steels (continued)
resistance welding ... **A6:** 848 **M6:** 527
tensile properties of ... **A1:** 864, 865
weldability of ... **A1:** 902-904
Precipitation-hardening stainless steels, specific types
13-8 PH, cup-and-cone tension overload fracture ... **A12:** 370
13-8 PH, high-cycle fatigue fracture ... **A12:** 371
13-8 PH, low-cycle fatigue fracture ... **A12:** 371
13-8 PH, SCC fracture ... **A12:** 372
Armco 15-5 PH, high-cycle fatigue fracture ... **A12:** 373
Armco 17-7 PH, brittle intergranular fracture ... **A12:** 374
Armco 17-7 PH, fracture by cross-check defect ... **A12:** 374
Precipitation-hardening steels
materials for die-casting dies ... **A18:** 629
Precipitation-strengthened alloys
postweld heat treatment ... **A6:** 572-574
Precision *See also* Reproducibility
and accuracy, compared ... **A10:** 524-525
analysis and automation electrogravimetry ... **A10:** 199
analysis, electrogravimetry as ... **A10:** 197
in corrosion testing ... **A13:** 195-196
defined ... **A8:** 10 **A10:** 679
experimental, replication as method of ... **A8:** 640-641
high-, electrometric titration for ... **A10:** 202
potentiometer, for thermocouple calibration ... **A8:** 314
of preform ... **A14:** 160-161
of radioanalysis ... **A10:** 246-247
of setup ... **A14:** 160
of single-crystal analysis ... **A10:** 352
of tension testing machine ... **A8:** 50
of tooling ... **A14:** 160
in UV/VIS absorption spectroscopy ... **A10:** 70
Precision boring machines ... **A16:** 161, 162, 168, 169, 171, 173
Precision capacitive bridges
as mercury displacement measure ... **A7:** 267-268
Precision cast hot-work wrought tool steels ... **A1:** 779
Precision casting *See also* Investment (lost wax) casting
of Alnico alloys ... **A15:** 737
defined ... **A15:** 9
furnaces, vacuum induction ... **A15:** 399-401
with gold-germanium alloys ... **A2:** 743
market trends ... **A15:** 44
titanium alloy ... **A15:** 824
titanium and titanium alloy castings ... **A2:** 634
types ... **A15:** 37
Precision casting, and powder forging
compared ... **A14:** 196
Precision castings *See also* Castings
computed tomography (CT) ... **A17:** 363
Precision ceramic molding processes
for patternmaking ... **A15:** 195
Precision extrusions
aluminum and aluminum alloys
Precision forging *See also* Forging; Precision casting ... **A14:** 158-185
advantages ... **A14:** 158
of aluminum alloys ... **A14:** 244, 251-254
applications ... **A14:** 158-159
defined ... **A14:** 10, 158
dies for ... **A14:** 51-52
equipment ... **A14:** 162-171
forming applications ... **A14:** 172-175
as new metalworking process ... **A14:** 17
and powder forging, compared ... **A14:** 196
process control ... **A14:** 160-162
process temperature, selection ... **A14:** 171-172
radial ... **A14:** 146-147
of spiral bevel gear ... **A14:** 173-175

Precision forging (continued)
through-die design ... **A14:** 17
tooling design ... **A14:** 159-160
Precision forgings
aluminum and aluminum alloys ... **A2:** 6
mechanically alloyed oxide
alloys ... **A2:** 948
dispersion-strengthened (MA ODS)
Precision forming
applications ... **A14:** 172-175
Precision machining
by scanning laser gage ... **A17:** 12
Precision matching study
fatigue striations in nickel ... **A12:** 205-206
Precision metal finishing
as interferometer application ... **A17:** 14-15
Precision molding *See also* Investment (lost wax) casting; Precision casting
types ... **A15:** 37
Precision optics
beryllium powder application ... **A7:** 169
Precision part ... **A7:** 9
Precision resistance alloys *See also* Electrical resistance alloys
properties and applications ... **A2:** 823, 825-826
Precision resistors
electrical resistance alloys ... **A2:** 822
Precision sintered part ... **A7:** 9
Precision warm forging ... **A14:** 158
Precision weighing
electric arc furnaces ... **A15:** 363
Precision-molded applications
critical properties ... **EM2:** 458
Precleaning *See also* Cleaning; Postcleaning; Surface preparation
for liquid penetrant inspection ... **A17:** 80
for urethane coatings ... **EL1:** 777
Precoated steel sheet *See* Steel sheet, precoated ... **A1:** 212-225
aluminum coatings ... **A1:** 218-220
base metal and formability ... **A1:** 219
coating weight ... **A1:** 218-219
corrosion resistance ... **A1:** 219
handling and storage ... **A1:** 220
heat reflection ... **A1:** 219-220
heat resistance ... **A1:** 219
mechanical properties ... **A1:** 218, 219
painting ... **A1:** 220
weldability ... **A1:** 220
aluminum-zinc alloy coatings ... **A1:** 220-221
organic composite coatings ... **A1:** 223
phosphate coatings ... **A1:** 222
prepainted sheet ... **A1:** 223-224
design considerations ... **A1:** 224
packaging and handling ... **A1:** 224
selection of paint system ... **A1:** 224
shop practices ... **A1:** 224
preprimed sheet ... **A1:** 222-223
formability ... **A1:** 222
zinc chromate primers ... **A1:** 222
zinc-rich primers ... **A1:** 222-223
terne coatings ... **A1:** 221-222
tin coatings ... **A1:** 221
zinc coatings ... **A1:** 212-218
chromate passivation ... **A1:** 214-215
coating tests and designations ... **A1:** 212-214, 215
corrosion resistance ... **A1:** 212, 581, 584
electrogalvanizing ... **A1:** 217
hot dip galvanizing ... **A1:** 216-217, 218
packaging and storage ... **A1:** 215-216
painting ... **A1:** 215
zinc alloy ... **A1:** 217-218
zinc spraying ... **A1:** 218
Zincrometal ... **A1:** 217
Precoated steels
for automotive industry ... **A13:** 1011-1015
weldability of ... **A1:** 609-610
Precoating *See also* Coatings
ceramic glass ... **A14:** 279
definition ... **A6:** 1212 **M6:** 13
metals, contour roll forming ... **A14:** 634
Precompaction
or "slugging" fabrication ... **A7:** 665
Precompression
effect on fatigue strength ... **A11:** 112
Preconditioning ... **EM3:** 23
defined ... **EM2:** 33
Precorrosion
and time-dependent subcritical crack propagation ... **A13:** 145
Precracked Charpy test ... **A8:** 267-268
Precracked specimens ... **A13:** 10, 253-260
Precracked test specimens
calculation of crack growth rates ... **A8:** 518-519
cantilever bend ... **A8:** 511
constant *K* ... **A8:** 515
crack configuration and orientation ... **A8:** 516
dimensional requirements ... **A8:** 515-516
double-beam ... **A8:** 513-515
loading arrangements and crack measurement ... **A8:** 518
machining ... **A8:** 516
modified compact ... **A8:** 512-513
preparation of ... **A8:** 515-517
SCC testing ... **A8:** 497-498, 510-519
testing procedure ... **A8:** 517-519
typical ... **A8:** 514
Precracking ... **A8:** 259, 382, 517
fatigue ... **A12:** 75, 236-237, 397
Precure *See also* Cure ... **EM3:** 23
defined ... **EM1:** 19 **EM2:** 33
Precursor
defined ... **EM2:** 33
Precursors
for carbon fibers ... **EM1:** 112, 868
carbon/carbon composite ... **EM1:** 917-918
defined ... **EM1:** 19
to graphite fiber MMCs ... **EM1:** 868-869
Prediction ... **A13:** 234, 316
accelerated life ... **EM2:** 788-795
of deformation under load ... **EM2:** 673-678
of isothermal oxidative stability ... **EM2:** 566
Prediction intervals ... **A8:** 626
Predictive modeling
thermoset resins ... **EM2:** 527
Predominance area diagrams
limitations ... **A13:** 64
Predrying
polyamide-imides (PAI) ... **EM2:** 132-133
Prefailure phenomenon
and durability ... **EM2:** 551-554
Preferential
attack ... **A13:** 323-324, 876
corrosion ... **A13:** 117, 363
dissolution ... **A13:** 17, 57
oxidation, in gaseous corrosion ... **A13:** 17
pitting, tantalum weldments ... **A13:** 345-346
weld corrosion, carbon steels ... **A13:** 363
Preferential alloy removal *See also* Dealloying
decarburization/selective oxidation as ... **A13:** 134
Preferential attack
in a palladium-based electrical contact material ... **A9:** 564
in electropolishing ... **A9:** 51
in stainless steels ... **A11:** 200
of the eutectic in magnesium alloys ... **A9:** 427
Preferential corrosion
tin-bronze alloy ... **A12:** 403
Preferential dissolution
in intergranular attack ... **A11:** 338
Preferential evaporation, of solute
in vacuum melting ultrapurification techniques ... **A2:** 1094
Preferential grain-boundary attack
ceramic fracture ... **A12:** 471
Preferential oxidation
at elevated temperatures ... **A11:** 405-406
of grain boundaries ... **A11:** 405
interdendritic ... **A11:** 406
as thermocouple failure mechanism ... **A2:** 881
Preferential sputtering ... **A18:** 854, 855, 857
Preferred crystallographic growth
SACP use to establish ... **A10:** 509
Preferred orientation *See also* Fiber; Fiber alignment; Fiber orientation; Orientation; Texture; Texture, crystallographic
in aluminum alloys, etchants for examination for ... **A9:** 355
beryllium ... **A7:** 756
in carbon fibers ... **EM1:** 50
crystallographic, measurement and analysis of ... **A10:** 357-364
defined ... **A9:** 14 **A11:** 8
Preferred orientation (continued)
Euler plots and orientation distribution ... **A10:** 360-361
from anisotropy in forging ... **A11:** 316
identification of, by polarized light etching ... **A9:** 59
metallurgical specification of ... **A10:** 359-360
oxide texture ... **A13:** 66
in plastic torsion ... **A8:** 143
pole figures ... **A10:** 360
polycrystalline, and property behavior ... **A10:** 358
RD-TD-ND system ... **A10:** 358, 359
specifying ... **A10:** 359
speci
in stereographic projection ... **A10:** 358-359
as XRPD source of error ... **A10:** 340-341
zirconium ... **A2:** 667
Preferred solder connection
defined ... **EL1:** 1153
Prefinishing
costs of ... **EM2:** 649-650
Prefit ... **EM3:** 23
defined ... **EM2:** 33
Prefits
defined ... **EM1:** 19
Prefixes, of SI units
names and symbols ... **A10:** 687
Preflow time
definition ... **M6:** 13
Preform *See also* Slury preforming ... **EM4:** 414-415
binder, defined ... **EM2:** 33
brazing filler metals available in this form ... **A6:** 119
defined ... **EL1:** 1153 **EM2:** 33
definition ... **A6:** 1212 **M6:** 13
Preform molding
as resin transfer molding ... **EM1:** 564
Preform patterns
Shaw process ... **A15:** 248-249
Preform(s) ... **A7:** 9
for brazing and soldering ... **A7:** 840
in Ceracon process ... **A7:** 537
compaction in commercial P/M forging ... **A7:** 415-417
cylindrical, press requirements ... **A7:** 545
deformation processing of ... **A7:** 531
design ... **A7:** 411-413, 539-540
hot formed, of automobile parts ... **A7:** 620
machine lens housing ... **A7:** 682
shape, fracture limits ... **A7:** 411
sintering ... **A7:** 416
spray-formed, parts production by ... **A7:** 530
Preformed butyl tapes ... **EM3:** 190
Preformed cores
ceramic ... **A15:** 9, 261
Preformed filler metal
induction brazing of steel ... **M6:** 971
resistance brazing ... **M6:** 982
Preformed resin-bonded mass finishing media ... **M5:** 135
Preforming ... **A7:** 9, 774
Preforms *See also* Fabrics; Multidirectionally reinforced preforms; P/M forging
3-D, fibers woven into ... **EM1:** 129
3-D polar, weaving machines ... **EM1:** 131
binder, defined ... **EM1:** 19
blanks, refractory metals ... **A14:** 788
by radial forging ... **A14:** 145, 147
closed-die forging ... **A14:** 77-78
continuous SiC fiber MMC ... **EM1:** 860-861
defined ... **A14:** 10 **EM1:** 19
design of ... **A14:** 50-51, 77-78, 154
eutectic die attach ... **EL1:** 214
experimental and modeling methods for ... **A14:** 51
explosive forming of ... **A14:** 636
fiber ... **EM1:** 529-532
geometry, computer design of ... **A14:** 411
hot upset, extrusion of ... **A14:** 306
hot-die/isothermal forging ... **A14:** 154
impressions, location of ... **A14:** 51
iron powder, reduction of ... **A14:** 194
materials/assembly ... **EM1:** 529-530
molding ... **EM1:** 564
multidirectional ceramic-ceramic composites ... **EM1:** 934-936
multidirectionally reinforced ... **EM1:** 129-131

Preforms (continued)
noncylindrical, weaving with shaping tool **EM1:** 131
porous, workability **A14:** 193
in powder forging **A14:** 191-192
for power spinning **A14:** 603
precision of **A14:** 160-161
for resin transfer molding **EM1:** 566-568
rigid epoxies **EL1:** 815
ring, powder forging **A14:** 192
solder, surface-mount soldering **EL1:** 700
surface conditions of **A14:** 160-161
three-dimensional **EM1:** 129-130
three-directional orthogonal construction **EM1:** 915
tooling for **A14:** 175
for tube spinning **A14:** 675
turbine shaft, by radial forging **A14:** 147
welded, explosive forming of **A14:** 642-643
woven multidirectional, carbon/graphite composite **EM1:** 915-917

Pregel
defined **EM1:** 19 **EM2:** 33

Pregelled, cationic corn starch
application or function optimizing powder treatment and green forming **EM4:** 49

Pregelling adhesives
automotive applications **EM3:** 553-554

Preheat **A7:** 9
definition **A6:** 1212
zone, furnace **A7:** 351

PREHEAT (software package) **A6:** 644

Preheat current
definition **M6:** 13

Preheat current (resistance welding)
definition **A6:** 1212

Preheat force **A6:** 888

Preheat temperature
definition **A6:** 1212 **M6:** 13

Preheat time
definition **M6:** 13

Preheating *See also* Die heating; Heat treatment; Heating; specific processes **EM3:** 23
aluminum alloys **A14:** 247
cast irons, for welding **A15:** 525-526
of castings **A15:** 525
cobalt-base alloys **A15:** 813-814
defined **EM1:** 19 **EM2:** 33
definition **M6:** 13
for drop hammer forming **A14:** 657
dryer for **A15:** 373
electric arc furnace **A15:** 360
HAZ, cast iron effect **A15:** 525
mold, and mold temperature **A15:** 283
and mold life **A15:** 281
for oxyacetylene welding **A15:** 530
for oxyfuel gas cutting **A14:** 722, 724
for plaster molding **A15:** 244
process, through-hole soldering **EL1:** 684-688
rigid epoxies **EL1:** 815
temperatures **A15:** 535
for through-hole soldering **EL1:** 684-685
titanium alloys, for forging **A14:** 278
of tool and die steels **A14:** 54
wave soldering **EL1:** 702
for welding aluminum alloys **A15:** 763

Preheating, substrate
for thermal spray coatings **A13:** 460

Prehistoric materials
radioanalytic dating of **A10:** 243

Preimpregnated materials
heat curing of **EM2:** 341

Preimpregnation *See also* Prepreg
defined **EM1:** 19 **EM2:** 33

Preinspection preparation *See* Surface preparation

Preionization structure
in typical K-shell ionization edge **A10:** 450

Preliminary laboratory examination
of failed parts **A11:** 173

Preliminary test
S-N data, for two-point program with up-and-down strategy **A8:** 705
with single fatigue test specimen **A8:** 705-706

Preliminary visual examination
for photography **A12:** 78
specimen **A12:** 72-73

Premature fracture *See also* Brittle fracture; Cracking; Fracture
causes of **A8:** 496

Premium engineered castings
as aluminum alloy specialty **A15:** 756-757
fatigue in **A15:** 764
mechanical property specifications **A15:** 757
quality assurance **A15:** 766

Premium quality aluminum alloy castings
high-strength, high-toughness **A2:** 149
mechanical properties **A2:** 147
properties **A2:** 127, 130

Premix
defined **EM1:** 19 **EM2:** 34

Premix burners
defined **A10:** 679
for flame source emission **A10:** 28
temperature of flames **A10:** 29

Premix chamber
AAS flame atomizers **A10:** 47, 48

Premixed flexible epoxy systems **EL1:** 818

Premixes *See also* Premixing
bronze alloys **A7:** 191, 279
copper-tin powders **A7:** 736
lubrication effects **A7:** 190
as noun, defined **A7:** 9
segregation **A7:** 190
as verb, defined **A7:** 9

Premixing **A7:** 186-189

Premolding **EM2:** 34, 85
defined **EM1:** 19

Preoxidation
as P/M oxide-dispersion technique **A7:** 719

Preoxidized-press-sinter-extrude process
for composite contact materials **A2:** 857

PREP *See* Plasma rotating electrode process

Prepaint processing
carbon steels **A13:** 528

Prepaint treatments **M5:** 476-478

Prepainted sheet **A1:** 223-224
design considerations **A1:** 224
packaging and handling **A1:** 224
select-on of paint system **A1:** 224
shop practices **A1:** 224

Prepainted steel sheet
applications **M1:** 176
bending **M1:** 176
crazing **M1:** 176
design considerations **M1:** 176
galvanized steel for coil coating **M1:** 169
packaging and handling **M1:** 176
selection of system **M1:** 176
shop practices **M1:** 176

Preparation *See also* Sand preparation; Surface preparation
cast for welding **A15:** 524-525
of corrosion tests **A13:** 193-196
of green sand **A15:** 225-226
metal, aluminum alloy casting **A15:** 753
and preservation, specimen **A12:** 72-77
of SEM specimens **A12:** 171-173
specimen **A13:** 194, 223, 253-260
of test specimens **A13:** 194

Preparation for cure
material types/functions **EM1:** 642-644

Preparative liquid chromatography
for obtaining purified compounds **A10:** 654

Prepared nitrogen-base atmospheres
advantages, disadvantages **M4:** 399
applications **M4:** 399-400

Prepared nitrogen-base atmospheres (continued)
generation **M4:** 399, 400, 401
generator maintenance **M4:** 402
maintenance **M4:** 402
molecular-sieve systems **M4:** 400-401
monoethanolamine system **M4:** 401
safety precautions **M4:** 402
types **M4:** 394, 399

Preplastication
defined **EM2:** 34

Preplied prepregs *See also* Multidirectional tape prepregs; Prepregs; Tape prepregs **EM1:** 146

Preplied tape *See also* Multidirectional tape prepregs **EM1:** 146

Preply *See also* Ply
defined **EM1:** 19 **EM2:** 34

Prepolishing *See also* Polishing
chemical or electrolytic, as FIM sample preparation **A10:** 585

Prepolymer **EM3:** 23, 110
defined **EM1:** 19

Prepolymer molding *See also* One-shot molding **EM3:** 23
defined **EM2:** 34

Prepolymer(s)
cyanates **EM2:** 232
defined **EM2:** 34
fiberglass reinforced BPADCy **EM2:** 232-233

Prepreg *See also* Impregnated fabric; Preimpregnation; Prepreg molding
defined **EL1:** 1153 **EM2:** 34
fiber content **EM2:** 338
fiber reinforcements **EM2:** 506
materials, manufacture **EL1:** 510
systems, resins and reinforcements **EL1:** 534
thermoset vs thermoplastic TPIs **EM2:** 178
urethane hybrids **EM2:** 270-271

Prepreg molding
molded-in color **EM2:** 306
processes **EM2:** 338-341
size and shape effects **EM2:** 291
surface finish **EM2:** 303-304
textured surfaces **EM2:** 305
tooling **EM2:** 341-343

Prepreg resins
aerospace applications **EM1:** 139-141
curatives **EM1:** 139-141
epoxy **EM1:** 139
formulation **EM1:** 141
future trends **EM1:** 142
lower performance applications **EM1:** 141-142

Prepreg tape *See also* Tape prepreg
for filament winding **EM1:** 138
requirements **EM1:** 105

Prepreg tape-laying machines
automated **EM1:** 631-635

Prepreg tow *See also* Prepregs; Tow **EM1:** 151-152
applications **EM1:** 151-152
carbon reinforced, for filament winding **EM1:** 138
manufacture **EM1:** 151
product forms **EM1:** 151
for winding **EM1:** 135

Prepregs *See also* Impregnated fabric; Preimpregnation; Prepreg resins; Tape prepregs
advantages **EM1:** 33
application **EM1:** 33
in bands, product forms from **EM1:** 152
boron fiber **EM1:** 58
cost guidelines **EM1:** 105
defined **EM1:** 19, 33
epoxy composite **EM1:** 71-73
glass-phenolic, as flame resistant **EM1:** 141
graphite tools from **EM1:** 587-588
mechanical property tests **EM1:** 737
near-net resin content **EM1:** 73
preplied quasi-isotropic, costs **EM1:** 146
properties analysis **EM1:** 737

Prepregs (continued)
resin, formulation **EM1:** 141
resins for **EM1:** 139-142
suppliers **EM1:** 141
tape and fabric, compared **EM1:** 602
tests for **EM1:** 291-292
thermoplastic, fabrication with **EM1:** 103-104
thermoplastic, potential of **EM1:** 99-100
tooling, properties **EM1:** 587
tow **EM1:** 151-152
uncured, ultrasonic ply cutting **EM1:** 615-618
useful life **EM1:** 144
woven fabric **EM1:** 148-150
Preprimed sheet **A1:** 222-223
formability of **A1:** 222
zinc chromate primers **A1:** 222
zinc-rich primers **A1:** 222-223
Preprimed steel sheet
primers for **M1:** 169, 175
Preproduction test **EM3:** 23
Prerinse
for liquid penetrant inspection **A17:** 82
Present worth (PW) method
economic analysis **A13:** 370-371
Present worth of future revenue requirements method
economic analysis **A13:** 370
Presentation of friction and wear data **A18:** 489-492
transition diagrams **A18:** 491
determination methodology **A18:** 491
tribographs **A18:** 489-491
dependence of tribodata on various parameters **A18:** 491
friction-time master curves **A18:** 489
wear-time master curves **A18:** 489-490
tribomaps **A18:** 491-492
wear regimes **A18:** 492
Preservation
and preparation, specimen **A12:** 72-77
techniques **A12:** 73
Preservation of specimens
carbon and alloy steels **A9:** 172
Preservation sample **A10:** 15, 16
Preservatives *See also* Solderability treatments
solderability, development/types **EL1:** 561-564
Preservice
environments, effect on stress-corrosion cracking **A11:** 211
testing, pipe failures in **A11:** 697-699
Presetting of springs **M1:** 290-291, 312-313
Preshadowed replica
defined **A9:** 14
Presintered blank **A7:** 9
Presintered density **A7:** 9
Presintering *See also* Sintering **A7:** 9
atmosphere, effect on P/M aluminum parts **A7:** 384
growth and shrinkage during **A7:** 480
of P/M ferrous powders **A7:** 683
of stainless steels **A7:** 368
Presolidification *See also* Liquid metal processing
fundamentals of **A15:** 49-97
Prespark period
defined **A10:** 679
Press
as consolidation technique **A7:** 719-720
deflection, in tooling design **A7:** 337
as noun, defined **A7:** 9
sinter aid, as P/M consolidation technique **A7:** 719
sinter re-press, P/M consolidation techniques **A7:** 719
tonnage and stroke capacity, in rigid tool compaction **A7:** 325-326
tools **A7:** 9, 286, 291
as verb, defined **A7:** 9
Press accessories **A14:** 497
Press and snap fits **EM2:** 713-721
Press bending *See also* Bending
accurate location/form, of holes **A14:** 531-532
accurate spacing, flanges **A14:** 532
bend orientation **A14:** 524-525
bendability and selection of steels **A14:** 523
compound dies **A14:** 527
of curved flanges **A14:** 530-531
of cylindrical parts **A14:** 529
Press bending (continued)
die construction **A14:** 525-527
edge bending **A14:** 529
of low-carbon steel **A14:** 523-532
lubrication **A14:** 528-529
minimum bend radius **A14:** 523-524
progressive dies **A14:** 527-528
separate dies **A14:** 528
single-operation dies **A14:** 527
springback control **A14:** 531
straight flanging **A14:** 529
transfer dies **A14:** 528
Press brake *See also* Bending brake; Press-brake forming
defined **A14:** 0
Press brakes
for bar bending **A14:** 662
capacity **A14:** 534
hybrid **A14:** 534
hydraulic **A14:** 534
length of stroke **A14:** 534
mechanical **A14:** 533-534
size **A14:** 534
vs punch press **A14:** 543-544
Press capacity
defined **A14:** 10
Press clave
defined **EM1:** 19 **EM2:** 34
Press columns
ultrasonic inspection **A17:** 232
Press consolidation process **A7:** 7, 749
Press, failed extrusion
EPMA-identified microconstituents in **A11:** 37
Press fits
in fatigue crack initiation **A8:** 371
Press fitting
of shafts **A11:** 459, 469-470
Press forging
defined **A14:** 10
and radial forging, compared **A14:** 148
Press forming
accuracy **A14:** 552-554
aluminum-coated steels **A14:** 561-562
annealed steel **A14:** 558-559
auxiliary operations **A14:** 554
blanking and piercing **A14:** 556-558
chromium-plated steels **A14:** 563-564
coated high-strength steels **A14:** 565
deburring blanks **A14:** 548
defined **A14:** 10, 504
dies **A14:** 504-507, 545-546
ferritic stainless steels **A14:** 764
galvanized steels **A14:** 560-561
high-carbon steel **A14:** 556-559
hole flanging **A14:** 559
hot forming **A14:** 554
large irregular shapes **A14:** 549
of low-carbon steels **A14:** 545-555
lubrication **A14:** 504-505, 545
multiple-slide forming **A14:** 559
nickel-plated steels **A14:** 563
organic-coated steels **A14:** 564-565
presses **A14:** 545
of pretempered steel **A14:** 558
process development **A14:** 548
process variables **A14:** 504
of ribs, beads, and bosses **A14:** 552
safety **A14:** 555
speed of **A14:** 545
stainless steels **A14:** 763-765
steel selection for **A14:** 546-547
and stretching, stainless steels **A14:** 764
surface finish **A14:** 547-548
tin-coated and terne-coated steels **A14:** 562-563
vs alternative methods **A14:** 554-555
vs swaging **A14:** 140
work metal thickness **A14:** 548-549
workpiece shape **A14:** 549
Press forming dies
cemented carbides **M3:** 492
chromium plating **M3:** 491-492
die life **M3:** 489, 491
galling resistance **M3:** 490-493
lower dies **M3:** 489, 490, 492, 493
nitriding **M3:** 491
powder metallurgy **M3:** 492
punches **M3:** 489-490
Press forming dies (continued)
steel-bonded carbides **M3:** 493
tool materials for **A14:** 504-505
upper dies **M3:** 489, 490
wear **M3:** 489, 491
wear and life **A14:** 505-507
Press frames
types **A14:** 492-493
Press load
defined **A14:** 10
Press quenching **M4:** 66
Press roll straightening
automatic **A14:** 687-688
Press slide *See* Slide
Press station
multiple-slide machines **A14:** 567-568
Press straightening **A14:** 682-684, 690-691
Press(es) *See* Hydraulic press; Hydromechanical press; specific press
Press-brake bending **A8:** 119
Press-brake forming **A14:** 533-544
aluminum alloys **A14:** 794-795
applicability **A14:** 533
defined **A14:** 533
dies and punches **A14:** 535-536
dimensional accuracy **A14:** 542-543
machine selection **A14:** 534-535
of magnesium alloys **A14:** 827
press brakes **A14:** 533-534
principles **A14:** 533
of refractory metal sheet **A14:** 786
rotary bending **A14:** 538-539
safety **A14:** 544
special dies and punches **A14:** 536-538
specific shapes, procedures for **A14:** 540-541
stainless steels **A14:** 759, 762-763
of titanium alloys **A14:** 844
tool material selection **A14:** 539-540
vs alternative processes **A14:** 543-544
work metal variables, effects **A14:** 541-542
Press-fit inserts
in magnesium alloy parts **A2:** 466-467
Press-fit interface
by liquid penetrant inspection **A17:** 86
Press-fit pin process
for mother boards **EL1:** 117
Press-sinter-extrude
of electrical contact composite materials **A2:** 857
Press-sintering *See also* Sintering
of composite electrical contact materials .... **A2:** 856, --857
Press-type resistance welding machines **M6:** 484
Press/hammer data base **A14:** 413
Pressed bar **A7:** 9
Pressed briquet XRS samples **A10:** 93
Pressed density **A7:** 9
Pressed-ceramic packages *See also* Ceramic packages
thermal performance of **EL1:** 409-410
Presses *See also* Hydraulic presses; Mechanical presses; specific presses
accessories **A14:** 497
accumulator-drive **A14:** 32-33
accuracy of **A14:** 495
as ancillary process, ring rolling **A14:** 123
ASEA Quintus fluid forming **A14:** 614-615
ASEA Quintus rubber-pad **A14:** 608
auxiliary operations in **A14:** 554
for blanking **A14:** 445-447, 451, 477
capacity **A14:** 495-496
characteristics **A14:** 490
closed-die forging in **A14:** 75-82
clutches and brakes **A14:** 496
coil handling equipment **A14:** 501-502
for coining **A14:** 180-181
for cold extrusion **A14:** 308
data base **A14:** 413
deep drawing **A14:** 577-579
defined **A14:** 10
die cushions **A14:** 497-499
die, or dieing machines **A14:** 502
direct drive **A14:** 32
direct-electric-drive **A14:** 33-34
double-action **A14:** 495
drive mechanisms **A14:** 31-33
eccentric one-point **A14:** 40
eccentric two-point **A14:** 40

Presses (continued)
feed mechanisms **A14:** 499-500
for fine-edge blanking and piercing **A14:** 473-474, 502
flexible-die forming **A14:** 503
fluid-forming, for deep drawing **A14:** 587
for forging **A14:** 25, 29-31
frame types **A14:** 492-493
friction drive **A14:** 33
Guerin process **A14:** 605
for high production **A14:** 502-503
for hot extrusion **A14:** 319-320
hot forging in **A14:** 36
hydraulic **A14:** 31-33, 490-491
identification **A14:** 489
mechanical **A14:** 29-31, 34, 489-490, 492-497
mechanical vs hydraulic **A14:** 492
mechanical/screw **A14:** 34
motor selection **A14:** 496
multiple-ram **A14:** 35
multiple-slide **A14:** 502
for open-die forging **A14:** 64
for piercing **A14:** 463-464, 477
for press forming **A14:** 545
safety of **A14:** 503
screw **A14:** 33-35
selection of **A14:** 491-492
sheet metal forming **A8:** 547 **A14:** 489-499
slide actuation in **A14:** 493-495
slides, number of **A14:** 495
speed, deep drawing **A14:** 582-583
for stainless steels **A14:** 227
straightening in **A14:** 682-684
transfer **A14:** 501-502
triple-action **A14:** 495
unloading **A14:** 500-501
Verson hydroform process **A14:** 612
Verson-Wheelon process **A14:** 609
vertical, die and die materials for **A14:** 43-58
wedge **A14:** 40
Presses, P/M **A7:** 329-338
cam-driven **A7:** 330
eccentric-driven **A7:** 330, 331
factors affecting dimensional change **A7:** 291
for hot pressing **A7:** 502
hydraulic **A7:** 330-332
mechanical **A7:** 329-330
production **A7:** 329-332
for secondary pressing operations **A7:** 338
stroke capacity **A7:** 325
types of **A7:** 334-335
Pressing *See* Compact
of blended elemental Ti-6Al-4V **A2:** 649
crack **A7:** 9
factors influencing dimensional change **A7:** 291
as noun, defined **A7:** 9
of powders, for structural ceramics **A2:** 1020
in rigid dies **A7:** 297, 298
skin **A7:** 9
tools **A7:** 9, 286, 291
as verb, defined **A7:** 9
viscoelastic properties required **EM4:** 116
Pressing tools **A7:** 9, 286, 291
Pressure *See also* Pressure die casting; Pressure systems; Pressure testing; Vapor pressure
abbreviation **A8:** 724
of abrasive waterjet cutting **A14:** 752
appearance effect, aluminum alloy **A15:** 459
appearance effect, copper alloys **A15:** 467
aqueous corrosion testing, laboratory **A13:** 226-227
assisted sintering **A7:** 309
average according to Hertzian contact theory **A18:** 40
balance by contained argon **A7:** 434
behavior of metal powders under **A7:** 297-304
blankholders, for magnesium alloys **A14:** 828
bonding **A7:** 9

Pressure (continued)
capacities, hot extrusion **A14:** 319
capacities, hot upset forging **A14:** 83
in chemical cleaning **A13:** 1142
clamping defined *See* Clamping pressure
compacting, beryllium powders **A7:** 171
control, autoclave **EM1:** 704
controlled in gas mass spectrometer **A10:** 151-152
and density **A7:** 298-300, 449, 538
duration, squeeze casting **A15:** 324
effects, kinetics gaseous corrosion **A13:** 70
effects on electrical conductivity **A7:** 608, 610
effects, produced fluids **A13:** 479
in erosion/cavitation testing **A13:** 312-313
extrusion **A14:** 312
flow under, in injection molding **EM1:** 164-167
fluid, conversion factors **A8:** 722 **A10:** 686
fluid hydrostatic **EM1:** 755-757
forging and ejection, at part full-density **A7:** 417
forging, and ejection force, powder forged parts **A14:** 193
forging, die temperature effects **A14:** 153
forging, low-carbon steel **A14:** 217
forging, prediction, closed-die forging **A14:** 79-80
forging, vs temperature **A14:** 216
friction during metal forming **A18:** 59, 60
in gating systems **A15:** 593-594
gradient, infiltration **A7:** 552-553
and green density **A7:** 298, 299
head, and change, in gating systems **A15:** 592
high, electroslag remelting under **A15:** 405
for hot extrusion **A14:** 322-323
hot forming, nickel-base alloys **A14:** 262
in hot pressing **A7:** 504
hydrodynamic, in liquid-erosion failures **A11:** 163-171
in hydrostatic extrusion **A14:** 327
impact extrusion **A14:** 312
intensifier, defined *See* Pressure intensifier
laminate, test tool for **EM1:** 756
and leakage measurement **A17:** 57-59
levels, squeeze casting **A15:** 324
limit, electromagnetic forming **A14:** 646
low static, liquid erosion in **A11:** 167
for magnesium alloy forgings **A14:** 260
mean free path **A17:** 59
measurement in HIP units **A7:** 423
metal, vs metal flow, die casting **A15:** 292
mold effects **A15:** 346
of nonmetallic inclusions, in steel **A11:** 70
peak, and explosivity **A7:** 196
physical chemistry principles **A15:** 50-52
ports, in hydraulic torsional system **A8:** 216
and powder flow direction **A7:** 300
profiling, superplastic metals **A14:** 866-867
proof, defined *See* under Proof pressure
rate increase, and explosivity **A7:** 196
and relative density **A7:** 298-300
roll **A14:** 344-345
SI unit/symbol for **A8:** 721
sintering **A7:** 9, 501, 800
straightening, control of **A14:** 690
stress, SI derived unit and symbol for **A10:** 685
in stress-corrosion cracking **A13:** 147
stressing, for optical holographic interferometry **A17:** 410
structural changes as function of **A10:** 420
temperature effects, magnesium alloys **A14:** 260
testing, acidified chloride solutions **A8:** 418
testing, back pressuring as **A17:** 65
theory, hydrogen damage **A13:** 164
thermodynamic, solidification effect **A15:** 84
time trace, iron powder explosion **A7:** 197
transducers **A8:** 368, 613-614
for tungsten forging **A14:** 238
uniform **A7:** 300
vs temperature, heat-resistant alloys **A14:** 233
vs upset reduction, stainless steels **A14:** 223-224

Pressure (continued)
waveform, electromagnetic forming **A14:** 652
Pressure and vacuum testing
soldered joints **A6:** 982
Pressure bag molding **EM3:** 23
defined **EM1:** 19 **EM2:** 34
Pressure bags
for epoxy composites **EM1:** 71
Pressure bar **A8:** 200-202
Pressure bonding *See* Diffusion welding
Pressure break
defined **EM2:** 34
Pressure bubble plug-assist vacuum thermoforming **EM2:** 401
Pressure calcintering **EM4:** 300
Pressure casting **EM4:** 34
Pressure coefficients of resistance
electrical resistance alloys **A2:** 824
Pressure components
steam-generator failures in **A11:** 603
Pressure contacts
as chip interconnect **EL1:** 232-233
Pressure densification **EM4:** 242, 296-301
techniques **EM4:** 296, 297-301
conditions **EM4:** 298
gas pressure sintering **EM4:** 298, 299-300, 301
hot forging **EM4:** 300, 301
hot isostatic pressing **EM4:** 298, 299, 300, 301
hot pressing **EM4:** 297-299
modified techniques **EM4:** 300-301
parameters **EM4:** 298
press forging **EM4:** 300
pressure calcintering **EM4:** 300
reactive hot pressing **EM4:** 300
superplasticity **EM4:** 300-301
uniaxial hot pressing **EM4:** 297-299, 301
theory **EM4:** 296, 297
deformation maps **EM4:** 297
dislocation creep **EM4:** 296
Pressure die casting *See also* Centrifugal casting; Cold chamber pressure casting Die casting; Injection molding; Squeeze casting
of metal-matrix composites **A15:** 845
Pressure die castings
aluminum and aluminum alloys **A2:** 5
zinc alloy **A2:** 528-529
Pressure dies
for bending **A14:** 666
Pressure dispensing
in surface-mount soldering **EL1:** 699
Pressure forming
defined **EM2:** 34
Pressure gage
in stress-relaxation compression tester **A8:** 325
Pressure gas welding
definition **M6:** 13
Pressure gas welding (PGW)
definition **A6:** 1212
Pressure granulation methods **EM4:** 100
Pressure intensifier **EM3:** 23
defined **EM1:** 19 **EM2:** 34
Pressure lubrication
defined **A18:** 15
Pressure, molding *See* Molding pressure
Pressure on clamped specimens **A9:** 28
Pressure pad *See* Hold-down plate
Pressure pipe **M1:** 315, 321, 324, 325
Pressure piping
codes governing **M6:** 824
ultrasonic inspection **A17:** 232
Pressure plate
defined **A14:** 10
Pressure plating **A18:** 843
Pressure reduction technique **EM3:** 762
Pressure regulators
oxyfuel gas welding **M6:** 584-585
Pressure requirements for thermoplastic mounting **A9:** 30
Pressure, saturation *See* Saturation pressure

Pressure sintering *See* Hot pressing
Pressure steel tubes **A1:** 327, 333-334
Pressure systems
leak testing, without tracer gases **A17:** 59-61
visual leak testing of **A17:** 66
Pressure terminations
flexible printed boards **EL1:** 591
Pressure testing **A15:** 264, 557
of aluminum alloy castings **A17:** 534
of brazed assemblies **A17:** 604
of powder metallurgy parts **A17:** 545-547
of soldered joints **A17:** 606
Pressure thermit welding **M6:** 694
Pressure tightness
gray cast iron **M1:** 21-22
in gray iron **A1:** 22-23
Pressure tightness tests
for investment castings **A15:** 264
Pressure transducers **A8:** 368, 613-614
Pressure tubes **M1:** 316, 321, 323, 324, 325
Pressure tubing *See* Steel tubular products
Pressure vessel
autoclave **EM1:** 645
field, for multiaxial testing **A8:** 344
and low-cycle fatigue **A8:** 347
nuclear, design code **A8:** 263-264
steel plate, changes Charpy V-notch properties for **A8:** 262
Pressure vessel plate **A1:** 234-235, 236, 237
Pressure vessel quality steel plate
ASTM specifications **M1:** 183-184
compositions **M1:** 185-187, 189
mechanical properties **M1:** 192-193, 195-196
Pressure vessels *See also* Boilers; Pressure vessels, failures of; Weld(s); Weldments **A6:** 377-381 **A7:** 419-420, 445 **A13:** 1064, 1076, 1083, 1088-1089
acoustic emission inspection **A17:** 290, 291, 642, 654-656
carbon steel, with stainless steel liner cracked **A11:** 656
codes governing **M6:** 823
codes, inspection methods **A17:** 641-644
deep drawing of **A14:** 587
eddy current inspection **A17:** 642
failed, fracture path of **A11:** 665
failures, statistical summary **A11:** 644
formulas **EM2:** 653-654
head, failures of **A11:** 644
head forgings **A14:** 71
hydrogen sulfide SCC in **A11:** 426
image systems **A17:** 654
in-service quantitative evaluation **A17:** 653-654
inspection during fabrication **A17:** 645-646
joining processes **M6:** 56
liquid penetrant inspection **A17:** 642
low-carbon, failure from caustic embrittlement by potassium hydroxide **A11:** 658-660
magnetic paint inspected **A17:** 128
magnetic particle inspection **A17:** 642
plates, forgings, tubes, inspection of **A17:** 644-645
radiographic inspection **A17:** 641-642, 647-649
replication microscopy **A17:** 642-644
shell, failures of **A11:** 644
standards for explosion welding **M6:** 711
thick-wall alloy steel, failure by weld HAZ cracks **A11:** 650-652
titanium alloy, brittle tensile fracture in **A11:** 77
ultrasonic inspection **A17:** 649-653
visual inspection **A17:** 47
Pressure vessels, failures of **A11:** 643-669
analysis, procedures for **A11:** 643-644
brittle fractures in **A11:** 663-666
of composite materials **A11:** 654-656
creep and stress rupture in **A11:** 666-668
ductile fractures in **A11:** 666
fabrication practices, effects of **A11:** 646-654
fatigue in **A11:** 668-669
hydrogen embrittlement in **A11:** 661-663
metallurgical discontinuities, effects of **A11:** 646
stress-corrosion cracking in **A11:** 656-661
unsuitable alloys, effects of **A11:** 644-646
Pressure wave method molding machines
green sand molding **A15:** 343
Pressure welding *See also* Forge welding
aluminum alloys **A6:** 739
Pressure welding processes
applications **EL1:** 238
Pressure-assisted sintering **EM4:** 288
Pressure-containing parts
cast steels for **M1:** 379
Pressure-controlled welding
definition **A6:** 1212 **M6:** 13
Pressure-feed powder flame spraying system **M5:** 540
Pressure-impregnation-carbonization (PIC) *See also* Hot isostatic press (HIP)
defined **EM1:** 19
Pressure-induced phase transformations
XRPD analysis **A10:** 333
Pressure-saturation **EM3:** 23
Pressure-sensitive adhesive
defined **EM1:** 19 **EM2:** 34
Pressure-sensitive adhesive (PSA) **EM3:** 23, 40, 74, 76
additives **EM3:** 83
advantages and limitations **EM3:** 75
applications **EM3:** 83-84
automotive applications **EM3:** 83, 84
bonding substrates and applications **EM3:** 85
characteristics **EM3:** 74, 83
chemical families **EM3:** 84-86
construction applications **EM3:** 83, 84
for corrosion protection **EM3:** 83, 84
for electrical applications **EM3:** 83, 84
for health care tapes **EM3:** 47
for hospital/first aid applications **EM3:** 83, 84
industrial applications **EM3:** 47
for labels **EM3:** 84
markets **EM3:** 47, 83
methods of application **EM3:** 47
for office/graphic arts applications **EM3:** 83, 84
for packaging **EM3:** 47, 83, 84
properties **EM3:** 82-83
providing faster processing cycles **EM3:** 143
shelf life **EM3:** 84
silicone-base **EM3:** 85
suppliers **EM3:** 86
for surface-mount technology bonding **EM3:** 570
for tapes **EM3:** 84
tapes and labels, as consumer products **EM3:** 47
Pressure-shear plate impact testing
anvil plate preparation **A8:** 234-235
basic concepts **A8:** 231-233
experimental record interpretation **A8:** 234-236
experimental results **A8:** 236-238
flyer plate **A8:** 231-232, 234-235
high strain rate **A8:** 187, 231-232
hydrostatic pressure **A8:** 237
normal stress-particle velocity **A8:** 232
plate impact facility **A8:** 233-234
requirements **A8:** 231
shear waves **A8:** 231
specimens **A8:** 230-232, 234, 237
for stress-strain curves **A8:** 230-238
Pressure-shear waves **A8:** 231
Pressure-temperature diagram
diamond and graphite **A2:** 1009
water **EL1:** 244
Pressure-temperature phase diagrams .... **A3:** 1 • 2
Pressure-vessel applications
steel plate for **M1:** 181-184
Pressure-viscosity coefficient **A18:** 79, 539, 540
defined **A18:** 15
nomenclature for lubrication regimes **A18:** 90
symbol and units **A18:** 544
Pressureless densification **EM4:** 242
Pressureless sintering **A7:** 9
alumina **EM4:** 516
TTZ **EM4:** 512
versus densification, parameters **EM4:** 296
Pressureless-sintered silicon nitride
joining non-oxide ceramics **EM4:** 526, 527-528
Pressurized gas lubrication
defined **A18:** 15
Pressurized loop **A8:** 411
Pressurized lubricating systems
failure of **A11:** 153
Pressurized water
chamber for fatigue crack growth testing **A8:** 427, 429
reactor **A8:** 423-424
Pressurized water reactors **A13:** 927, 937, 940, 948, 951
Preston equation **EM4:** 468, 469
Prestressed concrete
wire for **A1:** 283 **M1:** 264
Prestressing
effects on maraging steels **A11:** 218
Prestressing structures
corrosion of **A13:** 1310
Prestretching
effect on fatigue strength **A11:** 112
Pretempered steels *See also* Temper; Tempering
press forming of **A14:** 558
Pretinning
definition **A6:** 1212
Pretreatment
metal molds **A15:** 303-304
Prevention *See also* Corrosion prevention
airframe corrosion **A13:** 1035-1036
of cold-formed parts failures **A11:** 308-313
of crevice corrosion **A13:** 112-113
of erosion damage **A11:** 170-171
filiform corrosion **A13:** 107-108
of hydrogen damage **A11:** 250-251
intergranular corrosion, austenitic stainless steels **A13:** 124
of lubricant failures **A11:** 154
Prewash stations
liquid penetrant inspection **A17:** 78
Preweighed, packaged kits **EM3:** 687, 688, 689-690
advantages **EM3:** 690
barrier- and injection-style kits **EM3:** 689, 690
coaxial cartridge, static mix systems
divider bags (hinge pack) **EM3:** 687, 689
side-by-side (dual) syringes **EM3:** 687, 689
side-by-side, static mix systems **EM3:** 688, 689-690
Preweld and postweld heat treatments, effect of on weldability **A1:** 606
Preweld heat treatment
friction welding and **A6:** 152
Preweld interval
definition **M6:** 14
Price *See also* Costs
and density estimates, hybrids **EL1:** 250-252
Prices *See also* Cost(s); Pricing
quoting **EM2:** 86-87
Pricing *See also* Cost(s)
acrylonitrile-butadiene-styrenes (ABS) **EM2:** 109-110
of parts **EM2:** 83-85
Pricing history *See also* Cost(s)
of bismuth **A2:** 754
of indium **A2:** 751
Primacord
for explosive forming detonation **A14:** 636
Primary (x-ray)
defined **A9:** 14
Primary Adhesively Bonded Structure Technology program (PABST) **EM3:** 44, 523-524, 558
Primary alloy *See also* Secondary alloy
defined **A15:** 9
Primary alpha
defined **A9:** 14
Primary austenite
in cast iron **A15:** 173-174
Primary bone grafting
for unstable internal fixation **A11:** 673
Primary constituent .... **A3:** 1 • 20
Primary cracks
opening **A12:** 77
Primary creep *See* Creep
defined **A8:** 308 **A11:** 8 **A12:** 19
in elevated-temperature failures **A11:** 263
Primary crystals
defined **A9:** 14
Primary current distribution
defined **A13:** 10
Primary dendrite arm spacing
as an indicator of growth conditions in steel **A9:** 625
defined **A9:** 624

Primary dendrite arm spacing (continued)
and secondary dendrite arm spacing as a function of distance from chill surface **A9:** 626
Primary electrons
as x-ray source **A17:** 305
Primary etching
defined **A9:** 14
Primary excitation *See also* Excitation
AES, electron beams for **A10:** 550
Primary extinction
defined **A9:** 14
effect, in dynamic theory of diffraction **A10:** 366-367
as XRPD source of error **A10:** 341
Primary fabrication *See also* Fabrication; Fabrication characteristics; Secondary fabrication
hafnium **A2:** 662-664
metal-matrix composites **A2:** 903
wrought titanium alloys **A2:** 609-611
zirconium **A2:** 662-664
Primary ferrite (PF) **A6:** 76
grain boundary ferrite **A6:** 76
intragranular polygonal ferrite **A6:** 76
Primary forging
and ingot breakdown **A14:** 222-223
Primary graphite
in cast iron **A15:** 174
Primary ion bombardment
effects of **A10:** 611
Primary knock-on atom (PKA) **A18:** 851, 852
Primary leakage
defined **A18:** 15
Primary metastable precipitation
of beta **A15:** 128
Primary mills **M1:** 110, 114
Primary molding operations
costs **EM2:** 84-85
Primary nucleation **EM3:** 23
defined **EM2:** 34
Primary oxide compounds
for high-temperature superconductors **A2:** 1085-1086
Primary passive potential (passivation potential)
defined **A13:** 10
Primary phase
alloying **A13:** 46
Primary phases
cast iron, austenite **A15:** 173-174
cast iron, graphite **A15:** 174
Primary purification reaction **A1:** 986-987
Primary recrystallization *See also* Recrystallization
effect on crystallographic texture **A9:** 700
Primary shear angle **A18:** 610, 611
Primary shear zone **A18:** 609, 610, 611
Primary silicon particles
aluminum-silicon alloys **A15:** 165-166
Primary standard dosimetry system **EM3:** 23
defined **EM2:** 34
Primary standards
assay by electrometric titration **A10:** 202
Primary testing direction
for alloy systems **A8:** 667
long transverse **A8:** 672
Primary x-rays *See also* X-rays
defined **A10:** 679
Primary-beam transmitted electrons **A18:** 377
Prime Western zinc *See* Zinc, specific types, Prime Western
Primer *See also* Adhesion promoter; Paint and painting
characteristics and properties **M5:** 472, 475-477, 499-500
corrosion prevention with **M5:** 432
defined **EM1:** 19 **EM2:** 34
magnesium alloys, types used **M5:** 629, 647-648
stripping of **M5:** 19
Primer cup plate
spalled surface **A11:** 566-567
Primer(s) **A13:** 10, 912-914, 1015
Primers *See also* Paint **EM3:** 23, 42, 51-52, 254-258, 640-641
for acetal copolymer **EM3:** 278
aerospace industry applications **EM3:** 255
automotive industry applications **EM3:** 255
blackout for urethanes **EM3:** 554-555
for bonding urethane windshield sealant to metal **EM3:** 204
cast irons **M1:** 105
chromic acid anodization **EM3:** 261
coating system selection **EM3:** 640-641
combined with coupling agent **EM3:** 256-257
corrosion inhibitive compound for identification **EM3:** 249
corrosion-inhibiting adhesive (CIAPs) **EM3:** 807-808, 817, 819
cracked-lap-shear (CLS) specimens **EM3:** 450
definition **EM3:** 39
electrodeposition of **EM3:** 254
enhances cohesive bond strength to grit-blasted composite surfaces **EM3:** 841
epoxies and **EM3:** 101
for epoxy resins **EM3:** 277
examples **EM3:** 256
failure **EM3:** 640
formulation **EM3:** 640
function **EM3:** 640
for glass **EM3:** 281-282, 283
for missile radome-bonded joints **EM3:** 564
moisture effect **EM3:** 640-641
noncorrosion-inhibiting **EM3:** 808
for nylons **EM3:** 278
for plastics **EM3:** 277
for polyether silicones **EM3:** 229, 231
polyurethane filoform corrosion-resistant **EM3:** 808, 819
polyurethane, providing environmental durability **EM3:** 809
preprimed steel sheet **M1:** 169, 175
properties **EM3:** 256
suppliers **EM3:** 802
surface preparation for processing quality control **EM3:** 738
to improve wet strength durability **EM3:** 625-626
top coats **EM3:** 640-641
types **EM3:** 640
for urethanes **EM3:** 112, 206, 207
used in aircraft interior assembly **EM3:** 44
used with latexes **EM3:** 212
used with silicones (RTV) **EM3:** 218
Primitive space lattice ..... **A3:** 1 • 15
Principal strain
in ductility bending tests **A8:** 127
and principal stress, coincidence of **A8:** 343
Principal stress
biaxial **A8:** 10
and creep/creep rupture **A8:** 343
defined **A8:** 10 **A11:** 8
in ductility bending tests **A8:** 127
and principal strain, coincidence of **A8:** 343
triaxial **A8:** 10
uniaxial **A8:** 10
Principal stress (normal)
defined **A13:** 10-11
Principal stresses
and stress distribution **A10:** 382
Principio Furnace and Forge Company (Maryland) **A15:** 25
Principio furnaces
illustrated **A15:** 25
Principle of independent action (PIA) theory **EM4:** 700-703, 707
Principles
of liquid metal processing **A15:** 49
of physical chemistry **A15:** 50-54
of solidification **A15:** 99-185
of superconductivity **A2:** 1030-1042
Principles of Powder Metallurgy (Jones) **A7:** 18
Print applications screen and stencil **EL1:** 652-653
Print contrast
effect of photographic paper on **A9:** 85-87
Print resolution
solder paste **EL1:** 732
Printed board assembly
defined **EL1:** 1154
Printed board circuit
abbreviation **A8:** 725
Printed board coupon
metallographic evaluation **EL1:** 572-577
Printed board material systems **EM3:** 592
Printed board(s)
configurations, and rework **EL1:** 711-712
defined **EL1:** 1153-1154
failure analysis **EL1:** 1038-1040
parylene coating application **EL1:** 797-798
Printed circuit board *See* Circuit board; Printed wiring board
abbreviation for **A11:** 797
defined **EM1:** 19
Printed circuit board laminates
PCD tooling **A16:** 110
Printed circuit boards
waterjet machining **A16:** 522, 523, 525
Printed circuit boards (PCBs)
laser-beam welding **A6:** 263
Printed circuit technologies
development **EL1:** 89
Printed circuits
powders used **A7:** 573
shorts risk sites, identified **EL1:** 127
soldering **A6:** 631
vs conventional wiring **EL1:** 505
Printed circuits (radio and television)
powders used **A7:** 574
Printed wiring
defined **EL1:** 1154
substrates **EL1:** 15
Printed wiring assembly (PWA)
coefficient of thermal expansion, control of **EL1:** 77
complexities, and through-hole soldering **EL1:** 685-686
cure types **EL1:** 764-765
high-frequency digital systems **EL1:** 76
radiation curing of conformal coatings **EL1:** 854
solder joint inspection **EL1:** 735
Printed wiring board **EM2:** 34, 232-234
Printed wiring board (PWB) **A6:** 132, 134-136
printed wiring board (PWB) technology
current capabilities/limitations **EL1:** 507-512
etching **EL1:** 511
imaging **EL1:** 508-510
laminating **EL1:** 510
machining **EL1:** 508
plating **EL1:** 510-511
PWB development **EL1:** 507
PWB technological capabilities **EL1:** 508-511
Printed wiring boards **A13:** 1120
Printed wiring boards (PWBS) *See also* Ceramic printed wiring boards; Flexible printed wiring boards; Printed boards; Printed circuits; Printed wiring assembly (PWA); Printed wiring board (PWB) manufacture; Printed wiring board (PWB) technology; Rigid printed wiring boards **EM3:** 23, 588-591
advantages **EL1:** 7
air-cooled integral heat exchanger on **EL1:** 54-55
application, and design **EL1:** 516-517
area **EL1:** 20
assembly failures, types **EL1:** 943-944
assembly requirements, preservatives **EL1:** 562
ceramic, circuit construction **EL1:** 387
cleaning, for conformal coatings **EL1:** 763, 776-777
cleaning process, flow chart **EL1:** 777

Printed wiring boards (PWBS) (continued)
complexity ... **EL1:** 551-552
component attachment ... **EL1:** 388
computer-aided design ... **EL1:** 527-533
conduction enhanced circuit pack ... **EL1:** 47
conductive polymer interconnection for ... **EL1:** 16
conductor traces/ground-base radar system, corroded ... **EL1:** 1109-1110
conformal coatings/encapsulants for ... **EL1:** 759-761
connections ... **EL1:** 311
constrained fiber ... **EL1:** 984
construction, aramid fibers ... **EL1:** 616-617
construction, polymer-on-metal ... **EL1:** 620-622
construction, quartz fabric ... **EL1:** 619
cooling ... **EL1:** 47
defined ... **EL1:** 1154
design of ... **EL1:** 505, 513-526
development ... **EL1:** 513
dielectric materials for ... **EL1:** 597-610
dual-in-line packages, failure analysis ... **EL1:** 1109
electrical configurations typical ... **EL1:** 597-598
electronic/electrical corrosion failure analysis of ... **EL1:** 1109-1110
engineering design system for ... **EL1:** 127
environmental stress screening ... **EL1:** 878
environmental testing of ... **EL1:** 944
example (spreadsheet) ... **EL1:** 609
failure analysis ... **EL1:** 1038-1040
failure mechanisms in ... **EL1:** 1018-1030
fiberglass-epoxy, silicone conformal coatings for ... **EL1:** 822-824
flexible/rigid ... **EL1:** 505-506
flux contamination, failure analysis ... **EL1:** 1109
flux on board components, failure analysis ... **EL1:** 1110
functional densities ... **EL1:** 505
future design ... **EL1:** 506
geometrical structures, design of ... **EL1:** 40-41
heat sinks in ... **EL1:** 1129-1131
high-speed ... **EL1:** 608-609
instrumentation/testing ... **EL1:** 365
introduction ... **EL1:** 505-506
layout, by computer-aided design ... **EL1:** 528
lines-and-spaces reduction ... **EL1:** 253
manufacturability ... **EL1:** 608-609
manufacture, quality control in ... **EL1:** 505, 539, 608-609, 869-874
material properties ... **EL1:** 869
multilayer, design requirements ... **EL1:** 516
patent drawing ... **EL1:** 507
physical test methods ... **EL1:** 941-945
polymer matrix composite ... **EL1:** 1117-1118
precleaning ... **EL1:** 776-777
rigid, fabrication of ... **EL1:** 538-552
routing ... **EL1:** 115
solder masks ... **EL1:** 553-560
structures, analyses of ... **EL1:** 81-82
test methods ... **EL1:** 869, 941-945
thermal expansion mismatch ... **EL1:** 611
thermal stresses in ... **EL1:** 62
with through-hole and surface- mount technologies ... **EL1:** 670
water-soluble flux, corrosion from ... **EL1:** 1110
printed wiring boards (PWBS) manufacture
core processes ... **EL1:** 869-872
defect verification process ... **EL1:** 872-874
electroless copper plating ... **EL1:** 870
electrolytic copper plating ... **EL1:** 871-872
etching ... **EL1:** 872
low-CTE metal planes ... **EL1:** 627
photoprinting process ... **EL1:** 870-871
primary plated through-hole drilling ... **EL1:** 869-870
raw materials ... **EL1:** 869
Printed wiring laminate materials ... **A6:** 133
Printer hammer guide assembly ... **A7:** 669
Printers
for digital image enhancement ... **A17:** 462-463
Printing *See also* Phosphorus printing, Sulfur printing
conductor resistor, and via filling ... **EL1:** 463-464
defined ... **A9:** 14
as multilayer inner layer process ... **EL1:** 542
of photomicroscopy films ... **A9:** 85
postfired, ceramic packages ... **EL1:** 466
Printing (continued)
in thick-film process ... **EL1:** 332
Printing inks
aluminum flake for ... **A7:** 594-595
copper and gold bronze pigments for ... **A7:** 595
of copper-based powder metals ... **A7:** 734
Printing, magnetic *See* Magnetic printing
Printing wheels
for high-speed printing equipment ... **A7:** 669
Priopionaldehyde
physical properties ... **EM3:** 104
Prior art
in NDE reliability ... **A17:** 674
Prior-beta grain size
defined ... **A9:** 14
Prism
defined ... **A10:** 679
Prismatic bar
torsion testing of ... **A8:** 139-141
Pristine strength
glass fiber ... **EM1:** 46
Probabilistic design of ceramic components, NASA/CARES computer program ... **EM4:** 700-707
applications in industry ... **EM4:** 707
capability of program ... **EM4:** 700-702
code ... **EM4:** 700
cyclic symmetry of modeling ... **EM4:** 702
definition of CARES ... **EM4:** 700
description of program ... **EM4:** 700-702
design methodology ... **EM4:** 700
functions of program ... **EM4:** 700
input requirements ... **EM4:** 702
input categories ... **EM4:** 702
keyword driven ... **EM4:** 702
TEMPLET INP file ... **EM4:** 702
output information ... **EM4:** 702
corresponding element numbers ... **EM4:** 702
probability of failure and survival ... **EM4:** 702
reliability evaluation ... **EM4:** 702
risk-of-rupture intensity values ... **EM4:** 702
significance levels ... **EM4:** 702
PC-based version ... **EM4:** 700
reliability of components ... **EM4:** 701-702
theory ... **EM4:** 703-707
Probabilistic fracture mechanics
for prediction of crack size distributions ... **A8:** 683
Probability *See also* Probability of detection (POD); Reliability prediction; Statistical methods
conditional, in NDE discrimination ... **A17:** 675
and confidence values, unknown distribution ... **A8:** 666
curves, fatigue fracture ... **A8:** 253
defined ... **A8:** 624
density function ... **A8:** 624-625, 628-634
of failure ... **A8:** 627, 705
of false alarms (POFA) ... **A17:** 676
introduction to ... **EL1:** 895-896
and statistics ... **A8:** 624
Probability density distribution
for NDE reliability ... **A17:** 675-676
Probability density function
defined ... **A8:** 629
exponential distribution ... **A8:** 634-635
histogram of tensile yield strength with ... **A8:** 628
log normal distribution ... **A8:** 629, 630-631, 632
normal distribution ... **A8:** 629
normal, graph of ... **A8:** 629
of statistical distributions ... **A8:** 628-629
Weibull distribution ... **A8:** 632, 634
Probability density function (PDF) ... **A18:** 296
Probability mass function
binomial distribution ... **A8:** 635-636
Poisson distribution ... **A8:** 636-637
Probability of detection (POD) *See also* POD(A) function
analysis techniques ... **A17:** 664
for automatic eddy current inspection ... **A17:** 664
curves, for characteristic performance ... **A17:** 679
and demonstration program design ... **A17:** 664
and flaw size, as POD(A) function ... **A17:** 689
and fracture control/damage tolerance ... **A17:** 671
functions, for NDE reliability ... **A17:** 664
models, of eddy current inspection ... **A17:** 708
models, ultrasonic inspection ... **A17:** 706
Probability of detection (POD) (continued)
in NDE reliability modeling ... **A17:** 702
of true positive outcome, calculated ... **A17:** 676
Probability of escape (electron) ... **A18:** 451
Probability of false alarms (POFA)
calculated ... **A17:** 676
Probability parameter
as roughness parameter ... **A12:** 201
Probability sampling
in chemical analysis ... **A7:** 249
Probability statistics *See* Process capability; Quality control; Statistical quality design and control
Probability-stress-life-plot
from fatigue experiments at varied stresses ... **A8:** 700-701
Probe changers
coordinate measuring machines ... **A17:** 25
Probe coils
eddy current inspection ... **A17:** 176
Probe ion
defined ... **A10:** 679
Probe-flaw interaction models
eddy current inspection ... **A17:** 707-708
Probertite ... **EM4:** 380
Probes *See also* Coils; Conductors; Electron probe x-ray microanalysis; Pulsed laser atom probe; Search units; Sensors
absolute, eddy current inspection ... **A17:** 180-181
automatic calibration (CMMS) ... **A17:** 20
compensation, in CMMs ... **A17:** 26
differential, eddy current inspection ... **A17:** 180-181
edge-finder, coordinate measuring machine ... **A17:** 25
electric current perturbation ... **A17:** 137-140
electrical resistance ... **A13:** 199-200
electroanalytical, for process control ... **A10:** 197
for ferromagnetic heat exchanger tube inspection ... **A17:** 181-182
ferromagnetic resonance eddy current ... **A17:** 220-223
hydrogen ... **A13:** 201
microwave eddy testing ... **A17:** 218
monitoring, preferred locations ... **A13:** 202
operation, remote-field eddy current inspection ... **A17:** 196-197
reflect-on, eddy current inspection ... **A17:** 178
rotating, eddy current inspection ... **A17:** 183, 187
short-field, microwave inspection ... **A17:** 212
stripline ... **A17:** 219
surface analytical techniques ... **A7:** 251
three-electrode polarization ... **A13:** 200
types, coordinate measuring machine ... **A17:** 25
Probing
as specific-gas detector application ... **A17:** 64-65
Probit method ... **A8:** 702-703
Procedure
definition ... **M6:** 14
Procedure qualification
definition ... **M6:** 14
Procedure qualification record
definition ... **M6:** 14
Procedure qualification record (PQR)
definition ... **A6:** 1212
Proceedings of the Joint Military/Government Industry Symposium on Structural Adhesive Bonding ... **EM3:** 68
Process *See also* Process capability; Process control
behavior, over time ... **A17:** 723-724
capability assessment ... **A17:** 737-740
compatibility, connector ... **EL1:** 23
complexity, WSI ... **EL1:** 354
definition ... **A6:** 1212
improvement, defined ... **A17:** 740
performance factors, in quality design ... **A17:** 722
speed, human vs. machine vision ... **A17:** 30
test ... **EL1:** 373-374
verification ... **EL1:** 742
yield's, surface-mount soldering ... **EL1:** 708
Process annealing
and cold-heading properties ... **A14:** 293
defined ... **A9:** 14
Process capability *See also* Quality control; Statistical methods ... **EM3:** 786-790, 791
assessment, for quality control ... **A17:** 737-740
examples ... **A17:** 738-740
indices ... **A17:** 739
statistical process control ... **A17:** 739-740

Process capability (continued)
tolerances and control limits, compared ... **A17:** 739
vs. process control, compared ... **A17:** 738-739
Process control *See also* Mounting; Processes; Quality control; Soldering
adhesive-bonded joints ... **A17:** 632-636
of adhesives ... **EL1:** 672-674
audit, solder joint inspection as ... **EL1:** 735
by thermal inspection ... **A17:** 402
as coordinate measuring machine application ... **A17:** 20
in die casting ... **A15:** 286
directional solidification ... **A15:** 321
electric arc furnace ... **A15:** 361-362
electrometric titration for ... **A10:** 202
epoxy materials ... **EL1:** 833-836
of fluxing ... **EL1:** 683-684
human factors ... **A17:** 679
of inspection materials ... **A17:** 678
of leaded and leadless surface-mount joints ... **EL1:** 731-733
for liquid penetrant inspection ... **A17:** 88
magnetohydrodynamic casting ... **A15:** 331
and management personnel ... **A17:** 680
modeling for ... **EL1:** 447
NDE applications ... **A17:** 677-680
on-line, electrogravimetry ... **A10:** 199
preheating, through-hole soldering ... **EL1:** 686-688
probability of detection (POD) curves ... **A17:** 679
and product control, compared ... **A17:** 734
qualification, of inspection equipment ... **A17:** 679
qualification, of inspection processes ... **A17:** 678-679
real-time, plating thickness inspection ... **EL1:** 943
relative operating characteristics (ROC) curves ... **A17:** 679-680
requirements, green sand molds ... **A15:** 222
of sand reclamation ... **A15:** 352-353, 355
solder waves ... **EL1:** 690-691
of solid samples, XRS for ... **A10:** 83
statistical (SPC) ... **EL1:** 288, 696
statistical, and statistical tolerance model ... **A17:** 739-740
statistical, and variation ... **A17:** 723
tool, stress modeling as ... **EL1:** 446
ultrasonic inspection ... **A17:** 232, 254
use of ICP-AES in ... **A10:** 31
vs. process capability, in quality control ... **A17:** 738-739
worksheets, rework process ... **EL1:** 728-729
Process control monitor (PCM) testing
for electrical performance ... **EL1:** 951-952
Process design *See also* Computer-aided design; Process modeling
computer-aided ... **A14:** 21
data acquisition for ... **A14:** 439-442
method ... **A14:** 413-416
Process development, forging
controllable factors ... **A14:** 409
Process documentation
and qualification ... **A17:** 678-679
Process flow
CAD/CAM ... **EL1:** 128
for cofired ceramic multilayer packages ... **EL1:** 461
curing, conformal coating ... **EL1:** 786
material and process selection ... **EL1:** 112-113
parallel ... **EL1:** 136
sequential ... **EL1:** 132
Process flow diagram
gold-nickel-chromium system, thin-film hybrids ... **EL1:** 316
gold-tantalum nitride system ... **EL1:** 316
Process fluids
inhibitors in ... **M1:** 756
Process function variation
sources ... **A17:** 722-730
Process gas *See also* Gases
analytic methods for ... **A10:** 8, 11
inorganic, analytic methods for ... **A10:** 8
microcircuit, analyzed ... **A10:** 156-157
microcircuit, gas mass spectrometry for ... **A10:** 156-157
organic, analytic methods for ... **A10:** 11
Process industry applications *See also* Chemical industry applications
refractory metals and alloys ... **A2:** 558
tooling, structural ceramics ... **A2:** 1019
Process log ... **EM3:** 793
Process modeling *See also* Computer applications; Modeling; Process design ... **A14:** 911-927
analysis of large plastic incremental deformation ... **A14:** 425-431
analytical methods ... **A14:** 913-919
analytical, of forging operations ... **A14:** 425
deformation maps ... **A14:** 420-421
deformation mechanisms ... **A14:** 420-421
dynamic material modeling ... **A14:** 421-424
finite-element analysis methods ... **A14:** 919-924
future developments ... **EM1:** 501
generalized approach ... **A14:** 911-913
material, constitutive equations ... **A14:** 417-420
numerical ... **EM1:** 500
and optimization ... **EM1:** 499-502
physical modeling ... **A14:** 431-437
of rolling mills ... **A14:** 354
submodels of ... **EM1:** 500
techniques ... **A14:** 409-438
Process monitoring
FT-IR spectroscopy for ... **A10:** 112
Process parameters
controlling ... **A14:** 407-408
Process potential ... **EM3:** 786
Process reagents *See also* Reagents
trace impurities analysis in ... **A10:** 43
Process selection
for base material/insulators ... **EL1:** 113-114
for conductors ... **EL1:** 114
manufacturing ... **EM2:** 277-403
and materials ... **EL1:** 112-118
modeling for ... **EL1:** 447
for plating (PTH, vias, surface wiring) ... **EL1:** 113-115
for precious metal plating ... **EL1:** 116
process flow ... **EL1:** 112-113
for solder mask/protective coat ... **EL1:** 115-116
for soldering/interconnection ... **EL1:** 116-117
tool, stress modeling as ... **EL1:** 446
Process simulation *See also* Modeling; Process modeling; Simulation
of metalworking ... **A14:** 19-20
Process specification (traveler) ... **A14:** 414-416
Process streams
analysis ... **A13:** 201
as corrosive ... **A11:** 210
galvanic corrosion evaluation in ... **A13:** 236
in-service monitoring ... **A13:** 197, 201
UV/VIS on-line monitoring of species in ... **A10:** 60
Process tolerance
closed-die steel forgings ... **M1:** 365-367
Process variables
controlling workability ... **A14:** 367-370
Process verification
for reliability ... **EL1:** 742
Process water
corrosion by ... **M1:** 713
treatment of ... **M1:** 749-750
Process waters *See also* Rinse waters; Water
wet chemical analysis ... **A10:** 165
Process window
for microstructure control ... **A14:** 412
Processed wire
definition ... **M1:** 262
Processes *See* Manufacturing processes
Processibility
fiber effects ... **EM2:** 98-99, 253
Processing *See also* Fabrication; Manufacturing; Material processing; Metal processing of elemental cobalt
acrylics ... **EM2:** 106-107
acrylonitrile-butadiene-styrenes (ABS) ... **EM2:** 112-113
aminos ... **EM2:** 230-231
bismaleimides (BMI) ... **EM2:** 254-255
of castings ... **A15:** 502-565
of cold-formed parts ... **A11:** 307-308
conditions, electrical breakdown from ... **EM2:** 466-467
of continuous fiber reinforced composites ... **A11:** 733
costs ... **EM2:** 647-648
cyanates ... **EM2:** 236-238
design guidelines ... **EM2:** 710
developments, metal casting ... **A15:** 38
effects, ductile-to-brittle transition temperature structural steels ... **A11:** 68
effects, on molecular orientation ... **EM2:** 281-282
epoxies ... **EM2:** 241
as epoxy matrix selection parameter ... **EM1:** 76
of epoxy resin matrices ... **EM1:** 76
forging defects from ... **A11:** 317
germanium and germanium compounds ... **A2:** 735
high-density polyethylenes (HDPE) ... **EM2:** 165-166
high-impact polystyrenes (PS, HIPS) ... **EM2:** 198
homopolymer/copolymer acetals ... **EM2:** 101-102
hydrogen embrittlement in ... **A8:** 541
information sources on ... **EM1:** 40-42
ionomers ... **EM2:** 122-123
liquid crystal polymers (LCP) ... **EM2:** 181-182
melt, polyaryl sulfones (PAS) ... **EM2:** 146
methods, and reinforcement effects ... **EM2:** 283
methods, thermosetting resins ... **EM2:** 223-224
parameters, supplier data sheets ... **EM2:** 639
phenolics ... **EM2:** 244-245
polyamide-imides (PAI) ... **EM2:** 130, 132-137
polyamides (PA) ... **EM2:** 127
polyarylates (PAR) ... **EM2:** 140-141
polyaryletherketones (PAEK, PEK PEEK, PEKK) ... **EM2:** 142-143
polybenzimidazoles (PBI) ... **EM2:** 150
polybutylene terephthalates (PBT) ... **EM2:** 155
polycarbonates (PC) ... **EM2:** 152
polyether sulfones (PES, PESV) ... **EM2:** 161-162
polyether-imides (PEI) ... **EM2:** 158
polyetheretherketone (PEEK) ... **EM2:** 144
polyethylene terephthalates (PET) ... **EM2:** 175
polyphenylene ether blends (PPE PPO) ... **EM2:** 184-185
polyphenylene sulfides (PPS) ... **EM2:** 189
polysulfones (PSU) ... **EM2:** 201-202
polyurethanes (PUR) ... **EM2:** 262-264
as selection parameter ... **EM1:** 38-39
styrene-acrylonitriles (SAN, OSA ASA) ... **EM2:** 216
styrene-maleic anhydrides (S/MA) ... **EM2:** 220-221
thermoplastic fluoropolymers ... **EM2:** 119
thermoplastic injection molding ... **EM2:** 308
thermoplastic polyimides (TPI) ... **EM2:** 177
thermoplastic polyurethanes (TPUR) ... **EM2:** 207
titanium alloys ... **M3:** 361-371
ultrahigh molecular weight polyethylenes (UHMWPE) ... **EM2:** 169-170
of unalloyed uranium ... **A2:** 671-672
unsaturated polyesters ... **EM2:** 249-251
vinyl esters ... **EM2:** 274-275
zirconium and hafnium ... **A2:** 661-662
Processing additives ... **EM4:** 115-121
applications ... **EM4:** 115
binders ... **EM4:** 120-121
for aqueous system ... **EM4:** 120
for nonaqueous system ... **EM4:** 120
properties ... **EM4:** 120-121

Processing additives (continued)
viscosity grade at concentration **EM4:** 120
dispersants **EM4:** 118-120
deflocculants **EM4:** 119-120
DVLO theory **EM4:** 119
polyelectrolytes **EM4:** 119
work of dispersion **EM4:** 118
forming **EM4:** 116
handling characteristics **EM4:** 116
properties **EM4:** 116
required viscoelastic properties **EM4:** 116
selection of process **EM4:** 116
lubricants **EM4:** 121
plasticizers **EM4:** 120-121
postforming operations **EM4:** 121
processing scheme **EM4:** 115
purposes **EM4:** 42
sintering aids and dopants **EM4:** 115-116
solvents **EM4:** 116-117
total dissolved solids content **EM4:** 117
used in ceramics processing **EM4:** 117
surfactants **EM4:** 117-118, 121
wetting agents **EM4:** 118, 120
cation exchange capacity **EM4:** 118
cloud point **EM4:** 118
critical micelle concentration **EM4:** 118-119
effectiveness **EM4:** 118
Krafft point **EM4:** 118
Processing aids, effect
chemical susceptibility **EM2:** 572
Processing equipment
aluminum and aluminum alloys **A2:** 13
Processing for Adhesive Bonded Structures **EM3:** 68
Processing map
for aluminum **A8:** 572
with safe region for forming **A8:** 572-573
Processing maps
for deformation processing conditions **A14:** 365, 370-371
determined **A14:** 423
development and integration of **A14:** 440-441
for titanium alloy **A14:** 424
Processing of solid steel **A1:** 115-124
annealing **A1:** 122-123
cold rolling **A1:** 121-122, 123
conventional controlled rolling (CCR) **A1:** 117
cooling and coiling system **A1:** 118-119, 121
dynamic recrystallization controlled rolling (DRCR) **A1:** 117-118
hot rolling **A1:** 115
precipitation during cooling and coiling **A1:** 119-120, 123
precipitation of carbonitrides and sulfides **A1:** 115-116
recrystallization controlled rolling (RCR) **A1:** 117, 120
Stelco coil box **A1:** 118, 120
warm rolling **A1:** 120, 124
Processing operations
effect on crystallographic texture **A9:** 700-701
metal crystallographic textures developed by **A9:** 706
Processing quality control **EM3:** 735-742
application geometry **EM3:** 736-737
curing **EM3:** 739-742
film adhesives **EM3:** 739-741
single-part sealants **EM3:** 742
two-part adhesives and sealants **EM3:** 741
handling and exposure to contaminants **EM3:** 738
lay-up technique **EM3:** 739
material age-life history **EM3:** 735-736
material manufacturing consistency **EM3:** 735
moisture exposure **EM3:** 736
quality assurance **EM3:** 742
surface preparation **EM3:** 737-738
tooling verification **EM3:** 738-739
Processing, use of phase diagrams in .... **A3:** 1 • 26-27
Processing window
defined **EM1:** 19 **EM2:** 34
Procurement
of corrosion test materials **A13:** 193-194
Prod contacts
applications **A17:** 94
magnetization by **A17:** 130
magnetizing, advantages/limitations **A17:** 94
Prod-contact method
of generating magnetic fields **A17:** 94, 97, 130
Produced fluids
corrosivity factors **A13:** 479-480
Producer's risk **EM4:** 86
Product
development cycles, and computer-aided design **EL1:** 528
environment, and electrical testing **EL1:** 566
Product analysis
for classifying steels **A1:** 141
Product applications *See* Applications
Product control, and process control
compared **A17:** 734
Product design **EM3:** 36
for die casting **A15:** 286-288
thermoplastic resins **EM2:** 622-623
***Product Design with Plastics, A Practical Manual* (Dym)** **EM2:** 94
Product form
change to prevent corrosion **A11:** 194
magnesium alloy, effect on SCC **A11:** 223
Product forms *See also* Material forms
cobalt-base corrosion-resistant alloys **A2:** 453-454
cobalt-base high-temperature alloys **A2:** 451-452
cobalt-base wear-resistant alloys **A2:** 448-449
continuous reinforcing fibers **EM1:** 33
epoxy resin **EM1:** 66-67
magnesium alloys **A2:** 462-466
multidirectional tape prepregs **EM1:** 146
of prepreg tow **EM1:** 151
selection, tape prepreg **EM1:** 145
of unidirectional tape prepregs **EM1:** 143-145
Product forms, and multiple heats
creep-rupture testing **A8:** 330
Product function variation
sources **A17:** 722-738
Product hologram
microwave **A17:** 225
Product liability
of lubricants **A14:** 518
Product life cycle *See also* Life tests
of titanium **A2:** 587
Product performance, factors
in quality design **A17:** 722
Product specifications
SCC testing for **A8:** 496
Production *See also* Fabrication; Manufacturing; Manufacturing process
control, circle arc elongation test for **A8:** 556
control functions, computer-aided **EL1:** 132
copper metals **A2:** 237-238
costs, defined **EM1:** 423
high-volume, millimeter/microwave applications **EL1:** 757-758
of indium **A2:** 750-751
molybdenum and molybdenum alloys **A2:** 574-575
presses **A7:** 329-332
quality control, acoustic emission inspection **A17:** 289-290
rate, and demagnetization **A17:** 122
rate, as material selection parameter **EM1:** 38-39
refractory metals and alloys **A2:** 559-565
rhenium **A2:** 581
schedule **EL1:** 877
sintering atmospheres **A7:** 339-350
sintering equipment **A7:** 351-359
size vibratory mill **A7:** 67
stampings, hydraulic bulge test for **A8:** 559
statistics, aluminum and aluminum alloys
testing, acoustic emission inspection **A17:** 284
testing, Brinell machines for **A8:** 87
time order, and quality control **A17:** 723-724
tooling **A7:** 329, 332-337
trends, monitored by CMMs **A17:** 20
Production forgings
by hot-die/isothermal forging **A14:** 157
Production grinding methods and techniques **EM4:** 336-349
applications **EM4:** 338-349
ceramic grinding applications and methods **EM4:** 336-337, 343
cost-effective production grinding of ceramics **EM4:** 349
diamond grinding wheel construction **EM4:** 340, 341, 342
Production grinding methods and techniques (continued)
emerging ceramic machining technologies **EM4:** 349
optimizing precision grinding parameters **EM4:** 345-349
practical aspects of grinding machines and processes **EM4:** 339-345
Production volumes
acrylonitrile-butadiene-styrenes (ABS) **EM2:** 109-110
allyls (DAP ,DAIP) **EM2:** 226
amino molding compounds **EM2:** 231
cyanates **EM2:** 232
high-impact polystyrenes (PS, HIPS) **EM2:** 194
homopolymer/copolymer acetals **EM2:** 100
liquid crystal polymers (LCP) **EM2:** 180
phenolics **EM2:** 242
polyamides (PA) **EM2:** 125
polybenzimidazoles (PBI) **EM2:** 147
polybutylene terephthalates (PBT) **EM2:** 153
polycarbonates (PC) **EM2:** 151
polyethylene terephthalates (PET) **EM2:** 172
polyphenylene ether blends (PPE PPO) **EM2:** 183
polyurethanes (PUR) **EM2:** 258
polyvinyl chlorides (PVC) **EM2:** 209
silicones (SI) **EM2:** 265
styrene-acrylonitriles (SAN, OSA ASA) **EM2:** 214
styrene-maleic anhydrides (S/MA) **EM2:** 217
thermoplastic polyurethanes (TPUR) **EM2:** 204-205
thermosetting resins **EM2:** 222
unsaturated polyesters **EM2:** 246
Productivity, and quality
fundamentals of **A17:** 719-723
Products
P/M **A7:** 257, 295, 451-462, 569-574
Products of principal metalworking processes
brazed joints, failures of **A11:** 450-455
cold-formed parts, failures of **A11:** 307-313
forgings, failures of **A11:** 314-343
iron castings, failures of **A11:** 344-379
steel castings, failures of **A11:** 380-410
Products, specific *See* Nondestructive inspection of specific products
Proeutectic
aluminum-rich dendrites in welded joints of aluminum alloys **A9:** 579
Proeutectoid
defined **A9:** 14-15
Proeutectoid carbide
defined **A9:** 14
Proeutectoid cementite **A9:** 178
Proeutectoid constituent .... **A3:** 1 • 21
Proeutectoid ferrite *See* Free ferrite **A1:** 129 **A9:** 178-179
defined **A9:** 14
in the solidification structure of ferrous alloys **A9:** 579
Professional education and training *See* Information sources
Professional P/M societies **A7:** 19
Profile
defined **A13:** 11
Profile angular distributions
for fracture surface area **A12:** 202-204
Profile extrusion
process **EM2:** 386
of short-fiber composites **EM1:** 121
Profile gaging
by lasers **A17:** 12
Profile generation
from replicas **A12:** 199
metallographic sectioning methods **A12:** 198-199
nondestructive **A12:** 199
Profile imaging
schematic **A17:** 14
Profile meter
for part shape measurement **A8:** 549
Profile parameters
fracture surface roughness **A12:** 199-200, 212-215
Profile rolling
defined **A14:** 10
Profile roughness parameters **A12:** 212-215

Profiles *See also* Composition profiles; Depth profiles; Depth profiling
alloy steels **A12:** 214, 327
by light microscope.......... **A12:** 94, 99
damage depth, RBS analysis for.......... **A10:** 632-633
depth, FIM/AP surface analysis **A10:** 583
depth-composition, stainless steel.......... **A10:** 555
examining **A12:** 95, 100
fatigue fracture **A12:** 15, 22
for fractal analysis.......... **A12:** 212
fractal properties **A12:** 211-215
fracture **A12:** 15, 22, 95-96, 212, 214
of fracture surface roughness.......... **A12:** 199-205, 212-215
from replicas **A12:** 199
generation, quantitative fractography **A12:** 198-199
impurity, RBS analysis for.......... **A10:** 632
matching, fatigue striations in nickel **A12:** 205-206
modified fractal curve **A12:** 212
nondestructive **A12:** 199
roughness, parameters **A12:** 199-200
sections **A12:** 95-96
and surface roughness parameters relationship **A12:** 212
Profiling
in conjunction with milling **A16:** 322
Profilometer, stylus *See* Stylus profilometer
Profilometry
to determine surface roughness in ceramic powder characterization.......... **EM4:** 27
Programmability
very-large-scale integration (VLSI) **EL1:** 8
wafer-scale integration **EL1:** 363
Programmable calculator programs
for composite materials analysis **EM1:** 277-279
Programmable controllers *See also* Automation
with molding machines **A15:** 350
Programmable logic control (PLC) units
for liquid penetrant inspection **A17:** 80
Programmable logic controllers (PLC).......... **EM4:** 36
Programmable logic devices (PLDS).......... **EL1:** 167
Programmable machine tools **A16:** 4
Programmable read-only memories (PROM) **EL1:** 160
Programming
wafer-scale integration **EL1:** 362-363
Programs, computer *See* Computer programs
Progression
defined **A14:** 10
Progression marks *See* Beach marks
Progressive aging
defined **A9:** 15
Progressive block sequence
definition.......... **M6:** 14
Progressive dies **A14:** 10, 457, 480-481
for blanking.......... **A14:** 454-455
coining in.......... **A14:** 182
for electrical steel sheet **A14:** 478-479
multiple-slide forming.......... **A14:** 569
for piercing.......... **A14:** 466
for press bending **A14:** 527-528
for press forming.......... **A14:** 546
for sheet metal drawing.......... **A14:** 579
for stainless steels.......... **A14:** 765-766
vs separate dies **A14:** 528
vs simple dies, stainless steels **A14:** 765
Progressive field evaporation
in the atom probe.......... **A10:** 591
Progressive flame hardening **M4:** 485
Progressive forming
defined **A14:** 10
Progressive fracturing
fracture mechanics of **A8:** 439-440
rapid, resistance to crack extension in **A8:** 444
Progressive grinding
stainless steel.......... **M5:** 45
Progressive headers
for cold heading **A14:** 292
Progressive ply failures
laminates.......... **EM1:** 238-239
Progressive solidification
and directional solidification.......... **A15:** 778
permanent mold casting **A15:** 278
and riser design.......... **A15:** 578-579
Progressive-spinning flame hardening.......... **M4:** 486
Projected area *See also* Surface area
defined **A15:** 9
Projected area diameter
particle size measurement **A7:** 225-226
Projected images
assumption of randomness **A12:** 194-195
and fatigue striation spacing, correlated **A12:** 205
fracture path and microstructure correlated **A12:** 195-196
and fracture surface, parametric relationships **A12:** 202
quantitative fractography **A12:** 194-196
quantitative statistical treatment of **A9:** 134
single SEM fractograph.......... **A12:** 194
and spatial features, stereological relationships **A12:** 196
Projected views of a ternary diagram .... **A3:** 1 • 5
Projectile
impacts, to generate high strain rates **A8:** 190
for plate impact test.......... **A8:** 233
Projectile rotating bands
powders used.......... **A7:** 573
Projection
defined **A17:** 384
Projection data
defined **A17:** 384
Projection distance
defined **A9:** 15
Projection lens
defined **A9:** 15
Projection plane
basic quantities and relations.......... **A12:** 194-196
Projection stereology
imaging by.......... **A12:** 194
Projection topography
for defect imaging of crystals **A10:** 369
Projection weld
definition.......... **M6:** 14
Projection welding **A7:** 456 **M6:** 503-524
advantages.......... **M6:** 504
applicability.......... **M6:** 504-505
comparison to arc welding.......... **M6:** 522-523
comparison to spot welding.......... **M6:** 520
comparison to staking.......... **M6:** 523-524
control of weld quality.......... **M6:** 522
cooling of electrodes and dies **M6:** 512
cross wire welding.......... **M6:** 517-519
definition.......... **M6:** 14, 503
electrode force.......... **M6:** 507
electrode holders **M6:** 510-511
electrodes **M6:** 509-513
fixtures.......... **M6:** 509-510
heat balance.......... **M6:** 508
high-production welding.......... **M6:** 510
limitations.......... **M6:** 505
metallurgical effects **M6:** 506
metals welded **M6:** 506
process variables **M6:** 507-509
projections.......... **M6:** 513
quality control.......... **M6:** 505
weld formation **M6:** 503-504
weld time **M6:** 507-508
welding current **M6:** 507
welding dies.......... **M6:** 509-510
welding machines **M6:** 505-506
workpiece cleaning **M6:** 506-507
workpiece size **M6:** 508
Projection welding (PW) **A6:** 230-237
aluminum **A6:** 233
Projection welding (PW) (continued)
aluminum-base alloys.......... **A6:** 233
annular, projection designs and process requirements for thin-gage low-carbon steel **A6:** 235
annular projection welding **A6:** 230, 231, 235-236, 237
applications **A6:** 230-233
copper.......... **A6:** 233
copper-base alloys **A6:** 233
cross-wire welding.......... **A6:** 230-231, 235, 236, 237
definition.......... **A6:** 230, 1212
die geometries for intermediate-gage steels, spherical projections **A6:** 233
diffusion bonding.......... **A6:** 232-233
edge-to-sheet welds **A6:** 232
embossed-projection welding **A6:** 230, 231, 232, 235
equipment.......... **A6:** 233, 236
fast follow-up (low inertia) head (typical).......... **A6:** 231, 233
inductive "skin" effect.......... **A6:** 236
material effects.......... **A6:** 232-233
nickel-base alloys.......... **A6:** 233
nut welding **A6:** 232, 235
personnel.......... **A6:** 233-235
process requirements.......... **A6:** 235-237
heavy-gage low-carbon steels.......... **A6:** 232
intermediate-gage low-carbon steels.......... **A6:** 234
projection and die geometries for heavy-gage steels.......... **A6:** 232
pulsation welding schedules.......... **A6:** 235
resistance welding.......... **A6:** 230, 233, 235
solid-projection welding **A6:** 230, 231, 232, 233, 235-236
stainless steels **A6:** 233
steels **A6:** 232, 233
titanium alloys **A6:** 233
weld nuts **A6:** 232, 235
Projection welding of
aluminum **M6:** 506
aluminum alloys.......... **M6:** 503, 510-511
barstock to sheet metal.......... **M6:** 515-516
coated metals **M6:** 506
coated steel **M6:** 520-521
copper and copper alloys **M6:** 503, 548, 552-554
dissimilar metals **M6:** 519-520
free-machining steels **M6:** 506, 515-516
low-alloy steels **M6:** 506
low-carbon steels.......... **M6:** 506-507, 509, 513-514, 518, 520
Monel alloys.......... **M6:** 506
naval brass.......... **M6:** 506
nickel-copper alloys **M6:** 503, 506
powder metallurgy parts **M6:** 521-522
sheet metal parts **M6:** 513-514
intermediate thickness workpieces.......... **M6:** 514
thick workpieces **M6:** 514
thin workpieces.......... **M6:** 514
unequal thicknesses.......... **M6:** 514-515
stainless steel.......... **M6:** 503, 509, 530-531
austenitic stainless steels **M6:** 506
tube to sheet metal.......... **M6:** 516-517
electrode design **M6:** 517
heat balance **M6:** 517
projection design.......... **M6:** 516-517
weld strength.......... **M6:** 517
Projections
massive, as casting defects **A11:** 381
metallic, as casting defects **A11:** 381
with rough surfaces, as casting defects.......... **A11:** 381
Projections for projection welding.......... **M6:** 513
design **M6:** 513
spacing **M6:** 513
types **M6:** 512-513
Projective magnification
with microfocus x-ray sources.......... **A17:** 300
Promethium *See also* Rare earth metals
properties.......... **A2:** 1185

SUBJECTS OF THE INDEXED VOLUMES: ASM Handbook (designated by the letter "A"): **A1:** Properties and Selection: Irons, Steels, and High-Performance Alloys (1990); **A2:** Properties and Selection: Nonferrous Alloys and Special-Purpose Materials (1990); **A3:** Alloy Phase Diagrams (1992); **A4:** Heat Treating (1991); **A6:** Welding, Brazing, and Soldering (1993); **A7:** Powder Metallurgy (1984); **A8:** Mechanical Testing (1985); **A9:** Metallography and Microstructures (1985); **A10:** Materials Characterization (1986); **A11:** Failure Analysis and Prevention (1986); **A12:** Fractography (1987); **A13:** Corrosion (1987); **A14:** Forming and Forging (1988); **A15:** Casting (1988); **A16:** Machining (1989); **A17:** Nondestructive Testing and Quality Control (1989); **A18:** Friction, Lubrication, and Wear Technology (1992). **Metals Handbook, 9th Edition** (designated by the letter "M"): **M1:** Properties and Selection: Irons and Steels (1978); **M2:** Properties and Selection: Nonferrous Alloys and Pure Metals (1979); **M3:** Properties and Selection: Stainless Steels, Tool Materials and Special-Purpose Materials (1980); **M4:** Heat Treating (1981); **M5:** Surface Cleaning, Finishing, and Coating (1982); **M6:** Welding, Brazing, and Soldering (1983). **Engineered Materials Handbook** (designated by the letters "EM"): **EM1:** Composites (1987); **EM2:** Engineering Plastics (1988); **EM3:** Adhesives and Sealants (1990); **EM4:** Ceramics and Glasses (1991); **Electronic Materials Handbook** (designated by the letters "EL"): **EL1:** Packaging (1989).

Promethium (continued)
pure ... **M2:** 788
as rare earth ... **A2:** 720
Promoter *See also* Accelerator; Catalyst ... **EM3:** 23
adhesion ... **EM1:** 3, 122-123
defined ... **EM1:** 19 **EM2:** 34
Prompt gamma activation analysis ... **A10:** 239-240, 689
Prompt neutron-capture γ-rays
in NAA ... **A10:** 234
Prompt scrap ... **A1:** 1023
Proof *See also* Die proof ... **EM3:** 23
defined ... **A14:** 10 **EM1:** 19 **EM2:** 34
pressure, defined ... **EM1:** 19
Proof gold *See* Commercial fine gold
Proof load
defined ... **A14:** 10
Proof loading
composite materials ... **A8:** 717
followed by fatigue ... **A8:** 717-718
and minimum fatigue cycle relationship ... **A8:** 718
Proof pressure ... **EM3:** 23
defined ... **EM2:** 34
Proof strength in bending
for spring-tempered phosphor bronze strip ... **A8:** 136
Proof stress *See also* Offset yield strength
of bolt or stud ... **A1:** 296-297
defined ... **A8:** 10 **A14:** 10
of nut ... **A1:** 297
Proof stress, threaded fasteners ... **M1:** 273, 274, 277-278, 280-282
Proof testing ... **A8:** 717-718
of brazed assemblies
brazed joints ... **A6:** 1119
equipment, acoustic emission inspection ... **A17:** 289
as mechanical stress test ... **A7:** 490-491
soldered joints ... **A6:** 981
Proof-loading acceptance test ... **EM3:** 529-530
Proofstressing ... **M6:** 892
Propadiene
chemical bonding ... **M6:** 900
Propagation *See also* Crack propagation; Velocity, of programation; Wave propagation
crack ... **EM1:** 201
of decoupled modes ... **EL1:** 36-37
defined ... **EL1:** 1154
explosion ... **A7:** 196
phase, crevice corrosion ... **A13:** 303
of stress-corrosion cracking ... **A13:** 245-246
time ... **EL1:** 20-21, 601
Propagation constant
defined ... **EL1:** 1154
Propagation delay to signal rise
considered ... **EL1:** 521-522
Propagation velocities *See* Crack speed
Propane
chemical bonding ... **M6:** 900
fuel gas for oxyfuel gas cutting ... **A6:** 1157-1162
fuel gas for torch brazing ... **A6:** 328 **M6:** 950
in high-velocity oxyfuel powder spray process ... **A18:** 830
for oxyfuel gas cutting ... **A14:** 723
oxyfuel gas welding fuel gas ... **A6:** 281, 282, 283, 285
properties as fuel gas ... **M6:** 899
use in oxyfuel gas welding ... **M6:** 584
valve thread connections for compressed gas cylinders ... **A6:** 1197
Propane gas
as atmosphere ... **A7:** 341
Propellant gun
in flat plate impact test ... **A8:** 210
Propeller agitators ... **M4:** 46, 61, 62, 64
Propeller blade, cold-straightened aluminum alloy
fatigue fracture of ... **A11:** 125
Propeller blades
optical holography of ... **A17:** 424-425
Propeller bronze
properties and applications ... **A2:** 386
Proper fixturing ... **A16:** 404-410
clamping ... **A16:** 405-407
CNC automation ... **A16:** 407
cost factors ... **A16:** 409, 410
definition ... **A16:** 404
Proper fixturing (continued)
duplex fixtures ... **A16:** 404
fixture design ... **A16:** 404-405, 408
hydraulic clamping ... **A16:** 405-407
hydraulic/mechanical locking clamps ... **A16:** 407
manual clamping ... **A16:** 406
modular fixturing for limited production ... **A16:** 407-410
PLC automation ... **A16:** 407
pneumatic clamping ... **A16:** 405-406
positive-lock clamping ... **A16:** 407, 408
SAFE Modular Fixturing System ... **A16:** 409-410
vacuum clamping ... **A16:** 406
Properties *See also* Aggregate properties approach; Constituent materials; Fiber properties analysis; Laminate properties analysis; Material properties; Material properties analysis; Mechanical properties; Mechanical strength; Physical properties; Properties considerations; Refractory properties; Specific metals and alloys; specific properties; specific properties analysis
agency-related, data sheets ... **EM2:** 409
of composite materials ... **EM1:** 177-179
design guidelines ... **EM2:** 710
effects, filament winding ... **EM2:** 369-371
effects, long-term environmental factors ... **EM2:** 423-432
effects, RTM and SRIM, compared ... **EM2:** 346-349
important divergencies ... **EM2:** 655-658
material anisotropy effects on ... **EM2:** 405
modification, by additives ... **EM2:** 493-507
polyethylene terephthalates (PET) ... **EM2:** 172
in process selection ... **EM2:** 279-281
and rotational molding ... **EM2:** 361
size effects ... **EM2:** 658
thermoplastic process effects ... **EM2:** 282-286
thermosetting process effects ... **EM2:** 286-287
Properties considerations *See also* Properties; specific properties
aggregate properties approach, to design ... **EM2:** 407-411
electrical properties ... **EM2:** 460-480
introduction ... **EM2:** 405-406
long-term environmental factors ... **EM2:** 423-432
mechanical properties ... **EM2:** 433-438
modification, by additives ... **EM2:** 493-507
modification, by polymer-polymer mixtures ... **EM2:** 487-492
optical properties ... **EM2:** 481-486
thermal properties, engineering thermoplastics ... **EM2:** 445-459
thermal properties, engineering thermosets ... **EM2:** 439-444
thermoplastic structural forms ... **EM2:** 508-513
viscoelasticity ... **EM2:** 412-422
Properties, elevated temperature
aluminum alloys ... **M2:** 56, 57-58, 62
Properties, low-temperature
aluminum alloys ... **M2:** 62
Properties modification
by additives ... **EM2:** 493-507
by polymer-polymer mixtures ... **EM2:** 487-492
polymer selection for ... **EM2:** 489-490
Property
derived, defined ... **A8:** 667
static metallic material, design allowables for ... **A8:** 662-677
Properzi process
wheel-and-band machines ... **A15:** 314
Properzi system
for copper and copper alloy wire rod ... **A2:** 255
Propionic acid ... **A13:** 646, 1159-1160
Propionic acid ($CH_3CH_2CO_2H$)
as solvent used in ceramics processing ... **EM4:** 117
Proportional control
shape memory alloys ... **A2:** 900
Proportional integral differential (PID) control ... **EM4:** 253
Proportional limit *See also* Elastic limit; Hooke's law ... **EM3:** 24
defined ... **A8:** 10 **A10:** 679 **A11:** 8 **A14:** 10 **EM1:** 19 **EM2:** 34, 433
Proportional loading
in multiaxial testing ... **A8:** 344
in plastic train field ... **A8:** 447
Proportional-integral-derivative (PID) loop controller
preheaters ... **EL1:** 687
Proportional-integral-differential (PID) control systems ... **A6:** 1062
Proportioned kits ... **EM3:** 688-689
disadvantages ... **EM3:** 688-689
Proprietary gases
oxyfuel gas welding ... **A6:** 281, 282, 283
Proprietary methods
of magnetic particle inspection ... **A17:** 122-128
Proprietary phosphates
applications ... **EM4:** 47
composition ... **EM4:** 47
supply sources ... **EM4:** 47
Propylene
chemical bonding ... **M6:** 900
in high-velocity oxyfuel powder spray process ... **A18:** 830
for oxyfuel gas cutting ... **A14:** 723-724
properties as fuel gas ... **M6:** 899
valve thread connections for compressed gas cylinders ... **A6:** 1197
Propylene glycol ... **EM3:** 674
surface tension ... **EM3:** 181
Propylene glycol monomethyl ether
surface tension ... **EM3:** 181
Propylene plastics *See also* Plastics ... **EM3:** 24
defined ... **EM2:** 34
Prost patent (1970) ... **EM4:** 710
Prostheses
anchoring ... **A11:** 670-671
failed, from fatigue and stem loosening ... **A11:** 690-693
joint ... **A11:** 670-671
microhoning of ... **A16:** 491
types of ... **A11:** 670
Prosthetic devices ... **A13:** 1324-1335
artificial joints ... **A11:** 670-671
background ... **A13:** 1324-1329
biocompatibility of ... **A13:** 1328-1329
corrosion forms ... **A13:** 1330-1332
corrosion significance ... **A13:** 1328-1329
corrosion testing ... **A13:** 1332-1333
defined ... **A11:** 670
electrochemistry and corrosion ... **A13:** 1329-1330
metals/alloys ... **A13:** 1325-1328
titanium and titanium alloy ... **A2:** 589
tumor resections ... **A11:** 671
types of ... **A11:** 670-671
Prosthetics
carbon-carbon composite ... **EM1:** 923
powders used ... **A7:** 573, 754
Protactinium
M lines use for ... **A10:** 86
pure ... **M2:** 788, 832-833
Protactinium, as actinide metal
properties ... **A2:** 1194
Protection *See* Corrosion protection
Protection potential
pitting/crevice corrosion ... **A13:** 231
Protection tubes
powders used ... **A7:** 573
for thermocouples ... **A2:** 883-884
Protective atmosphere *See also* Atmospheres ... **A7:** 9
definition ... **A6:** 1212 **M6:** 14
Protective coatings *See also* Coatings; Coatings, specific coatings; Conformal coatings; Corrosion protection; Films; Plating
aluminum anodizing ... **A13:** 396-398
for chloride SCC ... **A13:** 327
compared ... **EL1:** 584
copper/copper alloys ... **A13:** 636
corrosion-resistant *See* Corrosion protection
electrochemical evaluation ... **A13:** 219-220
electroplated ... **A13:** 419
for elevated-temperature failures ... **A11:** 269-271
flexible printed boards ... **EL1:** 583-584
materials and processes selection ... **EL1:** 115-116
porcelain enamels ... **A13:** 446-452
for springs ... **A11:** 560
as surface preparation ... **EL1:** 675
tests for ... **EM2:** 425
uranium/uranium alloys ... **A13:** 818-821
vacuum coatings covered by ... **M5:** 402, 408-409
for wood/wood laminate patterns ... **A15:** 194

Protective film *See also* Film
atmospheric **A13:** 82
by alloying **A13:** 47-48
organic coatings and linings **A13:** 399-418
oxide, characteristics **A13:** 517-518
Protective gas **A7:** 9
Protective potential
defined **A13:** 11
range, defined **A13:** 11
Protective scale growth rate **A18:** 208
Protein
dynamics, MFS analysis for **A10:** 72
surface, IR analysis **A10:** 119-120
Protocol
sampling **A10:** 13, 15-16
Proton energy
abbreviation for **A10:** 690
Proton microprobes
of biological samples, compared **A10:** 107
capabilities **A10:** 102
particle-induced x-ray emission and **A10:** 107
Proton milliprobes
capabilities **A10:** 102
Proton-induced x-ray emission (PIXE) **EM3:** 237
for preliminary identification of heavier elements **EM3:** 644, 647, 648
Protons
defined **A10:** 680
Prototype **EM3:** 24
defined **EM1:** 19 **EM2:** 34
Prototype crystals .... **A3:** 1 • 16
Prototype development
blow molding **EM2:** 299, 356
compression molding **EM2:** 298
injection molding **EM2:** 296
molds, rotational molding **EM2:** 367
of plastic parts **EM2:** 80-81
Prototype facets
true area values **A12:** 203
Protuberances *See* Asperities
Prout-Tompkins
rate law expression **EM4:** 55
Proving ring *See also* Elastic proving ring
for Brinell test **A8:** 88
capacities **A8:** 614
elastic, as calibration device **A8:** 614-615
for load weighing systems **A8:** 48
screw-driven testing machine calibrated with **A8:** 615
for stress-corrosion testing **A8:** 507
Prow formation *See* Wedge formation
Proximity fuze cup
powder used **A7:** 573
PS *See* High-impact polystyrenes; Polystyrenes
PSD *See* Position-sensitive detector
Pseudarthrosis
and bone repair **A11:** 674
"Pseudo" alloys
silver powder use **A7:** 147
Pseudo-Kikuchi patterns *See also* Kikuchi patterns **A9:** 94
Pseudo-Kossel lines
in divergent-beam topography **A10:** 371
Pseudobinary .... **A3:** 1 • 5
Pseudobinary sections of a ternary diagram .... **A3:** 1 • 5
Pseudoboehmite **EM3:** 624
Pseudocolor
electronic image enhancement **A9:** 138, 152
Pseudocolor, in NDE
advantages/limitations of **A17:** 483-488
Pseudoelasticity
of shape memory alloys **A2:** 898, 901
Pseudomonas
biological corrosion by **A13:** 118-119
Pseudoplastic behavior
defined **A18:** 15
Pseudostatic tension test
strain rate ranges for **A8:** 40
Pseudothermoplastics **EM1:** 546-547
PSM System-100 .... **EM4:** 67
PST *See* Polycrystal scattering topography
PSU *See* Polysulfones
PSZ (CaO • MgO)
properties **A6:** 949
Pt-Rh (Phase Diagram) .... **A3:** 2 • 346
Pt-Si (Phase Diagram) .... **A3:** 2 • 347
Pt-Sn (Phase Diagram) .... **A3:** 2 • 347
Pt-Te (Phase Diagram) .... **A3:** 2 • 347
Pt-Ti (Phase Diagram) .... **A3:** 2 • 348
Pt-U (Phase Diagram) .... **A3:** 2 • 348
Pt-V (Phase Diagram) .... **A3:** 2 • 349
Pt-Zr (Phase Diagram) .... **A3:** 2 • 349
PTB
as synchrotron radiation source **A10:** 413
PTFE *See* Polytetrafluoroethylene
PTFE-fiberglass
for printed board materials systems **EM3:** 592
PTFE-Kevlar
for printed board materials systems **EM3:** 592
PTH *See* Plated-through hole (PTH)
Pu-Sc (Phase Diagram) .... **A3:** 2 • 349
Pu-U (Phase Diagram) .... **A3:** 2 • 350
Pu-Zn (Phase Diagram) .... **A3:** 2 • 350
Pu-Zr (Phase Diagram) .... **A3:** 2 • 350
Puckers **EM1:** 19, 1-6
defined **EM2:** 34
Puddle
definition **A6:** 1212
Puffed compact **A7:** 9
Puffing, laminate
during cure **EM1:** 662
Pull gun technique
definition **A6:** 1212
Pull rod
for creep test stand **A8:** 312
Pull-out and microdrop technique **EM3:** 402, 403
advantages **EM3:** 394
limitations **EM3:** 394
Pull-through tears
as fastener failure **A11:** 531
Pull-type lockbolts
materials and composite applications for **A11:** 530
Pulldown in forgings **M1:** 367-368
Pulldowns
in iron castings **A11:** 353
Pulled surface
defined **EM2:** 34
Pullucite
chemical system **EM4:** 870
Pulmonary disease
chronic, from cadmium toxicity **A2:** 1240
from aluminum toxicity **A2:** 1256
from beryllium toxicity **A2:** 1239
Pulp
aramid fiber **EM1:** 115
molding **EM1:** 19
Pulp and paper industry **A13:** 1186-1220
corrosion control, pulp bleach plants **A13:** 1190-1196
kraft pulping liquors corrosion **A13:** 1208-1214
mechanical pulping equipment corrosion **A13:** 1214-1218
nickel-base alloy applications **A13:** 654
paper machine corrosion **A13:** 1186-1188
pollution control **A13:** 1370
recovery boiler corrosion **A13:** 1198-1202
stainless steel corrosion **A13:** 561-562
suction roll corrosion **A13:** 1202-1208
sulfite pulping liquor corrosion **A13:** 1196-1198
Pulp and paper industry applications *See also* Paper and printing industry applications
cobalt-base wear-resistant alloys **A2:** 451
nickel and nickel alloys **A2:** 430
structural ceramics **A2:** 1019
Pulp bleach plants
corrosion control in **A13:** 1190-1196
Pulp digester vessel
neck liner removed from **A11:** 403
Pulp molding
defined **EM2:** 34
Pulping
thermomechanical and chemi-thermomechanical **A13:** 1217-1218
Pulsation
AISI/SAE alloy steels **A12:** 304
of electric current **A17:** 91
Pulse
attenuator, in explosively loaded torsional Kolsky bar **A8:** 227
definition **M6:** 14
filter **A8:** 224-227
loading, short duration, and flyer plate **A8:** 210-212
smoother, in explosively loaded torsional Kolsky bar **A8:** 224-225, 227
Pulse (resistance welding)
definition **A6:** 1212
Pulse fraction curves
for atom probe microanalysis **A10:** 594-595
Pulse generator, high-voltage
for atom probe microanalysis **A10:** 591
Pulse method
of measuring relaxation times **A10:** 258
Pulse NMR spectrometer **A10:** 283
Pulse plating deposition **A18:** 834, 835, 836
titanium alloys **A18:** 781
Pulse polarography
differential **A10:** 193
as improved voltammetry **A10:** 193
normal **A10:** 193
Pulse repetition rates (PRR) **A6:** 40
Pulse soldering
defined **EL1:** 1154
Pulse start delay time
definition **M6:** 14
Pulse time
definition **M6:** 14
Pulse width modulation (PWM) **A6:** 39
Pulse-echo method
nuclear magnetic resonance **A10:** 281
Pulse-echo methods *See also* Pulse-echo ultrasonic inspection
A-scan displays, interpreting **A17:** 244-246
data interpretation **A17:** 243-246
echo amplitude **A17:** 245-246
loss of back reflection **A17:** 246
and microwave inspection **A17:** 211
presentation **A17:** 241-244
principles **A17:** 241
ultrasonic inspection **A17:** 240-248
Pulse-echo ultrasonic inspection *See also* Pulse-echo methods
control system **A17:** 253-254
of forged shafts **A17:** 508
instrument **A17:** 253
Pulse-echo ultrasonics **EM1:** 770
Pulse-height analysis **A18:** 325
Pulse-inspection circuitry
energy-dispersive spectrometer **A10:** 519-520
Pulse-modulated reflection
microwave inspection **A17:** 206
Pulse-modulated transmission
microwave inspection **A17:** 206
Pulsed current flow
electromigration in **EL1:** 964
Pulsed gas-metal arc welding (GMAWP)
power source selected **A6:** 37
Pulsed laser atom probe
facilities built into ECAP **A10:** 598
schematic **A10:** 597
of ternary 3:5 semiconductor **A10:** 601-602
Pulsed laser beam welding **M6:** 656-657
Pulsed laser thermal detection system
schematic **EL1:** 369

Pulsed laser-beam welding
mechanically alloyed oxide dispersion-strengthened (MA ODS) alloys ... **A2:** 949
Pulsed leaky Lamb wave testing
C-scan ... **A17:** 253
Pulsed magnetic field
in high-energy-rate compacting ... **A7:** 306
Pulsed mode
sonic converters ... **A8:** 243
Pulsed optical preamplifier
x-ray spectrometers ... **A10:** 91
Pulsed power welding
definition ... **M6:** 14
Pulsed spectrometers
in acoustic ESR ... **A10:** 258
Pulsed spray welding
definition ... **M6:** 14
Pulsed xenon lamps
for radiation curing ... **EL1:** 864
Pulsed-arc welding
steel weldment soundness in gas-shielded processes ... **A6:** 409
to solve problems in joining thin sections by oxyfuel gas welding ... **A6:** 288
Pulsed-incident waves
microwave inspection ... **A17:** 202, 206
Pulsed-laser optical holographic interferometry ... **A17:** 407, 410-411
Pulser circuit
electronic ultrasonic inspection ... **A17:** 252
Pulses, threshold-crossing
acoustic emission inspection ... **A17:** 281-283
Pultrusion ... **EM1:** 533-543 **EM2:** 389-398
applications ... **EM1:** 533-534
defined ... **EM1:** 19, 72-73 **EM2:** 34
design guide lines ... **EM1:** 542-543
design, guidelines ... **EM2:** 396-398
of epoxy composites ... **EM1:** 72-73
of glass rovings ... **EM1:** 109
of graphite-reinforced MMC ... **EM1:** 870-872
material forming ... **EM1:** 536
materials ... **EM1:** 536-537 **EM2:** 392-393
molded-in color ... **EM2:** 306
orientation options ... **EM1:** 537-540 **EM2:** 393-395
process ... **EM1:** 533-536
process components ... **EM2:** 390-391
process description ... **EM2:** 389-390
product characteristics ... **EM2:** 397
properties ... **EM1:** 540-542 **EM2:** 395-396
resin formation ... **EM1:** 535-536
surface finish ... **EM2:** 304
of thermoplastic resin composites ... **EM1:** 549
thermosetting ... **EM2:** 303
tooling ... **EM1:** 536 **EM2:** 392
unsaturated polyesters ... **EM2:** 251
Pultrusion of resin-matrix composites ... **A9:** 591
Pulverization ... **A7:** 9
Pumice
abrasive in commercial prophylactic paste ... **A18:** 666, 668, 669
in dentifrices ... **A18:** 668
Pump casings, cast iron
coatings for ... **M1:** 105
Pump gear
economy in manufacture ... **M3:** 855, 856
Pump impeller, cast stainless steel
cavitation damage ... **A13:** 142
Pump wink ... **EM3:** 720
Pumping
schematics, vacuum leak testing ... **A17:** 67
speed, leak detection effects ... **A17:** 69-70
Pumping efficiency
defined ... **A18:** 15
Pumping equipment
for resin transfer molding ... **EM1:** 169
Pumping system, vacuum
SEM microscope ... **A10:** 491
Pumps *See* Circulation pump; Molten metal pump ... **A13:** 1139, 1294-1296, 1316
abrasive waterjet cutting ... **A14:** 744-746
bowl, graphitic corrosion ... **A11:** 372-373
impeller, bronze, cavitation damage failure ... **A11:** 167-168
impeller, cast iron, graphitic corrosion ... **A11:** 374-375
impeller, erosion-corrosion in ... **A11:** 402
Pumps (continued)
mud, bending fatigue fracture in pushrod of ... **A11:** 469
parts, wear of cast iron ... **A11:** 365-367
peristaltic, in ICP concentric nebulizers ... **A10:** 35
quenching medium agitation ... **M4:** 64, 67
quenching medium circulation ... **M4:** 67
rotary and diffusion vacuum, in gas mass spectrometer ... **A10:** 151-152
rotary, damaged austenitic cast iron impellers in ... **A11:** 355-356
sampling ... **A10:** 16
stainless steel, cavitating mercury damage ... **A11:** 165
Toepler ... **A10:** 152
Pumps, vacuum
for SEM systems ... **A12:** 171
Pumps, wear of ... **A18:** 593-600
abrasive wear ... **A18:** 595, 597-598, 599
cavitation erosion ... **A18:** 593, 597, 599-600
centrifugal pumps ... **A18:** 599-600
reciprocating pumps ... **A18:** 600
corrosive wear ... **A18:** 593
erosive wear ... **A18:** 593, 597-599
lubrication ... **A18:** 595, 597
mechanisms causing loss of surface material from component parts ... **A18:** 593
particulate erosion ... **A18:** 597-599
rubbing wear ... **A18:** 593-597
PUN *See* Phenotic urethane no-bake resin system
Punch *See* Force plug
Punch and die
method, blanking shear test as ... **A8:** 64
sheet metal forming ... **A8:** 547
Punch loading
aluminum, strain rate sensitivity ... **A8:** 230-231
for high shear testing ... **A8:** 228-230
for high strain rate shear testing ... **A8:** 187, 229-230
high-strength steel ... **A8:** 229-230
Kolsky bar apparatus ... **A8:** 229-230
Punch presses ... **A14:** 464, 523, 543-544
Punch, punches ... **A7:** 10
air-mounted lower outer punch ... **A7:** 323
arrangement, stress and density distribution with ... **A7:** 300-301
clamp rings, materials for ... **A7:** 337
component stress ... **A7:** 335-336
compression ... **A7:** 336, 337
designing ... **A7:** 335-337
and dies, tool steel applications ... **A7:** 792
for hot pressing ... **A7:** 502
materials for ... **A7:** 337
spring-mounted lower outer ... **A7:** 323
stationary and mountings ... **A7:** 335, 336
Punch shear
in piercing ... **A14:** 463
Punch-to-die clearance *See* Clearance; Die clearance
Punchability
of electrical steel sheet ... **A14:** 477
Punches *See also* Punchability, Punching
arbor-type ... **A14:** 537
back extrusion ... **A2:** 970-971
blanking and piercing, materials for ... **A14:** 484
bulging, fluid forming ... **A14:** 614
carburization cracking ... **A11:** 573-574
for cold extrusion ... **A14:** 309
deep-drawing, materials for ... **A14:** 508-509
defined ... **A14:** 10
design, cold extrusion ... **A14:** 302
expanding drawn workpieces with ... **A14:** 587
gooseneck ... **A14:** 536
high-pressure, cemented carbide ... **A2:** 972
for hot trimming, stainless steel forgings ... **A14:** 230
materials for ... **A14:** 483, 511, 580
for open-die forging ... **A14:** 62, 63
perforator, materials for ... **A14:** 485
for piercing ... **A14:** 464-465
powder compacting, cemented carbide ... **A2:** 971
for press-brake forming ... **A14:** 535-539
quench-crack failure of ... **A11:** 566
radii, deep drawing effects ... **A14:** 580-581
square-end, cutting force ... **A14:** 448
stamping, cemented carbide ... **A2:** 971
Punching *See also* Blanking; Punches
aluminum and aluminum alloys ... **A2:** 10
Punching (continued)
in conjunction with broaching ... **A16:** 195
defined ... **A14:** 10, 55
dies ... **A14:** 55-56
and gear manufacture ... **A16:** 330
high strain rate shear testing ... **A8:** 215
NC implemented ... **A16:** 613
nickel-titanium shape memory effect (SME) alloys ... **A2:** 899
refractory metals and alloys ... **A2:** 560-562
and shearing machines ... **A14:** 716
tungsten ... **A2:** 562
Punching technique for subdividing solids ... **A10:** 165
Punctual illumination
MOLE/Raman analysis ... **A10:** 129-130
Punty rod ... **EM4:** 394
PUR *See* Polyurethanes; Urethanes
Purchase scrap
storage of ... **A15:** 363
Purchased scrap ... **A1:** 1023
Pure adhesion term (μ_a) ... **A18:** 35
Pure aluminum *See also* Aluminum; Aluminum, specific types ... **A7:** 130
applications and properties ... **A2:** 62-65
direct-chill casting ... **A15:** 314
dynamic yield stress vs. strain rate in ... **A8:** 41
high-stacking fault energy material ... **A8:** 173
K-M CBEDP from ... **A10:** 441
for niobium-titanium superconducting materials ... **A2:** 1045
properties ... **A2:** 3, 1099-1100
rotor alloys ... **A2:** 127-129
solid-solution effects ... **A2:** 38
wrought series ... **A2:** 29
Pure bending
bent beam for ... **A8:** 505
Pure ceramics
electrical discharge machining ... **EM4:** 376
Pure chromium
oxide scale formation ... **A13:** 97
Pure cobalt *See also* Elemental cobalt
electrical and magnetic properties ... **A2:** 447
mechanical properties ... **A2:** 447
properties ... **A2:** 1109
Pure compression
plastic flow in ... **A8:** 576-577
Pure copper *See also* Copper ... **A15:** 314, 774
applications ... **A2:** 216, 224, 239-240
castability ... **A2:** 346
commercial, types ... **A2:** 223, 230, 234
cooling in ultrasonic testing ... **A8:** 247
corrosion on ... **A11:** 201
as electrical contact material ... **A2:** 843
mechanical working ... **A2:** 219
P/M parts ... **A2:** 397-398
melt treatment ... **A15:** 774
properties ... **A2:** 224, 1110
scrap, for recycling ... **A2:** 1215
Pure copper P/M parts ... **A7:** 735-736
Pure gold *See also* Commercial fine gold; Gold; Precious metals
properties ... **A2:** 1116
properties and applications ... **A2:** 692
Pure iron *See also* Iron; iron alloys; specific iron alloys ... **A7:** 480
damping capacity ... **M1:** 32
hydrogen damage ... **A13:** 166-169
properties ... **A2:** 1118-1127
soft magnetic properties ... **A2:** 762
Pure lead *See also* Corroding lead; Lead; Lead alloys
compositions and grades ... **A2:** 543-545
properties ... **A2:** 1129
Pure magnesium ... **A7:** 131
Pure materials
as primary standards ... **A10:** 162
Pure metal
resonant specimen lengths for ... **A8:** 249
workability ... **A8:** 574-575
Pure metal preparation methods
chemical vapor deposition ... **A2:** 1094
distillation ... **A2:** 1094
fractional crystallization ... **A2:** 1093
solid-state refining techniques ... **A2:** 1094-1095
vacuum melting ... **A2:** 1094
zone refining ... **A2:** 1093-1094

Pure metal(s) *See also* Actinide metals; High purity; Pure metals, specific types; Rare earth metals; Transplutonium actinide metals
aluminum, properties **A2:** 1099-1100
antimony, properties **A2:** 1100
arsenic, properties **A2:** 1101
barium, properties **A2:** 1101
beryllium, properties **A2:** 1102
boron, properties
cadmium, properties **A2:** 1104
calcium, properties **A2:** 1105
cesium, properties **A2:** 1107
chemical vapor deposition (CVD) **A2:** 1094
chromium, properties **A2:** 1107
cobalt, properties **A2:** 1109
columbium *See* Niobium
copper, properties **A2:** 1110
distillation **A2:** 1094
electrical resistance, properties **A2:** 823
fractional crystallization **A2:** 1093
gallium, properties **A2:** 1114
germanium, properties **A2:** 1115
gold, properties **A2:** 1116
as heating alloys **A2:** 829
impurity concentrations **A2:** 1096
indium, properties **A2:** 1117
iridium, properties **A2:** 1117
iron, properties **A2:** 1118-1127
lead, properties **A2:** 1129
lithium, properties **A2:** 1131
magnesium, properties **A2:** 1132
manganese, properties **A2:** 1135
mercury, properties **A2:** 1138
molybdenum, properties **A2:** 1140
nickel **A2:** 435, 437, 441
nickel, properties **A2:** 1143
niobium **A2:** 1144
osmium, properties **A2:** 1145
palladium, properties **A2:** 1146
platinum, properties **A2:** 1147
potassium, properties **A2:** 1148
praseodymium, properties **A2:** 1149
preparation and characterization **A2:** 1093-1097
preparation methods **A2:** 1093-1095
purity, six nines characterization **A2:** 1097
resistance-ratio test **A2:** 1096-1097
rhenium, properties **A2:** 1150
rhodium, properties **A2:** 1151
rubidium, properties **A2:** 1151
ruthenium, properties **A2:** 1153
selenium, properties **A2:** 1153
silicon, properties **A2:** 1154
silver, properties **A2:** 1156
sodium, properties **A2:** 1158
solid-state refining techniques **A2:** 1094-1095
strontium, properties **A2:** 1159
tantalum, properties **A2:** 1160
technetium, properties **A2:** 1163
tellurium, properties **A2:** 1165
thallium, properties **A2:** 1165
tin, properties **A2:** 1166
titanium, properties **A2:** 1169
trace element analysis **A2:** 1095-1096
tungsten, properties **A2:** 1170
vacuum melting **A2:** 1094
vanadium, properties **A2:** 1172
zinc, properties **A2:** 1174
zirconium, properties **A2:** 1175
zone refining **A2:** 1093-1094
Pure metals *See* index entries under individual elements **M2:** 709-713
allotropic transformations **A9:** 655
characteristics **A13:** 46
characterization **M2:** 711-712, 713
FIM images of defects in **A10:** 588-589
normal grain growth in **A9:** 697
preparation **M2:** 709-711
solidification structures of **A9:** 607-610
structures and thermal properties **A13:** 62-63
Pure metals, specific types
aluminum, by zone-refining technique **A2:** 1094
barium, by distillation **A2:** 1094
bismuth, by zone-refining technique **A2:** 1094
calcium, by distillation **A2:** 1094
chromium, by iodide/chemical vapor deposition **A2:** 1094
copper, by zone-refining technique **A2:** 1094
gallium, by fractional crystallization **A2:** 1093
gold, by zone-refining technique **A2:** 1094
hafnium, by chemical vapor deposition **A2:** 1094
lead, by zone-refining technique **A2:** 1094
lithium, by distillation **A2:** 1094
magnesium, by distillation and zone refinement **A2:** 1094
molybdenum, by chemical vapor deposition **A2:** 1094
molybdenum, by zone-refining technique **A2:** 1094
niobium, by chemical vapor deposition **A2:** 1094
niobium, by zone-refining technique **A2:** 1094
rare earth metals, by electrotransport purification **A2:** 1094-1095
silicon crystals, by zone-refining technique **A2:** 1094
silver, by zone-refining technique **A2:** 1094
sodium, by distillation **A2:** 1094
tantalum, by chemical vapor deposition **A2:** 1094
tantalum, by zone-refining technique **A2:** 1094
thorium, by chemical vapor deposition **A2:** 1094
tin, by zone-refining technique **A2:** 1094
titanium by external gettering **A2:** 1094
titanium, by iodide/chemical vapor deposition **A2:** 1094
titanium, by zone-refining technique **A2:** 1094
tungsten, by zone-refining technique **A2:** 1094
ultrapure gold, by fractional crystallization **A2:** 1093
ultrapure palladium, by fractional crystallization **A2:** 1093
ultrapure platinum, by fractional crystallization **A2:** 1093
ultrapure silver, by fractional crystallization **A2:** 1093
vanadium, by chemical vapor deposition **A2:** 1094
vanadium, by zone-refining technique **A2:** 1094
yttrium, by external gettering **A2:** 1094
zinc, by zone-refining technique **A2:** 1094
zirconium, by electrotransport purification **A2:** 1094-1095
zirconium, by external gettering **A2:** 1094
zirconium, by iodide/chemical vapor deposition **A2:** 1094
zirconium, by zone-refining technique **A2:** 1094
Pure nickel
effect of relative humidity on fretting **A13:** 140
K-absorption edge **A10:** 408
properties and characteristics **A2:** 435, 437, 441, 1143
Pure palladium *See also* Palladium
properties **A2:** 714-716, 1146
Pure plastic bending
geometry of deformation **A8:** 120, 122
radial stress and tangential stress ratio **A8:** 121-122
schematic of circumferential and radial stresses in plate **A8:** 121, 123
strain and stress states **A8:** 120, 122, 124
tangential and radial stress distribution **A8:** 122, 124
Pure plasticity theory **A18:** 435
Pure platinum *See also* Platinum
properties **A2:** 707, 1147
Pure plowing term (μ_p) **A18:** 35
Pure refractory metals *See also* Refractory metals and alloys **A7:** 766
applications **A2:** 557-558
mechanical and physical properties **A2:** 559
Pure shear
yield criteria **A8:** 577
Pure silicon *See also* Silicon
properties **A2:** 1154
Pure silver *See also* Fine silver; Silver
AES spectrum of **A11:** 34
properties **A2:** 1156
Pure tension
plastic flow in **A8:** 576-577
Pure tin *See also* Pure metals; Tin **A13:** 770-772
chemical properties **A2:** 518-519
corrosion behavior **A2:** 518-519
creep characteristics **A2:** 518
fatigue strength **A2:** 518
impact strength **A2:** 518
properties **A2:** 1166
purity **A2:** 518
solders, applications, specifications compositions **A2:** 521
tensile properties **A2:** 519
Pure titanium *See also* Pure metals; Titanium; Wrought titanium
alloying of **A2:** 596-597
blended elemental parts **A2:** 656
chemical reactivity **A2:** 592
commercial, effects of vacuum on fatigue **A12:** 48-49
commercial, fatigue striations **A12:** 20
corrosion resistance **A2:** 592
dimples, SEM stereo pair **A12:** 171
effect of strain rate on ductility in **A8:** 42
effect of stress intensity factor range on fatigue crack growth rate **A12:** 57
properties **A2:** 596, 1169
Pure titanium carbides **A7:** 158
Pure titanium powders **A7:** 165
Pure tungsten
effect of temperature on strength and ductility **A8:** 36
types and properties **A2:** 577-578, 1170
Pure water
high-temperature, aerated conditions **A8:** 420-422
high-temperature, deaerated conditions **A8:** 422-425
Pure zinc *See also* Zinc
properties **A2:** 1174
Purge (verb)
defined **A7:** 10
Purging gas
efficiency of **A15:** 86
Purification *See also* Pure metals; Purity; Ultrapurification
gallium **A2:** 744
of germanium and germanium compounds **A2:** 735
of melts **A15:** 71, 74-81
of molybdenum powders **A7:** 155
powders, refractory metals and alloys **A2:** 560
in precipitation from solution **A7:** 54
of rare earth metals **A2:** 720-721
and reduction reactions **A7:** 152
techniques, tungsten powder production **A7:** 152
tungsten powders **A7:** 152-154
Purified inert gas
brazing atmosphere sources **A6:** 628
Purifying hydrogen
powder used **A7:** 574
Purifying hydrogen catalysts
powders used **A7:** 572
Purity
of aluminum **A2:** 3-4
in aqueous environment synthesis **A8:** 416
characterization of **A2:** 1095-1097
commercial **A2:** 1093

Purity (continued)
of cyanates **EM2:** 234
determined by EFG **A10:** 212
effect on fatigue crack growth **A8:** 411
of gallium arsenide (GaAs) **A2:** 741
of indium **A2:** 750
of iron powder by electrolysis **A7:** 93
of lead **A2:** 543
of materials, NAA analysis for **A10:** 233
of metal, iron
of metals, preparation methods **A2:** 1094-1095
six nines characterization **A2:** 1097
of solvents and reagents, UV/VIS analysis **A10:** 68
tungsten **A2:** 577
unalloyed uranium, specifications **A2:** 672
of water atomized low-carbon iron **A7:** 84
wrought aluminum alloy, and fracture toughness **A2:** 42
Purple plague
defined **EL1:** 1154
as intermetallic-related failure **EL1:** 1042
as silicon transistor failure mechanism **EL1:** 960
"Purple plague," eliminating in solid-state electronics .... **A3:** 1 • 28
Push angle
definition **M6:** 14
Push feeds
of blanks **A14:** 500
Push weld
definition **M6:** 14
Push welding
definition **M6:** 14
Push-out device
for induction furnace linings **A15:** 372-373
Push-pull loading
in tests of cast steels **A8:** 348
Push-pull pins
failure in **A11:** 548
Push-up
as casting defect **A11:** 384
Pusher furnaces *See also* Furnaces ... **A7:** 10, 351, 354, 356
Pusher-type furnaces *See also* Furnaces **A7:** 10
Pusher-type seal
defined **A18:** 15
Pushrod, alloy steel
bending-fatigue fracture **A11:** 469
PV factor **A18:** 515, 518, 520
defined **A18:** 15
PV formula
for bearing load-carrying capacities **A7:** 709
PV limit **A18:** 239
PVA sealants
tensile strength affected by temperature **EM3:** 189
PVAC *See* Polyvinvl acetate; Polyvinyl acetates
PVAL *See* Polyvinyl alcohol
PVB *See* Polyvinyl butyral
PVC *See* Poly,(vinylchloride) (PVC); Polyvinyl chloride; Polyvinyl chlorides
PVC coating *See also* Nitrocellulose coating **A13:** 105, 108
PVC liquid coating
protection of terminals and wiring connections **EM3:** 611
PVC-coated aluminum foil
filiform corrosion **A13:** 105
PVD and CVD coatings *See also* Physical vapor deposition **A18:** 840-849
chemical vapor deposition processes **A18:** 840, 841, 846-848
advanced techniques **A18:** 848
applications **A18:** 846
classification of reactions **A18:** 846
complex reactions **A18:** 846
conventional (CCVD) **A18:** 840, 846-847
definition **A18:** 840, 846
excitation **A18:** 840
hot-filament CVD **A18:** 848
laser-induced (LCVD) **A18:** 846, 848
low-pressure (LPCVD) **A18:** 847
metal-organic (MOCVD) **A18:** 846
microwave excitation **A18:** 840, 847
organic-metallic (OMCVD) **A18:** 846
parameters **A18:** 841
photon excitation **A18:** 840
PVD and CVD coatings (continued)
plasma-assisted (PACVD) **A18:** 840, 846, 847-850
rate-limiting steps **A18:** 846
reactors **A18:** 846-847
thermal **A18:** 840, 846-847
classification of processes on basis of material deposited on substrate **A18:** 840
future outlook **A18:** 849
materials deposited by techniques **A18:** 848
modeling steps **A18:** 840
physical vapor deposition processes **A18:** 840-843
evaporation deposition **A18:** 840, 841, 842-843
sputter deposition **A18:** 840-842
unbalanced magnetron sputter deposition **A18:** 849
PVD techniques for deposition of metals, alloys, and compounds **A18:** 843-846
activated reactive evaporation (ARE) **A18:** 840, 841-845
alloy deposition **A18:** 843
ARE (BARE) process **A18:** 845
decomposition of compounds **A18:** 843-846
definition **A18:** 840
diode ion plating **A18:** 840, 844-845
direct evaporation **A18:** 844
hybrid PVD processes **A18:** 844-846
ion plating **A18:** 840, 844-845
laser ablation deposition **A18:** 844
plasma-assisted reactive evaporation **A18:** 840, 843-844
reactive evaporation process **A18:** 844
reactive ion plating (RIP) **A18:** 840, 844, 845-846
reactive PVD processes **A18:** 844
reactive sputtering (RS) **A18:** 840, 844
single-element species deposition **A18:** 843
PVD versus CVD **A18:** 840
wear applications **A18:** 848-849
cutting toot life improvements by coatings on tool substrates **A18:** 849
PVDC *See* Polyvinylidene chloride
PVDF *See* Polyvinylidene fluoride
PVF *See* Polyvinyl formal
PWA *See* Printed wiring assembly
PWA 90
electrochemical grinding **A16:** 547
grinding **A16:** 547
milling **A16:** 547
PWA 689
electrochemical grinding **A16:** 547
grinding **A16:** 547
milling **A16:** 547
PWA 1004
electrochemical grinding **A16:** 547
grinding **A16:** 547
milling **A16:** 547
PWA 1480
aging cycle **A4:** 812
composition **A4:** 795
PWA-682 Ti
broaching **A16:** 203, 209
PWB *See* Printed wiring boards
PY6 (GTE Laboratories)
processed by glass- encapsulated HIP **EM4:** 199
Pycnometer **A7:** 266
Pycnometry
as measure of density **A7:** 262, 265-266
to determine density of ceramic powders **EM4:** 27
Pyles Thermo-O-Flow System
hot-melt butyl applicator **EM3:** 200
Pyralin products **EM3:** 158
Pyramid-type three-roll forming machines **A14:** 617-618
Pyramidal Knoop indenter
with indentations in piece **A8:** 91
Pyre-ML-Polyimide
as thermocouple wire insulation **A2:** 882
Pyrene monomer
fluorescence emission spectra **A10:** 76
Pyrex
mechanical and physical properties **A18:** 813
Pyrex glass
applications, laboratory and process **EM4:** 1089, 1090
composition **EM4:** 1103
Pyrex glass (continued)
crystal structure **EM4:** 30
encapsulation in HIP conforming to body configuration **EM4:** 197
heat transfer coefficient compared **EM4:** 1090
properties **EM4:** 30, 330
tempered form **EM4:** 1103
Pyrex glass molds **A7:** 533
Pyridine
adsorbed, vibrational behavior **A10:** 134
SERS of **A10:** 136
Pyridylazonaphthaol
as metallochromic indicator **A10:** 174
Pyridylazorescorcinol
as metallochromic indicator **A10:** 174
Pyrite
Miller numbers **A18:** 235
Pyrite (FeS_2)
purpose for use in glass manufacture **EM4:** 381
Pyrocatechol violet
as metallochromic indicator **A10:** 174
Pyroceram **A18:** 532
aerospace applications **EM4:** 1017, 1018-1019
application **EM4:** 1
fortification **EM4:** 1019
properties **EM4:** 1018
seals to metals **EM4:** 499
tableware **EM4:** 1102
Pyrochlore
niobium deposits in **A7:** 160
Pyrochlore, in waste form simulant
EPMA analysis for **A10:** 532-535
Pyrochlore ore
recovery of **A2:** 1043-1044
Pyrociectric device
infrared detector as **A10:** 223
Pyrolysis **EM4:** 62
defined **EM1:** 19-20 **EM2:** 35
rice hull **EM1:** 64, 889, 896
of thin-film hybrids **EL1:** 313
Pyrolysis and consolidation as a fabrication process for carbon-carbon composites **A9:** 591
Pyrolysis gas chromatography/mass spectroscopy
analysis of rubber sheaths by **A10:** 648
polymer analysis by **A10:** 647-648
Pyrolysis products
analytic methods for **A10:** 11
Pyrolytic chemical vapor deposition (PCVD) process **A18:** 848
Pyrolytic gas chromatography (PGC)
peak areas of a brake lining resin **A18:** 571
Pyrolytic graphite (PG)
effect on SiC fiber production **EM1:** 858-859
Pyrolytic graphite used to redensify a carbon- carbon composite **A9:** 595-596
Pyromellitic dianhydride/oxydianiline (PMDA/ODA) polymers **EM1:** 290
Pyromet 31
composition **A4:** 794 **A6:** 573
Pyromet 860
composition **A4:** 794 **A6:** 573
Pyromet CTX-1
composition **A4:** 794
oxidation **A4:** 798
Pyromet CTX-3
cold-working effect on aging **A4:** 800
oxidation **A4:** 798
Pyromet CTX-909
cold-working effect on aging **A4:** 800
oxidation **A4:** 798
Pyromet CTX3M
precipitation strengthening and grain size **A4:** 800
Pyrometallurgy
of nickel and nickel alloys **A2:** 429
Pyrometers
for thermal inspection **A17:** 399
Pyrometric cone equivalent (PCE) **EM4:** 12-13
Pyrometric cones **EM4:** 36
Pyron iron powders **A7:** 83, 182
annealing **A7:** 182
chemical composition **A7:** 83
properties and uses **A7:** 83
Pyron process **A7:** 82-83
Pyrophoric
ignition **A7:** 194, 198-199

Pyrophoric (continued)
powder **A7:** 10
Pyrophoricity *See also* Explosion(s); Explosions; Explosive(s); Explosivity; Fires; Pyrometallurgy...... **A7:** 10, 24, 194-200
of aluminum powders **A7:** 125, 127, 130
degree of **A7:** 198-199
of manganese powders **A7:** 72
of metal powders **A7:** 24, 194, 198-200
prevention of pyrophoric reactions **A7:** 199-200
reaction parameters **A7:** 199
of uranium and uranium alloys **A2:** 670-671
Pyrophoricity of uranium **A9:** 477
Pyrophosphate copper plating *See* Copper pyrophosphate plating
Pyrophosphate, zirconium weighed as
in gravimetric analysis **A10:** 171
Pyrophosphoric acid
description **A9:** 68
as electrolyte **A9:** 54
Pyrophyllite **EM4:** 6, 44
in ceramic tiles **EM4:** 926
composition **EM4:** 6, 761, 932
fired properties **EM4:** 762
properties **EM4:** 760
refractory applications **EM4:** 901
sheet structure **EM4:** 759
structure **EM4:** 761, 762
in typical ceramic body compositions **EM4:** 5
Pyrophyllite, as refractory
core coatings **A15:** 240
Pyrosilicate
crystal structure **EM4:** 881
Pyrotechnic actuators
corrosion in **A10:** 510-511
Pyrotechnic materials
SEM analyses **A10:** 510-511
Pyrotechnics **A7:** 597, 600-604
as aluminum flake applications **A7:** 595
devices using metals as fuels **A7:** 600
magnesium powders for **A7:** 131
metals requirements **A7:** 600-601
powders used **A7:** 574
Pyrowear 53
nominal compositions **A18:** 726
Pyrowear Alloy 53
composition **A4:** 320
vacuum carburizing **A4:** 348
Pyroxenes
chain structure **EM4:** 759, 760
Pyrrhotite-grinding media system **A18:** 275-276
PZ 6400 (347) surfacing
plasma-MIG welding **A6:** 225
PZ 6410 (308) surfacing
plasma-MIG welding **A6:** 225
PZ 6415 (309) surfacing
plasma-MIG welding **A6:** 225
q
percentiles of the studentized range **A8:** 658
robing
direct current (dc) **EL1:** 946
testing problems **EL1:** 365
types **EL1:** 371-372

Q

Q *See* Fatigue notch sensitivity
defined **EL1:** 1154
Q, quality factor
defined **A17:** 217
Q value
defined **A7:** 10
Q-band microwave frequency
use in ESR **A10:** 255
Q-meter
for magabsorption measurement **A17:** 152
Q-meter method **EM3:** 431
Q-switched pulse mode
optical holographic interferometry **A17:** 411
QPL *See* Qualified products list
Quad flatpacks **EL1:** 485, 1154
Quadrant durometer **A8:** 107
Quadrature component
microwave inspection **A17:** 205
Quadrature-phase signals
iron carbonyl powders **A17:** 144
nuclear magnetic resonance **A17:** 144
Quadrivariant equilibrium
defined **A9:** 15
Quadrupole interaction
in Mössbauer effect **A10:** 290-293
Quadrupole magnets
for niobium-titanium superconducting materials **A2:** 1056
Quadrupole mass filter, schematic
of gas mass spectrometer **A10:** 153
Quadrupole mass spectrometer
for trace element analysis **A2:** 1095
Quadrupole mass spectrometers **A7:** 258
SIMS **A11:** 34-35
Quadrupole resonance
capabilities **A10:** 253
Qualification
of inspection equipment **A17:** 679
of inspection processes **A17:** 678-679
of personnel **A17:** 679
procedure, for reliability **EL1:** 742
procedure, underwater welding **M6:** 924
programs, environmental testing **EL1:** 502-503
responsibility, military applications **EL1:** 908
solder masking **EL1:** 554
standard, defined **A17:** 677
welder/diver **M6:** 924
Qualification test **EM3:** 24
Qualified product
defined **EL1:** 908
Qualified products list (QPL) **EM3:** 24, 63, 730
defined **EM2:** 35
military standardization documents **EL1:** 908
Qualitative analysis *See also* Qualitative analysis, methods for
of anions and cations, organic and inorganic **A10:** 658
and Auger emission probabilities **A10:** 550
classical wet chemistry **A10:** 167-168
defined **A10:** 680
direct-current plasma **A10:** 40
electrographic **A10:** 202
electrometric titration **A10:** 202-206
electron energy loss spectroscopy **A10:** 450
energy-dispersive spectrometry **A10:** 522-523
energy-dispersive x-ray spectrometry **A10:** 82-101
functional group **A10:** 654
infrared, factor analysis for **A10:** 116
of inorganic gases, analytic methods for **A10:** 8
of inorganic liquids and solutions, analytic methods for **A10:** 7
of inorganic solids, analytical methods for **A10:** 4-6
LEISS spectra in **A10:** 604-605
of major components, analytic methods for **A10:** 4
micro-, AEM-EDS for **A10:** 446-447
of minor components, analytic methods for **A10:** 4
of organic solids and liquids, analytic methods for **A10:** 9-10
resolution enhancement methods **A10:** 116-117
search/match methods as **A10:** 455
of surface phase on silicon **A10:** 341-342
of trace elements, analytic methods for **A10:** 4
of ultratrace components, analytic methods for **A10:** 4
wavelength-dispersive spectrometry **A10:** 523-524
Qualitative analysis, methods for *See also* Qualitative analysis
analytical transmission electron microscopy **A10:** 429-489
Auger electron spectroscopy **A10:** 549-567
electron probe x-ray microanalysis **A10:** 516-535
electron spin resonance **A10:** 253-266
field ion microscopy **A10:** 583-602
gas analysis by mass spectrometry **A10:** 151-157
gas chromatography/mass spectrometry **A10:** 639-648
infrared spectroscopy **A10:** 109-125
ion chromatography **A10:** 658-667
liquid chromatography **A10:** 649-659
low-energy ion-scattering spectroscopy **A10:** 603-609
molecular fluorescence spectrometry **A10:** 72-81
Mössbauer spectroscopy **A10:** 287-295
neutron diffraction **A10:** 420-426
optical emission spectroscopy **A10:** 21-30
particle-induced x-ray emission **A10:** 102-108
Raman spectroscopy **A10:** 126-138
scanning electron microscopy **A10:** 490-515
secondary ion mass spectroscopy **A10:** 610-627
spark source mass spectrometry **A10:** 141-150
ultraviolet/visible absorption spectroscopy **A10:** 60-71
x-ray diffraction **A10:** 325-332
x-ray photoelectron spectroscopy **A10:** 568-580
x-ray powder diffraction **A10:** 333-343
x-ray spectrometry **A10:** 82-101
Qualitative and quantitive properties
evaluating **EM2:** 405
Qualitative corrosion data
lead/lead alloys **A13:** 780
Qualitative data
optical holographic interferometry **A17:** 414-415
variables, in factorial designs **A17:** 746
Qualitative dynamic fracture tests **A8:** 259
Qualitative elemental analysis
crystallographic texture measurement and analysis **A10:** 357-364
dc arc emission spectroscopy **A10:** 25
Qualitative evaluation, turbine blades
by holography **A17:** 405
Qualitative inspection
defined **A17:** 39
Qualitative level
for fatigue experiments **A8:** 695
Qualitative SEM examination
of particles **A7:** 234-236
Quality *See also* Joint design and quality; Quality assurance; Quality control **EM3:** 429, 432
acceptable level, defined **EL1:** 1133
assembly/packaging for **EL1:** 441-448
assessment, by glass transition temperature **EM2:** 564-565
assessment criteria, attribute data **A17:** 734
assurance, in liquid penetrant inspection **A17:** 84-85
assurance, organic coatings and linings **A13:** 416-418
assurance testing **A8:** 662
of contour roll forming **A14:** 633
of cut, oxyfuel gas cutting **A14:** 720-721
of cut, plasma arc cutting **A14:** 731
defined **A17:** 720
design and control, statistical **A17:** 719-753
design, robust design approach **A17:** 721-722
and engineering specifications **A17:** 720
of fabrics, verification **EM1:** 126
of fastener holes **EM1:** 711
from computer-aided design **EL1:** 527-528
function deployment **A17:** 719
improvements, continuous casting **A15:** 313
of inspection techniques **A17:** 663-665
internal/surface, and gating, die casting **A15:** 289

Quality (continued)
loss function of **A17:** 720-721
of manual spinning **A14:** 600
as material selection parameter **EM1:** 38-39
problems, classified **A17:** 723
and productivity fundamentals **A17:** 719-722
of reclaimed sand, green sand molding **A15:** 225-226
and reproducibility, NDE **A17:** 676
of surface **A14:** 882
surface, determining **A15:** 544
surface, of castings **A17:** 512
of surface preparation and coating, thermal spray **A13:** 462
surface, sheet metal forming **A8:** 553
tension testing as index of **A8:** 19
in thermoplastic injection molding **EM2:** 317-318
uniform, of cut edges **A14:** 479
and variation reduction **A17:** 721
water, effect in water-recirculating systems **A13:** 489-494

Quality assurance *See also* Life cycle optimization; Life cycle testing; Process control; Quality control; Testing
epoxy materials **EL1:** 831-836
filament winding **EM2:** 368
introduction **EL1:** 867-868
in materials engineering **M3:** 826, 830
measurement **EM2:** 318
for powder forged parts **A14:** 203-205
for sampling **A10:** 17

Quality assurance, aluminum alloys
heat treating **M4:** 714-717

Quality assurance and quality control
in closed-die forgings **A1:** 339

Quality assurance and quality control of forgings **M1:** 351-352, 366

Quality conformance test circuitry
defined **EL1:** 1154

Quality control *See also* Control charts; Design; Joint evaluation and quality control; NDE reliability; NDE reliability applications; NDE reliability data analysis NDE reliability models; Process control; Production; Quality; Quality design; Statistical methods; Statistical process control **A7:** 295, 480-491 **EM1:** 729-763 **EM3:** 522, 523
aluminum alloy **A15:** 762
of boiler/pressure vessel fabrication **A17:** 641
of brazed joints **A6:** 1117-1123
overall quality systems **A6:** 1117-1118
by image analysis **A10:** 309, 316
Charpy test for **A8:** 262
closed-loop cure as **EM1:** 761-763
compacted graphite irons **A15:** 671
control charts, for individual measurements **A17:** 732-734
as coordinate measuring machine application **A17:** 20
copper casting **A15:** 778
core manufacturing processes **EL1:** 869-872
corrosion tests **A13:** 193, 242
of cure **EM1:** 745-760
defect verification process **EL1:** 872-874
defined **EM1:** 729-730
design of experiments **A17:** 740-743
design requirements for **EM1:** 182
and design, statistical **A17:** 719-753
as discipline **A17:** 719
Domfer iron powder production process **A7:** 91-92
in electroless copper plating **EL1:** 870
in electrolytic copper plating **EL1:** 871-872
electron beam melting **A15:** 412
electron-beam welding, gas turbine engine repairs **A6:** 866
elevated/low temperature tension testing as **A8:** 34-37
in etching **EL1:** 872
examination technique **A12:** 140-143
examples **A17:** 750-752
for exfoliation corrosion **A13:** 242
factorial designs **A17:** 740-750
fiber properties analysis **EM1:** 731-735
incoming, of solder paste **EL1:** 732
of inhibitors **A13:** 483
of integrated circuits **A10:** 122-123

Quality control (continued)
introduction **EL1:** 867 **EM1:** 729-730
laminate, factors **EM1:** 730
loss function concept **A17:** 720-721
mechanical properties and **A15:** 765
medical adhesives **EM3:** 575-576
of metallurgical products **A10:** 200
metalworking lubricants **A18:** 148-149
as minimizing variation **A17:** 720
of mixing and blending **A7:** 186-187, 189
network **EM3:** 703
and nondestructive evaluation **A15:** 664
nondestructive surface residual stress measurement for **A10:** 380
of nonferromagnetic heat exchanger tubesheet rolled joints **A17:** 180-181
on-line structural assessment, aluminum-silicon alloys **A15:** 166
out-of-control conditions **A17:** 731-732
p-chart, for fraction defective **A17:** 735-737
parameters **A8:** 27, 439
personnel, responsibilities **EM1:** 740-744
in photoprinting process **EL1:** 870-871
plan, objectives for CMMs **A17:** 27-28
premium engineered castings **A15:** 766
prepreg, factors **EM1:** 730
primary plated-through hole drilling **EL1:** 869-870
printed wiring boards manufacture **EL1:** 869-874
process capability assessment **A17:** 737-740
processing **EM3:** 735-742
application geometry **EM3:** 736-737
curing **EM3:** 739-742
handling and exposure to contaminants **EM3:** 738
lay-up technique **EM3:** 739
material age-life history **EM3:** 735-736
material manufacturing consistency **EM3:** 735
moisture exposure **EM3:** 736
quality assurance **EM3:** 742
surface preparation **EM3:** 737-738
tooling verification **EM3:** 738-739
production, acoustic emission inspection **A17:** 289-290
for purity or composition, NAA analysis as **A10:** 233
quality and productivity fundamentals **A17:** 719-722
raw materials **EL1:** 860 **EM3:** 729-734
acceptance alternatives **EM3:** 733-734
alternative techniques **EM3:** 732
product acquisition approach **EM3:** 730
product test phases **EM3:** 730-731
test techniques **EM3:** 731-732
testing considerations **EM3:** 732-733
of raw materials, sampling in **A10:** 12
of reinforcing material lay-up **EM1:** 740-744
resin properties analysis **EM1:** 736-737
reverberatory furnaces **A15:** 378-379
robust design, implementing **A17:** 750-752
in sampling **A10:** 17
Scleroscope for **A8:** 104
of semiconductor devices, by SEM **A10:** 490
of sheet metals, springback tests for **A8:** 564-565
Shewhart control charts, for attribute data **A17:** 734
solder paste **EL1:** 655
of soldered joints **A6:** 1124-1128
soldering **A6:** 112
sources of variation countermeasures **A17:** 722-730
squeeze casting **A15:** 324-326
tensile testing for **A12:** 101
testing equipment, electric arc furnace **A15:** 363
testing methods **EM3:** 37
for thermal spray coatings **A13:** 462
tool **A8:** 89
of tooling **EM1:** 738-739
total (TQC), in tape automated bonding (TAB) **EL1:** 288
transportion industry, magnetic particle inspection **A17:** 89
two-level fractional factorial designs **A17:** 746-750
u-chart, for number defects per unit **A17:** 737
verification **EM1:** 126, 744
visual inspection as **EL1:** 872

Quality control (continued)
in welding **A17:** 590
x-ray spectrometry of solid samples as **A10:** 83
zone rules, for control chart analysis **A17:** 730-732

Quality corridors
in ratio-analysis diagrams **A11:** 62

Quality descriptors
for carbon steels **A1:** 201, 203
for steel **A1:** 143-144, 146

Quality design *See also* Quality control; Robust design; Statistical methods
of boilers/pressure vessels **A17:** 641
and control, statistical **A17:** 719-753
parameter design stage **A17:** 722
product/process performance factors **A17:** 722
robust design approach **A17:** 721-722
systems design stage **A17:** 722
tolerance design stage **A17:** 722

Quality factor *See also* Tan delta **EM3:** 24
defined **EL1:** 1154 **EM2:** 35, 592

Quality factor, Q
defined **A17:** 217

Quality index **A6:** 1018

Quality of surface finish **A1:** 592-593

Quantitative analysis *See also* Quantitative analysis, methods for
AEM-EDS micro **A10:** 447
atom probe microanalysis, requirements for **A10:** 594-595
defined **A10:** 680
electron energy loss spectroscopy **A10:** 450
electron probe x-ray microanalysis, standards accuracy, and precision **A10:** 524-525
elemental, of industrial materials **A10:** 162-179
of field ion microscopy images **A10:** 590-591
of four titanate phases, waste form simulant **A10:** 534
functional group **A10:** 654
of hydroxyl and boron content in glass **A10:** 121-122
for impurities in LPCVD thin films **A10:** 624
of inorganic gases, analytic methods for **A10:** 8
of inorganic liquids and solutions, analytic methods for **A10:** 7
of inorganic solids, analytic methods for **A10:** 4-6
of major components, analytic methods for **A10:** 4
of minor components, analytic methods for **A10:** 4
of organic liquids and solutions, analytic methods for **A10:** 10
of organic solids, analytic methods for **A10:** 9
of oxygen in silicon wafers **A10:** 122-123
of phosphorus ion-implantation profile in silicon **A10:** 623-624
of risk sites **EL1:** 127-140
and separation, of metal ions **A10:** 200-201
stripping analysis **A10:** 202
surface **A7:** 251
of trace components, analytic methods for **A10:** 4
ultrasensitive destructive **A10:** 233
of ultratrace components, analytic methods for **A10:** 4
x-ray powder diffraction **A10:** 339-340
x-ray scanning electron microscopy **A9:** 93
of ZnO in calcite **A10:** 342

Quantitative analysis, methods for *See also* Quantitative analysis
analytical transmission electron microscopy **A10:** 429-489
atomic absorption spectrometry **A10:** 43-59
Auger electron spectroscopy **A10:** 549-567
classical wet analytical chemistry **A10:** 161-180
controlled-potential coulometry **A10:** 207-211
electrochemical analysis **A10:** 181-211
electrogravimetry **A10:** 197-201
electrometric titration **A10:** 202-206
electron probe x-ray microanalysis **A10:** 516-535
electron spin resonance **A10:** 253-266
elemental and functional group analysis **A10:** 212-220
gas analysis by mass spectrometry **A10:** 151-157

Quantitative analysis, methods for (continued)
- inductively coupled plasma atomic emission spectroscopy **A10:** 31-42
- infrared spectroscopy **A10:** 109-125
- ion chromatography **A10:** 658-667
- liquid chromatography **A10:** 649-659
- low-energy ion-scattering spectroscopy **A10:** 603-609
- molecular fluorescence spectrometry **A10:** 72-81
- Mössbauer spectroscopy **A10:** 287-295
- neutron activation analysis **A10:** 233-242
- neutron diffraction **A10:** 420-426
- optical emission spectroscopy **A10:** 21-30
- particle-induced x-ray emission **A10:** 102-108
- potentiometric membrane electrodes **A10:** 181-187
- Rutherford backscattering spectrometry **A10:** 628-636
- spark source mass spectrometry **A10:** 141-150
- ultraviolet/visible absorption spectroscopy **A10:** 60-71
- voltammetry **A10:** 188-196
- x-ray diffraction **A10:** 325-332
- x-ray photoelectron spectroscopy **A10:** 568-580
- x-ray spectrometry **A10:** 82-101

Quantitative chemical analysis **A1:** 1031-1032

Quantitative corrosion data
- lead/lead alloys **A13:** 780

Quantitative diffractometer studies **A18:** 469

Quantitative dynamic fracture tests **A8:** 259

Quantitative evaluation of microstructural geometry by image analysis with material
- contrast **A9:** 94

Quantitative fractography *See also* Fracture surface(s); Photography; Roughness parameters; Surface(s) **A12:** 193-210
- analytical procedures **A12:** 199-205
- angular distributions, profile **A12:** 201-204
- applications, example cases **A12:** 205-208
- area, fracture surface, estimating **A12:** 201-205
- defined **A12:** 193
- development **A12:** 193-194
- experimental techniques **A12:** 194-199
- geometrical methods **A12:** 198
- goal and history **A12:** 8
- information from **A11:** 56
- metallographic sectioning methods **A12:** 198-199
- nondestructive profiles **A12:** 199
- partially oriented surfaces **A12:** 201
- photogrammetric methods **A12:** 197-198
- profile generation **A12:** 198-199
- profile parameters **A12:** 199-200
- profiles from replicas **A12:** 199
- projected images **A12:** 194-196
- research, summary of **A12:** 211
- roughness parameters **A12:** 199-205
- statistical vs individual measurement **A12:** 207-208
- stereoscopic methods **A12:** 198
- surface parameters **A12:** 200-201
- triangular elements **A12:** 201-202

Quantitative fracture surface analysis **EM4:** 652-661
- crack branching **EM4:** 657-658
  - crossbending **EM4:** 657, 658
  - drumhead tension **EM4:** 657, 658
  - hoop stress at failure **EM4:** 658
  - internal pressure **EM4:** 657, 658
  - ratio of principal stresses at failure **EM4:** 658
  - stress at failure from branching patterns **EM4:** 658
  - torsion **EM4:** 657, 658
- fracture features **EM4:** 652-653
  - crack tip stress intensity **EM4:** 652
  - critical stress intensity **EM4:** 652
  - failure origin **EM4:** 652
  - fracture mechanics relations **EM4:** 652-653
  - Wallner lines **EM4:** 652
- fracture surface markings and their usefulness **EM4:** 652
- mirror radius **EM4:** 655-657
  - application of mirror radius/failure stress relation to polycrystalline materials **EM4:** 656-657
  - branching **EM4:** 655, 656
  - failure stress from measurement **EM4:** 655-656
  - hackle **EM4:** 655, 656
  - mist **EM4:** 655, 656
  - residual stresses **EM4:** 656
  - single-crystal fracture **EM4:** 657
  - x-ray microradiography **EM4:** 656
- mirror region of crack propagation **EM4:** 654-655
  - crack velocity determination **EM4:** 654-655
  - ultrasonic fractography **EM4:** 654
  - Wallner lines **EM4:** 654
- origin flaws **EM4:** 653-654
  - cone cracks **EM4:** 653-654
  - estimating failure stress from size **EM4:** 653
  - geometric characteristics of surface cracks caused by contract stresses **EM4:** 653-654
  - Hertzian cone flaws **EM4:** 653-654
  - identification **EM4:** 653
  - measurement of size **EM4:** 653
  - particle impact parameters **EM4:** 654
- stress-corrosion fracture **EM4:** 658-661
  - cavitation scarp **EM4:** 660, 661
  - critical flaw size **EM4:** 661
  - delayed failure **EM4:** 659
  - example **EM4:** 660-661
  - indirect fractographic evidence of water in field failures **EM4:** 659
  - intersection scarp **EM4:** 659-660
  - transition hackle **EM4:** 661
  - Wallner lines **EM4:** 661

Quantitative image analysis
- digital **A9:** 152-153

Quantitative image analysis technique **EM4:** 66

Quantitative image analyzers **A7:** 229

Quantitative inspection
- defined **A17:** 39

Quantitative level
- for fatigue experiments **A8:** 695

Quantitative metallography *See* Image analysis **A9:** 123-134
- application of basic equations **A9:** 125-126
- basic measurements **A9:** 123-125
- of cemented carbides **A9:** 275
- contrasting by interference layers **A9:** 60
- defined **A9:** 15
- of grain size **A9:** 129
- of oriented structures **A9:** 126-129
- particle relationships **A9:** 129-130
- particle-size distributions **A9:** 131-134
- of pore size distribution in green compacts **A7:** 299
- projected images used in **A9:** 134
- use of image analyzers in **A9:** 83

Quantitative microscopy for homogenization in multiphase systems **A7:** 316

Quantitative nondestructive evaluation *See also* NDE reliability
- fracture control philosophy **A17:** 666-673
- history of **A17:** 663-664
- introduction **A17:** 663-665
- models for predicting NDE reliability **A17:** 702-718
- NDE reliability data analysis **A17:** 689-701
- NDE systems reliability, applications **A17:** 674-688

Quantitative prediction, transformation hardening in steels **A4:** 20-31
- continuous cooling transformation (CCT) curves **A4:** 20, 21-25, 26, 27
- Creusot-Loire systematization of CCT curves **A4:** 21, 22-23, 28, 30, 31
- distortion predicted by finite element analysis program **A4:** 31, 32
- grain growth and size **A4:** 25, 27, 28, 30
- Grossmann factors **A4:** 27, 28, 29
- international set of experimental CCT curves **A4:** 21-22
- isothermal transformation (IT) diagrams **A4:** 21-25, 27
- IT-to-CCT transformation procedure **A4:** 25
- Jominy end-quench hardenability curves **A4:** 20, 21, 22, 23, 25-30
- Jominy end-quench hardenability test **A4:** 25
- Jominy hardenability data set **A4:** 25-30
- Jominy hardenability prediction **A4:** 27-28
- Jominy measurement reliability **A4:** 28-30
- Jominy predictor application **A4:** 30-31
- Jominy test precision and reproducibility **A4:** 26-27
- Lamont transformations **A4:** 25, 29
- mechanical properties **A4:** 20
- modeling of post-hardening treatments **A4:** 30-32
- residual stresses **A4:** 22, 31
- Rockwell C hardnesses (HRC) **A4:** 28
- Scheil-Avrami Rule **A4:** 24-25, 28
- Vickers hardness **A4:** 23

Quantitative reduced-pressure technique
- for hydrogen measurement **A15:** 458

Quantitative scanning electron microscopy image analysis **A9:** 96

Quantitative spectrographic analysis (QSA)
- chemical analysis of glass-quality sand **EM4:** 378
- to analyze minor contaminants in soda ash **EM4:** 380
- to analyze salt cake **EM4:** 380

Quantitative stereoscopy
- defined **A12:** 171

Quantitative stress measurement
- using test structures **EL1:** 444-446

Quantitative thermal analysis
- interpretation and use **A15:** 183

Quantitative thermal inspection
- methods **A17:** 400-402

Quantitative thin-film analysis
- by EPMA **A11:** 41

Quantity
- of electricity, SI unit/symbol for **A8:** 721
- of heat, SI unit/symbol for **A8:** 721
- symbol for change in **A8:** 726 **A10:** 692

Quantity loss
- leak testing by **A17:** 65-66

Quantity of electricity
- SI derived unit and symbol for **A10:** 685

Quantity of heat
- SI derived unit and symbol for **A10:** 685

Quantity vs cost
- permanent mold casting **A15:** 285

Quantum detection efficiency (QDE)
- defined **A17:** 298

Quantum efficiency
- defined **EL1:** 1154

Quantum mechanical band theory (solid state)
- of electrical phenomena **EL1:** 96-99
- insulators and dielectric materials **EL1:** 99-100
- magnetic materials **EL1:** 103
- resistive materials **EL1:** 100-101
- semiconductors **EL1:** 101-103

Quantum mechanical tunneling process
- field ionization as **A10:** 584

Quantum mechanics
- defined **A10:** 680
- of electron spin resonance **A10:** 254

Quantum mottle, defined
- radiography **A17:** 317

SUBJECTS OF THE INDEXED VOLUMES: ASM Handbook (designated by the letter "A"): **A1:** Properties and Selection: Irons, Steels, and High-Performance Alloys (1990); **A2:** Properties and Selection: Nonferrous Alloys and Special-Purpose Materials (1990); **A3:** Alloy Phase Diagrams (1992); **A4:** Heat Treating (1991); **A6:** Welding, Brazing, and Soldering (1993); **A7:** Powder Metallurgy (1984); **A8:** Mechanical Testing (1985); **A9:** Metallography and Microstructures (1985); **A10:** Materials Characterization (1986); **A11:** Failure Analysis and Prevention (1986); **A12:** Fractography (1987); **A13:** Corrosion (1987); **A14:** Forming and Forging (1988); **A15:** Casting (1988); **A16:** Machining (1989); **A17:** Nondestructive Testing and Quality Control (1989); **A18:** Friction, Lubrication, and Wear Technology (1992). **Metals Handbook, 9th Edition** (designated by the letter "M"): **M1:** Properties and Selection: Irons and Steels (1978); **M2:** Properties and Selection: Nonferrous Alloys and Pure Metals (1979); **M3:** Properties and Selection: Stainless Steels, Tool Materials and Special-Purpose Materials (1980); **M4:** Heat Treating (1981); **M5:** Surface Cleaning, Finishing, and Coating (1982); **M6:** Welding, Brazing, and Soldering (1983). **Engineered Materials Handbook** (designated by the letters "EM"): **EM1:** Composites (1987); **EM2:** Engineering Plastics (1988); **EM3:** Adhesives and Sealants (1990); **EM4:** Ceramics and Glasses (1991); **Electronic Materials Handbook** (designated by the letters "EL"): **EL1:** Packaging (1989).

Quantum noise (mottle)
and detective quantum efficiency **A17:** 371
Quantum number
defined **A10:** 680
Quantum number designations
defined **EL1:** 90
Quantum of energy
defined **EL1:** 90
Quantum theory *See also* Quantum mechanics
of scattered radiation **A10:** 126-127
Quantum yields, fluorescence
structural effects on **A10:** 74
Quantum-well laser
research **A2:** 747
Quartering **A7:** 10
and cone samples **A7:** 226-227
of samples **A10:** 17, 165
Quartz **EM4:** 44
blasting with **M5:** 84, 93-94
in ceramic tile **EM4:** 926, 928
chemical composition **A6:** 60
comminution **EM4:** 77
in dentifrices **A18:** 668
as extender **EM3:** 176
fabric, reinforced epoxy resin composites **EM1:** 399-400, 414-415
as fiber reinforcement **EL1:** 535, 605, 618-619
fibers, defined **EM1:** 29, 360
field-assisted bonding **EM4:** 479
hardness **A18:** 433
hydrofluoric acid as dissolution medium **A10:** 165
as internal reflection element **A10:** 113
JCPDS data for **A10:** 341
lapping **A16:** 499
laser cutting of **A14:** 742
on Mohs scale **A8:** 108
properties **EM4:** 330, 499
purpose for use in glass manufacture **EM4:** 381
silica sands as **A15:** 208
silica sols to bond aggregates **EM4:** 446
sintering agent for **A10:** 166
solid-particle impingement erosion of cobalt-base alloys **A18:** 768, 769
surfaces, water adsorption at **A15:** 212
ultrasonic machining ... **A16:** 530, 531 **EM4:** 359, 360
Quartz blanks
defects in **EL1:** 979
Quartz crystals
as piezoelectric elements **A17:** 254-255
Quartz glass
applications
electronics processing **EM4:** 1055
lighting **EM4:** 1032, 1034, 1036-1037
composition **EM4:** 742
when used in lamps **EM4:** 1033
crystal structure **EM4:** 879
effect on chippage and quality of cut **EM4:** 347
encapsulation in HIP conforming to body configuration **EM4:** 197
fractures encountered in cutoff of tubes **EM4:** 347
framework structure **EM4:** 759
maximum operating temperatures **EM4:** 1035
nozzle material for hot gas soldering **A6:** 361
properties **EM4:** 742, 1033, 1034
strength **EM4:** 755
thermal expansion coefficient **EM4:** 499
thermal properties **EM4:** 754
uses **EM4:** 742
Quartz infrared preheaters
wave soldering **EL1:** 685-686
Quartz laminates
advantages/ disadvantages **EL1:** 619
Quartz load washer **A8:** 193
Quartz piezoelectric device
to reduce load cell ringing **A8:** 193
Quartz refractor plate
in polychromators **A10:** 38
Quartz sensitive tint plate
placement **A9:** 138
Quartz substrates
physical characteristics **EL1:** 106
Quartz tube atomizer
characteristics **A10:** 48
Quartz-iodine light sources for microscopes **A9:** 72
Quartz-tube radiant furnace
for torsion testing **A8:** 159
Quartzite
Miller numbers **A18:** 235
refractory material composition **EM4:** 896
Quasi-brittle fatigue
polycarbonate sheet **A12:** 479
Quasi-cleavage
in micro-fracture mechanics **A8:** 465
Quasi-cleavage fracture
AISI/SAE alloy steels **A12:** 303, 305, 319
austenitic stainless steels **A12:** 357
in austenitized iron alloy **A12:** 26
brittle fracture by **A11:** 82
by hydrogen embrittlement **A12:** 31
defined **A11:** 8 **A12:** 20
ductile irons **A12:** 232, 235
facets, titanium alloys **A12:** 448
facets, unaged iron alloy **A12:** 458-459
iron alloy **A12:** 457-459
malleable iron **A12:** 238
precipitation-hardening stainless steels **A12:** 370-371
rosette tensile fractures with **A12:** 104
solute-depleted ASTM/ASME alloy steels **A12:** 350
stress state effect on appearance **A12:** 31, 40
tool steels **A12:** 26, 375, 379, 380
Quasi-flake graphite
in cast iron **M1:** 6, 7
Quasi-hydrodynamic lubrication **A18:** 80, 89, 94
defined **A18:** 15
in sheet metal forming **A14:** 512
Quasi-isotropic *See* Isotropic
Quasi-isotropic composites
joint design for **EM1:** 479
specific tensile strength vs. specific tensile modulus **EM1:** 28
Quasi-isotropic fiber-reinforced composites
properties **EL1:** 1123
Quasi-isotropic laminate *See also* Laminate(s) **EM3:** 24
defined **EM2:** 35
Quasi-isotropic laminates
defined **EM1:** 2
prepreg, costs **EM1:** 14
properties **EM1:** 223-22
Quasi-static loading
explosively loaded torsional Kolsky bar **A8:** 225
rapidly propagating cracks as dynamic fracture under **A8:** 259
vs. dynamic loading **A8:** 259
Quasi-static testing
direct-current differential transformers for **A8:** 223
extensions to higher loading rates of **A8:** 259
requirements, for rapid-load fracture testing **A8:** 260-261
stored-torque Kolsky bar for **A8:** 223
strain rate, Kolsky bar for **A8:** 219
torsional **A8:** 145-148
Wheatstone bridge in **A8:** 223
x-y recorder for **A8:** 223
Quasi-static torsional testing **A8:** 145-148
Quasi-stationary temperature distribution **A6:** 9-10
Quasi-striations
fatigue fracture **A12:** 15, 22
Quasibinary sections of a ternary diagram ..... **A3:** 1 • 5
Quaternary phase diagrams
defined **A13:** 46
Quaternary phosphonium salts **EM3:** 95
Quaternary system
defined **A9:** 15
Quaternary system or diagram ..... **A3:** 1 • 2
Quaternary systems
wrought aluminum alloys **A2:** 37
Quebec Metal Powder process (QMP) **A7:** 86-89
of annealing **A7:** 183
powder characteristics **A7:** 87-89
for welding-grade iron powders **A7:** 818
Quench *See also* Quench cracking; Quench cracks; Quenching **EM3:** 24
-age embrittlement, defined **A13:** 11
aging, defined **A13:** 11
Quench (continued)
-and-tempered steel, elongated dimples **A11:** 25
cracking, defined **A13:** 11
defined **A18:** 15 **EM2:** 35
hardening, defined **A13:** 11
rate, effect in forging failure **A11:** 340
water, low-alloy steel **A11:** 393
Quench aging
defined **A9:** 15 **A12:** 129
as embrittlement, examination/ interpretation **A12:** 129-130
Quench annealing
defined **A9:** 15
Quench cracking *See also* Quench **A1:** 698
alloy steel seamless tubing failure from **A11:** 334-335
as casting defect **A11:** 383
in coarse austenitic grain size **A11:** 96
from soft spots **A11:** 570
of steel, ductile and brittle fracture **A11:** 94-97
tool and die **A11:** 563, 566, 568-570
of tool steel ring forging **A11:** 570
Quench cracking in tool steels **A9:** 259
Quench cracking of aluminum alloys **A9:** 358
Quench cracks *See also* Quench
AISI/SAE alloy steels **A12:** 305
brittle intergranular fracture by **A12:** 335
by liquid penetrant inspection **A17:** 86
defined **A11:** 8
delayed **A11:** 95-96
effect on fatigue strength **A11:** 122
as embrittlement, examination/ interpretation **A12:** 130-132
in forging **A11:** 333-334
from nonuniform forging temperatures **A11:** 317
intergranular **A11:** 122
low-carbon steel **A12:** 251
magnetic particle detection **A17:** 103
in round steel bar **A11:** 94
service fracture by **A12:** 65
in shafts **A11:** 459
tool steels **A12:** 375
treatments for **A11:** 121
Quench hardening
defined **A9:** 15
plain carbon steels **A15:** 713
Quench modification
aluminum-silicon alloys **A15:** 162
Quench presses
effect on distortion **A11:** 141
Quench quality of solution-heat-treated aluminum alloys, etchants for examination for **A9:** 355
Quench time
definition **M6:** 14
"Quench welding" techniques **A6:** 75
Quench-age embrittlement **A1:** 692-693 **M1:** 684, 701, 704
for iron-nitrogen and iron-carbon alloys **A1:** 692-693
for low-carbon steels **A1:** 692
of steels **A11:** 98
Quench-cracked AISI 4340 steel threaded rod
overload failure of **A10:** 511-513
Quenched and tempered high-strength alloy steels **M6:** 282-292
composition **M6:** 282-285
electrode selection **M6:** 287
heat input **M6:** 288
joint design **M6:** 285
microstructure **M6:** 282-285
postweld heat treatment **M6:** 288-289
preheating **M6:** 287-288
selection of welding process **M6:** 285
Quenched and tempered steels
applications
automotive **A6:** 395
offshore structures **A6:** 384
classification and group description **A6:** 405, 406, 407
H-steels **A1:** 474-476, 480-481
heat flow in fusion welding **A6:** 13
hydrogen cracking **A6:** 384
low-alloy steels **A1:** 391-392, 396
martensitic stainless steels **A1:** 939-942

Quenched and tempered steels (continued)
microstructures and processing of **A1:** 136-137
structural carbon steels **A1:** 390-391
weldability **A6:** 421-424, 427
weldability of **A1:** 609
Quenched and tempered steels, specific types
A508, composition and carbon content **A6:** 406
A514 .... **A6:** 406
A517, composition and carbon content **A6:** 406
HY-80 .... **A6:** 406
HY-100 .... **A6:** 406
HY-130 .... **A6:** 406
Quenched steels
shielded metal arc welding **A6:** 176
Quenching *See also* Heat treatment; Plunge quenching; Quench; Rapid quenching; Subcritical quenching **A4:** 67-120 **A13:** 11, 47 **M1:** 182
after heat treatment **A8:** 502
agitation of media **A4:** 101, 103, 104, 107, 113-115 **M4:** 47, 48, 49, 60-61
equipment **A4:** 95, 110-116 **M4:** 46, 61, 62, 64
factors controlling agitation **A4:** 69, 75, 88-89, 113-116 **M4:** 60-61
impellers **A4:** 95, 110, 115, 118
measurement of velocity **A4:** 72-76, 88, 103, 114-115, 117 **M4:** 61
molten salt **A4:** 88-89, 105 **M4:** 49, 60
oil flow **A4:** 69, 74-76, 95-96 **M4:** 47, 48, 49, 60
propeller agitators **M4:** 46, 61, 62, 64
pumps **A4:** 89, 95, 100, 110, 113, 116-118 **M4:** 64, 67
turbulent agitation **A4:** 75, 88, 101, 115 **M4:** 61
variables affecting agitation **A4:** 69, 114 **M4:** 61, 62, 64-65
water and brine **A4:** 70, 75, 89 **M4:** 60
alloying elements in **A1:** 395, 456-457
aluminum alloys .... **A4:** 84 **A15:** 760 **M2:** 32-35, 40-41, 43 **M4:** 684 , 689, 690, 691, 692, 693, 694, 695, 713
applications, in gas quenching of steel **A4:** 105 **M4:** 45, 46, 59
austempering **A4:** 67, 68
austenitic stainless steels, effect on carbides **A9:** 283
beryllium-copper alloys **A2:** 406
brittle fracture from **A11:** 85
carburizing and carbonitriding **M1:** 533, 535-538, 539
cause of distortion **M1:** 469
characteristic temperature **A4:** 68
cold die **A4:** 106 **M4:** 66
continuous annealing process **A4:** 58
cooling curve analysis **A4:** 68-88, 90-93, 108-109, 114
cooling media **M4:** 63, 66
crack, defined **A11:** 8
cracking **A4:** 76-80, 84, 88, 91, 97, 98
defined, in atomic fluorescence spectrometry **A10:** 46
die-casting dies **A18:** 632
direct quenching **A4:** 67 **M4:** 31
distortion **A4:** 76-79, 88, 91, 97-99, 105, 106, 116
for ductile iron **A1:** 38, 41-42
effect of surface oxidation **A4:** 69, 75-76, 95, 97, 110 **M4:** 50, 61-62
cooling curves **A4:** 69 **M4:** 50, 62
high-speed motion-picture techniques **A4:** 76 **M4:** 62
magnetic testing **A4:** 75-76 **M4:** 61-62
effect on martensite in titanium alloys **A9:** 461
effect on tool and die failures **A11:** 565-566, 573
equipment **A4:** 95, 110, 113, 115, 118 **M4:** 63, 66-67
design of quench tanks **A4:** 89, 114-116, 117, 118 **M4:** 65
gas quenching unit **A4:** 89 **M4:** 58-59
gravity fall **A4:** 110
heaters **A4:** 110, 116, 117 **M4:** 67
impellers **A4:** 95, 110, 115, 118

Quenching (continued)
maintenance **A4:** 89, 100 **M4:** 67-68
propellers **A4:** 113-114, 116, 117
pumps **A4:** 89, 95, 100, 110, 113, 116, 117, 118 **M4:** 67
restraint fixtures **A4:** 110, 116 **M4:** 65
storage tanks **A4:** 89, 114-116, 117, 118 **M4:** 63, 66-67
support fixtures **A4:** 95, 110, 116 **M4:** 65
flame and induction hardened parts **A4:** 107
fluidized bed **A4:** 69, 106, 107
of fluorescence **A10:** 73, 79-80
fog **A4:** 67, 106 **M4:** 31-32, 60
gas quenching, recirculation of gases **A4:** 67, 69, 105 **M4:** 58
gear materials **A18:** 261
Grossmann number (quench severity factor) **A4:** 71-75, 84-88, 91, 94, 100-102, 104, 110
hardenability related to **M1:** 480-496
hardenable steels **M1:** 457-460
hardening of tool steel **A4:** 77-79, 105-106
hardening tool steel **M4:** 59
hardness of steel **A4:** 80-88, 105-106
hardness versus rewetting time **A4:** 71, 72, 74
heat removal stages **A4:** 68-71, 75
of high-chromium white irons **A15:** 684
of high-speed tool steels **A16:** 56, 57
of high-strength structural carbon steels **A1:** 389, 391
ideal critical diameter **A4:** 80-81
induction heat treating **A4:** 195
interrupted quenching **A4:** 67-68
isothermal quenching **A4:** 67, 68
Jominy equivalent **A4:** 82-83, 85, 86
Leidenfrost phenomenon **A4:** 68
Leidenfrost temperature **A4:** 68
of low-alloy steel sheet/strip **A1:** 209
low-carbon steel forgings **A4:** 38
macroetching to reveal soft spots from **A9:** 176
magnesium alloys **A4:** 903 **M4:** 748-749
malleable iron **M1:** 60-61, 68, 70
marquenching **A4:** 67, 68, 75-76, 91, 93
martempering of steel **A4:** 67, 68, 75-76, 91, 93 **M4:** 85
mechanisms **M4:** 32
media
air **A4:** 903
water **A4:** 903
media, selection **A11:** 94
medium for full hardness **A7:** 453
mediums **M1:** 458
microcracking from **A11:** 324
in nitrocarburizing **A7:** 455
notch toughness of steel, effect on **M1:** 706
oil quenching, induction heat treating **A4:** 195
oil, tool steel die crack during **A11:** 565
polymer solutions **A4:** 67, 69, 72-76, 88, 94, 100-104, 107, 209-210
control measures **A4:** 101-102
polyacrylates **A4:** 101, 102-104
polyalkylene glycols (PAG) **A4:** 69, 70, 74, 100-104, 107, 854-855, 867
polysodium acrylate **A4:** 100
polyvinyl alcohol (PVA) **A4:** 76, 100, 101, 104, 106, 195
polyvinyl/pyrrolidone (PVP) **A4:** 72, 100, 101, 102, 104, 854
sodium polyacrylate **A4:** 101, 102, 104, 105
press quenching **A4:** 106-107 **M4:** 66
process **A4:** 67-68, 110
quench factor analysis **A4:** 71, 84-85
in radiation curing **EL1:** 857
rapid, metallic glasses **A2:** 805-806
rate, uranium alloys **A2:** 679
Rushman-Lamont method **A4:** 87-88
safety precautions **A4:** 94, 95, 96, 115, 118-119 **M4:** 68
as secondary operation **A7:** 453

Quenching (continued)
section size effects **A4:** 75, 108, 110-115
selection of medium **A4:** 67, 74, 98
selective **A4:** 67 **M4:** 31
in selective heat treating **M1:** 529
severity (intensity) **A4:** 100-102, 104, 110
of shape memory alloys **A2:** 900
solutions **M4:** 55-58
control measures **M4:** 56
polyacrylates **M4:** 44, 57-58
polyalkylene glycols (PAG) **M4:** 55
polyvinyl alcohol (PVA) **M4:** 55
polyvinylpyrrolidone (PVP) **M4:** 56-57
spray **A4:** 67, 94, 107, 114, 195, 196 **M4:** 31
of steel plate **A1:** 231
storage tanks **A4:** 89, 110, 114-116, 117, 118 **M4:** 63, 66-67
subcritical, uranium alloys **A2:** 673
surface oxidation **A4:** 69, 75-76, 95, 97, 110
systems **A4:** 109-119
continuous quenching **A4:** 110 **M4:** 58, 63
oil quenching systems for continuous carburizers **A4:** 96, 110-113 **M4:** 59, 60, 63
special techniques **M4:** 61, 63-64
water quenching system **A4:** 110, 118 **M4:** 58, 63
and tempering **A7:** 453
tests for hardenability **A4:** 80-83, 90, 107-109
time quenching **A4:** 67 **M4:** 31
titanium **M4:** 766-767, 768, 769
titanium alloys **A4:** 917, 921
of tool and die steels **A14:** 54-55
tool steels **A4:** 77, 78, 79, 105-106, 744-745 **M4:** 572
of tool steels, effect on microstructure **A9:** 258-259
uniform **A11:** 325
of uranium alloys **A2:** 672-674
warping from **A11:** 141
Quenching crack
defined **A9:** 15
Quenching in of thermal vacancies
formation of point defects by **A9:** 116
Quenching media **A1:** 471-473 **M4:** 39
agitation of **A4:** 72-76, 85, 95-96, 101, 103-104, 107, 113-115 **M4:** 33, 47, 48, 49, 60-61
factors controlling agitation **A4:** 69, 75, 88-89, 113-116 **M4:** 60-61
measurement of velocity **A4:** 72-76, 88, 103, 114-115, 117 **M4:** 61
molten salt **A4:** 88-89, 105 **M4:** 49, 60
oil flow **A4:** 69, 74-76, 95-96 **M4:** 47, 48, 49, 60
turbulent agitation **A4:** 75, 88, 101, 115 **M4:** 61
variables affecting agitation **A4:** 69, 75, 88-89 **M4:** 61, 62, 64-65
water and brine **A4:** 70-75, 88-89 **M4:** 60
air **A4:** 67, 72, 75, 195, 903
aqueous *See* brine solutions
brine solutions **A4:** 69, 70, 72, 74-76, 88-93 **M4:** 39, 41-43
caustic solutions **A4:** 88, 90, 101, 115 **M4:** 39, 43
cold dies **A4:** 106
dry dies **M4:** 39, 66
electron-beam heat treating **M4:** 521
fluidized beds **A4:** 69, 106, 107
fog **A4:** 67, 106 **M4:** 39, 60
gases (still or moving) **A4:** 67, 69, 89, 105-106 **M4:** 39, 58-59
magnetic test **A4:** 75-76, 92, 93, 98, 109 **M4:** 61-62
molten metals **A4:** 83 **M4:** 39
molten salts **A4:** 67-69, 72, 83, 87-89, 91, 101, 104-105 **M4:** 39, 60
oils **A4:** 67-69, 72-76, 78, 87, 89-100, 195 **M4:** 43-54
polyacrylates **A4:** 101, 102-104
polyalkylene glycols (PAG) **A4:** 69, 70, 74, 100-104, 107, 854-855, 867

Quenching media (continued)
polymer solutions ... **A4:** 67, 69, 72-76, 88, 94, 100-104, 107, 209-210 **M4:** 39, 55-58
polyacrylates ... **M4:** 44, 57-58
polyalkylene glycols (PAG) ... **M4:** 55-56
polyvinyl alcohol (PVA) ... **M4:** 55
polyvinyl pyrrolidone (PVP) ... **A4:** 72, 100, 101, 102, 104, 854
polyvinylpyrrolidone (PVP) ... **M4:** 56-57
sodium polyacrylate (PA) ... **A4:** 101, 102, 104, 105
polysodium acrylate ... **A4:** 100
polyvinyl alcohol (PVA) ... **A4:** 76, 100, 101, 104, 106, 195
selection of ... **A4:** 67, 74, 98
severity ... **A4:** 71-74, 84-88, 90-92, 94, 100, 101
tests and evaluation ... **A4:** 71-76, 107-109 **M4:** 35-38
water ... **A4:** 67, 69, 72-75, 78, 87-89, 98-99, 113 **M4:** 39, 63
for magnesium alloys ... **A4:** 903
tank maintenance ... **A4:** 115, 118
Quenching of steel *See* Steel, quenching ... **M4:** 31-68
compared to conventional quenching and tempering ... **M4:** 104
Quenching oil
defined ... **A18:** 15
Quenching stresses
defined ... **EM2:** 751
Quick disconnect
defined ... **EL1:** 1154
Quick opening
in crack arrest testing ... **A8:** 444
Quick-break circuitry
for coils ... **A17:** 95
direct current ... **A17:** 98
Quick-coupling connectors
for magnetic particle inspection ... **A17:** 93
Quick-quiet basic oxygen process (Q-BOP) ... **A1:** 111
Quick-release clamp ... **A8:** 220-221
Quick-release pins
failure in ... **A11:** 548
Quick-stop devices
and orthogonal machining ... **A16:** 8
Quickline neutralization
plating wastes ... **M5:** 313
Quinary system or diagram ... **A3:** 1 • 2
Quinnipiac River Bridge
cracks in ... **A11:** 709
Quoting prices
guidelines ... **EM2:** 86

R

Recommendations for Storage and Handling of Aluminum Pigments and Powders **(Aluminum Association)** ... **A7:** 130
R *See* Radius of curvature; Rare earths; Stress ratio
R control charts *See also* Control charts; Shewhart control chart model
construction and interpretation ... **A17:** 725
interpretation ... **A17:** 728
setting up ... **A17:** 727-728
R phase
in austenitic stainless steels ... **A9:** 284
***r* space**
Fourier transform to ... **A10:** 412-413
***r* value** *See also* Anisotropy factor; Plastic strain ratio
determining, uniaxial tensile testing ... **A8:** 556-557
drawn cup with ears in direction of high ... **A8:** 550
role in strain distribution, sheet metal forming ... **A8:** 550
***R*-curve**
analysis, for fracture toughness ... **A11:** 64
applications ... **A8:** 451
defined ... **A8:** 11 **A11:** 8
format, use of crack tip opening displacement in ... **A8:** 457
in fracture mechanics ... **A8:** 451-453
from measurement of *K* and crack extension ... **A8:** 450
load vs. crack mouth opening displacement curve for ... **A8:** 461
schematic ... **A8:** 453
standard methods, aluminum alloys ... **A8:** 461
R-curve behavior ... **EM4:** 586
***R*-curve method**
for fracture toughness ... **A8:** 449
R-glass, high-strength
high-modulus ... **EM1:** 107
R-Monel, applications
protection tubes and wells ... **A4:** 533
R-output
defined ... **A17:** 303
high-energy sources ... **A17:** 307-308
tube current effect ... **A17:** 304
tube voltage effect ... **A17:** 304
of x-ray tube ... **A17:** 304
***R*-ratio**
residual stress as change in ... **A11:** 57
R.R. Moore rotating beam machine ... **A8:** 369-370
R_2O_3 group
elements precipitated in ... **A10:** 168
RA *See* Reduction in area; Reduction of area
RA 330
composition ... **A6:** 564 **M6:** 354
erosion test results ... **A18:** 200
RA 333
composition ... **A6:** 573
erosion test results ... **A18:** 200
mill annealing temperature range ... **A6:** 573
solution annealing temperature range ... **A6:** 573
RA-330
annealing ... **M4:** 655
composition ... **M4:** 651-652
stress relieving ... **M4:** 655
RA85H
characteristics ... **A4:** 512
composition ... **A4:** 512
recommended for carburizing and carbonitriding furnace parts ... **A4:** 513
RA330
composition ... **A4:** 512
radiant tube applications ... **A4:** 517
recommended for carburizing and carbonitriding furnace parts ... **A4:** 513
recommended for furnace parts and fixtures ... **A4:** 513
recommended for parts and fixtures for salt baths ... **A4:** 514
RA330HC
composition ... **A4:** 512
recommended for furnace parts and fixtures ... **A4:** 513
RA333
composition ... **A4:** 512, 794
mill annealing ... **A4:** 810
solution annealing ... **A4:** 810
Race (or raceway)
defined ... **A18:** 15
Raceway
bearing ... **A11:** 490, 498, 504
deformation, by ball and rollers ... **A11:** 510
fretting failure of ... **A11:** 498
outer-ring, bearing failure from improper heat treatment ... **A11:** 509-510
roller-bearing, spalling on ... **A11:** 504
surface, rolling-contact fatigue in ... **A11:** 503-504
Rachinger correction ... **A10:** 386
Rack cutting
and machining of helical gears ... **A16:** 339, 341
and machining of large spur gears ... **A16:** 341
used for gear manufacture ... **A16:** 333, 334-335
Rack gears
described ... **A11:** 587
Rack-and-panel connector
defined ... **EL1:** 1154
Rack-and-pinion feed mechanisms
for rotary swaging ... **A14:** 134
RAD *See* Ratio-analysis diagram
Rad units
radiography ... **A17:** 301
Radance
of particle shape ... **A7:** 242
Radar coolant-system assembly
brazed joint failure ... **A11:** 452-453
Radar modulator
connector corrosion failure analysis ... **EL1:** 1112
Radar, phased-array
with monolithic microwave integrated circuits ... **A2:** 740
Radar waves *See also* Microwave inspection; Scattered radiation
airborne ... **A17:** 228
defined ... **A17:** 202
side-looking ... **A17:** 227
Radial ball bearings ... **A18:** 499, 505, 506, 508
basic load rating ... **A18:** 505
f_o factor ... **A18:** 510
X and *Y* factors ... **A18:** 509
X_o and Y_o factors ... **A18:** 510
Radial brushes ... **M5:** 151-156
Radial clearance of bearing
nomenclature for Raimondi-Boyd design chart ... **A18:** 91
Radial contact ball bearings ... **A18:** 509
Radial crack
definition ... **EM4:** 633
Radial crushing strength ... **A7:** 10
of copper-based materials ... **A7:** 464
Radial cylindrical roller bearings ... **A18:** 500, 501, 502, 511
Radial displacement
for stress-strain determination ... **A8:** 210
Radial distribution function analysis ... **A10:** 393-401
applications ... **A10:** 393, 398-401
data reduction ... **A10:** 396-398
defined ... **A10:** 680
diffraction data ... **A10:** 394-395
estimated analysis time ... **A10:** 393
general uses ... **A10:** 393
instrumentation ... **A10:** 395-396
interpretation ... **A10:** 398
introduction ... **A10:** 393-394
limitations ... **A10:** 393
in liquid and amorphous materials ... **A10:** 420
related techniques ... **A10:** 393
samples ... **A10:** 393, 398-401
x-ray sources ... **A10:** 395
Radial draw forming ... **A14:** 10, 595-596
Radial drill presses
boring ... **A16:** 170
Radial field inspection *See* Flux leakage method; Magnetic field testing
Radial forging *See also* Forging; Rotary forging ... **A14:** 145-149
advantages ... **A14:** 146-147
application ... **A14:** 145, 147
defined ... **A14:** 10
equipment and process ... **A14:** 145-146
four-hammer machines capacities/sizes ... **A14:** 145, 147-148
machines ... **A14:** 16-17, 147-149, 227
as new metalworking process ... **A14:** 16-17
over mandrel ... **A14:** 146
parts formed by ... **A14:** 146
radial precision forging ... **A14:** 146-147
and rotary forging, compared ... **A14:** 176
of stainless steels ... **A14:** 227
two-hammer radial forging machines ... **A14:** 149
vs rotary (orbital) forging ... **A14:** 145
Radial fracture *See also* Radial marks
AISI/SAE alloy steels ... **A12:** 312
zone ... **A12:** 311
Radial friction welding ... **A6:** 318-320
applications ... **A6:** 320
equipment ... **A6:** 318-320
method ... **A6:** 318
process development ... **A6:** 318
prototype/production machine ... **A6:** 320
Radial hardness, variations
in swaging ... **A14:** 143
Radial lip seal
defined ... **A18:** 15
Radial marks *See also* Chevron pattern
AISI/SAE alloy steels ... **A12:** 294, 302, 312-314, 334
at fracture origin ... **A11:** 87, 95-9
and circular spall ... **A12:** 113, 123
cobalt alloy ... **A12:** 398
crack origin from ... **A11:** 83
curved, star or rosette fracture as ... **A12:** 103-104
defined ... **A8:** 10-11 **A11:** 8
fine/coarse ... **A12:** 102-103
and line spall ... **A12:** 114, 129-130
SEM fractographs ... **A12:** 169
studies ... **A12:** 3
tool steels ... **A12:** 377-378
Radial needle roller bearings ... **A18:** 500, 501-502

Radial precision forging **A14:** 146-147
Radial roll
defined **A14:** 10
Radial roller bearings **A18:** 500-501, 507
applicable load **A18:** 511
basic static radial load rating **A18:** 508-510
X and *Y* factors **A18:** 509
X_0 and Y_0 factors **A18:** 510
Radial rolling force
defined **A14:** 10
Radial sectioning
for plating thickness inspection **EL1:** 943
Radial segregation
in wrought heat-resistant alloys **A9:** 305
Radial spherical roller bearings **A18:** 500, 501, 502
Radial stress
during bending **A8:** 121, 123-124
Radial structure/texturing
carbon fibers **EM1:** 5
Radial wafer technique **A7:** 651-652
Radial wafer turbine blades **A7:** 651-652
Radial weaving **EM1:** 129, 13
Radial-axial horizontal rolling machines **A14:** 113
Radial-axial ring rolling
bead formation during **A14:** 115
machines, types and characteristics
mill, schematic **A14:** 110
Radial-contact ball bearings **A11:** 490, 503-504
Radial-load bearing
defined **A18:** 15
Radial-locking pins
failures in **A11:** 548
Radian **A10:** 685, 691
Radiance
SI derived unit and symbol for **A10:** 685
SI unit/symbol for **A8:** 721
Radiant continuous oven
paint curing process **M5:** 488
Radiant energy **A10:** 62, 680
Radiant flux
SI unit/symbol for **A8:** 721
Radiant intensity **A10:** 680, 685
SI unit/symbol for **A8:** 721
Radiant power
exponential decay, as function of concentration **A10:** 62
flux, defined **A10:** 680
intensity as **A10:** 62
as transfer of radiant energy **A10:** 62
Radiant tubes
corrosion failure **A11:** 293-29
Radiation *See also* Bremsstrahlung radiation; Radiation resistance; Radiation safety; specific radiation forms; Synchrotron radiation; X-rays
absorption and scatter, in x-ray spectrometry **A10:** 84
absorption, in neutron radiography **A17:** 390
alpha particles **EL1:** 965
beta particles **EL1:** 966
characteristics, gamma-rays **A17:** 308
continuum or bremsstrahlung, defined **A10:** 83
control techniques **A13:** 951
copper **A10:** 326
for curing coatings **EL1:** 854
damage **A10:** 253, 365, 583, 588
Damage, defined **A11:** 8 **A13:** 11
defined **A10:** 83
degradation factors **EM2:** 575-576
degradation, photooxidation as **EM2:** 424
diffraction, radiographic inspection **A17:** 345
divisions, electromagnetic spectrum **A17:** 202
effect, light water reactor corrosion **A13:** 947
effect, radioactive waste disposal **A13:** 973-974
effects, semiconductor chips **EL1:** 965-966
electromagnetic **A10:** 83
electromagnetic, and neutrons, compared **A17:** 390
electromagnetic, attenuation **A17:** 309-311
Radiation (continued)
electromagnetic, in x-ray spectrometry **A10:** 83
electromagnetic, microwaves as **A17:** 202
embrittlement **A11:** 69, 10
embrittlement effects **A12:** 127
energy, absorbance vs **A10:** 85-86
as environmental health hazard **A10:** 247
in ESR analysis **A10:** 254
exposure **A13:** 949
far-infrared, defined **A10:** 673
fields, corrosion influence on **A13:** 948-952
from tunable lasers **A10:** 142
gamma, and polyether-imides (PEI) **EM2:** 157
holographic **A17:** 405
-induced changes, ion-implantation **A10:** 484
infrared **A10:** 109, 114 **A11:** 76
infrared, evapograph-detected **A17:** 208
intensity, determined, radiographic inspection **A17:** 308
ionizing, defined **EL1:** 854
Kα **A10:** 326
leakage, radiographic inspection **A17:** 301
molecular polarization as source of **A10:** 127
monitoring **A17:** 301-302
monochromatic, in reflection topography **A10:** 366
neutron **A11:** 17, 6
penetration, in neutron radiography **A17:** 387
penetration into stress gradient, and measurement errors **A10:** 388
polychromatic **A10:** 97, 366
in Raman spectroscopy **A10:** 126-130
safety **A17:** 300-302
as scattered, defined **A17:** 205
scattered, shadow formation radiography **A17:** 313-314
secondary, in neutron radiography **A17:** 387
sources **A13:** 949 **A17:** 207-298, 368-369
sources, computed tomography (CT) **A17:** 368-369
sources, MFS **A10:** 76
sources, synchrotron radiation **A10:** 413
spectrum, x-ray absorption effect **A17:** 310
stability, polyamide-imides (PAI) **EM2:** 129
stem, defined **A17:** 305
susceptibility **EM2:** 575-580
test methods **EM2:** 576-580
thermal **A17:** 396
ultraviolet, defined **A10:** 683
units **A17:** 300-301
used in scanning electron microscopy **A9:** 90
UV/VIS, molecular absorption as requirement for fluorescence **A10:** 73
wavelength and energy theories of **A10:** 83
white, defined **A10:** 325-326
in x-ray range **A10:** 85-86
Radiation and vacuum effects on adhesives **EM3:** 644-649
evaluation methods **EM3:** 644-646
evaluation results **EM3:** 646-649
simulated space environment **EM3:** 646
space application adhesives **EM3:** 644
Radiation beam intensity modulation
in neutron radiography **A17:** 387
Radiation chemistry
defined **EL1:** 854
Radiation curing **EL1:** 854, 858
Radiation curve coatings **M5:** 486
Radiation damage *See* Neutron irradiation
transmission electron microscopy **A9:** 116-117
Radiation heat transfer
sintering **A7:** 341
Radiation output *See* R-output
Radiation properties **EM1:** 5
aramid fibers
glass fibers **EM1:** 4
Radiation resistance
polyether-imides (PEI) **EM2:** 157
thermoplastic polyimides (TPI) **EM2:** 177
Radiation safety
access control **A17:** 302
maximum permissible dose **A17:** 301
radiation monitoring **A17:** 301-302
radiation protection **A17:** 301
radiation units **A17:** 300-301
Radiation scattering
to analyze ceramic particle size **EM4:** 67
Radiation shielding
powder used **A7:** 573
Radiation units
rem and rad units **A17:** 301
sievert and gray units **A17:** 301
for x-rays and gamma-rays **A17:** 301
Radiation-assisted chemical vapor deposition **EM3:** 594
Radiation-enhanced diffusion **A18:** 854, 855
Radiation-induced segregation **A1:** 654 **A18:** 854, 855
Radiationless transition, and photoelectric effect
in XPS development **A10:** 568
Radiative heat flow equation **A6:** 1023
Radical *See* Free radicals **EM3:** 24
Radical ions **A10:** 263, 265
Radical production and decay
kinetics of **A10:** 265-266
Radical scavengers **A18:** 104- 105, 106
Radii
steel forging **M1:** 362-363, 370-371
Radii, internal/external
design of **EM2:** 615
Radii, sharp
as stress concentrators **A11:** 318, 31
Radio alloys *See also* Copper-nickel resistance alloys; Electrical resistance alloys
properties and applications **A2:** 823-825
Radio and television
applications of copper-based powder metals **A7:** 734
cores, powders used **A7:** 17, 574
Radio frequency
abbreviation for **A10:** 691
bridge, magabsorption measurement **A17:** 150
defined **A10:** 680
generators, analytic ICP systems **A10:** 34, 37
permeability, magabsorption theory **A17:** 148
signals, in magabsorption **A17:** 143, 147-152
spectrometers, defined **A10:** 680
Radio frequency (RF)
discrete transistors, electrical performance testing **EL1:** 946-951
fixturing **EL1:** 949
losses, in superconductors **A2:** 1040
magnetron sputtering, for high-temperature superconductors **A2:** 1087
packages **EL1:** 428-429
passive devices **EL1:** 1005
Radio frequency (RF) curing
of polymers **EL1:** 786
Radio frequency (RF) preheating
defined **EM2:** 35
Radio frequency (rf) sputter deposition technique **A18:** 842
Radio frequency (RF) sputtering **EM4:** 32
Radio frequency discrete transistors **EL1:** 946-951
Radio frequency generators
for induction heating **A8:** 414
Radio frequency heating *See* Induction soldering
Radio frequency induction
ion plating process **M5:** 418
Radio frequency sputtering **M5:** 414
Radio frequency switching relay
defined **EL1:** 1154
Radio frequency testing **EL1:** 946-949
Radio frequency welding *See also* Welding
defined **EM2:** 35
Radio tuning devices
iron powder cores for **A7:** 17

Radio-frequency spectroscopy
defined ... **A10:** 680
Radioactive decay
half-life, defined ... **A10:** 244
modes ... **A10:** 244-245
principles of ... **A10:** 244
scheme ... **A10:** 292
Radioactive decay spectrometry ... **A10:** 246
Radioactive isotopes
radioanalysis of ... **A10:** 243
specific activity of ... **A10:** 244
Radioactive materials *See also* Uranium
naturally occurring, radioanalysis of ... **A10:** 243
optical microscopy of ... **A9:** 83
replicas used to examine ... **A9:** 108
requirements for ... **A10:** 247
Radioactive metals
electropolishing procedures for ... **A9:** 50
electropolishing used in situ ... **A9:** 55
structures ... **A9:** 620
Radioactive soils, removal of
vapor degreasing ... **M5:** 54-55
Radioactive sources
for gamma-rays ... **A17:** 308
for neutrons ... **A17:** 389-390
Radioactive wastes
corrosion of containment materials for ... **A13:** 971-980
Radioactivity
decay modes ... **A10:** 244-245
defined ... **A10:** 680
detection and measurement ... **A10:** 245-246
elemental cobalt as ... **A2:** 446
of NAA samples ... **A10:** 233
remote analysis by electrometric titration for ... **A10:** 202
uranium ... **A7:** 666
of uranium and uranium alloys ... **A2:** 670
Radioactivity-induced silicon p-n junction failures ... **A11:** 784-78
Radioanalysis ... **A10:** 243-250
accuracy, precision, and sensitivity ... **A10:** 246-247
analytical procedure ... **A10:** 247-249
apparatus for, schematic ... **A10:** 247
applications ... **A10:** 243, 249-250
background and coincidence corrections ... **A10:** 248
chemical preparations ... **A10:** 247
defined ... **A10:** 680
detection and measurement of radioactivity ... **A10:** 245-246
equipment ... **A10:** 247-248
estimated analysis time ... **A10:** 243
general uses ... **A10:** 243
half-life ... **A10:** 244
introduction ... **A10:** 244
limitations ... **A10:** 243
radioactive decay modes ... **A10:** 244-245
radioactive measurement ... **A10:** 247-248
related techniques ... **A10:** 243
sample preparation ... **A10:** 248
samples ... **A10:** 243, 248-250
Radiochemical (destructive) TNAA elemental assay ... **A10:** 233, 238-239
for iridium determination in Cretaceous- Tertiary boundary ... **A10:** 240-241
Radiocobalt
toxicity effects ... **A2:** 1251
Radiofrequency (RF) curing method ... **EM3:** 577
Radiofrequency (RF) preheating ... **EM3:** 24
Radiofrequency (RF) shielding ... **EM3:** 52
Radiofrequency (RF) welding ... **EM3:** 24
Radiofrequency interference (RFI) protection/shielding ... **EM1:** 359, 728
Radiogallium, as medical therapy
toxic effects ... **A2:** 1257
Radiographic contrast
contrast sensitivity ... **A17:** 299
dynamic range ... **A17:** 299
subject contrast ... **A17:** 298-299
Radiographic coverage
defined ... **A17:** 347
Radiographic definition ... **A17:** 300
film unsharpness
projective magnification with microfocus x-ray sources ... **A17:** 300
screen unsharpness ... **A17:** 300
Radiographic equivalence
x-ray/gamma-ray absorption ... **A17:** 311
Radiographic film *See also* X-ray film
automatic processing ... **A17:** 353-355
feeding procedures ... **A17:** 354
fixer removal, tests for ... **A17:** 355-356
manual processing ... **A17:** 351-353
microfilming of radiographs ... **A17:** 356
processing ... **A17:** 351-356
Radiographic inspection *See also* NDE reliability; Radiography ... **A17:** 295-357 **M6:** 848-849 **EL1:** 369
acceptance standards ... **A17:** 347
of aluminum alloy castings ... **A17:** 534
applications ... **M6:** 827
arc welds ... **A17:** 334-335
of arc-welded nonmagnetic ferrous tubular products ... **A17:** 567
attenuation of electromagnetic radiation ... **A17:** 309-311
beam model for ... **A17:** 710
of boilers and pressure vessels ... **A17:** 641-642
of brazed assemblies ... **A17:** 604
of casting defects ... **A15:** 554-555, 557
of castings ... **A17:** 526-529
codes, standards, and specifications ... **M6:** 827
of complex shapes ... **A17:** 333-334
and computed tomography, compared ... **A17:** 295
detector model for ... **A17:** 710
of double submerged arc welded steel pipe ... **A17:** 565-566
explosion welds ... **M6:** 711
film radiography ... **A17:** 323-330
of forgings ... **A17:** 511
gamma-ray sources ... **A17:** 308-309
high-energy x-ray sources ... **A17:** 307-308
identification markers ... **A17:** 338-343
image conversion media ... **A17:** 314-317
magnesium alloys ... **A6:** 781-782
NDE reliability models of ... **A17:** 709-711
penetrameters (image-quality indicators) ... **A17:** 338-343
of pipeline girth welds ... **A17:** 580
plasma-MIG welding ... **A6:** 224
of powder metallurgy parts ... **A17:** 537-539
of pressure vessels ... **A17:** 647
process qualification ... **A17:** 678
radiograph appearance, of specific flaws ... **A17:** 348-351
radiographic film, processing of ... **A17:** 351-356
radiographs, interpretation of ... **A17:** 345-348
radiography, principles of ... **A17:** 297-302
radiography, uses of ... **A17:** 295-297
real-time radiography ... **A17:** 317-323
of resistance welds ... **A17:** 335
of resistance-welded steel tubing ... **A17:** 565
sample interaction model for ... **A17:** 710
scattered radiation, control of ... **A17:** 343-345
of seamless pipe ... **A17:** 579
of seamless steel tubular products ... **A17:** 571
selection of view ... **A17:** 330-338
shadow formation, principles of ... **A17:** 311-314
of simple shapes ... **A17:** 332-333
soldered joints ... **A6:** 981
to check brazing of stainless steels ... **A6:** 919
to detect subsurface slag inclusions ... **A6:** 1074
of tubular sections ... **A17:** 335-337
underwater welds ... **M6:** 924
of weldments ... **A17:** 334-335, 592-594
x-ray tubes ... **A17:** 302-307
Radiographic paper
as recording medium ... **A17:** 314
Radiographic screens *See* Screens
Radiographic sensitivity
detail perceptibility, of images ... **A17:** 300
radiographic definition ... **A17:** 299-300
Radiographic standard shooting sketch (RSSS) ... **A17:** 347-348
Radiographic testing ... **A6:** 1081, 1083, 1085, 1086, 1087, 1088
brazed joints ... **A6:** 1119, 1122
Radiographic view
selection of ... **A17:** 330-338
Radiographs
interpretation ... **A17:** 345-348
microfilming of ... **A17:** 356
thermal neutron ... **A17:** 391-395
Radiography *See also* Autoradiography; Digital radiography; Film radiography; In-motion radiography Microradiography; Microradiography; Neutron radiography Radiographic inspection; Real-time radiography; Xeroradiography ... **EM1:** 775-77
applicability ... **A17:** 296
for boilers and pressure vessels ... **A17:** 647-649
for brazing inspection ... **A11:** 451
carbon steels, cracking ... **A6:** 643
digital, color images by ... **A17:** 485
as failure analysis ... **A11:** 17-18
fitness for service evaluation ... **A6:** 376
gas-tungsten arc welding ... **A6:** 451
image conversion ... **A17:** 298
in-motion ... **A17:** 337-338
and industrial computed tomography compared ... **A17:** 361-362
of internal discontinuities, castings ... **A17:** 512
laminographic ... **EL1:** 738
limitations ... **A17:** 297
as nondestructive fatigue testing ... **A11:** 13
plasma arc welding ... **A6:** 198
principles of ... **A17:** 297-302
radiation sources ... **A17:** 297-298
selection of view ... **A17:** 330-338
to detect cracks, electron-beam welding ... **A6:** 1077-1078
to detect geometric weld discontinuities ... **A6:** 1075
to detect lack of fusion ... **A6:** 1075, 1078
to detect subsurface gas porosity ... **A6:** 1074
to detect tungsten inclusions ... **A6:** 1074
to detect weld discontinuities, electroslag welding ... **A6:** 1078
to detect weld metal and base metal cracks ... **A6:** 1075
to measure weld pool surface velocity ... **A6:** 1149
transmissive ... **EL1:** 738
uses of ... **A17:** 295-297
vs. ultrasonic inspection, for primary mill products ... **A17:** 267
x-ray ... **EL1:** 369
x-ray, for adhesive-bonded joints ... **A17:** 624
Radioisotope tracer test method
cleaning process efficiency ... **M5:** 20
Radioisotopes
analysis for ... **A10:** 233
decay rate ... **A10:** 235
defined ... **A10:** 234, 680
as neutron source ... **A17:** 389
properties, for thermal-neutron radiography ... **A17:** 390
testing ... **A17:** 63-64
Radiology. also Radiographic inspection; Radiography
defined ... **A17:** 295
Radiometers ... **EM3:** 762-763
for thermal inspection ... **A17:** 399
Radionuclide, computed tomography (CT)
defined ... **A17:** 384
Radionuclide methods ... **A18:** 319- 327
applications ... **A18:** 322-323
aerospace ... **A18:** 323
diagnostic condition monitoring ... **A18:** 322
laboratory wear characterization ... **A18:** 322
tracer monitoring ... **A18:** 322-323
considerations in planning SLA testing ... **A18:** 321-322
details for executing SLA measurements ... **A18:** 323-327
activation fundamentals ... **A18:** 323
attenuation and detection of gamma rays ... **A18:** 324-325
characteristics of sodium iodide spectrometry ... **A18:** 325
detector configuration ... **A18:** 325
detector mounting ... **A18:** 326-327
laboratory marker calibrations ... **A18:** 323-324
spectrum details ... **A18:** 325-326
statistical considerations ... **A18:** 326
tracer calibrations ... **A18:** 324
useful SLA radionuclides from common metals and their production methods ... **A18:** 323
future trends ... **A18:** 327

Radionuclide methods (continued)
radionuclides applied to tribology **A18:** 319-321
activation methods **A18:** 319-321
advantages of surface layer activation **A18:** 321
gamma detectors **A18:** 319, 320
markers **A18:** 319
tracers **A18:** 319
Radionuclides *See also* Radiosiotopes
activity of, SI unit and symbol for **A10:** 685
Radioscopy *See also* Real-time radiography
defined **A17:** 295
Radiotracer method
dental tissue wear plate measurement **A18:** 668
Radium containers
composite metals for **A7:** 17
Radius
defined **A14:** 10
Radius at root of groove welds **M6:** 64-65
Radius forming
for press-brake forming **A14:** 536
Radius of bend
defined **A8:** 11 **EM2:** 35
Radius of curvature
interference microscope measurement **A17:** 17
symbol for **A8:** 725
Radius of equivalent cylinder
nomenclature for lubrication regimes **A18:** 90
Radiusing **A16:** 33
abrasive flow
in conjunction with boring **A16:** 165
machining **A16:** 514-515, 517, 518, 519
Raffinal (super-purity aluminum)
applications and properties **A2:** 64-65
Rail shear test **EM1:** 30
Rail shear tests **EM3:** 822, 827-828
Rail steels
fracture modes **A12:** 115, 117, 135-137
Rail welding *See* Thermit welding
Railroad applications
acoustic emission inspection **A17:** 290
Barkhausen noise measurement **A17:** 160
car springs, steels for **M1:** 303
cast steels **M1:** 378, 379
copper-based powder metals **A7:** 734
crucible steel **A15:** 31
early, chilled iron **A15:** 30
hot rolled bars and shapes **M1:** 212-213
market trends **A15:** 44
steel rail, ultrasonic inspection **A17:** 232, 272
tank cars, magnetic paint inspection **A17:** 128
ultrasonic inspection **A17:** 232
Railroad cars
of aluminum and aluminum alloys **A2:** 11
Railroad construction
joining processes **M6:** 57
Railroad diesel oils
lubricant classification **A18:** 85
Railroad equipment **A6:** 395-398
Railroad rail head
residual stress distribution across a flash-butt welded **A10:** 391
Railroad rail, steel
transverse fracture in **A11:** 7
Railroad rolling stock
codes governing **M6:** 824
Rails
ultrasonic inspection **A17:** 232, 272
Railway tank car
brittle fracture from weld imperfections in **A11:** 93-94
Railway tank car applications
high-strength low-alloy steels for **A1:** 419-420
Raimondi-Boyd design chart **A18:** 90, 91
Rain erosion *See also* Erosion (erosive wear); Erosive wear
components affected **A11:** 163
defined **A18:** 15
Rainflow method
of damage analysis **A8:** 682
Raining
as centrifugal casting defect **A15:** 306-307
Rainwater *See also* Water
beryllium corrosion in **A13:** 809
chloride concentration **A13:** 909
corrosion in **A13:** 17
effects, zinc/zinc alloys and coatings **A13:** 757
GFAAS analysis for silver **A10:** 55
as water drop impingement corrosive **A13:** 142
Raised core
as casting defect **A11:** 381
Raised sand
as casting defect **A11:** 381
Raised-stress regimes (RSR) **EM3:** 651-652, 653-654
Raisers, stress *See* Stress raisers
Raj damage nucleation deformation map
example **A14:** 421-422
Rake
of straight-knife shears **A14:** 704-705
Rake angle **A16:** 121, 297 **A18:** 610, 611
Raleigh wave speed **A8:** 445
Ram *See* Force plug; Random access memory (RAM) **A7:** 10
in electrohydraulic testing machine **A8:** 160
in servo-hydraulic test frames **A8:** 192
Ram pressing **EM4:** 8
Ram tensile testing
explosion welds **M6:** 711
Ram travel
defined **EM2:** 35
Ram-and-inner-frame HERF machines **A14:** 29, 101
Raman band
defined **A10:** 680
Raman effect
experimental considerations **A10:** 128-129
fundamentals **A10:** 126-128
sampling **A10:** 129-130
Raman line
defined **A10:** 680
Raman microprobe analysis
capabilities **A10:** 516
of pyridine adsorbed on metal oxide surfaces **A10:** 134
Raman scattering
surface-enhanced **A10:** 135-136
Raman shift
defined **A10:** 680
Raman spectrometer
conventional **A10:** 128
with multichannel detector **A10:** 128
Raman spectroscopy **A10:** 126-138 **A13:** 1114-1115
as advanced failure analysis technique **EL1:** 1104
anti-Stokes radiation **A10:** 127
applications **A10:** 131-133
bulk materials analysis by **A10:** 126, 130-133
capabilities **A10:** 333
capabilities, compared with infrared spectroscopy **A10:** 109
defined **A10:** 680
estimated analysis time **A10:** 126
experimental utility **A10:** 127-128
and gas analysis by mass spectrometry compared **A10:** 151
for graphites **A10:** 132-133
in situ, laser excitation sources for **A10:** 135
information obtainable from **A10:** 130
and infrared spectroscopy, compared **A10:** 127
of inorganic liquids and solutions **A10:** 7
of inorganic solids **A10:** 4-6
introduction **A10:** 126
laser Raman microprobe **A10:** 129
lasers **A10:** 129
limitations **A10:** 126, 135, 136
for metal corrosion **A10:** 134-135
for metal oxides **A10:** 130-131
Raman spectroscopy (continued)
molecular optical laser examiner (MOLE) **A10:** 129
molecular vibration **A10:** 127
monochromator-detector assembly of **A10:** 128-129
of organic gases **A10:** 11
of organic liquids and solutions **A10:** 10
of organic solids **A10:** 9
polarization **A10:** 127
for polymers **A10:** 131-132
Pourbaix diagram **A10:** 135
Raman effect **A10:** 126-130
related techniques **A10:** 126
sampling **A10:** 126, 129-130
selection rules **A10:** 127
spectrometers for **A10:** 128
Stokes radiation in **A10:** 127
surface-enhanced Raman scattering **A10:** 135-137
Raman spectroscopy (RS)
for chemical analysis of polymer fibers **EM4:** 223
for phase analysis **EM4:** 561
to analyze the surface composition of ceramic powders **EM4:** 73
Raman spectrum
defined **A10:** 680
Ramaway
as casting defect **A11:** 387
Ramberg-Osgood equation **A11:** 50
modified version for stress-strain response of adhesives **EM3:** 515
Rammed graphite molding
titanium alloys **A15:** 825
titanium and titanium alloy castings **A2:** 635
Rammed graphite molds **A15:** 273-274
Rammer, air
jolt rollovers topped with **A15:** 29
Ramming **EM4:** 151
defined **A15:** 9
early practice **A15:** 28
effect, synthetic sands **A15:** 29
mixes, induction furnaces **A15:** 373
Ramming, soft or insufficient
as casting defect **A11:** 386
RAMOD-2 bilinear equation **EM3:** 515
Ramoff
as casting defect **A11:** 387
Ramp
defined **A12:** 59
rates, and mean stress **A12:** 63
Ramp rate *See* Thermal ramp rate
Rams *See also* Hammers; Presses
in counterblow hammers **A14:** 42
defined **A14:** 10
deflection, built-in die mismatch for **A14:** 50
displacement vs load, nonlubricated extrusion **A14:** 316
in electrohydraulic gravity-drop hammers **A14:** 42
in hammers **A14:** 25
in open-die forging **A14:** 61
in presses **A14:** 29, 31, 33, 35
speed, straight-knife shearing **A14:** 705
velocity, in high-energy-rate forging **A14:** 100
Ramsbottom method **A18:** 84
Randles equivalent circuit **EM3:** 436
Random (statistical) particle pattern **A7:** 186
Random access
defined **EL1:** 1154
Random access memory
abbreviation for **A11:** 797
Random access memory (RAM) **A10:** 92 **EL1:** 8, 160
Random alloys
EXAFS analysis **A10:** 407
Random copolymers **EM2:** 58
Random duplex aggregates **A9:** 604
Random errors
in sampling **A10:** 12

SUBJECTS OF THE INDEXED VOLUMES: ASM Handbook (designated by the letter "A"): **A1:** Properties and Selection: Irons, Steels, and High-Performance Alloys (1990); **A2:** Properties and Selection: Nonferrous Alloys and Special-Purpose Materials (1990); **A3:** Alloy Phase Diagrams (1992); **A4:** Heat Treating (1991); **A6:** Welding, Brazing, and Soldering (1993); **A7:** Powder Metallurgy (1984); **A8:** Mechanical Testing (1985); **A9:** Metallography and Microstructures (1985); **A10:** Materials Characterization (1986); **A11:** Failure Analysis and Prevention (1986); **A12:** Fractography (1987); **A13:** Corrosion (1987); **A14:** Forming and Forging (1988); **A15:** Casting (1988); **A16:** Machining (1989); **A17:** Nondestructive Testing and Quality Control (1989); **A18:** Friction, Lubrication, and Wear Technology (1992). **Metals Handbook, 9th Edition** (designated by the letter "M"): **M1:** Properties and Selection: Irons and Steels (1978); **M2:** Properties and Selection: Nonferrous Alloys and Pure Metals (1979); **M3:** Properties and Selection: Stainless Steels, Tool Materials and Special-Purpose Materials (1980); **M4:** Heat Treating (1981); **M5:** Surface Cleaning, Finishing, and Coating (1982); **M6:** Welding, Brazing, and Soldering (1983). **Engineered Materials Handbook** (designated by the letters "EM"): **EM1:** Composites (1987); **EM2:** Engineering Plastics (1988); **EM3:** Adhesives and Sealants (1990); **EM4:** Ceramics and Glasses (1991); **Electronic Materials Handbook** (designated by the letters "EL"): **EL1:** Packaging (1989).

Random homogeneous mixture (RHM) **EM4:** 96
Random intermittent welds
definition **A6:** 1212 **M6:** 14
Random mesh registration *See also* Registration
defined **EL1:** 1154
Random order
in experimental study **A17:** 743
Random orientation
in castings, methods to control **A9:** 701
defined **A9:** 15
determination of **A9:** 702
Random pattern
defined **EM1:** 2 **EM2:** 35
Random porosity
cast aluminum fracture surface **A12:** 66, 67
Random samples
defined **A10:** 12
Random sequence
definition **M6:** 14
Random variables
density function of **A8:** 624
statistical **A8:** 624
Random vibration environmental stress screening
parameters **EL1:** 878-880
Random wound
definition **M6:** 14
Randomization
defined **EM2:** 599-600
effect of **A8:** 640, 643-644
in fatigue data **A8:** 696
mechanical **A8:** 696
Randomized block experiments
analysis of **A8:** 640, 644, 646, 656-657
Randomized experimental designs **A8:** 643-650, 695
and block designs **A8:** 643-650
block plans **A8:** 644
complete **A8:** 643-644
incomplete blocks **A8:** 644-646
Randomly oriented surface
true area and length **A12:** 204
Randomness
in projected images **A12:** 194-195
of test observations **A8:** 624
Raney catalyst
powders used **A7:** 572
Range **EM3:** 791-793
sensing, machine vision **A17:** 43
setting, calibration for ultrasonic inspection **A17:** 267
Range of stress
defined **A8:** 11
symbol for **A11:** 797
Rank test
for comparing fatigue behavior **A8:** 707-709
Rank-ordering parameter
acceptability limit for **A13:** 316
Rankine-cycle steam systems
failures in **A11:** 602
Ranking
in computation for unknown distribution **A8:** 666
Rao-Raj deformation maps **A14:** 421
Raoult's law
defined **A15:** 51
Raoultian activity
defined **A15:** 51
Rap-jolt molding machines
green sand molding **A15:** 342-343
Rapeseed oil
as lubricant **A15:** 311
Rapi-Press composite consolidation
graphite-reinforced MMCs **EM1:** 87
Rapid burn-off zone
furnaces **A7:** 351
Rapid coalescence
intergranular fracture by **A12:** 349
Rapid cooling *See also* Cooling
grain refinement by **A15:** 476
Rapid filling *See also* Filling; Pouring; Risering
rates **A15:** 38
Rapid fracture
analysis, crack speed and control tendency of stress field as **A8:** 440-441
onset, toughness measurement as **A8:** 441
predicted by *R*-curves **A8:** 450
and stress-corrosion cracking **A8:** 498
Rapid loading
toughness control tests involving **A8:** 453
Rapid omnidirectional compaction **A7:** 542-546
and cold compaction **A7:** 544
and cold pressing/sintering, compared **A7:** 545
press requirements for cylindrical preforms **A7:** 545
process advantages and limitations **A7:** 545-546
Rapid omnidirectional consolidation
aluminum P/M alloys **A2:** 7, 203-204
Rapid oxidation *See also* Oxidation
aircraft powerplants **A13:** 1040
Rapid quenching *See also* Quenching
amorphous materials and metallic glasses **A2:** 809
beryllium powder **A2:** 685
metallic glasses **A2:** 805-806
neodymium-iron-boron permanent magnet materials **A2:** 791-792
titanium P/M products **A2:** 656
of uranium alloys **A2:** 673
Rapid solidification *See also* Solidification
Rapid run-arrest fracture
influence of dynamic effects **A8:** 285
Rapid solidification *See also* Rapid omnidirectional compaction; Rapid solidification rate; Solidification **A14:** 249
advantages for copper alloys **A9:** 640
for copier powders **A7:** 587
defined **A9:** 615
effect on microsegregation **A9:** 615-616
microsegregation in **A15:** 138
nucleation effect **A15:** 101
powders, in rapid omnidirectional compaction **A7:** 542
processing, eutectic growth in **A15:** 125
in rotating disk atomization **A7:** 45-46
solidification structures **A9:** 615-617
technology **A7:** 570
technology, Soviet **A7:** 692-693
in titanium powder production **A7:** 167
Rapid solidification (RS) alloys
high-strength aluminum P/M alloys **A2:** 200
wear-resistant **A2:** 205
Rapid solidification rate *See also* Rapid solidification; Solidification
Auger depth profile of **A7:** 255
microstructure **A7:** 47, 48
powder machine **A7:** 45, 46
technology **A7:** 255
Rapid solidification rate (RSR) **A1:** 973
Rapid solidification technology
P/M materials **A13:** 823
Rapid wear *See* Wear
Rapid-load plane-strain fracture toughness testing **A8:** 260-261
Rapid-*n* test
for sheet metals **A8:** 557
specimen **A8:** 557
Rapidly solidified powders
for copiers **A7:** 587
Rapping of pattern
as casting defect **A11:** 386
Rare earth aluminosilicate glasses
electrical properties **EM4:** 851
magnetic properties **EM4:** 854, 855
Rare earth compounds
as colorants **EM4:** 380
Rare earth dispersoids
in titanium **A7:** 167
Rare earth doped barium titanates **EM4:** 58
Rare earth elements
in metal powder-glass frit method **EM3:** 305
Rare earth elements in steel **M1:** 411
notch toughness, effect on **M1:** 694, 706-707
steel sheet, effect on formability **M1:** 556
Rare earth gallogermanate glasses
optical properties **EM4:** 854
Rare earth garnets **A9:** 538 **EM4:** 1162
Rare earth metal **A7:** 10
defined **A13:** 11
Rare earth metals *See also* Cerium; Dysprosium; Erbium; Europium; Gadolinium; Holmium; Lanthanum; Lutetium; Neodymium; Praseodymium; Promethium; Samarium; Terbium; Thulium; Ytterbium
as additions to steel to change sulfide inclusion morphology **A9:** 628
alloy additives **A2:** 727-729
alloy formation **A2:** 726-727
applications **A2:** 727-731
boiling points and sublimation energies **A2:** 723-724
cerium, properties **A2:** 720, 1178
chemical properties **A2:** 725
with cobalt, as permanent magnet materials **A2:** 787-788
compounds **A2:** 726
cuprates, high-temperature conductivity in **A2:** 1027
dysprosium, properties **A2:** 720, 1179
elastic and mechanical properties **A2:** 725
electronic configurations **A2:** 721-722
erbium, properties **A2:** 720, 1180
europium, properties **A2:** 720, 1180
gadolinium, properties **A2:** 720, 1181
holmium, properties **A2:** 720, 1181
hydrogen storage alloys **A2:** 731
immiscible liquids **A2:** 726-727
lanthanum, properties **A2:** 720, 1182
lighter flints **A2:** 729
in magnesium alloys **A9:** 427-428
magnetic material applications **A2:** 729-731
magnetic properties **A2:** 724-725
magnetooptical material applications **A2:** 731
melting/transformation temperatures **A2:** 723
metallography and surface passivation **A2:** 725
metals in magnets **A9:** 539
mischmetal, properties **A2:** 720, 1183
neodymium, properties **A2:** 720, 1184
oxide dispersion-strengthened (ODS) alloys **A2:** 729
physical properties **A2:** 721-725
praseodymium, properties **A2:** 720, 1185
preparation and purification **A2:** 720-721
promethium, properties **A2:** 720, 1185
research grade vs commercial grade **A2:** 720
samarium, properties **A2:** 720, 1186
scandium, properties **A2:** 720, 1186
structure, metallic radius, atomic volume and density **A2:** 722-723
terbium, properties **A2:** 720, 1187
thulium, properties **A2:** 720, 1188
ultrapure, by electrotransport purification **A2:** 1094-1095
ytterbium, properties **A2:** 720, 1188
yttrium, properties **A2:** 720, 1189
Rare earth orthoferrites **EM4:** 52-58
Rare earth oxide
effect on brake lining friction **A18:** 572
Rare earth oxides **EM4:** 56
Rare earths
copper oxide, for carbon monoxide, carbon dioxide removal **A10:** 222
as desulfurization reagent **A15:** 75
electron spin resonance and **A10:** 262
elements, laser-induced resonance ionization mass spectrometry for **A10:** 142
ESR analysis of **A10:** 262
fluoride separations **A10:** 169
as grain refiners, magnesium alloys **A15:** 481
ICP-MS analysis of **A10:** 40
oxygen produced by **A15:** 78
Rare-earth elements
effect of, on notch toughness **A1:** 742
in ferrite **A1:** 408
Rare-earth permanent magnets
bioapplication of **A13:** 1363
Rarefaction waves *See* Tensile waves
Rasberry-Heinrich model
calibration of x-ray spectrometry **A10:** 98
Raster pattern
SEM **A12:** 167
Rasters
with Auger spectrometer **A10:** 554
formation, and scan coils, SEM microscopes **A10:** 493
image analysis scanners **A10:** 310
Ratchet marks
defined **A8:** 11 **A11:** 8 **A12:** 112

Ratchet marks (continued)
in fatigue tests ... **A8:** 696
from multiple crack initiation ... **A11:** 77
martensitic stainless steels ... **A12:** 368
medium-carbon steels ... **A12:** 260, 269, 272
on shafts ... **A11:** 463
Ratcheting
creep, and stress relaxation, compared ... **A11:** 144
and distortion failure ... **A11:** 143-144
Rate of creep *See* Creep rate
Rate of flame propagation
definition ... **M6:** 14
Rate of oil flow ... **A7:** 10
Rate-dependent variables
affecting fatigue crack growth rate ... **A8:** 412
Rated tire load ... **A18:** 578
Rating and comparing structural adhesives ... **EM3:** 471-476
inverse skin-doubler coupon ... **EM3:** 471-472
rational design of adhesively bonded structural joints ... **EM3:** 474-476
Rating formability
for cold heading ... **A14:** 291
Rating life ... **A18:** 505
defined ... **A18:** 15
Ratio-analysis diagram (RAD) ... **A11:** 62, 797
Ratiometric instruments
MFS analysis ... **A10:** 77
Rational polynomial creep equation ... **A8:** 687-689
Rational sampling *See also* Samples: Sampling
common pitfalls ... **A17:** 729-730
concept ... **A17:** 729
importance ... **A17:** 728
size and frequency ... **A17:** 729
Rationalized erosion rate ... **A18:** 227-228
Rationalized incubation period ... **A18:** 227, 228
Rattail *See also* Buckle
as casting defect ... **A11:** 384
defined ... **A11:** 8 **A15:** 9
in iron castings ... **A11:** 353
Rattler test
for green strength ... **A7:** 288, 302
Raw materials *See also* Materials; Materials selection ... **EM4:** 43-51
acid steelmaking ... **A15:** 364
basic steelmaking ... **A15:** 360-367
for ductile iron ... **A15:** 647
for fiber-reinforced polymeric matrix composites ... **EM1:** 105-171
primary ceramics ... **EM4:** 49-51
alumina ... **EM4:** 49
metallurgical grade bauxite ... **EM4:** 49, 50
refractory grade bauxite ... **EM4:** 49
silica ... **EM4:** 49
zircon/zirconia ... **EM4:** 50-51
quality control in ... **EL1:** 869
selection of ceramics based on product applications ... **EM4:** 43-49
additives ... **EM4:** 44
advanced ceramics ... **EM4:** 45-49
antiscumming compounds ... **EM4:** 44
building clays ... **EM4:** 43-44
heavy clay products production, Europe ... **EM4:** 43
industrial and residential building products ... **EM4:** 43-44
refractories ... **EM4:** 44-45, 46
refractories, hazardous ... **EM4:** 45
whiteware ... **EM4:** 44, 45
validation, as quality control ... **EM1:** 72
wrought copper and copper alloy products ... **A2:** 241-242
Raw materials, industrial
sampling of ... **A10:** 12-18
Raw materials quality control ... **EM3:** 729-734
acceptance alternatives ... **EM3:** 733-734
alternative techniques ... **EM3:** 732
consistency controls ... **EM3:** 732
Raw materials quality control (continued)
shipping controls ... **EM3:** 732
product acquisition approach ... **EM3:** 730
product test phases ... **EM3:** 730-731
acceptance ... **EM3:** 730-731, 732-733
certification ... **EM3:** 730, 731, 732-733
design allowables ... **EM3:** 730
qualification test ... **EM3:** 730
raw material quality control ... **EM3:** 731
test techniques ... **EM3:** 731-732
flatwise tension ... **EM3:** 731-732
peel ... **EM3:** 732
shear ... **EM3:** 731
testing considerations ... **EM3:** 732-733
materials procedures, control ... **EM3:** 733
test temperatures ... **EM3:** 732-733
Raw materials/batching ... **EM4:** 378-384
batch formulation ... **EM4:** 681-682
batch sizes ... **EM4:** 382
collecting ... **EM4:** 384
cost/quality trade-offs ... **EM4:** 384
fining ... **EM4:** 384
melting ... **EM4:** 384
cullet ... **EM4:** 381
filling ... **EM4:** 384
important minerals ... **EM4:** 378-380
borate materials ... **EM4:** 380
feldspar ... **EM4:** 379
gypsum ... **EM4:** 380
lead oxides and silicates ... **EM4:** 380
limestone ... **EM4:** 379
materials and their purpose for use in glass manufacture ... **EM4:** 381
nepheline syenite ... **EM4:** 379
salt cake ... **EM4:** 380
silica sand ... **EM4:** 378-379
soda ash ... **EM4:** 379-380
sodium nitrate ... **EM4:** 380
minor materials ... **EM4:** 380-381
colorants ... **EM4:** 380-381
fining agents ... **EM4:** 380
melting accelerators ... **EM4:** 380
mixing ... **EM4:** 384
raw materials handling ... **EM4:** 382
segregation of materials during shipping, storage, and mixing ... **EM4:** 382-383
bin segregation ... **EM4:** 382-383
demixing ... **EM4:** 383
solid-solid mixing characteristics ... **EM4:** 382
weighing ... **EM4:** 384
wetting of glass batch ... **EM4:** 383-384
batch wetting with caustic soda ... **EM4:** 384
batch wetting with water ... **EM4:** 383-384
Ray diagram
for lensing action ... **A10:** 492
Rayleigh scattering ... **A18:** 409-410
and Compton scatter, x-rays ... **A10:** 85
defined ... **A10:** 127, 680
energy-level diagram ... **A10:** 127
from x-ray absorption ... **A10:** 84
in Raman spectroscopy ... **A10:** 126-128
Rayleigh scattering processes ... **EM4:** 853
Rayleigh step bearing ... **A18:** 523, 525
defined ... **A18:** 15
Rayleigh velocity ... **A18:** 407
Rayleigh waves
electromagnetic radiation ... **A17:** 309
in microwave inspection ... **A17:** 205
for optical holographic interferometry ... **A17:** 409
ultrasonic ... **A17:** 233-234
ultrasonic inspection ... **A17:** 233-234
Rayleigh-Ritz method
for elastic bending ... **A8:** 118
Rayon
as carbon fiber precursor ... **EM1:** 52, 11
Rayon production
powder used ... **A7:** 574
Razor blades
scanning laser gages for ... **A17:** 12
Rb-Sb (Phase Diagram) ... **A3:** 2•351
Rb-Se (Phase Diagram) ... **A3:** 2•351
Rb-Tl (Phase Diagram) ... **A3:** 2•351
Rb_2
single-crystal analysis of ... **A10:** 355
RBS *See* Rutherford backscattering spectrometry
RBSC *See* Reaction-bonded silicon carbide
RBSN *See* Reaction-bonded silicon-nitride
RDF *See* Radial distribution function analysis
RDS *See* Rheometric dynamic scanning
RDX (explosive)
friction coefficient data ... **A18:** 75
Re-pressing *See also* Hot re-pressing ... **A7:** 10 **M1:** 332, 339, 345, 346
aluminum and aluminum alloys ... **A2:** 6
aluminum P/M alloys ... **A2:** 211
defined ... **A14:** 11
forging, modes for ... **A7:** 414
powder forging ... **A14:** 190, 195
as secondary pressing operation ... **A7:** 337-338
Re-Ru (Phase Diagram) ... **A3:** 2•352
Re-Si (Phase Diagram) ... **A3:** 2•352
Re-Te (Phase Diagram) ... **A3:** 2•352
Re-U (Phase Diagram) ... **A3:** 2•353
Re-V (Phase Diagram) ... **A3:** 2•353
Re-X (GE)
seals to metals ... **EM4:** 499
Reacting hydrogen and oxygen catalysts
powder used ... **A7:** 572
Reaction bonding
of structural ceramics ... **A2:** 1020-1021
Reaction flux
definition ... **M6:** 14
Reaction flux (soldering)
definition ... **A6:** 1212
Reaction forming ... **EM4:** 124
Reaction hot pressing (RHP) ... **EM4:** 228
Reaction injection molding (RIM) *See also* Injection molding; Reinforced reaction injection molding (RRIM); Resin transfer molding (RTM)
application, automotive industry ... **EM1:** 833, 83
defined ... **EM1:** 2 **EM2:** 35
of epoxy composites ... **EM1:** 71-7
mechanical properties ... **EM2:** 261-263
for polyurethanes (PUR) ... **EM2:** 258-264
properties effects ... **EM2:** 287
size and shape effects ... **EM2:** 292
structural ... **EM1:** 56 **EM2:** 262
thermoset ... **EM2:** 320-321
as thermoset molding ... **EM1:** 55
types, and PUR properties ... **EM2:** 260-262
urethane hybrids ... **EM2:** 269
Reaction kinetic behavior
as quality control factor ... **EM1:** 730
Reaction kinetics
analyses ... **A10:** 109, 628
Reaction milling
aluminum P/M alloys ... **A2:** 202
Reaction products
identified ... **A10:** 212, 628
Reaction rates
equations ... **A13:** 67
linear oxidation ... **A13:** 66
logarithmic and inverse logarithmic ... **A13:** 66-67
Reaction sintering ... **A7:** 10 **EM4:** 291-295
additional reaction-bonding processes ... **EM4:** 293-295
chemical vapor infiltration ... **EM4:** 293-294
directed oxidation and nitridation processes ... **EM4:** 294
Oak Ridge National Laboratory process improvements ... **EM4:** 294
polymer-derived ceramics ... **EM4:** 294
self-propagating high-temperature synthesis ... **EM4:** 294
solid-solid exchange reactions ... **EM4:** 294
reaction-forming processes ... **EM4:** 291
Reaction soldering
definition ... **M6:** 14

SUBJECTS OF THE INDEXED VOLUMES: ASM Handbook (designated by the letter "A"): **A1:** Properties and Selection: Irons, Steels, and High-Performance Alloys (1990); **A2:** Properties and Selection: Nonferrous Alloys and Special-Purpose Materials (1990); **A3:** Alloy Phase Diagrams (1992); **A4:** Heat Treating (1991); **A6:** Welding, Brazing, and Soldering (1993); **A7:** Powder Metallurgy (1984); **A8:** Mechanical Testing (1985); **A9:** Metallography and Microstructures (1985); **A10:** Materials Characterization (1986); **A11:** Failure Analysis and Prevention (1986); **A12:** Fractography (1987); **A13:** Corrosion (1987); **A14:** Forming and Forging (1988); **A15:** Casting (1988); **A16:** Machining (1989); **A17:** Nondestructive Testing and Quality Control (1989); **A18:** Friction, Lubrication, and Wear Technology (1992). **Metals Handbook, 9th Edition** (designated by the letter "M"): **M1:** Properties and Selection: Irons and Steels (1978); **M2:** Properties and Selection: Nonferrous Alloys and Pure Metals (1979); **M3:** Properties and Selection: Stainless Steels, Tool Materials and Special-Purpose Materials (1980); **M4:** Heat Treating (1981); **M5:** Surface Cleaning, Finishing, and Coating (1982); **M6:** Welding, Brazing, and Soldering (1983). **Engineered Materials Handbook** (designated by the letters "EM"): **EM1:** Composites (1987); **EM2:** Engineering Plastics (1988); **EM3:** Adhesives and Sealants (1990); **EM4:** Ceramics and Glasses (1991); **Electronic Materials Handbook** (designated by the letters "EL"): **EL1:** Packaging (1989).

Reaction stress
definition **A6:** 1212 **M6:** 14
Reaction zone **A6:** 897
Reaction-based plastics
as binders **A15:** 211
Reaction-bonded alumina (RBAO)
infiltration **EM4:** 842
Reaction-bonded silicon carbide (RBSC) **EM4:** 239-240
applications **EM4:** 240, 964
wear **EM4:** 975
ceramic corrosion in the presence of combustion products **EM4:** 982
key features **EM4:** 676
mineral processing **EM4:** 961
mixing operations **EM4:** 98
properties **EM4:** 240, 677
reaction sintering **EM4:** 291, 293, 294
development **EM4:** 293
mechanical properties **EM4:** 293
processing **EM4:** 293
shapes **EM4:** 240
siliciding with SiO vapor **EM4:** 240
Reaction-bonded silicon nitride (RBSN)
applications, heat exchangers **EM4:** 984
ceramic corrosion in the presence of combustion products **EM4:** 982
compact thickness effect **EM4:** 237
control of reaction exotherm **EM4:** 238
corrosion resistance **EM4:** 238
erosion resistance **A18:** 205
fabrication **EM4:** 812
flexural strengths **EM4:** 237, 238
formation of shapes **EM4:** 237-238
green machining and cost-effectiveness of tool life **EM4:** 184
grinding **EM4:** 334
high-purity properties **EM4:** 237, 238
hot isostatically pressed **EM4:** 815, 818
influence of reaction variables **EM4:** 237
joining non-oxide ceramics **EM4:** 528
key features **EM4:** 676
made by a reaction-bonding technique **EM4:** 236-239
polysilazane infiltrants to fill voids **EM4:** 295
products **EM4:** 238
properties .... **EM4:** 330 , 677, 812, 815, 816, 187
reaction sintering **EM4:** 291-293
comparison of RBSN from silicon powders versus high-quality silicon powder from conventional processes **EM4:** 292
high-purity silicon powder processes **EM4:** 292-293
impurity levels and particle dimensions of silicon powders **EM4:** 292
silane gas source **EM4:** 292
rejected for turbocharger application **EM4:** 725, 726
strength retention **EM4:** 1001
strength-limiting defects **EM4:** 593
thermal properties **EM4:** 815-816, 818, 992
Reaction-bonded silicon-nitride (RBSN)- titanium nitride **EM4:** 292-293
Reaction-formed ceramics **EM4:** 236-240
Reaction-forming processes **EM4:** 236-240
reaction with a liquid phase **EM4:** 239-240
reaction-bonded silicon carbide **EM4:** 239-240
reaction with gas phases **EM4:** 236
reaction-bonded silicon nitride **EM4:** 236-239
compact thickness effect **EM4:** 237
control of reaction exotherm **EM4:** 238
formation of shapes **EM4:** 237-238
influence of reaction variables **EM4:** 237
products **EM4:** 238
properties **EM4:** 237, 238
silicon oxynitride **EM4:** 239
made by reaction-forming processes **EM4:** 239
properties **EM4:** 239
properties of ceramic bodies **EM4:** 239, 240
Reaction-infiltrated cermets
platelet reinforcement formation **A2:** 990
Reactive alloys, cleaning and finishing processes *See also* specific alloys by name **M5:** 650-668
Reactive contaminants **A7:** 178
Reactive diluent
defined **EM2:** 35
Reactive diluents **EM3:** 24, 94, 96
for epoxies **EM3:** 100
Reactive elements
mechanically alloyed oxide dispersion-strengthened (MA ODS) alloys **A2:** 943
Reactive evaporation vacuum coating **M5:** 391
Reactive gases
aluminum refining with **A15:** 80
Reactive hot pressing
pressure densification **EM4:** 300
Reactive ion plating **M5:** 418 **EM4:** 218
Reactive ion plating (RIP) process **A18:** 840, 844, 845-846
Reactive materials
metal powders for **A7:** 597
Reactive metals *See also* Hafnium; Titanium; Titanium alloys; Titanium and titanium alloys; Zirconium; Zirconium alloys **A16:** 844-857
abrasive cutoff sawing, Ti **A16:** 846
abrasive disk grinding, Ti **A16:** 846
band sawing, Ti **A16:** 851
boring, Zr **A16:** 852, 853, 856
chemical milling **A16:** 852, 853
chip formation **A16:** 853, 855, 856
chlorine-containing fluids **A16:** 845-846
climb milling, Ti **A16:** 845, 846, 848
climb milling, Zr **A16:** 854
continuous flow melting **A15:** 416
cutting fluids **A16:** 845, 850, 851, 853, 854, 855
cutting fluids, Ti **A16:** 846, 847, 848, 854, 856
cutting speeds **A16:** 845, 851
defined **A13:** 11
diffusion welding **A6:** 885
drilling **A16:** 845
drilling, Hf **A16:** 856
drilling, Ti **A16:** 846, 847, 848, 850, 851
drilling, Zr **A16:** 852, 854
ECM, Ti **A16:** 852
EDM, Zr **A16:** 852
electron beam drip melted **A15:** 413
electron-beam welding **A6:** 854, 855
face milling, Ti **A16:** 845, 846-847, 848, 850
fire hazard **A16:** 844, 846, 850, 851, 853, 854, 855, 856
friction welding **A6:** 890-891
galvanic corrosion **A13:** 85-86
gas-metal arc welding shielding gases **A6:** 66
gas-tungsten arc welding **A6:** 190
grinding, Ti **A16:** 844, 846, 847-851, 853, 854
grinding, Zr **A16:** 852, 854, 855, 856
hacksawing, Ti **A16:** 851, 854
hafnium as reactive metal **A16:** 844, 853, 855-857
health and safety, Hf **A16:** 856
health and safety, Zr **A16:** 855
high-frequency welding **A6:** 252
hydrogen fluoride/hydrofluoric acid corrosion **A13:** 1169
induction brazing **A6:** 947
laser beam machining, Ti **A16:** 852, 855
machinability **A16:** 844
machining guidelines, Ti **A16:** 844
milling, Hf **A16:** 856
milling, Ti **A16:** 845, 846-847
milling, Zr **A16:** 852, 853-854
peripheral end milling, Ti **A16:** 846-847, 848, 849
planing, Hf **A16:** 856
power band sawing, Ti **A16:** 846
power hacksawing, Ti **A16:** 846
rammed graphite molds of **A15:** 273
reaming **A16:** 845
reaming, Ti **A16:** 846, 847
reaming, Zr **A16:** 854
sawing, Zr **A16:** 852, 855
shaping, Hf **A16:** 856
surface finish **A16:** 844, 846, 851, 854, 856
tapping, Ti **A16:** 846, 847, 851
tapping, Zr **A16:** 854
threading, Ti **A16:** 851
titanium as reactive metal **A16:** 844-852, 853, 854
tool life, Ti **A16:** 846-847, 848, 851, 854, 856
turning, Hf **A16:** 856
turning, Ti **A16:** 845, 846, 847
turning, Zr **A16:** 852, 853, 856
ultrasonic welding **A6:** 894
Reactive metals and alloys
applications **M6:** 1049
Reactive metals and alloys (continued)
arc welding **M6:** 446-465
brazing **M6:** 1049-1054
applications **M6:** 1053-1054
atmospheres **M6:** 1053
base metals **M6:** 1049
fluxes **M6:** 1053
precleaning and surface preparation **M6:** 1052-1053
process and equipment **M6:** 1052
Reactive metals, brazing of **A6:** 941-947
alpha alloys **A6:** 944
alpha-beta alloys **A6:** 944
applications **A6:** 943
atmospheres **A6:** 941
beryllium alloys **A6:** 945, 947
beta alloys **A6:** 944
brazing procedures **A6:** 946-947
commercially pure alloys **A6:** 943-944
equipment **A6:** 947
filler metals **A6:** 941, 944, 945-946
fixturing **A6:** 946
health and safety rules **A6:** 946
joint designs **A6:** 946
physical properties **A6:** 941
restrictive titanium brazing attributes **A6:** 944
surface preparation **A6:** 946-947
titanium alloys **A6:** 943-944
torch brazing **A6:** 947
zirconium alloys **A6:** 944-945
Reactive milling **A7:** 64-65
Reactive plastisol adhesives
for body assembly **EM3:** 554
Reactive rubber adhesives **EM1:** 686-68
Reactive sputtered films for interference contrast layers **A9:** 59-60
Reactive sputtering
of interference films **A9:** 138, 148-150
Reactive sputtering (RS) **A18:** 840, 844
compounds and synthesized deposition rates **A18:** 848
Reactive sputtering system **M5:** 413-414
Reactive sputtering techniques **EM4:** 23
Reactivity
of titanium alloys **A15:** 824-825
Reactivity control applications **A7:** 666
Reactor
definition **M6:** 14
Reactor vessels
mining **A13:** 1297
Read camera
as glancing angle **A10:** 336
schematic **A10:** 336
Read-only memories (ROMs)
as circuit type **EL1:** 160
Readout instrumentation
eddy current inspection **A17:** 178-179
Readout methods
optical holographic interferometry **A17:** 413-414
Readout-type extensometers **A8:** 616
Reagent chemicals
defined **A10:** 680
Reagents *See also* Chemical reagents; Etchants
analytic methods for **A10:** 10
chemical, analytic methods for **A10:** 6-10
chemical, defined **A10:** 680
contaminated, controlling for **A10:** 12
defined **A10:** 680
effect on analyte extraction **A10:** 164
elimination by NAA use **A10:** 234
Karl Fischer **A10:** 204
process, AAS for trace impurities in **A10:** 43
pure, use in SSMS **A10:** 144
purity, in UV/VIS analysis **A10:** 68-69
recommended practices **A7:** 249
as solvent extractants **A10:** 170
suitability, classical wet analysis **A10:** 163
trace impurities analyzed **A10:** 31
use of excess, for UV/VIS interferences **A10:** 66
Reagents, corrosive
effects on tantalum **A13:** 734, 737-738
Real area of contact **A18:** 31, 40, 41, 343, 475
defined **A8:** 11 **A18:** 15
fretting wear **A18:** 245, 246, 247
Real leaks
defined **A17:** 57

Real time
in x-ray spectrometers **A10:** 92
Real-time
filmless x-ray image **EL1:** 369
process control **EL1:** 943
x-ray inspection **EL1:** 954
Real-time contouring
holographic **A17:** 427
Real-time imaging *See also* Radiography
media, radiography **A17:** 298
as neutron detection method **A17:** 391
Real-time inspection
radiographic **A11:** 17
Real-time interferometry *See also* Optical holography
as optical holographic interferometry **A17:** 408
Real-time laser confocal microscope (RLCM)
hardware configurations **A18:** 358
image acquisition **A18:** 359
Real-time radiography
advantages/limitations **A17:** 295, 321
applications **A17:** 321
background **A17:** 317-319
defined **A17:** 295, 317
digital radiography **A17:** 320
enlargement effect **A17:** 312
example **A17:** 321-323
and film radiography, compared **A17:** 295
with fluorescent screens **A17:** 319-320
image intensifiers **A17:** 318
image processing **A17:** 320-321
imaging system **A17:** 322
system layout and operating console **A17:** 323
system performance **A17:** 323
of weldments **A17:** 594
Real-time scanning optical microscope (RSOM)
hardware configurations **A18:** 358
Reamed and drifted pipe **M1:** 320-321
Reamers *See* Cutting tools
Reaming **A16:** 19, 239-248
aircraft engine components, surface finish requirements **A16:** 22
Al alloys **A16:** 766-767, 780, 782, 785
bushings and fixtures **A16:** 247-248
cast irons **A16:** 651, 655, 659, 660
compared to broaching **A16:** 194, 196, 201, 209
in conjunction with boring **A16:** 160, 162, 168
in conjunction with broaching **A16:** 195
in conjunction with drilling **A16:** 213-217, 219-223, 226, 228, 229
in conjunction with honing **A16:** 484
in conjunction with lapping **A16:** 494
in conjunction with tapping **A16:** 256, 261, 262
in conjunction with turning **A16:** 135
Cu alloys **A16:** 812-813, 814, 815
cutting fluids **A16:** 248
drilled hole size **A7:** 461
heat-resistant alloys **A16:** 750-752
in machining centers **A16:** 393
as machining process **A7:** 461
Mg alloys **A16:** 822, 823, 824-825
MMCs **A16:** 896, 899
multifunction machining **A16:** 366, 375, 379-380, 384
Ni alloys **A16:** 839
P/M high-speed tool steels **A16:** 65, 67
P/M materials **A16:** 889, 890, 891
and power feeding in chip removal operations **A16:** 33
process capabilities **A16:** 239
reamer design **A16:** 240-241
reamer materials **A16:** 240
reamers **A16:** 57, 58, 241-245, 253
recommended feeds **A7:** 461
refractory metals **A16:** 860, 862, 864, 865-867
selection of reamer **A16:** 241
special-purpose reamers, applications **A16:** 245
stainless steels **A16:** 690, 698, 702-703, 704, 705
step **A16:** 387
Reaming (continued)
surface alterations produced **A16:** 23
Ti alloys **A16:** 845, 846, 847, 852
tool adapters **A16:** 381
tool life **A16:** 239-244, 246, 247
tool steels **A16:** 719
in transfer machines **A16:** 395-397
workpiece material and hardness **A16:** 239
zirconium **A16:** 854
Zn alloys **A16:** 832-833
Rear surface mirror
for plate impact testing **A8:** 234
Réaumur process
for malleable iron **A15:** 31
Rebar protection
in breweries **A13:** 1224-1225
Rebar rolls
cemented carbide **A2:** 970
Reblending *See also* Blending
of molybdenum powders **A7:** 156
of tungsten powders **A7:** 154
Rebores, permissible
vertical centrifugal casting molds **A15:** 304
Rebound hardness test *See also* Dynamic hardness test **A8:** 71
Rebounding
effects in white iron **A12:** 239
Rebuilding, improper
tool steel cracking **A12:** 375
Recalescence
defined **A9:** 15
Recalibration, ultrasonic inspection
forgings **A17:** 506
Recarburizing
defined **A9:** 15
Recausticizing equipment
corrosion of **A13:** 1211-1213
Receiver-amplifier circuits
ultrasonic inspection **A17:** 253
Receiving inspection
magnetic particle inspection **A17:** 89
Receiving tubes
powders used **A7:** 574
Recess pressure
nomenclature for hydrostatic bearings with orifice or capillary restrictor **A18:** 92
Recessed heading tools
for upsetting with sliding dies **A14:** 90-91
Recesses
in coining **A14:** 185
Recessing
in conjunction with milling **A16:** 322
Cu alloys **A16:** 815-816
multifunction machining **A16:** 375
Recessing, deep
drop forming limits **A14:** 656
Reciprocal finishing **M5:** 133
Reciprocal lattice
defined **A9:** 15 **A10:** 680
Reciprocal lattice points
intensity distributions **A9:** 110-111
Reciprocal lattice points and double diffraction **A9:** 109
Reciprocal linear dispersion
defined **A10:** 680
Reciprocating engines
nickel alloy applications **A2:** 430
Reciprocating flow ion exchange recovery process
plating wastes **M5:** 317
Reciprocating screens **A7:** 177
Reciprocating screw injection molding machine **EM1:** 555-55
Reciprocating straight-line polishing and buffing machines **M5:** 121-122
Reciprocating-screw injection molding *See also* Injection molding
defined **EM2:** 35
Reciprocating-screw molding **EM2:** 35, 319-320
Reciprocation
and honing **A16:** 486, 491
and lapping **A16:** 497
Reciprocity failure
defined **A12:** 86
Reciprocity law or relation
in eddy current inspection modeling **A17:** 707
for radiographic exposure **A17:** 304
in ultrasonic inspection modeling **A17:** 704
x-ray tubes, radiography **A17:** 304
Recirculating bearing packs
coordinate measuring machines **A17:** 24
Recirculating steam generators **A13:** 937-938
Recirculation, and purification system
argon **A7:** 30
Reclaimed rubber **EM3:** 75, 86, 89
Reclaimed sand
for green sand molding **A15:** 225-226
Reclaimed scrap
for commercially pure titanium **A2:** 595
Reclamation *See also* Reclaimed sand; Sand reclamation
of chemically bonded sand **A15:** 351-354
of clay-bonded system sand **A15:** 354-355
effects, base sands **A15:** 355
thermal **A15:** 353-354
Recoil energy **A18:** 448
Recoil line *See* Impact line
Recoil-free fraction
as basis of Mössbauer spectroscopy **A10:** 287-288
Recombination-generation term/common emitter current gain
bipolar transistor analysis **EL1:** 152-153
Reconditioning
by metal spraying **A11:** 480
Reconfiguration, or configuration
of defective circuitry **EL1:** 9
Reconstructed glasses *See* Porous and reconstructed glasses
Reconstruction *See also* Reconstruction techniques
holographic **A17:** 407
LEED analysis **A10:** 536
in scanning acoustical holography **A17:** 440-441
surfaces, FIM/AP study of **A10:** 583
voxel, color image **A17:** 486
Reconstruction of bone
by orthopedic implants **A11:** 671
Reconstruction techniques
computed tomography (CT) **A17:** 379-382
direct Fourier **A17:** 380
filtered-backprojection **A17:** 380-382
iterative **A17:** 359, 382
projection data **A17:** 379-380
transform **A17:** 359
Recorder
analog-to-digital transient **A8:** 192
calibrator for **A8:** 618
drum-type *x-y* **A8:** 617
flatbed *x-y* **A8:** 617
load-elongation **A8:** 618
load-strain system **A8:** 617
Recording calipers **A7:** 229
Recording media
image conversion, radiographic inspection **A17:** 298
radiographic paper **A17:** 314-315
x-ray film **A17:** 314
xeroradiography **A17:** 315
Recording micrometer eyepieces **A7:** 229
Recording tapes
powder used **A7:** 573
Records
inspection **A11:** 134
photographic **A11:** 15-16
service of boilers and related equipment **A11:** 602
Recovery **EM3:** 24
after load removal **EM2:** 671-672
analysis of **A10:** 468-470

Recovery (continued)
defined ... **A9:** 15
effect on microstructure during creep ... **A8:** 305
thermal, and creep ... **A8:** 309
and work hardening, in creep ... **A8:** 301-302
Recovery, and recrystallization
in cold working ... **A7:** 61-62
Recovery boilers
wood pulp industry ... **A13:** 1198-1202
Recovery effect of annealing
effect on crystallographic texture ... **A9:** 700
Recovery hardness ... **A18:** 614, 615
Recovery in cold-worked metals during annealing ... **A9:** 692-694, 697
Recovery methods
of bismuth ... **A2:** 753-754
gallium ... **A2:** 742-744
for indium ... **A2:** 750
Recovery temperatures of aluminum alloys ... **A9:** 351
Recovery, waste *See* Waste recovery and treatment
Recreation and leisure
applications for stainless steels ... **A7:** 731
Recreation applications
acrylonitrile-butadiene-styrenes ... **EM2:** 111
blow molding ... **EM2:** 359
high-impact polystyrenes (PS HIPS) ... **EM2:** 194-195
of part design ... **EM2:** 616
polycarbonates (PC) ... **EM2:** 151
styrene-acrylonitriles (SAN, OSA ASA) ... **EM2:** 215
thermoplastic polyurethanes (TPUR) ... **EM2:** 205
urethane hybrids ... **EM2:** 268
Recreational equipment applications ... **EM1:** 845-84
Recrystallization *See also* Dynamic recrystallization; Postdynamic recrystallization; Static recrystallization ... **A6:** 163 **A7:** 10
in aluminum alloys, etchants for examination of ... **A9:** 355
annealing, for SCC control, copper/copper alloys ... **A13:** 615
cold-worked metals during annealing ... **A9:** 692, 694-697
and cooling, magnesium alloy forgings ... **A14:** 260
defined ... **A9:** 15 **A13:** 11 **A15:** 9
in duplex stainless steels ... **A9:** 286
effect of grain coarseness on ... **A9:** 697
effect on crystallographic texture ... **A9:** 700
in forging ... **A14:** 231
front, cementite particles pinning ... **A10:** 471
gold alloy ... **A9:** 562-563
line etching used to study ... **A9:** 62
of low-melting-point metals observed with optical microscopes ... **A9:** 82
measurement and analysis by crystallographic texture ... **A10:** 357-364
palladium ... **A9:** 563
palladium alloys ... **A9:** 564
and recovery, in cold working ... **A7:** 61-62
rolling temperatures above ... **A14:** 343
silver ... **A9:** 554-555
in situ ... **A9:** 694
solvent ... **EM2:** 773-774
structures, analysis of ... **A10:** 468-472
studied by x-ray topography ... **A10:** 365, 376
temperature, defined ... **A15:** 9
temperature, palladium ... **A2:** 716
textures, orientation relationships in ... **A10:** 358
tin and tin alloys as a result of working during specimen preparation ... **A9:** 449-450
titanium and titanium alloys ... **A9:** 460-461
tungsten ... **A2:** 562 **A9:** 442
in weldments ... **A13:** 344
in weldments, refractory metals and alloys ... **A2:** 563
zinc alloys ... **A9:** 489
zirconium ... **A2:** 666
Recrystallization annealing
defined ... **A9:** 15
wrought titanium alloys ... **A2:** 619, 620
Recrystallization controlled rolling (RCR) ... **A1:** 117, 120
Recrystallization kinetics of low-carbon steel
effect of penultimate grain size on ... **A9:** 697
Recrystallization nuclei, formation of
during recovery ... **A9:** 693
Recrystallization temperature
defined ... **A9:** 15
Recrystallization-anneal
cycle and microstructure ... **A6:** 510
Recrystallized grain size
defined ... **A9:** 15
Recrystallized grains
of single-phase palladium solid solution ... **A9:** 126
Recrystallized layers
in abraded zinc ... **A9:** 34, 37-38
Recrystallized structures
in magnesium alloys ... **A9:** 428-430
Recrystallized zone in ferrous alloy welded joints ... **A9:** 581
Rectangular chip resistors ... **EL1:** 178, 184
Rectangular plaque design
of penetrameters ... **A17:** 339
Rectangular pulses
mean current of a train of ... **A6:** 40
Rectangular wire
wrought copper and copper alloys ... **A2:** 252
Rectangular workpieces *See also* Shapes; Workpiece(s)
boxlike, drawing of ... **A14:** 586
Rectification
of alternating current ... **A17:** 91
Rectification systems ... **A6:** 41-42
Rectifiers ... **M6:** 470
for plasma arc welding ... **M6:** 216
silicon-controlled ... **M6:** 470
high-frequency welding ... **M6:** 764
stud arc welding ... **M6:** 732
Recuperative hot blast system
cupolas ... **A15:** 384
Recycling *See also* Aluminum recycling; Copper recycling; Electronic scrap recycling; Lead recycling; Magnesium recycling; Reclamation; Sand reclamation; Tin recycling; Titanium recycling; Zinc recycling ... **A1:** 1023-1033
of aluminum ... **A2:** 46, 1205-1213
aluminum-lithium alloys ... **A2:** 183-184
automobile scrap ... **A2:** 1211-1213
of carbon fiber scrap ... **EM1:** 153-15
copier powders ... **A7:** 582
of copper ... **A2:** 1213-1216
definition of ... **A1:** 1023
electronic scrap ... **A2:** 1228-1231
fluid dies ... **A7:** 543
gas, plasma melting/casting ... **A15:** 424
of home scrap ... **A1:** 1023
inclusions from ... **A15:** 95
of iron and steel scrap ... **A1:** 1023
factors influencing scrap demand ... **A1:** 1024
purchased scrap supply ... **A1:** 1024-1026
scrap use by industry ... **A1:** 1023-1024
of lead ... **A2:** 543, 1221-1223
of magnesium ... **A2:** 1216-1218
in materials engineering ... **M3:** 826, 829
of nonferrous alloys ... **A2:** 1205-1232
powders in magnetic separation ... **A7:** 591
and process selection ... **EM2:** 278
sand, green sand molding ... **A15:** 225-226
scrap processor ... **A1:** 1026
blending ... **A1:** 1027
collection ... **A1:** 1025, 1026
detinning ... **A1:** 1026-1027
incineration ... **A1:** 1027
separation and sorting ... **A1:** 1026
size reduction and compaction ... **A1:** 1026, 1027
of stainless steel and superalloy
blending ... **A1:** 1032
collection ... **A1:** 1029
degreasing ... **A1:** 1032
metallurgical wastes ... **A1:** 1032
processing stainless steel and superalloy scrap ... **A1:** 1028-1032
scrap ... **A1:** 1027-1028
scrap demand ... **A1:** 1028
scrap use by industry ... **A1:** 1028
secondary nickel refining ... **A1:** 1032
separation ... **A1:** 1028, 1029-1032
size reduction and compaction ... **A1:** 1032
of tin ... **A2:** 517, 1218-1220
of titanium ... **A2:** 595, 639, 1226-1228
titanium, by electron beam melting ... **A15:** 410
titanium scrap ... **A2:** 595, 639
Recycling (continued)
uranium scrap ... **A2:** 2903
of zinc ... **A2:** 1223-1226
Recycling of shipping containers
adhesives and sealants ... **EM3:** 693
Red brass *See also* Copper and copper alloys
for dezincification ... **A13:** 614
examination of lead particles ... **A9:** 400
ultimate shear stress ... **A8:** 148
Red brass (85% Cu; 9% Sm; 6% Zn)
thermal properties ... **A18:** 42
Red brass, 85%, microstructure of ... **A3:** 1•22
Red brasses *See also* Brasses; Leaded red brasses; Wrought coppers and copper alloys
applications and properties ... **A2:** 298-299
brazing ... **A6:** 629
dimensional tolerances ... **A2:** 350-351
gas-metal arc welding ... **A6:** 763
as low-shrinkage alloy ... **A2:** 346
properties and applications ... **A2:** 225
recycling ... **A2:** 1213
resistance spot welding ... **A6:** 850
salt-bath dip brazing ... **A6:** 922
weldability ... **A6:** 753
Red copper coloring solutions ... **M5:** 625
Red fuming nitric acid
titanium/titanium alloy SCC ... **A13:** 687
Red golds *See* Gold-silver-copper alloys
Red lead
as rust-inhibitor ... **A2:** 548
Red lead (Pb_3O_4)
purpose for use in glass manufacture ... **EM4:** 381
Red, methyl
as acid-base indicator ... **A10:** 172
Red mud *See* Cocoa
Red phosphorus
for flame retardance ... **EM3:** 179
Red rust *See* Corrosion products
Red shift
defined ... **A10:** 680
Red tracer material
as pyrotechnic application ... **A7:** 603
Reddish bronze-to-dark brown copper coloring solution ... **M5:** 625
Redox *See also* Oxidation; Oxidation-reduction (redox); Reduction
completeness of reaction ... **A10:** 164
endpoint detection ... **A10:** 164
titrations ... **A10:** 174-176
Redox number ... **EM4:** 381
Redox potential ... **A6:** 585
defined ... **A13:** 11
Redox reaction
for acrylics ... **EM3:** 119, 120
Redrawing
defined ... **A14:** 10
direct ... **A14:** 584-585
of magnesium alloys ... **A14:** 828
reverse ... **A14:** 585
Reduced cobalt powders *See also* Cobalt powders
properties ... **A7:** 145-146
Reduced gage section
cylindrical compression specimen with ... **A8:** 589
Reduced iron powders *See also* Iron powder alloys; Iron powders; Welding
apparent density ... **A7:** 297
business machine parts ... **A7:** 669
comminution processes ... **A7:** 615
for copier powders ... **A7:** 587
effect of tapping on loose powder density ... **A7:** 297
for food enrichment ... **A7:** 614-615
for welding ... **A7:** 817-818
Reduced lamina stiffness matrix
defined ... **EM1:** 21
Reduced oxide removal
heat-resistant alloys ... **M5:** 564
Reduced-oxide surfaces, nickel alloys
cleaning processes ... **M5:** 671-672
Reduced-pressure test
for hydrogen measurement ... **A15:** 457-458
Reducer rolls ... **A14:** 96-97
Reducing
atmosphere, early smelting ... **A15:** 15
fluxes, copper alloys ... **A15:** 449
heat, basic steelmaking ... **A15:** 366-367

Reducing acids
titanium/titanium alloy resistance **A13:** 678-679
Reducing agents **A13:** 11, 56, 1140
mixing with oxidizing agents **A9:** 69
precipitation by **A10:** 169
Reducing atmosphere
definition **A6:** 1212 **M6:** 14
Reducing atmospheres
heating-element materials **A2:** 834-835
Reducing flame **A6:** 352
definition **A6:** 1212 **M6:** 14
Reducing salt bath descaling **M5:** 97, 100
Reducing sections
welding of pipe **M6:** 591
Reduction *See also* Area reduction; Oxidation; Oxides; Reduction of copper oxides; Strain; Strain rate; Upset reduction
aramid fibers **EM3:** 285
in area, defined **A14:** 10
by isothermal forging **A14:** 150
calculation, for rotary swaging **A14:** 129-130
cold, by swaging **A14:** 129
of copper oxide **A7:** 105-111
defined **A10:** 680 **A11:** 8 **A13:** 11 **A14:** 10
of drawn shells **A14:** 586
effect, rotary swaging **A14:** 139
equipment **A7:** 65-70
finish-forging, nickel-base alloys **A14:** 262-263
forged low-alloy steel powders **A7:** 470
as forging factor **A14:** 231
forging ratio effects **A14:** 218
for full density nickel strip **A7:** 401
furnace, powder bed cross section **A7:** 155
of hydrodynamic intensity **A11:** 170
of iron oxide, ancient **A7:** 14
-oxidation reactions, classical wet chemistry analysis of **A10:** 163-164
of oxide **A7:** 10
of P/M and wrought titanium and alloys **A7:** 475
polarographic **A10:** 191
potential **A7:** 54
in precision forging **A14:** 159
rate, as forging factor **A14:** 231
ratio **A7:** 10
reversible, completeness of reaction as function of potential for **A10:** 209
of species soluble in solution **A10:** 208
of stress **A11:** 115
in tube swaging **A14:** 135-136
of tungsten powder, and particle growth **A7:** 153
uniform, in rolling **A14:** 344
upset, ductility effects **A14:** 219
in voltammetry **A10:** 189
Reduction (or Stockholm) convention
of electrode potentials **A13:** 21-22
Reduction and oxidation during electrochemical etching **A9:** 60-61
Reduction in area *See also* Ductility
at fracture, as ductility measure **A8:** 22
carbon and alloy steels, selected grades **M1:** 680
and cold rolling reduction in tension testing **A8:** 595-596
as creep-rupture, analysis of **A8:** 693
defined **A8:** 11
and elongation **A8:** 27
fatigue resistance, calculation from **M1:** 678
from hot tension test **A8:** 586
measurement, uniaxial tensile testing **A8:** 555
steel wire fabrication, relation to **M1:** 587-588, 593
as structure sensitive ductility parameter **A8:** 20, 27
symbols and unit **A8:** 662
vs. test temperature for Unitemp HN by on cooling testing **A8:** 587
Reduction in area (continued)
vs. test temperature from on heating testing **A8:** 586-587
and workability rating scale **A8:** 586
Reduction in forging **M1:** 353-354
Reduction of area
defined **A11:** 8
Reduction of area (RA) **EM3:** 24
defined **EM2:** 35
Reduction of copper oxide **A7:** 105-111
finished powders **A7:** 110
processing steps **A7:** 105
properties .... **A7:** 110, ill
of superalloys **A7:** 473
Reduction potentials, standard
magnesium/magnesium alloys **A13:** 740
Reduction ratio
at centerburst fracture in aluminum alloys **A8:** 577-578
Reductor
Jones **A10:** 175
Redundancy
in in-service monitoring **A13:** 202
in wafer scale integration **EL1:** 268
Redwood modulus of elasticity **A18:** 539
symbol and units **A18:** 544
Redwood viscosity
defined **A18:** 15
Reed wire **A1:** 852
Reed's vortex stabilization technique
use in ICP-AES **A10:** 32
Reentrant
abbreviation for **A10:** 691
Reference *See also* Reference blocks; Reference standards; Reference waves Standard reference blocks; Test blocks
discontinuities, eddy current inspection **A17:** 179
radiographs **A17:** 347
Reference block technique
ultrasonic inspection **A17:** 262-265
Reference blocks
standard **A17:** 263-265
Reference edge
defined **EL1:** 1155
Reference electrode *See also* Calomel electrode; Electrodes
anodic protection **A13:** 464
copper-saturated copper sulfate **A13:** 468
defined **A13:** 11
liquid junction potential **A13:** 23-24
low polarizability **A13:** 23
for marine corrosion **A13:** 921
operating conditions **A13:** 24
potential measurements with **A13:** 21-24
schematic **A13:** 23
Reference electrodes
defined **A10:** 680
for ISE **A10:** 185
schematic of **A10:** 186
Reference, frames of *See* Frames of reference
Reference gaging
visual **A17:** 10-11
Reference materials *See also* Controls
atmospheric-corrosion tests **A13:** 204
defined **A10:** 680
Reference plate
ultrasonic inspection **A17:** 264, 267
Reference resistors
of electrical resistance alloys **A2:** 823
Reference samples **A7:** 258
eddy current inspection **A17:** 179
Reference signal
microwave inspection **A17:** 205
Reference sphere, in stereographic projection
grain orientation **A10:** 358
Reference standards
adhesive-bonded joints **A17:** 634-636
durability **A17:** 677
Reference standards (continued)
in NDE engineering **A17:** 676-677
for quantitative defect evaluation **A17:** 676
requirements **A17:** 677
thermal inspection **A17:** 401
Reference stress
in fracture mechanics **A11:** 50-51
Reference tables
for thermocouples **A2:** 881
Reference temperature **A6:** 1132
Reference wave
holography **A17:** 224
microwave holography **A17:** 207
modulation as **A17:** 218
Refined soft lead
compositions **A2:** 544
Refinement *See also* Grain refinement; Refining effects, aluminum-silicon alloys **A15:** 753
of hypereutectic aluminum-silicon alloys **A15:** 753
and modification **A15:** 753
Refinery tubing
ASTM specifications for **M1:** 323
Refining *See also* Grain refinement; Refinement
in acid steelmaking **A15:** 364
aluminum, by evaporation treatment **A15:** 80
beryllium **A2:** 684
by continuous flow melting **A15:** 415-416
by electroslag remelting **A15:** 401
cold hearth, electron beam **A15:** 414-415
electrochemical, of aluminum melts **A15:** 80-81
filter bed **A15:** 471
final, lead alloys **A15:** 476
fire, effect on molten impurities **A15:** 450-451
fluxes **A15:** 446, 448
as function of circulation rate, VD **A15:** 433
gas injection, principles **A15:** 470-471
hafnium **A2:** 662
of heat, basic steelmaking **A15:** 366
of lead **A2:** 543 **A15:** 474-476
melt, copper alloys **A15:** 449
plasma **A15:** 710
secondary , argon oxygen decarburization as **A15:** 426-429
secondary, degassing procedures as **A15:** 426
zirconium **A2:** 662
Refining, and melting
Domfer process **A7:** 89-91
Refining fluxes
aluminum alloys **A15:** 446
magnesium alloys **A15:** 448
Refining industry
alloy steel corrosion in **A13:** 535
nickel-base alloy applications **A13:** 655-656
Reflectance
-absorption attachment, IRRAS spectroscopy **A10:** 114
defined **A10:** 680
specular infrared, for chemical surface studies **A10:** 177
Reflectance systems
for solder joint inspection **EL1:** 736-737
Reflected beam ultrasonic inspection **A17:** 248
Reflected light microscopy
for microstructural analysis **EM4:** 578
optical etching **A9:** 58-59
Reflected pulse
in torsional Kolsky bar dynamic test **A8:** 228-229
Reflected wave, and transmitted wave
timing between **A8:** 203
Reflected-light meter
for photomacrography **A12:** 85
Reflected-light optical micrograph
of polyethylene pipe fracture surface **A11:** 761
Reflected-light techniques for microscopic examination of lead and lead alloys **A9:** 416
Reflection
back, pulse-echo ultrasonic inspection **A17:** 246

SUBJECTS OF THE INDEXED VOLUMES: ASM Handbook (designated by the letter "A"): **A1:** Properties and Selection: Irons, Steels, and High-Performance Alloys (1990); **A2:** Properties and Selection: Nonferrous Alloys and Special-Purpose Materials (1990); **A3:** Alloy Phase Diagrams (1992); **A4:** Heat Treating (1991); **A6:** Welding, Brazing, and Soldering (1993); **A7:** Powder Metallurgy (1984); **A8:** Mechanical Testing (1985); **A9:** Metallography and Microstructures (1985); **A10:** Materials Characterization (1986); **A11:** Failure Analysis and Prevention (1986); **A12:** Fractography (1987); **A13:** Corrosion (1987); **A14:** Forming and Forging (1988); **A15:** Casting (1988); **A16:** Machining (1989); **A17:** Nondestructive Testing and Quality Control (1989); **A18:** Friction, Lubrication, and Wear Technology (1992). **Metals Handbook, 9th Edition** (designated by the letter "M"): **M1:** Properties and Selection: Irons and Steels (1978); **M2:** Properties and Selection: Nonferrous Alloys and Pure Metals (1979); **M3:** Properties and Selection: Stainless Steels, Tool Materials and Special-Purpose Materials (1980); **M4:** Heat Treating (1981); **M5:** Surface Cleaning, Finishing, and Coating (1982); **M6:** Welding, Brazing, and Soldering (1983). **Engineered Materials Handbook** (designated by the letters "EM"): **EM1:** Composites (1987); **EM2:** Engineering Plastics (1988); **EM3:** Adhesives and Sealants (1990); **EM4:** Ceramics and Glasses (1991); **Electronic Materials Handbook** (designated by the letters "EL"): **EL1:** Packaging (1989).

Reflection (continued)
defined ... **A17:** 204
degree, in ultrasonic inspection ... **A17:** 231
field, leaky Lamb wave testing ... **A17:** 252
laws, of microwaves ... **A17:** 204
in signal transmission ... **EL1:** 170-172
specular, effect in optical holography ... **A17:** 412
spurious, ultrasonic inspection ... **A17:** 246
theoretical indications ... **A17:** 213
in VHSIC technology ... **EL1:** 76
wave theory of ... **A10:** 83
Reflection (x-ray) *See* Diffraction
Reflection density
defined, radiography ... **A17:** 314
x-ray film ... **A17:** 323
Reflection electron diffraction ... **A18:** 387
Reflection EXAFS detection technique ... **A10:** 418
Reflection grating
defined ... **A10:** 680
Reflection high-energy electron diffraction
acronym ... **A10:** 689
with Auger electron spectroscopy ... **A10:** 554
capabilities ... **A10:** 536
effect of grazing incidence ... **A10:** 540
instrumentation ... **A10:** 539-540
introduction ... **A10:** 537
measurements ... **A10:** 539-540
Reflection high-energy electron diffraction (RHEED) ... **EM3:** 237
Reflection holographic systems
portable ... **A17:** 420-421
Reflection method
defined ... **A9:** 15
Reflection optical technique ... **A6:** 145
Reflection polariscope
for stress analysis ... **A17:** 451
Reflection spectrometers
FMR ... **A10:** 270
Reflection technique, thick-sample
for pole determination ... **A10:** 360
Reflection techniques
eddy current inspection ... **A17:** 183-184
fixed-frequency continuous-wave ... **A17:** 206
pulse modulated ... **A17:** 206
swept-frequency continuous-wave ... **A17:** 206
Reflection topography
applications ... **A10:** 366
asymmetric ... **A10:** 371
Bragg case ... **A10:** 366
camera for ... **A10:** 369
Laue case ... **A10:** 366
polychromatic and monochromatic ... **A10:** 366
Schulz and Berg-Barrett methods of ... **A10:** 368-369
Reflective (white) light ... **EL1:** 570
Reflectivity
of fracture surfaces, and photolighting ... **A12:** 84-88
of silver, atomic oxygen effects ... **A12:** 481
Reflectometers
continuous wave, microwave inspection ... **A17:** 212-213
frequency modulated ... **A17:** 213-214
microwave ... **A17:** 206
for thickness gaging ... **A17:** 211
Reflectors
as aluminum alloy application ... **A2:** 14
Reflex klystrons, defined
microwave inspection ... **A17:** 208
Reflow
Hot-gas, equipment ... **EL1:** 727-728
of leaded and leadless surface-mount joints ... **EL1:** 733
Reflow soldering
definition ... **A6:** 1212 **M6:** 14
Reflow soldering n design ... **EL1:** 520
flexible printed boards ... **EL1:** 590
infrared, outer lead bonding ... **EL1:** 286
of passive components ... **EL1:** 180
through-hole soldering ... **EL1:** 693-694
Reflowed solder plate
as preservation ... **EL1:** 563
Reflowing
definition ... **A6:** 1212
Reflux
defined ... **A10:** 680
Reformation
3-D surface, defined ... **A17:** 384
planar, defined ... **A17:** 384
Refractaloy 26
composition ... **A4:** 794 **A16:** 736
contour band sawing ... **A16:** 363
machining ... **A16:** 738, 741-743, 746-747, 749-757
Refraction
wave theory of ... **A10:** 83
Refraction laws
of microwaves ... **A17:** 204
Refractive index *See also* Optical properties ... **A10:** 680, 690
of a lens and focal length ... **A9:** 75
defined ... **A9:** 15
glass fibers ... **EM1:** 4
of interference films, role in interference effect ... **A9:** 136
optical testing ... **EM2:** 595-596
to determine concentration of microemulsions and micellar solutions ... **A18:** 144
Refractive indices of interference layers ... **A9:** 60
Refractometer
for emulsion concentration ... **A14:** 516
Refractories *See also* Monolithic and Fibrous Refractories; Refractory; Refractory metals ... **EM4:** 895-908
applications ... **EM4:** 985, 899-908
backup, Shaw process ... **A15:** 249
basic, for basic slag ... **A15:** 386
basic oxygen process furnace applications ... **EM4:** 896
ceramic shell mold, for investment casting ... **A15:** 258
chemical analysis of some typical raw materials ... **EM4:** 896
for crucibles ... **A15:** 31
definitions ... **EM4:** 895
deposit, reverberatory furnace ... **A15:** 378
fibrous ... **EM4:** 910-917
as inclusion forming ... **A15:** 90, 488
investment casting, compositions and properties ... **A15:** 258
investment casting, linear thermal expansion ... **A15:** 259
mixing operations ... **EM4:** 98
monolithic ... **EM4:** 910-917
physical properties ... **EM4:** 896-899
properties ... **EM4:** 896-900
as slag, compositions ... **A15:** 358
testing ... **EM4:** 547
types ... **EM4:** 895-896
for VIM crucible linings ... **A15:** 394
Refractories Institute, The (TRI)
as information source ... **EM1:** 41
Refractory *See also* Refractories
control, of inclusions ... **A15:** 90
defined ... **A15:** 9
Refractory alloy TZM
surface alterations from material removal processes ... **A16:** 27
Refractory alloys *See* Refractory metals and alloys; Refractory metals and alloys, specific types; specific refractory metals
liquid-metal embrittlement of ... **A11:** 234
susceptibility to hydrogen damage ... **A11:** 250, 338
susceptibility to LME ... **A11:** 234
Refractory base metals
ultrasonic welding ... **A6:** 894
Refractory brick inclusions in steel ... **A9:** 185
Refractory cermets *See also* Boride cermets; Carbide cermets ... **A7:** 813-814
carbonitride - and nitride-based cermets ... **A2:** 1004-1005
diamond-containing cermets ... **A2:** 1005
graphite-containing cermets ... **A2:** 1005
silicide cermets ... **A2:** 1005
Refractory coating inclusions
as casting defect ... **A11:** 387
Refractory coatings on cemented carbides
preparation for examination ... **A9:** 273
Refractory compounds
nitrous oxide-acetylene flame atomizer for ... **A10:** 48
Refractory fillers ... **EM4:** 1072
Refractory heavy metal (RHM) impurities
in sandstone deposits ... **EM4:** 378
Refractory inclusions
in rolling ... **A14:** 358
Refractory lining
cupolas ... **A15:** 386
Refractory linings
nonmetallic ... **A11:** 316
Refractory materials ... **EM4:** 32
applications ... **EM4:** 14
basic ... **EM4:** 13-14
composition ... **EM4:** 14
properties ... **EM4:** 14
chrome magnesite ... **EM4:** 13, 14
chromite ... **EM4:** 13, 14
classes ... **EM4:** 13
composition ... **EM4:** 14
deposition methods ... **EM4:** 202
fireclay ... **EM4:** 13, 14
high alumina ... **EM4:** 13, 14
high-duty oxides ... **EM4:** 13, 14
composition ... **EM4:** 14
properties ... **EM4:** 14
magnesite ... **EM4:** 13, 14
mullite ... **EM4:** 13, 14
properties ... **EM4:** 12, 13, 14
silica brick ... **EM4:** 13, 14
spalling ... **EM4:** 13
Refractory materials, high-silica
dissolution medium for ... **A10:** 165
Refractory metal ... **A13:** 11, 179, 1169
Refractory metal alloys
applications ... **A7:** 765
mechanical properties ... **A7:** 469
Refractory metal and oxide vacuum coating ... **M5:** 388-392, 400-401
Refractory metal carbides
properties of ... **A2:** 952
Refractory metal fiber reinforced composites
applications ... **A2:** 583-584
mechanical and thermal properties ... **A2:** 583
processing ... **A2:** 583
wires as reinforcement materials ... **A2:** 582
Refractory metals *See also* Molybdenum; Molybdenum and molybdenum alloys; Niobium; Niobium and niobium alloys; Refractories; Refractory cermets; Refractory metal alloys; specific refractory metals; Tantalum; Tantalum and tantalum alloysTungsten and tungsten alloys; Tungsten ... **A9:** 439-446 **A16:** 858-869
abrasive blasting ... **M5:** 652-653, 659, 663, 667
abrasive cutoff sawing ... **A16:** 867, 868
abrasive sawing ... **A16:** 867, 868
abrasives ... **A9:** 439
as additions to nickel-base heat-resistant casting alloys ... **A9:** 334
applications ... **A7:** 765
bars, mechanical properties ... **A7:** 469
-based composite structures infiltration ... **A7:** 555
batch-type vacuum sintering furnaces for ... **A7:** 357
blank forming ... **A14:** 788
boring ... **A16:** 162, 859, 862, 863
brazes used in ceramic/metal joining ... **EM4:** 479
brazing ... **A6:** 623
applications ... **A6:** 634
and soldering characteristics ... **A6:** 634-635
ceramic coating of ... **M5:** 532-533, 535, 537, 542
chemical blanking ... **A16:** 868
chemical machining ... **A16:** 868
chemical milling ... **A16:** 859
chucking and fixturing ... **A16:** 859
circular sawing ... **A16:** 867, 868
cleaning processes ... **M5:** 650-656, 659, 662-663, 667
climb milling ... **A16:** 861, 867
coating for dies ... **A18:** 644
for cofiring the metallization with the alumina ... **EM4:** 544-545
of commercial interest ... **A7:** 152
consolidation by hot isostatic pressing ... **A7:** 441-442
continuous flow melting ... **A15:** 416
for core coatings ... **A15:** 240
counterboring ... **A16:** 860, 863
creep rupture testing of ... **A8:** 302-303
cutting fluids ... **A16:** 125, 860-861, 863, 865, 867-869
defined ... **A7:** 10
descaling of ... **M5:** 650-654, 659, 662-663, 667
diffusion welding ... **A6:** 885
drilling ... **A16:** 860, 861, 863-865
ductile-to-brittle temperature ... **A6:** 634

Refractory metals (continued)
edge retention **A16:** 29
electrical and magnetic applications **A7:** 626-629
electrical discharge machining **A16:** 859, 865, 868
electrochemical machining **A16:** 868
electromechanical polishing of **A9:** 42-43, 45
electron beam drip melted **A15:** 413
electron-beam welding **A6:** 851, 855, 869-871
electroplating of **M5:** 658-661, 663-664, 668
end milling-slotting **A16:** 866, 867
enhanced sintering with **A7:** 317
etchants **A9:** 440
exothermic brazing **A6:** 345
face milling **A16:** 859, 865, 867
-faced welding electrodes **A7:** 629
field evaporation of **A10:** 586
finishing processes **M5:** 656-668
fire hazards **A16:** 862
forging characteristics **A14:** 237
forming of **A14:** 519, 785-788
friction welding **A6:** 890-891
for gas-lubricated bearings **A18:** 532
gas-tungsten arc welding **A6:** 192
grain-boundary embrittlement in, IAP studies **A10:** 599-600
grinding **A9:** 439 **A16:** 861, 862, 868-869
grinding parameters **A16:** 868-869
grinding ratios **A16:** 868-869
high-speed tool steels used **A16:** 59
as high-temperature materials **A7:** 765
hollow milling **A16:** 862, 867
honing **A16:** 477
induction brazing **A6:** 947
interstitial element contamination **A6:** 870
low-stress grinding procedures **A16:** 28
lubricants for **A14:** 519
lubrication of tool steels **A18:** 738
mechanical properties **A7:** 469, 476-478
microstructures **A9:** 441-442
milling **A16:** 860, 861, 867
molten salt bath descaling of **M5:** 653-654, 659
molybdenum machining **A16:** 858, 859, 860
mounting **A9:** 439
niobium machining **A16:** 858, 859, 860
nitric/hydrofluoric acid as dissolution medium **A10:** 166
oil hole or pressurized-coolant drilling **A16:** 861, 863, 865
oxidation protective coatings for **M5:** 375-376, 379-380
oxidation-resistant coatings, high-temperature **M5:** 661-662, 664-665
P/M history **A7:** 17
P/M process plus forging and rolling **A7:** 522
peripheral end milling **A16:** 865, 866, 867
pickling of **M5:** 654-655, 659, 663
polishing **A9:** 439-440
polishing and buffing **M5:** 655-656
powder or briquet sample preparation **A10:** 93
power band sawing **A16:** 867
power hacksawing **A16:** 867
pure, mechanical and physical properties **A7:** 766
radiation-damaged, FIM studies of **A10:** 588
reaming **A16:** 860, 862, 864, 865-867
resistance to corrosive media **A7:** 766
sectioning **A9:** 439
shaping **A16:** 190
sintering **A7:** 317, 389-393
sodium peroxide fusion **A10:** 167
solid-state bonding in joining non-oxide ceramics **EM4:** 525, 526
spade drilling **A16:** 861, 863, 864, 865
specimen preparation **A9:** 439
spotfacing **A16:** 860, 863
tantalum machining **A16:** 858, 859, 860
tapping **A16:** 860, 862, 865-867
techniques **A7:** 18, 522
thread milling **A16:** 867

Refractory metals (continued)
tool geometry **A16:** 859
tool life **A16:** 859
tools **A16:** 859-860, 861, 865
trepanning **A16:** 860, 862, 863
tungsten machining **A16:** 858, 859, 860
turning **A16:** 858-859, 861, 862-863
upset welding **A6:** 249
vibratory compacting **A7:** 306
wire EDM **A16:** 859

Refractory metals and alloys *See also* Refractory metals and alloys, specific types; specific metal by name **A2:** 557-585
applications **A2:** 557-560 **M3:** 314-315 **M6:** 1054
arc welding **M6:** 446-465
brazing **M6:** 1054-1060
cleaning methods **M6:** 1057
joining dissimilar metals **M6:** 1060
processes **M6:** 1055
and carbide-base composites, for electrical make-break contacts **A2:** 854-855
characteristics and properties **M6:** 1055
cleaning **A2:** 563
coatings **A2:** 564-565 **M3:** 319-320
compositions,for electrical contact materials **A2:** 854-855
definition **M6:** 1054
diffusion welding **M6:** 684-685
ductile-to-brittle transition **M6:** 1055
electron beam welding **M6:** 638-641
fabrication **A2:** 560, 561 **M3:** 314
forming **A2:** 562-563 **M3:** 316-318
friction welding **M6:** 722
gas metal arc welding **M6:** 153
gas tungsten arc welding **M6:** 182, 205
introduction **A2:** 557-565
joining **A2:** 563-564 **M3:** 318-319, 320
laser beam welding **M6:** 647
machining **A2:** 560-562 **M3:** 315-316
production **M3:** 315, 317
production of **A2:** 560-565
reactions with gas and carbon **M6:** 1055
recrystallization temperatures **M6:** 1055
shielded metal arc welding **M6:** 75
ultrasonic welding **M6:** 746

Refractory metals and alloys, specific types *See also* Refractory alloys; specific refractory alloys
molybdenum **A2:** 574-577
niobium **A2:** 565-572
refractory metal fiber reinforced composites **A2:** 582-584
rhenium **A2:** 581-582
tantalum **A2:** 571-574

Refractory Metals Association **A7:** 19

Refractory metals, brazing of **A6:** 941-947
alloy availability **A6:** 942
applications **A6:** 941
atmospheres **A6:** 941, 947
ductile-to-brittle transition temperature **A6:** 941-942
equipment **A6:** 947
filler metals **A6:** 941, 942-943
fixturing **A6:** 946
gas reactions **A6:** 942
joint designs **A6:** 946
physical properties **A6:** 941
surface preparation **A6:** 946-947
torch brazing **A6:** 947

Refractory metals, heat treating *See also* Heat-resistant alloys, heat treating; Heat-resisting alloys, heat treating **A4:** 815-819
annealing **A4:** 815-819 **M4:** 655, 670-671
atmospheres **A4:** 816-819 **M4:** 666, 671
hydrogen embrittlement **A4:** 818
nitriding **A4:** 816, 818
recrystallization **A4:** 815-816, 818, 819
stress relieving **M4:** 670
stress-relieving **A4:** 815, 816, 817

Refractory metals, heat treating (continued)
surface contamination **A4:** 816-818, 819 **M4:** 671

Refractory metals, special metallurgical welding considerations for **A6:** 580-582
brazing **A6:** 580, 581
diffusion bonding **A6:** 581
electron-beam welding **A6:** 500, 581, 582
explosion bonding **A6:** 580, 581
friction welding **A6:** 580
gas-tungsten arc welding **A6:** 580, 581
hot pressure welding **A6:** 581
laser-beam welding **A6:** 581
molybdenum alloys **A6:** 581
niobium alloys **A6:** 581
plasma arc welding **A6:** 581
resistance welding **A6:** 580, 581
rhenium alloys **A6:** 581
solid-state diffusion bonding **A6:** 580
tantalum alloys **A6:** 580-581
tungsten alloys **A6:** 581-582

Refractory oxides
in chemical flux cutting **A14:** 728-729
in mechanically alloyed oxide alloys **A2:** 943
dispersion-strengthened (MA ODS)
shell systems, lost-wax investment molding **A2:** 635-636

Refractory products **EM4:** 44-45, 46-47
additives **EM4:** 46-47
applications **EM4:** 46-47
composition **EM4:** 46-47
raw material origins **EM4:** 46-47

Refractory properties
of substrates **EL1:** 104-105

Refractory silicides
as ordered intermetallics **A2:** 934-935

Refractory supports
open resistance heaters **A2:** 830

Refrasil fiber **EM1:** 61

Refrax 20C
erosion test results **A18:** 200

Refresh rate
in acidified chloride solutions **A8:** 419

Refrigerants, magnetic
as rare earth application **A2:** 730-731

Refrigerated liquid chlorine **A13:** 1171-1172

Refrigeration
as casting market **A15:** 34

Refrigeration of adhesives **EM3:** 35

Refrigeration treatment
nickel-chromium white irons **A15:** 680

Refrigeration-grade high-impact polystyrenes (PS, HIPS) **EM2:** 199

Refrigerator liners
economy in manufacture **M3:** 852
porcelain enameled steel sheet for **M1:** 180

Refurbishment, casting
hot isostatic pressing **A15:** 544

REG *See* Rare-earth garnets

Registration *See also* Misregistration,- Random mesh registration
coupons **EL1:** 870
defects **EL1:** 1022
defined **EL1:** 1155
holes, creating **EL1:** 541-542
marks, defined **EL1:** 1155
multilayer hole-to-internal-feature **EL1:** 552
schemes, rigid printed wiring boards **EL1:** 541
stencil/board, and stencil printer **EL1:** 732
systems, panel **EL1:** 542
via, ceramic packages **EL1:** 464

Regression analysis **A8:** 653
Analysis of variance procedure **A8:** 669, 677
degrees of freedom for **A8:** 625
determining design allowables by **A8:** 664, 668-670
graphical display of F-test for **A8:** 669
lot-centered, for creep-rupture analysis **A8:** 691-693

Regression analysis (continued)
models of ... **A17:** 741
simple linear regression method for ... **A8:** 700
steps in ... **A8:** 669-670
Regression coefficient
bearing steel (ball bearings) ... **A18:** 728
Regrind
defined ... **EM2:** 35
use, thermoplastic injection molding ... **EM2:** 311
Regrinding
boring tools ... **A16:** 162
broaches ... **A16:** 203, 208, 415
cemented carbide tools ... **A16:** 84
dies ... **A16:** 284
and drills ... **A16:** 222, 233, 234
eliminated by use of indexable-insert cutters ... **A16:** 316
and multiple-operation machining ... **A16:** 366
and shaving cutters ... **A16:** 341
tool steel cutter die cracked and spalled after ... **A11:** 567, 569
Regular eutectics ... **A9:** 621-622
and irregular eutectics ... **A15:** 120
Regular quality
alloy steel sheet and strip ... **M1:** 164
steel plate ... **M1:** 182
Regular quality of low-alloy steel ... **A1:** 208
Regular reflection *See* Specular reflection
Regular transmittance ... **EM3:** 24
defined ... **EM2:** 35
Regulation
iron powders for food enrichment ... **A7:** 615
Regulators
current and voltage, for resistance spot welding ... **M6:** 472
definition ... **M6:** 14
Rehardened-and-tempered malleable iron ... **A1:** 79
Rehbinder effect ... **A18:** 189
defined ... **A18:** 15
Reheaders
for cold heading ... **A14:** 292
Reheat cracking
examination ... **A12:** 139-140
Reheat steam piping line
failure at power- generating station ... **A11:** 652-653
Reheat zone
weld characterization ... **A6:** 103
Reheat zone of a weldment
defined ... **A9:** 577
solid-state phase transformations in titanium alloys ... **A9:** 581
transformation behavior in ferrous alloys ... **A9:** 580-581
Reheaters
abbreviation for ... **A11:** 797
heat transfer control by ... **A11:** 605
heat transfer factors ... **A11:** 604
plasma ladle ... **A15:** 442
ruptured tubes from pendant-style ... **A11:** 610
steam/water-side boilers ... **A13:** 992
Reheating *See also* Heat treatment; Heating; Sintering
for austenitic manganese steel ... **A1:** 833
of copper and copper alloys ... **A14:** 257
of heat-resistant alloys ... **A14:** 235
in powder forging ... **A14:** 192-194
Rehydration
Antioch process ... **A15:** 246
Reiboxydation *See also* Fretting wear ... **A18:** 242
Reinforced composites *See also* Composites; Laminates; Metal-matrix composites; Subcomposites
flexible printed boards ... **EL1:** 583
Reinforced foam *See also* Foams
integral skin, polyurethanes (PUR) ... **EM2:** 261-262
size and shape effects ... **EM2:** 292
Reinforced molding compound *See also* Molding compound
defined ... **EM2:** 35
Reinforced nylon
thermal properties ... **A18:** 42
Reinforced phenolics *See* Phenolics
Reinforced plastics *See also* Fiber reinforcement; Laminate(s); Plastic; Plastics ... **EM3:** 24
defined ... **EM1:** 2 **EM2:** 35
Reinforced Plastics/Composites Institute (of SPI) ... **EM1:** 42
Reinforced polypropylene ... **EM3:** 24
Reinforced polypropylenes (PP) *See also* Polypropylenes (PP); Thermoplastic resins
applications ... **EM2:** 192-193
commercial forms ... **EM2:** 192
coupling agents, effects ... **EM2:** 193
defined ... **EM2:** 35
fillers and extenders ... **EM2:** 192
future trends ... **EM2:** 193
physical properties ... **EM2:** 192-193
Reinforced reaction injection molding (RRIM) *See also* Injection molding; Reaction injection molding
application, automotive industry ... **EM1:** 833, 835
defined ... **EM1:** 20 **EM2:** 35
with discontinuous fibers ... **EM1:** 121
Reinforced thermoset composites
epoxy resin cure in ... **EM1:** 654-656
Reinforced thermoset plastics
hydrochloric acid corrosion of ... **A13:** 1164
Reinforced-epoxy laminates
for base materials/insulators ... **EL1:** 113-114
Reinforcement *See also* Fiber(s); Filler; Reinforcement fibers ... **EM3:** 24
defined ... **EM1:** 20
forms ... **EM1:** 27
shielded metal arc welds ... **M6:** 93
in x-ray spectrometers ... **A10:** 88
Reinforcement (weld) ... **A6:** 27, 29
Reinforcement fibers *See also* Fiber(s)
chemical tests ... **EM1:** 285
comparative properties ... **EM1:** 58
mechanical tests ... **EM1:** 286-287
physical tests ... **EM1:** 285-286
Reinforcement of weld
definition ... **M6:** 14
Reinforcement(s) *See also* Fiber reinforcements; Fiber(s)
conductive ... **EM2:** 469-473
content, RTM and SRIM ... **EM2:** 347
continuous ... **EM2:** 393
defined ... **EM2:** 35
in engineering thermoplastics ... **EM2:** 98-99
and fillers, compared ... **EM2:** 72
limitations ... **EM2:** 281
mechanical properties ... **EM2:** 71-73
polyether sulfones (PES, PESV) ... **EM2:** 161
process, capabilities and properties ... **EM2:** 283
in pultrusions ... **EM2:** 390, 392-393
thermosetting resins ... **EM2:** 223
ultrahigh molecular weight polyethylenes (UHMWPE) ... **EM2:** 170
Reinforcement-to-resin ratio
defined ... **EL1:** 598
Reinforcements *See also* Composites; Fiber; Metal-matrix composites ... **EL1:** 535
for aluminum-matrix composites ... **A2:** 7, 904-907
amorphous materials and metallic glasses ... **A2:** 819
for copper-matrix composites ... **A2:** 908-909
dielectric, constraining ... **EL1:** 615-619
glass fibers, for base/insulators ... **EL1:** 113-114
inorganic, chemical compositions ... **EL1:** 604
for intermetallic-matrix composites ... **A2:** 909-911
for magnesium-matrix composites ... **A2:** 907-908
and matrix materials ... **EL1:** 1119-1121
for metal-matrix composites, types ... **A2:** 903
printed wiring board ... **EL1:** 604-605
properties
PWB, properties ... **EL1:** 604
for superalloy-matrix composites ... **A2:** 909
for titanium-matrix composites ... **A2:** 908
Reinforcing process
porcelain enameling ... **M5:** 519
Reinforcing steel
corrosion ... **A13:** 1309
Rejected take-off (RTO) stop ... **A18:** 583, 585, 586
Rejection *See also* Acceptance; Acceptance or rejection
of flaws, determination ... **A17:** 49
record, adhesive-bonded joints ... **A17:** 634
standards, liquid penetrant inspection ... **A17:** 88
Relational data base
KI SHELL mapped onto ... **A14:** 412
Relationship tree ... **A18:** 353-354
Relative bioavailability (RBV)
of iron powders for food enrichment ... **A7:** 614-615
Relative corrosivity ... **A13:** 204-205
Relative density *See also* Density
and forging modes ... **A7:** 414
and pressure ... **A7:** 298
Relative dielectric constant
defined ... **EL1:** 90
Relative displacement
crosshead displacement as ... **A8:** 41, 45
Relative erosion factor (REF) ... **A18:** 200, 201
Relative humidity *See also* Humidity ... **EM3:** 24
critical, defined ... **A13:** 82
defined ... **A13:** 11 **EM2:** 35
effect, atmospheric corrosion ... **A13:** 81-82
effect, in fretting ... **A13:** 140
effects on nickel-iron low-expansion alloys ... **A2:** 892
effects, zinc/zinc alloys and coatings ... **A13:** 757
fatigue strength as function of ... **A11:** 252
iron/magnesium corrosion rates in ... **A13:** 908
reduction, for filiform corrosion ... **A13:** 107-108
time of ... **A13:** 82
Relative hydrogen susceptibility
of metals ... **A8:** 541-542
Relative operating characteristic (ROC) curves ... **A17:** 679-680
Relative permittivity
defined ... **EL1:** 597
of glass fibers ... **EM1:** 46
Relative plate thickness ... **A6:** 12
Relative potency factor (RPF)
definition ... **A6:** 89
Relative pressure
friction during metal forming ... **A18:** 67
Relative ratings of resistance to SCC ... **A11:** 220
Relative rigidity ... **EM3:** 24
Relative sensitivity factor
abbreviation for ... **A10:** 691
LEISS ... **A10:** 606
use in SSMS ... **A10:** 145
Relative sintering temperature ... **A7:** 10
Relative sliding velocity
between two surfaces ... **A18:** 40
Relative standard deviation
defined ... **A10:** 681
Relative thermal index (RTI)
for service temperatures ... **EM2:** 569-570
Relative transmittance
defined ... **A10:** 681
Relative variability ... **A8:** 625
Relative viscosity ... **EM3:** 24
defined ... **EM2:** 35
Relativistic heavy ion collider (RHIC)
as niobium-titanium superconducting material application ... **A2:** 1055-1056
Relativistic voltage ... **A6:** 43
Relaxation *See* Strain relaxation; Stress relaxation
and creep, compared ... **A11:** 542
defined ... **EM1:** 190
effect, fastener performance at elevated temperatures ... **A11:** 542
modulus and creep compliance, viscoelasticity as ... **EM1:** 190-191
parameter, ferromagnetic materials ... **A10:** 275, 276
pulse method of measuring times of ... **A10:** 258
rate, factors affecting ... **A11:** 542
saturation method of measuring times of ... **A10:** 258
spin, rates of ... **A10:** 275-276
spin-spin ... **A10:** 257
in stress, due to layer removal ... **A10:** 388
as supplemental ESR technique ... **A10:** 257-258
surface atom, in single crystal ... **A10:** 628
as temperature effect on damping ... **EM1:** 215
time, defined ... **EM1:** 20
times, saturation and pulse methods of measuring ... **A10:** 258
vibrational, abbreviation for ... **A10:** 691
Relaxation at grain boundaries
models for describing ... **A9:** 119
Relaxation behavior
constructional steels for elevated temperature use ... **M1:** 642, 647, 649, 651, 656, 657
Relaxation curve
defined ... **A8:** 11

Relaxation curves, steel springs **M1:** 296-297, 298-300
Relaxation, in threaded steel fasteners **A1:** 295
effect of thread design on **A1:** 296
strengths **A1:** 631
Relaxation parameter
as ferromagnetic resonance application **A10:** 275-276
Relaxation, strain *See* Strain relaxation
Relaxation tests
and elevated-temperature service **A1:** 624, 631
Relaxation time **EM3:** 24
defined **EM2:** 35
Relaxation-time experiments
effects of temperature on **A10:** 257
Relaxed layer **A18:** 177
Relaxed modulus *See* Static modulus of elasticity
Relaxed stress *See also* Stress relaxation **EM3:** 24
defined **A8:** 11 **EM1:** 20 **EM2:** 35
Relay blades
oxide dispersion-strengthened copper **A2:** 401
Relay weld integrity
as gas analysis spectroscopy application **A10:** 156
Relay(s)
blades and contact supports **A7:** 715
parts magnets, powders used **A7:** 573
powders used **A7:** 573
Relays
failure mechanisms **EL1:** 981
as magnetically soft material application **A2:** 761
Release
methodology **EL1:** 127
organization **EL1:** 130-132
Release agent *See also* Mold release agent; Molding release agent **EM3:** 24
defined **EM1:** 20 **EM2:** 35
for sheet metal compounds **EM1:** 158
Release agents
for coremaking **A15:** 240
Release film *See also* Films; Separator **EM3:** 24
defined **EM1:** 20 **EM2:** 35
Release interface transmittals (RiTs) **EL1:** 130
Release paper **EM3:** 24
Relevant indications *See also* Nonrelevant indications
defined **A17:** 103
liquid penetrant inspection **A17:** 86
versus nonrelevant indications **A17:** 106-108
Reliability *See also* Dimensional accuracy; Failure analysis; Failure rate; Failure(s); NDE reliability; NDE reliability applications; NDE reliability data analysis; NDE reliability models Process control; prediction Reliability testing; Reliability; Reproducibility; Statistical methods; Testing
basis for **A8:** 624
ceramic multilayer packages **EL1:** 468
component interconnections **EL1:** 261
of connectors **EL1:** 21-23
of corrosion testing results **A13:** 195-196, 316-317, 322
criteria, by flaw size **A17:** 664
defined **EL1:** 1155
defined, in welding **A17:** 590
of designs **EL1:** 747
of electronics packaging **EL1:** 127
and end use **EL1:** 1
environmental, of ceramic hybrid circuits **EL1:** 381
environmental stress screening **EL1:** 876
exponential distribution for **A8:** 634
and failure analysis **EL1:** 957
and failure probability **A8:** 627
flexible printed boards **EL1:** 584
of flexible printed wiring **EL1:** 580
function, defined **EL1:** 896
of glass-to-metal seals **EL1:** 458-459
hardware **EL1:** 79

Reliability (continued)
improvement **EL1:** 133-135
of integrated circuits **A11:** 766, 771
of investment casting **A15:** 265
of level 1 packages **EL1:** 403
levels, and confidence levels **A8:** 627
of LSI/VLSI circuit chips **EL1:** 961
mathematics of **EL1:** 895
of nondestructive evaluation **A17:** 663-715
objectives (downtime) **EL1:** 128
phases, "bathtub" curve **EL1:** 740
of plastic packages **EL1:** 245-246, 479-480
plastic pin-grid arrays **EL1:** 480
plated-through hole **EL1:** 613-614
of polymer die attach **EL1:** 218-220
prediction, program factors **A17:** 664
preservation, environmental stress screening **EL1:** 876
solder joints **EL1:** 675
of surface-mount electronic assembly **EL1:** 730
tape automated bonding (TAB) **EL1:** 287-288, 480
of wafer-scale integration **EL1:** 263
wire bond **EL1:** 226-228
Reliability and wear, concepts of: failure modes **A18:** 493-495
dependence of failure rate on operating duration **A18:** 494-495
relationship between wear and reliability **A18:** 493
reliability characteristics **A18:** 493
reliability probability concepts **A18:** 493
statistical distributions of wear and reliability **A18:** 493-494
exponential distribution **A18:** 493
gamma distribution **A18:** 494
log normal distribution **A18:** 493
normal distribution **A18:** 493, 494
Weibull distribution **A18:** 493-494
wear and failure modes **A18:** 494
Reliability block diagrams
types **EL1:** 899
Reliability demonstration program
importance **A17:** 664
Reliability growth program
defined **EL1:** 903
Reliability of measurements
and sample quality **A10:** 12
Reliability prediction *See also* Reliability
estimating failure factors, from test data **EL1:** 900-903
failure density distributions **EL1:** 896-897
life cycle **EL1:** 897-899
mathematics of reliability **EL1:** 895-896
reliability growth **EL1:** 903-904
reliability of systems **EL1:** 899-900
Reliability testing
combined environments (CERT) **EL1:** 1103
electrical **EL1:** 567
structured, failure mechanisms **EL1:** 959-961
of VLSI **EL1:** 887
Relict **EM4:** 110, 111, 112, 113
Relief
in blanking die **A14:** 447
Relief (clearance) angle **A16:** 121, 195
Rem units
radiography **A17:** 301
Remaining life
of elevated-temperature materials
creep-rupture tests for **A8:** 339
estimation techniques **A8:** 338
from measurements of microstructure **A8:** 339
measurement of rupture properties after service **A8:** 338
Remaining stress *See also* Stress relaxation
defined **A8:** 11
Remalloy *See* Permanent magnet materials, specific types **A9:** 538
Remanence
magnetic **A7:** 643

Remelt hardening
ductile iron **A15:** 660
Remelting *See also* Heat treatment; Melting; Remelt hardening
early air furnace **A15:** 25
electroslag **A15:** 401-406
metal-matrix composites **A15:** 848-849
plasma arc **A15:** 424
of stainless steels **A14:** 222
under high pressure, as ESR process **A15:** 405
vacuum induction **A15:** 399-401
Remote crevice assemblies
electrochemical **A13:** 308-309
Remote-field eddy current inspection
See also Eddy current inspection **A17:** 195-201
applications **A17:** 199-201
current research **A17:** 197-198
examples **A17:** 199-201
flaw detection sensitivity, techniques **A17:** 198-199
flaw models **A17:** 198
instrumentation **A17:** 197
limitations **A17:** 197
no-flaw models **A17:** 197-198
theory **A17:** 195-197
Removal
component, and replacement **EL1:** 288
solvent, from surface-mount assemblies **EL1:** 666
Removal coefficient **A18:** 431, 434
Removers *See* Solvent removers
Renal effects *See also* Renal failure; Renal tubular dysfunction
from lead **A2:** 1244-1245
lesions, from gold toxicity **A2:** 1257
Renal failure *See also* Renal effects; Renal tubular dysfunction
from bismuth **A2:** 1256
from uranium toxicity **A2:** 1261
as gallium toxicity **A2:** 1257
Renal tubular dysfunction *See also* Renal effects; Renal failure
cadmium toxicity effects **A2:** 1240
from lead toxicity **A2:** 1244
lining cell, lead-induced inclusion bodies **A2:** 1245
René 41
aging **A4:** 796
aging cycle **A4:** 812 **M4:** 656
aging cycles **A6:** 574
aging, effect on properties **A4:** 800 **M4:** 659
aging precipitates **A4:** 808
annealing **A4:** 808-809 **M4:** 655
broaching **A16:** 204, 206, 208, 209, 743, 745-746
chemical milling **A16:** 584
composition **A4:** 794, 795 **A6:** 564, 573 **A16:** 736 **M4:** 651-652 **M6:** 354
contour band sawing **A16:** 363
drilling **A16:** 748, 749
electrochemical grinding **A16:** 547
electrochemical machining removal rates **A16:** 534
electron-beam welding **A6:** 869
flash welding **M6:** 557
grinding **A16:** 547
machining **A16:** 738, 741-743, 746-747, 749-757, 758
manufacturing rating **A16:** 739
material for jet engine components **A18:** 591
milling **A16:** 547, 755
overaging **M6:** 358
photochemical machining **A16:** 588
for press forging heat-resistant alloys, nickel-base alloys **A18:** 625
production time **A16:** 739
reheat treatment, effect on properties **M4:** 666
self-lubricating powder metallurgy composites sliding on **A18:** 120
shaping **A16:** 192

René 41 (continued)
shielded metal arc welding **M6:** 75
solution treating **M4:** 656
solution treatment **A6:** 574
solution-treating **A4:** 796, 806-807
stress relieving **M4:** 655
surface alterations from material removal processes **A16:** 27
thermomechanical processing **A4:** 809
thread grinding **A16:** 275
René 63
machining **A16:** 738, 741-743, 746-747, 749-757
thread grinding **A16:** 275
René 77
aging cycle **A4:** 812
composition **A4:** 795
machining **A16:** 737, 738, 741-743, 746-757
material for jet engine components **A18:** 588, 590
thread grinding **A16:** 275
René 80
aging cycle **A4:** 812
composition **A4:** 795 **A16:** 737 **M4:** 653
machining **A16:** 738, 741-743, 746-757, 758
material for jet engine components **A18:** 588
René 80 Hf
composition **A4:** 795
René 95
composition **A4:** 794 **A6:** 573 **A16:** 736
machining **A16:** 738, 741-743, 746-747, 749-757, 758
thermomechanical processing **A4:** 798
thread grinding **A16:** 275
René 100
composition **A4:** 794, 795 **A6:** 573 **A16:** 737
machining **A16:** 738, 741-743, 746-757
René 125
hydrogen fluoride cleaning **A6:** 926
machining **A16:** 757, 758
material for jet engine components **A18:** 588, 590
René alloys *See* Superalloys, nickel-base, specific types, René
René N4
aging cycle **A4:** 812
composition **A4:** 795
Reoxidation prevention
use of primers **EM3:** 101
REP *See* Rotating electrode process
Repair *See also* Maintainability; Maintenance; Reparability; Rework; Rework processes; Weld repair **EM3:** 37-38
of aluminum alloy forgings **A14:** 248
of cast steel castings **A15:** 531
of castings, by oxyacetylene welding **A15:** 529-530
in computer-aided manufacturing **EL1:** 131
of conventionally forged titanium alloys **A14:** 280
cost escalation **EL1:** 877
defects, adhesive-bonded joints **A17:** 615
design requirements for **EM1:** 183
nondestructive **EL1:** 710-711
of parylene coatings **EL1:** 798-799
of patterns **A15:** 196
rigid printed wiring boards **EL1:** 542-543, 547
of steel forgings defects **A17:** 503
of urethane-coated components **EL1:** 780
weld, of titanium castings **A15:** 833
Repair of advanced composite commercial aircraft structures **EM3:** 829-837
damage types **EM3:** 829, 830
inspection techniques used **EM3:** 829, 830
repair development **EM3:** 829-833
graphite-epoxy composite materials **EM3:** 830-832
graphite-polyimide composite materials **EM3:** 832-833
scarf repairs **EM3:** 829-830
repair durability **EM3:** 833-837
baseline results **EM3:** 835
exposure and test plan **EM3:** 834
exposure results **EM3:** 835-836
outdoor exposure test setup **EM3:** 834-835
program synopses **EM3:** 836-837
tabbed laminate specimen **EM3:** 834
Repair scheduling
by damage analyses **A8:** 682
Repair station soldering
as surface-mount soldering **EL1:** 707
Repair welding **A6:** 1103-1107
air-carbon arc cutting (CAC-A) **A6:** 1104, 1105
arc welds of
beryllium **M6:** 462
cast irons **M6:** 316-317
titanium and titanium alloys **M6:** 456
austenitic stainless steels **A6:** 1105-1106
base metal weldability **A6:** 1103
chemical analysis test **A6:** 1103
simulated weld tests **A6:** 1103
spark test **A6:** 1103
base-metal preparation **A6:** 1104
carbon steels **A6:** 1105
cast irons **A6:** 1105
categories **A6:** 1103
codes and standards **A6:** 1104
duplex stainless steels **A6:** 1107
electrodes **A6:** 1105, 1106
for stainless steels **A6:** 1106
electrogas welds **M6:** 243
electron beam welds **M6:** 634-637
explosion welds **M6:** 716
ferritic stainless steels **A6:** 1106
filler metals **A6:** 1105, 1106, 1107
gas metal arc welds of aluminum alloys **M6:** 388
guidelines for various base metals **A6:** 1104-1107
heat-affected zone **A6:** 1105, 1106, 1107
high-carbon steels **A6:** 1105
low-carbon steel **A6:** 1105
magnesium alloy castings **A6:** 779-782 **M6:** 432-435
martensitic stainless steels **A6:** 1106
medium-carbon steels **A6:** 1105
nature of failure categories **A6:** 1103-1104
nickel alloys **A6:** 579
nickel-base alloys **A6:** 591-592
oxyacetylene braze welds of cast irons **M6:** 599-600
oxyacetylene welds of cast irons **M6:** 601-602
oxyfuel gas welds **M6:** 592-593
precipitation-hardening stainless steels **A6:** 1106-1107
preliminary assessment **A6:** 1103-1104
stainless steels **A6:** 1105-1107
thermit welds **M6:** 698-702
titanium **A6:** 1107
welding process selection **A6:** 1104
Repairability *See also* Repair; Rework
Hybrid microcircuits **EL1:** 261
of silicone conformal coatings **EL1:** 822
solder masks **EL1:** 553
Repairing
defined **EL1:** 1155
Repairing aerospace components
powders used **A7:** 572
Repairs
by cold straightening, fatigue fracture from **A11:** 125
faulty, and distortion failure **A11:** 141-142
Repassivation
effect on SCC **A12:** 25
kinetics **A12:** 42
Repassivation potentials
titanium/titanium alloys **A13:** 684-685
Repeatability
defined **A8:** 11 **A17:** 18
of EMF processing **A14:** 646
in quantitative EPMA **A10:** 525
Repeated impact *See* Impingement
Repeated tension test
ductile to brittle fracture **A8:** 353-354
Rephosphorized steels *See also* Phosphorus in steel **A1:** 208 **M1:** 162
machinability ratings **M1:** 576
weldability **M1:** 563
Replacement
component **EL1:** 288
of defective circuitry **EL1:** 9
Replacement scheduling
by damage analyses **A8:** 682
Replenishment, developer
radiographic film processing **A17:** 352-354
Replica
defined **A9:** 15
Replica-stripping cleaning
of fractures **A12:** 74
Replicas *See also* Replication; Surface replicas
artifacts in **A12:** 184-185
cast, magnetic rubber inspection indications **A17:** 125
cellulose acetate, SEM **A12:** 171
extraction **A12:** 182-183 **A17:** 54
extraction for TEM **A10:** 452
fractographic, compared **A12:** 95, 100
grating **A12:** 167
for light microscopy **A12:** 94-95
profiles from **A12:** 199
sectioning **A12:** 199
shadowing **A12:** 7, 95, 100, 172, 183-184
single-stage **A12:** 179-182
two-stage **A12:** 182
Replicas used in transmission electron microscopy **A9:** 108-109
Replicast ceramic shell process **A15:** 270-272
flow chart for **A15:** 270
and investment casting, compared **A15:** 270
as special investment casting process **A15:** 267
Replicast full mold process **A15:** 270
Replicast process **A15:** 36, 270-272
advantages/applications **A15:** 37
applications **A15:** 271-272
ceramic coating and firing **A15:** 271
cleaning **A15:** 271
computer-integrated manufacturing system for **A15:** 569
core elimination in **A15:** 271
dimensional accuracy **A15:** 271
pattern assembly **A15:** 271
pattern production **A15:** 270-271
pouring **A15:** 271
as recent development **A15:** 37-38
Replicast CS (ceramic shell) **A15:** 270
Replicast FM (full mold) **A15:** 270
schematic **A15:** 36
surface finish **A15:** 271
Replicate
defined **A9:** 15
Replication *See also* Surface replication **A8:** 643, 696-697
for crack analysis **A17:** 54
effect on experimental bias **A8:** 640
in experimental design **A8:** 640-641
in experimental factorial design **A17:** 743
extraction **A17:** 52, 54
extraction replicas **A12:** 182-183
procedures, for light microscopy **A12:** 94-95
in sampling **A10:** 12
single-stage replicas **A12:** 7, 179-182
surface **A17:** 52-53
tape, cellulose acetate **A12:** 73
techniques **A12:** 7-8 **A17:** 52-54
two-stage replicas **A12:** 7, 182
Replication (experimental)
defined **EM2:** 600
Replication microscopy techniques (NDE) **A17:** 52-56
application **A17:** 643
of boilers and pressure vessels **A17:** 642-644
compared **A17:** 53
microstructural analysis **A17:** 54-55
replication techniques **A17:** 52-54
specimen preparation **A17:** 52
Replication techniques
in electropolishing **A9:** 55
one-step replicas **A12:** 7-8, 179-182
for preparing transmission electron microscopy specimens **A9:** 108-109
shadowing **A12:** 7-8, 95, 100, 172, 183-184
TEM **A12:** 7
two-step replicas **A12:** 7, 182
Report writing **A11:** 31-32
Reprecipitation and dissolution
as gravimetric sample preparation **A10:** 163
Reprecipitation and solution
effect of sintering time **A7:** 320
liquid-phase sintering **A7:** 320
Representative sample
defined **A10:** 13
Reprocessability
of thermoplastics **EM1:** 100

Reprocessed fibers
as aligned discontinuous **EM1:** 153-155
applications **EM1:** 153, 155-156
process **EM1:** 153
Reprocessed plastic *See also* Plastics **EM3:** 24
defined **EM2:** 35-36
Reprocessing
uranium **A7:** 666
Reproducibility *See also* Dimensional accuracy; Reliability
of corrosion tests **A13:** 194
defined **A8:** 11
dimensional **A15:** 614-623
in direct current electrical potential method **A8:** 389
of metal castings **A15:** 40
parameter, defined **A8:** 387
Reproducibility, of electrode materials
voltammetric monitoring for **A10:** 189
Reproductive effects
of arsenic toxicity **A2:** 1238
Request for quotation
automated **A14:** 409-410
Required life
sliding bearings **A18:** 515
Rerolling quality
defined **A14:** 11
Rerouting technologies
wafer-scale integration **EL1:** 268-269
Rescreening
in environmental stress screening **EL1:** 880
Research grade
rare earth metals **A2:** 720
Research reactor
as common source for low-energy TNAA neutrons **A10:** 234
Research-quality metallograph with projection screen **A9:** 74
Research-quality optical microscopes **A9:** 73
Resenes **EM3:** 24
defined **EM2:** 36
Reservoir effect
polymeric lubricants **A18:** 137
Reset
defined **A14:** 11
Resetting
and multiple-operation machining **A16:** 366
Reshaping
of round tubing **A14:** 631-632
Residual
defined **A8:** 699
Residual blinding
screen **A7:** 177
Residual compression strength
long-term exposure testing **EM1:** 825
thermosetting/thermoplastic systems **EM1:** 98
Residual core
materials, computed tomography (CT) **A17:** 361
turbine blades, neutron radiographic detection **A17:** 393-394
Residual dopants
tungsten powders **A7:** 153
Residual elements
defined **A9:** 15
in steels **A1:** 141
Residual elements in steel *See also* Impurity elements **M1:** 115-116, 121
tensile properties of steel plate effect on **M1:** 194, 197
Residual flexural modulus
long-term exposure testing **EM1:** 825
Residual flexural strength
long-term exposure testing **EM1:** 825
Residual gas analysis (RGA) *See also* Hermeticity; Hermeticity testing **EM3:** 24
application **EL1:** 1063-1064
data interpretation **EL1:** 1064-1066
defined **EM1:** 20 **EM2:** 36
and package-level test methods **EL1:** 927
Residual gas analysis (RGA) (continued)
test method **EL1:** 1064
Residual gas analyzer
in environmental test chamber **A8:** 411
Residual gas analyzers
as gas chromatographs **A17:** 64
as vacuum leak testing method **A17:** 68
Residual interparticle diffusion
sintered **A7:** 314
Residual magnetism *See also* Magnetization
defined **A17:** 100
magnetic field **A17:** 91
as magnetic particle inspection method **A17:** 110
Residual magnetization
application **A17:** 97
in leakage field testing **A17:** 130
Residual plots **A8:** 699
Residual resistance ratio (RRR), of copper
in NbTi superconductors **A2:** 1045
Residual short beam shear strength
from long-term exposure testing **EM1:** 825
Residual soil weight test method
cleaning process efficiency **M5:** 20
Residual strain *See also* Strain **EM3:** 24
defined **EM1:** 20 **EM2:** 36
Residual strain hardening *See also* Strain hardening
of zone-melted iron, changes during isothermal recovery **A9:** 693
Residual strength
after impact, modeling **EM1:** 435
analysis **EM1:** 203
of aramid fibers **EM1:** 36
calculated **A11:** 55-57
of composite failure **A8:** 716
and damage tolerances **EM1:** 265-266
defined **A17:** 669
fatigue **EM1:** 245-246
full-scale test for **EM1:** 351
graphite-epoxy composite material **A8:** 716
impact damage, resin effects **EM1:** 262
and static strength distributions **A8:** 717
Residual stress *See also* Engineering stress; Mean stress; Nominal stress; Normal stress; True stress **A16:** 22, 25, 26, 30 **EM1:** 20, 232 **EM3:** 24, 400, 401, 403
4340 steel surface milling **A16:** 34
and applied stress distribution **A8:** 124
arc-welded steel failure by **A11:** 417
arithmetically defined **A10:** 385
associated with failures caused by fatigue or stress corrosion **A10:** 380
Barkhausen noise measurement **A17:** 159-160
brittle fracture by **A11:** 85, 90
in ceramics **A11:** 751
in closed-die forgings **A1:** 342
closed-die steel forgings **M1:** 356
cold finished bars **M1:** 230-234
compressive **A11:** 97
defined **A8:** 11 **A10:** 681 **A11:** 8 **A13:** 11 **A14:** 11 **EM2:** 36
determined by neutron diffraction **A10:** 420
distribution, longitudinal, in welded railroad rail **A10:** 391-392
effect of stress relief, aluminum alloy **A13:** 591
effect on fatigue strength **A11:** 112
electrical resistance alloys **A2:** 824
evaluated by x-ray diffraction **A11:** 125-126
failures from **A11:** 97-98
fatigue behavior, effect on **M1:** 673-675
and fatigue resistance **A1:** 680-681, 682, 683
and fatigue strength **A8:** 374
ferromagnetic resonance **A10:** 267-276
fracture mechanics of **A11:** 57
from plastic deformation **A13:** 255
from surface grinding of D6ac steel **A16:** 33
gray iron **M1:** 26-28
hoop **A11:** 97-98
Residual stress (continued)
for induction heating stress improvement **A13:** 932
integral-finned stainless steel tubed cracked by **A11:** 636
local variations produced by surface grinding **A10:** 390-391
magabsorption measurement **A17:** 156-157
and magnetic hysteresis **A17:** 134
magnetically soft materials **A2:** 763
maximum, magnitude and direction produced by machining **A10:** 392
measured by magnetically induced velocity changes (MIVC) **A17:** 161-162
measurement, electromagnetic techniques **A17:** 159-163
measurement of **A10:** 425-426
NDE methods for **A17:** 51
neutron diffraction analysis of **A10:** 424, 425-426
nonlinear harmonics measurement **A17:** 160-161
and percent cold work distribution in belt-polished and formed Inconel 600 tubing **A10:** 390
and percent cold work distribution in tubing **A10:** 390
and percent cold work distributions caused by forming **A10:** 380
stress-relieving heat treatment or
profile **A16:** 27
profile using XRD technique **A16:** 24
profiles, minimum and maximum principal **A10:** 392
Rutherford backscattering analysis **A10:** 628-636
and SCC failure **A13:** 615
as source of stress-corrosion cracking **A8:** 502
specimens **A8:** 509-510
specimens, SCC testing **A13:** 253
and springback, in bending **A8:** 122-124
springs, steel **M1:** 290-291, 312-313
for steel springs **A1:** 309, 311
and stress concentration, iron castings **A11:** 362
stress-corrosion cracking **A13:** 145
study, methods for **A11:** 134
subsurface, and hardness distribution in induction-hardened steel shaft **A10:** 389-390
subsurface, in steels **A10:** 389-390
surface, and hardness, on raceway of ball and roller bearings **A10:** 380
surface, and shaft failures **A11:** 473-474
surface compressive, uniformity determined **A10:** 380
surface hardened steel **M1:** 528, 538, 540
surface, in fatigue **A13:** 295
surface, nondestructive measurement for quality control **A10:** 380
in swaging **A14:** 143-144
tensile, from manufacturing **A11:** 97
uranium alloys **A2:** 675
welding **A11:** 624 **A13:** 256, 344
x-ray diffraction techniques **A10:** 380-392
and yield stress **A11:** 57
Residual stresses *See also* Stress(es) **A6:** 1094-1102 **M6:** 856-892
alternate names **A6:** 1094
aluminum alloy castings **A15:** 762
analyses in weldments **A6:** 1095
austenitic stainless steels **A6:** 469
brazing of dissimilar materials **A6:** 623, 629
brittle fracture **M6:** 840-841
carbon steels **A6:** 643, 651
causes **M6:** 856
changes in **A6:** 1100-1101
classification of measurement techniques **A6:** 1095
combined effects of distortion **M6:** 886-887
corrosion of weldments **A6:** 1068
defined **A15:** 9
definition **A6:** 1212 **M6:** 14

Residual stresses (continued)
dissimilar metal joining **A6:** 826
distribution in weldments **A6:** 1096-1097 **M6:** 865-867
aluminum alloys **M6:** 867-868
high-strength steels **M6:** 866-867
plug welds **M6:** 865
titanium alloys **M6:** 867-868
welded pipe **M6:** 866
welded shapes and columns **M6:** 865-866
edge welds **M6:** 861
effect on service behavior **A6:** 1100 **M6:** 880-887
brittle fracture of welded structures **M6:** 881-883
changes caused by tensile loading **M6:** 881, 884
effect of stress relieving temperature on brittle fracture **M6:** 882-883, 885
fatigue fracture **M6:** 883
effect on stress-relieving treatments **A6:** 1101
effect on weld discontinuities **M6:** 840-841
effects **M6:** 856
effects on brittle fracture of welded structures **A6:** 1101
effects on fatigue fracture of welded structures **A6:** 1101- 1102
environmental effects **A6:** 1102
equilibrium condition of **A6:** 1095
fatigue failure **M6:** 841
formation **M6:** 856
compressive stresses **M6:** 856-857
mismatch of stress **M6:** 856-857
tensile stresses **M6:** 856-857
uneven distribution of nonelastic strains **M6:** 857-858
formation of **A6:** 1094
heat-affected zone **A6:** 1102
heavy weldments **M6:** 868-869, 871
lead frame strip **EL1:** 483
machinery and equipment **A6:** 393
macroscopic **M6:** 856
magnitude in weldments **A6:** 1096-1097
measurement **M6:** 862-865
Gunnert drilling technique **M6:** 862-864
Mathar Soete drilling technique **M6:** 862-864
photoelatic coating-drilling techniques **M6:** 864
sectioning **M6:** 862-864
stress-relaxation techniques **M6:** 863-864
x-ray diffraction **M6:** 864-865
measurement techniques **A6:** 1095-1096
mechanical stress relieving **A6:** 1100
microscopic **M6:** 856
nickel-base alloys **A6:** 927, 928
plastic upsetting **A6:** 1095
postweld heat treatment **A6:** 1102
preheat **A6:** 1102
reduction **M6:** 887
in solidification **A15:** 616
specimen length, effect of **M6:** 867-868, 870-871
specimen width, effect of **M6:** 860, 868
stainless steels **A6:** 625-626, 679
stress-corrosion cracking **A6:** 1102
techniques for analyzing **M6:** 858, 860-862
analytical simulation **M6:** 860-862
finite element method **M6:** 858, 862
incompatible strain method **M6:** 862
thermal stresses and metal movement during welding **A6:** 1094-1095
thermal treatments of weldments **A6:** 1102
underwater welding, weld-metals **A6:** 1012
welding sequence, effect of **M6:** 887
weldments **M6:** 859-860
Residual surface stresses
compressive **A8:** 374
tensile **A8:** 374
Residual-pattern test method
cleaning process efficiency **M5:** 20
Residue tack
defined **EL1:** 644
Residues
analytical techniques for **A10:** 177
assembly, sources **EL1:** 658
in capillary spaces, cleaning **EL1:** 666
ionic **EL1:** 660
measurement **EL1:** 667
metal and alloy, isolation of **A10:** 176
refinement of **A10:** 176-177

Residues (continued)
volatilization during ignition **A10:** 163
Resilience *See also* Elastic energy; Modulus of resilience; Strain energy; Toughness **EM3:** 24
beryllium copper alloys **A2:** 416-418
defined **A8:** 11, 22 **EM1:** 20 **EM2:** 36
modulus of *See* Modulus
Resin **EM3:** 24
liquid **EM3:** 24
Resin binder processes **A15:** 214-221
classification **A15:** 214
cold box processes **A15:** 219-221
hot box processes **A15:** 218
no-bake process **A15:** 214-217
oven-bake processes/core-oil binders **A15:** 218-219
shell (Croning) process **A15:** 217-218
warm box processes **A15:** 218
Resin bonds
bonded-abrasive grains **A2:** 1014
Resin coating
aluminum and aluminum alloys **M5:** 609-610
copper and copper alloys **M5:** 621, 626-627
Resin coatings
fused dry **A15:** 565
Resin compounds
acrylics, types **EM2:** 107-108
acrylonitrile-butadiene-styrenes (ABS) **EM2:** 114
cyanates **EM2:** 236
high-impact polystyrenes (PS HIPS) **EM2:** 198-199
ionomers **EM2:** 123
liquid crystal polymers (LCP) **EM2:** 181-182
polyamide-imides (PAI) **EM2:** 137
polyamides (PA) **EM2:** 127
polybutylene terephthalates (PBT) **EM2:** 155
polycarbonates (PC) **EM2:** 152
polyether sulfones (PES, PESV) **EM2:** 162
polyphenylene ether blends (PPE PPO) **EM2:** 185
polyurethanes (PUR) **EM2:** 264
polyvinyl chlorides (PVC) **EM2:** 212
styrene-acrylonitriles (SAN, OSA ASA) **EM2:** 216
styrene-maleic anhydrides (S/MA) **EM2:** 220-221
thermoplastic polyurethanes (TPUR) **EM2:** 207
ultrahigh molecular weight polyethylenes (UHMWPE) **EM2:** 170, 171
Resin content **EM3:** 24
for closed-loop cure **EM1:** 761
defined **EM1:** 20, 737
tested **EM1:** 737
Resin deoxygenating column
simulated pressurized water reactor water system **A8:** 423-424
Resin flask
for immersion testing **A13:** 222
Resin flow patterns
by computed tomography (CT) **A17:** 363
Resin fluxes **EL1:** 646
Resin impregnation
in pultrusion **EM2:** 390
Resin injection
and fiber performs **EM1:** 529-532
inspection **EM1:** 532
resin film infusion **EM1:** 530-531
resin transfer molding **EM1:** 532
Resin layers
interference films used to improve contrast **A9:** 59
Resin microflow
direction in composites **A11:** 742, 743
Resin pastes
sheet molding compounds **EM1:** 159-160
Resin pocket *See also* Resin-rich area
defined **EM1:** 20 **EM2:** 36
Resin producers *See* Suppliers
Resin properties analysis
chemical tests **EM1:** 736-737
component material tests **EM1:** 736-737
mixed resin system tests **EM1:** 737
prepreg tests **EM1:** 737
of resin/prepreg mechanical properties **EM1:** 737
Resin removal **EM1:** 604, 655
Resin smear **EL1:** 575, 1155
Resin system **EM3:** 25

Resin systems
defined **EM1:** 20 **EM2:** 36
effect, composite properties **EM1:** 162
for filament winding **EM1:** 504-505
flow **EM1:** 753-755
high-performance, for thermoplastics **EM1:** 544
high-temperature, types **EM2:** 443-444
low-temperature, types **EM2:** 439-440
medium-temperature, types **EM2:** 441-443
mixed, properties analysis **EM1:** 737
RTM and SRIM, compared **EM2:** 347
temperature requirements, changing **EM1:** 105
thermoplastic **EM1:** 544-546
Resin transfer molding **EM4:** 224
Resin transfer molding (RTM) **EM1:** 564-568
application, automotive industry **EM1:** 834
cost analysis **EM1:** 170-171
defined **EM1:** 20 **EM2:** 36
discontinuous fiber reinforced matrix **EM1:** 33
economic factors **EM2:** 299-301, 349
of epoxy composites **EM1:** 71, 72
equipment **EM1:** 565-566
for large parts **EM2:** 344
materials **EM1:** 168-171, 566-567
materials selection **EM1:** 168-171
mold design and construction **EM1:** 168-169
molded-in color **EM2:** 306
parts manufactured using **EM1:** 168
process **EM2:** 344-346
process variants **EM1:** 564-565
process/applications **EM1:** 567
properties effects **EM2:** 287
pumping/dispensing equipment **EM1:** 169
reinforcement selection **EM1:** 169-171
as resin injection process **EM1:** 532
resin selection **EM1:** 169-170
and structural reaction injection molding (SRIM), compared **EM2:** 344-351
surface finish **EM2:** 304
surface requirements **EM2:** 350
textured surfaces **EM2:** 305
tooling materials **EM1:** 169
unsaturated polyesters **EM2:** 250-251
urethane hybrids **EM2:** 269
Resin(s) *See also* Acetal resins; Acetals; Aminos; Constituent materials; Contact pressure resins; Epoxies (EP); Liquid resin; Low-profile resins; Matrices; Matrix resin tests; Phenolics; Phenoxy resins; Phenylsilane resins; Reinforcements; Resin compounds; Resin content; Resin injection; Resin properties analysis; Resin removal; Resin systems; Resole resin; Rigid resin; specific resins; Thermoplastic resins; Thermosetting resins; Tooling resin; Unsaturated polyesters **EL1:** 534-535
acrylated epoxy, for solder masks **EL1:** 555
amorphous **EM2:** 632-633, 636-637
as binders **A15:** 211, 214-221
coatings, for solderability retention **EL1:** 680
commercial, suppliers **EL1:** 534
for compression molding **EM1:** 559-560
content, defined **EM2:** 36
cost guidelines **EM1:** 105
defined **EL1:** 1155 **EM1:** 20, 135 **EM2:** 36
effect, laminate properties **EM1:** 287
effect on solder paste print resolution **EL1:** 732
effects, damage tolerances **EM1:** 262
engineering thermoset, compared **EM2:** 444
epoxy **EM1:** 66-77
epoxy novolac **EL1:** 241, 244
-fiber pullout, by machining/drilling **EM1:** 667-668
for filament winding **EM2:** 369-371
filament-winding **EM1:** 135-138, 505
flexible-epoxy **EL1:** 817-821
flow, testing **EM1:** 737
FR 4 type epoxy **EL1:** 606
hard, for wax patterns **A15:** 197
high-temperature, polyimides as **EM1:** 810-815
homogeneity, as manufacturing factor **EL1:** 82
interfaces, damage tolerance testing of **EM1:** 264
interlaminar fracture toughness **EM1:** 99
life cycle **EM1:** 748
matrix, thermal analysis techniques for **EM1:** 779
mechanical property tests **EM1:** 737
modulus vs. composite interlaminar fracture toughness **EM1:** 99

Resin(s) (continued)
molding, properties **EL1:** 241
polyimide **EL1:** 606 **EM1:** 78-89, 810-815
powder, for impregnation **EM1:** 102
prepreg **EM1:** 139-142
properties **EL1:** 534-537
properties analysis **EM1:** 736-737
properties tests for **EM1:** 289-294
PWB, alternative types **EL1:** 606
recession **EL1:** 575
and reinforcements **EL1:** 534-537
for resin transfer molding **EM1:** 169-170
for rotational molding **EM2:** 361-365
self-extinguishing **EM1:** 21
shear fracture **EM1:** 263
shrinkage effects, investment casting **A15:** 254
solids content, defined and tested **EM1:** 737
in space and missile applications **EM1:** 817-818
specialty **EL1:** 534-535
starvation **EM1:** 660-661
stiffness, and damage tolerance **EM1:** 262
synthetic polymer, as principal constituent **EM2:** 1
temperature ranges **EM2:** 439-444
thermoplastic **EM1:** 97-104
thermoset, properties/tests for **EM1:** 289-290
toughness, impact damage effects **EM1:** 262
treatment, for injection molding **EM1:** 164
types **EL1:** 534-535, 605-607
types, for investment casting waxes **A15:** 254
wet-lay up **EM1:** 132-134
Resin-base caulks **EM3:** 188
Resin-based photopolymer emulsions
screen printing **EM4:** 472
Resin-bonded abrasive wheels **A9:** 24-25
Resin-impregnated composite strands
mounting **A9:** 588
Resin-matrix composites
carbon epoxy wet specimen, local failure **A12:** 477
carbon epoxy wet specimen, tension tested **A12:** 477
carbon phenolic shear specimen **A12:** 478
cleaning **A9:** 588
fabrication methods **A9:** 591
fractographs **A12:** 474-478
fracture morphology **A12:** 477
fracture/failure causes illustrated **A12:** 217
grinding **A9:** 588
longitudinal carbon epoxy specimen, failed in flexure **A12:** 475-476
longitudinal carbon epoxy specimen, failed in tension **A12:** 476-477
matrix-rich regions **A9:** 591
microstructure **A9:** 592
mounting and mounting materials for **A9:** 588
overload tensile resin failure **A12:** 477
ply buckling **A12:** 478
polishing **A9:** 589-591
resin failure **A12:** 474
transverse carbon epoxy, fracture sequence **A12:** 474
transverse carbon epoxy room moisture specimen, failed in tension **A12:** 474
transverse carbon epoxy wet specimen, failed in tension **A12:** 475
voids **A9:** 591
woven carbon fabric phenolic prepreg panels fractures **A12:** 478
short beam shear, compression, and flexure
Resin-matrix composites, specific types
Fiberite 93 epoxy resin, fracture from intemal defect **A12:** 474
Fiberite 934 epoxy resin, fracture from intemal defect **A12:** 474
Resin-matrix graphite composites
material for jet engine components **A18:** 588
Resin-rich area *See also* also Resin pocket; Resin pocket **EM3:** 25
defined **EM1:** 20 **EM2:** 36
Resin-starved area **EM3:** 25
defined **EM1:** 20
Resin-to-fiber bond
aramid fibers **EL1:** 616
Resinking
of dies **A14:** 53
Resinography **EM3:** 24
defined **EM2:** 36
Resinoid **EM3:** 25
Resinoid bond
wheels for thread grinding **A16:** 271-272, 276
Resinoid-bonded abrasive wheels **A9:** 24
Resinous coatings
as corrosion control **A11:** 194
Resinous plasticizers, soft
for wax patterns **A15:** 197
Resinous precoat materials
vacuum coating **M5:** 397
Resins
alkyd **A13:** 400-403
analytic methods for **A10:** 9
auto-oxidative cross-linked **A13:** 400
crack resistance **A13:** 329
ion-exchange, as sampling substrates **A10:** 94
ion-exchange, defined **A10:** 675
organic **EM4:** 45
paint **M5:** 472-473, 483, 497-500, 503
polyimide, curing mechanism **A10:** 285
polymer **A10:** 164
principal coating **A13:** 401-402
thermoplastic **A13:** 403-406
Resins as mounting materials for electropolishing **A9:** 49
Resins, flour-filled
PCD tooling **A16:** 110
Resintering *See also* Postsintering; Sintering
aluminum and aluminum alloys, defined **A2:** 6
defined **A7:** 10
effect on copper powder conductivity **A7:** 116
Resist removal
rigid printed wiring boards **EL1:** 542, 546
Resistance *See also* Modulus of resistance, specific resistances
apparent, defined **EM2:** 593
calculations, cathodic protection applications **A13:** 470-477
capacitance **EL1:** 6, 30-31
change **EL1:** 588, 1002-1003
change, in magabsorption **A17:** 149
crack-growth, *R*-curve as **A11:** 64
curve measurements **A8:** 451-453
defined **A13:** 11 **EL1:** 89, 417, 1155 **EM2:** 460, 467
domain interpretation **A17:** 145
effective crack **A8:** 452
measuring **EL1:** 89
on-chip interconnection lines **EL1:** 6
relative ratings to SCC, aluminum alloys **A11:** 220
temperature dependence **EL1:** 139
term nation **EL1:** 5
to corrosion of chromium carbide-based cermets **A7:** 806
to crack extension measurement **A8:** 456-457
to oxidation of chromium carbide-based cermets **A7:** 806
to titanium carbide-based cermets **A7:** 808
welds, failures in **A11:** 440-442
Resistance (element) heating
as hot pressing setup **A7:** 505-507
for tungsten and molybdenum sintering **A7:** 392
Resistance alloys *See also* Copper alloys; Electrical resistance alloys; Electrical resistance alloys, specific types; Nickel alloys
atmospheres **A2:** 833-835
Resistance alloys (continued)
Constantan alloy **A2:** 825
copper-manganese-nickel resistance alloys (manganins) **A2:** 825
copper-nickel resistance alloys **A2:** 824-825
heating alloys **A2:** 827-829
introduction **A2:** 822
iron-chromium-aluminum alloys **A2:** 828-829
nickel-base **A2:** 433
nickel-chromium alloys **A2:** 825-826, 828
nickel-chromium-iron alloys **A2:** 828
nonmetallic materials **A2:** 829
open resistance heaters, design of **A2:** 829-830
open resistance heaters, fabrication of **A2:** 830-831
properties **A2:** 823
pure metals **A2:** 829
radio alloys, properties **A2:** 823
resistors **A2:** 822-824
service life of heating elements **A2:** 831-833
sheathed heaters **A2:** 831
thermostat metals **A2:** 826-827
types **A2:** 824-829, 835-839
Resistance, arc *See* Arc resistance
Resistance brazing **M6:** 976-988
alloy brazing **M6:** 988
applicability **M6:** 976
capacitor-discharge energy pulses **M6:** 985-987
cleaning **M6:** 982-983
cross-wire resistance brazing **M6:** 978
definition **M6:** 14
electrodes **M6:** 979-981
arrangement **M6:** 980-981
carbon **M6:** 979-980
design **M6:** 980
metal **M6:** 980
equipment **M6:** 976-979
controls **M6:** 977
hand-held tongs **M6:** 979
machine construction **M6:** 977
portable machines **M6:** 977
fast follow-up **M6:** 984-985
filler metals **M6:** 981-982
aluminum-silicon alloys **M6:** 981
application **M6:** 982
copper-phosphorous alloys **M6:** 981
forms **M6:** 982
silver alloys **M6:** 981
step brazing **M6:** 981-982
fluxes **M6:** 982-983
application **M6:** 982
selection **M6:** 982
joint design **M6:** 983-984
butt joints **M6:** 983
lap joints **M6:** 983
self-fixturing **M6:** 983-984
metals brazed **M6:** 976
overheating prevention **M6:** 987
overlapping spot resistance brazing **M6:** 987
plastic-coated wire technique **M6:** 985
power sources **M6:** 976-985
capacitors **M6:** 977
solidified joints **M6:** 987
step brazing **M6:** 984
filler metals **M6:** 981
Resistance brazing (RB) **A6:** 123, 339-342
advantages **A6:** 339
aluminum **A6:** 340
aluminum alloys **A6:** 342
applications **A6:** 339, 342
carbon electrodes **A6:** 340-341
contaminants **A6:** 340
copper **A6:** 339, 340, 342
copper and copper alloys **A6:** 339, 342, 935
definition **A6:** 339, 1212
electrodes **A6:** 339, 340-342
equipment **A6:** 339, 340
factors contributing to high quality in an RB joint **A6:** 339-340

Resistance brazing (RB) (continued)
filler metals **A6:** 340, 341-342
aluminum-silicon alloys **A6:** 342
composition and thermal properties **A6:** 341
copper-phosphorus alloys **A6:** 342
silver alloys **A6:** 341, 342
fluxes **A6:** 342
iron-base alloys **A6:** 341
limitations **A6:** 339
metal electrodes **A6:** 340-342
precious metals **A6:** 340, 936
process parameters **A6:** 339-340
setups, typical **A6:** 339
stainless steel **A6:** 341
steel **A6:** 341
system selection **A6:** 340
Resistance brazing of
aluminum **M6:** 976
aluminum alloys **M6:** 1030
copper and copper alloys **M6:** 976, 1045-1048
copper-tungsten **M6:** 976
low-carbon steels **M6:** 976
nickel alloys **M6:** 976
silver **M6:** 976
silver-graphite **M6:** 976
silver-molybdenum **M6:** 976
silver-tungsten **M6:** 976
stainless steels **M6:** 976
Resistance butt welding
definition **A6:** 1212
Resistance capacitance (RC) analysis
lumped **EL1:** 25, 30-31
Resistance, corrosion *See* Corrosion resistance
Resistance furnaces
high-frequency **A10:** 221-222
Resistance, heat *See* Heat resistance
Resistance heated furnace
dip brazing **A6:** 337
Resistance heaters
for steam generator corrosion **A13:** 942
Resistance heating
for optical holographic interferometry **A17:** 410
Resistance heating vaporization
ion beam plating **M5:** 418
Resistance induction torch welding *See also* Attachment methods; Joining; Welding
as attachment method, sliding contacts **A2:** 842
Resistance, insulation *See* Insulation resistance
Resistance projection welding
of TO packages **EL1:** 239
Resistance projection welding, applications
automotive **A6:** 393, 394-395
Resistance seam weld timer
definition **M6:** 14
Resistance seam welding **M6:** 494-502
advantages **M6:** 494
overlap **M6:** 494
seam width **M6:** 494
applications **M6:** 494
coated steel **M6:** 502
control of welding conditions **M6:** 498-499
quality control **M6:** 498-499
welding of a corner arc **M6:** 498
welding schedules **M6:** 498
definition **M6:** 14
electrode force, effect of **M6:** 499-500
electrodes **M6:** 496-497
bar electrodes **M6:** 497
cooling **M6:** 497
cup-shaped electrodes **M6:** 496-497
face contours **M6:** 496
maintenance of face contour **M6:** 497
wheel size **M6:** 496
heat and cool time, effect of **M6:** 500
joint overlap **M6:** 500
limitations **M6:** 494
fatigue life **M6:** 494
weld design **M6:** 494
machines **M6:** 494-496, 530
circular machines **M6:** 495
controls **M6:** 495
electrode force and support **M6:** 495
electrode or workpiece drives **M6:** 495
longitudinal machines **M6:** 495
portable machines **M6:** 495-496
power supplies **M6:** 495
Resistance seam welding (continued)
universal machines **M6:** 495
metals welded **M6:** 494
practices for series 300 stainless steels **M6:** 530
preweld cleaning **M6:** 494
stainless steel sheet **M6:** 530-531
types of welds *See* Seam welds
welding current, effect of **M6:** 499
welding methods **M6:** 497-498
continuous motion **M6:** 497-498
intermittent motion **M6:** 497-498
workpiece design, effect on electrode shape **M6:** 500
Resistance seam welding (RSEW) **A6:** 238-245
advantages **A6:** 239
aluminum **A6:** 241, 245
aluminum alloys **A6:** 241
applications **A6:** 238-239, 242, 243, 244
automotive **A6:** 393
butt seam welds **A6:** 239, 240, 243-244
carbon steels **A6:** 241
coated steels **A6:** 245
cooling the weld **A6:** 241
definition **A6:** 238
electrodes **A6:** 238, 239, 241, 242, 243, 244, 245
filler metals **A6:** 244
flange-joint lap seam welds **A6:** 242
fundamentals of lap-seam welding **A6:** 239-242
galvanized steel **A6:** 245
lap seam welds **A6:** 238-239, 242, 243
leak-tight seam welding **A6:** 238
limitations **A6:** 239, 240
low-alloy steels **A6:** 241, 245
low-carbon steels **A6:** 239, 240, 241, 243, 244, 245
magnesium alloys **A6:** 245
mash seam welds **A6:** 239, 242-243
metals welded **A6:** 241-242
mild steel **A6:** 243
nickel **A6:** 241
nickel alloys **A6:** 241, 245
nondestructive testing **A6:** 245
postweld heat treatments **A6:** 239
power sources **A6:** 244
processing equipment **A6:** 244-245
reinforced roll spot welding **A6:** 238
resistance seam welds **A6:** 245
roll spot welding **A6:** 238
stainless steels **A6:** 241, 243
terneplate **A6:** 244
tin-plated steel **A6:** 245
types of **A6:** 242-244
weld quality and process control **A6:** 245
weld time, speed, and current pulsation **A6:** 239-241
wheel dressing tools **A6:** 241
wheel geometry, weld force and wheel maintenance **A6:** 241
zirconium alloys **A6:** 787
Resistance seam welding of
alloy steels **M6:** 494
aluminum and aluminum alloys **M6:** 494, 542
coated steels **M6:** 494, 502
copper and copper alloys **M6:** 494-548
high-carbon steels **M6:** 494
high-strength low-alloy steels **M6:** 494
low-carbon steels **M6:** 494, 497-498, 502
magnesium alloys **M6:** 494
nickel and nickel alloys **M6:** 494
stainless steels **M6:** 494, 529-531
Resistance soldering
defined **EL1:** 1155
definition **M6:** 15
Resistance soldering (RS) **A6:** 357-358
aluminum **A6:** 357
aluminum alloys **A6:** 357
applications **A6:** 357
carbon steels **A6:** 357
copper **A6:** 357
copper alloys **A6:** 357
definition **A6:** 357, 1212
electrode configurations **A6:** 357
equipment **A6:** 357
limitations **A6:** 357
low-alloy steels **A6:** 357
nickel **A6:** 357
nickel alloys **A6:** 357
personnel training **A6:** 357
Resistance soldering (RS) (continued)
practice **A6:** 358
stainless steels **A6:** 357
steels **A6:** 357
Resistance spot welding **A7:** 624 **M6:** 469-492
applications **M6:** 469
clamping fixtures **M6:** 489
coated steels **M6:** 491-492
cycles **M6:** 484-485
definition **M6:** 15
destructive testing **M6:** 488-489
direct welding **M6:** 475-476
multiple-spot setups **M6:** 476, 478
single-spot setups **M6:** 475-477
dissimilar metals **M6:** 493
electrode design **M6:** 482
electrode follow-up **M6:** 532-533
electrode holders **M6:** 481-482
electrodes, effect of heat on **M6:** 482-486
equipment **M6:** 469-470
specifications **M6:** 470
tap switches **M6:** 473
windings **M6:** 472
feedback control **M6:** 489-491
heat **M6:** 476-477, 482-486
current flow **M6:** 476
resistances **M6:** 476
welding temperature **M6:** 477
heat-affected zone **M6:** 486-487
heating, effect on **M6:** 482-486
electrode composition **M6:** 482
electrode design **M6:** 482
surface finish requirements **M6:** 486
time **M6:** 484-485
weld spacing **M6:** 486
welding current **M6:** 483
welding force **M6:** 483-484
workpiece surface condition **M6:** 17
lobe curve **M6:** 478, 484, 486
machines **M6:** 470-476
mechanical properties **M6:** 487-488
microstructures **M6:** 486
monitoring **M6:** 489-491
acoustic emission **M6:** 490
electrical **M6:** 490-491
ultrasonic signal **M6:** 4900521
multiple-impulse welding **M6:** 473
nuggets **M6:** 469, 477-478, 483, 490
practices for series 300 stainless steels **M6:** 528
quality control **M6:** 488-489
destructive testing **M6:** 488-489
visual inspection **M6:** 488
recommended practice **M6:** 477-478
roll spot welding **M6:** 493
series welding **M6:** 475-476
multiple-spot setups **M6:** 476-478
spacings of welds **M6:** 486
spot weldability **M6:** 486-487
composition **M6:** 486
material processing **M6:** 486-487
surface conditions **M6:** 485
surface preparation **M6:** 477-478, 485
tests for mechanical properties **M6:** 487-488
visual inspection **M6:** 488
weld current **M6:** 478
weld time **M6:** 478, 484-485, 487
workpiece condition **M6:** 485
Resistance spot welding (RSW) *See also* Resistance welding; Welding **A6:** 226-229
advantages **A6:** 226, 393
aluminum alloy weldability **A6:** 229
aluminum alloys **A6:** 739
applications **A6:** 226, 833-836
automobiles **A6:** 393-394, 395
definition **A6:** 226, 1212
direct welding **A6:** 835-836, 837
electrical circuit **A6:** 226-227
electrodes for **A6:** 227-228
coolant parameters **A6:** 228, 229
shapes **A6:** 228
equipment **A6:** 226-227
heat sources **A6:** 1145-1146
joint design recommendations **A6:** 834
machine construction **A6:** 227
material selection **A6:** 833
mechanically alloyed oxide alloys **A2:** 949

Resistance spot welding (RSW) (continued)
dispersion-strengthened (MA ODS)
multiple spot welding machines **A6:** 227
multiwelders **A6:** 226
nickel-base corrosion-resistant alloys
containing molybdenum **A6:** 594
pedestal-type welding machines **A6:** 227
portable welding guns **A6:** 227, 229
power requirements **A6:** 835
press-type direct-acting machines **A6:** 227
pulsation welding **A6:** 834
push-pull welding **A6:** 227, 836
secondary impedance **A6:** 227
series **A6:** 227
series welding **A6:** 835-836, 837
sheet steel limitations **A6:** 834
single-weld configurations **A6:** 226
specifications for controls **A6:** 226
specifications for equipment **A6:** 226
steel weldability **A6:** 228-229
carbon content **A6:** 228
high-strength low-alloy **A6:** 228
low-carbon **A6:** 228
nitrogen content **A6:** 228
phosphorus content **A6:** 228
recommended practices **A6:** 228
sulfur content **A6:** 228
titanium content **A6:** 228
uncoated **A6:** 228
zinc-coated **A6:** 228-229
thickness parameters **A6:** 834, 836
two-weld configurations **A6:** 226-227
weld strength **A6:** 834-835
zirconium alloys **A6:** 787
Resistance spot welding machines **M6:** 470-476
direct-energy **M6:** 470-473
controls **M6:** 470-472
equipment **M6:** 472-473
secondary circuit **M6:** 473
single-phase **M6:** 470
three-phase **M6:** 470
multiple-electrode **M6:** 475
portable machines **M6:** 474-475
press-type machines **M6:** 474
rocker arm machines **M6:** 473-474
air operated **M6:** 473
foot operated **M6:** 473
motor operated **M6:** 473-474
roll spot welding **M6:** 475
Resistance spot welding of
aluminum alloys **M6:** 479-480, 538, 542
aluminum-coated steels **M6:** 472
electrodes **M6:** 492
welding conditions **M6:** 492
aluminum-plated steel **M6:** 480
carbon steel **M6:** 478, 486
chromium-plated **M6:** 491
coated steels **M6:** 479-480, 491
copper and copper alloys **M6:** 479-480, 548-552
high-strength low-alloy steels **M6:** 478, 484, 486-488
Inconel **M6:** 480
low-alloy steels **M6:** 477, 486
low-carbon steel **M6:** 477, 479-480, 486
magnesium alloys **M6:** 479-480
medium-carbon steels **M6:** 477, 486
Monel alloys **M6:** 480
nickel alloys **M6:** 479
nickel silver **M6:** 479
nickel-plated steel **M6:** 479
silicon bronze **M6:** 479
stainless steels **M6:** 480, 528-529
steels **M6:** 477-478, 480
terne-metal-plated steel **M6:** 480
tin-coated steel **M6:** 492-493
electrodes **M6:** 492-493
welding currents **M6:** 493
tin-plated steel **M6:** 479-480

Resistance spot welding of (continued)
tin-zinc coated steel **M6:** 491-493
electrodes **M6:** 492-493
welding currents **M6:** 493
titanium **M6:** 478
vanadium **M6:** 478
zinc-coated steel **M6:** 491-492
electrode composition **M6:** 491-492
electrode cooling **M6:** 492
electrode design **M6:** 491
welding conditions **M6:** 24
zinc-plated steel **M6:** 480
Resistance strain gage method
of stress analysis **A17:** 448-450
Resistance thermometers
electrical resistance alloys **A2:** 822-823
Resistance to decarburization in wrought tool steels **A1:** 777
Resistance weld(s)
acoustic emission inspection **A17:** 284
discontinuities from **A17:** 588
in pipe, inspection of **A17:** 579
radiographic inspection **A17:** 335
Resistance welding *See also* Attachment methods, Joining; Welding
aluminum alloys *See* Resistance welding of aluminum alloys
aluminum metal-matrix composites **A6:** 555, 558
applications **A6:** 340, 845
sheet metals **A6:** 398
as attachment method, sliding contacts **A2:** 842
capacitor discharge stud welding **A6:** 833
copper and copper alloys *See* Resistance welding of copper and copper alloys
copper metals **M2:** 441, 442
definition **A6:** 1213 **M6:** 15
dispersion-strengthened aluminum
alloys **A6:** 543
distortion **A6:** 834
electrodes **A6:** 833, 834, 835, 836, 837, 838, 839, 840, 843, 844, 846, 849
failure mechanisms **EL1:** 1043-1044
ferritic stainless steels **A6:** 448
flash welding **A6:** 833
heat-treatable aluminum alloys **A6:** 528
high frequency welding **M6:** 757-759
high-frequency resistance welding
(HFRW) **A6:** 833
limitations in choosing materials **A6:** 373
limitations on procedure qualification **A6:** 1093
niobium alloys **A6:** 581
oxide-dispersion-strengthened
materials **A6:** 1038, 1039-1040
power sources **A6:** 40, 41-42
precipitation-hardening stainless steels **A6:** 449, 489, 491, 492
processes, applications **EL1:** 238-239
projection welding (PW) **A6:** 833, 841, 842, 849
refractory metals and alloys **A2:** 563
resistance seam welding (RSEW) **A6:** 833, 836-837, 838
electrode wheels **A6:** 836-837, 838, 839
materials welded **A6:** 837
safety precautions **A6:** 1191, 1201-1202 **M6:** 58
seam welded joints **A9:** 581
as secondary operation **A7:** 456-457
spot welded joints **A9:** 581
stainless steels *See* Resistance welding of' stainless steels
standard procedure qualification test
weldments **A6:** 1090
tantalum alloys **A6:** 580
titanium alloys **A6:** 514, 521-522, 783, 784
to solve problems in joining thin sections by oxyfuel gas welding **A6:** 288
upset welding (UW) **A6:** 833
weld discontinuities **A6:** 1079
wrought martensitic stainless steels **A6:** 441
zirconium alloys **A6:** 787

Resistance welding (RW), procedure development and process considerations **A6:** 833-850
advantages **A6:** 837, 838
aluminum alloys **A6:** 848
applications **A6:** 837, 838, 841
capacitor discharge stud welding **A6:** 846-847
applications **A6:** 847
method variations **A6:** 846-847
sequence of operations **A6:** 847
copper **A6:** 849-850
copper alloys **A6:** 849-850
beryllium copper **A6:** 849
bronzes **A6:** 850
copper nickels **A6:** 850
high-zinc brasses **A6:** 849-850
low-zinc brasses **A6:** 849-850
nickel silvers **A6:** 850
description **A6:** 833
direct and series welding **A6:** 835-836
direct multiple-spot welding **A6:** 835
direct single-spot welding **A6:** 835
push-pull welding **A6:** 836
series multiple-spot welding **A6:** 835-836
electrode wire seam welding **A6:** 837, 839
electrodes **A6:** 836, 840, 843-844, 849
equipment **A6:** 836, 838-840, 842-844, 847, 849
flash welding **A6:** 840-845
foil butt seam welding **A6:** 837, 839
high-frequency resistance welding
(HFRW) **A6:** 845-846
applications **A6:** 846
continuous seam welding **A6:** 846
finite-length welding **A6:** 846
process variations **A6:** 846
welding parameters **A6:** 846
lap seam welding **A6:** 837, 839
limitations **A6:** 837, 838
mash seam welding **A6:** 837, 839
metal finish seam welding **A6:** 837, 839
projection welding **A6:** 837-840, 841, 842, 843
cross-wire welding **A6:** 838
welding machines **A6:** 838-840
resistance seam welding **A6:** 836-837
applications **A6:** 833-835
process variations **A6:** 836-837
seam welding machines **A6:** 836
types of **A6:** 836
resistance spot welding (RSW) **A6:** 833-836
stainless steels **A6:** 833, 837, 840, 841, 842, 847-848
coefficient of thermal expansion **A6:** 847
equipment **A6:** 847
factors affecting **A6:** 847
welding characteristics **A6:** 847-848
upset welding (UW) **A6:** 845
Resistance welding electrode *See also* Electrodes for resistance welding, specific types
definition **A6:** 1213 **M6:** 15
Resistance welding electrodes **A7:** 624-629
as dispersion-strengthened copper
application **A7:** 714-715
powders used **A7:** 573
Resistance welding gun
definition **A6:** 1213 **M6:** 15
Resistance welding of aluminum alloys
base-metal characteristics **M6:** 535-536
alclad alloys **M6:** 535
corrosion resistance **M6:** 536
weldability, effects on **M6:** 535-536
conductivity, electrical and thermal **M6:** 536
cross wire welding **M6:** 543
electrode holders **M6:** 537-538
electrodes **M6:** 537-538
cooling **M6:** 538
design **M6:** 537
electrode holders **M6:** 538
face contour **M6:** 537-538
maintenance **M6:** 538

Resistance welding of aluminum alloys (continued)
seam welding wheels **M6:** 538
flash welding **M6:** 543
grain structure **M6:** 540
inspection and testing **M6:** 543
machines **M6:** 537-540
multiple-electrode-force cycles **M6:** 537
single-phase direct-energy **M6:** 537, 540
slope control **M6:** 537
stored-energy **M6:** 537, 540
synchronous controls **M6:** 537
three-phase direct-energy **M6:** 537, 540
plastic range **M6:** 536
projection welding **M6:** 542-543
roll resistance spot welding **M6:** 542
seam welding **M6:** 542
shrinkage during cooling **M6:** 536
shunting **M6:** 536
spot weld spacing **M6:** 541
spot welding **M6:** 539-542
electrode force **M6:** 541
weld time **M6:** 541
welding current **M6:** 540-541
spot welding practice **M6:** 541-542
conditions for **M6:** 540-541
order of welding **M6:** 542
position for welding **M6:** 542
workpiece thickness **M6:** 541-542
surface oxide **M6:** 536-537
surface preparation **M6:** 538-539
cleaning **M6:** 539
contact-resistance test **M6:** 539
oxide removal **M6:** 539
seam sealants **M6:** 539
temperature **M6:** 536
weld defects **M6:** 543-544
burning of holes **M6:** 544
cracks and porosity **M6:** 543-544
electrode pickup **M6:** 543
expulsion of molten metal **M6:** 544
incomplete fusion **M6:** 544
indentation **M6:** 544
irregular shape **M6:** 544
sheet separation **M6:** 544
weld strength **M6:** 535
Resistance welding of copper and copper alloys
beryllium copper **M6:** 553-558
procedures **M6:** 551
projection welding **M6:** 551-554
spot welding **M6:** 551-552
brasses, low- and high-zinc
procedures **M6:** 554-555
projection welding **M6:** 555-556
bronzes **M6:** 556
cleaning, preweld **M6:** 548
copper nickels **M6:** 556
coppers **M6:** 548-551
controlling heating of workpieces **M6:** 550-551
plating composition **M6:** 551
procedures **M6:** 548
spot welding **M6:** 548-551
electrodes **M6:** 547-548
equipment **M6:** 545, 547
electrostatic stored-energy machines **M6:** 545
single-phase direct-energy machines **M6:** 545
three-phase direct-energy machines **M6:** 545
welding machine controls **M6:** 545, 547
nickel silvers **M6:** 556
safety **M6:** 556
selection of processes **M6:** 548
projection welding **M6:** 548
seam welding **M6:** 548
spot welding **M6:** 548
welding characteristics **M6:** 545, 547
contacting overlap **M6:** 543, 547
electrode force **M6:** 547
physical properties **M6:** 546-547
spot spacing recommendations **M6:** 545, 547
weld time **M6:** 547
welding current **M6:** 547
welding indexes **M6:** 546-547
Resistance welding of stainless steels **M6:** 525-534
austenitic stainless steels **M6:** 527
austenitic stainless steels
nitrogen-strengthened **M6:** 527
compositions of specific types **M6:** 526
cross wire welding **M6:** 531-532

Resistance welding of stainless steels (continued)
equipment **M6:** 525
ferritic stainless steels **M6:** 527
martensitic stainless steels **M6:** 527
multiple-impulse spot welding **M6:** 529
physical properties **M6:** 527
precipitation-hardening stainless steels **M6:** 527
projection welding **M6:** 530-531
resistance seam welding **M6:** 529-531
electrodes **M6:** 530
heat distortion **M6:** 530
machines **M6:** 530
use of mashing **M6:** 529-530
welding of sheet of dissimilar
thickness **M6:** 530-531
welding schedules **M6:** 530
roll resistance spot welding **M6:** 529-531
spot welding **M6:** 528-529
effect of heat on distortion **M6:** 529
electrode follow-up **M6:** 532-533
electrode force **M6:** 528
electrodes **M6:** 528-529
expulsion of molten metal **M6:** 529
spacing of spots **M6:** 528-529
weld time **M6:** 528
welding current **M6:** 528
surface preparation **M6:** 527-528
variables affecting welding **M6:** 525-526
coefficient of thermal expansion **M6:** 525-527
contact resistance **M6:** 526
electrical resistivity **M6:** 525
high strength **M6:** 526
melting temperatures **M6:** 525
thermal conductivity **M6:** 525
weld defects **M6:** 533-534
welding of dissimilar metals **M6:** 532-533
Resistance-inductance-capacitance (RLC) line
cross talk **EL1:** 361-362
homogeneous **EL1:** 357-359
inhomogeneous **EL1:** 359-361
model of **EL1:** 357
stretch factors **EL1:** 360
Resistance-ratio test for pure metals **M2:** 711-712
Resistance-ratio test, for trace elements
ultra-high purity metals **A2:** 1096
Resistive inks
thick-film systems **EL1:** 208, 346-347
Resistive losses (skin effect) **EL1:** 41-42, 603
Resistive materials
physical characteristics **EL1:** 107-108
quantum mechanical band theory
(solid state) **EL1:** 100-101
Resistivity *See also* Electrical resistivity **A6:** 365
EM3: 433-434
common metals and alloys **A17:** 168
copper **EL1:** 249
of copper, in niobium-titanium superconducting materials **A2:** 1045
defined **EL1:** 1155 **EM1:** 20 **EM2:** 36, 593
and dislocation density **A17:** 216
electrical **EM2:** 467
of filler volume, and polymer and
metal particle size ratio **A7:** 608-609
and magnetoresistance **A17:** 144
measured, in shape memory alloys **A2:** 899
and permeability, compared **A17:** 145
of polymers **A7:** 607, 608
pressure tube, as error, remote-field
eddy current inspection **A17:** 200
stable, of electrical resistance alloys **A2:** 822
temperature dependence of **EL1:** 95-96, 98-99
testing, of powder metallurgy parts **A17:** 541
Resistivity stream scanning
to analyze ceramic powder particle
sizes **EM4:** 67
Resistol *See* B-stage
Resistor and single-coil system
eddy current inspection **A17:** 177
Resistor capacitance (RC) elements **EL1:** 76
Resistor films, deposition of
vacuum coating process **M5:** 408
Resistor networks **EL1:** 178, 184
embrittlement in lead attachments of **A11:** 45-46
Resistor-transistor logic (RTL) **EL1:** 160, 1155
Resistors
ballast resistors **A2:** 823
carbon **EL1:** 178

Resistors (continued)
defined **EL1:** 1155
for electrical/electronic devices
classified **A2:** 822-824
fabrication **EL1:** 184-185
failure mechanisms **EL1:** 999-1003
fundamentals **EL1:** 999
germanium point-contact, packaging **EL1:** 958
grown-junction, packaging **EL1:** 958
as IC modification **EL1:** 249
implementation at microwave
frequency **EL1:** 178
materials selection for **EL1:** 182
metallic **EL1:** 178
monolithic, active analog components **EL1:** 144
in passive components **EL1:** 178
precision resistors **A2:** 822
rectangular chip **EL1:** 178
reference resistors **A2:** 823
removal methods and tools **EL1:** 724-727
resistance thermometers **A2:** 822-823
thick-film pastes **EL1:** 343-345
thin-film **EL1:** 316-320, 1099, 1114-1115
through-hole packages, failure
mechanisms **EL1:** 970-971
Resite **EM3:** 25
Resitol *See* B-stage **EM3:** 25
Resol resin **EM3:** 25
Resole *See* A-stage
Resole phenolic resin
properties/tests for **EM1:** 289-290
Resole resin *See also* Resin(s)
defined **EM2:** 36
Resols **EM3:** 25, 105
for binding electrical, industrial, and
decorative laminates **EM3:** 105
for bonding coated abrasives **EM3:** 105
chemistry **EM3:** 103
commercial forms **EM3:** 104
curing mechanism **EM3:** 104
Resolution
of acoustical holography **A17:** 439-441
of borescopes and fiberscopes **A17:** 8
calibration, ultrasonic inspection **A17:** 266-267
CCD videoscopes **A17:** 6
defined **A9:** 15 **A10:** 681
depth, AES **A10:** 556
depth, atom probe analysis **A10:** 595
detector (peak broadening) **A10:** 519
enhancement, methods for **A10:** 116-117
F-F-IR spectrometers **A10:** 112
of features, SEM **A10:** 490
field ion microscope **A10:** 588
high contrast, defined **A17:** 384
high SEM, use in Jominy bar analysis **A10:** 508
image, optical holographic
interferometry **A17:** 414
interferometers **A17:** 15
lateral, atom probe analysis **A10:** 595-596
limits, SEM and optical microscopes
compared **A10:** 495
liquid chromatography **A10:** 651
machine/human vision capabilities **A17:** 30
mass, LEISS analysis **A10:** 605
neutron diffraction **A10:** 422
photographic **A12:** 80-81
of scanning electron microscopes **A9:** 95
SEM **A10:** 494
spatial **A10:** 448, 525, 595-596
spectral, EDS and WDS compared **A10:** 521-522
of transmission electron microscopes **A9:** 103
of x-ray spectrometers **A10:** 91
Resolution (statistical)
and confounding **A17:** 748
of two-level fractional factorial designs **A17:** 747
Resolution enhancement methods
IR qualitative analysis **A10:** 116-117
Resolution in macrophotography **A9:** 87
Resolution of optical microscopes **A9:** 75-76
formula for determining **A9:** 75
relationship to numerical aperture for
four wavelengths of light **A9:** 78
Resolution range
of scanning electron microscopes **A9:** 89
Resolving power
defined **A9:** 15

Resonance *See also* Electron spin resonance
complex, appearance of **A10:** 255
displacement and strain in **A8:** 242
effect, RBS analysis **A10:** 635
electromechanical fatigue tester **A8:** 392
equations, for single crystals **A10:** 269
excitation, high-frequency, specimen geometries for crack growth measurements under **A8:** 251
field, defined **A10:** 267
longitudinal, frequency of specimens **A8:** 249
properties, acoustic extension horn materials **A8:** 245
and specimen lengths for pure metals **A8:** 249
systems **A8:** 241, 392-394
and ultrasonic testing **A8:** 241-242
Resonance methods
electron spin resonance **A10:** 253-266
ferromagnetic resonance **A10:** 267-276
Mössbauer spectroscopy **A10:** 287-295
nuclear magnetic resonance **A10:** 277-286
Resonance scattering **A18:** 409-410
Resonance testing
of powder metallurgy parts **A17:** 540
Resonance transitions
optical emission spectroscopy **A10:** 22
Resonant bar
lengths for pure metals **A8:** 249
uniform **A8:** 242
Resonant fatigue crack growth rate
specimen design **A8:** 251
Resonant fatigue tester
Amsler **A8:** 393
closed-loop **A8:** 393
computer-controlled **A8:** 394
Resonant forced vibration technique **EM3:** 35
defined **EM2:** 36
Resonant-type sensors
acoustic emission inspection **A17:** 280
Resorcinol **EM3:** 103, 104, 105
Resorcinol diglycidyl ether (RDE)
epoxy resin **EM1:** 67
Resorcinol formaldehyde **EM3:** 104
Respiratory disabilities and bronchitis
as gold toxic reaction **A7:** 205
Respiratory disease *See also* Pulmonary disease
from manganese dust **A2:** 1253
from platinum dust **A2:** 1258
from vanadium toxicity **A2:** 1262
Response *See also* Images; Signal response; Signal response analysis; Signal(s)
brightness, control **A17:** 678
curves **A8:** 11, 641-642, 702-703
discrimination **A17:** 675-676
experimental, defined **A8:** 640
as experimental end-property **A8:** 650
function, experimental **A8:** 650
NDE, defined **A17:** 674
parameters, new **A17:** 675
surface **A8:** 650
Response curve
least squares computations **A8:** 703
of Probit fatigue test data **A8:** 702
Response curve for *N* cycles
defined **A8:** 11
Rest potential *See* Corrosion potential
Restacked drilled billet method
of multifilamentary conductor assembly **A2:** 1048-1049
Restacked monofilament method
of multifilamentary conductor assembly **A2:** 1047-1048
Restoration
substructure due to **A10:** 469-470
Restoration, image *See* Image restoration
Restorations
dental **A13:** 1342-1344
Restraint cracks
in friction welds **A11:** 444
Restrictor rings
defined **A18:** 15
Restrike **A7:** 10
Restriking *See also* Sizing
coining as **A14:** 180
defined **A14:** 11
Restructure test
WSI **EL1:** 374, 377
Restructuring, or structuring
of defective circuitry **EL1:** 9
Resulfurized and rephosphorized steel
composition ranges and limits **M1:** 126
Resulfurized carbon and alloy steels
macroetching **A9:** 172
Resulfurized carbon steels, machinability of **A1:** 597-599
control an e of sulfide morphology **A1:** 598
economic **A1:** 598-599
manganese content **A1:** 597
Resulfurized steel
carbon steel wire rod **M1:** 254-255
cold drawing, effects of on tensile properties **M1:** 239-240
composition ranges and limits **M1:** 126
machinability ratings **M1:** 573-576
threaded fasteners **M1:** 274
weldability **M1:** 563
workability **A8:** 165, 575
Resulfurized steels for case hardening
compositions of **A9:** 219
Resulfurized steels, specific types
1117 bar, normalized by austenitizing and cooled in still air **A9:** 224
1117, carbonitrided and oil quenched **A9:** 227
Retained austenite *See also* Austenite
abrasion resistance affected by **M1:** 614-615
abrasion-resistant cast irons **M1:** 81-85
alloy cast irons **M1:** 79-80, 87
banded alloy segregation from **A11:** 121
in bearing materials **A11:** 509
carburized or carbonitrided steels **M1:** 534, 538
effect in distortion **A11:** 141
and forging failure **A11:** 326
gear train backlash affected by **M1:** 615, 618, 620
rapid wear of impact breaker bar due to **A11:** 367-368
tempering for **A11:** 122
Retainer *See* Cage
Retainer rings
steel **M1:** 290, 291
Retainer spring
failure of **A11:** 561-562
Retardation plate
defined **A9:** 15
Retarder *See* Inhibitor **EM3:** 25
Retarding field analyzers
for AES analyses **A10:** 554
Retention
parameters for defining **A10:** 651
time, defined **A10:** 681
Retentivity
magnetic **A17:** 91-92
Reticle
defined **EL1:** 951
Reticles
microscope eyepiece **A9:** 73-74
Reticular structure **A18:** 752
Reticulation **EM3:** 560-561, 562
Retirement-for-cause (RFC)
equipment **A17:** 687-688
of life management **A17:** 672-673
Retort furnace
furnace brazing **A6:** 121
Retracting die **A7:** 10
Retrieval, and storage systems
automatic **A15:** 570-571
Retro-Diels-Aider reaction **EM3:** 154
Retrofitting spark
imaging detectors for **A10:** 144
Retrogradation **EM3:** 25
Return bend
rupture by SCC and inclusions **A11:** 646
Return bend fittings
pipe welding **M6:** 591
Return dies
for blanking **A14:** 453-454
Reused sand *See also* Reclamation; Sand reclamation
for green sand molding **A15:** 225-226
Reverberations
required for stress equilibration **A8:** 191
Reverberatory furnaces *See also* Air furnace
aluminum alloy melting practice **A15:** 376-38
and crucible furnaces **A15:** 374-38
defined **A15:** 10
hearth, types of **A15:** 374-37
safety, operation and design of **A15:** 380-381
Revere, Paul
as early founder **A15:** 2
Reversal development
copier powders **A7:** 584
Reverse bias failures
passive devices **EL1:** 997
Reverse chill *See* Inverse chill
Reverse current cleaning *See* Anodic electrocleaning
Reverse Diels-Alder (RDA) reaction
as addition-type polyimide **EM1:** 79-80
during cure **EM1:** 83
PMR-15 polyimide, reaction scheme **EM1:** 82
Reverse drawing *See also* Drawing
defined **A14:** 11
Reverse engineering
as coordinate measuring machine application **A17:** 18
Reverse extrusion *See also* Extrusion
of aluminum alloys **A14:** 244
Reverse flange
defined **A14:** 11
Reverse helical winding
defined **EM2:** 36
Reverse impact test *See also* Impact test **EM3:** 25
defined **EM1:** 20 **EM2:** 36
Reverse osmosis recovery process
plating waste treatment **M5:** 318
Reverse polarity **A6:** 30, 31
definition **A6:** 1213 **M6:** 15
Reverse polarity air carbon arc cutting **A14:** 734
Reverse redrawing **A14:** 585
punch and die materials **A14:** 511
of thin-wall shells **A14:** 508
tooling for **A14:** 585
Reverse slip *See also* Slip
AISI/SAE alloy steels **A12:** 298
Reverse stress **A8:** 392
forced-displacement system **A8:** 392
Reverse yielding
fracture mechanics of **A11:** 57
Reverse-bend test
and axial push-pull tests, cycles to fracture vs cycle time **A8:** 351
combined creep-fatigue data for austenitic stainless steel sheet **A8:** 356
hold time effect **A8:** 351
Horger data for steel shafts in **A8:** 372
specimen size effect on fatigue limit of carbon steel in **A8:** 372
Reverse-current cleaning *See* Anodic cleaning
Reversed bending fatigue fracture
AISI/SAE alloy steels **A12:** 298
medium-carbon steels **A12:** 269
Reversed cyclic creep test
ductile to brittle fracture **A8:** 353-354
Reversed sigmoidal curves
linearization of **A12:** 213-214
Reversed stresses
crack nucleation under **A11:** 102
Reversed stressing
medium-carbon steel **A12:** 262

Reversed tension-compression testing
ultrasonics ... **A8:** 248
Reversed torsional fatigue fracture
alloy steels ... **A12:** 301
Reversed torsional loading
of shafts ... **A11:** 525
Reversed-bending fatigue
alloy steel lift pin ... **A11:** 77
axle shaft fracture by ... **A11:** 321
beach marks from ... **A11:** 321
in shafts ... **A11:** 462-463
of steel fan shaft ... **A11:** 476-477
of structural bolt ... **A11:** 322
Reversed-phase chromatography
defined ... **A10:** 681
as LC mode ... **A10:** 652-653
separation of cations and anions ... **A10:** 663
Reversed-phase ion chromatography
separation mode ... **A10:** 663
Reversible electrode reactions
in voltammetry ... **A10:** 190-191
Reversible permeability ... **A17:** 145-147
Reversible process ... **A3:** 1 • 7
Reversible temper embrittlement ... **A1:** 698
Revolution, American *See* American Revolution
Rework *See also* Repair; Rework processes
by desoldering ... **EL1:** 722
cost, environmental stress screening ... **EL1:** 876-877
hybrid microcircuits ... **EL1:** 261
processes ... **EL1:** 710-729
solderability defect analysis ... **EL1:** 1036-1037
tape automated bonding (TAB) ... **EL1:** 288
Rework processes *See also* Rework
component types and mounting techniques ... **EL1:** 712-715
materials and processes selection ... **EL1:** 117
printed board configurations ... **EL1:** 711-712
reliable ... **EL1:** 727-729
removal methods, common ... **EL1:** 715-718
removal techniques, surface-mount components ... **EL1:** 722-724
and repair problems ... **EL1:** 710-711
solder extraction system ... **EL1:** 718-722
tool and removal methods ... **EL1:** 724-727
universal repair ... **EL1:** 711
Reworked plastic *See also* Plastics
defined ... **EM2:** 36
Reworking
of magnesium alloys ... **A14:** 825
reamers ... **A16:** 241
of unacceptable flaws ... **A17:** 86
Rexalloy
honing stone selection ... **A16:** 476
Reyn
defined ... **A18:** 15
Reynold's equation ... **A18:** 79, 524
defined ... **A18:** 15
tribological behavior of fluid-film lubrication ... **A18:** 477
Reynold's numbers
and liquid flow ... **A15:** 591-59
Reynolds number ... **A13:** 34-35
Reynolds numbers ... **A6:** 162, 1133
RF *See* Radio frequency
rf excitation ... **A18:** 840
RF generator
ICP systems ... **A10:** 34, 37
RFC inspection equipment
as NDE reliability case study ... **A17:** 687-688
RFEC *See* Remote-field eddy current inspection
RGA *See* Residual gas analysis; Residual gas analysis (RGA)
Rh-Se (Phase Diagram) ... **A3:** 2 • 353
Rh-Ta (Phase Diagram) ... **A3:** 2 • 354
Rh-Ti (Phase Diagram) ... **A3:** 2 • 354
Rh-U (Phase Diagram) ... **A3:** 2 • 354
Rh-V (Phase Diagram) ... **A3:** 3 • 5
RHEED *See* Reflection high-energy electron diffraction
Rhenium *See also* Pure rhenium; Refractory metals and alloys; Refractory metals and alloys, specific types; Rhenium alloys; Rhenium alloys, specific types ... **A9:** 447-448
addition to filler metal ... **M6:** 464
addition to improve weldability of tungsten ... **A6:** 870
Rhenium (continued)
alloying effect on electron beam welding ... **M6:** 639
as an addition to tungsten alloys ... **A9:** 442
atomic interaction descriptions ... **A6:** 144
consumption ... **A2:** 557
corrosion resistance ... **A2:** 581-582
effect of alloying additions on ductility ... **A7:** 770
effect on weld ductility ... **M6:** 464
electrical resistivity vs. temperature for ... **A7:** 769
electrodes for resistance brazing ... **A6:** 340
evaporation fields for ... **A10:** 587
filament, gas mass spectrometer ... **A10:** 153
for heating elements for electrically heated furnaces ... **EM4:** 247
as high-temperature material ... **A7:** 769
highest oxide, titration of ... **A10:** 173
machining techniques ... **A2:** 561-562
mechanical and physical properties ... **A2:** 582
mechanical properties ... **A7:** 477
in nickel-base superalloys ... **A1:** 984
occurrence and production ... **A2:** 581
photochemical machining ... **A16:** 588
pure ... **M2:** 788-790
pure, properties ... **A2:** 1150
refractory metal brazing, filler metal ... **A6:** 942
in sintered metal powder process ... **EM3:** 304
solubility in molybdenum during etching ... **A9:** 447
species weighed in gravimetry ... **A10:** 172
sulfuric acid as dissolution medium ... **A10:** 165
in tungsten-rhenium thermocouples used in vacuum heat treating ... **A4:** 506, 507
welding ... **A2:** 563
Rhenium alloys *See also* Rhenium
electron-beam welding ... **A6:** 581, 582
gas-tungsten arc welding ... **A6:** 581
as high-temperature materials ... **A7:** 769
mechanical properties ... **A7:** 477
Rhenium powder ... **A9:** 447
Rhenium, zone refined
impurity concentration ... **M2:** 713
Rhenium-bearing alloys ... **A9:** 447-448
Rheocasting *See also* Semisolid metal casting and forging
alternative approaches ... **A15:** 330-331
defined ... **A15:** 10
as innovative ... **A15:** 3
key parameters ... **A15:** 329-33
operations flow chart for ... **A15:** 20
original method ... **A15:** 32
Rheodynamic lubrication
defined ... **A18:** 15
Rheological agent
effect on solder paste print resolution ... **EL1:** 732
Rheological behavior
models of ... **EL1:** 847-852
Rheological characterization *See also* Rheology
flows and material functions ... **EL1:** 838-842
material behavior, examples ... **EL1:** 842-847
rheological behavior, models ... **EL1:** 847-852
Rheological properties
measurement of ... **EM3:** 322-323
Rheology ... **EM3:** 25
defined ... **A18:** 15 **EL1:** 1155 **EM1:** 20 **EM2:** 36
in encapsulation ... **EL1:** 838
as epoxy quality assurance tool ... **EL1:** 835
melt, cone/plate/parallel geometries in ... **EM2:** 535-540
polyamide-imides (PAI) ... **EM2:** 132
as thermal testing ... **EM2:** 527
Rheology modifiers
with precious metal powders ... **A7:** 149
Rheometric dynamic scanning (RDS) analysis ... **EM1:** 704
Rheometry *See also* Rheology
dynamic mechanical ... **EM2:** 536-538
extensional ... **EM2:** 535
steady-shear ... **EM2:** 535
torque ... **EM2:** 534
Rheopectic material
defined ... **A18:** 15-16
Rheostat transducer
for torsion testing ... **A8:** 158
Rhodium *See also* Precious metals ... **A13:** 801-803, 805
annealing ... **A4:** 945, 946-947 **M4:** 760, 761
coatings, for sterling silver ... **A2:** 691
corrosion resistance ... **M2:** 669
determined by controlled-potential coulometry ... **A10:** 209
electrical circuits for electropolishing ... **A9:** 49
electroplated coating ... **A18:** 838
evaporation fields for ... **A10:** 587
gravimetric finishes ... **A10:** 171
hardness ... **A4:** 947
mechanical properties ... **A4:** 944
in medical therapy, toxic effects ... **A2:** 1258
overplating silver to prevent tarnishing ... **EM4:** 474
platinum-rhodium used in control thermocouples in vacuum heat treating ... **A4:** 506
as precious metal ... **A2:** 688
as prompt emission converter for thermal neutron radiography ... **EM3:** 759
pure ... **M2:** 790-791
pure, properties ... **A2:** 1151
in PWB manufacturing ... **EL1:** 510
refractory metal brazing, filler metal ... **A6:** 942
relative solderability ... **A6:** 134
as a function of flux types ... **A6:** 129
resources and consumption ... **A2:** 689-690
semifinished products ... **A2:** 694
special properties ... **A2:** 692, 694
thermal diffusivity from 20 to 100 °C ... **A6:** 4
thermal expansion coefficient ... **A6:** 907
toxicity ... **A7:** 207
as tube anode material, x-ray spectrometers ... **A10:** 90
tube, from iron and plastic, Compton scatter for ... **A10:** 99
vapor pressure, relation to temperature ... **A4:** 495 **M4:** 310
working of ... **A14:** 850
Rhodium coatings and films
soldering ... **A6:** 631
Rhodium electroplating
corrosion protection ... **M1:** 753, 754
Rhodium plate ... **A9:** 564
Rhodium plating ... **M5:** 290-291
applications ... **M5:** 290-291
barrel process ... **M5:** 290-291
copper and copper alloys ... **M5:** 623
low-stress solutions ... **M5:** 290-291
molybdenum ... **M5:** 660
nickel undercoatings used in ... **M5:** 291
phosphate process ... **M5:** 290
solution compositions ... **M5:** 290-291
specifications ... **M5:** 290-291
sterling silver ... **M5:** 290-291
sulfamate process ... **M5:** 290-291
Rhodium trichloride
toxic effects ... **A2:** 1258
Rhombohedral
defined ... **A9:** 15
Rhombohedral crystal system ... **A3:** 1 • 10, 15 **A9:** 706
Rhombohedral unit cells ... **A10:** 346-348
Rib
defined ... **EM1:** 20
Rib mark
definition ... **EM4:** 633
Rib markings
on fracture surface ... **EM2:** 810
Rib sink
in injection molding ... **EM1:** 166
Rib(s) ... **A14:** 11, 552
Rib-web forging
simulation of ... **A14:** 428-429
Ribbon, annealed
in open resistance heaters ... **A2:** 830
Ribbon cables
optical fiber ... **EL1:** 9
Ribbons, abraded
FMR study of ... **A10:** 274
Riboflavin
ESR studied ... **A10:** 264
Ribs
design of ... **EM2:** 615
polyamide-imides (PAI) ... **EM2:** 131

Ribs in forgings M1: 362, 363, 364
Rice hull pyrolysis EM1: 64, 889, 896
Rice J-integral
for crack-extension force in elastic deformation A8: 440
defined and applied A8: 447-450
J analysis based on A8: 446
Rich alloy *See* Hardener
Rich exothermic gases, composition
See also Exothermic gas A7: 342
Richards, Theodore .... A3: 1 • 7
Riddle
for sand screening A15: 3
Ridge shear failure model A18: 60, 67
Ridges *See also* Tear ridges; Tearing
AISI/SAE alloy steels A12: 304, 331
austenitic stainless steels A12: 360
by shear A12: 371
from sulfur-containing environments A12: 41, 53
high-carbon steels A12: 289
maraging steels A12: 385
precipitation-hardening stainless steels A12: 370-371
tear, AISI/SAE alloy steels A12: 293
tear, wrought aluminum alloys A12: 417
tool steels A12: 380
Ridges, in shafts
from torsional fatigue A11: 464
Ridging (wear)
defined A18: 16
Ridging-type Brinell indentation A8: 85
Rietveld method
$Nd_2(Co_{0.1}Fe_{0.9})_{14}B$ analysis by A10: 425-426
in neutron powder diffraction A10: 423, 425-426
of x-ray or neutron diffraction data A10: 344
Rietveld refinement methods EM4: 73
Rifflers
as blenders for samples A10: 17
chute and spinning A7: 226
Riffling EM4: 83, 85
of particulate samples A10: 165
Rifle receivers
exfoliation failures in A11: 338-342
Rifles, cylindrical receivers
economy in manufacture M3: 849-850
Rigging *See also* Gates; Risers
defined A15: 10, 19
Right-angle mold sections
monocrystal casting A15: 323
Right-hand rule
of magnetic field direction A17: 90
Rigid borescopes
types A17: 4-5
Rigid dies
compacting at high pressures in A7: 306
compaction and isostatic pressing densities compared A7: 298-299
effect of lubrication on density and stress distribution A7: 302
pressing of powders in A7: 297
pressure and density of pressed iron powders A7: 298
Rigid epoxies *See* Coatings; Conformal coatings; Epoxies; Epoxy; Epoxy materials; Flexible epoxies
encapsulant composition EL1: 812-813
encapsulation processes EL1: 815-816
epoxy characteristics EL1: 810-812
for mechanical support EL1: 810
performance-related properties EL1: 813-815
Rigid foam
polyurethane (PUR) EM2: 259
Rigid integral skin foam
polyurethane (PUR) EM2: 259
Rigid plastics *See also* Plastics; Semi-rigid plastic EM3: 25
defined EM2: 36
Rigid PWB fabrication techniques *See also* Printed wiring boards (PWBs)
artwork EL1: 548-549
buried/semiburied (blind) vias EL1: 550
gold plating EL1: 549-550
high aspect ratio holes EL1: 550-551
hot-air solder leveling (HASL) EL1: 550
manufacturing overview EL1: 539-540
mass lamination EL1: 551
materials EL1: 538-539
multilayer hole-to-internal-feature registration EL1: 552
printed circuit complexity EL1: 551-552
printed circuit manufacturing, in-depth EL1: 540-548
solder mask, over bare copper EL1: 550
types EL1: 539
Rigid resin *See also* Resins EM3: 25
defined EM2: 36
Rigid shear tool (Amsler)
for shear testing A8: 62-63
Rigid tool compaction A7: 322-328
for shape attainment A7: 295
systems, part sizes A7: 324
Rigid tool set A7: 323
Rigid-viscoplastic method
analytical modeling A14: 425
Rigidity
modulus of EM3: 322
Rigidity, modulus *See* Modulus of rigidity
Rigidity, modulus of *See* Modulus of rigidity
RIM *See* Reaction injection molding
Rimlock diamond cutting wheels A9: 25
Rimmed steel M1: 112, 123, 556
1008, coiled, cold rolled, effects of different process temperatures A9: 182
1008, orange peel A9: 182
1008, stretcher strains A9: 182
aged, transition between elastic and plastic regions A8: 554
as-rolled A9: 179
carbon steel wire rod M1: 255, 257
cold rolled *See* Cold rolled rimmed steel
corrosion in seawater M1: 741
engineering stress/strain curve for A8: 554
finish rolled, different rolling temperatures A9: 179-180
grain shape, steel sheet M1: 558
load-extension curves, sheet products M1: 548
manganese sulfide content and ductility in hot torsion tests A8: 166
mechanical properties related to formability M1: 548
notch toughness of M1: 694, 698
plastic strain ratio, typical M1: 548
sheet, strain-age embrittlement of *See also* Box-annealed rimmed steel; Steel sheet M1: 683-684
Rimmed steels A1: 6, 141, 143, 145, 578
sample dissolution for A10: 176
Ring
for flyer plate impact test A8: 211
standard, theoretical calibration curve for A8: 585
thin-wall, Rockwell hardness testing of A8: 81, 83
Ring assembly
arcing failure of A12: 488
Ring compression test A1: 583 A8: 585-586
for workability A14: 379-381
Ring diffraction patterns
SAD/ATEM A10: 436-437
Ring displacement A8: 210
Ring illumination
circular fluorescent-light tube A12: 83
Ring patterns
in wrought heat-resistant alloys A9: 305
Ring rolling *See also* Forging; Ring rolling machines; Ring rolling mills A14: 108-127
allowances, machining A14: 125-126
Ring rolling (continued)
and alternative processes A14: 126
of aluminum alloys A14: 244
ancillary operations A14: 123-124
application A14: 108-111
blank preparation A14: 122-123
blanking and rolling tools for A14: 62-63, 121, 124-125
and closed-die forging, combined A14: 125
and closed-die forging, compared A14: 122-123
contour A14: 117-120
of copper and copper alloys A14: 255-256
defined A14: 11, 108
of heat-resistant alloys A14: 231-232
machines A14: 111-115
product A14: 108-111
product and process technology A14: 115-122
relative displacement during A14: 112
rolling forces, power, and speeds A14: 120-122
slip fields from A14: 112
of stainless steels A14: 222
of titanium alloys A14: 273
tolerances, rolled ring A14: 125-126
work rolls A14: 124-125
wrought aluminum alloy A2: 34
Ring rolling machines *See also* Ring rolling; Ring rolling mills A14: 111-115
automatic radial-axial multiple-mandrel ring mills A14: 15
closed-die axial rolling A14: 114-115
history A14: 111-113
multiple-mandrel mills A14: 114
radial-axial horizontal A14: 113
vertical A14: 109, 112-113
Ring rolling mills A14: 109-111
Ring seal
defined A18: 16
Ring shear test A18: 404
Ring test
advantages A8: 210
expanding, for high-rate tensile testing A8: 210
specimen A8: 585
Ring tools
for open-die forging A14: 62-63
Ring(s) *See also* Ring rolling; Rolling
C-profiled, production stages A14: 117
forging of A14: 69
height and diameter, hyperbolic relationships A14: 116
preform, powder forging A14: 192
seamless, allowances and tolerances A14: 126-127
shapes, by ring rolling A14: 108-111
starter, closed-die forged and ring rolled compared A14: 122
tools A14: 62-63
weld-neck flange, closed-die forged and ring rolled, compared A14: 123
Ring-and-plug joints
properties EL1: 640-641
Ring-opening polymerization
molding compounds EL1: 805
Ring-rolled forgings
inspection of A17: 495
Ring-shaped parts
magnetizing A17: 94
Ring-up time
defined A8: 209
Ringdown counts *See* Threshold-crossing counts
Ringing
in aluminum alloys A8: 40, 44
effect in dynamic tests A8: 40
in high-speed torsional system A8: 216
in load cells A8: 193
low pass filter to reduce A8: 193
in servohydraulic testing systems A8: 260
in signal transmission EL1: 170-172
in torsional impact testing A8: 216
Rings
bearing, magnetic particle inspection A17: 117

Rings (continued)
enlargement, in bearing failures **A11:** 494
magnetizing **A17:** 90, 98
materials for **A11:** 515
mechanically alloyed oxide dispersion-strengthened (MA ODS) alloys **A2:** 948
retaining, brittle fracture by arc strikes **A11:** 97
steel, fretting failure **A11:** 498
Rings, solder *See* Solder rings
Rinse aids
as aqueous cleaners **EL1:** 664
Rinse time
off-chip **EL1:** 401
Rinse waters
acidity-basicity measured **A10:** 172
Rinsing
for chromate conversion coating **A13:** 389-390
of cold extruded parts **A14:** 304
final, phosphate coating **A13:** 387
in liquid penetrant inspection **A17:** 83-84
in surface cleaning **A13:** 380-382
trivalent chromium **A13:** 387
Rip-up routers
as automatic trace routing **EL1:** 532-533
Ripening
in creep testing **A8:** 306
Ripple formation (rippling)
defined **A18:** 16
Ripple mark
definition **EM4:** 633
Ripple voltage
defined **EL1:** 1156
Ripples
AISI/SAE alloy steels **A12:** 329
and cleavage, compared **A12:** 453
from slip step formation **A12:** 16
mechanism **A12:** 13
Rise time
classification **EL1:** 25
defined **EL1:** 1156 **EM1:** 20 **EM2:** 36
degradation, and signal attenuation **EL1:** 603
in explosively loaded torsional Kolsky bar **A8:** 224
of incident wave **A8:** 200-201
signal, and local physical performance **EL1:** 5
Rise time, defined
acoustic emission inspection **A17:** 283
Riser
broken casting at **A11:** 386
defined **A13:** 11
Riser block
cam plastometer **A8:** 195-196
Riser design
feed metal volume **A15:** 577-578
feeding aids **A15:** 586-588
liquid feed metal availability duration **A15:** 582-585
neck configurations **A15:** 588
optimum **A15:** 577-585
plain carbon steels **A15:** 711-712
riser configurations **A15:** 588
riser location **A15:** 578-582
riser necks and breaker cores **A15:** 587-588
riser size, factors **A15:** 586-587
Riser necks
and breaker cores **A15:** 587-588
Riser(s) *See also* Blind riser; Gates; Mold cavity; Rigging; Riser design; Risering
copper alloy casting **A15:** 779-781
defined **A15:** 10, 204
magnesium alloy **A15:** 806
nickel alloys **A15:** 822
as pattern feature **A15:** 192
Risering *See also* Riser(s)
aluminum alloys **A15:** 754-755
automated **A15:** 36
defined **A15:** 192
design of **A15:** 577-588
ductile iron **A15:** 651-652
modified **A15:** 192
principles **A15:** 754
Risers
ductile cast iron **M1:** 33
Risers, use
copper casting alloys **A2:** 346
Rising film evaporation
plating waste recovery process **M5:** 314, 316
Rising step-load test
for HY steel compositions **A8:** 540-541
for hydrogen embrittlement **A8:** 539-540 **A13:** 286-287
load-time record for **A8:** 540
loading frame for **A8:** 541
Ritchie-Knott-Rice (RKR) model, for cleavage fracture toughness **A8:** 466-467
Rivet, pin bearing testing of **A8:** 59-61
Risk analysis
for microbiological corrosion testing data **A13:** 315
Risk function
reliability of glass strength **EM4:** 744
Risk site
analysis, in computer-aided analysis **EL1:** 133
concept **EL1:** 127
shorts, forms of **EL1:** 127
River discharge
effects in seawater **A13:** 894
River marks
definition **EM4:** 633
River patterns *See also* Hackle
AISI/SAE alloy steels **A12:** 291
Alnico alloy **A12:** 461
in Armco iron **A12:** 18
in cleavage fractures **A12:** 13, 172, 175, 252, 397, 424
in composites **A11:** 742, 743
corrosion fatigue, aluminum alloy **A12:** 43, 54
defined **A11:** 8 **A12:** 13
in ductile irons **A12:** 230-231
and fracture origin **A11:** 80
and fracture temperature **A11:** 83, 84
generation **A11:** 22, 23, 25
in iron **A12:** 18, 219, 222-224
low-carbon steel **A12:** 252
metal-matrix composites **A12:** 466
mode I tension fractures, composites **A11:** 735
nickel alloys **A12:** 397
on brittle transgranular fracture surface **A11:** 77
on fracture surfaces **A12:** 13, 252
polymers **A12:** 479
precipitation-hardening stainless steels **A12:** 370
superalloys **A12:** 395
and thermal evaporation **A12:** 172
titanium alloys **A12:** 448
wrought aluminum alloys **A12:** 424
Rivers process **A1:** 819
Rivet shear tool
for shear testing **A8:** 62
for thick aluminum plate **A8:** 63
Riveted joints
crevice corrosion in **A11:** 184
failures in **A11:** 544-545
Riveters
electromagnetic **A14:** 648-649
Riveting
magnesium **M2:** 549
of magnesium and magnesium alloys **A2:** 1420-1421
mechanically alloyed oxide alloys **A2:** 949
dispersion-strengthened (MA ODS)
Riveting assembly
furnace brazing **M6:** 940
Rivets **EM3:** 34
bearing-surface failure **A11:** 544
blind **A11:** 530
defined **A11:** 529
failures in **A11:** 544-545
flexible printed boards **EL1:** 590
flush-head, loading and failure **A11:** 544-545
hole, delamination at **A11:** 551-553
mechanical fastening of **EM2:** 713
positioning **A11:** 544
shear in shank **A11:** 544
steel wire for **M1:** 265-266
stress-corrosion cracking **A11:** 544-545
types of **A11:** 544
Rivets, aluminum
eddy current inspection **A17:** 187
RKR *See* Ritchie-Knott-Rice (RKR) model
$\mathbf{R_L}$ *See* Roughness parameters
RLC line *See* Resistance-inductance- capacitance (RLC) line
RMI 0.2 Pd *See* Titanium alloys, specific types, Ti-Pd alloys
RMI 30 *See* Titanium alloys, specific types, Ti grade 1
RMI 40 *See* Titanium alloys, specific types, Ti grade 2
RMI 55 *See* Titanium alloys, specific types, Ti grade 3
RMI 70 *See* Titanium alloys, specific types, Ti grade 4
rms noise **A18:** 339
rms roughness **A18:** 334, 343
Road map **EL1:** 436-437, 1156
Road planing
with cemented carbide tools **A2:** 973
Roadarm weldment
brittle fracture of **A11:** 391-392
Roberts-Austen, William .... **A3:** 1 • 23
Robertson test **A8:** 259
compared with Esso (Feely) and Navy tear- test **A11:** 59-60
for notch toughness **A11:** 59-60
Robinson detector **A12:** 168
Robot-based noncontact measuring system
applied **A17:** 41
Robotic dispensing applications **EM3:** 699-700, 701-702
Robotic foundry applications
cleaning **A15:** 566-567
mold spraying **A15:** 568
mold venting **A15:** 568
pick and place operations **A15:** 568
riser cutting **A15:** 567-568
Robotic foundry operations
holding devices **A15:** 509-510
ladies, in automatic pouring systems **A15:** 498
transferring devices **A15:** 509-510
Robotic inspection systems *See also* Machine vision **A17:** 29-45
applications **A17:** 37-41
future outlook **A17:** 41-45
machine vision process **A17:** 30-37
Robotic systems
waterjet machining **A16:** 525, 526
Robotic ultrasonic inspection **EM2:** 845
Robotics **EM3:** 705, 716-725
advancements in dispensing technology **EM3:** 719-720
applications **EM3:** 720-724
automotive body shop robotic sealing **EM3:** 723-724
automotive door bonding **EM3:** 721-722
automotive interior seam sealing **EM3:** 720-721
automotive windshield bonding **EM3:** 723-724
for blind fastening **EM1:** 711
developing a robotic system **EM3:** 724-725
dispensing equipment for robotic applications **EM3:** 716-719
bead management methods **EM3:** 718-719
dispensing gun or valve **EM3:** 716, 717-718
header system **EM3:** 716, 717
pumping system **EM3:** 716-717
gas-metal arc welding applications **A6:** 180
laser-beam welding and **A6:** 266-267
material handling systems for plasma arc cutting **A6:** 1169
"operating window" of robot **A6:** 1062
portable welding guns used in resistance spot welding **A6:** 227, 229
robot accuracy **A6:** 1062
robot repeatability **A6:** 1062
WELDEXCELL system **A6:** 1057-1064
Robots *See* Industrial robots
Robots, and robotics
for surface-mount joints **EL1:** 731-732
Robust design *See also* Experimental design
approach, to quality design **A17:** 721-722
degree of control **A17:** 750
parameter design, analytical approaches **A17:** 750-751
ROC curves *See* Relative operating characteristic curves
Rochelle copper cyanide plating **M5:** 159, 161-163, 167, 169

Rochelle salt additions
bronze plating solutions **M5:** 288-289
Rock
ground by diamond wheels **A16:** 455
TNAA detection limits for **A10:** 237-238
Rock-candy fracture
Alnico alloy **A12:** 461
austenitic stainless steels **A12:** 351
as casting defect **A11:** 383
defined **A11:** 8
from stress corrosion **A11:** 28
as intergranular **A12:** 110-111
large grain size of **A11:** 75
low-alloy steel **A11:** 392
low-carbon steel **A12:** 249
Rocker arm drives
mechanical presses **A14:** 494
Rocker arm, forged steel
burning of **A11:** 119-120
Rocker arm resistance welding machines **M6:** 473-474
Rocker lever
failed malleable iron **A11:** 350-352
Rocker shear
cut-to-length lines **A14:** 711
Rocker-type dies
for press-brake forming **A14:** 537
Rocket fuel catalysts
powders used **A7:** 572
Rocket fuels
powder used **A7:** 572
Rocket launcher parts
powders used **A7:** 573
Rocket motor case
filament-wound **EM1:** 505-510
Rocket nozzles **A7:** 561
Rocket-motor case
brittle fracture **A11:** 95-96
Rocket-nozzle component
combined SEM/AES analysis of failed niobium alloy **A11:** 42
Rocking
refractory metals and alloys **A2:** 563
Rocking chair effect
fatigue resistance **A8:** 712
Rocking curves
defined **A10:** 681
gold single crystal **A10:** 373
grains, polycrystalline sample **A10:** 374
as intensity profiles **A10:** 372
polycrystal analysis **A10:** 371-373
profiles, for epitaxial films, differing thicknesses **A10:** 375
Rocking-die forge **A14:** 177-178
Rocking-die machines
capabilities **A14:** 178
Rockshaft servo cam and rockshaft cam follower
P/M farm machinery **A7:** 672
Rockwell A hardness numbers
Brinell hardness conversions, steel **A8:** 111
equivalent Rockwell B numbers, steel **A8:** 109-110
Rockwell C hardness conversion, steel **A8:** 110
Vickers hardness conversions, steel **A8:** 112-113
Rockwell B hardness numbers
approximate equivalents, steel **A8:** 109-110
Brinell hardness conversions, steel **A8:** 111
Rockwell C hardness conversions for steel **A8:** 110
Vickers hardness conversions, steel **A8:** 12-113
Rockwell C hardness **A7:** 451, 452
function of carbon content **A18:** 874
Rockwell C hardness numbers
equivalent hardness numbers for **A8:** 110
equivalent Rockwell B numbers steel **A8:** 109-110
Vickers hardness conversions, steel **A8:** 112-113
Rockwell C test
and Vickers hardness test **A8:** 91
Rockwell D hardness numbers
Brinell hardness conversions, steel **A8:** 111
Rockwell C hardness conversions, steel **A8:** 110
Vickers hardness conversions, steel **A8:** 112-113
Rockwell F hardness numbers, equivalent Rockwell B numbers
steel **A8:** 109-110
Rockwell hardness *See also* Hardness **EM3:** 25
abbreviation for **A10:** 690
defined **EM1:** 20 **EM2:** 36
diffraction-peak breadth at half height as function of **A10:** 387
of engineering plastics **EM2:** 245
number (HR), defined **A11:** 8-9
symbol for **A11:** 797
test, defined **A11:** 9
Rockwell hardness number
Brinell hardness conversions, steel **A8:** 111
defined **A8:** 11, 73 **A18:** 16
Rockwell hardness scales
applications **A8:** 76
as defined by indenter and load **A8:** 74-77
factors for selecting **A8:** 75-77
indenters **A8:** 76
specimen thickness for **A8:** 75-77
symbol **A8:** 76
values **A8:** 74
Rockwell hardness testing **A8:** 11, 74-83, 102
adjustments for specimen size and configuration **A8:** 80-81
applications **A8:** 102
at elevated temperatures **A8:** 81-82
and Brinell test, compared **A8:** 74
calibration testing machines for **A8:** 79
of castings **A17:** 521
correction factors for cylindrical workpieces **A8:** 82
depth **A8:** 102
depth of penetration **A8:** 75, 77
diagonal or diameter **A8:** 102
indenter, load, and scale selection **A8:** 74-77
indenters **A8:** 102
of large/long specimens **A8:** 80-81
load **A8:** 102
machines for **A8:** 77-78
method of measurement **A8:** 102
minimum work-metal hardness values for **A8:** 77
minor/major loads **A8:** 74
of rings, tubes, gears **A8:** 81, 83
scales for **A8:** 74-77
of specific materials **A8:** 82-83
specimen thickness **A8:** 75-77
standard hardness **A8:** 76
as static indentation test **A8:** 71
surface preparation **A8:** 102
techniques compared **A8:** 102
testing methodology **A8:** 79-80
Of workpieces with curved or inner surfaces **A8:** 81, 83
Rockwell superficial hardness number *See* Rockwell hardness number
defined **A11:** 9
Rockwell superficial hardness numbers
Brinell hardness conversions, steel **A8:** 111
equivalent Rockwell B numbers steel **A8:** 109-110
Rockwell C hardness conversions, steel **A8:** 110
Vickers hardness conversions, steel **A8:** 112-113
Rockwell superficial hardness test
applications **A8:** 102
correction factors for cylindrical workpieces **A8:** 82
defined **A8:** 11 **A18:** 16
depth **A8:** 102
indenters **A8:** 102
load **A8:** 74, 102
method of measurement **A8:** 102
Rockwell superficial hardness test (continued)
scales for **A8:** 76
techniques compared **A8:** 102
Rockwell testing machines **A8:** 77-79, 82
automatic **A8:** 79
bench-type **A8:** 78
correction factors for cylindrical workpieces **A8:** 82
dead weight or spring **A8:** 77-78
modified for elevated temperatures **A8:** 81-83
portable **A8:** 78, 79
production **A8:** 78
Rockwell-inch test **M1:** 457
Rod *See also* Connecting rods
annealed, in open resistance heaters **A2:** 830
beryllium **A2:** 683
beryllium-copper alloys **A2:** 403, 409, 411
brass *See* Brass, rod
brazing filler metals available in this form **A6:** 119
defined **A14:** 11
drawing of **A14:** 333-334
drawing, schematic **A14:** 330
forming, nickel-base alloy **A14:** 836
headers, for cold heading **A14:** 292
polyamide (PA) **EM2:** 125
preparation, for wiredrawing and wire stranding **A2:** 255-256
steel *See* Steel, rod
unalloyed uranium **A2:** 671
wrought aluminum alloy **A2:** 33
wrought beryllium-copper alloys **A2:** 409
zirconium **A2:** 663
Rod eutectic microstructure ..... **A3:** 1 • 20
Rod eutectic structures *See* Lamellar eutectic structures
Rod extensometers
for creep testing **A8:** 303
for low/elevated tension testing **A8:** 36
Rod gun flame spraying systems
ceramic coating processes **M5:** 540-542
thermal spray coating processes **M5:** 365-366
Rod impact (Taylor) test
analysis **A8:** 205-206
asymmetric rod impact test **A8:** 204
asymmetric rod impact test at elevated temperatures **A8:** 205
basic principles **A8:** 203-204
C-HEMP to determine flow curve **A8:** 205
classic Taylor test **A8:** 203
for high strain rates **A8:** 190, 203-206
symmetric rod impact test **A8:** 203-205
test procedure **A8:** 205
Rod mill rolls
cemented carbide **A2:** 969
Rod process
for assembly of A15 superconductors **A2:** 1065-1066
Rod, rolled and drawn aluminum
specimen locations **A8:** 60
Rod, steel
annealing **M4:** 19, 24-25, 26
Rod-pumped wells
oil/gas production **A13:** 1247
Rods
connecting, and shafts **A11:** 459
connecting, magnetic particle inspection methods **A17:** 117-119
explosion welding **M6:** 709
fiber textures in **A10:** 359
low-hydrogen welding **A11:** 251
nickel-plated aluminum, magabsorption measurement **A17:** 156-157
oxyacetylene welding of cast iron **M6:** 602-603
piston, and shafts **A11:** 459
preferred orientation in **A10:** 359
push, bending fatigue fracture of **A11:** 469
steel articulated, fatigue fracture from electroetched numeral in **A11:** 473

Rods (continued)
steel master connecting, fatigue cracking ... **A11:** 477
Roe formalism
and Bunge formalism, compared ... **A10:** 362
complete set of angles for ... **A10:** 359
notation, Euler space in ... **A10:** 361
ODF coefficient determined ... **A10:** 362
for specifying orientation in crystallographic measurement ... **A10:** 359-361
Roelands equation ... **A18:** 83
Roentgen
abbreviation for ... **A10:** 691
as radiation (rem) unit ... **A17:** 300-301
Rogers/Microtech process ... **EL1:** 621
Rogowski contour ... **EM4:** 539
Rolfe-Novak Charpy/fracture toughness correlation ... **A8:** 265
Roll
defined ... **A17:** 18 **EL1:** 1156
Roll bending ... **A8:** 119
of bar ... **A14:** 661
defined ... **A14:** 11
tube ... **A14:** 669
Roll bonding *See* Roll welding
Roll bonding processes
bearing materials ... **A18:** 755-756
Roll bonding technique
brazing with clad brazing materials ... **A6:** 347, 348
Roll calibration
Scleroscope hardness test ... **A8:** 104
Roll cladding *See* Roll welding
Roll coining ... **A14:** 185
Roll compacting *See also* Powder rolling ... **A7:** 10, 401-409
of cermets ... **A2:** 983-984
commercial production ... **A7:** 401-403
finishing in ... **A7:** 405-406
into sheet or strip ... **A7:** 297
powder feeding in ... **A7:** 403-404
production procedures ... **A7:** 403-406
recent developments ... **A7:** 408-409
and rigid tool compaction ... **A7:** 323
role diameter ... **A7:** 404-406
rolls, type and position ... **A7:** 403, 405
Soviet technologies for ... **A7:** 692
specialty applications ... **A7:** 402-403, 406-409
Roll compaction ... **A7:** 10, 323, 401-409
applications ... **EM4:** 164
compared to tape casting ... **EM4:** 164
Roll design ... **A14:** 352
parameters, contour roll forming ... **A14:** 629-630
passes ... **A14:** 347-350
tube rolling ... **A14:** 631
Roll dies
types ... **A14:** 97-99
Roll feeds *See also* Feed mechanisms ... **A14:** 134, 499-500
Roll flattening
defined ... **A14:** 11
Roll forging *See also* Forging ... **A14:** 96-99
of aluminum alloys ... **A14:** 244
applications ... **A14:** 96
defined ... **A14:** 11, 96
of heat-resistant alloys ... **A14:** 231
machines ... **A14:** 96-97
materials for ... **A14:** 97
operation, schematic ... **A14:** 97
roll dies ... **A14:** 97-99
of stainless steels ... **A14:** 222
of titanium alloys ... **A14:** 272-273
wrought aluminum alloy ... **A2:** 34
Roll forging machines ... **A14:** 96-97
Roll forming
aluminum alloys, lubricants for ... **A14:** 519
of copper and copper alloys, lubricants for ... **A14:** 519
defined ... **A14:** 11
lubricants for ... **A14:** 519
of stainless steels ... **A14:** 519, 759
of titanium alloys, lubricants for ... **A14:** 520
Roll groove geometry
for round wire ... **A14:** 396
Roll lubricants
for wrought copper and copper alloys ... **A2:** 245
Roll materials
cast iron ... **A14:** 352
Roll materials (continued)
cast steel ... **A14:** 353
chilled iron ... **A14:** 353
hardened forged steel ... **A14:** 353-354
sleeve ... **A14:** 354
Roll painting ... **A7:** 460
Roll pass designs
computer-aided ... **A14:** 347-350
Roll pressure
in strip rolling ... **A14:** 344-345
Roll resistance spot welding
aluminum alloys ... **M6:** 542
Roll spot resistance welding
definition ... **M6:** 15
Roll spot resistance welding machines ... **M6:** 475
Roll spot welding ... **M6:** 493
Roll straightening
defined ... **A14:** 11
Roll threading
alloy steel wire for ... **A1:** 287 **M1:** 269-270
defined ... **A14:** 11
steel fasteners ... **M1:** 274-276
Roll threading quality steel wire rod ... **M1:** 256
Roll torques, and roll forces for Nb-V steel compared ... **A8:** 180
Roll welding ... **M6:** 676
applications ... **M6:** 676
definition ... **M6:** 15
metals welded ... **M6:** 676
pack roll welding ... **M6:** 676
pressure during rolling ... **M6:** 676
procedures ... **M6:** 676
Roll welding (ROW) ... **A6:** 312-313
advantages ... **A6:** 312
"alligatoring" ... **A6:** 312
aluminum ... **A6:** 312-314
aluminum alloys ... **A6:** 312
applications ... **A6:** 312-314
cladding of metals by strip roll welding ... **A6:** 314
roll welded heat exchangers ... **A6:** 312-314
copper ... **A6:** 312-314
copper alloys ... **A6:** 312
definition ... **A6:** 312, 1213
high-alloy steels ... **A6:** 312
killed low-carbon steels ... **A6:** 312
Kirkendall diffusion ... **A6:** 312
limitations ... **A6:** 312
low-alloy steels ... **A6:** 312
low-carbon steels ... **A6:** 312
nickel ... **A6:** 312
nickel-base alloys ... **A6:** 312
niobium alloys ... **A6:** 312
nonpack rolling ... **A6:** 312
pack rolling ... **A6:** 312
postheat treatment ... **A6:** 312, 314
pressure hill effects ... **A6:** 312
process description ... **A6:** 312
semikilled low-carbon steel ... **A6:** 312
stainless steels ... **A6:** 312
steels ... **A6:** 312
strip roll welding ... **A6:** 314
tantalum alloys ... **A6:** 312
titanium ... **A6:** 312
titanium alloys ... **A6:** 312
zirconium alloys ... **A6:** 312
Roll(s) *See also* Roll design; Roll dies
aluminum bronze, contour roll forming ... **A14:** 624
angle of, tube straightening ... **A14:** 691, 692
bending ... **A14:** 11, 661
for cold swaging ... **A14:** 131
for contour roll forming ... **A14:** 630
deflection, in three-roll forming ... **A14:** 623
diameter, in roll compacting ... **A7:** 404-405
dies ... **A14:** 97-99
elastic deflection of ... **A14:** 346
formed, for bending ... **A14:** 666
forming, contour ... **A14:** 628-630
maintenance, three-roll forming machines ... **A14:** 618-619
in roll compacting ... **A7:** 403, 405
and roll materials ... **A14:** 352-354, 630
rolling Mill, principal parts ... **A14:** 353
separating force, and torque ... **A14:** 345
-separating, in plate rolling ... **A14:** 346
stress distribution ... **A14:** 344-345
Roll(s) (continued)
for swaging ... **A14:** 131
for three-roll forming machines ... **A14:** 618-619
Roll-compacted strip materials *See also* Roll compacting ... **A7:** 18, 403-404
Roll-separating force
estimating method for ... **A14:** 344
Rolled compact ... **A7:** 10
Rolled copper
α- and β-fibers in ... **A10:** 363
Rolled fcc materials
textures in ... **A10:** 363-364
Rolled plate
straight-beam top ultrasonic inspection ... **A17:** 268-269
Rolled products
steel normalizing ... **A4:** 37, 38, 40-41
Rolled shapes
ultrasonic inspection ... **A17:** 270-272
Rolled sheet materials
expected pole orientations of preferred orientations ... **A10:** 360
grain orientation ... **A10:** 358
preferred orientation in ... **A10:** 359
Rolled zinc
atmospheric corrosion ... **M2:** 647, 650
Rolled zinc alloy
properties ... **A2:** 541
Rolled-annealed (RA) copper ... **EL1:** 581
Rolled-in scale
as forging flaw ... **A17:** 493
Rolled-in slugs
tubular products ... **A17:** 568
Roller air analyzers ... **A7:** 10
for particle size distribution ... **A7:** 219-220
Roller bearing cup
fatigue life ... **A7:** 620
Roller bearing(s)
coordinate measuring machines ... **A17:** 24
magnetic particle inspection ... **A17:** 116-117
Roller bearings
alloy steel wire for ... **M1:** 269
and ball bearings, compared ... **A11:** 490
carburizing ... **A18:** 875
defined ... **A18:** 16
fatigue spalling life ... **A18:** 258
inner cone, galling in ... **A11:** 158
types of ... **A11:** 490
Roller burnishing ... **A16:** 252-254
Al alloys ... **A16:** 787
bearingizing ... **A16:** 254
in conjunction with boring ... **A16:** 168-169
Cu alloys ... **A16:** 813
fillet rolling ... **A16:** 254
P/M materials ... **A16:** 889, 890, 891
speed, feed, and lubrication ... **A16:** 253-254
tolerance and finish ... **A16:** 253
tool life ... **A16:** 253
tools ... **A16:** 252-253
workpiece requirements ... **A16:** 252
Roller chain components
wear of ... **M1:** 628-630
Roller coating
of lubricants ... **A14:** 515
Roller coating (machine process)
paint ... **M5:** 482, 494
Roller drum peel test ... **EM3:** 732
Roller hearth furnaces ... **A7:** 10, 351, 354, 357
Roller leveler breaks
defined ... **A14:** 11
Roller levelers ... **A14:** 502
Roller leveling ... **A1:** 693
of carbon steel sheet ... **A1:** 206
defined ... **A14:** 11
steel sheet ... **M1:** 157, 160
Roller supports, aircraft
economy in manufacture ... **M3:** 854-855
Roller(s)
as impression dies ... **A14:** 44
for power spinning of cones ... **A14:** 602-603
radius, for tube spinning ... **A14:** 678
for tube spinning ... **A14:** 677
Roller-path patterns
in failed bearings ... **A11:** 491-492
Rollers
banded alloy segregation in ... **A11:** 121

Rollers (continued)
carburized-and-hardened, fracture surface **A11:** 121
fretting damage **A11:** 341
rubber office-chair, failure of **A11:** 763-764
Rolling *See also* Ring rolling; Roll forging; Sliding; Spin **A8:** 593-596
basic processes **A14:** 343-344
of blended elemental Ti-6Al-4V **A2:** 649
as bulk forming process **A14:** 16
caliber **A14:** 347
defects **A8:** 593-595
defects in **A14:** 358-359
defined **A14:** 11
deformation zone parameter **A8:** 593
double-barreled deformation in **A8:** 593-594
effect of sliding **A11:** 465
end and side deformations in **A8:** 593, 595
flow stress of high-purity aluminum during **A8:** 179
heated-roll **A14:** 356-358
high-strength low-alloy steel in torsion **A8:** 179
inhomogeneous deformation **A8:** 593-595
instruments and controls **A14:** 354-355
loads **A14:** 357
materials for **A14:** 355-356
microstructure **A14:** 355
of mill shapes **A7:** 522
mills **A14:** 351-352
Nb-V steel, comparison of roll forces and roll torques **A8:** 180
of P/M billets **A7:** 522-529
as paint application technique **A13:** 415
painting process **M5:** 489, 493
plate, mechanics of **A14:** 346
primary objectives **A14:** 343-344
processes, classified **A14:** 16
pure, effects in pitting **A11:** 340
pure, shear stress in **A11:** 133-134, 465
pure, stress in
reduction, ODF along fiber lines as function of **A10:** 364
refractory metals and alloys **A2:** 562
refractory metals, and formability **A14:** 785
rolls and roll materials **A14:** 352-354
shape **A14:** 346-350
single-edge barreling in **A8:** 594, 596
of sleeve-type rings **A14:** 116
steel clad in aluminum by **M5:** 345-346
strategies, for ring types **A14:** 117
stress distribution in contacting surfaces from **A11:** 592
strip, theory of **A14:** 344-346
temper and flex, effect on Lüders lines **A8:** 553
textures, features in fcc materials **A10:** 363-364
tools **A14:** 121
tube **EM1:** 569-574
of unalloyed uranium **A2:** 671
unidirectional flux by **A7:** 315
universal **A14:** 347-348
vs sliding, effects on pitting **A11:** 340-341
of washer-type rings **A14:** 116
wear **A8:** 603
of weld-neck flange **A14:** 118
of whisker-reinforced MMCs **EM1:** 899
of wire, in Turk's head machines **A14:** 694
wire rod **A2:** 253-254
wrought titanium alloys **A2:** 610-611
Rolling bearings
laminar particles and spheres as wear indication **A18:** 303
Rolling bolster assemblies **A14:** 497
Rolling contact fatigue tester
for bearings **A8:** 370
Rolling direction
effect, blanking **A14:** 450
effect, formability **A14:** 780-781
in press-brake forming **A14:** 541
Rolling direction (in rolled metals) *See* Longitudinal direction
Rolling direction (of a sheet)
abbreviation for **A11:** 797
Rolling element **A18:** 499
Rolling friction coefficient **A18:** 478
Rolling friction force **EM3:** 510
Rolling lap
fracture from **A12:** 64
as planar flaw **A17:** 50
Rolling mandrel *See also* Mandrels
defined **A14:** 11
Rolling mill **A7:** 401, 408
Rolling mills **M1:** 110-111, 114
defined **A14:** 11
four-high **A14:** 351
instrumentation and controls **A14:** 354-355
planetary **A14:** 352
ring, computer program **A14:** 117
Sendzimir **A14:** 351-352
specialty **A14:** 351-352
three-high **A14:** 351
two-high **A14:** 351
for wrought copper and copper alloys **A2:** 244-245
Rolling oils **EM3:** 41
Rolling temperature *See* Temperature(s)
Rolling velocity *See* Sweep velocity
Rolling velocity of gear
symbol and units **A18:** 544
Rolling velocity of pinion
symbol and units **A18:** 544
Rolling-contact bearings **A18:** 741
Rolling-contact fatigue
in bearings **A12:** 115, 134
carburized/through-hardened P/F materials **A14:** 201
from powder forging **A14:** 203
in gears **A11:** 593-594
radial-contact ball bearing failure by **A11:** 503
in rolling-element bearings **A11:** 500-505
in shafts **A11:** 465
stresses in **A11:** 133-134
Rolling-contact fatigue (RCF) *See also* Rolling contact wear **A18:** 257, 258, 259
bearings in valve train assembly **A18:** 558
carburizing steels **A18:** 875
ceramic bearing **A18:** 260-261
defined **A18:** 16
jet engine components **A18:** 588, 590
mechanisms **A18:** 260
testing methods summary **A18:** 258, 259, 260, 261
Rolling-contact wear (RCW) *See also* Rolling-contact fatigue **A18:** 257-262
adhesive wear **A18:** 257
applications **A18:** 257
defined **A18:** 16, 257
first signs **A18:** 257
lubrication **A18:** 257, 259, 260, 261
magnitude of effects **A18:** 257
mechanisms **A18:** 259-262
process steps **A18:** 259-260
physical signs **A18:** 257-259
gears **A18:** 257-258
rolling-element bearings **A18:** 258-259
rolling-contact fatigue testing **A18:** 259
Rolling-element antifriction bearings
flux leakage inspection **A17:** 133
Rolling-element bearings *See also* Ball bearings; Bearings; Rolling-element bearings, failures of **A1:** 380
components of **A11:** 490
failures, characteristics and causes **A11:** 493-494
fretting failures **A11:** 497-498
low-alloy, failure from stray electric currents and moisture **A11:** 495, 497
lubrication of **A11:** 509-512
materials for **A11:** 490
microstructural alterations **A11:** 505
Rolling-element bearings (continued)
misalignment, types of **A11:** 507
types of failure effects **A11:** 492
wear failures **A11:** 493-496
Rolling-element bearings, failures of
See also Ball bearings; Rolling-element bearings **A11:** 490-513
bearing materials **A11:** 490
bearing-load ratings **A11:** 490-491
by corrosion **A11:** 498-499
by damage **A11:** 505-506
by fretting **A11:** 497-498
by plastic flow **A11:** 499-500
by rolling-contact fatigue **A11:** 500-505
by wear **A11:** 493-496
examination of **A11:** 491-492
fabrication practices, effects of **A11:** 506-508
and hardness of bearing components **A11:** 508-509
heat treatment and **A11:** 508-509
lubrication of rolling-element bearings **A11:** 509-512
types of **A11:** 492-493
Rolling-element bearings, friction and wear of **A18:** 499-513
bearing component materials **A18:** 502-503
cage materials **A18:** 503
rolling-contact component steels **A18:** 502-503
rolling-element ceramics **A18:** 503
bearing fatigue life **A18:** 507-510
basic static load rating **A18:** 508-510
contamination effect on fatigue life **A18:** 508
equivalent load **A18:** 508, 510
fatigue load limit **A18:** 508
nonstandard application conditions **A18:** 508
defined (rolling-element bearing) **A18:** 16
fatigue spalling life **A18:** 258
history and development of rolling-element bearings **A18:** 499
hybrid (ceramics and metals) **A18:** 261
load ratings **A18:** 504-507
basic load rating **A18:** 505-506
bearing endurance **A18:** 505
fatigue life dispersion **A18:** 505
median life **A18:** 505
standard bearing contact geometry **A18:** 506-507
standard bearing internal geometry **A18:** 506
standard roller geometry **A18:** 507
lubrication requirements and methods **A18:** 503-504
bath lubrication **A18:** 504
circulating-oil lubrication **A18:** 504, 505
grease lubrication **A18:** 503-504
properties determining applications **A18:** 261
rolling bearing friction **A18:** 510-511
bearing friction torque **A18:** 510-511
sources of sliding friction **A18:** 510
rolling-contact wear **A18:** 258-259
rubbing wear in centrifugal and axial flow pumps **A18:** 593, 595, 597
types of rolling-element bearings **A18:** 499-502
angular-contact ball bearings **A18:** 500, 505, 506, 511
angular-contact groove ball bearings **A18:** 509
ball bearings **A18:** 499-500, 505
Conrad-type ball bearings **A18:** 499-500
cylindrical roller bearings **A18:** 505, 511
cylindrical roller thrust bearings **A18:** 505
deep-groove ball bearings **A18:** 511
drawn cup needle roller bearings **A18:** 505
duplex angular-contact ball bearings **A18:** 500
filling slot ball bearings **A18:** 505
needle roller bearings **A18:** 511
needle roller with machined rings bearings **A18:** 505
radial ball bearings **A18:** 499, 505, 506, 508, 509, 510
radial contact ball bearings **A18:** 509

Rolling-element bearings, friction and wear of (continued)
radial cylindrical roller bearings...... **A18:** 500, 501, 502, 511
radial needle roller bearings...... **A18:** 500, 501-502
radial roller bearings.......... **A18:** 500-501, 507, 508, 510, 511
radial spherical roller bearings......... **A18:** 500, 501, 502
roller bearings.............................. **A18:** 500-502, 505
self-aligning ball bearings **A18:** 500, 505, 508, 509, 511
single-row radial contact separable ball bearings (magneto bearings) **A18:** 509, 607
spherical roller bearings **A18:** 505, 511
spherical roller thrust bearings **A18:** 502, 505, 511
split inner ring ball bearings **A18:** 500
tapered roller bearings...... **A18:** 500, 501, 502, 505, 511
thrust ball bearings........... **A18:** 500, 505, 506, 507, 508, 509, 510, 511
thrust cylindrical roller bearings...... **A18:** 502, 503, 505, 511
thrust needle roller bearings..... **A18:** 502, 505, 511
thrust roller bearings......... **A18:** 502, 503, 505, 507, 510, 511, 607
thrust spherical roller bearings **A18:** 511
thrust tapered roller bearings............ **A18:** 502, 505
in valve train assembly of internal combustion engine **A18:** 558
wear **A18:** 511-512
modes contributing to bearing failure **A18:** 511-512
wear control **A18:** 512-513
lubricant film thickness **A18:** 512-513
Rollover tests **EM3:** 53
Rollovers *See* Jolt rollovers
Rolls
ultrasonic inspection....................................... **A17:** 232
Rolls, metalworking *See* Metalworking rolls
Rondelles
nickel powder **A7:** 141-142
Roof bolt corrosion
mining industry.. **A13:** 1294
Roof coatings
aluminum flake pigments for **A7:** 594
powder used **A7:** 573
waterproofing, powders used......................... **A7:** 574
Roof structure
electric arc furnace **A15:** 357-358
Room temperature *See also* Ambient temperature; Temperature(s) **EM3:** 25
corrosion, beryllium-copper alloys **A2:** 422
defined **EM2:** 36
mechanical properties, sand cast magnesium alloys... **A2:** 497
mechanical properties, wrought magnesium alloys.................................. **A2:** 480-482
physical properties, nickel alloys **A2:** 441
tensile properties, lead alloys **A2:** 550
Room temperature systems
materials and processes selection.......... **EL1:** 112-118
Room-temperature cleaning processes **M5:** 12, 15
Room-temperature compression testing........ **A8:** 195, 201
Room-temperature cure *See also* Cure
adhesive, defined **EM1:** 20
filament winding... **EM1:** 135
of polyester resin systems **EM1:** 133
wet lay-up resins.................................. **EM1:** 132-134
Room-temperature curing adhesive
defined **EM2:** 36
Room-temperature out-time
defined and tested **EM1:** 737
Room-temperature properties
and composition **A7:** 628
Room-temperature structure
of gray iron.. **A1:** 14
Room-temperature vulcanizing (RTV) *See also* Silicones (room-temperature vulcanizing); Vulcanization **EM3:** 25
defined **EM1:** 20 **EM2:** 36
RTV sealant, compared to fluorocarbon sealants..................................... **EM3:** 226
as silicone conformal coating type.............. **EL1:** 823
Room-temperature vulcanizing (RTV) (continued)
of silicone-base coatings................................ **EL1:** 773
Room-temperature vulcanizing (rubber)
symbol for.. **A11:** 797
Room-temperature-curing adhesive **EM3:** 25
Room-temperature-setting adhesive.............. **EM3:** 25
Room-type blast cleaning machine **A15:** 508-509
Root
definition.. **A6:** 1213
Root bead
definition.. **A6:** 1213
Root bend tests... **A6:** 102
Root, bolt-threaded
as fatigue-initiation site................................ **A11:** 532
Root crack
definition... **M6:** 15
hydrogen-induced **A11:** 92, 93
Root edge
definition... **A6:** 1213 **M6:** 15
Root face
definition... **A6:** 1213 **M6:** 15
Root faces of groove welds **M6:** 64, 67-68
Root gap
definition.. **A6:** 1213
Root head
definition... **M6:** 15
Root node
broadcasting from **EL1:** 5
Root of joint
definition... **M6:** 15
Root of weld
definition... **M6:** 15
Root opening
definition... **A6:** 1213 **M6:** 15
Root opening of groove welds **M6:** 64-65
Root penetration *See also* Penetration
definition... **A6:** 1213 **M6:** 15
incomplete, radiographic appearance.......... **A17:** 350
Root radius
definition.. **A6:** 1213
notched-specimen testing **A8:** 315
Root reinforcement
definition... **A6:** 1213 **M6:** 15
Root surface
definition... **A6:** 1213 **M6:** 15
Root welds
magnetic particle inspection.......................... **A17:** 111
Root-mean-square composite surface roughness
symbol and units.. **A18:** 544
Root-mean-square end-to-end distance **EM3:** 25
defined **EM2:** 36
Root-mean-square surface roughness of gear
symbol and units.. **A18:** 544
Root-mean-square surface roughness of pinion
symbol and units.. **A18:** 544
Root-mean-square value.................................... **A18:** 28
Roots blower
development of... **A15:** 27
Rope
steel wire.. **A11:** 515-521
Rope pressure
steel wire rope **A11:** 516-517
Rope wire **A1:** 283-284, 851
steel **M1:** 265, 266, 271
Rope-lay stranded copper conductors........... **M2:** 266, 269, 270-271
Ropes
flux leakage inspection of............................. **A17:** 133
Roping
of tape prepregs.. **EM1:** 144
Rosenthal alkaline earth silicate
composition **EM4:** 1102
properties **EM4:** 1102
Rosenthal's equations
PTCT diagram applications.............................. **A6:** 73
Rosenthal-Norton sectioning technique....... **A6:** 1095
Rosette
defined **A8:** 11 **A9:** 15
Rosette (star) fractures
as radial marks **A12:** 103-104
Rosette graphite
defined **A9:** 15
Rosin.. **EM3:** 25
for cork and rubber gasket sealing **EM3:** 57-58
Rosin acids
chemical structure **EL1:** 645
Rosin mildly activated (RMA) flux.............. **EL1:** 731
Rosin mildly activated (RMA) fluxes
furnace soldering... **A6:** 354
noncorrosive (organic) soldering fluxes........ **A6:** 628
Rosin nonactivated
noncorrosive (organic) soldering fluxes........ **A6:** 628
Rosin(s)
coatings **EL1:** 680
effect on solder paste print resolution **EL1:** 732
flux, defined **EL1:** 1156
Rosin-base fluxes.. **EL1:** 645-646
Rotary actuator, and load cell
for torsional fatigue testing **A8:** 151
Rotary automatic polishing and buffing machines **M5:** 120-122
Rotary bending
of bar **A14:** 662
dies, for press bending................................... **A14:** 526
in press brake forming **A14:** 538-539
Rotary bending fatigue fracture
medium-carbon steel **A12:** 272
Rotary degassing *See also* Degassing; Gases
for hydrogen removal **A15:** 460-461
for inclusion removal **A15:** 490
Rotary dip test... **A6:** 136
as solderability wetting time test................ **EL1:** 677
test standards used to evaluate solderability **A6:** 136
Rotary drilling
with cemented carbides **A2:** 974
Rotary drum
for sand reclamation.............................. **A15:** 353-354
Rotary drum shears **A14:** 711-712
Rotary fluxers
through-hole soldering.................................. **EL1:** 682
Rotary forges
classified.. **A14:** 176
Rotary forging *See also* Forging; Orbital forging; Radial forging...................... **A14:** 176-179
advantages/limitations **A14:** 177
of aluminum alloys.. **A14:** 244
applications **A14:** 176-177
of bicycle hub bearing retainer..................... **A14:** 178
of carbon steel clutch hub **A14:** 178-179
of copper alloy seal fitting.............................. **A14:** 179
defined **A14:** 11
die arrangement **A14:** 17
die motion, examples **A14:** 178
dies **A14:** 17, 178
examples **A14:** 178-179
machines **A14:** 177-178
as new metalworking process........................... **A14:** 17
and radial forging, compared **A14:** 145, 176
of titanium alloys **A14:** 273
warm.. **A14:** 178-179
workpiece materials **A14:** 177
wrought aluminum alloy **A2:** 34
Rotary furnace
in tungsten powder production **A7:** 153
Rotary kilns... **A7:** 52, 79
Rotary piercing
of copper and copper alloy tube shells................ **A2:** 249-250
for steel tubular products............... **A1:** 328 **M1:** 316
Rotary presses.................................... **A7:** 10, 334, 335
Rotary pump
damaged austenitic cast iron impellers in **A11:** 355-356
Rotary pumps
gas mass spectrometry **A10:** 151-152
for SEM vacuum system................................ **A12:** 171
Rotary seal
defined **A18:** 16
Rotary shakeout device **A15:** 347-348, 503
Rotary shear
defined **A14:** 11
Rotary shearing
accuracy **A14:** 706
applicability... **A14:** 705-706
circle generation **A14:** 706
of plate and flat sheet.............................. **A14:** 705-707
rotary cutter adjustment **A14:** 706
Rotary straighteners
multiroll rotary **A14:** 686-687
two-roll rotary... **A14:** 686
Rotary swager
defined **A14:** 11

Rotary swaging *See also* Forging; Swaging
applicability **A14:** 128, 141-142
automatic swaging machines **A14:** 134
auxiliary tools **A14:** 133-134
defined **A14:** 11
die taper angle **A14:** 139
dimensional accuracy **A14:** 140
and drilling, combined **A14:** 141
feed **A14:** 139, 143-144
hot swaging **A14:** 142-143
lubrication **A14:** 139-140
machines **A14:** 129-132
material response **A14:** 143-144
metal flow during **A14:** 128-129
noise suppression **A14:** 144
reduction effects **A14:** 139
special applications **A14:** 141-142
surface contaminants **A14:** 39
surface finish **A14:** 140
swaging dies **A14:** 132-133
tube swaging, with mandrel **A14:** 137-139
tube swaging, without mandrel **A14:** 134-137
and turning, combined **A14:** 141
vs press forming **A14:** 140
vs spinning **A14:** 140-141
vs turning **A14:** 141
Rotary swaging machines **A14:** 129-136
Rotary table
magnetic particle inspection methods **A17:** 118
Rotary transducer
in torsion testing **A8:** 158, 216
Rotary tumbling
for deburring **A7:** 458
Rotary ultrasonic machining (RUM) **EM4:** 362
Rotary valve, high-temperature
expansion and distortion failure in **A11:** 374-376
Rotary vane
for hydraulic torsional system **A8:** 216
Rotary variable-differential transformer
for torsion testing **A8:** 158
Rotary welding
definition **M6:** 15
Rotary-arbor straighteners **A14:** 687
Rotate-only systems
computed tomography (CT) **A17:** 366-367
Rotating beam fatigue strength **A7:** 469
Rotating beam machine
fatigue test specimen **A8:** 368, 371
for fatigue testing **A8:** 369
grip ends **A8:** 371
R.R. Moore-type **A8:** 369-370
single-end **A8:** 370
and specimen stress distribution **A8:** 392-393
types **A8:** 369-370
Rotating cantilever beam
fatigue test specimen **A8:** 371
Rotating cylinder method
measuring angle of repose **A7:** 282, 283
Rotating dial
in cold extrusion **A14:** 303
Rotating disk atomization *See also* Atomization
equipment and production sequence **A7:** 46-47, 74-76
powder characteristics and process limitations **A7:** 47
Rotating disk atomizer **A7:** 74-76
Rotating eccentric mass machine
with direct-stress fixture **A8:** 369
Rotating electrode powder **A7:** 10
Rotating electrode process **A7:** 10, 34-37
atomized powders, Auger profiles **A7:** 254, 255
for copier powders **A7:** 586-587
equipment **A7:** 34-35
spherical powders **A7:** 586-587
Rotating electrode process (REP) **A1:** 973
Rotating equipment **A6:** 39
Rotating frequency
in forced-vibration systems **A8:** 391
Rotating hanger blast cleaning **A15:** 508
Rotating platinum microelectrode **A10:** 204, 691
Rotating probes
eddy current inspection **A17:** 183, 187
Rotating sample cell
Raman analysis of graphites **A10:** 132
Rotating shafts
with press-fitted elements **A11:** 470
Rotating tester
for gear fatigue **A8:** 370
Rotating-beam test **A1:** 861
Rotating-bending fatigue
in shafts **A11:** 463-464
surface of **A11:** 321
Rotating-die machines **A14:** 178-179
Rotation
effect on shafts **A11:** 463
Rotation axis
defined in crystal symmetry **A10:** 346
Rotation, molecular
in Raman spectroscopy **A10:** 127
Rotation, of images
digital image enhancement **A17:** 458
Rotation pattern
single-crystal diffraction **A10:** 330
Rotation rate
and particle size distribution **A7:** 41, 43
Rotation speed
horizontal centrifugal casting **A15:** 297-298
vertical centrifugal casting **A15:** 305
Rotational angle
control, in low-cycle torsional fatigue **A8:** 150
in cyclic torsional testing **A8:** 150
Rotational bending
and fatigue-crack propagation **A11:** 108-109
Rotational bending systems **A8:** 392-393
electromechanical fatigue tester **A8:** 392
Rotational casting
defined **EM2:** 36
properties effects **EM2:** 286
Rotational creep
in rolling-element bearings **A11:** 492, 496
Rotational moiré patterns **A9:** 110
Rotational molding
defined **EM2:** 36
equipment type/size **EM2:** 365-366
molded-in color **EM2:** 306
process **EM2:** 360
selection factors **EM2:** 361-365
size and shape effects **EM2:** 291
textured surfaces **EM2:** 305
tooling **EM2:** 366-367
Roto-percussive drilling
with cemented carbides **A2:** 974-975
Rotopeening
for steam generators **A13:** 942
Rotor
blade, ceramic, impact fracture of **A11:** 753
bowl, stainless steel, fatigue cracking in **A11:** 398
high-temperature, expansion and distortion failure **A11:** 374-376
microstructures **A11:** 376-377
oxidation and thermal fatigue cracking **A11:** 376
shaft, torsional fatigue failure **A11:** 464
shafts, brittle fracture from seams **A11:** 478
Rotor castings
aluminum alloy **A15:** 755-757
aluminum casting alloys **A2:** 127-129
Rotors
for elevated-temperature service **A1:** 620-621
Rototiller gear assemblies **A7:** 678
Rouge
batch size **EM4:** 382
Rouge buffing compounds **M5:** 117
Rough analysis tools
for forging process design **A14:** 411
Rough blank
defined **A14:** 11
Rough machining
effect on fatigue strength **A11:** 122
Rough threading
definition **M6:** 15
thermal spray coating process using **M5:** 367
for thermal spray coatings **A13:** 460
Rough-polishing process
defined **A9:** 15
Roughening, surface *See* Surface roughening
Roughing mechanical vacuum pump **A8:** 413
Roughing stand
defined **A14:** 11
Roughing tool
cracked after heat treatment **A11:** 571
Roughness *See also* Surface roughness **A7:** 242
defined **EL1:** 1156
fracture surface **A12:** 199-205, 276
index of surface **A12:** 201
light-section microscopy used to examine **A9:** 80-82
oblique illumination used to examine **A9:** 76
parameters 4340 steel **A12:** 213
parameters, quantitative fractography **A12:** 199-205
severe, as casting defect **A11:** 384
surface, as casting defect **A11:** 381
surface, fractal analysis **A12:** 211-215
Roughness average **A18:** 28, 334, 335, 340-341, 343
Roughness average (RA) **A6:** 883
Roughness parameters *See also* Surface roughness parameters
average peak-to-trough height **A12:** 200
fractal dimension, irregular planar curve **A12:** 200
fracture path preference index **A12:** 201
probability parameter **A12:** 201
profile configuration **A12:** 200
profile parameters **A12:** 199-200
quantitative fractography **A12:** 199-205
surface **A12:** 200-201
true length **A12:** 199-200
vertical **A12:** 200
Roughness profile **A18:** 518
Roughness-induced crack closure
and corrosion fatigue **A8:** 408-409
Round bar
dynamic notched testing with **A8:** 275-282
Round blanks
layout of **A14:** 449-450
Round bores
in gears and/or gear trains **A11:** 589
Round durometer **A8:** 107
Round hole bone plates
as internal fixation device **A11:** 671
Round particles
apparent density **A7:** 272
Round sections
computer-aided roll pass design for **A14:** 349
Round steel bars
eddy current inspection **A17:** 185-186
Round wire
wrought copper and copper alloys **A2:** 251-252
Round-headed bolts
economy in manufacture **M3:** 852, 853
Rounding of edges during abrasion and polishing **A9:** 44-45
Routability, of surface mounted chip carrier
vs DIP **EL1:** 15
Routers
channel **EL1:** 532
rip-up **EL1:** 532-533
shove-aside **EL1:** 533
Routing *See also* Rerouting
algorithms **EL1:** 81, 529
automatic trace, computer-aided design **EL1:** 529-533
daisy chain **EL1:** 83
digital systems **EL1:** 76
gridless **EL1:** 533

Routing (continued)
Manhattan-type patterns **EL1:** 76
PCD tooling inserts **A16:** 110
printed wiring boards **EL1:** 115
rigid wiring boards **EL1:** 547
signal path, rigid printed wiring boards **EL1:** 551-552
Roving *See also* Woven roving
cloth, defined **EM2:** 37
as continuous reinforcement
pultrusion **EM2:** 393
defined **EM2:** 36
Roving ball
defined **EM1:** 20 **EM2:** 36
Roving cloth
defined **EM1:** 20
Rovings *See also* Fabrics; Weaves
applications **EM1:** 109
aramid fiber **EM1:** 114-115
collimated, defined *See* Collimated roving
continuous strand, tests for **EM1:** 291
direct forming single-end, process
fiberglass **EM1:** 109
Kevlar, sizes **EM1:** 114
pultrusion **EM1:** 537
woven, and fiberglass mat **EM1:** 109
woven glass **EM1:** 109
woven/fabric, for resin transfer molding **EM1:** 169
yield (yd/lb), determined
Row nucleation **EM3:** 25
defined **EM2:** 37
Rowland circle mount
polychromators with **A10:** 23
Royal Brass Foundry Drawings **A15:** 20
R_p *See* Roughness parameters
RPDROD computer program
for roll pass design **A14:** 349
RPE *See* Rotating platinum microelectrode
RRIM *See* Reinforced reaction injection molding
RS *See* Raman spectroscopy; Surface roughness parameters
RSCs *See* Reversed sigmoidal curves
RSF *See* Relative sensitivity factor
RTI *See* Relative thermal index
RTM *See* Resin transfer molding (RTM); Resin transfer/molding
RTV *See also* Room-temperature vulcanizing; Room-temperature vulcanizing (RTV); Room-temperature vulcanizing (RTV) silicones
RTV silicone rubbers
applications **EM2:** 266
RTV-I (moisture-curing) silicone conformal coatings **EL1:** 823-824
RTV-II (addition curing) silicone conformal coatings **EL1:** 823-824
Ru-Si (Phase Diagram) .... **A3:** 2•355
Ru-Ta (Phase Diagram) .... **A3:** 2•355
Ru-Ti (Phase Diagram) .... **A3:** 2•356
Ru-U (Phase Diagram) .... **A3:** 2•356
Ru-V (Phase Diagram) .... **A3:** 2•356
Rub marks
by abrasion **A11:** 27
fatigue **A12:** 118-119
Rubber
abrasive blasting of **M5:** 91
analysis, nitrile sheath vs neoprene sheath **A10:** 648
-base adhesives **EM1:** 686-687
cross-linked, ductility dependence on temperature-compensated strain rate **A8:** 39, 43
defined **EM1:** 20
drilling **A16:** 229, 230, 237
ESR for **A10:** 263
honing **A16:** 476
liquid, for plastic replicas **A17:** 53
magnetic *See* Magnetic rubber inspection
modified polymers, toughness of **EM1:** 27
natural and synthetic, seal materials **A18:** 550
O-ring assemblies, neutron radiography of **A17:** 388
sliding and adhesive wear **A18:** 240
thermal energy method of deburring **A16:** 578
wheel, abrasive wear tester **A8:** 605
Rubber bag materials
natural latex **A7:** 447
Rubber bag materials (continued)
natural molded **A7:** 447
Rubber coatings
chlorinated **A13:** 404
Rubber compounds
for faying surfaces **EM3:** 604
for fillets **EM3:** 604
for rivets **EM3:** 604
Rubber forming
defined **A14:** 11
Rubber injection molding **A7:** 669
Rubber lattices
for packaging **EM3:** 45
Rubber, mold materials for *See* Plastics and rubber, mold materials for
Rubber network theory
molecular models from **EL1:** 849
Rubber office-chair roller
failure of **A11:** 763-764
Rubber padding
for drop hammer forming **A14:** 654
Rubber pads
for drop hammer forming **A14:** 657
in press brake dies **A14:** 539-540
Rubber phase associating plasticizers **EM3:** 183
Rubber phase associating resins **EM3:** 183
Rubber tape selective cadmium plating process **M5:** 268
Rubber tile
waterjet machinery **A16:** 522
Rubber wheel abrasion tests **M1:** 600-601
Rubber-backed carpet
waterjet machinery **A16:** 522
Rubber-base adhesives *See* Elastomeric adhesives
Rubber-bonded abrasive wheels **A9:** 24-25
Rubber-die flanging
failures in **A14:** 615
Rubber-modified styrene-acrylonitriles *See* Styrene-acrylonitriles
Rubber-pad forming *See also* Fluid forming; Fluid-cell process; Guerin process; Marforming process **A14:** 605-615
advantages/disadvantages **A14:** 605
aluminum alloys **A14:** 799-800
ASEA Quintus rubber-pad press **A14:** 608
of copper and copper alloys **A14:** 818-819
defined **A14:** 11, 605
drop hammer forming, with trapped rubber **A14:** 608
equipment **A14:** 5
failures, rubber-die flanging **A14:** 615
fluid forming **A14:** 611-614
fluid-cell forming **A14:** 608-611
Guerin process **A14:** 605-607
of magnesium alloys **A14:** 829-830
Marform process **A14:** 607-608
stainless steels **A14:** 773-774
of titanium alloys **A14:** 845-846
types **A14:** 605-608
Rubber-resin bonded abrasive wheels **A9:** 24-25
Rubber-wheel test
of abrasion-resistant cast iron **A1:** 97
Rubbers **EM3:** 25, 86
advantages and limitations **EM3:** 85
defined **EM2:** 37
effect of concentration on elastic modulus **EM3:** 321
electrical properties **EM2:** 589
heat-curable (HCR) **EM2:** 267
hot-applied pumpable for automotive applications **EM3:** 57
natural **EM3:** 75, 76, 82-84, 143
additives and modifiers **EM3:** 145-146
bonding substrates and applications **EM3:** 85
chemistry **EM3:** 145
commercial forms **EM3:** 145
cross-link density effect on tensile strength **EM3:** 412
markets **EM3:** 145
properties **EM3:** 143-144, 145, 146
for self-sealing tires **EM3:** 58
for tapes, labels, and protective films **EM3:** 84
reclaimed **EM3:** 75, 76, 82-83, 145
resins and plasticizers used in formulations **EM3:** 183
RTV silicone **EM2:** 266
Rubbers (continued)
sulfur-vulcanized natural, microstructural analysis **EM3:** 412
synthetic **EM3:** 75, 143
and synthetic latexes
applications **EM3:** 45
characteristics **EM3:** 45
vulcanized, microstructural analysis **EM3:** 412-413
wire splice protection **EM3:** 612
Rubbing
and abrasion **A11:** 26
in shafts **A11:** 463
Rubbing bearing
defined **A18:** 16
Rubidium
cations, in glasses, Raman analysis **A10:** 131
epithermal neutron activation analysis **A10:** 239
explosive reactivity in moisture **A7:** 194
in insect eggs, GFAAS analysis **A10:** 55
organic precipitant for **A10:** 169
pure **M2:** 791-792
as pyrophoric **A7:** 199
Raman vibrational behavior **A10:** 133
species weighed in gravimetry **A10:** 172
TNAA detection limits **A10:** 238
vapor pressure **A6:** 621
Rubidium, and liquid metals
safety of **A13:** 94-95
Rubidium monoxide-silicon dioxide (Rb_2O-SiO_2)
self-diffusion coefficients of alkali ions **EM4:** 461
Rubidium, pure
properties **A2:** 1151
Rubidium, vapor pressure
relation to temperature **A4:** 495 **M4:** 310
Ruby
ESR studied **A10:** 264
fusion with acidic fluxes **A10:** 167
ultrasonic machining **A16:** 530, 532
Ruby lasers
mounting **M6:** 667
operation **M6:** 649
for optical holographic interferometry **A17:** 418
Ruby pulsed lasers, parameters
laser-beam welding applications **A6:** 263
Ruby pulsed solid-state lasers **A6:** 266
Ruess model
composite Young's modulus **EM4:** 860
Rugged boundaries
fractals as descriptors **A7:** 243-244
Rugged systems
fractals as descriptors of **A7:** 243-245
Ruggedness elongation diagram
particle shape **A7:** 240
Rugosities *See* Asperities
Rule of mixtures **EM1:** 868
Ruled surface
true area and length **A12:** 204
Rules
behavioral model as **EL1:** 129
as form of CAD model definition **EL1:** 129
technology, of engineering design system (EDS) **EL1:** 128
Run length encoding
machine vision process **A17:** 34
Run-arrest cleavage behavior
crack arrest toughness and **A8:** 453
Runner box
defined **A15:** 10
Runner system
defined **EM2:** 37
Runner(s) *See also* Mold cavity
copper alloy casting **A15:** 777-778
defined **A15:** 10
design, gating system **A15:** 589-590, 592-593
in gating, die casting **A15:** 289
Runnerless thermoset molding *See* Warm-runner molding
Running Cracks
crack arrest testing and **A8:** 453
measuring toughness with **A8:** 284
Running-in coatings
valve train assembly components **A18:** 559
Running-in period **A18:** 489
Runoff tab **A6:** 88
Runoff weld tab **A6:** 51

Runout
as casting defect **A11:** 385
defined **A15:** 10
Runouts
in schematic of fatigue data collection **A8:** 701
treatment in fatigue testing **A8:** 701
Rupture *See also* Creep rupture; Dimple ruptures; Modulus of rupture; Rupture life; Rupture strength; Rupture stress; Rupture time; Stress rupture; Transverse rupture strength **EM3:** 25
by embrittlement **A11:** 612-614
caused by overheating **A11:** 603
characteristics, and particle lubrication **A7:** 288
and creep, with STAMP process **A7:** 549
defined **EL1:** 1156 **EM1:** 20 **EM2:** 37
external, in gears **A11:** 596-597
internal, in gears **A11:** 596
life **A11:** 265
of low-carbon steel boiler tubes, from overheating **A11:** 607-608
mechanical discontinuities as **A11:** 383
properties measurement after service **A8:** 338
rating in notched-bar upset test **A8:** 588-599
strength, defined **EM2:** 37
strength, transverse, cemented carbides **A2:** 956
sudden tube **A11:** 603
thick-lip, in steam-generator tubes **A11:** 605-606
thin-lip, in steam-generator tubes **A11:** 606
vs stress-rupture stress **A11:** 265
Rupture ductility **A1:** 933-934
Rupture life **A8:** 685
analyses of **A8:** 689-693
data, analysis of **A8:** 686
Inconel 751, notch effects **A8:** 316-318
Nimonic 80A, notch effect **A8:** 316-318
S-816, notch effect **A8:** 316-318
under constant stress, creep testing **A8:** 305
vs. stress for aluminum alloy **A8:** 332
Waspaloy, notch effect **A8:** 316-318
Rupture, modulus of *See* Modulus of rupture, in bending; Modulus of rupture, in torsion
Rupture strength **EM3:** 25
alloy cast irons **M1:** 95-96
of chromium-molybdenum steel **A8:** 340
ratio of notched and unnotched specimens determined by **A8:** 315
for steels **A8:** 330
Rupture stress
defined **A8:** 11 **A11:** 9
in gears **A11:** 596-597
in pressure vessels **A11:** 666
selection for extensive study **A8:** 339-340
stress **A11:** 265
Rupture stresses
maraging steels **M1:** 450
Rupture time
Discaloy **A8:** 316-317
heat treatment effects, Waspaloy **A8:** 316
Waspaloy, notch effect **A8:** 316
Rupture(s) *See also* Decohesive rupture(s); Dimple rupture(s); Fracture(s)
intergranular, and transcrystalline cleavage **A12:** 222
low-carbon steels **A12:** 240
materials illustrated in **A12:** 217
spontaneous, AISI/SAE alloy steels **A12:** 299
titanium alloys **A12:** 449
Rural atmospheres
contaminants in **A13:** 81
corrosion in **A11:** 193
galvanized coatings in **A13:** 440
magnesium/magnesium alloys in **A13:** 743
simulated service testing **A13:** 204
telephone cables in **A13:** 1127
Rust *See also* Corrosion; Oxidation; Rusting; White rust
defined **A11:** 9 **A13:** 11
Rust (continued)
derusters, alkaline cleaning process *See also* Alkaline derusting and descaling **M5:** 22
in marine atmospheres **A13:** 778
painting process affected by **M5:** 473-474, 499, 504-505
Pictorial Representation of Rust Classification **M5:** 499, 503
and porosity, in arc welds **A11:** 413
protection against *See also* Rust-preventive compounds **M5:** 459-470
cadmium plate on steel **M5:** 266
extent of protection, determining **M5:** 460
rust-proofing step, vapor degreasing **M5:** 47
shot-peened materials **M5:** 145
removal of, process types and selection **M5:** 11-14
resistance testing, tin coatings **A13:** 781
rusting rate, effects on **M5:** 460
sheared edges, fabricated steel sheet, aluminum coating process affected by **M5:** 334
staining, from galvanic corrosion **A13:** 86
Rust and corrosion inhibitors **A18:** 99, 105-106
applications **A18:** 106
formation of **A18:** 105-106
multifunctional nature **A18:** 111
tests **A18:** 106
types **A18:** 105
Rust and oxidation (R&O) inhibited oils
lubricants for rolling-element bearings **A18:** 133-134
Rust inhibitors **M5:** 435, 453
grease additives **A18:** 124
tapping **A16:** 263
Rust on etched specimens
prevention and removal **A9:** 172
Rust-preventive compounds **M5:** 459-470
applications **M5:** 460-466
applying, methods of **M5:** 466-467
chemical reactions with surface effects of **M5:** 466
costs **M5:** 468
dry-film types *See* Dry-film rust- preventive compounds
emulsion cleaners containing *See also* Emulsion rust-preventive compounds
film, physical characteristics **M5:** 466, 470
film thickness, control of **M5:** 467-468, 470
fingerprint removers *See* Fingerprint removers and neutralizers
grease used as **M5:** 460-464, 467-468
hard-film types **M5:** 460, 465-466, 468
lubrication oils used as **M5:** 460-464
material, selection of **M5:** 460-466
metal specimen mass effects of **M5:** 468-470
military specifications **M5:** 460-464
nonmetallic materials attacked by **M5:** 466
oil types *See* Oil rust-preventive compounds
petrolatum types *See* Petrolatum rust-preventive compounds
quality control **M5:** 469-470
removability and removal of **M5:** 45, 468
safety precautions **M5:** 470
solvent-cutback type *See* Solvent- cutback petroleum-based rust- preventive compounds
specific gravity effects **M5:** 468-470
storage conditions determining material selection **M5:** 460-466
surface preparation for **M5:** 470
temperature effects **M5:** 467-468, 470
types available **M5:** 459-464
viscosity, control of **M5:** 469
water-based types **M5:** 470
water-displacing polar types *See* Water-displacing polar rust-preventive compounds
withdrawal rate, effect of **M5:** 468-469
Rusting *See also* Corrosion; Oxidation
deliberate light **A15:** 18
effect of atmospheric pollution on **A13:** 511
as electrochemical **A13:** 18
erosion and, carbon steel **A13:** 136
ferrous scrap, as inclusion-forming **A15:** 90
passivators **A13:** 1259
scale, carbon steel **A13:** 520
tubercule, localized corrosion under **A13:** 488
Ruthenium *See also* Precious metals
annealing **A4:** 946-947 **M4:** 760
applications **A4:** 946
corrosion of **A13:** 805-806
determined by controlled-potential coulometry **A10:** 209
distillation **A10:** 169
electrical circuits for electropolishing **A9:** 49
evaporation fields for **A10:** 587
gravimetric finishes **A10:** 171
hardness **A4:** 947
in medical therapy, toxic effects **A2:** 1258
and palladium alloys, as electrical contact materials **A2:** 847
and platinum alloys, as electrical contact materials **A2:** 847
as precious metal **A2:** 688
pure **M2:** 792
pure, properties **A2:** 1153
resources and consumption **A2:** 689-690
special properties **A2:** 692, 694
substitution for titanium producing electrical conductivity **EM4:** 542
tensile strength **A4:** 944
thermal expansion coefficient **A6:** 907
toxicity **A7:** 207
vapor pressure, relation to temperature **A4:** 495 **M4:** 310
working of **A14:** 851
Rutherford backscattering spectrometry **A10:** 628-636
applications **A10:** 628, 631-636
backscattering **A10:** 629
capabilities **A10:** 610
capabilities, and FIM/AP **A10:** 583
channeling effect in **A10:** 630-631
collision kinematics **A10:** 629
defined **A10:** 681
energy loss **A10:** 630
equipment for **A10:** 631
estimated analysis time **A10:** 628
general uses **A10:** 628
of inorganic solids **A10:** 4-6
introduction and principles **A10:** 629
limitations **A10:** 628
related techniques **A10:** 628
resonance effect **A10:** 635
samples **A10:** 628, 632-636
scattering cross section **A10:** 629-630
sensitivity **A10:** 630
Rutherford backscattering spectroscopy (RBS) **EL1:** 926, 1092 **EM3:** 237
depth profiling **EM3:** 247
Rutherford cable, two-layer
for niobium-titanium superconducting materials **A2:** 1051-1052
Rutile
bright-field and dark-field images of annealing twin in **A10:** 443
in ceramic waste form simulant, EPMA analysis for **A10:** 532-535
chemical composition **A6:** 60
functions in FCAW electrodes **A6:** 188
Miller numbers **A18:** 235
sintering agent for **A10:** 166
RWMA group B refractory metal rod
bar and inserts **A7:** 629
Ryton thermoplastic **EM1:** 99
RZ powder **A7:** 10

SUBJECTS OF THE INDEXED VOLUMES: ASM Handbook (designated by the letter "A"): **A1:** Properties and Selection: Irons, Steels, and High-Performance Alloys (1990); **A2:** Properties and Selection: Nonferrous Alloys and Special-Purpose Materials (1990); **A3:** Alloy Phase Diagrams (1992); **A4:** Heat Treating (1991); **A6:** Welding, Brazing, and Soldering (1993); **A7:** Powder Metallurgy (1984); **A8:** Mechanical Testing (1985); **A9:** Metallography and Microstructures (1985); **A10:** Materials Characterization (1986); **A11:** Failure Analysis and Prevention (1986); **A12:** Fractography (1987); **A13:** Corrosion (1987); **A14:** Forming and Forging (1988); **A15:** Casting (1988); **A16:** Machining (1989); **A17:** Nondestructive Testing and Quality Control (1989); **A18:** Friction, Lubrication, and Wear Technology (1992). **Metals Handbook, 9th Edition** (designated by the letter "M"): **M1:** Properties and Selection: Irons and Steels (1978); **M2:** Properties and Selection: Nonferrous Alloys and Pure Metals (1979); **M3:** Properties and Selection: Stainless Steels, Tool Materials and Special-Purpose Materials (1980); **M4:** Heat Treating (1981); **M5:** Surface Cleaning, Finishing, and Coating (1982); **M6:** Welding, Brazing, and Soldering (1983). **Engineered Materials Handbook** (designated by the letters "EM"): **EM1:** Composites (1987); **EM2:** Engineering Plastics (1988); **EM3:** Adhesives and Sealants (1990); **EM4:** Ceramics and Glasses (1991); **Electronic Materials Handbook** (designated by the letters "EL"): **EL1:** Packaging (1989).

σ phase
orientation relationship in **A10:** 453-454
The Rauch Guide to the U.S. Adhesives and Sealants Industry **EM3:** 71

S

17-4 PH *See* Stainless steels
6150 steel **A1:** 437-438
heat treatments for **A1:** 438
properties of **A1:** 438, 439
noncontact methods **A8:** 193
in hold-time tests, stainless steel **A8:** 356
partitioning, in creep-fatigue interaction **A8:** 357-358
Strain range **A8:** 357
and strain rate, in ultrasonic fatigue testing **A8:** 241
vs. cycles-to-failure for low-cycle fatigue **A8:** 364
vs. fatigue-life reduction factor **A8:** 353
vs. hold period in compression **A8:** 353
vs. hold periods in tension **A8:** 353
vs. mean stress **A8:** 353-354
S basis
of design values **A8:** 662
S glass
properties **EM4:** 849, 1057
seals to metals **EM4:** 499
S polarization *See* Polarization
S-2 glass **EM1:** 29, 43
S-590
composition **M4:** 651-652
S-816
aging **A4:** 796
aging cycle **M4:** 656
annealing **M4:** 655
broaching **A16:** 209
composition **A4:** 794, 795 **A6:** 564, 573, 929 **A16:** 736 **M4:** 651-652 **M6:** 354
grinding **A16:** 759
machining **A16:** 738, 741-743, 746-747, 749-758
solution treating **M4:** 656
solution-treating **A4:** 796
stress relieving **M4:** 655
S-band microwave frequency
use in ESR analysis **A10:** 255
S-basis **EM3:** 25
defined **EM2:** 37
S-glass
defined **EM2:** 38
as PWB reinforcement **EL1:** 604
S-glass fibers
applications **EM1:** 45, 107
composition **EM1:** 45, 107
defined **EM1:** 21, 29
effect, polyester resins **EM1:** 92
effect, vinyl ester resins **EM1:** 92
pristine strength **EM1:** 46
properties **EM1:** 175
reinforced epoxy resin composites **EM1:** 400
in space and missile applications **EM1:** 817
tensile strength **EM1:** 107
thermal stability **EM1:** 107
S-gun magnetron sputtering technique **A18:** 841, 842
***S-N* curve** **A8:** 376
for 50% survival **A8:** 12
for constant amplitude and sinusoidal loading **A8:** 364
defined **A8:** 12
fatigue crack initiation **A8:** 364
fatigue limit **A8:** 364
for *p*% survival **A8:** 12
***S-N* curves** *See* Fatigue data **A13:** 292 **EM1:** 201, 441-442
cast steels **M1:** 389, 397
defined **A11:** 9
and fatigue resistance **A1:** 674-675
from fatigue-crack initiation tests **A11:** 102-103
for reversed-bending stress and temperature **A11:** 130
springs, steel **M1:** 292, 297, 304
***S-N* diagram** **EM3:** 26
defined **A13:** 11 **EM1:** 22 **EM2:** 39
***S-N* relation**
bolted joint fatigue **EM1:** 440
mean stress **EM1:** 438-439
notch **EM1:** 439-440
S-Se (Phase Diagram) ..... **A3:** 2 • 357
S-Sn (Phase Diagram) ..... **A3:** 2 • 357
S-Te (Phase Diagram) ..... **A3:** 2 • 358
S-Ti (Phase Diagram) ..... **A3:** 2 • 358
S. Jarvis Adams Company *See* Mackintosh-Hemphill Company
S.W.G. System *See* Steel wire gage system
S/MA *See* Styrene-maleic anhydrides
S1, S2, S3, etc *See* Tool steels, specific types
SAAB rubber-diaphragm method
of fluid forming **A14:** 611, 614
SAAS 3M computer program for structural analysis **EM1:** 268, 272
SAAS III computer program for structural analysis **EM1:** 268, 272-273
Sabotage
of pipelines **A11:** 704
SAC (surface area center) test **A1:** 150-151, 152-153, 454
SAC hardenability test *See* Surface-area -center test **M1:** 473-474, 477
Saccharin
use in nickel plating **M5:** 205
Saccharin (benzoic sulfimide) **EM3:** 114
Saccharin salt + alpha hydroxy sulphone
generating free radicals for acrylic adhesives **EM3:** 120
Sachs (slab) analytical process method **A14:** 425
SACMA *See* Suppliers of Advanced Composite Materials Association
SACP *See* Selected-area channeling pattern
Sacrificial anode
on steel ship hulls **A8:** 537, 540
Sacrificial anodes *See also* Cathodic protection
aluminum **A13:** 469
corrosion protection in seawater **M1:** 745
energy characteristics **A13:** 921
and impressed-current anodes, compared **A13:** 922
magnesium **A13:** 468-469
for marine corrosion **A13:** 920-921
soil corrosion prevented by **M1:** 731
for stray-current corrosion **A13:** 87
systems, cathodic protection **A13:** 467-469
zinc **A13:** 469
Sacrificial coatings
for cast irons **A13:** 570-571
corrosion prevention with **M5:** 432-433
Sacrificial metal systems
for cladding **A13:** 888-889
Sacrificial protection
defined **A13:** 11
SAD *See* Selected-area diffraction
Saddle
forging **A14:** 62-63, 69-70
open-die forging of ring with **A14:** 127
supports, for open-die forging **A14:** 62
Saddle forging **A14:** 62-63, 69-70
SADP *See* Selected-area diffraction pattern
SAE *See* Society of Automotive Engineers
SAE aerospace material specifications (AMS) **EM2:** 91
SAE Aerospace Materials Division, Composites Committee
as information source **EM1:** 41
SAE alloy 13 ..... **A9:** 419
SAE alloy 14 ..... **A9:** 419
SAE alloy steels *See* AISI/SAE alloy steels
SAE Handbook, Aerospace Index **EM3:** 72
SAE Nonmetallics Committee **EM3:** 62
SAE specifications *See also* AISI-SAE specifications; listings in data compilations for individual alloys
bearing materials **M3:** 813-815, 817-819
composition ranges and limits for steels **M1:** 131-134
designation system for steels **M1:** 124-127, 132
ductile iron **M1:** 34, 35, 36
experimental steels, discussion of **M1:** 127, 132
ferrous P/M materials **M1:** 332, 333
gray iron **M1:** 16, 18, 19
HSLA steels, discussion of **M1:** 132, 403, 408-409
SAE specifications (continued)
J435c specification requirements **M1:** 377, 378, 380
superhard tool materials **M3:** 448-449, 453
SAE strength grades
threaded fasteners **M1:** 274, 275, 277
SAE-AISI designations
cross-referenced to international steel specifications
British (BS) steel specifications **A1:** 166-174
French (AFNOR) steel specifications **A1:** 166-174
German (DIN) steel specifications **A1:** 166-174
Italian (UNI) steel specifications **A1:** 166-174
Japanese (JIN) steel specifications **A1:** 166-174
Swedish (SS) steel specifications **A1:** 166-174
for ductile iron **A1:** 35, 36
for former standard steels **A1:** 155
for free-cutting (resulfurized) steels **A1:** 150
for free-cutting steels **A1:** 151
for high-strength low-alloy steels **A1:** 154
for low-alloy (alloy) steels **A1:** 152, 153
for merchant quality steels **A1:** 150
for potential standard steels **A1:** 153
for steel castings **A1:** 364, 365, 366
steels, formerly listed **A1:** 151, 155-156
for threaded fasteners **A1:** 289, 290, 292, 296
for threaded steel fasteners **A1:** 289, 290, 292, 296
Safe forming
in austenitic stainless steel **A8:** 154
Safe region
in aluminum processing map **A8:** 572
Safety *See also* Explosions; Fires; Management; Operators; Personnel; Precautions; Pyrophoricity; Toxicity
abrasive waterjet cutting **A14:** 755
alarm, cupola **A15:** 30
aluminum melting, reverberatory furnace **A15:** 377
of aluminum-lithium alloys **A2:** 182-184
aspects. of damage tolerance **EM1:** 259
autoclave **EM1:** 648
of beryllium **A2:** 687
in beryllium forming **A14:** 806
beryllium-containing alloys **A2:** 426-427
in blanking operations **A14:** 458
of chromate conversion coating **A13:** 395
of cleaning methods **A17:** 81-82
in deep drawing **A14:** 590
with die and die materials **A14:** 58
in discontinuous ceramic fiber MMC production **EM1:** 903
in electromagnetic forming **A14:** 650
in electropolishing **A9:** 51-55
in explosive forming **A14:** 643, 781
in fine-edge blanking and piercing **A14:** 474
with forging hammers/presses **A14:** 35
in hand lay-up technique **EM1:** 132
in handling and mixing etchants **A9:** 68-69
in handling electrolytes **A9:** 51-55
in hot forging **A14:** 58
in hydrogen sulfide use **A10:** 169
information, bibliography of **A14:** 35
with liquid metals **A13:** 94-97
of lubrication **A14:** 517-518
in machining magnesium and magnesium alloys **A2:** 475-476
with magnesium alloys **A14:** 826
in open-die forging **A14:** 73
in piercing **A14:** 471
with polyamide-imides (PAI) **EM2:** 137
of pouring devices **A15:** 27
in press bending **A14:** 532
in press forming **A14:** 555
in press-brake forming **A14:** 544
of presses **A14:** 474, 503
radiation **A17:** 300-302
salt bath explosions **A2:** 182
in sample dissolution **A10:** 165-167
of semisolid metalworking processes **A15:** 338
in shearing **A14:** 707
structural, fracture control for **A17:** 666
in three-roll forming **A14:** 623
toxicity, of additives **EM2:** 494
of ultrasonic cutting **EM1:** 618
of vertical centrifugal casting **A15:** 305-307
with water-suspended magnetic particles **A17:** 101-102

Safety equipment
of polyarylates (PAR) **EM2:** 138
of polycarbonates (PC) **EM2:** 151
Safety factor rule
laminate ranking **EM1:** 452-453
Safety glass **EM4:** 453
recommended waterjet cutting speeds **EM4:** 366
Safety ladle
development of **A15:** 28
geared, development **A15:** 33
Safety pin wire **A1:** 286 **M1:** 269
Safety precautions **A6:** 1189-1205 **EM3:** 685-686
abrasive blasting **M5:** 21, 96
acid cleaning **M5:** 21, 65
adhesive bonding **A6:** 1204-1205
alkaline cleaning **M5:** 21, 453-454
with aluminum powders **A7:** 130
arc welding **A6:** 1192-1193
arc welding and cutting **A6:** 1201
austempering of steel **M4:** 116
belt grinding **M5:** 557
brazing **A6:** 1202
buffing *See* Safety precautions, polishing and buffing
cadmium plating **M5:** 267
chlorofluorocarbon (CFC) solvents **EL1:** 667
chromic acid handling **M5:** 454
cleaning materials **EL1:** 667
clothing **M6:** 57-58
cyanide tarnish removal solutions **M5:** 613
dip brazing **A6:** 1202
electrical installations **M6:** 58
electrical safety **A6:** 1199-1200
cables **A6:** 1200
connections **A6:** 1200
electric shock **A6:** 1199
equipment installation **A6:** 1199
equipment selection **A6:** 1199
grounding **A6:** 1199-1200
maintenance **A6:** 1200
modifications **A6:** 1200
multiple arc welding operations **A6:** 1200
operations **A6:** 1200
pacemakers, wearers of **A6:** 1199
perception threshold **A6:** 1199
personnel training **A6:** 1199
prevention of fires **A6:** 1200
shock mechanism **A6:** 1199
shock sources **A6:** 1199
electrolytic cleaning **M5:** 31, 38
electron-beam welding **A6:** 1202, 1203
emulsion cleaning **M5:** 21, 38-39
explosion welding **A6:** 1203
exposure factors **A6:** 1193-1194
eyes and face **M6:** 57
fire protection **M6:** 58
fire safety requirements, NFPA **M5:** 20-21
flame hardening **M4:** 497-498
fluxes **A6:** 1202
friction welding **A6:** 1203
fuel gases **A6:** 1200-1201
fumes and gases **A6:** 1192
furnaces **M4:** 378-385
gas carburizing **M4:** 139-141
gas cylinders **M6:** 58
gas-tungsten arc welding **A6:** 1202
general welding safety **A6:** 1189-1192
burns **A6:** 1191
explosion **A6:** 1191
eye and face protection **A6:** 1191
fire **A6:** 1190-1191
general housekeeping **A6:** 1190
hot work permit system **A6:** 1191
machinery guards **A6:** 1192
Materials Safety Data Sheet (MSDS) **A6:** 1189-1190
noise **A6:** 1192
permissible exposure limits (PEL) **A6:** 1196
protection in the general area **A6:** 1190

Safety precautions (continued)
protective clothing **A6:** 1191-1192
public demonstrations **A6:** 1190
Threshold Limit Value (TLV) **A6:** 1189, 1192, 1196
handling of compressed gases **A6:** 1196-1199
cryogenic cylinders and tanks **A6:** 1197-1198
fuel gases **A6:** 1198-1199
gas cylinders and containers **A6:** 1196-1197
manifolds **A6:** 1198
oxygen **A6:** 1198
regulators **A6:** 1198
shielding gases **A6:** 1199
hard chromium plating **M5:** 178-179, 184, 186
health requirements, OSHA **M5:** 20-21
hot dip tin coating process **M5:** 355
laser-beam cutting **A6:** 1203
laser-beam welding **A6:** 1203
with liquid carbonyl compounds **A7:** 138
magnesium alloy cleaning and finishing processes **M5:** 648-649
magnesium alloys **M4:** 751-753
with magnesium powder production **A7:** 132-133
martempering of steel **M4:** 88
mass finishing **M5:** 136
measurement of exposure **A6:** 1196
with metals as fuels **A7:** 598
in nuclear applications **A7:** 666
optical testing **EL1:** 571
oxyfuel gas cutting **A6:** 1193, 1200-1201
oxyfuel gas welding **A6:** 1193, 1200-1201
for P/M toxicity, regulatory agencies and standards **A7:** 201-202
painting **M5:** 472, 494
parylene coatings **EL1:** 800
passivation processes **M5:** 560
personnel protection **M6:** 57
phosphate coating **M5:** 453-454
pickling **M5:** 21, 81, 613-614, 669
plasma arc cutting **A6:** 1201
polishing and buffing **M5:** 648-649
power brushing **M5:** 150
precautionary labeling **A6:** 68
processes **A6:** 1200-1205
quenching **M4:** 68
recommended practices **A7:** 249
resistance welding **A6:** 1201-1202
respiratory **M6:** 58
rust-preventive compounds **M5:** 470
selective plating **M5:** 298
with sintering atmospheres **A7:** 348-349
soldering **A6:** 1202
solvent cleaning **M5:** 21, 41-42, 44
thermal spray coating **M5:** 372
thermal spraying **A6:** 1204
thermite welding **A6:** 1203-1204
toxic materials **M6:** 58
training **M6:** 58
tumbling **M5:** 21
ultrasonic welding **A6:** 1203
with urethane coatings **EL1:** 780-781
vapor degreasing **M5:** 21, 45-46, 50, 53, 57
ventilation **A6:** 1194-1196
general **A6:** 1194
local **A6:** 1194-1195
respiratory protective equipment **A6:** 1195
special situations **A6:** 1195-1196
Wave soldering **EL1:** 689-690
Safety precautions for
arc welding of beryllium **M6:** 462
brazing of aluminum alloys **M6:** 1032
brazing of beryllium **M6:** 1052
dip brazing of steels **M6:** 995
electron beam welding **M6:** 646
explosion welding **M6:** 716-717
furnace brazing **M6:** 945-949
burnout method of, purging **M6:** 948-949
carbon monoxide poisoning and suffocation hazards **M6:** 947

Safety precautions for (continued)
cold chambers **M6:** 946
critical periods **M6:** 946
flammable and nonflammable gases **M6:** 945-946
ignition temperatures **M6:** 946
interruption of flammable gas flow **M6:** 946-947
leaky gas valves **M6:** 947
leaky retorts or muffles **M6:** 947
purging **M6:** 946
purging with nonflammable gas **M6:** 947-948
gas metal arc welding **M6:** 180-181
gas tungsten arc welding **M6:** 213
high frequency welding **M6:** 767-768
laser beam welding **M6:** 669-670
oxyfuel gas cutting **M6:** 914-915
oxyfuel gas welding **M6:** 593-594
percussion welding **M6:** 745
plasma arc cutting **M6:** 917
resistance welding of copper and copper alloys **M6:** 556
shielded metal arc welding **M6:** 95
soldering **M6:** 1097-1100
assessing lead exposure **M6:** 1099-1100
cleaning agents **M6:** 1097-1098
fluxes **M6:** 1097
solder constituents **M6:** 1098-1099
specific processes **M6:** 58-59
submerged arc welding **M6:** 137
Safety valves, pressure vessels
failures of **A11:** 644
Saffil fibers **EM1:** 63
Sag
defined **A15:** 10 **EM2:** 37
Sag, cylindrical/spherical
measured **A17:** 17
Sag resistance
epoxies **EM3:** 98
of steel sheet for porcelain enameling **M1:** 177-180
Sagging **EM3:** 25
Sagittal plane
defined **A17:** 384
Saha equation **A10:** 24
Sailboat hardware
powders used **A7:** 574
Sailors-Corten Charpy/fracture toughness correlation **A8:** 265
Saint-Venant's principle **EM3:** 485
Saline groundwaters
copper alloy corrosion rates **A13:** 621
Saline solutions *See also* Salt solutions; Salts
cast iron resistance **A13:** 570
Saliva
and dental alloy tarnish/corrosion **A13:** 1341-1342
Salt annealing
steel wire **M1:** 262
Salt atmosphere
effect, glass-to-metal seals **EL1:** 459
tests **EL1:** 494, 499-500
Salt bath carburizing **A4:** 262 **A18:** 873
diffusion of carbon **A18:** 873
Salt bath descaling **A14:** 230 **M5:** 97-103
applications **M5:** 101-103
aqueous caustic descaling process
refractory metals **M5:** 654, 659
carbon removal process **M5:** 99-100
cast iron **M5:** 98-100, 102
castings **M5:** 98-102
electrolytic process *See* Electrolytic salt bath descaling
equipment **M5:** 100-101
hafnium alloys **M5:** 667
heat process scale removal **M5:** 100-102
heat-resisting alloys **M5:** 565-568
high-temperature baths **M5:** 653-654
low-temperature baths **M5:** 654

Salt bath descaling (continued)
molybdenum **M5:** 659
nickel alloys **M5:** 672
organic coating removal **M5:** 102-103
oxidizing process *See* Oxidizing salt bath descaling
pickling process **M5:** 97-99, 102
concentration and temperature of acids **M5:** 97-98
process *See also* specific processes by name **M5:** 4, 12-13, 97-101
reducing process **M5:** 97, 100
refractory metals **M5:** 653-654, 659
rust and scale removal by **M5:** 12-14
safety precautions **M5:** 3-21
sand removal process **M5:** 98-99
sodium hydride process *See* Sodium hydride salt bath descaling
stainless steel **M5:** 97-98, 101-102, 553-554
strip **M5:** 101-102
wire **M5:** 98, 102
superalloys **M5:** 97, 101-102
titanium and titanium alloys **M5:** 97-98, 102, 653-654
tungsten **M5:** 659
zirconium alloys **M5:** 667
Salt bath equipment **A4:** 726-728
applications
distortion control **A4:** 475
selection of a salt **A4:** 475
surface protection **A4:** 475
furnaces
air-quality assurance **A4:** 481-482
automatic **A4:** 482-483, 727, 728 **M4:** 297-298
efficiency **A4:** 475
electrical resistance **A4:** 477, 730 **M4:** 233, 293
electrodes, service life **A4:** 478
externally heated **A4:** 476-477 **M4:** 233, 293-294
gas fired **A4:** 476-477
immersed-electrode **A4:** 477-480, 727 **M4:** 294-296
isothermal quench **A4:** 481-482, 730-731 **M4:** 298
oil-fired **A4:** 476
pot service life **A4:** 477
refractories, service life **A4:** 478
safety precautions **A4:** 477
semiautomatic **A4:** 482-483 **M4:** 297-298
steel pot **A4:** 479, 480 **M4:** 295-296
submerged-electrode **A4:** 477, 478, 480-481, 727 **M4:** 296-297
salt pots **A4:** 476 **M4:** 293-294
Salt bath explosions
aluminum-lithium alloys **A2:** 182
Salt bath hardening
of sintered high- speed steels **A7:** 54, 55
Salt bath nitrided cast iron **A9:** 229
Salt bath nitrided steels **A9:** 229
Salt bath nitriding **A4:** 263
Salt bath prepickling process, nickel alloys **M5:** 672
electrolytic type **M5:** 672
oxidizing type **M5:** 672
Salt bath(s)
compositions **A4:** 726
processing temperature ranges **A4:** 726
Salt baths
cleaning for cast irons **M6:** 600
dip brazing of copper **M6:** 1048
dip brazing of stainless steels **M6:** 1012
liquid carburizing **M4:** 229
of molten barium chloride **A11:** 277
preparation for oxyacetylene braze welding **M6:** 598
Salt brine
Miller numbers **A18:** 235
Salt cake **EM4:** 381
batch size **EM4:** 382
as by-product **EM4:** 380
chemical analysis **EM4:** 380
as melting accelerator **EM4:** 380
mining techniques **EM4:** 380
purity **EM4:** 380
purpose for use in glass manufacture **EM4:** 381
sources **EM4:** 380
Salt deposits
high-level waste disposal in **A13:** 976-978
Salt flux grain refining
aluminum **A15:** 477
Salt fog test **A13:** 11, 224-226, 1113-1114
as environmental failure analysis **EL1:** 1102-1103
Salt fog tests
shot peening corrosion resistance improvement **M5:** 145
Salt precursors **EM4:** 113
Salt solution
precipitation by **A7:** 54, 55, 375
Salt solutions
dilute, uranium/uranium alloys in **A13:** 815
electroplated chromium deposit resistance to **A13:** 874
magnesium/magnesium alloys in **A13:** 742
niobium resistance in **A13:** 722
titanium/titanium alloy resistance to **A13:** 679-680
Salt spray
corrosion, chromium-plated parts **A13:** 871-872
data, chromate conversion coatings **A13:** 393
magnesium/magnesium alloys in **A13:** 745, 746
performance, galvanized coatings **A13:** 440
Salt spray (fog) testing **A13:** 11, 224-226, 1113-1114
application and types **A13:** 225
cabinets **A13:** 225-226
neutral, acetic, and copper-accelerated acetic **A13:** 225
Salt spray test **EM3:** 659, 661, 664, 669-671
Salt water *See also* Seawater; Water
corrosion, copper/copper alloys **A13:** 622-625
corrosion fatigue in **A8:** 255
high-strength steel SCC behavior in **A8:** 527, 528
zirconium/zirconium alloy resistance **A13:** 708-709
Salt water corrosion
copper alloys **M2:** 471-472
Salt-bath brazing **A6:** 121, 122
safety precautions **A6:** 1191
Salt-bath dip brazing
copper and copper alloys **A6:** 935
Salt-coated magnesium particles **A7:** 132
Salt-spray test
chromium plating **M5:** 198
magnesium alloy finishes **M5:** 638, 646
mechanical coatings **M5:** 300-301
paint **M5:** 492
rust-preventive compounds **M5:** 469-470
shot peening corrosion resistance improvement **M5:** 145
Salting
as AAS flame modification **A10:** 54
Salting up
with concentric nebulizers **A10:** 35
with total consumption burners **A10:** 28
Salts
as colorants **EM4:** 380
copper/copper alloy corrosion in **A13:** 630
corrosive (inorganic) soldering fluxes **A6:** 628
dissolved, in water **A13:** 490
dissolved, zinc corrosion in **A13:** 763
effect on tantalum **A13:** 732-733
filler metal for molten-salt dip brazing **A6:** 338
gold corrosion in **A13:** 799
hot dry chloride, and SCC **A11:** 223
hot, SCC testing of titanium alloys in **A13:** 273-275
immersion, of magnesium/magnesium alloys **A13:** 745
molten **A11:** 276-277 **A12:** 24
molten, corrosion in **A13:** 17
nickel alloys, corrosion **M3:** 174
nonstoichiometric, single-crystal analysis **A10:** 355
osmium corrosion in **A13:** 807
palladium corrosion in **A13:** 804
plating **A11:** 45
platinum corrosion in **A13:** 802
rhodium corrosion in **A13:** 805
silver corrosion in **A13:** 796
in sinters and fusions **A10:** 166
solution, as alloy corrosive environment **A11:** 78-79
stainless steel corrosion in **A13:** 559
tantalum resistance to **A13:** 727
zirconium, corrosion in **M3:** 785-786
Salts (continued)
zirconium/zirconium alloy resistance to **A13:** 716-717
Salts, copper alloys
corrosion rate **M2:** 477-479
Salts, dissolved *See* Dissolved salts
Salts for dip brazing **M6:** 990-991
addition of fluxing agents **M6:** 991
carburizing and cyaniding salts **M6:** 991
neutral salts **M6:** 990-991
Salts used in etchants **A9:** 66-68
Saltwater *See also* Seawater; Water
crack growth of pipeline steel in **A11:** 54
effect on shafts and pistons **A11:** 467
heat-exchanger condenser tube, pitting of **A11:** 631
Saltwater corrosion resistance *See also* Marine
of magnesium alloys **A2:** 456
nickel alloys **A2:** 429
Saltykov equations
for three-dimensional planar figures **A9:** 132
for two-dimensional planar figures **A9:** 131
Saltykov method for determining the surface- to-volume ratio of discrete particles **A9:** 125
Salvage, scrap
by hot isostatic pressing **A15:** 543
SAM *See* Scanning acoustic microscopy (SAM); Scanning Auger microscopy
Samarium *See also* Rare earth metals
epithermal neutron activation analysis (ENAA) **A10:** 239
pure **M2:** 792-793
as rare earth metal, properties **A2:** 720, 1186
TNAA detection limits **A10:** 237
Samarium cobalt ($SmCo_5$)
hot pressing applications **EM4:** 192
Samarium in garnets **A9:** 538
Samarium-cobalt magnets **A9:** 539, 548-549
Samarium-cobalt permanent magnets
as rare earth metal application **A2:** 729-730
Samarium-praseodymium-cobalt permanent magnets **A7:** 643
SAMPE *See* Society for the Advancement of Material and Process Engineering
SAMPE Journal **EM3:** 67
as information source **EM2:** 92
SAMPE Quarterly **EM3:** 67
as information source **EM2:** 92-93
Sample *See also* Specimen
average **A8:** 11
defined **A8:** 11, 624
degrees of freedom for **A8:** 625
for direct computation of normal distribution **A8:** 664
median **A8:** 11
number and accuracy of test results **A8:** 623-624
percentage **A8:** 11
of SCC test materials **A8:** 501
size **A8:** 696-697
standard deviation **A8:** 11, 709-711
in statistical distributions **A8:** 628
variance, defined **A8:** 11
Sample cavity
ESR spectrometer **A10:** 256
Sample cell
rotating **A10:** 132
Sample dissolution *See also* Dissolution
mediums for **A10:** 165, 166, 168
Sample geometry
and properties divergencies **EM2:** 655
Sample interaction model
for radiographic inspection **A17:** 710
Sample matrix **A10:** 83, 162, 168-170
Sample mean **A18:** 481
Sample means control charts *See also* Control charts
construction and interpretation **A17:** 725, 728
setting up **A17:** 727-728
Sample preparation *See also* Samples; Sampling
analytical electron microscopy **A10:** 450-453
for classical wet analytical chemistry **A10:** 165-167
elemental and functional group analysis **A10:** 213
field ion microscopy **A10:** 584-586
gas chromatography/mass spectrometry **A10:** 644-645

Sample preparation (continued)
for image analysis **A10:** 309-310, 313-314
microbeam analysis **A10:** 529
for microstructural analysis of ion-implanted alloys **A10:** 484
techniques for selected materials **A10:** 586
x-ray diffraction residual stress techniques **A10:** 384-385
x-ray photoelectron spectroscopy **A10:** 574-576
for x-ray spectrometry **A10:** 93-95
as XRPD source of error **A10:** 341
Sample selection *See also* Sample(s)
for analysis **A11:** 15-16
of fracture surfaces **A11:** 19-20
for heat-exchanger failed parts **A11:** 628-629
of metallographic sections **A11:** 23-24
for pipeline failure investigation **A11:** 696-697
Sample size **EM3:** 791
defined, in NDE reliability data analysis **A17:** 694
effect, NDE reliability **A17:** 689
and flaw size, in NDE reliability experiments **A17:** 694
for hit/miss data **A17:** 694
for POD(A) function analysis **A17:** 694
Sample standard deviation **A18:** 481
Sample standard deviations
defined **A8:** 11
different, difference between two mean populations **A8:** 711
similar, differences between two population means **A8:** 709-711
Sample tube, pure quartz
in ESR analysis **A10:** 256
Sample(s) *See also* Objects; Sample selection; Sample size; Sampling; Specimen(s); Surface preparation
chamber, SEM installation **A7:** 235
contamination, in microanalytical techniques **A11:** 36
definition **A17:** 728
density, electron penetration depth as function of accelerating voltage and **A11:** 41
electrical conductivity of **A11:** 36
handling, in tool and die failures **A11:** 568
for microanalyses **A11:** 36
mixing **A17:** 730
for plane-strain fracture toughness **A11:** 61
and populations, compared **A17:** 738
preparation, scanning electron microscope **EL1:** 1095
preparation, solder joints **EL1:** 1038-1039
preparation, uniform corrosion exposure tests **A13:** 229-230
protection of **A11:** 173
reference **A7:** 258
selection, as rational **A17:** 727
selection of **A11:** 15-16
splitter **A7:** 10, 212
standard reference, eddy current inspection **A17:** 179
stratification, in rational sampling **A17:** 730
thief **A7:** 10, 212, 213
variable sizes, for p-chart **A17:** 735-736
Samples *See also* Sample preparation; Sampling
acidity/basicity measured in **A10:** 172
aqueous, optical emission spectroscopy for **A10:** 21
briquets, for x-ray spectrometry **A10:** 93-94
bulk, Rutherford backscattering spectrometry **A10:** 631
charging, AES **A10:** 556
composite **A10:** 13
defined **A10:** 681
dissolution **A10:** 4, 56-57, 163, 165-167
as error source in image analysis **A10:** 313
fluorescence of **A10:** 387
fusion of **A10:** 94

Samples (continued)
geometry, effect on XRD analysis **A10:** 388
gross, defined **A10:** 674
high vapor pressure, AES **A10:** 556
holders, MFS analysis **A10:** 76-77
identification **A10:** 17
infinitely thick/thin, x-ray spectrometry **A10:** 93
labeling and recording of **A10:** 13
laboratory, defined **A10:** 676
matrix **A10:** 83, 162, 168-170
mechanical strength of **A10:** 587
metallurgical, precision analysis by electrogravimetry **A10:** 197
minimum number **A10:** 14
minimum size **A10:** 14
multiphase **A10:** 516, 529-530
NAA, radioactive contamination of **A10:** 236
particulate, preparation of **A10:** 165
positioning, XRD analysis **A10:** 385
powder, fluxes for fusing **A10:** 167
powders, for x-ray spectrometry **A10:** 93-94
preservation **A10:** 12, 15, 16
pretreatment **A10:** 12, 15
quality of **A10:** 12
radioactive, remote analysis by electrometric titration **A10:** 202
representative **A10:** 13
rotation of **A10:** 130, 132
sectioning, for OM analysis **A10:** 300
size and form, importance in classical wet chemistry **A10:** 162
size and homogeneity of **A10:** 13
small diameter, surface stress measurement of **A10:** 385
solid, for x-ray spectrometry **A10:** 93
storage **A10:** 12, 15
systematic **A10:** 12-13
temperature control of **A10:** 257
thin-film, x-ray spectrometry **A10:** 95
types of **A10:** 12-13
unknown, AAS for **A10:** 46
variability of **A10:** 12
Sampling *See also* Rational sampling; Sample preparation; Sample(s); Samples **A7:** 212-213 **A10:** 12-18 **A18:** 294
accessories, infrared spectroscopy **A10:** 112
blending and mixing analyses **A7:** 187
of bulk materials **A10:** 13-14
by Grimm glow discharge source **A10:** 27
by sputtering **A10:** 27
for chemical analysis **A7:** 246, 248-249
composite samples **A10:** 13
copper and copper alloys for chemical composition **A7:** 249
of corrosion products **A11:** 602
costs of **A10:** 15
in dc arc **A10:** 25
defined **A10:** 12, 681
during discharge in continuous stream **A7:** 213
and excitation, in emission sources **A10:** 25
experimental **EM2:** 606-607
for fiber properties analysis **EM1:** 731, 733-734
field sampling **A10:** 16
filter, applications **A10:** 94
for fractal analysis **A12:** 212
from packaged containers **A7:** 213
isokinetic, devices for **A10:** 16
materials in discrete units **A10:** 14
mechanical cutting and flame cutting, compared methods **A7:** 187, 212-213, 249
model of operation for **A10:** 13
on-site, of failed parts **A11:** 173
for particle sizing **A7:** 226-227
plan **A10:** 13-15 **A12:** 212
point **A12:** 196-198
polymer, milligram quantities **EM2:** 836-837
practical aspects of **A10:** 15-16
precautions **A11:** 173, 602

Sampling (continued)
preliminary considerations **A10:** 12
procedures **A7:** 212-213
of process output, historical **A17:** 719
protocol **A10:** 13, 15-16
quality assurance for **A10:** 17
random **A10:** 12
rational **A17:** 728-730
representative samples **A10:** 13
resources, optimizing **A10:** 15
sample contamination **A10:** 16
sample damage **A10:** 15-17
sample discrimination **A10:** 16
sample preservation **A10:** 16
sampling protocol **A10:** 15-16
for SCC failure analysis **A11:** 212
scoop **A7:** 227
segregated (stratified) materials **A10:** 14
size **A7:** 212
specific materials **A10:** 17
statistical methods **A7:** 212-213
statistics **A10:** 12-15
for steam equipment **A11:** 602
strategy, microbeam analysis **A10:** 529-530
subsampling **A10:** 13, 15
substrates, for x-ray spectrometry **A10:** 94
tables **A7:** 226, 227
techniques, infrared spectroscopy **A10:** 112-116
techniques, microscopy and image analysis **A7:** 225-230
and the Raman effect **A10:** 129-130
thief **A7:** 212, 213
of tool steels **A9:** 256
uncertainty **A10:** 12, 15, 17
volume, by neutron diffraction **A10:** 423
vs environmental stress screening **EL1:** 885
Sampling boat
for flame AAS **A10:** 49
SAN *See* Styrene-acrylonitrile
Sand
angle of repose **A7:** 283
batch size **EM4:** 382
as carrier core, copier powders **A7:** 584
fuel-pump shaft wear from **A11:** 464
hole, defined **A11:** 9
inclusions, as casting defect **A11:** 387
in iron castings **A11:** 353
spots, slag streaks as **A11:** 322
Sand and sand fill
Miller numbers **A18:** 235
Sand, and SiC wear abrasives
compared **A8:** 603
Sand, as alloying element
and impurity specifications **A2:** 16
Sand blasting
aluminum and aluminum alloys **M5:** 572
health and safety precautions **M5:** 96
magnesium alloys **M5:** 628-629
properties, characteristics, and effects of sand **M5:** 84, 87
stainless steel **M5:** 553-555
Sand casting *See also* Castings, Foundry products; Sand molding
aluminum alloys **M2:** 144, 145-146, 147
aluminum casting alloys **A2:** 139-140
aluminum-silicon alloy, characteristics of **A15:** 159
of copper alloys **A2:** 346, 348 **M2:** 384
defined **A15:** 10
magnesium alloy compositions for **A15:** 798-799
magnesium alloys **A2:** 456-459, 494-516
mechanization **A15:** 32
of metal-matrix composites **A15:** 844
procedure **A15:** 203
tolerances **A15:** 617-619
vs permanent mold casting **A15:** 285
zinc alloys **A15:** 797
Sand compaction machines
for dry sand molding **A15:** 228

Sand conditioning *See also* Sand preparation; Sand(s)
mechanization **A15:** 32
Sand expansion defects **A15:** 210
Sand grains *See also* Grain size; Grain size distribution; Grains; Shape
distribution, defined **A15:** 10
rounded, sizes of **A15:** 223
shape and distribution of **A15:** 208-209
Sand iron rolls
rolling **A14:** 353
Sand mixes
carbonaceous additions **A15:** 211
cellulose **A15:** 211
cereals **A15:** 211
Sand mold *See also* Mold(s)
schematic **A15:** 189
vertical centrifugal casting **A15:** 301
Sand molding *See also* Sand casting; Sand(s) **A15:** 222-237
aluminum alloys, heat treatments for **A15:** 758-759
baked, Alnico alloys **A15:** 736
bonded sand molds **A15:** 222-230
chemically bonded and green sand compared **A15:** 341
chemically bonded self-setting **A15:** 37
dry sand molding **A15:** 228
equipment, development **A15:** 35
flow chart for **A15:** 203
green sand molds **A15:** 222-228
historical use **A15:** 22
loam molding **A15:** 228-229
phosphate bonded molds **A15:** 229-230
production, flow diagram for **A15:** 203
requirements **A15:** 208
resin binder processes **A15:** 222
silicate bonded molds **A15:** 229-230
skin-dried molds **A15:** 228
unbonded sand molds **A15:** 230-237
vs permanent mold process **A15:** 34
Sand preparation *See also* Preparation; Sand conditioning; Sand(s)
complete plant for **A15:** 32
for coremaking **A15:** 238-239
green sand and chemically bonded compared **A15:** 341
green sand molding **A15:** 225-226, 341, 344-345
intensive mixer **A15:** 344, 346
mechanization of **A15:** 32
plaster molding, conventional **A15:** 244
Shaw process **A15:** 249
of silica sands **A15:** 209
Sand processing *See also* Green sand molding; Reclamation; Sand molding; Sand reclamation; Sands
green sand molding, equipment/processing **A15:** 341-351
sand reclamation **A15:** 351-355
Sand reclamation *See also* Reclaimed sand; Reclamation; Recycled sand; Sand(s) **A15:** 351-355
of chemically bonded sands **A15:** 351-354
of clay-bonded system sand **A15:** 354-355
defined **A15:** 10
dry scrubbing/attrition **A15:** 227
dry systems **A15:** 351-353
effect, green sand appearance **A15:** 227
green sand molding **A15:** 225-228
multiple-hearth furnace/vertical shaft furnace **A15:** 227-228
wet systems **A15:** 351
wet washing/scrubbing **A15:** 227
Sand, removal of
pickling process **M5:** 69, 72
salt bath descaling process **M5:** 98-99
Sand slingers
development of **A15:** 28-29
for dry sand molding **A15:** 228
molding machines, green sand molding **A15:** 342-343
Sand tempering *See also* Temper point; Temper water
defined **A15:** 10
Sand(s) *See also* Alumina; Bank sand; Blended sand; Burned sand; Molding sands; Olivine sands; Sand casting; Sand conditioning; Sand grains; Sand mixes; Sand mold(s); Sand molding; Sand preparation; Sand processing; Sand reclamation; Sand slingers; Sand tempering; Sandblasting; Silica; Silica sands; Zircon
aluminum silicates **A15:** 209-210
base, reclamation effects **A15:** 355
chromate, reclamation **A15:** 355
chromite **A15:** 209
compounding **A15:** 32
control, green sand molds **A15:** 222-224
densities and mold hardnesses, development of **A15:** 29
grinding **A15:** 32
high-temperature, recovery of **A15:** 348-349
for lost foam casting **A15:** 233
metal separation from **A15:** 350
mixing of **A15:** 32, 211
mulling of **A15:** 32
natural vs synthetic **A15:** 208
olivine **A15:** 209
processing **A15:** 341-355
properties, controlling of **A15:** 222-224
reclamation of **A15:** 208, 320-355
screening machines for **A15:** 32
silica **A15:** 208-209
synthetic, development of **A15:** 29
systems, zirconium alloys **A15:** 836-837
testing of **A15:** 35, 345
zircon **A15:** 209
Sand-cast
ductile iron brake drum, brittle fracture **A11:** 371
gray iron flanged nut, brittle fracture **A11:** 369-370
gray iron pump bowl, graphitic corrosion **A11:** 373
oil-pump gears, brittle fracture **A11:** 344
steel axle housing, fatigue fracture **A11:** 388-389
Sand-cast gray iron
flaking fracture **A12:** 225
Sand-cast white iron
structure **M1:** 85
Sand-to-sand partings
early practice **A15:** 28
Sandblasting *See also* Shotblasting
compared to abrasive jet machining **A16:** 511
defined **A15:** 10
development of **A15:** 33
as surface preparation **A17:** 52, 591
Sanding
surface alterations produced **A16:** 23
Sandpaper test
as cure test **A13:** 418
Sandwich beam flexure testing
polybenzimidazoles **EM3:** 170
Sandwich braze **A6:** 335
definition **M6:** 15
Sandwich column compression tests
of polybenzimidazoles **EM3:** 170, 171
Sandwich construction **EM3:** 25
defined **EM2:** 37
Sandwich constructions *See also* Hybrid composites; Hybrid laminates
beam, four-point load test **EM1:** 328
concept **EM1:** 726-727
defined **EM1:** 21
honeycomb **EM1:** 726-728
laminate, damping, analysis **EM1:** 214
Sandwich edgewise compression testing
polybenzimidazoles **EM3:** 170-171
Sandwich flatwise tension testing
polybenzimidazoles **EM3:** 170
Sandwich heating **EM3:** 25
defined **EM2:** 37
Sandwich injection molding **EM1:** 557
Sandwich materials
roll compacting **A7:** 406-408
Sandwich mechanical coatings **M5:** 301
Sandwich molding
properties effects **EM2:** 284-285
size and shape effects **EM2:** 290
surface finish **EM2:** 304
Sandwich-peel testing
polybenzimidazoles **EM3:** 170
Sanford anodizing process
aluminum and aluminum alloys **M5:** 592
Sanitary ware *See* Fluid handling applications
Sanitaryware
composition **EM4:** 45
glazes **EM4:** 1061
properties of fired ware **EM4:** 45
SANS *See* Small-angle neutron scattering
SAP 4-5 computer program for structural analysis **EM1:** 268, 273
Saponification cleaning
copper and copper alloys **M5:** 617
mechanism of action **M5:** 3-4, 23
Saponification number
defined **A18:** 16
Saponifiers
as aqueous cleaners **EL1:** 664-665
Saponify
defined **A18:** 16
Sapphire **EM4:** 18, 19
effect on chippae and quality of cut **EM4:** 347
friction coefficient data **A18:** 75
fusion with acidic fluxes **A10:** 167
hardness **EM4:** 806
as internal reflection element **A10:** 113
nozzles for abrasive jet machining **A16:** 511-512
subcritical crack growth **EM4:** 695
ultrasonic machining ... **A16:** 529, 530, 532 **EM4:** 359
wire guide for wire EDM **A16:** 562
Sapphire substrates
physical characteristics **EL1:** 106
Sapphire, synthetic
lapping **A16:** 493
for orifice of waterjet nozzle **A16:** 521, 522
Sapphire window
autoclave **A8:** 422
Sarcoidosis **A7:** 202
SAS *See* Small-angle scattering
Satellite drops formation **A7:** 40-41, 43
Satellite imaging systems
beryllium parts for **A2:** 686-687
Satellite lines **A10:** 86, 521
Satellite radiator panels
graphite fiber reinforced copper-matrix composites for **A2:** 922
Satellite spots
as fine structure effect **A10:** 438
Satellites **A18:** 447
shake-up **A10:** 572
Satellites in diffraction patterns
cause of **A9:** 653
Satellites, space
corrosion of **A13:** 1101-1105
Satin **EM3:** 25
defined **EM1:** 21 **EM2:** 37
Satin (crowfoot) weave
fiberglass fabric
five-harness **EM1:** 125
five-harness, damping in **EM1:** 213
unidirectional/two-directional fabrics **EM1:** 125
for woven fabric prepregs **EM1:** 148, 148
Satin finishing
aluminum and aluminum alloys **M5:** 574-575, 577
magnesium alloys **M5:** 632
nickel alloys **M5:** 674-675
Satin weave *See* Harness satin
Saturable reactors **A6:** 38
Saturated backward coupled noise
defined **EL1:** 35
Saturated calomel
electrode, defined **A13:** 11
use, galvanic series **A13:** 235
Saturated calomel electrode **A8:** 416
abbreviation for **A11:** 797
in amperometric titration **A10:** 204
defined **A10:** 681
as reference electrode **A10:** 189
Saturated gun
defined **A9:** 15
Saturation *See* Degree of saturation; Wet strength
effect in compression-hold-only test **A8:** 348-349
effect in no-hold-test **A8:** 348-349
effect in symmetrical-hold testing **A8:** 348-349
magnetic **A9:** 534
Saturation, aluminum
in niobium-titanium superconducting materials **A2:** 1045

Saturation and relaxation
in ESR **A10:** 257-258
Saturation, defined *See* Wet strength
Saturation encapsulation process **EM3:** 586
Saturation induction
iron, alloying effects **A2:** 763
in magnetic hysteresis **A2:** 782
magnetically soft materials **A2:** 761
Saturation, magnetic
and hysteresis **A17:** 99-100
Saturation magnetization
in ferrites **EM4:** 1161, 1163
microstructural dependence **A17:** 131-132
Saturation mode
MOSFET in **EL1:** 966
Saturation pressure **EM3:** 25
defined **EM2:** 37
Saturation vapor pressure **EM4:** 130
Saugus Iron works
as US casting birthplace **A15:** 24
Sauter mean diameter
defined **A18:** 16
Saw burn
defined **EM2:** 37
Saw cutting
of specimens **A12:** 76
wrought aluminum alloys **A12:** 429, 435
Saw quality
alloy steel sheet and strip **M1:** 164
Saw tips, and corrosion
cemented carbides **A13:** 856
Sawcutting
notch root radius **A8:** 376
Sawing **A16:** 356-365
as a sectioning method **A9:** 23
as a sectioning method for tool steels **A9:** 256
aluminum alloy forgings **A14:** 247
aluminum and aluminum alloys **A2:** 10
applications
band **A16:** 29, 356-359
band, Al alloys **A16:** 770
band, Mg alloys **A16:** 827, 828
band, of titanium alloys **A14:** 841
band sawing machines **A16:** 356-357
band, vs. planing **A16:** 186
billet preparation by **A14:** 164
circular **A16:** 356, 365
circular, Al alloys **A16:** 770, 794
circular, cemented carbide tools **A16:** 84
circular, Cu alloys **A16:** 817-818
circular, heat-resistant alloys **A16:** 756
circular, MMCs **A16:** 894
circular, refractory metals **A16:** 867, 868
circular, tool steels **A16:** 725
circular, Zr **A16:** 855
in conjunction with milling **A16:** 308
contour band **A14:** 458 **A16:** 356-364
of copper and copper alloys **A14:** 257
Cu alloys **A16:** 805, 808
cutoff band **A16:** 360, 361
cutting fluids used **A16:** 125, 358, 360-363
fixtures and attachments **A16:** 357
hacksawing **A16:** 356, 365
Mg alloys **A16:** 827, 828
MMCs **A16:** 897
PCD tooling inserts **A16:** 110
safety **A16:** 364-365
saws **A16:** 1, 57, 58, 59
stack **A16:** 358
stainless **A16:** 705
of thermoplastic composite **EM1:** 552
of titanium alloys **A14:** 841
tool life **A16:** 356, 358, 359, 365
vs. planing **A16:** 186
zirconium **A16:** 852
Zn alloys **A16:** 834
Saws
as sampling tools **A10:** 16
Saws, ditching
of cemented carbides **A2:** 974
Sawtooth edge
in shear testing **A8:** 68
SAXS *See* Small-angle x-ray scattering
Saybold Universal Viscosity
defined **A18:** 16
Saybolt universal second (SUS)
viscosity measure of lubrication oil **A16:** 254
Sb-Se (Phase Diagram) .... **A3:** 2 • 358
Sb-Si (Phase Diagram) .... **A3:** 2 • 359
Sb-Sm (Phase Diagram) .... **A3:** 2 • 359
Sb-Sn (Phase Diagram) .... **A3:** 2 • 359
Sb-Sr (Phase Diagram) .... **A3:** 2 • 360
Sb-Tb (Phase Diagram) .... **A3:** 2 • 360
Sb-Te (Phase Diagram) .... **A3:** 2 • 360
Sb-Ti (Phase Diagram) .... **A3:** 2 • 361
Sb-U (Phase Diagram) .... **A3:** 2 • 361
Sb-Y (Phase Diagram) .... **A3:** 2 • 361
Sb-Zn (Phase Diagram) .... **A3:** 2 • 362
SBS *See* Short beam shear; Styrene-butadiene-styrene
$SbSbS_4$
effectiveness changing with temperature range **A18:** 824
thermogravimetric analysis **A18:** 823
wear factor **A18:** 824
SC *See* Single-crystal casting
Sc-Ti (Phase Diagram) .... **A3:** 2 • 362
Sc-Y (Phase Diagram) .... **A3:** 2 • 362
Sc-Zr (Phase Diagram) .... **A3:** 2 • 363
Scab
defined **A15:** 10
Scabbing
defined **A18:** 16
as dynamic fracture by stress waves **A8:** 287
Scabs
alloy steel fracture by **A12:** 337
in billets, magnetic particle inspection **A17:** 115-116
defined **A11:** 9
in iron castings **A11:** 353
as mechanical damage **A12:** 284
in rolling **A14:** 358
in steel bar and wire **A17:** 549
tubular products **A17:** 568
Scale *See also* Descaling; Scaling; Scating **EM3:** 25
angular deflection **A8:** 132
conditioning, heat-resistant alloys **M5:** 564-566
coordinate measuring machines **A17:** 24-25
corrosion and hardness, in boiler tubes **A11:** 616
defined **A9:** 15 **A14:** 57 **EM2:** 37
deposits, intemal, in boilers **A11:** 604
and die life **A14:** 57
embedded, in steel bar and wire **A17:** 549
of eutectic structures **A15:** 121-122
forged-in, as surface defect **A11:** 327
heat-treat, blast cleaning of **A15:** 506
intemal buildup, in steam-generator tubes **A11:** 605
Moment **A8:** 132
on scrap, as inclusion-forming **A15:** 90
oxide, as casting defect **A11:** 385
removal, investment castings **A15:** 264
removal of *See also* Salt bath descaling
cast iron **M5:** 13-14
ceramic coating process **M5:** 537-538
copper and copper alloys **M5:** 611-612
descaling time, effects on, pickling operations **M5:** 74-80
heat-resisting alloys **M5:** 12, 564-568
nickel and nickel alloys **M5:** 671-672
process types and selection **M5:** 65-66, 68-69, 72-80, 86, 92, 304
process variables affecting, pickling procedures **M5:** 72-80
reactive and refractory alloys **M5:** 650-654, 659, 662-663, 667
Scale (continued)
stainless steel **M5:** 553 554
steel **M5:** 12-14, 17
rolled-in, as forging flaw **A17:** 493
as rolling defect **A14:** 358
tempering, quench crack in steel containing **A11:** 94, 95
thickness, as function of Larson-Miller parameter **A11:** 605
thickness, determination of steel **M5:** 73-77
water-side, removal from steam equipment **A11:** 616
Scale breaking
use in pickling operations **M5:** 74, 76, 78
Scale dipping
solution composition **M5:** 611-612
Scale dips
copper and copper alloy forgings **A14:** 258
Scale erosion rate **A18:** 208, 210
Scale formation and control
steam systems **M1:** 748, 749
Scale growth rate **A18:** 210
Scale sensitivity factor **EM3:** 653
Scale spalling **A18:** 208-209, 210
Scale-up
in tumble-type blenders **A7:** 189
Scaler analysis
FMR eddy current probes **A17:** 221, 223
Scales *See also* Oxide scale; Oxides; Rockwell hardness scales
control, gas/oil production **A13:** 1247
corrosive, on weathering steel structure **A13:** 1305
deposition, water-formed **A13:** 490-494
forming minerals **A13:** 1137
mill **A13:** 524
in phosphate coatings **A13:** 384, 388
rusting, carbon steel **A13:** 520
and solvents **A13:** 1140
surface, cleaning of **A13:** 380-381
Scaling *See also* Descaling; Scale **A13:** 11, 97
in boilers and steam equipment **A11:** 614-621
as casting defect **A11:** 385
compacted graphite irons **A15:** 673
defined **A15:** 10 **A18:** 16
ductile iron **A15:** 663
factors, quality design **A17:** 722
as function of temperature **A14:** 162
in gray iron **A1:** 27, 100, 101-102
heat-resistant cast irons **M1:** 91-94
image, digital image enhancement **A17:** 458
oxide, in oil-well waterflood flow line **A11:** 191
removal, titanium alloys
resistance, in heat exchangers **A11:** 628
in superheaters **A11:** 616
temperature effects **A14:** 174
of titanium alloys **A14:** 838, 842
in turbines **A11:** 616
water-side **A11:** 616
Scaling coefficients
determination **EL1:** 747
Scalping
wrought copper and copper alloys **A2:** 243-244
Scan
defined **A17:** 385
Scan coils and raster formation
SEM microscopes **A10:** 493
Scan lines
image analysis scanners **A10:** 310
Scandium
fluoride separation **A10:** 169
pure **M2:** 793-794
species weighed in gravimetry **A10:** 172
TNAA detection limits **A10:** 238
weighed as the fluoride **A10:** 171
Scandium, as rare earth metal
properties **A2:** 720, 1186
Scandium, vapor pressure
relation to temperature **A4:** 495 **M4:** 310

Scanners
image analyzers **A10:** 310-311
infrared imaging **A17:** 398
touch, coordinate measuring machines **A17:** 25
Scanning
definition **M6:** 632
electronic systems **M6:** 633-634
equipment, ultrasonic inspection **A17:** 261
geometries, computed tomography (CT) **A17:** 365-368
light beams, laser inspection **A17:** 12
longitudinally **A17:** 5
optical systems **M6:** 633
orbital **A17:** 4-5
procedure in electron beam welding **M6:** 632-633
as rigid borescopes **A17:** 5
system, industrial computed tomography **A17:** 359
techniques **M6:** 633
Scanning acoustic microscope (SAM) **EL1:** 370, 1070-1071
Scanning acoustic microscopy
to detect weld discontinuities in diffusion welding **A6:** 1080
Scanning acoustic microscopy (SAM) **A18:** 406-409 **EM4:** 625
applications **A18:** 408-409
defined **A17:** 465
evaluation of surface conditions caused by machining **A18:** 408-409
operating principles **A17:** 467-468
parameters/techniques, compared **A17:** 470
of powder metallurgy parts **A17:** 540
of soldered joints **A17:** 607
thin-film thickness measurements **A18:** 408
Scanning acoustical holography *See also* Acoustical holography
for adhesive-bonded joints **A17:** 631
commercial equipment **A17:** 441-443
and liquid-surface acoustical holography compared **A17:** 4
object size **A17:** 441
reconstruction **A17:** 11, 440-441
sensitivity and resolution **A17:** 441
Scanning Auger microprobe (SAM) **EM3:** 237, 238-240
combined with ion sputtering **EM3:** 244
depth profiling by ball cratering **EM3:** 245
failure analysis **EM3:** 248-249
process **EM3:** 238-240
surface behavior diagrams **EM3:** 247
titanium adherends **EM3:** 266
Scanning Auger microscope
quantification **EM3:** 242-243
vacuum chamber on **A11:** 35
Scanning Auger microscopes
with secondary electron or x-ray detectors **A10:** 501
and SEM, compared **A10:** 509-510
Scanning Auger microscopy **A7:** 251-252, 254 **A18:** 376
defined **A10:** 681
magnetostrictive alloy analysis by **A10:** 510
Scanning Auger microscopy (SAM)
for surface analysis **EM4:** 25
Scanning capacitance microscope (SCAM) **A18:** 397
Scanning chemical potential microscopy (SCPM) **A6:** 145
Scanning coil
in SEM imaging **A12:** 167
Scanning electron beam instruments
compared **A10:** 501
sample volume **A10:** 499-500
signals generated by **A10:** 497-499
Scanning electron micrograph
of AISI 4130 steel **A8:** 478
of cyclic cleavage of Fe-4Si **A8:** 484-485
of Fe-2.5Si cycled near stress-intensity factor range threshold **A8:** 491
of Fe-4Si **A8:** 480
Inconel 718 fracture surface **A8:** 482, 848
Scanning electron micrographs
titanium adherends **EM3:** 266
Scanning electron microscope *See also* Scanning electron microscopy
basic configuration **A18:** 379
Scanning electron microscope (continued)
current **A9:** 89
defined **A9:** 15
design **A9:** 89-90
electronic image analysis system **A9:** 138
imaging and analysis **A18:** 382-385
backscattered electron images **A18:** 383
secondary electron signal **A18:** 382-383
thermal wave imaging **A18:** 385
x-ray microanalysis and mapping **A18:** 383-384
for secondary electron imaging **A9:** 89
stereomicroscopy **A18:** 396, 397
used in structural analysis of lead alloys **A9:** 417
used to examine case hardened steels **A9:** 217
voltages **A9:** 89
Scanning electron microscope (SEM)
for plastic replicas **A17:** 53
Scanning electron microscope(s)
applied to fractography **A12:** 8
for dimples **A12:** 96
fractographs, compared with dark-field fractographs **A12:** 92-100
illumination and light-field illumination
geometry of image formation **A12:** 196
history **A12:** 7-8
literature **A12:** 8
schematic cross section **A12:** 166
vacuum system **A12:** 171
Scanning electron microscopes
area channeling analysis with **A10:** 357
basic components **A10:** 491
depth of field **A10:** 496-497
detectors and image formation in **A10:** 493-494
double-deflection system **A10:** 493
electron column **A10:** 431
electron gun for **A10:** 491-492
electron optical column, compared with TEM column **A10:** 431
with electron probe microanalyzer **A10:** 517
image contrast in **A10:** 500-504
as input devices for image analyzers **A10:** 310
lenses **A10:** 492
for particle image analyses **A10:** 318
resolution limits **A10:** 495-496
and SAM, compared **A10:** 509-510
scan coils and raster formation **A10:** 493
with secondary electron and x-ray detectors **A10:** 501
vacuum pumping system **A10:** 491
Scanning electron microscopy **A9:** 89-102 **A10:** 490-515 **A12:** 166-178 **A13:** 1117
acceptable degree of edge rounding in samples **A9:** 44
advantages **A10:** 494-497
of aluminum/iron and aluminum/brass interfaces **A10:** 531-532
applications **A9:** 89, 99-101 **A10:** 490, 508-514
backscattered electrons **A9:** 92
capabilities **A10:** 309, 365, 402, 429, 536
capabilities, and FIM/AP **A10:** 583
capabilities, compared with optical metallography **A10:** 299
capabilities, defined **A12:** 166
cathodoluminescence **A10:** 507
and chip formation process **A16:** 8-9
compositional analysis of welds **A6:** 104
contrast enhancement **A9:** 95
dark-field, and light-field illumination fractographs, compared **A12:** 92-100
defined **A10:** 681
depth of region below surface from which information is obtained **A9:** 90
detected signals **A9:** 90
detectors **A9:** 90
display system **A12:** 169-171
of ductile fractures **A12:** 173-174
electron beam induced current **A10:** 507
electron channeling patterns and contrast **A10:** 504-506
and electronic image analysis **A9:** 152-153
etching for **A9:** 57
of fiber composites **A9:** 592
fractographs **A12:** 92-100, 173-176, 185-192, 216
fractographs, types of **A12:** 216
fractography **A12:** 173-176
general uses **A10:** 490
Scanning electron microscopy (continued)
illuminating/imaging system **A12:** 167-168
image and dark-field light microscope fractograph, compared **A12:** 92
image contrast **A10:** 500-504
image morphology **A10:** 309
image quality **A9:** 90
information system **A12:** 168-169
of inorganic solids, types of information from **A10:** 6
instrumentation **A12:** 166-171
integrated circuit problems solved by **A10:** 513-514
of intergranular brittle fracture **A12:** 174-175
introduction **A10:** 491
limitations **A10:** 490
microscope for **A10:** 491-494
mounting of specimens **A9:** 97
operation modes **A9:** 90-91
of organic solids, information from **A10:** 9
of powder metallurgy materials **A9:** 508
preparation, of fractographs **A12:** 185, 192
preparation, of specimens **A12:** 171-173
related techniques **A10:** 490
replication procedures **A12:** 94-95
resolution **A9:** 95
and SAM **A10:** 509-510
samples **A10:** 490, 508-514
scanning electron beam instruments **A10:** 497-500
secondary electrons **A9:** 92
selection of mounting materials for specimens **A9:** 28-29, 32
special techniques with **A10:** 506-508
specimen demagnetization for **A12:** 77
specimen preparation **A9:** 97-98
stereo pair, flat cleavage fracture, irons **A12:** 219
stereo, tilt method **A12:** 171
for stereo viewing **A12:** 192
study of magnetic domains **A9:** 536-537
surface texture measurements **A16:** 27
and TEM fractographs, compared **A12:** 185-192
of tin and tin alloy coatings **A9:** 451
of transgranular fracture modes **A12:** 175-176
types of electron-beam-excited electrons **A9:** 90
types of radiation .... **A9:** go
used to examine aluminum alloy powders **A9:** 358
used to examine tin plate **A9:** 198
used to reveal titanium alloy subgrain boundaries **A9:** 461
vacuum system **A12:** 171
voltage contrast **A10:** 506-507
of wrought heat-resistant alloys **A9:** 307-308
with x-ray analysis, advantage **A12:** 168
x-rays **A9:** 92-93
Scanning electron microscopy (SEM) **A8:** 412, 476, 481-483 **EM1:** 285, 771 **EM3:** 237
abrasive wear and lubricant analysis **A18:** 308
as advanced failure analysis technique **EL1:** 1106
and AES analysis, of failed niobium alloy rocket-nozzle component **A11:** 41-42
of composites **A11:** 741
conductive coating techniques not practical with Auger electron spectroscopy **A18:** 456
as contactless probing **EL1:** 371-372
for descriptive fractography **EM4:** 640, 641, 643, 648
with EDS attachment, capabilities description **EL1:** 1094-1095
with energy-dispersive x-ray analyzer **A11:** 38
examination example **EL1:** 1099-1100
failure analysis **EM3:** 249
of failure surface, rocker lever **A11:** 350
fractographs, T-hook **A11:** 369
fracture mode identification chart for **A11:** 80
for fracture surface analysis of optical fibers **EM4:** 663-667
of fracture surfaces **A11:** 22
fracture-surface analysis, failed hip prosthesis **A11:** 692-693
impact wear measurements **A18:** 263
for instrumentation/testing **EL1:** 367-368
for intermetallic-related failures **EL1:** 1043
micrograph, failed low-alloy casting **A11:** 392

Scanning electron microscopy (SEM) (continued)
- micrograph, integrated circuit using voltage contrast ... **A11:** 768
- micrographs ... **EL1:** 1096, 1097
- for microstructural analysis ... **EM4:** 25, 26, 570, 578
- mode of operation ... **A18:** 379
- nanoindentation attachment for equipment ... **A18:** 419
- ofshafts ... **A11:** 460
- for particle shape ... **A7:** 234-236
- for particle sizing ... **A7:** 228
- for phase analysis ... **EM4:** 25
- phosphoric acid anodized surfaces ... **EM3:** 249
- photographs for failure analysis ... **EM4:** 630
- process ... **EM3:** 241-242
- sample preparation and examination ... **EL1:** 1095
- schematic ... **EL1:** 1094
- secondary electron image ... **A11:** 358
- specimen preparation ... **A18:** 380-381
- stages detected in sliding wear of metal-matrix composites ... **A18:** 809
- studies of erosion of ceramics ... **A18:** 206
- summary and future outlook ... **A18:** 391
- surface analysis ... **EM3:** 259, 260, 261
- for surface examination ... **A18:** 291
- and TEM, contrasted ... **EL1:** 1101
- for temperature-time control in polymer removal techniques ... **EM4:** 137
- to analyze ceramic powder particle size ... **EM4:** 67
- to analyze polymer-derived ceramic fiber ... **EM4:** 225
- to analyze strength and fracture phenomena ... **EM4:** 586, 593
- to determine surface roughness in ceramic powder characterization ... **EM4:** 27
- to examine fracture surfaces of green ceramics ... **EM4:** 96
- to view fracture surface of a compact pressed from spray-dried powder ... **EM4:** 107
- use with integrated circuits ... **A11:** 768
- as wafer-level physical test method ... **EL1:** 918-920

Scanning electron microscopy energy-dispersive systems ... **A9:** 92

Scanning infrared treatment ... **EM3:** 35

Scanning laser acoustic microscope (SLAM) ... **EL1:** 370, 1070-1071

Scanning laser acoustic microscopy (SLAM) ... **A18:** 406, 410-412 **EM4:** 625
- color image ... **A17:** 487
- defined ... **A17:** 465
- operating principles ... **A17:** 465-466
- parameters/techniques, compared ... **A17:** 470
- of powder metallurgy parts ... **A17:** 540
- principles ... **A18:** 410-411
- reconstruction of images by holography ... **A18:** 411-412
- of soldered joints ... **A17:** 607

Scanning laser beam technique
- inspection by ... **A17:** - - 12

Scanning laser gage ... **A17:** 12-13

Scanning light photomacrography ... **A12:** 81-82

Scanning log ratio
- analysis by ... **A10:** 144

Scanning low-energy electron microscopy (SLEEM or LEEM) ... **A6:** 145

Scanning monochromators
- radiation detection ... **A10:** 38

Scanning transmission electron microscopes ... **A10:** 501

Scanning transmission electron microscopes (STEMs) ... **EM3:** 242

Scanning transmission electron microscopy ... **A9:** 104
- capabilities ... **A10:** 490
- defect analysis by ... **A10:** 464-468
- defined ... **A10:** 681
- lenses for ... **A10:** 432

Scanning transmission electron microscopy (continued)
- profiles, diffusion interfaces ... **A10:** 478

Scanning transmission electron microscopy (STEM) ... **A18:** 380, 386, 387, 390-391
- basic configuration ... **A18:** 379
- dedicated STEM ... **A18:** 380

Scanning tunneling microscopy (STM) ... **A6:** 145 **A18:** 376, 393-397, 402
- adhesion interfacial analysis ... **A6:** 144
- applications ... **A18:** 395-396
- data acquisition and analysis ... **A18:** 395
- equipment ... **A18:** 393
- future trends ... **A18:** 397
- historical perspective ... **A18:** 393
- limitations and solutions ... **A18:** 396-397
- motion control ... **A18:** 394
- principle of STM imaging ... **A18:** 393-394
- technique comparison ... **A18:** 396
- tip preparation ... **A18:** 395
- variants ... **A18:** 393, 397
- versus other surface roughness measurement techniques ... **A18:** 397
- vibration isolation ... **A18:** 394-395

Scanning tunnelling microscopy (STM) ... **EM3:** 237

Scanning ultrasonic inspection
- of weldments ... **A17:** 594-596

Scans
- daughter ion ... **A10:** 646
- in ESR spectrometers ... **A10:** 255
- natural loss, GC/MS ... **A10:** 646-647
- parent ion ... **A10:** 646

Scarf joint *See also* Joints; Lap joint ... **EM3:** 25, 487, 491, 492, 537
- brazing ... **A6:** 120
- defined ... **EM1:** 21 **EM2:** 37
- definition ... **A6:** 1213

Scarf joints
- brazing of aluminum alloys ... **M6:** 1025
- definition ... **M6:** 1025

Scars
- as casting defects ... **A11:** 384

Scatter
- allowance for in creep-rupture testing ... **A8:** 340
- band of fracture toughness, stainless steel ... **A8:** 479-481
- bands, from multiple heat rupture tests ... **A8:** 330
- Compton ... **A10:** 84
- in creep measurement ... **A8:** 339
- of creep-rupture test data ... **A8:** 329-330
- from heterogeneities ... **A8:** 329-330
- in machinability ratings ... **A1:** 593
- in periodic overstrain data ... **A8:** 701
- quantifying ... **A8:** 625
- ratio of Compton-to-Rayleigh ... **A10:** 84
- wave theory of ... **A10:** 83
- x-ray, defined ... **A10:** 84

Scatter, data
- effect on fracture mechanics accuracy ... **A11:** 55

Scatter radiation
- computed tomography (CT) ... **A17:** 377

Scattered electrons ... **A18:** 385

Scattered radiation
- at high photon energies ... **A17:** 345
- back scatter, protection against ... **A17:** 344
- collimators ... **A17:** 344
- control, computed tomography (CT) ... **A17:** 377
- diffraction (mottling) ... **A17:** 345
- filtration ... **A17:** 345
- lead screens ... **A17:** 344
- masks and diaphragms ... **A17:** 344

Scattering *See also* Backscattering; Scatter
- in AFS atomizer ... **A10:** 46
- anti-Stokes ... **A10:** 126-128
- approximations, in NDE reliability models ... **A17:** 705-706
- at high photon energies ... **A17:** 345
- atomic absorption spectrometry ... **A10:** 51-52

Scattering (continued)
- binary collisions ... **A10:** 604
- Compton ... **A17:** 309
- contributions to scattered intensities ... **A10:** 604
- conversion electron Mössbauer ... **A10:** 293-294
- cross section, in Rutherford backscattering spectrometry ... **A10:** 629
- cross section, symbol for ... **A10:** 692
- defined ... **A10:** 681
- effect in MFS analysis ... **A10:** 77-78
- elastic, ATEM ... **A10:** 432-433
- electron, defined ... **A10:** 672
- electron, volume of ... **A10:** 434
- factor, in effect absorption, x-rays ... **A17:** 310-311
- factor, of atom ... **A10:** 349
- inelastic ... **A10:** 433, 540, 674
- infrared ... **A10:** 117
- internal, shadow formation, radiography ... **A17:** 313
- kinetic energy in ... **A17:** 390
- light ... **A10:** 126-130
- of microwaves ... **A17:** 204-205
- molecular ... **A10:** 126-130
- multiple-, in EXAFS ... **A10:** 410
- neutron and x-ray, compared ... **A10:** 421
- pathological electron, effects in microbeam analysis ... **A10:** 530
- phase differences from different electrons within an atom ... **A10:** 328
- photoelectron, in EPMA energy-dispersive spectrometer ... **A10:** 518
- principles, LEISS ... **A10:** 604
- process, in Auger electron production ... **A10:** 550
- Raman ... **A10:** 135-136
- Rayleigh ... **A10:** 126-128 **A17:** 309
- reduction, in shadow formation ... **A17:** 312
- in reflection and refraction ... **A17:** 204
- side scatter, shadow formation radiography ... **A17:** 313
- small-angle x-ray and neutron ... **A10:** 402-406
- Stokes ... **A10:** 126-128
- techniques, microwave inspection ... **A17:** 206-207
- types of curves ... **A10:** 404
- of ultrasonic beams ... **A17:** 238-239
- ultrasonic wave attenuation by ... **A17:** 231
- x-ray anomalous ... **A10:** 407
- x-rays, defined ... **A17:** 385

Scattering (x-ray)
- defined ... **A9:** 15

Scattering cross section
- symbol for ... **A10:** 692

Scavenging
- chemical ... **A10:** 46
- oxygen ... **A13:** 482, 485

SCb-291
- aging and ductility loss ... **A6:** 581

SCC *See* Stress corrosion cracking; Stress-corrosion cracking; Stress-corrosion cracking (SCC); Stress-corrosion craking

SCE *See* Saturated calomel electrode

SCF 19 *See* Stainless steels, specific types, S21000

Schaeffler (constitution) diagram ... **A14:** 225

Schaeffler constitution diagram for stainless steels ... **A6:** 431

Schaeffler diagram ... **A6:** 82, 83, 100, 457, 458, 461, 500, 501, 677, 678, 679, 685, 686 **M6:** 40, 322, 341
- microstructure prediction ... **A6:** 825
- stainless steel weld metal ... **A6:** 809
- tool for prediction in overlaying ... **M6:** 808, 810
- weld cladding prediction ... **A6:** 817-818, 819, 821

Schaeffler diagrams
- to predict microstructures of welded joints in steels ... **A9:** 582

Schallamach waves ... **A18:** 36

Schallamach-Turner equation ... **A18:** 579

Schallamach-Turner model
- wear of tire treads ... **A18:** 579

Scheelite
- sintering agent for ... **A10:** 166

SUBJECTS OF THE INDEXED VOLUMES: ASM Handbook (designated by the letter "A"): **A1:** Properties and Selection: Irons, Steels, and High-Performance Alloys (1990); **A2:** Properties and Selection: Nonferrous Alloys and Special-Purpose Materials (1990); **A3:** Alloy Phase Diagrams (1992); **A4:** Heat Treating (1991); **A6:** Welding, Brazing, and Soldering (1993); **A7:** Powder Metallurgy (1984); **A8:** Mechanical Testing (1985); **A9:** Metallography and Microstructures (1985); **A10:** Materials Characterization (1986); **A11:** Failure Analysis and Prevention (1986); **A12:** Fractography (1987); **A13:** Corrosion (1987); **A14:** Forming and Forging (1988); **A15:** Casting (1988); **A16:** Machining (1989); **A17:** Nondestructive Testing and Quality Control (1989); **A18:** Friction, Lubrication, and Wear Technology (1992). **Metals Handbook, 9th Edition** (designated by the letter "M"): **M1:** Properties and Selection: Irons and Steels (1978); **M2:** Properties and Selection: Nonferrous Alloys and Pure Metals (1979); **M3:** Properties and Selection: Stainless Steels, Tool Materials and Special-Purpose Materials (1980); **M4:** Heat Treating (1981); **M5:** Surface Cleaning, Finishing, and Coating (1982); **M6:** Welding, Brazing, and Soldering (1983). **Engineered Materials Handbook** (designated by the letters "EM"): **EM1:** Composites (1987); **EM2:** Engineering Plastics (1988); **EM3:** Adhesives and Sealants (1990); **EM4:** Ceramics and Glasses (1991); **Electronic Materials Handbook** (designated by the letters "EL"): **EL1:** Packaging (1989).

Scheelite (continued)
tungsten-bearing ore ... **A7:** 152
Scheil equation ... **A6:** 56 **A9:** 614
Schelleng, R.D
compacted graphite irons ... **A15:** 667
Scherzer focus ... **A18:** 389
Schielern ... **EM3:** 25
Schlapfer and Bukowski method
for free lime content ... **A10:** 179
Schlieren
defined ... **EM2:** 37
Schlieren photography ... **A6:** 34
Schmidt number ... **A13:** 35
Schnadt notched-bar impact test ... **A11:** 57, 58
Schnadt specimen
for testing ship plate ... **A11:** 57, 58
Schoefer diagram
austenitic iron-chromium-nickel alloy castings ... **A13:** 577
Schöniger combustion ... **A10:** 167, 681
Schöniger flask method
for common elements ... **A10:** 215
for elemental analysis ... **A10:** 215
estimated analysis time ... **A10:** 215
limitations ... **A10:** 215
for sample dissolution ... **A10:** 167
use in ion chromatography ... **A10:** 664
Schott alkaline earth silicate
composition ... **EM4:** 1102
properties ... **EM4:** 1102
Schott glass
glass/metal seal combinations ... **EM4:** 497
glass/metal seals ... **EM4:** 494-495
Schottky
barrier diodes ... **EL1:** 201, 960
barrier field-effect transistors (FETS) ... **EL1:** 201
defects, defined ... **EL1:** 93
FAST advanced logic, design guidelines ... **EL1:** 82
Schottky barrier contacts
epitaxial precipitation of silicon in ... **A11:** 778
Schottky barrier technique
EBIC arrangement using ... **A10:** 507
Schottky barriers ... **EM3:** 581-582
Schottky defect
ionic oxide ... **A13:** 65
Schrödinger's equation ... **A18:** 388
Schulz method
reflection topography ... **A10:** 368-369
Scientific and Technical Aerospace Reports (STAR) ... **EM1:** 41
Scientific glass era ... **EM4:** 21
Scientific products ... **EM4:** 1007-1013
Scientific products industry applications
structural ceramics ... **A2:** 1019
Scintillation
counter, diffractometers ... **A10:** 351
crystal-photodiode array detectors ... **A17:** 371
crystals, radiographic inspection ... **A17:** 298
defined ... **EM2:** 593
detectors ... **A10:** 88-89
detectors, computed tomography (CT) ... **A17:** 370
Scintillation crystals
production ... **A2:** 743
Scintillation failures ... **EL1:** 995-997
Scintillator
defined ... **A17:** 385
in neutron radiography ... **A17:** 390-391
Scissor shear
of composites ... **EM1:** 98
Scleroscope hardness
number (HSc), defined ... **A11:** 9
number (HSd), defined ... **A11:** 9
test, defined ... **A11:** 9
Scleroscope hardness numbers
Brinell hardness conversions, steel ... **A8:** 111
equivalent Rockwell B numbers steel ... **A8:** 109-110
HSc or HSd, defined ... **A8:** 11
Rockwell C hardness conversions, steel ... **A8:** 110
Vickers hardness conversions, steel ... **A8:** 112-113
Scleroscope hardness testers, Models C D, and D digital ... **A8:** 104-105
Scleroscope hardness testing ... **A8:** 104-106
calibration HFRSc/HFRSd ... **A8:** 104-105
defined ... **A8:** 11
as dynamic hardness testing ... **A8:** 71
Scleroscope hardness testing (continued)
procedure ... **A8:** 105
SCM420
nominal compositions ... **A18:** 725
Scoop sampling ... **A7:** 227
Scoops, plastic
as sampling devices ... **A10:** 16
Scorches
cracks from ... **A11:** 89
Scoring *See also* Die line; Galling; Pickup ... **A18:** 296, 715
in cold-formed parts ... **A11:** 308
damage, by chipping ... **A11:** 157
defined ... **A8:** 11 **A9:** 15-16 **A11:** 9 **A14:** 11 **A18:** 16
gears ... **A18:** 535
ofshafts ... **A11:** 466
in sheet metal forming ... **A8:** 548
Scotch-yoke drive mechanism
mechanical presses ... **A14:** 31-32
Scott spray chamber
for analytic ICP systems ... **A10:** 35
Scott volumeter
for apparent density ... **A7:** 274, 275
Scouring abrasion *See* Abrasion
Scragging of springs *See* Presetting of springs
Scrap *See also* Recycling; Reprocessed plastic
for aluminum melts ... **A15:** 79-81
aluminum-lithium alloys ... **A2:** 183-184
automobile, recycling technology ... **A2:** 1211-1213
blow molding, costs ... **EM2:** 299
burners, electric arc furnace ... **A15:** 360
cast iron/steel, as metal charge ... **A15:** 388
classification, copper recycling ... **A2:** 1213-1215
compression molding, costs ... **EM2:** 297
electronic, recycling ... **A2:** 1228-1231
gallium sources ... **A2:** 745-746
in injection molding, costs ... **EM2:** 294
lead, sources ... **A2:** 1221
magnesium, sources ... **A2:** 1216-1217
percentages, as allowances ... **EM2:** 84
purchase, storage of ... **A15:** 363
pure copper ... **A2:** 1215
salvage, hot isostatic pressing ... **A15:** 543
selection ... **A15:** 74
streams, aluminum recycling ... **A2:** 1207-1208
thermoforming, costs ... **EM2:** 301
tin, types ... **A2:** 1218-1219
titanium, recycling ... **A2:** 639
titanium, sources ... **A2:** 1227
zinc alloy ... **A15:** 787-789
zinc, sources ... **A2:** 1223-1224
Scrap burners
electric arc furnace ... **A15:** 360
Scrap, carbon fiber
recycling of ... **EM1:** 153-156
Scrap disposal
in blanking and piercing ... **A14:** 478
uranium ... **M3:** 778-779
Scrap metal, silver
analysis of ... **A10:** 41
Scrap recovery
uranium ... **A7:** 666
Scraper
defined ... **A18:** 16
Scrapers
for induction furnaces ... **A15:** 372-373
Scraping
of samples ... **A10:** 575
Scraping artifacts
TEM replicas ... **A12:** 185
Scrapless nut quality carbon steel wire rod ... **M1:** 254-255
Scrapless nut quality rod ... **A1:** 273
Scratch
defined ... **A9:** 16 **A18:** 16
definition ... **EM4:** 633
Scratch adhesion test ... **A18:** 434
Scratch brushing
copper and copper alloys ... **M5:** 615-617
Scratch cross-sectional area ... **A18:** 432, 436
Scratch hardness ... **A18:** 432, 433
Scratch hardness test ... **A8:** 11, 71, 104, 107-108
defined ... **A18:** 16
Scratch starting ... **A6:** 200
Scratch testing ... **A18:** 404, 430-436
fundamentals of scratching
deformation ... **A18:** 431-433
contact geometry ... **A18:** 431
relation of friction to scratching deformation ... **A18:** 432-433
removal coefficient ... **A18:** 431
specific grooving energy ... **A18:** 431-432
objectives ... **A18:** 430
scratch adhesion testing of thin hard coatings ... **A18:** 434-436
parameters affecting $F_{N,C}$ ... **A18:** 435-436
test equipment and procedures ... **A18:** 434-435
scratch testing of monolithic solids ... **A18:** 433-434
abrasion resistance ... **A18:** 434
scratch hardness ... **A18:** 433, 434
size effects ... **A18:** 433-434
types of scratch test devices ... **A18:** 430-431
apparatus classification ... **A18:** 430
in situ scratching devices ... **A18:** 430-431
measured quantities ... **A18:** 430
quick-stop devices ... **A18:** 431
scratching elements ... **A18:** 430
Scratch trace
defined ... **A9:** 16
Scratch-repassivation testing method
for localized corrosion ... **A13:** 217
Scratch-resistant coatings
definition ... **EM4:** 633
Scratches
adhesive-bonded joints ... **A17:** 613
visual inspection ... **A17:** 3
Scratching *See also* Abrasion; Plowing; Plowing (ploughing); Ridging wear ... **A18:** 259
defined ... **A8:** 11 **A11:** 9 **A18:** 16
magnesium powder production by ... **A7:** 131
Scratching abrasion *See also* Abrasvie wear; Low-stress abrasion ... **M1:** 599, 600
Scratching speed ... **A18:** 432, 436
Screeding
of molds ... **A15:** 191
Screen *See also* Screening ... **A7:** 10
analysis ... **A7:** 10, 25
classification ... **A7:** 10
defined ... **A15:** 10
Screen analysis *See also* Sieve analysis
loam molding sand ... **A15:** 229
of reclaimed sands ... **A15:** 354-355
Screen distribution test
for sand reclamation ... **A15:** 355
Screen print application
solder pastes ... **EL1:** 652
Screen printing
capabilities/limitations ... **EL1:** 508-509
flexible printed boards ... **EL1:** 583
for particulate depositions in metallizing ... **EM4:** 542, 543
solder masks ... **EL1:** 555
in surface-mount soldering ... **EL1:** 699
thick films on ceramic substrates ... **EL1:** 206-208
of thick-film circuits ... **EL1:** 249
viscoelastic properties required ... **EM4:** 116
Screen printing dispensing technique
general solder paste parameters ... **A6:** 988
Screen-printed solder mask
advantages/disadvantages ... **EL1:** 555
Screening *See also* Sand(s); Screen ... **A7:** 10, 176-177
as adhesive application method ... **EL1:** 672
Charpy V-notch impact tests as ... **A8:** 263
DSC/TGA ... **EM2:** 560-561
in encapsulation hot isostatic pressing ... **A7:** 431
equipment ... **A7:** 177
high-cycle fatigue properties, by ultrasonic testing ... **A8:** 240
as inclusion control ... **A15:** 90
key operations ... **A7:** 177
of platinum powders ... **A7:** 150
of precious metal powders ... **A7:** 149
screen types ... **A15:** 350
sieve sizes, and related particle sizes ... **A7:** 176
surfaces ... **A7:** 176-177
TGA method ... **EM2:** 565-566
of uranium dioxide pellets ... **A7:** 665
vs testing ... **EL1:** 875
Screens
filtration, radiography ... **A17:** 298
fluorescent intensifying ... **A17:** 316-317

Screens (continued)
fluorescent, real-time radiography with **A17:** 319-320
fluorometallic, radiography **A17:** 317
intensifying, radiography **A17:** 298
lag **A17:** 317
mottle, defined **A17:** 317
radiography, unsharpness **A17:** 300
speeds **A17:** 316-317
Screw
-action grips **A8:** 51
dislocations, in ultrasonic testing **A8:** 256
micrometer **A8:** 619
se, f-drilling tapping, hydrogen embrittlement in **A8:** 541-542
set, for stored-torque Kolsky bar **A8:** 220
-threaded copper current, input connections **A8:** 389
threads, in fatigue crack initiation **A8:** 371
Screw coining press **A7:** 15
Screw conveyors for seed separation **A7:** 589
Screw dislocation
defined **A13:** 45-46
Screw dislocations defined **A9:** 719
diffraction contrast **A9:** 114
effect of low temperature and high strain rate on **A9:** 688
in iron **A9:** 690
Screw machine tests **A1:** 591, 593
Screw machining *See also* Multiple-operation machining
cutting fluid flow recommendations **A16:** 127
cutting fluids used **A16:** 125
susceptibility to flaking **A16:** 281
Screw plasticating injection molding *See also* Injection molding
defined **EM2:** 37
Screw press
defined **A14:** 11
Screw presses **A14:** 33-35
for aluminum alloys **A14:** 245
drives, variations in **A14:** 41
energy vs load diagram **A14:** 41
as energy-restricted machines **A14:** 25, 37
load and energy **A14:** 40-41
for precision forging **A14:** 169
time-dependent characteristics **A14:** 41
for titanium alloys **A14:** 274-275
Screw threads
die cutting speeds **A16:** 301
Screw-driven machines
balanced beam universal **A8:** 613
constant strain rate test on **A8:** 43
conventional load frame **A8:** 192
hydraulic **A8:** 613
tension, components **A8:** 48
Screw-micrometer ocular **A9:** 74
Screwed-in inserts
in magnesium alloy parts **A2:** 467
Screws *See also* Extruder; specific types by name
alloy steel for **M1:** 256
carbon steel for **M1:** 254
construction, extruder **EM2:** 379-380
design **EM2:** 380-381
injection molding **EM1:** 165-166
injection molding, polyvinyl chlorides (PVC) **EM2:** 211-212
mechanical fastening of **EM2:** 711-712
zones, extruders **EM2:** 380
Screws, hip
austenitic stainless steels **A12:** 359-364
Scribe line
for flow localization in torsion **A8:** 169
in tubular specimen **A8:** 224
Scribing
gage marks **A8:** 548
Scribing tool, metal
with coordinate measuring machine **A17:** 24
Scrim *See also* B-stage; Glass cloth **EM3:** 25
defined **EM1:** 21 **EM2:** 37
Script *See* Chinese script
Scrubber systems **A13:** 1368, 1370
Scrubbing
dry **A15:** 227
high-energy wet, cupolas **A15:** 387
mechanical, sand reclamation by **A15:** 352
pneumatic, sand reclamation by **A15:** 352
of steel wire rope **A11:** 519
and wet washing, for sand reclamation **A15:** 227
Scrubbing, mechanical *See* Mechanical scrubbing
Scuff marks
liquid penetrant inspection of **A17:** 86
Scuff resistance
in gray iron **A1:** 24-25
Scuffing **A18:** 296, 715
defined **A8:** 11-12 **A11:** 9 **A18:** 16
gears **A18:** 535, 537-538
internal combustion engine parts **A18:** 555, 556, 557, 559
ofshafts **A11:** 466
Scuffing, resistance, gray cast iron *See also* Adhesive wear **M1:** 24-25
steel wire for **M1:** 265-266
Scuffing temperature
symbol and units **A18:** 544
Scuffing/adhesion
surface property effect **A18:** 342, 343
Sculpture
metalworking of **A15:** 20-22
SDL-1
nickel aluminide alloy sliding wear data with this lubricant **A18:** 776
SE *See* Secondary electrons
Se-Sn (Phase Diagram) ..... **A3:** 2•363
Se-Sr (Phase Diagram) ..... **A3:** 2•363
Se-Te (Phase Diagram) ..... **A3:** 2•364
Se-Tl (Phase Diagram) ..... **A3:** 2•364
Se-Tm (Phase Diagram) ..... **A3:** 2•364
Se-U (Phase Diagram) ..... **A3:** 2•365
Sea bottom
Miller numbers **A18:** 235
Seacoal
as carbon mold addition **A15:** 211
Seal
for acidified chloride solution fatigue testing **A8:** 418-419
for specimen in aqueous solutions at ambient temperatures **A8:** 415
Seal coat
definition **A6:** 1213 **M6:** 15
Seal design techniques **EM4:** 532-541
chemical integrity **EM4:** 539-541
processing effects **EM4:** 540-541
survival in the environment **EM4:** 539-540
electrical integrity **EM4:** 538-539
factors affecting design **EM4:** 532
finite-element stress-analytic techniques **EM4:** 535-538
ceramic/metal seals **EM4:** 538
general approach **EM4:** 535-536
glass/metal seals **EM4:** 536-538
mechanical integrity **EM4:** 532-535
ceramic fracture **EM4:** 532-533, 535
coefficient of thermal expansion **EM4:** 533-535, 540
mechanical behavior of interfaces ... **EM4:** 533-534
proof testing **EM4:** 533
residual stresses **EM4:** 534-535
static fatigue **EM4:** 533
stress intensity factor **EM4:** 532, 533
stress minimization **EM4:** 534-535
philosophy **EM4:** 532
Seal flank closing force **A18:** 550
Seal materials
wear particles **A18:** 305
Seal nose
defined **A18:** 16
Seal rings
cemented carbide **A2:** 972
Seal surface
stainless steel parts as **A7:** 662
Seal weld
definition **A6:** 1213 **M6:** 15
Seal(s) *See also* Glass-to-metal seals; Hermetic seals; Hermeticity
ceramic-to-metal **EL1:** 678
glass-to metal **EL1:** 455-459, 678, 958
lid, test procedures **EL1:** 953-954
reliability/testing **EL1:** 459
Seal-swell agents **A18:** 110
applications **A18:** 110
functions **A18:** 110
materials **A18:** 110
in nonengine lubricant formulations **A18:** 111
Sealant
defined **EM2:** 37
Sealant design considerations **EM3:** 545-550
accessibility **EM3:** 546
applications **EM3:** 547-550
elastic recovery **EM3:** 546
functions of sealants **EM3:** 545
materials considerations **EM3:** 545-547
movement capability **EM3:** 545, 546, 547, 549
strain **EM3:** 545
Sealant Waterproofers, and Restoration Institute (SWRI) **EM3:** 188
Sealants *See also* Adhesives and sealants, specific types; Faying surface sealing **EM3:** 25, 48-55
for ablative barriers **EM3:** 605
additives **EM3:** 674
advantages and disadvantages **EM3:** 675
for aerodynamic smoothing **EM3:** 677
aerospace industry market **EM3:** 58-59
appliance market **EM3:** 59
application methods **EM3:** 607, 642
application procedures guide **EM3:** 188
applications **EM3:** 56-60, 604-612
asphaltic **EM3:** 675
automotive market **EM3:** 608-610
aviation applications **EM3:** 611
bituminous, properties **EM3:** 677
body assembly applications **EM3:** 608
for cabin pressure sealing **EM3:** 58
for canopy sealing **EM3:** 605
for carbon-carbon composites **EM1:** 920
casting technique **EM3:** 59
for channel sealing **EM3:** 677
chemical properties **EM3:** 52
chemistry **EM3:** 49-51
commonly used polymer systems **EM3:** 673
compared to adhesives **EM3:** 56
compared to caulks and putties **EM3:** 187
component functions **EM3:** 674
as conductive sealants **EM3:** 604-605
construction application areas **EM3:** 608
for construction joints **EM3:** 605
for construction market **EM3:** 605-608
for contraction joints **EM3:** 605
for corrosion protection **EM3:** 604
cure mechanism **EM3:** 674
cure properties **EM3:** 51
cure times **EM3:** 189
for cure-in-place gasketing **EM3:** 548
defined **EM1:** 21
design of sealing systems **EM3:** 52-54
distribution categorized by polymeric binders **EM3:** 673
double **EM3:** 578
electrical and electronics market **EM3:** 610-612
electrical properties **EM3:** 52
electronics market **EM3:** 59-60
encapsulating technique **EM3:** 59
for engine sealing **EM3:** 609

SUBJECTS OF THE INDEXED VOLUMES: ASM Handbook (designated by the letter "A"): **A1:** Properties and Selection: Irons, Steels, and High-Performance Alloys (1990); **A2:** Properties and Selection: Nonferrous Alloys and Special-Purpose Materials (1990); **A3:** Alloy Phase Diagrams (1992); **A4:** Heat Treating (1991); **A6:** Welding, Brazing, and Soldering (1993); **A7:** Powder Metallurgy (1984); **A8:** Mechanical Testing (1985); **A9:** Metallography and Microstructures (1985); **A10:** Materials Characterization (1986); **A11:** Failure Analysis and Prevention (1986); **A12:** Fractography (1987); **A13:** Corrosion (1987); **A14:** Forming and Forging (1988); **A15:** Casting (1988); **A16:** Machining (1989); **A17:** Nondestructive Testing and Quality Control (1989); **A18** Friction, Lubrication, and Wear Technology (1992). **Metals Handbook, 9th Edition** (designated by the letter "M"): **M1:** Properties and Selection: Irons and Steels (1978); **M2:** Properties and Selection: Nonferrous Alloys and Pure Metals (1979); **M3:** Properties and Selection: Stainless Steels, Tool Materials and Special-Purpose Materials (1980); **M4:** Heat Treating (1981); **M5:** Surface Cleaning, Finishing, and Coating (1982); **M6:** Welding, Brazing, and Soldering (1983). **Engineered Materials Handbook** (designated by the letters "EM"): **EM1:** Composites (1987); **EM2:** Engineering Plastics (1988); **EM3:** Adhesives and Sealants (1990); **EM4:** Ceramics and Glasses (1991); **Electronic Materials Handbook** (designated by the letters "EL"): **EL1:** Packaging (1989).

Sealants (continued)
for environmental barriers **EM3:** 608, 609
environmental considerations **EM3:** 673-679
chemistry and technology of sealants **EM3:** 673-674
degradation **EM3:** 678-679
features of sealant types **EM3:** 78
performance requirements increased **EM3:** 673
for environmental seals **EM3:** 604
for expansion joints **EM3:** 605, 607, 608
for faying surface sealing **EM1:** 719-720
for faying surfaces **EM3:** 604
form considerations **EM3:** 56
forms and types **EM3:** 674
for fuel tank sealing **EM3:** 677
functions **EM3:** 33, 36, 48-49, 187
for glazing **EM3:** 606
global standards **EM3:** 188-189
high-performance, shelf life **EM3:** 188
history **EM3:** 48
hydrolysis **EM3:** 679
for industrial applications **EM3:** 58
for integral fuel tanks **EM3:** 58
for jet turbine blades **EM3:** 605
joints between roadway concrete slabs **EM3:** 188
major classification types **EM3:** 189-192
markets **EM3:** 56-60, 605-612
methods of application **EM3:** 58, 610
no adverse effect by water-displacing corrosion inhibitors **EM3:** 641
nonsag **EM3:** 56
oleoresinous **EM3:** 674---675, 677
packaging **EM3:** 56
performance **EM3:** 188, 674
physical properties **EM3:** 51-52, 188-189
pit and fissure (dental) **A18:** 672
pot life **EM3:** 604
potting technique **EM3:** 59
price of high-performance sealants **EM3:** 51
processing techniques **EM3:** 59
properties **EM3:** 191, 605, 677
reservoirs and canals **EM3:** 188
for rivet heads **EM3:** 604
self-leveling **EM3:** 56
service life **EM3:** 678
shelf life, high-performance **EM3:** 188
specifications **EM3:** 188-189
for structural integrity **EM3:** 604
suppliers **EM3:** 57, 58, 60
technology basics **EM3:** 48-49
for thermal barriers **EM3:** 605
thermal properties **EM3:** 52
for thermoset housings **EM3:** 548
thread fitting **EM3:** 610
three-dimensional application method **EM3:** 609
turbocharged engine requirements **EM3:** 609
types and uses **EM3:** 56, 188-191, 642, 674
vacuum impregnation process **EM3:** 609
well-defined standards **EM3:** 611
for window glazing **EM3:** 545
for windshield sealing **EM3:** 605
Sealants, bare copper assembly
furnace soldering **A6:** 355
Sealants, effect
eddy current inspection **A17:** 191-192
Sealants in Construction **EM3:** 70
Sealed cans
corrosion resistance **A13:** 779
Sealed chrome pickle
magnesium alloys **M5:** 632, 634, 636, 640-642
Sealing
aluminum spray coatings **M5:** 344-345
anodic coatings **M5:** 590-591, 594-596, 603-604, 606
dual sealing treatments **M5:** 594-595
electrolytic brightened finishes **M5:** 580-581
glass, selection of **EL1:** 456-458
glasses, nuclear radiation induced device failures **EL1:** 1056
heat welding and **EM2:** 724-725
hermetic **EL1:** 734
low-viscosity sealers **M5:** 369
moisture monitoring after **EL1:** 953
packages, methods **EL1:** 237-243
plastic pin-grid arrays **EL1:** 476
thermal spray coatings **M5:** 369
of thermal sprayed coatings **A13:** 461, 907-909
Sealing (continued)
vacuum impregnation method **M5:** 369
vitreous dielectrics for **EL1:** 109
weathering steels **A13:** 519
Sealing face
defined **A18:** 16
Sealing glasses **EM4:** 22, 1069-1072
applications **EM4:** 22
common problems **EM4:** 1071
disposal **EM4:** 1072
firing cycle stages **EM4:** 1070-1071
crystallizing sealing cycle **EM4:** 1071
vitreous sealing cycle **EM4:** 1070-1071
health issues **EM4:** 1072
refractory fillers **EM4:** 1072
respirators for workers **EM4:** 1072
safety issues **EM4:** 1072
seal geometry and stresses **EM4:** 1072
solder glass technology **EM4:** 1071-1072
solder glasses **EM4:** 1069
application methods **EM4:** 1069-1070
bask glasses **EM4:** 1071-1072
disannealing **EM4:** 1069
hot dipping processes **EM4:** 1070
physical forms **EM4:** 1069-1070
types **EM4:** 1069
uses **EM4:** 1069
vehicles **EM4:** 1069-1070
vitreous **EM4:** 1069, 1070
suppliers **EM4:** 1072
Sealing, heat *See* Heat sealing
Seals
for explosive forming **A14:** 638-639
indium and indium-based **A2:** 752
and shaft failures **A11:** 466
shear fracture studies of Kovar-glass **A10:** 577-578
Seals, friction and wear of **A18:** 546-552
defined (seal) **A18:** 16
design considerations **A18:** 549-550
chemical compatibility **A18:** 550
environmental and operational conditions **A18:** 549-550
operational factors **A18:** 550
greases used **A18:** 129
material selection **A18:** 550-551
methods of reducing wear **A18:** 551
monitoring **A18:** 552
seal friction and wear problems **A18:** 547-549
abrasive wear **A18:** 549
adhesive wear **A18:** 549
cavitation **A18:** 549-550
corrosion **A18:** 549-550
erosion **A18:** 549-550
properties of hard mating face materials **A18:** 548
thermoelastic instability **A18:** 548-549
seal specifications based on application **A18:** 548
surface coatings used **A18:** 551
types of seals **A18:** 546-547
applications **A18:** 546
dynamic seals **A18:** 546-547, 550
face seals **A18:** 546-547, 548-549
gaskets **A18:** 546, 550
hybrid seals **A18:** 547
labyrinth seals **A18:** 547
lip seals **A18:** 546, 547, 550, 551
materials **A18:** 546, 547
O-rings **A18:** 546, 549, 550
packing **A18:** 546, 547, 550-551
piston ring seals **A18:** 547
quad rings **A18:** 546, 550
static seals **A18:** 547
wear testing **A18:** 551-552
Seam
defined **A9:** 16 **A15:** 10
Seam weld
definition **A6:** 1213 **M6:** 15
Seam welded joints, resistance
metallography and microstructures **A9:** 581
Seam welding
definition **M6:** 15
high frequency welds **M6:** 759
laser beam welds **M6:** 657
resistance *See* Resistance seam welding
ultrasonic welds **M6:** 746, 749
Seam welds
butt **M6:** 501
flange-joint lap **M6:** 501
foil butt **M6:** 501-502
lap **M6:** 501
in magnesium and magnesium alloys **A2:** 473
mash **M6:** 501
radiographic inspection **A17:** 335
Seamless mechanical tubing **A1:** 335 **M1:** 324-325
Seamless pipe
inspection of **A17:** 579
Seamless processes
for steel tubular products **A1:** 328 **M1:** 316
Seamless tubes
magnetic particle inspection **A17:** 111
steel, inspection of **A17:** 567-571
ultrasonic inspection **A17:** 272
Seamless tubing
as encapsulation material **A7:** 429
quench cracking in **A11:** 335
Seams **M1:** 182
AISI/SAE alloy steels **A12:** 326, 335
in billets, magnetic particle inspection **A17:** 115
in bolts **A12:** 131, 149
brittle fracture of rotor shaft splines from **A11:** 478
brittle intergranular fracture from **A12:** 335
as casting defects **A11:** 384, 388
in crane hook, magnetic particle inspection **A17:** 107
defined **A11:** 9
as discontinuities, defined **A12:** 64-65
eddy current inspection of **A17:** 164
effect on cold-formed part failure **A11:** 307
in fasteners **A11:** 530
fatigue failure initiated at **A11:** 551
forged stainless steel forceps, fracture by **A11:** 329, 331
as forging defect **A11:** 317, 327, 329, 331
as forging process flaws **A17:** 493
high-carbon steels **A12:** 281, 287
in hot rolled bars **M1:** 199-200
in hot-rolled steel bars **A1:** 240
as planar flaw **A17:** 50
quench cracking from **A12:** 131, 149
radiographic methods **A17:** 296
in rolled steel bar, surface indications **A11:** 329, 331
in rolling **A14:** 358
as spring failure origin **A11:** 555
in spring wire **M1:** 290, 301
in steel bar and wire **A17:** 550
in steel plate **A1:** 230
tool and die failure from **A11:** 574
in tool steel coil spring, cracked during heat treatment **A11:** 574, 579
tubular products **A17:** 568
walls, alloy steels **A12:** 335
wire springs fracture from **A12:** 63
Seams in rolled steel
revealed by macroetching **A9:** 176
Search units *See also* Probe(s); Sensors
contact-type **A17:** 257-258
focused **A17:** 259-261
immersion-type **A17:** 258-259
ultrasonic inspection **A17:** 252, 256-261
Search/match methods
for unknown phases or particles **A10:** 455-459
Seasalt brine
line-pipe steel exposure to **A11:** 299
Season cracking *See also* Stress-corrosion cracking
defined **A11:** 9 **A13:** 11
SCC of brass as **A12:** 28
Seasoning
for residual stresses **A15:** 616
Seawater *See also* Marine atmosphere; Marine corrosion; Ocean water; Pacific Ocean; Salt water; Saltwater; Seawater corrosion; Water
aerated, corrosion of carbon steels **A13:** 512
aluminum coating corrosion resistance **A13:** 435
aluminum-zinc alloy coatings **A13:** 436
aluminum/aluminum alloys in **A13:** 597-598, 603, 607
applications, nickel-base alloys **A13:** 653-654
biological organisms, influences **A13:** 900-902
boron determined in **A10:** 205
carbon steel in **A13:** 1256

Seawater (continued)
cavitation-erosion resistance ratings in **A11:** 190
consistency, and major Ions **A13:** 893-894
corrosion **A13:** 433-434
corrosion fatigue crack growth rate **A8:** 407
corrosion, low-carbon steel **A12:** 250
corrosion protection in **M1:** 751
as corrosive environment **A12:** 24
crevice corrosion in **A13:** 108-112, 303, 309, 556
effect, galvanized coatings **A13:** 440
effect of velocity on steel corrosion rate **A11:** 188
filtered, crevice corrosion in stainless steels **A13:** 556
flowing, copper alloy galvanic couple data **A13:** 624
flowing, corrosion potentials **A13:** 557
flowing, galvanic series **A13:** 675
galvanic series in **A11:** 185 **A13:** 83, 235, 324, 488, 613, 718, 876, 1329
gas solubility **A13:** 1256
general corrosion, titanium/titanium alloys **A13:** 676-677
general properties **A13:** 893
marine corrosion in **A13:** 893-902
metal matrix composites in **A13:** 861-863
pollutants, effects of **A13:** 898-900
salinity **A13:** 894, 900
and SCC in titanium and titanium alloys **A11:** 224
for SCC of titanium alloys **A8:** 531
specific conductance **A13:** 896
stress ratio effect on corrosion fatigue of steel in **A8:** 407
synthetic **A13:** 309
tin/tin alloys in **A13:** 772
variability, and minor ions **A13:** 894-898
velocity effect **A13:** 516
zinc anodes composition in **A13:** 765
zinc/zinc alloys and coatings in **A13:** 762
Seawater corrosion
basic rates **M1:** 741-742
cast irons **A13:** 570
clad metal **A13:** 889
corrosion, aluminum alloy weldments **A13:** 345
of couple members **A13:** 545
desalination plant equipment **M1:** 744-745
factors, carbon/alloy steels immersed in **A13:** 539
fresh water corrosion compared with **M1:** 740, 741
as function of velocity **A13:** 333
of galvanic couples **A13:** 624, 773 **M1:** 740-741, 742-743
mill scale, effect of **M1:** 739
oxygen concentration effect of **M1:** 739-740, 743
pilings, carbon and steel **M1:** 742-744
pitting **M1:** 739, 742, 744-745
preventive measures **M1:** 745
profile **A13:** 539
stainless steel **A13:** 303, 556
stainless steels **M3:** 70-78, 79, 80
titanium/titanium alloy **A13:** 676-677
tropical waters, rates in **M1:** 742
velocity, effect of **M1:** 740-742
zones of **M1:** 740, 742-744
Secant method
of calculating crack growth rates **A8:** 378, 415, 518, 680
Secant modulus *See also* Modulus of elasticity; Tangent modulus; Young's modulus **EM3:** 25
defined **A8:** 12 **A11:** 9 **A14:** 11 **EM1:** 21 **EM2:** 37, 434
Second, as SI base unit
symbol for **A10:** 685
Second friction force **A6:** 888
Second Law of Thermodynamics .... **A3:** 1 • 6-7 **A13:** 61
Second order rate law expression **EM1:** 55
Second phases effect of dendritic structure on **A9:** 639
effect on dislocation distribution **A9:** 693
Second phases in beryllium **A9:** 390
Second-degree blocking **EM3:** 26
Second-layer package **EL1:** 113-117
Second-level package
cross section **EL1:** 112
Second-order polynomial
fit of intensity vs concentration **A10:** 97
Second-phase cleavage fracture
in titanium alloy **A8:** 485-486
Second-phase constituents
of stainless steels **A9:** 284
wrought aluminum alloy **A2:** 37, 42, 43
Second-phase distribution
SIMS analysis **A10:** 610
Second-phase imaging **A10:** 309, 490
Second-phase inclusions
analytical transmission electron microscopy **A10:** 429-489
electron probe x-ray microanalysis **A10:** 516-535
field ion microscopy; **A10:** 583-602
scanning electron microscopy **A10:** 490-515
small-angle x-ray and neutron scattering **A10:** 402-406
x-ray diffraction **A10:** 325-332
Second-phase inhomogeneity
definition **EM4:** 633
Second-phase particles
brittle fracture, wrought aluminum alloys **A12:** 423
cleaved, Alnico alloy **A12:** 451
effect on fatigue strength **A11:** 119
effect on steel tensile ductility **A14:** 364
and field evaporation **A10:** 587
hard, in aluminum alloys **A11:** 84
iron sulfide **A12:** 219
large, effect on striation **A12:** 16
microvoid coalescence at **A12:** 12
as particle-mix decohesion **A11:** 82
rock-candy fracture from **A11:** 75
volume fraction, by autocorrelational analysis **A10:** 594
volume fraction effect on steel tensile ductility **A8:** 571-572
wrought aluminum alloys **A12:** 428
Second-phase particles in aluminum alloys **A9:** 358-360
effect on boundary migration **A9:** 696-697
effect on diffraction patterns **A9:** 118
extraction replicas used to study **A9:** 108
Second-phase particles in plate steels
examination for **A9:** 203
Second-phase precipitates
transmission electron microscopy **A9:** 117-118
Second-phase precipitation **A6:** 622, 625
brazeability of base metal **A6:** 622, 625
Second-phase regions
quantitative metallography of **A9:** 129
Second-phase strengthening
ion implantation strengthening mechanisms **A18:** 855, 856, 858
Second-phase testing
isolation **A10:** 176-177
methods **A10:** 177
purification **A10:** 177
residue refinement and analysis **A10:** 176-177
Second-phase volume fraction (SPVF) **A18:** 206, 207
Secondary alloy *See also* Primary alloy
defined **A15:** 10
producers, aluminum scrap from **A15:** 79
Secondary aluminum
alloying effects **A2:** 46
Secondary arm coarsening
in dendritic structure **A15:** 117
Secondary atomization **A7:** 28, 29, 31
Secondary bonding *See also* Binder systems; Binders; Bonding; Bonds; Co-curing **EM2:** 37, 59 **EM3:** 25
in cast irons **A15:** 238
defined **EM1:** 21
Secondary circuit
definition **A6:** 1213 **M6:** 15
Secondary coolant *See* Water spraying
Secondary corrosion attack
in stainless steel implants **A11:** 683, 688
Secondary cracking
AISI/SAE alloy steels **A12:** 293-294, 299, 301, 306, 308, 327, 334
angular, titanium alloys **A12:** 448
ASTM/ASME alloy steels **A12:** 346, 348
at elevated temperatures, titanium alloys **A12:** 442
at machining marks **A12:** 251
austenitic stainless steels **A12:** 353
circumferential, precipitation-hardening stainless steels **A12:** 370
copper alloys **A12:** 403
of corrosion products **A12:** 29
deep intergranular, iron alloy **A12:** 459
in dimpled surface, titanium alloys **A12:** 450
ductile irons **A12:** 232
and fatigue striations **A12:** 20
intergranular **A12:** 47, 401, 425, 448
iridium **A12:** 462
low-carbon steel **A12:** 244
in low-melting metal embrittlement **A12:** 30, 38
nickel alloys **A12:** 397
OFHC copper **A12:** 401
opening **A12:** 77
in oxygen-embrittled iron **A12:** 222
parallel, maraging steels **A12:** 386
precipitation-hardening stainless steels **A12:** 372
as splitting **A12:** 68
stress-corrosion **A12:** 27
titanium alloys **A12:** 441, 448-450, 452
tool steels **A12:** 377, 380, 382
transgranular **A12:** 449, 452
transverse, AISI/SAE alloy steels **A12:** 301
wrought aluminum alloys **A12:** 414, 424, 425, 434
Secondary cracks
opening of **A11:** 19-20
visual macroanalysis **A12:** 72
Secondary creep *See also* Creep
defined **A12:** 19
in elevated-temperature failures **A11:** 263
Secondary creep rate
from transient data **A8:** 336
Secondary crystallization **EM3:** 25
Secondary dendrite arm spacing
as a function of distance from chill surface in 4340 steel **A9:** 625
as an index of solidification conditions in steel **A9:** 624-625
in copper alloy ingots **A9:** 637
defined **A9:** 624
as in indicator of degree of homogenization in steel **A9:** 625
and primary dendrite arm spacing as a function of distance from chill surface **A9:** 626
Secondary disintegration **A7:** 29, 33
Secondary electron detectors
for AES analysis **A10:** 554
image contrast with SEM **A10:** 501-502
sample configurations **A10:** 495
with scanning electron beam instruments **A10:** 501
types of electrons detected **A10:** 502
Secondary electron emission
as low-energy **A10:** 433-434
surface, AES analysis for **A10:** 549
Secondary electron imaging **A9:** 91-92
compared to backscattering electron imaging **A9:** 92

SUBJECTS OF THE INDEXED VOLUMES: ASM Handbook (designated by the letter "A"): **A1:** Properties and Selection: Irons, Steels, and High-Performance Alloys (1990); **A2:** Properties and Selection: Nonferrous Alloys and Special-Purpose Materials (1990); **A3:** Alloy Phase Diagrams (1992); **A4:** Heat Treating (1991); **A6:** Welding, Brazing, and Soldering (1993); **A7:** Powder Metallurgy (1984); **A8:** Mechanical Testing (1985); **A9:** Metallography and Microstructures (1985); **A10:** Materials Characterization (1986); **A11:** Failure Analysis and Prevention (1986); **A12:** Fractography (1987); **A13:** Corrosion (1987); **A14:** Forming and Forging (1988); **A15:** Casting (1988); **A16:** Machining (1989); **A17:** Nondestructive Testing and Quality Control (1989); **A18:** Friction, Lubrication, and Wear Technology (1992). **Metals Handbook, 9th Edition** (designated by the letter "M"): **M1:** Properties and Selection: Irons and Steels (1978); **M2:** Properties and Selection: Nonferrous Alloys and Pure Metals (1979); **M3:** Properties and Selection: Stainless Steels, Tool Materials and Special-Purpose Materials (1980); **M4:** Heat Treating (1981); **M5:** Surface Cleaning, Finishing, and Coating (1982); **M6:** Welding, Brazing, and Soldering (1983). **Engineered Materials Handbook** (designated by the letters "EM"): **EM1:** Composites (1987); **EM2:** Engineering Plastics (1988); **EM3:** Adhesives and Sealants (1990); **EM4:** Ceramics and Glasses (1991); **Electronic Materials Handbook** (designated by the letters "EL"): **EL1:** Packaging (1989).

Secondary electron imaging (continued)
contrast enhancement ... **A9:** 95
resolution ... **A9:** 95
scanning electron microscope ... **A9:** 89
voltage of primary electron beam ... **A9:** 91
Secondary electron yield in scanning electron microscopy
as a function of atomic number of the electron beam ... **A9:** 92
material interacting with the primary contrast ... **A9:** 93
effect of angle between incident beam and surface ... **A9:** 92
Secondary electron-coupled (SEC) vidicons *See also* Vidicons
in optical sensors ... **A17:** 10
Secondary electrons *See also* Electrons ... **A18:** 377, 379, 380
abbreviation for ... **A10:** 691
Auger ejection ... **A10:** 86
and backscattered electrons, compared ... **A12:** 168
and backscattered electrons, yields for ... **A10:** 502
defined ... **A10:** 681
emission ... **A10:** 433-434, 549
escape depths ... **EL1:** 1095
image, coarse-grain iron alloy ... **A12:** 93-94
images of ... **A10:** 554, 558
measuring instruments ... **A10:** 554
micrographs, beryllium copper alloy ... **A10:** 559
peak intensity ... **A10:** 498-499
production and x-ray photon emission ... **A10:** 86
in SEM imaging ... **A12:** 168
signal generation ... **A10:** 500
STEM mode, signal detector ... **A10:** 435
Secondary electrons in scanning electron microscopy energy spectrum ... **A9:** 92
and surface potentials ... **A9:** 95
Secondary etching
defined ... **A9:** 16
Secondary explosion ... **A7:** 194
Secondary extinction
defined ... **A9:** 16
Secondary fabrication *See also* Fabrication; Primary fabrication
hafnium ... **A2:** 664-665
zinc ... **A2:** 530-531
zirconium ... **A2:** 664-665
Secondary flow
nickel alloys ... **A12:** 396
Secondary hardening
austenitizing effects ... **A12:** 341
during tempering ... **A1:** 396, 641
in elevated-temperature service ... **A1:** 637-638, 640, 641
Secondary ignition sources ... **A7:** 197
Secondary ion
defined ... **A10:** 681
imaging, schematic diagrams ... **A10:** 622
Secondary ion mass spectrometry (SIMS) ... **A18:** 456, 458-460, 461 **EM3:** 237
as advanced failure analysis ... **EL1:** 1107
advantages and limitations ... **EM3:** 239
applications ... **A18:** 458, 459-460 **EL1:** 1085-1088
carbon depth profile, thin-film hybrids ... **EL1:** 319
depth profiling by ball cratering ... **EM3:** 245
dynamic ... **A18:** 459
equipment ... **A18:** 459
fundamentals ... **A18:** 459 **EL1:** 1083-1084
instrumentation ... **EL1:** 1084-1085
process ... **EM3:** 240-241
spectrum ... **A18:** 459
static ... **A18:** 459
for surface analysis ... **EL1:** 1083-1088
versus AES and XPS ... **A18:** 458, 459
as wafer-level physical test method ... **EL1:** 924-926
Secondary ion mass spectroscopy ... **A7:** 257-259 **A10:** 610-627 **A13:** 1118
apparatus ... **A7:** 257, 258
application ... **A7:** 258-259
applications ... **A10:** 610, 622-626
artifacts, effects of ... **A10:** 619
with Auger electron spectroscopy ... **A10:** 554
capabilities ... **A10:** 516, 517, 549, 568, 603
capabilities, and FIM/AP ... **A10:** 583
capabilities, compared with classical wet analytical chemistry ... **A10:** 161
defined ... **A10:** 681
Secondary ion mass spectroscopy (continued)
depth profiles ... **A10:** 617-620
detection limits ... **A10:** 622
elemental in-depth concentration profiling by ... **A10:** 610
and EPMA, compared ... **A10:** 516, 517
estimated analysis time ... **A10:** 610
general uses ... **A10:** 610
hydrogen analysis by ... **A10:** 610
images, schematic diagrams ... **A10:** 622
of inorganic solids ... **A10:** 4-6
instrumentation ... **A10:** 613
introduction ... **A10:** 611
ion imaging ... **A10:** 621
isotope abundances by ... **A10:** 610
limitations ... **A10:** 610
principles, schematic ... **A10:** 611
quantitative analysis ... **A10:** 620-621
related techniques ... **A10:** 610
samples ... **A10:** 610, 622-626
secondary ion emission ... **A10:** 612
secondary ion mass spectra ... **A10:** 615-617
sensitivity enhanced ... **A10:** 618
sputtering ... **A10:** 611-612
surface compositional analysis by ... **A10:** 610
system components ... **A10:** 613-615
testing parameters ... **A7:** 251
Secondary ion mass spectroscopy (SIMS) ... **EM1:** 285
development of ... **A11:** 33-35
hydrogen embrittlement study by ... **A11:** 45-46
for phase analysis ... **EM4:** 557
for surface analysis ... **EM4:** 25
uses for ... **A11:** 37
Secondary knock-on atom (SKA) ... **A18:** 851
Secondary loads
pipeline failure from ... **A11:** 704
Secondary magnesium
properties ... **A2:** 1218
Secondary manufacturing processes *See also* Manufacturing process selection; Manufacturing processes; Processing
injection molding, costs ... **EM2:** 296
molding operations, costs ... **EM2:** 84-86
thermoplastic injection molding ... **EM2:** 310
types, and design ... **EM2:** 277
Secondary manufacturing techniques
and heat exchanger failures ... **A11:** 629
Secondary metallurgy *See also* Decarburization; Metallurgy; Refining
converter ... **A15:** 426-431
freeboard requirements ... **A15:** 433
ladle ... **A15:** 432-444
Secondary milling ... **A7:** 56
Secondary neutral mass spectroscopy (SN MS) ... **EM3:** 237
advantages and limitations ... **EM3:** 239
process ... **EM3:** 240-241
Secondary nucleation ... **EM3:** 25
Secondary operations ... **A7:** 10, 451-462
powder forging ... **A14:** 197-198
pressing ... **A7:** 337-338
zinc alloys ... **A2:** 530-531
Secondary oxidation *See also* Deoxidation; Oxidation; Oxygen
inclusion formation by ... **A15:** 91
Secondary phase alloying
defined ... **A13:** 46
Secondary phases in iron-chromium-nickel heat-resistant casting alloys, identification of ... **A9:** 332-333
Secondary phases in nickel-base heat-resistant casting alloys ... **A9:** 334
Secondary processing methods
for composites ... **EM1:** 35
Secondary radiation *See also* Radiation; Scattered radiation
filtration, lead screens ... **A17:** 315
in neutron radiography ... **A17:** 387
Secondary recovery
of gallium ... **A2:** 745-746
Secondary recrystallization *See also* Grain growth ... **A9:** 697-698
effect on crystallographic texture ... **A9:** 700-701
in Fe-3Si ... **A9:** 698
in type 304 stainless steel ... **A9:** 698
Secondary recycling *See also* Recycling
aluminum-lithium alloys ... **A2:** 183-184
Secondary seal
defined ... **A18:** 16-17
Secondary structure ... **EM3:** 26
defined ... **EM1:** 21 **EM2:** 37
Secondary tension test
hole and slot ... **A8:** 584-585
Secondary tooling
blanking and piercing ... **A14:** 484-485
Secondary x-rays
defined ... **A9:** 16 **A10:** 681
Secondary-phase structures
for rolling direction ... **A12:** 422
Secondary-standard dosimetry system ... **EM3:** 25
defined ... **EM2:** 37
Secondary-target excitation
x-ray spectrometers ... **A10:** 89
Secondary-tension test ... **A1:** 583
for workability ... **A14:** 379
Section area measurements
use in calculating particle-size distribution ... **A9:** 131-133
Section sensitivity
gray cast iron ... **M1:** 11, 13-15
of gray iron ... **A1:** 14-16
Section size *See also* Section thickness
ductile iron ... **A15:** 656
effect on carbon steel casting microstructures ... **A9:** 231
and mass effects, low-alloy steels ... **A15:** 717-718
and mass effects, plain carbon steel ... **A15:** 704
strength effect, CG iron ... **A15:** 673
Section size, effect of
in compacted graphite iron ... **A1:** 60, 61, 62
in gray iron ... **A1:** 15
Section thickness *See also* Casting section thickness
die castings ... **A15:** 288
in gray iron ... **A15:** 633-635
zinc alloy ... **A15:** 792
Section-mass changes
and tool and die failure ... **A11:** 564
Sectioning ... **A9:** 23-27
aluminum alloys ... **A9:** 351-352
austenitic manganese steel castings ... **A9:** 237-238
beryllium ... **A9:** 389
beryllium-copper alloys ... **A9:** 392
beryllium-nickel alloys ... **A9:** 392
carbon and alloy steels ... **A9:** 165-166
carbon steel casting specimens ... **A9:** 230
carbonitrided steels ... **A9:** 217
carburized steels ... **A9:** 217
cast irons ... **A9:** 242-243
cemented carbides ... **A9:** 273
defined ... **A9:** 23
ferrites and garnets ... **A9:** 533
fiber composites ... **A9:** 587
of fracture surfaces ... **A11:** 19
hafnium ... **A9:** 497
lead and lead alloys ... **A9:** 415
low-alloy steel casting samples ... **A9:** 230
for macroscopic examination ... **A12:** 92
magnesium alloys ... **A9:** 425
medium-carbon steels ... **A12:** 268
metallographic, methods ... **A12:** 198-199
metallographic, of weldments ... **A11:** 412
metallographic, preparation and analyses ... **A11:** 23-24
nitrided steels ... **A9:** 218
permanent magnet alloys ... **A9:** 533
planar ... **A12:** 198-199
powder metallurgy materials ... **A9:** 503-504
printed board coupons ... **EL1:** 572-577
profile parameters from ... **A12:** 194
replica ... **A12:** 199
rhenium and rhenium-bearing alloys ... **A9:** 447
sampling plan ... **A12:** 212
serial, profile from ... **A12:** 198
sleeve bearing materials ... **A9:** 565
tin and tin alloys ... **A9:** 449
titanium and titanium alloys ... **A9:** 458, 461
tool steels ... **A9:** 256
transmission electron microscopy specimens ... **A9:** 104
transverse ... **A12:** 179
uranium ... **A9:** 477-478

Sectioning (continued)
vertical, for true fracture surface area **A12:** 198-199, 211-212
wrought heat-resistant alloys **A9:** 305
wrought stainless steels **A9:** 279
zinc and zinc alloys **A9:** 488
zirconium and zirconium alloys **A9:** 497
Sectioning technique using electric resistance strain gages **A6:** 1095
Sections
fracture profile **A12:** 95-96
for metallographic analysis **A10:** 300, 303
mounting of **A9:** 35
recommended size **A9:** 35
taper **A12:** 96
thickness, effect on fracture surface **A12:** 105
Sections used for metallographic examination of welded joints **A9:** 578-579
Sector analyzers
AES analysis **A10:** 554
Sedigraph
to study turbidity of a constantly falling column **EM4:** 26
Sedimentary deposits
high-level waste disposal in **A13:** 978
Sedimentation **A7:** 124, 218-220 **EM4:** 87
in equiaxed grain growth **A15:** 132
to measure particle size **EM4:** 66
versus weighting factor **EM4:** 85
Sedimentation field flow fractionation
to analyze ceramic powder particle sizes **EM4:** 67, 69
Sedimentation methods
to analyze ceramic powder particle size **EM4:** 67
Sedimentation potential **EM4:** 74
Seebeck coefficient **EM4:** 251, 748
Seebeck effect **EM3:** 711, 712
Seebeck emf *See* Electromotive force; Thermocouple materials
Seed(s)
cleaning, powder used **A7:** 572
coating, powder used **A7:** 572
conditioning **A7:** 589
lots, purity **A7:** 589
magnetic separation **A7:** 589-592
Seeded gel
alumina abrasive **A16:** 432
grinding **A16:** 421
Seeding
in monocrystal casting **A15:** 323
Seeman-Bohlin x-ray diffractometer **A10:** 337
Segment
defined **A10:** 681
Segment die **A7:** 10
Segmented die **A7:** 10
Segmented dies *See also* Split die
design, for physical modeling **A14:** 435
expanding drawn workpieces with **A14:** 587
Segments **A18:** 569
Segregated (stratified) materials
sampling **A10:** 14
Segregated impurities and pipe in ingots **A9:** 174
Segregated network
in metal-filled polymers **A7:** 607-608
Segregated particle pattern **A7:** 186
Segregation *See also* Inverse segregation; Macrosegregation; Microsegregation; Normal segregation **A6:** 1073 **A7:** 10
alloy **A11:** 121
of alloy and compound constituents to surface **A10:** 603
of alloy elements and impurities to dislocations and interfaces **A10:** 583
antimony **A12:** 350
of binary/ternary trace elements in solids **A10:** 544
in blending and premixing **A7:** 187-189

Segregation (continued)
brittle fracture from **A11:** 85
carbon **A15:** 405
in cast or wrought product **A11:** 314-315
centerline **A15:** 325
channel **A15:** 140-141, 156
chemical, ingot **A17:** 491
chemical-element **A11:** 121
in cold-formed parts **A11:** 307
composition, in solidification **A15:** 102
in decohesive rupture **A12:** 18
defined **A11:** 9 **A15:** 10 **EM2:** 37
as discontinuity, defined **A12:** 67
during solidification **A15:** 109
effect in creep strengthening **A10:** 598
effect in fastener failure **A11:** 529-530
effect on warping **A11:** 141
extrusion **A15:** 325
fractured forging die by **A11:** 324-326
grain-boundary **A10:** 481-484, 610 **A13:** 156-157
grain-boundary, phosphorus **A12:** 29
gravity **A15:** 139
high-alloy steels **A15:** 731
hydrogen, during solidification **A15:** 82
interfacial, in molybdenum **A10:** 599
intergranular embrittlement by **A12:** 110
inverse **A15:** 16
of lead to surface of tin-lead solder **A10:** 607-608
LEED analysis of **A10:** 536
liquid flow induced **A15:** 139
low-gravity, and dendritic growth **A15:** 153-156
macroetching of cast iron for **A11:** 344
manganese, dual phase steel **A10:** 483
metal particle, in polymers **A7:** 606
micro, in weldments **A13:** 344
microstructural, metallographic sectioning **A11:** 24
of nonmetallic inclusions **A11:** 477-478
P/M techniques for eliminating **A7:** 18
phosphorus, tool steel **A12:** 375
radiographic appearance **A17:** 349
rate, and austenite **A13:** 47
and solid phase movement **A15:** 141
and solubility **A15:** 109-110
solute, AES analysis **A10:** 549
steel plate **M1:** 182
and steel plate imperfections **A1:** 230
sulfur **A12:** 349
surface **A10:** 564-566, 583, 593
in titanium ingot **A2:** 596
transverse fracture from **A12:** 142, 163
in uranium isotopes, radioanalytic Measurement **A10:** 243
in weldments **A17:** 582
Segregation (coring) etching
defined **A9:** 16
Segregation banding
as centrifugal casting defect **A15:** 306
defined **A9:** 16
resemblance to artifact banding **A9:** 38
Segregation in aluminum alloys
etchants for examination of **A9:** 355
Segregation in carbon and alloy steels
macroetching to reveal **A9:** 173
Segregation in steel
during dendritic growth **A9:** 625-626
of nonmetallic inclusions **A9:** 625-626
Segregation of alloying elements *See also* Microsegregation
defined **A9:** 16
Seignette salt brightening
aluminum and aluminum alloys **M5:** 582
Seize **A7:** 10
Seizing
of deformation workpiece **A8:** 575-576
and distortion **A11:** 141-142
in fasteners at elevated temperatures **A11:** 542
and lubrication **A7:** 190, 192
of shafts **A11:** 466

Seizing (continued)
in spool-type hydraulic valve **A11:** 141
Seizure
defined **A18:** 17
lead-base alloys **A18:** 749
of pistons in internal combustion engines **A18:** 557
sliding bearing materials **A18:** 744, 747
in tin-base bearing alloys **A18:** 748
tool steels **A18:** 734
Seizure resistance **M1:** 611, 612
Selected complexation **A10:** 65-66
Selected Research in Microfiche (SRIM) (NTIS) **EM1:** 41
Selected-area channeling patterns
arrangement for **A10:** 506
for pearlite growth **A10:** 508-509
for preferred crystallographic growth **A10:** 509
as SEM technique **A10:** 505-506
Selected-area channeling patterns (SACP) **A6:** 145
Selected-area diffraction
defined **A10:** 681
Kikuchi patterns **A10:** 437-438
patterns, and CBEDP, compared **A10:** 439, 441
ring patterns **A10:** 436-437
spot patterns **A10:** 437, 438, 440
TEM **A10:** 436-438
Selected-area diffraction (SAD) **A18:** 386, 387
Selected-area diffraction pattern
defined **A10:** 681
***Selecting Thermoplastics for Engineering Applications* (MacDermott)** **EM2:** 94
Selection *See also* Economic process selection factors; Material selection; Material(s); Materials selection; Processes; Sample selection; Sample(s)
of copper alloy castings **A2:** 346-355
of copper alloy, for minimum-draft forgings **A14:** 258
of die materials **A14:** 45-47
electrical contact materials **A2:** 840
of flux **EL1:** 650
of forging equipment **A14:** 36-42
gold and aluminum wire **EL1:** 110-111
of lubricant **A14:** 518-520
of machine size **A14:** 83-85
of manufacturing process **EM2:** 277-403
material, for closed-die forging **A14:** 75-76
materials **EL1:** 112-118 **EM2:** 611-637
of NDE method **A17:** 49, 51, 561
of packaging **EL1:** 128
of press **A14:** 491-492
of process, hot-die/isothermal forging **A14:** 152-153
of process temperature, precision forging **A14:** 171
processes **EL1:** 112-118
of product form, magnesium alloys **A2:** 462-466
of roll forging machine **A14:** 96-97
of steel, for press forming **A14:** 546-547
of technology **EL1:** 128
of titanium alloy forging method **A14:** 282
Selection guides **EM3:** 35
Selection of adhesives **EM3:** 35-36
Selection of metals
case depth in case hardened parts **M1:** 627
shafting, factors in selection **M1:** 606
wear resistance **M1:** 603, 606-611
Selection rules
effect on x-radiation K lines **A10:** 86
Mössbauer spectroscopy **A10:** 288
Selective attack
by molten salts **A13:** 50
Selective block sequence
definition **M6:** 15
Selective coating
hot dip tin **M5:** 354-355
Selective combustion **A10:** 223-224

Selective corrosion
in aluminum alloy weldments **A13:** 345
Selective dissolution
in molten salts and liquid metals **A13:** 134
Selective epitaxy
GaAs-silicon wafer production method **A2:** 747
Selective film coatings *See also* Coatings; Conformal coatings
conformal coatings **EL1:** 764
Selective forced cooling
for stress and distortion **A15:** 616
Selective leaching *See also* Corrosion; Dealloying; Decarburization; Decobaltification; Dezincification; Graphitic corrosion **A13:** 131-134
alloys and environments subject to **A11:** 178
in brasses, as dezincification **A11:** 633
by corrosion **A11:** 178-180
cast irons **A13:** 568
copper alloys **A13:** 334
in copper-nickel alloys, as denickelification **A11:** 633
defined **A11:** 9 **A13:** 11
detection of **A11:** 179-180
dezincification **A11:** 178, 633
in heat exchangers **A11:** 628, 633-634
of iron, effects of **A11:** 373-374
as localized erosion, copper alloys **A11:** 633
material selection for **A13:** 333-334
mechanisms of **A11:** 178
mining industry **A13:** 1296
as molten-salt corrosion **A13:** 89
space shuttle orbiter **A13:** 1069, 1080
Selective oxidation
of chromium **A13:** 91, 134
as dealloying **A13:** 134
in gaseous corrosion **A13:** 17
oxide scales **A13:** 73-74
with two reactive elements **A13:** 73-74
Selective plating **M5:** 292-299
anode **M5:** 294-295
anode-cathode motion, control of **M5:** 295
applications **M5:** 298-299
automated operation **M5:** 298-299
brush plating compared to **M5:** 292
cadmium **M5:** 267-268, 298
chromium **M5:** 187, 194
current densities **M5:** 295
electrochemical metallizing **M5:** 298
equipment **M5:** 293-297
flame spray or plasma metallizing compared to **M5:** 292-293
flow plating process **M5:** 295-297
gold **M5:** 283-284
limitations of **M5:** 298-299
manual operation **M5:** 292-294
masking *See* Masking
power pack, features of **M5:** 293-294
process **M5:** 292-294
safety precautions **M5:** 298
solution compositions and operating
bonding solutions **M5:** 296-297
buildup solutions **M5:** 297
characteristics **M5:** 296-297
preparatory solutions **M5:** 296
specifications and standards **M5:** 299
steel **M5:** 298
stopping-off *See* Stopping-off
tank plating compared to **M5:** 292-293, 298
thickness, control of **M5:** 297-298
welding compared to **M5:** 292-293
Selective quenching **M4:** 31
Selective surface hardening
of pearlitic malleable iron **A1:** 84 **A15:** 697
Selective transfer
defined **A18:** 17
Selective vaporization
in arc sources **A10:** 25
Selective-ion potentiometry
to analyze the bulk chemical composition of starting powders **EM4:** 72
Selectivity
coefficient, defined **A10:** 165
in controlled-potential coulometry analysis **A10:** 208
defined **A10:** 681
determining electrode **A10:** 182-183
of fluorescence analysis **A10:** 76
Selectivity (continued)
UV VIS **A10:** 68
of x-ray spectrometry **A10:** 96
Selectivity coefficient
defined **A10:** 165
Selenic acid
use in rhodium plating solutions **M5:** 290
Selenic acid in color etchants **A9:** 142
Selenium **A13:** 182, 728
alloying, copper and copper alloys **A2:** 236
as an addition to Alnico alloys **A9:** 539
as an alloying addition to austenitic stainless steels **A9:** 284
in austenitic stainless steels **A6:** 468
biologic effects and toxicity **A2:** 1254-1255
as chalcogen **A2:** 1077
content additions to P/M materials **A16:** 885
content in stainless steels **A16:** 682-683, 684 **M6:** 320
in copper alloys **A6:** 753
deposited by color etching **A9:** 141
determined in natural waters **A10:** 41
distillation **A10:** 169
effect of, on machinability of carbon steels **A1:** 599
effects, electrolytic tough pitch copper **A2:** 270
embrittlement effect of **A11:** 236
embrittlement of iron by **A1:** 691
epithermal neutron activation analysis **A10:** 239
as essential metal **A2:** 1250, 1254-1255
in free-machining metals **A16:** 389
gaseous hydride, for ICP sample introduction **A10:** 36
gravimetric finishes **A10:** 171
heat-affected zone fissuring in nickel-base alloys **A6:** 588
machinability additives **A16:** 685, 686, 687, 688
and machine turning of resulfurized steels **A16:** 673, 677
poisoning **A2:** 1254
pure **M2:** 794
pure, properties **A2:** 1153
quartz tube atomizers for **A10:** 49
redox titration **A10:** 174, 175
reduction, by iodimetric titration **A10:** 174
sample modification, GFAAS analysis **A10:** 55
in semiconductor alloys, electrometric titration for **A10:** 206
sulfuric acid as dissolution medium **A10:** 165
TNAA detection limits **A10:** 238
toxicity **A6:** 1195
as trace element **A15:** 394
vapor pressure **A6:** 621
volatilizing **A10:** 166
Selenium (Se)
as colorant **EM4:** 380
purpose for use in glass manufacture **EM4:** 381
volatilization losses in melting **EM4:** 389
Selenium in copper **M2:** 242-243
Selenium in steel
improved machinability **M1:** 576
Selenium oxychloride
as industrial hazard **A2:** 1255
Selenium pink **EM4:** 380
Selenium poisoning
acute **A2:** 1254
Selenium sulfide
toxic effects **A2:** 1255
Selenium-coated plates
radiography **A17:** 315
Self-absorption
defined **A10:** 681
emission profile of **A10:** 22
in glow discharges **A10:** 28
Self-acting bearing *See* Gas bearing; Self-lubricating bearing
Self-actuation **A18:** 574
Self-aligning ball bearings **A18:** 500, 505, 508, 509
basic load rating **A18:** 505
f_v factors for lubrication methods **A18:** 511
z and y factors **A18:** 511
Self-aligning bearing
defined **A18:** 17
Self-cleaning effect
sputtering **M5:** 414
Self-cured solvent-base alkyl silicates **A13:** 412
Self-curing **EM3:** 26
Self-curing water-base alkali silicates **A13:** 411-412
Self-deconvolution, Fourier
as method of resolution enhancement **A10:** 116
Self-diffusion activation energy **A8:** 309
Self-electrode
defined **A10:** 681
Self-extinguishing resin *See also* Flame resistance **EM3:** 26
defined **EM1:** 21 **EM2:** 37
Self-fluxing alloys
definition **M6:** 15
Self-fluxing alloys (thermal spraying)
definition **A6:** 1213
Self-formed cores
ceramic **A15:** 261
Self-ionization
of water **A10:** 203
Self-jigging joints **A6:** 130, 132
Self-locating parts
design of **EL1:** 124
Self-lubricating bearing
defined **A18:** 17
Self-lubricating bearings *See also* Bearings; Lubricated bearings **A7:** 10, 16, 18, 319, 704-709
Self-lubricating material *See also* Solid lubricant
defined **A18:** 17
Self-lubricating parts
of copper-based powder metals **A7:** 734
Self-lubricating sintered bronze bearings **A2:** 394-396
Self-lubrication
of engineering plastics **EM2:** 1
Self-propagating high-temperature synthesis (SHS) **EM4:** 227-230
combustion modes **EM4:** 229
combustion theory **EM4:** 228-229
disadvantages **EM4:** 230
experimental procedures **EM4:** 229-230
experimental parameters **EM4:** 229-230
facilities **EM4:** 229
thermodynamic considerations **EM4:** 229
functionally graded materials **EM4:** 230
gas-pressure sintering **EM4:** 230
gas-solid combustion **EM4:** 228-229
historical development process **EM4:** 227-228
microstructures **EM4:** 230
potential applications **EM4:** 230
process advantages **EM4:** 227
process characteristics **EM4:** 227
properties **EM4:** 230
cast parts **EM4:** 230
dense products **EM4:** 230
powders **EM4:** 230
solid-solid combustion **EM4:** 228, 229
Self-resistance heating
in elevated/low temperature tension testing **A8:** 36
in tungsten and molybdenum sintering **A7:** 389, 391-392
Self-reversal
defined **A10:** 681
in emission spectroscopy **A10:** 22, 25
Self-shielded flux-cored arc welding
shielding gases **A6:** 68-69
Self-similitude
in rough planar curves **A12:** 211
Self-skinning foam **EM3:** 26
defined **EM1:** 21 **EM2:** 37
Self-tapping inserts **EM2:** 723
Self-tempering
9 Ni-4Co steels **M1:** 440
Self-test *See also* Built-in self-test (BIST)
design for **EL1:** 374-376
Self-threading screws
mechanical fastening **EM2:** 712-713
Self-tumbling
aluminum and aluminum alloys **M5:** 573
Self-vulcanizing **EM3:** 26
Selvage **EM1:** 21, 125-126
defined **EM2:** 37
SEM *See* Scanning electron microscopy
SEM microscopes *See* Scanning electron microscopes
SEM special techniques
cathodoluminescence **A10:** 507
electron beam induced current **A10:** 507

SEM special techniques (continued)
magnetic contrast **A10:** 506
signal processing **A10:** 507-508
specimen current detectors **A10:** 506
voltage contrast **A10:** 506-507
Semi-killed steels **M1:** 112, 123-124
carbon steel wire rod **M1:** 257
notch toughness of **M1:** 694, 698
plate **M1:** 181
Semiapochromatic objective lenses **A9:** 73
Semiaustenitic precipitation-hardenable stainless steels *See also* Wrought stainless steels; Wrought stainless steels, specific types **A9:** 285
Semiaustenitic steels
advantages **A6:** 797
applications **A6:** 797
Semiaustenitic steels, advantages and applications of materials for surfacing
build-up, and hardfacing **A18:** 650
Semiautogenous grinding (SAG) mill
grinding wear dependent on force **A18:** 273
Semiautomatic arc welding
definition **A6:** 1213 **M6:** 15
Semiautomatic brazing
definition **M6:** 15
Semiautomatic digital image analyzer **A10:** 310
Semiblind joint
definition **M6:** 15
Semibright plating additives
duplex nickel plating **M5:** 207
Semiburied vias
rigid printed wiring boards **EL1:** 550
Semicentrifugal casting
defined **A15:** 300
Semicoherent interface
between matrix and precipitate **A9:** 648
defined **A9:** 647
Semiconducting ceramics **EM4:** 17
Semiconducting compounds
as gallium application **A2:** 739
Semiconductive adhesives
for bonding electrical wires and devices **EM3:** 45
Semiconductor applications
thermoplastic fluoropolymers **EM2:** 117
Semiconductor caps
pure nickel **A7:** 404
Semiconductor chips *See also* Chips
chemical effects **EL1:** 965
electromigration on **EL1:** 963-964
electrostatic discharge (ESD) **EL1:** 965-967
failures on **EL1:** 963-967
fracture **EL1:** 978
fracture, as failure mechanism **EL1:** 978-979
hot carriers **EL1:** 965
oxide breakdown **EL1:** 965
radiation effects **EL1:** 965-966
stress cracking of **EL1:** 964-965
Semiconductor device
defined **EL1:** 1156
Semiconductor devices
defect analysis and quality control **A10:** 490
metal contact **A11:** 778
silicon, failure analysis of **A11:** 766-792
Semiconductor flaws
radiographic appearance **A17:** 350-351
Semiconductor integrated circuit (die)
polyimide formulations **EM3:** 159
Semiconductor lasers
as gallium compound application **A2:** 739
Semiconductor materials *See also* Semiconductors; Semiconductors, characterization of
diffused or ion-implanted, SIMS analysis of **A10:** 610
interdiffusion analyzed **A10:** 583
metallization **A10:** 583
oxidation, FIM/AP study of **A10:** 583
PIXE analysis of **A10:** 102
Semiconductor materials (continued)
pulsed laser atomic probe analysis of **A10:** 597
reconstruction of surfaces **A10:** 536
selenium and tellurium in, by electrometric titration **A10:** 206
SSMS analysis of high-purity silicon for **A10:** 141
Semiconductor oxides **A13:** 65-66
Semiconductor packaging *See also* Discrete semiconductor packages; Integrated semiconductor packages
components of **EL1:** 397
defined **EL1:** 397
Semiconductor packaging road map
as trend **EL1:** 436-437
Semiconductor technology
effect, passive component fabrication **EL1:** 178
gallium arsenide **EL1:** 160
Semiconductor(s)
development, and failure mechanisms **EL1:** 958
integrated circuits, in hybrids **EL1:** 249
leakage detection **EL1:** 1089-1090
metrology/inspection, optical vs SEM for **EL1:** 367
parylene coatings **EL1:** 799
plastic packaged, structure **EL1:** 241
quantum mechanical band theory (solid state) **EL1:** 101-103
vs electron tubes **EL1:** 958
water cooling of **EL1:** 310
Semiconductors *See also* Semiconductor materials; Semiconductors, characterization of
analytic methods for **A10:** 4
compound, ECAP/PLAP analysis of **A10:** 598, 601-602
field evaporation for **A10:** 590
FIM images of **A10:** 589-590
of germanium and germanium compounds **A2:** 733, 735-737
gold plating of **M5:** 283
of indium **A2:** 752-753
intrinsic, sample preparation for FIM **A10:** 584
laser-enhanced etching **A16:** 576
powder used **A7:** 573
rocking curve analyses of **A10:** 371
substrates, of germanium and germanium compounds **A2:** 743
ternary 3:5, local composition fluctuations in **A10:** 601-602
x-ray topographic analysis **A10:** 366
Semiconductors, characterization of *See also* Semiconductor materials; Semiconductors
analytical transmission electron microscopy **A10:** 429-489
atomic absorption spectrometry **A10:** 43-59
Auger electron spectroscopy **A10:** 549-567
classical wet analytical chemistry **A10:** 161-180
controlled-potential coulometry **A10:** 207-211
electrochemical analysis **A10:** 181-211
electrogravimetry **A10:** 197-201
electrometric titration **A10:** 202-206
electron probe x-ray microanalysis **A10:** 516-535
electron spin resonance **A10:** 253-266
extended x-ray absorption fine structure **A10:** 407-419
field ion microscopy **A10:** 583-602
inductively coupled plasma atomic emission spectroscopy **A10:** 31-42
infrared spectroscopy **A10:** 109-125
low-energy electron diffraction **A10:** 536-545
low-energy ion-scattering spectroscopy **A10:** 603-609
neutron activation analysis **A10:** 233-242
neutron diffraction **A10:** 420-426
optical emission spectroscopy **A10:** 21-30
particle-induced x-ray emission **A10:** 102-108
potentiometric membrane electrodes **A10:** 181-187
Raman spectroscopy **A10:** 126-138
Semiconductors, characterization of (continued)
Rutherford backscattering spectrometry **A10:** 628-636
scanning electron microscopy **A10:** 490-515
secondary ion mass spectroscopy **A10:** 610-627
single-crystal x-ray diffraction **A10:** 344-356
spark source mass spectrometry **A10:** 141-150
ultraviolet/visible absorption spectroscopy **A10:** 60-71
voltammetry **A10:** 188-196
x-ray diffraction **A10:** 325-332
x-ray photoelectron spectroscopy **A10:** 568-580
x-ray powder diffraction **A10:** 333-343
x-ray spectrometry **A10:** 82-101
x-ray topography **A10:** 365-379
Semiconductors, doped
conditions for growth **A9:** 612
Semiconductors, friction and wear of **A18:** 685-689
abrasive wear **A18:** 685
application of cathodoluminescent signal **A18:** 378
and doping **A18:** 688-689
dynamic friction coefficient in silicon and gallium arsenide as a function of temperature
lubrication **A18:** 685, 686, 687
machining operations **A18:** 685-686
mechanical damage at silicon surfaces caused by dicing **A18:** 686
mechanical damage at silicon surfaces caused by wafering **A18:** 685-686
microcracking **A18:** 685
simulation of dicing damage by single-point diamond scratching of silicon **A18:** 686-688
Semicontinuous casting
direct-chill as **A15:** 313-314
wrought copper and copper alloys **A2:** 243
Semicontinuous castings
unsoundness in **A9:** 643
Semicontinuous vacuum coating **M5:** 397-398, 403-404
Semicrystalline *See also* Crystalline plastic; Crystallinity **EM3:** 26
defined **EM1:** 21 **EM2:** 37
Semicrystalline plastic *See* Crystalline plastic
Semicrystalline polymers
defined **A11:** 758
immiscible blends **EM2:** 633-635
Semicrystalline resins
miscible blends **EM2:** 636-637
Semicylindrical dies
roll forging **A14:** 97
Semiductile cast iron *See* Compacted graphite iron
Semifinisher
defined **A14:** 11
Semifractal plot *See also* Fractal plot **A12:** 211
Semigraphical creep analysis **A8:** 688
Semiguided bend *See also* Free bend
defined **A8:** 12
Semihollow shapes
wrought aluminum alloy **A2:** 33-34
Semikilled low-carbon steel
roll welding **A6:** 312
Semikilled steel **A1:** 142-143, 226, 227
Semikilled steels
sample dissolution for **A10:** 176
Semilogarithmic plot
of stress vs. log rupture time **A8:** 333
Semimet brake linings *See* Semimetallic brake linings
Semimetallic (resin-bonded metallic, or semimet) brake linings **A18:** 569, 570-571, 572, 573, 575, 576
debris **A18:** 574
frictional behavior **A18:** 576
Semimetallic elements
partitioning oxidation states in **A10:** 178

SUBJECTS OF THE INDEXED VOLUMES: ASM Handbook (designated by the letter "A"): **A1:** Properties and Selection: Irons, Steels, and High-Performance Alloys (1990); **A2:** Properties and Selection: Nonferrous Alloys and Special-Purpose Materials (1990); **A3:** Alloy Phase Diagrams (1992); **A4:** Heat Treating (1991); **A6:** Welding, Brazing, and Soldering (1993); **A7:** Powder Metallurgy (1984); **A8:** Mechanical Testing (1985); **A9:** Metallography and Microstructures (1985); **A10:** Materials Characterization (1986); **A11:** Failure Analysis and Prevention (1986); **A12:** Fractography (1987); **A13:** Corrosion (1987); **A14:** Forming and Forging (1988); **A15:** Casting (1988); **A16:** Machining (1989); **A17:** Nondestructive Testing and Quality Control (1989); **A18:** Friction, Lubrication, and Wear Technology (1992). **Metals Handbook, 9th Edition** (designated by the letter "M"): **M1:** Properties and Selection: Irons and Steels (1978); **M2:** Properties and Selection: Nonferrous Alloys and Pure Metals (1979); **M3:** Properties and Selection: Stainless Steels, Tool Materials and Special-Purpose Materials (1980); **M4:** Heat Treating (1981); **M5:** Surface Cleaning, Finishing, and Coating (1982); **M6:** Welding, Brazing, and Soldering (1983). **Engineered Materials Handbook** (designated by the letters "EM"): **EM1:** Composites (1987); **EM2:** Engineering Plastics (1988); **EM3:** Adhesives and Sealants (1990); **EM4:** Ceramics and Glasses (1991); **Electronic Materials Handbook** (designated by the letters "EL"): **EL1:** Packaging (1989).

Semipermanent mold
defined ... **A15:** 10
Semipermanent mold casting
defined ... **A15:** 275
and permanent mold casting, compared ... **A15:** 275
Semipermanent pins, machine
failures in ... **A11:** 545
Semipositive mold *See also* Mold(s)
defined ... **EM2:** 37
Semiprecision resistance alloys *See also* Electrical resistance alloys; Precision resistance alloys
properties and applications ... **A2:** 826
Semiquantitative analysis *See also* Semiquantitative analysis, methods for
dc arc emission spectroscopy ... **A10:** 25
electrographic ... **A10:** 202
of inorganic gases ... **A10:** 8
of inorganic liquids and solutions, methods for ... **A10:** 7
of inorganic solids, applicable analytical methods ... **A10:** 4-6
of organic solids and liquids, techniques for ... **A10:** 9, 10
Semiquantitative analysis, methods for *See also* Semiquantitative analysis
analytical transmission electron microscopy ... **A10:** 429-489
Auger electron spectroscopy ... **A10:** 549-567
electron probe x-ray microanalysis ... **A10:** 516-535
electron spin resonance ... **A10:** 253-266
field ion microscopy ... **A10:** 583-602
gas analysis by mass spectrometry ... **A10:** 151-157
gas chromatography/mass spectrometry ... **A10:** 639-648
infrared spectroscopy ... **A10:** 109-125
ion chromatography ... **A10:** 658-667
liquid chromatography ... **A10:** 649-659
low-energy ion-scattering spectroscopy ... **A10:** 603-609
molecular fluorescence spectrometry ... **A10:** 72-81
Mössbauer spectroscopy ... **A10:** 287-295
neutron diffraction ... **A10:** 420-426
optical emission spectroscopy ... **A10:** 21-30
particle-induced x-ray emission ... **A10:** 102-108
scanning electron microscopy ... **A10:** 490-515
secondary ion mass spectroscopy ... **A10:** 610-627
spark source mass spectrometry ... **A10:** 141-150
ultraviolet/visible absorption spectroscopy ... **A10:** 60-71
x-ray diffraction ... **A10:** 325-332
x-ray photoelectron spectroscopy ... **A10:** 568-580
x-ray powder diffraction ... **A10:** 333-343
x-ray spectrometry ... **A10:** 82-101
Semired brasses
applications and properties ... **A2:** 225
Semirefractory constituents in silver-base electrical contact materials ... **A9:** 551-552
Semirigid cast elastomers
polyurethane (PUR) ... **EM2:** 259
Semirigid molded foams
polyurethane (PUR) ... **EM2:** 258-259
Semirigid plastic *See also* Plastics; Rigid plastic ... **EM3:** 26
defined ... **EM2:** 37
Semisolid metal casting and forging
See also Semisolid metalworking ... **A15:** 327-338
applications, forging ... **A15:** 334-336
automated ... **A15:** 35
history/benefits ... **A15:** 327-328
magnetohydrodynamic casting ... **A15:** 331
market effects ... **A15:** 44
metal forming ... **A15:** 332-333
of metal-matrix composites ... **A15:** 338
metalworking processes ... **A15:** 328-333
as permanent mold process ... **A15:** 34
quality control ... **A15:** 336-337
raw material production, casting ... **A15:** 329-331
rheocasting ... **A15:** 207, 329-331
semisolid forging ... **A15:** 332-336
SIMA process ... **A15:** 332
wrought processes ... **A15:** 332-333
zinc alloys ... **A15:** 797
Semisolid metalworking *See also* Semisolid casting and forging
forging advantages/limitations ... **A15:** 333
Semisolid metalworking (continued)
forging applications of ... **A15:** 334
magnetohydrodynamic casting ... **A15:** 331
processes ... **A15:** 328-333
raw material production, casting ... **A15:** 329-331
rheocasting ... **A15:** 329-331
wrought ... **A15:** 332-333
Semisolid polishing and buffing compounds
removal of ... **M5:** 5, 10
Semisolid-metal processing
aluminum casting alloys ... **A2:** 142
Semisynthetic fluids ... **A18:** 144
Semivitreous earthenware ... **EM4:** 3, 4
Semiwidth of Hertzian contact band ... **A18:** 539, 540
symbol and units ... **A18:** 544
Sendzimir aluminum dip coating method ... **M5:** 335
Sendzimir mill rolls
cemented carbide ... **A2:** 969-970
Sendzimir mills
for wrought copper and copper alloys ... **A2:** 244
Sendzimir process
zinc-base coatings ... **A13:** 526
Sendzimir rolling mill ... **A14:** 11, 351-352
Sensing (transduction)
as electronic function ... **EL1:** 89
Sensing zone
versus weighting factor ... **EM4:** 85
Sensitive tint ... **A9:** 138
Sensitive tint plate
defined ... **A9:** 16
placement ... **A9:** 138
used to enhance coloration with crossed- polarized light ... **A9:** 72, 78
Sensitivity *See also* Radiographic sensitivity
acoustic emission inspection ... **A17:** 280-281
of acoustical holography ... **A17:** 439, 441
atomic, empirical factors ... **A10:** 574
contrast, and noise ... **A17:** 374-375
contrast, radiography ... **A17:** 299
defined, in welding ... **A17:** 590
depth, microwave inspection ... **A17:** 202
detection, AES ... **A10:** 556
and detection limits, UV/VIS ... **A10:** 70
of electric current perturbation ... **A17:** 137, 139
of electric currents ... **A17:** 101
elemental, LEISS analysis ... **A10:** 605-606
of half-wave current ... **A17:** 91
image quality as ... **A17:** 338
increasing, remote-field eddy current inspection ... **A17:** 198-199
of leak testing method ... **A17:** 70
levels, demonstration program for ... **A17:** 663
levels, of liquid penetrant methods ... **A17:** 77
magabsorption ... **A17:** 148-149
neutron radiography ... **A17:** 391-395
noise adjustments, acoustic emission inspection ... **A17:** 285
notch ... **A13:** 294
of radioanalysis ... **A10:** 246-247
radiographic, and image quality ... **A17:** 298-300
ranges, of leak testing methods ... **A17:** 59
of RBS ... **A10:** 630
relative, LEISS ... **A10:** 606
remote-field eddy current inspection ... **A17:** 195
of SIMS, enhanced by oxygen primary ion beam ... **A10:** 618
surface, LEISS ... **A10:** 605
surface, XPS analysis ... **A10:** 569-570
ultrasonic inspection ... **A17:** 231, 267
wavelength-dispersive spectrometer ... **A10:** 521
Sensitivity analysis
experimental ... **EM2:** 605-606
Sensitivity factors ... **EM3:** 242, 243, 247
Sensitivity parameter
defined ... **A8:** 387
Sensitivity, surface
of AES ... **A11:** 33
Sensitization ... **A6:** 622, 626
alloy, electrochemical testing ... **A13:** 218-219
in austenitic stainless steels ... **A1:** 706-707, 912
chromium nitride, P/M stainless steels ... **A13:** 828, 830
defined ... **A11:** 9, 400-401 **A13:** 11, 199
effects, stainless steel corrosion ... **A13:** 551
in ferritic stainless steels ... **A1:** 707-708
of test coupons ... **A13:** 199
Sensitization (continued)
time/temperature, curves ... **A13:** 154
to intergranular attack ... **A13:** 48
weld, and design ... **A13:** 342
Sensitization, chromium
in Inconel ... **A10:** 483, 600
Sensitizing heat treatment *See also* Sensitization
defined ... **A13:** 11
Sensitometric curve
radiography ... **A17:** 324
Sensor coils *See also* Coils; Probes; Sensors
in electric current perturbation ... **A17:** 136
Sensor response
acoustic emission inspection ... **A17:** 280
Sensor systems
acoustic emission ... **A6:** 1063
airborne acoustics ... **A6:** 1063
charge-coupled device (CCD) video camera ... **A6:** 1063
eddy current seam locator (probe) ... **A6:** 1064
emission spectroscopic ... **A6:** 1063
infrared thermal area-type cameras ... **A6:** 1064
laser scanning ... **A6:** 1063
low voltage "touch" ... **A6:** 1064
through-the-arc voltage current ... **A6:** 1063
ultrasonic ... **A6:** 1063
Sensor(s) *See also* Coils; Image sensors; Optical sensors; Probes
acoustic emission ... **A17:** 280-281
acoustic waveguide, applied ... **A17:** 281
configuration, and sensitivity, remote-field eddy current inspection ... **A17:** 198
coordinate measuring machine ... **A17:** 25
innovative configurations ... **A17:** 45
in leakage field testing ... **A17:** 130-131
response, acoustic emission inspection ... **A17:** 280
RFEC probes ... **A17:** 196
sensitivity, remote-field eddy current inspection ... **A17:** 197
Sensors
for closed-loop cure ... **EM1:** 762
for fatigue testing machines ... **A8:** 368
load cell ... **A8:** 368
pressure transducer ... **A8:** 368
Sentry holes
monitoring technique ... **A13:** 201
Separable component part
defined ... **EL1:** 1156
Separate-application adhesive ... **EM3:** 26
Separates
ASTM standards ... **EM3:** 61
Separation
advanced ... **A8:** 440, 443
automated ... **A10:** 199
beam-to-chip ... **EL1:** 287
beam-to-substrate ... **EL1:** 287
by complexation, for UV/VIS interferences ... **A10:** 65
by distillation ... **A10:** 169
of cadmium and lead, by internal electrolysis ... **A10:** 201
cavities, copper alloys ... **A12:** 402
circumferential, wrought aluminum alloys ... **A12:** 415
cupferron ... **A10:** 169
direct chemical, for UV/VIS interferences ... **A10:** 65
in ductile hole joining fracture mode ... **A8:** 450
effectiveness, measure of ... **A10:** 164
efficiency measured by radioanalysis ... **A10:** 243
final, titanium alloys ... **A12:** 441
grain boundary, as casting defect ... **A15:** 548
grain-boundary ... **A11:** 605, 675, 681
hackle-, in composites ... **A11:** 737
of interfering elements, in high-temperature combustion ... **A10:** 222
intergranular, high-purity copper ... **A12:** 399
ion exchange ... **A10:** 164-165
matrix, effect of crack tip ... **A12:** 16
mechanism of progressive fracturing as advanced ... **A8:** 440
of metals in electrogravimetry, emf conditions for ... **A10:** 197-198
nonmetallic inclusions by ... **A11:** 316
oblique-shear ... **A8:** 440
of praseodymium and neodymium ... **A10:** 249-250

Separation (continued)
precipitation ... **A10:** 168
and precipitation (Sherritt Gordon process) ... **A7:** 54
in precipitation from solution ... **A7:** 54
principles, ion chromatography ... **A10:** 658-659
and quantitative determination of metal ions ... **A10:** 200-201
of RDF into pair distribution functions ... **A10:** 397-398
sink/float density ... **A10:** 177
techniques, classical wet chemistry ... **A10:** 165, 168-170
techniques, for inclusion control ... **A15:** 90-91
types, tape automated bonding ... **EL1:** 173
Separation margin *See* Signal-to-noise ratio
Separation science
defined ... **A10:** 164
Separation techniques
classical wet chemistry ... **A10:** 162, 165, 168-170
hydroxide ... **A10:** 168
ion-exchange chromatography ... **A10:** 168
magnetic ... **A10:** 177
paper chromatography ... **A10:** 168
for solids ... **A10:** 165
Separations
by computed tomography (CT) ... **A17:** 361
microwave inspection ... **A17:** 212
Separator *See also* Release film ... **A18:** 499
as cure processing material ... **EM1:** 21, 642-643
defined ... **A18:** 17 **EM2:** 37-38
Separators
metal powders cleaning ... **A7:** 178-180
Septenary system or diagram ... **A3:** 1 • 2
Sequence timer
definition ... **M6:** 15
Sequence VE testing
oxidation inhibitors ... **A18:** 105
Sequence weld timer
definition ... **M6:** 15
Sequencing
shielded metal arc welding ... **M6:** 92
Sequential load/unload cycling
crack arrest fracture toughness ... **A8:** 292-293
Sequential multielement analysis
direct-current plasma ... **A10:** 40
monochromators for ... **A10:** 37, 38
scanning monochromator for ... **A10:** 38
and simultaneous multielement analysis, combined ... **A10:** 38
Sequential multifrequency techniques
eddy current inspection ... **A17:** 174
Sequential observation technique
erosion mechanisms ... **A18:** 202
Sequential process board
cross section ... **EL1:** 135
Sequential wavelength-dispersive x-ray spectrometers
automated ... **A10:** 87
Sequestrants
use in alkaline etching ... **M5:** 583
Serial access
defined ... **EL1:** 1156
Serial sectioning *See also* Sectioning
defined ... **A9:** 16
for fractal analysis ... **A12:** 212-213
for fracture surface area ... **A12:** 194
profile, fractured aluminum-copper alloy ... **A12:** 198
used to determine shapes of eutectic structures ... **A9:** 620
Series damping *See* Series termination
Series multiple-spot welding setups ... **M6:** 476
Series resistance
defined ... **EL1:** 1156
Series submerged arc welding
definition ... **M6:** 15
Series termination ... **EL1:** 171-172, 522
for reflection ... **EL1:** 171-172
Series welding
definition ... **A6:** 1213 **M6:** 15
Serpentine
Miller numbers ... **A18:** 235
Serpentine glide
defined ... **A12:** 13
formation, in copper ... **A12:** 17
from slip step ... **A12:** 16
wrought aluminum alloys ... **A12:** 434
Serrated steel rollers
for SMC compaction ... **EM1:** 160
Serrating
in conjunction with boring ... **A16:** 174
multifunction machining ... **A16:** 386
SERS *See* Surface-enhanced Raman scattering
Serthi and Wright's model
corrosion-affected erosion ... **A18:** 210
Serum glutamate pyruvate transaminase
reaction rate analysis ... **A10:** 70
Service *See also* Service conditions; Service failures
conditions, anomalous, as failure cause ... **EM1:** 767
history ... **A11:** 158, 460
life, of forged products, and design ... **A11:** 17-19
records, of boilers and related equipment ... **A11:** 602
static, allowable stresses ... **A11:** 136
stress sources in ... **A11:** 206
temperatures, and material selection ... **EM1:** 38
Service conditions
abnormal ... **A11:** 16
anomalous, of continuous fiber reinforced composites ... **A11:** 733
brittle tensile fracture under ... **A11:** 77
data collection of ... **A11:** 15
effect on stress-corrosion cracking ... **A11:** 208-211
in failure analysis ... **A11:** 747
forging failures ... **A11:** 317, 338-342
influence on tool and die failures ... **A11:** 575-577
of iron castings ... **A11:** 367-378
for shaft failures ... **A11:** 467-468
spring failures in ... **A11:** 550, 560-562
Service environment
correlation of accelerated SCC test mediums with ... **A8:** 523
and field testing, SCC evaluation ... **A13:** 263-265
outdoor, exposure of coatings to ... **A13:** 395
SCC of aluminum alloy ... **A13:** 266
Service failures
by fatigue cracking ... **A11:** 129
by low impact resistance and grinding burns ... **A11:** 90-92
ductile, as overload failures ... **A11:** 85
from brittle fracture ... **A11:** 69-71, 85
from high-cycle fatigue ... **A11:** 106
marine-air ... **A11:** 309-310
of pipes ... **A11:** 699-704
of steel wing slat track ... **A11:** 140-141
Service fractures
AISI/SAE alloy steels ... **A12:** 297, 317, 319
cast aluminum alloys ... **A12:** 409-412
fatigue, low-carbon steel ... **A12:** 241, 251
macroscopic examination ... **A12:** 91-93
mating fracture ... **A12:** 251
rotary bending fatigue fracture, medium-carbon steels ... **A12:** 272
superalloys ... **A12:** 390, 391
tool steels ... **A12:** 376
wrought aluminum alloys ... **A12:** 417
Service life *See also* Life tests
aircraft parts ... **A13:** 1020-1022
changes in test temperature or load effect on ... **A8:** 337-338
corrosion testing for ... **A13:** 193
electrical contacts, as selection criterion ... **A2:** 840
evaluating creep damage and remaining ... **A8:** 337-339
factors influencing ... **A13:** 321
Service life (continued)
hazards, effects, electrical contact materials ... **A2:** 840
of heating elements, electrical resistance alloys ... **A2:** 831-833
hot dip galvanized steel ... **A13:** 440
life tests, electrical contact materials ... **A2:** 858-861
prediction ... **A13:** 278-279, 316
specimens ... **A8:** 337
steam turbine materials ... **A13:** 956-958
zinc/zinc alloys and coatings ... **A13:** 762
Service loading
as cause of stress-corrosion cracking ... **A8:** 496
estimates from crack extension behavior ... **A8:** 439
Service pipe *See also* Pipe
lead and lead alloy ... **A2:** 552-553
Service qualification tests
constructional steels for elevated temperature use ... **M1:** 643
Service temperature *See also* Temperature(s)
defined ... **EM2:** 569
determined, by thermal degradation ... **EM2:** 568-570
polyamides (PA) ... **EM2:** 126
range, of polyamide-imides (PAI) ... **EM2:** 129
test program ... **EM2:** 569-570
Service temperatures *See also* Temperature(s) ... **EM3:** 36
heating elements, electrical resistance alloys ... **A2:** 831
thermocouples ... **A2:** 883
Serviceability
of SCC-tested materials ... **A8:** 503
Servo control
electrical discharge grinding ... **A16:** 565
method of operation for EDM ... **A16:** 557, 562
relation to adaptive control ... **A16:** 618
sawing ... **A16:** 357
Servo-controller
features ... **A8:** 399
in servohydraulic fatigue testing system ... **A8:** 398-399
in torsion testing ... **A8:** 158
Servo-valve
in hydraulic torsional system ... **A8:** 215-216
in servohydraulic testing system ... **A8:** 398-399
in torsion testing ... **A8:** 158
Servohydraulic closed-loop system ... **A8:** 368-369
Servohydraulic fatigue-testing machine
closed-loop ... **A11:** 278
Servohydraulic shear testing machine ... **A8:** 215
Servohydraulic test frame ... **A8:** 187, 190, 192
Servohydraulic testing systems
closed-loop ... **A8:** 43, 368-369
components ... **A8:** 396-400
constant strain rate tests on ... **A8:** 43
determining critical *J*-values and *J*-resistance curves using ... **A8:** 261
for fatigue testing ... **A8:** 395-400
rapid-load plane-strain fracture toughness tests with ... **A8:** 260-261
specimens ... **A8:** 400
Servohydraulic torsional fatigue testing machine ... **A8:** 369-370
Servomechanical electromechanical fatigue tester ... **A8:** 392
Servomechanical systems ... **A8:** 392, 394-395
Sessile dislocations
effect on crack extension ... **A8:** 439
Set ... **EM3:** 26
defined ... **EM1:** 11
Set (mechanical) ... **EM3:** 26
defined ... **EM2:** 38
Set (polymerization) ... **EM3:** 26
Set up ... **EM3:** 26
defined ... **EM1:** 21 **EM2:** 38
Set-up wheel polishing *See* Wheel polishing

Setdown
definition ... **A6:** 1213
Setting temperature ... **EM3:** 26
Setting time ... **EM3:** 26
Setting-down of springs *See* Presetting of springs
Settling
in encapsulation hot isostatic pressing ... **A7:** 431
Settling time
defined ... **EL1:** 7
Setups
fractographic ... **A12:** 78
Severe cold forming applications
carbon steel for ... **M1:** 255
Severe oxidational wear theory ... **A18:** 288
Severe wear ... **A8:** 603
defined ... **A18:** 17
Severity index (SI) ... **A18:** 308
Severity index gradient ... **A18:** 578, 580
Severity of quench
evaluation ... **M4:** 34, 35, 38-39
Severn gage ... **A6:** 461
Sewage
digested, Miller numbers ... **A18:** 235
potentiometric membrane electrode analysis of ... **A10:** 181
raw, Miller numbers ... **A18:** 235
Sewer pipes
ductile iron for ... **M1:** 99
linings for ... **M1:** 100
Sexinary system or diagram ... **A3:** 1 • 2
SFC *See* Supercritical fluid chromatography
SG *See* Spin glass
SG iron ... **A1:** 56
Shackle, dragline bucket
failure of ... **A11:** 399-400
Shading, and image processing
radiography ... **A17:** 320
Shadow angle
defined ... **A9:** 16
Shadow cast *See* Shadowing
Shadow cast replica
defined ... **A9:** 16
Shadow cure
of epoxidized silicones ... **EL1:** 824
Shadow effect
UV curing ... **EL1:** 786-787
Shadow formation
distortion ... **A17:** 312-313
enlargement ... **A17:** 311-312
geometric unsharpness ... **A17:** 313
intensity, and inverse-square law ... **A17:** 313
in neutron radiography ... **A17:** 387
principles, radiography ... **A17:** 311-314
scattered radiation ... **A17:** 313-314
Shadow mask
definition ... **A6:** 1213 **M6:** 15
Shadow microscope
defined ... **A9:** 16
Shadow optic method of caustics
with precracked Charpy test ... **A8:** 268, 269
Shadow(s) *See also* Shadow formation
projection, scanning laser gages ... **A17:** 12
radiographic, view of ... **A17:** 331-332
Shadowed carbon-platinum replicas ... **A9:** 108
Shadowgraphs
for part shape measurement ... **A8:** 549
Shadowing
defined ... **A9:** 16
effect of ... **EM1:** 747
as film deposition ... **A12:** 172
methods ... **A12:** 183-184
of replicas ... **A12:** 95, 100, 183-184
single-stage plastic replicas ... **A12:** 184
in TEM replication ... **A12:** 7
two-stage plastic-carbon replicas ... **A12:** 184
Shaeffler diagram
cast stainless steels ... **A13:** 574-575
Shaft
rotation, record from dynamic torsion test ... **A8:** 216-217
and shaft collar, compression test fixture ... **A8:** 198
torsion tests for ... **A8:** 139
in torsional hydraulic actuator ... **A8:** 217
Shaft boring
with cemented carbide tools ... **A2:** 974
Shaft run-out
defined ... **A18:** 17
Shaft whirl
effect on bearings ... **A11:** 484-485
Shaft(s)
disk or gear on, inspection ... **A17:** 113
drive-pinion, magnetic particle inspection ... **A17:** 113
forged, ultrasonic inspection of ... **A17:** 506-510
straightening processes, acoustic emission inspection ... **A17:** 289
ultrasonic inspection ... **A17:** 232, 271
Shafts *See also* Crankshafts; Shafts, failures of alignment and deflection, bearing failure and ... **A11:** 507
applications for closed-die steel forgings ... **M1:** 360
automotive drive, EMF assembly ... **A14:** 648
axle, fracture at journal from reversed-bending fatigue ... **A11:** 321
and bearing, overheating failure from misalignment ... **A11:** 507-508
bent, and fatigue fracture of steering knuckle ... **A11:** 342
by radial forging ... **A14:** 147
cast iron coatings for ... **M1:** 104
changes of diameter in ... **A11:** 468
chloride SCC in ... **A11:** 660
common stress raisers ... **A11:** 467-468
compressor, peeling-type fatigue cracking ... **A11:** 471
corrosion in ... **A11:** 467
cross-travel, service failure ... **A11:** 525
damaged during assembly ... **A11:** 475
defined ... **A11:** 459
distortion of ... **A11:** 467
effect on gears and gear trains ... **A11:** 590
exciter, fatigue failure of ... **A11:** 425
failed, examination of ... **A11:** 459-460, 463
failures of ... **A11:** 459-483, 524-525
fan, developing inspection criteria for ... **A11:** 107-108
fan, reversed-bending fatigue ... **A11:** 476
fatigue-crack growth behavior ... **A11:** 108
fillets, effects on bearing material failure ... **A11:** 506-507
forged, allowances and tolerances ... **A14:** 72
forged steel, ductile fracture ... **A11:** 481-482
fractured, macrograph of ... **A11:** 123
fuel-pump drive, wear failure ... **A11:** 465
journals on ... **A11:** 89
large steel, torsional-fatigue fracture ... **A11:** 464
main hoist, fatigue cracking ... **A11:** 525
materials for ... **A11:** 515
mechanical conditions of ... **A11:** 459-460
microscopic examination ... **A11:** 460
misalignment of ... **A11:** 475
misapplication of material in ... **A11:** 459
mixer paddle, SCC in ... **A11:** 403-404
motorcycle-transmission, high-cycle fatigue in ... **A11:** 106
plunger, fatigue fracture from sharp fillet ... **A11:** 319-320
repair by welding ... **A11:** 480-481
as roller-bearing raceway, flaking damage in ... **A11:** 502
as roller-bearing raceway, fretting damage ... **A11:** 498
rotating, with press-fitted elements ... **A11:** 470
rotor, brittle fracture from seams ... **A11:** 478
shoulder heights, effect on bearings ... **A11:** 506-507
size and mass of ... **A11:** 472
splined alloy steel, corrosion fatigue ... **A11:** 261
splined, fatigue fracture ... **A11:** 122-123
stationary ... **A11:** 459
steel coal pulverizer, fatigue failure ... **A11:** 468
steel crane, fatigue fracture ... **A11:** 524-525
steel cross-travel, fatigue fracture ... **A11:** 525
steel fan, fatigue fracture of ... **A11:** 476
steel main hoist, fatigue cracking ... **A11:** 525
steel pump, bending-fatigue fracture ... **A11:** 109
stepped, by cold extrusion ... **A14:** 305-306
stress concentrators in hot or cold forming ... **A11:** 459
stress systems acting on ... **A11:** 460-461
stub-, ductile fracture ... **A11:** 481-482
surface coatings ... **A11:** 480-482
Shafts (continued)
surface, schematic stress distribution ... **A11:** 123
tool steel, unidirectional-bending fatigue failure ... **A11:** 462
torsional stresses in ... **A11:** 115
types of contact and wear failures ... **A11:** 465-466
types of fatigue failures ... **A11:** 461-464
Shafts, failures of *See also* Shafts ... **A11:** 459-482
brittle fracture ... **A11:** 466
changes in shaft diameter ... **A11:** 468-472
common stress raisers ... **A11:** 467-468
contact fatigue ... **A11:** 465
distortion and corrosion ... **A11:** 467
ductile fracture ... **A11:** 466-467
examination of ... **A11:** 459-460
fabricating practices ... **A11:** 472-477
fatigue failures ... **A11:** 461-464
fracture origins ... **A11:** 459
metal fatigue as common cause of ... **A11:** 459
metallurgical factors ... **A11:** 477-480
stress systems in ... **A11:** 460-461
surface coatings, effects of ... **A11:** 480-482
wear ... **A11:** 465-466
Shake-and-bake method, of powder precursor preparation
high-temperature superconductors ... **A2:** 1086
Shake-down (of surface layers)
defined ... **A18:** 17
Shake-up lines ... **A18:** 446
Shake-up satellites ... **A18:** 447
in XPS analysis ... **A10:** 572
Shakeout *See also* Knockout
austenitic ductile irons ... **A15:** 7(10
defined ... **A15:** 10, 502
green sand molding ... **A15:** 347-348
high-chromium white irons ... **A15:** 683-684
high-silicon irons ... **A15:** 699
magnesium alloy castings ... **A15:** 806
nickel-chromium white irons ... **A15:** 680
operation ... **A15:** 502-503
rapid, for stress/distortion ... **A15:** 616
Shakeout, early
as casting defect ... **A11:** 386
Shakeout practice
effect on gray iron ... **A1:** 28
gray cast iron ... **M1:** 28
Shaker-hearth furnace
for carbonitriding ... **A7:** 454
Shaking
loose powder filling by ... **A7:** 431
screens ... **A7:** 177
Shale
ground by diamond wheels ... **A16:** 455
Miller numbers ... **A18:** 235
Shales
defined ... **A9:** 16
Shallow drawing
by Guerin process ... **A14:** 606
Shallow parts
by Guerin process ... **A14:** 607
Shank
defined ... **A14:** 11
Shank, shear in
rivet failures from ... **A11:** 544
Shanked ladles *See also* Ladles
development ... **A15:** 27
Shape *See also* Particle shape; Shaping
accuracy ... **A7:** 11
attainment ... **A7:** 295
discontinuities in, shafts ... **A11:** 467
distortion, defined ... **A11:** 136
effect, filament winding ... **EM2:** 372
effect on corrosion control ... **A13:** 339-340
effect, rotational molding ... **EM2:** 363-364
factor, defined ... **EM2:** 38
fundamentals, in rigid tool compaction ... **A7:** 322-328
hardening of steel related to ... **M1:** 481-482
incorrect, as casting defect ... **A11:** 386-387
making, with titanium powders ... **A7:** 749-750
multidimensional ... **A7:** 240-241
planar ... **A7:** 240
as process selection factor ... **EM2:** 288-292
and rigid toot compaction ... **A7:** 295, 322-328
RTM/SRIM, compared ... **EM2:** 348
stereology of ... **A7:** 238-240
stress-wave, effect on corrosion-fatigue ... **A11:** 255

Shape (continued)
terms and definitions ... **A7:** 239
Shape casting processes *See also* Continuous casting
grain refiners in ... **A15:** 478
types, classified ... **A15:** 204
vacuum induction ... **A15:** 399-401
Shape, compact
as material selection parameter ... **EM1:** 38-39
Shape distortion *See also* Springback
as formability problem ... **A8:** 548
and material properties ... **A8:** 552-553
Shape factor ... **EM3:** 26
Shape index used to express elongation ... **A9:** 128
Shape measurements
sheet metal forming ... **A8:** 548-549
Shape memory alloys (SMA) ... **A2:** 897-902
alloys having effect ... **A2:** 897
applications ... **A2:** 900-901
characterization methods ... **A2:** 898-899
commercial copper-base shape memory alloys ... **A2:** 899-900
commercial shape memory effect (SME) alloys ... **A2:** 899-900
crystallography ... **A2:** 898
defined ... **A2:** 897
future prospects ... **A2:** 901
general characteristics ... **A2:** 897
history ... **A2:** 897
nickel alloy ... **A2:** 433
nickel-titanium alloys ... **A2:** 899
thermomechanical behavior ... **A2:** 898
Shape memory effects and martensitic structures ... **A9:** 674
Shape resolution
defined ... **A9:** 16
Shape rolling ... **A14:** 346-350
of airfoil sections ... **A14:** 348-349
computer-aided roll pass design ... **A14:** 347-349
elongation, estimated ... **A14:** 347
finite-element modeling ... **A14:** 350
Shape(s) *See also* Shape casting processes
alloy extrudability ... **A2:** 35
aluminum casting alloys for ... **A15:** 744-745
casting, inspection effects ... **A17:** 529-530
classification ... **A2:** 34-35
of coils, eddy current inspection ... **A17:** 176-177
complex, radiographic inspection ... **A17:** 333-334
complex, ultrasonic inspection ... **A17:** 504
defined ... **A2:** 33
dendritic solid/liquid interface ... **A15:** 155-156
design of ... **A2:** 34-36
difficult-to-inspect, magnetic rubber inspection of ... **A17:** 124-125
echo, ultrasonic inspection ... **A17:** 245
factor, in feed metal availability ... **A15:** 582
of flaws ... **A17:** 50
of hysteresis loop, effects ... **A17:** 100
of inclusions ... **A15:** 91, 709
incorrect, as casting defect ... **A17:** 518-519
interface, effect, insoluble particles ... **A15:** 144
L-sections ... **A15:** 599, 604-610
metal casting to ... **A15:** 40
mold life effect ... **A15:** 281
and mold temperature, permanent molds ... **A15:** 282
object, NDE method selection by ... **A17:** 51
rolled, ultrasonic inspection ... **A17:** 270-272
simple, radiographic inspection ... **A17:** 332-333
and size, wrought aluminum alloy ... **A2:** 35
T-sections ... **A15:** 599-604, 607, 609
thermal, changing ... **A15:** 606-610
of titanium carbide particles, steel-bonded cermet ... **A2:** 997
wrought aluminum alloy ... **A2:** 33-34
X-sections ... **A15:** 599, 604, 606-607, 609-610
"Shaped charge" effects
liquid impact erosion ... **A18:** 224
Shaped ends
by multiple-slide forming ... **A14:** 569-570
Shaped tube electrolytic machining (STEM) ... **A16:** 509, 554-556
acid electrolytes ... **A16:** 554, 555, 556
advantages ... **A16:** 554
applications ... **A16:** 554
CNC machines ... **A16:** 555
compared to electrostream and capillary drilling ... **A16:** 551
equipment ... **A16:** 554
limitations ... **A16:** 554
power supply ... **A16:** 555
process capabilities ... **A16:** 554
process parameters ... **A16:** 556
tooling ... **A16:** 555-556
Shapers ... **A16:** 1, 187-188
for bar ... **A14:** 663
crank-driven horizontal ... **A16:** 187, 188
horizontal ... **A16:** 187-188, 193
hydraulic horizontal ... **A16:** 187
speeds by type ... **A16:** 189
vertical ... **A16:** 188, 190, 193
Shapes *See also* Complex shapes; Near-net shape; Net shape; Rolling; Shape rolling; Workpiece(s)
of blanks ... **A14:** 449-450
by deep drawing ... **A14:** 575
by three-roll forming ... **A14:** 616
complex, drawing of ... **A14:** 337
of cylindrical compression specimen ... **A14:** 381
extruded, characterized ... **A14:** 322
extrusion processes for ... **A14:** 304
forging, classified ... **A14:** 77-78
of forgings, closed-die forging ... **A14:** 75
forming ... **A14:** 622
heated-roll rolling ... **A14:** 357-358
for hot upset forging ... **A14:** 89
ingot, stainless steel ... **A14:** 222
internal, by swaging ... **A14:** 138-139
irregular, hot upset forging ... **A14:** 89, 549-552
large irregular, for press forming ... **A14:** 549-552
with locked-in metal ... **A14:** 549
measurements ... **A14:** 879
nesting, for gas cutting ... **A14:** 728
of open-die forgings ... **A14:** 61
press-brake bending ... **A14:** 540-541
rectangular, of superplastic metals ... **A14:** 863-865
of rolled ring ... **A14:** 110-111
semicircular, press-brake forming ... **A14:** 540
severely formed, press forming ... **A14:** 549
spin-forged aluminum alloy ... **A14:** 245
straightening of ... **A14:** 680-689
workpiece, for press forming ... **A14:** 549
Shapes, geometric
unified-shapes checker for ... **EL1:** 133
Shapes, hot rolled *See* Hot rolled bars
Shaping ... **A7:** 461 **A16:** 187-193
Al alloys ... **A16:** 778
carbon and alloy steels ... **A16:** 676
carbon and low-alloy steel gears ... **A16:** 347
cast irons ... **A16:** 659-660
and chip formation ... **A16:** 8
compared to broaching ... **A16:** 209
compared to sawing ... **A16:** 363-364
cutting fluids used ... **A16:** 125, 191
electrochemical machining ... **A16:** 527
external or internal contours ... **A16:** 192-193
flat surfaces ... **A16:** 192
form cutting ... **A16:** 193
and gear manufacture ... **A16:** 193, 330, 333-335, 339-340, 342-344
grooves, slots, and keyways ... **A16:** 193
hafnium ... **A16:** 856
heat-resistant alloys ... **A16:** 742-743
Mg alloys ... **A16:** 820, 821-822, 823
and milling ... **A16:** 329
Ni alloys ... **A16:** 837
process capabilities ... **A16:** 187
ram stroke and clearance ... **A16:** 190-191
shapers, and die threading ... **A16:** 297
shapers, and dimensional control ... **A16:** 192
Shaping (continued)
shapers, and form cutting ... **A16:** 193
shapers, high-speed tool steels used ... **A16:** 58
speed, feed, and depth of cut ... **A16:** 191
and spur gears ... **A16:** 338
surface roughness arithmetic average extremes ... **A18:** 340
tool life and slotting ... **A16:** 187, 190, 191
tool steels ... **A16:** 714
workholding devices ... **A16:** 188
and worm gears ... **A16:** 340
zirconium ... **A16:** 855
Shaping tool, for weaving
noncylindrical preforms ... **EM1:** 131
Sharks teeth
definition ... **EM4:** 633
Sharp corners
effect on tool and die failure ... **A11:** 564-566
spring failure at ... **A11:** 558-559
Sharp radii
in cold-formed parts ... **A11:** 307
Sharp-notch strength
defined ... **A8:** 12
Sharpening
defined ... **A17:** 385
Sharpness severity
fracture toughness and ... **A8:** 450
of initial crack ... **A8:** 452
in plane-strain fracture toughness ... **A8:** 450-451
Shatter crack *See* Flake
Shatter cracks *See* Flakes; Flaking
as hydrogen damage ... **A13:** 164
Shaver's disease
as aluminum poisoning ... **A7:** 206
Shaving
after blanking ... **A14:** 457-458
after piercing ... **A14:** 470
allowance ... **A14:** 457
die ... **A14:** 458
in progressive die ... **A14:** 457
punch and die materials for ... **A14:** 485
setups ... **A14:** 457-458
Shaw process *See also* Ceramic molding
all-ceramic molding casting ... **A15:** 249
burn-off, defined ... **A15:** 252
composite ceramic mold ... **A15:** 250
patterns ... **A15:** 248-249
and Unicast process, compared ... **A15:** 250-251
Shear *See also* Interlaminar shear strength; Shear bands; Shear deformation; Shear dimples; Shear failure; Shear fracture(s); Shear lips; Shear modulus; Shear strength; Shear stress ... **EM3:** 26
adiabatic ... **A12:** 31-33, 42-43
at elevated temperatures and low strain rates ... **A9:** 690
axial, as failure mode ... **EM1:** 198
bands, flow localization by ... **A14:** 385
blades ... **A8:** 68
of composites ... **EM1:** 98
crack, in compressed 4340 steel ... **A8:** 58
defined ... **A13:** 11 **A14:** 11-12 **EM1:** 21 **EM2:** 38
dies with, cutting force ... **A14:** 448
ductile tensile fracture ... **A11:** 82
edge ... **EM1:** 21
edgewise loaded ... **EM1:** 331-333
effect of crystal structure on ... **A9:** 684
effect, thermoplastics ... **EM1:** 100
elongated dimple formation ... **A12:** 13
elongated voids, irradiated stainless steels ... **A12:** 365
as fatigue modes ... **A12:** 12
final fast fracture, austenitic stainless steels ... **A12:** 353
fracture and failure ... **A8:** 12, 541-542, 548, 574
fractures, Mode II in-plane ... **A11:** 736-738
of gear tooth ... **A11:** 595
failures, in fasteners ... **A11:** 531
in-plane, ply ... **EM1:** 238

Shear (continued)
longitudinal, and damping **EM1:** 207
modulus **A8:** 12, 139, 724
on secondary slip systems, furrow-type fracture **A12:** 44, 54
overload failure by **A11:** 398-399
pin, in torsional impact machine **A8:** 217
plane, influence on shear strength **A8:** 64
plate, in torsional impact machine **A8:** 217
punch, and force, in piercing **A14:** 463
pure, delamination resistance in **EM1:** 264
resulting in deformation twins **A9:** 687
reverse direction, gear and pinions **A11:** 595
ridges, precipitation-hardening stainless steels **A12:** 371
sample **A8:** 155, 180-181
in shank, rivet failure by **A11:** 544
short beam, resin-matrix composites **A12:** 478
single crystals in, Kolsky bar testing **A8:** 219
step, in iron **A12:** 224
strength, defined **A13:** 11 **A14:** 12
stress, copper/copper alloys in seawater **A13:** 624
stress, defined **A14:** 12
and surface tilt, in formation of martensite plate **A9:** 668
test methods **EM1:** 299-300
testing **A14:** 888-889
types, on piercing punches **A14:** 464
void coalescence by **A14:** 393
yield strength, from torsion shear test **A8:** 64-65
zone, in low-carbon steel specimen **A8:** 229-230
Shear (or friction) factor **A18:** 60
Shear and torsional strength
of malleable irons **A1:** 82
Shear angles **A16:** 14
Shear band tearing
ductile fracture **A8:** 572
Shear bands *See also* Shear-deformation bands
in 70-30 brass **A9:** 686
adiabatic **A12:** 31-33, 42-43
along slip planes **A12:** 32
in cold rolled steel using torsional Kolsky bar **A8:** 222
crack during high-energy-rate forging **A8:** 573
defined **A9:** 16, 693 **A11:** 9
development of **A9:** 693
during deformation **A8:** 155
effect on high strain rate deformation **A8:** 45
in electrolytic magnesium, cold rolled 50% **A9:** 687
formation and fracture, titanium alloys **A12:** 444-445
fractures **A8:** 574
in high-energy-rate forming **A8:** 170
in iron **A9:** 686
in isothermal forging **A8:** 170
as nucleation sites **A9:** 694
orientation **A12:** 445
thermal **M6:** 707-708
in thin-walled tubular specimen **A8:** 222
in torsion and flow localization **A8:** 170
torsional, austenitic stainless steels **A12:** 359
Shear blades *See also* Cutting; Shearing; Shears
for bar **A14:** 716-717
blade life **A14:** 717
conforming **A14:** 717
design and production **A14:** 717-718
materials, hot shearing **A14:** 716
profile **A14:** 716-717
Shear compaction **A7:** 56
in polymers **A7:** 606
Shear compliance, defined *See* Compliance
Shear cracks
as forging flaws **A17:** 494
Shear cutting
and gear manufacture **A16:** 330, 333, 334
and internal gears **A16:** 339
and spur gears **A16:** 338
Shear damping
in unidirectional composites **EM1:** 207-208
Shear deformation
instability effect **EM1:** 447
low-carbon steel **A12:** 244
mechanical damage during, tool steels **A12:** 380
ridge formation during, tool steels **A12:** 380
Shear dimples
maraging steels **A12:** 383
martensitic stainless steels **A12:** 369
titanium alloys **A12:** 455
Shear edge **EM3:** 26
defined **EM2:** 38
Shear failure
continuous fiber composites **EM1:** 786, 789-790
discontinuous fiber composites **EM1:** 797
Shear fatigue, solder
analytical model of **EL1:** 6
Shear flow stress **A8:** 577
calculated for plate impact test **A8:** 236
Shear fracture *See also* Shear stress **EM3:** 26
defined **A11:** 9 **EM2:** 38
as fatigue fracture plane-stress mode **A11:** 105
of Kovar-glass seals **A10:** 577-578
and material properties **A14:** 881
stainless steel implant screw with **A11:** 676, 682
Shear fractures
AISI/SAE alloy steels **A12:** 341
effect of striation **A12:** 16
elongated dimples **A12:** 12, 15-16
maraging steels **A12:** 383
martensitic stainless steels **A12:** 367
precipitation-hardening stainless steels **A12:** 371
Shear front-lamella structure **A16:** 8
Shear hackle
definition **EM4:** 633
Shear lag **EM1:** 33, 243-244 **EM3:** 478
Shear ledges *See* Radial marks
Shear lip
crack origin determined from **A11:** 83
defined **A8:** 12 **A11:** 9
development, *J*-controlled fracture testing **A8:** 449
as indication of toughness and failure mode **A11:** 396
in quench crack, round steel bar **A11:** 94
of slant fracture **A11:** 20
zone, shear dimples in **A11:** 76
Shear lips
alloy steels **A12:** 294, 302, 306-307, 311, 314-315, 318, 332-334
ductile, wrought aluminum alloys **A12:** 414
low-carbon steel **A12:** 244
maraging steels **A12:** 383
martensitic stainless steels **A12:** 366, 369
medium-carbon steels **A12:** 253
on crater **A12:** 318
precipitation-hardening stainless steels **A12:** 370
toot steels **A12:** 379
wrought aluminum alloys **A12:** 414, 425
Shear load transfer
through adhesively bonded joints **EM1:** 480-481
through mechanical fasteners **EM1:** 488
Shear mixing
short fiber effect **EM1:** 120
statistical analysis **A7:** 187
Shear modulus *See also* Modulus of rigidity; Modulus of rigidity, Torsional modulus, and Elastic properties **EM3:** 26, 319-320, 321
aluminum oxide-containing cermets **A7:** 804
defined **A11:** 9 **EM1:** 21 **EM2:** 38
glass fiber reinforced epoxy resin **EM1:** 406
Kevlar 49 fiber/fabric reinforced epoxy resin **EM1:** 408
medium-temperature thermoset matrix composites **EM1:** 384
Shear plane **A16:** 7, 8, 10, 11
Shear properties *See also* Shear strength **A1:** 61
wrought aluminum and aluminum alloys **A2:** 92-94
Shear rate
defined **EM2:** 38
Shear rates **EM3:** 26
adhesive-backed tape application by hand **EM3:** 40
Shear reactions
observation with a hot-stage microscope **A9:** 82
Shear spinning *See also* Power spinning; Spinning **A14:** 600-604
refractory metals and alloys **A2:** 562
Shear spinning tools
mandrels **M3:** 500, 501
rollers **M3:** 500, 501
Shear spinning tools (continued)
surface finish **M3:** 501
Shear stability *See also* Penetration (of a grease)
defined **A18:** 17
Shear storage modulus
resin systems **EM2:** 274
Shear strain **EM3:** 26
control, in low-cycle torsional fatigue **A8:** 150
defined **A8:** 12 **A11:** 9 **EM1:** 21 **EM2:** 38
and effective stress and strain, in torsional loading **A8:** 142-143
from torsion **A8:** 327
rate, for plate impact test **A8:** 236
rate in torsion **A8:** 158
symbols for **A8:** 726
vs. cyclic and monotonic shear stress in steel **A8:** 151
vs. life, in steel **A8:** 151
vs. shear stress curves, torsion testing **A8:** 141
Shear strain rate **A6:** 997
Shear strain rate within lubricant **A18:** 60, 63
Shear strains
calculated **A17:** 449
Shear strength *See also* Mechanical properties; Shear; Shear properties; Shear yield strength; Shearing; Ultimate shear strength **A16:** 4 **EM3:** 26, 395
aluminum casting alloys **A2:** 153-177
aluminum oxide-containing cermets **A7:** 804
defined **A8:** 12 **A11:** 9 **EM1:** 21 **EM2:** 38
for
ductile iron **M1:** 38, 41
epoxy resin matrices **EM1:** 73
of magnesium alloys **A2:** 461
single-shear tests **A8:** 63-64
spot welds, in magnesium alloys **A2:** 477
variables affecting **A8:** 68
wrought aluminum and aluminum alloys **A2:** 77, 81, 96, 98, 101
Shear strength tests
butyl tapes **EM3:** 202
Shear stress *See also* Shear **A16:** 11, 14-16 **A18:** 60, 610 **EL1:** 445-446 **EM3:** 26, 395-402, 491-493
adhesive **EM1:** 683
alternating **A11:** 465
at specimen-anvil interface, remelted 4340 steel **A8:** 235-236
constant maximum **A8:** 72
cyclic and monotonic, vs. shear strain **A8:** 151
defined **A8:** 12 **A11:** 9 **EM1:** 21 **EM2:** 38
derivations for arbitrary flow laws **A8:** 182-184
differential testing **A8:** 182-183
distribution during noncylindrical bending **A8:** 119, 121
and effective stress and strain, in torsional loading **A8:** 142-143
and free-surface transverse velocity vs. time profiles and stress-strain curves **A8:** 235-236
from dynamic torsion test **A8:** 216-217
interlaminar **EM3:** 402
and mean axial stress dependence on strain fixed-end high-temperature torsion tests **A8:** 157
multiple test piece method **A8:** 182-184
in pure rolling **A11:** 133-134
in shafts **A11:** 460-461
subsurface, by cylindrical roller **A11:** 500-501
surface, converted to torque **A8:** 142
symbol for **A8:** 726
torque and angle of elastic unloading **A8:** 183-184
in torsional hydraulic actuator **A8:** 217
torsional Kolsky bar **A8:** 228
transverse, particle velocity for flyer, anvil specimen **A8:** 232
vs. shear strain curves, torsion testing **A8:** 141
Shear stress, and viscosity
semisolid alloys **A15:** 328
Shear Stress Hypothesis **A18:** 476
Shear stress-strain
curve, for copper **A8:** 231
curve, for OFHC copper **A8:** 216-217
reduction to effective stress and strain **A8:** 161
Shear stress-strain curve
static and dynamic, for hot rolled steel **A8:** 225
Shear testing **A8:** 62-68, 187 **A14:** 888-889
application **A8:** 65-66

Shear testing (continued)
blanking shear, mill products **A8:** 64
data, applications **A8:** 62
double-notch **A8:** 228-229
double-shear, fasteners **A8:** 66
double-shear tests, mill products **A8:** 62-63
of fasteners **A8:** 65-68
high strain rate **A8:** 215-239
lap-joint shear, fasteners **A8:** 67
machines **A8:** 67-68, 215
Marciniak in-plane sheet torsion test **A8:** 559
of mill products **A8:** 62-65
Miyauchi shear test **A8:** 559-560
procedures, fasteners **A8:** 68
report **A8:** 65
of sheet metals **A8:** 559-560
single-shear, fasteners **A8:** 66-67
single-shear, mill products **A8:** 63-64
torsion-shear, mill products **A8:** 64-65
Shear tests **EM3:** 387-388, 389, 502-503
Shear thickening *See also* Shear thinning
defined **A18:** 17
Shear thinning *See also* Shear thickening
defined **A18:** 17
of thermoplastics **EM1:** 102
Shear ultimate strength **A8:** 662-663, 667, 672
Shear velocity **A18:** 610, 611
Shear wave **A8:** 231-232
Shear wave ultrasonic inspection **A17:** 233, 506
Shear yield strength *See also* Mechanical properties; Shear strength; Shearing; Yield; Yield strength **A18:** 34, 610
wrought aluminum and aluminum alloys **A2:** 85, 90, 97
Shear zone **A16:** 7-8
Shear-area transition temperature
symbol for **A11:** 797
Shear-deformation bands
defined **A11:** 9
in polymers **A11:** 760
width as function of crack length, polyethylene pipe **A11:** 762
Shear-face tensile fractures
by plane strain **A11:** 75
by plane stress **A11:** 75
in ductile materials **A11:** 76
Shear-out, edge
as fastened sheet failure **A11:** 531
Shearing *See also* Blanking; Cutting; Shear blades; Shears
as a sectioning method **A9:** 23
accuracy **A14:** 714
aluminum alloy forgings **A14:** 247
aluminum and aluminum alloys **A2:** 10
of angles, blades for **A14:** 718
of bars **A14:** 714-719
for billet separation, precision forging **A14:** 162
of carbon and alloy steels **A9:** 165
closed-die steel forgings **M1:** 367-368
of coiled sheet and strip **A14:** 708-713
of copper and copper alloys **A14:** 256-257
defined **A14:** 12
of flat sheet **A14:** 701-707
in header **A14:** 311
hot **A14:** 716
in hot upset forging **A14:** 83
machines **A14:** 714-716, 718-719
in metallic bond formation **A7:** 303
of nickel-base alloys **A14:** 520, 832
of nickel-titanium shape memory effect (SME) alloys **A2:** 899
of organic-coated steels **A14:** 565
of plate **A14:** 701-707
in press brake **A14:** 538
and punching machines **A14:** 716
refractory metals and alloys **A2:** 560
rotary **A14:** 705-707
safety **A14:** 707
sheet, process modeling of **A14:** 913-915

Shearing (continued)
straight-knife **A14:** 701-705
of titanium alloys **A14:** 840-841
tungsten **A2:** 562
Shearing and slitting tools
cold shearing, blade materials **M3:** 478-479
hardness, effect on wear **M3:** 479
hot shearing **M3:** 479-480, 481
machine knives **M3:** 481-483
rotary slitting, blade materials **M3:** 478-479
shear-blades, service data **M3:** 480, 481
slitting tools **M3:** 481
Shearing lines *See* Cut-to-length lines
Shearing machines **A14:** 714-716, 718-719
Shearography **A6:** 1150
Shears *See also* Cutting; Shear blades; Shearing
alligator, for bar **A14:** 714-715
guillotine, for bar **A14:** 715
rotary drum **A14:** 711-712
for sheet or plate **A14:** 701-702
Sheath **A7:** 11
Sheath, sliver-lead alloy cable
creep-rate/time curves **A8:** 321, 323
Sheathed heaters
of electrical resistance alloys **A2:** 831
Sheave grooves
steel wire ropes **A11:** 516
Shed
in textile looms **EM1:** 127-128
Shedding
aramid composites **EM1:** 667
Sheet
aluminum and aluminum alloys **A2:** 5, 10, 33, 53
aluminum-lithium alloys, fatigue of **A2:** 195
brazing of aluminum alloys **M6:** 1023-1024
cast, acrylic **EM2:** 103
electrical steel, properties **A2:** 769
embossed, tempering **A2:** 26
flash welding **M6:** 558
formability, of magnesium alloys **A2:** 467-468
high molecular weight **EM2:** 165
high-impact polystyrenes (PS, HIPS) **EM2:** 195
laser beam welding **M6:** 663-664
lead, applications **A2:** 551-552
long terne steel **A2:** 554-555
materials, for compression molding **EM1:** 559-560
mechanically alloyed oxide alloys **A2:** 949
dispersion-strengthened (MA ODS)
molybdenum, cutting **A2:** 561
niobium and tantalum, forming **A2:** 562
overlay, defined *See* Overlay sheet, defined
oxyfuel gas welding **M6:** 589, 592
pewter, mechanical properties **A2:** 523
plastic, thermoforming **EM2:** 399-403
polyamides (PA) **EM2:** 125
polycarbonates (PC) **EM2:** 151
polyphenylene sulfides (PPS) **EM2:** 190
polysulfones (PSU) **EM2:** 202
reinforced polypropylenes (PP) **EM2:** 193
resistance spot welding **M6:** 469, 486, 491
spunlaced, aramid staple fiber **EM1:** 115
stainless steel *See* Stainless steel, sheet
steel *See* Steel, sheet
stop-off materials, hard chromium plating **M5:** 187
take-up, for joining **EM1:** 710-711
titanium alloy, superplastic forming **A2:** 590-591
for tube rolling **EM1:** 573
twin-sheet forming, size and shape effects **EM2:** 290
unalloyed uranium **A2:** 671
wrought aluminum alloy **A2:** 33, 53
wrought copper **A2:** 241-248
of wrought magnesium alloys **A2:** 459-460, 467-468, 484-490
wrought titanium alloys **A2:** 610-611
Sheet and plate
magnesium **M2:** 529, 541-543, 544

Sheet and strip
steel, annealing **M4:** 21-23
steel, normalizing **M4:** 12-13
Sheet bar can **A7:** 424, 425
Sheet bending *See also* Sheet
finite-method analysis **A14:** 921-922
process modeling of **A14:** 913-914
Sheet, defined *See also* forming; Sheet bending; Sheet forming; Sheet metal; Sheet metal(s); Sheet rolling; Sheet shearing; Sheet stretching **A14:** 12, 343
Sheet extrusion *See also* Extrusion
acrylonitrile-butadiene-styrenes (ABS) **EM2:** 113-114
products **EM2:** 384-385
Sheet finishes **A1:** 885-886
Sheet formability of steel *See also* Formability **A1:** 573-580, 888-889
circle grid analysis **A1:** 575-576
correlation between microstructure and formability **A1:** 578-579
grain shape **A1:** 579
grain size **A1:** 578
microconstituents **A1:** 579
effect of metallic coatings on formability **A1:** 579-580
effects of steel composition on
aluminum **A1:** 577
carbon **A1:** 576
cerium **A1:** 577
chromium, nickel, molybdenum, and vanadium **A1:** 577
copper **A1:** 577
formability **A1:** 576-577
manganese **A1:** 576
niobium **A1:** 577
nitrogen **A1:** 577
oxygen **A1:** 577
phosphorus and sulfur **A1:** 577
silicon **A1:** 577
titanium **A1:** 577
effects of steelmaking practices on
aluminum-killed steels **A1:** 578
cold-rolled steel **A1:** 577-578
formability **A1:** 577-578
hot-rolled steel **A1:** 577
interstitial-free steel **A1:** 578
rimmed steels **A1:** 578
surface finish **A1:** 578
mechanical properties and formability **A1:** 573-575
planar anisotropy **A1:** 575
plastic strain ratio **A1:** 575
strain-hardening exponent **A1:** 575
total elongation **A1:** 573-574
uniform elongation **A1:** 574
yield point elongation **A1:** 574-575
yield strength **A1:** 573
selection of steel sheet **A1:** 580
simulative forming tests **A1:** 576
of stainless steel **A1:** 888-889
Sheet formability testing **A8:** 547-570
Sheet forming *See also* Bulk forming; Sheet
CAD/CAM applications in **A14:** 903-910
defined **A14:** 12, 15
new processes for **A14:** 20
process modeling and simulation **A14:** 911-927
processes, classified **A14:** 16
Sheet liner cladding **M6:** 804
Sheet martensite **A9:** 672
Sheet materials *See also* Heat exchangers, failures o
fastened, types of failures in **A11:** 531
flat-face fractures **A11:** 109-110
heat exchangers, application **A11:** 628
high-strength, tests for **A11:** 61
thin, buckling failure in riveted **A11:** 544
thin, final fracture in **A11:** 105
use of *R*-curve with **A11:** 64

Sheet metal *See also* Aluminum sheet; Sheet formability testing; Sheet metal forming; Sheet steel, coated; Sheet steel specific types; Tin plate
aluminum alloy, effects of lubricants and cleaners on bearing strength **A8:** 60
annealed, computation procedures for design allowables **A8:** 672-677
austenitic stainless steel, hardness conversion tables **A8:** 109
automotive industry application **EM1:** 834
bend testing **A8:** 117, 145, 547
bending test specimens **A8:** 126
beryllium.. **A7:** 509, 759
biaxial stretch testing of **A8:** 558-559
circle arc elongation test **A8:** 556
containment, for part shape in HIP processing **A7:** 425, 427
deformed, strained state **A8:** 549
eddy current inspection **A17:** 187
encapsulation **A7:** 427, 428
envelope hot pressing of beryllium **A7:** 509
fully reversed loading in ultrasonic testing **A8:** 242
Knoop minimum thickness chart for **A8:** 96, 101
lubricants for **A8:** 567-568
major strain/minor strain combinations **A8:** 549
maximum strain levels **A8:** 551
minimum thickness for Scleroscope hardness testing **A8:** 105
Modul-*r* test for **A8:** 557
mounting of specimens **A9:** 198
orientation, in bending **A8:** 547
plane-strain tensile testing of **A8:** 557-558
powder cans **A7:** 431
primary testing direction, various alloys **A8:** 667
product fabrication, high-energy-rate compacting in **A7:** 305
properties of **A8:** 549-553
quasi-static torsional testing of **A8:** 145
radiographic methods **A17:** 296
rapid-*n* test **A8:** 556-667
roll compacting **A7:** 297
rolling **A8:** 158
shear data standards for **A8:** 62
shear testing of **A8:** 559-560
single-shear slotted, for shear testing **A8:** 64
specimen, constant-load testing **A8:** 314
specimens for pin bearing testing **A8:** 59
stampings, laser triangulation sensors for **A17:** 13
as tantalum mill product **A7:** 770-771
tensile test specimen for **A8:** 553
thin, buckling in axial compression **A8:** 56
thin, soft, Rockwell scale for **A8:** 76
typical tensile properties **A8:** 555
uniaxial tensile testing of **A8:** 553-557
wide, bending of **A8:** 552
Sheet metal forming
biaxial stretch testing **A8:** 558
circle grid analysis **A8:** 566-677
combined types **A8:** 548
deformation measurement in **A8:** 548-549
drawbead forces **A8:** 547, 567
effect of material properties in **A8:** 549-553
effect of temperature **A8:** 553
formability problems **A8:** 548
formability tests **A8:** 553
forming limit diagrams **A8:** 566
hardness testing **A8:** 560
high- and low-temperature **A8:** 553
intrinsic tests **A8:** 553-560
lubricant selection and use **A14:** 512-520
lubricants **A8:** 567-568
materials, lubricants for **A14:** 518-520
plane-strain tensile testing **A8:** 557-558
shear testing **A8:** 559-560
simulative tests **A8:** 553, 560-566
as tension source for stress-corrosion cracking **A8:** 502
testing in **A8:** 547-570
types of **A8:** 547-548
uniaxial tensile testing **A8:** 550-557
Sheet metal materials
determining pole figures in **A10:** 360
diffraction in **A10:** 360
Sheet metal materials (continued)
effects of steel surface carbons on paint adherence **A10:** 224
texture of **A10:** 363
Sheet metal, molds
fabricated **EM2:** 367
Sheet metal(s) *See also* Sheet; Superplastic sheet forming
bending of **A14:** 835-836, 877
beryllium, formability **A14:** 805
for blanking **A14:** 449
coiled, slitting and shearing of **A14:** 708-713
copper and copper alloy formability **A14:** 809-811
corrugated **A14:** 541
cut-to-length lines (shearing lines) **A14:** 711-713
deep drawing of **A14:** 575-590
defined **A14:** 12, 343
electrical steel, blanking and piercing **A14:** 476-482
explosive forming of **A14:** 640-641
flat, shearing of **A14:** 701-707
flatteners and levelers for **A14:** 713
flow strength **A14:** 576
formability testing of **A14:** 877-899
forming **A14:** 512-520
forming, presses and auxiliary equipment for **A14:** 489-503
heated-roll rolling **A14:** 356-357
magnesium alloy **A14:** 825-826
nickel-base alloys **A14:** 835-836
preforms, explosive forming of **A14:** 642-643
refractory, compositions **A14:** 785
refractory, forming **A14:** 787-788
thickness, for drop hammer forming **A14:** 656
thickness, for press forming **A14:** 504
thickness, in deep drawing **A14:** 584
Sheet metals **A6:** 398-400
shear bands in **A9:** 693
Sheet molding compounds (SMC) *See also* Molding compounds..... **EM1:** 157-160 **EM3:** 97, 294
application, automotive industry **EM1:** 832-835
bonding parts to metals **EM3:** 255
catalysts **EM1:** 157-158
compaction **EM1:** 160
in compression molding/stamping **EM2:** 324-333
costs, compared **EM1:** 170
defined **EM1:** 21 **EM2:** 38
designing with **EM2:** 327-331
discontinuous fiber reinforced fibers for **EM1:** 33
fillers **EM1:** 158
flame retardants **EM1:** 158
glass roving production for **EM1:** 109
high-strength, properties effects **EM2:** 286
as lower performance prepreg res-in application **EM1:** 141-142
machines **EM1:** 159-160
material components **EM1:** 157-158
maturation room environments **EM1:** 160
mixing techniques, for resin pastes **EM1:** 159
output and feed requirements **EM1:** 160
paste metering **EM1:** 159-160
physical properties **EM1:** 158
pigments **EM1:** 158
prepreg resins **EM1:** 141-142
process, urethane hybrids **EM2:** 269
processing machine **EM1:** 157
properties, compared **EM2:** 325
properties effects **EM2:** 286
release agents **EM1:** 158
short fiber for **EM1:** 121
size and shape effects **EM2:** 291
suppliers **EM1:** 142
surface finish **EM2:** 303
take-up **EM1:** 160
thermoplastic polymers **EM1:** 158
thickeners **EM1:** 158
ultraviolet (UV) absorbers **EM1:** 158
vs steels, costs compared **EM2:** 334
Sheet, normal direction
abbreviation for **A10:** 690
Sheet resistivity **EL1:** 89, 1156
Sheet rolling
torsion testing **A14:** 373
Sheet separation
definition **M6:** 16
Sheet separation (resistance welding)
definition **A6:** 1213
Sheet shearing
process modeling of **A14:** 913-915
Sheet steel *See* Steel sheet
annealed, Lüders bands **A9:** 693
chromized, color etched **A9:** 156
coated **A9:** 197-201
tint etched to color ferrite grains **A9:** 180
Sheet steel, specific types
1006, chromized **A9:** 200
1006, electrogalvanized **A9:** 199
1006, hot-dip galvanized **A9:** 199
1008, hot-dip galvanized **A9:** 199
1008, tin-plated **A9:** 200
1008, with type 1 hot-dip aluminum coating **A9:** 200
1008, with type 2 hot-dip aluminum coating **A9:** 200
1010, Galvalume coated **A9:** 199
Sheet steels *See also* Carbon steel sheet and strip
aluminum-coated **A13:** 527
cold-rolled HSLA steel **A1:** 420
copper-bearing, failure time **A13:** 519
crevice corrosion **A13:** 10
forming properties of various types **A1:** 398, 418
frits for **A13:** 446
HSLA sheet steels, tensile properties of **A1:** 411
interstitial-free steels **A1:** 112-113, 131-132, 405
composition **A1:** 417
deep-drawing properties of **A1:** 398
production of **A1:** 112-113, 131-132, 578
linings **A13:** 400
mechanical properties and formability *See also* Sheet formability of steel
planar anisotropy **A1:** 575
plastic-strain ratio **A1:** 575
strain-hardening exponent **A1:** 575
total elongation **A1:** 573-574
uniform elongation **A1:** 574
yield point elongation **A1:** 574-575
yield strength **A1:** 573
precoated *See* Precoated steel sheet
selection of **A1:** 580
simulative forming tests **A1:** 576
stainless steel **A1:** 888-889
Sheet stretching
finite-element analysis of **A14:** 922-923
Sheet surfaces
mounting with clamps **A9:** 28
Sheet take-up
in blind fastening **EM1:** 710-711
Sheet texture in crystalline lattices **A9:** 701
pole-figure technique used to examine **A9:** 702-706
Sheet, zinc
galvanized **M2:** 651, 653
Sheet-metal bearings **A18:** 741
Sheet-metal forming
tool steels **A18:** 737-378
Sheet-polishing mills **M5:** 124
Sheeting
defined **EM2:** 38
Sheetmaking
porous parts for **A7:** 698
Sheets
examination by electropolishing **A9:** 55
Shelf, defined
as ingot flaw **A17:** 492
Shelf level
of packaging **EL1:** 13
Shelf life *See also* Storage life **EM3:** 26
age-life history **EM3:** 735-736
compromised when kits are used **EM3:** 688
defined **EM1:** 21 **EM2:** 38
Shelf roughness
defined **A8:** 12
Shelf-life identification tags **EM3:** 685
Shell
cupolas **A15:** 384
electric arc furnaces **A15:** 358
and head cracking, gray cast iron rolls **A11:** 653-654
liner, rapid wear by severe abrasion **A11:** 375
pressure vessel, failures of **A11:** 644

Shell (continued)
and stainless steel liner, weld cracking in ... **A11:** 655-656
Shell char
methods used for synthesis ... **A18:** 802
Shell cracks
high-carbon steel ... **A12:** 288
Shell flour
as extender ... **EM3:** 176
Shell mold casting
aluminum casting alloys ... **A2:** 140
Shell mold casting, aluminum alloys
See also Castings; Foundry products ... **M2:** 146
Shell molding *See also* Alpha process; Croning process; Dip coating
Alnico alloys ... **A15:** 736
ceramic, for investment casting ... **A15:** 257-261
as conventional process ... **A15:** 37
as coremaking system ... **A15:** 238
defined ... **A15:** 10
novolac shell-molding binders ... **A15:** 217
producing cores and molds ... **A15:** 217-218
as resin binder system ... **A15:** 217
tolerances ... **A15:** 622
zirconium alloys ... **A15:** 837
Shell panel
instability of ... **EM1:** 448-449
Shell process *See* Croning process; Shell molding
Shell reamers ... **A16:** 239, 241, 243-244
Shell tooling ... **EM3:** 26
defined ... **EM1:** 21
Shell-cast bearings ... **A18:** 741
Shelling *See* Spalling
defined ... **A9:** 16 **A18:** 17
Shells
boxlike, drawing of ... **A14:** 585
reducing drawn ... **A14:** 586
structural analysis of ... **EM1:** 461
Shells, electron
defined ... **EL1:** 90
Shellvest system
as special investment casting process ... **A15:** 266-267
Shepherd fracture grain size technique
austenite grain size evaluated by ... **A11:** 563
Shepherd fracture grain size technique used to examine tool steels ... **A9:** 258
Shepherd P-F test
for toot steels ... **A12:** 141, 162
Sherritt Gordon process ... **A7:** 54, 134, 138-142
ammonia leach process ... **A7:** 139
cobalt refining process ... **A7:** 144
of rolling nickel powders ... **A7:** 401
soluble cobaltic pentammine process ... **A7:** 144-145
Shetty's mixed mode equation ... **EM4:** 701, 706
Shewart variable control charts ... **EM3:** 793, 795, 796
Shewhart control chart model
attribute data ... **A17:** 734
operational definitions ... **A17:** 734
for quality control/design ... **A17:** 719, 725-728
Shewhart control charts ... **EM4:** 85
Shielded carbon arc welding
definition ... **M6:** 16
Shielded metal arc (SMA) weld deposition ... **A18:** 653
Shielded metal arc cutting
definition ... **M6:** 16
Shielded metal arc cutting (SMAC)
definition ... **A6:** 1213
Shielded metal arc welding ... **A7:** 456 **M6:** 75-95
accessibility for electrodes ... **M6:** 87
alternating current ... **M6:** 79
arc blow ... **M6:** 78-79, 87-88
arc starting ... **M6:** 78-79
capabilities ... **M6:** 75
of cast irons ... **A15:** 523-524
comparison to flux cored arc welding ... **M6:** 107-108, 112-113

Shielded metal arc welding (continued)
comparison to other processes ... **M6:** 95
cost ... **M6:** 95
definition ... **M6:** 16
direct current ... **M6:** 78-79
distortion ... **M6:** 91-92
thick sections ... **M6:** 92
thin sections ... **M6:** 92
electrode classification ... **M6:** 81-82
electrode coverings ... **M6:** 76
electrode deposition rates ... **M6:** 84-85
electrode holders ... **M6:** 79
electrode orientation and manipulation ... **M6:** 87
electrodes ... **M6:** 78-79, 81-85
coverings ... **M6:** 81-82
fillet welds ... **M6:** 76, 85, 88-89
fixtures ... **M6:** 79-81
groove welds ... **M6:** 76, 85, 89-90
ground clamps ... **M6:** 79
hardfacing ... **M6:** 783-784
heat flow calculations ... **M6:** 31
induction brazing as replacement ... **M6:** 974-975
jigs ... **M6:** 79-81
joint quality ... **M6:** 75
limitations ... **M6:** 75
metals welded ... **M6:** 75
moisture in electrode coverings ... **M6:** 82-83
polarity ... **M6:** 78-79
positioners ... **M6:** 79-81
positions for welding ... **M6:** 76
power supplies ... **M6:** 76-78
constant-current output ... **M6:** 77
cost ... **M6:** 77-78
deposition rate ... **M6:** 78
melting rate ... **M6:** 78
motor-generator units ... **M6:** 77-78
selection factors ... **M6:** 77
transformer-rectifier units ... **M6:** 77-78
transformers ... **M6:** 77-78
prevention of weld defects ... **M6:** 92-94
rating of weldability ... **M6:** 94-95
recommended grooves ... **M6:** 69-71
safety ... **M6:** 58, 95
sections of unequal thickness ... **M6:** 91
selection of electrode class ... **M6:** 83-84
cost ... **M6:** 84
material composition ... **M6:** 83-84
mechanical properties ... **M6:** 83
position of welding ... **M6:** 84
quality ... **M6:** 84
selection of electrode size ... **M6:** 84-85
sequencing ... **M6:** 92
sheet metal welding ... **M6:** 78-79
surface condition, effect on weld quality ... **M6:** 95
thick sections ... **M6:** 79, 90-91
thin sections ... **M6:** 90
electrodes ... **M6:** 90
fit-up ... **M6:** 90
tack welds ... **M6:** 90
welding speed and current ... **M6:** 90
underwater welding ... **M6:** 922
weld defects ... **M6:** 92-94, 830
arc strikes ... **M6:** 93
cracking ... **M6:** 92-93
gaps from incomplete fusion ... **M6:** 92-93
microfissuring ... **M6:** 93
overlapping ... **M6:** 93-94
oxidation ... **M6:** 93
porosity ... **M6:** 92
sink or concavity ... **M6:** 93
slag inclusions ... **M6:** 92
undercuts ... **M6:** 92-93
wagon tracks ... **M6:** 92
weld craters ... **M6:** 93
weld spatter ... **M6:** 94
weld overlaying ... **M6:** 807
weld properties ... **M6:** 84-85
weld spatter ... **M6:** 79

Shielded metal arc welding (continued)
welding arc ... **M6:** 86-87
arc length ... **M6:** 86
control to reduce porosity ... **M6:** 86
striking, maintaining, and breaking ... **M6:** 86-87
welding procedures ... **M6:** 88
welding speed ... **M6:** 85-86
Shielded metal arc welding (SMAW) ... **A6:** 175-179
advantages ... **A6:** 175
all-weld-metal chemical compositions for martensitic stainless steel filler metals ... **A6:** 439
aluminum alloys ... **A6:** 738
aluminum bronzes ... **A6:** 754, 765-766
applications ... **A6:** 175, 176
sheet metals ... **A6:** 398
shipbuilding ... **A6:** 384
austenitic stainless steels ... **A6:** 461, 1018
base-metal thicknesses ... **A6:** 175
carbon steels ... **A6:** 651, 652, 653, 654, 656-657
cast irons ... **A6:** 716-718
cast iron electrodes ... **A6:** 717
composition of electrodes ... **A6:** 717
copper-base electrodes ... **A6:** 718
nickel-base electrodes ... **A6:** 717-718
stainless steel electrodes ... **A6:** 717
steel electrodes ... **A6:** 717
copper ... **A6:** 760
copper alloys ... **A6:** 752, 755, 762, 763, 765-766, 768-769
copper-nickel alloys ... **A6:** 768-769
cryogenic service ... **A6:** 1017
definition ... **A6:** 175, 1213
deposition rate ... **A6:** 177, 178, 179
discontinuities, types ... **A17:** 582
dissimilar metal joining ... **A6:** 824, 827, 828
duplex stainless steels ... **A6:** 476, 477, 480
electrode amperage ranges ... **A6:** 179
electrode choice determining hydrogen pickup sources ... **A6:** 1069
electrode holder size and capacity ... **A6:** 176
electrode holders ... **A6:** 176
electrodes ... **A6:** 176-177
for carbon steels ... **A6:** 654, 656-657
ferritic stainless steels ... **A6:** 446, 447
low-alloy steel covered and their suffix symbols and compositions ... **A6:** 178
mild and low-alloy steels ... **A6:** 57
for nickel alloys ... **A6:** 746
for stainless steels ... **A6:** 699, 700, 701
equipment ... **A6:** 175-176
features of process ... **A6:** 175
ferritic stainless steels ... **A6:** 446-447, 450, 451
filler metals ... **A6:** 447
filter lens shades recommended for use ... **A6:** 179
firecracker welding ... **A6:** 178
fluxes ... **A6:** 60, 61, 62
gravity welding ... **A6:** 178
hardfacing alloy consumable form ... **A6:** 796
hardfacing alloys ... **A6:** 797, 799, 800, 801-802, 803
heat input ... **A6:** 427
heat sources ... **A6:** 1144
heat-treatable low-alloy steels ... **A6:** 669-670
high-strength low-alloy quench and tempered structural steels ... **A6:** 666
high-strength low-alloy steels ... **A6:** 663-664
high-strength low-alloy structural steels ... **A6:** 663-664
limitations ... **A6:** 175
low-alloy metals for pressure vessels and piping ... **A6:** 668
low-alloy steels ... **A6:** 662, 663-664, 666, 668, 669-670, 674, 676
low-alloy steels for pressure vessels and piping ... **A6:** 667
low-carbon steels ... **A6:** 10
magnetic particle inspection ... **A17:** 114-115

Shielded metal arc welding (SMAW) (continued)
matching filler metal specifications A6: 394
metallurgical discontinuities A6: 1073
mild steels and low-alloy steels, chemical composition of coverings used in electrodes A6: 61
multi-arc gravity-fed welding A6: 179
multipass SMAW/flux-cored arc weld on pipe steel, industrial application A6: 102-104
nickel alloys...... A6: 740, 742, 744, 745, 746-747, 749, 751
nickel alloys to dissimilar alloys A6: 751
nickel-base corrosion-resistant alloys containing molybdenum A6: 594
oxide-dispersion-strengthened materials A6: 1039
parameters used to obtain multipass weld in X-65 steel pipe A6: 101
phosphor bronzes A6: 754, 755, 764
power source selected A6: 36, 37, 38, 39, 41
precipitation-hardening stainless steels A6: 483, 489, 490
pressure vessel manufacture A6: 379
process selection guidelines for arc welding A6: 653
rating as a function of weld parameters and characteristics A6: 1104
for repair of high-carbon steels A6: 1105
for repair welding A6: 1103, 1107
safety considerations A6: 179
safety precautions A6: 1192-1193, 1200
silicon bronzes A6: 754, 766
slipping and binding agents for electrodes A6: 60
special applications A6: 178-179
stainless steel casting alloys A6: 496
stainless steels A6: 688, 693, 694, 698, 699-700, 701
steel weldment soundness...... A6: 408, 409, 413, 414
suggested viewing filter plates A6: 1191
to prevent hydrogen-induced cold cracking A6: 436
tool and die steels A6: 674, 676
tungsten carbides A18: 761
underwater welding A6: 178-179, 1010, 1012
variations of process A6: 178
vs. flux-cored arc welding A6: 186, 187
vs. gas-metal arc welding A6: 180
weld cladding A6: 819
weld procedures A6: 177-178
weld quality A6: 175
weld schedules A6: 177
welding circuit A6: 175

Shielded metal arc welding of
alloy steels M6: 297-298
electrode selection M6: 298
aluminum alloys M6: 75, 398-399
aluminum bronzes M6: 425
brasses M6: 425
carbon steels M6: 75, 83
carbon steels, hardenable M6: 265-266
cast irons M6: 75, 310
copper alloys M6: 75
copper nickels M6: 426
copper to aluminum M6: 425
coppers M6: 425
ductile iron M6: 315, 317-318
gray iron M6: 315
heat-resistant alloys M6: 75
cobalt-based alloys M6: 370
iron-nickel-chromium and iron-chromium-nickel alloys M6: 367
nickel-based alloys M6: 362-363
high-strength alloy steels M6: 75
high-strength low-alloy steels M6: 95
lead M6: 75
low-alloy steels M6: 75
low-carbon steels M6: 75
nickel alloys M6: 75, 440-442
phosphor bronzes M6: 425
reactive metals M6: 75
refractory metals M6: 75
silicon bronzes M6: 425-426
stainless steels M6: 75
stainless steels, austenitic M6: 324-326
electrode coatings M6: 324-325

Shielded metal arc welding of (continued)
electrodes M6: 324-325
stainless steels, ferritic M6: 347
stainless steels, martensitic M6: 348-349
stainless steels, nitrogen-strengthened austenitic M6: 345
electrodes M6: 345
tin M6: 75
zinc M6: 75

Shielded metal arc welding of specific materials
aluminum A6: 176
cadmium A6: 179
cast irons A6: 176
copper A6: 179
copper alloys A6: 176, 179
high-alloy steels A6: 176
high-strength steels A6: 176
lead A6: 179
low-alloy steels A6: 176
low-carbon steels A6: 176
mild steels A6: 176
nickel A6: 176
nickel alloys A6: 176
quenched steels A6: 176
stainless steels A6: 176
steels A6: 176
tempered steels A6: 176
zinc A6: 179

Shielded metal-arc welding (SMAW)
of nickel alloys A2: 445

Shielding
defined A9: 16
electromagnetic interference A7: 609-610, 612
flexible printed boards EL1: 588
radiation, computed tomography (CT) A17: 364-365
for radiography A17: 301

Shielding (gamma ray)
powders used A7: 573

Shielding (neutron)
powders used A7: 573

Shielding (nuclear engineering)
powder used A7: 573

Shielding gas
definition M6: 16
effect on weld metallurgy M6: 40-41
equipment for gas metal arc welding M6: 161

Shielding gases A6: 64-69
accuracy of gas blends A6: 65-66
basic properties A6: 64-65
dissociation A6: 64
gas density A6: 64, 65
gas purity A6: 65
ionization potential A6: 64
reactivity/oxidation potential A6: 64
recombination A6: 64
surface tension A6: 64-65
thermal conductivity A6: 64
blend component characteristics A6: 65
capacitor discharge stud welding, for aluminum A6: 222
definition A6: 1213
effect on weld metallurgy A6: 64
flux composition for CO_2 shielded FCAW electrodes A6: 61
from fluxes A6: 58
fume generation A6: 68
function A6: 64
influence on weld mechanical properties A6: 68
safety precautions A6: 1192, 1199
selection of A6: 65
steel weldment soundness A6: 408
tungsten inclusions after gas-metal arc welding as result of improper choice A6: 1074

Shielding gases for
arc welding of
beryllium M6: 462
copper and copper alloys M6: 402
magnesium alloys M6: 428
stainless steel, austenitic M6: 331-332, 334
titanium and titanium alloys M6: 447-448
zirconium and hafnium M6: 458
electrogas welding M6: 240-241
electron beam welding M6: 621-622
flux cored arc welding M6: 102-104

Shielding gases for (continued)
gas metal arc welding M6: 163-164
of aluminum alloys M6: 380
of nickel alloys M6: 439-440
of nickel-based heat-resistant alloys M6: 362
gas tungsten arc welding M6: 197-200
argon versus helium M6: 197-198
argon-helium mixture M6: 198
argon hydrogen mixtures M6: 198-199
effect of nitrogen M6: 199
flow M6: 199-200
oxygen-bearing argon mixtures M6: 199
purity M6: 199
supply and control M6: 199-200
gas tungsten arc welding of
alloys M6: 365
aluminum alloys M6: 390-391
aluminum bronzes M6: 412
copper nickels M6: 414
copper-zinc alloys M6: 409
coppers M6: 405
iron-nickel-chromium and iron- chromium-nickel heat-resistant
nickel alloys M6: 438
nickel-based heat-resistant alloys M6: 358-360
phosphor bronzes M6: 410
plasma arc welding M6: 217

Shielding gases for specific applications
aluminum metal-matrix composites A6: 555
aluminum-lithium alloys A6: 550
austenitic stainless steels A6: 468
electrogas welding A6: 270, 275
low-alloy steels A6: 662
electron-beam welding A6: 857, 868
high-strength alloy steels A6: 867
flux-cored arc welding A6: 64, 67, 186-187, 189
of carbon steels A6: 657
low-alloy steels A6: 662
gas-metal arc welding A6: 64, 65, 66-67, 181, 183, 185
aluminum alloy A6: 738
cast irons A6: 718, 719
ferritic stainless steels A6: 446, 449
hardfacing alloys A6: 803
low-alloy steels A6: 662
nickel alloys A6: 743-745
of silicon bronzes A6: 766
stainless steels A6: 705, 706, 707
gas-tungsten arc welding A6: 65, 67-68, 193, 488-489
aluminum alloys A6: 736
of aluminum bronzes A6: 764
of cast irons A6: 720
copper alloys A6: 756
coppers A6: 758
dispersion-strengthened aluminum alloys A6: 543
ferritic stainless steels A6: 445-446, 449, 452, 453
of hardfacing alloys A6: 803-804
low-alloy steels A6: 662
nickel alloys A6: 742
in space and low-gravity environments A6: 1022
stainless steels A6: 703, 704
of titanium alloys A6: 783, 786
laser hardfacing A6: 807
laser-beam welding A6: 878
aluminum alloys A6: 739
magnesium alloys A6: 772, 778
plasma arc cutting A6: 1167
plasma arc welding A6: 65, 67, 68
aluminum alloys A6: 735
low-alloy steels A6: 662
plasma-MIG welding A6: 224
self-shielded flux-cored arc welding A6: 68-69
submerged arc welding A6: 58
titanium alloys A6: 784, 786
tungsten alloy welding A6: 582
ultrahigh-strength low-alloy steels A6: 673
welding of molybdenum alloys A6: 581
zirconium alloys A6: 787-788

Shields
in torsional testing A8: 146

Shift
as casting defect A11: 387
defined A15: 10

Shift tolerance *See* Mismatch tolerance
Shifted core
as casting defect ... **A11:** 387
Shim *See* Pin or Mandrel
brazing filler metals available in this form ... **A6:** 119
defined ... **A14:** 12
Shim, liquid *See* Liquid shim
Shimming
in press-brake forming ... **A14:** 540
Shimmy trimming
of blanks ... **A14:** 446-447
Shine rolling ... **M5:** 134
Shingling ... **A6:** 820
Ship, and submarine
corrosion ... **A13:** 543-544, 546, 919, 924
Ship hulls
composite ... **EM1:** 837-838
Ship P/M applications ... **A7:** 574
Ship steel
wide-plate tests of ... **A8:** 440
Ship steel, grades A
B, C, laser-beam welding ... **A6:** 264
Ship-bottom paints
powders used ... **A7:** 574
Shipbuilding ... **A6:** 381-385
codes ... **M6:** 824
high-strength low-alloy steels for ... **A1:** 419
joining processes ... **M6:** 56
Shipbuilding applications
magnetic painting ... **A17:** 128
magnetic particle inspection methods ... **A17:** 114-115
Shipment tonnages
metal casting ... **A15:** 41-42
Shipping
effects on tubing ... **A11:** 630
environments, stress-corrosion cracking in ... **A11:** 211-212
first-level packages ... **EL1:** 991
of heat-exchanger tubing, effect on corrosion resistance ... **A11:** 630
in winter, fracture from ... **A11:** 97-98
SHKH15-SHD
nominal compositions ... **A18:** 725
Shock
effect in creep and stress-rupture testing ... **A8:** 312
fronts, high strain rates in ... **A8:** 190
loads, defined ... **A8:** 12
Shock absorber fluids ... **A18:** 98-99
extreme-pressure agents used ... **A18:** 101
friction modifiers ... **A18:** 104
Shock absorbers
powders used ... **A7:** 572
Shock, and vibration
effects ... **EL1:** 62-65, 589
Shock fronts ... **A6:** 161
Shock load
defined ... **A11:** 9
Shock loading *See also* Loading; Shock load
distortion from ... **A11:** 138
of oil-pump gear ... **A11:** 344-345
of steel wire rope, vibrational fatigue in ... **A11:** 518-519
Shock loads
spring steels for ... **M1:** 284, 285
Shock pressure
liquid impingement erosion ... **A18:** 223
Shock resistance
titanium carbide-steel cermets ... **A7:** 810
Shock-resisting cold-work tool steels
forging temperatures ... **A14:** 81
Shock-resisting steels
composition limits ... **A18:** 735
service temperature of die materials in forging ... **A18:** 625
Shock-resisting tool steels *See* Tool Steels, shock-resisting ... **A1:** 766-767
for hot-forging dies ... **A18:** 625
Shock-wave transmission
explosive forming ... **A14:** 640
Shockley partial used to eliminate a stacking fault area ... **A9:** 116
Shoe
defined ... **EM1:** 21
Shoe-type pinch-roll forming machines ... **A14:** 617
Shoefer diagram
for ferrite content ... **A15:** 725
Shoes
electroslag welding ... **M6:** 227-228
Shore A scale ... **EM3:** 189, 190
butyl tapes ... **EM3:** 202
urethane sealant values ... **EM3:** 205
Shore hardness *See also* Hardness ... **EM3:** 26
defined ... **EM1:** 21 **EM2:** 38
Shore hardness test *See* Scleroscope hardness test
Shore OO scale ... **EM3:** 189
Short
defined ... **EM2:** 38
Short bar machine ... **A7:** 34
plasma rotating electrode process ... **A7:** 39, 41
Short beam shear
resin-matrix composites ... **A12:** 478
Short beam shear (SBS) ... **EM3:** 26
defined ... **EM1:** 21
strength, long-term exposure testing ... **EM1:** 825
tests ... **EM1:** 287
Short blanks
multiple-slide forming ... **A14:** 570
Short circuit(s)
defect causing ... **EL1:** 1018
defined ... **EL1:** 1156
as failure sites ... **EL1:** 127
from solder hairs ... **EL1:** 561
as hard defect ... **EL1:** 568
solder ... **EL1:** 691
in wound film capacitors ... **EL1:** 999
Short circuiting arc welding ... **M6:** 329-330
Short circuiting gas metal arc welding
to solve problems in joining thin sections by oxyfuel gas welding ... **A6:** 288
Short circuiting transfer
definition ... **M6:** 16
Short core
adhesive-boned joints ... **A17:** 614
Short crack
behavior ... **A8:** 379-380
defined ... **A8:** 379
threshold stress intensity ... **A8:** 379
Short distances
RDF parameters defining ... **A10:** 396-397
Short fiber *See* Chopped fiber; Discontinuous fibers; Short fiber composites; Short fibers; Staple fiber
Short fiber composites *See also* Discontinuous fiber composites
defined ... **EM1:** 27
fiber lengths ... **EM1:** 119
with recovered carbon fibers ... **EM1:** 155-156
Short fibers *See also* Reinforcements ... **EM1:** 119-121
aligned, production facility ... **EM1:** 154
for aluminum metal-matrix composites ... **A2:** 7
properties ... **EM1:** 119
Short lines
for reflection ... **EL1:** 170
Short shot *See also* Sink mark
defined ... **EM1:** 21
Short terne coating
steel plate ... **M1:** 173
Short terne coatings ... **M5:** 358
Short weld sequence
cast iron welding ... **A15:** 526
Short-beam shear (SBS)
carbon fiber reinforced composites ... **EM2:** 237
defined ... **EM2:** 38
Short-beam shear test ... **EM3:** 392, 402, 403
Short-chain branching ... **EM3:** 26
defined ... **EM2:** 38
Short-circuit diffusion
high-temperature gaseous corrosion ... **A13:** 69
Short-circuiting
in direct current electrical potential method ... **A8:** 389
Short-circuiting transfer ... **A6:** 26
Short-crack corrosion-fatigue ... **A11:** 254
Short-field probes
microwave inspection ... **A17:** 212
Short-pulse-duration tests
dynamic fracture testing ... **A8:** 282-283
Short-range order ... **A10:** 407, 691
Short-run blanking dies
steel-rule ... **A14:** 451-452
subpress ... **A14:** 452-453
template ... **A14:** 452
Short-shot *See* Short
Short-term etching
defined ... **A9:** 16
Short-term loads
metals vs plastics ... **EM2:** 75
Short-time annealing of steel to reveal as-cast solidification structures ... **A9:** 624
Short-time tensile tests
for high-temperature properties measurement ... **A12:** 121, 139
Short-transverse direction, of grain structure
SCC and ... **A8:** 501
Short-trim dense wire-filled radial brushes ... **M5:** 153-154
Shorting pin
in flyer plate impact test ... **A8:** 210-211
Shortness ... **EM3:** 26
defined ... **A9:** 16
Shortness, cold *See* Cold shortness
Shortness, hot *See* Hot shortness
Shorts *See also* Short circuits
opens, low- and high-voltage ... **EL1:** 565
risk site analysis, methodology ... **EL1:** 130
Shot ... **A7:** 11
capacity, defined ... **EM1:** 21
defined ... **A15:** 10
discontinuous oxide fiber ... **EM1:** 62-63
in Domfer process ... **A7:** 89-91
in injection molding ... **EM1:** 166
size, and abrasive particles, cleaning efficiency by ... **A15:** 518-519
size, specifications ... **A15:** 514
Shot blast cleaning
core knockout ... **A15:** 506
Shot blasting
abrasive wear and material property effects ... **A18:** 186
aluminum and aluminum alloys ... **M5:** 571
hot dip galvanized coating process ... **M5:** 326, 329
properties, characteristics and effects, shot ... **M5:** 83-87, 91
size specifications, cast and cut steel wire shot ... **M5:** 83-84
stainless steel ... **M5:** 553
Shot capacity
defined ... **EM2:** 38
Shot cleaning abrasive finishing
powders used ... **A7:** 572
Shot, lead or iron
and shotgun barrel distortion ... **A11:** 139-140
Shot peening *See also* Peening ... **A16:** 21, 27, 34, 35 **A18:** 253 **M5:** 138-149
Almen test strip ... **M5:** 139-140, 148-149
aluminum alloys ... **A14:** 803
aluminum and aluminum alloys ... **M5:** 141-142, 145
applications ... **M5:** 141, 145-147
limitations ... **M5:** 146-147
problems and corrections ... **M5:** 147-148
benefits for lubricated wear ... **M1:** 637
brass ... **M5:** 145-146

Shot peening (continued)
breakdown of shot **M5:** 143
cast iron shot **M5:** 141-145
cast shot size specifications **M5:** 84
cast steel shot **M5:** 141
compressive residual stresses **A10:** 380
compressive residual stresses by **A11:** 125
costs **M5:** 148
cycling of shot **M5:** 142-143
dies **A18:** 643
effect on AISI 4340 steel **A16:** 35
effect on carburized steels **A4:** 371
effect on dross inclusion, wrought aluminum alloys **A12:** 422
effect on fatigue resistance **M1:** 674-675, 682
electrical discharge machining **A16:** 559
electrochemical machining **A16:** 539
equipment **M5:** 143-145
exposure time, coverage related to **M5:** 138-139
fatigue strength improved by **M5:** 138, 141-142, 145-146
forgings **M5:** 146-147
glass bead **A7:** 669
glass bead shot **M5:** 142-145
dry method **M5:** 144
wet method **M5:** 144-145
hardness of shot **M5:** 142
heat-resisting alloys **M5:** 566-567
impingement angle effects of **M5:** 143, 147
improvement of fatigue strength **A16:** 26-27
inadequate, alloy steel failure by **A12:** 327
incomplete, in helicopter-blade spindle **A11:** 126
Inconel 718 fatigue strength **A16:** 36
limitations-workpiece, surface and temperature **M5:** 146-147
mechanism of action **M5:** 138
peening intensity **M5:** 139-149
selection of **M5:** 139-140
physical and mechanical properties affected by **M5:** 308
process variables, control of **M5:** 139-142
processing after peening **M5:** 148-149
production peening, problems and corrections **M5:** 147-148
propulsion of shot-wheel, air blast, and gravitational force methods **M5:** 139
for SCC control **A13:** 327-328, 942
screening tolerances, shot **M5:** 141
of shafts **A11:** 459
shot peen testing of adhesion **M5:** 143, 146
silver plate adhesion, testing of **M5:** 146
size of shot **M5:** 138-149
standard specifications **M5:** 140-141
of springs **A1:** 313-314
stainless steel **M5:** 145-147
stainless steels **A18:** 715, 716, 723
steel **M5:** 140-142, 145-149, 186
steel springs **M1:** 293, 296, 297, 312-313
stop-offs (masking) **M5:** 144
strain peening **M5:** 138
stress corrosion resistance improvement of **M5:** 145-146
superficial, effect in wrought aluminum alloys **A12:** 420
surface conditions, effects of **M5:** 138-139
surface coverage (saturation) **M5:** 138-139
exposure time related to **M5:** 138-139
measuring-visual, Straub, Peensean, and Valentine methods **M5:** 138-140
surface stresses **A10:** 383
temperature effects **M5:** 146-147
testing **M5:** 144, 146, 149
to reduce thermal and mechanical fatigue **A18:** 640
types of shot **M5:** 140-142
valve spring failure from **A11:** 554
velocity of shot **M5:** 142-143
work handling mechanisms **M5:** 144
Shotblasting *See also* Sandblasting
defined **A15:** 10
as weldment cleaning **A17:** 591
of zirconium castings **A15:** 838
Shotgun barrel
bulging of **A11:** 139-140
Shotguns
powders used **A7:** 574
Shotting **A7:** 11
and atomization, defined **A7:** 25
of copper **A7:** 106
Shoulder
definition **A6:** 1213
Shoulder design
for torsion specimen **A8:** 155-156
Shoulder joint prosthesis
total **A11:** 670
Shouldered end specimen
for constant-load testing **A8:** 314
Shove-aside router **EL1:** 533
Shoving furnace
for continuous annealing **A15:** 31
Shrink control agents **EM3:** 41
Shrink etching
defined **A9:** 16
Shrink fixture *See* Cooling fixture
Shrink tape debulking
in tube rolling **EM1:** 573-574
Shrink-fit inserts
in magnesium alloy parts **A2:** 466-467
Shrinkage *See also* Casting shrinkage; Cavities; Liquid shrinkage; Macroshrinkage; Microshrinkage; Mold shrinkage; Porosity; Shrinkage cavities; Shrinkage cavity; Shrinkage porosity; Solid shrinkage; Solidification; Solidification shrinkage; Voids...... **A7:** 11 **A9:** 238 **A14:** 12, 838 **M6:** 859 **EM1:** 21, 158
in activated sintering **A7:** 318
allowance, improper, as casting defect **A11:** 386
allowances **A15:** 303, 805
aluminum alloy **A15:** 768
in aluminum alloys **A9:** 358
aluminum casting alloys **A2:** 146-147
apparent, filler effects **EM2:** 280-281
area, iron **A12:** 223
axial **A11:** 382
boring mill effects **A15:** 20
casting, defined *See* Casting shrinkage
cavities, radiographic appearance **A17:** 349
cavity, defined *See* Shrinkage cavities
centerline **A11:** 2, 315, 382
centerline, defined **A15:** 2
centerline, radiographic appearance **A17:** 349
in cold isostatic pressing **A7:** 444
in compacted graphite iron **A1:** 57
compacted graphite irons **A15:** 671
of compacts **A7:** 309
condition, graphite structure in **A11:** 369
in container welding **A7:** 431
control, by chills **A11:** 354
of copper alloys **A2:** 346, 348
of copper powder compacts **A7:** 309
core **A11:** 382
corner **A11:** 382
cracks, by liquid penetrant inspection **A17:** 86
cracks, defined **A15:** 10
cracks, radiographic inspection **A17:** 296
defined **EM2:** 38
and densification during sintering dilatometer-measured **A7:** 310
as directionally solidified defect **A15:** 321
dispersed, as casting defect **A11:** 382
during delubrication, presintering and sintering **A7:** 480
feeding of, die casting **A15:** 291-292
of flexible epoxies **EL1:** 817
from solidification **A15:** 109
from thermoplastic processing **EM2:** 280
from thermosetting processes **EM2:** 280
as function of green density **A7:** 310
as function of metal structure **A7:** 310
as function of sintering temperature **A7:** 310
as function of sintering time **A7:** 310
fusible alloys **A2:** 756
ingot pipe **A11:** 315
with injection molding **A7:** 495
internal, defined *See* Internal shrinkage
internal or blind **A11:** 382
internal, radiographic methods **A17:** 296
liquid, and feed metal volume **A15:** 577
mold, rammed graphite **A15:** 274
open or external, as casting defect **A11:** 382
patternmaker's, cast copper alloys **A2:** 356-391
patternmaker's, defined *See* Patternmaker's shrinkage
Shrinkage (continued)
polyamide-imides (PAI) **EM2:** 132, 135
polyaryl sulfones (PAS) **EM2:** 146
porosity **A12:** 405-406
porosity, aluminum casting alloys **A2:** 136
porosity, radiographic appearance **A17:** 348
and pouring temperature **A15:** 283
in sintering of tungsten powders **A7:** 390
solid, defined *See* Casting shrinkage
substrate **EL1:** 462
tears, in all-ceramic mold castings **A15:** 249
thermoplastic injection molding **EM2:** 313
void, cast aluminum alloy **A12:** 67
voids, copper alloys, inspection for **A15:** 558
voids, in aluminum alloy castings **A17:** 535
voids, in weldments **A17:** 582
volume, and size of rigid tool set **A7:** 322-323
as volumetric flaw **A17:** 50
Shrinkage allowance *See also* Patternmakers' rules **M1:** 30-31, 67
Shrinkage cavities *See also* Cavities; Dimple(s)
cast aluminum alloys **A12:** 409
in copper alloy ingots **A9:** 643
defined **A15:** 10
as gray iron defect **A15:** 640-641
iron **A12:** 220
low-carbon steel **A12:** 249
maraging steels **A12:** 384
revealed by macroetching **A9:** 174
solidification in **A12:** 140, 160
and white spot, iron **A12:** 220
zirconium alloys **A15:** 838
Shrinkage cavity *See also* Shrinkage
defined **A11:** 9
fatigue fracture from **A11:** 120
from forging **A11:** 315
Shrinkage, mold *See* Mold shrinkage
Shrinkage of compression-mounting epoxies **A9:** 29
Shrinkage porosity *See also* Shrinkage
at bolt-boss sites, ductile iron cylinder head **A11:** 354-355
brittle fracture of cast low-alloy steel from **A11:** 389-391
cast aluminum alloys **A12:** 405-406
as casting internal discontinuity **A11:** 354
coolant leakage by **A11:** 355, 357
as die casting defect **A15:** 294-295
effect on fatigue strength **A11:** 120
fatigue fracture from **A11:** 120
in iron castings **A11:** 354
permanent mold casting **A15:** 279
in semisolid alloys **A15:** 337
tensile fracture from **A11:** 389-390
Shrinkage stress
definition **A6:** 1213
Shrinkage stresses
effect on plated specimens **A9:** 28
Shrinkage void
definition **A6:** 1073, 1213 **M6:** 16
Shroyer, H.F
lost foam casting **A15:** 230
Shunt error
in tungsten-rhenium thermocouples **A2:** 876
Shunting
resistance spot welding **M6:** 472, 475, 529
resistance welding of aluminum alloys **M6:** 536
Shurtronics harmonic bond tester **EM3:** 527, 528
Shurtronics Mark I harmonic bond tester
for adhesive-bonded joints **A17:** 621
Shut height *See also* Daylight
defined **A14:** 12
Shuttle extractors
for unloading presses **A14:** 500-501
Shuttles
in textile looms **EM1:** 127-128
SI *See* Silicones; Système International d'Unités
SI units
base, supplementary, and derived **A10:** 685
guide for **A10:** 685-687
prefixes, names and symbols **A10:** 687
Si(Li) detectors *See* Lithium-drifted silicon detectors
SI-based sievert (SV) unit
radiography **A17:** 301
Si-Sn (Phase Diagram) .... **A3:** 2 • 365
Si-Sr (Phase Diagram) .... **A3:** 2 • 365
Si-Ta (Phase Diagram) .... **A3:** 2 • 366

SI-Te (Phase Diagram) ... A3: 2 • 366
Si-Th (Phase Diagram) ... A3: 2 • 366
Si-Ti (Phase Diagram) ... A3: 2 • 367
Si-U (Phase Diagram) ... A3: 2 • 367
Si-V (Phase Diagram) ... A3: 2 • 367
Si-Zn (Phase Diagram) ... A3: 2 • 368
Si-Zr (Phase Diagram) ... A3: 2 • 368
SiAlON *See also* Silicon nitride ... A16: 100-103, 106 EM4: 47, 49
abrasive machining ... EM4: 317, 319
applications ... A18: 812 EM4: 48
automotive ... EM4: 960
heat exchangers ... EM4: 982
behavior diagram ... EM4: 814
brazing with glasses ... EM4: 519
cemented carbides for machining ... A16: 87
ceramic cutting tools ... EM4: 966
creep behavior ... EM4: 818
cutting of Ni-base superalloys ... A16: 109
cutting speed and work material relationship ... A18: 616
cutting tool material for high removal rate machining ... A16: 608
for cutting tools ... EM4: 959
development ... EM4: 814
formation ... EM4: 814
formed during active metal brazing ... EM4: 524
forms ... EM4: 814
heat treatments ... A18: 814
and high removal rate machining ... A16: 608
high-speed machining ... A16: 602
insert for turning fiber FP Al MMCs ... A16: 898
joining non-oxide ceramics ... EM4: 528
key product properties ... EM4: 48
mechanical and physical properties ... A18: 813
properties ... EM4: 814, 815, 816, 817
raw materials ... EM4: 48
reactive hot pressing for pressure densification ... EM4: 300
sintering ... EM4: 814, 817
sintering aids ... EM4: 814
thermal shock resistance ... EM4: 818
tool material properties ... A16: 107
tool materials for cast iron machining ... A16: 656
turning heat-resistant alloys ... A16: 740
types ... EM4: 814
Sialon-titanium nitride
material removal rates with electrical discharge machining ... EM4: 376
Sialons, specific types
composition and properties ... EM4: 873
SiC whisker-reinforced aluminum metal-matrix composites ... A14: 20
Side clearance *See* Clearance
Side corings
defined ... EM2: 38
Side cutting edge angle (SCEA) ... A16: 19, 20, 143, 151, 192
Side deformation
in rolling ... A8: 593, 595
Side draw pins *See* Side corings
Side gating
permanent mold casting ... A15: 279
Side gears
economy in manufacture ... M3: 847
Side grooves
in aqueous solution test specimen ... A8: 417
Side groups
of mer ... EM2: 57-58
Side loading
preventing ... A8: 48
Side milling
surface roughness arithmetic average extremes ... A18: 340
Side rails
overloading to distortion failure of ... A11: 137
Side rake angle (SR) ... A16: 142, 143, 151, 162
Side relief angle (SRF) ... A16: 143
Side scatter, shadow formation
radiography ... A17: 313
Side slip, as buckling
axial compression testing ... A8: 56
Side thrust
defined ... A14: 12
Side-action brushes ... M5: 155
Side-and-bottom brushes ... M5: 155
Side-brazed packages ... EL1: 205, 213-217
Side-egress flatpack/quadpack
as surface-mount option ... EL1: 77
Side-stream (bypass) loops
in-service monitoring ... A13: 201
Sidebands in diffraction patterns
cause of ... A9: 653
Sideplate ferrite (SPF) ... A6: 1011
Sidepressing
isothermal plane-strain ... A8: 172-173
test ... A8: 588-589
Sidepressing test ... A1: 583-585
for forgeability ... A14: 215, 383
Siderocapsa
biological corrosion by ... A13: 117
Siderosis ... A7: 203
Siemans
symbol for ... A8: 726
Siemens
abbreviation for ... A10: 691
defined ... A10: 681 EL1: 89
as IC unit of conductance ... A10: 659
as SI derived unit, symbol for ... A10: 685
Siemens, Sir William
as inventor ... A15: 32
Siemens-Martin open hearth furnace
development ... A15: 32
Sieve analysis *See also* Particle size distribution; Screening; Sub-sieve analysis ... A7: 11
AFS standard ... A15: 209
definition ... M6: 16
equipment ... A7: 215
fireclay use ... A15: 210
as particle size and size distribution measure ... A7: 214-216
sieving problems ... A7: 216
of stainless steel powder ... A7: 100
standard test method ... A7: 215-216
standard U.S. sieve series ... A7: 215
of tin powder ... A7: 123
of water-atomized high-speed tool steels ... A7: 103
Sieve classification *See* Sieve analysis ... A7: 11
Sieve fraction *See also* Sub-sieve fraction ... A7: 11
Sieve shaker ... A7: 11
Sieve underside ... A7: 11
Sieve(s) *See also* Screen; Screening; Sieve analysis; Sub-sieve analysis ... A7: 11, 215
agglomeration problems ... A7: 216
agitators ... A7: 215
blinded ... A7: 216
classification ... A7: 11
damaged ... A7: 216
defined ... A7: 11
equipment ... A7: 215
fraction ... A7: 11
as inclusion control ... A15: 90
for irregularly shaped particles ... A7: 216
overloaded ... A7: 216
shaker ... A7: 11
types ... A7: 216
underside ... A7: 11
Sievert units
radiography ... A17: 301
Sievert's law ... A6: 543
hydrogen solubility ... A13: 329
Sieving
of particulate samples ... A10: 165
in sampling ... A10: 16
Sieving techniques ... EM4: 87
versus weighting factor ... EM4: 85
Sieving, wet and dry
to analyze ceramic powder particle sizes ... EM4: 66, 67
Sifter screens ... A7: 177
Sigma blade mixer
for bulk molding compounds ... EM1: 161
Sigma bonding
defined ... A10: 681
Sigma formation ... A6: 587
Sigma phase ... A13: 11, 125, 127
in 18-8 stainless steel, polarization curve for potentiostatic etching ... A9: 145
in austenite, detection by phase contrast etching ... A9: 59
defined ... A9: 16 A12: 132 A17: 55
effect of delta ferrite on formation in austenitic stainless steels ... A9: 136-137
embrittlement ... A12: 132-133, 150
extraction replicas ... A17: 55
in ferritic chromium steel, detection by phase contrast etching ... A9: 59
orientation relationship in ... A10: 453-454
revealed through color etching ... A9: 136
in rhenium-bearing alloys ... A9: 448
in stainless steel casting alloys, role of ... A9: 297
in stainless steels, potentiostatic etching ... A9: 145-146
in type 304 stainless steel, magnetic etching used to study ... A9: 65
Sigma phase embrittlement ... A1: 708-709
in austenitic and ferritic stainless steels ... A1: 709-711
in duplex stainless steels ... A1: 711
Sigma phase in iron-chromium-nickel heat- resistant casting alloys
formation of ... A9: 334
identification of ... A9: 332-333
preparation for examination ... A9: 330-331
Sigma phase in wrought heat-resistant alloys ... A9: 309, 312
specimen preparation for identifying ... A9: 307
Sigma phase in wrought stainless steels
in austenitic grades ... A9: 284
in duplex grades ... A9: 286
etching ... A9: 281-282
in ferritic grades ... A9: 285
Sigma precipitation
in HAZs ... A13: 350
Sigma-phase embrittlement *See also* Sensitization ... A13: 11, 1265 M1: 686
in steels ... A11: 99
Sigmoidal curves
reversed ... A12: 212-215
Sign convention
of electrode potentials ... A13: 21-22
Signal *See also* Small signal
averaging, for fluorescence noise ... A10: 130
conditioning, during medium stress and strain measurement ... A8: 191
conditioning equipment ... A8: 197
defined ... EL1: 1156
delay ... EL1: 5
detectors ... A10: 434-435
generation , electron beam ... A10: 498-500
path, routing, rigid printed wiring boards ... EL1: 551-552
processing ... A10: 507-508, 631
processing, electronic, in dynamic testing machines ... A8: 40-41
scanning electron beam instrument ... A10: 501
-to-noise ratios ... A10: 68, 409, 413, 681
transmission, in digital integrated circuits ... EL1: 168-172
vias ... EL1: 112
Signal amplitude
in ECP flaw characterization ... A17: 140-141
as proportional to flaw size ... A17: 131

SUBJECTS OF THE INDEXED VOLUMES: ASM Handbook (designated by the letter "A"): **A1:** Properties and Selection: Irons, Steels, and High-Performance Alloys (1990); **A2:** Properties and Selection: Nonferrous Alloys and Special-Purpose Materials (1990); **A3:** Alloy Phase Diagrams (1992); **A4:** Heat Treating (1991); **A6:** Welding, Brazing, and Soldering (1993); **A7:** Powder Metallurgy (1984); **A8:** Mechanical Testing (1985); **A9:** Metallography and Microstructures (1985); **A10:** Materials Characterization (1986); **A11:** Failure Analysis and Prevention (1986); **A12:** Fractography (1987); **A13:** Corrosion (1987); **A14:** Forming and Forging (1988); **A15:** Casting (1988); **A16:** Machining (1989); **A17:** Nondestructive Testing and Quality Control (1989); **A18:** Friction, Lubrication, and Wear Technology (1992). **Metals Handbook, 9th Edition** (designated by the letter "M"): **M1:** Properties and Selection: Irons and Steels (1978); **M2:** Properties and Selection: Nonferrous Alloys and Pure Metals (1979); **M3:** Properties and Selection: Stainless Steels, Tool Materials and Special-Purpose Materials (1980); **M4:** Heat Treating (1981); **M5:** Surface Cleaning, Finishing, and Coating (1982); **M6:** Welding, Brazing, and Soldering (1983). **Engineered Materials Handbook** (designated by the letters "EM"): **EM1:** Composites (1987); **EM2:** Engineering Plastics (1988); **EM3:** Adhesives and Sealants (1990); **EM4:** Ceramics and Glasses (1991); **Electronic Materials Handbook** (designated by the letters "EL"): **EL1:** Packaging (1989).

Signal amplitude (continued)
remote-field eddy current inspection **A17:** 198
shapes, ECP inspection **A17:** 137
Signal attenuation
causes **EL1:** 603
design considerations **EL1:** 41-42
Signal conditioning
ultrasonic inspection **A17:** 253
Signal cross talk *See* Cross talk
Signal detectors
analytical transmission electron microscopy **A10:** 434-436
positioning, in AEM microscope column **A10:** 434
secondary electrons (STEM mode) **A10:** 435
transmitted and scattered electrons (STEM mode) **A10:** 435
transmitted and scattered electrons (TEM mode) **A10:** 434-435
Signal display
A-scans, ultrasonic inspection **A17:** 242
B-scans **A17:** 243
C-scans
control **A17:** 254
transmission ultrasonic inspection **A17:** 249
Signal edge speed degradation
high-speed digital systems **EL1:** 80
Signal line
apparent, impedance **EL1:** 37
characteristic impedance **EL1:** 29-30
density **EL1:** 389
digital, characteristics **EL1:** 169-170
impedance, choice of **EL1:** 30
isolated **EL1:** 28-34
parallel **EL1:** 34-36
Signal line resistance
effect, low end systems **EL1:** 27
Signal plane
design improvements **EL1:** 131
printed circuit, design features **EL1:** 127
Signal propagation
accurate solution of **EL1:** 35
velocity, maximum **EL1:** 5
Signal response analysis **A17:** 697-700
confidence bound calculation **A17:** 698
multiple inspections per flaw **A17:** 698-699
parameter estimation **A17:** 698
response analysis **A17:** 697-700
Signal rise
classification **EL1:** 25
times, and local physical performance **EL1:** 5
Signal transformation
interpreting images produced by **A9:** 96
Signal transmission lines *See also* Line(s); Transmission line
branch lines **EL1:** 361
high-frequency effects **EL1:** 362
resistance-inductance-capacitance (RLC) line, homogeneous **EL1:** 357-359
resistance-inductance-capacitance (RLC) line, inhomogeneous **EL1:** 359-361
Signal(s) *See also* Response; Signal amplitude; Signal display; Signal response; Signal response analysis; Signal-to-noise ratio
conditioning **A17:** 253
detection, acoustic emission inspection **A17:** 281-282
factors, quality design **A17:** 722
measurement, acoustic emission inspection **A17:** 282-283
as NDE response, and NDE reliability **A17:** 674
processing, remote-field eddy current inspection **A17:** 198-199
Signal-to-noise ratio **A18:** 295-296
defined **A10:** 681
effect on secondary electron imaging **A9:** 95
EXAFS **A10:** 409, 413
of inspection materials **A17:** 678
in NDE engineering **A17:** 676
in NDE reliability **A17:** 675
as performance measure **A17:** 750
UV/VIS **A10:** 68
Signal-to-noise ratio (SNR) **EL1:** 738
Signature
defined **A17:** 51
Signature analysis
as NDE area **A17:** 49, 51
Signature, thermal *See* Thermal signature
Significance
level **A8:** 12, 627
statistical, for factorial experiments **A8:** 654
statistical, in comparative experiments **A8:** 642-643
tests, computations for **A8:** 711
Significant
defined **A8:** 12
Sigran
composition **EM4:** 873
manufacturer **EM4:** 873
properties **EM4:** 873
thermal conductivity **EM4:** 873
Silahydrocarbons (SiHCs) **A18:** 151
high-vacuum lubricant applications **A18:** 156, 157
molecular structures **A18:** 156
Silal *See* Gray cast iron, medium silicon gray iron **A1:** 103
Silane adhesion promoters
for urethane sealants **EM3:** 206
Silane coupling agents .... **EM3:** 40 , 42, 49, 181-182, 674
for aircraft canopies and windshield priming aluminum adherends **EM3:** 263
for epoxies **EM3:** 99-100
for glass **EM3:** 281-283
polyaramid **EM3:** 286
for priming glass and glass fibers **EM3:** 101
reactive **EM3:** 52
in siliconized acrylics **EM3:** 50
to improve joint durability **EM3:** 626
to promote adhesion in sealants **EM3:** 188
Silane(s)
as coupling agents **EM1:** 29, 32
organofunctional **EM1:** 123
as sizing **EM1:** 123
Silanols
decomposition increasing viscosity **EM4:** 212
Silcoro 60
brazing, composition **A6:** 117
wettability indices on stainless steel base metals **A6:** 118
Silcoro 75
brazing, composition **A6:** 117
wettability indices on stainless steel base metals **A6:** 118
Silcrome alloy *See* Valve alloys, specific types, Silcrome
Silica *See also* Sand(s); Silica sands; Silica-base bonds; Silicon
as abrasive **A14:** 747-748
abrasive wear **A18:** 188
-aluminas **A10:** 130
background fluorescence of **A10:** 130
chemical composition **A6:** 60
composition in flux **M6:** 229
corrosion fatigue test specification **A8:** 423
defined **A15:** 10
electrode coatings **A6:** 60
as extender **EM3:** 175, 176
as filler
for conductive adhesives **EM3:** 76
for polysulfides **EM3:** 139
fracture toughness **A18:** 192
from aluminum silicates **A15:** 209-210
fumed **EM3:** 176, 177-178
filler for urethane sealants **EM3:** 205
function and composition for mild steel SMAW
fused **EM3:** 176
as grease thickener **A18:** 126
as lining material, induction furnaces **A15:** 372
manufacturing processes **EM3:** 176
material for surface force apparatus **A18:** 402
as mold refractory, investment casting **A15:** 258
and molten iron, slag compound from **A15:** 208
physical properties **A18:** 192
platinized, for removal of carbon monoxide/dioxide in high-temperature combustion **A10:** 222
precipitated, filler for urethane sealants **EM3:** 205
resistance spot welding **A6:** 850
slip-cast fused, rain erosion applications **A18:** 222
synthetic **EM3:** 176
thermal expansion of **A15:** 224
Silica (SiO_2) *See also* Engineering properties of single oxides; Silicon dioxide **EM4:** 32, 44, 49
as addition during batch melting **EM4:** 386
applications **EM4:** 46
composition **EM4:** 46
density at various pressures **EM4:** 583, 584
as filler **EM4:** 6
for glass fibers, sol-gel processed **EM4:** 450
glass forming ability **EM4:** 494
impurity found in gypsum **EM4:** 380
mechanical properties **EM4:** 850
melting point **EM4:** 494
melting/fining **EM4:** 391
as optical fiber glass material **EM4:** 413, 414, 415
porous glass beads commercially marketed **EM4:** 421
specific properties imparted in CTV tubes **EM4:** 1039
subcritical crack growth **EM4:** 695
superduty, refractory physical properties **EM4:** 897, 898, 899
supply sources **EM4:** 46
viscosity **EM4:** 848
viscosity at melting point **EM4:** 494
Silica abrasive blasting
aluminum and aluminum alloys **M5:** 571-572
properties and characteristics of silica **M5:** 84, 93-94
Silica brick
applications, refractory **EM4:** 899, 903, 906-907, 912
refractory compositions **EM4:** 896
spalling resistance **EM4:** 897
Silica dust, abrasive jet machining
health hazard **A16:** 512
Silica fibers
and carbon fibers **EM1:** 129
Silica fibers, vitrified
as thermocouple wire insulation **A2:** 882
Silica flour
defined **A15:** 10
Silica flour filled epoxy resin (SFFER) **A16:** 108, 109
Silica glass **EM4:** 49
applications
aerospace **EM4:** 1017
electronic processing **EM4:** 1055, 1056
lighting **EM4:** 1032
consolidated 96%, properties **EM4:** 430, 431
fatigue **EM4:** 744
RDF analysis **A10:** 395-399
x-ray diffraction patterns for **A10:** 395, 398-399
Silica grinding balls **A7:** 58
Silica impurities in ferrites **A9:** 538
Silica in a complex mixture **A9:** 185
Silica removal
from boiler tubes **A11:** 616
Silica sand **EM4:** 378-379
for abrasive blasting **A7:** 458
abrasive mixed with high-pressure waterjet **A16:** 521
applications, dental **EM4:** 1093
chemical analysis **EM4:** 378
chemical properties **EM4:** 855
chrome content **EM4:** 378
coloration **EM4:** 378-379
density **EM4:** 846
deposits in U.S **EM4:** 378
elastic modulus **EM4:** 849, 850
electrical properties **EM4:** 851
heat capacity **EM4:** 847
hydrofluoric acid effect **EM4:** 1061
iron content **EM4:** 378
mining techniques **EM4:** 378
nickel content **EM4:** 378
particle size distribution analysis **EM4:** 378
properties **EM4:** 851
purpose for use in glass manufacture **EM4:** 381
refractory heavy metal deposits **EM4:** 378
typical oxide compositions of raw materials **EM4:** 550
viscosity **EM4:** 849
Silica sand(s) *See also* Sand(s); Silica; Silica-base bonds
composition **A15:** 208
dried, synthetic sands as **A15:** 29

Silica sand(s) (continued)
grains, shape and distribution **A15:** 208-209
preparation of **A15:** 209
reclamation **A15:** 354-355
subangular-to-round shaped **A15:** 208
types **A15:** 208
Silica-alumina glasses
chemical strengthening by ion exchange **EM4:** 462
Silica-base bonds
clay-water **A15:** 212
colloidal silica **A15:** 212
ethyl silicate **A15:** 212
sodium silicate **A15:** 212-213
Silica-phosphate glass-ceramic
bonding to bone **EM4:** 1010
Silica-silica composites *See also* composites
densification process **EM1:** 935
machining of **EM1:** 936
tensile strength **EM1:** 937-938
Silicate
carburizing affected by content in steels **A18:** 875
electroplated coatings **A18:** 838
Silicate alkaline cleaners **M5:** 23-24, 28, 35, 576-577
Silicate bonded molds *See also* Mold(s)
sodium silicate/carbon dioxide system **A15:** 229
Silicate cement
wheel polishing process using **M5:** 208-209
Silicate ceramic coatings **M5:** 533-534
Silicate ceramics
chemical etching **EM4:** 575
Silicate ester
properties **A18:** 81
Silicate glasses **EM4:** 21-22
categories **EM4:** 742
compositions **EM4:** 741, 742
crack velocity versus stress-intensity relation **EM4:** 658
indirect fractographic evidence of water in field failure **EM4:** 659
properties **EM1:** 45-47 **EM4:** 742
raw materials **EM4:** 741
uses **EM4:** 742
viscosity **EM4:** 567
Silicate inclusions
in electrical steels **A9:** 537
Silicate scale (porous)
thermal conductivity **A11:** 604
Silicate-type inclusions
defined **A9:** 16
Silicate/ester-catalyzed no-bake binder system **A15:** 215-216
Silicates
chemical-setting **A13:** 453-455
as filler for sealants **EM3:** 674
fluoboric acid as dissolution medium **A10:** 165
glasses, bonding topologies in **A10:** 393
hydrofluoric acid as dissolution medium for **A10:** 165
identification of **EM3:** 243
as inclusions **A12:** 65
for packaging **EM3:** 45
postcured water-base inorganic **A13:** 411
scales, water-formed **A13:** 491
self-cured solvent-base alkyl **A13:** 412
self-curing water-base alkali **A13:** 411-412
sintering **A10:** 166
Silicates in steel
effect of hot rolling on **A9:** 628
formation of **A9:** 631-626
randomly distributed in **A9:** 626
Silicic acid sols **EM4:** 446
Silicide
chemical vapor deposition of **M5:** 381
Silicide ceramic coatings **M5:** 535-537, 542-545
Silicide cermets
application and properties **A2:** 1005
Silicide coating
molybdenum **M5:** 661-662
niobium **M5:** 664-666
tantalum **M5:** 664-666
tungsten **M5:** 661-662
Silicide diffusion coatings **M5:** 380
Silicide formation
RBS analysis **A10:** 628
Silicide-based cermets **A7:** 813
Silicides *See also* Ordered intermetallics; Trialuminides
Fe_3Si alloys, properties **A2:** 933-934
$MoSi_2$ alloys **A2:** 934-935
Ni_3Si alloys, properties **A2:** 933
refractory **A2:** 934-935
as refractory cermet **A7:** 813
Silicocarbides
in high-alloy graphitic irons **A15:** 698
Silicon *See also* Aluminum-silicon alloys; Binary iron-base alloys; Iron-base alloys; SiAlON; Silica; Silicon carbide; Silicon modifiers; Silicon nitride
as a beta stabilizer in titanium alloys **A9:** 458
in active fluxes for submerged arc welding **A6:** 204
activity, in iron melt **A15:** 62
added to reduce solubility product in austenite **A4:** 245
as addition to aluminum alloys **A4:** 842, 843, 844, 845, 846
addition to aluminum-base alloys **A18:** 752
as addition to brazing filler metals **A6:** 904
addition to low-alloy steels for pressure vessels and piping **A6:** 667
addition to solid-solution nickel alloys **A6:** 575
alloying addition to heat-treatable aluminum alloys **A6:** 528
alloying effect in titanium alloys **A6:** 508
alloying effect on nickel-base alloys **A6:** 589, 591
alloying effects on copper alloys **M6:** 402
alloying in aluminum alloys **M6:** 373
alloying, in cast irons **A13:** 566-567
alloying, in microalloyed uranium **A2:** 677
alloying, magnetic property effects **A2:** 762
alloying, nickel-base alloys **A13:** 641
alloying, of magnetically soft materials **A2:** 766
alloying, wrought aluminum alloy **A2:** 54-55
in aluminum alloys **A15:** 746
aluminum coating affected by **M5:** 334, 336, 338, 345
-aluminum contacts, interdiffusion at **A11:** 777-778
in aluminum powder metallurgy alloys microstructure **A9:** 511
aluminum-silicon alloys, resistance brazing filler metals **A6:** 342
aluminum-silicon-lead alloys, mixed bearing microstructure **A18:** 744
aluminum-silicon-tin alloys ,mixed bearing microstructure **A18:** 744
as an addition to austenitic manganese steel castings **A9:** 239
as an addition to nickel-iron alloys **A9:** 538
as an addition to wrought heat-resistant alloys **A9:** 312
as an alloying addition to austenitic stainless steels **A9:** 283-284
analysis of phosphorus ion-implantation profile in **A10:** 623-624
with arsenic, ESR studies **A10:** 263
at elevated-temperature service **A1:** 640
in austenitic manganese steel **A1:** 822, 824
in austenitic stainless steels **A6:** 457, 458, 461, 463, 465
autocorrelation functions for surface texture **A18:** 336, 337
boule, alignment for cutting along crystallographic planes **A10:** 342
brazing of cast irons, effect on **M6:** 996
Silicon (continued)
brittle-to-ductile transition **A18:** 688
bump technology, and flip-chip assembly **EL1:** 440
in cast iron **A1:** 5, 6, 86-87, 88, 100
in cast iron composition **A18:** 695
cast iron heat treating, effect on **A4:** 667, 669
in ceramics and cermets **A18:** 813
chemical vapor deposition of **M5:** 381
chip, thermomechanical design **EL1:** 415-416
coefficient of friction as a function of temperature and doping **A18:** 688-689
composition in laser claddings **M6:** 797-798
composition, wt% (maximum) liquation cracking **A6:** 568
composition-depth profile **A7:** 256
contamination, of platinum thermocouple **A11:** 296
content, CG iron, optimum **A15:** 668
content effect on alloy solidification cracking **A6:** 89-90
content effect on fluid flow phenomena **A6:** 21
content, effect on sintering **A7:** 372
content, effect on solubility and equilibrium temperatures **A15:** 63
content in Al alloys and machinability **A16:** 761, 767-770, 775-777, 779, 785, 789, 797
content in HSLA Q&T steels **A6:** 665
content in HTLA steels **A6:** 670
content in MnS P/M powders **A16:** 885
content in stainless steels **A16:** 682-683 **M6:** 320, 322
content in tool and die steels **A6:** 674
content in ultrahigh-strength low-alloy steels **A6:** 673
content loss in electrodes after rebaking **A6:** 415
content of weld deposits **A6:** 675
control, ductile iron **A15:** 647-648
-controlled rectifier system **A7:** 423
in copper alloys **A6:** 753
cracking geometries **A18:** 687
cracking sensitivity in stainless steel casting alloys **A6:** 497
crystal, lithium-doped, for EPMA **A10:** 519
crystals, spin-dependent recombination analysis of **A10:** 258
crystals, ultrapurification by zone-refining technique **A2:** 1094
in cupolas **A15:** 388, 390
Czochralski crystal growth process **EL1:** 191
deoxidizing, copper and copper alloys **A2:** 236
depth profiles for LPCVD thin films on **A10:** 624
determined by 14-MeV FNAA **A10:** 239
diffused, temperature effect **EL1:** 958
diffusion in brazing sheet **M6:** 1023
diffusivity in solid aluminum **A11:** 777
dislocations, strong-beam and weak-beam images compared **A9:** 114
distribution in pearlite **A9:** 661
in ductile iron **A1:** 43
in ductile irons **A4:** 686, 689, 692
effect, atmospheric corrosion **A13:** 514
effect, equilibrium temperature, cast irons **A15:** 65
effect, malleable iron **A15:** 31
effect, nitrogen solubility **A15:** 82
effect of, on corrosion resistance **A1:** 912, 913
effect of, on hardenability **A1:** 393, 395
effect of, on notch toughness **A1:** 740-741
effect on aluminum alloy soldering **A6:** 628
effect on base metal color matching in aluminum alloys **A6:** 730
effect on carburization resistance in iron-alloys **A9:** 333
chromium-nickel heat-resistant casting effect on crack formation **M6:** 833

Silicon (continued)
effect on Curie point **A4:** 187
effect on ferrite formation in iron-alloys **A9:** 333
chromium-nickel heat-resistant casting
effect on hardness of tempered martensite **A4:** 124, 128-129
effect on oxidation resistance in cobalt-base heat-resistant casting alloys **A9:** 334
effect on precipitation-hardening stainless steels **A6:** 490
effect on SCC of copper **A11:** 221
effect on sigma formation in ferritic stainless steels **A9:** 285
effect, ternary iron-base alloys **A15:** 64-65
effect, wrought/cast aluminum alloys **A13:** 586
effects during water atomization **A7:** 256
in electrical steel sheet **A14:** 476
in electrical steels **A9:** 537
electrochemical grinding **A16:** 543
and electrochemical machining **A16:** 535
electroslag welding, reactions **A6:** 273, 274
in enamel cover coats **EM3:** 304
in enameling ground coat **EM3:** 304
enriched water-atomized surfaces **A7:** 252
epitaxial precipitation, in Schottky barrier contacts **A11:** 778
erosion in ceramics **A18:** 205
eutectic, growth of **A15:** 163
evaporation fields for **A10:** 587
in ferrite **A1:** 401, 406
as ferrite stabilizer **A13:** 47
filler metals for brazing of aluminum alloys **M6:** 1022-1023
FIM sample preparation of **A10:** 586
formation, wafer preparation **EL1:** 191
free energy of reaction, and gas porosity **A15:** 82
FT-IR spectra of **A10:** 123
functions in FCAW electrodes **A6:** 188
gold-silicon eutectic **EM3:** 584
gold-silicon phase diagram **EM3:** 585
in gray irons **A4:** 670, 671, 673, 675, 677, 678
growth kinetics of **A15:** 79
as H-WSI substrate **EL1:** 88
in hardfacing alloys **A18:** 759, 762, 764, 765
in heat-resistant alloys **A4:** 510, 511, 512, 514
in high-alloy white irons **A15:** 679-680
high-energy neutron irradiation of **A10:** 233
high-purity, under oxygen bombardment, ion microscope **A10:** 615
in high-speed tool steels **A16:** 52
hot dip aluminizing bath **M1:** 172
influence of, on microstructure of white iron **A1:** 94
integrated circuits, failure analysis of **A11:** 766-792
as internal reflection element **A10:** 113
inversion, due to contamination **A11:** 781
for laser alloying **A18:** 866
in laser cladding material **A18:** 867
laser-enhanced etching **A16:** 576
laser-induced CVD for synthesis **A18:** 848
laser-induced CVD for synthesis (polycrystalline) **A18:** 848
loss effect on welding parameters **A6:** 68
lubricant indicators and range of sensitivities **A18:** 301
in magnesium alloys **A9:** 428
as major element, gray iron **A15:** 629-630
in maraging steels **A4:** 222, 224
vapor pressure, relation to temperature **A4:** 495
metal-oxide semiconductors (MOS) **EL1:** 2
microalloying of **A14:** 220
modification **A15:** 161-162
modules as failure mechanism **EL1:** 1016-1017
n-type, scratch morphology **A18:** 688
nickel plating bath contaminated by **M5:** 208
in nickel-chromium white irons **A15:** 679-680
nodules, after etching away aluminum, SEM micrographs of **A11:** 774
oxidized, as substrate **EL1:** 106
oxygen cutting, effect on **M6:** 862
p-n junction failures **A11:** 782-786

Silicon (continued)
p-type, groove surface morphology and temperature effect **A18:** 687
particles, primary **A15:** 165-166
photometric analysis methods **A10:** 64
pickup in submerged arc welding **M6:** 116
polyimide adhesion **EM3:** 157
powders fused after flame spraying of cast irons **A6:** 720
presence in cast irons **M6:** 307-308
prompt gamma activation analysis of **A10:** 240
properties **EM4:** 1
pullout, as TAB mechanical failure **EL1:** 287
pure **M2:** 796-797
pure, properties **A2:** 1154
qualitative analysis of surface phase on **A10:** 341-342
RBS profiles of arsenic in **A10:** 632
reactions for formation of films **EM3:** 593
recovery from selected electrode coverings **A6:** 60
reductant of metal oxides **M6:** 692, 694
removal, by oxygen blowing **A15:** 78
semiconductor devices, failure analysis of **A11:** 766-792
significance in failure analysis programs **A18:** 311
as silicon modifier **A15:** 161
in simple steels, partitioning oxidation states in **A10:** 178
SIMS analysis of phosphorus ion-implantation profile in **A10:** 623
single-crystal (001), uniform hardness making it a control specimen **A18:** 424
single-crystal, growth **EL1:** 191
species weighed in gravimetry **A10:** 172
spectrometric metals analysis **A18:** 300
SSMS analysis of impurities in high-purity **A10:** 141
in stainless steels **A18:** 716, 720, 721
in steel **A1:** 14f-142, 145, 577
in steel weldments **A6:** 418, 419
submerged arc welding **M6:** 115
content in fluxes **M6:** 140-141
transfer due to flux content **M6:** 124-125
substrates matched to **EL1:** 441
surface mechanical damage of semiconductors caused by dicing **A18:** 686
surface mechanical damage of semiconductors caused by wafering **A18:** 685-686
surface segregation, temperature dependence of **A10:** 565
tantalum corrosion at elevated temperatures **A13:** 728
thermal diffusivity from 20 to 100 °C **A6:** 4
in thermal spray coating materials **A18:** 832
tilt boundaries studied by high resolution electron microscopy **A9:** 121
to fabricate integrated circuits **EM3:** 378
to harden and strengthen steel **A16:** 667
in tool steels **A16:** 52, 53 **A18:** 734, 735-736
ultrasonic machining ... **A16:** 532 **EM4:** 359, 360, 361
use in flux cored electrodes **M6:** 103
for valve springs for reciprocating compressors **A18:** 604
Vickers and Knoop microindentation hardness numbers **A18:** 416
volume resistivity and conductivity **EM3:** 45
in wafer processing **EM3:** 580-583
wafers, analysis of oxygen in **A10:** 122
weld-metal content, underwater welding **A6:** 1010-1011
in white iron composition **A18:** 698
in zinc alloys **A15:** 788
[111# single-crystal, Kikuchi diffraction patterns from **A10:** 438

Silicon (electrical) steels
argon oxygen decarburization **A15:** 42

Silicon bipolar drivers
in mixed technologies **EL1:** 8

Silicon boride (SiB_6)
pressure densification **EM4:** 298

Silicon brass
dezincification **A13:** 128

Silicon brasses *See also* Cast copper alloys; Leaded silicon brass
corrosion rating **A2:** 353-354
foundry properties for sand casting **A2:** 348
as high-shrinkage foundry alloy **A2:** 346
melt treatment **A15:** 775
nominal composition **A2:** 347
properties and applications **A2:** 226, 372-373

Silicon bronze *See also* Cast copper alloys
corrosion ratings **A2:** 353-354
foundry properties for sand casting **A2:** 348
galling **A18:** 723
as high-shrinkage foundry alloy **A2:** 346
nominal composition **A2:** 347
properties and applications **A2:** 226, 334-335, 371-372

Silicon bronzes *See also* Copper-silicon alloys **A6:** 752, 753
for art casting **A15:** 22
brazeability **M6:** 1034
brazing **A6:** 630, 931
composition and properties **M6:** 401
corrosion in various media **M2:** 468-469
desiliconification **A13:** 133
filler metals **A6:** 756
gas metal arc welding **M6:** 421
gas tungsten arc welding **M6:** 413-414
gas-metal arc butt welding **A6:** 760
gas-metal arc welding **A6:** 755, 766
gas-tungsten arc welding **A6:** 192, 767
to high-carbon steels **A6:** 827, 878
to low-alloy steels **A6:** 827, 828
to low-carbon steel **A6:** 827, 828
to medium-carbon steel **A6:** 827, 828
to stainless steel **A6:** 827, 828
globular-to-spray transition currents for electrodes **A6:** 182
melt treatment **A15:** 775
plasma arc welding **A6:** 754
resistance spot welding **A6:** 850 **M6:** 479
shielded metal arc welding **A6:** 754, 755, 763, 766 **M6:** 425-426
surface condition **A6:** 754
weld cladding **A6:** 822
weld overlay for hardfacing alloys **A6:** 820
weld overlay material **M6:** 816
weldability **A6:** 753

Silicon carbide *See also* Electrical resistance alloys **M3:** 646, 647, 655
abrasive disks, for surface preparation **A17:** 52
in abrasive flow machining **A16:** 517
abrasive for abrasive jet machining **A16:** 512, 513
abrasive for cast irons **A16:** 661
abrasive for electrochemical grinding **A16:** 545
abrasive in honing cast irons **A16:** 664
in abrasive slurry for ultrasonic machining **A16:** 529
for abrasive slurry in ultrasonic machining **A16:** 529
abrasive wear **A18:** 189
for abrasive wear of nickel aluminide alloys **A18:** 773-774
abrasives **A16:** 102, 430-432, 434, 444, 450, 453
applications **A18:** 812, 814
blasting with **M5:** 84, 94
ceramic mechanical seals for pumps **A18:** 598
coating for laser cladding materials **A18:** 868
coating for titanium alloys **A18:** 781, 782
coatings for composites **EM3:** 293-294
contact bridge when brass "contacts" **A18:** 236
as CTE-matched material **EL1:** 306
defined **EM1:** 21
as erodent in solid-particle erosion of nickel aluminide alloys **A18:** 773
erosion in ceramics **A18:** 205
erosion in metals **A18:** 202, 203, 204
erosion test results **A18:** 200
fibers, acoustic emission inspection **A17:** 288
fretting wear **A18:** 248, 250
friction coefficient data **A18:** 72
for gas-lubricated bearings **A18:** 532
and gear manufacture **A16:** 350, 351, 353
grinding **A16:** 421
for grinding carbide cutters **A16:** 317
grinding wheels for Al alloys **A16:** 801
grinding wheels for carbon and alloy steels **A16:** 676

Silicon carbide (continued)
grinding wheels for CPM 10V tool -steel ... **A16:** 735
grinding wheels for Cu alloys ... **A16:** 819
grinding wheels for Ni alloys ... **A16:** 843
grinding wheels for refractory metals ... **A16:** 869
grinding wheels for stainless steels ... **A16:** 705
grinding wheels for thread grinding ... **A16:** 271, 272
grinding wheels for Ti alloys ... **A16:** 848, 850, 851
grinding wheels for tool steels ... **A16:** 728
grinding wheels for tools for Mg alloys ... **A16:** 821
grinding wheels for W ... **A16:** 859
grinding wheels for Zr ... **A16:** 854, 855, 856
honing Al alloys ... **A16:** 802
in honing stones ... **A16:** 475, 476, 477, 478, 484, 490
incorporated in nickel electroplating ... **A18:** 836, 837
lapping ... **A16:** 492-493, 501-502
metallization ... **EL1:** 306-307
methods used for synthesis ... **A18:** 802
and Mg alloys ... **A16:** 827
normal and abnormal flaws in ... **A11:** 751
particle, properties ... **EL1:** 1118
particles, effect on solid/liquid interface ... **A15:** 145
polishing with ... **M5:** 108
properties ... **A18:** 192, 801, 803, 813
seal material ... **A18:** 551
siliconized microstructure ... **A9:** 148, 158
sintered, fracture surface ... **A11:** 745
solid particle erosion ... **A18:** 210
solid particle impingement erosion of cobalt-base alloys ... **A18:** 768, 769
thermal properties ... **A18:** 42
tool material for cast irons ... **A16:** 651, 652
tramp impurity in semimet brake lining ... **A18:** 572
vacuum deposition of an interference film ... **A9:** 148
Vickers and Knoop microindentation hardness numbers ... **A18:** 416
wheels, thread grinding ... **A16:** 271, 272
whiskers ... **A15:** 88, 84

Silicon carbide (SiC) *See also* Engineering properties of single oxides ... **EM4:** 17-20, 25, 32-33, 47
as abrasive ... **EM4:** 324
abrasive for truing ... **EM4:** 347
as abrasive for ultrasonic machining ... **EM4:** 360
as abrasive grains or cutting tool tips for grinding or machining ... **EM4:** 329
abrasive property control ... **EM4:** 332
applications ... **EM4:** 1, 46, 48, 331, 960
aerospace ... **EM4:** 1003, 1005
heat exchangers ... **EM4:** 981, 982, 983
refractory ... **EM4:** 900, 902, 915
wear ... **EM4:** 974-975, 976, 977
automotive applications ... **EM4:** 960
bar fractures analyzed by CARES program compared to IEA Annex II agreement results ... **EM4:** 704-705
bond type ... **EM4:** 331
bondability ... **EM4:** 332
brazing ... **A6:** 635-636
ceramic component development ... **EM4:** 718
ceramic dies ... **EM4:** 187
characteristics ... **EM4:** 976
chemical etching ... **EM4:** 575
composition ... **EM4:** 46
crack growth ... **EM4:** 586
crystal structure ... **EM4:** 30
densification resisted in HIP process ... **EM4:** 197
diluent for boriding ... **A4:** 441
electrical properties ... **EM4:** 807-808
as embedding agent ... **EM4:** 572
eutectic joining ... **EM4:** 526
fabrication ... **EM4:** 806-807

Silicon carbide (SiC) (continued)
fiber ... **EM4:** 224
gas turbine engine blade foreign object damage ... **EM4:** 719
gas turbine engine blade strength at high temperatures and speeds ... **EM4:** 719
grades available ... **EM4:** 676
grain-growth inhibitor ... **EM4:** 188
as grinding abrasive before microstructural analysis ... **EM4:** 572, 574
for handling and processing equipment ... **EM4:** 959
hardness ... **EM4:** 351, 806, 807, 808
for heat treating furnace equipment ... **A4:** 468, 472, 473
as heating material ... **A2:** 829
joining method and wear application ... **EM4:** 974
in joining non-oxide ceramics ... **EM4:** 528
key product properties ... **EM4:** 48
lapping abrasive ... **EM4:** 351, 352
magnetic properties ... **EM4:** 808
manufacture ... **EM4:** 806
manufacturing processes ... **EM4:** 331
material selection for structural ceramics ... **EM4:** 29
matrix material for ceramic-matrix composites ... **EM4:** 840
mechanical properties ... **EM4:** 316, 331, 807, 808
metallic binder phase, effects ... **A2:** 1008
as monoxide refractory material ... **EM4:** 14
for non-oxide ceramic heating elements of electrically heated furnaces ... **EM4:** 248-249, 250
non-oxide ceramic joining ... **EM4:** 480
optical properties ... **EM4:** 808
phase analysis ... **EM4:** 25
physical properties ... **EM4:** 30, 191, 316
pressure densification
pressure ... **EM4:** 301
technique ... **EM4:** 301
temperature ... **EM4:** 301
properties ... **A6:** 629, 949 **EM4:** 1, 332, 351, 808, 974, 976
adiabatic engine use ... **EM4:** 990
raw materials ... **EM4:** 48
refractory composition ... **EM4:** 896
refractory physical properties ... **EM4:** 897, 898, 899
reinforcements, aluminum metal-matrix composites ... **A2:** 7
rejected for turbocharger application ... **EM4:** 725, 726
scuffing temperatures and coefficients of friction between ring and cylinder liner materials ... **EM4:** 991
silicon-infiltrated applications ... **EM4:** 1003
siliconized, for gas turbine scrolls ... **EM4:** 720
sintering aid ... **EM4:** 188
solid-state bonding ... **EM4:** 525
solid-state sintering ... **EM4:** 273, 278
as structural ceramic, applications and properties ... **A2:** 1022
structures ... **EM4:** 806, 807
as superhard material ... **A2:** 1008
supply sources ... **EM4:** 46
thermal etching ... **EM4:** 575
thermal expansion ... **EM4:** 685-686
thermal expansion coefficient ... **A6:** 907
thermal properties ... **EM4:** 30, 191, 316, 331, 807, 808, 974
thermal shock resistance ... **EM4:** 1003
thermostructural ceramic for aerospace applications ... **EM4:** 1003
ultrasonic machining ... **EM4:** 359-360
ultrasonic measurements ... **EM4:** 622, 623, 624
Vickers hardness ... **EM4:** 974
whisker-reinforced alumina ... **A2:** 1023-1024

Silicon carbide (SiC) fibers *See* also Continuous silicon carbide fiber MMCs; Silicon carbide (SiC) whiskers ... **EM1:** 58-59
ceramic, thermal stability ... **EM1:** 64
continuous ... **EM1:** 31, 63-64
costs ... **EM1:** 859
in discontinuous ceramic fiber MMCs ... **EM1:** 903-910
importance ... **EM1:** 43
production, for continuous SiC fiber MMCs ... **EM1:** 858-859
properties ... **EM1:** 58, 175
for short fiber reinforced composites ... **EM1:** 120
tennis racket application ... **EM1:** 31
tensile strength ... **EM1:** 59
types ... **EM1:** 58-59
variations ... **EM1:** 859

Silicon carbide (SiC) reinforced ceramics ... **EM1:** 941-944
applications ... **EM1:** 943
crack growth failure, low stress ... **EM1:** 943
creep resistance ... **EM1:** 943
fracture toughness/flexural strength ... **EM1:** 942-943
mechanical properties ... **EM1:** 942
thermal conductivity/expansion ... **EM1:** 943
thermal shock response ... **EM1:** 943

Silicon carbide (SiC) whisker reinforced aluminum alloys ... **EM1:** 890-894

Silicon carbide (SiC) whiskers
ceramic composite reinforcement by ... **EM1:** 941-944
characteristics ... **EM1:** 941
in metal matrix composites (MMCs) ... **EM1:** 889-902

Silicon carbide + 2 magnesium oxide (SiC + 2MgO)
adiabatic temperatures ... **EM4:** 229
synthesized by SHS process ... **EM4:** 229

Silicon carbide abrasives
for hand polishing ... **A9:** 35

Silicon carbide aluminum composites
properties ... **EM1:** 862-863

Silicon carbide as an abrasive for nonferrous metals and nonmetals ... **A9:** 24

Silicon carbide ceramic fibers
thermal stability ... **EM1:** 64

Silicon carbide ceramic/metal composite
applications ... **EM4:** 963, 964
property comparison, mineral processing ... **EM4:** 962

Silicon carbide ceramics
confocal microscopy application to analyze cracks from wear ... **A18:** 360, 361

Silicon carbide fiber metal-matrix composites ... **A2:** 904-905

Silicon carbide, hot pressed *See* Hot pressed silicon carbide

Silicon carbide metalloid cermets
application and properties ... **A2:** 1002

Silicon carbide particle reinforced aluminum
as application ... **EL1:** 1127
applications ... **EL1:** 1122-1125
coefficient of thermal expansion (CTE) ... **EL1:** 1124
as heat sink ... **EL1:** 1130-1131
microwave packages ... **EL1:** 1127-1128
properties ... **EL1:** 1123

Silicon carbide, reaction-bonded *See* Reaction-bonded silicon carbide

Silicon carbide Refel
properties ... **A18:** 548

Silicon carbide, sintered *See* Sintered silicon carbide

Silicon carbide whisker
properties ... **A18:** 803

Silicon carbide whisker-reinforced alumina
heat treatments ... **A18:** 814
as structural ceramic ... **A2:** 1023-1024

Silicon carbide whisker-reinforced alumina ($Al_2O_3 \bullet SiC_w$)
properties ... **A6:** 949

SUBJECTS OF THE INDEXED VOLUMES: ASM Handbook (designated by the letter "A"): **A1:** Properties and Selection: Irons, Steels, and High-Performance Alloys (1990); **A2:** Properties and Selection: Nonferrous Alloys and Special-Purpose Materials (1990); **A3:** Alloy Phase Diagrams (1992); **A4:** Heat Treating (1991); **A6:** Welding, Brazing, and Soldering (1993); **A7:** Powder Metallurgy (1984); **A8:** Mechanical Testing (1985); **A9:** Metallography and Microstructures (1985); **A10:** Materials Characterization (1986); **A11:** Failure Analysis and Prevention (1986); **A12:** Fractography (1987); **A13:** Corrosion (1987); **A14:** Forming and Forging (1988); **A15:** Casting (1988); **A16:** Machining (1989); **A17:** Nondestructive Testing and Quality Control (1989); **A18:** Friction, Lubrication, and Wear Technology (1992). **Metals Handbook, 9th Edition** (designated by the letter "M"): **M1:** Properties and Selection: Irons and Steels (1978); **M2:** Properties and Selection: Nonferrous Alloys and Pure Metals (1979); **M3:** Properties and Selection: Stainless Steels, Tool Materials and Special-Purpose Materials (1980); **M4:** Heat Treating (1981); **M5:** Surface Cleaning, Finishing, and Coating (1982); **M6:** Welding, Brazing, and Soldering (1983). **Engineered Materials Handbook** (designated by the letters "EM"): **EM1:** Composites (1987); **EM2:** Engineering Plastics (1988); **EM3:** Adhesives and Sealants (1990); **EM4:** Ceramics and Glasses (1991); **Electronic Materials Handbook** (designated by the letters "EL"): **EL1:** Packaging (1989).

Silicon carbide-copper composites
properties ... **EM1:** 863
Silicon carbide-magnesium composites
properties ... **EM1:** 863
Silicon carbide-titanium composites
properties ... **EM1:** 863
Silicon carbide/aluminum composites ... **A13:** 860
Silicon carbide/aluminum metal-matrix composites ... **A2:** 905
Silicon carbides
effect of carbon KVV lineshapes on ... **A10:** 553
Silicon carboxynitride ... **EM4:** 19
Silicon ceramics
properties ... **A6:** 992
Silicon chlorate ($SiCl_4$)
chemical vapor deposition ... **EM4:** 445
Silicon circuit board (SCB)
bare, testing of ... **EL1:** 362-363
defined ... **EL1:** 354
Silicon crystal
dislocation lines in ... **A9:** 127
Silicon dioxide ... **EM3:** 592-594
abrasive in commercial prophylactic paste ... **A18:** 666
as dielectric medium ... **EL1:** 88
direct evaporation ... **A18:** 844
for multichip structures ... **EL1:** 301-302
properties ... **A18:** 801
to fabricate integrated circuits ... **EM3:** 378
Vickers and Knoop microindentation hardness numbers ... **A18:** 416
in wafer processing ... **EM3:** 582, 583, 584
Silicon dioxide (SiO_2) *See also* Engineering properties of single oxides; Silica
additive used to attain requisite viscosity for blowing ... **EM4:** 21
component in photochromic ophthalmic and flat glass composition ... **EM4:** 442
component in photosensitive glass composition ... **EM4:** 440
composition ... **EM4:** 14
in composition of glass-ceramics ... **EM4:** 499
in composition of leachable alkali-borosilicate glasses ... **EM4:** 428
in composition of textile products ... **EM4:** 403
in composition of wool products ... **EM4:** 403
content effect in fluxes ... **A6:** 57, 58
corrosion resistance of refractories ... **EM4:** 391
in drinkware compositions ... **EM4:** 1102
electrical/electronic applications ... **EM4:** 1105
in glaze composition for tableware ... **EM4:** 1102
in ovenware compositions ... **EM4:** 1103
pressure densification
pressure ... **EM4:** 301
technique ... **EM4:** 301
temperature ... **EM4:** 301
properties ... **EM4:** 14, 424
properties (vitreous) ... **A6:** 629
role in glazes ... **EM4:** 1062
solderable and protective finishes for substrate materials ... **A6:** 979
in tableware compositions ... **EM4:** 1101
Silicon dioxide as an interference film ... **A9:** 147
Silicon dioxide, as refractory
core coatings ... **A15:** 24
Silicon dioxide films
PVD applications ... **EM4:** 319
Silicon dioxide glass *See* Fused silica fibers
Silicon grinding balls ... **A7:** 58
Silicon in cast iron *See also* High silicon cast irons; Medium silicon cast irons
content ... **M1:** 3-4
corrosion resistance, effect on ... **M1:** 88-91
depth of chill, effect on ... **M1:** 77-78
ductile iron ... **M1:** 37-41, 49, 51, 53
elevated-temperature properties effect on ... **M1:** 91-94
gray iron ... **M1:** 11-12, 21-22, 28-31
malleable iron ... **M1:** 58-60, 64, 73
structure, influence on ... **M1:** 85
Silicon in copper ... **M2:** 242
Silicon in steel ... **M1:** 115, 410, 417
atmospheric corrosion affected by ... **M1:** 717, 721-722
castings, effect on ... **M1:** 378, 399
Silicon in steel (continued)
constructional steels for elevated temperature use, effect on ... **M1:** 647
effect on hot dip galvanized coatings ... **M1:** 170
graphitization, effect on ... **M1:** 686
modified low-carbon steels ... **M1:** 162
notch toughness, effect on ... **M1:** 693
relation to deoxidation practice ... **M1:** 121, 123-124
seawater corrosion, effect on ... **M1:** 741
steel sheet, effect on formability ... **M1:** 554
temper embrittlement, effect on ... **M1:** 684
Silicon intermetallic compounds in uranium alloys ... **A9:** 477
Silicon irons *See also* Cast irons, Magnetic materials ... **A9:** 245-246
Silicon microelectronic die
silicone conformal coatings for ... **EL1:** 822
Silicon modification
aluminum-silicon alloys ... **A15:** 161-162
Silicon modifiers
characteristics ... **A15:** 161
effects ... **A15:** 162-165
sodium as ... **A15:** 162-165
strontium as ... **A15:** 162-165
types ... **A15:** 7
Silicon monoxide films
PVD applications ... **EM4:** 219
Silicon nitride *See also* Ceramics; SiAlON ... **EM3:** 592, 593, 594
abrasive wear ... **A18:** 189
applications ... **A18:** 812-813, 814, 815
internal combustion engine parts ... **A18:** 558, 559
brazing with Cusil-ABA in Ar-O_2 atmosphere ... **A6:** 957
coatings ... **A16:** 103
crack initiation site ... **A11:** 745
diffusion welding ... **A6:** 886
erosion test results ... **A18:** 200
fretting wear ... **A18:** 248, 250
friction coefficient data ... **A18:** 72, 815
for gas-lubricated bearings ... **A18:** 532
gray iron metal removal rates ... **A16:** 652
high-speed machining ... **A16:** 602
hot-filament chemical vapor deposition ... **A18:** 848
machinability ... **A16:** 639, 640, 643, 644
matrix composites ... **A2:** 1024
on copper-titanium base filler metals ... **A6:** 115-116
primary applications ... **A16:** 639
properties ... **A18:** 42, 801, 812-813
properties (Si_3N) ... **A6:** 629, 949
reaction-sintered, fracture mirrors ... **A11:** 745
rolling contact fatigue of advanced ceramics ... **A18:** 260, 261
rolling-element ceramics for bearings ... **A18:** 503
SIMS analysis (in water) ... **A18:** 459
solderable and protective finishes for substrate materials (Si_3N_4) ... **A6:** 979
as structural ceramic, application and properties ... **A2:** 1022
as superhard material ... **A2:** 1008
thickener for grease high-vacuum application lubricants ... **A18:** 157, 158
to fabricate integrated circuits ... **EM3:** 378
tools for cast irons ... **A16:** 656
wettability (Si_3N_4) ... **A6:** 115
Silicon nitride (Si_3N_4) *See also* Engineering properties of nitrides ... **EM4:** 19, 32, 47, 49
as abrasive grains or cutting tool tips for grinding or machining ... **EM4:** 32
active metal brazing ... **EM4:** 523-524, 525
adiabatic temperatures ... **EM4:** 229
applications ... **EM4:** 48, 200, 230, 960
automotive ... **EM4:** 529, 960
diesel engines ... **EM4:** 677, 678
electrical/electronic ... **EM4:** 1105
turbomachinery ... **EM4:** 677-678
wear ... **EM4:** 975
bar fractures analyzed by CARES program compared to IEA Annex II agreement results ... **EM4:** 704-705
brazing with glasses ... **EM4:** 519-520
CARES computer program for preliminary design of rotor applications ... **EM4:** 707
for ceramic blade materials preventing foreign object damage ... **EM4:** 719
Silicon nitride (Si_3N_4) (continued)
ceramic component development ... **EM4:** 718
ceramic cutting tools ... **EM4:** 966
chemical etching ... **EM4:** 575
chemical vapor deposition ... **EM4:** 217
comminution ... **EM4:** 77
complementary flexure and tensile strength rupture data ... **EM4:** 595
crack growth ... **EM4:** 586, 587, 593
crystal structure ... **EM4:** 30
densified ... **EM4:** 1000, 1001
eutectic joining ... **EM4:** 526
gas turbine engine blade strength at high temperatures and speeds ... **EM4:** 719-720
grades available ... **EM4:** 676
grain-growth inhibitor ... **EM4:** 188
high-cycle fatigue, and slow crack growth ... **EM4:** 685
high-temperature strength of joints ... **EM4:** 527, 528
hot isostatic pressing ... **EM4:** 812, 998
hot pressing ... **EM4:** 812
improved refractory resulting from bond with SiC ... **EM4:** 14
joined to metals, automotive turbocharger application ... **EM4:** 724
key product properties ... **EM4:** 48
machining flaw ... **EM4:** 643
material removal rates with electrical discharge machining ... **EM4:** 376
material selection for structural ceramics ... **EM4:** 29
material-surface degradation ... **EM4:** 685
matrix material for ceramic-matrix composites ... **EM4:** 840
mechanical properties ... **EM4:** 30, 191, 316
non-oxide ceramic joining ... **EM4:** 480
particle rearrangement process ... **EM4:** 297
physical properties ... **EM4:** 30, 191, 316
pressure densification ... **EM4:** 298-301
pressure ... **EM4:** 301
technique ... **EM4:** 301
temperature ... **EM4:** 301
properties ... **EM4:** 1, 30
properties, adiabatic engine use ... **EM4:** 990
properties as a function of temperature ... **EM4:** 678
raw materials ... **EM4:** 48
recrystallization ... **EM4:** 190
scuffing temperatures and coefficients of friction between ring and cylinder liner materials ... **EM4:** 991
sintering aid ... **EM4:** 188
solid-state bonding in joining non-oxide ceramics ... **EM4:** 525
spherical media composition for wet-milling ... **EM4:** 78
strength-limiting defects ... **EM4:** 593
stress-rupture life ... **EM4:** 1001
superplasticity ... **EM4:** 301
synthesized by SHS process ... **EM4:** 229
thermal etching ... **EM4:** 575
thermal expansion ... **EM4:** 685-686
thermal properties ... **EM4:** 30, 191
as thermostructural ceramic for aerospace applications ... **EM4:** 1003
ultrasonic machining ... **EM4:** 359, 360
ultrasonic measurements ... **EM4:** 624
Weibull plot of strength ... **EM4:** 31
zirconia as additive ... **EM4:** 777
Silicon nitride fibers ... **EM1:** 118
Silicon nitride films
PVD applications ... **EM4:** 219
Silicon nitride, hot pressed *See* Hot-pressed silicon nitride
mechanical properties ... **A7:** 516
Silicon nitride matrix composites
as structural ceramics ... **A2:** 1024
Silicon nitride powders
densification by polymers ... **EM4:** 226
Silicon nitride, reaction-bonded *See* Reaction-bonded silicon nitride
Silicon nitride, sintered *See* Sintered silicon nitride (SSN)

Silicon nitride-beryllium nitride [$Si_3N_4(Be_3N_2)$#, pressure densification
pressure ... **EM4:** 301
technique ... **EM4:** 301
temperature ... **EM4:** 301
Silicon nitride-bonded silicon carbide (SNBSC)
applications ... **EM4:** 964
mineral processing ... **EM4:** 961
property comparison, mineral processing ... **EM4:** 962
Silicon nitride-titanium nitride
electrical discharge machining ... **EM4:** 374
Silicon oxide ... **A7:** 252, 256, 802
ion sputtering effect ... **EM3:** 245
vacuum heat-treating support fixture material ... **A4:** 503
Silicon oxide (glass)
thermal properties ... **A18:** 42
Silicon oxide cermets
applications and properties ... **A2:** 992
Silicon oxide failures *See also* Silicon oxide interface, failures
charge trapping in ... **A11:** 780
defect-related dielectric breakdown ... **A11:** 778-779
integrated circuits ... **A11:** 778-781
intrinsic dielectric breakdown ... **A11:** 779-780
Silicon oxide inclusions in steel ... **A9:** 185
Silicon oxide interface failures *See also* Silicon oxide failures
charge ... **A11:** 782
integrated circuits ... **A11:** 781-782
interface traps ... **A11:** 782
ion-induced threshold drift ... **A11:** 782
ionic contamination ... **A11:** 781
surface-charge accumulation ... **A11:** 781
Silicon oxide vacuum coatings ... **M5:** 390-391, 394-395, 401
Silicon oxynitride ... **EM4:** 819
corrosion resistance as a cryolite container ... **EM4:** 239
flexural strength ... **EM4:** 239
made by a reaction-bonding technique ... **EM4:** 236
made by reaction-forming processes ... **EM4:** 239
properties of ceramic bodies ... **EM4:** 239, 240
properties when reaction bonded ... **EM4:** 240
Silicon oxynitride film
to increase fatigue resistance of glass by coating ... **EM4:** 744
Silicon *p-n* junction failures
alpha particle induced ... **A11:** 782-784
integrated circuits ... **A11:** 782-786
latch-up in CMOS ... **A11:** 785-786
radioactivity-induced ... **A11:** 784-785
Silicon polyester coating for steel sheet ... **M1:** 176
Silicon red brass *See also* Brasses; Red brass; Wrought coppers and copper alloys
applications and properties ... **A2:** 337
Silicon rubber ... **EM3:** 594
Silicon rubber compound
for surface replicas ... **A17:** 53
Silicon rubber insulation
wrought copper and copper alloy products ... **A2:** 258
Silicon semiconductor devices *See also* Integrated circuits; Integrated circuits, failure analysis of; Metal-oxide semiconductor (MOS) devices
failure analysis of ... **A11:** 766-792
failure mechanisms ... **EL1:** 959
time-dependent failure mechanisms ... **A11:** 767
Silicon semiconductors
field-assisted bonding ... **EM4:** 479
Silicon steel
hot dip galvanized coating of ... **M5:** 324, 327-330
particles of Fe_3C at grain boundaries of ... **A9:** 127
Silicon steel transformer sheets
metallographic texture control ... **A9:** 62-63
Silicon steels *See also* Magnetically soft materials ... **M3:** 1
bainitic structures ... **A9:** 663-665
corrosion resistance ... **A2:** 778
flat-rolled products, as magnetically soft materials ... **A2:** 766-769
heat treatment ... **A2:** 762
nonoriented ... **A14:** 476, 480
nonoriented, grades ... **A2:** 766-769
oriented ... **A2:** 767-769 **A14:** 476-477, 480
Silicon substrates
advanced, for hybrid circuits ... **EL1:** 8
epitaxial GaAs on ... **EL1:** 200-201
large-aspect ratio ... **EL1:** 8
multichip structures ... **EL1:** 304-305
Silicon thyristors
acoustic microscopy ... **A17:** 481
Silicon transistors
failure mechanisms ... **EL1:** 959-961
Silicon, vapor pressure
relation to temperature ... **M4:** 309, 310
Silicon VLSI memory and logic
in mixed technologies ... **EL1:** 8
Silicon wafers
optoelectronics on ... **EL1:** 8
oxygen determined in ... **A10:** 122-123
Silicon yellow brass *See also* Cast copper alloys; Yellow brasses
nominal composition ... **A2:** 347
properties and applications ... **A2:** 373-374
Silicon-aluminum-oxynitride (SiAlON)
applications and properties ... **A2:** 1022
Silicon-based polymers ... **EM4:** 62
Silicon-bronze alloys
contact-finger retainer, failed ... **A11:** 310
corrosion in ... **A11:** 201
Silicon-carbide refractory, applications
protection tubes and wells ... **A4:** 533
Silicon-carbide-glass ceramic composite ... **A9:** 597
Silicon-chromium cast irons
scaling resistance ... **M1:** 93-94
Silicon-controlled rectifier power supply
infrared soldering ... **A6:** 136
Silicon-controlled rectifiers (SCR) contactors, flash welding ... **A6:** 247, 248
power sources ... **A6:** 38-39, 42, 43
Silicon-copper alloys
pickling of ... **M5:** 612-613
Silicon-epoxies ... **EM3:** 594
Silicon-infiltrated silicon carbide (Si/SiC)
scanning acoustic microscopy wear studies ... **A18:** 409
Silicon-iron (oriented)
permeability ... **EM4:** 1162
resistivity ... **EM4:** 1162
saturation flux density ... **EM4:** 1162
Silicon-iron alloys, bar and heavy strip
as magnetically soft ... **A2:** 769
Silicon-iron electrical steels *See* Electrical steels
Silicon-iron-bronze
cage material for rolling-element bearings ... **A18:** 503
Silicon-modified polyimides ... **EM3:** 594
Silicon-modifying additions
wrought aluminum alloy ... **A2:** 44
Silicon-on-silicon hybrids ... **EL1:** 372-373
Silicon-oxynitride ... **EM3:** 592, 593, 594
Silicon-silicon carbide composite
radiant tube applications ... **A4:** 517-518
Silicone
as bag material ... **A7:** 447
as defoamers for metalworking lubricants ... **A18:** 142
as foam inhibitors ... **A18:** 108
properties ... **A18:** 81
Silicone adhesives ... **EM1:** 684, 687
Silicone alkyd coatings ... **M5:** 474, 500, 501
Silicone caulks
water-base ... **EM3:** 192
Silicone conformal coatings *See also* Conformal coatings; Silicone-base coatings; Silicones
advantages ... **EL1:** 773
application methods ... **EL1:** 773-774
property ranges ... **EL1:** 773
ultraviolet-curable ... **EL1:** 824
Silicone faying surface sealants ... **EM1:** 719
Silicone fluids
lubricants for rolling-element bearings ... **A18:** 135-136
polymer additives ... **A18:** 154, 157
Silicone gel ... **EM3:** 594
Silicone oils
high-vacuum lubricant applications ... **A18:** 155
surface tension ... **EM3:** 181
Silicone phenolics ... **EM3:** 104
Silicone plastics
defined ... **EM1:** 21
Silicone polymers ... **EM3:** 49
Silicone resins
applications ... **EM3:** 278
surface preparation ... **EM3:** 278
Silicone resins and coatings ... **M5:** 473-475, 621
thermosetting ... **M5:** 621
Silicone rubber
contact-angle testing ... **EM3:** 277
for formed-in-place gaskets ... **EM3:** 57
lap-shear strength ... **EM3:** 276
medical applications ... **EM3:** 576 **EM4:** 1009
for platen seals ... **EM3:** 694
principal stress pattern ... **EM3:** 327
for Thermex process press tooling ... **EM3:** 711
Silicone vapor
atmospheric effect causing dusting ... **A18:** 684
Silicone-base coatings *See* also Coatings; Conformal coatings; Silicone conformal coatings; Silicones
application methods ... **EL1:** 773-774
chemical reactions ... **EL1:** 774
materials ... **EL1:** 773
physical properties ... **EL1:** 773
Silicone-base sealers
aluminum coatings ... **M5:** 345
Silicone-polyimides
for coating and encapsulation ... **EM3:** 580
Silicone-rubber encapsulated integrated-circuit chip ... **A11:** 775
Silicones *See also* Coatings; Conformal coatings; Silicon conformal coatings; Silicone-base coatings
chemical structure ... **EL1:** 244
as coating/encapsulant, introduction ... **EL1:** 759-760
for coating/encapsulation ... **EL1:** 242
as conformal coating ... **EL1:** 763
as encapsulants, ILB chips ... **EL1:** 283
major suppliers ... **EL1:** 823
properties ... **EL1:** 822-823
types of silicone coatings ... **EL1:** 823-824
ultraviolet-curable silicone coatings ... **EL1:** 824
and urethane, acrylic, epoxy, compared ... **EL1:** 775
Silicones (SI) *See also* Thermosetting resins ... **EM3:** 26, 44, 74-75, 82-83, 133-137
acetoxy ... **EM3:** 77
advantages and disadvantages ... **EM3:** 675
acrylated ... **EM3:** 598
addition cure mechanism ... **EM3:** 598, 599
additives and modifiers ... **EM3:** 135
for adhesive bonding of aircraft canopies and windshields ... **EM3:** 563-564
advantages and limitations ... **EM3:** 79, 85
for aerodynamic smoothing ... **EM3:** 59
aerospace industry applications ... **EM3:** 218, 220, 559
for aircraft and device encapsulation ... **EM3:** 677
alkoxy-cured ... **EM3:** 59
appliance market applications ... **EM3:** 59
for attaching stone, metal, or glass to a building side ... **EM3:** 188

Silicones (SI) (continued)
for automotive electronics bonding **EM3:** 553
bonding substrates and applications **EM3:** 56, 57, 79, 85
catalysts **EM3:** 598, 599
for caulking **EM3:** 607
characteristics **EM2:** 267 **EM3:** 53
chemical properties **EM3:** 52
chemistry **EM2:** 66, 265 **EM3:** 49, 50, 79, 133-134
for circuit protection **EM3:** 592
for coating and encapsulation **EM3:** 580
coating for aramid fibers **EM3:** 285
compared to epoxies **EM3:** 98
compared to polysulfides **EM3:** 195
competing adhesives **EM3:** 134
competing with urethane sealants **EM3:** 205
for component carrier tapes **EM3:** 134
component covers **EM3:** 611
component terminal sealant **EM3:** 612
as conductive adhesives **EM3:** 76
conformal overcoat **EM3:** 592
construction industry applications **EM3:** 218-220
contaminant for aluminum adherends **EM3:** 264
as contaminants in a metal bond clean-room environment **EM3:** 738
cost factors **EM3:** 134
costs and production volume **EM2:** 265
critical surface tensions **EM3:** 180
cross-linking **EM3:** 133, 135, 137
cure mechanisms **EM3:** 79, 133, 134, 136, 137
cure properties **EM3:** 51
for curtain wall construction **EM3:** 549
defined **EM2:** 38
dielectric sealants **EM3:** 611
durability superior to solvent acrylics **EM3:** 190
electrical contact assemblies **EM3:** 611
electrical insulator protection **EM3:** 612
electrical/electronics applications, silicones **EM3:** 44
for electrically conductive bonding **EM3:** 572
for electronic general component bonding **EM3:** 573
electronic packaging applications **EM3:** 594, 597-599
electronics industry applications **EM3:** 218, 220
engineering adhesives family **EM3:** 567
epoxidized **EM3:** 599
for exterior mirrors bonding **EM3:** 553
for exterior seals in construction **EM3:** 56
failure analysis role **EM3:** 249
films and electrical properties after curing **EM3:** 136
for formed-in-place gaskets **EM3:** 50, 57-58
for gasket replacements **EM3:** 547
heat-curable rubber (HCR) **EM2:** 267
for high-temperature insulation and seals for plasma spray masking **EM3:** 134
for high-temperature sealing of aircraft assemblies **EM3:** 808
for highway construction joints **EM3:** 57
for in-house glazing **EM3:** 58
as injection-moldable **EM2:** 322
for insulated glass (IG) construction **EM3:** 46, 50, 58, 196
IPN polymers **EM3:** 602
for jet engine firewall sealing **EM3:** 58, 59
for jet window sealing **EM3:** 59
markets **EM3:** 134
for masking of printed circuit boards **EM3:** 134
for medical bonding (class IV approval) **EM3:** 576
as medium-temperature resin system **EM2:** 442-443
for metal building construction **EM3:** 57
for mirror assemblies **EM3:** 553
modified **EM3:** 675
moisture cure mechanisms **EM3:** 598, 599
moisture exposure **EM3:** 736
molds, rotational molding **EM2:** 367
for mounting wires and circuitry **EM3:** 137
for organic junction coatings **EM3:** 587
packaging of **EM3:** 135-136
peel strengths **EM3:** 136, 137
performance **EM3:** 674
photoinitiators **EM3:** 598
as polyimide toughening agent **EM3:** 161
pot life **EM3:** 79, 135-136

Silicones (SI) (continued)
for potting **EM3:** 585
prebond treatment **EM3:** 35
predicted 1992 sales **EM3:** 77
pressure-sensitive properties **EM3:** 133, 134, 135, 137
primer requirements **EM3:** 136
product forms/applications **EM2:** 266-267
properties **EM2:** 267 **EM3:** 49-50, 134-137
property ranges of coatings **EM3:** 599
for protecting electronic components **EM3:** 59
resin functionality **EM3:** 133-134
room-temperature-vulcanizing (RTV) **EM3:** 50, 215-222
acetoxy-cured **EM3:** 59
aerospace window sealant **EM3:** 220
applications **EM3:** 606, 608
for automotive circuit devices **EM3:** 610
automotive circuit devices sealant **EM3:** 610
automotive gasketing **EM3:** 220
basic and acid catalysts **EM3:** 215
body assembly sealants **EM3:** 609
building and highway expansion joints **EM3:** 218-219
chemical gasketing applications **EM3:** 609
chemistry **EM3:** 215-217
competing with anaerobics **EM3:** 116
condensation catalysts used **EM3:** 216, 217
consumer market **EM3:** 221
cost factors **EM3:** 220, 221
cross-linking system **EM3:** 215-216, 217
cure characteristics **EM3:** 218
cure mechanisms **EM3:** 215-216, 217, 219
dielectric sealants **EM3:** 611
for eight-cavity mold used in fabricating thermoset test coupons **EM3:** 399
electrical properties **EM3:** 218
for engine gasketing **EM3:** 553
for faying surfaces **EM3:** 604
fillers **EM3:** 215, 216, 217
for fillets **EM3:** 604
flame-retardant fire stop sealant **EM3:** 219-220
forms **EM3:** 217-218
general glazing for weatherproofing **EM3:** 219
general industrial sealant **EM3:** 220-221
heating and air conditioning duct sealing and bonding **EM3:** 610
industrial cost factors **EM3:** 221
for lighting subcomponent bonding **EM3:** 552
medical applications **EM3:** 576
methods of application **EM3:** 218
plumbing sealant **EM3:** 608
primers used **EM3:** 218
properties **EM3:** 217-218, 219
recreational vehicle modifications **EM3:** 610, 611
resistance to ultraviolet radiation and ozone **EM3:** 217-218
for rivets **EM3:** 604
sanitary mildew-resistant sealant ... **EM3:** 219, 220
silanol-terminated polydimethylsiloxanes **EM3:** 215, 216
structural glazing **EM3:** 219
suppliers **EM3:** 220, 221
thermal aging properties **EM3:** 217
water-dispersed **EM3:** 221
RTV silicone rubbers **EM2:** 266-267
sealants
for aerospace applications **EM3:** 192
appliance applications **EM3:** 192
automotive applications **EM3:** 192
characteristics (wet seals) **EM3:** 57, 188
chemical resistance **EM3:** 642
for construction **EM3:** 192
for consumer do-it-yourself **EM3:** 192
for fuel containment **EM3:** 192
for highway construction **EM3:** 192
properties **EM3:** 191-192, 677
for structural glazing **EM3:** 192
tensile strength affected by temperature **EM3:** 189
secondary seal for polyisobutylene **EM3:** 190
service temperature **EM3:** 621
shelf life **EM3:** 79, 135-136
for shelving and window construction **EM3:** 45
silicone fluids **EM2:** 266

Silicones (SI) (continued)
for structural glazing **EM3:** 606, 607
suppliers **EM2:** 267 **EM3:** 58, 59, 79, 134
surface contamination effect **EM3:** 638
for surface-mount technology bonding **EM3:** 570
thermal properties **EM3:** 52
thermal resistance and shear strength **EM3:** 98
for thermally conductive bonding **EM3:** 571
thermoset **EM2:** 630-631
to facilitate gun-dispensible application of sealants **EM3:** 188
UV-curing **EM3:** 568
for wall construction **EM3:** 56
for wave solder masking **EM3:** 134
for window glazing and mounting **EM3:** 577
for window sealing **EM3:** 56
for wrapping and bundling of wires **EM3:** 134
for wrapping transformers **EM3:** 134
Silicones/methacrylates
for solar panel construction **EM3:** 578
Siliconized silicon carbides (Si/SiC) **EM4:** 19
Silicosis **A7:** 202
from gold toxicity **A7:** 205
Silk fibers
friction coefficient data **A18:** 75
Silk screen process
for solder mask/protective coat **EL1:** 115
Silk screen surfaces **A7:** 176
Silky fracture
defined **A8:** 12 **A11:** 9
Silky fractures **A12:** 2
Sillimanite
as aluminum silicate molding sand **A15:** 20
applications, refractory **EM4:** 906
density **EM4:** 762
island structure **EM4:** 758
structure **EM4:** 762
versus mullite **EM4:** 762
Silo hoist
destructive wear in steel worm used in **A11:** 598
Siloxanes
failure analysis role **EM3:** 249
identification of **EM3:** 243
Siltemp fiber **EM1:** 61
Silumin (Al-Si: 86.5% Al; 1% Cu)
thermal properties **A18:** 42
Silver *See also* Fine silver; Nickel silver; Precious metal powders; Precious metals; Pure silver; Silver alloys; Silver alloys, specific types; Silver coatings; Silver contact alloys; Silver powders; Silver-base composites; Silver-graphite powders; Silver/cadmium oxides; Sterling silver; Tumbaga **A13:** 793-798
(liquid) contact angles on beryllium at various test temperatures in argon and vacuum atmospheres **A6:** 116
as a conductive coating for scanning electron microscopy specimens **A9:** 97
absorptivity **A6:** 265
in acids **A13:** 795
addition to plasma-sprayed coatings **A18:** 119
adhesion measurement of fcc metals **A6:** 144
adhesion to nickel **A6:** 144
adhesion to silver **A6:** 144
as alloyable coating **EL1:** 679
alloying **A13:** 794-795
alloying, aluminum casting alloys **A2:** 132
alloying, wrought aluminum alloy **A2:** 55
in aluminum alloys **A15:** 74
annealing **M4:** 760-761
anode x-ray tube **A10:** 90
anodes **A7:** 148
applications **A2:** 699, 1259-1260 **A7:** 205
atomic interaction descriptions **A6:** 144
brazing, composition **A6:** 117
in carbon-graphite materials **A18:** 816
in clad and electroplated contacts **A2:** 848
in cloud seeded rainwater, GFAAS analysis **A10:** 55
coinability of **A14:** 183
color buffing of **M5:** 306
colors obtained from various thicknesses of silver iodide interference films **A9:** 136
commercially pure **M2:** 671-673
commercially pure, properties **A2:** 699-700

Silver (continued)
compatibility in bearing materials **A18:** 743
Compton-scattered **A10:** 84
as conductive thick-film material **EL1:** 249
conductor ink **EL1:** 208
corrosion applications **A13:** 795
corrosion resistance **A13:** 793-794, 797 **M2:** 670
current-time curves in coulometric determinations of **A10:** 210
deformation twinning **A9:** 554-555
in dental amalgam **A18:** 669
dental solders **A2:** 698
deposition **A10:** 199
determined by controlled-potential coulometry **A10:** 209
diffusion bonding **A6:** 159
diffusion in an electronic circuit **A9:** 101
in diffusion welding welds **A6:** 884-885, 886
dislocation pairs **A9:** 115
dispersion-strengthened **A7:** 716-720
effect on thermal conductivity **EM3:** 621
effects, electrolytic tough pitch copper **A2:** 269
for electrical and thermal conductivity **EM3:** 584
electrical and thermal conductivity of **A2:** 840-841, 843-845
electrical contacts, use in **M3:** 666-668
electrical resistance applications **M3:** 641
as electrically conductive filler **EM3:** 45, 76, 130, 178, 572, 596, 620
electrochemical grinding **A16:** 543, 545
electrodes, oxidation in alkaline environments **A10:** 135
in electronic scrap recycling **A2:** 1228
electroplating of bearing materials **A18:** 756, 838
electropolishing of **M5:** 306
embrittlement by **A11:** 236 **A13:** 182
enamels **EM3:** 302
in eutectic alloys **A6:** 127
evaporation fields for **A10:** 587
explosion welding **A6:** 896 **M6:** 710
exposure limits **A7:** 205
fabrication **A13:** 793
filler for polymers **A7:** 606
filler metal for beryllium alloys **A6:** 946
film, grown on mica, analysis of grain size in **A10:** 543-544
film grown on mica, grain size of **A10:** 543-544
foil, solid-state welding **A6:** 169
friction coefficient data **A18:** 71, 72
gas-tungsten arc welding **A6:** 192
gaseous corrosion **A13:** 798
as gold alloy **A2:** 690
gold plating, uses in **M5:** 282-283
gravimetric finishes **A10:** 171
in halogens **A13:** 795
heat-affected zone fissuring in nickel-base alloys **A6:** 588
high-purity, SSMS analysis **A10:** 144
high-vacuum lubricant application **A18:** 153
honing stone selection **A16:** 476
in hydrochloric acid **A13:** 794
impregnation effects on typical carbon-graphite base material **A18:** 817
impregnation effects on typical graphite-base material **A18:** 817
-infiltrated tungsten, properties **A7:** 562
as interlayer metal for solid-state welding **A6:** 165-166, 168, 169, 170
internal electrolysis of **A10:** 200
ion implantation **A18:** 856
lap welding **M6:** 673
liquid-metal embrittlement of **A11:** 234
in liquid-phase metallizing **EM3:** 306
low-melting temperature indium-base solder compatibility **A2:** 752
low-temperature solid-state welding **A6:** 300
lubricant indicators and range of sensitivities **A18:** 301

Silver (continued)
metal filler for polyimide-base adhesives **EM3:** 159
in metal powder-glass frit method **EM3:** 305
migration, electronics industry **A13:** 1110
migration, polymer die attach **EL1:** 218
as minor toxic metal , biologic effects **A2:** 1259-1260
in molten sodium chloride **A13:** 50
mutual solubility **A11:** 452
not a polyphase alloy **A18:** 743
in organic compounds **A13:** 797
oxidation **A13:** 794
percussion welding **M6:** 740
photochemical machining etchant **A16:** 590
plating, uses **EL1:** 679
and platinum, cold heading **A14:** 850
powders, catalytic oxidation by SERS analysis **A10:** 136
as precious metal **A2:** 688
precoating **A6:** 131
price per pound **A6:** 964
production **M2:** 659-660
properties **A2:** 699-700 **A13:** 793
as protective finish in electronic applications **A6:** 990, 991
pure **M2:** 794-796
pure (99.9%), thermal properties **A18:** 42
pure, AES spectrum of **A11:** 34
pure, properties **A2:** 1156
purest, thermal properties **A18:** 42
recommended impurity limits of solders **A6:** 986
reflecting layer coating on outer substrate surface of surface force apparatus (SFA) **A18:** 400
refractory metal brazing, filler metal **A6:** 942
reinforcement of glass ionomers as dental elements **A18:** 673
relative solderability **A6:** 134
as a function of flux type **A6:** 129
resistance brazing **M6:** 976
resistance to specific corroding agents **A2:** 699-721
resistance welding **A6:** 833
resources and consumption **A2:** 689
rhodium plating of **M5:** 290
roll welding **A6:** 314
rosin flux use **A6:** 129
safety standards for soldering **M6:** 1099
in salts **A13:** 796
scrap metal **A10:** 41
semifinished products **A2:** 694
as SERS metal **A10:** 136
as solder impurity **EL1:** 638
solderability **A6:** 978
in solid solution alloys **A6:** 127
solution potential **M2:** 207
special properties **A2:** 691 **M2:** 662
species weighed in gravimetry **A10:** 172
spectrometric metals analysis **A18:** 300
in stainless steel brazing filler metals **A6:** 911-913, 915, 920
sterling **A14:** 184
suitability for cladding combinations **M6:** 691
tantalum corrosion by **A13:** 735
texture orientations **A10:** 360
thermal diffusivity from 20 to 100 °C **A6:** 4
thermal expansion coefficient **A6:** 907
TNAA detection limits **A10:** 238
toxicity **A6:** 1195, 1196 **A7:** 204-205
trade practices **A2:** 690-691
transferred patches on a copper specimen in TEM **A18:** 385
in trimetal bearing material systems **A18:** 748
TWA limits for particulates **A6:** 984
ultrapure, by fractional crystallization **A2:** 1093
ultrapure, by zone refining **A2:** 1094
ultrasonic welding **M6:** 746

Silver (continued)
vapor pressure, relation to temperature **M4:** 309, 310
Vickers and Knoop microindentation hardness numbers **A18:** 416
Volhard titration of **A10:** 173
volume resistivity and conductivity **EM3:** 45
volumetric procedures for **A10:** 175
weighed as the chloride **A10:** 171
wettability indices on stainless steel base metals **A6:** 118
x-ray tube emission, spectrum of **A10:** 90

Silver (Ag)
for brazing in ceramic/metal seals **EM4:** 504, 505, 506
component in photochromic ophthalmic and flat glass composition **EM4:** 442
component in photosensitive glass composition **EM4:** 440
in filler metal used for direct brazing **EM4:** 517-518, 519
in filler metals for active metal brazing **EM4:** 523-524, 526, 530
joining **EM4:** 487
materials for conductors **EM4:** 1141
metallic decorating material **EM4:** 474
for metallizing **EM4:** 443, 542, 543
variation in the experimentally determined melting point **EM4:** 251

Silver alloys *See also* Fine silver; Pure metals; Pure silver; Silver; Silver alloys, specific types; Silver contact alloys; Silver-base composites; specific sliver alloys
applications **A2:** 700, 702-704
brazing, definition **A6:** 1213
dental amalgam, properties **A2:** 703-704
as filler metal for dip brazing **A6:** 338
microstructures of **A9:** 551-552
properties **A2:** 699-704
recommended gap for braze filler metals **A6:** 120
resistance brazing filler metals **A6:** 341, 342

Silver alloys, for brazing
composition and properties **A7:** 838

Silver alloys, specific types *See also* Sleeve bearing materials, specific types
70Ag-30Cu, graphite-silver copper composite **A9:** 595, 597
72Ag-28Cu foil **A6:** 146
75Ag- 19.5Cu-5Cd-0.5Ni **A9:** 555
75Ag-24.5Cu-0.5Ni **A9:** 555
85Ag-15Cd **A9:** 555
90Ag-10Cu **A9:** 555
96.5Sn-3.5Ag **A6:** 351
98.58Ag-0.22MgO-0.2Ni **A9:** 559
Ag-0.25Mg-0.20Ni, as electrical contact materials **A2:** 845
Ag-3Pd, as electrical contact materials **A2:** 845
Ag-5% Cu, weld microstructures **A6:** 52
Ag-5.5Cd-0.2Ni-7.5Cu, as electrical contact materials **A2:** 845
Ag-5Cu, electron beam melted and resolidified **A9:** 615-616
Ag-10Au, as electrical contact materials **A2:** 845
Ag-15Cd, as electrical contact materials **A2:** 845
Ag-15Cu, electron beam melted and resolidified **A9:** 615
Ag-22.6Cd-0.4Ni, as electrical contact materials **A2:** 845
Ag-23Pd-12Cu-5Ni, as electrical contact materials **A2:** 845
Ag-24.5Al (at.%), single crystal from massive transformation **A9:** 657
Ag-24.5Cu-0.5Ni, as electrical contact materials **A2:** 845
Ag-28Cu, melt spun **A9:** 616

Silver alloys, specific types (continued)
Ag-Cu-Zn (silver-copper-zinc) alloys, brazing
available product forms of filler metals **A6:** 119
joining temperatures **A6:** 118
Ag-Cu-Zn-Cd (silver-copper-zinc-cadmium) alloys, brazing
available product forms of filler metals **A6:** 119
joining temperatures **A6:** 118
Silver alloys, use for bearings *See also* Bearings, sliding **M3:** 820
Silver and silver alloys
as addition to aluminum alloys **A4:** 843
aging **A4:** 941
annealing **A4:** 939-941
applications **A4:** 939, 941, 946
brazing, furnace atmosphere **A4:** 548, 559
cold working **A4:** 940
facing for split inductor coils **A4:** 178
hardness **A4:** 941, 945
mechanical properties **A4:** 940, 941, 946
probe material for quenching cooling curve analysis **A4:** 68
sleeve bearing liners **A9:** 567
tensile properties **A4:** 939, 940
vapor pressure of silver, relation to temperature **A4:** 495
Silver azide (AgN_3) (explosive)
friction coefficient data **A18:** 75
Silver brazing
method for attaching ultrasonic machining tool to toolholder **EM4:** 360
Silver brazing alloy
cladding material for brazing **A6:** 347
Silver brazing filler metals
properties **A2:** 702
Silver bromide
as internal reflection element **A10:** 113
Silver carbonate **A7:** 148
Silver chloride
as internal reflection element **A10:** 113
stain used for amber color in glass **EM4:** 474
Silver chloride electrode
for acidified chloride solutions **A8:** 419
Silver coatings
applications **A2:** 691
Silver coatings and films
soldering **A6:** 631
Silver, coin *See* Silver-copper alloys
Silver contact alloys *See also* Silver
electrical and thermal conductivity **A2:** 843
fine silver **A2:** 844-845
limitations **A2:** 844
multi-component alloys **A2:** 845
silver-cadmium alloys **A2:** 845
silver-copper alloys **A2:** 845
silver-gold alloys **A2:** 845
silver-palladium alloys **A2:** 845
silver-platinum alloys **A2:** 845
types **A2:** 843-845
Silver filling *See* Dental amalgam
Silver flake **A7:** 147-148, 596 **EM3:** 33
Silver halide **EM4:** 22
Silver halide grains
in radiography **A17:** 326
Silver in copper **M2:** 242, 243
Silver in fusible alloys **M3:** 799
Silver iodide interference films on silver
colors obtained at various thicknesses **A9:** 136
Silver matrix composite with nickel fibers **A9:** 100
Silver migration
electronics industry **A13:** 1110
Silver migration failures
tantalum capacitors **EL1:** 998
Silver nitrate in water as an etchant for tin- lead alloys **A9:** 450
Silver nitrate, stainless steels
corrosion resistance **M3:** 87-88
Silver nitrite
biologic effects **A2:** 1260
Silver oxides by thermal reduction **A7:** 148
Silver plating **M5:** 279-280, 566
aluminum and aluminum alloys **M5:** 605-606
applications **M5:** 279
copper and copper alloys **M5:** 617, 621-623
Silver plating (continued)
cyanide and noncyanide systems **M5:** 279-280
electroless process **M5:** 606, 621-622
heat-resistant alloys **M5:** 566
immersion process **M5:** 622
magnesium alloys, stripping of **M5:** 647
solution compositions and operating conditions **M5:** 279-280, 622-623
impurities **M5:** 280
specifications **M5:** 280
stripping of **M5:** 647
Silver powders *See also* Precious metal powders; Silver
chemical processes **A7:** 147-148
electrochemical reduction and processes **A7:** 148
electrolytic **A7:** 71, 72
irregularly shaped galvanic **A7:** 148
reducing agents **A7:** 147-148
thermal reduction **A7:** 148
Silver, pure
effect of atomic oxygen **A12:** 481
Silver salts **A7:** 147-148
Silver sand
angle of repose **A7:** 282
Silver scrap metal
analysis of **A10:** 41
Silver soldering
definition **A6:** 1213
Silver solders *See* Silver-base brazing alloys; Silver-base brazing filler metals
chemical analysis and sampling **A7:** 249
Silver, stacking-fault tetrahedra **A9:** 117
effect on lead-tin microstructures **A9:** 417
incipient melting from electrical overload **A9:** 555
melting in **A9:** 555
recrystallized **A9:** 554-555
sheet, cold rolled **A9:** 554-555
Silver, sterling *See* Silver-copper alloys
Silver streaking *See* Splay marks
Silver striking
aluminum and aluminum alloys **M5:** 605
Silver sulfide
biologic effects **A2:** 1260
Silver tungsten
electrical contact applications **A7:** 631
electrode material for EDM **A16:** 559, 560
tongs for manual resistance brazing **A6:** 340
Silver-base alloys
as bearing alloys **A18:** 748, 753
applications **A18:** 753
mechanical properties **A18:** 753
fibers for reinforcement **A18:** 803
fretting wear **A18:** 248, 250
processing techniques **A18:** 803
properties **A18:** 803
torch brazing filler metals **A6:** 328
Silver-base brazing alloys **M2:** 675-677
Silver-base brazing filler metals
properties **A2:** 702
Silver-base composites
with dispersed oxides **A2:** 856
for electrical contact materials **A2:** 855-856
multiple-component **A2:** 856
with pure element or carbide **A2:** 855
silver-cadmium oxide group **A2:** 856
silver-iron **A2:** 856
silver-nickel **A2:** 855
silver-tin oxide **A2:** 856
silver-tungsten/silver molybdenum **A2:** 855
silver-zinc oxide **A2:** 856
Silver-base filler alloys
corrosion resistance **A13:** 880-883
Silver-base powder metallurgy materials, specific types
40Ag-60WC **A9:** 559-560
50Ag-50Mo **A9:** 559
50Ag-50Ni **A9:** 558
50Ag-50W **A9:** 559
50Ag-50WC **A9:** 559
60Ag-40Ni **A9:** 558
65Ag-35WC **A9:** 559
80Ag-20Ni **A9:** 558
85Ag-15CdO **A9:** 556-557
85Ag-15Ni **A9:** 558
88Ag-10Ni-2graphite **A9:** 558
Silver-base powder metallurgy materials, specific types (continued)
89Ag-11CdO **A9:** 556
90Ag-10CdO **A9:** 555-557
90Ag-10graphite **A9:** 558
90Ag-10W **A9:** 559
95Ag-5graphite **A9:** 558
97Ag-3graphite **A9:** 558
Silver-bearing copper alloys
characteristics **A2:** 230, 234
Silver-bearing tough pitch copper *See* Copper alloys, specific types, C11300, C11400, C11500 and C11600
applications and properties **A2:** 274-275
Silver-cadmium alloys
as electrical contact materials **A2:** 845 **A9:** 551
feathery structures in **A9:** 656
Silver-cadmium oxide
percussion welding **M6:** 740, 745
Silver-cadmium oxide composites
as electrical contact materials **A2:** 851, 856
Silver-cadmium oxide materials
microexamination of **A9:** 550
polishing of **A9:** 550
Silver-copper alloys *See also* Silver alloys; Sterling silver **A13:** 775, 878 **M2:** 673-675
as electrical contact materials **A2:** 845 **A9:** 551
properties **A2:** 700-702
weld microstructure **A6:** 52, 53
Silver-copper-nickel alloys
as electrical contact materials **A9:** 551
Silver-copper-phosphorus alloys
torch brazing filler metals **A6:** 328
Silver-copper-titanium (Ag-Cu-Ti)
used to braze silver metallization onto alumina **EM4:** 544
Silver-filled epoxy
volume resistivity and conductivity **EM3:** 45
Silver-glass
bonding **EL1:** 349
die attach, for hermetic packages **EL1:** 215-216
Silver-gold alloys
as electrical contact materials **A2:** 845
Silver-graphite
microexamination of **A9:** 550
resistance brazing **M6:** 976
Silver-graphite composites, properties
for electrical make-break contacts **A2:** 852
Silver-graphite mixtures
applications **A7:** 147
electrical contact applications **A7:** 631
electrical/magnetic applications **A7:** 633
Silver-indium alloys
dental **A13:** 1362
Silver-iron composites, properties
for electrical make-break contacts **A2:** 852
Silver-iron mixtures
applications **A7:** 147
Silver-lead alloy cable sheath
creep-rate/time curves **A8:** 321, 323
Silver-lead solders
microstructures of **A9:** 417, 422
Silver-magnesium-nickel alloys **M2:** 677-678
properties **A2:** 702
Silver-mercury alloys (dental amalgam)
properties **A2:** 703-704
Silver-molybdenum
resistance brazing **M6:** 965
Silver-molybdenum mixtures
applications **A7:** 147
electrical contact applications **A7:** 631
Silver-nickel
microexamination of **A9:** 550
Silver-nickel composites, properties
for electrical make-break contacts **A2:** 852
Silver-nickel mixtures **A7:** 632-633
applications **A7:** 147
electrical contact applications **A7:** 631
microstructural analysis **A7:** 488-489
Silver-palladium alloys **A13:** 1360
as electrical contact materials **A2:** 845
termination of multilayer ceramic capacitors **EM4:** 1117
Silver-platinum alloys
as electrical contact materials **A2:** 845

Silver-reinforced glass ionomer
composite restorative material (dental) combined with **A18:** 670
Silver-tin alloys
peritectic reactions **A9:** 676
Silver-tin oxide composites
as electrical contact materials **A2:** 852, 856
Silver-tungsten
percussion welding **M6:** 740, 745
resistance brazing **M6:** 965
Silver-zinc oxide composites
as electrical contact materials **A2:** 852, 856
Silver/cadmium oxide
applications **A7:** 147, 631, 633, 720
contacts **A7:** 718, 720
electrical/magnetic applications **A7:** 633
materials, compared **A7:** 718
Silver/refractory metals **A7:** 633-634
Silver/silver chloride electrode **A8:** 416
Silver/tungsten carbide mixtures **A7:** 147, 631
applications **A7:** 631
Silverware
coining dies for **M3:** 509-510
coining of **A14:** 182
SIMA process
brass application **A15:** 33
as semisolid metalworking **A15:** 33
Simple (lattices)
defined **A9:** 16
Simple bending
as springback **A8:** 552
tests, simulative, for sheet metals **A8:** 560
Simple cubic array
density and coordination number **A7:** 296
Simple hydrostatic extrusion **A14:** 327-328
Simple linear regression **A8:** 700
Simple set-point control systems **A6:** 1062
Simple shear **A8:** 155, 180-181
Simple space lattice .... **A3:** 1 • 15
Simple upsetting **A14:** 87-88
Simple-beam theory
bending **A8:** 118-119
Simple-SIMS instrument
for qualitative analysis **A10:** 613
Simplex (experimental) designs **EM2:** 601
Simplified impulse and inductive inert gas fusion furnaces **A10:** 228
Simplified Marker-and-Cell program **A15:** 86
SIMS *See* Secondary ion mass spectrometry; Secondary ion mass spectroscopy; Secondary, ion mass spectroscopy,
Simulated pressurized water reactor **A8:** 423-425
Simulated service testing *See also* Corrosion testing; Evaluation; In-service monitoring; Monitoring **A13:** 204-211
corrosion testing **A13:** 208-210
corrosion testing, atmospheric **A13:** 204-206, 226
corrosion testing in water **A13:** 207-208
for intergranular corrosion **A13:** 239
of magnesium/magnesium alloys **A13:** 743-747
Simulated-bolthole test
for gas-turbine components **A11:** 279
Simulated-service tests *See also* Simulation
conditions for **A11:** 214
for corrosion **A11:** 174
of failures **A11:** 31
for stress-corrosion cracking **A11:** 214
Simulation *See also* Analysis; Computer applications; Mathematical modeling; Model simulation; Model(s); Modeling; NDE reliability models; Physical modeling; Process modeling; Simulated-service tests **A14:** 911-927
analytical methods **A14:** 913-919
finite-element analysis methods **A14:** 919-924
and fracture mechanics **A8:** 262
generalized approach **A14:** 911-913
High-frequency digital systems **EL1:** 77-81
laboratory, failure analysis **EM2:** 818-819
Simulation (continued)
of LME mechanism **A11:** 724-725
of metalworking **A14:** 19-20
and models, fatigue testing by **A11:** 135
models, for high-frequency digital system design **EL1:** 81
mold fill **EM2:** 312
output, interpreting **A15:** 863
procedure, for parallel signal lines **EL1:** 35
software, for design **EL1:** 419
of solidification **A15:** 36, 863-865
of tool and die failures **A11:** 563
variable mesh (VMS) **EL1:** 129
Simulations *See also* Computers
in high resolution electron microscopy **A9:** 121
Simulative formability tests **A14:** 883, 889-893
Simulative forming tests **A8:** 560-566
for steel sheet **A1:** 576
Simulative tests
bending **A8:** 560-561
drawing **A8:** 562-563
forming **A8:** 553, 560-566
for lubricant evaluation in forming operations **A8:** 568
and material properties **A8:** 565
springback **A8:** 564-565
stretch-drawing **A8:** 563
stretching **A8:** 561-562
types of **A8:** 560
for wear testing **A8:** 604-606
wrinkling and buckling **A8:** 563-564
Simulators
interconnection network electrical performance (AT&T) **EL1:** 14
Simultaneous carburization **A7:** 158
Simultaneous engineering
design for **EM1:** 835-836
and NDE reliability models **A17:** 702
Simultaneous multielement analysis
direct-current plasma **A10:** 40
direct-current plasma atomic emission spectroscopy **A10:** 43
inductively coupled plasma atomic emission spectroscopy **A10:** 31-42, 43
neutron activation analysis **A10:** 233-242
and sequential multielement analysis combined **A10:** 38
Sin^2 ψ technique
plane-stress elastic model **A10:** 384
six-angle, residual stress pattern **A10:** 386
Sine law
in mechanics of cone spinning **A14:** 601
Single bending impact load
medium-carbon steel fracture **A12:** 270
Single coil tests
eddy current **A17:** 177, 543-544
Single crystal
iron alloy, fracture surface **A8:** 479-480
specimen, aluminum, arrangement to Kolsky bar **A8:** 222, 224
Single crystal spheres
oxidized **A9:** 137
Single crystalline graphite
Raman analysis **A10:** 132
Single crystals *See also* Crystals
of aluminum, spot diffraction pattern from **A10:** 437
defined **A10:** 681
diffraction experiments with **A10:** 330
for ESR study **A10:** 256
EXAFS analysis **A10:** 410
experiment with monochromatic beams **A10:** 330
GaAs, growth methods **A2:** 744
garnet, gallium gadolinium (GGG) **A2:** 740
growth of **A9:** 656-657
heat tinting **A9:** 136
high-purity gallium **A2:** 739
lattice location of impurities in **A10:** 628
Single crystals (continued)
NaCl, rotation pattern for **A10:** 330
neutron diffraction **A10:** 424
orientation by XRPD analysis **A10:** 333
orientations determined **A10:** 357
resonance equations for **A10:** 269
strain-anneal growing technique **A9:** 696
study of near-surface defects in **A10:** 633
surface, density of steps determined **A10:** 544
surfaces, EXAFS for orientation adsorbed molecules on **A10:** 407
unit mesh size and shape of overlayer adsorbed on **A10:** 544
Single cyclic stress
analysis **EM1:** 201-202
Single fibers
mechanical testing of **EM1:** 731-732
Single fractures
in milling **A7:** 59
Single impulse welding
definition **A6:** 1213
Single metal powders **A7:** 309-314
Single monochromators
stray light rejection for **A10:** 129
Single oxides, engineering *See* Engineering properties of single oxides
Single particles **A7:** 59-60
Single pressure bar
coaxial capacitor to measure strain **A8:** 199
high-speed photography to measure strain **A8:** 199
Hopkinson **A8:** 198-199
Single scattering approximation
in EXAFS analysis **A10:** 409-410
Single scattering formalism
in EXAFS analysis **A10:** 407, 417
Single spread **EM3:** 26
Single stage direct carbon replicas
formation **A12:** 181
Single strain
fatigue resistance at **A8:** 706-712
Single stress
fatigue resistance at **A8:** 706-712
Single transducer system
ultrasonic testing **A8:** 244, 245
Single-action press **A7:** 11
Single-action tooling systems **A7:** 332-333
Single-angle technique
plane-stress elastic model **A10:** 383-384
Single-bevel groove welds
applications **M6:** 61
arc welding of nickel alloys **M6:** 437
combinations with fillet welds **M6:** 66-67
comparison to fillet welds **M6:** 64
flux cored arc welding **M6:** 105-106
gas metal arc welding of commercial coppers **M6:** 403
gas tungsten arc welding of silicon bronzes **M6:** 413
preparation **M6:** 67-68
recommended proportions **M6:** 69-72
Single-bevel-groove weld
definition **A6:** 1213
Single-chip package
thermal control **EL1:** 47
Single-circuit winding
defined **EM1:** 21 **EM2:** 38
Single-coil and resistor system
eddy current inspection **A17:** 177
Single-column chromatography anion
for suppression of background conductivity **A10:** 660
Single-column chromatography ion
with conductivity detection **A10:** 660-661
Single-compound-angle groove welds **M6:** 442-443
Single-crystal
materials, crystal pulling for **EL1:** 959
silicon, growth of **EL1:** 191

SUBJECTS OF THE INDEXED VOLUMES: ASM Handbook (designated by the letter "A"): **A1:** Properties and Selection: Irons, Steels, and High-Performance Alloys (1990); **A2:** Properties and Selection: Nonferrous Alloys and Special-Purpose Materials (1990); **A3:** Alloy Phase Diagrams (1992); **A4:** Heat Treating (1991); **A6:** Welding, Brazing, and Soldering (1993); **A7:** Powder Metallurgy (1984); **A8:** Mechanical Testing (1985); **A9:** Metallography and Microstructures (1985); **A10:** Materials Characterization (1986); **A11:** Failure Analysis and Prevention (1986); **A12:** Fractography (1987); **A13:** Corrosion (1987); **A14:** Forming and Forging (1988); **A15:** Casting (1988); **A16:** Machining (1989); **A17:** Nondestructive Testing and Quality Control (1989); **A18:** Friction, Lubrication, and Wear Technology (1992). **Metals Handbook, 9th Edition** (designated by the letter "M"): **M1:** Properties and Selection: Irons and Steels (1978); **M2:** Properties and Selection: Nonferrous Alloys and Pure Metals (1979); **M3:** Properties and Selection: Stainless Steels, Tool Materials and Special-Purpose Materials (1980); **M4:** Heat Treating (1981); **M5:** Surface Cleaning, Finishing, and Coating (1982); **M6:** Welding, Brazing, and Soldering (1983). **Engineered Materials Handbook** (designated by the letters "EM"): **EM1:** Composites (1987); **EM2:** Engineering Plastics (1988); **EM3:** Adhesives and Sealants (1990); **EM4:** Ceramics and Glasses (1991); **Electronic Materials Handbook** (designated by the letters "EL"): **EL1:** Packaging (1989).

Single-crystal (continued)
structure, defined **EL1:** 93
Single-crystal casting *See also* Monocrystal casting
and equiaxed casting, compared **A15:** 39
furnaces, vacuum induction remelting/shape casting **A15:** 400-401
market effects **A15:** 4
nickel alloy **A15:** 817, 819, 823
superalloys **A15:** 41
Single-crystal diffraction
basic principles of **A10:** 350
Single-crystal neutron diffraction **A10:** 424-425
Single-crystal superalloys *See also* Superalloys **A1:** 995-996, 998-1006
casting of **A1:** 998-1000
chemistry control **A1:** 998-999
compositions of **A1:** 996, 997
development and characteristics **A2:** 429
fatigue properties **A1:** 1002, 1004, 1005, 1006
heat treatment, effect of **A1:** 1005
hot isostatic pressing, effect of **A1:** 1005
heat treatment **A1:** 1000-1002
hot corrosion **A1:** 1004
microstructure **A1:** 1000-1002
heat treatment, effect of **A1:** 1000, 1003
precipitate formation **A1:** 1001
oxidation of **A1:** 1004, 1006
protective coatings **A1:** 1006
Single-crystal topography
as x-ray diffraction radiography **A10:** 330-331
Single-crystal x-ray diffraction **A10:** 344-356
accuracy and precision **A10:** 352
applications **A10:** 344, 353-355
assumptions of **A10:** 352
capabilities **A10:** 333, 393
crystal diffraction **A10:** 346
crystal structure definition **A10:** 348
crystal symmetry **A10:** 346
crystallographic problems **A10:** 352-353
diffraction intensities **A10:** 348-349
estimated analysis time **A10:** 344
experimental procedure **A10:** 351-352
general uses **A10:** 344
introduction and principles **A10:** 345-346
limitations **A10:** 344, 352-353
phase problem **A10:** 349-351
related techniques **A10:** 344
samples **A10:** 344, 351, 353-355
space groups **A10:** 347
unit cells **A10:** 346-347
Single-cylinder engine tests
Caterpillar IK/IH2 .... **A18:** 101
Single-detector translate-rotate systems
computed tomography (CT) **A17:** 365-366
Single-edge barreling
edge cracking in cast alloy after **A8:** 594, 596
in rolling **A8:** 594, 596
Single-edge cracked specimens
geometry for **A8:** 251
Single-edge notched specimen
calibration curve **A8:** 386
and current input and potential measurement probes **A8:** 386
grips **A8:** 382
for vacuum and oxidizing fatigue testing **A8:** 414-415
Single-edged
notched three-point bend (SENB) specimens **EM3:** 447, 448
Single-embedded fiber test **EM3:** 402, 403
Single-embedded-fiber technique **EM3:** 394-399, 402
Single-end rotating cantilever testing machine **A8:** 370
Single-end rovings
applications **EM1:** 109
Single-exposure technique
plane-stress elastic model **A10:** 383
Single-extension dies **A14:** 132
Single-fiber fragmentation method **EM3:** 403
Single-fiber fragmentation test **EM3:** 394-397, 398, 399, 400
Single-flare-groove weld
definition **A6:** 1213
Single-flare-V-groove weld
definition **A6:** 1213
Single-fluid atomization **A7:** 75, 76
Single-impulse welding
definition **M6:** 16
Single-in-line package (SIP)
defined/outline **EL1:** 210
die attachments **EL1:** 217-221
DRAM, memory density **EL1:** 439
package configurations **EL1:** 485
as package family **EL1:** 404
resistor networks **EL1:** 178
resistors, in design layout **EL1:** 514
vertical orientation of substrates **EL1:** 444
Single-J-groove weld
definition **A6:** 1213
Single-J-groove welds
applications **M6:** 61
recommended proportions **M6:** 69-71
Single-lap joints
failure **EM3:** 473, 474
finite-element analysis **EM3:** 480, 481
normal stress **EM3:** 487, 491, 492
stress singularities **EM3:** 477-478
stresses in the bond line **EM3:** 489
Single-lap shear specimens
dynamic mechanical analysis **EM3:** 645-649
Single-lap specimen **EM3:** 26
Single-lap-shear test
with and without an adhesive fillet ... **EM3:** 493-495
Single-lead heating and pulling method
of component removal **EL1:** 717
Single-lens-reflex cameras
35-mm **A12:** 78-79
Single-operation dies
for blanking **A14:** 453-454
for piercing **A14:** 465
for press bending **A14:** 527
for press forming **A14:** 546
for sheet metal drawing **A14:** 579
Single-overload fracture behavior
shafts **A11:** 460-461
Single-pass lubrication systems **A18:** 133
Single-pass soldering (SPS) **EL1:** 694-695
Single-pass weldments **A1:** 603, 604
Single-peak flow curve
grain refinement **A8:** 175
Single-phase alloys
alloy composition **A15:** 11
defined **A13:** 46
heat flow conditions **A15:** 114-11
instability **A15:** 12
interface morphologies, types **A15:** 114-11
interface velocity, above critical velocity **A15:** 116-11
interface velocity, below critical velocity **A15:** 114-11
normal grain growth in **A9:** 697
phase diagram **A15:** 11
Single-phase ceramics **EM4:** 47
Single-phase direct-energy resistance welding machines **M6:** 537, 540, 545
Single-phase full-wave direct current
defined **A17:** 91
Single-phase material
dislocations **A8:** 173
dynamic recovery and recrystallization in **A8:** 173-177
grain boundaries **A8:** 173
intergranular fracture **A8:** 486
torsion test for occurrence of recovery/recrystallization **A8:** 176
transgranular fracture **A8:** 486
workability **A8:** 574
Single-phase materials
EPMA compositional analysis **A10:** 516
image analysis **A10:** 309
Single-phase microstructure
dimpled rupture **A8:** 476
Single-phase microstructures **A9:** 602-603
Single-piece heaters
modified for elevated/low temperature tension testing **A8:** 36
Single-point bonding
innerlead **EL1:** 278
outer lead bonding **EL1:** 285
Single-point cutting tools **M3:** 470-472
"Single-point" machining **A7:** 425
Single-point tooling
in green machining **EM4:** 183
Single-port nozzle
definition **A6:** 1213 **M6:** 16
Single-pot tinning
cast iron and steel **M5:** 353
Single-punch opposing ram presses **A7:** 334-335
Single-punch withdrawal presses **A7:** 335
Single-row radial contact separable ball bearings (magneto bearings) **A18:** 509
Single-shear test
and blanking shear test, compared **A8:** 64, 65
and double-shear test, for fasteners compared **A8:** 66
failure modes **A8:** 63-64
fasteners **A8:** 66-67
for ultimate shear strength **A8:** 63-64
Single-sided boards
rework processes **EL1:** 711
rigid printed wiring **EL1:** 539
Single-sided gage
microwave thickness gaging **A17:** 212
Single-spindle drawing machines **A14:** 336
Single-spindle or multiple-spindle bar machines **A16:** 136, 140, 367-369, 371-374, 376-378
Single-spot welding setups **M6:** 475-477
Single-square-groove weld
definition **A6:** 1213
Single-stage extraction replicas
for small-particle analysis **A10:** 452
Single-stage method
U-bend stressing **A8:** 512
Single-stage replicas
conversion oxide film **A12:** 181-182
direct carbon **A12:** 181
production methods **A12:** 179
technique, schematic **A12:** 180
thick plastic **A12:** 180-181
thin plastic **A12:** 179-180
Single-stage surface replicas
defined **A17:** 53
Single-stand mill
defined **A14:** 12
Single-station dies **A14:** 478-479
Single-station tooling
cold extrusion **A14:** 303
Single-step drawing
copper and copper alloys **A14:** 815-816
Single-stroke open-die headers
for cold heading **A14:** 292
Single-stroke solid-die headers
for cold heading **A14:** 292
Single-taper dies
standard **A14:** 132
Single-tube attachments
for flame cutting **A7:** 843
Single-U-groove butt joint, welding of titanium and titanium alloys
joint dimensions **A6:** 785
Single-U-groove weld
definition **A6:** 1213
Single-U-groove welds
arc welding of austenitic stainless steels **M6:** 328
oxyfuel gas welding **M6:** 590-591
recommended proportions **M6:** 69-72
submerged arc welding of nickel alloys **M6:** 442-443
Single-V butt joints
plasma arc welding **A6:** 198
Single-V-groove butt joint, welding of titanium and titanium alloys
joint dimensions **A6:** 785
Single-V-groove weld
definition **A6:** 1213
Single-V-groove welds
applications **M6:** 61
arc welding of austenitic stainless steels **M6:** 328
electroslag welding **M6:** 242
flux cored arc welding **M6:** 105
gas metal arc welding of
commercial coppers **M6:** 403
coppers and copper alloys **M6:** 416-417
gas tungsten arc welding of
magnesium alloys **M6:** 431-432
silicon bronzes **M6:** 413
preparation **M6:** 67-68

Single-V-groove welds (continued)
recommended proportions **M6:** 69-72
submerged arc welding **M6:** 114
of nickel alloys **M6:** 442-443
Single-valued functions **A18:** 346
Single-wall, single-image radiographic technique
for tube **A17:** 337
Single-welded joint
definition **A6:** 1213 **M6:** 16
Singularity zone size
K-dominated **A8:** 450-451
Sink
shielded metal arc welds **M6:** 93
Sink mark *See also* Short shot
defined **EM1:** 21 **EM2:** 38
Sink marks
as casting defect **A11:** 384
Sink/float density separations
for high-carbon steels **A10:** 177
Sinking
defined **A14:** 12
Sinking-type Brinell indentation **A8:** 85
Sinter
(noun) defined **A7:** 11
(verb) defined **A7:** 11
Sinter bonding
and adhesive bonding **A7:** 457
Sinter forging *See also* Powder forging; Sintering
defined **A14:** 188
Sinter-aluminum-pulver (SAP) technology
aluminum P/M alloys **A2:** 202
Sinter-hot isostatic pressing (sinter-HIP) **EM4:** 194, 197-199
pressure densification **EM4:** 300
Sinterability **A7:** 211
in milling of single particles **A7:** 59
Sintercarburizing process
for P/M automotive parts **A7:** 620
Sintered (porous) P/M stainless steels *See also* Stainless steels; Stainless steels, specific types
application and selection **A13:** 824-825
commercial, compositions **A13:** 824
corrosion resistance **A13:** 825-832
iron contamination effects **A13:** 826-827
Sintered alpha-silicon carbide
applications, heat exchanger **EM4:** 984
ceramic corrosion in the presence of combustion products **EM4:** 982
Sintered alumina
applications **EM4:** 963
fretting wear **A18:** 248, 249
mineral processing **EM4:** 961
property comparison, mineral processing **EM4:** 962
Sintered aluminum powder (SAP) **A18:** 263
Sintered austenitic stainless steels
corrosion resistance **A13:** 831
Sintered bronze bearings **A7:** 705
self-lubricating **A2:** 394-396
Sintered carbide
cobalt-enhanced cutting ability **A7:** 144
for cold extrusion tools **A18:** 628
for deep-drawing dies **A18:** 634
die materials for sheet metal forming **A18:** 628
for shallow forming dies **A18:** 633
Sintered compacts
tin and tin alloy **A2:** 519-520
Sintered compacts composition-depth profiles *See also* Compact(s); Compacting; Compaction **A7:** 252
Sintered density *See also* Density; Sintering **A7:** 11, 309
effect on copper powder conductivity **A7:** 116
ratio **A7:** 11
sintering time, and sintering temperature **A7:** 314, 372
variation during homogenization **A7:** 314
Sintered density, corrosion resistance effects
P/M stainless steel **A13:** 830-832
Sintered density ratio **A7:** 11
Sintered friction materials **A7:** 701-702, 739
Sintered high-carbon steel
friction welding **A6:** 153
Sintered high-speed tool steels
powder metallurgy **A1:** 786
Sintered iron-base P/M parts **A13:** 823-824
Sintered materials *See* Powder metallurgy materials
Sintered metal friction materials
in aircraft brakes **A18:** 582
Sintered metals
contrasting by interference layers **A9:** 60
Sintered nickel electrode
in alkaline batteries **A13:** 1318
Sintered parts
aluminum P/M alloys **A2:** 213
properties **A13:** 825
Sintered polycrystalline cubic boron nitride *See also* Cubic boron nitride
properties of **A2:** 1012
Sintered polycrystalline diamond *See also* Diamond; Synthetic diamond
with a metallic second phase **A2:** 1011-1012
properties of **A2:** 1011-1012
Sintered reaction-bonded silicon nitride (RBSN)
additives **EM4:** 813
formation **EM4:** 813
properties **EM4:** 813
reaction forming as a post-sintering step **EM4:** 236
sintering **EM4:** 813
Sintered silicon carbide (SSC)
applications **EM4:** 964
boron-doped, strength tests **EM4:** 710, 711
fabrication **EM4:** 807
fast fracture flexure strength **EM4:** 1000
grinding **EM4:** 334
key features **EM4:** 676
mechanical properties **EM4:** 807
mineral processing **EM4:** 961
properties **EM4:** 330, 677
property comparison, mineral processing **EM4:** 962
strength retention **EM4:** 1001
turbine scroll failures **EM4:** 720
Sintered silicon nitride (SSN) **EM4:** 813
additives **EM4:** 813
applications **EM4:** 984
creep behavior **EM4:** 818
development **EM4:** 813
key features **EM4:** 676
low-pressure, piston material for adiabatic diesel engine **EM4:** 993
production **EM4:** 813
properties **EM4:** 330, 677, 815
sintering **EM4:** 813
strength **EM4:** 816, 817
superior for turbocharger wheel applications **EM4:** 723, 725-726
thermal shock resistance **EM4:** 818
turbine scroll failures **EM4:** 720
Sintered strength
and lubrication **A7:** 190-192
Sintered tungsten *See* Tungsten
Sintering *See also* also; Diffusion bonding; Heat treatment; Presintering **A7:** 295 **A16:** 101 **EM4:** 35-36, 242
activated **A7:** 316-321
activators, tungsten and molybdenum sintering **A7:** 391
additives and their effect on joining non-oxide ceramics **EM4:** 527-528
additives for S13N **A16:** 100
of alloy steels **A7:** 366
Alnico alloys **A2:** 786
alumina, for medical applications **EM4:** 1008
aluminum and aluminum alloys **A2:** 6
aluminum P/M alloys **A2:** 211
aluminum P/M parts **A7:** 743
Sintering (continued)
applications, specialty **A7:** 340-341
in argon and beryllium powders **A7:** 171
atmospheres *See* Sintering atmospheres
atmospheric pressure **A7:** 376
of beryllium **A2:** 685-686 **A16:** 870
of blended elemental Ti-6AI-4V **A2:** 649
boron carbide **EM4:** 805
of brass **A7:** 378-381
and calcination temperature and time **EM4:** 113
calcium phosphate ceramics **EM4:** 1011-1012
cemented carbides **A2:** 951 **A7:** 385-389 **A16:** 74, 78, 79
ceramic grain resistance to **A7:** 539
ceramics **A16:** 98, 100
of cermets **A2:** 985-986 **A7:** 799-800
of compacts **A7:** 309-314
complete homogenization during **A7:** 308
of composite bearings **A7:** 408
in consolidation **A7:** 295
contact area between particles during **A7:** 312
continuous furnaces for **A7:** 352
control of carbon content during **A7:** 370
of copper powders **A7:** 735
copper-base structural parts **A2:** 397
of copper-based materials **A7:** 376-381
cycles **A7:** 11, 367, 369-370, 394
defined **A14:** 12 **EM1:** 21 **EM2:** 38
and densification in CAP process **A7:** 533
density **M4:** 793
diamond and **A16:** 105-106
dimensional change on **A7:** 115, 291, 292, 480-481
in dissociated ammonia, and corrosion resistance **A7:** 256-257
effect on copper powder dimensional change **A7:** 115
effect on copper powder tensile strength **A7:** 116
effect on strength and fracture toughness **EM4:** 587
as energy-efficient **A7:** 569
enhanced, use with refractory metals **A7:** 317
equipment *See also* Furnaces; Sintering furnaces **A1:** 804-805 **A7:** 351-359, 391-392
expansion during **A7:** 311
factors affecting **A7:** 291, 371-373
of ferrous materials **A7:** 360-368, 683
ferrous P/M materials **M1:** 329-346
and ferrous powder metallurgy materials **A1:** 803-804
full-dense **A7:** 23, 525
furnaces *See also* furnaces **A7:** 339, 352, 381-383
glass-ceramics as ceramic substrates **EM4:** 1111
and hardness values **A7:** 189
and heat treating **A7:** 339
of high-speed steels **A7:** 370-376
high-temperature **A7:** 366-368, 482
high-temperature superconductors **A2:** 1086
history **A7:** 14-16
homogenization during **A7:** 308, 314-315
and injection molding **A7:** 497
of iron powders **A7:** 361-362
iron-base compacts, alloying additions **M4:** 796-797
of iron-copper powders **A7:** 365-366
of iron-graphite powder **A7:** 362-365
and isostatic pressing **A7:** 435
and joining **A7:** 457
with joining oxide ceramics **EM4:** 511
liquid-phase **A7:** 309, 316-317, 319-321 **A16:** 98, 100 **M4:** 797
liquid-phase, transient **A7:** 319
loose powders **A7:** 296
mechanisms **A7:** 371
and melting/fining **EM4:** 391
mercury porosimetry curves for **A7:** 270
microstructures **A7:** 375-376
of molybdenum **A7:** 389-392
molybdenum billets **A14:** 237

Sintering (continued)
multilayer alumina substrates **EM4:** 1109
of nickel alloy powders **A7:** 395-398
of nickel silvers **A7:** 378-381
as noun, defined **A7:** 11
optical fibers **EM4:** 414-415
overpressure **A2:** 989
in oxide reduction **A7:** 52
P/M materials **A16:** 886, 887, 888, 889
pellet, oxygen removal during **A7:** 665
of polycrystalline cubic boron nitride and diamond **A2:** 1009
polymer compacts **A7:** 607
in powder forging **A14:** 192-194
and powder forging, compared **A14:** 196
prealloyed powders **M4:** 797
pressure assisted **A7:** 309
pressureless **A16:** 100
in production of cemented carbides **A16:** 72
production practices **A7:** 360-400
of refractory metals **A7:** 389-393
self-lubricating bronze bearings **A2:** 1394
silica gels **EM4:** 449
silicon nitride-based ceramics **EM4:** 812-814
silicon-nitride ceramic cutting tools **EM4:** 966
sinter, HIP process, cemented carbides **A16:** 72
solid phase **A7:** 308
of stainless steel powders **A7:** 368-370, 729-730
and strength **A7:** 190-192
structural ceramics **A2:** 1020
and surface behavior **A7:** 262
of tantalum **A7:** 393
techniques **A1:** 804
temperatures *See* Sintering temperatures
and the FULDENS process **A16:** 65
time *See* Sintering time
time and temperature **M4:** 796
of titanium **A7:** 393-395
to high density plus hot working **A7:** 525
of tool steels **A7:** 370-376
transient liquid-phase, self-lubricating bearings **A7:** 319
tungsten **A7:** 389-393
tungsten alloys **A2:** 579
of tungsten carbide **A7:** 774
of tungsten heavy alloys **A7:** 393-394
two spheres together **A7:** 313
in vacuum **A7:** 171
as verb, defined **A7:** 11
vs. wrought **A7:** 375
zinc oxide varistors **EM4:** 1150-1151, 1153, 1154
zirconia **EM4:** 775, 776, 777, 779
Sintering, and electron beam melting
compared **A15:** 41
Sintering atmospheres *See also* Atmospheres; Sintering **A7:** 11, 339-349, 360-361 **M4:** 794-796
in activated sintering **A7:** 319
Al and Al alloy powders **A7:** 383-384
delubing in **A7:** 340
dissociated ammonia **M4:** 795
effect of chemical composition **A7:** 246
endothermic gas **M4:** 795
exothermic gas **M4:** 795-796
hydrogen **M4:** 795
for nickel and nickel alloys **A7:** 397-398
nitrogen-base **M4:** 796
physical properties of gases and liquids used **A7:** 341
powder metallurgy products **M4:** 795
requirements **A7:** 339-341
safety precautions in **A7:** 341, 348-349
for stainless steels **A7:** 368
vacuum **M4:** 796
Sintering cemented carbides **A18:** 796
self-lubricating powder metallurgy composites **A18:** 120
steel brakes **A18:** 583
Sintering cycles **A7:** 11, 367, 369-370, 394
Sintering equipment *See also* Furnaces; Sintering; Sintering equipment
for aluminum and aluminum alloys **A7:** 381-383
continuous **A7:** 352
for steel **A7:** 339
for tungsten heavy alloys **A7:** 392
Sintering fundamentals **EM4:** 260-268
definition and description of process **EM4:** 260-261
densification **EM4:** 261, 262, 263, 264
final density **EM4:** 260-261
neck size ratio **EM4:** 260-262
pore elimination process **EM4:** 261
surface area measurement **EM4:** 260-261, 264
effect on compact properties **EM4:** 265-266
effect on pore structures **EM4:** 264-265
densification **EM4:** 265
Ostwald ripening **EM4:** 265
enhanced solid-state sintering **EM4:** 266-268
activated sintering **EM4:** 266, 267, 268
liquid-phase sintering **EM4:** 260, 266, 267-268
phase stabilization **EM4:** 266, 267
reactive sintering **EM4:** 266, 267
supersolidus liquid-phase sintering **EM4:** 268
kinetics of sintering **EM4:** 261-264
initial stage sintering equation **EM4:** 263
isothermal neck growth determined by initial stage sintering model **EM4:** 262
transport mechanisms **EM4:** 261-262, 263
mixed-powder sintering **EM4:** 266
homogenization **EM4:** 266
mixed-phase sintering **EM4:** 266
modification results **EM4:** 267
rate-controlled sintering **EM4:** 264
sintering diagrams **EM4:** 265, 266
stages **EM4:** 260, 263, 264, 265, 268
Sintering furnaces
mesh-belt conveyers **M4:** 794
pusher-type **M4:** 794
roller-hearth **M4:** 794
vacuum **M4:** 794
walking-beam **M4:** 794
Sintering studied by scanning electron microscopy **A9:** 99-100
Sintering temperature **A7:** 11
of brass **A7:** 379-380
for bronze **A7:** 378
and compact hardness **A7:** 312
density as function of compacting pressure **A7:** 310
effect of **A7:** 367
effect on apparent hardness **A7:** 367
effect on dimensional change **A7:** 367, 369
effect on elongation **A7:** 369
effect on porosity of iron compacts **A7:** 362
effect on tensile and yield strengths **A7:** 369
effect on transverse rupture strength **A7:** 367
for nickel slivers **A7:** 379-380
shrinkage as function of **A7:** 310
and shrinkage of compacts **A7:** 309
of single metal powders **A7:** 308
and sintered density **A7:** 314, 372
and time *See also* Sintering; Sintering time **A7:** 372
Sintering time **A7:** 11
for bronze **A7:** 378
and densities, tensile properties and velocities **A7:** 484, 485
density of copper powder compacts as function of **A7:** 309
effect on elongation and dimensional change **A7:** 370
effect on tensile and yield strength **A7:** 370
for iron-graphite powders **A7:** 364
shrinkage as function of **A7:** 310
and sintering density and temperature **A7:** 372
and temperature **A7:** 372
and temperature, microstructural changes as function of **A7:** 311
variation of sintered density with **A7:** 314
versus density, liquid-phase sintering **A7:** 320
Sintering zone
furnaces **A7:** 351-352
Sintering-compacting combination
of cermets **A2:** 988-989
Sinters
solid sample digestion by **A10:** 166-167
Sintrate **A7:** 11, 564
Sinusoidal excitation test
methods **EM2:** 551
Sinusoidal loading
S-N curve **A8:** 364
S-N curves for **A11:** 103
Sinusoids
phasor representation of **A17:** 167
SIP *See* Single-in-line package (SIP)
Siphon
use in electrogravimetry **A10:** 200
Sisal buffs **M5:** 118-119
Sister hooks
materials for **A11:** 522
Site symmetry
determined **A10:** 407
Six nines characterization *See also* Pure metals of purity **A2:** 1097
Size *See also* Device size; Dimension; Dimensional; Particle size; Shapes; Sizing; Workpiece(s) **EM3:** 26
component, by metal casting **A15:** 4
component, SRIM/RTM compared **EM2:** 348
of crystal, as x-ray diffraction analysis **A10:** 325
decreasing, as trend **EL1:** 12
defined **EM1:** 21
definition, extruded section **A14:** 322
device, minimum for conventional logic **EL1:** 2
die, and IC complexity **EL1:** 400
drawing to **A14:** 294
as driving force **EL1:** 438
effect, defined **A8:** 12
effect, filament winding **EM2:** 372
effect of fatigue test specimen **A8:** 372-373
effect, on properties divergencies **EM2:** 658
effect, rotational molding **EM2:** 363-364
effects, in wave soldering **EL1:** 520
factors, laminates **EM1:** 236
of forgings, closed-die forging **A14:** 75
hardening of steel related to **M1:** 481-482
hybrid vs single-chip packages **EL1:** 451
of match plate patterns **A15:** 24
measurements with image analysis **A10:** 315
of open-die forgings **A14:** 61
as process selection factor **EM2:** 288-292
reduction, by hybrid circuitry **EL1:** 8
riser, and feeding aids **A15:** 586-58
of rolled rings **A14:** 108-110
of rounded sand grains **A15:** 223
workpiece, for coining **A14:** 180
workpiece, in hot upset forging **A14:** 83
Size, as treatment
defined **EM2:** 38
Size distortion
defined **A11:** 136
Size effect
erosion mechanisms **A18:** 203
probabilistic design of ceramic components **EM4:** 700, 707
Size fraction **A7:** 11
Size of weld
definition **M6:** 16
Size reduction equipment **A7:** 69-70
Size scale, nucleation
casting effects **A15:** 101
Size tolerances
cold finished bars **M1:** 217, 218
Size(s) *See also* Cracks; Discontinuities; Flaws; Sizing
of ceramic component, in cermets **A2:** 978
code/standard/requirement for **A17:** 49
of coils **A17:** 176-177
detector, x-ray **A17:** 369
difficult-to-inspect, magnetic rubber inspection of **A17:** 124-125
discontinuity, magnetic effect **A17:** 100
of flat-rolled products, magnesium alloys **A2:** 472
of flaws **A17:** 50
limitations, copper alloy casting **A2:** 346-347
of micron diamond powders **A2:** 1013
of object, NDE methods for **A17:** 51
part, and demagnetization **A17:** 122
and partial volume effect **A17:** 374
and shape, wrought aluminum alloy **A2:** 35
of synthetic diamond abrasive grains **A2:** 1010-1011
Size-distribution of particles **A9:** 131-134
comparison of methods for obtaining **A9:** 132
expressed as numerical parameters **A9:** 133
Size-exclusion chromatography **A10:** 654, 681

Sizing *See also* also Coatings; Coupling agent; Fiber size; Fiber sizing; Restriking **A7:** 11 **EM3:** 26
after derodding, of tube **A14:** 691
agents, functions/types **EM1:** 122
aluminum and aluminum alloys, defined **A2:** 6
blocks, for open-die forging **A14:** 62
butadiene-acrylonitride elastomers as **EM1:** 124
of carbon fiber **EM1:** 113, 868
for carbon fibers/thermoplastics **EM1:** 103
classifications/t-unctions **EM1:** 122, 123
as coining **A14:** 180, 185
and composite mechanical properties **EM1:** 122
computer-aided, through crack tip diffraction **A17:** 654
content, measured **EM1:** 21-22, 285
of cracks, microwave inspection **A17:** 203
defect/flaw, NDE capabilities **A17:** 677
defined **A14:** 12 **EM1:** 122
die **A7:** 11
dimensional change during **A7:** 291, 480
ferrous P/M materials **M1:** 331
final, of niobium-titanium superconducting materials **A2:** 1051
of flaws, magnetic characterization **A17:** 131
glass fibers **EM1:** 108
knockout **A7:** 11
of particulate samples **A10:** 165
as protective coating **EM1:** 123
punch **A7:** 11
removal **EM1:** 123
in resin transfer molding **EM1:** 170
as secondary pressing operation **A7:** 337-338
self-lubricating sintered bronze bearings **A2:** 394-395
stripper **A7:** 11
and surface treatment, compared **EM1:** 122
tube/pipe rolling **A14:** 631
wrought titanium alloys **A2:** 619
Sizing content
defined **EM2:** 38
Skein
defined **EM1:** 22 **EM2:** 38
Skeletal points
beam **A8:** 326-327
Skeletal system
cadmium toxicity effects **A2:** 1241
gallium toxicity **A2:** 1257
Skeletons **A7:** 11
austenitic stainless steel **A7:** 559
compact powder matrix as **A7:** 551
Skewed populations **A8:** 629
Skewness **A18:** 334-335, 336, 343, 348
of particle shape **A7:** 242
SKF plasmadust process
for zinc recycling **A2:** 1225
Skid polishing **A9:** 42
defined **A9:** 16
Skidding
defined **A18:** 17
jet engine components **A18:** 591
Skids **A18:** 337-338, 341
Skids, for street sweepers
of cemented carbide **A2:** 973
Skim gate
defined **A15:** 10
Skimmers, mechanical
induction furnaces **A15:** 374
Skimming
defined **A15:** 10
for inclusion control **A15:** 90
ladles, for aluminum alloys **A15:** 95
Skin **EM3:** 26
defined **EM2:** 38
Skin curing **EM3:** 712
Skin damage from castable resins **A9:** 31
Skin depth **A6:** 365
defined **EL1:** 1157
Skin depth (continued)
effect, microwave vs. eddy current **A17:** 218
and microwave inspection **A17:** 203
in stress corrosion detection, microwave inspection **A17:** 217
test frequency, eddy current inspection **A17:** 182
Skin effect
by alternating current (ac) **A17:** 91
as cause, signal attenuation **EL1:** 603
as eddy current variation **A17:** 169
and operating frequency selection **A17:** 165
resistance, design considerations **EL1:** 41-42
Skin effects
of beryllium toxicity **A2:** 1239
Skin sections
eddy current inspection **A17:** 187
Skin-doubler specimen **EM3:** 475-476
Skin-dried molds, or skin drying
defined **A15:** 10, 228
Skinning butyls
residential construction applications **EM3:** 188
Skins **EM1:** 104
Skip defects
solder **EL1:** 647
Skips **A6:** 366, 367
Skirt
defined **EM1:** 22 **EM2:** 38-39
Skull **A6:** 118
definition **A6:** 1213 **M6:** 16
Slag inclusions **A6:** 1073, 1074
Slag viscosity **A6:** 59-60
Slot weld, definition **A6:** 1213
Skyscraper
early cast iron supported **A15:** 33
Sl units, base
supplementary and derived **A11:** 193
Slab
defined **A14:** 12, 343
Slab bearings **A18:** 741
Slab milling *See* Milling, peripheral
Slab zinc
composition **A13:** 760
Slabbing
defined **A14:** 12
Slabs
ultrasonic inspection **A17:** 267-268
Slabs, hot rolled
hardness of **M1:** 203
Slag *See also* Fluxes
acid, compositions **A15:** 391
acid, iron oxide content **A15:** 357
basic, composition effects **A15:** 384, 390-391
basic, compositions **A15:** 384, 391
basic, iron oxide content **A15:** 357
basic, refractories for **A15:** 386
as casting defect **A11:** 382, 384
cobalt-base casting **A15:** 813
as crevice corrosion site **A13:** 348-349
defined **A9:** 16 **A15:** 10
effect, brass melting loss **A15:** 449
effect on nickel alloy weld metal **M6:** 443
electro-, failure origins **A11:** 440
of equilibrium thickness **A11:** 618
ESR, compositions **A15:** 402
floating, teapot ladle for **A15:** 90
function in flux cored arc welding **M6:** 103
inclusion, defined **A15:** 10
Inclusions **A11:** 322-323, 440
lead removal by **A15:** 452
modifiers, as coating material **A7:** 817
property control, as electrode coating function **A7:** 816
reduction, in argon oxygen decarburization **A15:** 428
removal from
arc welds of nickel alloys **M6:** 442
shielded metal arc welds **M6:** 442
submerged arc welds **M6:** 130, 135
removal, in gating **A15:** 589
Slag (continued)
volumetric flaws in **A17:** 50
Slag bath
in electroslag remelting **A15:** 401
Slag entrapment in welded joints **A9:** 581
Slag inclusion *See also* Inclusions
defined **A15:** 10
Slag inclusions **M6:** 837-838, 843-944
definition **M6:** 16
radiographic appearance **A17:** 350
radiographic methods **A17:** 296
in steel pipe **A17:** 565
structural importance **M6:** 837-838
in weldments **A17:** 582-584
Slag inclusions in
shielded metal arc welds **M6:** 92
wagon tracks **M6:** 92
submerged arc welds **M6:** 130
Slag stringers
as woody fracture pattern **A12:** 224
Slags
analysis of oxidation states in **A10:** 162
analytic methods applicable **A10:** 6
partitioning oxidation states in **A10:** 178
sodium peroxide fusion for **A10:** 167
Slagsitalls (slagsital) **EM4:** 23
composition **EM4:** 873
manufacturer **EM4:** 873
properties **EM4:** 873
thermal conductivity **EM4:** 876
SLAM *See* Scanning laser acoustic microscopy (SLAM)
Slant fracture
defined **A8:** 12 **A11:** 9
surface of **A11:** 20
Slant-shear fractures *See* Shear-face tensile fractures
Slat track, aircraft wing
bending distortion in **A11:** 140-141
Sled friction tester **A18:** 47
Sled test **A18:** 47
Sledge-hammer head
forging lap failure **A11:** 574, 580
Sleeve bearing *See also* Sliding bearing
defined **A18:** 17
powder rolled **A7:** 407-408
Sleeve bearing materials **A9:** 565-576
chemical compositions **A9:** 566
etchants **A9:** 565
etching **A9:** 567
grinding **A9:** 565
manufacturing methods **A9:** 567
microstructures **A9:** 567
mounting **A9:** 565
polishing **A9:** 565-567
sectioning **A9:** 565
specimen preparation **A9:** 565-567
Sleeve bearing materials, specific types
98Ag-2Pb, electroplated on steel **A9:** 576
AMS 4815, unalloyed silver electroplated on steel **A9:** 576
AMS 4825, high-leaded tin bronze, gravity cast against inner surface of steel shell **A9:** 569
SAE 12, tin-base babbitt, continuously cast onto steel backing strip **A9:** 568
SAE 12, tin-base babbitt, overlaid on copper-lead **A9:** 572
SAE 12, tin-base babbitt, overlaid on leaded tin bronze **A9:** 572
SAE 13, lead-base babbitt, continuously cast onto steel backing strip **A9:** 568
SAE 14, lead-base babbitt, centrifugally cast against inside of steel shell **A9:** 568
SAE 14, lead-base babbitt, continuously cast onto steel backing strip **A9:** 568

Sleeve bearing materials, specific types (continued)
SAE 15, lead-base babbitt, continuously cast onto steel backing strip **A9:** 568
SAE 48, copper-lead alloy, continuously cast onto steel backing strip **A9:** 569
SAE 48, copper-lead alloy, gravity cast against inner wall of steel shell **A9:** 569
SAE 49, copper-lead alloy, continuously cast onto steel backing strip **A9:** 569
SAE 49, copper-lead alloy, gravity cast against inside wall of steel shell... **A9:** 568-569
SAE 49, copper-lead alloy, prealloyed powder, sintered, cold rolled resintered **A9:** 570
SAE 49, copper-lead alloy, sintered **A9:** 570-571
SAE 49, copper-lead alloy, with lead-indium overlay **A9:** 575
SAE 49, high-leaded tin bronze continuously cast onto steel backing strip **A9:** 570
SAE 49, high-leaded tin bronze, gravity cast against inside surface of steel shell **A9:** 570
SAE 191, electroplate **A9:** 570
SAE 192, lead-tin copper, electroplated cadmium alloy, and steel backing **A9:** 575
overlay on rolled aluminum-silicon
SAE 192, lead-tin copper, electroplated overlay on copper-lead **A9:** 572
SAE 192, lead-tin copper, electroplated overlay on Cu-40Pb-5.5Ag **A9:** 572
SAE 192, lead-tin copper, electroplated overlay on high-leaded tin bronze **A9:** 572
SAE 192, lead-tin copper, electroplated overlay on low-tin aluminum alloy **A9:** 575-576
SAE 192, lead-tin-copper, electroplate **A9:** 571
SAE 193, electroplate **A9:** 571
SAE 194, lead-indium alloy, electroplated overlay on cast copper-lead alloy **A9:** 575
SAE 485, high-leaded tin bronze, prealloyed powder, sintered, cold rolled resintered **A9:** 570
SAE 770, low-tin aluminum alloy permanent mold cast **A9:** 573
SAE 780, low-tin aluminum alloy strip, clad rolling **A9:** 573, 575-576
to steel backing by warm
SAE 781, aluminum-silicon alloy, rolled to steel **A9:** 575
SAE 781, aluminum-silicon alloy strip, clad to steel backing by warm rolling **A9:** 573
SAE 783, formation of liner **A9:** 567
SAE 783, high-tin aluminum alloy strip clad to nickel-electroplated steel by warm rolling **A9:** 573
SAE 783, high-tin aluminum alloy strip clad to unalloyed aluminum, then steel **A9:** 573
SAE 784, aluminum-silicon alloy strip, clad to low-carbon steel by warm rolling **A9:** 574
SAE 785, aluminum alloy strip, clad to low- carbon steel by warm rolling **A9:** 574
SAE 786, aluminum-tin alloy strip, clad to unalloyed aluminum bonding layer by warm rolling, then to steel **A9:** 574
SAE 787, aluminum-lead alloy strip, clad to low-carbon steel by warm rolling **A9:** 574
SAE 787, aluminum-lead prealloyed powder, roll compacted and sintered, clad to steel by warm rolling **A9:** 574-575
SAE 791, leaded tin bronze, strip, cold rolled and annealed **A9:** 569
SAE 792, leaded tin bronze, centrifugally cast against inner wall of steel shell **A9:** 569

Sleeve bearing materials, specific types (continued)
SAE 792, leaded tin bronze, gravity cast against inner wall of steel shell **A9:** 569
SAE 792, leaded tin bronze, prealloyed powder, sintered, cold rolled resintered **A9:** 570
SAE 793, leaded tin bronze, continuously cast onto steel backing strip **A9:** 569
SAE 793, leaded tin bronze, prealloyed powder, sintered, cold rolled resintered **A9:** 570
SAE 794, high-leaded tin bronze, continuously cast onto steel backing strip **A9:** 570
SAE 794, high-leaded tin bronze, gravity cast against inside wall of steel shell **A9:** 569
SAE 794, high-leaded tin bronze, prealloyed powder, sintered, cold rolled resintered **A9:** 570
SAE 795, commercial bronze, strip, hot- rolled **A9:** 569
tin-bronze infiltrated by lead-base babbitt **A9:** 571
tin-bronze infiltrated by Teflon **A9:** 571
Sleeve, roll-assembly
brittle fracture of **A11:** 327
Sleeve rolls **A14:** 354
Sleeve-bearing stock
acoustical holography inspection of **A17:** 446-447
Sleeving
for steam generators **A13:** 942
Slenderness ratio **EM3:** 26
defined **A8:** 12 **EM1:** 22 **EM2:** 39
Slice
defined **A17:** 385
Slice thickness
defined **A17:** 385
Slide
actuation, in mechanical presses **A14:** 493-495
adjustment, defined **A14:** 12
defined **A14:** 12
feeds **A14:** 499
number in presses **A14:** 495
Slide forming wire **A1:** 852
Slide welding **M6:** 674
Slide-roll ratio *See* Slide-sweep ratio
Slide-sweep ratio
defined **A18:** 17
Slide-to-roll ratios **A18:** 474
Slider blade blast wheels **M5:** 86-87
Slider-crank drive mechanisms
kinematics of **A14:** 37-38
Slides
for die casting **A15:** 287
Sliding *See also* Rolling; Specific sliding; Spin
defined **A18:** 17
grain-boundary **A12:** 19, 25
imposed on rolling, stresses in **A11:** 134
low-stress, cracks in copper crystal **A8:** 603
material transfer during **A10:** 566
seal, high-temperature aerated water testing **A8:** 421
smearing from **A11:** 496
stress distribution in contacting surfaces from **A11:** 592
stresses, in shafts **A11:** 460-461
superimposed on rolling, effect of **A11:** 465
wear **A8:** 601-603
Sliding and adhesive wear **A18:** 236-241
characteristics and examples **A18:** 236
prevention of adhesive wear **A18:** 241
primary parameters in the wear of metals, polymers, and ceramics **A18:** 237-241
ceramics **A18:** 237, 240-241
metals **A18:** 237-239, 241
polymers **A18:** 237, 239-240, 241
sliding surface **A18:** 236-237
conformity of contacting surfaces **A18:** 237
material-dependent bond strength **A18:** 236-237
standard conditions for sliding of three classes of materials **A18:** 236
wear equations, design criteria, and material selection **A18:** 237

Sliding, and deformation
in loose powder compaction **A7:** 298
Sliding bearing materials, friction and wear of **A18:** 741-756
bearing alloys **A18:** 748-754
aluminum-base alloys **A18:** 748, 752-753
cemented carbides **A18:** 748, 754
copper-base alloys **A18:** 748, 750-752
gray cast irons **A18:** 748, 754
laminated phenolics) **A18:** 748, 754
lead-base alloys **A18:** 748, 749-750
nonmetallic materials (nylon, PTFE, carbon-graphite, wood, rubber, and
silver-base alloys **A18:** 748, 753
tin-base alloys **A18:** 748-749
zinc-base alloys **A18:** 748, 753-754
bearing material selection **A18:** 756
bearing material systems **A18:** 745-748
aluminum alloys **A18:** 746
bimetal systems **A18:** 747
copper alloys **A18:** 746
porous metal bushings **A18:** 746-747
single-metal systems **A18:** 745, 747
trimetal systems **A18:** 747-748
zinc alloys **A18:** 746
casting processes **A18:** 754-755
babbitt centrifugal casting **A18:** 754-755
bearing performance properties **A18:** 754
bimetal systems **A18:** 754-755
bronze centrifugal casting **A18:** 755
bronze gravity casting **A18:** 755
bronze strip and slab casting **A18:** 755
single-metal systems **A18:** 754
trimetal systems **A18:** 755
definition of sliding bearing **A18:** 17, 741
designations by tenus that describe their application **A18:** 741
electroplating processes **A18:** 756
fatigue life **A18:** 744
history and development **A18:** 741
lubrication **A18:** 744, 747, 750
powder metallurgy (P/M) processes **A18:** 755
applications of bearing materials **A18:** 755
bimetal and trimetal systems **A18:** 755
continuous sintering process **A18:** 755
impregnation and infiltration **A18:** 755
powder rolling **A18:** 755
single-metal systems **A18:** 755
properties of bearing materials **A18:** 741-745
bearing material microstructures **A18:** 743-744
compatibility **A18:** 743
conformability and embeddability versus hardness and fatigue strength **A18:** 743
corrosion resistance **A18:** 743, 744-745
hardness **A18:** 743
heat and temperature effects **A18:** 745
load capacity **A18:** 745
measurement and testing **A18:** 742-743
microstructures **A18:** 743-744
surface and bulk properties **A18:** 741-742
roll bonding processes **A18:** 755-756
wear damage mechanisms **A18:** 741, 742, 743
Sliding bearings *See also* Sliding bearings, failures of
abrasion damage **A11:** 488
cavitation damage **A11:** 488
classification of **A11:** 483-489
fretting in **A11:** 487
pairs, friction of **A11:** 486
surfaces, fatigue failure **A11:** 487
Sliding bearings, failures of *See also* Bearings; Lubrication; Sliding bearings **A11:** 483-489
bearing failures **A11:** 486-489
bearing materials **A11:** 483-484
classification of sliding bearings **A11:** 483
contaminants **A11:** 485, 486
debris **A11:** 486
elasto-hydrodynamics **A11:** 485
failure analysis procedure **A11:** 486
fluid-film lubrication **A11:** 484
grooves **A11:** 485
load-carrying capacity **A11:** 485
lubricants **A11:** 485
shaft whirl **A11:** 484-485
squeeze-film lubrication **A11:** 485
surface roughness **A11:** 485

Sliding bearings, friction and wear of **A18:** 515, 521
defined (sliding bearings) **A18:** 17
dynamic loads **A18:** 519-520
friction and wear under mixed-film lubrication **A18:** 518-519
full-film lubricated bearings **A18:** 516-518
asperity height (shaft roughness) effect **A18:** 517, 518, 519
friction effect **A18:** 518
low h_{min} value significance **A18:** 518
minimum oil film thickness permissible **A18:** 517
oil additives and surface treatments **A18:** 519
operation conditions, effect on lubricated bearing performance **A18:** 520-521
corrosion **A18:** 520
dirty environment **A18:** 520
dynamic loading **A18:** 520
geometric imperfections **A18:** 520
inadequate lubrication **A18:** 520-521
PV factor effect on rubbing bearing performance **A18:** 515
wear rate for dry rubbing **A18:** 515-516
metallic bearings **A18:** 515-516
nomenclature and units **A18:** 516
polymer bearings **A18:** 516
Sliding bearings, materials for *See* Bearings, sliding
Sliding contact wear test **A8:** 605-606
Sliding contacts *See also* Electrical contact materials
and arcing contacts, compared **A2:** 841-842
and arcing contacts, defined **A2:** 841
brush contacts **A2:** 842
brush materials **A2:** 841
interdependence factors **A2:** 842
for power circuits, recommended materials **A2:** 862
Sliding dies
upsetting with **A14:** 90-91
Sliding friction **A16:** 15, 16
Sliding knee joint prosthesis **A11:** 670
Sliding traction coefficient **A18:** 93
Sliding velocity
defined **A18:** 17
Sliding wear **A8:** 601-603
alumina **A18:** 389
aluminum-silicon alloys **A18:** 788
cobalt-base alloys **A13:** 663 **A18:** 768-770
galling **A18:** 768-769, 770
oxide control **A18:** 769-770
of cobalt-base wear-resistant alloys **A2:** 447-448
jet engine components **A18:** 588, 591
metal-matrix composites **A18:** 803-804, 806, 807, 809-810
nonferrous hardfacing alloys **A18:** 762, 764-765
thermoplastic composites **A18:** 820-822
Slime film, effect
biological corrosion **A13:** 88, 907
Slip *See also* Flow; Slip planes; Slip reversal; Slip steps; Slip traces **EM3:** 26
as a result of electric discharge machining **A9:** 27
as acoustic emission source **A17:** 287
at crack tip **A12:** 15, 21
at low temperatures **A12:** 33
bands, persistent, in fatigue cracking **A11:** 102
and cleavage, in hcp and bcc metals **A11:** 75
cross **A8:** 34
cross-slip, cobalt alloy **A12:** 398
in crystals **A9:** 719-720
defined **A8:** 12 **A9:** 16 **A11:** 9 **A13:** 11
in deformation **A8:** 188
dissolution, crack initiation by **A13:** 149
distance, in particle fracture model **A12:** 207
during simple shear deformation **A8:** 180-181
effect in liquid erosion **A11:** 165
effect of temperature **A8:** 34
effect on texturing **A10:** 358
enamel **A13:** 47-49

Slip (continued)
environmental effects **A12:** 35
fatigue as process of **A12:** 35
in FIM samples **A10:** 587
formation, stages **A13:** 45-46
as glide of dislocations **A8:** 34
lattice rotation by **A9:** 700
light microscopy for **A12:** 106
lines, in irons **A12:** 219
lines, topographic methods **A10:** 368
localized, titanium alloys **A12:** 450
multiple **A8:** 34
plane, defined **A13:** 45
-plane fracture, crack-initiation by **A11:** 104
in plastically deformed metals **A9:** 684-693
progression, by DIC illumination **A12:** 121
revealed by differential interference contrast **A9:** 152
rolling contact wear **A18:** 257, 260
severe deformation, wrought aluminum alloys **A12:** 426
side, as buckling in axial compression testing **A8:** 56
and stacking fault energy **A9:** 693
in torsional testing **A8:** 147
Slip aids **EM3:** 41
Slip angle **A18:** 578, 579
defined **EM1:** 22 **EM2:** 39
Slip band
below plane-strain fracture surface strain-induced transformation **A8:** 481-482
defined **A8:** 12
due to creep deformation **A8:** 306
Slip bands
in a single crystal of Co-8Fe **A9:** 689
AISI/SAE alloy steels **A12:** 298
in Armco iron **A12:** 224
in austenitic manganese steel castings **A9:** 239
austenitic stainless steels **A12:** 360
cracks **A12:** 224, 298
defined **A9:** 16, 687 **EM1:** 201
formation, in fatigue **A12:** 117
persistent **A12:** 117
superalloy fracture surface along **A12:** 392
wrought aluminum alloys **A12:** 420
Slip behavior
in cobalt-base alloys **A18:** 766
Slip bonding **EM4:** 181
Slip casting **A7:** 11, 296 **EM4:** 33, 34, 35, 123-124, 151
advanced ceramics **EM4:** 49
applications **EM4:** 153
binders **EM4:** 157
centrifugal casting **EM4:** 153
as cermet forming technique **A7:** 800
of cermets **A2:** 984
comminution techniques **EM4:** 157
constituents of powder formulation **EM4:** 126
definition **EM4:** 153
demixing **EM4:** 95, 98
dispersants **EM4:** 155
drain casting **EM4:** 153, 154
factors influencing ceramic forming process selection **EM4:** 34
filtration kinetics model **EM4:** 156
fugitive wax slip casting **EM4:** 153
for gas turbine engine component fabrication **EM4:** 718, 720
gypsum molds **EM4:** 157-158
mechanical consolidation **EM4:** 125, 126, 127, 128
mechanics **EM4:** 156
plasticizers **EM4:** 157
advantages **EM4:** 158
equipment **EM4:** 158
process considerations **EM4:** 156-158
gypsum molds **EM4:** 157-158
slip control **EM4:** 156-157
in reaction sintering **EM4:** 292
release agents **EM4:** 157

Slip casting (continued)
rheology **EM4:** 153-156, 157
colloidal phase equilibria **EM4:** 153, 155-156
electrostatic stabilization **EM4:** 153, 154
electrosteric stabilization **EM4:** 153, 155
polymeric stabilization **EM4:** 155
steric stabilization **EM4:** 153, 155
surface forces **EM4:** 153-154
size fractionation methods **EM4:** 157
solid casting **EM4:** 153
surface chemistry **EM4:** 153-156
colloidal phase equilibria **EM4:** 155-156
electrostatic stabilization **EM4:** 154
electrosteric stabilization **EM4:** 153, 155
polymeric stabilization **EM4:** 155
steric stabilization **EM4:** 153, 155
surface forces and rheology **EM4:** 153-154
to form shapes of reaction-bonded silicon carbide **EM4:** 240
to make compacts of silicon **EM4:** 237
to make silicon oxynitride shapes by reaction bonding **EM4:** 239
tungsten-reinforced nickel-base superalloys by **EM1:** 885
vacuum casting **EM4:** 153
viscoelastic properties required **EM4:** 116
Slip crack **A7:** 11
Slip direction
defined **A9:** 16
Slip factor **A18:** 264, 266, 267
Slip flask
defined **A15:** 10
Slip forming
defined **EM2:** 39
Slip Line
defined **A8:** 12
Slip lines **A9:** 686-687
defined **A9:** 16
examination by phase contrast etching **A9:** 59
in plastically deformed hafnium **A9:** 500
in plastically deformed zirconium and zirconium alloys **A9:** 500
in type 304 stainless **A9:** 66
Slip pack oxidation-resistant coating **M5:** 664-666
Slip planes **A9:** 684
cracking **A12:** 14
defined **A9:** 16
and dislocation cell walls **A9:** 693
displacement, on dimples **A12:** 13
fracture, in fatigue **A12:** 117
shear bands along **A12:** 32, 32
Slip reversal
effect of oxidation **A12:** 15
effect on crack propagation **A12:** 35
partial **A12:** 15, 21
in vacuum **A12:** 46
Slip ring-brush assemblies
recommended contact materials **A2:** 866
Slip rings, miniature
microcontact materials for **A2:** 866
Slip steps
formation, and serpentine glide, ripples from **A12:** 16
in iron **A12:** 224
on grain-boundary cavities **A12:** 219
wrought aluminum alloys **A12:** 426
Slip systems
contrast of dislocations **A9:** 116
Slip traces
with fracture striations **A12:** 15
and striations, compared **A12:** 119
Slip-in rack
coupon testing **A13:** 199
Slip-line field analysis **A18:** 34, 35
Slip-line field analytical process modeling **A14:** 425
Slip-line field solution
in indentation testing **A8:** 71-72
vs. elastic theory of hardness **A8:** 72

Slip-line fields
for double indentation **A14:** 394
for dynamic material modeling **A14:** 423
from ring rolling **A14:** 112
for rolling **A14:** 395
Slip/cross slip
cobalt alloys **A12:** 398
Slippage **EM3:** 26
Slipping, of components
distortion from **A11:** 142-143
Slips
forming structural ceramics from **A2:** 1020
Slit scanning method
for particle sizing **A7:** 229
Slit width
UV/VIS **A10:** 68
Slit-island technique
for surface roughness **A12:** 200
Slit-scan radiography *See also* Radiography
defined **A17:** 385
Slitter knives
cemented carbide **A2:** 970
Slitters
edge-trim **A14:** 712
Slitting
burrs **A14:** 708
camber **A14:** 709-710
capacity **A14:** 710
clearances **A14:** 708-709
of coiled sheet and strip **A14:** 708-713
defined **A14:** 12
fracture during **A8:** 548
in hot upset forging **A14:** 83
knives **A14:** 708
lines **A14:** 708
paper stuffing **A14:** 710
re-coiling of stock **A14:** 710
speed **A14:** 710-711
of titanium alloys **A14:** 841
of woven fabric prepreg **EM1:** 150
wrought copper and copper alloys **A2:** 247-248
Slitting tools *See* Shearing and slitting tools
Sliver *See also* Strand
defined **EM1:** 22 **EM2:** 39
Slivers
defined **A9:** 16
as forging flaws **A17:** 293
in hot-rolled steel bars **A1:** 240-241
as PTH failure mechanism **EL1:** 1024-1025
in rolling **A14:** 358
in steel bar and wire **A17:** 549
Slot extension
defined **EM2:** 39
Slot geometry
in secondary-tension test **A8:** 584-585
Slot welds
definition **M6:** 16
oxyacetylene braze welding **M6:** 597
recommended groove proportions **M6:** 70
Slot-headed screws
carbon steel for **M1:** 254
Slots
measured by CMMs **A17:** 18
as stress concentrator **A11:** 318
Slotting **A16:** 184, 187-193
Al alloys **A16:** 790
cemented carbides used **A16:** 75
in conjunction with milling **A16:** 304, 308, 320, 321, 322
machines **A16:** 187-188
Mg alloys **A16:** 827, 828
multifunction machining **A16:** 374
process capabilities **A16:** 187
ram stroke and clearance **A16:** 190-191
refractory metals **A16:** 866, 867
and tube piercing **A14:** 470
workholding devices **A16:** 188
Slow axial flow (carbon dioxide) laser **A14:** 735-736
Slow cooling
of aluminum bronzes **A2:** 350
Slow cooling zone
furnaces **A7:** 352
Slow crack growth **A18:** 403
Slow crack growth (SCG)
caused by surface roughness of ceramic powders **EM4:** 27
Slow monotonic fracture
ductile irons **A12:** 230-232
Slow neutron capture **A10:** 234
Slow neutrons *See* Thermal neutrons
Slow opening
crack **A8:** 454
Slow strain rate embrittlement *See* Hydrogen embrittlement
Slow strain rate technique
defined **A13:** 11
Slow strain rate test
for stress-corrosion cracking **A1:** 725-726
Slow strain rate testing
of aluminum alloys **A8:** 524-525
apparatus **A13:** 262
apparatus for **A8:** 519, 520
as dynamic loading **A13:** 260-263
effect of low strain rates **A8:** 498, 499
load-deflection curve interpreted in **A8:** 498-499
magnitude of strain rate in **A8:** 498
of nickel-based alloy **A8:** 530
potentiostatic tensile **A13:** 288
stress-corrosion cracking **A13:** 268
for stress-corrosion testing **A8:** 496, 498-499, 519-520
tensile, for hydrogen embrittlement **A13:** 288
test specimen selection **A8:** 519
Slow sweep mode
eddy current inspection **A17:** 190
Slow-bend fracture
iron alloy **A12:** 457
maraging steels **A12:** 387
toughness test, LME in **A12:** 30, 38
Slow-bend tests
correlations between **A11:** 60
with impact testing **A8:** 453
Lehigh **A11:** 59
Slow-rate tests **EM3:** 448
Slow-strain-rate embrittlement *See also* Hydrogen embrittlement **M1:** 687
Slow-wave structure
microwave eddy current measurement **A17:** 223
SLR cameras *See* Cameras
Sludge
defined **A18:** 17
dewatering **M5:** 313-314
disposal of *See* Plating waste disposal and recovery
formation, in acid cleaning, control of **M5:** 63-65
as inclusion-producing **A15:** 96
in phosphate coating **A13:** 384, 388
Sludge treatment
constant conditions maintained by electrometric titration **A10:** 202
Slug **A7:** 11
Slug(s) *See also* Blank
copper and copper alloy **A14:** 257
defined **A14:** 12
preparation, cold extrusion of copper/copper alloy parts **A14:** 310
preparation, impact extrusion **A14:** 311-312
shape, by cold extrusion **A14:** 303
stock for **A14:** 309
Slug-casting metal
as lead and lead alloy application **A2:** 549
Slugging **A6:** 1183
definition **A6:** 1213 **M6:** 16
in uranium dioxide fabrication **A7:** 665
Slump test
porcelain enameling **M5:** 523, 531
Slumpability
defined **A18:** 17
Slurries
as core coatings **A15:** 240
ethyl silicate, as pH sensitive **A15:** 212
foamed plaster, mixing of **A15:** 247
formation, investment casting **A15:** 260
mixing, for plaster molding **A15:** 244
Shaw ceramic **A15:** 250
solid-sample Babington-style nebulizers for **A10:** 36
structural ceramics from **A2:** 1020
for Unicast process **A15:** 251-252
zircon, formulations and properties **A15:** 260
Slurry
defined **A18:** 17 **EL1:** 1157
Slurry (continued)
magnetic paint **A17:** 126-127
in spray drying **A7:** 73-78
in wet magnetic compaction **A7:** 327-328
Slurry abrasion response (SAR) **A18:** 235
Slurry abrasion response (SAR) number
defined **A18:** 17
Slurry abrasivity
defined **A18:** 17
Slurry casting processes **EM4:** 34-35
Slurry coating
powder used **A7:** 573
Slurry coating processes
aluminum coating, steel **M5:** 341-343
procedures and equipment **M5:** 342-343
slurry compositions **M5:** 342-343
fusion *See* Fused slurry process
oxidation protective coating, superalloys and refractory metals **M5:** 379-380
Slurry erosion **A18:** 233-235
cobalt-base alloys **A18:** 768, 770
defined **A18:** 17, 233
dry abrasivity **A18:** 235
effects of wear **A18:** 234
nickel-base alloys **A18:** 768, 770
slurry wear modes **A18:** 233-234
abrasion-corrosion wear **A18:** 233, 234, 235
cavitation **A18:** 234
crushing and grinding **A18:** 234, 235
high-velocity erosion **A18:** 234, 235
low-velocity erosion **A18:** 234
saltation wear **A18:** 234
scouring wear **A18:** 233-234, 235
stainless steels **A18:** 768, 770
Slurry handling systems
corrosive wear **A18:** 271
Slurry infiltration **EM4:** 35
Slurry method
ceramic coatings for adiabatic diesel engines **EM4:** 992
Slurry preforming *See also* Preform
defined **EM1:** 22 **EM2:** 39
Slurry wear test
Bureau of Mines **A18:** 274, 275
Slurry wet abrasive blasting systems **M5:** 93-96
Slurry-sinter oxidation-resistant coating **M5:** 664-666
Slush casting
defined **A15:** 10, 34
development of **A15:** 34-35
Slush casting zinc alloys
gravity castings **A2:** 530
properties **A2:** 538-539
Slush molding
defined **EM2:** 39
Slushing compounds
for oil/gas piping **A13:** 11, 1259
Slushing oil
defined **A18:** 17
Sly cleaning machine
development of **A15:** 33
SM solder attachments *See* Solder attachments
Sm-Sn (Phase Diagram) ..... **A3:** 2 • 368
Sm-Tl (Phase Diagram) ..... **A3:** 2 • 369
Sm-Zn (Phase Diagram) ..... **A3:** 2 • 369
SMAC software program **A15:** 867
Small angle x-ray scattering **EM4:** 87
Small Engine Components Technology Studies **EM4:** 716
Small parts *See also* Part(s); Specimen(s); Workpiece(s)
demagnetizing **A17:** 121
magnetic particle inspection **A17:** 117
magnetizing **A17:** 94
microhardness testing for **A8:** 96
Small signal *See also* Signal; Small signal packages
equivalent circuit **EL1:** 155, 159
frequency response, model **EL1:** 153-154
parameters, junction field effect transistors **EL1:** 156-159
Small signal equivalent circuit **EL1:** 155, 159
Small systems
future trends **EL1:** 393
Small tool welding *See* Lap welding
Small-angle boundaries
in crystals **A9:** 719-720

Small-angle neutron scattering *See also* Small-angle x-ray and neutron scattering **A10:** 402-406
analysis of ceramics **A10:** 405
analysis of glasses **A10:** 405
analysis of metals **A10:** 405
analysis of polymers **A10:** 405
Small-angle neutron scattering (SANS) **EM4:** 87
to measure pore sizes in range of 1 to 10^4 nm **EM4:** 71, 72
Small-angle scattering *See* Small-angle x-ray and neutron scattering
Small-angle x-ray and neutron scattering **A10:** 402-406
applications **A10:** 402, 405
estimated analysis time **A10:** 402
experimental aspects **A10:** 402-403
general uses **A10:** 402
introduction **A10:** 402
related techniques **A10:** 402
samples **A10:** 402
theoretical aspects **A10:** 403-405
Small-angle x-ray scattering **A10:** 402-406
analysis of ceramics **A10:** 405
analysis of glasses **A10:** 405
analysis of metals **A10:** 405
analysis of polymers **A10:** 405
of organic solids, information from **A10:** 9
Small-bubble artifact
in replicas **A12:** 184
Small-outline integrated circuit (SOIC)
die attachments **EL1:** 217-221
as package family **EL1:** 404
removal methods **EL1:** 724-727
as surface-mount package option **EL1:** 77
thermal resistance **EL1:** 409-410
Small-outline packages (SOPs)
and substrates **EL1:** 209
Small-outline transistors
removal methods **EL1:** 724
Small-particle examination, sample preparation
ATEM **A10:** 452
Small-rotation assumption
and beams **EM2:** 692-694
and plates **EM2:** 694-696
Small-sample fatigue testing procedures **A8:** 706
Small-scale integration (SSI) *See also* Small-outline integrated circuits (SOIC)
development **EL1:** 160
tape automated bonding (TAB) for **EL1:** 274
Small-scale integration (SSI) devices **A11:** 766, 797
Small-signal packages
construction details **EL1:** 432
defined **EL1:** 422, 452
leaded plastic **EL1:** 424
metal-body devices **EL1:** 422-423
multiple-terminal **EL1:** 432
plastic body, and surface mount **EL1:** 423-424
plastic-bodied **EL1:** 425
surface mounted **EL1:** 424
"Smart" materials **EM4:** 17
SMC *See* Sheet molding compound (SMC); Sheet molding compounds
SMC-C compression molded sheet
composition/properties **EM1:** 560
SMC-C/R compression molded sheet
composition/properties **EM1:** 560
SMC-D compression molded sheet
composition/properties **EM1:** 560
SMC-R compression molded sheet
composition/properties **EM1:** 560
Smear
Removal, rigid printed wiring boards **EL1:** 544
resin, printed board coupon **EL1:** 575
Smearing *See also* Transfer
in bearing failures **A11:** 493
cause from line source **A10:** 403
Smearing (continued)
cumulative material transfer as **A11:** 496
defined **A18:** 17
Smelt
defined **A13:** 11
Smelting
history of **A15:** 15
of lead **A2:** 543
and spray drying **A7:** 76
SMIE *See* Solid-metal embrittlement
Smith forging *See* Handforge; Open-die forging
Smith/Hieftje system
AAS spectrometers **A10:** 52
Smoke *See also* Flammability
emission, and flame spread, polyester systems **EM1:** 95
liberation, of composites **EM1:** 35
Smoke bombs
leak detection by **A17:** 66
Smoke candles
leak detection by **A17:** 66
Smokes
metal fuel pyrotechnic device **A7:** 600, 602, 603
Smooth specimens
elastic strain, for SCC testing **A8:** 503-508
plastic strain, SCC testing **A8:** 508-509
for SCC evaluation **A13:** 246-253, 265-266, 275
for stress-corrosion cracking tests **A8:** 496-497, 503-510
surface preparation of **A8:** 510
vs. precracked specimen, in SCC testing **A8:** 498
Smoothing
defined **A17:** 385
Smoothing contour function **EM3:** 33
SMT *See* Surface mount technology (SMT)
Smudge remover **M5:** 582
Smut removal
aluminum and aluminum alloys **M5:** 580-581, 584-585, 590-591
surface cleaning **A13:** 381
SN-N-X (whisker)
properties **A18:** 803
Sn-Te (Phase Diagram) ..... **A3:** 2 • 370
Sn-Ti (Phase Diagram) ..... **A3:** 2 • 370
Sn-U (Phase Diagram) ..... **A3:** 2 • 371
Sn-Y (Phase Diagram) ..... **A3:** 2 • 371
Sn-Yb (Phase Diagram) ..... **A3:** 2 • 371
Sn-Zn (Phase Diagram) ..... **A3:** 2 • 372
Sn-Zr (Phase Diagram) ..... **A3:** 2 • 369, 372
Snag grinding
of crankshaft **A11:** 472-473
Snake skins
closed feedwater heaters **A13:** 990
Snap fits **EM2:** 713-714
Snap flasks
defined **A15:** 10
development **A15:** 28
green sand molding **A15:** 341
Snap temper
for ductility **A11:** 95
Snapoff
as solder paste parameter **EL1:** 732
Snapping
of interlocking extrusions **A2:** 36
Sneddon's relation **A18:** 422
Snell's law **EM4:** 1050, 1077
critical angle of internal reflection of optical fibers **EM4:** 409
index of refraction of glass **EM4:** 565
of light refraction **A10:** 113
Snorkels **A6:** 273
Snowflakes *See* Flaking
Snowplow blades
cemented carbide **A2:** 973-974
SNS process
polyaramid fibers **EM3:** 286
Snubber-type wire grip
for axial fatigue testing **A8:** 369
Soak cleaning *See* Alkaline cleaning, soak process; Emulsion cleaning, soak process
Soaking
effects on alloy segregation **A11:** 121
Soaking procedures
steam generators **A13:** 944
Soap **A18:** 125, 126
chemical structures (thickeners) **A18:** 126
defined **A18:** 17
grease incompatibilities **A18:** 129
lubricants for rolling-element bearings **A18:** 136, 137
reasons for use **A18:** 126
for tool steel lubrication **A18:** 737-738
types of thickeners **A18:** 126
Society for Automotive Engineers (SAE) **A7:** 19, 463
Society for the Advancement of Material and Process Engineering (SAMPE)
as information source **EM1:** 41
Society for the Advancement of Material/Process Engineering (SAMPE) **EM2:** 92-93, 95
Society of Automotive Engineers **A15:** 514
(SAE) **A8:** 469, 726
materials specifications of **A13:** 322
Society of Automotive Engineers (SAE)
See SAE-AISI designations **EM3:** 61
Aerospace Materials Specifications **EM3:** 62
Aerospace Recommended Practice (ARP) for aircraft wheels and brakes (ARP 597) **A18:** 586
Crankcase Classification System **A18:** 163, 164
engine oil specifications and viscosity grades **A18:** 162-163, 164
as information source **EM1:** 41, 701
performance specifications for engine lubricants **A18:** 98
performance specifications for nonengine lubricants **A18:** 98
performance testing of engine oils **A18:** 170
two-stroke cycle engine oil service classifications **A18:** 166-167
viscosity grades of lubricants **A18:** 85, 98, 99
Society of Automotive Engineers (SAE) standards
brake codes for heavy commercial vehicles (J880; J9781) **A18:** 577
brake lining friction rating specification (J866a) **A18:** 576
performance testing of engine oils (JI83) **A18:** 170
quality control test procedure for friction materials test machine (J661a) **A18:** 576
viscosity classification systems for engine oils (J300; JI536) **A18:** 163, 164, 166
Society of Automotive Engineers, Japan
performance testing of engine oils **A18:** 170
Society of Automotive Engineers Lubricants Review Institute (LRI)
performance testing of engine oils **A18:** 170
Society of Die Casting Engineers **A15:** 34
Society of Manufacturing Engineers **A7:** 19
classification characteristics of structural adhesives **EM3:** 74
Society of Plastics Engineers (SPE) **EM2:** 93, 95
as information source **EM1:** 41
Society of the Plastics industry (SPI)
as information source
Society of the Plastics Industry, Inc **EM2:** 94
Socket spanner head
forging fold cracking **A11:** 329, 331
Socket-head screws
alloy steel for **M1:** 256
Socketing, as connection
multichip packaging **EL1:** 311
Sockets
desoldering **EL1:** 721-722
Soda ash **A13:** 1178-1179

Soda ash (Na_2CO_3)
batch size ... **EM4:** 382
chemical analysis ... **EM4:** 380
composition ... **EM4:** 379, 380
decalhydrates ... **EM4:** 380
heat evolved by hydration of ... **EM4:** 384
heptahydrates ... **EM4:** 380
mining techniques ... **EM4:** 380
process for making ... **EM4:** 379
process plants ... **EM4:** 379
purpose for use in glass manufacture ... **EM4:** 381
sources ... **EM4:** 379-380
trona ... **EM4:** 380
Soda ash addition
fluxes ... **A15:** 389
Soda ash tailings
Miller numbers ... **A18:** 235
Soda borosilicate
chemical corrosion ... **EM4:** 1047
properties, non-CRT applications ... **EM4:** 1048-1049
Soda feldspar ($NaAlSi_3O_8$) ... **EM4:** 6, 7
Soda-alumina-silica glass
Corning glass codes 0313, 0315, 0319 derived ... **EM4:** 463
Soda-barium-silicate glass
lighting applications ... **EM4:** 1036
Soda-lime ... **EM4:** 460
coefficient of thermal expansion ... **EM4:** 1102
composition ... **EM4:** 1102
material to which crystallizing solder glass seal is applied ... **EM4:** 1070
material to which vitreous solder glass seal is applied ... **EM4:** 1070
softening point ... **EM4:** 1102
Soda-lime glass ... **EM4:** 460
advantages ... **EM4:** 1101
applications
electronic processing ... **EM4:** 1055, 1056
glass containers ... **EM4:** 1082, 1083, 1084
information display ... **EM4:** 1046
lighting ... **EM4:** 1032, 1034, 1035, 1036
solar-cell glass covers ... **EM4:** 1019
coloration ... **EM4:** 1101
composition ... **EM4:** 566, 742, 1033, 1083, 1088, 1101
disadvantages ... **EM4:** 1101
for drinkware ... **EM4:** 1102
durability ... **EM4:** 1101
electrical resistivity ... **EM4:** 404
glass-contact and fuzed AZS refractories ... **EM4:** 904
heat transfer to batch materials ... **EM4:** 386, 387
not strengthened by ion-exchange ... **EM4:** 462
for ovenware ... **EM4:** 1103
proof testing ... **EM4:** 745
properties ... **EM4:** 742, 863, 1034, 1083, 1088, 1101
refractive index ... **EM4:** 566
regenerative heat exchanger applications ... **EM4:** 906
softening point ... **EM4:** 566
strength ... **EM4:** 851
superstructure and crown refractories applications ... **EM4:** 906
tempering ... **EM4:** 1101
thermal properties ... **EM4:** 566, 1101
uses ... **EM4:** 742
Young's modulus ... **EM4:** 566
Soda-lime tableware
defect inclusion levels ... **EM4:** 392
Soda-lime-borosilicate
as C-glass composition ... **EM1:** 45
Soda-lime-silica glasses
applications ... **EM4:** 1015
dental ... **EM4:** 1096
glass containers ... **EM4:** 1085
information display ... **EM4:** 1045
laboratory and process ... **EM4:** 1087
optical glass products ... **EM4:** 1076
batch formulation ... **EM4:** 381
chemical corrosion ... **EM4:** 1047
elastic modulus ... **EM4:** 849
fatigue ... **EM4:** 744
fining system ... **EM4:** 380
float process used ... **EM4:** 377
glass-to-metal seals ... **EM3:** 302
ion-exchange ... **EM4:** 460
Soda-lime-silica glasses (continued)
MgO substituted for CaO to increase meltability ... **EM4:** 379
properties, non-CRT applications ... **EM4:** 1048-1049
soda ash in composition ... **EM4:** 379
strength ... **EM4:** 850
stress-corrosion failure ... **EM4:** 658, 659, 660-661
versus bioactive glasses ... **EM4:** 1010
Soda-lime-silica sheet glass
forming ... **EM4:** 399
Soderberg's law ... **A11:** 111, 112
mean stress effect on fatigue strength ... **A8:** 374
and static yield strength ... **A8:** 374
Sodium *See also* Liquid sodium; Molten salts; Salts ... **A13:** 515
acid-base titration ... **A10:** 173
as addition to aluminum-silicon alloys ... **A18:** 788
alloying, aluminum casting alloys ... **A2:** 132
in aluminum alloys ... **A15:** 746
cations, in glasses, Raman analysis ... **A10:** 131
corrosion fatigue test specification ... **A8:** 423
deoxidizing, copper and copper alloys ... **A2:** 236
as desulfurization reagent ... **A15:** 75
detected by Auger electron spectroscopy ... **A7:** 251, 254
effect, embrittlement ... **A13:** 182-183
effect on glass substrate bonding ... **EM3:** 283
in enamel cover coats ... **EM3:** 304
in enameling ground coat ... **EM3:** 304
flame emission sources for ... **A10:** 30
as flux ... **A10:** 167
fusion, crucibles for ... **A10:** 167
as inoculant ... **A15:** 105
ions, exchanged in water softeners ... **A10:** 658-659
lubricant indicators and range of sensitivities ... **A18:** 301
as modifier addition ... **A15:** 484
molten, applications ... **A13:** 56
pressurized water reactor specification ... **A8:** 423
pure ... **M2:** 797-799
pure, properties ... **A2:** 1158
as pyrophoric ... **A7:** 199
as sample contaminant ... **A10:** 236
as silicon modifier ... **A15:** 79, 161-165
species weighed in gravimetry ... **A10:** 172
spectrometric metals analysis ... **A18:** 300
and strontium, in aluminum-silicon alloys compared ... **A15:** 167
thermal diffusivity from 20 to 100 °C ... **A6:** 4
TNAA detection limits ... **A10:** 237
ultrapure, by distillation ... **A2:** 1094
use in flux cored electrodes ... **M6:** 103
used to make detergents ... **A18:** 100
vapor pressure ... **A6:** 621
volatilization losses in melting ... **EM4:** 389
Sodium acid pyrophosphate cleaning process
iron and steel ... **M5:** 60
Sodium alloys, resistance of
to liquid-metal corrosion ... **A1:** 635
Sodium aluminum borosilicate glass
properties ... **EM4:** 1057
Sodium antimonate ($2Na_2O{\cdot}2Sb_2O_5{\cdot}H_2O$)
purpose for use in glass manufacture ... **EM4:** 381
Sodium bentonite *See also* Western bentonite
and calcium bentonite, blending effect ... **A15:** 210
Sodium bicarbonate
abrasive for abrasive jet machining ... **A16:** 512
for explosion prevention ... **A7:** 197
Sodium bisulfate
as acidic flux ... **A10:** 167
Sodium bisulfate cleaning process
iron and steel ... **M5:** 60
Sodium borate
electrochemical grinding ... **A16:** 545
Sodium borohydride
as reductant ... **A10:** 49
Sodium borohydride electroless nickel plating process ... **M5:** 221-223, 229-231
coatings, properties of ... **M5:** 229-231
solution composition and operating conditions ... **M5:** 221-223
Sodium borosilicate glass
composite strength as a function of weight gain ... **EM4:** 867
effect of interfacial reaction in ceramic-matrix composites ... **EM4:** 866
Sodium borosilicate glass (continued)
effect of volume fraction and size of dispersed phase ... **EM4:** 866
Sodium carbonate
cadmium cyanide plating bath content ... **M5:** 257, 266-267
carburizing role ... **A4:** 325
in cathodic cleaning ... **A12:** 75
ECDG electrolyte ... **A16:** 548
ECG electrolyte ... **A16:** 545
used to neutralize etching acids ... **A9:** 172
zinc cyanide plating bath content ... **M5:** 248
Sodium carbonate (Na_2CO_3) ... **EM4:** 379
Sodium carbonate cleaner ... **M5:** 24, 28, 35
Sodium carbonate electrolytic brightening
aluminum and aluminum alloys ... **M5:** 580-581
Sodium chlorate
ECG electrolyte ... **A16:** 545
ECM electrolyte ... **A16:** 533, 535, 536
Sodium chloride ... **A13:** 50, 271-272, 893
alternate immersion test, aluminum alloys ... **A8:** 523
as analyzing crystal ... **A10:** 88
aqueous, crack growth in ... **A8:** 403
boiling, Al alloys in continuous immersion in ... **A8:** 523
detected by Auger electron spectroscopy ... **A7:** 251, 254
ECG electrolyte ... **A16:** 545
electrolyte for ECM ... **A16:** 533, 535, 536, 538, 540, 541
electrolyte for Ti alloys, ECM ... **A16:** 852
as fining agent ... **EM4:** 380
mediums for accelerating SCC in Al alloys ... **A8:** 524
photochemical machining etchant ... **A16:** 591
and sodium sulfate in boiling water vs
alloy ... **A8:** 427, 429
steel ... **A8:** 427, 429
stress-intensity factor for titanium
stress-intensity range for stainless
solution, tension SCC testing in ... **A8:** 508
solution, testing high-strength steels in ... **A8:** 527
and stress-corrosion cracking ... **A16:** 27
Sodium chromate and glacial acetic acid, as an electrolyte for austenitic manganese steel
casting specimens ... **A9:** 238
Sodium cyanide
in cathodic cleaning ... **A12:** 75
used to electrolytically etch heat-resistant casting alloys ... **A9:** 331-333
and water as an electrolyte for platinum alloys ... **A9:** 551
Sodium cyanide plating
cadmium plating bath content ... **M5:** 257, 266
copper plating baths, high- efficiency ... **M5:** 159-165, 167-169
zinc cyanide plating bath content ... **M5:** 249-250
Sodium dichromate
passivation of zinc coating with ... **M1:** 169
Sodium diethyidithiocarbamate
as extractant ... **A10:** 170
Sodium dodecyl sulfate ... **A18:** 143
Sodium fluoride
photo-nucleated phase ... **EM4:** 440, 441
Sodium fluoroborate glasses
electrical properties ... **EM4:** 852
optical properties ... **EM4:** 854
Sodium germanate glasses
density ... **EM4:** 846
Sodium hydride cycle
for descaling stainless steels ... **A14:** 230
Sodium hydride salt bath descaling
heat-resisting alloys ... **M5:** 566-567
process ... **M5:** 100, 654
Sodium hydroxide
as an electrolyte for tungsten ... **A9:** 440
as an etchant for wrought stainless steels ... **A9:** 281-282
cadmium cyanide plating bath content ... **M5:** 257, 266
ECG electrolyte ... **A16:** 545
electrolyte for ECM ... **A16:** 533, 535
epoxy synthesis and ... **EM3:** 94
etchant for chemical milling of MMCs ... **A16:** 896
as ferrous cleaning agent ... **A12:** 75
photochemical machining etchant ... **A16:** 589

Sodium hydroxide (continued)
as precipitate ... **A10:** 168-169
safety hazards ... **A9:** 69
solutions, in cathodic cleaning ... **A12:** 75
used in chemical treatment before disposal ... **A16:** 131
zinc cyanide plating bath content ... **M5:** 246, 248
Sodium hydroxide alkaline etching
aluminum and aluminum alloys ... **M5:** 583
Sodium hydroxide anodization (SHA)
of polyphenylquinoxalines ... **EM3:** 166, 167
Sodium hydroxide cleaner ... **M5:** 24, 28, 35
Sodium hydroxide corrosion ... **A13:** 1174-1178, 1269
Sodium hydroxide, removal
adhesion failure and ... **A11:** 43
Sodium hydroxide, stainless steels
corrosion resistance ... **M3:** 87, 88
Sodium hypochlorite ... **A13:** 1179-1180
Sodium hypophosphite electroless nickel plating process ... **M5:** 220-231
coatings, properties of ... **M5:** 223-231
fatigue strength affected by ... **M5:** 231
solution composition and operating conditions ... **M5:** 220-223
Sodium iodide doped with thallium
as scintillator ... **A17:** 371
Sodium ion exchange resins
used in water treatment of cutting fluids ... **A16:** 130
Sodium lamps
as indium application ... **A2:** 752
Sodium lead silicate
properties ... **EM4:** 1057
Sodium metabisulfite as a tint etchant ... **A9:** 170
Sodium metaphosphate, insoluble
dentifrice abrasive ... **A18:** 665
Sodium metasilicate
as SCC inhibitor ... **A13:** 327
Sodium metasilicate cleaner ... **M5:** 24, 28, 35
Sodium molybdate
ECM electrolyte ... **A16:** 535
use in color etching ... **A9:** 142
Sodium nitrate
ECG electrolyte ... **A16:** 544, 545
ECM electrolyte ... **A16:** 533, 535, 536, 541
ECM electrolyte for Ni alloys ... **A16:** 843
laser-enhanced etching neutral salt solution ... **A16:** 576
Sodium nitrate ($NaNO_3$)
as a colorant ... **EM4:** 380-381
as fining agent ... **EM4:** 380
function ... **EM4:** 380
hygroscopic ... **EM4:** 380
impurities ... **EM4:** 380
ion-exchange in melts ... **EM4:** 461
as oxidizing agent ... **EM4:** 380
purpose for use in glass manufacture ... **EM4:** 381
sources ... **EM4:** 380
specific properties imparted in CTV tubes ... **EM4:** 1040-1041
Sodium nitrate, apparent threshold stress values
low-carbon steel ... **A8:** 526
Sodium nitrate-potassium nitrate, impact treatment bath for photochromic glasses ... **EM4:** 462
application or function optimizing powder treatment and green forming ... **EM4:** 49
Sodium nitrite ... **A18:** 141, 144
fracture surfaces of copper specimen in ... **A8:** 490
Sodium oleate ... **A18:** 143
Sodium oxide
in binary phosphate glasses ... **A10:** 131
Sodium peroxide
fusions ... **A10:** 166-167
as sintering agent ... **A10:** 166
Sodium phosphate
and ECG ... **A16:** 545
Sodium picrate as an etchant for silicon steel transformer sheets ... **A9:** 62-63
Sodium polyelectrolyte
application or function optimizing powder treatment and green forming ... **EM4:** 49
Sodium reduction ... **A7:** 161, 162, 165
Sodium silicate
applications ... **EM4:** 47
binding agent ... **A6:** 61
chemical composition ... **A6:** 60
composition ... **EM4:** 47
function and composition for mild steel SMAW electrode coatings ... **A6:** 60
supply sources ... **EM4:** 47
Sodium silicate/CO_2 *See* Carbon dioxide process
Sodium silicates
carbon dioxide cold box process ... **A15:** 238
carbon dioxide resin binder system ... **A15:** 221
competing with anaerobics ... **EM3:** 116
liquid, as binder, ceramic shell molds ... **A15:** 259
for porosity sealing in castings ... **EM3:** 51
sand molding, as chemically bonded self-setting ... **A15:** 37
as silica-base bond, characteristics ... **A15:** 212-213
Sodium soap ... **A18:** 126
Sodium stannate tin plating process ... **M5:** 271
Sodium sulfate
as melting accelerators ... **EM4:** 380
Miller numbers ... **A18:** 235
photochemical machining etchant ... **A16:** 593
Sodium sulfide, stainless steels
corrosion resistance ... **M3:** 88
Sodium sulfite ... **A13:** 1244
Sodium tellurite
biologic effects ... **A2:** 1260
Sodium tetraborate
as flux ... **A10:** 167
glass-forming fusions with ... **A10:** 94
Sodium tetraborate ($Na_2O{\cdot}2B_2O_3{\cdot}10H_2O$)
purpose for use in glass manufacture ... **EM4:** 381
Sodium tetrahydroborate
for hydride generation ... **A10:** 36
Sodium tetraphenylborate
as narrow-range precipitant ... **A10:** 169
Sodium thiosulfate as an etchant base for line etching ... **A9:** 62
Sodium tripolyphosphate cleaner ... **M5:** 24, 28, 35
Sodium tungstate
and ECM ... **A16:** 535
Sodium, vapor pressure
relation to temperature ... **A4:** 495 **M4:** 310
Sodium-aluminosilicate glass
applications, military ... **EM4:** 1020
properties ... **EM4:** 1020
Sodium-ammonium chloride zinc plating system ... **M5:** 251-252
Sodium-arc light sources for microscopes ... **A9:** 72
Sodium-borate glasses
optical properties ... **EM4:** 853, 854
Sodium-desilicate glass, Fe/CoO-containing
joining ... **EM4:** 488
Sodium-potassium alloys ... **A13:** 56, 515
Sodium-potassium alloys, resistance of
to liquid-metal corrosion ... **A1:** 635
Sodium-potassium nitrate, molten
corrosion rates ... **A13:** 1314
Sodium-potassium-aluminum silicate
abrasive in commercial prophylactic paste ... **A18:** 666, 668
Sodium-reduced
potassium tantalum fluoride ... **A7:** 161
sponge fines ... **A7:** 165
tantalum powders ... **A7:** 161, 162
Sodium-resistant borate glass
lighting applications ... **EM4:** 1036
Sodium-silicate glasses
density ... **EM4:** 846
Sodium-silicate glasses (continued)
electrical properties ... **EM4:** 851, 852
properties ... **EM4:** 1057
Sodium/sulfur batteries ... **A13:** 1320
Soft bronze
properties and applications ... **A2:** 382
Soft elastic systems ... **A18:** 193
Soft error
alpha-radiation induced ... **EL1:** 808-809
integrated circuits ... **A11:** 783-784
Soft failure
as environmental failure mechanism ... **EL1:** 493-494
Soft glass
friction coefficient ... **A18:** 32
Soft honing ... **A16:** 21
Soft joints
testing used ... **EM3:** 382-390
Soft magnetic alloy ... **A7:** 11
Soft magnetic alloys *See also* Magnetically soft materials
amorphous materials and metallic glasses ... **A2:** 818-819
nickel alloy ... **A2:** 433
Soft magnetic materials ... **A7:** 639-641
Soft magnetic parts
powders used ... **A7:** 573
Soft materials *See also* Plasticity; Soft magnetic materials ... **A7:** 11, 56, 299, 302, 639-641
and green strength ... **A7:** 302
terminal density ... **A7:** 299
Soft metal bearings
materials for ... **A11:** 483
Soft metals
in electroplated coatings ... **A18:** 838
indentation hardness ... **A8:** 109
Soft solder
definition ... **A6:** 1213
Soft solder (70-30) tin alloy
application and composition ... **A2:** 521
Soft solder structures ... **A9:** 453
Soft solders
tin/tin alloys ... **A13:** 772
Soft spots
effect on carbon tool steel ... **A11:** 570
Soft tool bending test devices ... **A8:** 125-126
Soft undesilverized lead *See* Leads and lead alloys, specific types, corroding lead
Soft vacuum plasma spraying ... **EM4:** 205
Soft water *See also* Water
defined ... **A13:** 12
Soft x-rays
defined ... **A10:** 83
Soft-temper wire ... **A1:** 850
Softening ... **A9:** 694
of copper alloy carrier, sliding electrical contacts ... **A2:** 842
cyclic strain, defined ... **A11:** 144
during creep ... **A8:** 302
iron, historical ... **A15:** 26
resistance, copper and copper alloys ... **A2:** 234
of rolling-element bearings ... **A11:** 500
to refine lead ... **A15:** 475
Softening point *See* Softening range; Vicat softening point ... **EM1:** 47, 736 **EM4:** 424
defined ... **EL1:** 1157
in drinkware compositions ... **EM4:** 1102
in tableware compositions ... **EM4:** 1101
tool materials ... **A16:** 601
Softening range ... **EM3:** 27
defined ... **EM1:** 22 **EM2:** 39
Softening resistance
hot-work tool steels ... **A14:** 46
lead frame alloys ... **EL1:** 491
Softness
lead and lead alloys ... **A2:** 545
Software *See also* also Computers; CADICAM; Computer-aided Design (CAD); Computers
for automated solder joint inspection ... **EL1:** 738

SUBJECTS OF THE INDEXED VOLUMES: ASM Handbook (designated by the letter "A"): **A1:** Properties and Selection: Irons, Steels, and High-Performance Alloys (1990); **A2:** Properties and Selection: Nonferrous Alloys and Special-Purpose Materials (1990); **A3:** Alloy Phase Diagrams (1992); **A4:** Heat Treating (1991); **A6:** Welding, Brazing, and Soldering (1993); **A7:** Powder Metallurgy (1984); **A8:** Mechanical Testing (1985); **A9:** Metallography and Microstructures (1985); **A10:** Materials Characterization (1986); **A11:** Failure Analysis and Prevention (1986); **A12:** Fractography (1987); **A13:** Corrosion (1987); **A14:** Forming and Forging (1988); **A15:** Casting (1988); **A16:** Machining (1989); **A17:** Nondestructive Testing and Quality Control (1989); **A18:** Friction, Lubrication, and Wear Technology (1992). **Metals Handbook, 9th Edition** (designated by the letter "M"): **M1:** Properties and Selection: Irons and Steels (1978); **M2:** Properties and Selection: Nonferrous Alloys and Pure Metals (1979); **M3:** Properties and Selection: Stainless Steels, Tool Materials and Special-Purpose Materials (1980); **M4:** Heat Treating (1981); **M5:** Surface Cleaning, Finishing, and Coating (1982); **M6:** Welding, Brazing, and Soldering (1983). **Engineered Materials Handbook** (designated by the letters "EM"): **EM1:** Composites (1987); **EM2:** Engineering Plastics (1988); **EM3:** Adhesives and Sealants (1990); **EM4:** Ceramics and Glasses (1991); **Electronic Materials Handbook** (designated by the letters "EL"): **EL1:** Packaging (1989).

Software (continued)
by computer-aided engineering (CAE) **EL1:** 81
for calibration, x-ray spectrometers **A10:** 98
casting design **A15:** 610-611
for composite materials analysis **EM1:** 275-281
control, for electronic packaging **EL1:** 12
for coordinate measuring machines **A17:** 20, 26
design **EL1:** 419
electrical simulation **EL1:** 419
empirical correction, x-ray spectrometry **A10:** 100
for failure analysis and fracture mechanics **A11:** 55
finite-element method **A15:** 860-861
for GaAs applications **EL1:** 390
for ICP-AES computer systems **A10:** 39
library, of object location and recognition algorithms **A17:** 37
MAC **A15:** 867, 871-872
MARC program **A15:** 861
Michigan solidification simulator **A15:** 861
MITAS-II, for heat transfer **A15:** 861
model, simulator exercising of **EL1:** 129
program evaluation **EM1:** 69-274
real-time radiography **A17:** 323
simulator exercising of **EL1:** 129
SMAC **A15:** 867
SOLA program **A15:** 867
SOLSTAR **A15:** 611
standardized, algorithms **A17:** 44
for statistical/data analysis **EM2:** 607-608
SWIFT **A15:** 611
system structure constraints by **EL1:** 127
thermal analysis (ADINAT) **EL1:** 446-447
unified -shapes checking (USC), for testing **EL1:** 135
Software engineering
in process design **A14:** 409
Software tools *See also* Modeling
CAD/CAM **A14:** 905
detail analysis tools **A14:** 411-412
geometry representation **A14:** 410
knowledge-based expert systems **A14:** 409-410
rough analysis tools **A14:** 411
Software-controlled image processing in scanning electron microscopy **A9:** 95
Softwood
carbides for machining **A16:** 75
SOIC *See* Small-outline integrated circuits
Soil composition
as corrosive to buried metals **A11:** 192
Soil conditioning
powders used **A7:** 572
Soil corrosion *See also* Moisture; Soils **M1:** 725-731
M5: 430-431
adobe, pit depths **A13:** 517
aluminum alloys **M2:** 225
of aluminum coatings **A13:** 435
of aluminum/aluminum alloys **A13:** 598-599
Arctic soils **M1:** 731
bacterial action, effects of **M1:** 726-727
of carbon steels **A13:** 512-513
of cast irons **A13:** 570
of cast steels **A1:** 376 **M1:** 400
causes **M1:** 725, 731
of copper, iron, lead, zinc **A13:** 621
differential aeration, effects of **M1:** 725, 731
driven steel pilings **M1:** 730-731
of galvanized steels **A13:** 434
of lead/lead alloys **A13:** 789
of magnesium/magnesium alloys **A13:** 743
microbiological **A13:** 314
oxygen content of soil, effects of **M1:** 725, 731
preventive measures **M1:** 731
relationship to electrical resistivity **A13:** 762
soil characteristics, effects of **M1:** 725-726, 728-731
of zinc/zinc alloys and coatings **A13:** 762-763
Soil stabilization
cemented carbide tools for **A2:** 973
Soil types, effects of
selection of cleaning process **M5:** 3-15
Soils *See also* Soil corrosion
aluminum-zinc alloy coatings in **A13:** 436
characteristics, for corrosion testing **A13:** 208-209
copper corrosion resistance **M2:** 467, 470
dissimilar, pipeline corrosion in **A13:** 1288
Soils (continued)
galvanic series in neutral **A13:** 1288
galvanized coatings in **A13:** 440
molecular structure and orientation determined in **A10:** 109
potentiometric membrane electrode analysis **A10:** 181
resistance, of porcelain enamels **A13:** 449
simulated service testing in **A13:** 208-210
steam cleaning for **A13:** 414
TNAA detection limits for **A10:** 237
Sol-gel glasses
Brunauer-Emmett-Teller equation for surface area measurements **EM4:** 27
Sol-gel process **EM4:** 32, 124, 209-213, 445-451
for advanced ceramics **EM4:** 47
advantages **EM4:** 213, 445
alkoxide-derived gels **EM4:** 210-211
aerogels **EM4:** 210-211, 212-213
ceramers **EM4:** 210, 211
cryogels **EM4:** 211
diphasic gels **EM4:** 211
drying problem **EM4:** 210
ormocers **EM4:** 210, 211
oxyfluoride gels **EM4:** 210, 211
oxynitride gels **EM4:** 210, 211
Scherer drying model **EM4:** 210
sonogels **EM4:** 211
vapogels **EM4:** 211
xerogels **EM4:** 211, 212, 213
applications **EM4:** 450-451
ceramic coatings for adiabatic diesel engines **EM4:** 992
ceramic preforms **EM4:** 211-212
applications **EM4:** 212
bulk shapes **EM4:** 212
characteristics **EM4:** 212
coatings **EM4:** 211-212
compositions **EM4:** 212
fibers **EM4:** 212
films **EM4:** 211-212
monoliths **EM4:** 212
ceramic-matrix composites **EM4:** 840, 842
for chemical synthesis **EM4:** 62
for complex glass shapes **EM4:** 741-742
compositions **EM4:** 450-451
densification **EM4:** 449-450
development **EM4:** 377
disadvantages **EM4:** 445
droplet generation method of porous glass beads **EM4:** 421
drying **EM4:** 449-450
of ferrites **EM4:** 1163
filler materials for dental applications **EM4:** 1092
forms **EM4:** 450-451
fibers **EM4:** 450
monoliths **EM4:** 450-451
multicomponent glasses **EM4:** 451
thin films **EM4:** 450
gels produced by hydrolysis and polycondensation of metal alkoxides **EM4:** 448-449
acidic gels **EM4:** 448
monolithic gels **EM4:** 448-449
multicomponent gels **EM4:** 449
gels produced by sol destabilization, aggregation, and aging **EM4:** 446-448
dispersion processing **EM4:** 447
gelation of aqueous silica sols **EM4:** 446
monolithic silica gels **EM4:** 446-447
multicomponent glass compositions **EM4:** 447
nonaqueous solvents **EM4:** 447-448
silica sol formation **EM4:** 446
glass manufacture **EM4:** 741
glass-free mullite preparation **EM4:** 763
limitations **EM4:** 213
mixed oxide ceramics with zirconia **EM4:** 777
for oxynitride glasses **EM4:** 22
polycrystalline fiber production **EM4:** 912
powder-free process **EM4:** 209
processing diagnostics **EM4:** 212-213
formulation of the solution or sol **EM4:** 212
gel drying **EM4:** 213
gelation **EM4:** 213
stabilizing heat treatment **EM4:** 213
properties **EM4:** 450-451
quality control procedures **EM4:** 212-213
Sol-gel process (continued)
sol formulations **EM4:** 445-446
solutions versus sols **EM4:** 209-210, 211
multicomponent solutions and sols **EM4:** 209
one-component solutions and sols **EM4:** 209
sol-gel transition **EM4:** 209-210
Sol-gels
for coating and encapsulation **EM3:** 580
discontinuous oxide fiber from **EM1:** 63
SOLA computer program **A15:** 867
Solar cells
as gallium compound application **A2:** 740, 747, 749
of indium phosphide/indium-copper-diselenide/cadmium **A2:** 753
Solar collector sealants **EM3:** 678
Solar Gemini turbine engine **EM4:** 716
Solar heat welding
in space and low-gravity environments **A6:** 1022-1023
Solar spectrum
and AAS **A10:** 43
Solarization **EM4:** 439
Solder *See also* Lead-tin, Silver-lead solders, microstructures **A9:** 422
analysis of **A10:** 179
back-scattered electron imaging **A9:** 451
cadmium-silver solders **M6:** 1073
cadmium-zinc solders **M6:** 1073
compositions **M6:** 1069-1070
tin-lead solders **M6:** 1070
copper/copper alloy corrosion in **A13:** 635, 637
definition **M6:** 16
embrittlement **A13:** 12, 183
forms **M6:** 1074
fusible solders **M6:** 1074
gold/silver **A13:** 1362
high-temperature, elemental mapping of **A10:** 532
impurities **M6:** 1071-1072
indium-based solders **M6:** 1073-1074
lead-silver solders **M6:** 1073
lead-silver-tin solders **M6:** 1073
melting points **M6:** 1070
precious metal solders **M6:** 1074
properties **M6:** 1091-1093
segregation of lead to surface **A10:** 607-608
soft **A13:** 772
tin-antimony solder **M6:** 1072
tin-lead alloy **A13:** 780-781
tin-lead antimony solders mechanical properties **M6:** 1092
tin-lead solder **M6:** 1070-1072
mechanical properties **M6:** 1091-1093
physical properties .... **M6:** logo
tin-silver solders **M6:** 1072-1073
tin-zinc solders **M6:** 1073
zinc-aluminum solder **M6:** 1073
Solder alloys *See also* Soldering; Solders
dental, of precious metals **A2:** 696-697
development **EL1:** 631
for hybrid packages **EL1:** 454
lead, compositions **A2:** 544
for millimeter/microwave applications **EL1:** 754-758
selection, for passive components **EL1:** 183-184
Solder attachments
leaded **EL1:** 744-745
leadless **EL1:** 744
reliability of **EL1:** 740-742
Solder ball test **EL1:** 655
Solder balls
condensation (vapor-phase) soldering **EL1:** 704
as contaminants **EL1:** 661
from excess solder **EL1:** 691
from rapid solder heating **EL1:** 694
as sealed package particles **EL1:** 679
Solder bars
defined **EL1:** 639
Solder blister
as laminate thermal property **EL1:** 536-537
Solder blocks as mounting materials for tungsten **A9:** 441
Solder bump(s)
defined **EL1:** 1157
techniques, VHSIC technology **EL1:** 76

Solder coating
flexible printed boards **EL1:** 583
Solder, cracked
in electronic materials **A12:** 481-482
Solder creams
defined **EL1:** 639, 651-657
Solder die attach
for hermetic packages **EL1:** 216-217
for plastic-encapsulated devices **EL1:** 221
Solder embrittlement
defined **A13:** 12
Solder extraction system **EL1:** 718-722
defined **EL1:** 718-719
desoldering **EL1:** 719-722
devices **EL1:** 718
tools **EL1:** 719
Solder float
as component- and board-level physical testing **EL1:** 944
Solder flow bath
for component removal **EL1:** 718
Solder fluxes, resistance
of flexible epoxies **EL1:** 821
Solder fountain process
for reworking **EL1:** 117
Solder glasses **EM4:** 22, 1069-1070
application methods **EM4:** 1069-1070
applications **EM4:** 22, 1072
base glasses **EM4:** 1071-1072
crystallizing-type **EM4:** 1069
advantages **EM4:** 1069
commercially available **EM4:** 1070
disadvantages **EM4:** 1069
material to which seal is applied **EM4:** 1070
disannealing **EM4:** 1069
physical forms **EM4:** 1069-1070
suppliers **EM4:** 1072
types **EM4:** 1069
uses **EM4:** 1069
vehicles **EM4:** 1069
vitreous **EM4:** 1069, 1070
advantages **EM4:** 1069
commercially available **EM4:** 1070
disadvantages **EM4:** 1069
hot dipping processes **EM4:** 1070
material to which seal is applied **EM4:** 1070
for microelectronic package sealing **EM4:** 1072
Solder hairs
as defect **EL1:** 561
Solder icicles *See* Icicles
Solder ingots
defined **EL1:** 639
Solder interface
definition **A6:** 1213
Solder joint inspection *See also* Joints; Solder joints; Solder(s)
automation **EL1:** 738-739
future considerations **EL1:** 739
inspection criteria **EL1:** 735-736
laminographic radiography **EL1:** 738
methods **EL1:** 736-738
overview **EL1:** 735
as physical testing **EL1:** 942
reflectance systems **EL1:** 736-737
thermal systems **EL1:** 737-738
transmissive radiography **EL1:** 738
x-ray systems **EL1:** 738
Solder joint(s) *See also* Joints; Solder joint inspection; Solder(s); Soldering
attachments, reliability **EL1:** 740-742
configuration, first-level package **EL1:** 991
design **EL1:** 632
evaluation criteria **EL1:** 737
failure analysis **EL1:** 1034-1040
failure mechanisms **EL1:** 1031-1032
fatigue **EL1:** 743-744
heating factors **EL1:** 715
inspection **EL1:** 735-739
leadless chip carrier **EL1:** 987
Solder joint(s) (continued)
for millimeter/microwave applications **EL1:** 754-758
reliability **EL1:** 626-627, 675
reliability, aramid fibers **EL1:** 616
reliability, metal cores **EL1:** 620
reliability, prediction of **EL1:** 751-752
reliability, quartz fiber reinforcement **EL1:** 618-619
reliable, defined **EL1:** 735
sample preparation **EL1:** 1038-1039
strain model **EL1:** 984
strain relationships **EL1:** 611
surface-mount **EL1:** 117
surface-mount technology effects **EL1:** 631
tall **EL1:** 985
thermal failures **EL1:** 60-62
thermal fatigue of **EL1:** 640-641
through-hole failure mechanisms **EL1:** 969-970
yield **EL1:** 691
Solder leveling
defined **EL1:** 1157
Solder masks
applying/imaging, rigid printed wiring boards **EL1:** 546-547
described **EL1:** 553-554
designs **EL1:** 558-559
dry film **EL1:** 555-556, 584
future trends **EL1:** 559
liquid photoimageable **EL1:** 556-557
materials **EL1:** 115, 555
nomenclature **EL1:** 559
over bare copper **EL1:** 550
processes **EL1:** 115, 554-558
requirements **EL1:** 554-555
rigid printed wiring boards **EL1:** 548
screen printing **EL1:** 555
temporary **EL1:** 559
Solder metal business
development **EL1:** 631
Solder pastes **EL1:** 654-657
application **EL1:** 656-657, 698-700
condensation soldering **EL1:** 703
defined **EL1:** 639
distribution **EL1:** 652
flux **EL1:** 653-654
functional parameters **EL1:** 732
powder for **EL1:** 651-653
quality control **EL1:** 655
reflow soldering of **EL1:** 693
set, surface-mount soldering **EL1:** 699-700
and solder creams **EL1:** 651-657
as solderability preservative **EL1:** 563
surface-mount soldering **EL1:** 698-700
testing **EL1:** 655-656
usage **EL1:** 732
Solder plate
immersion, as solderability preservation **EL1:** 564
reflowed, as preservation **EL1:** 563
relative solderability **A6:** 134
Solder pot
dynamic (flowing) **EL1:** 731
Solder powder
mesh, paste print resolution effects **EL1:** 732
shape, paste print resolution effect **EL1:** 732
for solder pastes **EL1:** 651-653
Solder preforms
defined **EL1:** 639
leaded and leadless surface-mount joints **EL1:** 732
in surface-mount soldering **EL1:** 700
Solder, specific types
63Sn-37Pb, soldered joint **A9:** 457
Sn-30Pb, different magnifications compared **A9:** 453
Sn-31Pb-18Cd, structure **A9:** 455
Sn-37Pb, effect of cooling rate **A9:** 453
Sn-40Pb, structure **A9:** 453
Solder, specific types (continued)
Sn-40Pb, wave-soldered circuit board joint **A9:** 455-456
Sn-50Pb, lamellar eutectic **A9:** 453
Solder station
wave soldering **EL1:** 702
Solder wafer bumping
technology **EL1:** 278
Solder waves
process controls **EL1:** 690-691
process requirements **EL1:** 688
soldering/tinning oils **EL1:** 689-690
through-hole soldering **EL1:** 688-691
wave configurations **EL1:** 688-689
Solder wetting analysis
as SIMS application **EL1:** 1085-1086
Solder wires
defined **EL1:** 639
Solder(s) *See also* Solder alloys; Solder attachment; Solder balls; Solder extraction system; Solder joint inspection; Solder joints; Solder masks; Solder paste; Solder plate; Solder preforms; Solderability; Solderability treatment; Soldering; Soldering in electronic applications; Tin-lead alloys; Wave solder
63Sn-37Pb **A6:** 113
effect of filler reinforcement on yield strength **A6:** 113
torch soldering composition **A6:** 351
vapor-phase soldering **A6:** 369
bismuth **A6:** 995
bismuth alloys **A6:** 985, 986
bridging, and fluxes **EL1:** 647
bumps **EL1:** 278
cadmium **A6:** 968
characteristics **A6:** 905
cohesively bonded, removal **EL1:** 692
compositions **EL1:** 633-637
corrosion resistance **A6:** 628, 632
defined **EL1:** 1157
definition **A6:** 1213
development **EL1:** 633
for electronic applications **A6:** 985-988
forms and properties **EL1:** 2
"fusible" **A6:** 985, 986
as fusible finish **EL1:** 679
gold-germanium **A6:** 968, 977
gold-silicon **A6:** 968, 977
gold-tin **A6:** 968, 977
hidden, inspection of **EL1:** 942
high-melting-point **A6:** 628
high-purity **EL1:** 642
impurities **A6:** 986
impurity effects **EL1:** 637-639, 642
indium **A6:** 968, 976
indium-tin **A6:** 985, 986
intermediate-melting-point **A6:** 628
joint defects, and solder impurities **EL1:** 642
lead-base tin-lead alloys **A6:** 985
lead-indium **A6:** 991, 995
lead-indium alloys **A6:** 985, 986
lead-silver **A6:** 964-967
low-melting-point **A6:** 628
paste **A6:** 111
plated, as preservation **EL1:** 562-563
plating, uses **EL1:** 679
precious-metal solders **A6:** 968, 976, 977
rapid solidification technology **A6:** 905
reflow failures **EL1:** 995
rings **EL1:** 117
rosin-base **A6:** 988
sealing, schematic **EL1:** 240
shear fatigue, analytical model of **EL1:** 743-746
skip defects, and fluxes **EL1:** 647
specifications **EL1:** 633-634
sucker, defined **EL1:** 117
tin-antimony **A6:** 967, 972, 999
tin-antimony alloys **A6:** 985-986

Solder(s) (continued)
tin-antimony-silver **A6:** 964-967, 968, 969, 972, 973
tin-base tin-lead **A6:** 985
tin-bismuth **A6:** 995
tin-cadmium
composition **A6:** 632
properties **A6:** 632
tin-lead **A6:** 964-967, 968, 969, 970, 971, 972, 981, 989, 996, 999
composition **A6:** 632
properties **A6:** 632
tin-lead-silver **A6:** 964-967, 995
tin-silver **A6:** 967-968, 973, 974, 975, 995, 999
tin-silver alloys **A6:** 985
tin-zinc **A6:** 968, 975
composition **A6:** 632
properties **A6:** 632
wave soldering **EL1:** 701
zinc
composition **A6:** 632
properties **A6:** 632
zinc-aluminum **A6:** 968, 975
zinc-cadmium
composition **A6:** 632
properties **A6:** 632
Solderability
board, leaded and leadless surface-mount joints **EL1:** 731
component- and board-level physical testing of **EL1:** 943-944
component/board, leaded and leadless surface-mount joints **EL1:** 731
defect analysis **EL1:** 1035-1037
defined **EL1:** 675
definition **M6:** 16
dewetting mechanism **EL1:** 676
first-level packaging **EL1:** 989-991
glass-to-metal seals **EL1:** 459
lead frame alloys **EL1:** 490-491
in mass soldering **EL1:** 631
and materials **EL1:** 676-677
mechanisms **EL1:** 675-676, 1032-1034
meniscograph **EL1:** 954
nonwetting mechanism **EL1:** 676
plated-through hole **EL1:** 1026
preservative treatments **EL1:** 561-564
retention **EL1:** 680
and solder joint failure analysis **EL1:** 1034-1040
and surface preparation **EL1:** 675
testing **EL1:** 677-678, 954
of tin coatings **A13:** 781
wetting mechanism **EL1:** 675-676
Solderability defect analysis (SDA) **EL1:** 1035-1038
Solderability preservatives *See* Solderability treatments
Solderability testing **EL1:** 677-678, 965
Solderability treatments
assembly requirements **EL1:** 562
development **EL1:** 561-562
hot air solder leveling (HASL) **EL1:** 563
immersion solder plating **EL1:** 564
plated solder **EL1:** 562-563
preservation options **EL1:** 562-564
reflowed solder plate **EL1:** 563
solder paste **EL1:** 563
Soldered joints *See also* Joint(s); Weldment(s)
cleanliness **A17:** 60
corrosion of **A17:** 608-60
cracked, causes **A17:** 608
destructive inspection **A17:** 605
flaw types **A17:** 605-60
laser inspection **A17:** 60
micrographs of **A9:** 422-423
nondestructive inspection **A17:** 606-60
pressure/vacuum testing **A17:** 60
visual inspection **A17:** 605-60
Soldering *See also* Condensation (vapor phase) soldering; Conductive (hot bar) soldering; Conductive belt soldering; Desoldering; Hand-held iron soldering; Infrared (IR) soldering; Laser soldering; Repair station soldering; Solder alloys; Solder extraction; Solder joint; Solder preforms; Solder waves; Solder(s); Solderability; Solderability treatments; Soldering in electronic applications; Solders; Through-hole soldering **A6:** 110-113, 964-984 **A7:** 837-841 **M6:** 1069-1101
adhesives for surface mounting **A6:** 111
advantages **A6:** 110, 137 **M6:** 1069
airborne lead concentrations by operation **A6:** 984
alloy formation and phase diagrams **A6:** 127
alloy powders, composition and properties **A7:** 840
aluminum **A6:** 628, 631, 632 **M2:** 200-201
aluminum alloys **A6:** 628, 631, 632, 739
application of solder **A6:** 133
applications **A6:** 964, 965, 968
aerospace **A6:** 387
automotive **A6:** 393, 395
aqueous cleaning **A6:** 112
automation of **A6:** 110
available solder-metal forms **A6:** 968-969
base and coated metal
base metals **M6:** 1074-1075
coated metals **M6:** 1075-1076
solderability **M6:** 1074-1076
solderability testing **M6:** 1074
beryllium-copper alloys **A2:** 414
blood lead levels in workers by job description **A6:** 984
cadmium plating, solderability **M5:** 264
cadmium-containing solders **A6:** 968
cascade soldering **A6:** 135
case histories **M6:** 1100-1101
as casting defect **A11:** 384
chronology **A6:** 111
circuit board soldering **A6:** 977
coatings **A6:** 971, 979, 981, 988
as component attachment **EL1:** 347-348
compositions of solders **A6:** 965 **M6:** 1069-1070
lead-silver **A6:** 965
tin-lead **A6:** 965
tin-lead solders **M6:** 1071
tin-lead-antimony **A6:** 965
tin-lead-silver **A6:** 965
condensation soldering **A6:** 112, 136
condensation/vapor-phase reflow **A6:** 130
contact angle **A6:** 128, 129, 134-135
copper **A6:** 630-631
copper alloys **A6:** 630-631
copper metals **M2:** 443-449, 450
copper plating, solderability of **M5:** 168
current technology **A6:** 112
defects **EL1:** 556, 1034
defined **EL1:** 1157
definition **A6:** 1213 **M6:** 16
destructive testing **A6:** 982-983
development **EL1:** 631
die, as die casting defect **A15:** 294-295
as die attachment method **EL1:** 213
dip soldering (DS) **A6:** 135
dip technique **A6:** 977
dissimilar metal joining **A6:** 822
dissimilar metals **A6:** 981
drag soldering **A6:** 111-112, 135
drag technique **A6:** 977
electrical resistance alloys **A2:** 822, 824
electroless nickel plating solderability **M5:** 228
environmental, safety, and health issues **A6:** 983-984
equipment **A6:** 134-136
failure mechanisms **EL1:** 1031-1040
fatigue life, equation **A6:** 975
fixture material **A6:** 978
flexible printed boards **EL1:** 590
flux classification scheme **A6:** 622, 628
flux evaluation **A6:** 130
flux safety precautions **A6:** 1202
flux selection guidelines **A6:** 129
flux specifications **A6:** 972
flux technology **A6:** 111
flux types **A6:** 129-130
fluxes **A6:** 135, 971-974 **A7:** 840 **M6:** 1081-1085
characteristics **M6:** 1081-1083
corrosive general-purpose fluxes **M6:** 1083-1084
halide-free **A6:** 973
inorganic content **M6:** 1082
inorganic-acid **A6:** 971, 972, 973-974, 980, 983
intermediate fluxes **M6:** 1084
noncorrosive fluxes **M6:** 1084
organic content **M6:** 1082
organic-acid **A6:** 971, 972, 973, 983
rosin-base **A6:** 971, 972-973, 976, 977, 983
vehicular content **M6:** 1082
fluxes electronic **M6:** 1084
organic fluxes **M6:** 1084-1085
rosin fluxes **M6:** 1085
fluxes for **EL1:** 647
fluxes, removal of **M5:** 57
fluxing effects on joint formation **M6:** 1089
focused infrared soldering **A6:** 977
fundamentals **A6:** 126-129
furnace (infrared or convection), reflow technique **A6:** 977
furnace soldering **A6:** 135
future outlook **A6:** 137
future trends **A6:** 112-113
gold-germanium solders **A6:** 968, 977
of gold-palladium powders **A7:** 151
gold-silicon solders **A6:** 968, 977
gold-tin solders **A6:** 968, 977
heat rate recognition **EL1:** 714-715
history **A6:** 126
hot gas reflow technique **A6:** 977
hot-gas soldering **A6:** 136
"immersion" platings **A6:** 971
impurities in solders **M6:** 1071-1072
aluminum **M6:** 1072
antimony **M6:** 1072
arsenic **M6:** 1072
bismuth **M6:** 1072
cadmium **M6:** 1072
copper **M6:** 1072
iron and nickel **M6:** 1072
phosphorus and sulfur **M6:** 1072
zinc **M6:** 1072
indium-containing solders **A6:** 968
properties **A6:** 976
induction method **A6:** 130
induction soldering (IS) **A6:** 135
induction soldering technique **A6:** 977
infrared heating **A6:** 112
infrared soldering (IRS) **A6:** 136
inspection **A6:** 980-983
inspection and testing **M6:** 1089-1091
destructive testing **M6:** 1090-1091
nondestructive testing .... **M6:** logo
repair .... **M6:** log
visual inspection **M6:** 1089-1090
interconnections, passive components **EL1:** 180
intermetallic compounds **A6:** 127, 128, 129, 134
introduction **EL1:** 631-632
joint design **A6:** 130-131
joint designs **M6:** 1077-1079
joints, mechanical properties **A7:** 841
laser soldering **A6:** 112
laser soldering technique **A6:** 977
lasers to provide heat **A6:** 112
leaching **A6:** 132
lead regulations **A6:** 984
lead-silver solders **A6:** 964-967
low-alloy steels **A6:** 624
low-carbon steels **A6:** 624
low-melting fusible alloy solders **A6:** 968
mechanical test data **A6:** 976
physical and mechanical properties **A6:** 976
low-temperature **EL1:** 686, 692-693
machine **EL1:** 633
machine soldering **A6:** 111
mass **A6:** 110-111
materials and processes selection **EL1:** 116-117
melting and solidification of pure metals **A6:** 126-127
metallurgy of solder joints **M6:** 1095-1097
methods, flexible printed boards **EL1:** 592
of microwave/millimeter wave assemblies **EL1:** 754-758
and mounting technology **EL1:** 632-633
"no-clean" setups **A6:** 112
nondestructive inspection **A6:** 981-982
nonwetting **A6:** 128, 129
in outer lead bonding **EL1:** 285-286
oven method **A6:** 130
as package sealing method **EL1:** 237-239

Soldering (continued)
passivation characteristic of the base metal **A6:** 127-128
pipe and tube soldering **M6:** 1093-1095
planar soldering **A6:** 135
postassembly cleaning procedures **A6:** 978-980
powders used **A7:** 573
precious-metal solders **A6:** 968, 976
physical and mechanical properties **A6:** 977
precleaning **A6:** 131
degreasing **A6:** 131
mechanical cleaning **A6:** 131
pickling **A6:** 131
precleaning and surface preparation **M6:** 1079-1081
acid cleaning **M6:** 1079-1080
degreasing **M6:** 1079
mechanical preparation with abrasives **M6:** 1081
preforms **A6:** 133
principles of joining **M6:** 1069-1070
process and equipment **M6:** 1086-1089
flame or torch soldering **M6:** 1086
furnace soldering **M6:** 1086
hot dip soldering **M6:** 65-26
induction heating **M6:** 1086
infrared heating **M6:** 1086
resistance heating **M6:** 1086
soldering iron or bit **M6:** 1086
ultrasonic soldering **M6:** 1086-1087
process overview **A6:** 126
process parameters **A6:** 133-134
properties of solders and solder joints **M6:** 1091-1093
mechanical properties of solders **M6:** 1091-1092
physical properties of tin-lead solders **M6:** 1091
soldered joint **M6:** 1092-1093
tensile properties of bulk solders **M6:** 1091
properties of solders for the electronics industry **A6:** 126
protective layer **A6:** 988
quality control **A6:** 112, 136-137
resistance heating **A6:** 112
resistance method **A6:** 130
resistance soldering (RS) **A6:** 136
safety **M6:** 1097-1100
assessing lead exposure **M6:** 1099-1100
cleaning agents **M6:** 1097-1098
fluxes **M6:** 1097
solder constituents **M6:** 1098-1099
safety precautions **A6:** 1191, 1202
sheet metals **A6:** 398-399
single-pass (SPS) **EL1:** 694-695
slump **A6:** 986
solder (filler-metal) alloys **A6:** 964-969
solder joint assembly **A6:** 974-977
solder metals **A6:** 111
solder paste **A6:** 133, 986, 987, 988
solder pot materials **A6:** 986
solder resists **A6:** 133
solderability **A6:** 128-129, 134
solderability testing **M6:** 1076-1077
accelerated aging **M6:** 1077
applications **M6:** 1077
area-of-spread tests **M6:** 1076
capillary penetration tests **M6:** 1076-1077
globule test **M6:** 1076
rotary dip test **M6:** 1076
surface tension balance test **M6:** 1076
vertical dip test **M6:** 1076
wave soldering test **M6:** 1076
solderable and protective finishes **A6:** 979
solderable layer **A6:** 988
soldering iron technique **A6:** 977
soldering irons **A6:** 134-135
solders for **EL1:** 633
spalling **A6:** 965
spray-gun soldering **A6:** 136

Soldering (continued)
standards **M6:** 1101
strengthening of solder materials **A6:** 112-113
stress-corrosion cracking **A6:** 130
substrate materials **A6:** 969-971
cleaning solutions for **A6:** 978
preassembly cleaning procedures **A6:** 969-971
surface preparation **A6:** 131
electrodeposition **A6:** 131
hot dipping **A6:** 131
immersion coatings **A6:** 131
precoating **A6:** 131
surface-mount **EL1:** 647, 697-709
tackiness **A6:** 986
techniques **A6:** 977-978
temperatures/dwell time **EL1:** 676
tensile properties of bulk solders as a function of testing temperature **A6:** 968
terne coating for **M1:** 173-174
test methods **A6:** 967, 970
through-hole **EL1:** 681-696
time-weighted average (TWA) limits for organic substances used in solder processing **A6:** 983
tin coating for **M1:** 173
tin-antimony solders **A6:** 967
compositions **A6:** 972
properties **A6:** 967
tin-antimony-silver solders **A6:** 967
creep strength **A6:** 973
fatigue strength **A6:** 973
properties **A6:** 972
tensile strength (bulk) **A6:** 972
tin-lead solders **A6:** 964-967, 970 **M6:** 1070-1072
at cryogenic temperatures **A6:** 967
creep life data **A6:** 969
creep-rupture strength **A6:** 970, 971
fatigue life **A6:** 969, 971
physical and mechanical properties **A6:** 965, 966, 967, 968, 969
physical properties **A6:** 970
reaction zone determination **A6:** 981
shear strength **A6:** 969, 971
tensile strength **A6:** 969, 972
test methods for tensile strength of adhesive joints **A6:** 966, 970
tin-lead-antimony solders **A6:** 964-967
creep life data **A6:** 969
creep-rupture data **A6:** 972
hardness after room-temperature aging **A6:** 969
mechanical properties **A6:** 968
tin-lead-silver solders **A6:** 964-967
tin-silver solders **A6:** 967-968
composition **A6:** 973, 974
creep-rupture test data **A6:** 975
fatigue strength **A6:** 975
physical properties **A6:** 967
tensile strength (bulk) **A6:** 975
tin-zinc solders **A6:** 968
physical properties **A6:** 975
tool steels **A6:** 625
torch method **A6:** 130
torch soldering (TS) **A6:** 135
torch technique **A6:** 977
total strain **A6:** 975
TWA limits for inorganic acids and alkalines **A6:** 983
TWA limits for metal particulates **A6:** 984
ultrasonic soldering **A6:** 136
ultrasonic soldering (bath)
equipment **A6:** 982
technique **A6:** 977
vapor phase soldering **A6:** 112, 136
vapor-phase (condensation) reflow technique **A6:** 977
vs. other joining technologies **A6:** 110-111

Soldering (continued)
wave soldering **A6:** 111, 135 **M6:** 1087-1089
combination of automatic lead cutting **M6:** 1088
conveyance **M6:** 1088
developments **M6:** 1088
flux **M6:** 1088
new technology **M6:** 1089
operating unit **M6:** 1087-1088
operation **M6:** 1087
preheat **M6:** 1088
supplementary operations **M6:** 1088-1089
wave soldering technique **A6:** 977
wettability **A6:** 128-129, 134
wettability of metals by solder **A6:** 127-128
wetting phenomena effect on solderability **A6:** 128-129
wrought aluminum alloys **A2:** 30-32
zinc alloys **A15:** 795
zinc-aluminum solders **A6:** 968
physical properties **A6:** 975

Soldering alloys
dental **A13:** 1357

Soldering, bonds
ultrasonic inspection **A17:** 23

Soldering consumables, selection criteria **A6:** 903-905
joint considerations **A6:** 903-904
process stages **A6:** 903
product forms **A6:** 903-904
solders **A6:** 905

Soldering gun
definition **A6:** 1213 **M6:** 16

Soldering in electronic applications
See also Solder(s); Soldering **A6:** 132-133, 985-1000
aluminum and aluminum alloys **A6:** 990
base materials **A6:** 988-991
ceramic materials **A6:** 991
ceramic substrates **A6:** 132
coatings **A6:** 990, 994, 999
composite laminates **A6:** 132
copper alloys **A6:** 988-990, 998
corrosion **A6:** 990, 991
finishes **A6:** 988-991
fluxes **A6:** 985, 986-988
hot-air leveling technique **A6:** 989
iron-base alloys **A6:** 990-991
metal substrates **A6:** 133
nickel and nickel-base alloys **A6:** 990
precious metals **A6:** 991, 994-995
solder joint design **A6:** 991-999
coefficients of thermal expansion **A6:** 992-993, 996, 997
connector technology **A6:** 991, 998-999
controlled collapse chip connection (C^4) **A6:** 997-998
mixed technology **A6:** 991, 994-997
PWB assemblies (organic laminate and ceramic) **A6:** 994-997
surface-mount technology **A6:** 991, 994-998
tape automated bonding (TAB) **A6:** 991, 997
through-hole technology **A6:** 991, 992-994
solderability testing in electronics applications **A6:** 999-1000
solders **A6:** 985-988
solidus/liquidus temperatures of solders **A6:** 985
storage **A6:** 991
storage/corrosion issues **A6:** 988-991
substrates for electronic components **A6:** 132

Soldering iron
definition **A6:** 1213 **M6:** 16

Soldering, joint evaluation and quality control **A6:** 1124- 1128
automated inspection techniques **A6:** 1126
laser inspection **A6:** 1126
structured-light, three-dimensional vision system **A6:** 1126

Soldering, joint evaluation and quality control (continued)
x-ray laminography **A6:** 1126
destructive evaluation **A6:** 1126-1128
surface-mount assemblies **A6:** 1124-1125
through-hole assemblies **A6:** 1124, 1125
visual inspection **A6:** 1124-1126
Soldering of
aluminum **M6:** 1075
beryllium copper **M6:** 1075
copper **M6:** 1075
iron **M6:** 1075
nickel **M6:** 1075
precious metals **M6:** 1075
Soldering oils
in wave soldering **EL1:** 689-690
Solders *See also* Solder alloys; Soldering; Solders, specific types; specific types, such as soft solder and tin-silver solder, under Tin alloys, specific types **EM3:** 40
applications **M2:** 201, 445
bismuth-base **A2:** 756-757
compositions for soldering aluminum **M2:** 201
compositions for soldering copper **M2:** 445
dental appliances, gold **M2:** 686, 687
embrittlement by **A11:** 236
gold/silver dental **A2:** 698
indium-base and indium-alloyed **A2:** 753
lead and lead alloy **A2:** 553
lead-base **M2:** 497, 505-506
low-melting temperature indium-base **A2:** 751-752
melting range **M2:** 201, 445
recycling **A2:** 1218
tin alloy **M2:** 614
tin and tin alloy **A2:** 520-522
Solenoid valve
for hydraulic torsional system **A8:** 216
Solenoids
as magnetically soft material application **A2:** 761
powder used **A7:** 573
Solid ^{13}C nuclear magnetic resonance (NMR) spectroscopy
on polyimides **EM3:** 157
Solid aluminum
diffusivity of silicon in **A11:** 777
Solid angle
SI base unit and symbol for **A10:** 685
SI unit/symbol for **A8:** 721
Solid cadmium *See also* Cadmium
effect on crack depth in titanium alloys **A11:** 239-241
induced crack morphology, titanium alloy **A11:** 241
Solid casting **EM4:** 9, 34, 35
Solid crystalline membrane electrodes **A10:** 182
Solid cylinders *See also* Cylinders; Tubing; Tubular products
diameter over 75 mm (3 in.), eddy current inspection **A17:** 185
diameter under 75 mm (3. in.), eddy current inspection **A17:** 184-185
eddy current inspection **A17:** 184-185
radiographic inspection **A17:** 332-333
tubes on, inspection machines **A17:** 185
Solid cylindrical bar
eddy current impedance **A17:** 171-172
Solid degassing *See also* Degassing
fluxes, copper alloys **A15:** 467
vs nitrogen purging **A15:** 468
Solid density **A7:** 11
Solid deposits
effect on crevice corrosion **A11:** 184
Solid die headers
for cold heading **A14:** 292
Solid dies **A14:** 292-293, 639
Solid diffusion
chemical kinetics of **A15:** 52
Solid diffusion rates
effect on dendritic structures **A9:** 613-614
Solid friction **A18:** 27
Solid graphite molds
as machined permanent molds **A15:** 285
Solid lubricants *See also* Lubricants **A13:** 960-964 **A18:** 89, 113-120
characteristics of materials **A18:** 113
Solid lubricants (continued)
composition and use **A14:** 515
defined **A18:** 17
electron microscopy as analytical tool **A18:** 376
electroplated coatings **A18:** 838
electroplating with nickel **A18:** 836
extreme-temperature **A18:** 118-120
ceramic-bonded fluorides **A18:** 118-119
comparison of ceramic-bonded and fused fluoride coatings **A18:** 119
fused fluoride coatings **A18:** 119
plasma-sprayed coatings **A18:** 119-120
self-lubricating powder metallurgy composites **A18:** 120
jet engine components **A18:** 591
layer lattice **A18:** 113-118
bearing tests of GFRPI **A18:** 117
bulk properties of some hard coat materials **A18:** 115
graphite **A18:** 114-115, 117
graphite fluoride **A18:** 116
molybdenum disulfide and other dichalcogenides **A18:** 114
physical vapor deposition (PVD) of tribological coatings **A18:** 115-116
polyimide and polyimide bonded CF_x coatings **A18:** 116-117
polymer composites **A18:** 117
mainshaft bearings of jet engines **A18:** 590
rolling-element bearings **A18:** 137-138, 261
titanium alloys **A18:** 781-783
and wear **A11:** 152
Solid lubrication
bearing designs using **A11:** 153
Solid magnesium
comminution of **A7:** 131
Solid material fractures
progressive fracturing of **A8:** 439-440
Solid materials
categories of **EL1:** 1119
Solid mechanics, foundations
and formulas **EM2:** 653
Solid metal
dissolved, effect on pressure vessels **A11:** 656
environments, embrittlement by **A11:** 239-244
and liquid metal, interactions **A11:** 718
Solid metal induced embrittlement *See* Solid-metal embrittlement
Solid metal induced embrittlement (SMIE) **A11:** 239-244
characteristics of **A11:** 240
defined **A11:** 239
delayed failure and mechanism of **A11:** 242-244
investigations of **A11:** 240-242
and liquid-metal embrittlement **A11:** 240
of metals **A11:** 239-244
occurrence in nonferrous alloys **A11:** 243
in steels **A11:** 243, 244
Solid missile fuel
powders used **A7:** 573
Solid oxides *See also* Oxides
defect structure **A13:** 61, 65
Solid particle erosion (SPE) **A18:** 199-210
abrasion **A18:** 199, 203-204
coarse two-phase microstructures, erosion of **A18:** 206-207
edge effect **A18:** 206-207
tungsten carbide-cobalt cermets **A18:** 207
void nucleation **A18:** 206
copper **A18:** 388-389
corrosion-affected erosion regime **A18:** 210
definition **A18:** 199
erosion **A18:** 199-201
rate **A18:** 199
erosion of ceramics **A18:** 204-206
particle hardness **A18:** 205-206
erosion-enhanced corrosion regime **A18:** 210
erosion/corrosion (E/C) **A18:** 207-210
comparison with other E/C classifications **A18:** 209-210
corrosion-affected erosion (CAE) **A18:** 207, 209, 210
erosion-enhanced corrosion (EEC) **A18:** 207-208, 209, 210
pure corrosion **A18:** 208-209
pure erosion **A18:** 207, 209
Solid particle erosion (SPE) (continued)
time- versus mass-based E/C rates **A18:** 209
manifestations in service **A18:** 199
metals, erosion of **A18:** 201-204
embedding of erodent fragments **A18:** 203
mechanisms **A18:** 201-203
micromachining **A18:** 202, 203
particle flux **A18:** 204
particle hardness **A18:** 203-204
particle shape **A18:** 203
particle size **A18:** 203
platelet mechanism **A18:** 202-203
rate **A18:** 199-203
resistance **A18:** 201, 202
sequential observation technique **A18:** 202
temperature **A18:** 204
scale flaking or spalling **A18:** 210
steels, erosion of **A18:** 204
versus liquid impingement erosion **A18:** 222-223
Solid particle retention
and permeability in porous parts **A7:** 696
Solid patterns *See also* Loose patterns
described **A15:** 195
Solid phases *See also* Phases
fraction, as function of time **A15:** 183
movement, and segregation **A15:** 141
Solid polishing and buffing compounds
removal of **M5:** 5, 10-11
Solid propellants **A7:** 598-600
aluminum specifications for **A7:** 599-600
applications **A7:** 599
grain configuration **A7:** 599
grain shapes **A7:** 598
Solid reamers **A16:** 239, 243-244, 246
Solid sample analysis **A10:** 36, 113
Solid shapes *See also* Shapes
wrought aluminum alloy **A2:** 33
Solid shrinkage *See* Casting shrinkage
defined **A11:** 9
Solid solubility
schematic binary phase diagrams **A9:** 612
Solid solution **A6:** 127
defined **A13:** 12
Solid solution, hydrogen
in aluminum alloys **A15:** 748
Solid solution nickel-base wrought heat- resistant alloys **A9:** 309
Solid solution solidification structures **A9:** 611-617
Solid solutions
decomposition during precipitation reactions **A9:** 650
defined **A9:** 16
Solid steel
processing of **A1:** 114, 118
solubility of carbonitrides in austenite **A1:** 407
Solid(s)
alloy additions as **A15:** 72-74
inhomogeneous distribution, and segregation **A15:** 140-141
-liquid interface, insoluble particles at **A15:** 142-147
macroscopic ,solidification of **A15:** 101-102
microscopic, solidification of **A15:** 102-103
nucleation of **A15:** 103
-solid contraction, in solidification **A15:** 598-599
state, solidification of **A15:** 109-110
Solid-electrode voltammetry
capabilities **A10:** 207
Solid-film lubrication **A14:** 512-513, 515
defined **A18:** 17
Solid-film polymers
for plastic-encapsulated devices **EL1:** 220-221
Solid-liquid embrittlement couples
compiled **A11:** 238
Solid-liquid mixing
mixed oxide ceramics with zirconia **EM4:** 777
Solid-metal embrittlement *See also* Liquid-metal embrittlement **A1:** 721 **A13:** 12, 145, 184-187, 689
defined **A12:** 29-30
delayed failure and mechanisms **A13:** 186-187
as environmentally induced cracking **A13:** 145, 184-187
in iron-base alloys **A1:** 721
in leaded alloy steels **A1:** 719, 721-722
in leaded carbon and alloy steels **A1:** 722

Solid-metal embrittlement (continued)
occurrence in steels ... **A13:** 184
titanium/titanium alloy SCC ... **A13:** 689
Solid-particle erosion
of cobalt-base wear-resistant alloys ... **A2:** 450
Solid-particle impingement erosion
ceramics ... **A18:** 814
cobalt-base alloys ... **A18:** 767-768
Solid-phase chemical dosimeter ... **EM3:** 27
defined ... **EM2:** 39
Solid-phase embrittlement
of brazed joints ... **A13:** 879
Solid-phase forming
defined ... **EM2:** 39
Solid-phase method
composite processing ... **A2:** 583
Solid-phase sintering ... **A7:** 308
Solid-solid contraction
in solidification ... **A15:** 598-599
Solid-solution alloys
tungsten ... **A14:** 238
Solid-solution copper alloys ... **A2:** 1019
Solid-solution coring in aluminum alloys
etchants for examination of ... **A9:** 355
Solid-solution hardening
nickel aluminides ... **A2:** 916-917
of nickel and nickel alloys ... **A2:** 429
Solid-solution hardening of nickel-base heat- resistant casting alloys, effect on
strength ... **A9:** 334
Solid-solution mechanisms ... **A3:** 1 • 15, 16-17
Solid-solution melting in aluminum alloys ... **A9:** 358
Solid-solution nickel-base alloys
electron-beam welding ... **A6:** 869
Solid-solution strengthening
aluminum P/M alloys ... **A2:** 202
copper ... **A2:** 234-235
copper and copper alloys ... **A14:** 809-810
effect of alloying elements on ... **A1:** 400
effect on fatigue strength ... **A11:** 119
in elevated-temperature service ... **A1:** 637, 639
wrought aluminum alloy ... **A2:** 38
Solid-solution-strengthened alloys
postweld heat treatment ... **A6:** 572-574
Solid-state
migration, thick-film pastes ... **EL1:** 342
miniature power source ... **EL1:** 103
quantum mechanical band theory of ... **EL1:** 96-103
rheology, for epoxies ... **EL1:** 835-836
Solid-state amorphization
amorphous materials and metallic glasses ... **A2:** 807
Solid-state bonding *See* Diffusion welding
in space and low-gravity environments ... **A6:** 1023
Solid-state cameras
machine vision process ... **A17:** 32-33
Solid-state devices
microwave inspection ... **A17:** 209
Solid-state diffusion
gaseous corrosion ... **A13:** 67-69
sintering- caused ... **A7:** 314
Solid-state diffusion bonding
tantalum alloys ... **A6:** 580
Solid-state diffusion welding
titanium-matrix composites ... **A6:** 527
Solid-state electronic device materials
x-ray topographic studies of ... **A10:** 365, 376
Solid-state infrared detection system ... **A10:** 223
Solid-state lasers
welding suitability ... **M6:** 651
Solid-state nuclear magnetic resonance (NMR) ... **EM4:** 52
Solid-state phase transformations
characterized ... **A10:** 333
Solid-state phased-array jammers
as gallium arsenide MMIC application ... **A2:** 740
Solid-state power supplies
induction brazing ... **M6:** 967
Solid-state precipitation ... **A3:** 1 • 21-22
Solid-state reduction *See also* Oxide reduction ... **A7:** 734
Solid-state refining techniques *See also* Pure metal preparation techniques
electrotransport purification ... **A2:** 1094-1095
external gettering ... **A2:** 1094
solid-state vacuum degassing ... **A2:** 1094
Solid-state resistance welding *See* Forge welding
Solid-state sintering ... **A7:** 11 **EM4:** 270-282, 285-287
binders ... **EM4:** 275
chemical effects of dopant addition ... **EM4:** 275-279
enhancement of lattice diffusion ... **EM4:** 276
preservation of ideal power characteristics ... **EM4:** 276
prevention of exaggerated grain growth ... **EM4:** 276-278
chemical effects of sintering atmosphere ... **EM4:** 278-281
gas solubility ... **EM4:** 278
reactions with dopants ... **EM4:** 278
reactions with powder ... **EM4:** 278-281
conclusions ... **EM4:** 281-282
kinetic effects of firing schedule on preparation and sintering processes ... **EM4:** 279-281
cooling rate effect on second-phase evolution ... **EM4:** 280-281
heating rate effect on densification ... **EM4:** 279
kinetics of powder preparation ... **EM4:** 279
lubricants used in ceramic processing ... **EM4:** 274
objectives ... **EM4:** 270
physical characteristics of powder compacts ... **EM4:** 274-275
compact inhomogeneities effects ... **EM4:** 274-275
compaction ... **EM4:** 274
physical characteristics of powders ... **EM4:** 270-274
agglomerates ... **EM4:** 270-272
aggregates ... **EM4:** 270-272
microstructural changes ... **EM4:** 273-274
particle shape ... **EM4:** 272
particle size ... **EM4:** 270
particle size distribution ... **EM4:** 273
of tungsten heavy alloys ... **A7:** 392
Solid-state stamping *See* Solid-phase forming
Solid-state transformation structures
outline ... **A9:** 602
Solid-state transformations in weldments ... **A6:** 70-86
aluminum alloys ... **A6:** 83
continuous-cooling transformation (CCT) diagrams ... **A6:** 70, 71, 72, 73, 77
fusion zone in multipass weldments ... **A6:** 81-82
fusion zone of a single-pass weld ... **A6:** 75-80
relating weld metal toughness to the microstructure ... **A6:** 78-79
titanium oxide steels ... **A6:** 79
transformation effect on transient weld stresses ... **A6:** 79-80
transformations in single-pass weld metal ... **A6:** 75-78
heat-affected zone ... **A6:** 70, 71, 72, 73, 74, 75, 79, 80-81, 82, 83, 84
heat-affected zone in multipass weldments ... **A6:** 80-81
GMAW to limit the size of local brittle zones ... **A6:** 81
GTAW for temper-bead procedure ... **A6:** 81
postweld heat treatments ... **A6:** 81
heat-affected zone of a single-pass weld ... **A6:** 70-75
continuous heating transformation (CHT) diagrams ... **A6:** 73
peak temperature-cooling time diagrams ... **A6:** 70-73
precipitate stability and grain boundary pinning ... **A6:** 73-74
Solid-state transformations in weldments (continued)
unmixed and partially melted zones in a weldment ... **A6:** 74-75
hydrogen-induced cracking ... **A6:** 79, 80
Newton's law of cooling ... **A6:** 70
nickel-base superalloys ... **A6:** 83-84
special factors affecting transformation behavior in a weldment ... **A6:** 70
stainless steels ... **A6:** 82-83
titanium alloys ... **A6:** 84-86
Solid-state vacuum degassing
for ultrapurification of metals ... **A2:** 1094
Solid-state welding
advantages ... **M6:** 672-673
applications of deformation welding ... **M6:** 686-691
cold welding of aluminum tubing ... **M6:** 691
metal cladding by strip roll welding ... **M6:** 689-690
roll welded heat exchangers ... **M6:** 689
seal welds ... **M6:** 686-687
thermocompression welding ... **M6:** 686
applications of diffusion welding ... **M6:** 682-686
composites ... **M6:** 684
iron-based alloys ... **M6:** 682-683
nickel-based alloys ... **M6:** 683-684
other metals ... **M6:** 685-686
refractory metals ... **M6:** 684
titanium and titanium alloys ... **M6:** 682
cladding of metals ... **M6:** 689-690
copper metals ... **M2:** 441, 442
definition ... **M6:** 16
deformation welding ... **M6:** 678-679, 686-691
cold welding ... **M6:** 673-674
extrusion welding ... **M6:** 676-677
forge welding ... **M6:** 675-676
roll welding ... **M6:** 676
thennocompression welding ... **M6:** 674-675
diffusion welding ... **M6:** 677-679, 682-686
fundamentals of welding ... **M6:** 677
Interlayers ... **M6:** 680-681
metallurgical considerations ... **M6:** 679-680
intermetallic systems ... **M6:** 680
pure elements ... **M6:** 679
two-phase systems ... **M6:** 680
pressure-temperature combinations ... **M6:** 678
problems in welding ... **M6:** 681-682
corrosion failure ... **M6:** 681
porosity ... **M6:** 681
selection of processes ... **M6:** 672-673
Solid-state welding (SSW)
definition ... **A6:** 141, 1213
soft interlayer mechanical properties ... **A6:** 165-171
application methods ... **A6:** 165
environmentally induced failure of interlayers ... **A6:** 171
interlayer fabrication method effect ... **A6:** 168-169
interlayer strain ... **A6:** 167
interlayer thickness effect on stress ... **A6:** 166-167
microstructure of interlayer welds ... **A6:** 165
multiaxial loading ... **A6:** 170-171
shear loading ... **A6:** 169-170
tensile loading of soft-interlayer welds ... **A6:** 165-169
time-dependent failure ... **A6:** 167-168
Solid-surface velocities
laser measurement ... **A17:** 16-17
Solid-tool machining and drilling ... **EM1:** 667-672
of composites, problems ... **EM1:** 667
drilling techniques ... **EM1:** 667-672
machining techniques ... **EM1:** 667
Solid/liquid interface
insoluble particles at ... **A15:** 142-147
silicon carbide particle effect ... **A15:** 145
Solidification *See also* Bridging; Directional solidification; Freezing; Multidirectional solidification; Rapid solidification; Solidification heat transfer; Solidification modeling; Solidification rate;

Solidification (continued)
Solidification sequence; Solidification shrinkage ... **A3:** 1 • 19 **A13:** 501-504, 823
aluminum casting alloys ... **A2:** 146
aluminum Castings ... **M2:** 150-151
aluminum, hydrogen solubility during ... **A15:** 85
of aluminum-silicon alloys ... **A15:** 159-168
of cast iron ... **A15:** 168-181
in casting design ... **A15:** 598-599
columnar to equiaxed transition during ... **A15:** 130-135
continuous flow melting ... **A15:** 415
contours, and riser placement ... **A15:** 779
control, copper casting alloys ... **A2:** 349
cooling curves, interpretation and use ... **A15:** 182-185
copper casting alloys ... **M2:** 385
cracking, ASTM/ASME alloy steels ... **A12:** 345
and crystal growth ... **A15:** 109-113
of cylinders ... **A15:** 606
defined ... **A15:** 10, 598
delayed ... **A11:** 350
dendritic ... **A15:** 102, 145-146
dendritic growth during ... **A11:** 354
direction, coating effect ... **A15:** 281
directional ... **A15:** 5, 174-175
distortion by ... **A15:** 615-616
of ductile iron ... **A15:** 651-652
enthalpy/heat capacity and ... **A15:** 50
equiaxed ... **A15:** 146, 887-890
of eutectic alloys ... **A15:** 159-181
of eutectic structures ... **A15:** 121-122
of eutectics ... **A15:** 119-125
evolution of gases during ... **A15:** 87
in filament winding ... **EM1:** 35
of flake graphite ... **A15:** 174
of flat plates ... **A15:** 606
and fluidity, relationship ... **A15:** 767
front movement, computed ... **A15:** 863
gas-liquid reactions during ... **A15:** 83-84
gray ... **A15:** 67
of gray iron ... **A1:** 13-14 **A12:** 226 **A15:** 630-631
and growth, principles of ... **A15:** 109-158
heat release during ... **A15:** 109
historical study ... **A12:** 2
impact, by carbon equivalence ... **A15:** 629
ingot, by electroslag remelting ... **A15:** 403-404
in iron-chromium-aluminum alloy ... **A12:** 140, 160
isothermal ... **A15:** 174
liquid state ... **A15:** 109-110
local, insoluble particles and ... **A15:** 142
low-gravity effects during ... **A15:** 147-158
of macroscopic solids ... **A15:** 101-102
of malleable iron ... **A1:** 72
maraging steels ... **A12:** 384
mass and heat transport ... **A15:** 111-113
metal-matrix composites ... **A2:** 903
metallurgical aspects ... **A15:** 298-299
of microscopic solids ... **A15:** 102-103
mode, and riser location ... **A15:** 579-580
modeling/computer applications ... **A15:** 857, 887-890
multidirectional ... **A15:** 175-180
nonequilibrium ... **A15:** 136-137
nucleation during ... **A15:** 103-105
nucleation kinetics of ... **A15:** 101-108
of peritectics ... **A15:** 125-129
physical properties relevant to ... **A15:** 109-110
plane front, defined ... **A15:** 112-113
porosity, wrought aluminum alloys ... **A12:** 431
principles of ... **A15:** 99-185
progressive ... **A15:** 278
in shrinkage cavity ... **A12:** 140, 160
shrinkage crack, defined ... **A11:** 9
shrinkage, defined ... **A11:** 9
of single-phase alloys ... **A15:** 114-119
solid/liquid interface ... **A15:** 110-111
in spray wet lay-up technique ... **EM1:** 132
stresses from ... **A15:** 615-616
structure, control of ... **A15:** 61
studies of ... **A10:** 365
temperature (undercooling), low-gravity ... **A15:** 150
temperature, onset/offset ... **A15:** 101
thermal analysis of ... **A15:** 182-185
thermal aspects ... **A15:** 298
and thermal stress ... **EM2:** 752

Solidification (continued)
thermodynamic pressure effect ... **A15:** 84
thermodynamics of ... **A15:** 101-103
ultra-rapid ... **A7:** 18
undercooling required for ... **A15:** 105
uniform, copper casting ... **A15:** 782

Solidification cracking ... **A6:** 409
aluminum metal-matrix composites ... **A6:** 555
austenitic stainless steels ... **A6:** 456, 458-459, 461, 462, 463-464, 467
carbon steels ... **A6:** 641, 649-651, 657
duplex stainless steels ... **A6:** 474, 476, 477-478
electron-beam welding ... **A6:** 851
heat-treatable aluminum alloys ... **A6:** 530-531, 533
precipitation-hardening stainless steels ... **A6:** 484, 487, 490
stainless steels ... **A6:** 82
submerged arc welding ... **A6:** 208-209
titanium alloys ... **A6:** 516

Solidification cracks *See* Hot cracks

Solidification effects in welded joints
sections used to study ... **A9:** 578

Solidification heat transfer
computational system ... **A15:** 861-862
data base ... **A15:** 862-863
data interpretation ... **A15:** 863
future simulation ... **A15:** 863-865
geometric description and discretization ... **A15:** 858-861
modeling of ... **A15:** 858-866

Solidification hot cracking
stainless steel casting alloys ... **A6:** 497, 498

Solidification isotherms of welded joints
mapping of ... **A9:** 579

Solidification modeling *See also* Computer applications; Modeling; Simulation
application ... **A15:** 889-890
industry use ... **A15:** 883
software, for patterns ... **A15:** 198

Solidification processes
in pure metal castings ... **A9:** 608-610

Solidification range
defined ... **A9:** 16

Solidification rate ... **A6:** 49-50, 51, 52, 53
coating effect ... **A15:** 281
effect, inclusion-forming ... **A15:** 95
effect on dendritic structures in copper alloy ingots ... **A9:** 637-638, 640
effect on eutectic structure of tin-lead alloys ... **A9:** 452
effect, particle size ... **A15:** 146
eutectic growth at ... **A15:** 125
under multidirectional solidification ... **A15:** 145

Solidification rate in cast aluminum alloys
determination by examining dendrite arm spacing ... **A9:** 357
effect on phases ... **A9:** 360

Solidification sequence
graphs ... **A15:** 600-604
L-sections ... **A15:** 604-606
liquid-solid contraction, casting by ... **A15:** 599
T-sections ... **A15:** 599-604
X-sections ... **A15:** 604

Solidification shrinkage *See also* Casting shrinkage
aluminum melts ... **A15:** 79
defined *See* Casting shrinkage
and feed metal volume ... **A15:** 577
liquid and solid state ... **A15:** 109
plastics effect ... **A15:** 254
of semisolid materials ... **A15:** 328

Solidification shrinkage crack
defined ... **A9:** 16

Solidification structures
of aluminum alloy ingots ... **A9:** 629-636
of copper alloy ingots ... **A9:** 637-645
of eutectic alloys ... **A9:** 618-622
of pure metals ... **A9:** 607-610
of solid solutions ... **A9:** 611-617
of steel ... **A9:** 623-628
in welded joints ... **A9:** 578-580

Solids *See also* Solids, characterization of
acid mediums for digesting ... **A10:** 166
dissolution, in liquid metals ... **A13:** 56-60
dynamic mechanical properties ... **EM2:** 538-539
EPMA elemental analysis ... **A10:** 516
IC analysis of ... **A10:** 663-664
inorganic, analytic methods for ... **A10:** 4

Solids (continued)
meltable, as IR samples ... **A10:** 112-113
organic compounds, ESR studied ... **A10:** 263
phase composition determined ... **A10:** 126
in production and quality control, XRS analysis for ... **A10:** 83
pure ... **A10:** 129
subdividing, wet chemical analysis techniques for ... **A10:** 165
suspended, in water ... **A13:** 489
transition series elements identified ... **A10:** 253-266
XRS analysis of ... **A10:** 93

Solids, characterization of *See also* Solids
analytical transmission electron microscopy ... **A10:** 429-489
atomic absorption spectrometry ... **A10:** 43-59
Auger electron spectroscopy ... **A10:** 549-567
classical wet analytical chemistry ... **A10:** 161-180
controlled-potential coulometry ... **A10:** 208-211
crystallographic texture measurement and analysis ... **A10:** 357-364
electrochemical analysis ... **A10:** 181-211
electrogravimetry ... **A10:** 197-201
electrometric titration ... **A10:** 202-206
electron probe x-ray microanalysis ... **A10:** 516-535
electron spin resonance ... **A10:** 253-266
elemental and functional group analysis ... **A10:** 212-220
extended x-ray absorption fine structure ... **A10:** 407-419
ferromagnetic resonance ... **A10:** 267-276
field ion microscopy ... **A10:** 583-602
gas chromatography/mass spectrometry ... **A10:** 639-648
inductively coupled plasma atomic emission spectroscopy ... **A10:** 31-42
infrared spectroscopy ... **A10:** 109-125
ion chromatography ... **A10:** 658-667
liquid chromatography ... **A10:** 649-659
low-energy electron diffraction ... **A10:** 536-545
low-energy ion-scattering spectroscopy ... **A10:** 603-609
molecular fluorescence spectrometry ... **A10:** 72-81
Mössbauer spectroscopy ... **A10:** 287-295
neutron activation analysis ... **A10:** 233-242
neutron diffraction ... **A10:** 420-426
nuclear magnetic resonance ... **A10:** 277-286
optical emission spectroscopy ... **A10:** 21-30
optical metallography ... **A10:** 299-308
particle-induced x-ray emission ... **A10:** 102-108
potentiometric membrane electrodes ... **A10:** 181-187
radial distribution function analysis ... **A10:** 393-401
Raman spectroscopy ... **A10:** 126-138
Rutherford backscattering spectrometry ... **A10:** 628-636
scanning electron microscopy ... **A10:** 490-515
secondary ion mass spectroscopy ... **A10:** 610-627
single-crystal x-ray diffraction ... **A10:** 344-356
spark source mass spectrometry ... **A10:** 141-150
ultraviolet/visible absorption spectroscopy ... **A10:** 60-71
voltammetry ... **A10:** 188-196
x-ray diffraction ... **A10:** 325-332
x-ray photoelectron spectroscopy ... **A10:** 568-580
x-ray powder diffraction ... **A10:** 333-343
x-ray spectrometry ... **A10:** 82-101
x-ray topography ... **A10:** 365-379

Solids contact clarifier ... **M5:** 314

Solids content ... **EM3:** 27

Solidus *See also* Freezing range; Liquidus; Melting range ... **A3:** 1 • 2 **A6:** 127
defined ... **A15:** 10
definition ... **M6:** 16
lines, single-phase alloys ... **A15:** 114
temperature, determined ... **A15:** 184-185

Solidus composition (T_s) ... **A6:** 89

Solidus line ... **A6:** 46

Solidus temperature *See also* Thermal properties
aluminum casting alloys ... **A2:** 153-177
cast copper alloys ... **A2:** 356-391
defined ... **A9:** 16, 611
wrought aluminum and aluminum alloys ... **A2:** 62-122

Solithane 60 adhesive
glass adherends showing crack profile displacements ... **EM3:** 453

Soller collimator
WDS spectrometers **A10:** 87
Soller slit
use in topographic methods **A10:** 370, 374, 375
SOLSTAR casting design software **A15:** 611
Solubility **A7:** 552, 801
carbon, cast irons/carbon steels **A13:** 46-47
of cermets **A2:** 990-991
extended, of iron in aluminum **A10:** 294-295
as function of temperature **A15:** 61
gas, in cast iron **A15:** 82-85
gases, copper alloys **A15:** 464-465
in gravimetric analysis **A10:** 163
hydrogen and nitrogen, in cast iron **A15:** 82
hydrogen, during aluminum solidification **A15:** 85-86
of hydrogen, in aluminum **A15:** 79, 456, 747
of hydrogen in copper **A15:** 86, 466
hydrogen in magnesium alloys **A15:** 462-465
hydrogen, temperature effect **A15:** 86
of inclusions **A15:** 90
iron-carbon system, silicon effects **A15:** 63
of lead **A13:** 786
of liquid/solid states **A15:** 109-110
nitrogen, in austenitic stainless steels **A13:** 827
nitrogen, in cementite **A15:** 82
nitrogen, sulfur effect **A15:** 83
of oxygen, in seawater **A13:** 900
of plasticizers **EM2:** 496
of polymers **EM2:** 61
solid, of carbon in austenitic stainless steels **A13:** 827
temperature, and plasticizers **EM2:** 496
of thermoplastics **EM1:** 103
third element factors, ternary iron-base alloys **A15:** 66
Solubility lines
calculated **A15:** 61-70
Solubility, mutual
in brazing **A11:** 452
Soluble gas atomization *See also* Vacuum atomization **A7:** 25, 26, 37-39
Soluble oil
defined **A18:** 17
Soluble oils
as lubricant **A14:** 696
Soluble phases in aluminum alloys **A9:** 358-360
Soluble-gas process **A1:** 972
Solutal convection
and temperature gradient **A15:** 148
Solute
defined **A13:** 12 **A15:** 10
species, concentration, activity, in SCC **A13:** 147
third element, effect in microsegregation **A15:** 138
Solute concentration
phase diagram **A9:** 611
Solute content, effect
melting and solution temperatures **A14:** 367
Solute diffusion
Brody-Flemings model **A15:** 883
Solute diffusion in precipitation reactions **A9:** 647
Solute redistribution
and convection, low gravity effects **A15:** 148-149
in nonequilibrium solidification **A15:** 136-137
Solute-solvent equilibria
voltammetric analysis **A10:** 188
Solutes *See also* Solution analysis; Solutions
defined **A9:** 16 **A10:** 682
effect on boundary migration **A9:** 696-697
effect on dendritic structures in copper alloy ingots **A9:** 638
effect on plastic deformation **A9:** 693
effect on stacking fault energy **A9:** 693
elemental PIXE analysis **A10:** 102
segregation in **A10:** 549
-solute equilibria **A10:** 188
Solutes, transverse distribution
ingot **A11:** 324-325
Solution
activity in, calculated **A15:** 55
behavior, by phase diagram **A15:** 56-57
composition, in aqueous environment synthesis **A8:** 416
containment, acidified chloride **A8:** 418-419
defined **A9:** 16
effect, partial molar thermal properties **A15:** 55
precipitation from **A7:** 52, 54, 55
temperature, in aqueous environment synthesis **A8:** 416
Solution Algorithm program **A15:** 867
Solution analysis *See also* Solutes; Solutions
ATR spectroscopy **A10:** 113
chemical, by controlled-potential coulometry **A10:** 207
electrometric titration **A10:** 202-206
element transition series identified by **A10:** 253-266
elemental, AAS as **A10:** 44
gravimetric, of reagents **A10:** 163
ICP-AES as **A10:** 34
of structure **A10:** 188
to determine bonding distance, coordination neighbors **A10:** 407
Solution and reprecipitation **A7:** 309, 320
Solution annealing *See also* Annealing **A4:** 224-225 **A7:** 690
of austenitic stainless steels, effect on carbides **A9:** 283
beryllium-copper alloys **A2:** 405-406 **A9:** 395
of beryllium-nickel alloys **A9:** 395-396
of heat-resistant alloys **A14:** 780
nonferrous high-temperature materials **A6:** 572-574
stainless steel casting alloys **A6:** 497
Solution annealing, of austenitic stainless steels **A1:** 898-899, 912, 945
effect on creep-rupture strength **A1:** 945
effect on intergranular corrosion **A1:** 912, 945
Solution chemisty
polycrystalline fiber production **EM4:** 912
Solution hardening .... **A3:** 1 • 17
aluminum alloys **A13:** 592
Solution heat treatment *See also* Heat treatment
for age-hardenable alloys **A11:** 122
aluminum alloys **A6:** 83 **A15:** 759-760 **M2:** 31, 32, 35, 38-40
cast copper alloys **A2:** 357
defined **A9:** 16 **A13:** 12, 931
effect on beta flecks in titanium and titanium alloys **A9:** 459-460
permanent magnet materials **A2:** 786
of uranium alloys **A2:** 673-674
wrought titanium alloys **A2:** 619-620
Solution potentials **A13:** 12, 584-585
aluminum alloys **M2:** 206-207
Solution purification **A7:** 134, 139, 140, 145
Solution resistance
in galvanic corrosion **A13:** 234
in laboratory corrosion tests **A13:** 214
modeling of **A13:** 234
Solution spinning
of silicon nitride fiber **EM1:** 118
Solution temperature *See also* Fabrication characteristics; Heat treatment; Solution heat treatment; Temperature(s)
aluminum casting alloys **A2:** 153-177
cast copper alloys **A2:** 356-391
wrought aluminum and aluminum alloys **A2:** 82, 103
Solution treat and age
cycle and microstructure **A6:** 510
Solution treating
titanium **M4:** 766, 767, 768
Solution treating and quenching
nonferrous high temperature materials **A6:** 572, 574
Solution viscosity
for molecular weight **EM2:** 533
Solution-annealed wrought stainless steels
carbide precipitation **A9:** 284
etching to reveal grain boundaries **A9:** 281
Solution-gelation process *See* Sol-gel process
Solution-heat-treated (T4) temper of aluminum alloys
identification of **A9:** 358
Solution-heat-treated aluminum alloys
etchants for examination of **A9:** 355
Solution-heat-treated and artificially aged (T6) temper of aluminum alloys, identification of **A9:** 358
Solution-precipitation **EM4:** 267
Solutions .... **A3:** 1 • 8
aqueous, containing nickel and cobalt ions spectra compared **A10:** 65
aqueous, use in ion chromatography **A10:** 658-667
colored, electrometric titration analysis of **A10:** 202
defined **A10:** 682 **A13:** 12
inorganic, analytic methods for **A10:** 7
liquid, analytic methods for **A10:** 7
as lubricant form **A14:** 513
molal, abbreviation for **A10:** 690
normal, abbreviation for **A10:** 690
organic, analytic methods for **A10:** 10
preparation of known **A10:** 162
turbid, electrometric titration for **A10:** 202
very dilute, electrometric titration for **A10:** 202
Solvation **EM3:** 27
defined **EM1:** 22 **EM2:** 39
Solvay process **EM4:** 379
nickel powder production **A7:** 134
Solvent
chlorinated, for chemical cleaning **A15:** 561
coating, for woven fabric prepregs **EM1:** 149
for condensation polyimides **EM1:** 79
defined **A9:** 16 **A15:** 10
effect, thermoplastics **EM1:** 101
impregnation vs. solvent resistance, as composite problem **EM1:** 102
removal, by curing **EM1:** 655
resistance, of thermoplastics **EM1:** 33, 100, 293-294
Solvent acrylics **EM3:** 190, 208-209
for air conditioner perimeter sealing **EM3:** 209
applications **EM3:** 209
for brick and stone masonry pointing **EM3:** 209
characteristics **EM3:** 53, 209
chemistry **EM3:** 208
compared to butyl compounds **EM3:** 209
for control joints **EM3:** 209
cure mechanisms **EM3:** 208
for exterior panel joints **EM3:** 209
formulation **EM3:** 208
for glass-to-mullion joints **EM3:** 209
for glazing of insulating glass panels **EM3:** 209
for heel filler beads for glazing **EM3:** 209
properties **EM3:** 208-209
sealant characteristics (wet seals) **EM3:** 57
sealants **EM3:** 57, 188
suppliers **EM3:** 209
tensile testing **EM3:** 209
Solvent adhesive **EM3:** 27, 36
Solvent alcohols
use in etchants **A9:** 67-68
Solvent cement **EM3:** 567
applications **EM3:** 45
characteristics **EM3:** 45
for composite panel construction **EM3:** 46
as consumer product **EM3:** 47
for packaging **EM3:** 45
for woodworking **EM3:** 46

Solvent cleaners *See also* Removers; Solvent removers; Solvent(s)
blends **EL1:** 663
for liquid penetrant inspection **A17:** 75-76
organic **EL1:** 662-663
Solvent cleaning *See also* Cleaning **A13:** 413-414
M5: 40-58
agitation, use of **M5:** 41-42, 44
aluminum and aluminum alloys **M5:** 576-577
applications **M5:** 43-44
chips and cutting fluids removed by **M5:** 3-10
cleanness requirements and testing **M5:** 41, 43-44
cold cleaning **M5:** 40-44
copper and copper alloys **M5:** 617-618
drying process **M5:** 41-42
emulsifiable solvents *See* Emulsifiable solvents
enameling **EM3:** 303
equipment **M5:** 42-43
grinding, honing, and lapping compounds removed by **M5:** 3-15
limitations of **M5:** 43-44
magnesium alloys **M5:** 629
mechanism of action **M5:** 3-4
methods, liquid penetrant inspection **A17:** 81, 82
molybdenum **M5:** 659
niobium **M5:** 663
pigmented drawing compounds removed by **M5:** 3-8
polishing and buffing compounds removed by **M5:** 3-10
process variables, control of **M5:** 41-42
safety precautions **M5:** 3-21, 41-42, 44
for soldering/ interconnection **EL1:** 117
solvent compositions and operating conditions **M5:** 40-41
solvent flash points **M5:** 40-41, 44
solvent reclamation **M5:** 41
spray process **M5:** 3-15, 42
tantalum **M5:** 663
temperature, operating **M5:** 41
tungsten **M5:** 659
ultrasonic vibration used in **M5:** 40, 44
unpigmented oils and greases removed by **M5:** 6, 8-9
vapor degreasing *See* vapor degreasing
workpiece size and shape and work quantity affecting **M5:** 43
zinc alloy die castings **M5:** 676-677
Solvent extraction
by methylisobutyl ketone **A10:** 169
classical wet chemical analyses **A10:** 164
separations, common **A10:** 170
techniques **A10:** 170
using ethyl ether **A10:** 169
Solvent extraction process **A7:** 54
copper powders **A7:** 120
tungsten powder **A7:** 153
Solvent molding
defined **EM2:** 39
Solvent removal process
urethane coatings **EL1:** 780
Solvent removers *See also* Cleaning; Solvent cleaning
for liquid penetrant inspection **A17:** 75-76
Solvent resistance *See also* Chemical resistance
of acetals **EM2:** 100
of parylene coatings **EL1:** 796-797
of silicone conformal coatings **EL1:** 822
Solvent spraying
for liquid penetrant inspection **A17:** 81
Solvent wipes **EM3:** 34, 42
Solvent wiping
for liquid penetrant inspection **A17:** 81, 82
Solvent(s)
bonding **EM2:** 725
cleaning methods **A17:** 81, 82
as flux component **EL1:** 644
halogenated, properties **EL1:** 662
leaching, of additives **EM2:** 774
nonhalogenated, properties **EL1:** 665
polymeric thick-film systems **EL1:** 346
stability, of organic solvent blends **EL1:** 663
stripping, for urethane coating removal **EL1:** 780
ultrasonic immersion with **A17:** 81
Solvent-activated adhesive **EM3:** 27, 35
Solvent-base resins
competing with anaerobics **EM3:** 116
Solvent-base sealants
applications **EM3:** 177
Solvent-borne resin dispersion
as silicone coating **EL1:** 823
Solvent-cutback petroleum-base compounds
as fracture preservatives **A12:** 73
Solvent-cutback petroleum-based rust preventive compounds **M5:** 459-463, 466-470
applying, methods of **M5:** 467
duration of protection **M5:** 469
safety precautions **M5:** 470
Solvent-dispersed adhesives **EM3:** 36
Solvent-release butyl sealants **EM3:** 190
Solvent-removable fluorescent penetrant method
of liquid penetrant inspection **A17:** 77
Solvent-removable liquid penetrant inspection **A17:** 77-78, 502
Solvent-removable penetrants **A17:** 84
Solvent-removable visible penetrant method
of liquid penetrant inspection **A17:** 78
Solvent-suspendible nonaqueous developer (form D) application **A17:** 83
Solvents
analytic methods for **A10:** 10
decontamination **A13:** 951
defined **A10:** 682 **A13:** 12
effect in MFS **A10:** 77
effects in flame emission spectroscopy **A10:** 29
evaporation, as IR sample **A10:** 112
ICP-AES for trace impurities in **A10:** 31
for ion chromatography sample preparation **A10:** 663
liquid chromatography analysis for low-level organic contaminants **A10:** 649
nonpolar, effect in extraction **A10:** 164
purity, in UV/VIS analysis **A10:** 68-69
role in etchants for wrought stainless steels **A9:** 281
rub, as cure test **A13:** 418
sample preparation, x-ray spectrometry **A10:** 95
and scales **A13:** 1140
-solute equilibria, voltammetric analysis **A10:** 188
surface tensions **EM3:** 181
Solvus ..... **A3:** 1 • 3 **A6:** 128
defined **A9:** 16
Sommerfeld number *See also* Hershey number; Ocvirk number **A18:** 90, 516, 518, 519
and coefficient of friction **A11:** 485
friction during metal forming **A18:** 59, 60, 62, 63, 67-68
nomenclature for Raimondi-Boyd design chart **A18:** 91
Sonar domes
glass-reinforced plastic composite **EM1:** 83
Sonar reflection plate
as ordnance application **A7:** 680
Sonic and ultrasonic properties **A1:** 67-68, 69
Sonic bond tests
defect detection **EM3:** 751
Sonic converters
lead-zirconium-titanate ..... **A8:**)44
pulsed mode vs. continuous cycling mode **A8:** 243
single and double **A8:** 244-245
for ultrasonic testing **A8:** 243-245
Sonic flow *See* Choked flow
Sonic velocity **A6:** 161
Sonic waves *See* Ultrasonic inspection; Ultrasonic waves
Sonogels
alkoxide-derived gels **EM4:** 210, 211
Sonotrode **A6:** 324, 325, 326
Soot blowers
for ash removal **A11:** 619
Soot deposition **EM4:** 414, 415
Sooting **A7:** 342, 347
SOR Ring
as synchrotron radiation source **A10:** 413
Sorbite
defined **A9:** 17
Sorel-metal **A7:** 86
Soret effect
in microsegregation **A15:** 138
Sorption pump
in environmental test chamber **A8:** 411
Sorting
by laser dimensional measurement **A17:** 15-16
of steel bar and wire **A17:** 556-557
of tubular products, nondestructive inspection for **A17:** 561
Sorting, and inspection
automatic **A15:** 571-572
Sound alarms, as readout
eddy current inspection **A17:** 178
Sound conduction method
ultrasonic inspection **A17:** 241
Sound control materials
lead and lead alloy **A2:** 556
Sound damping
P/M metals **A7:** 670
Sound deadening plastic
powders used **A7:** 573
Sound, speed of
and ultrasonic fatigue testing specimens **A8:** 249
Sound velocity
in split Hopkinson pressure bar test **A8:** 202
Sound waves
in ultrasonic inspection **A17:** 231
Sound-control materials **M2:** 498, 499
Soundness
casting, of aluminum-silicon alloys **A15:** 167
ductile iron castings **A15:** 663
from ceramic molding **A15:** 248
from Unicast process **A15:** 251
Sour gas **A13:** 12, 654-655, 1257
Sour gas environments **A11:** 298-303
definition of **A11:** 301
failures in **A11:** 298-303
Sour gas environments, failures in
field environments **A11:** 302-303
hydrogen-induced cracking **A11:** 299
and materials selection **A11:** 300-301
stress-corrosion cracking **A11:** 299-300
sulfide-stress cracking **A11:** 298-299
test methods **A11:** 301-302
weight-loss corrosion **A11:** 300
Sour gas systems
defined **A11:** 301
Sour multiphase systems
defined **A11:** 301
Sour water **A13:** 12, 1267
Source (x-rays)
defined **A9:** 17
Source function analysis
acoustic emission inspection **A17:** 278-280
Source location analysis
acoustic emission inspection **A17:** 279-280
South America
early metalworking in **A15:** 19
Southern bentonite, as molding clay
characteristics **A15:** 210
Southwell plot
boron-epoxy panel **EM1:** 334
Southwest Research Institute **A8:** 202
Southwire casting system
continuous casting **A15:** 314
Southwire continuous rod system
for wire rod **A2:** 254-255
Soviet steel, 30Kh2N2M
hydrogen-induced cracking **A6:** 414
Soviet Union
P/M technology in **A7:** 691-694
Sow block
defined **A14:** 12
Soxhlet apparatus
use in removing fluids from powder metallurgy materials **A9:** 504
Soybean-oil storage tank
brittle fracture of **A11:** 416
Space and missile systems
advanced composite applications **EM1:** 816-822
carbon-carbon composite applications **EM1:** 922-924
carbon-carbon composite components **EM1:** 921
Space boosters
corrosion of **A13:** 1101-1105
Space charge **A6:** 30-31

Space groups
230, relation to crystal symmetry and crystal systems ... **A10:** 348
and atomic symmetry ... **A10:** 347
atoms residing in ... **A10:** 348
for defining crystal structure ... **A10:** 348
identification for single-crystal analysis ... **A10:** 351
of recrystallized organic polysulfide ... **A10:** 353
Space lattice *See* Lattice
defined ... **EL1:** 93
Space lattices ... **A3:** 1 • 10, 15
Space markets
for hybrids ... **EL1:** 254
Space motors
as advanced composite application ... **EM1:** 817, 819-821
Space satellites
corrosion of ... **A13:** 1101-1105
Space shuttle orbiter ... **A13:** 1058-1075
Space shuttle program
as NDE reliability case study ... **A17:** 685-686
Space-charge aberration
defined ... **A9:** 17
Space-domain subtraction function
thermal inspection ... **A17:** 400
Space-group notations ... **A3:** 1 • 16 **A9:** 707
for simple metallic crystals ... **A9:** 716-718
Spacecraft *See* Aerospace industry applications; Space shuttle program
codes governing ... **M6:** 824-825
Spacecraft industry *See* Aerospace applications
Spacer bar edge preparations ... **M6:** 68
Spacer strip
definition ... **A6:** 1213
Spacers ... **EM3:** 547
economy in manufacture ... **M3:** 849
Spacers used in mechanical mounts ... **A9:** 28
Spacerstrip
definition ... **M6:** 16
Spacing
conductor, flexible printed boards ... **EL1:** 587
conductor, selection criteria ... **EL1:** 518-519
fatigue striation, and stress-intensity range related ... **A8:** 482, 484
of flanges ... **A14:** 532
of holes, in piercing ... **A14:** 468
of indentation, Scleroscope hardness test ... **A8:** 105
of indentations, Brinell test ... **A8:** 85, 88
of indentations, Rockwell hardness testing ... **A8:** 80
of interconnects ... **EL1:** 417
and platelet thickness in transformed microstructure ... **A8:** 480
violation, as near defect ... **EL1:** 568
Spacing (lattice planes) *See* Interplanar distance
Spacing, dendritic
and gravity ... **A15:** 154-155
Spade drilling
Al alloys ... **A16:** 777, 782
refractory metals ... **A16:** 861, 863, 864, 865
Spall ring ... **A8:** 210-211
Spall stress ... **A8:** 211-212
Spallation neutron sources
for neutron diffraction ... **A10:** 421
Spalling *See also* Fatigue wear; Galling; Scabbing; Shelling ... **A13:** 12, 98, 1305, 1222 **A18:** 260, 296
(splitting) test, green sand ... **A15:** 345
AISI/SAE alloy steels ... **A12:** 322, 329
carburizing steels ... **A18:** 875
circular, in hardened steel roll ... **A11:** 89
defined ... **A8:** 12, 211 **A11:** 9 **A18:** 17
as dynamic fracture by stress waves ... **A8:** 287
erosion-enhanced corrosion ... **A18:** 208-209, 210
as fatigue failure ... **A12:** 113-115, 329
fatigue, in bearings ... **A11:** 491
in flyer plate impact tests ... **A8:** 211
Spalling (continued)
in forged hardened steel rolls ... **A12:** 113
in forging ... **A11:** 341
free surface velocity data ... **A8:** 212
from bearing overloading ... **A11:** 501
from contact fatigue testing ... **A11:** 133-134
from improper lubrication ... **A11:** 130
in gear tooth ... **A11:** 600
gears ... **A18:** 258
initiated at true-brinelling indentations ... **A11:** 500
medium-carbon steels ... **A12:** 275
micro-, in tapered-roller bearing ... **A11:** 502
mold, from thermal expansion ... **A15:** 208
on shaft as roller-bearing raceway ... **A11:** 504
porcelain enamels ... **M5:** 527, 529
pumps ... **A18:** 595
in rolling-element bearings ... **A11:** 500, 504 **A18:** 258
strength, of sands ... **A15:** 345
subcase fatigue as ... **A11:** 134
in tool steel primer cup plate ... **A11:** 566-567
tool steels ... **A18:** 737
in white iron ... **A12:** 239
Spangle
galvanized steel ... **M1:** 170, 171
Spangles
as formability problem in zinc-coated steels ... **A8:** 548
SPAR (Struc Perf Analy and Redesign) computer program for structural analysis ... **EM1:** 268, 273
Sparging
dissolved oxygen removal by ... **A10:** 204
Spark
defined ... **A10:** 682
Spark discharge spraying
ceramic coatings for adiabatic diesel engines ... **EM4:** 992
Spark erosion *See* Electrical pitting
Spark machining *See* Electric discharge machining
Spark plug (body)
powder used ... **A7:** 572
Spark plug (corrosion protection)
powder used ... **A7:** 572
Spark plug shells
cold-formed ... **A7:** 621
Spark sintering ... **A7:** 11
Spark source mass spectrometry ... **A10:** 141-150
applications ... **A10:** 141, 146-150
basis of technique ... **A10:** 141-142
defined ... **A10:** 682
depth profiling ... **A10:** 142
electric and magnetic sectors ... **A10:** 143
estimated analysis time ... **A10:** 141
general elemental surveys ... **A10:** 144-145
general uses ... **A10:** 141
of inorganic solids ... **A10:** 4-6
instrumentation ... **A10:** 142-143
internal standardization techniques ... **A10:** 145
introduction ... **A10:** 141
ion detection methods ... **A10:** 143-144
isotope dilutions ... **A10:** 145-146
limitations ... **A10:** 141
mass spectra ... **A10:** 144
neutron activation analysis and compared ... **A10:** 233
quantitative elemental measurement ... **A10:** 145-146
related techniques ... **A10:** 141, 142
samples ... **A10:** 141, 146-150
Spark sources
applications ... **A10:** 29
controlled waveform ... **A10:** 25, 26
high-voltage, defined ... **A10:** 25
parameters ... **A10:** 26
Spark spectrometer
calibration ... **A10:** 26
Spark testing ... **A1:** 1030
Spark volatilization
for solid sample analysis ... **A10:** 36
Spark-ignition (Otto cycle) engines ... **A18:** 553
Sparking off ... **A10:** 26
Spatial distribution
of constituents, EPMA mapping of ... **A10:** 525-529
of elemental species, SIMS analysis for ... **A10:** 610
Spatial features
and projected images, stereological relationships ... **A12:** 196
stereometry/profile analysis for ... **A12:** 194
Spatial filtering
image processing and enhancement ... **A17:** 459
in thermal inspection ... **A17:** 400
Spatial filters
for optical holography ... **A17:** 418
Spatial grain size
defined ... **A9:** 17
Spatial resolution
atom probe analysis ... **A10:** 595-596
computed tomography (CT) ... **A17:** 372-373
defined ... **A17:** 385
factors affecting ... **A17:** 373
modulation transfer function (MTF) ... **A17:** 373
point spread function (PSF) ... **A17:** 372-373
quantitative EMPA ... **A10:** 525
real-time radiography ... **A17:** 319
surface analytical techniques ... **A7:** 251
test patterns ... **A17:** 372
x-ray, effect in microanalysis ... **A10:** 448
Spatter ... **A6:** 27-28, 38, 1073
argon-helium shielding gas ... **A6:** 67
argon/carbon dioxide shielding gas ... **A6:** 66
definition ... **M6:** 16
in shielded metal arc welds ... **M6:** 79, 94
in weldments ... **A17:** 582
Spatter loss
definition ... **M6:** 16
Spatter, weld *See* Weld spatter
SPEAR *See* Stanford Position Electron Accelerator Ring
Special brasses
corrosion in various media ... **M2:** 468-469
gas-metal arc butt welding ... **A6:** 760
gas-metal arc welding ... **A6:** 763
to high-carbon steel ... **A6:** 828
to low-alloy steels ... **A6:** 828
to low-carbon steel ... **A6:** 828
to medium-carbon steel ... **A6:** 828
to stainless steels ... **A6:** 828
Special cast iron ... **A1:** 11
Special cause ... **EM3:** 785
Special die drawing
cold finished bars ... **M1:** 243-249, 250-251
Special engineering topics
recycling, nonferrous alloys ... **A2:** 1205-1232
toxicity of metals ... **A2:** 1233-1269
Special equipment applications
refractory metals and alloys ... **A2:** 558
Special helical profiles
thread grinding ... **A16:** 278
Special imaging techniques ... **EL1:** 1067-1073
Special interface transmittals (SITS) ... **EL1:** 130
Special pipe ... **M1:** 315
Special ply termination tests ... **EM3:** 822, 823
Special quality hot rolled bars ... **M1:** 205-206
Special quality hot-rolled carbon steel bars ... **A1:** 244-245
Special roll-turning lathes *See also* High removal rate machining ... **A16:** 153
***Special Technical Publications* (STPS) ASTM** ... **EM2:** 95
Special-killed low-carbon steel sheet and strip ... **M1:** 155
Special-purpose alloys, heat treating *See* Uranium, tantalum, niobium
Special-purpose fasteners
failures in ... **A11:** 548-549

SUBJECTS OF THE INDEXED VOLUMES: ASM Handbook (designated by the letter "A"): **A1:** Properties and Selection: Irons, Steels, and High-Performance Alloys (1990); **A2:** Properties and Selection: Nonferrous Alloys and Special-Purpose Materials (1990); **A3:** Alloy Phase Diagrams (1992); **A4:** Heat Treating (1991); **A6:** Welding, Brazing, and Soldering (1993); **A7:** Powder Metallurgy (1984); **A8:** Mechanical Testing (1985); **A9:** Metallography and Microstructures (1985); **A10:** Materials Characterization (1986); **A11:** Failure Analysis and Prevention (1986); **A12:** Fractography (1987); **A13:** Corrosion (1987); **A14:** Forming and Forging (1988); **A15:** Casting (1988); **A16:** Machining (1989); **A17:** Nondestructive Testing and Quality Control (1989); **A18:** Friction, Lubrication, and Wear Technology (1992). **Metals Handbook, 9th Edition** (designated by the letter "M"): **M1:** Properties and Selection: Irons and Steels (1978); **M2:** Properties and Selection: Nonferrous Alloys and Pure Metals (1979); **M3:** Properties and Selection: Stainless Steels, Tool Materials and Special-Purpose Materials (1980); **M4:** Heat Treating (1981); **M5:** Surface Cleaning, Finishing, and Coating (1982); **M6:** Welding, Brazing, and Soldering (1983). **Engineered Materials Handbook** (designated by the letters "EM"): **EM1:** Composites (1987); **EM2:** Engineering Plastics (1988); **EM3:** Adhesives and Sealants (1990); **EM4:** Ceramics and Glasses (1991); **Electronic Materials Handbook** (designated by the letters "EL"): **EL1:** Packaging (1989).

Special-purpose fatigue testing machines **A8:** 369-370
Special-purpose materials
cemented carbides **A2:** 950-977
cermets **A2:** 978-1007
dispersion-strengthened nickel-base and iron-base alloys **A2:** 943-949
electrical contact materials **A2:** 840-868
electrical resistance alloys **A2:** 822-839
low-expansion alloys **A2:** 889-896
magnetically soft materials **A2:** 761-781
metal-matrix composites **A2:** 903-912
metallic glasses **A2:** 804-821
ordered intermetallics **A2:** 913-942
permanent magnet materials **A2:** 782-803
shape-memory alloys **A2:** 897-902
structural ceramics **A2:** 1019-1024
superabrasives and ultrahard tool materials **A2:** 1008-1018
thermocouple materials **A2:** 869-888
Special-purpose systems
machine vision for **A17:** 41
trends **A17:** 44
Specially denatured alcohols **A9:** 68
Specialty applications
roll compacting **A7:** 402-403, 406-409
Specialty castings
aluminum alloy **A15:** 755-757
steels, plasma melting/casting **A15:** 420
Specialty glasses
applications **EM4:** 379, 1015
composition **EM4:** 741
defects and cost of losses **EM4:** 392
fining **EM4:** 387
Specialty P/M strip **A7:** 402-403
Specialty polymers **A7:** 606-613
Specialty resins *See* Resins
Specialty rolling mills **A14:** 351-352
Specialty spherical shaped powders **A7:** 40
Specialty steels
hot-workability ratings **A14:** 381
Speciation
of inorganic gases, analytic methods for **A10:** 8
of inorganic solids, analytic methods to determine **A10:** 5-6
of organic solids and liquids, techniques for **A10:** 9, 10
population, in sampling **A10:** 13
Species concentration
in iron corrosion **A13:** 37-38
Specific adhesion **EM3:** 27
Specific crosshead rate
and machine stiffness effects **A8:** 41
Specific cutting force **A18:** 432
Specific damping capacity *See also* Damping; Damping properties analysis
for beams **EM1:** 210
defined **EM1:** 206
fiber orientation effect **EM1:** 215
variation with temperature **EM1:** 214-215
vs. stress, ferrous/nonferrous metals **EM1:** 216
Specific density of specimens, effect on depth of information in x-ray scanning electron
microscopy **A9:** 93
Specific energy **A8:** 721
conformal coating materials **EL1:** 783
SI derived unit and symbol for **A10:** 685
Specific entropy **A8:** 721
Specific film thickness **A18:** 539, 542, 544
symbol and units **A18:** 544
Specific gravity *See also* Density **A7:** 11 **EM1:** 22, 107, 158 **EM3:** 27
cemented carbides **A2:** 957
control, of fluxes **EL1:** 648-649
data sheet information **EM2:** 410
defined **EL1:** 1157 **EM2:** 39
of fillers **EM2:** 84
of fluxes, wave soldering **EL1:** 683
of gases, defined **A10:** 682
ordered intermetallics **A2:** 935
of solids and liquids, defined **A10:** 682
of thermoset molding compounds **EM2:** 84
Specific grooving energy **A18:** 431-432, 434
Specific heat *See also* Thermal properties **A8:** 722 **EM3:** 27
aluminum casting alloys **A2:** 153-177
Specific heat (continued)
aluminum oxide-containing cermets **A7:** 804
of C-glass **EM1:** 47
carbon fiber/fabric reinforced epoxy resin **EM1:** 412
cast copper alloys **A2:** 356-391
and coefficient of thermal expansion **EM2:** 455-456
conversion factors **A10:** 686
defined **A17:** 396 **EM1:** 22 **EM2:** 39
ductile iron **M1:** 49
of E-glass **EM1:** 47
epoxy resin system composites **EM1:** 403
glass fabric reinforced epoxy resin **EM1:** 405
glass fiber reinforced epoxy resin **EM1:** 407
of glass fibers **EM1:** 47
graphite fiber reinforced epoxy resin **EM1:** 414
high-temperature thermoset matrix composites **EM1:** 376
Kevlar 49 fiber/fabric reinforced epoxy resin **EM1:** 409
low-temperature thermoset matrix composites **EM1:** 397
and material selection **EM1:** 38
medium-temperature thermoset matrix composites **EM1:** 384, 387, 390, 391
para-aramid fibers **EM1:** 56
of polyester resins **EM1:** 92
selected steel grades **M1:** 149
SI derived unit and symbol for **A10:** 685
thermoplastic matrix composites **EM1:** 367, 369, 372
various materials **EM2:** 456
white cast iron **M1:** 83
wrought aluminum and aluminum alloys **A2:** 62-122
Specific heat capacity **A8:** 721
Specific heat per unit mass
symbol and units **A18:** 544
Specific humidity **EM3:** 27
defined **EM2:** 39
Specific impulse **A7:** 598
Specific interfacial free energy **A7:** 312
Specific ion effect
stress-corrosion cracking **A11:** 207-208
Specific metals and alloys
aluminum and aluminum alloys, alloy and temper designation systems **A2:** 15-28
aluminum and aluminum alloys introduction **A2:** 3-14
aluminum foundry products **A2:** 123-151
aluminum mill and engineered wrought products **A2:** 29-61
aluminum-lithium alloys **A2:** 178-199
beryllium **A2:** 683-687
beryllium-copper and other beryllium-containing alloys **A2:** 403-427
cast aluminum alloys, properties **A2:** 152-177
cast copper alloys, properties **A2:** 356-391
cobalt and cobalt alloys **A2:** 446-454
copper alloy castings, selection and application **A2:** 346-355
copper alloys, introduction **A2:** 216-240
gallium and gallium compounds **A2:** 739-749
germanium and germanium compounds **A2:** 733-738
high-strength aluminum P/M alloys **A2:** 200-215
indium and bismuth **A2:** 750-757
lead and lead alloys **A2:** 543-556
magnesium alloys, properties **A2:** 480-516
magnesium and magnesium alloys, selection and application **A2:** 455-479
nickel and nickel alloys **A2:** 428-445
precious metals **A2:** 699-719
rare earth metals **A2:** 720-732
refractory metals and alloys **A2:** 557-585
tin and tin alloys **A2:** 517-526
titanium and titanium alloy castings **A2:** 634-646
titanium and titanium alloys introduction **A2:** 586-591
titanium P/M products **A2:** 647-660
uranium and uranium alloys **A2:** 670-682
wrought aluminum and aluminum alloys properties **A2:** 62-122
wrought copper and copper alloy products **A2:** 241-264
Specific metals and alloys (continued)
wrought copper and copper alloys, properties **A2:** 265-345
wrought titanium and titanium alloys **A2:** 592-633
zinc and zinc alloys **A2:** 527-542
zirconium and hafnium **A2:** 661-669
Specific modulus
beryllium **A2:** 683
defined **A2:** 683
Specific pressure **A7:** 11
Specific properties **EM3:** 27
defined **EM1:** 22 **EM2:** 39
Specific resistance *See* Electrical resistility
defined **EL1:** 89
Specific sliding
defined **A18:** 18
Specific surface **A7:** 11
Specific surface area *See also* Specific surface; Surface area
BET method of measuring **A7:** 262-263
copier powders **A7:** 585
gas adsorption as measure of **A7:** 262-263
low, of palladium powders **A7:** 151
permeametry as measure of **A7:** 262-265
Specific surface free energy **A7:** 312
Specific thrust **A7:** 598
Specific viscosity *See also* Viscosity **EM3:** 27
defined **EM2:** 39
Specific volume **A7:** 265 **A8:** 721
SI derived unit and symbol for **A10:** 685
Specific wear **A18:** 708
Specific wear rate **A18:** 515
defined **A18:** 18
Specification compliance
image analysis for **A10:** 309
Specification limits **EM3:** 786-787
Specification wire **M1:** 262-263
Specifications *See also* AMS specifications; API specifications, ASME, ASTM; ASTM specifications; entries under issuing organization: ASTM specifications; Federal specifications; Foreign specifications; Listing under issuing agency; MIL specifications; Reference; SAE specifications; Standards **EM3:** 61-64
A5.2, "Iron and Steel Gas Welding Rods" **A6:** 285
A36, "Specification for Structural Steel" **A6:** 375, 376
A242, HSLA as-rolled pearlitic structural steels **A6:** 662
A440, HSLA as-rolled pearlitic structural steels **A6:** 662
A441, HSLA as-rolled pearlitic structural steels **A6:** 662
A500, "Specification for Structural Steel" **A6:** 375, 376
A501, "Specification for Structural Steel" **A6:** 375, 376
A514, "Specification for Structural Steel" (HSLA Q&T structural steels) **A6:** 375, 376, 664
A516, "Specification for Structural Steel" **A6:** 376
A517, HSLA Q&T structural steels **A6:** 664
A543, HSLA Q&T structural steels **A6:** 664
A572, "Specification for Structural Steel" (HSLA as-rolled pearlitic structural steels) **A6:** 375, 376, 662
A588, "Specification for Structural Steel" (HSLA as-rolled pearlitic structural steels) **A6:** 375, 376, 662
A606, HSLA as-rolled pearlitic structural steels **A6:** 662
A607, HSLA as-rolled pearlitic structural steels **A6:** 662
A618, HSLA as-rolled pearlitic structural steels **A6:** 662
A633, HSLA as-rolled pearlitic structural steels **A6:** 662
A656, HSLA as-rolled pearlitic structural steels **A6:** 662
A690, HSLA as-rolled pearlitic structural steels **A6:** 662
A709, "Specification for Structural Steel" **A6:** 376

Specifications (continued)
A710, HSLA as-rolled pearlitic structural steels ... **A6:** 662
A715, HSLA as-rolled pearlitic structural steels ... **A6:** 662
A852, "Specification for Structural Steel" ... **A6:** 375
adhesive-bonded joints ... **A17:** 633-634
adhesives ... **EM1:** 689-701
Aerospace Material Specification 4779 standard, powders used for flame spraying ... **A6:** 715
Aerospace Material Specifications (AMS)
aerospace sealants and governed products (3266-3268) ... **EM3:** 195
property data comparison (fluorosilicone) (3357) ... **EM3:** 196
aluminum casting alloys ... **A2:** 125
American National Standards Institute (ANSI) mat-formed wood particleboard (A208.1) ... **EM3:** 105
American Society for Testing and Materials ... **EM3:** 62
butyl rubber-based solvent release sealants (C 1085) ... **EM3:** 190
creep testing method described (D 2990) ... **EM3:** 316-317
elastomeric sealants ... **EM3:** 185, 189, 195, 212, 213
insulating glass units, performance of (E 774) ... **EM3:** 195
latex sealing compounds standard (C 834) ... **EM3:** 190, 212, 213
rubber properties in tension test methods (D 412) ... **EM3:** 189
shear property measurement of low-modulus adhesives (D 3983, D 4027) ... **EM3:** 315-316
short-beam shear test (D 2344) ... **EM3:** 402
stress relaxation test method (D 2991) ... **EM3:** 318
tear strength measurement (D 624) ... **EM3:** 189
tensile test (D 638) ... **EM3:** 315
tensile test, details of (D 3518) ... **EM3:** 401
viscosity measurement (D 1084, D 2556) ... **EM3:** 322
wedge-crack propagation test (D 3672) ... **EM3:** 261, 262, 263, 267, 268, 269
AMS-2642, "blue etch" technique ... **A4:** 919
AMS-2770, polymer quenchant application ... **A4:** 853, 855
AMS-3025, polymer quenchant application ... **A4:** 853
ANSI Z49.2, "Fire Prevention in Use of Welding and Cutting Processes" ... **A6:** 1176
ANSI Z117.1, "Safety Requirements for Working in Tanks and Other Confined Spaces" ... **A6:** 1195
ANSI Z136.1, "Safe Use of Lasers" ... **A6:** 267
ANSI/ASC Z49.1, "Safety in Welding and Cutting" ... **A6:** 1199, 1202
ANSI/ASTM A 263, simple bend and shear test ... **A6:** 163
ANSI/ASTM A 264, simple bend and shear test ... **A6:** 163
ANSI/ASTM A 265, simple bend and shear test ... **A6:** 163
ANSI/ASTM B 432, simple bend and shear test ... **A6:** 163
ANSI/AWS D3.6-83, weld specification for underwater welding ... **A6:** 1012, 1014
ANSI/AWS Z49.1, "Safety in Welding and Cutting" ... **A6:** 179, 185, 283, 1165, 1176
ANSI/NEMA EW-1, "Electrical Arc Welding Apparatus" ... **A6:** 1199
ASTM A 236, locomotive-axle forgings ... **A4:** 39
ASTM A 255, Jominy end-quench test ... **A4:** 73-74, 81-84, 87-89, 107, 108, 229

Specifications (continued)
ASTM A 255-89, Jominy test ... **A4:** 26
ASTM A 297, mechanical properties of heat-resistant alloys ... **A4:** 510
ASTM A 436, grades of austenitic gray iron alloys ... **A4:** 697, 698
ASTM A 439, grades of austenitic ductile irons ... **A4:** 698
ASTM A 518M, high-silicon iron alloys, defined ... **A4:** 700
ASTM A 532, composition and hardness of white iron grades ... **A4:** 700, 701, 703
ASTM A 897, ASTM A 897M, minimum grades ... **A4:** 684
values of austempered ductile iron
ASTM B 154, test method for residual stress and stress-relief effectiveness ... **A4:** 885
ASTM B 310, class A, powder metallurgy part carbonitriding, carbon and nitrogen penetration ... **A4:** 386
ASTM Code A 255-89 appendices, Grossmann factors ... **A4:** 27
ASTM D 92, gravity of quenching and martempering oils ... **A4:** 91, 93
ASTM D 94, saponification of quenching and martempering oils ... **A4:** 91, 93
ASTM D 95, percent of water in quenching and martempering oils ... **A4:** 91, 93, 98
ASTM D 97, pour point of quenching and martempering oils ... **A4:** 91, 93
ASTM D 287, gravity of quenching and martempering oils ... **A4:** 91, 93
ASTM D 445, viscosity of quenching and martempering oils ... **A4:** 91, 93, 98
ASTM D 482, percent of ash in quenching and martempering oils ... **A4:** 91, 93
ASTM D 1210, pigment particle size for thermal spray coatings ... **A6:** 1006
ASTM D 1533, percent of water in quenching and martempering oils ... **A4:** 91, 93
ASTM D 1835, special-duty propane ... **A4:** 312
ASTM D 2161, viscosity of quenching and martempering oils ... **A4:** 91, 93, 98
ASTM D 2273, sludge test of quenching oils ... **A4:** 98
ASTM D 3520, magnetic quenchometer test for oil classification ... **A4:** 92, 109
ASTM E 384, microhardness measurements after liquid nitriding ... **A4:** 419
AWS 5.16, "Specification for Titanium and Titanium Alloy Bare Welding Rods and Electrodes" ... **A6:** 522
AWS A4.3-86, procedure for measuring hydrogen gas volume escaping from a test weld ... **A6:** 413
AWS A5.20, "Specification for Carbon Steel Electrodes for Flux Cored Arc Welding" ... **A6:** 188-189
AWS A5.22, "Specification for Flux Cored Corrosion Resisting Chromium and Chromium-Nickel Steel Electrodes" ... **A6:** 189
AWS A5.29, "Specification for Low Alloy Steel Electrodes for Flux Cored Arc Welding" ... **A6:** 188-189
AWS A6.3-69, "Recommended Safe Practices for Plasma Arc Cutting" ... **A6:** 1170
AWS D10.6, "Recommended Practices for Gas-Tungsten Arc Welding of Titanium Pipe and Tubing" ... **A6:** 522
AWS D14.1, "Welding Industrial and Mill Cranes and Other Material Handling Equipment" ... **A6:** 393

Specifications (continued)
AWS D14.2, "Welding Industrial and Mill Cranes and Other Material Handling Equipment" ... **A6:** 393
AWS D14.3, "Welding Industrial and Mill Cranes and Other Material Handling Equipment" ... **A6:** 393
AWS D14.4, "Welding Industrial and Mill Cranes and Other Material Handling Equipment" ... **A6:** 393
AWS D14.5, "Welding Industrial and Mill Cranes and Other Material Handling Equipment" ... **A6:** 393
AWS D14.6, "Welding Industrial and Mill Cranes and Other Material Handling Equipment" ... **A6:** 393
AWS F2.1, "Recommended Safe Practices for Electron Beam Welding and Cutting" ... **A6:** 1202
AWS F4.1, "Recommended Safe Practices for the Preparation for Welding and Cutting of Containers and Piping That Have Held Hazardous Substances" ... **A6:** 1195
BMS aerospace sealant specifications ... **EM3:** 185
Canadian specifications, property data comparison, polysulfide B-2 (CS-3204) ... **EM3:** 196
cast copper alloys ... **A2:** 356-391
conformal coatings ... **EL1:** 765-766
corrosion-resistant fasteners ... **M3:** 184, 185
defined ... **EM1:** 689, 700
definition ... **M1:** 119-120
and distortion failure ... **A11:** 138-142
ductile iron ... **M1:** 34, 35, 36
engineering, and quality ... **A17:** 720
Federal specifications ... **EM3:** 63
(MMA-A-132A) ... **EM3:** 731
butyl caulks (TT-S-001657) ... **EM3:** 291
construction sealant and governed products (SS-S-200E, TT-S-00227E) ... **EM3:** 195
heat-resistant airframe structural, metal-to-metal adhesives
joint movement capability (TT-S-00230C) ... **EM3:** 206, 212, 213
lap-shear test (Federal Test Method Standard No. 175, method 1033) ... **EM3:** 773, 774
lap-shear test (MMM-A-132) ... **EM3:** 460, 466, 646, 647, 773, 774, 807
latex (AAMA 802.3) ... **EM3:** 213
polysulfides (TT-SS-00230) ... **EM3:** 193
windshield bonding (Federal Motor Vehicle Safety standards) ... **EM3:** 231
filler metal specifications, welding of low-alloy steels ... **A6:** 662
film, optical holographic interferometry ... **A17:** 414
functional, as system definition ... **EL1:** 128
for germanium and germanium compounds ... **A2:** 736
gray cast iron ... **M1:** 16-19
high-purity uranium ... **A2:** 672
International Explosion Metalworking Association (IEMA), explosion welds ... **A6:** 305
IPC-A-600C, "Guidelines for Acceptability of Printed Boards" ... **A6:** 998
IPC-AJ-820, Assembly-Joining Handbook ... **A6:** 998
IPC-CM-770, Printed Board Component Mounting ... **A6:** 992
IPC-D-275, Design Standard for Rigid Printed Boards and Rigid Printed Board Assemblies ... **A6:** 992
IPC-D-279, Design Guidelines for Reliable Surface Mount Technology ... **A6:** 992
IPC-D-300G, Printed Board Dimensions and Tolerances ... **A6:** 992

Specifications (continued)
IPC-D-322, Guidelines for Selecting Printed Wiring Board Sizes Using Standard Panels **A6:** 992
IPC-D-330, Design Guide **A6:** 992
IPC-MC-324, Performance Specifications for Metal Core Boards **A6:** 992
IPC-PD-325, Electronic Packaging Handbook **A6:** 992
IPC-S-804, Solderability Test Methods for Printed Wiring Boards **A6:** 998, 999
IPC-S-805, Solderability Tests for Component Leads and Terminations **A6:** 998, 999
IPC-S-815A, General Requirements for Soldering Electronic Interconnections **A6:** 989, 992
IPC-S-816, SMT Process Guideline and Checklist **A6:** 992
IPC-SM-780, Component Packaging and Interconnecting with Emphasis on Surface Mounting **A6:** 992
IPC-SM-782, Surface Mount Land Patterns (Configurations and Design Rules) **A6:** 992
IPC-SM-785, Guidelines for Accelerated Reliability Testing of Surface Mount Solder Attachments **A6:** 992
IPC-TM-650, Test Methods Manual **A6:** 989, 998
J-STD-001, Requirements for Soldered Electrical and Electronic Assemblies **A6:** 992
J-STD-002, Solderability Tests for Component Leads, Terminations, Lugs, Terminals and Wires **A6:** 992
J-STD-003, Solderability Tests for Printed Boards **A6:** 992
Japanese Institute of Steel, high-performance construction industry sealants (JIS A 5757) **EM3:** 188
lamination **EL1:** 523
for liquid penetrant inspection **A17:** 87-88
malleable cast irons **M1:** 63-64, 68
material **A13:** 322
MIL-10699, salts for heat-treating furnaces **A4:** 475
MIL-A-21180, premium-quality casting **A4:** 867
MIL-C-55302, Connectors, Printed Circuit Subassembly, and Accessories **A6:** 992
MIL-D-3464E, "Dessicants, Activated, Bagged, Packaging Use and Static Dehumidification" **A6:** 980
MIL-F-14256, "Flux Soldering, Liquid (Rosin Base)" **A6:** 137, 989
MIL-G-45204C, gold electroplated layers **A6:** 971, 990
MIL-H-6088
commercial equipment difference between surface and center temperatures **A4:** 848
maximum quench delays **A4:** 851-852
probe check on furnace monthly recommendation **A4:** 874
water-immersion quenching **A4:** 852, 854
MIL-H-6088C, distance restriction between sensing element and working zone in furnace **A4:** 874
MIL-H-6875, loading locations prescribed for temperature uniformity surveys **A4:** 622
MIL-H-81200, temperature-control equipment for heat treating titanium alloys **A4:** 920
MIL-M-6857, heat treating of magnesium castings **A4:** 901
MIL-M-10699A (Ordnance), salt bath compositions and operating temperatures **A4:** 127-128
MIL-P-50804, dip-and-look solderability test **A6:** 999
MIL-P-50884C, printed wiring (Flexible and Rigid-Flex) **A6:** 998
MIL-P-55110, General Specification for Printed Wiring Boards (dip-and-look solderability test) **A6:** 998, 999

Specifications (continued)
MIL-STD-150D, Sampling Procedures and Tables for Inspection by Attributes **A6:** 998
MIL-STD-202F method 208F, Solderability **A6:** 998
MIL-STD-750C method 2026.5, Solderability **A6:** 998
MIL-STD-883C method 2022.2, Wetting Balance Solderability **A6:** 998, 999
MIL-STD-883C method 5005.11, Qualification and Quality Conformance Procedures **A6:** 998
MIL-STD-883D, method 2003.7, dip-and-look solderability test **A6:** 999
MIL-STD-1276D, gold electroplate thickness **A6:** 990
MIL-STD-2000, solderability testing **A6:** 1000
MIL-STD-2000A, Standard Requirements for Soldered Electrical and Electronic Assemblies (solderability testing) **A6:** 989, 998, 999
Military specifications
adhesive-bonded metal faced sandwich structures, acceptance criteria (MIL-A-83376) **EM3:** 767-768
adhesive-bonded structures, NDT process in evaluating quality (MIL-1-6780) **EM3:** 767
aerospace sealants and governed products (MIL-S-8516, MIL-S-29574, MIL-S-81733, MIL-S-83430) **EM3:** 195, 196
aerospace sealants and governed products (MIL-S-8802) **EM3:** 195, 808, 812-813
bonded structures, preparation of acceptance/rejection criteria (MIL-A-83377) **EM3:** 767, 768
core material, aluminum, for sandwich construction (MIL-C-7438F) **EM3:** 733
flatwise tensile specimen preparation (MIL-STD-401) **EM3:** 772-773
nondestructive testing of panels (MIL-STD-860) **EM3:** 755, 773
nondestructive testing, personnel requirements (MIL-STD-410) **EM3:** 768
MR-01-75, National Association of Corrosion Engineers (NACE) materials and fabrication requirements for H_2S service **A6:** 378
P/M steels **M1:** 333, 335-336
pipe and tubing, stainless steel **M3:** 16
QQ-S-571, "Solder, Electronic" **A6:** 137
and quality control **EM1:** 740
requirements, of NDE methods **A17:** 663
requirements, tubular products **A17:** 561
SAE J406, Grossmann factors **A4:** 27
SAE J423
case depth measurement **A4:** 454
liquid nitriding (microscopic method) **A4:** 418, 419
SAE J1268, hardenability data **A4:** 319
SAE J1868, hardenability data **A4:** 319
Society of Automotive Engineers specifications **EM3:** 62
solder **EL1:** 633-634
sources for **EM1:** 701
titanium castings **A2:** 637
TM-01-77 (test solution) NACE, carbon steels for use in wet H_2S service **A6:** 378
TM-02084 (test method) NACE, carbon steels for use in wet H_2S service **A6:** 378
W48.5-M1990 (Canadian), "Carbon Steel Electrodes for Flux- and Metal-Cored Arc Welding **A6:** 188
weld procedure qualifications **A6:** 1089
wrought aluminum and aluminum alloys **A2:** 62-122
wrought copper and copper alloys **A2:** 265-345
Specifications and standards
ASTM standards **EM2:** 90
defense program **EM2:** 89
International Organization for Standardization **EM2:** 91
military and federal **EM2:** 89-90
SAE aerospace material **EM2:** 91
Underwriters' Laboratories (UL) standards **EM2:** 91

Specified limiting values
recommended practices **A7:** 249
Specimen *See also* Compact Specimen; Cylindrical specimen; Plate specimen; Sample; Sample(s); Smooth specimens; Specimen size; Specimens; Test specimen; Workpiece
for acidified chloride solution testing **A8:** 420
-anvil interface, shear stress at **A8:** 235-236
bent-beam **A8:** 503
for biaxial fatigue test **A8:** 370
buckling, in axial compression testing **A8:** 55-56
buckling, in pin bearing testing **A8:** 59
C-ring **A8:** 505-506
in cam plastometer **A8:** 195-196
for cantilever beam bend test **A8:** 132-133, 537
cantilever bend **A8:** 511
chevron-notched **A8:** 469-475
complex geometries, ultrasonic fatigue testing **A8:** 250
compression test **A8:** 195-197, 578-579
constant *K* **A8:** 515
conventional fatigue vs. high-frequency resonant **A8:** 242
copper, in torsional hydraulic actuator **A8:** 217
for crack arrest testing **A8:** 454
for creep experiments **A8:** 302
defined **A8:** 12
deformation during split Hopkinson pressure bar test **A8:** 199
deformed, Marciniak in-plane sheet torsion test **A8:** 559
design for cold upset testing **A8:** 578-579
design for liquid metal environments **A8:** 426
design for resonant fatigue crack growth rate **A8:** 251
design for torsion testing **A8:** 155-157
design, ultrasonic fatigue testing **A8:** 248-252
different sized, elongation measurement of **A8:** 26
displacement measurement in drop tower compression testing **A8:** 197
double-beam **A8:** 505, 513
double-cantilever beam **A8:** 513-515
double-notch **A8:** 228-229
for ductility bending tests **A8:** 125-127
dynamic notched round bar **A8:** 277-278
elastic train **A8:** 503-508
in electrohydraulic testing machine **A8:** 160
fatigue crack growth **A8:** 379-382, 384
for fatigue testing **A8:** 368, 370-371
fatigue-fractured, scanning electron microscopy for **A8:** 412
flat, grips for **A8:** 50-51
four-point loaded **A8:** 505
fracture, single-edge notched **A8:** 441
fully supported **A8:** 505
grips, ultrasonic fatigue testing **A8:** 252
heating and temperature control in vacuum and oxidizing environments **A8:** 414
for high-temperature aerated water testing **A8:** 422
holder in cantilever beam bend test **A8:** 132-133
hollow dumbbell design for ultrasonic testing **A8:** 247
in hydraulic torsional system **A8:** 215-216
imperfections controlling flow localization **A8:** 169-170
Izod, for impact cantilever bend testing **A8:** 262
large, Rockwell hardness testing of **A8:** 80
length, inclination effect of mean burgers vector on **A8:** 180-181
load requirements for **A8:** 58
in load train **A8:** 368
loading, in constant-load testing **A8:** 314
long, Rockwell hardness testing of **A8:** 80-81
Miyauchi shear test **A8:** 560
modified compact **A8:** 512-513
modified double-beam **A8:** 505
for multiaxial testing **A8:** 344
normal stress-particle velocity **A8:** 232
notched **A8:** 7, 315-318
O-ring **A8:** 506
for one-point bend test **A8:** 274
orientation, double-shear test **A8:** 62-63
for pin bearing testing **A8:** 59-60

Specimen (continued)
for plane-strain fracture toughness **A8:** 451
for plane-strain tensile testing **A8:** 558
for plane-stress fracture toughness
plastic deformation **A8:** 510
plastic strain **A8:** 508-509
plate, pressure-shear impact testing **A8:** 231-232
-platen assembly, cam plastometer **A8:** 195-196
precracked, classification of **A8:** 514
precracked test **A8:** 510-519
preparation, bending ductility tests **A8:** 117
preparation, fatigue testing **A8:** 371
preparation for notched-bar upset forgeability test **A8:** 588
preparation for pressure-shear impact testing **A8:** 234
preparation for three- and four-point bend tests **A8:** 134-135
preparation, in wear test **A8:** 604, 606
profile, effect on free-surface strain **A8:** 580
for *R*-curve method **A8:** 452
rapid -*n* test **A8:** 557
for rapid-load fracture testing **A8:** 260-261
residual stress **A8:** 509-510
round bar, with grid jig **A8:** 279
round, grips for **A8:** 50
round notched tensile, for fracture toughness testing **A8:** 460
for SCC testing **A8:** 503-519
seals, for aqueous solutions at ambient temperatures **A8:** 415
shear, planes of shear and loading directions **A8:** 63
shear stress/transverse particle velocity **A8:** 232
for sheet tensile test **A8:** 553, 554
in single pressure bar configuration **A8:** 199
single-crystal, for torsional Kolsky bar test **A8:** 222
single-shear slotted-sheet **A8:** 64
smooth, for stress-corrosion cracking tests **A8:** 496-497
smooth test **A8:** 503-510
smooth vs. precracked, for SCC testing **A8:** 498
in split Hopkinson pressure bar configuration **A8:** 199
stainless steel, deformation of **A8:** 191
statically deformed double-notch shear **A8:** 229
for steam or boiling water with contaminants testing **A8:** 427, 429
in stored-torque Kolsky bar **A8:** 220
in stress wave propagation **A8:** 191
in subpress assembly for medium strain rate testing **A8:** 192-193
surface preparation, Rockwell hardness testing **A8:** 80
tension, SCC testing **A8:** 506-508
for tension testing **A8:** 19
thickness, for Rockwell hardness testing **A8:** 75-77
thin sheet, buckling in axial compression **A8:** 56
three-point loaded **A8:** 505
tolerances, axial compression testing **A8:** 55
for torsion-shear test **A8:** 65
for torsional impact testing **A8:** 216-217
for torsional Kolsky bar tests **A8:** 221-222
tuning fork **A8:** 508
two-point loaded **A8:** 503-505
U-bend **A8:** 508-509
ultrasonic fatigue testing **A8:** 247-252
uniform deformation conditions in split Hopkinson bar test **A8:** 200
uniform deformation in, stress wave propagation **A8:** 191
V-notch, for bar impact testing **A8:** 261
vapor-deposited **A8:** 237
for wedge test **A8:** 587
weld **A8:** 510

Specimen chamber
defined **A9:** 17
Specimen charge
defined **A9:** 17
Specimen configuration
magnetically soft materials **A2:** 763
Specimen contamination
defined **A9:** 17
Specimen current **A18:** 378
Specimen current detection
scanning electron microscopy **A9:** 90
Specimen current detectors
as SEM special technique **A10:** 506-508
Specimen current imaging **A9:** 91
Specimen damage from
fracturing **A9:** 23
sawing **A9:** 23
shearing **A9:** 23
Specimen distortion
defined **A9:** 17
Specimen grid
defined **A9:** 17
Specimen holder
defined **A9:** 17
Specimen life (time to failure)
in SCC **A13:** 275-276
Specimen preparation *See also* Specimens; Surface preparation
for replication microscopy **A17:** 52, 643
Specimen preservation
carbon and alloy steels **A9:** 172
Specimen screen
defined **A9:** 17
Specimen size
and configuration, for Rockwell hardness testing **A8:** 80-83
and creep-rupture **A8:** 329
effect of **A8:** 372-373
effect on fatigue limit of steel, reversed bending **A8:** 372
Specimen stage
defined **A9:** 17
Specimen strain
defined **A9:** 17
Specimen surface
depth of region from which information is obtained **A9:** 90
Specimen(s) *See also* Fracture specimens; Sample(s); Samples; Specimen preparation; Surface preparation; Test specimen(s)
assembled, crevice corrosion **A13:** 564
atmospheric **A13:** 205
for atmospheric galvanic corrosion testing **A13:** 237-238
bent-beam **A13:** 248-249
burial, in soil **A13:** 209
C-ring **A13:** 249
cantilever bend **A13:** 254
cast and wrought **A13:** 193-194
conductivity, for SEM imaging **A12:** 171
constant K_1 ..... **A13:** 256
for corrosion testing in soil **A13:** 209
corrosion testing, in water **A13:** 207
cylindrical **A13:** 293
design, for atmospheric corrosion tests **A13:** 206
double-beam **A13:** 255-256
elastic strain **A13:** 248-252
electrolytic polishing **A17:** 52
evaluation, radiographic inspection **A17:** 348
evaluation techniques for atmospheric corrosion **A13:** 207
exposure, for galvanic corrosion evaluation **A13:** 236-238
fatigue test **A13:** 292-293
flat sheet **A13:** 293
flatwise tensile, adhesive-bonded joints **A17:** 638
fracture sources illustrated **A12:** 217

Specimen(s) (continued)
fracture toughness, ultrasonic cleaning of **A12:** 74
galvanic-corrosion **A13:** 236-238
geometry, effects in SEM **A12:** 168
immersion test **A13:** 223
ISO wire helix **A13:** 205
marine atmosphere **A13:** 904-905
mechanical polishing of **A17:** 52
modified compact **A13:** 254-255
O-ring **A13:** 249-250
orientation and fracture plane identification **A13:** 259
plastic deformation **A13:** 253
plastic strain **A13:** 252-253
plate **A13:** 293
preparation **A13:** 194, 223, 253-260
preparation/ preservation **A12:** 6-7, 72-77, 171-173, 179
radioactive, neutron radiography **A17:** 391
replication **A13:** 195
residual stress **A13:** 253
retrieval, from soil **A13:** 210
ring-loaded wedge-opening, test setup **A13:** 261
for SCC and hydrogen embrittlement in soil **A13:** 210
sectioning and cutting **A12:** 76-77
for SEM **A12:** 171-173
size, effects in fatigue testing **A13:** 293
slow strain rate testing **A13:** 261
smooth, for SCC evaluation **A13:** 246-253, 265-266, 275
support, total immersion tests **A13:** 222
for TEM **A12:** 6-7, 179
tension **A13:** 250-251
thin-foil, direct observation of **A12:** 179
tilt, effects in SEM imaging **A12:** 168
tuning fork **A13:** 251-252
U-bend **A13:** 252-253
wedge-opening load **A13:** 261
weld **A13:** 253
Specimens *See also* Notched specimens; Shapes; Testing; Workpiece(s)
center crack **EM1:** 255-256
circular hole **EM1:** 256-257
compression test **A14:** 391
damage tolerance **EM1:** 264
preparation, exposure during **EM1:** 295-296
tensile test **EM1:** 297
Speckle metrology **A17:** 432-437
electron microscopy speckle method **A17:** 435
fatigue determined by **A17:** 435-436
future outlook **A17:** 436-437
laser speckle patterns **A17:** 432
measurement, of in-plane deformation **A17:** 432-434
measurement, of out-of-plane deformation **A17:** 434
plastic strain determined by **A17:** 435-436
speckle displacement, recording delineation **A17:** 432-434
surface roughness determined by **A17:** 435-436
white light methods **A17:** 434-435
Spectra *See also* Spectrum
characteristic **A10:** 325-326
corrected, for MFS **A10:** 77
electromagnetic, x-radiation in high-energy region of **A10:** 83
energy-dispersive, sum and escape peaks in **A10:** 520
ESR absorption **A10:** 260
excitation and emission, MFS analysis **A10:** 74-75
gamma-ray **A10:** 235
infrared, defined **A10:** 674
LEISS, in qualitative analysis **A10:** 604-605
Raman accessibility to low-frequency regions of **A10:** 133
resonance **A10:** 254
of silver x-ray tube emission **A10:** 90

Spectra (continued)
wavelength-dispersive, stainless steel **A10:** 87, 88
x-ray, EPMA measurement **A10:** 518-522
Spectra polyethylene fibers
properties **EM1:** 54, 56
Spectral analysis
ultrasonic inspection **A17:** 240-241
Spectral background
defined **A10:** 682
Spectral distribution curve, defined **A10:** 682
of synchrotron radiation from SPEAR **A10:** 411
Spectral interferences *See also* Interferences
with flame emission sources **A10:** 30
in flame spectroscopy **A10:** 29
in ICP-AES **A10:** 33-34
Spectral line
defined **A10:** 682
Spectral lineshapes
in infrared spectra **A10:** 116
Spectral order
defined **A10:** 682
Spectral peak overlap
Auger electron spectroscopy **A10:** 556
optical emission spectroscopy **A10:** 22
Spectral resolution, EDS and WDS
compared **A10:** 521
Spectral response
human eye vs. vidicon camera **A17:** 30
Spectral sensitivity
x-ray film **A17:** 326-327
Spectral shift
surface analytical techniques **A7:** 251
Spectral-stripping techniques
as IR qualitative analysis **A10:** 116
Spectrochemical analysis
atomic absorption spectrometry as **A10:** 43
atomic emission as **A10:** 44
defined **A10:** 682
Spectrofluorometer
double-beam **A10:** 77
Spectrogram
defined **A10:** 682
Spectrograph
defined **A10:** 682
Spectrographic analysis
of residues **A10:** 177
use in carbon control **M4:** 436
Spectrometer equipment
for testing **A15:** 390
Spectrometer, mass *See* Mass spectrometer(s)
Spectrometers
AAS double-beam **A10:** 50
atomic absorption **A10:** 43, 45, 50-52
Auger, cylindrical mirror analyzer in **A10:** 554
continuous-wave NMR **A10:** 283
curved crystal, x-ray spectrum of **A10:** 517
defined **A10:** 682
defocusing, effects of **A10:** 527
dispersive and FT-IR, infrared reflec-
tion- absorption by **A10:** 114
echelle, for direct-current plasma **A10:** 40
electron, for AES analysis **A10:** 554
for elemental mapping **A10:** 527-528
emission, defined **A10:** 673
energy-dispersive x-ray **A10:** 89-93, 519-520
ferromagnetic antiresonance **A10:** 271
Fourier transform **A10:** 38-39
FT-IR **A10:** 112
gas mass **A10:** 151-156
high-resolution, and spectral
interferences **A10:** 33
inert gas-purged **A10:** 37
infrared **A10:** 117
magnetic and detector, for EELS **A10:** 435
microwave **A10:** 254-255, 270-271
monochromators **A10:** 38-39
photodiode arrays **A10:** 38
polychromators **A10:** 37-38
pulse NMR **A10:** 283
Raman **A10:** 128
reflection, for FMR **A10:** 270
scanning monochromator **A10:** 38
simultaneous x-ray fluorescence **A10:** 162
spark source mass **A10:** 142
time-of-flight mass **A10:** 142, 591
with Triplemate device **A10:** 129
vacuum **A10:** 29, 41
Spectrometers (continued)
wavelength-dispersive x-ray **A10:** 87, 89-93,
527-528
WDS vs EDS **A10:** 527-528
x-ray **A11:** 32-33, 37-42
Zeeman-corrected **A10:** 52
zero-dispersion double **A10:** 129
Spectrometric oil analysis program
(SOAP) **A18:** 299, 300
Spectrometry **EM3:** 27
defined **EM2:** 39
gas analysis by mass **A10:** 151-157
gas chromatography/mass **A10:** 639-648
molecular fluorescence **A10:** 72-81
Rutherford backscattering **A10:** 628-636
spark source mass **A10:** 141-150
x-ray **A10:** 82-101
Spectrophotometer
paint color testing **M5:** 490-491
Spectrophotometers
defined **A10:** 682
dispersive single-beam **A10:** 67
double-beam, block diagram **A10:** 67
dual-beam dispersive **A10:** 67-68
Spectrophotometric detection **A10:** 661, 663
ion chromatography with **A10:** 661
Spectrophotometric titrations **A10:** 70
Spectrophotometry
indirect **A10:** 70
Spectroscope **EM3:** 27
Spectroscopes
defined **A10:** 682
Spectroscopic testing **A1:** 1030-1031, 1032
Spectroscopies *See* Chemical analysis techniques
Spectroscopies, infrared
visible, and ultraviolet **A13:** 1114
Spectroscopy
attenuated total reflectance **A10:** 113-114
Auger electron **A10:** 549-567
categories of **A10:** 264
defined **EM2:** 39
diffuse reflectance **A10:** 114
electron or x-ray methods of **A10:** 549-580
inductively coupled plasma atomic
emission **A10:** 31-42
infrared **A10:** 109-125
low-energy ion-scattering **A10:** 603-609
mass **A10:** 141-157
molecular, methods of **EM2:** 825-828
Mössbauer **A10:** 287-295
optical and x-ray **A10:** 21-138
optical emission **A10:** 21-30
Raman **A10:** 126-138
secondary ion mass **A10:** 610-627
ultraviolet/visible absorption **A10:** 60-71
x-ray photoelectron **A10:** 568-580
Spectrum *See also* Spectra
defined **A10:** 681
-fitting programs, x-ray spectrometers **A10:** 91
radiation, and x-ray absorption **A17:** 310
x-ray, defined **A17:** 383
x-ray, radiography **A17:** 303
Spectrum shifter
in polychromators **A10:** 38
Specular, and matte
surface finish, aluminum alloys **A15:** 762-763
Specular finishes
aluminum and aluminum alloys **M5:** 574, 576,
580-581, 609-610
Specular gloss
defined **EM2:** 598
Specular reflectance
defined **A10:** 115
infrared instruments **A10:** 177
as IR technique **A10:** 115
Specular reflection **A18:** 191
defined **A9:** 17
effect in optical holography **A17:** 412
Specular transmittance
defined **A10:** 682
Speculum metal
copper and copper alloys plated with **M5:** 623
polishing of **A9:** 451
Speculum plating **M5:** 288-289
Speed *See also* Devic-e speed; Processing speed
of abrasive waterjet cutting **A14:** 748-751
change drives **A14:** 497
Speed (continued)
contour roll forming **A14:** 625
cutting, bar shearing **A14:** 714
of eddy current inspection **A17:** 563
electromagnetic forming **A14:** 644
of electronic polarization **EL1:** 100
extrusion **A14:** 317
formability effects, heat-resistant alloys **A14:** 781
of forming **A14:** 623
as IC performance parameter **EL1:** 400
magnesium alloy forming **A14:** 828
of manual spinning **A14:** 600
of power spinning **A14:** 603
of press forming **A14:** 545
press, in deep drawing **A14:** 582-583
rotary shearing **A14:** 706
of screens, radiography **A17:** 316-317
slitting **A14:** 710-711
of sound, and ultrasonic fatigue test-
ing specimens **A8:** 249
as specification **EL1:** 128
of straightening **A14:** 687
of strain rate tension testing machines **A8:** 208
of testing, ASTM prescribed **A8:** 39
of testing, range **A8:** 47
of tube spinning **A14:** 678
of ultrasonic inspection **A17:** 564
of wire forming **A14:** 694
Speed cracks
extrusion **A11:** 87, 91
Speed of rotation
horizontal centrifugal casting **A15:** 297-298
vertical centrifugal casting **A15:** 305
Speed-of-light delay
as communication delay limit **EL1:** 2
SPF *See* Superplastic forming
Sphalerite
as gallium source **A2:** 742
Spher-a-Caps
material loss on abrasion of dental
amalgams **A18:** 669
Spheres
limits for grouped planar sections of **A9:** 133
properties of **A9:** 133
sintering together **A7:** 313
Spherical aberration
defined **A9:** 17 **A10:** 682
SEM illuminating/imaging **A12:** 167-168
Spherical aberration in lenses **A9:** 75, 77
Spherical aberration of the objective
lens of a transmission electron
microscope **A9:** 103
Spherical bearing
defined **A18:** 18
Spherical bearings
burrs on hole drilled in **A11:** 489
corrosion fatigue failure **A11:** 488
Spherical boring machines **A16:** 166, 167
Spherical bronze bearings **A7:** 707
Spherical crystals *See also* Crystal growth; Growth
free growth **A15:** 112
Spherical dilational waves
in elastic pressure bars **A8:** 199
Spherical domes
of superplastic metals **A14:** 862-863
Spherical eutectic microstructure .... **A3:** 1 • 20
Spherical fixed-catalyst bed reactor
failed after long service **A11:** 664
Spherical grains *See also* Grains; Sand(s)
equiaxed **A15:** 130-135
Spherical harmonics
in series ODF method **A10:** 362-363
Spherical particles *See also* Grains; Particle(s);
Powder
from atomization **A2:** 684-685
radiographic inspection **A17:** 341
Spherical particles, apparent density of
See also Spherical powders **A7:** 272
Spherical powders **A7:** 11, 233, 234
aluminum **A7:** 125
copier **A7:** 586-587
green strength **A7:** 288
for hardfacing **A7:** 824
Spherical projection
defined **A9:** 17
Spherical roller bearing **A18:** 511
basic load rating **A18:** 505

Spherical roller bearings
defined **A18:** 18
f_v factors for lubrication method **A18:** 511
Spherical roller thrust bearings **A18:** 502, 511
basic load rating **A18:** 505
Spherical sag
interference microscope measurement **A17:** 17
Spherical seat
compression test fixture **A8:** 198
with drop tower compression system **A8:** 196
Spherical-roller bearings
described **A11:** 490
Spheroidal graphite
defined **A9:** 17
eutectic growth **A15:** 176-178
solid particle effects **A15:** 142
Spheroidal graphite cast iron *See* Ductile iron
Spheroidal graphite iron *See* Ductile iron
Spheroidal powders **A7:** 11
Spheroidal-graphite cast iron *See* Ductile cast iron
Spheroidite
defined **A9:** 17 **A13:** 12
Spheroidization
at elevated-temperature service **A1:** 642, 644
effect in carbon and low-alloy steels **A11:** 613
Spheroidization as a result of heat in aluminum alloys **A9:** 359
Spheroidize annealing **A1:** 272, 280
of low-alloy steel sheet/strip **A1:** 209
steel wire **M1:** 262
steel wire rod **M1:** 253, 255
Spheroidize-annealed tool steels *See* Tool steels
Spheroidized structure
defined **A9:** 17
Spheroidizing
alloy steel sheet and strip **M1:** 164-165
cold finished bars **M1:** 234
for cold heading **A14:** 293
defined **A9:** 17
effect on fatigue behavior **M1:** 675, 677
steel wire, improved fabrication.... **M1:** 590-591, 593
ultra-high strength steels **M1:** 423-425, 427, 430, 432-435, 438
Spherometer
stress-relaxation bend test **A8:** 326
Spherulite
defined **EM2:** 39
Spherulite size in semicrystalline, thermoplastic-matrix composites
determination of **A9:** 592
Spherulites **EM3:** 409-410, 417, 418
Spherulitic graphite *See* Nodular graphite
Spherulitic-graphite cast iron *See* Ductile cast iron
SPI *See* Society of the Plastics Industry
SPI Facts and Figures of the U.S. Plastics Industry (1986) **EM2:** 94
SPICE analyses
analog, and CAE software **EL1:** 81
for complex interconnect structures **EL1:** 80
defined **EL1:** 1158
of high-frequency digital systems **EL1:** 80-81, 86-87
simulations, controlled impedance **EL1:** 86
Spider
defined **EM2:** 39
Spider cracks
in macrostructure of continuous- cast copper ingot **A10:** 302
Spike forging
simulation of **A14:** 428-429
Spike weld or spot weld
laser-beam welding **A6:** 879, 880
Spikes **EM3:** 581
Spiking **A6:** 23
definition **M6:** 16
Spiking defect
effects of **A11:** 350, 446
Spiking method
XRPD analysis **A10:** 340
Spin *See also* Rolling; Sliding
defined **A18:** 18
Spin blocks *See* Mandrels
Spin coating
polyimide thin-film hybrids **EL1:** 326
Spin decouplers
in ESR analysis **A10:** 258
Spin echo
schematic **A10:** 282
spectra of Ni_3, room-temperature zero-field **A10:** 285
technique **A10:** 258
Spin forging *See also* Forging
of aluminum alloy **A14:** 244
of titanium alloys **A14:** 273
wrought aluminum alloy **A2:** 34
Spin friction coefficient **A18:** 478
Spin glass
abbreviation for **A10:** 691
defined **A10:** 682
identification of magnetic states in **A10:** 253
Spin lattice relaxation time
in ESR analysis **A10:** 257
Spin plating
to make ferrite films **EM4:** 1163
to make garnet films **EM4:** 1163
Spin relaxation rates
determined **A10:** 275-276
Spin resonances
in ferromagnetic resonance (FMR) **A17:** 220
Spin riming **EM4:** 83
Spin test and rig
for creep testing **A11:** 279-280
Spin tests
for gas turbine disks **A8:** 344
Spin wave
defined **A10:** 682
NMR studies in ferromagnetic materials **A10:** 277
resonance **A10:** 268, 273, 275
Spin welding **EM2:** 725
Spin-dependent recombination
for analysis of change in photo-induced conductivity **A10:** 258
Spin-spin relaxation
ESR **A10:** 257
Spindle
open-die forging of **A14:** 68
in spring-material test apparatus **A8:** 134-135
Spindle defects
magnetic particle inspection **A17:** 102
Spindle finishing **M5:** 131-132
nickel alloys **M5:** 673-674
Spindle, helicopter blade
fatigue fracture of **A11:** 126
Spindle oil
defined **A18:** 18
Spindle wall thickness
flaw detection through **A17:** 139
Spindle-type mandrel
for tube swaging **A14:** 137
Spindles
economy in manufacture **M3:** 851
Spinel ($MgAl_2O_4$) **EM4:** 18, 61
applications **EM4:** 46, 766
cation and anion variety of selected compounds **EM4:** 765
composition **EM4:** 46
electrical properties **EM4:** 765
electrochemical characteristics of rechargeable lithium cells **EM4:** 767
general characteristics **EM4:** 765
hot pressing **EM4:** 191
insertion electrodes for lithium batteries **EM4:** 766
lithium ion effect **EM4:** 765-766
magnetic properties **EM4:** 765
precipitation process **EM4:** 59, 60
properties **EM4:** 759, 765
Spinel ($MgAl_2O_4$) (continued)
protective coating for solid-electrolyte sensors **EM4:** 1137
structure **EM4:** 765
supply sources **EM4:** 46
Spinels
affecting slag removal in welds **A6:** 61
defined **A13:** 76
Spinels, multiphase
as inclusions **A11:** 322
Spinnability
tube **A14:** 679
Spinneret
for aramid fiber **EM1:** 114
defined **EM2:** 39
Spinning **A14:** 599-604
aluminum alloys **A14:** 519, 797-798
assembly by **A14:** 604
of beryllium **A14:** 807-808
cone **A14:** 601
of copper and copper alloys **A14:** 818
defined **A14:** 12
effects, work metal properties **A14:** 604
equipment **A14:** 599-600
fire-extinguisher case failure from overheating during **A11:** 649
of heat-resistant alloys, lubricants for **A14:** 519
of hemispheres **A14:** 603-604
lubricants for **A14:** 519
machines **A14:** 601-602
of magnesium alloys **A14:** 520, 828-829
manual **A14:** 599, 828-829
of nickel-base alloys **A14:** 520, 833-834
power **A14:** 600-604, 829
of refractory alloys, lubricant for **A14:** 519
refractory metals and alloys **A2:** 562
of stainless steels **A14:** 759, 771-773
strength and hardness effects **A14:** 604
of titanium alloys **A14:** 520, 845
tools **A14:** 602-603
tube **A14:** 673, 675-679
vs swaging **A14:** 140-141
of zirconium **A2:** 665
Spinning assembly
furnace brazing **M6:** 52-12
Spinning brass
applications and properties **A2:** 300-302
Spinning flame hardening **M4:** 485-486
Spinning rifflers
for particle sizing sampling **A7:** 226
Spinodal alloy
nominal composition **A2:** 347
Spinodal curve
defined **A9:** 17
Spinodal decomposition **A9:** 652-653
of Alnico alloys **A9:** 539
Alnico permanent magnet materials **A2:** 786-787
composition profile of **A10:** 593
copper alloys **A12:** 402 **M2:** 259-260
of Fe-base magnet alloy **A10:** 598-600
formation of Guinier-Preston zones during -phase mixture precipitation reactions **A9:** 650-651
formation of two **A9:** 652
hardening by **A2:** 236
resultant microstructure **A9:** 653-654
resultant phase **A9:** 604
SAS techniques for **A10:** 405
Spinodal structures **A9:** 652-654
defined **A9:** 17
Spiral bevel gears
carburized, fatigue fracture **A11:** 599
contact wear **A11:** 596
described **A11:** 587
gear-tooth contact in **A11:** 588
internal rupture **A11:** 597
and pinion set, reverse shear **A11:** 595
precision forming of **A14:** 173-175
tooth **A11:** 592, 594

Spiral bevel gears (continued)
tooth, case crushing **A11:** 595
Spiral bevel pinion
fatigue tooth breakage **A11:** 601
rippled surface and pitting **A11:** 600
Spiral cracks **A14:** 365
Spiral flow
polyaryl sulfones (PAS) **EM2:** 146
styrene-maleic anhydrides (S/MA) **EM2:** 220
temperature effects, polysulfones (PSU) **EM2:** 201
test, defined **EM2:** 39
thermoplastic polyurethanes (TPUR) **EM2:** 207
Spiral gouges
high-carbon steels **A12:** 280
Spiral groove journal bearings **A18:** 528
Spiral mold cooling
defined **EM2:** 39
Spiral pinion
tooth **A11:** 593
Spiral power springs
distortion in **A11:** 140
Spiral wrapping
for composite tube **EM1:** 571-572
Spiral-flow test **EM3:** 27
Spiral-flute chucking reamers **A16:** 241, 242
Spiral-point taps **A7:** 462
Spiral-weld steel pipe
inspection of **A17:** 567
Spit
definition **A6:** 1213
Spit pipe backing
definition **A6:** 1213
Spitting
as preheating defect **EL1:** 687
Splash lubrication
defined **A18:** 18
Splash pockets **A7:** 323, 325
Splash revealed by macroetching **A9:** 173
in alloy steel ingot **A9:** 175
Splash zone
marine structures **A13:** 542-544
Splashing
corrosivity **A13:** 339
Splat cooling
aluminum P/M alloys **A2:** 201-202
Splat powder **A7:** 11
Splat quenching **A7:** 11
Splay
defined **EM1:** 22 **EM2:** 39
Splay marks
defined **EM2:** 39
Splice
defined **EM1:** 22
Spline function fit
for *da/dN* vs. stress intensity **A8:** 681
Splined bores
effect in gears and gear trains **A11:** 589-590
Splined drums
nitriding for wear resistance **M1:** 630
Splines
brittle fracture from seams **A11:** 478
drive gear and coupling, distortion in **A11:** 143
as notches **A11:** 85
roots, machining marks at **A11:** 122
Splines, mangled
alloy steels **A12:** 333
Split die **A7:** 11
compacted parts **A7:** 325
defined **A14:** 12
systems **A7:** 326, 327
Split furnace
on lever-arm creep testing machine **A8:** 313
Split Hopkinson bar in tension
for high strain rate testing **A8:** 212-214
Split Hopkinson pressure bar **A8:** 198-203
Split inner ring ball bearings **A18:** 500
Split mull
for IR samples **A10:** 113
Split pipe backing
definition **M6:** 16
Split punch **A7:** 11
Split saddle tensile testing **A7:** 684
Split seal
defined **A18:** 18
Split shoulder
in split Hopkinson bar tension test **A8:** 213
Split-ring mold *See also* Mold(s)
defined **EM2:** 39
Splits
as forging flaw **A17:** 494
Splitters
as impression dies **A14:** 44
Splitting **A8:** 591, 594
blankholder pressure in **A8:** 548
diagnosis by circle grid analysis **A8:** 566
fiber, metal-matrix composites **A12:** 467
maraging steels **A12:** 387
multiplet **A10:** 572
in sheet metal forming **A8:** 548
in tensile specimens **A12:** 104-107
in unfavorable grain flow **A12:** 67-68
Spodumene
flame emission sources for **A10:** 29-30
Spodumene ($LiO_2{\cdot}Al_2O_3{\cdot}4SiO_2$)
chain structure **EM4:** 759
composition **EM4:** 932
crystal structure **EM4:** 881
as melting accelerator **EM4:** 380
purpose for use in glass manufacture **EM4:** 381
Sponge *See also* Sponge iron; Sponge iron powders
commercial-purity P/M titanium mechanical properties **A7:** 475
direct deposition of **A7:** 71-72
fines titanium powder **A7:** 164, 165
iron process, Swedish **A7:** 79-82
powders **A7:** 52
Sponge effect *See also* Squeeze effect
defined **A18:** 18
Sponge indium
purity of **A2:** 750
Sponge iron *See also* Sponge; Sponge iron powders **A7:** 11, 14, 18
annealing **A7:** 81, 182
by ground magnetite reduction **A7:** 8
green density and green strength **A7:** 289
particles **A7:** 81, 243
Sponge iron powders *See also* Sponge; Sponge iron; Spongy **A7:** 11
compressibility curve **A7:** 287
for copier powders **A7:** 587
pressure-density in rigid dies **A7:** 298
shape **A7:** 243-244
for welding **A7:** 818
Sponge irons
microstructures **A9:** 509
polishing to open pores **A9:** 507
Sponge metal *See also* Sponge indium; Titanium sponge
zirconium and hafnium **A2:** 661-662
Spongy **A7:** 11
Spontaneous fission
as neutron source for NAA **A10:** 234
Spontaneous rupture
AISI/SAE alloy steels **A12:** 299
Spool
definition **A6:** 1213 **M6:** 16
Spool (birdcage) rack
for test coupons **A13:** 199
Spool specimen test racks
crevice corrosion evaluation **A13:** 304
Spooling *See also* Winding
carbon fiber **EM1:** 112, 113
creel war supply, textile equipment **EM1:** 128
of graphite epoxy tape **EM1:** 143
Sporting darts
powder used **A7:** 574
Sports applications *See* Recreation applications
Sports equipment applications **EM1:** 163, 845-847
Spot (stationary) flame hardening **M4:** 485
Spot and seam welding
precipitation-hardening stainless steels **A6:** 489
Spot diffraction patterns **A10:** 437, 438, 440
Spot drilling
combined with countersinking **A16:** 249-250
Spot, high-pressure *See* Resin-starved area
Spot market
as material price component **EM2:** 84
Spot metering
photographic **A12:** 79, 85
Spot polishing and buffing
copper and copper alloys **M5:** 616
Spot pulsing
laser **EL1:** 358
Spot resistance brazing
overlapping **M6:** 977
Spot resistance welded joints
metallography and microstructures **A9:** 581
Spot scanning devices
for particle counting **A7:** 229
Spot size
effects in SEM imaging **A12:** 167
Spot test kits
for alloy identification and sorting **A10:** 168
Spot tests
chemical analysis **A11:** 30-31
Spot tests, single and triple
terne coatings **M5:** 359
Spot weld
definition **A6:** 1213 **M6:** 16
laser-beam welding **A6:** 879
Spot welding **A7:** 506, 624
of blanks **A14:** 450
comparison to projection welding **M6:** 520
definition **M6:** 16
gas metal arc welding **M6:** 174-175
of aluminum alloys **M6:** 388-390
gas tungsten arc welding **M6:** 210-212
of austenitic stainless steels **M6:** 339
laser beam welds **M6:** 647, 657
applications **M6:** 647, 657
equipment **M6:** 657
ultrasonic **EM2:** 722
ultrasonic welds **M6:** 746, 749
Spot welds *See also* Weld(s)
heat-treatment fatigue failure **A11:** 310-311
in magnesium and magnesium alloys **A2:** 473
radiographic inspection **A17:** 335
resistance, poor fit-up failure **A11:** 441
Spot-type meter
for photomacrography **A12:** 85
Spot-welded potential probes **A8:** 389
Spotfacing **A16:** 249-251
cast irons **A16:** 660
in conjunction with drilling **A16:** 213, 215, 219, 235
heat-resistant alloys **A16:** 751, 752
refractory metals **A16:** 860, 863
speeds and feeds **A16:** 251
tools **A16:** 250
Zn alloys **A16:** 833
Spotswood, Colonel Alexander
as early founder **A15:** 25
Spotty particle oxide inclusions
powder forging **A14:** 191
Spout
electric arc furnace **A15:** 359
Spragging
defined **A18:** 18
Spray chambers
corrosion-resistant **A10:** 35
for ICP-MS **A10:** 35, 40
Spray cleaning processes
acid *See* Acid cleaning, spray process
alkaline *See* Alkaline cleaning, spray process
emulsion *See* Emulsion cleaning, spray process
solvent *See* Solvent cleaning, spray process
vapor-spray-vapor degreasing process **M5:** 46-48, 50, 53, 55-56
Spray coating *See also* Application methods; Coatings; Spraying
of conformal coatings **EL1:** 763-764
equipment, schematic **EL1:** 779
powders used **A7:** 573
of urethane coatings **EL1:** 778-779
Spray coating processes
aluminum *See* aluminum coating, spray process
ceramic coatings *See* Ceramic coating, spray processes
flame spraying *See* Flame spraying
paint *See* Paint and painting, spraying process
porcelain enameling **M5:** 517-518, 520-521, 523
thermal *See* Thermal spray coating
Spray deposit
definition **A6:** 1213 **M6:** 16
Spray deposit density ratio (thermal spraying)
definition **A6:** 1214
Spray deposition *See also* Deposition **A7:** 530-532
aluminum and aluminum alloys **A2:** 7
Spray dryers **A7:** 73, 75

Spray drying A7: 11, 54, 73-78
agglomeration technique for dry pressing EM4: 146
applications A7: 76-77
atomization mechanism A7: 27
closed-system A7: 73, 76
of composite powders A7: 173
of copier powders A7: 587
demixing EM4: 95, 98
equipment A7: 73-75
explosive hazards A7: 197
resulting in hollow agglomerates EM4: 128
Spray fluxers
through-hole soldering EL1: 682
Spray forming
of carbon steel sheet and strip A1: 210, 211
Spray forming techniques
aluminum P/M alloys A2: 204
Spray lay-up *See also* Spray molding; Spray-up EM1: 132-134
defined EM1: 22
fabrication, of rovings EM1: 109
procedures, for epoxy composites EM1: 71
processes EM2: 338-341
properties effects EM2: 287
size and shape effects EM2: 291
tooling EM2: 341-343
Spray molding *See also* Spray lay-up
processes EM2: 338-340
Spray nozzle A7: 11
Spray nozzles
liquid penetrant inspection A17: 74
Spray paint stripping method M5: 18-19
Spray painting A7: 459
types A13: 415-416
Spray phosphate coating systems M5: 434-436, 438, 441-442, 446-449, 452-454
equipment M5: 446-449, 452-454
safety precautions M5: 453-454
spray cabinets M5: 446-447
Spray phosphating
for conversion coatings A13: 387
Spray quenching M4: 31
Spray rate
definition M6: 16
Spray rinsing
liquid penetrant inspection A17: 84
Spray transfer
definition M6: 16
Spray transfer (arc welding)
definition A6: 1214
Spray-and-fuse hardfacing process A7: 830-832
Spray-dry granulation
viscoelastic properties required EM4: 116
Spray-formed preforms A7: 530-532
Spray-up
defined EM2: 39-40
unsaturated polyesters EM2: 249
Sprayed metal coatings
ceramic coating applications M5: 539
for corrosion control A11: 195
Sprayed metal molds
defined EM1: 22
Sprayed-metal molds *See also* Mold(s)
defined EM2: 39
Spraying *See also* Application methods
of chromate conversion coatings A13: 390-391
electrostatic powder A13: 447-448
as flux application method EL1: 648
fountain A7: 73
of porcelain enamels A13: 447
primer, automotive A13: 1015
as silicone conformal coating application EL1: 773
surface cleaning by A13: 381-382
Spraying for hardfacing M6: 787-793
detonation gun spraying M6: 792-793
advantages and disadvantages M6: 793
coating deposition M6: 793
Spraying for hardfacing (continued)
workpiece surface preparation M6: 793
plasma spraying M6: 790-792
advantages and disadvantages M6: 791
coating deposition M6: 791
under low pressure M6: 791-792
workpiece surface preparation M6: 791
spray-and-fuse process M6: 788-790
advantages and disadvantages M6: 790
fusion M6: 789-790
spraying M6: 789
workpiece surface preparation M6: 789
Spraying of
carbon steels M6: 789
chromium M6: 789
chromium-vandium M6: 789
copper M6: 789
irons M6: 789
manganese M6: 789
molybdenum M6: 789
nickel M6: 789
nickel-chromium-molybdenum M6: 789
nickel-chromium-vanadium M6: 789
nonferrous metals and alloys M6: 789
stainless steels M6: 789
Spraying sequence
definition M6: 16
Spread EM3: 27
in plate rolling A14: 346
Spreadability *See also* Wettability
of fluoropolymer coatings EL1: 782-783
Spreader *See also* Gutterway
defined A18: 18 EM2: 40
Spreader pockets
defined A18: 18
Spreading test A6: 136-137
Spreading thermal resistances
in component package design EL1: 412-413
Spring brass
applications and properties A2: 300-302
Spring clips
as special-purpose fasteners A11: 548-549
Spring closing force A18: 550
Spring coiling
of wire A14: 695
Spring constant EM3: 27
defined EM1: 22 EM2: 40
Spring ejectors
for rotary swaging A14: 134
Spring material
of connectors EL1: 23
Spring materials
copper and copper alloys A14: 819-820
Spring model
of viscoelasticity EM2: 414
Spring rate, in torsion
defined A8: 145
Spring steels, blanking
die materials for M3: 485, 486, 487
Spring wire A1: 851
coating lubricant for A14: 697
steel M1: 266-268, 269-270
Spring-material test apparatus A8: 134-135
Spring-mounted lower outer punch A7: 323
Spring-steel fasteners EM2: 713
Spring-temper wire A1: 850
Springback *See also* Shape distortion A7: 11
aftereffects of bending and stretching A8: 565
at molding, dimensional changes during A7: 480
of beam, in simple bending A8: 552
as casting defect A11: 387
in contour roll forming A14: 633
control of A14: 531
copper and copper alloys A14: 820
defined A8: 12 A14: 12
in drop hammer forming A14: 656-657
effect of stretching on A8: 553
in electromagnetic forming A14: 646
Springback (continued)
examples of A8: 552-553
as formability problem A8: 548
of magnesium alloys A14: 825
and material properties A8: 552-553 A14: 881-882
as molding problem A15: 346
in multiple-slide forming A14: 567
in press-brake forming A14: 541, 762
and residual stress, in bending A8: 122-124
simulative tests, for sheet metals A8: 564
stainless steel, in stretch forming A14: 776
steel wire fabrication M1: 588
tester for yield strength A8: 565
tests A14: 893-894
of titanium alloys A14: 838, 846
in wire forming A14: 694
Springback, elastic
beryllium-copper alloys A2: 412
Springs *See also* Compression springs; Helical springs; Hot wound springs; Leaf springs; Motor springs; Springs, failures of M1: 283-313
automotive valve, distortion failure of A11: 138-139
cadmium plating M5: 261, 263-264
carbon steel, split wire A11: 556
chromium-silicon steel M1: 306
chromium-vanadium steel M1: 306
coatings for M1: 291, 313
coil, seam cracking during heat treatment A11: 574, 579
counterbalance, fatigue failure A11: 558
deflection M1: 296-297, 306, 308, 311, 312
design M1: 303-304, 306-311
design stress M1: 284-286, 288, 301, 303, 304-307
elevated temperature effects M1: 296-300
elimination of EL1: 123
fabrication of steel springs M1: 588
failures, from sharp-edged pitted area A11: 555
fatigue properties M1: 291-296, 297, 303, 304, 312
finishes M1: 313
flat, failure origin A11: 559
hard drawn spring wire M1: 306
hardenability requirements M1: 297, 301-303, 311-312
helical, fatigue failure A11: 559
hydrogen embrittlement M1: 291
Inconel X-750, SCC failure A11: 560
landing-gear A11: 560
locomotive, fatigue fracture A11: 552
materials
applications M1: 284-286
carbides in M1: 287
costs M1: 303, 305
decarburization M1: 290, 301
grades and specifications M1: 284-286, 288-290, 311-312
hardness measurement M1: 283, 286
hardness vs. yield strength M1: 301
heat treatment M1: 287-288
mechanical properties M1: 283-288, 300, 301, 312
stress relieving M1: 291, 293
thickness, effect on mechanical properties M1: 283, 286-288, 303
music wire M1: 290, 291, 292, 293, 296, 300, 306
music-wire, fractured A11: 558
nickel alloy, cleaning of M5: 673
oil-tempered wire M1: 296, 297, 306
oxide coated steel wire for M1: 262
pawl A11: 553
peening M1: 293, 296, 297, 312, 313
phosphor bronze, premature failure A11: 556
plating M1: 291
presetting M1: 290-291, 312-313
relaxation M1: 296-298, 300
residual stresses M1: 290-291, 312-313
S-N curves M1: 291-292, 297, 304
seams M1: 290, 301

Springs (continued)
shot peening of **M5:** 147
spiral power, compared for distortion **A11:** 140
steel valve, fatigue fracture **A11:** 120
steel wire for **M1:** 266-268, 270
stress relieving temperatures and treatments **M1:** 290-291, 293
stress-relaxation tests **A8:** 327-328
surface quality **M1:** 290, 301
titanium alloy, surface fracture **A11:** 551
toggle-switch, fatigue fracture **A11:** 557
tool steel safety-valve, corrosion-fatigue fracture in moist air **A11:** 261
valve, distortion in **A11:** 139
valve-seat retainer **A11:** 561
wiper **A11:** 559
wire **A11:** 555
Springs, failures of **A11:** 550-562
alloy steel valve **A11:** 551
carbon steel **A11:** 555
carbon steel counterbalance **A11:** 558
carbon steel pawl **A11:** 551-553
carbon steel wiper **A11:** 558-559
causes of **A11:** 550-551
common failure mechanisms **A11:** 550
corrosion-caused **A11:** 559-560
during high-stress fatigue **A11:** 553-555
flat-spring **A11:** 559
from design errors **A11:** 551
from fabrication **A11:** 555-559
from fracture at transverse wire defect **A11:** 554
from material defects **A11:** 551-553
from operating conditions **A11:** 550, 560-562
from surface defect **A11:** 554-555
from weld spatter **A11:** 559
and inclusions **A11:** 554
inconel X-750 ..... **A11:** 559
landing-gear flat **A11:** 560-561
locomotive suspension **A11:** 551
music-wire **A11:** 557
originating at seam **A11:** 555
phosphor bronze **A11:** 555-557
retainer spring **A11:** 561-562
toggle-switch **A11:** 557
valve, from grinding and shot peening **A11:** 554
Sprocket assemblies
farm planters **A7:** 674
Sprocket gears **A7:** 617, 619
Sprockets
wear of **M1:** 628-630
Sprue *See also* Mold cavity; Pouring basin; Runner **EM1:** 22, 166
copper alloy casting **A15:** 776-777
defined **A15:** 11 **EM2:** 40
in gating system, die casting **A15:** 289
historic use **A15:** 16
Sprue hooks
design of **EM2:** 615
Spun roving
defined **EM1:** 22 **EM2:** 40
Spunlaced sheets
aramid fiber **EM1:** 115
Spur gears
described **A11:** 586
economy in manufacture **M3:** 848
external rupture **A11:** 597
failure modes **A11:** 595
gear-tooth contact in **A11:** 588
thermal fatigue cracking in **A11:** 594
tooth **A11:** 592, 595
Spur pinion
tooth bending fatigue **A11:** 591
Sputter analysis
vs. depth-profiling analysis **A7:** 251
Sputter chambers
low-pressure **A10:** 54
Sputter coating
SEM specimens **A12:** 173
Sputter deposition **A18:** 840-842
cylindrical magnetron **A18:** 841, 842
definition **A18:** 840
ion beam deposition **A18:** 842
magnetron deposition **A18:** 841-842
magnetron limitations **A18:** 842
parameters **A18:** 841
planar diode glow discharge deposition **A18:** 841
Sputter deposition (continued)
planar magnetron **A18:** 841, 842
primary components **A18:** 841
radio frequency (rf) deposition **A18:** 842
S-gun magnetron **A18:** 841, 842
sputter target erosion rate **A18:** 840, 841
sputtering yield **A18:** 841
titanium alloys **A18:** 778, 779
unbalanced magnetron **A18:** 841
Sputter etching **A10:** 556, 558
to remove surface oxide layer for solid-state welding **A6:** 165
Sputter ion gun **A7:** 251 **A10:** 554
Sputtered coatings for contrast enhancement of high-speed steel scanning electron
microscopy specimens **A9:** 99
Sputtering **A7:** 11 **A13:** 12, 456 **M5:** 412-416 **EM4:** 414
amorphous materials and metallic glasses **A2:** 806
applications of **M5:** 415
approximate thickness by **A7:** 256
artifacts, AES **A10:** 556
cathode systems **M5:** 413-414
coating properties and structure **M5:** 414-415
copper deposition **EL1:** 303
defined **A9:** 17
deposition rate **M5:** 412-414
differential, as artifact **A10:** 556
as effect of primary ion bombardment **A10:** 611
equipment **M5:** 414
as glow discharge effect **A10:** 27
high-rate system **M5:** 413
inert gas ion, use with LEISS **A10:** 603
ion beam, for AES analysis **A10:** 550
ion plating process **M5:** 418
limitations of **M5:** 414
magnetron systems **M5:** 413-414
metallization process **EL1:** 511
for metallizing ceramics **EM4:** 542, 543
oxidative protective coating process **M5:** 379
planar electrode systems **M5:** 413-415
process control **M5:** 412-414
process variations **M5:** 413
profiles, Auger Astroloy powders **A7:** 255
radio frequency (RF) magnetron **A2:** 1087
radio frequency process **M5:** 414
rates **A7:** 255
reactive system **M5:** 413-414
sampling by **A10:** 27
self-cleaning effect **M5:** 414
species, schematic diagram **A10:** 612
sputtering rate **M5:** 412-414
substrate, contoured coating of **M5:** 413-414
substrate temperature control of **M5:** 413-415
for thin-film hybrids **EL1:** 313
to make ferrite films **EM4:** 1163
to make garnet films **EM4:** 1163
types, for superconducting thin film materials **A2:** 1081-1082
uniform **A7:** 258
vacuum coating process **M5:** 387
Sputtering films
crystallographic texture in **A9:** 700
Sputtering in ion-beam thinning of transmission electron microscopy
specimens **A9:** 107
Sputtering yield **A9:** 107 **A18:** 852, 857
effect of ion energy on **A9:** 107
variation with the angle of ion incidence **A9:** 107
SQ brazing **EM4:** 52
Square butt joints
plasma arc welding **A6:** 198
Square fractures *See* Flat-face tensile fractures
Square grids *See* Open square grids
Square groove welds
definition, illustration **M6:** 60-61
flux cored arc welding **M6:** 105
gas metal arc welding of commercial coppers **M6:** 403
Square mesh cloth **A7:** 176
Square root compensation **A6:** 43
Square wire
wrought copper and copper alloys **A2:** 252
Square-end punches
cutting force **A14:** 448
Square-groove butt joints
arc welding of
austenitic stainless steels **M6:** 328-329, 334-339, 343
cobalt-based alloys **M6:** 367-368
heat-resistant alloys **M6:** 356-357, 360, 363, 367-368
nickel-based alloys **M6:** 356-357, 360, 363
titanium and titanium alloys **M6:** 446
electrogas welding **M6:** 242
electron beam welds **M6:** 615
electroslag welding **M6:** 225
gas metal arc welding of
commercial coppers **M6:** 403
coppers and copper alloys **M6:** 416-417
gas tungsten arc welding **M6:** 201
of aluminum alloys **M6:** 395-396
of heat-resistant alloys **M6:** 356-357, 360
of magnesium alloys **M6:** 431-432
of silicon bronzes **M6:** 413
oxyfuel gas welding **M6:** 589
plasma arc welding **M6:** 218
recommended proportions **M6:** 69, 72
shielded metal arc welding of nickel-based heat-resistant alloys **M6:** 363
submerged arc welding of nickel alloys **M6:** 442
Square-groove butt joints, welding of titanium and titanium alloys
joint dimensions **A6:** 785
Square-groove joints
radiographic inspection **A17:** 334
Square-groove weld
definition **A6:** 1214
electron-beam welding **A6:** 260
Square-loop ferrites
as magnetically soft materials **A2:** 776
Squaring arms
straight-knife shearing **A14:** 703
Squaring shears
for plate and flat sheet **A14:** 701
Squeegee
defined **EL1:** 1158
pressure, effect on stencil printers **EL1:** 732
Squeeze action
swaging by **A14:** 131
Squeeze casting *See also* Pressure casting **A15:** 323-327
advantages **A15:** 323
aluminum casting alloys **A2:** 141-142
defects in **A15:** 324-326
defined **A15:** 11, 323
as innovative **A15:** 37
market effects **A15:** 44
of metal-matrix composites **A15:** 845-847
microstructure **A15:** 326
as permanent mold process **A15:** 34, 277
process description **A15:** 323-325
process variables **A15:** 324
quality control **A15:** 324-326
zinc alloys **A15:** 797
Squeeze effect (sponge effect)
defined **A18:** 18
Squeeze, hydraulic
and sand blowing **A15:** 29
Squeeze time
definition **M6:** 17
Squeeze-out **EM3:** 27
Squeezer machines
development **A15:** 28
Squirt-on paint stripping method **M5:** 3-19
Squirter techniques
ultrasonic inspection **A17:** 248
Sr-Te (Phase Diagram) ..... **A3:** 2 • 372
Sr-Ti (Phase Diagram) ..... **A3:** 2 • 373
Sr-Zn (Phase Diagram) ..... **A3:** 2 • 373
SRI algorithms
as machine vision technique **A17:** 34, 36
SRIM *See* Structural reaction injection molding
SRM *See* Standard Reference Materials
SRO *See* Short-range order
SRS
as synchrotron radiation source **A10:** 413
SSC *See* Sulfide-stress cracking
SSMS *See* Spark source mass spectrometry
St Venant maximum normal strain **A8:** 344
St.Georg total ringer joint prosthesis **A11:** 670

St.Georg total shoulder joint prosthesis **A11:** 670
Stability *See* Dimensional stability; Instability
of aluminum alloy castings **A15:** 762
chemical thermodynamic **A15:** 50-52
defined **A2:** 824 **A15:** 762
dimensional, of molding materials **A15:** 208-209
of electrical resistance alloys **A2:** 824
heat treatment, austenitic ductile irons **A15:** 700-701
iron-carbon alloys **A15:** 61
metallurgical, electrical resistance alloys **A2:** 824
morphological, low gravity **A15:** 153
of niobium-titanium superconducting materials **A2:** 1044
permanent magnet materials **A2:** 794-795
phase diagram as map of **A15:** 57
Stability analysis *See also* Quality control
of planar interface growth **A15:** 116
and statistical control **A17:** 742
Stability constants
as voltammetric information **A10:** 193
Stability limit **A6:** 52
Stabilization **EM3:** 27
adiabatic **A2:** 1038
in carbon fiber conversion **EM1:** 112
cryogenic **A2:** 1037-1038
defined **EM1:** 22 **EM2:** 40
dynamic **A2:** 1038-1039
niobium-titanium superconducting materials **A2:** 1051
permanent magnet materials **A2:** 794
in superconductors **A2:** 1036-1039
Stabilization annealing
wrought titanium alloys **A2:** 619
Stabilization bake
as temperature-induced stress test **EL1:** 499
Stabilized zirconia (ZrO_2)
electrochemical sensors **EM4:** 252
for oxide ceramic heating elements of electrically heated furnaces **EM4:** 249
Stabilizer(s)
effect on blending in Fe-Al mixtures **A7:** 188
effect on iron-iron contact formation **A7:** 187
Stabilizers *See* Heat stabilizers; Ultraviolet (UV) stabilizers **EM3:** 27, 49
for ceramics **A18:** 814
defined **EM2:** 40
heat **EM2:** 494-495
hot-melt adhesives **EM3:** 80
light **EM2:** 495
ultraviolet (UV) **EM2:** 572
Stabilizers, effect
ternary iron-base alloys **A15:** 65
Stabilizers/initiators
formulation **EM3:** 123
Stabilizing **A1:** 259
Stabilizing heat treatments
effect on austenitic stainless steels **A9:** 284
Stabilizing treatment
defined **A13:** 12
Stable crack growth
beach marks from **A11:** 87
and fracture **A11:** 75
and subcritical crack growth, compared **A11:** 63
Stable emulsion cleaners
use of **M5:** 33-35
Stable equilibrium ..... **A3:** 1 • 1
Stable free radical hydrazyl
ESR analysis of **A10:** 265
Stack cutting
definition **A6:** 1214 **M6:** 17
Stack molding
defined **A15:** 11
Stack-up, multilayer
rigid printed wiring boards **EL1:** 543
Stacker crane
bending-fatigue failure of **A11:** 518
Stackers
as transfer equipment **A14:** 501
Stacking
of laminations **A14:** 478
of weathering steels **A13:** 518
Stacking fault energy
and cross slip **A9:** 693
distribution of dislocations **A9:** 693
and slip **A9:** 693
Stacking faults **A18:** 387, 388, 389
in 18Cr-8Ni stainless steel **A9:** 686
in brass with bainitic structure **A9:** 666
as carbide precipitation sites in austenitic stainless steels **A9:** 284
in crystals **A9:** 719
defined **A13:** 45
and dislocation loops **A9:** 116-117
effect in topographs **A10:** 369-370
in fcc cobalt-base alloy **A10:** 466
FIM/AP study of point defects in **A10:** 583
imaged by x-ray topography **A10:** 365
in nonferrous martensite **A9:** 672-673
Stacking faults as stored energy sites in
cold-worked metals **A9:** 684
transmission electron microscopy **A9:** 118-119
Stacking sequence **EM3:** 27
defined **EM1:** 22 **EM2:** 40
effect, tensile strengths **EM1:** 260
lamina **EM1:** 209-210
and strength **EM1:** 230
Stacking-fault energy
cavitation erosion of metals and alloys **A18:** 215, 217
in cobalt-base alloys **A18:** 766, 768
liquid impingement erosion **A18:** 228
stainless steels **A18:** 715
Stacking-fault tetrahedra **A9:** 116-117
Stadimetry
defined **A17:** 34
Stadiums
corrosion in **A13:** 1299, 1308-1309
Stage
defined **A9:** 17
micrometer, for microhardness testing **A8:** 92-93
movable, bench-mounted tester **A8:** 92
Stage I
fatigue fracture **A12:** 14, 19
Stage II
fatigue fracture **A12:** 14, 19-21
Stage III
fatigue fracture **A12:** 14, 16
Stage of an optical microscope **A9:** 74
Staggered intermittent weld
definition **A6:** 1214
Staggered intermittent welds
definition **M6:** 17
Staging
defined **EM2:** 40
Stagnation
corrosive effects **A13:** 339
STAGS
finite-element analysis code **EM3:** 480
STAGSC-1 computer program for structural analysis **EM1:** 268, 273
Stained glass
recommended waterjet cutting speeds **EM4:** 366
Staining
defined **A9:** 17
of integrated circuits **A11:** 769
Staining etchants for heat-resistant casting alloys **A9:** 330-332
Staining techniques
for IA samples **A10:** 313
Staining to render isotropic metals optically active **A9:** 78
Stainless steel *See also* Austenitic stainless steel; Cast stainless steels; Non-heat-treatable stainless steels; Wrought stainless steels
abrasive blasting of **M5:** 552-553
Stainless steel (continued)
acid etching and acid dipping of **M5:** 561
alkaline cleaning of **M5:** 561
aluminum coating of **M5:** 342-343, 345
annealing of **M5:** 102, 553-554
austenitic sheet, creep-fatigue data in reversed bending **A8:** 356
bar, mill finishes **M5:** 552
base metal solderability **EL1:** 677
buffing *See* Stainless steel, polishing and buffing of
buffing compounds **M5:** 117
cadmium plating of **M5:** 264
ceramic coating of **M5:** 537-538
chemical polishing of **M5:** 559
chemical resistance **M5:** 4
chromic acid cleaning of **M5:** 59
cleaning processes **M5:** 552-554
sequence of processes **M5:** 553-554
composite graph for Gill-Goldhoff correlation for **A8:** 337
copper plating of **M5:** 163, 168
corrosion resistance, effects of polishing and buffing on **M5:** 558
critical strain rate for SCC in **A8:** 519
effect of electrochemical factors in SCC initiation **A8:** 499
electrolytic cleaning of **M5:** 561
electroplating of **M5:** 561-562
precleaning for **M5:** 561-562
electropolishing of **M5:** 305-309, 559
for environmental test chambers **A8:** 411
etching of **M5:** 560-561
fatigue life vs. hold-period time **A8:** 349
for fatigue test chamber **A8:** 412
finishing processes **M5:** 551-552, 554-562
fluxer tanks **EL1:** 681
gage length microstructure to measure flow localization **A8:** 169
galvanic corrosion with magnesium **M2:** 607
grinding of **M5:** 555-559
belt **M5:** 556
belt life **M5:** 556-557
equipment and procedures **M5:** 558-559
mechanized belt **M5:** 556
progressive **M5:** 555-556
rough surfaces **M5:** 555
safety precautions **M5:** 557
solid wheels **M5:** 555
weld beads **M5:** 555
wheel **M5:** 555, 559
wheel speeds **M5:** 555, 559
high-chromium, workability **A8:** 165
hold period in tension and hold period in compression results **A8:** 349-350
ion plating of **M5:** 420-421
mass finishing of **M5:** 554-555
applications, specific **M5:** 554-555
posttreatments **M5:** 554
mill finishes **M5:** 551-552, 558
grade limitations **M5:** 552
matching **M5:** 558
preservation of **M5:** 552
nickel plating of **M5:** 204, 216, 232
electroless **M5:** 232
nickel striking of **M5:** 232
non-heat-treatable, SCC testing of **A8:** 527-529
organic acid cleaning of **M5:** 66
passivation of **M5:** 306, 558-560
precleaning for **M5:** 560
safety precautions **M5:** 560
solution composition and operating conditions **M5:** 560
passivity, corrosion resistance and **M5:** 431-432
phosphate coating of **M5:** 437-438
pickling of **M5:** 72-73, 76
polishing and buffing of **M5:** 108, 112-114, 120-121, 557-559
belt polishing **M5:** 557
cleaning and passivation after **M5:** 558

Stainless steel (continued)
color buffing ... M5: 558
corrosion resistance affected by ... M5: 558
equipment and procedures ... M5: 558-559
hard buffing ... M5: 557
wheel polishing ... M5: 557-559
wheel speeds ... M5: 557-559
porcelain enameling of ... M5: 512-513
precipitation-hardenable, SCC environments ... A8: 526
precipitation-hardenable, workability ... A8: 165
precipitation-hardened, stress-intensity factor range effect on fracture ... A8: 486
rust and scale removal ... M5: 12
salt bath descaling of ... M5: 97-98, 101-102, 553-554
selenium, workability ... A8: 165, 575
sheet
annealing and scale removing operations, mill processing ... M5: 553-554
grinding, polishing, and buffing ... M5: 108, 112-114, 556, 559
mass finishing ... M5: 554
mill finishes ... M5: 551-552
shot peening of ... M5: 145-147
with soluble carbides or nitrides workability ... A8: 165
solution potential ... M2: 207
specimen, deformation of ... A8: 191
strip
aluminum-clad, processing of ... M5: 345
color buffing ... M5: 557-558
grinding, polishing, and buffing ... M5: 557-559
mill finishes ... M5: 552
salt bath descaling of ... M5: 101-102
superplasticity of ... A8: 553
surface activation of ... M5: 232, 561
surface-conditioning operations special ... M5: 559
tubing, mill finishes ... M5: 552
vapor degreasing of ... M5: 54-55
water system, crack growth rate ... A8: 421-422
wick test for ... A8: 529
wire brushing of ... M5: 559
wire, salt bath descaling of ... M5: 98, 102
Stainless steel alloys
chemical resistance ... EM3: 639
enamels ... EM3: 303
etching procedure ... EM3: 272
glass-to-metal seals ... EM3: 301, 302
interleaves protected by elastomer cover ... EM3: 636
lap-shear strength, storage, and weathering effect ... EM3: 658
medical applications ... EM3: 576
phenolic bond properties ... EM3: 106
polysulfides as sealants ... EM3: 196
reinforcing interleaves on deep-sea compliant oil platform ... EM3: 634
seawater exposure effect on adhesives ... EM3: 632
for sporting goods manufacturing ... EM3: 576
Stainless steel casting alloys *See* Cast stainless steels, selection ... A9: 297-304
classification of microstructures ... A9: 298
compositions of ... A9: 298
effect of carbon content on ... A9: 298
effect of composition on microstructure ... A9: 298
effect of heat treatment on microstructure ... A9: 298
etchants for ... A9: 297
etching ... A9: 297
grinding ... A9: 297
microstructures ... A9: 297-298
polishing ... A9: 297
preparation of specimens ... A9: 297
role of sigma phase ... A9: 297
Stainless steel casting alloys, specific types
440C, with dendritic structure and interdendritic carbide network ... A9: 304
CA-6NM, as-cast ... A9: 299
CA-6NM, martensitic ... A9: 298
Ca-6NM, normalized and tempered ... A9: 299
CA-15, as-cast ... A9: 299
CA-15, austenitized, showing effect of section thickness on structure ... A9: 300
CA-15, martensitic ... A9: 298
CA-15, normalized and tempered ... A9: 299
CB-7Cu-1, as-cast ... A9: 300
Stainless steel casting alloys, specific types (continued)
CB-7Cu-1, austenitized ... A9: 300
CB-7Cu-1, precipitation hardening ... A9: 298
CD-4MCu, as-cast ... A9: 300
CD-4MCu, duplex phase ... A9: 298
CD-4MCu, effect of homogenization ... A9: 300
CF types, structures of ... A9: 298
CF-3, as-cast ... A9: 301
CF-3, solution treated and water quenched ... A9: 300-301
CF-3M, as-cast ... A9: 301
CF-3M, solution treated and water quenched ... A9: 301
CF-3M, with sigma phase ... A9: 301
CF-8, as-cast ... A9: 302
CF-8, as-cast, solution treated and water quenched ... A9: 302
CF-8, solution treated ... A9: 301
CF-8, solution treated and water quenched ... A9: 302
CF-8C, solution treated, water quenched stabilized, with niobium carbide ... A9: 302
CF-8M, solution treated and water quenched ... A9: 301
CF-8M, solution treated, water quenched and air cooled ... A9: 302
CF-16F, as-cast ... A9: 302
CF-16F, solution treated and water quenched ... A9: 302-303
CF-20, as-cast ... A9: 303
CK-20, as-cast, with inclusions ... A9: 303
CK-20, solution treated and water quenched ... A9: 303
CN-7M, as-cast, with precipitated chromium carbide ... A9: 303
CN-7M, solution treated and water quenched, with inclusions ... A9: 303-304
Stainless steel castings ... A4: 785-792
austenitic alloys ... A4: 786-787, 790 M4: 639, 642-645
cleaning, prior to heat treating ... A4: 787
ferritic alloys ... A4: 785, 786-787, 790 M4: 639, 642-645
homogenization ... A4: 786 M4: 639-642
martensitic alloys ... A4: 785-786, 787-790 M4: 642, 645
precipitation-hardening alloys ... A4: 787, 790-792 M4: 642, 643, 645-646
Stainless steel corrosion-resistant casting alloys, specific types
CA-6N
composition ... A6: 496
microstructure ... A6: 496
CA-6NM
composition ... A6: 496
microstructure ... A6: 496
special welding considerations ... A6: 497
CA-15
composition ... A6: 496, 684
filler metals for ... A6: 684
microstructure ... A6: 496
properties ... A6: 684
special welding considerations ... A6: 497
CA-15M
composition ... A6: 496
microstructure ... A6: 496
CA-28MWV
composition ... A6: 496
microstructure ... A6: 496
CA-40
composition ... A6: 496
microstructure ... A6: 496
special welding considerations ... A6: 497
CA-40F
composition ... A6: 496
microstructure ... A6: 496
CB-7Cu
composition ... A6: 496
microstructure ... A6: 496-497
CB-7Cu-1
composition ... A6: 496
microstructure ... A6: 496
special welding considerations ... A6: 497
CB-7Cu-2
composition ... A6: 496
Stainless steel corrosion-resistant casting alloys, specific types (continued)
microstructure ... A6: 496
special welding considerations ... A6: 497
CB-30
composition ... A6: 496
microstructure ... A6: 496
CC-50
composition ... A6: 496
microstructure ... A6: 496
CD-4MCu
composition ... A6: 496
microstructure ... A6: 496
CE-30
composition ... A6: 496
microstructure ... A6: 496
CF-3
composition ... A6: 496
microstructure ... A6: 496, 498
CF-3M
composition ... A6: 496
microstructure ... A6: 496, 498
CF-3MN
composition ... A6: 496
microstructure ... A6: 496
CF-8(e)
composition ... A6: 496
microstructure ... A6: 496
CF-8C
composition ... A6: 496
microstructure ... A6: 496
CF-8M
composition ... A6: 496, 498
microstructure ... A6: 496-497
nomenclature ... A6: 495
welding defects ... A6: 497
CF-10
composition ... A6: 496
microstructure ... A6: 496
CF-10M
composition ... A6: 496
microstructure ... A6: 496
CF-10MC
composition ... A6: 496
microstructure ... A6: 496
CF-10SMnN
composition ... A6: 496
microstructure ... A6: 496
CF-12M
composition ... A6: 496
microstructure ... A6: 496
CF-16F
composition ... A6: 496
microstructure ... A6: 496
CF-20
composition ... A6: 496
microstructure ... A6: 496
CG-6MMN
composition ... A6: 496
microstructure ... A6: 496
CG-8M
composition ... A6: 496
microstructure ... A6: 496
CG-12
composition ... A6: 496
microstructure ... A6: 496
CH-8
composition ... A6: 496
microstructure ... A6: 496
CH-10
composition ... A6: 496
microstructure ... A6: 496
CH-20
composition ... A6: 496
microstructure ... A6: 496
CK-3MCuN
composition ... A6: 496
microstructure ... A6: 496
CK-20
composition ... A6: 496
microstructure ... A6: 496
CN-3M
composition ... A6: 496
microstructure ... A6: 496
CN-7M
composition ... A6: 496
hot cracking ... A6: 497

Stainless steel corrosion-resistant casting alloys, specific types (continued)
microstructure ... **A6:** 496
CN-7MS
composition ... **A6:** 496
microstructure ... **A6:** 496
CT-15C
composition ... **A6:** 496
microstructure ... **A6:** 496
Stainless steel fibers
as reinforcements ... **EM2:** 472-473
Stainless steel flake pigments ... **A7:** 596
Stainless steel heat-resistant casting alloys, specific types
HC, composition ... **A6:** 497
HD, composition ... **A6:** 497
HE, composition ... **A6:** 497
HF, composition ... **A6:** 497
HH, composition ... **A6:** 497
HI, composition ... **A6:** 497
HK, composition ... **A6:** 497
HK30, composition ... **A6:** 497
HK40
composition ... **A6:** 497
hot-cracking sensitivity ... **A6:** 497
HL, composition ... **A6:** 497
HN, composition ... **A6:** 497
HP, composition ... **A6:** 497
HP-50WZ, composition ... **A6:** 497
HT
composition ... **A6:** 497
hot cracking ... **A6:** 497
HT30, composition ... **A6:** 497
HU, composition ... **A6:** 497
HW, composition ... **A6:** 497
HX, composition ... **A6:** 497
Stainless steel, sintering
time and temperature ... **M4:** 796
Stainless steel, specific types
16-25-6, electroless nickel plating of ... **M5:** 237
17-4PH, corrosion fatigue ... **A8:** 253, 254
17-4PH, effect of notch ... **A8:** 254
17-4PH, fatigue performance ... **A8:** 253, 254
17-4PH, notch and plain bar fatigue properties ... **A8:** 255
18-8, electropolishing of ... **M5:** 306
200, series, cleaning and finishing processes ... **M5:** 552, 560
201, mill finishes ... **M5:** 552
202, mill finishes ... **M5:** 552
300, series, cleaning and finishing processes ... **M5:** 72-73, 112-114, 305, 552, 554, 556-560
301, mill finishes ... **M5:** 552
302
electropolishing of ... **M5:** 305
grinding ... **M5:** 556
mill finishes ... **M5:** 552, 559
302, ultimate shear stress ... **A8:** 148
302B, mill finishes ... **M5:** 552
303, electropolishing of ... **M5:** 305
303S, 303Se, mill finishes ... **M5:** 552
304
electropolishing of ... **M5:** 305
mill finishes ... **M5:** 552
304, axial tests on ... **A8:** 351-352
304, creep analyses ... **A8:** 686, 689-692
304 creep strain/time behavior ... **A8:** 691, 692
304, creep-fatigue interaction plot for hold-time data ... **A8:** 356
304, creep-fatigue tests ... **A8:** 347-348
304, cycles to fracture vs. cycle period ... **A8:** 351
304, ductility from hot torsion tests ... **A8:** 165-166
304, effect of hold-period in tension-hold-only testing on fatigue resistance ... **A8:** 349
304, effect of temperature on strength and ductility ... **A8:** 36

Stainless steel, specific types (continued)
304, fatigue and stress-rupture data in tension-hold-only test ... **A8:** 349
304, heat transfer effect on flow localization during torsion ... **A8:** 172-173
304, hold period and strain waveform effect on ... **A8:** 347-348
304, low-cycle fatigue data in tension-hold-only test ... **A8:** 351
304, pitting potentials on unstrained specimens ... **A8:** 418
304, plastic-strain fatigue resistance ... **A8:** 348
304, push-pull fatigue ... **A8:** 352
304, re-annealed, multiple heats ... **A8:** 330
304, spread in creep elongation ... **A8:** 336
304, stress amplitude vs. time-to-fracture ... **A8:** 349-350
304, threshold stress intensity ... **A8:** 256
304, time-to-fracture vs. tension-hold-only test ... **A8:** 350
304L, Crockcroft and Latham criterion ... **A8:** 168-169
304L, flow curves from torsion tests ... **A8:** 161-162
304L, flow localization during torsion ... **A8:** 169
304L, grain size ... **A8:** 174-175
304L, micrographs from torsion tests ... **A8:** 170
304L, mill finishes ... **M5:** 552
304L, strain rate effect on torsional ductility ... **A8:** 166-167
304L, torsion flow stress data and compared ... **A8:** 162, 164
compression tension data
304L, torsion flow stress data at various temperatures ... **A8:** 162, 164
304L, torsion tests at hot working temperatures ... **A8:** 163
304L, twisted at various strain rates and temperatures ... **A8:** 175
305, mill finishes ... **M5:** 552
309, 309S, mill finishes ... **M5:** 552
310
grinding ... **M5:** 556
mill finishes ... **M5:** 552
316
ion plating of ... **M5:** 421
mass finishing ... **M5:** 554-555
mill finishes ... **M5:** 552
316, autoclave construction ... **A8:** 424
316, constant-stress creep curves ... **A8:** 320-321
316, creep-rupture ... **A8:** 691
316, initial and secondary creep rates ... **A8:** 336-337
316, multiple heats ... **A8:** 330
316, stress amplitude vs. time-to-fracture ... **A8:** 349-350
316L, mill finishes ... **M5:** 552
321
aluminum coating, tensile strength affected by ... **M5:** 342
buffing ... **M5:** 557-558
mill finishes ... **M5:** 552
347
buffing ... **M5:** 557-558
grinding ... **M5:** 556
mill finishes ... **M5:** 552
347, autoclave construction ... **A8:** 424
347, low-cycle fatigue curve ... **A8:** 367
348, mill finishes ... **M5:** 552
400, series, cleaning and finishing processes ... **M5:** 72-73, 305, 552
403, alloying addition effects for three heats ... **A8:** 479-481
403, corrosion fatigue ... **A8:** 253, 254
403, effect of cathodic polarization on corrosion fatigue ... **A8:** 254
403, fatigue performance ... **A8:** 253-254
403, fracture toughness ... **A8:** 479-480
403, hydrazine effect on near-threshold fatigue crack propagation ... **A8:** 427, 429

Stainless steel, specific types (continued)
403, mill finishes ... **M5:** 552
403, pH effect on near-threshold fatigue crack growth rate ... **A8:** 427, 429
403, sodium chloride vs. threshold curve for ... **A8:** 427, 429
stress-intensity range average regression
410
electroless nickel plating of ... **M5:** 237
mill finishes ... **M5:** 552
410, ductility from hot torsion tests ... **A8:** 165-166
410, hydrogen embrittlement ... **A8:** 537
414, mill finishes ... **M5:** 552
416, 416S, 416Se, mill finishes ... **M5:** 552
420, mill finishes ... **M5:** 552
430
color buffing of ... **M5:** 557-558
electropolishing of ... **M5:** 305
mill finishes ... **M5:** 552
431, mill finishes ... **M5:** 552
440A, B, C, mill finishes ... **M5:** 552
446, mill finishes ... **M5:** 552
Stainless steel spring wire
characteristics of ... **A1:** 308
Stainless steel weld metal
different etchants to reveal delta ferrite ... **A9:** 289
different etchants to reveal sigma phase ... **A9:** 289
Stainless steel weld metals, specific types
E308, color etching after creep rupture testing ... **A9:** 137
E308, role of sigma phase in creep rupture and separation of phases ... **A9:** 137
Stainless steel(s) *See also* Austenitic stainless steels; Stainless steels, specific types; Steel(s)
arc-welded, failures in ... **A11:** 426-433
austenitic, intergranular corrosion of ... **A11:** 180
ball bearings, pitting failure ... **A11:** 495, 497
carbide precipitation in ... **A11:** 451
composition failures in ... **A11:** 391
effect of carburization on ... **A11:** 271-272
elevated-temperature ductility ... **A11:** 265
forceps, forging seam fracture ... **A11:** 329, 331
fuel-control lever, fractured ... **A11:** 388
general corrosion in ... **A11:** 200
hot-salt corrosion in ... **A11:** 200
for implants ... **A11:** 672
lever, fatigue fracture in ... **A11:** 113-114
liquid erosion resistance ... **A11:** 167
microvoid coalescence in ... **A11:** 85, 86
oxide stability ... **A11:** 452
pitting attack in ... **A11:** 200
preferential attack, types of ... **A11:** 200
sensitization ... **A11:** 200
service failure due to coarse grain size ... **A11:** 70
in sour gas environments ... **A11:** 300
susceptibility to hydrogen damage ... **A11:** 249
wrought, maximum operating temperatures for ... **A11:** 271
Stainless steel-europium oxide
in military pressurized water reactors ... **A7:** 666
Stainless steels *See also* Austenitic stainless steels; Duplex stainless steels; Ferritic stainless steels; High-alloy steels; Magnetic materials; Martensitic stainless steels; Precipitation hardening stainless steels; Sintered (porous) P/M stainless steels; Stainless steels, specific types; Steels; Steels, specific types ... **A6:** 677-707 **A7:** 728-732 **A13:** 547-565 **A14:** 222-230, 759-778 **A16:** 681-707
abrasive jet machining ... **A16:** 706
abrasive waterjet machining ... **A16:** 704, 706
absorption/enhancement effects in ... **A10:** 97
adaptive control implemented ... **A16:** 618
air-carbon arc cutting ... **A6:** 1172, 1176
AISI numbering system ... **A16:** 681
Al_2O_3 effect on stainless steels ... **A16:** 688
alloy selection ... **A14:** 759-761
annealing ... **A7:** 185

Stainless steels (continued)
apparent density ... **A7:** 297
for appliance parts ... **A7:** 622, 731-732
applications ... **A6:** 377-381 **A7:** 730-731 **M3:** 3-4, 38-39
aerospace ... **A6:** 385, 387
engine oil coolers ... **A6:** 961
arc welding *See* Arc welding of stainless steels
arc welding with nickel alloys ... **M6:** 443
ASTM XM designations ... **A16:** 681
atomized, effect of particle size on apparent density ... **A7:** 273
austenitic classification ... **A16:** 681, 682, 683, 684
austenitic, dislocation interaction ... **A10:** 469
austenitic grades ... **A7:** 100
hydrogen embrittlement of ... **M1:** 687
neutron embrittlement of ... **M1:** 686
sigma-phase embrittlement of ... **M1:** 686
austenitic, properties ... **M3:** 7, 17
austenitic, sintering ... **A7:** 308
austenitic stainless steels ... **A6:** 686-695
base metals ... **A6:** 689-693
engineering for use in as-welded condition ... **A6:** 693-694
engineering for use in postweld heat-treated condition ... **A6:** 694-695
metallurgy ... **A6:** 686
primary austenite solidification ... **A6:** 686
primary ferrite solidification ... **A6:** 686
austenitic, ultrasonic inspection ... **A17:** 569-570
austentic ... **A14:** 224-226
automotive applications ... **A7:** 732
backing bars ... **M6:** 382
bar ... **M3:** 12-13
bar and tube, die materials for drawing ... **M3:** 525
barnacle corrosion ... **A13:** 115
base metal, clad brazing material applications ... **A6:** 961, 962, 963
binary iron-chromium equilibrium phase diagram ... **A6:** 678, 681
biological corrosion ... **A13:** 117-118
blanking ... **A14:** 761-762
blanking, die materials for ... **M3:** 485, 486, 487
boriding ... **A4:** 440, 445
brazed joints ... **A13:** 880
brazing *See* Brazing of stainless steels
brazing and soldering characteristics ... **A6:** 625-626
brazing properties ... **M6:** 966
brazing with clad brazing materials ... **A6:** 347
broaches for stainless steels ... **A16:** 704
broaching ... **A16:** 206, 700-701, 704
in business machines ... **A7:** 732
calcium deoxidation ... **A16:** 688
capacitor discharge stud welding ... **A6:** 221 **M6:** 730, 736
carbide precipitation ... **A6:** 695
carbon concentration ... **A15:** 431
and carbon steels, formabilities compared ... **A14:** 760
for casings in air-fuel gas burners ... **A4:** 274
for casings of high-velocity convection burners ... **A4:** 274
cast *See* Corrosion-resistant steel castings
cast, corrosion of ... **A13:** 574-582
cast structures, influence on properties ... **M3:** 32
CBN as abrasive for honing ... **A16:** 476
cemented carbide machining applications ... **A16:** 86, 87, 88
ceramic cutting tools ... **A16:** 103
ceramic molding products ... **A15:** 248
cermet tools applied ... **A16:** 92, 96
in chemical environments ... **A13:** 556
chemical flux cutting of ... **A14:** 729
chemical integrity of seals ... **EM4:** 541
chemical milling ... **A16:** 579, 585
chip formation ... **A16:** 696
chlorine corrosion ... **A13:** 1171
chromium levels ... **A15:** 431
clad aluminum alloy, galvanic corrosion prevention ... **A13:** 1017
classification ... **A16:** 681-685
cleaning ... **A14:** 230 **M3:** 6-8, 38-39, 52-55
cleaning solutions for substrate materials ... **A6:** 978
coinability of ... **A14:** 183
cold cracking ... **A6:** 677

Stainless steels (continued)
cold heading of ... **A14:** 291
cold isostatic pressing dwell pressures ... **A7:** 449
cold reduced products, influence on properties ... **M3:** 33
cold reduction swaging effects ... **A14:** 129
cold working and austenitic alloys ... **A16:** 689
commercial P/M grades ... **A7:** 100
compact, isolated residual pores in ... **A7:** 435, 436
compactibility ... **A7:** 729
compacting grade, properties ... **A7:** 728
compacting pressure ... **A7:** 729
comparisons ... **A16:** 691
composition ... **A16:** 682-683
composition control in weld
cross-wire projection welding ... **M6:** 518
overlays ... **M6:** 809-811
composition effects ... **A13:** 550-551
compositions, standard/nonstandard grades ... **A13:** 548-549
in compound and progressive dies ... **A14:** 765-766
compressibility and green strength ... **A7:** 101
constitution diagrams ... **A6:** 677-678
contamination source for niobium electron-beam welding ... **A6:** 871
contour band sawing ... **A16:** 362
contour roll forming ... **A14:** 634, 775-776
copier parts ... **A7:** 732
corrosion *See also* Stainless steels, corrosion resistance ... **M3:** 6, 7, 11-12, 56-93
corrosion forms ... **A13:** 553-554
corrosion in aerospace applications ... **A6:** 385
corrosion in specific environments ... **A13:** 554-559
corrosion in various applications ... **A13:** 559-563
corrosion resistance ... **A6:** 585 **A7:** 254, 728
corrosion resistance mechanism ... **A13:** 550
corrosion testing ... **A13:** 563-564
covering for welding electrodes ... **A6:** 176
cracking ... **A6:** 677, 678, 679, 680, 681, 682, 687, 693
crevice corrosion ... **A13:** 110, 303, 564
cryogenic temperature behavior ... **A16:** 683
cryogenic treatment ... **A4:** 205
cutoff band sawing ... **A16:** 360
cutting fluids used ... **A16:** 125, 691-698, 700-701, 703-705
cutting precautions ... **A6:** 1196
cutting speeds ... **A16:** 685, 700
deep drawing ... **A14:** 767-771
deep drawn parts, materials for drawing tools ... **M3:** 494-496
definition ... **A6:** 677
degassing procedures ... **A15:** 427-428
delta ferrite role in weld deposits ... **A6:** 1066-1067
as dental alloys ... **A13:** 1352
deoxidation practice ... **A16:** 688
design ... **A13:** 552-553
determination of Cr, Ni, and Mn in ... **A10:** 146-147
die forgings, materials for forging tools ... **M3:** 529, 536, 532, 534
die lubrication ... **A14:** 229
die threading ... **A16:** 698, 700-701
dies ... **A14:** 227-228
diffusion brazing ... **A6:** 343
diffusion welding ... **A6:** 884, 886
dilution effects ... **A6:** 677, 680
dimensional change ... **A7:** 292, 730
dip brazing ... **A6:** 338
dissimilar metal joining ... **A6:** 821, 824, 825
distortion ... **A6:** 626
drilling ... **A16:** 220-221, 226-227, 229-230, 690, 697-698, 699
drilling test ... **A16:** 686, 687, 690, 691
drop hammer forming of ... **A14:** 656, 774
duplex *See* Duplex alloys
duplex, by 885 °F embrittlement ... **A12:** 155
duplex classification ... **A16:** 681, 682, 683, 684
duplex ferritic-austenitic stainless steels ... **A6:** 697-699
base metals ... **A6:** 697-698
distortion ... **A6:** 699
engineering for use in the as-welded condition ... **A6:** 698- 699
engineering for use in the postweld heat-treated condition ... **A6:** 699
metallurgy ... **A6:** 697

Stainless steels (continued)
properties ... **A6:** 697
effect of coarse and fine particle mixture ... **A7:** 273, 296
effect of lubricant an compacting pressure on green strength ... **A7:** 729
effect of sintering atmosphere on mechanical properties ... **A7:** 729
electrical discharge grinding ... **A16:** 566
electrical discharge machining ... **A16:** 706
electrochemical grinding ... **A16:** 542, 546
electrochemical machining ... **A16:** 706
electrodes for flux-cored arc welding ... **A6:** 189
electrolytic inclusion and phase isolation in ... **A10:** 176
electron beam drilling ... **A16:** 570, 571
electron beam machining ... **A16:** 705, 706
electron beam welding ... **M6:** 638
electron-beam welding ... **A6:** 679, 688, 698, 699, 828, 851, 853, 868-869, 870
electronic applications ... **A6:** 998
electropolishing ... **M3:** 36
electroslag welding ... **A6:** 278 **M6:** 226
electrostream and capillary drilling ... **A16:** 551
electrostream and shaped tube electrolytic machining ... **A16:** 706
embrittlement ... **A6:** 686
embrittlement of ... **M1:** 685-686
end milling ... **A16:** 326, 703
enhanced-machining alloys ... **A16:** 684-685, 689
environments that cause stress-corrosion cracking ... **A6:** 1101
etchants ... **A9:** 281-282
-europium oxide, ordnance application ... **A7:** 666
exothermic brazing ... **A6:** 345
explosion welding ... **A6:** 162, 163, 303-304, 896 **M6:** 710-711
extralow-interstitial (ELI) alloys ... **A16:** 683
fabrication *See also* Stainless steels, wrought, fabrication ... **M3:** 6-8
families ... **A6:** 677
families of ... **A13:** 547-550
fasteners, use in ... **M3:** 183-184
fatigue at subzero temperatures ... **M3:** 752, 4-755, 756, 764, 765
fatigue strength ... **M3:** 30, 32
ferrite analysis ... **A6:** 1059
Ferrite Number (FN) ... **A6:** 677, 678
ferritic ... **A13:** 325, 355-358 **A14:** 226
ferritic, 885 °F (475 °C) embrittlement ... **A12:** 136, 155
ferritic classification ... **A16:** 681-685
ferritic grades ... **A7:** 100
400 to 500 °C, embrittlement of ... **M1:** 685-686
embrittlement of ... **M1:** 685-686
sigma-phase embrittlement of ... **M1:** 686
ferritic, properties ... **M3:** 8, 17, 24-25
ferritic stainless steels ... **A6:** 682-686
base metals ... **A6:** 683
engineering for use in as-welded condition ... **A6:** 683-686
engineering for use in the postweld heat-treated condition ... **A6:** 686
metallurgy ... **A6:** 682-683
filler for polymers ... **A7:** 606
filler metals ... **A6:** 679, 681, 683, 686, 689, 692, 693, 694, 695-696, 697, 698, 699, 703-704, 705
for fine-edge blanking and piercing ... **A14:** 472
finishes ... **M3:** 8, 33, 36-38
flake pigments ... **A7:** 596
flame hardening ... **A4:** 284
flash welding ... **M6:** 558
fluid flow phenomena ... **A6:** 22
flux-cored arc welding ... **A6:** 186, 188, 680, 681, 688, 693, 698, 699, 705-706
foil ... **M3:** 12
forgeability ... **A14:** 223-224
forging equipment for ... **A14:** 227
forging, ferritic-austenitic properties ... **A7:** 549
forging methods ... **A14:** 222
forging of ... **A14:** 222-230
formability ... **A14:** 759-761
forming of ... **A14:** 519, 759-778
forming vs machining ... **A14:** 778
fracture toughness at subzero temperatures ... **M3:** 752, 763
free-machining alloys classification ... **A16:** 681, 684

Stainless steels (continued)
free-machining and
non-free-machining stainless
steels correspondence **A16:** 684
friction welding **A6:** 152, 153, 154, 890 **M6:** 721
fume generation from shielding gases **A6:** 68
fusion welding to carbon steels **A6:** 826, 827
fusion welding to low-alloy steels **A6:** 826, 827
galvanic corrosion **A6:** 625 **A13:** 84-85
gas metal arc welding **M6:** 153
gas nitriding **A4:** 424
gas tungsten arc welding **M6:** 182, 203, 205
gas-metal arc welding **A6:** 180, 677, 680, 688, 693, 694, 697, 698, 699, 705, 706, 707
of aluminum bronze **A6:** 828
of copper nickels **A6:** 828
of coppers **A6:** 828
of high-zinc brasses **A6:** 828
of low-zinc brasses **A6:** 828
of phosphor bronzes **A6:** 828
shielding gases **A6:** 67
of silicon bronzes **A6:** 828
of special brasses **A6:** 828
of tin brasses **A6:** 828
gas-tungsten arc welding **A6:** 20, 190, 191, 192, 677, 679, 686, 688, 693, 697, 698, 699, 703-705, 870
of aluminum bronzes **A6:** 827
of copper nickels **A6:** 827
of coppers and copper-based alloys **A6:** 827
of phosphor bronzes **A6:** 827
shielding gas selection **A6:** 67, 68
of silicon bronzes **A6:** 827
for gastight shell of cold-wall vacuum
furnace **A4:** 498
general guidelines for minimizing difficulties in machining **A16:** 690-691
globular-to-spray transition currents
for electrodes **A6:** 182
grain-boundary precipitation in **A13:** 155-156
green strength **A7:** 729
grinding **A16:** 436, 705
and hardenability **A7:** 728
hardfaced, erosion **A13:** 137
hardfacing **A6:** 789, 807
hardness level **A16:** 688-689
heat treating **A16:** 690, 704
heat-affected zone **A6:** 679, 681, 683, 686, 689, 695, 697, 698-699
heat-affected-zone cracks **A6:** 92
heating for forging **A14:** 228-229
heating of dies **A14:** 229
HERF forgeability **A14:** 104
high-alloy, hydrogen fluoride/hydrofluoric acid corrosion **A13:** 1168
high-frequency welding **A6:** 252, 253
high-temperature solid-state welding **A6:** 298, 299
higher-alloyed P/M **A13:** 832
hone forming **A16:** 488
honing **A16:** 476, 477
hot cracking **A6:** 677, 693, 695, 696, 699
hot cracking in **A12:** 123
hot isostatic pressing, effects **A15:** 541
hot processing, influence on properties **M3:** 32-33
humpback furnace for sintering **A7:** 355
hydroxide melt corrosion **A13:** 91
indentification systems **A13:** 547
induction brazing **A6:** 335
induction heating energy requirements
for metalworking **A4:** 189
induction heating temperatures for
metalworking processes **A4:** 188
induction soldering, physical
properties **A6:** 364
ingot breakdown **A14:** 222-223
inorganic fluxes **A6:** 980
inorganic fluxes used **A6:** 129
intergranular corrosion **A13:** 239-240, 324, 562
interim protection during shipping **M3:** 38

Stainless steels (continued)
ion-implanted, AEM analysis **A10:** 484
joined to aluminum alloys **A6:** 739
joint, high-temperature corrosion **A13:** 878
lapping **A16:** 499
laser beam machining **A16:** 575, 706
laser beam welding **M6:** 647, 662
penetration in welds **M6:** 654, 656
welding rates **M6:** 658
laser cutting **A14:** 741-742
laser-beam welding **A6:** 262, 263, 679, 688, 697, 698, 699
limiting draw ratios **A14:** 575
liquid nitriding **A4:** 412
low-temperature solid-state welding **A6:** 300, 301
lubricant effect **A7:** 729
lubrication **A14:** 519, 761
machinability **A16:** 1, 645, 685, 737
machinability additives **A16:** 685-688
machinability of austenitic alloys **A16:** 689, 691
machinability of duplex alloys **A16:** 689-690, 691
machinability of ferritic and martensitic alloys **A16:** 89
machinability of PH alloys **A16:** 690
machinability test matrix **A16:** 639-640
machining characteristics **A16:** 681
machining of **A14:** 778
macroetchants **A9:** 281
magabsorption measurement **A17:** 152
magnetic applications **M3:** 605, 607
as magnetically soft materials **A2:** 776-778
manual spinning **A14:** 771-772
marine corrosion **A13:** 555
martensitic **A14:** 226
martensitic classification **A16:** 681, 682, 683, 684
martensitic grade **A7:** 100
martensitic grades neutron embrittlement of **M1:** 686-687
martensitic, properties **M3:** 8, 17-25, 26-28
martensitic stainless steels **A6:** 678-682
base metals **A6:** 679
engineering for use after postweld
heat treatment **A6:** 680-682
engineering for use in the as-welded
condition **A6:** 679- 680
metallurgy **A6:** 678-679
maximum service temperatures, air **A13:** 558
mechanical properties **M3:** 6, 18-22, 23-25, 26-28
mechanical properties of medium-density **A7:** 464, 468
mechanical properties of nearly dense
and fully dense **A7:** 464-466, 471
mechanical properties of product **A7:** 728
as metallic implants/prosthetic
devices **A13:** 1325-1326
metallurgical effects on corrosion **A13:** 124-127
microduplex, superplasticity **A14:** 871
microstructural analysis **A7:** 488
microstructure **A6:** 677, 678
mill finishes **M3:** 36-38
milling **A16:** 312-314, 327, 699, 701, 703-704
in moist chlorine **A13:** 1173
molten salt corrosion **A13:** 89
multiple-slide forming **A14:** 766-767
nearly dense P/M **A7:** 471
nickel alloys, welding to **A6:** 578
nitric/hydrochloric acid as dissolution
medium **A10:** 166
nitrided, SCC in **A13:** 933
nitrogen-strengthened **A6:** 462 **A14:** 225-226
no phase transformation in stabilized
ferritic **A6:** 84
non-free-machining alloys
classification **A16:** 681-684
nonheat-treatable, SCC testing **A13:** 272-273
nonmagnetic characteristics **A7:** 728
nonstandard types **M3:** 4, 9-10
notch toughness **M3:** 28-30, 32
nozzle material for hot gas soldering **A6:** 361

Stainless steels (continued)
nuclear grade **A13:** 931
Osprey atomizing of **A7:** 531
oxidation **M1:** 93-94
oxidation resistance of **A7:** 728
oxyacetylene welding **A6:** 281
oxyfuel cutting **M6:** 903
oxyfuel gas cutting **A6:** 1155 **A14:** 725 **M6:** 897, 912, 914
oxyfuel gas welding **A6:** 281, 285
P/M *See* Sintered (porous) P/M stainless steels
P/M materials, hardness and density **A16:** 882
P/M parts **A7:** 728-732
parameter selection **A6:** 699-707
particle shape **A7:** 728
percussion welding **M6:** 740
peritectic structures **A9:** 679-680
in petroleum refining and petrochemical operations **A13:** 1263
PH alloys classification **A16:** 681, 683, 684
in pharmaceutical production facilities **A13:** 1226-1227
phases **A13:** 47, 48
photochemical machining **A16:** 587, 590, 591
physical properties **M3:** 33, 34-35
piercing **A14:** 761-762
pipe **M3:** 16
plasma (ion) nitriding **A4:** 420, 421, 423
plasma arc cutting **A6:** 1167, 1169, 1170 **A14:** 731-732 **M6:** 914, 916-917
plasma arc machining **A16:** 706
plasma arc welding **A6:** 197, 199, 698, 699 **M6:** 214
manufacture of tubing **M6:** 220-221
welding of foil **M6:** 221
plasma-MIG welding **A6:** 224, 225
plate **M3:** 5-6, 8-11
plunge machining test **A16:** 687
porous **A7:** 699, 731
postweld heat treatments **A6:** 677, 680-682, 686, 688, 695, 697
power hacksawing **A16:** 704
power requirement **A16:** 690
power spinning **A14:** 772-773
precipitate identification by
light-element analysis **A10:** 459-461
precipitation-hardening **A14:** 226-227 **M3:** 25-28, 30-31
precipitation-hardening stainless steels **A6:** 695-697
prediction of microstructure **M6:** 40
preheating **A6:** 679
press forming **A14:** 763-765
press-brake forming **A14:** 762-763
pressure-density relationships **A7:** 299
probe material for quench-cooling
curve analysis **A4:** 68
procedures for weld overlays **M6:** 811-816
processing sequence **A7:** 728-730
processing/fabrication/external treatment effects **A13:** 551-553
and product physical appearance **A7:** 728
production **A7:** 100-101 **M3:** 3-4
products, semifinished **M3:** 3-4
projection welding **A6:** 233 **M6:** 503, 509
properties, commercial P/M grades **A7:** 100, 731
properties, elevated temperatures **M3:** 30-31
properties, influence of product form **M3:** 31-33
quenching **A4:** 67
reaction with graphite hearths in vacuum heat treating **A4:** 503
reaming **A16:** 249, 690, 698, 702-703, 704-705
recommended guidelines for selecting
PAW shielding gases **A6:** 67
recommended shielding gas selection
for gas-metal arc welding **A6:** 66
reductions by cold swaging **A14:** 128
relative solderability **A6:** 134

SUBJECTS OF THE INDEXED VOLUMES: ASM Handbook (designated by the letter "A"): **A1:** Properties and Selection: Irons, Steels, and High-Performance Alloys (1990); **A2:** Properties and Selection: Nonferrous Alloys and Special-Purpose Materials (1990); **A3:** Alloy Phase Diagrams (1992); **A4:** Heat Treating (1991); **A6:** Welding, Brazing, and Soldering (1993); **A7:** Powder Metallurgy (1984); **A8:** Mechanical Testing (1985); **A9:** Metallography and Microstructures (1985); **A10:** Materials Characterization (1986); **A11:** Failure Analysis and Prevention (1986); **A12:** Fractography (1987); **A13:** Corrosion (1987); **A14:** Forming and Forging (1988); **A15:** Casting (1988); **A16:** Machining (1989); **A17:** Nondestructive Testing and Quality Control (1989); **A18:** Friction, Lubrication, and Wear Technology (1992). **Metals Handbook, 9th Edition** (designated by the letter "M"): **M1:** Properties and Selection: Irons and Steels (1978); **M2:** Properties and Selection: Nonferrous Alloys and Pure Metals (1979); **M3:** Properties and Selection: Stainless Steels, Tool Materials and Special-Purpose Materials (1980); **M4:** Heat Treating (1981); **M5:** Surface Cleaning, Finishing, and Coating (1982); **M6:** Welding, Brazing, and Soldering (1983). **Engineered Materials Handbook** (designated by the letters "EM"): **EM1:** Composites (1987); **EM2:** Engineering Plastics (1988); **EM3:** Adhesives and Sealants (1990); **EM4:** Ceramics and Glasses (1991); **Electronic Materials Handbook** (designated by the letters "EL"): **EL1:** Packaging (1989).

Stainless steels (continued)
relative solderability as a function of
flux type **A6:** 129
relative weldability ratings, resistance
spot welding **A6:** 834
repair welding **A6:** 1105-1107
Replicast process for **A15:** 271-272
residual stress **M3:** 50-51
residual stresses **A6:** 625-626, 679
resistance brazing **A6:** 341 **M6:** 976
resistance seam welding **A6:** 241, 243 **M6:** 494
resistance soldering **A6:** 357
resistance spot welding **M6:** 480
resistance welding *See* Resistance
welding of stainless steels.... **A6:** 833, 837, 840,
841, 842, 847-848
roll welding **A6:** 312, 314
rubber-pad forming **A14:** 773-774
rupture life **M1:** 95
sawing **A16:** 358, 705
SCC test result distribution **A13:** 276
SCC under thermal insulation **A13:** 1146-1147
SCC/general corrosion resistance **A13:** 1023
Schaeffler diagram **A6:** 431
schematic anodic polarization
diagrams **A13:** 48
screw machine test **A16:** 689
second-phase constituents **A9:** 284
selection factors **M3:** 4-8
sensitization **A6:** 688 **M3:** 50, 60-62
shaped tube electrolytic machining **A16:** 554
sheet **M3:** 5-6, 11-12
sheet metals **A6:** 399-400
shielded metal arc welding **A6:** 176, 688, 693,
694, 698, 699-700, 701 **M6:** 75
sigma phase in **A17:** 55
SIMS analysis of surface composition
effects in **A10:** 622-623
sintered properties **A7:** 731
sintering **A7:** 340-341, 368-370, 729
sintering atmospheres for **A7:** 341, 729
slab milling **A16:** 324
slag removal from weldments **A6:** 61
sodium hydroxide corrosion **A13:** 1175
soft-magnetic modified versions **A16:** 684
solderable and protective finishes for
substrate materials **A6:** 979
soldering **A6:** 127, 631
solid-state transformations in
weldments **A6:** 82-83
solidification cracking **A6:** 82
spark source mass spectrometry for **A10:** 146-147
spinning **A14:** 771-773
spraying for hardfacing **M6:** 789
STAMP processed **A7:** 548-549
standard types **M3:** 4-8, 11-12
straightening, for cold heading **A14:** 687
stress-corrosion cracking **A6:** 625, 626 **M3:** 63-64
stress-relief heat treating **A4:** 34
stretch forming **A14:** 776-777
strip **M3:** 12
structure, principles **A13:** 47
stud arc welding..... **A6:** 210, 211, 212, 213, 214, 216,
218-219 **M6:** 730, 733
stud material **M6:** 730-731, 735
submerged arc welding **A6:** 203, 680, 681-682,
688, 693, 694, 698, 699, 700-703 **M6:** 116
suction shell alloys **A13:** 1204
suitability for cladding combinations **M6:** 691
"super ferritics" **A6:** 683, 686
surface composition effects during
laser treatment of **A10:** 622-623
surface finish **A16:** 685, 686, 687, 689, 691, 693,
697, 705
surface finishing **M3:** 33-36, 55
and surface integrity **A16:** 22
tapping **A16:** 263, 698-700
taps for stainless steels **A16:** 699, 700
tempering **A4:** 124
tensile properties **A7:** 729, 730 **M3:** 17-28, 29
tensile properties at subzero
temperatures **M3:** 748, 751-752, 757-762
texture orientations **A10:** 360
thermal properties **A6:** 17
thread grinding **A16:** 270, 274
thread milling **A16:** 269
thread rolling **A16:** 282, 293, 701-703

Stainless steels (continued)
threading **A16:** 299
three-roll forming **A14:** 774-775
tin containing **A7:** 254
tool for ECM **A16:** 533, 536, 537
tool life **A16:** 681, 685-686, 688-689, 691, 695, 701,
703
tool life test **A16:** 690
torch brazing **A6:** 328
torch soldering **A6:** 351, 352
transition temperature **M3:** 28-30, 32
trimming **A14:** 229-230
tube, optical micrograph with precipitates and cracks **A10:** 459
tube stock **A14:** 671
tubing **M3:** 16-17
tubing, aircraft grades **M3:** 16-17
tubing, bending of **A14:** 777
tubing, forming of **A14:** 671, 777-778
turning **A16:** 144, 146, 147, 690, 696-697
ultrasonic inspection **A17:** 650-652
ultrasonic welding **A6:** 327, 893, 894
umpire analysis of **A10:** 178-179
Unicast molding of **A15:** 251
Unified Numbering System (UNS) **A16:** 681
use in breweries **A13:** 1221-1222
vacuum heat treating **A4:** 494
vacuum sintering atmospheres for **A7:** 345
for valves and piping used in
carbonitriding **A4:** 376, 383
water corrosion **A13:** 555-556
water-atomized, effect of oxide films
on green strength **A7:** 303
wear resistance of **A7:** 728
weld cladding **A6:** 814
by submerged arc welding **A6:** 813
composition control of weld
overlays **A6:** 817-820
filler metals **A6:** 809
welding parameters **A6:** 817
"weld cold" rule **A6:** 683
weld cracking **M3:** 51-52
weld decay, corrosion of weldments **A6:** 1066
weld metal, measurement of δ-ferrite
in **A10:** 287
weld overlay material **M6:** 806
"weld ugly" **A6:** 694, 696
weldability **A15:** 535-537
welding **M3:** 48-52
welding process selection **A6:** 699-707
welding to copper and copper alloys **A6:** 769
wire **M3:** 13-15
wire, die materials for drawing **M3:** 522
for wire-drawing dies **A14:** 336
X-ray photoelectron spectroscopy **A7:** 256, 257
x-ray spectrometry of **A10:** 100

Stainless steels, ACI specific types
CA-6N, composition and
microstructure **A4:** 775
CA-6NM
annealing **A4:** 786, 790
applications **M3:** 94
composition **A4:** 772 **M3:** 95, 104-105
composition and microstructure **A4:** 775
corrosion resistance **M3:** 94
hardening **A4:** 787, 790
impact strength **A4:** 787
intergranular attack not a problem **A4:** 787
machining speeds and feeds **M3:** 100
mechanical properties **A4:** 786
property data **M3:** 104-105
reaustenitizing **A4:** 787
stress-relieving **A4:** 786
tempering temperature effect **A4:** 787-789
tempering temperature effect on
hardness **A4:** 789, 791
welding conditions **M3:** 101
CA-6NM, tempering temperature,
effect on properties **M4:** 644
CA-15
annealing **A4:** 790
applications **M3:** 94
composition **M3:** 95, 105
composition and microstructure **A4:** 775
corrosion resistance **M3:** 94
hardening **A4:** 790
heat treatment **A4:** 785 **M4:** 642

Stainless steels, ACI specific types (continued)
heat treatment methods, effect on
mechanical properties **A4:** 790
impact strength **A4:** 787
machining speeds and feeds **M3:** 100
mechanical properties **A4:** 790 **M3:** 96-97
mechanical properties, effect of heat
treating **M4:** 641
property data **M3:** 107-108
replaced by CA-6NM **A4:** 786
tempering temperature effect on
hardness **A4:** 789, 791
tempering temperature effect on
mechanical properties **A4:** 791
tempering temperature, effect on
properties **M4:** 644, 645
welding conditions **M3:** 101
CA-15M, composition and
microstructure **A4:** 775
CA-28MWV, composition and
microstructure **A4:** 775
CA-40
annealing **A4:** 790
composition **M3:** 94, 106
composition and microstructure **A4:** 775
hardening **A4:** 790
heat treatment **A4:** 785
machining speeds and feeds **M3:** 100
property data **M3:** 106-107
welding conditions **M3:** 101
CA-40, heat treatment **M4:** 642
CA-40F, composition and
microstructure **A4:** 775
CB-7Cu
applications **M3:** 95-96
composition **M3:** 95, 107
corrosion resistance **M3:** 95-96
mechanical properties **M3:** 97
property data **M3:** 107-108
welding conditions **M3:** 101
CB-7Cu-1, composition and
microstructure **A4:** 775
CB-7Cu-2, composition and
microstructure **A4:** 775
CB-30
annealing **A4:** 774, 786, 790
applications **M3:** 96-97
composition **M3:** 95, 108
composition and microstructure **A4:** 775
corrosion **M3:** 109
machining speeds and feeds **M3:** 98
property data **M3:** 108-109
welding conditions **M3:** 101
CB-30, annealing **M4:** 639
CC-50
annealing **A4:** 774, 786, 790
composition **M3:** 95, 109
composition and microstructure **A4:** 775
machining speeds and feeds **M3:** 100
property data **M3:** 109-110
welding conditions **M3:** 101
CC-50, annealing **M4:** 639
CD-4MCu
applications **M3:** 96
composition **M3:** 95, 110
corrosion **M3:** 96, 111
property data **M3:** 110, 111-112
welding conditions **M3:** 101
CD-4MCu, composition and
microstructure **A4:** 775
CE-30
annealing **A4:** 774, 790
composition **M3:** 95, 112
composition and microstructure **A4:** 775
machining speeds and feeds **M3:** 100
property data **M3:** 112
welding conditions **M3:** 101
CE-30, annealing **M4:** 639
CF-3
annealing **A4:** 774, 790
composition **M3:** 95, 112-113
composition and microstructure **A4:** 775
corrosion chart **M3:** 113
property data **M3:** 113
welding conditions **M3:** 101
CF-3, annealing **M4:** 639
CF-3, corrosion resistance **A4:** 787

Stainless steels, ACI specific types (continued)
CF-3A
composition **M3:** 113
property data **M3:** 113
CF-3M
annealing **A4:** 774, 790
applications **M3:** 95
composition **M3:** 95, 113
composition and microstructure **A4:** 775
corrosion resistance **A4:** 787 **M3:** 95
property data **M3:** 113-114
welding conditions **M3:** 101
CF-3M, annealing **M4:** 639
CF-3MA
composition **M3:** 113
property data **M3:** 114
CF-3MN, composition and microstructure **A4:** 775
CF-8
applications **M3:** 95
composition **M3:** 95, 114, 756
corrosion **M3:** 81, 93, 100, 115
machining speeds and feeds **M3:** 100
mechanical properties **M3:** 96, 103
property data **M3:** 114, 115
tensile properties at subzero temperatures **M3:** 762
welding conditions **M3:** 101
CF-8, annealing **A4:** 774, 790 **M4:** 639
composition and microstructure **A4:** 775
CF-8A
composition **M3:** 114
property data **M3:** 114
CF-8C
annealing **A4:** 774, 790
composition **M3:** 95, 116
composition and microstructure **A4:** 775
machining speeds and feeds **M3:** 100
property data **M3:** 116
stabilizing treatment **A4:** 787
welding conditions **M3:** 101
CF-8C, annealing **M4:** 639
CF-8M
annealing **A4:** 774, 790
applications **M3:** 95
composition **M3:** 95, 116, 756
composition and microstructure **A4:** 775
corrosion **M3:** 95, 100, 117-118
corrosion in chemical solutions **M3:** 81, 84, 86, 89, 90
corrosion in foods **M3:** 93
machining speeds and feeds **M3:** 100
property data **M3:** 116-118
tensile properties at subzero temperatures **M3:** 762
welding conditions **M3:** 101
CF-8M, annealing **M4:** 639
CF-10M, composition and microstructure **A4:** 775
CF-10MC, composition and microstructure **A4:** 775
CF-10SMnN, composition and microstructure **A4:** 775
CF-12M
annealing **A4:** 774, 790
composition **M3:** 95, 116
composition and microstructure **A4:** 775
corrosion charts **M3:** 117-118
property data **M3:** 116-118
welding conditions **M3:** 101
CF-12M, annealing **M4:** 639
CF-16F
annealing **A4:** 774, 790
composition **M3:** 95, 118
composition and microstructure **A4:** 775
machining speeds and feeds **M3:** 100
property data **M3:** 118-119
welding conditions **M3:** 101
CF-16F, annealing **M4:** 639

Stainless steels, ACI specific types (continued)
CF-20
annealing **A4:** 774, 790
composition **M3:** 95, 119
composition and microstructure **A4:** 775
machining speeds and feeds **M3:** 100
property data **M3:** 119
welding conditions **M3:** 101
CF-20, annealing **M4:** 639
CG-6MMN, composition and microstructure **A4:** 775
CG-8M
applications **M3:** 95
composition **M3:** 95, 119
corrosion resistance **M3:** 95
property data **M3:** 119-120
welding conditions **M3:** 101
CG-8M, composition and microstructure **A4:** 775
CG-12, composition and microstructure **A4:** 775
CH-8, composition and microstructure **A4:** 775
CH-10
composition **M3:** 120
corrosion in silver nitrate **M3:** 87
property data **M3:** 120-121
CH-10, composition and microstructure **A4:** 775
CH-20
annealing **A4:** 774, 790
composition **M3:** 95, 120
composition and microstructure **A4:** 775
corrosion in silver nitrate **M3:** 87
machining speeds and feeds **M3:** 100
property data **M3:** 120-121
welding conditions **M3:** 101
CH-20, annealing **M4:** 639
CHF20-8M
annealing **A4:** 774, 790
composition and microstructure **A4:** 775
CK-3MCuN, composition and microstructure **A4:** 775
CK-20
annealing **A4:** 774, 790
composition **M3:** 95, 121
composition and microstructure **A4:** 775
machining speeds and feeds **M3:** 100
mechanical properties **M3:** 96
property data **M3:** 121
welding conditions **M3:** 101
CK-20, annealing **M4:** 639
CN-3M, composition and microstructure **A4:** 775
CN-7M
annealing **A4:** 774, 790
applications **M3:** 96
composition **M3:** 95, 121
composition and microstructure **A4:** 775
corrosion **M3:** 96, 100, 122
corrosion in chemical solutions **M3:** 80-81, 84-86, 89
machining speeds and feeds **M3:** 100
mechanical properties **M3:** 96
property data **M3:** 121-123
welding conditions **M3:** 101
CN-7M, annealing **M4:** 639
CN-7MS
applications **M3:** 96
composition **M3:** 95
corrosion resistance **M3:** 96
CN-7MS, composition and microstructure **A4:** 775
CT-15C, composition and microstructure **A4:** 775
Stainless steels, ASTM specific types
A297, grade HF, rupture life **M1:** 95
Stainless steels, austenitic **A4:** 769-776
annealing **A4:** 769-772, 773, 774, 775-776 **M4:** 623-624, 639

Stainless steels, austenitic (continued)
boriding **A4:** 441
bright annealing **A4:** 773 **M4:** 625-626
conventional composition **A4:** 769 **M4:** 623-624
cooling curves **A4:** 114
hardness and oil quenching **A4:** 99
high-nitrogen alloys **A4:** 769, 772, 773 **M4:** 623, 624, 626
highly alloyed grades **A4:** 769, 773 **M4:** 623, 624, 626
intergranular attack **A4:** 769-773, 776
low-carbon **A4:** 769-772 **M4:** 623, 624
magnetic permeability **A4:** 772, 773 **M4:** 625
plasma (ion) nitriding **A4:** 424
stabilized compositions **A4:** 769 **M4:** 623, 624-625
stress relieving **M4:** 647-649
stress-relieving **A4:** 772, 773, 774-776, 777
for trays and grids **A4:** 514
Stainless steels, cast *See* Cast stainless steels, selection
Stainless steels, cast, specific types *See also* Nickel alloys, specific types; Stainless steels, ACI specific types
Illium P
composition **M3:** 123
property data **M3:** 123, 124
Illium PD
composition **M3:** 123, 124
property data **M3:** 123-124
Kromarc 55
composition **M3:** 756
tensile properties at subzero temperatures **M3:** 762
Stainless steels, corrosion resistance
architectural applications **M3:** 67
atmospheric corrosion **M3:** 65-68
biofouling **M3:** 76
cathodic protection **M3:** 73-74, 78
cavitation erosion **M3:** 77
chromium carbide precipitation **M3:** 61-62
cold work **M3:** 59, 60
composition effects **M3:** 57, 70-71, 72, 73
corrosion fatigue **M3:** 64-65
corrosion testing **M3:** 65
crevice corrosion **M3:** 62, 72-73, 74, 75, 76
design, effect of **M3:** 59-61
dissimilar-metal couples **M3:** 62, 72-74
fabrication, effect of **M3:** 59-60
foods **M3:** 92-93
galvanic corrosion **M3:** 62-63
heat treatment, effect of **M3:** 57-59, 60, 64
impingement attack **M3:** 76-77, 79
intergranular corrosion **M3:** 57, 58, 60-62, 63
localized corrosion **M3:** 56, 57
passivity **M3:** 56-57
pharmaceuticals **M3:** 91-92
pitting **M3:** 58, 61
pollution **M3:** 76
pulp and paper **M3:** 92
seawater environments **M3:** 70-80
ship propellers **M3:** 77-78, 80
stress-corrosion cracking **M3:** 63-64
transportation equipment **M3:** 68-70
welding **M3:** 59-60
Stainless steels, dissimilar welds **A6:** 500-504
cladding austenitic stainless steel to carbon steels **A6:** 502- 504
cladding austenitic stainless steels to low-alloy steels **A6:** 502-504
filler metals **A6:** 500, 501-502
hot cracking **A6:** 504
submerged arc welding **A6:** 502
welding austenitic stainless steels to carbon steels **A6:** 500- 501, 502
welding austenitic stainless steels to low-alloy steels **A6:** 500-501
welding austenitic-stainless-clad carbon or low-alloy steels **A6:** 501-502

Stainless steels, dissimilar welds (continued)
welding dissimilar austenitic stainless steels **A6:** 500
welding ferritic stainless steels to carbon steels **A6:** 501
welding ferritic stainless steels to low-alloy steels **A6:** 501
welding martensitic stainless steels to carbon steels **A6:** 501
welding martensitic stainless steels to low-alloy steels **A6:** 501
Stainless steels, duplex
annealing **A4:** 777, 778
yield strengths **A4:** 777
Stainless steels, ferritic **A4:** 776-777
annealing **A4:** 776 **M4:** 625
embrittlement **A4:** 776-777 **M4:** 629
plasma (ion) nitriding **A4:** 424
Stainless steels, heat treating *See also* Stainless steels, austenitic; Stainless steels, ferritic **M4:** 623-646
Stainless steels, martensitic **A4:** 777-782
annealing **A4:** 778, 780 **M4:** 628, 633-634
austenitizing **A4:** 778-779, 781 **M4:** 625, 627, 631
cleaning prior to heat treating **A4:** 778 **M4:** 630
hardness **A4:** 778-779, 780 **M4:** 625, 630
hydrogen embrittlement **A4:** 781-782 **M4:** 635
impact strength **A4:** 779, 780
martempering **A4:** 779
preheating **A4:** 778 **M4:** 630-631
protective atmospheres **A4:** 781 **M4:** 634-635
quenching **A4:** 779 **M4:** 632
reheating **A4:** 780 **M4:** 627, 629, 631, 632, 633, 634, 635
retained austenite **A4:** 779 **M4:** 632
salt baths **A4:** 780-781 **M4:** 634
soaking time **A4:** 779 **M4:** 626, 627, 628, 631-632
subzero cooling **A4:** 779 **M4:** 632-633
tempering **A4:** 778, 779, 780, 781
Stainless steels, powder metallurgy materials
etching **A9:** 509
microstructures **A9:** 511
Stainless steels, precipitation-hardening **A4:** 782-785
bright annealing **A4:** 783
cleaning prior to heat treating **A4:** 782 **M4:** 636
furnace atmospheres **A4:** 782-783 **M4:** 636-637
furnace(s) **A4:** 782
furnaces **M4:** 636
heat treating cycles, variations **M4:** 638, 640, 641, 642, 643
heat treating procedures **M4:** 635-639
heat-treating cycles, variations **A4:** 785
heat-treating procedures **A4:** 783-785
scale removal **A4:** 785 **M4:** 638-639
Stainless steels, selection of **A6:** 431
alloying additions **A6:** 431
chromium addition **A6:** 431
chromium equivalents **A6:** 431
classification scheme **A6:** 431
nickel equivalents **A6:** 431
Stainless steels, specific types *See also* Stainless steels
0.30C-13Cr, wear resistance **A18:** 705
2RK65
annealing **A4:** 773
composition **A4:** 771
2RK65, annealing **M4:** 624
3RE60
annealing **A4:** 778
composition **A4:** 772
7L4, annealing **A4:** 773
7MoPlus
annealing **A4:** 778
composition **A4:** 772
12Cr, chromic acid corrosion resistance **M3:** 82
12SR, chemical composition **A6:** 444
12SR, composition **A4:** 772
13-8, chemical milling **A16:** 584
13-8 PH, hydrogen embrittlement **A12:** 30
13-8Mo alloys, heat-treating procedures **A4:** 784
13-8PH, gas nitriding **A4:** 387
14Cr, crevice corrosion **M3:** 58
15-5 PH
composition **M6:** 350

Stainless steels, specific types (continued)
resistance welding **M6:** 527
15-5 PH, seizure resistance **M1:** 611
15-5 PH, wrought heat-resistant **A16:** 738
15-5Ni alloys, heat-treating procedures **A4:** 783-784
15-5PH
composition **A6:** 483, 484 **M3:** 5-6, 190-191
composition effect **A6:** 487
electron-beam welding **A6:** 491, 869
heat treatment **A6:** 485
mechanical properties **A6:** 486 **M3:** 30-31
physical properties **M3:** 34-35
precipitation-hardening phases **A6:** 483
recommended filler metals for welding **A6:** 1107
resistance to environments **M3:** 11-12
weldability **A6:** 484, 487
15-5PH, composition **A4:** 770
15-5PH, gas nitriding **A4:** 387
15-7 PH Mo, and polybenzimidazoles **EM3:** 171
15Cr-5Ni, composition **M6:** 526
16-8-2H
composition **A6:** 691
filler metals for **A6:** 693
properties **A6:** 693
16Cr-18Ni, cold rolling, effect on tensile strength **M3:** 33
17-4 PH
composition **M6:** 350, 526
flash welding **M6:** 557
resistance welding **M6:** 527
17-4 PH (630) bolt, hydrogen embrittlement **A11:** 249
17-4 PH CD, seawater corrosion resistance **M3:** 74
17-4 PH, intergranular SCC failure **A11:** 23
17-4 PH, intergranular SCC-caused decohesive fracture **A12:** 24
17-4 PH, machining **A16:** 203, 269, 360, 362, 534, 584, 588, 738
17-4 PH power-plant gate-valve stem, SCC failure in high-purity water **A11:** 218
17-4 PH steam-turbine blade, corrosion fatigue failure by corrosion pitting **A11:** 257
17-4PH
aging **A4:** 791
composition **A4:** 770 **A6:** 483 **M3:** 5-6, 190-191
composition effect **A6:** 487
cutoff band sawing with bimetal blades **A6:** 1184
electron-beam welding **A6:** 484, 490, 491, 869
fasteners **M3:** 184, 185
fracture toughness **M3:** 32
furnace brazing **A6:** 920-921
gas nitriding **A4:** 387, 401
gas-metal arc welding **A6:** 483, 489
gas-tungsten arc welding **A6:** 483, 489
hardening temperature, effect on properties **A4:** 789 **M4:** 643
heat treating procedures **M4:** 635-639
heat treatment **A6:** 485
heat-treating procedures **A4:** 783, 785
homogenization **A4:** 790
laser weld joining, YAG **A6:** 88
machinability **M3:** 45
mechanical properties **A4:** 791 **A6:** 483, 486, 489 **M3:** 30-31, 200, 202
mechanical properties, effect of aging time **M4:** 643, 645-646
M_f temperature **A6:** 482
microstructure **A6:** 488
physical properties **M3:** 34-35
plasma arc welding **A6:** 490
postweld heat treatment **A6:** 490
precipitation-hardening phases **A6:** 483
recommended filler metals for welding **A6:** 1107
resistance to environments **M3:** 11-12
shielded metal arc welding **A6:** 483, 489
temperature, effect on aging **M4:** 642, 645-646
weldability **A6:** 483, 484
17-7 PH *See* Stainless steels, specific types
composition **A6:** 483 **M6:** 350, 526
composition effect **A6:** 487

Stainless steels, specific types (continued)
cutoff band sawing with bimetal blades **A6:** 1184
electron-beam welding **A6:** 869
filler metals **A6:** 488
furnace brazing **A6:** 918-919, 920-921
gas-metal arc welding **A6:** 489
gas-tungsten arc welding **A6:** 488-489, 492
heat treatment **A6:** 485
mechanical properties **A6:** 486
M_f temperature **A6:** 482
microstructure **A6:** 488
M_s temperature **A6:** 482
postweld heat treatment **A6:** 488, 489
precipitation-hardening phases **A6:** 483
recommended filler metals for welding **A6:** 1107
resistance welding **A6:** 489 **M6:** 527
spot and seam welding **A6:** 489
submerged arc welding **A6:** 489
ultrasonic welding **A6:** 326, 327
weldability **A6:** 488
17-7 PH annealed, and polybenzimida-zoles **EM3:** 170
17-7 PH annealed, and polybenzimidazoles **EM3:** 170
17-7 PH, distortion in **A11:** 140
17-7 PH, machining **A16:** 360, 362, 584, 588, 738
17-7PH
annealing **A4:** 787
annealing temperature variation, effect on properties **M4:** 640
austenite-conditioning temperature, effect of variations on properties **M4:** 641
austenitizing temperature **A4:** 789
composition **A4:** 770 **M3:** 5-6, 190-191
fasteners **M3:** 185
fracture toughness **M3:** 32
gas nitriding **A4:** 387, 401
hardening temperature and time, effect of variations on properties **M4:** 642
hardening temperature and time variations **A4:** 789
heat treating procedures **M4:** 635-639
heat treatment **M3:** 202
heat-treating procedures **A4:** 783, 785
isochronous stress-strain curves **M3:** 192
machinability **M3:** 45
mechanical properties **A4:** 787, 789 **M3:** 30-31, 202, 204
physical properties **M3:** 34-35
resistance to environments **M3:** 11-12
transformation treatment temperature and time variations **A4:** 787
transformation treatment temperature, effect of variations on properties **M4:** 640
17-7PH, die wear and die life **A18:** 633
17-7PH, intergranular fracture **A13:** 1046
17-7PH RH950, stress-corrosion failure **A13:** 1102
17-10 P
composition **M3:** 9-10
mechanical properties **M3:** 30-31
17-10 P, composition **M6:** 350
17-10 P, not electron-beam welded **A6:** 869
17-14-4LN, composition **A4:** 771 **A6:** 458
17-14CuMo
composition **M3:** 190-191
seawater corrosion resistance **M3:** 74
17Cr, crevice corrosion **M3:** 58
18 SR, wrought heat-resistant **A16:** 738
18% Cr + 8% Ni (18-8), wear resistance **A18:** 651
18-2FM (XM-34), composition **A4:** 771
18-2FM, composition **M3:** 9-10
18-2Mn, composition **A6:** 458
18-3Mn, composition **A6:** 458
18-8 *See also* 302, 304; Stainless steels, specific types
acetic acid corrosion resistance **M3:** 81
cooling curves **A4:** 96
cooling rate with quenching **A4:** 76
corrosion resistance, marine atmospheres **M3:** 67
fasteners **M3:** 183, 184
fatty acids corrosion resistance **M3:** 83

Stainless steels, specific types (continued)
plasma (ion) nitriding A4: 404, 423
quenching A4: 96
seawater corrosion resistance M3: 70, 72, 73
stress-rupture, affected by sigma phase M3: 226, 229
sulfuric acid corrosion resistance M3: 88-89
18-8, drilling A16: 219
18-8, polarization curves A9: 144-145
18-8 properties A6: 992
18-8, radiographic absorption A17: 311
18-8Mo *See also* 316
chromic acid corrosion resistance M3: 82
18-8Ti, chromic acid corrosion resistance M3: 82
18-9, phosphoric acid corrosion resistance M3: 86
18-18 Plus
annealing A4: 773
composition A4: 771 M3: 9-10
mechanical properties M3: 23-24
18-18 Plus, composition A6: 458 M6: 526
18-18-2
composition M3: 9-10
mechanical properties M3: 18-22
18-18-2 (XM-15), composition A4: 771 A6: 458
18Cr, ammonium sulfate corrosion resistance M3: 81
18Cr-2Mo
acetic acid corrosion resistance M3: 78-81
ammonia corrosion resistance M3: 81
composition M3: 190-191
18Cr-2Mo, wrought heat-resistant A16: 738
18Cr-2Ni-12Mn, annealing A4: 773
18Cr-2Ni-12Mn, composition M6: 526
18Cr-8Ni, ammonium sulfate corrosion resistance M3: 81
18Cr-8Ni, stacking faults and mechanical twins A9: 686
18Cr-9Ni, Mn content effect A16: 687
18Cr-9Ni, S content effect A16: 685
18Cr-9Ni-3Mn, effect of C and N on machinability A16: 690
18Cr-12Ni- high-purity austenitic, atom probe analysis A10: 595
18Ni maraging steel, composition A6: 1017
18SR
annealing A4: 776
composition A4: 772 M3: 9-10, 190-191
mechanical properties M3: 24-25
oxidation resistance M3: 196
tensile properties M3: 194
18SR, chemical composition A6: 444
19-9 DL
composition M3: 190-191
stress-rupture M3: 205, 230
tensile properties M3: 205
19-9 DX
composition M3: 190-191
stress-rupture M3: 205
tensile properties M3: 205
19-9DL
composition A6: 458
electron-beam welding A6: 869
19-9DX, wrought heat-resistant A16: 738
20Cb, as filler metal for weld cladding applications A6: 809
20Cb-3
annealing A4: 773
composition A4: 771 A6: 458
mechanical properties A6: 468
20Cb-3 tools for shaped tube electrolytic machining A16: 555
20Cr, crevice corrosion M3: 58
20Mo-4
composition A6: 458
mechanical properties A6: 468
20Mo-4, composition A4: 771

Stainless steels, specific types (continued)
20Mo-6
composition A6: 458
mechanical properties A6: 468
20Mo-6, composition A4: 771
21-6-9
composition M3: 190-191, 756
fatigue-crack-growth rate M3: 764
friction welding A6: 153, 154
gas-tungsten arc welding and fluid flow phenomena A6: 20, 21, 22
stress-rupture M3: 205
tensile properties M3: 205, 760, 762
trace element impurity effect on GTA weld penetration A6: 20
wettability of commercial braze filler metals A6: 116
21-6-9, wrought heat-resistant A16: 738
21-6-9LC, composition A4: 771 A6: 458
21Cr-6Ni-9Mn, annealing A4: 773
21Cr-6Ni-9Mn, composition M6: 526
21Cr-6Ni-9Mn, liquid-copper penetration A11: 431
22Cr-13Ni-5Mn, annealing A4: 773
22Cr-13Ni-5Mn, composition M6: 526
25-6Mo
composition A6: 691
filler metals for A6: 693
properties A6: 693
25Cr, crevice corrosion M3: 58
25CR-12Ni cast, sigma-phase embrittlement A12: 150
26-1 Ti
composition M3: 9-10, 190-191
mechanical properties M3: 24-25
tensile properties M3: 194
26-1-Ti, chemical composition A6: 446 29-4
composition A6: 687
filler metals for A6: 887
properties A6: 687
26-1S, pulp and paper corrosion resistance M3: 92
26-1Ti, wrought heat-resistant A16: 738
26-1X, pulp and paper corrosion resistance M3: 92
26Cr-1Mo, corrosion in chemical solutions M3: 79-81, 84, 86
27Cr, chromic acid corrosion resistance M3: 82
29-4
composition M3: 9-10
mechanical properties M3: 24-25
29-4-2
composition M3: 9-10
mechanical properties M3: 24-25
29Cr-4Mo
composition M3: 190-191
corrosion in acids M3: 81, 84, 86, 87
corrosion in foods M3: 92
corrosion resistance, pulp and paper industry M3: 92
29Cr-4Mo, wrought heat-resistant A16: 738
29Cr-4Mo-2Ni
corrosion resistance, phosphoric acid M3: 87
corrosion resistance, pulp and paper industry M3: 92
30Cr, crevice corrosion M3: 58
44LN, composition A4: 772
200 series A16: 93
200, turning with cermet tools A16: 93
201
annealing A4: 773
composition A4: 770 A6: 457, 690 M3: 5-6
corrosion resistance, marine atmospheres M3: 67
cutoff band sawing with bimetal blades A6: 1184
filler metals for A6: 692
mechanical properties M3: 23-24

Stainless steels, specific types (continued)
metallurgy A6: 688
physical properties M3: 34-35
properties A6: 692
resistance to environments M3: 11-12
similar/dissimilar welds A6: 500
transportation equipment M3: 68-70
201, annealing M4: 624
202
annealing A4: 773
composition A4: 770 A6: 457, 690 M3: 5-6, 190-191, 756
corrosion resistance, marine atmospheres M3: 67
cutoff band sawing with bimetal blades A6: 1184
filler metals for A6: 692
mechanical properties M3: 23-24, 760
physical properties M3: 34-35
properties A6: 692
resistance to environments M3: 11-12
similar/dissimilar welds A6: 500
transportation equipment M3: 69
202, annealing M4: 624
202, machining A16: 144-147, 179, 207, 274, 323, 324, 326, 360, 362, 738
203 EZ (XM-1), composition A4: 771
203, mechanical cutting A6: 1180
203EZ (XM-1), composition A6: 458
205
composition A6: 457, 690 M3: 5-6
filler metals for A6: 692
mechanical properties M3: 23-24
physical properties M3: 34-35
properties A6: 692
resistance to environments M3: 11-12
205, composition A4: 770
215, photochemical machining A16: 588
216
composition M3: 9-10, 190-191
mechanical properties M3: 23-24
tensile properties M3: 205
216 (XM-17), composition A4: 771 A6: 458
216, wrought heat-resistant A16: 738
216L (XM-18), composition A4: 771 A6: 458
223MA, composition A6: 458
250, surface alterations A16: 27
250, turning with cermet tools A16: 93
253 MA, composition A4: 771
254 SMO
annealing A4: 773
composition A4: 771
254 SMO, annealing M4: 624
254Mo
composition A6: 690
filler metals for A6: 692
properties A6: 692
255
composition A6: 698
filler metals for A6: 698
properties A6: 698
300
ion-nitriding atmospheres A4: 565
plasma (ion) nitriding A4: 424
pot material to hold noncyanide carburizing process A4: 331
for vacuum chamber construction A4: 503, 504
300, machining A16: 231, 547
300, press-formed parts, materials for forming tools M3: 492, 493
300 series A16: 57, 93, 95, 266, 267, 281
stud arc welding A6: 215
yield strength vs. temperature A6: 1016
300 series, nitriding of powder metallurgy materials A9: 503
300/400 series, analysis of A10: 100
300/400 series, data reduction and standardization in x-ray spectrometry in A10: 100

Stainless steels, specific types (continued)
300M
composition A4: 207
heat treatment temperatures A4: 208
heat treatment(s) A4: 210-211
martempered to full hardness A4: 140
mechanical properties A4: 211
plasma (ion) nitriding A4: 424
pot material to hold noncyanide carburizing process A4: 331
properties, effect of mass on A4: 211
for vacuum chamber construction A4: 503, 504
301
annealing A4: 773
arc-welding, filler metals for M3: 49
atmospheric corrosion M3: 67-68
cold rolling, effect on tensile strength M3: 33
composition A4: 770 A6: 457, 690 M3: 56, 756
fatigue life M3: 765
filler metals for use in arc welding A6: 692, 1106
forging temperature M3: 42
gas nitriding A4: 401
machinability M3: 45
mechanical properties A6: 468 M3: 18-22, 759, 762
metallurgy A6: 688
physical properties M3: 34-35
plasma (ion) nitriding A4: 424
pot material to hold noncyanide carburizing process A4: 331
properties A6: 692
resistance to environments M3: 11-12
similar/dissimilar welds A6: 500
stress-strain curves M3: 17
thermal diffusivity from 20 to 100 °C A6: 4
transportation equipment M3: 68, 69
ultrasonic welding A6: 326
for vacuum chamber construction A4: 503, 504
301, annealing M4: 624
301, bonding to glass-phenolic laminate with nitrile phenolics EM3: 107
301, composition M6: 526
301, effect of stress intensity factor range on fatigue crack growth rate A12: 57
301, hydrogen-stress cracking A13: 1133
301 springs, strip for M1: 285
302 *See also* 18-8
annealing A4: 773
arc welding, filler metals for M3: 49
architectural applications M3: 67
cold rolling, effect on tensile strength M3: 33
composition A4: 770 A6: 457, 690 M3: 5-6 M6: 526
content, effects on weldability M6: 320
corrosion M3: 11-12, 66-68, 72-76, 77, 78, 79, 83, 87
corrosion fatigue M3: 64-65
cutoff band sawing with bimetal blades A6: 1184
electron-beam welding A6: 868, 869
fasteners M3: 183, 184
filler metals for joining A6: 692, 693
filler metals for use in arc welding A6: 1106
forging temperature M3: 42
friction welding A6: 153
gas nitriding A4: 401, 402
hardness gradients A4: 402
machinability M3: 45
mechanical cutting A6: 1179
mechanical properties A6: 468 M3: 18-22
metal powder cutting M6: 912
metallurgy A6: 688
modulus of rigidity M1: 300
physical properties M3: 34-35
plasma (ion) nitriding A4: 424
pot material to hold noncyanide carburizing process A4: 331
properties A6: 689, 692
proven application for borided ferrous materials A4: 445
resistance welding M6: 527
similar/dissimilar welds A6: 500

Stainless steels, specific types (continued)
springs M1: 284-285, 287, 290, 291, 294, 295, 298, 300, 305, 307
stud arc welding A6: 211
tensile strength, spring wire M3: 14-15
ultrasonic welding A6: 326
ultrasonic welding power requirements M6: 750
for vacuum chamber construction A4: 503, 504
302, annealing M4: 624
302, back reflection intensity A17: 238
302, effect of stress intensity factor range on fatigue crack growth rate A12: 57
302B
annealing A4: 773
arc-welding, filler metals for M3: 49
composition A4: 770 A6: 457, 690 M3: 5-6
filler metals for A6: 692
filler metals for use in arc welding A6: 1106
forging temperature M3: 42
machinability M3: 45
mechanical properties M3: 18-22
physical properties M3: 34-35
properties A6: 692
resistance to environments M3: 11-12
similar/dissimilar welds A6: 500
302B, annealing M4: 624
302B, thread grinding A16: 274
302Cu, composition A4: 770
302Cu, mechanical properties M3: 18-22
302Cw, composition A6: 457
302HQ, stud arc welding A6: 211
303 *See also* 18-8
annealing A4: 773
arc-welding, filler metals for M3: 49
composition A4: 770 A6: 457, 690 M3: 5-6, 756
corrosion in seawater M3: 80
corrosion resistance, marine atmospheres M3: 68
cutoff band sawing with bimetal blades A6: 1184
fasteners M3: 183, 184
filler metals for A6: 692
filler metals for use in arc welding A6: 1106
forging temperature M3: 42
furnace brazing A6: 916, 919
gas nitriding A4: 401
machinability M3: 44-46
mechanical properties A6: 468 M3: 18-22, 757
no stud arc welding A6: 211
physical properties M3: 34-35
plasma (ion) nitriding A4: 424
pot material to hold noncyanide carburizing process A4: 331
properties A6: 692
resistance to environments M3: 11-12
similar/dissimilar welds A6: 500
for vacuum chamber construction A4: 503, 504
303, annealing M4: 624
303, aqueous corrosion resistance A13: 834
303, cutting performance A7: 536
303, effect of content on weldability M6: 320
303 Plus X (XM-5), composition A4: 771 A6: 458
303 Plus X, composition M3: 9-10
303, sulfide inclusions in A9: 127
303, temperature effect, nitric acid corrosion A13: 221
303 valve, corrosion of A11: 182-183
303(Se) wire-rope terminal A11: 521
303F, contour band sawing A16: 362
303F, cutoff band sawing A16: 360
303F, cutoff band sawing with bimetal blades A6: 1184
303L, applications A7: 730-731
303L, dimensional change A7: 292
303S, seawater corrosion resistance M3: 74, 76
303Se
annealing A4: 773
composition A4: 770 A6: 457, 690 M3: 5-6
filler metals for A6: 692
filler metals for use in arc welding A6: 1106
mechanical properties A6: 468
properties A6: 692
resistance to environments M3: 11-12
303Se, effect of content on weldability M6: 320

Stainless steels, specific types (continued)
304 *See also* 18-8
absorptivity A6: 265
annealing A4: 773
applications A6: 395
applications, protection tubes and wells A4: 533
arc welding, filler metals for M3: 49
atmospheric corrosion M3: 66-68
brazing A6: 920
capacitor discharge stud welding A6: 222
carbide precipitation A6: 689
as cladding A6: 502
cladding for stainless steels A6: 502
composition A4: 770 A6: 69, 457 M3: 5-6, 190-191, 756 M6: 526
content, effect on weldability M6: 320
corrosion A6: 467
corrosion in chemical solutions M3: 78-91
corrosion in fertilizer M3: 83
corrosion in foods M3: 92-93
corrosion in seawater M3: 70, 71, 72, 73-74, 75, 76, 77, 79, 80
corrosion of weldments, fusion zone A6: 1065
corrosion resistance, pulp and paper industry M3: 92
creep-rupture behavior, solid-state-welded interlayers A6: 168-169, 170, 171
creep-rupture properties M3: 195, 205
cross-wire welding M6: 531-532
cutoff band sawing with bimetal blades A6: 1184
deep-penetration electron beam and laser welding A6: 22, 23
delta ferrite role in weld deposits A6: 1067
differences in space-based (Skylab) and earth-based weld samples A6: 1024
dissimilar metal joining A6: 825
electron-beam brazing A6: 922-923
electron-beam welding A6: 853, 858, 868, 869, 870
electron-beam welding in a space environment A6: 1023- 1025
fasteners M3: 183, 184
fatigue M3: 32, 733, 764
ferrite analysis A6: 1059
as filler metal for weld cladding applications A6: 809
filler metals for joining A6: 692, 693
filler metals for use in arc welding A6: 1106
flash welding M6: 557
fluid flow phenomena A6: 20, 21
flux-cored arc welding A6: 188
forging temperature M3: 42
furnace brazing A6: 918-919
gas nitriding A4: 401
gas-tungsten arc welding A6: 703, 870, 1038
heat flow in fusion welding A6: 15, 17
heat-resistant alloy applications A4: 515, 516
laser beam weld rates M6: 658
laser melt/particle inspection M6: 802
laser-beam welding A6: 262, 264
low-temperature solid-state welding A6: 300, 301
machinability M3: 44-46
mechanical properties A6: 468 M3: 18-22, 205, 751, 755, 757, 759
metallurgy A6: 688
oil quenching A4: 95
oil quenching and water contamination A4: 99
oxidation resistance M3: 196
penetration in laser beam welds M6: 654
physical properties M3: 34-35
plasma (ion) nitriding A4: 424
polyvinyl alcohol solution quenching A4: 106
pot material to hold noncyanide carburizing process A4: 331
precipitation A6: 689
probe material for quenching cooling curve analysis A4: 72
properties A6: 587, 629, 689, 692, 697
properties and compositions studied in M512 melting experiments A6: 1024
resistance to environments M3: 11-12

Stainless steels, specific types (continued)
resistance welding **M6:** 527
roll welding **A6:** 313
salt-bath dip brazing **A6:** 922
sealing compounds, effect on crevice corrosion **M3:** 75
similar/dissimilar welds **A6:** 500
stress-corrosion cracking **A6:** 477 **M3:** 63, 64, 82
stress-relieving **A4:** 776 **A6:** 469
stress-strain curves **M3:** 17
stud arc welding **A6:** 211
submerged arc welding **A6:** 703
temperature measurement by OSRLR **A6:** 1150
tensile strength, spring wire **M3:** 14-15
thermal diffusivity from 20 to 100 °C **A6:** 4
time-temperature-sensitization curves in a mixture of $CuSO_4$ and HSO_4 containing copper **A6:** 1067
torch brazing **A6:** 914
trace element impurity effect on GTA weld penetration **A6:** 20
transportation equipment **M3:** 68-70
for vacuum chamber construction **A4:** 503, 504
wear data **M3:** 590
weld cladding **A6:** 822
weld microstructure **M6:** 37
weld-metal ferrite **A6:** 462
weldability **A6:** 500, 501
wettability of commercial braze filler metals **A6:** 116
304, annealing **M4:** 624
304, AOD refining parameters **A15:** 428
304, aqueous corrosion resistance **A13:** 834
304 austenitic, SCC of **A11:** 299
304, average crack propagation rate **A13:** 154
304, biological pitting **A13:** 119
304 bone screw, pitting corrosion **A11:** 687, 691
304, chloride SCC **A13:** 354
304 clad, corrosion control **A13:** 888-889
304, corrosive attack in HAZ of weld in **A11:** 428
304, crack initiation defects **A13:** 148
304, crack propagation rate/strain rate **A13:** 160
304, crevice corrosion **A13:** 109
304, cyclic potentiodynamic polarization curves **A13:** 218
304, density with high-energy compacting **A7:** 305
304, depth-composition profiles **A10:** 555
304, die life in upset forging **A14:** 228
304, dissolved oxygen and electrochemical potential **A13:** 932
304, effect of frequency and wave form on fatigue properties **A12:** 59, 60
304 foil, surface segregation in **A10:** 564-566
304, frequencies for eddy current inspection **A17:** 182
304, galvanic corrosion **A13:** 1297
304, grain-boundary segregation measurement **A13:** 157
304 HN (XM-21), composition **A6:** 458
304, hot-rolled and recrystallized **A10:** 470
304 integral-finned tube, cracked from chlorides and residual stresses **A11:** 636
304, intergranular SCC in **A11:** 204
304, intergranular/transgranular cracking **A13:** 152
304, liquid lithium corrosion **A13:** 93
304, microbial intergranular cracking **A13:** 315
304, molten salt corrosion **A13:** 53
304 pipe, SCC by residual welding stresses **A11:** 624-625
304 sensitized cerclage wire, intercrystalline corrosion on **A11:** 676, 681
304, sensitized, grain boundaries **A13:** 930
304 sensitized, intergranular attack **A11:** 180, 181
304, shielded metal arc weld **A9:** 583

Stainless steels, specific types (continued)
304 tee fitting, low-cycle thermal fatigue **A11:** 622
304, time-temperature-sensitization curves **A13:** 551
304 tubing, ultrasonic inspection **A17:** 565
304, weight loss, by intergranular corrosion **A13:** 123
304 wire, SCC failure **A12:** 133, 152
304-L, refined, composition **A15:** 427
304B4, composition **A4:** 771
304BI, composition **A6:** 458
304Cb, weldability **A6:** 501
304H
composition **A6:** 457, 690 **M3:** 5-6
filler metals for **A6:** 692
properties **A6:** 692
resistance to environments **M3:** 11-12
304H, composition **A4:** 770
304HN
composition **M3:** 9-10
mechanical proper-ties **M3:** 23-24
304HN (XM-21), composition **A4:** 771
304L
annealing **A4:** 773
arc welding, filler metals for **M3:** 49
brazing **A6:** 944
brazing with copper cladding **A6:** 347
carbide precipitation resistance **M6:** 321
clad brazing material, application of **A6:** 962
cladding **A6:** 502
composition **A4:** 770 **A6:** 457, 690 **M3:** 5-6, 190-191, 756 **M6:** 526
content, effect on weldability **M6:** 320
corrosion in nitric acid **M3:** 84-85
diffusion brazing **A6:** 343
electron-beam welding **A6:** 462, 464
explosion welding **A6:** 303
explosion welding, interface failure of trilayer **A6:** 163, 164
fatigue at subzero temperatures **M3:** 764, 765
fatigue strength for gas-tungsten arc welds **A6:** 1018
filler metals for joining **A6:** 692, 693
filler metals for use in arc welding **A6:** 1106
gas-tungsten arc welding **A6:** 19
gas-tungsten arc welding and fluid flow phenomena **A6:** 21
machinability **M3:** 45
mechanical properties **A6:** 468 **M3:** 18-22, 757, 759
physical properties **M3:** 34-35
plasma arc welding **A6:** 199
properties **A6:** 692
resistance to environments **M3:** 11-12
resistance welding **M6:** 527
solid-state-welded interlayers **A6:** 169
stress-relieving **A6:** 469
thermal cycling effects **M6:** 712
to avoid carbide formation and weld decay **A6:** 1066
torch brazing **A6:** 914
upset welding **A6:** 250
weld (vertical) characterization **A6:** 99
weld cladding **A6:** 809, 818
weld overlays **M6:** 810
weld-metal ferrite **A6:** 462
weldability **A6:** 468
wettability of commercial braze filler metals **A6:** 116
304L (P/M), machining **A16:** 883-886, 890
304L, annealing **M4:** 624
304L, applications **A7:** 730-731
304L, copper-containing, corrosion resistance of **A13:** 832
304L, corrosion resistance **A13:** 829
304L, dimensional changes **A7:** 730
304L, flow rate through Hall and Carney funnels **A7:** 279

Stainless steels, specific types (continued)
304L, galvanic/intergranular corrosion, space shuttle orbiter **A13:** 1068
304L, insert for pump impeller **A7:** 622
304L, microbiological corrosion **A13:** 355
304L, molten lithium corrosion **A13:** 52-54
304L P/M, weight loss and corrosion time **A13:** 834
304L, similar/dissimilar welds **A6:** 500
304L tin-modified, corrosion resistance **A13:** 829
304L vacuum-sintered, copper alloying effect **A13:** 832
304L, water-atomized **A7:** 728
304LN
annealing **A4:** 773
composition **A4:** 770 **A6:** 457, 690 **M3:** 5-6
filler metals for **A6:** 692
mechanical properties **M3:** 18-22
properties **A6:** 689, 692
304LSi, applications **A7:** 731
304N
annealing **A4:** 773
composition **A4:** 770 **A6:** 457, 690 **M3:** 5-6, 190-191
filler metals for **A6:** 692
mechanical properties **M3:** 18-22, 23-24
physical properties **M3:** 34-35
properties **A6:** 692
resistance to environments **M3:** 11-12
304N, annealing **M4:** 624
304N, wrought heat-resistant **A16:** 738
305 *See also* 18-8
annealing **A4:** 773
arc welding, filler metals for **M3:** 49
cold rolling, effect on tensile strength **M3:** 33
composition **A4:** 770 **A6:** 457, 690 **M3:** 5-6
fasteners **M3:** 184
filler metals for joining **A6:** 692, 693
filler metals for use in arc welding **A6:** 1106
forging temperature **M3:** 42
furnace brazing **A6:** 916
mechanical properties **A6:** 468 **M3:** 18-22
physical properties **M3:** 34-35
plasma (ion) nitriding **A4:** 424
pot material to hold noncyanide carburizing process **A4:** 331
properties **A6:** 692
resistance to environments **M3:** 11-12
similar/dissimilar welds **A6:** 500
stud arc welding **A6:** 211
tensile strength, spring wire **M3:** 14-15
for vacuum chamber construction **A4:** 503, 504
305, annealing **M4:** 624
305L, dimensional change **A7:** 292
306
composition **A6:** 690
filler metals for **A6:** 692
properties **A6:** 692
308
annealing **A4:** 773
arc welding, filler metals for **M3:** 49
atmospheric corrosion **M3:** 67
cladding **A6:** 503
cold rolling, effect on tensile strength **M3:** 33
composition **A4:** 770 **A6:** 690 **M3:** 5-6, 9-10
corrosion in seawater **M3:** 71
cutoff band sawing with bimetal blades **A6:** 1184
electrodes for flux-cored arc welding **A6:** 189
Ferrite Number **A6:** 677
ferrite percentage **A6:** 809
filler metal for oxide-dispersion-strengthened materials **A6:** 1038
as filler metals **A6:** 692, 695-696
filler metals for use in arc welding **A6:** 1106
forging temperature **M3:** 42

Stainless steels, specific types (continued)
gas nitriding ... **A4:** 401
gas-tungsten arc welding ... **A6:** 465, 466
mechanical properties ... **A6:** 468 **M3:** 18-22
physical properties ... **M3:** 34-35
plasma (ion) nitriding ... **A4:** 424
plasma-MIG welding ... **A6:** 224
pot material to hold noncyanide carburizing process ... **A4:** 331
properties ... **A6:** 692
resistance to environments ... **M3:** 11-12
similar/dissimilar welds ... **A6:** 500
stud arc welding ... **A6:** 211
for vacuum chamber construction ... **A4:** 503, 504
weld morphology ... **A6:** 462
weldability ... **A6:** 502
308, annealing ... **M4:** 624
308, machining ... **A16:** 144-147, 179, 207, 274, 323, 324, 326, 360, 362
308L
cladding ... **A6:** 504
composition ... **M3:** 9-10
Ferrite Number ... **A6:** 677
in ferritic base metal-filler metal combinations ... **A6:** 449
filler metal for stainless steel casting alloys ... **A6:** 496
mechanical properties ... **M3:** 18-22
weld cladding ... **A6:** 818
308L, pitting underalloyed weld metal ... **A13:** 349
309
annealing ... **A4:** 773
arc welding, filler metals for ... **M3:** 49
buttering material ... **A6:** 501
cladding ... **A6:** 503
cold rolling, effect on tensile strength ... **M3:** 33
composition ... **A4:** 770 **A6:** 457, 690 **M3:** 5-6, 190-191 **M6:** 526
corrosion ... **M3:** 11-12, 67, 70, 71, 78-81, 84-87
creep-rupture properties ... **M3:** 195, 205
cutoff band sawing with bimetal blades ... **A6:** 1184
dissimilar metal joining ... **A6:** 827
fasteners ... **M3:** 184
Ferrite Number ... **A6:** 677, 678
ferrite percentage ... **A6:** 809
filler metal ... **A6:** 443, 500-501, 502, 679, 692, 695-696
as filler metal for weld cladding applications ... **A6:** 809
filler metals for use in arc welding ... **A6:** 1106
flash welding ... **M6:** 557
forging temperature ... **M3:** 42
gas nitriding ... **A4:** 401
heat-resistant alloy applications ... **A4:** 515, 516, 517
machinability ... **M3:** 45
mechanical properties ... **A6:** 468 **M3:** 18-22
oxidation resistance ... **M3:** 196
physical properties ... **M3:** 34-35
plasma (ion) nitriding ... **A4:** 424
plasma-MIG welding ... **A6:** 224
pot material to hold noncyanide carburizing process ... **A4:** 331
properties ... **A6:** 692, 693
similar/dissimilar welds ... **A6:** 500
stress-relieving ... **A4:** 774
stud arc welding ... **A6:** 211
tensile properties ... **M3:** 205
for vacuum chamber construction ... **A4:** 503, 504
weld cladding ... **A6:** 818
weldability ... **A6:** 500, 501, 502
309, annealing ... **M4:** 624
309 Cb+Ta, composition ... **M3:** 9-10
309, elevated temperature behavior ... **M1:** 93-94
309 lead bath pan, premature failure ... **A11:** 274-275
309, x-ray spectrometric results ... **A10:** 100
309C, annealing ... **M4:** 624
309Cb
composition ... **A6:** 690
ferrite percentage ... **A6:** 809
filler metals for ... **A6:** 501-502, 692
properties ... **A6:** 692
309Cb, stress-relieving ... **A4:** 774

Stainless steels, specific types (continued)
309H
composition ... **A6:** 690
filler metals for ... **A6:** 692
properties ... **A6:** 692
309HCb
composition ... **A6:** 690
filler metals for ... **A6:** 692
properties ... **A6:** 692
309L
cladding ... **A6:** 504
in ferritic base metal-filler metal combinations ... **A6:** 449
filler metal ... **A6:** 500-502, 679
309LSi
in ferritic base metal-filler metal combinations ... **A6:** 449
309Mo
as filler metal ... **A6:** 501-502
weldability ... **A6:** 501, 502
309MoL, weldability ... **A6:** 501
309S
annealing ... **A4:** 773
arc welding, filler metals for ... **M3:** 49
composition ... **A4:** 512, 770 **A6:** 457, 690 **M3:** 5-6, 9-10
filler metals for ... **A6:** 692
filler metals for use in arc welding ... **A6:** 1106
machinability ... **M3:** 45
mechanical properties ... **A6:** 468
precipitation ... **A6:** 689
properties ... **A6:** 692
resistance to environments ... **M3:** 11-12
similar/dissimilar welds ... **A6:** 500
309S (Nb) evaporator tube, defective weld seam ... **A11:** 432
309S, annealing ... **M4:** 624
309S Cb, composition ... **A4:** 771
309S-Cb
composition ... **M3:** 10
corrosion in nitric acid ... **M3:** 85
309SCb, composition ... **A6:** 458
310
acetic acid corrosion resistance ... **M3:** 79
annealing ... **A4:** 773
arc welding, filler metals for ... **M3:** 49
atmospheric corrosion, industrial sites ... **M3:** 66, 67
cladding ... **A6:** 503
cold rolling, effect on tensile strength ... **M3:** 33
composition ... **A4:** 770 **A6:** 457, 690 **M3:** 5-6, 190-191, 756 **M6:** 526
corrosion in seawater ... **M3:** 70, 71, 79
corrosion in silver nitrate ... **M3:** 75-76, 90
creep-rupture properties ... **M3:** 195, 205
cutoff band sawing with bimetal blades ... **A6:** 1184
delta ferrite role in weld deposits ... **A6:** 1067
dissimilar metal joining ... **A6:** 827
fasteners ... **M3:** 184
fatigue behavior, rotation-beam ... **M3:** 32
fatigue life ... **M3:** 765
ferrite percentage ... **A6:** 809
as filler metal for weld cladding applications ... **A6:** 809
filler metals for ... **A6:** 491, 501-502
filler metals for use in arc welding ... **A6:** 1106
flash welding ... **M6:** 557
forging temperature ... **M3:** 42
heat-resistant alloy applications ... **A4:** 515
laser beam welding ... **M6:** 662
laser-beam welding ... **A6:** 876
materials for parts and fixtures in nitriding furnaces ... **A4:** 398
mechanical properties ... **A6:** 468
nitric acid corrosion resistance ... **M3:** 85
oxidation resistance ... **M3:** 196
phenol corrosion resistance ... **M3:** 87
physical properties ... **M3:** 34-35
plasma (ion) nitriding ... **A4:** 424
Poisson's ratio ... **M3:** 755
pot material to hold noncyanide carburizing process ... **A4:** 331
precipitation ... **A6:** 689
properties ... **A6:** 692, 693
resistance to environments ... **M3:** 11-12

Stainless steels, specific types (continued)
similar/dissimilar welds ... **A6:** 500
stress-relieving ... **A4:** 774
stud arc welding ... **A6:** 211
tensile properties ... **M3:** 205, 757, 759, 762
to avoid carbide formation and weld decay ... **A6:** 1066
for vacuum chamber construction ... **A4:** 503, 504
water quenching required ... **A4:** 769
weld cladding ... **A6:** 809, 818
weld morphologies ... **A6:** 462
Young's modulus ... **M3:** 751
310 (ELC), in ferritic base metal-filler metal combination ... **A6:** 449
310, annealing ... **M4:** 624
310, ringlike forging ... **A14:** 225
310Cb
composition ... **A6:** 458, 690
filler metals for ... **A6:** 692
mechanical properties ... **A6:** 468
properties ... **A6:** 692
310Cb, composition ... **A4:** 771
310H
composition ... **A6:** 690
filler metals for ... **A6:** 692
properties ... **A6:** 692
310HCb
composition ... **A6:** 690
filler metals for ... **A6:** 692
properties ... **A6:** 692
310Mo, hydrochloric acid corrosion resistance ... **M3:** 83-84
310MoLN
composition ... **A6:** 690
filler metals for ... **A6:** 692
properties ... **A6:** 692
310S
annealing ... **A4:** 773
arc welding, filler metals for ... **M3:** 49
composition ... **A4:** 512, 770 **A6:** 457, 690 **M3:** 5-6, 756
fatigue-crack-growth rate ... **M3:** 764
filler metals for ... **A6:** 692
filler metals for use in arc welding ... **A6:** 1106
fracture toughness ... **M3:** 763
machinability ... **M3:** 45
mechanical properties ... **A6:** 468
properties ... **A6:** 692
resistance to environments ... **M3:** 11-12
similar/dissimilar welds ... **A6:** 500
tensile properties ... **M3:** 758, 762
310S, annealing ... **M4:** 624
312
composition ... **A6:** 474 **M3:** 9-10
ferrite percentage ... **A6:** 809
as filler metal ... **A6:** 501-502, 679, 680
laser-beam welding ... **A6:** 463
mechanical properties ... **M3:** 18-22
pitting resistance equivalent values ... **A6:** 474
plasma (ion) nitriding ... **A4:** 424
pot material to hold noncyanide carburizing process ... **A4:** 331
properties ... **A6:** 475
stress-relieving ... **A4:** 774
for vacuum chamber construction ... **A4:** 503, 504
weld cladding ... **A6:** 818
weldability ... **A6:** 500, 501, 502
312 weld metal, sigma-phase embrittlement ... **A12:** 151
314
composition ... **A4:** 770 **A6:** 457, 690 **M3:** 5-6 **M6:** 526
cutoff band sawing with bimetal blades ... **A6:** 1184
filler metals for ... **A6:** 692
flash welding ... **M6:** 557
forging temperature ... **M3:** 42
heat-resistant alloy applications ... **A4:** 516
mechanical properties ... **M3:** 18-22
physical properties ... **M3:** 34-35
plasma (ion) nitriding ... **A4:** 424
pot material to hold noncyanide carburizing process ... **A4:** 331
properties ... **A6:** 692
resistance to environments ... **M3:** 11-12
similar/dissimilar welds ... **A6:** 500

Stainless steels, specific types (continued)
for vacuum chamber construction...... **A4:** 503, 504
314, annealing **M4:** 624
314, contour band sawing **A16:** 362
314, cutoff band sawing **A16:** 360
314, thread grinding **A16:** 274
315
composition **A6:** 474
corrosion **A6:** 477
pitting resistance equivalent values **A6:** 474
properties **A6:** 475
316 *See also* 18-8Mo
annealing **A4:** 773
applications **A6:** 395
applications, protection tubes and wells **A4:** 533
arc welding, filler metals for **M3:** 49
cladding **A6:** 502, 503
composition **A4:** 770 **A6:** 457, 690, 1017 **M3:** 5-6, 190-191, 756 **M6:** 526
content, effect on weldability **M6:** 320
corrosion in atmospheres **M3:** 66, 67
corrosion in chemical solutions **M3:** 79-90
corrosion in foods **M3:** 92-93
corrosion in paper industry **M3:** 92
corrosion in pharmaceuticals **M3:** 91
corrosion in seawater **M3:** 70, 71, 72, 74, 76, 77, 78, 79
corrosion in silver nitrate **M3:** 87
creep-rupture properties **M3:** 195, 205
crevice corrosion **M3:** 58
cryogenic treatment **A4:** 205
cutoff band sawing with bimetal blades **A6:** 1184
dissimilar metal joining **A6:** 823
dissolved by 1% HCl solution **A6:** 585
electrodes for flux-cored arc welding **A6:** 189
fasteners **M3:** 184
fatigue life **M3:** 765
fatigue-crack-growth rate **M3:** 764
ferrite analysis **A6:** 1059
Ferrite Number **A6:** 677
ferrite percentage **A6:** 809
as filler metal for weld cladding applications **A6:** 809
filler metals for **A6:** 692
filler metals for use in arc welding **A6:** 1106
flash welding **M6:** 557
flux-cored arc welding **A6:** 188
forging temperature **M3:** 42
gas nitriding **A4:** 401
gas-tungsten arc welding **A6:** 703
hardfacing coating material **A6:** 808
heat-resistant alloy applications **A4:** 516
laser beam welding **M6:** 662
Laves phase, effect on impact strength **M3:** 229
machinability **M3:** 44-46
mechanical properties ... **A6:** 468 **M3:** 14-15, 18-22, 205, 758
physical properties **M3:** 34-35
plasma (ion) nitriding **A4:** 424
Poisson's ratio **M3:** 755
pot material to hold noncyanide carburizing process **A4:** 331
precipitation **A6:** 689
properties **A6:** 692
proven applications for borided ferrous materials **A4:** 445
resistance to environments **M3:** 11-12
resistance welding **M6:** 527
similar/dissimilar welds **A6:** 500
stress-corrosion cracking **A6:** 477 **M3:** 63-64
stress-relieving **A4:** 776 **A6:** 469
stud arc welding **A6:** 211
temperature measurements by OSRLR **A6:** 1150
thermal diffusivity from 20 to 100 °C........ **A6:** 4
thermal spray processing **A6:** 806

Stainless steels, specific types (continued)
trace element impurity effect on GTA weld penetration **A6:** 20
ultrasonic welding **A6:** 326
for vacuum chamber construction...... **A4:** 503, 504
wear data **M3:** 590
wettability of commercial braze filler metals **A6:** 116
yield strength vs. fracture toughness **A6:** 1017
Young's modulus **M3:** 751
316, annealing **M4:** 624
316, apparent density **A7:** 272
316, calcium chloride pitting **A13:** 113
316, controlled spray deposition **A7:** 532
316, conventional SADP and ZOLZ-CBEDP in **A10:** 441
316, crevice corrosion **A13:** 109
316, ductile-to-brittle transition **A11:** 84, 86
316, effect of frequency and wave form on fatigue properties **A12:** 59, 60
316, effect of stress intensity factor range on fatigue crack growth rate **A12:** 57
316, effect of vacuum on fatigue **A12:** 48
316, fatigue fracture in air/vacuum **A12:** 48, 55
316 heat-exchanger shell, hot shortness failure **A11:** 640
316, in molten lithium, weight changes **A13:** 52
316, liquid lead corrosion **A13:** 96
316, liquid lithium corrosion **A13:** 94
316, liquid sodium corrosion **A13:** 90
316 low-carbon remelted, intramedullary tibia nail **A11:** 671, 680
316, mass transfer, liquid-metal corrosion **A13:** 57
316 piping, SCC failure at welds **A11:** 216
316, preferential biological corrosion **A13:** 117
316, preferential weldment attack, HAZ attack **A13:** 347
316, shallow crevice corrosion
316, springs **M1:** 284, 285, 290
316 tubing, SCC failure from chloride-contaminated steam condensate **A11:** 625-626
316Cb
composition **A6:** 458, 690
ferrite percentage **A6:** 809
filler metals for **A6:** 692
filler metals for use in arc welding **A6:** 1106
properties **A6:** 692
316Cb, composition **A4:** 771
316F
composition **M3:** 5-6
mechanical properties **M3:** 18-22
resistance to environments **M3:** 11-12
316F, composition **A4:** 770 **A6:** 457
316H
composition **A6:** 457, 690 **M3:** 5-6
filler metals for **A6:** 692
properties **A6:** 692
resistance to environments **M3:** 11-12
316H, composition **A4:** 770
316HQ, composition **A4:** 771 **A6:** 458
316L
annealing **A4:** 773
arc welding, filler metals for **M3:** 49
brazing **A6:** 943
carbide precipitation resistance **M6:** 321
composition **A4:** 770 **A6:** 457, 690 **M3:** 5-6, 190-191 **M6:** 526
content, effect on weldability **M6:** 320
corrosion **A6:** 465, 466, 467
corrosion in chemical solutions **M3:** 81, 85
corrosion in foods **M3:** 92-93
explosion welding, wave morphology of trilayer **A6:** 162
fatigue strength for gas-tungsten arc welds **A6:** 1018

Stainless steels, specific types (continued)
filler metal for stainless steel casting alloy **A6:** 496
as filler metal for weld cladding applications **A6:** 809
filler metals for **A6:** 692
filler metals for use in arc welding **A6:** 1106
gas-tungsten arc welding **A6:** 686
laser-beam welding **A6:** 463
machinability **M3:** 45
mechanical properties **A6:** 468 **M3:** 18-22
physical properties **M3:** 34-35
plasma arc welding **A6:** 199
properties **A6:** 689, 692, 694, 697
resistance to environments **M3:** 11-12
resistance welding **M6:** 527
sigma-phase formation susceptibility **A4:** 772
similar/dissimilar welds **A6:** 500
stress-relieving **A4:** 776 **A6:** 469
temperature measurement, validation strategies **A6:** 1149
time-temperature-precipitation diagram **M3:** 226, 228
to avoid carbide formation and weld decay **A6:** 1066
316L (P/M), drilling **A16:** 886
316L, annealing **M4:** 624
316L, applications **A7:** 730-731
316L, Auger spectra **A7:** 251-255
316L, composition-depth profiles **A7:** 252
316L, dimensional change **A7:** 292
316L, effect of sintering temperature on elongation and dimensional change **A7:** 369
316L, effect of sintering time on tensile and yield strengths **A7:** 370
316L, ejector pad for refrigerator automatic icemaker **A7:** 731
316L, fracture by hot brine exposure **A11:** 432
316L, injection molded **A7:** 495
316L, nitrogen atomized **A7:** 101
316L P/M, Auger composition depth profile **A13:** 830
316L P/M, sintering temperature effects on tensile/yield strengths, apparent hardness **A13:** 825
316L, pitting **A13:** 561
316L, powder compacts, lubricant effect **A7:** 191
316L, SCC striations **A11:** 28
316L, shear strength as function of porosity **A7:** 697
316L sintered, microstructure **A13:** 827
316L, sintering atmosphere **A7:** 729
316L, sintering time effects **A13:** 826
316L, tensile properties **A7:** 729, 730
316L, tin-modified type **A7:** 257
316L vacuum-sintered, density effect on corrosion **A13:** 831
316L wrought/sintered, corrosion resistance **A13:** 825
316L-P (porous), tensile and shear strength as function of density **A7:** 696
316LM
composition **A6:** 458, 691
filler metals for **A6:** 692
properties **A6:** 692
316LMN
composition **A6:** 691
filler metals for **A6:** 693
properties **A6:** 693
316LN
annealing **A4:** 773
composition **A4:** 770 **A6:** 458, 691 **M3:** 5-6
cryogenic service **A6:** 1017
filler metals for **A6:** 693
gas-tungsten arc welding **A6:** 1018
mechanical properties **M3:** 18-22
properties **A6:** 693

Stainless steels, specific types (continued)
quality index vs. inclusion spacing **A6:** 1018
weldability **A6:** 466
316LR bone plate, fretting and fretting corrosion **A11:** 688, 691-692
316LR bone plate, top surface **A11:** 680, 684
316LR bone screw hole, with fretting and fretting corrosion **A11:** 688, 691-692
316LR bone screws, fatigue failure **A11:** 679, 682
316LR cold-worked, effect of load on fatigue-surface damage **A11:** 684, 689
316LR intramedullary tibia nail, fatigue striations and corrosion pits **A11:** 683, 688
316LR, *S-N* fatigue curves for cold-worked **A11:** 683, 688-689
316LR screw, with shearing fracture **A11:** 676, 682
316LR straight bone plate, fatigue initiation **A11:** 679, 684-686
316LR test specimen, development of fatigue- surface damage **A11:** 684, 688
316LSi, applications **A7:** 730-731
316N
annealing **A4:** 773
composition **A4:** 770 **A6:** 457, 690 **M3:** 5-6, 190-191
filler metals for **A6:** 692
magnetic permeability **A4:** 773
mechanical properties **M3:** 23-24
physical properties **M3:** 34-35
properties **A6:** 692
resistance to environments **M3:** 11-12
316N, annealing **M4:** 624
316N, wrought heat-resistant **A16:** 738
316Ti
boriding **A4:** 445
composition **A4:** 771 **A6:** 458, 690
filler metals for **A6:** 692
properties **A6:** 692
317
annealing **A4:** 773
arc welding, filler metals for **M3:** 49
atmospheric corrosion **M3:** 66, 67
composition **A4:** 770 **A6:** 457, 691 **M3:** 5-6, 190-191 **M6:** 526
content, effect on weldability **M6:** 320
corrosion in chemical solutions **M3:** 79, 80, 81-82, 83-84, 89, 90
corrosion in paper industry **M3:** 92
corrosion in seawater **M3:** 71
cutoff band sawing with bimetal blades **A6:** 1184
fasteners **M3:** 184
Ferrite Number **A6:** 677
ferrite percentage **A6:** 809
filler metal for weld cladding applications **A6:** 809
filler metals for **A6:** 692
filler metals for use in arc welding **A6:** 1106
forging temperature **M3:** 42
mechanical properties **A6:** 468 **M3:** 18-22
physical properties **M3:** 34-35
plasma (ion) nitriding **A4:** 424
pot material to hold noncyanide carburizing process **A4:** 331
properties **A6:** 692
resistance to environments **M3:** 11-12
resistance welding **M6:** 527
shielded metal arc welding **A6:** 746
similar/dissimilar welds **A6:** 500
for vacuum chamber construction **A4:** 503, 504
317, annealing **M4:** 624
317L
annealing **A4:** 773
arc welding, filler metals for **M3:** 49
composition **A4:** 770 **A6:** 457, 691 **M3:** 5-6
as filler metal for weld cladding applications **A6:** 809
filler metals for **A6:** 692
filler metals for use in arc welding **A6:** 1106
mechanical properties **A6:** 468 **M3:** 18-22
physical properties **M3:** 34-35
properties **A6:** 689, 692
resistance to environments **M3:** 11-12
sigma-phase formation susceptibility **A4:** 772

Stainless steels, specific types (continued)
similar/dissimilar welds **A6:** 500
317L, annealing **M4:** 624
317L plus, annealing **A4:** 773 **M4:** 624
317L, selective chloride SCC attack **A13:** 353
317L4, annealing **M4:** 624
317LM
annealing **A4:** 773
composition **A4:** 771 **M3:** 9-10
mechanical properties **M3:** 18-22
317LM, annealing **M4:** 624
317LMO, annealing **A4:** 773 **M4:** 624
317LN, composition **A4:** 771
317LX, annealing **A4:** 773 **M4:** 624
318
ferrite percentage **A6:** 809
filler metals for use in arc welding **A6:** 1106
Schaeffler diagram **A6:** 809
318, annealing **M4:** 624
318, arc welding, filler metals for **M3:** 49
318, effect of content on weldability **M6:** 320
320
cladding **A6:** 503
composition **A6:** 691
filler metals for **A6:** 693
gas-metal arc welding **A6:** 694
properties **A6:** 693
321
annealing **A4:** 773
arc welding, filler metals for **M3:** 49
atmospheric corrosion **M3:** 67-68
composition **A4:** 770 **A6:** 457, 691 **M3:** 5-6, 190-191, 756 **M6:** 526
content, effect on weldability **M6:** 320
corrosion in chemical solutions **M3:** 79-80, 81, 83
corrosion in seawater **M3:** 71
corrosion resistance **A6:** 466
creep-rupture properties **M3:** 193
cutoff band sawing with bimetal blades **A6:** 1184
fasteners **M3:** 184
as filler metal for weld cladding applications **A6:** 809
filler metals for **A6:** 693
filler metals for use in arc welding **A6:** 1106
flash welding **M6:** 557
forging temperature **M3:** 42
gas nitriding **A4:** 401, 402
hardness gradients **A4:** 402
induction brazing **A6:** 321
laser-beam welding **A6:** 264
machinability **M3:** 45
mechanical properties **A6:** 468
oxidation resistance **M3:** 196
physical properties **M3:** 34-35
plasma (ion) nitriding **A4:** 424
pot material to hold noncyanide carburizing process **A4:** 331
for pots for molten salt baths of aluminum alloys **A4:** 873
properties **A6:** 689, 693
resistance to environments **M3:** 11-12
resistance welding **M6:** 527
similar/dissimilar welds **A6:** 500
solid-state-welded interlayers **A6:** 169
stress-relieving **A6:** 469
stud arc welding **A6:** 211
tensile properties **M3:** 205, 758
to avoid carbide formation and weld decay **A6:** 1066
for vacuum chamber construction **A4:** 503, 504
321 aircraft freshwater storage tank, pitting failure **A11:** 177
321, annealing **M4:** 624
321 bellows expansion joint, fatigue cracking **A11:** 131-133
321, brazed joint failure, from inadequate cleaning **A11:** 452-453
321, carbide precipitation in **A11:** 451
321, containing titanium carbonitride blocky inclusions **A11:** 30
321, effect of frequency and wave form on fatigue properties **A12:** 59, 60
321 elbow assembly, weld cracking **A11:** 117
321 fuel-nozzle-support assembly, cracking in **A11:** 429

Stainless steels, specific types (continued)
321 fuel-nozzle-support, weld-repair cracking **A11:** 430
321 heat-exchanger bellows, fatigue failure **A11:** 640
321 radar coolant system assembly, brazed joint failure in **A11:** 452-453
321, SCC of brazed joint **A11:** 453
321, superheater tube, thick-lip stress rupture **A11:** 605-606
321 type welded liners, fatigue fracture **A11:** 118
321H
composition **A6:** 457, 691 **M3:** 5-6
filler metals for **A6:** 693
mechanical properties **A6:** 468
properties **A6:** 693
resistance to environments **M3:** 11-12
321H, composition **A4:** 770
325, seawater corrosion resistance **M3:** 71
327, to avoid carbide formation and weld decay **A6:** 1066 329
composition **A6:** 474, 698
filler metals for **A6:** 698
metallurgy **A6:** 697
pitting resistance equivalent values **A6:** 474
properties **A6:** 475, 698
329
annealing **A4:** 778
composition **A4:** 770 **M3:** 5-6
corrosion in hydrochloric acid **M3:** 83-84
corrosion in seawater **M3:** 71
mechanical properties **M3:** 18-22
physical properties **M3:** 34-35
plasma (ion) nitriding **A4:** 424
pot material to hold noncyanide carburizing process **A4:** 331
resistance to environments **M3:** 11-12
stress-relieving **A4:** 774
for vacuum chamber construction **A4:** 503, 504
329, electroslag-remelted **A7:** 549
329, ferritic-austenitic, STAMP-processed **A7:** 548-549
330
composition **A4:** 770 **A6:** 457, 691 **M3:** 5-6
cutoff band sawing with bimetal blades **A6:** 1184
filler metals for **A6:** 693
forging temperature **M3:** 42
materials for parts and fixtures in nitriding furnaces **A4:** 398
mechanical properties **A6:** 468 **M3:** 18-22
physical properties **M3:** 34-35
plasma (ion) nitriding **A4:** 424
pot material to hold noncyanide carburizing process **A4:** 331
properties **A6:** 693
resistance to environments **M3:** 11-12
similar/dissimilar welds **A6:** 500
for vacuum chamber construction **A4:** 503, 504
330, contour band sawing **A16:** 362
330, cutoff band sawing **A16:** 360
330, thread grinding **A16:** 274
330HC
composition **M3:** 9-10
mechanical properties **M3:** 18-22
332
composition **A4:** 771 **A6:** 458 **M3:** 9-10
mechanical properties **A6:** 468 **M3:** 18-22
plasma (ion) nitriding **A4:** 424
pot material to hold noncyanide carburizing process **A4:** 331
for vacuum chamber construction **A4:** 503, 504
347
annealing **A4:** 773
annealing temperature **M4:** 624
applications **A4:** 402
arc welding, filler metals for **M3:** 49
atmospheric corrosion **M3:** 66, 67
cladding **A6:** 502, 503, 504
composition **A4:** 770 **A6:** 457, 691 **M3:** 5-6, 190-191, 756 **M6:** 526
content, effect on weldability **M6:** 320
corrosion in chemical solutions **M3:** 78-88
corrosion in seawater **M3:** 70, 71, 74
corrosion resistance **A6:** 466
creep-rupture properties **M3:** 193, 205

Stainless steels, specific types (continued)
cutoff band sawing with bimetal blades ... **A6:** 1184
ductility dip cracking ... **A6:** 465
fasteners ... **M3:** 184
fatigue life ... **M3:** 765
Ferrite Number ... **A6:** 677
ferrite percentage ... **A6:** 809
as filler metal for weld cladding applications ... **A6:** 809
filler metals for ... **A6:** 693
filler metals for use in arc welding ... **A6:** 1106
flash welding ... **M6:** 557
flux-cored arc welding ... **A6:** 188
forging temperature ... **M3:** 42
furnace brazing ... **A6:** 915, 916-917, 919
gas nitriding ... **A4:** 401, 403
gas-tungsten arc welding ... **A6:** 705
HAZ liquation cracking ... **A6:** 464
heat-resistant alloy applications ... **A4:** 516
interface formation ... **A6:** 146
machinability ... **M3:** 45
mechanical properties ... **A6:** 468
oxidation resistance ... **M3:** 196
physical properties ... **M3:** 34-35
plasma (ion) nitriding ... **A4:** 424
plasma-MIG welding ... **A6:** 224
pot material to hold noncyanide carburizing process ... **A4:** 331
for pots for molten salt baths of aluminum alloys ... **A4:** 873
properties ... **A6:** 689, 693
resistance to environments ... **M3:** 11-12
resistance welding ... **M6:** 527
roll welding ... **A6:** 313
similar/dissimilar welds ... **A6:** 500
stress relieving ... **M4:** 648
stress-corrosion cracking ... **M3:** 63-64
stress-relieving ... **A4:** 774
stud arc welding ... **A6:** 211
tensile properties ... **M3:** 205, 758
for vacuum chamber construction ... **A4:** 503, 504
weld cladding ... **A6:** 820
weldability ... **A6:** 500
347, carbide precipitation in ... **A11:** 451
347, effect of frequency and wave form on fatigue properties ... **A12:** 59, 60
347, for steam generator tube support structures ... **A13:** 938
347 inlet header, poor welding crack ... **A11:** 430
347 pipeline, grain-boundary embrittlement ... **A11:** 132
347, pressure-probe housing, brazed joint failure ... **A11:** 454
347 shaft, in hydrogen-bypass valve, chloride SCC in ... **A11:** 660
347, wavelength-dispersive spectrum of ... **A10:** 87, 88
347F, machinability ... **M3:** 45
347H
composition ... **A6:** 457, 691 **M3:** 5-6
filler metals for ... **A6:** 693
properties ... **A6:** 693
resistance to environments ... **M3:** 11-12
347H, composition ... **A4:** 770
348
annealing ... **A4:** 773
arc welding, filler metals for ... **M3:** 49
composition ... **A4:** 770 **A6:** 457, 691 **M3:** 5-6 **M6:** 526
content, effect on weldability ... **M6:** 320
filler metals for ... **A6:** 693
filler metals for use in arc welding ... **A6:** 1106
mechanical properties ... **A6:** 468
plasma (ion) nitriding ... **A4:** 424
pot material to hold noncyanide carburizing process ... **A4:** 331
properties ... **A6:** 693
resistance to environments ... **M3:** 11-12

Stainless steels, specific types (continued)
resistance welding ... **M6:** 527
similar/dissimilar welds ... **A6:** 500
for vacuum chamber construction ... **A4:** 503, 504
348, annealing ... **M4:** 624
348, machining ... **A16:** 144-147, 179, 207, 274, 323, 324, 326
348H
composition ... **A6:** 457, 691 **M3:** 5-6
filler metals for ... **A6:** 693
properties ... **A6:** 693
resistance to environments ... **M3:** 11-12
348H, composition ... **A4:** 770
350, turning with cermet tools ... **A16:** 93
370
composition ... **A4:** 771
plasma (ion) nitriding ... **A4:** 424
pot material to hold noncyanide carburizing process ... **A4:** 331
for vacuum chamber construction ... **A4:** 503, 504
370, composition ... **A6:** 458
380L, gas-tungsten arc welding and fluid flow phenomena ... **A6:** 22
384
composition ... **A4:** 770 **M3:** 5-6
corrosion in seawater ... **M3:** 77
mechanical properties ... **M3:** 18-22
physical properties ... **M3:** 34-35
plasma (ion) nitriding ... **A4:** 424
pot material to hold noncyanide carburizing process ... **A4:** 331
resistance to environments ... **M3:** 11-12
for vacuum chamber construction ... **A4:** 503, 504
384, composition ... **A6:** 457
385
composition ... **M3:** 9-10
mechanical properties ... **M3:** 18-22
385, machining ... **A16:** 144-147, 179, 207, 274, 323, 324, 326
400, machining ... **A16:** 93, 231, 547
400 series ... **A16:** 88, 93, 95, 266
403
annealing ... **A4:** 780 **M4:** 624
arc welding, filler metals for ... **M3:** 49
austenitizing ... **A4:** 781 **M4:** 624
brazing ... **A6:** 626
chemical composition ... **A6:** 432
composition ... **A4:** 770 **A6:** 684 **M3:** 5-6, 190-191 **M6:** 526
content, effect on weldability ... **M6:** 348
corrosion in seawater ... **M3:** 71
filler metals for ... **A6:** 684
filler metals for use in arc welding ... **A6:** 439, 1106
flash welding ... **M6:** 557
forging temperature ... **M3:** 42
gas-metal arc welding parameters ... **A6:** 439
hardening ... **A4:** 778, 779, 780 **M4:** 624
hardness ... **M3:** 25, 47
hydrogen embrittlement ... **A4:** 782
impact strength ... **A4:** 779, 780
laser welding ... **A6:** 441
machinability ... **M3:** 45
mechanical properties ... **M3:** 26-28, 199
mechanical properties in various conditions ... **A6:** 434
preheating ... **A4:** 778
properties ... **A6:** 684
resistance to environments ... **M3:** 11-12
resistance welding ... **A6:** 848 **M6:** 527
rupture life ... **M3:** 194
specific welding recommendations, filler metals ... **A6:** 439
tempering ... **A4:** 778 **M4:** 624
403, anodic polarization curves ... **A13:** 956
403, forgeability ... **A14:** 380
403 modified, steam-turbine blades, effect of Stellite erosion shield ... **A11:** 170-171
403, single-origin fatigue crack ... **A11:** 257

Stainless steels, specific types (continued)
403 steam turbine blade, fracture surface ... **A11:** 258
404
composition ... **M3:** 9-10
mechanical properties ... **M3:** 26-28
405
408Cb, composition ... **A4:** 772
annealing ... **A4:** 776
arc welding, filler metals for ... **M3:** 49
capacitor discharge stud welding ... **A6:** 222
chemical composition ... **A6:** 444
composition ... **A4:** 770, 772 **A6:** 687 **M3:** 5-6, 190-191 **M6:** 526
content, effect on weldability ... **M6:** 346
corrosion ... **A6:** 450
corrosion in seawater ... **M3:** 70, 71
in ferritic base metal-filler metal combination ... **A6:** 449
filler metals for ... **A6:** 687
filler metals for use in arc welding ... **A6:** 1106
forging temperature ... **M3:** 42
machinability ... **M3:** 45
mechanical properties ... **M3:** 24-25, 194
metallurgy ... **A6:** 683
physical properties ... **M3:** 34-35
properties ... **A6:** 443, 683, 687
resistance to environments ... **M3:** 11-12
resistance welding ... **A6:** 848 **M6:** 527
stud arc welding ... **A6:** 215
405, annealing ... **M4:** 624
405, for steam generators ... **A13:** 940
405Nb, in ferritic base metal-filler metal combinations ... **A6:** 449
406, chemical composition ... **A6:** 444
406, composition ... **M3:** 190-191
406, wrought heat-resistant ... **A16:** 738
409
annealing ... **A4:** 776
applications ... **A6:** 443, 685
arc welding ... **A6:** 685
brazing with copper cladding ... **A6:** 347
chemical composition ... **A6:** 444
composition ... **A4:** 770 **A6:** 687 **M3:** 5-6, 190-191
corrosion in acids ... **M3:** 82, 86
corrosion in fertilizer ... **M3:** 83
embrittlement tendency minimal ... **A4:** 777
in ferritic base metal-filler metal combination ... **A6:** 449
filler metals ... **A6:** 448, 451, 687
gas-metal arc welding ... **A6:** 446
gas-tungsten arc welding ... **A6:** 448
hydrogen embrittlement ... **A4:** 781-782
mechanical properties ... **M3:** 24-25, 194
metallurgy ... **A6:** 682-683
oxidation resistance ... **M3:** 196
physical properties ... **M3:** 34-35
properties ... **A6:** 683, 687
resistance to environments ... **M3:** 11-12
resistance welding ... **A6:** 443
transportation equipment ... **M3:** 68, 69, 70
weldability ... **A6:** 445
409, annealing ... **M4:** 625
409, effect of content on weldability ... **M6:** 346
409, for steam generators ... **A13:** 940
409Cb, chemical composition ... **A6:** 444
409Nb, in ferritic base metal-filler metal combination ... **A6:** 449
410
annealing ... **A4:** 780 **M4:** 628
applications ... **A4:** 402 **A6:** 443
arc welding ... **A6:** 679-680
arc welding, filler metals for ... **M3:** 49
atmospheric corrosion ... **M3:** 65, 66
austenitizing ... **A4:** 781
austenitizing temperature ... **A4:** 781 **M4:** 628, 629
capacitor discharge stud welding ... **A6:** 222
chemical composition ... **A6:** 432

Stainless steels, specific types (continued)
composition........ A4: 770 A6: 684 M3: 5-6, 190-191, M6: 526
constitution diagram A6: 685
content, effect on weldability M6: 348
corrosion in chemical solutions.......... M3: 81, 82
corrosion in seawater.... M3: 70, 73, 74, 76, 77, 78, 79
creep-rupture properties.......... M3: 195
cutoff band sawing with bimetal blades A6: 1184
design curves, total deformation M3: 198
electrodes for flux-cored arc welding A6: 189
fasteners M3: 184
filler metals for A6: 684
filler metals for use in arc welding A6: 1106
filler metals used in arc welding this steel A6: 439
flame hardening A4: 283
flash welding M6: 557, 578-579
forging temperate M3: 42
friction welding A6: 153
furnace brazing A6: 917
gas nitriding A4: 401, 402
hardening A4: 778 M4: 627
hardness M3: 25, 47
hydrogen embrittlement A4: 782
isothermal transformation diagram.......... A6: 438, 678-679, 683
liquid nitriding A4: 411 M4: 252
machinability M3: 45
martempered to full hardness A4: 140
martempering in salt applications A4: 148
martensitic start temperature A6: 437
mechanical properties M3: 26-28, 197, 199
mechanical properties in various conditions A6: 434
nitriding time A4: 388 M4: 193
oxidation resistance M3: 196
physical properties M3: 34-35
physical properties in the annealed condition A6: 435
preheat and interpass temperatures A6: 680
preheating A4: 778
properties A6: 629, 679, 684
proven applications for borided ferrous materials A4: 445
resistance to environments M3: 11-12
resistance welding A6: 848 M6: 527
specific welding recommendations, filler metals A6: 439- 440
stress rupture M3: 227
stress-relieving A4: 777
stud arc welding A6: 215
tempering A4: 778, 781 M4: 627, 629
tempering temperatures, influence on properties M3: 29
thermal diffusivity from 20 to 100 °C A6: 4
transition behavior M3: 32
410, arc strikes damage A11: 414
410 bars, magabsorption measurement A17: 155
410, brazing and heat treating of A11: 451
410, cracks in HAZs A11: 428
410, for steam generators A13: 938
410, hardness and magnetic hysteresis A17: 134
410, hydrogen-stress cracking A13: 1132-1133
410, liquid-erosion resistance A11: 167
410, stress effects on hysteresis loops A17: 161
410 tube, pitted by chloride ions in flush water A11: 630
410Cb
composition M3: 9-10
mechanical properties M3: 26-28
410Cb (XM-30), chemical composition A6: 432
410Cb (XM-30), composition A4: 772
410Cb, mechanical properties in various conditions A6: 434
410L, applications A7: 730-731
410L, dimensional change A7: 292
410L, garbage disposal part A7: 731
410L, sintered properties A7: 731
410NiMo
composition A6: 684
filler metals for A6: 684
properties A6: 684
410S
chemical composition A6: 432

Stainless steels, specific types (continued)
composition A6: 684 M3: 9-10
filler metals for A6: 684
mechanical properties A6: 434 M3: 26-28
properties A6: 684
410S, composition A4: 772
414
annealing A4: 780 M4: 628
austenitizing A4: 781 M4: 627, 630
austenitizing temperature A4: 782
chemical composition A6: 432
composition........ A4: 770 A6: 684 M3: 5-6 M6: 526
content, effect on weldability M6: 348
filler metals for A6: 684
filler metals for use in arc welding A6: 439
flame hardening A4: 283
forging temperature M3: 42
hardening A4: 778 M4: 627
hardness M3: 25
hydrogen embrittlement A4: 782
machinability M3: 45
mechanical properties M3: 26-28
mechanical properties in various conditions A6: 434
no response to full annealing A4: 780
physical properties M3: 34-35
physical properties in the annealed condition A6: 435
properties A6: 684
resistance to environments M3: 11-12
resistance welding A6: 848 M6: 527
specific welding recommendations, filler metals A6: 440
tempering A4: 778, 782 M4: 627, 630
414, properties A7: 499-500
414 stud, service fracture A11: 428
414L
composition M3: 9-10
mechanical properties M3: 26-28
molds for plastics and rubber, use for M3: 547
416
annealing A4: 780 M4: 628
arc welding, filler metals for M3: 49
austenitizing A4: 781 M4: 627, 631
austenitizing temperature A4: 783
chemical composition A6: 432
composition....... A4: 770 A6: 684 M3: 5-6, 190-191, 756
corrosion in seawater M3: 70-71, 73, 74, 76
cutoff band sawing with bimetal blades A6: 1184
fasteners M3: 184, 185
filler metals for A6: 684
filler metals for use in arc welding A6: 439, 1106
flame hardening A4: 283
forging temperature M3: 42
hardening A4: 778 M4: 627
hardness M3: 25, 47
induction brazing A6: 921
machinability M3: 44-46
mechanical properties M3: 26-28, 761
mechanical properties in various conditions A6: 434
not recommended for welding A6: 439
physical properties M3: 34-35
physical properties in the annealed condition A6: 435
plasma (ion) nitriding A4: 421
preheating A4: 778
properties A6: 684
resistance to environments M3: 11-12
tempering A4: 778, 783 M4: 627, 631
416, effect of content on weldability M6: 348
416, examples of preferential detection, image analysis A10: 311
416 plus X
chemical composition A6: 432
composition M3: 9-10
mechanical properties M3: 26-28
mechanical properties in various conditions A6: 434
416 Plus X (XM-6), composition A4: 772
416Se
annealing A4: 780
chemical composition A6: 432

Stainless steels, specific types (continued)
composition A4: 770 A6: 684 M3: 5-6
filler metals for A6: 684
filler metals for use in arc welding A6: 1106
hardening A4: 778
mechanical properties M3: 26-28
mechanical properties in various conditions A6: 434
not recommended for welding A6: 439
properties A6: 684
resistance to environments M3: 11-12
tempering A4: 778
416Se, effect of content on weldability M6: 348
418
chemical composition A6: 432
mechanical properties in various conditions A6: 434
418 (Greek Ascoloy), composition A4: 772
418, mechanical properties M3: 26-28
420
annealing A4: 780 M4: 628
applications A4: 402 A6: 679
arc welding, filler metals for M3: 49
austenitizing A4: 781 M4: 627, 632
austenitizing temperature A4: 784
chemical composition A6: 432
composition A4: 770 A6: 684 M3: 5-6
content, effect on weldability M6: 348
cutoff band sawing with bimetal blades A6: 1184
fasteners M3: 185
filler metals for A6: 684
filler metals for use in arc welding A6: 439, 1106
flame hardening A4: 283
forging temperature M3: 42
gas nitriding A4: 402
hardening A4: 778, 779, 780 M4: 627
hardness M3: 25
impact strength A4: 779, 780
induction hardening M4: 470
machinability M3: 44-46
martempering, forming after A4: 145
mechanical properties M3: 26-28
mechanical properties in various conditions A6: 434
molds for plastics and rubber, use for M3: 547
nitriding M4: 193
nitriding time A4: 388
physical properties M3: 34-35
physical properties in the annealed condition A6: 435
plasma (ion) nitriding A4: 404, 423
preheat and interpass temperatures A6: 680
preheating A4: 778
properties A6: 684
proven applications for borided ferrous materials A4: 445
resistance to environments M3: 11-12
resistance welding A6: 848 M6: 527
specific welding recommendations, filler metals A6: 440
submerged arc welding A6: 681-682
tempering A4: 778, 784 A6: 682 M4: 627, 632
tempering temperatures, influence on properties M3: 29
420F
chemical composition A6: 432
composition A6: 684
cutoff band sawing with bimetal blades A6: 1184
filler metals for A6: 684
hardness M3: 25
machinability M3: 45
not recommended for welding A6: 439
properties A6: 684
resistance to environments M3: 11-12
420F, composition A4: 770
420F Se, composition A4: 772
420FSe
composition A6: 684
filler metals for A6: 684
properties A6: 684
422
applications A4: 402
chemical composition A6: 432

Stainless steels, specific types (continued)
composition A4: 770 M3: 5-6, 190-191
gas nitriding A4: 402
mechanical properties M3: 26-28, 197, 199
mechanical properties in various conditions A6: 434
physical properties M3: 34-35
physical properties in the annealed condition A6: 435
resistance to environments M3: 11-12
tempering in service M3: 198
422, by STAMP process A7: 549
422, creep-rupture relationship A7: 549
422, ferritic STAMP processed chemical composition A7: 548
422, flash welding M6: 557
422, machining A16: 144-147, 179, 207, 738
429
chemical composition A6: 443
composition A6: 687 M3: 5-6
in ferritic base metal-filler metal combination A6: 449
filler metals for A6: 687
mechanical properties M3: 24-25
physical properties M3: 34-35
properties A6: 687
resistance to environments M3: 11-12
429, broaching A16: 207
429, composition A4: 770
429, trepanning A16: 179
429FSe, chemical composition A6: 432
430
annealing A4: 776
arc welding, filler metals for M3: 49
atmospheric corrosion M3: 65-68
austenite-martensite embrittlement A4: 776
capacitor discharge stud welding A6: 222
chemical composition A6: 443
cold rolling, effect on tensile strength M3: 33
composition A4: 770 M3: 5-61, 190-191 M6: 526
content, effect on weldability M6: 346
corrosion A6: 450
corrosion in chemical solutions M3: 81, 82, 84, 85, 87
corrosion in seawater M3: 71, 73, 74, 76-79
creep-rupture properties M3: 195
cross-wire welding M6: 531
cutoff band sawing with bimetal blades A6: 1184
ductile-to-brittle transition temperature A6: 444
fasteners M3: 184, 185
in ferritic base metal-filler metal combination A6: 449
filler metal A6: 443, 448, 451
filler metals for A6: 687
filler metals for use in arc welding A6: 1106
flash welding M6: 557
forging temperature M3: 42
gas nitriding A4: 401, 402
gas-tungsten arc welding A6: 448
hardness gradients A4: 402
heat-resistant alloy applications A4: 516
low-cyanide liquid nitriding equipment A4: 414
machinability M3: 45
mechanical properties M3: 24-25, 194, 199
metallurgy A6: 682
microstructure A6: 686
oxidation resistance M3: 196
physical properties M3: 34-35
postweld heat treatment A6: 686
preheating A6: 450
properties A6: 683, 687
resistance to environments M3: 11-12
resistance welding A6: 848 M6: 527
roll welding A6: 313
shielded metal arc welding A6: 450, 451

Stainless steels, specific types (continued)
stud arc welding A6: 215
thermal diffusivity from 20 to 100 °C A6: 4
transportation equipment M3: 68, 69
weldability A6: 445
430, annealing M4: 625
430, development of banded structure A9: 627
430, potentiostatic passive anodic polarization A13: 218
430F
annealing A4: 776
arc welding, filler metals for M3: 49
composition A4: 770 A6: 687 M6: 526
content, effect on weldability M6: 346
corrosion in seawater M3: 73, 74, 76
cutoff band sawing A6: 1184
filler metals for A6: 687
filler metals used in arc welding A6: 1106
forging temperature M3: 42
machinability M3: 44-46
mechanical properties M3: 24-25
physical properties M3: 34-35
properties A6: 687
resistance to environments M3: 11-12
430F, annealing M4: 625
430F Se composition A4: 770
430F-Se, effect of content on weldability M6: 346
430FSe
chemical composition A6: 443
composition A6: 687 M3: 5-6
filler metals for A6: 687
filler metals used in arc welding A6: 1106
properties A6: 687
resistance to environments M3: 11-12
430L vacuum-sintered ferritic, carbon effects A13: 827
430Se, spade drilling A16: 225
430Ti
composition M3: 9-10
mechanical properties M3: 24-25
430Ti, composition A4: 771
430Ti, in ferritic base metal-filler metal combination A6: 449
431
annealing A4: 780 M4: 628
arc welding, filler metals for M3: 49
austenitizing A4: 781 M4: 627, 633
austenitizing temperature A4: 785
chemical composition A6: 432
composition A4: 770 A6: 684 M3: 5-6, 190-191 M6: 526
content, effect on weldability M6: 348
corrosion in seawater M3: 74, 76
filler metals for A6: 684
filler metals for use in arc welding A6: 439, 1106
flame hardening A4: 283
forging temperature M3: 42
hardening A4: 778, 779, 780 M4: 627
hardness M3: 25
hydrogen embrittlement A4: 781, 782
impact strength A4: 779, 780
Izod impact properties A4: 779
machinability M3: 45
mechanical properties M3: 26-28, 199
mechanical properties in various conditions A6: 434
no response to full annealing A4: 780
oxidation resistance M3: 196
physical properties M3: 34-35
physical properties in the annealed condition A6: 435
preheating A4: 778
properties A6: 684
resistance to environments M3: 11-12
resistance welding A6: 848 M6: 527
retained austenite A4: 779

Stainless steels, specific types (continued)
specific welding recommendations, filler metals A6: 440
tempering A4: 778, 785 M4: 627, 633
tempering temperatures, influence on properties M3: 29
431, springs M1: 299
434
annealing A4: 776
austenite-martensite embrittlement A4: 776
chemical composition A6: 443
composition A4: 770 M3: 5-6, 190-191
corrosion A6: 450
mechanical properties M3: 24-25
physical properties M3: 34-35
preheating A6: 450
resistance to environments M3: 11-12
roll welding A6: 313
transportation equipment M3: 68, 69
weldability A6: 445
434, annealing M4: 625
434L, applications A7: 730-731
434L, compactibility A7: 184
434L, dimensional change A7: 292
434L, ferritic, green strength A7: 184, 185
436
composition M3: 5-6
mechanical properties M3: 24-25
physical properties M3: 34-35
resistance to environments M3: 11-12
436, broaching A16: 207
436, chemical composition A6: 443
436, composition A4: 770
436, trepanning A16: 179
439
annealing A4: 776
austenite-martensite embrittlement avoided A4: 776
chemical composition A6: 444
composition A4: 770 A6: 687 M3: 190-191
in ferritic base metal-filler metal combination A6: 449
filler metals for A6: 687
properties A6: 687
tensile properties M3: 194
439, annealing M4: 625
439 calibration tube, eddy current inspection A17: 182
439, wrought heat-resistant A16: 738
440
composition M3: 5-6
flame hardening A4: 283
forging temperature M3: 42
preheating A4: 778
440 FSe
chemical composition A6: 432
composition A6: 684
filler metals for A6: 684
not recommended for welding A6: 439
properties A6: 684
440A
annealing A4: 780 M4: 628
applications A6: 433, 679
arc welding A6: 680
austenitizing M4: 627
chemical composition A6: 432
composition A4: 770 A6: 684 M3: 5-6 M6: 526
content, effect on weldability M6: 348
cutoff band sawing with bimetal blades A6: 1184
filler metals for A6: 684
filler metals for use in arc welding this steel A6: 439
hardening A4: 778 M4: 627
hardness M3: 25
machinability M3: 45
mechanical properties M3: 26-28
mechanical properties in various conditions A6: 435

SUBJECTS OF THE INDEXED VOLUMES: ASM Handbook (designated by the letter "A"): **A1:** Properties and Selection: Irons, Steels, and High-Performance Alloys (1990); **A2:** Properties and Selection: Nonferrous Alloys and Special-Purpose Materials (1990); **A3:** Alloy Phase Diagrams (1992); **A4:** Heat Treating (1991); **A6:** Welding, Brazing, and Soldering (1993); **A7:** Powder Metallurgy (1984); **A8:** Mechanical Testing (1985); **A9:** Metallography and Microstructures (1985); **A10:** Materials Characterization (1986); **A11:** Failure Analysis and Prevention (1986); **A12:** Fractography (1987); **A13:** Corrosion (1987); **A14:** Forming and Forging (1988); **A15:** Casting (1988); **A16:** Machining (1989); **A17:** Nondestructive Testing and Quality Control (1989); **A18:** Friction, Lubrication, and Wear Technology (1992). **Metals Handbook, 9th Edition** (designated by the letter "M"): **M1:** Properties and Selection: Irons and Steels (1978); **M2:** Properties and Selection: Nonferrous Alloys and Pure Metals (1979); **M3:** Properties and Selection: Stainless Steels, Tool Materials and Special-Purpose Materials (1980); **M4:** Heat Treating (1981); **M5:** Surface Cleaning, Finishing, and Coating (1982); **M6:** Welding, Brazing, and Soldering (1983). **Engineered Materials Handbook** (designated by the letters "EM"): **EM1:** Composites (1987); **EM2:** Engineering Plastics (1988); **EM3:** Adhesives and Sealants (1990); **EM4:** Ceramics and Glasses (1991); **Electronic Materials Handbook** (designated by the letters "EL"): **EL1:** Packaging (1989).

Stainless steels, specific types (continued)
physical properties **M3:** 34-35
physical properties in the annealed condition **A6:** 435
properties **A6:** 684
resistance to environments **M3:** 11-12
resistance welding **A6:** 848 **M6:** 527
specific welding recommendations, filler metals **A6:** 440
tempering **A4:** 778 **M4:** 627
440A, applications **A7:** 731
440B
annealing **A4:** 780 **M4:** 628
applications **A6:** 433, 679
austenitizing **M4:** 627
chemical composition **A6:** 432
composition **A4:** 770 **A6:** 684 **M3:** 5-6
cutoff band sawing with bimetal blades **A6:** 1184
filler metals for **A6:** 684
hardening **A4:** 778 **M4:** 627
hardness **M3:** 25
machinability **M3:** 45
mechanical properties **M3:** 26-28
mechanical properties in various conditions **A6:** 435
properties **A6:** 684
resistance to environments **M3:** 11-12
specific welding recommendations, filler metals **A6:** 440
tempering **A4:** 778 **M4:** 627
440B, effect of content on weldability **M6:** 348
440C
annealing **A4:** 780 **M4:** 628
applications **A6:** 433, 679
austenitizing **A4:** 781 **M4:** 627, 634
austenitizing temperature **A4:** 786
chemical composition **A6:** 432
composition **A4:** 770 **A6:** 684 **M1:** 610 **M3:** 5-6
cutoff band sawing with bimetal blades **A6:** 1184
filler metals for **A6:** 684
friction welding **A6:** 441
hardening **A4:** 778 **M4:** 627
hardness **M3:** 25
hydrogen embrittlement **A4:** 781
ion implantation **A4:** 266
machinability **M3:** 44-46
mechanical properties **M3:** 26-28
mechanical properties in various conditions **A6:** 435
physical properties **M3:** 34-35
physical properties in the annealed condition **A6:** 435
properties **A6:** 684
resistance to environments **M3:** 11-12
retained austenite **A4:** 779
seizure resistance **M1:** 611
shafting for mining or off-road construction machinery **M1:** 606
specific welding recommendations, filler metals **A6:** 440
tempering **A4:** 778, 786 **M4:** 627, 634
440C, cracking **A13:** 1056
440C, effect of content on weldability **M6:** 348
440C, for bearings **A11:** 490
440C radial-contact ball bearings, rolling-contact fatigue in **A11:** 503-504
440CM, composition **M1:** 610
440F
annealing **A4:** 780 **M4:** 628
austenitizing **M4:** 627
chemical composition **A6:** 432
composition **A4:** 772 **A6:** 684
cutoff band sawing with bimetal blades **A6:** 1184
filler metals for **A6:** 684
hardening **A4:** 778 **M4:** 627
hardness **M3:** 25
machinability **M3:** 45
not recommended for welding **A6:** 439
properties **A6:** 684
tempering **A4:** 778 **M4:** 627
440F Se, composition **A4:** 772
441, chemical composition **A6:** 444
441, composition **A4:** 771

Stainless steels, specific types (continued)
442
chemical composition **A6:** 443
composition **M3:** 5-6 **M6:** 526
in ferritic base metal-filler metal combination **A6:** 449
forging temperature **M3:** 42
mechanical properties **M3:** 24-25
preheating **A6:** 450
resistance to environments **M3:** 11-12
resistance welding **A6:** 848 **M6:** 527
weldability **A6:** 445
442, composition **A4:** 770
443, cutoff band sawing with bimetal blades **A6:** 1184
444
annealing **A4:** 776
austenite-martensite embrittlement avoided **A4:** 776
chemical composition **A6:** 446
composition **A4:** 770 **A6:** 687 **M3:** 9-10
ductility loss **A6:** 444
filler metals for **A6:** 687
gas-tungsten arc welding **A6:** 686
mechanical properties **M3:** 24-25
metallurgy **A6:** 683
physical properties **M3:** 34-35
properties **A6:** 687
weldability **A6:** 454
444, annealing **M4:** 624
444, resistance welding **M6:** 527
446
annealing **A4:** 776
applications, protection tubes and wells **A4:** 533
arc welding, filler metals for **M3:** 49
atmospheric corrosion, industrial sites **M3:** 66
chemical composition **A6:** 443
composition **A4:** 770 **A6:** 687 **M3:** 5-6, 190-191 **M6:** 526
content, effect on weldability **M6:** 346
corrosion **A6:** 450
corrosion in nitric acid **M3:** 86
creep-rupture properties **M3:** 195
cutoff band sawing with bimetal blades **A6:** 1184
in ferritic base metal-filler metal combination **A6:** 449
filler metals for **A6:** 687
filler metals for use in arc welding **A6:** 1106
forging temperature **M3:** 42
furnace brazing **A6:** 919
gas nitriding **A4:** 401, 402
hardness gradients **A4:** 402
hardness vs temperature **M3:** 199
heat-resistant alloy applications **A4:** 517
liquid nitriding equipment **A4:** 415
machinability **M3:** 44-46
mechanical properties **M3:** 24-25, 199
metallurgy **A6:** 682
oxidation resistance **M3:** 196
physical properties **M3:** 34-35
preheating **A6:** 450
properties **A6:** 683, 687
recommended for parts and fixtures for salt baths **A4:** 514
resistance to environments **M3:** 11-12
resistance welding **A6:** 848 **M6:** 527
weldability **A6:** 445
446, annealing **M4:** 625
446, embrittlement of **M1:** 686-687
450 (custom), hot isostatic pressing **A15:** 541
500 series, cemented carbides **A16:** 88
501
composition **M3:** 5-6
mechanical properties **M3:** 26-28
resistance to environments **M3:** 11-12
501, machining **A16:** 144-147, 179, 207
501, thermal diffusivity from 20 to 100 °C **A6:** 4
501A, composition **M3:** 5-6
501B, composition **M3:** 5-6
502
composition **M3:** 5-6
mechanical properties **M3:** 26-28

Stainless steels, specific types (continued)
resistance to environments **M3:** 11-12
502, machining **A16:** 144-147, 179, 207
503
composition **M3:** 5-6
resistance to environments **M3:** 11-12
504
composition **M3:** 5-6
resistance to environments **M3:** 11-12
600, STAMP processed **A7:** 549
615
composition **A6:** 684
filler metals for **A6:** 684
properties **A6:** 684
616
composition **A6:** 684
filler metals for **A6:** 684
properties **A6:** 684
619
composition **A6:** 684
filler metals for **A6:** 684
properties **A6:** 684
630
arc welding **A6:** 696, 697
composition **A6:** 696
filler metals for **A6:** 696
properties **A6:** 696
630, liquid-erosion resistance **A11:** 167
630 type 17-4 PH, poppet-valve stem, fracture of **A11:** 320-321
631
composition **A6:** 696
filler metals for **A6:** 696
properties **A6:** 696
631 (17-7PH) Belleville washers, distortion from heat treatment **A11:** 140
631, liquid-erosion resistance **A11:** 167
631, springs **M1:** 284, 285, 298, 291
632
composition **A6:** 696
filler metals for **A6:** 696
heat treatment **A6:** 695
properties **A6:** 696
semiaustenitic PH stainless steel **A6:** 695
632, composition **M3:** 9-10
633
composition **A6:** 696
filler metals for **A6:** 696
properties **A6:** 696
633, composition **M3:** 9-10
634
composition **A6:** 696
filler metals for **A6:** 696
properties **A6:** 696
634, composition **M3:** 9-10
635
composition **A6:** 696
filler metals for **A6:** 696
properties **A6:** 696
635, composition **M3:** 9-10
660 (A286)
composition **A6:** 696
filler metals for **A6:** 696
properties **A6:** 695, 696
662
composition **A6:** 696
filler metals for **A6:** 696
properties **A6:** 696
830, applications **A7:** 731
904L
annealing **A4:** 773
composition **A4:** 771 **A6:** 458, 691 **M3:** 9-10
filler metals for **A6:** 693
mechanical properties **A6:** 468 **M3:** 18-22
properties **A6:** 693
904L, annealing **M4:** 624
2205
applications **A6:** 697
composition **A6:** 698
filler metals for **A6:** 698
properties **A6:** 698
2205, composition **A4:** 772
2205 duplex, chemical compositions **A13:** 359
2209, as filler metal **A6:** 698
2304
composition **A6:** 698
filler metals for **A6:** 698

Stainless steels, specific types (continued)
properties ... **A6:** 698
2304, composition ... **A4:** 772
2507
composition ... **A6:** 698
filler metals for ... **A6:** 698
properties ... **A6:** 698
2553, as filler metal ... **A6:** 698
4340, brazing ... **A6:** 930
4340, die life in upset forging ... **A14:** 228
9310, die life in upset forging ... **A14:** 228
A-286
aging ... **A6:** 482
composition ... **A6:** 483, 564
composition effect ... **A6:** 487
constitutional liquation in multicomponent systems ... **A6:** 568
electron-beam welding ... **A6:** 491, 492, 493, 865, 869
filler metals ... **A6:** 491
flash butt welding ... **A6:** 492
gas-tungsten arc welding ... **A6:** 492, 493
heat treatment ... **A6:** 485
hot cracking ... **A6:** 696
mechanical properties ... **A6:** 487
microstructure ... **A6:** 488
M_s temperature ... **A6:** 482
postweld heat treatment ... **A6:** 492
precipitation-hardening phases ... **A6:** 483
resistance welding ... **A6:** 491, 492
ultrasonic welding ... **A6:** 326
upset welding ... **A6:** 249
weldability ... **A6:** 490
ACI CN-7 cast pump impeller ... **A13:** 142
AF22
composition ... **A4:** 772
corrosion resistance ... **A4:** 777
AISI 4340, hydrogen embrittlement ... **A12:** 31
AL 29-4
chemical composition ... **A6:** 445
in ferritic base metal-filler metal combination ... **A6:** 449
weldability ... **A6:** 449
AL 29-4-2
chemical composition ... **A6:** 445
in ferritic base metal-filler metal combination ... **A6:** 449
weldability ... **A6:** 449
AL 29-4-2, annealing ... **M4:** 625
AL 29-4C (S44735)
chemical composition ... **A6:** 446
in ferritic base metal-filler metal combination ... **A6:** 449
hydrogen embrittlement ... **A6:** 450
AL 29-4C, annealing ... **M4:** 625
AL-4X, annealing ... **A4:** 773 **M4:** 624
AL-6X
annealing ... **A4:** 773
composition ... **A4:** 771 **M3:** 9-10
corrosion in foods ... **M3:** 92
corrosion resistance, pulp and paper industry ... **M3:** 92
mechanical properties ... **M3:** 18-22
AL-6X, annealing ... **M4:** 624
AL-6X, mechanical properties ... **A6:** 468
AL-6XN
composition ... **A6:** 458, 691
filler metals for ... **A6:** 693
properties ... **A6:** 693
AL-6XN, composition ... **A4:** 771
AL29-4-2
annealing ... **A4:** 776
composition ... **A4:** 772
AL29-4C
annealing ... **A4:** 776
composition ... **A4:** 772
AL433, chemical composition ... **A6:** 444
AL446, chemical composition ... **A6:** 444
AL468, chemical composition ... **A6:** 444

Stainless steels, specific types (continued)
ALFA IV, composition ... **A4:** 772
Almar 363, composition ... **M3:** 190-191
Almar 363, wrought heat-resistant ... **A16:** 738
AM 350, composition ... **M6:** 350
AM 355
composition ... **M6:** 350
flash welding ... **M6:** 557
AM-350
composition ... **A4:** 772 **A6:** 483 **M3:** 9-10, 190-191
composition effect ... **A6:** 487
electron-beam welding ... **A6:** 489
filler metals ... **A6:** 490
gas nitriding ... **A4:** 387
gas-metal arc welding ... **A6:** 489
gas-tungsten arc welding ... **A6:** 489
heat treatment ... **A6:** 485 **M3:** 202
heat-treating procedures ... **A4:** 784, 785, 791-792
laser-beam welding ... **A6:** 489
mechanical properties ... **A6:** 486 **M3:** 30-31, 202, 204
precipitation-hardening phases ... **A6:** 483
recommended filler metals for welding ... **A6:** 1107
resistance welding ... **A6:** 489, 490
shielded metal arc welding ... **A6:** 490
stress-rupture ... **M3:** 203
submerged arc welding ... **A6:** 489
ultrasonic welding ... **A6:** 326
weldability ... **A6:** 489
welding sequence ... **A4:** 791-792
AM-350, chemical milling ... **A16:** 584
AM-350, wrought heat-resistant ... **A16:** 738
AM-355
composition ... **A4:** 772 **A6:** 483 **M3:** 9-10, 190-191
composition effect ... **A6:** 487
electron-beam welding ... **A6:** 489
filler metals ... **A6:** 490
gas nitriding ... **A4:** 387
gas-metal arc welding ... **A6:** 489
gas-tungsten arc welding ... **A6:** 489
heat treatment ... **A6:** 485 **M3:** 202
heat-treating procedures ... **A4:** 784, 785, 791-792
laser-beam welding ... **A6:** 489
mechanical properties ... **A6:** 488 **M3:** 30-31, 202, 203
microstructure ... **A6:** 488, 490
postweld heat treatment ... **A6:** 490
precipitation-hardening phases ... **A6:** 483
recommended filler metals for welding ... **A6:** 1107
resistance welding ... **A6:** 489, 490
shielded metal arc welding ... **A6:** 490
submerged arc welding ... **A6:** 489
tempering temperature effect on properties ... **A4:** 792
ultrasonic welding ... **A6:** 326
weldability ... **A6:** 489
welding sequence ... **A4:** 791-792
AM-355, chemical milling ... **A16:** 584
AM-355, wrought heat-resistant ... **A16:** 738
AM-363
composition ... **M3:** 9-10
mechanical properties ... **M3:** 30-31
AM355
heat treating procedures ... **M4:** 635-639
mechanical properties, effect of tempering temperatures ... **M4:** 643, 646
mechanical properties, effect of welding sequence ... **M4:** 643, 646
AMS-5616 Greek Ascoloy flash welding ... **M6:** 557
ASTM A313 *See* Stainless steels, specific types
ASTM A693 *See* Stainless Steels
CA-15, suction roll shell, muriatic acid corrosion ... **A13:** 1204
CA-15M
composition ... **A6:** 684
filler metals for ... **A6:** 684

Stainless steels, specific types (continued)
properties ... **A6:** 684
CA-28MWV
composition ... **A6:** 684
filler metals for ... **A6:** 684
properties ... **A6:** 684
CA-40
composition ... **A6:** 684
filler metals for ... **A6:** 684
properties ... **A6:** 684
CA-40F
composition ... **A6:** 684
filler metals for ... **A6:** 684
properties ... **A6:** 684
CA6N
composition ... **A6:** 684
filler metals for ... **A6:** 684
properties ... **A6:** 684
CA6NM
application ... **A6:** 679
austenite-start temperature ... **A6:** 680
composition ... **A6:** 684
filler metals for ... **A6:** 684
mechanical properties in various conditions ... **A6:** 434
preheat and interpass temperatures ... **A6:** 680
properties ... **A6:** 679, 684
Carpenter 18-18 Plus ... **A16:** 738
composition ... **M3:** 190-191
tensile properties ... **M3:** 205
Carpenter 18-18 plus, annealing ... **M4:** 624
Carpenter 18-18 Plus, magnetic permeability ... **A4:** 773
Carpenter 18Cr-2Ni-12Mn, annealing ... **M4:** 624
Carpenter 20 Cb-3
composition ... **M3:** 9-10, 172, 210
corrosion in chemical solutions ... **M3:** 80, 81, 82, 83-84, 88, 89, 173
corrosion in pharmaceuticals ... **M3:** 92
corrosion in seawater ... **M3:** 70
fasteners, use for ... **M3:** 185
mechanical properties ... **M3:** 18-22
physical properties ... **M3:** 217
Carpenter 20Cb-3, annealing ... **M4:** 624
Carpenter 20Cb-3, applications ... **A4:** 769
Carpenter 21Cr-6Ni-9Mn, annealing ... **M4:** 624
Carpenter 22Cr-13Ni-5Mn, annealing ... **M4:** 624
Carpenter H-46
composition ... **M3:** 190-191
mechanical properties ... **M3:** 197
tempering in service ... **M3:** 198
CB-30
composition ... **A6:** 687
filler metals for ... **A6:** 687
properties ... **A6:** 687
CC-50
composition ... **A6:** 687
filler metals for ... **A6:** 687
properties ... **A6:** 687
CD-4MCw
composition ... **A6:** 698
filler metals for ... **A6:** 698
metallurgy ... **A6:** 697
properties ... **A6:** 698
CE-30
composition ... **A6:** 691
filler metals for ... **A6:** 693
properties ... **A6:** 693
CF-3
composition ... **A6:** 691
filler metals for ... **A6:** 693
properties ... **A6:** 693
CF-3M
composition ... **A6:** 691
filler metals for ... **A6:** 693
properties ... **A6:** 693
CF-3MN
composition ... **A6:** 691
filler metals for ... **A6:** 693

SUBJECTS OF THE INDEXED VOLUMES: ASM Handbook (designated by the letter "A"): **A1:** Properties and Selection: Irons, Steels, and High-Performance Alloys (1990); **A2:** Properties and Selection: Nonferrous Alloys and Special-Purpose Materials (1990); **A3:** Alloy Phase Diagrams (1992); **A4:** Heat Treating (1991); **A6:** Welding, Brazing, and Soldering (1993); **A7:** Powder Metallurgy (1984); **A8:** Mechanical Testing (1985); **A9:** Metallography and Microstructures (1985); **A10:** Materials Characterization (1986); **A11:** Failure Analysis and Prevention (1986); **A12:** Fractography (1987); **A13:** Corrosion (1987); **A14:** Forming and Forging (1988); **A15:** Casting (1988); **A16:** Machining (1989); **A17:** Nondestructive Testing and Quality Control (1989); **A18:** Friction, Lubrication, and Wear Technology (1992). **Metals Handbook, 9th Edition** (designated by the letter "M"): **M1:** Properties and Selection: Irons and Steels (1978); **M2:** Properties and Selection: Nonferrous Alloys and Pure Metals (1979); **M3:** Properties and Selection: Stainless Steels, Tool Materials and Special-Purpose Materials (1980); **M4:** Heat Treating (1981); **M5:** Surface Cleaning, Finishing, and Coating (1982); **M6:** Welding, Brazing, and Soldering (1983). **Engineered Materials Handbook** (designated by the letters "EM"): **EM1:** Composites (1987); **EM2:** Engineering Plastics (1988); **EM3:** Adhesives and Sealants (1990); **EM4:** Ceramics and Glasses (1991); **Electronic Materials Handbook** (designated by the letters "EL"): **EL1:** Packaging (1989).

Stainless steels, specific types (continued)
properties A6: 693
CF-8
composition A6: 691
filler metals for A6: 693
properties A6: 693
CF-8C
composition A6: 691
filler metals for A6: 693
properties A6: 693
CF-8M
composition A6: 691
filler metals for A6: 693
properties A6: 693
CF-8M, radiographic inspection A17: 333-334
CF-16F
composition A6: 691
filler metals for A6: 693
properties A6: 693
CF-20
composition A6: 691
filler metals for A6: 693
properties A6: 693
CF10SMnN
composition A6: 691
filler metals for A6: 693
properties A6: 693
CG-8M
composition A6: 691
filler metals for A6: 693
properties A6: 693
CG-12
composition A6: 691
filler metals for A6: 693
properties A6: 693
CG6MMN
composition A6: 691
filler metals for A6: 693
properties A6: 693
CH-20
composition A6: 691
filler metals for A6: 693
properties A6: 693
CK-3MCuN
composition A6: 691
filler metals for A6: 693
properties A6: 693
CK-20
composition A6: 691
filler metals for A6: 693
properties A6: 693
CN-3M
composition A6: 691
filler metals for A6: 693
properties A6: 693
CN-7M
composition A6: 691
filler metals for A6: 693
properties A6: 693
CN-7MS
composition A6: 691
filler metals for A6: 693
properties A6: 693
Cr-Ni-Mo-Ti, liquid lithium corrosion A13: 93
Cronifer 1815 LCSi, composition A6: 458
Cronifer 1815LCSi, composition A4: 771
Cronifer 1925 hMo, composition A6: 458
Cronifer 1925hMo, composition A4: 771
Cronifer 2328, composition A4: 771 A6: 458
Crutemp 25
composition M3: 9-10
mechanical properties M3: 18-22
Cryogenic Tenelon (XM-14),
composition A4: 771 A6: 458
Cryogenic Tenelon, composition M3: 9-10
Custom 450
composition A6: 483, 487 M3: 9-10, 190-191
composition effect A6: 487
filler metals A6: 487
fracture toughness M3: 32
gas-metal arc welding A6: 487
gas-tungsten arc welding A6: 487
heat treatment A6: 485
mechanical properties A6: 486 M3: 30-31, 200
precipitation-hardening phases A6: 483

Stainless steels, specific types (continued)
weldability A6: 487
Custom 450 (XM-25)
composition A4: 772
heat-treating procedures A4: 784
Custom 450, composition M6: 350, 526
Custom 450, wrought heat-resistant A16: 738
Custom 455
composition A6: 483 M3: 9-10, 190-191
composition effect A6: 487
filler metals A6: 487
fracture toughness M3: 32
gas-metal arc welding A6: 487
gas-tungsten arc welding A6: 485
mechanical properties A6: 486 M3: 30-31, 200
microstructure A6: 483
precipitation-hardening phases A6: 483
Custom 455 (XM-16)
composition A4: 772
heat-treating procedures A4: 784
Custom 455, composition M6: 350, 526
Custom 455, wrought heat-resistant A16: 738
DP-3
annealing A4: 778
composition A4: 772
corrosion resistance A4: 777
E Brite 26-1
composition M3: 9-10, 190-191
corrosion in sodium hydroxide M3: 88
crevice corrosion M3: 58
mechanical properties M3: 24-25
oxidation resistance M3: 196
tensile properties M3: 194
E-4, mechanical properties in various
conditions A6: 434
E-Brite 26-1
chemical composition A6: 445
in ferritic base metal-filler metal
combination A6: 449
weldability A6: 448, 449, 453
E-BRITE 26-1, composition A4: 771
E-Brite 26-1, wrought heat-resistant A16: 738
E-Brite alloy
corrosion A6: 451-452
hydrogen embrittlement A6: 450
weldability A6: 446
E-BRITE, annealing A4: 776 M4: 625
E4, composition A4: 772
EN 56, flash welding M6: 557
EN 58, flash welding M6: 557
EN58J, wear factors in orthopedic
implants A18: 659, 660
Esshete 1250, composition A4: 771 A6: 458
Fe-21Cr-6Ni-9Mn-0.3N, composition A6: 1017
Fe-22Mn, composition A6: 1017
Fe-27.7Cr, potentiostatic etching A9: 145-146
Fe-28Cr-5Mo, gas-tungsten arc
welding A6: 445, 448
Ferralium 255
annealing A4: 778
composition A4: 772
corrosion resistance A4: 777
Ferralium alloy 255, preferential corro-
sion ferrite phase A13: 360
FV 448, flash welding M6: 557
FV 535, flash welding M6: 557
Gall-Tough, composition A4: 771 A6: 458
Greek Ascoloy
composition M3: 190-191
mechanical properties M3: 197, 198
Greek Ascoloy, wrought heat-resistant A16: 738
H-20 Mod, corrosion resistance, pulp
and paper industry M3: 92
H-46, wrought heat-resistant A16: 738
Hastelloy B, corrosion in sulfuric acid M3: 88
Hastelloy C, corrosion in sulfuric acid M3: 88
HK-40 *See also* 310
corrosion in ammonia M3: 81
HNM
composition M3: 9-10
mechanical properties M3: 30-31
HNM, not electron-beam welded A6: 869
HT9 (12Cr-1Mo-0.3V)
applications A6: 433
chemical composition A6: 432
gas-tungsten arc welding A6: 435, 436
laser welding A6: 441

Stainless steels, specific types (continued)
microstructure A6: 435
orientation and PWHT effect A6: 437
specific welding recommendations,
filler metals A6: 440
tempering behavior A6: 440
JBK-75
composition A6: 483, 492
composition effect A6: 487
cracking A6: 492-493
electron-beam welding A6: 493
gas-tungsten arc welding and fluid
flow phenomena A6: 493
mechanical properties A6: 493
postweld heat treatment A6: 493
precipitation-hardening phases A6: 483
trace element impurity effect on
GTA weld penetration A6: 20
Jessops G 88, flash welding M6: 557
Jessops G 183, flash welding M6: 557
Jethete M-152, composition M3: 190-191
Jethete M-152, wrought heat-resistant A16: 738
JS-77
annealing A4: 773
composition A4: 771
JS-700
annealing A4: 773
composition A4: 771 A6: 458 M3: 9-10
corrosion in foods M3: 93
corrosion resistance, pulp and paper
industry M3: 92
mechanical properties M3: 18-22
JS-777
composition M3: 9-10
corrosion resistance, pulp and paper
industry M3: 92
mechanical properties M3: 18-22
JS700, annealing M4: 624
JS777, annealing M4: 624
Kromarc 58
composition M3: 756
fatigue-crack-growth rate M3: 764
fracture toughness M3: 763
tensile properties M3: 760-762
Lapelloy
chemical composition A6: 432
mechanical properties in various
conditions A6: 434
Lapelloy, composition A4: 772
Lescalloy BG 42, composition M1: 610
M-152 .... A16: 22
Moly Ascoloy
composition M3: 190-191
mechanical properties M3: 197
Moly Ascoloy, wrought heat-resistant A16: 738
Monel 400 tube, eddy current
inspection A17: 182-183
Monit
composition M3: 9-10
mechanical properties M3: 24-25
MONIT (25-4-4), composition A4: 771
MONIT, annealing A4: 776 M4: 625
MVMA, composition A4: 771 A6: 458
N08020, composition A16: 682, 683
N08020, machining A16: 96, 698-700, 703, 704
N08330, tapping A16: 695, 699
N08367, weldability A6: 449
NBS SRM-442 standard A10: 147
Nitronic 30
gas nitriding A4: 387
magnetic permeability A4: 773
Nitronic 32
annealing A4: 773
composition A4: 771 M3: 9-10, 190-191
magnetic permeability A4: 773
mechanical properties M3: 23-24
Nitronic 32 (18-2Mn), composition A6: 458
Nitronic 32, annealing M4: 624
Nitronic 32, wrought heat-resistant A16: 738
Nitronic 33
annealing A4: 773
composition A4: 771 M3: 9-10, 190-191
magnetic permeability A4: 773
mechanical properties M3: 23-24, 205
Nitronic 33 (18-3Mn), composition A6: 458
Nitronic 33, annealing M4: 624
Nitronic 33, wrought heat-resistant A16: 738

Stainless steels, specific types (continued)
Nitronic 40
annealing A4: 773
composition A4: 771 M3: 9-10, 756
gas nitriding A4: 387
magnetic permeability A4: 773
mechanical properties M3: 23-24, 760
Nitronic 40 (XM-10), composition A6: 458
Nitronic 40, annealing M4: 624
Nitronic 50
annealing A4: 773
composition A4: 771 M3: 9 10, 190-191
gas nitriding A4: 387
magnetic permeability A4: 773
mechanical properties M3: 23-24, 205
stress-rupture M3: 205
Nitronic 50 (XM-19), composition A6: 458
Nitronic 50, annealing M4: 624
Nitronic 50, wrought heat-resistant A16: 738
Nitronic 60
annealing A4: 773
composition A4: 771 M3: 9-10, 190-191, 756
gas nitriding A4: 387
magnetic permeability A4: 773
mechanical properties M3: 23-24, 205, 760
stress-rupture M3: 205
Nitronic 60, annealing M4: 624
Nitronic 60, composition A6: 458
Nitronic 60, wrought heat-resistant A16: 738
Nitronic alloys, gas-tungsten arc welding A6: 465
NuMonit (S44635), chemical composition A6: 446
PH 13-8 Mo
composition M3: 5-6, 190-191 M6: 350
fatigue behavior, constant-life M3: 32
fracture toughness M3: 32
mechanical properties M3: 30-31, 200, 202
physical properties M3: 34-35
resistance to environments M3: 11-12
resistance welding M6: 527
PH 13-8 Mo, composition A4: 770
PH 13-8 Mo, wrought heat-resistant A16: 738
PH 14-8 Mo, composition M6: 350
PH 14-8 Mo, electron-beam welding A6: 869
PH 15-7, chemical milling A16: 584
PH 15-7, flash welding M6: 557
PH 15-7 Mo
aging A6: 482
composition A4: 772 A6: 483 M3: 190-191
composition effect A6: 487
fasteners, use for M3: 185
filler metals A6: 488
fracture toughness M3: 32
furnace brazing A6: 916
gas-metal arc welding A6: 489
gas-tungsten arc welding A6: 488-489, 492
heat treatment A6: 485 M3: 202
heat-treating procedures A4: 783, 785
mechanical properties A6: 486 M3: 30-31, 202
postweld heat treatment A6: 489
precipitation-hardening phases A6: 483
recommended filler metals for welding A6: 1107
resistance welding A6: 489
spot and seam welding A6: 489
submerged arc welding A6: 489
ultrasonic welding A6: 326
weldability A6: 488
PH 15-7 Mo, composition M6: 350, 526
PH 15-7 Mo, machining A16: 738, 739
PH 15-7, photochemical machining A16: 588
PH13-8Mo
composition A6: 483
composition effect A6: 487
electron-beam welding A6: 487, 491
gas-metal arc welding A6: 487
gas-tungsten arc welding A6: 487
heat treatment A6: 485

Stainless steels, specific types (continued)
mechanical properties A6: 486
microstructure A6: 488
precipitation-hardening phases A6: 483
weldability A6: 487
Pyromet 350, heat-treating procedures A4: 784
Pyromet 355, heat-treating procedures A4: 784
Pyromet 538
composition M3: 756
fatigue-crack-growth rate M3: 764
fracture toughness M3: 763
tensile properties M3: 760, 762
RA 85 H, composition A4: 771
RA 85H
composition A6: 458, 690
filler metals for A6: 692
properties A6: 692
RH 950 bonded using polybenzimidazoles
RR517, flash welding M6: 557
S110, flash welding M6: 557
S129, flash welding M6: 557
S130, flash welding M6: 557
S13800 (XM-13), composition A16: 683, 684
S13800, machining A16: 692-694, 696, 698-700, 703, 704
S15500 (AISI 15-5PH), chromium-plated versus plasma spray coated Ti-10V-2Fe-3Al A18: 780
S15500 (XM-12), composition A16: 683
S15500, machining A16: 692-694, 696, 698-700, 703, 704
S17400 (630), composition A16: 683, 684
S17400 (AISI 17-4PH)
corrosive wear A18: 719
material for jet engine components A18: 588, 591
microstructure A18: 712
precipitates formed A18: 712
temperature effect on adhesive wear A18: 721
S17400, machining A16: 691-694, 696, 698-700, 703, 704
S17700 (631), composition A16: 683, 684
S17700 (AISI PH17-7), microstructure A18: 712
S17700, machining A16: 692-694, 696, 698-700, 703, 704
S18200 (XM-34), composition A16: 682
S18200, machining A16: 684, 692-694, 696, 698-700, 702, 704
S18235, composition A16: 682
S18235, machining A16: 686, 692-694, 696, 698-700, 702, 704
S20100 (201), composition A16: 682, 683
S20100, machining A16: 144-147, 179, 207, 225, 274, 323, 324, 326, 360, 362, 692-694, 696, 698-700, 703, 704
S20300 (XM-1), composition A16: 682, 684
S20300, machining A16: 684, 686, 689, 692-694, 696, 698-700, 702, 704
S20910 (XM-19), composition A16: 682, 683
S20910, machining A16: 689-694, 696, 698-700, 703, 704
S21800 (AISI Nitronic 60)
cavitation erosion rate A18: 774
galling threshold load A18: 595
sliding wear A18: 769
temperature effect on adhesive wear A18: 721
S21904 (XM-11), composition A16: 682, 683
S21904, machining A16: 692-694, 696, 698-700, 703, 704
S24100 (XM-28), composition A16: 682, 683
S24100, machining A16: 692-694, 696, 698-700, 703, 704
S28200, composition A16: 682, 683
S28200, machining A16: 689, 692-694, 696, 698-700, 703, 704
S30000 series
galling resistance with various material combinations A18: 596

Stainless steels, specific types (continued)
hardfacing alloys based on A18: 762
S30100 (301), composition A16: 682
S30100, machining A16: 144-147, 179, 207, 225, 323, 324, 326, 584, 588, 692-694, 696, 698-700, 703, 704
S30200 (302), composition A16: 682, 683
S30200, machining A16: 147, 179, 207, 225, 274, 301, 323, 324, 326, 360, 588, 682-684, 690, 692-694, 696, 698-700, 703, 704
S30300 (303), composition A16: 682
S30300, (P/M), drilling A16: 885, 886
S30300, (P/M), machinability A16: 880
S30300, machining A16: 58, 225, 237, 269, 282, 301, 360, 362, 566, 684-687, 689, 692-694, 696, 698, 702, 704
S30310 (XM-5), composition A16: 682
S30310, machining A16: 684, 686, 689, 692-694, 696, 698-700, 702, 704
S30323 (303Se), composition A16: 682
S30323, machining A16: 225, 684, 692-694, 696-700, 702, 704
S30330 (303Cu), composition A16: 682, 684
S30330, machining A16: 689, 692-694, 696, 698-700, 702, 704
S30345 (XM-2), composition A16: 682, 684
S30345, machining A16: 692-694, 696, 698-700, 702, 704
S30360 (XM-3), composition A16: 682
S30360, machining A16: 684, 692-694, 696, 698-700, 702, 704
S30400 (304), composition A16: 682, 683, 684
S30400 (AISI 304)
cavitation erosion rate A18: 774
corrosive wear A18: 715, 719
erosion test results A18: 200
friction coefficient data A18: 71
ion implantation A18: 857-858
laser melt/particle injection A18: 869
metallographic sections to detect subsurface deformation A18: 374, 375
service life of coal-handling equipment A18: 719
temperature effect on adhesive wear A18: 721
wear rates for test plates in drag conveyor bottoms A18: 720
S30400, machining A16: 58, 144-148, 179, 204, 207, 209, 223, 225, 237, 269, 274, 301, 323, 324, 326, 529, 566, 581, 584, 588, 684, 685, 689, 690, 692-696, 698-700, 703, 704, 738
S30403 (AISI 304L), sliding wear A18: 774
S30403, composition A16: 682, 683
S30403, machining A16: 144-147, 174, 179, 207, 225, 274, 323, 324, 326, 692-694, 696, 698-700, 703, 704, 738
S30403, stress-corrosion cracking A6: 477
S30430
composition M3: 5-6
physical properties M3: 34-35
resistance to environments M3: 11-12
S30430 (XM-7), composition A16: 682
S30430, machining A16: 684, 692-694, 696, 698-700, 703, 704
S30431, composition A16: 682, 684
S30431, machining A16: 42-694, 696, 698-700, 702, 704
S30452 (XM-21), composition A16: 682, 683
S30452, machining A16: 692-694, 696, 698-700, 703, 704
S30500 (305), composition A16: 682
S30500, machining A16: 144-147, 179, 207, 225, 588, 692-694, 696, 698-700, 703, 704
S30800 (stickweld overlay), cavitation erosion rate A18: 774
S30900 (309), composition A16: 682, 683
S30900, machining A16: 225, 274, 360, 362, 689, 692-694, 696, 698-700, 703, 704, 738
S30908 (309S), composition A16: 682, 683

Stainless steels, specific types (continued)
S30908, machining A16: 225, 274, 692-694, 696, 698-700, 703, 704
S31000 (310), composition A16: 682, 683
S31000, machining A16: 225, 274, 360, 362, 692-694, 696, 698-700, 703, 704, 738
S31008 (310S), composition A16: 682
S31008, machining A16: 225, 274, 692-694, 696, 698-700, 703, 704
S31254, pitting A6: 477
S31260
composition A6: 474, 475
pitting resistance equivalent values A6: 474, 475
properties A6: 475
S31600 (316), composition A16: 682, 683, 684
S31600 (AISI 316)
corrosive wear A18: 719
electro-spark deposition A18: 645
erosion resistance, liquid impingement erosion A18: 228
erosion test results A18: 200
galling A18: 10
polarization curves, slurry wear testing A18: 274, 275, 276
sliding wear A18: 774
thermal spray coating material A18: 832
wear rates for test plates in drag conveyor bottoms A18: 720
wear resistance relation to toughness A18: 707
S31600, machining A16: 65, 225, 236, 237, 274, 282, 360, 362, 513, 537, 540, 584, 588, 684, 689-696, 698-700, 703, 704, 738, 876-877
S31603 (316L), composition A16: 682
S31603 (AISI 316L)
annealed hardness A18: 768
applications, femoral components of hip and knee replacements A18: 657, 658, 661
erosive wear A18: 768
fretting wear A18: 250
S31603, machining A16: 225, 274, 692-694, 696-698, 703, 704, 738
S31603, stress-corrosion cracking A6: 477
S31620 (316F), composition A16: 682
S31620, machining A16: 225, 684, 692-694, 696, 698-700, 702, 704
S31700 (317), composition A16: 682, 683
S31700, machining A16: 225, 274, 360, 362, 692-694, 696, 698-700, 703, 704, 738
S31703 (317L), composition A16: 682
S31703, machining A16: 225, 692-694, 696, 698-700, 703, 704
S31803
applications A6: 477
composition A6: 474, 475
physical properties A6: 476
pitting A6: 477
pitting resistance equivalent values A6: 474, 475
properties A6: 475
stress-corrosion cracking A6: 477
S31803, composition A16: 682, 684
S31803, machining A16: 94, 696, 698-700, 703, 704
S32100 (321), composition A16: 682, 683
S32100 (AISI 321), work material for ion implantation A18: 858
S32100, machining A16: 144-147, 179, 207, 225, 274, 301, 323, 324, 326, 360, 584, 588, 689, 690, 692-694, 696, 698-700, 703, 704, 738
S32304
applications A6: 477
composition A6: 474, 475
pitting resistance equivalent values A6: 474, 475
properties A6: 475
stress-corrosion cracking A6: 477
S32550
applications A6: 477
composition A6: 474, 475
physical properties A6: 476
pitting A6: 477
pitting resistance equivalent values A6: 474, 475
properties A6: 475
S32550, composition A16: 682

Stainless steels, specific types (continued)
S32550, machining A16: 225, 692-694, 696, 698-700, 703, 704
S32750
applications A6: 477
composition A6: 474
physical properties A6: 476
pitting A6: 477
pitting resistance equivalent values A6: 474
properties A6: 475
stress-corrosion cracking A6: 477
S32760
applications A6: 477
composition A6: 474, 475
pitting resistance equivalent values A6: 474, 475
properties A6: 475
S32900 (329), composition A16: 682, 683, 684
S32900, machining A16: 225, 692-694, 696, 698-700, 703, 704
S32950
composition A6: 474
pitting resistance equivalent values A6: 474
properties A6: 475
S32950, composition A16: 682, 684
S32950, machining A16: 689-694, 696, 698-700, 703, 704
S34700 (347), composition A16: 682, 683, 684
S34700, machining A16: 144-147, 179, 203, 207, 225, 274, 323, 324, 326, 588, 684, 693, 694, 696, 698-700, 703, 704, 738
S34720 (347F), composition A16: 682
S34720, machining A16: 225, 684, 692-694, 696, 698-700, 702, 704
S34723 (347FSe), composition A16: 682
S34723, machining A16: 225, 684, 692-694, 696, 698-700, 702, 704
S35000 (633), composition A16: 683
S35000, machining A16: 690, 692-694, 696, 698-700, 703, 704
S35500 (634), composition A16: 683, 684
S35500, machining A16: 690, 692-694, 696, 698-700, 703, 704
S38400 (384), composition A16: 682
S38400, machining A16: 144-147, 179, 207, 225, 274, 323, 324, 326, 692-694, 696, 698-700, 703, 704
S40000 series (hard), galling resistance with various material combinations A18: 596
S40000 series (soft), galling resistance with various material combinations A18: 596
S40300 (403), composition A16: 682
S40300 (AISI 403), liquid impingement erosion A18: 221, 225
S40300, machining A16: 144-147, 179, 203, 204, 693, 694, 696, 698-700, 702, 704, 738
S40400, turning A16: 692
S40500 (405), composition A16: 682, 683
S40500, machining A16: 179, 207, 225, 688, 692-694, 696, 698-700, 702, 704
S40900 (409), composition A16: 682, 683
S40900 (AISI 409)
corrosive wear A18: 719
modified, corrosive wear A18: 715
S40900, machining A16: 179, 207, 225, 692-694, 696, 698-700, 703, 704, 738
S41000 (410), composition A16: 682, 683, 684
S41000 (AISI 410)
cavitation erosion A18: 763
corrosive wear A18: 719
service life of coal-handling equipment A18: 719
temperature effect on adhesive wear A18: 721
wear rates for test plates in drag conveyor bottoms A18: 720
S41000, machining A16: 27, 29, 58, 144-147, 204, 207, 225, 301, 360, 362, 588, 684, 688, 689, 692-694, 696, 698-700, 702, 704, 737, 738
S41400 (414), composition A16: 682, 683
S41400, machining A16: 225, 689, 692-694, 696, 698-700, 702, 704
S41600 (416), composition A16: 682
S41600, machining A16: 203, 225, 237, 269, 282, 301, 360, 362, 684-686, 688, 689, 692-694, 696, 698-700, 702, 704, 738
S41610 (XM-6), composition A16: 682

Stainless steels, specific types (continued)
S41610, machining A16: 684, 686, 692-694, 696, 698-700, 702, 704
S41623 (416Se), composition A16: 682
S41623, machining A16: 684, 692-694, 696, 698-700, 702, 704
S42000 (420), composition A16: 682, 683, 684
S42000, machining A16: 144-147, 179, 207, 225, 588, 684, 692-694, 696, 698-700, 702, 704
S42010, composition A16: 682, 683
S42010, machining A16: 692-694, 696, 698-700, 702, 704
S42020 (420F), composition A16: 682
S42020, machining A16: 225, 301, 360, 362, 684, 692-694, 696, 698-700, 702, 704
S42023 (420FSe), composition A16: 682
S42023, machining A16: 225, 684, 692-694, 696, 698-700, 702, 704
S43000 (430), composition A16: 682, 683, 684
S43000 (AISI 430), erosion test results A18: 200
S43000, machining .... A16: 179, 225, 282, 360, 362, 588 , 692-694, 696, 698-700, 702, 704, 738
S43020 (430F), composition A16: 682
S43020, machining A16: 225, 301, 360, 362, 684, 692-694, 696, 698-700, 702, 704
S43023 (430FSe), composition A16: 682
S43023, machining A16: 225, 684, 692-694, 696, 698-700, 702, 704
S43100 (431), composition A16: 682, 683
S43100, machining A16: 204, 225, 689, 692-694, 696, 698-700, 702, 704, 738
S43400 (434), composition A16: 682, 683
S43400, machining A16: 179, 207, 692-694, 696, 698-700, 702, 704, 738
S43735, weldability A6: 449
S44002 (440A), composition A16: 682, 683
S44002, machining A16: 225, 302, 360, 361, 692-694, 696, 698-700, 702, 704
S44003 (440B), composition A16: 682, 683
S44003, machining A16: 225, 360, 361, 692-694, 696, 698-700, 702, 704
S44004 (440C), composition A16: 682, 683, 684
S44004 (AISI 440C)
ceramic coatings and retarding degradation of polytetrafluoroethylene (PTFE) A18: 156
contact fatigue in rolling-element bearings A18: 260
endurance lives when coated with molybdenum disulfide solid lubricant A18: 115, 116
friction coefficient data A18: 74
graphite fluoride as lattice layer lubricant A18: 116
high-resolution electron microscopy to study sliding wear A18: 389
metal wear generated by sliding plastics A18: 240
part material for ion implantation A18: 858
pin-on-disk bench tests A18: 117
polyimide solid lubricants A18: 116, 117
rolling-contact component steels A18: 503
surface, modification with and without lubrication, high-vacuum applications A18: 158
versus titanium alloys modified by evaporation A18: 780, 781
wear data with unlubricated tool steel A18: 737
S44004, machining A16: 225, 360, 362, 479, 684, 692-694, 696, 698-700, 702, 704
S44020 (440F), composition A16: 682
S44020, machining A16: 684, 692-694, 696, 698-700, 702, 704
S44023 (440FSe), composition A16: 682
S44023, machining A16: 225, 684, 692-694, 696, 698-700, 702, 704
S44200 (442), composition A16: 682, 683
S44200, machining A16: 179, 207, 692-694, 696, 698-700, 702, 704
S44300 (443), composition A16: 682
S44300, machining A16: 225, 360, 362, 692-694, 696, 698-700, 702, 704
S44400 (444), composition A16: 682, 684
S44400, machining A16: 225, 684, 692-694, 696, 698-700, 702, 704
S44600 (446), composition A16: 682, 683

Stainless steels, specific types (continued)
S44600, machining A16: 179, 207, 225, 360, 362, 688, 692-694, 696, 698-700, 702, 704, 738
S45000 (XM-25), composition A16: 683, 684
S45000, machining A16: 692-694, 696, 698-700, 703, 704
S45500 (XM-16), composition A16: 683, 684
S45500, machining A16: 692-694, 696, 698-700, 703, 704
S66286 (660), composition A16: 683, 684
S66286 (AISI A286)
material for jet engine components A18: 588, 591
microstructure A18: 712
S66286, machining A16: 690, 692-694, 696, 698-700, 703, 704
SAF 2205
annealing A4: 778
corrosion resistance A4: 777
Sanicro 28
annealing A4: 773
composition A4: 771 A6: 458, 691
filler metals for A6: 693
properties A6: 693
Sanicro 28, annealing M4: 624
Sea-Cure (S44660), chemical composition A6: 446
Sea-Cure (SC-1)
annealing A4: 776
composition A4: 771
SEA-CURE SC-1, annealing M4: 625
Sea-cure/SC-1
composition M3: 9-10
mechanical properties M3: 24-25
Sealmet 1, composition A4: 772
seawater corrosion resistance M3: 74
SHOMAC 30-2, chemical composition A6: 445
specific types
SS-303, mechanical properties A7: 468
SS-303L-12, hardness and density A16: 882
SS-303N1-25, hardness and density A16: 882
SS-303N2-35, hardness and density A16: 882
SS-304L-13, hardness and density A16: 882
SS-304N1-30, hardness and density A16: 882
SS-304N2-33, hardness and density A16: 882
SS-316, mechanical properties A7: 468
SS-316L-15, hardness and density A16: 882
SS-316N2-33, hardness and density A16: 882
SS-316NI-25, hardness and density A16: 882
SS-410, mechanical properties A7: 468
SS-410-90HT, hardness and density A16: 882
Stainless W
composition M3: 9-10
mechanical properties M3: 30-31
SUS 304, effect of cyclic load on fatigue crack rate A12: 62
SUS 304, effect of frequency and wave form on fatigue properties A12: 62
T316LN
composition A6: 690
filler metals for A6: 692
properties A6: 692
Tenelon
composition M3: 9-10
mechanical properties M3: 23-24
Tenelon (XM-31), composition A4: 771 A6: 458
TP304
composition A6: 690
filler metals for A6: 692
properties A6: 692
TP304H
composition A6: 690
filler metals for A6: 692
properties A6: 692
TP304LN
composition A6: 690
filler metals for A6: 692

Stainless steels, specific types (continued)
properties A6: 692
TP316
composition A6: 690
filler metals for A6: 692
properties A6: 692
TP316H
composition A6: 690
filler metals for A6: 692
properties A6: 692
TP316LN
filler metals for A6: 692
properties A6: 692
TP321
composition A6: 691
filler metals for A6: 693
properties A6: 693
TP321H
composition A6: 691
filler metals for A6: 693
properties A6: 693
TP347
composition A6: 691
filler metals for A6: 693
properties A6: 693
TP347H
composition A6: 691
filler metals for A6: 693
properties A6: 693
TP348
composition A6: 691
filler metals for A6: 693
properties A6: 693
TrimRite
chemical composition A6: 432
mechanical properties in various conditions A6: 434
TrimRite, composition A4: 772
type 303, pressed and sintered A9: 523-524
type 304, cold rolled and annealed A9: 687
type 304, cross rolled A9: 684
type 304, delta ferrite stringers A9: 65
type 304, delta ferrite to sigma phase transformation A9: 65-66
type 304, recrystallized to cube texture A9: 698
type 304, strain induced martensite A9: 66
type 316, exposed to air at high temperatures A9: 158
type 316, gas-atomized powder A9: 514
type 316, pressed and sintered A9: 523
type 316, vacuum deposition of an interference film A9: 147-148
type 316L, gas-atomized powder A9: 529
type 316L, rotating electrode processed powder A9: 514
type 316L, water-atomized powder A9: 529
type 410, pressed and sintered A9: 524
Ultimet 04, mechanical properties A7: 471
Ultimet 16, mechanical properties A7: 471
Ultimet 40C, mechanical properties A7: 471
Ultimet 304, mechanical properties A7: 471
Ultimet 316, mechanical properties A7: 471
Ultimet 440C, mechanical properties A7: 471
UNS S13800
precipitation hardening M4: 635-639
solution annealing M4: 635-639
UNS S13800, precipitation hardening A4: 788
UNS S15500
precipitation hardening M4: 635-639
solution annealing M4: 635-639
UNS S15500, precipitation hardening A4: 788
UNS S15700, heat treating procedures M4: 635-639
UNS S15700, precipitation hardening A4: 788
UNS S17400 *See* 17-4PH
UNS S17400, precipitation hardening A4: 788
UNS S17700 *See* 17-7PH
UNS S17700, precipitation hardening A4: 788

Stainless steels, specific types (continued)
UNS S35000, heat treating procedures M4: 635-639
UNS S35500 *See* AM 355
UNS S35500, precipitation hardening A4: 788
UNS S45000
precipitation hardening M4: 635-639
solution annealing M4: 635-639
UNS S45000, precipitation hardening A4: 788
UNS S45500
precipitation hardening M4: 635-639
solution annealing M4: 635-639
UNS S45500, precipitation hardening A4: 788
UNS S65000, precipitation hardening A4: 788
Uranus 50, composition A4: 772
XM-1
composition A6: 690
filler metals for A6: 692
properties A6: 692
XM-2
composition A6: 690
filler metals for A6: 692
properties A6: 692
XM-3
composition A6: 690
filler metals for A6: 692
properties A6: 692
XM-5
composition A6: 690
filler metals for A6: 692
properties A6: 692
XM-6
composition A6: 684
filler metals for A6: 684
properties A6: 684
XM-7
composition A6: 690
filler metals for A6: 692
properties A6: 692
XM-9
composition A6: 696
filler metals for A6: 696
properties A6: 696
XM-10
composition A6: 458, 690
filler metals for A6: 692
properties A6: 692
XM-11
composition A6: 690
filler metals for A6: 692
properties A6: 692
XM-12
composition A6: 696
filler metals for A6: 696
properties A6: 696
XM-13
composition A6: 696
filler metals for A6: 696
properties A6: 696
XM-14
composition A6: 690
filler metals for A6: 692
properties A6: 692
XM-15
composition A6: 458, 691
filler metals for A6: 693
properties A6: 693
XM-16
composition A6: 696
filler metals for A6: 696
properties A6: 696
XM-17
composition A6: 690
filler metals for A6: 692
properties A6: 692
XM-18
composition A6: 690
filler metals for A6: 692

Stainless steels, specific types (continued)
properties ... **A6:** 692
XM-19
composition ... **A6:** 458, 690
filler metals for ... **A6:** 692
properties ... **A6:** 692
XM-21
composition ... **A6:** 690
filler metals for ... **A6:** 692
properties ... **A6:** 692
XM-25
composition ... **A6:** 696
filler metals for ... **A6:** 696
properties ... **A6:** 696
XM-26
composition ... **A6:** 698
filler metals for ... **A6:** 698
properties ... **A6:** 698
XM-27
composition ... **A6:** 687
corrosion ... **A6:** 452
filler metals for ... **A6:** 687
properties ... **A6:** 687
XM-28
composition ... **A6:** 690
filler metals for ... **A6:** 692
properties ... **A6:** 692
XM-29
composition ... **A6:** 690
filler metals for ... **A6:** 692
properties ... **A6:** 692
XM-30
composition ... **A6:** 684
filler metals for ... **A6:** 684
properties ... **A6:** 684
XM-31
composition ... **A6:** 458, 690
filler metals for ... **A6:** 692
properties ... **A6:** 692
XM-32
composition ... **A6:** 684
filler metals for ... **A6:** 684
properties ... **A6:** 684
XM-33
composition ... **A6:** 687
filler metals for ... **A6:** 687
properties ... **A6:** 687
XM-34
composition ... **A6:** 687
filler metals for ... **A6:** 687
properties ... **A6:** 687
YUS 190L, chemical composition ... **A6:** 445
YUS436S, chemical composition ... **A6:** 444
Stainless steels, wear of ... **A18:** 693, 710-723
abrasive wear ... **A18:** 718, 719, 720, 723, 767, 774
adhesive wear resistance ... **A18:** 720, 721
applications ... **A18:** 710, 712, 713, 715, 722-723
femoral components of hip and knee replacements ... **A18:** 657, 658
austenitic ... **A18:** 710-712 , 713-714, 715, 719, 721-723
applications ... **A18:** 723
cavitation erosion ... **A18:** 217-218
cavitation resistance ... **A18:** 600
classification and composition of hardfacing alloys ... **A18:** 652
composition ... **A18:** 711
erosion resistance in liquid impingement erosion ... **A18:** 228
fretting wear ... **A18:** 248, 250, 251
galling ... **A18:** 721-722
microstructure ... **A18:** 710
not suitable for die-casting applications ... **A18:** 630
properties ... **A18:** 710
reference specimens for optical reading of microindentation testing ... **A18:** 414
relation of wear resistance to toughness ... **A18:** 707
specific types ... **A18:** 650, 651
spray material for oxyfuel wire spray process ... **A18:** 829
wear resistance ... **A18:** 651
work-hardening ... **A18:** 712
classification of stainless steels ... **A18:** 710
compositions ... **A18:** 658, 710, 711, 712, 726

Stainless steels, wear of (continued)
corrosive wear ... **A18:** 719, 723
dependence on slurry ... **A18:** 273
cutting tool materials and cutting speed relationship ... **A18:** 616
damage dominated by fatigue fracture ... **A18:** 180, 181
design considerations ... **A18:** 723
duplex ... **A18:** 712, 713
applications ... **A18:** 712
composition ... **A18:** 712
galling ... **A18:** 721
microstructure ... **A18:** 712
properties ... **A18:** 712
endodontic instruments manufactured from ... **A18:** 666, 675
erosion/corrosion ... **A18:** 208
extension of tool life via ion implantation, examples ... **A18:** 643
families of stainless steels ... **A18:** 710-712
austenitic ... **A18:** 710-712, 713-714, 715, 719, 721, 723
duplex ... **A18:** 712, 713, 721
ferritic ... **A18:** 710, 713, 716, 721
martensitic ... **A18:** 712, 713-714, 715, 716, 719, 721, 723
precipitation-hardenable (PH) ... **A18:** 712, 713, 721
ferritic ... **A18:** 710, 713, 716, 721
advantages ... **A18:** 710
applications ... **A18:** 710
composition ... **A18:** 710
galling ... **A18:** 721
microstructure ... **A18:** 710
properties ... **A18:** 710
ferrography application to identify wear particles ... **A18:** 305
fretting wear ... **A18:** 247, 248, 250, 251
friction coefficient data ... **A18:** 71, 74
for gears, resistant to scuffing ... **A18:** 538
high-silicon alternate material hardfacing alloys ... **A18:** 762, 764
erosive wear ... **A18:** 762
galling ... **A18:** 763
properties ... **A18:** 762
hot forging ... **A18:** 625
laser alloying ... **A18:** 866
laser cladding ... **A18:** 867
laser melt/particle injection ... **A18:** 869
laser melting ... **A18:** 864, 866
martensitic ... **A18:** 712-716, 719
applications ... **A18:** 723
classification and composition of hardfacing alloys ... **A18:** 652
composition ... **A18:** 712
galling ... **A18:** 721-722
microstructure ... **A18:** 712
properties ... **A18:** 712
specific types ... **A18:** 222
spray material for oxyfuel wire spray process ... **A18:** 829
material for jet engine components ... **A18:** 588
material for surface force apparatus ... **A18:** 402
metal forming lubricants ... **A18:** 148
orthodontic wires ... **A18:** 666, 675-676
precipitation-hardenable (PH) ... **A18:** 712, 713
classification ... **A18:** 712
galling ... **A18:** 721-722
microstructure ... **A18:** 712
properties ... **A18:** 712
wear resistance and cost effectiveness ... **A18:** 706
properties ... **A18:** 659, 710, 712-713, 714, 715, 716, 721
seal materials ... **A18:** 550
sliding wear ... **A18:** 769
slurry erosion ... **A18:** 768, 770
for shallow forming dies ... **A18:** 633
threshold galling stress results ... **A18:** 721, 722
for valve plates for reciprocating compressors ... **A18:** 604
for valve springs for reciprocating compressors ... **A18:** 604
wear and galling ... **A18:** 713-716, 723
abrasive wear ... **A18:** 713-714, 716, 717
adhesive wear ... **A18:** 715, 716, 717
corrosive wear ... **A18:** 715, 716, 717

Stainless steels, wear of (continued)
factors affecting ... **A18:** 715-716
fatigue wear ... **A18:** 715
fretting wear ... **A18:** 714-715
galling ... **A18:** 715, 721, 722
wear and galling tests, commonly used ... **A18:** 716-718
block-on-ring test ... **A18:** 717
button-on-block test ... **A18:** 718, 720, 721
corrosive wear testing ... **A18:** 717
crossed-cylinder test ... **A18:** 717
dry sand/rubber wheel test ... **A18:** 716-717
galling test ... **A18:** 718
pin-on-disk test ... **A18:** 717-718
ring-on-ring test ... **A18:** 718
wear compatibility of dissimilar-mated stainless steels ... **A18:** 720
wear data ... **A18:** 718-722
abrasive/corrosive wear ... **A18:** 718-719
adhesive wear ... **A18:** 719-720
galling resistance ... **A18:** 721-722
wear effect on polarization curves ... **A18:** 274
weld overlays for resistance to cavitation erosion ... **A18:** 217
welding factor ... **A18:** 541
work-hardening rates ... **A18:** 712
Stainless steels, wrought *See* Wrought stainless steels
Stainless steels, wrought, fabrication
annealing ... **M3:** 46-47
atmospheres ... **M3:** 48
brazing ... **M3:** 49-50
cold drawing ... **M3:** 44
cold heading ... **M3:** 43-44
cold working ... **M3:** 43-44
constitution diagram, weld metal ... **M3:** 51
extrusion, cold ... **M3:** 44
forging, design ... **M3:** 42-43
forming ... **M3:** 41-42
heat treating ... **M3:** 46-48
hot heading ... **M3:** 43
machining ... **M3:** 44-46
passivation ... **M3:** 46, 53, 54
plating ... **M3:** 55
riveting ... **M3:** 44
soldering ... **M3:** 50
stress relieving ... **M3:** 47-48
welding ... **M3:** 48-49, 50-52
Stainless welding wire ... **A1:** 852
Stainless-clad sheet steels
color etching ... **A9:** 141-142, 145
passive potential ranges
specimen preparation ... **A9:** 197-198
Stains
weathering steels ... **A13:** 520
Stains, surface
LEISS determined ... **A10:** 603
Stair step fracture
cobalt alloy ... **A12:** 398
Stair-rod dislocations
connecting stacking- fault tetrahedra ... **A9:** 117
Staircase method
for defining fatigue strength ... **A8:** 703-704, 706
Staking
comparison to projection welding ... **M6:** 523-524
furnace brazing of steels ... **M6:** 940
ultrasonic ... **EM2:** 721-722
Stamp mills
aluminum flake ... **A7:** 125
for metallic flake powders ... **A7:** 593
STAMP process ... **A7:** 547-550 **A16:** 60
for new P/M alloys ... **A7:** 549
processing sequence ... **A7:** 547-548
Stamped identification marks
effects on tools and dies ... **A11:** 568-570
fatigue fracture by ... **A11:** 130
Stamping *See also* Compression stamping; Thermoplastic stamping ... **A14:** 12, 804, 896
aluminum alloy sheet ... **M2:** 180-186
beryllium-copper alloys ... **A2:** 411
dies, cemented carbide ... **A2:** 971
in green machining ... **EM4:** 183
large, sheet metal forming of ... **A8:** 547
lead frame ... **EL1:** 484, 486
punches, cemented carbide ... **A2:** 971
size and shape effects ... **EM2:** 290
stretching in ... **A8:** 547

Stamping (continued)
of thermoplastic composite **EM1:** 552, 562
thermoset, properties effects **EM2:** 287
thermoset, size and shape effects **EM2:** 292
twin-sheet, size and shape effects **EM2:** 290
Stampings
aluminum alloy **M2:** 13-14
brass, mass finishing of................................... **M5:** 615
copper alloy *See* Copper alloys, stampings
inspection fixtures for...................................... **M3:** 557
pigmented drawing compounds removed from.. **M5:** 4, 6
steel *See* Steel, stampings
Stand
defined **A14:** 12
Stand linseed oil **M5:** 498
Stand-alone controller
for creep test stand **A8:** 312
Standard addition
defined **A10:** 682
method, XRPD analysis **A10:** 340
Standard beryllium-copper casting alloy
properties and applications **A2:** 360-362
Standard calomel electrode (SCE) **A18:** 250
Standard Charpy V-notch impact test
for dynamic fracture testing **A8:** 259, 262-264
Standard deviation **EM3:** 786-797
computed for normal distribution **A8:** 664
defined **A8:** 12, 625 **EM1:** 22
of fatigue test groups **A8:** 707-708
and random error... **A10:** 12
relative, defined **A10:** 681
symbol **A8:** 629
Standard deviation of the asperity height distribution **A18:** 33
Standard electrode potential
defined **A10:** 682 **A13:** 12
Standard error of estimate
for stress-life equations **A8:** 700
Standard four-part ASTM system
alloy/temper designations for magnesium alloys... **A2:** 456
Standard free energy
of metal oxide formation **A7:** 53
Standard free energy of formation
defined **A15:** 51
Standard Gibbs free energies
liquid aluminum... **A15:** 59
Standard grain-size micrograph
defined **A9:** 17
Standard grip
for axial fatigue testing **A8:** 369
Standard Malaysian Rubber system **EM3:** 145
Standard mount dimensions **A9:** 28
Standard normal distribution
in statistical quality control **A17:** 738
Standard parts
and design for manufacturability (DFM).. **EL1:** 122-123
Standard pipe **A1:** 331
Standard reference blocks *See also* Reference; Reference standards
area-amplitude **A17:** 264
ASTM blocks **A17:** 264
ASTM reference plate **A17:** 265
distance-amplitude... **A17:** 264
IIW **A17:** 264
miniature angle-beam.............................. **A17:** 264-265
ultrasonic inspection **A17:** 263-265
Standard reference materials
abbreviation for **A10:** 691
defined **A10:** 682
for environmental/industrial effluent waters **A10:** 95
Standard Reference Materials (SRM) **A18:** 415
Standard reference sample
eddy current inspection **A17:** 179
Standard state
hypothetical, thermal properties for **A15:** 56-57
Standard state (continued)
thermodynamic, defined **A15:** 50-51
Standard Test and Evaluation Bottle (STEB) **EM1:** 511, 514
Standard test methods **EM3:** 321-322
Standard test-bus interfaces
system test **EL1:** 376-377
Standard thermocouples *See also* Thermocouple materials, specific types
properties **A2:** 871
types **A2:** 871-873
Standard weights
verification method **A8:** 611
Standardization
ASTM/international **A13:** 322
by ion chromatography................................... **A10:** 664
of common titrants.. **A10:** 172
defined **A10:** 682
of hydrogen embrittlement tests **A13:** 284
techniques, internal, in SSMS....................... **A10:** 145
of tests, electrochemical crevice corrosion... **A13:** 308-309
of tests, stress-corrosion cracking................ **A13:** 246
Standardization documents
government/industry guide to............... **EM1:** 40-42
military **EL1:** 906-911
Standardless method
XRPD analysis.. **A10:** 340
Standardless ratio (Cliff-Lorimer) microanalytic technique **A10:** 447
Standards *See also* Reference; Specifications; Specifications and standards............. **EM3:** 61-64
86C, National Fire Protection Association, safety considerations for conveyor belt furnaces........................... **A4:** 547
acceptance/rejection, liquid penetrant inspection **A17:** 88
American Petroleum Institute (API), 1104 (pipeline welding specification)... **A6:** 98
American Society for Testing and Materials
aluminum surface preparation for structural adhesives bonding (D 3933).. **EM3:** 733
Boeing wedge test (D 3762) **EM3:** 332-333, 351-352, 353
bond line corrosion dependent on exposure time in a salt spray chamber (B 117)................................. **EM3:** 664
bonding permanency of water- or solvent-soluble liquid adhesives for labeling glass bottles (D 1581)... **EM3:** 657
butt joint testing (napkin ring test) (E 229) **EM3:** 331, 361
cleavage peel test specimen, DCB type (D 3807)............................. **EM3:** 353, 442
climbing drum test (D 1781)........... **EM3:** 332, 442
complex modulus test (Oberst Bar) (E 756) **EM3:** 556
complex permittivity (D 2520)................. **EM3:** 432
corrosion resistance of bonded steel joints (B 117)....................................... **EM3:** 670
creep test (D 1780, D 2293, D 2294, D 2919, D 3160) **EM3:** 333
creep test (D 2918) **EM3:** 333, 657
direct tensile stressed-durability testing (D 897, D 1344, D 2095) ... **EM3:** 351, 442
double-cantilever-beam specimen load transfer (D 3433)...... **EM3:** 442, 445, 614
double-overlap shear configuration (D 3528)..................................... **EM3:** 442, 731
dry arc flashover resistance of surfaces (D 495).. **EM3:** 438
electrical contact and resistance measurements (D 257)............ **EM3:** 429, 433
fringe corrections for electrode shapes and dimensions (D 150)...... **EM3:** 429
Standards (continued)
irradiation procedures B and C (D 1879)... **EM3:** 646
lap joint test (D 3165)........................ **EM3:** 330, 442
maximum electric field calculation (D 3756)... **EM3:** 439
napkin ring test (D 3658).................. **EM3:** 361, 442
peel test (D 903) **EM3:** 442, 657
residual static strength in composite
resistance measurements, tracking (D 2303)... **EM3:** 439
resistance of adhesives to cyclic lab oratory aging conditions (D 1183).. **EM3:** 657
roller drum peel test (D 3167) **EM3:** 332, 442, 732
single-lap joint test (D 1002) **EM3:** 325, 333, 350, 353, 355, 442, 471-473, 535-536, 731
for substrates in block form (D 229, D 816, D 905, D 906, D 1062, D 2094, D 2295, D 2339, D 2557, D 2558, D 3164, D 4027) **EM3:** 442
for substrates in block form (D 429) **EM3:** 422, 442
for substrates in block form, standard impact test (D 950) **EM3:** 333, 442
surface corona flashover resistance of surfaces (D 2275) **EM3:** 438
T-peel test (D 1876) **EM3:** 332, 442
tension lap-shear strength of adhesives (D 3163)..................... **EM3:** 442, 644-645
tension specimens (D 3039)............................ **EM3:** 834
tension test of flat sandwich construction in flatwise plane (C 297)... **EM3:** 731, 732
thick-adherend test (D 3983)........... **EM3:** 330, 442
volatile content of coatings, test method for (D 2369) **EM3:** 732
ANSI H35.1, temper designations for heat-treatable aluminum alloys **A4:** 878-879
ANSI Z49.1 "Safety in Welding and Cutting" **A6:** 68
ASTM A 263, explosion welds....................... **A6:** 305
ASTM A 264, explosion welds........................ **A6:** 305
ASTM A 265, bond shear strength measurement, explosion welds............ **A6:** 305
ASTM A 273, ultrasonic testing..................... **A6:** 253
ASTM A 426, eddy current testing **A6:** 253
ASTM A 450, hydrostatic pressure testing **A6:** 253
ASTM A 578, ultrasonic inspection, explosion welds **A6:** 305
ASTM E 213, ultrasonic testing **A6:** 253
ASTM E 309, eddy current testing................ **A6:** 253
ASTM E 390, steel weld reference radiographs.. **A6:** 98
ASTM E 399, fracture toughness measurements **A6:** 101
ASTM E 570, flux leakage examination **A6:** 253
B 432, explosion welds **A6:** 305
bodies, American-based **EL1:** 734
D 897-78, standard test method for tensile properties of adhesive bonds **A6:** 970
D 1002-72, standard test method for tensile properties of adhesive bonds **A6:** 970
D 1876-72, standard test method for tensile properties of adhesive bonds **A6:** 970
D 2294-69, standard test method for tensile properties of adhesive bonds **A6:** 970
D 3166-73, standard test method for tensile properties of adhesive bonds **A6:** 970
D 3528-76, standard test method for tensile properties of adhesive bonds **A6:** 970
defined **EM1:** 700

Standards (continued)
E 8-89, standard test method for tensile properties of adhesive bonds **A6:** 970
E 143-87, standard test method for tensile properties of adhesive bonds **A6:** 970
for EPMA **A10:** 524, 530
F 1044-87, standard test method for tensile properties of adhesive bonds **A6:** 970
German standards
durability of joints (DIN 54 456) **EM3:** 661
resistance of adhesives to cyclic laboratory aging conditions (DIN 53 295) **EM3:** 657
IPC-S-815B, industry soldering standard **A6:** 349
for liquid penetrant inspection **A17:** 87-88
in materials engineering **M3:** 826, 830
MIL-STD-1946, ballistic shock tests **A6:** 540
MIL-STD-2000A, military soldering standard **A6:** 349
National Fire Protection Association (NFPA), 51B **A6:** 185
for physical test methods **EL1:** 941
primary, assay by electrometric titration **A10:** 202
pure-element, x-ray scan across **A10:** 527
quantitative LEISS analysis **A10:** 606
solderability **EL1:** 677
sources for **EM1:** 701
sources of **EM3:** 63, 64
surface mount technology **EL1:** 734
U.S. Pharmacopoeia (USP) class VI tests, adhesive compatibility with human tissue and blood **EM3:** 575
ultrasonic inspection **A17:** 261-265
of units and measures, SI **A10:** 685
verification by controlled-potential coulometry **A10:** 207
Standards and Specifications **EM3:** 72
Standards organizations **EM4:** 40
Standards Search **EM3:** 72
Standing waves
acoustic, for optical holographic interferometry
continuous-beam ultrasonic inspection **A17:** 249
microwave **A17:** 204
system, surface crack detection **A17:** 214-215
techniques **A17:** 206, 211
for thickness gaging **A17:** 211
Standoff, component
and lead preparation **EL1:** 731
Standoff distance
definition **A6:** 1214 **M6:** 17
with laser triangulation sensors **A17:** 13
Stands
microscope **A9:** 74
Stanford Position Electron Accelerator Ring (SPEAR)
spectral of synchrotron radiation from **A10:** 411
Stanford Research Institute algorithms *See* SRI algorithms
Stanford Synchrotron Radiation Laboratory
EXAFS experimental apparatus at **A10:** 521
as EXAFS radiation source **A10:** 411-413
Stannate bronze plating **M5:** 288-289
Stannate tin plating **M5:** 270-272
Stannates **EM4:** 60
Stannic chloride, stainless steels
corrosion resistance **M3:** 88
Stannite
in bronze plating **M5:** 288
Stannous chloride
as reducing agent **A10:** 169
Stannous fluoborate tin plating **M5:** 272
Stannous fluoborate tin-lead plating **M5:** 276-278
Stannous fluoride, stainless steels
corrosion resistance **M3:** 88
Stannous octoate **EM3:** 95
Stannous sulfate tin plating **M5:** 271-272
Staple fibers *See also* Chopped fibers; Discontinuous fibers; Fiber(s); Short fibers **A7:** 49
aramid **EM1:** 115
defined **EM1:** 22 **EM2:** 40
Staple yarns
aramid fiber **EM1:** 115
STAR *See* Scientific and Technical Aerospace Reports
Star (rosette) fractures
as radial marks **A12:** 103-104
Star craze *See also* Crazing
defined **EM2:** 40
Star marks
alloy steels **A12:** 301
Star polymers
as viscosity improvers **A18:** 109
Starch
as binding agent for samples **A10:** 94
Starch blended with dry colloidal aluminosilicate
application or function optimizing powder treatment and green forming **EM4:** 49
Starch conversions
for packaging **EM3:** 45
Starch-plastic blends
biodegradation of **EM2:** 786-787
Starch/starch derivatives
as sizing **EM1:** 122
Starches
industrial applications **EM3:** 567
STARDYNE computer program for structural analysis **EM1:** 268, 273
Stark effect
defined **A10:** 682
splittings **A10:** 264
Stark line broadening
in emission spectroscopy **A10:** 22
Start current
definition **M6:** 17
Start time
definition **M6:** 17
Start voltage
definition **M6:** 17
Start-up/shutdown
chemical processing plant **A13:** 1135-1136
Starting feeds and speeds
PCBN tools **A2:** 1017
Starting motors
as magnetically soft materials application **A2:** 779
Starting torque
defined **A18:** 18
Starting troughs
electrogas welding **M6:** 241
Starve infiltration **A7:** 564
Starved area **EM3:** 27
defined **EM1:** 22 **EM2:** 40
Starved joint **EM3:** 27
defined **EM1:** 22 **EM2:** 40
State, change of
as x-ray diffraction analysis **A10:** 325
State movement
automated **A10:** 310
State variables **A3:** 1 • 1
Static
defined **A11:** 9
electrical discharges, ball bearing pitting failure from **A11:** 495, 497
fatigue **A11:** 28
impact fracture toughness (K_{tc}), as function of test temperature **A11:** 54
service, allowable stresses for **A11:** 136
tensile loading, fracture under **A11:** 390
Static and dynamic fatigue testing **EM3:** 349-370
analysis of test results **EM3:** 353-361
behavior of the interphase **EM3:** 367-368
Boeing wedge test **EM3:** 351-352
components of an adhesive joint **EM3:** 350
crack opening displacement **EM3:** 366-367
diffusion mechanism of water **EM3:** 363-364
dynamic fatigue **EM3:** 367
energy balance **EM3:** 366
factors affecting joint durability **EM3:** 349-350
general test program **EM3:** 353
joint durability **EM3:** 349
joint specimen-test result relationship **EM3:** 353-361
maximum principal stress **EM3:** 366-367
mixed-mode fracture criteria **EM3:** 366-367
moisture ingression and joint performance **EM3:** 361-364, 367, 368
recommendations
qualitative accuracy **EM3:** 368-370
Static and dynamic fatigue testing (continued)
quantitative accuracy **EM3:** 368-370
sealant durability **EM3:** 368
single-lap geometry **EM3:** 350-351
strain-energy density **EM3:** 366
stress and free volume effect on sorption behavior **EM3:** 362-363
stress whitening effect on fluid ingression **EM3:** 363
stress-concentration factors for single-lap specimens **EM3:** 359-361
temperature-dependent delayed-failure behavior of bonded single-lap joints **EM3:** 355-361
tensile (butt) joint geometry **EM3:** 351
test specimen geometries for quantitative
Arcan specimen **EM3:** 366, 370
bonded cantilever beam specimen **EM3:** 366
cracked lap-shear specimen **EM3:** 366
independently loaded mixed-mode specimen **EM3:** 366, 370
Iosipescu specimen **EM3:** 365-366, 370
testing **EM3:** 365-366
thick-adherend specimen **EM3:** 352-353
viscoelasticity considerations for DCB fracture testing **EM3:** 364-365
viscoelasticity-viscoplasticity considerations **EM3:** 353-355
Static babbitting **M5:** 356-357
Static cast copper alloy ingots
grain structures **A9:** 641
Static charge **EM3:** 27
defined **EM2:** 40
Static coefficient of friction **A18:** 27, 46, 47, 476
defined **A18:** 18
Static cold pressing
as cermet forming technique **A7:** 800
Static creep
rupture life, prior fatigue effect on **A8:** 353, 355
test, ductile to brittle fracture **A8:** 354
test results **A8:** 353-354
Static decay rate
defined **EM2:** 461
Static elimination
testing for **EM2:** 475-476
Static equilibrium viscosity **A18:** 93
Static equivalent load
defined **A18:** 18
Static equivalent radial load **A18:** 510
Static fatigue *See also* Fatigue; Fatigue loading; Hydrogen embrittlement; Hydrogen stress cracking; Hydrogen induced delayed cracking **M1:** 687 **EM3:** 27
defined **EM1:** 22 **EM2:** 40
described **EM2:** 702
hydrogen embrittlement as **A12:** 124
tests **EM2:** 703-704
Static fatigue strength *See* Creep rupture strength
Static fracture mechanics
short-pulse duration tests and **A8:** 282-283
Static friction *See* Limiting static friction
Static friction coefficients for selected materials **A18:** 70-75
factors affecting relative contributions **A18:** 70
Static friction factor value **A18:** 67, 68
Static hot pressing **A7:** 11
as cermet forming technique **A7:** 800
Static indentation test
durometer **A8:** 104, 106-107
as hardness test **A8:** 71
Static joint lap shear tests
for joint/fastener strength **EM1:** 706, 710
Static leak testing
defined **A17:** 61
Static load rating
defined **A18:** 18
Static load strain surveys **EM3:** 543
Static loading
damage development under **A8:** 714
delamination during **A8:** 714
stress-corrosion cracking tests by **A8:** 496-498, 501-503
tests to predict stress-corrosion performance **A8:** 501-503
Static loading tests
bending vs uniaxial tension **A13:** 247-248
constant-strain vs. constant-load **A13:** 246-247

Static loading tests (continued)
of corrosion inhibitors ... **A13:** 483
elastic-strain specimens ... **A13:** 248-252
plastic-strain specimens ... **A13:** 252-253
of SCC smooth specimens ... **A13:** 246-253
Static magnetic parameters
FMR quantitative determination of ... **A10:** 267
Static metallic material properties
design allowables for ... **A8:** 662-677
Static modulus ... **EM3:** 27
defined ... **EM1:** 22 **EM2:** 40
Static modulus of elasticity
vs. dynamic modulus ... **A8:** 249
Static pressing
cold, of cermets ... **A2:** 980-981
hot, of cermets ... **A2:** 986
Static proof test
truncated fatigue life by ... **A8:** 718
Static random-access memories (SRAMs)
soft errors ... **EL1:** 805
Static random-access memory
as digital GaAs application ... **A2:** 740
Static recrystallization ... **A9:** 690-691
in aluminum ... **A9:** 691
in Fe-3.25Si ... **A9:** 691
in oxygen-free copper, reduced 86% ... **A9:** 691
Static secondary ion mass spectroscopy (SSIMS) ... **EM2:** 811, 817
Static shear stress-shear strain
curve ... **A8:** 225
Static sliding friction coefficient ... **A18:** 478
Static strength ... **EM1:** 432-435
distribution ... **A8:** 716, 717
and fatigue life, related ... **A8:** 717-718
lamina ... **EM1:** 432-434
stress concentrations/damage ... **EM1:** 434-435
structural composite laminate ... **EM1:** 433-434
Static stress ... **EM3:** 27
defined ... **EM1:** 22 **EM2:** 40
Static tensile stress
and fretting wear ... **A18:** 248
Static tensile stresses
role of in stress-corrosion cracking ... **A1:** 724
Static tension loading ... **A8:** 717-718
Static test
full-scale ... **EM1:** 347-350
Static testing ... **EM3:** 315-316
Static uniaxial compaction
of cermets ... **A2:** 979
Static uniaxial load
in tension testing ... **A8:** 19
Static viscosity *See* Viscosity
Static yield strength
and Soderberg's law ... **A8:** 374
Station-function approach
to Astroloy data ... **A8:** 334
Stationary crucible furnaces ... **A15:** 381-383
Stationary cutting machines ... **M6:** 907
Stationary equipment
for liquid penetrant inspection ... **A17:** 78-79
magnetic particle inspection ... **A17:** 93
Stationary oxyfuel gas cutting machine ... **A14:** 728
Stationary phase
defined ... **A10:** 682
Stationary punch and mountings
designing ... **A7:** 335, 336
Stationary shears
cut-to-length lines ... **A14:** 711
Stationary-spindle swagers ... **A14:** 130-131
Statistic
defined ... **A8:** 12
Statistical (random) distribution particle pattern ... **A7:** 186
Statistical analysis *See also* Data analysis; Quality control; Statistical methods; Statistical process control
approach ... **A14:** 928
blending and premixing ... **A7:** 187
of casting defects ... **A17:** 523
Statistical analysis (continued)
control charting ... **A14:** 928-931
of data ... **EM2:** 599-609
deformation control ... **A14:** 932-937
of dimensional inspection data ... **A15:** 560-561
experimental design ... **A14:** 937-939
of fiber properties ... **EM1:** 731, 733-734
of forming processes ... **A14:** 928-939
historical tracking ... **A14:** 728-937
image interpretation, machine vision ... **A17:** 35-36
of mechanical properties ... **EM1:** 302-307
multiple batches ... **EM1:** 304-306
single-batch ... **EM1:** 303-304
Statistical concepts
basic ... **A8:** 623-624
central tendency ... **A8:** 624-625
confidence limits ... **A8:** 626
degrees of freedom ... **A8:** 625
hypothesis testing ... **A8:** 626-627
intervals ... **A8:** 626
probability ... **A8:** 624
random variables ... **A8:** 624
reliability ... **A8:** 627
variability ... **A8:** 625
Statistical distribution ... **A8:** 628-638
selecting ... **A8:** 637-638
verified by goodness-of-fit test ... **A8:** 637-638
Statistical measurement, and individual measurement
compared ... **A12:** 207-208
Statistical methods
confidence interval method ... **A17:** 746
estimating failure factors, from test data ... **EL1:** 900-903
failure density distributions ... **EL1:** 896-987
hypothesis testing ... **A17:** 746
for process capability ... **A17:** 738-739
quality design and control ... **A17:** 719-753
reliability growth ... **EL1:** 903-904
reliability life cycle ... **EL1:** 897-899
reliability mathematics ... **EL1:** 895-896
reliability of systems ... **EL1:** 899-900
for reliability prediction ... **EL1:** 895-905
solder joint fatigue ... **EL1:** 745-746
time-to-failure ... **EL1:** 888-889
Statistical methods, sampling *See also* Sample(s); Sampling ... **A7:** 212-213
Statistical planning
and analysis ... **A13:** 316-317
Statistical precision
assumptions of ... **A10:** 525
Statistical procedures
parametric and nonparametric ... **A8:** 653
Statistical process control *See also* Process control; Statistical methods ... **A14:** 928-937
and stability analysis, experimental study ... **A17:** 742
and statistical tolerance model ... **A17:** 739-740
Statistical process control (SPC) *See also* Automation; Process control; Quality control; Statistical methods ... **EM3:** 701, 735, 785-798
algorithmic ... **EM4:** 87
application ... **EM3:** 797-798
application to ceramics ... **EM4:** 85-86, 87
attribute control charts ... **EM3:** 795-797
control charts for individual measurements ... **EM3:** 794-795
control charts for measurements ... **EM3:** 795
DOE-ATTAP program ... **EM4:** 998
effect on rework, tape automated bonding ... **EL1:** 288
monitoring process output over time ... **EM3:** 790-791
polished components ... **EM4:** 469
process capability and product quality ... **EM3:** 785-790
process control ... **EM3:** 791-794
testing normality ... **EM3:** 790
wave soldering ... **EL1:** 696
Statistical process control of operations ... **A4:** 620-637
examples ... **A4:** 623-624
integration of SPC and SQC ... **A4:** 636-637
economic considerations ... **A4:** 637
modeling and feedback for process refinement ... **A4:** 636-637
process analysis ... **A4:** 624-636
characterization plan ... **A4:** 624-625
computerization of SPC/SQC systems ... **A4:** 635-636
experiment design ... **A4:** 625-628
monitor/control decisions of furnace atmospheres ... **A4:** 628-635
noise factors ... **A4:** 626-628
process factors ... **A4:** 626-628, 629
signal-to-noise ratio analysis ... **A4:** 626-628, 629
process capabilities ... **A4:** 622-624
process deterioration ... **A4:** 621-622
processing method considerations ... **A4:** 621
product capabilities ... **A4:** 622-624
SPC/SPQ nomenclature ... **A4:** 620-621
versus traditional control ... **A4:** 620-622
Statistical process control procedures ... **A15:** 664
Statistical significance
in comparative experiments ... **A8:** 642
Statistical summary report
of castings ... **A17:** 523
Statistical theory of rubber elasticity ... **EM3:** 412
Statistical tolerancing
defined ... **A17:** 739
Statistical weight
plot vs. true probability of failure ... **A8:** 705
Statistician
role in planning experiments ... **A8:** 639
Statistics
defined ... **A8:** 624
descriptive ... **A8:** 624
for image analysis ... **A10:** 313
and probability ... **A8:** 624
sampling ... **A10:** 12-15
Statuary bronze copper coloring solutions ... **M5:** 625-626
Statuary, metal *See* Art founding
Statue of Liberty
galvanic corrosion ... **A13:** 86
Stave bearing
defined ... **A18:** 18
Staying ... **EM3:** 27
Steadite
in cast irons, magnifications to resolve ... **A9:** 245
defined ... **A9:** 17
in gray iron ... **A15:** 633
in gray iron, effect of different etchants ... **A9:** 246
Steadite, in gray cast iron
effect on wear resistance ... **M1:** 24
Steady component of stress *See* Mean stress
Steady humidity-temperature test
under bias ... **EL1:** 495
Steady loads
defined ... **A8:** 12
Steady shear flows
in mold filling ... **EL1:** 838-839
Steady shear viscosity
and normal stresses ... **EL1:** 842-843
Steady-rate creep *See* Creep
Steady-shear rheometry ... **EM2:** 535
Steady-state creep ... **A8:** 301-302, 304, 306
defined ... **A12:** 19
effect of work hardening ... **A8:** 310
stress dependence of ... **A8:** 309
temperature dependence of ... **A8:** 309
Steady-state creep rate ... **A8:** 304, 306
characteristics ... **A8:** 308
effect of loading direction ... **A8:** 302
time-to-rupture as function of ... **A8:** 306
Steady-state life tests ... **EL1:** 494, 498
Steady-state scale thickness ... **A18:** 207, 208, 209, 210

Steady-state strain rate
high-purity copper **A12:** 399-400
Steady-state thermal inspection methods **A17:** 400
Steady-state wave model **A18:** 59, 60, 66-68
Steady-state wear period **A18:** 613
Steam *See also* Elevated temperatures; Moisture; Water; Water vapor
-bypass system, in boilers **A11:** 615
in chemical processing plant **A13:** 1135
chemistry **A13:** 992-993
cleaning, of surfaces **A13:** 414, 1139
condensate, copper/copper alloys **A13:** 622
condensate, corrosiveness of **M1:** 735
with contaminants, fatigue crack growth testing **A8:** 426-430
corrosion, casting alloys **A13:** 575
corrosion, copper/copper alloys **A13:** 328, 622
flow, in steam-cooled tubes **A11:** 606
heat transfer from **A11:** 628
high-temperature, alloy steel resistance...... **A13:** 574
and moist air, corrosion fatigue crack propagation in steel in **A8:** 409
superheated, P/M parts oxidated with....... **A13:** 823
systems
carbon dioxide, effects of **M1:** 733
chemical cleaning of **M1:** 748-749
cooling tower water, treatment of **M1:** 749-750
corrosion **M1:** 747-751
feedwater treatment **M1:** 747-748
scale formation and control **M1:** 748
tantalum resistance to **A13:** 731
treating of ferrous P/M materials......... **M1:** 339-342, 343
zirconium/zirconium alloys corrosion in **A13:** 708
Steam accumulator
weld penetration failure in **A11:** 647-648
Steam blackening
dimensional change during **A7:** 480, 481
Steam blackening of powder metallurgy materials **A9:** 512
bainitic structures **A9:** 662-665
color etching **A9:** 141-142
columnar grains in the macrostructure of **A9:** 623
copper-infiltrated powder metallurgy materials, microstructures **A9:** 510
corrosion test coupon, x-ray map **A9:** 162
effect of hot rolling on solidification structures **A9:** 626-628
equiaxed grains in the macrostructure of **A9:** 623
eutectoid structures **A9:** 658-661
low carbon sheet, resistance spot weld......... **A9:** 586
low carbon-manganese plate, CCT weld metal curve **A9:** 580
macrostructure of an ingot or casting **A9:** 623
martensitic structures **A9:** 668-672
microstructures **A9:** 177-179
nonmetallic inclusions in **A9:** 625-626
orange peel **A9:** 688
peritectic structures **A9:** 679-680
plastically deformed, resulting in orange peel **A9:** 688
powder metallurgy materials, etching **A9:** 508-509
powder metallurgy materials, macroexamination **A9:** 507
revealing as-cast solidification structures **A9:** 624
solidification structures of **A9:** 623-628
Steam bronze *See also* Leaded tin bronzes
properties and applications **A2:** 375-376
Steam bubbles
as casting internal discontinuity **A11:** 354
in iron castings **A11:** 357
Steam cleaning
high pressure **A17:** 81-82
Steam condensate
copper/copper alloys **A13:** 622
Steam corrosion
copper alloys **M2:** 471
Steam distillation
in determination of nitrogen **A10:** 173
Steam engine
effect on foundry industry **A15:** 27
Steam equipment *See also* Boilers
cause of failure in **A11:** 602-603
condensers and feedwater heaters, corrosion of **A11:** 615
corrosion of **A11:** 615
corrosion protection **A11:** 615-616
dissimilar-metal welds in **A11:** 620-621
fire-side corrosion **A11:** 616-620
generator failures, causes of **A11:** 603
generator tubes, thick-lip ruptures in................ **A11:** 605-606
overheating ruptures in **A11:** 603-614
reformer components, elevated-temperature failures in....... **A11:** 290-292
turbine blade, liquid-erosion shield for **A11:** 170-179
turbine components, elevated-temperature failures in....... **A11:** 282-288
water-side corrosion **A11:** 614-616
Steam generators **A13:** 937-945
design/corrosion problems **A13:** 938-945
inside surface (primary-side) SCC **A13:** 941
once-through **A13:** 937-938
recirculating **A13:** 937-940
units affected by **A13:** 939
Steam hammer **A14:** 12, 25, 234
Steam line bellows, Inconel 600
EDX/AES failure analysis **A11:** 38-41
Steam molding
defined **EM2:** 40
Steam power plants
failure analysis procedures for **A11:** 602
Steam process
alkaline cleaning **M5:** 24
Steam surface condenser
corrosion of **A13:** 986-989
Steam tempering **A7:** 453
Steam treating
effects on density and apparent hardness of P/M materials **A7:** 466
of P/M business machine parts **A7:** 669
processing sequence **A7:** 453
as secondary operation **A7:** 453
Steam treatment **A7:** 11
Steam turbine power plants
nickel alloy applications
Steam turbines *See also* Turbines
design **A13:** 995
environment **A13:** 994
erosion-corrosion **A13:** 993-994
low-pressure disks, SCC **A13:** 993
materials **A13:** 994
SCC in **A13:** 952-959
Steam, zirconium
corrosion in **M3:** 786-787
Steam-injected cleaning **A13:** 1139
Steam-lift forging hammers
capacities **A14:** 25
Steam-treating
of P/M part **A13:** 823
Steam/water-side boilers
corrosion of **A13:** 990-993
Stearic acid
copper/copper alloy resistance **A13:** 629
effect on sintered strength **A7:** 192
as lubricant **A7:** 190-193
as mold release agent **EM1:** 158
Stearic acid, as binding agent for samples
x-ray spectrometry **A10:** 94
Stearic acids
as investment casting wax **A15:** 254
Stearone, as wax
investment casting **A15:** 254
Steatite
as ceramic thermocouple wire insulation **A2:** 883
material to which crystallizing solder glass seal is applied **EM4:** 1070
material to which vitreous solder glass seal is applied **EM4:** 1070
in porcelain compositions **EM4:** 4, 5
stone molds of **A15:** 15
thermal expansion coefficient **A6:** 907
STEB *See* Standard Test and Evaluation Bottle
Steel *See also* Ferritic steel; High-carbon spring steel; High-strength steels; Low-carbon steel; Martensitic steel; Stainless steel; Steel, specific types; Steels; Structural steel
50% aluminum-zinc alloy coating **M5:** 348-350
abrasive blasting of **M5:** 91, 94
acid cleaning of **M5:** 59-67
acid dipping of **M5:** 16-17
acid etching of **M5:** 16-18
aged rimmed, Luders bands in **A8:** 548
alkaline cleaning of........ **M5:** 16-18, 24, 27-30, 34-37, 70-71, 81
aluminum coating of **M5:** 333-347
aluminum-killed draw quality, tensile properties **A8:** 555
aluminum-killed, strain measurements and forming limit diagram for............. **A8:** 566
applications, cobalt-base wear-resistant alloys **A2:** 451
approximate equivalent hardness numbers for Rockwell B hardness **A8:** 109-110
arc welding to cast irons **M6:** 307
austempering **M4:** 104-116
austenitizing temperatures **A4:** 961-962
backing bars **M6:** 382
bake-hardening **A4:** 61
bar
abrasive blasting of **M5:** 91
aluminum coating of **M5:** 334
pickling of **M5:** 69, 71, 76-77
rust-preventive compounds used on **M5:** 465-466
base metal solderability **EL1:** 677
in bending or axial fatigue loading notch-sensitivity **A8:** 373
billet, pickling of **M5:** 69
brass plating of **M5:** 285
bridge, toughness control, method for.......... **A8:** 453
Brinell test for **A8:** 84, 88, 89
bronze plating of **M5:** 288
buffing *See* Steel, polishing and buffing of
buffing compounds **M5:** 117
cadmium plating of **M5:** 256, 261-264, 266-268
cadmium-plated, galvanic corrosion with magnesium **M2:** 607
carbon *See* Carbon Steel
castings *See* Cast steels
aluminum coating of **M5:** 339
pickling of **M5:** 72
ceramic/metal seals **EM4:** 508
characteristics **EM4:** 976
Charpy specimens, impact response curves for **A8:** 269
Charpy V-notch test for quality assurance in **A8:** 263
Charpy/fracture toughness correlations for **A8:** 265
chemical resistance **M5:** 4, 7
chemical vapor deposition coating of **M5:** 382-384
chemical vapor deposition of tools............ **EM4:** 217
chips and cutting fluids removed from **M5:** 5, 9-10
chromium plating, decorative................ **M5:** 189-191, 196-197
chromium plating, hard................ **M5:** 171-172, 180, 184-186
removal of plate **M5:** 184-185
chromium/molybdenum in hydrogen, at high and low stress-intensity ranges **A8:** 409-410
coil, pickling of **M5:** 71, 76, 80
cold rolled *See* Cold rolled Steel
cold rolled aluminum-killed *See* Cold rolled aluminum-killed steel
cold rolled aluminum-killed, *r* value **A8:** 550
cold rolled, porcelain enameling of..... **M5:** 512, 517, 527
cold rolled rimmed *See* Cold rolled rimmed steel
cold-rolled, glass-to-metal seals **EL1:** 455
columbium-stabilized, porcelain enameling of **M5:** 512-513
commercial quality (CQ)............ **A4:** 51, 56, 57, 59-61
control of surface carbon content in heat treating............ **M4:** 418, 419, 421, 424-431
cooling curves *See* Cooling curves
copper plating of **M5:** 160-161, 163

Steel (continued)
corner castings, aluminum coating of **M5:** 339
corrosion of **M5:** 431
crack growth **A8:** 678-679
decarburized, porcelain enameling of **M5:** 509-510, 512, 514-515, 521
deep case-hardened, Rockwell scale for **A8:** 76
deep-drawing-quality (DDQ/DQSK) **A4:** 59-61
dip brazing *See* Dip brazing of steels in molten salt
drawing quality (DQ) **A4:** 51-52, 57, 59-61
drawing quality special killed (DQSK) **A4:** 52
drawing-quality, special-killed, porcelain enameling of **M5:** 513
electrolytic cleaning of **M5:** 16-18, 27-30, 34-37
electroplating of *See* Steel plating of
electropolishing of **M5:** 305-308
electroslag welding **M6:** 226
emulsion cleaning of **M5:** 35
enamelability, porcelain enameling in **M5:** 512-513
equivalent hardness numbers for Rockwell C hardness **A8:** 110
equivalent hardness numbers, Vickers hardness **A8:** 112-113
etching of **M5:** 16-18, 180
fabricability, effects of aluminum coatings on **M5:** 336-337
fabrications
aluminum coating of **M5:** 334
electrocleaning of **M5:** 29, 36
fasteners
abrasive blasting of **M5:** 91
aluminum coating of **M5:** 334, 337
cadmium plating of **M5:** 36, 261
electrocleaning of **M5:** 29, 36
ion plating of **M5:** 420
pickling of **M5:** 36
zinc plating of **M5:** 255
fatigue strength, coatings affecting **M5:** 201, 231-232, 325
ferritic, brittle fracture in **A8:** 262
file hardness test for **A8:** 107-108
fluxing of **M5:** 353
forgings, pickling of **M5:** 72, 80
formability
hot dip galvanizing and **M5:** 325
porcelain enameling and **M5:** 512
formed, terne coating of **M5:** 358
forming limit diagram **A8:** 551
forming of, phosphate coating aiding **M5:** 436-437
fully annealed, hardened steel ball indenters for **A8:** 74
furnace brazing *See* Furnace brazing of steels
galvanized *See* Galvanized steel
galvanized, phosphate coating of **M5:** 438, 441
hard, thin-gage, springback tests for **A8:** 564-565
hardened, diamond indenter for **A8:** 74
hardening, effects of **M5:** 325
hardness
aluminum coating affecting **M5:** 339, 343-344
chrome plated, hydrogen embrittlement and **M5:** 185-186
hardness affected by carbon concentration **A4:** 77, 78, 79, 80
hardness and yield strength in **A8:** 560
high- and low-toughness, stress-strain curves for **A8:** 22
high-production parts, aluminum coating of **M5:** 339-341
high-strength low-alloy *See* HSLA steel
high-strength, punch loading **A8:** 229-230
hot dip galvanizing of **M5:** 323-332
silicon steels **M5:** 324, 327-330
hot dip lead alloy coating of **M5:** 358-360
hot dip tin coating of **M5:** 351-355
hot rolled *See* Hot rolled steel

Steel (continued)
hot rolled and normalized, cyclic and monotonic shear stress vs. shear strain of **A8:** 151
hot rolled and normalized, shear strain vs life **A8:** 151
hot workability ratings for **A8:** 586
hot-rolled, porcelain enameling of **M5:** 513-514
hydrogen embrittlement notch tensile testing of **A8:** 27
hydrogen embrittlement of **M5:** 185-186, 237, 268-269, 325
impact toughness, hot dip galvanizing affecting **M5:** 325
induction brazing *See* Induction brazing of steels
interstitial-free (IF) **A4:** 61
interstitial-free, porcelain enameling of **M5:** 512-513
interstitial-free, tensile properties **A8:** 555
ion plating of **M5:** 420-421
k-gradient effect on near-threshold fatigue crack growth rates, *K*-decreasing method **A8:** 379-380
killed, porcelain enameling of **M5:** 512-514
lead plating stripped from **M5:** 275
macrocracks in **A8:** 57
with manganese, fracture surface tested in hydrogen gas **A8:** 487
maraging, for pressure bar construction **A8:** 200
maximum strain level **A8:** 551
mechanical coating of **M5:** 300, 302
mechanical properties, effect of hot dip galvanizing on **M5:** 324-325
medium-strength, load-time response **A8:** 266
nickel plating of **M5:** 199-205, 216-217, 230-232, 238-240
electroless **M5:** 230-232, 238-240
nickel-plated, chrome plating removed from **M5:** 185
non-oxide ceramic joining **EM4:** 480
nonaustenitic, hardness conversion numbers for **A8:** 106
notched high-strength, surface residual stress **A8:** 374
oxyacetylene braze welding *See* Oxyacetylene braze welding of steel and cast irons
painting of **M5:** 473-477, 481, 499, 503-505
phosphate coating of **M5:** 434-439, 441-443, 449
pickling of **M5:** 16-18, 36, 68-82, 352-353, 358, 438
phosphate coating after **M5:** 438
pilings *See* Pilings
pipe *See also* Steel tubular products
ASTM specifications **M1:** 317, 318, 320-322, 324, 325
casing **M1:** 320-321
compositions **M1:** 318-319, 322-324
conduit **M1:** 318
drill pipe **M1:** 319-320
drive pipe **M1:** 320
line pipe **M1:** 315, 319
nickel plating of **M5:** 201, 203
nipples, pipe for **M1:** 318-319
oil country tubular goods **M1:** 315, 319-320
pickling of **M5:** 69
piling pipe **M1:** 318
porcelain enameling of **M5:** 513-514
pressure pipe **M1:** 315, 321, 324, 325
reamed and drifted pipe **M1:** 320-321
special pipe **M1:** 315
standard pipe **M1:** 315, 318
tensile properties **M1:** 320-321, 325
transmission pipe **M1:** 319
types and uses **M1:** 315, 317, 318-321
water main pipe **M1:** 319
water well pipe **M1:** 315, 320-321
plain carbon in reversed bending, fatigue limit **A8:** 372

Steel (continued)
plate, porcelain enameling of **M5:** 513-514
plate, strain-age embrittlement of **M1:** 683-684
plating, preparation for **M5:** 16-18
polishing and buffing **M5:** 108-109, 112-114, 123
poppet valves, aluminum coating of **M5:** 335, 339-341
porcelain enameling of **M5:** 509-531
design parameters **M5:** 524-525
evaluation of enameled surfaces **M5:** 527, 529
frits, composition **M5:** 509-510
methods **M5:** 516-520
porcelain enameling of **M5:** 509-531
sag characteristics **M5:** 512-513, 523-525, 531
selecting, factors in **M5:** 512-514
surface preparation for **M5:** 514-515
types used **M5:** 512
power brushing of **M5:** 151
primary testing direction **A8:** 667
properties **EM4:** 677, 976
quenched and tempered, crack growth rate **A8:** 403
quenching **M4:** 31-68
rare earth alloy additives **A2:** 727
recommended machining specifications for rough and finish turning with HIP metal-oxide ceramic insert cutting tools **EM4:** 969
recovery-annealed **A4:** 62
resistance spot welding **M6:** 477-478, 480
resistance to hydrogen embrittlement, test for **A8:** 539
resulfurized, workability **A8:** 165
rimmed *See* Rimmed steel
rimmed, engineering stress-strain curve for **A8:** 554
rimmed, tensile properties **A8:** 555
Rockwell C and B scales for **A8:** 74
Rockwell hardness scale for **A8:** 76
rod
abrasive blasting of **M5:** 91
aluminum cladding by rolling **M5:** 345-346
pickling of **M5:** 68, 71-72
rust-preventive compounds used on **M5:** 465-466
rust and scale removal **M5:** 12-14, 17
rust-preventive compounds for *See* Rust-preventive compounds
selective plating of **M5:** 298
self-drilling tapping screws, hydrogen cracking in **A8:** 541
shapes, used in mass finishing **M5:** 135
sheet
55% aluminum-zinc alloy coating of **M5:** 348-350
aluminum cladding by rolling **M5:** 345-346
aluminum coating of **M5:** 333-334, 345
chemical vapor deposition coating of **M5:** 383
painting of **M5:** 474
phosphate coating of **M5:** 434-437, 439
pickling of **M5:** 69, 72-77, 80
rust-preventive compounds used on **M5:** 465-466
terne coating of **M5:** 358-360
shot, blasting with **M5:** 83-84, 86-87, 91
shot peening of **M5:** 140-142, 145-149, 186
soft, Rockwell scale for **A8:** 76
spring-tempered, preparation for plating **M5:** 17-18
springs, mechanical coating of **M5:** 300
stampings, pickling and electrocleaning of **M5:** 36
strength
aluminum coating affecting **M5:** 336-339, 342
porcelain enameling affecting **M5:** 513-514, 526-527
strengths of ultrasonic welds **M6:** 752
strip
aluminum coating of **M5:** 333-337
pickling of **M5:** 69, 72-76, 78-79

Steel (continued)
rust-preventive compounds used on **M5:** 465-466
terne coating of **M5:** 358-360
stripping methods **M5:** 3-19
structural quality (SQ).. **A4:** 52
structures, red lead rust inhibitors **A2:** 548
substrate for thermoreactive deposition/diffusion process **A4:** 449
suitability for cladding combinations.......... **M6:** 691
superplasticity of **A8:** 553
tapered, for torsional Kolsky bar high-temperature test..................... **A8:** 222, 225
tensile ductility, volume fraction second-phase particle effect **A8:** 571-572
tensile strength, effects of coatings on **M5:** 324-325, 336-337, 342
terne coating of... **M5:** 358-360
thin, Rockwell scale for.................................... **A8:** 76
tin plating of **M5:** 270
titanium-stabilized, porcelain enameling of... **M5:** 512-513
torch brazing *See* Torch brazing of steels
for torsional Kolsky bar strain rate testing **A8:** 227
toughness as function of crack velocity **A8:** 284
tubing
pickling of **M5:** 69
porcelain enameling of **M5:** 513
rust-preventive compounds used on **M5:** 465-466
typical tensile properties................................ **A8:** 555
ultrahigh-strength steels **A4:** 218
valves, aluminum coating of......... **M5:** 335, 339-341, 345-346
vapor degreasing of **M5:** 45, 53-55
weldments, aluminum coating of **M5:** 334-335
wire
55% aluminum-zinc alloy coating of................ **M5:** 348-349
aluminum cladding by rolling **M5:** 346
aluminum coating of................... **M5:** 335-337, 346
brass plating of... **M5:** 285
cut steel wire shot, blasting with........... **M5:** 83-84
fabrications, pickling and electrocleaning of... **M5:** 36
nickel plating of **M5:** 201
pickling of **M5:** 36, 68-69, 71-72
workability, sulfide effect on **A8:** 166
yield stress dependence on strain rate in.. **A8:** 38, 41
zinc plating of **M5:** 254-255
zinc plated, galvanic corrosion with magnesium... **M2:** 607
Steel, AISI-SAE, specific types
1010
electroless nickel plating of........................ **M5:** 237
wet abrasive blasting of.......................... **M5:** 94-95
1020
aluminum coating of.................................. **M5:** 335
electroless nickel plating of........................ **M5:** 237
4130
electroless nickel plating of........................ **M5:** 237
electropolishing of **M5:** 305
4140, electropolishing of **M5:** 305
4340, electroless nickel plating of................. **M5:** 237
8620, rust and scale removal......................... **M5:** 3-13
Steel alloys *See also* Steel alloys, specific types; Steel(s); Steels
adherends
conversion coating treatments.......... **EM3:** 272-273
nitric acid anodization **EM3:** 272
nitric-phosphoric acid etch.......................... **EM3:** 271
no surface preparation............................... **EM3:** 273
phosphoric acid-alcohol.................... **EM3:** 271, 272
adhesion-dominated durability **EM3:** 666, 667
aluminum 2024-T3 bonding, strain-energy release rate **EM3:** 352
aluminum 2219-T81 bonding, polybenzimidazoles............................. **EM3:** 171
analysis for copper in, neocuproine method.. **A10:** 65
analysis for oxide inclusions in **A10:** 162
anodized... **EM3:** 417
arc-welded, failures in............................ **A11:** 423-426
atom probe composition profile, for heat- treatment responses in **A10:** 594
Steel alloys (continued)
brittle fracture **A11:** 522
catalytic effect on oxidation of polyolefins.. **EM3:** 418
chemical pipe sealants for plumbing.......... **EM3:** 608
chemical resistance... **EM3:** 639
corrosion resistance of bonded joints......... **EM3:** 670
durability of adhesive Joints during storage.. **EM3:** 658
corrosion-dominated durability of shotblasted surfaces **EM3:** 669
delayed failure in SMIE and LME systems **A11:** 243
enamels **EM3:** 302, 303
as lap plate material **EM4:** 353
optical emission spectroscopy......................... **A10:** 21
polyethylene coatings................................... **EM3:** 412
polyurethane shear strength when bonded to polycarbonate **EM3:** 663, 664
Russell's process **EM3:** 271
shear strength.. **EM3:** 670
substrate cure rate and bond strength for cyanoacrylates.............................. **EM3:** 129
sulfuric acid/dichromate anodization........ **EM3:** 272
sulfuric acid/dichromatic etch **EM3:** 272
surface parameters **EM3:** 41
wedge testing **EM3:** 667, 668
wedge-crack propagation tests **EM3:** 272
wedge-test specimens.................................. **EM3:** 271
zinc-phosphate treatment **EM3:** 272-273
Steel alloys, specific types *See also* Steel alloys; Steels
202, SIMS depth profiles................................ **A10:** 623
1010, effect of improper polishing **A10:** 301
1070 shaft, residual stress and correction for surface removal........................ **A10:** 389
4140 steel hook, flow lines in forged........... **A10:** 303
4340 ground, effect of stress gradient correction on measurement of near-surface stresses for.. **A10:** 388
4340, optical micrograph of fracture surface.. **A10:** 512
4340, overload failure of quench-cracked threaded rod **A10:** 511-513
52100 bearing, high-resolution Jominy bar analysis.................................... **A10:** 508-509
commercial chromium-molybdenum, atom probe mass spectrum for........... **A10:** 592
NBS reference, positive SIMS spectra under oxygen bombardment.............. **A10:** 616
tempered 2.25Cr-Mo, atom probe mass spectrum, carbide particle................... **A10:** 592
Steel and aluminum sheet
explosively welded **A9:** 386
Steel, annealing.. **A4:** 42-55
annealing cycles........... **A4:** 42-45, 49, 51, 52, 53 **M4:** 14-15
atmospheres, furnace................ **A4:** 45-46, 50 **M4:** 21
austenitizing time...................... **A4:** 43, 44-45 **M4:** 16
bar **A4:** 54-55 **M4:** 19, 25-26
batch (box) annealing **A4:** 51, 52, 54
compared to normalizing **A4:** 35
cooling after transformation..... **A4:** 51, 52, 53, 54, 55 **M4:** 16
critical temperature **M4:** 14, 15
critical temperatures **A4:** 42, 43, 45, 46
cycle annealing **A4:** 45, 46, 49, 50 **M4:** 20
dead-soft steel **A4:** 44-45 **M4:** 16
decomposition of austenite..... **A4:** 42, 43, 44, 53 **M4:** 16
forgings **A4:** 53-54
furnace atmospheres **A4:** 548, 552
furnaces............... **A4:** 45-46, 49-50, 51, 53 **M4:** 20-21
guidelines... **A4:** 45 **M4:** 16-17
high-strength cold rolled **A4:** 50-53 **M4:** 23
induction heating energy requirements........ **A4:** 189
induction heating temperatures **A4:** 188
lamellar annealing........ **A4:** 43, 44, 45, 49, 53 **M4:** 20
machining, structures for......... **A4:** 49, 50, 53-54 **M4:** 20, 21
plate **A4:** 50-51, 52, 55 **M4:** 26
prior structure, effect of................ **A4:** 46, 48 **M4:** 16
process annealing.......... **A4:** 47-48, 50, 51 **M4:** 18-19
rod **A4:** 50-51, 54-55 **M4:** 19, 25-26
sheet **A4:** 50-51
spheroidizing ... **A4:** 43, 46-47, 48, 49, 50, 54 **M4:** 18, 19
Steel, annealing (continued)
strip **A4:** 50-51, 52
temperatures **A4:** 42-43, 45-46, 48, 52, 53, 54 **M4:** 17-18, 19
terminology **A4:** 42, 52 **M4:** 26-27
thermal cycles *See* Annealing cycles; annealing cycles
tin mill products **A4:** 52
tubular products....................... **A4:** 47, 49, 55 **M4:** 26
ultrahigh-strength steels **A4:** 208-216, 218
wire **A4:** 48, 54-55 **M4:** 19, 25-26
Steel, ASTM, specific types
A203, porcelain enameling of **M5:** 513
A225, porcelain enameling of **M5:** 513
A285, porcelain enameling of **M5:** 513
A387, porcelain enameling of **M5:** 513
A446, aluminum-zinc coated sheet production.. **M5:** 349-350
A783, aluminum-zinc wire production **M5:** 349-350
A784, aluminum-zinc coated wire production... **M5:** 349
A785, aluminum-zinc coated wire production... **M5:** 350
Steel, austempering
ultrahigh-strength steels **A4:** 213
Steel, austenitic
physical properties, related to thermal stresses **A4:** 605
Steel auto body panel
galvanic corrosion **A13:** 86
Steel balls
magnetic particle inspection of................. **A17:** 98-99
Steel bands
development.. **A15:** 28
Steel, bar *See also* Bar(s)
annealing... **A4:** 870
eddy current inspection **A17:** 553-555
electromagnetic inspection methods..... **A17:** 552-555
flaw detection.. **A17:** 557
flaws, types of... **A17:** 549-550
inspection methods **A17:** 550-554
liquid penetrant inspection **A17:** 550-551
magnetic particle inspection......................... **A17:** 550
magnetic permeability systems **A17:** 555-556
NDE equipment requirements..................... **A17:** 557
residual stresses **A4:** 605, 606
round, eddy current inspection **A17:** 185-186
sorting procedure **A17:** 557
tempering.. **A4:** 132, 133
ultrasonic inspection............... **A17:** 271-272, 551-552
Steel bar and tube, normalizing
alloy steels **A4:** 39-40 **M4:** 8
furnaces **A4:** 39-40 **M4:** 8
Steel bars *See* Alloy steel bars; Carbon steel bars; Cold-finished steel bars; Hot-rolled steel bars and shapes
Steel bars, extruded
centerbursts in **A8:** 595
Steel bearing alloys, specific types
52100, bar, different heat treatments compared **A9:** 195-196
52100, bar, different magnifications compared **A9:** 195
52100, damaged by an abrasive cutoff wheel... **A9:** 196
52100, rod, austenitized and slack quenched in oil **A9:** 196
52100, roller, crack from a seam in bar stock **A9:** 196
Steel black copper coloring solution **M5:** 625
Steel bridge
toughness criteria **A8:** 264-265
Steel can method
hot extrusion of powder mixtures **A2:** 988
Steel, carburizing... **A4:** 363-373
alloying effects **A4:** 366-367
austenite.................... **A4:** 364-366, 369, 370, 371, 373
boost-diffuse method **A4:** 364
carbon gradient... **A4:** 364
carbon potentials **A4:** 364
carbon profiles **A4:** 363-364, 370
case depth measurement **A4:** 363-364, 365, 370
effect on fatigue cracking................................ **A4:** 370
eta (η)-carbide **A4:** 366
excessive retained austenite and massive carbides **A4:** 369-370

Steel, carburizing (continued)
fatigue mechanisms **A4:** 372-373
grain size **A4:** 365, 366-367, 369
hardness profile **A4:** 363-364, 365, 368-369
intergranular fracture at austenite grain boundaries **A4:** 367-369
lath martensite **A4:** 366
martensite **A4:** 364-366, 369, 370, 371, 373
microcracking **A4:** 369
microstructures **A4:** 363, 364-366, 367-368, 372
plasma (ion) nitriding **A4:** 423
residual stresses **A4:** 370 371
surface and internal oxidation **A4:** 371-372
Steel castings *See also* Austenitic manganese steel castings; Carbon steel castings; Cast steel; Low-alloy steel castings; Steels **A1:** 363-379, 483
classifications and specifications **A1:** 363-364, 365, 366
development **A15:** 31
engineering properties **A1:** 376-378
corrosion resistance **A1:** 376
elevated-temperature properties **A1:** 376-377
low-temperature toughness **A1:** 377-378
machinability **A1:** 378
soil corrosion **A1:** 376
wear resistance **A1:** 376
weldability **A1:** 378
high-carbon cast steels **A1:** 372
history of **A15:** 31
low-alloy cast steels **A1:** 367, 372-374, 375
low-carbon cast steels **A1:** 364, 371-372
markets for **A15:** 42
mechanical properties **A1:** 365, 367-371
ductility **A1:** 365
fatigue properties **A1:** 369-370
heat treatment **A1:** 367, 370-371
section size and mass effects **A1:** 370, 373
specimens **A1:** 370
tensile and yield strengths **A1:** 365
toughness and impact resistance **A1:** 365, 367-369
medium-carbon cast steels **A1:** 364, 372, 373
nondestructive inspection **A1:** 378-379
olivine sands for **A15:** 209
physical properties **A1:** 374, 376
density **A1:** 374
elastic constants **A1:** 374
electrical properties **A1:** 374
magnetic properties **A1:** 374
volumetric changes **A1:** 374, 376
production, flow diagram of **A15:** 203
Steel castings, failures of *See also* Steel(s) **A11:** 380-410
by corrosion **A11:** 401-405
casting defects **A11:** 380-391
composition and **A11:** 391-392
in elevated temperatures **A11:** 405-408
improper heat treatment and **A11:** 392-396
overload of **A11:** 396-399
related to hydrogen-assisted cracking **A11:** 408-410
related to welding **A11:** 399-401
Steel castings, normalizing
cooling **A4:** 40 **M4:** 12
heating **A4:** 40 **M4:** 12
loading **A4:** 40 **M4:** 12
loading temperatures **A4:** 40 **M4:** 12
normalizing **A4:** 40
soaking **A4:** 40 **M4:** 12
Steel castings, specific types
C5, normalizing **A4:** 40
C12, normalizing **A4:** 40
WC9, normalizing **A4:** 40
Steel check-valve poppet
brittle fracture of **A11:** 70-71
Steel chips, analysis for silver
lead, and cadmium in **A10:** 55
Steel, cold treating
advantages **A4:** 204 **M4:** 118
equipment **A4:** 204 **M4:** 118
stress relief **A4:** 203-204 **M4:** 118
transformation *See* Steel, transformation
Steel, corrosion-resistant
liquid nitriding **A4:** 419
Steel crane hooks
materials and failures of **A11:** 522-524
Steel, cryogenic treatment **A4:** 203, 204-206
case studies **A4:** 204-205
direct spray system **A4:** 206
equipment **A4:** 206
heat-exchanger system **A4:** 206
kinetics **A4:** 204
treatment cycles **A4:** 204
Steel dies
for cold upset testing **A8:** 579
Steel, ferrite *See* Ferrite steels
Steel, forged
use in metalworking rolls **M3:** 503, 506-507
Steel, forging
induction heating energy requirements **A4:** 189
induction heating temperatures **A4:** 188
Steel forgings
flaws, and inspection methods **A17:** 495-496
liquid penetrant inspection **A17:** 501-503
Steel forgings, annealing
cold forming **M4:** 24
hardness after annealing **M4:** 25, 26
machinability, annealing for **M4:** 24-25
pearlitic microstructures **M4:** 24-25
Steel forgings, normalizing
axle-shaft **M4:** 10
furnaces **M4:** 8
low-carbon **M4:** 10
mechanical properties, effect on **M4:** 10-11
multiple treatments **M4:** 11
structural stability **M4:** 10
Steel Founder's Society of America **A15:** 34
Steel grit blasting
aluminum and aluminum alloys **M5:** 571
ceramic coating preparation **M5:** 539
Steel, hardenability
Climax Molybdenum calculator **A4:** 81
ultrahigh-strength steels **A4:** 208-216, 218
United States Steel (USS) calculator **A4:** 81
Steel, heat-treating principles **A4:** 3-18
carbon content effect **A4:** 3, 4, 5, 10, 11, 16
case hardening **A4:** 15-16
composition **A4:** 3, 4
computer simulation of transformation diagrams **A4:** 7-9
computer-controlled spray cooling **A4:** 13
continuous cooling transformation diagrams (CCT) **A4:** 7, 8, 9, 10, 12, 13
continuous heating transformation (CHT) diagrams **A4:** 6-7, 9
cooling media **A4:** 11-13, 17
cracking and distortion due to hardening **A4:** 18
critical diameter (D_o) **A4:** 9
dilatometer curves **A4:** 7
Grossmann hardenability test **A4:** 9-10
Grossmann number (quench severity concept) **A4:** 13
hardenability **A4:** 9-11
ideal diameter (D_i) **A4:** 9, 10
induction hardening **A4:** 16-18
isothermal cooling (IT) diagrams **A4:** 5, 6, 7, 8, 17
isothermal transformation (ITh) diagrams **A4:** 5-6, 7, 8, 9, 17, 18
Johnson-Mehi-Avrami expression **A4:** 8
Jominy end-quench test **A4:** 10-11
microconstituents **A4:** 3, 4, 5, 10
phase transformations **A4:** 4, 13-18
phases **A4:** 3-4, 13-18
press hardening (fixtures) **A4:** 18
quench intensity **A4:** 11-13
Steel, heat-treating principles (continued)
residual stresses **A4:** 13-16, 17, 18
Scheil-Avrami additivity rule **A4:** 8, 9
temper embrittlement **A4:** 11
tempered martensite embrittlement **A4:** 11
tempering principles **A4:** 11
thermal stresses during the residual stresses after heat treatment **A4:** 13-18
through hardening **A4:** 14-15
transformation temperatures **A4:** 4, 5, 7
TTT diagrams **A4:** 8, 10
Steel, high carbon
tempering artifacts in **A9:** 36, 37
Steel, induction heat treating **A4:** 164-202
advantages **A4:** 164
aging **A4:** 194
annealing with induction heating **A4:** 193-194
applications **A4:** 164, 188, 189, 191, 196-200
armor-piercing projectiles **A4:** 199
automotive **A4:** 199-200
axle shafts **A4:** 197
bar stock **A4:** 200
crankshafts **A4:** 196-197
gears **A4:** 198, 199
hand tools **A4:** 199
miscellaneous **A4:** 199-200
pipe-mill products **A4:** 200
railroad rails **A4:** 199
rolling mill rolls **A4:** 199
structural members **A4:** 200
surface-hardening **A4:** 196-200
through-hardening **A4:** 200
transmission shafts **A4:** 197
valve seats **A4:** 198, 199
control equipment **A4:** 183-184
coupling **A4:** 166-167
definition **A4:** 164
distortion **A4:** 201
eddy current characteristics **A4:** 165-166, 201
equipment **A4:** 167-171, 201
flux shaping **A4:** 166-167
grain refinement **A4:** 194
heat generation **A4:** 166, 201
impedance matching **A4:** 182-183
capacitors **A4:** 182, 183
kVAR ability **A4:** 183
transformers **A4:** 182-183, 184
tuning of fixed-frequency systems **A4:** 182, 183
tuning of variable-frequency systems **A4:** 182
induction hardening **A4:** 184-191
austenitizing temperatures **A4:** 184-186
case depth **A4:** 188, 189
Curie point **A4:** 187, 188
electrical properties **A4:** 186-187, 201
energy requirements **A4:** 189
frequency selection **A4:** 188-189
heating parameters **A4:** 187-188
induction heating temperatures for metalworking processes **A4:** 188
induction tempering **A4:** 186, 191, 194
magnetic properties **A4:** 186-187, 201
operating conditions for through hardening **A4:** 190, 192
power density and heating time **A4:** 189-191
power ratings for surface hardening **A4:** 190
residual stresses **A4:** 185-186, 202
temperatures required **A4:** 188
time-temperature relations **A4:** 184-186
inductor coils **A4:** 171-182
butterfly coils **A4:** 178
coil characterization **A4:** 174-176
coil construction **A4:** 180
coil cooling **A4:** 181-182
coil design **A4:** 171-174, 180-182, 201
coil inserts **A4:** 177
coil insulation **A4:** 180-181
conveyor/channel coils **A4:** 179
coupling distance **A4:** 175-176, 178
electrical characteristics **A4:** 180, 201

Steel, induction heat treating (continued)
flux concentrators **A4:** 179-180
flux diverters (robbers) **A4:** 176, 177
induction scanning **A4:** 174, 175, 177-178
irregularities of parts, effect on heating patterns **A4:** 176
longitudinal flux mode **A4:** 171
master work coils **A4:** 177
mechanical design considerations **A4:** 180
multiturn coils **A4:** 174, 181
offsetting of coil turns **A4:** 174, 175
progressive hardening (scanning) coils **A4:** 177-178
series/parallel coil construction **A4:** 179
single-shot process **A4:** 174-175
single-turn coils **A4:** 174, 177, 181
split coils **A4:** 178, 179
split-return inductors **A4:** 178
transverse flux mode (proximity heating) **A4:** 171
transverse-flux coils **A4:** 178-179, 180
malleable iron **A4:** 695-696
mechanical properties **A4:** 196-200
power requirements **A4:** 167-168
power supplies **A4:** 167-171
constant-current (load-resonant) inverters **A4:** 169-170
constant-voltage (swept-frequency) inverters **A4:** 169, 170
core-type induction heaters **A4:** 168
frequency multipliers **A4:** 167, 169, 182
function **A4:** 167
line-frequency systems **A4:** 167, 168-169, 182
MOSFET (metal-oxide semiconductor field-effect transistor) output devices **A4:** 171
motor-generators **A4:** 167, 168, 169, 170, 182, 184
power ratings as a function of frequency **A4:** 167
power supplies **A4:** 167-171
radio-frequency (RF) power supplies **A4:** 167, 170-171
solid-state (static) inverters **A4:** 167, 168, 169-170, 182, 183, 184
solid-state RF power supplies ... **A4:** 171, 182, 183, 184
spark-gap converters **A4:** 167, 171, 182, 184
power supplies, vacuum tube RF generators **A4:** 168, 170, 171, 184
precipitation hardening **A4:** 194
principles **A4:** 164-167
process control considerations **A4:** 200-202
products **A4:** 164
quality control considerations **A4:** 200-202
quench systems **A4:** 194-196, 197, 199, 202
cracking **A4:** 196, 202
distortion **A4:** 196, 202
quench control **A4:** 195-196
skin effect **A4:** 165, 188, 189, 190-191
stress-relieving **A4:** 193
surface hardening by induction **A4:** 190-191
selective **A4:** 191
volume **A4:** 191
tempering **A4:** 192-193, 197
through hardening with induction heating **A4:** 191-192
tooth-by-tooth hardening **A4:** 198, 199
workhandling equipment **A4:** 183
Steel ingot
schematic of macrostructure **A9:** 623
Steel, low carbon
effect of load in vibratory polishing **A9:** 44
effect of suspending liquid in vibratory polishing **A9:** 43
Steel, martempering
advantages **A4:** 137-138
agitation quenching **A4:** 142, 143, 144, 146, 147, 150, 151 **M4:** 93
applications **A4:** 143-151 **M4:** 96, 97, 98
austenitizing temperatures **A4:** 141, 142, 147 **M4:** 91, 92
bath temperature **A4:** 138, 142, 151 **M4:** 91
cooling **A4:** 143 **M4:** 85, 93
dimensional control **A4:** 143-145 **M4:** 93-97, 99
equipment, austenitizing **A4:** 146-147 **M4:** 97-98
equipment maintenance **A4:** 149-151 **M4:** 100-102
Steel, martempering (continued)
equipment, martempering **A4:** 138, 147-149 **M4:** 98-99, 100, 101, 102
fixturing **A4:** 151 **M4:** 102-103
fluidized beds **A4:** 139, 147
martempering media **A4:** 138-139 **M4:** 86-88
oil ... **A4:** 138, 139, 140, 141, 143, 148, 150, 151 **M4:** 87-88
salt **A4:** 138-140, 141, 143, 146, 148, 149, 151 **M4:** 86, 87
material selection, suitability **A4:** 140-141 **M4:** 88-90, 91
microstructure **A4:** 137
modified **A4:** 138 **M4:** 86
quenching **A4:** 137, 138, 139, 142, 143-144, 146, 147, 151 **M4:** 85, 86
quenching bath time **A4:** 142 **M4:** 91-92
racking **A4:** 151 **M4:** 102-103
safety precautions **A4:** 139-140 **M4:** 88
salt contamination **A4:** 139, 141-142, 147 **M4:** 91
tempering **A4:** 137
ultrahigh-strength steels **A4:** 213
washing the work **A4:** 151
water addition to salt **A4:** 138, 139-140, 142-143 **M4:** 92
Steel, medium carbon
tempering artifacts in **A9:** 36, 37
Steel mill products
U.S., net shipments **A13:** 509-510
Steel molds *See also* Mold(s)
horizontal centrifugal casting **A15:** 296
permanent, water cooling **A15:** 304
Steel mud drum tubing
remote-field eddy current inspection **A17:** 200-201
Steel, nitriding
plasma (ion) nitriding **A4:** 423
ultrahigh-strength steels **A4:** 214, 215
Steel, normalizing **A4:** 35-41
alloy steels **A4:** 35-41 **M4:** 7, 8
applications **A4:** 38 **M4:** 8
carbon steels **A4:** 35-37, 38, 39 **M4:** 7, 8, 9-10
double **A4:** 39
induction heating energy requirements **A4:** 189
induction heating temperatures **A4:** 188
normalizing range, carbon steel **A4:** 36 **M4:** 6
properties after treatment **A4:** 36, 37, 39
temperatures **A4:** 35, 36, 37, 38, 39 **M4:** 6, 7
ultrahigh-strength steels **A4:** 208-216, 218
uses **A4:** 35, 36, 39-40 **M4:** 6-8, 9-10, 11
Steel, overaging
ultrahigh-strength steels **A4:** 218
Steel parts *See also* Parts; Steel bars; Steels; Steels, specific types
cold-extruded, ultrasonic inspection **A17:** 271
Steel, pearlitic *See* Pearlitic steels
Steel, pearlitic wire
curly lamellar structure **A9:** 688
Steel pipe *See also* Pipe; Tubular products
continuous butt-welded **A17:** 567
double submerged arc welded **A17:** 565-566
galvanic corrosion **A13:** 86
spiral-weld **A17:** 567
Steel pipe, specific types *See also* Steel tubing, specific types; Steel tubular products
API 5A, Grade K55, seamless, as-rolled **A9:** 211
API 5AC, Grade C-90, seamless, austenitized, quenched and tempered **A9:** 212
API 5AX, Grade N-80, seamless, austenitized, quenched and tempered **A9:** 211
API 5AX, Grade P-110, seamless, austenitized, quenched and tempered **A9:** 211
API 5L, Grade A, continuous welded, as- rolled **A9:** 212
API 5L, Grade X46, resistance weld **A9:** 212
API 5L, Grade X52, seamless, as-rolled **A9:** 211
API 5L, Grade X52, submerged arc welded, as-rolled **A9:** 212
API 5L, Grade X60, electric resistance welded, as-rolled **A9:** 212
API 5L, Grade X60, gas metal arc weld **A9:** 212
ASTM A106, Grade A, seamless, as hot drawn **A9:** 212
Steel pipe, specific types (continued)
ASTM A106, Grade A, seamless, normalized by austenitizing **A9:** 212
ASTM A106, Grade B, seamless, different sections and heat treatments **A9:** 212-213
ASTM A335, Grade P2, seamless, cold drawn and stress relieved **A9:** 213
ASTM A335, Grade P5, seamless, annealed **A9:** 213
ASTM A335, Grade P7, seamless, annealed **A9:** 213
ASTM A335, Grade P11, seamless, annealed **A9:** 214
ASTM A335, Grade P22, seamless, hot drawn and annealed **A9:** 214
ASTM A381, Class Y52, gas metal arc welded, annealed **A9:** 214
Steel pipelines
nondestructive evaluation **A17:** 578-579
Steel plate *See* Plate steels **A1:** 226-239
applications of **A1:** 226
fabrication considerations **A1:** 238-239
formability **A1:** 238
machinability **A1:** 238
weldability **A1:** 238-239
heat treatment of **A1:** 230-232
normalizing **A1:** 231
quenching **A1:** 231
stress relieving **A1:** 231-232
tempering **A1:** 231
high-strength low-alloy steel plate
compositions **A1:** 401, 406, 410
controlled rolling of **A1:** 408-409, 586-587
mechanical properties of **A1:** 401, 409, 411, 586-587
normalized HSLA steel plate **A1:** 409-410
specifications **A1:** 399, 401
imperfections **A1:** 230
decarburization **A1:** 230
seams **A1:** 230
segregation **A1:** 230
mechanical properties **A1:** 237-238
directional properties **A1:** 238
elevated-temperature properties **A1:** 238
fatigue strength **A1:** 238
low-temperature impact energy **A1:** 238
static tensile properties **A1:** 227, 228, 229, 230, 232-233, 235-236, 237-238
platemaking practices **A1:** 228, 230
quality of **A1:** 226, 234-235, 236-237, 741
quenched and tempered carbon steel **A1:** 391
low-alloy steel compositions **A1:** 392
tensile properties **A1:** 391, 396
steelmaking practices **A1:** 226-228
austenitic grain size **A1:** 227-228
deoxidation practice **A1:** 226-227
melting practices **A1:** 228
types of **A1:** 226, 227-228, 232-237
aircraft quality **A1:** 237
carbon **A1:** 232-233
forging quality **A1:** 237
high-strength low-alloy **A1:** 235-236
low-alloy **A1:** 233
pressure vessel **A1:** 237
regular quality **A1:** 236
structural quality **A1:** 236
Steel processing technology **A1:** 107-125
basic oxygen process (BOP) **A1:** 110, 111, 112
Kawasaki basic oxygen process **A1:** 111
quick-quiet basic oxygen process **A1:** 112
controlled rolling of microalloyed steels **A1:** 115, 117-118, 130-131, 408-409, 587-588
conventional controlled rolling **A1:** 117, 409
dynamic recrystallization controlled rolling **A1:** 117-118, 409
recrystallization controlled rolling **A1:** 117, 409
furnaces **A1:** 108-109, 110
basic oxygen furnace **A1:** 110
blast furnace **A1:** 109
hot metal desulfurization **A1:** 109
current technology **A1:** 108, 109
ironmaking **A1:** 107-109
blast furnace stove use **A1:** 107-108
cokemaking **A1:** 107

Steel processing technology (continued)
current blast furnace technology **A1:** 108-109
liquid processing **A1:** 107-114
desulfurization **A1:** 109-110
future technology for **A1:** 114
ironmaking **A1:** 107-109
steelmaking **A1:** 110-114
of solid steel **A1:** 114-123
annealing **A1:** 122-123, 132-133
cold rolling **A1:** 121-122, 132-133
hot rolling **A1:** 115-120
warm rolling **A1:** 120
steelmaking **A1:** 110-114, 226-228, 930
effects on formability **A1:** 577-588
electric furnace steelmaking **A1:** 111
ferroalloy/deoxidizer additions **A1:** 111-112
first-stage refining **A1:** 110, 111, 112
ladle steelmaking **A1:** 112-113
mold metallurgy **A1:** 114
second-stage refining and technology advances **A1:** 111
temperature-time schedules for various steel processing technologies **A1:** 130, 131
third-stage refining **A1:** 113
tundish metallurgy and continuous casting **A1:** 113-114

Steel production
raw steel production by type of furnace steel grade, and casting technique **A1:** 147
raw steel production by various countries **A1:** 154, 165

Steel, quenching **A4:** 67-120
agitation ... **A4:** 69, 72-76, 85, 88, 95-96, 101, 103-104, 107, 113-115 **M4:** 47, 48, 49, 60-61
factors controlling agitation **A4:** 69, 75, 88-89, 113-116 **M4:** 60-61
measurement of velocity **A4:** 72-76, 88, 103, 114-115, 117 **M4:** 61
molten salt **A4:** 88-89, 105 **M4:** 49, 60
oil flow **A4:** 69, 74-76, 95-96 **M4:** 47, 48, 49, 60
turbulent agitation **A4:** 75, 88, 101, 115 **M4:** 61
water and brine **A4:** 70, 75, 88-89 **M4:** 60
agitation equipment **A4:** 95, 110, 113, 115, 118 **M4:** 46, 61, 62, 64
gravity fall **A4:** 110 **M4:** 64
impellers **A4:** 95, 110, 115, 118
movement of the workpiece **A4:** 110, 113 **M4:** 64
propellers **A4:** 113-114, 116, 117 **M4:** 46, 61, 62, 64
pumps ... **A4:** 89, 95, 100, 110, 113, 116-118 **M4:** 64
air **A4:** 67, 72, 75
austempering **A4:** 67-68
brine solutions **A4:** 69-70, 72, 74-76, 88-93, 101, 115 **M4:** 36, 37, 41-43
advantages and disadvantages **A4:** 89-90, 115 **M4:** 36, 41
cooling rates **A4:** 69, 89, 90, 93 **M4:** 41-42
effect of brine concentration **A4:** 89-90 **M4:** 36, 37, 42
effect of brine temperature **A4:** 89, 90 **M4:** 36, 37, 42-43
effect of contamination **A4:** 90 **M4:** 43
maintenance schedules for quenching systems **A4:** 118
caustic solutions **A4:** 88, 90, 101, 115 **M4:** 43
cold die quenching **A4:** 106 **M4:** 66
cooling curves **A4:** 68-88, 90-93, 96, 108-109, 114 **M4:** 32-34
agitation **A4:** 69, 101, 103 **M4:** 33
applications **A4:** 74-76
calculation methods **A4:** 70, 80-83
convective stage **A4:** 69, 71, 88, 91, 92, 97, 102
effects of immersion **A4:** 68-69, 74 **M4:** 32, 33
initial liquid contact stage **M4:** 32, 33
Leidenfrost phenomenon **A4:** 68
Leidenfrost temperature **A4:** 68

Steel, quenching (continued)
liquid cooling stage **M4:** 32, 33
nucleate boiling stage **A4:** 69, 70-71, 88, 91, 102
significance of **A4:** 69-70 **M4:** 33
temperature of quenchant **A4:** 68-75, 101 **M4:** 33-34
vapor blanket cooling stage **A4:** 68-71, 88, 91, 102 **M4:** 32, 33
vapor transport cooling stage **M4:** 32, 33
workpiece temperature **M4:** 34
cooling systems **A4:** 117-118 **M4:** 63, 66
maintenance costs **A4:** 117 **M4:** 66
selection of **A4:** 117
shell-and-tube heat exchanger **A4:** 117 **M4:** 66
cracking **A4:** 76-80, 84, 88, 91, 97, 98
definition **A4:** 67 **M4:** 31
design of quench tanks **M4:** 65
distortion **A4:** 76-79, 88, 91, 97-99, 105-106, 116
evaluation of severity **M4:** 34, 35, 38-39
examples **A4:** 98-100, 102-104, 105-106, 116
fixtures **A4:** 95, 110, 116 **M4:** 65
restraint fixturing **A4:** 110, 116 **M4:** 65
support fixtures **A4:** 95, 110, 116 **M4:** 65
fluidized bed quenching **A4:** 69, 106, 107
fog quenching **A4:** 67, 106 **M4:** 60
gas quenching **A4:** 67, 69, 89, 105-106 **M4:** 45, 46, 58-59
applications **A4:** 105 **M4:** 45, 46, 59
gas quenching unit **A4:** 105 **M4:** 58-59
hardening tool steel **A4:** 77, 78-79, 105-106 **M4:** 59
recirculation **A4:** 67, 69, 105 **M4:** 58
Grossmann number **A4:** 71-75, 84-88, 91, 94, 100-102, 104, 110
hardness correlation **A4:** 71, 72, 74, 80-88
heaters **A4:** 95, 117 **M4:** 67
ideal critical diameter **A4:** 80-81
induction heat treating **A4:** 195
isothermal quenching **A4:** 67, 68
Jominy equivalent **A4:** 82-83, 85, 86
maintenance of quenching installations **A4:** 118 **M4:** 67-68
major variables **A4:** 67, 94 **M4:** 31-32
direct quenching **A4:** 67 **M4:** 31
fog quenching **A4:** 67 **M4:** 31-32
interrupted quenching **A4:** 67
selective quenching **A4:** 67 **M4:** 31
spray quenching **A4:** 67, 94, 107, 114, 195, 196 **M4:** 31
time quenching **A4:** 67 **M4:** 31
marquenching **A4:** 67, 68, 75-76, 91, 93
mass and section size .... **A4:** 75, 108, 110-115 **M4:** 51, 52 , 54, 56, 57, 62
mechanical conditioning of quenching oils **A4:** 100 **M4:** 67
mechanism of quenching **A4:** 75 **M4:** 32
metallurgical aspects **A4:** 68, 75, 77-88 **M4:** 33, 34-35
carbon content **A4:** 77, 78, 79, 80, 82, 83 **M4:** 34-35
cooling rates **A4:** 75, 82-89, 91-94, 101, 105 **M4:** 35
hardenability **A4:** 75, 80-88, 107-109 **M4:** 34-35
martensite **A4:** 68, 80, 83 **M4:** 33, 34-35
oil quenching **A4:** 67, 69, 72, 74-76, 78, 89-100, 109 **M4:** 38, 39, 40, 41, 42, 43-54
control of quenching oils **A4:** 91, 93, 94, 97-98 **M4:** 53
conventional quenching oils **A4:** 90, 92-95, 97 **M4:** 44
cooling characteristics **A4:** 69, 91-93 **M4:** 38, 39, 40, 43, 44, 45-47
emulsions **A4:** 90, 91, 93-94, 96, 97 **M4:** 40, 45
fast quenching oils **A4:** 72, 76, 91-95, 97-100, 109, 114, 115 **M4:** 44
hot quenching oils *See* martempering **M4:** 44-45
induction heat treating **A4:** 195

Steel, quenching (continued)
maintenance schedules for quenching systems **A4:** 118
martempering **A4:** 75-76, 91 **M4:** 44-45
oil temperature **A4:** 69, 92, 93-96 **M4:** 40, 47-49
quench loads **A4:** 86, 91, 95 **M4:** 49
recirculation **A4:** 113
safety precautions **A4:** 118-119
selection of quenching oil **M4:** 42, 53-54
staining **A4:** 97
surface conditions of quenched work **A4:** 92 **M4:** 53
water contamination **A4:** 96-97, 99 **M4:** 41, 49-53
polymer solutions **A4:** 67, 69, 72-75, 100-104, 107, 209-210
control measures **A4:** 101-102
cooling characteristics **A4:** 101, 102, 104, 106
polyacrylates **A4:** 101, 102-104
polyalkylene glycols (PAG) **A4:** 69, 70, 74, 100-102, 103, 104, 107
polysodium acrylate **A4:** 100
polyvinyl alcohol (PVA) **A4:** 76, 100, 101, 104, 106, 195
polyvinylpyrrolidone (PVP) **A4:** 72, 100, 101, 102, 104
sodium polyacrylate (PA) ... **A4:** 101, 102, 104, 105
press quenching **A4:** 106-107 **M4:** 66
pumps **A4:** 89, 95, 100, 110, 113, 116-118 **M4:** 67
quench factor analysis **A4:** 71, 84-85
quenching media **A4:** 88-105 **M4:** 39
air **A4:** 67, 72, 75, 195, 903
aqueous *See* brine solutions
brine solutions **A4:** 72, 74, 75, 87-90 **M4:** 39, 41-43
caustic solutions **A4:** 88, 90 **M4:** 39, 43
cold dies **A4:** 106
dry dies **M4:** 39, 66
fog quenching **A4:** 67, 106 **M4:** 39, 60
gases (still or moving) **A4:** 67, 69, 89, 105-106 **M4:** 39, 58-59
molten metals **A4:** 83 **M4:** 39
molten salts **A4:** 67-69, 72, 83, 87-89, 91, 101, 104-105 **M4:** 39, 60
oils **A4:** 72-74, 87-100 **M4:** 43-54
polymer solutions **A4:** 67, 69, 72-76, 88, 94, 100-104, 107, 209-210 **M4:** 39, 55-58
selection of **A4:** 67, 74, 98
water **A4:** 72-74, 87-89, 903 **M4:** 39, 63
quenching of flame heated parts **M4:** 68
quenching of flame-heated parts **A4:** 107
quenching systems **A4:** 67, 109-119 **M4:** 58, 59, 60, 61, 62-63
continuous quenching **A4:** 110 **M4:** 58, 63
oil quenching systems for continuous carburizers **A4:** 96, 110-113 **M4:** 59, 60, 63
special techniques **M4:** 61, 63-65
water quenching system **A4:** 110, 118 **M4:** 58, 63
Rushman-Lamont method **A4:** 87-88
safety precautions, and extinguishing oil fires **A4:** 96, 115, 118-119
safety precautions, extinguishing oil fires **M4:** 68
severity (intensity) **A4:** 71-75, 84-88, 90-92, 94, 100-102, 104, 110
solutions **M4:** 35, 42, 43, 44, 55-58
control measures **M4:** 56
cooling characteristics **M4:** 35, 42, 43, 55-58
polyacrylates **M4:** 44, 57-58
polyalkylene glycols (PAG) **M4:** 55-56
polyvinyl alcohol (PVA) **M4:** 55
polyvinylpyrrolidone (PVP) **M4:** 56-57
storage or supply tanks, design **A4:** 67, 89, 95, 110, 114-116, 117, 118 **M4:** 63, 66-67
surface oxidation **A4:** 69, 75-76, 95, 97, 110 **M4:** 50, 61-62
cooling curves **A4:** 75 **M4:** 50, 62

Steel, quenching (continued)
high-speed motion-picture techniques ... **A4:** 75-76 **M4:** 62
magnetic testing ... **A4:** 75-76 **M4:** 61-62
tank design (quench) ... **A4:** 67, 95, 114-116
testing and evaluation of quenching
cooling curve test ... **A4:** 98, 108-109
cooling power test ... **A4:** 108-109
crackle test ... **A4:** 98
GM Quenchometer (nickel ball) test ... **A4:** 90, 98, 109
hardening power tests ... **A4:** 107-108
hot wire test ... **A4:** 98, 109
immersion quench test ... **A4:** 107-108
internal test ... **A4:** 109
Jominy end-quench test (ASTM A 255) ... **A4:** 73-74, 81-84, 87-89, 107, 108
magnetic test ... **A4:** 75-76, 92, 93, 98, 109
media ... **A4:** 71-76, 107-109
nickel ball (GM Quenchometer) test ... **A4:** 90, 98, 109
sludge test ... **A4:** 98
viscosity test ... **A4:** 98
water test ... **A4:** 98
tests and evaluation of
cooling curve test ... **M4:** 35-37
cooling power tests ... **M4:** 35
hardening power tests ... **M4:** 35
hot wire test ... **M4:** 37-38
immersion quench test ... **M4:** 35
internal test ... **M4:** 38
jominy end-quench test ASTM A255 ... **M4:** 35
magnetic test ... **M4:** 37
quenching media ... **M4:** 35-38
time-temperature-property (TTP) function ... **A4:** 84
titanium alloys ... **A4:** 917, 921
tool steels ... **A4:** 77, 78, 79, 105-106
ultrahigh-strength steels ... **A4:** 218
variables affecting agitation ... **A4:** 113-116 **M4:** 61, 62, 64-65
effect of velocity ... **A4:** 114 **M4:** 64
number of agitators ... **A4:** 114 **M4:** 64
relation of tank design to agitation ... **A4:** 114-116 **M4:** 64-65
water ... **A4:** 67, 69, 72-75, 78, 87-89, 98, 99, 113, 115 **M4:** 35, 36, 40-41
agitation ... **A4:** 75, 88 **M4:** 40-41
contamination ... **A4:** 88-89, 98, 99 **M4:** 41
maintenance schedule for system ... **A4:** 118
Steel rails *See* Rails
Steel reinforcement
flux leakage inspection methods ... **A17:** 133
Steel, rimmed *See* Rimmed steel
Steel rolls
ceramic cutting tools ... **A16:** 98
Steel scrap, recycling of ... **A1:** 1023
factors influencing scrap demand ... **A1:** 1024
purchased scrap supply ... **A1:** 1024-1026
scrap use by industry ... **A1:** 1023-1024
Steel shafts
economy in manufacture ... **M3:** 856
Steel sheet *See also* Alloy steels, sheet and strip; Electrical steel sheet; Low-carbon steels, sheet and strip; Sheet; Sheet metals; Sheet steel
alloy steel *See* Alloy steel, sheet and strip
aluminum coatings ... **M1:** 171-173
bending of ... **M1:** 552-555
carbon steel *See* Low-carbon steel, sheet and strip
coatings for ... **M1:** 167-176
effect of material on economy in manufacture ... **M3:** 851
formability ... **M1:** 545-560
forming limit diagram ... **M1:** 549-553
galvanized ... **M1:** 167-171
grain shape, ferrite, effect on formability ... **M1:** 557-559
grain size, effect on formability ... **M1:** 557-558
load-extension curves ... **M1:** 548
metallic coatings ... **M1:** 167-174
phosphate coatings ... **M1:** 174-175
porcelain enameling of *See* Porcelain enameling
precoated, for appliances ... **M1:** 167, 169, 173-176
prepainted ... **M1:** 175-176
preprimed ... **M1:** 175
quality descriptors ... **M1:** 546-547
selection for formed parts ... **M1:** 559
Steel sheet (continued)
strain-age embrittlement of ... **M1:** 683-684
temper rolling, effect on formability ... **M1:** 548, 556
terne coatings ... **M1:** 173-174
thickness, effect on bending ... **M1:** 553, 554
tin coatings ... **M1:** 173
zinc coatings ... **M1:** 167-172
Steel, sheet and strip
normalizing ... **A4:** 40-41
Steel sheet and strip, annealing
cold rolled plain carbon ... **M4:** 23
hot dip galvanized products ... **M4:** 23
open-coil ... **M4:** 23
properties ... **M4:** 23
tin mill products ... **M4:** 23
Steel sheet and strip, normalizing
furnaces ... **M4:** 13
catenary ... **M4:** 13
conveyor-type ... **M4:** 13
equipment ... **M4:** 13
heating ... **M4:** 13
heating ... **M4:** 13
processing ... **M4:** 12
Steel shot
as abrasive ... **A15:** 504, 510-511
Steel shot blasting
aluminum and aluminum alloys ... **M5:** 571
Steel shot peening ... **A16:** 34, 35
Steel, specific types
0.55C-2.40Mn ... **A9:** 194
2.25Cr-1Mo plate, electron beam weld ... **A9:** 583
300M, effect of stress-intensity factor on fracture ... **A8:** 486
301, tensile properties ... **A8:** 555
302, surface conditions effect on fatigue properties ... **A8:** 373
321 HB, surface conditions effect on fatigue properties ... **A8:** 373
409, tensile properties ... **A8:** 555
1008 rimmed, Lüders bands on ... **A8:** 22
1008 sheet, minimum bend radii ... **A8:** 131
1010, microstructure from slow cooling ... **A9:** 624
1010 sheet, minimum bend radii ... **A8:** 131
1017, as strand cast, with columnar dendrites ... **A9:** 624
1017, sulfur print of ... **A9:** 625
1017, with manganese sulfide inclusions ... **A9:** 626
1017, with silicate inclusions ... **A9:** 626
1018, static and dynamic fracture toughness ... **A8:** 283
1020 cold rolled, load-time and displacement-time curves ... **A8:** 281
1020, finite element analysis of ... **A8:** 282
1020, load drop vs. crack length ... **A8:** 278, 279
1020, oxygen contamination effect ... **A8:** 408
1020, static and dynamic fracture toughness ... **A8:** 283
1020, static and dynamic shear stress/shear strain curves ... **A8:** 223, 225
1020, ultimate shear stress ... **A8:** 148
1022, banding from hot rolling ... **A9:** 626
1040, ductility from hot torsion tests ... **A8:** 165-166
1041, with carbon-rich bands ... **A9:** 626
1045 cold finished, fracture loci upset test specimens of ... **A8:** 580-581
1045, hot rolled and normalized ... **A8:** 150-151
1052, wear behavior ... **A8:** 604
1090, lamellar pearlite ... **A9:** 129
1095, ultimate shear stress ... **A8:** 148
1213, manganese sulfide stringers in ... **A9:** 627
1524, microstructure from rapid cooling ... **A9:** 624-625
2340, *S-N* curve ... **A8:** 364
4130, effect of environment on fatigue crack propagation ... **A8:** 404
4130, electrode potential effect on corrosion fatigue crack propagation ... **A8:** 407
4130, fayed to aluminum 7075-T651 ... **A9:** 387
4130, fracture surface with sulfide stringers ... **A8:** 477-478
4140, ultimate shear stress ... **A8:** 148
4340, compressed ... **A8:** 57-58
4340, corrosion fatigue crack growth ... **A8:** 405-406
4340, ductility from hot torsion tests ... **A8:** 165-166
Steel, specific types (continued)
4340, effect of annealing temperature on homogenization, calculated for diffusion of nickel and manganese in dendrites ... **A9:** 626
4340, effect of heat treatment and cracking ... **A8:** 539, 540
environment on hydrogen stress
4340, effect of hot rolling on dendritic pattern ... **A9:** 628
4340, for pressure bars ... **A8:** 200
4340, for symmetric rod impact tests ... **A8:** 168-169
4340, free-surface transverse velocity and shear stress ... **A8:** 235-236
4340, high strain rate pressure-shear tests ... **A8:** 236
4340, hydrogen embrittlement ... **A8:** 537
4340, hydrogen embrittlement crack growth rate as function of stress intensity ... **A8:** 539
4340, hydrogen stress incubation ... **A8:** 539, 540
4340, interferometer records ... **A8:** 235
4340, metallographic section strained before fracture ... **A8:** 479
4340, spacing between dendrite arms as a function of distance from chill surface ... **A9:** 625-626
4340, stress-strain curve in shear ... **A8:** 236-238
4340, tension, and torsion effective fracture strains ... **A8:** 168
4340, testing for hydrogen embrittlement from processing ... **A8:** 541
4360, electrode potential effect on corrosion fatigue crack propagation ... **A8:** 407
4620, fracture loci for elevated-temperature tests on ... **A8:** 583
A-36 plate, shielded metal arc weld ... **A9:** 584-585
A-710 plate, laser butt weld ... **A9:** 586
A-710 plate, submerged arc weld ... **A9:** 583, 585
A36, bridge, Charpy toughness requirements ... **A8:** 265
A36, stress intensity values ... **A8:** 453
A36, yield strength ... **A8:** 453
A242, bridge, Charpy toughness requirements ... **A8:** 265
A286, threshold stress intensity ... **A8:** 256
A440, bridge, Charpy toughness requirements ... **A8:** 265
A441, bridge, Charpy toughness requirements ... **A8:** 265
A471, in moist air or steam, crack closure ... **A8:** 409
A514 bridge, Charpy toughness requirements ... **A8:** 409
A533 B-1, shelf toughness ... **A8:** 467
A533, dynamic *J*-integral, *J*-resistance curves ... **A8:** 261
A533B1 , fatigue crack growth behavior ... **A8:** 377
A572, bridge, Charpy toughness requirements ... **A8:** 265
A588 bridge, Charpy toughness requirements ... **A8:** 265
AF 1410, heat treated, quenched, tempered color etched ... **A9:** 156
AISI 4340, closed-die forging, flow lines ... **A9:** 687
AISI 4340, solidification structures of welded joints ... **A9:** 579
AISI 4340, submerged arc weld ... **A9:** 584, 586
C100W1, fatigue crack propagated from hard surface coating ... **A9:** 96
copper infiltrated ... **A9:** 519
D-6 AC, cyclic microvoid ... **A8:** 484
D-6 AC, high-cycle fatigue ... **A8:** 485
D-6 AC, incubation period for hydrogen cracking ... **A8:** 539, 540
Distaloy, microstructure ... **A9:** 510
En 2A, tension and torsion effective fracture strains ... **A8:** 168
En 2D, tension and torsion effective fracture strains ... **A8:** 168
En 9, tension and torsion effective fracture strains ... **A8:** 168
EN-24, effect of stress-intensity factor range on fracture ... **A8:** 486
Fe-0.06C-0.35Mn-0.4SI-0.40Ti, sheet, color etched ... **A9:** 156

Steel, specific types (continued)
Fe-0.8C, pearlite structures **A9:** 658-660
Fe-0.8C, pressed and sintered **A9:** 527
Fe-0.8C, pressed and sintered, effect of polishing on pore opening **A9:** 506-507
Fe-0.8C, pressed and sintered, Knoop indenter mark for gaging rate of material removal **A9:** 507
Fe-0.8C, pressed and sintered, mounted for edge retention **A9:** 505
Fe-0.8C, pressed and sintered, steam blackened **A9:** 528
Fe-0.22C-0.88Mn-0.55Ni-0.50Cr-0.35Mn from a Jominy bar **A9:** 186
Fe-1.0C, different mounting techniques compared **A9:** 167
Fe-1.0C, etched to reveal cathodic cementite **A9:** 156
Fe-1.86C, color etched **A9:** 156
Fe-3.2Si, indentations in mechanical twin bands **A9:** 690
Fe-3.5Si, sheet, magnetic contrast **A9:** 94
Fe-3.25Si, compressed subgrains separated by small angle boundaries **A9:** 690
Fe-3.25Si, static recrystallization **A9:** 691
Fe-3.25Si, strain markings near a crack tip **A9:** 687
Fe-3lSi, water-atomized master alloy **A9:** 529
Fe-3SI, sheet, etch pit **A9:** 100
Fe-10Ni-8Co-#Mo, crack growth versus constant-amplitude stress cycles **A8:** 379
HY-80 plate, gas shielded flux core weld **A9:** 586
HY-130, rising step-load test in **A8:** 540, 541
HY-180, rising step-load test in **A8:** 540, 541
HY80, electrode potential effect on corrosion fatigue crack propagation **A8:** 407
Lukens Frostline plate, submerged arc bead- on-plate weld **A9:** 585
Lukens Frostline plate, submerged arc weld **A9:** 583
Nb-V microalloyed, ferrite structure **A8:** 180
Nb-V microalloyed, roll forces and roll torques compared **A8:** 180
René 95, fatigue-life reduction factor vs. strain range **A8:** 353
René 95, hold-time results **A8:** 353
René 95, mean stress vs. strain range **A8:** 353-354
René 95, strain range vs. hold period in compression **A8:** 353
René 95, strain range vs. hold period in tension **A8:** 353
SAE 4340, strain-hardening exponent and true stress values for **A8:** 24
X65 linepipe, cathodic potential and corrosion fatigue crack growth rate **A8:** 417

Steel, spheroidizing
ultrahigh-strength steels **A4:** 209, 210, 211, 212, 213

Steel springs *See* Springs **A1:** 302-326
characteristics of spring steel grade
annealed spring wire **A1:** 307-308
stainless steel spring wire **A1:** 308
valve-spring quality (VSQ) wire **A1:** 307
compression springs **A1:** 319-320, 322
active coils **A1:** 320
modulus change, effect of **A1:** 320
solid heights **A1:** 320, 322
costs **A1:** 317-318, 320
design **A1:** 318
life **A1:** 319, 321
stress range **A1:** 308, 309, 310, 319

Steel springs (continued)
Wahl corrections **A1:** 318-319, 320, 321
extension springs **A1:** 320-321
end hooks **A1:** 321
fatigue **A1:** 307, 312
shot peening **A1:** 313-314
stress range **A1:** 308, 309, 310, 311, 312-313
hot-wound springs **A1:** 315
hardenability requirements **A1:** 315-316, 317
surface quality **A1:** 316, 318
leaf springs **A1:** 321-324
mechanical prestressing **A1:** 325-326
mechanical properties **A1:** 325
steel grades **A1:** 322, 325
surface finishes and protective coatings **A1:** 326
for vehicle suspension **A1:** 322, 325
mechanical properties **A1:** 302-305
flat springs **A1:** 305, 306, 307
plating of springs **A1:** 311-312
hydrogen relief treatment **A1:** 312
mechanical plating **A1:** 312
residual stresses for **A1:** 309, 311
stress relieving **A1:** 307, 311
temperature, effect of **A1:** 303-304, 312, 313, 314-315
types of **A1:** 302
wire quality **A1:** 303-304, 308
decarburization **A1:** 308-309
magnetic particle and eddy current testing **A1:** 309
seams **A1:** 308

Steel, stainless type ...... **A3:** 1 • 27

Steel stamps
identification **A11:** 473

Steel storage tanks
cathodic protection system **A13:** 471-472

Steel strip *See also* Alloy steels, sheet and strip; Low-carbon steels, sheet and strip

Steel, surface hardening **A4:** 259-266
approaches to methods **A4:** 259
with arc lamps **A4:** 266
austenitic nitrocarburizing **A4:** 264
boriding **A4:** 264
carbonitriding **A4:** 264, 266, 282
carburizing **A4:** 261-263, 281, 282
diffusion methods **A4:** 259-264
electron beam (EB) hardening **A4:** 265
ferritic nitrocarburizing **A4:** 264
flame hardening **A4:** 264-265, 266
induction hardening **A4:** 281, 282
induction heating **A4:** 265, 266
ion implantation **A4:** 265-266
laser surface heat treatment **A4:** 265
nitriding **A4:** 263-264, 266, 281, 282
process selection **A4:** 266
selective carburizing **A4:** 266
selective surface hardening **A4:** 264-266
specialized diffusion methods **A4:** 264
titanium carbide **A4:** 264
tool steels **A4:** 745
Toyota diffusion process **A4:** 264

Steel tank car
brittle fracture from weld imperfections **A11:** 93-94

Steel technology
advances in **A1:** 665-666, 668, 669

Steel tempering *See* Tempering of steel

Steel, thermomechanical processing (TMP) **A4:** 237-253
applications to heat treat as-rolled microalloyed steels **A4:** 247-252
applications to heat treat low-alloy steels **A4:** 252
Broken Hill Proprietary (BHP) high toughness rolling **A4:** 251-252
conventional cold rolling (CCR) **A4:** 247
conventional controlled rolling **A4:** 237, 239, 240, 248-249, 250, 251, 252

Steel, thermomechanical processing (TMP) (continued)
conventional hot rolling (CHR) **A4:** 247, 249, 250
ductile-to-brittle transition temperature (DBTT) evaluation **A4:** 238, 239
dynamic recrystallization **A4:** 248-249
ferrite-pearlite **A4:** 237-240
fracture-appearance transition temperature (FATT) **A4:** 239, 240
fundamentals **A4:** 240-247
high-strength low-alloy (HSLA) steel **A4:** 238
hot rolling **A4:** 239, 240, 242, 246
HSLA steels **A4:** 252
intensified controlled rolling (ICR) **A4:** 250, 251
intragranular planar defects (IPD) **A4:** 242
mechanical properties, effect of microstructure **A4:** 237-240, 241
metadynamic recrystallization **A4:** 248-249
microalloying additions in austenite **A4:** 243, 245, 251
microalloying elements (MAE) **A4:** 237, 239, 240
effect on critical temperatures of austenite **A4:** 245-247
microstructure **A4:** 237-240, 241, 242
multiphase steels **A4:** 252
nil-ductility temperature (NDT) **A4:** 240
particle pinning **A4:** 243, 249-250, 251
physical metallurgy **A4:** 240-243
precipitation in austenite **A4:** 243-245, 246, 247, 249
recrystallization controlled rolling (RCR) **A4:** 242, 247-248, 250-251, 252
solubility products in austenite **A4:** 243-245
solute drag **A4:** 243, 246, 251
static recrystallization **A4:** 249-250, 251
Sumitomo high toughness rolling (SHT) **A4:** 249, 251
versus microalloyed (MA) steels **A4:** 237, 238, 239, 240

Steel, transformation
cold treating vs tempering **M4:** 117
hardness testing, cold treating **M4:** 117
precipitation-hardening **M4:** 118
process limitations, cold treating **M4:** 117
shrink fits, cold treating **M4:** 118

Steel tricone drill bits
application **A2:** 976

Steel tubes *See also* Steel tubing
ASTM specifications **M1:** 321, 323, 324, 325
compositions **M1:** 323, 324
pressure tubes **M1:** 316, 321, 323, 324, 325
square, rectangular and special-shape sections **M1:** 325-326
tensile properties **M1:** 325

Steel tubing *See also* Steel tubes
ASTM specifications **M1:** 323, 325, 326
compositions **M1:** 326
double-wall brazed tubing **M1:** 321
mechanical tubing **M1:** 323-326
structural tubing **M1:** 323, 325, 326
tensile properties **M1:** 326
Turk's head shaping **M1:** 325-326

Steel tubing, specific types
1015, resistance welded, different sections and heat treatments **A9:** 215-216
1018, welded and normalized **A9:** 216
1025, cold drawn, aluminate inclusion **A9:** 216
1215, cold drawn, sulfide inclusion **A9:** 216
4140, annealed **A9:** 216
4140, austenitized, quenched and tempered **A9:** 216
4620, silicate and sulfide inclusions **A9:** 216
5048, seamless, decarburization **A9:** 216
8620, silicate inclusion **A9:** 216
ASTM A161, seamless, hot drawn **A9:** 214
ASTM A200, Grade T5, seamless, annealed different magnifications **A9:** 214-215
ASTM A213, Grade T5c, hot finished **A9:** 215

Steel tubing, specific types (continued)
ASTM A254, Class 1, copper brazed joints **A9:** 215
Steel tubular products *See also* Steel pipe, Steel tubes, Steel tubing.... **A1:** 327-336 **A9:** 210-216 **M1:** 315-326
cold finishing **M1:** 316-317, 324-325
cold finishing for **A1:** 328-329
commercial classifications **A9:** 210
common types of pipe **A1:** 331-333
conduit pipe **A1:** 331
oil country tubular goods..... **A1:** 329, 330, 332-333
Piling pipe **A1:** 331
pipe nipples **A1:** 331
standard pipe **A1:** 331
transmission or line pipe **A1:** 331
water main pipe **A1:** 331-332
water well pipe **A1:** 329, 332, 333
compositions **A9:** 211 **M1:** 318-319, 322, 323, 324, 326
etchants **A9:** 211
etching **A9:** 211
maximum-use temperatures of boiler tube steels **A1:** 617
mechanical tubing **A1:** 334-336
continuous-welded cold-finished mechanical tubing **A1:** 335
seamless mechanical tubing **A1:** 335
square, rectangular, and special-shape sections **A1:** 335-336
welded mechanical tubing **A1:** 334-335
nondestructive evaluation **A17:** 561-581
pipe sizes and specifications for.... **A1:** 328, 329-331, 333
pressure tubes **A1:** 327, 333-334
double-wall brazed tubing **A1:** 333-334
structural tubing **A1:** 334
product classification **A1:** 327 **M1:** 315-316
production by welding **M1:** 316
seamless processes **M1:** 316
seamless processes for **A1:** 328
cupping and drawing **A1:** 328
hot extrusion **A1:** 328
rotary piercing **A1:** 328
specimen preparation **A9:** 210-211
welding processes
continuous welding **A1:** 328
double submerged arc welding **A1:** 328
electric resistance welding **A1:** 327-328
fusion welding **A1:** 328
Steel, ultrahigh-strength *See* Ultrahigh- strength steels
Steel, welding high-speed to low alloy .... **A3:** 1 • 26-27
Steel weldment soundness, influence of welding
common defects associated with arc welds **A6:** 408-409
hot cracks **A6:** 409
inclusions **A6:** 409
incomplete fusion **A6:** 408-409
lamellar tearing **A6:** 409
porosity **A6:** 408
rollover **A6:** 409
undercut **A6:** 409
gas-metal arc welding **A6:** 408, 409, 413
heat-affected zone **A6:** 408, 409, 410, 411, 412
high-strength low-alloy (HSLA) steels **A6:** 408, 411
hydrogen-induced cracking (HIC) **A6:** 408, 410-415
pulsed-arc welding **A6:** 409
shielded metal arc welding **A6:** 408, 409, 413, 414
shielding gases **A6:** 408
submerged arc welding **A6:** 409, 413, 414
Steel weldments, influence of welding on properties **A6:** 416-428
carbon steels **A6:** 417, 424
carbon-manganese steels **A6:** 420, 421, 422, 423, 424, 426-427
chromium-molybdenum steels **A6:** 420-421
continuous cooling transformation diagrams **A6:** 416, 417, 418, 419
corrosion **A6:** 424-425
dilution **A6:** 417-418, 419
electrode size effect **A6:** 418
embrittlement **A6:** 420, 421, 423
fatigue strength of joints **A6:** 425
Steel weldments, influence of welding on properties (continued)
filler metals **A6:** 417-418, 425
heat-affected zone ... **A6:** 416-417, 418, 419, 420, 421, 423-424, 425, 426-427
high-strength, low-alloy (HSLA) steels **A6:** 418-419
hydrogen cracking **A6:** 416, 424
hydrogen-induced cracking **A6:** 424
low-alloy steels **A6:** 420, 424
low-carbon steels **A6:** 419
microalloyed carbon-manganese steels **A6:** 420
microstructure **A6:** 418-424
mild steels **A6:** 419
postweld heat treatment **A6:** 416, 419, 420, 421, 422, 423, 424, 426-427
quenched and tempered steels **A6:** 421-424, 427
single-pass vs. multipass welding **A6:** 418, 424
steel types and weldability **A6:** 418-424
temper bead techniques **A6:** 420, 424
thermomechanically controlled process steels **A6:** 419, 420
underbead cold cracking **A6:** 423
weld metals **A6:** 417-418
welding procedure effect on properties **A6:** 425-428
heat input **A6:** 425-428
interpass temperature **A6:** 426
postweld heat treatment **A6:** 426-427
preheat temperature **A6:** 425-426
weldment considerations **A6:** 424-425
Steel wire *See also* Steel bar; Wire; Wire drawing **A1:** 277-288
applications **M1:** 587, 588
bend radii **M1:** 588
cold extrusion **M1:** 591-592
cold heading **M1:** 589-591
configurations and sizes **A1:** 277
E50100, tempering **A4:** 132
fabrication characteristics **M1:** 587-593
inspection of **A17:** 549-556
mechanical properties of round wire versus flat wire **A1:** 287-288
metallic coated wire **A1:** 281-282
for packaging and container applications **A1:** 282
quality descriptions and commodities **A1:** 282
alloy wire **A1:** 286-287
aluminum conductor steel reinforced wire **A1:** 283
for electrical or conductor applications **A1:** 283
fine steel wire **A1:** 286
low-carbon steel wire for general usage **A1:** 282
mechanical spring wire for general use **A1:** 284-285
mechanical spring wire for special applications **A1:** 285
for packaging and container applications **A1:** 282
for prestressed concrete **A1:** 283
structural applications (not prestressed concrete) **A1:** 282
upholstery spring construction wire **A1:** 285
wire for fasteners **A1:** 284
wire for other specific applications **A1:** 285-296
specification wire **A1:** 281
swaging **M1:** 593
wiremaking practices
cleaning and coating **A1:** 280
lubricants **A1:** 279
thermal treatments **A1:** 280-281
welds **A1:** 279
zinc galvanized **M2:** 652-654
Steel wire gage (SWG) system **A1:** 277
Steel Wire Gage system **M1:** 259-260
Steel wire rod **A1:** 272-276
alloy steel rod **A1:** 275
qualities and commodities for **A1:** 275
special requirements for **A1:** 275-276
carbon steel rod **A1:** 272
mechanical properties **A1:** 275-276
special requirements for **A1:** 274
cleaning and coating **A1:** 272
configurations and sizes of **A1:** 277
heat treatment of **A1:** 272
Steel wire rod (continued)
wiremaking practices **A1:** 277
wiredrawing **A1:** 277-279
Steel wire rope
abrasion and crushing failure **A11:** 519
components of **A11:** 515
corrosion **A11:** 518
drums **A11:** 517
eye terminal failure **A11:** 521
failure by overheating **A11:** 519-520
failure by shock loading **A11:** 518-519
failure of wires in **A11:** 519
failures of **A11:** 515-521
grooved drums **A11:** 517-518
hoisting, bending-fatigue failure **A11:** 518
rope pressure **A11:** 516-517
scrubbing **A11:** 519
sheave grooves **A11:** 516
sheave material **A11:** 517
shock loading of **A11:** 518
strength and stretch **A11:** 515-516
Steel(s) *See also* AISI/SAE alloy steels; Arc-welded low carbon steel; Arc-welded stainless steels; ASTM/ASME alloy steels; Austenitic stainless steels; Austenitic steels; Austentic- manganese steels; Carbon steels; Ferritic steels; Heat-resistant alloys; High-carbon steels; High-temperature alloys; Low-alloy steels; Low-carbon steels; Maraging steels; Martensitic stainless steels; Medium-carbon steels; Plain carbon steels; Precipitation-hardening stainless steels; Stainless steels; Stainless steels, specific types; Steel alloys; Steel castings, failures of; Steels, specific types; Tool steels; Tool steels, specific types; Ultrahigh-strength steels; Wrought carbon steels; Wrought cobalt-base alloys; Wrought heat-resisting alloys; Wrought steels
35Kh3N3M (Russian specification), tensile tests **A6:** 80
austenitic manganese, microstructural wear **A11:** 161
brazing with clad brazing materials **A6:** 347
bushings, in shafts **A11:** 470
carbide-containing, microstructural wear effects **A11:** 161
carbon arc welding **A6:** 200
carbon-manganese, lamellar tearing in **A11:** 92
carburized, microstructure **A11:** 326-327
carburized, surface-origin pitting from contact fatigue testing **A11:** 133
casting defects in **A11:** 380-391
chromium, feather markings **A12:** 18
cleaning solutions for substrate materials **A6:** 978
coarse-grained, quench cracking of **A11:** 96
coating, exposing and cleaning **A12:** 73
coextrusion welding **A6:** 311
distortion **A6:** 1098-1099
effect of temperature on dimple size **A12:** 34, 46
electrodes, nylon liners **A6:** 183-184
electron-beam welding **A6:** 851, 857, 860, 866
embrittlement by low-melting alloys **A12:** 29
embrittlement of **A11:** 98-101
environments that cause stress-corrosion cracking **A6:** 1101
explosion welding **A6:** 162-163, 303
fatigue striations in **A11:** 105
ferritic, ductile-to-brittle fracture transition in **A11:** 66
flash welding **A6:** 247
forge welding **A6:** 306
fractography of hydrogen-embrittled **A11:** 28-29
fracture types **A12:** 1-3
fracture-transition data, Charpy V-notch tests **A11:** 67
friction surfacing **A6:** 323
friction welding **A6:** 152
full-hard strip, mechanical cutting **A6:** 1180
fully killed, electron-beam welding **A6:** 866
fusion welding to aluminum-base alloys **A6:** 828
fusion welding to cobalt-base alloys **A6:** 828
fusion welding to copper-base alloys **A6:** 828
fusion welding to nickel-base alloys **A6:** 827-828
galvanized, brittle fracture **A11:** 100

Steel(s) (continued)
grade selection, and tool and die failure ... A11: 564, 566
grain refiners ... A6: 53
high-carbon, figure numbers for ... A12: 216
high-strength, effect of high temperature on overload fracture ... A12: 35, 50
high-strength, SCC failure in ... A11: 27
hot shortness ... A12: 126-127
hydrogen cracking in ... A11: 410
hydrogen embrittlement ... A12: 23, 25, 31, 124
hydrogen flaking ... A12: 125
hydrogen-damage failure ... A11: 336-337
hydrogen-induced flake formation in ... A11: 79
induction brazing ... A6: 333, 334
intergranular dimple rupture ... A12: 14
joined to aluminum alloys ... A6: 739
laser-beam welding ... A6: 263
LME failures in ... A11: 719-723
low-carbon copper-flashed, capacitor discharge stud welding ... A6: 222
low-carbon, figure numbers for ... A12: 216
medium-carbon, figure numbers for ... A12: 216
nickel alloys, welding to ... A6: 578
nitrided, decrease in nitrogen concentration ... A11: 43
nonmetallics occurring in ... A11: 316
overheating ... A12: 127-129
oxyfuel gas cutting ... A6: 1158
oxyfuel gas welding ... A6: 281, 282, 284, 285, 288
pearlitic eutectoid, TTS fracture ... A12: 28
phase transformations ... A6: 84
pipe, parameters used to obtain multipass weld in 1.07 m diameter X-65 steel pipe ... A6: 101-102
plane-strain fracture toughness ... A11: 54
plasma-MIG welding ... A6: 224
plate, hydrogen flaking ... A12: 141
precoated before soldering ... A6: 131
projection welding
heavy-gage steels ... A6: 232
hot- and cold-drawn wires ... A6: 236
intermediate-gage low-carbon steels ... A6: 234
process requirements ... A6: 235-237
process requirements for heavy-gage low-carbon steels ... A6: 232
thin-gage low-carbon steels ... A6: 235
quality, tool and die failure and ... A11: 564
quench cracking of ... A11: 94-97
R-curve for ... A11: 64
ratio-analysis diagram for ... A11: 62
rehardened high-speed, brittle fracture of ... A11: 574
relative solderability as a function of flux type ... A6: 129
relative weldability ratings, resistance spot welding ... A6: 834
rephosphorized, resulfurized, breakage elimination in ... A11: 70
resistance brazing ... A6: 341
resistance soldering ... A6: 357
resistance spot welding ... A6: 228-229
resistance welding ... A6: 840
roll welding ... A6: 312, 314
S-N curves ... A11: 103
SCC of ... A12: 27
semikilled, electron-beam welding ... A6: 806
shielded metal arc welding ... A6: 176
solderability ... A6: 978
soldering ... A6: 631
structural, ductile-to-brittle transition temperature ... A11: 68-69
sulfide inclusions ... A12: 14
thermal conductivity ... A11: 604
torch brazing ... A6: 328
torch soldering ... A6: 351, 352
ultrasonic welding ... A6: 324
welding to copper and copper alloys ... A6: 769

Steel(s), specific types
1.6Mn-5Cr forged gear and pinion, carburization failure of ... A11: 336
1.25Cr-0.5Mo reheater tube rupture ... A11: 610
1Cr-4Ni-0.2Mo high-strength cast, effect of stress-intensity rate on plane strain fracture toughness ... A11: 54
2.25Cr-1Mo superheater tube, creep failure ... A11: 610-612
5B41 connecting rod, fatigue fracture ... A11: 328-329
15 B28 induction-hardened, replica fractographs of fatigue fracture, compared ... A12: 95, 100
15 B28, photo-illumination effects ... A12: 85
15B41 forged connecting-rod cap, fatigue fracture ... A11: 120
21-2 valve stem, fracture surface ... A11: 288
300M failed flash-welded joint in ... A11: 442
300M jackscrew drive pins, SCC failure ... A11: 546-548
1008 drawn container, fatigue cracking ... A11: 307-308
1008 strip, photo effects of lighting ... A12: 87
1015 semi-killed forged hook ... A11: 524
1020 C-hook, design of ... A11: 523
1020 cold-worked, inelastic cycle buckling ... A11: 144
1020 crane hook, fatigue fracture ... A11: 523
1020, hydrogen blister ... A11: 247
1020 stop-block guide ... A11: 526-527
1025 tube post, fatigue failure ... A11: 424
1030 crane shaft ... A11: 524-525
1030 pinion shaft, fatigue fracture surface ... A11: 524
1030 tube, erosion-corrosion of ... A11: 342
1035 automobile stub axle, slag inclusions in ... A11: 323
1035 cap screws, fatigue fracture ... A11: 533, 535
1039 fatigue specimen, photolighting effects ... A12: 84
1040, cleavage fractures ... A12: 175
1040, cleavage-crack nucleation ... A11: 23
1040 coil hook, fatigue fracture ... A11: 523-524
1040 crankshaft, fatigue cracking from nonmetallic inclusions ... A11: 477-478
1040 fan shaft, reversed-bending fatigue ... A11: 476
1040 hot-rolled, cleavage fracture ... A11: 22
1040 hot-rolled, shear dimples in shear-lip zone ... A11: 76
1040 main hoist shaft, fatigue cracking ... A11: 525
1040 main-bearing journal, fatigue failure ... A11: 478
1040 shaft for amusement ride, crack propagation and final fast fracture ... A11: 418
1040 splined shaft, fatigue fracture ... A11: 122-123
1040, tongues in cleavage ... A11: 22
1041, SMIE in ... A11: 243
1045 U-bolts, fatigue fracture ... A11: 533-535
1048 diesel truck crankshaft, fatigue fracture ... A11: 78
1050 axle shaft, fracture surface ... A11: 21
1050 crankshaft, fatigue fracture ... A11: 324
1055 crane-bridge wheel ... A11: 527-528
1055 stripper crane wheel, fatigue failure ... A11: 526
1055 wire strands, effects of heat-treatment conditions on tensile strength and elongation ... A11: 444
1065 oval wire, corrosion-fatigue cracking ... A11: 258
1070 hardened-and-tempered clamps, distortion in ... A11: 140
1085, stereo-pair photographs ... A12: 89
1095 nickel-plated pawl spring, fatigue fracture ... A11: 553
1095, occurrence of SMIE ... A11: 243

Steel(s), specific types (continued)
1095 yam eyelet, matrix hardness ... A11: 160
1113 worm, destructive wear ... A11: 598
1117, valve spool cylinder, distortion in ... A11: 141-142
1138 shotgun barrel, distortion in ... A11: 139
1151 induction-hardened, brittle fracture of ... A11: 478-479
3340, occurrence of SMIE ... A11: 243
4130 exciter shaft, fatigue failure ... A11: 425
4130 permanent mold, overheating failure of ... A11: 276
4130 rolled bar, seam in ... A11: 329, 331
4130 shaft, fatigue-fracture surface ... A11: 104
4140, brittle cracks from liquid zinc embrittlement ... A11: 237
4140, cadmium-plated, arc striking hard spot in ... A11: 97
4140 cross-travel shaft, service failure ... A11: 525
4140, effect of temperature on embrittlement ... A11: 239
4140, embrittlement by liquid metals ... A11: 226
4140, embrittlement systems ... A11: 242, 244
4140 forged crankshaft, fatigue fracture ... A11: 472-473
4140 in indium, crack propagation ... A11: 244
4140, liquid zinc-induced cracking ... A11: 237
4140 locking collar, microstructural fibering or banding in ... A11: 320
4140, occurrence of SMIE ... A11: 243
4140, quenched-and-tempered, elongated dimples on ... A11: 25
4140 radar-antenna bearing, heat-treatment failure ... A11: 509-510
4140 slat track, distortion in ... A11: 140-141
4140, tire tracks on ... A11: 27
4145, occurrence of SMIE ... A11: 243
4150 chuck jaw, brittle fracture from white-etching layer ... A11: 573, 576
4150 overhead crane drive axle, fatigue fracture ... A11: 117
4150 plunger shaft, fatigue fracture ... A11: 319
4150 pump shaft, bending-fatigue fracture ... A11: 109
4330V part, fatigue-fracture surface ... A11: 105
4337 forging, rod fracture surface ... A11: 474
4337 master connecting rod, fatigue cracking ... A11: 477
4340 cadmium-plated, embrittlement ... A11: 242
4340 compressor shaft, peeling-type fatigue cracking ... A11: 471
4340, copper-induced liquid-metal embrittlement ... A11: 232, 233
4340, influence of lead on fracture morphology of ... A11: 226
4340, intergranular corrosion-fatigue fracture ... A11: 260
4340, light fractographs ... A12: 83
4340, occurrence of SMIE ... A11: 243
4340 pressure vessel, liquid lead induced brittle fracture ... A11: 225
4340 rotor shaft, torsional fatigue in ... A11: 464
4340 steering knuckle, fracture surface ... A11: 342
4340 wing-attachment bolt, cracked along seam ... A11: 530, 532
4520 wheel studs, fatigue fracture ... A11: 531-537
4615, chain link failure ... A11: 521-522
6150 coal pulverizer shaft, fatigue failure ... A11: 468
6150 landing-gear spring, fracture during hard landing ... A11: 560
6260 gear, adhesive wear ... A11: 596
6263 impeller drive gear, pitting and wear ... A11: 598-599
6470 knuckle pins, fatigue fracture ... A11: 128-129
8617 carbonitrided bushing, fatigue fracture ... A11: 121
8620, brittle fracture ... A11: 90-92
8620, for bearings ... A11: 490

Steel(s), specific types (continued)
8620, photo-illumination effects **A12:** 86
8620 valve spool, seizing of **A11:** 141-142
8640 fuel shaft, fatigue fracture **A11:** 476
8740 cadmium-plated, burning failure from forging **A11:** 332-334
8740 cadmium-plated fasteners, hydrogen embrittlement **A11:** 541
8740 cadmium-plated, intergranular fracture **A11:** 29
9310 gear, fracture surface, electron beam weld **A11:** 447
52100 bearing, hardness, grain size, and retained austenite variations in **A11:** 509
52100 bearing, microstructural alterations **A11:** 505
52100 bearing, overheating failure from misalignment **A11:** 507-508
52100, for bearings **A11:** 490
52100 steel rings, fretting failure of raceways on **A11:** 498
A-4, occurrence of SMIE **A11:** 243
A36 structural, fracture surface **A11:** 89
A186 double-flange trailer wheel, fatigue failure **A11:** 130
A293, fatigue-crack growth behavior **A11:** 108
A356 turbine casing, failed by cracking **A11:** 287-288
A533 type B, fracture-transition data **A11:** 67
A533 type B, transition temperatures **A11:** 67
API grade X-60 line pipe, sinusoidal fracture path **A12:** 116
API-5LX grade X42 carbon-manganese on fatigue crack growth rate **A12:** 56
pipeline, effect of stress intensity factor range
D-6ac alloy, brittle fracture from delayed quench cracking **A11:** 95-96
D-6ac, occurrence of SMIE **A11:** 243
D-6ac structural member, fatigue fracture **A11:** 116
Grade X42 pipeline, effect of hydrogen on fatigue fracture appearance **A12:** 37, 51
Halmo bearing, hardness, grain size, and retained austenite variations in **A11:** 509
HY-130 bainitic, TTS fracture **A12:** 21, 28
HY-180, stress-corrosion fracture **A12:** 27, 34
Nitralloy, LAMMA analysis and spectra **A11:** 42-43
RQC-90 steel plate, cold cracks **A12:** 156

Steel-bonded cermets
hardening of **A2:** 997

Steel-bonded titanium carbide, and other wear-resistant materials
compared **A2:** 997

Steel-bonded titanium carbide cermets
applications and properties **A2:** 996-998
hardening **A2:** 997
machining and grinding **A2:** 997-998
manufacturing **A2:** 997

Steel-bonded tungsten carbide cermets
applications and properties **A2:** 1000

Steel-carbon fiber composite
adhesive material properties used in stress analysis **EM3:** 316

Steel-cutting cemented carbides
compositions and microstructures **A2:** 952-953

Steel-mill billets
magnetic particle inspection of **A17:** 119-120

Steel-polypropylene joints
weathering and long-term testing **EM3:** 659-662

Steel/acrylonitrile-butadiene-styrene
acrylic properties **EM3:** 122

Steel/glass
acrylic properties **EM3:** 122

Steel/glass-reinforced plastics
acrylic properties **EM3:** 122

Steel/natural rubber
acrylic properties **EM3:** 122

Steel/nylon
acrylic properties **EM3:** 122

Steel/polystyrene
acrylic properties **EM3:** 122

Steel/polyvinyl chloride
acrylic properties **EM3:** 122

Steel/steel (clean)
acrylic properties **EM3:** 122

Steelmaking *See also* Steel processing technology
effect of practices on formability **A1:** 577-578
electric furnace steelmaking **A1:** 111
ferroalloy/deoxidizer additions **A1:** 111-112
first-stage refining **A1:** 110, 111, 112
of HSLA steel **A1:** 405-406
ladle steelmaking **A1:** 112-113
mold metallurgy **A1:** 114
second-stage refining and technology advances **A1:** 111
stainless steels **A1:** 930
steel plate **A1:** 226-228
third-stage refining **A1:** 113
tundish metallurgy and continuous casting **A1:** 113-114

Steelmaking furnaces **M1:** 109-114

Steelmaking industry
acid melting practice **A15:** 363-364
basic melting practice **A15:** 366
horizontal centrifugal casting in **A15:** 300

Steelmaking practices **M1:** 109-116
$2^1/_4$ Cr-1Mo steel **M1:** 654
formability, effects on **M1:** 556-557

Steels *See also* Alloy steels; ASP steels; Carbon steels; Cast steels; Ferrous casting alloys; High-alloy steels; High-strength steels; Low-alloy steels; Low-carbon steels; Maraging steels; Plain carbon steels; Silicon steels; specific types; Stainless steels; Steel alloys; Steel alloys, specific types; Steels, AISI-SAE; Steels, AMS; Steels, ASME; Steels, ASTM; Tool steels; Wrought steels, specific steels **A7:** 100-104
abrasion resistance **A18:** 490, 491
abrasive wear materials **A18:** 190
additives **A7:** 105
Al-killed, analysis of **A10:** 231
alloy production, powders used **A7:** 572
alloying element partitioning in, FIM/AP study of **A10:** 583
alloying elements in **A10:** 56
aluminized **A13:** 434-435
aluminum coating **A13:** 458
aluminum-caused inclusions in **A15:** 90
aluminum-zinc alloy coated **A13:** 435-436
analysis, as ratioed with the intensity of iron **A10:** 26
applications
cylinder liner material for pistons **A18:** 556
internal combustion engine parts **A18:** 553, 561
ASTM specifications **M3:** 739-740
atmospheric corrosion **A13:** 82, 205, 542, 1299-1301
bearing, grinding **A16:** 437
bearing, ground by CBN wheels **A16:** 455
bendability and selection, for press bending **A14:** 523
in bimetal bearing material systems **A18:** 747
biological corrosion **A13:** 116-117
bonded to aluminum-silicon-tin or aluminum- silicon-lead alloys **A18:** 744
boring **A16:** 163, 164
boring bars **A16:** 163, 171
broachability constant **A16:** 200
carburizing **A10:** 380
as carrier core, copier powders **A7:** 584
case hardening for rolling-contact component steels **A18:** 502
case-hardening steel **A16:** 67
cast and wrought, weldability compared **A15:** 535
chemical analysis and sampling **A7:** 249
chemical milling **A16:** 579, 582, 585, 586
chemical-setting ceramic linings for **A13:** 454
chloride salt corrosion **A13:** 90
chromium-molybdenum, atom probe analysis **A10:** 592
classified **A15:** 628
coated, filiform corrosion **A13:** 104-107
coated, press forming **A14:** 560-566
coinability of **A14:** 183
cold form tapping **A16:** 266
cold-rolled, corrosion resistance **A10:** 556-557
complex inclusions in **A15:** 92-93
composition **A7:** 464
constructional, hobs for **M3:** 477

Steels (continued)
continuous casting of **A15:** 308-313
contour band sawing **A16:** 364
and conventional steel, microstructures compared **A7:** 785
corrosion, Evans diagram **A13:** 49
corrosion fatigue **A13:** 764
corrosion under insulation **A13:** 1144-1146
counterboring **A16:** 251
crack growth kinetics **A13:** 279
crucible, development of **A15:** 31
cutting fluids used **A16:** 125
cutting tool materials and cutting speed relationship **A18:** 616
damage dominated by fatigue fracture **A18:** 181
damage dominated by surface cracking **A18:** 178, 179
debris effect on wear **A18:** 249
dependence of wear on load and velocity **A18:** 490
desulfurization **A15:** 75
determination of alloying elements by flame AAS **A10:** 56
die block **A14:** 230
die cutting speeds **A16:** 301
dissolution, in iron-carbon melts **A15:** 73-74
drilling **A16:** 229
drop hammer forming of **A14:** 655-656
edge retention **A16:** 29
effect of sintering atmosphere on carbon content **A7:** 340, 341
effects of density on elastic modulus Poisson's ratio, and thermal expansion **A7:** 466
electrical discharge machining **A16:** 558, 559, 560
electrochemical discharge grinding **A16:** 549
electrochemical machining **A16:** 535, 539
electrogalvanized **A13:** 1012
electromagnetic techniques for **A17:** 159-163
electron beam drip melted **A15:** 413-414
electroslag remelting **A15:** 401-403
elements implanted to improve wear and friction properties **A18:** 858
EPMA analysis of inclusions **A10:** 516
erosion **A18:** 200, 201, 204
relative erosion factor **A18:** 200, 201
exogenous inclusions **A15:** 93-94
extrudability of **A14:** 300-301
fatigue, magnetic rubber inspection monitoring **A17:** 125
fatigue-crack-growth rates **M3:** 741, 755
FIM sample preparation of **A10:** 586
flow stress relations **A14:** 166-168
forged **A18:** 654
fracture toughness **M3:** 740-741, 754
free-machining, cold extrusion **A14:** 301
fretting wear **A18:** 252
fretting wear of ropes **A18:** 250
friction coefficient data **A18:** 74, 75
fully dense, STAMP process **A7:** 547
fully-killed, cleanliness assayed **A10:** 176
galling threshold loads **A18:** 595
galvanic corrosion **A13:** 84
galvanized **A13:** 432-434
glass-lined **A13:** 1228
grade 250 maraging EDG **A16:** 567
grinding rod materials **A18:** 654
grit-blast descaling for **A7:** 435
guide shoe material for honing operation **A16:** 478
hardened and tempered, erosive attack on die surface **A18:** 630
heat treated, properties **A18:** 813
high removal rate machining **A16:** 607
high-alloy **A15:** 722-735
high-resolution energy-compensated atom probe analysis of **A10:** 597
high-speed machining **A16:** 598, 600
high-strength, hydrogen embrittlement **A13:** 1039
high-strength, precleaning embrittlement **A17:** 81
high-strength, stress-corrosion crack initiation **A13:** 149
hone forming **A16:** 488
honing stone selection **A16:** 476
hot dip coating **A13:** 432-445
hot dip galvanized **A13:** 1012

Steels (continued)
hot extrusion, billet temperatures for **M3:** 537
hot extrusion of **A14:** 322
hot forming temperatures for **A14:** 620
hot-rolled commercial-quality, for press forming **A14:** 546-547
hydrogen traps, by size **A13:** 167
inclusions in **A15:** 91-94
induction-hardened shaft, subsurface residual in **A10:** 389-390
stress and hardness distributions
infiltrated, composition and properties **A7:** 564
infiltrated, microstructural analysis **A7:** 487-488
iron-based, AAS analysis of **A10:** 55
isolation of inclusions in **A10:** 176
Izod impact testing on hot forged **A7:** 410
lapping **A16:** 499
laser alloying **A18:** 866
lead and sulfur content and flaking **A16:** 281
low-strength, stepwise cracking **A13:** 170
low-stress grinding procedures **A16:** 28
machinability **A16:** 643, 644, 646
mechanical properties **A18:** 774
melts, purification of **A15:** 75-79
metallic coated **A13:** 526-527
microhoning **A16:** 490
microstructural analysis **A7:** 487, 488
mild **A18:** 71, 73, 246, 251, 549
mill scale breaks, galvanic corrosion from **A13:** 84
milling **A16:** 307, 312, 313, 314
in moist chlorine **A13:** 1172-1173
molten, surface protection **A15:** 311
Mössbauer measurement of retained austenite in **A10:** 287
nickel **A7:** 488
nitrides as inclusions **A15:** 93
nonmetallic elements determined in **A10:** 178
operating limits, to avoid hydrogen attack **A13:** 331
optical emission spectroscopy **A10:** 21
organic-coated **A13:** 528-529
oxides as inclusions **A15:** 91-92
oxyfuel resistance **A14:** 722
oxygen removal **A15:** 74
part material for ion implantation **A18:** 858
pearlitic, atom probe composition profile of **A10:** 593
pH effect on corrosion rate **A13:** 991
phase stability in **A10:** 583
phase transformation in **A10:** 583
photochemical machining **A16:** 590, 591
physical properties **A18:** 192
planing **A16:** 184, 185, 186
prealloyed, microstructural analysis **A7:** 488
precision forming of **A14:** 158-175
precoated **A13:** 1011-1015
prepainted **A13:** 528-529
quantitative determination of carbon and sulfur by high-temperature combustion in **A10:** 223-224
radiographic absorption **A17:** 311
radiographic film selection **A17:** 328
reaction with cementitious materials **A13:** 1299, 1301-1303, 1306-1308
reaming **A16:** 239, 245, 247, 248
resulfurized or leaded **A16:** 149
reverse redrawing **A14:** 511
roll bonding processes **A18:** 755-756
roller burnishing **A16:** 252
rolling of **A14:** 343, 355
sand castings, noncritical tolerances **A15:** 619
scanning acoustic microscopy for wear studies **A18:** 408, 409
scrap, as charge **A15:** 388
second-phase particle effect, tensile ductility **A14:** 364
semikilled and rimmed, sample dissolution of **A10:** 176

Steels (continued)
shaping **A16:** 191
sheet and strip, powder used **A7:** 574
simple, partitioning silicon oxidation states n **A10:** 178
simulated, range of Kα doublet blending for **A10:** 385
sintering furnace **A7:** 339
slugs, for cold extrusion **A14:** 303-304
sodium hydroxide corrosion **A13:** 1174
soft-annealed, erosive attack on the die surface **A18:** 630, 631
solid-metal embrittlement **A13:** 184
for space boosters/satellites **A13:** 1102
spade drilling **A16:** 225
spark emission sources for **A10:** 29
specialty, hot workability ratings **A14:** 381
specialty, plasma melting/casting **A15:** 420
spotfacing **A16:** 251
steel parts, steam treating **A7:** 453
steel-copper systems, galvanic corrosion prediction **A13:** 236
strip, multiple-slide forming of **A14:** 567-574
strip-backed main connecting rod bearings, automotive **A7:** 565
structural, atmospheric corrosion **A13:** 542
structure, principles **A13:** 46-47
sulfides as inclusions **A15:** 92
sulfur removal **A15:** 74
surface condition effects on fatigue **A13:** 294
surface decarburization measured **A17:** 134
and surface integrity **A16:** 22
surface preparation, for porcelain enameling **A13:** 447
tempered, microhardness **A7:** 489
tensile properties at subzero temperatures **M3:** 740, 743, 748
thermal energy method of deburring **A16:** 578
thermal properties **A18:** 42
thread grinding **A16:** 271
thread milling **A16:** 269
thread rolling **A16:** 282
threading **A16:** 299
threading, tangential chasers **A16:** 297
tin coatings on **A13:** 775-776
to collimate or shield sources for storage and shipment **A18:** 325
trace metals AAS analysis of **A10:** 55
in trimetal bearing material systems **A18:** 748
tube stock **A14:** 671
tuberculation in **A13:** 121
tubing, as stock **A14:** 671
turning **A16:** 135, 144
turning machinability ratings **A16:** 668-669
use in breweries **A13:** 1221
for valve seats and guards for reciprocating compressors **A18:** 604
wear rates **A18:** 272
weathering **A13:** 515-521
weathering, Raman analysis **A10:** 135
welding of **A15:** 520, 531-537
work material for ion implantation **A18:** 858
zinc-alloy coated **A13:** 1014

Steels, AISI specific types
601, composition **M1:** 649
602
composition **M1:** 649
elevated temperature properties **M1:** 649, 651
603
composition **M1:** 649
elevated temperature properties **M1:** 640, 642, 649, 651
610 composition **M1:** 649

Steels, AISI-SAE
classification **M6:** 289
electrode selection **M6:** 291
filler metals **M6:** 289-290
postheating **M6:** 291
preheating **M6:** 291

Steels, AISI-SAE specific types
10B20, applications, austempered parts **A4:** 157
10B21
notch toughness **M1:** 692, 1022
10B53
applications, austempered parts **A4:** 155
heat treatment effect on fracture appearance **A4:** 156
10L18, broaching **A16:** 207
12L13
bushings, machining cost vs. 1213 steel **M1:** 578, 580
machinability rating **M1:** 570
recommended machining conditions **M1:** 569
12L13, arc welding **A6:** 646
12L13, machining **A16:** 179, 207, 342, 345, 347, 349
12L14
arc welding **A6:** 646
composition **A6:** 641 **M1:** 126
machinability **M1:** 236, 237
machinability rating **M1:** 576
tensile properties, cold drawn bars **M1:** 243
12L14, machining **A16:** 164, 179, 207, 251, 263, 342, 345, 347, 349, 364, 391, 667, 668, 672, 673
12L15, machining **A16:** 179, 207, 342, 345, 347, 349
14B35H
tempering with induction heating **A4:** 193
through-hardening by induction heating **A4:** 192
15B21H
composition **M1:** 127
hardenability curve **M1:** 499
15B35
annealing **A4:** 53, 54
hardness **A4:** 54
15B35H
composition **M1:** 127
hardenability curve **M1:** 500
15B37H
composition **M1:** 127
hardenability curve **M1:** 501
15B41H
composition **M1:** 127
hardenability curve **M1:** 501
15B48H
composition **M1:** 127
hardenability curve **M1:** 502
15B62H
composition **M1:** 127
hardenability curve **M1:** 502
41L40
machinability rating **M1:** 582
41L40, machining **A16:** 164, 225, 246, 247, 263, 269, 342, 345, 347, 349, 364, 669
41L45, machining **A16:** 342, 345, 347, 349
41L47, machining **A16:** 342, 345, 347, 349
41L50, machining **A16:** 81, 342, 345, 347, 349
43L40, machining **A16:** 342, 345, 347, 349
50B40
annealing temperatures **A4:** 46 **M4:** 18
austenitizing temperature **A4:** 961 **M4:** 29
composition **M1:** 128
hardenability equivalence **M1:** 483-484
machinability rating **M1:** 582
normalizing temperatures **M4:** 7
50B40, machining **A16:** 144-147, 179, 207, 274, 281, 343, 346, 348, 349, 669
50B40H
composition **M1:** 130
hardenability curve **M1:** 511
50B44
annealing temperatures **A4:** 46 **M4:** 18
austenitizing temperature **A4:** 961 **M4:** 29
composition **M1:** 128
machinability rating **M1:** 582

Steels, AISI-SAE specific types (continued)
normalizing temperatures **M4:** 7
50B44, machining **A16:** 144-147, 179, 207, 274, 343, 346, 348, 349, 669
50B44H
composition **M1:** 130
hardenability curve **M1:** 511
50B46
annealing temperatures **A4:** 46 **M4:** 18
austenitizing temperature **A4:** 961 **M4:** 29
composition **M1:** 128
machinability rating **M1:** 582
normalizing temperatures **M4:** 7
50B46, machining **A16:** 144-147, 179, 207, 274, 343, 346, 348, 349, 669
50B46H
composition **M1:** 130
hardenability curves **M1:** 512
50B50
annealing temperatures **A4:** 46 **M4:** 18
austenitizing temperature **A4:** 961 **M4:** 29
composition **M1:** 128
machinability rating **M1:** 582
normalizing temperatures **M4:** 7
50B50, machining **A16:** 207, 274, 343, 346, 348, 349, 669
50B50H
composition **M1:** 130
hardenability curve **M1:** 512
50B60
annealing temperatures **A4:** 46 **M4:** 18
austenitizing temperature **A4:** 961 **M4:** 29
composition **M1:** 128
machinability rating **M1:** 582
50B60, machining **A16:** 207, 274, 343, 346, 348, 349, 669
50B60H
composition **M1:** 130
hardenability curve **M1:** 512
springs, hot wound **M1:** 302
51B60
annealing temperatures **A4:** 46 **M4:** 18
austenitizing temperature **A4:** 961 **M4:** 29
composition **M1:** 128
51B60, machining **A16:** 200, 207, 225, 274, 343, 346, 348, 349, 669
51B60H
composition **M1:** 130
hardenability curve **M1:** 516
springs, hot wound **M1:** 302
51L32, machining **A16:** 342, 345, 347, 349
52L100, machining **A16:** 342, 345, 347, 349
60B60, normalizing temperatures **M4:** 7
81B45
annealing temperatures **A4:** 46 **M4:** 18
austenitizing temperature **A4:** 961 **M4:** 29
composition **M1:** 128
machinability rating **M1:** 582
normalizing temperatures **M4:** 7
81B45, machining **A16:** 144-147, 179, 207, 225, 274, 343, 346, 348, 349, 669
81B45H
composition **M1:** 130
hardenability curve **M1:** 517
86B30H
composition **M1:** 131
hardenability curve **M1:** 519
86B45
annealing temperatures **A4:** 46 **M4:** 18
austenitizing temperature **A4:** 961 **M4:** 29
composition **M1:** 128
machinability rating **M1:** 582
normalizing temperatures **M4:** 7
86B45, machining **A16:** 144-147, 179, 207, 225, 274, 343, 346, 348, 349, 669
86B45H
composition **M1:** 131
hardenability curve **M1:** 521
hardenability variation **M1:** 479
86L20
machinability rating **M1:** 582
86L20, machining **A16:** 342, 346, 347, 349, 669
86L40, machining **A16:** 342, 345, 347, 349
94B15
composition **M1:** 129

Steels, AISI-SAE specific types (continued)
machinability rating **M1:** 582
94B15, machining **A16:** 144-147, 179, 207, 274, 342, 346, 347, 349, 669
94B15, normalizing temperatures **M4:** 7
94B15H
composition **M1:** 131
hardenability curve **M1:** 524
94B17
composition **M1:** 129
machinability rating **M1:** 582
94B17, machining **A16:** 144-146, 179, 207, 274, 342, 346, 347, 349, 669
94B17, normalizing temperatures **M4:** 7
94B17H
composition **M1:** 131
hardenability curve **M1:** 524
94B30
annealing temperatures **A4:** 46 **M4:** 18
austenitizing temperature **A4:** 961 **M4:** 29
composition **M1:** 129
hardenability equivalence **M1:** 483-484
machinability rating **M1:** 582
normalizing temperatures **M4:** 7
94B30, machining ... **A16:** 144-147, 207, 274, 323-325, 343, 346, 348, 349, 669
94B30H
composition **M1:** 131
hardenability curve **M1:** 525
94B40
annealing temperatures **A4:** 46 **M4:** 18
austenitizing temperature **A4:** 961 **M4:** 29
normalizing temperatures **M4:** 7
95B17, turning **A16:** 147
98BV40, turning **A16:** 144-147
300M, properties, effect of mass on **M4:** 123
410, martempering in oil, applications **M4:** 98
1000 to 6000 (hard), galling resistance with various material combinations **A18:** 596
1000 to 6000 (soft), galling resistance with various material combinations **A18:** 596
1005
composition **M1:** 125
machinability rating **M1:** 576
1005, machining **A16:** 144-147, 179, 207, 274, 323-325, 342, 345, 349, 668
1006
capacitor discharge stud welding **A6:** 222
composition **A6:** 641 **M1:** 125
machinability rating **M1:** 576
stud arc welding **A6:** 215
1006, machining **A16:** 144-147, 179, 207, 274, 323-325, 342, 345, 347, 349, 668
1007
capacitor discharge stud welding **A6:** 222
stud arc welding **A6:** 215
1008
annealing **A4:** 53
arc welding **A6:** 645
brazing with copper cladding **A6:** 347
capacitor discharge stud welding **A6:** 222
carbonitriding **A4:** 376
composition **M1:** 125
cutoff band sawing with bimetal blades **A6:** 1184
flash welding schedule **M6:** 577
liquid carburizing **A4:** 339
machinability rating **M1:** 576
projection welding **M6:** 511
resistance welding **A6:** 835
stud arc welding **A6:** 215
1008, gas carburizing **M4:** 163
1008, machining **A16:** 144-147, 179, 207, 274, 323-325, 342, 345, 347, 349, 360, 361, 388, 668, 672
1009
capacitor discharge stud welding **A6:** 222
composition **M1:** 125
cutoff band sawing with bimetal blades **A6:** 1184
stud arc welding **A6:** 215
1009, machining **A16:** 144-147, 179, 207, 274, 323-325, 342, 345, 347, 349, 360, 361
1009, projection welding **M6:** 511
1010
applications, austempered parts **M4:** 108

Steels, AISI-SAE specific types (continued)
arc welding **A6:** 645
austenitizing temperature **M4:** 30
capacitor discharge stud welding **A6:** 222
carbonitriding **M4:** 183, 186
composition **A6:** 641 **M1:** 125
composition and carbon equivalent **A6:** 406
critical temperatures, annealing **M4:** 15
cutoff band sawing with bimetal blades **A6:** 1184
electrogas welding process, selected steel grades **A6:** 656
flux-cored arc welding **A6:** 187
laser-beam welding **A6:** 264
machinability rating **M1:** 576
mechanical properties **M4:** 26
resistance seam welding **A6:** 840
resistance spot welding after SAW **A6:** 228
resistance welding **A6:** 842
specific heat vs. temperature **M1:** 149
stud arc welding **A6:** 215
tensile properties, hot rolled bars **M1:** 204
1010 (also EN32)
applications, austempered parts **A4:** 157
austenitic nitrocarburizing **A4:** 431
austenitizing temperature **A4:** 962
carbonitriding **A4:** 376, 378, 379, 380, 381, 383-384
case depth, measurement **A4:** 457-458
composition **A4:** 320
critical temperatures **A4:** 43
distortion in heat treatment **A4:** 614
flame hardening **A4:** 283
liquid carburizing **A4:** 345
liquid nitriding **A4:** 410
liquid nitrocarburizing **A4:** 417
mechanical properties after annealing **A4:** 54
shim stock analysis test strip, carbon potential **A4:** 589
1010, cage material for rolling-element bearings **A18:** 503
1010, machining **A16:** 144-147, 179, 207, 209, 274, 282, 291, 293, 323-325, 342, 345, 347, 360, 361, 388, 584, 592, 668, 708
1010, projection welding **M6:** 508-509
1011
arc welding **A6:** 645
capacitor discharge stud welding **A6:** 222
carbonitriding **A4:** 376, 380
composition **M1:** 125
cutoff band sawing with bimetal blades **A6:** 1184
flame hardening **A4:** 283
machinability rating **M1:** 576
stud arc welding **A6:** 215
1011, machining **A16:** 207, 274, 342, 345, 347, 349, 360, 361
1012
arc welding **A6:** 645
austenitizing temperature **A4:** 962
capacitor discharge stud welding **A6:** 222
composition **M1:** 125
cutoff band sawing with bimetal blades **A6:** 1184
flame hardening **A4:** 283
machinability rating **M1:** 576
stud arc welding **A6:** 215
1012, austenitizing temperature **M4:** 30
1012, flash welding **M6:** 574
1012, machining **A16:** 144-147, 179, 207, 274, 323-325, 342, 345, 347, 349, 360, 361, 668
1013
arc welding **A6:** 645
capacitor discharge stud welding **A6:** 222
composition **M1:** 125
cutoff band sawing with bimetal blades **A6:** 1184
machinability rating **M1:** 576
stud arc welding **A6:** 215
1013, flame hardening **A4:** 283
1013, machining **A16:** 207, 274, 342, 345, 347, 349, 360, 361, 668
1014
capacitor discharge stud welding **A6:** 222
stud arc welding **A6:** 215
1014, flame hardening **A4:** 283

Steels, AISI-SAE specific types (continued)
1015
- arc welding **A6:** 645
- austenitizing temperature **A4:** 962 **M4:** 30
- capacitor discharge stud welding **A6:** 222
- carbonitriding **A4:** 376, 381
- case depth, measurement **A4:** 457-458
- composition **A4:** 300 **M1:** 125
- cutoff band sawing with bimetal blades **A6:** 1184
- electron beam hardening treatment **A4:** 300
- electron beam hardening treatment (as C15 steel) **A4:** 308
- flame hardening **A4:** 283
- hardness **A4:** 407
- liquid carburizing **A4:** 334
- liquid nitriding **A4:** 412 **M4:** 253
- machinability rating **M1:** 576
- mass, effect on hardness **A4:** 39 **M4:** 11
- nitrogen diffusion **M4:** 255
- nominal compositions and applications **A18:** 703
- normalizing temperatures **A4:** 36 **M4:** 7
- properties, various heat treating conditions **M4:** 9-10
- properties, various heat-treating conditions
- relative erosion factor **A18:** 200
- stud arc welding **A6:** 215
- vacuum nitrocarburizing **A4:** 406, 407

1015, machining **A16:** 144-147, 174, 179, 207, 262, 274, 323-325, 342, 345, 347, 349, 360-362, 668

1016
- arc welding **A6:** 645
- austenitizing temperature **A4:** 962
- capacitor discharge stud welding **A6:** 222
- carbonitriding **A4:** 262, 376, 380
- cold work, effect on mechanical properties **M1:** 225-228
- composition **M1:** 125
- cutoff band sawing with bimetal blades **A6:** 1184
- flame hardening **A4:** 283
- hardfacing **A6:** 807
- machinability **M1:** 236, 238
- machinability rating **M1:** 576
- stud arc welding **A6:** 215

1016, austenitizing temperature **M4:** 30
1016, laser cladding **A18:** 867
1016, machining **A16:** 207, 274, 291, 342, 345, 347, 349, 360, 361, 668, 670

1017
- arc welding **A6:** 645
- austenitizing temperature **A4:** 962
- capacitor discharge stud welding **A6:** 222
- composition **M1:** 125
- cutoff band sawing with bimetal blades **A6:** 1184
- flame hardening **A4:** 283
- machinability rating **M1:** 576
- stud arc welding **A6:** 215

1017, austenitizing temperature **M4:** 30
1017, machining **A16:** 144-147, 179, 207, 225, 274, 321, 323-325, 342, 345, 347, 349, 360, 361, 668

1018
- arc welding **A6:** 645
- austenitizing temperature **A4:** 962 **M4:** 30
- bars **M1:** 223, 241, 245-246
- capacitor discharge stud welding **A6:** 222
- carbonitriding **A4:** 262, 376, 377, 379, 381, 382, 385 **M4:** 179, 180
- carburizing temperature, effect on grain size **M4:** 158
- case depth, measurement **A4:** 457-458
- composition **A4:** 320 **M1:** 125
- cutoff band sawing with bimetal blades **A6:** 1184
- damping capacity **M1:** 41, 45
- deep drawing dies, use for **M3:** 499
- die-casting dies, use in **M3:** 543
- electrogas welding process, selected steel grades **A6:** 656
- erosion rate and particle flux effect **A18:** 204
- erosion rate and particle size effect **A18:** 203
- flame hardening **A4:** 283
- friction welding **A6:** 153 **M6:** 721
- full annealing **A4:** 46
- hardfacing **A6:** 807
- ion-carburizing atmospheres **A4:** 565
- laser alloying **M6:** 793-794
- laser alloying with chromium **A18:** 866
- laser beam welding **M6:** 658
- laser cladding **A18:** 867 **M6:** 800
- laser surface hardening **A4:** 286
- laser transformation hardening **A18:** 862
- liquid carburizing **A4:** 345
- machinability **M1:** 236, 238
- machinability rating **M1:** 576
- magnetic-comparator electromagnetic test specimen **A4:** 590
- martempering in oil, applications **A4:** 148 **M4:** 98
- mechanical properties **M4:** 26
- mechanical properties after annealing **A4:** 54
- mechanical properties, cold drum
- nominal compositions and applications **A18:** 703
- stud arc welding **A6:** 215
- surface laser alloying **A18:** 866
- temperatures and cooling cycles, annealing **M4:** 17

1018, machining **A16:** 27, 58, 207, 225, 262, 274, 282, 285, 301, 342, 345, 347, 349, 360, 361, 668

1019
- arc welding **A6:** 645
- austenitizing temperature **A4:** 962
- capacitor discharge stud welding **A6:** 222
- carbonitriding **A4:** 262, 376
- Composition **A4:** 320 **M1:** 125
- cutoff band sawing with bimetal blades **A6:** 1184
- flame hardening **A4:** 283
- machinability rating **M1:** 576
- stud arc welding **A6:** 215
- wristpins, tool life for planing vs. 8620 steel **M1:** 572-573, 575

1019, austenitizing temperature **M4:** 30
1019, machining **A16:** 207, 225, 274, 342, 345, 347, 349, 360, 361, 668

1020
- abrasive wear data **A18:** 706
- annealing **A4:** 47
- arc welding **A6:** 645
- austenitizing temperature **A4:** 962 **M4:** 30
- cage material for rolling-element bearings **A18:** 503
- capacitor discharge stud welding **A6:** 222
- carbon gradient **M4:** 440
- carbon gradients and surface carbon content **A4:** 592, 593
- carbon restoration **A4:** 599 **M4:** 447
- carbonitrided, effect of nitrogen on case hardenability **M1:** 536
- carbonitriding **A4:** 376, 377, 378, 379, 381, 385 **M4:** 180, 182
- case hardened, hardness profiles **M1:** 632
- composition **A4:** 320 **A6:** 641 **M1:** 125
- composition and carbon equivalent **A6:** 406
- critical temperatures **A4:** 43
- critical temperatures, annealing **M4:** 15
- cutoff band sawing with bimetal blades **A6:** 1184
- die-casting dies, use in **M3:** 543
- electrical resistivity **A6:** 587
- electrogas welding **A6:** 660
- electrogas welding process, selected steel grades **A6:** 656

Steels, AISI-SAE specific types (continued)
- erosion rate **A18:** 206
- flame hardening **A4:** 283
- friction coefficient data **A18:** 71
- friction welding **A6:** 187
- full annealing **A4:** 46
- hardness **A18:** 706
- laser-beam welding **A6:** 877
- liquid carburizing **A4:** 330-331, 334, 340, 345
- liquid nitriding **A4:** 410
- machinability **M1:** 236, 238
- machinability rating **M1:** 576
- mass, effect on hardness **A4:** 39 **M4:** 11
- mating material used to demonstrate friction of different types of carbon **A18:** 818
- mating material used to demonstrate wear of different types of carbon **A18:** 818
- molds for plastics and rubber, use for **M3:** 547, 548
- nominal compositions and applications **A18:** 703
- normalizing temperatures **A4:** 36 **M4:** 7
- pack carburizing applications **A4:** 326
- plasma (ion) carburizing **A4:** 355, 356, 358
- press forming dies, use for **M3:** 491
- properties **A6:** 587
- properties, various heat treating conditions **M4:** 9-10
- properties, various heat-treating conditions **A4:** 37
- proven applications for borided ferrous materials **A4:** 445
- recommended upsetting pressures for flash welding **A6:** 843
- stud arc welding **A6:** 215
- surface carbon variability **A4:** 595, 596
- temperatures and cooling cycles, annealing **M4:** 17
- tempering **A4:** 122
- tensile properties, hot rolled bars **M1:** 204
- toughness **A18:** 706
- trimming tools, use for **M3:** 532
- ultrasonic welding **A6:** 326
- wear data with unlubricated tool steel **A18:** 737
- wear resistance and cost effectiveness **A18:** 706
- work material for ion implantation **A18:** 858

1020, flash welding schedule **M6:** 577
1020, machining **A16:** 15, 58, 144-151, 164, 179, 180, 203, 207-209, 225, 246, 247, 251, 261, 263, 264, 269, 274, 282, 291, 323-325, 342, 345, 347, 349, 360, 361, 363, 364, 541, 584, 668, 854
1020 spheroidized, composition **A18:** 806

1021
- arc welding **A6:** 645
- capacitor discharge stud welding **A6:** 222
- carbonitriding **A4:** 376
- composition **A4:** 320 **M1:** 125
- cutoff band sawing with bimetal blades **A6:** 1184
- machinability rating **M1:** 576
- stud arc welding **A6:** 215

1021, machining **A16:** 207, 225, 274, 342, 345, 347, 349, 360, 361, 668

1022
- arc welding **A6:** 645
- austenitizing temperature **A4:** 962 **M4:** 30
- capacitor discharge stud welding **A6:** 222
- carbon gradient **M4:** 441
- carbon gradients after carburizing **A4:** 592, 594, 595
- carbonitriding **A4:** 262, 376, 381
- carburized and ground cams, wear affected by grinding burns **M1:** 634, 636
- carburizing temperature, effect on grain size **M4:** 158

SUBJECTS OF THE INDEXED VOLUMES: ASM Handbook (designated by the letter "A"): **A1:** Properties and Selection: Irons, Steels, and High-Performance Alloys (1990); **A2:** Properties and Selection: Nonferrous Alloys and Special-Purpose Materials (1990); **A3:** Alloy Phase Diagrams (1992); **A4:** Heat Treating (1991); **A6:** Welding, Brazing, and Soldering (1993); **A7:** Powder Metallurgy (1984); **A8:** Mechanical Testing (1985); **A9:** Metallography and Microstructures (1985); **A10:** Materials Characterization (1986); **A11:** Failure Analysis and Prevention (1986); **A12:** Fractography (1987); **A13:** Corrosion (1987); **A14:** Forming and Forging (1988); **A15:** Casting (1988); **A16:** Machining (1989); **A17:** Nondestructive Testing and Quality Control (1989); **A18:** Friction, Lubrication, and Wear Technology (1992). **Metals Handbook, 9th Edition** (designated by the letter "M"): **M1:** Properties and Selection: Irons and Steels (1978); **M2:** Properties and Selection: Nonferrous Alloys and Pure Metals (1979); **M3:** Properties and Selection: Stainless Steels, Tool Materials and Special-Purpose Materials (1980); **M4:** Heat Treating (1981); **M5:** Surface Cleaning, Finishing, and Coating (1982); **M6:** Welding, Brazing, and Soldering (1983). **Engineered Materials Handbook** (designated by the letters "EM"): **EM1:** Composites (1987); **EM2:** Engineering Plastics (1988); **EM3:** Adhesives and Sealants (1990); **EM4:** Ceramics and Glasses (1991); **Electronic Materials Handbook** (designated by the letters "EL"): **EL1:** Packaging (1989).

Steels, AISI-SAE specific types (continued)
composition ... A4: 320 M1: 125
cutoff band sawing with bimetal blades ... A6: 1184
full annealing ... A4: 46
liquid carburizing ... A4: 345
machinability ... M1: 236, 238
machinability rating ... M1: 576
mass, effect on hardness ... A4: 39 M4: 11
mechanical properties ... M4: 26
mechanical properties after annealing ... A4: 54
normalizing temperatures ... A4: 36 M4: 7
pack carburizing applications ... A4: 326
properties, various heat treating conditions ... M4: 9-10
properties, various heat-treating conditions ... A4: 37
sprocket, wear of ... M1: 628
statistical process control of operations ... A4: 624
stud arc welding ... A6: 215
temperatures and cooling cycles, annealing ... M4: 17
test bar material for carburizing facility's quality control ... A4: 599-600
1022, machining ... A16: 58, 207, 225, 274, 290, 342, 345, 347, 349, 360, 361, 389, 672, 668
1023
arc welding ... A6: 645
capacitor discharge stud welding ... A6: 222
composition ... M1: 125
cutoff band sawing with bimetal blades ... A6: 1184
machinability rating ... M1: 576
1023, machining ... A16: 179, 207, 225, 274, 323-325, 342, 345, 347, 349, 360, 361, 668
1024
arc welding ... A6: 645
austenitizing temperature ... A4: 142 M4: 92
capacitor discharge stud welding ... A6: 222
carbon and hardness gradients ... M4: 442
carbon gradient effect on case properties ... A4: 594, 596
carbonitriding ... A4: 376, 381-382
carburizing temperature, effect on grain size ... M4: 158
cutoff band sawing with bimetal blades ... A6: 1184
electron-beam welding ... A6: 866
gas carburizing ... M4: 163
martempering temperature ... A4: 142 M4: 92
1025
arc welding ... A6: 645
austenitizing temperature ... A4: 961 M4: 29
capacitor discharge stud welding ... A6: 222
carbonitriding ... A4: 376
composition ... M1: 125
cutoff band sawing with bimetal blades ... A6: 1184
flame hardening ... A4: 283
full annealing ... A4: 46
induction heat treating ... A4: 200
machinability rating ... M1: 576
mechanical properties, cold drawn bars ... M1: 223
normalizing temperatures ... A4: 36 M4: 7
pack carburizing applications ... A4: 326
temperatures and cooling cycles, annealing ... M4: 17
thermal properties ... M4: 511
1025, machining ... A16: 144-147, 179, 207, 225, 274, 323-325, 342, 345, 347, 349, 360, 361, 668
1026
arc welding ... A6: 645
capacitor discharge stud welding ... A6: 222
carbonitriding ... A4: 376
composition ... M1: 125
cutoff band sawing with bimetal blades ... A6: 1184
flame hardening ... A4: 283
machinability rating ... M1: 576
1026, flash versus friction welding ... M6: 720
1026, machining ... A16: 207, 208, 225, 274, 342, 345, 347, 349, 360, 361, 668, 672
1027
arc welding ... A6: 645
capacitor discharge stud welding ... A6: 222
carbonitriding ... A4: 376
cutoff band sawing with bimetal blades ... A6: 1184
flame hardening ... A4: 283
1028
carbonitriding ... A4: 376
flame hardening ... A4: 283
1028, capacitor discharge stud welding ... A6: 222
1029
arc welding ... A6: 645
capacitor discharge stud welding ... A6: 222
carbonitriding ... A4: 376
composition ... M1: 125
cutoff band sawing with bimetal blades ... A6: 1184
flame hardening ... A4: 283
machinability rating ... M1: 576
1029, machining ... A16: 207, 225, 274, 342, 345, 347, 349, 360, 361, 668
1030
arc welding ... A6: 645
austenitizing temperature ... A4: 961 M4: 29
capacitor discharge stud welding ... A6: 222
carbonitriding ... A4: 376, 381
composition ... A6: 641 M1: 125
composition and carbon equivalent ... A6: 406
critical temperatures ... A4: 43
critical temperatures, annealing ... M4: 15
cutoff band sawing with bimetal blades ... A6: 1184
flame hardening ... A4: 283
flux-cored arc welding ... A6: 187
full annealing ... A4: 46
hardness after tempering ... M4: 71
induction surface hardening ... A4: 191
machinability rating ... M1: 576
mass, effect on hardness ... A4: 39 M4: 11
mechanical properties ... M4: 26
mechanical properties after annealing ... A4: 54
mechanical properties vs. processing variables ... M1: 462-463
normalizing temperatures ... A4: 36 M4: 7
pack carburizing applications ... A4: 326
properties, various heat treating conditions ... M4: 9-10
properties, various heat-treating conditions ... A4: 37
recommended machining conditions ... M1: 569
temperatures and cooling cycles, annealing ... M4: 17
tempering ... A4: 122
tensile properties, hot rolled bars ... M1: 204
1030, machining ... A16: 144-147, 163, 165, 179, 207, 225, 274, 323-325, 342, 345, 347, 349, 360, 361, 668
1030, shear blades, service data ... M3: 481
1031
capacitor discharge stud welding ... A6: 222
carbonitriding ... A4: 376
cutoff band sawing with bimetal blades ... A6: 1184
flame hardening ... A4: 283
1032
capacitor discharge stud welding ... A6: 222
carbonitriding ... A4: 376
cutoff band sawing with bimetal blades ... A6: 1184
flame hardening ... A4: 283
1032, friction coefficient data ... A18: 71, 73, 74
1033
capacitor discharge stud welding ... A6: 222
carbonitriding ... A4: 376
composition ... M1: 125
cutoff band sawing with bimetal blades ... A6: 1184
flame hardening ... A4: 283
1033, machining ... A16: 144-147, 179, 207, 225, 274, 323-325, 342, 345, 347, 349, 360, 361
1034
austempering, suitability ... A4: 154 M4: 106
capacitor discharge stud welding ... A6: 222
carbonitriding ... A4: 376
cutoff band sawing with bimetal blades ... A6: 1184
flame hardening ... A4: 283
hardenability ... A4: 140, 141
time-temperature transformation diagram ... M4: 89
time-temperature transformation diagrams ... A4: 140, 141
1035
arc welding ... A6: 645
austenitizing temperature ... A4: 961 M4: 29
capacitor discharge stud welding ... A6: 222
carbon restoration ... A4: 598 M4: 447
carbonitriding ... A4: 376
composition ... M1: 125
corrosion protection ... A4: 160 M4: 112
cutoff band sawing with bimetal blades ... A6: 1184
flame hardening ... A4: 283, 284
friction welding ... A6: 152-153
full annealing ... A4: 46
machinability ... M1: 236, 238
machinability rating ... M1: 576
mechanical properties, cold drawn bars ... M1: 223
normalizing temperatures ... A4: 36 M4: 7
pack carburizing applications ... A4: 326
SAC hardenability ... M1: 477, 478
shafts, induction hardened, fatigue life of ... M1: 675
temperatures and cooling cycles, annealing ... M4: 17
tempering ... A4: 131
1035, machining ... A16: 144-148, 179, 207, 225, 226, 246, 247, 274, 323-325, 342, 345, 347, 349, 360, 361, 363, 668
1035 mod
tempering with induction heating ... A4: 193
through-hardening by induction heating ... A4: 192
1035, trimming tools, use for ... M3: 532
1036
arc welding ... A6: 645
capacitor discharge stud welding ... A6: 222
carbonitriding ... A4: 376
cutoff band sawing with bimetal blades ... A6: 1184
flame hardening ... A4: 284
through-hardening by induction heating ... A4: 192
1037
arc welding ... A6: 645
austenitizing temperature ... A4: 961
automobile axle shafts; induction hardened, fatigue life ... M1: 674, 675
bolts, hardenability compared to 1541 ... M1: 277, 279
capacitor discharge stud welding ... A6: 222
carbonitriding ... A4: 376
composition ... M1: 125
cutoff band sawing with bimetal blades ... A6: 1184
distortion occurring during quenching ... A4: 79
flame hardening ... A4: 284
machinability rating ... M1: 576
notch toughness ... M1: 692
threaded fastener applications ... M1: 277-281
1037, austenitizing temperature ... M4: 29
1037, machining ... A16: 144-148, 179, 204, 207, 209, 225, 274, 323-325, 342, 345, 347, 349, 360, 361, 668
1037 mod
tempering with induction heating ... A4: 193
through-hardening by induction heating ... A4: 192
1038
arc welding ... A6: 645
austenitizing temperature ... A4: 961 M4: 29
capacitor discharge stud welding ... A6: 222
carbon restoration ... A4: 599 M4: 446
carbonitriding ... A4: 376
cutoff band sawing with bimetal blades ... A6: 1184
flame hardening ... A4: 284
mechanical properties ... M4: 26
mechanical properties after annealing ... A4: 54
tempering ... A4: 134
tempering with induction heating ... A4: 193

Steels, AISI-SAE specific types (continued)
through-hardening by induction
heating A4: 192
1038, machining A16: 144-147, 179, 207, 225, 274, 287, 288, 291, 293, 294, 323-325, 342, 345, 347, 349, 360, 361, 668
1038H
composition M1: 127
hardenability curve M1: 497
1039
arc welding A6: 645
austenitizing temperature A4: 961
capacitor discharge stud welding A6: 222
carbonitriding A4: 376
composition M1: 125
cutoff band sawing with bimetal
blades A6: 1184
flame hardening A4: 284
machinability rating M1: 576
1039, austenitizing temperature M4: 29
1039, machining A16: 144-147, 179, 207, 225, 274, 323-325, 342, 345, 347, 349, 360, 361, 668
1040
annealing A4: 49
arc welding A6: 645
austenitic nitrocarburizing (as EN 8) A4: 432
austenitizing temperature A4: 961 M4: 29
auxiliary tools, hot upset forging,
use for M3: 518
bolts, fatigue life M1: 280
capacitor discharge stud welding A6: 222
carbonitriding A4: 376, 381
cold extrusion tools, use for M3: 518
composition A6: 641 M1: 125
critical temperatures A4: 43
critical temperatures, annealing M4: 15
cutoff band sawing with bimetal
blades A6: 1184
flame hardening A4: 275, 277, 283, 284
forging without use of a lubricant A18: 639
full annealing A4: 46
hardenability, process variability
and experiment design (statistical process control) A4: 626-628
hardness after tempering M4: 71
hardness profile A4: 404
ion nitriding A4: 404
Jominy curves A4: 650
laser surface transformation
hardening A4: 293
laser transformation hardening A4: 265
machinability M1: 237, 238
machinability rating M1: 576
mass, effect on hardness A4: 39 M4: 11
mechanical properties, cold drawn
bars M1: 223, 228-229
mesh belting application A4: 516
normalizing temperatures A4: 36 M4: 7
plasma (ion) nitriding A4: 423
plasma nitrocarburizing A4: 433
properties, various heat treating
conditions M4: 9-10
properties, various heat-treating
conditions A4: 37
SAC hardenability variation M1: 479
seizure resistance M1: 611
shallow-hardening steel A18: 706
spheroidizing A4: 46, 48, 49
temperatures and cooling cycles,
annealing M4: 17
tempering A4: 122, 134
tensile properties, hot rolled bars M1: 204
1040, hardness in a flash weld M6: 578
1040, machining A16: 144-147, 179, 207, 225, 274, 285, 291, 302, 323-325, 342, 345, 347, 360, 361, 668, 672
1041
arc welding A6: 645
austempering, suitability A4: 153

Steels, AISI-SAE specific types (continued)
austenitizing temperature A4: 961
capacitor discharge stud welding A6: 222
carbonitriding A4: 376, 386
cutoff band sawing with bimetal
blades A6: 1184
flame hardening A4: 283, 284
martempering A4: 140
tempering with induction heating A4: 193
through-hardening by induction
heating A4: 192
1041, machining A16: 225, 264, 285, 360, 361
1042
arc welding A6: 645
capacitor discharge stud welding A6: 222
composition M1: 125
cutoff band sawing with bimetal
blades A6: 1184
machinability rating M1: 576
1042 (DIN Ck 45)
boriding A4: 445
carbonitriding A4: 376
flame hardening A4: 283, 284
induction hardening A4: 184, 185
proven applications for borided ferrous materials A4: 445
1042, austenitizing temperature M4: 29
1042, machining A16: 144-147, 179, 207, 225, 274, 323-325, 342, 345, 347, 349, 360, 361, 668
1043
arc welding A6: 645
capacitor discharge stud welding A6: 222
composition M1: 125
cutoff band sawing with bimetal
blades A6: 1184
machinability rating M1: 576
1043 (DIN C 45)
austenitizing temperature A4: 961
boriding A4: 437, 438, 439
carbonitriding A4: 376
flame hardening A4: 283, 284
proven applications for borided ferrous materials A4: 445
tempering with induction heating A4: 193
through-hardening by induction
heating A4: 192
1043, austenitizing temperature M4: 29
1043, machining A16: 144-147, 179, 207, 225, 274, 323-325, 342, 345, 347, 349, 360, 361, 668
1044
arc welding A6: 645
capacitor discharge stud welding A6: 222
carbonitriding A4: 376
composition M1: 125
cutoff band sawing with bimetal
blades A6: 1184
flame hardening A4: 283, 284
1044, machining A16: 144-147, 179, 207, 225, 274, 323-325, 342, 345, 347, 349, 360, 361
1045
arc welding A6: 645
austenitizing temperature A4: 961 M4: 29
capacitor discharge stud welding A6: 222
carbonitriding A4: 376
composition A4: 300 M1: 125
cooling curves A4: 138 M4: 86
cutoff band sawing with bimetal
blades A6: 1184
electron beam hardening treatment A4: 300
electron beam hardening treatment
(as C 45 steel) A4: 299, 300, 302-308
erosion resistance versus hardness A18: 202, 204
flame hardening A4: 274, 275, 282, 283, 284
friction welding A6: 153
full annealing A4: 46
hardness A4: 71
hardness, and oil quenching A4: 91, 94
hardness vs. tempering temperature M1: 468

Steels, AISI-SAE specific types (continued)
heavy-draft drawing, suitability M1: 243
hot forging, die wear A18: 635
induction hardened, heating time vs.
depth of hardening M1: 529
induction hardened, residual stress
in M1: 528
induction heat treating A4: 196, 198
induction surface hardening A4: 191
laser hardening M4: 513
laser heat treating M4: 514
laser surface transformation
hardening A4: 295
laser transformation hardening A18: 862, 863
liquid nitriding A4: 410
liquid nitrocarburizing A4: 417
machinability M1: 237, 238
machinability rating M1: 576
martempering A4: 141, 142
martempering in salt, applications A4: 148 M4: 96
martempering time M4: 92
mechanical properties M4: 26
mechanical properties after
annealing A4: 54
mechanical properties, cold drawn
bars M1: 223, 243
non-martensitic transformation
products A4: 141
normalizing temperatures A4: 36 M4: 7
recommended upsetting pressures
for flash welding A6: 843
residual stress patterns, cold drawn
bars M1: 235
residual stresses A4: 608
residual stresses and quenching A4: 15, 16
springs, hot wound M1: 303
sprocket, wear of M1: 628
temperatures and cooling cycles,
annealing M4: 17
tempering A4: 11, 131
thermal properties M4: 511
wheel spindles, machining time
compared with 1141 steel M1: 575, 579
1045, friction welding M6: 721
1045, machining A16: 58, 81, 87, 88, 90, 92, 144-148, 164, 165, 172, 179, 191, 207, 209, 225, 246, 247, 251, 263, 274, 285, 323-325, 342, 345, 347, 349, 360, 361, 364, 388, 640, 643, 668, 673, 674, 676, 886
1045, shear blades, service data M3: 481
1045H
composition M1: 127
hardenability curve M1: 497
1046
arc welding A6: 645
austenitizing temperature A4: 961 M4: 29
capacitor discharge stud welding A6: 222
carbonitriding A4: 376
composition M1: 125
cutoff band sawing with bimetal
blades A6: 1184
flame hardening A4: 283, 284
hardness A4: 138
hardness, effect of quenchant and
agitation on M4: 87
heat treated, machinability compared with 1144 steel M1: 575-578
machinability rating M1: 576
SAC hardenability variation M1: 479
tempering A4: 131, 133-134
1046, machining A16: 144-147, 179, 207, 225, 274, 323-325, 342, 345, 347, 349, 360, 361, 668
1047
carbonitriding A4: 376
flame hardening A4: 283, 284
1047, capacitor discharge stud welding A6: 222
1048
arc welding A6: 645
capacitor discharge stud welding A6: 222

Steels, AISI-SAE specific types (continued)
carbonitriding ... **A4:** 376
cutoff band sawing with bimetal blades ... **A6:** 1184
flame hardening ... **A4:** 283, 284
1049
arc welding ... **A6:** 645
capacitor discharge stud welding ... **A6:** 222
carbonitriding ... **A4:** 376
composition ... **M1:** 125
cutoff band sawing with bimetal blades ... **A6:** 1184
flame hardening ... **A4:** 283, 284
machinability rating ... **M1:** 576
normalizing ... **A4:** 38
1049, machining ... **A16:** 23-325, 342, 345, 347, 349, 360, 361, 668
1050
applications, austempered parts ... **A4:** 157 **M4:** 108
arc welding ... **A6:** 645
austempering ... **A4:** 157, 162
austenitizing temperature ... **A4:** 961 **M4:** 29
capacitor discharge stud welding ... **A6:** 222
carbonitriding ... **A4:** 98, 376
composition ... **A6:** 641 **M1:** 125
critical temperatures ... **A4:** 43
critical temperatures, annealing ... **M4:** 15
cutoff band sawing with bimetal blades ... **A6:** 1184
flame hardening ... **A4:** 283, 284
full annealing ... **A4:** 46
hardness ... **A4:** 186
hardness after tempering ... **M4:** 71
hardness, and oil quenching ... **A4:** 98-99
heavy-draft drawing, suitability ... **M1:** 250
induction hardening ... **A4:** 186
laser surface transformation hardening ... **A4:** 293
laser transformation hardening ... **A4:** 265
machinability ... **M1:** 237, 238
machinability rating ... **M1:** 576
mass, effect on hardness ... **A4:** 39 **M4:** 11
mechanical properties ... **A4:** 123
mechanical properties, cold drawn bars ... **M1:** 224
normalizing temperatures ... **A4:** 36 **M4:** 7
notch toughness ... **M1:** 705, 706
oil quenching ... **M4:** 53
properties affected by composition variations ... **M1:** 243, 456, 457, 463, 467
properties, various heat treating conditions ... **M4:** 9-10
properties, various heat-treating conditions ... **A4:** 37
residual stress patterns, cold drawn bars ... **M1:** 233-234
section size, austempered parts ... **A4:** 155 **M4:** 107
springs ... **M1:** 285, 287
surface finish, drilled and reamed part, leaded vs. standard steel ... **M1:** 578
temperatures and cooling cycles, annealing ... **M4:** 17
tempering ... **A4:** 98, 122, 123, 131, 134, 186
tempering temperature, effect on properties ... **M4:** 72
tensile properties, hot rolled bars
wear rate versus load ... **A18:** 239
weld cladding ... **A6:** 811, 819
work material for ion implantation ... **A18:** 858
1050, machining ... **A16:** 58, 150, 179, 207, 225, 274, 323-325, 342, 345, 347, 349, 360, 361, 668
1051, cutoff band sawing with bimetal blades ... **A6:** 1184
1052
arc welding ... **A6:** 645
carbonitriding ... **A4:** 376
cutoff band sawing with bimetal blades ... **A6:** 1184
flame hardening ... **A4:** 275, 282, 284
1053
arc welding ... **A6:** 645
composition ... **M1:** 125
cutoff band sawing with bimetal blades ... **A6:** 1184

Steels, AISI-SAE specific types (continued)
machinability rating ... **M1:** 576
1053, machining ... **A16:** 144-147, 179, 207, 225, 274, 323-325, 342, 345, 347, 349, 360, 361, 668
1054, cutoff band sawing with bimetal blades ... **A6:** 1184
1055
austenitizing temperature ... **A4:** 961
carbonitriding ... **A4:** 376
composition ... **M1:** 125
flame hardening ... **A4:** 283
machinability rating ... **M1:** 576
mesh belting application ... **A4:** 516
1055, austenitizing temperature ... **M4:** 29
1055, cutoff band sawing with bimetal blades ... **A6:** 1184
1055, erosion rate ... **A18:** 204
1055, machining ... **A16:** 144-147, 179, 207, 225, 274, 323-325, 342, 345, 347, 349, 360, 361, 607, 668, 677
1056
carbonitriding ... **A4:** 376
flame hardening ... **A4:** 283
1056, cutoff band sawing with bimetal blades ... **A6:** 1184
1057
carbonitriding ... **A4:** 376
flame hardening ... **A4:** 283
1057, cutoff band sawing with bimetal blades ... **A6:** 1184
1058
carbonitriding ... **A4:** 376
flame hardening ... **A4:** 283
1058, cutoff band sawing with bimetal blades ... **A6:** 1184
1059
carbonitriding ... **A4:** 376
composition ... **M1:** 125
flame hardening ... **A4:** 283
machinability rating ... **M1:** 576
1059, band sawing ... **A16:** 360, 361
1059, cutoff band sawing with bimetal blades ... **A6:** 1184
1059, turning machinability rating ... **A16:** 668
1060
applications, austempered parts ... **A4:** 157, 161 **M4:** 108
austenitizing temperature ... **A4:** 961 **M4:** 29
carbonitriding ... **A4:** 376
cold work, effect on mechanical properties ... **M1:** 228-229
composition ... **A6:** 641 **M1:** 125
corrosion protection ... **A4:** 160 **M4:** 112
critical temperatures ... **A4:** 43
critical temperatures, annealing ... **M4:** 15
cutoff band sawing with bimetal blades ... **A6:** 1184
flame hardening ... **A4:** 283
full annealing ... **A4:** 46
hardness after tempering ... **M4:** 71
liquid nitriding ... **A4:** 412
machinability rating ... **M1:** 576
mass, effect on hardness ... **A4:** 39 **M4:** 11
mechanical properties ... **M4:** 26
mechanical properties after annealing ... **A4:** 54
normalizing temperatures ... **A4:** 36 **M4:** 7
properties, various heat treating conditions ... **M4:** 9-10
properties, various heat-treating conditions ... **A4:** 37
recommended machining conditions ... **M1:** 569
temperatures and cooling cycles, annealing ... **M4:** 17
tempering ... **A4:** 122, 134
tensile properties, hot rolled bars ... **M1:** 204
1060, machining ... **A16:** 225, 274, 323-325, 360, 361, 668
1061
carbonitriding ... **A4:** 376
flame hardening ... **A4:** 283
machinability rating ... **M1:** 576
1061, cutoff band sawing with bimetal blades ... **A6:** 1184
1062
carbonitriding ... **A4:** 376

Steels, AISI-SAE specific types (continued)
flame hardening ... **A4:** 275, 276, 277, 282, 283
1062, cutoff band sawing with bimetal blades ... **A6:** 1184
1063
carbonitriding ... **A4:** 376
flame hardening ... **A4:** 283
1063, band sawing ... **A16:** 360, 361
1063, broaching ... **A16:** 203, 209
1063, cutoff band sawing with bimetal blades ... **A6:** 1184
1064
carbonitriding ... **A4:** 376
composition ... **M1:** 125
flame hardening ... **A4:** 283
hardness, and oil quenching ... **A4:** 99
machinability rating ... **M1:** 576
1064, band sawing ... **A16:** 360, 361
1064, cutoff band sawing with bimetal blades ... **A6:** 1184
1064, thread grinding ... **A16:** 274
1064, turning machinability rating ... **A16:** 668
1065
applications, austempered parts ... **A4:** 157 **M4:** 108
austenitizing temperature ... **A4:** 961 **M4:** 29
carbonitriding ... **A4:** 376
composition ... **M1:** 125
cutoff band sawing with bimetal blades ... **A6:** 1184
flame hardening ... **A4:** 283
machinability rating ... **M1:** 576
martempering in oil, applications ... **A4:** 148 **M4:** 98
mechanical properties ... **M4:** 26
mechanical properties after annealing ... **A4:** 54
recommended pressures for flash welding ... **A6:** 843
section size, austempered parts ... **A4:** 155 **M4:** 107
1065, band sawing ... **A16:** 360, 361
1065, corrosive wear ... **A18:** 719
1065, thread grinding ... **A16:** 274
1065, turning machinability rating ... **A16:** 668
1066
carbonitriding ... **A4:** 376
flame hardening ... **A4:** 283
machinability rating ... **M1:** 576
section size, austempered parts ... **A4:** 155
1066, band sawing ... **A16:** 360, 361
1066, cutoff band sawing with bimetal blades ... **A6:** 1184
1066, section size, austempered parts ... **M4:** 107
1066, turning machinability rating ... **A16:** 668
1067
carbonitriding ... **A4:** 376
flame hardening ... **A4:** 283
1067, cutoff band sawing with bimetal blades ... **A6:** 1184
1068
carbonitriding ... **A4:** 376
flame hardening ... **A4:** 283
1068, cutoff band sawing with bimetal blades ... **A6:** 1184
1069
carbonitriding ... **A4:** 376
composition ... **M1:** 125
flame hardening ... **A4:** 283
machinability rating ... **M1:** 576
1069, band sawing ... **A16:** 360, 361
1069, cutoff band sawing with bimetal blades ... **A6:** 1184
1069, thread grinding ... **A16:** 274
1069, turning machinability rating ... **A16:** 668
1070
annealing ... **A4:** 54
applications ... **A4:** 104
applications, austempered parts ... **A4:** 157 **M4:** 108
austenitizing temperature ... **A4:** 142, 961 **M4:** 29, 92
carbon restoration ... **A4:** 599 **M4:** 446
carbonitriding ... **A4:** 376
composition ... **A4:** 300 **A6:** 641 **M1:** 125
critical temperatures ... **A4:** 43
critical temperatures, annealing ... **M4:** 15

Steels, AISI-SAE specific types (continued)
cutoff band sawing with bimetal blades ... A6: 1184
electron beam hardening treatment ... A4: 300
electron beam hardening treatment (as Ck 67 steel) ... A4: 304, 308
flame hardening ... A4: 283
full annealing ... A4: 46
hardening in an exothermic-endothermic based atmosphere ... A4: 563, 564
induction hardened, effect of prior structure on hardness profile ... M1: 531
induction surface hardening ... A4: 191
laser surface transformation hardening ... A4: 293
laser transformation hardening ... A4: 265
machinability rating ... M1: 576
martempering temperature ... A4: 142 M4: 92
springs, helical ... M1: 299
temperatures and cooling cycles, annealing ... M4: 17
tensile properties, hot rolled bars ... M1: 204
1070, machining ... A16: 209, 225, 274, 360, 361, 668, 672
1070, wear rate ... A18: 705
1071
applications ... A4: 104
carbonitriding ... A4: 376
flame hardening ... A4: 283
1071, cutoff band sawing with bimetal blades ... A6: 1184
1072
applications ... A4: 104
carbonitriding ... A4: 376
flame hardening ... A4: 283
1072, cutoff band sawing with bimetal blades ... A6: 1184
1073
applications ... A4: 104
carbonitriding ... A4: 376
flame hardening ... A4: 283
1073, cutoff band sawing with bimetal blades ... A6: 1184
1074
applications ... A4: 104
austenitizing temperature ... A4: 961
carbonitriding ... A4: 376
composition ... M1: 125
flame hardening ... A4: 283
machinability rating ... M1: 576
springs ... M1: 285, 287
1074, austenitizing temperature ... M4: 29
1074, band sawing ... A16: 360, 361
1074, cutoff band sawing with bimetal blades ... A6: 1184
1074, thread grinding ... A16: 274
1074, turning machinability rating ... A16: 668
1075
applications ... A4: 104
applications, austempered parts ... A4: 157
carbonitriding ... A4: 376
composition ... M1: 125
flame hardening ... A4: 283
machinability rating ... M1: 576
1075, applications, austempered parts ... M4: 108
1075, band sawing ... A16: 360, 361
1075, cutoff band sawing with bimetal blades ... A6: 1184
1075, erosion rate and resistance ... A18: 204
1075, thread grinding ... A16: 274
1075, turning machinability rating ... A16: 668
1076, cutoff band sawing with bimetal blades ... A6: 1184
1077, cutoff band sawing with bimetal blades ... A6: 1184
1078
applications ... A4: 104

Steels, AISI-SAE specific types (continued)
austenitizing temperature ... A4: 961 M4: 29
carbonitriding ... A4: 376
composition ... M1: 125
induction heat treating ... A4: 199
laser surface transformation hardening ... A4: 290
machinability rating ... M1: 576
thermal properties ... M4: 511
1078, band sawing ... A16: 360, 361
1078, cutoff band sawing with bimetal blades ... A6: 1184
1078, erosion rate ... A18: 204
1078, thread grinding ... A16: 274
1078, turning machinability rating ... A16: 668
1079, cutoff band sawing with bimetal blades ... A6: 1184
1080
annealing ... A4: 184, 185
applications ... A4: 104
applications, austempered parts ... A4: 157 M4: 108
austempering ... A4: 162, 163
austempering equipment ... M4: 111
austempering equipment requirements ... A4: 159
austempering, suitability ... A4: 153, 154-155 M4: 106
austenitizing temperature ... A4: 961 M4: 29
carbonitriding ... A4: 376
composition ... A6: 641 M1: 125
critical temperatures ... A4: 43
critical temperatures, annealing ... M4: 15
cutoff band sawing with bimetal blades ... A6: 1184
drawing temperature, effect on mechanical properties ... M1: 246-247
erosion rate ... A18: 204
flame hardening ... A4: 283
full annealing ... A4: 46
hardness after tempering ... M4: 71
induction hardening ... A4: 184, 185
induction heat treating ... A4: 199
machinability rating ... M1: 576
mass, effect on hardness ... A4: 39 M4: 11
normalizing temperatures ... A4: 36 M4: 7
properties ... A18: 713, 714
properties, various heat treating conditions ... M4: 9-10
properties, various heat-treating conditions ... A4: 37
temperatures and cooling cycles, annealing ... M4: 17
tempering ... A4: 122
tensile properties, hot rolled bars ... M1: 204
time-temperature transformation ... M4: 115
wear resistance and cost effectiveness ... A18: 705
1080, band sawing ... A16: 360, 361
1080 spheroidized, composition ... A18: 806
1080, thread grinding ... A16: 274
1080, turning machinability rating ... A16: 668
1081
applications ... A4: 104
carbonitriding ... A4: 376
flame hardening ... A4: 283
1081, cutoff band sawing with bimetal blades ... A6: 1184
1082
applications ... A4: 104
carbonitriding ... A4: 376
flame hardening ... A4: 283
1082, band sawing ... A16: 360, 361
1082, cutoff band sawing with bimetal blades ... A6: 1184
1083
applications ... A4: 104
carbonitriding ... A4: 376

Steels, AISI-SAE specific types (continued)
flame hardening ... A4: 283
1083, cutoff band sawing with bimetal blades ... A6: 1184
1084
applications ... A4: 104
austenitizing temperature ... A4: 961 M4: 29
carbonitriding ... A4: 376
composition ... M1: 125
flame hardening ... A4: 283
machinability rating ... M1: 576
section size, austempered parts ... A4: 155 M4: 107
1084, band sawing ... A16: 360, 361
1084, cutoff band sawing with bimetal blades ... A6: 1184
1084, thread grinding ... A16: 274
1084, turning machinability rating ... A16: 668
1085
applications ... A4: 104
austenitizing temperature ... A4: 961 M4: 29
carbonitriding ... A4: 376
composition ... M1: 125
distortion, and oil quenching ... A4: 99
flame hardening ... A4: 283
hardness ... A4: 99, 100
machinability rating ... M1: 576
oil quenching ... M4: 54
1085, abrasive wear data ... A18: 705
1085, band sawing ... A16: 360, 361
1085, cutoff band sawing with bimetal blades ... A6: 1184
1085, thread grinding ... A16: 274
1085, turning machinability rating ... A16: 668
1086
applications ... A4: 104
austenitizing temperature ... A4: 961 M4: 29
carbonitriding ... A4: 376
composition ... M1: 125
flame hardening ... A4: 283
machinability rating ... M1: 576
section size, austempered parts ... A4: 155 M4: 107
tempering ... A4: 133
1086, band sawing ... A16: 360, 361
1086, cutoff band sawing with bimetal blades ... A6: 1184
1086, thread grinding ... A16: 274
1086, turning machinability rating ... A16: 668
1087
applications ... A4: 104
carbonitriding ... A4: 376
flame hardening ... A4: 283
1087, cutoff band sawing with bimetal blades ... A6: 1184
1088
applications ... A4: 104
carbonitriding ... A4: 376
flame hardening ... A4: 283
1088, cutoff band sawing with bimetal blades ... A6: 1184
1089
applications ... A4: 104
carbonitriding ... A4: 376
flame hardening ... A4: 283
1089, cutoff band sawing with bimetal blades ... A6: 1184
1090
applications ... A4: 104
applications, austempered parts ... A4: 157 M4: 108
austenitizing temperature ... A4: 961 M4: 29
carbonitriding ... A4: 376
composition ... M1: 125
flame hardening ... A4: 283
full annealing ... A4: 46
machinability rating ... M1: 576
martempered to full hardness ... A4: 140

Steels, AISI-SAE specific types (continued)
mechanical properties, austempered compared to oil quenched **A4:** 155 **M4:** 108
normalizing temperatures **A4:** 36 **M4:** 7
section size, austempered parts **A4:** 155 **M4:** 107
temperatures and cooling cycles, annealing **M4:** 17
tempering **A4:** 134
tensile properties, hot rolled bars **M1:** 204
time-temperature transformation diagram **M4:** 89
time-temperature transformation diagrams **A4:** 140, 141
1090, abrasive wear data **A18:** 705
1090, band sawing **A16:** 360, 361
1090, cutoff band sawing with bimetal blades **A6:** 1184
1090, thread grinding **A16:** 274
1090, turning machinability rating **A16:** 668
1090, wear data **M3:** 590
1091
carbonitriding **A4:** 376
flame hardening **A4:** 283
1091, cutoff band sawing with bimetal blades **A6:** 1184
1092
carbonitriding **A4:** 376
flame hardening **A4:** 283
1092, cutoff band sawing with bimetal blades **A6:** 1184
1093
carbonitriding **A4:** 376
flame hardening **A4:** 283
1093, cutoff band sawing with bimetal blades **A6:** 1184
1094
carbonitriding **A4:** 376
flame hardening **A4:** 283
1094, cutoff band sawing with bimetal blades **A6:** 1184
1095
annealing **A4:** 49
austempering time, effect on hardness **A4:** 160 **M4:** 112
austenitizing temperature **A4:** 961 **M4:** 29
carbon restoration **A4:** 599 **M4:** 446
carbonitriding **A4:** 376
composition **A6:** 641 **M1:** 125
core properties of carburized versus induction-hardened components **A18:** 725
cutoff band sawing with bimetal blades **A6:** 1184
dimensional control, austempered parts **A4:** 161
flame hardening **A4:** 283
full annealing **A4:** 46
gas nitriding **A4:** 402
gas quenching **A4:** 105 **M4:** 59
grinding rod material **A18:** 654
hardness **A18:** 704
hardness, after quenching **A4:** 106
hardness after tempering **M4:** 71
heat treatments **A4:** 138
machinability rating **M1:** 576
martempering **A4:** 138
mass, effect on hardness **A4:** 39 **M4:** 11
mechanical properties **A4:** 138
mechanical properties, heat treated **M4:** 85, 105
mechanical properties, heat-treated **A4:** 152
normalizing temperatures **A4:** 36 **M4:** 7
properties, various heat treating conditions **M4:** 9-10
properties, various heat-treating conditions **A4:** 37
quenching **A4:** 76
section size, austempered parts **A4:** 155 **M4:** 107
springs **M1:** 285, 287, 288, 297, 299, 301-303
temperatures and cooling cycles, annealing **M4:** 17
tempering **A4:** 122, 133, 138
time-temperature-transformation diagram **A18:** 704, 705

Steels, AISI-SAE specific types (continued)
wear resistance and cost effectiveness **A18:** 706
1095, machining **A16:** 274, 282, 315, 360, 361, 668
1108
arc welding **A6:** 645
banding or microalloy segregation **A4:** 623
carbonitriding **A4:** 376
composition **A6:** 641
cutoff band sawing with bimetal blades **A6:** 1184
flame hardening **A4:** 283
1108, band sawing **A16:** 360, 361
1108, broaching **A16:** 207
1109
arc welding **A6:** 645
austenitizing temperature **A4:** 962
banding or microalloy segregation **A4:** 623
carbonitriding **A4:** 376
cutoff band sawing with bimetal blades **A6:** 1184
flame hardening **A4:** 283
1109, austenitizing temperature **M4:** 30
1109, band sawing **A16:** 360, 361
1109, broaching **A16:** 207
1110
arc welding **A6:** 645
banding or microalloy segregation **A4:** 623
carbonitriding **A4:** 376
composition **M1:** 126
cutoff band sawing with bimetal blades **A6:** 1184
flame hardening **A4:** 283
1111
banding or microalloy segregation **A4:** 623
carbonitriding **A4:** 376
flame hardening **A4:** 283
1112
banding or microalloy segregation **A4:** 623
carbonitriding **A4:** 376, 378
cutoff band sawing with bimetal blades **A6:** 1184
flame hardening **A4:** 283
recommended upsetting pressures for flash welding **A6:** 843
1112, machining **A16:** 15, 148, 149, 203, 204, 209, 246, 247, 251, 269, 282, 286, 318, 360, 361, 362, 364, 717
1113
banding or microalloy segregation **A4:** 623
carbonitriding **A4:** 376, 379
flame hardening **A4:** 283
liquid carburizing **A4:** 334, 345
1113, cutoff band sawing with bimetal blades **A6:** 1184
1113, multifunctional machining **A16:** 391
1113, reaming **A16:** 239, 360-362
1113, sawing **A16:** 360-362
1114
banding or microalloy segregation **A4:** 623
carbonitriding **A4:** 376
flame hardening **A4:** 283
1114, cutoff band sawing with bimetal blades **A6:** 1184
1115
austenitizing temperature **A4:** 962
banding or microalloy segregation **A4:** 623
carbonitriding **A4:** 376
flame hardening **A4:** 283
1115, austenitizing temperature **M4:** 30
1115, broaching **A16:** 207
1115, cutoff band sawing with bimetal blades **A6:** 1184
1115, sawing **A16:** 360-362
1116
arc welding **A6:** 645
banding or microalloy segregation **A4:** 623
carbonitriding **A4:** 376
cutoff band sawing with bimetal blades **A6:** 1184
flame hardening **A4:** 283
1116, machining **A16:** 179, 207, 225, 342, 345, 347, 349, 360-362
1117
applications, austempered parts **A4:** 157 **M4:** 108
arc welding **A6:** 645

Steels, AISI-SAE specific types (continued)
austenitizing temperature **A4:** 962 **M4:** 30
banding or microalloy segregation **A4:** 623
blanking and piercing dies, use for **M3:** 487
carbonitriding **A4:** 376, 379, 380, 381-382 **M4:** 181
carburizing temperature, effect on grain size **M4:** 158
composition **A4:** 320 **M1:** 126
consecutive cuts analysis of carbon control **A4:** 588
cutoff band sawing with bimetal blades **A6:** 1184
die-casting dies, use in **M3:** 543
flame hardening **A4:** 283
liquid carburizing .... **A4:** 332 , 333, 334, 340, 345
machinability **M1:** 236, 237
machinability compared with 1119 steel **M1:** 579-580
machinability rating **M1:** 576
martempering in oil, applications **A4:** 148 **M4:** 98
mass, effect on hardness **A4:** 39 **M4:** 11
mechanical properties, cold drawn bars **M1:** 223, 242
normalizing temperatures **A4:** 36 **M4:** 7
properties, various heat treating conditions **M4:** 9-10
properties, various heat-treating conditions **A4:** 37
1117, machining **A16:** 164, 179, 207, 225, 246, 247, 251, 263, 274, 282, 342, 345, 347, 349, 360-362, 364, 668
1117L, martempering in salt, applications **A4:** 148 **M4:** 96
1118
arc welding **A6:** 645
austenitizing temperature **A4:** 962 **M4:** 30
banding or microalloy segregation **A4:** 623
carbonitriding **A4:** 376, 381-382
composition **M1:** 126
cutoff band sawing with bimetal blades **A6:** 1184
flame hardening **A4:** 283
liquid carburizing **A4:** 345
machinability **M1:** 236, 237
machinability rating **M1:** 576
mass, effect on hardness **A4:** 39 **M4:** 11
mechanical properties, cold drawn bars **M1:** 223
properties, various heat treating conditions **M4:** 9-10
properties, various heat-treating conditions **A4:** 37
1118, cross-wire projection welding **M6:** 518
1118, machining **A16:** 179, 207, 225, 246, 247, 342, 345, 347, 349, 360, 361, 668
1118, nominal compositions and applications **A18:** 703
1118, plug gaging of, wear of gage materials **M3:** 554
1119
arc welding **A6:** 645
banding or microalloy segregation **A4:** 623
carbonitriding **A4:** 376
cutoff band sawing with bimetal blades **A6:** 1184
flame hardening **A4:** 283
liquid carburizing **A4:** 345
machinability **M1:** 236, 237
machinability compared with 1117 steel **M1:** 579-580
1119, machining **A16:** 179, 207, 225, 342, 345, 347, 349, 360, 361
1120, cutoff band sawing with bimetal blades **A6:** 1184
1121, cutoff band sawing with bimetal blades **A6:** 1184
1122, cutoff band sawing with bimetal blades **A6:** 1184
1123, cutoff band sawing with bimetal blades **A6:** 1184
1124, cutoff band sawing with bimetal blades **A6:** 1184
1125
banding or microalloy segregation **A4:** 623
carbonitriding **A4:** 376

Steels, AISI-SAE specific types (continued)
flame hardening ... A4: 283
1125, cutoff band sawing with bimetal blades ... A6: 1184
1126
banding or microalloy segregation ... A4: 623
carbonitriding ... A4: 376
flame hardening ... A4: 283
1126, cutoff band sawing with bimetal blades ... A6: 1184
1127
banding or microalloy segregation ... A4: 623
carbonitriding ... A4: 376
flame hardening ... A4: 283
1127, cutoff band sawing with bimetal blades ... A6: 1184
1128
banding or microalloy segregation ... A4: 623
carbonitriding ... A4: 376
flame hardening ... A4: 283
1128, cutoff band sawing with bimetal blades ... A6: 1184
1129
banding or microalloy segregation ... A4: 623
carbonitriding ... A4: 376
flame hardening ... A4: 283
1129, cutoff band sawing with bimetal blades ... A6: 1184
1130
banding or microalloy segregation ... A4: 623
carbonitriding ... A4: 376
flame hardening ... A4: 283
1130, cutoff band sawing with bimetal blades ... A6: 1184
1131
banding or microalloy segregation ... A4: 623
carbonitriding ... A4: 376
flame hardening ... A4: 283
1131, cutoff band sawing with bimetal blades ... A6: 1184
1132
arc welding ... A6: 646
banding or microalloy segregation ... A4: 623
carbonitriding ... A4: 376
cutoff band sawing with bimetal blades ... A6: 1184
flame hardening ... A4: 283
tempering ... A4: 134
1132, cutoff band sawing with bimetal blades ... A6: 1184
1132, machining A16: 179, 207, 225, 342, 345, 347, 349, 360, 361
1133
banding or microalloy segregation ... A4: 623
carbonitriding ... A4: 376
flame hardening ... A4: 283
1134
banding or microalloy segregation ... A4: 623
carbonitriding ... A4: 376
flame hardening ... A4: 283
1135
banding or microalloy segregation ... A4: 623
carbonitriding ... A4: 376
flame hardening ... A4: 283
1136
banding or microalloy segregation ... A4: 623
carbonitriding ... A4: 376
flame hardening ... A4: 283
1137
arc welding ... A6: 646
austenitizing temperature ... A4: 961 M4: 29
banding or microalloy segregation ... A4: 623
carbonitriding ... A4: 376
composition ... M1: 126
cutoff band sawing with bimetal blades ... A6: 1184
flame hardening ... A4: 283
hardness after tempering ... M4: 71
heavy-draft drawing, suitability ... M1: 242

Steels, AISI-SAE specific types (continued)
machinability ... M1: 236, 238
machinability rating ... M1: 576
mass, effect on hardness ... A4: 39 M4: 11
mechanical properties, cold drawn bars ... M1: 224, 226, 242
normalizing temperatures ... A4: 36 M4: 7
properties, various heat treating conditions ... M4: 9-10
properties, various heat-treating conditions ... A4: 37
shafts, induction hardened, fatigue life ... M1: 675
tempering ... A4: 122
1137, machining A16: 164, 179, 207, 225, 246, 247, 251, 263, 342, 345, 347, 349, 360, 361, 364, 668
1137, V-blocks run against by Cr-Mo/Ti-6Al-4V specimens ... A18: 781
1138
austenitizing temperature ... A4: 961
banding or microalloy segregation ... A4: 623
carbonitriding ... A4: 376
flame hardening ... A4: 283
proven applications for borided ferrous materials ... A4: 445
1138, austenitizing temperature ... M4: 29
1138, cutoff band sawing with bimetal blades ... A6: 1184
1139
arc welding ... A6: 646
banding or microalloy segregation ... A4: 623
carbonitriding ... A4: 376
composition ... A6: 641 M1: 126
cutoff band sawing with bimetal blades ... A6: 1184
flame hardening ... A4: 283
1139, machining A16: 179, 207, 225, 342, 345, 347, 349, 360, 361
1140
arc welding ... A6: 646
austenitizing temperature ... A4: 961
banding or microalloy segregation ... A4: 623
carbonitriding ... A4: 376
composition ... M1: 126
cutoff band sawing with bimetal blades ... A6: 1184
flame hardening ... A4: 283
machinability rating ... M1: 576
mechanical properties, cold drawn bars ... M1: 223
1140, austenitizing temperature ... M4: 29
1140, machining A16: 179, 207, 225, 342, 345, 347, 349, 360, 361, 668
1141
arc welding ... A6: 646
austenitizing temperature ... A4: 961 M4: 29
banding or microalloy segregation ... A4: 623
carbonitriding ... A4: 376
composition ... M1: 126
cutoff band sawing with bimetal blades ... A6: 1184
flame hardening ... A4: 283
hardness after tempering ... M4: 71
heavy-draft drawing, suitability ... M1: 239
machinability ... M1: 236, 238
machinability rating ... M1: 576
martempered to full hardness ... A4: 140
martempering ... A4: 140
martempering in oil, applications ... A4: 148 M4: 98
mass, effect on hardness ... A4: 39 M4: 11
mechanical properties, cold drawn bars ... M1: 223
normalizing temperatures ... A4: 36 M4: 7
properties, various heat treating conditions ... M4: 9-10
properties, various heat-treating conditions ... A4: 37
quench cracking ... A4: 610

Steels, AISI-SAE specific types (continued)
quench severity ... A4: 90
Rushman-Lamont quenching technique ... A4: 87-88
tempering ... A4: 122, 134
universal joint spline shafts, wear vs. hardness ... M1: 631
wheel spindles, machining time compared with 1045 steel ... M1: 575, 579
1141, machining A16: 176, 179, 207, 225, 342, 345, 347, 349, 360, 361, 668, 882
1142
banding or microalloy segregation ... A4: 623
carbonitriding ... A4: 376
flame hardening ... A4: 283
1142, cutoff band sawing with bimetal blades ... A6: 1184
1143
banding or microalloy segregation ... A4: 623
carbonitriding ... A4: 376
flame hardening ... A4: 283
1143, cutoff band sawing with bimetal blades ... A6: 1184
1144
arc welding ... A6: 646
austenitizing temperature ... A4: 961 M4: 29
banding or microalloy segregation ... A4: 623
carbonitriding ... A4: 376
cold drawn, machinability compared with heat treated ... M1: 575, 578
composition ... M1: 126
cutoff band sawing with bimetal blades ... A6: 1184
drawing temperature, effect of ... M1: 245, 247-249
flame hardening ... A4: 283
hardness after tempering ... M4: 71
heavy-draft drawing ... M1: 245-248
machinability ... M1: 236, 238, 240
machinability rating ... M1: 576
mass, effect on hardness ... A4: 39 M4: 11
mechanical properties, cold drawn bars
normalizing temperatures ... A4: 36 M4: 7
properties, various heat treating conditions ... M4: 9-10
properties, various heat-treating conditions ... A4: 37
quench cracking ... A4: 610
tempering ... A4: 122, 134
1144, machining A16: 179, 207, 225, 282, 342, 345, 347, 349, 360, 361, 668
1145
arc welding ... A6: 646
austenitizing temperature ... A4: 961
banding or microalloy segregation ... A4: 623
carbonitriding ... A4: 376
cutoff band sawing with bimetal blades ... A6: 1184
machinability, comparison to malleable iron ... M1: 71
mechanical properties, cold drawn bars ... M1: 223
1145, austenitizing temperature ... M4: 29
1145, machining A16: 179, 207, 209, 225, 323-325, 342, 345, 347, 349, 360, 361
1146
arc welding ... A6: 646
austenitizing temperature A4: 142, 961 M4: 29, 92
banding or microalloy segregation ... A4: 623
carbonitriding ... A4: 376
composition ... M1: 126
cutoff band sawing with bimetal blades ... A6: 1184
flame hardening ... A4: 283
machinability rating ... M1: 576
martempering temperature ... A4: 142 M4: 92
1146, machining A16: 179, 207, 225, 342, 345, 347, 349, 360, 361, 668, 673

Steels, AISI-SAE specific types (continued)
1147
banding or microalloy segregation **A4:** 623
carbonitriding **A4:** 376
flame hardening **A4:** 283
1147, cutoff band sawing with bimetal blades **A6:** 1184
1148
banding or microalloy segregation **A4:** 623
carbonitriding **A4:** 376
flame hardening **A4:** 283
1148, cutoff band sawing with bimetal blades **A6:** 1184
1149
banding or microalloy segregation **A4:** 623
carbonitriding **A4:** 376
flame hardening **A4:** 283
1149, cutoff band sawing with bimetal blades **A6:** 1184
1150
banding or microalloy segregation **A4:** 623
carbonitriding **A4:** 376
flame hardening **A4:** 283
1150, cutoff band sawing with bimetal blades **A6:** 1184
1151
arc welding **A6:** 646
austenitizing temperature **A4:** 961
banding or microalloy segregation **A4:** 623
carbonitriding **A4:** 376
composition **A6:** 641 **M1:** 126
cutoff band sawing with bimetal blades **A6:** 1184
flame hardening **A4:** 283
machinability rating **M1:** 576
mechanical properties, special-die-drawn bars **M1:** 243
1151, austenitizing temperature **M4:** 29
1151, machining **A16:** 179, 207, 225, 342, 345, 347, 349, 360, 361, 668
1211
arc welding **A6:** 646
composition **A6:** 641 **M1:** 126
machinability **M1:** 236, 237
1211, machining **A16:** 179, 207, 225, 342, 345, 347, 349
1212
arc welding **A6:** 646
composition **M1:** 126
cutoff band sawing with bimetal blades **A6:** 1184
machinability **M1:** 236, 237
machinability rating **M1:** 576
1212, gages, use for **M3:** 555
1212, machining **A16:** 179, 207, 225, 342, 345, 347, 349, 360, 361, 668
1213
arc welding **A6:** 646
banding or microalloy segregation **A4:** 623
bushings, machining cost vs. 12LI3 steel **M1:** 578, 580
carbonitriding **A4:** 376, 381
composition **M1:** 126
cutoff band sawing with bimetal blades **A6:** 1184
machinability **M1:** 236, 237
machinability rating **M1:** 576
1213, machining **A16:** 207, 225, 360, 361, 387, 668
1215
composition **M1:** 126
machinability rating **M1:** 576
1215, arc welding **A6:** 646
1215, broaching **A16:** 207
1215, turning machinability rating **A16:** 668
1315, recommended upsetting pressures for flash welding **A6:** 843
1320
annealing **A4:** 47
carbonitriding **A4:** 376
1320, temperatures and time cycles **M4:** 19
1330
annealing temperatures **A4:** 46 **M4:** 18
arc welding **A6:** 646
austenitizing temperature **A4:** 142, 961 **M4:** 29, 92
carbonitriding **A4:** 376
composition **A6:** 670 **M1:** 127, 129
cutoff band sawing with bimetal blades **A6:** 1184
hardenability, affected by hot work and location in bar **M1:** 478, 480
hardness after tempering **M4:** 71
machinability rating **M1:** 582
martempering **A4:** 141
martempering temperature **A4:** 142 **M4:** 92
normalizing temperatures **A4:** 36 **M4:** 7
oxyfuel gas welding **A6:** 290
shielded metal arc welding **A6:** 670
tempering **A4:** 122
welding preheat and interpass temperatures **A6:** 672
1330, machining **A16:** 144-147, 207, 225, 274, 323-325, 343, 346, 348, 349, 360, 361, 669
1330H
composition **M1:** 130
hardenability curve **M1:** 498
1331, cutoff band sawing with bimetal blades **A6:** 1184
1332, cutoff band sawing with bimetal blades **A6:** 1184
1333, cutoff band sawing with bimetal blades **A6:** 1184
1334, cutoff band sawing with bimetal blades **A6:** 1184
1335
annealing temperatures **A4:** 46 **M4:** 18
arc welding **A6:** 646
austenitizing temperature **A4:** 961 **M4:** 29
carbonitriding **A4:** 376
composition **M1:** 127, 129
cutoff band sawing with bimetal blades **A6:** 1184
induction heat treating **A4:** 198
machinability rating **M1:** 582
martempering **A4:** 141
normalizing temperatures **A4:** 36 **M4:** 7
recommended upsetting pressures for flash welding **A6:** 843
1335, machining **A16:** 144-147, 207, 225, 274, 285, 323-325, 343, 346, 348, 349, 360, 361, 669
1335H
composition **M1:** 130
hardenability curve **M1:** 498
1336, cutoff band sawing with bimetal blades **A6:** 1184
1337, cutoff band sawing with bimetal blades **A6:** 1184
1338, cutoff band sawing with bimetal blades **A6:** 1184
1339, cutoff band sawing with bimetal blades **A6:** 1184
1340
annealing **A4:** 47
annealing temperatures **A4:** 46 **M4:** 18
arc welding **A6:** 646
austenitizing temperature **A4:** 961 **M4:** 29
carbonitriding **A4:** 376
composition **A4:** 300 **A6:** 670 **M1:** 127, 129
critical temperatures **A4:** 43
critical temperatures, annealing **M4:** 15
cutoff band sawing with bimetal blades **A6:** 1184
electron beam hardening treatment **A4:** 300
electron beam hardening treatment (as 42 Mn V 7 steel) **A4:** 308
flame hardening **A4:** 283
hardenability **M1:** 489, 494
hardness after tempering **M4:** 71
machinability rating **M1:** 582
martempering **A4:** 141
mass, effect on hardness **A4:** 39 **M4:** 11
non-martensitic transformation products **A4:** 141
normalizing temperatures **A4:** 36 **M4:** 7
properties, various heat treating conditions **M4:** 9-10
properties, various heat-treating conditions **A4:** 37
shielded metal arc welding **A6:** 670
temperatures and time cycles **M4:** 19
tempering **A4:** 122, 134
welding preheat and interpass temperatures **A6:** 672
1340, machining **A16:** 144-147, 179, 203, 204, 207, 209, 225, 274, 343, 346, 347, 349, 360, 361, 669
1340, plug gaging of, wear of gage materials **M3:** 554, 556
1340H
composition **M1:** 130
hardenability curve **M1:** 494, 498
1341
carbonitriding **A4:** 376
flame hardening **A4:** 283
martempering **A4:** 141
1341, cutoff band sawing with bimetal blades **A6:** 1184
1342
carbonitriding **A4:** 376
flame hardening **A4:** 283
martempering **A4:** 141
1342, cutoff band sawing with bimetal blades **A6:** 1184
1343
carbonitriding **A4:** 376
flame hardening **A4:** 283
martempering **A4:** 141
1343, cutoff band sawing with bimetal blades **A6:** 1184
1344
carbonitriding **A4:** 376
flame hardening **A4:** 283
martempering **A4:** 141
1344, cutoff band sawing with bimetal blades **A6:** 1184
1345
annealing temperatures **A4:** 46 **M4:** 18
arc welding **A6:** 646
austenitizing temperature **A4:** 961 **M4:** 29
carbonitriding **A4:** 376
composition **M1:** 127, 129
cutoff band sawing with bimetal blades **A6:** 1184
flame hardening **A4:** 283
machinability rating **M1:** 582
martempering **A4:** 141
1345, machining **A16:** 144-147, 179, 203, 207, 225, 274, 343, 346, 347, 349, 360, 361, 669
1345, wear rate compared with those of squeeze-cast composites **A18:** 809
1345H
composition **M1:** 130
hardenability curve **M1:** 499
1345H, quench cracking **A4:** 610
1350
carbonitriding **A4:** 376
section size, austempered parts **A4:** 155
1350, section size, austempered parts **M4:** 107
1513
arc welding **A6:** 646
composition **A6:** 641 **M1:** 126
1513 annealing **A4:** 53
austenitizing temperature **A4:** 962
carbonitriding **A4:** 376
case depth, measurement **A4:** 458
1513, machining **A16:** 207, 225, 274, 342, 345, 347, 349
1518
austenitizing temperature **A4:** 962
carbonitriding **A4:** 376
case depth, measurement **A4:** 458
composition **M1:** 126
1518, arc welding **A6:** 646
1518, machining **A16:** 207, 225, 274, 342, 345, 347, 349
1522
austenitizing temperature **A4:** 962
carbonitriding **A4:** 376
case depth, measurement **A4:** 458
composition **M1:** 126
1522, arc welding **A6:** 646
1522, machining **A16:** 207, 225, 274, 342, 345, 347, 349
1522H
composition **M1:** 127
hardenability curve **M1:** 499
1524
annealing **A4:** 53

Steels, AISI-SAE specific types (continued)
austenitizing temperature ... A4: 962
carbonitriding ... A4: 376
case depth, measurement ... A4: 458
composition ... A4: 320 M1: 125, 126
hardenability as a basis for selection ... M1: 493
machinability rating ... M1: 576
mechanical properties after annealing ... A4: 54
plasma (ion) carburizing ... A4: 359
surface hardening ... A4: 262
1524, arc welding ... A6: 647
1524, carburizing and hardenability ... A18: 875
1524, machining ... A16: 207, 225, 274, 342, 345, 347, 349
1524, mechanical properties ... M4: 26
1524H
composition ... M1: 127
hardenability curve ... M1: 500
1525
austenitizing temperature ... A4: 962
carbonitriding ... A4: 376
case depth, measurement ... A4: 458
composition ... M1: 126
1525, arc welding ... A6: 646
1525, machining ... A16: 144-147, 179, 207, 225, 274, 323-325, 342, 345, 347, 349
1526
austenitizing temperature ... A4: 962
carbonitriding ... A4: 376
case depth, measurement ... A4: 458
composition ... M1: 126
1526, arc welding ... A6: 647
1526, machining ... A16: 144-147, 179, 207, 225, 274, 323-325, 342, 345, 347, 349
1526H
composition ... M1: 127
hardenability curve ... M1: 500
1527
arc welding ... A6: 647
austenitizing temperature ... A4: 962
carbonitriding ... A4: 376
case depth, measurement ... A4: 458
composition ... A4: 320 A6: 641 M1: 125, 126
heavy-draft drawing, suitability ... M1: 250
machinability rating ... M1: 576
1527, machining ... A16: 144-147, 179, 207, 225, 274, 323-325, 342, 345, 347, 349
1536
austenitizing temperature ... A4: 961
carbonitriding ... A4: 376
case depth, measurement ... A4: 458
composition ... M1: 125, 126
heavy-draft drawing, suitability ... M1: 250
machinability rating ... M1: 576
1536, arc welding ... A6: 647
1536, machining ... A16: 207, 225, 274, 342, 345, 347, 349
1541
austenitizing temperature ... A4: 961
bolts, hardenability compared to 1038 ... M1: 276, 279
carbonitriding ... A4: 376
case depth, measurement ... A4: 458
composition ... A6: 641 M1: 125, 126
continuous cooling transformation diagram ... A6: 70, 71
flame hardening ... A4: 284
heavy-draft drawing, suitability ... M1: 250
induction hardened, hardness profile ... M1: 532
machinability rating ... M1: 576
mechanical properties after annealing ... A4: 54
1541, carburizing and hardenability ... A18: 875
1541, machining ... A16: 207, 225, 274, 342, 345, 347, 349
1541, mechanical properties ... M4: 26
1541H
composition ... M1: 127

Steels, AISI-SAE specific types (continued)
hardenability curve ... M1: 501
1547
composition ... M1: 126
machinability rating ... M1: 576
1547, arc welding ... A6: 647
1547, machining ... A16: 207, 225, 274, 342, 345, 347, 349
1548
austenitizing temperature ... A4: 961
carbonitriding ... A4: 376
case depth, measurement ... A4: 458
composition ... M1: 125, 126
machinability rating ... M1: 576
1548, arc welding ... A6: 647
1548, machining ... A16: 207, 225, 274, 342, 345, 347, 349
1551
composition ... M1: 126
1551, machining ... A16: 207, 225, 274, 342, 345, 347, 349
1552
austenitizing temperature ... A4: 961
carbonitriding ... A4: 376
case depth, measurement ... A4: 458
composition ... M1: 125, 126
flame hardening ... A4: 284
machinability rating ... M1: 576
1552, machining ... A16: 207, 225, 274, 342, 345, 347, 349
1561
composition ... M1: 126
1566
austenitizing temperature ... A4: 961
carbonitriding ... A4: 376
case depth, measurement ... A4: 458
composition ... M1: 126
1566, composition ... A6: 641
1572
composition ... M1: 126
1855, honing with CBN ... A16: 479
2317, pack carburizing applications ... A4: 326
2325, pack carburizing applications ... A4: 326
2330, hardness after tempering ... M4: 71
2330, tempering ... A4: 122
2340, annealing ... A4: 47
2340, temperatures and time cycles ... M4: 19
2345, annealing ... A4: 47
2345, temperatures and time cycles ... M4: 19
2512, carbon gradient ... M4: 440
3115, pack carburizing applications ... A4: 326
3115-4615, broachability constant ... A16: 200
3120, annealing ... A4: 47
3120, temperatures and time cycles ... M4: 19
3130, hardening after tempering ... M4: 71
3130, tempering ... A4: 122
3135, normalizing temperatures ... A4: 36 M4: 7
3135, recommended upsetting pressures for flash welding ... A6: 843
3135, thread rolling ... A16: 285
3138, residual stress and quenching ... A4: 16
3140
annealing ... A4: 47
annealing temperatures ... A4: 46 M4: 18
applications, austempered parts ... A4: 157 M4: 108
austenitizing temperature ... A4: 961 M4: 29
critical temperatures ... A4: 43
critical temperatures, annealing ... M4: 15
flame hardening ... A4: 283
hardness after tempering ... M4: 71
mass, effect on hardness ... A4: 39 M4: 11
normalizing temperatures ... A4: 36 M4: 7
properties, various heat treating conditions ... M4: 9-10
properties, various heat-treating conditions ... A4: 37
quenching ... A4: 12
temperatures and time cycles ... M4: 19

Steels, AISI-SAE specific types (continued)
tempering ... A4: 122
3141, flame hardening ... A4: 283
3142, flame hardening ... A4: 283
3143, flame hardening ... A4: 283
3144, flame hardening ... A4: 283
3145, flame hardening ... A4: 283
3150, annealing ... A4: 47
3150, temperatures and time cycles ... M4: 19
3310
annealing ... A4: 47
applications ... A4: 397 A18: 704
austenitizing temperature ... A4: 962 M4: 30
composition ... A4: 320
flame hardening ... A4: 283
gas carburizing ... A4: 319
gas nitriding ... A4: 397
hardness, and oil quenching ... A4: 99
mass, effect on hardness ... A4: 39 M4: 11
nominal compositions ... A18: 704, 725
normalizing temperatures ... A4: 36, 38 M4: 7
oil quenching ... M4: 54
surface carbon content and shim stock analysis ... A4: 589
surface carbon content, variations ... M4: 444
surface carbon variability ... A4: 597
surface hardening ... A4: 262
temperatures and time cycles ... M4: 19
3310, broaching ... A16: 209
3312
austenitizing temperature ... A4: 142 M4: 92
carbon gradient ... M4: 440
liquid carburizing ... A4: 331, 334
martempered to full hardness ... A4: 140
martempering in salt, applications ... A4: 148 M4: 96
martempering temperature ... A4: 142 M4: 92
normalizing temperatures ... A4: 38
3350
flame hardening ... A4: 283
normalizing temperatures ... A4: 38
4012
carbonitriding ... A4: 376
composition ... M1: 127
machinability rating ... M1: 582
martempering ... A4: 141
4012, machining ... A16: 144-147, 179, 207, 342, 346, 347, 349
4022, nominal compositions and applications ... A18: 703
4023
carbonitriding ... A4: 376
carburizing ... A4: 262
carburizing and hardenability ... A18: 875
composition ... A4: 320 A6: 670 M1: 127
cutoff band sawing with bimetal blades ... A6: 1184
hardenability, selection for ... M1: 492
machinability rating ... M1: 582
martempering ... A4: 141
nominal compositions and applications ... A18: 703
shielded metal arc welding ... A6: 670
surface hardening ... A4: 262
welding preheat and interpass temperatures ... A6: 672
4023, machining ... A16: 144-147, 179, 207, 225, 274, 342, 346, 347, 349, 360, 361, 669, 672
4024
composition ... M1: 127
machinability rating ... M1: 582
4024, cutoff band sawing with bimetal blades ... A6: 1184
4024, machining ... A16: 144-147, 179, 207, 225, 274, 342, 346, 347, 349, 360, 361, 669
4024, martempering, equipment ... M4: 99-100
4025, cutoff band sawing with bimetal blades ... A6: 1184

Steels, AISI-SAE specific types (continued)
4026, cutoff band sawing with bimetal blades ... A6: 1184
4027
carbon gradients after gas carburizing ... A4: 592-593, 595
carbonitriding ... A4: 376
composition ... A4: 320 M1: 127
critical temperatures ... A4: 43
critical temperatures, annealing ... M4: 15
machinability rating ... M1: 582
martempering ... A4: 141
mass, effect on hardness ... A4: 39 M4: 11
normalizing temperatures ... A4: 36 M4: 7
surface carbon variability ... A4: 596, 597
through hardened, fatigue limits ... M1: 677
zone control, effect on carbon gradient ... M4: 442
4027, cutoff band sawing with bimetal blades ... A6: 1184
4027, machining ... A16: 144-147, 207, 225, 274, 281, 323-325, 343, 346, 347, 349, 360, 361, 669
4027, nominal compositions and applications ... A18: 703
4027H
composition ... M1: 130
4028
carbonitriding ... A4: 376
composition ... A6: 670 M1: 127
cutoff band sawing with bimetal blades ... A6: 1184
machinability rating ... M1: 582
martempering ... A4: 141
martempering, equipment ... M4: 99-100
martempering equipment requirements ... A4: 148, 149
normalizing temperatures ... A4: 36 M4: 7
shielded metal arc welding ... A6: 670
welding preheat and interpass temperatures ... A6: 672
4028, machining ... A16: 144-147, 207, 225, 274, 323-325, 343, 346, 348, 349, 360, 361, 669
4028H
composition ... M1: 130
hardenability curve ... M1: 502
4029, cutoff band sawing with bimetal blades ... A6: 1184
4030, cutoff band sawing with bimetal blades ... A6: 1184
4031, cutoff band sawing with bimetal blades ... A6: 1184
4032
carbonitriding ... A4: 376
composition ... M1: 127
machinability rating ... M1: 582
martempering ... A4: 141
normalizing temperatures ... A4: 36
through hardened, fatigue limits ... M1: 677
4032, cutoff band sawing with bimetal blades ... A6: 1184
4032, machining ... A16: 144-147, 207, 225, 274, 323-325, 343, 346, 347, 349, 360, 361, 363, 669
4032, normalizing temperatures ... M4: 7
4032H
composition ... M1: 130
hardenability curve ... M1: 503
4033, cutoff band sawing with bimetal blades ... A6: 1184
4034, cutoff band sawing with bimetal blades ... A6: 1184
4035, cutoff band sawing with bimetal blades ... A6: 1184
4036, cutoff band sawing with bimetal blades ... A6: 1184
4037
annealing ... A4: 54
annealing temperatures ... A4: 46 M4: 18
austenitizing temperature ... M4: 29
austenitizing temperatures ... A4: 961
bolts, fatigue life ... M1: 280
carbonitriding ... A4: 376
composition ... M1: 127
machinability rating ... M1: 582
martempering ... A4: 141
normalizing temperatures ... A4: 36 M4: 7

Steels, AISI-SAE specific types (continued)
statistical process control of heat treatment ... A4: 622, 623
4037, cutoff band sawing with bimetal blades ... A6: 1184
4037, machining ... A16: 144-147, 207, 225, 274, 323-325, 343, 346, 348, 349, 360, 361, 669
4037H
composition ... M1: 130
hardenability curve ... M1: 503
4038, cutoff band sawing with bimetal blades ... A6: 1184
4039, cutoff band sawing with bimetal blades ... A6: 1184
4040, cutoff band sawing with bimetal blades ... A6: 1184
4041, cutoff band sawing with bimetal blades ... A6: 1184
4042
annealing ... A4: 47
annealing temperatures ... A4: 46 M4: 18
austenitizing temperature ... A4: 961 M4: 29
carbonitriding ... A4: 376
composition ... M1: 127
critical temperatures ... A4: 43
critical temperatures, annealing ... M4: 15
machinability rating ... M1: 582
martempering ... A4: 141
normalizing temperatures ... A4: 36 M4: 7
temperatures and time cycles ... M4: 19
4042, carburizing and hardenability ... A18: 875
4042, cutoff band sawing with bimetal blades ... A6: 1184
4042, machining ... A16: 144-147, 179, 207, 225, 274, 343, 346, 348, 349, 360, 361, 669
4042H
composition ... M1: 130
hardenability curve ... M1: 503
4043, cutoff band sawing with bimetal blades ... A6: 1184
4044, cutoff band sawing with bimetal blades ... A6: 1184
4045, cutoff band sawing with bimetal blades ... A6: 1184
4046, cutoff band sawing with bimetal blades ... A6: 1184
4047
annealing ... A4: 47
annealing temperatures ... A4: 46 M4: 18
austenitizing temperature ... A4: 961 M4: 29
carbonitriding ... A4: 376, 381
composition ... A6: 670 M1: 127
cutoff band sawing with bimetal blades ... A6: 1184
machinability rating ... M1: 582
martempering in salt, applications ... A4: 148 M4: 96
normalizing temperatures ... A4: 36 M4: 7
shielded metal arc welding ... A6: 670
temperatures and time cycles ... M4: 19
tempering ... A4: 134
welding preheat and interpass temperatures ... A6: 672
4047, machining ... A16: 144-147, 179, 207, 225, 274, 343, 346, 348, 349, 360, 361, 669
4047H
composition ... M1: 130
hardenability curve ... M1: 504
through hardened, effect of hardness on fatigue limit ... M1: 676
through hardened, fatigue limits
4053
carbonitriding ... A4: 376
tempering ... A4: 133
4062
annealing ... A4: 47
carbonitriding ... A4: 376
4062, temperatures and time cycles ... M4: 19
4063
annealing temperatures ... A4: 46 M4: 18
austenitizing temperature ... A4: 142, 961 M4: 29, 92
carbonitriding ... A4: 376
flame hardening ... A4: 276, 278, 283
martempering temperature ... A4: 142 M4: 92
mass, effect on hardness ... A4: 39 M4: 11
normalizing temperatures ... A4: 36 M4: 7

Steels, AISI-SAE specific types (continued)
section size, austempered parts ... A4: 155 M4: 107
tempering ... A4: 134
4068
applications, austempered parts ... A4: 157
carbonitriding ... A4: 376
Jominy testing ... A4: 27
4068, applications, austempered parts ... M4: 108
4118
applications ... A18: 703
carbonitriding ... A4: 376
carburizing ... A4: 373
carburizing and fabricability ... A18: 876
composition ... A4: 320 A6: 670 M1: 127, 129
gas carburizing ... M4: 163
hardenability, selection for ... M1: 492
machinability rating ... M1: 582
martempering ... A4: 141
mass, effect on hardness ... A4: 39 M4: 11
nominal compositions ... A18: 703, 725
normalizing temperatures ... A4: 36 M4: 7
plasma (ion) nitriding ... A4: 423
quench cracking ... A4: 610
shielded metal arc welding ... A6: 670
surface hardening ... A4: 262
welding preheat and interpass temperatures ... A6: 672
4118, machining ... A16: 144-147, 179, 207, 225, 274, 342, 346, 347, 349, 669
4118H
composition ... M1: 130
hardenability curve ... M1: 504
4120, turning machinability rating ... A16: 669
4121
carbonitriding ... A4: 376
carburizing ... A4: 365, 368, 369, 370
martempering ... A4: 141
plasma (ion) nitriding ... A4: 423
quench cracking ... A4: 610
4130
adhesive wear resistance ... A18: 721
annealing ... A4: 47
annealing temperatures ... A4: 46 M4: 18
applications ... A4: 395
austenitizing temperature ... A4: 142, 961 M4: 29, 92
carbonitriding ... A4: 376
carburizing production of carbide networks in prior austenite grain boundaries ... A18: 874
CCT diagrams ... A4: 7, 8, 10, 83, 87
chemical composition limits ... A4: 85, 86
coarse primary carbides produced by carburizing ... A18: 874
composition ... A4: 207 A6: 670 M1: 127, 129, 422
critical temperatures ... A4: 43
critical temperatures, annealing ... M4: 15
cutoff band sawing with bimetal blades ... A6: 1184
flame hardening ... A4: 283
flux-cored arc welding ... A6: 188, 670
friction surfacing ... A6: 323
gas nitriding ... A4: 395
gas quenching ... A4: 105, 106
gas-metal arc welding ... A6: 671
hardness ... A4: 74, 81
hardness after tempering ... M4: 71
hardness, annealed sheet and strip ... M1: 165
hardness from Jominy test method ... A4: 11
hardness in casting positions measured ... A4: 85, 87, 88
heat treatment ... M1: 423 M4: 120-121
heat treatment(s) ... A4: 208-209
heat-treatment temperatures ... A4: 208
IT diagrams ... A4: 7, 8
Jominy hardenability testing ... A4: 81, 84
laser cladding ... A18: 868
laser-beam welding ... A6: 264
machinability rating ... M1: 582
martempered to full hardness ... A4: 140
martempering ... A4: 141
martempering in salt, applications ... A4: 148 M4: 96
martempering temperature ... A4: 142 M4: 92
mass, effect on hardness ... A4: 39 M4: 11
mass effects on typical properties ... A4: 209

Steels, AISI-SAE specific types (continued)
mechanical properties A4: 209 M1: 422-423 M4: 120
mechanical properties, normalized A4: 39, 40 M4: 11
nitrided, hardness profile M1: 540
nitriding time A4: 388 M4: 193
normalizing temperatures A4: 36 M4: 7
notch toughness M1: 705, 706
oxyfuel gas welding A6: 286
plasma (ion) nitriding A4: 423
properties, effect of mass on M4: 121
properties, various heat treating conditions M4: 9-10
properties, various heat-treating conditions A4: 37
quench cracking A4: 610
residual stresses A6: 1097
shielded metal arc welding A6: 669-670
temperatures and time cycles M4: 19
tempering A4: 122, 134, 135
tempering with induction heating A4: 193
through-hardening by induction heating A4: 192
TTP curves A4: 85, 86, 88
TTT diagram A4: 83, 86
welding preheat and interpass temperatures A6: 672
work material for ion implantation A18: 858
4130, die-casting dies, use in M3: 543
4130, flash welding M6: 557
4130, machining A16: 144-148, 159, 186, 207, 209, 220, 225, 230, 232, 233, 246, 247, 274, 323-325, 329, 343, 346, 348, 349, 360, 361, 427, 428, 584, 669, 708, 738, 862
4130H
composition M1: 130
hardenability curve M1: 504
4131
carbonitriding A4: 376
flame hardening A4: 283
martempering A4: 141
plasma (ion) nitriding A4: 423
quench cracking A4: 610
4131, cutoff band sawing with bimetal blades A6: 1184
4132
carbonitriding A4: 376
flame hardening A4: 283
martempering A4: 141
plasma (ion) nitriding A4: 423
quench cracking A4: 610
tempering A4: 134
4132, cutoff band sawing with bimetal blades A6: 1184
4133
carbonitriding A4: 376
flame hardening A4: 283
martempering A4: 141
plasma (ion) nitriding A4: 423
quench cracking A4: 610
4133, cutoff band sawing with bimetal blades A6: 1184
4134
carbonitriding A4: 376
flame hardening A4: 283
martempering A4: 141
plasma (ion) nitriding A4: 423
quench cracking A4: 610
4134, cutoff band sawing with bimetal blades A6: 1184
4135
annealing temperatures A4: 46 M4: 18
austenitizing temperature A4: 961
carbonitriding A4: 376
composition M1: 127, 129
flame hardening A4: 283
machinability rating M1: 582
martempering A4: 141

Steels, AISI-SAE specific types (continued)
microstructure and hardness, effect on tool life in machining M1: 571, 574
normalizing temperatures A4: 36 M4: 7
oil quenching A4: 98 M4: 53
plasma (ion) nitriding A4: 423
quench cracking A4: 610
tempering A4: 134
4135, cutoff band sawing with bimetal blades A6: 1184
4135, machining A16: 92, 93, 144-147, 207, 225, 274, 323-325, 343, 346, 348, 349, 360, 361, 669, 672
4135H
composition M1: 130
hardenability curve M1: 505
4135H, flame hardening A4: 284
4136, cutoff band sawing with bimetal blades A6: 1184
4137
annealing temperatures A4: 46 M4: 18
austenitizing temperature A4: 961 M4: 29
carbonitriding A4: 376
composition M1: 127, 129
hardenability equivalence M1: 483-484
machinability rating M1: 582
martempering A4: 141
minimum hardenability vs. M1: 491
normalizing temperatures A4: 36 M4: 7
plasma (ion) nitriding A4: 423
quench cracking A4: 610
4137, cutoff band sawing with bimetal blades A6: 1184
4137, machining A16: 144-147, 207, 225, 274, 323-325, 343, 346, 348, 349, 360, 361, 669
4137H
composition M1: 130
hardenability curve M1: 505
4138, cutoff band sawing with bimetal blades A6: 1184
4139, cutoff band sawing with bimetal blades A6: 1184
4140
abrasive wear data A18: 705
annealing A4: 47
annealing temperatures A4: 46 M4: 18
applications A4: 394, 397
applications, austempered parts A4: 157 M4: 108
applications, machinery and equipment A6: 391
austempering, suitability A4: 153
austenitizing temperature A4: 142, 961 M4: 29, 92
blanking and piercing dies, use for M3: 485, 486, 487
boriding A4: 445
carbonitriding A4: 376
case depth during double-stage nitriding A4: 389
case depth, nitriding M4: 195
CCT diagram A4: 9
CHT diagram A4: 7
composition A4: 207, 300 A6: 670 M1: 127, 129, 163, 422 M3: 490 M4: 120
composition and carbon content A6: 406
critical temperatures A4: 43
critical temperatures, annealing M4: 15
cutoff band sawing with bimetal blades A6: 1184
deep drawing dies, use for M3: 498, 499
die material for sheet metal forming A18: 628
die wear in shallow forming dies A18: 633, 634
die-casting dies, use in M3: 542, 543
distortion of a Jominy specimen A4: 31
drawing temperature, effect on mechanical properties M1: 248, 250
electron beam hardening treatment A4: 300
electron beam hardening treatment (as 42 Cr Mo 4 steel) A4: 308

Steels, AISI-SAE specific types (continued)
erosion resistance A18: 204
flame hardening A4: 280, 283
friction welding A6: 153
gages, use for M3: 555
gas nitriding A4: 388, 389, 392, 394, 397
gas-metal arc welding A6: 671
gear materials, surface treatment, and minimum surface hardness A18: 261
hardenability M1: 489, 494
hardenability variation M1: 479
hardness after tempering M4: 71
hardness, and oil quenching A4: 91, 94
hardness, annealed sheet and strip
hardness gradients, nitriding M4: 256
hardness vs. tempering temperature M1: 469
heat treatment M1: 424
heat treatment(s) A4: 209
heat treatments M4: 120-121
heat-treatment temperatures A4: 208
hot extrusion tools, use for M3: 538, 540
impact wear A18: 268
induction hardening A4: 17
ion-nitriding atmosphere A4: 565
isothermal heating diagram A4: 6
IT cooling diagram A4: 7
Jominy end-quench hardenability curves A4: 20, 27, 650
laser heat treating M4: 515
laser surface transformation hardening A4: 293-294
laser transformation hardening A18: 863
liquid nitriding A4: 411, 413 M4: 253, 256
machinability, free-machining vs. standard grades M1: 570, 581, 583
machinability rating M1: 582
martempered to full hardness A4: 140
martempering temperature A4: 142 M4: 92
mass, effect on hardness A4: 39 M4: 11
mass effects on typical properties A4: 209
mechanical properties A4: 98, 99, 209 M1: 423-424 M4: 121
metal flow in forging M1: 354
nitrided, case structure M1: 540
nitrided, hardness profile M1: 632
nitriding time A4: 388 M4: 193
normalizing temperatures A4: 36 M4: 7
notch toughness A4: 123
notch toughness, tempering temperature M4: 73
notch toughness vs. tempering temperature M1: 469
oil quenching A4: 98, 99 M4: 53
plasma (ion) nitriding A4: 420, 423
press forming dies, use for M3: 489, 490
properties, effect of mass on M4: 121
properties, various heat treating conditions M4: 9-10
properties, various heat-treating conditions A4: 37
proven applications for borided ferrous materials A4: 445
quench cracking A4: 610
recommended upsetting pressures for flash welding A6: 843
residual stresses A4: 17
shafts, through hardened, fatigue life M1: 675
shear spinning tools, use for M3: 501
shielded metal arc welding A6: 669-670
temperatures and time cycles M4: 19
tempering A4: 122, 123, 132, 134, 135, 136
thermal properties M4: 511
through hardened, fatigue limits M1: 676, 677
welding preheat and interpass temperatures A6: 672
work material for ion implantation A18: 858
4140, flash welding M6: 557
4140, machining A16: 95, 144-148, 164, 176, 178, 179, 203, 204, 207-209, 225, 236, 246, 247, 251, 263,

Steels, AISI-SAE specific types (continued)
269, 274, 282, 285, 315, 342, 343, 345-349, 360, 361, 364, 618, 640, 669, 674-676, 721
4140 mod
nitrided, hardness profile ... **M1:** 540
4140+S machining ... **A16:** 164, 246, 247, 251, 263, 364
4140H
composition ... **M1:** 130
hardenability ... **M1:** 488, 489, 493, 494
hardenability curve ... **M1:** 505
mechanical properties of castings ... **M1:** 399
4140H, austenitizing temperature ... **M4:** 29
4140H flame hardening ... **A4:** 284
quench cracking ... **A4:** 610
4140Se, machining ... **A16:** 342, 345, 347, 349
4141 carbonitriding ... **A4:** 376
CCT diagram ... **A4:** 9
flame hardening ... **A4:** 283
plasma (ion) nitriding ... **A4:** 423
quench cracking ... **A4:** 610
4142
austenitizing temperature ... **M4:** 29
composition ... **M1:** 127, 129, 163
hardness, annealed sheet and strip ... **M1:** 165
machinability rating ... **M1:** 582
minimum hardenability vs. ... **M1:** 491
nitriding time ... **M4:** 193
normalizing temperatures ... **M4:** 7
tool wear for screw machining tellurium-containing vs. standard grades ... **M1:** 576, 579
4142 applications ... **A4:** 394
austenitizing temperature ... **A4:** 961
carbonitriding ... **A4:** 376
CCT diagram ... **A4:** 9
flame hardening ... **A4:** 283
gas nitriding ... **A4:** 394
nitriding time ... **A4:** 388
normalizing temperatures ... **A4:** 36
plasma (ion) nitriding ... **A4:** 423
quench cracking ... **A4:** 610
tempering curves ... **A4:** 11
4142, machining ... **A16:** 144-147, 179, 207, 274, 343, 346, 348, 349, 669
4142H
composition ... **M1:** 130
hardenability curve ... **M1:** 506
4142Te, machining ... **A16:** 342, 345, 347, 349
4143
carbonitriding ... **A4:** 376
flame hardening ... **A4:** 283
plasma (ion) nitrocarburizing ... **A4:** 423
quench cracking ... **A4:** 610
4144
carbonitriding ... **A4:** 376
flame hardening ... **A4:** 283
plasma (ion) nitrocarburizing ... **A4:** 423
quench cracking ... **A4:** 610
4145
annealing temperatures ... **A4:** 46 **M4:** 18
austenitizing temperature ... **A4:** 961 **M4:** 29
carbonitriding ... **A4:** 376
composition ... **M1:** 127, 129, 163
flame hardening ... **A4:** 283
hardness, annealed sheet and strip ... **M1:** 165
hardness vs. tempering temperature ... **M1:** 468
machinability rating ... **M1:** 582
normalizing temperatures ... **A4:** 36 **M4:** 7
plasma (ion) nitriding ... **A4:** 423
quench cracking ... **A4:** 610
4145, machining ... **A16:** 144-147, 179, 207, 225, 274, 343, 346, 348, 349, 669
4145H
composition ... **M1:** 130
hardenability curve ... **M1:** 506
4145H, quench cracking ... **A4:** 610
4145Se, machining ... **A16:** 342, 345, 347, 349
4147
annealing temperatures ... **A4:** 46 **M4:** 18
austenitizing temperature ... **A4:** 961 **M4:** 29
carbonitriding ... **A4:** 376
composition ... **M1:** 127
flame hardening ... **A4:** 283
machinability rating ... **M1:** 582
normalizing temperatures ... **A4:** 36 **M4:** 7
plasma (ion) nitriding ... **A4:** 423

Steels, AISI-SAE specific types (continued)
quench cracking ... **A4:** 610
4147, machining ... **A16:** 144-147, 179, 207, 209, 225, 274, 343, 346, 348, 349, 669
4147H
composition ... **M1:** 130
hardenability curve ... **M1:** 506
4147Te, machining ... **A16:** 345, 347, 349
4148
carbonitriding ... **A4:** 376
flame hardening ... **A4:** 283
plasma (ion) nitriding ... **A4:** 423
quench cracking ... **A4:** 610
4149
carbonitriding ... **A4:** 376
flame hardening ... **A4:** 283
plasma (ion) nitriding ... **A4:** 423
quench cracking ... **A4:** 610
4150
annealing ... **A4:** 47
annealing temperatures ... **A4:** 46 **M4:** 18
applications ... **A4:** 394
austenitizing temperature ... **A4:** 961 **M4:** 29
boriding ... **A4:** 445
carbonitriding ... **A4:** 376
composition ... **A6:** 670 **M1:** 127, 163
critical temperatures ... **A4:** 43
critical temperatures, annealing ... **M4:** 15
flame hardening ... **A4:** 280, 282, 283
flash welding ... **M6:** 557
gas nitriding ... **A4:** 394
hardness after tempering ... **M4:** 71
hardness, annealed sheet and strip ... **M1:** 165
hardness in a flash weld ... **M6:** 578
for hot extrusion tools ... **A18:** 627
hot extrusion tools, use for ... **M3:** 538, 540
hot upset forging tools, use for ... **M3:** 534
laser transformation hardening ... **A18:** 863
machinability rating ... **M1:** 582
machining data ... **M1:** 566-567, 580, 583
martempered to full hardness ... **A4:** 140
mass, effect on hardness ... **A4:** 39 **M4:** 11
nitriding time ... **M4:** 193
non-martensitic transformation products ... **A4:** 141
normalizing temperatures ... **A4:** 36 **M4:** 7
plasma (ion) nitriding ... **A4:** 423
properties, various heat treating conditions ... **M4:** 9-10
properties, various heat-treating conditions ... **A4:** 37
proven applications for borided ferrous materials ... **A4:** 445
quench cracking ... **A4:** 610
section size, austempered parts ... **A4:** 155 **M4:** 107
shielded metal arc welding ... **A6:** 670
temperatures and time cycles ... **M4:** 19
tempering ... **A4:** 122, 134, 135
welding preheat and interpass temperatures ... **A6:** 672
4150, machining ... **A16:** 207, 225, 274, 342, 343, 345-349, 429, 640, 641, 643, 669
4150 mod, recommendations for die-casting dies and die inserts ... **A18:** 629
applications ... **A18:** 704
nominal compositions ... **A18:** 704, 725
tapered roller bearing material ... **A18:** 502
4150H
composition ... **M1:** 130
hardenability curve ... **M1:** 507
4150H, quench cracking ... **A4:** 610
4161
annealing temperatures ... **A4:** 46 **M4:** 18
austenitizing temperature ... **A4:** 961 **M4:** 29
carbonitriding ... **A4:** 376
composition ... **M1:** 127
machinability rating ... **M1:** 582
plasma (ion) nitriding ... **A4:** 423
quench cracking ... **A4:** 610
springs, hot wound ... **M1:** 302
4161, machining ... **A16:** 207, 274, 343, 346, 348, 349, 669
4161H
composition ... **M1:** 130

Steels, AISI-SAE specific types (continued)
hardenability curve ... **M1:** 507
4317
boriding ... **A4:** 445
normalizing temperatures ... **A4:** 38
plasma (ion) nitriding ... **A4:** 423
proven applications for borided ferrous materials ... **A4:** 445
4320
annealing ... **A4:** 47
austenitizing temperature ... **A4:** 142, 962 **M4:** 30, 92
carburized, fracture toughness ... **M1:** 536
carburized sealing rings wear of ... **M1:** 624-625
composition ... **A4:** 320 **A6:** 670 **M1:** 127
cutoff band sawing with bimetal blades ... **A6:** 1184
hardenability, selection for ... **M1:** 493
machinability rating ... **M1:** 582
martempering, equipment ... **M4:** 100, 102
martempering temperature ... **A4:** 142 **M4:** 92
mass, effect on hardness ... **A4:** 39 **M4:** 11
normalizing temperatures ... **A4:** 36, 38 **M4:** 7
notch toughness vs. temperature ... **M1:** 536
plasma (ion) nitriding ... **A4:** 423
properties, various heat treating conditions ... **M4:** 9-10
properties, various heat-treating conditions ... **A4:** 37
shafts, carburized, fatigue life of ... **M1:** 675
shielded metal arc welding ... **A6:** 670
surface carbon variability ... **A4:** 596, 597
surface hardening ... **A4:** 262
temperatures and time cycles ... **M4:** 19
tempering ... **A4:** 133
welding preheat and interpass temperatures ... **A6:** 672
4320, machining ... **A16:** 144-148, 179, 207, 225, 274, 342, 346, 347, 349, 360, 361, 669
4320H
composition ... **M1:** 130
hardenability curve ... **M1:** 507
4322, gas carburizing ... **M4:** 163
4330
martempering applications, equipment requirements ... **A4:** 149, 150
normalizing temperatures ... **A4:** 38
notch toughness ... **M1:** 706, 709
plasma (ion) nitriding ... **A4:** 423
4330, machining ... **A16:** 166, 247, 329, 360, 361
4330 mod
forgings, mechanical properties vs. orientation ... **M1:** 358
4330 Mod steel
composition ... **A4:** 207
heat treatment(s) ... **A4:** 209
heat-treatment temperatures ... **A4:** 208
properties after heat treatment ... **A4:** 210
4330V, turning ... **A16:** 144-147
4330V, yield strength ... **A4:** 208
4335
mechanical properties, normalized ... **A4:** 39, 40
normalizing temperatures ... **A4:** 38
plasma (ion) nitriding ... **A4:** 423
4335, mechanical properties, normalized ... **M4:** 11
4335V, gas-tungsten arc welding ... **A6:** 671
4335V, yield strengths ... **A4:** 208
4337
annealing temperatures ... **A4:** 46 **M4:** 18
austenitizing temperature ... **A4:** 961 **M4:** 29
flame hardening ... **A4:** 283
gas nitriding ... **A4:** 392
normalizing temperatures ... **A4:** 36, 38 **M4:** 7
plasma (ion) nitriding ... **A4:** 423
4337, adhesive wear resistance ... **A18:** 721
4337, flash welding ... **M6:** 557
4338
flame hardening ... **A4:** 283
normalizing temperatures ... **A4:** 38
plasma (ion) nitriding ... **A4:** 423
4339
flame hardening ... **A4:** 283
normalizing temperatures ... **A4:** 38
plasma (ion) nitriding ... **A4:** 423
4340
abrasive wear data ... **A18:** 705, 706, 719

Steels, AISI-SAE specific types (continued)
annealing ... **A4:** 47
annealing temperatures ... **A4:** 46 **M4:** 18
applications, machinery and equipment ... **A6:** 391
austenitizing temperature ... **A4:** 142, 961 **M4:** 29, 92
blanking and piercing dies, use for ... **M3:** 486
cage material for rolling-element bearings ... **A18:** 503
carbonitriding ... **A4:** 376
cold extrusion tools, use for ... **M3:** 518
composition ... **A4:** 207 **A6:** 670 **M1:** 127, 129, 163, 422 **M4:** 120
composition and carbon content ... **A6:** 406
constant lifetime fatigue data ... **M1:** 667-669, 681
correlation between wear resistance and toughness ... **A18:** 706-707
corrosive wear ... **A18:** 719
crankshafts, nitrided or shot peened fatigue behavior ... **M1:** 674
critical temperatures ... **A4:** 43
critical temperatures, annealing ... **M4:** 15
electron beam welding ... **M6:** 617-618
electron-beam welding ... **A6:** 866, 867
fatigue behavior affected by inclusions ... **M1:** 672-674
fatigue life, effect of overstrain ... **M1:** 681
fatigue life vs. fatigue ductility ... **M1:** 671
fatigue life vs. fatigue strength ... **M1:** 672
fatigue life vs. total strain ... **M1:** 672
flame hardening ... **A4:** 280, 283
flash welding ... **M6:** 557
forging, mechanical properties vs. orientation ... **M1:** 358
fracture toughness ... **A4:** 211 **M4:** 123
friction welding ... **A6:** 578
gas nitriding ... **A4:** 392
gas-tungsten arc welding ... **A6:** 866
gear materials, surface treatment, and minimum surface hardness ... **A18:** 261
hardenability ... **M1:** 489, 495
hardness ... **A18:** 706
hardness after tempering ... **M4:** 71
hardness, annealed sheet and strip ... **M1:** 165
hardness gradients, nitriding ... **M4:** 256
hardness, variation with tempering temperature ... **M4:** 122
heat treatment ... **M1:** 424-425
heat treatment(s) ... **A4:** 210
heat treatments ... **M4:** 121
heat-treatment temperatures ... **A4:** 208
hot extrusion tools, use for ... **M3:** 538, 540
hot upset forging tools, use for ... **M3:** 534
hydrogen-induced cracking ... **A6:** 411
isothermal transformation diagram ... **A6:** 671, 672
Jominy depth of hardening ... **A4:** 26, 29
laser melt/particle injection ... **A18:** 869
laser surface transformation hardening ... **A4:** 265, 293
laser transformation hardening ... **A18:** 862
laser-beam welding ... **A6:** 263
liquid nitriding ... **A4:** 413
M_{90} temperature ... **A6:** 676
machinability rating ... **M1:** 582
machining data ... **M1:** 569, 570, 580-582
martempered to full hardness ... **A4:** 140
martempering in salt, applications ... **A4:** 148 **M4:** 96
martempering temperature ... **A4:** 142 **M4:** 92
mass, effect on hardness ... **A4:** 39 **M4:** 11
mass, effect on mechanical properties ... **A4:** 210
mechanical properties ... **A4:** 123, 210 **M1:** 424-428 **M4:** 121-122, 123
mechanical properties, normalized ... **A4:** 39 **M4:** 11
nitrided, hardness profile ... **M1:** 540, 632
nitriding time ... **A4:** 388 **M4:** 193
nominal compositions and applications ... **A18:** 703
non-martensitic transformation products ... **A4:** 141
normalizing ... **A4:** 39, 40
normalizing temperatures ... **A4:** 36, 38 **M4:** 7
notch toughness ... **A4:** 211 **M1:** 690, 703-707 **M4:** 123
oxyfuel gas welding ... **A6:** 286
plasma (ion) nitriding ... **A4:** 408, 423
postweld heat treatment ... **A6:** 675-676
processing ... **M1:** 424
properties, effect of mass on ... **M4:** 122
properties, various heat treating conditions ... **M4:** 9-10
properties, various heat-treating conditions ... **A4:** 37
recommended upsetting pressures for flash welding ... **A6:** 843
residual stresses ... **A4:** 608
seizure resistance ... **M1:** 611
shafting, aircraft applications ... **M1:** 606
shear blades, service data ... **M3:** 480
shielded metal arc welding ... **A6:** 670
temperatures and time cycles ... **M4:** 19
tempering ... **A4:** 122, 123, 134
thermal expansion and contraction ... **A4:** 78, 80
through hardened, fatigue limits ... **M1:** 677, 678
time-temperature transformation diagram ... **M4:** 90
time-temperature transformation diagrams ... **A4:** 140, 141
toughness ... **A18:** 706
welding preheat and interpass temperatures ... **A6:** 672
4340, machining ... **A16:** 44-45, 90-91, 144-147, 167, 177, 179, 183, 184, 203, 207, 209, 225, 230, 232, 233, 247, 274, 282, 314, 329, 343, 346, 348, 349, 360, 361, 423, 425, 534, 538, 567, 584, 598-601, 640, 669, 738
4340, machining effects on properties ... **A16:** 24, 25, 27, 29-31, 33-36
4340H
composition ... **M1:** 130
hardenability curve ... **M1:** 508
4340H, flame hardening ... **A4:** 284
4340M, martempered to full hardness ... **A4:** 140
4340Si, turning ... **A16:** 144-147
4347
flame hardening ... **A4:** 283
normalizing temperatures ... **A4:** 38
plasma (ion) nitriding ... **A4:** 423
4350
austenitizing temperature ... **A4:** 142 **M4:** 92
martempering in salt, applications ... **A4:** 148 **M4:** 96
martempering temperature ... **A4:** 142 **M4:** 92
normalizing temperatures ... **A4:** 38
plasma (ion) nitriding ... **A4:** 423
4350, for hot extrusion tools ... **A18:** 627
4350, machining ... **A16:** 186, 299
4350 mod, for hot extrusion tools ... **A18:** 627
4360, bainite ... **A16:** 667
4360, liquid nitriding ... **M4:** 256
4365
normalizing temperatures ... **A4:** 38
plasma (ion) nitriding ... **A4:** 423
section size, austempered parts ... **A4:** 155
4419
composition ... **M1:** 127
machinability rating ... **M1:** 582
notch toughness vs. temperature ... **M1:** 536
4419, machining ... **A16:** 144-147, 179, 207, 225, 274, 342, 346, 347, 349, 669
4419, mass, effect on hardness ... **A4:** 39 **M4:** 11
4419H
composition ... **M1:** 130
4422
composition ... **M1:** 127
gear cutting, comparison with 8620 steel ... **M1:** 579, 581
machinability rating ... **M1:** 582
4422, machining ... **A16:** 144-147, 179, 207, 225, 274, 342, 346, 347, 349, 669
4422, martempering ... **A4:** 141
4427
composition ... **M1:** 127
machinability rating ... **M1:** 582
4427, machining ... **A16:** 144-147, 207, 225, 274, 323-325, 343, 346, 347, 349, 669
4427, martempering ... **A4:** 141
4520
martempering ... **A4:** 141
normalizing temperatures ... **A4:** 36
4520, normalizing temperatures ... **M4:** 7
4600
carbonitriding ... **A4:** 376
hardenability ... **A4:** 232, 233
high-temperature sintering ... **A4:** 232
normalizing temperatures ... **A4:** 38
prealloyed powders similar ... **A4:** 231
4615
austenitizing temperature ... **A4:** 142, 962 **M4:** 30, 92
carbon gradient ... **M4:** 440
carbonitriding ... **A4:** 376
carburizing temperature, effect on grain size ... **M4:** 158
composition ... **M1:** 127, 129
critical temperatures ... **A4:** 43
critical temperatures, annealing ... **M4:** 15
flame hardening ... **A4:** 283
liquid carburizing ... **A4:** 331, 334, 340-341
machinability rating ... **M1:** 582
martempering temperature ... **A4:** 142 **M4:** 92
normalizing temperatures ... **A4:** 38
4615, machining ... **A16:** 144-147, 179, 207, 225, 274, 342, 346, 347, 349, 669
4616
carbonitriding ... **A4:** 376
flame hardening ... **A4:** 283
normalizing temperatures ... **A4:** 38
4617
austenitizing temperature ... **A4:** 962 **M4:** 30
carbonitriding ... **A4:** 376
carburizing temperature, effect on grain size ... **M4:** 158
composition ... **M1:** 128, 129
flame hardening ... **A4:** 283
machinability rating ... **M1:** 582
normalizing temperatures ... **A4:** 38
pack carburizing applications ... **A4:** 326
4617, machining ... **A16:** 144-147, 179, 207, 225, 274, 342, 346, 347, 349, 669
4618
carbonitriding ... **A4:** 376
flame hardening ... **A4:** 283
normalizing temperatures ... **A4:** 38
4619
carbonitriding ... **A4:** 376
flame hardening ... **A4:** 283
normalizing temperatures ... **A4:** 38
4620
annealing ... **A4:** 47, 49
applications ... **A4:** 598
austenitizing temperature ... **M4:** 30
austenitizing temperatures ... **A4:** 962
carbonitriding ... **A4:** 376
carburized, fracture toughness ... **M1:** 536
carburizing ... **A4:** 367
composition ... **A4:** 320 **A6:** 670 **M1:** 128, 129
flame hardening ... **A4:** 283
gas carburizing ... **A4:** 321

SUBJECTS OF THE INDEXED VOLUMES: ASM Handbook (designated by the letter "A"): **A1:** Properties and Selection: Irons, Steels, and High-Performance Alloys (1990); **A2:** Properties and Selection: Nonferrous Alloys and Special-Purpose Materials (1990); **A3:** Alloy Phase Diagrams (1992); **A4:** Heat Treating (1991); **A6:** Welding, Brazing, and Soldering (1993); **A7:** Powder Metallurgy (1984); **A8:** Mechanical Testing (1985); **A9:** Metallography and Microstructures (1985); **A10:** Materials Characterization (1986); **A11:** Failure Analysis and Prevention (1986); **A12:** Fractography (1987); **A13:** Corrosion (1987); **A14:** Forming and Forging (1988); **A15:** Casting (1988); **A16:** Machining (1989); **A17:** Nondestructive Testing and Quality Control (1989); **A18:** Friction, Lubrication, and Wear Technology (1992). **Metals Handbook, 9th Edition** (designated by the letter "M"): **M1:** Properties and Selection: Irons and Steels (1978); **M2:** Properties and Selection: Nonferrous Alloys and Pure Metals (1979); **M3:** Properties and Selection: Stainless Steels, Tool Materials and Special-Purpose Materials (1980); **M4:** Heat Treating (1981); **M5:** Surface Cleaning, Finishing, and Coating (1982); **M6:** Welding, Brazing, and Soldering (1983). **Engineered Materials Handbook** (designated by the letters "EM"): **EM1:** Composites (1987); **EM2:** Engineering Plastics (1988); **EM3:** Adhesives and Sealants (1990); **EM4:** Ceramics and Glasses (1991); **Electronic Materials Handbook** (designated by the letters "EL"): **EL1:** Packaging (1989).

Steels, AISI-SAE specific types (continued)
hardenability, selection for ... M1: 492
liquid carburizing ... A4: 334
machinability rating ... M1: 582
martempered to full hardness ... A4: 140
martempering in oil, applications ... M4: 98
mass, effect on hardness ... A4: 39 M4: 11
nominal compositions ... A18: 725
normalizing temperatures ... A4: 36, 38 M4: 7
properties, various heat treating conditions ... M4: 9-10
properties, various heat-treating conditions ... A4: 37
roller chain pins, carburized wear compared to Nitralloy N ... M1: 628
shielded metal arc welding ... A6: 670
surface carbon content, variations ... M4: 446
surface carbon variability ... A4: 597, 598
surface hardening ... A4: 262
tapered roller bearing material ... A18: 502
temperatures and time cycles ... M4: 19
welding preheat and interpass temperatures ... A6: 672
4620, machining ... A16: 144-147, 179, 207, 225, 274, 342, 346, 347, 349, 669
4620H
composition ... M1: 130
hardenability curve ... M1: 508
martempering ... A4: 146
martempering in oil, applications ... A4: 148
4620H, martempering in oil, applications ... M4: 98
4621
austenitizing temperature ... A4: 962 M4: 30
carbonitriding ... A4: 376
composition ... M1: 128
machinability rating ... M1: 582
normalizing temperatures ... A4: 36, 38 M4: 7
4621, machining ... A16: 144-147, 179, 207, 225, 274, 342, 346, 347, 349, 669
4621H
composition ... M1: 130
hardenability curve ... M1: 508
4626
austenitizing temperature ... A4: 962
carbonitriding ... A4: 376
composition ... M1: 128
machinability rating ... M1: 582
normalizing temperatures ... A4: 38
4626, austenitizing temperature ... M4: 30
4626, machining ... A16: 144-147, 207, 225, 274, 323-325, 343, 346, 348, 349, 669
4626H
composition ... M1: 130
hardenability curve ... M1: 509
4640
annealing ... A4: 47
carbonitriding ... A4: 376
flame hardening ... A4: 280, 283
hardness after tempering ... M4: 71
martempered to full hardness ... A4: 140
normalizing temperatures ... A4: 38
P/M, mechanical properties vs. density ... M1: 345-346
recommended upsetting pressures for flash welding ... A6: 843
shielded metal arc welding ... A6: 670
temperatures and time cycles ... M4: 19
tempering ... A4: 122, 134
welding preheat and interpass temperatures ... A6: 672
4650
P/M, densification ... M1: 346
4718
austenitizing temperature ... A4: 962 M4: 30
composition ... M1: 128
hardness, and oil quenching ... A4: 99
machinability rating ... M1: 582
normalizing temperatures ... A4: 36 M4: 7
4718, machining ... A16: 144-147, 179, 207, 274, 342, 346, 347, 349, 669
4718H
composition ... M1: 130
hardenability curve ... M1: 509
4720
austenitizing temperature ... A4: 142, 962 M4: 30, 92

Steels, AISI-SAE specific types (continued)
composition ... M1: 128
machinability rating ... M1: 582
martempering temperature ... A4: 142 M4: 92
normalizing temperatures ... A4: 36 M4: 7
notch toughness vs. toughness ... M1: 536
4720, machining ... A16: 144-147, 179, 207, 274, 342, 346, 347, 349, 669
4720, nominal compositions and applications ... A18: 704
4720H
composition ... M1: 130
hardenability curve ... M1: 509
4815
austenitizing temperature ... A4: 962 M4: 30
carbon gradients after gas carburizing ... A4: 593-594, 596
carburizing and fabricability ... A18: 876
composition ... A4: 320 M1: 128
cutoff band sawing with bimetal blades ... A6: 1184
gas carburizing ... M4: 157
hardfacing ... A6: 807
hardness, and oil quenching ... A4: 99
laser cladding ... A18: 867, 868
liquid carburizing ... A4: 334
machinability rating ... M1: 582
martempering in salt, applications ... A4: 148 M4: 96
nominal compositions and applications ... A18: 704
normalizing temperatures ... A4: 36 M4: 7
oil quenching ... M4: 54
surface hardening ... A4: 262
4815, machining ... A16: 144-147, 179, 207, 274, 342, 346, 347, 360, 361, 669
4815H
composition ... M1: 130
hardenability curve ... M1: 510
4816, cutoff band sawing with bimetal blades ... A6: 1184
4817
austenitizing temperature ... A4: 962 M4: 30
composition ... M1: 128
hardenability, selection for ... M1: 493
machinability rating ... M1: 582
normalizing temperatures ... M4: 7
surface hardening ... A4: 262
4817, cutoff band sawing with bimetal blades ... A6: 1184
4817, machining ... A16: 144-147, 179, 207, 274, 342, 346, 347, 349, 360, 361, 669
4817H
composition ... M1: 130
hardenability curve ... M1: 510
4817H, martempering in oil, applications ... A4: 148 M4: 98
4818, cutoff band sawing with bimetal blades ... A6: 1184
4819, cutoff band sawing with bimetal blades ... A6: 1184
4820
annealing ... A4: 47
austenitizing temperature ... A4: 962 M4: 30
carbon profile determination using computer prediction ... A4: 652, 653
composition ... A4: 320 M1: 128
machinability rating ... M1: 582
martempering in oil, applications ... A4: 148 M4: 98
mass, effect on hardness ... A4: 39 M4: 11
normalizing temperatures ... M4: 7
notch toughness vs. temperature ... M1: 536
properties, various heat treating conditions ... M4: 9-10
properties, various heat-treating conditions ... A4: 37
surface carbon variability ... A4: 596, 597
temperatures and time cycles ... M4: 19
4820, cutoff band sawing with bimetal blades ... A6: 1184
4820, machining ... A16: 144-147, 156-157, 179, 207, 274, 342, 346, 347, 349, 360, 361, 669
4820H
composition ... M1: 130
hardenability curve ... M1: 510
5010, turning machinability rating ... A16: 669

Steels, AISI-SAE specific types (continued)
5015
composition ... M1: 128
machinability rating ... M1: 582
5015, machining ... A16: 144-147, 179, 207, 274, 342, 346, 347, 349, 669
5015, martempering ... A4: 141
5040, martempering, equipment ... M4: 100
5040, martempering equipment requirements ... A4: 149
5045, annealing ... A4: 47
5045, temperatures and time cycles ... M4: 19
5046
annealing temperatures ... A4: 46 M4: 18
austenitizing temperature ... A4: 961 M4: 29
composition ... M1: 128
critical temperatures ... A4: 43
critical temperatures, annealing ... M4: 15
machinability rating ... M1: 582
martempering ... A4: 141
normalizing temperature ... M4: 7
5046, cutoff band sawing with bimetal blades ... A6: 1184
5046, machining ... A16: 144-147, 179, 207, 274, 343, 346, 348, 349, 360, 361, 669
5046H
composition ... M1: 130
hardenability curve ... M1: 511
5060
composition ... M1: 128
machinability rating ... M1: 582
5060, machining ... A16: 207, 274, 343, 346, 348, 349, 669
5110, turning machinability rating ... A16: 669
5115
carbonitriding ... A4: 376
composition ... M1: 128
liquid nitriding ... A4: 412
machinability rating ... M1: 582
plasma (ion) nitriding ... A4: 423
proven applications for borided ferrous materials ... A4: 445
5115, machining ... A16: 144-147, 179, 207, 274, 342, 346, 347, 349, 669
5117
composition ... M1: 128
5120
annealing ... A4: 47
applications ... A18: 703
carbonitriding ... A4: 376
carburizing ... A4: 262
carburizing and hardenability ... A18: 875
composition ... A4: 320 A6: 670 M1: 128
critical temperatures ... A4: 43
critical temperatures, annealing ... M4: 15
cutoff band sawing with bimetal blades ... A6: 1184
distortion in heat treatments ... A4: 614
gas carburizing ... M4: 163
machinability rating ... M1: 582
martempered to full hardness ... A4: 140
nominal compositions ... A18: 703, 725
normalizing temperatures ... M4: 7
plasma (ion) nitriding ... A4: 423
shielded metal arc welding ... A6: 670
surface hardening ... A4: 262
temperatures and time cycles ... M4: 19
welding preheat and interpass temperatures ... A6: 672
5120, cutoff band sawing with bimetal blades ... A6: 1184
5120, flash versus friction welding ... M6: 720
5120, machining ... A16: 144-148, 179, 200, 207, 225, 274, 342, 346, 347, 349, 674, 669
5120H
composition ... M1: 130
hardenability curve ... M1: 513
5121, cutoff band sawing with bimetal blades ... A6: 1184
5122, cutoff band sawing with bimetal blades ... A6: 1184
5123, cutoff band sawing with bimetal blades ... A6: 1184
5124, cutoff band sawing with bimetal blades ... A6: 1184
5125, cutoff band sawing with bimetal blades ... A6: 1184

Steels, AISI-SAE specific types (continued)
5126, cutoff band sawing with bimetal blades ... A6: 1184
5127, cutoff band sawing with bimetal blades ... A6: 1184
5128, cutoff band sawing with bimetal blades ... A6: 1184
5129, cutoff band sawing with bimetal blades ... A6: 1184
5130
annealing temperatures ... A4: 46 M4: 18
austenitizing temperature ... A4: 961 M4: 29
carbonitriding ... A4: 376
composition ... A4: 320 M1: 128
hardness after tempering ... M4: 71
machinability rating ... M1: 582
normalizing temperatures ... M4: 7
plasma (ion) nitriding ... A4: 423
tempering ... A4: 122, 134
thermal properties ... M4: 511
5130, cutoff band sawing with bimetal blades ... A6: 1184
5130, machining ... A16: 144-147, 200, 207, 225, 274, 323, 346, 348, 349, 669
5130H
composition ... M1: 130
hardenability curve ... M1: 513
5131, cutoff band sawing with bimetal blades ... A6: 1184
5132
annealing ... A4: 47
annealing temperatures ... A4: 46 M4: 18
austenitizing temperature ... A4: 961 M4: 29
carbonitriding ... A4: 376
composition ... M1: 128
machinability rating ... M1: 582
normalizing temperatures ... M4: 7
plasma (ion) nitriding ... A4: 423
temperatures and time cycles ... M4: 19
5132, cutoff band sawing with bimetal blades ... A6: 1184
5132, machining ... A16: 144-147, 200, 207, 225, 274, 323-325, 343, 346, 348, 349, 669
5132H
composition ... M1: 130
hardenability curve ... M1: 513
5133, cutoff band sawing with bimetal blades ... A6: 1184
5134, cutoff band sawing with bimetal blades ... A6: 1184
5135
annealing temperatures ... A4: 46 M4: 18
austenitizing temperature ... A4: 961 M4: 29
carbonitriding ... A4: 376
composition ... M1: 128
machinability rating ... M1: 582
normalizing temperatures ... M4: 7
plasma (ion) nitriding ... A4: 423
5135, cutoff band sawing with bimetal blades ... A6: 1184
5135, machining ... A16: 144-147, 200, 207, 225, 274, 323, 346, 348, 349, 669
5135H
composition ... M1: 130
hardenability curve ... M1: 514
5136, cutoff band sawing with bimetal blades ... A6: 1184
5137, cutoff band sawing with bimetal blades ... A6: 1184
5138, cutoff band sawing with bimetal blades ... A6: 1184
5139, cutoff band sawing with bimetal blades ... A6: 1184
5140
annealing ... A4: 47
annealing temperatures ... A4: 46 M4: 18
austempering, suitability ... A4: 153, 154 M4: 106
austenitizing temperature ... A4: 961 M4: 29
carbonitriding ... A4: 376, 378, 379, 381 M4: 183

Steels, AISI-SAE specific types (continued)
carburizing and hardenability ... A18: 875
composition ... M1: 163
critical temperatures ... A4: 43
critical temperatures, annealing ... M4: 15
die material for sheet metal forming ... A18: 628
hardenability ... M1: 489, 495
hardness, annealed sheet and strip ... M1: 165
Jominy curves ... A4: 650
machinability rating ... M1: 582
martempered to full hardness ... A4: 140
mass, effect on hardness ... A4: 39 M4: 11
microstructural transformations determinations using computer simulation ... A4: 648, 649
nitrided, hardness profile ... M1: 633
normalizing temperatures ... M4: 7
plasma (ion) nitriding ... A4: 423
properties, various heat treating conditions ... M4: 9-10
properties, various heat-treating conditions ... A4: 37
section size, austempered parts ... A4: 155 M4: 107
temperatures and time cycles ... M4: 19
tempering ... A4: 134
time-temperature transformation diagram ... M4: 90
time-temperature transformation diagrams ... A4: 140, 141
5140, cutoff band sawing with bimetal blades ... A6: 1184
5140, machining ... A16: 144-147, 179, 199, 200, 203, 204, 207, 209, 225, 274, 343, 346, 348, 349, 360, 361, 669
5140H
composition ... M1: 130
hardenability curve ... M1: 514
5140H, carbonitriding ... A4: 381
5141, cutoff band sawing with bimetal blades ... A6: 1184
5142, cutoff band sawing with bimetal blades ... A6: 1184
5143, cutoff band sawing with bimetal blades ... A6: 1184
5144, cutoff band sawing with bimetal blades ... A6: 1184
5145
annealing temperatures ... A4: 46 M4: 18
austenitizing temperature ... A4: 961 M4: 29
carbonitriding ... A4: 376
composition ... A6: 670 M1: 128
cutoff band sawing with bimetal blades ... A6: 1184
hardness vs. tempering temperature ... M1: 468
machinability rating ... M1: 582
normalizing temperatures ... M4: 7
plasma (ion) nitriding ... A4: 423
shielded metal arc welding ... A6: 670
welding preheat and interpass temperatures ... A6: 672
5145, machining ... A16: 144-147, 179, 200, 207, 225, 274, 343, 346, 348, 349, 360, 361, 669
5145H
composition ... M1: 130
hardenability curve ... M1: 514
5146, cutoff band sawing with bimetal blades ... A6: 1184
5147
annealing temperatures ... A4: 46 M4: 18
austenitizing temperature ... A4: 961 M4: 29
carbonitriding ... A4: 376
composition ... M1: 128
machinability rating ... M1: 582
normalizing temperatures ... M4: 7
plasma (ion) nitriding ... A4: 423
5147, cutoff band sawing with bimetal blades ... A6: 1184

Steels, AISI-SAE specific types (continued)
5147, machining ... A16: 144-147, 179, 200, 207, 225, 274, 343, 346, 348, 349, 360, 361, 669
5147H
composition ... M1: 130
hardenability curve ... M1: 515
5148, cutoff band sawing with bimetal blades ... A6: 1184
5149, cutoff band sawing with bimetal blades ... A6: 1184
5150
annealing ... A4: 47
annealing temperatures ... A4: 46 M4: 18
austenitizing temperature ... A4: 961 M4: 29
carbonitriding ... A4: 376
composition ... M1: 128, 163
hardenability equivalence ... M1: 483-484
hardness after tempering ... M4: 71
hardness, annealed sheet and strip ... M1: 165
machinability rating ... M1: 582
mass, effect on hardness ... A4: 39 M4: 11
normalizing temperatures ... M4: 7
plasma (ion) nitriding ... A4: 423
properties, various heat treating conditions ... M4: 9-10
properties, various heat-treating conditions ... A4: 37
resistance to scuffing wear ... M1: 25
temperatures and time cycles ... M4: 19
tempering ... A4: 122
5150, cutoff band sawing with bimetal blades ... A6: 1184
5150, machining ... A16: 200, 207, 225, 274, 343, 346, 348, 349, 360, 361, 669
5150H
composition ... M1: 130
hardenability curve ... M1: 515
springs, hot wound ... M1: 302
5151, cutoff band sawing with bimetal blades ... A6: 1184
5152, cutoff band sawing with bimetal blades ... A6: 1184
5153, cutoff band sawing with bimetal blades ... A6: 1184
5154, cutoff band sawing with bimetal blades ... A6: 1184
5155
annealing temperatures ... A4: 46 M4: 18
austenitizing temperature ... A4: 961 M4: 29
carbonitriding ... A4: 376
composition ... M1: 128
machinability rating ... M1: 582
normalizing temperatures ... M4: 7
plasma (ion) nitriding ... A4: 423
5155, cutoff band sawing with bimetal blades ... A6: 1184
5155, machining ... A16: 200, 207, 225, 274, 343, 346, 348, 349, 360, 361, 669
5155H
composition ... M1: 130
hardenability curve ... M1: 515
5156, cutoff band sawing with bimetal blades ... A6: 1184
5157, cutoff band sawing with bimetal blades ... A6: 1184
5158, cutoff band sawing with bimetal blades ... A6: 1184
5159, cutoff band sawing with bimetal blades ... A6: 1184
5160
annealing ... A4: 49, 50, 53
annealing temperatures ... A4: 46 M4: 18
austenitizing temperature ... A4: 961 M4: 29
carbonitriding ... A4: 376
composition ... M1: 128, 129, 163
core properties of carburized versus induction-hardened components ... A18: 725
critical temperatures ... A4: 43 M4: 15

Steels, AISI-SAE specific types (continued)
deep-hardening steel ... **A18:** 706
hardenability variation ... **M1:** 479
hardness ... **A4:** 155
hardness, annealed sheet and strip ... **M1:** 165
machinability rating ... **M1:** 582
machining ... **M4:** 19
mass, effect on hardness ... **A4:** 39 **M4:** 11
plasma (ion) nitriding ... **A4:** 423
properties, various heat treating conditions ... **M4:** 9-10
properties, various heat-treating conditions ... **A4:** 37
section size, austempered parts ... **A4:** 155 **M4:** 107
springs, wire for ... **M1:** 305
5160, cutoff band sawing with bimetal blades ... **A6:** 1184
5160, machining ... **A16:** 200, 207, 225, 274, 343, 346, 348, 349, 360, 361, 669
5160H
composition ... **M1:** 130
hardenability curve ... **M1:** 516
springs, hot wound ... **M1:** 302
torsion bars, fatigue life distribution ... **M1:** 677
5210, spheroidal carbides dispersed ... **A16:** 675-676
5210, turning machinability rating ... **A16:** 669
6118
carbonitriding ... **A4:** 376
composition ... **M1:** 128
machinability rating ... **M1:** 582
martempering ... **A4:** 141
plasma (ion) nitriding ... **A4:** 423
6118, cutoff band sawing with bimetal blades ... **A6:** 1184
6118, machining ... **A16:** 144-147, 179, 200, 207, 225, 274, 342, 346, 347, 349, 360, 361, 669
6118, normalizing temperatures ... **M4:** 7
6118H
composition ... **M1:** 130
hardenability curve ... **M1:** 516
hardenability equivalence ... **M1:** 483, 486
6119, cutoff band sawing with bimetal blades ... **A6:** 1184
6120
carbonitriding ... **A4:** 376
carburizing ... **A4:** 262, 266
martempering ... **A4:** 141
plasma (ion) nitriding ... **A4:** 423
6120, cutoff band sawing with bimetal blades ... **A6:** 1184
6120, normalizing temperatures ... **M4:** 7
6121, cutoff band sawing with bimetal blades ... **A6:** 1184
6122, cutoff band sawing with bimetal blades ... **A6:** 1184
6123, cutoff band sawing with bimetal blades ... **A6:** 1184
6124, cutoff band sawing with bimetal blades ... **A6:** 1184
6125, cutoff band sawing with bimetal blades ... **A6:** 1184
6126, cutoff band sawing with bimetal blades ... **A6:** 1184
6127, cutoff band sawing with bimetal blades ... **A6:** 1184
6128, cutoff band sawing with bimetal blades ... **A6:** 1184
6129, cutoff band sawing with bimetal blades ... **A6:** 1184
6130, cutoff band sawing with bimetal blades ... **A6:** 1184
6131, cutoff band sawing with bimetal blades ... **A6:** 1184
6132, cutoff band sawing with bimetal blades ... **A6:** 1184
6133, cutoff band sawing with bimetal blades ... **A6:** 1184
6134, cutoff band sawing with bimetal blades ... **A6:** 1184
6135, cutoff band sawing with bimetal blades ... **A6:** 1184
6136, cutoff band sawing with bimetal blades ... **A6:** 1184
6137, cutoff band sawing with bimetal blades ... **A6:** 1184

Steels, AISI-SAE specific types (continued)
6138, cutoff band sawing with bimetal blades ... **A6:** 1184
6139, cutoff band sawing with bimetal blades ... **A6:** 1184
6140, cutoff band sawing with bimetal blades ... **A6:** 1184
6141, cutoff band sawing with bimetal blades ... **A6:** 1184
6142, cutoff band sawing with bimetal blades ... **A6:** 1184
6143, cutoff band sawing with bimetal blades ... **A6:** 1184
6144, cutoff band sawing with bimetal blades ... **A6:** 1184
6145
austempering, suitability ... **A4:** 153
carbonitriding ... **A4:** 376
plasma (ion) nitriding ... **A4:** 423
6145, cutoff band sawing with bimetal blades ... **A6:** 1184
6145, die casting of copper alloys and erosive wear of dies ... **A18:** 629
6146, cutoff band sawing with bimetal blades ... **A6:** 1184
6147, cutoff band sawing with bimetal blades ... **A6:** 1184
6148, cutoff band sawing with bimetal blades ... **A6:** 1184
6149, cutoff band sawing with bimetal blades ... **A6:** 1184
6150
annealing ... **A4:** 47
annealing temperatures ... **A4:** 46 **M4:** 18
applications, austempered parts ... **A4:** 157 **M4:** 108
austenitizing temperature ... **A4:** 961 **M4:** 29
carbonitriding ... **A4:** 376
composition ... **A4:** 207 **A6:** 670 **M1:** 128, 129, 163, 422 **M4:** 120
critical temperatures ... **A4:** 43
critical temperatures, annealing ... **M4:** 15
cutoff band sawing with bimetal blades ... **A6:** 1184
deep drawing dies, use for ... **M3:** 499
die-casting dies, use in ... **M3:** 543
distortion in heat treatment ... **A4:** 614
electron beam hardening treatment (as 50 Cr V 4 steel) ... **A4:** 308
electron-beam welding ... **A6:** 867
flame hardening ... **A4:** 283
hardness after tempering ... **M4:** 71
hardness and impact energy ... **A4:** 213
hardness, annealed sheet and strip ... **M1:** 165
heat treatment ... **M1:** 431-432
heat treatments ... **A4:** 212-213 **M4:** 124
heat-treatment temperatures ... **A4:** 208
liquid nitriding ... **A4:** 413 **M4:** 256
machinability rating ... **M1:** 582
martempered to full hardness ... **A4:** 140
martempering applications equipment requirements ... **A4:** 149, 150
martempering, equipment ... **M4:** 100, 101
martempering in salt, applications ... **A4:** 148 **M4:** 96
mass, effect on hardness ... **A4:** 39 **M4:** 11
mass effects on typical properties ... **A4:** 213
mechanical properties ... **M1:** 432
normalizing temperatures ... **M4:** 7
plasma (ion) nitriding ... **A4:** 423
processing ... **M1:** 431
properties, effect of mass on ... **M4:** 126
properties, various heat treating conditions ... **M4:** 9-10
properties, various heat-treating conditions ... **A4:** 37
shear blades, service data ... **M3:** 481
springs, wire for ... **M1:** 305
temperatures and time cycles ... **M4:** 19
tempering ... **A4:** 122, 134
tempering temperature ... **M4:** 125
tensile properties ... **A4:** 213
tensile properties, heat treated ... **M4:** 124, 125
trimming tools, use for ... **M3:** 532
6150, for hot extrusion tools ... **A18:** 627
6150, machining ... **A16:** 200, 207, 225, 274, 343, 346, 348, 349, 358, 360, 361, 669

Steels, AISI-SAE specific types (continued)
6150H
composition ... **M1:** 130
hardenability curve ... **M1:** 517
6150H, flame hardening ... **A4:** 284
6152, proven applications for borided ferrous materials ... **A4:** 445
6411, composition ... **A4:** 207
7140
austenitizing temperature ... **M4:** 191
case depth ... **A4:** 391
composition ... **M4:** 191
gas nitriding ... **A4:** 387, 390, 391
hardness gradients ... **A4:** 388, 390, 391 **M4:** 195
hardness gradients, nitriding ... **M4:** 256
liquid nitriding ... **A4:** 413
tempering temperature ... **M4:** 191
7140, interface considerations for die-casting die wear ... **A18:** 632
8115
austenitizing temperature ... **A4:** 962 **M4:** 30
composition ... **M1:** 128
critical temperatures ... **A4:** 43
critical temperatures, annealing ... **M4:** 15
machinability rating ... **M1:** 582
8115, machining ... **A16:** 144-147, 179, 207, 225, 274, 285, 342, 346, 347, 349, 669
8615
austenitizing temperature ... **A4:** 962
carbonitriding ... **A4:** 376
composition ... **M1:** 163
flame hardening ... **A4:** 283
liquid carburizing ... **A4:** 334
machinability rating ... **M1:** 582
plasma (ion) nitriding ... **A4:** 423
quench cracking ... **A4:** 610
8615, austenitizing temperature ... **M4:** 30
8615, cutoff band sawing with bimetal blades ... **A6:** 1184
8615, sawing ... **A16:** 360, 361
8615, trepanning ... **A16:** 178
8615, turning machinability rating ... **A16:** 669
8615H, liquid carburizing ... **A4:** 340
8616
carbonitriding ... **A4:** 376
flame hardening ... **A4:** 283
plasma (ion) nitriding ... **A4:** 423
quench cracking ... **A4:** 610
8617
austenitizing temperature ... **A4:** 142, 962 **M4:** 30, 92
carbonitriding ... **A4:** 376, 381
carburized, residual stress ... **M1:** 538
composition ... **A4:** 320 **M1:** 128, 129
flame hardening ... **A4:** 283
machinability rating ... **M1:** 582
martempering, equipment ... **M4:** 100
martempering equipment requirements ... **A4:** 149
martempering temperature ... **A4:** 142 **M4:** 92
normalizing temperatures ... **M4:** 7
plasma (ion) nitriding ... **A4:** 423
quench cracking ... **A4:** 610
tempering ... **A4:** 133
8617, machining ... **A16:** 144-147, 179, 203, 207, 225, 274, 342, 346, 347, 349, 360, 361, 669
8617H
carbonitriding ... **A4:** 381
composition ... **M1:** 130
hardenability curve ... **M1:** 517
martempering in oil, applications ... **A4:** 148
8617H, martempering in oil, applications ... **M4:** 98
8618
carbonitriding ... **A4:** 376
flame hardening ... **A4:** 283
plasma (ion) nitriding ... **A4:** 423
quench cracking ... **A4:** 610
8619
carbonitriding ... **A4:** 376
flame hardening ... **A4:** 283
plasma (ion) nitriding ... **A4:** 423
quench cracking ... **A4:** 610
8620
annealing ... **A4:** 47, 53
applications ... **A4:** 397 **A18:** 704

Steels, AISI-SAE specific types (continued)
applications, austempered parts A4: 157 M4: 108
austenitizing temperature A4: 142, 962 M4: 30, 92
carbon gradients M4: 271, 440
carbon gradients and surface carbon content A4: 592, 593
carbonitriding A4: 376, 377, 381, 382 M4: 179
carburized, fracture toughness M1: 536
carburizing A4: 365-370
carburizing and fabricability A18: 876
carburizing and hardenability A18: 875
carburizing effect on carbon concentration gradients and case hardness A18: 873-874
carburizing in a fluidized bed A4: 486, 487
carburizing temperature, effect on grain size M4: 158
case depth variation A4: 591, 592, 593
case hardened, hardness profiles M1: 633
case-depth measurements M4: 438, 439
center-line hardness and microstructure of steel bar A4: 27
cold extrusion tools, use for M3: 518
composition A4: 29, 320 A6: 670 M1: 163
consecutive cuts analysis of carbon control A4: 588
core properties of carburized versus induction-hardened components A18: 725
critical cooling rates and microstructure A4: 24
critical temperatures A4: 43
critical temperatures, annealing M4: 15
cutoff band sawing with bimetal blades A6: 1184
determination of case hardenability M1: 473
distortion, and oil quenching A4: 99, 100
electron-beam welding A6: 867
fatigue life versus film thickness in bearing steels A18: 727, 728
flame hardening A4: 282, 283
gages, use for M3: 555
gas carburizing A4: 315, 319, 363, 364 M4: 163
gas nitriding A4: 397
gear cutting, comparison with 4422 steel M1: 579, 581
hardenability, selection for M1: 492
hardness A4: 99, 100, 486, 487
hardness after tempering A4: 31
hardness as a function of bar diameter A4: 23, 24
isothermal transformation diagram A6: 671, 672
Jominy curve A4: 29
Jominy end-quench hardenability curves A4: 20, 27
laser transformation hardening A18: 862
liquid carburizing A4: 331, 334, 340, 345
M_{90} temperature A6: 672
machinability rating M1: 582
machining compared with 4620 steel M1: 568, 579
martempered to full hardness A4: 140
martempering in oil, applications A4: 148 M4: 98
martempering in salt, applications A4: 148 M4: 96
martempering temperature A4: 142 M4: 92
mass, effect on hardness A4: 39 M4: 11
M_s temperature A6: 671, 672
nominal compositions A18: 704, 725
normalizing temperatures M4: 7
oil quenching M4: 54
pack carburizing applications A4: 326
plasma (ion) carburizing A4: 356
plasma (ion) nitriding A4: 423

Steels, AISI-SAE specific types (continued)
process variability (statistical process control) A4: 625
properties, various heat treating conditions M4: 9-10
properties, various heat-treating conditions A4: 37
quench cracking A4: 610
recommended upsetting pressures for flash welding A6: 843
shielded metal arc welding A6: 670
surface carbon variability A4: 596, 597
surface hardening A4: 262
tapered roller bearing material A18: 502
temperatures and time cycles M4: 19
vacuum carburizing A4: 351
welding preheat and interpass temperatures A6: 672
wrist pins, tool life for planing vs. 1019 steel M1: 572-573, 575
8620, flash welding schedule M6: 577
8620, machining A16: 58, 144-148, 179, 200, 203, 207, 208, 246, 247, 251, 263, 274, 282, 285, 326, 347, 349, 360, 361, 364, 640, 669, 672
8620H
carbonitriding A4: 382, 383
carburizing A4: 364, 365
case-depth values for a carbon-gradient plot A4: 455, 456
composition M1: 130
consecutive cuts analysis of carbon control A4: 588
gas carburizing process parameters A4: 653
hardenability curve M1: 518
hardness A4: 454
martempering, dimensional changes A4: 144, 145
process variability (statistical process control) A4: 624-625
quench cracking A4: 611
8620H, dimensional control, martempering M4: 95
8621, cutoff band sawing with bimetal blades A6: 1184
8622
austenitizing temperature A4: 962 M4: 30
carbonitriding A4: 376, 381
composition M1: 128, 129
hardenability variation M1: 479
machinability rating M1: 582
normalizing temperatures M4: 7
plasma (ion) nitriding A4: 423
quench cracking A4: 610
8622, cutoff band sawing with bimetal blades A6: 1184
8622, machining A16: 144-148, 157, 179, 207, 225, 274, 342, 346, 347, 349, 360, 361, 669
8622H
composition M1: 130
hardenability curve M1: 518
8623, cutoff band sawing with bimetal blades A6: 1184
8624, cutoff band sawing with bimetal blades A6: 1184
8625
austenitizing temperature A4: 962 M4: 30
carbonitriding A4: 376
composition M1: 128, 129
dimensional control, martempering M4: 94
machinability rating M1: 582
martempering in oil, applications A4: 148 M4: 98
martempering to reduce distortion A4: 144, 146, 147
normalizing temperatures M4: 7
plasma (ion) nitriding A4: 423
quench cracking A4: 610
8625, cutoff band sawing with bimetal blades A6: 1184

Steels, AISI-SAE specific types (continued)
8625, machining A16: 144-147, 207, 225, 274, 323-325, 343, 346, 348, 349, 360, 361, 669
8625H
composition M1: 131
hardenability curve M1: 518
8625H, dimensional control, martempering M4: 95
8625H, martempering, dimensional changes A4: 144, 145
8626, cutoff band sawing with bimetal blades A6: 1184
8627
annealing temperatures A4: 46 M4: 18
austenitizing temperature A4: 962 M4: 30
carbonitriding A4: 376
composition M1: 128, 129
machinability rating M1: 582
normalizing temperatures M4: 7
plasma (ion) nitriding A4: 423
quench cracking A4: 610
8627, cutoff band sawing with bimetal blades A6: 1184
8627, machining A16: 144-147, 207, 225, 274, 323-325, 343, 346, 348, 349, 360, 361, 669
8627H
composition M1: 131
hardenability curve M1: 519
8628, cutoff band sawing with bimetal blades A6: 1184
8629, cutoff band sawing with bimetal blades A6: 1184
8630
annealing A4: 47
annealing temperatures A4: 46 M4: 18
austenitizing temperature A4: 961 M4: 29
carbonitriding A4: 376
composition A6: 670 M1: 128, 129, 163
cooling rate and quenching A4: 80, 82
cutoff band sawing with bimetal blades A6: 1184
fatigue properties of castings M1: 389, 397
flame hardening A4: 283
flash welding M6: 557
friction welding M6: 721
hardness after tempering M4: 71
hardness, annealed sheet and strip M1: 165
machinability rating M1: 582
martempered to full hardness A4: 140
mass, effect on hardness A4: 39 M4: 11
mechanical properties, cold finished bars M1: 222, 230-231
non-martensitic transformation products A4: 141
normalizing temperatures M4: 7
oxyfuel gas welding A6: 286
plasma (ion) nitriding A4: 423
properties, various heat treating conditions M4: 9-10
quench cracking A4: 610
recommended upsetting pressures for flash welding A6: 843
shielded metal arc welding A6: 669-670
temperature measurement by OSRLR A6: 1150
temperatures and time cycles M4: 19
tempering A4: 122, 134
trace element impurity effect on GTA weld penetration A6: 20
welding preheat and interpass temperatures A6: 672
yield strengths A4: 208
8630, hot upset forging tools, use for M3: 535
8630, machining A16: 144-147, 207, 225, 274, 323-325, 343, 346, 348, 349, 360, 361, 669
8630 mod Charpy impact properties of castings M1: 392, 398
8630H
composition M1: 131

Steels, AISI-SAE specific types (continued)
hardenability curve **M1:** 519
hardenability variation **M1:** 479
8631
carbonitriding **A4:** 376
flame hardening **A4:** 283
plasma (ion) nitriding **A4:** 423
quench cracking **A4:** 610
8631, cutoff band sawing with bimetal blades **A6:** 1184
8632
carbonitriding **A4:** 376
flame hardening **A4:** 283
plasma (ion) nitriding **A4:** 423
quench cracking **A4:** 610
tempering **A4:** 134
8632, cutoff band sawing with bimetal blades **A6:** 1184
8633
carbonitriding **A4:** 376
flame hardening **A4:** 283
plasma (ion) nitriding **A4:** 423
quench cracking **A4:** 610
8633, cutoff band sawing with bimetal blades **A6:** 1184
8634
carbonitriding **A4:** 376
flame hardening **A4:** 283
plasma (ion) nitriding **A4:** 423
quench cracking **A4:** 610
8634, cutoff band sawing with bimetal blades **A6:** 1184
8635
carbonitriding **A4:** 376
flame hardening **A4:** 283
plasma (ion) nitriding **A4:** 423
quench cracking **A4:** 610
8635, cutoff band sawing with bimetal blades **A6:** 1184
8636
carbonitriding **A4:** 376
flame hardening **A4:** 283
plasma (ion) nitriding **A4:** 423
quench cracking **A4:** 610
8636, cutoff band sawing with bimetal blades **A6:** 1184
8637
annealing temperatures **A4:** 46 **M4:** 18
austenitizing temperature **A4:** 961 **M4:** 29
carbonitriding **A4:** 376
composition **M1:** 128, 129
flame hardening **A4:** 283
machinability rating **M1:** 582
normalizing temperatures **M4:** 7
plasma (ion) nitriding **A4:** 423
quench cracking **A4:** 610
8637, cutoff band sawing with bimetal blades **A6:** 1184
8637, machining **A16:** 144-147, 207, 225, 274, 323-325, 343, 346, 348, 349, 360, 361, 669
8637H
composition **M1:** 131
hardenability curve **M1:** 520
8638
carbonitriding **A4:** 376
flame hardening **A4:** 283
plasma (ion) nitriding **A4:** 423
quench cracking **A4:** 610
8638, cutoff band sawing with bimetal blades **A6:** 1184
8639
carbonitriding **A4:** 376
flame hardening **A4:** 283
plasma (ion) nitriding **A4:** 423
quench cracking **A4:** 610
8639, cutoff band sawing with bimetal blades **A6:** 1184
8640
annealing **A4:** 47
annealing temperatures **A4:** 46 **M4:** 18
applications, austempered parts **A4:** 157 **M4:** 108
austenitizing temperature **A4:** 961 **M4:** 29
carbonitriding **A4:** 376
carburizing and hardenability **A18:** 875
composition **A4:** 207 **A6:** 670 **M1:** 128, 129, 163, 422 **M4:** 120

Steels, AISI-SAE specific types (continued)
critical temperatures **A4:** 43
critical temperatures, annealing **M4:** 15
cutoff band sawing with bimetal blades **A6:** 1184
die material for sheet metal forming **A18:** 628
fatigue properties **M1:** 389, 397
flame hardening **A4:** 283
gas nitriding **A4:** 392
hardenability equivalence **M1:** 483-484
hardness, annealed sheet and strip **M1:** 165
heat treatment **M1:** 433
heat treatments **A4:** 213-214 **M4:** 124
heat-treatment temperatures **A4:** 208
machinability rating **M1:** 582
martempered to full hardness **A4:** 140
mass, effects on typical properties **A4:** 214
mechanical properties **A4:** 214 **M1:** 433-435 **M4:** 126
nitrided, hardness profile **M1:** 633
normalizing temperatures **M4:** 7
plasma (ion) nitriding **A4:** 423
processing **M1:** 432
properties, effect of mass on **M4:** 126
quench cracking **A4:** 610
shielded metal arc welding **A6:** 670
temperatures and time cycles **M4:** 19
welding preheat and interpass temperatures **A6:** 672
8640, flash welding schedule **M6:** 577
8640, machining **A16:** 144-147, 179, 207, 225, 274, 343, 346, 348, 349, 360, 361, 669
8640H
composition **M1:** 131
hardenability curve **M1:** 520
hardenability variation **M1:** 479
8640H, flame hardening **A4:** 284
8641, cutoff band sawing with bimetal blades **A6:** 1184
8642
annealing temperatures **A4:** 46 **M4:** 18
austenitizing temperature **A4:** 961 **M4:** 29
carbonitriding **A4:** 376
composition **M1:** 128
flame hardening **A4:** 283
hardenability equivalence **M1:** 483-484
machinability rating **M1:** 582
normalizing temperatures **M4:** 7
plasma (ion) nitriding **A4:** 423
quench cracking **A4:** 610
8642, cutoff band sawing with bimetal blades **A6:** 1184
8642, machining **A16:** 144-147, 179, 207, 225, 274, 343, 346, 348, 349, 360, 361, 669
8642H
composition **M1:** 131
hardenability curve **M1:** 520
8642H, flame hardening **A4:** 284
8643
carbonitriding **A4:** 376
flame hardening **A4:** 283
plasma (ion) nitriding **A4:** 423
quench cracking **A4:** 610
8643, cutoff band sawing with bimetal blades **A6:** 1184
8644
carbonitriding **A4:** 376
flame hardening **A4:** 283
plasma (ion) nitriding **A4:** 423
quench cracking **A4:** 610
8644, cutoff band sawing with bimetal blades **A6:** 1184
8645
annealing temperatures **A4:** 46 **M4:** 18
austenitizing temperature **A4:** 961 **M4:** 29
carbonitriding **A4:** 376
composition **M1:** 128, 163
flame hardening **A4:** 283
hardness, annealed sheet and strip **M1:** 165
machinability rating **M1:** 582
normalizing temperatures **M4:** 7
plasma (ion) nitriding **A4:** 423
quench cracking **A4:** 610
8645, cutoff band sawing with bimetal blades **A6:** 1184
8645, machining **A16:** 144-147, 179, 207, 225, 274, 343, 346, 348, 349, 360, 361, 669

Steels, AISI-SAE specific types (continued)
8645H
composition **M1:** 131
hardenability curve **M1:** 521
hardenability variation **M1:** 479
8646
carbonitriding **A4:** 376
flame hardening **A4:** 283
plasma (ion) nitriding **A4:** 423
quench cracking **A4:** 610
8647
carbonitriding **A4:** 376
flame hardening **A4:** 283
plasma (ion) nitriding **A4:** 423
quench cracking **A4:** 610
8648
carbonitriding **A4:** 376
flame hardening **A4:** 283
plasma (ion) nitriding **A4:** 423
quench cracking **A4:** 610
8649
carbonitriding **A4:** 376
flame hardening **A4:** 283
plasma (ion) nitriding **A4:** 423
quench cracking **A4:** 610
8650
annealing **A4:** 47
annealing temperatures **A4:** 46 **M4:** 18
austenitizing temperature **A4:** 961 **M4:** 29
carbonitriding **A4:** 376
composition **M1:** 128
flame hardening **A4:** 283
hardenability **M1:** 471, 472
hardness after tempering **M4:** 71
machinability rating **M1:** 582
mass, effect on hardness **A4:** 39 **M4:** 11
normalizing temperatures **M4:** 7
plasma (ion) nitriding **A4:** 423
properties, various heat treating conditions **M4:** 9-10
properties, various heat-treating conditions **A4:** 37
quench cracking **A4:** 610
springs, hot wound **M1:** 304
temperatures and time cycles **M4:** 19
tempering **A4:** 122, 134
8650, machining **A16:** 207, 225, 274, 343, 346, 348, 349, 669
8650H
composition **M1:** 131
hardenability curve **M1:** 521
8651
carbonitriding **A4:** 376
flame hardening **A4:** 283
plasma (ion) nitriding **A4:** 423
quench cracking **A4:** 610
8652
carbonitriding **A4:** 376
flame hardening **A4:** 283
plasma (ion) nitriding **A4:** 423
quench cracking **A4:** 610
8653
carbonitriding **A4:** 376
flame hardening **A4:** 283
plasma (ion) nitriding **A4:** 423
quench cracking **A4:** 610
8654
carbonitriding **A4:** 376
flame hardening **A4:** 283
plasma (ion) nitriding **A4:** 423
quench cracking **A4:** 610
8655
annealing temperatures **A4:** 46 **M4:** 18
austenitizing temperature **A4:** 961 **M4:** 29
carbonitriding **A4:** 376
composition **M1:** 128, 129
flame hardening **A4:** 283
machinability rating **M1:** 582
normalizing temperatures **M4:** 7
plasma (ion) nitriding **A4:** 423
quench cracking **A4:** 610
8655, machining **A16:** 207, 225, 274, 343, 346, 348, 349, 669
8655H
composition **M1:** 131
hardenability curve **M1:** 522

Steels, AISI-SAE specific types (continued)
springs, hot rolled ... M1: 303
8656
carbonitriding ... A4: 376
flame hardening ... A4: 283
plasma (ion) nitriding ... A4: 423
quench cracking ... A4: 610
8657
carbonitriding ... A4: 376
flame hardening ... A4: 283
plasma (ion) nitriding ... A4: 423
quench cracking ... A4: 610
8658
carbonitriding ... A4: 376
flame hardening ... A4: 283
plasma (ion) nitriding ... A4: 423
quench cracking ... A4: 610
8659
carbonitriding ... A4: 376
flame hardening ... A4: 283
plasma (ion) nitriding ... A4: 423
quench cracking ... A4: 610
8660
annealing ... A4: 47
annealing temperatures ... A4: 46 M4: 18
austenitizing temperature ... A4: 961 M4: 29
carbonitriding ... A4: 376
composition ... M1: 128
flame hardening ... A4: 283
machinability rating ... M1: 582
normalizing temperatures ... M4: 7
plasma (ion) nitriding ... A4: 423
quench cracking ... A4: 610
springs, hot wound ... M1: 304
temperatures and time cycles ... M4: 19
8660, machining ... A16: 207, 225, 274, 343, 346, 348, 349, 669
8660H
composition ... M1: 131
hardenability curve ... M1: 522
8719
carbonitriding ... A4: 376
carburizing ... A4: 372
gas carburizing ... A4: 366
plasma (ion) nitriding ... A4: 423
8720
annealing ... A4: 47, 53
austenitizing temperature ... A4: 962 M4: 30
carbonitriding ... A4: 376
composition ... A4: 320 M1: 128
gas carburizing ... A4: 319
hardenability variation ... M1: 479
machinability rating ... M1: 582
martempering in salt, applications ... A4: 148 M4: 96
normalizing temperatures ... M4: 7
notch toughness vs. temperature ... M1: 536
plasma (ion) nitriding ... A4: 423
temperatures and time cycles ... M4: 19
8720, band sawing ... A16: 360, 361
8720, carburizing and hardenability ... A18: 875
8720 to 8740, cutoff band sawing with bimetal blades ... A6: 1184
8720, turning machinability rating ... A16: 669
8720H
composition ... M1: 131
hardenability curve ... M1: 522
8735
austempering time, effect on hardness ... A4: 160
carbonitriding ... A4: 376
plasma (ion) nitriding ... A4: 423
8735, austempering time, effect on hardness ... M4: 112
8740
annealing ... A4: 47
annealing temperatures ... A4: 46 M4: 18
austenitizing temperature ... A4: 142, 961 M4: 29, 92
carbonitriding ... A4: 376
composition ... M1: 128
fatigue limit variability ... M1: 676, 679
hardenability equivalence ... M1: 483-484
hardness after tempering ... M4: 71
machinability rating ... M1: 582
martempered to full hardness ... A4: 140
martempering in salt, applications ... A4: 148 M4: 96
martempering temperature ... M4: 92
mass, effect on hardness ... A4: 39 M4: 11
plasma (ion) nitriding ... A4: 423
properties, various heat treating conditions ... M4: 9-10
properties, various heat-treating conditions ... A4: 37
temperatures and time cycles ... M4: 19
tempering ... A4: 122, 134
8740, carburizing and hardenability ... A18: 875
8740, flash welding ... M6: 557
8740, machining ... A16: 144-147, 179, 207, 225, 274, 343, 346, 348, 349, 360, 361, 669
8740H
composition ... M1: 131
hardenability curve ... M1: 523
8742
annealing temperatures ... A4: 46 M4: 18
austenitizing temperature ... A4: 961 M4: 29
carbonitriding ... A4: 376
composition ... M1: 129
flame hardening ... A4: 275
normalizing temperatures ... M4: 7
plasma (ion) nitriding ... A4: 423
8742, machining ... A16: 144-147, 179, 207, 225, 274, 343, 346, 348, 349
8745
carbonitriding ... A4: 376
martempered to full hardness ... A4: 140
plasma (ion) nitriding ... A4: 423
8750
annealing ... A4: 47
austempering time, effect on hardness ... A4: 160 M4: 112
carbonitriding ... A4: 376
hardness after tempering ... M4: 71
plasma (ion) nitriding ... A4: 423
section size, austempered parts ... A4: 155 M4: 107
temperatures and time cycles ... M4: 19
tempering ... A4: 122
8822
austenitizing temperature ... A4: 962 M4: 30
carbonitrided, effect of nitrogen on case hardenability ... M1: 536
composition ... A4: 320 M1: 129
machinability rating ... M1: 582
normalizing temperatures ... M4: 7
8822, machining ... A16: 144-147, 179, 207, 274, 342, 346, 347, 349, 669
8822H
composition ... M1: 131
hardenability curve ... M1: 523
9112, gas carburizing ... M4: 163
9254
composition ... M1: 129
machinability rating ... M1: 582
springs ... M1: 285, 300
9254, austenitizing temperature ... A4: 961 M4: 29
9254, machining ... A16: 207, 274, 342, 346, 348, 349, 669
9255
austenitizing temperature ... A4: 961 M4: 29
composition ... M1: 129
machinability rating ... M1: 582
mass, effect on hardness ... A4: 39 M4: 11
normalizing temperatures ... M4: 7
properties, various heat treating conditions ... M4: 9-10
properties, various heat-treating conditions ... A4: 37
9255, machining ... A16: 207, 274, 343, 346, 348, 349, 669
9260
annealing ... A4: 47
annealing temperatures ... A4: 46 M4: 18
austenitizing temperature ... A4: 961 M4: 29
composition ... M1: 129
critical temperatures ... A4: 43
critical temperatures, annealing ... M4: 15
machinability rating ... M1: 582
martempering in salt, applications ... A4: 148 M4: 96
normalizing temperatures ... M4: 7
temperatures and time cycles ... M4: 19
9260, machining ... A16: 207, 274, 343, 346, 349, 669
9260H
composition ... M1: 131
hardenability curve ... M1: 523
9261, austempering, suitability ... A4: 154 M4: 106
9262, normalizing temperatures ... M4: 7
9310
annealing ... A4: 47
austenitizing temperature ... A4: 142, 962 M4: 30, 92
carburized, toughness ... M1: 536
carburizing and fabricability ... A18: 876
carburizing and its effect on case hardness ... A18: 874
composition ... A4: 320 M1: 129
cutoff band sawing with bimetal blades ... A6: 1184
electron-beam welding ... A6: 867
flame hardening ... A4: 282
gas carburizing ... A4: 319 M4: 157
hardenability, selection for ... M1: 493
liquid carburizing ... A4: 345
machinability rating ... M1: 582
martempered to full hardness ... A4: 140
martempering in salt, applications ... A4: 148 M4: 96
martempering temperature ... A4: 142 M4: 92
mass effect on hardness ... A4: 39 M4: 11
nominal compositions and applications ... A18: 704
normalizing temperatures ... M4: 7
plasma (ion) nitriding ... A4: 423
properties, various heat treating conditions ... M4: 9-10
properties, various heat-treating conditions ... A4: 37
surface carbon content and shim stock analysis ... A4: 589
surface carbon content, variations ... M4: 444
surface carbon variability ... A4: 597
surface hardening ... A4: 262
temperatures and time cycles ... M4: 19
vacuum carburizing ... A4: 351
9310, machining ... A16: 144-147, 175, 179, 203, 204, 207, 209, 274, 342, 346, 347, 349, 360, 361, 669
9310H
composition ... M1: 131
hardenability curve ... M1: 524
9315, carburizing temperature, effect on grain size ... M4: 158
9400, quenching with molten salts ... A4: 105
9440, austempering, suitability ... A4: 153
9445
cooling rates ... A4: 95-96, 97
oil quenching ... A4: 95-96, 97
quenching temperature ... A4: 97
9620, gear shaping ... A16: 348
9840
annealing ... A4: 47
annealing temperatures ... A4: 46 M4: 18
austenitizing temperature ... A4: 961 M4: 29
normalizing temperatures ... M4: 7

Steels, AISI-SAE specific types (continued)
plasma (ion) nitriding **A4:** 423
temperatures and time cycles **M4:** 19
tempering **A4:** 134
9840, broaching **A16:** 203, 209
9850
annealing **A4:** 47
hardness after tempering **M4:** 71
normalizing temperatures **M4:** 7
plasma (ion) nitriding **A4:** 423
temperatures and time cycles **M4:** 19
tempering **A4:** 122, 134
50100 *See* following 50B60
annealing temperatures **A4:** 46 **M4:** 18
austenitizing temperature **A4:** 961 **M4:** 29
composition **M1:** 128, 609
machinability rating **M1:** 582
section size, austempered parts **A4:** 155 **M4:** 107
50100, machining **A16:** 179, 207, 274, 360, 361
50100 to 52100, cutoff band sawing with bimetal blades **A6:** 1184
51100 *See* following 51B60
annealing temperatures **A4:** 46 **M4:** 18
austenitizing temperature **A4:** 961 **M4:** 29
composition **M1:** 128, 609
machinability rating **M1:** 582
surface finish, influence on wear of lubricated surfaces **M1:** 634, 636
51100, machining **A16:** 179, 207, 274, 360, 361
52100 *See* following 51100
abrasive wear **A18:** 774
abrasive wear coefficient **A18:** 491
agitation effect on surface hardness **A4:** 143
analysis **A4:** 589
annealing **A4:** 47, 49, 53, 55
annealing temperatures **A4:** 46 **M4:** 18
application **A18:** 158
application in gas turbine mainshaft bearings **A18:** 590
applications **A4:** 757
austenitizing temperature ... **A4:** 141, 142, 961 **M4:** 29, 92
bearing life versus specific film thickness in bearing steels **A18:** 727, 728
carburizing **A4:** 366
carburizing and effect on its microstructure **A18:** 873, 875
composition **M1:** 128, 609
contact fatigue in rolling-element bearings **A18:** 260
core properties of carburized versus induction-hardened components **A18:** 725
critical temperatures **A4:** 43
critical temperatures, annealing **M4:** 15
cutoff band sawing with bimetal blades **A6:** 1184
damage by microstructural changes **A18:** 177-178
electron beam hardening treatment (as 100 Cr 6 hypereutectoid steel) **A4:** 302, 303, 305
electron-beam welding **A6:** 867, 868
endurance lifetime with molybdenum disulfide coating **A18:** 782
flame hardening **A4:** 275, 280, 283
flash welding **M6:** 557
friction coefficient data **A18:** 71, 73
friction coefficient, tribotesting **A18:** 485
friction welding **A6:** 153 **M6:** 721
grain size, martempering **M4:** 91
grinding rod material **A18:** 654
hardness after annealing **M4:** 26
hardness, and oil quenching **A4:** 99
heat treatment for use in bearings **A18:** 702
high-resolution electron microscopy to study sliding wear **A18:** 389
ion implantation **A4:** 266
laser surface transformation hardening **A4:** 293
laser transformation hardening **A4:** 265 **A18:** 862
machinability rating **M1:** 582
martempered in hot salt **M4:** 93
martempered to full hardness **A4:** 140

Steels, AISI-SAE specific types (continued)
martempering applications, equipment requirements **A4:** 149, 150, 151
martempering, equipment **M4:** 100, 102
martempering for distortion control **A4:** 144, 146
martempering in oil, applications **A4:** 148 **M4:** 98
martempering in salt, applications **A4:** 148 **M4:** 96
martempering temperature **A4:** 142 **M4:** 92
microstructural changes from rolling contact fatigue **A18:** 389
microstructure **M1:** 612
nominal compositions **A18:** 725
non-martensitic transformation products **A4:** 141
normalizing **A4:** 38
part material for ion implantation **A18:** 858
promotes degradation of perfluoropolyalkylether (PFPE) **A18:** 156
rolling contact fatigue **A18:** 260, 389
rolling-contact component steel material for rolling-element bearings **A18:** 502, 503
shafting for mining or off-road construction machinery **M1:** 606
sliding wear **A18:** 774
springs, helical **M1:** 299
surface carbon content and shim stock
surface modification with and without lubrication and high-vacuum
temperatures and time cycles **M4:** 19
tempering **A4:** 134
VAMAS round-robin sliding wear tests **A18:** 486
Vickers and Knoop microindentation hardness numbers **A18:** 416
52100, coining dies, use for **M3:** 508-509
52100, machining **A16:** 177, 179, 203, 207, 209, 466, 498, 541, 608
A36
weld compositional analysis **A6:** 101
weld macrostructure **A6:** 99, 101
A537, weld macrostructure **A6:** 99, 101
B1111, arc welding **A6:** 646
B1112, arc welding **A6:** 646
B1112, machinability **A16:** 643
B1112, scatter in machinability rating **M1:** 567
B1112, spade drilling **A16:** 225
B1113, arc welding **A6:** 646
B1113, broaching **A16:** 203, 204
B1113, scatter in machinability rating **M1:** 567
B1113, spade drilling **A16:** 225
B1212, carbonitriding **A4:** 385
C-45, applications, laser transformation hardening **A18:** 863
C-1080
composition similar to P/M steel **A4:** 229
porosity and hardenability **A4:** 229
C1008, roll welding **A6:** 313
C1018
carbon penetration **A4:** 231
case depth **A4:** 232
composition similar to P/M steels **A4:** 232
hardness **A4:** 232
porosity effect on case depth **A4:** 230
C1095 (spring steel), impact wear **A18:** 268
carbonitriding **A4:** 376
Ck 45N, gas nitriding effect on compound layer concentration versus distance from surface **A18:** 878, 879
Class 45010, flame hardening **A4:** 283
Class 50007, flame hardening **A4:** 283
Class 53004, flame hardening **A4:** 283
Class 60003, flame hardening **A4:** 283
Class 80002, flame hardening **A4:** 283
D-6a, residual stresses **A6:** 1097
D-6ac, electron-beam welding **A6:** 867
E 7024, weld (horizontal) characterization **A6:** 99
E4340
composition **M1:** 127, 129

Steels, AISI-SAE specific types (continued)
machinability rating **M1:** 582
E4340H
composition **M1:** 130
E52100
composition **A4:** 300
electron beam hardening treatment **A4:** 300
proven applications for borided ferrous materials **A4:** 445
En3l, oxidational wear **A18:** 286, 287, 288
En8, oxidational wear **A18:** 283, 286
EN40B, plasma nitrocarburizing **A4:** 432, 433
EN40C, vacuum nitrocarburizing **A4:** 407
EN41, vacuum nitrocarburizing **A4:** 407
EX24-type steels, carburizing **A4:** 368, 370
M1008
composition **M1:** 126
M1010
composition **M1:** 126
M1012
composition **M1:** 126, 205
M1015
composition **M1:** 126, 205, 1016
M1017
composition **M1:** 126, 205
M1020
composition **M1:** 126, 205
M1023
composition **M1:** 126, 205
M1025
composition **M1:** 126, 205
M1031
composition **M1:** 126, 205
M1044
composition **M1:** 126, 205
martempering **A4:** 141
martempering, equipment requirements **A4:** 148, 149
Steels, AMS, specific types *See also* Nonferrous alloys, AMS specific types
5010E **M1:** 140
5020 .... **M1:** 140
5022G **M1:** 140
5024D **M1:** 140
5032B **M1:** 140
5040F **M1:** 140
5041 .... **M1:** 140
5042F **M1:** 140
5044D **M1:** 140
5045C **M1:** 140
5047A **M1:** 140
5050F **M1:** 140
5053C **M1:** 140
5060C **M1:** 140
5061B **M1:** 140
5062B **M1:** 140
5069A **M1:** 140
5070C **M1:** 140
5075B **M1:** 140
5077B **M1:** 140
5080D **M1:** 140
5082A **M1:** 140
5085A **M1:** 140
5110B **M1:** 140
5112E **M1:** 140
5115C **M1:** 140
5120F **M1:** 140
5121C **M1:** 140
5122C **M1:** 140
5132D **M1:** 140
6242C **M1:** 141
6250F **M1:** 141
6260
applications **A4:** 397
gas nitriding **A4:** 397
6260G **M1:** 141
6263D **M1:** 141
6264D **M1:** 141
6265C **M1:** 141
6266C **M1:** 141
6267A **M1:** 141
6270G **M1:** 141
6272E **M1:** 141
6274G **M1:** 141
6275B **M1:** 141
6276C **M1:** 141
6277A **M1:** 141

Steels, AMS, specific types (continued)
6280E **M1:** 141
6281C **M1:** 141
6282D **M1:** 141
6290C **M1:** 141
6292C **M1:** 141
6294C **M1:** 141
6299A **M1:** 141
6300 ..... **M1:** 141
6302, composition **M1:** 649
6302B **M1:** 141
6303, composition **M1:** 649
6303A **M1:** 141
6304, composition **M1:** 649
6304, normalizing and tempering **A4:** 39
6304C **M1:** 141
6312A **M1:** 141
6317B **M1:** 141
6320F **M1:** 141
6321A **M1:** 141
6322F **M1:** 141
6323D **M1:** 141
6324C **M1:** 141
6325D **M1:** 141
6327D **M1:** 141
6328E **M1:** 141
6330A **M1:** 141
6342D **M1:** 141
6350D **M1:** 141
6351A **M1:** 141
6352B **M1:** 141
6354 ..... **M1:** 141
6355G **M1:** 141
6356A **M1:** 141
6357D **M1:** 141
6358B **M1:** 141
6359B **M1:** 141
6360F **M1:** 141
6361 ..... **M1:** 141
6362 ..... **M1:** 141
6365E **M1:** 141
6370F **M1:** 141
6371D **M1:** 141
6372D **M1:** 141
6373A **M1:** 141
6378 ..... **M1:** 142
6379 ..... **M1:** 142
6381B **M1:** 142
6382, gas nitriding **A4:** 392
6382G **M1:** 142
6385, composition **M1:** 649
6385B **M1:** 142
6386A **M1:** 142
6390A **M1:** 142
6395 ..... **M1:** 142
6406A **M1:** 142
6407B **M1:** 142
6411 ..... **M1:** 142
6412, gas nitriding **A4:** 392
6412F **M1:** 142
6413D **M1:** 142
6414A **M1:** 142
6415G **M1:** 142
6416 ..... **M1:** 142
6417 ..... **M1:** 142
6418C **M1:** 142
6419 ..... **M1:** 142
6421A **M1:** 142
6422C **M1:** 142
6423A **M1:** 142
6426A **M1:** 142
6427D **M1:** 142
6428A **M1:** 142
6429A **M1:** 142
6430A **M1:** 142
6431A **M1:** 142
6432 ..... **M1:** 142
6433A **M1:** 142
6434 ..... **M1:** 421-422

Steels, AMS, specific types (continued)
6434A **M1:** 142
6435A **M1:** 142
6436, composition **M1:** 649
6436A **M1:** 142
6437, composition **M1:** 649
6437A **M1:** 142
6438A **M1:** 142
6440E **M1:** 142
6441D **M1:** 142
6442C **M1:** 142
6443B **M1:** 142
6444B **M1:** 142
6445A **M1:** 142
6446 ..... **M1:** 142
6447 ..... **M1:** 142
6448C **M1:** 142
6450C **M1:** 142
6455, springs, strip for **M1:** 285
6455C **M1:** 142
6458, composition **M1:** 649
6470
  applications **A4:** 397
  austenitizing temperature **A4:** 387 **M4:** 192
  composition **A4:** 387 **M4:** 192
  gas nitriding **A4:** 388, 390, 397
  hardness gradients **A4:** 390
  nitriding time **A4:** 388 **M4:** 193
  tempering temperature **A4:** 387 **M4:** 192
6470, composition and heat treatment **M1:** 649
6470F **M1:** 142
6471 ..... **M1:** 142
6472 ..... **M1:** 142
6475
  austenitizing temperature **M4:** 192
  composition **M4:** 192
  hardness gradient **M4:** 196
  nitriding time **M4:** 193
  tempering temperature **M4:** 192
6475 (Nitralloy N)
  applications **A4:** 395
  austenitizing temperature **A4:** 387
  composition **A4:** 387
  gas nitriding **A4:** 388, 390, 395
  hardness gradients **A4:** 388, 390
  liquid nitriding **A4:** 413
  nitriding time **A4:** 388
  tempering temperature **A4:** 387
6475, composition and heat treatment **M1:** 649
6475C **M1:** 142
6485, composition **M1:** 649
6485B **M1:** 143
6487, composition and heat treatment **M1:** 649
6487C **M1:** 143
6488 ..... **M1:** 143
6490B **M1:** 143
6512 ..... **M1:** 143
6514 ..... **M1:** 143
6520 ..... **M1:** 143
6521 ..... **M1:** 143
6526 ..... **M1:** 143
6530E **M1:** 143
6535D **M1:** 143
6540A **M1:** 143
6541A **M1:** 143
6542A **M1:** 143
6545A **M1:** 143
6546A **M1:** 143
6550E **M1:** 143

Steels, and sheet molding compounds
  costs compared **EM2:** 334

Steels, ASME specific types *See also* Steels, ASTM specific types
  SA106, grades A and B, composition and mechanical properties **M1:** 648
  SA182, grade F22, mechanical properties **M1:** 654
  SA199, grade T22, mechanical properties **M1:** 654

Steels, ASME specific types (continued)
  SA204, grade A, composition and mechanical properties **M1:** 648
  SA213, grade T22, mechanical properties **M1:** 654
  SA217
    grade C12, composition and mechanical properties **M1:** 648
    grade WC1, mechanical properties **M1:** 654
    grade WC6, composition and mechanical properties **M1:** 648
  SA285, grade A, composition and mechanical properties **M1:** 648
  SA299, composition and mechanical properties **M1:** 648
  SA302, grade A, composition and mechanical properties **M1:** 648
  SA333, grade P22, mechanical properties **M1:** 654
  SA335, grade P12, composition and mechanical properties **M1:** 648
  SA336
    class P22, mechanical properties **M1:** 654
    class P22a, mechanical properties **M1:** 654
  SA369, grade FP22, mechanical properties **M1:** 654
  SA387
    grade 5, class 2, mechanical properties **M1:** 648
    grade 5, composition **M1:** 648
    grade 22, class 1, composition and mechanical properties **M1:** 648, 654
    grade 22, class 2, composition and mechanical properties **M1:** 648, 654
  SA426, grade CP22, mechanical properties **M1:** 654
  SA517, grade F, composition and mechanical properties **M1:** 648
  SA533, type B, class 2, composition and mechanical properties **M1:** 648
  SA542
    class 1, mechanical properties **M1:** 654
    class 2, mechanical properties **M1:** 654

Steels, ASTM specific types *See also* Nonferrous alloys, ASTM specific types; Steels, ASME specific types
  A 553, grade II, composition **M3:** 754
  A7, notch toughness **M1:** 417
  A27, specification requirements **M1:** 378
  A36
    composition **M1:** 136, 185
    seawater corrosion **M1:** 745
    structural bars and shapes, composition and properties **M1:** 207
    tensile properties **M1:** 190
  A36, laser beam welding **M6:** 662
  A53 ..... **M1:** 317
    composition **M1:** 318
    tensile properties **M1:** 320
  A106 ..... **M1:** 317
    composition **M1:** 318
    tensile properties **M1:** 320
  A109
    characteristics **M1:** 154
    composition **M1:** 135
    mechanical properties **M1:** 155
    standard sizes **M1:** 154
  A113
    composition **M1:** 185
    properties **M1:** 207
    tensile properties **M1:** 190
  A120 ..... **M1:** 317
  A131
    composition **M1:** 185, 188
    structural bars and shapes, composition and properties **M1:** 207, 210, 211
    tensile properties **M1:** 190
  A134 ..... **M1:** 317

Steels, ASTM specific types (continued)

A135 .... **M1:** 317
composition **M1:** 318
tensile properties **M1:** 320
A139 .... **M1:** 317
composition **M1:** 318
tensile properties **M1:** 320
A148, specification requirements **M1:** 378
A155 .... **M1:** 317
A161
composition **M1:** 323, 324
tensile properties **M1:** 325
A178
composition **M1:** 323
tensile properties **M1:** 325
A179, composition **M1:** 323
A192
composition **M1:** 323
tensile properties **M1:** 325
A199 .... **M1:** 323
composition **M1:** 324
tensile properties **M1:** 325
A200 .... **M1:** 323
composition **M1:** 324
tensile properties **M1:** 325
A202
composition **M1:** 139, 186
tensile properties **M1:** 192
A203, composition **M1:** 139, 186
A203, grade E, fatigue-crack-growth rate **M3:** 755
A203, grades A thru D, compositions **M3:** 754
A204
composition **M1:** 139, 186
tensile properties **M1:** 192
A209 .... **M1:** 323
composition **M1:** 324
tensile properties **M1:** 325
A210
composition **M1:** 323
tensile properties **M1:** 325
A211 .... **M1:** 317
A213 .... **M1:** 323
composition **M1:** 324
tensile properties **M1:** 325
A214, composition **M1:** 323
A216
grade WCB, machinability compared to ductile iron **M1:** 56
grades WCB and WC1, service life in steam systems **M1:** 747
A217 grades WC6 and WC9, service life in steam systems **M1:** 747
A225
composition **M1:** 139, 185
tensile properties **M1:** 192
A226
composition **M1:** 323
tensile properties **M1:** 325
A227 *See also* Hard drawn spring wire
springs, helical **M1:** 294, 298
springs, wire for **M1:** 284
A228 *See also* Music wire
springs, helical **M1:** 292-295, 298, 299, 300
springs, wire for **M1:** 284, 287, 296
A229 *See also* Oil-tempered wire
springs, helical **M1:** 292, 295, 297, 298
springs, wire for **M1:** 284, 287, 297
A230 *See also* Valve-spring wire
springs, helical **M1:** 300
springs, wire for **M1:** 284, 287, 289
A231 *See also* Chromium-vanadium spring steel
springs, helical **M1:** 298, 299
springs, wire for **M1:** 284
A232 *See also* Chromium-vanadium steel spring wire
springs, helical **M1:** 300
springs, wire for **M1:** 284, 287, 289
A242
atmospheric corrosion **M1:** 722, 723
composition **M1:** 136, 188, 407
description **M1:** 403, 405, 409
mechanical properties **M1:** 406
seawater corrosion **M1:** 740, 742, 745
structural bars and shapes, composition and properties **M1:** 210, 211

Steels, ASTM specific types (continued)

tensile properties **M1:** 191
A250 .... **M1:** 323
composition **M1:** 324
tensile properties **M1:** 325
A252 .... **M1:** 317
composition **M1:** 318
tensile properties **M1:** 320
A254 .... **M1:** 321, 323
composition **M1:** 323
tensile properties **M1:** 325
A283
composition **M1:** 136, 185
tensile properties **M1:** 190
A284
composition **M1:** 136, 185
tensile properties **M1:** 190
A285
composition **M1:** 138, 185, 195, 196
tensile properties **M1:** 192
A288, composition **M1:** 138
A299
composition **M1:** 185
tensile properties **M1:** 192
A302
composition **M1:** 139, 187
tensile properties **M1:** 192
A313 *See also* Stainless steels specific types, 302
springs, helical **M1:** 294, 295
springs, wire for **M1:** 284
A333 .... **M1:** 317
composition **M1:** 318, 322
tensile properties **M1:** 320
A334
composition **M1:** 323, 324
tensile properties **M1:** 325
A335 .... **M1:** 317
composition **M1:** 324
tensile properties **M1:** 325
A336
composition **M1:** 135
standard sizes **M1:** 154
A352, Charpy impact properties **M1:** 381, 389, 392, 398
A353
composition **M1:** 139, 187 **M3:** 754
tensile properties **M1:** 192
tensile properties at subzero temperatures **M3:** 754
A381 .... **M1:** 317
composition **M1:** 318
tensile properties **M1:** 320
A387
composition **M1:** 139, 187
grade 22, seawater corrosion **M1:** 745
tensile properties **M1:** 192, 193
A387, laser cladding **M6:** 797-798
A401 *See also* Chromium-silicon spring steel
springs **M1:** 284, 289, 294, 297-299
A405 .... **M1:** 317
composition **M1:** 324
tensile properties **M1:** 325
A414, composition **M1:** 135
A423 .... **M1:** 323
composition **M1:** 324
tensile properties **M1:** 325
A434, composition and properties **M1:** 209
A440
composition and properties **M1:** 136, 185, 207, 405, 407
fatigue life vs. total strain **M1:** 672
mechanical properties **M1:** 406
tensile properties **M1:** 190
A441, composition and properties **M1:** 136, 188, 191, 210, 211, 405-407
A442, composition and properties **M1:** 138, 185, 192
A455, composition and properties **M1:** 138, 185, 192, 209
A487, specification requirements **M1:** 379
A500, composition and properties **M1:** 325, 326
A501, composition and properties **M1:** 325, 326
A512 .... **M1:** 325
A513 .... **M1:** 325
A514
atmospheric corrosion **M1:** 722, 723

Steels, ASTM specific types (continued)

composition and properties **M1:** 136-137, 186, 190
A515, composition and properties **M1:** 138, 185, 192, 195
A516, composition and properties **M1:** 138, 185, 192, 195
A517, composition and properties **M1:** 139, 187, 193
A519 .... **M1:** 325
A523, composition and properties **M1:** 317, 318, 320
A524, composition and properties **M1:** 317, 318, 320
A529, composition and properties **M1:** 136, 185, 190, 207
A533, composition and properties **M1:** 139, 187, 193, 209
A537, composition and properties **M1:** 138, 185, 192
A538, composition and properties **M1:** 140, 189, 193
A539, composition and properties **M1:** 323, 325
A542, composition and properties **M1:** 140, 187, 193
A543, composition and properties **M1:** 140, 187, 193
A553, composition and properties **M1:** 140, 187, 193
A553, grade I
composition **M3:** 754
fatigue-crack-growth rate **M3:** 755
tensile properties at subzero temperatures **M3:** 754
A556, composition and properties **M1:** 323, 325
A557, composition and properties **M1:** 323, 325
A562, composition and properties **M1:** 140, 185, 192
A569
characteristics **M1:** 154
composition **M1:** 135
standard sizes **M1:** 154
A570
characteristics **M1:** 154
composition **M1:** 135
mechanical properties **M1:** 155
standard sizes **M1:** 154
A572
composition **M1:** 137, 188, 407
composition and properties, structural bars and shapes **M1:** 210, 211
description **M1:** 405
mechanical properties **M1:** 406
tensile properties **M1:** 191
A573
composition **M1:** 136, 185
tensile properties **M1:** 190
A587 .... **M1:** 317
composition **M1:** 318
tensile properties **M1:** 320
A588
atmospheric corrosion **M1:** 722, 723
composition **M1:** 137-138, 188, 407
composition and properties, structural bars and shapes **M1:** 210, 211
description **M1:** 405
mechanical properties **M1:** 406
tensile properties **M1:** 191
A589 .... **M1:** 317
composition **M1:** 318
tensile properties **M1:** 320
A590
composition **M1:** 140, 189
tensile properties **M1:** 193
A595 .... **M1:** 325
composition **M1:** 326
tensile properties **M1:** 326
A605
composition **M1:** 140, 189
tensile properties **M1:** 193
A606
composition **M1:** 135, 407
description **M1:** 405, 409
mechanical properties **M1:** 406
A607
composition **M1:** 135, 407
description **M1:** 405, 409

Steels, ASTM specific types (continued)
mechanical properties ... **M1:** 406
A611
characteristics ... **M1:** 154
composition ... **M1:** 135
mechanical properties ... **M1:** 155
standard sizes ... **M1:** 154
A612
composition ... **M1:** 138, 185
properties, concrete-reinforcing bars ... **M1:** 212
tensile properties ... **M1:** 192
A616 properties, concrete-reinforcing bars ... **M1:** 212
A617 properties, concrete-reinforcing bars ... **M1:** 212
A618 ... **M1:** 325
composition ... **M1:** 326, 407
description ... **M1:** 405
mechanical properties ... **M1:** 406
tensile properties ... **M1:** 326
A619
characteristics ... **M1:** 154
composition ... **M1:** 135
standard sizes ... **M1:** 154
A620
characteristics ... **M1:** 154
composition ... **M1:** 135
standard sizes ... **M1:** 154
A621
characteristics ... **M1:** 154
composition ... **M1:** 135
standard sizes ... **M1:** 154
A622
characteristics ... **M1:** 154
composition ... **M1:** 135
standard sizes ... **M1:** 154
A633
composition ... **M1:** 138, 189, 407
composition and properties, structural bars and shapes ... **M1:** 210, 211
description ... **M1:** 405
mechanical properties ... **M1:** 406
tensile properties ... **M1:** 191
A635
characteristics ... **M1:** 154
standard sizes ... **M1:** 154
A645
composition ... **M1:** 140, 187 **M3:** 754
fatigue-crack-growth rate ... **M3:** 755
tensile properties ... **M1:** 193
tensile properties at subzero temperatures ... **M3:** 754
A656
composition ... **M1:** 138, 189, 407
description ... **M1:** 405
mechanical properties ... **M1:** 406
tensile properties ... **M1:** 191
A658
composition ... **M1:** 187
tensile properties ... **M1:** 193
A659
characteristics ... **M1:** 154
standard sizes ... **M1:** 154
A662
composition ... **M1:** 138, 185
tensile properties ... **M1:** 192
A671 ... **M1:** 317
A672 ... **M1:** 317
A678
composition ... **M1:** 136, 185
tensile properties ... **M1:** 190
A679 *See also* High-tensile hard drawn wire springs, wire for ... **M1:** 284
A682 springs, strip for ... **M1:** 285
A690
composition ... **M1:** 407
composition and properties, structural bars and shapes ... **M1:** 210-211
description ... **M1:** 406

Steels, ASTM specific types (continued)
mechanical properties ... **M1:** 406
A691 ... **M1:** 317
A699
composition ... **M1:** 138, 186
properties, concrete-reinforcing bars ... **M1:** 212
tensile properties ... **M1:** 190
A709
composition ... **M1:** 185, 186, 189
composition and properties, structural bars and shapes ... **M1:** 207, 210, 211
tensile properties ... **M1:** 190, 191
A710
composition ... **M1:** 138, 186
composition and properties, structural bars and shapes ... **M1:** 209
tensile properties ... **M1:** 191
A714 ... **M1:** 317
composition ... **M1:** 322
tensile properties ... **M1:** 321
A715
bend test radii ... **M1:** 419
composition ... **M1:** 135, 407
description ... **M1:** 406
mechanical properties ... **M1:** 406
A722, properties, concrete-reinforcing bars ... **M1:** 212
A724
composition ... **M1:** 138, 185
tensile properties ... **M1:** 192
A734
composition ... **M1:** 140, 187, 189
tensile properties ... **M1:** 193
A735
composition ... **M1:** 140, 187
tensile properties ... **M1:** 193
A736
composition ... **M1:** 140, 187
tensile properties ... **M1:** 193
A737
composition ... **M1:** 140, 189
tensile properties ... **M1:** 193
A738
composition ... **M1:** 185
tensile properties ... **M1:** 192

Steels, continuous annealing ... **A4:** 50, 51-54, 56-65
advantages ... **A4:** 63-64
antiaging ... **A4:** 56, 57, 58, 59, 61
automotive applications ... **A4:** 59-64
cooling ... **A4:** 57-59, 64
deep-drawing-quality special-killed (DQSK or DDQ) ... **A4:** 57
enameling applications ... **A4:** 65
metallurgical advantages ... **A4:** 56
overaging ... **A4:** 62, 63
process description ... **A4:** 56-59
temperatures ... **A4:** 56-58, 59, 60, 61, 63, 64
tinplate applications ... **A4:** 64-65
versus batch annealing ... **A4:** 56, 63, 65

Steels, DIN, specific types
16MnCr5 steel, shot peening effect on residual stresses ... **A4:** 371
20MnCr5, carburizing ... **A4:** 372
20MnCr5B, boron added to improve toughness ... **A4:** 366
22CrMo44, residual stresses ... **A4:** 606, 607
42CrMo4 steel (AISI-SAE 4140), electron beam hardening treatment ... **A4:** 308
42MnV7 steel (AISI-SAE 1340), electron beam hardening treatment ... **A4:** 308
50CrV4 steel (AISI-SAE 6150), electron beam hardening treatment ... **A4:** 308
55Crl steel, electron beam hardening treatment ... **A4:** 306
90MnV8 steel
austenite grain growth diagrams ... **A4:** 643
electron beam hardening treatment ... **A4:** 304, 307

Steels, DIN, specific types (continued)
St 37, boriding ... **A4:** 440, 444, 445
100Cr6 hypereutectoid steel (AISI-SAE 52100), electron beam hardening treatment ... **A4:** 302, 303, 305
C15 steel (AISI-SAE 1015), electron beam hardening treatment ... **A4:** 308
C45 steel (0.45% C), boriding ... **A4:** 437, 438, 439
C45 steel (AISI-SAE 1045), electron beam hardening treatment ... **A4:** 299, 300, 302-308
C100W1 steel (tool steel WI), electron beam hardening treatment ... **A4:** 306
Ck45, boriding ... **A4:** 445
Ck67 steel (AISI-SAE 1070), electron beam hardening treatment ... **A4:** 304, 308

Steels for case hardening
compositions of ... **A9:** 219

Steels, hardened
chilled cast iron ... **EM4:** 972
oxide-based ceramic inserts for cutting tools ... **EM4:** 967
turning and milling recommended ceramic grade inserts for cutting tools ... **EM4:** 972

Steels, magnet *See* Permanent magnet materials, specific types

Steels, microstructures of ... **A3:** 1•24

Steels, MPIF specific types *See* P/M steels

Steels, SAE experimental
EX 9, composition ... **M1:** 131
EX 10, composition ... **M1:** 131
EX 11, composition ... **M1:** 131
EX 12, composition ... **M1:** 131
EX 13, composition ... **M1:** 131
EX 14, composition ... **M1:** 131
EX 15, composition ... **M1:** 131
EX 16, composition ... **M1:** 131
EX 17, composition ... **M1:** 131
EX 18, composition ... **M1:** 131
EX 19
composition ... **M1:** 131
hardenability of carburized bar ... **M1:** 473
EX 20, composition ... **M1:** 131
EX 21, composition ... **M1:** 131
EX 24, composition ... **M1:** 131
EX 27, composition ... **M1:** 131
EX 29, composition ... **M1:** 131
EX 30, composition ... **M1:** 131
EX 31, composition ... **M1:** 131
EX 32, composition ... **M1:** 131
EX 33, composition ... **M1:** 131
EX 34, composition ... **M1:** 131
EX 35, composition ... **M1:** 131
EX 36, composition ... **M1:** 131
EX 37, composition ... **M1:** 131
EX 38, composition ... **M1:** 132
EX 39, composition ... **M1:** 132
EX 40, composition ... **M1:** 132
EX 41, composition ... **M1:** 132
EX 42, composition ... **M1:** 132
EX 43, composition ... **M1:** 132
EX 44, composition ... **M1:** 132
EX 45, composition ... **M1:** 132
EX 46, composition ... **M1:** 132
EX 47, composition ... **M1:** 132
EX 48, composition ... **M1:** 132
EX 49, composition ... **M1:** 132
EX 50, composition ... **M1:** 132
EX 51, composition ... **M1:** 132
EX 52, composition ... **M1:** 132
EX 53, composition ... **M1:** 132
EX 54, composition ... **M1:** 132
EX 55, composition ... **M1:** 132
EX 56, composition ... **M1:** 132
EX1, composition ... **M1:** 131

Steels, silicon
photochemical machining etchant ... **A16:** 590

Steels, specific types *See also* ASP steels; Carbon steels; Low- carbon steel powders; Low-alloy steels; Steels; Tool steels; Tool steels, specific types
0.25% C, indigenous sulfide inclusions **A15:** 89
3% Ni, gaseous hydrogen damage **A13:** 164
15-7PH Mo, and polybenzimidazoles **EM3:** 171
17-7PH, and polybenzimidazoles **EM3:** 170
1010, coining of **A14:** 185-186
1015 tube, wall reduction and mechanical properties **A14:** 678
1018 steel bars, magabsorption measurement **A17:** 155
1025, flow stresses **A14:** 241
1030, drive gears **A7:** 670
1040, fracture surface **A13:** 1053
1040 hot rolled bar, mechanical properties **A14:** 172
1045 crane hook, magnetic particle inspection **A17:** 107
1045, fracture locus **A14:** 392
1050, welded, acoustic emission inspection **A17:** 289
1130, coining of **A14:** 186
4118 steel shaft, ultrasonic inspection **A17:** 508-509
4130, fretting fatigue **A13:** 140
4130, tempering temperature effects on sulfide fracture toughness **A13:** 535
4135, yield strength effect, critical stress and sulfide fracture toughness **A13:** 534
4137, hydrogen-assisted SCC failure **A13:** 332
4137H, hardenability **A7:** 451, 452
4140, fibrous structure (flow lines) **A14:** 218
4140, forged, flow lines in **A14:** 367
4140, indium embrittlement **A13:** 186
4140, solid-metal embrittlement **A13:** 186
4330 aircraft bracket, magnetic rubber inspection **A17:** 124
4340, aircraft powerplant failure **A13:** 1041
4340 bars, magabsorption measurement **A17:** 155
4340 cadmium-plated, corrosion products **A13:** 1050
4340 cadmium-plated, solid-metal embrittlement **A13:** 186
4340, coercive force and hardness **A17:** 134
4340, corrosion fatigue **A13:** 1036
4340, delayed failure from hydrogen cathodic charging **A13:** 535, 536
4340 forged shaft, pulse-echo ultrasonic inspection **A17:** 508
4340, forging pressures **A14:** 217
4340, gaseous hydrogen damage **A13:** 164
4340 high-strength, hydrogen embrittlement **A13:** 289
4340, hydrogen embrittlement incubation **A13:** 286
4340, hydrogen embrittlement testing **A13:** 284
4340, pitting corrosion/fatigue failure **A13:** 1026
4340 plate, effect of yield strength on crack growth **A13:** 169
4340, stress-corrosion cracking **A13:** 1030
4340 ultrahigh-strength, corrosion fatigue **A13:** 143
4600, effect of admixed lubricant on green strength **A7:** 303
4600, prealloyed, compactibility **A7:** 683
4620, axial fatigue for forging deformation **A7:** 416
4620, Izod impact testing **A7:** 410
4640, isothermal transformation diagram for **A7:** 686
4640, response to heat treatment **A7:** 686
4650, deformation of preforms during consolidation **A7:** 538
4650, properties, Ceracon process **A7:** 540
4680, shrinkage during sintering **A7:** 480
5046 crankshaft, magabsorption measurement **A17:** 155
7075-T6 landing gear trunnion, ultrasonic inspection **A17:** 509-510
8620 hot rolled bar, mechanical properties **A14:** 173
8822 billet, ultrasonic inspection **A17:** 507-508

Steels, specific types (continued)
A514
microcrystalline conversion coatings **EM3:** 273
phosphoric-acid-etched **EM3:** 271, 272
A516, inclusion shape modification in **A15:** 91
A536, ductile iron, mechanical properties **A15:** 665
A606
microcrystalline conversion coatings **EM3:** 273
phosphoric-acid-etched **EM3:** 271, 272
A633C, calcium treatment as inclusion control **A15:** 92
CA-15, mechanical properties **A15:** 727
CA-40, mechanical properties **A15:** 727
CB-30, mechanical properties **A15:** 727
CF-8 high-alloy, elevated temperature effects **A15:** 728
CF-8 high-alloy, ferrite effects **A15:** 725
D-6ac aircraft part, magnetic rubber inspection **A17:** 124
D6-AC, hydrogen embrittlement testing **A13:** 286
HK-40, creep rupture properties **A15:** 730
Hoeganaes 1000B, composition-depth profiles **A7:** 266
HP-50WZ, creep and stress rupture properties **A15:** 729-730
SAE 1010 cold-rolled
compared to FPL aluminum substrates **EM3:** 271
high-performance acrylic adhesive bonding when oiled **EM3:** 122
X38CrMoV51, hydrogen/nitrogen removal, VIM processed **A15:** 395-296
Steels, spring
photochemical machining etchant **A16:** 590
Steels, ultrahigh strength *See* Ultrahigh-strength steels
300M, machining **A16:** 144-147, 320
D6ac, machining **A16:** 27, 33, 34, 144-147, 567
Steels, wrought abrasive wear data **A18:** 706
hardness **A18:** 706
nominal compositions and applications **A18:** 703
toughness **A18:** 706
Steering gears, automotive
economy in manufacture **M3:** 850
Steering knuckle
fracture surface **A11:** 342
Steering potentiometer **A13:** 1121
Steering potentiometers
corrosion failure analysis **EL1:** 1111
Stefan-Boltzmann constant **A18:** 441 **EM4:** 615
Stefan-Boltzmann law **A18:** 441 **EM4:** 251
Steiner Tunnel Test
for burning rate/smoke generation **EM1:** 96
Stelco coil box **A1:** 118, 120
Stellite
applications, pump components **A18:** 598
broaching **A16:** 203
diagnostic condition monitoring by radionuclide methods **A18:** 322
electrical discharge machining **A16:** 560
electrochemical grinding **A16:** 545, 546
erosion **A18:** 200, 206
erosion in ceramics **A18:** 205
erosion resistance **A18:** 228
friction coefficient data **A18:** 71
furnace brazing **A6:** 920-921
galling resistance with various material combinations **A18:** 596
hardfacing **A6:** 807
honing **A16:** 476
laser cladding **A18:** 867, 868
laser cladding components and techniques **A18:** 869
liquid impingement erosion and hardness **A18:** 228
seal adhesive wear **A18:** 549
seal material **A18:** 551
Stellite 3
erosion test results **A18:** 200
properties, sand cast **A18:** 548
Stellite 6
composition **A18:** 806
friction surfacing **A6:** 322, 323
laser cladding **A18:** 867, 868

Stellite 6 (continued)
laser cladding components and techniques **A18:** 869
Stellite 6B
adhesive wear resistance **A18:** 721
as coating to protect against liquid impingement erosion **A18:** 221, 222
composition **A6:** 564 **M6:** 354
erosion mechanisms and embedding of particle fragments **A18:** 203
erosion test results **A18:** 200
explosion welding **M6:** 710
fatigue wear **A18:** 715
galling test use **A18:** 718
liquid impingement erosion **A18:** 225
relative erosion factor **A18:** 200, 201
replaced by stainless steels **A18:** 723
temperature effect on adhesive wear **A18:** 721
Stellite 6K
erosion test results **A18:** 200
Stellite 21 (stick)
cavitation erosion rate **A18:** 774
Stellite 31
broaching **A16:** 209
Stellite alloy 31
composition **A4:** 795
Stellite alloys *See* Cobalt alloys, specific types; Haynes Stellite; Superalloys, cobalt-base, specific types
effects of eutectic melting in **A11:** 122
friction surfacing **A6:** 321, 322, 323
liquid-erosion resistance **A11:** 167
as liquid-erosion shield **A11:** 170-171
Osprey process for **A7:** 530
Stellite SF **A18:** 6
laser cladding **A18:** 867
laser cladding components and techniques **A18:** 869
Stellite SF6
hardfacing **A6:** 807
STEM *See* Scanning transmission electron microscopy; Shaped tube electrolytic machining
Stem radiation, x-ray tubes
radiography **A17:** 305
Stencil printing
as adhesive application method **EL1:** 672
application, solder pastes **EL1:** 653
equipment **EL1:** 732
for surface-mount components **EL1:** 732
in surface-mount soldering **EL1:** 699
Stencil printing dispensing technique
general solder paste parameters **A6:** 988
Step aging
defined **A9:** 17
Step bearing *See also* Rayleigh step bearing; Stepped bearing
defined **A18:** 18
Step brazing
definition **M6:** 17
filler metals for resistance brazing **M6:** 981
production application **M6:** 981-982
use in resistance brazing **M6:** 984
Step cooling
to evaluate temper embrittlement **M1:** 684-685
Step excitation test
methods **EM2:** 549-551
Step fracture
definition **EM4:** 633
Step shear rate
rheology **EL1:** 840, 845-846
Step soldering
definition **M6:** 17
Step strain
rheology **EL1:** 840, 846
Step wedge penetrameters
radiographic inspection **A17:** 341
Step-down creep test **A8:** 324
Step-down tension testing **A8:** 324
Step-loading curve
for modulus of elasticity **A8:** 315
Step-scan **A7:** 258
Step-stress analysis
screening stress levels by **EL1:** 879-880
Stepdown test
defined **A9:** 17
STEPLAP computer procedure **A6:** 1041, 1042-1043, 1044

Stepped bearing *See also* Step bearing
defined **A18:** 18
Stepped compact **A7:** 12
Stepped pockets **A18:** 551
Stepped shafts
cold forming **A14:** 305-306
Stepped-lap joints
design details **EM3:** 493
maximum principal stresses **EM3:** 493
spacing distance **EM3:** 490, 491, 492, 494
Stepped-up joints
normal stress **EM3:** 487, 491, 492, 493
Stepper motors **A13:** 1121
electronic/electrical corrosion failure analysis of **EL1:** 1111
Steps
in ceramics **A11:** 745
from rough machining **A11:** 122
multiple, on fracture surface **A11:** 398
in shafts **A11:** 467
shear, as fatigue indicators **A11:** 258, 259
Stepwise cracking *See* Hydrogen-induced cracking (HIC)
from hydrogen damage **A13:** 332
Steradian, as SI supplementary unit
symbol for **A10:** 685
Stereo angle
defined **A9:** 17
Stereo imaging *See also* Imaging
fractographic **A12:** 87-88
parallax determination **A12:** 197
SEM display systems **A12:** 171
Stereo microscopy
defined **EL1:** 1067
Stereo viewing
TEM and SEM fractographs **A12:** 192
Stereo vision *See also* Three-dimensional images
in machine vision process **A17:** 34
Stereo zoom optical microscope
fractographic **EM2:** 805
Stereo-pair photographs **A12:** 87-89
Stereo-pair viewer
effects **A12:** 87
Stereochemistry, and conformation
molecular .... **A10:** log
Stereogrammetry *See* Stereophotogrammetry
Stereographic projection
by XRPD **A10:** 333
construction
crystal axes using **A10:** 359
preferred crystallographic orientations in **A10:** 358-359
Stereoimaging **A6:** 145
Stereoisomer *See also* Isomer **EM3:** 27
defined **EM2:** 40
as molecular structure **EM2:** 58
of polypropylene **EM2:** 58
Stereological measurements
use of image analyzers in making **A9:** 83
use of semi-automatic tracing devices in making **A9:** 83
Stereological techniques used for quantitative metallography of cemented carbides **A9:** 275
Stereology *See* Quantitative metallography **A10:** 310, 316
of particle shape **A7:** 237-241
in quantitative fractography **A12:** 193-196
Stereometry
for three-dimensional spatial analysis **A12:** 194
Stereomicroscopes *See also* Microscopes **A12:** 78-79
Stereomicroscopes used in macrophotography **A9:** 86-88
Stereomicroscopy
fractographs by **A11:** 56
scanning electron microscopy **A9:** 97
Stereophotogrammetry *See also* Photogrammetry
carpet plot, titanium alloy **A12:** 198
Stereophotogrammetry (continued)
contour map and profiles by **A12:** 197
SEM **A12:** 171
Stereoscan
development of **A12:** 8
Stereoscopes
Hilger-Watts **A12:** 171
Stereoscopic imaging
fracture examination **A12:** 92
methods **A12:** 196-198
in quantitative fractography **A12:** 196-198
Stereoscopic micrographs
defined **A9:** 17
Stereoscopic specimen holder
defined **A9:** 17
Stereospecific plastics *See also* Plastics **EM3:** 27
defined **EM2:** 40
Stereotype metal **A9:** 424
Stereoviewing
in fracture studies by scanning electron microscopy **A9:** 99
Sterling, Lord
as early founder **A15:** 26
Sterling silver *See also* Silver; Silver-copper alloys; Sliver-copper alloys
applications and properties **A2:** 691
properties **A2:** 700
rhodium plating of **M5:** 290-291
Sterling silver, coining of *See also* Silver **A14:** 184
Stern-tube bearing
defined **A18:** 18
Stevenson clamp design
for torsional Kolsky bar **A8:** 221
Stibine
as antimony toxic gas **A2:** 1259
"Stick droopers" **A6:** 37
Stick electrode
definition **A6:** 1214
Stick electrode welding
definition **A6:** 1214
Stick welding *See* Shielded metal arc welding
Stick-shift, automobile transmission
failure of **A11:** 763
Stick-slip *See also* Spragging **A18:** 47, 48, 56, 57
defined **A18:** 18
Sticker
as casting defect **A11:** 381
Sticking
in squeeze casting **A15:** 326
Sticking, platinum group metals
as electrical contact materials **A2:** 846
Stickout
definition **A6:** 1214
Stiction **A18:** 576
defined **A18:** 18
Stiff adherend test **EM3:** 443, 444
Stiff pastes
forming structural ceramics from **A2:** 1020
Stiffness *See also* Effective stiffness; Strength; Stress-strain **EM3:** 27
of acetals **EM2:** 100
composite, and fiber direction **EM1:** 179
defined **A8:** 12 **EM1:** 22 **EM2:** 40
degradation, measured **EM1:** 775
environmental effects, elasticity **EM1:** 188
and failure analysis **EM1:** 775
of fiber-reinforced polymers **EM1:** 36
of fibers **EM1:** 29
in heat-exchanger tubing **A11:** 628, 640
lamina **EM1:** 219-220
loading frame **A8:** 50
machine, experimental determination of **A8:** 44
moisture/temperature effects **EM1:** 226
of multidirectional tape prepregs **EM1:** 146
nomenclature for hydrostatic bearings with orifice or capillary restrictor **A18:** 92
polymer, as property **EM2:** 61
Stiffness (continued)
prediction **EM1:** 432
resin, and damage tolerance **EM1:** 262
of selected structural metals **A2:** 478
of SiC fibers **EM1:** 58
in stress-corrosion cracking testing **A8:** 502
of testing machine, effects on strain rate **A8:** 41-42, 45
testing machine, experimental values of **A8:** 42, 43
torsional, improved **EM1:** 36
-weight ratios, composites and metals compared **EM1:** 178
Stiffness reduction (fatigue) model **EM1:** 246-247
Stigmators
as SEM lens **A12:** 167
Still tank cadmium plating systems **M5:** 256-259
Still tank nickel plating *See* Nickel plating, still tank process
Stippled area
definition **EM4:** 633
Stir casting *See* Gircast process; Rheocasting; Semi-solid metal casting and forging
Stirring
argon **A15:** 368
effect in electrogravimetry **A10:** 197, 200
electromagnetic **A15:** 369
ladle, inductive/inert gas injection **A15:** 432
in LF/V D-VAD **A15:** 437
microstructural effects **A15:** 329
mode, VIM process **A15:** 398
as semisolid variable **A15:** 327-331
in stress-corrosion cracking **A13:** 147
Stitch bonding *See* Thermocompression welding
Stitching **EM1:** 171, 529
Stitching, as flaw
defined **A17:** 562
STN International computer network system **EM4:** 692
Stock *See also* Forging stock; Stock preparation; Stock reels
defined **A14:** 12
feed mechanism, multiple-slide forming **A14:** 567
forging, titanium alloy **A14:** 277-278
slit, re-coiling of **A14:** 710
for slugs **A14:** 309
straighteners, multiple-slide machines **A14:** 567
thick, piercing of **A14:** 467
thickness, electrical steel sheet **A14:** 479-480
thin, confined explosive forming of **A14:** 636
thin, piercing of **A14:** 467-468
Stock piping
corrosion of **A13:** 1186
Stock preparation
aluminum alloys **A14:** 247
copper and copper alloy **A14:** 256
heat-resistant alloys **A14:** 235
titanium alloys **A14:** 278
Stock reels
for coil handling **A14:** 501
contour roll forming **A14:** 627
Stock removal
and gear manufacture **A16:** 335, 350
and gear shaving **A16:** 341-343
gear-tooth honing **A16:** 487
honing **A16:** 482, 483, 484, 485
lapping **A16:** 495, 496, 499, 503, 505
microhoning vs. honing **A16:** 489
Stockholm convention
of electrode potentials **A13:** 21-22
Stoichiometric tungsten carbide **A7:** 156-157
Stoichiometry **A6:** 146
and adhesion **A6:** 143, 144, 145
in basic chemical equilibria and analytical chemistry **A10:** 162
Stoke (centistoke)
defined **A18:** 18

Stoke's law
for particle flotation **A15:** 79
Stokes radiation *See* Stokes scattering
Stokes Raman line
defined **A10:** 682
Stokes scattering
energy-level diagram **A10:** 127
in Raman spectroscopy **A10:** 126-128
Stokes' law **EM4:** 66, 67, 68
Stokes' lines **EM4:** 561
Stoking **A7:** 12
Stomatitis
as gold toxic effect **A2:** 1257
Stone
carbides for machining **A16:** 75
ground by diamond wheels **A16:** 455
PCD tooling **A16:** 110
Stone molds
Bronze Age **A15:** 15-16
Stoneware
definition **EM4:** 3
subclassifications **EM4:** 3
Stoody Deloro HG1
cavitation erosion rate **A18:** 774
Stop
defined **A14:** 12
Stop blocks
and compression fixture, drop tower **A8:** 197
with drop tower compression system **A8:** 196
Stop cock
use in electrogravimetry **A10:** 200
Stop rods
for rotary swaging **A14:** 134
Stop-block guide
brittle fracture of **A11:** 526-527
Stop-off coating
lead wires in acidified chloride solutions **A8:** 420
Stopoff
definition **A6:** 1214 **M6:** 17
furnace brazing of steels **M6:** 938-939
Stopper rod
defined **A15:** 11
Stopping distance **A18:** 583
PIXE analysis **A10:** 103
Stopping off
defined **A15:** 11
Stopping-off *See also* Masking
cadmium plating **M5:** 267-268
chromium plating, media for **M5:** 187
decorative chromium plating processes **M5:** 194
shot peening, masks for **M5:** 144-145
Stops
defined **EM1:** 22 **EM2:** 40
Storage
of aluminum/aluminum alloys **A13:** 602-603
automatic **A15:** 570-571
of beryllium **A13:** 810
energy, silicon metal-oxide semiconductor effect **EL1:** 2
failure mechanisms during **EL1:** 961
and mold life **A15:** 281
of patterns **A15:** 196
of rammed graphite molds **A15:** 274
scrap and alloy, electric arc furnace **A15:** 363
of weathering steels **A13:** 518
Storage and loss moduli **EM3:** 318-319, 320
Storage containers for chemicals **A9:** 69
Storage control (SC) **EL1:** 128, 991
Storage environments
stress-corrosion cracking in **A11:** 211-212
Storage life *See also* Shelf life **EM3:** 27, 35
defined **EM1:** 22 **EM2:** 40
Storage modulus **EM3:** 27
for closed-loop cure **EM1:** 761
defined **EM2:** 40
Storage of adhesives and sealants **EM3:** 683-686
aging effects **EM3:** 683
facilities **EM3:** 683-685
ambient- or room-temperature **EM3:** 685
freezer **EM3:** 684
refrigerated **EM3:** 684-685
formulation constituents **EM3:** 683
handling considerations **EM3:** 685-686
inventory control mechanisms **EM3:** 685
Storage of mounted specimens **A9:** 32
Storage reservoirs
liquids from P/M porous parts **A7:** 700
Storage stability
defined **A18:** 18
Storage tank construction
codes governing **M6:** 827
joining processes **M6:** 56
Storage tanks
acoustic emission inspection **A17:** 290
aluminum and aluminum alloys **A2:** 9
fiberglass, acoustic emission inspection **A17:** 291
Storage vessels
anodic protection of **A13:** 465
Stored energy in crystals as a result of cold working **A9:** 684
Stored energy welding
definition **M6:** 17
Stored-energy resistance
welding machines **M6:** 537-540, 545
Stored-torque torsional Kolsky bar **A8:** 219-221
Stovetops
porcelain enameled sheet for **M1:** 179
Straddle milling *See* Milling
Straight bevel gears
described **A11:** 587
Straight bone plate, stainless steel
fatigue initiation **A11:** 579, 684-686
Straight flanging **A14:** 529
Straight Küntscher intramedullary femoral nail
as internal fixation device **A11:** 671
Straight lead
as lead formation **EL1:** 733-734
Straight leaded devices
small-signal plastic bodied **EL1:** 423-424
Straight notch with rounded adherends (SNRA) **EM3:** 444
Straight notch with sharp adherends (SNSA) **EM3:** 444
Straight polarity **A6:** 30, 32
definition **A6:** 1214 **M6:** 17
Straight WC-CO, uncoated
machining applications **A2:** 967
Straight-beam contact-type ultrasonic search units **A17:** 257
Straight-beam edge ultrasonic inspection
of rolled plate **A17:** 269-270
Straight-beam top ultrasonic inspection
of rolled plate **A17:** 268-269
Straight-flute chucking reamers **A16:** 241, 242
Straight-knife shearing
accessory equipment **A14:** 703
accuracy **A14:** 705
applicability **A14:** 701
capacity **A14:** 702
knives **A14:** 703-705
machines **A14:** 701-703
power requirements **A14:** 702-703
ram speed in **A14:** 705
straight shear knives **A14:** 703-705
Straight-line depreciation method **A13:** 371-372
Straight-line polishing and buffing machines **M5:** 121-124
Straight-side presses **A14:** 464, 493
Straight-through attachments
flexible printed boards **EL1:** 590
Straight-through drawing machines **A14:** 334
Straighteners **A14:** 502
parallel-roll **A14:** 685-686
rotary **A14:** 686-687
rotary-arbor **A14:** 687
Straightening
automatic press roll **A14:** 687-688
of bars **A14:** 680-689
by heating **A14:** 681-682
camber **A14:** 680
cold **A11:** 125-126
cold-finished bars **M1:** 224, 235
tolerances **M1:** 219
in continuous casting **A15:** 312
defined **A14:** 12
effect on fatigue strength **A11:** 125
epicyclic **A14:** 689
equipment, contour roll forming **A14:** 628
fracture from **A11:** 88
growth during **A14:** 687
hot-rolled bars and shapes **M1:** 203
of investment castings **A15:** 264
Straightening (continued)
of long parts **A14:** 680-689
manual **A14:** 680-681
material displacement **A14:** 680-681
moving-insert **A14:** 688
for multiple-slide forming **A14:** 567
of nickel-base alloys **A14:** 837
parallel-rail **A14:** 688-689
parallel-roll **A14:** 684-686
in presses **A14:** 682-684
rotary straighteners **A14:** 686-687
rough **A14:** 690
of shapes **A14:** 680-689
speed **A14:** 687
stainless steel, for cold heading **A14:** 687
of titanium alloy forgings **A14:** 281-282
total indicator reading **A14:** 680
tube/pipe rolling **A14:** 631
of tubing **A14:** 690-693
wrought titanium alloys **A2:** 619
Straightening, shaft
acoustic emission inspection **A17:** 289
Straightness
in contour roll forming **A14:** 632-633
tolerances **A14:** 680-689
Strain *See also* Axial strain; Critical strain; Effective strain; Engineering strain; Initial strain; Linear strain; Reduction; Residual strain; Shear strain; Strain aging; Strain distribution; Strain rate; Strain rate distribution; Stress; Stress-strain; Transverse strain; True strain **EM3:** 27
in 300 series stainless steels, effect on the formation of martensite **A9:** 66
accumulation, in load-controlled continuous-cycling tests **A8:** 353, 355
-age embrittlement **A13:** 12
aging, defined **A13:** 12
along acoustic wave train **A8:** 244
amplitude, defined **EM2:** 40
amplitude, effect on material fatigue resistance **A8:** 712-713
antinode, in ultrasonic test specimen, cooling of **A8:** 247
asymmetrical, effect on dimple size **A12:** 12, 16
at failure, PAN precursor carbon fibers **EM1:** 52
autographic recording of **A8:** 58
axial **EM3:** 27
Barkhausen noise dependence **A17:** 160
in bulk forming processes **A14:** 389
calculation, three-dimensional grid for **A14:** 436-437
calculations for torsional Kolsky bar **A8:** 225-226
causes of nonuniform application **A9:** 686
in composite structure analysis **EM1:** 459-460
compressive, to failure **A8:** 58
as controlling workability **A14:** 363, 367-368
and CTE vs temperature, copper-Invar-copper **EL1:** 622
curvature, in bending **A8:** 118
cycles, in fatigue testing **A11:** 102
cycles, niobium-titanium (copper) superconducting materials **A2:** 1053-1054
cycling, creep relaxation during hold periods **A8:** 347
defined **A8:** 12 **A11:** 10 **A13:** 12 **A14:** 12 **EM1:** 22 **EM2:** 40, 433
definition **EM4:** 633
in deformation process **A8:** 578-579
depth profiles **A18:** 468, 469
and displacement, in ultrasonic testing **A8:** 242
distribution, by simple-beam theory **A8:** 118-119
distribution, elastic plastic bending **A8:** 119, 121
drum camera for recording **A8:** 217
effect on deformation cell structure **A9:** 693
elastic, effect on fatigue **M1:** 668, 670, 672
elastic-plastic, as *J*-integral parameter **A8:** 261
energy **A11:** 49-50, 797
engineering bending **A8:** 118-119
for engineering stress-strain curve **A8:** 20
fields, defined **EM1:** 186
fixed critical **A8:** 449
in flows and material functions **EL1:** 838
-hardening relief, mechanism of **A12:** 42
increasing to fracture, malleable iron **A12:** 238
indicator, high-quality **A17:** 450

Strain (continued)
-induced martensitic transformation and SCC **A12:** 26
initial **EM3:** 27
laminar **EM1:** 218
levels, maximum, forming limit diagrams for **A8:** 555
limiting, determined by Marciniak biaxial stretching test **A8:** 558
local, in low-cycle torsional fatigue **A8:** 150-151
localization, zone of **A12:** 42
localized **A14:** 389-390
maximum levels **A14:** 880
measurement, of single fibers **EM1:** 732
measurements, constraining substrate **EL1:** 612
nondimensional profile **A8:** 209
nonlinear functions **EL1:** 848-849
nonuniform, effect in axial compression testing **A8:** 55-58
normalized, in a bar **A8:** 209
optical holographic measurement **A17:** 405
path, and fracture limit line, compared **A8:** 581
peak, effect of hold periods on fatigue life of René 95 ..... **A8:** 352
plastic, determined, by speckle metrology **A17:** 435
plastic, effect on fatigue **M1:** 665, 668, 670-672
pure plastic bending **A8:** 120, 122
range **A11:** 103
rate, effect in liquid-metal embrittlement **A11:** 230
rate, effect on failure mode **EM2:** 684-687
representation in sheet metal forming **A8:** 549
representation of **A14:** 879
residual **EM3:** 27
role in fatigue **M1:** 665, 668, 670-672
-sensing methods **A17:** 51
sensitivity, of A15 compounds **A2:** 1061
shear **EM3:** 27
shear, calculated **A17:** 449
shear stress and mean axial stress dependence on, fixed end high temperature torsion tests **A8:** 157
softening, compressive flow stress curve with **A8:** 583
state **A14:** 389, 879
stress to produce in creep testing **A8:** 304
symbol for **A8:** 726
-time equation, for creep analysis **A8:** 686-687
to failure, and strain rate, cross-linked rubber **A8:** 43
to fracture, stress state influence **A14:** 369
to fracture, stress state influence on **A8:** 576
total, at aluminum alloy forging temperatures **A14:** 241
transducer, linear variable-differential transformer **A8:** 313
transverse **EM3:** 27
true **EM3:** 27
true bending **A8:** 118-119
true compressive hoop **A8:** 564
true local necking **A8:** 24-25
true strain at fracture, selected steels **M1:** 680
for two-dimensional flow, computation of **A14:** 433-434
ultimate, and expanding ring test **A8:** 210
values, relation to stress values **A17:** 51
vs. deflection **A8:** 58
vs. interlaminar fracture toughness thermoplastics **EM1:** 100
waveform **A8:** 347-348
and workability **A8:** 572
and workability, bulk forming processes **A14:** 388-389
zero **A11:** 57
Strain aging **A14:** 12, 547
in bcc materials **A8:** 256-257
in carbon steel sheet and strip **A1:** 204-205

Strain aging (continued)
in carbon steels at cold working temperatures **A8:** 179
defined **A8:** 12 **A9:** 17
effect of diffusion **A8:** 35
as embrittlement, examination/interpretation **A12:** 129-130
embrittlement, in steels **A11:** 98
Lüders bands from **A12:** 129, 148
of rimmed steel tube, brittle fracture after **A11:** 648
stretcher-strain formation by **A12:** 129, 148
Strain amplitude **EM3:** 27
Strain analysis of sheet during forming **M1:** 545-546
Strain annealing **A1:** 259, 280
Strain contrast
in aluminum-copper due to precipitates **A9:** 117
described by dynamical diffraction in imperfect crystals **A9:** 112
of second-phase precipitates **A9:** 117
of stair-rod dislocations bounding stacking- fault tetrahedra **A9:** 117
in strong-beam images **A9:** 112
structure-factor contrast **A9:** 113
Strain control cyclic torsional testing **A8:** 149
data, linear model dummy variable approach for **A8:** 712
test, for low-cycle fatigue testing **A8:** 696
Strain distribution
as ALPID result **A14:** 427
in cube **A14:** 437
homogeneous, in tubular specimen **A8:** 224
material properties determining **A14:** 879-880
nonhomogeneous, in tubular specimen **A8:** 222, 224
sheet formability testing **A8:** 550
Strain energy *See also* Elastic energy; Modulus of resilience; Resilience; Toughness defined **A8:** 12-13
Strain energy of the critical nucleus
in precipitation hardening **A9:** 646
Strain energy remaining from working
effect on tin-antimony alloys **A9:** 450
Strain etching
defined **A9:** 17
Strain, expression of **A9:** 117
microstructural deformation modes as a function of, for alpha-brass and copper **A9:** 686
Strain gage *See also* Rosette **EM3:** 27
for amplitude detection, ultrasonic testing **A8:** 246
applications **A8:** 617
bonded resistance, fatigue test specimen with **A8:** 618
bridge, four-arm electric resistance **A8:** 220
for creep testing **A8:** 303
defined **A8:** 13 **EM1:** 22 **EM2:** 40
for deformation measurement, sheet metal forming **A8:** 548-549
for drop tower specimen measurement **A8:** 197
for dynamic notched round bar testing **A8:** 277
electrical resistance, for strain measurement **A8:** 209
for elevated temperatures **A8:** 36
and extensometers, sheet metal forming **A8:** 549
extensometers using **A8:** 49, 50, 617
for fatigue testing machines **A8:** 368
foil bonded resistance **A8:** 618
for high strain rate tests **A8:** 208
installation on pressure bar **A8:** 202
lead wire attachment during testing **A8:** 202
load cell **A8:** 400, 613-614
measurements and stress-strain behavior related **A8:** 198-199

Strain gage (continued)
for measuring medium stress and strain rates **A8:** 191
for multiaxial testing **A8:** 344
output voltages, from torsional Kolsky bar dynamic tests **A8:** 228
outputs, dynamic notched round bar testing **A8:** 277
paperback **A8:** 201
records, dynamic toughness from **A8:** 274-275
with single pressure bar **A8:** 199
with split Hopkinson pressure bar **A8:** 198-199, 201, 213
surveys **EM3:** 543
to measure strain in pressure bars **A8:** 201
for torsional impact system **A8:** 216-217
for torsional Kolsky bar **A8:** 218-219
in torsional testing machine **A8:** 220
types **A8:** 618
typical **A8:** 48
Strain gage amplifier **A8:** 95
Strain gages **A6:** 1150 **EL1:** 445, 446, 612
and Barkhausen noise measurement **A17:** 160
for deformation measurement **A14:** 879
for residual stress study **A11:** 134
as strain-sensing method **A17:** 51
in stress analysis **A17:** 448-450
Strain hardening *See also* Residual strain hardening; Work hardening
annealing for removing **A7:** 185
changes due to recovery **A9:** 693
and compressibility in loose powder compaction **A7:** 298
defined **A8:** 13 **A9:** 17 **A11:** 10 **A13:** 12 **A14:** 12
effect of deformation rate on **A8:** 38, 39
effect of strain rate on **A8:** 38, 39
effect of temperature on **A8:** 38, 40
exponent, defined **A14:** 12
exponent, symbol for **A11:** 797
and limit analysis **A11:** 136-137
mechanisms, effect on failure **A8:** 34
in multiaxial creep **A8:** 343
parameter, defined **A8:** 45
rate, effect of environment on **A11:** 225
rate, in contour roll forming **A14:** 624
rate of, and strain-hardening exponent **A8:** 24
and strain rate, in structural aluminum **A8:** 38, 40
temper designations **A2:** 21, 25
wrought aluminum alloy **A2:** 38-39
Strain impact
SIMA aluminum alloy microstructure **A15:** 330
Strain increment **A6:** 1138
Strain induced, melt activated process *See* SIMA process
Strain life
curve **A8:** 696-698
Langer equation **A8:** 698
level, fatigue resistance of **A8:** 712
material response, statistical characterization of **A8:** 697-701
testing, replication and sample size **A8:** 696-697
Strain markings **A9:** 686-687
defined **A9:** 17
in Fe-3.25SI alloy steel **A9:** 687
Strain measurement *See also* Stress analysis **A8:** 208
at medium rates **A8:** 193
brittle-coating method **A17:** 453
in cold upset testing **A8:** 579
in creep testing **A8:** 303
in drop tower compression testing **A8:** 197
during increasing strain rate **A8:** 191-192
during rupture tests **A8:** 337
in elevated/low temperature tension testing **A8:** 35-36
grids on upset cylinders **A8:** 579
and high-rate tensile testing **A8:** 209-210
machines/equipment **A8:** 49-50, 191, 193, 211

Strain measurement (continued)
photoelastic coating method ... **A17:** 450-453
resistance strain gage method ... **A17:** 448-450
for stress analysis ... **A17:** 448-453
Strain peening *See* Stress peening ... **M5:** 138
Strain point ... **EM4:** 424
Strain point, glass fiber
viscosity at ... **EM1:** 47
Strain rate *See also* Reduction rate; Strain ... **A6:** 1138
analysis of ... **A8:** 44-45
ASTM defined ... **A8:** 39
beryllium, formability effects ... **A14:** 805
calculations for torsional Kolsky bar ... **A8:** 225-226
changes, stress-strain curves for tests with ... **A8:** 177
elastic and plastic ... **A8:** 41-42
constant ... **A8:** 171
control during deformation ... **A8:** 178
as controlling workability ... **A14:** 363, 368
and crack propagation rate ... **A13:** 160
in creep ... **A8:** 301-302
critical ... **A13:** 260-262
defined ... **A8:** 13, 190, 575 **A13:** 12
deformation, as cause of shear bands ... **A8:** 45
deformation mode change from ... **A8:** 188
as deformation rate ... **A8:** 44
-dependent material, ultrasonic testing of ... **A8:** 255-256
distribution, as ALPID result ... **A14:** 427
effect in polymers ... **A8:** 38, 43
effect in torsion testing ... **A8:** 141-142
effect, liquid-metal embrittlement ... **A13:** 175-177
effect of nonhomogenous deformation ... **A8:** 44
effect on creep crack propagation, austenitic stainless steels ... **A12:** 354
effect on dimple rupture ... **A12:** 31-33
effect on ductility ... **A8:** 19, 38, 42-43
effect on embrittlement ... **A12:** 29
effect on fatigue life, stainless steels ... **A12:** 59
effect on flow properties ... **A8:** 38-45
effect on flow stress of alpha iron and copper ... **A8:** 178, 179
effect on flow stress, titanium alloys ... **A14:** 270
effect on forgeability, carbon/alloy steels ... **A14:** 215-216
effect on fracture appearance, steel alloy ... **A12:** 31, 41
effect on intergranular creep rupture ... **A12:** 19
effect on strength ... **A8:** 19 **A12:** 121
effect on torsional ductility ... **A8:** 166-168
effect on true yield stress ... **A8:** 38, 39
effect on yield strength and deformation rate ... **A8:** 38, 39
effect, SCC and hydrogen-induced cracking ... **A13:** 261
effective, disk forging simulation ... **A14:** 431
flow stress dependence on ... **A8:** 236-238
flow stress relationships, process modeling ... **A14:** 419
fracture strain as function of ... **A13:** 330
as function of frequency and strain range, I ultrasonic fatigue testing ... **A8:** 241
high, compression testing ... **A8:** 190-207
high, pressure-shear impact testing ... **A8:** 231-232
high, shear testing ... **A8:** 215-239
high, tension testing ... **A8:** 208-214
history, effect on workability in torsion testing ... **A8:** 178
increased at ultrasonic frequency ... **A8:** 240, 241
increases, data acquisition during ... **A8:** 191
increases, stress and strain measurement during ... **A8:** 191-192
increasing or decreasing, effect on copper flow stress in torsion ... **A8:** 177-178
inertia effects on ... **A8:** 42-43
influence on deformation and failure mode high strain rate testing ... **A8:** 188
limitations, torsional Kolsky bar ... **A8:** 226-228
with lower yield stress, double-notch shear testing ... **A8:** 229
magnitude in slow strain rate testing ... **A8:** 498
medium, conventional load frames for compression testing at ... **A8:** 192-193
moderately high, effects ... **A12:** 31
occurrences ... **A8:** 208
Strain rate (continued)
or temperature change effect on flow stress cubic metals ... **A8:** 223, 226
regimes ... **A8:** 190-191
sensitivity parameter, determining ... **A14:** 439-440
steady-state, effect in high-purity copper ... **A12:** 400
and stress ... **A8:** 343
stress-corrosion crack growth and ... **A13:** 160
stress-log, for titanium alloys ... **A8:** 214
symbol for ... **A8:** 726
and temperature, and tensile fracture ... **A14:** 363
and temperature, combined effect, process modeling ... **A14:** 420
and temperature effect on alpha parameter ... **A8:** 172
and temperature, effect on flow stress ... **A14:** 151
and temperature, effect on yield strength ... **A8:** 38, 40
and temperature function, tensile fracture modes ... **A8:** 571
testing at constant crosshead speeds ... **A8:** 43-44
testing, incremental, with Kolsky bar ... **A8:** 223-224
testing machine stiffness and ... **A8:** 41-42
tests at constant ... **A8:** 42-43
and time relationship during constant-load creep test ... **A8:** 331
to failure, average ... **A8:** 693
twist reversal on aluminum after deformation and ... **A8:** 174
in ultrasonic fatigue testing ... **A8:** 240-241
very high, effects ... **A12:** 31-33
very high, Kolsky bar apparatus for double-notch shear testing ... **A8:** 229
very low, effects ... **A12:** 31
vs flow stress, aluminum alloy forging ... **A14:** 242
vs specific energy, heat-resistant alloys ... **A14:** 233
vs. cycles to fracture ... **A8:** 349-350
vs. time-to-fracture ... **A8:** 349-350
and wave propagation ... **A8:** 40
and workability ... **A8:** 178, 572, 575
Strain rate and temperature
effect on plastic deformation ... **A9:** 688-691
Strain rate as a function of mean free distance between particles in copper-aluminum
dispersion alloys ... **A9:** 131
Strain rate sensitivity *See also* m value
aluminum ... **A8:** 230-231
defined ... **A8:** 13, 45
of ductile iron ... **A1:** 47
of engineering alloys, verification ... **A8:** 251
of explosive LX-04-M ... **A8:** 38, 42
in incremental testing ... **A8:** 223
methods of determining, sheet metals ... **A8:** 557
negative ... **A8:** 38, 40
polycrystalline aluminum ... **A8:** 237-238
role in strain distribution, sheet metal forming ... **A8:** 550
of sheet metals ... **A8:** 555, 556
vacuum arc remelted 4340 steel and 1100-O aluminum, compared ... **A8:** 238
vs. percent elongation ... **A8:** 42
Strain relaxation ... **EM3:** 27
defined ... **EM1:** 20 **EM2:** 40
Strain relieving ... **A1:** 259
defined
Strain wave form *See* Waveform
Strain-age cracking
nickel-base superalloys ... **A6:** 83
nonferrous high-temperature materials ... **A6:** 566-567
Strain-age embrittlement ... **M1:** 683-684
in carbon and alloy steels ... **A1:** 693-694, 695
defined ... **A8:** 12
dynamic strain aging ... **A1:** 694
low-carbon steels ... **A1:** 693-694
Strain-aging
low-carbon steel sheet and strip ... **M1:** 154, 157, 162
notch toughness, effect on ... **M1:** 701
Strain-anneal technique of growing single crystals ... **A9:** 696
Strain-based approach to fatigue ... **A1:** 677-678, 679
Strain-controlled fatigue test ... **A8:** 346-347
Strain-energy release rate ... **A18:** 403
Strain-energy release rate (G) ... **EM3:** 444, 446, 447, 449-450, 451
Strain-energy release rates ... **A6:** 147
Strain-gauge pressure transducer ... **A7:** 423
Strain-hardening coefficient *See* n value; Strain-hardening exponent; Work hardening
role in strain distribution sheet metal forming ... **A8:** 550
Strain-hardening exponent ... **A1:** 575
correlation with seizure resistance ... **M1:** 611
Strain-hardening exponent defined ... **A8:** 13, 24
range of values ... **A8:** 24
and rate of strain hardening ... **A8:** 24
symbol for ... **A8:** 725
values for metals at room temperature ... **A8:** 24
Strain-hardening rate
solid-state-welded interlayers ... **A6:** 169
Strain-induced martensite
in steel ... **A9:** 178
in wrought stainless steels, magnetic etching ... **A9:** 282
Strain-rate sensitivity (m value)
defined ... **A14:** 12
Strain-sensing methods
types ... **A17:** 51
Strain-to-failure of interlayers ... **A6:** 167, 168, 170
Strained-layer superlattices, RBS analysis
of ... **A10:** 634
Strainer core
defined ... **A15:** 11
in gating system ... **A15:** 596
Straining
compression ... **A8:** 357
directions ... **A8:** 357
slow, as isothermal ... **A8:** 44
tension ... **A8:** 357
Strains
ferromagnetic resonance analysis ... **A10:** 275
fields of ... **A10:** 365
frozen-in ... **A10:** 268
interfacial, evaluated by x-ray topography ... **A10:** 365
magnitude of ... **A10:** 325
measurement, in lattices ... **A10:** 633-635
measurement of ... **A10:** 275
relaxation of, stress measurement ... **A10:** 385
surface, and corrosion products ... **A10:** 607
Straits tin *See* Tin alloys, specific types, commercially pure tins
Strand *See also* Fiber(s); Fibers; Filament(s); Filaments; Sliver
defined ... **EM1:** 22 **EM2:** 40
integrity, defined ... **EM1:** 22
test methods ... **EM1:** 23, 732
Strand annealing ... **A1:** 280
Strand annealing of steel wire ... **M1:** 262
Strand casting ... **M1:** 114
Strand count *See also* Count
defined ... **EM1:** 22 **EM2:** 40
Strand integrity
defined ... **EM2:** 40
Strand tensile test
defined ... **EM1:** 23
Strand wire
coating weight ... **M1:** 263
description ... **M1:** 264
Stranded copper conductors ... **M2:** 266, 268-273
Stranded electrode
definition ... **A6:** 1214 **M6:** 17
Stranded wire
production ... **A2:** 257-258
wrought copper and copper alloys ... **A2:** 252-253
Stranded-wire springs ... **M1:** 304
Strap-type clamp
stress-corrosion failure ... **A11:** 309
Strapping wire ... **A1:** 282 **M1:** 264
Strategic materials availability and supply ... **A1:** 1009-1022
COSAM program approach ... **A1:** 1013
advanced processing ... **A1:** 1018
alternate materials ... **A1:** 1018-1020
results ... **A1:** 1020-1021
substitution ... **A1:** 1014-1018
reserves and resources ... **A1:** 1009, 1010
strategic materials ... **A1:** 1009-1011
index method ... **A1:** 1010
survey method ... **A1:** 1011
superalloys ... **A1:** 1011-1013

Strategic missile systems
composite components **EM1:** 816
Stratification
of samples **A17:** 730
Stratified material
sampling **A10:** 14
Straub measurement method
shot peen coverage **M5:** 139
Straube-Pfeiffer test
for hydrogen measurement **A15:** 457-458
Strauss, B
as metallurgist **A15:** 32
Strauss coupling reaction
polyimides **EM3:** 157
Stray current
defined **A13:** 12
electrolysis **A13:** 87
in lead/lead alloys **A13:** 788-789
Stray current corrosion **M5:** 432
Stray radiation *See* Cross talk
Stray-current corrosion *See also* Corrosion **A13:** 87
in carbon and low-alloy steels **A11:** 199
defined **A11:** 10 **A13:** 12
designing for **A13:** 342
and galvanic corrosion, compared **A13:** 87
identifying **A13:** 87
in oil/gas production **A13:** 1234
sources **A13:** 87
in telephone cables **A13:** 1128
Streak cameras
for ring displacement **A8:** 210
Streaking
alloy steels **A12:** 346
Stream atmospheres
processing considerations **M4:** 411
steam treating effects of **M4:** 411
Stream inoculation *See also* Inoculation
of gray iron **A15:** 638-639
Stream sampling
chemical analysis **A7:** 246
Stream scanning methods
to analyze ceramic powder particle sizes **EM4:** 67
Stream tin
historical use **A15:** 16
Streaming potential **EM4:** 74
Street sweepers
cemented carbide skids for **A2:** 973
Strength *See also* Adhesive strength; Bearing ultimate strength; Bearing yield strength; Bond strength; Burst strength; Cohesive strength; Compressive strength; Compressive yield strength; Dielectric strength; Dry strength; Elongation; Fatigue strength; Flexural strength; Green strength; Impact strength; Offset yeild strength; Peel strength Residual strength; Reduction in area; Shear strength; Shear ultimate strength; Strength analysis; Tensile strength; Tensile strengths; Tensile ultimate strength; Tensile yield strength; Ultimate compressive strength; Ultimate shear strength; Ultimate strength; Ultimate tensile strength, Wet strength; Wet strength; Yield strength **EM3:** 27
of aluminum P/M alloys **A7:** 747
aluminum-lithium alloys **A2:** 185-186, 192
analysis **EM1:** 192-204
assessment of **EM2:** 551-554
of austempered ductile irons **A15:** 35
bearing **EM1:** 314-316
bending, tests for **A8:** 117, 132-136
beryllium-copper alloys **A2:** 409
of boiler tubes **A11:** 603
of compact infiltrated copper alloys **A7:** 565
compressive **EM3:** 27
compressive, to failure **A8:** 57-58
computer program for **EM1:** 276-277
copper and copper alloys **A2:** 216
Strength (continued)
and corrosion resistance **A13:** 48
corrosion-resistant high-alloy **A15:** 726
and damping **EM1:** 216
data, three-parameter Weibull distribution **A8:** 645
as decreasing with crack size **A11:** 55
defined **A8:** 13
degradation model **A8:** 716-717
dielectric **EM2:** 62
dry **EM3:** 27
of ductile iron **A15:** 647
and ductility **A7:** 319, 414
effect of grain size **A11:** 307
effect of SSC on **A11:** 298-299
effect of strain rate **A8:** 19
effect of temperature **A8:** 19, 34, 670 **A12:** 121
and electrical conductivity, beryllium-copper alloys **A2:** 417
epoxy resin matrices **EM1:** 73
excessive, effect in AISI/SAE alloy steels **A12:** 318
fatigue **EM1:** 436-444
fatigue, bridge components **A11:** 707-708
fatigue surface, wrought aluminum alloys **A12:** 420
of fibers **EM1:** 29, 36, 46, 58, 193-194
flexural **EM3:** 27
fracture and flow, related **A11:** 138
galvanizing effect, steels **A13:** 438
and hardness, carbon steels **A15:** 702
and hardness, low-alloy steels **A15:** 716
of heat-exchanger tubes **A11:** 628
in high strain rate testing **A8:** 188
improvement, wrought aluminum alloy **A2:** 36, 37-41
lamina laminate **EM1:** 227, 230-235, 276-277, 432-434
lead and lead alloys **A2:** 545
long-term, polyether sulfones (PES PESV) **EM2:** 161
mean **A8:** 626
of mechanical fasteners **EM1:** 706
mechanical, polyamide-imides (PAI) **EM2:** 129
of plastics, design guidelines **EM2:** 709
ply **EM1:** 137-238
of polymers **EM2:** 61
prediction **EM1:** 275, 432
pristine, glass fiber **EM1:** 46
residual **A11:** 55-57, 731
shear **EM3:** 27
SiC fibers **EM1:** 58
static **EM1:** 432-435
static, effect of temperature on **A11:** 130
and stress distribution **A8:** 627
stress rupture, tungsten-reinforced MMCs **EM1:** 880
stress-rupture, defined **A11:** 131
sustained stress, exposure testing **EM1:** 825
tensile **EM3:** 27
and toughness **A8:** 23
and tungsten content **A7:** 691
ultimate, defined **A13:** 13
under combined stress **EM1:** 198-201
universal testing machine for **A8:** 612
and weight **A11:** 325
-weight ratios, composites and metals compared **EM1:** 178
wet **EM3:** 27
of work metal **A14:** 620
and workability **A8:** 581
of woven fabric prepreg, float effect **EM1:** 150
yield **EM3:** 28
zinc alloys **A15:** 786
Strength analysis *See also* Strength
crack-closure scheme for **EM1:** 248-250
and damage accumulation **EM1:** 246-247
failure modes in **EM1:** 240-244
fatigue failure in **EM1:** 244-246
Strength analysis (continued)
finite-element procedure for **EM1:** 247-250
of laminates **EM1:** 236-251
ply failure criteria, point stress **EM1:** 238
ply strength properties **EM1:** 237-238
progressive ply failure in **EM1:** 238-239
stress approaches **EM1:** 236-237, 239-240
three-dimensional stress in **EM1:** 239-240
Strength and proof testing **EM4:** 585-596
high-temperature strength test methods for isotropic ceramics **EM4:** 594-595
creep and stress rupture **EM4:** 594-595
fast fracture **EM4:** 594
proof testing **EM4:** 596
room-temperature strength test methods for isotropic ceramics **EM4:** 588-593
Brazilian disk test **EM4:** 589
compression tests **EM4:** 593, 594
diametral compression test **EM4:** 589
elastic modulus **EM4:** 588, 590, 595
flaw size distributions **EM4:** 592
flexure testing **EM4:** 588-589, 590, 594, 595-596
fractography **EM4:** 592, 594
interpretation of uniaxial strength **EM4:** 590
maximum likelihood estimation scheme **EM4:** 592
multiaxial strength **EM4:** 593, 594
threshold stress **EM4:** 591, 592
uniaxial compression strength **EM4:** 592-593
uniaxial tensile strength **EM4:** 588-590
Weibull strength distribution **EM4:** 590, 592, 593
strength and fracture phenomena **EM4:** 585-588
crack growth **EM4:** 586-587
creep **EM4:** 586-587
defect management in ceramic fabrication **EM4:** 587-588
dynamic fatigue **EM4:** 587
Griffith and fracture mechanics approaches **EM4:** 585, 586, 592, 593
R-curve behavior **EM4:** 586, 592
slow crack growth **EM4:** 587
static fatigue testing **EM4:** 587
stress corrosion **EM4:** 587
stress intensity factor **EM4:** 586, 587
strength of continuous fiber reinforced composites **EM4:** 595-596
Strength coefficient *See* Strain-hardening exponent
Strength level
and fatigue resistance **A1:** 674, 678, 679, 681
Strength reduction factors (SRFs)
and fretting wear **A18:** 242
Strengthening mechanisms
from second-phase constituents, wrought aluminum alloy **A2:** 38
heat-resistant alloys **A14:** 779
heat-treatable wrought aluminum alloy **A2:** 36, 39-41
non-heat treatable wrought aluminum alloy **A2:** 36-39
Strescon *See* Copper alloys, specific types, C19500
applications and properties **A2:** 294
Stress *See also* Applied stress; Effective stress; Engineering stress; Mean stress; Nominal stress; Normal stress; Residual stress; Stress concentration; Stress relaxation; Stress relief; Stress relieving; Stress rupture; Stress- corrosion cracking; Stress-corrosion cracking (SCC); Tensile stress; True stress **EM3:** 28
analysis, software for **A14:** 412
analytical model **EM3:** 395-396
applied **A8:** 124, 363-364
applied, effect on embrittlement **A12:** 29
applied, examining **A12:** 92
at equatorial surface of upset test specimen **A8:** 580
average axial **A7:** 301
bar history **A8:** 209
-based failure prediction theories **A8:** 343

SUBJECTS OF THE INDEXED VOLUMES: ASM Handbook (designated by the letter "A"): **A1:** Properties and Selection: Irons, Steels, and High-Performance Alloys (1990); **A2:** Properties and Selection: Nonferrous Alloys and Special-Purpose Materials (1990); **A3:** Alloy Phase Diagrams (1992); **A4:** Heat Treating (1991); **A6:** Welding, Brazing, and Soldering (1993); **A7:** Powder Metallurgy (1984); **A8:** Mechanical Testing (1985); **A9:** Metallography and Microstructures (1985); **A10:** Materials Characterization (1986); **A11:** Failure Analysis and Prevention (1986); **A12:** Fractography (1987); **A13:** Corrosion (1987); **A14:** Forming and Forging (1988); **A15:** Casting (1988); **A16:** Machining (1989); **A17:** Nondestructive Testing and Quality Control (1989); **A18:** Friction, Lubrication, and Wear Technology (1992). **Metals Handbook, 9th Edition** (designated by the letter "M"): **M1:** Properties and Selection: Irons and Steels (1978); **M2:** Properties and Selection: Nonferrous Alloys and Pure Metals (1979); **M3:** Properties and Selection: Stainless Steels, Tool Materials and Special-Purpose Materials (1980); **M4:** Heat Treating (1981); **M5:** Surface Cleaning, Finishing, and Coating (1982); **M6:** Welding, Brazing, and Soldering (1983). **Engineered Materials Handbook** (designated by the letters "EM"): **EM1:** Composites (1987); **EM2:** Engineering Plastics (1988); **EM3:** Adhesives and Sealants (1990); **EM4:** Ceramics and Glasses (1991); **Electronic Materials Handbook** (designated by the letters "EL"): **EL1:** Packaging (1989).

Stress (continued)
biaxial, effect on dimple rupture **A12:** 31, 39
buckling, axial compression testing **A8:** 55
calculations for torsional Kolsky bar **A8:** 225-226
circumferential, during bending **A8:** 121, 123
combined, fatigue test specimens **A8:** 368
compressive, and workability **A8:** 576
conversion factors **A8:** 722-723
and creep rates **A8:** 302, 308, 343, 345
in creep/creep-rupture analyses **A8:** 302-304, 685
criteria, mixed and unmixed **A8:** 344
decreasing, effects in AISI/SAE alloy steels **A12:** 315
defined **A8:** 13, 308 **A13:** 12 **A14:** 12
definition **EM4:** 633
and deformation **A8:** 308
and density distribution in pressed powder compacts **A7:** 300-302
dependence of steady-state creep **A8:** 309
as dependent variable **A8:** 698
direction, effects, aluminum SCC **A13:** 590-591
distribution, defined **A10:** 382
distribution, elastic plastic bending **A8:** 119, 121
distribution, elastic-perfect plastic material **A8:** 120-121
effect, aluminum SCC **A13:** 590
effect in dimple rupture **A12:** 30-31
effect in high-purity copper **A12:** 399-400
effect in visual examination **A12:** 72
effect, liquid-metal embrittlement **A13:** 177
effect on cracking in elevated-temperature water **A8:** 421-422
effect, produced fluids **A13:** 479
effects, permanent magnet materials **A2:** 801
elastic-plastic, as *J* integral parameter **A8:** 261
in electroplated hard chromium deposits **A13:** 871
end **EM3:** 34
engineering, defined in tension testing **A8:** 20
equilibration, in wave propagation **A8:** 191
fatigue caused by **M1:** 665, 668
fatigue-test stresses described **M1:** 666
field, control tendency of, effect in rapid fracturing **A8:** 440
force per unit area, conversion factors **A10:** 686
for fracture of particles **A7:** 59
gradient, fatigue specimen size effect from **A8:** 372-373
gradients, subsurface, effects on measurement **A10:** 388-389
grain interaction, neutron diffraction analysis **A10:** 424
initial **EM3:** 28
internal **EM3:** 34
levels **A8:** 679, 702-706
linear, beam distribution **A8:** 118-119
local **EM3:** 33-34
-log strain rate, for titanium alloys **A8:** 214
macroscopic ,defined **A10:** 676
macroshear, effect in medium-carbon steels **A12:** 253
mean, effect on fatigue testing **A8:** 374
measurement, by x-ray diffraction **A10:** 381-382
measurement devices **A8:** 191
measurement during increasing strain rate **A8:** 191-192
methods, sustained service tension **A13:** 246
micro **A10:** 380, 386-387
microscopic, defined **A10:** 676
-modified critical strain model, and RKR compared **A8:** 467
critical stress model for cleavage
Mohr's circle for, x-ray diffraction stress measurement **A10:** 381
-moment equations, in bending **A8:** 118
multiaxial, effect on creep and creep rupture **A8:** 343-345
nominal **EM3:** 28
nonuniform, effects in axial compression testing **A8:** 55-58
normal **EM3:** 28
as not directly measurable **A10:** 381
in oxide scales **A13:** 71
-particle velocity, normal **A8:** 232
peak **EM3:** 34
plane strain bending **A8:** 121, 123
in plate rolling, prediction **A14:** 346

Stress (continued)
principal **A10:** 382
principal, and creep/creep rupture **A8:** 343
principal, direction, and yield criteria **A14:** 370
principal, direction effects on dimple shape **A12:** 30
as producing dynamic strain **A8:** 498
pulse, stress-intensity histories for cracks subjected to **A8:** 284
pure plastic bending **A8:** 120, 122
radial, in plate during bending **A8:** 121, 123
raisers **A8:** 13, 364, 371
raisers, defined **A13:** 12 **A14:** 12
range, vs. time-to-fracture **A8:** 349-350
rate, during elastic deformation **A8:** 39
relaxed **EM3:** 28
-relief cracking **A12:** 139-140
residual, and springback **A8:** 122-124
residual stresses, effect on fatigue **M1:** 673-675, 682
reverse, medium-carbon steel fracture by **A12:** 262
scanning electron microscopy examination of specimens **A9:** 97
in SCC **A13:** 145, 247
schematic **A13:** 247
shear **EM3:** 28
shear, constant maximum **A8:** 72
SI unit/symbol for **A8:** 721
sources, hydrogen embrittlement **A13:** 284
state, effect on dimple shape **A12:** 12
state, tensile specimen **A12:** 104
states **A14:** 368-369, 388-389
in steam turbine SCC **A13:** 952
and strain rate **A8:** 343
and strength distributions **A8:** 627
stress ratio, defined **M1:** 666
design of forgings **M1:** 359-360
Stress-concentration factor, use in
stress-concentration factor, defined **M1:** 667
in substrates **A13:** 431
surface residual **A13:** 295
and surface tension, LaPlace equation for **A7:** 312
symbol for **A10:** 692
system, compressive **A8:** 576
tangential **A8:** 121-122
and temperature dependence, of creep equations **A8:** 688-689
-temperature plots, for isothermal high strain rate flow curves **A8:** 161
tensile **EM3:** 28
test **A8:** 503
time-to-rupture as function of **A8:** 304
to produce strain, in creep testing **A8:** 304, 498
torsional **EM3:** 28
in torsional testing **A8:** 146
torsional, vs. life in high-cycle regime wrought aluminum alloy **A8:** 150
total **A10:** 385
triaxial, from necking **A8:** 25
true **EM3:** 28
true, at maximum load **A8:** 24
uniformity of **A7:** 300
vs. cycles-to-failure, high-cycle fatigue testing **A8:** 367
and workability **A14:** 363, 388-389
Stress alloying *See* Liquid-metal embrittlement; Solid-metal embrittlement
Stress amplitude **A8:** 374 **A13:** 150, 295 **EM3:** 28
defined **A8:** 13 **A11:** 10 **EM2:** 41
effect of mean stress on **A11:** 112
effect on corrosion fatigue **A8:** 375
effect on fatigue resistance **A8:** 712-713
effect on fatigue strength **A11:** 112
fatigue life as function of **A8:** 253
and sinusoidal loading, *S-N* curves for **A11:** 103
and stress-intensity factor range, effects on corrosion-fatigue **A11:** 255
vs. time-to-fracture **A8:** 349-351
Stress analysis *See also* Stress modeling; Stress tests; Stress(es) **EM1:** 771-773
applications **A17:** 448
brittle-coating method **A17:** 453
color images by **A17:** 488
of composites **A11:** 741-742
computer program for **EM1:** 277

Stress analysis (continued)
and damage tolerance design **A17:** 702
and die design **A14:** 414
experimental **A11:** 18
laminate **EM1:** 236-237, 239-240
of laminates **EM1:** 227-230
micro/mini/macro scales **EM1:** 432
nonlinear, of laminates **EM1:** 230
part rejection determined by **A17:** 49
photoelastic coating method **A17:** 450-453
resistance strain gage method **A17:** 448-450
strain measurement for **A17:** 448-453
of temperature cycling **EL1:** 961
thermoplastic injection molding **EM2:** 312
three-dimensional **EM1:** 239-240
Stress buffers
polyimides **EL1:** 770
Stress coefficients
measurement of **EM3:** 322, 323
Stress concentration **EM3:** 28
at cutouts **EM1:** 462
by forming operations, wrought aluminum alloys **A12:** 414
cold shut as **A11:** 352-353
in corrosion fatigue testing **A13:** 293-294
crack formation at **A12:** 111
deburring drum fatigue at **A11:** 346-348
defined **A11:** 10 **EM1:** 23
effect of fillet radius size, in shafts **A11:** 468
effect of stress raisers on **A11:** 113
effect on fatigue cracking **A11:** 102-103 **A12:** 322
effect on fatigue strength **A11:** 113-115
effect on fatigue-crack propagation **A11:** 109-110
in elevated-temperature failures **A11:** 265-266
factor, defined **A11:** 10 **A13:** 12 **EM1:** 23
factors, laminate circular hole **EM1:** 234
fatigue fracture from **A11:** 122-123
in forgings **A11:** 318
in fracture analysis, laminates **EM1:** 252-253
as fracture origin, shafts **A11:** 459
from visual examination **A12:** 72
from welding defects, failures from **A11:** 400
local, design details for **A13:** 343
in notched cylinder in tension **A11:** 318
and residual stresses, castings **A11:** 362
spring failure from **A11:** 558
and static strength **EM1:** 434-435
Stress concentration factor
and fatigue resistance **A1:** 675
Stress corrosion **A10:** 380 **A16:** 21 **EM3:** 28
at sustained load **A11:** 53
cracks, branching and propagation of **A11:** 81
defined **EM1:** 23 **EM2:** 41
failure, heat-resistant alloys **A11:** 309
failure, high-strength aluminum alloy **A11:** 28
fracture mechanics and **A11:** 47
from cutting fluid traces **A16:** 35
improved by shot peening **A16:** 26
intergranular, in stainless steel bolts **A11:** 536-537
microwave measurement **A17:** 215-218
radiographic methods **A17:** 296
resistance of maraging steels to **A1:** 799-800
retarding cracking **A16:** 27
rock candy from **A11:** 28
test data **A11:** 53-54
woody **A11:** 27
Stress corrosion cracking *See also* Cracking; Stress(es)
acoustic emission inspection **A17:** 287
definition **M6:** 17
eddy current inspection **A17:** 573
effect of boundary precipitation and solute segregation in **A10:** 549
fracture surfaces **A10:** 562-564
and low ductility **A7:** 254
near-crack tensile sample geometry for **A10:** 563
propagation **A17:** 50, 54, 86
three brass alloy environment for **A10:** 563-564
Stress crack *See also* Crazing **EM3:** 28
defined **EM1:** 23 **EM2:** 41
Stress cracking *See also* Thermal stress cracking
corrosion **EL1:** 56-59, 1006-1007
defined **EM1:** 23 **EM2:** 67
environmental, high-impact polystyrenes (PS, HIPS) **EM2:** 197-198
environmental, of polymers **A11:** 761
failure, defined **EM2:** 41

Stress cracking (continued)
in integrated circuits **A11:** 774
semiconductor chips **EL1:** 964-965
thermally induced **EL1:** 416
Stress crazing
environmental **EM2:** 796-803
and uniaxial tensile creep **EM2:** 667
Stress creep, constant
experiments **EL1:** 840, 846-847
Stress cycle
defined **A8:** 13
Stress cycles
defined **A11:** 10
and fatigue life **A11:** 102
Stress cycles endured n **A8:** 13, 725
Stress decay *See* Stress relaxation
Stress dependence
of Barkhausen noise **A17:** 160
of magnetically induced velocity changes (MIVC) **A17:** 162
of nonlinear harmonics **A17:** 161
Stress distribution **A7:** 301-302
as ALPID result **A14:** 427
angle of forking as indicator of **A11:** 744
at neck **A8:** 25-26
at time of fracture, and Wallner lines **A11:** 745
in contacting surfaces, from rolling, sliding **A11:** 592
defined **A10:** 382
effects of stress raisers on **A11:** 113
from machine notch and tensile loading **A11:** 115
in rotating shafts with press-fitted elements **A11:** 470
schematic, of shaft surfaces **A11:** 123
specimen, rotating-beam mechanism **A8:** 392-393
in strip rolling **A14:** 344-346
torsion testing **A8:** 140
Stress equalizing
nickel alloys **A4:** 907, 908, 911 **M4:** 755, 756, 758
Stress field analysis
crack front simplified for **A8:** 442
fracture mechanics **A8:** 441-443
Stress fracture *See* Fracture stress
Stress fracture criteria
average stress **EM1:** 254-255
and center notch experimental data compared **EM1:** 256
and critical stress-intensity factor compared **EM1:** 256
and fracture toughness, compared **EM1:** 256
point stress **EM1:** 254
Stress in cladding **EM4:** 424
Stress intensity *See also* Stress-intensity factor
applied, and crack growth rate **A11:** 107
calibration, defined **A8:** 13
concepts, ultrasonic fatigue testing specimens **A8:** 251-252
and crack growth, by incremental polynomial method **A8:** 678-679
crack tip **A13:** 147
crack velocity as function of **A13:** 170
for damage analyses **A8:** 681-682
effect in maraging steels **A11:** 218
effect of time, crack extension and load hydrogen embrittlement testing **A8:** 538
effect on fracture in wrought beryllium-copper alloys **A9:** 393
effects, SCC kinetics **A13:** 246
factor, defined **A13:** 12
and stress amplitude factor range, effect on corrosion-fatigue **A11:** 255
of surface flaw **A11:** 52
Stress intensity, effective
and FSS **A12:** 205
Stress intensity factor (*K*)
defined **A12:** 15-16, 54, 56
effect on fracture modes **A12:** 441
Stress intensity factor range **A13:** 12, 143, 297-299
defined **A12:** 54, 56
effect on fatigue **A12:** 54-58
effect on fatigue crack growth rate **A12:** 56-58
wrought aluminum alloys **A12:** 417, 418
Stress level
controlling, in fatigue crack growth rate testing **A8:** 679
increments in staircase method **A8:** 703-704
required to cause specimen failure **A8:** 702
selection **A8:** 705-706
Stress life
curve **A8:** 696-698
data, for two-point test **A8:** 705
data, tolerance limits on **A8:** 700
diagram, log normal life distribution stresses **A8:** 699
level, fatigue resistance of **A8:** 712
material response, statistical characterization of **A8:** 697-701
results, ultrasonic fatigue testing **A8:** 252
specimen design, ultrasonic fatigue testing **A8:** 249-250
testing, replication and sample size **A8:** 696-697
Stress measurement
and acoustic emission inspection **A17:** 286
by strain **A17:** 448
magabsorption **A17:** 153-158
in nonferromagnetic materials, by magabsorption **A17:** 155-156
ultrasonic inspection **A17:** 276
Stress modeling *See also* Modeling; Stress(es)
as process control/selection tool **EL1:** 446
Stress peening of leaf springs **M1:** 312
Stress perturbation
as failure origin **EM1:** 194
Stress raisers *See also* Stress concentration
arc-welded steel failures from **A11:** 417
as cause of shaft failures **A11:** 459
corrosion pits as **A11:** 637
created during hot trimming, fatigue fracture from **A11:** 472
and design, iron castings **A11:** 346
flow stress associated with **A11:** 113
in forging **A11:** 334
from corrosive environment **A11:** 134
from welding-produced surface defects **A11:** 127
in gray iron bearing cap **A11:** 347-348
inclusions as **A15:** 88
keyways as **A11:** 115
mechanical, and shaft failure **A11:** 459
metallurgical, and shaft failure **A11:** 459
in pressure vessels, effect of **A11:** 647-648
produced by rough machining **A11:** 122
seams as **A11:** 478
as service stress **A11:** 206
stamp marks as **A11:** 130
as stress source **A11:** 205
types in shafts **A11:** 467-468
Stress ratio *See also* Load ratio **A13:** 12, 143, 292, 298
A or *R*, defined **A8:** 13
for aluminum alloys **A8:** 697
and composite material age **A8:** 716
and corrosion fatigue behavior **A8:** 408
defined **A11:** 10
effect on corrosion fatigue crack propagation **A8:** 406-407
and fatigue crack growth rate **A8:** 415
fatigue crack initiation **A8:** 363
and fatigue resistance **A1:** 674
in fatigue-crack initiation **A11:** 102
increase effect **A8:** 406-407
symbol for **A8:** 724, 725
Stress relaxation **EM3:** 28, 52
compression testing **A8:** 325-328
copper and copper alloys **M2:** 484-490
and creep **EM2:** 659-678
Stress relaxation (continued)
and creep, as time-dependent **A11:** 144
defined **A8:** 13 **EL1:** 1158 **EM1:** 23 **EM2:** 41, 673
in ductile materials, effect on SCC **A8:** 502
during strain-controlled fatigue cycle **A8:** 347
effect of stationary cracks on **A8:** 439
elastic strain **A8:** 323-324
failure analysis of **EM2:** 730
high-impact polystyrenes (PS, HIPS) **EM2:** 197
microstructural crystallinity **EM3:** 411
models **EM2:** 659-666
moisture effects **EM2:** 763-765
polyether sulfones (PES, PESV) **EM2:** 161
in polymers **A11:** 758
rheology **EL1:** 846
step excitation test methods **EM2:** 549-551
stress-strain diagram for **A8:** 323
test **EM3:** 318, 319
test, defined **A8:** 307
tests **EM2:** 435
in thermoplastic fluoropolymers **EM2:** 119
time-dependent compressive **EM2:** 673, 675
and viscoelasticity **EM2:** 414
wrought copper and copper alloys **A2:** 260-263
Stress relief *See also* Stress relief anneal; Stress relieving
copper casting alloys **A2:** 348
cracking, defined **A13:** 12
defined **EL1:** 1158
electrical resistance alloys **A2:** 824
embrittlement, of steels **A11:** 98-99
heat treatment, of pressure vessels **A11:** 664-666
for internal stresses, cast alloys **A11:** 345
oxide scales **A13:** 71-72
for residual stress **A11:** 98
for steam generators **A13:** 944
temperature, aluminum casting alloys **A2:** 169-170
treatments **A13:** 327, 942
uranium alloys **A2:** 675
wrought copper and copper alloys **A2:** 247
Stress relief annealing (SRA) **A6:** 411
Stress relief cracking *See also* Cracking
definition **A6:** 1214 **M6:** 17
Stress relief heat treatment *See also* heat treatment for specific processes
definition **A6:** 1214 **M6:** 17
Stress relieving *See also* Annealing; Heat Treatment; Stress relief
after heat treatment **A4:** 616
annealing process **A4:** 53
austenitic ductile irons **A4:** 698-699 **A15:** 700
in beryllium forming **A14:** 806
cast irons **A4:** 668, 669 **A6:** 714
cold finished bars **M1:** 223-233, 234, 239-240, 250-251
copper alloys **M2:** 255-256
copper and copper alloys **A2:** 216
copper metals **M2:** 462-463
for ductile iron **A1:** 41 **A15:** 657 **M1:** 37
effects, aluminum SCC **A13:** 590
gray cast iron **M1:** 27, 28
gray iron **A15:** 643 **M4:** 540, 541-543
gray irons **A4:** 679-681
high-chromium white irons **A4:** 708
high-silicon irons **A4:** 700
for hot-rolled steel bars **A1:** 241
induction hardening heat treatment **A4:** 193
of magnesium alloys **A2:** 472-473 **A14:** 825
nickel alloys **A4:** 907, 908, 911 **M4:** 755, 756, 758
nonferrous high-temperature materials **A6:** 572-574
parameters, copper alloys **A13:** 615
plain carbon steels **A15:** 714
postweld, zirconium alloys **A15:** 838
for powder metallurgy high-speed tool steels **A1:** 783
refractory metals **A4:** 815, 816, 817
springs **M1:** 290-291, 293

Stress relieving (continued)
stainless steels **A4:** 772, 773, 774-776 **M3:** 47-48
steam atmospheres **A4:** 562
of steel plate **A1:** 231-232
temperature, cast copper alloys **A2:** 356-391
times and temperatures, magnesium alloy arc welds **A2:** 475
titanium **M4:** 764, 765
of tool and die steels **A14:** 54
tool steels **A4:** 735, 737-738, 739-740, 743, 757, 758, 765
ultrahigh-strength steels **A4:** 208, 210, 212, 214, 215, 217 **M1:** 425, 430, 435, 439, 441
welded cast steels **A15:** 534-535
wrought titanium alloys **A2:** 618
Stress relieving, austenitic stainless steel **M4:** 647-649
annealing **M4:** 649
inadequate stress relief **M4:** 648, 649
intergranular corrosion **M4:** 649
metallurgical characteristics, influence on selection **M4:** 647-648
stress corrosion, prevention **M4:** 649
treatment selection **M4:** 647-648, 649
Stress residual **EM3:** 28
cure **EM1:** 761
Stress rupture *See also* Creep-rupture strength; Stress rupture fractures; Stress-relief cracking
of carbon and low-alloy steels **A1:** 620, 622, 623-624, 629
of consolidated superalloys **A7:** 439, 472
crack, in welds of low-alloy steel pipe **A11:** 427
cracking **A12:** 139-140
cracking, from defective material **A11:** 603
and creep, determination of **A11:** 29
curves **A11:** 265
data, cobalt-base superalloys **A2:** 452
defined **A11:** 9
of dispersion-strengthened copper **A7:** 717
ductility **A11:** 265
in elevated-temperature failures **A11:** 264-266
external **A11:** 596-597
failure **A11:** 75, 590, 607
fracture **A11:** 264-265
in gears **A11:** 590, 596
of heat-resistant alloys **A11:** 290-292 **A15:** 729
of infiltrated graded bucket **A7:** 562
internal **A11:** 596
intergranular cracks in **A11:** 264
of metal borides and boride-based cermets **A7:** 812
platinum dispersion-strengthened alloys **A7:** 721
polyamide-imides (PAI) **EM2:** 129
polyvinyl chlorides (PVC) **EM2:** 210
in pressure vessels **A11:** 666
resistance, mechanically alloyed oxide alloys **A2:** 943-947
dispersion-strengthened (MA ODS)
as solder joint failure mechanism **EL1:** 1031
of stainless steels **A1:** 932-933, 934, 937, 938, 939, 942, 945, 946
steam equipment failure by **A11:** 602
strength **A11:** 131, 603
stress, vs rupture life **A11:** 265
superheater tube failure from **A11:** 609-610
tests, for high-temperature effects **A12:** 121
zirconium alloys **A2:** 668
Stress rupture fracture(s)
examination **A12:** 139-140
metal-matrix composites **A12:** 468
superalloys **A12:** 391
Stress rupture strength
tungsten-reinforced MMCs **EM1:** 880
Stress screening
environmental **EL1:** 944
levels, through step-stress analysis **EL1:** 879-880
nonuniversality (temperature) **EL1:** 876
sequence **EL1:** 880
temperature chamber **EL1:** 881
Stress shear
growth **EL1:** 846
Stress shock
UHMWPE resistance **EM2:** 168
Stress singularities in bonded dissimilar materials **EM3:** 497-500
Stress singularity methods **EM3:** 506
Stress state
at free-surface of compressed specimen **A8:** 580
classifying **A8:** 576
in forging, billet shape and enclosure effect on **A8:** 588
influence on strain to fracture **A8:** 576
and workability **A8:** 572
yield criteria **A8:** 577
Stress tests
burn-in **EL1:** 497-498
endurance life **EL1:** 498
high-temperature storage life **EL1:** 498
highly accelerated stress testing (HAST) **EL1:** 495-497
with hot air knives **EL1:** 692
humidity-induced **EL1:** 495-497
power-temperature cycling **EL1:** 499
stabilization bake **EL1:** 499
steady-state life **EL1:** 498
temperature cycling **EL1:** 498
temperature-induced **EL1:** 497-499
Stress wave
deformation by **A8:** 40, 44
dynamic fracture by **A8:** 287
gages, for short-pulse-duration tests **A8:** 283
propagation **A8:** 40-41, 190-191
velocity, generated by impacts **A8:** 191
Stress wave form
and fatigue **A13:** 295
Stress wave propagation
liquid impact erosion **A18:** 224
Stress whitening
defined **EM2:** 667
Stress(es) *See also* Alternatitig stress; Applied stresses; Bearing stress; Compression; Compressive stress; Corrosion; Critical longitudinal stress, Glass stress; Hoop stress Initial (instantaneous) stress; Initial stress; Nominal stress; Nominal stress Normal stress, Relaxed stress; Normal stress; Relaxed stress; Residual stress; Residual stress Shear stress, Stress relaxation; Shear stress; Static stress; Strain; Stress analysis Stress corrosion cracking; Stress cracking; Stress measurement; Stress modeling; Stress relaxation; Stress screening; Stress tests; Tensile stress; Tensile stress Torsional stress; Tension; Testing; Torsional stress; True stress
in accelerated testing **EL1:** 887
allowable, for static service **A11:** 136
alternating **A11:** 461
amplitude, in corrosion-fatigue **A11:** 255
applied, in fatigue-crack initiation **A11:** 102
and applied stress **A11:** 48
at a point. theory of **EM1:** 238
averaging nonuniform **EM1:** 177
biaxial **A11:** 124
butterflies, in steel bearing ring **A11:** 505
by mechanical loads, laminate **EM1:** 227-228
by quantitative fractography **A11:** 56
casting **A11:** 345-346, 362-365
in chip metallization **EL1:** 445
in chip metallization/passivation **EL1:** 443-444
combined, strength under **EM1:** 198-201
complex **A11:** 112-113
in composites **A11:** 741-742
cooling **EM2:** 751
crack-tip **A11:** 47
critical longitudinal, defined *See* Critical longitudinal stress
cyclic, combined **EM1:** 202-203
in damage tolerance design **A17:** 702
defined **A11:** 10 **EM1:** 23 **EM2:** 41
direction, effect on SCC **A11:** 220
and distortion ratios, from overloading **A11:** 137
distribution of **A11:** 113
edge, laminate **EM1:** 230-231
effect, magnetic domain orientation **A17:** 145, 153
effect on fatigue strength **A11:** 110-112
effect on hysteresis loops, stainless steels **A17:** 161
effect on SSC resistance **A11:** 299
fiber **A11:** 137
fields, defined **EM1:** 186
fields, torsional-fatigue cracks and **A11:** 471
flow, and stress raisers **A11:** 113
in flows and material functions **EL1:** 838
Stress(es) (continued)
fluctuating **A11:** 461
formulas for **EM2:** 652-653
free-body system of **A11:** 460-461
from encapsulation **EL1:** 45
functions, nonlinear, of rheological behavior **EL1:** 849
gear **A11:** 589
heat-treating **A11:** 98
high static tensile **A11:** 537
-induced crystallization, defined **EM2:** 41
-induced plastic-package failures, integrated circuits **A11:** 789
intensity factor, effect in corrosion-fatigue **A11:** 255
intensity, threshold **A11:** 205
for interferometry **A17:** 408-410
interlaminar **EM1:** 229-230
internal, plastic encapsulants **EL1:** 806-808
level, as environmental factor **EM2:** 70
lowering, to prevent hydrogen damage **A11:** 251
magabsorption amplitude signals as function of **A17:** 155
and magnetic properties, in electromagnetic techniques **A17:** 159
maximum compressive **A11:** 102
maximum contact, subsurface **A11:** 340-341
maximum, criterion **EM1:** 238
maximum, schematic stress distribution of **A11:** 123
maximum tensile **A11:** 102
mean **A11:** 110-111, 255
measurements, quantitative **EL1:** 444-446
mechanical, and chip failure, plastic packages **EL1:** 480
minimum **A11:** 102
modeling of **A17:** 702
and moment resultants, structural analysis **EM1:** 460
near-surface, Barkhausen noise measurement **A17:** 160
negative tensile **A11:** 102
nominal, in shafts **A11:** 461
on conformal-coated components **EL1:** 765
operating, and corrosion-fatigue **A11:** 259
orientation **EM2:** 751-752
and package cracking **EL1:** 480
plastic, minimization **EL1:** 444
in polyimides **EL1:** 768-769
principal (normal), defined **A11:** 8
quenching **EM2:** 751
reduction of **A11:** 115
residual **A11:** 8, 97-98, 112, 473
reversed **A11:** 102
reversed-bending **A11:** 462
in rolling-contact fatigue **A11:** 133-134
rupture, defined **A11:** 9
shear, test structures for **EL1:** 445
shock, UHMWPE resistance **EM2:** 168
single cyclic **EM1:** 201-202
sources in manufacture **A11:** 205
subsurface, Barkhausen noise measurement **A17:** 160
sustained, strength/modulus effects **EM1:** 825
symbol for **A11:** 797
system, terms describing **A11:** 75
systems acting on shafts **A11:** 460-461
systems, and fatigue **EM2:** 701-702
temperature/moisture, in laminates **EM1:** 228-229
testing, zero-risk **EL1:** 136
thermal, residual/induced **EL1:** 45
threshold **A11:** 204
values, relation to strain values **A17:** 51
-wave shape **A11:** 255
Stress-annealed pyrolytic graphite
Raman analysis **A10:** 132-133
Stress-based approach to fatigue **A1:** 675, 676, 677
Stress-concentration factor **A8:** 13, 364, 371-372 **EM2:** 41, 709 **EM3:** 28
Stress-corrosion cracking *See also* Corrosion; Corrosion fatigue; Corrosive environment; Hydrogen embrittlement; Stress-corrosion cracking evaluation; Stress-corrosion fracture(s) **A6:** 374 **A13:** 145-163 **M1:** 687
accelerated testing media **A13:** 194

Stress-corrosion cracking (continued)
aircraft ... **A13:** 1026, 1043
AISI/SAE alloy steels ... **A12:** 27, 299, 305, 320
alloy composition ... **A13:** 615
alloys with high/moderate/low resistance to ... **A13:** 1103-1104
of aluminum ... **A12:** 28
aluminum alloys ... **A6:** 727 **M2:** 210-211, 212-218
aluminum-magnesium alloys ... **A6:** 622
in aluminum/aluminum alloys ... **A13:** 590-594
in amorphous metals ... **A13:** 868-869
anodic ... **A13:** 245-246
austenitic stainless steels ... **A6:** 466
behavior, criteria ... **A13:** 275-278
in beryllium-copper alloys ... **A9:** 393
in boiling water reactors ... **A13:** 927-936
of brass ... **A12:** 28
brass soldering ... **A6:** 130
in brazed joints ... **A13:** 878
brazing and ... **A6:** 117
in brewing piping ... **A13:** 1221
of carbon steel weldments ... **A13:** 365
in cast irons ... **A13:** 568-569
cathodic ... **A13:** 245-246
chloride, austenitic stainless steels ... **A12:** 354
chloride, prevention ... **A13:** 327
cleavage fracture, titanium alloys ... **A12:** 453
in closed feedwater heaters ... **A13:** 989-990
of cobalt-base alloys ... **A13:** 662
as cold-working effect ... **A13:** 49
compressive-stress ... **A12:** 25
conditions leading to ... **A13:** 615
copper ... **M2:** 239-240
copper alloys ... **A12:** 403
copper metals ... **M2:** 459, 462-463
of copper-zinc-tin alloys, and dezincification ... **A13:** 132
of copper/copper alloys ... **A13:** 614-616
and corrosion fatigue ... **A13:** 143-144
and corrosion fatigue cracking, compared ... **A13:** 291
and corrosion fatigue, unified theory ... **A12:** 42
corrosion testing for ... **A13:** 194
of corrosion-resistant cast steels ... **A13:** 582
crack initiation ... **A13:** 148-150
crack propagation process ... **A13:** 150-162
data, precision of ... **A13:** 278
decohesion fracture, aluminum alloy forging ... **A12:** 18, 25
decohesive rupture from ... **A12:** 18, 24
defined ... **A12:** 24 **A13:** 12, 145-146
dissimilar metal joining ... **A6:** 826
duplex stainless steels ... **A6:** 471, 477, 479
effect on dimple rupture ... **A12:** 24-29
electrochemical testing ... **A13:** 218
electrochemical/mechanical factors ... **A13:** 246
end grains, effect on resistance to ... **M1:** 355-356
environment-alloy combinations for ... **A13:** 326
environmental causes for selected metals and alloys ... **A6:** 1101
as environmentally induced cracking ... **A13:** 145-147
examination/interpretation ... **A12:** 133-134
external, prevention of ... **A13:** 1147
facets, austenitic stainless steels ... **A12:** 354, 357
failures, space boosters ... **A13:** 1102
ferritic stainless steel weldments ... **A13:** 361
flutes and cleavage from ... **A12:** 28
fracture ... **A13:** 148
fracture mode change by ... **A12:** 26-27
in generators ... **A13:** 1006-1007
grain-boundary separation by ... **A12:** 174
heat-treatable aluminum alloys ... **A6:** 534-535
hot cell ... **A13:** 936
initiation ... **A13:** 245-246
inside surface (primary side) tube, steam generators ... **A13:** 941
intergranular ... **A12:** 373, 357
intergranular, austenitic stainless steels ... **A13:** 124

Stress-corrosion cracking (continued)
and intergranular cracking ... **A13:** 123, 614
interlayer/base-metal interfaces and solid-state welding ... **A6:** 165, 171
iron-nickel alloy soldering ... **A6:** 130
irradiation-assisted ... **A13:** 935-936
kinetics ... **A13:** 153-155
of manganese alloys ... **A13:** 535
in manned spacecraft ... **A13:** 1082-1087
maraging steels ... **M1:** 450-451
material selection to avoid/minimize ... **A13:** 325-329
materials exhibiting ... **A13:** 145
mechanisms ... **A13:** 145, 147, 614-615
mechanisms of ... **A12:** 25-27
metal matrix composites ... **A13:** 860-861
micromechanisms ... **A13:** 145
nickel alloys ... **A6:** 740, 749
nickel alloys, resistance to ... **M3:** 171, 172, 174
of nickel-base alloys ... **A6:** 928 **A13:** 648-650
in nitrided stainless steel ... **A13:** 933
parameters, controlling ... **A13:** 147-148
in petroleum refining and petrochemical operations ... **A13:** 1274-1277
phenomenon of ... **A13:** 146-147
plateau crack velocities ... **A13:** 268
precipitation-hardening stainless steels ... **A12:** 372-373
precracked specimens, classification ... **A13:** 257
premature fracture by ... **A13:** 245
pressure vessels ... **A6:** 379
propagation ... **A13:** 245-246
of radioactive waste containers ... **A13:** 971
rate-determining steps ... **A13:** 147
ratings, aluminum alloys ... **A13:** 591-592
ratings, wrought products, high-strength aluminum alloys ... **A13:** 593
residual stresses ... **A6:** 1102
resistance, costs of ... **A13:** 327
revealed by differential interference contrast ... **A9:** 152
schematic ... **A13:** 151
in soil ... **A13:** 210
in space boosters/satellites ... **A13:** 1103-1105
in space shuttle orbiter ... **A13:** 1060
stages of ... **A13:** 146-147
stainless steel ... **M3:** 63, 64
in stainless steel casting alloys, role of ferrite in ... **A9:** 297
of stainless steels ... **A6:** 625, 626 **A13:** 554, 564, 933
in steam surface condensers ... **A13:** 988-989
in steam turbine materials ... **A13:** 952-959
steel weldments ... **A6:** 424-425
of steels ... **A12:** 27
strain rate effect, schematic ... **A13:** 261
stress sources ... **A13:** 615
in substrates ... **A13:** 421
sulfide, alloy steels ... **A12:** 299
summary ... **A13:** 162
in superheaters/reheaters ... **A13:** 992
technique ... **A6:** 1095, 1096
in telephone cable ... **A13:** 1130
temperature effects, high-strength steels ... **A13:** 533
testing, titanium/titanium alloys ... **A13:** 675
thermal stress, in brazed joint ... **A13:** 879
titanium ... **M3:** 415-416
titanium alloys ... **A12:** 28-29, 453
titanium-base corrosion-resistant alloys ... **A6:** 599
of titanium/titanium alloys ... **A13:** 674-675, 686-690
under thermal insulation ... **A13:** 1146-1147
of uranium/uranium alloys ... **A13:** 817-818
velocity ... **A13:** 277-278
welds ... **A6:** 101, 104
wrought aluminum alloys ... **A12:** 414, 433-436
of zinc/zinc alloys and coatings ... **A13:** 763-764
zirconium ... **M3:** 784
of zirconium/zirconium alloys ... **A13:** 718

Stress-corrosion cracking (SCC) *See also* Corrosion ... **A1:** 299-300, 723-728 **A11:** 203-224
accelerated testing mediums for ... **A8:** 522-532
accelerated testing of ... **A8:** 496-501, 522-532
and alloying element additions ... **A8:** 487, 489
alloys susceptible to ... **A11:** 206-207
aluminum alloy C-ring fracture by ... **A11:** 78-79
in aluminum alloys ... **A8:** 523-525 **A11:** 218-220
aluminum P/M alloys ... **A2:** 204-206
atmospheric environments contributing to ... **A11:** 208
in austenitic stainless steels ... **A11:** 215-217
bending vs. uniaxial tension ... **A8:** 503
in boilers and steam equipment ... **A11:** 624-626
in brazed joints ... **A11:** 451-453
brittle fracture from ... **A11:** 85
cast copper alloys ... **A2:** 361-362
of cast stainless steels ... **A1:** 917
in casting, with weld-metal deposits ... **A11:** 404
as cause of premature cracking ... **A8:** 496
causes and conditions for ... **A8:** 495-496
change in net section stress with onset of ... **A8:** 502
of cobalt-base corrosion-resistant alloys ... **A2:** 453
in copper alloys ... **A8:** 525-526
copper and copper alloys ... **A2:** 216 **A11:** 220-223
copper-zinc alloy failed by ... **A11:** 222
and corrosion fatigue ... **A8:** 495, 499
crack growth rate and ... **A11:** 53
crack initiation ... **A11:** 203
as craze cracking ... **A11:** 451
defined ... **A8:** 13, 495 **A11:** 10
determined ... **A11:** 27
of die-cast zinc alloy nut ... **A11:** 538-539
as distortion, in shafts ... **A11:** 467
effect of electrochemical factors ... **A8:** 499-500
effect of high-purity zinc on ... **A11:** 539
effect of *K* in ... **A8:** 497-498
effects of temper embrittlement on ... **A11:** 335
elastic strain specimens for ... **A8:** 503-508
electrochemical tests for ... **A8:** 533
environmental effects on ... **A11:** 207-212
environmental factors ... **A8:** 499-500
evaluation in stainless steels ... **A1:** 724, 725-728, 873
failure analysis of ... **A11:** 212-214
in ferritic stainless steels ... **A11:** 217
from caustic embrittlement by potassium hydroxide ... **A11:** 659
from hot chlorides, in pressure vessels ... **A11:** 660
general features of ... **A11:** 27, 203-204
in heat exchangers ... **A11:** 635-636
in high-strength steels ... **A8:** 526-527
and hydrogen embrittlement, compared ... **A8:** 537
hydrogen sulfide ... **A11:** 426
of Inconel 600 safe-end on a reactor nozzle ... **A11:** 660-661
of Inconel X-750 springs ... **A11:** 559-560
incubation period ... **A11:** 718
initiation method ... **A8:** 527
ions and substances causing ... **A11:** 207
and liquid-metal embrittlement, compared ... **A11:** 718
in low-carbon steels ... **A8:** 526
in magnesium alloys ... **A8:** 529-530 **A11:** 223
in maraging steels ... **A11:** 218
in marine-air environment ... **A11:** 309-310
in martensitic stainless steels ... **A11:** 217-218
mechanisms ... **A8:** 496
mechanisms of ... **A11:** 203
metal susceptibility to ... **A11:** 206-207
in mixer paddle shafts ... **A11:** 403-404
in neck liner,pulp digester vessel ... **A11:** 403
nickel alloys ... **A2:** 432-433 **A8:** 530
in nickel and nickel alloys ... **A11:** 223
in non-heat-treatable stainless steels ... **A8:** 527-529

Stress-corrosion cracking (SCC) (continued)
of nuclear steam-generator vessel **A11:** 656-657
of oil-well production tubing **A11:** 298
parameters affecting **A1:** 724
of pipeline **A11:** 701-703
plastic strain specimens for **A8:** 508-509
in precipitation-hardening stainless steels **A11:** 217-218
precracked specimens for **A8:** 497-498, 510-519
in pressure vessels **A11:** 656-661
propagation rate as function of *K* **A8:** 497
properties and conditions producing **A1:** 723-724
residual stress specimens **A8:** 509-510
resistance, of nickel alloys **A8:** 530
resistance ratings, wrought commercial aluminum alloys **A11:** 220
in rivets **A11:** 544-545
sampling of test materials for **A8:** 501
secondary **A8:** 520
service environment and **A8:** 522
in shafts **A11:** 467
slow strain rate testing of **A8:** 496-499, 519-520
smooth test specimens for **A8:** 503-510
in sour gas environments **A11:** 299-300
sources of stress **A11:** 204-206
sources of sustained tension **A8:** 501-502
of stainless steel bolts **A11:** 536-537
in stainless steel, by chloride-contaminated steam condensate **A11:** 625-626
of stainless steel eye terminal **A11:** 521
of stainless steel integral-finned tube **A11:** 635
of stainless steel pipe, by residual welding stress **A11:** 624-625
of stainless steel T-bolt **A11:** 538
static loading tests for **A8:** 496-498, 501-503
steam equipment failure by **A11:** 602
of steel castings **A11:** 402-403
of steel jackscrew drive pins **A11:** 546-548
in steel wire rope **A11:** 521
in steels **A11:** 101
stress-relief heat treating to reduce **A4:** 33, 34
striations **A11:** 8
and subcritical fracture mechanics **A11:** 53
sustained-load failure **A8:** 486
test environment selection **A8:** 521-522
testing specimens **A8:** 514
tests for **A8:** 495-536
in threaded fasteners **A11:** 536
in titanium alloys **A8:** 530-532
in titanium and titanium alloys **A11:** 223-224
tube sheet failed by **A11:** 210
verification procedures **A1:** 724-725
weldment testing **A8:** 520-521
worst case for **A8:** 499
wrought aluminum alloys **A2:** 30-32
in wrought carbon and low-alloy steels **A11:** 214-215
Stress-corrosion cracking evaluation **A13:** 245-282
of aluminum alloys **A13:** 265-268
of copper alloys **A13:** 268-270
dynamic loading: slow strain rate testing **A13:** 260-263
of high-strength steels **A13:** 270-272, 638-639
initiation and propagation **A13:** 245-246
interpretation of results **A13:** 275-279
of nonheat-treatable stainless steels **A13:** 272-273
of precracked (fracture mechanics) specimens **A13:** 146-147, 253-260
selection of test environments **A13:** 263-265
slow strain rate **A13:** 147
standardized tests **A13:** 246
static loading, smooth specimens **A13:** 146, 246-253
surface preparation, smooth specimens **A13:** 275
test coupons **A13:** 198
test method selection **A13:** 279
tests, magnesium/magnesium alloys **A13:** 273, 745
of titanium alloys **A13:** 273-275
of weldments **A13:** 275
Stress-corrosion cracking test **A1:** 611
Stress-corrosion cracking, zone 2
package exterior **EL1:** 1007
Stress-corrosion failures
in magnesium alloys **A9:** 426
Stress-corrosion fracture(s) *See also* Stress- corrosion cracking
austenitic stainless steel **A12:** 34
in brass **A12:** 36
path, effect of electrochemical potential **A12:** 27, 35
in steel **A12:** 34
Stress-corrosion threshold (K_{Iscc})
described **A11:** 53
Stress-cracking failure **EM3:** 28
Stress-induced atomic diffusion at elevated temperatures and low strain rates **A9:** 690
Stress-induced crystallization **EM3:** 28
Stress-induced martensite in titanium alloys **A9:** 461
Stress-intensity factor **A6:** 143, 147
Stress-intensity factor (*K*) *See also* Dynamic stress-intensity factor; Minimum stress- intensity factor
by specimen geometry, ultrasonic testing **A8:** 252
calculation, in crack growth propagation analysis **A8:** 680
closed form solutions for **A8:** 682
computations and estimates related to **A8:** 443-444
and corrosion fatigue **A8:** 405
and corrosion fatigue in steel in hydrogen **A8:** 409-410
and crack extension, resistance curves from **A8:** 450
and crack growth rates **A8:** 242, 365, 403-404
and crack propagation rate **A8:** 678-682
and crack tip stress fields **A8:** 441-443
defined **A11:** 10, 708
-dominated singularity zone size **A8:** 450-451
dynamic critical, from dynamic notched round bar testing **A8:** 275
in dynamic notched round bar testing **A8:** 282
effect in rack driving tendency **A8:** 439
effect on cleavage **A8:** 485-486
effect on crack propagation of aluminum alloy 7079-T651 .... **A8:** 408
effect on crack-extension force **A8:** 439
effect on ductile rupture **A8:** 485-486
effect on intergranular fracture **A8:** 485-486
effect on mixed fracture modes **A8:** 485-486
effect on plane-strain fracture toughness **A11:** 54
estimates for part-through surface cracks **A8:** 444
and fatigue crack growth rates **A8:** 403, 415
in fatigue crack propagation **A11:** 103
and fatigue striation spacing related **A8:** 482, 484
in fracture mechanics **A11:** 48, 55
in fracture-toughness testing **A11:** 60-61
and frequency effect on corrosion fatigue **A8:** 405-406
high, fatigue crack growth rate **A8:** 376-377
and high loading rates **A8:** 259-260
history, dependence on one-point bend test parameters **A8:** 272-273
history, for fracture test with long time-to-fracture **A8:** 275
hydrogen embrittlement crack growth rate as function of **A8:** 539
and loading frequency effect on crack growth in ultrahigh-strength steel **A8:** 405-406
with precracked specimens **A8:** 514
range **A8:** 13, 403, 405, 408-409, 485-486
range, symbol for **A11:** 797
rates, for dynamic fracture **A8:** 259
relation to crack opening, crack-extension force, *J*-integral **A8:** 440
relation to crack speed, semi-brittle material **A8:** 441
and residual stress **A11:** 57
solution, range of application **A8:** 379-380
strength ratio correlation, hydrogen embrittlement testing **A8:** 541
and stress-corrosion cracking **A8:** 497-498
subscripts, defined **A11:** 10
symbol for **A8:** 725 **A11:** 797
for three-dimensional problems **A8:** 444
Stress-intensity factor (*K*) (continued)
threshold, and hydrogen embrittlement **A8:** 537
threshold, and minimum crack growth rate **A8:** 254, 256
threshold, design significance **A8:** 538
threshold, short cracks **A8:** 379
and toughness, compared **A11:** 49
as value for onset of cleavage, rapid-load testing **A8:** 453
Stress-intensity factors **EM3:** 367, 483, 503, 504
Stress-range histograms
for Yellow Mill Pond Bridge **A11:** 708
Stress-related failure mechanisms
microelectronic devices **EL1:** 1052-1054
Stress-relaxation testing **A8:** 311, 322-328
applications **A8:** 322
bend **A8:** 326-328
clamp spring **A8:** 327
curve, defined **A8:** 13
curve for continuous tensile **A8:** 324-325
equipment **A8:** 311
helical compression spring **A8:** 328
system for step-down tension testing **A8:** 324
tension **A8:** 323-325
torsion **A8:** 327
Stress-relief anneal *See also* Annealing; Fabrication characteristics
wrought aluminum and aluminum alloys **A2:** 110-111, 118
Stress-relief annealing *See* Annealing
Stress-relief cracking **A1:** 607
examination/interpretation **A12:** 139-140
metallographic example **A12:** 159
Stress-relief heat treating **A4:** 33-34 **M4:** 3-5
annealing aluminum alloys **A4:** 870-871
cold treating **A4:** 203-204
cooling **A4:** 33, 34 **M4:** 5
creep **A4:** 33-34 **M4:** 5
induction heating energy requirements **A4:** 189
induction heating temperatures **A4:** 188
residual stress **A4:** 33-34 **M4:** 3-5
temperatures **A4:** 34 **M4:** 4
time-temperature relation **A4:** 33, 34 **M4:** 4
welding, effect on residual stress **A4:** 33 **M4:** 3, 4
yield strength, relation to temperature **A4:** 34 **M4:** 4-5
Stress-relief heat treatment
mechanical property effects **A16:** 25
Stress-relieved high-carbon steel wire **M1:** 264
Stress-relieved uncoated high-carbon wire **A1:** 283
Stress-relieving
before hard chromium plating **M5:** 185-186
Stress-rupture properties *See also* Creep behavior; Elevated temperature properties; specific material type
constructional steels for elevated temperature use **M1:** 640, 642, 646, 649-652, 655-656, 658
D-6a steel **M1:** 432
ductile iron **M1:** 47, 49, 50, 51, 95
malleable iron **M1:** 65-66, 70-72
Stress-rupture strength *See* Creep rupture strength; Creep-rupture strength
Stress-rupture testing
constant-load **A8:** 313-318
constant-stress **A8:** 318-321
data presentation **A8:** 315
and ductility, influence of triaxiality factor on **A8:** 345
equipment **A8:** 311-313
of Inconel 718 .... **A8:** 333, 335
method for master Larson-Miller curve in **A8:** 333, 335
notch effects **A8:** 315-316
stress amplitude vs. time to fracture **A8:** 349-350
temperature control and measurement in furnace **A8:** 312-313
and tension-hold-only, compared **A8:** 348, 349
Stress-sorption theory
and stress-corrosion cracking **A11:** 203
Stress-strain *See also* Stiffness **EM3:** 28
beam distributions **A8:** 119, 121
behavior, as mechanical properties measurement **EM2:** 433-435
behavior, superplastic alloys **A14:** 855-857

Stress-strain (continued)
curve **EM1:** 23, 296
curves, defined **EM2:** 41
curves, metal and plastic **EM2:** 74
defined **EM1:** 23 **EM2:** 41
diagram, defined **A14:** 13, 19
of fibers **EM1:** 175
field, effect on crack extension **A8:** 439
from wave propagation **A8:** 209
in lamina **EM1:** 218-220, 231
loop, for constant-strain cycling **A8:** 367
of matrices **EM1:** 176
ply **EM1:** 146, 459
pure plastic bending **A8:** 118-124
rate behavior, superplastic alloys **A14:** 854
relations, forming, stainless steels **A14:** 760
relationships in bending **A8:** 118-124
relationships in cantilever beam bend test **A8:** 133
and strain gage measurements related **A8:** 198-199
tensile test configuration schematic **A8:** 208
tests, compressive **A8:** 197
vs. life approach, in torsional fatigue testing **A8:** 149
Stress-strain curve *See also* Engineering stress-strain curve; Stress-strain diagram **EM3:** 28
for aluminum **A8:** 236-237
in bending **A8:** 117, 133-134
in cantilever beam bend test **A8:** 133
cantilever bending normalized **A8:** 134
for carbon steel **A8:** 174
dynamic, of steel in shear **A8:** 236
for elastic plastic bending **A8:** 119, 121
for electrolytic tough pitch (ETP) copper in tension **A8:** 213-214
engineering **A8:** 20-27
from high strain rate shear testing **A8:** 215
from incremental strain rate testing **A8:** 223
for iron **A8:** 174
for nickel **A8:** 174
niobium, yield strength and deformation rate **A8:** 38
pure plastic bending **A8:** 120, 122
and rate changes **A8:** 177
serrations in **A8:** 35
shear, for copper **A8:** 229, 231
shear, for OFHC copper **A8:** 216-217
for torsional Kolsky bar dynamic testing **A8:** 228
for vacuum arc remelted steel **A8:** 237-238
Stress-strain curves
cold-finished bars **M1:** 221, 244, 245-249
gray cast iron **M1:** 18, 20
malleable iron **M1:** 65
Stress-strain diagram **A8:** 13-14, 34-35, 124, 323
defined **A10:** 682 **A11:** 10
nonlinear **A11:** 47
Stress-strain relationships **EM3:** 392-394
Stress-time curve *See* Stress-relaxation curve
in stress-corrosion cracking **A13:** 276
Stress-wave shape
effect on corrosion fatigue **A11:** 255
Stresscoating
for residual stress study **A11:** 134
Stresses
imperfection-caused **A15:** 88
in solidification, distortion from **A15:** 615-616
weldment, weld beads for **A15:** 526
Stresses, residual **A4:** 603-610
after surface hardening **A4:** 606-607
in boriding **A4:** 607, 608
in carburizing **A4:** 606-607, 608, 611
cast irons **A4:** 681
compressive **A4:** 604, 605, 606, 607-608
control in heat-treated parts **A4:** 609
development in processed parts **A4:** 604-609
effects **A4:** 603-604
Stresses, residual (continued)
in induction hardening **A4:** 607
measurement **A4:** 609-610
in nitriding **A4:** 607, 611
in nitrocarburizing **A4:** 607
in nonferrous heat-treated alloys **A4:** 608-609
quench cracking **A4:** 610-612
tensile **A4:** 604, 605, 606, 607-608
thermal contraction **A4:** 605
thermal evaluation for residual stress analysis (TERSA) **A4:** 610
vapor pressure, relation to temperature **A4:** 495
volumetric changes **A4:** 605
Stressing frames
SCC testing **A8:** 507
Stretch
automotive chain wear **A18:** 566
Stretch bending
of bar **A14:** 661
Stretch draw forming
form-block method **A14:** 593
lancing **A14:** 593
mating-die method **A14:** 593
Stretch factors, critical distance
RLC line **EL1:** 360
Stretch forming **A14:** 591-598
accuracy **A14:** 596-597
advantages **A14:** 591-592
aluminum alloys **A14:** 798
applicability **A14:** 591-592
of beryllium **A14:** 807
compression forming **A14:** 594-595
of copper and copper alloys **A14:** 814-818
defined **A14:** 13
draw forming **A14:** 593-594
equalizing **A14:** 777
fundamentals **A14:** 591
of large parts **A14:** 596
limitations **A14:** 592
machines and accessories **A14:** 592, 596
magnesium **M2:** 543, 545-546
magnesium alloys **A2:** 469-471 **A14:** 830
methods **A14:** 591
operating parameters **A14:** 598
radial draw forming **A14:** 595-596
stainless steels **A14:** 776-777
stretch wrapping **A14:** 594
surface finish **A14:** 597-598
of titanium alloys **A14:** 846
vs alternative methods **A14:** 598
Stretch forming machines and accessories **A14:** 593, 596
Stretch straightening **A14:** 681
Stretch wrapping **A14:** 594
Stretch-bending test **A8:** 14, 560-561
Stretch-drawing tests, simulative
for sheet metals **A8:** 563
Stretchability
measure of **A8:** 561
Stretched zone
measured **A11:** 56
Stretcher leveling
of carbon steel sheet **A1:** 206
defined **A14:** 13
Stretcher leveling of steel sheet **M1:** 160
Stretcher straightening
defined **A14:** 13
Stretcher strains *See also* Deformation bands; Lüders lines **A1:** 574 **A14:** 13, 547
appearance **A9:** 687
in carbon steel sheet and strip **A1:** 204
in deep drawn low-carbon steel aerosol can **A9:** 688
defined **A9:** 17
as formability problem **A8:** 548
Stretcher strains, low-carbon sheet and strip *See also* Luders lines **M1:** 157
Stretcher-strain formation
by strain aging **A12:** 129, 148
Stretching
balanced biaxial **A8:** 547
and bending, as springback **A8:** 553
biaxial test **A8:** 558-559
defined **A14:** 13
effect on springback **A8:** 553, 565
from creep, nickel-base alloy **A11:** 283
m and *n* values for **A8:** 550-551
of sheet metals **A14:** 877
tests **A14:** 890-891
tests, simulative, for sheet metals **A8:** 561-562
titanium alloys **A12:** 453
wrought aluminum alloys **A12:** 434
zone, toughness calculated by size of **A11:** 56
Stria *See* Weld line
Striae *See* Flow line
Striation *See also* Beach marks **A13:** 12, 937
defined **A8:** 14
definition **EM4:** 633
ductile intergranular **A8:** 487
Striation spacing *See* Fatigue striation spacings
Striations *See* Fatigue striations
by high-cycle fatigue **A11:** 78
by microscopy **A11:** 105
concentric circular, in polymers **A11:** 762
count, by quantitative fractography **A11:** 56
in deburring drum SEM fractograph **A11:** 347
defined **A11:** 10
fatigue **A11:** 75, 78, 105, 347, 657, 683, 688
in fatigue fractures **A11:** 75
fatigue-crack growth, in bridge components **A11:** 708
as indicator of fatigue **A11:** 258, 259
-like rub marks **A11:** 27
in nuclear steam-generator wall **A11:** 657
SCC, in stainless steel **A11:** 28
stage II **A11:** 22, 23
in stainless steel implant **A11:** 683, 687
surface, from fatigue **A11:** 80-81
and Wallner lines, compared **A11:** 26
Stribeck curve **A18:** 60, 66, 89
coefficient of friction of sliding bearings **A18:** 518, 519
defined **A18:** 18
lubrication regimes of tribosystems **A18:** 476
Stribeck-Hersey curve **A18:** 48, 49
Striker bar
for explosively loaded torsional Kolsky bar **A8:** 227
for incident wave generation **A8:** 201
with single pressure bar **A8:** 199
in split Hopkinson pressure bar **A8:** 198-199
Strikes
as surface plating **EL1:** 679
Striking surface
defined **A14:** 13
Stringer
in ultrahigh-strength steel **A8:** 477-478
Stringer bead
definition **A6:** 1214 **M6:** 17
Stringer bead welding
cast irons **A6:** 710
nickel-base corrosion-resistant alloys containing molybdenum **A6:** 594
Stringers *See also* Sulfide stringers
deep shell crack from **A12:** 288
defined **A9:** 17 **A11:** 10 **A12:** 65
as inclusion **A12:** 65, 67
inclusion, in failed extrusion press **A11:** 38
as linear element in quantitative metallography **A9:** 126
in magnesium alloys **A9:** 429
nonmetallic, high-carbon steels **A12:** 285
slag **A11:** 316
of slag, as woody fracture **A12:** 224
sulfide, AISI/SAE alloy steels **A12:** 294
Stringiness **EM3:** 28

Strip *See also* Carbon steel sheet and strip; Cobalt powder strip; Nickel powder strip; Roll compacted strip; Roll compacting **A7:** 12, 297, 401-409
bend testing **A8:** 117, 132
bending moment-deflection data **A8:** 134
bending, nickel-base alloys **A14:** 835-836
beryllium-copper alloys **A2:** 403, 413
for blanking **A14:** 449
brazing filler metals available in this form **A6:** 119
coiled, slitting and shearing of **A14:** 708-713
cold-rolled **A1:** 846
defined **A14:** 13
flash welding **M6:** 558
flatteners and levelers for **A14:** 713
heat treatment **A1:** 846, 848
heated-roll rolling **A14:** 356-357
heavy, silicon-iron, as magnetically soft materials **A2:** 769
high-strength low-alloy steel strip
cold-forming strip **A1:** 417
compositions **A1:** 406, 417
forming properties **A1:** 418, 420
specifications **A1:** 399, 401
yield and tensile strengths **A1:** 417
hot-rolled **A1:** 846
lamination, iron-cobalt alloy, properties **A2:** 777
lockseaming/can seaming **A14:** 572
stainless steel *See* Stainless steel, strip **A1:** 846, 848
steel *See* Steel, strip
steel, multiple-slide forming of **A14:** 567-574
thickness for hardness testing **A8:** 105
wrought copper **A2:** 241-248
wrought high-strength beryllium-copper alloys **A2:** 409
Strip chart recorder
for fatigue testing machines **A8:** 368
Strip chart recorders
as eddy current inspection readout **A17:** 179
Strip, continuous galvanized
chromating of **A13:** 390
Strip, copper
temper designations **M2:** 248-249
Strip finishes **A1:** 886
Strip line
defined **EL1:** 1158
Strip liner cladding **M6:** 804
Strip plastic zone model
fracture tests using **A8:** 449
Strip rolling
theory **A14:** 344-346
Strip-back method **A6:** 591
Strip-type bearings **A18:** 741
Strip-type wide face brushes **M5:** 152-153
Stripline probes
microwave **A17:** 219
Stripline properties
impedance models **EL1:** 602
Stripped die method **A7:** 12
Stripper **EM3:** 28
defined **A14:** 13
guided, for piercing **A14:** 465
punch, defined **A14:** 13
Stripper crane wheel
fatigue failure **A11:** 526
Stripper plate *See* Hold-down plate
Stripper punch **A7:** 12
Stripping
of carbon replicas **A12:** 181
defined **A15:** 11
force, in blanking **A14:** 448
mechanism, multiple-slide forming **A14:** 569
plate, contoured **A15:** 28
solvent **EL1:** 780
Stripping analysis, electrometric titration and
compared **A10:** 202
Stripping methods
aluminum and aluminum alloys **M5:** 3-19, 218
cadmium plating **M5:** 267-268, 647
chrome pickle **M5:** 648
chromium **M5:** 218
chromium plating **M5:** 184-185, 198, 646
coatings, various **M5:** 3-19
copper alloys **M5:** 218
Stripping methods (continued)
copper plate **M5:** 567-568, 646
dichromate coatings **M5:** 648
gold plate **M5:** 283, 647
iron-base alloys **M5:** 218
lead deposits from steel **M5:** 275
magnesium alloys, plated deposits on **M5:** 646-647
nickel plate **M5:** 218, 567-568, 646-647
paint *See* Paint and painting, stripping methods
phosphate coating weight determination **M5:** 451
primers **M5:** 647-648
silver plate **M5:** 647
steel **M5:** 3-19
tin plate **M5:** 647
vinyl-base enamels **M5:** 648
zinc alloys **M5:** 218
zinc plate **M5:** 646
Stripping pressure **A7:** 190-191
Stripping voltammetry **A10:** 192-193
Strips **A18:** 569
Stroh formalism **A6:** 146
Stroke
defined **A14:** 13
Stroke capacity, and press tonnage
rigid tool compaction **A7:** 325-326
Stroke-restricted machines **A14:** 25, 37
Stromeyer, Wilhelm
as metallurgical physicist **A15:** 29
Strong acid number **A18:** 84
Strong base number **A18:** 84
Strong-beam images **A9:** 111
compared to weak-beam images **A9:** 114
of curved dislocation segment **A9:** 114
strain contrast in **A9:** 112
Strongbacks
electroslag welding **M6:** 230
Strontium
as addition to aluminum-silicon alloys **A18:** 788
alloying, aluminum casting alloys **A2:** 132
alloying, wrought aluminum alloy **A2:** 55
in aluminum alloys **A15:** 746
epithermal neutron activation analysis **A10:** 239
as modifier addition **A15:** 484
pure **M2:** 799
pure, properties **A2:** 1159
as silicon modifier **A15:** 79, 161-165
and sodium, in aluminum-silicon alloys compared **A15:** 167
species weighed in gravimetry **A10:** 172
sulfate ion separation **A10:** 169
sulfuric acid as dissolution medium **A10:** 165
TNAA detection limits **A10:** 237
vapor pressure **A6:** 621
Strontium ferrites **A9:** 539
thermal etching **EM4:** 575
Strontium oxide
in binary phosphate glasses **A10:** 131
Strontium oxide (SrO)
glass-ceramic/metal seals **EM4:** 499
glass/metal seals **EM4:** 494
in glaze composition for tableware **EM4:** 1102
specific properties imparted in CTV tubes **EM4:** 1039
tooling for pressed ware **EM4:** 398
Strontium titanate **EM4:** 60, 542
application **EM4:** 542
Strontium titanium oxide ($SrTiO_3$)
dielectric properties **EM4:** 877
gas pressure sintering for pressure densification **EM4:** 299
Strontium, vapor pressure, relation
to temperature **M4:** 310
Structural film adhesives
aerospace applications **EM3:** 44
curing methods **EM3:** 44
resin system types **EM3:** 44
shelf life **EM3:** 44
Structural adhesive *See also* Adhesives
defined **EM1:** 23 **EM2:** 41
Structural Adhesive Joints in Engineering **EM3:** 70-71
Structural adhesives
addition of secondary phases to suppress viscoelasticity **EM3:** 513
advantages and limitations **EM3:** 75, 77
Structural adhesives (continued)
aerospace applications **EM3:** 44
automotive, advantages **EM3:** 555
characteristics **EM3:** 74
chemical families **EM3:** 77-80
polyetheretherketone (PEEK), fracture toughness **EM3:** 509, 510
properties **EM3:** 78
selection **EM3:** 76
shear properties **EM3:** 321
suppliers **EM3:** 510, 513
for woodworking **EM3:** 46
Structural Adhesives and Bonding **EM3:** 69
Structural Adhesives Bonding **EM3:** 68
Structural Adhesives: Developments in Resins and Primers **EM3:** 71
Structural Adhesives—Chemistry and Technology **EM3:** 69
Structural alloys
effect of temperature on fatigue life of **A11:** 130
fatigue crack propagation in liquid metal environments **A8:** 425-426
overload failure mechanism **A12:** 12
Structural aluminum alloys
R-curve measurements **A8:** 452
strain rate sensitivity of **A8:** 38, 40
Structural analysis *See also* Comosite structural analysis; Composite structures; Structural analysis and design; Structure; Structure determinations; Substructure; Surface structure
by CBED **A10:** 461-464
of change, by neutron diffraction **A10:** 420
computer programs for **EM1:** 268-274
crystal, applicable analytical methods **A10:** 4
of defects, applicable analytical methods **A10:** 4
electronic **A10:** 60, 277
elemental, applicable analytical methods **A10:** 4
extended x-ray absorption fine structure **A10:** 407-419
in situ, of active sites in catalysts **A10:** 407
in vivo, of active sites in metalloproteins **A10:** 407
inert gas fusion, in titanium **A10:** 231
of inorganic solids, applicable analytical methods **A10:** 4-6
of milligram quantities **EM2:** 836-837
molecular spectroscopy **EM2:** 825-828
molecular weight **EM2:** 828-830
phase distribution, applicable analytical methods **A10:** 4
phase identification, applicable analytical methods **A10:** 4
of polymers **EM2:** 835-836
problem solving **EM2:** 824-825
in solutions, voltammetric analysis **A10:** 188
substructure, due to cold **A10:** 468-469
substructure, due to hot deformation and restoration **A10:** 469-470
thermal, methods of **EM2:** 830-835
x-ray diffraction (XRD) analysis **EM2:** 835
Structural analysis and design *See also* Structural analysis; Structure
assembly methods **EM2:** 711-725
creep and stress relaxation **EM2:** 659-678
design guidelines, general **EM2:** 707-710
engineering formulas, use of **EM2:** 652-654
fatigue loading **EM2:** 701-706
impact loading **EM2:** 679-700
introduction **EM2:** 651
properties divergencies **EM2:** 655-658
Structural anomalies
as casting defects **A11:** 387-388 **A17:** 519-520
Structural anomalies, as casting defects
types **A15:** 552-553
Structural applications *See also* Honeycomb structures
aircraft components, eddy current inspection **A17:** 189-194
casting inspections **A17:** 530-531
characterization **A17:** 50-51
component assessment testing program **A17:** 686
components, cemented carbide **A2:** 971-972
computed tomography (CT) **A17:** 364
ductile iron **M1:** 36

Structural applications (continued)
hot rolled bars and shapes **M1:** 207, 209, 211-213
lead and lead alloy **A2:** 555-556
magnetic paint inspection **A17:** 128
malleable iron **M1:** 9
nickel aluminides **A2:** 918
parts, copper-base **A2:** 396-398
plate **M1:** 181
steel, red lead rust inhibitors **A2:** 548
Structural applications for technical, engineering, and advanced ceramics **EM4:** 959-960
advantages gained by substituting ceramics for metals **EM4:** 959
application areas **EM4:** 959
automotive applications **EM4:** 960
ceramic heat exchangers **EM4:** 960
engineering ceramics market growth chart **EM4:** 959
Structural assessment
aluminum-silicon alloys **A15:** 166
Structural bond **EM3:** 28
defined **EM1:** 23 **EM2:** 41
Structural carbon steels **A1:** 389
high-strength **A1:** 390-391
hot-rolled carbon-manganese structural steels **A1:** 390, 391
mild steels **A1:** 390
Structural ceramic composites
carbon fiber reinforcement **EM1:** 929-930
ceramic fiber reinforcement **EM1:** 930-931
fabrication **EM1:** 926-927
metal reinforcement **EM1:** 927-929
Structural ceramics *See also* Ceramics **EM4:** 18-20
alumina ceramics **A2:** 1021
aluminum titanate **A2:** 1021-1022
applications **A2:** 1019, 1021-1024 **EM4:** 18, 19
aerospace **EM4:** 1005
automotive components **EM4:** 19
biomedical applications **EM4:** 19
boron carbide **A2:** 1022
composite ceramics **A2:** 1023-1024
defined **A2:** 1019
design practices in automotive turbocharger wheels **EM4:** 722-726
fabrication processes **EM4:** 18, 19
finishing of **A2:** 1021
forming and fabrication **A2:** 1020
load-bearing applications **EM4:** 18
materials for tribological applications **EM4:** 19
materials selection **EM4:** 19-20
power generation systems **EM4:** 19
primary properties of interest **EM4:** 19
processing **A2:** 1019-1021
processing equipment **EM4:** 19
properties **A2:** 1019, 1021-1024 **EM4:** 18, 19, 29
properties needed in key design areas **EM4:** 691
raw material preparation **A2:** 1019-1020
silicon carbide **A2:** 1022
silicon nitride **A2:** 1022
silicon-aluminum-oxynitride (SiAlON) **A2:** 1022
thermal treatment **A2:** 1020-1021
toughened ceramics **A2:** 1022-1023
uses **A2:** 1019
zirconia **A2:** 1022
Structural Ceramics Database (SCD) project
National Institute of Standards and Technology (NIST) **EM4:** 692
Structural ceramics, design practices in gas turbine engines *See* Design practices for structural ceramics in gas turbine engines
Structural ceramics in gasoline engines, design practices *See* Design practices for structural ceramics in gasoline engines
Structural clay products **EM4:** 943-951
aggregates **EM4:** 950
chemically resistant units **EM4:** 950-951
common properties **EM4:** 943
facing materials **EM4:** 943-946
Structural clay products (continued)
glazes **EM4:** 1061
load-bearing products **EM4:** 946-949
passive solar brick **EM4:** 951
paving units **EM4:** 949-950
product applications **EM4:** 943, 944
roofing tile **EM4:** 950
terra cotta **EM4:** 951
vitrified clay pipe **EM4:** 951
Structural components
use of tool materials for **M3:** 558-559
Structural composites *See also* Composites
cyanates **EM2:** 232-233
Structural compression molding **EM1:** 561-562
Structural configuration, effect
damage tolerance **EM1:** 262-264
Structural design
categories of **A11:** 115
corrosion protection by **A13:** 379
Structural diagrams
Laplanche diagram **A15:** 68-70
Maurer diagram **A15:** 68-69
Structural evolution
EXAFS described **A10:** 407
Structural failure
predicting **A8:** 682
Structural foams *See also* Foams **EM3:** 28
defined **EM2:** 41
thermoplastic **EM2:** 508-513
Structural grades
beryllium **A2:** 686
Structural grades of beryllium **A9:** 390
Structural gradients **A9:** 602
Structural HSLA steels
applications **A1:** 399, 401, 420
compared with carbon steel **A1:** 389
compositions **A1:** 401, 406, 410, 420
normalized **A1:** 410
tensile properties
cold-rolled **A1:** 420
hot-rolled **A1:** 411
Structural liquid and paste adhesives
aerospace applications **EM3:** 44
shelf life **EM3:** 44
Structural parts
bronze and copper **A7:** 105
Structural plastics
chemistry **EM2:** 65-66
types **EM2:** 65-66
Structural quality
hot rolled bars **M1:** 204, 207-213
plate
ASTM specifications **M1:** 183
compositions **M1:** 185-189
mechanical properties **M1:** 190-191
sheet and strip **M1:** 154, 155
formability **M1:** 547
Structural quality carbon steels **A1:** 202
Structural reaction injection molding (SRIM) *See also* Injection molding; Reaction injection molding (RIM); Resin transfer molding (RTM)
defined **EM2:** 41
economic factors **EM2:** 349-350
for large parts **EM2:** 344
for polyurethanes (PUR) **EM2:** 262
process **EM2:** 344-346
and resin transfer molding compared **EM2:** 344-351
surface finish **EM2:** 304
surface requirements **EM2:** 350
textured surfaces **EM2:** 305
urethane hybrids **EM2:** 269-270
Structural reaction injection molding (structural RIM) **EM1:** 564
Structural sections
computer-aided roll pass design **A14:** 349-350
Structural shapes *See* Shapes
Structural stability
mechanically alloyed oxide dispersion-strengthened (MA ODS) alloys **A2:** 947
Structural steel *See also* As-rolled structural steels; HSLA steels
atmospheric corrosion of **M1:** 723
fatigue crack growth in **A1:** 663-664, 665, 666
fracture toughness characteristics of **A1:** 664, 666-667, 669
high-strength low-alloy steels for **A1:** 419
hot-rolled shapes **M1:** 207, 209, 212-213
low-temperature properties of **A1:** 662-672
notch toughness of **A1:** 412
specifications **A1:** 664-665, 667, 668
tubing **M1:** 323, 325, 326
Structural steel tubing **A1:** 334, 335
Structural steels
applications
offshore structures **A6:** 384
shipbuilding and offshore structures **A6:** 381-385
atmospheric corrosion **A13:** 532
capacitor discharge stud welding **A6:** 222
Charpy V-notch impact test for **A8:** 263
corrosion failures **A13:** 1308-1309
crack arrest testing of **A8:** 453-455
ductile-to-brittle transition temperature **A11:** 68-69
dynamic fracture toughness for **A8:** 269
electroslag welding **A6:** 277 **M6:** 226
fracture testing, effects of cleavage **A8:** 458
gas tungsten arc welding **M6:** 203-204
general biological corrosion **A13:** 87, 88
impact stress-intensity testing of **A8:** 453
marine corrosion **A13:** 540, 544
plane-strain fracture toughness **A8:** 450
resilience of **A8:** 23
ship/submarine **A13:** 546
submerged arc welding **M6:** 116
Structural steels, specific types *See also* Low-alloy steels, specific types
15B22M, electron-beam welding **A6:** 1077
A36
composition **A6:** 642
dissimilar metal joining **A6:** 825
electrogas welding **A6:** 656, 660
flux-cored arc welding **A6:** 187
mechanical properties **A6:** 642
recommended preheat and interpass temperatures **A6:** 644
A53, recommended preheat and interpass temperatures **A6:** 644
A106, recommended preheat and interpass temperatures **A6:** 644
A108, stud arc welding **A6:** 211
A131
composition **A6:** 642
electrogas welding **A6:** 656
mechanical properties **A6:** 642
recommended preheat and interpass temperatures **A6:** 644
A131-C, electrogas welding **A6:** 660
A139, recommended preheat and interpass temperatures **A6:** 644
A203
electrogas welding **A6:** 660
flux-cored arc welding **A6:** 187
A204-A, electroslag welding **A6:** 660
A242
electrogas welding **A6:** 656
recommended preheat and interpass temperatures **A6:** 644
A283
composition **A6:** 642
electrogas welding **A6:** 656
mechanical properties **A6:** 642
A284
composition **A6:** 642

Structural steels, specific types (continued)
mechanical properties ... **A6:** 642
A285
composition ... **A6:** 642
corrosion of weldments ... **A6:** 1066
electrogas welding ... **A6:** 656
flux-cored arc welding ... **A6:** 187
mechanical properties ... **A6:** 642
A302-B, electroslag welding ... **A6:** 660
A381, recommended preheat and interpass temperatures ... **A6:** 644
A387
flux-cored arc welding ... **A6:** 187
laser hardfacing ... **A6:** 807
A387-C, electroslag welding ... **A6:** 660
A387-D, electroslag welding ... **A6:** 660
A387A, weld cladding ... **A6:** 818
A441
electrogas welding ... **A6:** 660
recommended preheat and interpass temperatures ... **A6:** 644
A442
composition ... **A6:** 642
mechanical properties ... **A6:** 642
A500, recommended preheat and interpass temperatures ... **A6:** 644
A501, recommended preheat and interpass temperatures ... **A6:** 644
A514
flux-cored arc welding ... **A6:** 188
recommended preheat and interpass temperatures ... **A6:** 644
A515
composition ... **A6:** 642
electrogas welding ... **A6:** 656
flux-cored arc welding ... **A6:** 187
mechanical properties ... **A6:** 642
A515 GR70, electroslag welding ... **A6:** 660
A516
composition ... **A6:** 642
electrogas welding ... **A6:** 656, 660
flux-cored arc welding ... **A6:** 187
mechanical properties ... **A6:** 642
recommended preheat and interpass temperatures ... **A6:** 644
A517
flux-cored arc welding ... **A6:** 188
recommended preheat and interpass temperatures ... **A6:** 644
A524, recommended preheat and interpass temperatures ... **A6:** 644
A529, recommended preheat and interpass temperatures ... **A6:** 644
A537
composition ... **A6:** 642
electrogas welding ... **A6:** 656
mechanical properties ... **A6:** 642
recommended preheat and interpass temperatures ... **A6:** 644
weld compositional analysis ... **A6:** 101
A537-1, electrogas welding ... **A6:** 660
A570, recommended preheat and interpass temperatures ... **A6:** 644
A572
electrogas welding ... **A6:** 656, 660
recommended preheat and interpass temperatures ... **A6:** 644
A572-50, electrogas welding ... **A6:** 660
A573
composition ... **A6:** 642
electrogas welding ... **A6:** 656
mechanical properties ... **A6:** 642
recommended preheat and interpass temperatures ... **A6:** 644
A588
electrogas welding ... **A6:** 656, 660
flux-cored arc welding ... **A6:** 187
A588, recommended preheat and interpass temperatures ... **A6:** 644
A595, recommended preheat and interpass temperatures ... **A6:** 644
A606, recommended preheat and interpass temperatures ... **A6:** 644
A607, recommended preheat and interpass temperatures ... **A6:** 644
A618, recommended preheat and interpass temperatures ... **A6:** 644

Structural steels, specific types (continued)
A633
electrogas welding ... **A6:** 660
recommended preheat and interpass temperatures ... **A6:** 644
A662
composition ... **A6:** 642
mechanical properties ... **A6:** 642
A709, recommended preheat and interpass temperatures ... **A6:** 644
A710
flux-cored arc welding ... **A6:** 187
recommended preheat and interpass temperatures ... **A6:** 644
A808, recommended preheat and interpass temperatures ... **A6:** 644
ABS, recommended preheat and interpass temperatures ... **A6:** 644
API 2H, recommended preheat and interpass temperatures ... **A6:** 644
API 5L, recommended preheat and interpass temperatures ... **A6:** 644
Structural test applications
acoustic emission inspection ... **A17:** 290-292
Structural testing
element and subcomponent ... **EM1:** 313-345
Structural tests
of VLSI/ULSI/WSI devices ... **EL1:** 377
Structural welding
joining processes ... **M6:** 56
Structural-element testing
data analysis and reporting ... **EM1:** 327-329
in-plane loaded ... **EM1:** 320-325
joint elements ... **EM1:** 325-326
methodology ... **EM1:** 326-327
normal and bending loaded ... **EM1:** 325
Structurally reinforced carbon-carbon composites ... **EM1:** 922-924
Structure *See also* Structural analysis; Structural analysis and design; Structure determinations; Substructure; Surface structure
actinide metals ... **A2:** 1189-1198
amorphous materials and metallic glasses ... **A2:** 809-812
atomic-scale, FIM/AP for ... **A10:** 584
banded, defined ... **A11:** 1
benefits, aluminum P/M alloys ... **A2:** 200-201
beryllium-copper alloys ... **A2:** 286-290
cartridge brass ... **A2:** 302
columnar, defined
commercial bronze ... **A2:** 297
crystal, of organic solids, analytic methods for ... **A10:** 9
defined ... **A9:** 17
and degradation, of plasma-polymerized hexamethyidisiloxane ... **A10:** 285-286
design guidelines ... **EM2:** 710
electrical resistance alloys ... **A2:** 836-839
fibrous, defined ... **A11:** 4
as form of model definition ... **EL1:** 129
gilding metal ... **A2:** 295
of inorganic liquids and solutions, analytic methods ... **A10:** 7
of inorganic solids, analytic methods applicable ... **A10:** 6
layered, RBS analysis of ... **A10:** 628
low brass ... **A2:** 299-300
of materials, and materials characterization ... **A10:** 1
molecular, of organic solids, analytic methods for ... **A10:** 9
molecular, of thermoplastic resins ... **EM2:** 619-621
of organic solids and liquids, techniques for ... **A10:** 10
of organic solids, methods for ... **A10:** 9
palladium-silver alloys ... **A2:** 716
phase distribution/morphology, of organic solids, analytic methods for ... **A10:** 9
platinum ... **A2:** 709
platinum alloys ... **A2:** 709-714
platinum-iridium alloys ... **A2:** 710
polymer ... **EM2:** 571-572
polymer, and electrical properties ... **EM2:** 462-464
polymer, and properties ... **EM2:** 57-59
polymer, thermoset resins ... **EM2:** 626

Structure (continued)
-property relations, polyether-imides (PEI) ... **EM2:** 156-157
pure metals ... **A2:** 1099-1178
rare earth metals ... **A2:** 722, 1178-1189
SAS techniques for ... **A10:** 405
static, aluminum and aluminum alloys ... **A2:** 9-10
sub-boundary *See* Grain boundary
ternary molybdenum chalcogenides (chevrel phases) ... **A2:** 1078
vs resistance to attack ... **EM2:** 426
yellow brass ... **A2:** 303-304
Structure amplitude
effect on structure-factor contrast ... **A9:** 113
Structure and properties of glasses ... **EM4:** 564-568
glass characterization ... **EM4:** 564-565
ASTM definition ... **EM4:** 564
glass modifiers ... **EM4:** 564
Pauling's rules ... **EM4:** 564
Zachariasen's rules ... **EM4:** 564
properties of glass ... **EM4:** 565-568
annealing of glass ... **EM4:** 567-568
chemical durability ... **EM4:** 566
density ... **EM4:** 566
dielectric constant of glass ... **EM4:** 566
dispersion ... **EM4:** 565
elastic properties ... **EM4:** 566
electrical conductivity ... **EM4:** 566
index of refraction ... **EM4:** 565, 568
optical ... **EM4:** 565
photoelasticity ... **EM4:** 565
static fatigue ... **EM4:** 567
strength ... **EM4:** 566-567
thermal expansion ... **EM4:** 568
transmission of light ... **EM4:** 565
viscosity ... **EM4:** 567
Structure control phases
heat-resistant alloys ... **A14:** 266
Structure determinations *See also* Structural analysis; Structure; Substructure; Surface structure
analytical transmission electron microscopy ... **A10:** 429-489
crystallographic texture measurement and analysis ... **A10:** 357-364
electron spin resonance ... **A10:** 253-266
extended x-ray absorption fine structure ... **A10:** 407-419
ferromagnetic resonance ... **A10:** 267-276
field ion microscopy ... **A10:** 583-602
infrared spectroscopy ... **A10:** 109-125
Mössbauer spectroscopy ... **A10:** 287-295
neutron diffraction ... **A10:** 420-426
nuclear magnetic resonance ... **A10:** 277-286
Raman spectroscopy ... **A10:** 126-138
single-crystal x-ray diffraction ... **A10:** 344-356
small-angle x-ray and neutron scattering ... **A10:** 402-406
x-ray diffraction ... **A10:** 325-332
x-ray powder diffraction ... **A10:** 333-343
Structure factor
complete equation for ... **A10:** 352
defined ... **A9:** 17 **A10:** 349, 682-683
equation, cells, and diffraction in unit cell ... **A10:** 329
F, derivation of ... **A10:** 350
Structure image
formed by transmission electron microscopy ... **A9:** 104
Structure maps
for ordered intermetallics ... **A2:** 929-930
Structure prototypes ... **A3:** 1•16
Structure prototypes of crystals ... **A9:** 707-708, 711-718
Structure, secondary *See* Secondary structure
Structure-factor calculations for ordered structures ... **A9:** 109
Structure-factor contrast in imperfect crystals ... **A9:** 112-113
Structure-insensitive electrical properties
defined ... **A2:** 761-762
Structure-sensitive electrical properties
defined ... **A2:** 761-762
Structured light method
of object orientation ... **A17:** 34
Structured query language (SQL) ... **EM4:** 692
Structured reliability testing
oxide contamination effect ... **EL1:** 959-960

Structured reliability testing (continued)
purple plague ... **EL1:** 960
temperature-cycling-sensitive designs ... **EL1:** 961
temperature-sensitive mechanisms ... **EL1:** 959-961
Structured-light
three-dimensional vision system ... **A6:** 1126
Structures *See* Composite structures
corrosion in ... **A13:** 1299-1310
formation of ... **A9:** 602
observation methods related to size of structure ... **A9:** 605
Structuring, and restructuring
of defective circuitry ... **EL1:** 9
Strukturbericht designations ... **A3:** 1•16
Strukturbericht symbols ... **A9:** 706-707
conversion to Pearson symbols ... **A9:** 707
Strut, composite
design of ... **EM1:** 183-184
Stub
definition ... **A6:** 1214
Stub axles
slag inclusions in ... **A11:** 322-323
Stub lathes ... **A16:** 153, 156, 157
Stud arc welding *See also* Stud welding
definition ... **M6:** 17
Stud arc welding (SW) ... **A6:** 210-220
alloy steel ... **A6:** 210, 211, 212, 213, 214
aluminum ... **A6:** 210, 211, 212, 213, 214, 216, 218-219
applications ... **A6:** 210, 213, 214, 216
automotive ... **A6:** 393-395
austenitic stainless steels ... **A6:** 213
base plate material and thickness ... **A6:** 213-214
brass ... **A6:** 210, 218-219
carbon steel(s) ... **A6:** 210, 211, 212, 213, 214, 216, 652, 653, 659-660
counterbore and countersink weld flash clearance dimensions ... **A6:** 215
definition ... **A6:** 1214
equipment ... **A6:** 210, 211, 212, 214, 217, 219-220
failure mode ... **A6:** 218
fixturing ... **A6:** 215-216, 218, 219
"flash" ... **A6:** 210, 213, 217
gas-shielded arc welding ... **A6:** 214
inspection ... **A6:** 216-219
plating or contaminant removal ... **A6:** 218-219
power supplies ... **A6:** 210
power/control ... **A6:** 210, 211, 212
process overview ... **A6:** 210-215
process selection guidelines for arc welding ... **A6:** 210, 653
process variations ... **A6:** 214
qualification ... **A6:** 216-219
quality control ... **A6:** 216-219
recommended base metal thicknesses ... **A6:** 215
safety precautions ... **A6:** 219-220
short-cycle stud arc welding ... **A6:** 214
special equipment ... **A6:** 214
stainless steel ... **A6:** 210, 211, 212, 213, 214, 216, 218-219
stud strength ... **A6:** 214
testing methods ... **A6:** 219
tooling ... **A6:** 215-216
types of ... **A6:** 210
typical settings ... **A6:** 214
vs. capacitor discharge stud welding ... **A6:** 210
Stud arc welding of
alloy steels ... **M6:** 730
aluminum alloys ... **M6:** 730, 733
carbon steels ... **M6:** 730, 733
copper alloys ... **M6:** 730
galvanized steel ... **M6:** 733
high-carbon steels ... **M6:** 733
high-strength low-alloy steels ... **M6:** 733
low-carbon steel ... **M6:** 733
stainless steel ... **M6:** 730, 733
Stud steels-temperature ... **A1:** 292
Stud welding ... **M6:** 729-738
of aluminum alloys ... **M6:** 399
Stud welding (continued)
applications ... **M6:** 729-730
capacitor discharge stud welding ... **M6:** 729, 734-738
automatic feed systems ... **M6:** 737
drawn arc method ... **M6:** 735
guns ... **M6:** 736-737
initial contact method ... **M6:** 735
initial gap method ... **M6:** 735
inspection and quality control ... **M6:** 738
metals welded ... **M6:** 738
portable system ... **M6:** 734-735
power/control units ... **M6:** 737
studs ... **M6:** 730, 735-736
weld current/weld time relationships ... **M6:** 737
comparison to percussion welding ... **M6:** 739
definition ... **A6:** 1214 **M6:** 17
equipment ... **M6:** 732, 736-737
fillet dimensions ... **M6:** 730
inspection and quality control ... **M6:** 734
bend tests ... **M6:** 734
limitations ... **M6:** 730
metals welded ... **M6:** 733, 738
process selection ... **M6:** 730
base metal ... **M6:** 730
base-metal thickness ... **M6:** 730
weld-base diameter ... **M6:** 730
standard procedure qualification test weldments ... **A6:** 1090
stud arc welding ... **M6:** 730-734
automatic feed systems ... **M6:** 732
guns ... **M6:** 732
heat-affected zone ... **M6:** 733
metals welded ... **M6:** 733
power/control units ... **M6:** 732
studs and ferrules ... **M6:** 731-732
weld current/weld time relationships ... **M6:** 733
Stud-mount devices
three-terminal ... **EL1:** 429
two-terminal ... **EL1:** 431-432
Student's *t* statistic
value, for confidence level ... **A8:** 699
values ... **A8:** 711
Studies, fracture
history of ... **A12:** 1-8
Studs *See also* Threaded fasteners ... **M6:** 731-732
alloy steel for ... **M1:** 256
for capacitor-discharge stud welding ... **M6:** 730, 735-736
composition ... **M1:** 274
description ... **M1:** 274
designs ... **M6:** 730-732, 735-736
materials ... **M6:** 730-731, 735-736
measured by coordinate measuring machines ... **A17:** 18
mechanical properties ... **M1:** 274, 278-280
proof stress ... **M1:** 273, 274, 277-278
roll threading ... **M1:** 274-276
selection of steel for ... **M1:** 274-276
strength grades and property classes ... **M1:** 273-277
for stud arc welding ... **M6:** 730-732
Stuffing box
fatigue fracture of ... **A11:** 346-347
Stulen equivalent stress parameter
Goodman diagram plot for ... **A8:** 713
Stump-type lockbolts
materials and composite applications for ... **A11:** 530
Sturtevant blower
as cyclone type ... **A15:** 27
Stylus profilometer
for plating thickness testing ... **EL1:** 943
Stylus profilometry ... **A18:** 396, 397
Styrene
copolymerized with methyl methacrylate (MMA) ... **EM1:** 94
Styrene (continued)
as diluent, sheet molding compounds ... **EM1:** 142
effect, flexural strength, glass-polyester composite ... **EM1:** 93
oxidative degradation ... **EM1:** 94
as polyester diluent ... **EM1:** 132-133
polymerization of ... **A10:** 132
Styrene acrylonitrile (SAN) ... **EM3:** 28
surface preparation ... **EM3:** 279-280
Styrene block copolymer
advantages and disadvantages ... **EM3:** 675
Styrene copolymers *See also* Copolymer(s)
as engineering thermoplastics ... **EM2:** 446-447
environmental effects ... **EM2:** 427
Styrene polyesters
prebond treatment ... **EM3:** 35
Styrene polymers
environmental effects ... **EM2:** 427
Styrene-acrylonitriles (SAN, OSA, ASA) *See also* Thermoplastic resins
applications ... **EM2:** 215
characteristics ... **EM2:** 215-216
commercial forms ... **EM2:** 214
competitive materials ... **EM2:** 215
copolymer compatibilities ... **EM2:** 216
costs and production volume ... **EM2:** 214
defined ... **EM2:** 41
design considerations ... **EM2:** 216
processing ... **EM2:** 216
properties ... **EM2:** 215
suppliers ... **EM2:** 216
Styrene-butadiene ... **EM3:** 51
applications ... **EM3:** 56
properties ... **EM3:** 677
Styrene-butadiene block copolymer
used as modifier ... **EM3:** 121
Styrene-butadiene copolymers ... **EM3:** 75
Styrene-butadiene rubber
wear index and test severity ... **A18:** 580
Styrene-butadiene rubber (SBR) ... **EM3:** 76, 82-83, 86, 89
additives and modifiers ... **EM3:** 148
for auto body sealing ... **EM3:** 57
for auto headliners ... **EM3:** 147
for automotive padding and insulation ... **EM3:** 147
for body sealing and glazing materials ... **EM3:** 57
for book binding ... **EM3:** 147
for ceramic and tile bonding compounds ... **EM3:** - - 147
characteristics ... **EM3:** 90
chemistry ... **EM3:** 147
commerical forms ... **EM3:** 147
for construction ... **EM3:** 147
cross-link density effect on tensile strength ... **EM3:** 412
laminating bond ... **EM3:** 148
markets ... **EM3:** 147-148
for nonwoven goods binder ... **EM3:** 147-148
for orthopedic devices ... **EM3:** 147
for packaging ... **EM3:** 147
properties ... **EM3:** 143-144, 147
for tirecord coating ... **EM3:** 147
Styrene-butadiene rubber/polyethylene terephthalate (PET)
fracture energy versus debond rate ... **EM3:** 383, 385
Styrene-butadiene-styrene (SBS) ... **EM3:** 677
thermoplastic elastomers ... **EM3:** 51
Styrene-diene polymers
as viscosity improvers ... **A18:** 109, 110
Styrene-ester polymers
hydrocarbon moiety of dispersants ... **A18:** 99
Styrene-ethylene/butylene-styrene (SEBS) ... **EM3:** 677
Styrene-isoprene
infrared spectroscopy data ... **A18:** 301
Styrene-isoprene-styrene (SIS) ... **EM3:** 677
Styrene-maleic anhydride (SMA) ... **EM3:** 28

Styrene-maleic anhydrides (S/MA) *See also* Thermoplastic resins
alloys and blends **EM2:** 217
applications **EM2:** 217-219
characteristics **EM2:** 219-221
commercial forms **EM2:** 217
competitive materials **EM2:** 218-219
copolymer chemistry **EM2:** 66
costs and production volume **EM2:** 217
defined **EM2:** 41
design considerations **EM2:** 219
extrusion **EM2:** 220
filled vs unfilled **EM2:** 220-221
injection molding **EM2:** 220
and polycarbonate alloy, properties **EM2:** 221
processing **EM2:** 220
properties **EM2:** 219-221
resin compound types **EM2:** 220-221
Suppliers **EM2:** 221
Styrene-methyl methacrylate copolymer syrup
typical formulation **EM3:** 121
Styrene-polyester (STPE) polymers
as dispersants **A18:** 111
as pour-point depressants **A18:** 111
as viscosity improvers **A18:** 109, 110, 111
Styrene-rubber plastics **EM3:** 28
defined **EM2:** 41
Styrene-thinned polyesters
competing with anaerobics **EM3:** 116
Styrofoam pattern
defined **A15:** 11
St‡$$‡Adablein successive milling technique **A6:** 1095
Sub-boundary structure
defined **A11:** 10
Sub-seabed sediments
high-level waste disposal in **A13:** 978-980
Sub-sieve analysis *See also* Fisher sub-sieve analysis; Sieve analysis; Sieve(s) **A7:** 12
Sub-sieve fraction *See also* Fisher sub-sieve analysis; Sieve fraction **A7:** 12
Sub-sieve size **A7:** 12
Sub-sow block
defined **A14:** 13
Subassembly adhesives **EM3:** 37
Subboundaries
in crystals **A9:** 601, 719-720
formation in pure, single-crystal metals **A9:** 607
in germanium **A9:** 608
in high-purity tin **A9:** 607
Subboundary structure
defined **A9:** 17-18
Subcase fatigue
bearing components **A11:** 505
as spalling **A11:** 134
Subcase fatigue fracture
alloy steel **A12:** 322-323
Subcomponent testing
bolted-joint **EM1:** 336-337
bonded-joint **EM1:** 337-339
combined in-plane/normal loaded **EM1:** 334-336
experimental procedure/data/reports **EM1:** 339-340
in-plane loaded **EM1:** 329-334
normal loaded **EM1:** 333-334
structural **EM1:** 313, 329-334
Subcomposites
fabricating process steps **EL1:** 134
Subcritical annealing
defined **A9:** 18
Subcritical assemblies
as neutron sources **A17:** 389-390
Subcritical crack growth **A13:** 283, 1085
curve **A11:** 55
and stable crack growth, compared **A11:** 63
Subcritical fracture mechanics (SCFM)
abbreviation for **A11:** 797
crack-propagation calculated by **A11:** 55
defined **A11:** 47
in failure analyses **A11:** 52-53
fatigue and **A11:** 52-53
stress-corrosion cracking (SCC) and **A11:** 53
Subcritical heat treatment
white irons **A15:** 684
Subcritical quenching
of uranium alloys **A2:** 673
Subcritically reheated grain-coarsened (SCGC) zone **A6:** 80-81
Subgrain
defined **A9:** 18 **A11:** 10
Subgrain boundaries in titanium alloys
method of revealing **A9:** 461
Subgrain boundary
effect on cleavage fracture **A12:** 17
Subgrain growth in annealed cold-worked specimens
at grain boundaries **A9:** 696
observation of **A9:** 694
as recovery mechanism **A9:** 697
Subgrains *See also* Grains
caused by dislocations in pure metals **A9:** 607
defined **A9:** 18
formation **A9:** 601
formation and growth during recovery **A9:** 693-694
as result of deformation at elevated temperatures and high strain rates **A9:** 690
size and shape **A10:** 365
topographic methods for **A10:** 368
Subgroup **EM3:** 791
Subject contrast
radiography **A17:** 298-299
Sublaminates
in laminate ranking method **EM1:** 450-456
Sublattice ordering in intermetallic compounds
NMR analysis **A10:** 283-284
Sublimation curves ..... **A3:** 1 • 2
Sublimation energies
of rare earth metals **A2:** 723-724
Sublimation for purifying metals **M2:** 710
Sublimation point
in reduction reactions **A7:** 53
Submarine, and ship
corrosion **A13:** 543-544, 546
Submarine structures
composite **EM1:** 838
Submerged arc welding
advantages **M6:** 115
arc starting **M6:** 134-135
automatic welding **M6:** 132
of cast irons **A15:** 523-524
cleaning, initial and interpass **M6:** 134
comparison to flux cored arc welding **M6:** 113
consumables **M6:** 119-127
cooling rate, effect of **M6:** 117-118
cracking **M6:** 128-130
hydrogen induced **M6:** 129-130
solidification **M6:** 128-129
weld bead profile effects **M6:** 127
definition **M6:** 17
distortion **M6:** 130
effect of base-metal and electrode compositions **M6:** 117
effect of flux composition on
basicity **M6:** 125
control of element transfer **M6:** 126
weld-metal composition **M6:** 124-126
weld-metal oxygen content **M6:** 125
welding parameters **M6:** 125-126
electrical relationships **M6:** 135
oscillographic voltage traces **M6:** 135
electrode wire **M6:** 119-122
composition **M6:** 119-121
current range **M6:** 121-122
melting rate **M6:** 122
packaging **M6:** 119-120
wire surface **M6:** 120
failure origins **A11:** 415
flux formulation and deposit **M6:** 117
fluxes **M6:** 122-127
agglomerated **M6:** 123
bonded **M6:** 122-123
classification system **M6:** 122-123
composition **M6:** 122-124
composition, effect on weld hydrogen content **M6:** 124-125
effect of flux composition on content **M6:** 124-125
handling and storage **M6:** 123
low-oxygen potential types **M6:** 123
Submerged arc welding (continued)
melting rate **M6:** 126-127
prefused **M6:** 122
type of welding current **M6:** 126
viscosity and conductivity **M6:** 127
hardfacing **M6:** 784
heat flow calculations **M6:** 31
heat input, effect of **M6:** 117-119
heat-affected zone microstructure and toughness **M6:** 118-119
grain coarseness **M6:** 118-119
relationship of heat input **M6:** 119
jigs, fixtures, and booms **M6:** 134
design principles **M6:** 134
tack welds **M6:** 134
joint design **M6:** 133-134
avoiding entrapment of slag **M6:** 141-142
backing rings or bars **M6:** 134, 143
horizontal edge-flange weld **M6:** 143-144
reducing cost and distortion **M6:** 142-143
use of an offset **M6:** 143
limitations **M6:** 115
metallurgical considerations **M6:** 116
heat-affected zone toughness **M6:** 116
oxygen potential **M6:** 116
weld-metal properties **M6:** 116
multiple-electrode welding **M6:** 132-133
electrode position **M6:** 133
number of electrodes **M6:** 132
operation, principles of **M6:** 114-115
machines **M6:** 114-115
setup **M6:** 114
porosity in welds **M6:** 127-128
arc blow **M6:** 127-128
contaminants **M6:** 127
insufficient flux coverage **M6:** 127
prevention **M6:** 128
postheating **M6:** 135
powder metal joint fill **M6:** 133
power supplies **M6:** 130-132
alternating current **M6:** 131
direct current **M6:** 131
volt-ampere curves **M6:** 131-132
preheating **M6:** 135
production examples **M6:** 137-152
meeting toughness and hardness specifications **M6:** 138-139
service in corrosive environments **M6:** 140-141
weld-bead quality **M6:** 140
recommended grooves **M6:** 72
safety precautions **M6:** 58, 137
semiautomatic welding **M6:** 133
slag inclusions **M6:** 130
slag removal **M6:** 135
strip electrode welding **M6:** 133
weld discontinuities **M6:** 830
weld overlaying **M6:** 807
weld size and shape, effects on **M6:** 135-137
current **M6:** 135-138
electrode-wire diameter **M6:** 137
travel speed **M6:** 136-138
voltage **M6:** 136
weld-bead fusion area **M6:** 137
weld-metal microstructure **M6:** 116-118
high-strength low-alloy steels **M6:** 116-117
weldability of steels **M6:** 115-116
welding position **M6:** 134
welding tabs **M6:** 135
wire-feed systems **M6:** 132
Submerged arc welding (SAW) **A6:** 202-209
additions to fluxes **A6:** 61
advantages **A6:** 202
all-weld-metal chemical compositions for martensitic stainless steel filler metals **A6:** 439
alloy steels **A6:** 202, 203
applications **A6:** 62, 204
automatic welding **A6:** 202, 205
carbon steels **A6:** 202, 203, 204, 642, 643, 647, 648, 649, 650, 652, 653, 654, 658
cast irons **A6:** 716, 720, 721
chromium **A6:** 206
cobalt **A6:** 206
copper alloys **A6:** 752, 755
cryogenic service **A6:** 1017
definition **A6:** 202, 1214
deposition rate **A6:** 202, 206-207

Submerged arc welding (SAW) (continued)
discontinuities from **A17:** 582
dissimilar metal joining **A6:** 824
duplex stainless steels **A6:** 477, 480
electrodes
electrode selection **A6:** 202, 203, 204-205, 209
equipment **A6:** 202, 203-205
for stainless steels **A6:** 700, 702
ferritic stainless steels **A6:** 448
filler metals **A6:** 748
fluid flow phenomena **A6:** 24
flux classification **A6:** 203-204
relative to basicity index (BI) **A6:** 204
relative to effect on alloy content of weld deposit **A6:** 203-204
relative to production method **A6:** 203
flux-to-wire ratio **A6:** 204, 206, 207
fluxes **A6:** 202, 203, 224, 700-701
acid **A6:** 204
active **A6:** 204
alloy **A6:** 204
basic **A6:** 204
bonded **A6:** 203
forms of **A6:** 62
fused **A6:** 203
neutral **A6:** 204
types of **A6:** 62
used for applications **A6:** 62
fume/consumable removal systems **A6:** 1063
hardfacing alloy consumable form **A6:** 796
hardfacing alloys **A6:** 796, 798, 799, 800, 801, 802
heat sources **A6:** 1144
heat-affected zone **A6:** 202-203
heat-treatable low-alloy steels **A6:** 669, 670-671
heat-treated steels **A6:** 202
high-strength low-alloy structural steels **A6:** 663, 664
hydrogen cracking **A6:** 209
insufficient fusion **A6:** 207-208
joint configurations **A6:** 203
limitations **A6:** 202
low-alloy metals for pressure vessels and piping **A6:** 668
low-alloy steels **A6:** 204-205, 662, 663, 664, 668, 669, 670-671, 674, 676
for pressure vessels and piping **A6:** 667
low-carbon steels **A6:** 10
manganese **A6:** 206
matching filler metal specifications **A6:** 394
melting and solidification sequence **A6:** 202, 203
metallurgical discontinuities **A6:** 1073
mild steel **A6:** 204
nickel **A6:** 206
nickel alloys **A6:** 740, 743, 748, 749
nickel-based electrodes **A6:** 720
nonferrous alloys **A6:** 203
not for HSLA Q&T structural steels **A6:** 666
not recommended for nickel-base corrosion-resistant alloys containing molybdenum **A6:** 594
oxygen content in weld metal effect on flux basicity index **A6:** 59
parameters **A6:** 206-209
electrical stickout **A6:** 207
flux layer depth **A6:** 207
travel speed **A6:** 207
weld current **A6:** 206-207
weld voltage **A6:** 207
personnel considerations **A6:** 205-206
personnel training **A6:** 202
pipe inspection **A17:** 578-579
porosity **A6:** 209
postweld heat treatment **A6:** 426
power source selected **A6:** 36, 37
power sources **A6:** 203
power supplies **A6:** 203-205
precipitation-hardening stainless steels **A6:** 489
pressure vessel manufacture **A6:** 379
principles of operation **A6:** 202

Submerged arc welding (SAW) (continued)
process applications **A6:** 202-203
process selection guidelines for arc welding **A6:** 653
railroad equipment **A6:** 396
rating as a function of weld parameters and characteristics **A6:** 1104
for repair of high-carbon steels **A6:** 1105
for repair welding **A6:** 1103, 1107
safety precautions **A6:** 206, 1191, 1192-1193
semiautomatic welding **A6:** 202, 205
shielding gases **A6:** 58
shipbuilding **A6:** 384
slag **A6:** 202, 203
slag entrapment **A6:** 207-208
slag viscosity **A6:** 60
solidification cracking **A6:** 208-209
stainless steel dissimilar welds **A6:** 502
stainless steels **A6:** 203, 680, 681-682, 688, 693, 694, 698, 699, 700-703
steel weldment soundness **A6:** 409, 413, 414
titanium alloys **A6:** 521, 783
twin arc process **A6:** 202
vanadium **A6:** 206
vs. flux-cored arc welding **A6:** 186
vs. gas-metal arc welding **A6:** 180
wear-resistant steel **A6:** 203
weld cladding **A6:** 816, 817, 820-821, 822
weld cladding of stainless steels **A6:** 813
welding conditions **A6:** 425
Submerged arc welding of
alloy steels **M6:** 301-303
filler metals **M6:** 301
fluxes **M6:** 301-302
low-temperature applications **M6:** 303
carbon steels, hardenable **M6:** 266-268
carbon-manganese steels **M6:** 119
cast irons **M6:** 310
copper and copper alloys **M6:** 426
heat-resistant alloys **M6:** 367
high-strength low-alloy steels **M6:** 115-116
low-carbon steels **M6:** 115
malleable iron **M6:** 316-317
medium-carbon steels **M6:** 115-116
nickel alloys **M6:** 442
stainless steels **M6:** 1116
stainless steels, austenitic **M6:** 326-329
circumferential welding **M6:** 329
current **M6:** 327
electrode wires **M6:** 327-328
filler metals **M6:** 327
flux **M6:** 327
joint design **M6:** 328
weld backing **M6:** 327-328
weld quality **M6:** 328
stainless steels, ferritic **M6:** 347
stainless steels, martensitic **M6:** 349
stainless steels, nitrogen-strengthened austenitic **M6:** 345
structural steels **M6:** 116
Submerged arc welding process **A7:** 821-822 **A18:** 644, 653
Submerged cutting *See* Submerged sectioning
as a sectioning method for tool steels **A9:** 256
Submerged sectioning **A9:** 24
Submerged wires
to measure mercury displacement **A7:** 268
Submerged zone
marine structures **A13:** 542-544
Submersibles
for commercial operations **EM1:** 838-839
Submicron powder **A7:** 12
Submicron spatial resolution failure analysis
by SIMS **EL1:** 1088
Submicron tungsten carbide-cobalt alloys
compositions and structure **A2:** 952
Submicroscopic
defined **A9:** 18

Submonolayers
effect of oxygen adsorbed on metal surfaces **A10:** 552
Subsample
defined **A10:** 683
Subsampling **A10:** 13, 15-17
Subsoil zone
marine structures **A13:** 542-544
Substance, amount of
SI base unit and symbol for **A10:** 685
SI unit/symbol for **A8:** 721
Substitute steels
effect on economy in manufacture **M3:** 847
Substitution of elements in aluminum alloy phases **A9:** 359
Substitutional element
defined **A9:** 18
Substitutional solid solution .... **A3:** 1 • 15, 17
defined **A9:** 18 **A13:** 46
Substitutional solid-solution strengthening
ion implantation strengthening mechanisms **A18:** 855-856, 858
Substrate **EM3:** 28, 39
defined **A9:** 18 **EM2:** 41
definition **A6:** 1214 **M6:** 17
effects of grain size on surface kinetics **A10:** 560
heavily TaC-coated WC + cobalt, XPS analysis of **A10:** 576-577
high-resolution SIMS spectra for phosphorus- doped silicon **A10:** 623
high-surface-area, infrared-transparent, DRS for **A10:** 114
light, diffusion in **A10:** 628
lighter elements, surface impurities of heavy elements on **A10:** 628
perfection of **A10:** 375
platinum, extent of coverage, Ni-P film on **A10:** 608
platinum, LEISS spectra from Ni-P film on **A10:** 609
rocking curve analyses of **A10:** 371
semiconductor, topographic methods for **A10:** 368
silicon, organometallic silicate film deposited on **A10:** 617
tin-nickel, composition vs depth of passive i film on **A10:** 608-609
WC-Co tool, vapor-deposited multilayer structure, by AES **A10:** 561
Substrate considerations
selection of cleaning process **M5:** 4-5, 7-10, 14
Substrate intensity attenuation method
for thin-film samples **A10:** 95
Substrate interconnection technology
for HWSI systems **EL1:** 88
Substrate(s) *See also* Passive substrates
adhesion promotion **EL1:** 624
advanced ceramic and silicon **EL1:** 8
alternative electronic **EL1:** 306
available materials for **EL1:** 105-107
bonding, low-CTE metal planes **EL1:** 627
card edge **EL1:** 624
ceramic, medical and military applications **EL1:** 386
for ceramic packages **EL1:** 203-206
ceramic, thick-film hybrid technology **EL1:** 206-208
ceramic, world market **EL1:** 460
cleaning, for conformal (urethane) coating **EL1:** 776-778
compensation, layer artwork and **EL1:** 623-624
cooling **EL1:** 308-309
copper, multichip structures **EL1:** 307
costs **EL1:** 612
CTE target for **EL1:** 612-613
defined **EL1:** 1158
dielectric, flexible printed boards **EL1:** 582-583
double-sided rigid (DSR) **EL1:** 15
epitaxial effects **EL1:** 104

Substrate(s) (continued)
flexible/rigid ... **EL1:** 505
glass ... **EL1:** 338
hard-wired, by rigid epoxies ... **EL1:** 810
high density ... **EL1:** 439, 441
low-expansion, characteristics ... **EL1:** 611-613
matched to silicon ... **EL1:** 441
materials ... **EL1:** 318, 538-539
metal core ... **EL1:** 337-338
metallization, in eutectic bonding ... **EL1:** 214
multichip structures ... **EL1:** 304-307, 441
multilayer thin-film hybrid ... **EL1:** 323
organic, aramids as ... **EL1:** 535
orientation ... **EL1:** 444
and packages ... **EL1:** 203-212
passive ... **EL1:** 8
physical characteristics ... **EL1:** 104-106
plastic packages ... **EL1:** 209-212
plastic pin-grid arrays ... **EL1:** 476
polymeric ... **EL1:** 338-339
porcelain-coated steel ... **EL1:** 337
preparation ... **EL1:** 195-197
programmable thin-film silicon ... **EL1:** 301
properties ... **EL1:** 802-803
separate interconnection, with HWSI assemblies ... **EL1:** 76
silicon, epitaxial GaAs on ... **EL1:** 200
specific types ... **EL1:** 105-106
thermal-mechanical effects ... **EL1:** 740-753
thermomechanical properties ... **EL1:** 612
thick-film hybrid technology ... **EL1:** 206-208, 334-339
for thick-film screening ... **EL1:** 207
for thin-film hybrids ... **EL1:** 315-316
Substrate-integrated communication circuitry ... **EL1:** 8
Substrates
beryllium ... **A2:** 684
corrosion of ... **A13:** 108, 420-422
formability ... **A14:** 560
of germanium and germanium compounds ... **A2:** 743
for SEM specimens ... **A12:** 172
for SiC fibers ... **EM1:** 858
specific types ... **EL1:** 105-106
of superconducting thin-film materials ... **A2:** 1081
Substrates, moving
tape casting ... **EM4:** 162
Substructure
due to cold deformation ... **A10:** 468-469
due to hot deformation and restoration ... **A10:** 469-470
Substructure of metals ... **A9:** 604
Subsurface
blowholes, as casting defects ... **A11:** 382
contact fatigue, cracking from ... **A11:** 134
defects ... **A11:** 127, 330-331
defined ... **A18:** 376
discontinuities ... **A11:** 120-121
inclusions, fatigue fracture from ... **A11:** 323
-initiated fatigue, in bearing components ... **A11:** 504
nonmetallic inclusions, rolling;contact fatigue from ... **A11:** 504
residual stress and hardness distributions in induction-hardened steel shaft ... **A10:** 389-390
shear stresses, cylindrical roller ... **A11:** 500-501
Subsurface corrosion *See also* Subsurface flaws
aircraft, eddy current inspection ... **A17:** 193
defined ... **A13:** 12
Subsurface cracking
in cleavage, low-carbon steel ... **A12:** 117
high-carbon steels ... **A12:** 280
medium-carbon steels ... **A12:** 261
Subsurface fatigue ... **A18:** 259
Subsurface flaws
by thermal inspection ... **A17:** 396
casting defects, inspection methods ... **A17:** 512
castings, by ultrasonic inspection ... **A17:** 530
cracks, electric current perturbation inspection of ... **A17:** 136
cracks, microwave inspection ... **A17:** 203
cracks, phase-sensitive CRT instrumentation ... **A17:** 194
from electron beam welding ... **A17:** 586-587
from plasma arc welding ... **A17:** 587
Subsurface flaws (continued)
magnetic particle inspection ... **A17:** 105
ultrasonic inspection ... **A17:** 231
in weldments ... **A17:** 593-594
Subsurface residual stress distributions ... **A10:** 380, 388-390
Subsurface stress *See also* Subsurface cracks; Subsurface flaws
Barkhausen noise measurement ... **A17:** 160
Subsurface stress gradients
effect on subsurface measurement ... **A10:** 388
Subsurface(s)
cracking ... **A12:** 117, 261, 280
fracture, ductile irons ... **A12:** 236
thermal-wave imaging of ... **A12:** 166, 169
Subtraction, image *See* Image subtraction
Subtraction techniques ... **A10:** 183
Subtractive processing
defined ... **EL1:** 82
for printed wiring boards ... **EL1:** 505
for rigid printed wiring boards ... **EL1:** 539-540, 543-547
Suburban environments
galvanized coatings in ... **A13:** 440
Subzero temperatures
alloys for structural applications ... **M3:** 721-772
Succinate electroless nickel plating process ... **M5:** 222-224
Succinonitrile-5.5 mole% acetone, directional
solidification in ... **A9:** 612
Suck-in
as casting defect ... **A11:** 384
Sucker rod corrosion
oil/gas production ... **A13:** 1248
Suction bell, cast iron
cavitation erosion of ... **A11:** 624
Suction roll corrosion
pulp and paper industry ... **A13:** 1202-1208
Suction rolls
alloys ... **A13:** 1203-1204
failures ... **A13:** 1205-1206
shells, manufacturing methods ... **A13:** 1206-1207
Suction systems
dry blasting ... **M5:** 87-88
Suction-specific speed ... **A18:** 599
Suffix symbols
for welding electrodes ... **A6:** 176-177
Sugar
copper/copper alloy resistance ... **A13:** 631
Suh's delamination theory of wear ... **A18:** 260
SUJ 1
nominal compositions ... **A18:** 725
Sulfamate plating
lead ... **M5:** 273-275
nickel ... **M5:** 200-201, 216
rhodium ... **M5:** 290-291
Sulfamic acid
as chemical cleaning solution ... **A13:** 1140-1141
Sulfate ions
as narrow-range precipitant ... **A10:** 169
Sulfate liquor, pickling process
killing of ... **M5:** 81
Sulfate plating
chromium, hard ... **M5:** 171-176, 181
copper ... **M5:** 160, 165-167
nickel ... **M5:** 200-204, 207-208
tin ... **M5:** 271-272
Sulfate-reducing bacteria
corrosion by ... **A13:** 116-117, 482-483
Sulfates
anions, separation by ion chromatography ... **A10:** 659
calcination ... **EM4:** 111
calibration curve for glass microballoons ... **A10:** 667
copper/copper alloy corrosion in ... **A13:** 636
as fining agents ... **EM4:** 380
ion chromatography analysis of geological waters for ... **A10:** 665-666
as melting accelerators ... **EM4:** 380
molten ... **A13:** 50, 91
as salt precursors ... **EM4:** 113
titanium/titanium alloy resistance ... **A13:** 683
weighing as the, gravimetric analysis ... **A10:** 171
Sulfidation *See also* Hot corrosion; Sulfide corrosion
aircraft powerplants ... **A13:** 1038-1040
defined ... **A11:** 10 **A13:** 12
Sulfidation (continued)
heat-resistant cast alloys ... **A13:** 576
high-temperature ... **A13:** 99-100
metal-processing equipment ... **A13:** 1313
-oxidation, test for resistance to hot corrosion by ... **A11:** 280
resistance, P/M superalloys ... **A13:** 839
in steel castings at elevated temperatures ... **A11:** 405
superalloys failure by ... **A12:** 388
tests, metal-processing equipment ... **A13:** 1314
Sulfidation and elevated service ... **A1:** 630-632, 636, 637
Sulfidation attack in heat-resistant casting alloys
effect of chromium on ... **A9:** 333-334
effect of nickel on ... **A9:** 333
Sulfidation resistance
cobalt-base high-temperature alloys ... **A2:** 452
mechanically alloyed oxide alloys ... **A2:** 946-947
dispersion-strengthened (MA ODS)
Sulfide
effect on workability of steels ... **A8:** 166
Stress cracking, of high-strength steels ... **A8:** 527
Sulfide concentrates
mineralogical and chemical compositions ... **A7:** 138-139
Sulfide corrosion *See also* Hot corrosion; Sulfide stress cracking; Sulfides ... **A13:** 1142-1143
in petroleum refining and petrochemical operations ... **A13:** 1270
steam surface condensers ... **A13:** 987
Sulfide films
deposited by color etching ... **A9:** 141-142
deposited by potentiostatic etching ... **A9:** 146-147
Sulfide inclusions
in ductile dimple fracture ... **A11:** 83
effect in locomotive axle failures ... **A11:** 723
fatigue cracking from ... **A11:** 477-478
in forging ... **A11:** 322
as nonmetallic in steels ... **A11:** 316
sulfur printing ... **A9:** 176
in type 303 stainless steel ... **A9:** 127
Sulfide inclusions in steel
effect on machinability ... **M1:** 573-575, 577
Sulfide ions
corrosion caused by ... **M1:** 726-727
elements precipitated by ... **A10:** 168
ion chromatography analysis of ... **A10:** 661
Sulfide ores
nitric acid as dissolution medium for ... **A10:** 166
Sulfide spheroidization
defined ... **A9:** 18
Sulfide stress cracking *See also* Environmental cracking; Hydrogen embrittlement ... **M1:** 687
acceptability criteria development ... **A13:** 316
defined ... **A13:** 12
from petroleum industry corrosion ... **A13:** 533-538
in high-strength steels ... **A13:** 270-271
in low-alloy steel ... **A13:** 532
Sulfide stringers in
manganese oxide ... **A9:** 184
Monel R-405 ... **A9:** 437
Sulfide stringers, managanese
in wrought steel ... **A11:** 317
Sulfide treatment
as surface treatment in wrought tool steels ... **A1:** 779
Sulfide-stress cracking (SSC)
defined ... **A11:** 10
in sour gas environments ... **A11:** 101, 298-300
Sulfide-type inclusions
defined ... **A9:** 18
Sulfides *See also* Inclusions
in alloy steels ... **A12:** 14, 291, 294, 299
anaerobic, in aqueous corrosion ... **A13:** 43
effect, copper/copper alloys in seawater ... **A13:** 625
effect in hydrogen embrittlement ... **A12:** 125
as electrical conductor ... **A13:** 65
as electrode ... **A10:** 185
in gray cast iron ... **A15:** 94
as inclusions ... **A10:** 176 **A12:** 14, 65, 239
inclusions, effects ... **A13:** 48
inclusions formed in fluxes ... **A6:** 56

Sulfides (continued)
as inclusions in steel **A15:** 92
as indigenous inclusion **A15:** 89
intergranular cracking by **A12:** 123
of iron and manganese, mixed **A9:** 184
Leforte aqua regia as dissolution medium **A10:** 166
in malleable iron **A12:** 239
in mating fracture, medium-carbon steels **A12:** 263
particles, effect on ridge formation **A12:** 41, 53
plasma-assisted physical vapor deposition **A18:** 848
precipitation, on grain boundaries **A12:** 349
SCC by, alloy steels **A12:** 299
as seawater pollutant **A13:** 898-899
in silicon-iron electrical steels **A9:** 537
sintering agents for **A10:** 166
solutions, copper/copper alloy corrosion in **A13:** 636
solutions, pure tin resistance **A13:** 772
stringers, AISI/SAE alloy steels **A12:** 294
weighing as the, gravimetry analysis **A10:** 171
Sulfides in steel
control of, with additions **A9:** 628
formation of **A9:** 625
Sulfides in wrought stainless steels
microstructure in austenitic grades **A9:** 284
sulfur printing **A9:** 279
Sulfides, precipitation of
in processing of solid steel **A1:** 115-116
Sulfite pulping liquors **A13:** 1196-1198
Sulfochlorinated lubricant
defined **A18:** 18
Sulfonate-base coatings
for oil/gas pipes **A13:** 1259
Sulfonates
as detergents **A18:** 111
as rust and coffosion inhibitors **A18:** 111
Sulfonation products, stainless steels
corrosion resistance **M3:** 88, 89
Sulfonium metallic hexafluorides
cationic curing **EL1:** 862
Sulfur *See also* Desulfurizing
accumulated by slag **M6:** 443
addition improving machinability **A16:** 125
addition to Ni alloys for free-machining **A16:** 840
addition to P/M tool steels **A4:** 765
additive for copper alloys **M6:** 400
additive to aid milling of carbon and alloy steels **A16:** 676
in alloy cast irons **A1:** 109
alloying effect on nickel-base alloys **A6:** 590
alloying, wrought aluminum alloy **A2:** 55
in Alnico alloys **A9:** 539
and aluminum, ductility/impact strength effects **A15:** 93
as an addition to tool steels **A9:** 258
asphalt, fracture surface **A12:** 473
at elevated-temperature service **A1:** 640
atomic interactions and adhesion **A6:** 144
attack, effect on fatigue crack growth rate **A12:** 41
in austenitic manganese steel **A1:** 826
in austenitic stainless steels **A6:** 457, 458, 463, 468
bacterial effects on **A12:** 245
in bearing steels, abrasive wear **A18:** 733
behavior in electroslag remelting **A15:** 403
cause of cracking in nickel alloy welds **M6:** 443
cause of hot cracks **A6:** 409
cement, bonding **A12:** 472
as chalcogen **A2:** 1077
chemistry at surfaces, AES analysis of **A10:** 553
in commercial CPM tool steel compositions **A16:** 63
composition, wt% (maximum), liquation cracking **A6:** 568- 569, 570
compounds, bacteria-transforming **A13:** 41-42

Sulfur (continued)
compounds, copper/copper alloy resistance **A13:** 631
concentration, cupolas **A15:** 390
concrete, fracture surfaces **A12:** 472-473
-containing atmospheres, ridges and striations **A12:** 41, 53
contaminant of aluminum alloys **A4:** 848, 850
content effect, electron-beam welding **A6:** 851
content effect on gas-tungsten arc welding and fluid flow phenomena **A6:** 20, 21, 22
content effect on nickel alloy heat treating **A4:** 910
content effect on solidification cracking **A6:** 90
content in carbon and alloy steels **A16:** 669, 670, 671, 672
content in HSLA Q&T steels **A6:** 665
content in P/M materials **A16:** 884, 885, 887, 888
content in stainless steels **A16:** 682-683, 684, 685
M6: 320, 322
content in tool steels affecting grindability **A16:** 727, 728, 732
in copper alloys **A6:** 753
corrosion, oil/gas production **A13:** 1232-1233
cracking sensitivity in stainless steel casting alloys **A6:** 497
crystalline, in concrete **A12:** 472
for curing butyl rubber **EM3:** 146
for curing styrene-butadiene rubber **EM3:** 147
cycle, biological corrosion **A13:** 41-42
determination by high-temperature combustion **A10:** 221-225
determination from iron x-radiation **A10:** 94
determination in oil **A10:** 101
determination in petroleum products, by XRS **A10:** 82
determined by iodimetric titrations **A10:** 174
detrimental effect in thermit welds **M6:** 693-694
detrimental to welding of alloy systems **A6:** 89
distribution and morphology, in asphalt **A12:** 473
in ductile iron **A15:** 648
in duplex stainless steels **A6:** 472
effect, gas dissociation **A15:** 83
effect, nitrogen solubility **A15:** 83
effect of, on machinability of carbon steels **A1:** 597-599
effect of, on notch toughness **A1:** 740, 741
effect of, on steel composition and formability **A1:** 577
effect on Cu alloy machinability **A16:** 805-808
effect on hardenability **A4:** 25
effect, ultimate tensile strength **A14:** 200
effects, electrolytic tough pitch copper **A2:** 269
effects, nickel-base alloys **A14:** 779
effects, silver-copper alloys **A2:** 700
in electrical steels **A9:** 537
in electroplated nickel **A18:** 836
electroslag welding reactions **A6:** 273, 274
embrittlement in nickel alloys **M6:** 437
embrittlement of iron by **A1:** 690-691
as embrittler, and material selection **A13:** 335
as embrittler of fcc metals **A12:** 123
enrichment, crack tip region **A12:** 41
in ferritic stainless steels **A6:** 454
forming lower melting point eutectics **A6:** 588
and free-cutting grades of carbon or low-alloy steels **A16:** 149
in free-machining metals **A16:** 389
in free-machining powder metallurgy steels **A9:** 510
free-machining steel additive **A16:** 672, 674
gas, solubility in copper alloys **A15:** 465
glow discharge to determine **A10:** 29
grain boundary adhesion **A6:** 144

Sulfur (continued)
heat-affected zone fissuring in nickel-base alloys **A6:** 588, 589
in high-speed tool steels **A16:** 53
hot/cold shortness effects **A15:** 29
ICP-determined in natural waters **A10:** 41
impurity in solders **M6:** 1072
as Impurity, magnetic effects **A2:** 762
as inclusion, steels **A15:** 92
infrared detection, high-temperature combustion **A10:** 223
in inorganic solids, applicable analytical methods **A10:** 4, 6
interaction coefficient, ternary iron-base alloys **A15:** 62
as ion chromatography application **A10:** 658, 664
lubricant indicators and range of sensitivities **A18:** 301
machinability additives **A16:** 685-686, 687, 688
maps **A12:** 349, 388
measured, in lubricants **A14:** 516
microalloying of **A14:** 220
Miller numbers **A18:** 235
as minor element, gray iron **A15:** 630
in most widely used EP additives **A16:** 123, 124, 125, 126
in oil, EDS determination **A10:** 101
oil-ash corrosion with **A11:** 618
oxygen cutting, effect on **M6:** 898
in P/M alloys **A1:** 810
as poison **A13:** 330
in polysulfides **EM3:** 50
primary curative of natural rubber **EM3:** 145
recovery unit **A13:** 365-366
relationship to hot cracking **M6:** 38, 833
removal, from ferrous melts **A15:** 74-78, 366
resistance spot welding of steels and content effect **A6:** 228
segregation **A12:** 349
segregation and solid friction **A18:** 28
as solder impurity **EL1:** 638, 642
as solid lubricant inclusion for stainless steels **A18:** 716
species weighed in gravimetry **A10:** 172
spectrometric metals analysis **A18:** 300
in steel **A1:** 144-145
in steel, effect on nonmetallic inclusions **A9:** 179
in steel weldments **A6:** 418
and stress corrosion **A16:** 35
submerged arc welding
effect on cracking **M6:** 128-129
influence on weld-metal toughness **M6:** 117
surface segregation during heating **A10:** 564-565
tantalum resistance to **A13:** 727
tantalum sulfide formation in **A13:** 728
in thread grinding oils **A16:** 273
as tin solder impurity **A2:** 520-521
as tramp element **A8:** 476
use in resistance spot welding **M6:** 486
use of copper accelerators with **A10:** 222
vapor pressure **A6:** 621
volatilization losses in melting **EM4:** 389
volumetric procedures for **A10:** 175
Sulfur copper *See* Copper alloys, specific types, C14700
Sulfur corrosion
copper metals **M2:** 463, 464
Sulfur dioxide
as atmospheric contaminant **A13:** 81
copper/copper alloy corrosion in **A13:** 632, 635-636
for core curing **A15:** 240
effect, carbon steel corrosion rate **A13:** 909
electrometric titration for **A10:** 205
oxidation, SERS analysis of **A10:** 136
formation, wood pulp superheaters **A13:** 1200
furan resin binder process **A15:** 219-221
in marine atmospheres **A13:** 903-904

Sulfur dioxide (continued)
oxi
plain carbon steel corrosion in **A13:** 81
reduction with **A7:** 54
removal by manganese dioxide in high- temperature combustion........... **A10:** 222
service, pollution control **A13:** 1370
use, oil/gas production...................... **A13:** 1245
use to determine sulfur in high-temperature combustion...... **A10:** 221-225
Sulfur dioxide, and ammonia reactions
as detection method...................... **A17:** 61
Sulfur dioxide reducing method
chromic acid wastes...................... **M5:** 186-187
Sulfur dioxide test method
anodic coating sealing...................... **M5:** 596
Sulfur embrittlement
brazing and **A6:** 117
Sulfur hexafluoride detectors
of gas/leaks...................... **A17:** 62
Sulfur in cast iron
alloy effect **M1:** 77, 78
ductile iron **M1:** 37
malleable iron **M1:** 58
Sulfur in steel *See also* Resulfurized steel...................... **M1:** 116
castings...................... **M1:** 399
maraging steels...................... **M1:** 447
notch toughness...................... **M1:** 692
P/M materials...................... **M1:** 338
sheet, effect on formability...................... **M1:** 554
Sulfur, nonsulfate
in Sherritt nickel powder production........... **A7:** 140
Sulfur oxides
effect on atmospheric corrosion **M1:** 717, 718
Sulfur print
defined **A9:** 18
Sulfur print technique, for revealing distribution of manganese sulfides in
wrought stainless steels **A9:** 279
Sulfur print test
used for macroexamination of tool steels...................... **A9:** 256
Sulfur printing
carbon and alloy steels...................... **A9:** 176
to reveal as-cast solidification structures in steel **A9:** 624-625
Sulfur removal *See* Desulfurization; Sulfur
Sulfur segregation
macroetching to reveal...................... **A9:** 176
Sulfur trioxide
formation **A13:** 1200
removal of...................... **A10:** 222
Sulfur-bearing copper *See* Copper alloys, specific types, C14700
applications and properties...................... **A2:** 278-280
Sulfuric acid *See also* Sulfuric acid corrosion
in alcohol as an electrolyte for refractory metals **A9:** 440
as an etchant for wrought stainless steels...................... **A9:** 281
cemented carbides resistance to **A13:** 852
as chemical cleaning solution **A13:** 1140
chemical milling etchant...................... **A16:** 873
corrosion inhibitors used in **M1:** 755-756
electrolyte for ECM...................... **A16:** 536
electrolyte for Ni alloy ECM...................... **A16:** 843
formation, by liquor decomposition.......... **A13:** 1197
formation, by oxidation...................... **A13:** 1197
with inorganic acids as electrolyte...................... **A9:** 54
mounting materials for use with...................... **A9:** 54
nickel alloys, corrosion...................... **M3:** 172-173
in organic solvent (Group IV electrolytes)...................... **A9:** 52-54
in petroleum refining and petrochemical operations **A13:** 1269
photochemical machining etchant........ **A16:** 591, 593
residue isolation using **A10:** 176
safety hazards **A9:** 69
as sample dissolution medium...................... **A10:** 165
service, pollution control **A13:** 1370
stainless steels, corrosion resistance **M3:** 88-91
used in chemical treatment before disposal...................... **A16:** 131
in water (Group IV electrolytes) **A9:** 52-54
Sulfuric acid and hydrochloric acid
as an etchant for carbon and alloy steels...................... **A9:** 171
Sulfuric acid anodizing process
aluminum and aluminum alloys........... **M5:** 586-589, 592-595
voltage requirements **M5:** 588-589
Sulfuric acid cleaning process
iron and steel **M5:** 60
Sulfuric acid corrosion *See also* Sulfuric acid **A13:** 1148-1154
alloy steels **A13:** 544
austenitic stainless steels...................... **A13:** 1150-1151
brick linings...................... **A13:** 1153
carbon steel...................... **A13:** 1148-1149
cast iron resistance **A13:** 569
cast irons **A13:** 1149-1150
cast stainless steels **A13:** 1151
copper/copper-base alloys **A13:** 627, 1153
higher austenitic stainless steels........ **A13:** 1151-1152
lead/lead alloys...................... **A13:** 788, 1153
lined pipe...................... **A13:** 1153-1154
mechanisms **A13:** 1148
nickel-base alloy resistance.......... **A13:** 643-644, 1152
polyvinyl chloride...................... **A13:** 1153
stainless steels **A13:** 557-558
tantalum resistance to **A13:** 725, 1153
titanium...................... **A13:** 1153
zirconium/zirconium alloys................. **A13:** 709-710, 1152-1153
Sulfuric acid electropolishing solutions **M5:** 303-305, 308
Sulfuric acid etching bath **M5:** 180
Sulfuric acid, hydrochloric acid, and nitric acid, as an etchant for wrought heat-resistant alloys...................... **A9:** 307
Sulfuric acid, hydrofluoric acid and nitric acid
as an etchant for beryllium **A9:** 390
Sulfuric acid, inhibited
in cathodic cleaning...................... **A12:** 75
Sulfuric acid pickling
copper and copper alloys **M5:** 611-614
hot dip galvanized coating process **M5:** 325-326, 328
iron...................... **M5:** 68-82
magnesium alloys **M5:** 629, 640-641
process variables affecting scale removal...................... **M5:** 72-77
solution compositions and operating conditions.... **M5:** 12, 68-70, 73-74, 611-614, 655
stainless steel...................... **M5:** 553-554
steel...................... **M5:** 68-82
titanium a titanium alloys **M5:** 654-655
waste recovery and treatment **M5:** 81-82
Sulfuric acid/potassium dichromate solution...................... **EM3:** 35
Sulfuric acid/sodium dichromate solution...................... **EM3:** 35
Sulfuric anodize...................... **A13:** 396-397
Sulfuric-oxalic anodizing process
aluminum and aluminum alloys...................... **M5:** 592
Sulfuric-phosphoric acid as electrolyte
current- voltage relation **A9:** 48-49
Sulfuric-phosphoric-chromic acid electrolytic brightening
aluminum and aluminum alloys........... **M5:** 580-581
Sulfurized lubricant
defined **A18:** 18
Sulfurous acid
as reducing agent **A10:** 169
SCC by **A13:** 327, 558
Sulfurous acid, stainless steels
corrosion resistance...................... **M3:** 89-91
Sum peaks
defined **A10:** 92, 683
in energy-dispersive spectra **A10:** 520
and escape peaks...................... **A10:** 520
Sumitomo top and bottom blowing (SFB) process...................... **A1:** 111
Summing
in real-time radiography...................... **A17:** 320
Sump lubrication...................... **A18:** 132-133
Sunlight
effects in marine atmospheres **A13:** 905
Sunlight, as source
photolytic degradation **EM2:** 776-782
Sunshine open-flame carbon arc weatherometer **EM2:** 578
Super heterodyne detection
use for low- temperature ESR studies......... **A10:** 257
Super Z300 *See* Superplastic zinc
Super-alpha titanium alloys...................... **A9:** 458
Super-Invar *See also* Low-expansion alloys
composition, properties, applications..... **A2:** 894-895
as low-expansion alloy...................... **A2:** 889
Super-purity aluminum
applications and properties...................... **A2:** 64-65
Super-Z 300 *See* Zinc alloys, specific types, superplastic zinc alloy
Superabrasive grains *See also* Superabrasives
bonded-abrasive grains **A2:** 1013-1015
commercially available...................... **A2:** 1011
loose abrasive grains **A2:** 1012-1013
Superabrasives *See also* Cubic boron nitride; Diamond; Grinding equipment and processes; Superabrasive grains **A16:** 453-471
abrasive stick dressing method **A16:** 468
abrasive-jet dressing method **A16:** 469
applications **A16:** 454-456 **EM4:** 331
automation by multiaxis CNC machines, as a variable...................... **A16:** 460
balancing, as operational factor...................... **A16:** 465
batch production dressing methods **A16:** 469
batch production truing methods **A16:** 467
bond type...................... **EM4:** 331
conditioning **A16:** 469-470
conditioning, as operational factor **A16:** 466
coolants **A16:** 463, 464, 465, 466, 467
cubic boron nitride (CBN), properties of...................... **A2:** 1010-1011
cubic boron nitride (CBN), synthesis of...................... **A2:** 1008-1009
diamond and CBN compared............... **A16:** 453-454
diamond, properties of...................... **A2:** 1009-1010
diamond, synthesis of **A2:** 1008-1009
dressing, as operational factor.............. **A16:** 465-466
dressing methods **A16:** 468-469
form truing methods for production grinding...................... **A16:** 467
high-pressure waterjet dressing method...................... **A16:** 469
mechanical properties...................... **EM4:** 331
metals ground/machined with...................... **A2:** 1013
modulus of resilience **EM4:** 331
powered truing methods for production grinding...................... **A16:** 467
properties...................... **EM4:** 331, 332, 333
sintered polycrystalline cubic boron nitride, properties of **A2:** 1012
sintered polycrystalline diamond, properties of **A2:** 1011-1012
slurry dressing method **A16:** 469
stationary tool truing method for production grinding...................... **A16:** 466-467
superabrasive grains **A2:** 1012-1015
superhard materials of commercial interest **A2:** 1008
thermal properties...................... **EM4:** 331
truing, as operational factor................. **A16:** 466, 467
truing parameters **A16:** 468
and ultrahard tool materials **A2:** 1008, 1015-1017
wheel application and machine tool variables **A16:** 460
wheel applications **A16:** 458-471
wheel bond systems **A16:** 457-458, 461, 462-463
wheel selection...................... **A16:** 460-465
wheel truing objectives **A16:** 467
wheels...................... **A16:** 456-471
Superalloy nickel eutectic **A9:** 619
Superalloy scrap, recycling of **A1:** 1027-1028
blending **A1:** 1032
by industry **A1:** 1028
collection **A1:** 1029
degreasing...................... **A1:** 1032
demand **A1:** 1028
of metallurgical wastes...................... **A1:** 1032
processing **A1:** 1028-1032
secondary nickel refining...................... **A1:** 1032
separation and sorting................ **A1:** 1028, 1029-1032
size reduction and compaction.................... **A1:** 1032
Superalloy(s)
directionally solidified, development........... **A2:** 429
dispersion-strengthened, development **A2:** 429

Superalloy(s) (continued)
ground/machines with superabrasives/ultrahard tool materials **A2:** 1013
metal-matrix composites **A2:** 909
nickel alloy, P/M development **A2:** 429
rare earth alloy additives **A2:** 727-728
single-crystal, development **A2:** 429
Superalloy-matrix composites
development and fabrication **A2:** 909
Superalloys *See also* Cobalt-base superalloys; Fiber-reinforced superalloys (FRS); Heat-resistant alloys; Heat-resistant castings; Nickel alloys, specific types; Nickel-based superalloys; specific types; Superalloys, specific types; Wrought heat-resistant alloys **A18:** 623 **M3:** 207-268, 271
aerospace applications **EM4:** 1005
alloy segregation **A7:** 468
alloy stability **M3:** 208, 219-229
aluminum coating of **M5:** 335, 339, 342
application, gas turbine engine components **A18:** 813
applications **M3:** 207-208, 213-214
arc-welded, heat-resisting **A11:** 433-434
argon oxygen decarburization **A15:** 426
argon-atomized, Auger composition-depth profile **A7:** 254
ASP **A16:** 61
atomization cooling rates **A7:** 33, 36
boundary contamination in **A7:** 428, 439
cast cobalt-base superalloys **A1:** 983, 985-987
cast nickel-base superalloys **A1:** 981-985, 986-994
chemical milling **A16:** 579, 584, 586
coatings **M3:** 208
cobalt-base **M3:** 208, 210-211, 213
erosion resistance for pump components **A18:** 598
for hot-forging dies **A18:** 626
material for jet engine components **A18:** 588
cobalt-base, phases **A15:** 812
cobalt-enhanced high-temperature properties **A7:** 144
commercial, oxygen levels **A7:** 434
compositions **A7:** 524, 647 **M3:** 210-211
consolidation by hot isostatic pressing **A7:** 439-441
contaminant types **A7:** 178
continuous flow electron beam melted **A15:** 417
creep behavior **A8:** 331
creep damage **M3:** 225-226
directionally solidified superalloys **A1:** 995, 996-998
dispersion-strengthened **A7:** 722-723
effects of hot pressing **A7:** 512
electron beam investment casting **A15:** 417-418
electron-beam welding **A6:** 851, 865-866
elements implanted to improve wear and friction properties **A18:** 858
fatigue crack propagation **A8:** 409
fatigue properties **A7:** 439
fractographs **A12:** 388-395
fracture/failure causes illustrated **A12:** 217
friction welding **A6:** 154
high-speed machining tool systems **A16:** 602
high-speed tool steels used **A16:** 59, 63
hot corrosion **M3:** 207-209
hot extrusion plus forging **A7:** 523-524
hot isostatic pressing plus forging **A7:** 522-523
hot isostatically pressed **A7:** 18
hot workability ratings for **A8:** 586
hydrogen fluoride/hydrofluoric acid corrosion **A13:** 1168
iron-base **M3:** 209-213
iron-base, for hot-forging dies **A18:** 626
iron-base superalloys **A1:** 950, 958-962, 965
for isothermal/hot-die forging **A14:** 151-152
jet engine applications **A7:** 646-652
laser alloying **A18:** 866
Laves phase formation **A8:** 479

Superalloys (continued)
material utilization through HIP **A7:** 440, 443
materials for dies and molds **A18:** 622, 625, 626
mechanical properties **A7:** 441, 468, 473
melting methods **M3:** 214
melting point **A16:** 601
microstructure and strength **A8:** 408
microstructure in rapid solidification **A7:** 47, 48
nickel-base **M3:** 208-211, 213-214
hot forging **A18:** 625, 626
material for jet engine components **A18:** 588, 590
resistance to plastic deformation **A18:** 626
service temperature of die materials in forging **A18:** 625
nickel-base, eutectic joining **EM4:** 526
nickel-base, for rolling **A14:** 356
nickel-base, hot isostatic pressing effects **A15:** 539
nickel-based *See also* Nickel-based superalloys **A7:** 33, 36, 47, 48, 134, 178, 439-441
nominal compositions **A7:** 473
notch effects in stress rupture **M3:** 230-234
for notched-specimen testing **A8:** 315
Osprey process for **A7:** 530
oxidation protection coatings for **M5:** 375-379
oxide dispersion-strengthened **A7:** 440
oxide stability **A11:** 452
oxide-dispersion-strengthened alloys **M3:** 209, 212, 216
oxygen effect on rupture life **A15:** 396
P/M **A13:** 834-838
P/M processing **M3:** 207, 214-216
physical properties **M3:** 217
plasma melting/casting **A15:** 420
plasma rotating electrode particles **A7:** 42, 44
powder metallurgy (P/M) cobalt-base alloys **A1:** 977-980
powder metallurgy (P/M) superalloys **A1:** 972-976
processed from hot work CAP stock **A7:** 536
processing **M3:** 214-216
product forms **M3:** 216, 218
roughing and finishing with whisker-reinforced alumina ceramic insert cutting tools **EM4:** 972
rupture strengths/compositions **EM1:** 881
salt bath descaling of **M5:** 97, 101-102
selection **M3:** 218-219
semisolid casting/forging of **A15:** 327
short-time tensile properties **M3:** 218-221, 223, 225
single-crystal superalloys **A1:** 995-996, 998-1006
solidification structures of welded joints **A9:** 580
specifications, aircraft gas turbine requirements **M3:** 234
spherical powder, contaminant removal **A7:** 178-179
strengthening mechanisms **M3:** 209, 212
stress-rupture curves **M3:** 277, 279
stress-rupture data **M3:** 221-223, 227, 229-235, 237-241
superplastic **A7:** 18
testing of **M3:** 229-237
tungsten-reinforced nickel-base **EM1:** 885
upset welding **A6:** 249
vacuum degassing **A7:** 435
vacuum induction melting **A15:** 393
versus ceramics for turbine nozzles **EM4:** 995
workability **A8:** 165, 575
wrought cobalt-base superalloys **A1:** 950, 962-968
wrought nickel-base superalloys **A1:** 950-959, 968-972
Superalloys, cobalt-base, specific types *See also* Cobalt alloys, specific types
AiResist 13
composition **M3:** 257, 271

Superalloys, cobalt-base, specific types (continued)
property data **M3:** 257
stress-rupture curve **M3:** 279
AiResist 213
composition **M3:** 211, 257, 271
property data **M3:** 257
AiResist 215
composition **M3:** 258, 271
property data **M3:** 258
stress-rupture curve **M3:** 279
FSX-414
composition **M3:** 258
property data **M3:** 258
Haynes 21 .... **M3:** 278
composition **M3:** 271
stress-rupture curve **M3:** 279
Haynes 25 .... **M3:** 590-591
composition **M3:** 210, 258, 271, 590
hot corrosion **M3:** 208
incipient melting temperature **M3:** 224
mechanical properties **M3:** 218, 223
oxidation resistance **M3:** 259
physical properties **M3:** 217
property data **M3:** 258-259
stress-rupture curve **M3:** 279
wear data **M3:** 590
Haynes 88, notch effects in stress rupture **M3:** 230, 231
Haynes 151
composition **M3:** 271
stress-rupture curve **M3:** 279
Haynes 188
composition **M3:** 210, 259
hot corrosion **M3:** 208
incipient melting temperature **M3:** 224
mechanical properties **M3:** 218
microstructure **M3:** 223, 226
oxidation resistance **M3:** 259
physical properties **M3:** 217
property data **M3:** 259-261
Haynes 263, hot corrosion **M3:** 208
J-1650, composition **M3:** 271
L-605 *See* Haynes
MAR-M 302
composition **M3:** 263, 271, 279
property data **M3:** 263
stress-rupture data **M3:** 263, 277
MAR-M 322
composition **M3:** 263, 279
property data **M3:** 263
stress-rupture curve **M3:** 279
MAR-M 509 .... **M3:** 279
composition **M3:** 263, 271
property data **M3:** 263-364
stress-rupture data **M3:** 264, 279
MAR-M 918, composition **M3:** 271
MP-35N, composition **M3:** 211
MP-159, composition **M3:** 211
S-816
composition **M3:** 210, 271
mechanical properties **M3:** 210
notches, effect on fatigue and stress rupture **M3:** 232, 233, 235
stress-rupture curves **M3:** 232, 233, 279
S816, cleaning and finishing processes **M5:** 567
UMCo-50
composition **M3:** 210, 266
physical properties **M3:** 217
property data **M3:** 266-267
V-36, composition **M3:** 271
X-40 .... **M3:** 278
composition **M3:** 268, 271
property data **M3:** 267, 268
stress-rupture data **M3:** 268, 279
X-45
composition **M3:** 268
property data **M3:** 268

Superalloys, heat treating *See also* Heat-resistant alloys, heat treating **A4:** 793-813
aging **A4:** 795-798, 799, 800, 808, 812, 813
aging cycles **A4:** 799
aging precipitates **A4:** 795-796
alloy depletion **A4:** 798, 813
annealing.. **A4:** 793, 808, 809-810
applications **A4:** 807, 808, 812, 813
atmosphere for aging **A4:** 799
carbide precipitation **A4:** 809-810, 811
carbon pickup **A4:** 798
cast **A4:** 812-813
cast, compositions **A4:** 795
cold working effect on recrystallization and grain growth **A4:** 808, 810
cold working effect on response **A4:** 800-801
constituents observed **A4:** 797
direct aging.. **A4:** 804
double aging **A4:** 796, 798, 801, 802
endothermic atmospheres for protection **A4:** 799
exothermic atmospheres for protection......... **A4:** 799
fixtures for support or restraint..................... **A4:** 799
furnace equipment **A4:** 799
gas furnace quench (GFQ).. **A4:** 813
grain growth **A4:** 799-800, 808, 810, 811
hardness **A4:** 800-801
hot isostatic pressing (HIP) **A4:** 812, 813
hydrogen for protection.. **A4:** 799, 813
inert gas for protection.. **A4:** 798, 813
mechanical properties.. **A4:** 808
mill annealing **A4:** 809-810, 811
miscellaneous contaminants.. **A4:** 798
oxidation **A4:** 798, 813
precipitation-strengthened nickel-base superalloys **A4:** 805-809
precipitation-strengthened nickel-iron-base superalloys **A4:** 799-805
quenching **A4:** 793-795
reheating for hot working **A4:** 793
solid-solution-strengthened iron-, nickel-, and cobalt-base **A4:** 809-811
solid-solution-strengthened iron/nickel-, nickel-, and cobalt-base **A4:** 813
solution annealing.. **A4:** 810, 811
solution treating **A4:** 793, 800, 808, 810, 811, 812-813
stabilization **A4:** 804, 806, 813
stress relieving **A4:** 793, 799, 809-810, 811, 812, 813
tensile properties **A4:** 801
thermomechanical processing................ **A4:** 798, 801
vacuum atmosphere for protection........ **A4:** 798, 813
vacuum brazing.. **A4:** 811
welding **A4:** 812, 813
wrought.. **A4:** 799-812
wrought, compositions.. **A4:** 794

Superalloys, iron-base, specific types *See also* Iron-base alloys
16-25-6, composition **M3:** 210
17-14CuMo, composition **M3:** 210
19-9 DL, cleaning and finishing processes **M5:** 567-568
19-9 DL, composition.. **M3:** 210
A-286
composition.. **M3:** 210, 538, 756
fasteners.. **M3:** 185
fatigue at subzero temperatures **M3:** 764, 765
fracture toughness **M3:** 763
Poisson's ratio.. **M3:** 755
tensile properties.. **M3:** 761, 762
Young's modulus.. **M3:** 751
A-286, cleaning and finishing processes **M5:** 567
Discaloy
composition.. **M3:** 210
hardness, effect on stress rupture....... **M3:** 232-233
Haynes.. **M3:** 556
composition.. **M3:** 210, 261
physical properties **M3:** 217
property data.. **M3:** 261-262
Incoloy 800
composition.. **M3:** 210
incipient melting temperature..................... **M3:** 224
oxidation resistance **M3:** 196
physical properties **M3:** 217

Superalloys, iron-base, specific types (continued)
rupture strength.. **M3:** 221
Incoloy 801
composition.. **M3:** 210
physical properties **M3:** 217
rupture strength.. **M3:** 221
Incoloy 802
composition.. **M3:** 210
rupture strength.. **M3:** 221
Incoloy 825
corrosion resistance, pulp and paper industry **M3:** 92
incipient melting temperature..................... **M3:** 224
physical properties **M3:** 217
Incoloy 903, composition **M3:** 210
Incoloy, creep-rupture properties **M3:** 195
Incoloy MA 956, oxidation resistance **M3:** 212
N-155
composition.. **M3:** 210
hot corrosion.. **M3:** 208
rupture strength.. **M3:** 222
Pyromet CTX-1, composition........................ **M3:** 211
RA-330, composition.. **M3:** 210
V-57
composition.. **M3:** 211, 538
hot extrusion tools, use for **M3:** 538, 540
W-545, composition **M3:** 211

Superalloys, nickel-base *See* Nickel-base superalloys

Superalloys, nickel-base, specific types *See also* Nickel alloys, specific types
Astroloy
composition.. **M3:** 211
mechanical properties.. **M3:** 218
B-1900; B-1900 + Hf **M3:** 278
composition.. **M3:** 242, 271
property data **M3:** 242-243
stress-rupture data.. **M3:** 242, 277
D-979
composition.. **M3:** 211
mechanical properties.. **M3:** 218
D-979, cleaning and finishing processes **M5:** 567-568
Hastelloy B, composition.. **M3:** 210
Hastelloy B-2
composition.. **M3:** 210
physical properties **M3:** 217
Hastelloy C
composition.. **M3:** 210
hot corrosion.. **M3:** 208
Hastelloy C-4
composition.. **M3:** 210
physical properties **M3:** 217
Hastelloy C-276
composition.. **M3:** 210
hot corrosion.. **M3:** 208
physical properties **M3:** 217
Hastelloy N
composition.. **M3:** 210
physical properties **M3:** 217
Hastelloy S
composition.. **M3:** 210
hot corrosion.. **M3:** 208
physical properties **M3:** 217
Hastelloy W
composition.. **M3:** 210
physical properties **M3:** 217
Hastelloy X
composition.. **M3:** 210, 271
degradation of strength by overtemperature exposure **M3:** 222-223, 225
hot corrosion.. **M3:** 208
incipient melting temperature..................... **M3:** 224
mechanical properties.. **M3:** 218
microstructure **M3:** 225, 227
physical properties **M3:** 217
HDA 8077, oxidation resistance **M3:** 212
IN-100.. **M3:** 278
composition.. **M3:** 211, 243, 271
mechanical properties.. **M3:** 216
property data **M3:** 243
stress-rupture data.. **M3:** 243, 277
IN-102
composition.. **M3:** 211
mechanical properties.. **M3:** 218
IN-162
composition.. **M3:** 244

Superalloys, nickel-base, specific types (continued)
property data **M3:** 243, 244
IN-731
composition.. **M3:** 244
property data **M3:** 244
IN-738.. **M3:** 278
composition.. **M3:** 244, 271
hot tensile properties **M3:** 216
property data **M3:** 244-245
stress-rupture data.. **M3:** 244, 277
IN-738 + Y_2O_3, oxidation resistance............ **M3:** 212
IN-792.. **M3:** 278
composition.. **M3:** 245, 271
property data **M3:** 245
IN-939
composition.. **M3:** 245
proper data **M3:** 245-246
Incoloy 901, composition **M3:** 211
Inconel
corrosion in sulfuric acid.. **M3:** 88-89
creep-rupture properties.. **M3:** 195
Inconel 617
composition.. **M3:** 210
hot corrosion.. **M3:** 208
incipient melting **M3:** 224
physical properties **M3:** 217
rupture strength.. **M3:** 221
Inconel 713C.. **M3:** 278
composition.. **M3:** 246, 271
property data **M3:** 246-247
Inconel 600
composition.. **M3:** 210
hot corrosion.. **M3:** 208
mechanical properties **M3:** 218
physical properties **M3:** 217
rupture strength.. **M3:** 221
Inconel 601
composition.. **M3:** 210
hot corrosion.. **M3:** 208
mechanical properties **M3:** 219
oxidation resistance **M3:** 196
rupture strength.. **M3:** 221
Inconel 604, composition **M3:** 210
Inconel 625
composition.. **M3:** 210
hot corrosion.. **M3:** 208
incipient melting temperature.................. **M3:** 224
mechanical properties **M3:** 219
rupture strength.. **M3:** 222
Inconel 671, physical properties **M3:** 217
Inconel 690, physical properties **M3:** 217
Inconel 706
composition.. **M3:** 211
mechanical properties **M3:** 219
Inconel 713LC **M3:** 278
composition.. **M3:** 247, 271
property data **M3:** 247-248
Inconel 718
composition.. **M3:** 211, 271, 538
hot corrosion.. **M3:** 208
hot extrusion tools, use for **M3:** 538, 540
master Larson-Miller curve................ **M3:** 239-240
mechanical properties **M3:** 219
rupture strength.. **M3:** 222
Inconel 751
composition.. **M3:** 211
rupture strength.. **M3:** 222
Inconel MA 754, oxidation resistance.......... **M3:** 212
Inconel MA 6000E, oxidation resistance **M3:** 212
Inconel X750
composition.. **M3:** 211, 271
hot corrosion.. **M3:** 208
incipient melting temperature.................. **M3:** 224
mechanical properties **M3:** 219
notch effects in stress rupture **M3:** 231
physical properties **M3:** 217
rupture strength.. **M3:** 222
solution treatment.. **M3:** 224
tensile strength affected by rate of heating.. **M3:** 229, 230
M-22
composition.. **M3:** 248
property data **M3:** 248
M-252
composition.. **M3:** 211, 271

Superalloys, nickel-base, specific types (continued)
mechanical properties ... M3: 219
M-252, cleaning and finishing processes ... M5: 564
MAR-M 200; MAR-M 200 + Hf ... M3: 278
composition ... M3: 248, 271
property data ... M3: 248-249
stress-rupture data ... M3: 248-249, 277
MAR-M 246 ... M3: 278
composition ... M3: 249, 271
property data ... M3: 249-250
MAR-M 247 ... M3: 278
composition ... M3: 250, 271
property data ... M3: 250
MAR-M 421
composition ... M3: 250
property data ... M3: 250-251
MAR-M 432
composition ... M3: 251
property data ... M3: 251
MC-102
composition ... M3: 251
property data ... M3: 251-252
NA-224, composition ... M3: 210
Nimocast 75
composition ... M3: 252
property data ... M3: 252
Nimocast 80
composition ... M3: 252
property data ... M3: 252
Nimocast 90
composition ... M3: 252
property data ... M3: 252, 253
Nimocast 242
composition ... M3: 252
property data ... M3: 252-253
Nimocast 263
composition ... M3: 253
property data ... M3: 253
Nimocast 738, hot corrosion ... M3: 209
Nimocast 739, hot corrosion ... M3: 209
Nimonic 75
composition ... M3: 210
mechanical properties ... M3: 219
physical properties ... M3: 217
rupture strength ... M3: 222
Nimonic 80A
composition ... M3: 211
hot corrosion ... M3: 209
incipient melting temperature ... M3: 224
mechanical properties ... M3: 219
physical properties ... M3: 217
rupture strength ... M3: 222
stress-rupture curves ... M3: 233
Nimonic 81, hot corrosion ... M3: 209
Nimonic 90
composition ... M3: 211
hot corrosion ... M3: 209
incipient melting temperature ... M3: 224
mechanical properties ... M3: 220
physical properties ... M3: 217
rupture strength ... M3: 222
solution treatment ... M3: 224
Nimonic 91, hot corrosion ... M3: 209
Nimonic 95, composition ... M3: 211
Nimonic 100
composition ... M3: 211
physical properties ... M3: 217
Nimonic 101, hot corrosion ... M3: 209
Nimonic 105
composition ... M3: 211
hot corrosion ... M3: 209
incipient melting temperature ... M3: 224
mechanical properties ... M3: 220
physical properties ... M3: 217
reheat treatment, for healing creep damage ... M3: 225, 228
rupture strength ... M3: 222

Superalloys, nickel-base, specific types (continued)
solution treatment ... M3: 224
Nimonic 115
composition ... M3: 211
hot corrosion ... M3: 209
mechanical properties ... M3: 220
rupture strength ... M3: 222
Nimonic 263
composition ... M3: 211
rupture strength ... M3: 222
NX 188
composition ... M3: 253, 271
proper data ... M3: 253, 254
Pyromet 860
composition ... M3: 211
mechanical properties ... M3: 220
RA-333
composition ... M3: 210
hot corrosion ... M3: 208
Refractaloy 26
composition ... M3: 211
notch rupture strength ... M3: 231
rupture strength ... M3: 222
René 41
composition ... M3: 211
hot corrosion ... M3: 208
incipient melting temperature ... M3: 224
mechanical properties ... M3: 220
physical properties ... M3: 217
René 41, cleaning and finishing processes ... M5: 564-565
René 77
composition ... M3: 253, 271
property data ... M3: 253, 254
René 80 ... M3: 278
composition ... M3: 253, 271
property data ... M3: 253, 254
René 95
composition ... M3: 211
mechanical properties ... M3: 220
René 100
composition ... M3: 211, 254, 271
property data ... M3: 254
TAZ-8A
composition ... M3: 254
property data ... M3: 254, 255
TRW 1900, stress-rupture curve ... M3: 277
TRW-NASA VIA
composition ... M3: 254, 271
property data ... M3: 254-255
Udimet 500
composition ... M3: 211, 255, 271
hot corrosion ... M3: 209
incipient melting temperature ... M3: 224
mechanical properties ... M3: 220
physical properties ... M3: 217
property data ... M3: 255
solution treatment ... M3: 224
Udimet 520
composition ... M3: 211
mechanical properties ... M3: 220
Udimet 630, composition ... M3: 211
Udimet 700 ... M3: 278
composition ... M3: 211, 255, 271
incipient melting temperature ... M3: 224
mechanical properties ... M3: 216, 220
microstructure ... M3: 223, 226
physical properties ... M3: 217
property data ... M3: 255
solution treatment ... M3: 224
stress-rupture affected by sigma phase ... M3: 224, 227
stress-rupture data ... M3: 255, 277
Udimet 710
composition ... M3: 211, 256, 271
mechanical properties ... M3: 221
property data ... M3: 256
Unitemp AF2-1DA
composition ... M3: 211

Superalloys, nickel-base, specific types (continued)
mechanical properties ... M3: 221
Waspaloy
composition ... M3: 211, 271
hot corrosion ... M3: 208
incipient melting temperature ... M3: 224
mechanical properties ... M3: 221
physical properties ... M3: 217
solution treatment ... M3: 224
stress rupture ... M3: 232, 233
WAZ-20 (DS)
composition ... M3: 256
property data ... M3: 256
WAZ-20, composition ... M3: 271
Superalloys, special metallurgical welding considerations ... A6: 575-579
clad metals and overlays ... A6: 578-579
outside-diameter clad tubing ... A6: 579
overaging ... A6: 575
Superalloys, specific types *See also* Nickel-base superalloys; Superalloys
713C, service failure ... A12: 390
713LC, service fracture ... A12: 391
718, trace element effects ... A15: 395
A-286 iron-nickel-base (UNS S66286)
effect of neutron irradiation on fracture mode/toughness ... A12: 388
A-286 nickel-base in air and vacuum, fatigue crack growth rates ... A8: 413
A-286, notch sensitivity ... A8: 316
AF-115, composition ... A7: 524, 647
AF-115, nominal compositions and mechanical properties ... A7: 473
Ancorsteel 1000B, ultrasonic velocity and strength for ... A7: 484
Astroloy, Auger sputtering profiles ... A7: 255
Astroloy, impurities in hot isostatically pressed ... A7: 428
Astroloy, jet turbine disks in ... A7: 649-650
Astroloy, mechanical properties ... A7: 441
Astroloy, nominal compositions and mechanical properties ... A7: 473
Astroloy powder, particles produced by argon gas atomization ... A7: 428
Discaloy, notch sensitivity ... A8: 316
Discaloy, rupture time variations ... A8: 316-317
Haynes 88, notch sensitivity ... A8: 315-316
IN-100, composition ... A7: 647
IN-100, for turbine F-100 disks ... A7: 649
IN-100, mechanical properties ... A7: 650
IN-100, nominal compositions and mechanical properties ... A7: 473
IN718, hot isostatic pressing ... A15: 543
IN738 nickel-base, hot isostatic pressing ... A15: 539
Incoloy MA 956 ... A7: 726
Inconel 600, molten fluoride corrosion ... A13: 90
Inconel 706, liquid sodium corrosion ... A13: 91
Inconel 718, deformation twins in threshold regime ... A8: 488
Inconel 718, ductile striations ... A8: 482, 484
Inconel 718, mechanical properties ... A7: 536
Inconel 750, notch-rupture strength ratio vs temperature ... A8: 316
Inconel 751, notch effect on rupture life ... A8: 316-318
Inconel MA 754 ... A7: 725-726
Inconel MA 754, turbine vanes ... A7: 650-651
Inconel MA 6000 ... A7: 726
Inconel MA 6000E, small aircraft gas turbine blades ... A7: 651
Inconel, molten salt corrosion resistance ... A13: 51
low-carbon Astroloy ... A7: 524
low-carbon Astroloy, hot isostatic pressing compared with rapid omnidirectional compaction ... A7: 545
MA 754, composition ... A7: 524, 647
MA 956, composition ... A7: 524, 647

Superalloys, specific types (continued)
MA 6000, composition ... **A7:** 524, 647
MA754, creep fracture ... **A12:** 393
MA754, departure side pinning ... **A12:** 393
MERL 76, composition ... **A7:** 524, 647
MERL 76, nominal compositions and mechanical properties ... **A7:** 473
MERL 76, turbine disks for Turbofan jet engines ... **A7:** 650
NASA IIB-7, elevated-temperature fatigue crack propagation ... **A8:** 405-406
Nimonic 80A, notch effect on rupture life ... **A8:** 316-318
Nimonic 115, fatigue fracture ... **A12:** 394
PA-101, composition ... **A7:** 524, 647
PA-101, microstructure, compared with cast alloy C103 ... **A7:** 652
RA-330 iron-nickel-base (UNS N08330) sulfidation failure ... **A12:** 388
René 95, aerospace applications ... **A7:** 646-650
René 95, composition ... **A7:** 524, 647
René 95, nominal compositions and mechanical properties ... **A7:** 473
René 95, sphericity and surface quality ... **A7:** 42, 44
René 95, T-700 engine hardware ... **A7:** 648
S-816, grain size effect on notch-rupture strength ... **A8:** 316-317
S-816, notch effect on rupture life ... **A8:** 316-318
S-816, notched and unnotched, creep-rupture strength comparison ... **A8:** 317, 319
single-crystal PWA 1480, effects of thermal cycling ... **A12:** 394
Stellite 31, atomizing variables ... **A7:** 36
Stellite 31, composition ... **A7:** 524, 647
Stellite 31, turbine blade dampers ... **A7:** 651
Udimet 100, nominal compositions and mechanical properties ... **A7:** 473
Udimet 700, ductility in hot torsion tests ... **A8:** 165-166
Udimet 700, mechanical properties ... **A7:** 473
Udimet 700, microstructure, in torsion and extrusion ... **A8:** 176-178
Udimet 700, temperature effect on torsional ductility ... **A8:** 166-167
Udimet 700, tension and torsion effective fracture strains ... **A8:** 168
Waspaloy, ductility in hot torsion tests ... **A8:** 165-166
Waspaloy, flow curves for ... **A8:** 162-163
Waspaloy, for high-temperature grips ... **A8:** 159
Waspaloy, grain size effect on notch-rupture strength ... **A8:** 316-317
Waspaloy, heat treatment effects on rupture time ... **A8:** 316
Waspaloy, notch effect on rupture life ... **A8:** 316-318

SUPERB computer program for structural analysis ... **EM1:** 268, 273
Superbeta npn bipolar junction transistor ... **EL1:** 147
Supercomponent concept
microwave ... **A17:** 210
Superconducting alloy, multifilament
color etched ... **A9:** 157
Superconducting applications *See also* Superconducting materials
A15 superconductors ... **A2:** 1070-1074
high-energy physics (HEP) ... **A2:** 1055-1056
magnetic confinement for thermonuclear fusion ... **A2:** 1056-1057
magnetic energy storage ... **A2:** 1057
magnetic resonance imaging (MRI) ... **A2:** 1054-1055
niobium-titanium superconductors ... **A2:** 1043-1044, 1054-1057
in power industry ... **A2:** 1057
ternary molybdenum chalcogenides (chevrel phases) ... **A2:** 1079-1080
thin-film superconductors ... **A2:** 1083
Superconducting ceramics
properties ... **EM4:** 1
Superconducting magnetic energy storage (SMES)
materials for ... **A2:** 1057
Superconducting materials *See also* A15 superconductors; Conductor(s); High-temperature superconductors for wire and tape; Niobium-titanium superconductors; Superconducting applications; Superconductivity; Superconductors; Ternary molybdenum chalcogenides (chevrel phases); Thin-film materials ... **A7:** 636-638
A15 superconductors ... **A2:** 1060-1076
BSCCO, properties ... **A2:** 1082
high-temperature superconductors for wires and tape ... **A2:** 1085-1089
introduction ... **A2:** 1027-1029
low and high temperature ... **A2:** 1082-1083
niobium-titanium superconductors ... **A2:** 1043-1059
plasma-assisted physical vapor deposition ... **A18:** 848
principles of superconductivity ... **A2:** 1030-1042
TBCCO, properties ... **A2:** 1082-1083
ternary molybdenum chalcogenides (chevrel phases) ... **A2:** 1077-1080
thin-film ... **A2:** 1081-1084
thin-film materials ... **A2:** 1081-1084
YBCO, properties ... **A2:** 1082
Superconducting quantum interference devices (SQUID) ... **EM4:** 17
Superconducting supercollider (SSC) ... **A2:** 1027, 1044, 1056
Superconduction ... **A7:** 636-638
Superconductivity *See also* Superconducting applications; Superconducting materials; Superconductor composites; Superconductor filaments; Superconductors
30 K, discovery, and thin-film materials ... **A2:** 1081
alternating current losses ... **A2:** 1039-1040
of amorphous materials/metallic glasses ... **A2:** 806, 816-817
critical parameters ... **A2:** 1033-1034
cryogenic stability ... **A2:** 1037-1038
defined ... **A2:** 1030
double domain ... **A2:** 1077
dynamic stability ... **A2:** 1038-1039
eddy current losses ... **A2:** 1040
flux pinning ... **A2:** 1034-1035
as future trend ... **EL1:** 395
high-temperature, thin-film materials ... **A2:** 1083
history ... **A2:** 1027-1028
hysteresis losses ... **A2:** 1039
Josephson effects ... **A2:** 1040-1041
low temperature ... **A2:** 1082-1083
magnetic properties ... **A2:** 1035-1036
penetration losses ... **A2:** 1039-1040
principles of ... **A2:** 1030-1042
radio frequency effects ... **A2:** 1039, 1040
stabilization ... **A2:** 1036-1039
ternary molybdenum chalcogenides (chevrel phases) ... **A2:** 1077-1080
theoretical background ... **A2:** 1031-1034
weak link ... **A2:** 1088
of wire filaments, ternary molybdenum chalcogenides ... **A2:** 1079
Superconductor composites
assembly techniques ... **A2:** 1046-1049
billet cleanliness ... **A2:** 1046
cabling ... **A2:** 1051-1052
extrusion ... **A2:** 1049-1050
isostatic compaction ... **A2:** 1049
monofilamentary conductors ... **A2:** 1046-1047
multifilamentary conductors ... **A2:** 1047-1049
stabilizing ... **A2:** 1051
twisting and final sizing ... **A2:** 1051
welding of ... **A2:** 1049
wire drawing ... **A2:** 1050
Superconductor filaments
geometric uniformity ... **A2:** 1052-1053
heat treatment ... **A2:** 1053-1054
properties ... **A2:** 1052-1054
strain cycles ... **A2:** 1053-1054
Superconductor rotor windings
niobium-titanium alloy ... **A2:** 1057
Superconductors *See also* A15 superconductors; Conductors; Drawing; High-temperature superconductors; Niobium-titanium superconductors; Superconducting materials; Superconductivity; Ternary molybdenum chalcogenides (chevrel phases); Thin-film materials
A15, development, processing and applications ... **A2:** 1060-1074
amorphous materials ... **A2:** 816-817
Superconductors (continued)
applications, niobium-titanium superconductors ... **A2:** 1043-1044, 1054-1057
billet stacking manufacture method ... **A14:** 338-341
commercial, manufacture of ... **A14:** 338-342
freeze drying ... **EM4:** 62
high-temperature, for wires and tapes ... **A2:** 1085-1089
history ... **A2:** 1027-1028
modified jelly-roll manufacture method ... **A14:** 341-342
niobium-titanium ... **A2:** 1043-1059
and normal metals ... **A7:** 636
processing and properties, A15 types ... **A2:** 1065-1070
Type I ... **A2:** 1032-1033
Type II ... **A2:** 1032-1033
Superconductors, ceramic
applications ... **EM4:** 1106
Supercooling *See also* Constitutional supercooling; Undercooling ... **A6:** 45, 46-48, 49, 51, 52, 53
constitutional ... **A9:** 611-612
defined ... **A9:** 18 **A15:** 11, 101
diagram, constitutional ... **A15:** 115
as driving mechanism, equiaxed nucleation ... **A15:** 130-131
effect on pure metal solidification structures ... **A9:** 609-610
influence, calculated ... **A15:** 102
in planar interface growth ... **A15:** 114-116
Supercritical drying ... **EM4:** 446, 440
Supercritical fluid chromatography ... **A10:** 116
Supercritically reheated grain-refined (SCGR) zone ... **A6:** 81
Superelastic applications
shape memory alloys ... **A2:** 900-901
Superelements ... **EM3:** 485
Superficial hardness test *See* Rockwell superficial hardness test
Superfine particle separator ... **A7:** 179
Superfines ... **A7:** 12
Superhard coatings
low-pressure synthesis of ... **A2:** 1009
Superhard materials *See also* Superabrasives; Ultrahard tool materials
principal types ... **A2:** 1008
Superhard tool materials *See* Tool materials, superhard
Superheat
big bang mechanism and ... **A15:** 131
defined ... **A15:** 11
effect, grain growth ... **A15:** 135
effect on grain size of austenitic manganese steel castings ... **A9:** 237
effect on pure metal solidification structures ... **A9:** 608-610
in equiaxed grain growth ... **A15:** 132
temperatures, effect, alloy additions ... **A15:** 71-72
Superheated tubes
ASTM specifications ... **M1:** 323
Superheaters
abbreviation for ... **A11:** 797
ash deposits ... **A13:** 1201
corrosion of ... **A13:** 998-999
heat transfer in ... **A11:** 604, 605
outlet header, interligament cracking in ... **A11:** 667
recovery boilers ... **A13:** 1200-1201
scaling in ... **A11:** 616
steam/water-side boilers ... **A13:** 992
thermal fatigue and stress rupture in ... **A11:** 626
tubes, circumferential corrosion-fatigue cracks ... **A11:** 79
tubes, coal ash corrosion of ... **A11:** 617-618
tubes, ruptured by overheating ... **A11:** 609
tubes, thick-lip stress rupture ... **A11:** 606
Superheating *See also* Heat treatment
defined ... **A9:** 18 **A15:** 11
for magnesium alloy grain refinement ... **A15:** 480
titanium alloys ... **A15:** 831-832
titanium and titanium alloy castings ... **A2:** 642
Superlattice diffraction vectors
determination of ... **A9:** 109
Superintendent of Documents (U.S. Government Printing Office)
source of standards ... **EM3:** 64

Superlattice *See also* Ordered crystal structures
defined ... **A9:** 708
long-period ... **A9:** 710, 719
in ordered intermetallics ... **A2:** .913-914
Superlattice reflections in transmission electron microscopy diffraction patterns ... **A9:** 109
Superlattices *See also* Lattices; Ordered structure ... **A3:** 1•10
defined ... **A10:** 683
diffraction pattern from ... **A10:** 540
element location in planes of ... **A10:** 599
as impeding determination of atomic structure ... **A10:** 344, 353
and interface studies ... **A10:** 634-635
strain measurement ... **A10:** 628
Superpicral as an etchant for wrought stainless steels ... **A9:** 281
Superplastic alloys ... **A7:** 18
characterized ... **A14:** 853-857
Superplastic duplex stainless steels ... **A9:** 286
Superplastic ferrous alloys
outlook for ... **A14:** 871
Superplastic formed titanium ... **EM1:** 35, 862
Superplastic forming
aluminum alloys ... **A14:** 800
blow-forming method, illustrated ... **A14:** 18
as new metalworking process ... **A14:** 18
titanium alloy ... **A15:** 824
of titanium alloys ... **A14:** 842-844
Superplastic forming (SPF) *See also* Forming ... **EM1:** 23, 862
aluminum P/M alloys ... **A2:** 210
defined ... **EM2:** 41
titanium alloy sheet ... **A2:** 590-591
wrought titanium alloys ... **A2:** 616
Superplastic forming/diffusion bonding ... **A14:** 844, 857, 859-860
Superplastic forming/diffusion bonding (SPF/DB)
titanium alloys ... **A6:** 156
Superplastic forming/diffusion welding (SPF/DW) techniques ... **A6:** 884, 885
Superplastic metals
defined ... **A8:** 45
Superplastic properties in zinc alloys
effect of aluminum on ... **A9:** 489
Superplastic sheet forming ... **A14:** 852-873
cavitation ... **A14:** 867
forming equipment and tools ... **A14:** 860-861
hypereutectoid plain carbon steels ... **A14:** 869
hypoeutectoid/eutectoid plain carbon steels ... **A14:** 869
iron-base alloys, superplasticity in ... **A14:** 868-872
low-alloy steels ... **A14:** 871
manufacturing ... **A14:** 867-868
medium alloy steels ... **A14:** 871
microduplex stainless steels ... **A14:** 871
processes ... **A14:** 857-860
superplastic alloys ... **A14:** 853-857
superplastic ferrous alloys ... **A14:** 871-872
superplasticity requirements ... **A14:** 852-853
thinning characteristics ... **A14:** 861-867
white cast irons ... **A14:** 869
Superplastic titanium alloys ... **A14:** 842-844
Superplastic zinc alloy
properties ... **A2:** 542
Superplasticity
in aluminum alloys ... **A14:** 800
defined ... **A14:** 13, 852
as effect of temperature on formability ... **A8:** 553
of extruded Alloy 100 ... **A14:** 152
m values for metals in ... **A8:** 550
pressure densification ... **EM4:** 301
requirements, for superplastic sheet forming ... **A14:** 852-853
Supersaturated
defined ... **A15:** 11
Supersaturation
and nucleation ... **A15:** 52-53
Supersonic vibrations
for compacting iron powders ... **A7:** 306
Superstone 40 *See* Cast copper alloys, specific types (C95700); Copper alloys, specific types, C95700
Superstructures of crystals ... **A9:** 708
Supplemental operation ... **A7:** 12
Supplementary SI units
guide for ... **A10:** 685
Supplementary units
Système International d'Unités (SI) ... **A11:** 793
Supplier data sheets *See also* Data sheets
interpreting ... **EM2:** 638-645
properties ... **EM2:** 638-645
structure ... **EM2:** 638
Suppliers
of acrylic acid and vinyl-acrylic polymer emulsion ... **EM3:** 211
of acrylic rubber phenolic adhesives ... **EM3:** 104
of acrylics ... **EM2:** 108 **EM3:** 124
of acrylonitrile-butadiene-styrenes (ABS) ... **EM2:** 114
of aircraft sealants ... **EM3:** 59
of allyls (DAP, DAIP) ... **EM2:** 229
of amino molding compounds ... **EM2:** 231
of anaerobics ... **EM3:** 79, 116
of asphalt ... **EM3:** 59
of butyls ... **EM3:** 199, 202
of conductive adhesives ... **EM3:** 76
of cresol novolacs ... **EM3:** 94
of cresol-base epoxy-novolac resins ... **EM3:** 104
of cyanates ... **EM2:** 238
of cyanoacrylates ... **EM3:** 79, 127-128, 129
engineering/development by ... **EM1:** 35
of epoxidized phenols ... **EM3:** 94
of epoxies ... **EM2:** 241 **EM3:** 78, 98, 139
of etchants ... **EM3:** 802, 803
of EVA copolymers ... **EM3:** 82
of film adhesives ... **EM3:** 76
of fluoroelastomer sealants ... **EM3:** 227
of fluorosilicones ... **EM3:** 59
of glazing sealants ... **EM3:** 58
of high-density polyethylenes (HDPE) ... **EM2:** 166
of high-impact polystyrenes (PS, HIPS) ... **EM2:** 199
of high-temperature adhesives ... **EM3:** 80
of high-temperature structural adhesives ... **EM3:** 508
of homopolymer/copolymer acetals ... **EM2:** 102
of hot-melt adhesives ... **EM3:** 82
of insulated glass ... **EM3:** 58
of ionomers ... **EM2:** 123
of latex ... **EM3:** 211
of liquid crystal polymers (LCP) ... **EM2:** 182
of metal building sealant tapes ... **EM3:** 58
of mixing and dispensing equipment ... **EM3:** 604
of modified acrylics ... **EM3:** 78
of neoprene phenolics ... **EM3:** 104
of nitrile phenolics ... **EM3:** 104
of phenol ... **EM3:** 104
of phenolic silicones ... **EM3:** 105
of phenolics ... **EM2:** 245 **EM3:** 80, 104
of plastisol ... **EM3:** 59
of polyamides (PA) ... **EM2:** 127
of polyarylates (PAR) ... **EM2:** 141
of polybenzimidazoles ... **EM3:** 169
of polybenzimidazoles (PBI) ... **EM2:** 150
of polybutene ... **EM3:** 59, 82
of polybutylene terephthalates (PBT) ... **EM2:** 155
of polycarbonates (PC) ... **EM2:** 152
of polyether silicones ... **EM3:** 232-233
of polyether sulfones (PES, PESV) ... **EM2:** 162
of polyethylene ... **EM3:** 82
of polyethylene terephthalates (PET) ... **EM2:** 176
of polyimides ... **EM3:** 158-160, 161
of polyphenylene ether blends (PPE PPO) ... **EM2:** 185
of polyphenylene sulfides (PPS) ... **EM2:** 190
of polyphenylquinoxalines ... **EM3:** 164
of polypropylenes ... **EM3:** 59
of polysulfides ... **EM3:** 138, 193, 195
Suppliers (continued)
of polyurethanes ... **EM3:** 78, 111
of polyurethanes (PUR) ... **EM2:** 264
of polyvinyl acetal ... **EM3:** 82
of polyvinyl chlorides (PVC) ... **EM2:** 212-213
of pressure-sensitive adhesives ... **EM3:** 86
pricing of product ... **EM3:** 730
of primers ... **EM3:** 802
products, and material selection ... **EM1:** 38
of Pyralin products ... **EM3:** 158
of sealants ... **EM3:** 57, 58, 60
of silicones ... **EM2:** 267 **EM3:** 59, 79, 134
of silicones (RTV) ... **EM3:** 220
of solvent acrylics ... **EM3:** 209
of structural adhesives ... **EM3:** 510, 513
of styrene-acrylonitriles (SAN, OSA ASA) ... **EM2:** 216
of styrene-maleic anhydrides (S/MA) ... **EM2:** 221
testing of materials ... **EM3:** 735
testing of product for quality control ... **EM3:** 730, 733-734
of thermoplastic fluoropolymers ... **EM2:** 119
of thermoplastic polyurethanes (TPUR) ... **EM2:** 207
of ultrahigh molecular weight polyethylenes (UHMWPE) ... **EM2:** 171
of urethane hybrids ... **EM2:** 271
of urethane sealants ... **EM3:** 204-205
of UV/EB-cured adhesives ... **EM3:** 92
of vinyl esters ... **EM2:** 275
of water-base adhesives ... **EM3:** 90
Suppliers of Advanced Composite Materials Association (SACMA)
as information source ... **EM1:** 41
Supply pressure
nomenclature for hydrostatic bearings with orifice or capillary restrictor ... **A18:** 92
Support arm, front-end loader
brittle fracture of ... **A11:** 69
Support blocks
in spring-material test apparatus ... **A8:** 134-135
Support plate
defined ... **A14:** 13
Support wire ... **A1:** 283
Supported plumbum materials
applications ... **A2:** 555
Supporting electrode
defined ... **A10:** 683
Suppressed chromatography
anion, reaction of ... **A10:** 659
cation, reaction of ... **A10:** 659-660
Suppressors
columns ... **A10:** 660
fiber ... **A10:** 660
hollow fiber anion ... **A10:** 660
role in ion chromatography ... **A10:** 659
schematic, membrane-type ... **A10:** 660
SURF 11
as synchrotron radiation source ... **A10:** 413
Surface *See also* Surface analysis; Surface analysis and characterization; Surface composition; Surface contamination; Surface roughness; Surface segregation; Surface sensitivity; Surface structure; Surface-sensitive analytical techniques; Surfacefilms
adsorbates, identified on metal electrodes ... **A10:** 126, 134
adsorbed or chemisorbed species on ... **A7:** 258
adsorption ... **A10:** 109, 114, 126, 134, 407, 583
angle of test, Rockwell hardness testing ... **A8:** 80
anti-static ... **A7:** 610-611
catalysts ... **A10:** 114, 253
chemical analyses of ... **A10:** 177-178, 591
chemical bonding on ... **A7:** 257
coatings, effects on explosivity *See also* Coatings ... **A7:** 194
coatings, identified ... **A10:** 168
composition of atomized powders ... **A7:** 25

Surface (continued)
compressive residual stresses **A10:** 380
condition, effect on fatigue properties of steel **A8:** 373
condition, effect on powder compact **A7:** 211
condition, specimen, in cold upset testing **A8:** 579
contaminants *See also* Contaminants **A7:** 255
cracking **A8:** 591-594
crystallography of **A10:** 536-538
crystallography, vocabulary for **A10:** 537-538
curved, Rockwell hardness testing with **A8:** 81, 83
decarburization, tested by magnetic bridge sorting **A7:** 491
deeply etched, SEM analysis **A10:** 490
defects, detection by magnetic particle inspection **A7:** 575-579
defined **A18:** 376
depth profiling through **A10:** 583
dielectric properties **A10:** 136
diffraction from, principles **A10:** 538-539
diffusion **A10:** 583
discontinuities, detected by magnetic particle inspection **A7:** 576
dynamic processes of, LEED analysis of **A10:** 536
effects and fatigue **A8:** 373-374
effects of, determined **A10:** 273-274
electrode, XPS spectrum showing elements present in **A10:** 578
electroplating and fatigue limit **A8:** 373
elemental analysis by XPS **A10:** 568
enhancement **A10:** 136
etching **A10:** 575
fatigue specimen, factors affecting **A8:** 373
finishing **A10:** 93
flat metal, IRRAS analysis **A10:** 114
forces, microcompact **A7:** 58
fracture **A10:** 490, 497
free-radical reactions, ESR studied **A10:** 253
friction-velocity curve with hydrodynamic lubrication **A8:** 605
grinding **A10:** 287, 390-391
hardening, microhardness testing for monitoring **A8:** 96
inhomogeneity **A7:** 260
inner, Rockwell hardness testing of **A8:** 81, 83
interaction of lubricants with, AES analysis **A10:** 565
layers of glass, analysis of **A10:** 624-625
LEISS segregation of lead to **A10:** 607-608
materials, resource evaluations by NAA **A10:** 233
metal **A10:** 118, 583, 586
microstructure by LEED **A10:** 536
modification of frictional **A10:** 565
near-, effect of stress gradient correction on Measurement of **A10:** 388
of nontransparent samples characterized **A10:** 70-71
oriented, structural information **A10:** 407
-phase detection, Mössbauer effect **A10:** 287, 293
phase, qualitative XRPD analysis on silicon **A10:** 341
preparation **A8:** 80, 93, 96 **A10:** 93
quality, sheet metal forming **A8:** 553
reactions, FIM/AP study of **A10:** 583
reconstruction **A10:** 536, 583
residual stresses **A8:** 373-374
roughness **A8:** 373
screening **A7:** 176-177
segregation **A10:** 544, 564-566, 583, 593, 607-608
shear, determined by quasi-static torsional testing **A8:** 145
shear stress **A8:** 142, 145
of sheet metals, effect on hardness **A8:** 560
single-crystal **A10:** 407
solid **A10:** 198, 603
species **A10:** 133, 134
of stainless steel, SIMS analysis of effects of laser treatment on **A10:** 622-623
stains and corrosion, LEISS identified **A10:** 607
strains of **A10:** 607
stress-corrosion crack fracture, AES analysis **A10:** 562-564

Surface (continued)
tension, and stress, LaPlace equation for **A7:** 312
tension, SI derived unit and symbol for **A10:** 685
test, Brinell hardness **A8:** 85, 88
texture, in milling of single particles **A7:** 59
texture, undesirable, as formability problem **A8:** 548
topography **A10:** 595
topography, tin powders **A7:** 124
UV/VIS characterizations of **A10:** 70-71
x-ray diffraction stress measurement confined to **A10:** 382

Surface abrasion **EM3:** 35
Surface acoustic wave (SAW) **A18:** 408
Surface activation **A7:** 174
copper alloys **M5:** 232-233
electroless nickel plating processes **M5:** 232-233
magnesium alloys **M5:** 642, 644, 646
nickel plate **M5:** 198, 216
stainless steel **M5:** 232, 561

Surface adsorption theory
of graphite growth **A15:** 178
hydrogen damage **A13:** 164

Surface alloying
and fatigue resistance **A1:** 680

Surface analysis *See also* Surface; Surface analysis and characterization; Surface finish; Surface(s)
advanced failure analysis techniques **EL1:** 1107
amorphous solar cell formation **EL1:** 1080-1081
Auger electron spectroscopy **EL1:** 1077-1080
Auger electron spectroscopy (AES) **EM2:** 811, 816-817
by EPMA electron beam **A10:** 517, 529-530
chemical, AES high lateral resolution for **A10:** 549, 556-566
electron spectroscopy **EL1:** 1074-1080
electron spectroscopy, applications **EL1:** 1080-1083
elemental, inorganic solids, analytical methods **A10:** 4-6
elemental, of organic solids, methods for **A10:** 9
and failure analysis **EL1:** 1074
failure analysis case histories **EM2:** 817-822
of failures, types **EM2:** 811
graphite, Raman spectroscopy **A10:** 132
imaging photoemission microscopy **EL1:** 1077
of inorganic solids, applicable analytical methods **A10:** 4-6
ion scattering techniques **EL1:** 1090-1092
laser microprobe mass spectroscopy **EL1:** 1088-1090
methods, capabilities **EL1:** 1074
molecular/compound, of inorganic solids methods for **A10:** 4-6
molecular/compound, of organic solids methods for **A10:** 9
of organic solids, methods for **A10:** 9
photoemission spectroscopy instrumentation **EL1:** 1076-1077
Raman spectroscopy **A10:** 126, 133-137
secondary ion mass spectroscopy **EL1:** 1083-1088
smooth, Raman spectroscopy for **A10:** 135-136
static secondary ion mass spectroscopy (SSIMS) **EM2:** 811, 817
techniques **EM2:** 811-817
x-ray photoelectron spectroscopy **EM2:** 811, 812-816

Surface analysis and characterization *See also* Surface; Surface analysis
Auger electron spectroscopy **A10:** 549-567
infrared spectroscopy **A10:** 109-125
low-energy electron diffraction **A10:** 536-545
low-energy ion-scattering spectroscopy **A10:** 603-609
Mössbauer spectroscopy **A10:** 287-295
particle-induced x-ray emission **A10:** 102-108
Raman spectroscopy **A10:** 126-138
Rutherford backscattering spectrometry **A10:** 628-636
secondary ion mass spectroscopy **A10:** 610-627
x-ray photoelectron spectroscopy **A10:** 568-580

Surface analysis techniques *See also* Surface chemical analysis **A7:** 250-261
compared **EM1:** 285

Surface analysis techniques and applications **EM3:** 236-251
adsorption of hydration inhibitors **EM3:** 250-251
data analysis by chemical state information **EM3:** 243-244
data analysis by depth distribution
ball cratering **EM3:** 245-246
comparison with less surface-sensitive I inelastic scattering ratio **EM3:** 247
information **EM3:** 244, 247
ion sputtering **EM3:** 244-245, 246
multiple anodes/different transitions **EM3:** 247
techniques **EM3:** 247
variable take-off angle **EM3:** 246-247
data analysis by depth profiling **EM3:** 249
data analysis by quantification **EM3:** 242-243
failure analysis application to adhesive
adhesive failure **EM3:** 248
bonding **EM3:** 248-249
cohesive failure **EM3:** 248
interfacial failure **EM3:** 248, 249
mixed-mode failure **EM3:** 248
hydration of phosphoric-acid-anodized aluminum **EM3:** 249-250
near-surface-sensitive techniques **EM3:** 237
surface behavior diagrams (SBDS) **EM3:** 247-248
surface-sensitive techniques **EM3:** 236-242

Surface analysis with laser ionization (SALI) **EM3:** 237

Surface area *See also* Projected area; Specific surface area
of atomized aluminum powder **A7:** 125, 129
defined **A15:** 11
and degree of pyrophoricity **A7:** 198-199
effect, clay-water bonds **A15:** 212
and explosivity **A7:** 195
and green strength **A7:** 288
of metal powders **A7:** 262
methods to determine **A7:** 262-270
microcompact **A7:** 58
and oxygen content, aluminum powder **A7:** 130
sand, molding effects **A15:** 208
and sintering behavior **A7:** 262
and steady-state flow conditions **A7:** 264-265
of tin powders **A7:** 124
volume-to-, ratio **A15:** 611

Surface area fraction **A18:** 465
Surface area measurements **EM4:** 583-584
BET theory **EM4:** 583-584
dynamic (continuous flow) method **EM4:** 584
experimental methods **EM4:** 584
single-point method **EM4:** 584
vacuum volumetric technique **EM4:** 584

Surface area per unit volume
equation for **A9:** 125-126

Surface behavior diagrams
adsorption of hydration inhibitors **EM3:** 250

Surface blisters *See* Blistering

Surface blow
as semisolid alloy defect **A15:** 337

Surface carbon content control *See also* Carbon control, evaluation **A4:** 573-586 **M4:** 417-431
carbon potential, control of **A4:** 573-574, 575, 576, 577, 583 **M4:** 417-419
control-system features **A4:** 575
gas carburizing, process control **A4:** 575-577 **M4:** 420-422
gas reactions **A4:** 573 **M4:** 417
gas-with-gas reactions **A4:** 574 **M4:** 419
gas-with-solid reactions **A4:** 574 **M4:** 419
kinetics **A4:** 574-575 **M4:** 419-420

Surface carbon content control, analyzers **A4:** 577-586
atmosphere samples for **A4:** 577-579 **M4:** 422-424
dew point **A4:** 581-583, 584 **M4:** 426-428
gas chromatography **A4:** 585-586 **M4:** 430-431
hot-wire **A4:** 584-585 **M4:** 429-430
infrared **A4:** 579-581, 584 **M4:** 424-426
Orsat **A4:** 584 **M4:** 429
oxygen probes **A4:** 583-584 **M4:** 428-429

Surface carbons
carbide or graphitic identified **A10:** 568
contamination, effect on painted, cold-rolled steel **A10:** 556
determined **A10:** 223-224

Surface carbons (continued)
effects on paint adherence on metal cabinetry **A10:** 224
Surface chemical analysis *See also* Surface analysis techniques **A7:** 250-261 **A18:** 445-461
Auger electron spectroscopy (AES) **A18:** 445, 446-447, 449, 450-456
applications **A18:** 454-455
data analysis **A18:** 452-453
effects of chemical environment on AES spectrum **A18:** 454
electron escape depth **A18:** 451-452
equipment **A18:** 452
fundamentals **A18:** 450-451
ion etching (sputtering) **A18:** 453-454, 455
limitations **A18:** 455-456
platinized titania application **A18:** 449-450, 451
spectrum **A18:** 452-453
versus SIMS **A18:** 458, 459
by Auger electron spectroscopy **A7:** 250-255
by ion scattering spectroscopy **A7:** 259-260
by secondary ion mass spectroscopy **A7:** 257-259
by x-ray photoelectron spectroscopy **A7:** 255-257
electron-stimulated desorption (ESD) **A18:** 456-458
applications **A18:** 457-458
detection limit **A18:** 456
equipment **A18:** 456-457
fundamentals **A18:** 456
spectrum **A18:** 457
infrared (IR) spectroscopy **A18:** 460-461
ion-scattering spectrometry (ISS) **A18:** 448-450
applications **A18:** 449-450
equipment **A18:** 449
fundamentals **A18:** 448-449
neutralization of ions **A18:** 449
resolution **A18:** 449
sensitivity **A18:** 449
spectrum **A18:** 449
secondary ion mass spectrometry (SIMS) **A18:** 456, 458-460, 461
applications **A18:** 458, 459-460
dynamic **A18:** 459
equipment **A18:** 459
fundamentals **A18:** 459
spectrum **A18:** 459
static **A18:** 459
versus AES and XPS **A18:** 458, 459
techniques **A7:** 250-261
X-ray photoelectron spectroscopy (XPS) **A18:** 445-448, 449, 452, 456
data analysis **A18:** 447-448
development **A18:** 445
equipment **A18:** 446-447
fundamentals **A18:** 445-446
spectrum **A18:** 447, 448
versus SIMS **A18:** 458
Surface chemistry, powder **A7:** 250-261
and bonding mechanisms **A7:** 260
Surface cleanliness
measurement of **EM3:** 322
Surface coating *See* Coatings, Fiber sizing; Sizing
Surface coatings *See also* Coatings; Metallic Coatings; Surface(s)
antireflective, for photography **A12:** 83
effect in shaft failures **A11:** 480-482
effects, visual examination of **A12:** 72
for fasteners **A11:** 530
for fracture preservation **A12:** 73
gas-lubricated bearings **A18:** 532
seals **A18:** 551
and sectioning **A12:** 77
sputter and thermal evaporation **A12:** 173
Surface cold shuts
in copper alloy ingots **A9:** 642
Surface composition
analysis by SIMS **A10:** 610
Surface composition (continued)
effects during laser treatment of stainless steel **A10:** 622-623
variation during heating **A10:** 564
Surface condition
forgings, effect on fatigue **M1:** 355
hot rolled bars **M1:** 199-200
Surface conditioners
for polysulfides **EM3:** 196-197
Surface conductivity test **A6:** 130
Surface considerations for joining ceramics and glasses **EM3:** 298-310
bonding fundamentals **EM3:** 298-301
chemical bonding **EM3:** 300
mechanical bonding **EM3:** 300
physical bonding **EM3:** 299-300
wetting by sessile drop method **EM3:** 299
ceramic and glass bonding **EM3:** 308-309
ceramic chemistry and microstructure **EM3:** 298
ceramic-metal bond stresses, elastic modulus effect **EM3:** 301
ceramic-to-metal joints and seals **EM3:** 304-308
ceramic films and coatings on metals **EM3:** 306-308
ceramic metallization and joining **EM3:** 304-306
enamels, joining ceramics and glasses **EM3:** 301-304
glass chemistry and microstructure **EM3:** 298
glass-to-metal seals **EM3:** 301-302
glasses compared to ceramics **EM3:** 298
organic bonding, surface preparation of ceramics and glasses **EM3:** 309-310
stresses in ceramic-metal bonds **EM3:** 300-301
Surface contact points **A7:** 12
Surface contamination
AES analysis **A10:** 549
anion determination **A10:** 658
FMR analysis **A10:** 268
identification **A10:** 168
RBS analysis **A10:** 628
Surface contamination considerations **EM3:** 845-847
composite substrates **EM3:** 845-847
thermoplastics **EM3:** 845, 846-847
thermosets **EM3:** 845-846
contamination definition and description **EM3:** 845
interlaminar failure **EM3:** 846
jet fuel (JP-4) contamination **EM3:** 846
metallic substrates **EM3:** 845
moisture presence **EM3:** 846
Surface cracks *See also* Cracks; Subsurface flaws; Surface(s)
casting **A17:** 512
laser-detected **A17:** 17
microstructure and replica, compared **A17:** 54
microwave detection **A17:** 214-215
in pipeline girth welds **A17:** 581
radiographic methods **A17:** 296
ultrasonic inspection **A17:** 250
visual inspection **A17:** 3
Surface damage **A18:** 176-183
classes of **A18:** 176-177
cumulative, fatigue crack development from **A11:** 684, 688-689
defined **A18:** 18
diagnosis **A18:** 177
effects of the surface damage **A18:** 177
properties of the surface layer **A18:** 177
relative importance of rival wear mechanisms **A18:** 177
as failure mechanism **A11:** 75
intensity, as dependent upon load **A11:** 685, 689
tribography as a tool **A18:** 183
tribological examples **A18:** 177-182
damage by interaction between corrosion and fracture **A18:** 181-182
damage dominated by brittle fracture **A18:** 180
Surface damage (continued)
damage dominated by chip formation **A18:** 179-180
damage dominated by contact-generated corrosion **A18:** 182-183
damage dominated by dissolution or diffusion **A18:** 181
damage dominated by extrusion **A18:** 179
damage dominated by fatigue fracture **A18:** 180-181
damage dominated by microstructural changes **A18:** 177-178
damage dominated by plastic deformation **A18:** 178
damage dominated by shear fracture **A18:** 178-179
damage dominated by surface cracking **A18:** 178
damage dominated by tearing **A18:** 180
wear failure from **A11:** 156
Surface damage contribution factor **A18:** 266, 268
Surface decarburization in martensitic stainless steels **A9:** 285
Surface defects *See* Defect(s); Surface cracks; Surface flaws; Surface(s)
casting **A11:** 353-354
cold-formed parts **A11:** 307
from welding **A11:** 127
spring failure from **A11:** 554-555
in tire-mold casting **A11:** 358
Surface delamination
by machining/drilling **EM1:** 667
Surface diffusion **A7:** 12, 313
Surface digitizers
noncontacting **A8:** 549
Surface discontinuities
effect on fatigue strength **A11:** 119
from burning **A11:** 120
on iron castings **A11:** 352-353
on shafts **A11:** 459, 467, 472, 478
Surface distress
defined **A18:** 18
Surface energy **A6:** 143 **A18:** 399-400
parylene coatings **EL1:** 795-796
Surface EXAFS detection technique **A10:** 418
Surface examination **A18:** 290- 292
analysis of data collected **A18:** 292
first level visual inspection **A18:** 290-291
matters of scale **A18:** 292
planning **A18:** 290
second level electron microscopy **A18:** 291
selection of chemical analysis instruments **A18:** 291-292
Surface exposure time (SET) **EM3:** 277
Surface expulsion
by liquid penetrant inspection **A17:** 86
definition **A6:** 1214
Surface extended x-ray absorption fine structure (SEXAFS) **EM3:** 237
Surface extension parameter **A6:** 308
Surface fatigue **A18:** 259
nitrided surfaces **A18:** 879, 880, 881-882
sliding bearings **A18:** 742, 743
Surface fatigue wear, thermal spray coating applications **A18:** 831, 833
performance factors **A18:** 833
Surface films **A7:** 12
AES in-depth compositional evaluation **A10:** 549
heterogeneous, AES analysis of **A10:** 565-566
magnesium-oxide-enriched **A7:** 254
on electrical contacts, XRS analysis **A10:** 578
oxidation states of metal atoms, determined in **A10:** 568
Surface finger oxides **A14:** 205
Surface finish *See also* Coatings; Finish; Finishes; Finishing; Surface preparation; Surface(s) **A16:** 19-36
abrasive flow machining **A16:** 517, 518, 519

SUBJECTS OF THE INDEXED VOLUMES: ASM Handbook (designated by the letter "A"): **A1:** Properties and Selection: Irons, Steels, and High-Performance Alloys (1990); **A2:** Properties and Selection: Nonferrous Alloys and Special-Purpose Materials (1990); **A3:** Alloy Phase Diagrams (1992); **A4:** Heat Treating (1991); **A6:** Welding, Brazing, and Soldering (1993); **A7:** Powder Metallurgy (1984); **A8:** Mechanical Testing (1985); **A9:** Metallography and Microstructures (1985); **A10:** Materials Characterization (1986); **A11:** Failure Analysis and Prevention (1986); **A12:** Fractography (1987); **A13:** Corrosion (1987); **A14:** Forming and Forging (1988); **A15:** Casting (1988); **A16:** Machining (1989); **A17:** Nondestructive Testing and Quality Control (1989); **A18:** Friction, Lubrication, and Wear Technology (1992). **Metals Handbook, 9th Edition** (designated by the letter "M"): **M1:** Properties and Selection: Irons and Steels (1978); **M2:** Properties and Selection: Nonferrous Alloys and Pure Metals (1979); **M3:** Properties and Selection: Stainless Steels, Tool Materials and Special-Purpose Materials (1980); **M4:** Heat Treating (1981); **M5:** Surface Cleaning, Finishing, and Coating (1982); **M6:** Welding, Brazing, and Soldering (1983). **Engineered Materials Handbook** (designated by the letters "EM"): **EM1:** Composites (1987); **EM2:** Engineering Plastics (1988); **EM3:** Adhesives and Sealants (1990); **EM4:** Ceramics and Glasses (1991); **Electronic Materials Handbook** (designated by the letters "EL"): **EL1:** Packaging (1989).

Surface finish (continued)
abrasive waterjet cutting **A14:** 746-748
abrasives and grinding **A16:** 431, 433, 434, 441, 446
and adaptive control **A16:** 618, 621, 623, 624
after cold heading **A14:** 296
Al alloys **A16:** 761, 764-765, 768-769, 773-780, 784-785, 795-797
aluminum alloy **A15:** 762-763
Be alloys **A16:** 872
boring **A16:** 162-164, 169
of Brinell test workpiece **A8:** 86, 88
broaching **A16:** 194, 197, 203, 206, 208, 209
and burnishing **A16:** 210
by abrasives, compared **A15:** 514
by CBN grinding wheels **A16:** 462-463, 464, 467, 469
by metal casting **A15:** 40
by plaster molding **A15:** 242
CAD/CAM analysis **A16:** 628
carbon and alloy steels **A16:** 670, 672, 673, 675
carbonaceous additions for **A15:** 211
cast irons **A16:** 657, 661, 662, 663, 664, 665
with ceramic molding **A15:** 248
chemical milling **A16:** 579, 581-586
classifications **A16:** 22
in coining **A14:** 186
in contour roll forming **A14:** 633-634
copper and copper alloy forgings **A14:** 258
Cu alloys **A16:** 805, 808, 812, 815, 817, 819
designations **A16:** 20
dimensional tolerances **A16:** 19
disk-pressure test for hydrogen embrittlement in **A8:** 540-541
and drilling **A16:** 222
for durometer testing **A8:** 106
effect on fatigue life **A8:** 711
effect, optical holographic interferometry **A17:** 412
electrical discharge grinding **A16:** 565-566
electrical discharge machining **A16:** 558, 561
electrochemical discharge grinding **A16:** 550
electrochemical grinding **A16:** 543, 545, 546
electrochemical machining **A16:** 534-541
filament winding **EM2:** 373-374
flaws **A16:** 19
of forged heat-resistant alloys **A14:** 236
and gear manufacture **A16:** 344
and gear shaving **A16:** 341-343
gear-tooth honing **A16:** 486-487
gray cast iron **M1:** 12, 25, 26
grinding **A16:** 421-424, 426, 427
and grinding of gears **A16:** 350, 354-355
and grit size in honing **A16:** 483
hafnium **A16:** 844, 856
heat-resistant alloys **A16:** 740, 744, 745, 755, 756
in HERF processing **A14:** 105
high removal rate machining **A16:** 608, 609
and high-speed machining **A16:** 601, 602
and hobbing of gears **A16:** 340
honing **A16:** 473, 476, 481, 482, 484
improved by distribution of cutting forces **A16:** 299
lapping **A16:** 492, 496, 503, 504
lay **A16:** 19, 20
and machinability **A16:** 646-647
machined surfaces **M1:** 566-567, 571-572, 574, 576-579
Mg alloys **A16:** 820, 821, 822, 823, 824, 825, 826, 827
microhoning **A16:** 472, 489
milling **A16:** 305, 316, 319-321, 326, 327, 329
and NC programming **A16:** 614
Ni alloys **A16:** 835
P/M tool steels **A16:** 735
permanent mold casting **A15:** 284-285
photochemical machining **A16:** 587
plateau finish from honing **A16:** 487
of powder forgings **A14:** 204
precision, as interferometer application **A17:** 14-15
and press forming **A14:** 547-548
reaming **A16:** 240, 246
with Replicast process **A15:** 37, 271
requirements, and process selection **EM2:** 302-207
roller burnishing **A16:** 253

Surface finish (continued)
rotary swaging **A14:** 140
roughness **A16:** 19-21
roughness, arithmetic averages **A16:** 26
roughness, maximum peak-to-valley **A16:** 26
RTM and SRIM **EM2:** 350
sawing **A16:** 356, 358, 359
for Scleroscope hardness testing **A8:** 105
shaping **A16:** 190, 191
slotting **A16:** 190, 191
stainless steels **A16:** 681
steel sheet **M1:** 557
with stretch drawing **A14:** 597-598
substrate **EL1:** 104
and superabrasive grinding wheels **A16:** 458, 460-461, 462
symbols **A16:** 19
tapping **A16:** 263, 264
thread grinding **A16:** 271
thread rolling **A16:** 280, 291-292, 293, 294
titanium **A16:** 844, 851
titanium castings **A15:** 831
treatments, investment castings **A15:** 264
trepanning **A16:** 175
tube spinning **A14:** 678
turning **A16:** 152, 153
types **EM2:** 303-306
ultrasonic machining **A16:** 529
uranium alloys **A16:** 874, 875, 876-877
Verson hydroform process **A14:** 613
visual inspection **A17:** 3
waviness **A16:** 19
wear resistance, effect on **M1:** 636-637
of wire **A14:** 694
workpiece, and liquid penetrant inspection **A17:** 81
zirconium **A16:** 844, 854
Zn alloys **A16:** 831, 832, 834
Surface finishes and protective coatings
for leaf springs
Surface finishing
and fatigue strength **A11:** 122
in green machining **EM4:** 183
improper, by carbon flotation **A11:** 358-359
of stainless steel **A1:** 884-885
as stress source **A11:** 205
Surface flaws *See also* Cracks; Flaws; Subsurface flaws; Surface cracks
electron beam welding **A17:** 585-586
in flat plate, electric current perturbation inspection **A17:** 137-138
forging **A17:** 494
from plasma arc welding **A17:** 587
laser inspection **A17:** 17
magnetic particle inspection **A17:** 499
NDE detection methods **A17:** 50
on weldments **A17:** 593
shallow, in plate, eddy current inspection **A17:** 187
small, electric current perturbation for **A17:** 137
in thread roots **A17:** 138-139
visual inspection **A17:** 3
Surface folds
as casting defects **A11:** 383
Surface force apparatus (SFA) **A18:** 400, 401, 402
Surface forces and adhesion, measurement of **A18:** 399-405
atomic force microscopy **A18:** 402
basic concepts **A18:** 399-400
interfacial energy **A18:** 399-400
surface energy and surface forces **A18:** 399-400
work of adhesion **A18:** 400, 403, 404, 405
measuring adhesion **A18:** 402-404
adhesion between curved surfaces **A18:** 403
fracture experiments **A18:** 403
fundamental adhesion measurements **A18:** 403
history dependence and sample preparation **A18:** 403
modes of separation **A18:** 404
practical adhesion measurements **A18:** 403-404
rate-dependent effects **A18:** 403
measuring surface forces **A18:** 400-402
Derjaguin approximation **A18:** 400, 401
environments **A18:** 402
instrumental requirements **A18:** 400

Surface forces and adhesion, measurement of (continued)
other measurements (adsorption, friction, refractive index, viscosity) **A18:** 402
preparation of surfaces and fluids **A18:** 402
pull-off force **A18:** 401, 403
substrate materials **A18:** 401
techniques **A18:** 400-401
state of the art **A18:** 404-405
Surface free energy change **A6:** 114-115
Surface grinding *See* Grinding
Surface hardening *See also* Hardening **A11:** 333, 395
of ductile iron **A1:** 42 **A15:** 659-660 **M1:** 37-38
and fatigue resistance **A1:** 680-681
for hot-rolled steel bars **A1:** 241
malleable iron **M1:** 73
of maraging steels **A1:** 797-798
selective, pearlitic malleable iron **A15:** 697
steels **M1:** 527-542
carburizing and carbonitriding **M1:** 528, 533-540
characteristics of **M1:** 528
fatigue resistance improvement **M1:** 665, 674, 675
hardened zone, effect of depth on wear **M1:** 607
material requirements **M1:** 529-532, 533, 535-538, 540-541
nitriding **M1:** 540-542
selective **M1:** 527-532
Surface impedance
modulation **A17:** 217
Surface imperfections
in porcelain enameled steel sheet **M1:** 179
Surface insulation resistance
adhesive **EL1:** 674
from flux residues **EL1:** 643
test boards **EL1:** 648
testing, as cleanliness measurement **EL1:** 667-668
Surface integrity **A16:** 19-36
abrasive processes **A16:** 31-33
chip removal operations **A16:** 33
controls **A16:** 30
effects in material removal processes compared **A16:** 28
and electrical, chemical, and thermal material processes **A16:** 30-31
and electrical, chemical, and thermal removal processes **A16:** 33
inspection practices to meet requirements **A16:** 35-36
measuring effects **A16:** 27
microstructural alterations detected **A16:** 28
postprocessing guidelines **A16:** 35
principal causes of surface alterations **A16:** 23
typical problems **A16:** 22
Surface inversion/mobile ions
in accelerated testing **EL1:** 892
Surface layer activation (SLA) **A18:** 319
Surface methods
versus weighting factor **EM4:** 85
Surface modification **A13:** 498-505
defined/advantages and disadvantages **A13:** 498
ion implantation **A13:** 498-500
laser surface processing **A13:** 501-503
Surface Mount Council (SMC) **EL1:** 734
Surface Mount Equipment Manufacturers Association (SMENA) **EL1:** 734
Surface Mount Technology Association (SMTA) **EL1:** 734
Surface mounting
adhesives for **EL1:** 670-674
defined **EL1:** 1158
Surface nitriding
to improve abrasive wear resistance **A18:** 639
Surface oxidation **A7:** 252, 587
Surface oxidation of stored specimens
prevention **A9:** 32
Surface oxide films *See also* Oxide films; Surface oxides
annealing to remove **A7:** 182
effects on green strength **A7:** 302-303
Surface oxide layers
abrasion procedures for **A9:** 46

Surface oxide layers (continued)
cleaning of specimens to be examined for **A9:** 28
polishing procedures for **A9:** 46
Surface oxides
determined **A10:** 177
electron spectroscopy of **A7:** 252
enhancement by bombarding ion **A7:** 259
penetration, tested by magnetic bridge sorting **A7:** 491
and pyrophoricity **A7:** 199
stripped in sintering process **A7:** 340
Surface peak
RBS analysis for **A10:** 633
Surface phase
detection **A10:** 293
qualitative analysis, on silicon **A10:** 341-342
Surface pitting **A11:** 133-134
Surface plasmon effects
cathodoluminescence used to detect **A9:** 91
Surface porosity *See also* Porosity
effect of grinding **A7:** 462
finishing operations affecting **A7:** 451
open or shut **A7:** 486, 487
Surface potentials
effect on voltage contrast **A9:** 94
Surface preparation *See also* Cleaning; Coatings; Film(s); Postcleaning; Precleaning; Solderability; specific processes; Surface(s) **EL1:** 675-680 **EM3:** 28, 34-35, 42, 704
for adhesive bonding **EM1:** 681-682, 687
for automotive painting **A13:** 1015
of billets **A17:** 558
of carbon steels, for coatings **A13:** 522
chemical cleaning **EL1:** 678
chromic-acid-anodization, wet peel testing **EM3:** 668
codeposited organics **EL1:** 678-680
for cold extrusion **A14:** 309
cold extrusion of copper/copper alloy parts **A14:** 310-311
composites, adhesive-bonded repair **EM3:** 840-844
CSA process compared to FPL process (etching) **EM3:** 667
defined **EM1:** 23
definition **M6:** 17
for drawing **A14:** 332
drying laminates that contain absorbed moisture **EM3:** 844
equipment and procedures prior to sealant **EM3:** 642
Forest Products Laboratory (FPL) etching process **EM3:** 259, 260, 261
adsorption of NTMP **EM3:** 250-251
aluminum adherends **EM3:** 242, 263-264, 267
compared to PAA process **EM3:** 625
processing quality control **EM3:** 738
wet peel testing **EM3:** 668
functions **EM3:** 235
fusible finish **EL1:** 679
grit blasting **EM3:** 841, 842, 843
for hot dip galvanizing **A13:** 436-437
hot solder dipping (tinning) **EL1:** 678-679
for liquid penetrant inspection **A17:** 74
of magnesium/magnesium alloys **A13:** 749-750
for magnetic rubber inspection **A17:** 122
materials and solderability **EL1:** 676-677
methods **EL1:** 678-679
for optical holography **A17:** 412
for optical testing **EL1:** 571
organic coatings **EL1:** 680
for organic coatings and linings **A13:** 412-415, 912-913
peel plies **EM3:** 841-843
phosphoric acid nontank anodizing procedure evaluation **EM3:** 803-804, 805
plating for **EL1:** 679
for porcelain enameling **A13:** 447
Surface preparation (continued)
pretreatment variations for wedge testing **EM3:** 668
processing quality control **EM3:** 737-738
pumice/abrasive cleaner scrubbing **EM3:** 841, 843
for replication microscopy **A17:** 643
scuff-sand and solvent **EM3:** 843-844
solderability mechanisms **EL1:** 675-676
solderability testing **EL1:** 677-678
solvent wipe cleaning of surface contamination **EM3:** 843-844
steel slugs **A14:** 304
for surface-mount soldering **EL1:** 697-698
techniques **A13:** 413-415
thermal inspection **A17:** 397
for thermal spray coatings **A13:** 460, 462
thin immersion plates **EL1:** 679
thin-film hybrids **EL1:** 326
to prevent weak bonds **EM3:** 521
for zinc plating **A13:** 767
Surface preparation methods
chemical cleaning **EL1:** 678
hot solder dipping (tinning) **EL1:** 678-679
Surface preparation of composites **EM3:** 281-295
abrasion **EM3:** 292, 293, 294
acid etch **EM3:** 293, 294
adherend pretreatment **EM3:** 293
adhesive bonding of polymeric materials **EM3:** 290-292
argon treatment **EM3:** 291, 294
bonding of composites to composites **EM3:** 292-295
chlorine exposure plus UV irradiation **EM3:** 290-291
chromic acid (dichromate/sulfuric acid) **EM3:** 290-291
corona discharge treatments **EM3:** 290-291, 292, 293, 294
cross linking by activated species of inert gases (CASING) **EM3:** 290-291
electrical discharge treatment **EM3:** 291
ESCA/XPS analysis **EM3:** 291, 292, 293, 294
fiber-matrix adhesion **EM3:** 281-290
flame **EM3:** 290-291, 294
gas plasma treatment **EM3:** 285-286
hot chlorinated solvents **EM3:** 290-291
IR laser treatment (laser ablation) **EM3:** 294
mold release agents and/or release cloth components **EM3:** 294-295
nitrogen treatment **EM3:** 291
plasma treatment **EM3:** 291, 292-293, 294
solvent cleaning **EM3:** 292, 293, 294
surface treatments **EM3:** 294
UV irradiation only **EM3:** 290-291
Surface preparation of metallographic specimens **A9:** 33-47
Surface preparation of metals **EM3:** 259-273
alkaline peroxide (AP solution), titanium adherends **EM3:** 265, 266
alkaline-chlorite etch, copper adherends **EM3:** 269-270
chromic acid anodization (CAA) **EM3:** 259-260, 261, 262, 264
processing steps **EM3:** 260
titanium adherends **EM3:** 264-270
chromic-sulfuric acid etching procedures **EM3:** 259, 260
copper adherends **EM3:** 269-270
Ebonol C etch, copper adherends **EM3:** 269-270
ferric chloride-nitric acid solution, copper adherends **EM3:** 269-270
Forest Products Laboratory (FPL) etching procedure **EM3:** 259, 260, 261
low-carbon steels **EM3:** 271
NaTES: etching procedure, titanium adherends **EM3:** 265, 266, 267, 269
nitric acid anodization, steel adherends **EM3:** 272
Surface preparation of metals (continued)
no surface preparation, steel adherends **EM3:** 273
P2 etching procedure **EM3:** 259
phosphoric acid anodization (PAA) **EM3:** 259, 260, 261
processing steps **EM3:** 260
plasma spraying, titanium adherends **EM3:** 265, 266, 267, 268, 269, 270
Russell's process, steel adherends **EM3:** 271
sodium hydroxide anodization (SHA), titanium adherends **EM3:** 265, 266, 267, 268, 270
steel adherends **EM3:** 270-273
nitric-phosphoric acid etch **EM3:** 271
phosphoric acid-alcohol etch **EM3:** 271, 272
sulfuric acid/dichromate anodization, steel adherends **EM3:** 272
tartaric acid-anodized (TAA) **EM3:** 261
titanium adherends
classification **EM3:** 267
durability **EM3:** 266-269
Surface preparation of plastics **EM3:** 276-280
abrasion **EM3:** 278, 279, 280
cleaning procedures **EM3:** 276-277
chemical treatment **EM3:** 277
intermediate cleaning methods **EM3:** 276-277
solvent cleaning methods **EM3:** 277
evaluating preparation effectiveness
contact-angle test **EM3:** 277
postbonding tests **EM3:** 277
surface exposure time effect **EM3:** 277
water-break test **EM3:** 277
gas plasma treatment **EM3:** 278, 279, 280
oxidizing flame **EM3:** 279, 280
thermoplastic materials preparation **EM3:** 278-280
thermosetting materials **EM3:** 277-278
Surface Preparation Techniques for Adhesive Bonding **EM3:** 70
Surface relief
relationship to martensite **A9:** 668
topographic methods for **A10:** 365, 368
Surface replicas *See also* Replicas
direct, single-stage **A17:** 53
indirect **A17:** 53
plastic **A17:** 53
Surface replication
defined **A17:** 52-53
as nondestructive test technique **EM1:** 775
techniques **A17:** 52-65
Surface resistance
defined **EM2:** 41, 461
Surface resistivity *See also* Resistivity
defined **EL1:** 1158
electrical, defined **EM2:** 41, 461
and electrical testing **EM2:** 585-586
and microwave inspection **A17:** 216
Surface roughening
definition **M6:** 17
electroplating pretreatment **M5:** 601
nickel alloys **M5:** 674
thermal spray coating process **M5:** 367-368
Surface roughness *See also* Roughness; Surface(s) **A18:** 436
and apparent density **A7:** 273
and Auger spectra interpretation **A7:** 255
bearings **A11:** 485
casting **A11:** 384
as casting defect **A15:** 549
correction factors **A11:** 116
determined, by speckle metrology **A17:** 435-436
diffusion bonding **A6:** 158
effect in AES analysis **A10:** 553
effect in x-ray diffraction residual stress techniques **A10:** 387
effect on cracking **A11:** 57
effect on fatigue strength **A11:** 122
and film thickness **A11:** 152

SUBJECTS OF THE INDEXED VOLUMES: ASM Handbook (designated by the letter "A"): **A1:** Properties and Selection: Irons, Steels, and High-Performance Alloys (1990); **A2:** Properties and Selection: Nonferrous Alloys and Special-Purpose Materials (1990); **A3:** Alloy Phase Diagrams (1992); **A4:** Heat Treating (1991); **A6:** Welding, Brazing, and Soldering (1993); **A7:** Powder Metallurgy (1984); **A8:** Mechanical Testing (1985); **A9:** Metallography and Microstructures (1985); **A10:** Materials Characterization (1986); **A11:** Failure Analysis and Prevention (1986); **A12:** Fractography (1987); **A13:** Corrosion (1987); **A14:** Forming and Forging (1988); **A15:** Casting (1988); **A16:** Machining (1989); **A17:** Nondestructive Testing and Quality Control (1989); **A18:** Friction, Lubrication, and Wear Technology (1992). **Metals Handbook, 9th Edition** (designated by the letter "M"): **M1:** Properties and Selection: Irons and Steels (1978); **M2:** Properties and Selection: Nonferrous Alloys and Pure Metals (1979); **M3:** Properties and Selection: Stainless Steels, Tool Materials and Special-Purpose Materials (1980); **M4:** Heat Treating (1981); **M5:** Surface Cleaning, Finishing, and Coating (1982); **M6:** Welding, Brazing, and Soldering (1983). **Engineered Materials Handbook** (designated by the letters "EM"): **EM1:** Composites (1987); **EM2:** Engineering Plastics (1988); **EM3:** Adhesives and Sealants (1990); **EM4:** Ceramics and Glasses (1991); **Electronic Materials Handbook** (designated by the letters "EL"): **EL1:** Packaging (1989).

Surface roughness (continued)
interference microscope measurement **A17:** 17
in moil point **A11:** 573, 578
monitoring, by optical sensors **A17:** 10
monitoring, line projection schematic **A17:** 8
secondary ion mass spectroscopy **A7:** 258
standards, by visual inspection **A17:** 9
and surface electronic structure, SERS and **A10:** 136
Surface roughness parameters **A12:** 212-215
defined **A12:** 212
for fractal analysis **A12:** 212
and profile roughness parameter calculated **A12:** 212
surface roughness **A12:** 200-201
surface volume **A12:** 201
topographic index **A12:** 201
Surface scratches as a result of grinding **A9:** 37-39
Surface segregation
analysis of **A10:** 536, 564-566, 583, 593, 607-608
Surface selection rule **A10:** 119
Surface sensitivity *See also* Sensitivity; Surface; Surface-sensitive analytical techniques
dependence on takeoff angle **A10:** 574
ferromagnetic resonance **A10:** 268
LEISS analysis **A10:** 605
of x-ray photoelectron spectroscopy **A10:** 569-570, 573
Surface structure
analysis of **A10:** 133-134
EXAFS electron detection studies adsorbates **A10:** 418
peak, RBS analysis for **A10:** 633
study by channeling and blocking **A10:** 633
Surface temperature measurement **A18:** 438-444
metallographic techniques **A18:** 438-439, 444
radiation detection techniques **A18:** 441-444
infrared detectors **A18:** 442-443
optical and infrared (IR) photography **A18:** 441-442, 443
photography **A18:** 441
photon collection **A18:** 443
photon detection **A18:** 441
pyrometry **A18:** 441, 442
thermal imaging **A18:** 441
total emissivity of solid surfaces at 25°C, representative values **A18:** 441
thermocouples and thermistors **A18:** 439-441
contact thermocouples **A18:** 439-440
dynamic thermocouples **A18:** 440
embedded subsurface thermocouples **A18:** 440
thin-film temperature sensors **A18:** 440-441
Surface tension *See also* Surfactant **EM3:** 28, 181
balance, eutectic growth **A15:** 121
defined **EM2:** 41
defined, for component removal **EL1:** 715
and fluidity **A15:** 766
gas porosity by **A15:** 82
liquid **A18:** 400
in reflow **EL1:** 733
SI derived unit and symbol for **A10:** 685
SI unit/symbol for **A8:** 721
Surface tension, effect
liquid penetrant inspection **A17:** 71
Surface tension temperature coefficient **A6:** 20, 21, 22
Surface texture **A18:** 334-344
applications **A18:** 341-344
automotive engine components surface roughness specifications **A18:** 342
lip seals **A18:** 344
magnetic storage **A18:** 343
metalworking **A18:** 341-343
surface parameter effect on component performance **A18:** 342
surface property effect on component failure causes **A18:** 342
tribology and wear **A18:** 342-343
definition **A18:** 334
design **A18:** 336, 344
experimental issues for stylus instruments **A18:** 336-341
bandwidth limits for surface metrology **A18:** 338-339
examples of roughness measurement results **A18:** 340-341
Surface texture (continued)
height resolution and range **A18:** 336-338
lateral resolution and range **A18:** 338
other distortions **A18:** 339-340
stylus load and surface deformation **A18:** 339
instrument calibration **A18:** 341
comparison of roughness parameters **A18:** 341
comparison of surface profiles **A18:** 341
general calibration issues **A18:** 341
lubrication role **A18:** 336, 342-343
profiling techniques **A18:** 334
profile methods **A18:** 334
raster area methods **A18:** 334
statistical functions **A18:** 335-336
amplitude density function **A18:** 335
autocorrelation function (ACF) **A18:** 335-336, 337
autocovariance function **A18:** 335-336
bearing area curve **A18:** 335
height distribution **A18:** 335
power spectral density (PSD) **A18:** 335, 336
surface parameters **A18:** 334-335
height parameters **A18:** 334
hybrid parameters **A18:** 334, 335
shape parameters **A18:** 334-335
wavelength parameters **A18:** 334
surface statistics, introduction **A18:** 334
versus roughness **A18:** 334
Surface tilting
relationship to martensitic structure **A9:** 668-669
Surface topography **A16:** 21
and color metallography **A9:** 138
Surface topography and image analysis (area) **A18:** 346-354
computing differences between two traces or surfaces **A18:** 350-352
conventions **A18:** 346-347
curvature **A18:** 352
Hough transforms **A18:** 352
pattern matching **A18:** 352
definitions **A18:** 346-347
estimation and combination of intensity and topographic images **A18:** 347-348
rendering and combining images **A18:** 348
transforming an intensity image **A18:** 347-348
fractals, trees, and future investigations **A18:** 352-354
implementation on personal computers and data bases **A18:** 347
lessons from two-dimensional analysis **A18:** 349
point spacing and image compression **A18:** 347
potential pitfalls **A18:** 347
relating two- and three-dimensional parameters **A18:** 348-349
selecting an appropriate coordinate system **A18:** 349-350
Surface treatment *See also* Finishes; Sizing
of carbon fiber **EM1:** 112-113
defined **EM1:** 23, 122
hot rolled bars **M1:** 200
maraging steels **M1:** 448-451
and sizing, compared **EM1:** 122
Surface treatments
aluminum alloys **A13:** 588
for beryllium **A13:** 811
cemented carbides **A13:** 857
and coatings in copier powders **A7:** 585
identifying compositional changes from **A7:** 260
protective, for powder stability **A7:** 182
standardized **A13:** 194
tool steels **M3:** 446
Surface treatments for wrought tool steels **A1:** 779
carburizing **A1:** 779
nitriding **A1:** 779
oxide coatings **A1:** 779
plating **A1:** 779
sulfide treatment **A1:** 779
titanium nitride **A1:** 779
Surface veils *See also* Veil
for fiberglass mats **EM1:** 109
for resin transfer molding **EM1:** 169
Surface void
definition **EM4:** 633
Surface wave angle-beam technique
ultrasonic inspection **A17:** 247
Surface wave velocity (v_R) **A18:** 408
Surface waves *See also* Rayleigh waves
for optical holographic interferometry **A17:** 409
ultrasonic **A17:** 233-234
Surface wetting *See* Wetting
Surface wiring
materials and processes selection **EL1:** 113-115
methods, compared **EL1:** 115
wiring sources **EL1:** 115
Surface(s) *See also* Cleaning; Coatings; Cracks; Defects; Discontinuities; Film; Film thickness; Films; Finish; Flaws; Fracture surface(s); Oxides; Specimen(s); Subsurface cracking; Subsurface flaws Surface cracks; Subsurface(s); Sulfide modification; Surface analysis; Surface coatings; Surface defects; Surface finish; Surface flaws; Surface modification; Surface preparation; Surface preparation Surface replicas; Surface replication Surface roughness; Surface resistance; Surface resistivity; Surface roughness parameters; Surface tension; Surface treatment; Surface waves; Surface wiring; Surface-mount; Thick films; Thin films
abrasion, from mechanical damage **A11:** 342
adhesive bonding preparation **EM1:** 681-682, 687
alloy **A13:** 46
aluminum **A2:** 3
analysis **A13:** 1117-1118
of beryllium **A2:** 683
blowholes, as casting defects **A11:** 382
carbon fiber **EM1:** 868
carburization **A11:** 406
of castings, inspection methods **A17:** 512
-charge accumulation, in silicon oxide interface failures **A11:** 781
chemical analyses **A11:** 30, 160
chemical tests for **EM1:** 285
chemistry **EL1:** 104
cleaning, before coating **A15:** 561
cleaning, for conversion coatings **A13:** 380-382
cleanliness **A13:** 382, 417
coated, magnetic rubber inspection of **A17:** 123-124
compression, effect on fatigue strength **A11:** 125-126
condition, and atmospheric corrosion **A13:** 81-82
condition, effects in stainless steels **A13:** 552
condition, for brazing **A11:** 450-451
condition, for optical holographic interferometry **A17:** 412
configuration **A11:** 159-160
-contact fatigue, in gears **A11:** 592-593
contamination **A14:** 139, 785
contamination, and conformal coatings **EL1:** 763
contamination of **EM2:** 597-598
contouring, by acoustical holography **A17:** 447
conversion, cleaning for **A13:** 380-382
corroded, microscopic examination **A11:** 173
cracks, flaw shape parameter curves for **A11:** 108
crystal effects **EL1:** 104
damage **A11:** 75, 156, 684, 688, 689
debris, cleaning techniques for **A12:** 73-76
decarburization **A14:** 204
defective, castings **A17:** 515-517
defects **A11:** 127, 307, 353-354, 358, 554-555 **A13:** 150
defects, hot isostatic pressing effects **A15:** 542
defects, on inclusions **A12:** 220
defects, types **A15:** 548-550
delamination, by cutting **EM1:** 667
diffusion from **A15:** 82
discontinuities in **A11:** 119
discontinuity, crack initiation at **A13:** 148-149
dislocation at **A13:** 47
drilled hole, fatigue fracture **A11:** 21
effect of lubrication on **A11:** 154
effects, and fatigue **A13:** 294
effects, surgical implants **A13:** 1329
electrical leakage **EL1:** 660
embrittlement, in polymers **A11:** 761
energy, effects **EM2:** 773
evaluation, by magnetic rubber inspection **A17:** 125
of failed bearing, cap **A11:** 349
faying, defined *See* Faying surface

Surface(s) (continued)
fibrillations, liquid crystal polymers (LCP) **EM2:** 181
film, stability, gaseous corrosion **A13:** 17
films, passive, as SCC mechanism **A12:** 25
finger oxides **A14:** 204
finish **A13:** 303, 344
finish, extrusions, aluminum and aluminum alloys **A2:** 5-6
finishing, as stress source **A11:** 122, 205, 358-359
of flash welds **A11:** 442
flaws, ceramics **A11:** 750-751
fractal analysis of **A12:** 211-215
fracture, features of **EM2:** 807-810
fractured, photography of **A12:** 78-90
hardening **A11:** 333, 395
hardness, of bearing steels **A11:** 490
-initiated fatigue, of bearing components **A11:** 501-502
inside mold line (IML) **EM1:** 168
internal, blast cleaning of **A15:** 520
interpretation by machine vision **A17:** 42
inversion in metal-oxide semiconductor (MOS) devices **A11:** 766
irregular, fractal plots **A12:** 211-215
irregular, RSC profiles **A12:** 213
irregularities, radiographic appearance **A17:** 349
irregularity, optical effects **EM2:** 597-598
joint, heat exchangers **A11:** 639
laser inspection **A17:** 17
layers, as source, acoustic emissions **A17:** 287
liquid penetrant inspection **A17:** 71-88
measurement, specialized, by coordinate measuring machines **A17:** 19
metal **A13:** 45-46
microcracking **A11:** 760, 761, 764
modification **A13:** 498-505
moisture ,header assembly **EL1:** 958
mold **EM2:** 614
mold, defined *See* Mold surface
nondestructive inspection effects **A12:** 77
outside mold line (OML) **EM1:** 168
oxide growth **A13:** 71
partially oriented, parametric relationships for **A12:** 201
passivation, of rare earth metals **A2:** 725
perfectly oriented, true area and length **A12:** 204
phenomena **EL1:** 103-104
pickling, titanium/titanium alloys **A13:** 696
pinholes, as surface defects **A11:** 382
pitting fatigue, in carburized steel **A11:** 133-134
of preforms **A14:** 160-161
preparation *See* Surface preparation
preparation, to prevent hydrogen damage **A11:** 251
properties, change **A13:** 294-295
prototype faceted, true profile length values **A12:** 200
quality, and gating, die casting **A15:** 289
quality, determining **A15:** 544
quality of **A14:** 882
randomly oriented, true area and length **A12:** 204
reactions, as SCC parameter **A13:** 147
residual stresses, in shafts **A11:** 473
residual stresses, x-ray diffraction for **A17:** 51
resistivity (electrical), defined **EM1:** 23
rolling, compressive residual stress by **A11:** 133-134
rough, and swaging **A14:** 143
roughness *See* Surface-roughness
roughness, and fatigue **A13:** 294
ruled, true area and length **A12:** 204
sensitivity, of- AES **A11:** 33
shape, and atmospheric corrosion **A13:** 81-82
smooth **EM2:** 303-304
smoothness, of ceramics **EL1:** 336
-structure ,gradients, in iron castings **A11:** 359

Surface(s) (continued)
substrate **EL1:** 104
temperature, thermal inspection **A17:** 396
tension, aluminum casting alloys **A2:** 145-146
tension, effect, liquid penetrant inspection **A17:** 71
tension, effect on bearing lubrication **A11:** 484
textured **EM2:** 304-305
three-dimensional laser gaging **A17:** 12, 16
topography, and microstructure **A12:** 393
treatment, as multilayer inner layer process **EL1:** 543
treatment, defined **EM2:** 41
treatments, electrolytic and chemical **A11:** 195
volume, as surface roughness parameter **A12:** 201
wetting on **EL1:** 1031-1035
Surface-activated acrylics
performance of **EM3:** 124
Surface-active agents
in milling environment **A7:** 63
Surface-area-center test *See also* Rockwell-inch test **M1:** 457
Surface-contact fatigue
in gears **A11:** 592-593
Surface-enhanced Raman scattering **A10:** 135-136
Surface-hardened steels and irons
CBN for machining **A16:** 105
Surface-mount *See* Surface-mount assemblies; Surface-mount components (SMC); Surface-mount devices (SMD); Surface-mount joints; Surface-mount packages; Surface-mount soldering; Surface-mount technology (SMT)
Surface-mount assemblies
cleaning of **EL1:** 666-667
integrated semiconductor packages **EL1:** 437-438
Surface-mount components (SMC)
construction **EL1:** 178
removal **EL1:** 722-724
soldering methods **EL1:** 180
Surface-mount devices (SMD)
assembly **EL1:** 437-438
construction **EL1:** 178
failure mechanisms **EL1:** 1046-1047
schematic **EL1:** 431
small-signal plastic-bodied products as **EL1:** 423-424
solder masking **EL1:** 558
thermal expansion mismatch **EL1:** 611
two-terminal **EL1:** 430-432
vs plated-through hole designs **EL1:** 559
Surface-mount devices (SMDs) **EM3:** 569-571
Surface-mount joints
leaded and leadless **EL1:** 730-734
properties **EL1:** 641-642
Surface-mount soldering *See also* Joints; Solder joints; Soldering; Surface-mount technology (SMT)
adhesives **EL1:** 700
component placement **EL1:** 700-701
condensation (vapor phase) soldering **EL1:** 702-704
conductive (hot bar) soldering **EL1:** 705-706
conductive belt and hot platen soldering **EL1:** 706
defluxing **EL1:** 707-708
development **EL1:** 697
fluxes for **EL1:** 647
hand-held iron/repair station soldering **EL1:** 707
infrared (IR) soldering **EL1:** 704-705
laser soldering **EL1:** 706-707
process yields **EL1:** 708
solder paste application **EL1:** 698-700
solder processes **EL1:** 701-707
surface preparation **EL1:** 697-698
wave soldering **EL1:** 701-702

Surface-mount technology (SMT) *See also* Mixed technology (MT); Plated-through hole (PTH); Through-hole soldering **A6:** 133
adhesives **EL1:** 671
advantages **EL1:** 730
chip carriers with **EL1:** 76
design guidelines, leaded and leadless surface-mount joints **EL1:** 733-734
discrete semiconductor packages **EL1:** 434-435
effect, passive component fabrication **EL1:** 178
effect, soldering/mounting **EL1:** 631-632
as emerging package option **EL1:** 76-77
future trends **EL1:** 391-392
hybrid vs polymeric **EL1:** 252
leaded and leadless surface-mount joints **EL1:** 730-734
nonwetting, causes **EL1:** 1033
packaging requirements **EL1:** 15-16
removal techniques **EL1:** 722-724
solder joint **EL1:** 117
standards **EL1:** 734
tape automated bonding (TAB) with **EL1:** 274
thermal-mechanical effects **EL1:** 740-753
with VHSIC, modeling/simulation requirements **EL1:** 77-81
Surface-mount technology (SMT) adhesives **EM3:** 569-571
application methods **EM3:** 570-571
cure method **EM3:** 570-571
pot life **EM3:** 570
Surface-mounted packages
bottom-brazed flatpack solder joint failures **EL1:** 992-993
failure mechanisms in **EL1:** 892-993
first-level, failure mechanisms **EL1:** 989-992
leaded package technology **EL1:** 982-983
leadless packaging technology **EL1:** 983-989
plastic, fabrication **EL1:** 470-482
Surface-sensitive analytical techniques *See also* Surface; Surface sensitivity
Auger electron spectroscopy **A10:** 549-567
secondary ion mass spectroscopy **A10:** 610-627
x-ray photoelectron spectroscopy **A10:** 568-580
Surface-to-volume ratio of discrete particles **A9:** 124-125
Chalkley method **A9:** 125
Saltykov method **A9:** 125
Surfacing
cast irons **A6:** 720-721
definition **A6:** 1214 **M6:** 17
Surfacing mat *See also* Overlay sheet
defined **EM1:** 23 **EM2:** 41
for resin transfer molding **EM1:** 169
Surfacing material
definition **A6:** 1214
Surfacing materials
cast irons **A6:** 720-721
cast iron alloys **A6:** 721
ceramic materials **A6:** 721
copper alloys **A6:** 720-721
hardfacing alloys **A6:** 721
high-nickel alloys **A6:** 721
stainless steels **A6:** 721
Surfacing metal
definition **A6:** 1214
Surfacing weld
definition **A6:** 1214 **M6:** 17
Surfacing welding electrodes **A6:** 176
Surfactant *See also* Rehbinder effect; Surface tension **EM3:** 28, 674
defined **A18:** 18 **EM2:** 41
for metalworking lubricants **A18:** 141, 142, 143, 144
Surfactant molecules, in water
structural changes in **A10:** 118
Surfactants **A13:** 12, 380, 1140
acid cleaning process **M5:** 59-60, 64
alkaline cleaners **M5:** 35
as aqueous cleaners **EL1:** 665-666

Surfactants (continued)
chemical brightening process **M5:** 579-580
effect in blending and premixing **A7:** 188
effect on material flow **A7:** 188
nickel plating process **M5:** 200, 206, 209
pickling process **M5:** 69, 71
Surge
current, failures **EL1:** 997
suppressors, failure mechanisms **EL1:** 974
Surgical gauzes and pins
powders used **A7:** 573
Surgical implants
metals/alloys used **A13:** 1325
powders used **A7:** 573
titanium and titanium alloy **A2:** 589
Surveillance applications
of beryllium **A2:** 684
of germanium **A2:** 743
Suspended solids
for metalworking lubricants **A18:** 141
in water **A13:** 489
Suspendible developer baths
liquid penetrant inspection **A17:** 79
Suspending agents
for slurry in spray drying **A7:** 75
Suspending liquids
for magnetic particles **A17:** 100-104
Suspension bridge piers
compression testing **A8:** 55
Suspension system
in core coatings **A15:** 240
Suspensions
as lubricant form **A14:** 514
Sustained fluid flow
corrosivity **A13:** 339-340
Sustained load
cracking, abbreviated **A8:** 726
microscopic fracture mode **A8:** 477
Sustained-load failure **A8:** 476, 486-488
Suutala diagram **A6:** 463
Swabbing
defined **A9:** 18
Swage
defined **A14:** 13
Swage dies **A14:** 43, 61, 132-133
Swageability
of tapers **A14:** 137
Swagers
alternate-blow **A14:** 131
capacity **A14:** 131-132
creeping-spindle swaging **A14:** 131
die-closing **A14:** 131
standard rotary **A14:** 130
stationary-spindle **A14:** 130-131
Swaging *See also* Rotary swaging **A7:** 12 **M1:** 316
alternate blow **A14:** 131
by squeeze action **A14:** 131
cold work, in ordnance applications **A7:** 691
of copper and copper alloys **A14:** 819
creeping-spindle **A14:** 131
dies **A14:** 43, 61, 132-133
and drilling, combined **A14:** 141
ferule from tube stock **A14:** 140
hot **A14:** 142-143
in hot upset forging **A14:** 83
in irons **A12:** 219
metal flow during **A14:** 128-129
problems and solutions **A14:** 143-144
and shrinkage in sintering **A7:** 310
tube types for **A14:** 135
and turning, combined **A14:** 141
of unalloyed uranium **A2:** 671
vs press forming **A14:** 140
vs spinning **A14:** 140-141
vs turning **A14:** 141
of wire **A14:** 694
wire fabrication **M1:** 593
Swaging assembly
furnace brazing of steels **M6:** 940
Swaging of rods and wires
preferred orientation during **A10:** 359
Swaging, rotary *See* Rotary swaging
Swaging transfer machine
multistation automatic **A14:** 136
Swarf **A7:** 12
defined **A9:** 26
Sweat soldering
definition **A6:** 1214 **M6:** 17
Sweat, tin *See* Tin sweat
Sweating **A7:** 12
defined **A18:** 18
Sweco vibratory grinding mills **A7:** 68
Swedish sponge iron process **A7:** 79-82
Swedish standards for steels **A1:** 159
compositions of **A1:** 193, 194
cross-referenced to SAE-AISI steels **A1:** 166-174
Sweep
defined **A15:** 11
Sweep blasting
before painting **M5:** 332
Sweep velocity
defined **A18:** 18
Sweeping
of molds **A15:** 191
Sweet gas **A13:** 1256-1257
Swelling
coefficients of **EM1:** 188
and dissolution **EM2:** 771-773
and kinetics **EM2:** 771
laminate **EM1:** 226-227
in liquid phase sintering **A7:** 320
moisture **EM1:** 188-190
Swells
as casting defect **A15:** 546
as casting defects **A11:** 381
Swept-frequency continuous-wave reflection
microwave inspection **A17:** 206
Swept-frequency continuous-wave transmission
microwave inspection **A17:** 205-206
SWIFT casting design software **A15:** 611
Swift cup test **A1:** 576 **A8:** 14, 565
defined **A14:** 13
Swift flat-bottomed cup test
as drawing test **A8:** 563
for lubricant evaluation **A8:** 568
Swift round-bottomed cup test
correlation with material properties **A8:** 565
for lubricant evaluation **A8:** 568
for sheet metals **A8:** 563
Swift-moving water
corrosion in **A11:** 188
Swimming
defined **EL1:** 1158
Swirling tank reactor
for degassing **A15:** 461, 463
Switches, electrical snap-action
contact materials **A2:** 864
Switching
energy, defined **EL1:** 2
failure mechanisms **EL1:** 980
false, avoidance of **EL1:** 34
intrinsic, silicon metal-oxide semiconductor effect **EL1:** 2
matrix, for electrical testing **EL1:** 567
speeds **EL1:** 76, 128
systems, as hybrid telecommunications application **EL1:** 382-383
time, performance range classifications **EL1:** 25
Syenite
purpose for use in glass manufacture **EM4:** 381
Symbols **A4:** 968-970
and abbreviations **A8:** 724-726 **A11:** 796-798 **A12:** 492-494 **A13:** 1375-1377 **A14:** 944-945 **A15:** 896-897 **A17:** 758-760 **EL1:** 1166-1168 **EM1:** 948-950 **EM2:** 850-852 **EM3:** 852-853
in geometric dimensioning and tolerancing **A15:** 623
Greek, for liquid metal processing terms **A15:** 49
SI prefixes **A11:** 795
Symbols, abbreviations, and tradenames **A1:** 1038-1041 **A2:** 1273-1277
Symbols and abbreviations **A10:** 689-692
Symbols for quantitative metallography **A9:** 123
Symmetric rod impact test **A8:** 203-206
Symmetrical laminate *See also* Balanced laminate; Laminate(s) **EM3:** 28
defined **EM1:** 23 **EM2:** 41
properties of **EM1:** 222-224
Symmetrical-hold-only test
and compression-hold-only test **A8:** 347
saturation effect **A8:** 348-349
Symmetry
center of **A10:** 346
x-ray diffraction analysis and **A10:** 325
Synchronous excitation spectroscopy **A10:** 78
Synchronous initiation
definition **M6:** 17
Synchronous loading device
SCC testing **A8:** 507
Synchronous system modules
data transfer between **EL1:** 7
Synchronous timing (resistance welding)
definition **A6:** 1214
Synchrotron
defined **A10:** 683
Synchrotron radiation *See also* Radiation
and bremsstrahlung output, compared **A10:** 411
defined **A10:** 683
dynamic processes monitored **A10:** 375
effect on EXAFS as atomic probe **A10:** 408
EXAFS scan of nickel using **A10:** 408
from SPEAR, spectral distribution of **A10:** 411
radiation sources **A10:** 412
as x-ray source for EXAFS **A10:** 407, 411-412
and x-ray topography **A10:** 365, 374-376
Syndiotactic stereoisomerism **EM3:** 28
defined **EM2:** 41
Syneresis **EM3:** 28
defined **EM2:** 41
Syneresis (of a grease) *See* Bleeding
Synergistic component of damage from wear **A18:** 274
Synergy
in SCC crack propagation **A13:** 145
Synovial joints **A18:** 656, 662
boundary lubrication **A18:** 656, 662
friction coefficients **A18:** 656, 662
schematic **A18:** 656
wear factors **A18:** 656
Syntactic cellular plastics **EM3:** 28
defined **EM2:** 41
Syntactic foams
defined **EM1:** 23
Syntactics **EM3:** 176, 561, 566
for aerospace industry **EM3:** 176
physical properties **EM3:** 562
Syntectic reaction .... **A3:** 1 • 5
Synthesis
of cubic boron nitride (CBN) and diamond **A2:** 1008-1009
Synthesis of caffeine
powder used **A7:** 574
Synthesis of hydrocarbons
powder used **A7:** 574
Synthesized stoneware **EM4:** 3
Synthetic activated fluxes **EL1:** 647
Synthetic adhesives **EM3:** 86, 89-90
Synthetic diamond *See also* Diamond; Polycrystalline diamond; Superabrasives; Ultrahard tool materials
as an abrasive for wire sawing **A9:** 26
cube-, cubooctahedron-, octahedron-shaped **A2:** 1010-1011
metal impurities in **A10:** 417
synthesis of diamond grit **A2:** 1008-1009
synthesis of polycrystalline diamond **A2:** 1009
as ultrahard material **A2:** 1008
Synthetic equilibrium
defined **A9:** 18
Synthetic fiber(s)
inorganic, forms **EM1:** 175
and natural fibers, properties compared **EM1:** 117
organic, forms **EM1:** 175
reinforcement **EM1:** 117-118
Synthetic fluids
in lubricants **A14:** 514
Synthetic frits
functions in FCAW electrodes **A6:** 188
Synthetic hydrocarbon fluids (SHC) **A18:** 132, 134
Synthetic hydroxylapatites
biomedical applications **EM4:** 19
Synthetic marble
PCD tooling **A16:** 110
Synthetic materials
analysis of new **A10:** 353-354
five-component, MFS analysis **A10:** 78

Synthetic materials (continued)
high-temperature combustion analysis of ... **A10:** 224
Synthetic methanol
grades ... **A9:** 67
Synthetic nitrogen-based atmospheres ... **A7:** 345-346
Synthetic oil
defined ... **A18:** 18
as metalworking lubricants ... **A18:** 142, 144, 146
Synthetic organic waxes ... **A7:** 192
Synthetic Polaroid filters ... **A9:** 76
Synthetic polyimide fiber
as thermocouple wire insulation ... **A2:** 882
Synthetic polymers
electrically conductive ... **A7:** 607
Synthetic random mass finishing media ... **M5:** 135
Synthetic resins
compatibility with steels ... **A18:** 743
Synthetic rubbers ... **EM3:** 75, 86
Synthetic sands *See also* Sands; Silica sands
defined ... **A15:** 29
development of ... **A15:** 29
vs natural sands ... **A15:** 209
Synthetic stone
PCD tooling ... **A16:** 110
Synthetic waxes ... **A7:** 190-193
Synthetics ... **EM3:** 86-87
Syphilis
bismuth as treatment for ... **A2:** 1256
Syringe dispensing
as adhesive application ... **EL1:** 671-672
of flexible epoxy systems ... **EL1:** 817
Syrup, 25% polymer in methyl methacrylate (carboxylated butadiene acrylonitrile)
formulation ... **EM3:** 122
Syrup, baumé
for rammed graphite molds ... **A15:** 273-274
System cabling
as level 4 components ... **EL1:** 76
System clocks
and mounting technology ... **EL1:** 143
System noise
experimental study ... **A17:** 742
System(s)
advanced, thermal issues ... **EL1:** 80
environmental stress screening ... **EL1:** 878
-level products, software impacts ... **EL1:** 12
macrofabrication, and device scaling ... **EL1:** 10
requirements, interconnection ... **EL1:** 12-17
test, wafer-scale integration ... **EL1:** 374
System-generated electromagnetic pulse ... **A13:** 1109
System-level test issues ... **EL1:** 373-377
Systematic error
in corrosion testing ... **A13:** 195-196
Systematic names
of polymers ... **EM2:** 53
Systematic samples
defined ... **A10:** 12-13
Système International d'Unités
abbreviation ... **A8:** 726
abbreviation for ... **A11:** 797
guide for ... **A12:** 489-491
metric and conversion data ... **A8:** 721-723
prefixes, names and symbols ... **A11:** 795
Systems ... **A3:** 1 • 1-2
Systems design stage
quality design ... **A17:** 722
Systems, disordered and ordered
EXAFS analysis of ... **A10:** 407
Systolic arrays ... **EL1:** 8, 9
The Science of Adhesive Joints ... **EM3:** 71

T

1,1,1-Trichloroethane solvent cleaners ... **M5:** 40-41, 44-48, 57
cold solvent cleaning, use in ... **M5:** 40-41
flash point ... **M5:** 40
vapor degreasing, use in ... **M5:** 44-48, 57
1,1,1-Trichloroethane, surface tension ... **EM3:** 181
300M steel ... **A1:** 435-436
heat treatments for ... **A1:** 435
properties of ... **A1:** 435-436
μ-μ diffractometers ... **A10:** 337
T *See* Temperature; Thickness; Time
***t* distribution** ... **A8:** 655
T grades, pressure tubes
composition ... **M1:** 324
tensile properties ... **M1:** 325
T prime phase in zinc-copper alloys ... **A9:** 489
t statistic
values ... **A8:** 709, 711
T temper
solution heat-treated temper, defined ... **A2:** 21
T1 through T10 temper, defined ... **A2:** 26-27
variations, additional, defined ... **A2:** 27
T-1 structural steel
composition and mechanical properties ... **M1:** 621
T-3 lamps
for wave soldering preheating ... **EL1:** 684-685, 687
T-50
brazing, composition ... **A6:** 117
wettability indices on stainless steel base metals ... **A6:** 118
T-51
brazing, composition ... **A6:** 117
wettability indices on stainless steel base metals ... **A6:** 118
T-52
brazing, composition ... **A6:** 117
wettability indices on stainless steel base metals ... **A6:** 118
T-bolts
SCC of ... **A11:** 538
T-fittings
pipe welding ... **M6:** 591
T-joint weld
magnetic particle inspection ... **A17:** 109
nonrelevant indications in ... **A17:** 106-108
T-joint(s)
definition ... **A6:** 1214
electron-beam welding ... **A6:** 260
hydrogen-induced cracking ... **A6:** 436
lamellar tearing ... **A6:** 95
laser-beam welding ... **A6:** 264, 879, 880
oxyfuel gas welding ... **A6:** 286, 287
T-joints
arc welding of
cobalt-based alloys ... **M6:** 368
heat-resistant alloys ... **M6:** 356-357, 360, 363, 368
nickel alloys ... **M6:** 437
nickel-based alloys ... **M6:** 360, 363
stainless steels, austenitic ... **M6:** 334
definition, illustration ... **M6:** 18, 60, 61
electron beam welds ... **M6:** 616-618
electroslag welding ... **M6:** 225
flux cored arc welding ... **M6:** 112-113
gas metal arc welding of aluminum alloys ... **M6:** 384
gas tungsten arc welding ... **M6:** 202
of aluminum alloys ... **M6:** 396
of commercial coppers ... **M6:** 403
of heat-resistant alloys ... **M6:** 356-357, 360
of magnesium alloys ... **M6:** 431-432
laser beam welding ... **M6:** 664
laser brazing ... **M6:** 1065
magnetic particle inspection ... **A17:** 114
oxyfuel gas welding ... **M6:** 589-590
T-joints (continued)
radiographic inspection ... **A17:** 334
shielded metal arc welding of nickel-based heat-resistant alloys ... **M6:** 356, 363
submerged arc welds ... **M6:** 127
***t-n* diagram**
for creep-fatigue interaction analysis ... **A8:** 358
T-peel strength ... **EM3:** 29
T-peel testing
polybenzimidazoles ... **EM3:** 170
T-plate for the humeral and tibial head
as internal fixation device ... **A11:** 671
T-rail ... **A6:** 397
T-sections
as basic casting shape ... **A15:** 599-604, 607, 609
T-wave ... **A6:** 367
T-X diagram
defined ... **A9:** 19
T1, T2, T3, etc *See* Tool steels, specific types
T1 tool steel springs *See* High-speed tool steel springs
Ta-Th (Phase Diagram) ... **A3:** 2 • 373
Ta-Ti (Phase Diagram) ... **A3:** 2 • 374
Ta-U (Phase Diagram) ... **A3:** 2 • 374
Ta-V (Phase Diagram) ... **A3:** 2 • 374
Ta-W (Phase Diagram) ... **A3:** 2 • 375
Ta-Zr (Phase Diagram) ... **A3:** 2 • 375
Tab bonding
as specimen exposure ... **EM1:** 295
Taber abrader ... **A18:** 670, 673
used for abrasion tests of dental amalgam ... **A18:** 669
weight loss resulting from abrasive wear on electroplated coatings ... **A18:** 835
Taber abraser test ... **EM2:** 642
Taber abrasion test
anodic coating resistance ... **M5:** 596
paint abrasion resistance ... **M5:** 490-491
Table-type blast cleaning machines ... **A15:** 508
Table-type dry blasting machine ... **M5:** 88
Tables
hardness conversion ... **A8:** 109-113
Tableware
coining practice ... **A14:** 183-189
glazes ... **EM4:** 1061, 1063
Tabor criterion ... **A18:** 195
Tabs *See also* Doubler
defined ... **EM1:** 23 **EM2:** 41
submerged arc welding ... **M6:** 135
Tabular punch
effect of axial compressive force ... **A7:** 335, 336
tensile stresses ... **A7:** 336
TAC process
for alkali metal removal ... **A15:** 470
Tachometer
for fatigue testing machines ... **A8:** 368
Tack ... **EM3:** 28
defined ... **EM1:** 23, 139 **EM2:** 41
life, defined ... **EM1:** 139
of prepregs ... **EM1:** 33, 58
range, defined ... **EM1:** 23
as selection criterion ... **EM1:** 77
testing ... **EM1:** 737
of unidirectional tape prepreg ... **EM1:** 143-144
Tack range ... **EM3:** 28
defined ... **EM2:** 42
Tack weld
definition ... **A6:** 1214
Tack welds
arc welding of heat-resistant alloys ... **M6:** 366
definition ... **M6:** 17
electron beam welds ... **M6:** 630
furnace brazing of steels ... **M6:** 940
gas tungsten arc welding of titanium ... **M6:** 454-455
shielded metal arc welding ... **M6:** 90
submerged arc welding ... **M6:** 135
use in welding of pipe ... **M6:** 591

Tacker
definition **A6:** 1214
Tackifiers **EM3:** 85, 89-90
added to natural rubber latex **EM3:** 88
Tacky-dry **EM3:** 28
Taconite rock
severe abrasion by **A11:** 375
Tactical missile system
composite components **EM1:** 816
Tacticity
cis/trans isomers and **EM2:** 64
Tafel
diagram, defined **A13:** 12-13
extrapolation, electrochemical corrosion testing **A13:** 213-214
line, defined **A13:** 12-13
method, corrosion rate measurement **A13:** 32-33
polarization plot **A13:** 215
slope, defined **A13:** 12-13
Taft cokeless cupola **A15:** 392
Taguchi (experimental) designs **EM2:** 602
Taguchi methods *See also* Quality control
of orthogonal arrays **A17:** 748-750
of parameter design **A17:** 750-751
Taguchi techniques
of experimental design **A14:** 938-939
Taguchi's method **EM4:** 87
Tail stock
in torsional impact machine **A8:** 217
Tail wire
aerial telephone cable clamp **A13:** 1132
Tailings (all types)
Miller numbers **A18:** 235
Tailoring
coefficient of thermal expansion (CTE) **EL1:** 611, 614-628, 684-685, 687
Take-up device **EM1:** 108, 160
Talc **A6:** 60 **EM4:** 6, 32
in ceramic tiles **EM4:** 926
chemical composition **A6:** 60
composition **EM4:** 932
as filler/extender material **EM2:** 192, 500
hardness **A18:** 433
on Mohs scale **A8:** 108
properties **A18:** 801
in typical ceramic body compositions **EM4:** 5
typical oxide compositions of raw materials **EM4:** 550
Talc (hydrated magnesium silicate) **EM3:** 175
as filler **EM3:** 177, 178
as filler for polysulfides **EM3:** 139
Talc, as refractory
core coatings **A15:** 240
Talc earthenware **EM4:** 3, 4
Tampico brushing
stainless steel **M5:** 559
Tampico-filled brushes **M5:** 155-156
Tampon-type probe for local electropolishing **A9:** 55
Tan delta *See also* Quality factor **EM3:** 28
for closed-loop cure **EM1:** 761
defined **EM2:** 42
Tandem arc equipment
for heavy structural steels **A7:** 821
Tandem die *See* Follow die
Tandem mill
defined **A14:** 13
Tandem scanning microscope (TSM) **A18:** 357
hardware configurations **A18:** 358
Tandem scanning reflected-light microscope (TSRLM)
hardware configurations **A18:** 358
image acquisition **A18:** 359
simplified optical diagram **A18:** 358
Tandem seal
defined **A18:** 19
Tangent bending *See also* Wiper forming
defined **A14:** 13
Tangent diameter
as basic figure quantity **A12:** 194
Tangent modulus *See also* Modulus of elasticity; Secant modulus **EM3:** 28
defined **A8:** 14 **A11:** 10 **A14:** 13 **EM1:** 23 **EM2:** 42
Tangential friction force **A18:** 432
Tangential stress *See also* Shear stress
and radial pure plastic bending **A8:** 124
Tangential stress (continued)
and radial stress ratio during bending **A8:** 121-122
Tank cars
magnetic paint inspection **A17:** 128
Tanks *See also* Storage tanks
corrosion of **A13:** 1296, 1314-1315
electrogas welding **M6:** 243
failure, acoustic emission inspection **A17:** 291
for quenching **M4:** 64-65
Tannate **EM4:** 113
Tannin and ammonium hydroxide
as precipitant **A10:** 169
Tantalite **A7:** 160
Tantalum *See also* Pure tantalum; Refractory metals; Refractory metals and alloys; Refractory metals and alloys, specific types; Tantalum alloy; Tantalum alloys; Tantalum alloys, specific types; Tantalum powders **A13:** 725-739
as a beta stabilizer in titanium alloys **A9:** 458
acid effects **A13:** 729
alloy, high-cycle fatigue fracture **A12:** 464
alloying effect in titanium alloys **A6:** 508
alloying, nickel-base alloys **A13:** 641
alloys, workability **A8:** 165, 575
in amorphous metals **A13:** 868
as an addition to cobalt-base heat-resistant casting alloys **A9:** 334
as an addition to niobium alloys **A9:** 441
annealing **M3:** 324
applications **A2:** 557-559, 571-572 **M3:** 323-324
arc welding *See* Arc welding of tantalum
atomic interaction descriptions **A6:** 144
in austenitic stainless steels **A6:** 457
boriding **A4:** 437, 438
brazing **M6:** 1058-1059
filler metals and their properties **M6:** 1058
precleaning and surface preparation **M6:** 1057-1059
processes and equipment **M6:** 1059
brazing and soldering characteristics **A6:** 634
cans used for making material sample's of silicon nitride and boron **EM4:** 196
capacitor powder anodes, green strength **A7:** 162, 163
capacitor, solid electrolyte metal case **A7:** 770
cathodic protection **A13:** 735-736
cleaning **A2:** 563 **M3:** 324
cleaning processes **M5:** 662-663
coatings **M3:** 324
in cobalt-base alloys **A1:** 985
coextrusion welding **A6:** 311
compositions **M6:** 460
consumption **A2:** 557
corrosion **A2:** 573
corrosion, in specific media **A13:** 725-735
corrosion resistance **M3:** 324
corrosion resistance mechanism **A13:** 725
corrosive reagents effects **A13:** 734
creep rupture testing of **A8:** 302
differences in space-based (Skylab) and earth-based weld samples **A6:** 1024
diffraction pattern **A9:** 109
diffusion bonding **A2:** 564 **A6:** 156
diffusion welding **M6:** 677
effect of hydrogen content on ductility **A11:** 338
effect on anodic dissolution of phases in wrought heat-resistant alloys **A9:** 308
effect on maraging steels **A4:** 222
electrical discharge machining **A2:** 561
electrical resistance applications **M3:** 641, 646, 647, 655
electron beam drip melted **A15:** 413
electron beam welding **M6:** 641
electron-beam welding **A6:** 870, 871
electron-beam welding in a space environment **A6:** 1023- 1025
electronic applications **M3:** 314-315
electroplating of **M5:** 663-664
embrittled by mercury **A11:** 234
embrittlement **A13:** 179
embrittlement, and formability **A14:** 785
epithermal neutron activation analysis **A10:** 232
equipment, applications for **A13:** 726-728
erosion rate dependent on static hardness **A18:** 201
evaporation fields for **A10:** 587
Tantalum (continued)
explosion welding **A6:** 303, 896 **M6:** 710, 713
explosively bonded to aluminum alloy 5052 .... **A9:** 445
explosively bonded to Nickel 201 .... **A9:** 445
explosively bonded to phosphorus-deoxidized copper **A9:** 445
extrusion welding **M6:** 677
fabrication **A2:** 573
filament, gas mass spectrometer **A10:** 152-153
filler metals for brazing of **A6:** 943
finishing processes **M5:** 663-666
forging of **A14:** 237-238
forming **M3:** 324
friction welding **A6:** 152 **M6:** 722
galvanic effects **A13:** 735-736
and gamma double prime in wrought heat- resistant alloys **A9:** 309
gas-tungsten arc welding **A6:** 190, 193
glass-to-metal seals **EM3:** 302
grains, striations **A12:** 20-21
heat-exchanger, fatigued **A12:** 20
in heat-resistant alloys **A4:** 512
heating element, use in vacuum furnace **A4:** 500 **M4:** 316
for heating elements and hot furnace structures **A4:** 497
for heating elements for electrically heated furnaces **EM4:** 247, 248, 249
vapor pressure **EM4:** 249
high-purity, SSMS analysis **A10:** 144
as high-temperature material **A7:** 769-771
hydrochloric acid corrosion **A13:** 1164
hydrogen damage **A13:** 171
hydrogen embrittlement **A13:** 735
as implant material **A11:** 673
joining **M3:** 325
lamp filaments **A7:** 16
liquid-metal embrittlement of **A11:** 234
machinability **M3:** 324
in MAR-M 247 .... **A1:** 1016-1018
mechanical and physical properties **A2:** 573-574
microstructures **A9:** 441-442
mill products: wire, sheet, foil **A7:** 770-771
molten metals effect on **A13:** 736
in nickel-base superalloys **A1:** 984
and niobium, production flowchart **A7:** 161
ores, fusion flux for **A10:** 167
ores, hydrofluoric acid as dissolution medium **A10:** 165
oxidation-resistant coating of **M5:** 665-666
percussion welding **M6:** 740
photometric analysis methods **A10:** 64
polish-etching **A9:** 440-441
polishing **A9:** 440
processing characteristics **M3:** 325
production **A2:** 572-573
properties and compositions studied in M512 melting experiments **A6:** 1024
pure **M2:** 799-804
pure, properties **A2:** 1160
as pyrophoric **A7:** 199
reactions with gases and carbon **M6:** 1055
recrystallization temperatures **M6:** 1055
as refractory metal, commercial uses **A7:** 17
room-temperature bend strength of silicon nitride metal joints **EM4:** 526
salt effects **A13:** 732-733
separation from niobium **A7:** 160
sheath, in ternary molybdenum chalcogenides (chevrel phases) **A2:** 1079
sheet, forming **A2:** 562 **A14:** 787
shim for Ti-6Al-4V weld **A6:** 102, 104-106
in sintered metal powder process **EM3:** 304
sintering conditions **M4:** 796
solid-state bonding in joining non-oxide ceramics **EM4:** 525
sources and applications **A7:** 206
species weighed in gravimetry **A10:** 172
suitability for cladding combinations **M6:** 691
in sulfuric acid **A13:** 723
sulfuric acid corrosion **A13:** 1153
and tantalum alloys **A2:** 571-574
temperatures, mill processing **M3:** 317
thermal diffusivity from 20 to 100 °C **A6:** 4
thermal expansion coefficient **A6:** 907
thermal spray coating of **M5:** 362

Tantalum (continued)
TNAA detection limits **A10:** 238
to promote hardness **A4:** 124
toxicity and exposure limits **A7:** 206-207
transition temperatures **M6:** 1055
tubing, production techniques **A2:** 562-563
ultrapure, by chemical vapor deposition **A2:** 1094
ultrapure, by zone-refining technique **A2:** 1094
ultrasonic welding **A6:** 326, 327
unalloyed powder metallurgy sheet annealed **A9:** 443
unalloyed sheet, cold reduced and annealed **A9:** 443
used as heating elements in vacuum furnaces **A4:** 499, 500
vacuum heat-treating support fixture material **A4:** 503
vapor pressure, relation to temperature **A4:** 495 **M4:** 310
weldments **A13:** 345-347
welds, ductility **A2:** 564
wire, applications **A2:** 558-559

Tantalum alloys *See also* Tantalum; Tantalum alloys, specific types **A7:** 771 **M3:** 320, 324-325
annealing **A4:** 817-819
anodizing **A9:** 142
applications **A2:** 557-559
arc oscillations effect **A6:** 581
brazing **A6:** 580
cleaning **A4:** 819
cleaning processes **M5:** 662-663
corrosion rates, concentrated sulfuric acid **A13:** 737
corrosion resistance of **A13:** 736-738
density and melting temperature **M5:** 380
diffusion welding **A6:** 886
ductility **A6:** 580
electron-beam welding **A6:** 580-581
electroplating of **M5:** 663-664
explosion bonding **A6:** 580
finishing processes **M5:** 663-666
forging of **A14:** 237-238
friction welding **A6:** 580
FS61 (KBI-6), annealing **A4:** 816
FS63 (KBI-6), annealing **A4:** 816
furnaces **A4:** 817-819
fusion zones **A6:** 580, 581
gas-tungsten arc welding **A6:** 580, 581
heat-affected zone **A6:** 581
hydrogen damage **A13:** 171
interstitial impurities effect **A6:** 581
joint design **A2:** 563-564
machining **A2:** 560
microstructure effect **A6:** 580
oxidation protective coating of **M5:** 379-380
oxidation-resistant coating of **M5:** 665-666
physical properties **A6:** 941
plasma arc welding **A6:** 197
preferential pitting **A13:** 345-346
refractory metal brazing, filler metal **A6:** 942
resistance welding **A6:** 580
roll welding **A6:** 312
sintered, ultrasonic welding **A6:** 327
solid-state diffusion bonding **A6:** 580
T111, annealing **A4:** 816
T222, annealing **A4:** 816
Ta, annealing **A4:** 816
Ta-10W (FS60, KBI-10), annealing **A4:** 816
tantalum and **A2:** 571-574
tensile properties **A6:** 580
welding conditions effect **A6:** 581
weldments **A13:** 345-347

Tantalum alloys, specific types *See also* Tantalum
"61" metal
composition **M3:** 343
property data **M3:** 343
"63" metal
composition **M3:** 316, 343
property data **M3:** 326, 343-344
90Ta-10W
annealing **M4:** 655
stress relieving **M4:** 655
Astar 811C **M3:** 325
ASTAR 811C, machining **A16:** 858-869
commercially pure Ta **M3:** 326
Fansteel 60, corrosion resistance **A13:** 736
Fansteel 65, corrosion resistance **A13:** 736
FS60, annealing **M4:** 788
FS61, annealing **M4:** 788
FS63, annealing **M4:** 788
T-111 .... **M3:** 325
annealing treatment, postweld **M3:** 319
composition **M3:** 316, 345
property data **M3:** 326, 345, 346
welding conditions **M3:** 326
T-111 (Ta-8.0W-2.5Hf-0.003C), electron-beam welding **A6:** 871
T-111, machining **A16:** 858-869
T-222 .... **M3:** 325
annealing treatment, postweld **M3:** 319
composition **M3:** 316, 345
property data **M3:** 326, 345-347
temperatures, mill processing **M3:** 317
welding conditions **M3:** 319
T-222 (Ta-9.64W-2.4Hf-0.01C), electron-beam welding **A6:** 871
T-222, machining **A16:** 858-869
T111, annealing **M4:** 788
T222, annealing **M4:** 788
Ta-2.5W, chemical composition and applications **A2:** 573
Ta-2.5W-0.15Nb **M3:** 325
Ta-7.5W **M3:** 325-326
Ta-7.5W, compositional limits, applications **A2:** 573
Ta-8W-2Hf, physical properties **A6:** 941
Ta-10W **M3:** 325
annealing treatment, postweld **M3:** 319
composition **M3:** 316, 344
electron-beam welding **A6:** 871
physical properties **A6:** 941
property data **M3:** 326, 344-345, 346
temperatures, mill processing **M3:** 317
welding conditions **M3:** 319
Ta-10W, chemical composition, applications **A2:** 573-574
Ta-10W, fully recrystallized structure and banding **A9:** 444
Ta-10W, gas tungsten arc weld to Nb-10Hf-1Ti plate **A9:** 155
Ta-10W, machining **A16:** 858-869
Ta-40Nb, sheet, forged, cold rolled and annealed **A9:** 444
Ta-Hf, machining **A16:** 858-869
Ta63, machining **A16:** 858-869
Tantaloy '63' welded tube **A9:** 444
Tantaloy 63, corrosion resistance **A13:** 736

Tantalum and tantalum alloys
addition to complex metal carbonitrides **A16:** 91-92
in cast Co alloys **A16:** 69
in cemented carbide coatings **A16:** 80-83
in cemented carbides **A16:** 73-74, 78-80
content in tools machining uranium **A16:** 875-876
electrochemical machining **A16:** 534
grinding **A16:** 868-869
machining **A16:** 858-868
niobium carbides used in cermets **A16:** 91, 92, 97
nontraditional machining **A16:** 868
photochemical machining **A16:** 588
thermal fatigue resistance **A16:** 76
thermal shock resistance **A16:** 74
W-Ti-Ta(Nb) carbides **A16:** 71

Tantalum borides in wrought heat-resistant alloys **A9:** 312

Tantalum capacitors *See also* Capacitors
chip, fabrication **EL1:** 187
electrolytic, types **EL1:** 179
failure mechanisms **EL1:** 972-973
hermetically sealed **EL1:** 995
hermetically sealed solid **EL1:** 995-997
wet-slug **EL1:** 997-998

Tantalum carbide **A13:** 846-847
in cemented carbides **A18:** 795
properties **A18:** 795
thermal expansion coefficient **A6:** 907

Tantalum carbide (TaC)
adiabatic temperatures **EM4:** 229
pressure densification
pressure **EM4:** 301
technique **EM4:** 301
temperature **EM4:** 301
sprayed using IPS method with cryogenic cooling **EM4:** 207
synthesized by SHS process **EM4:** 229
Young's modulus **EM4:** 30

Tantalum carbide cermets
applications and properties **A2:** 1001-1002

Tantalum carbide in superalloy nickel matrix **A9:** 619

Tantalum carbide powders **A7:** 158
production **A7:** 158
tap density **A7:** 277

Tantalum carbides *See* Tantalum and tantalum alloys

Tantalum carbides in wrought heat-resistant alloys **A9:** 311

Tantalum nitride (TaN)
synthesized by SHS process **EM4:** 227

Tantalum oxide **EM3:** 592
reference material for ion etching with Auger electron spectroscopy **A18:** 453
sputter-induced reduction of oxides **EM3:** 245

Tantalum powder metallurgy products, specific types
Ta-100 ppm V, capacitor foil **A9:** 444
Ta-150 ppm Si, capacitor wire, fully recrystallized **A9:** 443
Ta-250 ppm Y, effect of annealing on banding **A9:** 443
'61' metal spring wire, warm rolled and cold drawn **A9:** 444

Tantalum powders *See also* Tantalum; Tantalum alloys
applications **A7:** 206-207
capacitance **A7:** 771
electron-beam melted degassed-hydride **A7:** 161-162
feedstock **A7:** 160
manufacture **A7:** 160-162
particle shape **A7:** 160, 161
platelet particle shapes **A7:** 161
powder shape **A7:** 162
production **A7:** 160-162
sintering **A7:** 393
sodium-reduced **A7:** 161

Tantalum screens
radiography **A17:** 316

Tantalum, tantalum alloys, heat treating
annealing **M4:** 788
cleaning **M4:** 788
furnaces **M4:** 788-789

Tantalum wire
mechanical properties **A7:** 476

Tantalum, zone refined
impurity concentration **M2:** 713

Tantalum-base corrosion-resistant alloys, selection of **A6:** 589, 599
applications **A6:** 599
cost effectiveness and cladding **A6:** 599
ductile-to-brittle transition temperature **A6:** 599

Tantalum-base corrosion-resistant alloys, selection of (continued)
microstructure ... **A6:** 599
oxidation and reaction ... **A6:** 599
welding characteristics ... **A6:** 599
Tantalum-copper-niobium composite superconducting wire, electronic digital
scan image ... **A9:** 152, 161
Tantalum-molybdenum alloys
corrosion resistance ... **A13:** 738
Tantalum-niobium alloys
corrosion resistance ... **A13:** 738
Tantalum-titanium alloys
corrosion resistance ... **A13:** 738
Tantalum-tungsten alloys ... **A13:** 736-738
Tantalum/niobium
as dopant for tungsten carbide ... **A18:** 795
Tantalum/titanium/niobium carbides
in cemented carbides ... **A18:** 795
Tantalus I
as synchrotron radiation source ... **A10:** 413
Tantung 144 ... **M3:** 461
Tantung 144 (Co) alloy ... **A16:** 69-70
Tantung G ... **M3:** 461, 462
Tantung G (Co) alloy ... **A16:** 69-70
Tap density ... **A7:** 12, 276-277
and apparent density ... **A7:** 297
of atomized aluminum powders ... **A7:** 129
change with differently shaped particles ... **A7:** 188, 189
equipment and test procedures ... **A7:** 276-277
as packed density ... **A7:** 297
Tap drill charts ... **A7:** 462
Tap hole
electric arc furnace ... **A15:** 359
Tap switches
resistance spot welding ... **M6:** 472-473
Tap test
for adhesive-bonded joints ... **A17:** 626-627
Tap testing ... **EM3:** 750, 751, 777
Tap tooth
showing hardness variations during grinding ... **A8:** 98, 101
Tap water
as corrosive ... **A12:** 24, 320
zinc corrosion in ... **A13:** 761
Tap-and-charge operation
induction furnaces ... **A15:** 374
Tap-out pits
electric arc furnace ... **A15:** 359
Tap-Pak volumeter ... **A7:** 276
Tape *See also* Prepregs; Tape Tape prepregs
basic structures ... **EL1:** 228-229
defined ... **EM1:** 23 **EM2:** 42
fabrication process ... **EL1:** 278
high-temperature superconductors for ... **A2:** 1085-1089
for tape automated bonding ... **EL1:** 276, 477-478
types ... **EL1:** 275-276
Tape automated bonding (TAB) *See also* Tape automated bonding (TAB) technology ... **A6:** 133 **EM3:** 584, 585
assembly, plastic packages ... **EL1:** 476-479
at level ... **EL1:** 76
chip bumping ... **EL1:** 477
with chip-on-board technology ... **EL1:** 452
defined ... **EL1:** 1158
development ... **EL1:** 274
as electrical connection ... **EL1:** 228-231
encapsulation ... **EL1:** 479
failure mechanisms ... **EL1:** 977-978, 1047
inner lead bonding ... **EL1:** 228, 478-479
mounted package ... **EL1:** 984
outer lead bonding ... **EL1:** 228, 479
reliability ... **EL1:** 480
tape ... **EL1:** 276, 477-478
tape construction ... **EL1:** 477
thermal management ... **EL1:** 286-287
types of ... **EL1:** 274-275
Tape automated bonding (TAB) technology *See also* Tape automated bonding
advantages/disadvantages ... **EL1:** 274
bum-in on tape-on-chip ... **EL1:** 281-282
encapsulation ... **EL1:** 282-283
failure mechanisms ... **EL1:** 1047
inner lead bonding ... **EL1:** 278-281
outer lead bonding ... **EL1:** 283-286
Tape automated bonding (TAB) technology (continued)
overview of ... **EL1:** 274-296
recommendations ... **EL1:** 289-290
reliability ... **EL1:** 287-288
rework ... **EL1:** 288-289
TAB, types of ... **EL1:** 274-275
tape, types of ... **EL1:** 275-276
testing ... **EL1:** 281-282
thermal management, TAB ... **EL1:** 286-287
wafer bumping technology ... **EL1:** 276-278
Tape bonding
defined ... **EL1:** 1158
Tape cable
defined ... **EL1:** 1158
Tape casting ... **EM4:** 33, 34-35, 123, 124, 161-164
advanced ceramics ... **EM4:** 49
alternate thin-sheet forming methods ... **EM4:** 164
aluminum nitride ... **EM4:** 819
applications ... **EM4:** 164
ceramic packages ... **EL1:** 462
comminution ... **EM4:** 161
compared to extrusion ... **EM4:** 164
compared to roll compaction ... **EM4:** 164
compared to slip casting ... **EM4:** 161
constituents of powder formulation ... **EM4:** 126
critical materials parameters ... **EM4:** 163
defined ... **EL1:** 1158
development ... **EM4:** 161
factors influencing ceramic forming process selection ... **EM4:** 34
formulations used in oxidizing sintering atmospheres ... **EM4:** 163
material parameters ... **EM4:** 161
mechanical consolidation ... **EM4:** 125, 126, 127
process ... **EM4:** 161-163
actual tape forming ... **EM4:** 161-162
ball milling ... **EM4:** 161
de-airing ... **EM4:** 161
deflocculation/dispersion ... **EM4:** 161, 163
filtering ... **EM4:** 161
inorganic powders ... **EM4:** 161
mixing ... **EM4:** 161
tape-casting machines ... **EM4:** 162-163
processing variables ... **EM4:** 161
slurry drying ... **EM4:** 133
substrates, moving ... **EM4:** 162
tape applications ... **EM4:** 163-164
multilayer process ... **EM4:** 163-164
viscoelastic properties required ... **EM4:** 116
Tape drapability *See* Drape
Tape laying
application, automotive industry ... **EM1:** 833-835
contoured ... **EM1:** 631-635
flat ... **EM1:** 624-630
machines, mechanically assisted lay-up ... **EM1:** 605-606
of thermoplastic resin composite ... **EM1:** 549
Tape methods
selective cadmium plating ... **M5:** 268
selective chromium plating ... **M5:** 187
Tape prepregs *See* also Multidirectional tape prepregs; Prepregs
boron-copper, continuous reentrant fill ... **EM1:** 127
boron-epoxy preimpregnated ... **EM1:** 58
braiding ... **EM1:** 519
chemical tests for ... **EM1:** 291
cross-plied, properties ... **EM1:** 147
dimensions ... **EM1:** 143
fastener holes, techniques/tools for ... **EM1:** 713-714
form selection, factors affecting ... **EM1:** 145
graphite, properties ... **EM1:** 139
machine, typical ... **EM1:** 143
multidirectional ... **EM1:** 146-147
natural fiber ... **EM1:** 117
tape laying machines ... **EM1:** 631-635
thermoplastic matrix ... **EM1:** 546
unidirectional ... **EM1:** 143-145
unidirectional, chemical tests for ... **EM1:** 291
unidirectional, for tubes ... **EM1:** 569-570
vs. fabric, cost ... **EM1:** 105
vs. fabric, impact resistance ... **EM1:** 262
woven curved ... **EM1:** 127
Tape process
General Electric ... **A7:** 636
Tape recorders, magnetic
for eddy current inspection ... **A17:** 179
Tape replica method
defined ... **A9:** 18
Tape test
for adhesion failures by thin-film contaminants ... **A11:** 43
Tape tests
cadmium plate adhesion ... **M5:** 269
Tape wrapped
defined ... **EM1:** 23 **EM2:** 42
Tape wrapping
application ... **EM1:** 355, 356
Tape-on-chip
bum-in ... **EL1:** 281-282
Taper *See* Draft
bar specimen, ultrasonic fatigue testing ... **A8:** 250
upset test specimen, fracture loci ... **A8:** 580-581
Taper delay time
definition ... **M6:** 17
Taper pin
fatigue fracture of ... **A11:** 545-546
Taper reamers ... **A16:** 241, 245
Taper sections
defined ... **A9:** 18
of fractures ... **A12:** 96
Taper sections of tinplate
components of ... **A9:** 450
steps in preparation ... **A9:** 451
Taper sections used to examine porcelain enameled sheet steel ... **A9:** 198
Taper time
definition ... **M6:** 17
Taper-point dies ... **A14:** 132, 139
Tapered bonded joints ... **EM3:** 475, 476
Tapered bores
effect in gears and gear trains ... **A11:** 589
Tapered dies
strains in compression between ... **A7:** 412
Tapered land bearing
defined ... **A18:** 19
Tapered roller bearing ... **A18:** 500, 501, 502
application in automotive manual transmissions ... **A18:** 565
basic load rating ... **A18:** 505
defined ... **A18:** 19
f_1 factors ... **A18:** 511
f_v factors for lubrication methods ... **A18:** 511
Tapered section
defined ... **A18:** 19
Tapered sections
by metal casting ... **A15:** 40
Tapered-box structure, graphite-epoxy
compression fracture of ... **A11:** 742-743
Tapered-roller bearings
burnup with plastic flow in ... **A11:** 500-501
cone, bulk damage by gross impact loading ... **A11:** 506
described ... **A11:** 490
electrical pitting in ... **A11:** 494, 496
raceway deformation by ... **A11:** 510
Tapers
swaging of ... **A14:** 136-139
Tapes
acetate, for plastic replicas ... **A17:** 53
electrical-resistance heating ... **A17:** 409
indicator, as gas/leak detection devices ... **A17:** 61
TAPPI *See* Technical Association of the Pulp and Paper Industry
Tapping ... **A16:** 255-267
accuracy and finish factors ... **A16:** 261
adjustable taps ... **A16:** 265
Al alloys ... **A16:** 764, 766-767, 769, 787-791
apparatus ... **A7:** 276
automatic nut tappers ... **A16:** 260, 264
automatic tapping machines ... **A16:** 257
carbon and alloy steels ... **A16:** 676
cast irons ... **A16:** 649, 651, 655, 660, 661
cold form tapping ... **A16:** 265-267
collapsible taps ... **A16:** 258-262, 265
compared to thread rolling ... **A16:** 289
in conjunction with boring ... **A16:** 160
in conjunction with drilling ... **A16:** 212, 213, 214, 218, 221
in conjunction with turning ... **A16:** 135
Cu alloys ... **A16:** 813-815, 816
cutting fluids used ... **A16:** 125, 262-264, 392

Tapping (continued)
effect on loose powder density **A7:** 297
expansion taps **A16:** 258
gang machines **A16:** 256
heat-resistant alloys **A16:** 751-753
inserted-chaser taps **A16:** 258
loose powder filling by **A7:** 431
in machining centers **A16:** 393
metal composition and hardness factor **A16:** 260-261
Mg alloys **A16:** 822, 825-826
MMCs **A16:** 896, 897
monitoring systems **A16:** 412, 414, 415, 416
multifunction machining **A16:** 366, 368, 370, 375, 380, 386
multifunction machining and vertical machines **A16:** 379
multiple-spindle automatic chucking
multiple-spindle machines **A16:** 259
multiple-spindle tapping machines **A16:** 256, 259
NC implemented **A16:** 614
Ni alloys **A16:** 835, 837, 839
P/M high-speed tool steels applied **A16:** 65, 67
P/M materials **A16:** 880, 881, 889, 891
of P/M parts, as machining process **A7:** 462
refractory metals **A16:** 860, 862, 865-867
sectional taps **A16:** 258
selection of tap features **A16:** 259-260
self-reversing attachments **A16:** 255
shell taps **A16:** 258
single-spindle tapping machines **A16:** 256
solid taps **A16:** 256-258, 259, 261
speed factor **A16:** 262
stainless steels **A16:** 694, 695, 696, 698-700, 701
standard taper pipe threads (NPT) **A16:** 265
surface alterations produced **A16:** 23
surface treatment of taps **A16:** 259
tap classification system **A16:** 256
tap life **A16:** 259, 260, 261, 262, 264, 266, 267
tap materials **A16:** 259
taper pipe thread tapping **A16:** 265
tapping of blind holes as a factor **A16:** 262
thread grinding of taps **A16:** 270
Ti alloys **A16:** 846, 847, 851
TiN coating for taps **A16:** 57, 58
to fill flexible envelopes for isostatic pressing **A7:** 297
tool steels **A16:** 720
torque **A16:** 264-265
two-spindle tapping machines **A16:** 261
uranium alloys **A16:** 875
vertical tapping machines **A16:** 261
workpiece size and shape factors **A16:** 261
zirconium **A16:** 854
Zn alloys **A16:** 833
Tapping, intermittent/continuous
cupolas **A15:** 384
Taps *See* Cutting tools
definition **M6:** 17
TAR
for highway construction joints **EM3:** 57
as metallochromic indicator **A10:** 174
Tar sand
Miller numbers **A18:** 235
Tar sands
GC/MS analysis of volatile compounds in **A10:** 639
Tare weight **A6:** 395-396
Target
defined **A10:** 683
in divergent-beam topography **A10:** 370
Target (x-ray)
defined **A9:** 18
Target load/life test method
for production designs **A11:** 135
Target plate
in flyer plate impact test **A8:** 211
Target poisoning **A18:** 844
Target population
defined **A10:** 12
Tarnish
and corrosion, compared **A13:** 1360
defined **A13:** 13, 97
of dental alloys **A13:** 1336-1366
lead **A9:** 416
microstructural effects **A13:** 1359-1360
Tarnish removal
copper and copper alloys **M5:** 613, 619
heat-resisting alloys **M5:** 472
nickel alloys **M5:** 670-671
titanium and titanium alloys **M5:** 654
Tarnish rupture model
crack propagation **A13:** 160-161, 163
Tarnishing
defined **A13:** 97
Tarnishing, of copper powder
effect on green strength ..... **A7:** 109, Ill
Tars/pitches applications **EM4:** 47
composition **EM4:** 47
supply sources **EM4:** 47
Tartaric acid
copper/copper alloy resistance **A13:** 629
Tartaric acid solution and ammonium persulfate solution as an etchant for lead-antimony-tin alloys **A9:** 417
TASS computer program for structural analysis **EM1:** 268, 273
Taylor constant **A1:** 591
Taylor impact test **A8:** 187, 203
Taylor series expansion **A10:** 414
for interaction coefficient **A15:** 55
ternary iron-base alloys **A15:** 62
Taylor vertices
defined **A18:** 19
Taylor's tool life equation **A16:** 42
Taylor-von Karman theory **A8:** 200
Taylor-von Mises criterion **A11:** 26
Taylor-Wharton Iron and Steel Company (New Jersey) **A15:** 25, 32
Tb-Tl (Phase Diagram) ..... **A3:** 2 • 375
TBCCO superconducting materials
properties **A2:** 1082
TBS-9
nominal composition **A18:** 725
TCARES program predicting time-independent reliability of whisker-toughened ceramic components **EM4:** 733
TCMP steels **A1:** 666
strength in **A1:** 666
TCP *See* Topologically close-packed
TCP phases **A1:** 956
TD nickel
composition **A16:** 736
machining **A16:** 738, 741-743, 746-747, 749-757
upset welding **A6:** 249
TDI *See* Toluene diisocyanate
Te-Tl (Phase Diagram) ..... **A3:** 2 • 376
Te-U (Phase Diagram) ..... **A3:** 2 • 376
Te-Yb (Phase Diagram) ..... **A3:** 2 • 376
Te-Zn (Phase Diagram) ..... **A3:** 2 • 377
TEA *See* Thermogravimetric analysis
Teach-in system
automatic pouring **A15:** 501
Teapot ladles *See also* Ladles
automatic **A15:** 497
defined **A15:** 11
for inclusion control **A15:** 90
for plain carbon steels **A15:** 710
Tear *See* Tear dimples; Tear fracture(s); Tearing
Tear dimples **A12:** 12, 15
alloy steels **A12:** 304
maraging steels **A12:** 387
titanium alloys **A12:** 452
Tear dropping
as casting defect **A11:** 385
Tear fractures
effect of striations **A12:** 16
Tear fractures (continued)
elongated dimples **A12:** 12, 15
surface (Mode 1), dimple shape **A12:** 12, 14
Tear resistance
flexible printed boards **EL1:** 589
Tear ridges *See also* Ridges; Tearing
AISI/SAE alloy steels **A12:** 304
in aluminum alloy **A12:** 19, 20
in Armco iron **A12:** 224
as fatigue indicators **A11:** 258, 259
and fluting **A12:** 20-21
in iron **A12:** 26
maraging steels **A12:** 387
titanium alloys **A12:** 453
in TTS fractures **A12:** 21, 28
wrought aluminum alloys **A12:** 417
Tear strength
measurement of **EM3:** 189
Tear test
defined **EL1:** 1159
Tear-out, edge
as fastened sheet failure **A11:** 531
Tear-test toughness of steel plate **M1:** 701
Tearing *See also* Hot tearing; Microtearing; Ridges; Tear dimples; Tear ridges
by decohesion, in welds **A12:** 139
by machining, AISI/SAE alloy steels **A12:** 296
as casting defect **A15:** 548
cold, as casting defect **A11:** 383
complex **A12:** 28
in ductile fracture **A12:** 173
ductile fracture as **A11:** 82
in ductile irons **A12:** 230-236
historical study **A12:** 2
hot, in aluminum alloy castings **A17:** 535
lamellar **A12:** 138-139, 158
lamellar, carbon-manganese steel **A11:** 92
lamellar, in weldments **A17:** 582, 585
polymers **A12:** 479, 480
rapid, AISI/SAE alloy steels **A12:** 327
resin-matrix composites **A12:** 474
shear fractures **A11:** 697
tensile, elongated dimples by **A11:** 76
transgranular **A12:** 305
Tearing instability
analysis, *J-R* measurement for **A8:** 457
conditions for **A8:** 452, 457
elastic-plastic analysis for **A8:** 446
Tearing topography surface fracture
appearance **A12:** 28-29
defined **A12:** 21-22
Tears
brittle fracture from **A11:** 85
pull-through, as fastener failure **A11:** 531
radiographic inspection **A17:** 297, 349
Teart disease (cattle and sheep)
from molybdenum toxicity **A2:** 1253
Technetium
determined **A10:** 209
pure **M2:** 804-806
Technetium, pure
properties **A2:** 1163
Technical Association of the Pulp and Paper Industry (TAPPI) **EM3:** 61
standards **EM3:** 62
Technical ceramics **EM4:** 16
applications **EM4:** 16
materials included **EM4:** 16
properties **EM4:** 16
Technical meetings
on nondestructive evaluation **A17:** 51
Technical whiteware ceramics **EM4:** 4
absorption **EM4:** 4
applications **EM4:** 4
compositions **EM4:** 4
glazing **EM4:** 4
products **EM4:** 4
Technically vitreous stoneware **EM4:** 3, 4

Techniques, nondestructive evaluation *See* Nondestructive evaluation methods Nondestructive evaluation techniques specific techniques
Technology
cost effectiveness curves **EL1:** 390
dilemma of **EL1:** 390
integrated circuits, trends **EL1:** 399-401
selection, in computer-aided design **EL1:** 128
Technology, adhesives and sealants
failure analysis **EM3:** 1-2
fundamentals **EM3:** 1-2
systems approach **EM3:** 1
Technology, embedded *See* Embedded technology
Technology rules
of engineering design system **EL1:** 128
Technology Transfer Society (TTS)
as seminar information source **EM1:** 42
Technora HM-50 fabric
as reinforcement **EL1:** 535
Tectyl 506 surface coating **A12:** 73
Teeth **EM3:** 28
discoloration of **A13:** 1337
lead toxicity **A2:** 1246
Teflon **A7:** 606 **A18:** 144, 152
as additive for liquid sealants in construction industry **EM3:** 608
as bearing material **A18:** 754
bearings, for stored-torque Kolsky bar **A8:** 219-220
coating, business machine parts **A7:** 669
as coating for bonding fixture **EM3:** 707
coating, lead wires in acidified, chloride solutions **A8:** 420
development of **EM3:** 223
friction coefficient data **A18:** 73, 74
for low temperature tension testing **A8:** 36
material for surface force apparatus **A18:** 402
porous, abradable seal material **A18:** 589
sheet, for friction in axial compression testing **A8:** 56-57
surface pores trapping oils **EM3:** 264
thermal properties **A18:** 42
as thermocouple wire insulation **A2:** 882
use in metal-on-polyer total hip replacements **A18:** 657
Teflon in sleeve bearings **A9:** 567
Teflon-based materials *See also* Polytetrafluoroethylene (PTFE)
for base material/insulators **EL1:** 114
for printed wiring boards **EL1:** 607
Telecommunication applications *See also* Communications applications
acrylonitrile-butadiene-styrenes (ABS) **EM2:** 111
high-density polyethylenes (HDPE) **EM2:** 163
liquid crystal polymers (LCP) **EM2:** 180
polybutylene terephthalates (PBT) **EM2:** 153
Telecommunications and related uses **EM4:** 1050-1054
applications of high-silica fibers for communications **EM4:** 1050-1054
cost **EM4:** 1053
dispersion **EM4:** 1050-1051, 1052
erbium-doped fiber amplifiers **EM4:** 1054
fatigue **EM4:** 1052-1053
fiber amplifiers and lasers **EM4:** 1054
fiber design **EM4:** 1050-1051, 1052
fiber processing **EM4:** 1053
glass couplers, splitters, and components **EM4:** 1053-1054
graded refractive index cores **EM4:** 1051
lasers **EM4:** 1050, 1051
light guidance **EM4:** 1050-1051
light-emitting diode **EM4:** 1050, 1051
mechanical strength **EM4:** 1052-1053
numerical aperture **EM4:** 1050, 1051
optical fiber types **EM4:** 1051
optical transmission loss **EM4:** 1051-1052
polarization-maintaining fibers **EM4:** 1053
Rayleigh scattering **EM4:** 1052
refractive index **EM4:** 1051
future directions/summary **EM4:** 1054
Telecommunications applications
of commercial hybrids **EL1:** 382-385
copper and copper alloys **A2:** 239
equipment, recommended contact materials **A2:** 864-865
Telegas instrument
for hydrogen measurement **A15:** 458-459
Telegraph and telephone wire **M1:** 255, 264-265
Telegrapher's equation **EL1:** 29
Telegraphing **EM3:** 28
Telephone and telegraph wire **A1:** 283
Telephone cable plants **A13:** 1127-1133
case histories **A13:** 1132-1133
corrosion causes **A13:** 1128 1130
corrosion prevention **A13:** 1130-1131
and environment **A13:** 1127-1128
metallic components **A13:** 1127
Telephone components
powders used **A7:** 573
Telephony communication networks **EL1:** 8-9
Television
thick-film hybrid applications **EL1:** 385
Television (TV) glasses
material to which crystallizing solder glass seal is applied **EM4:** 1070
material to which vitreous solder glass seal is applied **EM4:** 1070
Television applications
high-impact polystyrenes (PS, HIPS) **EM2:** 195
Polyphenylene ether blends (PPE, PPO) **EM2:** 183
polyurethanes (PUR) **EM2:** 259
Television cameras *See also* Cameras
in pulsed video thermography **A17:** 512
radiographic inspection **A17:** 318-319
real-time radiography **A17:** 319
use, neutron radiography **A17:** 391
with vidicon tubes **A17:** 10
Television monitors used in optical microscopy **A9:** 83-84
Television scanners, conventional
with image analyzers **A10:** 310
Tellurites
toxicity of **A2:** 1260
Tellurium
additive for copper alloys **M6:** 400
additive for enhanced machinability of carbon and alloy steels **A16:** 673, 674, 675, 676
in alloy cast irons **A1:** 112
alloying, copper and copper alloys **A2:** 236
in Alnico alloys **A9:** 539
applications **A2:** 1260
in cast iron **A1:** 8
as chalcogen **A2:** 1077
content additions to P/M materials **A16:** 885
in copper alloys **A6:** 753
Cu-Ni-Te alloys, microdrilling **A16:** 238
determined by 14-MeV FNAA **A10:** 239
-doped gamma-ray detectors **A10:** 235
effect, gas dissociation **A15:** 83
effect in copper **A12:** 401
effect, iron-base alloy **A15:** 83
effect of, on machinability of carbon steels **A1:** 599
effect on Cu alloy machinability **A16:** 805-808
effects, electrolytic tough pitch copper **A2:** 270
embrittlement by **A13:** 183
embrittlement of **A11:** 236
embrittlement of iron by **A1:** 691
as embrittler in steels **A11:** 226-227
in flake graphite composition **A18:** 699
in free-machining metals **A16:** 389
gaseous hydride, for ICP sample introduction **A10:** 36
gravimetric finishes **A10:** 171
HAZ fissuring in nickel-base alloys **A6:** 588
as inoculant **A15:** 105
machinability additive **A16:** 685, 687
in malleable iron **A1:** 10
as minor toxic metal, biologic effects **A2:** 1260
pure **M2:** 806
pure, properties **A2:** 1165
quartz tube atomizers with **A10:** 49
radiochemical, destructive TNAA for **A10:** 238
in semiconductor alloys, electrometric titration for **A10:** 206
tantalum corrosion in **A13:** 728, 735
toxicity, during retrograde pyelorgraphy **A2:** 1260
as trace element **A15:** 394
as tramp element **A8:** 476
Tellurium (continued)
vapor pressure **A6:** 621
volatilization losses in melting **EM4:** 389
Tellurium in cast iron
carbide-inducing effect **M1:** 80
malleable cast irons **M1:** 58
Tellurium in copper **M2:** 242-243
Tellurium-copper alloy
as electrode for EDM **A16:** 559
Telomer **EM3:** 28
defined **EM2:** 42
TEM *See* Transmission electron microscope(s); Transmission electron microscopy
TEM direct carbon replicas
iron **A12:** 224
TEM p-c replicas
iron **A12:** 224
Temper *See also* Alloy designation systems; Green sand; T temper; Temper designation system; Tempering; Thermal conductivity **A1:** 281
for aluminum and aluminum alloys **A2:** 8
aluminum/aluminum alloys, corrosion resistance ratings **A13:** 586-588
beryllium-copper alloys **A2:** 409-411
color, defined **A13:** 13
color, of tools and dies **A11:** 563
defined **A13:** 13 **A15:** 11
definition **EM4:** 633
designations, copper and copper alloys **A2:** 223
effects, AISI/SAE alloy steels **A12:** 341
embrittlement **A12:** 134-135
embrittlement, defined **A13:** 13
as measure of hardness **A2:** 219
of nickel-base alloys **A14:** 831
over-, AISI/SAE alloy steels **A12:** 301
snap **A11:** 95-96
straightening **A14:** 682
T1 through T10, defined **A2:** 26-27
temperature, effects on sulfide fracture toughness **A13:** 535
temperature vs. yield strength, low-alloy steels **A13:** 954
transition-temperature shift from **A11:** 69
Temper annealing
ferrous alloys **A7:** 185
Temper brittleness **A1:** 698
defined **A8:** 14 **A11:** 10
Temper carbon *See also* Nodular graphite **M1:** 6, 9, 57-61, 63-64, 67
defined **A9:** 18
Temper designation system *See also* Alloy designation systems; specific tempers
additional tempers **A2:** 25-26
aluminum and aluminum alloys **A2:** 16, 21-22, 25-27
for annealed products **A2:** 27
basic designations **A2:** 16, 21
copper and copper alloys **A2:** 223
F, as fabricated, defined **A2:** 21
of foreign tempers **A2:** 27
H, strain-hardened (wrought products only) defined **A2:** 21
for heat-treatable alloys **A2:** 26-27
magnesium and magnesium alloys **A2:** 456
O, annealed, defined **A2:** 21
for strain-hardened products **A2:** 21, 25
T, solution heat-treated, defined **A2:** 21
of unregistered tempers **A2:** 27
W, solution heat-treated, defined **A2:** 21
Temper designations
aluminum and aluminum alloys **M2:** 24-27
Temper embrittlement *See also* Embrittlenient **M1:** 468-469, 684-685, 701-704
deterioration by **A11:** 69
in forging **A11:** 335
intergranular, in brittle fracture **A11:** 75
low-alloy steels **A15:** 721
of shafts **A11:** 479-480
of steels **A11:** 99
transition-temperature shift from tempering range of **A11:** 69
Temper embrittlement in alloy steels
See also Tempered martensite embrittlement **A1:** 698-703
composition, effect of **A1:** 699-700
grain size, effect of **A1:** 700-702

Temper embrittlement in alloy steels (continued)
microstructure, effect of ... **A1:** 702
Temper mill scale breaking
use in pickling operations ... **M5:** 74-76, 78
Temper point
in clay-water bonds ... **A15:** 212
Temper rolling ... **A1:** 693
of carbon steel sheet ... **A1:** 205-206
defined ... **A9:** 18
effect on formability on steel sheet ... **M1:** 548, 556
effect on Lüders lines ... **A8:** 553
galvanized steel sheet ... **M1:** 170
Temper straightening ... **A14:** 682
Temper time
definition ... **M6:** 17
Temper water
defined ... **A15:** 212
"Temper-bead" procedures ... **A6:** 81
Temper-embrittled steel
microscopic models for ... **A8:** 466
Temperature *See also* Ambient temperature; Elevated temperature; Elevated temperatures; Elevated-temperature failures; Heat treating; Heat treatments; High temperature; Hot pressing; Low temperature; Room temperature; Sintering; Sintering temperatures; Temperature control; Transient temperature; Transition temperature
abbreviation for ... **A10:** 691
abnormal increase, in bearing failures ... **A11:** 494
activated sintering ... **A7:** 318
and alloying effect on torsional ductility ... **A8:** 164-166
ambient, symmetric rod impact test at ... **A8:** 204-205
of austenite formation, symbol for ... **A8:** 724
for austenite relationships, symbols for ... **A11:** 796
of austenite to pearlite, symbol for ... **A11:** 796
austenitizing, effect in tool steels ... **A11:** 571
axial force dependence on ... **A8:** 181
of brittle cracking, by Robertson test ... **A11:** 59-60
brittle-to-ductile transition ... **A11:** 28
bulk, effects on chemical-reaction rates ... **A11:** 160
of cementite precipitation from austenite on cooling ... **A8:** 724
changes during creep-rupture testing ... **A8:** 337-339
and chevron patterns in low-alloy steel ... **A11:** 77
conducive to SCC in titanium alloys ... **A11:** 223
control, in ion selective electrode measurement ... **A10:** 185-186
-control, tool and die failure and ... **A11:** 573
conversion factors ... **A8:** 723 **A10:** 686
copper oxide reduction ... **A7:** 108, 109
and corrosion ... **A11:** 156, 199, 255-256
and corrosion rates ... **A11:** 175-176, 199
in creep/creep-rupture analyses ... **A8:** 685
cryogenic, effect on fatigue ... **A12:** 52-53
Curie, abbreviation for ... **A10:** 691
cycling, shape distortions from ... **A11:** 266
dependence of copper texture on ... **A8:** 182
dependence, of polycrystalline materials ... **A11:** 138
dependence of steady-state creep ... **A8:** 309
dependence of toughness ... **A8:** 262
differential-, cells ... **A11:** 184-185
and distortion failure ... **A11:** 138
dry-bulb ... **EM3:** 28
ductile-to-brittle transition, defined ... **A12:** 34
-ductility behavior variability ... **A8:** 34, 36
ductility transition, defined ... **A11:** 59
during creep-rupture tests ... **A8:** 330
effect, austenitic stainless steel ... **A12:** 50-52
effect, dimple rupture ... **A12:** 33-35
effect, dimple size ... **A12:** 34, 46
effect, ductile iron fracture ... **A12:** 231
effect, embrittlement rate ... **A12:** 29
effect, high-purity copper ... **A12:** 399-400
effect in AAS emission signals ... **A10:** 44

Temperature (continued)
effect in changing failure mode of metal ... **A8:** 188
effect in emission sources ... **A10:** 24
effect in MFS analysis ... **A10:** 77
effect in Rietveld method ... **A10:** 423
effect in tension testing ... **A8:** 49
effect in ultrasonic testing ... **A8:** 247
effect, intergranular creep rupture ... **A12:** 19
effect, medium-carbon steels ... **A12:** 254
effect of chemical composition ... **A7:** 246
effect of high, on corrosion-resistant cast steel ... **A1:** 917-918
effect of, on austenitic manganese steel ... **A1:** 836-837
effect of, on steel springs ... **A1:** 303-304, 312, 313, 314-315
effect of, on stress-corrosion cracking ... **A1:** 724-725
effect on abrasive wear ... **A18:** 188
effect on cast heat-resistant alloy ... **A11:** 407
effect on corrosion of carbon and low-alloy steels ... **A11:** 199
effect on corrosion-fatigue ... **A11:** 255-256
effect on crack growth and toughness ... **A11:** 54, 75
effect on crack growth rate ... **A11:** 75
effect on creep rate ... **A8:** 301, 308
effect on ductility ... **A8:** 19, 34, 262
effect on dynamic friction coefficient in silicon and gallium arsenide ... **A18:** 688-689
effect on embrittlement in steels ... **A11:** 239
effect on eutectoid transformation parameters ... **A9:** 661
effect on fatigue crack growth ... **A8:** 411
effect on fatigue strength and static strength ... **A11:** 130
effect on flow stress of alpha iron and copper ... **A8:** 178
effect on formability of sheet metals ... **A8:** 553
effect on homogenization kinetics ... **A7:** 315
effect on interlamellar spacing in pearlite ... **A9:** 660
effect on magnetic saturation ... **A9:** 534
effect on pearlite growth ... **A9:** 660
effect on rates of phase transformations, in metals ... **A10:** 318
effect on reactive sputtering of interference films ... **A9:** 149-150
effect on SCC in aluminum alloys ... **A11:** 220
effect on SCC in titanium and titanium alloys ... **A11:** 223-224
effect on sintered iron-graphite powders ... **A7:** 364
effect on slip ... **A8:** 34
effect on strength ... **A8:** 19, 34
effect on the amount of martensite ... **A9:** 669
effect on torsional ductility of U-700 ... **A8:** 166-167
effect on torsional fatigue testing ... **A8:** 152
effect on torsional flow curve of carbon pearlitic alloy ... **A8:** 176
effect on toughness, schematic ... **A11:** 66
effect on upper bainite ... **A9:** 663
effect, slip ... **A12:** 33
effect, strength ... **A12:** 121
effects in electrogravimetry ... **A10:** 198
electron ... **A10:** 24
equicohesive ... **A12:** 121
equilibrium transformation, steel, symbol for ... **A11:** 796
eutectic melting ... **A11:** 122
exceeding normal operating effects of ... **A11:** 282, 374-376
explosion characteristics ... **A7:** 133, 196
as fatigue environment ... **A12:** 35
of ferrite transformation to austenite, symbol for ... **A8:** 724
and flow localization ... **A8:** 170-171
fracture-strain dependence on ... **A8:** 167-168

Temperature (continued)
gas kinetic, and kinetic energies of heavy particles ... **A10:** 24
hazardous ignition, magnesium powder ... **A7:** 133
high, effect on dimple rupture ... **A12:** 34-35
high, FMR probe for ... **A10:** 270
high-purity oxygenated water ... **A8:** 420
history, effect on workability, torsion testing ... **A8:** 178
homologous, defined ... **A8:** 34
in hot pressing ... **A7:** 504
impact fracture toughness as function of ... **A11:** 54
influence in deformation ... **A8:** 188
influence in stress-corrosion cracking ... **A8:** 499
interval, conversion factors ... **A8:** 723
in iron castings, effects of ... **A11:** 374
and liquid-metal embrittlement ... **A11:** 230
in liquid-phase sintering ... **A7:** 320
low, effect on dimple rupture ... **A12:** 33-34
low, probe for FMR measurement at ... **A10:** 270
low- and elevated, design properties for ... **A8:** 670-671
of lubricant breakdown, effect on bearings ... **A11:** 490
and lubrication, rolling-element bearings ... **A11:** 510-511
maximum, for operation of wrought stainless steels ... **A11:** 271
and median time-to-failure of aluminum and aluminum alloys ... **A11:** 772
melting, symbol for ... **A8:** 726
minimum metal, for flue-gas corrosion ... **A11:** 619
modulus of elasticity values at different ... **A8:** 23
of nonuniform forging, effects ... **A11:** 317
operating, of gas-turbine engine ... **A11:** 283
or strain rate change effect on flow stress ... **A8:** 226
and oxidation ... **A12:** 35
oxide reduction ... **A7:** 52, 108, 109
plane-strain fracture toughness as function of ... **A8:** 450
of premix flames ... **A10:** 29
room, corrosion in heat-resisting alloys ... **A11:** 201
room variations ... **A10:** 172
-sensitive paint, for ultrasonic testing ... **A8:** 247
of specimen during deformation ... **A8:** 191
and SSC resistance ... **A11:** 298
and strain rate effect on alpha parameter ... **A8:** 172
and strain rate, effects on yield strength ... **A8:** 38, 40
and strain rate function, tensile fracture modes ... **A8:** 571
and strength ... **A8:** 670
and stress dependence, creep equations ... **A8:** 688-689
structural changes as function of ... **A10:** 420
symbol for ... **A8:** 726 **A11:** 798
tempering ... **A7:** 374, 453, 454 **A8:** 27
tempering, effect on microstructural failure ... **A11:** 325
test ... **A8:** 586-587
thermodynamic ... **A8:** 721
thermodynamic, SI base unit and symbol for ... **A10:** 685
and time for sintering tungsten and molybdenum ... **A7:** 389
-time parameters, for creep-rupture analysis ... **A8:** 690
-time plot, pearlite decomposition ... **A11:** 613
of transformation to ferrite or ferrite and cementite, symbol for ... **A8:** 724
transition ... **A7:** 12 **A11:** 66-68, 154
variable, ESR spectrometers ... **A10:** 257
variation of field-evaporation rate with ... **A10:** 594
various, for stress-corrosion cracking ... **A11:** 207

Temperature (continued)
vs. notch-rupture strength ratio, Inconel X-750 .... **A8:** 316
wear failure from corrosion and **A11:** 156
and workability **A8:** 572, 574-575
Temperature and high strain rate
effect on plastic deformation **A9:** 688-691
Temperature coefficient
defined **EL1:** 1159
Temperature coefficient of resistance (TCR)
electrical resistance alloys **A2:** 822
nonmetallic materials **A2:** 829
resistance alloys **A2:** 824-826
wrought aluminum and aluminum alloys **A2:** 122
Temperature control **A4:** 529-541 **M4:** 345-360
basic control loop **A4:** 529, 530, 538, 540 **M4:** 345-346
of constant-load test specimen **A8:** 314-315
control instruments **A4:** 529, 537-541 **M4:** 354, 355-359
in creep-rupture testing **A8:** 330
distributed control systems (DCSs) **A4:** 540-541
effect on coating thickness in thermoreactive deposition/diffusion process **A4:** 449
elevated/low temperature tension testing **A8:** 36
emissivity values for materials at 0.65 μm wavelengths **A4:** 538
and measurement in creep furnace **A8:** 312-313
measurement instruments **A4:** 529, 537-541 **M4:** 354-355
noncontact sensors **A4:** 534-537 **M4:** 350-355
precautions **A4:** 530, 531
process characteristics **A4:** 537, 538, 539-540, 541 **M4:** 359-360
product characteristics **A4:** 537, 539, 540, 541 **M4:** 360
resistance temperature detectors **M4:** 350
resistance temperature detectors (RTDs) **A4:** 533-534
set-point programmers ... **A4:** 538, 539, 540 **M4:** 356, 357, 358-359
and specimen heating in vacuum and oxidizing environments **A8:** 414
in steam or boiling water with contaminants **A8:** 427
for step-down tension testing **A8:** 324
temperature scales **A4:** 529
thermocouples **A4:** 529-532, 535, 536 **M4:** 346-350, 351, 352
Temperature cycling *See also* Thermal cycling
aluminum bonding pad **EL1:** 893
beam-lead/bump-contact designs **EL1:** 961
as component- and board-level physical testing **EL1:** 944
environmental stress screening parameters **EL1:** 880-884
as intermittent failure test **EL1:** 1060
package-level testing **EL1:** 937
and power cycling, compared **EL1:** 961
profiles **EL1:** 881
in semiconductor chips **EL1:** 964
-sensitive designs, new **EL1:** 960-961
as stress test **EL1:** 498-499
test conditions **EL1:** 498
tests **EL1:** 494, 498-499
wear-out modes **EL1:** 287
Temperature effects **EM3:** 420-427
Eyring formulation **EM3:** 421-422
governing equations **EM3:** 425-526
kinetics of cure **EM3:** 423-425
physical aging **EM3:** 422
relaxation time concept **EM3:** 421
Struik concept **EM3:** 422
thermoset adhesive processing **EM3:** 422-425
time-temperature relations in polymers **EM3:** 420-422
Temperature gradient
ahead of interface **A15:** 144
during forging **A14:** 161
effect, scaled ingots **A14:** 66
and solutal convection **A15:** 148
and thermal convection **A15:** 147-148
Temperature gradients **A6:** 46-50, 51, 52, 53
Temperature overshoot **EM3:** 701
Temperature profiles
infrared (IR) soldering **EL1:** 705
of PWAs **EL1:** 685
Temperature resistance *See also* Temperature(s); Thermal analysis
glass transition temperature, determined **EM2:** 564-565
high-performance polymers **EM2:** 559
and polymer composition **EM2:** 566-567
quality assessment method **EM2:** 564-565
silicones (SI) **EM2:** 267
techniques and parameters **EM2:** 559-564
thermal analysis **EM2:** 566-567
thermogravimetric analysis, as screening **EM2:** 565-566
thermoplastic polyurethanes (TPUR) **EM2:** 206
Temperature scales
in thermocouple calibration **A2:** 878-879
Temperature sensors
use in temperature control **A4:** 529-537, 540 **M4:** 346-355
Temperature(s) *See also* Aging temperature; Casting temperatures; Cooling; Dry bulb temperature; Elevated temperature; Elevated temperatures; Equilibrium temperatures; Finishing temperature; Glass transition temperature; Heat treatment; Heating; High temperature; High-temperature; High-temperature service; Liquidus temperature; Low temperature; Low temperature flexibility; Melting temperature; Microwave thermography; Mold temperature; Pouring temperature; Room temperature; Service temperature; Solidus temperature; Solution temperature; Temper; Temperature coefficient of resistance (TCR); Temperature gradient; Temperature resistance; Tempering; Thermal; Thermal analysis; Thermal conductivity; Thermal inspection; Thermal mismatching; Thermal neutrons; Thermal wave imaging Thermography; Thermalization; Thermocouple materials; Transformation temperature; Transition temperature
accurate measurement of **A2:** 869-871
for adhesives, structural **EM1:** 684
aging **EM2:** 569, 751
aging, effects, uranium alloys **A2:** 680
aging, uranium alloys **A2:** 680
of air, maximum service, for stainless steels **A13:** 558
for alloy steel forgings **A14:** 81, 215-216, 367
and aluminum alloy forgeability **A14:** 241-242, 247
annealing, beryllium-copper alloys **A2:** 405
annealing, cast copper alloys **A2:** 367
application, creep at **EM2:** 1
beryllium, formability effects **A14:** 805
beta transus, wrought titanium alloys **A2:** 623
billet, for hot extrusion **A14:** 315
bonding **EL1:** 224
brazing **A13:** 881-883
calibration source, thermal inspection **A17:** 400
of carbon solubility in austenite **A15:** 66
for carbon steel working **A14:** 81, 215-216
casting, lead and lead alloys **A2:** 547-548
in chemical cleaning **A13:** 1142
coefficient of resistance, electrical resistance alloys **A2:** 822
compensation, magnetic, alloys for **A2:** 773-774
composition, and free energy composition **A15:** 53
and conductivity **EL1:** 98
constant, and time effects, permanent magnet materials **A2:** 798
control, in sand reclamation **A15:** 352-353
control, in SiC fiber production **EM1:** 858
control, radiographic film processing **A17:** 352
control, wave soldering **EL1:** 690-691
control, zinc casting **A15:** 790
as controlling workability **A14:** 363, 368
copper/copper alloy, blanking/piercing effects **A14:** 812
crack growth rate and **A13:** 164
for crevice corrosion **A13:** 305
critical, for A15 superconducting compounds **A2:** 1061
critical ordering, ordered intermetallic compounds **A2:** 913-914
critical pitting **A13:** 347, 646
Temperature(s) (continued)
cure, filament winding **EM1:** 135
curing, epoxy resins classified by **EM1:** 654
data, from hot compression testing **A14:** 439
defined, thermal inspection **A17:** 396
deflection, under load *See* Deflection temperature under load
deformation effects, titanium alloys **A14:** 268-269
deformation, of iron powder preforms **A14:** 194
dependence, and glass transition **EM2:** 452
dependence, nickel aluminide yield strength **A2:** 915-916
dependence, of coefficient of linear thermal expansion **EL1:** 814
dependence, of modulus **EM2:** 453
dependence, of resistivity **EL1:** 95-99, 139
dependence, of viscoelastic properties **EM1:** 191
-dependent corrosion, in accelerated testing **EL1:** 891
as design parameter **EM2:** 670-671
die, control of **A14:** 81-82
die, in die casting **A15:** 289
and die life **A14:** 57
die, zinc alloy casting **A15:** 790
differential, distortion and stress effects **A15:** 616
display, color image **A17:** 488
drop, for boiling cooling **EL1:** 364
drying, Antioch process **A15:** 246-247
dynamic viscosity and **A15:** 110
effect, adhesives **EL1:** 673
effect, atmospheric corrosion **A13:** 81-82
effect, copper alloys in seawater **A13:** 624
effect, corrosion rates of iron, zinc, copper **A13:** 910
effect, diffused germanium/silicon **EL1:** 958-959
effect, erosion/cavitation testing **A13:** 313
effect, flexible epoxies **EL1:** 820
effect, high-impact polystyrenes (PS, HIPS) **EM2:** 197
effect, hydrogen absorption, low-carbon steel **A13:** 330
effect, in fretting **A13:** 140
effect, kinetics gaseous corrosion **A13:** 70
effect, laminate strength **EM1:** 227
effect, liquid penetrant inspection **A17:** 71-74
effect, liquid-metal embrittlement **A13:** 175-177
effect, marine atmospheres **A13:** 905
effect, nuclear reactor erosion-corrosion **A13:** 965
effect, on damping **EM1:** 215-216
effect on formability **A14:** 882-883
effect on pressure, magnesium alloys **A14:** 260
effect on pressure, titanium alloys **A14:** 269-270
effect on strength properties, aluminum oxide-chromium cermets **A2:** 994
effect, pitting resistance **A13:** 114
effect, pollution control **A13:** 1367
effect, polyether sulfones (PES, PESV) **EM2:** 160, 161
effect, produced fluids **A13:** 479
effect, pulp bleach plants **A13:** 1193
effect, SCC in steam generators **A13:** 942
effect, SCC of high-strength steels **A13:** 533
effect, seawater corrosivity **A13:** 894-895, 899-900
effect, tensile strength, platinum wire **A13:** 800
effect, water quality **A13:** 489
effect, zinc alloy castings **A2:** 533
effect, zinc corrosion in distilled water **A13:** 760-761
effects, coordinate measuring machines **A17:** 27
effects, permanent magnet materials **A2:** 797-799
effects, polyurethanes (PUR) **EM2:** 260
electrical breakdown from **EM2:** 465-466
electrical effects **EM2:** 584
embrittler-melting **A13:** 186
and emf relationship **A2:** 878
equilibrium, in ternary iron-base systems **A15:** 65-67
equilibrium liquidus, and nucleation **A15:** 101
extrusion **A14:** 317-319
failure mode effects **EM2:** 684-687
failures, thermal analysis techniques for **EM1:** 779-780
fiber/matrix CTEs as function of **EM1:** 189
finishing, for steels **A14:** 620

Temperature(s) (continued)
firing, rammed graphite molds **A15:** 273-274
and flexural strength, unsaturated polyesters **EM2:** 248
flow stress and workability as function of **A14:** 166-170
-flow stress relations, process modeling **A14:** 419-420
flux **EL1:** 682
forging, for wrought aluminum alloy **A2:** 34
forging, of various alloys **A14:** 75
in forming of magnesium alloys **A14:** 825
in full-scale static tests **EM1:** 348
and furnace life, reverberatory furnaces **A15:** 378-379
and gating, die casting **A15:** 289
Gibbs energy of formation, oxides **A13:** 63
glass transition, defined *See* Glass transition temperature
gradient, effect, insoluble particles **A15:** 144
and gradient mass transfer **A13:** 51
and heat transfer, aqueous corrosion **A13:** 39-40
heat-deflection **EM2:** 68
for heat-resistant alloys **A14:** 232-234, 266
high, as inclusion-forming **A15:** 90, 488
high, ceramic resistance **EL1:** 335
for high-temperature alloys **A14:** 224
histories, during thermal cycling **EL1:** 881-882
homogeneity, with circulation pump **A15:** 455
horizontal centrifugal casting **A15:** 297
hot forming, for steel **A14:** 620
hot-working, beryllium-copper alloys **A2:** 415
hot/wet in-service, organic matrix materials **EM1:** 33
-humidity-bias testing **EL1:** 1050
hydrogen solubility and **A15:** 82, 86
ideal scale for **A2:** 878
indexes, supplier data sheets **EM2:** 642
-induced stress testing **EL1:** 497-499
of intermittent solution tests **A13:** 223
interpass, cast iron arc welding **A15:** 525-526
in iron corrosion **A13:** 37-38
junction, thermal conductivity effects **EL1:** 814
in leakage measurement **A17:** 57-59
liquation, defined *See* Liquation temperature
liquidus, by thermal analysis **A15:** 184-185
liquidus, cast copper alloys **A2:** 356-391
liquidus, glass fiber **EM1:** 107
long-term resistance **EM2:** 68
low, as degradation factor **EM2:** 576
low, for semisolid casting/forging **A15:** 328
for magnesium alloys **A14:** 259, 825
magnetic, rare earth metals **A2:** 724
as manufacturing factor **EL1:** 82
maximum upper use, as composite property **EM1:** 43
measurement, by thermocouple **A2:** 870
melt, effect on alloy addition **A15:** 72
melting, nickel-base alloys **A14:** 265
melting, of indium and bismuth **A2:** 750
metal, of pours **A15:** 498
and molecular structure, properties **EM2:** 436
for nickel-base alloys **A14:** 263
operating, closed-die forging **A14:** 82
operating, failure rate dependence **EL1:** 23
of peritectic reaction **A15:** 127
and permeability, magnetically soft materials **A2:** 773
and permittivity **EL1:** 600
in phase diagrams **A15:** 57
polyimides categorized by **EM1:** 78
of porcelain enameling **A13:** 448
postcure, of polyimide resins **EM1:** 663
pouring **A15:** 94, 281-283, 639, 651
pouring, of lead and lead alloys **A2:** 545
and power requirements, three-roll forming **A14:** 621
precision forging **A14:** 171-172
preheating, cast steel welding **A15:** 535

Temperature(s) (continued)
properties effects, epoxy resin system composites **EM1:** 401-415
properties effects, high-temperature thermoset matrix composites **EM1:** 373-380
properties effects, low-temperature thermoset matrix composites **EM1:** 394-398
properties effects, medium-temperature thermoset matrix composites **EM1:** 383-392
properties effects, thermoplastic matrix composites **EM1:** 365-372
range, silicone conformal coatings **EL1:** 822
ranges, resin systems **EM2:** 439-444
recrystallization, defined *See* Recrystallization temperature
refractory metals, and formability **A14:** 786-787
in reliability testing **EL1:** 742
resistance, structural ceramics **A2:** 1019
rolling, vs deformation resistance, in steels **A14:** 119
sand, and sand/casting recovery **A15:** 348-349
scales, thermocouple calibration **A2:** 878-879
scaling as function of **A14:** 162, 174
for SCC in titanium alloys **A13:** 274
seasonal changes, chemical processing effects **A13:** 1136
sensors, contact **A17:** 399
sensors, noncontact, thermal inspection **A17:** 398-399
sintering, effects in P/M stainless steels **A13:** 825
soldering **EL1:** 590, 676, 678, 688
solidification (undercooling), low-gravity **A15:** 150
solidification, onset/offset **A15:** 101
solidus, by thermal analysis **A15:** 184-185
solidus, cast copper alloys **A2:** 356-391
solubility **EM2:** 496
solubility as function of **A15:** 61
of solution, total immersion tests **A13:** 221
specific, coatings for **A13:** 461
specific damping capacity variation with **EM1:** 214-215
spinning, refractory metals and alloys **A2:** 562
in squeeze casting **A15:** 324
stability, and unbalanced bridge eddy current inspection **A17:** 177
for stainless steel heat treatment **A14:** 229
for steel working **A14:** 81, 119, 216, 620
and strain rate, and tensile fracture **A14:** 363
and strain rate, combined effect, process modeling **A14:** 420
and strain rate, effect on flow stress **A14:** 151
and strain rate, uranium alloys **A2:** 678
stress, defined **EL1:** 1159
in stress-corrosion cracking **A13:** 147
stresses, in laminates **EM1:** 228-229
for superconductivity **A2:** 1030, 1033-1034
surface, in thermal inspection **A17:** 396
tempering **A14:** 55, 198, 682
tensile effect, polysulfones (PSU) **EM2:** 200
and tensile properties, CG/SG irons **A15:** 674
thermal management, by design **EL1:** 45-55
for thermal shock testing **EL1:** 500
-time development, radiographic film **A17:** 351
and time of cure, viscosity effects **EM1:** 649
-time relation, wave soldering **EL1:** 685
for titanium alloy working **A14:** 268-270, 278
for tool steel forgings **A14:** 81
tooling **A15:** 324
tooling, precision forming **A14:** 162
and toughness **A14:** 162
and tracer-gas concentrations **A17:** 67
transition, defined **A13:** 13
transition, refractory metals **A14:** 786
uranium, effect on mechanical properties **A2:** 671

Temperature(s) (continued)
vs deformation resistance, carbon/alloy steels **A14:** 216
vs ductility, heat-resistant alloys **A14:** 234
vs dynamic hot hardness, as forgeability **A14:** 226
vs formability, drop hammer forming **A14:** 657
vs hardness, lead frame materials **EL1:** 492
vs mechanical properties **EM2:** 752
vs modulus, polyvinyl chlorides (PVC) **EM2:** 210
vs modulus, reinforcement effects **EM2:** 281-282
vs pressure, heat-resistant alloys **A14:** 233
vs pressure, steels **A14:** 216
vs strain/CTE, copper-Invar-copper **EL1:** 622
vs volume resistivity, allyls (DAP, DAIP) **EM2:** 227
vs. tensile strength **EM1:** 362
warm forging **A14:** 168, 171-172
work metal, variables affecting **A14:** 36
and workability, alloy systems **A14:** 367
workpiece, as forging factor **A14:** 231
yield, fusible alloys **A2:** 756

Temperature-composition phase diagrams .... **A3:** 1 • 2
Temperature-humidity-bias (THB) test **EM3:** 434
Temperature-induced phase transformations
XRPD analysis **A10:** 333
Temperature-sensitive failure mechanisms **EL1:** 959-961
Temperature-time curves
for thermal analysis **A15:** 182-185
Tempered borosilicate glass
coefficient of thermal expansion **EM4:** 1103
composition **EM4:** 1103
Tempered glass *See* Annealed and tempered glass
Tempered layer
defined **A9:** 18
Tempered martensite
defined **A9:** 18
Tempered martensite embrittlement **A1:** 703-706
activating mechanism **A1:** 704-706
defined **A13:** 13
Tempered martensitic embrittlement **A12:** 135-136
Tempered soda-lime glass
coefficient of thermal expansion **EM4:** 1103
composition **EM4:** 1103
Tempered steels
microhardness **A7:** 489
shielded metal arc welding **A6:** 176
Tempered zone in ferrous alloy welded joints **A9:** 581
Tempered-martensite embrittlement
See also Embrittlement, 500 °F embrittlement **M1:** 685
of steels **A11:** 99
Tempering *See also* Heat treatment; Self-tempering; Temper **M1:** 182
after quenching **A7:** 453
alloy steel sheet and strip **M1:** 165
aluminum alloys, designations/practices **A15:** 758
of carbonitriding parts **A7:** 454
carburized and carbonitrided parts **M1:** 536, 539
cast steels **M1:** 379, 382, 383, 384, 385, 386-389
defined
for delayed cracking, alloy steels **A11:** 122
of dual-phase steels **A1:** 427-428
for ductile iron **A1:** 41-42 **A15:** 658-659
effect on machining marks **A11:** 89-90
effect on martensite in steel **A9:** 178
effects of alloys on **A1:** 393-394, 395, 396
effects of chromium on **A1:** 641
forced air convection-type furnaces for **A7:** 453
from grinding, macroetching to reveal **A9:** 176
gear materials **A18:** 261
grinding cracks prevented by **A11:** 567-568
hardenable steels **M1:** 463, 466-469
for hardened steels **A1:** 458-459, 462
heat checking of die surfaces **A18:** 631

Tempering (continued)
of high-chromium white irons **A15:** 684
of high-strength structural carbon steels **A1:** 389, 391
of hot-rolled steel bars **A1:** 241
for hydrogen-damage control **A11:** 251
of low-alloy steel sheet/strip **A1:** 209
malleable cast iron **M1:** 60-63, 68-71
of martensite **A1:** 134-136, 137 **M1:** 564
of martensitic stainless steels **A9:** 285
medium and temperatures **A7:** 453
nitrided parts, prior to nitriding **M1:** 540, 542
notch toughness of steels, effect on **M1:** 701-703, 706, 709
plain carbon steels **A15:** 713-714
of powder metallurgy high-speed tool steels **A1:** 783
response, powder forging **A14:** 202
scale, quench cracking from **A11:** 94-95
as secondary operation **A7:** 453
selective tempering **M1:** 527, 529
selectively hardened parts **M1:** 529
springs, steel **M1:** 287-288, 301
steam **A7:** 453
of steel plate **A1:** 231
steel wire **M1:** 262
temperature and carbon content, hardness effects **A14:** 198
temperature, effect on microstructural failure **A11:** 325
temperature, hardness effects **A14:** 55
temperature, heating below, for straightening **A14:** 682
temperature, property variations in P/M steels F-0005-T and FN-0205-T **M1:** 344
temperatures for carbonitrided parts **A7:** 454
temperatures of high-speed steels **A7:** 374
of tool and die steels **A14:** 55
and tumbling or deburring **A7:** 454
ultrahigh-strength steels **M1:** 423-427, 430, 431, 433, 435, 438, 441
Tempering artifacts in carbon steel **A9:** 38
Tempering, gray iron **M4:** 535-537
Tempering of steel *See also* Tempering of steel, equipment **A4:** 121-136 **M4:** 70-84
alloy content, effect of **A4:** 121, 124, 128-129 **M4:** 72-75, 77
autotempering **A4:** 121
blue brittleness **A4:** 135
bulk processing **A4:** 125, 133
carbon content, effect of **M4:** 72, 74-75, 76, 77
carbon content, effect on **A4:** 121, 123-124, 127
carburized components **A4:** 130 **M4:** 81
computer simulation **A4:** 645-646
cooling rate **A4:** 121, 122-123 **M4:** 71-72
cracking in processing **A4:** 134-135 **M4:** 83-84
cyclic heating and cooling **A4:** 125
dimensional changes **A4:** 121
embrittlement **A4:** 122-123, 124, 134 **M4:** 83-84
examples **A4:** 121, 123, 133-134
fixture use
flame tempering **A4:** 125, 126
frequency for various applications **A4:** 189
furnace atmosphere **A4:** 548
hardenability **A4:** 124, 127-132, 134, 135
hydrogen embrittlement **A4:** 136 **M4:** 84
induction heating energy requirements **A4:** 189
induction heating temperatures **A4:** 188
induction tempering **A4:** 125, 126, 130-132, 134 **M4:** 81-83
multiple tempering **A4:** 125, 133 **M4:** 82-83
nonmartensitic structures **A4:** 130 **M4:** 81, 82, 83
power density **A4:** 131, 132
power source selection **A4:** 189
procedures **A4:** 124-125 **M4:** 75-76
process control **A4:** 129, 131, 132 **M4:** 79, 80-81
quench tempering **A4:** 121
residual elements, effect on **A4:** 121, 124
safety precautions **A4:** 128, 129
secondary hardening **A4:** 121, 124
selective tempering **A4:** 125, 129, 132-133 **M4:** 81-82, 83
snap draw treatment **A4:** 134
stages of tempering **A4:** 121
steam atmosphere **A4:** 562
temper embrittlement **A4:** 135

Tempering of steel (continued)
temperature **A4:** 121-122, 124, 125-126, 127, 129, 130 **M4:** 70-71, 72-73, 77
tempered martensite embrittlement (TME) **A4:** 122, 135-136
time **A4:** 121, 122, 124, 127 **M4:** 71, 73
toughness **A4:** 122-123
ultrahigh-strength steels **A4:** 209-216
Tempering of steel, equipment **A4:** 125-129, 130
convection furnaces **A4:** 125, 126 **M4:** 76-78
molten metal baths **A4:** 125, 126, 128-129 **M4:** 80
oil baths **A4:** 125, 126, 128 **M4:** 79-80
salt bath compositions **A4:** 127
salt bath furnaces **A4:** 125, 127, 128
salt-bath compositions **M4:** 78-79
salt-bath furnaces **M4:** 78
selection **A4:** 129 **M4:** 80
temperature control **A4:** 126-127, 128, 129 **M4:** 80
vapor pressure, relation to temperature **A4:** 495
Tempering of tool steels
effects on microstructure **A9:** 259
Tempering times **A1:** 80
Tempers
galvanized wire **M1:** 263
steel sheet, bend limitations **M1:** 555
steel strip, mechanical properties for various **M1:** 155
Tempers of aluminum alloys **A9:** 358
Template **EM1:** 23, 144
defined **A14:** 13 **EM2:** 42
Template matching
in machine vision process **A17:** 36
Templates
for part shape measurement **A8:** 549
Templet *See* Template
Temporary coatings
stripping methods **M5:** 19
Temporary solder masks
application **EL1:** 559
Temporary viscosity loss **A18:** 84, 109-110
Temporary weld
definition **M6:** 17
Tenacity
defined **EM1:** 23 **EM2:** 42
Tennis racquets **EM1:** 31, 845
Tensile
-compressive systems **A8:** 576
data, reliability and utility of **A8:** 19
ductility, volume fraction second-phase particle effect on **A8:** 571-572
elongation **A8:** 36
fracture **A8:** 156, 571
hysteresis energy, for creep fatigue interaction analysis **A8:** 358
load, and compliance **A8:** 383-386
normal stress, in delamination **A8:** 714
plastic flow stress **A8:** 452
residual surface stresses, and fatigue strength **A8:** 374
strain **A8:** 353, 355
stress systems **A8:** 576
stress-relaxation curve **A8:** 324-325
and torsion fracture strains correlated **A8:** 168-169
waves, flyer plate impact test **A8:** 211
Tensile and hardness tests **A1:** 242
Tensile compliance, defined *See* under Compliance
Tensile creep *See also* Creep
high-impact polystyrenes (PS, HIPS) **EM2:** 197
of polyarylates (PAR) **EM2:** 139
polyether sulfones (PES, PESV) **EM2:** 160
polyether-imides (PEI) **EM2:** 157
thermoplastics **EM2:** 113
Tensile ductility
effect of second-phase particles **A14:** 364
loss by hydrogen damage **A13:** 164
Tensile ductility, loss of
and hydrogen-stress cracking **A1:** 712-717
Tensile elastic modulus
glass fiber reinforced epoxy resin **EM1:** 405
high-temperature thermoset matrix composites **EM1:** 375
Tensile elongation
carbon fiber/fabric reinforced epoxy resin **EM1:** 405, 410, 412

Tensile elongation (continued)
medium-temperature thermoset matrix composites **EM1:** 383, 386
percent, epoxy resin system composites **EM1:** 401
thermoplastic matrix composites **EM1:** 365, 368, 371
Tensile failure/fracture
continuous fiber composites **EM1:** 787-789, 791-792
discontinuous fiber composites **EM1:** 795-796
fiber mode **EM1:** 200, 202
matrix mode **EM1:** 200
Tensile failures
in magnesium alloys **A9:** 426
Tensile fatigue
pure **A12:** 342
Tensile fracture
as fatigue fracture plane-strain mode **A11:** 105
flat-face **A11:** 76
from shrinkage porosity, cast low-alloy steel connector **A11:** 389-390
in graphite-epoxy lay-ups **A11:** 734
shear-face **A11:** 76
Tensile fracture modes
and temperature/strain rate **A14:** 363
Tensile fracture surfaces of beryllium-copper alloys **A9:** 393
Tensile fractures *See also* Tensile-test fracture
cup-and-cone, studies **A12:** 3
shear, elongated dimple formation **A12:** 13
shear, martensitic stainless steels **A12:** 368-369
Tensile fractures of austenitic manganese steel castings **A9:** 237
Tensile impact
and yield point **EM2:** 556
Tensile linear strain *See* Linear strain
Tensile load
deformation bands as a result of **A9:** 684
vs. elongation curves for yielding **A9:** 684
Tensile loading
pure **A11:** 746
Tensile modulus *See also* Modulus of elasticity; Young's modulus **EM3:** 28
carbon-epoxy laminates **EM1:** 147
of engineering thermoplastics **EM2:** 98
fiber reinforcements, compared **EM1:** 113
glass addition effect **EM2:** 72
glass fabric reinforced epoxy resin **EM1:** 404
glass fibers **EM1:** 107
para-aramid fibers **EM1:** 54
polyester resins **EM1:** 91, 92
RTM materials, E-glass reinforcement **EM1:** 567
sheet molding compounds **EM1:** 158
thermoplastics **EM1:** 544
vs. bulk resistivity, graphite fiber **EM1:** 113
vs. temperature, Kevlar aramid fiber **EM1:** 362
Tensile opening mode
determining **EM1:** 264
Tensile Poisson's ratio *See also* Poisson's ratio
carbon fiber/fabric reinforced epoxy resin **EM1:** 410
glass fiber reinforced epoxy resin **EM1:** 406
graphite fiber reinforced epoxy resin **EM1:** 413
Kevlar 49 fiber/fabric reinforced epoxy resin **EM1:** 407
low-temperature thermoset matrix composites **EM1:** 395
medium-temperature thermoset matrix composites **EM1:** 386
Tensile properties *See also* Mechanical properties; specific material type; Tensile strength
aluminum casting alloys **A2:** 143-144, 152-177
aluminum-silicon castings **A15:** 160
cast copper alloys **A2:** 356-391
CG iron **A15:** 672, 673
correlation with simulative tests, sheet metals **A8:** 565
corrosion-resistant cast irons **M1:** 89
density effects **A14:** 202-203
design stresses, ductile iron **A15:** 662
ductile iron **M1:** 35-38, 41, 47, 52
effect of density in P/M copper **A7:** 736
elevated temperature, alloy cast irons **M1:** 94
fatigue strength related to **M1:** 665
heat-resistant cast irons **M1:** 92-94
high-impact polystyrenes (PS, HIPS) **EM2:** 196

Tensile properties (continued)
lead alloys ... **A2:** 550
low-expansion alloys ... **A2:** 892
malleable iron ... **M1:** 64-66, 68-71
measurement ... **EM2:** 433-434
mechanically alloyed oxide alloys ... **A2:** 946
dispersion-strengthened (MA ODS)
palladium ... **A2:** 715
powder forgings ... **A14:** 202-203
prealloyed titanium P/M compacts ... **A2:** 651-652
pure tin ... **A2:** 519
of sheet metals ... **A8:** 555
stainless steel powders ... **A7:** 729, 730
tension testing of ... **A8:** 19
titanium and titanium alloy castings ... **A2:** 637, 641
of titanium-based powders, HIP, rotating electrode processed ... **A7:** 41, 42
uranium alloys ... **A2:** 672, 677
urethane hybrids ... **EM2:** 269
vinyl esters ... **EM2:** 274
white cast irons ... **M1:** 86-87
wrought aluminum and aluminum alloys ... **A2:** 62-122
wrought titanium alloys ... **A2:** 621-623, 630
Tensile pull-through
for fastener/joint evaluation ... **EM1:** 710
Tensile residual stresses
and fatigue resistance ... **A1:** 681
Tensile residual surface stress
effect on fatigue strength ... **A11:** 112
Tensile shearing ... **A12:** 13, 368-369
Tensile strength *See also* Strength; Tensile properties; Ultimate strength; Ultimate tensile strength; Yield strength ... **EM3:** 51, 400-401
aligned short fibers ... **EM1:** 154, 155
aluminum and aluminum alloys ... **A2:** 8
in aluminum-killed steels ... **A10:** 231
aluminum-lithium alloys ... **A2:** 184
axial, analysis ... **EM1:** 192-194
boron fiber ... **EM1:** 58
composites vs. metals ... **EM1:** 259-260
copper and copper alloys ... **A2:** 219
copper casting alloys ... **A2:** 348-350
defined ... **A8:** 14, 20 **A11:** 10 **A13:** 13 **A14:** 13 **EM1:** 23 **EM2:** 42
and deformation ... **A8:** 168-169
ductile iron ... **A15:** 654, 660, 662
effect, fiber orientation ... **EM1:** 120
effect, hydrogen embrittlement, nickel-base alloys ... **A13:** 651-652
as engineering stress-strain parameter ... **A8:** 20-21
of engineering thermoplastics ... **EM2:** 98, 618
epoxy resin matrices ... **EM1:** 73
ferrite effect, high-alloy steels ... **A15:** 725
fiber ... **EM1:** 113, 193-194
fiber, measurement techniques ... **EM1:** 286
fiber reinforcements, compared ... **EM1:** 113
of filament, test method effect ... **EM1:** 286
of flexible epoxies ... **EL1:** 821
glass addition effect ... **EM2:** 72
of glass fibers ... **EM1:** 46, 107, 192, 362, 566
graphite-epoxy laminates ... **EM1:** 233
gray iron ... **A15:** 643
heat-treated copper casting alloys ... **A2:** 355
hot/wet conditions, p-aramid fibers ... **EM1:** 55
and hydrogen porosity ... **A15:** 747
Kevlar aramid fiber ... **EM1:** 55
laminate, test method effect ... **EM1:** 286
of laminates ... **EM1:** 286, 311
loss, as function of time ... **A13:** 911
loss, by atmospheric corrosion ... **A13:** 600-602, 775
low-temperature thermoset matrix composites ... **EM1:** 394
magnesium alloys ... **A2:** 460
and material selection ... **EM1:** 38
measurement of ... **EM3:** 189
and modified Goodman law ... **A8:** 374
notch, high-strength steel ... **A13:** 169
Tensile strength (continued)
and notch sensitivity ... **A8:** 372
p-aramid fibers ... **EM1:** 54-55
palladium wire, temperature effects ... **A13:** 804
with PAN precursor ... **EM1:** 112
pearlitic malleable iron ... **A15:** 695
plane-strain fracture toughness varies with ... **A11:** 54
platinum wire, temperature effect ... **A13:** 800
polyaryl sulfones (PAS) ... **EM2:** 146
polyaryletherketones (PAEK, PEK PEEK, PEKK) ... **EM2:** 143
polybenzimidazoles (PBI) ... **EM2:** 148
polyester resins ... **EM1:** 91, 92
polyether-imides (PEI) ... **EM2:** 157
polysulfones (PSU) ... **EM2:** 200
RTM materials, E-glass reinforcement ... **EM1:** 566
of sheet metals ... **A8:** 555
sheet molding compounds ... **EM1:** 158
SiC fibers ... **EM1:** 59
single-filament, glass ... **EM1:** 192
sintering temperature effects ... **A13:** 825
in spinning ... **A14:** 604
of steel wire strands, effect of heat treatment ... **A11:** 444
tape prepregs ... **EM1:** 143, 144
tests ... **EM1:** 286, 298
thermoplastic polyurethanes (TPUR) ... **EM2:** 206
thermoplastics ... **EM1:** 544
tow, test method effect ... **EM1:** 286
ultimate ... **EM3:** 29
in uniaxial tensile testing ... **A8:** 555
vs. material form, multidirectional tape prepreg ... **EM1:** 146
vs. temperature, glass fiber ... **EM1:** 362
welded gray iron ... **A15:** 527
wrought aluminum alloy ... **A2:** 57, 58
wrought magnesium alloys ... **A2:** 483, 485
zinc alloys ... **A2:** 532
Tensile strength, ultimate *See* Ultimate tensile strength
Tensile strength(s) *See also* Strength; Ultimate tensile strength
of aluminum alloys ... **A7:** 474
of aluminum oxide-containing cermets ... **A7:** 804
of consolidated superalloys ... **A7:** 430
of copper-based P/M materials ... **A7:** 470
effect of sintering temperature ... **A7:** 369
effect of sintering time ... **A7:** 370
electrolytic copper powder ... **A7:** 115, 116
of ferrous P/M materials ... **A7:** 465, 466
of injection molded materials ... **A7:** 471, 499
of metal borides and boride-based cermets ... **A7:** 812
of molybdenum and molybdenum alloys ... **A7:** 476
of nickel-based, cobalt-based alloys ... **A7:** 472
of P/M and wrought titanium and alloys ... **A7:** 475
of P/M forged low-alloy steel powders ... **A7:** 470
of P/M stainless steels ... **A7:** 468
of porous uninfiltrated and silver-infiltrated ... **A7:** 562
ratio in composite/matrix ... **A7:** 612
of rhenium and rhenium-containing alloys ... **A7:** 477
of superalloys ... **A7:** 430, 473
of tantalum wire ... **A7:** 478
testing, and mechanical testing ... **A7:** 489-491
of titanium carbide-based cermets ... **A7:** 808
titanium P/M parts ... **A7:** 752
Tensile stress *See also* Compressive stress ... **A13:** 13, 145, 929-931, 941 **EM3:** 29
criterion, as fracture model ... **A14:** 395-396
defined ... **A8:** 14 **A11:** 10 **A14:** 13 **EM1:** 23 **EM2:** 42-43
in fatigue testing ... **A11:** 102
high, static, in threaded fasteners ... **A11:** 537
Tensile stress (continued)
initial, effect on time-to-fracture by SCC ... **A11:** 220
negative ... **A11:** 102
pulse, for dynamic notched round bar testing ... **A8:** 276-277
sustained, for stress-corrosion cracking ... **A8:** 495
systems ... **A8:** 576
systems, as controlling workability ... **A14:** 369
Tensile stress-strain curve
as design requirement ... **EM2:** 407
Tensile tearing
elongated dimples by ... **A11:** 76
Tensile test *See also* Tension testing
of bars prestrained in torsion ... **A8:** 155
for hot-rolled steel bars ... **A1:** 242, 243
resistance spot welds ... **M6:** 487-488
schematic ... **A8:** 208
validity, and wave propagation effects ... **A8:** 209
Tensile test bar
machined and unmachined ... **A7:** 490
Tensile test, strand *See* Strand tensile test
Tensile testing *See also* Tension testing ... **EM3:** 315, 387-388
of ductile-to-brittle fracture transition, in ferritic steel ... **A11:** 66
limitations of ... **A11:** 18-19
plane-strain ... **A14:** 887
uniaxial ... **A14:** 883-887
uniaxial, superplastic metals ... **A14:** 861-862
Tensile testing used for fracturing ... **A9:** 23
Tensile tests
modulus-directed ... **EM2:** 546-547
strength-directed ... **EM2:** 547
toughness-directed ... **EM2:** 547-548
wrought martensitic stainless steels ... **A6:** 441
Tensile ultimate strength
computation of derived ... **A8:** 667
direct computation of design allowables for ... **A8:** 663
examples of computational procedures ... **A8:** 672-676
symbols and unit ... **A8:** 662
Tensile yield strength
and compressive yield strength ... **A14:** 20
from pin bearing testing ... **A8:** 61
histogram, with probability density functions ... **A8:** 628
Tensile-compressive stress states
as controlling workability ... **A14:** 368-369
Tensile-impact failures
in magnesium alloys ... **A9:** 426
Tensile-tensile fiber failure mode ... **EM1:** 202
Tensile-test fracture *See also* Tensile fracture
carbon steel casting ... **A12:** 105
ductile and brittle ... **A12:** 102
interpretation of ... **A12:** 98-105
iron ingot ... **A12:** 219
low-carbon, high oxygen iron ... **A12:** 223
maraging steels ... **A12:** 384
titanium alloys ... **A12:** 454
types, fcc metals ... **A12:** 100
Tension *See also* Stress; Surface tension; Tension testing ... **A12:** 3, 13 **A13:** 13, 246, 250-251
band fixation ... **A11:** 676, 681
-compression hold period ... **A8:** 353
-creep curve, with creep stages ... **A8:** 331
defined ... **A8:** 14 **A11:** 10 **A14:** 13
effect, magabsorption measurement ... **A17:** 154
effect on resistance, domain interpretation ... **A17:** 145
effective fracture strain ... **A8:** 168
effective stress-strain curves for stainless steel ... **A8:** 162, 164
failures, in fasteners ... **A11:** 531
far-field ... **A8:** 442
in forced-displacement system ... **A8:** 392
forced-vibration system ... **A8:** 392

Tension (continued)
-hold-only test **A8:** 347-352
horizontal, deformation under **A14:** 38
as in-plane failure mode **EM1:** 781, 782-783
instability in **A8:** 25
interlaminar **EM1:** 787-789
load cell, in electrohydraulic testing machine **A8:** 160
-loaded elements **EM1:** 321-323
loaded subcomponents **EM1:** 331
-loading techniques, for split Hopkinson bar in tension test **A8:** 212-213
longitudinal, and damping **EM1:** 207-208
and magnetic field, resistivity effects **A17:** 144
measurement, by Barkhausen noise **A17:** 160
mechanical behavior of materials under **A8:** 19, 20-27
mode I fractures in composites **A11:** 735-736
monotonic, failure in **A11:** 20
repeated, test results **A8:** 353-354
resonance system **A8:** 392
rotational bending system **A8:** 392
servomechanical system **A8:** 392
simple, in shafts **A11:** 460-461
as stress, on shafts **A11:** 460
stress-relaxation curve in **A8:** 323-324
stress-relaxation test in **A8:** 323
-tension failure cycles **A8:** 715-716
test, for workability **A14:** 37
testing, aluminum alloys **A14:** 79
and torsion flow curves at room temperature for copper **A8:** 162-164
translaminar **EM1:** 791-792
transverse, ply **EM1:** 237-238
types, in fatigue fracture **A11:** 75
uniaxial, ply **EM1:** 237
universal testing machines for **A8:** 612
Tension effective fracture strain **A8:** 168
Tension knuckle-drive forging **A14:** 173-17
Tension lap-shear tests **EM3:** 646-649
Tension leveling
of carbon steel sheet **A1:** 206
steel sheet **M1:** 160
Tension overload fractures
AISI/SAE alloy steels **A12:** 304, 311-315, 334
copper alloys **A12:** 402
equiaxed dimples, precipitation-hardening stainless steels **A12:** 372
free-machining copper **A12:** 401
iron-base alloy **A12:** 460
martensitic stainless steels **A12:** 367
precipitation-hardening stainless steels **A12:** 370, 372
titanium alloys **A12:** 449
tool steels **A12:** 379
wrought aluminum alloys **A12:** 423, 425, 437
Tension specimens
method to decrease SCC susceptibility **A8:** 508
ring-stressed **A8:** 507
SCC testing **A8:** 506-508
transverse, effective fracture strain **A8:** 168
Tension, surface
SI derived unit and symbol for **A10:** 685
Tension test **A1:** 581-582
for gray iron **A1:** 16
for hot-rolled steel bars **A1:** 242, 243
Tension testing *See also* Tension; Testing Test(s)
advantages **A8:** 19, 34
for bulk workability assessment **A8:** 577-578
correlation of cold rolling reduction with reduction in area **A8:** 595-596
defined **A8:** 14, 19, 20 **A11:** 10
ductility measurement in **A8:** 26-27
elevated temperature **A8:** 34-37
equipment **A8:** 19
explosion-bulge test, drop-weight test and compared **A11:** 58
fracture regimes determined by **A8:** 154
for full-size parts **A8:** 19
fully reversed, ultrasonics **A8:** 248
high strain rate **A8:** 208-214
high-temperature hydrodynamic **A11:** 281
Hopkinson bar **A8:** 198
hot **A8:** 586-587
instrumentation **A8:** 47-51
interpretation and limitations **A8:** 19
low temperature **A8:** 34-37
Tension testing (continued)
machine **A8:** 45, 47-51, 472
mechanical behavior of materials under **A8:** 20-27
necking in **A8:** 578
rationalizing strain distribution in **A8:** 26
specimen, and upset test specimen orientation **A8:** 581
speed of **A8:** 39, 47
speed of deformation in **A8:** 38
step-down **A8:** 324
tensile strength from **A8:** 20
and torsion testing **A8:** 163, 165
and upset test fracture strains, compared **A8:** 581
Tension testing machine **A8:** 45, 47-51, 472
Tension-hold-only test **A8:** 347-352
Tension-shear testing
explosion welds **M6:** 711
Tent lighting
for reflective parts **A12:** 88
Tenth-scale vessel
defined **EM1:** 23
Tenting
defined **EL1:** 1159
Teratogenicity
of arsenic **A2:** 1238
Terbium
pure **M2:** 806-807
TNAA detection limits for **A10:** 238
Terbium, as rare earth metal
properties **A2:** 720, 1187
Terbium basic carbonate
sol-gel processing **EM4:** 447
Terbium in garnets **A9:** 538
Terephthalic polyester resin *See also* Unsaturated polyesters
properties **EM2:** 246
Terfenol
as rare earth magnetic application **A2:** 730
Terminal creep
defined **A12:** 19
Terminal erosion rate **A18:** 228
defined **A18:** 19
Terminal period
defined **A18:** 19
Terminal phases .... **A3:** 1 • 3, 18
Terminal pins, and solder
elemental mapping of **A10:** 532
Terminal solid solution
defined **A9:** 18
Terminal(s)
desoldering **EL1:** 721
number, in IC packages **EL1:** 251
three-, discrete semiconductor **EL1:** 422-429
Terminals **A14:** 142, 82
Terminals (electrical)
powder used **A7:** 573
Termination(s)
density, comparisons **EL1:** 730
design **EL1:** 521-522
failures, ceramic capacitors **EL1:** 994
failures, passive devices **EL1:** 994
lead, common methods **EL1:** 713
mechanical crimp **EL1:** 590-591
parallel and series **EL1:** 170-172, 522
pressure **EL1:** 591
resistance, and local physical performance **EL1:** 5
Terminations in eutectic microstructures **A9:** 620
Terminology *See also* Categorization; Classification; Glossary; Nomenclature; Notation
of composite materials **EM1:** 176-177
electrical **EM2:** 461-462, 590-593
Terms *See also* Categorization; Classification; Nomenclature; Terminology
and definitions **A8:** 1-15
experimental design **A8:** 640
glossary of **A15:** 1-12 **EL1:** 1133-1162 **EM2:** 2-47
for liquid metal processing **A15:** 49
Terms related to phase diagrams .... **A3:** 1 • 2
Ternary
alloys, tantalum, corrosion resistance **A13:** 738
phase diagram, defined **A13:** 46
Ternary alloy (copper-tin) plating **M5:** 288
Ternary alloys
implantation in **A10:** 486
Ternary iron-base alloys *See also* Iron-base alloys
carbon solubility **A15:** 66-68
equilibrium temperatures **A15:** 65-67
Fe-C-Mn system, thermodynamics **A15:** 65
Fe-C-P system, thermodynamics **A15:** 65
Fe-C-Si system, thermodynamics **A15:** 64-65
interaction coefficients for alloying elements **A15:** 62
multicomponent **A15:** 68-69
solubility factors **A15:** 66
thermodynamics of **A15:** 62-64
third element effects **A15:** 67-68
Ternary iron-cobalt-chromium alloys
FIM/AP analysis **A10:** 598-599
Ternary molybdenum chalcogenides (chevrel phases) *See also* Superconducting materials
applications, potential **A2:** 1079-1080
cold processing (niobium/tantalum sheaths) **A2:** 1079
development **A2:** 1077
fabrication technology **A2:** 1077-1079
hot processing (molybdenum sheath) **A2:** 1077-1079
properties **A2:** 1077
structure and bonding **A2:** 1078
wire filaments, superconducting properties **A2:** 1079
Ternary mullite
maximum use temperature **EM4:** 875
Ternary system
defined **A9:** 18
Ternary system or diagram .... **A3:** 1 • 2, 4-5
Ternary-alloy phase diagrams .... **A3:** 3 • 5-58
Terne
defined **A13:** 13
Terne coating *See also* Tin-lead plating **M5:** 358-360
air-knife system **M5:** 359-360
cleaning processes **M5:** 358-359
continuous system **M5:** 359-360
long **M5:** 358-359
semicontinuous system **M5:** 359
short **M5:** 358
specifications **M5:** 358-359
spot tests, single and triple **M5:** 359
steel **M5:** 358-360
formed **M5:** 358
sheet **M5:** 358-359
strip **M5:** 359
thickness **M5:** 358-359
Terne coatings **A1:** 221-222 **M1:** 171, 173-174
applications **M1:** 173-174
coating weights **M1:** 174
corrosion resistance **M1:** 174
designations **M1:** 174
formability **M1:** 173-174
handling and storage **M1:** 174
lead and lead alloys **A2:** 554-555
mechanical properties **M1:** 171, 174
painting of **M1:** 174
specifications for **M1:** 174
toxicity, precautions when welding **M1:** 174
weldability **M1:** 174
Terne, long and short
defined **A2:** 555
Terne metal **M2:** 497-498
micrograph **A9:** 422
Terne-coated steels
press forming of **A14:** 562-563
Terne-metal-coated steels
resistance spot welding **M6:** 479-480, 491
Terne-metal-plated steel
coating for resistance seam welding **M6:** 502
resistance spot welding **M6:** 480
Terneplate **A1:** 221
relative weldability ratings, resistance spot welding **A6:** 834
resistance seam welding **A6:** 244
Terpenes
as cleaning solvents **EL1:** 663
Terpineol (80%) (terpine alcohol)/20% styrene resin
as vehicle for solder glass powder **EM4:** 1070
Terpolymer *See also* Polymers **EM1:** 23, 78 **EM3:** 29
defined **EM2:** 42

Terrace cracks
by ductile shearing ... **A12:** 139
"Tertiarium" mixtures ... **A6:** 126
Tertiary amide polymer
application or function optimizing powder treatment and green forming ... **EM4:** 49
Tertiary amines ... **EM3:** 96, 99, 100
for accelerating epoxy curing reactions ... **EM3:** 95
for core curing ... **A15:** 240
Tertiary creep
defined ... **A8:** 308 **A12:** 19
in elevated-temperature failures ... **A11:** 263-264
onset ... **A8:** 693
time to ... **A8:** 690
Tesla
abbreviation for ... **A10:** 691
charge ... **A10:** 32
as SI derived unit, symbol for ... **A10:** 685
Test *See also* Testing
bending strength ... **A8:** 132-136
chamber, fatigue ... **A8:** 412-413
dry-sand rubber wheel ... **A8:** 605
environment, effect on fatigue and creep-fatigue results ... **A8:** 354
equipment, for fracture toughness testing ... **A8:** 19, 470
flat plate impact ... **A8:** 210
interval ... **A8:** 374
light, in spring-material test apparatus ... **A8:** 134-135
limitations, split Hopkinson pressure bar tests ... **A8:** 202-203
matrix, for creep testing ... **A8:** 686
nonstandard torsional ... **A8:** 147
selection of quasi-static torsional ... **A8:** 145
stand ... **A8:** 311-313
standard methods for torsional testing ... **A8:** 147
stress, SCC testing ... **A8:** 503
surface, Brinell test ... **A8:** 85, 88
temperature ... **A8:** 586-587
torsional creep ... **A8:** 147
validity ... **A8:** 190-191
Test and maintenance (TM) ... **EL1:** 376
Test bar properties
in gray iron ... **A1:** 16-19
Test bars ... **A7:** 489-491
Test block ... **A8:** 374
Brinell standardized ... **A8:** 85
verification by ... **A8:** 88
Test blocks *See also* Reference standards; Standard reference blocks
acoustical holography ... **A17:** 444-445
with artificial flaws ... **A17:** 262
with natural flaws ... **A17:** 262
reference ... **A17:** 262-265
for ultrasonic inspection standards ... **A17:** 261-265
Test board
defined ... **EL1:** 1159
Test chips, as design tools ... **EL1:** 419
Test coupon *See also* Coupon
defined ... **EL1:** 1159
printed board, metallographic evaluation ... **EL1:** 572-577
Test coupons
closed-die steel forgings ... **M1:** 351-352
Test equipment
fixture for pin bearing testing ... **A8:** 60-61
Test exposures
fractographic ... **A12:** 86-87
Test frequencies *See also* Frequencies
and flaw size, eddy current inspection ... **A17:** 194
Test frequency
magnetically soft materials ... **A2:** 763
Test geometry evaluation ... **EM3:** 441-454
impact performance of bonded joints ... **EM3:** 447-449
in situ testing of bonded joints ... **EM3:** 442-444, 445
Test geometry evaluation (continued)
mixed-mode bond line fracture characterization ... **EM3:** 444-447
tools for assessing specimen performance ... **EM3:** 449-454
Test grids used in quantitative metallography ... **A9:** 124-125
Test lines used in quantitative metallography ... **A9:** 124
Test loop
gas/oil production monitoring ... **A13:** 1251
Test methods *See also* Analytical methods; Life tests
for germanium and germanium compounds ... **A2:** 736
magnetic, magnetically soft materials ... **A2:** 763-778
mechanical, aluminum casting alloys ... **A2:** 147-148
Test module
zero risk stress testing ... **EL1:** 136, 140
Test Monitoring Center (TMC)
performance testing of engine oils ... **A18:** 170
Test object *See also* Object(s); Specimen(s); Workpieces
thermal active, thermal inspection ... **A17:** 397
Test phantom *See* Phantom
Test program design *See also* Statistical analysis; Test(s); Testing
data analysis ... **EM2:** 602-606
data analysis software ... **EM2:** 607-608
economic factors ... **EM2:** 606-607
experimental ... **EM2:** 599-602
selection ... **EM2:** 599
terminology ... **EM2:** 599-600
for variability sources ... **EM2:** 606
Test solution
for acidified chloride environments ... **A8:** 419
Test specimen *See also* Specimen
fatigue crack propagation ... **A8:** 379-382, 678
for pin bearing testing ... **A8:** 59-60
for ultrasonic fatigue testing ... **A8:** 248-252
Test specimen(s) *See also* Fracture specimens; Specimens
fracture sources illustrated ... **A12:** 217
preparation/preservation ... **A12:** 72-77
Test structures
quantitative stress measurement using ... **EL1:** 444-446
Test(s) *See also* Mechanical testing; Nondestructive evaluation (NDE); Nondestructive testing (NDT); Statistical analysis; Test program design; Testing; Testing and characterization; Ultrasonic nondestructive analysis
accelerated, for corrosion ... **A11:** 174
for cannon tubes ... **A11:** 281
carousel-type thermal fatigue ... **A11:** 278-279
cascade ... **A11:** 280-281
Charpy V-notch impact ... **A11:** 2, 57-60, 67-69, 76, 84-88, 705-706, 796
correlations and comparisons between ... **A11:** 60
crack-opening displacement ... **A11:** 62
and data, fracture toughness ... **A11:** 53-55
differential scanning calorimetry (DSC) ... **EM2:** 523-525, 540
drop-weight (DWT) ... **A11:** 57-58
dynamic mechanical analysis (DMA) ... **EM2:** 526-527
dynamic tear (DT) ... **A11:** 61-62
electrical properties ... **EM2:** 461-462, 475-478, 581-587
electrochemical ... **A11:** 174
for elevated-temperature failures ... **A11:** 278-281
for environmental attack, polymers ... **EM2:** 425-426
Esso (Feely) ... **A11:** 60
explosion-bulge ... **A11:** 58-59
fatigue loading ... **EM2:** 703-706
ferris-wheel disk ... **A11:** 279
flexed-beam impact, and notch sensitivity ... **EM2:** 556-557
flexed-plate impact ... **EM2:** 557
fluidized-bed thermal fatigue ... **A11:** 278
Test(s) (continued)
for gas-turbine components ... **A11:** 278-281
high-temperature hydrodynamic tension ... **A11:** 281
impact ... **EM2:** 554-556, 687-688
instrumented-impact ... **A11:** 64
Izod ... **A11:** 57
J testing ... **A11:** 62-64
laboratory cracking ... **A11:** 250
Lehigh bend ... **A11:** 59
mechanical ... **A11:** 18-19
methods, for magnesium alloys ... **A11:** 223
Navy tear ... **A11:** 60
notched slow-bend ... **A11:** 59
oxidation-corrosion ... **A11:** 279-280
peel ... **A11:** 43
plane-strain fracture-toughness ... **A11:** 60-61
preservice, of pipeline ... **A11:** 697-699
for protective coatings ... **EM2:** 425
R-curve analysis ... **A11:** 64
for radiation degradation ... **EM2:** 576-580
ratio-analysis diagram (RAD) ... **A11:** 62
resistance to hot corrosion by sulfida- tion- oxidation ... **A11:** 280
rheological analysis ... **EM2:** 527
Robertson ... **A11:** 59-60
Schnadt specimen ... **A11:** 57
simulated-bolthole ... **A11:** 279
simulated-service ... **A11:** 31, 174, 214
spectroscopic evaluation ... **EM2:** 426
spin ... **A11:** 279
spot, chemical analysis ... **A11:** 30-31
static fatigue ... **EM2:** 703-704
tape ... **A11:** 43
tensile, limitations of ... **A11:** 18-19
thermal-shock ... **A11:** 281
thermomechanical analysis (TMA) ... **EM2:** 525-526, 540-541
thermomechanical fatigue ... **A11:** 278
for weather aging ... **EM2:** 576-580
Test-bar properties
gray cast iron ... **M1:** 16-19
Test-bus standardization ... **EL1:** 376
Testability *See also* Testing
design for ... **EL1:** 374
of level 1 packages ... **EL1:** 403
Tester friction
effect on microhardness readings ... **A8:** 96
Testicular tumors
from zinc toxicity ... **A2:** 1255-1256
Testing *See also* Accelerated testing; Board-level physical test methods; Chemical analysis; Component-level physical test methods; Corrosion testing; Electrical performance testing; Environmental testing; Evaluation; Fiber properties analysis; Formability testing; In-process testing; In-service monitoring; Inspection; Instrumentation; Laminate properties analysis; Life cycle testing; Material properties; Material properties analysis Statistical analysis; Mechanical testing; Model(s); Modeling; Monitoring; Nondestructive evaluation; Performance; Physical test methods; Reliability; Reliability prediction; Reliability testing; Simulated service testing; Specific testing techniques; Test; Test program design; Test(s); Testing and characterization; Tests; Ultrasonic nondestructive analysis; Visual inspection; Workability tests ... **EM1:** 283-351 **EM4:** 547-548
accelerated ... **EL1:** 887-894 **EM2:** 789-790
ad hoc optical ... **EM2:** 598
of adhesives, specifications ... **EM1:** 696-699
Alnico alloys ... **A15:** 738-739
aluminum coatings ... **M1:** 172
aluminum stamping alloys ... **M2:** 180-182
ASTM test methods ... **EM2:** 334
at elevated temperatures ... **A8:** 202
atmospheric corrosion ... **M1:** 717-720
axial compression ... **A8:** 55-58

Testing (continued)
Brinell hardness ... **A8:** 84-89
bulk workability ... **A8:** 571-597
and burn-in tape-on-chip ... **EL1:** 281-282
cam plastometer ... **A8:** 195-196
cast steels ... **M1:** 378-381, 400-402
of casting defects ... **A15:** 544-561
chemical, reinforcement fibers ... **EM1:** 285
closed-loop system ... **A8:** 151
complete WSI modules ... **EL1:** 363
compression, by conventional load
constant tensile load ... **EM2:** 802
constant-load vs constant-strain ... **EM2:** 801
constant-strain ... **EM2:** 802
constructional steels for elevated temperature use ... **M1:** 639-643, 657-659
control, for cupolas ... **A15:** 390
corrosion, graphite-to-aluminum joints ... **EM1:** 717-718
cost drivers in ... **EM1:** 421
creep ... **A8:** 311-328
cyclic torsional ... **A8:** 149-153
cyclic-stress ... **A11:** 102
for damage tolerances ... **EM1:** 264
for design allowables ... **EM1:** 308-312
digital signal processing (DSP) ... **EL1:** 378
direction, primary ... **A8:** 667
documentation ... **EM1:** 300
of ductile iron ... **A15:** 652-654 **M1:** 37, 54-56
ductile iron pipe ... **M1:** 98
dynamic tests ... **EM2:** 418-419
electrical ... **EL1:** 372-378, 565-571, 946-972 **EM2:** 581-593
electrical properties ... **EM2:** 78, 461-462, 475-478
electrical resistivity ... **EM2:** 475
electromagnetic interference (EMI) shielding ... **EM2:** 476-478
electronic materials, special procedures ... **EL1:** 953-955
of elements and subcomponents ... **EM1:** 313-345
environmental ... **EL1:** 493-503
environmental attack on polymers ... **EM2:** 425-426
environmental exposure ... **EM1:** 295-301
for environmental stress crazing ... **EM2:** 801-803
environments, stress-corrosion cracking in ... **A11:** 211
of extrusions
fabric strength ... **EM1:** 732-733
fatigue crack propagation ... **A8:** 376-378
fatigue loading ... **EM2:** 703
fatigue, standards and practices for ... **A8:** 375
ferrous P/M materials ... **M1:** 327-330
of flexible epoxies ... **EL1:** 820
fluidity ... **A15:** 767-768
fracture toughness ... **EM2:** 739-740
frames ... **A8:** 192-193
full-scale ... **EM1:** 346-351
of gases, copper alloys ... **A15:** 465-466
of glass-to-metal seals ... **EL1:** 458-459
hardenability ... **M1:** 457, 471-479
hardness ... **A8:** 71-73
hermeticity ... **EL1:** 1062-1063
high strain rate ... **A8:** 215-239
high strain rate tension ... **A8:** 208-214
of honeycomb ... **EM1:** 724-726
hot rolled bars and shapes ... **M1:** 201-202
for hydrogen, aluminum alloys ... **A15:** 457-459
hypotheses ... **A8:** 626
information sources for ... **EM2:** 92-95
and instrumentation ... **EL1:** 365-380
introduction ... **EM1:** 283
of investment castings ... **A15:** 264
leadless packaging ... **EL1:** 988-989
for life cycle ... **EL1:** 135-140
of lubricants ... **A14:** 896-89
machine capability ... **A8:** 58
of matrix resin properties ... **EM1:** 289-294
mechanical ... **EM2:** 544-558
of mechanical properties ... **EM1:** 95-307, 731-735
medium strain rate compression ... **A8:** 192-193
mediums, accelerated, for SCC testing ... **A8:** 522-523
metallization integrity ... **EL1:** 953
methods, ASTM standard, plastics ... **EM2:** 90
methods, for hydrogen embrittlement ... **A8:** 537-541

Testing (continued)
methylene chloride, for chemical resistance ... **EL1:** 536
microhardness ... **A8:** 90-103
monocrystal casting ... **A15:** 322-323
multifilament yarns ... **EM1:** 732
natural environmental ... **EM2:** 578
nondestructive ... **A11:** 16-18
optical ... **EM2:** 594-598
for oxides, in aluminum ... **A15:** 749
physical test methods ... **EL1:** 365-372
pin bearing ... **A8:** 59-61
plain carbon steels ... **A15:** 702
plane-strain fracture toughness ... **EM2:** 739-740
pressure-shear plate impact ... **A8:** 230-238
printed board coupons ... **EL1:** 574
and prototyping, plastic parts ... **EM2:** 80-81
quantitative stress, using test structures ... **EL1:** 444-446
quasi-static torsional ... **A8:** 145-148
reclaimed sands ... **A15:** 355
of reinforcement fibers ... **EM1:** 285-288
reliability, ceramic multilayer packages ... **EL1:** 468
rigid printed wiring boards ... **EL1:** 542-543, 547
solder pastes ... **EL1:** 655-656
spalling (splitting) strength, of green sands ... **A15:** 345
standard mechanical, for strain rate regime compression testing ... **A8:** 190
static elimination ... **EM2:** 475-476
statistical analysis, of mechanical properties ... **EM1:** 302-307
steel forgings ... **M1:** 351-359
steel sheet for formability ... **M1:** 547-554
stress-relaxation ... **A8:** 311-328
structural. element and subcomponent ... **EM1:** 313-345
subcomponent ... **EM1:** 313, 329-334
substrates, ceramic packages ... **EL1:** 467
surface insulation resistance ... **EL1:** 667-668
ten-year worldwide ground-based exposure ... **EM1:** 823-825
tensile ... **EM2:** 546-548
tin coatings ... **M1:** 173
torsion, for workability ... **A8:** 154-184
torsional impact ... **A8:** 216-218
transient tests ... **EM2:** 417-418
uniaxial tensile creep ... **EM2:** 666-667
vs screening ... **EL1:** 875
wafer-scale integration ... **EL1:** 362-363
water absorption test, for chemical resistance ... **EL1:** 536-537
wear ... **A8:** 601-608
weight ... **A15:** 363, 545
wire bonds ... **EL1:** 226-228
zinc coatings ... **M1:** 168-169
Testing and analysis ... **EM3:** 313-314
Testing and characterization *See also* Testing; Tests; Ultrasonic-nondestructive analysis
chemical analysis, thermoplastic resins ... **EM2:** 533-543
chemical analysis, thermoset resins ... **EM2:** 517-532
chemical susceptibility ... **EM2:** 571-574
electrical testing ... **EM2:** 581-593
introduction ... **EM2:** 515-516
mechanical testing ... **EM2:** 544-558
optical testing ... **EM2:** 594-598
physical analysis, thermoplastic resins ... **EM2:** 533-543
physical analysis, thermoset resins ... **EM2:** 517-532
temperature resistance ... **EM2:** 559-567
test program design and statistical analysis ... **EM2:** 599-609
thermal analysis, thermoplastic resins ... **EM2:** 533-543
thermal analysis, thermoset resins ... **EM2:** 517-532
thermal degradation, and service temperatures ... **EM2:** 568-570
weather aging and radiation susceptibility ... **EM2:** 575-580
Testing direction
primary ... **A8:** 667
Testing equipment axial fatigue ... **A8:** 369
bending fatigue ... **A8:** 369
for Brinell hardness test ... **A8:** 86-88
calibration of ... **A8:** 611-619

Testing equipment axial fatigue (continued)
constant-load ... **A8:** 311
constant-stress ... **A8:** 311
creep ... **A8:** 311-313
maintenance ... **A8:** 88
multiaxial fatigue ... **A8:** 370
for Rockwell hardness testing ... **A8:** 77-78
for shear testing ... **A8:** 67-68
special-purpose fatigue ... **A8:** 369-370
stand ... **A8:** 311-313
stiffness, experimental values ... **A8:** 42-43
stress-relaxation ... **A8:** 311
stress-rupture ... **A8:** 311-312
torsional fatigue ... **A8:** 369
universal ... **A8:** 612-614
verification of ... **A8:** 611-612
Testing machine *See also* Testing equipment
defined ... **A8:** 14
Testing methods nondestructive ... **A7:** 491-492, 575-579
Testing of adhesive joints ... **EM3:** 37
Testing of wrought tool steels ... **A1:** 771-778
fabrication ... **A1:** 774-778
distortion and safety in hardening ... **A1:** 777
grindability ... **A1:** 775
hardenability ... **A1:** 775-777
machinability ... **A1:** 774-775
resistance to decarburization ... **A1:** 778
weldability ... **A1:** 775
performance in service ... **A1:** 771-774
Testpiece *See* Specimens; Workpiece(s)
Testpieces for examination
carbonitrided steels ... **A9:** 217
carburized steels ... **A9:** 217
nitrided steels, preparation of ... **A9:** 217-218
Tetrabromethane
separation medium for heavy mineral analysis of glass-quality sand ... **EM4:** 378
Tetrabromobisphenol A
in epoxy resin manufacture ... **EM1:** 66
Tetracalcium aluminoferrite
in composition of portland cement ... **EM4:** 11
Tetraethylorthosilicate (TEOS)
hydrolysis stages ... **EM4:** 209, 211
polymerization stages ... **EM4:** 210
sol-gel processing ... **EM4:** 445, 448, 449
sol-gel transition ... **EM4:** 209, 211
Tetrafluoroethylene oxide phenylquinoxaline elastomer (FEX) ... **EM3:** 677
Tetraglycidyl methylene dianiline (TGMDA) ... **EM2:** 240 241
Tetraglycidylmethylenedianiline (TGMDA)
epoxy resin ... **EM1:** 67, 69
Tetragonal
defined ... **A9:** 18
Tetragonal crystal system ... **A3:** 1•10, 15 **A9:** 706
Tetragonal crystals under polarized light ... **A9:** 77
Tetragonal forms
in uranium ... **A9:** 476
Tetragonal unit cells ... **A10:** 346-348
Tetragonal zirconia polycrystal (TZP) ... **EM4:** 756
joining oxide ceramics ... **EM4:** 512
properties ... **EM4:** 512, 756
Tetrahydrofuran ... **EM3:** 35
Tetrahydrofuran (C_4H_8O)
as solvent used in ceramics processing ... **EM4:** 117
Tetramethyl butane diamine (TMBDA)
catalyst ... **EM3:** 590
Tetramethylammonium hydroxide (TMAH) ... **EM4:** 57, 59
Tetramethylorthosilicate (TMOS)
sol-gel processing ... **EM4:** 449, 450
sol-gel transition ... **EM4:** 209
Tetrathiofulvalenium tetracyanoquinodimethane (TTF-TCNQ) ... **EM3:** 436
Tetronics plasma process
for EAF dust zinc recycling ... **A2:** 1225
Tevatron
as largest superconducting device ... **A2:** 1027
Tex
defined ... **EM1:** 23 **EM2:** 42
Texas Instrument process ... **EL1:** 621
TEXGAP84 2nd-3d texcap for contact computer program for structural analysis ... **EM1:** 268, 273

TEXLESP (Texas Large Elas Stn Pg) computer program for structural analysis **EM1:** 268, 273

Textile applications
silicones (SI) **EM2:** 266
ultrahigh molecular weight polyethylenes (UHMWPE) **EM2:** 168

Textile equipment
creel warp supply spools **EM1:** 128
fly-shuttle loom, unidirectional fabric **EM1:** 127
rapier loom, plain-weave carbon fabric **EM1:** 127
for unidirectional/two-directional fabrics **EM1:** 127-128

Textile fiberglas
defect inclusion levels **EM4:** 392
defects and cost of losses **EM4:** 392

Textile fibers *See also* Fabric; Fiber(s); Yarns
defined **EM1:** 23 **EM2:** 42

Textile industry applications
aluminum and aluminum alloy equipment **A2:** 13-14
cobalt-base wear-resistant alloys **A2:** 451

Textile machine
spiral power spring distortion in **A11:** 140

Textile machine wear plate
material selection **M1:** 612, 630

Textile oil
defined **A18:** 19

Textile yarn *See also* Yarns
fiberglass **EM1:** 110-111

Textural fractal dimension **A18:** 353

Texture *See also* Crystallographic information; Crystallographic texture measurement and analysis; Fiber; Texturing
analysis **A10:** 357-364
of body-centered cubic metals in torsion **A8:** 181, 183
by x-ray topography **A10:** 365
of composite fractures **A11:** 735
copper, dependence on temperature **A8:** 181-182
in copper twisted at high temperature **A8:** 181
crystallographic, as measure of average grain orientation **A10:** 358
crystallographic, measurement and analysis of **A10:** 357-364
crystallographic, superplastic titanium alloys **A14:** 84
defined **A9:** 18 **A10:** 683 **A11:** 11
for degradation detection **EM2:** 574
determined by neutron diffraction **A10:** 420, 423-424, 425
development, in torsion **A8:** 180-181
dimples as **A12:** 399-400
of ductile fractures **A11:** 82-83
of face-centered cubic metals **A8:** 181-182
fiber, defined **A10:** 359
fracture, photographing **A12:** 82-85
of fracture surfaces **A11:** 20, 75
high-purity copper, temperature and stress effects **A12:** 399
magnetic measurement of **A17:** 129
and material behaviors, correlated **A10:** 357-358
nondestructive measurement **A10:** 423
orientation distribution function as measurement of **A10:** 360-361
of oxides **A13:** 66
preferred orientation **A10:** 358-361
rolling, in fcc material **A10:** 363-364
surface **A14:** 87
symmetrical variations **A10:** 363
and texture gradients **A10:** 420
true volume-averaged, by neutron diffraction **A10:** 357

Texture, crystallographic **A9:** 700-706
changes due to high-angle boundary migration **A9:** 694
characterization of **A9:** 701-706
control of **A9:** 701
defined **A9:** 700

Texture, crystallographic (continued)
development of **A9:** 700-701
during normal grain growth **A9:** 689
during secondary recrystallization **A9:** 690
effect on anisotropic properties **A9:** 700
pole figures used to describe **A9:** 703-706
in specimens deformed to large strains **A9:** 693
types **A9:** 701

Textured aramid
process **EM1:** 115

Textured surfaces
processes for **EM2:** 304-305

Textures
metal, developed by various processing operations **A9:** 706
relationship of lattice orientation to grain structure **A9:** 602

Texturing
glass yarn **EM1:** 111
radial, carbon fibers **EM1:** 51

Texturing, defined
in forging **A11:** 319

T_g *See* Glass transition temperature; Glass-transition temperature

TGA *See* Thermogravimetric analysis

Th-Ti (Phase Diagram) ..... **A3:** 2 • 377
Th-Tl (Phase Diagram) ..... **A3:** 2 • 377
Th-Zn (Phase Diagram) ..... **A3:** 2 • 378
Th-Zr (Phase Diagram) ..... **A3:** 2 • 378

Thailand
bronze casting in **A15:** 22

Thallium
adhesion and solid friction **A18:** 33
determined by controlled-potential coulometry **A10:** 209
effect in embrittlement **A11:** 236
embrittlement by **A13:** 183
as minor toxic metal, biologic effects **A2:** 1260-1261
photometric analysis methods **A10:** 64
pure **M2:** 807
pure, properties **A2:** 1165
radionuclide methods, use in **A18:** 325
species weighed in gravimetry **A10:** 172
toxicity, biologic effects **A2:** 1260-1261
vapor pressure **A6:** 621
weighed as the chromate **A10:** 171

Thallium in fusible alloys **M3:** 799

Thallium, vapor pressure
relation to temperature **M4:** 310

Thallium-base 125 K oxide superconductor
processing of **A2:** 1085-1086

Thallium-doped sodium iodide crystal
in scintillation detectors **A10:** 89

Theorem of Le Châtelier ..... **A3:** 1 • 7
Theorems of limit analysis of plasticity **EM1:** 198
Theoretical calibration curve **A8:** 585
Theoretical density **A7:** 12

Theoretical properties
graphite and diamond **EM1:** 49

Theoretical stress concentration factor
symbol for **A8:** 725

Theoretical stress-concentration factor *See* Stress concentration

Theoretical throat
definition **A6:** 1214

Theory
elastic, Hertz **A8:** 72
of elasticity **A8:** 71-72
of plasticity **A8:** 71-72
of slip-line field solution **A8:** 71-72

Theory of elasticity
and fracture mechanics **A11:** 47

Thermal *See also* Thermal aging; Thermal analysis; Thermal conductivity; Thermal cycling; Thermal design considerations; Thermal design methodologies; Thermal environment; Thermal expansion; Thermal expansion properties; Thermal fatigue; Thermal humidity and cycling test;

Thermal (continued)
Thermal management; Thermal mechanical effects; Thermal mismatching; Thermal performance, Thermal properties; Thermal resistance; Thermal shock; Thermal stress; Thermal vias
degradation, from moisture and carbon dioxide **EL1:** 1066
design considerations **EL1:** 52-55
effects, rheological behavior **EL1:** 849-850
electromotive forces **A8:** 389
endurance, parylene coatings **EL1:** 794
expansion, conversion factors **A8:** 723
failures, in wire bonds **EL1:** 62
fatigue, defined **A8:** 14
history, cooling **A8:** 178
inspection systems, for solder joints **EL1:** 737-738
issues, of advanced systems **EL1:** 80
linkage, desoldering **EL1:** 714
load, and mounting technology **EL1:** 143
properties, and flow localization **A8:** 169
recovery, and creep **A8:** 309
sensors, noncontact **EL1:** 687
stability, of flexible epoxies **EL1:** 821

Thermal (electrical) breakdown
defined **EM2:** 464-465

Thermal (hot) plasma melting and casting **A15:** 419-425

Thermal agglomeration *See also* Agglomeration
of niobium powder **A7:** 163
of powder blends **A7:** 173
of tantalum powder **A7:** 162

Thermal aging *See also* Aging; Heat aging
defined **EL1:** 1159
of polyphenylene sulfides (PPS) **EM2:** 188

Thermal analysis *See also* Coefficient of thermal expansion (CTE); Cooling curve analysis; Cooling curves; Modeling; Temperature resistance; Temperature(s); Testing; Thermal analysis (thermoplastic resins); Thermal analysis (thermoset resins); Thermal expansion; Thermal expansion properties; Thermal image analysis; Thermal inspection; Thermal properties; Thermal shock; Thermal stress; Thermogravimetric; Thermogravimetric analysis (TGA); Thermomechanical
color images by **A17:** 488
for composition effects **EM2:** 566-567
for degree of modification **A15:** 482-483
deviation from equilibrium **A15:** 183
differential **A15:** 183-184
expansion analysis **EM1:** 188-190
of high-performance polymers **EM2:** 559
of laminates **EM1:** 226-227
liquidus and solidus temperature determined **A15:** 184-185
model **EL1:** 14
modeling **EL1:** 50-52
on-line, aluminum-silicon alloys **A15:** 166
oriented plastic **EM2:** 70
quantitative **A15:** 182-183
of residues **A10:** 177
software ADINAT **EL1:** 446
structural, methods of **EM2:** 830-835
techniques, master alloy processing **A15:** 108
thermography **EM1:** 777
thermoplastic polyurethanes (TPUR) **EM2:** 206

Thermal analysis (thermoplastic resins) **EM2:** 533, 540-541

Thermal analysis (thermoset resins) **EM2:** 517, 523-527
differential scanning calorimetry (DSC) **EM2:** 523-525
dynamic mechanical analysis (DMA) **EM2:** 526-527
rheological analysis **EM2:** 527
thermomechanical analysis (TMA) **EM2:** 525-526

Thermal and humidity cycling test
as environmental failure analysis **EL1:** 1102

Thermal annealing
as failure mechanism **EL1:** 1012
Thermal barrier ceramic coatings **M5:** 541-542
Thermal chemical vapor deposition **A18:** 841
Thermal cleaning **A13:** 1143
Thermal coefficient of resistance (TCR) **EM4:** 543
Thermal color images
examples **A17:** 488
Thermal conductivity *See also* Conduction; Conductivity; Thermal; Thermal properties **A1:** 58, 64, 66-68, 69 **EM3:** 29, 33
alloy cast irons **M1:** 88
aluminum and aluminum alloys **A2:** 3, 9
aluminum casting alloys **A2:** 153-177
of aluminum oxide-containing cermets **A7:** 804
of aluminum P/M parts **A7:** 742
ASTM test methods **EM2:** 334
at cryogenic temperatures, beryllium-copper alloys **A2:** 420
and average flash temperature **A18:** 41
beryllium **A2:** 683-684
beryllium-copper alloys **A2:** 409
carbon determined **A10:** 221-225
carbon fiber/fabric reinforced epoxy resin **EM1:** 412
carbon fibers **EM1:** 52, 412, 924
carbon-carbon composites **EM1:** 924
cast copper alloys **A2:** 356-391
ceramic multilayer packages **EL1:** 468
of ceramics **EL1:** 335
of cermets **A2:** 978
CG irons **A15:** 675
of chromium carbide-based cermets **A7:** 806
of common packaging materials **EL1:** 454
of compacted graphite iron **A1:** 64, 66-67
constructional steels for elevated temperature use **M1:** 652, 653
conversion factors **A8:** 723 **A10:** 686
copper and copper alloys **A2:** 216, 219
of copper casting alloys **A2:** 355
of copper-filled epoxy **A7:** 608, 610
of copper/copper alloys **A13:** 610
of cubic boron nitride (CBN) **A2:** 1010
defined **EM1:** 23 **EM2:** 42
defined, thermal inspection **A17:** 396
as detector for C and S in high-temperature combustion **A10:** 221-222
of diamond **A2:** 1010
and diffusivity **EM2:** 451
of dispersion-strengthened copper **A7:** 712
ductile iron **M1:** 53
effect, insoluble particles **A15:** 143
effect, thermal performance **EL1:** 411
electrical contact materials **A2:** 840
of engineering plastics **EM2:** 68-69
epoxy resin system composites **EM1:** 402, 405, 408, 412, 414
equivalent physical quantities **EM1:** 191
and fatigue strength, beryllium-copper alloys **A2:** 419
and forging cracks **A11:** 317
gages, as gas/leak detection **A17:** 62-63, 68
gages, for vacuum testing **A17:** 62-63, 68
of gases **A10:** 223, 230
glass addition effect **EM2:** 70
glass fabric reinforced epoxy resin **EM1:** 405
of glass fibers **EM1:** 47
graphite fiber reinforced epoxy resin **EM1:** 414
of gray iron **A1:** 31 **A15:** 644-645 **M1:** 31, 53
in heat exchangers **A11:** 628
high-temperature thermoset matrix composites **EM1:** 376
impact, junction temperature **EL1:** 814
of internal scale deposits **A11:** 604
Kevlar 49 fiber/fabric reinforced epoxy resin **EM1:** 408
in laminates **EM1:** 226
of lead frame alloys **EL1:** 490
lead frame materials **EL1:** 489
low-temperature thermoset matrix composites **EM1:** 397
maraging steels **M1:** 451
and material selection **EM1:** 38
materials, compared **EM2:** 452
and maximum flash temperature **A18:** 40
Thermal conductivity (continued)
and mean free path **A17:** 59
medium-temperature thermoset matrix composites **EM1:** 384, 388, 390, 391
of metal borides and boride-based cermets **A7:** 812
metal-matrix composites **EL1:** 1124-1125
of plastic encapsulants **EL1:** 808
of polyester resins **EM1:** 92
of polyesters/polymers **EM2:** 452
polymer, as property **EM2:** 60
polymer die attach **EL1:** 218
polyurethanes (PUR) **EM2:** 262
relationship between electrical conductivity and, of ductile iron **A1:** 50-51, 52
selected steel grades **M1:** 148
sheet molding, compounds **EM1:** 158
SI derived unit and symbol for **A10:** 685
SI unit/symbol for **A8:** 721
silver contact alloys **A2:** 843
of specialty polymers **A7:** 606
of substrates **EL1:** 104, 317
thermoplastic matrix composites **EM1:** 366, 369, 371
tungsten **A2:** 562
of tungsten-reinforced composites **EM1:** 884
values, for two bare surfaces **EL1:** 414
for various materials **EL1:** 52
wrought aluminum and aluminum alloys **A2:** 62-122
wrought magnesium alloys **A2:** 483
Thermal conductivity gages **A4:** 507-508
Thermal contact coefficients **A18:** 40-41, 540
symbol and units **A18:** 544
Thermal contact coefficients ratio **A18:** 44
Thermal contraction **A7:** 480
in precision forging **A14:** 16
in spiral bevel gear forging **A14:** 17
Thermal contraction overload
brittle fracture of ductile iron brake drum by **A11:** 370-371
Thermal control plate
radioanalysis **A10:** 248
Thermal convection
and temperature gradient **A15:** 147-148
Thermal convection loop
molten salts **A13:** 51
Thermal cracks
as forging defects **A11:** 317
Thermal curing **EM3:** 554
Thermal cutting **A14:** 720-73
air carbon arc cutting **A14:** 73
air carbon arc cutting and gouging **M6:** 918-920
arc cutting **A14:** 720, 729-73
chemical flux cutting **A14:** 728-72
definition **M6:** 17
exo-process **A14:** 73
gouging **A14:** 732-73
and mechanical cutting, compared **A14:** 72
metal powder cutting **A14:** 72
oxyfuel gas cutting **A14:** 720-72 **M6:** 896-915
oxyfuel gas gouging **M6:** 913-914
oxygen arc cutting **A14:** 73 **M6:** 920
oxygen cutting **A14:** 720-72
plasma arc cutting **M6:** 915-918
Thermal cutting (TC)
definition **A6:** 1214
Thermal cycle simulator (TCS) **A6:** 70
Thermal cycles
steel **M4:** 14
Thermal cycling *See also* Temperature; Temperature cycling; Thermal **EM1:** 296, 886
defined **EL1:** 1159
prediction **EL1:** 883
refractory metals and alloys **A2:** 563
as screening environment **EL1:** 875
superalloys **A12:** 394
temperature histories **EL1:** 881-882
Thermal cycling test **A13:** 1113
Thermal decomposition **A7:** 12, 52, 54-55
of cobalt oxide **A7:** 145-146
in nickel carbonyl formation **A7:** 136, 137
polymer **EM2:** 60
Thermal deformation
of cemented carbides **A7:** 779
Thermal degradation *See also* Degradation **EM1:** 33, 153
and chemical susceptibility **EM2:** 573
mechanisms **EM2:** 423
polyolefins **EM2:** 423
processes, kinetic curves **A13:** 96
service temperature determined by **EM2:** 568-570
Thermal design considerations *See also* Design; Temperature(s); Thermal
cooling techniques **EL1:** 413-414
die attach **EL1:** 411-412
enhancing thermal performance **EL1:** 410-411
key-parameter trends **EL1:** 409
package thermal resistance, defined **EL1:** 409
spreading and interface thermal resistances **EL1:** 412-413
thermal performance **EL1:** 409-410
Thermal design methodologies
introduction **EL1:** 45-46
of level 1 packages **EL1:** 402
modeling **EL1:** 50-52
state-of-the-art technologies **EL1:** 47-50
thermal control **EL1:** 46
thermal design considerations **EL1:** 52-55
thermal management, components and systems **EL1:** 46-55
Thermal desorption mass spectroscopy **EM1:** 285
Thermal diffusion *See also* Diffusion **EM4:** 131
effect on dendritic growth in pure metals **A9:** 610
and surface modification, compared **A13:** 498
Thermal diffusivity **A6:** 3-4
ceramics **A18:** 814-815
defined **A17:** 396 **EM2:** 752
Thermal diffusivity, constructional steels for elevated temperature use **M1:** 643, 652, 653
Thermal effects on adhesive joints **EM3:** 616-621
interfacial effects **EM3:** 616-617
chemical bonding theory **EM3:** 616
differential thermal expansion **EM3:** 617
diffusion theory **EM3:** 616
electrostatic theory **EM3:** 616
interfacial tension **EM3:** 17
mechanical interlocking theory **EM3:** 616
physical absorption theory **EM3:** 616
weak boundary layer theory **EM3:** 616
temperature limits of adhesives **EM3:** 621
thermal conductivity of adhesives **EM3:** 620
thermal degradation of adhesives **EM3:** 619-620
depolymerization **EM3:** 619-620
oxidative degradation **EM3:** 620
thermal transitions in adhesives **EM3:** 617-619
glass transition **EM3:** 617-618
melting of semicrystalline polymers **EM3:** 618
temperature effects **EM3:** 618-619
viscoelastic properties of adhesives **EM3:** 618
Thermal electromotive force
defined **A13:** 13
Thermal embrittlement *See also* Embrittlement
defined **A13:** 13
examination/interpretation **A12:** 136
of maraging steels **A1:** 697-698
Thermal emf *See* Electromotive force; Thermocouple materials
Thermal endurance **EM3:** 29
defined **EM1:** 23-24 **EM2:** 42
Thermal energy method (TEM) of deburring **A16:** 509, 577-578
Thermal environment
leadless packaging **EL1:** 985-986
stress screening **EL1:** 882-884
Thermal equilibrium model
local (ion calculation) **A7:** 258
Thermal etching **A9:** 61-62
defined **A9:** 18
iron-nickel alloys **A9:** 532
Thermal evaporation
of amorphous materials and metallic glasses **A2:** 806
Thermal evaporation films
SEM specimens **A12:** 172-173
Thermal excitation
in optical emission spectroscopy **A10:** 24
thermal inspection **A17:** 397-398

Thermal expansion *See also* Coefficient of linear thermal expansion; Coefficient of thermal expansion (CTE); Thermal stress **A1:** 67, 68 **EM3:** 37, 300-301
alloy cast irons **M1:** 88
of aluminum oxide-containing cermets **A7:** 804
of aluminum silicates **A15:** 209
analysis, of fiber composites **EM1:** 188-190
ASTM test methods **EM2:** 334
beryllium-copper alloys **A2:** 408
of carbon-carbon composites **EM1:** 913
of ceramics **EL1:** 336
CG irons **A15:** 675
of chromium carbide-based cermets **A7:** 806
coated carbide tools **A2:** 960
coefficient, ductile iron **A15:** 663
coefficient of **EM2:** 69
coefficient of, defined *See* Coefficient of thermal expansion
coefficients, aniosotropic, XRPD determined **A10:** 333
coefficients, for ferritic and austenitic steels **A11:** 620
coefficients, tooling materials **EM1:** 428
of composite materials **A11:** 732
of composites **EM1:** 36
constructional steels for elevated temperature use **M1:** 653
conversion factors **A10:** 686
correction, for composite tooling **EM1:** 580-581
cyclic differential, as solder joint failure mechanism **EL1:** 740
defined **A15:** 11 **EL1:** 1159 **EM2:** 752
and density, in P/M steels **A7:** 466
and density, low-alloy ferrous parts **A7:** 463
distortion, steam equipment failure by **A11:** 602
ductile iron **A15:** 663 **M1:** 53
epoxies and **EM3:** 101
of ferritic steels **A1:** 647, 651, 652
fiber volume fraction effects **EM1:** 190
of fibers, testing for **EM1:** 286
friction welding affected by **A6:** 154
glasses versus ceramics **EM3:** 300
of gray iron **A15:** 645 **M1:** 31
in impact extrusion **A14:** 31
in laminates **EM1:** 224-226
lead and lead alloys **A2:** 545, 547
leadless packaging **EL1:** 983-984
linear **A13:** 71
linear coefficient of **EL1:** 217
linear, investment casting refractories **A15:** 259
in magnesium alloys **A14:** 82
malleable irons **M1:** 67
maraging steels **M1:** 451
of metal borides and boride-based cermets **A7:** 812
mismatch problem **EL1:** 611, 742-743, 747, 957
mold spalling from **A15:** 208
molding methods **EM1:** 590-591
nickel alloys **A2:** 433-435
olivine **A15:** 209
P/M materials **M1:** 344
of polymers **EM2:** 60
properties **EL1:** 611-629
sand, cellulose for **A15:** 211
selected steel grades **M1:** 146-147
silica sand **A15:** 208, 224
of thermoplastics **EM1:** 294
of titanium carbide-based cermets **A7:** 808
of titanium carbide-steel cermets **A7:** 810
in tungsten-copper contact materials **A7:** 560
volumetric, filler material effects investment casting **A15:** 255
white cast iron **M1:** 83
zircon **A15:** 209
Thermal expansion coefficient **A6:** 113
solid-state-welded interlayers **A6:** 170, 171
Thermal expansion coefficients **A18:** 575
hardfacing deposition **A18:** 653-654
Thermal expansion coefficients (continued)
heat checking **A18:** 630
mismatch a factor in composite restorative materials (dental) **A18:** 672
Thermal expansion, differential
in soldered joints **A17:** 608
Thermal expansion molding **EM1:** 24, 590-591 **EM3:** 29
defined **EM2:** 42
Thermal expansion properties
CTE tailoring, structures available for **EL1:** 700-704
dielectric materials, constraining **EL1:** 614
dielectric reinforcements, constraining **EL1:** 615-619
low-CTE metal planes, constraining **EL1:** 625-628
low-expansion substrate characteristics **EL1:** 611-613
metal cores, constraining **EL1:** 619-625
plated-through hole reliability **EL1:** 699-700
thick-film ceramic multilayer interconnect boards **EL1:** 15
wiring density **EL1:** 613
Thermal exposure and aging
at elevated-temperature service **A1:** 638, 641
Thermal exposure, effect
thermoplastics **EM1:** 100
Thermal failure
as fatigue failure mechanism **EM2:** 743-744
Thermal fatigue *See also* Fatigue; Fatigue failure **A11:** 133
in boilers and steam equipment **A11:** 623
in cast low-alloy steel ingot tub **A11:** 408
as cemented carbide tool wear mechanism **A2:** 954-955
and coal-ash corrosion, circumferential grooves from **A11:** 618
of compacted graphite iron **A1:** 6344, 66, 67
control, in precision forging **A14:** 16
cracking, crazed pattern **A11:** 623
cracking, in spur gear **A11:** 594
cracks, nickel-base alloy **A11:** 284
defined **A11:** 11, 133
as die failure cause **A14:** 4
effect of section size on resistance to **A11:** 273
in elevated-temperature failures **A11:** 266
and elevated-temperature service **A1:** 626
in engine valves **A11:** 289
failure of main steam line by **A11:** 668-669
fracture, in elevated-temperature failures **A11:** 266
fracture, steam equipment failure by **A11:** 602
in gears **A11:** 594-595
in heat-resistant alloys **A1:** 927-928
in iron castings **A11:** 371
in microelectronics **EL1:** 58-59
and oxidation, in cast ductile iron rotor **A11:** 376
resistance, mechanically alloyed oxide dispersion-strengthened (MA ODS) alloys **A2:** 945
resistance, of low-melting temperature indium-base solders **A2:** 752
of solder joints **EL1:** 1
of steel castings at elevated temperatures **A11:** 407-408
in superheater tube **A11:** 626
testing, carousel-type rig for **A11:** 279
in tungsten-reinforced composites **EM1:** 881-882
turbine vane, failure by **A11:** 285-286
valve guttering as **A11:** 289
Thermal fatigue resistance
of heat resistant alloys **A15:** 729
Thermal fatigue testing
explosion welds **M6:** 711
Thermal gradient
effect on control and size of eutectic structures **A9:** 618
effect on grain growth in weldments **A9:** 579
Thermal gradient (continued)
effect on twinned columnar growth in aluminum alloy ingots **A9:** 631
in tooling **A14:** 16
Thermal gradient mass transfer
liquid-metal circuit **A13:** 57
Thermal gradients
and gating **A15:** 589
for tungsten and molybdenum sintering **A7:** 390-391
Thermal gravimetric analyzer unit **A7:** 197
Thermal image analysis *See also* Thermal analysis
for instrumentation/testing **EL1:** 368-369
Thermal imaging
color **A17:** 488
Thermal inspection *See also* High temperature; Microwave thermography; Temperature(s) Temperature sensors; Thermal wave imaging Thermography; Thermalization Thermal neutrons **A17:** 396-404
of adhesive-bonded joints **A17:** 628-630
applications **A17:** 402-403
of castings **A17:** 512
color images by **A17:** 488
defined **A17:** 396
equipment **A17:** 398-400
heat flow, establishing **A17:** 397-398
inspection methods **A17:** 399
noncontact temperature sensors **A17:** 398-399
of powder metallurgy parts **A17:** 541
principles **A17:** 396
quantitative methods **A17:** 400-402
recording equipment **A17:** 399-400
surface preparation **A17:** 397
Thermal insulation
as application **EM2:** 1
corrosion under **A11:** 184 **A13:** 1144-1147, 1230-1231
Thermal management
analysis **EL1:** 286-287
defined **EL1:** 45
design **EL1:** 287
for digital integrated circuits **EL1:** 174-176
heat sinks as **EL1:** 1129-1131
tape automated bonding (TAB) **EL1:** 286-287
thermal model **EL1:** 174-175
Thermal mismatching *See also* Coefficient of thermal expansion (CTE); CTE-matched materials
encapsulant effects **EL1:** 67
as failure mechanism **EL1:** 957
Thermal model
digital integrated circuits **EL1:** 174-175
Thermal molding
for replicas **A12:** 181
Thermal motion, atomic
to determine crystal structure **A10:** 352
Thermal neutron activation analysis
automated systems **A10:** 238
calibration for **A10:** 236
detection limits **A10:** 238
detection limits, rock and soil **A10:** 237-238
nondestructive **A10:** 234-238
radiochemical, destructive **A10:** 238-239
sample handling **A10:** 236
of zirconium **A10:** 234
Thermal neutron capture
for radioisotope production **A10:** 234
Thermal neutron cross section
various metals **A2:** 664
Thermal neutron irradiation
and epithermal irradiation, compared **A10:** 234
neutron sources for **A10:** 234
Thermal neutron radiography **EM3:** 753
defect detection **EM3:** 749, 750, 751
Thermal neutron(s) *See also* Neutron radiography; Neutron(s)
capture cross section **A17:** 390
defined **A17:** 387
for neutron radiography **A17:** 387-390

Thermal neutron(s) (continued)
sources ... **A17:** 388
Thermal noise *See also* Noise; Signal-to-noise ratio
defined ... **A10:** 683
and flaw detection, thermal inspection ... **A17:** 400
Thermal oxidation *See also* Oxidation ... **A13:** 499, 694, 696
and additives ... **EM2:** 494
degradation, and chemical susceptibility ... **EM2:** 573
Thermal oxide *See also* Oxide contamination
effect, structured reliability testing ... **EL1:** 959-960
Thermal performance ... **EL1:** 409-411
Thermal polymerization of styrene
Raman analysis ... **A10:** 131
Thermal processing *See also* Heat treatment
applications, of structural ceramics ... **A2:** 1019
and homogenization ... **A7:** 308
as stress source ... **A11:** 205
of structural ceramics ... **A2:** 1020
Thermal properties *See also* Coefficient of linear thermal expansion; Electrical properties; Fiber properties analysis; Laminate properties analysis; Liquidus temperature; Material properties; Material properties analysis; Mechanical properties; specific materials; specific properties by name; Temperature(s); Thermal analysis; Thermal conductivity; Thermal degradation; Thermal expansion; Thermal stability; Thermal stress ... **EM3:** 420-427
acrylonitrile-butadiene-styrenes (ABS) ... **EM2:** 111-113
actinide metals ... **A2:** 1189-1198
and activities, aluminum and copper alloys ... **A15:** 55-60
aluminum casting alloys ... **A2:** 153-177
aluminum-silicon alloys ... **A15:** 57
ASTM test methods ... **EM2:** 334
cast copper alloys ... **A2:** 356-391
cast magnesium alloys ... **A2:** 492-516
ceramics ... **EL1:** 335-336
coefficient of thermal expansion ... **EM2:** 69
commercially pure tin ... **A2:** 518-519
constructional steels for elevated temperature use ... **M1:** 643, 647, 649, 652-653
copper-aluminum alloys ... **A15:** 58
copper-clad E-glass laminates ... **EL1:** 536
copper-zinc alloys ... **A15:** 59
data sheet, typical ... **EM2:** 409
of ductile iron ... **A1:** 49, 50
effect, optical properties ... **EM2:** 481-482
electrical resistance alloys ... **A2:** 836-839
engineering plastics ... **EM2:** 68-69
of engineering thermoplastics ... **EM2:** 445-459
engineering thermoset resins compared ... **EM2:** 444
of engineering thermosets ... **EM2:** 439-444
of germanium ... **A2:** 734
of glass fibers ... **EM1:** 47
gold and gold alloys ... **A2:** 705
heat-deflection temperature ... **EM2:** 68
high-impact polystyrenes (PS, HIPS) ... **EM2:** 196
high-temperature resin systems ... **EM2:** 443-444
integral, aluminum-magnesium alloys ... **A15:** 57
intrinsic ... **EM2:** 451-456
lead frame materials ... **EL1:** 488
liquid copper-nickel alloys ... **A15:** 58
liquid copper-tin alloys ... **A15:** 59
liquid crystal polymers (LCP) ... **EM2:** 181
long-term temperature resistance ... **EM2:** 68
low-temperature resin systems ... **EM2:** 439-440
of make-break arcing contacts ... **A2:** 841
medium-temperature resin systems ... **EM2:** 441-443
niobium alloys ... **A2:** 567-571
on supplier data sheets ... **EM2:** 642
packaging materials ... **EL1:** 454
palladium and palladium alloys ... **A2:** 573-576
para-aramid fibers ... **EM1:** 56
parameters ... **EM2:** 613
parylene coatings ... **EL1:** 794
phenolics ... **EM2:** 245
of plastics ... **EM2:** 69
platinum and platinum alloys ... **A2:** 708
polyamide-imides (PAI) ... **EM2:** 133
polyaryl sulfones (PAS) ... **EM2:** 145
polyarylates (PAR) ... **EM2:** 138-139
polyether sulfones (PES, PESV) ... **EM2:** 160
Thermal properties (continued)
polyether-imides (PEI) ... **EM2:** 158
of polyimides ... **EL1:** 768-769
polymeric substrates ... **EL1:** 338
of polymers ... **EM2:** 59-60
polysulfones (PSU) ... **EM2:** 201
pure metals ... **A2:** 1099-1178 **A13:** 62-63
rare earth metals ... **A2:** 1178-1189
refractory metal fiber reinforced composites ... **A2:** 583
of reinforcements ... **EL1:** 1120
of selected oxides ... **A13:** 64
silicone conformal coatings ... **EL1:** 822
silver and silver alloys ... **A2:** 699-704
specialty HIPS ... **EM2:** 199
structural ceramics ... **A2:** 1019, 1021-1024
styrene-acrylonitriles (SAN, OSA ASA) ... **EM2:** 215
thermal conductivity ... **EM2:** 68-69
thermal expansion effects ... **EM3:** 420
thermoplastic polyurethanes (TPUR) ... **EM2:** 207
tin solders ... **A2:** 522
ultrahigh molecular weight polyethylenes (UHMWPE) ... **EM2:** 169
unsaturated polyesters ... **EM2:** 247
wrought aluminum and aluminum alloys ... **A2:** 62-122
wrought copper and copper alloys ... **A2:** 265-345
wrought magnesium alloys ... **A2:** 480-491
of zinc alloys ... **A2:** 532-542
Thermal radiation *See also* Thermal analysis; Thermal image analysis
factors ... **EL1:** 368
Thermal ramp rate
wave soldering ... **EL1:** 685
Thermal ratcheting
defined ... **A11:** 144
Thermal reclamation
of sands ... **A15:** 353-354
Thermal reduction
silver powders ... **A7:** 148
Thermal relaxation failure
microelectronic ... **EL1:** 56
Thermal resistance
defined ... **EL1:** 409, 1159
effect of thermal via area ... **EL1:** 54
epoxies ... **EM3:** 98
package, schematic ... **EL1:** 409
of packages ... **EL1:** 409, 412-413
resin systems ... **EL1:** 534-535
silicone-base coatings ... **EL1:** 773
spreading and interface ... **EL1:** 412-413
and thermal conductivity, lead frame materials ... **EL1:** 489
Thermal resistivity
defined ... **EL1:** 1159
"Thermal runaway" ... **A6:** 239
Thermal sampling
and Grimm emission source ... **A10:** 27
Thermal Severity Number (TSN) ... **A6:** 94, 606
Thermal shape
changing ... **A15:** 606-510
Thermal shock *See also* Thermal shock failures
ceramic design process overview ... **EM4:** 686
in ceramics ... **A11:** 751-753
coating effect ... **A15:** 281
defined ... **A11:** 11 **EL1:** 741, 1159
definition ... **EM4:** 633
as environmental test ... **EL1:** 494, 499
failure ... **EL1:** 56-59, 740-742
failure, pharmaceutical production ... **A13:** 1229
glass-to-metal seals ... **EL1:** 459
as physical testing ... **EL1:** 944
resistance, heat-resistant alloys ... **A15:** 729-730
resistance, of aluminum silicates ... **A15:** 209
resistance, porcelain enamels ... **A13:** 451
testing ... **EL1:** 468, 944
tests, for gas turbine components ... **A11:** 281
Thermal shock failures
fracture, lack of crack branching in ... **A11:** 744
fracture surface ... **A11:** 744, 752
initiation at material flaw ... **A11:** 753
testing machine ... **A11:** 281
turbine blade crack by ... **A11:** 279
Thermal shock resistance
aluminum oxide-containing cermets ... **A7:** 804
cemented carbides ... **A2:** 958
Thermal shock resistance (continued)
in heat-resistant alloys ... **A1:** 928
package-level testing ... **EL1:** 937
Thermal shock tests ... **A8:** 285-286, 454
constructional steels for elevated temperature use ... **M1:** 643
Thermal signature
as preheating process control ... **EL1:** 687
Thermal spray coating ... **M5:** 361-374
aluminum on steel ... **M5:** 343-346
applications ... **M5:** 379, 343-346, 361-363
bonding of ... **M5:** 361-362, 372
testing ... **M5:** 372
burnishing and polishing process ... **M5:** 370-371
cleaning and degreasing ... **M5:** 367
corrosion resistance ... **M5:** 361
electric arc process ... **M5:** 365-366, 368
equipment ... **M5:** 363-368
finishing treatments *See also* specific processes by name ... **M5:** 368-370
surface ... **M5:** 369
flame-spraying process ... **M5:** 365-368
grinding process ... **M5:** 371
grit blasting process ... **M5:** 367-368
hardness testing ... **M5:** 371
inert atmosphere chamber process ... **M5:** 364-365
machining process ... **M5:** 369-370
mechanism of action ... **M5:** 361-362
metallographic examination and evaluation of coatings ... **M5:** 371-372
oxidation protective coatings ... **M5:** 362, 379
plasma-arc process *See* Plasma-arc thermal spray coating
processes *See also* specific processes by name ... **M5:** 361, 363-368
quality assurance ... **M5:** 370-372
repair, coating ... **M5:** 370
rough threading process ... **M5:** 367
safety precautions ... **M5:** 372
sealing process ... **M5:** 369
sectioning and mounting process ... **M5:** 371
surface preparation ... **M5:** 366-368
surface roughening process ... **M5:** 367
tantalum ... **M5:** 362
terminology, glossary of ... **M5:** 373-374
thickness ... **M5:** 367-368
wear resistance ... **M5:** 363
Thermal spray coatings *See also* Flame spraying; Plasma spraying ... **A13:** 459-462 **A18:** 644, 829-833
application method for jet engine component coatings ... **A18:** 590, 591
applications ... **A13:** 462 **A18:** 832
for atmospheric/immersion environments ... **A13:** 460-461
for carbon steel ... **A13:** 523
coating selection ... **A13:** 460-461
coatings for friction and wear applications ... **A18:** 831-833
abrasive wear ... **A18:** 831, 832-833
adhesive wear ... **A18:** 831, 832, 833
cavitation ... **A18:** 832, 833
corrosive wear ... **A18:** 832, 833
suitability criteria ... **A18:** 831
surface fatigue ... **A18:** 831, 832, 833
deposition methods ... **A13:** 459-460
hardfacing cobalt-base alloys by ... **A13:** 664-665
hardfacing deposition in mining and mineral industries ... **A18:** 653
for high-temperature environments ... **A13:** 461
inspection and quality control ... **A13:** 462
lubrication ... **A18:** 832
for marine corrosion ... **A13:** 907-909
material groups ... **A18:** 829
process application area categories ... **A18:** 829
process definition ... **A18:** 829
process fundamentals ... **A13:** 459
process operational mechanisms ... **A18:** 829
process parameters ... **A18:** 830-831
bond coat materials ... **A18:** 831
coating finishing method ... **A18:** 831
coating thickness limitations ... **A18:** 831
deposition rate ... **A18:** 831
surface preparation ... **A18:** 830-831
sealing of ... **A13:** 461, 907-909
surface preparation ... **A13:** 460

Thermal spray coatings (continued)
thermal spray methods commercially available ... **A18:** 829-830, 832
electric arc wire (EAW) spray ... **A18:** 829-830, 832
high-velocity oxyfuel (HVOF) powder spray ... **A18:** 829, 830, 831, 832
oxyfuel powder (OFP) spray ... **A18:** 829, 830, 832
oxyfuel wire (OFW) spray ... **A18:** 829, 830, 832
plasma arc (PA) powder spray ... **A18:** 829, 830, 831, 832
zinc ... **A13:** 767-768
Thermal spray coatings (TSC) ... **A6:** 1004-1009
advantages ... **A6:** 1003
aluminum and zinc
corrosion protection ... **A6:** 1004, 1006, 1008
service life information ... **A6:** 1006, 1007
aluminum metal-matrix service life ... **A6:** 1006
aluminum, service life ... **A6:** 1006, 1007
applications ... **A6:** 1004, 1007-1008
classification of processes relative to method of heat generation and feedstock form ... **A6:** 1004
corrosion protection ... **A6:** 1004, 1005-1006, 1007
deposition rates ... **A6:** 1008-1009
description of spraying processes ... **A6:** 1004
galvanic properties ... **A6:** 1006
industrial process instruction for applications on steel ... **A6:** 1008
parameters of selected processes ... **A6:** 1005
porosity ... **A6:** 1006
process parameters (typical) ... **A6:** 1008
processes ... **A6:** 1004-1005
arc spraying ... **A6:** 1005, 1008-1009
flame spraying ... **A6:** 1004, 1008-1009
plasma spraying ... **A6:** 1005
properties ... **A6:** 1004
safety precautions ... **A6:** 1008
sealants, role of ... **A6:** 1006-1008
selection to preserve integrity of steels ... **A6:** 1005-1008
testing, evaluation and analysis of service and life cycle cost (LCC) ... **A6:** 1008
thermite sparking hazard ... **A6:** 1006
topcoats, role of ... **A6:** 1006-1008
zinc ... **A6:** 1007, 1008
Thermal spray powders ... **A7:** 12
as composite powder application ... **A7:** 174-175
nickel-chromium/chromium carbide as ... **A7:** 174
spray drying applications ... **A7:** 77
Thermal spray processes ... **A6:** 789, 796, 797, 798, 799, 803
air-cooled continuous combustion HVOF process ... **A6:** 806
applications ... **A6:** 808, 809, 810, 811, 814-816
buildup ... **A6:** 803
coating material ... **A6:** 808
coating selection criteria ... **A6:** 814
coatings used for elevated-temperature service ... **A6:** 808
coatings used for hardfacing applications ... **A6:** 808
comparison of processes ... **A6:** 804
definition ... **A6:** 807-808
detonation gun (D-gun) HVOF process ... **A6:** 804, 806
electric arc wire (EAW) process ... **A6:** 804, 805, 808, 809, 810
finishing treatment ... **A6:** 813-814
high-velocity oxyfuel powder spray (HVOF) process ... **A6:** 806, 807, 808, 812-813
continuous-combustion ... **A6:** 806, 812-813
pulsed-combination ... **A6:** 812
materials ... **A6:** 814-816
corrosion resistance ... **A6:** 815
dimensional restorative coatings ... **A6:** 816
electrically conductive coatings ... **A6:** 816
medically compatible coatings ... **A6:** 816
Thermal spray processes (continued)
oxidation protection ... **A6:** 815-816
thermal barrier coatings ... **A6:** 816
thermal insulation ... **A6:** 816
wear coatings ... **A6:** 815
method selection criteria ... **A6:** 808-809
methods ... **A6:** 808-813
oxyfuel powder spray (OFP) process ... **A6:** 804, 805, 808, 810, 811
applications ... **A6:** 810
flame spray and fuse ... **A6:** 810
oxyfuel wire spray (OFW) process ... **A6:** 804, 805, 808, 809-810, 811
advantages ... **A6:** 809
applications ... **A6:** 809, 811
equipment ... **A6:** 809
fuel gases ... **A6:** 809
materials ... **A6:** 809
percentage of market captured ... **A6:** 808
plasma arc spray (PA) process ... **A6:** 804, 805, 808, 809, 810-812
safety precautions ... **A6:** 1204
surface preparation ... **A6:** 813
tungsten carbide-cobalt coating compared to ... **A6:** 807
tungsten carbide-cobalt coating particle velocity compared to ... **A6:** 807
water-cooled continuous combustion HVOF process ... **A6:** 806
Thermal spraying *See also* Flume spraying; Plasma spraying; Thermat spray coatings ... **M6:** 804
cast irons ... **A6:** 720
defined ... **A13:** 13
definition ... **A6:** 1214 **M6:** 17
weld overlaying ... **M6:** 807-808
Thermal spraying gun
definition ... **A6:** 1214 **M6:** 18
Thermal sprays
as coatings ... **A15:** 563
Thermal stability *See also* Stability; Thermal properties ... **A18:** 84
beryllium-copper alloys ... **A2:** 416-417
of composite tooling ... **EM1:** 587
of composites ... **EM2:** 372
of glass-polyester composites ... **EM1:** 93
initial, and thermo-oxidative stability ... **EM2:** 565-566
of molding materials ... **A15:** 208
para-aramid fibers ... **EM1:** 56
phenolics ... **EM2:** 242, 244
of polyester resins ... **EM1:** 93
of polyimide resins ... **EM1:** 43
polyphenylene sulfides (PPS) ... **EM2:** 186
polysulfones (PSU) ... **EM2:** 200
and process selection ... **EM2:** 277
S-glass ... **EM1:** 107
SiC ceramic fibers ... **EM1:** 64
structural foams ... **EM2:** 509
unsaturated polyesters/fiberglass composites ... **EM2:** 249
wrought titanium alloys ... **A2:** 627-628
Thermal straining
low-cycle fatigue cracking from ... **A11:** 284
Thermal stress *See also* Aging; Stress relief; Stress relieving; Stress(es); Thermal properties
in adhesively bonded joints ... **EL1:** 57
alloy steels ... **A12:** 291
defined ... **A11:** 11 **EL1:** 55-56
definition ... **A6:** 1214 **EM4:** 633
distribution ... **EM2:** 753
effect on SCC of T-bolt ... **A11:** 538
effect, transistor junction ... **EL1:** 56-57
in electronic packaging ... **EL1:** 56-59
evaluation ... **EM2:** 754
failure, pharmaceutical production ... **A13:** 1229
failures ... **EL1:** 55-62
failures, in microelectronics ... **EL1:** 55-62
and fracture mechanics ... **A11:** 57
Thermal stress (continued)
in heat exchangers ... **A11:** 636
heat sinks for ... **EL1:** 1129
measurement ... **EM2:** 754
mismatch ... **EM2:** 753-754
overload, in iron castings ... **A11:** 370
oxide scales ... **A13:** 71
and physical aging ... **EM2:** 751-760
in PWBs ... **EL1:** 62
relief, for SCC, plain carbon/low-alloy steels ... **A13:** 328
in TAB assemblies ... **EL1:** 287-288
Thermal stress cracking *See also* Stress cracking; Stress-cracking failure ... **EM3:** 29
defined ... **EM1:** 24 **EM2:** 42
Thermal stresses *See also* residual stresses
causes ... **M6:** 856-858
definition ... **M6:** 18
during welding ... **M6:** 858-859
metal movement ... **M6:** 858-859
Thermal stressing
for optical holographic interferometry ... **A17:** 409-410
Thermal taper *See* Thermal wedge
Thermal tempering
laminated glass ... **EM4:** 423
Thermal transfer inspection
brazed joints ... **A6:** 1119-1120
Thermal transport *See* Heat transport
amorphous materials and metallic glasses ... **A2:** 813
Thermal treatment *See* Heat treatment ... **M6:** 889-892
code requirements ... **M6:** 889-890
postweld ... **M6:** 890-891
preheating ... **M6:** 890
reduction of residual stresses ... **M6:** 887
low-temperature stress relieving ... **M6:** 891
peening ... **M6:** 891-892
Thermal treatments
effect on phase structure in aluminum alloys ... **A9:** 360
ion-implanted alloys ... **A10:** 486
for steel wire ... **A1:** 280-281
Thermal treatments, of aluminum P/M alloys *See also* Heat treating; Heat treatment ... **A7:** 381
Thermal vias *See also* Via(s)
area, effect on thermal resistance ... **EL1:** 54
effect on multichip modules ... **EL1:** 53-54
Thermal vibration
analysis of ... **A10:** 536
Thermal volumetric strain rate ... **A6:** 1137
Thermal wave imaging
of powder metallurgy parts ... **A17:** 541
Thermal wave interferometer systems
thermal inspection ... **A17:** 399
Thermal wear
defined ... **A18:** 19
Thermal wedge
defined ... **A18:** 19
Thermal-activation energy
in integrated circuits ... **A11:** 766, 767
Thermal-calcining
for sand reclamation ... **A15:** 227
Thermal-conductive detection
of carbon and sulfur ... **A10:** 223
cell ... **A10:** 230
inert gas fusion ... **A10:** 229-230
Thermal-dry scrubbing
for sand reclamation ... **A15:** 227
Thermal-energy dissipation ... **A18:** 611
Thermal-mechanical analysis (TMA)
of flexible epoxies ... **EL1:** 821
Thermal-mechanical effects
accelerated fatigue reliability testing ... **EL1:** 748-751
design for reliability ... **EL1:** 746-748
prediction, of SM solder joint reliability ... **EL1:** 751-752

Thermal-mechanical effects (continued)
reliability overview EL1: 740-742
solder shear fatigue, analytical model EL1: 743-746
thermal expansion mismatch problem EL1: 742-743
Thermal-mechanical processing A14: 19, 26
Thermal-mechanical-controlled processing (TMCP) steels
classification and group description A6: 405, 406, 407
Thermal-mechanical-controlled processing steels, specific types
A841, composition and carbon content A6: 406
Thermal-neutron radiographs
examples A17: 391-395
Thermal-stress cracking
in copper alloy ingots A9: 642
Thermal-wave imaging A9: 91
SEM A12: 169
Thermalization, and collimation
in neutron radiography A17: 389
Thermalloy 40B
weld repair M6: 367
Thermally activated dislocation glide A8: 34, 169
Thermally activated growth
and isothermal phase transformations A10: 317
Thermally active/passive test objects
thermal inspection A17: 397
Thermally aged adhesion
defined EL1: 1159
Thermally conductive adhesives EM3: 571-572
application methods EM3: 571-572
pot life EM3: 571
Thermally induced embrittlement
of steels A11: 98-100
Thermally induced failures
passive devices EL1: 1000-1001
Thermally induced porosity A7: 181
Thermally labile species
Raman analysis of A10: 129
Thermally quenched phosphors
brazed joint inspection A17: 604-605
in thermal inspection A17: 399
Thermally-induced radiation
in thermal inspection A17: 396
Thermid 600 resins EM1: 83, 84, 89
Thermid IP-600 resins EM1: 83, 89
Thermid MC-600 resins EM1: 83, 84, 89
Thermionic cathode gun
defined A9: 18
Thermionic emission A6: 30, 44
defined A9: 18
Thermistors
microwave detection by A17: 208
for thermal inspection A17: 399
Thermistors and related sensors EM4: 1145-1149
applications EM4: 1145
barrier layers EM4: 1145-1146
common features linking oxide thermistors and gas sensors EM4: 1145
electrical contacts EM4: 1145
gas sensors EM4: 1147-1148
humidity sensors EM4: 1148-1149
negative temperature coefficient thermistors EM4: 1145, 1146
positive temperature coefficient resistors EM4: 1145, 1146, 1147
Thermit crucible
definition A6: 1214 M6: 18
Thermit mixture
definition A6: 1214 M6: 18
Thermit mold
definition A6: 1214 M6: 18
Thermit reaction
definition A6: 1214 M6: 18
Thermit reactions A7: 52, 55
Thermit welding M6: 692-704
electrical connections M6: 703-704
fusion welding M6: 694
metallurgical parameters M6: 693-694
metallurgical principles M6: 692
free energy of oxide formation M6: 692
powder used A7: 573
pressure welding M6: 694
principles of welding M6: 692-693
aluminothermic compounds M6: 692-694
Thermit welding (continued)
oxide reactions M6: 692
variables affecting reductants M6: 692-693
rail welding M6: 695-698
grain structure M6: 697
metallurgical aspects of welds M6: 697
testing of welds M6: 697-699
welding with preheat M6: 695-696
welding without preheat M6: 696
reinforcement bar welding M6: 702-703
repair welding M6: 698-702
applications M6: 701-702
applying the mold M6: 699-700
changing the crucible M6: 701
joint preparation M6: 699
molding sand characteristics M6: 700-701
preheating M6: 701
safety precautions M6: 58
Thermit welding (TW)
definition A6: 1214
Thermite
flame AAS analysis of A10: 56, 57
Thermite reactions A7: 194
Thermite welding (TW) A6: 291-293
applications A6: 291-293
railroad equipment A6: 397
definition A6: 291
heat-affected zone (HAZ) A6: 291, 292
metallurgy A6: 291
principles A6: 291
rail welding A6: 291
safety precautions A6: 293, 1203-1204
vs. exothermic brazing A6: 345
Thermo-forming
as superplastic forming A14: 857-85
Thermo-oxidative stability
and thermal stability EM2: 565-566
Thermochemical conversion surface treatments
titanium alloys A18: 779, 780-781
Thermochemical processing (TCP) *See also* Thermal processing
prealloyed titanium P/M compacts A2: 654
Thermochemistry *See also* Physical chemistry; Thermodynamics
gas-metal A7: 295
of inclusion-forming reactions A15: 89
Thermochromic coatings
for adhesive-bonded joints A17: 629
Thermocompression ball bonds
defect between gold wire/vacuum-deposited aluminum A12: 484
intermetallic formation effects A12: 485
Thermocompression bonding *See* Thermocompression welding EM3: 584, 586
defined EL1: 1159
definition A6: 1214
as electrical interconnection EL1: 224-225
failure mechanisms EL1: 62, 1042-1043
as inner lead process EL1: 278-279
leaded and leadless surface-mount joints EL1: 734
of leads, low-temperature cofired ceramic EL1: 467
thermal failures EL1: 62
Thermocompression welding M6: 674-675
applications M6: 674, 686
equipment M6: 675
metals joined M6: 675
procedures M6: 675
resistance heating M6: 675
Thermocouple
classifications and comparisons A8: 414
defined A8: 14 A9: 18 A13: 13
in temperature control of tension testing A8: 36
as transducer in creep testing A8: 312-313
-type temperature controller A8: 414
-type vacuum gage A8: 414
use in ultrasonic testing A8: 247
Thermocouple alloys
percussion welding M6: 740
Thermocouple calibration *See also* Thermocouple materials
comparison method A2: 880-881
direct emf measurement vs platinum A2: 880
freezing-point calibration A2: 879-880
Thermocouple calibration (continued)
International Temperature Scale of 1990 (ITS-90) A2: 879
methods A2: 879
temperature scales A2: 878-879
Thermocouple extension wires *See also* Thermocouple materials
circuitry A2: 876-877
color coding A2: 878
error analysis A2: 877-878
properties A2: 878
selection criteria A2: 876
Thermocouple materials *See also* Thermocouple calibration; Thermocouple extension wires; Thermocouple materials, specific types; Thermocouple thermometers; Thermocouple wires
calibration, change during service A2: 881-882
color coding of thermocouple wires/extension wires A2: 878
industrial applications, selection criteria A2: 885
insulation and protection A2: 882-884
nonstandard thermocouples, types A2: 874-876
selection criteria A2: 885
standard thermocouples, types A2: 871-873
thermocouple assemblies A2: 884-885
thermocouple calibration A2: 878-881
thermocouple extension wires A2: 876-878
thermocouple maintenance A2: 886-887
thermocouple materials, types A2: 871-876
thermocouple practice A2: 885-886
thermocouple thermometers, principles A2: 869-871
Thermocouple materials, specific types *See also* Thermocouple materials
19 alloy/20 alloy, for elevated temperatures A2: 874
iridium-rhodium thermocouples A2: 874
nonstandard thermocouples A2: 874-876
Platinel, types and application A2: 875-876
platinum 67, as reference standard A2: 870
platinum-molybdenum, under neutron radiation A2: 874-875
standard thermocouples A2: 871-873
tungsten-rhenium, types and application A2: 876
type B, still air/inert atmospheres A2: 873
type E, thermoelectric power A2: 873
type J, versatility and cost A2: 871-872
type K, industrial applications A2: 872
type N, Nicrosil/Nisil thermocouple A2: 873
type R, stability of A2: 873
type S, as calibration standard A2: 873
type T, cryogenic applications A2: 872-873
Thermocouple thermometers
measurement of temperature by A2: 870
preparation of measuring junction A2: 870-871
Thermocouple wires *See also* Thermocouple extension wires; Thermocouple materials
color coding A2: 878
wire insulation, types A2: 882-883
Thermocouple(s) *See also* Thermocouple calibration; Thermocouple extension wires; Thermocouple materials; Thermocouple materials, specific types
conventional, assemblies A2: 885
defined A2: 869
metal-sheathed, assemblies A2: 884-885
reference tables/calibration A2: 881
Thermocouples A11: 296-297 M3: 696-720
for a practical temperature measurement EM4: 251
assemblies, conventional M3: 715, 716
assemblies, metal-sheathed M3: 715-716
calibration M3: 701, 708-714
change of during service M3: 712-714
methods of M3: 709-712
tolerances, initial M3: 701, 712
color coding of thermocouple wires and extension wires M3: 708, 709
cryogenic applications M3: 717
extension wires M3: 706-708, 709
furnace brazing A6: 330
insulation and protection M3: 714-715, 716
in magnesium melting A15: 801
maintenance M3: 718-719
manganese cast alloy for A15: 32

Thermocouples (continued)
materials, nonstandard ... **M3:** 702-706
composition ... **M3:** 702-704
mechanical properties ... **M3:** 703, 704, 707, 708
physical properties ... **M3:** 702, 703, 704, 705, 707, 708
temperature limits ... **M3:** 702-704, 707, 708
materials, standard ... **M3:** 699-702
composition ... **M3:** 699-702
physical properties ... **M3:** 699, 700
temperature limits ... **M3:** 699-702
powders used ... **A7:** 573
reference tables ... **M3:** 712
selection, criteria for ... **M3:** 716-717
for temperature control with surface carbon content control ... **A4:** 576, 579
for thermal inspection ... **A17:** 399
thermocouple thermometers ... **M3:** 696-699
use in temperature control ... **A4:** 506-508, 529-532 **M4:** 346-350, 351, 352
Thermodynamic effect
in liquid erosion ... **A11:** 164
Thermodynamic equilibrium
and physical aging ... **EM2:** 755-756
Thermodynamic modeling of phase diagrams ... **A3:** 1•18
Thermodynamic principles ... **A3:** 1•5-7
Thermodynamic properties *See also* Thermodynamics
activities and thermal properties ... **A15:** 55
aluminum-base alloys ... **A15:** 55-60
amorphous materials and metallic glasses ... **A2:** 812-813
of binary iron-base systems ... **A15:** 61-62
copper-base alloys ... **A15:** 55-60
interaction coefficients ... **A15:** 55-56
of iron-base alloys ... **A15:** 61-70
of multicomponent Fe-C systems ... **A15:** 68-69
structural diagrams ... **A15:** 69-70
of ternary iron-base alloys ... **A15:** 62-64
ternary iron-carbon systems ... **A15:** 64-68
thermal properties, hypothetical standard state ... **A15:** 56-57
Thermodynamic scale
as ideal temperature scale ... **A2:** 878
Thermodynamic stability of cold-worked metals ... **A9:** 692
Thermodynamic temperature ... **A8:** 721
SI base unit and symbol for ... **A10:** 685
Thermodynamic values
types of invariant point ... **EM4:** 883
Thermodynamics *See also* Thermodynamic properties
of aqueous corrosion ... **A13:** 18-28, 32
of argon oxygen decarburization ... **A15:** 426
chemical ... **A15:** 50-52
corrosion protection by ... **A13:** 377
first principle, half-cell potentials from ... **A10:** 164
of high-temperature corrosion in gases ... **A13:** 61-64
of hydrogen removal ... **A15:** 457
of multicomponent Fe-C systems ... **A15:** 68-69
of particle behavior, interfacial ... **A15:** 142-143
principles, of corrosion ... **A13:** 17
of solidification ... **A15:** 101-103
of stress-corrosion cracking ... **A13:** 151
of superconducting to normal transition ... **A2:** 1028
of surface or grain-boundary segregation LEED analysis ... **A10:** 544
of ternary FE-C-X systems ... **A15:** 64-68
Thermoelastic coefficient
of Invar ... **A2:** 891
Thermoelastic deformation
and mechanical durability ... **EL1:** 55
Thermoelastic instability (TEI)
ceramics ... **A18:** 814, 815
defined ... **A18:** 19
Thermoelastic martensite
in shape memory alloys ... **A2:** 897
Thermoelasticity, laminate
computer program for ... **EM1:** 276
Thermoelectric devices
for thermal inspection ... **A17:** 399
Thermoelectric effect ... **A6:** 42
Thermoelectric potential versus copper
electrical resistance alloys ... **A2:** 822
Thermoelectric power
defined ... **A2:** 870
Thermoelectric testing ... **A1:** 1031
Thermoelement, negative and positive
defined ... **A2:** 869-871
Thermoformers
types ... **EM2:** 401-403
Thermoforming *See also* Air-assist vacuum forming; Bubble forming; Drop forming; Forming; Matched-mold forming; Plug-and-ring forming; Plug-assist forming; Ridge forming; Scrapless thermoforming; Slip forming; Twin-sheet thermoforming; Vacuum forming ... **EM2:** 399-403
acrylics ... **EM2:** 107
air-slip ... **EM2:** 401
defined ... **EM1:** 24 **EM2:** 42, 303
economic factors ... **EM2:** 301, 403
high-impact polystyrenes (PS, HIPS) ... **EM2:** 198
machines ... **EM2:** 401-403
matched-mold ... **EM2:** 400
molded-in color ... **EM2:** 306
pressure bubble plug-assist vacuum ... **EM2:** 401
process ... **EM2:** 399-400
properties effects ... **EM2:** 285
sheet, reinforced polypropylenes (PP) ... **EM2:** 193
size and shape effects ... **EM2:** 290-291
surface finish ... **EM2:** 304
techniques ... **EM2:** 400-401
textured surfaces ... **EM2:** 305
trapped-sheet, contact heat, pressure ... **EM2:** 401
vacuum snapback ... **EM2:** 401
Thermogalvanic corrosion *See also* Galvanic corrosion
defined ... **A13:** 13
Thermographs
microwave holography ... **A17:** 225
Thermography *See also* Microwave thermography
high-resolution ... **EL1:** 954-955
microstructure effect ... **A17:** 51
as nondestructive test technique ... **EM1:** 777
pulsed video ... **A17:** 512
Thermogravimetric analysis (TGA) *See also* Thermal analysis ... **EM2:** 525, 540, 834-835 **EM3:** 29
defined ... **EM1:** 24 **EM2:** 42
of epoxy resin system composites ... **EM1:** 403
of high-temperature polymers ... **EM1:** 810
of leaded and leadless surface-mount joints ... **EL1:** 733
of medium-temperature thermoset matrix composites ... **EM1:** 384
for phase analysis ... **EM4:** 561, 562
polysiloxane preceramic polymer ... **EM4:** 223
for temperature resistance ... **EM2:** 559-566
thin-film hybrids ... **EL1:** 324
to study evaporation of low-boiling point organics during spray drying ... **EM4:** 107
Thermogravimetry
for outgassing ... **A7:** 434
Thermogravimetry (TG) ... **EM4:** 52, 53
to measure relative or time differential weight change ... **EM4:** 27
Thermoluminescent coatings
for adhesive-bonded joints ... **A17:** 629
Thermomagnetometry (TM) ... **EM4:** 52, 54
Thermomechanical
design considerations ... **EL1:** 414-416
performance, rigid epoxies ... **EL1:** 814
Thermomechanical (continued)
properties, of substrates ... **EL1:** 612
simulation, design tools for ... **EL1:** 419
Thermomechanical analysis
abbreviation for ... **A11:** 798
Thermomechanical analysis (TMA) ... **EM1:** 779-780 **EM2:** 525-526, 540-541 **EM3:** 420
defined ... **EM2:** 832-833
for temperature resistance ... **EM2:** 559-564
Thermomechanical analyzer (TMA)
for substrates ... **EL1:** 612
Thermomechanical effects
aluminum-lithium alloys ... **A2:** 180
shape memory alloys ... **A2:** 898
Thermomechanical fatigue (TMF) tests ... **A1:** 626
Thermomechanical fatigue tests
for gas-turbine components ... **A11:** 278-281
Thermomechanical processing *See also* Controlled rolling
accelerated cooling of high-strength low-alloy steels ... **A1:** 398
interpass cooling ... **A1:** 409
of microalloyed steels ... **A1:** 585-589
temperature-time schedules ... **A1:** 130, 131
Thermomechanical processing (TMP)
titanium alloys ... **A6:** 84, 85
Thermomechanical processing (TMP), in mechanical alloying ... **A7:** 725
oxide reduction by ... **A7:** 256
Thermomechanical properties
polymer composition effects ... **EM2:** 566-567
Thermomechanical pulping equipment ... **A13:** 1217-1218
Thermomechanical treatments
ultrahigh-strength steels ... **M1:** 421
Thermomechanical working
defined ... **A14:** 13
Thermomechanically controlled process (TMCP) steels
weldability ... **A6:** 419, 420
Thermomechanically processed cobalt- chromium-molybdenum alloy
for orthopedic implants ... **A7:** 658
Thermometers *See also* Thermocouple materials
resistance, of electrical resistance alloys ... **A2:** 822-823
thermocouple, principles of ... **A2:** 869-871
types ... **A2:** 869
Thermonuclear fusion
magnetic containment of ... **A2:** 1056-1057
Thermophysical properties ... **EM4:** 610-616
emissivity ... **EM4:** 615-616
calorimetric determinations ... **EM4:** 616
direct measure of emitted radiant flux ... **EM4:** 616
reflectivity measurements ... **EM4:** 616
heat capacity ... **EM4:** 614-615
calorimetry ... **EM4:** 614
differential scanning calorimetry ... **EM4:** 614
differential thermal analysis ... **EM4:** 615
specific heat ... **EM4:** 614, 615
thermal conductivity ... **EM4:** 612-614
guarded hot plate technique ... **EM4:** 613
laser flash method ... **EM4:** 613-614
radial heat flow method ... **EM4:** 613
thermal diffusivity ... **EM4:** 612-614
thermal expansion ... **EM4:** 610-612
ASTM standards ... **EM4:** 611
coatings ... **EM4:** 612
coefficient of thermal expansion ... **EM4:** 610
dilatometry ... **EM4:** 610-611
films ... **EM4:** 612
interferometry ... **EM4:** 611
measuring microscopy ... **EM4:** 611-612
x-ray diffraction ... **EM4:** 611
Thermophysical properties data base
die/workpiece ... **A14:** 41
Thermopiles
for thermal inspection ... **A17:** 399

Thermoplastic
defined **EL1:** 1159 **EM1:** 24
Thermoplastic acrylic **A13:** 405
Thermoplastic coal-tar-based coating **M1:** 176
Thermoplastic composites, friction and wear of **A18:** 820-826
abrasive wear **A18:** 821
adhesive wear **A18:** 821
advantages **A18:** 820
coefficients of friction **A18:** 820, 821, 822, 824, 825, 826
composition of injection-moldable thermoplastic composites **A18:** 821
conclusions **A18:** 826
corrosive wear **A18:** 820
lubricant effect on friction **A18:** 825-826
lubricant effect on wear factor **A18:** 822-825
lubrication **A18:** 820, 822-826
material selection **A18:** 822-826
PEEK composites, friction and wear properties **A18:** 821
properties **A18:** 820
sliding wear **A18:** 820-822
wear tests **A18:** 822
coefficient of friction **A18:** 822
volume wear **A18:** 822
wear factor **A18:** 822
Thermoplastic compression molding **EM1:** 562-563
Thermoplastic, defined *See also* Thermoplastics **EM2:** 42
Thermoplastic elastomer
as thermocouple wire insulation **A2:** 882
Thermoplastic elastomers (TEs) *See also* Elastomers
chemistry **EM2:** 66
thermal properties **EM2:** 450
Thermoplastic extrusion *See also* Extrusion; Thermoplastic processes **EM2:** 378-388
competing processes **EM2:** 387
defined **EM2:** 303
extruder parameters **EM2:** 378-383
multiple-screw extruders **EM2:** 383
process/product parameters
additional **EM2:** 386-387
products, types of **EM2:** 383-386
proving the system **EM2:** 387
size and shape effects **EM2:** 290
Thermoplastic fluoropolymers *See also* Fluoroplastics; Thermoplastic resins **EM2:** 115-119
applications **EM2:** 117-118
characteristics **EM2:** 118-119
chlorotrifluoroethylene **EM2:** 116
commercial forms **EM2:** 115-117
ethylene chlorotrifluoroethylene **EM2:** 115-118
ethylene-tetrafluoroethylene **EM2:** 115-118
fluorinated
perfluoroethylene-propylene **EM2:** 116
perfluoro alkoxy alkane **EM2:** 116
polytetrafluoroethylene **EM2:** 115-116
polyvinylidene fluoride **EM2:** 115-119
suppliers **EM2:** 119
thermal properties **EM2:** 448
Thermoplastic injection molding *See also* Injection molding **EM2:** 308-318
defined **EM2:** 42
economic analysis **EM2:** 308-310
injection molds and process **EM1:** 556-557
materials **EM2:** 310-311
mechanical properties **EM2:** 311-312
mold construction **EM2:** 313-317
quality assurance **EM2:** 317-318
reciprocating screw injection molding machine for **EM1:** 555-557
technical analysis **EM2:** 312
tolerance capability **EM2:** 312-313
two-stage injection/accumulator machine **EM1:** 557
variants **EM1:** 557
Thermoplastic matrix composites *See also* Thermoplastic matrix processing; Thermoplastic resins **A9:** 592 **EM1:** 363-372
discontinuous fiber **EM1:** 121
interlaminar fracture toughness **EM1:** 100
polyester resin and fiber-resin **EM1:** 363
polysulfone resin and fiber-resin **EM1:** 363
Thermoplastic matrix composites (continued)
properties **EM1:** 363-372
testing **EM1:** 97-98
Thermoplastic matrix processing **EM1:** 544-553
comparative assessment **EM1:** 142
economics of **EM1:** 552-553
fabrication **EM1:** 547-552
product forms **EM1:** 546-547
thermoplastic resin systems **EM1:** 544-546
Thermoplastic media
for decorating **EM4:** 475
Thermoplastic molding compounds **EM1:** 164-166
Thermoplastic polyesters *See also* Polybutylene terephthalate (PBT); Polyesters; Polyethylene terephthalate (PET); Thermosetting polyesters; Unsaturated polyesters
defined **EM1:** 24 **EM2:** 42
as engineering plastic **EM2:** 429
Thermoplastic polyimides **EM1:** 78-79
Thermoplastic polyimides (TPI) *See also* Polyimides (PI); Thermoplastic resins
applications **EM2:** 177
chemistry and properties **EM2:** 177
defined **EM2:** 42
processibility **EM2:** 177
processing **EM2:** 177-178
vs thermoset polyimides **EM2:** 177
Thermoplastic polymers *See also* Polymer(s); Thermoplastic resins; Thermoplastics
chemical structures **EM2:** 54
glass transition temperatures **EM2:** 54
heterochain, glass transition temperature **EM2:** 53
hydrocarbon **EM2:** 50
nonhydrocarbon carbon-chain **EM2:** 51-52
in sheet molding compounds **EM1:** 158
Thermoplastic polyurethanes (TPUR) *See also* Polyurethanes (PUR); Thermoplastic resins
alloys, elastomeric alloys **EM2:** 204
applications **EM2:** 205-206
chain extenders **EM2:** 203
characteristics **EM2:** 206-207
chemistry **EM2:** 203
commercial forms **EM2:** 203-204
competitive materials **EM2:** 206
costs and production volume **EM2:** 204-205
defined **EM2:** 42
design considerations **EM2:** 206-207
diisocyanates **EM2:** 203
glass-reinforced **EM2:** 205
Polyols **EM2:** 203
processing **EM2:** 207
properties **EM2:** 203, 205
as PUR commercial form **EM2:** 258
special compounds **EM2:** 207
suppliers **EM2:** 207
Thermoplastic prepreg
fabrication with **EM1:** 103-104
potential of **EM1:** 99-100
Thermoplastic processes *See also* Thermoplastic resins; Thermoplastics
blow molding, size and shape effects **EM2:** 290
compression molding, properties effects **EM2:** 285
compression molding, size and shape effects **EM2:** 290
conventional blow molding, properties effects **EM2:** 285
effects, on properties **EM2:** 282-286
extrusion, size and shape effects **EM2:** 290
filament winding, properties effects **EM2:** 285-286
filament winding, size and shape effects **EM2:** 291
foam injection molding, properties effects **EM2:** 284
foam injection molding, size and shape effects **EM2:** 290
hollow injection molding, properties effects **EM2:** 284
hollow injection molding, size and shape effects **EM2:** 290
injection blow molding, properties effects **EM2:** 285
injection compression molding properties effects **EM2:** 283-284
Thermoplastic processes (continued)
injection molding, properties effects **EM2:** 282-284
injection molding, size and shape effects **EM2:** 288-289
injection-compression molding, size and shape effects **EM2:** 289-290
rotational casting, properties effects **EM2:** 286
rotational casting, size and shape effects **EM2:** 291
sandwich molding, properties effects **EM2:** 284-285
sandwich molding, size and shape effects **EM2:** 290
shrinkage from **EM2:** 280
size and shape effects **EM2:** 288-291
stamping, size and shape effects **EM2:** 290
thermoforming, properties effects **EM2:** 285
thermoforming, size and shape effects **EM2:** 290-291
and thermoset processes, compared **EM2:** 289
twin-sheet forming, size and shape effects **EM2:** 290
twin-sheet stamping, size and shape effects **EM2:** 290
Thermoplastic processing
in superalloys **A7:** 440
Thermoplastic resin bonds
bonded-abrasive grains **A2:** 1014
Thermoplastic resins *See also* Compression- mounting materials; Engineering plastics; Engineering thermoplastics; Resins; Thermoplastic matrix composites; Thermoplastics **A13:** 329, 403-406 **EM1:** 97-104 **EM2:** 98-221, 618-625
acrylics **EM2:** 103-108
acrylonitrile-butadiene-styrenes (ABS) **EM2:** 109-114
advanced composites **EM2:** 621-622
for aerospace application **EM1:** 32-33
amorphous **EM2:** 620-621
application, automotive industry **EM1:** 834
applications **EM2:** 623-625
carbon fiber treatment for **EM1:** 52
characterization tests **EM1:** 294
chopped fiber reinforced grades **EM1:** 43
commodity, elevated-temperature behavior **EM1:** 32
cost considerations **EM2:** 622
cost elements **EM1:** 97
creep **EM1:** 293
creep data **EM2:** 621
crystalline, injection molding screws for **EM1:** 166
and cyanates, compatibility **EM2:** 234
damage tolerance **EM1:** 97-98, 293
electrical applications **EM2:** 591
electrical properties **EM2:** 589
engineering-grade, for resin transfer molding **EM1:** 169
environmental resistance **EM1:** 293
fabrication with thermoplastic prepreg **EM1:** 103-104
for filament winding **EM1:** 138
as fully reacted **EM1:** 142
future trends **EM1:** 142
grades of **EM2:** 98
high-density polyethylenes (HDPE) ... **EM2:** 163-166
high-impact polystyrenes (PS HIPS) **EM2:** 194-199
for high-modulus composites suitability **EM1:** 98-100
high-performance **EM1:** 43, 100-101
high-temperature **EM1:** 138, 142
homopolymer and copolymer acetals (AC) **EM2:** 100-102
impact strength, tests **EM1:** 292-293
impregnation problems **EM1:** 101-103
introduction **EM2:** 98-99
ionomers **EM2:** 120-123
linearity/reprocessability **EM1:** 100
liquid crystal polymers (LCP) **EM2:** 179-182
low-performance, for resin transfer molding **EM1:** 169
melt impregnation **EM1:** 101-102
melt values **EM1:** 294
melt viscosity **EM1:** 102
military specifications **EM2:** 90

Thermoplastic resins (continued)
moisture effects **EM2:** 767-768
molecular structure **EM2:** 619-621
molecular weights **EM1:** 101
neat form **EM2:** 98
physical, chemical, thermal analysis **EM2:** 533-543
polyamide-imides (PAI) **EM2:** 128-137
polyamides (PA) **EM2:** 124-127
polyaryl sulfones (PAS) **EM2:** 145-146
polyarylates (PAR) **EM2:** 138-141
polyaryletherketones (PAEK, PEK PEEK, PEKK) **EM2:** 142-144
polybenzimidazoles (PBI) **EM2:** 147-150
polybutylene terephthalates (PBT) **EM2:** 153-155
polycarbonates (PC) **EM2:** 151-152
polyether sulfones (PES, PESV) **EM2:** 159-162
polyether-imides (PEI) **EM2:** 156-158
polyethylene terephthalates (PET) **EM2:** 172-176
polyphenylene ether blends (PPE PPO) **EM2:** 183-185
polyphenylene sulfides (PPS) **EM2:** 186-191
polysulfones (PSU) **EM2:** 200-202
polyvinyl chlorides (PVC) **EM2:** 209-213
problems **A9:** 30
product design **EM2:** 622-623
programs, government-funded **EM1:** 103
properties **A9:** 29 **EM1:** 101, 292, 364 **EM2:** 618-619
for pultrusion matrices **EM2:** 394-395
reinforced polypropylenes (PP) **EM2:** 192-193
solvent/chemical resistance **EM1:** 293-294
in space and missile applications **EM1:** 817
styrene-acrylonitriles (SAN, OSA ASA) **EM2:** 214-216
styrene-maleic anhydrides (S/MA) **EM2:** 217-221
target properties **EM1:** 544
tensile modulus **EM1:** 544
tensile strengths **EM1:** 544
tests for **EM1:** 292-294
thermal expansion **EM1:** 294
thermoplastic fluoropolymer **EM2:** 115-119
thermoplastic polyimides (TPI) **EM2:** 177-178
thermoplastic polyurethanes (TPUR) **EM2:** 203-208
thermoset resins, trade-offs for commercial aircraft composites **EM1:** 100
and thermosetting resins compared **EM2:** 222, 225
toughness tests **EM1:** 293
ultrahigh molecular weight polyethylenes (UHMWPE) **EM2:** 167-171
viscoelasticity **EM2:** 625
viscosity **EM1:** 294
Thermoplastic stamping **EM1:** 562-563
molded-in color **EM2:** 306
surface finish **EM2:** 303
textured surfaces **EM2:** 305
Thermoplastic structural foam injection molding **EM1:** 557
Thermoplastic structural foams *See also* Foams; Injection molding
aesthetic concerns **EM2:** 512-513
blow agents **EM2:** 510-512
foam cell effects **EM2:** 508-509
part performance **EM2:** 508-510
properties **EM2:** 508-513
Thermoplastics *See* Thermoplastic matrix composites; Thermoplastic resins **EM3:** 29, 35, 74, 151
100% solid **EM3:** 75
abrasive wear resistance **A18:** 490
for adhesive bonding of aircraft canopies and windshields **EM3:** 563
aromatic polyarylates (PARS) **EM2:** 66
aromatic polysulfones (PSU) **EM2:** 66
compared to epoxies **EM3:** 98
as composites **A11:** 731
for die attach in device packaging **EM3:** 584
Thermoplastics (continued)
for die attachment and interconnection **EM3:** 580
elastomers **EM3:** 85
additives **EM3:** 85
bonding substrates and applications **EM3:** 85
properties **EM3:** 82, 85
for tapes and labels **EM3:** 85
elastomers, chemistry **EM2:** 66
for electronic packaging applications **EM3:** 601
epoxy bonding **EM3:** 96
fillers for epoxies **EM3:** 99
film adhesives **EM3:** 35
fluoroplastics **EM2:** 66
fluoropolymers **EM3:** 29
friction welding **A6:** 317
glassy, stress crazing **EM2:** 797-800
hot melts **EM3:** 44
for electronic general component bonding **EM3:** 573
headliner adhesives **EM3:** 552
for surface-mount technology bonding **EM3:** 570
hydrocarbon resins as tackifiers **EM3:** 182
hydrochloric acid corrosion **A13:** 1164
for industrial sealants **EM3:** 58
inorganic polymers **EM2:** 66
for insulated double-pane window construction **EM3:** 46
liquid crystal polymers **EM2:** 66
military specifications **EM2:** 90
nondestructive testing **A6:** 1087
nonreactive **EM3:** 97
in oil/gas production **A13:** 1243-1244
polyesters **EM3:** 29, 576
surface preparation **EM3:** 291
polyetheretherketone (PEEK), fracture analysis **EM3:** 345
polyetherketone **EM2:** 66
polyimides (TPI) **EM3:** 29, 152-153
polyimidesulfone **EM3:** 355-359, 513
mechanical properties **EM3:** 360
overlap edge stress concentration for single-lap joints **EM3:** 360-361
polymer chemistry of **EM2:** 63-67
polymer, prebond treatment **EM3:** 35
polymers as **EM2:** 48
polyphenylene sulfide (PPS) **EM2:** 66
polyurethanes (TPUR) **EM3:** 29, 121
properties **EM3:** 74, 601
styrene-maleic anhydride (SMA) **EM2:** 66
surface preparation **EM3:** 278-280
abrasion **EM3:** 278-279
chromic acid etching **EM3:** 278-279
satinizing **EM3:** 278
tensile creep **EM2:** 113
as toughener **EM3:** 185
vs thermosets, properties **EM2:** 437
Thermoplastics and Thermosets
Plastics (9th) (D.A.T.A) **EM2:** 94
Thermoplastics: Materials Engineering (Mascia) **EM2:** 94
Thermoreactive deposition/diffusion process (TRD) **A4:** 448--453
carbide coatings **A4:** 451-453
chromium effects **A4:** 449, 451
compared to chemical vapor deposition **A4:** 448
niobium effects **A4:** 451
process characteristics **A4:** 448-451
temperature effect on coating thickness **A4:** 449
time effect on coating thickness **A4:** 449
tooling applications **A4:** 451-452
Thermoset
defined **EL1:** 1159 **EM1:** 24 **EM2:** 42
epoxies, hardness **EL1:** 817
materials, for base/insulators **EL1:** 113
Thermoset (TS) processing
comparative assessment **EM1:** 142
Thermoset engineering plastics *See also* Thermosetting resins
types, volumes, costs **EM2:** 222
Thermoset injection molding *See also* Injection molding **EM1:** 558
defined **EM2:** 42, 319
injection-moldable thermoset resins **EM2:** 321-322
molding methods **EM2:** 319-321
reaction injection molding, reinforced (RRIM) **EM2:** 320-321
reciprocating-screw molding **EM2:** 319-320
thermoset selection **EM2:** 322-323
transfer molding **EM2:** 319-320
Thermoset matrix composites
high-temperature, properties **EM1:** 373-380
low-temperature, properties **EM1:** 392-398
medium-temperature **EM1:** 399-415
medium-temperature, properties **EM1:** 381-391
process modeling of **EM1:** 500-501
Thermoset molding compounds *See also* Bulk molding compounds (BMC)
bulk properties **EM1:** 161-162
curing **EM1:** 167
for injection **EM1:** 165
shear/heat effects **EM1:** 164
specific gravity **EM2:** 84
Thermoset plastic
surface preparation **EM3:** 291
Thermoset polyester
for bonded auto lens assembly **EM3:** 575
Thermoset polymers
prebond treatment **EM3:** 35
Thermoset resins
bismaleimide **EM1:** 32, 289
cyanate esters **EM1:** 290
epoxies **EM1:** 32, 289
phenolic **EM1:** 32, 289-290
polyesters **EM1:** 290
polyimides **EM1:** 32, 290
properties and tests for **EM1:** 289-290
rheological behavior, dynamic mechanical spectroscopy for **EM1:** 654-656
and thermoplastic resins, trade-offs for commercial aircraft composites **EM1:** 100
Thermoset stamping *See also* Compression stamping; Stamping
properties effects **EM2:** 287
size and shape effects **EM2:** 292
Thermoset-based resin systems
for filament winding **EM1:** 138
Thermosets *See also* Thermosetting **EM3:** 29, 74, 76-78, 151, 182-183
bismaleimides (BMIs) **EM3:** 154-156
bisnadimides (BNIs) **EM3:** 154
chemical cross-linking reaction **EM3:** 151-157
as composite materials **A11:** 731
cross-linking reactions **EM3:** 423
curing methods **EM3:** 74
for die attach in device packaging **EM3:** 584
for die attachment and interconnection **EM3:** 580
epoxy bonding **EM3:** 96
microstructure **EM3:** 406-407
polymers as **EM2:** 48
processing and temperature effects **EM3:** 422-425
properties **EM3:** 74
reactive oligomer end groups **EM3:** 154
starting materials **EM2:** 55
surface preparation **EM3:** 277-278
toughening methods **EM3:** 185
vs thermoplastics, properties **EM2:** 437
Thermosetting **EM3:** 29, 35
polyesters **EM3:** 29
polyimides **EM3:** 153-158
polyurethanes **EM3:** 278
Thermosetting acrylics **A13:** 405
Thermosetting coatings
cross-linked **A13:** 406-410
epoxies **A13:** 407
urethane **A13:** 408-410

Thermosetting diallyl phthalate
as a mounting for hafnium **A9:** 497
as a mounting for zirconium and zirconium alloys **A9:** 497
Thermosetting epoxy resins as mounting materials for
carbon and alloy steels **A9:** 166-167
carbonitrided steels **A9:** 217
carburized steels **A9:** 217
Thermosetting lacquers
use of **M5:** 626-627
Thermosetting molding materials
electrical properties **EM2:** 590
Thermosetting polyamide resins **EM2:** 124
Thermosetting polyesters *See also* Liquid resins; Polyesters; Solid resins; Unsaturated polyesters
defined **EM1:** 24 **EM2:** 42-43
as low-temperature Fesin system **EM2:** 440
Thermosetting processes
cold press molding, properties effects **EM2:** 287
combinations of **EM2:** 287
compression molding, properties effects **EM2:** 286-287
effects, on properties **EM2:** 286-287
filament winding, properties effects **EM2:** 287
hand lay-up, properties effects **EM2:** 287
high-speed resin injection, properties effects **EM2:** 287
injection molding, properties effects **EM2:** 287
powder compression molding, size and shape effects **EM2:** 291
properties effects **EM2:** 286-287
reaction injection molding (RIM), properties effects **EM2:** 287
resin transfer molding (RTM), properties effects **EM2:** 287
sheet molding compound (SMC), size and shape effects **EM2:** 291
shrinkage from **EM2:** 280
spray lay-up, properties effects **EM2:** 287
and thermoplastic processes, compared **EM2:** 289
Thermosetting pultrusion *See also* Pultrusion
defined **EM2:** 303
Thermosetting resins *See also* Chemical analysis (thermoset resins); Compression- mounting materials; Engineering plastics; Engineering plastics families; Polymer families; Polymer(s); Resins **A13:** 329 **EM2:** 222-225, 626-631
allyls (DAP, DAIP) **EM2:** 226-229
aminos **EM2:** 230-231
applications **EM2:** 223-224
bismaleimides (BMI) **EM2:** 252-256
characteristics **EM2:** 222-223
chemical structures **EM2:** 55
classification **EM2:** 626
cross linking/characteristics **EM2:** 626-627
and cross-linked thermoplastics **EM2:** 631
cure monitoring **EM2:** 528
cyanates **EM2:** 232-239
defined **EM2:** 222
electrical applications **EM2:** 589
electrical properties **EM2:** 589
for elevated temperature service **EM2:** 319
epoxies (EP) **EM2:** 240-241
future trends **EM2:** 224-225
injection-moldable **EM2:** 321-322
introduction **EM2:** 222-225
material development **EM2:** 626
materials selection **EM2:** 627-631
matrix properties **EM2:** 235
military specifications **EM2:** 90
moisture effects, mechanical properties **EM2:** 766-767
phenolics **EM2:** 242-245
physical, chemical, thermal analysis **EM2:** 517-532
polymer chemistry of **EM2:** 63-64
polyurethanes (PUR) **EM2:** 257-264
problems **A9:** 30
processing **EM2:** 223-224
properties **A9:** 29
research and development of **EM2:** 631
selection, for injection molding **EM2:** 322-323
silicones (SI) **EM2:** 265-267
thermal properties **EM2:** 439-444
Thermosetting resins (continued)
and thermoplastic resins compared **EM2:** 222, 225
unsaturated polyesters **EM2:** 246-251
urethane hybrids **EM2:** 268-271
vinyl esters **EM2:** 272-275
Thermosetting silicone and polyester styrene resins **M5:** 621
Thermosonic bonding **A6:** 325 **EM3:** 584-585
defined **EL1:** 1159
as electrical interconnection **EL1:** 225-226
wire, thermal failures **EL1:** 62
Thermostat bi-metal **A6:** 962
Thermostat metals **M3:** 640, 645, 646
electrical resistance alloys **A2:** 826-827
Thermotropic liquid crystal **EM3:** 29
defined **EM2:** 43
Thiazoline derivatives
as biocides **A18:** 110
Thiazolyazoresorcinol
as metallochromic indicator **A10:** 154
Thick film circuits **EM4:** 1140-1144
applications **EM4:** 1140, 1144
compositions and processes
for dielectrics **EM4:** 1143-1144
for thick film conductors **EM4:** 1141-1142
for thick film resistors **EM4:** 1142-1143
general compositions for thick films **EM4:** 1140-1141
metallo-organic deposition **EM4:** 1141
thick film tape **EM4:** 1141
Thick film printing
viscoelastic properties required **EM4:** 116
Thick film vacuum coatings for high-temperature protection **M5:** 395, 397, 399-400, 402-403, 409
Thick film(s) *See also* Film(s); Thick-film circuits; Thick-film hybrids; Thick-film pastes; Thick-film technology
ceramic wiring boards, thermal expansion **EL1:** 614-615
chip inductors, fabrication **EL1:** 187-188
metallization, ceramic packages **EL1:** 462-463
properties/advantages **EL1:** 321-322
resistors **EL1:** 343-345
vs thin films, for hybrids **EL1:** 321-322
Thick films *See also* Films; Thin films
analysis of **A10:** 561
LEISS analyses for **A10:** 603
RBS analysis for **A10:** 631
Thick, infinitely
XRS samples **A10:** 93
Thick samples
and light absorption **A10:** 61
PIXE analysis **A10:** 102
Thick single-stage plastic replicas
formation **A12:** 180-181
Thick stock
piercing of **A14:** 46
Thick-film applications
gold powders **A7:** 149-150
nonprecious metals in **A7:** 151
precious metal powders **A7:** 151
Thick-film circuits
in hybrid IC systems **EL1:** 249
hybrid microcircuit packages, types **EL1:** 451-454
phases **EL1:** 249
screen printing **EL1:** 206
Thick-film dielectrics
physical characteristics **EL1:** 108-109
Thick-film hybrids
ceramic substrates **EL1:** 206-208, 334-338, 386
cermet paste systems **EL1:** 339-345
circuitry **EL1:** 206
component attachment technology **EL1:** 347-351
development **EL1:** 332
industrial, defined **EL1:** 381
medical and military applications **EL1:** 386-389
multilayer, as VLSI packaging approach **EL1:** 269-270
polymeric paste systems **EL1:** 345-347
polymeric substrates **EL1:** 338-339
substrates **EL1:** 334-339
thick-film multilayer interconnects **EL1:** 347
thick-film pastes **EL1:** 339-347
thick-film process **EL1:** 332-334
Thick-film inks
for hybrid microcircuits **EL1:** 207
Thick-film lubrication *See also* Thin-film lubrication **A18:** 89-94
defined **A18:** 19
for sheet metal forming **A14:** 51
Thick-film market
commercial applications **EL1:** 381
Thick-film multilayer interconnects
hybrid **EL1:** 347
Thick-film pastes
for ceramic multilayer packages **EL1:** 460, 462-463
cermet systems **EL1:** 339-345
polymeric systems **EL1:** 345-347
Thick-film printing process **EL1:** 332
Thick-film resistors
properties **EL1:** 343-345
Thick-film technology
Hybrid microcircuits **EL1:** 255-256, 258
of hybrids **EL1:** 255-256, 258, 332-353
vs thin-film **EL1:** 255-257
Thick-lip ruptures
in steam-generator tubes **A11:** 605-606
Thick-plate equation **A6:** 14
Thick-sample reflection technique
pole figures determined by **A10:** 360
Thickener
defined **A18:** 19
Thickeners
as lubricant additives **A14:** 51
for sheet metal compounds **EM1:** 158
Thickness *See also* Casting section thickness; Section thickness; Thickness gaging; Thickness measurement; Wall thickness
abbreviation for **A10:** 691
aluminum-zinc alloy coatings **A13:** 436
of anodized coatings **A13:** 397
of blanking work metal **A14:** 45
case, for Scleroscope hardness testing **A8:** 105
castings with various **A15:** 581-582
center-cracked tension specimen **A8:** 381
of chromate conversion coatings **A13:** 392
compact-type specimen **A8:** 381
conductor, design effects **EL1:** 517
conformal coating **EL1:** 761
of corrosion product, gas-dryer piping **A11:** 631
of crack growth specimen **A8:** 381
distribution, interlaminar shear stress **EM1:** 240
dry-film **A13:** 417
for durometer testing **A8:** 106
effect in polycarbonate sheet **A11:** 762
effect in rocking curve profile for epitaxial films **A10:** 375
effect in wrinkling **A8:** 551
effect on crack growth **A11:** 54
effect on critical stress-intensity factor **A11:** 54
effect on plane-strain **A8:** 551
effect on plastic buckling by overloading **A11:** 137
effect on toughness **A11:** 51, 54
effect on x-ray energy in crystals **A10:** 367-370
effect, radiographic inspection **A17:** 295
effective, woven E-glass cloth
effects, abrasive waterjet cutting **A14:** 748-74
for electrodeposited bright nickel-chromium coatings **A13:** 428
of electrolyte layer **A13:** 82
electroplated hard chromium **A13:** 871-872
fiber calculation for filament winding **EM1:** 508
gaging, microwave inspection **A17:** 211-212
galvanized steel/aluminized steel coatings **A13:** 435
intermetallic compounds **EL1:** 635
of laminates
limit, oxyfuel gas cutting **A14:** 72
loci, by eddy current inspection **A17:** 172
loss, marine corrosion **A13:** 542
measurements **A14:** 879, 929-93
measurements, sheet metal forming **A8:** 548-549
measurements, uniaxial tensile testing **A8:** 554-555
metal, and press bending **A14:** 52
metal, for three-roll forming **A14:** 61
for metal powder cutting **A14:** 72
metal, with ultrasonic thickness data analysis **A13:** 202

Thickness (continued)
and NDE method selection ... **A17:** 51
nonconductive coatings, eddy current inspection ... **A17:** 164
nonmagnetic metal coating on magnetic material ... **A17:** 164
oxide scales ... **A13:** 70
of plated coatings ... **A13:** 425
plating, inspection of ... **EL1:** 942-943
porcelain enamel ... **A13:** 448, 451
printed board coupon ... **EL1:** 575-576
of pultruded product, compared ... **EM1:** 541
radiographic methods ... **A17:** 296
radomes, microwave inspection ... **A17:** 202
reduction, atmospheric corrosion ... **A13:** 518
reduction, in rolling ... **A14:** 34
sample, and light absorption ... **A10:** 61
of selected structural metals ... **A2:** 478
of shapes, wrought aluminum alloy ... **A2:** 35
sheet, for press forming ... **A14:** 50
sheet, in deep drawing ... **A14:** 58
specimen, for EELS analysis ... **A10:** 450
of specimens ... **EM1:** 297
steel, for drop hammer forming ... **A14:** 65
stress and strain, true ... **A8:** 559
symbol for ... **A8:** 724 **A11:** 797
of thin films, determined ... **A10:** 100, 631-632
through-, cracks as ... **A11:** 51
tin/tin alloy coatings ... **A13:** 775-776
-to-width ratio, fatigue cracked specimens ... **A8:** 377
of tube scale vs temperature, for values of heat flux ... **A11:** 609
ultrasonic inspection ... **A17:** 240
ultrasonic measurement ... **A13:** 200
for ultrasonic ply cutting ... **EM1:** 616
vs theoretical intensity, single-element x-ray spectrometry ... **A10:** 100
wall, flaw detection through ... **A17:** 139
wall, for tube spinning ... **A14:** 67
of walls, in heat exchangers ... **A11:** 628
weathering steel curtain walls ... **A13:** 520
wet-film ... **A13:** 417
work metal, for press forming ... **A14:** 548-54
of workpiece, Brinell test ... **A8:** 85-86, 88
workpiece, in deep drawing ... **A14:** 57
zinc coating ... **A13:** 756, 759

Thickness blocks
ultrasonic inspection ... **A17:** 263

Thickness, dielectric plastic
and breakdown ... **EM2:** 466

Thickness gaging *See also* Thickness; Thickness measurement
microwave inspection ... **A17:** 211-212
ultrasonic inspection ... **A17:** 273
ultrasonic welding inspection ... **A17:** 597

Thickness measurement *See also* Thickness; Thickness gaging
blocks, ultrasonic inspection ... **A17:** 263
boilers/pressure vessels, by ultrasonic inspection ... **A17:** 649-650
neutron radiography ... **A17:** 390
ultrasonic inspection ... **A17:** 232, 240, 273-274

Thickness measurements *See also* Thickness
optical metallography ... **A10:** 299-308
Rutherford backscattering spectrometry ... **A10:** 628-636
scanning electron microscopy ... **A10:** 490-515
x-ray spectrometry ... **A10:** 82-101

Thickness, steel plate, effect on mechanical properties ... **M1:** 188-189, 194, 196

Thickness-to-diameter ratio (t/d)
soft interlayer for solid-state welding ... **A6:** 166, 167

Thief ... **A7:** 12
defined ... **EL1:** 1159

Thieves
as sampling devices ... **A10:** 16

Thin film
analyses ... **A11:** 36, 41
contaminants, adhesion failures caused by ... **A11:** 43
lubrication, friction and wear in ... **A11:** 150
lubrication, of rolling-element bearings ... **A11:** 510

Thin film diffractometers
XRPD analysis ... **A10:** 337

Thin film, theories of *See also* Film ... **A13:** 67

Thin film(s) *See also* Thin-film capacitors; Thin-film chip resistors; Thin-film circuits Thin-film conductors; Thin-film hybrids; Thin-film inductors
conformal coatings as ... **EL1:** 759
dielectrics, physical characteristics ... **EL1:** 108-109, 323-324
metallization, as new multichip technology ... **EL1:** 299
multichip modules, as future trend ... **EL1:** 391
package, defined ... **EL1:** 1159
technology, hybrid microcircuits ... **EL1:** 257
vs thick films ... **EL1:** 255-257, 321-322

Thin films *See also* Films; Thick films; Thin samples; Ultrathin films
carbon ... **A12:** 173
changes in nickel on silicon ... **A10:** 631, 632
characterization of ... **A10:** 559-561
with columnar growth morphology ... **A10:** 544
composition and layer thickness of ... **A10:** 631-632
composition profiles by XPS ... **A10:** 568
compositional AES analysis ... **A10:** 549, 559-561
defects revealed by differential interference contrast ... **A9:** 59
diffractometer ... **A10:** 337
FIM/AP study of local composition variation ... **A10:** 583
impurity analysis in LPCVD ... **A10:** 624
molecular structure and orientation in ... **A10:** 109
nucleation and growth ... **A10:** 583
passive ... **A10:** 557-558
rocking curve analyses of ... **A10:** 371
Rutherford backscattering spectrometry analysis ... **A10:** 628
sample preparation for ATEM ... **A10:** 452
as samples, x-ray spectrometry ... **A10:** 95, 100
SIMS analysis of surface layers ... **A10:** 610
solvent evaporation for infrared analysis ... **A10:** 112
thickness determined ... **A10:** 100
transmission electron microscopy used to study ... **A9:** 103-122
x-ray intensity vs thickness, single element analysis ... **A10:** 100

Thin foil
microhardness testing of ... **A8:** 96

Thin foils
used in scanning electron microscopy ... **A9:** 95

Thin immersion plates
as surface preparation ... **EL1:** 679

Thin, infinitely
XRS samples ... **A10:** 93

Thin plastic components
design/analysis of ... **EM2:** 691-700

Thin samples
PIXE analysis of ... **A10:** 102

Thin sections *See also* Section thickness
by ceramic molding ... **A15:** 248
by plaster molding ... **A15:** 242

Thin sheet specimens
buckling in axial compression ... **A8:** 56

Thin sheet tin and tin alloy coated specimens
preparation of ... **A9:** 450-451

Thin single-stage plastic replicas
formation ... **A12:** 179-180

Thin slab casting ... **A1:** 211

Thin stock
confined explosive forming of ... **A14:** 63
electrical sheet, blanking and piercing ... **A14:** 479-48
piercing of ... **A14:** 467-46

Thin strip casting ... **A1:** 210, 211

Thin wire
diffraction pattern techniques for ... **A17:** 13

Thin-film capacitors
for hybrids ... **EL1:** 320-321

Thin-film chip resistors
chromium, corrosion failure analysis ... **EL1:** 1114-1115
defined ... **EL1:** 178
fabrication ... **EL1:** 185
failure mechanisms ... **EL1:** 1003

Thin-film circuits
defined ... **EL1:** 1159
in hybrid integrated circuitry ... **EL1:** 249
hybrid microcircuits, package types ... **EL1:** 451-454

Thin-film conductors
films ... **EL1:** 318-320
multilevel, hybrid structures ... **EL1:** 324

Thin-film hybrids ... **EL1:** 313-331
dielectrics, multilayer ... **EL1:** 323-324
fabrication ... **EL1:** 313-316, 326-330
film preparation ... **EL1:** 313-316
generic microcircuit ... **EL1:** 313
multilayer, as VLSI packaging approach ... **EL1:** 270
multilevel hybrid structures ... **EL1:** 322-324
polyimide chemistry and properties ... **EL1:** 324-326
processing thin-film hybrid structures ... **EL1:** 326-330
substrate-conductor/resistor system ... **EL1:** 315
thin films vs thick films ... **EL1:** 321-322
thin-film capacitors/inductors ... **EL1:** 320-321
thin-film resistor/conductor design ... **EL1:** 316-320

Thin-film inductors
for hybrids ... **EL1:** 320-321
types ... **EL1:** 323

Thin-film lubrication *See also* Thick-film lubrication ... **A18:** 89, 94-96
asperity lubrication modes ... **A18:** 94-96
micro-EHL and friction polymer films ... **A18:** 94-95
oxide film ... **A18:** 94, 96
physically adsorbed and other surface films ... **A18:** 94, 95-96
defined ... **A18:** 19
for sheet metal forming ... **A14:** 51

Thin-film materials *See also* Superconducting materials
applications ... **A2:** 1083
electron-beam coevaporation ... **A2:** 1081
future outlook ... **A2:** 1083
in situ film growth ... **A2:** 1082
sputtering techniques ... **A2:** 1081-1082
substrates and buffer layers ... **A2:** 1081
superconducting materials ... **A2:** 1081
superconducting properties ... **A2:** 1082-1083
thin-film deposition techniques ... **A2:** 1081-1083

Thin-film resistors
design ... **EL1:** 316-318
nichrome, light microscope photograph ... **EL1:** 1099

Thin-film resistors (electrical)
powders used ... **A7:** 573

Thin-film resistors (electronic)
powders used ... **A7:** 573

Thin-film scratch test ... **A18:** 421

Thin-film thermocouples (TFTCS) ... **A18:** 440-441

Thin-foil specimens
direct observation ... **A12:** 179
electropolishing of magnetic materials ... **A9:** 534
Lorentz microscopy ... **A9:** 536
stainless steel electropolishing ... **A9:** 283
wrought heat-resistant alloys ... **A9:** 307-308

Thin-foil specimens, sectioning of ... **A9:** 25
used to observe cell structure development ... **A9:** 693-694

Thin-foil transmission electron microscopy
observation of $AuCu_3$ Structures using ... **A9:** 682
used to monitor Lüders front ... **A9:** 685

Thin-foil transmission microscopy
and surface replication ... **A17:** 52
Thin-layer chromatography
of epoxies ... **EL1:** 834
Thin-layer chromatography (TIC)
defined ... **EM1:** 24
Thin-layer chromatography (TLC) ... **EM2:** 43, 520-521 **EM3:** 29
Thin-lip ruptures, as transgranular
tensile fractures ... **A11:** 606
Thin-metal holding devices
for microhardness test specimens ... **A8:** 96
Thin-plate equation ... **A6:** 13-14, 15
Thin-plate theory
for laminates ... **EM1:** 220
Thin-section polarized light microscopy
of fiber composites ... **A9:** 592
Thin-sheet specimens
mounting ... **A9:** 31
Thin-wall casting
by FM process ... **A15:** 38
defined ... **A15:** 11
Thin-wall tube
inspection ... **A17:** 169-170, 172
mechanically alloyed oxide dispersion-strengthened (MA ODS) alloys ... **A2:** 949
Thin-wall tubes
bending of ... **A14:** 671-67
Thinner ... **EM3:** 29
"Thinnest outer sheet" (TOS) ... **A6:** 240
Thinning
carbon steel ... **A17:** 200-201
corrosion ... **A17:** 50, 195
electrochemical vs electrojet ... **A10:** 451
in polymers ... **A12:** 479
r value as measure of resistance to ... **A8:** 553
radiographic methods ... **A17:** 296
as sample preparation technique ... **A10:** 450-452
with superplastic metals ... **A14:** 861-867
of tube walls ... **A11:** 606-607, 617
Thinning of transmission electron microscopy specimens ... **A9:** 105-108
Thiobacillus bacteria
low-carbon steel fracture from ... **A12:** 245
Thiobacillus thiooxidans
biological corrosion by ... **A13:** 119
Thiokol liquid polymer ... **EM3:** 96
Thiols
reaction with epoxies ... **EM3:** 96
Thiosulfate pitting corrosion ... **A13:** 349, 352, 1190, 1205
Third Law of Thermodynamics ... **A3:** 1 • 7
Third-level package
materials/process selection ... **EL1:** 116-117
Third-particle abrasive wear
in bearings ... **A11:** 494-495
Thixocasting *See also* Rheocasting; Semisolid metal casting and forging
defined ... **A15:** 328
and rheocasting, compared ... **A15:** 328
Thixotrope
sheet metal compounds ... **EM1:** 141
Thixotropic
agent, paste print resolution effect ... **EL1:** 732
defined ... **EL1:** 1159 **EM1:** 24 **EM2:** 43
loop, rheological characterization ... **EL1:** 840-842, 847
Thixotropic additives
for sealants ... **EM3:** 674
Thixotropic paste adhesives ... **EM3:** 35
Thixotropic properties
of semisolid metals ... **A15:** 327-328
Thixotropy *See also* Rheopectic material ... **EM3:** 29
defined ... **A18:** 19
of fillers ... **EM3:** 177, 178
Thoma number ... **A18:** 599
Thomas Register ... **EM3:** 70
Thomson effect ... **A6:** 31
Thomson, William ... **A3:** 1 • 7
Thoria
property data ... **EM4:** 863
safety precaution ... **A6:** 191
thermal expansion coefficient ... **A6:** 907
thermionic emission production ... **A6:** 30
Thoriam insulation
for thermocouples ... **A2:** 883
Thoriated tungsten ... **A9:** 442
Thoriated tungsten wire ... **A7:** 153
annealed ... **A9:** 446
Thorium
as actinide metal, properties ... **A2:** 1195
classification in tungsten alloy electrodes for GTAW ... **A6:** 191
content in magnesium alloys ... **M6:** 427
detector mounting in radionuclide methods ... **A18:** 326
epithermal neutron activation analysis ... **A10:** 239
M lines for ... **A10:** 86
in magnesium alloys ... **A9:** 427-428
plutonium diffused into ... **A10:** 249
powder metallurgy ... **A7:** 18
pure ... **M2:** 807-810, 823-833
as pyrophoric ... **A7:** 199
species weighed in gravimetry ... **A10:** 172
thermal expansion coefficient ... **A6:** 907
TNAA detection limits ... **A10:** 238
ultrapure, by chemical vapor deposition ... **A2:** 1094
vacuum heat-treating support fixture material ... **A4:** 503
vapor pressure, relation to temperature ... **A4:** 495
weighed as the fluoride ... **A10:** 171
Thorium dioxide
dispersion-strengthened nickel (TD nickel) ... **A7:** 55
effect in volatile tungsten compounds ... **A7:** 154
in tungsten powders for welding electrodes ... **A7:** 153
Thorium fluoride-based glasses
electrical properties ... **EM4:** 853
Thorium nitrate
doping of tungsten oxide ... **A7:** 153
Thorium oxide cermets
applications and properties ... **A2:** 993
Thorium oxide-containing cermets ... **A7:** 803-804
Thorium, vapor pressure
relation to temperature ... **M4:** 310
Thorium-magnesium eutectic
effect on tantalum ... **A13:** 735
Thornel 50
properties ... **A18:** 803
Thornel 300
properties ... **A18:** 803
Thread *See* Fiber(s)
Thread chasers ... **A16:** 290-299
circular ... **A16:** 302
for Cu alloys ... **A16:** 814-815
dovetail ... **A16:** 726
nitrided high speed steel used ... **A16:** 302
operating details ... **A16:** 300
radial ... **A16:** 302
stainless steels ... **A16:** 697, 700-701
standard high-speed steel used ... **A16:** 299
tangential ... **A16:** 302
tapping ... **A16:** 258-259
thread grinding application ... **A16:** 278
tool life ... **A16:** 299, 300, 302
WC used ... **A16:** 302
for Zn alloys ... **A16:** 834
Thread count *See also* Count
defined ... **EM1:** 24 **EM2:** 43
Thread cutting dies
thread grinding application ... **A16:** 278
Thread defects
AISI/SAE alloy steels ... **A12:** 318
Thread grinding ... **A16:** 296, 270-279
applications ... **A16:** 270
centerless grinding of threads ... **A16:** 276-277
compared to thread rolling ... **A16:** 295
crush dressing of wheels ... **A16:** 273, 275, 276, 277
cylindrical grinding of threads ... **A16:** 273-276, 277
diamond dressing of wheels ... **A16:** 272-273, 275, 276, 277
grinding fluids ... **A16:** 273
grinding speed ... **A16:** 273
high-volume applications ... **A16:** 277-279
multirib wheel grinding ... **A16:** 275-276
production practice ... **A16:** 277-279
thread cutting taps ... **A16:** 278
thread gaging ... **A16:** 266, 267, 270, 278
thread grinding machine categories ... **A16:** 279
tolerances ... **A16:** 270-271
Thread grinding (continued)
truing grinding wheels ... **A16:** 272 273
wheel selection ... **A16:** 271-272
Thread inspection
automated/magnetic ... **A17:** 133
Thread joining
furnace brazing ... **M6:** 940
Thread milling ... **A16:** 268-269
carbide metal cutting tools ... **A2:** 965
compared to thread grinding ... **A16:** 277, 279
compared to thread rolling ... **A16:** 295
heat-resistant alloys ... **A16:** 751-754
machines, product-on thread mills ... **A16:** 268
multiple form thread milling cutter ... **A16:** 268, 269
NC machines ... **A16:** 269
planetary thread mills ... **A16:** 268
refractory metals ... **A16:** 867
single-form thread milling cutter ... **A16:** 268, 269
universal thread mills ... **A16:** 268
Thread milling cutters
thread grinding application ... **A16:** 278
Thread rolling ... **A16:** 280-295, 296
capacities and limitations ... **A16:** 280
compared to tapping ... **A16:** 265, 267
continuous rolling ... **A16:** 288
die life ... **A16:** 280-282, 284, 285, 287-291, 293, 294
end-feed rolling ... **A16:** 286-287, 289, 292
factors affecting die life ... **A16:** 289-291
flaking ... **A16:** 281
flat traversing die rolling ... **A16:** 283, 289, 290, 292, 293
fluids ... **A16:** 294-295
internal thread rolling ... **A16:** 288-289
load requirements and penetration rate ... **A16:** 292
MMCs ... **A16:** 896
planetary thread rolling ... **A16:** 287-288, 289, 293
preparation and feeding of work blanks ... **A16:** 281-283
radial-infeed rolling (cylindrical die) ... **A16:** 283-286, 288, 290-292, 294, 295
selection of rolling method ... **A16:** 289
stainless steels ... **A16:** 701-703
surface speed ... **A16:** 292
tangential rolling ... **A16:** 285-286, 292
thin-wall parts threading ... **A16:** 293
thread form effect on processing ... **A16:** 291
through-feed rolling ... **A16:** 286-287, 289, 290, 292, 293, 294
vs. alternative processes ... **A16:** 295
warm rolling ... **A16:** 292-293
work-hardening materials threading ... **A16:** 293
Thread rolling dies
as tool steels application ... **A7:** 792
Thread roots
surface flaws by ECP detection ... **A17:** 138-139
Thread-rolling dies
circular dies ... **M3:** 551, 552, 553
die life ... **M3:** 551, 552, 553
flat dies ... **M3:** 551, 552, 553
grinding ... **M3:** 552
hardness ... **M3:** 551, 552, 553
materials for ... **M3:** 551-553
Threaded end closures
hot isostatic pressure vessels ... **A7:** 420
Threaded fasteners *See also* specific types by name
case hardening ... **M1:** 276, 279
clamping forces ... **M1:** 280-282
compositions ... **M1:** 275, 277
corrosion in ... **A11:** 535-541
defined ... **A11:** 529
elimination of ... **EL1:** 123
fabrication ... **M1:** 274, 277, 280
failure types and origins ... **A11:** 529-531
fatigue in ... **A11:** 531-534
fatigue strength ... **M1:** 275-276, 279-280, 282
forging to prevent failure ... **A11:** 532, 534
grade markings and identification code ... **A11:** 529, 531
hardness ... **M1:** 278-281
heat treatment ... **M1:** 276-277
ISO property classes ... **M1:** 274, 275, 277
mechanical properties ... **M1:** 274, 278-282
proof testing ... **M1:** 277-278
roll threading ... **M1:** 274-276
SAE strength grades ... **M1:** 274-275, 277
selection of steel for ... **M1:** 273-277

Threaded fasteners (continued)
steels for, cost comparison **M1:** 276
steels for, hardenability comparisons **M1:** 276
testing of **M1:** 277-278
Threaded grip ends
for torsion specimens **A8:** 156-157
Threaded joints
design **EM3:** 53
Threaded steel fasteners **A1:** 289-301
clamping forces of **A1:** 300-301
corrosion protection for **A1:** 291
aluminum coatings **A1:** 295
cadmium coatings **A1:** 295
zinc coating **A1:** 295
fabrication **A1:** 300
platings and coatings **A1:** 300
fastener performance at elevated temperatures **A1:** 295
bolt steels for elevated temperatures **A1:** 296, 620
coatings for elevated temperatures **A1:** 296
effect of thread design on relaxation **A1:** 296
relaxation strengths **A1:** 631
time- and temperature-related factors **A1:** 295-296
fastener tests **A1:** 296
proof stress of a bolt or stud **A1:** 296-297
proof stress of nut **A1:** 297
wedge tensile test of bolts **A1:** 296
mechanical properties **A1:** 291, 297-300, 301
fatigue failures **A1:** 297-299, 300, 301
hardness versus tensile strength **A1:** 297
strengths with static loads **A1:** 297
stress-corrosion cracking (SCC) **A1:** 299-300
specifications and selection **A1:** 289, 290
steels for **A1:** 290-291, 292, 293
bolt steels **A1:** 290-291, 294
nut steels **A1:** 291, 292-294, 295
selection of steel for bolts and studs **A1:** 292
stud steels **A1:** 292
strength grades and property classes **A1:** 289-290
Threading
Al alloys **A16:** 766-767, 769-791
carbide metal cutting tools **A2:** 965
cemented carbides used **A16:** 75, 85, 87, 95
cermet tools applied **A16:** 92, 95, 96, 97
coated carbides used **A16:** 95
in conjunction with turning **A16:** 135, 154
in conjunction with ultrasonic machining **A16:** 530
Cu alloys **A16:** 813-815
cutting fluids used **A16:** 125
in machining centers **A16:** 393
MMCs **A16:** 896
multifunction machining **A16:** 366, 376, 379
Ni alloys **A16:** 835, 839
PCBN cutting tools used **A16:** 114, 115, 116
single-point **A16:** 296
stainless steels **A16:** 154, 155, 695
Ti alloys **A16:** 851
tool life **A16:** 302
and turning **A16:** 158
uranium alloys **A16:** 875
of wire **A14:** 69
Threading and knurling
definition **M6:** 18
Threading, rough *See* Rough threading
Threads *See also* Fiber(s)
internal/external, design of **EM2:** 616
molded-in, mechanical fastening **EM2:** 711
as notches or stress concentrator **A11:** 85, 318
polyamide-imides (PAI) **EM2:** 131
Three-blow headers
for cold heading **A14:** 29
Three-body abrasion **A18:** 537
Three-dimensional circuits **EL1:** 8, 10
Three-dimensional defect analysis
ATEM **A10:** 466
Three-dimensional grains and particles
equations for particle-size distribution **A9:** 132
Three-dimensional images
by coordinate measuring machines **A17:** 18
color **A17:** 486
holograms as **A17:** 405
interpretation **A17:** 42
pseudo, digital image enhancement **A17:** 463
reformation, computed tomography (CT) **A17:** 378
surface gaging, holographic **A17:** 12, 16
Three-dimensional packaging **EL1:** 79, 441, 444-445
Three-dimensional preforms **EM1:** 129-130
Three-dimensional reinforcements **EM1:** 129-131
Three-dimensional scanning electron microscopy images **A9:** 96-97
Three-dimensional stress analysis
laminates **EM1:** 239-240
Three-flute taps **A7:** 462
Three-high rolling mills **A14:** 35
Three-leaded power devices
discrete semiconductor **EL1:** 424-428
Three-parameter point stress criterion **EM1:** 255
Three-phase ac plasma ladle furnace
heating/degassing **A15:** 440-442
Three-phase direct-energy resistance welding machines **M6:** 537, 540, 545
Three-phase equilibrium .... **A3:** 1 • 3
Three-phase power supplies
flash welding **M6:** 559
Three-phase rectifiers
gas tungsten arc welding **M6:** 187-188
Three-point bend bars
for hydrogen embrittlement testing **A8:** 539-540
Three-point bend test **A8:** 132-135, 505 **A14:** 37
Three-point bending *See also* V-bend die
defined **A14:** 13
nickel alloys **A12:** 396
Three-roll bending
of beryllium **A14:** 80
Three-roll forming **A14:** 616-62
alternative processes **A14:** 62
of beryllium **A14:** 80
blank preparation **A14:** 61
cold vs hot forming **A14:** 619-620
and contour-roll forming **A14:** 623
and deep drawing **A14:** 623
diameter and width **A14:** 616
hot-forming temperatures, steel **A14:** 620
of large cylinders **A14:** 621
machines **A14:** 616-618
metal thicknesses **A14:** 616
metals formed **A14:** 616
power requirements **A14:** 620-621
roll deflection **A14:** 623
rolls **A14:** 618-619
safety **A14:** 623
shapes produced **A14:** 616
of small cylinders **A14:** 621
speed of **A14:** 623
stainless steels **A14:** 774-775
of titanium alloys **A14:** 846
of truncated cones **A14:** 621
of two halves **A14:** 623
Three-roll forming machines
conventional pinch-type **A14:** 616-617
pyramid-type **A14:** 617-618
shoe-type pinch-roll **A14:** 617
Three-sector journal bearing **A18:** 527-528
Three-terminal devices *See also* Multiple terminals; Two-terminal devices
discrete semiconductors **EL1:** 422-429
performance **EL1:** 423
power **EL1:** 427-428
radio-frequency (RF) packages **EL1:** 29
stud-mount **EL1:** 429
Three-zone furnace
for step-down tension testing **A8:** 324
Threshold
regime, cyclic crack growth rate testing in **A8:** 378-379
regime, ultrasonic fatigue testing **A8:** 241
stress-corrosion, correlation of smooth/precracked specimens **A8:** 498
stresses **A8:** 14, 499
Threshold crack tip stress intensity factor
effect on SCC **A12:** 24-25
Threshold energy
EXAFS analysis **A10:** 409
Threshold galling stress (TGS) **A18:** 718, 722
Threshold stress **A13:** 13, 276
effects of composition and SCC on **A11:** 204
intensity **A13:** 12, 276-277
Threshold stress intensity *See also* Stress-intensitv, factor (K)
conventional vs. ultrasonic resonance test methods **A8:** 256
in hydrogen embrittlement testing **A8:** 537
for hydrogen stress cracking **A8:** 537-541
and minimum crack growth rate **A8:** 254, 256
range, vs. sodium chloride for stainless steel and titanium alloys **A8:** 427, 429
for stress-corrosion cracking **A8:** 537-541
symbol for **A8:** 725
Threshold stress intensity (K_{th})
defined **A11:** 10
factor, symbol for **A11:** 797
and stress-corrosion cracking **A11:** 204-205, 797
Threshold-crossing pulses
in acoustic emission inspection **A17:** 281-283
Thresholding
with digital image enhancement **A17:** 457
gray-level, as feature detection mode, image analyzers **A10:** 311
use in image analysis **A10:** 311-312
Throat clearance
machine size selection by **A14:** 8
Throat depth **A6:** 41
definition **M6:** 18
Throat gap **A6:** 41
Throat height
definition **M6:** 18
Throat of a fillet weld
definition **A6:** 1214 **M6:** 18
Throat size of fillet welds **M6:** 63, 66-67
Throttle control linkage
hydrogen embrittlement of bolts in **A11:** 540-541
Through connection
defined **EL1:** 1159
Through hardening *See* Induction hardening; Through hardening
Through transmission system
eddy current inspection **A17:** 178
Through-die design
precision forging **A14:** 17
Through-hardened materials
for bearings **A11:** 490
Through-hardened steel
welding factor **A18:** 541
Through-hardening *See also* Hardening
gear materials **A18:** 261
in powder forging **A14:** 202
Through-hardening alloy steels
machinability of **A1:** 600-601
Through-hole circuit boards
design steps for **EL1:** 515-518
Through-hole drilling, mechanical
computer-aided **EL1:** 131
Through-hole mounting
defined **EL1:** 1159
Through-hole packages
capacitors **EL1:** 971-973
crystals **EL1:** 979
diodes **EL1:** 973-974
electromechanical components **EL1:** 979-981
failure mechanisms **EL1:** 969-981
integrated circuits (ICs) **EL1:** 975-979

Through-hole packages (continued)
integrated semiconductor packages...... **EL1:** 437-438
mounted plastic...... **EL1:** 210
resistors...... **EL1:** 970-971
transistors...... **EL1:** 974-975
Through-hole soldering
combined processes...... **EL1:** 694-695
computer control and data logging...... **EL1:** 695-696
debridging hot air knives...... **EL1:** 691-693
evolution...... **EL1:** 507
fluxing process...... **EL1:** 681-684
joint inspection...... **EL1:** 735
preheating process...... **EL1:** 684-688
reflow soldering techniques...... **EL1:** 693-694
solder waves...... **EL1:** 91
as wave soldering...... **EL1:** 681
Through-substrate electrical connections
optical...... **EL1:** 10
Through-substrate plated-through holes (TSPTH)
thermal expansion...... **EL1:** 624
Through-thickness center cracks *See* Center crack
Through-thickness cracks
fracture mechanics of...... **A11:** 51
Through-transmission ultrasonics
abbreviation for...... **A11:** 798
Through-transmission ultrasonics (TTU)...... **EM1:** 770
Through-wall cracking
austenitic stainless steels...... **A12:** 354
Through-wall examination
by electric current perturbation...... **A17:** 139
by remote-field eddy current inspection...... **A17:** 195
Throughput...... **A7:** 12
Throughput rate
and clock skew...... **EL1:** 7
Throw
defined...... **A14:** 13
Throwing power *See also* Covering power
defined...... **A13:** 13
Thrust ball bearings...... **A18:** 500, 505, 506, 507, 508
applicable load...... **A18:** 511
basic load rating...... **A18:** 505
carburizing...... **A18:** 875
f_v factors for lubrication method...... **A18:** 511
static equivalent axial load...... **A18:** 510
X and Y factors...... **A18:** 509
z and y factors...... **A18:** 511
Thrust bearing...... **A18:** 741
for centrifugal compressors...... **A18:** 607
defined...... **A18:** 19
Thrust bearings
applications...... **A11:** 490
electrical wear...... **A11:** 487
Thrust cylindrical roller bearings...... **A18:** 502, 503, 505
f_1 factors...... **A18:** 511
f_v factors for lubrication method...... **A18:** 511
Thrust loading
in bearings...... **A11:** 492
Thrust needle roller bearings...... **A18:** 502
basic load rating...... **A18:** 505
f_v factors for lubrication method...... **A18:** 511
Thrust roller bearings...... **A18:** 502, 503, 505, 507
applicable load...... **A18:** 511
static equivalent axial load...... **A18:** 510
Thrust spherical roller bearings
f_1 factors...... **A18:** 511
f_v factors for lubrication method...... **A18:** 511
Thrust tapered roller bearings...... **A18:** 502
basic load rating...... **A18:** 505
Thulium *See also* Rare earth metals
pure...... **M2:** 810-811
as rare earth metal, properties...... **A2:** 720, 1188
Thulium in garnets...... **A9:** 538
Thymol blue
as acid-base indicator...... **A10:** 172
Thymolphthalein
as acid-base indicator...... **A10:** 172
Thyratron tubes
resistance spot welding...... **M6:** 470
Thyristors...... **A6:** 37, 38-39, 40
Thyristors, silicon
acoustic microscopy of...... **A17:** 481
Ti Code 12 *See* Titanium alloys, specific types, Ti-0.3Mo-0.8Ni
Ti Tech 0.2 Pd *See* Titanium alloys, specific types, Ti-Pd alloys
Ti-0.2Pd *See* Titanium alloys, specific types, Ti-Pd alloys
Ti-3-2½ *See* Titanium alloys, specific types, Ti-3Al-2.5V
Ti-6-2-4-2 *See* Titanium alloys, specific types, Ti-6Al-2Sn-4Zr-2Mo
Ti-6-2-4-6 *See* Titanium alloys, specific types, Ti-6Al-2Sn-4Zr-6Mo
Ti-6-6-2 *See* Titanium alloys, specific types, Ti-6Al-6V-2Sn
Ti-6-22-22-S *See* Titanium alloys, specific types, Ti-6Al-2Sn-2Zr-2Cr-2Mo-0.25Si
Ti-6Al-4V *See* Titanium alloys, specific types; Titanium casting alloys, specific types; Titanium P/M products
Ti-8-1-1 *See* Titanium alloys, specific types, Ti-8Al-1Mo-1V
Ti-10-2-3 *See* Titanium alloys, specific types, Ti-10V-2Fe-3Al
Ti-17 *See* Titanium alloys, specific types, Ti-5Al-2Sn-2Zr-4Mo-4Cr
Ti-35A *See* Titanium alloys, specific types, Ti grade 1
Ti-50A *See* Titanium alloys, specific types, Ti grade 2
Ti-65A *See* Titanium alloys, specific types, Ti grade 3
Ti-75A *See* Titanium alloys, specific types, Ti grade 4
Ti-621/0.8 *See* Titanium alloys, specific types, Ti-6Al-2Nb-1Ta-0.8Mo
Ti-679 *See* Titanium alloys, specific types, Ti-2.25Al-11Sn-5Zr-1Mo
Ti-811 *See* Titanium alloys, specific types, Ti-8Al-1Mo-1V
Ti-5522S *See* Titanium alloys, specific types, Ti-5Al-5Sn-2Zr-2Mo-0.25Si
Ti-6242 *See* Titanium alloys, specific types, Ti-6Al-2Sn-4Zr-2Mo
Ti-6246 *See* Titanium alloys, specific types, Ti-6Al-2Sn-4Zr-6Mo
Ti-U (Phase Diagram)...... **A3:** 2•378
Ti-V (Phase Diagram)...... **A3:** 2•379
Ti-W (Phase Diagram)...... **A3:** 2•379
Ti-Y (Phase Diagram)...... **A3:** 2•379
Ti-Yb (Phase Diagram)...... **A3:** 2•380
Ti-Zn (Phase Diagram)...... **A3:** 2•380
Ti-Zr (Phase Diagram)...... **A3:** 2•380
Ticusil
brazing, composition...... **A6:** 117
wettability indices on stainless steel base metals...... **A6:** 118
wetting behavior and joining...... **EM4:** 492
Tidal zone
marine structures...... **A13:** 542-544
Tide marks *See* Beach marks
Tie bar
defined...... **A15:** 11
Tie lines...... **A3:** 1•8
Tie triangles...... **A3:** 1•8
Tied-arch floor beams
bridge...... **A11:** 714
Tier charts...... **EM3:** 791, 792
Ties, and anchors
in masonry walls...... **A13:** 1306-1310
Tifran...... **A18:** 781, 782
TIG welding *See* Gas tungsten arc welding
definition...... **A6:** 1214
TIGER (multilayer electron transport program)
energy deposition per incident electron...... **EM3:** 646
Tight binding method...... **A6:** 143
Tight flasks
green sand molding...... **A15:** 341
Tile whiteware
alternative/competitive materials...... **EM4:** 929
applications...... **EM4:** 927-928
ceramic tile types used for building in the United States...... **EM4:** 926
characteristics of ceramic tile...... **EM4:** 926
classification of ceramic tiles...... **EM4:** 925-926
common shapes and sizes for floor and wall applications...... **EM4:** 928
definitions...... **EM4:** 925
ceramic mosaic tile...... **EM4:** 925
decorative wall tile...... **EM4:** 925
individual tile whiteware grades...... **EM4:** 925
paver tile...... **EM4:** 925
porcelain tile...... **EM4:** 925
Tile whiteware (continued)
quarry tile...... **EM4:** 925
wall tile...... **EM4:** 925
development of the industry...... **EM4:** 925
disadvantages of tile...... **EM4:** 929
emerging materials...... **EM4:** 929
flat ceramic mosaic tile...... **EM4:** 972
flat glazed wall tile...... **EM4:** 927
flat paver tile...... **EM4:** 927
flat quarry tile...... **EM4:** 927
floor tiles...... **EM4:** 928
glazes...... **EM4:** 926-927, 1061
porcelain tiles...... **EM4:** 928-929
processing steps...... **EM4:** 926
properties...... **EM4:** 926
tile types...... **EM4:** 927
cottoforte...... **EM4:** 927, 929
decorated wall tiles...... **EM4:** 927
exterior wall tiles...... **EM4:** 927, 928
faience tiles...... **EM4:** 927
porous single-fired ("Monoporosa") tiles...... **EM4:** 927, 929
stove tiles (majolica, earthenware)...... **EM4:** 927, 929
water absorption characteristics...... **EM4:** 928
Tilghman, R.E
as early founder...... **A15:** 33
Tilt *See also* Tilt boundary; Tilt-twist boundary
angles across subgrain boundaries, evaluated by x-ray topography...... **A10:** 365
in area measurement...... **A12:** 208
casting, as permanent mold method...... **A15:** 275-276
control, electric arc furnace...... **A15:** 358-359
controlled, for phase/particle confirmation...... **A10:** 458-459
defined...... **A10:** 683
effect in determining orientation relationships...... **A10:** 453-454
effect in XPS depth analysis...... **A10:** 573
experiment, for unknown phase or particle confirmation...... **A10:** 458
method, stereo SEM...... **A12:** 171
specimen, effect in SEM imaging...... **A12:** 168
Tilt boundaries
defined...... **A9:** 719
direct imaging by high resolution electron microscopy...... **A9:** 121
schematic...... **A9:** 720
Tilt boundary
in Armco iron...... **A12:** 18
defined...... **A12:** 13
schematic...... **A12:** 17
Tilt boundary in gold...... **A9:** 609
Tilt furnace...... **A7:** 12
Tilt method
stereo SEM 1866...... **A12:** 171
Tilt pin
copper...... **A8:** 233-234
Tilt-twist boundary
defined...... **A12:** 13
in iron...... **A12:** 17
Tilting
furnaces, crucible...... **A15:** 382
ladle...... **A15:** 498
Tilting stages
photographic...... **A12:** 88
Tilting stages for microscopes...... **A9:** 82
Tilting table method
angle of repose measurement...... **A7:** 282
Tilting-pad bearing...... **A18:** 527
for centrifugal compressors...... **A18:** 607
defined...... **A18:** 19
Tilting-pad journal bearing...... **A18:** 527, 528
pivot circle clearance...... **A18:** 527
pivot design...... **A18:** 527
yaw stability...... **A18:** 527
Tilting-pad thrust bearing...... **A18:** 523, 525
Time *See also* Assembly time; Curing time; Duration; Gelation time; Get time; Heat treatment; Heating; Heating time; Relaxation time; Rise time; Temperature time curves; Time-temperature equivalency principle; Time-temperature super-position; Zero time
(at) temperature maximums, for magnesium alloys...... **A2:** 473
abbreviation for...... **A10:** 691
assembly, defined *See* Assembly time

Time (continued)
at reflow temperatures, wave soldering ... **EL1:** 694
base, calibration, ultrasonic inspection ... **A17:** 266
-based measurement, multiple, thermal inspection ... **A17:** 401
-based thermal inspection methods ... **A17:** 400
board, in solder wave ... **EL1:** 688
charge-up *See* Charge-up time
compression, in ultrasonic fatigue testing ... **A8:** 240-241
control, radiographic film processing ... **A17:** 352
corrosion rate changes as function of ... **A13:** 911
and creep or strain rate ... **A11:** 264
as creep/creep-rupture variable ... **A8:** 303
curing, in coremaking ... **A15:** 240
cycle, reduction, and costs ... **EL1:** 448
dead, live, and real, in x-ray spectrometry ... **A10:** 92
delay, squeeze casting ... **A15:** 324
dependence, polymer ... **EM2:** 405, 657-658
-dependent characteristics, mechanical presses ... **A14:** 39
-dependent characteristics, screw presses ... **A14:** 41
-dependent corrosion fatigue cracking ... **A8:** 405-406
-dependent deformation, in pressure vessels ... **A11:** 666-668
-dependent factors, galvanic corrosion ... **A11:** 187
-dependent failures ... **EL1:** 887
-dependent failures, in semiconductor devices ... **A11:** 767
-dependent strain, distortion failure by ... **A11:** 138
-dependent variables, affecting crack growth rate ... **A8:** 412
developing, liquid penetrant inspection ... **A17:** 83
dwell, of penetrants ... **A17:** 82
effect, in marine atmospheres ... **A13:** 906
effect in stress-corrosion cracking ... **A8:** 495
effect on liquid-erosion rate ... **A11:** 165
effect on thermodynamics and kinetics ... **A15:** 50
effect, weight loss, polyphenylene sulfides (PPS) ... **EM2:** 187
and effective modulus, plastics ... **EM2:** 412
effects, at constant temperature, permanent magnet materials ... **A2:** 798
and electrical breakdown ... **EM2:** 466
emulsification ... **A17:** 83
fail, data on ... **EL1:** 903
failures as function of ... **A8:** 635
freezing, compared ... **A15:** 243
-function, for determining dynamic fracture toughness ... **A8:** 271
gamma-ray spectrum changes as function of ... **A10:** 236
hardening, in multiaxial creep ... **A8:** 343
in hot pressing ... **A7:** 504
immersion, effects ... **A15:** 72
for injection molds ... **EM2:** 296
local relative, and clock skew ... **EL1:** 7
LS unit/symbol for ... **A8:** 721
for machine flat tape lay-up ... **EM1:** 629
melting ... **A15:** 72-74
for microanalytical techniques ... **A11:** 36
and money, engineering economy ... **A13:** 369
order, of production, and quality control ... **A17:** 723-724
out, defined *See* Out time, defined
and oxide breakdown ... **EL1:** 965
of pour, and flow rate ... **A15:** 500
power dissipation and, chip ... **EL1:** 46
processing, with machine vision ... **A17:** 42
profile, defined ... **EM2:** 43
propagation, interconnection effect ... **EL1:** 20-21
related characteristics, forming machines ... **A14:** 16
rinse, liquid penetrant inspection ... **A17:** 83

Time (continued)
rise, defined in acoustic emission inspection ... **A17:** 283
set, match plate patterns ... **A15:** 245
SI base unit and symbol for ... **A10:** 685
for silicone conformal coatings ... **EL1:** 824
sintering, effects, in P/M stainless steel ... **A13:** 826
solid phase fraction as function of ... **A15:** 183
solidification, Chvorinov's rule ... **A15:** 601
and strain rate relationship ... **A8:** 331, 686-687
symbol for ... **A11:** 797
-temperature development, radiographic film ... **A17:** 351
and temperature for sintering tungsten and molybdenum ... **A7:** 389
-temperature plot, of- pearlite decomposition ... **A11:** 613
-temperature relation, in wave soldering ... **EL1:** 685
-temperature superposition, shift factors ... **EL1:** 850
temperature-atmosphere, requirements for consolidation ... **A7:** 295
and tensile strength loss ... **A13:** 911
to complete ultrasonic fatigue tests at differing frequencies ... **A8:** 241
to failure, multiaxial testing ... **A8:** 344
to perforation, as atmospheric-exposure data ... **A13:** 510
-to-blister, as laminate thermal property ... **EL1:** 536
-to-failure, data for ... **A11:** 56
-to-failure, polyethylene pipes ... **A11:** 760
-to-fracture, fracture mechanics of ... **A11:** 53
-to-market cycle reduction ... **EL1:** 448-449
total computation (settling), defined ... **EL1:** 7
total cycle, vertical centrifugal casting ... **A15:** 304
of transit, ultrasonic waves ... **A17:** 231
-varying deflection control, fatigue tests ... **A8:** 368
-varying force control, fatigue tests ... **A8:** 368
vs flow rate, gating system ... **A15:** 597
vs potential, potentiometric membrane electrodes ... **A10:** 186
vs. boric acid concentrations for pressurized water reactor ... **A8:** 423
vs. creep deformation curve, creep stages ... **A8:** 331
of wetness/relative humidity, effect on atmospheric corrosion ... **A13:** 82
Time dependence
of polymer mechanical behavior ... **EM2:** 657-658
properties effects ... **EM2:** 405
Time dependence of current density during interference film formation ... **A9:** 145
Time of flight (TOF)
as propagation time ... **EL1:** 25
Time profile ... **EM3:** 29
Time quenching ... **M4:** 31
Time series ... **A18:** 294
Time to failure
effect in stress-corrosion cracking ... **A8:** 497
measured in multiaxial testing ... **A8:** 344
in stress-corrosion cracking ... **A13:** 275-276
Time-average interferometry *See also* Optical holography
as optical holographic interferometry ... **A17:** 408, 416
Time-delay refractometry
for impedance ... **EL1:** 522
Time-delayed embrittlement *See* Hydrogen embrittlement
Time-delayed stress
as testing factor ... **EL1:** 893
Time-dependent dielectric breakdown (TDDB) ... **EL1:** 965

Time-dependent fatigue behavior
methods for predicting ... **A1:** 629
Time-domain subtraction functions
thermal inspection ... **A17:** 400
Time-of-arrival pins
for flyer impact test ... **A8:** 210-211
Time-of-flight mass spectrometer ... **A11:** 35-36
Time-of-flight measurement
atom probe analysis ... **A10:** 591
field ion microscopy ... **A10:** 584
imaging atom probe analysis ... **A10:** 596
spark source mass spectrometry ... **A10:** 142
Time-of-flight powder diffractometer
neutron diffraction ... **A10:** 422
Time-of-flight single-crystal diffractometer
at pulsed neutron source ... **A10:** 424
Time-of-flight spectrometers ... **A10:** 142, 591
Time-temperature curve
defined ... **A9:** 18
Time-temperature equivalency principle ... **EM2:** 412, 415, 454
Time-temperature parameters ... **A8:** 333-335, 690; **M3:** 237-241
Time-temperature shift factor ... **EM3:** 421
Time-temperature superposition principle (TTSP) ... **EM3:** 354, 355
Time-temperature transformation ... **A8:** 726
Time-temperature transformation diagrams ... **M6:** 25
Time-temperature-precipitation (TTP) curves
austenitic stainless steels ... **A6:** 466, 467
Time-temperature-stress-moisture-superposition (TTSMSP) procedures ... **EM2:** 788
Time-temperature-superposition principle (TTSP) ... **EM2:** 788
for deformation prediction ... **EM2:** 676-677
Time-temperature-transformation (TTT) diagram ... **EM3:** 414, 423
Time-temperature-transformation (TTT) isothermal cure diagram
for phase-separating epoxy system ... **EM3:** 514
Time-temperature-transformation diagrams
isothermal solidification ... **A15:** 172
Time-to-failure *See also* Accelerated testing
modeling ... **EL1:** 887-888
statistics ... **EL1:** 888-889
Time-to-fracture *See also* Fracture energy,
long/short, one-point bend test ... **A8:** 275
measured, dynamic fracture toughness low-alloy steel ... **A8:** 272
measurement, impact response curves ... **A8:** 270
measurement, precracked Charpy specimen ... **A8:** 270
and stress corrosion cracking ... **A8:** 499
in tension-hold-only low-cycle fatigue tests ... **A8:** 351-352
vs. hold period in tension ... **A8:** 350-351
vs. strain rate ... **A8:** 349-350
vs. stress range or stress amplitude ... **A8:** 349-350
vs. tension-hold-only test ... **A8:** 350
Time-to-rupture
in creep ... **A8:** 302-304, 306
Timed solder rise test
for solderability ... **EL1:** 677
test standards used to evaluate solderability ... **A6:** 136
Timer
for creep test stand ... **A8:** 312
for fatigue testing machines ... **A8:** 368
Timing controls
ultrasonic inspection ... **A17:** 254
Timing marks ... **EM4:** 638
Timken 17-22AS
for opposing surfaces of steel aircraft brakes ... **A18:** 584
Timken tests
metalworking fluids ... **A18:** 101

Tin *See also* Tin alloys; Tin brass; Tin bronzes; Tin coatings; Tin powders; Tin recycling; Tin silver; Tin solders; Tin-base bearing alloys **M2:** 613-616
as addition to aluminum alloys **A4:** 843
addition to aluminum-base bearing alloys **A18:** 752
as addition to brazing filler metals **A6:** 904
addition to lead alloys providing corrosive resistance **A18:** 744
adhesion and solid friction **A18:** 33
alloying, aluminum casting alloys **A2:** 132
alloying effect in titanium alloys **A6:** 508, 509
as alloying element **A15:** 16
alloying, in wrought titanium alloys **A2:** 599
alloying, wrought aluminum alloy **A2:** 55
alloys, for soft metal bearings **A11:** 483
in aluminum alloy bearing material systems **A18:** 746
in aluminum alloys **A15:** 746
aluminum-silicon-tin alloys, mixed bearing microstructure **A18:** 744
as an addition to titanium alloys **A9:** 458
as an addition to zinc alloys **A9:** 490
as an addition to zirconium **A9:** 497
bearing alloys **A2:** 522-525
boiling point **M5:** 270
bright ac-d **EL1:** 679
bronze plating bath content **M5:** 288-289
-cadmium induced LME failure, of nose landing gear socket **A11:** 226, 229
cans as work material for ion implantation **A18:** 858
in cast iron **A1:** 6
catalyst for sealants **EM3:** 674
catalyst for urethane sealants **EM3:** 204
causing embrittlement **A4:** 124, 135
chemical analysis and sampling **A7:** 248
chemical compositions per ASTM specifications **A6:** 787
chemical resistance **M5:** 8
in chemicals **A2:** 520
chip combustion accelerators **A10:** 222
coating for valve train assembly components **A18:** 559
in coatings **A2:** 517-518
in compacted graphite iron **A1:** 57, 59
compatibility in bearing materials **A18:** 743
in composition, effect on ductile iron **A4:** 686
in composition, effect on gray irons **A4:** 671
in compounds providing flame retardance and smoke suppression **EM3:** 179
conformability and embeddability **A18:** 743
content in solders **M6:** 1069-1073
in copper alloys **A6:** 752
in copper-base alloys **A18:** 750
corrosion of **A13:** 770-783
in dental amalgam **A18:** 669
determined by controlled-potential coulometry **A10:** 209
diffusion in sleeve bearing liners **A9:** 567
distillation **A10:** 169
effect, gas dissociation **A15:** 83
effect of number of grains on elongation **A9:** 125
effect of number of grains on ultimate tensile strength **A9:** 125
effect on cast iron machinability **A16:** 652, 654
effect on Cu alloy machinability **A16:** 808
effect on dealloying **A13:** 132, 614
effect on macrosegregation in copper alloys **A9:** 639
effects of, on notch toughness **A1:** 742
effects on SCC of copper **A11:** 221
electrochemical grinding **A16:** 543
electroplated coatings for bearings **A18:** 838
electroplating **A2:** 518
electropolishing with alkali hydroxides **A9:** 54
embrittlement by **A11:** 236 **A13:** 183-184
in enamel cover coats **EM3:** 304
in enameling ground coat **EM3:** 304
evaporation fields for **A10:** 587
fiber for reinforcement **A18:** 803
in filler metal used for direct brazing **EM4:** 517-518, 519
friction coefficient data **A18:** 71

Tin (continued)
galvanic corrosion **A13:** 85
gas-atomized **A7:** 32
gaseous hydride, for ICP sample introduction **A10:** 36
gold-tin eutectic **EM3:** 584
gold-tin, for solder sealing **EM3:** 585
as gray iron alloying element **A15:** 639
heat-affected zone fissuring in nickel-base alloys **A6:** 588
hone forming **A16:** 488
hot dip coatings **A2:** 518
hot extrusion of **A14:** 321
in hydrogen peroxide, GFAAS analysis of **A10:** 57-58
ingot, grades **M5:** 354
in intermetallic compounds **A6:** 127
iodimetric titration for **A10:** 174
in iron-base alloys, flame AAS analysis of **A10:** 56
as lead additive **A2:** 519
in lead-base alloys **A18:** 750
as low-melting embrittler **A12:** 29
lubricant indicators and range of sensitivities **A18:** 301
melting point **M5:** 270
mill products, annealing **A4:** 52
as minor element, ductile iron **A15:** 648
as minor toxic metal, biologic effects **A2:** 1261
moderate explosivity class **A7:** 196
-modified stainless steel **A7:** 253
ores, dissolution of inorganic materials in **A10:** 167
penetration, x-ray elemental dot map of **A11:** 720
pewter **A2:** 522
photochemical machining etchant **A16:** 590
photometric analysis methods **A10:** 64
plain carbon steel resistance to **A13:** 515
plate **A13:** 778-780
plating, uses **EL1:** 679
powders **A2:** 519-520
precoating **A6:** 131
price per pound **A6:** 964
processing technique **A18:** 803
production and consumption **A2:** 517
properties **A18:** 803
PSG as diffusion barrier **EM3:** 583
pure **A2:** 518-520 **A13:** 770-772 **M2:** 811-814
pure, properties **A2:** 1166
pure, thermal properties **A18:** 42
quartz tube atomizers with **A10:** 49
recovery from brass **A10:** 200
recycling **A2:** 1218-1220
relative solderability as a function of flux type **A6:** 129
resistance of, to liquid-metal corrosion **A1:** 636
resistance to corroding agents **A2:** 520
safety standards for soldering **M6:** 1098
segregation and solid friction **A18:** 28
shielded metal arc welding **M6:** 75
shrinkage allowance **A15:** 303
soft solders **A13:** 772-774
solder characteristics **M6:** 1070-1071
as solder impurity **EL1:** 638-639
solderability **A6:** 978
solders **A2:** 520-522
solid lubricant inclusion for stainless steels **A18:** 716
solution potential **M2:** 207
solvent extractant for **A10:** 170
species weighed in gravimetry **A10:** 172
spectrometric metals analysis **A18:** 300
in steel weldments **A6:** 420
stream **A15:** 16
subboundaries in high-purity **A9:** 607
suitability for cladding combinations **M6:** 691
tap density **A7:** 277
thermal diffusivity from 20 to 100 °C **A6:** 4
thermal expansion coefficient **A6:** 907
thermal spray coating material **A18:** 832
tinplate **A2:** 517-518
tinplate applications, continuous annealing **A4:** 64-65
toxicity and exposure limits **A7:** 206
trace, analysis in hydrogen peroxide by GFAAS **A10:** 57-58

Tin (continued)
as trace element, cupolas **A15:** 388
trace levels in H20-1, GFAAS determined **A10:** 57-58
as tramp element **A8:** 476
TWA limits for particulates **A6:** 984
ultrapure, by zone-refining technique **A2:** 1094
unalloyed **M2:** 614
unalloyed, applications **A2:** 519
usage, PWB manufacturing **EL1:** 510
vapor pressure, relation to temperature **A4:** 495
volatilizing **A10:** 166
volumetric procedures for **A10:** 175
in zinc alloys **A15:** 788

Tin alloys *See also* Tin; Tin coatings
battery grid alloys **A2:** 526
battery-grid **M2:** 615
bearing **A13:** 774
bearing alloys **A2:** 522-525
bearing materials **M2:** 614-615
cast iron, addition to **M2:** 615
cast irons **A2:** 526
cast tin bronze, flux effect **A15:** 448
casting alloys **A2:** 525
chemical analysis and sampling **A7:** 248
in coatings **A2:** 517-518 **A13:** 775-778
collapsible tubes and tin foil **A2:** 525
and copper alloys **A2:** 526
copper-tin bronzes **M2:** 615
corrosion of **A13:** 770-783
dental alloys **A2:** 526 **M2:** 615
effect of number of grains on elongation **A9:** 125
effect of number of grains on ultimate tensile strength **A9:** 125
electroplating **A2:** 518
fusible alloys **A2:** 525
hard tin, applications **A2:** 525
hot dip coating **A2:** 518
melting heat for **A15:** 376
for organ pipes **A2:** 525
pewter **A2:** 522 **A13:** 774 **M2:** 614
powders **A2:** 519-520
production and consumption **A2:** 517
sample dissolution medium **A10:** 166
Sn-0.2Pb, LEISS spectra **A10:** 608
solder **A13:** 772-774
soldering **A6:** 631
solders **A2:** 520-522 **M2:** 614
solid-state phase transformations in welded joints **A9:** 581
thermal expansion coefficient **A6:** 907
tin foil **A2:** 525-526
tin-copper **A13:** 774-775
tin-silver **A13:** 775
tinplate **A2:** 517
titanium, addition to **M2:** 615
and titanium alloys **A2:** 526
type metals **M2:** 615
use in organ pipes **M2:** 614
wetting behavior and joining **EM4:** 491
zirconium, addition to **M2:** 615-616
and zirconium alloys **A2:** 526

Tin alloys, specific types *See also* Sleeve bearing materials, specific types
40Sn-60Pb **A6:** 351
63Sn-37Pb solder **A6:** 113
phase diagram **A6:** 128
63Sn-37Pb solder, embrittlement in **A11:** 45-46
90Sn-10Pb solder plating, peeling of **A11:** 43-45
95Sn-5Sb **A6:** 351
96.5Sn-3.5Ag **A6:** 351
antimonial tin solder **M2:** 619
bearing alloy **M2:** 617, 623
casting alloy **M2:** 617, 623
commercially pure tins **M2:** 617, 618-619
hard tin **M2:** 619
pewter **M2:** 617, 624-625
Sn-0.05Pb, liquid-solid interface structures **A9:** 613
Sn-0.6Cd, directionally solidified **A9:** 612
Sn-15Pb, semisolid viscosity/shear stress **A15:** 328
Sn-17Co, peritectic and peritectoid structures **A9:** 679

Tin alloys, specific types (continued)
Sn-20Pb, cell formation, intercellular eutectic ... **A15:** 116
Sn-44Pb, directionally solidified dendritic structure ... **A9:** 613
soft solder, 60Sn-40Pb ... **M2:** 620-621
soft solder, 63Sn-37Pb ... **M2:** 620
soft solder, 70Sn-30Pb ... **M2:** 620
Ti-6Al-2Sn-4Zr-2Mo, electron-beam welding ... **A6:** 865
tin babbitt alloy 1 ... **M2:** 617, 621-622
tin babbitt alloy 2 ... **M2:** 617, 622
tin babbitt alloy 3 ... **M2:** 617, 622, 623
tin die-casting alloy ... **M2:** 623
tin foil ... **M2:** 623-624
tin-silver solder ... **M2:** 619
white metal ... **M2:** 624
Tin alloys, use for bearings *See also* Bearings, sliding ... **M3:** 813-814
Tin and tin alloy coatings *See also* Tin and tin alloys; Tinplate ... **A9:** 450-457
electrical terminal coating ... **A9:** 457
etching ... **A9:** 451
grinding ... **A9:** 450
intermetallic layer formed beneath ... **A9:** 456
mounting ... **A9:** 450
polishing ... **A9:** 450-451
removal of, for electron microscopy ... **A9:** 451
specimen preparation ... **A9:** 450-451
Tin and tin alloys *See also* Tin and tin alloy coatings ... **A9:** 449-457
abrasives ... **A9:** 449
electrodeposited on copper substrate ... **A9:** 456
etchants ... **A9:** 450
etching ... **A9:** 450
eutectics ... **A9:** 452
grinding ... **A9:** 449
high-purity tin ... **A9:** 453
microstructures ... **A9:** 452
polishing ... **A9:** 449
pure tin with twinned grains and recrystallized grains ... **A9:** 453
sectioning ... **A9:** 449
specimen preparation ... **A9:** 450
Tin and tin alloys, specific types
50Sn-50In, structure ... **A9:** 455
63Sn-37Pb, soldered joint ... **A9:** 457
Sn-0.4Cu, structure ... **A9:** 453
Sn-3lPb-18Cd, structure ... **A9:** 455
Sn-4.5Sb-4.5Cu, structure ... **A9:** 454
Sn-5Ag, structure ... **A9:** 455
Sn-5Sb, effect of cooling rate ... **A9:** 454
Sn-6Sb-2Cu, cold-rolled ... **A9:** 456
Sn-6Sb-2Cu, structure ... **A9:** 454
Sn-7Sb-3.5Cu, structure ... **A9:** 454
Sn-8Sb-8Cu, structure ... **A9:** 454
Sn-9Sb-4Cu, structure ... **A9:** 454
Sn-12Sb-10Pb-3Cu, structure ... **A9:** 454
Sn-13Sb-5Cu, structure ... **A9:** 454
Sn-15Sb-18Pb-2Cu, structure ... **A9:** 454
Sn-18Ag-15Cu, coated with a platinum oxide layer ... **A9:** 61
Sn-28Zn-2Cu, structure ... **A9:** 455
Sn-30Pb, different magnifications compared ... **A9:** 453
Sn-30Sb, structure ... **A9:** 454
Sn-30Zn, structure ... **A9:** 455
Sn-37Pb, effect of cooling rate ... **A9:** 453
Sn-40Pb, structure ... **A9:** 453
Sn-40Pb, wave-soldered circuit board joint ... **A9:** 455-456
Sn-48Pb-2Sb, cast, contraction cavity ... **A9:** 456
Sn-50Pb, lamellar eutectic ... **A9:** 453
Sn-57Bi, chill cast ... **A9:** 456
Tin babbitt
in bearing alloys ... **A18:** 748
bearing material microstructure ... **A18:** 743, 744
in bimetal bearing material systems ... **A18:** 747
casting processes ... **A18:** 754-755
Tin babbitt (continued)
in trimetal bearing material systems ... **A18:** 748
Tin brass
applications and properties ... **A2:** 315
Tin brasses
corrosion resistance ... **A13:** 610-611
gas-metal arc butt welding ... **A6:** 760
gas-metal arc welding ... **A6:** 763
to high-carbon steel ... **A6:** 828
to low-alloy steel ... **A6:** 828
to low-carbon steel ... **A6:** 828
to medium-carbon steel ... **A6:** 828
to stainless steels ... **A6:** 828
thermal expansion coefficient ... **A6:** 907
weldability ... **A6:** 753
Tin bronze
bearing material systems ... **A18:** 746-747
applications ... **A18:** 746
bearing performance characteristics ... **A18:** 746
load capacity rating ... **A18:** 746
inverse segregation ... **A9:** 644
Tin bronze filters ... **A7:** 736
Tin bronzes *See also* Cast copper alloys
applications and properties ... **A2:** 226
composition/melt treatment ... **A15:** 772, 776
corrosion ratings ... **A2:** 353-354
development of ... **A15:** 16
foundry properties for sand casting ... **A2:** 348
freezing range ... **A2:** 348
as general-purpose copper casting alloys ... **A2:** 352
microstructural wear ... **A11:** 161
nominal compositions ... **A2:** 347
occurrence of SMIE in ... **A11:** 243
properties and applications ... **A2:** 374
shrinkage allowances ... **A15:** 303
Tin bronzes (cast)
thermal expansion coefficient ... **A6:** 907
Tin cans ... **A13:** 778-779
Tin chemicals
applications ... **M2:** 616
as metallic tin application ... **A2:** 520
Tin chip combustion accelerators ... **A10:** 222
Tin coating, hot dip *See* Hot dip tin coating
Tin coatings ... **A1:** 221 **A13:** 775-778
cast irons ... **M1:** 102, 103
copper/copper alloys ... **A13:** 636
corrosion protection ... **M1:** 752-754
electrodeposited ... **A13:** 427
electroplating ... **M2:** 614
hot dip ... **M2:** 614
immersion ... **A13:** 776
nails ... **M1:** 271
on nonferrous metals ... **A13:** 776
on steel ... **A13:** 775-776
steel sheet ... **M1:** 173
applications ... **M1:** 173
coating weights ... **M1:** 173
designations ... **M1:** 173
specifications for ... **M1:** 173
surface finish and texture ... **M1:** 173
testing ... **M1:** 173
steel wire ... **M1:** 1-63
thicknesses for ... **A13:** 426
tin-cadmium alloy ... **A13:** 776
tin-cobalt ... **A13:** 776-777
tin-copper ... **A13:** 777
tin-lead ... **A13:** 777
tin-nickel ... **A13:** 777
tin-zinc ... **A13:** 777-778
tinplate ... **M2:** 613-614
Tin foil
tin-base alloy ... **A2:** 525-526
Tin grain size test
for tinplate ... **A13:** 782
Tin in copper ... **M2:** 242
Tin in fusible alloys ... **M3:** 799
Tin in steel ... **M1:** 116
Tin in titanium ... **M3:** 356
Tin mechanical coating ... **M5:** 300-302
Tin melt stock ... **A7:** 124
Tin oxide
ESDIED spectrum ... **A18:** 457, 458
Tin oxide as an interference film ... **A9:** 147
Tin oxides (SnO and SnO_2) purpose for use in glass manufacture ... **EM4:** 381
Tin plate
acid-bath plated ... **A9:** 200
alkaline-bath plated ... **A9:** 200
specimen preparation ... **A9:** 198
Tin plating ... **M5:** 270-272
acid process ... **M5:** 271-272
alkaline process ... **M5:** 270-271
aluminum and aluminum alloys ... **M5:** 604-605
anodes ... **M5:** 270-272
filming ... **M5:** 270-271
applications ... **M5:** 270
copper and copper alloys ... **M5:** 581-584
copper-tin plating, magnesium alloys ... **M5:** 646-647
stripping of ... **M5:** 647
current density ... **M5:** 271-272
electroless process ... **M5:** 621
fluoborate process ... **M5:** 272
rinsewater recovery ... **M5:** 318
solution compositions and
acid electrolytes ... **M5:** 271-272
alkaline electrolytes ... **M5:** 270-271
operating conditions ... **M5:** 270-272
stannate process ... **M5:** 270-272
steel ... **M5:** 270
stripping of ... **M5:** 647
sulfate process ... **M5:** 271-272
temperature ... **M5:** 271-272
throwing power ... **M5:** 272
Tin powders *See also* Tin; Tin alloys
applications ... **A7:** 123 **M2:** 616
atomizing apparatus ... **A7:** 123
commercially available ... **A7:** 123
particles, air-atomized ... **A7:** 124
production and properties ... **A7:** 123-124
shipment tonnage ... **A7:** 24
Tin recycling
facts and figures ... **A2:** 1219
recycled metal behavior ... **A2:** 1220
scrap tin materials, types ... **A2:** 1218-1219
technology ... **A2:** 1219-1220
trends and future ... **A2:** 1220
Tin salts
additive to silicones ... **EM3:** 598
Tin soap
condensation catalyst for silicones ... **EM3:** 216, 218
Tin solders
applications, specifications, compositions ... **A2:** 521
general purpose ... **A2:** 520
impurities ... **A2:** 520-521
tin-zinc solders ... **A2:** 520
Tin sweat
from inverse segregation ... **A15:** 16
Tin tetraiodide
dissolution to ... **A10:** 167
Tin, vapor pressure
relation to temperature ... **M4:** 310
Tin whiskers *See also* Whiskers ... **A13:** 770, 1110, 1126
corrosion failure analysis ... **EL1:** 1116
growth, and dendrite formation ... **EL1:** 969-970
growth, from code posted organics ... **EL1:** 679
as reinforcement ... **EL1:** 1119-1121
Tin-alloy microstructures
tin-indium (50-50) ... **A3:** 1 • 19
tin-lead ... **A3:** 1 • 20
Tin-aluminum alloys
not directly roll bonded to steel ... **A18:** 756
Tin-antimony
filler metal for torch soldering ... **A6:** 352
heat treating ... **M4:** 775

Tin-antimony alloys
microstructures ... **A9:** 452
strength of ... **EL1:** 636
Tin-antimony-copper alloys
microstructures ... **A9:** 452
Tin-antimony-copper-lead alloys
microstructures ... **A9:** 452
Tin-base alloys
as bearing alloys ... **A18:** 748-749
composition ... **A18:** 749
designations ... **A18:** 749
mechanical properties ... **A18:** 749
microstructure ... **A18:** 749
in bimetal bearing material systems ... **A18:** 747
bonding providing corrosion resistance ... **A18:** 744
heat and temperature effects on strength retention ... **A18:** 745
Tin-base bearing alloys *See also* Tin; Tin alloys
compositions ... **A2:** 523-524
lead-base (lead-base babbitts) ... **A2:** 523-524
material selection ... **A2:** 522-523
properties ... **A2:** 524
Tin-base casting alloys
types and applications ... **A2:** 525
Tin-bismuth
heat treating ... **M4:** 775
Tin-bismuth alloys (eutectic)
for solders ... **EL1:** 637
Tin-bronze alloys
preferential corrosion ... **A12:** 403
Tin-cadmium alloy coatings ... **A13:** 776
Tin-cadmium alloys
peritectic structures ... **A9:** 679
Tin-coated copper
surface damage by corrosion and plastic deformation ... **A18:** 183
Tin-coated steel
sheet metal forming with liquid lubrication for tool steels ... **A18:** 738
soldering ... **A6:** 631
Tin-coated steels
press forming of ... **A14:** 562-563
resistance spot welding ... **M6:** 479-480, 491-493
Tin-coated wire
wrought copper and copper alloys ... **A2:** 253
Tin-containing brasses
brazing ... **A6:** 629-630
Tin-copper alloy plating
copper and copper alloys ... **M5:** 623
Tin-copper alloys
microstructures ... **A9:** 452
types ... **A13:** 774
Tin-free aluminum alloys ... **A18:** 752-753, 756
Tin-gold alloys
strength of ... **EL1:** 636
Tin-gold eutectic ... **EM3:** 584
Tin-indium alloys
microstructures ... **A9:** 452
Tin-lead
heat treating ... **M4:** 775
precoating ... **A6:** 131
Tin-lead alloy coatings *See* Terne coating
Tin-lead alloys *See also* Solder; Tin-lead alloys, specific types
development ... **EL1:** 631
equilibrium diagram ... **EL1:** 634
microstructures ... **A9:** 452
for plating rigid wiring boards ... **EL1:** 546
for solder sealing ... **EM3:** 585
as solders ... **EL1:** 633-634
usage, PWB manufacturing ... **EL1:** 510
Tin-lead alloys, specific types
60Sn-40Pb eutectic, for machine soldering ... **EL1:** 633
60Sn-40Pb, intermetallic compound layer growth ... **EL1:** 635
63Sn-37Pb eutectic, for machine soldering ... **EL1:** 633
Tin-lead coatings
microstructures ... **A9:** 456
Tin-lead phase diagram ... **A2:** 552
Tin-lead plating *See also* Terne coatings ... **M5:** 276-278
addition agents used in ... **M5:** 276-277
agitation ... **M5:** 277-278
anodes, composition of ... **M5:** 277-278
Tin-lead plating (continued)
applications ... **M5:** 276
boric acid used in ... **M5:** 276-278
carbon treatment ... **M5:** 278
copper and copper alloys ... **M5:** 623-624
current densities ... **M5:** 277-278
equipment, construction materials for ... **M5:** 278
filtration process ... **M5:** 278
metallic impurity removal ... **M5:** 278
peptone used in ... **M5:** 277-278
solution compositions and operating conditions ... **M5:** 276-278
temperature effects ... **M5:** 277
Tin-lead solder
LEISS segregation of lead to surface of ... **A10:** 607-608
Tin-lead solder(s)
dip soldering ... **A6:** 356
electronic packaging applications ... **A6:** 618
wave soldering ... **A6:** 367
wetting of steel ... **A6:** 114
Tin-lead solders ... **EM3:** 584
Tin-lead solders, specific types
40Sn-60Pb, torch soldering composition ... **A6:** 351
60Sn-40Pb, properties ... **A6:** 992
63Sn-37Pb ... **A6:** 113, 351
furnace soldering ... **A6:** 354
phase diagram ... **A6:** 128
Tin-lead-antimony alloys
for solders ... **EL1:** 636
Tin-lead-cadmium alloys *See also* Solder
microstructures ... **A9:** 452
Tin-lead-copper alloys
for solders ... **EL1:** 636
Tin-lead-gold alloys
equilibrium diagram ... **EL1:** 636
Tin-lead-silver alloys
for solders ... **EL1:** 636
Tin-nickel
relative solderability ... **A6:** 134
Tin-nickel base metal
solderability ... **EL1:** 677
Tin-nickel substrate
composition vs depth of passive film on ... **A10:** 608-609
Tin-plated nickel
intermetallic compound ... **A9:** 456
Tin-plated steel
resistance seam welding ... **A6:** 245
resistance spot welding ... **M6:** 480
Tin-rich alloys ... **A4:** 924
aging ... **A4:** 924
annealing ... **A4:** 924
binary alloys ... **A4:** 924 **M4:** 775
heat treating ... **A4:** 924 **M4:** 775-776
pewter ... **A4:** 924 **M4:** 776
solution-treating ... **A4:** 924
tempering ... **A4:** 924
ternary alloys ... **A4:** 924 **M4:** 775-776
tin-antimony ... **A4:** 924 **M4:** 775
tin-bismuth ... **A4:** 924 **M4:** 775
tin-lead ... **A4:** 924 **M4:** 775
tin-silver ... **A4:** 924 **M4:** 775
Tin-rich alloys, specific types
Sn-3Cd-7Sb, tensile strength ... **A4:** 924
Sn-9Sb-1.5Cd, tensile strength ... **A4:** 924
Tin-silver
filler metal for torch soldering ... **A6:** 352
heat treating ... **M4:** 775
Tin-silver alloys
atmospheric corrosion ... **A13:** 775
microstructures ... **A9:** 452
Tin-silver eutectic alloy solders
applications and composition ... **A2:** 521
Tin-silver solders
applications and compositions ... **A2:** 521
Tin-silver-lead alloys *See also* Lead; Lead alloys
as solder ... **A2:** 553
Tin-zinc alloys
microstructures ... **A9:** 452
Tin-zinc solders
applications ... **A2:** 520
Tin-zinc-copper
microstructures ... **A9:** 452
Tin/lead alloys
ferrographic application to identify wear particles ... **A18:** 305
strain-rate behaviors (Sn-38% Pb) ... **A18:** 426
Tinned wire ... **A1:** 282
Tinning *See also* Hot solder dipping
defined ... **A13:** 13 **EL1:** 1159
definition ... **A6:** 1214
for solderability ... **EL1:** 678-679
of surface-mount components ... **EL1:** 731
Tinning oils
in wave soldering ... **EL1:** 689-690
Tinplate ... **A13:** 778-782
components of a taper section ... **A9:** 450
intermetallic layer ... **A9:** 456
recycling ... **A2:** 1218
steel-tin interface ... **A9:** 455
Tinplate, relative weldability ratings
resistance spot welding ... **A6:** 834
Tint etchants *See also* Color etchants
enhancement of colors using polarized light ... **A9:** 78
for heat-resistant casting alloys ... **A9:** 330-332
used with carbon and alloy steels ... **A9:** 70
Tint etching *See also* Color etching
of austenitic manganese steel casting specimens ... **A9:** 239
wrought stainless steels ... **A9:** 281
Tinting *See* Heat tinting
Tipping solder
micrograph ... **A9:** 422
Tire bead wire ... **A1:** 286 **M1:** 269
Tire rolling radius ... **A18:** 566
Tire studs
of cemented carbides ... **A2:** 974
powder used ... **A7:** 572
Tire tracks
AISI/SAE alloy steels ... **A12:** 298
defined ... **A12:** 18
as fatigue indicators ... **A11:** 258, 259
on fatigue fracture surface ... **A12:** 23
on quenched-and-tempered steel ... **A12:** 23
on steel ... **A11:** 27
polymer ... **A12:** 479
and striations, compared ... **A12:** 119
Tire-mold casting
internal discontinuities in ... **A11:** 358
Tires, friction and wear of ... **A18:** 578-581
abrasion and wear ... **A18:** 578
basic mechanism ... **A18:** 578
technology related to tire wear and their definitions ... **A18:** 578
tire axis system ... **A18:** 578, 579
effects of various factors, experimental design study ... **A18:** 580
tire-related factors, effect on tire wear rate ... **A18:** 580
fleet testing ... **A18:** 579, 580
future outlook ... **A18:** 581
load and velocity effects ... **A18:** 579-580
resilience ... **A18:** 579
standard wear rate ... **A18:** 579
test severity ... **A18:** 580
pavement surface severity ... **A18:** 580
tire force severity ... **A18:** 580
weather severity ... **A18:** 580
tire conceptual model ... **A18:** 580-581
carcass foundation stiffness ... **A18:** 580
coefficient of friction of rubber ... **A18:** 580-581
flexural rigidity ... **A18:** 580
tire wear models ... **A18:** 579
empirical ... **A18:** 579
theoretical ... **A18:** 579
trailer testing ... **A18:** 579-580
Tissues, body
tantalum resistance to ... **A13:** 728
Titanate phases
ceramic nuclear waste forms ... **A10:** 532-535
Titanates ... **EM3:** 674 **EM4:** 47
applications ... **EM4:** 47
as coupling agents ... **EM3:** 181, 182
Titania
chemical composition ... **A6:** 60
content effect in fluxes ... **A6:** 57, 58
flux composition, CO_2 shielded FCAW electrodes ... **A6:** 61
in opaque cover coat enamels ... **EM4:** 953

Titania (continued)
semiconductor sensors **EM4:** 1137-1138
Titania porcelain enamels
composition of **M5:** 510
Titania slag.. **A7:** 86
Titania-doped silica glass
aerospace applications................................ **EM4:** 1016
Titania/silica (TiO_2/SiO_2) gel compositions **EM4:** 446-447
Titanium *See also* Advanced titanium materials; Introduction to titanium and titanium alloys; Niobium-titanium superconductors; Pure titanium; Reactive metals; Titanium alloy castings; Titanium alloy precision forgings; Titanium alloys; Titanium alloys, specific types; Titanium carbides; Titanium castings; Titanium P/M products; Titanium powders; Titanium recycling; Titanium sponge; Titanium-aluminum-vanadium alloys; Wrought titanium; Wrought titanium alloys; Wrought titanium alloys, specific types **A13:** 669-706
abrasive blasting of................................. **M5:** 652-653
abrasive wear **A18:** 189
active brazing process in ceramic/metal seals............................. **EM4:** 504-505, 506
in active metal process **EM3:** 305
added to nuclear waste, EPMA analysis **A10:** 532-535
addition to
ferritic stainless steels **A6:** 444, 445
filler metals for ceramic materials **A6:** 951, 954
fluxes affecting ionization process.......... **A6:** 57-58
molybdenum for electron-beam welding **A6:** 870-871
precipitation-hardenable nickel alloys **A6:** 575, 576
underwater welds, microstructural development....................................... **A6:** 1011
adhesion and solid friction....................... **A18:** 32, 33
aging **A4:** 913, 914, 916-919, 921, 922-923
aging furnaces................................... **A4:** 921 **M4:** 772
air-carbon arc cutting **A6:** 1176
alloy compositions **M3:** 357
alloy systems.. **M3:** 355-357
alloy types **A4:** 913 **M3:** 359-360
alloying, aluminum casting alloys **A2:** 132
alloying effect on nickel-base alloys **A6:** 589
alloying elements' effects on (X-p transformation... **A4:** 913
alloying for corrosion resistance **M3:** 414-415
alloying, of cast irons **A13:** 567
alloying, of nickel-base alloys **A13:** 641
alloying, of uranium/uranium alloys **A13:** 814
in Alnico alloys... **A9:** 539
alpha case removal **A15:** 264
alpha structure.. **A9:** 156
in aluminum alloys......................... **A15:** 95, 105, 746
aluminum pretreatment process **M5:** 457
in ambient seawater...................................... **A13:** 677
as an addition to austenitic manganese steel castings.. **A9:** 239
as an addition to beryllium-nickel alloys.. **A9:** 395
as an addition to cobalt-base heat-resistant casting alloys.................. **A9:** 334
as an addition to niobium alloys **A9:** 441
as an alloying addition to austenitic stainless steels **A9:** 283-284
as an alloying addition to wrought heat- resistant alloys **A9:** 309-311
annealing............ **A4:** 913, 914, 915-916, 917, 920 **M4:** 765-766
anodized.. **EM3:** 416-417
applications **A2:** 587-590, 1261 **M3:** 353-355
applications, sheet metals............................... **A6:** 400
in aqueous oxalate solutions as a function of pH **EM4:** 59
arc welding *See* Arc welding of titanium and titanium alloys

Titanium (continued)
arc-welded **A11:** 437-439
atmospheres **A4:** 920, 921 **M4:** 771
atomic interaction descriptions...................... **A6:** 144
in austenitic stainless steels........................... **A6:** 457
back reflection intensity **A17:** 238
bainitic-like microstructures.................... **A9:** 665-666
basis for organometallic coupling agents.. **EM3:** 182
Bauschinger effect **A4:** 914
belt grinding of... **M5:** 652
beta transus temperatures **A4:** 913
bonding preparations necessary................. **EM3:** 703
bonding with advanced composites using polyphenylquinoxalines **EM3:** 163
bone screw head, wear and fretting at plate hole.. **A11:** 689, 692
boriding.. **A4:** 437, 438
and boron, as inoculants................................ **A15:** 105
boron-implanted, nanoindentation **A18:** 426
brazing.. **M6:** 1049-1051
applications **M6:** 1053
brazeability.. **M6:** 1049-1050
filler metals **M6:** 1050-1051
preparation... **M6:** 1052-1053
brazing and soldering characteristics **A6:** 633-634
brazing to aluminum.................................... **M6:** 1031
brazing with clad brazing materials............. **A6:** 347
capacitor discharge stud welding **M6:** 738
as carbide-stabilizing element...................... **A13:** 325
carbonitride blocky inclusions, microanalysis.. **A11:** 38
carburizing affected by content in steels.. **A18:** 875
castings............................. **M3:** 354, 355, 360, 407-412
cathodic attack of **A11:** 202
cavitation resistance...................................... **A18:** 600
chemical conversion coating of **M5:** 656-658
chemical integrity of seals **EM4:** 540-541
chemical resistance... **M5:** 4
chrome plating, hard **M5:** 180-181
cleaning processes **M5:** 650-656
coatings, bonded-abrasive grains **A2:** 1015
cobalt laser cladding materials not used **A18:** 866-867
coextrusion welding... **A6:** 311
commercial grades **M3:** 357-360
commercially pure *See* Titanium alloys, specific types **A2:** 592-597 **A12:** 16, 20, 48-49, 57, 65, 171 **A14:** 129, 838-83
commercially pure, bone screw with shear fracture................................ **A11:** 677, 682
commercially pure, corrosion resistance **A6:** 585
commercially pure, ductile fracture surface... **A11:** 685, 689
commercially pure, for encapsulation........... **A7:** 428
commercially pure, hex nuts.......................... **A7:** 682
commercially pure implant, fracture surface with mixed morphology **A11:** 686, 689
concentration effect on shear strength **A6:** 116
concentration effect on wettability................ **A6:** 116
consolidation **A7:** 748-749
content effect on interstitial-free steels........... **A4:** 61
content in stainless steels.............................. **M6:** 320
content of military airframes **A7:** 748
continuous flow electron beam melted............. **A15:** 416-417
copper plating of **M5:** 658
corrosion **M3:** 354, 413-417
corrosion of... **A13:** 669-706
cracking in seawater **M3:** 416
crevice corrosion.............. **A13:** 323, 671-672, 681-683
crevice corrosion and pitting in.................... **A11:** 202
current technology **A2:** 586
cutting tool material selection based on machining operation **A18:** 617

Titanium (continued)
cutting tool materials and cutting speed relationship **A18:** 616
for cutting tools **A18:** 612
debris effect on wear **A18:** 249
deoxidation, low-alloy steels........................ **A15:** 715
descaling process...................................... **M5:** 650-654
problems **M5:** 650-651
procedures... **M5:** 651-654
determination in paint, enhancement and absorption effects............................ **A10:** 98
determined by controlled-potential coulometry **A10:** 209
diffusion bonding... **A6:** 145
diffusion brazing **A6:** 343
diffusion welding..... **A6:** 884, 885, 886 **M6:** 677, 682
effect, amorphous metals **A13:** 868
effect of stress intensity factor range on fatigue crack growth rate **A12:** 57
effect of vacuum on fatigue **A12:** 48-49
effect on brake lining friction....................... **A18:** 592
effect on carbon fiber..................................... **EM1:** 52
effect on case depth in carbonitriding........... **A4:** 376
electron beam investment casting **A15:** 418
electron-beam welding **A6:** 854
electroplating of....................................... **M5:** 658-659
electroslag welding **A6:** 278
electroslag welding, reactions **A6:** 273, 274
elongated dimples on shear fracture surface... **A12:** 16
embrittlement **A13:** 179
evaporation fields for **A10:** 587
explosion welding..... **A6:** 162-163, 303-304, 896 **M6:** 707, 710, 713
explosive-bonded to zirconium **A9:** 157
extrusion welding **M6:** 677
FASIL adhesion excellent............................ **EM3:** 678
as fastener material **A11:** 530
fasteners, use in **M3:** 184, 185
fatigue levels **EM3:** 504
fatigue striations **A12:** 20
ferrite formation **M6:** 346
as ferrite stabilizer... **A13:** 47
in ferritic stainless steels **A6:** 451, 454
in filler metals used for active metal brazing........................... **EM4:** 523-524, 526, 529
in filler metals used for direct brazing **EM4:** 517-518, 519
finishing processes **M5:** 656-659
fixtures.................................. **A4:** 916, 921, 922 **M4:** 772
foil and film adhesive, to repair advanced composite structures.......... **EM3:** 821
forged, microstructure **A14:** 271
forgings **M3:** 358, 360
fretting in **A13:** 140
fretting wear.. **A18:** 250
friction coefficient.................................... **A18:** 71, 72
friction welding **A6:** 152, 153 **M6:** 722
friction-welded to zirconium **A9:** 157
functions in FCAW electrodes **A6:** 188
furnace atmospheres **A4:** 920 **M4:** 771
furnaces **A4:** 920-921
galvanic corrosion with magnesium............ **M2:** 607
galvanic effects and discontinuities **M5:** 651
gas absorption by **M5:** 651
gas metal arc welding **M6:** 753
gas-metal arc welding, shielding gases........... **A6:** 66
gas-tungsten arc welding......................... **A6:** 190, 193
glass coatings referred to as enamels **EM3:** 301-302
glass-to-metal seals **EM3:** 302
gold plating of **M5:** 659
for grain refinement, aluminum alloys **A15:** 95
harmful alloying elements and corrosive substances...................................... **A11:** 202
heat treating **M4:** 763-774
heat treating practices **M4:** 772
heat treatment... **M3:** 359-360
in heat-resistant alloys...................................... **A4:** 512

SUBJECTS OF THE INDEXED VOLUMES: ASM Handbook (designated by the letter "A"): **A1:** Properties and Selection: Irons, Steels, and High-Performance Alloys (1990); **A2:** Properties and Selection: Nonferrous Alloys and Special-Purpose Materials (1990); **A3:** Alloy Phase Diagrams (1992); **A4:** Heat Treating (1991); **A6:** Welding, Brazing, and Soldering (1993); **A7:** Powder Metallurgy (1984); **A8:** Mechanical Testing (1985); **A9:** Metallography and Microstructures (1985); **A10:** Materials Characterization (1986); **A11:** Failure Analysis and Prevention (1986); **A12:** Fractography (1987); **A13:** Corrosion (1987); **A14:** Forming and Forging (1988); **A15:** Casting (1988); **A16:** Machining (1989); **A17:** Nondestructive Testing and Quality Control (1989); **A18:** Friction, Lubrication, and Wear Technology (1992). **Metals Handbook, 9th Edition** (designated by the letter "M"): **M1:** Properties and Selection: Irons and Steels (1978); **M2:** Properties and Selection: Nonferrous Alloys and Pure Metals (1979); **M3:** Properties and Selection: Stainless Steels, Tool Materials and Special-Purpose Materials (1980); **M4:** Heat Treating (1981); **M5:** Surface Cleaning, Finishing, and Coating (1982); **M6:** Welding, Brazing, and Soldering (1983). **Engineered Materials Handbook** (designated by the letters "EM"): **EM1:** Composites (1987); **EM2:** Engineering Plastics (1988); **EM3:** Adhesives and Sealants (1990); **EM4:** Ceramics and Glasses (1991); **Electronic Materials Handbook** (designated by the letters "EL"): **EL1:** Packaging (1989).

Titanium (continued)
heat-treating ... **A4:** 913-923
heat-treating principles ... **A4:** 921
heater tube, embrittlement of ... **A11:** 640-642
high frequency resistance welding ... **M6:** 760
high-frequency welding ... **A6:** 252, 253
history ... **A2:** 586
hot isostatic pressing effects ... **A15:** 539-540
hydrochloric acid corrosion ... **A13:** 1164
hydrogen embrittlement ... **A12:** 23
impurity concentrations ... **A2:** 1097
incident-ion energy ... **A18:** 851, 852, 854
indirect brazing of PSZ for joining oxide ceramics ... **EM4:** 517
induction heating energy requirements for metalworking ... **A4:** 189
induction heating temperatures for metalworking processes ... **A4:** 188
ingot characteristics ... **M3:** 362-364
as inoculant ... **A15:** 105
inoculant for aluminum ... **A6:** 53
intensity of K lines in ... **A10:** 97
intergranular corrosion of ... **A11:** 182
interlayer material ... **M6:** 681
interstitial-element content, effect on properties ... **M3:** 361
investment casting, by vacuum arc skull melting ... **A15:** 409-410
investment-cast, hot isostatic pressing of ... **A15:** 263
ion plating of ... **M5:** 425-426
ion removal from ... **A10:** 200
ion sputtering of adherend surfaces ... **EM3:** 244
iron levels ... **A4:** 913, 921
in iron-base alloys, flame AAS analysis of ... **A10:** 56
joined to aluminum alloys ... **A6:** 739
Jones reductor for ... **A10:** 176
lap welding ... **M6:** 673
lap-shear coupon and finite-element analysis ... **EM3:** 485
laser beam welding ... **M6:** 647, 662-663
weld properties ... **M6:** 662-663
laser cladding ... **A18:** 867
laser-beam welding ... **A6:** 263-264
liquid-metal embrittlement ... **M3:** 416
lubricant indicators and range of sensitivities ... **A18:** 301
as major structural metal ... **A2:** 647
in maraging steels ... **A4:** 220, 222, 223, 224
market data ... **M3:** 355
market development ... **A2:** 587
martensitic structures ... **A9:** 673-674
mass finishing of ... **M5:** 655-656
materials, types ... **A14:** 838-83
matrix composites, acoustic emission inspection ... **A17:** 288
matrix, ultrasonic inspection ... **A17:** 250
mechanical properties ... **A4:** 913 **A7:** 468-469, 475
metal characteristics, general ... **A2:** 586
in metal powder-glass frit method ... **EM3:** 305
microalloying of ... **A14:** 22
as minor element, ductile iron ... **A15:** 648
as minor toxic metal, biologic effects ... **A2:** 1260
neutron and x-ray scattering, and absorption compared ... **A10:** 421
new developments ... **A2:** 590-591
and niobium welds, neutron radiography of ... **A17:** 393
nitric acid corrosion ... **A13:** 1156
as nitride-forming ... **A15:** 93
nitride-forming element ... **A18:** 878
non-oxide ceramic joining ... **EM4:** 480
nonimplanted, nanoindentation ... **A18:** 426
optical micrograph ... **A8:** 476-477
ores, commercial, fusion with acidic fluxes ... **A10:** 167
oxidation potentials and bonding ... **EM4:** 482
oxidational wear ... **A18:** 287
oxygen determined in ... **A10:** 231
oxygen levels ... **A4:** 913, 921
P/M technology ... **A7:** 748-755
passivation and corrosion inhibition ... **M3:** 413-414
petroleum refining and petrochemical operations ... **A13:** 1263
in pharmaceutical production facilities ... **A13:** 1227
photometric analysis methods ... **A10:** 64

Titanium (continued)
physical properties related to thermal stresses ... **A4:** 605
pickling of ... **M5:** 654-655
pin bearing testing of ... **A8:** 59
plasma and shielding gas compositions ... **A6:** 197
plasma arc cutting ... **M6:** 916
platinized titania ISS and AES applications ... **A18:** 449-450, 451
platinum plating of ... **M5:** 658-659
polishing and buffing ... **M5:** 655-656
and polybenzimidazoles ... **EM3:** 170
polyimidesulfone (thermoplastic) adhesive ... **EM3:** 358
polyphenylquinoxaline bonding ... **EM3:** 166, 167
polysulfides as sealants ... **EM3:** 196
Pourbaix diagram ... **A13:** 1330
powder metallurgy materials, etching ... **A9:** 509
powder metallurgy materials microstructures ... **A9:** 511
powder metallurgy, Soviet ... **A7:** 693-694
precipitate stability and grain boundary pinning ... **A6:** 73
in precipitation-hardening steels ... **M6:** 350
precleaning embrittlement ... **A17:** 81
processing, effect on properties ... **M3:** 361-371
product forms ... **M3:** 354, 355, 358
product life cycle ... **A2:** 587
production examples, heat treating processes ... **M4:** 772-774
production examples, heat-treating processes ... **A4:** 921-923
production, for niobium-titanium superconductors ... **A2:** 1044
protective coatings, effects on descaling ... **M5:** 651
pure ... **M2:** 814-816
pure, effect of strain rate on ductility in ... **A8:** 42
pure, furrow-type fatigue fracture ... **A12:** 44, 54
pure, hydrogen entry ... **A13:** 329
pure, optical micrograph ... **A8:** 476-477
pure, properties ... **A2:** 1169
pure, sponge fines ... **A7:** 165
as pyrophoric ... **A7:** 199
quenching media ... **A4:** 917, 921 **M4:** 772
radiographic absorption ... **A17:** 311
rare earth alloy additives ... **A2:** 729
raw materials ... **M3:** 362
as reactive material, properties ... **A7:** 597
recommended shielding gas selection for gas-metal arc welding ... **A6:** 66
recovery from selected electrode coverings ... **A6:** 60
recrystallized implant, fatigue fracture surfaces ... **A11:** 685, 689
recycling ... **A2:** 1226-1228
relative solderability ... **A6:** 134
relative solderability as a function of flux type ... **A6:** 129
repair welding ... **A6:** 1107
resistance spot welding ... **M6:** 478
resistance spot welding of steels and content effect ... **A6:** 228
resistance welding ... **A6:** 847
Rockwell scale for ... **A8:** 76
roll welding ... **A6:** 312
for rolling ... **A14:** 35
S-N curve ... **A11:** 103
salt bath descaling of ... **M5:** 97-98, 102
scrap, plasma melting/casting ... **A15:** 420
for screws and plate nuts to repair Gr-Ep laminates ... **EM3:** 835
seal adhesive wear ... **A18:** 549
seal materials ... **A18:** 550
seawater exposure effect on adhesives ... **EM3:** 632
selection factors ... **M3:** 353-355
sheet, preparation for forming ... **A14:** 839-84
shielding gas purity ... **A6:** 65
in sintered metal powder process ... **EM3:** 304
in skins for which NDT methods used ... **EM3:** 767
solution-treating ... **A4:** 913, 914, 915-919, 920, 921, 922-923
species weighed in gravimetry ... **A10:** 172
specifications ... **M3:** 354, 357-358, 360
spectrometric metals analysis ... **A18:** 300

Titanium (continued)
sponge ... **M3:** 353, 361, 362
in stainless steels ... **A18:** 710, 712, 713
in steel weldments ... **A6:** 417, 418
stereo pair dimples ... **A12:** 171
strain-age cracking in precipitation-strengthened alloys ... **A6:** 573
strength of ultrasonic welds ... **M6:** 752
stress corrosion, microwave inspection ... **A17:** 215
stress relieving ... **M4:** 764-765
stress-corrosion cracking ... **M3:** 415-416
stress-corrosion cracking in ... **A11:** 223-224
stress-relieving ... **A4:** 913, 914-915, 920, 923
structural bonding with aluminum, primers ... **EM3:** 254
structural problems analyzed by inert gas fusion ... **A10:** 231
submerged arc welding
additions to restrict grain growth ... **M6:** 116
content in fluxes ... **M6:** 124
influence on heat-affected zone toughness ... **M6:** 119-120
substrate for polyimides ... **EM3:** 161
suitability for cladding combinations ... **M6:** 691
sulfuric acid corrosion ... **A13:** 1153
in superalloys, solution-treating ... **A4:** 799
superplasticity of ... **A8:** 553
surface preparation ... **EM3:** 264-269, 799
as surgical implants ... **A13:** 1327-1328
susceptibility to hydrogen damage ... **A11:** 249-250
tantalum/titanium/niobium carbides in cemented carbides ... **A18:** 795
tarnish removal ... **M5:** 654-655
thermal diffusivity from 20 to 100 °C ... **A6:** 4
thermal expansion coefficient ... **A6:** 907
Ti 387 eV, Auger electron spectroscopy ... **A18:** 454
and titanium alloys ... **A15:** 824-835
in titanium oxide, determined by controlled- potential coulometry ... **A10:** 207
TNAA detection limits ... **A10:** 237
to enhance case hardness ... **A4:** 263
to promote hardness ... **A4:** 124
to reduce electromigration ... **EM3:** 581, 582
trace amounts affecting induction hardening ... **A4:** 185
as trace element, cupolas ... **A15:** 388
twin bands in plastically deformed ... **A9:** 689
ultrapure, by external gettering ... **A2:** 1094
ultrapure, by iodide/chemical vapor deposition ... **A2:** 1094
ultrapure, by zone refining ... **A2:** 1094
ultrasonic welding ... **A6:** 326, 327 **M6:** 746
unalloyed, as implant material ... **A11:** 672
unalloyed, general corrosion data ... **A13:** 701-705
unalloyed grades, corrosion resistance ... **A13:** 328
unalloyed, hydrided ... **A13:** 674
as unknown particle ... **A10:** 457
upset welding ... **A6:** 249
use in flame atomizers ... **A10:** 48
use in flux cored electrodes ... **M6:** 103
use in resistance spot welding ... **M6:** 486
vacuum deposition of interference films ... **A9:** 148
vacuum heat-treating support fixture material ... **A4:** 503
vapor degreasing of ... **M5:** 8
vapor pressure, relation to temperature ... **A4:** 495 **M4:** 309-310
in vapor-phase metallizing ... **EM3:** 306
Vickers and Knoop microindentation hardness numbers ... **A18:** 416
volumetric procedures for ... **A10:** 175
weld fracture from incomplete fusion ... **A12:** 65
in wet chlorine ... **A13:** 1173
wire, brushing of ... **M5:** 656
wire, copper plated ... **M5:** 658
wrought ... **A2:** 592-633
yield strength vs. fracture toughness ... **A6:** 1017
in zinc alloys ... **A9:** 489
in zinc/zinc alloys and coatings ... **A13:** 759
Titanium alloy castings *See also* Titanium; Titanium alloys; Titanium castings
alloy properties ... **A2:** 636
as-cast and cast + HIP microstructures ... **A2:** 638
cast microstructure ... **A2:** 637-638
casting design ... **A2:** 640-642

Titanium alloy castings (continued)
chemical milling **A2:** 643
compared **A2:** 637
fatigue and fatigue crack growth rate **A2:** 640
heat treatment **A2:** 643-644
hot isostatic pressing **A2:** 401, 638
introduction **A2:** 634
lost-wax investment molding **A2:** 635-636
mechanical properties **A2:** 639-640
melting and pouring practice **A2:** 642
microstructure **A2:** 637-639
microstructure modification **A2:** 639
molding methods **A2:** 635-636
new alloys **A2:** 636-637
oxygen influence **A2:** 639
porosity **A2:** 638-639
product applications **A2:** 644-645
rammed graphite molding **A2:** 635
specifications **A2:** 636
superheating **A2:** 642
technology, history **A2:** 634-635
and titanium castings **A2:** 634-646
tolerances **A2:** 641-642
vacuum consumable electrode melting **A2:** 642
weld repair **A2:** 643
Titanium alloy precision forgings **A14:** 283-287
design criteria **A14:** 28
forging processing **A14:** 285-28
technology development effectiveness **A14:** 286-28
tooling and design **A14:** 284-285
Titanium alloys *See also* Advanced titanium materials; Advanced titanium-base allloys, Reactive metals; specific types; Titanium; Titanium alloy castings; Titanium alloy precision forging; Titanium alloys; Titanium alloys, specific types; Titanium carbides; Titanium sponge; Titanium-aluminum-vanadium powders; Wrought titanium; Wrought titanium alloys; Wrought titanium alloys, specific types **A6:** 507-523 **A13:** 669-706 **A14:** 267-287
abrasive blasting of **M5:** 652-653
advanced materials, forging **A14:** 282-283
aerospace applications **A2:** 587-588
aged, general corrosion **A13:** 682
aging **A4:** 913, 914, 916-919, 921, 922-923 **A6:** 509, 510, 512, 515, 518 **M4:** 767
allotropic forms **A14:** 26
alloy composition effects **A13:** 670
alloy development **A15:** 824-827
alloy types **A2:** 586-587 **A4:** 913
alloying element effects **A6:** 508
alloying elements' effects on α-βtransformation **A4:** 913
alpha + beta alloys **A2:** 586
alpha alloys **A2:** 586 **A6:** 508, 509, 518, 783, 786
alpha case removal **A15:** 264
alpha-beta alloys
alpha-beta forging **A14:** 271
annealing **A4:** 913, 914, 915-916, 917, 920 **M4:** 765-766
stability **A4:** 913, 914, 916 **M4:** 766
strength during **A4:** 915 **M4:** 766
anodic pitting **A13:** 683-685
anodizing **A9:** 142
applications **A2:** 587-590 **A6:** 84, 507, 522, 944 **A15:** 833-834
arc welding *See* Arc welding of titanium and titanium alloys
arc-welded **A11:** 437-439
architecture applications **A2:** 589-590
atmospheres, oxidizing heat treating **A4:** 920, 921 **M4:** 769
Auger spectra **A7:** 254
automotive components **A2:** 589
bar and tube, die materials for drawing **M3:** 525
Bauschinger effect **A4:** 914 **A14:** 83
belt grinding of **M5:** 652

Titanium alloys (continued)
beta alloys **A2:** 586-587
beta forging, example **A14:** 271-27
beta transus temperatures **A4:** 913
blank preparation **A14:** 840-841
blended elementalP/M, ductility **A7:** 254
brazed with an aluminum-silicon brazing alloy **A9:** 158
brazing **M6:** 1033-1035
brazing and soldering characteristics **A6:** 633-634
capacitor discharge stud welding **M6:** 738
carbon dioxide **A4:** 920
carbon monoxide **A4:** 920
casting design **A15:** 829-831
cathodic charging **A13:** 673
chemical conversion coating of **M5:** 656-658
chemical milling **A15:** 832
chemical resistance **EM3:** 639
chlorides contamination **A4:** 920 **M4:** 770
chrome plating, hard **M5:** 180-181
classes **A14:** 26
classification of **A6:** 508-512
classifications according to phase **A6:** 84
cleaning **A6:** 784-785
cleaning prior to heat treating **A4:** 919-920 **M4:** 770
cleaning processes **M5:** 650-656
cold forming **A14:** 841
commercially pure (CP) (unalloyed) titanium **A6:** 508-509
composition **M3:** 357, 755-756, 766
composition and impurity limits **M6:** 1049
consumer goods **A2:** 590
containing tin **A2:** 526
containment autoclaves **A8:** 420
contamination cracking **A6:** 516
continuous cooling transformation (CCT) diagrams **A6:** 84
continuous-drive welding **A6:** 522
contour roll forming **A14:** 84
conventional forging process **A14:** 271, 27
cooling in ultrasonic testing **A8:** 247
corrosion **A6:** 509
corrosion applications **A2:** 588-589
corrosion fatigue **A13:** 676
corrosion forms **A13:** 671
corrosion resistance enhancement **A13:** 670, 693-696
corrosion resistance mechanism **A13:** 669-670
crack depth as function of exposure time to solid cadmium in **A11:** 239-241
creep forming **A14:** 846-84
crevice corrosion **A13:** 671-672, 681-683
cryogenic service **A6:** 1017
deep drawing **A14:** 844-845
deformation differences **A8:** 583-584
dental **A13:** 1352, 1361
descaling process **M5:** 650-654
problems **M5:** 650-651
procedures **M5:** 651-654
designations/nominal compositions **A13:** 669
die specifications **A14:** 274-27
diffraction techniques, elastic constants, and bulk values for **A10:** 382
diffusion bonding **A6:** 156, 157-158, 159
diffusion welding **A6:** 522, 884, 885, 886 **M6:** 682
dimpling **A14:** 84
disk-forming process, simulation of **A14:** 43
distortion **A6:** 522
drop hammer forming **A14:** 657, 84
ductile-to-brittle temperature **A6:** 633
effect of hot pressing **A7:** 512-513
effect of milling time on density and flowability **A7:** 59
effect of strain rate on ductility in **A8:** 38, 42
effect of vacuum **A12:** 47-48
electric current perturbation inspection **A17:** 136, 138-139
electrochemical polarization **A8:** 531

Titanium alloys (continued)
electron beam welding **M6:** 643
electron-beam welding **A6:** 85, 512, 513, 514, 516, 517, 518, 519, 520, 521, 522, 865, 872
electroplating of **M5:** 658-659
electroslag welding **A6:** 783
elevated-temperature **A2:** 625-626
embrittled by mercury **A11:** 234
embrittled by zinc **A11:** 234
embrittlement by low-melting alloys **A12:** 29
environments that cause stress-corrosion cracking **A6:** 1101
equipment for gas-tungsten arc welding **A6:** 785-786
erosion-corrosion **A13:** 676, 692-693
evaluation/certification **A15:** 833
explosive forming **A14:** 84
explosive welding **A6:** 522
extra-low interstitial (ELI) grades **A6:** 520
extrusion **M3:** 368-369, 537
fatigue at subzero temperatures **M3:** 763, 764, 765, 769, 770
fatigue striations in **A12:** 176
filler metals **A6:** 510, 783, 784
FIM sample preparation of **A10:** 586
finishing processes **M5:** 656-659
fixtures **A4:** 921, 922
flash-butt welding **A6:** 522
flow stress and workability as function of temperature **A14:** 17
flux-cored arc welding **A6:** 783
fluxes **A6:** 521
forge welding **A6:** 306
forgeability **A14:** 267-27
forging **M3:** 366-369
forging equipment **A14:** 273-274
forging methods **A14:** 270-273, 277-278
forging temperature ranges **A14:** 26
forgings, flaws and inspection methods **A17:** 497-498
forming **M3:** 369
forming, lubricants for **A14:** 52
fractographs **A12:** 441-455
fracture topography **A12:** 441-443
fracture toughness **M3:** 763, 769
fracture/failure causes illustrated **A12:** 217
friction welding **A6:** 152, 153, 783, 784 **M6:** 722
fully dense **A7:** 435
furnaces **A4:** 920-921
fusion zone **A6:** 513-514, 515, 516, 521
galvanic corrosion **A13:** 675-676, 690-692
galvanic couples **A13:** 673
gas metal arc welding **M6:** 153
gas porosity in electron beam welds of **A11:** 445
gas tungsten arc welding **M6:** 182, 203, 206
gas-metal arc welding **A6:** 512, 513, 521, 522
gas-tungsten arc welding **A6:** 192, 513, 514, 515, 516, 518, 520, 521, 522
general corrosion **A13:** 671, 705-706
general welding considerations **A6:** 512
growth during heat treatment **A4:** 920 **M4:** 771
heat treating **A4:** 913-923
heat treatment **A15:** 833
heat-affected zone **A6:** 85, 509, 510, 513, 514, 516, 519, 521, 783, 784, 786
heat-treating principles **A4:** 921
HERF forgeability **A14:** 10
high-temperature alkaline conditions **A13:** 673
HIP temperatures **A7:** 437
historical perspective **A15:** 825
history **A2:** 586
hot forming **A14:** 841-84
hot isostatic pressing **A15:** 832
hot sizing of **A14:** 84
hot wire welding **A6:** 786
hydrogen damage **A13:** 170-171, 673-674, 685-686
hydrogen damage in **A11:** 338
hydrogen embrittlement **A6:** 516-517 **A12:** 23, 32, 124

Titanium alloys (continued)
hydrogen entry **A13:** 329
hydrogen pickup **A4:** 919, 920, 921 **M4:** 770
hydrogen pickup, composition affecting **M5:** 655
hydrogen testing **A13:** 673-674
implants **A7:** 657
induction heating for **A8:** 159
inertia friction welding **A6:** 522
influence of oxygen content **A8:** 480-481
investment casting, by vacuum arc skull melting **A15:** 409-410
iron levels **A4:** 913, 923
for isothermal/hot-die forging **A14:** 152
joggling **A14:** 847
joining **M3:** 368, 369-370
joint preparation **A6:** 784
laser beam welding **M6:** 662-663
weld properties **M6:** 662-663
laser melt/particle inspection **M6:** 801
laser-beam welding **A6:** 85, 263-264, 512, 513, 514, 516, 517, 518, 519, 520, 521, 783, 784
liquid-metal embrittlement of **A11:** 234
low-cycle fatigue, eddy current inspection **A17:** 190
low-temperature applications **M3:** 755
machining **A2:** 966
macrosegregation **A6:** 515-516
market development **A2:** 587
mass finishing of **M5:** 655-656
mechanical deformation from shearing **A9:** 159
mechanical properties **A4:** 913 **A7:** 468-469 **A15:** 828-829
melting and pouring practice **A15:** 831-832
metal removal after thermal exposure **A4:** 919
metastable beta alloys **A6:** 508, 509, 511-512, 514-515, 516, 518-519, 783
microsegregation **A6:** 516
microstructural effects **A8:** 476-477
microstructure **A6:** 507-508, 509, 512-519, 520, 943-944 **A15:** 827-828
modulus of elasticity at different temperatures **A8:** 23
molding methods **A15:** 825-826
molten salt bath descaling of **M5:** 653-654
near-alpha alloys **A6:** 508, 509, 515, 517
new developments **A2:** 590
nitric acid cleaning **A12:** 75
nitrogen contamination **A4:** 920 **M4:** 770
nitrogen-implanted **A10:** 485
nondestructive testing **A6:** 1086
nonisothermally upset, axial cross sections **A8:** 589
nonmetallic inclusions in **A11:** 316
occurrence of SMIE **A11:** 243
optical micrograph **A8:** 476-477
oxidation rates **A4:** 919, 920 **M4:** 770
oxygen content effect on near-threshold fatigue crack growth rate **A8:** 427, 430
oxygen levels **A4:** 913, 921
oxygen pickup **A4:** 919 **M4:** 770
oxygen rates **A4:** 919-920 **M4:** 769, 770
phase transformations **A6:** 84, 514
physical metallurgy **A6:** 507-508
pickling **A6:** 785
pickling of **M5:** 654-655
pitting **A13:** 672-673
plasma arc welding **A6:** 197, 198, 512, 513, 514, 516, 519, 520, 521, 522 **M6:** 214
polishing and buffing **M5:** 655-656
polishing powder metallurgy materials **A9:** 507
postweld heat treatments **A6:** 85-86, 508, 509, 510, 512, 514, 515, 517, 518-519, 786
powder metallurgy **M3:** 370-371
powder metallurgy alloys **A2:** 590
powder metallurgy materials microstructures **A9:** 511
powder production techniques **A7:** 438
power spinning **A14:** 845
precautions **A6:** 522
precision forgings **A14:** 283-287
press-brake forming **A14:** 844
press-formed parts, materials for forming tools **M3:** 492, 493
pressed, DGV percentage for **A8:** 590

Titanium alloys (continued)
pressure vessel, brittle tensile fracture in **A11:** 77
primary testing direction **A8:** 667
principal ASTM specifications for weldable nonferrous sheet metals **A6:** 400
processing, effect on properties **M3:** 361-371
production examples, heat-treating processes **A4:** 921-923
projection welding **A6:** 233
properties **A6:** 507, 510, 519
properties, compared **A15:** 826
quenching **A4:** 917, 921 **M4:** 766-767, 768, 769
rammed graphite molds of **A15:** 273
ratio-analysis diagrams **A11:** 62-63
reaction with molybdenum hearths in vacuum heat treating **A4:** 503
repair welding **A6:** 786
requirements **A6:** 522
residual stresses **A4:** 608 **A6:** 50, 518
resistance welding **A6:** 514, 521-522, 783, 784, 847
roll welding **A6:** 312
for rolling **A14:** 356 **M3:** 365-366
rubber-pad forming **A14:** 845-846
SCC, data representations **A13:** 676
SCC environments/temperatures **A13:** 274
SCC of **A12:** 28-29
SCC testing of **A8:** 530-532
SCC-conducive environments and temperatures **A11:** 223
scrap, recycling of **A2:** 755
secondary phases and martensitic transformations **A6:** 508
selection, forging method **A14:** 282
semisolid casting/forging of **A15:** 327
sheet, crevice corrosion testing **A13:** 674
sheet, preparation for forming **A14:** 839-840
sheet, roll compacted **A7:** 408
shielding gases **A6:** 784, 786
solid-state phase transformations in welded joints **A9:** 581
solid-state transformations in weldments **A6:** 84-86
solidification cracking **A6:** 516
solidification structures of welded joints **A9:** 579
solution treating **M4:** 766-767, 768
solution-treating **A4:** 913, 914, 916-919, 920, 921, 922-923
specific types **A7:** 475
springs, surface fracture **A11:** 551
stress intensity factor range effect on fracture **A8:** 486
stress relieving **A6:** 786 **M4:** 764-765
stress-corrosion cracking **A13:** 273-275, 674-676, 686-690
stress-corrosion cracking in **A11:** 223-224
stress-log strain rate data for **A8:** 214
stress-relieving **A4:** 913, 914-915, 920, 923
stretch forming **A14:** 846
submerged arc welding **A6:** 521, 783
subsolidus cracking **A6:** 517
super-alpha alloys **A6:** 509
superplastic forming **A14:** 842-844
superplastic forming and diffusion bonding **A2:** 590-591
superplastic forming/diffusion bonding (SPF/DB) **A6:** 156
surface preparation **EM3:** 522
as surgical implants **A13:** 1327-1328
susceptibility to hydrogen damage **A11:** 249-250
sustained-load cracking in inert environments **A8:** 531
tarnish removal **M5:** 654-655
tensile properties at subzero temperatures **M3:** 758, 762-763, 767
tension and torsion effective fracture strains **A8:** 168
test methods **A13:** 273-275, 671-676
testing in water and aqueous solutions **A8:** 531
thermal expansion coefficient **A6:** 907
thermal properties **A6:** 17
thermal treatments, special **A4:** 918-919 **M4:** 768-769
thermomechanical processing **A6:** 84, 85
three-roll forming **A14:** 846

Titanium alloys (continued)
Ti-6Al-4V
aerospace applications **EM3:** 264
anodization procedures **EM3:** 265, 266
bonded with polyphenylquinoxalines **EM3:** 166, 167
chromic acid anodization **EM3:** 266, 267, 269, 270
wedge-crack propagation test **EM3:** 268, 269
tin addition **M2:** 615
titanium weldment structure/property relationships **A6:** 519- 520
titanium-base intermetallic compounds **A2:** 590
titanium-matrix composites **A2:** 590
tool materials/lubricants for forming **A14:** 840
transformation temperatures **A4:** 919 **M4:** 766, 768
ultrasonic welding **A6:** 522, 894
unalloyed **A6:** 508-509
unalloyed titanium **A6:** 783, 784, 786
Unicast process for **A15:** 251
used as grain refiners in aluminum alloys **A9:** 630
vacuum forming **A14:** 847
weld design criteria **A6:** 522
weld design limitations **A6:** 522
weld metal porosity **A6:** 517-518
weld repair **A15:** 833
weldability **A6:** 507, 509, 510
welding defects **A6:** 515-518
welding process application **A6:** 520-523
welding specifications **A6:** 522-523
weldments, stress relieving **A4:** 914-915, 921-922 **M4:** 764, 765
wide-beam Auger spectrum **A7:** 251
with Widmanstatten structure **A8:** 476-477
Widmanst‡$$‡Adatten structure **A6:** 85, 86
wire brushing of **M5:** 656
wire, die materials for drawing **M3:** 522
for wire-drawing dies **A14:** 336
wrought **A2:** 592-633

Titanium alloys, friction and wear of **A18:** 693, 778-783
advantages of **A18:** 778
applications **A18:** 780, 781
femoral components of hip and knee replacements **A18:** 657, 658
carbonitriding **A18:** 866
carburizing **A18:** 866
disadvantages **A18:** 778
elements implanted to improve wear and friction properties **A18:** 858
fretting wear **A18:** 248, 251, 252
friction coefficient **A18:** 778, 779, 780, 781, 782
galling **A18:** 715
hot forging **A18:** 625, 626
for laser alloying **A18:** 866
laser melt/particle injection **A18:** 869, 870-871
laser melting **A18:** 864
lubrication **A18:** 778, 780, 781-783
of tool steels **A18:** 738
machined by high-speed steel cutting tools **A18:** 615
material for jet engine components **A18:** 588, 590, 591
metal forming lubricants **A18:** 147
nitriding **A18:** 866
part material for ion implantation **A18:** 858
protection against liquid impingement erosion **A18:** 222
surface modification treatments **A18:** 778-783
activated reactive evaporation (ARE) **A18:** 779, 780
surface modification treatments, advantages and limitations **A18:** 779
boriding **A18:** 779, 780-781
carburizing **A18:** 779, 780-781
erosive wear **A18:** 781
evaporation **A18:** 779, 780
fatigue life **A18:** 779
ion implantation **A18:** 778-780
laser surface treatment **A18:** 781
nitriding **A18:** 779, 780-781
physical vapor deposition (PVD) **A18:** 778-780
plasma spray coatings **A18:** 778, 780
plating **A18:** 778, 779
solid lubrication **A18:** 778-783

Titanium alloys, friction and wear of (continued)
sputtering **A18:** 778, 779, 782
thermochemical conversion surface treatments **A18:** 779, 780-781
versus S44004 (AISI 440C) steel when disk modified by evaporation **A18:** 780
Titanium alloys, specific powder metallurgy products
Ti-6Al-4V, compact, hot isostatically pressed fatigue specimen **A9:** 471
Ti-6Al-4V, pressed and sintered **A9:** 525
Titanium alloys, specific types *See also* Advanced titanium-base alloys, specific types; Titanium; Titanium alloy castings; Titanium alloys; Titanium castings
99%, contour band sawing **A16:** 363
99%, electrochemical machining **A16:** 537
230, flash welding **M6:** 558
317, flash welding **M6:** 558
318A, flash welding **M6:** 558
684, flash welding **M6:** 558
7075-T6, low-melting metal embrittlement fracture **A12:** 30, 38
CermeTi, as blended elemental titanium compacts **A2:** 650
commercially pure *See also* Ti grades 1 to 4
castings, applications **M3:** 407
tensile properties of castings **M3:** 409
commercially pure, diffraction techniques elastic constants, and bulk values for **A10:** 382
commercially pure, machining **A16:** 847, 849-852, 854
commercially pure Ti, 0.25 max O_2, brazing **A6:** 634
commercially pure Ti, 0.40 max O_2, brazing **A6:** 634
commercially pure titanium
annealing **A4:** 916 **M4:** 765
beta transformation temperatures **A4:** 914
oxide, thickness **M4:** 771
stress relief treatments **M4:** 764
stress-relief treatments **A4:** 915
contact angles on beryllium at various test temperatures in argon and vacuum atmospheres **A6:** 116
CORONA 5
electron-beam welding **A6:** 516, 520
gas-tungsten arc welding **A6:** 519
plasma arc welding **A6:** 516, 520
postweld heat treatment **A6:** 516
properties **A6:** 516
toughness vs. yield strength plotted **A6:** 85
extra-low-interstitial (ELI) grade **A6:** 783
Hylite 50, flash welding **M6:** 558
IMI 155 (British commercially pure), effect of stress intensity factor range on fatigue crack growth rate **A12:** 57
IMI 550, fretting wear **A18:** 251
IMI-685, effect of frequency and wave form on fatigue properties **A12:** 60-62
T1-6Al-6V-2Sn, fracture surface of a tension test bar **A9:** 472
Ti Code 12
composition **M3:** 357
mechanical properties **M3:** 357
Ti, grade 1
fasteners **M3:** 184
forging temperatures **M3:** 364
property data **M3:** 368, 372-374
rolling temperatures **M3:** 365
tensile properties, sheet **M3:** 362
Ti, grade 2
fasteners **M3:** 184
forging temperatures **M3:** 364
property data **M3:** 362, 368, 374-375
rolling temperatures **M3:** 365
Ti, grade 3
forging temperatures **M3:** 364

Titanium alloys, specific types (continued)
property data **M3:** 362, 368, 376-377
rolling temperatures **M3:** 365
Ti, grade 4
fasteners **M3:** 184
forging temperatures **M3:** 364
property data **M3:** 362, 368, 377-379
rolling temperatures **M3:** 365
Ti, grade 5 *See* Ti-6Al-4V
Ti, grade 7 *See* Ti-0.2Pd
Ti-0.2Pd
fasteners **M3:** 184
property data **M3:** 379-380
Ti-0.2Pd, sheet, hot rolled, annealed **A9:** 462
Ti-0.3Mo-0.8Ni
compositions **A6:** 510
properties **A6:** 510
stress-relieving **A6:** 515
Ti-0.3Mo-0.8Ni (Ti code 12)
beta transformation temperatures **A4:** 914
stress-relief treatments **A4:** 915
Ti-0.3Mo-0.8Ni, property data **M3:** 380-382
Ti-0.3Mo-0.8Ni, stress-relief treatments **M4:** 764
Ti-0.8Ni-0.3Mo, brazing **A6:** 634
Ti-0.15Pd, stress relieving **M6:** 456
Ti-0.350, fluting and cleavage **A12:** 27
Ti-1.5Fe-2.5Cr, cutoff band sawing **A16:** 361
Ti-1.5Fe-2.5Cr, cutoff band sawing with bimetal blades **A6:** 1184
Ti-2.5Cu (IMI 230)
aging **A4:** 917
annealing **A4:** 916
beta transformation temperatures **A4:** 914
solution-treating **A4:** 917
stress-relief treatments **A4:** 915
weldability **A4:** 913
Ti-2.5Cu (IMI 230), stress relieving **A6:** 515
Ti-2.25Al-11Sn-5Zr-1Mo
composition **M3:** 357
compositions **A6:** 510
mechanical properties **M3:** 357
properties **A6:** 510
property data **M3:** 387
Ti-2Fe-2Cr-2Mo, band sawing **A16:** 361, 363
Ti-2Fe-2Cr-2Mo, cutoff band sawing with bimetal blades **A6:** 1184
Ti-3.9Mn, tensile fracture **A12:** 454
Ti-3.9Mn, voids with particles **A12:** 455
Ti-3Al-2.5V
annealing **A4:** 916 **M4:** 765
beta transformation temperatures **A4:** 914
brazing **A6:** 634
classification **A6:** 633
composition **M3:** 357
compositions **A6:** 510
mechanical properties **M3:** 357, 368
properties **A6:** 510
property data **M3:** 399-400
stress-relief treatments **A4:** 915 **M4:** 764
stress-relieving **A6:** 515
stress-relieving times and temperatures **A6:** 786
Ti-3Al-2.5V, tube, cold drawn, stress relieved **A9:** 473
Ti-3Al-2.5V, tube, vacuum annealed **A9:** 473
Ti-3Al-8V-6Cr-2Zr-4Mo, capabilities **A15:** 827
Ti-3Al-8V-6Cr-4Mo
physical properties **A6:** 941
stress relieving **A6:** 515
weldability **A6:** 783
Ti-3Al-8V-6Cr-4Mo-4Zr **M3:** 357
brazing **A6:** 634
composition
compositions **A6:** 510
mechanical properties **M3:** 357
properties **A6:** 510
property data **M3:** 404-405
Ti-3Al-8V-6Cr-4Mo-4Zr (Beta C)
aging **A4:** 917

Titanium alloys, specific types (continued)
annealing **A4:** 916
beta transformation temperatures **A4:** 914
solution-treating **A4:** 917
stress-relief treatments **A4:** 915
tensile strength, relation to size **A4:** 918
Ti-3Al-8V-6Cr-4Mo-4Zr, machining **A16:** 847, 849-852
Ti-3Al-8V-6Cr-4Zr-4Mo
aging **M4:** 767, 769
annealing **M4:** 765
solution treating **M4:** 766-767
stress-relief treatments **M4:** 764
tensile strength, relation to size **M4:** 769
Ti-3Al-8V-6Cr-4Zr-4Mo, composition and properties **A2:** 637
Ti-3Al-8V-6Cr-4Zr-4Mo, rod, cold drawn solution treated, aged **A9:** 473
Ti-3Al-8V-6Cr-4Zr-4Mo weldability **M6:** 446
Ti-3Al-10V-2Fe, classification **A6:** 633
Ti-3Al-13V-11Cr, solid cadmium induced crack morphology for **A11:** 241
Ti-3Al-13V-11Cr, weldability **A6:** 783 **M6:** 446
Ti-4.5Al-5Mo-1.5Cr (CORONA 5)
compositions **A6:** 510
gas-tungsten arc welding **A6:** 515
properties **A6:** 510
Ti-4.5Al-5Mo-1.5Cr, as blended elemental compacts **A2:** 650
Ti-4Al-3Mo-1V, rolling temperatures **M3:** 365
Ti-4Al-4Mn, band sawing **A16:** 361, 363
Ti-4Al-4Mn, cutoff band sawing with bimetal blades **A6:** 1184
Ti-4Al-4Mo-2Sn-0.5Si (IMI 550)
aging **A4:** 917
annealing **A4:** 916
beta transformation temperatures **A4:** 914
quenching **A4:** 917
solution-treating **A4:** 917
stress-relief treatments **A4:** 915
Ti-4Al-4Mo-2Sn-0.5Si (IMI 550), stress relieving **A6:** 515
Ti-4Al-4Mo-2Sn-0.5Si (IMI 551)
aging **A4:** 917
annealing **A4:** 916
beta transformation temperatures **A4:** 914
solution-treating **A4:** 917
stress-relief treatments **A4:** 915
Ti-4Al-4Mo-4Sn-0.5Si (IMI 551), stress relieving **A6:** 515
Ti-4Mo-8V-6Cr-4Zr-3Al, classification **A6:** 633
Ti-4Ni, bainitic-like microstructures **A9:** 665-666
Ti-5.5Al-3.5Sn-3Zr-1Nb-0.3Mo-0.3Si (IMI 829)
aging **A4:** 917
annealing **A4:** 916
beta transformation temperatures **A4:** 914
solution-treating **A4:** 917
stress-relief treatments **A4:** 915
Ti-5.5Al-3.5Sn-3Zr-1Nb-0.3Mo-0.3Si (IMI 829), stress-relieving **A6:** 515
Ti-5.8Al-4Sn-3.5Zr-0.7Nb-0.5Mo-0.3Si (IMI 834)
aging **A4:** 917
annealing **A4:** 916
beta transformation temperatures **A4:** 914
solution-treating **A4:** 917, 918-919
stress-relief treatments **A4:** 915
Ti-5.8Al-4Sn-3.5Zr-0.7Nb-0.5Mo-0.3Si (IMI 834), stress-relieving **A6:** 515
Ti-5.8Mn, tensile brittle fracture **A12:** 454
Ti-5Al-2.5Sn
annealing **A4:** 916 **M4:** 765
beta transformation temperatures **A4:** 914
brazing **A6:** 634
casting applications **M3:** 407
classification **A6:** 633
composition **M3:** 357, 766
compositions **A6:** 510

Titanium alloys, specific types (continued)
fatigue at subzero temperatures M3: 765, 769, 770
filler metals for A6: 784
flash welding M6: 558
forging temperature M3: 364
fracture toughness M3: 769
gas metal arc welding conditions M6: 455
heat treating A4: 921, 922
mechanical properties M3: 357, 362, 368, 370, 409, 758, 763, 767
physical properties A6: 941
properties A6: 510
property data M3: 382-384
rolling temperatures M3: 365
stress relieving M6: 456
stress-relief treatments M4: 764
stress-relieving A4: 915, 922 A6: 515
stress-relieving times and temperatures A6: 786
thermal diffusivity from 20 to 100 °C A6: 4
ultrasonic welding A6: 326
weldability A4: 913 M6: 446
Ti-5Al-2.5Sn, composition and applications A2: 637
Ti-5Al-2.5Sn, effect of cooling rate on A9: 460
Ti-5Al-2.5Sn, effect of frequency and wave form on fatigue properties A12: 60-62
Ti-5Al-2.5Sn, forged, annealed, air cooled A9: 463
Ti-5Al-2.5Sn, forged, lap or fold, alpha case A9: 463
Ti-5Al-2.5Sn gas-turbine fan duct, repair weld contamination failure A11: 439
Ti-5Al-2.5Sn, hot worked, annealed different cooling rates compared A9: 463
Ti-5Al-2.5Sn, hydrogen-embrittled A12: 23, 32
Ti-5Al-2.5Sn, machining A16: 847, 849-852
Ti-5Al-2.5Sn, relative hydrogen susceptibility A8: 542
Ti-5Al-2.5Sn, strain-induced porosity A9: 463
Ti-5Al-2.5Sn, stress-corrosion cracks A9: 463
Ti-5Al-2.5Sn-ELI
compositions A6: 510
properties A6: 510
Ti-5Al-2.5Sn-ELI, machining A16: 847, 849-852
Ti-5Al-2Cr-1Fe, as blended elemental compacts A2: 650
Ti-5Al-2Sn, turning A16: 845
Ti-5Al-2Sn-2Zr-4Cr-4Mo, beta-process forging A9: 473
Ti-5Al-2Sn-2Zr-4Mo-4Cr
composition M3: 357
mechanical properties M3: 357
property data M3: 396-397
Ti-5Al-2Sn-2Zr-4Mo-4Cr (Ti 17)
aging A4: 917
annealing A4: 916
beta transformation temperatures A4: 914
designed for strength in heavy sections A4: 913
quenching A4: 917
solution-treating A4: 917
stress-relief treatments A4: 915
tensile strength, relation to size A4: 918
Ti-5Al-2Sn-4Cr-4Mo-2Zr
classification A6: 633
compositions A6: 510
properties A6: 510
stress-relieving A6: 515
Ti-5Al-2Sn-4Mo-2Zr-4Cr
aging M4: 767
annealing M4: 765
solution treating M4: 767, 769
stress-relief treatments M4: 764
tensile strength, relation to size M4: 769
Ti-5Al-4V, chemical machining A16: 25
Ti-5Al-5Sn, turning A16: 845
Ti-5Al-5Sn-2Zr-2Mo
composition M3: 357
compositions A6: 510
mechanical properties M3: 357
properties A6: 510
weldability A6: 783
Ti-5Al-5Sn-2Zr-2Mo, weldability M6: 446
Ti-5Al-5Sn-2Zr-2Mo-0.25Si, property data M3: 388

Titanium alloys, specific types (continued)
Ti-5Al-5Sn-5Zr, stress relieving M6: 456
Ti-5Al-6Sn-2Zr-1Mo-2.5Si, reduced 75% by temperatures A9: 465
upset forging, different forging
Ti-5Mo-4.5Al-1.5Cr, tensile ductile fracture A12: 455
Ti-6A-4V, effect of heat treatment and microstructure on fracture appearance A12: 32
Ti-6A-4V, hydrogen embrittled A12: 32
Ti-6Al-6V-2Sn, electrochemical machining A16: 852
Ti-6Al-2.5Sn bracket-brace assembly, fatigue failure A11: 439
Ti-6Al-2Cb-1Ta-0.8Mo
annealing M4: 765
stress-relief treatments M4: 764
Ti-6Al-2Cb-1Ta-1Mo, weldability M6: 446
Ti-6Al-2Nb-1Ta, physical properties A6: 941
Ti-6Al-2Nb-1Ta-0.8Mo
annealing A4: 916
beta transformation temperatures A4: 914
brazing A6: 634
composition A6: 510
properties A6: 510
stress-relief treatments A4: 915
stress-relieving A6: 515
subsolidus cracking A6: 517
Ti-6Al-2Nb-1Ta-0.8Mo, effect of temperature on ductility A12: 35, 48
Ti-6Al-2Nb-1Ta-0.8Mo, plate, hot rolled A9: 463
Ti-6Al-2Nb-1Ta-0.80Mo machining A16: 847, 849-852
Ti-6Al-2Nb-1Ta-1Mo
gas-tungsten arc welding A6: 520
weldability A6: 783
Ti-6Al-2Nb-1Ta-1Mo laser beam weld, ductile fracture A12: 443
Ti-6Al-2Nb-1Ta-1Mo plate, gas metal arc weld A9: 584-585
Ti-6Al-2Nb-1Ta-1Mo plate, gas tungsten arc weld A9: 584-585
Ti-6Al-2Nb-1Ta-IMo
high fracture toughness A4: 913
stress corrosion resistant A4: 913
Ti-6Al-2Sn, properties, after postweld heat treatment M3: 369
Ti-6Al-2Sn-2Zr-2Cr-2Mo
composition M3: 357
mechanical properties M3: 357
Ti-6Al-2Sn-2Zr-2Cr-2Mo-0.25Si, property data M3: 397
Ti-6Al-2Sn-2Zr-2Mo-2Cr
compositions A6: 510
properties A6: 510
Ti-6Al-2Sn-2Zr-2Mo-2Cr-0.25 Si
aging M4: 767
annealing M4: 765
solution treating M4: 767
stress-relief treatments M4: 764
Ti-6Al-2Sn-2Zr-2Mo-2Cr-0.25Si
aging A4: 917
annealing A4: 916
beta transformation temperatures A4: 914
solution-treating A4: 917
stress-relief treatments A4: 915
Ti-6Al-2Sn-2Zr-2Mo-2Cr-0.25Si, stress-relieving A6: 515
Ti-6Al-2Sn-4Zr-2Mo
aging A4: 917 M4: 767
annealing A4: 916 M4: 765
beta annealing A4: 918
beta transformation temperatures A4: 914
classification A6: 633
compositions A6: 510
designed for creep resistance A4: 913
electron-beam welding A6: 865
mechanical properties M3: 357, 369, 409, 410
properties A6: 510
property data M3: 385-386
rolling temperatures M3: 365
solution treating M4: 767
solution-treating A4: 917
stress-relief treatments A4: 915 M4: 764
stress-relieving A6: 515

Titanium alloys, specific types (continued)
thermal treatment M4: 768
weldability A6: 783
Ti-6Al-2Sn-4Zr-2Mo, application A15: 826
Ti-6Al-2Sn-4Zr-2Mo, application in jet engine nozzle assemblies A18: 591
Ti-6Al-2Sn-4Zr-2Mo, composition and applications A2: 637
Ti-6Al-2Sn-4Zr-2Mo, diffraction techniques elastic constants, and bulk values for A10: 382
Ti-6Al-2Sn-4Zr-2Mo, disk-forming process simulated A14: 430-431
Ti-6Al-2Sn-4Zr-2Mo, effect of forging deformation on A9: 460
Ti-6Al-2Sn-4Zr-2Mo, forged ingot, air cooled, reduced 15% by upset forging A9: 465
Ti-6Al-2Sn-4Zr-2Mo, forgings, reheated to different temperatures, air cooled A9: 465
Ti-6Al-2Sn-4Zr-2Mo, hot isostatic pressing A15: 542
Ti-6Al-2Sn-4Zr-2Mo, initial microstructures A14: 418
Ti-6Al-2Sn-4Zr-2Mo, machining A16: 323, 324, 326, 586, 602-603, 846, 847, 850-852
Ti-6Al-2Sn-4Zr-2Mo, processing map for A14: 424
Ti-6Al-2Sn-4Zr-2Mo, tree rings in macroslice A9: 465
Ti-6Al-2Sn-4Zr-2Mo, weldability M6: 446
Ti-6Al-2Sn-4Zr-2Mo-0.1Si, fatigue crack growth fracture topography A12: 441-443
Ti-6Al-2Sn-4Zr-2Mo-0.1Si, gas-tungsten arc welding A6: 515, 519
Ti-6Al-2Sn-4Zr-2Mo-0.08Si
casting applications M3: 407
forging temperatures M3: 364
tensile properties, variation with section size M3: 365
Ti-6Al-2Sn-4Zr-2Mo-0.25Si machining A16: 323, 324, 326, 847, 849-852
Ti-6Al-2Sn-4Zr-4Mo, composition and applications A2: 637
Ti-6Al-2Sn-4Zr-6Mo
aging A4: 917 M4: 767, 769
annealing A4: 916 M4: 765
beta transformation temperatures A4: 914
classification A6: 633
composition M3: 357
compositions A6: 510
designed for strength in heavy sections A4: 913
mechanical properties M3: 357
properties A6: 510, 519
property data M3: 395-396
quenching A4: 917
solution treating M4: 767, 769
solution-treating A4: 917
stress-relief treatments A4: 915 M4: 764
stress-relieving A6: 515
tensile strength, relation to size A4: 918 M4: 769
weldability A6: 511
Ti-6Al-2Sn-4Zr-6Mo , application A15: 826-827
Ti-6Al-2Sn-4Zr-6Mo, as blended elemental compacts A2: 650
Ti-6Al-2Sn-4Zr-6Mo, effect of frequency and wave form on fatigue properties A12: 62
Ti-6Al-2Sn-4Zr-6Mo, flash welding M6: 558
Ti-6Al-2Sn-4Zr-6Mo, forged bar, solution treated, water quenched A9: 471
Ti-6Al-2Sn-4Zr-6Mo, forged billet annealed A9: 471
Ti-6Al-2Sn-4Zr-6Mo, forged, solution compared A9: 471
treated, different temperatures
Ti-6Al-2Sn-4Zr-6Mo, forging, solution treated above the beta transus, quenched in water A9: 471
Ti-6Al-2Sn-4Zr-6Mo, high-temperature effect overload fracture A12: 35, 49
Ti-6Al-2Sn-4Zr-6Mo machining A16: 323, 324, 326, 847, 849-852

Titanium alloys, specific types (continued)
Ti-6Al-2Sn-6V, classification ... **A6:** 633
Ti-6Al-4V ... **A6:** 511
aging ... **A4:** 917 **M4:** 767, 769
annealed, yield strength vs. temperature ... **A6:** 1016
annealing ... **A4:** 916 **A18:** 780-781 **M4:** 765
applications ... **A18:** 781, 782
applications, femoral components of hip and knee replacements ... **A18:** 657
belt grinding of ... **M5:** 652
beta annealing ... **A4:** 918
beta transformation temperatures ... **A4:** 914
boriding ... **A4:** 445
brazing ... **A6:** 634
butt weld, characterization example ... **A6:** 102, 104-106
carbon ion implantation effect on resistance to polishing wear ... **A18:** 424
carburizing ... **A18:** 780
casting applications ... **M3:** 407
classification ... **A6:** 633
composition ... **M3:** 357, 766
compositions ... **A6:** 510
contact corrosive wear with UHMWPE improved by ion implantation ... **A18:** 779
cryogenic service ... **A6:** 1017
cutoff band sawing with bimetal blades ... **A6:** 1184
diffusion welding ... **A6:** 522, 885
electron beam welding ... **M6:** 643
electron-beam welding ... **A6:** 257, 518, 865, 872
erosion test results ... **A18:** 200
fasteners ... **M3:** 184
fatigue at subzero temperatures ... **M3:** 765, 767, 769, 770
filler metals for ... **A6:** 784
flash welding ... **M6:** 558, 579
flash welding schedule ... **M6:** 577
forging ... **M3:** 366, 367
fracture toughness ... **M3:** 410, 769
friction coefficient data ... **A18:** 71, 73
friction coefficient with no lubricant in air environment ... **A18:** 778
friction coefficient with PFPE lubricant ... **A18:** 778
friction surfacing ... **A6:** 323
friction welding ... **A6:** 153
gas metal arc welding conditions ... **M6:** 455
gas-metal arc welding ... **A6:** 521
gas-tungsten arc welding ... **A6:** 512, 514, 516, 521
growth during heat treatment ... **A4:** 920 **M4:** 771
heat treating ... **A4:** 922
high strength at low-to-moderate temperatures ... **A4:** 913
hydrogen contamination ... **A4:** 921
laser alloying ... **A18:** 866
laser cladding ... **A18:** 868
laser melt/particle injection ... **A18:** 868-869, 870, 871
laser melting ... **A18:** 864
laser nitriding ... **A18:** 781
laser-beam welding ... **A6:** 263, 264, 511, 517, 521, 876
mechanical properties ... **M3:** 357, 362, 365, 366, 368, 370, 409, 410, 758, 763
microstructure ... **A6:** 509, 512, 514
molybdenum disulfide coating effect on endurance lifetimes ... **A18:** 781, 782-783
nitriding ... **A18:** 780-781
nitrogen ion implantation versus carbon ion implantation ... **A18:** 779
notch fatigue strength of castings ... **M3:** 411
plasma arc welding ... **A6:** 520, 521
postweld heat treatment ... **A6:** 510
properties ... **A6:** 510, 511, 941
property data ... **M3:** 388-391
pulse plating deposition ... **A18:** 781

Titanium alloys, specific types (continued)
quench delay effect ... **A4:** 917, 918
quenching, effect on tensile properties ... **M4:** 768, 769
recrystallization annealing ... **A4:** 918
relative erosion factor ... **A18:** 200
resistance spot welding ... **A6:** 522
rolling temperatures ... **M3:** 365
scale removal ... **M5:** 651, 653, 660
solid-state cracking ... **A6:** 511
solid-state-welded interlayers ... **A6:** 169
solidification cracking ... **A6:** 516
solution treating ... **M4:** 767, 769
solution treating and overaging ... **A4:** 918
solution-treating ... **A4:** 917, 921, 922, 923
stress relieving ... **M6:** 456
stress-relief treatments ... **M4:** 764
stress-relieving ... **A4:** 914, 915 **A6:** 515
stress-relieving times and temperatures ... **A6:** 786
surface treatments ... **A18:** 780-782
tensile properties ... **A4:** 917, 918
tensile properties at subzero temperatures ... **M3:** 768
tensile strength, relation to size ... **A4:** 918 **M4:** 769
thermal diffusivity from 20 to 100 °C ... **A6:** 4
thermal treatment ... **M4:** 768-769
toughness vs. yield strength plotted ... **A6:** 85
ultrasonic welding ... **A6:** 326, 327
unimplanted versus carbon ion-implanted and fatigue ... **A18:** 780
untreated versus nitrogen-implanted wear volume loss ... **A18:** 779
weldability ... **A6:** 518, 519, 520, 783 **M6:** 446
Ti-6Al-4V aircraft component, fracture surface ... **A11:** 260
Ti-6Al-4V, alpha-beta billet, high aluminum defect ... **A9:** 470
Ti-6Al-4V, alpha-beta billet, high interstitial defect ... **A9:** 470
Ti-6Al-4V, alpha-beta, different illumination modes compared ... **A9:** 160
Ti-6Al-4V, as implant material ... **A11:** 672
Ti-6Al-4V, as-cast ... **A9:** 466
Ti-6Al-4V, as-forged ... **A9:** 468
Ti-6Al-4V, bar, annealed and air cooled ... **A9:** 467
Ti-6Al-4V, bar, held above the beta transus different cooling methods compared ... **A9:** 467
Ti-6Al-4V, bar, held below the beta transus different cooling methods compared ... **A9:** 467
Ti-6Al-4V, beta-annealed fatigued plate ... **A9:** 470
Ti-6Al-4V, brittle fracture surface ... **A12:** 448
Ti-6Al-4V, chip formation (high-speed machining) ... **A16:** 598, 600
Ti-6Al-4V, cleavage facets ... **A9:** 470
Ti-6Al-4V, closed-die forging techniques ... **A14:** 273
Ti-6Al-4V, compositional and microstructural effects on toughness ... **A8:** 480-481
Ti-6Al-4V, corrosion fatigue ... **A8:** 253-254 **A13:** 142
Ti-6Al-4V, cost comparisons for conventional vs hot-die forging ... **A14:** 154
Ti-6Al-4V, crack morphology ... **A11:** 240, 241
Ti-6Al-4V, cup-and-cone fracture ... **A12:** 451
Ti-6Al-4V, die temperature effects on forging pressure ... **A14:** 153
Ti-6Al-4V, different forging temperatures compared ... **A9:** 468
Ti-6Al-4V, diffraction techniques, elastic constants, and bulk values for ... **A10:** 382
Ti-6Al-4V, diffusion bonded to boron-aluminum composite ... **A9:** 595
Ti-6Al-4V, dimple size and secondary cracking ... **A12:** 451

Titanium alloys, specific types (continued)
Ti-6Al-4V, effect of beta flecks on ... **A9:** 459
Ti-6Al-4V, effect of biaxial tension on dimple rupture ... **A12:** 31, 39
Ti-6Al-4V, effect of cooling rate on ... **A9:** 460
Ti-6Al-4V, effect of frequency and wave form on fatigue properties ... **A12:** 62
Ti-6Al-4V, effect of stress intensity factor range on fatigue crack growth rate ... **A12:** 57
Ti-6Al-4V, effect of temperature on strength and ductility ... **A8:** 36
Ti-6Al-4V, electron beam welds in ... **A11:** 446
Ti-6Al-4V ELI, composition and applications ... **A2:** 637
Ti-6Al-4V ELI, ductile overload fracture ... **A12:** 443
Ti-6Al-4V ELI, fretting wear ... **A18:** 250
Ti-6Al-4V, extrusion, heated and air cooled ... **A9:** 467
Ti-6Al-4V, fatigue performance ... **A8:** 253-254
Ti-6Al-4V, fatigue strength (machining effects) ... **A16:** 26, 31, 35
Ti-6Al-4V, feathery fracture surface ... **A12:** 450
Ti-6Al-4V, forged above the beta transus ... **A9:** 468
Ti-6Al-4V, forging, different cooling methods compared ... **A9:** 468
Ti-6Al-4V, forging, gas tungsten arc weld titanium hydride ... **A9:** 469
Ti-6Al-4V, forging, oxide inclusion ... **A9:** 469
Ti-6Al-4V, forging, transgranular stress- corrosion cracks ... **A9:** 469
Ti-6Al-4V, forgings, gas tungsten arc welds different sections ... **A9:** 469
Ti-6Al-4V, fracture toughness and fraction of transformed structure ... **A8:** 480, 482
Ti-6Al-4V, grain-boundary cracking ... **A12:** 449
Ti-6Al-4V ground, diffraction peak location methods compared ... **A10:** 386
Ti-6Al-4V, high-cycle fatigue fracture ... **A12:** 452
Ti-6Al-4V, hot isostatic pressing ... **A15:** 540
Ti-6Al-4V, hydrazine effect on near-threshold fatigue crack propagation ... **A8:** 427, 429
Ti-6Al-4V, hydrogen embrittlement, manned spacecraft ... **A13:** 1086
Ti-6Al-4V, laser surface alloying of molybdenum into ... **A13:** 504
Ti-6Al-4V, low-cycle fatigue fracture ... **A12:** 452
Ti-6Al-4V, machining ... **A16:** 65, 67, 209, 323, 324, 326, 361, 363, 540, 580, 598, 600, 603, 608, 844-853, 858
Ti-6Al-4V, martensite needles, color etched ... **A9:** 157
Ti-6Al-4V, material savings by hot-die/isothermal forging ... **A14:** 151
Ti-6Al-4V, mechanical properties ... **A2:** 639-640 **A15:** 828-829
Ti-6Al-4V, microstructure ... **A2:** 637-639 **A15:** 827-828
Ti-6Al-4V, nitrogen-implanted ... **A10:** 485
TI-6Al-4V, pH effect on near-threshold fatigue crack growth rate ... **A8:** 427, 429
Ti-6Al-4V, plate, diffusion-bonded joint with bond-line contamination ... **A9:** 467
Ti-6Al-4V, plate, gas tungsten arc weld ... **A9:** 584-585
Ti-6Al-4V, plate, rolled, annealed, air cooled ... **A9:** 466
Ti-6Al-4V, prealloyed compacts, microstructure ... **A2:** 652
Ti-6Al-4V, prealloyed compacts, tensile/fracture toughness properties ... **A2:** 653

Titanium alloys, specific types (continued)
Ti-6Al-4V, press forging, different reductions and forging temperatures compared ... **A9:** 468
Ti-6Al-4V, radial fractures ... **A12:** 446
Ti-6Al-4V, recrystallized, creep rupture tested ... **A9:** 156
Ti-6Al-4V, recrystallized-annealed ... **A9:** 466
Ti-6Al-4V, relative hydrogen susceptibility ... **A8:** 542
Ti-6Al-4V, second-phase cleavage ... **A8:** 485, 488
Ti-6Al-4V, secondary cracks ... **A12:** 448-449
Ti-6Al-4V, shear band fracture ... **A12:** 444-445
Ti-6Al-4V shear fastener, LME failed ... **A11:** 228
Ti-6Al-4V, sheet, rolled, annealed, furnace cooled ... **A9:** 466
Ti-6Al-4V STA alloy rocket motor, adiabatic shear bands ... **A12:** 43
Ti-6Al-4V, stress-corrosion cracking ... **A12:** 452
Ti-6Al-4V, surface alterations ... **A16:** 22, 27
Ti-6Al-4V, temperature and strain rate effects on flow stress ... **A14:** 151
Ti-6Al-4V, tensile overload fracture ... **A12:** 449, 450
Ti-6Al-4V, tensile properties and fracture toughness ... **A2:** 641
Ti-6Al-4V, tensile properties/fracture toughness ... **A15:** 830
Ti-6Al-4V, tool mark fracture ... **A12:** 447
Ti-6Al-4V, TTS fracture ... **A12:** 29
Ti-6Al-4V-ELI
beta annealing ... **A4:** 918
compositions ... **A6:** 510
high fracture toughness properties ... **A4:** 913
recrystallization annealing ... **A4:** 918
stress corrosion resistant ... **A4:** 913
Ti-6Al-4V-ELI machining ... **A16:** 323, 324, 326, 847, 849-852
Ti-6Al-4V-ELI, thermal treatment ... **M4:** 768-769
Ti-6Al-4Zr-2Mo-2Sn, brazing ... **A6:** 634
Ti-6Al-5Zr-0.5Mo-0.2Si (IMI 685)
aging ... **A4:** 917
annealing ... **A4:** 916
beta transformation temperatures ... **A4:** 914
designed for creep resistance ... **A4:** 913
solution-treating ... **A4:** 917
stress-relief treatments ... **A4:** 915
Ti-6Al-5Zr-0.5Mo-0.2Si (IMI 685), stress-relieving ... **A6:** 515
Ti-6Al-5Zr-0.5Mo-0.5Si, forged, solution treated, oil quenched, aged and air cooled ... **A9:** 464
Ti-6Al-5Zr-0.5Mo-0.25Si, striation spacing ... **A12:** 60, 61
Ti-6Al-5Zr-4Mo-1Cu-0.2Si, as-cast, and solution treated in argon ... **A9:** 466
Ti-6Al-5Zr-4Mo-1Cu-0.2Si, forging annealed ... **A9:** 466
Ti-6Al-6Mo-4Zr-2Sm, brazing ... **A6:** 634
Ti-6Al-6Sn-4Zr-2Mo, peripheral end milling ... **A16:** 846, 849
Ti-6Al-6V-2Sn ... **A6:** 510
aging ... **M4:** 767, 769
annealing ... **A4:** 916 **M4:** 765
brazing ... **A6:** 634
casting applications ... **M3:** 407
composition ... **M3:** 357
compositions ... **A6:** 510
forging ... **M3:** 364, 369
gas-tungsten arc welding ... **A6:** 514
high strength at low-to-moderate temperatures ... **A4:** 913
mechanical properties ... **M3:** 357, 365, 368, 370
properties ... **A6:** 510, 519, 941
property data ... **M3:** 391-392
rolling temperatures ... **M3:** 365
solidification cracking ... **A6:** 516
solution treating ... **M4:** 767, 769
stress-relief treatments ... **M4:** 764
tensile strength, relation to size ... **M4:** 769
weldability ... **A6:** 511, 520
Ti-6Al-6V-2Sn (Cu + Fe)
aging ... **A4:** 917
annealing ... **A4:** 916
beta transformation temperatures ... **A4:** 914
solution-treating ... **A4:** 917
stress-relief treatments ... **A4:** 915

Titanium alloys, specific types (continued)
tensile strength, relation to size ... **A4:** 918
Ti-6Al-6V 2Sn (Cu ‡m+ Fe), stress-relieving ... **A6:** 515
Ti-6Al-6V-2Sn, alpha-beta forged billet macroscopic appearance of beta flecks ... **A9:** 472
Ti-6Al-6V-2Sn, as-extruded ... **A9:** 471
Ti-6Al-6V-2Sn, billet forged below the beta transus ... **A9:** 471
Ti-6Al-6V-2Sn, cleaved alpha grains ... **A8:** 490
Ti-6Al-6V-2Sn, effect of beta flecks on ... **A9:** 459
Ti-6Al-6V-2Sn, effect of elevated temperature on overload fracture ... **A12:** 35, 50
Ti-6Al-6V-2Sn, effect of stress intensity factor range on fatigue crack growth rate ... **A12:** 57
Ti-6Al-6V-2Sn, forging, solution treated quenched, aged ... **A9:** 472
Ti-6Al-6V-2Sn, fracture surface ... **A11:** 105
Ti-6Al-6V-2Sn, hand forging, solution treated, water quenched, aged, air cooled ... **A9:** 472
Ti-6Al-7Nb (IMI 367)
annealing ... **A4:** 916
beta transformation temperatures ... **A4:** 914
stress-relief treatments ... **A4:** 915
Ti-6Al-7Nb (IMI 367), stress-relieving ... **A6:** 515
Ti-6Al-25N-4Zr-2Mo, flash welding ... **M6:** 558
Ti-6Al-Nb-1Ta-0.8Mo
composition ... **M3:** 357
mechanical properties ... **M3:** 357, 368
property data ... **M3:** 386
Ti-6Be
spreading coefficients ... **A6:** 115
test conditions effect on interfacial energies ... **A6:** 117
wetting of beryllium ... **A6:** 115
Ti-7Al-2Mo-1V, plate, solution treated ... **A9:** 466
Ti-7Al-2Nb 1Ta, stress-corrosion cracking ... **A12:** 453
Ti-7Al-2Nb-1Ta, cleavage fracture ... **A12:** 453
Ti-7Al-2Nb-1Ta, staining and large dimples ... **A12:** 453
Ti-7Al-4Mo
annealing ... **A4:** 916 **M4:** 765
beta transformation temperatures ... **A4:** 914
brazing ... **A6:** 634
composition ... **M3:** 357
compositions ... **A6:** 510
forging temperatures ... **M3:** 364, 369
mechanical proper-ties ... **M3:** 357
properties ... **A6:** 510
property data ... **M3:** 393-394
rolling temperatures ... **M3:** 365
stress-relief treatments ... **A4:** 915 **M4:** 764
stress-relieving ... **A6:** 515
Ti-7Al-4Mo, aircraft powerplant, cracking ... **A13:** 1041
Ti-7Al-12Zr, stress relieving ... **M6:** 456
Ti-7Al-IMo-IV, fracture surface with tear ridge ... **A12:** 453
Ti-8.5Mo-0.5Si, twinned athermal alpha double-prime martensite ... **A9:** 474
Ti-8Al with 1800 ppm O_2, sheet, aged ... **A9:** 462
Ti-8Al-1Mo, flutes and cleavage ... **A12:** 27
Ti-8Al-1Mo-1V
aging ... **A4:** 917 **M4:** 767
annealing ... **A4:** 916 **M4:** 765
beta transformation temperatures ... **A4:** 914
brazing ... **A6:** 634
casting applications ... **M3:** 407
classification ... **A6:** 633
composition ... **M3:** 357
compositions ... **A6:** 510
electron-beam welding ... **A6:** 865
flash welding ... **M6:** 558
forging temperatures ... **M3:** 364, 369
mechanical properties ... **M3:** 357, 365, 368
properties ... **A6:** 510, 941
property data ... **M3:** 384-385
rolling temperatures ... **M3:** 365
solution treating ... **M4:** 767
solution-treating ... **A4:** 917
stress relieving ... **M6:** 456
stress-relief treatments ... **A4:** 915 **M4:** 764
stress-relieving ... **A6:** 515

Titanium alloys, specific types (continued)
stress-relieving times and temperatures ... **A6:** 786
weldability ... **M6:** 446
Ti-8Al-1Mo-1V, as-forged, ingot void ... **A9:** 464
Ti-8Al-1Mo-1V, effect of forging temperature ... **A9:** 460
Ti-8Al-1Mo-1V, electrical discharge grinding ... **A16:** 567
Ti-8Al-1Mo-1V, example of alpha-beta alloy ... **A9:** 458
Ti-8Al-1Mo-1V, fluting ... **A12:** 27, 28
Ti-8Al-1Mo-1V, forged at different starting temperatures ... **A9:** 464
Ti-8Al-1Mo-1V, forging, solution treated ... **A9:** 464
Ti-8Al-1Mo-1V, hydrogen-embrittled ... **A12:** 23, 32
Ti-8Al-1Mo-1V, martensitic structures ... **A9:** 674
Ti-8Al-1Mo-1V, pulse plating deposition ... **A18:** 781
Ti-8Al-1Mo-1V, SEM and TEM fractographs compared ... **A12:** 189-191
Ti-8Al-1Mo-1V, sheet, annealed ... **A9:** 464
Ti-8Al-1Mo-1V, sheet, duplex annealed ... **A9:** 464
Ti-8Al-1Mo-1V, sheet, solution treated ... **A9:** 464
Ti-8Al-2.8Sn-5.4Hf-3.6Ta-1Y-0.2Si, laser-beam welding ... **A6:** 514, 516
Ti-8Al-IMo-IV, effect, sustained-load cracking and SCC ... **A13:** 275
TI-8Al-IMo-IV, electrochemical polarization cracking ... **A8:** 531
Ti-8Al-IMo-IV, SCC crack initiation ... **A13:** 274
Ti-8Mn
annealing ... **A4:** 916 **M4:** 765
beta transformation temperatures ... **A4:** 914
brazing ... **A6:** 634
composition ... **M3:** 357
compositions ... **A6:** 510
forging temperatures ... **M3:** 364
mechanical properties ... **M3:** 357
properties ... **A6:** 510
property data ... **M3:** 393
rolling temperatures ... **M3:** 365
stress-relief treatments ... **A4:** 915 **M4:** 764
stress-relieving ... **A6:** 515
ultrasonic welding ... **A6:** 316, 327
Ti-8Mn, chemical milling ... **A16:** 585
Ti-8Mo-8V-2Fe-3Al
classification ... **A6:** 633
composition ... **M3:** 357
compositions ... **A6:** 510
mechanical properties ... **M3:** 357
properties ... **A6:** 510
property data ... **M3:** 403-404
weldability ... **A6:** 512, 783
Ti-8Mo-8V-2Fe-3Al, machining ... **A16:** 52
Ti-8Mo-8V-2Fe-3Al, weldability ... **M6:** 446
Ti-9Mo, martensitic structure ... **A9:** 674
Ti-10V-2Fe-3Al
aging ... **A4:** 917 **M4:** 767, 769
annealing ... **A4:** 916 **M4:** 765
beta transformation temperatures ... **A4:** 914
composition ... **M3:** 357
compositions ... **A6:** 510
high strength at low-to-moderate temperatures ... **A4:** 913
laser melting ... **A18:** 864
lubrication bearing test results after plasma spray coating ... **A18:** 780
mechanical properties ... **M3:** 357
plasma spray coating applications ... **A18:** 780
properties ... **A6:** 510
property data ... **M3:** 397-399
solution treating ... **M4:** 767, 769
solution-treating ... **A4:** 917
stress-relief treatments ... **A4:** 915 **M4:** 764
stress-relieving ... **A6:** 515
tensile strength, relation to size ... **A4:** 918 **M4:** 769
Ti-10V-2Fe-3Al, beta solution treated, water induced alpha double prime ... **A9:** 475
quenched, strained 5%, deformation-
Ti-10V-2Fe-3Al, carpet plot by stereophotogrammetry ... **A12:** 198
Ti-10V-2Fe-3Al, deformed and recrystallized, thermally etched ... **A9:** 474

Titanium alloys, specific types (continued)
Ti-10V-2Fe-3Al, fractured, SEM stereo
pair carpet plot, and contour plot **A12:** 172
Ti-10V-2Fe-3Al, machining **A16:** 847, 849-852
Ti-10V-2Fe-3Al, near-beta alloy **A9:** 458
Ti-10V-2Fe-3Al, pancake forging **A9:** 474
Ti-11.5Mo-6Zr-4.5Sn
aging **M4:** 767, 769
annealing **M4:** 765
brazing **A6:** 634
classification **A6:** 633
composition **M3:** 357
compositions **A6:** 510
mechanical properties **M3:** 357
properties **A6:** 510
property data **M3:** 408-409
solution treating **M4:** 767, 769
stress-relief treatments **M4:** 764
stress-relieving **A6:** 515
tensile strength, relation to size **M4:** 769
weldability **A6:** 783
Ti-11.5Mo-6Zr-4.5Sn (Beta III)
aging **A4:** 917
annealing **A4:** 916
beta transformation temperatures **A4:** 914
solution-treating **A4:** 917
stress-relief treatments **A4:** 915
tensile strength, relation to size **A4:** 918
Ti-11.5Mo-6Zr-4.5Sn, machining **A16:** 847, 849-852
Ti-11.5Mo-6Zr-4.5Sn, sheet, solution
treated and water quenched,
aged **A9:** 473
Ti-11.5Mo-6Zr-4.5Sn, weldability **M6:** 446
Ti-11Cr-13V-3Al, brazing **A6:** 943
Ti-13V-11Cr-3Al
aging **A4:** 917 **M4:** 767, 769
annealing **A4:** 916 **M4:** 765
beta transformation temperatures **A4:** 914
brazing **A6:** 634
classification **A6:** 633
composition **M3:** 357
compositions **A6:** 510
forging temperatures **M3:** 364, 369
mechanical properties **M3:** 357, 368
postweld heat treatment **A6:** 786
properties **A6:** 510
property data **M3:** 400-403
rolling temperatures **M3:** 365
solution treating **M4:** 767, 769
solution-treating **A4:** 917
stress-relief treatments **A4:** 915 **M4:** 764
stress-relieving **A6:** 515
stress-relieving times and
temperatures **A6:** 786
tensile strength, relation to size **A4:** 918 **M4:** 769
weldability **A6:** 512
Ti-13V-11Cr-3Al, beta rich alloy **A9:** 458
Ti-13V-11Cr-3Al, machining **A16:** 847, 849-852
Ti-13V-11Cr-3Al, sheet, rolled, solution
treated, air cooled, aged **A9:** 473
Ti-13V-11Cr-3Al, stress relieving **M6:** 456
Ti-15Cu-15Ni, reactive metal brazing,
filler metal **A6:** 945
Ti-15Mo
laser alloying **A18:** 866
laser melting **A18:** 864
Ti-15Mo-2.7Nb-3Al-0.2Si
compositions **A6:** 510
properties **A6:** 510
Ti-15Ni-15Cu, reactive metal brazing,
filler metal **A6:** 945
Ti-15V-3Al-3Cr-3Sn
aging **A4:** 917 **M4:** 767
annealing **A4:** 916 **M4:** 765
beta transformation temperatures **A4:** 914
solution treating **M4:** 767
solution-treating **A4:** 917

Titanium alloys, specific types (continued)
stress-relief treatments **A4:** 915 **M4:** 764
Ti-15V-3Al-3Cr-3Sn, capabilities **A15:** 827
Ti-15V-3Al-3Cr-3Sn, composition and
properties **A2:** 637
Ti-15V-3Cr-3Al-3Sn
classification **A6:** 633
compositions **A6:** 510
electron-beam welding **A6:** 520
gas-tungsten arc welding **A6:** 513, 515, 520, 521
laser-beam welding **A6:** 520
properties **A6:** 510
stress-relieving **A6:** 515
weldability **A6:** 512, 518, 519, 783
Ti-15V-3Cr-3Al-3Sn, cold-rolled strip
decorative aging **A9:** 475
Ti-15V-3Cr-3Al-3Sn, cold-rolled strip
progression of aging **A9:** 475
Ti-15V-3Cr-3Al-3Sn, weldability **M6:** 446
Ti-17Al, aged and plastically deformed **A9:** 689
Ti-24V, true profile length values **A12:** 200
Ti-25Cr-3Be, refractory metal brazing,
filler metal **A6:** 942
Ti-25Cr-13Ni, brazing **A6:** 943
Ti-25Cr-21V
brazing **A6:** 943
refractory metal brazing, filler metal **A6:** 942
Ti-28V, true profile length values **A12:** 200
Ti-28V-4Be, brazing **A6:** 943
Ti-28Zr-8Ge, brazing **A6:** 943
Ti-30V, brazing **A6:** 943
Ti-30V-4Be, brazing **A6:** 943
Ti-40 at.% Nb, beta solution heat
treated water quenched, aged,
with beta prime **A9:** 475
Ti-48Zr-4Be
brazing **A6:** 943, 945
reactive metal brazing, filler metal **A6:** 945
Ti-75A
composition **M3:** 766
tensile properties at subzero
temperatures **M3:** 767
Ti-1100
compositions **A6:** 510
properties **A6:** 510
Ti-1100, composition and properties **A2:** 637
Ti-commercially pure, flash welding **M6:** 558
Ti-Cr-V, brazing **A6:** 943, 944
Ti-Mo-Zr-Sn, orthodontic wires **A18:** 666
Ti-Ni-Cu, brazing **A6:** 944
Ti-O-N, surface treatment material for
titanium alloys **A18:** 779
$TiAl_3$ constituent, in aluminum-silicon
alloys **A15:** 160
unalloyed Ti *See* Ti, grades 1 to 4
Titanium alloys, welding of **A6:** 783-786
electron-beam welding **A6:** 783, 784
gas-metal arc welding **A6:** 783, 784, 786
gas-tungsten arc welding **A6:** 783-784, 785-786
plasma arc welding **A6:** 783, 784, 786
process selection **A6:** 783-784
weldability **A6:** 783
Titanium alloys, wrought
composition **A18:** 658
physical and mechanical properties **A18:** 659
Titanium aluminides *See also* Ordered intermetallics
alpha-2 alloys **A2:** 926-927
application **A2:** 925
forging of **A14:** 283
gamma alloys **A2:** 927-929
Titanium and titanium alloys *See also*
Reactive metals; Titanium alloys,
specific types **A9:** 458-475 **A16:** 844-857
abrasive waterjet machining **A16:** 527
aged structures **A9:** 461
alloy classes **A9:** 458
alloyed with uranium for hardness **A16:** 874
alpha microstructures **A9:** 460
alpha stabilizers **A9:** 458

Titanium and titanium alloys (continued)
beta flecks **A9:** 459
beta flecks, macroscopic appearance **A9:** 472
beta microstructures **A9:** 461
beta stabilizers **A9:** 458
broaching **A16:** 200, 203, 206
chemical milling **A16:** 579-583, 584, 586
commercial-purity, bar, annealed, and
water quenched **A9:** 462
commercial-purity, different levels of
hydrogenation compared **A9:** 462
commercial-purity, sheet, different
anneals compared **A9:** 462
content in stainless steels **A16:** 682-683, 684, 689
contour band sawing **A16:** 364
cutoff band sawing **A16:** 360, 361
cutting fluids used **A16:** 125
drilling **A16:** 222, 227, 229, 230
drilling/countersinking **A16:** 899
edge retention **A16:** 29
electrochemical grinding **A16:** 543, 547
electrochemical machining **A16:** 534, 535
electron beam machining **A16:** 570
end milling **A16:** 326
grinding **A9:** 458-459 **A16:** 437, 547
high aluminum defects **A9:** 459
high aluminum defects, appearance **A9:** 470
high interstitial defects **A9:** 459
high interstitial defects, appearance **A9:** 470
high-purity sheet, cold-rolled and
annealed **A9:** 462
high-speed machining **A16:** 597, 598, 600, 602-605
high-speed tool steels used **A16:** 58, 59
honing **A16:** 476, 477
laser beam machining **A16:** 575
low-stress grinding procedures **A16:** 28, 31
machinability **A16:** 1, 645
macroexamination **A9:** 459-460
martensite microstructures **A9:** 460-461
melting point **A16:** 601
milling **A16:** 307, 312, 313, 314, 317, 321, 547
mounting **A9:** 458
omega phase **A9:** 461, 474
peck drilling **A16:** 899
phase transformation temperature **A9:** 458
photochemical machining **A16:** 588, 590
polishing **A9:** 459
sectioning **A9:** 458
shear stresses and HPs **A16:** 15
slab milling **A16:** 324
spade drilling **A16:** 225
specimen preparation **A9:** 458-459
surface integrity **A16:** 22
thread grinding **A16:** 271
thread rolling **A16:** 282
tool life **A16:** 844-848
Titanium, ASTM specific types
grade 1
composition **M3:** 357
mechanical properties **M3:** 357, 360
grade 2
composition **M3:** 357
mechanical properties **M3:** 357, 360
grade 3
composition **M3:** 357
mechanical properties **M3:** 357, 360
grade 4
composition **M3:** 357
mechanical properties **M3:** 357, 360
grade 5, yield strength **M3:** 360
grade 6, yield strength **M3:** 360
grade 7
composition **M3:** 357
mechanical properties **M3:** 357, 360
grade 8, yield strength **M3:** 360
grade 9, yield strength **M3:** 360
grade 10, yield strength **M3:** 360
grade 11, yield strength **M3:** 360
Titanium, binary systems with .... **A3:** 1•23

Titanium boride
direct evaporation **A18:** 844
to improve cladding microhardness **A18:** 868
Titanium boride (TiB_2)
adiabatic temperatures **EM4:** 229
applications **EM4:** 230, 801
binary phase diagram **EM4:** 792
Hall-Petch relationship **EM4:** 800
hardness **EM4:** 799
properties **EM4:** 794-796, 798, 799
strength **EM4:** 800
synthesized by SHS process **EM4:** 229, 230
Young's modulus **EM4:** 799
Titanium boride cermets
application and properties **A2:** 1004
Titanium boride-based cermets **A7:** 811-812
Titanium carbide **A13:** 846-847
carbide coatings **A16:** 80, 81-82, 83, 87-88
cemented carbides **A16:** 73, 74, 78-80
in ceramics **A16:** 98, 101
in cermets **A16:** 90-94
chemical vapor deposition process **M5:** 382-384
coating for ceramic tools **A16:** 101
coating for high-speed tool steels **A16:** 51, 57
grain refiner in steels **A6:** 53
ground by diamond wheels **A16:** 462
and hard-phase Ni alloys **A16:** 835
honing stone selection **A16:** 476
inclusions affecting tool wear in stainless steels **A16:** 690
inserts, drilling **A16:** 237
ion plating process **M5:** 421
laser hardfacing **A6:** 806
powder production **A7:** 158
properties **A6:** 629
and stainless steels **A16:** 688
tap density **A7:** 277
thermal expansion coefficient **A6:** 907
Titanium carbide (TiC)
activated reactive evaporation process **A18:** 844
adiabatic temperatures **EM4:** 229
applications **EM4:** 230
in cemented carbides **A18:** 795
coatings
for cutting tool materials **A18:** 614
for dies **A18:** 643, 646
for jet engine components **A18:** 592
for titanium alloys **A18:** 779, 780
for tool steels **A18:** 739
for valve train assembly components **A18:** 559
friction coefficient data (on 440C stainless steel) **A18:** 74
for gas-lubricated bearings **A18:** 532
for hot-forging dies **A18:** 627
in joining non-oxide ceramics **EM4:** 529
for laser melt/particle injection **A18:** 869, 870-871
liquid impingement erosion **A18:** 227
methods used for synthesis **A18:** 802
pressure densification **EM4:** 298
properties **A18:** 812, 813
properties, adiabatic engine use **EM4:** 990
synthesized by SHS process **EM4:** 229, 230
thermal properties **A18:** 42
Vickers and Knoop microindentation hardness numbers **A18:** 416
Titanium carbide cermet
scuffing temperatures and coefficients of friction between ring and cylinder liner materials **EM4:** 991
thermal expansion coefficient **A6:** 907
Titanium carbide cermets **A7:** 802
as engineering materials **A2:** 978
gas turbine components **A7:** 807
hardening of **A2:** 997
hardness **A2:** 996
hardness, with ferrous metal binder **A7:** 810
impact and stress-rupture properties **A7:** 809
jet engine, turbine components **A7:** 563
manufacturing, hardening, machining, and grinding **A2:** 997-998
microstructure **A2:** 991
nickel-bonded, applications and properties **A2:** 995
sintered, effect of temperature on mechanical properties **A7:** 809
steel-bonded **A2:** 996-998
Titanium carbide cermets (continued)
steel-bonded, and other wear-resistant materials, compared **A2:** 997
Titanium carbide coating, on steel
fatigue crack **A9:** 96
Titanium carbide particulate-reinforced silicon nitride
fracture toughness **EM4:** 586
Titanium carbide-alumina (TiC-Al_2O_3)
erosion test results **A18:** 200
Titanium carbide-based cermets *See also* Titanium carbide cermets; Titanium carbide-based cermets, specific types **A7:** 806-810
composition and properties **A7:** 808
impact resistance of **A7:** 808
Titanium carbide-based cermets, specific types *See also* Titanium carbide cermets; Titanium carbide-based cermets
aluminum, composition **A7:** 808
chromium, composition **A7:** 808
cobalt, composition **A7:** 808
iron, composition **A7:** 808
molybdenum, composition **A7:** 808
nickel, composition **A7:** 808
titanium carbide, composition **A7:** 808
tungsten, composition **A7:** 808
Titanium carbide-nickel alloys
corrosion resistance **A18:** 800
Titanium carbide-steel cermets *See also* Titanium carbide-steel cermets, specific types **A7:** 810
Titanium carbide-steel cermets, specific types
aluminum, composition **A7:** 810
carbon, composition **A7:** 810
chromium, composition **A7:** 810
cobalt, composition **A7:** 810
iron, composition **A7:** 810
molybdenum, composition **A7:** 810
nickel, composition **A7:** 810
steel matrix, composition **A7:** 810
titanium carbide, composition **A7:** 810
titanium, composition **A7:** 810
Titanium carbide/nitride
applications **A18:** 812
Titanium carbides
in austenitic stainless steels **A9:** 284
in gray iron **A15:** 633
in wrought heat-resistant alloys **A9:** 311
Titanium carbohydride (TiC_xH_y)
synthesized by SHS process **EM4:** 227
Titanium carbonitride
chemical vapor deposition process **M5:** 383-384
Titanium carbonitride cermets
as engineering materials **A2:** 978
Titanium carbonitrides
in austenitic manganese steel castings **A9:** 239
in austenitic stainless steels **A9:** 284
Titanium castings *See also* Titanium; Titanium alloy castings; Titanium alloys
alloys **A2:** 636-637
applications **M3:** 407, 408
casting design **A2:** 640-642
chemical milling **A2:** 643
cost comparisons **M3:** 411, 412
electrode composition **A2:** 642
fatigue properties **M3:** 409, 411
final evaluation and certification **A2:** 644
fracture toughness **M3:** 409, 410
heat treatment **A2:** 643-644
hot isostatic pressing **A2:** 643
hot isostatic processing **M3:** 410-412
impact strength **M3:** 409
introduction **A2:** 634
lost-wax investment molding **A2:** 635-636
melting and pouring practice **A2:** 642
molding methods **A2:** 635-636
product applications **A2:** 644-645
production **M3:** 408
rammed graphite molding **A2:** 635
specifications **A2:** 637
superheating **A2:** 642
technology, historical perspective **A2:** 634-635
technology, history **A2:** 634-635
tensile properties **M3:** 409, 410
and titanium alloy castings **A2:** 634-646
tolerances **A2:** 641-642
vacuum consumable electrode melting **A2:** 642
Titanium castings (continued)
weld repair **A2:** 643 **M3:** 409-410
Titanium, commercially pure (unalloyed) *See* Titanium alloys, specific types; Unalloyed Titanium
Titanium diboride
thermal properties **A18:** 42
Titanium diboride (TiB_2)
erosion by thermal spalling in electrical discharge machining **EM4:** 375
grinding **EM4:** 334-335
pressure densification **EM4:** 298
synthesized by SHS process **EM4:** 227
as tooling for uniaxial hot pressing **EM4:** 298
Titanium dioxide
biologic effects **A2:** 1261
filler for elastomeric adhesives **EM3:** 150
function and composition for mild steel SMAW electrode coatings **A6:** 60
as pigment **EM3:** 179
Titanium dioxide (TiO_2)
medical applications **EM4:** 1009
properties **A18:** 801
Vickers and Knoop microindentation hardness numbers **A18:** 416
Titanium dioxide as an interference film **A9:** 147-148
Titanium dioxide/magnesium oxide (TiO_2/MgO)
methods used for synthesis **A18:** 802
Titanium in austenitic manganese steel **A1:** 826
in cast iron **A1:** 8
in cobalt-base alloys **A1:** 985
in compacted graphite iron **A1:** 56
effect of, on hardenability **A1:** 395, 413, 470
effect of, on notch toughness **A1:** 741-742
in ferrite **A1:** 404, 408
in malleable iron **A1:** 10
in maraging steels **A1:** 794-795
in microalloy steel **A1:** 359
in nickel-base superalloys **A1:** 983-984
in steel **A1:** 146, 577
Titanium in steel **M1:** 115, 411, 417
400 to 500 °C embrittlement, effect on susceptibility of high-chromium ferritic stainless steels **M1:** 686
maraging steels **M1:** 446, 447
notch toughness, effect on **M1:** 694
steel sheet, effect on formability **M1:** 555
temper embrittlement reduced by **M1:** 685
Titanium metal-matrix composites
forging of **A14:** 283
Titanium nitride **A16:** 95, 98
activated reactive evaporation process **A18:** 844
cermets **A16:** 91
chemical vapor deposition process **M5:** 383-384
coating for broaches for improved tool life **A16:** 59
coating for carbide tools for machining uranium alloys **A16:** 876
coating for carbide tools machining carbon steels **A16:** 673
coating for carbides **A16:** 639, 646, 656
coating for drills **A16:** 58, 710
coating for high-speed tool steels **A16:** 51, 57, 58
coating for milling cutters **A16:** 58, 59, 314
coating for saw bands **A16:** 358
coating for taps **A16:** 259
coatings
for cutting tool materials **A18:** 613, 614
for dies **A18:** 643
for jet engine components **A18:** 592
for sintered carbide tools to improve endurance **A18:** 812
for titanium alloys **A18:** 779, 780, 783
for tool steels **A18:** 739
for valve train assembly components **A18:** 559
drilling inserts **A16:** 236
friction coefficient data (on 440C stainless steel) **A18:** 74
grain refiner in steels **A6:** 53
HAZ microstructure and toughness **A6:** 79
ion plating process **M5:** 421
magnagold, friction coefficient data **A18:** 73, 74
maximum stability **A6:** 74
physical vapor deposition **A18:** 645
properties **A18:** 801
as surface treatment in wrought tool steels **A1:** 779

Titanium nitride (continued)
and tapping ... A16: 266
and tertiary wear mechanisms ... A16: 40, 41
to improve cladding microhardness ... A18: 868
Vickers and Knoop microindentation hardness numbers ... A18: 416
Titanium nitride (TiN) ... EM4: 20
as abrasive grains or cutting tool tips for grinding or machining ... EM4: 329
applications ... EM4: 230
in joining non-oxide ceramics ... EM4: 529
non-oxide ceramic joining ... EM4: 480
synthesized by SHS process ... EM4: 229
Titanium nitride coating
on high-speed steel ... A9: 99
Titanium nitride in plate steels
examination for ... A9: 203
Titanium nitride-coated carbides
cutting speed and work material relationship ... A18: 616
Titanium nitride-coated HSS (roughing grade)
cutting speed and work material relationship ... A18: 616
Titanium nitride-coated tool steel
part material for ion implantation ... A18: 858
Titanium nitride-coated WC
part material for ion implantation ... A18: 858
Titanium nitrides
in austenitic stainless steels ... A9: 284
in wrought heat-resistant alloys ... A9: 312
Titanium oxide *See also* Engineering properties of single oxides ... EM3: 592
alloyed with aluminum oxide for tough abrasives ... EM4: 332
coating formation in molten particle deposition ... EM4: 206
component in photochromic ophthalmic and flat glass composition ... EM4: 442
controlling reflectivity of coated architectural glass ... EM4: 450
depth profiles ... EM3: 249
effect on color of ceramics ... EM4: 5
HAZ microstructure and toughness ... A6: 79
hydration of oxide layers ... EM3: 625
metal brazing ... EM4: 490
solid-state sintering ... EM4: 272, 273, 280-281
specific properties imparted in CTV tubes ... EM4: 1039, 1042
sputter-induced reduction of oxides ... EM3: 245, 246
surface preparation ... EM3: 235, 259
thermal etching ... EM4: 575
as thixotrope ... EM3: 178
to improve thermal conductivity ... EM3: 178
Titanium oxide (TiO_2)
Knotek-Feibelman (KF) model for ESD ... A18: 456, 457
material transfer from Ti-6Al-4V to UHMWPE ... A18: 779
Titanium oxide and rubber
work material for ion implantation ... A18: 858
Titanium oxide steels
microstructure ... A6: 79
Titanium P/M products
applications ... A2: 647, 654-655
blended elemental (BE) products ... A2: 654-655
blended elemental compacts (BE P/M) ... A2: 647-651
blended elemental Ti-6Al-4V ... A2: 648-649
broken-up structure (BUS) heat treatment ... A2: 653-654
crack propagation ... A2: 650-653
fatigue strength ... A2: 650, 652-653
fracture toughness ... A2: 648-649, 651-653
future technology trends ... A2: 655-656
introduction ... A2: 647
mechanical properties ... A2: 647-654
postcompaction treatments ... A2: 653-654
Titanium P/M products (continued)
prealloyed (PA) products ... A2: 655
prealloyed compacts (PA P/M) ... A2: 651-653
tensile properties ... A2: 648-652
thermochemical processing (TCP) ... A2: 654
Ti-6Al-4V, as most commonly used alloy ... A2: 647-653
Titanium powder metallurgy materials
forging of ... A14: 283
Titanium powders *See also* Titanium; Titanium alloys; Titanium carbides; Titanium sponge; Titanium-aluminum- vanadium alloys; Wrought titanium ... A7: 748-755
aerospace applications ... A7: 653-655
applications ... A7: 164, 206, 653-655, 680-682, 752-755
blended elemental production ... A7: 164-165
chemical analysis and sampling ... A7: 249
cold isostatic pressing ... A7: 449-450
compacts, properties of electric spark-activated hot pressed elemental ... A7: 512
compacts, sintering ... A7: 393
cost reduction ... A7: 164
developing processes ... A7: 167
effects of hot pressing ... A7: 512-513
encapsulation of refractory metals ... A7: 428
keel splice ... A7: 682
mechanical properties ... A7: 468-469, 475, 752
for ordnance applications ... A7: 680-682
P/M parts *See also* Titanium powders, applications ... A7: 507, 680-682, 752-755
practical vacuum degassing cycle ... A7: 435
production ... A7: 164-168
as pyrophoric ... A7: 199
as reactive material, properties ... A7: 597
sintering ... A7: 393-395
Soviet ... A7: 693-694
specific types ... A7: 475
static properties of pressed-and-sintered ... A7: 395
toxicity and exposure limits ... A7: 206
vacuum degassing ... A7: 435
vacuum hot pressing ... A7: 507
vacuum sintering atmospheres for ... A7: 345
Titanium recycling
development ... A2: 1226-1227
fire hazards ... A2: 1227
melters ... A2: 1227
sacrificial uses ... A2: 1228
scrap ... A2: 639
trends ... A2: 1228
Titanium scrap
recycling of ... A2: 639
Titanium sponge *See also* Titanium; Titanium alloys ... A7: 164
bulk chemical analysis ... A7: 254
for commercially pure titanium ... A2: 594-595
fines, screen analysis and composition ... A7: 165, 393
plasma melting/casting ... A15: 420
production ... A2: 590
Titanium sponge fines *See also* Powder(s); Titanium sponge
for blended elemental compacts ... A2: 647-648
Titanium sulfides
in austenitic stainless steels ... A9: 284
in wrought heat-resistant alloys ... A9: 312
Titanium tetrachloride
chemical vapor deposition process ... M5: 383
Titanium tungsten
to form chemical bonds with alumina substrate in metallizing ... EM4: 545
Titanium vacuum arc skull melting furnaces ... A15: 409-410
Titanium-alloy phase diagrams
titanium-aluminum ... A3: 1 • 24
titanium-chromium ... A3: 1 • 24
titanium-vanadium ... A3: 1 • 24
Titanium-aluminum alloys
types and usage ... A2: 586
Titanium-aluminum phase diagram ... A6: 525
Titanium-aluminum-vanadium alloy, wrought
composition ... A18: 658
physical and mechanical properties ... A18: 659
Titanium-aluminum-vanadium alloys ... A7: 164, 165
aerospace applications ... A7: 654
by rotating electrode process ... A7: 41
in cold isostatic pressing ... A7: 435, 449
compacts, composition ... A7: 165
compacts, properties of electric spark activated hot pressed ... A7: 513
compacts, tensile properties ... A7: 752
complex shape from near-net shape ... A7: 41, 44
effect of milling time on particle shape change ... A7: 60
for encapsulation ... A7: 428
fatigue behavior ... A7: 41, 44
from mixed chlorides ... A7: 167
impeller of blended elemental powder ... A7: 749
keel splice preforms ... A7: 439
martensitic microstructure ... A7: 749
microstructure during sintering ... A7: 394
pores in ... A7: 165
processing conditions ... A7: 438-439
roll compacted ... A7: 408
shape reproducibility ... A7: 755
specific types ... A7: 475
tensile properties ... A7: 41, 42, 752
Titanium-base corrosion-resistant alloys, selection of ... A6: 598, 599
corrosion resistance ... A6: 599
heat-affected zone ... A6: 599
iron contamination ... A6: 599
mechanical properties ... A6: 599
microstructure ... A6: 599
postweld heat treatment ... A6: 599
stress-corrosion cracking ... A6: 599
weldability characteristics ... A6: 599
Titanium-base intermetallic compounds
as new technology ... A2: 590
Titanium-carbon refractory compound
synthesized by SHS process ... EM4: 228
Titanium-coated carbides
cutting speed and work material relationships ... A18: 616
Titanium-iron alloys, hydrided
phase analysis of ... A10: 293-294
Titanium-iron binary alloy
with omega phase ... A9: 474
Titanium-iron intermetallic compounds
phase analysis of ... A10: 293-294
Titanium-matrix composites
development and production ... A2: 589, 908
Titanium-matrix composites with boron fibers
grinding ... A9: 588
polishing ... A9: 591
Titanium-matrix composites with silicon- carbide fibers
grinding ... A9: 591
polishing ... A9: 591
Titanium-microalloyed steels ... A1: 403-404
Titanium-niobium microalloyed steels ... A1: 404
Titanium-palladium alloys
nitric acid corrosion in ... A13: 1156
Titanium-platinum-gold
interconnect and contact system ... EL1: 961
Titanium-stabilized steel
porcelain enameling of ... M5: 512-513
Titanium-stabilized wrought stainless steels
etching ... A9: 282
Titanocene
biologic/toxic effects ... A2: 1261
Titer
in volumetric work ... A10: 162
Titer technique
for analyte determination ... A10: 172

SUBJECTS OF THE INDEXED VOLUMES: ASM Handbook (designated by the letter "A"): **A1:** Properties and Selection: Irons, Steels, and High-Performance Alloys (1990); **A2:** Properties and Selection: Nonferrous Alloys and Special-Purpose Materials (1990); **A3:** Alloy Phase Diagrams (1992); **A4:** Heat Treating (1991); **A6:** Welding, Brazing, and Soldering (1993); **A7:** Powder Metallurgy (1984); **A8:** Mechanical Testing (1985); **A9:** Metallography and Microstructures (1985); **A10:** Materials Characterization (1986); **A11:** Failure Analysis and Prevention (1986); **A12:** Fractography (1987); **A13:** Corrosion (1987); **A14:** Forming and Forging (1988); **A15:** Casting (1988); **A16:** Machining (1989); **A17:** Nondestructive Testing and Quality Control (1989); **A18:** Friction, Lubrication, and Wear Technology (1992). **Metals Handbook, 9th Edition** (designated by the letter "M"): **M1:** Properties and Selection: Irons and Steels (1978); **M2:** Properties and Selection: Nonferrous Alloys and Pure Metals (1979); **M3:** Properties and Selection: Stainless Steels, Tool Materials and Special-Purpose Materials (1980); **M4:** Heat Treating (1981); **M5:** Surface Cleaning, Finishing, and Coating (1982); **M6:** Welding, Brazing, and Soldering (1983). **Engineered Materials Handbook** (designated by the letters "EM"): **EM1:** Composites (1987); **EM2:** Engineering Plastics (1988); **EM3:** Adhesives and Sealants (1990); **EM4:** Ceramics and Glasses (1991); **Electronic Materials Handbook** (designated by the letters "EL"): **EL1:** Packaging (1989).

Titrants
common standardizations ... **A10:** 172
unstable ... **A10:** 202
Titration
acid-base ... **A10:** 172-173
amperometric ... **A10:** 204
analytical, by potentiometric membrane electrodes ... **A10:** 181
automated, biamperometry or bipotentiometry use with ... **A10:** 204
back ... **A10:** 173
biamperometric ... **A10:** 204
bipotentiometric ... **A10:** 204
buret ... **A10:** 205
chelometric ... **A10:** 164
of chloride in silver nitrate solution ... **A10:** 164
complexometric ... **A10:** 164, 201
conductometric ... **A10:** 203
coulometric ... **A10:** 197, 202-205
dead-stop end-point ... **A10:** 204
defined ... **A10:** 683
EDTA ... **A10:** 173
electrochemical ... **A10:** 202
electrometric ... **A10:** 202-206, 672
for emulsion concentration ... **A14:** 516
iodimetric ... **A10:** 174
Karl Fischer, for surface oxides ... **A10:** 177
methods, with ion-selective membrane electrodes ... **A10:** 183
microwave inspection ... **A17:** 215
Mohr, defined ... **A10:** 164
oscillometric (high-frequency) ... **A10:** 203-204
permanganate, for chromium and vanadium ... **A10:** 176
potentiometric ... **A10:** 204
precipitation ... **A10:** 173
redox, miscellaneous ... **A10:** 174-176
spectrophotometric ... **A10:** 70
vessel, oscillometric (high-frequency) ... **A10:** 203
Vohhard, defined ... **A10:** 164
Titrimetric potentiometry
and ion-selection electrodes ... **A10:** 204
Titrimetry
acid-base ... **A10:** 172-173
amperometric ... **A10:** 204
buret ... **A10:** 205
classical ... **A10:** 205
complexation ... **A10:** 173
described ... **A10:** 162
iodimetric ... **A10:** 174
of metal alloys, by potentiometric membrane electrodes ... **A10:** 181
precipitation ... **A10:** 173
redox ... **A10:** 174-176
TL600 three-dimensional weaving machine ... **EM1:** 130-131
TL1000 three-dimensional weaving machines ... **EM1:** 130-131
TL1250 three-dimensional weaving machine ... **EM1:** 130-131
TLC *See* Thin-layer chromatography; Thin-layer chromatography (TLC)
TLW ... **M2:** 3-833 **M3:** 1-882 **M4:** 1-800 **M5:** 1-715 **M6:** 1-1112
TNAA *See* Thermal neutron activation analysis
TNT
aluminum additives ... **A7:** 600
TO packages *See also* Packages
3, discrete semiconductor ... **EL1:** 435
3, steel header ... **EL1:** 426
39, discrete semiconductor, header ... **EL1:** 423
92, discrete semiconductor ... **EL1:** 435
220, discrete semiconductor ... **EL1:** 435
220, lead frame assembly ... **EL1:** 427
220, metal body part ... **EL1:** 426
resistance projection welding ... **EL1:** 239
sealing of ... **EL1:** 237
Tobermorite gel ... **EM4:** 12
Tobin bronze
thermal spray coating material ... **A18:** 832
Toe and root cracks
in weldments ... **A17:** 585
Toe crack *See also* Cracking
definition ... **A6:** 1214 **M6:** 18
Toe of weld
definition ... **M6:** 18
Toepler pumps ... **A10:** 152
Toggle mechanisms
in mechanical presses ... **A14:** 494-495
Toggle press ... **A7:** 12
defined ... **A14:** 13
horizontal ... **A7:** 15
Toggle-switch springs
fatigue failure ... **A11:** 557
Tokamak fusion test reactor ... **A6:** 618
Tokamaks
fusion-containment devices ... **A6:** 433
as fusion-containment machines ... **A2:** 1056-1057
Tokamax (whisker)
properties ... **A18:** 803
Tolansky multiple-beam interferometer ... **A9:** 80
Tolerance *See also* Damage tolerance; Permissible variation ... **EM3:** 29
calibration accuracy ... **A8:** 612
close, of composites ... **EM1:** 36
damage, evaluating ... **A8:** 683
defined ... **A8:** 14 **EM1:** 24
interval ... **A8:** 626
specimen, in axial compression testing ... **A8:** 55
Tolerance design stage
quality design ... **A17:** 722
Tolerance limits
defined ... **A8:** 14
on fatigue data ... **A8:** 700
on stress-life data ... **A8:** 700
one-sided ... **A8:** 664-665, 700
Tolerance model, statistical
and process control ... **A17:** 739-740
Tolerance, to yielding
in limit analysis ... **A11:** 136-138
Tolerance(s) *See also* Damage tolerance; Dimensional tolerances; NDE reliability; Statistical tolerancing Tolerance model
analysis, flexible printed boards ... **EL1:** 595-596
calculations, high-speed PWBs ... **EL1:** 609
calibration, thermocouples ... **A2:** 881
and control limits, in process capability assessment ... **A17:** 739
defined ... **EM2:** 43
design stage, quality design ... **A17:** 722
and dimensions ... **EL1:** 595
fault, wafer-scale integration ... **EL1:** 9
flexible printed boards ... **EL1:** 592-595
of impurities, casting vs wrought copper alloys ... **A2:** 346
manufacturing/production, optical ... **EM2:** 482-483
thermoplastic injection molding ... **EM2:** 312-313
titanium and titanium alloy castings ... **A2:** 641-642
Tolerances *See also* Accuracy; Allowances; Dimensional accuracy ... **A14:** 6
alloy steel sheet and strip ... **M1:** 163
aluminum alloy closed-die forging ... **A14:** 243
for aluminum alloy precision forgings ... **A14:** 251-252
as-cast dimensions, die castings ... **A15:** 289
assignment, hot upset forging ... **A14:** 93-94
blanking and piercing high-carbon steels ... **A14:** 557-558
close, in cold extrusion ... **A14:** 307
cold finished bars ... **M1:** 217-219
contour roll forming ... **A14:** 632-634
defined ... **A14:** 73 **A15:** 11
die casting ... **A15:** 619-620
dimensional ... **A15:** 614-623
dimensional, in coining ... **A14:** 186
dimensional, of investment casting ... **A15:** 265
dimensional, of P/M materials and parts ... **A7:** 482
dimensioning and tolerancing ... **A15:** 622-623
for drop hammer forming ... **A14:** 657
effect on cost, hot upset forging ... **A14:** 94-95
evaporative foam castings ... **A15:** 622
explosive plate forming ... **A14:** 641
for explosive sheet forming ... **A14:** 640-641
explosive tube forming ... **A14:** 641
forging of steel ... **M1:** 362, 364-368, 370, 371, 373-375
of heat-resistant castings ... **M3:** 280, 283
high-temperature sintering ... **A7:** 482
hot rolled bars and shapes ... **M1:** 199
investment castings ... **A15:** 621-622
loose ... **A14:** 307
magnesium alloy die casting ... **A15:** 809
medium ... **A14:** 307
Tolerances (continued)
no-bake molds ... **A15:** 619
noncritical, steels ... **A15:** 619
on slug volume, cold extrusion ... **A14:** 309
open ... **A14:** 307
in open-die forging ... **A14:** 71-73
parting line, die casting ... **A15:** 289
permanent mold castings ... **A15:** 620-621
plaster mold castings ... **A15:** 622
of powder forged parts ... **A7:** 416 **A14:** 204
for precision forging ... **A14:** 160, 165
press bending ... **A14:** 552
and process capability ... **A15:** 615
requirements, process variables ... **A15:** 614-615
rolled ring ... **A14:** 125-127
sand castings ... **A15:** 617-619
for seamless rolled rings ... **A14:** 126-127
shell molded castings ... **A15:** 622
springs, steel ... **M1:** 288, 303, 304
in straightening ... **A14:** 680
for swaging bar and tubing ... **A14:** 140
titanium alloy castings ... **A15:** 830-831
titanium alloy forged aircraft part ... **A14:** 274
titanium alloy precision forging ... **A14:** 284
upsetting pipe and tubing ... **A14:** 91-92
variations ... **A14:** 307
Tolerancing
and dimensioning ... **A15:** 622-623
Toluene
as organic cleaning solvent ... **A12:** 74
surface tension ... **EM3:** 181
Toluene ($C_6H_5CH_3$)
as solvent used in ceramics processing ... **EM4:** 117
Toluene diisocyanate (TDI) ... **EM3:** 108, 203
in polyurethanes ... **EM2:** 257
Toluhydroquinone
typical formulation ... **EM3:** 121
Toluol/hexane
solvent used to help apply ferrography to grease-lubricated bearings ... **A18:** 307
Tolyltriazole ... **A18:** 141
Tombasil *See also* Cast copper alloys; Copper alloys, specific types, C87500 and C87800
properties and applications ... **A2:** 372-373
Tombstoning ... **A6:** 994, 995 **EL1:** 694, 704
Tomkeieff equations
for three-dimensional grains and particles ... **A9:** 132
for two-dimensional planar figures ... **A9:** 131
Tomographic plane
defined ... **A17:** 385
Tomography
defined ... **A17:** 385
Toner
in copier powders ... **A7:** 573, 582
for electrostatic copying machine powder used ... **A7:** 573, 582
Tong hold ... **A14:** 13, 87
Tongue and groove bonding ... **EM3:** 561
Tongue design
for press-brake forming ... **A14:** 536
Tongues
AISI/SAE alloy steels ... **A12:** 341
and cleavage facets, compared ... **A12:** 424
formation ... **A12:** 13
and hairline indications ... **A11:** 79
iron ... **A12:** 219, 224, 457
low-carbon steels ... **A12:** 252
medium-carbon steels ... **A12:** 267
micro-, in ductile irons ... **A12:** 232
on cleavage fracture surface ... **A12:** 13, 252
on steel weld metal ... **A12:** 17
pyramid-shaped, nickel alloys ... **A12:** 397
steel, in cleavage ... **A11:** 22
Tonnages
metal casting shipments ... **A15:** 41-42
Tool and die materials *See also* Die materials; Tool materials
for hot forging ... **A14:** 43
Tool and die steels ... **A6:** 662, 674-676
composition ... **A6:** 674, 675
cracking ... **A6:** 675
description of steels ... **A6:** 674
filler metals ... **A6:** 674-675
flux-cored arc welding ... **A6:** 674, 676
gas-metal arc welding ... **A6:** 674, 676

Tool and die steels (continued)
gas-tungsten arc welding **A6:** 674, 676
heat-affected zones **A6:** 675
plasma arc welding **A6:** 674
postweld heat treatment **A6:** 675-676
preheating **A6:** 675, 676
repair practices **A6:** 675, 676
shielded metal arc welding **A6:** 674, 676
welding applications **A6:** 674
welding procedures and practices **A6:** 675-676
welding processes **A6:** 674
Tool and die(s)
powders used **A7:** 574
steels, cobalt powders in **A7:** 144
Tool bits
high-speed tool steels **A16:** 57
Tool center point (TCP) velocity (robot) **EM3:** 718
Tool chips
as stress concentrator **A11:** 89
Tool condition monitoring systems **A16:** 411-417
bearing transducers **A16:** 412, 414
design consideration **A16:** 414
feed force transducers **A16:** 412
force transducers **A16:** 415
future developments **A16:** 417
load-sensitive transducers **A16:** 414
measuring plate transducers **A16:** 412
microprocessing **A16:** 412, 415
NC machines **A16:** 411, 412
signal amplification **A16:** 412
tapping transducers **A16:** 412
tension-measuring transducers **A16:** 412, 413
torque transducers **A16:** 415
variable monitored **A16:** 411-414
Tool design **A14:** 197, 474
Tool designer
information for **EM2:** 1
Tool dulling
effect in piercing **A14:** 461
Tool failures *See also* Tools; Tools and dies, failures
analytical approach **A11:** 563
causes of **A11:** 564-577
heat treatment, influence of **A11:** 564, 567-574
influence of design **A11:** 564-566
machining and **A11:** 564, 566-567
service conditions and **A11:** 575-577
steel grade and quality, influence of **A11:** 564, 566, 574-575
types of **A11:** 564
Tool inserts, carbide
vapor forming **M5:** 382-383
Tool joints
by radial forging **A14:** 145
Tool life **A1:** 591-592 **A14:** 139, 463 **A18:** 617
ductile cast iron, machining of **M1:** 54, 55
in machining aluminum **M2:** 187-189
steel, machining of **M1:** 566, 571, 572, 574, 581, 583
Tool life tests **A1:** 591
Tool marks **A14:** 13, 840
AISI/SAE alloy steels **A12:** 296
cobalt alloy **A12:** 398
effect on cold-formed part failures **A11:** 308
as fracture origin, medium-carbon steels **A12:** 258
magnetic rubber inspection **A17:** 125
spring failures from **A11:** 555-557
as stress concentrators **A11:** 318
titanium alloys **A12:** 446-447
wrought aluminum alloys **A12:** 419
Tool materials *See* Cemented carbides; Material selection; Tool Steels; Ultrahard tool materials
blanking and piercing dies **A14:** 484-485
for blanking/piercing high-carbon steels **A14:** 556
cold extrusion **A14:** 309
for cold heading **A14:** 293
for conventional blanking dies **A14:** 453
for drawing **A14:** 336
Tool materials (continued)
for drop hammer forming, steels **A14:** 656
for hot extrusion **A14:** 320-321
for hot upset forging **A14:** 86
for hydrostatic extrusion **A14:** 329
for piercing **A14:** 465
for press forming dies **A14:** 504-505
for press-brake forming **A14:** 539-540
for titanium alloy forming **A14:** 840
for trimming stainless steels **A14:** 230
Tool materials, structural components
use for **M3:** 558-559
Tool materials, superhard *See also* Cemented carbides **M3:** 448-465
abrasion resistance **M3:** 453, 455, 456, 457-458, 461, 464
applications **M3:** 449, 450, 451, 452, 460, 461-462, 463-464
cast Co-Cr-W-Nb-C alloys **M3:** 461-462
cemented carbides **M3:** 449-461
ceramics **M3:** 462-463, 464
classification **M3:** 448-449, 450, 451
composition **M3:** 460, 461
corrosion resistance **M3:** 456, 461
cubic boron nitride **M3:** 463-464
diamond **M3:** 464
hardness **M3:** 453, 455, 456, 457-458, 459, 460, 461, 463, 464
microstructure **M3:** 452-453, 454, 459, 460, 462, 463
Tool set **A7:** 12
Tool side
defined **EM1:** 24
Tool steel
borided, for tape casting, blanking, and hole fabrication **EM4:** 164
hardened, for shear testing **A8:** 68
hardness conversion tables for **A8:** 109
ion implantation of **M5:** 425
Knoop indentations in microconstituents of **A8:** 97, 101
parameters for machining with HIP metal-oxide composite grade ceramic insert cutting tools **EM4:** 968
properties **EM4:** 316, 974
Tool steel alloys, specific types *See also* ASP steels; High-speed tool steels; Tool steels **A7:** 471, 472
01, chemical composition **A7:** 525
A2, chemical composition **A7:** 525
ASP 23, chemical composition **A7:** 525
ASP 23, mechanical properties and compositions **A7:** 471
ASP 30, chemical composition **A7:** 525
ASP 30, mechanical properties and compositions **A7:** 471
ASP 60, chemical composition **A7:** 525
ASP 60, mechanical properties and compositions **A7:** 471
CPM Rex 20, chemical composition **A7:** 525
CPM Rex 20, mechanical properties and compositions **A7:** 471
CPM Rex 25, mechanical properties and compositions **A7:** 471
CPM Rex 76, chemical composition **A7:** 525
CPM Rex 76, mechanical properties and compositions **A7:** 471
CPM-10V, chemical composition **A7:** 525
CPM-10V, mechanical properties and compositions **A7:** 471
D2, as-sintered and wrought **A7:** 378
D2, chemical composition **A7:** 525
D2, mechanical properties and compositions **A7:** 472
D6, chemical composition **A7:** 525
FC-0200, tensile strengths **A7:** 499
FC-0400, tensile strengths **A7:** 499
FC-0600, tensile strengths **A7:** 499
FC-0800, tensile strengths **A7:** 499
Tool steel alloys, specific types (continued)
FN-0200, tensile strengths **A7:** 499
FN-0405 MPIF, tensile strengths **A7:** 499
FN-0500, tensile strengths **A7:** 499
FN-0600, tensile strengths **A7:** 499
FN-0605 MPIF, tensile strengths **A7:** 499
FN-0800, tensile strengths **A7:** 499
M2, chemical composition **A7:** 525
M2 CPM, mechanical properties and compositions **A7:** 471
M2, mechanical properties and compositions **A7:** 103, 472
M2, microstructure as-sintered **A7:** 378
M2, sintering data **A7:** 373
M2, water-atomized **A7:** 32
M3, composition and properties **A7:** 103
M4 CPM, mechanical properties and compositions **A7:** 471
M4, mechanical properties and compositions **A7:** 472
M42, chemical composition **A7:** 525
M42 CPM, mechanical properties and compositions **A7:** 471
M42 CPM Rex, chemical composition **A7:** 525
M42, mechanical properties and compositions **A7:** 103, 472
M42, replacing **A7:** 144
S7, chemical composition **A7:** 525
T15, chemical composition **A7:** 525
T15 CPM, mechanical properties and compositions **A7:** 471
T15 CPM, microstructures **A7:** 788
T15, mechanical properties and compositions **A7:** 103, 472
T15, microstructures ... **A7:** 38, 39, 103, 104, 535, 788
T15, particles, carbide enlargement **A7:** 184
T15, replacing **A7:** 144
T15, sintered microstructure **A7:** 371
Tool steel for case hardening
composition of **A9:** 219
Tool steel(s) *See* Tool steels
Tool steels *See also* ASP steels; High speed tool steels; Tool failures; Tool steel alloys, specific types; Tool steels, high-speed; Tool steels, specific types; Tools; Ultrahard tool materials **A7:** 784-793 **A9:** 256-272 **A16:** 708-732 **M3:** 421-447
abrasion resistance **M3:** 583
abrasive flow machining **A16:** 519
abrasive waterjet machining **A16:** 527
air-hardened, die failure from severe wear **A11:** 575, 582
air-hardening, composition of weld deposits **A6:** 675
air-hardening medium-alloy cold work **M3:** 425, 427
annealing **A4:** 35
anti-segregation process **A7:** 784-787
antiwear applications **A7:** 621
applications **A7:** 789-793
arc welding **M6:** 294-297
filler metals **M6:** 295-296
postweld heat treating **M6:** 296
preheating **M6:** 296
repair welding **M6:** 296-297
welding conditions **M6:** 295
welding processes **M6:** 295
argon oxygen decarburization **A15:** 426
boriding **A4:** 264, 437, 445
boring **A16:** 716
brazing **A6:** 908
brazing and soldering characteristics **A6:** 624-625
by STAMP process **A7:** 548
CAP-produced **A7:** 535-536
carbon, hydrogen flaking in **A11:** 574
causes of distortion **M3:** 466-467
CBN for machining **A16:** 105
centerless grinding **A16:** 730
ceramic molding of **A15:** 248

Tool steels (continued)
cermet tools applied **A16:** 92
chemical compositions **A7:** 525
chromium hot work **M3:** 426-427
circular sawing **A16:** 725
classification **M4:** 561, 562-564
classification for grindability **A16:** 726
cleanliness **A7:** 104
CNC grinding **A16:** 730
Co content **A16:** 57
cobalt enhanced cutting ability **A7:** 144
for coining **A14:** 182
cold compacted **A7:** 467
cold reduction swaging effects **A14:** 129
cold work **M3:** 427-428
cold-work, steel type and composition **A6:** 674
composition **M3:** 422-423, 424 **M4:** 561, 562-564
composition effect on cost **A16:** 728
composition effect on grindability **A16:** 727-728
composition limits **A16:** 709-710
consolidated by atmospheric pressure (CAP) **A7:** 535-536
contour band sawing **A16:** 362, 364
conventionally wrought and ASP tool, compared **A7:** 785
crucible particle metallurgy process (CPM) for **A7:** 787-788
cryogenic treatment **A4:** 204-205
cutoff band sawing **A16:** 360
cutting fluids **A16:** 125, 710, 716, 717, 726
cylindrical grinding **A16:** 730
decarburization resistance **M3:** 443
distortion **M3:** 443, 466-469
for drawing dies **A14:** 336-337
drilling **A16:** 229-231, 237, 710-711, 715-716, 718, 721, 727
early fractographs **A12:** 5
EDM procedures for **A11:** 91-92
effect of composition on microstructure **A9:** 258
electrical discharge machining **A16:** 527, 539, 557, 558, 560, 708
electrochemical discharge grinding **A16:** 548, 549, 550
electrochemical grinding **A16:** 547
electrochemical machining **A16:** 535, 539
electron beam welding **M6:** 638
preheating **M6:** 613
electron-beam welding **A6:** 258-259, 867
end milling **A16:** 722, 726
etchants for **A9:** 257
fabrication **M3:** 442-443
face milling **A16:** 715, 716
flame hardening **A4:** 284
flash welding **M6:** 558
flute grinding **A16:** 730-731
forging temperatures **A14:** 81
fractographs **A12:** 375-382
fracture tests for **A12:** 141, 162
fracture/failure causes illustrated **A12:** 217
friction welding **A6:** 152, 153
FULDENS process **A7:** 788-789
gas atomized, compositions **A7:** 103
gas nitriding **A4:** 392-393
gear machining **A16:** 723
grindability **M3:** 442, 443
grinding **A16:** 547, 722, 723-732
grinding conditions **A16:** 729-730
grinding fluids **A16:** 730-731
grinding ratios **A16:** 724, 726-727, 728, 730, 731, 732
grinding wheel selection **A16:** 728-730
grooving with cermets **A16:** 95
ground by CBN wheels **A16:** 455, 461
hardenability **M3:** 443
hardening by a molten salt bath treatment **A4:** 475-476
hardening by quenching **A4:** 77, 78, 79
hardfacing **A6:** 798
hardness effect on grindability **A16:** 727-728
hardness effect on surface finish **A16:** 728, 729
hardness scale, relation of Vickers and Rockwell **M3:** 440
heat treating of **A14:** 53-55
heat treatment **M3:** 425, 433-437, 438-440
high speed **M3:** 424-426
high-carbon **A12:** 141, 162

Tool steels (continued)
high-carbon, high-chromium cold work **M3:** 427
high-speed
composition of weld deposits **A6:** 675
steel type and composition **A6:** 674
high-speed molybdenum, for bearings **A11:** 490
high-speed, wear resistance influenced by carbides **M1:** 612
history **A7:** 18
for hot forging **A14:** 43, 46, 54-56
hot isostatically pressed gas- atomized **A7:** 467
for hot upsetting dies **A14:** 86
hot work **M3:** 426-427
composition of weld deposits **A6:** 675
steel type and composition **A6:** 674
hot-work, hot hardnesses **A14:** 46
hot-work, runner block wear from **A11:** 576, 583
hydrogen-stress cracking and loss of tensile ductility **A1:** 715
injection molded, tensile strengths **A7:** 499
internal grinding **A16:** 730
ion implantation **A4:** 266
for ironing punches and dies **A14:** 511
lapping **A16:** 492, 494, 499
laser surface transformation hardening **A4:** 293
laser transformation hardening **A4:** 265
liquid nitrocarburizing **A4:** 418, 419
liquid-erosion resistance **A11:** 167
low-alloy, quench cracking fracture **A12:** 375
low-alloy, special-purpose **M3:** 430-431
low-stress grinding guidelines (LSG) **A16:** 731
machinability **M3:** 442-443
machining allowances **M3:** 443-444
macroetching **A9:** 256
macroetching to reveal hardened zones **A9:** 175
macroexamination **A9:** 256
maraging steels, use for **M3:** 446-447
mechanical properties **A7:** 467-468, 471
mechanical testing of **A11:** 564
microetching **A9:** 257-258
microexamination **A9:** 256-258
microstructure **A16:** 708, 711, 715
microstructure of as-sintered **A7:** 377
microstructures **A9:** 258-259
mill products, ingot metallurgy vs. CAP production **A7:** 535
milling **A16:** 312-313, 547, 715-716, 718-719, 721, 723, 726-727
milling with PCBN tools **A16:** 112
Mn content **A16:** 52
mold steels **M3:** 430-431
mold steels, steel type and composition **A6:** 674
molybdenum high speed **M3:** 424-425
molybdenum hot work **M3:** 427
mounting **A9:** 256-257
nominal compositions of **A9:** 259
oil hardening, composition of weld deposits **A6:** 675
oil-hardening cold work **M3:** 427-428, 429-430
parts, effects of overaustenizing **A11:** 570
PCBN tooling used **A16:** 111, 112
peripheral milling **A16:** 721
phosphorus content **A16:** 52
photochemical machining **A16:** 590
physical properties **M3:** 434, 441, 442
plasma (ion) nitrocarburizing **A4:** 423
powder metallurgy **M3:** 441, 442, 443, 444-445
advantages over conventional tool steels **A1:** 780
applications of high-speed tool steels **A1:** 785-786
classification **A1:** 781-792
cold-work tool steels **A1:** 786-789
heat treatment of high-speed tool steels **A1:** 782-783
high-speed tool steels **A1:** 781-786
hot-work tool steels **A1:** 789-790
powder metallurgy materials, etching **A9:** 509
powder metallurgy materials microstructures **A9:** 511
power band sawing **A16:** 714
power hacksawing **A16:** 724
for precision aluminum forgings **A14:** 253
precision-cast hot work tools **M3:** 446

Tool steels (continued)
preparation of specimens **A9:** 256-258
production of **A7:** 100, 103-104
quench cracking in **A12:** 130
reaming **A16:** 719
residual stresses **M3:** 467
resistance welding **A6:** 84
rigid tool compaction of **A7:** 322
SAE grades **M3:** 432
sawing **A16:** 358-360, 363-364
sectioning **A9:** 256
selection **A16:** 708, 712
service performance **M3:** 425, 434-442
shape distortion **M3:** 467, 468
shaping **A16:** 191, 714
Shepherd P-F test for **A12:** 141, 162
shock-resisting **M3:** 428-430, 431
shock-resisting, steel type and composition **A6:** 674
sintering **A7:** 370-376
size distortion **M3:** 467-468
soldering **A6:** 625
spade drilling **A16:** 225
special classes **A4:** 734-760 **M4:** 581-613
special purpose, steel type and composition **A6:** 674
specifications **M3:** 424
stabilization **M3:** 469
sulfurization effect on grindability **A16:** 728, 729
surface finish caused **A16:** 708, 721, 724, 728, 732
surface grinding **A16:** 730
surface hardening **A4:** 259
tapping **A16:** 720
tempering **A4:** 124, 134
testing **M3:** 434
thermoreactive deposition/diffusion process **A4:** 448
thermoreactive deposition/diffusion process, applications **A4:** 451-452
thread grinding **A16:** 730
threading with cermets **A16:** 95
tool life **A16:** 708, 710, 720, 721, 726
for tooling in thermoreactive deposition/diffusion process **A4:** 449
tungsten high speed **M3:** 425-426
tungsten hot work **M3:** 427
turning **A16:** 150, 708-710, 711, 713, 714, 715, 718, 726
turning with cermets **A16:** 93
ultrasonic inspection **A17:** 232
Unicast process for **A15:** 251
vacuum sintering atmospheres for **A7:** 345, 467
water-atomized, composition and properties **A7:** 103, 467
water-hardening **M3:** 431-434
composition of weld deposits **A6:** 675
steel type and composition **A6:** 674
weldability **M1:** 563 **M3:** 442
wrought **A1:** 757-779
classification and characteristics **A1:** 757-771
cold-work steels **A1:** 763-766
fabrication of **A1:** 774-778
high-speed steels **A1:** 759-762
hot-work steels **A1:** 762-763
low-alloy special-purpose steels **A1:** 767
machining allowances **A1:** 778-779
mold steels **A1:** 767-768
precision cast hot-work **A1:** 779
shock resisting steels **A1:** 766-767
surface treatments **A1:** 779
testing of **A1:** 771-778
typical heat treatments and properties **A1:** 771
water-hardening steels **A1:** 768-771

Tool steels, carbon-tungsten special purpose **A4:** 757-758
annealing **A4:** 758 **M4:** 609, 610
austenitizing **A4:** 758 **M4:** 609, 610
composition **A4:** 758 **M4:** 562-563
normalizing **A4:** 758 **M4:** 609, 610
quenching **A4:** 758 **M4:** 610
stress relieving **M4:** 610
stress-relieving **A4:** 758
tempering **A4:** 758 **M4:** 610, 611

Tool steels, defects and distortion in heat-treated parts **A4:** 601-618
burning, detection and effects **A4:** 602

Tool steels, defects and distortion in heat-treated parts (continued)
control of distortion ... **A4:** 616-617
distortion after heat treatment ... **A4:** 617
distortion and its control in heat-treated aluminum alloys ... **A4:** 617
distortion in heat treatment ... **A4:** 612-616
examples of ... **A4:** 613-615
methods of preventing ... **A4:** 615-616
precautions ... **A4:** 615
types of ... **A4:** 612-613
etching characteristics ... **A4:** 602, 603
importance of design ... **A4:** 617-618
overheating and burning of low-alloy steels ... **A4:** 601-603
overheating detection ... **A4:** 601-602
overheating, factors affecting ... **A4:** 602-603
overheating, prevention ... **A4:** 603
quench cracking ... **A4:** 610-612
reclamation of overheated steel ... **A4:** 603
residual stresses ... **A4:** 603-610

Tool steels, distortion ... **A4:** 761-766
causes ... **A4:** 761-762 **M4:** 614-616
control ... **A4:** 761-762, 764-765 **M4:** 618
fixtures ... **A4:** 765
irreversible changes ... **A4:** 765 **M4:** 614
maraging steels ... **A4:** 765-766 **M4:** 620
metallurgical structure ... **A4:** 761-762 **M4:** 615-616
powder metallurgy steels ... **A4:** 765, 766 **M4:** 619
reversible changes ... **A4:** 761 **M4:** 614
shape ... **A4:** 761, 762-764 **M4:** 614, 615, 616
shape distortion control ... **A4:** 764-765 **M4:** 618
size ... **A4:** 761, 762 **M4:** 614, 615, 616
tempering ... **A4:** 762

Tool steels, friction and wear of ... **A18:** 693, 734-739
applications ... **A18:** 734, 736
boriding ... **A18:** 739
classification system ... **A18:** 734, 735-736
cold work (A, O, D) tool steels ... **A18:** 734, 735, 739
high-speed (M, T) tool steels ... **A18:** 734-736, 739
hot-work (H) tool steels ... **A18:** 734, 735, 739
low-alloy special purpose (L) tool steels ... **A18:** 736
low-carbon (P) mold steels ... **A18:** 736
shock-resisting (S) tool steels ... **A18:** 734, 735, 739
water-hardening (W) tool steels ... **A18:** 734, 736
compositions ... **A18:** 726, 734, 735-736, 737
extension of tool life via ion implantation, examples ... **A18:** 643
friction coefficient ... **A18:** 737, 738, 739
galling ... **A18:** 737, 739
heat treatments ... **A18:** 734
high-speed, erosion rate and resistance ... **A18:** 204
ion plating ... **A18:** 849
laser cladding ... **A18:** 868
laser melt/particle injection ... **A18:** 869
laser melting ... **A18:** 864, 865
lubrication ... **A18:** 736, 737-739
cutting ... **A18:** 738-739
hot extrusion ... **A18:** 738
hot forging ... **A18:** 738
sheet metal forming ... **A18:** 737-738
magnetron sputtering ... **A18:** 849
manufacturing methods ... **A18:** 734
mass loss study by National Institute of Standards and Technology (NIST) ... **A18:** 363
metallurgical aspects of wear ... **A18:** 736-737
abrasive wear ... **A18:** 190, 736-737, 738, 805
adhesive wear ... **A18:** 237-238, 736, 737-738
contact fatigue wear ... **A18:** 736, 737
corrosive wear ... **A18:** 736, 737
nitriding ... **A18:** 642, 739
part material for ion implantation ... **A18:** 858
properties ... **A18:** 734, 737, 739
as shield against liquid impingement erosion ... **A18:** 222
surface treatments ... **A18:** 739

Tool steels, heat treating ... **A4:** 711-725 **M4:** 561-574
annealing ... **A4:** 711, 714-717, 721 **M4:** 563-564, 565-566
applications ... **A4:** 711
austenitizing ... **A4:** 717, 720-721, 722 **M4:** 566-569
carburizing ... **A4:** 714, 723-724 **M4:** 574
chromium plating ... **A4:** 724
cold working ... **A4:** 723-725 **M4:** 574
electroless nickel plating ... **A4:** 724
equipment ... **A4:** 721, 723
full annealing ... **A4:** 714-717 **M4:** 574
grain growth ... **A4:** 714
hardening ... **A4:** 711, 717, 720-721, 722 **M4:** 567-569, 570-572
isothermal annealing ... **A4:** 717 **M4:** 564
lead baths ... **A4:** 721
martempering ... **A4:** 721-722 **M4:** 572
material selection ... **A4:** 711 **M4:** 561
nitriding ... **A4:** 724 **M4:** 574
normalizing ... **A4:** 714 **M4:** 563, 565-566
oxide coatings ... **A4:** 724
physical properties ... **A4:** 711, 714, 720, 721 **M4:** 563, 573, 574
preheating before austenitizing ... **A4:** 721 **M4:** 569-572
processing ... **A4:** 711, 714, 715-717, 718-719 **M4:** 562-563, 565-566, 567-569, 570-572
quenching ... **A4:** 721-722 **M4:** 572
salt baths ... **A4:** 721
service characteristics ... **A4:** 711, 714, 715-717, 718-719 **M4:** 562-563, 565-566, 567-569, 570-572
stress relieving ... **M4:** 561-566
stress-relieving ... **A4:** 717
sulfide treatment ... **A4:** 724
surface treatment ... **A4:** 723-725
tempering ... **A4:** 722-723 **M4:** 567-569, 572-574
titanium nitride coatings ... **A4:** 724

Tool steels, heat treating equipment
atmospheres ... **M4:** 579-580
austenitizing ... **M4:** 566-569
immersed-electrode furnaces ... **M4:** 576, 577
submerged-electrode furnaces ... **M4:** 576-578
tempering ... **M4:** 572-574

Tool steels, heat treating processes
automatic ... **M4:** 578-579, 580
molten salts ... **M4:** 575, 576
rectification of salt baths ... **M4:** 576, 579

Tool steels, heat-treating equipment ... **A4:** 726-733
atmospheres ... **A4:** 728-729
austenitizing ... **A4:** 726-728
fluidized-bed furnaces ... **A4:** 732-733
immersed-electrode furnaces ... **A4:** 727
submerged-electrode furnaces ... **A4:** 727
tempering ... **A4:** 732
vacuum furnaces ... **A4:** 729-732

Tool steels, heat-treating processes ... **A4:** 726-728
automatic ... **A4:** 727, 728
molten salts ... **A4:** 726-727
rectification of salt baths ... **A4:** 727-728

Tool steels, high speed ... **A4:** 749-757
annealing ... **A4:** 750 **M4:** 600
austenitizing ... **A4:** 750-751, 752, 754 **M4:** 599, 601
bainitic hardening ... **A4:** 751 **M4:** 602
boriding ... **A4:** 438
carburizing ... **A4:** 755 **M4:** 607
fracture toughness ... **A4:** 751
furnaces ... **A4:** 751, 752
heat treating practices ... **M4:** 582, 599-600
heat-treating practices ... **A4:** 749-757
machinability, partial hardening to improve ... **A4:** 751 **M4:** 602
martempering ... **A4:** 756, 757
nitriding ... **A4:** 752-754 **M4:** 604-606
pack annealing ... **A4:** 750
preheating ... **A4:** 750 **M4:** 600-601
quenching ... **A4:** 751 **M4:** 601-602
refrigeration treatment ... **A4:** 752 **M4:** 604
shape distortion control ... **A4:** 764

Tool steels, high speed (continued)
specific tools, hardening ... **A4:** 755-757
bearing components ... **A4:** 756 **M4:** 607, 609-610
broaches ... **A4:** 755 **M4:** 607
chasers ... **A4:** 751, 755-756 **M4:** 607
drills ... **A4:** 756 **M4:** 607
form tools ... **A4:** 756 **M4:** 608
hobs ... **A4:** 756 **M4:** 608
milling cutters ... **A4:** 756 **M4:** 607
reamers ... **A4:** 756 **M4:** 608
taps ... **A4:** 751, 755, 756 **M4:** 607-608
thread rolling dies ... **A4:** 756 **M4:** 608
threading dies ... **A4:** 756 **M4:** 608
tool bits ... **A4:** 756 **M4:** 608-609
steam treating ... **A4:** 754-755 **M4:** 606-607
tempering ... **A4:** 751-752, 753, 754, 756 **M4:** 602-604, 605, 606

Tool steels, high-carbon high-chromium cold work *See* Tool steels, medium-alloy air hardening cold work

Tool steels, high-speed *See also* Steels ... **A16:** 51-59, 60-68, 717-720
for Al alloys ... **A16:** 765, 766, 767
applications ... **A16:** 57-59, 639
band sawing, Ti alloys ... **A16:** 851
boring ... **A16:** 163, 716
boring, Al alloys ... **A16:** 771, 777, 778
boring, heat-resistant alloys ... **A16:** 742
boring, refractory metals ... **A16:** 859
broaching ... **A16:** 202, 203, 206, 207, 208
broaching, Al alloys ... **A16:** 774, 779
broaching, heat-resistant alloys ... **A16:** 743
broaching, stainless steels ... **A16:** 700, 701, 704
broaching, tool steels ... **A16:** 717
carbon content ... **A16:** 57
for cast iron ... **A16:** 650, 651, 652, 653, 654-655, 656
circular sawing ... **A16:** 725
circular sawing, Al alloys ... **A16:** 794, 800
circular sawing, refractory metals ... **A16:** 868
Co content ... **A16:** 52-53
coating to increase tool life ... **A16:** 58, 59
compared to cast Co alloy tools ... **A16:** 70
compared to P/M processed materials ... **A16:** 60, 61, 64
composition ... **A16:** 52
contour sawing, Al alloys ... **A16:** 800
counterboring, cast iron ... **A16:** 660
counterboring, refractory metals ... **A16:** 860
for Cu alloys ... **A16:** 809, 810, 811, 812, 813, 814, 816, 817, 818
for dressing EDG wheels ... **A16:** 565
drilling ... **A16:** 217, 219, 230, 233-234
drilling, Al alloys ... **A16:** 769, 775, 776, 781, 783, 784
drilling, cast iron ... **A16:** 657
drilling, heat-resistant alloys ... **A16:** 748
drilling, P/M materials ... **A16:** 887, 890
drilling, refractory metals ... **A16:** 860, 861, 863
drilling, Ti alloys ... **A16:** 850
drilling, tool steels ... **A16:** 718
effect of alloying elements ... **A16:** 51
end milling, Al alloys ... **A16:** 786-787, 790
end milling, cast irons ... **A16:** 663
end milling, heat-resistant alloys ... **A16:** 754
end milling, tool steels ... **A16:** 722
end milling-slotting, refractory metals ... **A16:** 866
face milling, Al alloys ... **A16:** 788-789, 794
face milling, cast iron ... **A16:** 662
face milling, refractory metals ... **A16:** 859
face milling, Ti alloys ... **A16:** 850
flank wear scars ... **A16:** 44
and gear cutting ... **A16:** 343
grinding ... **A16:** 428
grinding ratios ... **A16:** 732
hacksawing, Ti alloys ... **A16:** 851
for hafnium ... **A16:** 855
hardness property ... **A16:** 53, 54, 55
heat treatment ... **A16:** 55-57
hobbing ... **A16:** 345-346

Tool steels, high-speed (continued)
hollow milling, refractory metals **A16:** 862
for honeycomb structures **A16:** 900
honing **A16:** 476, 477
hot hardness property **A16:** 53, 54, 57, 58, 59
machinability ratings **A16:** 643, 644
for Mg alloys **A16:** 821, 823, 825, 827, 828
milling **A16:** 311-315, 317-318, 320-326, 329
milling, Al alloys **A16:** 793
milling, cast iron **A16:** 660-661
milling, refractory metals **A16:** 864, 867
milling, stainless steels **A16:** 703
milling, Ti alloys **A16:** 846
milling, tool steels **A16:** 718-719, 723, 728
multifunction machining **A16:** 380, 381, 384, 388
multipoint cutting tools **A16:** 58
for Ni alloys **A16:** 837, 838, 839, 841
for P/M materials **A16:** 881
peripheral end milling, refractory metals **A16:** 866
peripheral end milling, Ti alloys **A16:** 849
peripheral milling, tool steels **A16:** 721
planing **A16:** 185-186
planing, Al alloys **A16:** 773, 778
planing, cast irons **A16:** 657, 660
planing, heat-resistant alloys **A16:** 743
for planing tools **A16:** 183
powder metallurgy (P/M) **A16:** 53, 60-68
power band sawing, Al alloys **A16:** 795, 801
power band sawing, heat-resistant alloys **A16:** 755
power band sawing, tool steels **A16:** 723
power band sawing, Zr **A16:** 855
power hacksawing, Al alloys **A16:** 796, 800
power hacksawing, heat-resistant alloys **A16:** 757
power hacksawing, MMCs **A16:** 895
power hacksawing, tool steels **A16:** 724
power hacksawing, Zr **A16:** 855
reaming **A16:** 245-246
reaming, Al alloys **A16:** 781
reaming, cast iron **A16:** 659, 660
reaming, heat-resistant alloys **A16:** 750
reaming, refractory metals **A16:** 862, 865-867
reaming, stainless steels **A16:** 702-703, 705
reaming, Ti alloys **A16:** 847, 852
reaming, tool steels **A16:** 719
rough and finish cutting, bevel gears **A16:** 349
sawing **A16:** 357, 358, 360, 362, 364
sawing, stainless steels **A16:** 705
selection factors **A16:** 59
shaping, Al alloys **A16:** 778
slotting, Al alloys **A16:** 790
softening point **A16:** 601
spade drilling, Al alloys **A16:** 777
spotfacing, cast iron **A16:** 660
spotfacing, refractory metals **A16:** 860
for stainless steels **A16:** 691
sulfur content **A16:** 53
surface treatment **A16:** 57
tapping **A16:** 256, 258, 259
tapping, Al alloys **A16:** 782, 783, 791
tapping, cast irons **A16:** 661
tapping, refractory metals **A16:** 862, 864
tapping, Ti alloys **A16:** 847, 851
tapping, tool steels **A16:** 720
thread chasers **A16:** 299
thread grinding **A16:** 270, 274
thread milling **A16:** 269
thread milling, refractory metals **A16:** 867
threading chasers, stainless steel die heads **A16:** 700
for titanium **A16:** 844
tool grinding **A16:** 450
tool life **A16:** 57
toughness properties **A16:** 53, 54, 55
trepanning, refractory metals **A16:** 860
turning **A16:** 144, 147, 150, 151, 428
turning, Al alloys **A16:** 770, 774, 775
turning, carbon and alloy steels **A16:** 668
turning, heat-resistant alloys **A16:** 739, 740, 741
turning, Ni alloys **A16:** 837
turning, refractory metals **A16:** 858
turning, stainless steels **A16:** 692, 693
turning, Ti alloys **A16:** 846, 847
turning, tool steels **A16:** 708-710, 715, 718
turning, with PCBN tools **A16:** 112

Tool steels, high-speed (continued)
for uranium alloys **A16:** 875
wear resistance property **A16:** 53, 54, 57, 58, 59, 108
for zirconium **A16:** 853
for Zn alloys **A16:** 831, 832, 833, 834

Tool steels, high-speed, coatings for contrast enhancement under scanning electron microscopy **A9:** 98-99
heat tinting **A9:** 136

Tool steels, hot work **A4:** 742-749
annealing **A4:** 743 **M4:** 593, 594
austenitizing **A4:** 743-744 **M4:** 593-595
heat treating procedures **M4:** 597-599
heat-treating procedures **A4:** 742-749
normalizing **A4:** 742-743 **M4:** 593, 594
preheating **A4:** 743 **M4:** 593, 594
quenching **A4:** 744-745 **M4:** 594, 595
salt baths **A4:** 745
stress relieving **M4:** 593
stress-relieving **A4:** 743
surface hardening **A4:** 745 **M4:** 595-597
tempering **A4:** 745 **M4:** 595, 596

Tool steels, low-alloy special purpose
annealing **A4:** 757 **M4:** 608, 610
austenitizing **A4:** 757 **M4:** 608, 610
defects and distortion **M4:** 601-603
normalizing **A4:** 757 **M4:** 582, 608, 610
quenching **A4:** 757 **M4:** 610
stress relieving **M4:** 610
stress-relieving **A4:** 757
tempering **A4:** 757 **M4:** 610

Tool steels, medium-alloy air-hardening cold work **A4:** 739-742
annealing **A4:** 739 **M4:** 589
atmospheres **A4:** 741
austenitizing **A4:** 740-741 **M4:** 590-591
nitriding **A4:** 742 **M4:** 592, 593
normalizing **A4:** 739 **M4:** 589
preheating **A4:** 740 **M4:** 589-590
quenching **A4:** 741-742 **M4:** 590, 591
salt baths **A4:** 741
secondary hardening **A4:** 742
stress relieving **M4:** 589
stress-relieving **A4:** 739-740
tempering **A4:** 742 **M4:** 591-593

Tool steels, mold steels **A4:** 758-760
annealing **A4:** 759 **M4:** 611
applications **A4:** 758
composition **A4:** 758 **M4:** 562-563, 610
heat treatments, variations for specific steels **A4:** 758, 759-760 **M4:** 582, 611-613
pack annealing **A4:** 759
tempering **A4:** 760

Tool steels, oil-hardening cold work **A4:** 737-739
annealing **A4:** 737 **M4:** 585-586, 589
austenitizing **A4:** 738-739 **M4:** 586
cycle annealing **A4:** 737
martempering **A4:** 739 **M4:** 587, 590
normalizing **A4:** 737, 739 **M4:** 585
pack annealing **A4:** 737
preheating **A4:** 738, 739 **M4:** 586
quenching **A4:** 739 **M4:** 586-587
salt baths **A4:** 739
stress relieving **M4:** 586
stress-relieving **A4:** 737-738
tempering **A4:** 739, 740 **M4:** 587, 591

Tool steels, shock-resisting **A4:** 736-737
annealing **A4:** 736 **M4:** 584, 585
austenitizing temperatures **A4:** 737 **M4:** 585
normalizing **A4:** 736 **M4:** 584
pack annealing **A4:** 736-737 **M4:** 584
proprietary paints **A4:** 737 **M4:** 584
quenching **A4:** 739
salt baths **A4:** 737
stress relieving **M4:** 584-585
stress-relieving **A4:** 737
surface treatments **A4:** 737 **M4:** 585
tempering **A4:** 737, 738, 739 **M4:** 585, 586, 587

Tool steels, specific types
6F, for hot-forging dies **A18:** 622-623
6F2
annealing **A4:** 743, 744 **M4:** 594
auxiliary tools, hot upset forging, use for **M3:** 535
composition **A4:** 713 **M3:** 526, 533 **M4:** 562-564
die blocks and inserts, use for **M3:** 529, 530

Tool steels, specific types (continued)
die steel ratings **A18:** 624
for hammer forging **A18:** 625
hardening **A4:** 744 **M4:** 594
heat treatments **A4:** 749
for hot-forging dies **A18:** 623, 624, 625, 635
preheating not require **A4:** 743
for press forging **A18:** 625
tempering **A4:** 745
trimming tool materials, use for **M3:** 532
wear resistance **A18:** 637
6F3
annealing **A4:** 743, 744 **M4:** 594
catastrophic die failure/plastic deformation **A18:** 641
composition **A4:** 713 **M4:** 562-564
die steel ratings **A18:** 624
for hammer forging carbon steels **A18:** 625
hardening **A4:** 744 **M4:** 594
heat treatments **A4:** 749
for hot-forging dies **A18:** 623, 624
preheating not required **A4:** 743
for press forging **A18:** 625
tempering **A4:** 745
thermal fatigue in dies **A18:** 639
6F3 auxiliary tools, hot upset forging
composition **M3:** 526, 533
die blocks and inserts, use for **M3:** 529, 530
use for **M3:** 535
6F4
annealing **A4:** 743, 744 **M4:** 594
austenitizing **A4:** 743
composition **A4:** 713 **M4:** 562-564
die steel ratings **A18:** 624
hardening **A4:** 744 **M4:** 594
heat treatments **A4:** 749
for hot-forging dies **A18:** 622-623, 624
tempering **A4:** 745
6F4, composition **M3:** 533
6F5
annealing **A4:** 744 **M4:** 594
cold extrusion tools, use for **M3:** 518
composition **A4:** 713 **M4:** 562-564
die steel ratings **A18:** 624
hardening **A4:** 744 **M4:** 594
for hot-forging dies **A18:** 623, 625
molds for plastics and rubber, use for **M3:** 548
preheating not required **A4:** 743
6F6
annealing **A4:** 744 **M4:** 594
composition **A4:** 713 **M4:** 562-564
hardening **A4:** 744 **M4:** 594
6F6, molds for plastics and rubber, use for **M3:** 548
6F7
annealing **A4:** 743, 744 **M4:** 594
catastrophic die failure/plastic deformation **A18:** 641
composition **A4:** 713 **M4:** 562-564
die steel ratings **A18:** 624
hardening **A4:** 744 **M4:** 594
for hot-forging dies **A18:** 623, 625
normalizing **A4:** 742-743
normalizing temperature **M4:** 594
6G
annealing **A4:** 744 **M4:** 594
auxiliary tools, hot upset forging, use for **M3:** 535
composition **A4:** 713 **M3:** 526, 533 **M4:** 562-564
die blocks and inserts, use for **M3:** 529, 530
die steel ratings **A18:** 624
for hammer forging **A18:** 625
hardening **A4:** 744 **M4:** 594
hot upset forging tools, use for **M3:** 534
for hot-forging dies **A18:** 622-623, 625, 635
preheating not required **A4:** 743
for press forging **A18:** 625
tempering **A4:** 745
6H, for hot-forging dies **A18:** 622-623
6H1
annealing **A4:** 743, 744 **M4:** 594
composition **A4:** 713 **M3:** 533 **M4:** 562-564
die steel ratings **A18:** 624
hardening **A4:** 744 **M4:** 594
hot upset forging tools, use for **M3:** 534

Tool steels, specific types (continued)
for hot-forging dies ... **A18:** 635
6H2
annealing ... **A4:** 744 **M4:** 594
composition ... **A4:** 713 **M3:** 533 **M4:** 562-564
hardening ... **A4:** 744 **M4:** 594
hot upset forging tools, use for ... **M3:** 534
6H2, die steel ratings ... **A18:** 624
234, quasi-cleavage fracture ... **A12:** 26
234 saw disk, dimple rupture ... **A12:** 14
A group, grinding of ... **A16:** 729, 731
A group, machining ... **A16:** 710, 714-716
A1, gear materials, surface treatment and minimum surface hardness ... **A18:** 261
A2
annealing ... **A4:** 715, 741 **M4:** 565-566, 589
annealing temperatures ... **M3:** 433-434
applications ... **A4:** 756 **A6:** 930
austenitizing ... **A4:** 720-721, 722
blades, cold shearing, use for ... **M3:** 478
blades, rotary slitting, use for ... **M3:** 479
blanking and piercing dies, use for ... **M3:** 485, 486, 487
boriding ... **A4:** 438, 445
coining dies, use for ... **M3:** 509, 510
for cold extrusion ... **A18:** 627
cold extrusion tools, use for ... **M3:** 515, 516, 517, 518, 519
composition ... **A4:** 712 **M3:** 422-423, 490 **M4:** 562-564
composition of weld deposits ... **A6:** 675
cutoff band sawing with bimetal blades ... **A6:** 1184
cycle annealing ... **A4:** 740
deep drawing dies, use for ... **M3:** 496, 498, 499
for deep-drawing dies ... **A18:** 634
density ... **A4:** 720, 763 **M3:** 441 **M4:** 617
description ... **A6:** 930
die steel ratings ... **A18:** 624
die-casting dies, use in ... **M3:** 543
dimensional changes ... **M4:** 616
distortion, linear dimensions ... **M3:** 468
gages, use for ... **M3:** 555, 556
gas nitriding ... **A4:** 387
gas quenching ... **A4:** 105-106 **M4:** 59
gear materials, surface treatment, and minimum surface hardness ... **A18:** 261
hardening ... **A4:** 613, 716, 741, 742, 762 **M4:** 567-569, 570-572, 589, 590
hardening and tempering ... **M3:** 425, 435-440
for hot-forging dies ... **A18:** 623
machinability rating ... **M3:** 443
molds for plastics and rubber, use for ... **M3:** 547, 548
nitriding ... **A4:** 742
normalizing ... **A4:** 715
preheating ... **A4:** 740
press forming dies, use for ... **M3:** 489, 491, 492, 493
processing ... **A4:** 719
recommended for backward extrusion of two parts ... **A18:** 628
secondary hardening ... **A4:** 745
service characteristics ... **A4:** 719 **M3:** 438-440 **M4:** 570-572
for shallow forming dies ... **A18:** 633
shear blades, service data ... **M3:** 480
shear spinning tools, use for ... **M3:** 501
stress-relieving ... **A4:** 740
structural components, use for ... **M3:** 558, 559
tempering ... **A4:** 613, 716, 742, 762 **M4:** 567-569, 570-572
thermal expansion ... **A4:** 720, 763 **M3:** 441 **M4:** 617
thread-rolling dies, use for ... **M3:** 551, 552, 553
welding preheat and interpass temperatures ... **A6:** 675

Tool steels, specific types (continued)
working hardness ... **M3:** 510
A2 chromium-plated blanking,, die, carburization effects ... **A11:** 572-573, 575
A2, composition ... **A16:** 709
A2, decarburization cracking ... **A13:** 132, 134
A2, for thread rolling tools ... **A16:** 290, 291
A2, machining ... **A16:** 274, 360, 362, 708, 711-725, 728
A2 rolling tool mandrel, fatigue fracture ... **A11:** 474
A2 scoring die spalled from carbide bands ... **A11:** 575, 582
A3
annealing ... **A4:** 715, 741 **M4:** 565-566, 589
annealing temperatures ... **M3:** 433-434
composition ... **A4:** 712 **M3:** 422-423 **M4:** 562-564
hardening ... **A4:** 716, 741 **M4:** 567-569, 570, 572, 589
hardening and tempering ... **M3:** 435-440
machinability rating ... **M3:** 443
normalizing ... **A4:** 715
preheating ... **A4:** 740
processing ... **A4:** 719
service characteristics ... **A4:** 719 **M3:** 438-440 **M4:** 570-572
tempering ... **A4:** 716 **M4:** 567-569, 570-572
A3, composition ... **A16:** 709
A3, composition of weld deposits ... **A6:** 675
A3, gear materials, surface treatment and minimum surface hardness ... **A18:** 261
A3, machining ... **A16:** 274, 713-725, 728
A4
annealing ... **A4:** 715, 741 **M4:** 565-566, 589
annealing temperatures ... **M3:** 433-434
austenitizing ... **A4:** 740-741
composition ... **A4:** 712 **M3:** 422-423, 490 **M4:** 562-564
hardening ... **A4:** 716, 741, 742 **M4:** 567-569, 570-572, 589, 590
hardening and tempering ... **M3:** 435-440
machinability rating ... **M3:** 443
normalizing ... **A4:** 715
press forming dies, use for ... **M3:** 491
processing ... **A4:** 719
service characteristics ... **A4:** 719 **M3:** 438-440 **M4:** 570-572
stress-relieving ... **A4:** 740
tempering ... **A4:** 716, 742 **M4:** 567-569, 570-572
A4, composition ... **A16:** 709
A4, gear materials, surface treatment and minimum surface hardness ... **A18:** 261
A4, machining ... **A16:** 274, 713-725, 728
A4, spalling, from EDM hole ... **A11:** 566
A5
annealing ... **A4:** 741 **M4:** 589
applications ... **A6:** 930
austenitizing ... **A4:** 740-741
composition ... **A4:** 712 **M4:** 562-564
description ... **A6:** 930
hardening ... **A4:** 741, 742 **M4:** 589, 590
stress-relieving ... **A4:** 740
tempering ... **A4:** 742
A5, composition ... **M3:** 424
A5, gear materials, surface treatment and minimum surface hardness ... **A18:** 261
A6
annealing ... **A4:** 715, 741 **M4:** 565-566, 589
annealing temperatures ... **M3:** 433-434
austenitizing ... **A4:** 740-741
boriding ... **A4:** 445
boriding of dies ... **A18:** 642
coining dies, use for ... **M3:** 509
composition ... **A4:** 712 **M3:** 422-423 **M4:** 562-564
cycle annealing ... **A4:** 740
density ... **A4:** 720, 763 **M3:** 441 **M4:** 617
die steel ratings ... **A18:** 624
gas nitriding ... **A4:** 387

Tool steels, specific types (continued)
hardening ... **A4:** 6, 741, 742 **M4:** 567-569, 570-572, 589, 590
hardening and tempering ... **M3:** 435-440
machinability rating ... **M3:** 443
normalizing ... **A4:** 715
processing ... **A4:** 719
service characteristics ... **A4:** 719 **M3:** 438-440 **M4:** 570-572
stress-relieving ... **A4:** 740
tempering ... **A4:** 716, 742 **M4:** 567-569, 570-572
thermal expansion ... **A4:** 720, 763 **M3:** 441 **M4:** 617
A6, composition ... **A16:** 709
A6, machining ... **A16:** 274, 712-726
A6 shaft, unidirectional-bending fatigue failure ... **A11:** 462
A6, shattered during finish grinding ... **A11:** 569
A7
annealing ... **A4:** 715, 741 **M4:** 565-566, 589
annealing temperatures ... **M3:** 433-434
composition ... **A4:** 712 **M3:** 422-423 **M4:** 562-564
density ... **A4:** 720, 763 **M3:** 441 **M4:** 617
hardening ... **A4:** 716, 741 **M4:** 567-569, 570-572, 589
hardening and tempering ... **M3:** 435-440
machinability rating ... **M3:** 443
nitriding ... **A4:** 742
normalizing ... **A4:** 715
preheating ... **A4:** 740
press forming dies, use for ... **M3:** 492
processing ... **A4:** 719
service characteristics ... **A4:** 719 **M3:** 438-440 **M4:** 570-572
stress-relieving ... **A4:** 740
structural components, use for ... **M3:** 559
tempering ... **A4:** 716, 742 **M4:** 567-569, 570-572
thermal expansion ... **A4:** 720, 763 **M3:** 441 **M4:** 617
A7, composition ... **A16:** 709
A7, for hot-forging dies ... **A18:** 623
A7, machining ... **A16:** 713-728, 731
A8
annealing ... **A4:** 715, 741 **M4:** 565-566, 589
annealing temperatures ... **M3:** 433-434
composition ... **A4:** 712 **M3:** 422-423 **M4:** 562-564
density ... **A4:** 720, 763 **M3:** 441 **M4:** 617
die steel ratings ... **A18:** 624
hardening ... **A4:** 716, 741 **M4:** 567-569, 570-572, 589
hardening and tempering ... **M3:** 435-440
for hot-forging dies ... **A18:** 623
machinability rating ... **M3:** 443
normalizing ... **A4:** 715
preheating ... **A4:** 740
processing ... **A4:** 719
service characteristics ... **A4:** 719 **M3:** 438-440 **M4:** 570-572
tempering ... **A4:** 716 **M4:** 567-569, 570-572
thermal expansion ... **A4:** 720, 763 **M3:** 441 **M4:** 617
A8, composition ... **A16:** 709
A8, machining ... **A16:** 274, 713-725, 728
A8 ring forging cracked after forging and annealing ... **A11:** 574, 580
A9
annealing ... **A4:** 715, 741 **M4:** 565-566, 589
annealing temperatures ... **M3:** 433-434
blades, cold shearing, use for ... **M3:** 478
blades, rotary, use for ... **M3:** 479
composition ... **A4:** 712 **M3:** 422-423 **M4:** 562-564
density ... **A4:** 720, 763 **M3:** 441 **M4:** 617
die steel ratings ... **A18:** 624
hardening ... **A4:** 716, 741 **M4:** 567-569, 570-572, 589
hardening and tempering ... **M3:** 435-440
for hot-forging dies ... **A18:** 623
machinability rating ... **M3:** 443
normalizing ... **A4:** 715

Tool steels, specific types (continued)
preheating ... A4: 740
processing ... A4: 719
service characteristics ... A4: 719 M3: 438-440 M4: 570-572
tempering ... A4: 716 M4: 567-569, 570-572
thermal expansion ... A4: 720, 763 M3: 441 M4: 617
A9, composition ... A16: 709
A9, machining ... A16: 274, 713-725, 728
A10
annealing ... A4: 715, 741 M4: 565-566, 589
annealing temperatures ... M3: 433-434
austenitizing ... A4: 740-741
blanking and piercing dies, use for ... M3: 487
composition ... A4: 712 M3: 422-423 M4: 562-564
density ... A4: 720
dimensional changes ... M4: 616
distortion in heat treatment ... A4: 612
distortion, linear dimensions ... M3: 468
hardening ... A4: 613, 716, 741, 762 M4: 567-569, 570-572, 589
hardening and tempering ... M3: 435-440
machinability rating ... M3: 443
normalizing ... A4: 714, 715, 739
normalizing temperature ... M4: 589
normalizing temperatures ... M3: 433-434
processing ... A4: 719
service characteristics ... A4: 719 M3: 438-440 M4: 570-572
size distortion in heat treatment ... A4: 613
tempering ... A4: 613, 716, 762 M4: 567-569, 570-572
thermal expansion ... A4: 720
A10, composition ... A16: 709
A10, machining ... A16: 728
AISI 01, as-rolled ... A9: 260
AISI 01, influence of austenitizing temperature ... A9: 266
AISI 01, influence of tempering temperature ... A9: 268
AISI 02, oil quenched and tempered ... A9: 269
AISI 06, effect of carbon on hardening ... A9: 258
AISI 06, spheroidize annealed ... A9: 262
AISI A2, as-rolled ... A9: 260
AISI A2, influence of austenitizing temperature ... A9: 266
AISI A4N mating fracture surface, fibrous zones ... A12: 376
AISI A5, partially spheroidized ... A9: 263
AISI A6, air quenched and tempered ... A9: 271
AISI A6, spheroidize annealed ... A9: 263
AISI A7, as received ... A9: 261
AISI A7, box annealed ... A9: 263
AISI A10, as received ... A9: 263
AISI AHT, use of vapor-deposited zinc selenide to accentuate features ... A9: 265
AISI D2, air quenched and tempered ... A9: 271
AISI D2, brittle in-service failure ... A12: 376
AISI D2, influence of etchant on revealed martensite ... A9: 265
AISI D2, quenched and tempered ... A9: 265
AISI D3, oil quenched and tempered ... A9: 271
AISI F2, water quenched and tempered ... A9: 269
AISI H11, air quenched and double tempered ... A9: 271
AISI H11, high-cycle fatigue fracture ... A12: 381
AISI H11, hydrogen embrittlement ... A12: 381-382
AISI H11, low-cycle fatigue fracture ... A12: 380
AISI H11, tension overload fracture ... A12: 379
AISI H13, air quenched and double tempered ... A9: 271
AISI H13, gas nitrided ... A9: 228
AISI H13, spheroidize annealed ... A9: 263
AISI H21, oil quenched and tempered ... A9: 271
AISI H23, annealed by austenitizing ... A9: 263
AISI H26, annealed by austenitizing ... A9: 263
AISI L6, as-rolled ... A9: 260
AISI L6 high-nickel, fracture surfaces ... A12: 377
AISI L6, impact fracture ... A12: 377
AISI L6 low-carbon, cup-and-cone fracture ... A12: 377
AISI L6, oil quenched and tempered ... A9: 269
AISI LI, as-rolled ... A9: 260
AISI LI, spheroidize annealed ... A9: 262

Tool steels, specific types (continued)
AISI M2, carbide segregation ... A9: 264
AISI M2, cracking from improper rebuilding ... A12: 375
AISI M2, directionally solidified, peritectic austenite ... A9: 680
AISI M2, heat treated and oil quenched ... A9: 266
AISI M2, heat treated, oil quenched and double tempered ... A9: 266
AISI M2, oil quenched and double tempered ... A9: 271
AISI M2, oil quenched and triple tempered ... A9: 271
AISI M2, quenched and tempered ... A9: 266
AISI M2, spheroidize annealed ... A9: 263
AISI M2, water-atomized, vacuum annealed ... A9: 529
AISI M3, impact fracture ... A12: 378
AISI M4, oil quenched and double tempered ... A9: 271
AISI M7, cracking from improper rebuilding ... A12: 375
AISI MI, oil quenched and triple tempered ... A9: 271
AISI P2, used for hobbing ... A9: 258
AISI P5, heat treated ... A9: 270
AISI P20, hardness of ... A9: 258
AISI P20, water quenched and tempered ... A9: 270
AISI S1, effect of carbides ... A9: 259
AISI S1, low-cycle fatigue fracture ... A12: 377
AISI S1, oil quenched and tempered ... A9: 269
AISI S2, spheroidize annealed ... A9: 262
AISI S2, water quenched and tempered ... A9: 269
AISI S4, as-rolled ... A9: 260
AISI S5, as-rolled ... A9: 260
AISI S5, austenitized and isothermally transformed ... A9: 267
AISI S5, effect of carbides ... A9: 259
AISI S5, oil quenched and tempered ... A9: 270
AISI S5, spheroidize annealed ... A9: 262
AISI S5, used for hobbing ... A9: 258
AISI S7, air quenched and tempered ... A9: 270
AISI S7, continuous cooling transformations ... A9: 267
AISI S7, effect of carbides ... A9: 259
AISI S7, influence of austenitizing temperature ... A9: 265
AISI S7, spheroidize annealed ... A9: 263
AISI T1, carbide segregation ... A9: 264
AISI T1, directionally solidified, peritectic envelopes ... A9: 680
AISI T1, hammer-burst fracture ... A12: 378
AISI T1, impact fracture ... A12: 378
AISI T2 high-speed, TEM and SEM fractographs, compared ... A12: 186
AISI T15, powder made ... A9: 272
AISI T15, powdered ... A9: 511
AISI T15, pressed and sintered ... A9: 524
AISI T15, wrought ... A9: 524
AISI W1, as received ... A9: 262
AISI W1, as-rolled ... A9: 260
AISI W1, austenitized, brine quenched and tempered ... A9: 268
AISI W1, brine quenched ... A9: 268
AISI W1, case hardness ... A9: 176
AISI W1, fatigue fracture surfaces ... A12: 376
AISI W1, hardened zone ... A9: 177
AISI W1, influence of etchant on revealing as-quenched martensite ... A9: 264
AISI W1, influence of starting structure on spheroidization ... A9: 262
AISI W1, simulated service fractures ... A12: 376
AISI W2, catastrophic failure surface ... A12: 377
AISI W2, spheroidize annealed ... A9: 261
AISI W4, as received ... A9: 261
BM42, boriding ... A4: 438
C2, proven applications for borided ferrous materials ... A4: 445
CBM Rex 60, refrigeration treatment ... A4: 752
CBS-600, composition ... A4: 320
CBS-1000M
composition ... A4: 320

Tool steels, specific types (continued)
gas carburizing ... A4: 319
CPM 10V
blades, cold shearing, use for ... M3: 478
blades, rotary slitting, use for ... M3: 479
blanking for piercing dies, use for ... M3: 485, 486, 487, 488
coining dies, use for ... M3: 511
cold heading tools, use in ... M3: 512
press forming dies, use for ... M3: 492, 493
CPM 10V, machinability ... A16: 734, 735
CPM M ... A16: 65
CPM M42, temper resistance ... A16: 65
CPM Rex ... A16: 60, 63
CPM Rex 20, temper resistance ... A16: 65
CPM Rex 45 HS ... A16: 63
CPM Rex 76 HS ... A16: 63
CPM Rex M ... A16: 63
CPM Rex M2HCHS ... A16: 63
CPM Rex M3HCHS ... A16: 63
CPM Rex M4 HS ... A16: 63
CPM Rex M35HCHS ... A16: 63
CPM Rex T ... A16: 63
CPM Rex T15 HS ... A16: 63
CPM T ... A16: 63
D group, grindability ... A16: 728
D group machining ... A16: 710, 713-716, 721, 722, 729, 731
D1
annealing ... A4: 741 M4: 589
composition ... A4: 712 M3: 424 M4: 562-564
damage dominated by fatigue fracture ... A18: 180, 181
damage dominated by shear fracture typical of adhesive wear ... A18: 179
hardening ... A4: 717, 741, 742 M4: 567-569, 589, 590
hardening and tempering ... M3: 435-437
nitriding ... A4: 742
preheating ... A4: 740
stress-relieving ... A4: 740
tempering ... A4: 717, 742 M4: 567-569
D2
abrasive wear ... A18: 718-719, 767, 797
adhesive wear resistance ... A18: 721
annealed hardness ... A18: 768
annealing ... A4: 715, 741 M4: 565-566, 589
annealing temperatures ... M3: 433-434
applications ... A4: 756 A6: 930
austenitizing ... A4: 741
austenitizing procedure ... M4: 590
auxiliary tools, hot upset forging, use for ... M3: 535
blades, cold shearing, use for ... M3: 478
blades, rotary slitting, use for ... M3: 479
blanking and piercing dies, use for ... M3: 485, 486, 487
boriding ... A4: 45
coining dies, use for ... M3: 509, 510
for cold extrusion tools ... A18: 627
cold extrusion tools, use for ... M3: 515, 516, 517, 518
cold heading tools, use in ... M3: 512
composition ... A4: 712 M3: 422-423, 490 M4: 562-564
composition of weld deposits ... A6: 675
cutoff band sawing with bimetal blades ... A6: 1184
cycle annealing ... A4: 740
deep drawing dies, use for ... M3: 496, 498, 499
for deep-drawing dies ... A18: 634
density ... A4: 720, 763 M3: 441 M4: 617
description ... A6: 930
die materials for sheet metal forming ... A18: 628
die wear and die life ... A18: 633
dimensional changes ... M4: 616
distortion, linear dimensions ... M3: 468
electron-beam welding ... A6: 867
endothermic-atmosphere dew point for hardening ... A4: 729 M4: 579
extension of tool life via ion implantation, examples ... A18: 643
friction surfacing ... A6: 323
gages, use for ... M3: 555, 556
gas nitriding ... A4: 387, 392
hardening ... A4: 613, 741, 742, 762, 763 M4: 570-572, 589, 590

Tool steels, specific types (continued)
hardening and tempering M3: 438-440
laser cladding A18: 868
liquid nitriding A4: 411, 413 M4: 252
machinability rating M3: 443
microconstituents A4: 613 M4: 615
microconstituents after hardening A4: 762 M3: 468
molds for plastics and rubber, use for M3: 547, 548
nitrided, for shallow forming dies A18: 633
nitriding A4: 742
normalizing A4: 715
part material for ion implantation A18: 858
preheating A4: 740
press forming dies, use for M3: 489-493
processing A4: 719
proven applications for borided ferrous materials A4: 445
recommended for backward extrusion of two parts A18: 628
secondary hardening A4: 745
service characteristics A4: 719 M3: 438-440 M4: 570-572
for shallow forming dies A18: 633
shear, blades, service data M3: 480
shear spinning tools, use for M3: 501
stress-relieving A4: 740
structural components, use for M3: 558
tempering A4: 613, 742, 743, 762 M4: 570-572, 592
test duration of dry sand/rubber wheel test A18: 716
thermal expansion A4: 720, 763 M3: 441 M4: 617
thermoreactive deposition/diffusion process A4: 450, 451
thread-rolling dies, use for M3: 551, 552, 553
trimming tool materials, use for M3: 532
wear, cold shearing blades M3: 479, 481
wear resistance and cost effectiveness A18: 706
welding preheat and interpass temperatures A6: 675
working hardness M3: 510
D2, composition A16: 709
D2, CVD titanium carbide coating of M5: 384
D2, electron beam welding M6: 638
D2 flange edge, poor carbide morphology A11: 575, 582
D2, for thread rolling dies A16: 290, 291
D2, grinding cracks from failure to temper A11: 568
D2, machining A16: 191, 192, 360, 362, 563, 712-725, 728
D2, microstructure M1: 612
D2, surface alterations produced A16: 27
D3
annealing A4: 715, 741 M4: 565-566, 589
annealing temperatures M3: 433-434
applications A6: 930
coining dies, use for M3: 509
cold extrusion tools, use for M3: 518
composition A4: 712 M3: 422-423, 490 M4: 562-564
cutoff band sawing with bimetal blades A6: 1184
decarburization bands when held in a fluidized bed A4: 486
deep drawing dies, use for M3: 496, 499
for deep-drawing dies A18: 634
density A4: 720, 763 M3: 441 M4: 617
description A6: 930
dimensional changes M4: 616
distortion, linear dimensions M3: 468
endothermic-atmosphere dew point for hardening A4: 729 M4: 579
gas nitriding A4: 387

Tool steels, specific types (continued)
hardening A4: 486, 487, 613, 717, 741, 762 M4: 567-569, 570-572, 589
hardening and tempering M3: 435-440
heat transfer in fluidized beds M4: 301
isothermal quenching A4: 486, 487
machinability rating M3: 443
nitriding A4: 742
normalizing A4: 715
preheating A4: 740
press forming dies, use for M3: 490
processing A4: 719
proven applications for borided ferrous materials A4: 445
service characteristics A4: 719 M3: 438-440 M4: 570-572
shear spinning tools, use for M3: 501
stress-relieving A4: 740
tempering M4: 567-569, 570-572, 592
thermal expansion A4: 720, 763 M3: 441 M4: 617
TiN coating by physical vapor deposition A18: 739
D3, composition A16: 709
D3, machining A16: 115, 116, 360, 362, 479, 712, 717-720, 723, 724, 726, 728
D4
annealing A4: 715, 741 M4: 565-566, 589
annealing temperatures M3: 433-434
applications A6: 930
blanking and piercing dies, use for M3: 485, 486, 487
coining dies, use for M3: 509
cold extrusion tools, use for M3: 518
composition A4: 712 M3: 422-423 M4: 562-564
density A4: 720, 763 M3: 441 M4: 617
description A6: 930
dimensional changes M4: 616
distortion, linear dimensions M3: 468
endothermic-atmosphere dew point for hardening A4: 729 M4: 579
hardening A4: 613, 717, 741, 762 M4: 567-569, 570-572, 589
hardening and tempering M3: 435-440
machinability rating M3: 443
nitriding A4: 742
normalizing A4: 715
preheating A4: 740
press forming dies, use for M3: 493
processing A4: 719
service characteristics A4: 719 M3: 438-440 M4: 570-572
stress-relieving A4: 740
tempering A4: 613, 717, 742, 762 M4: 567-569, 570-572
thermal expansion A4: 720, 763 M3: 441 M4: 617
D4, composition A16: 709
D4, for deep-drawing dies A18: 634
D4, machining A16: 717-720, 723-725, 728
D5
annealing A4: 715, 741 M4: 565-566, 589
annealing temperatures M3: 433-434
composition A4: 712 M3: 422-423, 490 M4: 562-564
density A4: 720, 763 M3: 441 M4: 617
dimensional changes M4: 616
distortion, linear dimensions M3: 468
hardening A4: 613, 717, 741, 742, 762 M4: 567-569, 570-572, 589, 590
hardening and tempering M3: 435-440
machinability rating M3: 443
nitriding A4: 742
normalizing A4: 715
preheating A4: 740
press forming dies, use for M3: 490, 491
processing A4: 719
service characteristics A4: 719 M3: 438-440 M4: 570-572

Tool steels, specific types (continued)
stress-relieving A4: 740
tempering A4: 613, 717, 742, 762 M4: 567-569, 570-572
thermal expansion A4: 720, 763 M3: 441 M4: 616
D5, composition A16: 709
D5 cross-recessed die, segregation fracture A11: 324-326
D5, machining A16: 717-720, 723-725, 728
D6
boriding A4: 445
composition A4: 712 M4: 562-564
endothermic-atmosphere dew point for hardening A4: 729 M4: 579
nitriding A4: 742
preheating A4: 740
proven applications for borided ferrous materials A4: 445
stress-relieving A4: 740
tempering A4: 742
D6, composition M3: 424
D6, gas-tungsten arc welding A6: 671
D6, part material for ion implantation A18: 858
D7
annealing A4: 715, 741 M4: 565-566, 589
annealing temperatures M3: 433-434
composition A4: 712 M3: 422-423, 490 M4: 562-564
hardening A4: 717, 741 M4: 567-569, 570-572, 589
hardening and tempering M3: 435-440
machinability rating M3: 443
nitriding A4: 742
preheating A4: 740
press forming dies, use for M3: 490, 491
processing A4: 719
service characteristics A4: 719 M3: 438-440 M4: 570-572
stress-relieving A4: 740
tempering A4: 717, 742 M4: 567-569, 570-572
D7, composition A16: 709
D7, cutoff band sawing with bimetal blades A6: 1184
D7, die wear and die life A18: 633
D7, machining A16: 360, 362, 479, 714, 717-720, 723-725, 727, 728, 731
F group, grindability A16: 728, 729, 732
F group, machining A16: 710, 713-716, 721, 722
F1
annealing A4: 715, 758 M4: 565-566, 609
annealing temperatures M3: 433-434
composition A4: 713 M3: 424 M4: 562-564
hardening A4: 717, 758 M4: 567-569, 609
hardening and tempering M3: 435-437
normalizing A4: 715
normalizing temperature M4: 565-566, 609
normalizing temperatures M3: 433-434
tempering A4: 717 M4: 567-569
F1, machining A16: 717-721, 723-725
F2
annealing A4: 715, 758 M4: 565-566, 609
annealing temperature M3: 433-434
composition A4: 713 M3: 424 M4: 562-564
endothermic-atmosphere dew point for hardening A4: 729 M4: 579
hardening A4: 717, 758 M4: 567-569, 609
hardening and tempering M3: 435-437
normalizing A4: 715
normalizing temperature M4: 565-566, 609
normalizing temperatures M3: 433-434
tempering A4: 717 M4: 567-569
F2, machining A16: 238, 717-721, 723-725
F3
annealing A4: 758 M4: 609
composition A4: 713 M4: 562-564
endothermic-atmosphere dew point for hardening A4: 729 M4: 579
hardening A4: 757, 758 M4: 609

Tool steels, specific types (continued)
normalizing temperature **M4:** 609
F3, composition **M3:** 424
F3, for drills **A16:** 238
F3, machinability **A16:** 721
H group, grindability **A16:** 728, 731, 732
H group, machinability **A16:** 717
H-11, pitting corrosion **A13:** 1054
H10
annealing **A4:** 715, 744 **M4:** 565-566, 594
annealing temperatures **M3:** 433-434
austenitizing **A4:** 743
boriding **A4:** 445
composition **A4:** 712 **M3:** 422-423 **M4:** 562-564
density **A4:** 720, 763 **M3:** 441 **M4:** 617
die steel ratings **A18:** 624
hardening **A4:** 716, 744 **M4:** 567-569, 570-572, 594
hardening and tempering **M3:** 435-440
for hot-forging dies **A18:** 623, 624
machinability rating **M3:** 443
normalizing **A4:** 715
processing **A4:** 718
proven applications for borided ferrous materials **A4:** 445
service characteristics **A4:** 718 **M3:** 438-440
tempering **A4:** 716 **M4:** 567-569, 570-572
thermal expansion **A4:** 720, 763 **M3:** 441 **M4:** 617
thermal fatigue in dies **A18:** 639
H10, composition **A16:** 709
H10, machining **A16:** 274, 710, 712-725, 731, 797
H11
annealing **A4:** 715, 744 **M4:** 565-566, 594
annealing temperatures **M3:** 433-434
applications **A4:** 397
austenitizing **A4:** 743
auxiliary tools, hot upset forging, use for **M3:** 535
blanking and piercing dies, use for **M3:** 486
boriding **A4:** 445
catastrophic die failure/plastic deformation **A18:** 641
clamping dies **M6:** 562
coining dies, use for **M3:** 509, 510
composition **A4:** 712 **M3:** 422-423 **M4:** 562-564
composition of weld deposits **A6:** 675
cutoff band sawing with bimetal blades **A6:** 1184
density **A4:** 720, 763 **M3:** 441 **M4:** 617
die blocks and inserts, use for **M3:** 529, 530
for die casting **A18:** 632
die steel ratings **A18:** 624
dimensional changes **M4:** 616
distortion, linear dimensions **M3:** 468
electron beam welding .... **M6:** 624, -625, 638
electron-beam welding **A6:** 867, 868
endothermic-atmosphere dew point for hardening **A4:** 729 **M4:** 579
gas nitriding **A4:** 387, 392, 397
for hammer forging **A18:** 625
hardening **A4:** 613, 716, 744, 762 **M4:** 567-569, 570-572, 594
hardening and tempering **M3:** 435-440, 442
hardness gradients, nitriding **M4:** 256
heat treatments **A4:** 745-747
for hot extrusion tools **A18:** 627
hot extrusion tools, mandrel life **M3:** 538, 540
hot extrusion tools, use for **M3:** 538, 540
hot upset forging tools, use for **M3:** 534
for hot-forging dies **A18:** 623, 624, 626
interface considerations for die-casting die wear **A18:** 632
liquid nitriding **A4:** 413
machinability rating **M3:** 443
martempering, forming after **A4:** 145
normalizing **A4:** 715
for press forging **A18:** 625
press forming dies, use for **M3:** 490
processing **A4:** 718
proven applications for borided ferrous materials **A4:** 445
recommended for die-casting dies and die inserts **A18:** 629
recommended machining conditions **M1:** 569
resistance to plastic deformation **A18:** 626

Tool steels, specific types (continued)
service characteristics **A4:** 718 **M3:** 438-440 **M4:** 570-572
shear, blades, service data **M3:** 480
structural components, use for **M3:** 559
tempering **A4:** 124, 130, 613, 716, 762 **M4:** 567-569, 570-572
thermal conductivity **A4:** 721 **M3:** 442 **M4:** 573
thermal expansion **A4:** 720, 763 **M1:** 653 **M3:** 441 **M4:** 617
thermal fatigue of dies **A18:** 640
wear resistance relation to toughness **A18:** 707
wear vs. toughness **M1:** 607
welding preheat and interpass temperatures **A6:** 675
H11, composition **A16:** 709
H11, for thread rolling totals **A16:** 293
H11, machining **A16:** 274, 321, 710, 712-725
H11 Mod
composition **A4:** 207 **M1:** 422, 649 **M4:** 120
ductility **M4:** 127
heat treatment **M1:** 434-436
heat treatments **A4:** 214, 215 **M4:** 126-127
longitudinal mechanical properties **A4:** 215
mechanical properties **M1:** 435-437 **M4:** 126
processing **M1:** 434
tempering temperature **M4:** 127
tensile strength **M4:** 127
transverse strength and ductility **A4:** 215
H11 Mod, structural components, use for **M3:** 559
H12
annealing **A4:** 715, 744 **M4:** 565-566, 594
annealing temperatures **M3:** 433-434
applications **A6:** 930
austenitizing **A4:** 743
ceramic coatings of dies **A18:** 643
coining dies, use for **M3:** 509, 510
composition **A4:** 712 **M3:** 422-423 **M4:** 562-564
composition of weld deposits **A6:** 675
cutoff band sawing with bimetal blades **A6:** 1184
description **A6:** 930
die blocks and inserts, use for **M3:** 529, 530
for die casting **A18:** 632
die steel ratings **A18:** 624
endothermic-atmosphere dew point for hardening **A4:** 729 **M4:** 579
gas nitriding **A4:** 387, 392
for hammer forging **A18:** 625
hardening **A4:** 712, 716, 744 **M4:** 567-569, 570-572, 594
hardening and tempering **M3:** 435-440
heat treatments **A4:** 749
for hot extrusion tools **A18:** 627
hot extrusion tools, use for **M3:** 538, 540
hot upset forging tools, use for **M3:** 534
for hot-forging dies **A18:** 623, 624, 626
liquid nitriding **A4:** 413 **M4:** 256
normalizing **A4:** 715
for press forging **A18:** 625
processing **A4:** 718
service characteristics **A4:** 718 **M3:** 438-440 **M4:** 570-572
structural components, use for **M3:** 559
surface hardening **A4:** 745
tempering **A4:** 716 **M4:** 567-569, 570-572
thermal fatigue in dies **A18:** 639
Toyota Diffusion (TD) process **A18:** 643
wear, hot shearing blades **M3:** 481
wear resistance **A18:** 636
welding preheat and interpass temperatures **A6:** 675
H12 coil spring, cracked from seam during heat treatment **A11:** 574, 579
H12, composition **A16:** 709
H12, machining **A16:** 274, 360, 362, 710, 712-725
H12, nitrided, hardness profile **M1:** 633
H13
annealing **A4:** 715, 744 **M4:** 565-566, 594
annealing temperatures **M3:** 433-434
applications **A6:** 930
austenitizing **A4:** 743
boriding **A4:** 438
coining dies, use for **M3:** 509, 510
cold heading tools, use in **M3:** 512

Tool steels, specific types (continued)
composition **A4:** 207, 712 **M1:** 422 **M3:** 422-423 **M4:** 562-564
composition of weld deposits **A6:** 675
cutoff band sawing with bimetal blades **A6:** 1184
density **A4:** 720, 763 **M3:** 441 **M4:** 617
description **A6:** 930
for die casting **A18:** 632
die steel ratings **A18:** 624
die-casting dies, use in **M3:** 542, 543
dimensional changes **M4:** 616
distortion, linear dimensions **M3:** 468
elevated temperatures, resistance to softening **M3:** 426
endothermic-atmosphere dew point for hardening **A4:** 729 **M4:** 579
fracture toughness **A4:** 748, 749
friction surfacing **A6:** 323
gas nitriding **A4:** 387, 392
for hammer forging **A18:** 625
hardening **A4:** 613, 716, 744, 762 **M4:** 567-569, 570-572, 594
hardening and tempering **M3:** 435-440
heat treatment **M1:** 438-439
heat treatments **A4:** 214-215, 747-749 **M4:** 127-128
for hot extrusion tools **A18:** 627
hot extrusion tools, mandrel life **M3:** 538, 540
hot extrusion tools, use for **M3:** 538, 540
hot hardness, comparison for cast and wrought **M3:** 446
for hot-forging dies **A18:** 623, 624, 626, 635
interface considerations for die-casting die wear **A18:** 632
longitudinal mechanical properties **A4:** 216
machinability rating **M3:** 443
mechanical properties **M1:** 437-441 **M4:** 127
molds for plastics and rubber, use for **M3:** 547, 548
normalizing **A4:** 715
part material for ion implantation **A18:** 858
for press forging **A18:** 625
press forming dies, use for **M3:** 490
processing **A4:** 718 **M1:** 437-438
proven applications for borided ferrous materials **A4:** 445
recommended for die-casting dies and die inserts **A18:** 629
refractory-metal coatings **A18:** 644
residual stresses **A6:** 1097
service characteristics **A4:** 718 **M3:** 438-440 **M4:** 570-572
structural components, use for **M3:** 559
tempering **A4:** 613, 716, 762 **M4:** 567-569, 570-572
thermal conductivity **A4:** 721 **M3:** 442 **M4:** 573
thermal expansion **A4:** 720, 763 **M3:** 441
TiN coating for surface treatments **A18:** 739
Toyota Diffusion (TD) process **A18:** 643
welding preheat and interpass temperatures **A6:** 675
H13, composition **A16:** 709
H13, machining **A16:** 362, 479, 710, 712-726
H13 mandrel, surface chemistry changes **A11:** 576, 583
H13 nozzle, erosion failure **A11:** 576, 584
H13 shear knives, service condition failures **A11:** 575, 582-583
H14
annealing **A4:** 715, 744 **M4:** 565-566, 594
annealing temperatures **M3:** 433-434
austenitizing **A4:** 743
composition **A4:** 712 **M3:** 422-423 **M4:** 562-564
density **A4:** 720, 763 **M3:** 441 **M4:** 617
die steel ratings **A18:** 624
hardening **A4:** 716, 744 **M4:** 567-569, 570-572, 594
hardening and tempering **M3:** 435-440
for hot extrusion tools **A18:** 627
hot extrusion tools, use for **M3:** 538, 540
for hot-forging dies **A18:** 623, 624
machinability rating **M3:** 443
normalizing **A4:** 715
processing **A4:** 718
service characteristics **A4:** 718 **M3:** 438-440 **M4:** 570-572

Tool steels, specific types (continued)
tempering **A4:** 716 **M4:** 567-569, 570-572
thermal expansion **A4:** 720, 763 **M3:** 441 **M4:** 617
H14, composition **A16:** 709
H14, machining....................... **A16:** 274, 710, 712-725
H15
composition............................. **A4:** 712 **M4:** 562-564
hardness gradients, nitriding..................... **M4:** 256
liquid nitriding **A4:** 413
H15, composition **M3:** 424
H15, for hot-forging dies **A18:** 623, 624
H16
annealing **A4:** 744 **M4:** 594
austenitizing.. **A4:** 743
composition............................. **A4:** 712 **M4:** 562-564
hardening **A4:** 744 **M4:** 594
H16, composition **M3:** 424
H16, description **A6:** 930
H16, for hot-forging dies **A18:** 623, 624
H16, machining............... **A16:** 710, 712-716, 721, 722
H17, for hot-forging dies **A18:** 623
H18, for hot-forging dies **A18:** 623
H19
annealing **A4:** 715, 744 **M4:** 565-566
annealing temperatures **M3:** 433-434
austenitizing.. **A4:** 743
composition....... **A4:** 712 **M3:** 422-423 **M4:** 562-564
density..................... **A4:** 720, 763 **M3:** 441 **M4:** 617
hardening **A4:** 716, 744 **M4:** 567-569, 570-572
hardening and tempering **M3:** 435-440
hot extrusion tools, use for **M3:** 538, 540
machinability rating **M3:** 443
normalizing... **A4:** 715
processing.. **A4:** 718
service characteristics..... **A4:** 718 **M3:** 438-440 **M4:** 570-572
tempering **A4:** 716 **M4:** 567-569, 570-572
thermal expansion **A4:** 720, 763 **M3:** 441 **M4:** 617
H19 catastrophic die failure/plastic
deformation **A18:** 641
die steel ratings **A18:** 624
for hot-forging dies **A18:** 623, 626
wear resistance **A18:** 636
H19, composition **A16:** 709
H19, machining.............................. **A16:** 274, 712-725
H20
annealing **A4:** 744 **M4:** 594
austenitizing.. **A4:** 743
coining dies, use for **M3:** 510
composition............. **A4:** 712 **M3:** 424 **M4:** 562-564
die-casting dies, use in................................ **M3:** 542
hardening **A4:** 744 **M4:** 594
for hot-forging dies **A18:** 623
recommended for die-casting dies
and die inserts **A18:** 629
H20, machining............... **A16:** 710, 712-716, 721, 722
H21
annealing **A4:** 715, 744 **M4:** 565-566, 594
annealing temperatures **M3:** 433-434
applications **A6:** 930
austenitizing.. **A4:** 743
auxiliary tools, hot upset forging,
use for **M3:** 535
coining dies, use for **M3:** 510
composition....... **A4:** 712 **M3:** 422-423 **M4:** 562-564
cutoff band sawing with bimetal
blades **A6:** 1184
density..................... **A4:** 720, 763 **M3:** 441 **M4:** 617
description.. **A6:** 930
die-casting dies, use in................................ **M3:** 542
elevated temperatures, resistance to
softening **M3:** 426
hardening **A4:** 716, 744 **M4:** 567-569, 570-572, 594
hardening and tempering **M3:** 435-440
heat treatments.. **A4:** 745
hot extrusion tools, die life **M3:** 538-540

Tool steels, specific types (continued)
hot extrusion tools, use for **M3:** 538, 540
machinability rating **M3:** 443
normalizing... **A4:** 715
processing.. **A4:** 718
service characteristics..... **A4:** 718 **M3:** 438-440 **M4:** 570-572
shear blades, service life............................. **M3:** 481
tempering **A4:** 716 **M4:** 567-569, 570-572
thermal conductivity **A4:** 721 **M3:** 442 **M4:** 573
thermal expansion **A4:** 720, 763 **M3:** 441 **M4:** 617
H21 catastrophic die failure/plastic
deformation **A18:** 641
die steel ratings **A18:** 624
hot extrusion tools **A18:** 627
for hot-forging dies **A18:** 623, 626
recommended for die-casting dies
and die inserts **A18:** 629
thermal fatigue of dies................................ **A18:** 640
H21, composition **A16:** 709
H21, machining....... **A16:** 274, 360, 362, 710, 712-725
H21 safety valve spring, corrosion
fatigue fracture from moist air........... **A11:** 261
H22
annealing **A4:** 715, 744 **M4:** 565-566, 594
annealing temperatures **M3:** 422-423
austenitizing.. **A4:** 743
composition............................. **A4:** 712 **M4:** 562-564
density..................... **A4:** 720, 763 **M3:** 441 **M4:** 617
die-casting dies, use in................................ **M3:** 542
hardening **A4:** 716, 744 **M4:** 567-569, 570-572, 594
hardening and tempering **M3:** 435-440
for hot-forging dies **A18:** 623
machinability rating **M3:** 443
normalizing... **A4:** 715
processing.. **A4:** 718
recommended for die-casting dies
and die inserts **A18:** 629
service characteristics..... **A4:** 718 **M3:** 438-440 **M4:** 570-572
tempering **A4:** 716 **M4:** 567-569, 570-572
thermal expansion **A4:** 720, 763 **M3:** 441 **M4:** 617
H22, composition **A16:** 709
H22, machining....................... **A16:** 274, 710, 712-725
H23
annealing **A4:** 715, 744 **M4:** 565-566, 594
annealing temperatures **M3:** 433-434
applications **A6:** 930
austenitizing.. **A4:** 743
composition....... **A4:** 712 **M3:** 422-423 **M4:** 562-564
cutoff band sawing with bimetal
blades **A6:** 1184
description.. **A6:** 930
die steel ratings **A18:** 624
elevated temperatures, resistance to
softening **M3:** 426
hardening **A4:** 716, 744, 745 **M4:** 567-569, 570-572, 594
hardening and tempering **M3:** 435-440
hot extrusion tools **A18:** 627
for hot-forging dies **A18:** 623
normalizing... **A4:** 715
processing.. **A4:** 718
quenching... **A4:** 745
recommended for die-casting dies
and die inserts **A18:** 629
secondary hardening..................................... **A4:** 745
service characteristics..... **A4:** 718 **M3:** 438-440 **M4:** 570-572
tempering **A4:** 716 **M4:** 567-569, 570-572
H23, composition **A16:** 709
H23, machining....................... **A16:** 274, 710, 712-725
H24
annealing **A4:** 715, 744 **M4:** 565-566, 594
annealing temperatures **M3:** 433-434
austenitizing.. **A4:** 743

Tool steels, specific types (continued)
composition....... **A4:** 712 **M3:** 422-423 **M4:** 562-564
die steel ratings **A18:** 624
hardening **A4:** 716, 744 **M4:** 567-569, 570-572, 594
hardening and tempering **M3:** 435-440
for hot-forging dies **A18:** 623
machinability rating **M3:** 443
normalizing... **A4:** 715
processing.. **A4:** 718
service characteristics..... **A4:** 718 **M3:** 438-440 **M4:** 570-572
tempering **A4:** 716 **M4:** 567-569, 570-572
H24, composition **A16:** 709
H24, machining....................... **A16:** 274, 710, 712-725
H25
annealing **A4:** 715, 744 **M4:** 565-566, 594
annealing temperatures **M3:** 433-434
austenitizing.. **A4:** 743
composition....... **A4:** 712 **M3:** 422-423 **M4:** 562-564
hardening **A4:** 716, 744 **M4:** 567-569, 570-572, 594
hardening and tempering **M3:** 435-440
machinability rating **M3:** 443
normalizing... **A4:** 715
processing.. **A4:** 719
service characteristics..... **A4:** 719 **M3:** 438-440 **M4:** 570-572
shear blades, service data........................... **M3:** 480
tempering **A4:** 716 **M4:** 567-569
H25, composition **A16:** 709
H25, for hot-forging dies **A18:** 623
H25, machining....................... **A16:** 274, 710, 712-725
H26
annealing **A4:** 715, 744 **M4:** 565-566, 594
annealing temperatures **M3:** 433-434
austenitizing.. **A4:** 743
ceramic coatings of dies **A18:** 643
composition....... **A4:** 712 **M3:** 422-423 **M4:** 562-564
density..................... **A4:** 720, 763 **M3:** 441 **M4:** 617
die blocks and inserts, use for........... **M3:** 529, 530
die steel ratings **A18:** 624
elevated temperatures, resistance to
softening **M3:** 426
for hammer forging alloy and stain-
less steel **A18:** 625
hardening **A4:** 716, 744 **M4:** 567-569, 570-572, 594
hardening and tempering **M3:** 435-440
for hot extrusion tools................................ **A18:** 627
for hot-forging dies **A18:** 623, 625, 626, 635
machinability rating **M3:** 443
normalizing... **A4:** 715
for press forging heat-resistant
alloys, nickel-base alloys.................. **A18:** 625
processing.. **A4:** 718
service characteristics..... **A4:** 718 **M3:** 438-440 **M4:** 570-572
tempering **A4:** 716 **M4:** 567-569, 570-572
thermal expansion **A4:** 720, 763 **M3:** 441 **M4:** 617
H26, composition **A16:** 709
H26 exhaust-valve punch, fatigue
failure.. **A11:** 576, 585
H26, machining.............. **A16:** 274, 710, 712, 717-720, 723-725, 731
H41
annealing **A4:** 715, 744 **M4:** 565-566, 594
annealing temperatures **M3:** 433-434
austenitizing.. **A4:** 743
composition............. **A4:** 712 **M3:** 424 **M4:** 562-564
hardening **A4:** 716, 744 **M4:** 567-569, 594
hardening and tempering **M3:** 435-437
for hot-forging dies **A18:** 623
normalizing... **A4:** 715
preheating **A4:** 743
resistance to thermal fatigue..................... **A18:** 640
tempering **A4:** 717 **M4:** 567-569
H41, machining............... **A16:** 710, 713-716, 721, 722

Tool steels, specific types (continued)
H42
annealing A4: 715, 744 M4: 565-566, 594
annealing temperatures M3: 433-434
austenitizing A4: 743
composition A4: 712 M3: 422-423 M4: 562-564
density A4: 720, 763 M3: 441 M4: 617
hardening A4: 716, 744 M4: 567-569, 570-572, 594
hardening and tempering M3: 435-440
for hot-forging dies A18: 623, 626
machinability rating M3: 443
normalizing A4: 715
processing A4: 719
resistance to thermal fatigue A18: 640
service characteristics A4: 719 M3: 438-440 M4: 570-572
tempering A4: 716 M4: 567-569, 570-572
thermal expansion A4: 720, 763 M3: 441 M4: 617
H42, composition A16: 709
H42, machining A16: 710, 721-725
H43
annealing A4: 715, 744 M4: 565-566, 594
annealing temperatures M3: 433-434
austenitizing A4: 743
composition A4: 712 M3: 424 M4: 562-564
hardening A4: 716, 744 M4: 567-569, 594
hardening and tempering M3: 435-437
for hot-forging dies A18: 623
normalizing A4: 715
resistance to thermal fatigue A18: 640
tempering A4: 716 M4: 567-569
H43, machining A16: 710, 712-716, 721, 722
L group, grindability A16: 728, 732
L group, machinability A16: 721
L group, machinability rating A16: 710
L1
annealing A4: 757 M4: 608
composition A4: 713 M4: 562-564
hardening A4: 757 M4: 608
normalizing temperature M4: 608
wear resistance A4: 757
L1, composition M3: 424
L2
annealing A4: 715, 757 M4: 565-566, 608
annealing temperatures M3: 433-434
composition A4: 713 M3: 422-423 M4: 562-564
density A4: 720, 763 M3: 441 M4: 617
hardening A4: 717, 757 M4: 567-569, 570-572, 608
hardening and tempering M3: 435-440
machinability rating M3: 443
mechanical properties M3: 431
normalizing A4: 715
normalizing temperature M4: 565-566, 608
normalizing temperatures M3: 433-434
processing A4: 719
service characteristics A4: 720 M3: 438-440 M4: 570-572
structural components, use for M3: 559
tempering A4: 717 M4: 567-569, 570-572
thermal expansion A4: 720, 763 M3: 441 M4: 617
L2, composition A16: 710
L2, machining A16: 717-720, 723-725
L3
annealing A4: 715, 757 M4: 565-566, 608
annealing temperatures M3: 433-434
composition A4: 713 M3: 424 M4: 562-564
hardening A4: 717, 757 M4: 567-569, 608
hardening and tempering M3: 435-437
microconstituents A4: 613 M4: 615
microconstituents after hardening A4: 762 M3: 468
normalizing A4: 715
normalizing temperature M4: 565-566, 608
normalizing temperatures M3: 433-434
tempering A4: 717 M4: 567-569
wear resistance A4: 757
L4
composition A4: 713
hardenability A4: 757
wear resistance A4: 757
L4, composition M3: 424 M4: 562-564
L5, composition A4: 713 M3: 424 M4: 562-564

Tool steels, specific types (continued)
L6
annealing A4: 715, 757 M4: 565-566, 608
annealing temperatures M3: 433-434
coining dies, use for M3: 509, 510
composition A4: 713 M3: 422-423 M4: 562-564
density A4: 720, 763 M3: 441 M4: 617
hardening A4: 717, 757 M4: 567-569, 570-572, 608
hardening and tempering M3: 435-440
machinability rating M3: 443
mechanical properties M3: 431
normalizing A4: 715
normalizing temperature M4: 565-566, 608
normalizing temperatures M3: 433-434
processing A4: 719
proven applications for borided ferrous materials A4: 445
service characteristics A4: 719 M3: 438-440 M4: 570-572
shear blades, service data M3: 480
structural components, use for M3: 559
tempering A4: 717 M4: 567-569, 570-572
thermal expansion A4: 720, 763 M3: 441 M4: 617
L6, clamping dies M6: 562
L6, composition A16: 710
L6, cutoff band sawing with bimetal blades A6: 1184
L6, machining A16: 274, 717-721, 723-725
L6, tool steel selection A16: 712
L7
annealing A4: 757 M4: 608
applications A4: 757
composition A4: 713 M3: 424 M4: 562-564
gages, use for M3: 555, 556
hardening A4: 757 M4: 608
normalizing temperature M4: 608
wear resistance A4: 757
L7, machining A16: 274, 717-721, 723-725
M group, grindability A16: 729, 732
M1
annealing A4: 715, 750 M4: 565-566, 600
annealing temperatures M3: 433-434
applications A6: 930
case depth effect on spline size distortion A4: 628
cold heading tools, use in M3: 512
composition A4: 713 M3: 422-423 M4: 562-564
composition of weld deposits A6: 675
cutting tools, use for M3: 472, 473, 474, 476
density A4: 720, 763 M3: 441 M4: 617
description A6: 930
distortion after heat treatment A4: 617
endothermic-atmosphere dew point for hardening A4: 729 M4: 579
grinding ratio M3: 443
hardening A4: 716, 750 M4: 567-569, 570-572, 600
hardening and tempering M3: 435-440
normalizing A4: 715
processing A4: 718
service characteristics A4: 718 M3: 438-440 M4: 570-572
tempering A4: 716 M4: 567-569, 570-572
thermal expansion A4: 720, 763 M3: 441 M4: 617
thread-rolling dies, use for M3: 551, 552
welding preheat and interpass temperatures A6: 675
M1, composition A16: 52, 709 M1: 609
M1, contact fatigue in rolling-element bearings A18: 260
M1, for die threading dies A16: 698
M1, for drills A16: 58, 218, 219, 658, 718, 747-749, 776, 781, 850, 863
M1, for end mills A16: 58
M1, for milling cutters A16: 314
M1, for reamers A16: 58, 246, 659, 719, 750, 751, 781, 785, 852, 862
M1, for taps A16: 259, 260, 261, 263, 661, 696, 720, 753, 782, 787, 813, 825, 851, 862, 867
M1, grindability A16: 728, 732
M1, grinding ratios A16: 732
M1, hot hardness A16: 53
M1, machining A16: 716-725
M1, tool bit applications A16: 51, 52, 53, 57

Tool steels, specific types (continued)
M1, tool steel selection A16: 712
M2
annealing A4: 715, 750 M4: 565-566, 600
annealing temperatures M3: 433-434
anvil tip for ultrasonic welding A6: 326
applications A6: 930
austenitizing A4: 751, 755
auxiliary tools, hot upset forging, use for M3: 535
blades, rotary slitting, use for M3: 479
blanking and piercing dies, use for M3: 486, 487
case depth effect on spline size distortion A4: 628
cold extrusion tools, use for M3: 515, 516, 517, 518, 519
cold heading tools, use in M3: 512
composition A4: 713 M3: 422-423, 490 M4: 562-564
contact fatigue in rolling-element bearings A18: 260
cutoff band sawing with bimetal blades A6: 1184
cutting tools, use for M3: 471, 472, 473, 474, 476, 477
density A4: 720, 763 M3: 441 M4: 617
description A6: 930
dimensional changes M4: 616
distortion after heat treatment A4: 617
distortion, linear dimensions M3: 468
electron-beam welding A6: 867
extension of tool life via ion implantation, examples A18: 643
fracture toughness A4: 751
friction welding A6: 153
gages, use for M3: 555
gas nitriding A4: 387
gas pressure effect on cooling A4: 731, 732
grinding ratio M3: 443
hardening A4: 613, 716, 750, 751, 752, 762 M4: 567-569, 570-572, 599, 600, 601, 602, 606
hardening and tempering ... M3: 425, 426, 435-440
hardening, furnace atmosphere A4: 565
hot extrusion tools, use for M3: 538, 540
impact strength M4: 601
ion-nitriding atmospheres A4: 565-566
laser melting A18: 865
machinability rating M3: 443
mechanical properties M3: 445
microconstituents A4: 613 M4: 615
microconstituents after hardening A4: 762 M3: 468
normalizing A4: 715
out-of-roundness distortion in bars A4: 766
part material for ion implantation A18: 858
plasma nitriding A4: 404, 405
press forming dies, use for M3: 490, 492
processing A4: 718
recommended for backward extrusion of two parts A18: 628
service characteristics A4: 718 M3: 438-440 M4: 570-572
shear spinning tools, use for M3: 501
structural components, use for M3: 558, 559
tempering A4: 613, 716, 751-752, 755, 762 M4: 567-569, 570-572
tempering resistance A4: 749
thermal conductivity A4: 721 M3: 442 M4: 573
thermal expansion A4: 720, 763 M3: 441
thread-rolling dies, use for M3: 551, 552
welding preheat and interpass temperatures A6: 675
M2 bearing, hardness, grain size, and retained austenite variations in A11: 509
M2, composition A16: 52, 359, 709, 733 M1: 609
M2, CPM compositions A16: 63
M2, cutting edge wear of tools A16: 61
M2, electron beam welding M6: 638
M2, for boring tools A16: 164, 716, 742, 771, 813, 859
M2, for broaches A16: 59, 201, 203, 207-209, 700, 743, 746, 774, 779, 823
M2, for circular sawing A16: 725, 794, 817, 868
M2, for counterbores A16: 251, 660, 752, 860
M2, for cutoff tools, for plunge test A16: 678
M2, for drills A16: 218, 224, 232, 811

Tool steels, specific types (continued)
M2, for end mills.... A16: 58, 325, 326, 663, 664, 722, 790
M2, for face mills ... A16: 323, 662, 788-789
M2, for gear cutting tools ... A16: 342, 343
M2, for hobs ... A16: 345, 346
M2, for milling cutters ... A16: 59, 699, 718, 727, 792, 840
M2, for multipoint cutting tools ... A16: 58, 59
M2, for peripheral end mills ... A16: 721, 786-787, 846, 849, 866
M2, for planing tools ... A16: 183, 657, 743, 773
M2, for reamers ... A16: 58, 240, 242-243, 246, 659, 719, 751, 781, 785, 839, 852, 862
M2, for sawing ... A16: 358, 359, 360
M2, for shaping tools ... A16: 190, 347, 348
M2, for slab mills ... A16: 324
M2, for spotfacing tools ... A16: 251, 660, 752, 860
M2, for taps ... A16: 259, 261, 695, 700
M2, for thread milling cutters ... A16: 752, 867
M2, for trepanning tools ... A16: 179, 742
M2, for turning tools ... A16: 144, 147, 653, 715, 726, 740, 741, 771, 811, 846
M2, grindability ... A16: 728, 732
M2 high-speed, brittle fracture of rehardened ... A11: 574, 579
M2, honing with cubic boron nitride ... A16: 479
M2, hot hardness ... A16: 53
M2, localized surface melting ... A11: 573, 579
M2, machining ... A16: 713-721, 723-727
M2, microstructure ... A16: 708, 711
M2, powdered ... A9: 511
M2, pressed and sintered ... A9: 524
M2 roughing tool, cracked from heat treatment ... A11: 571
M2, tool bit applications ... A16: 51, 53, 56, 57, 58
M2, tool steel selection ... A16: 712
M3
annealing ... A4: 715, 750 M4: 565-566, 600
annealing temperatures ... M3: 433-434
applications ... A6: 930
case depth effect on spline size distortion ... A4: 628
composition ... A4: 713 M3: 422-423 M4: 562-564
cutoff band sawing with bimetal blades ... A6: 1184
cutting tools, use for ... M3: 473, 476
density ... A4: 720, 763 M3: 441 M4: 617
description ... A6: 930
hardening ... A4: 716, 750 M4: 567-569, 570-572, 600
hardening and tempering ... M3: 426, 435-440
machinability rating ... M3: 443
normalizing ... A4: 715
processing ... A4: 718
refrigeration treatment ... A4: 752
service characteristics ... A4: 718 M3: 438-440 M4: 570-572
size distortion in heat treatment ... A4: 613
tempering ... A4: 716 M4: 567-569, 570-572
thermal expansion ... A4: 720, 763 M3: 441 M4: 617
M3 class 1, composition ... A16: 52, 709
M3 class 1, machining ... A16: 710, 713-725
M3 class 2, composition ... A16: 52, 709
M3 class 2, for multipoint cutting tools ... A16: 59
M3 class 2, grindability ... A16: 728
M3 class 2, machining ... A16: 716-725
M3 class 2, wear resistance ... A16: 54, 57
M3, composition ... A16: 52, 709, 733
M3, CPM composition ... A16: 63
M3, for boring tools ... A16: 742, 771, 813, 859
M3, for broaches ... A16: 203, 743-745, 823
M3, for counterbores ... A16: 660, 752, 860
M3, for drills ... A16: 218, 224, 836
M3, for end mills ... A16: 325, 326, 663, 664, 754, 790, 866
M3, for gear cutters ... A16: 342, 343
M3, for hobs ... A16: 345, 346
M3, for milling cutters ... A16: 320, 755
M3, for multipoint cutting tools ... A16: 58, 59
M3, for peripheral end mills ... A16: 786-787, 849, 866
M3, for planing tools ... A16: 657, 773
M3, for reamers ... A16: 240
M3, for shaping tools ... A16: 192, 347, 348
M3, for spotfacing tools ... A16: 660, 752, 860
M3, for taps ... A16: 259, 261, 263
M3, for thread chasers ... A16: 299, 720
M3, for trepanning tools ... A16: 179
M3, for turning tools ... A16: 144, 147, 653, 740-741, 811
M3, grindability ... A16: 729
M3, microstructure ... A16: 734
M3, tool steel selection ... A16: 712
M4
annealing ... A4: 715, 750 M4: 565-566, 600
annealing temperatures ... M3: 433-434
coining dies, use for ... M3: 510-511
cold extrusion tools, use for ... M3: 515, 516, 517, 518, 519
composition ... A4: 713 M3: 422-423, 492 M4: 562-564
cutting tools, use for ... M3: 471, 476, 477
density ... A4: 720, 763 M3: 441 M4: 617
dimensional changes ... M4: 616
gas nitriding ... A4: 387
hardening ... A4: 716, 750 M4: 567-569, 570-572, 600
hardening and tempering ... M3: 435-440
normalizing ... A4: 715
press forming dies, use or ... M3: 491, 492
processing ... A4: 718
service characteristics ... A4: 719 M3: 438-440 M4: 570-572
shear spinning tools, use for ... M3: 150
tempering ... A4: 716 M4: 567-569, 570-572
thermal expansion ... A4: 720, 763 M3: 441 M4: 617
M4, composition ... A16: 52, 709, 733
M4, CPM composition ... A16: 63
M4, cutoff band sawing with bimetal blades ... A6: 1184
M4, die wear and die life ... A18: 633
M4, for broaches ... A16: 203
M4, for drills ... A16: 218, 224
M4, for multipoint cutting tools ... A16: 58, 59
M4, for reamers ... A16: 240, 248
M4, for taps ... A16: 259, 753
M4, for thread chasers ... A16: 299
M4, grindability ... A16: 728
M4, honing with cubic boron nitride ... A16: 479
M4, hot hardness ... A16: 53
M4, machining ... A16: 360, 362, 713-725
M4, sulfur content effect on machinability ... A16: 734
M4, tool steel selection ... A16: 712
M4, wear resistance ... A16: 52, 54, 57
M6
annealing ... A4: 715, 750 M4: 565-566, 600
annealing temperatures ... M3: 433-434
composition ... A4: 713 M3: 422-423 M4: 562-564
hardening ... A4: 716, 750 M4: 567-569, 570-572, 600
hardening and tempering ... M3: 435-440
normalizing ... A4: 715
processing ... A4: 718
service characteristics ... A4: 718 M3: 438-440 M4: 570-572
tempering ... A4: 716 M4: 567-569, 570-572
M6, composition ... A16: 52, 53, 709
M6, for drills ... A16: 218
M6, for planing tools ... A16: 183
M6, for shaping tools ... A16: 190
M6, machining ... A16: 710, 713-715, 718-724
M7
annealing ... A4: 715, 750 M4: 565-566, 600
composition ... A4: 713 M3: 422-423 M4: 562-564
cutting tools, use for ... M3: 472, 473, 474, 476
density ... A4: 720, 763 M3: 441 M4: 617
hardening ... A4: 716, 750 M4: 567-569, 570-572, 600
hardening and tempering ... M3: 435-440
magnetron sputtering ... A18: 849
normalizing ... A4: 715
part material for ion implantation ... A18: 858
processing ... A4: 718
service characteristics ... A4: 718 M3: 438-440 M4: 570-572
tempering ... A4: 716 M4: 567-569, 570-572
thermal expansion ... A4: 720, 763 M3: 441 M4: 617
M7, composition ... A16: 52, 53, 709
M7, for broaches ... A16: 207, 219, 700, 743, 774
M7, for circular sawing ... A16: 725, 756, 794, 817, 868
M7, for drill force testing ... A16: 678-679
M7, for drills ... A16: 58, 658, 677, 718, 726, 747, 749, 776, 781, 850, 851
M7, for end mills ... A16: 58, 325, 326, 663, 664, 722, 754, 790, 866
M7, for face mills ... A16: 323, 662, 788-789
M7, for gear cutting ... A16: 343
M7, for hobs ... A16: 345, 346
M7, for milling cutters ... A16: 314, 699
M7, for peripheral end mills ... A16: 721, 786-787, 849, 866
M7, for reamers ... A16: 58, 240, 246, 659, 702, 719, 750, 751, 781, 785, 852, 862
M7, for shaping tools ... A16: 347
M7, for slab milling cutters ... A16: 324
M7, for taps ... A16: 259, 261, 263, 661, 696, 720, 782, 787, 813, 825, 851, 862, 867
M7, for thread milling cutters ... A16: 752, 867
M7, grindability ... A16: 728
M7, hot hardness ... A16: 53
M7, machining ... A16: 713-721, 723-725
M8, composition ... A4: 713 M3: 424 M4: 562-564
M10
activated reactive evaporation ... A18: 849
annealing ... A4: 715, 750 M4: 565-566, 600
annealing temperatures ... M3: 433-434
applications ... A6: 930
auxiliary tools, hot upset forging, use for ... M3: 535
composition ... A4: 713 M3: 422-423 M4: 562-564
composition of weld deposits ... A6: 675
contact fatigue in rolling-element bearings ... A18: 260
cutoff band sawing with bimetal blades ... A6: 1184
cutting tools, use for ... M3: 472, 473, 474, 476
density ... A4: 720, 763 M3: 441 M4: 617
description ... A6: 930
hardening ... A4: 716, 750 M4: 567-569, 570-572, 600
hardening and tempering ... M3: 435-437
hot upset forging tools, use for ... M3: 534
normalizing ... A4: 715
processing ... A4: 718
service characteristics ... A4: 718 M3: 438-440 M4: 570-572
tempering ... A4: 716 M4: 567-569, 570-572
thermal expansion ... A4: 720, 763 M3: 441 M4: 617
welding preheat and interpass temperatures ... A6: 675
M10 bearing, hardness, grain size, and retained austenite variations in ... A11: 509
M10, composition ... A16: 51, 52, 709 M1: 609
M10, flash welding schedule ... M6: 577
M10, for boring tools ... A16: 742, 771, 813
M10, for drills ... A16: 58, 218, 658, 718, 747, 749, 776, 781, 850, 863
M10, for end mills ... A16: 58, 664

Tool steels, specific types (continued)
M10, for milling cutters A16: 314, 714, 840
M10, for planing tools A16: 657, 773
M10, for reamers A16: 58, 239, 240, 839
M10, for taps A16: 259, 260, 261, 263, 661, 696, 720, 782, 787, 813, 825, 851, 862, 867
M10, for turning tools A16: 145, 653, 718, 726, 741, 811
M10, grindability A16: 728
M10, machining A16: 713-725
M10, tool steel selection A16: 712
M15
applications A6: 930
composition M3: 424
cutoff band sawing with bimetal blades A6: 1184
description A6: 930
machinability rating M3: 443
M15, composition A4: 713 A16: 52, 54 M4: 562-564
M15, for taps A16: 259
M15, grindability A16: 728
M15, hardness A16: 708
M15, ion plating A18: 849
M15, machining A16: 715, 718, 722
M15, wear resistance A16: 52, 54
M25
distortion M4: 619
hardening, comparison between powder metallurgy and conventional M4: 619
M25, distortion M3: 444
M25, hardness A4: 766
M30
annealing A4: 715, 750 M4: 565-566, 600
annealing temperatures M3: 433-434
composition A4: 713 M3: 422-423 M4: 562-564
density A4: 720, 763 M3: 441 M4: 617
hardening A4: 716, 750 M4: 567-569, 570-572, 600
hardening and tempering M3: 435-440
normalizing A4: 715
processing A4: 718
service characteristics A4: 718 M3: 438-440 M4: 570-572
tempering A4: 716 M4: 567-569, 570-572
thermal expansion A4: 720, 763 M3: 441 M4: 617
M30, composition A16: 51, 52, 54, 709
M30, for face mills A16: 323, 662, 753, 788-789
M30, for gear cutters A16: 343
M30, for planing tools A16: 657, 773
M30, for shaping tools A16: 192
M30, for turning tools A16: 145, 653, 741
M30, hardness A16: 54
M30, machining A16: 713-725
M33
annealing A4: 715, 750 M4: 565-566, 600
annealing temperatures M3: 433-434
composition A4: 713 M3: 422-423 M4: 562-564
cutting tools, use for M3: 472, 473, 474
density A4: 720, 763 M3: 441 M4: 617
hardening A4: 716, 750 M4: 567-569, 570-572, 600
hardening and tempering M3: 435-440
normalizing A4: 715
processing A4: 718
service characteristics A4: 719 M3: 438-440 M4: 570-572
tempering A4: 716 M4: 567-569, 570-572
thermal expansion A4: 720, 763 M3: 441 M4: 617
M33, composition A16: 52, 53, 709
M33, for broaches A16: 207, 743
M33, for broaches, for heat-resistant alloys A16: 743
M33, for counterbores A16: 752, 860
M33, for drills A16: 58, 218, 233, 718, 747, 749, 860
M33, for end mills A16: 58, 325, 326, 722, 754
M33, for face mills A16: 323, 863
M33, for peripheral end mills A16: 721, 866
M33, for slab milling cutters A16: 324
M33, for spotfacing tools A16: 752, 860
M33, for taps A16: 261
M33, for trepanning tools A16: 179, 860
M33, for turning tools A16: 144, 741, 858
M33, hot hardness A16: 53
M33, machining A16: 713-725
M34
annealing A4: 715, 750 M4: 565-566, 600
annealing temperatures M3: 433-434
composition A4: 713 M3: 422-423 M4: 562-564
hardening A4: 716, 750 M4: 567-569, 570-572, 600
hardening and tempering M3: 435-440
normalizing A4: 715
processing A4: 718
service characteristics A4: 718 M3: 438-440 M4: 570-572
tempering A4: 716 M4: 567-569, 570-572
M34, composition A16: 52, 709
M34, for drills, for heat-resistant alloys A16: 747-748
M34, for shaping tools A16: 192, 233
M34, machining A16: 713-724
M34, specimens used in metallographic techniques A18: 438
M35
laser melting A18: 865
part material for ion implantation A18: 858
M35, and gear manufacturing A16: 66, 67
M35, composition A16: 52, 733 M3: 424 M4: 562-564
M35, CPM composition A16: 63
M35, machining A16: 713-717, 721, 722, 725
M36
annealing A4: 715, 750 M4: 565-566, 600
annealing temperatures M3: 433-434
composition A4: 713 M3: 422-423 M4: 562-564
density A4: 720, 763 M3: 441 M4: 617
hardening A4: 716, 750 M4: 567-569, 570-572, 600
hardening and tempering M3: 435-440
normalizing A4: 715
processing A4: 718
service characteristics A4: 718 M3: 438-440
tempering A4: 716 M4: 567-569, 570-572
thermal expansion A4: 720, 763 M3: 441 M4: 617
M36, composition A16: 52, 53, 709
M36, for machining heat-resistant alloys A16: 740, 741, 748
M36, for shaping tools A16: 190, 233
M36, hot hardness A16: 53
M36, machining A16: 710, 713-725
M41
annealing A4: 715, 750 M4: 565-566, 600
annealing temperatures M3: 433-434
composition A4: 713 M3: 422-423 M4: 562-564
density A4: 720, 763 M3: 441 M4: 617
dimensional changes M4: 616
distortion, linear dimensions M3: 468
grinding ratio M3: 443
hardening A4: 613, 716, 750, 762 M4: 567-569, 570-572, 600
hardening and tempering M3: 435-440
normalizing A4: 715
processing A4: 718
service characteristics A4: 718 M3: 438-440 M4: 570-572
size distortion in heat treatment A4: 613
tempering A4: 613, 716, 762 M4: 567-569, 570-572
thermal expansion A4: 720, 763 M3: 441 M4: 617
M41, composition A16: 51, 52, 709
M41, for boring tools A16: 164, 716, 742
M41, for broaches A16: 207, 743
M41, for counterbores A16: 752, 860
M41, for drills A16: 718, 749, 860
M41, for end mills A16: 325, 326, 722, 754, 866
M41, for face mills A16: 323, 863
M41, for slab milling cutters A16: 324
M41, for spotfacing tools A16: 752, 860
M41, for trepanning tools A16: 179, 860
M41, for turning tools A16: 144, 715, 741, 858
M41, grindability A16: 732
M41, machining A16: 713-724
M42
annealing A4: 715, 750 M4: 565-566, 600
annealing temperatures M3: 433-434
composition A4: 713, 722 M3: 422-423 M4: 562-564
cutting tools, use for M3: 471, 472, 474, 476, 477
density A4: 720, 763 M3: 441 M4: 617
grinding ratio M3: 443
hardening A4: 716, 750 M4: 567-569, 570-572, 600
hardening and tempering M3: 435-440
normalizing A4: 715
processing A4: 718
refrigeration treatment A4: 752
service characteristics A4: 718 M3: 438-440 M4: 570-572
structural components, use for M3: 559
tempering A4: 716 M4: 567-569, 570-572
thermal expansion A4: 720, 763 M3: 441 M4: 617
thread-rolling dies, use for M3: 552
M42, activated reactive evaporation A18: 849
M42 bearing, hardness, grain size, and retained austenite variations in A11: 509
M42, composition A16: 51, 52, 653, 659, 709, 733
M42, CPM composition A16: 63, 64
M42, for boring tools A16: 164, 716, 742
M42, for broaches A16: 207, 717, 700, 743
M42, for counterbores A16: 251, 752, 860
M42, for drills A16: 658, 718, 749, 850, 860
M42, for end mills A16: 58, 325, 326, 722, 754
M42, for face mills A16: 323, 753, 863
M42, for hobs A16: 345, 346
M42, for multipoint cutting tools A16: 59
M42, for peripheral end mills A16: 848, 866
M42, for reamers A16: 702-703, 750, 751, 852
M42, for sawing A16: 358, 359, 360, 362
M42, for shaping tools A16: 347
M42, for slab milling cutters A16: 324
M42, for spotfacing tools A16: 251, 752, 860
M42, for taps A16: 259, 261
M42, for tools for single-point turning test A16: 679
M42, for trepanning tools A16: 179, 860
M42, for tSubhead chasers A16: 299
M42, for turning tools A16: 144, 653, 715, 741, 858
M42, grindability A16: 732
M42, machining A16: 713-721, 723-725
M42, tool steel selection A16: 712
M43
annealing A4: 715, 750 M4: 565-566, 600
composition A4: 713 M4: 562-564
hardening A4: 716, 750 M4: 567-569, 570-572
normalizing A4: 715
processing A4: 718
service characteristics A4: 718 M4: 570-572
tempering A4: 717 M4: 567-569, 570-572
M43 annealing temperatures M3: 433-434
composition M3: 422-423
grinding ratio M3: 443
hardening and tempering M3: 435-440
service characteristics M3: 438-440
M43, composition A16: 709
M43, for boring tools A16: 164, 716, 742
M43, for broaches A16: 207, 743
M43, for broaches, for heat-resistant alloys A16: 743
M43, for counterbores A16: 752, 860
M43, for drills A16: 718, 749, 860
M43, for end mills A16: 325, 326, 722, 754
M43, for face mills A16: 323, 863
M43, for slab milling cutters A16: 324
M43, for spotfacing tools A16: 752, 860
M43, for trepanning tools A16: 179, 860
M43, for turning tools A16: 144, 715, 744, 858
M43, grindability A16: 726, 728, 732
M43, machining A16: 713-721, 723, 724
M44
annealing A4: 715, 750 M4: 565-566, 600
composition A4: 713 M4: 562-564
hardening A4: 716, 750 M4: 567-569, 570-572, 600
normalizing A4: 715
processing A4: 718
service characteristics A4: 718 M4: 570-572

Tool steels, specific types (continued)
tempering A4: 716 M4: 567-569, 570-572
M44 annealing temperatures.................. M3: 433-434
composition.. M3: 422-423
grinding ratio... M3: 443
hardening and tempering M3: 435-440
service characteristics........................... M3: 438-440
M44, composition A16: 709
M44, for boring tools.................... A16: 164, 716, 742
M44, for broaches................................... A16: 207, 743
M44, for counterbores A16: 752, 860
M44, for drilling A16: 718, 749, 860
M44, for end mills................. A16: 325, 326, 722, 754
M44, for face mills A16: 323, 863
M44, for planing tools A16: 183
M44, for slab milling cutters A16: 324
M44, for spotfacing tools A16: 752, 860
M44, for trepanning tools..................... A16: 179, 860
M44, for turning tools A16: 144, 715, 740, 741, 858
M44, grindability A16: 726, 728, 732
M44, machining A16: 713-721, 723-725
M44, tool steel selection A16: 712
M45, composition M3: 424
M45, for broaches................................... A16: 207, 743
M45, for counterbores A16: 752, 860
M45, for drills A16: 718, 749, 860
M45, for end mills.......................... A16: 325, 326, 754
M45, for face mills A16: 323, 863
M45, for slab milling cutters A16: 324
M45, for spotfacing tools A16: 752, 860
M45, for trepanning tools..................... A16: 179, 860
M45, for turning tools A16: 144, 741, 858
M46
annealing A4: 715, 750 M4: 565-566, 600
composition............................. A4: 713 M4: 562-564
density..................................... A4: 720, 763 M4: 617
hardening .... A4: 716, 750 M4: 567-569 , 570-572, 600
normalizing A4: 715
processing.. A4: 718
service characteristics A4: 719 M4: 570-572
tempering A4: 716 M4: 567-569, 570-572
thermal expansion A4: 720, 763 M4: 617
M46 annealing temperatures.................. M3: 433-434
composition.. M3: 422-423
density... M3: 441
hardening and tempering M3: 435-440
service characteristics........................... M3: 438-440
thermal expansion M3: 441
M46, composition A16: 52, 709
M46, for broaches................................... A16: 207, 743
M46, for counterbores A16: 752, 860
M46, for drills A16: 718, 749, 860
M46, for end mills.......................... A16: 325, 326, 754
M46, for face mills A16: 323, 863
M46, for slab milling cutters A16: 324
M46, for spotfacing tools A16: 752, 860
M46, for trepanning tools..................... A16: 179, 860
M46, for turning tools A16: 144, 741, 858
M46, machining A16: 717-720, 723-725
M47
annealing A4: 715, 750 M4: 565-566, 600
annealing temperatures M3: 433-434
composition....... A4: 713 M3: 422-423 M4: 562-564
density..................... A4: 720, 763 M3: 441 M4: 617
hardening A4: 716, 750 M4: 567-569, 570-572, 600
hardening and tempering M3: 435-440
normalizing A4: 715
processing.. A4: 718
service characteristics A4: 718 M3: 438-440 M4: 570-572
tempering A4: 716 M4: 567-569, 570-572
thermal expansion A4: 720, 763 M3: 441 M4: 617
M47, composition A16: 709
M47, for broaches................................... A16: 207, 743
M47, for counterbores A16: 752, 860
M47, for drills A16: 718, 749, 760
M47, for end mills.......................... A16: 325, 326, 754
M47, for face mills A16: 323, 863
M47, for slab milling cutters A16: 324
M47, for spotfacing tools A16: 752, 860
M47, for trepanning tools..................... A16: 179, 860
M47, for turning tools A16: 144, 741, 858
M47, machining A16: 717-720, 723-725
M48, composition A16: 52, 733
M48, CPM composition.................................. A16: 63
M50
annealing A4: 750 M4: 600
application in gas turbine mainshaft bearings.. A18: 590
applications A4: 749-750, 756
bath temperatures and cycle times............ A4: 757
composition.......................... A4: 713, 756 M4: 562-564
contact fatigue in rolling-element bearings.. A18: 260
friction coefficient data A18: 71, 72
hardening A4: 750 M4: 600
heat treatments....................................... A4: 756-757
liquid nitriding A4: 413 M4: 256
nominal compositions A18: 726
rolling-contact component steels A18: 503
tempering A4: 757
TT-F diagram... M4: 607
TTT diagram A4: 756
M50 bearing,, hardness, grain size, and retained austenite variations in... A11: 509
M50, composition A16: 52 M1: 609
M50, for drills A16: 58
M50NiL
application in gas turbine mainshaft bearings.. A18: 590
contact fatigue in rolling-element bearings.. A18: 260
rolling-contact component steels A18: 503
M52
annealing A4: 750 M4: 600
applications A4: 749-750
composition............................ A4: 713 M4: 562-564
hardening A4: 750 M4: 600
M52, composition A16: 52
M52, for drills A16: 58
M62, composition A16: 52, 733
M62, CPM composition.................................. A16: 63
O group machining........ A16: 710, 713-716, 721, 722, 728
O1
annealing A4: 715, 739 M4: 565-566, 600
annealing temperatures M3: 433-434
applications A6: 930
austenitizing.. A4: 739
blanking and piercing dies, use for.......... M3: 485, 486, 487
coining dies, use for M3: 509, 510
for cold extrusion tools.............................. A18: 627
cold extrusion tools, use for M3: 515, 516, 518
composition............ A4: 712 M3: 422-423, 490 M4: 562-564
composition of weld deposits...................... A6: 675
cutoff band sawing with bimetal blades A6: 1184
cycle annealing A4: 737, 740
decarburization bands when held in a fluidized bed................................... A4: 486
deep drawing dies, use for M3: 496, 498, 499
density..................... A4: 720, 763 M3: 441 M4: 617
description... A6: 930
dimensional changes M4: 590, 616
distortion, linear dimensions M3: 468
endothermic-atmosphere dew point for hardening A4: 729 M4: 579
erosion resistance versus hardness........... A18: 204
gages, use for.............................. M3: 554, 555, 556
hardenability bands... M3: 428
hardening....... A4: 613, 716, 739, 762 M4: 567-569, 570-572, 586
hardening and tempering M3: 429-430, 435-440
machinability rating M3: 443
martempering A4: 739, 740
molds for plastics and rubber, use for.. M3: 347, 548
normalizing A4: 715
normalizing temperature............ M4: 565-566, 586
normalizing temperatures M3: 433-434
press forming dies, use for M3: 489, 492
processing... A4: 719
recommended for backward extrusion of two parts A18: 628
service characteristics..... A4: 719 M3: 438-440 M4: 570-572
structural components, use for.................. M3: 558
tempering A4: 121, 123, 613, 716, 740, 763 M4: 567-569, 570-572
thermal expansion A4: 720, 763 M3: 441 M4: 617
trimming tool materials, use for M3: 531
welding preheat and interpass temperatures A6: 675
working hardness M3: 510
O1, clamping dies M6: 562
O1, composition... A16: 709
O1 die, hydrogen flaking after heat treatment A11: 574, 581
O1 fixture, oil quench cracking A11: 565
O1, for hand reamers..................................... A16: 246
O1, machining....... A16: 274, 360, 362, 710, 714, 717, 719, 723-725
O1 ring forging, quench cracked................. A11: 570
O1 tool steel die, oil quenching failure....... A11: 565
O1, tool steel selection................................... A16: 712
O2
annealing A4: 715, 739 M4: 565-566, 586
annealing temperatures M3: 433-434
applications A6: 930
austenitizing.. A4: 739
boriding A4: 445
composition............ A4: 300, 712 M3: 422-423 M4: 562-564
cutoff band sawing with bimetal blades A6: 1184
density..................... A4: 720, 763 M3: 441 M4: 617
description... A6: 930
electron beam hardening treatment A4: 300
endothermic-atmosphere dew point for hardening A4: 729 M4: 579
hardenability bands.. M3: 428
hardening A4: 716, 739 M4: 567-569, 570-572, 586
hardening and tempering M3: 429-430, 435-440
machinability rating M3: 443
normalizing A4: 715
normalizing temperature............ M4: 565-566, 586
normalizing temperatures M3: 433-434
processing... A4: 719
proven applications for borided ferrous materials A4: 445
service characteristics..... A4: 719 M3: 438-440 M4: 570-572
structural components, use for.................. M3: 559
tempering A4: 717, 740 M4: 567-569, 570-572
thermal expansion A4: 720, 763 M3: 441 M4: 617
use for M3: 555
O2, composition... A16: 709
O2, grindability...................................... A16: 726, 728
O4, tool steel selection................................... A16: 712
O6
annealing A4: 715, 739 M4: 565-566, 586
annealing temperatures M3: 433-434
blanking and piercing dies, use for........... M3: 487
composition....... A4: 712 M3: 422-423 M4: 562-564
composition of weld deposits...................... A6: 675

SUBJECTS OF THE INDEXED VOLUMES: ASM Handbook (designated by the letter "A"): **A1:** Properties and Selection: Irons, Steels, and High-Performance Alloys (1990); **A2:** Properties and Selection: Nonferrous Alloys and Special-Purpose Materials (1990); **A3:** Alloy Phase Diagrams (1992); **A4:** Heat Treating (1991); **A6:** Welding, Brazing, and Soldering (1993); **A7:** Powder Metallurgy (1984); **A8:** Mechanical Testing (1985); **A9:** Metallography and Microstructures (1985); **A10:** Materials Characterization (1986); **A11:** Failure Analysis and Prevention (1986); **A12:** Fractography (1987); **A13:** Corrosion (1987); **A14:** Forming and Forging (1988); **A15:** Casting (1988); **A16:** Machining (1989); **A17:** Nondestructive Testing and Quality Control (1989); **A18:** Friction, Lubrication, and Wear Technology (1992). **Metals Handbook, 9th Edition** (designated by the letter "M"): **M1:** Properties and Selection: Irons and Steels (1978); **M2:** Properties and Selection: Nonferrous Alloys and Pure Metals (1979); **M3:** Properties and Selection: Stainless Steels, Tool Materials and Special-Purpose Materials (1980); **M4:** Heat Treating (1981); **M5:** Surface Cleaning, Finishing, and Coating (1982); **M6:** Welding, Brazing, and Soldering (1983). **Engineered Materials Handbook** (designated by the letters "EM"): **EM1:** Composites (1987); **EM2:** Engineering Plastics (1988); **EM3:** Adhesives and Sealants (1990); **EM4:** Ceramics and Glasses (1991); **Electronic Materials Handbook** (designated by the letters "EL"): **EL1:** Packaging (1989).

Tool steels, specific types (continued)
density **A4:** 720
dimensional changes **M4:** 616
distortion, linear dimensions **M3:** 468
gages, use for **M3:** 554, 555, 556
hardenability bands **M3:** 428
hardening **A4:** 613, 716, 739, 762 **M4:** 567-569, 570-572, 586
hardening and tempering **M3:** 435-440
machinability rating **M3:** 443
normalizing **A4:** 715
normalizing temperature **M4:** 565-566, 586
normalizing temperatures **M3:** 433-434
processing **A4:** 719
service characteristics **A4:** 719 **M3:** 438-440 **M4:** 570-572
structural components, use for **M3:** 559
tempering **A4:** 613, 716, 740, 762 **M4:** 567-569, 570-572
thermal expansion **A4:** 720
welding preheat and interpass temperatures **A6:** 675
O6, composition **A16:** 709
O6, extension of tool life via ion implantation, examples **A18:** 643
O6 graphitic punch, service failure **A11:** 571
O6, machining **A16:** 274, 714, 717, 718, 720, 723, 724
O7
annealing **A4:** 715, 739 **M4:** 565-566, 586
annealing temperatures **M3:** 433-434
composition **A4:** 712 **M3:** 422-423 **M4:** 562-564
density **A4:** 720, 763 **M3:** 441 **M4:** 617
endothermic-atmosphere dew point for hardening **A4:** 729 **M4:** 579
hardening **A4:** 716, 739 **M4:** 567-569, 570-572, 586
hardening and tempering **M3:** 435-440
machinability rating **M3:** 443
normalizing **A4:** 715
normalizing temperature **M4:** 565-566, 586
normalizing temperatures **M3:** 433-434
processing **A4:** 719
service characteristics **A4:** 719 **M3:** 438-440 **M4:** 570-572
tempering **A4:** 716, 740 **M4:** 567-569, 570-572
thermal expansion **A4:** 720, 763 **M3:** 441 **M4:** 617
O7, composition **A16:** 709
O7, machining **A16:** 274, 714, 717, 718, 720, 723-725
P group, grindability **A16:** 732
P group, machinability **A16:** 722
P1
annealing **A4:** 759 **M4:** 611
applications **A4:** 759
composition **A4:** 713 **M3:** 424 **M4:** 562-564
hardening **A4:** 759 **M4:** 611
heat treatments **A4:** 759
molds for plastics and rubber, use for **M3:** 548
P1, grindability **A16:** 728, 732
P1, machining **A16:** 710, 713-716, 721, 722
P2
annealing **A4:** 715, 759 **M4:** 565-566, 611
annealing temperatures **M3:** 433-434
composition **A4:** 713 **M3:** 422-423 **M4:** 562-564
density **A4:** 720, 763 **M3:** 441 **M4:** 617
hardening **A4:** 716, 759 **M4:** 567-569, 570-572, 611
hardening and tempering **M3:** 435-440
heat treatment **A4:** 759
machinability rating **M3:** 443
normalizing **A4:** 715
processing **A4:** 719
service characteristics **A4:** 719 **M3:** 438-440 **M4:** 570-572
tempering **A4:** 716 **M4:** 567-569, 570-572
thermal expansion **A4:** 720, 763 **M3:** 441 **M4:** 617
P2, composition **A16:** 709
P2, grindability **A16:** 728, 732
P2, machining **A16:** 274, 710, 713-725
P2, tool steel selection **A16:** 712
P3
annealing **A4:** 715, 759 **M4:** 565-566, 611
annealing temperatures **M3:** 433-434
composition **A4:** 713 **M3:** 422-423 **M4:** 562-564
hardening **A4:** 716, 759 **M4:** 567-569, 570-572, 611
hardening and tempering **M3:** 435-440
heat treatment **A4:** 759
machinability rating **M3:** 443
molds for plastic and rubber, use for **M3:** 548
normalizing **A4:** 715
processing **A4:** 719
service characteristics **A4:** 719 **M3:** 438-440 **M4:** 570-572
tempering **A4:** 716, 759 **M4:** 567-569, 570-572
P3, composition **A16:** 710
P3, grindability **A16:** 728, 732
P3, machining **A16:** 710, 713-716, 721, 722
P3, tool steel selection **A16:** 712
P4
annealing **A4:** 715, 759 **M4:** 565-566, 611
annealing temperatures **M3:** 433-434
applications **A4:** 758, 759
composition **A4:** 713 **M3:** 422-423 **M4:** 562-564
hardening **A4:** 716, 759 **M4:** 567-569, 570-572, 611
hardening and tempering **M3:** 435-440
heat treatments **A4:** 759
machinability rating **M3:** 443
molds for plastics and rubber, use for **M3:** 547, 548
normalizing **A4:** 715
processing **A4:** 719
service characteristics **A4:** 719 **M3:** 438-440 **M4:** 570-572
tempering **A4:** 716, 759 **M4:** 567-569, 570-572
P4, composition **A16:** 710
P4, grindability **A16:** 732
P4, machining **A16:** 710, 713-725
P4, tool steel selection **A16:** 712
P5
annealing **A4:** 715, 759 **M4:** 565-566, 611
annealing temperatures **M3:** 433-434
composition **A4:** 713 **M4:** 562-564
density **A4:** 720, 763 **M3:** 441 **M4:** 617
hardening **A4:** 716, 759 **M4:** 567-569, 570-572, 611
hardening and tempering **M3:** 435-440
heat treatments **A4:** 759
machinability rating **M3:** 443
molds for plastics and rubber, use for **M3:** 548
normalizing **A4:** 715
processing **A4:** 719
service characteristics **A4:** 719 **M3:** 438-440 **M4:** 570-572
tempering **A4:** 716 **M4:** 567-569, 570-572
thermal expansion **A4:** 720, 763 **M3:** 441 **M4:** 617
P5, composition **A16:** 710
P5, grindability **A16:** 728, 732
P5, machining **A16:** 274, 710, 713-725
P6
annealing **A4:** 715, 759 **M4:** 565-566, 611
annealing temperatures **M3:** 433-434
composition **A4:** 713 **M3:** 422-423 **M4:** 562-564
density **A4:** 720, 763 **M3:** 441 **M4:** 617
hardening **A4:** 716, 759 **M4:** 567-569, 570-572, 611
hardening and tempering **M3:** 435-440
heat treatments **A4:** 759, 760
machinability rating **M3:** 443
molds for plastic and rubber, use for **M3:** 547
normalizing **A4:** 715
processing **A4:** 719
service characteristics **A4:** 719 **M3:** 438-440 **M4:** 570-572
tempering **A4:** 716, 759, 760 **M4:** 567-569, 570-572
thermal expansion **A4:** 720, 763 **M3:** 441 **M4:** 617
P6, composition **A16:** 710
P6, grindability **A16:** 728, 732
P6, machining **A16:** 274, 710, 713-725
P10, grindability **A16:** 728
P20
annealing **A4:** 715, 759 **M4:** 565-566, 611
annealing temperature **M3:** 433-434
applications **A4:** 758, 760
composition **A4:** 713 **M3:** 422-423 **M4:** 562-564
density **A4:** 720, 763 **M3:** 441 **M4:** 617
die-casting dies, use in **M3:** 542
extension of tool life via ion implantation, examples **A18:** 643
hardening **A4:** 716, 759 **M4:** 567-569, 570-572, 611
hardening and tempering **M3:** 435-440
heat treatments **A4:** 760
machinability rating **M3:** 443
molds for plastics and rubber, use for **M3:** 547
normalizing **A4:** 715
normalizing temperature **M4:** 565-566, 611
normalizing temperatures **M3:** 433-434
processing **A4:** 719
recommended for die-casting dies and die inserts **A18:** 629
service characteristics **A4:** 719 **M3:** 438-440 **M4:** 570-572
tempering **A4:** 716, 760 **M4:** 567-569, 570-572
thermal expansion **A4:** 720, 763 **M3:** 441 **M4:** 617
P20, composition **A16:** 710
P20, grindability **A16:** 728, 732
P20, machining **A16:** 710, 713-725
P20 mold, effects of- carburization **A11:** 571-572
P20, tool steel selection **A16:** 712
P21
annealing **A4:** 715, 759 **M4:** 565-566, 611
annealing temperatures **M3:** 433-434
applications **A4:** 758
composition **A4:** 713 **M3:** 422-423 **M4:** 562-564
hardening **A4:** 716, 759 **M4:** 567-569, 570-572, 611
hardening and tempering **M3:** 435-440
machinability rating **M3:** 443
molds for plastics and rubber, use for **M3:** 547
normalizing **A4:** 715
normalizing temperature **M4:** 565-566, 611
normalizing temperatures **M3:** 433-434
processing **A4:** 719
service characteristics **A4:** 719 **M3:** 438-440 **M4:** 570-572
tempering **A4:** 716 **M4:** 567-569, 570-572
P21, composition **A16:** 710
P21, grindability **A16:** 728, 732
P21, machining **A16:** 274, 710, 713-725
P30, grindability **A16:** 728
P40, grindability **A16:** 728
S group machining **A16:** 710, 713-716, 721, 722, 728, 731
S1
annealing **A4:** 715, 736 **M4:** 565-566, 585
annealing temperatures **M3:** 433-434
applications **A6:** 930
austenitizing **A4:** 737
boriding **A4:** 445
carbonitriding **A4:** 737
coining dies, use for **M3:** 509-510
cold extrusion tools, use for **M3:** 518
composition **A4:** 712, 736 **M3:** 422-423, 490 **M4:** 562-564, 587-588
composition of weld deposits **A6:** 675
deep drawing dies, use for **M3:** 498
density **A4:** 720, 763 **M3:** 441 **M4:** 617
description **A6:** 930
endothermic-atmosphere dew point for hardening **A4:** 729 **M4:** 579
hardening **A4:** 716, 736 **M4:** 567-569, 570-572, 585
hardening and tempering **M3:** 435-440
machinability rating **M3:** 443
mechanical properties **M3:** 431
normalizing **A4:** 715
pack annealing **A4:** 737
press forming dies, use for **M3:** 492
processing **A4:** 719
proven applications for borided ferrous materials **A4:** 445
quenching temperature **M4:** 587-588
service characteristics **A4:** 719 **M3:** 438-400 **M4:** 570-572
shear blades, service data **M3:** 480

Tool steels, specific types (continued)
tempering A4: 716, 738, 739 M4: 567-569, 570-572
thermal expansion A4: 720, 763 M3: 441 M4: 617
time allowable between quenching and tempering M4: 586
welding preheat and interpass temperatures A6: 675
S1, composition A16: 709
S1, cracking and spalling after regrinding A11: 569
S1, machining A16: 274, 712, 717-720, 723, 724
S2
annealing A4: 715, 736 M4: 565-566, 585
annealing temperatures M3: 433-434
austenitizing A4: 737
blades, cold shearing, use for M3: 478
coining dies, use for M3: 510
composition...... A4: 712 M3: 422-423 M4: 562-564, 587-588
density A4: 720, 763 M3: 441 M4: 617
endothermic-atmosphere dew point for hardening A4: 729 M4: 579
hardening A4: 716, 736 M4: 567-569, 570-572, 585
hardening and tempering M3: 425, 435-440
machinability rating M3: 443
normalizing A4: 715
pack annealing A4: 737
processing A4: 719
quenching temperature M4: 587-588
service characteristics..... A4: 719 M3: 438-440 M4: 570-572
tempering A4: 716, 738, 739 M4: 567-569, 570-572
thermal expansion A4: 720, 763 M3: 441 M4: 617
time allowable between quenching and tempering M4: 586
wear, cold shearing blades M3: 479
S2, composition A16: 709
S2, cutoff band sawing with bimetal blades A6: 1184
S2, for band sawing A16: 360, 362
S2, for broaches A16: 207, 717, 743, 774
S2, for circular sawing A16: 725, 756, 794, 868
S2, for drills A16: 658, 718, 746, 749, 776
S2, for end mills A16: 325, 326, 663, 664, 790, 866
S2, for face mills A16: 323, 662, 788
S2, for gear shaving A16: 342, 343
S2, for hobs A16: 345, 346
S2, for peripheral end mills A16: 786-787, 866
S2, for reamers A16: 659, 719, 750, 751, 781, 862
S2, for shaping tools A16: 347, 348
S2, for slab milling cutters A16: 324
S2, for taps A16: 661, 720, 782, 862
S2, for thread grinding tools A16: 274
S2, for thread milling cutters A16: 752, 867
S2, for trepanning tools A16: 179
S2, machining A16: 719, 720, 723, 724
S3
composition............. A4: 712 M4: 562-564, 587-588
quenching temperature M4: 587-588
tempering A4: 738, 739
time allowable between quenching and tempering M4: 586
S3, composition M3: 424
S3, for drills A16: 658, 718, 747, 749, 776, 777
S3, for reamers A16: 659, 719, 750, 751, 781, 862
S3, for taps A16: 661, 720, 782, 862
S4
annealing A4: 736 M4: 585
austenitizing A4: 737
cold extrusion tools, use for M3: 518
composition............ A4: 712 M3: 424 M4: 562-564, 587-588
hardening A4: 736 M4: 585
pack annealing A4: 737

Tool steels, specific types (continued)
quenching temperature M4: 587-588
tempering A4: 738, 739
time allowable between quenching and tempering M4: 586
wear, cold shearing blades M3: 479
S4, for boring tools A16: 771, 813, 859
S4, for broaches A16: 207, 717, 743, 774
S4, for circular sawing A16: 725, 756, 794, 868
S4, for counterbores A16: 660, 752, 860
S4, for end mills A16: 325, 326, 663, 790, 866
S4, for face mills A16: 323, 662, 788-789
S4, for gear shaving A16: 342, 343
S4, for hobs A16: 345, 346
S4, for peripheral end mills A16: 786-787, 866
S4, for planing tools A16: 657, 773
S4, for reamers A16: 575, 659, 719, 781, 862
S4, for shaping tools A16: 347, 348
S4, for slab milling cutters A16: 324
S4, for spotfacing tools A16: 6, 752, 860
S4, for thread milling cutters A16: 752, 867
S4, for trepanning tools A16: 179
S4, for turning tools A16: 144, 147, 653, 741, 811
S5
annealing A4: 715, 736 M4: 565-566, 585
annealing temperatures M3: 433-434
applications A6: 930
austenitizing A4: 737
blades, cold shearing, use for M3: 478
blades, rotary, use for M3: 479
coining dies, use for M3: 509-510
cold extrusion tools, use for M3: 518
composition...... A4: 712 M3: 422-423 M4: 562-564, 587-588
composition of weld deposits A6: 675
cutoff band sawing with bimetal blades A6: 1184
density A4: 720, 763 M3: 441 M4: 617
description A6: 930
hardening A4: 717, 736 M4: 567-569, 570-572, 585
hardening and tempering M3: 435-440
machinability rating M3: 443
mechanical properties M3: 431
normalizing A4: 715
pack annealing A4: 737
press forming dies, use for M3: 492
processing A4: 719
quenching temperature M4: 587-588
service characteristics..... A4: 719 M3: 438-440 M4: 570-572
shear blades, service data M3: 480
structural components, use for M3: 559
tempering A4: 715, 738, 739 M4: 567-569, 570-572
thermal expansion A4: 720, 763 M3: 441 M4: 617
time allowable between quenching and tempering M4: 586
welding preheat and interpass temperatures A6: 675
S5, composition A16: 709
S5, cracked during heat treatment A11: 570
S5, for boring tools A16: 771, 813, 859
S5, for contour band sawing A16: 362
S5, for counterbores A16: 660, 752, 860
S5, for cutoff band sawing A16: 360
S5, for end mills A16: 325, 326, 663, 664, 790, 866
S5, for gear shaving A16: 342, 343
S5, for hobs A16: 345, 346
S5, for peripheral end mills A16: 786-787, 866
S5, for planing tools A16: 657-773
S5, for shaping tools A16: 347, 348
S5, for spotfacing tools A16: 660, 752, 860
S5, for trepanning tools A16: 179
S5, for turning tools A16: 144, 147, 653, 741, 811
S5, fractured pin and gripping cani..... A11: 573, 576
S5, machining A16: 718-720, 723, 724, 861

Tool steels, specific types (continued)
S6
annealing A4: 736 M4: 585
blades, cold shearing, use for M3: 478
blades, rotary, use for M3: 479
coining dies, use for M3: 509-510
composition....... A4: 712 M3: 422-423 M4: 562-564
density A4: 720, 763 M3: 441 M4: 617
hardening A4: 736 M4: 570-572, 585
hardening and tempering M3: 438-440
machinability rating M3: 443
processing A4: 719
service characteristics..... A4: 719 M3: 438-440 M4: 570-572
tempering M4: 570-572
thermal expansion A4: 720, 763 M3: 441 M4: 617
S6, composition A16: 709
S6, machining A16: 274, 717-720, 723-725
S7
annealing A4: 715, 736 M4: 565-566, 585
annealing temperatures M3: 433-434
austenitizing A4: 737
blades, cold shearing, use for M3: 478
blades, rotary, use for M3: 479
cold extrusion tools, use for M3: 518
composition....... A4: 712 M3: 422-423 M4: 562-564
composition of weld deposits A6: 675
density A4: 720, 763 M3: 441 M4: 617
gas nitriding A4: 387
hardening A4: 716, 736 M4: 567-569, 570-572, 585
hardening and tempering M3: 435-440
machinability rating M3: 443
mechanical properties M3: 431
molds for plastics and rubber, use for M3: 547
normalizing A4: 715
press forming dies, use for M3: 492
processing A4: 719
service characteristics..... A4: 719 M3: 438-440 M4: 570-572
structural components, use for M3: 559
tempering A4: 716 M4: 567-569, 570-572
thermal expansion A4: 720, 763 M3: 441 M4: 617
welding preheat and interpass temperatures A6: 675
S7, composition A16: 709
S7, cracked plastic mold die A11: 568
S7, ductile fracture, torsional overload A11: 85, 90
S7 jewelry striking die, effects of excessive carburization A11: 571-572
S7, machining A16: 274, 712, 717-720, 723-725
S7 punch, excessive carburization of.... A11: 572-574
S7, quench cracked A11: 566
S7, quench cracking from stamp mark A11: 570
S9, for broaches A16: 207, 743
S9, for counterbores A16: 752, 860
S9, for drills A16: 658, 718, 749
S9, for end mills A16: 325, 326, 866
S9, for face mills A16: 323, 753, 863
S9, for slab milling cutters A16: 324
S9, for spotfacing tools A16: 752, 860
S9, for trepanning tools A16: 179, 860
S9, for turning tools A16: 144, 653, 741, 858
S9, machining A16: 717, 719
S10, for counterbores A16: 752, 860
S10, for drills A16: 718, 749, 860
S10, for end mills A16: 325, 326
S10, for face mills A16: 323, 863
S10, for slab milling cutters A16: 324
S10, for spotfacing tools A16: 752, 860
S10, for trepanning tools A16: 179, 860
S10, for turning tools A16: 144, 741, 858
S10, machining A16: 718-719
S11, for broaches A16: 207, 717, 743
S11, for counterbores A16: 752, 860

Tool steels, specific types (continued)
S11, for drills **A16:** 658, 749, 860
S11, for end mills **A16:** 325, 326
S11, for face mills **A16:** 323, 753, 863
S11, for hobs **A16:** 345, 346
S11, for spotfacing tools **A16:** 752, 860
S11, for trepanning tools **A16:** 179, 860
S11, for turning tools **A16:** 144, 653, 741, 858
S11, machining **A16:** 717, 719
S12, for counterbores **A16:** 752, 860
S12, for drills **A16:** 718, 749, 860
S12, for end mills **A16:** 325, 326
S12, for face mills **A16:** 323, 863
S12, for slab milling cutters **A16:** 324
S12, for spotfacing tools **A16:** 752, 860
S12, for trepanning tools **A16:** 179, 860
S12, for turning tools **A16:** 144, 741, 858
S12, machining **A16:** 717, 719
SAE J438b: type W108, composition **A16:** 710
SAE J438b: type W108, machining **A16:** 274, 717-720, 723-725
SAE J438b: type W109, composition **A16:** 710
SAE J438b: type W109, machining **A16:** 274, 717-720, 723-725
SAE J438b: type W110, composition **A16:** 710
SAE J438b: type W110 machining **A16:** 274, 717-720, 723-725
SAE J438b: type W112, composition **A16:** 710
SAE J438b: type W112 machining **A16:** 274, 717-720, 722-725
SAE J438b: type W209, composition **A16:** 710
SAE J438b: type W209 machining **A16:** 274, 717-720, 723-725
SAE J438b: type W210, composition **A16:** 710
SAE J438b: type W210 machining **A16:** 274, 717-720, 723-725
SAE J438b: type W310, composition **A16:** 710
SAE J438b: type W310 machining **A16:** 274, 717-720, 723-725
T group, grindability **A16:** 729, 732
T1
annealing **A4:** 715, 750 **M4:** 565-566, 600
annealing temperatures **M3:** 433-434
applications **A6:** 930
carbon content after nitriding **A4:** 753
carbon content, nitrided **M4:** 604
composition **A4:** 712 **M3:** 422-423 **M4:** 562-564
composition of weld deposits **A6:** 675
cutoff band sawing with bimetal blades **A6:** 1184
cutting tools, use for **M3:** 473, 476, 477
density **A4:** 720, 763 **M3:** 441 **M4:** 617
description **A6:** 930
distortion after heat treatment **A4:** 617
endothermic-atmosphere dew point for hardening **A4:** 729 **M4:** 579
gas quenching **A4:** 105, 106 **M4:** 59
hardening **A4:** 716, 750, 753, 755 **M4:** 567-569, 570-572, 600
hardening and tempering **M3:** 435-440
hardness gradients **M4:** 606
hot extrusion tools, use for **M3:** 538, 540
liquid nitriding **A4:** 752-753
machinability rating **M3:** 443
mechanical properties **A4:** 752
nitriding time, effect on nitrogen content **M4:** 604
normalizing **A4:** 715
processing **A4:** 718
service characteristics **A4:** 718 **M3:** 438-440 **M4:** 570-572
tempering **A4:** 716, 752 **M4:** 567-569, 570-572
tempering, effect on mechanical properties **M4:** 602
thermal conductivity **A4:** 721 **M3:** 442 **M4:** 573
thermal expansion **A4:** 720, 763 **M3:** 441 **M4:** 617
welding preheat and interpass temperatures **A6:** 675
T1 bearing, hardness, grain size, and retained austenite variations in **A11:** 509
T1, composition **A16:** 52, 709
T1, for cutoff band sawing **A16:** 360
T1, for drills **A16:** 218
T1, for planing tools **A16:** 183
T1, for reamers **A16:** 58, 240
T1, for shaping tools **A16:** 190
T1, for taps **A16:** 261
T1, for turning tools **A16:** 148, 150, 740
T1, for-milling cutters **A16:** 314
T1, grinding **A16:** 727, 728
T1, hot hardness **A16:** 53
T1, machining **A16:** 710, 716-718, 721, 723-725
T1, tool bit applications **A16:** 57, 59
T1, tool steel selection **A16:** 712
T1, wear resistance **A16:** 57
T2
annealing **A4:** 715, 750 **M4:** 565-566, 600
annealing temperatures **M3:** 433-434
composition **A4:** 712 **M3:** 422-423 **M4:** 562-564
composition of weld deposits **A6:** 675
cutoff band sawing with bimetal blades **A6:** 1184
density **A4:** 720, 763 **M3:** 441 **M4:** 617
hardening **A4:** 716, 750 **M4:** 567-569, 570-572, 600
hardening and tempering **M3:** 435-440
normalizing **A4:** 715
processing **A4:** 718
service characteristics **A4:** 718 **M3:** 438-440 **M4:** 570-572
tempering **A4:** 716 **M4:** 567-569, 570-572
thermal expansion **A4:** 720, 763 **M3:** 441 **M4:** 617
welding preheat and interpass temperatures **A6:** 675
T2, composition **A16:** 709
T2, for broaches **A16:** 203
T2, for cutoff band sawing **A16:** 360
T2, machining **A16:** 713-718, 720-725
T3, composition **A4:** 712 **M3:** 424 **M4:** 562-564
T3, grindability **A16:** 728
T4
annealing **A4:** 715, 750 **M4:** 565-566, 600
annealing temperatures **M3:** 433-434
composition **A4:** 712 **M3:** 422-423 **M4:** 562-564
composition of weld deposits **A6:** 675
cutting tools, use for **M3:** 471
density **A4:** 720, 763 **M3:** 441 **M4:** 617
hardening **A4:** 716, 750 **M4:** 567-569, 570-572, 600
hardening and tempering **M3:** 435-440
machinability rating **M3:** 443
normalizing **A4:** 715
processing **A4:** 718
service characteristics **A4:** 718 **M3:** 438-440 **M4:** 570-572
tempering **A4:** 716 **M4:** 567-569, 570-572
thermal expansion **A4:** 720, 763 **M3:** 441 **M4:** 617
welding preheat and interpass temperatures **A6:** 675
T4, composition **A16:** 51, 52, 709
T4, for broaches, for heat-resistant alloys **A16:** 743
T4, grindability **A16:** 728
T4, hot hardness **A16:** 57-58
T4, machining **A16:** 710, 713-721, 723
T4, wear resistance **A16:** 57-58
T5
annealing **A4:** 715, 750 **M4:** 565-566, 600
composition **A4:** 713 **M3:** 422-423 **M4:** 562-564
cutting tools, use for **M3:** 471
density **A4:** 720, 763 **M3:** 441 **M4:** 617
hardening **A4:** 716, 750 **M4:** 567-569, 570-572, 600
hardening and tempering **M3:** 435-440
normalizing **A4:** 715
processing **A4:** 718
service characteristics **A4:** 718 **M3:** 438-440 **M4:** 570-572
tempering **A4:** 716 **M4:** 567-569, 570-572
thermal expansion **A4:** 720, 763 **M3:** 441 **M4:** 617
T5, composition **A16:** 52, 709
T5, for boring tools **A16:** 164, 716
T5, for broaches **A16:** 203, 743, 744
T5, for turning tools **A16:** 715, 770, 838
T5, hot hardness **A16:** 57-58
T5, machining **A16:** 710, 713-725, 727
T5, tool steel selection **A16:** 712
T5, wear resistance **A16:** 57-58
T6
annealing **A4:** 715, 750 **M4:** 565-566, 600
annealing temperatures **M3:** 433-434
composition **A4:** 713 **M3:** 422-423 **M4:** 562-564
density **A4:** 720, 763 **M3:** 441 **M4:** 617
hardening **A4:** 716, 750 **M4:** 567-569, 570-572, 600
hardening and tempering **M3:** 435-440
normalizing **A4:** 715
processing **A4:** 719
service characteristics **A4:** 719 **M3:** 438-440 **M4:** 570-572
tempering **A4:** 716 **M4:** 567-569, 570-572
thermal expansion **A4:** 720, 763 **M3:** 441 **M4:** 617
T6, composition **A16:** 52, 709
T6, cutoff band sawing with bimetal blades **A6:** 1184
T6, for contour band sawing **A16:** 362
T6, for cutoff band sawing **A16:** 360
T6, for planing tools **A16:** 183
T6, machining **A16:** 713-725
T7, composition **A4:** 713 **M3:** 424 **M4:** 562-564
T7, machining **A16:** 713-716, 721, 722
T8
annealing **A4:** 715, 750 **M4:** 565-566, 600
annealing temperatures **M3:** 433-434
composition **A4:** 713 **M3:** 422-423 **M4:** 562-564
density **A4:** 720, 763 **M3:** 441 **M4:** 617
hardening **A4:** 716, 750 **M4:** 567-569, 570-572, 600
hardening and tempering **M3:** 435-440
normalizing **A4:** 715
processing **A4:** 718
service characteristics **A4:** 718 **M3:** 438-440 **M4:** 570-572
tempering **A4:** 716 **M4:** 567-569, 570-572
thermal expansion **A4:** 720, 763 **M3:** 441 **M4:** 617
T8, composition **A16:** 52, 709
T8, cutoff band sawing with bimetal blades **A6:** 1184
T8, for contour band sawing **A16:** 362
T8, for cutoff band sawing **A16:** 360
T8, hot hardness **A16:** 57-58
T8, machining **A16:** 713-725
T8, wear resistance **A16:** 57-58
T9, composition **A4:** 713 **M3:** 424 **M4:** 562-564
T9, grindability **A16:** 728
T9, machining **A16:** 713-715, 718, 721
T15
annealing **A4:** 715, 750 **M4:** 565-566, 600
annealing temperatures **M3:** 433-434
applications **A6:** 930
blades, cold shearing, use for **M3:** 478
cold extrusion tools, use for **M3:** 516, 519
composition **A4:** 713 **M3:** 422-423 **M4:** 562-564
cutoff band sawing with bimetal blades **A6:** 1184
cutting tools, use for **M3:** 471, 472, 473, 474, 475, 476
density **A4:** 720, 763 **M3:** 441 **M4:** 617
description **A6:** 930
gardening and tempering **M3:** 435-440, 442
grinding ratio **M3:** 443
hardening **A4:** 716, 750 **M4:** 567-569, 570-572, 600
machinability rating **M3:** 443
mechanical properties **M3:** 445
normalizing **A4:** 715
press forming dies, use for **M3:** 492
processing **A4:** 718
service characteristics **A4:** 718 **M3:** 438-440 **M4:** 570-572
structural components, use for **M3:** 558
tempering **A4:** 716 **M4:** 567-569, 570-572
thermal conductivity **A4:** 721 **M3:** 442 **M4:** 573
thermal expansion **A4:** 720, 763 **M3:** 441 **M4:** 617
T15, composition **A16:** 51, 52, 54, 709, 726, 733
T15, CPM composition **A16:** 63, 64
T15, for boring tools **A16:** 164, 716, 742
T15, for broaches **A16:** 207, 700, 743-746
T15, for counterbores **A16:** 251, 752, 860
T15, for cutoff band sawing **A16:** 360
T15, for deep-drawing dies **A18:** 634

Tool steels, specific types (continued)
T15, for drills ... **A16:** 58, 218, 224, 232, 658, 718, 746, 747, 850, 860
T15, for end mills ... **A16:** 325, 326, 722, 726, 754, 866
T15, for face mills ... **A16:** 323, 753, 863
T15, for gear cutters ... **A16:** 343
T15, for milling cutters ... **A16:** 314, 721, 864
T15, for multipoint cutting tools ... **A16:** 58, 59
T15, for peripheral end mills ... **A16:** 721, 849, 866
T15, for planing tools ... **A16:** 183
T15, for reamers ... **A16:** 240, 246, 702-703, 750-751, 852
T15, for shaping tools ... **A16:** 190
T15, for slab milling cutters ... **A16:** 324
T15, for spotfacing tools ... **A16:** 251, 752, 860
T15, for taps ... **A16:** 259, 261, 753, 825, 867
T15, for thread chasers ... **A16:** 299
T15, for trepanning tools ... **A16:** 179, 860
T15, for turning tools ... **A16:** 144, 653, 740-741, 846, 847, 858
T15, grindability ... **A16:** 726, 732
T15, grinding ratios ... **A16:** 732
T15, hardness ... **A16:** 54, 708
T15, hot hardness ... **A16:** 53
T15, machining ... **A16:** 710 , 713-715, 717-721, 723-725
T15, tool steel selection ... **A16:** 712
T15, wear resistance ... **A16:** 52, 54
W group, grindability ... **A16:** 728, 732
W group, machinability ... **A16:** 710
W1
- annealing ... **A4:** 715 **M4:** 565-566
- annealing temperatures ... **M3:** 433-434
- applications ... **A6:** 930
- blanking and piercing dies, use for ... **M3:** 485, 486, 487
- coining dies, use for ... **M3:** 509, 510
- cold extrusion tools, use for ... **M3:** 515, 516, 517, 518
- for cold heading dies ... **A18:** 627
- cold heading tools, use in ... **M3:** 512
- composition ... **A4:** 300, 712 **M3:** 422-423 **M4:** 562-564
- composition of weld deposits ... **A6:** 674
- cutoff band sawing with bimetal blades ... **A6:** 1184
- deep drawing dies, use for ... **M3:** 496, 498, 499
- for deep-drawing dies ... **A18:** 634
- density ... **A4:** 720, 763 **M3:** 441 **M4:** 617
- description ... **A6:** 930
- electron beam hardening treatment ... **A4:** 300
- electron beam hardening treatment (as C 100 W1 steel) ... **A4:** 306
- extension of tool life via ion implantation, examples ... **A18:** 643
- gages, use for ... **M3:** 555, 556
- hardening ... **A4:** 717, 764 **M4:** 567-569
- hardening and tempering ... **M3:** 426, 435-440
- hardening atmospheres ... **A4:** 735
- hot upset forging tools, use for ... **M3:** 534
- machinability rating ... **M3:** 443
- microconstituents ... **A4:** 613 **M4:** 615
- microconstituents after hardening ... **A4:** 762 **M3:** 468
- normalizing ... **A4:** 715
- normalizing temperature ... **M4:** 565-566
- normalizing temperatures ... **M3:** 433-434
- press forming dies, use for ... **M3:** 490
- processing ... **A4:** 719
- proven applications for borided ferrous materials ... **A4:** 445
- recommended for backward extrusion of two parts ... **A18:** 628
- service characteristics ... **A4:** 719 **M3:** 438-440 **M4:** 570-572
- structural components, use for ... **M3:** 558
- tempering ... **A4:** 134, 717 **M4:** 567-569, 570-572
- thermal conductivity ... **A4:** 721 **M3:** 442 **M4:** 573

Tool steels, specific types (continued)
- thermal deposition/diffusion process ... **A4:** 449
- thermal expansion ... **A4:** 720, 763 **M3:** 441 **M4:** 617
- welding preheat and interpass temperatures ... **A6:** 675
- working hardness ... **M3:** 510

W1, composition ... **A16:** 710
W1, for contour band sawing ... **A16:** 362
W1, for cutoff band sawing ... **A16:** 360
W1, for hand reamers ... **A16:** 240
W1 header die, failure from improper quenching ... **A11:** 577-578
W1, machining ... **A16:** 717-719, 720-725
W1, microstructure ... **A16:** 708, 711
W1 rough scaled surface after heat treatment ... **A11:** 573, 578
W1, seizure resistance ... **M1:** 611
W1 tool steel concrete roughers, failed ... **A11:** 565
W1, tool steel selection ... **A16:** 712
W2
- annealing ... **A4:** 715 **M4:** 565-566
- annealing temperatures ... **M3:** 433-434
- applications ... **A6:** 930
- cold extrusion tools, use for ... **M3:** 518
- cold heading tools, use in ... **M3:** 512
- composition ... **A4:** 712 **M3:** 422-423 **M4:** 562-564
- composition of weld deposits ... **A6:** 675
- density ... **A4:** 720, 763 **M3:** 441 **M4:** 617
- description ... **A6:** 930
- endothermic-atmosphere dew point for hardening ... **A4:** 729 **M4:** 579
- furnace atmosphere, effect on carbon content ... **A4:** 735 **M4:** 583
- hardening ... **A4:** 714, 764 **M4:** 567-569, 570-572
- hardening and tempering ... **M3:** 435-440
- hardening atmospheres ... **A4:** 735
- machinability rating ... **M3:** 443
- normalizing ... **A4:** 715
- normalizing temperature ... **M4:** 565-566
- normalizing temperatures ... **M3:** 433-434
- processing ... **A4:** 719
- service characteristics ... **A4:** 719 **M3:** 438-440 **M4:** 570-572
- shear blades, service data ... **M3:** 480
- stress relieving ... **A4:** 735
- tempering ... **A4:** 717 **M4:** 567-569, 570-572
- thermal expansion ... **A4:** 720, 763 **M3:** 441 **M4:** 617
- wear, cold shearing blades ... **M3:** 479
- welding preheat and interpass temperatures ... **A6:** 675

W2 carbon steel, quench cracked from soft spots ... **A11:** 570
W2, composition ... **A16:** 710
W2 die insert, fracture from rehardening ... **A11:** 575, 581
W2, for cold heading dies ... **A18:** 627
W2, for drills ... **A16:** 726
W2, for shaping tools ... **A16:** 192
W2, machining ... **A16:** 712, 714-720, 723, 724
W2, quench crack at base of threads ... **A11:** 566
W2 threaded parts, cracked during, quenching ... **A11:** 566
W2, tool steel selection ... **A16:** 712
W3
- composition ... **A4:** 712 **M3:** 424 **M4:** 562-564
- endothermic-atmosphere dew point for hardening ... **A4:** 729 **M4:** 579
- hardening ... **A4:** 717 **M4:** 567-569
- hardening and tempering ... **M3:** 435-437
- tempering ... **A4:** 717 **M4:** 567-569

W4, composition ... **A4:** 712 **M3:** 424 **M4:** 562-564
W5
- annealing ... **A4:** 715 **M4:** 565-566
- annealing temperatures ... **M3:** 433-434
- composition ... **A4:** 712 **M3:** 422-423 **M4:** 562-564
- hardening ... **M4:** 567-569, 570-572
- machinability rating ... **M3:** 443

Tool steels, specific types (continued)
- normalizing ... **A4:** 715
- normalizing temperature ... **M4:** 565-566
- normalizing temperatures ... **M3:** 433-434
- processing ... **A4:** 719
- service characteristics ... **A4:** 719 **M3:** 438-440 **M4:** 570-572
- tempering ... **M4:** 567-569, 570-572

W5, composition ... **A16:** 710
W5, for cold heading dies ... **A18:** 627
W5, machining ... **A16:** 717, 719, 720, 723-725
W6, composition ... **A4:** 712 **M3:** 424 **M4:** 562-564
W7, composition ... **A4:** 712 **M3:** 424 **M4:** 562-564

Tool steels, surface treatment
- carburizing ... **M4:** 619
- nitriding ... **M4:** 619
- oxide coatings ... **M4:** 619
- plating ... **M4:** 619
- sulfide treatment ... **M4:** 619

Tool steels, surface treatments
- carburizing ... **M3:** 446
- nitriding ... **M3:** 446
- oxide coatings ... **M3:** 446
- plating ... **M3:** 446
- sulfide ... **M3:** 446

Tool steels, W1
- case-hardened layer as OM macrograph ... **A10:** 303

Tool steels, water hardening ... **A4:** 734-736
- annealing ... **A4:** 734, 735
- atmospheres ... **A4:** 735, 736
- austenitizing temperatures ... **A4:** 734, 735, 736
- embrittlement ... **A4:** 736
- fluidized-bed furnaces ... **A4:** 735
- hardening ... **A4:** 734, 736
- lead baths ... **A4:** 735
- normalizing ... **A4:** 734-735
- quenching ... **A4:** 735-736
- salt baths ... **A4:** 735
- stress relieving ... **A4:** 735
- tempering ... **A4:** 736

Tool steels, water-hardening
- annealing ... **M4:** 582
- atmospheres ... **M4:** 583
- austenitizing temperatures ... **M4:** 582, 583
- hardening ... **M4:** 581, 582
- lead baths ... **M4:** 583
- normalizing ... **M4:** 581-582
- quenching ... **M4:** 582, 583
- salt baths ... **M4:** 583, 585
- stress relieving ... **M4:** 582-583
- tempering ... **M4:** 583-584

Tool wear
- chip motion strain ... **A16:** 38
- chip velocity ... **A16:** 38-39
- coatings improving high-speed tool steels ... **A16:** 51
- and cutting fluids ... **A16:** 392
- cutting fluids and wear zone temperatures ... **A16:** 39
- and cutting speed relationship ... **A16:** 645, 646
- cutting velocity ... **A16:** 38-39
- and machinability ... **A16:** 640, 642, 643, 644, 645
- machine, cutting tool, and tool wear interactions ... **A16:** 41-42
- monitoring systems ... **A16:** 411-417
- primary shear zone ... **A16:** 39
- secondary shear zone ... **A16:** 38
- shear velocity ... **A16:** 38-39
- surface shear stresses ... **A16:** 38
- Taylor's tool life tests ... **A16:** 644
- temperatures in the wear zones ... **A16:** 39
- tool life equation ... **A16:** 45-46
- tool life testing ... **A16:** 43-47
- variations in cross section affecting ... **A16:** 390
- volumetric chip removal rate ... **A16:** 38
- wear environment ... **A16:** 38-39, 41-42
- wear mechanisms ... **A16:** 39-40, 45

Tool wear in cold/hot compacting **A7:** 480
and lubrication **A7:** 190
Tool wear mechanisms
attrition wear/built-up edge **A2:** 954
cemented carbides **A2:** 954-955
crater wear **A2:** 954
depth-of-cut notching **A2:** 955
flank/abrasive wear **A2:** 954
thermal fatigue **A2:** 954-955
Tool(s) *See also* Tool and die(s); Tool steel alloys; Tool steels; Tooling
applications, cutting, cermets for **A2:** 978
applications, microcrystalline alloys for **A7:** 796
carbide metal cutting **A2:** 962-965
of cermets **A2:** 978
coal mining, cemented carbide **A2:** 975-976
components, rigid, lubrication and **A7:** 190
cutoff **A2:** 965
diamond cutting **A2:** 976
ditching, cemented carbide **A2:** 974
drilling, cemented carbide **A2:** 974
drills, carbide metal cutting **A2:** 964-965
edge preparation **A2:** 964
end mills **A2:** 964-965
forestry, cemented carbide **A2:** 974
geometry, polycrystalline cubic boron nitride (PCBN) tools **A2:** 1016-1017
grooving **A2:** 965
indexable carbide inserts **A2:** 963
for infiltration **A7:** 563
life, coated carbide tools **A2:** 960
machining, for refractory metals and alloys **A2:** 560-562
motions, advanced **A7:** 327-328
nose deformation versus vanadium carbide content, cutting tool materials **A2:** 999
performance, of CAP material **A7:** 535
thread milling **A2:** 965
threading **A2:** 965
trenching, cemented carbide **A2:** 973
Tool-chip interface mean temperature **A18:** 612
Toolholding
carbide metal cutting tools **A2:** 962-965
chipbreaking **A2:** 963-964
cutoff **A2:** 965
drills **A2:** 964-965
edge preparation **A2:** 964
end mills **A2:** 964-965
indexable carbide inserts **A2:** 963
threading **A2:** 965
Tooling *See also* Equipment; Tools **A7:** 12
adapters **A7:** 336, 337
alignment **A14:** 160
for aluminum alloy extrusion **A14:** 308-309
for aluminum alloy precision forgings **A14:** 252-253
with aluminum honeycomb structure **EM1:** 728
for autoclave molding **EM1:** 578-581
for beryllium forming **A14:** 805-806
blow molding, costs **EM2:** 299
clearance and finish materials **A7:** 337
cold extruded copper/copper alloy parts **A14:** 310
for cold extrusion **A14:** 302-303, 310
for cold isostatic pressing **A7:** 445-447
for composite fastener holes **EM1:** 712-715
composite, properties **EM1:** 703
compression molding, costs **EM2:** 298
for compression molding/stamping **EM2:** 331-333
configuration, radial forging **A14:** 17
for contour roll forming **A14:** 628-630
contoured, tape prepreg machine **EM1:** 145
costs **EM2:** 82
in curing process **EM1:** 702-703
data base **A14:** 413
deflection analysis **A7:** 336
design **A7:** 335-337, 668
design effects **EM1:** 430
diamond **EM1:** 295
for dimensional change **A7:** 480
for drop hammer forming **A14:** 654-655
effects on design process **EM1:** 428-431
for efficiency **A14:** 556
ejection, dimensional changes **A7:** 480

Tooling (continued)
elastic deflection **A14:** 161
elastomeric **EM1:** 590-601
elastomeric, defined *See* Elastomeric tooling
electroformed nickel **EM1:** 582-585
for electromagnetic forming **A14:** 646
of epoxy resin matrices **EM1:** 76, 77
for explosive plate forming **A14:** 641
for explosive sheet forming **A14:** 640
for explosive tube forming **A14:** 641
for fabrication with rapid omnidirectional compaction **A7:** 546
feature, defined **EL1:** 1159
fiberglass-epoxy-laminated **EM1:** 582-585
filament winding **EM2:** 374-377
flat, tape prepreg machine **EM1:** 145
graphite-epoxy **EM1:** 586-589
hand lay-up, spray-up, prepreg molding **EM2:** 341-343
heat transfer to **A14:** 161
for hemispherical dome tests **A8:** 561
holes, defined **EL1:** 1159
hot extrusion **A14:** 320-321
in hot isostatic pressing unit **A7:** 423
for hot upset forging **A14:** 88-94
hydrostatic extrusion **A14:** 329
impact extrusion **A14:** 311
injection molding, costs **EM2:** 295-296
layout **A7:** 335
manufacturing effects **EM1:** 430
master models **EM1:** 738-739
materials **A7:** 337
materials, coefficients of thermal expansion **EM1:** 428
materials, performance **EM1:** 586
materials, thermal characteristics **EM1:** 578
for multiple-slide rotary forming machines **A14:** 574
multiple-station **A14:** 303
pattern, investment casting **A15:** 256-257
personnel, interfaces with design/manufacturing personnel **EM1:** 428-431
for powder forging **A14:** 207
for precision aluminum forgings **A14:** 252
precision, tolerance bands **A14:** 160
for preform **A14:** 175
prepregs, properties **EM1:** 587
and process selection **EM2:** 278
production **A7:** 329, 332-337
prototype and temporary, investment casting **A15:** 265
pultrusion **EM1:** 536 **EM2:** 392
quality control **EM1:** 738-739
for redrawing **A14:** 585
for Replicast process **A15:** 270-271
resin transfer molding **EM1:** 169, 566
resin transfer molding, costs **EM2:** 301
rotational molding **EM2:** 367
for round tube reshaping **A14:** 632
setup, physical modeling **A14:** 435
setups, cold extrusion **A14:** 303
for superplastic forming **A14:** 860
support adapters **A7:** 337
for Swift-flat-bottomed cup test **A8:** 563
systems, of high-production P/M compacting presses **A7:** 332-334
temperature, precision forging **A14:** 162
temperatures **A15:** 324
thermal gradient **A14:** 162
thermoforming, costs **EM2:** 301
thermoplastic injection molding, costs **EM2:** 309
for transverse-rupture test **A7:** 290
wear, precision forging **A14:** 159
Tooling data base **A14:** 413
Tooling marks **A14:** 13, 840
Tooling quality control
composite tools **EM1:** 739
hand-faired master models **EM1:** 738
machined master models **EM1:** 738
second-generation patterns **EM1:** 738-739
Tooling resin *See also* Epoxy plastic; Resins; Silicone plastics
defined **EM1:** 24 **EM2:** 43
Toolmakers' microscope
described **A17:** 10
Toolroom lathes **A16:** 153

Tools *See also* Auxiliary equipment; Closed-die forging tools; Equipment; Gages; specific types: reamers, drills, etc; Tool failures; Tool steels; Tool steels, types; Tooling; Tools and dies, failures of; Wiredrawing dies
assembly components, cold extrusion **A14:** 302
for bar bending **A14:** 663
for bending **A14:** 663, 665-666
characteristics **A11:** 563
chemical vapor deposition coating of **M5:** 382-384
for cold heading **A14:** 292-293
for component removal **EL1:** 724-727
design **A14:** 197, 474 **EL1:** 419
and dies, failures of **A11:** 563-585
dulling, piercing effects **A14:** 461
for fine-edge blanking and piercing **A14:** 474-475
for Guerin process **A14:** 605-606
hard chromium plating of **M5:** 170-171, 177
for heat-resistant alloy forming **A14:** 781
hot extrusion, materials and hardnesses of **A14:** 320
lead frame stamping **EL1:** 486
life **A14:** 139, 463
for liquid penetrant inspection **A17:** 86
for manual spinning **A14:** 600
multiple-slide forming **A14:** 571
for nickel-base alloy forming **A14:** 832
operation, and failure **A11:** 564
for piercing **A14:** 464-465
for power spinning of cones **A14:** 602
process control, stress modeling as **EL1:** 446
rehardened high-speed steel **A11:** 574
rolling **A14:** 121
for rotary swaging **A14:** 133-134
rust-preventive compounds for **M5:** 465
setup, and failure **A11:** 564
shape, cold worked **A14:** 38
size, and clearance, in piercing **A14:** 462-463
steel cutter die, grinding damage **A11:** 567, 569
for tube piercing **A14:** 470
for tube spinning **A14:** 676
upsetter heading, types **A14:** 85-86
Verson-hydroform process **A14:** 612
warm heading **A14:** 297
Tools and dies, failures of- *See also* Dies; Tools **A11:** 563-585
Tools, farm
testing of **A8:** 605
Tooth (gear)
adapter, overload failure of **A11:** 398
-bending fatigue **A11:** 590-592
-bending impact, gear failures from **A11:** 595
chipping, gear failure from **A11:** 595
shear, gear failure from **A11:** 595
Tooth and bone implants
porous materials for **A7:** 700
Tooth involute form (TIF) **A18:** 565
Tooth tissues
thermal expansion rate **A7:** 611
Tooth-spur gear
military vehicular equipment **A7:** 687
Top gating
permanent mold casting **A15:** 279
"Top hat" tensile test **A6:** 163
Top inspection *See* Ultrasonic inspection
Top-braze flatpack/quadpack
as surface mount option **EL1:** 77
Top-cooling heat sink
multichip structures **EL1:** 309
Top-hat tensile test apparatus
and punch-die tooling components **A6:** 162
Topaz
hardness **A18:** 433
on Mohs scale **A8:** 108
TOPAZ forging process analysis tool **A14:** 412
Topogically close-packed phases ... **M3:** 209, 223-229, 278
Topographic contrast
scanning electron microscopy **A9:** 93, 99
wrought stainless steels phase identification **A9:** 282
Topographic details revealed under interference-contrast illumination **A9:** 79
Topographic image **A18:** 346, 348, 351-352
Topographic index
as surface roughness parameter **A12:** 201

Topographs
effect of surface relief in ... **A10:** 369
Lang, of dislocations ... **A10:** 370
projection or traverse ... **A10:** 369
Topography *See also* Microtopography
defined ... **A10:** 366
Lang section ... **A10:** 368
neutron, capabilities ... **A10:** 365
projection ... **A10:** 369
reflection ... **A10:** 368-369
single-crystal and x-ray ... **A10:** 330-331
surface, by magnetic rubber inspection ... **A17:** 125
transmission ... **A10:** 369-370
x-ray ... **A10:** 365-379
Topologically close-packed
abbreviation for ... **A10:** 691
Topologically close-packed (TCP) phases ... **A6:** 572
Topologically close-packed (TCP)-type phases ... **A1:** 952
Topologically close-packed phases in wrought heat-resistant alloys
anodic dissolution to extract ... **A9:** 308
electrolytic extraction and x-ray diffraction ... **A9:** 308
Topothesy ... **A18:** 352, 353
Topper-Sandor equivalent stress parameter
Goodman diagram for ... **A8:** 713
Toppets
materials for ... **M1:** 630
Topside coating systems
marine corrosion ... **A13:** 913-914
Torch
cutting, for sampling ... **A10:** 16
Fassel, for analytic ICP systems ... **A10:** 36-37
inductively coupled plasma ... **A10:** 34, 36-37
mini-, for the ICP ... **A10:** 37
Torch (flame)
relative rating of brazing process heating method ... **A6:** 120
Torch brazing
of aluminum alloys ... **M6:** 1029-1030
of copper and copper alloys ... **M6:** 1037-1042
definition ... **M6:** 18
of stainless steels ... **M6:** 1010-1012
of steels *See* Torch brazing of steels
Torch brazing (TB) ... **A6:** 328-329
advantages ... **A6:** 328
aluminum ... **A6:** 328
aluminum alloys ... **A6:** 939
applications ... **A6:** 328
brass ... **A6:** 328
carbides ... **A6:** 328
copper ... **A6:** 328
copper alloys ... **A6:** 328
copper and copper alloys ... **A6:** 933-934
definition ... **A6:** 328, 1214
equipment ... **A6:** 328-329
filler metals ... **A6:** 328, 329
fluxes ... **A6:** 328
fuel gases ... **A6:** 328
heat-resistant alloys ... **A6:** 328
limitations ... **A6:** 328
manual ... **A6:** 121, 122
precious metals ... **A6:** 936
safety precautions ... **A6:** 329, 1191
stainless steel ... **A6:** 328
steel ... **A6:** 328
suggested viewing filter plates ... **A6:** 1191
techniques ... **A6:** 329
Torch brazing of steels ... **M6:** 950-964
comparison to braze welding ... **M6:** 952
equipment for automatic torch brazing ... **M6:** 958-960
paste feeders ... **M6:** 959
production examples ... **M6:** 959-960
Torch brazing of steels (continued)
wire feeders ... **M6:** 959
equipment for machine torch brazing ... **M6:** 955-958
conveyor belts ... **M6:** 955
production examples ... **M6:** 955-958
turntables ... **M6:** 955
equipment for manual torch brazing ... **M6:** 952-954
gas savers ... **M6:** 954
gas-fluxing equipment ... **M6:** 954
mixing-chamber design ... **M6:** 952-953
multiple-flame tips ... **M6:** 953
torch tips ... **M6:** 953
torches ... **M6:** 952
feeding filler metal ... **M6:** 951
filler metals ... **M6:** 950, 961
copper-zinc ... **M6:** 961
silver alloy ... **M6:** 961
fixtures for manual torch brazing ... **M6:** 954-955
flux removal after brazing ... **M6:** 963-964
boric acid ... **M6:** 963
fluoride fluxes ... **M6:** 963-964
mixed borax and boric acid fluxes ... **M6:** 963
preparation for plating ... **M6:** 964
fluxes ... **M6:** 961-963
flux constituents ... **M6:** 962
gas fluxing ... **M6:** 962-963
types ... **M6:** 962
fluxing ... **M6:** 950-951
fuel gases ... **M6:** 950
heating ... **M6:** 951
inspection ... **M6:** 951-952
joint design ... **M6:** 950
prebraze cleaning ... **M6:** 950
principles and techniques ... **M6:** 950-952
quality control ... **M6:** 951-952
Torch nozzle
swaging of ... **A14:** 136
Torch, plasma *See* Plasma torch
Torch soldering
definition ... **M6:** 18
Torch soldering (TS) ... **A6:** 351-352
advantages ... **A6:** 351
aluminum ... **A6:** 351
applications ... **A6:** 351
brass ... **A6:** 351
common solders for ... **A6:** 351
copper ... **A6:** 351
copper alloys ... **A6:** 351, 352
definition ... **A6:** 1214
description ... **A6:** 351
equipment ... **A6:** 351
filler metals ... **A6:** 352
fluxes required ... **A6:** 351
fuel gases ... **A6:** 351-352
gold ... **A6:** 351
heating techniques, basic ... **A6:** 352
limitations ... **A6:** 351
stainless steel ... **A6:** 351, 352
steel ... **A6:** 351, 352
suggested viewing filter plates ... **A6:** 1191
Torch tip
definition ... **A6:** 1214
Torch velocity ... **A6:** 49, 50
Torch weaving ... **A6:** 54
Torches for
gas tungsten arc welding ... **M6:** 188-190
manual brazing ... **M6:** 952
oxyfuel gas welding ... **M6:** 585-586
plasma arc welding ... **M6:** 216-217
Toroidal-sector electrostatic lens
Poschenrieder analyzer as ... **A10:** 597
Toroids
defined ... **EL1:** 1160
Torpedoes
powders used ... **A7:** 573
Torpex explosives
aluminum powder containing ... **A7:** 601
Torque
balance equation ... **A8:** 226
calibration, Wheatstone-bridge circuit ... **A8:** 158
cell adapter ... **A8:** 160
conversion factors ... **A8:** 722 **A10:** 686
defined ... **A18:** 19
effect of twist rate on ... **A8:** 141-142
elastic springback method ... **A8:** 327
per unit length, conversion factors ... **A8:** 722
pulley ... **A8:** 219-220
-radius, of electrolytic tough pitch copper ... **A8:** 183-184
and roll-separating force ... **A14:** 345
sensor ... **A8:** 158
in slider-crank mechanism ... **A14:** 39
in torsional testing ... **A8:** 146
transducer ... **A8:** 158
-twist reduction, torsion testing ... **A8:** 160
Torque arm assembly, aircraft landing gear
fatigue fracture of ... **A11:** 114
Torque converter stall ratio ... **A18:** 566
Torque joints
by electromagnetic forming ... **A14:** 648
Torque rheometry
for molecular weight ... **EM2:** 534
Torque testing ... **EL1:** 954, 1160
Torque-coil magnetometer
defined ... **A10:** 683
Torque-twist curve ... **A8:** 141
for hot torsion tests, Ti alloy ... **A8:** 169-171
to measure flow localization ... **A8:** 169
in torsion, annealed ETP copper ... **A8:** 183
Torsion *See also* Torsional fatigue; Torsional loadingTorsional stress; Torsional vibration ... **EM3:** 29
applied to determine workability ... **A8:** 154-184
circular columns, formulas for ... **EM2:** 654
defined ... **A8:** 14 **A11:** 11 **A13:** 13 **A14:** 13 **EM1:** 24 **EM2:** 43
deformation, alloy steels ... **A12:** 304
ductility at various temperatures ... **A8:** 164-166
effective stress-strain curves for stainless steel in ... **A8:** 162, 164
equipment ... **A8:** 157-160, 179-180
and extrusion, microstructure of Udimet 700 in ... **A8:** 176-178
flow ... **A8:** 162-164
flow localization in ... **A8:** 169
flow softening measured in ... **A8:** 117
high-cycle fatigue, AISI/SAE alloy steels ... **A12:** 296
historical studies ... **A12:** 3
modulus of rupture in ... **A11:** 7
overload fractures failed in ... **A11:** 399
plastic, anisotropy in ... **A8:** 143
pure, elastic-stress distribution ... **A11:** 461
of round bar, inclusions ... **A8:** 155
shear strain from ... **A8:** 327
of solid circular prismatic bar ... **A8:** 139
specimen *See* Torsion specimen
as stress, fatigue fracture from ... **A11:** 75
as stress, on shafts ... **A11:** 460-461
study, of Widmanstatten alpha microstructure ... **A8:** 178
and tensile fracture strains correlated ... **A8:** 168-169
tensile tests on bars prestrained in ... **A8:** 155
and tension flow curves, for copper ... **A8:** 162-164
test, for workability ... **A14:** 373-374
of thin-walled tube, and simple shear ... **A8:** 155
ultimate shear stress in ... **A8:** 148
Torsion bars ... **M1:** 301, 303
fatigue life distribution ... **M1:** 677
Torsion effective fracture strain
for various materials ... **A8:** 168
Torsion equipment ... **A8:** 157-160, 179-180
to assess workability ... **A8:** 157-160, 179-180
Torsion flow ... **A8:** 162-164
Torsion fractures
brittle, medium-carbon steels ... **A12:** 273

Torsion fractures (continued)
elongated dimples **A12:** 16
fatigue, alloy steels **A12:** 323
fatigue, high-carbon steels **A12:** 282
fatigue, medium-carbon steels **A12:** 253
low-carbon steel **A12:** 250
medium-carbon steels **A12:** 253, 268
overload, alloy steels **A12:** 330
overload, austenitic stainless steels **A12:** 359
overload, high-carbon steels **A12:** 278
Torsion, modulus of rupture in *See* Modulus of rupture, in torsion
Torsion snap joints **EM2:** 718-719
Torsion specimen
buckling avoidance **A8:** 156
designs **A8:** 155-156
experimental and theoretical torque-twist curves for **A8:** 170
geometries for workability **A8:** 156
grip design **A8:** 155-157
with inclusions before and after twisting **A8:** 156
solid, effective fracture strain **A8:** 168
stainless steel, flow localization in torsion **A8:** 169
tubular, effective fracture strain **A8:** 168
Torsion springs **A1:** 302 **M1:** 290, 291, 293
wire for **M1:** 289
Torsion test **A1:** 582
resistance spot welds **M6:** 487-488
Torsion testing *See also* Fixed-end torsion testing; Free-end torsion testing; High-temperature torsion testing; Torsion; Torsional; Torsional fatigue testing **A8:** 139-144 **EM3:** 315
application of axial line on gage section surface **A8:** 157
applied to determine workability **A8:** 154-184
assumptions **A8:** 139
axial effects and alternate analysis methods in **A8:** 180-184
buckling in **A8:** 156
for bulk workability assessment **A8:** 577-578
control and data acquisition system for **A8:** 158
defined **A8:** 14
deformation heating in **A8:** 161
deformation-temperature-time of microalloyed steels during **A8:** 179
effect of strain rate **A8:** 141-142
effective stress-strain curves using Tresca criterion **A8:** 163
equipment **A8:** 155, 157-160
equipment, to access workability **A8:** 157-160
fixed end **A8:** 157, 180
flow curves **A8:** 161-162, 175
flow curves for stainless steel **A8:** 161-162
flow localization, stainless steel specimen **A8:** 169
for flow softening in two-phase alloys **A8:** 177
for flow stress **A8:** 154, 160-163
flow stress data **A8:** 160-163, 172
flow-localization-controlled failures **A8:** 154-155
fracture data, interpreted **A8:** 163-169
for fracture limits **A8:** 154, 163
and fracture-controlled failures **A8:** 154
fracture-strain data, stainless steel **A8:** 167-168
free-end vs. fixed-end **A8:** 181-182
gage length-to-radius ratio, effect on effective strain to failure **A8:** 165
hot, manganese sulfide content and ductility in **A8:** 166
hot, torque-twist behavior for Ti alloy **A8:** 169
hot, with stainless steel and aluminum **A8:** 163
of iron **A8:** 171
with Kolsky bar **A8:** 224-225
load-train design **A8:** 159-160
of prismatic bars **A8:** 139-141
processing history **A8:** 178-180
for recovery/recrystallization in single-phase materials **A8:** 176
reduction of shear stress/shear strain to effective stress and strain **A8:** 161
shear stress derivations for arbitrary flow laws **A8:** 182-184
for simulation die chilling effect on workability, multiphase alloys **A8:** 180
specimen design **A8:** 155-157

Torsion testing (continued)
of stainless steel **A8:** 175
stress distributions **A8:** 140
stress-relaxation **A8:** 327
study of multiphase alloy microstructures **A8:** 178
and temperature **A8:** 176, 178
and tension testing **A8:** 163, 165
and texture **A8:** 180-183
torque and **A8:** 158, 160, 183-184
torsional rotation rates in metalworking **A8:** 157-158
Torsion-dynamometer measurement
hard chromium plate thickness **M5:** 181
Torsion-shear test
for shear yield strength and modulus of rigidity **A8:** 64-65
specimen **A8:** 65
Torsional braid analysis (TBA) **EM3:** 319-320, 321, 423
Torsional deformation **A8:** 139
Torsional ductility **A8:** 164-169
Torsional fatigue *See also* Torsion
cracks, stress fields and **A11:** 471
fracture, steel valve spring **A11:** 120
in shafts **A11:** 464
steel rotor shaft failure by **A11:** 464
Torsional fatigue testing **A8:** 149-152, 371
Torsional flow
curve **A8:** 175, 176
stress data, correlation for carbon steel with Zener-Holloman parameter **A8:** 162-163
Torsional fracture
data interpretation **A8:** 163-169
strain **A8:** 155
Torsional hydraulic actuator **A8:** 216-217
Torsional impact
for high strain rate shear testing **A8:** 187
loading **A8:** 215
machine **A8:** 216-217
testing **A8:** 216-218
Torsional impact testing **A8:** 216-218
Torsional indicator
deflections in torsional testing **A8:** 147
Torsional Kolsky bar **A8:** 218-229
Torsional loading *See also* Torsion
and axial loading, testing machine for **A8:** 160
complications **A8:** 151
control, in low-cycle torsional fatigue **A8:** 150
dimensional changes **A8:** 143
effect of strain rate **A8:** 141-142
and fatigue-crack propagation **A11:** 109
fundamentals of **A8:** 139-144
Hopkinson bar **A8:** 198
overload, tool steel, ductile fracture by **A11:** 85, 90
reversed, of shafts **A11:** 525
Torsional modulus *See* Modulus of rigidity
Torsional pendulum **EM3:** 29, 319-320
defined **EM2:** 43
Torsional properties
of ductile iron **A1:** 42, 45 **A15:** 660 **M1:** 38, 41
gray iron **A15:** 644 **M1:** 18, 19
malleable iron **M1:** 70
malleable irons **A15:** 697
steel springs **M1:** 290
Torsional pulse
in Kolsky bar **A8:** 219
smoother **A8:** 224-225, 227
Torsional rotation rates **A8:** 157-158
Torsional shear strength
of gray iron **A1:** 18, 20
Torsional stiffness *See also* Stiffness. of high-modulus fiber-reinforced composites **EM1:** 36
Torsional straining
anisotropy theory applied to **A8:** 143
Torsional stress *See also* Stress(es). defined; Torsion **EM1:** 24 **EM3:** 29
defined **A8:** 14 **A11:** 11 **A13:** 13 **A14:** 13 **EM2:** 43
in shafts **A11:** 115
Torsional testing
cyclic *See* Cyclic torsional testing
fixturing examples in **A8:** 146
machine **A8:** 146
Poisson's ratio effect **A8:** 218

Torsional testing (continued)
quasi-static *See* Quasi-static torsional testing
Torsional testing machines
high-speed hydraulic **A8:** 215-216
Torsional vibration
retainer spring failure from **A11:** 561-562
Tortuosity
crack path **A12:** 38, 206
Tortuosity factor **EM4:** 70
Total *a* vs *N* curve method
crack propagation rate **A8:** 680
Total absorption
attenuation of electromagnetic radiation **A17:** 309-310
Total acid number (TAN) **A18:** 84, 300
lubricant indicators and range of sensitivities **A18:** 301
Total acid value tests
phosphate coating solutions **M5:** 442-443
Total base number (TBN) **A18:** 84, 100, 165, 166
corrosive wear and lubricant analysis **A18:** 310
Total carbon
defined **A13:** 13
Total carbon content
abbreviation for **A11:** 798
Total combustion **A10:** 223-224
Total consumption burners **A10:** 28
Total contact temperature
gears **A18:** 538, 539
Total dissolved gas
pressurized water reactor specification **A8:** 423
Total elbow joint prosthesis **A11:** 670
Total elongation *See also* Elongation; Percent elongation
components of **A8:** 26
defined **A8:** 14 **A14:** 13
of sheet metals **A8:** 555-556
to failure, analysis of **A8:** 693
Total enthalpy
defined **A15:** 50
Total friction coefficient **A18:** 35
Total frictional force
rubber **A18:** 580
Total hip and knee replacements **A7:** 657
Total hip joint prostheses
fractures of **A11:** 692-693
Total immersion tests *See also* Immersion tests **A13:** 221-222, 231
Total indicator reading
abbreviation for **A11:** 798
Total ion chromatogram
typical **A10:** 644
Total luminescence spectroscopy **A10:** 78
Total material loss **A18:** 274, 275, 276
Total nuclear magnetization
defined **A10:** 280
Total quality concept **EM3:** 785
Total quality control (TQC)
in tape automated bonding (TAB) **EL1:** 288
Total quantity of heat (Q)
resistance welding **A6:** 40, 41
Total ringer joint prosthesis **A11:** 670
Total shoulder joint prosthesis **A11:** 670
Total strain-energy release rate theory **EM4:** 700, 701
Total transfer of heat **A6:** 25
Total transmittance
defined **A10:** 683
Total unsharpness, defined
radiography **A17:** 300
Total wear rate **A18:** 275
Total-extension-under-load yield strength *See* Yield strength
Touch scanners
coordinate measuring machines **A17:** 25
Touch up
cost/efficacy of **EL1:** 695
Touch-trigger probes
coordinate measuring machines **A17:** 25
Touchdown development
of copier powders **A7:** 583
Tough pitch copper *See also* Electrolytic tough pitch copper
alloy types **A2:** 223, 230, 234
applications and properties **A2:** 269-276
SCC resistance **A13:** 615

Tough pitch copper with silver *See* Copper alloys, specific types, C11300, C11400, C11500 and C11600
Tough pitch coppers
brazing ... **A6:** 623, 628-629
no brazing ... **A6:** 931, 934
Tough-brittle transition
and impact tests ... **EM2:** 554-556
Toughened ceramics *See also* Ceramics; Structural ceramics
transformation-toughened zirconia ... **A2:** 1023
zirconia-toughened alumina (ZTA) ... **A2:** 1022-1023
Toughness *See also* Crack arrest toughness; Elastic energy; Fracture toughness; Fracture toughness of steel; Interlaminar fracture toughness; Modulus of toughness; notch toughness of steel; Plane-stress fracture toughness; Resilience; Strain energy ... **EM3:** 29
of acetals ... **EM2:** 100
aluminum-lithium alloys ... **A2:** 185-186, 192-193
ASP steels ... **A7:** 784-785
assessment of ... **EM2:** 554-557
calculated, by stretched zone size ... **A11:** 56
of cermets ... **A2:** 978
correlation with impact properties ... **A11:** 54-55
correlation with wear resistance ... **M1:** 607
corrosion-resistant high-alloy ... **A15:** 728
crack arrest, testing of ... **A8:** 453-455
and crack growth, parameters for ... **A11:** 54
and crack-tip stresses ... **A11:** 48
crazing effects on ... **EM2:** 737
and CTOD, correlated ... **A11:** 56
curves, as J_r curves ... **A11:** 64
cyanates ... **EM2:** 234
and damage tolerances ... **EM1:** 62
defined ... **A8:** 14, 22 **A13:** 13 **EM1:** 24 **EM2:** 43
and degree of crystallinity ... **EM1:** 101
of die materials ... **A14:** 46-47
and ductility ... **EM2:** 554
effect of temperature, schematic on ... **A11:** 66
effect of thickness ... **A11:** 51
elastic-plastic fracture, test for ... **A8:** 455-456
and engineering stress-strain curve ... **A8:** 22
of epoxy resin matrices ... **EM1:** 77
estimates from crack extension behavior ... **A8:** 439
evaluation, use of *J*-concept for ... **A8:** 457-458
fracture, evaluation of ... **A8:** 450
of fracture materials ... **A8:** 466-467
and gloss, high-impact polystyrenes (PS, HIPS) ... **EM2:** 195
high, fracture energy of ... **A11:** 51
high-impact polystyrenes (PS, HIPS) ... **EM2:** 195
HSLA steels ... **M1:** 403, 409-410, 414-415, 417-418
impact fracture ... **A11:** 54
of implant wrought cobalt alloy ... **A11:** 685, 689
inadequate, ASTM/ASME alloy steels ... **A12:** 347
influence of loading rate, ductile materials ... **A8:** 261
in *J*-integral terms ... **A8:** 261
low-, fracture energy of ... **A11:** 51
low-alloy steels ... **A15:** 717
maraging steels ... **A12:** 385
material, defined ... **A11:** 60
notch testing and evaluation ... **A11:** 57-60
opening mode, in high strain rate testing ... **A8:** 188
para-aramid fibers ... **EM1:** 55
plane stress/plane strain and ... **A11:** 51
of polymers ... **EM2:** 61
polyphenylene ether blends (PPE PPO) ... **EM2:** 184
polyvinyl chlorides (PVC) ... **EM2:** 209
and process selection ... **EM2:** 277, 279-280
requirements, bridge steels ... **A8:** 265
resin, effect impact damage residual strength ... **EM1:** 62
of shafts ... **A11:** 478-479
shear lips as indicator of ... **A11:** 396
Toughness (continued)
of sintered polycrystalline diamond ... **A2:** 1011
and strength, carbon steels ... **A15:** 702
and stress-intensity factor (*K*), compared ... **A11:** 49
and temperature ... **A14:** 162
temperature dependence ... **A8:** 262
test, and Charpy test, compared ... **A11:** 55
of thermoplastics ... **EM1:** 97-98, 293
ultrahigh molecular weight polyethylenes (UHMWPE) ... **EM2:** 167
and weight ... **A11:** 325
wrought aluminum alloy ... **A2:** 42
Toughness and impact resistance
of austenitic stainless steel after aging ... **A1:** 947
comparison of mild steel, HSLA steel, and heat-treated low-alloy steel ... **A1:** 389
of nickel-chromium-molybdenum steel ... **A1:** 396-397
of steel castings ... **A1:** 365, 367-369
Toughness, fracture
conversion factors ... **A10:** 686
Tow *See also* Prepreg tow; Towpreg
braiding ... **EM1:** 519
carbon fiber ... **EM1:** 51, 113
defined ... **EM1:** 24 **EM2:** 43
prepreg ... **EM1:** 151-152
size, determining ... **EM1:** 105
tensile testing ... **EM1:** 286-287
thermoplastic matrix ... **EM1:** 546
Towbar, trailer
fatigue fracture surface of ... **A11:** 77, 78
Towers
aluminum and aluminum alloys ... **A2:** 9
Towpreg ... **EM1:** 151-152
Toxic chemicals atmospheric constituents (carbon monoxide, ammonia, and methanol) ... **A7:** 349
Toxic elements
pollution studies by NAA for ... **A10:** 233
SSMS analysis of natural waters for ... **A10:** 141
Toxic metals
carcinogenesis ... **A2:** 1235
chelation ... **A2:** 1235-1237
chromium ... **A2:** 1242
defined ... **A2:** 1233-1234
dose-effect relationships ... **A2:** 1234
essential metals with potential for toxicity ... **A2:** 1250-1256
factors influencing ... **A2:** 1234-1235
ligands preferred for removal of ... **A2:** 1236
mercury ... **A2:** 1247-1250
metals with toxicity related to medical therapy ... **A2:** 1256-1258
minor toxic metals ... **A2:** 1258-1262
sources and standards ... **A2:** 1233
toxic metals with multiple effects ... **A2:** 1237-1250
Toxicity *See also* Health; Occupational metal toxicity; Precautions; Safety; Safety precautions; Toxic metals; Toxicology ... **A7:** 24, 201-208, 211
of additives ... **EM2:** 494
of aluminum, types ... **A2:** 1256
antimony ... **A2:** 1258-1259
arsenic ... **A2:** 1237-1239
barium ... **A2:** 1259
beryllium ... **A2:** 687, 1238-1239
bismuth ... **A2:** 1256-1257
cadmium ... **A2:** 1240
chromium ... **A2:** 1242
cobalt ... **A2:** 1251
copper ... **A2:** 1251-1252
exposure limits ... **A7:** 201-208
gallium ... **A2:** 1257
germanium and germanium compounds ... **A2:** 736
gold ... **A2:** 1257
indium ... **A2:** 1259
iron ... **A2:** 1252
lead ... **A2:** 1242-1247
Toxicity (continued)
lithium ... **A2:** 1257-1258
of lubricants ... **A14:** 517
magnesium ... **A2:** 1259
manganese ... **A2:** 1252-1253
mercury ... **A2:** 1247-1250
of metallic wires ... **EM1:** 118
molybdenum ... **A2:** 1253-1254
of nickel tetracarbonyl vapor ... **A7:** 136-137
particulate behavior in ... **A7:** 201
platinum ... **A2:** 1258
regulatory agencies and standards for ... **A7:** 202
selenium ... **A2:** 1254-1255
silver ... **A2:** 1259-1260
tellurium ... **A2:** 1260
thallium ... **A2:** 1260
tin ... **A2:** 1261
titanium ... **A2:** 1261
uranium ... **A2:** 670-671, 1261-1262
vanadium ... **A2:** 1263
water-base versus organic-solvent-base adhesives properties ... **EM3:** 86
zinc ... **A2:** 1255-1256
Toxicity of metals ... **A2:** 1233-1269
Toxicology
of arsenic ... **A2:** 1237-1238
of chromium ... **A2:** 1242
of germanium and germanium compounds ... **A2:** 736
of nickel ... **A2:** 1250
PIXE analysis in ... **A10:** 102
of thallium ... **A2:** 1260-1261
of tin ... **A2:** 1261
of titanium ... **A2:** 1261
Toyota Diffusion (TD) coating process ... **A4:** 448
Toyota Diffusion (TD) process ... **A18:** 642-643
Toys *See* Recreation applications
TPI *See* Thermoplastic polyimides; Thermoplastics (polyimides)
TPUR *See* Thermoplastic polyurethanes; Thermoplastics (polyurethanes)
Trabon grease test ... **A18:** 128
Trace analysis *See also* Trace analysis, methods of; Ultratrace analysis
14-MeV fast neutron activation analysis ... **A10:** 239
of alkali metals ... **A10:** 29
dc arc excitation for impurities in $CaWO_4$ ... **A10:** 29
habit plane determination as ... **A10:** 453-455
inductively coupled plasma atomic emission of inorganic gases, analytic methods for ... **A10:** 8
spectroscopy ... **A10:** 31-42
of inorganic liquids and solutions, applicable methods ... **A10:** 7
of inorganic solids, applicable analytical methods ... **A10:** 4-6
of metals ... **A10:** 44, 46
of organic solids and liquids, techniques for ... **A10:** 10
of organic solids, methods for ... **A10:** 9
sampling quality assurance for ... **A10:** 17
of toxic elements in ground water ... **A10:** 148-149
Trace analysis, methods of *See also* Trace analysis; Ultratrace analysis
atomic absorption spectrometry ... **A10:** 43-59
classical wet analytical chemistry ... **A10:** 161-180
controlled-potential coulometry ... **A10:** 207-211
electrochemical analysis ... **A10:** 181-211
electrogravimetry ... **A10:** 197-201
electrometric titration ... **A10:** 202-206
electron probe x-ray microanalysis ... **A10:** 516-535
electron spin resonance ... **A10:** 253-266
gas analysis by mass spectrometry ... **A10:** 151-157
gas chromatography/mass spectrometry ... **A10:** 639-648
inductively coupled plasma atomic emission spectroscopy ... **A10:** 31-42

Trace analysis, methods of (continued)
infrared spectroscopy **A10:** 109-125
ion chromatography **A10:** 658-667
liquid chromatography **A10:** 649-659
molecular fluorescence spectrometry **A10:** 72-81
neutron activation analysis **A10:** 233-242
particle-induced x-ray emission **A10:** 102-108
potentiometric membrane electrodes **A10:** 181-187
Raman spectroscopy **A10:** 126-138
Rutherford backscattering spectrometry **A10:** 628-636
secondary ion mass spectroscopy **A10:** 610-627
spark source mass spectrometry **A10:** 141-150
ultraviolet/visible absorption spectroscopy **A10:** 60-71
voltammetry **A10:** 188-196
Trace element analysis *See also* Pure metals
applications **A2:** 1093
techniques **A2:** 1095-1096
Trace elements
chemical analysis in pure metals **M2:** 711
control, cupolas **A15:** 388
effect, superalloys **A15:** 395
environmental routes **A2:** 1233
gas-tungsten arc welding affected **A6:** 20
in gray iron **A15:** 630-631
removal, vacuum induction furnace **A15:** 394-395
and steel casting failure **A11:** 392
Trace routing, automatic
types **EL1:** 529-533
Trace-element analysis for pure metals **M2:** 711
Traceability, of processing
materials **EM1:** 741
Tracer
defined **EM2:** 43
defined) **EM1:** 4
fibers, in woven fabric prepregs **EM1:** 150
radioanalysis for **A10:** 243
and shaping **A16:** 193
yarns, in woven fabric prepregs **EM1:** 149
Tracer gases
detection techniques **A17:** 64
in detector probe accumulation method **A17:** 65
as detectors, in leak testing systems **A17:** 61-65
diffusion rate, and system response **A17:** 69
diffusion rates, compared **A17:** 67
helium, in vacuum systems **A17:** 7
for leak testing, pressure systems **A17:** 61-65
leak testing without **A17:** 1
Tracer lathes **A16:** 1, 155
boring **A16:** 167
turning **A16:** 154, 156, 157
Tracer stripe
defined **EL1:** 1160
Tracer testing
curve **EL1:** 1059-1060
Tracers for cutting machines
magnetic **M6:** 907-908
manual **M6:** 907-908
Tracers with metal fuels **A7:** 600, 603
powders used **A7:** 573
Tracing devices, semi-automatic
used for quantitative metallography **A9:** 83
Track
defined **A18:** 19
Tracking
defined **EM2:** 593
Tracking pattern
defined **A18:** 19
Tracking resistance
defined **EM2:** 593
Traction
defined **A18:** 19
Tractive force
defined **A18:** 19
Tractor fitting
fatigue fracture of U-bolts on **A11:** 533-535
Tractor fluids
pour-point depressants **A18:** 108
Tractor hydraulic fluids
additives in formulation **A18:** 111
antisquawk additives **A18:** 104
antiwear agents used **A18:** 101
detergents used **A18:** 101
friction modifiers **A18:** 104
Tractor-trailer steel drawbar
fatigue fracture **A11:** 127-128
Tractors
as casting market **A15:** 34
Tractors P/M parts for **A7:** 671-673
two-wheel drive row crop **A7:** 671
Trade associations
powder metallurgy **A7:** 19
Trade magazines
as information source **EM2:** 92-93
Trade-off analyses, for steels
Al and Ti alloys **A11:** 62-63
Trade-off studies
procedure for **EM1:** 426-427
Trade-offs (design)
and driving forces, in design **EL1:** 408
leadless packaging **EL1:** 985-986
in level 1 packages **EL1:** 403
Tradenames **A4:** 970-971
Tradenames, abbreviations, and symbols **A1:** 1038-1041 **A2:** 1273-1277
Traditional ceramics **EM4:** 3-14, 893-894
characterization **EM4:** 4-5
color **EM4:** 5
compositions **EM4:** 5
etymology of ceramics **EM4:** 3
fine ceramics **EM4:** 3-4
china **EM4:** 3-4
cookware **EM4:** 4
earthenware **EM4:** 4
fine china **EM4:** 4
hotelware **EM4:** 4
porcelain **EM4:** 4
pottery **EM4:** 3
stoneware **EM4:** 3, 4
technical whiteware ceramics **EM4:** 4
vitreous china **EM4:** 4
whiteware **EM4:** 3, 4
forming processes **EM4:** 7-9
fabrication processes **EM4:** 8
moisture content **EM4:** 8
pressing **EM4:** 8, 9
pressure ranges **EM4:** 8
slip casting **EM4:** 8, 9
soft plastic forming **EM4:** 8
stiff plastic forming **EM4:** 8-9
glazes **EM4:** 9-10
application process **EM4:** 9
applications **EM4:** 10
Bristol **EM4:** 10
crystalline **EM4:** 10
definition **EM4:** 9
fritted **EM4:** 10
leadless **EM4:** 10
luster **EM4:** 10
porcelain **EM4:** 10
raw **EM4:** 10
raw lead **EM4:** 10
salt **EM4:** 10
slip **EM4:** 10
special **EM4:** 10
zinc-containing **EM4:** 10
mineralogy **EM4:** 5-7
clay minerals **EM4:** 5
commercial clays **EM4:** 6
heat effects **EM4:** 6-7
nonclay minerals **EM4:** 6
particle size distribution alterations **EM4:** 5
portland cement **EM4:** 12
air-entraining **EM4:** 12
applications **EM4:** 12
ASTM specifications **EM4:** 12
blast-furnace slag **EM4:** 12
cement chemistry **EM4:** 11
color **EM4:** 11, 12
composition **EM4:** 11
definition **EM4:** 12
development **EM4:** 10
expanding cement **EM4:** 12
hydration **EM4:** 11-12
manufacturing process **EM4:** 10-11
masonry cement **EM4:** 12
microstructure **EM4:** 11
oil well cement **EM4:** 12
properties **EM4:** 12
types **EM4:** 11, 12
Traditional ceramics (continued)
white **EM4:** 12
properties **EM4:** 5
raw materials **EM4:** 32
refractory materials **EM4:** 12-14
applications **EM4:** 14
basic **EM4:** 13-14
chrome magnesite **EM4:** 13, 14
chromite **EM4:** 13, 14
classes **EM4:** 13
composition **EM4:** 14
fireclay **EM4:** 13, 14
high alumina **EM4:** 13, 14
high-duty oxides **EM4:** 13, 14
magnesite **EM4:** 13, 14
mullite **EM4:** 13, 14
properties **EM4:** 12-13, 14
silica brick **EM4:** 13, 14
spalling **EM4:** 13
rheological behavior **EM4:** 5
shaping and finishing **EM4:** 313
structural clay products **EM4:** 9
applications **EM4:** 9, 10
classification **EM4:** 9
colors **EM4:** 9
compositions **EM4:** 9
materials **EM4:** 9
properties **EM4:** 9, 10
types **EM4:** 9
testing **EM4:** 547
Traditional glasses **EM4:** 1, 21
applications **EM4:** 21
fusing of glass objects **EM4:** 21
history **EM4:** 21
Trailer towbar
fatigue fracture surface of **A11:** 77, 78
Training *See also* Certification; Information sources; Management; Operators; Personnel
of coordinate measuring machine personnel **A17:** 28
of liquid penetrant inspection personnel **A17:** 85-86
Tram-rail assembly
fracture of **A11:** 526
Tramp element *See also* Trace elements
defined **A15:** 11
Tramp elements
and hydrogen **A8:** 487
segregation **A8:** 476
Tramp metal
separation and screening of **A15:** 350
Trans isomers
chemistry **EM2:** 64
Trans stereoisomer **EM3:** 29
defined **EM2:** 43
Transcrystalline *See* Intracrystalline
Transcrystalline cleavage, and intergranular rupture
iron **A12:** 222
Transcrystalline cracking *See* Transgranular cracking
defined **A9:** 18
Transcrystalline fractures
in magnesium alloys **A9:** 426
Transcrystalline layer (TCL) **EM3:** 417-418
Transducer
conditioner **A8:** 158
crack-opening displacement **A8:** 384
in drop tower compression test **A8:** 197
magnetostrictive **A8:** 240, 243-245
optical encoder **A8:** 158
piezoelectric **A8:** 240, 243-245
rheostat **A8:** 158
rotary capacitance **A8:** 216
rotary variable-differential transformer **A8:** 158
torque, for torsion testing **A8:** 158
Transducer elements
bandwidth **A17:** 255
barium titanate **A17:** 255
electromagnetic-acoustic (EMA) **A17:** 255-256
lead metaniobate **A17:** 255
lithium sulfate **A17:** 255
magnetostriction **A17:** 256
piezoelectric **A17:** 254-255
polarized ceramics **A17:** 255
quartz **A17:** 254-255
ultrasonic inspection **A17:** 254-256

Transducer(s) *See also* Transducer elements
electronic, in leak detection **A17:** 60
paintbrush **A17:** 258
ultrasonic inspection **A17:** 231
Transfer *See also* Selective transfer
defined **A18:** 19
feeds, of blanks **A14:** 500
headers, for cold heading **A14:** 292
in multiple-slide forming **A14:** 570
presses **A14:** 303, 502, 545
Transfer (xerography) and plain paper electrostatic copying **A7:** 582
Transfer dies
for blanking **A14:** 455
for piercing **A14:** 466
for press bending **A14:** 528
for press forming **A14:** 546
Transfer equipment
types **A14:** 292, 500-501, 570
Transfer film **A18:** 237
Transfer films **A8:** 604
in fabric prepreg **EM1:** 150
Transfer gear
single-piece **A7:** 668
Transfer ladle *See also* Ladles
defined **A15:** 11
permanent mold casting **A15:** 283
Transfer layer **A18:** 29
Transfer lines **A16:** 309
Transfer lines, flexible *See* Flexible transfer lines
Transfer machines **A16:** 395-397
Transfer, material
AES analysis of sliding during **A10:** 566
Transfer method
of neutron detection **A17:** 391, 625-626
Transfer molding **EM2:** 43, 319-320
of bulk molding compounds **EM1:** 161
for coating/encapsulation **EL1:** 240
defined **EM1:** 24
encapsulation by **EL1:** 473
of plastic packages
rigid epoxies **EL1:** 815-816
Transfer pattern, and metallization
thin-film hybrids **EL1:** 329-330
Transfer presses **A14:** 303, 502, 545
Transfer printing with sublimable dyes **EM4:** 475
Transference
defined **A13:** 13
Transferred arc
definition **M6:** 18
Transferred arc (plasma arc welding)
definition **A6:** 1214
Transferred arc plasma **A18:** 644
Transferred bump tape automated bonding **EL1:** 276
Transferred plasma-arc thermal spray coating **M5:** 363-364
Transform reconstruction algorithm
computed tomography (CT) **A17:** 359, 380-382
Transformation behavior
of the heat-affected zone in ferrous alloys **A9:** 580-581
of the reheat zone in ferrous alloys **A9:** 580-581
Transformation hardening
quantitative prediction **A4:** 20-31
Transformation of tetragonal zirconia
specialty zirconia refractories **EM4:** 907
Transformation phase diagram
engineering **EM1:** 750
Transformation ranges
defined **A9:** 18
Transformation, steel
cold treating **M4:** 117-118
Transformation stress cracks
service fracture by **A12:** 65
Transformation stresses
oxide scales **A13:** 71
Transformation stresses in tool steels **A9:** 259
Transformation temperature *See also* Melting temperature
$2^1/_2$Cr-1 Mo steel **M1:** 654
defined **A9:** 18
of rare earth metals **A2:** 723
Transformation temperatures
definitions **A4:** 4
Transformation-induced plasticity
ductile/cleavage mixed modes **A8:** 481
Transformation-induced plasticity steels
effects of plastic deformation on **A9:** 686
Transformation-toughened zirconia
as structural ceramic **A2:** 1023
Transformation-toughened zirconias (TTZ) **A6:** 949
Transformation-toughened zirconium oxide (TTZ)
applications **EM4:** 959, 960, 976, 977
characteristics **EM4:** 976
direct brazing with joining oxide ceramics **EM4:** 517
fabrication processes **EM4:** 512
properties **EM4:** 976
wear resistance **EM4:** 959
Weibull modulus increase **EM4:** 698
Transformations
allotropic, in pure metals **A9:** 655
amorphous to crystalline, EXAFS determined **A10:** 407
bainitic **A9:** 655
in cold-worked metals during annealing **A9:** 692-699
duplex growth **A9:** 656
equilibrium decomposition **A9:** 655
in situ, studied by x-ray topography and synchrotron radiation **A10:** 365
material, and crystal kinetics **A10:** 376
scanning electron microscopy used to study **A9:** 101
Widmanstatten precipitate growth **A9:** 655
Transformations in steel
bainite **M4:** 33, 34
martensite **M4:** 33, 34-35
pearlite **M4:** 33, 34
Transformed beta
defined **A9:** 18-19
Transformed microstructure fraction
and fracture toughness **A8:** 480, 482
Transformer
direct-current differential **A8:** 223
linear variable-differential transformer **A8:** 191
Transformer cores
analysis of **A10:** 224
Transformer rectifiers
gas tungsten arc welding **M6:** 187
of titanium **M6:** 453
shielded metal arc welding **M6:** 77-78
stud arc welding **M6:** 732
submerged arc welding **M6:** 131
Transformer sheet
effect of texture on anisotropic properties **A9:** 700
Transformer windings
resistance spot welding **M6:** 472
Transformers
aluminum and aluminum alloys **A2:** 13
flash welding **M6:** 560
gas tungsten arc welding **M6:** 188
high frequency welding **M6:** 764
as magnetically soft material application **A2:** 779-780
with niobium-titanium superconducting materials **A2:** 1057
passive devices **EL1:** 1005
percussion welding **M6:** 740
shielded metal arc welding **M6:** 77-78
submerged arc welding **M6:** 131
through-hole packages, failure mechanism **EL1:** 979-980
Transfusional siderosis
from iron toxicity **A2:** 1252
Transgranular *See also* Intergranular; Intracrystalline
cracking, defined **A13:** 13
defined **A11:** 11 **A13:** 13
fracture, defined **A13:** 13
Transgranular acicular carbides
precipitation of **A1:** 829
Transgranular cleavage
acoustic emission inspection **A17:** 287
fracture, in ferritic steel **A8:** 465
fracture surfaces **A8:** 487
micro-fracture mechanics morphology **A8:** 465
Transgranular cleavage fracture(s) *See also* Cleavage fracture(s); Transgranular fracture(s) **A12:** 175
AISI/SAE alloy steels **A12:** 301
cemented carbides **A12:** 470
high-carbon steel **A12:** 290
low-carbon steel **A12:** 240, 243
molybdenum alloy **A12:** 464
nickel alloys **A12:** 396
titanium alloys **A12:** 453
Transgranular cracking
brittle fracture by **A11:** 82
defined **A11:** 11
in stainless steel **A11:** 28
Transgranular cracking (or fracture)
due to creep deformation **A8:** 306, 486
Transgranular fracture
in beryllium-copper alloys **A9:** 393
brittle, cleavage facets of **A11:** 22
brittle determined **A11:** 25
brittle, fractography of **A11:** 25
cleavage, hairline indications in **A11:** 79
corrosion-fatigue cracking to **A11:** 467
defined **A11:** 11
definition **EM4:** 633
in stainless steel pressure tube **A11:** 453
Transgranular fracture(s)
at low temperatures **A12:** 121
austenitic stainless steels **A12:** 355
austenitizing effects **A12:** 339
chloride SCC in austenitic stainless steels **A12:** 357
copper alloys **A12:** 403
engineering alloys **A12:** 12
fatigue **A12:** 175-176
high-purity copper **A12:** 399-400
radial marks, SEM fractographs **A12:** 169
stress-corrosion, cleavage steps **A12:** 18
superalloys **A12:** 389
thumbprint, superalloys **A12:** 395
Transgranular stress corrosion
cleavage fracture by **A12:** 18
Transgranular stress-corrosion cracking **A13:** 148, 151-152, 157-158
Transgranular stress-corrosion cracking (TGSCC) **A6:** 379
Transgranular-intergranular fracture transition
and elevated-temperature failures **A11:** 266
Transient creep **A8:** 308
Transient film analysis **A18:** 91
Transient flow measurements **A7:** 264-265
Transient liquid-phase bonding *See also* Diffusion brazing
aluminum metal-matrix composites **A6:** 555, 557-558
titanium-matrix composites **A6:** 527
Transient liquid-phase sintering **A7:** 319, 320
Transient recorder **A8:** 197, 228
Transient state flow measurements **A7:** 264-265
Transient testing **EM3:** 316-318
Transient thermal impedance
defined **EL1:** 1160
Transistor board testers **EL1:** 873-874
Transistor outline metal can package **EL1:** 404, 1160

Transistor(s) *See also* Field effect transistors (FETs); Junction field effect transistors; MOS transistors
alloy-junction, development **EL1:** 958
application, ultrapure germanium for **A2:** 1093
ballistic **A2:** 747
bipolar junction transistor technology **EL1:** 195-196
chip Junction failure, microelectronic **EL1:** 56
defined **EL1:** 1160
double-diffused pnp **EL1:** 147
failure mechanisms **EL1:** 974-975
future trends **EL1:** 390
heterojunction bipolar (HBT) **A2:** 747
high-electron-mobility (HEMT) **A2:** 747
as IC modification **EL1:** 249
junction, thermal stress effect **EL1:** 56
n-type, development **EL1:** 958
pnp **EL1:** 146-147, 958
radio frequency discrete, electrical performance testing **EL1:** 946-951
surfaces, atmosphere for **EL1:** 958
Transistor-transistor logic (TTL)
as circuit interface **EL1:** 160
defined **EL1:** 1160
development **EL1:** 160-161
for digital ICs **EL1:** 161-162
for digital system implementation **EL1:** 76
and ECL, CMOS, compared **EL1:** 165-166
parts, in design layout **EL1:** 513
Transistors, field-effect
microwave inspection **A17:** 209
Transition
first-order **EM3:** 29
Transition alumina **EM4:** 111, 112
Transition band
defined **A9:** 685
Transition band nucleation **A9:** 694
Transition behavior *See* Nil ductility transition temperature; Notch toughness
"Transition current" **A6:** 181
Transition diagram
defined **A18:** 19
Transition elements compound analysis of **A10:** 262
on periodic table **A10:** 688
as system favorable for ESR **A10:** 262
Transition fatigue life **A8:** 712
Transition, first order *See* First-order transition
Transition, first-order *See* First-order transition
Transition, glass *See* Glass transition
Transition group metals
ESR analysis of hyperfine splitting in **A10:** 260
Transition ion content, ESR analysis for **A10:** 253
content, of fossil fuels **A10:** 253
in solids, local crystal environments around **A10:** 253-266
Transition load **A18:** 715
Transition metal
defined **A13:** 13
for galvanic corrosion **A13:** 87
-metal binary alloys, corrosion resistance **A13:** 866
-metalloid alloys, corrosion resistance **A13:** 866-867
systems, for cladding **A13:** 888-889
Transition metals
d-valence bond character and adhesive friction **A18:** 32
ion chromatography separation and detection of **A10:** 660-661
as source of background fluorescence in vibrational spectroscopy **A10:** 130
susceptibility to hydrogen damage **A11:** 250
Transition phases
defined **A9:** 19
in titanium alloys **A9:** 461
Transition point
in reduction reactions **A7:** 53
Transition scarp
definition **EM4:** 633
Transition structure
defined **A9:** 19
Transition temperature *See also* Glass transition temperature; Temperature; Temperature(s) **A7:** 12 **M1:** 696, 699 **EM3:** 29
aging, effect of **M1:** 704
Charpy test measure of **A8:** 262
cold finished bars **M1:** 228-231
correlations of tests for **A11:** 60
criteria for defining **M1:** 691-692
defined **A8:** 14 **A11:** 66 **A13:** 13 **EM1:** 24 **EM2:** 43
ductile iron **M1:** 40-41
ductile-to-brittle **A8:** 262 **A11:** 68-69 **A12:** 34
from Charpy V-notch impact tests, in steels **A11:** 67
grain size, effect of **M1:** 695
lubricant failure from **A11:** 154
nil-ductility *See* Nil-ductility transition temperature
refractory metals **A14:** 786
regimes, for Charpy/fracture toughness correlations **A8:** 265
in shafts **A11:** 478-479
shift, from tempering **A11:** 69
for structural steels **A11:** 68-69
Transition-element ions
ESR identification of valence states of **A10:** 253-266
Transition-metal ions
ESR analysis of **A10:** 254
Transitional flow
in leaks **A17:** 58
Transitions
Auger electron via $KL_{2,3}$ and $L_{2,3}$ **A10:** 550
identification of elements in **A10:** 253-266
interband, effect on Auger electron **A10:** 551
Translaminar fracture **EM1:** 786, 790-792
compression **EM1:** 792
tension **EM1:** 791-792
Translaminar fractures
in composites **A11:** 738-739
compression **A11:** 739-740
schematic **A11:** 734
tension **A11:** 738-739
Translaminar-tension fractures
graphite-epoxy composite **A11:** 738
Translation interfaces
transmission electron microscopy **A9:** 118-119
Translation, of images
digital image enhancement **A17:** 458
Translation vector **A9:** 118-119
Transmission
as electronic function **EL1:** 89
media, explosive forming **A14:** 639-640
optical testing of **EM2:** 594
shock-wave, explosive forming **A14:** 640
Transmission (pipe)lines
high-pressure long- distance **A11:** 695
Transmission anomalous
topography **A10:** 369-370
in x-ray topography **A10:** 367
Transmission cable
defined **EL1:** 1160
Transmission density
x-ray film **A17:** 323
Transmission electron fractography
for failure mode **A8:** 476
Transmission electron micrographs **A8:** 481-484
Transmission electron microscope **A18:** 376-377
basic configuration **A18:** 379
Transmission electron microscope(s)
history **A12:** 5-6
literature **A12:** 7
replication techniques **A12:** 7
specimens for **A12:** 6-7
Transmission electron microscopes **A9:** 103-104
components **A9:** 103
defined **A9:** 19
electron column **A10:** 431
as input device for image analyzers **A10:** 310
modes of operation **A9:** 103-104
resolution **A9:** 103
Transmission electron microscopes (TEM)
for plastic replicas **A17:** 53
Transmission electron microscopy *See also* Analytical transmission electron microscopy; Thin-foil transmission electron microscopy **A9:** 103-122 **A12:** 179-192 **A13:** 1117
(TEM) **A8:** 477
artifacts in replicas **A12:** 184-185
capabilities, and FIM/AP **A10:** 583
changes of beam amplitude after passing through a crystal **A9:** 118 119
cleaning fracture surfaces **A12:** 179-183
and color enhancement by electronic imaging **A9:** 153
defect analysis by **A10:** 464-468
defined **A10:** 683
deformation, recovery, and recrystallization analysis by **A10:** 468-470
diffract on patterns **A9:** 109
diffraction contrast theory **A9:** 110-113
diffraction studies of grain boundaries **A9:** 119-120
of dislocations **A9:** 113-116
effect of abrasion damage on samples for **A9:** 39
of fiber composites **A9:** 592
and FIM images, IN 939 nickel-base superalloy **A10:** 598
fractographs, and SEM fractographs compared **A12:** 185-192
fractographs, for stereo viewing **A12:** 192
fractographs, method of preparation **A12:** 185, 192
of inorganic solids, types of information from **A10:** 4-6
magnetic material specimens, preparation of **A9:** 534
material types **A12:** 216
of microstructure and magnetic properties, ductile permanent magnets **A10:** 599
of organic solids, information from **A10:** 9
of permanent magnet alloys **A9:** 533
point defect agglomerates **A9:** 116-117
radiation damage **A9:** 116-117
and Raman analysis, for intercalated graphites **A10:** 133
replicas **A9:** 108-109
replication, fracture surface preparation **A12:** 179-183
replication procedures **A12:** 94-95
resolution **A9:** 103
second-phase precipitates **A9:** 117
sectioning **A9:** 104
and SEM fractographs, compared **A12:** 185-192
shadowing of replicas **A12:** 183-184
specimen preparation **A9:** 104-108 **A12:** 179
specimen thickness **A9:** 103
of tin and tin alloy coatings **A9:** 451
to study solid-state transformations in weldments **A6:** 77
transmitted wave amplitudes **A9:** 119
used to examine case hardened steels **A9:** 217
used to examine plate steels **A9:** 203
used to examine tin plate **A9:** 198
weld characterization **A6:** 104-105
of wrought heat-resistant alloys **A9:** 307-308, 311
Transmission electron microscopy (TEM) **EM1:** 771 **EM3:** 237 **EM4:** 87
as advanced failure analysis technique **EL1:** 1106
of composites **A11:** 741
embedding of erodent fragments in nickel observed **A18:** 203
erosion of ceramics **A18:** 205
of fracture surfaces **A11:** 20-22
imaging and analysis **A18:** 385-391
amplitude contrast **A18:** 387
basic imaging and diffraction modes **A18:** 385
diffraction contrast **A18:** 387-389
EDS analysis in the TEM **A18:** 387, 389-390
electron diffraction **A18:** 385-387
electron energy loss spectrometry **A18:** 390
image contrast **A18:** 387
mode of operation **A18:** 379
phase contrast **A18:** 387
phase contrast imaging **A18:** 389
scanning transmission electron microscopy **A18:** 390-391
for microchemical analysis **EM4:** 25

Transmission electron microscopy (TEM) (continued)
for microstructural analysis **EM4:** 26, 570
ofshafts **A11:** 460
for particle sizing **A7:** 227-228
for phase analysis **EM4:** 560
and SEM, contrasted **EL1:** 1101
specimen preparation **A18:** 381-382
debris specimens **A18:** 382
specimens from worn surfaces **A18:** 381
surface replicas **A18:** 381-382
stripped oxides **EM3:** 242
summary and future outlook **A18:** 391
surface analysis **EM3:** 261
for surface examination **A18:** 291
to analyze ceramic powder particle size **EM4:** 66-67
as wafer-level physical test method **EL1:** 920-922
Transmission electron microscopy, specimens for *See also* Thin-foil specimens. preparation by electric discharge machining **A9:** 26
Transmission flex plates
economy in manufacture **M3:** 851
Transmission fluids **A18:** 98
additives in formulation **A18:** 111
antisquawk additives **A18:** 104
antiwear agents used **A18:** 101
demulsifiers **A18:** 107
detergents used **A18:** 101
dispersants used **A18:** 100
dyes **A18:** 110
extreme-pressure agents used **A18:** 101
friction modifiers **A18:** 104
oxidation inhibitors **A18:** 105
pour-point depressants **A18:** 108
seal-swell agents **A18:** 110
viscosity improvers used **A18:** 110
Transmission grating
defined **A10:** 683
Transmission line *See also* Line(s); Signal transmission line
abbreviation for **A11:** 798
analysis **EL1:** 25, 32
approximation **EL1:** 31-32
coupled, signal cross talk **EL1:** 34-37
defined **EL1:** 1160
design, WSI **EL1:** 354, 357-362
environments **EL1:** 601
fundamentals **EL1:** 28-30
lossy single, analytical solution **EL1:** 41-42
structures **EL1:** 419
Transmission method
defined **A9:** 19
Transmission, off-highway
cam plastometer **A8:** 194-195
Transmission oil
defined **A18:** 19
Transmission oil pump gears
farm tractors **A7:** 672
Transmission or line pipe **A1:** 331
Transmission parts
powders used **A7:** 572
Transmission pinhole camera
schematic **A10:** 334
Transmission pinhole photographs of iron- nickel alloy wires **A9:** 702
Transmission pipe **M1:** 319
Transmission pipelines **A13:** 1292
Transmission shaft
by radial forging **A14:** 147
Transmission shafts
economy in manufacture **M3:** 851, 852
fatigue life of **M1:** 675
Transmission stick-shift, automobile
failure of **A11:** 763
Transmission techniques *See also* Transmission ultrasonic inspection
eddy current inspection **A17:** 183-184
fixed-frequency continuous-wave **A17:** 205
Transmission techniques (continued)
microwave inspection **A17:** 205-206
pulse-modulated **A17:** 206
swept-frequency continuous-wave **A17:** 205-206
ultrasonic inspection **A17:** 240, 248-252
Transmission topography
Berg-Barrett transmission arrangement **A10:** 369
configurations for **A10:** 369
defect imaging with **A10:** 370
Transmission tower legs
weathering steel **A13:** 520
Transmission ultrasonic inspection *See also* Transmission techniques
applications **A17:** 249-250
continuous-beam testing **A17:** 249
displays **A17:** 249
examples **A17:** 249-250
Lamb wave testing **A17:** 250-251
pitch-catch testing **A17:** 249
ultrasonic inspection **A17:** 240, 248-252
Transmissive radiography
as x-ray solder joint inspection **EL1:** 738
Transmittance **A10:** 62-63, 683
Transmitted gage
for stored-torque Kolsky bar **A8:** 220
Transmitted light
defined **EL1:** 1067
Transmitted power
symbol and units **A18:** 544
Transmitted pulse
in torsional Kolsky bar dynamic tests **A8:** 228-229
Transmitted specimen current density
scanning electron microscopy **A9:** 90
Transmitted wave, and reflected wave
timing between **A8:** 203
Transmitted-beam images
used for quantitative metallography **A9:** 134
Transmitter bar
in torsional Kolsky bar dynamic test **A8:** 228
Transparency, electronic *See* Electronic transparency
Transparent electronic graticule **A7:** 229
Transparent frits and glazes
typical oxide compositions **EM4:** 550
Transparent layers
interference films used to improve contrast **A9:** 59
Transparent mounting materials **A9:** 29-30
Transpassive
dissolution, and corrosion resistance **A13:** 48
region, defined **A13:** 13
state, defined **A13:** 13
Transpassive anodic potential ranges **A9:** 144-145
Transpiration cooling
porous parts **A7:** 700
Transplutonium actinide metals *See also* Americium; Berkelium; Californium; Curium; Einsteinium; Fermium
properties **A2:** 1198-1201
Transport *See* Interphase mass transport; Mass transport; Transference
Transport aircraft components
flight service evaluation **EM1:** 826-831
Transportation applications *See also* Aerospace applications; Aircraft applications; Automotive applications
acrylonitrile-butadiene-styrenes (ABS) **EM2:** 111
blow molding **EM2:** 359
critical properties **EM2:** 458
liquid crystal polymers (LCP) **EM2:** 180
polyamides (PA) **EM2:** 125
polybenzimidazoles (PBI) **EM2:** 147
polycarbonates (PC) **EM2:** 151
unsaturated polyesters **EM2:** 246
Transportation engine project
Japan **EM4:** 717
Transportation equipment
use of stainless steels **M3:** 68-70
Transportation glass **EM4:** 1021-1024
aircraft window glazings **EM4:** 1023-1024
Transportation glass (continued)
automotive glass **EM4:** 1021-1023
Transportation industry applications
aluminum and aluminum alloys **A2:** 10-12
aluminum-lithium alloys **A2:** 182
cemented carbides **A2:** 973-974
copper and copper alloys **A2:** 239-240
of magnetic painting **A17:** 128
magnetic particle inspection **A17:** 89
Transportation, mold
green sand molding **A15:** 347
Transportation vessels
anodic protection of **A13:** 465
Transporting, and placement
outer lead bonding **EL1:** 285
Transverse
compression **EM1:** 198, 238
cracks, in weldments **A17:** 585
defined **A8:** 14
flaws, in steel bar and wire **A17:** 550
properties, carbon fibers **EM1:** 52
strain, defined **EM1:** 24
tension, ply **EM1:** 198, 237-238
waves, ultrasonic inspection **A17:** 233
Transverse bend test
for green strength **A7:** 288
Transverse breaking load
corrosion-resistant cast irons **M1:** 89
gray iron **M1:** 16-19
heat-resistant cast irons **M1:** 92
white cast irons **M1:** 78, 80, 87
Transverse connection plate
bridge girder web crack **A11:** 713
Transverse crack
definition **A6:** 1214
Transverse defects in copper alloy ingots **A9:** 642
Transverse direction
of a sheet, abbreviation for **A11:** 798
defined **A9:** 19 **A11:** 11
Transverse displacement interferometer **A8:** 233-235
Transverse distribution of solute
in ingots **A11:** 324-325
Transverse electric (TE) mode
microwave holography **A17:** 207-211
Transverse electric (TE) modes **A10:** 256, 691
Transverse electromagnetic (TEM) wave **EM3:** 429
Transverse fractures
cast aluminum alloys **A12:** 411-413
of crankshaft **A11:** 472-473
martensitic stainless steels **A12:** 369
shear, high-strength steel, fractograph **A12:** 174
in steel railroad rail **A11:** 79
Transverse Kerr effect **A9:** 535
Transverse magnetic (TM) mode
microwave holography **A17:** 207
Transverse magnetic field
overall critical current density as function of **A7:** 638
Transverse metallographic sections
titanium alloy bars **A8:** 173
Transverse properties
gray iron **A15:** 643
joint, ductile iron weldments **A15:** 528
Tread and flange, chilled iron **A15:** 30
Transverse radius of curvature of gear
symbol and units **A18:** 544
Transverse radius of curvature of pinion
symbol and units **A18:** 544
Transverse resistance seam welding
definition **M6:** 18
Transverse rupture strength
cemented carbides **A2:** 956
cemented carbides, compared **A2:** 989
Transverse sectioning
for TEM **A12:** 179

Transverse shrinkage
butt welds ... M6: 861, 870-875, 880
effect of restraint ... M6: 870-871, 873-875
effect of welding sequence ... M6: 875-876
mechanisms ... M6: 871-872
multipass welding ... M6: 872-874
rotational distortion ... M6: 870-873
fillet welds ... M6: 875
Transverse strain ... EM3: 29
defined ... A8: 14 EM2: 43
Transverse strain-measuring extensometers ... A8: 49
Transverse stretch-forming machines ... A14: 596
Transverse tension specimen
effective fracture strain in ... A8: 168
Transverse tension test ... A6: 101, 103
Transverse test
for gray iron ... A1: 16
Transverse velocity
at free surface of anvil ... A8: 235-236
Transverse wire defect
spring failure from ... A11: 554
Transverse yield strength
mortar tube alloys ... A11: 295
Transverse-flow (carbon dioxide) laser ... A14: 736
Transverse-rupture strength ... A7: 12, 311
of aluminum oxide-containing cermets ... A7: 804
of chromium carbide-based cermets ... A7: 806
effect of sintering temperature ... A7: 367
and green strength ... A7: 288
of iron, copper and graphite powder compacts ... A7: 266
of metal borides and boride-based cermets ... A7: 812
of sintered steel, combined carbon effect ... A7: 362
testing ... A7: 288, 290, 312, 490
of titanium carbide-based cermets ... A7: 808
of titanium carbide-steel cermets ... A7: 810
of tungsten carbide/cobalt ... A7: 775-776
of water-atomized nickel ... A7: 397
Transversely isotropic ... EM3: 29
defined ... EM1: 24 EM2: 43
Tranverse strength
of abrasion-resistant cast iron ... A1: 96
of gray iron ... A1: 18
Trapezoidal loading theory ... EM3: 607
Trapped rubber forming
drop hammer ... A14: 654-655
Trapped-sheet, contact heat
pressure thermoforming ... EM2: 401
Trapping
in impact milling ... A7: 57
Trapping, charge
in silicon oxide failures ... A11: 780-781
Trapping off ... EM3: 740
Traps
lead and lead alloy ... A2: 536-537
Traps, cold/weld
molding ... EM2: 615
Traptometer
for deflections in torsional testing ... A8: 147
Travel angle
definition ... M6: 18
Travel speed
effect on weld attributes ... A6: 182
Travel start delay time
definition ... M6: 18
Travel stop delay time
definition ... M6: 18
Travel trailers
aluminum and aluminum alloys ... A2: 11
Traveling acoustic waves
for optical holographic interferometry ... A17: 409
Traveling laser welds ... A6: 20
Traveling wave tube (TWT)
microwave inspection ... A17: 209
Traveling wave tube magnets
powders used ... A7: 573
Tray set, modular
for multiple small parts ... A7: 423
Trays, heat-resistant alloys
applications ... M4: 329, 330-331
life expectancy ... M4: 327-328, 330-331
materials recommended ... M4: 328
Tread life ... A18: 579
Tread loss ... A18: 578
Tread rubber abradability ... A18: 579
Tread wear index ... A18: 578, 580
Treatise on Adhesion and Adhesives ... EM3: 70
Treatment *See also* Ligands; Toxic metals; Toxicity; Toxicology
of arsenic toxicity ... A2: 1238
of bismuth toxicity ... A2: 1257
of cadmium toxicity ... A2: 1241-1242
combination, defined for factorial experiments ... A8: 641
effects, analyzing ... A8: 656-660
experimental, defined ... A8: 640
of iron toxicity ... A2: 1252
of lead toxicity ... A2: 1246-1247
levels, defining for fatigue experiments ... A8: 695-696
of lithium toxicity ... A2: 1257-1258
of manganese toxicity ... A2: 1253
of mercury toxicity ... A2: 1249-1250
of tellurium toxicity ... A2: 1260
Treatment transfer
defined ... EL1: 1160
Treatment, waste *See* Waste recovery and treatment
Tree ring pattern, as defect
vacuum arc remelting ... A15: 407
Tree rings in titanium alloys ... A9: 460
in Ti-6Al-2Sn-4Zr-2Mo forged billet ... A9: 465
Trenching tools
cemented carbide ... A2: 973
Trend identification
as quality control method ... A17: 731-732
Trend removal
digital image enhancement ... A17: 460
Trends *See also* Future trends
casting market ... A15: 43-45
in electrical design ... EL1: 416
finer lead pitch ... EL1: 438
interconnection system requirements modeling ... EL1: 12-17
key-parameter, in thermal design ... EL1: 409
lead frame materials ... EL1: 491-492
in level 1 packages ... EL1: 405
Trepanning ... A16: 175-180
abrasive wear and material property effects ... A18: 186
by ND-YAG laser ... A14: 737
in conjunction with drilling ... A16: 216
in conjunction with milling ... A16: 308
cutting fluid flow recommendations ... A16: 127
deep holes ... A16: 176-180
of fracture surfaces ... A11: 19
heat-resistant alloys
hot ... A14: 63
refractory metals ... A16: 863
tool life ... A16: 175, 176, 180
Trepanning methods
used with ultrasonic machining ... EM4: 361
Tresca criterion ... A8: 163, 343-344, 576 A18: 476
TRI *See* The Refractories Institute
Tri-N-butyl phosphate
solvent extractions with ... A10: 169
Tri-N-octylphosphine oxide
as solvent extractant ... A10: 170
Trial
defined for factorial experiments ... A8: 641
Trialuminides *See also* Ordered intermetallics
and cleavage fracture ... A2: 930-931
Co_3Ti, types and properties ... A2: 929-930
Co_3V, types and properties ... A2: 929
$L1_2$-ordered ... A2: 929-930
Ni_3X alloys, properties ... A2: 931-932
Zr_3Al, properties ... A2: 932
Triangular elements
for fracture surface area ... A12: 201-202
Triangular grip ends
for torsion specimen ... A8: 157
Triangularity
of particle shape ... A7: 242
Triangulation *See also* Laser triangulation
defined ... A17: 34
sensors, laser ... A17: 13
Triaxial compression ... A7: 304
Triaxial stress *See* Principal stress ... A8: 25, 576
effect on fracture mode ... A12: 30-31
in stainless steel ... A12: 31, 40
Triaxial tension
in fatigue fracture ... A11: 75
Triaxiality factor
(TF) ... A8: 344 345
Triazine resins *See* Cyanates
Triazines
as biocides ... A18: 110
Tribaloy 800
coating compositions for LPT blade interlocks ... A18: 590
material for jet engine components ... A18: 591
Tribaloy alloy
hardfacing ... A6: 807
Tribaloy alloys *See* Cobalt alloys, specific types, Tribaloy alloy
Tribaloy PWA 694
laser cladding components and techniques ... A18: 869
Tribaloy T-400
laser cladding ... A18: 868
Tribaloy T-800
laser cladding ... A18: 867
laser hardfacing ... A6: 807
Tribo-
defined ... A18: 19
Tribo-oxidation
nitrides surfaces ... A18: 879, 880, 881
Tribochemistry
defined ... A18: 19
Tribocontact parameter ... A18: 476, 477, 480
Tribocracks ... A18: 178
Triboelectric charging ... A7: 584, 585
Triboelement
defined ... A18: 19
Tribography ... A18: 176
Tribological parameters, basic ... A18: 473-479
data sheet, basic tribological parameters ... A18: 478-479
functional purpose of tribosystems, categories ... A18: 473
interaction parameters ... A18: 474-477
contact area/wear-track ratio ... A18: 476
contact deformation modes ... A18: 475-476
contact stresses ... A18: 476
interface forces and energies ... A18: 475
lubrication modes of tribosystems ... A18: 476-477
operational parameters ... A18: 474
kinematics ... A18: 474
type of motion of components of tribosystem for sliding and sliding and rolling ... A18: 475
velocities of tribosystem components for sliding and sliding and rolling ... A18: 475
structural parameters ... A18: 473-474
examples (components) of common tribosystems ... A18: 474
tribometric characteristics ... A18: 477-478
friction parameters ... A18: 478
wear parameters ... A18: 478
Tribology ... A8: 14, 601
AES analyses in ... A10: 566
defined ... A11: 11 A18: 20
Tribometer
defined ... A18: 20
Tribophysics
defined ... A18: 20
Tribopolymers ... A18: 29
Triboscience
defined ... A18: 20
Tribosurface
defined ... A18: 20
Tribosystem
defined ... A18: 20
Tribotechnology
defined ... A18: 20
Tribotesting ... A18: 480-485
categories ... A18: 480-481
chi-square (χ^2) distribution ... A18: 482, 485
conditions ... A18: 480-481
evaluation ... A18: 481-482
simulative ... A18: 481
Tributyl phosphate
surface tension ... EM3: 181
Tricalcium aluminate
in composition of portland cement ... EM4: 11, 12
Tricalcium phosphate
dentifrice abrasive ... A18: 665

Tricalcium phosphate (TCP)
resorbable biomaterial **EM4:** 1007
Tricalcium silicate
in composition of portland cement **EM4:** 11
Trichloroethane (TCE)
solvent resistance/repairability **EL1:** 822
Trichloroethylene **A6:** 119
as carcinogenic cleaning agent **A12:** 74
for SCC of titanium alloys **A8:** 531
Trichloroethylene (C_2HCl_3)
as solvent used in ceramics processing **EM4:** 117
Trichloroethylene solvent cleaners **M5:** 40, 44-48
cold solvent cleaning process, use in **M5:** 40
flash point **M5:** 40
magnesium alloy cleaning **M5:** 629
vapor degreasing, use in **M5:** 44-48
Trichlorofluoroethane
for SCC of titanium alloys **A8:** 531
Trichlorofluoroethane solvent cleaner **M5:** 40-41
flash point **M5:** 40
Trichromatic printing **EM4:** 473
Triclinic
defined **A9:** 19
Triclinic crystal system .... **A3:** 1 • 10, 15 **A9:** 706
Triclinic crystal systems
unit cells as **A10:** 346-348
Tricresyl phosphate
surface tension **EM3:** 181
Tricresylphosphate (TCP)
lubrication of adiabatic engines **EM4:** 991
Tridymite **EM4:** 754
crystal structure **EM4:** 879
framework structure **EM4:** 759
as quartz transition **A15:** 208
thermal expansion coefficients **EM4:** 499
Triethanolamine (TEA)
and chemical milling of Al alloys **A16:** 583, 584
Trigged capacitor discharge
defined **A10:** 683
Trigger pulse mode
eddy current inspection **A17:** 190
Trigger rod
for torsional impact system **A8:** 216-217
Triglycidyl ether of triphenyl methane (TGETPM)
epoxy resin **EM1:** 67, 69
Triglycidyl *p*-amino phenol (MY 0500) **EM3:** 94, 97
Triglycidyl *p*-aminophenol (TGAP), for epoxy adhesives **EM1:** 67
Trigonal crystal system .... **A3:** 1 • 10 **A9:** 706
Trigonal unit cells **A10:** 346-348
Trim **EL1:** 448, 1160
lines, defined **EL1:** 1160
patterns, thin-film **EL1:** 321
Trim and wipe-down
of blanks **A14:** 447
Trim tabs
as test coupons **EM1:** 744
Trimerized fatty acids
epoxidized **EL1:** 818
Trimetal bearing
defined **A18:** 20
Trimetal bearings **A9:** 567
Trimmer
defined **A14:** 13
Trimmer blade
defined **A14:** 13
Trimmer die
defined **A14:** 13
Trimmer punch
defined **A14:** 13
Trimming
of aluminum alloy forgings **A14:** 248
of blanks **A14:** 446-447
for closed-die forging **A14:** 82
in coining **A14:** 180
of copper and copper alloy forgings **A14:** 257
in deep drawing operations **A14:** 588-589
defined **A14:** 14, 55
Trimming (continued)
in die casting **A15:** 287
dies, material for **A14:** 257
fracture in **A8:** 548
hot **A14:** 8, 230, 257
in hot upset forging **A14:** 83
magnesium alloy forgings **A14:** 260
multiple-slide forming **A14:** 567
pinch **A14:** 446
shimmy **A14:** 446-447
of stainless steel forgings **A14:** 229-230
in thermoforming **EM2:** 400
of titanium alloy forgings **A14:** 279-280
trim and wipe-down **A14:** 447
and upsetting, tooling setup for **A14:** 88
vs developed blanks, deep drawing **A14:** 589
Trimming dies **A14:** 55-56, 257
Trimming, of composites **EM1:** 668
Trimming press
defined **A14:** 14
Trimming tolerance
forgings **M1:** 366, 367
Trimming tools
closed-die forgings, materials for **M3:** 531-532
hot upset forging, materials for **M3:** 535-536
TrimRite *See* Stainless steels, specific types, S42010
Triode-style electron guns **A6:** 254, 260-261
TRIP steels *See* Transformation-induced plasticity steels
Triple action press **A7:** 12
Triple curve .... **A3:** 1 • 2
defined **A9:** 19
Triple monochromators
stray light rejection for **A10:** 129
Triple point .... **A3:** 1 • 2
defined **A9:** 19
Triple point crack **A8:** 572, 574
Triple point of water
in thermocouple calibration **A2:** 879
Triple-action press *See also* Hydraulic press; Mechanical press; Slide
defined **A14:** 14
slides in **A14:** 495
Triple-coated carbides (TiC/Al_2O_3/TiN)
cutting speed and work material relationship **A18:** 616
Triple-point cracking *See also* Cracking; Fracture; Wedge cracks **A14:** 19, 364
austenitic stainless steels **A12:** 364
in fatigue **A12:** 119
intergranular creep rupture by **A12:** 19, 25
Triplemate
use in Raman spectroscopy **A10:** 129
Triplet states
ESR analysis of **A10:** 254
Triplex annealing
wrought titanium alloys **A2:** 619
Tripod
photographic **A12:** 89
Tripoli buffing compound **M5:** 117
Tripoly cleaner *See* Sodium tripolyphosphate cleaner
Trisodium phosphate **A6:** 119
Tristelle TS-1
composition **A18:** 806
Tristelle TS-2
composition **A18:** 806
Tristelle TS-3
composition **A18:** 806
Trithermal weld **A6:** 606
Tritium, molten
applications **A13:** 56
Triton
defined **A10:** 683
as surfactant, and adhesion of polymers on silver **A10:** 136
-X-100, as maximum suppressor voltammetry **A10:** 191
Tritonal explosives
aluminum powder containing **A7:** 601
Trivalent chromium
formation and oxidation to hexavalent chromium **M5:** 191
hard chromium plating baths contaminated by **M5:** 174-175
hexavalent chromium reduced to **M5:** 302, 311
Trivalent chromium plating **M5:** 196-198
solution compositions and operating conditions **M5:** 196-197
Trodaloy
resistance welding **A6:** 833-834
Trommels **A7:** 177
Trona **EM4:** 380
Trona mining tools
cemented carbide **A2:** 976
Troostite
defined **A9:** 19
Tropenas converter
development of **A15:** 32-33
Trowel coating process
ceramic coating **M5:** 535, 545
Truck and truck trailers
aluminum and aluminum alloys **A2:** 11
Truck engine
connecting rod fatigue fracture from forging lap **A11:** 328-329
Truck signal flares
powder used **A7:** 573
Truck-gear forging
simulation **A14:** 429
Trucks *See also* Bucket trucks
as casting market **A15:** 34
jumbo tube, acoustic emission inspection **A17:** 291
Trucks, diesel
beach marks on crankshaft fracture **A11:** 76, 77
True area *See also* Area
of fracture surface, importance **A12:** 211
True brinelling
and false brinelling, compared in bearings **A11:** 499-500
indentations, spalling initiated at **A11:** 500
rolling-contact fatigue from **A11:** 503-504
True centrifugal casting
vertical **A15:** 300
True density
by pycnometry **A7:** 265-266
True elastic limit *See also* Elastic limit
defined **A8:** 21
True fracture strain
defined **A8:** 24
True length *See also* Length
defined **A12:** 199-200
fractal analysis **A12:** 212
True local necking strain
defined **A8:** 24-25
True rake angle
milling **A16:** 317, 318, 319, 328, 329
True spacing equation for lamellar structures **A9:** 128
True strain *See also* Engineering strain; Linear strain; Strain; True stress/true strain curve **EM3:** 29
and area reduction **A8:** 575
in bending **A8:** 118-119
compression tests, cam plastometer **A8:** 194
curves, uniaxial tensile testing **A8:** 554
defined **A8:** 14, 194 **EM1:** 24 **EM2:** 43
determining **A8:** 23
engineering strain, area reduction and **A14:** 368
-flow stress, equations **A14:** 418
in forming refractory metal sheet **A14:** 786
rate compression, cam plastometer **A8:** 194
rate, in hot compression testing **A8:** 582
rate-flow stress, equations **A14:** 419
in SCC testing **A13:** 254
in sheet metal forming **A8:** 549

True strain (continued)
vs. engineering strain, in creep testing **A8:** 305
True stress *See also* Engineering stress; Mean stress; Nominal stress; Normal stress; Residual stress; Stress(es); True stress/true strain curve **EM3:** 29
at maximum load, as true stress/true strain curve parameter **A8:** 24
at maximum load, as true tensile strength **A8:** 24
curves, uniaxial tensile testing **A8:** 554
defined **A8:** 14 **EM1:** 24
determining **A8:** 23
and engineering stress **A8:** 23
in SCC testing **A13:** 254
symbol for **A8:** 726
values, for metals at room temperature **A8:** 24
True stress/true strain curve **A8:** 23-25
True tensile strain
relationship to Bridgman correction factor **A8:** 26
True uniform strain
defined **A8:** 24
True yield
in indentation tests **A8:** 71
stress, at various strains **A8:** 38-39
Truncated cones
forming **A14:** 621-622
indentation test, for forgeability **A14:** 384
Truncated fatigue life
by static proof test **A8:** 718
Truncated octohedron
properties of **A9:** 133
Truncated Weibull distribution **A8:** 717
Truncated-cone indentation test **A1:** 584
Trunnion bearing
defined **A18:** 20
Trunnions
materials for **A11:** 515
TRW-NASA VIA
composition **A16:** 737
machining **A16:** 738, 741-743, 746-758
Tryout
defined **A14:** 14
TSAAS computer program for structural analysis **EM1:** 268, 273
Tsai-Wu criterion **EM1:** 200, 433
TTL *See* Transistor-transistor logic
TTS fractures *See* Tearing topography surface fracture(s)
Tub vibratory finishing **M5:** 130
Tuballoy *See* Uranium
Tube *See also* Tubing; Tubular products
alloys, feedwater heaters **A13:** 990
aluminum and aluminum alloys **A2:** 5
beryllium-copper alloys **A2:** 403
beryllium-copper alloys, properties **A2:** 411
collapsible, tin-base alloy **A2:** 525
crevice corrosion **A13:** 110
cylindrical **EM1:** 569-570, 572
explosion welding **M6:** 709, 715
filament winding for **EM1:** 135
finished, production of **A2:** 250
flash welding **M6:** 558, 577
friction welding **M6:** 720-721, 726
fully reversed loading in ultrasonic testing **A8:** 242
gas metal arc welding of copper-nickel **M6:** 421-422
gas tungsten arc welding **M6:** 338-340
for gas/oil production **A13:** 1236-1237, 1258
graphite fiber MMC pultruded **EM1:** 870-871
high frequency resistance welding **M6:** 759, 760, 762
oxyfuel gas welding **M6:** 592
plasma arc welding **M6:** 220-221, 342-343
polyamides (PA) **EM2:** 125
primary testing direction, various alloys **A8:** 667
projection welding **M6:** 516-517
properties **A2:** 249
quasi-static torsional testing of **A8:** 145
recovery boiler **A13:** 1199
reducing, wrought copper and copper alloys **A2:** 250
Rockwell hardness testing of **A8:** 81
-sheet interface, crevice corrosion **A13:** 110
Tube (continued)
shells, production of **A2:** 249-250
soldering **M6:** 1093-1095
specimens **A8:** 168, 221-224
support, in torsional impact machine **A8:** 217
tapered **EM1:** 570-572
test specimens **EM1:** 326-327
testing **A8:** 250
thermoplastic polyurethanes (TPUR) **EM2:** 205
thick-walled, in torsional fatigue testing **A8:** 152
thin-walled for workability **A8:** 156
thin-walled, torsion and simple shear **A8:** 155
wastage, steam generators **A13:** 939
wrought aluminum alloy **A2:** 33
wrought beryllium-copper alloys **A2:** 409
wrought copper and copper alloys **A2:** 248-250
Tube bending *See also* Tube(s); Tubing
lubrication **A14:** 672-673
machines **A14:** 668-669
with mandrel **A14:** 666-668
of nickel-base alloys **A14:** 834-835
of titanium alloys **A14:** 848
tools **A14:** 665-666
without mandrel **A14:** 668
of zirconium **A2:** 664-665
Tube brass
applications and properties **A2:** 306
Tube, copper
temper designations **M2:** 249-251
Tube designs
borescopes **A17:** 3-4
Tube drawing *See also* Tube(s); Tubing
dies **A14:** 337
ironing operation in **A14:** 334-336
schematic **A14:** 330
without mandrel (tube sinking) **A14:** 330
Tube drawing dies
cemented carbide **M3:** 523-525
chromium plating **M3:** 525
diamond **M3:** 521, 523, 525
mandrels **M3:** 521, 524-525
polishing **M3:** 525
sectional, adjustable **M3:** 524
tool breakage **M3:** 525
tool steels **M3:** 523, 524, 525
Tube extensometers
for elevated/low temperature tension testing **A8:** 32
Tube extrusion
process **EM2:** 385
Tube forming *See also* Tube(s); Tubing
contour roll forming **A14:** 673
dies, for press brake forming **A14:** 537
nosing **A14:** 673
press methods **A14:** 673
spinning **A14:** 673
Tube furnace **A7:** 12
Tube length
effect on objective lenses **A9:** 72
Tube mills
vibratory **A7:** 66-68
Tube piercing **A14:** 470
Tube rating, x-ray tubes
radiography **A17:** 305-306
Tube reducing **M1:** 316 **M2:** 264
Tube reducing and swaging
for steel tubular products **A1:** 328
Tube rolling *See also* Tube(s); Tubing **EM1:** 569-574
contour roll forming **A14:** 630-632
convolute wrapping, cylindrical tubes **EM1:** 569-570
convolute wrapping, tapered tube **EM1:** 570-571
material forms **EM1:** 569
production methods **EM1:** 572-574
roll design **A14:** 631
spiral wrapping **EM1:** 571-572
Tube shear test **EM3:** 732, 733
Tube shells
extrusion **A2:** 249
rotary piercing **A2:** 249-250
Tube sinking *See* Tube drawing
Tube spinning *See also* Tube(s); Tubing **A14:** 673, 675-679
applicability **A14:** 675
backward **A14:** 675-676
Tube spinning (continued)
defined **A14:** 675
finish, of tube-spun parts **A14:** 678
forward **A14:** 676
lubricants **A14:** 678
machines **A14:** 676-678
methods **A14:** 675-676
preform requirements **A14:** 675
speeds and feeds **A14:** 678-679
tools **A14:** 676-677
tube spinnability **A14:** 679
tube wall thickness, limitations **A14:** 677
work metal properties, effects **A14:** 679
Tube steels
AISI compositions **A9:** 211
ASTM compositions **A9:** 211
Tube stock
defined **A14:** 14
Tube straightening *See also* Tube(s); Tubing **A14:** 690-693
multiple-roll rotary **A14:** 691-693
ovalizing, in rotary straighteners **A14:** 693
parallel-roll straightening **A14:** 691
press straightening **A14:** 690-691
pressure, control of **A14:** 690
tube material effects **A14:** 690
two-roll rotary **A14:** 691
Tube swaging **A14:** 128-144
Tube(s) *See also* Tube bending; Tube drawing; Tube forming; Tube piercing; Tube rolling; Tube spinning; Tube straightening; Tube swaging; Tubing
bending, nickel-base alloys **A14:** 834-835
by radial forging **A14:** 145
computing ovality in **A14:** 133
and cups, drawing with moving mandrel **A14:** 335-336
deflection **A14:** 692
drawing of **A14:** 335-336
electromagnetic forming of **A14:** 646
explosive forming of **A14:** 641-642
extruding, by Hydrafilm process **A14:** 329
internal shapes in **A14:** 138
piercing, and slotting **A14:** 470
rotary swaging of **A14:** 128-144
spinnability **A14:** 679
spinning **A14:** 675-679
stock, for bending **A14:** 671
swaging **A14:** 128-144
swaging, with mandrel **A14:** 137-139
swaging, without mandrel **A14:** 134-137
types, for swaging **A14:** 135
wall thickness **A14:** 135-136
Tubelets
flexible printed boards **EL1:** 590
Tuberculation **M1:** 734, 736
defined **A11:** 11 **A13:** 13
electrochemical/microbial processes **A13:** 122
localized biological corrosion **A13:** 119-120
tubercule formation **A13:** 43
Tubes *See also* Cylindrical parts; Heat exchangers, failures of Tubing; Holes; Photomultiplier tubes; Steel tubing; Tubing; Tubular products
with bends, remote-field eddy current inspection **A17:** 195
blockage, in boilers **A11:** 606
boiler, corrosion-fatigue cracks **A11:** 79
boiler/pressure vessel, inspection **A17:** 644-645
calandria, remote-field eddy current inspection **A17:** 199-200
connector, aluminum alloy, blowout failure **A11:** 312-313
copper alloys **M2:** 472
with diameter changes, remote-field eddy current inspection **A17:** 195
dimension, as test variable **A17:** 175
eddy current inspection of **A17:** 179-184
effect of exceeding service temperatures in **A11:** 290
erosion-corrosion of **A11:** 342
examination by electropolishing **A9:** 55
-expanding tool, fatigue fracture of **A11:** 474-475
and fin assembly, from economizer **A11:** 432
function in heat exchangers **A11:** 628
heat exchanger, eddy current inspection **A17:** 181-182
heat-flow path through **A11:** 603

Tubes (continued)
horizontal centrifugal casting of **A15:** 299-300
for image analyzer scanners **A10:** 310
integral-finned, SCC in **A11:** 635
leaks in **A11:** 606
mortar **A11:** 294-296
mounting **A9:** 31
Nessler, color comparison **A10:** 66
outside diameter over 75 mm (3 in.)
eddy current inspection **A17:** 182-183
outside diameter under 75 mm (3 in.)
eddy current inspection **A17:** 179-182
overrolling of **A11:** 629
pressure, SCC of brazed joint in **A11:** 453
pure quartz sample, ESR spectrometer **A10:** 256
quench crack failure of **A11:** 334-335
radiographic methods **A17:** 296
ruptures, by overheating and embrittlement **A11:** 604-614
seamless, quench crack failure **A11:** 334-335
seamless, ultrasonic inspection **A17:** 272
stainless steel, after carburizing and oxidizing **A11:** 272
steam, corroded inner surface **A11:** 176
steam-coated, internal scale thickness **A11:** 604
steam-generator, thick-lip ruptures in **A11:** 605-606
steam-generator, thin-lip ruptures in **A11:** 606
steel boiler **A11:** 603-614
steel macroscopic examination **A11:** 21
sudden rupture in **A11:** 603
superheater, circumferential corrosion fatigue cracks **A11:** 79
u-shaped, eddy current inspection **A17:** 184-185
uniform corrosion in **A11:** 175
wall-thinning in **A11:** 606
Waspaloy spray-manifold, fatigue fracture in **A11:** 454-455
wastage, creep failure from **A11:** 612
x-ray, and Z element determination by XRS **A10:** 101
x-ray, for x-ray spectrometry **A10:** 87-90
x-ray, molybdenum **A10:** 88
Zircaloy, liquid-metal embrittlement in **A11:** 234
Tubes, protection
for thermocouples **A2:** 883-884
Tubesheet interface
crevice corrosion **A13:** 110
Tubesheet rolled joints
eddy current inspection **A17:** 180-181
Tubing *See also* Heat exchangers, failures of Tube(s); Hole(s); Tube; Tube(s); Tubular products
bending, titanium alloy **A14:** 848
bending, with mandrel **A14:** 666-668
bending, without mandrel **A14:** 668
beryllium **A2:** 683
boxing, of **A11:** 630
brass *See* Brass, tubing
carbon steel, remote-field eddy current inspection **A17:** 200-201
characteristics of **A11:** 628
conducting, remote-field eddy current inspection **A17:** 195-201
copper, (111) pole figures from **A10:** 363
copper, ODF using Euler plots method **A10:** 361
cupro-nickel, pickling of **M5:** 612
dies and die materials for drawing **A14:** 337
distortion **A14:** 690
eddy current inspection of **A17:** 166, 179-184
extended surface (finned) **A11:** 628
ferromagnetic, flux leakage method **A17:** 132
fluorescent magnetic particle inspection **A17:** 105
impedance of **A17:** 169-170
inconel 600, residual stress and percent cold work distributions **A10:** 390
inside surface measured **A10:** 384
magnetic rubber inspection of **A17:** 123
magnetizing **A17:** 94

Tubing (continued)
material, straightening effects **A14:** 690
midwall, pole figures from **A10:** 362
nickel-base alloy, expanding of **A14:** 836
nonferrous, inspection of **A17:** 572-574
nonuniformity of texture in **A10:** 363-364
oil well, magnetic particle inspection of **A17:** 111-112
oil-well production, SCC of **A11:** 298
oval, bending of **A14:** 667
radiographic inspection **A17:** 335-337
refractory metals and alloys **A2:** 562-563
round, reshaping of **A14:** 631-632
round, straightening **A14:** 691
SCC in **A11:** 624-626
square and rectangular, straightening of **A14:** 691
stainless steel *See* Stainless steel, tubing
stainless steel, eddy current inspection **A17:** 182
stainless steel, precipitate identification in **A10:** 459-461
steel *See* Steel, tubing
steel, magnetic flaw characterization **A17:** 131
straightening of **A14:** 690-693
surface stress measurement in **A10:** 390
thickness, by eddy current inspection **A17:** 172
tool materials for drawing **A14:** 336
upsetting of **A14:** 91-93
welded, eddy current weld inspection **A17:** 186
zirconium, extrusion **A2:** 663
"Tubing" alloy *See also* Titanium alloys, specific types, Ti-3Al-2.5V
Tubular electrodes for arc welding
powders used **A7:** 573
Tubular heat exchangers
applications **A11:** 628
Tubular joint stresses **EM3:** 490
Tubular parts
cold extruded **A14:** 305
Tubular porous filters **A7:** 698
Tubular products *See also* Holes; Tube; Tube(s); Tubing; Wrought copper tubular products **A17:** 561-581
aluminum and aluminum alloys, as furniture **A2:** 13
arc-welded nonmagnetic ferrous **A17:** 566-567
characteristics **A17:** 561
classified **A17:** 561
in commercial applications **A17:** 574-575
continuous butt-welded steel pipe **A17:** 567
copper **M2:** 261-264
double submerged arc welded steel pipe **A17:** 565-566
duplex tubing **A17:** 572
eddy current inspection **A17:** 562-563
finned tubing **A17:** 571-572
flux leakage inspection **A17:** 563-564
in-service inspection **A17:** 574-581
inspection method, selection **A17:** 561-562
liquid penetrant inspection **A17:** 565
magnetic particle inspection **A17:** 565
nondestructive evaluation **A17:** 561-565
nonferrous tubing **A17:** 572-574
in oil and gas distribution ustry **A17:** 577-578
pipeline girth welds **A17:** 579-581
radiographic inspection **A17:** 565
resistance-welded steel **A17:** 562-565
seamless steel **A17:** 567-571
sections, radiographic inspection **A17:** 335-337
spiral-weld steel pipe **A17:** 567
steel, annealing **A4:** 39
steel, normalizing **A4:** 39-40
steel pipelines **A17:** 578-579
ultrasonic inspection **A17:** 564-565
wrought aluminum alloy **A2:** 33
wrought copper **A2:** 248-250
Tubular steel product *See* Steel tubular products
Tubular woven fabric
by fly-shuttle loom **EM1:** 128

Tuff
high-level waste disposal in **A13:** 975-976
Tuffriding
for valve train assembly components **A18:** 559
Tukon tester
for microhardness testing **A8:** 91
Tumbaga
as historic gold-silver-copper alloy **A15:** 19
Tumble coating
paint **M5:** 483
Tumble-type blenders **A7:** 189
Tumbler ball mills **A7:** 66
Tumbler mills **A7:** 66
Tumbling *See* Barrel finishing **EM4:** 100
abrasive blasting systems **M5:** 88-89, 93, 95
barrel finishing *See* Barrel finishing
defined **A15:** 11 **EM2:** 43
dry *See* Dry barrel finishing
finish for nails **M1:** 271
mass finishing *See* Mass finishing
mechanical coating *See* Mechanical coating
mill, development of **A15:** 33
rust and scale removed by **M5:** 12-13
safety precautions **M5:** 21
self- *See* Self-tumbling
wet *See* Wet barrel finishing; Wet tumbling
zinc alloys **A2:** 530
Tumbling, abrasive
for liquid penetrant inspection **A17:** 81-82
Tumbling, aircraft engine components
surface finish requirements **A16:** 22
Tumor resections
as prosthetic devices **A11:** 670-671
Tumorgenisis
and dental alloys **A13:** 1339
Tunable infrared lasers
applications **A10:** 112
Tundish
in atomization process **A7:** 25
in ferrous continuous casting **A15:** 310-311, 313
Tungstate solutions
copper/copper alloy corrosion in **A13:** 636
Tungsten *See also* Pure tungsten; Refractory metals; Refractory metals and alloys; Refractory metals and alloys, specific types; Specific tungsten materials; Tungsten alloys; Tungsten alloys, specific types; Tungsten carbide; Tungsten carbide powders; Tungsten fibers; Tungsten heavy metals; Tungsten oxide; Tungsten powders; Tungsten wire
additions to martensitic stainless steels **M6:** 348
adhesion and solid friction **A18:** 32
adhesion measurements **A6:** 144
AKS-doped **A2:** 578
alloying, nickel-base alloys **A13:** 641
alloys, workability of **A8:** 165, 575
-aluminum metal-matrix composites, splitting fracture **A12:** 467
as an addition to cobalt-base heat-resistant casting alloys **A9:** 334
as an addition to copper-base powder metallurgy electrical contacts **A9:** 551
as an addition to nickel-base heat-resistant casting alloys **A9:** 334
as an addition to niobium alloys **A9:** 441
as an addition to permanent magnet alloys **A9:** 538
as an addition to sliver-base switchgear materials **A9:** 552
as an addition to tantalum alloys **A9:** 442
annealing, effect on electrical resistivity **M3:** 327, 328
applications **A2:** 557-560
arc welding *See* Arc welding of molybdenum and tungsten
at elevated-temperature service **A1:** 641
atomic interaction descriptions **A6:** 144
back reflection intensity **A17:** 238
for base metal conductors **EM4:** 1142

Tungsten (continued)
brazing ... **M6:** 1059-1060
applications ... **M6:** 1059-1060
filler metals and their properties ... **M6:** 1059
fluxes and atmospheres ... **M6:** 1059
precleaning and surface preparation ... **M6:** 1057, 1059
process and equipment ... **M6:** 1059
brazing and soldering characteristics ... **A6:** 634-635
in cast iron ... **A1:** 6
characteristics and weldability ... **M6:** 462-463
chemical vapor deposition of ... **M5:** 382-383, 385
chromium plating of ... **M5:** 660
cleaning processes ... **M5:** 659-660
in cobalt-base alloys ... **A1:** 985 **A18:** 766
codeposited with nickel in electroplating ... **A18:** 836
for cofiring the metallization with the alumina ... **EM4:** 544
combustion accelerators ... **A10:** 222
commercial grade/undoped P/M ... **A2:** 577-578
commercially pure, infiltrated with copper ... **A9:** 446
commercially pure, pressed from powder ... **A9:** 446
in composition, effect on ductile iron ... **A4:** 686
composition similar to ceramics and glasses ... **EM3:** 300
consumption ... **A2:** 557
content in nickel-base and cobalt-base high-temperature alloys ... **A6:** 573
content in tool and die steels ... **A6:** 674
content of weld deposits ... **A6:** 675
creep rupture testing of ... **A8:** 302
crystal, cleavage planes ... **A12:** 462
dichalcogenides ... **A18:** 113
diffusion bonding ... **A6:** 156
effect, amorphous metals ... **A13:** 868
effect of, on hardenability ... **A1:** 395, 413
effect of temperature on strength and ductility of ... **A8:** 34, 36
effect on anodic dissolution of phases in wrought heat-resistant alloys ... **A9:** 308
effect on borided steels ... **A4:** 441
effect on maraging steels ... **A4:** 222
effect, Stellite alloys ... **A13:** 658
as electrical contact materials ... **A2:** 848-849
electrical contacts, use in ... **M3:** 671-672, 673, 674, 675
electrical discharge machining ... **A2:** 561
electrical properties ... **A2:** 580-581
electrical resistance applications ... **M3:** 641, 646, 647, 655
electrochemical polishing of ... **A9:** 45
electrodes for gas welding ... **M6:** 190-195
electrolytes for ... **A9:** 440
electrolytic etching ... **A9:** 440
electromechanical polishing ... **A9:** 441
electron beam drip melted ... **A15:** 413
electron beam welding ... **M6:** 639-640
electron-beam welding ... **A6:** 870
electroplating of ... **M5:** 660-661
electropolishing with alkali hydroxides ... **A9:** 54
embrittlement sources ... **A12:** 123
epithermal neutron activation analysis ... **A10:** 239
erosion resistance ... **A18:** 201
erosion test results ... **A18:** 200
etch-attack on ... **A9:** 45
fabrication ... **M3:** 314, 326
fabrication techniques ... **A2:** 562
field evaporation of ... **A10:** 586, 587
filament, gas mass spectrometer ... **A10:** 152-153
filler metals for ... **M3:** 320
filler metals for brazing of ... **A6:** 943
FIM sample preparation of ... **A10:** 586
finishing processes ... **M5:** 659-662
forge welding ... **A6:** 306 **M6:** 676
forging of ... **A14:** 237-238
friction coefficient data ... **A18:** 71
friction welding ... **M6:** 722
glass-to-metal seals ... **EM3:** 302
glass/metal seals ... **EM4:** 1037
grain boundary, FIM image ... **A10:** 589
gravimetric finishes ... **A10:** 171
grinding ... **A9:** 441
in heat-resistant alloys ... **A4:** 512

Tungsten (continued)
heating element, use in vacuum furnace ... **A4:** 500 **M4:** 316
for heating elements and hot furnace structures ... **A4:** 497
for heating elements for electrically heated furnaces ... **EM4:** 247, 248, 249
vapor pressure ... **EM4:** 249
high-resolution spectrum by ECAP analysis ... **A10:** 597
high-temperature solid-state welding ... **A6:** 298
hot swaging of ... **A14:** 142
hydrogen damage in ... **A11:** 338
inclusions ... **A17:** 50, 582, 584
joining ... **A2:** 564
material to which crystallizing solder glass seal is applied ... **EM4:** 1070
mechanical properties ... **M3:** 328-329
melted and resolidified ... **A9:** 561
methods for metallizing alumina ceramics ... **EM4:** 542
microstructure of wire ... **M3:** 327
microstructures ... **A9:** 442
Monte Carlo electron trajectories in ... **A12:** 167
mounting ... **A9:** 441
neutron and x-ray scattering, and absorption compared ... **A10:** 421
nickel plating of ... **M5:** 660
in nickel-base superalloys ... **A1:** 984
ore, flowchart of chemical processing ... **A7:** 152
oxidation-resistant coatings high-temperature ... **M5:** 662
oxygen cutting, effect on ... **M6:** 898
photometric analysis methods ... **A10:** 64
physical properties ... **A6:** 941
plasma-MIG welding ... **A6:** 224
polish-etching ... **A9:** 441
polishing of ... **A9:** 550
powder metallurgy ... **M3:** 326
powder metallurgy materials, polishing ... **A9:** 507
production ... **A2:** 577-578
properties ... **A6:** 629
pure ... **M2:** 816-821
pure, properties ... **A2:** 1170
purity ... **M3:** 326, 327
reactions with gases and carbon ... **M6:** 1055
recommended glass/metal seal combinations ... **EM4:** 497
recrystallization ... **M3:** 328-329
recrystallization and thermal conductivity ... **A2:** 562
recrystallization temperatures ... **M6:** 1055
relative erosion factor ... **A18:** 200
rolling ... **A2:** 562
sheet, blanking characteristics ... **M3:** 318
sheet forming ... **A14:** 787-788
single crystals uniformly hard, as control specimen ... **A18:** 424
sintered, intergranular fracture ... **A12:** 462
in sintered metal powder process ... **EM3:** 304
sintering ... **A7:** 389-392
sintering, time and temperature ... **M4:** 796
sintering/hot pressing ... **A2:** 579
solderability ... **A6:** 978
solid-state bonding in joining non-oxide ceramics ... **EM4:** 525
solubility of rhenium in, during etching ... **A9:** 447
-steel metal-matrix composites, splitting fractures ... **A12:** 467
stereographic projection ... **A10:** 585
temperatures, mill processing ... **M3:** 317
thermal diffusivity from 20 to 100 °C ... **A6:** 4
thermal expansion coefficient ... **A6:** 907
thermal properties ... **A18:** 42
in thermal spray coating materials ... **A18:** 832
TNAA detection limits ... **A10:** 237
to collimate or shield sources for storage and shipment ... **A18:** 325
to promote hardness ... **A4:** 124
tongs for manual resistance brazing ... **A6:** 340
in tool steels ... **A18:** 734, 735-736
toxicity and exposure limits ... **A7:** 206
transition temperatures ... **M6:** 1055
tubing ... **A2:** 563
and tungsten alloys ... **A2:** 577-581

Tungsten (continued)
typical field ion micrograph ... **A10:** 585
ultrapure, by zone-refining technique ... **A2:** 1094
undoped ... **A2:** 579-580
upset welding ... **A6:** 249
used as heating elements in vacuum furnaces ... **A4:** 499
vacuum heat-treating support fixture material ... **A4:** 503
vapor pressure, relation to temperature ... **A4:** 495 **M4:** 310
wear resistance of die material ... **A18:** 635-636
welds, ductility ... **A2:** 564
wire, die materials for drawing ... **M3:** 522
wire, x-ray tubes ... **A17:** 302
in wrought heat-resistant alloys ... **A9:** 311
x-ray characterization of surface wear results for various microstructures ... **A18:** 469

Tungsten alloys *See also* Tungsten
alloy production technique effect ... **A6:** 582
annealing ... **A4:** 816-817
applications ... **A2:** 557-560 **A7:** 469
brazing ... **A6:** 581
cleaning ... **A4:** 817
cleaning processes ... **M5:** 659-660
DBTT ... **A6:** 582
density and melting temperature ... **M5:** 380
diffusion bonding ... **A6:** 581
electrical properties ... **A2:** 580-581
electron-beam welding ... **A6:** 581, 582
electroplating of ... **M5:** 660-661
finishing processes ... **M5:** 659-662
forging of ... **A14:** 238
furnaces ... **A4:** 816-817
gas-tungsten arc welding ... **A6:** 580, 581
heat-affected zone ... **A6:** 582
heavy-metal, classes ... **A2:** 578-579
heavy-metal, manufacturing processes ... **A2:** 578-579
as high temperature materials ... **A7:** 766-767
interstitial impurities effect ... **A6:** 582
machinability ... **M3:** 330-332
main types ... **A2:** 578
mechanical properties ... **A7:** 469-476
microstructure effect ... **A6:** 582
oxidation protective coating of ... **M5:** 379-380
oxidation-resistant coatings high-temperature ... **M5:** 662
penetrators, flowchart of fabrication ... **A7:** 689
production ... **A2:** 578-581
property data ... **M3:** 348-349
stress-relieving ... **A4:** 817
structure ... **M3:** 326-327
tensile strength ... **A6:** 580
tool steels ... **A14:** 43, 54, 56, 81
tungsten and ... **A2:** 577-581
for wire-drawing dies ... **A14:** 336

Tungsten alloys electrolytes for ... **A9:** 440
mounting ... **A9:** 441

Tungsten alloys, specific types
2 thoria, boring ... **A16:** 859
2 thoria, end milling-slotting ... **A16:** 866
2 thoria, internal grinding ... **A16:** 869
2 thoria, spade drilling ... **A16:** 861
2 thoria, turning ... **A16:** 858
85% density, abrasive cutoff sawing ... **A16:** 868
85% density, boring ... **A16:** 859
85% density, counterboring ... **A16:** 860
85% density, drilling ... **A16:** 860
85% density, end milling-slotting ... **A16:** 866
85% density, face milling ... **A16:** 863
85% density, internal grinding ... **A16:** 869
85% density, oil hole or pressurized-coolant drilling ... **A16:** 861
85% density, reaming ... **A16:** 862
85% density, spade drilling ... **A16:** 861
85% density, spotfacing ... **A16:** 860
85% density, turning ... **A16:** 858
93% density, abrasive cutoff sawing ... **A16:** 868
93% density, boring ... **A16:** 859
93% density, counterboring ... **A16:** 860
93% density, drilling ... **A16:** 860
93% density, end milling-slotting ... **A16:** 866
93% density, internal grinding ... **A16:** 869
93% density, oil hole or pressurized-coolant drilling ... **A16:** 861

Tungsten alloys, specific types (continued)
93% density, reaming **A16:** 862
93% density, spade drilling **A16:** 861
93% density, spotfacing **A16:** 860
93% density, turning **A16:** 858
96% and 100% density, abrasive cutoff sawing **A16:** 868
96% and 100% density, boring **A16:** 859
96% and 100% density, drilling **A16:** 860
96% and 100% density, end milling-slotting **A16:** 866
96% and 100% density, internal grinding **A16:** 869
96% and 100% density, oil hole or pressurized-coolant drilling **A16:** 861
96% and 100% density, reaming **A16:** 862
96% and 100% density, spade drilling **A16:** 861
96% and 100% density, turning **A16:** 858
Anviloy 1100, circular sawing **A16:** 868
Anviloy 1150, circular sawing **A16:** 868
Anviloy 1200, circular sawing **A16:** 868
Doped W
composition **M3:** 316
wire, microstructure **M3:** 327
T-111, electron-beam welding **A6:** 871
T-222, electron-beam welding **A6:** 871
Ta-10W, electron-beam welding **A6:** 871
W-25Re, liquid-metal embrittlement of **A11:** 234
W-25Re, physical properties **A6:** 941
W-250_s, refractory metal brazing filler metal **A6:** 942
W-Mo **M3:** 326
composition **M3:** 316
W-Ni-Cu, class 1
composition **M3:** 330-331
mechanical properties **M3:** 330-332
W-Ni-Fe, class 1
composition **M3:** 330-331
mechanical properties **M3:** 330-332
W-Ni-Fe, class 3
composition **M3:** 330, 331
mechanical properties **M3:** 330-332
W-Ni-Fe class 4
composition **M3:** 330, 331
mechanical properties **M3:** 330-332
W-Re **M3:** 326
composition **M3:** 316
W-ThO_2 ..... **M3:** 326
composition **M3:** 316
WC-3Co cemented carbide, brittle fracture **A11:** 26
Tungsten alloys, specific types W-3Re, wire, non-sag, doped lamp grade **A9:** 446
W-10Ni, selected-area electron-channeling pattern **A9:** 94
WC-15Ti, micrograph **A9:** 46
Tungsten and tungsten alloys
boring **A16:** 859
in cast Co alloys **A16:** 69
electrical discharge machining **A16:** 558, 560
electrochemical grinding **A16:** 543, 545, 547
electrochemical machining **A16:** 533-535, 536, 537, 541
electrode material for EDM **A16:** 559
electron beam machining **A16:** 570
in high-speed tool steels **A16:** 52
in P/M high-speed tool steels **A16:** 61
photochemical machining **A16:** 588, 590
surface alterations **A16:** 27
Tungsten arc welding *See* Gas-tungsten arc welding
Tungsten bar
wrought **A7:** 627
Tungsten carbide *See also* Carbide tools; Carbides; Cemented carbides **A16:** 71 **A18:** 758, 760-761
in 75W-25Ag powder metallurgy material **A9:** 561
abrasive wear **A18:** 760-761
alloyed grades **A2:** 953-954

Tungsten carbide (continued)
as an addition to silver-base switchgear materials **A9:** 552
applications **A16:** 87, 88 **A18:** 758, 761
ball indenters **A8:** 84
bearing blocks, in axial compression testing **A8:** 57
in cemented carbides **A18:** 795, 796
cermets, steel-bonded **A2:** 1000
classification **A16:** 75
CO-bonded carbides **A16:** 70-71, 73-75, 77-79
coarse granular tube rod, advantages and applications of materials for surfacing build-up, and hardfacing **A18:** 650
coating for jet engine components **A18:** 592
coatings for **A16:** 79-82
cobalt-bonded, and heat-treatable steel-bonded carbides, compared **A2:** 996
compared with cermets **A16:** 93
composition **A18:** 758, 761
composition in laser cladding **M6:** 798, 800
compositions **A16:** 72, 73
cutting speed and work material relationship **A18:** 616
in diamond cutting tools **A2:** 976-977
diamond for machining **A16:** 105
electrical discharge grinding **A16:** 566
electrical discharge machining **A16:** 558, 559, 560
electrochemical grinding **A16:** 542, 547
fine-grain tubular rods, advantages and applications of materials for surfacing build-up, and hardfacing **A18:** 650
friction coefficient data **A18:** 72
for gas-lubricated bearings **A18:** 532
granules or inserts, advantages and applications of materials for surfacing, build-up and hardfacing **A18:** 650
ground by diamond wheels **A16:** 107
highly alloyed grades **A2:** 953
for hot-forging dies **A18:** 627
for laser melt/particle injection **A18:** 869, 870-871
manufacture **A18:** 760, 761
material yield strength **A16:** 39
methods used for synthesis **A18:** 802
microstructure **A18:** 760-761, 799
microstructures **A16:** 72-75
nozzle tips for abrasive jet machining **A16:** 511-512
part material for ion implantation **A18:** 858
PCD tooling **A16:** 110
as plasma-spray coating for titanium alloys **A18:** 780
polishing of **A9:** 550
powder, preparation of **A2:** 950
properties **A8:** 234 **A16:** 72, 77-79 **A18:** 761, 795, 796, 801, 812, 813
Rockwell scale for **A8:** 77
seal material **A18:** 551
service temperature of die materials in forging **A18:** 625
skeletons, composites with **A2:** 855
softening point **A16:** 601
straight grades **A2:** 954
submicron WC-CO alloys **A16:** 73, 74
substrate and diamonds **A16:** 106
support shims for PCBN inserts **A16:** 111
thermal properties **A18:** 42
thermal spray coating material **A18:** 832
thermal spray coating material **A18:** 832
tool wear mechanisms **A16:** 5-7
use in laser melt/particle inspection **M6:** 798
used to cut hydroxyapatite blocks **A18:** 668
Vickers and Knoop microindentation hardness numbers **A18:** 416
Tungsten carbide (WC) **EM4:** 808-810
abrasive machining **EM4:** 325, 326

Tungsten carbide (WC) (continued)
applications **EM4:** 810, 977
wear **EM4:** 975
for banding wheels used in decorating method **EM4:** 472
cemented carbides **EM4:** 808-810
crystal properties **EM4:** 810
ditungsten carbide **EM4:** 810
pressure densification
pressure **EM4:** 301
technique **EM4:** 301
temperature **EM4:** 301
properties **EM4:** 326, 806, 810, 974
Tungsten carbide + alumina (WC + Al_2O_3)
adiabatic temperatures **EM4:** 229
synthesized by SHS process **EM4:** 229
Tungsten carbide cermet
thermal expansion coefficient **A6:** 907
Tungsten carbide metal-cutting insert
wear mark **A9:** 99
Tungsten carbide phase in cemented carbides **A9:** 274-275
Tungsten carbide powders **A7:** 156-158
alloyed **A7:** 773
cemented, liquid-phase sintering **A7:** 320
cobalt-covered **A7:** 173
cold isostatic pressing dwell pressures **A7:** 449
compaction **A7:** 774
contact angle with mercury **A7:** 269
die inserts **A7:** 337
fine and coarse, tap densities **A7:** 277
grades **A7:** 776, 777
high-energy milling **A7:** 69
particles, ball milled with cobalt powder **A7:** 145
production of **A7:** 152, 156-158, 773-774
shipment tonnage **A7:** 24
sintering and preforming **A7:** 774
spray-dried, rigid tool compaction of **A7:** 322
tantalum carbides in **A7:** 158
Tungsten carbide precipitates in 99W-1Ni powder metallurgy material **A9:** 562
Tungsten carbide, sintering
time and temperature **M4:** 796
Tungsten carbide tricone drill bits **A2:** 976
Tungsten carbide(s)
advantages **A6:** 797
applications **A6:** 797
overlayed on cast irons **A6:** 721
properties **A6:** 629
Tungsten carbide-based cermets **A7:** 804-805
Tungsten carbide-cobalt (WC-Co)
sintering **EM4:** 268
Tungsten carbide-cobalt (WC-Co) cemented carbides **A13:** 846-848
Tungsten carbide-cobalt (WC-Co) composites
electrical discharge machining **EM4:** 374, 375
Tungsten carbide-cobalt alloys
coatings for titanium alloy jet engine components **A18:** 590, 591
for cold heading dies **A18:** 627
compositions and microstructures **A2:** 951-952
corrosion resistance **A18:** 796
laser cladding components and techniques **A18:** 869
mechanical and physical properties **A18:** 813
microstructures **A18:** 797, 799
plasma-spray coating for pistons **A18:** 556
properties **A18:** 796-797
relative erosion factor **A18:** 200
Tungsten carbide-copper composites, properties
for electrical make-break contacts **A2:** 853
Tungsten carbide-iron powder
laser cladding **A18:** 867
Tungsten carbide-nickel alloys
in cemented carbides **A18:** 795
corrosion resistance **A18:** 800

Tungsten carbide-silver composites, properties
for electrical make-break contacts **A2:** 853
Tungsten carbide/cobalt powders
atmosphere conditions for neutral carburizing potentials **A7:** 388
cold isostatic pressing applications **A7:** 450
density-temperature relationship **A7:** 388
effect of sintering on grain growth **A7:** 389
grades **A7:** 774-776
high-energy milling **A7:** 69
liquid-phase sintering **A7:** 319, 320
microstructure after liquid-phase sintering **A7:** 320
microstructure during sintering **A7:** 388
spray dried **A7:** 76, 77
tools, by containerless hot isostatic pressing **A7:** 441
Tungsten carbide/tantalum carbide/cobalt grades
properties **A7:** 776
Tungsten carbide/titanium carbide/cobalt grades
properties **A7:** 776
Tungsten carbonyl
chemical vapor deposition process **M5:** 382-383
Tungsten combustion accelerators **A10:** 222
Tungsten composites
polishing of **A9:** 550
Tungsten disulfide **A18:** 114
coating for gears, lubrication **A18:** 541
effect of testing parameters **A18:** 806
high-vacuum lubricant applications **A18:** 154, 159
rolling-element bearing lubricant **A18:** 138
Tungsten disulfide, powdered
for compression testing **A8:** 195
Tungsten electrode *See also* Electrodes and Gas tungsten arc welding
definition **M6:** 18
Tungsten fibers *See also* Continuous tungsten fiber MMCs; Fibers; Tungsten; Tungsten wire reinforced composites
brittle transgranular failure **A12:** 466
ductile-to-brittle transition effects **A12:** 467
embrittlement **A12:** 466
metal-matrix composites **A12:** 466
microstructures **EM1:** 878
splitting fracture **A12:** 467
Tungsten filament precursors for boron fibers **A9:** 592, 595
Tungsten hairpin filament electron gun **A10:** 492
Tungsten heavy metals **A7:** 17-18, 469
as high-temperature materials **A7:** 767
mechanical properties during sintering **A7:** 393
sintering **A7:** 392-393
Tungsten hexachloride
chemical vapor deposition process **M5:** 382-383
Tungsten hexafluoride
chemical vapor deposition process **M5:** 382-383
Tungsten high-speed steels
composition limits **A18:** 735
Tungsten high-speed tool steels
forging temperatures **A14:** 81
Tungsten hot-work steels
catastrophic die failure/plastic deformation **A18:** 641
composition limits **A18:** 735
for hot-forging dies **A18:** 623, 625
resistance to heat checking **A18:** 639
service temperature of die materials in forging **A18:** 625
Tungsten inclusions **A6:** 1073 **M6:** 839
Tungsten inert gas (TIG) welding
application method for interlock coatings **A18:** 590
Tungsten inert gas welding (TIG) *See* Gas-tungsten arc welding
Tungsten machinable alloys, manufacturing
finishing **M3:** 332
hot pressing **M3:** 331-332
machining **M3:** 332
metal powders **M3:** 331
sintering **M3:** 331
Tungsten nitride
coating for jet engine components **A18:** 592
Tungsten oxide
as lubricant **A14:** 238
Tungsten oxide reduction **A7:** 52-53, 153
Tungsten oxides **A7:** 153, 273, 623
Tungsten powder spheres
wetting by liquid copper during sintering **A9:** 100
Tungsten powders *See also* Tungsten; Tungsten alloys; Tungsten carbide powders; Tungsten carbide/cobalt powders; Tungsten heavy metals; Tungsten oxides; Tungsten wire, specific tungsten materials
activated sintering **A7:** 318
aluminum oxide, potassium, and silicon dioxide doping **A7:** 153
apparent tap density and particle shape **A7:** 189
chemical analysis **A7:** 154-155, 249
cold isostatic pressing dwell pressures **A7:** 449
compacts **A7:** 318, 389-390
contact angle with mercury **A7:** 269
electrode **A7:** 26
explosive isostatic compaction of **A7:** 305
filaments, as electrical/magnetic application **A7:** 16, 629-630
fine and coarse, tap densities **A7:** 277
finishing **A7:** 154
as high-temperature material **A7:** 765-767
hydrogen reduced, shipment tonnage **A7:** 24
lamp filaments **A7:** 16
melting point and density **A7:** 152
mesh heating and induction **A7:** 389
microstructures **A7:** 155, 318, 555
particle size distribution **A7:** 153, 154, 602
particle sizes **A7:** 153, 154
pressure and green density **A7:** 298
processing sequence flowchart **A7:** 767
production **A7:** 152-158
properties **A7:** 153-154
purity .... **A7:** 153— 154, 389-390
pyrotechnics **A7:** 597, 601
reduced **A7:** 632
as refractory metal, commercial uses **A7:** 17
relative density vs. sintering time **A7:** 390
sampling **A7:** 249
with selenium and molybdenum **A7:** 206
sintering **A7:** 389-392
toxicity and exposure limits **A7:** 206
Tungsten selenide
high-vacuum lubricant applications **A18:** 154
Tungsten steel
thermal properties **A18:** 42
Tungsten steels
composition of tool and die steel groups **A6:** 674
Tungsten tip plasma torch **A15:** 419-420
Tungsten tool steels
high-speed steels **A1:** 759-762
hot-work steels **A1:** 762-763
Tungsten trioxide
analysis of calcination and activation of **A10:** 133
Tungsten wire **A7:** 630, 767
annealed **A9:** 446
applications **A2:** 558, 582-584
non-sag doped lamp grade **A9:** 446
Tungsten wire reinforced composites **EM1:** 879, 880, 882, 886
Tungsten wires
mounting **A9:** 441
Tungsten, zone refined
impurity concentration **M2:** 713
Tungsten-base electrical contact materials
microstructures of **A9:** 552
Tungsten-base hot-work tool steels
composition **A14:** 43
forging temperatures **A14:** 81
heat treating **A14:** 54
tempering temperature effects **A14:** 56
Tungsten-base powder metallurgy materials, specific types
51W-49Ag **A9:** 562
55W-45Cu **A9:** 560
65W-35Ag **A9:** 561-562
68W-32Cu **A9:** 560
70W-30Cu **A9:** 561
72W-28Ag **A9:** 562
75W-25Ag **A9:** 561
75W-25Cu **A9:** 560
80W-20Cu **A9:** 560
81W-19Ag **A9:** 561
Tungsten-base powder metallurgy materials, specific types (continued)
87W-13Cu **A9:** 560
90W-6Ni-4Cu, as sintered **A9:** 446
90W-10Ag **A9:** 561
90W-10Cu **A9:** 560
98W-2ThO$_2$, pressed and sintered **A9:** 446
99W-1Ni **A9:** 562
Tungsten-copper composites, properties
for electrical make-break contacts **A2:** 853
Tungsten-copper contact materials **A7:** 560, 561
Tungsten-copper powders **A7:** 319, 555
Tungsten-copper-nickel
heavy metal compositions **A7:** 17-18
Tungsten-graphite-silver composites, properties
for electrical make-break contacts **A2:** 853
Tungsten-halogen filament lamps for microscopes **A9:** 72
Tungsten-inert gas weld
weld microstructure **A6:** 53
Tungsten-iodide lamp
for continuum source background correction **A10:** 51
Tungsten-nickel-copper, sintering
time and temperature **M4:** 796
Tungsten-nickel-copper-iron powders **A7:** 690
Tungsten-nickel-iron systems **A7:** 319, 320
Tungsten-rhenium
annealing practices for microexamination **A9:** 447-448
metallographic techniques for **A9:** 447-448
for thermocouples used in vacuum heat treating **A4:** 506, 507
Tungsten-rhenium alloy
mechanical properties **A7:** 477
Tungsten-rhenium thermocouples
insulation **A2:** 883
types/properties/application **A2:** 876
Tungsten-silicon carbide composite shell
vapor forming of **M5:** 382-383
Tungsten-silver composites, properties
for electrical make-break contacts **A2:** 853
Tungsten-silver powders **A7:** 319, 555, 561
Tungsten-titanium-tantalum (niobium) carbides
manufacture of **A2:** 950-951
Tungsten/titanium carbide powders **A7:** 158, 273
Tuning
in ultrasonic testing **A8:** 242-243
Tuning fork specimens
SCC testing **A13:** 251-252
Tunnel boring
with cemented carbide tools **A2:** 1002
Tunnel junction structures
metal/solid surface analysis by SERS **A10:** 136
study of molecules by SERS in **A10:** 137
Tunneling **A6:** 1078
crack-front **A11:** 20
electron, rate in field ionization **A10:** 585
field ionization as quantum mechanical process of **A10:** 584
Tup impact test *See also* Impact test
defined **EM2:** 43
Turbidimeter
schematic **A7:** 219
Turbidimetry
light and X-ray **A7:** 219
Turbine
blades **A11:** 29, 285
deposition in **A11:** 616
spacer **A11:** 285
vane **A11:** 285
Turbine alloys
alloying element partitioning **A10:** 583
phase stability and transformations FIM/AP **A10:** 583
Turbine blades *See also* Turbine(s) **A7:** 651
holographic inspection **A17:** 16
integrated, as NDE reliability case study **A17:** 686-687
optical holography of **A17:** 405, 424-425
residual core detection, by neutron radiography **A17:** 393-394
scanning laser gages for **A17:** 12
Turbine blades and vanes, superalloy
aluminum coating process **M5:** 335, 339
Turbine casing
for elevated-temperature service **A1:** 621

Turbine disks ... **A7:** 522-523, 646-650
Turbine engine parts
vacuum coating of ... **M5:** 395, 404, 406, 409, 411
Turbine oil
applications ... **A18:** 541
defined ... **A18:** 20
lubricant classification ... **A18:** 86
oxidation inhibitors ... **A18:** 105
Turbine rotor
open-die forging of ... **A14:** 669
Turbine rotor steels ... **A1:** 619 , 620-621, 937, 938
Turbine shaft preform
by radial forging ... **A14:** 147
Turbine wheel forging
contour ... **A14:** 71
Turbine wheels
fir-tree or dovetail slots ... **A16:** 196
Turbine(s)
borescope inspection ... **A17:** 9
flaws, through flexible fiberscope ... **A17:** 7
forgings, ultrasonic inspection ... **A17:** 232
gas engines, as NDE reliability case study ... **A17:** 681-684
Turbines
blades, hot corrosion of ... **A13:** 1000-1001
combustion, corrosion of ... **A13:** 999-1001
fretting wear ... **A18:** 243
steam ... **A13:** 993-995
Turbines, superalloy
oxidation protective coatings for ... **M5:** 375-377
Turbo-supercharger, for aircraft engines
and nickel alloy development ... **A2:** 429
Turbocharger turbine wheels
design practices for structural ceramics ... **EM4:** 722-726
Turbochargers ... **A18:** 567, 568
Turbomolecular pumps
for SEM vacuum system ... **A12:** 171
Turbostratic graphite
defined ... **EM1:** 49
Turbulence in the melt to effect grain refinement in aluminum alloy ingots ... **A9:** 629
Turbulent flow
in leaks ... **A17:** 58
Turbulent fluid flow ... **A6:** 162
Turbulent waves
in wave soldering ... **EL1:** 689
Turk's head machine
for wire rolling ... **A14:** 694
Turk's head shaping
steel mechanical tubing ... **M1:** 325-326
Turnbull, William
as early founder ... **A15:** 26
Turner equation
thermal expansion behavior of composite bodies ... **EM4:** 858
Turner, Joseph
as early founder ... **A15:** 25
Turning ... **A16:** 139-159
adaptive control implemented ... **A16:** 620, 622, 623-624, 625
after cold heading ... **A14:** 294
aircraft engine components, surface finish requirements ... **A16:** 22
Al alloys ... **A16:** 766-768, 769, 770
and arithmetic roughness average ... **A16:** 26
automatic bar machines ... **A16:** 141-142
automatic turning machines ... **A16:** 137, 140-141
axial ... **A16:** 383
bar-type machines ... **A16:** 138
basic lathe components ... **A16:** 136-137
Be alloys ... **A16:** 872
bench lathes ... **A16:** 137-138
carbon and alloy steels ... **A16:** 668-673, 675, 676
cast irons ... **A16:** 112, 648, 651, 652, 653, 654, 656-658
cemented carbides used ... **A16:** 75
ceramic tools ... **A16:** 101, 146

Turning (continued)
cermet tools ... **A16:** 90, 92, 93, 95-97
and chip formation ... **A16:** 8, 17
and chip removal for surface integrity ... **A16:** 33
compared to broaching ... **A16:** 205
compared to drilling ... **A16:** 234
compared to grinding ... **A16:** 426, 427, 428-429
compared to milling ... **A16:** 316
compared to thread grinding ... **A16:** 270
in conjunction with boring ... **A16:** 160, 164, 165, 168, 174
in conjunction with EDM ... **A16:** 560
in conjunction with tapping ... **A16:** 263
Cu alloys ... **A16:** 809-810, 811, 812
cutting fluid flow recommendations ... **A16:** 127
cutting fluids used ... **A16:** 125, 159
diamond (PCD) tooling ... **A16:** 110
dimensional accuracy ... **A16:** 155-157
duplicating lathes ... **A16:** 138
engine lathes ... **A16:** 136-138, 140, 142
equipment capacity ... **A16:** 154
flow, refractory metals and alloys ... **A2:** 562
gap-frame (gap-bed) lathes ... **A16:** 138
hafnium ... **A16:** 856
hand-screw machine ... **A16:** 139
heat-resistant alloys ... **A16:** 739-742
high-speed steel, carbide cutoff and form tools ... **A16:** 147
high-speed steel flank wear limits ... **A16:** 43
hollow-spindle (oil-country) lathes ... **A16:** 138
lathe classification system ... **A16:** 137
and ledge wear ... **A16:** 603
in machining centers ... **A16:** 393
maximum peak-to-valley roughness (height) ... **A16:** 26
Mg alloys ... **A16:** 820, 822-823
MMCs ... **A16:** 894, 896-898
multifunction machining ... **A16:** 366, 367, 368, 369, 370, 375, 376, 379, 384
multispindle automatic lathes ... **A16:** 141
NC implemented ... **A16:** 613, 614
Ni alloys ... **A16:** 837, 838, 839, 840, 841
P/M materials ... **A16:** 880, 881, 882, 884, 886, 889
PCBN cutting tools ... **A16:** 112
power consumption ... **A16:** 17
process capabilities ... **A16:** 135-159
radial tangential ... **A16:** 380
refractory metals ... **A16:** 858-859, 861, 862-863
residual stress distributions ... **A10:** 392
samples, for chemical surface studies ... **A10:** 177
screw machine ... **A16:** 139, 141
single-point cutting tools ... **A16:** 141, 142-148
speed lathes ... **A16:** 137
stainless steels ... **A16:** 155, 681, 690, 691, 692, 693, 696-697
surface alterations produced ... **A16:** 23
surface finish ... **A16:** 157, 158-159
surface finish requirements ... **A16:** 21
surface integrity effects ... **A16:** 28
surface roughness arithmetic average extremes ... **A18:** 340
and swaging, combined ... **A14:** 141
Swiss-type automatic screw machines ... **A16:** 141
theoretical surfaces produced ... **A16:** 22, 23
Ti alloys ... **A16:** 845, 846
tool geometries ... **A16:** 136
tool life ... **A16:** 143, 148-153, 158, 159
tool monitoring systems ... **A16:** 414, 416
tool theoretical surface roughness ... **A16:** 24
toolroom lathes ... **A16:** 137, 138
tracer lathes ... **A16:** 143
and transfer machines ... **A16:** 394, 397
tungsten and tungsten alloys ... **A2:** 560
turret lathes ... **A16:** 136, 138-139, 141, 142
uranium alloys ... **A16:** 874
vertical turret lathes ... **A16:** 139-140
vs swaging ... **A14:** 141
wheel lathes ... **A16:** 137, 138
workpiece configuration ... **A16:** 153-154

Turning (continued)
zirconium ... **A16:** 852, 853, 856
Zn alloys ... **A16:** 831-832
Turning, or boring
as machining processes ... **A7:** 461
Turning speed, steel machining *See* Cutting speed
Turnover/turnaround devices
between presses ... **A14:** 501
Turns per inch (TPI) *See also* Twist
defined ... **EM1:** 4 **EM2:** 43
Turnstile hanger blast cleaning machine ... **A15:** 508
Turntable vise
for microhardness test specimens ... **A8:** 93, 96
Turntables
for permanent mold casting ... **A15:** 276
Turquoise and metatorbernite
ESR analysis of ... **A10:** 265
Turret lathes ... **A16:** 1, 367, 368
automatic, tapping ... **A16:** 256, 259
boring ... **A16:** 160, 164, 168, 169, 170
Cu alloys machined ... **A16:** 815
die threading, dimensional control ... **A16:** 300
and drilling ... **A16:** 220, 229
horizontal ... **A16:** 168, 169
machine selection ... **A16:** 384, 385-386
manual ... **A16:** 369-371
mounting of tools ... **A16:** 383
reaming ... **A16:** 239-240, 242-245
selection of operations ... **A16:** 386
tapping ... **A16:** 256
trepanning ... **A16:** 176, 177
turning ... **A16:** 154, 155
workpiece fragility and shape ... **A16:** 383
Turret punch press
for piercing ... **A14:** 464
Tuyeres
adjustable ... **A15:** 30
cupolas ... **A15:** 385
defined ... **A15:** 11
double rows ... **A15:** 30
TV picture tubes *See* CRTs and TV picture tubes
Twaron aramid fibers ... **EM1:** 54
Twill weave *See also* Fabric(s); Weaves ... **EM1:** 111, 148
defined ... **EM2:** 43
Twin
defined ... **A9:** 19 **A11:** 11
Twin bands
defined ... **A9:** 19 **A11:** 11
in iron deformed 5% ... **A9:** 690
in plastically deformed titanium ... **A9:** 689
Twin boundaries
austenitic stainless steels ... **A12:** 351, 356
effect on magnetic domains ... **A9:** 535
in wrought stainless steels, etching to reveal ... **A9:** 281
Twin boundary energies ... **A6:** 143
Twin carbon arc brazing
definition ... **A6:** 1214 **M6:** 18
Twin carbon arc welding
definition ... **M6:** 18
Twin enclosed carbon arc weatherometer ... **EM2:** 577-578
Twin martensite ... **A6:** 76
Twin roll casting
continuous ... **A15:** 315
Twin shell blenders ... **A7:** 189
Twin-fluid atomization
copper powders ... **A2:** 392
Twin-matrix interface
iron cleavage fracture along ... **A12:** 224
Twin-on voltage instability
MOSFET ... **EL1:** 159
Twin-sheet
forming, size and shape effects ... **EM2:** 290
stamping, size and shape effects ... **EM2:** 290
thermoforming, defined ... **EM2:** 43

Twin-trace length as a function of angle for Mo-35Re single crystal........... **A9:** 128
Twinned columnar growth *See also* Columnar growth
in aluminum alloy 1100 .... **A9:** 631
in aluminum alloy 3003 .... **A9:** 631
in aluminum alloy ingots **A9:** 630
Twinned martensite **A9:** 673-674
effect of crystal structure on **A9:** 684
lattice rotation by **A9:** 700
in magnesium alloys.. **A9:** 427
in metals with medium to high stacking fault energies **A9:** 685
in molybdenum alloys...................................... **A9:** 127
in rhenium and rhenium-bearing alloys **A9:** 447
Twinning *See also* Mechanical twinning; Twinning, characterization of................ **A18:** 224
analytic methods for **A10:** 3
austenitic stainless steels............................... **A12:** 355
in cobalt-base alloys.. **A18:** 766
deformation-induced, effect on diffraction pattern **A10:** 440
effect on reorienting slip systems **A8:** 35
effect on texturing **A10:** 358
effect, unalloyed uranium................................ **A2:** 671
formation, low-carbon steel............................ **A12:** 252
imaged by x-ray topography **A10:** 366
light microscopy for... **A12:** 106
liquid impingement erosion............................ **A18:** 228
in master alloy processing.............................. **A15:** 108
mechanical, and cleavage **A12:** 4
parting, austenitic stainless steels **A12:** 356
in rutile, bright- and dark-field images of annealing **A10:** 443
in shape memory alloys.................................. **A2:** 897
subtle, as impeding determination of atomic structure **A10:** 344, 352-353
Twinning, as source
acoustic emissions **A17:** 287
Twinning, characterization of *See also* Twinning
analytical transmission electron microscopy.. **A10:** 429-489
optical metallography............................ **A10:** 299-308
x-ray diffraction.. **A10:** 380-392
Twins *See also* Deformation twins
in austenitic manganese steel castings **A9:** 239
in cadmium copper.. **A9:** 553
in crystals.. **A9:** 719
examination by phase contrast etching.......... **A9:** 59
in hafnium **A9:** 497, 499, 502
in pure metals **A9:** 610
quantitative metallography **A9:** 126
in silver **A9:** 556
in silver-cadmium alloy **A9:** 556
transmission electron microscopy **A9:** 118-119
in zirconium and zirconium alloys........ **A9:** 497, 499
Twins as a result of electric discharge machining **A9:** 27
Twist *See also* Balanced twist; Bend; Defect; Lay; Turns per inch (tpi); Turns per inch (TPI)
in contour roll forming **A14:** 633
defined **EL1:** 1160 **EM1:** 24 **EM2:** 43
drills, torsion tests for **A8:** 139
and flow localization **A8:** 170-171
glass textile yarns **EM1:** 110
measurement, in torsion testing **A8:** 158
number, determined **EM1:** 286
per unit length, in torsion, testing **A8:** 139
rate **A8:** 169-171, 726
reversal, effect on super-purity aluminum after deformation and strain........ **A8:** 174
symbol for... **A8:** 726
Twist, balanced *See* Balanced twist
Twist boundaries **A9:** 719
diffraction studies **A9:** 121
Twist boundary
cleavage steps from **A12:** 319
defined **A12:** 13
low-carbon steel.. **A12:** 252
molybdenum alloy **A12:** 464
schematic.. **A12:** 17
Twist boundary in gold **A9:** 609
Twist drills
of high-speed tool steels **A16:** 58
Twist hackle
in ceramics... **A11:** 745
Twist hackle (continued)
definition.. **EM4:** 633
Twist-compression bonding method
modifications.. **A6:** 144
Twist-off failures
in locomotive axles **A11:** 715
Twisting, multifilamentary wire
NbTi superconductors......................... **A2:** 1043, 1051
Two-angle technique
plane-stress elastic model **A10:** 384
Two-body abrasion **A18:** 537
Two-body collision process............................ **A18:** 448
Two-body wear **A18:** 263
Two-component adhesive **EM3:** 29, 51
Two-component dynamic hip screw plate
as internal fixation device.............................. **A11:** 671
Two-cycle oils
lubricant classification **A18:** 85
Two-dimensional defect analysis **A10:** 466
Two-dimensional images *See also* Three-dimensional images
human vs. machine vision.............................. **A17:** 30
Two-dimensional metal flow
strain computation for.......................... **A14:** 433-434
Two-dimensional planar figures
equations for particle-size distribution **A9:** 131
Two-directional fabrics **EM1:** 125-128
design **EM1:** 125-127
textile equipment for **EM1:** 127-128
Two-fluid atomization.......................... **A7:** 25, 75, 76
designs... **A7:** 29
metal buildup.. **A7:** 29
Two-frequency laser interferometer.......... **A17:** 14-15
schematic... **A17:** 15
Two-group rank test **A8:** 707
Two-hammer radial forging machines **A14:** 149
Two-high mill *See also* Cluster mill; Four-high mill
defined **A14:** 14
Two-high rolling mills **A14:** 351
Two-level fractional factorial designs **A17:** 746-750
Two-parameter Weibull distribution **A8:** 716-718
Two-part acrylic
performance of.. **EM3:** 124
Two-part adhesives
pot life **EM3:** 741
Two-part epoxy
performance of.. **EM3:** 124
Two-part urethane
performance of.. **EM3:** 124
Two-pattern size castings
uniform mold hardness for **A15:** 29
Two-phase alloys **A8:** 177-178
Two-phase field
relationship to massive transformations **A9:** 655
Two-phase instability
dendritic/eutectic **A15:** 122-123
Two-phase materials
atomic number contrast in analysis of **A10:** 508
FIM images of **A10:** 589, 590
Two-phase microstructures, coarse
erosion of **A18:** 206-207
Two-phase mixtures
formed by spinodal decomposition **A9:** 652-653
Two-plane curvature
by compression forming **A14:** 595
Two-point loaded specimens
SCC testing **A8:** 503-505
Two-point strategy
for defining fatigue strength **A8:** 704-705
Two-post loading frame
cam plastometer **A8:** 194
Two-pot tinning
cast iron and steel **M5:** 353-354
Two-ram HERF machines **A14:** 29, 101
Two-roll rotary straighteners **A14:** 686, 691
Two-shaft regenerative gas turbine engine **EM4:** 717
Two-sided alternative hypotheses **A8:** 626
Two-sphere model **A7:** 313
Two-stage plastic-carbon replicas **A9:** 108
Two-stage replicas
technique.. **A12:** 7, 182
Two-stage welding **A6:** 316
Two-step plastic-carbon technique
TEM replication **A12:** 7
Two-step temper embrittlement........................ **A1:** 698
Two-terminal devices *See also* Discrete semiconductor packages; Multiple-terminal devices; Three-terminal devices
compression mount **EL1:** 432
custom assemblies.. **EL1:** 432
as diodes **EL1:** 429-432
lead mount **EL1:** 429-430
opto devices **EL1:** 432
performance **EL1:** 423
stud mount **EL1:** 431-432
surface-mount **EL1:** 430-431
Two-way shape memory
shape memory alloys................................ **A2:** 897-989
Tx51 temper
defined **A2:** 27
Tx52 temper
defined **A2:** 27
Tx54 temper
defined **A2:** 27
Tx510 temper
defined **A2:** 27
Tx511 temper
defined **A2:** 27
Tying wire **A1:** 282
Type 6 nylon *See* Caprolactam
Type B thermocouples
properties and applications **A2:** 871, 873
Type E thermocouples
properties and applications **A2:** 871, 873
Type I magnetic contrast.................................. **A9:** 536
Type II magnetic contrast **A9:** 536-537
Type J thermocouples
properties and applications **A2:** 871-872
Type K thermocouples
properties and applications **A2:** 871-872
Type metals **M2:** 496
lead and lead alloy **A2:** 549-550
Type N thermocouples
Nicrosil/Nisil **A2:** 873
Type R thermocouples
properties and application......................... **A2:** 871, 873
Type S thermocouples
properties and application......................... **A2:** 871, 873
Type T thermocouples
properties and cryogenic application **A2:** 871, 873
Types I and II embrittlement
zinc **A13:** 184
Types of adhesives **EM3:** 35
Typical basis *See also* A-basis; B-basis; Design allowables; S-basis **EM3:** 29
defined **EM1:** 24
Typical-basis
defined **EM2:** 43
Tyranno **EM4:** 223
composition **EM4:** 225
mechanical properties, room-temperature............................... **EM4:** 225
Tyranno fibers **EM1:** 63
TZC *See* Molybdenum alloys, specific types **A16:** 858-864, 866-869
TZM *See* Molybdenum alloys, specific types **A16:** 858-864, 866-869
annealing.. **M4:** 655
composition **M4:** 651-652
electron-beam welding **A6:** 581
stress relieving **M4:** 655
TZM alloy *See* Molybdenum alloys, specific types, TZM
TZM molybdenum
for hot-forging dies **A18:** 627
materials for dies and molds **A18:** 622, 625
service temperature of die materials in forging **A18:** 625
TZM molybdenum alloy **A7:** 768-769
cold seal pressure vessel.................................. **A7:** 769
strength-to-density ratio, compared **A7:** 769
as tooling for uniaxial hot pressing **EM4:** 298
UNS R03630, properties **EM4:** 503
TZP (Y_2O_3, CeO_2)
properties.. **A6:** 949

U

u chart.. **EM3:** 796, 797

U-500
- contour band sawing **A16:** 363
- electrochemical machining removal rates **A16:** 534
- flash welding **M6:** 557
- grinding **A16:** 759
- milling **A16:** 314

U-700
- electrostream drilling **A16:** 539

U-bend heat-exchanger tubes
- corrosion fatigue **A11:** 637

U-bend specimens
- for SCC testing **A8:** 508-509 **A13:** 252-253
- single-stage stressing method **A8:** 512
- true stress/true strain **A8:** 511
- typical **A8:** 510

U-bend testing
- to verify stress-corrosion cracking **A1:** 725

U-bending dies **A14:** 14, 526

U-bends
- press-brake forming of **A14:** 540-541

U-bolts *See also* Studs
- fatigue fracture of **A11:** 533-535
- roll threading **M1:** 274-276
- selection of steel for **M1:** 274-276
- strength grades and property classes **M1:** 273-277

u-chart *See also* Control charts
- for number of defects per unit **A17:** 737

U-groove joints
- radiographic inspection **A17:** 334

U-groove weld
- definition **A6:** 1214

U-groove welds
- arc welding of heat-resistant alloys **M6:** 356-357, 360, 367-368
  - cobalt-based alloys **M6:** 367-368
  - nickel-based alloys **M6:** 360
- arc welding of nickel alloys **M6:** 437, 440, 442-443
- definition, illustration **M6:** 60-61
- gas metal arc welding of aluminum alloys **M6:** 384-385
- gas tungsten arc welding of heat-resistant alloys **M6:** 356-357, 360
- preparation **M6:** 66-68

U-shaped heat exchanger tubes
- eddy current inspection **A17:** 184-185

U-Zr (Phase Diagram) **A3:** 2•381

U.S. Army
- applications of ceramics in portable power pack **EM4:** 716

U.S. Government specifications
- ductile iron **M1:** 35
- QQ-N-286
  - springs, strip for **M1:** 286
  - springs, wire for **M1:** 284
- QQ-S-681 specification requirements **M1:** 377-378
- QQ-W-390, springs, wire for **M1:** 285
- zinc chromate primers for steel sheet **M1:** 175

U.S. government standards **EM3:** 61

U.S. military and government specifications
- liquid penetrant inspection **A17:** 87

U.S. Military and original equipment manufacturers (OEM)
- performance specifications of engine lubricants **A18:** 98
- performance specifications of nonengine lubricants **A18:** 99

U.S. Navy
- Garrett axial flow turbine engine demonstration **EM4:** 716

U.S. Steel
- grease mobility testing **A18:** 127

U.S.S. W.G. system *See* United States Steel Wire Gage system

UBE SNE-10 (high-purity silicon nitride powder)
- processed by glass-encapsulated HIP **EM4:** 199

Udel thermoplastic **EM1:** 99

Udimet 400
- composition **A6:** 573

Udimet 500
- aging **A4:** 796
- aging cycle **A4:** 812 **M4:** 656
- aging cycles **A6:** 574
- annealing **M4:** 655
- composition **A4:** 794, 795 **A6:** 573 **A16:** 736, 737 **M4:** 651-652
- creep strength **A4:** 807
- current densities **A16:** 543
- double aging **A4:** 798
- electrochemical grinding **A16:** 547
- grinding **A16:** 547, 759, 760
- for hot-forging dies **A18:** 626
- intermediate aging, effect of elimination on properties **M4:** 668
- machining **A16:** 738, 741-743, 746-758
- mechanical properties **A4:** 809
- milling **A16:** 547
- solution treating **M4:** 656
- solution treatment **A6:** 574
- solution-treating **A4:** 793, 796
- stress relieving **M4:** 655
- tensile properties **A4:** 807
- thread grinding **A16:** 275

Udimet 520
- composition **A4:** 794 **A6:** 573

Udimet 630
- composition **A4:** 794 **A6:** 573 **A16:** 736
- machining **A16:** 738, 741-743, 746-747, 749-758

Udimet 700
- aging **A4:** 796
- aging cycle **A4:** 812 **M4:** 656
- aging precipitates **A4:** 796
- annealing **M4:** 655
- applications **A4:** 807
- composition **A4:** 794, 795 **A6:** 564, 573 **A16:** 736, 737 **M4:** 651-652, 653 **M6:** 354
- electrochemical grinding **A16:** 547
- electrochemical machining removal rates **A16:** 534
- electron-beam welding (wrought) **A6:** 869
- grinding **A16:** 547
- heat treatments **A4:** 807-808
- for hot-forging dies **A18:** 626
- machinability **A16:** 737
- machining **A16:** 738, 741-743, 746-758
- milling **A16:** 547
- solution treating **M4:** 656
- solution-treating **A4:** 793, 796
- stress relieving **M4:** 655
- thread grinding **A16:** 275

Udimet 710
- aging precipitates **A4:** 796
- applications **A4:** 807
- composition **A4:** 794, 795 **A6:** 573 **A16:** 736
- heat treatments **A4:** 807-808
- machining **A16:** 738, 741-743, 746-747, 749-757
- thread grinding **A16:** 275

Udimet 720
- composition **A6:** 573

Udimet alloys *See* Nickel alloys, specific types, Udimet
- photochemical machining **A16:** 588

Uhlig (thin-film) theory **A13:** 67

UHMWPE *See* Ultrahigh molecular weight polyethylene

UHV **EM3:** 238

UKAEA Refel 1 (RBSC)
- properties **EM4:** 240

UL 94 flame classes **EM2:** 77

UL flammability ratings *See* Flammability

Ulexite **EM4:** 380

Ultem thermoplastic **EM1:** 99

Ultimate bearing strength, calculated
- pin bearing testing **A8:** 61

Ultimate compressive strength *See also* Compressive strength
- carbon fiber/fabric reinforced epoxy resin **EM1:** 411
- epoxy resin system composites **EM1:** 401, 404, 406, 408, 413
- glass fabric reinforced epoxy resin **EM1:** 404
- glass fiber reinforced epoxy resin **EM1:** 406
- graphite fiber reinforced epoxy resin **EM1:** 413
- high-temperature thermoset matrix composites **EM1:** 375, 379
- Kevlar 49 fiber/fabric reinforced epoxy resin **EM1:** 408
- low-temperature thermoset matrix composites **EM1:** 393, 395-396
- medium-temperature thermoset matrix composites **EM1:** 383, 386, 389, 390
- quartz fabric reinforced epoxy resin **EM1:** 414
- thermoplastic matrix composites **EM1:** 365, 368, 371

Ultimate elongation *See also* Elongation **EM3:** 29
- defined **EM1:** 24 **EM2:** 43

Ultimate resilience **A18:** 228

Ultimate shear strength *See also* Shear strength
- carbon fiber/fabric reinforced epoxy resin **EM1:** 411
- epoxy resin system composites **EM1:** 402
- glass fiber reinforced epoxy resin **EM1:** 406
- high-temperature thermoset matrix composites **EM1:** 376, 380
- influence of shear plane and loading direction **A8:** 64
- Kevlar 49 fiber/fabric reinforced epoxy resin **EM1:** 408
- low-temperature thermoset matrix composites **EM1:** 393, 396
- medium-temperature thermoset matrix composites **EM1:** 387, 389, 391
- single-shear test for **A8:** 63-64
- thermoplastic matrix composites **EM1:** 366, 369, 371
- torsion tests for **A8:** 139
- variables affecting **A8:** 68

Ultimate shear stress **A8:** 148

Ultimate static strength
- graphite-epoxy composite **A8:** 715

Ultimate strain
- and expanding ring test **A8:** 210

Ultimate strength *See also* Yield **A8:** 14
- defined **A11:** 11 **A13:** 13 **A14:** 14 **EM2:** 434
- effect on toughness and crack growth **A11:** 54
- from pin bearing testing **A8:** 61

Ultimate tensile strength *See also* Rupture; Tensile properties; Tensile strength; Tensile stress **A7:** 312 **A18:** 810 **EM3:** 29
- by constant crosshead speed testing **A8:** 43
- carbon fiber/fabric reinforced epoxy resin **EM1:** 410
- of crack growth specimen **A8:** 381
- defined **A8:** 20 **EM1:** 24 **EM2:** 43
- as engineering stress-strain parameter **A8:** 20-21
- epoxy resin system composites **EM1:** 401, 404-405, 407, 412, 414
- from results of two heats **A8:** 623
- glass fabric reinforced epoxy resin **EM1:** 404
- glass fiber reinforced epoxy resin **EM1:** 405
- graphite fiber reinforced epoxy resin **EM1:** 412
- high-temperature thermoset matrix composites **EM1:** 375, 378
- Kevlar 49 fiber/fabric reinforced epoxy resin **EM1:** 407
- low-temperature thermoset matrix composites **EM1:** 394
- medium-temperature thermoset matrix composites **EM1:** 383, 385, 388, 391
- quartz fabric reinforced epoxy resin **EM1:** 414
- sulfur/carbon effects **A14:** 200
- symbol for **A11:** 798
- tension testing machines for **A8:** 47
- thermoplastic matrix composites **EM1:** 365, 370
- titanium PA P/M alloy compacts **A2:** 654

Ultimate tensile strength (continued)
vs. temperature, Kevlar aramid fiber **EM1:** 362
Ultimate tensile strength (UTS) **A6:** 389
solid-state-welded interlayers **A6:** 165-167
Ultimate tensile strength, effect of number of grains on
in tin and tin alloys **A9:** 125
Ultimet
composition **A6:** 598
gas-metal arc welding **A6:** 598
microstructure **A6:** 598
tensile properties relative to wrought and cast products **A6:** 599
Ultra-high strain rate (dynamic) compaction
aluminum and aluminum alloys **A2:** 7
Ultraclean powders
by rotating electrode process **A7:** 36
generation and clean-room processing **A7:** 41-42, 44
Ultradry hydrogen atmospheres, for brazing **A11:** 450
Ultrafine grinding
of brittle and hard materials **A7:** 59
Ultrafine powders
by high-energy milling **A7:** 70
Ultrahard tool materials *See also* Cubic boron nitride; Diamond; Superhard materials; Tool(s)
applications **A2:** 1015
cubic boron nitride (CBN), properties of **A2:** 1010-1011
cubic boron nitride (CBN), synthesis of **A2:** 1008-1009
for cutting tool materials **A18:** 617
diamond, properties of **A2:** 1009-1010
diamond, synthesis of **A2:** 1008-1009
metals ground/machines with **A2:** 1013
polycrystalline cubic boron nitride (PCBN) tool blanks **A2:** 1016-1017
polycrystalline diamond tool blanks **A2:** 1015-1016
sintered polycrystalline cubic boron nitride properties of **A2:** 1012
sintered polycrystalline diamond, properties of **A2:** 1011-1012
superabrasive grains **A2:** 1012-1015
and superabrasives **A2:** 1008, 1015-1017
Ultrahigh molecular weight polyethylene **A7:** 607
Ultrahigh molecular weight polyethylene (UHMWPE) **EM3:** 29
contact corrosive wear with ion-implanted Ti-6Al-4V **A18:** 779
interfacial zone shear and solid friction **A18:** 36
tensile properties at selected strain rates **A18:** 659
tribological characteristics **A18:** 658-662
friction coefficients **A18:** 662
in vivo assessment of total joint replacement performance **A18:** 661-662
joint simulators **A18:** 660-661
pin-on-disk experiments **A18:** 659, 660, 661
pin-on-plate (reciprocating) experiments **A18:** 659-660
use in metal-on-polymer total hip replacements **A18:** 657, 658-662
wear equation **A18:** 661
wear factors **A18:** 659, 661
wear properties **A18:** 241
wear rate **A18:** 662
Ultrahigh molecular weight polyethylenes (UHMWPE) *See also* Polyethylenes
abrasion resistance **EM2:** 167
applications **EM2:** 167-168
characteristics **EM2:** 168-171
chemistry **EM2:** 167
costs **EM2:** 167
defined **EM2:** 43
processing **EM2:** 169-170
properties **EM2:** 169
resin compound types, properties **EM2:** 170-171
suppliers **EM2:** 171
Ultrahigh vacuum
abbreviation for **A10:** 691
atom probe microanalysis **A10:** 591
in Auger electron spectroscopy **A7:** 251
EXAFS surface structure detection in **A10:** 418
Ultrahigh-carbon steels *See also* Carbon steel
definition of **A1:** 148
Ultrahigh-modules pitch-base carbon fibers
properties **EL1:** 1122
Ultrahigh-purity metals
mass spectroscopy trace element analysis **A2:** 1095
Ultrahigh-speed photography
with symmetric rod impact test **A8:** 204
Ultrahigh-strength low-alloy steels **A6:** 662, 673-674
filler metals **A6:** 673
gas-metal arc welding **A6:** 673
gas-tungsten arc welding **A6:** 673, 674
heat-affected zones **A6:** 673-674
microstructure **A6:** 673-674
plasma arc welding **A6:** 673
postweld heat treatment **A6:** 674
preheating **A6:** 673-674
properties **A6:** 673
specifications **A6:** 673
thermal expansion coefficient **A6:** 907
welding procedures and practices **A6:** 673
welding processes **A6:** 673
Ultrahigh-strength low-alloy steels, specific types
300M
composition **A6:** 673
welding preheat and interpass temperatures **A6:** 673
AF1410
composition **A6:** 673
postweld heat treatments **A6:** 674
welding preheat and interpass temperatures **A6:** 673
AMS 6434
composition **A6:** 673
welding preheat and interpass temperatures **A6:** 673
D-6a
composition **A6:** 673
welding preheat and interpass temperatures **A6:** 673
H11 mod
composition **A6:** 673
M_s temperature **A6:** 673
welding preheat and interpass temperatures **A6:** 673
H13
composition **A6:** 673
M_s temperature **A6:** 673
welding preheat and interpass temperatures **A6:** 673
HP9-4-20
composition **A6:** 673
postweld heat treatments **A6:** 674
welding preheat and interpass temperatures **A6:** 673
HP9-4-30
composition **A6:** 673
postweld heat treatments **A6:** 674
welding preheat and interpass temperatures **A6:** 673
HY-180
composition **A6:** 673
oxygen content and welding **A6:** 673
postweld heat treatments **A6:** 674
welding preheat and interpass temperatures **A6:** 673
Ultrahigh-strength steel
Ault-Wald-Bertolo Charpy/fracture toughness correlation for **A8:** 265
stress-intensity range and loading frequency effects **A8:** 405-406
Ultrahigh-strength steels *See also* Maraging steels; specific types **A1:** 430-448 **M1:** 421-443
applications **M1:** 423, 424, 427, 429, 431-434, 437, 441
compositions **A4:** 207
corrosion fatigue crack growth **A13:** 299
environmental hydrogen cracks in **A11:** 410
fatigue data **M1:** 428, 432, 439-441
forging of **M1:** 422-424, 427, 429, 431, 432, 434, 437-438, 441
fracture toughness **M1:** 426-431, 437, 439, 441, 442
Ultrahigh-strength steels (continued)
heat treating **A4:** 207-218 **M4:** 119-129
heat treatment **M1:** 423-425, 427, 429-430, 431-432, 433, 434-436, 438-439, 441
heat-treatment temperatures **A4:** 208
high fracture toughness steels **A1:** 444
AF1410 steel **A1:** 431, 445, 446-447
HP-9-4-30 steel **A1:** 431, 444-446
machining and machinability ratings **M1:** 422, 424, 427, 429, 431, 432, 434, 438, 441
mechanical properties **M1:** 422-442
medium-alloy air-hardening **A4:** 214-215 **M4:** 125-129
medium-alloy air-hardening steels **A1:** 431, 439
H11 modified **A1:** 439-441
medium-carbon low-alloy **A4:** 208-214 **M4:** 119-125
medium-carbon low-alloy steels **A1:** 430-431
300M steel **A1:** 434-436
4130 steel **A1:** 431-432
4140 steel **A1:** 432
4340 steel **A1:** 432-434
6150 steel **A1:** 437-438, 439
8640 steel **A1:** 438-439
D-6a and D-6ac steel **A1:** 436-437, 438
H13 steel **A1:** 431, 441-444
plate **M1:** 189, 193
processing **M1:** 422-424, 426-427, 429, 431, 432, 434, 437-438, 441
stress relief **A4:** 208, 210, 212, 214, 215, 217
thermomechanical treatments **M1:** 421
types **A4:** 207-208
welding of **M1:** 422, 423, 424, 427, 429, 431, 432, 434, 438, 441
Ultrahigh-strength steels, specific types
9Ni-4Co steels **A4:** 215-217
AF1410 .... **A4:** 217-218
composition **A4:** 207
heat treatments **A4:** 216, 217-218
heat treatments, effect on impact energy **A4:** 218
mechanical properties, effect of reaustenitizing and aging
mechanical properties, heat-treatment effects **A4:** 217
temperatures **A4:** 216
tensile strength **A4:** 217, 218
yield strength **A4:** 217
Ultralarge-scale integration (ULSI) **EL1:** 2, 377-378
Ultralong depth-of-field optical microscope **EL1:** 954
Ultralow carbon bainite steels
low-temperature toughness **A4:** 252
Ultralow-carbon steel
temperature effect on fracture mode **A12:** 34, 46
Ultramicroscopic *See* Submicroscopic
Ultramicrotome
replica cutting by **A12:** 199
Ultramicrotomy
diamond knife **A12:** 179
Ultrapure metals *See also* Metal(s); Pure metals
applications **A2:** 1093
characterization **A2:** 1095-1097
purity, six nines purity **A2:** 1096
Ultrapure water
GFAAS trace metal analysis for **A10:** 57-58
Ultrapurification techniques
of metals **A2:** 1093-1095
Ultrarapid solidification processes **A7:** 18, 47-49
Ultrasonic
bonding, defined **EM2:** 43
C-scans **EM2:** 838-845
spot welding **EM2:** 722
staking **EM2:** 721-722
testing, defined **EM2:** 43
welding **EM2:** 721
Ultrasonic abrasive machining (UAM) *See* Ultrasonic machining
Ultrasonic agitation
organic solvents **A12:** 74
Ultrasonic and sonic techniques
defect detection **EM3:** 746
Ultrasonic assembly
types **EM2:** 721-722
Ultrasonic atomization **A7:** 25, 26, 48

Ultrasonic attenuation
to analyze ceramic powder particle size **EM4:** 67
Ultrasonic attenuation technique **A6:** 1095, 1096
Ultrasonic beams *See also* Ultrasonic inspection; Ultrasonic waves
absorption **A17:** 238
acoustic impedance **A17:** 238
attenuation, overall **A17:** 240
beam diameter/spreading **A17:** 240
diffraction **A17:** 239
near-field and far-field effects **A17:** 239
scattering of **A17:** 238-239
Ultrasonic bond testers
for adhesive-bonded joints **A17:** 619-624
Bondascope **A17:** 622-623
Fokker **A17:** 619-620
NDT **A17:** 620
NovaScope **A17:** 623
Shurtronics Mark I harmonic **A17:** 621
Ultrasonic bonding **EL1:** 225, 350 **EM3:** 29, 584, 585, 587
Ultrasonic bonds
defined **EL1:** 1160
Ultrasonic C-scan **EM1:** 8, 262, 776
Ultrasonic C-scan techniques
defect detection **EM3:** 746-749, 750
tooling verification **EM3:** 739
Ultrasonic cleaning
after grinding **A7:** 462
copper and copper alloys **M5:** 619
of corroded surface **A12:** 74
extraction replica **A12:** 183
of fatigue precrack region **A12:** 75
fiber composites **A9:** 588
fracture surfaces **A11:** 19
mechanism of action **M5:** 4, 15-17
organic solvents **A12:** 74
of specimens to be mounted **A9:** 28
for surface coating removal **A12:** 73
Ultrasonic coupler
definition **M6:** 18
Ultrasonic coupler (ultrasonic soldering and ultrasonic welding)
definition **A6:** 1214
Ultrasonic data recording and processing system (UDRPS) **A17:** 654
Ultrasonic degreasing, as secondary
cleaning operation **A7:** 459
Ultrasonic fatigue testing **A8:** 240-258
amplitude detection **A8:** 245-246
application **A8:** 240, 252-255
construction **A8:** 243-249
cooling systems **A8:** 246-248
effect of frequency **A8:** 255-256
electrochemical potential during **A8:** 254
environmental **A8:** 241, 248
frequency in **A8:** 240-241
for high strain rate fatigue testing **A8:** 187
infrared specimen thermogram **A8:** 247
machine, schematic **A8:** 244
strain rates **A8:** 240-241
test specimens **A8:** 249-252
testing equipment and methods **A8:** 243-249
time compression and **A8:** 240-241
uniform resonant bar **A8:** 242
Ultrasonic fractography
to assess local crack velocity **EM4:** 654
to observe crack propagation in brittle composites **EM4:** 863
Ultrasonic hardness testing **A8:** 90-103
test and tester for **A8:** 99-102
Ultrasonic imaging
C-scan, of powder metallurgy parts **A17:** 540
scanning acoustic microscopy **A17:** 540
scanning laser acoustic microscopy **A17:** 540
to evaluate gas turbine ceramic components **EM4:** 718
Ultrasonic immersion *See also* Immersion ultrasonic inspection
with solvents **A17:** 81
Ultrasonic impact grinding *See* Ultrasonic machining **A16:** 528, 529, 530, 531-532
Ultrasonic impedance analysis principle **EM3:** 757
Ultrasonic inspection *See also* Immersion ultrasonic inspection; Longitudinal wave ultrasonic inspection; NDE reliability Shear wave ultrasonic inspection Ultrasonic waves; Nondestructive evaluation; Nondestructive testing; Ultrasound transmission **A11:** 17 **A17:** 231-277
and acoustic emission inspection **A17:** 286
acoustic, of powder metallurgy parts **A17:** 539-541
of adhesive-bonded joints **A17:** 617-624
advantages/disadvantages **A17:** 231-232
of aluminum alloy castings **A17:** 534
applicability **A17:** 232
applications **M6:** 827
of arc-welded nonmagnetic ferrous tubular products **A17:** 567
area-amplitude curves, determined **A17:** 266
attenuation, ultrasonic beams **A17:** 238-240
beam models of **A17:** 705
of boilers and pressure vessels **A17:** 642
of bonded joints **A17:** 272-273
of brazed assemblies **A17:** 604
brazed joints **A6:** 1119, 1122
in brazing **A11:** 451
with CAD system **A17:** 712-713
calibration **A17:** 266-267
carbon steels, cracking **A6:** 643
of castings **A17:** 267, 529-531
of closed-die and upset forgings **A17:** 495
codes, boilers/pressure vessels **A17:** 642
codes, standards, and specifications **M6:** 827
color images by **A17:** 484-486, 488
control systems **A17:** 254
corrosion monitoring by **A17:** 275-276
couplants **A17:** 256
crack monitoring by **A17:** 273
defined **A17:** 231
detection of
lack of fusion **A6:** 1078
subsurface porosity **A6:** 1074
subsurface slag inclusions **A6:** 1074
tungsten inclusions **A6:** 1074
weld discontinuities, electroslag welding **A6:** 1078
weld discontinuities, friction welding **A6:** 1078
weld metal and base metal cracks **A6:** 1075
detection of subsurface cracking **A18:** 369
dimension-measurement by **A17:** 273-274
distance-amplitude curves determined **A17:** 265-266
of double submerged arc welded steel pipe **A17:** 566
of duplex tubing **A17:** 572
and eddy current inspection simultaneous **A17:** 272
electrogas welds **M6:** 244
electronic equipment **A17:** 252-254
electroslag welds **M6:** 234
equipment **A17:** 231, 252-254, 261
examples **A17:** 249-250
explosion welds **M6:** 710-711
of extrusions **A17:** 270-272
of flat-rolled products **A17:** 268-270
flaw detection **A17:** 267
of forgings **A17:** 268, 504-510
immersion, of nonferrous tubing **A17:** 574
for in-service inspection, tubular products **A17:** 574
inspection methods **A17:** 240-252
Ultrasonic inspection (continued)
inspection standards **A17:** 261-263
of internal discontinuities, castings **A17:** 512
of journal bearings **A11:** 717
in leak testing **A17:** 59-60
manual, of pressure vessels **A17:** 649
measurement model of **A17:** 704
of microstructure **A17:** 274-275
microstructure effect **A17:** 51
and microwave inspection, compared **A17:** 202
models, applications **A17:** 706-707, 712-713
NDE reliability models of **A17:** 703-707
of niobium-titanium superconducting materials **A2:** 1044
as nondestructive fatigue testing **A11:** 134
of open-die forgings **A17:** 495
orifice diameter effects **A17:** 60
of pipeline girth welds **A17:** 580-581
plasma arc welding **A6:** 198
plasma-MIG welding **A6:** 224
of powder metallurgy parts **A17:** 539
of pressure vessels **A17:** 649-653
of primary-mill products **A17:** 267-268
probability of detection (POD) models **A17:** 706
process qualification **A17:** 678
pulse-echo methods **A17:** 241-248
for residual stresses **A17:** 51
of resistance-welded steel tubing **A17:** 564-565
of ring-rolled forgings **A17:** 495
of rolled shapes **A17:** 270-272
scanning equipment **A17:** 261
of seamless pipe **A17:** 579
of seamless steel tubular products **A17:** 568-570
search units **A17:** 256-261
soldered joints **A6:** 981
of spiral-weld steel pipe **A17:** 567
standard reference blocks **A17:** 263-265
of steel bar and wire **A17:** 551-552
stress measurement **A17:** 276
of surface cracks/delamination, metal matrix **A17:** 250
of titanium-matrix composite panels **A17:** 250
transducer elements **A17:** 254-256
transmission methods **A17:** 240, 248-252
ultrasonic waves, characteristics **A17:** 232-234
underwater welds **M6:** 924
variables, major **A17:** 234-238
of welded joints **A17:** 272
of weldments **A17:** 272, 594-598
Ultrasonic machining (USM) **A16:** 509, 528-532 **EM4:** 313, 314, 359-362
abrasive grit size effect on surface finish **EM4:** 360
advantages **EM4:** 361
CNC of all axes **A16:** 529, 531
coolants **A16:** 528
equipment **A16:** 528 **EM4:** 361-362
impact grinders **EM4:** 361-362
rotary ultrasonic machining **EM4:** 362
grit sizes
key components for USM installation **EM4:** 359
machined materials **EM4:** 359
material removal rates **EM4:** 361
process **EM4:** 359
process capabilities **EM4:** 361
properties compared for selected ceramics after processing **EM4:** 360
rotary ultrasonic machining **A16:** 528-529, 530, 531-532
selection of abrasives **EM4:** 360-361
tooling requirements **EM4:** 359
versus laser beam machining **EM4:** 361
Ultrasonic methods
capabilities of **A10:** 380
of thickness measurement **A8:** 549
Ultrasonic microhardness testing **A8:** 98-103
applications **A8:** 99, 102-103
piezoelectric converter **A8:** 101

Ultrasonic nebulizers
for atomic absorption spectrometry **A10:** 55
for ICP sample introduction **A10:** 36
Ultrasonic nondestructive analysis **EM2:** 838-846
C-scans, gating techniques for **EM2:** 840-845
complex geometry inspection **EM2:** 845
future trends **EM2:** 845-846
system description **EM2:** 838-840
Ultrasonic ply cutting **EM1:** 615-618
cutting medium **EM1:** 615-616
cutting system features **EM1:** 617-618
cutting variables **EM1:** 616-617
reliability and safety **EM1:** 618
theory/operation principles **EM1:** 615
Ultrasonic resonance test
for threshold stress intensity **A8:** 256
Ultrasonic scanning of fiber composites **A9:** 591
Ultrasonic soldering
aluminum alloys **A6:** 628
definition **A6:** 1214 **M6:** 18
Ultrasonic techniques **A6:** 1095
Ultrasonic testing **A6:** 1081, 1083-1084, 1085-1086, 1087, 1088 **EM1:** 24, 776 **EM3:** 29
for casting defects **A15:** 555, 557
fitness for service evaluation **A6:** 376
flaws in electron-beam welding **A6:** 860
for inclusions **A15:** 493
of investment castings **A15:** 264
resistance seam welds **A6:** 245
to detect weld discontinuities in diffusion welding **A6:** 1080
for weld characterization **A6:** 97, 98-99, 100, 102
weld microstructures **A6:** 54
Ultrasonic thickness measurement and data analysis **A13:** 200, 202, 323
Ultrasonic transducer
in cleaning techniques **A7:** 459
Ultrasonic vapor degreasing system **M5:** 47
Ultrasonic velocity
measurement **A15:** 664
and mechanical properties **A7:** 484
and strength **A7:** 484
Ultrasonic vibration
copper plating baths **M5:** 162
solvent cleaning processes **M5:** 40, 44
Ultrasonic vibratory energy
theory and principles **EM1:** 615
Ultrasonic waves *See also* Ultrasonic inspection; Ultrasound transmission
in acoustical holography **A17:** 405
characteristics **A17:** 232-234
defined **A17:** 231
interference effects, adhesive-bonded joints **A17:** 637-638
Lamb **A17:** 234
longitudinal **A17:** 233
magnetically induced velocity changes (MICV) for **A17:** 161-162
surface (Rayleigh) **A17:** 233-234
transverse **A17:** 233
wave propagation **A17:** 233
Ultrasonic welding **M6:** 746-756 **EM2:** 721
applications **M6:** 747, 753-756
electrical and electronic assemblies **M6:** 754-755
foil and sheet splicing **M6:** 755
packaging **M6:** 755
solar energy systems **M6:** 755
structural applications **M6:** 755-756
definition **M6:** 18
equipment for welding **M6:** 747-750
anvils **M6:** 748
automated production equipment **M6:** 749-750
coupling systems **M6:** 747-748
force applications systems **M6:** 748
frequency converters **M6:** 747
machines for welding **M6:** 749
transducers **M6:** 747
welding tips **M6:** 748
metal combinations **M6:** 746
procedures for welding **M6:** 750-751
interaction of welding variables **M6:** 750-751
operating variables **M6:** 750
sur-face preparation **M6:** 751
workpiece resonance control **M6:** 751
safety precautions **M6:** 59
seam welding, continuous **M6:** 746
spot welding **M6:** 746
Ultrasonic welding (continued)
weld properties **M6:** 751-753
surface characteristics **M6:** 751-752
weld microstructure **M6:** 752-753
weld strength **M6:** 752
Ultrasonic welding (USW) **A6:** 324-327
advantages **A6:** 324
aluminum **A6:** 324, 326, 327
aluminum alloys **A6:** 739
anvils **A6:** 326
applications **A6:** 324
brass **A6:** 326
continuous seam welds **A6:** 325
copper **A6:** 324, 326
definition **A6:** 324, 893, 1214
dissimilar metal joining **A6:** 822
equipment **A6:** 324-325, 326
heat-affected zone **A6:** 327
limitations **A6:** 324-325
line welds **A6:** 325
mechanical properties **A6:** 327
microelectronic welds **A6:** 325
molybdenum **A6:** 327
nickel **A6:** 327
noise level **A6:** 325
parameters **A6:** 324-325
personnel **A6:** 325
procedures **A6:** 325-327
process mechanism **A6:** 324
process variations **A6:** 324-325
ring welds **A6:** 325, 327
safety precautions **A6:** 1203
sheet metals **A6:** 399
special atmospheres **A6:** 326
special considerations **A6:** 326
spot welds **A6:** 324-325
steel **A6:** 324
tantalum **A6:** 327
tips **A6:** 326
titanium **A6:** 326, 327
titanium alloys **A6:** 522
to solve problems in joining thin sections by oxyfuel gas welding **A6:** 288
tooling **A6:** 326
weld quality **A6:** 326-327
weldable materials **A6:** 894
Ultrasonic welding (USW), procedure development and practice considerations **A6:** 893-895
aluminum alloys **A6:** 894, 895
applications **A6:** 893, 894, 895
carbon steels **A6:** 893
copper alloys **A6:** 894-895
difficult-to-weld alloys **A6:** 893-894
dissimilar metal (nonferrous-to-nonferrous) combination **A6:** 895
heat-affected zone **A6:** 895
high-strength steels **A6:** 893
Inconel X **A6:** 895
Kovar **A6:** 895
low-alloy steels **A6:** 893
molybdenum alloys **A6:** 894
nickel-base alloys **A6:** 895
precious metals **A6:** 895
reactive metals **A6:** 894
refractory base metals **A6:** 894
stainless steels **A6:** 327, 893, 894
titanium alloys **A6:** 894
weld strength **A6:** 893
Ultrasonic welding of
aluminum alloys **M6:** 746
copper alloys **M6:** 746
gold **M6:** 746
iron **M6:** 746
nickel **M6:** 746
palladium **M6:** 746
platinum **M6:** 746
refractory metals **M6:** 746
silver **M6:** 746
titanium **M6:** 746
zirconium **M6:** 746
Ultrasonically installed inserts **EM2:** 722
Ultrasonics
measured by acoustic ESR **A10:** 258
Ultrasonics, detection of cracks
electron-beam welding **A6:** 1077-1078
Ultrasound
in acoustic microscopy **EL1:** 369
Ultrasound transmission *See also* Ultrasound inspection
in green compacts **A17:** 539
in sintered parts **A17:** 539-540
Ultrathin films
LEISS coverage analysis **A10:** 603
Ultrathin window
abbreviation for **A10:** 691
-EDS light-element analysis **A10:** 459-461
energy-dispersive spectrometer **A10:** 519
Ultratrace analysis
graphite furnace atomic absorption spectrometry **A10:** 57
of inorganic gases, applicable methods for **A10:** 8
of inorganic liquids and solutions, applicable methods for **A10:** 7
of inorganic solids, applicable methods for **A10:** 4-6
of organic solids and liquids, applicable methods for **A10:** 9, 10
uranium, determined by laser-induced fluorescence spectroscopy **A10:** 80
Ultraviolet (UV) **EM3:** 29
absorbers, in sheet molding compounds **EM1:** 158
defined **EM1:** 24 **EM2:** 43-44
degradation, long-ten-n exposure testing **EM1:** 823-825
resistance, of polyester resins **EM1:** 94-95
stabilizer, defined **EM1:** 24
Ultraviolet (UV) conformal coatings
advantages **EL1:** 788
development **EL1:** 785
equipment **EL1:** 787
future of **EL1:** 788
masking **EL1:** 787
materials **EL1:** 785-786
performance data **EL1:** 787-788
processing techniques **EL1:** 786-787
Ultraviolet (UV) curing **EL1:** 773, 785-788, 854, 864
acrylates **EL1:** 555
of adhesives **EL1:** 672
of coatings **EL1:** 773, 785-788, 854, 864
of conformal coatings **EL1:** 765
silicone coatings **EL1:** 824
Ultraviolet (UV) fluorescence
for optical imaging **EL1:** 1068
Ultraviolet (UV) radiation resistance **EL1:** 773
Ultraviolet (UV) spectra
parylene coatings **EL1:** 795
Ultraviolet (UV) spectroscopy **EL1:** 1103
Ultraviolet (UV) technology
future of **EL1:** 788
process flexibility **EL1:** 786
UV curing **EL1:** 785-788
Ultraviolet (UV)/visible spectrophotometry
for chemical analysis **EM4:** 553-554
Ultraviolet binders
for multicolor screening applications **EM4:** 475
Ultraviolet degradation **EM3:** 554-555
defined **EL1:** 1160
Ultraviolet illumination
effects **A12:** 84-85
Ultraviolet laser treatment **EM3:** 35
Ultraviolet light
for cold shut detection **A17:** 101
in fluorescent magnetic particle inspection **A7:** 578-579
in liquid penetrant inspection **A17:** 73
for magnetic particle inspection **A17:** 102-103
Ultraviolet light, as source
photolytic degradation **EM2:** 776-782
Ultraviolet light exposure
for curing **EM3:** 35
Ultraviolet photoelectron spectroscopy
of molybdena catalysts **A10:** 134
Ultraviolet photoelectron spectroscopy (UPS) **A18:** 445, 447 **EM3:** 237
Ultraviolet radiation
cross-linking effect on polyimides **EM3:** 157
defined **A10:** 683
effect on modified silicones **EM3:** 191
effect on two-part manually mixed polysulfide sealants **EM3:** 190

Ultraviolet radiation (continued)
effect on urethanes **EM3:** 190-191
Ultraviolet radiation resistance
as degradation factor **EM2:** 575-576
of engineering plastics **EM2:** 1
polyether-imides (PEI) **EM2:** 157
ultrahigh molecular weight polyethylenes (UHMWPE) **EM2:** 169, 171
Ultraviolet resistance **EM3:** 52
Ultraviolet spectra
DRS and ATR analysis **A10:** 114
Ultraviolet spectroscopy tests **EM2:** 426
Ultraviolet stability
polyamide-imides (PAI) **EM2:** 129
Ultraviolet stabilizer **EM3:** 30
Ultraviolet stabilizers
as additive, effects **EM2:** 425
defined **EM2:** 44
effect, chemical susceptibility **EM2:** 572
in engineering thermoplastics **EM2:** 98
as polymer additives **EM2:** 67
Ultraviolet wave energy curing **EM3:** 567-568
Ultraviolet-curable acrylates
for lighting subcomponent bonding **EM3:** 552
Ultraviolet-curable acrylics
for circuit protection **EM3:** 592
Ultraviolet-curable epoxies
liquid crystal display potting and sealing **EM3:** 612
Ultraviolet-curable resins
wire tacking on motor assemblies **EM3:** 612
Ultraviolet-curing adhesives **EM3:** 74
for electronic general component bonding **EM3:** 573
medical applications **EM3:** 576
surface-mount technology adhesives **EM3:** 570-571
for wire-winding operation **EM3:** 573, 574
Ultraviolet-curing methacrylates
medical applications **EM3:** 576
for medical bonding (class IV approval) **EM3:** 576
Ultraviolet-curing urethane-acrylates
for medical bonding (class IV approval) **EM3:** 576
Ultraviolet-visible radiation
contamination by **EL1:** 45
Ultraviolet/electron beam cured adhesives **EM3:** 90-92
advantages and limitations **EM3:** 75
applications **EM3:** 91
automotive applications **EM3:** 91
characteristics **EM3:** 74, 91
chemistry **EM3:** 91
components **EM3:** 91
cross-linking **EM3:** 90, 91
curing methods **EM3:** 90-91
electronics applications **EM3:** 91
for fiber optics **EM3:** 91
laminating adhesives **EM3:** 91
limitations **EM3:** 91
for magnetic media **EM3:** 91
markets **EM3:** 91
medical adhesives **EM3:** 91
for metal finishing **EM3:** 91
for package laminations **EM3:** 91
pressure-sensitive adhesives (PSA) **EM3:** 91
for release coatings **EM3:** 91
suppliers **EM3:** 92
UV-curable structural adhesives **EM3:** 91
Ultraviolet/visible absorption spectroscopy *See also* Optical and x-ray spectroscopy **A10:** 60-71
absorbance **A10:** 62
adapted for inorganic solids **A10:** 4-6
advantages **A10:** 60
applications **A10:** 60, 70-71
Beer's law **A10:** 61-63
capabilities **A10:** 181

Ultraviolet/visible absorption spectroscopy (continued)
capabilities, MFS compared **A10:** 72
color comparison kits **A10:** 66-67
defined **A10:** 683
effect of complexing agent **A10:** 64
estimated analysis time **A10:** 60
experimental parameters **A10:** 68-70
filter photometers **A10:** 67
general uses **A10:** 60
for inorganic liquids and solutions **A10:** 7
instrumentation **A10:** 66-68
introduction and principles **A10:** 61-63
limitations **A10:** 60
molecular fluorescence **A10:** 61-62
monochromators **A10:** 66
NMR, ESR, and IR compared with **A10:** 265
of organic liquids and solutions, information from **A10:** 10
of organic solids, information from **A10:** 9
Planck's constant **A10:** 61
quantitative analysis **A10:** 63-66
related techniques **A10:** 60
samples **A10:** 60, 70-71
sensitivity **A10:** 63-65
spectral region of interest **A10:** 61
terms and symbols used in **A10:** 62
Ultraviolet/visible solution spectrophotometry
AAS spectrometers and **A10:** 51
Umbra
defined for radiographic definition **A17:** 313
UMCo-50
composition **A6:** 564, 573
Umpire analysis
gravimetry for **A10:** 170
of stainless steel alloy **A10:** 178-179
Unalloyed aluminum *See also* Aluminum; Pure aluminum; Pure metals
compositions **A2:** 22-25
Unalloyed cast irons **A13:** 567
Unalloyed Ti-0.15Pd
stress-relieving times and temperatures **A6:** 786
Unalloyed tin *See also* Pure metals; Pure tin; Tin
applications **A2:** 519
Unalloyed titanium *See* Titanium alloys, specific types, Ti grades 1 to 4
Unalloyed uranium *See* Natural uranium; Pure uranium; Uranium
Unaltered grain-coarsened (UAGC) zone **A6:** 81
Unary system
defined **A9:** 19
Unary system or diagram .... **A3:** 1 • 2
Unbalanced magnetron sputter deposition process **A18:** 841, 849
Unbiased autoclave test
for humidity-induced stress **EL1:** 495
Unbond **EM3:** 30
defined **EM1:** 24-25 **EM2:** 44
Unbonded sand molds *See also* Bonded sand molding; Sand molding **A15:** 230-237
lost foam casting **A15:** 230-234
magnetic molding **A15:** 234-235
vacuum molding **A15:** 235-236
Unbonds
in adhesive-bonded joints **A17:** 612
holographic inspection **A17:** 423
optical holography of **A17:** 405
Unbreakable Metal *See* Zinc alloys, specific types, zinc-base slush-casting alloys
Uncertainties
sampling **A10:** 12, 15, 17
Uncertainty
defined **A8:** 14 **A10:** 683
Unclinching leads
conditions and methods **EL1:** 722
Uncoated alloyed carbide grades
machining applications **A2:** 967

Uncoated straight WC-CO grades
machining applications **A2:** 967
Uncoilers
as coil-handling equipment **A14:** 501
Unconsolidated interiors
tool and die failures from **A11:** 574-575
Unconstrained interconnection methods
surface-mounted **EL1:** 985
Unconventional metallurgy **A7:** 570
Unctuous
defined **A18:** 20
Under-deposit corrosion *See* Crevice corrosion
Underage treatment
beryllium-copper alloys **A2:** 407
Underbead cold cracking
steel weldments **A6:** 423, 425
Underbead crack
definition **M6:** 18
weld metal **A12:** 155
Underbead cracking *See also* Hydrogen-induced cracking **A6:** 410, 1068 **M6:** 830
in austenitic stainless steels **A1:** 898
carbon steels **A6:** 642
definition **A6:** 12, 14
pressure vessels **A6:** 379
in shielded metal arc welds **M6:** 92-93
sources **M6:** 44
in submerged arc welds **M6:** 130
underwater welding **A6:** 1011, 1012
Underbead cracking, welded steel
estimating risk of **M1:** 561
Underbead cracks
in girth weld **A11:** 699
in pipeline girth welds, inspection **A17:** 581
in weldments **A17:** 585
Undercoating, soft
for circuit encapsulation **EL1:** 241
Undercooling *See also* Cooling; Supercooling **A6:** 45, 52
data, low gravity **A15:** 150
defined **A15:** 101
of dendritic structures **A15:** 118-119
effect on cast iron bearing caps **A11:** 350
effect on eutectic structures **A9:** 619
effect on pure metal solidification structures **A9:** 610
evolution, low gravity **A15:** 150
Gibbs-Thomson **A15:** 110-111
gray-to-white transition by **A15:** 180
and lattice disregistry **A15:** 105
and microsegregation **A15:** 138
in peritectic reactions **A9:** 676
required to activate solidification **A15:** 105
in solidification modeling **A15:** 884-890
total growth, eutectic interface **A15:** 124
transcrystalline layer **EM3:** 418
Undercure *See also* Cure **EM3:** 30
defined **EM1:** 25 **EM2:** 44
Undercut **A6:** 1073, 1075, 1164, 1215 **M6:** 836-837, 844
by metal casting **A15:** 40
defined **A15:** 11 **EM1:** 25 **EM2:** 44
definition **M6:** 18
in gas metal arc welds **M6:** 172
in permanent molds **A15:** 277
polyamide-imides (PAI) **EM2:** 131
reduction by beam oscillation **M6:** 634
in shielded metal arc welds **M6:** 92-93
weld overlays **M6:** 817
Undercuts
in arc-welded aluminum alloys **A11:** 435
Undercutting
in conjunction with turning **A16:** 138, 158
defined **EL1:** 1160
during etching, effects **EL1:** 82
printed board coupons **EL1:** 577
weld **A17:** 350, 582
Underfill *See also* Nonfill **A6:** 1073, 1215 **M6:** 837
defined **A14:** 14

SUBJECTS OF THE INDEXED VOLUMES: ASM Handbook (designated by the letter "A"): **A1:** Properties and Selection: Irons, Steels, and High-Performance Alloys (1990); **A2:** Properties and Selection: Nonferrous Alloys and Special-Purpose Materials (1990); **A3:** Alloy Phase Diagrams (1992); **A4:** Heat Treating (1991); **A6:** Welding, Brazing, and Soldering (1993); **A7:** Powder Metallurgy (1984); **A8:** Mechanical Testing (1985); **A9:** Metallography and Microstructures (1985); **A10:** Materials Characterization (1986); **A11:** Failure Analysis and Prevention (1986); **A12:** Fractography (1987); **A13:** Corrosion (1987); **A14:** Forming and Forging (1988); **A15:** Casting (1988); **A16:** Machining (1989); **A17:** Nondestructive Testing and Quality Control (1989); **A18:** Friction, Lubrication, and Wear Technology (1992). **Metals Handbook, 9th Edition** (designated by the letter "M"): **M1:** Properties and Selection: Irons and Steels (1978); **M2:** Properties and Selection: Nonferrous Alloys and Pure Metals (1979); **M3:** Properties and Selection: Stainless Steels, Tool Materials and Special-Purpose Materials (1980); **M4:** Heat Treating (1981); **M5:** Surface Cleaning, Finishing, and Coating (1982); **M6:** Welding, Brazing, and Soldering (1983). **Engineered Materials Handbook** (designated by the letters "EM"): **EM1:** Composites (1987); **EM2:** Engineering Plastics (1988); **EM3:** Adhesives and Sealants (1990); **EM4:** Ceramics and Glasses (1991); **Electronic Materials Handbook** (designated by the letters "EL"): **EL1:** Packaging (1989).

Underfill (continued)
definition M6: 18
reduction by beam oscillation M6: 634
as weldment defect A17: 582
Underfilm corrosion
defined A13: 13
Underground ducts
lead/lead alloys corrosion in A13: 787-789
Underground installations
ductile iron pipe M1: 99
Underground mining, metallic ores
with cemented carbide tools A2: 975
Underground plant
telephone cables A13: 1127-1128
Underground zinc anodes A13: 765
Underlayer, exposed *See* Exposed underlayer
Underpouring
ladle A15: 497, 500
Undersintered powder metallurgy materials A9: 512
Undersize powder A7: 12
Undervoltage relay
defined EL1: 1160
Underwater applications A6: 374
Underwater cutting M6: 914, 921-925
depth limitations M6: 922
environmental effects on processes M6: 923
processes M6: 922
arc cutting M6: 922
gas cutting M6: 922
mechanical cutting M6: 922
remote cutting M6: 922
welder/diver qualification M6: 924
Underwater installations
ductile iron pipe M1: 99-100
Underwater welding A6: 1010-1014 M6: 921-925
applications, maintenance and repair examples A6: 1014
carbon monoxide reaction A6: 1010-1011, 1012
depth limitations M6: 922
environmental effects M6: 922-923
on design M6: 923
on processes M6: 923
on welder M6: 923
on welds M6: 922-923
fatigue as a function of porosity A6: 1013
flux-cored arc welding A6: 1010, 1012
gas-metal arc welding A6: 1010
gas-tungsten arc welding A6: 1010, 1014
heat flow in fusion welding A6: 9
heat sources A6: 1013-1014
heat-affected zone A6: 1010, 1011-1012
hydrogen content in weld-metals A6: 1011-1012
hydrogen cracking A6: 1010, 1011-1012, 1014
hydrogen mitigation A6: 1011-1012
hyperbaric welding A6: 1010, 1011, 1014
inspection M6: 924
isotopic radiography M6: 924
ultrasonic M6: 924
visual M6: 924
microstructural development of underwater welds A6: 1011
procedure qualification M6: 924
process selection M6: 924
processes M6: 921-922
chamber M6: 921
dry-box M6: 921
dry-spot M6: 921
habitat M6: 921
remote M6: 921
wet or open-water M6: 921
pyrometallurgy A6: 1010-1011
residual stress reduction practices A6: 1012
shielded metal arc welding A6: 1010, 1012
specifications A6: 1012, 1014
temper bead practice A6: 1012
time to cool value A6: 1010
underbead cracking A6: 1011, 1012
water pressure effect A6: 1013
weld-metal carbon A6: 1010, 1011, 1012
weld-metal manganese A6: 1010-1011
weld-metal oxygen content A6: 1010-1011
weld-metal porosity A6: 1012-1013
weld-metal silicon A6: 1010-1011
welder/diver qualification M6: 924
welding parameter space A6: 1011
wet welding A6: 1010, 1011
Underwriters' Laboratories (UL) standards EM2: 91, 461
Undiluted weld metal
plasma-MIG welding A6: 225
Undisturbed substrate limit A18: 465
Unfavorable grain flow *See* Grain flow
Unfilled resins
acrylic, friction as described by various properties A18: 669
diacrylate, friction as described by various properties A18: 669
Unfused chapiet
as casting defect A11: 383
Unfused chaplets
radiographic appearance A17: 349
UNI (Italian) standards for steels A1: 159
compositions of A1: 190-193
cross-referenced to SAE-AISI steels A1: 166-174
UNI programmable calculator program
for composite material analysis EM1: 277-279
Uni-type fill
fly-shuttle loom selvage with EM1: 128
Uniaxial
compression, ply EM1: 237
load, defined EM1: 25
orientation, defined *See* Oriented materials, defined
tension EM1: 237, 239
Uniaxial compacting A7: 12
Uniaxial compression A7: 304
Uniaxial compression in dies
carbon-graphite materials A18: 816
Uniaxial compression testing
creep specimens for A8: 302
flow localization A8: 171
Uniaxial dry pressing EM4: 123
constituents of powder formulation EM4: 126
mechanical consolidation EM4: 125, 126
Uniaxial extension
and compression EL1: 839, 844-845
Uniaxial flow
in indentation tests A8: 71
Uniaxial hot pressing A7: 309, 501 EM4: 186-190
additives EM4: 187, 188
brazing for joining non-oxide ceramics EM4: 528
comminution EM4: 188
constituents of powder formulation EM4: 126
die assembly EM4: 187
hot press densification EM4: 189
hot pressed product properties EM4: 189-190
hot pressing
environments EM4: 186
furnace EM4: 186-187
temperatures EM4: 186
mechanical consolidation EM4: 125, 126
powder loading EM4: 188-189
powder selection EM4: 187-188
pressure densification EM4: 297-299
conditions EM4: 298
parameters EM4: 298
pressure EM4: 301
temperature EM4: 301
Uniaxial load EM3: 30
defined EM2: 44
to failure A8: 19
Uniaxial pressing EM4: 9, 126, 127
Uniaxial strain *See also* Axial strain
defined A8: 210
Uniaxial stress *See* Principal stress
effect on fracture mode A12: 30-31
Uniaxial stress system A8: 576
Uniaxial tensile creep
testing of EM2: 666-667
Uniaxial tensile test
for welds A6: 101
Uniaxial tensile testing A14: 861-862, 883-887
determining *n* and *r* values A8: 556-557
engineering and true stress/strain curves A8: 554
as intrinsic formability test A8: 553-557
material properties A8: 555-556
rate of A8: 554-555
of sheet metals A8: 553-557
specimens for A8: 302
test procedure A8: 554-555
Uniaxial tension A8: 22, 503
in fatigue fracture A11: 75
Uniaxial tension tests
SCC evaluation A13: 247-248
Uniaxial tension-biaxial compression system A8: 576
Uniaxial tension-uniaxial compression system A8: 576
Uniaxially cold pressed beryllium powder A7: 171
Uniaxis assembly
design for EL1: 121-122
Unicast process A15: 250-252
alloys cast A15: 251
and Shaw process, compared A15: 250-251
Unidirectional bending
fatigue, in shafts A11: 461-462
and fatigue-crack propagation A11: 108
Unidirectional compacting A7: 12
Unidirectional composites (UDCS)
anisotropy of EM1: 218
aramid fiber reinforced, strength of EM1: 35
carbon fiber reinforced, strength of EM1: 35
damping analysis of EM1: 207-209
longitudinal shear, and damping in EM1: 207
longitudinal tension/compression and damping in EM1: 207-208
moisture effects EM1: 188-190
physical properties EM1: 185-192
properties EM1: 185
Unidirectional fabrics EM1: 125-128
design EM1: 125-127
maximum directional properties for minimum thickness EM1: 126
textile equipment for EM1: 127-128
Unidirectional flux A7: 315
Unidirectional laminate *See also* Bidirectional laminate; Laminate(s); Unidirectional composites (UDCs)
defined EM1: 25 EM2: 44
Unidirectional lubrication theory EM4: 176
Unidirectional Material Properties (UNI) programmable calculator program
for composite materials analysis EM1: 277-279
Unidirectional radial loads
ball bearings A11: 491-492
Unidirectional tape prepregs EM1: 143-145
applications EM1: 144-145
automatic machine lay-up EM1: 145
chemical tests for EM1: 291
dimensions EM1: 143
drape EM1: 144
fiber bundle dimensions EM1: 143
flow EM1: 144
future trends EM1: 145
gel time EM1: 144
hand lay-up EM1: 144
machine-cut patterns EM1: 145
manufacture/product forms EM1: 143
mechanical properties EM1: 143
properties EM1: 143-144
tack EM1: 143-144
vs. multidirectional tape prepregs EM1: 146
Unidirectional waves
in wave soldering EL1: 688
Unidirectionally solidified eutectics, quantitative metallography of parallel
rods in A9: 127
Unified life cycle engineering
and NDE reliability models A17: 702
Unified Numbering System (UNS)
for copper-base castings A2: 346-347
cross-referencing system A2: 16, 17-25
Unified Numbering System (UNS) designations *See also* SAE/AISI
designation A1: 151, 153, 842
Unified-shapes checking (USC) EL1: 127-129
Uniform corrosion *See also* Corrosion; General corrosion
automotive industry A13: 1011
in carbon steel pipe A11: 300
of cobalt-base corrosion-resistant alloys A2: 453
defined A13: 13
electrochemical testing methods A13: 213-215
evaluation of A13: 229-230
exposure tests A13: 229-230
and localized corrosion A13: 48
material selection to avoid/minimize A13: 323

Uniform corrosion (continued)
of metals ... **A11:** 174-176
in mining/mill applications ... **A13:** 1294-1295
rates, measurement ... **A13:** 229-230
in soil, measurement ... **A13:** 209
stainless steels ... **A13:** 553
of steel castings ... **A11:** 402
of steel tubes ... **A11:** 175
test coupons ... **A13:** 198
Uniform elongation *See also* Elongation ... **EM3:** 30
correlation with stretch-forming ... **A8:** 22
defined ... **A8:** 14 **EM2:** 44
effect on percent elongation ... **A8:** 27
n value defined in region of ... **A8:** 550
of sheet metals ... **A8:** 555-556
in uniaxial tens-le testing ... **A8:** 555-556
Uniform extension
in ductility measurement ... **A8:** 26
Uniform loading
effect on cracking ... **A12:** 176
Uniform quasi-static torsional testing ... **A8:** 145
Uniform quenching
in forging ... **A11:** 325
Uniform solidification
of copper alloy castings ... **A15:** 782
Uniform strain
defined ... **A8:** 14
true ... **A8:** 24
Uniformity
of blending and premixing ... **A7:** 186-189
Uniformity coefficients ... **EM3:** 444
Uniformity, product
with isothermal/hot-die forging ... **A14:** 150
Uniformly distributed impact test *See* Distributed impact test
Unilay stranded copper conductors ... **M2:** 266
Unimeric ... **EM3:** 30
defined ... **EM2:** 44
Uninhibited leaded Muntz metal
applications and properties ... **A2:** 311
Uninhibited naval brass
applications and properties ... **A2:** 319-320
Uninhibited pickling solutions
use of ... **M5:** 70
Union Furnace *See* Taylor-Wharton Iron and Steel Company
Unipolarity operation
definition ... **M6:** 18
Unit cell
defined ... **A13:** 45
Unit cell (crystallographic)
defined ... **EL1:** 93
Unit cell of a crystal *See also* Crystal structure ... **A3:** 1 • 10
defined ... **A9:** 19, 706
Unit cells
arrangement of atoms within ... **A10:** 345
copper tetrahedra and molybdenum octahedra in ... **A10:** 354
cubic ... **A10:** 346-348
defined ... **A10:** 683
described ... **A10:** 346-347
diffraction, and structure factor equation for ... **A10:** 329
dimensions, for defining crystal structure ... **A10:** 348
geometry of ... **A10:** 326-327
hexagonal ... **A10:** 346-348
identification, by single-crystal x-ray diffraction ... **A10:** 344, 346-347, 351
monoclinic ... **A10:** 346-348
orthorhombic ... **A10:** 346-348
tetragonal ... **A10:** 346-348
triclinic ... **A10:** 346-348
trigonal (rhombohedral) ... **A10:** 346-348
use to identify unknown crystalline phase ... **A10:** 353
Unit level, of interconnections
defined ... **EL1:** 13
Unit mesh size and shape
LEED analysis ... **A10:** 544
United Kingdom Atomic Energy Authority (UKAEA)
development of RBSC ... **EM4:** 293
United States
foundry history of ... **A15:** 24-36
telecommunication industry structure ... **EL1:** 384
United States Public Health Service (USPHS)
criteria for classifying loss of anatomical form of posterior restorations (dental) ... **A18:** 671
United States steel wire gage (USSWG) ... **A1:** 277
United States steel wire gage system ... **M1:** 259
Unitemp 1753
thread grinding ... **A16:** 275
Unitemp AF2-1DA
composition ... **A4:** 794 **A6:** 573
Unitemp AF2-1DA6
composition ... **A6:** 573
Units
of analysis, part tolerances ... **A17:** 18
environmental stress screening ... **EL1:** 878
of measure ... **A10:** 691
radiation ... **A17:** 300-301
SI standardized ... **A10:** 685
Units and measures
metric and conversion data for ... **A8:** 721-723
Univariant equilibrium ... **A3:** 1 • 2
defined ... **A9:** 19
Universal clamp and leveling vise
for microhardness specimens ... **A8:** 96
Universal fluxes
various ... **A10:** 167
Universal lead frame
defined ... **EL1:** 1160
Universal measuring machines *See also* Coordinate measuring machines (CMMs)
column coordinate ... **A17:** 21
Universal open-front holder
for axial fatigue testing ... **A8:** 369
Universal pinion
by precision forging ... **A14:** 163
Universal repair
techniques ... **EL1:** 711
Universal rolling ... **A14:** 347-348
Universal Slopes equation
for creep-fatigue predictions ... **A8:** 354
Universal straight-line polishing and buffing machines ... **M5:** 122
Universal Strip Cote ... **EM4:** 464
Universal testing machine ... **A8:** 612-616
calibration ... **A8:** 612-616
flat-face tensile fracture ... **A11:** 76
Fracjack test system and ... **A8:** 470
hydraulic ... **A8:** 612-613
load application systems ... **A8:** 612
with LVDT extensometer ... **A8:** 616
screw-driven ... **A8:** 613
Universal tilting stages for microscopes ... **A9:** 82
Universal unit
immersion ultrasonic scanning ... **A17:** 261
Universe *See* Population
Unknown crystalline phase identification
by single-crystal diffraction ... **A10:** 353
Unknown distribution
direct computation for ... **A8:** 666-667
example of computational procedures ... **A8:** 676
ranks of observations for ... **A8:** 666
Unknown phase identification, by electron diffraction/EDS
analysis of metallized ceramic ... **A10:** 457-458
assessment, experimental/reference data ... **A10:** 455-456
confirming ... **A10:** 457
data base use ... **A10:** 456
diffraction patterns obtained ... **A10:** 458
strategy of analysis ... **A10:** 456-457
Unleaded tin bronzes
applications ... **A18:** 750, 751
composition ... **A18:** 751
designations ... **A18:** 751
mechanical properties ... **A18:** 750, 752
product form ... **A18:** 751, 752
Unloading
of presses ... **A14:** 500-501
Unloading compliance technique
for crack closure evaluation ... **A8:** 391
Unlubricated sliding
defined ... **A18:** 20
Unmelted electrodes, ingot
as forging defect ... **A17:** 492
Unmixed stress criteria ... **A8:** 344
Unmixed zone ... **A6:** 749
nickel alloys ... **A6:** 588, 589
nickel-chromium alloys ... **A6:** 588, 589
nickel-chromium-iron alloys ... **A6:** 588, 589
nickel-copper alloys ... **A6:** 588, 589
Unmixed zones
in welded stainless steels ... **A13:** 344, 353
Unnotched axial specimen
for fatigue testing ... **A8:** 696
Unnotched Charpy impact strength
testing ... **A7:** 490
Unnotched fatigue limits
of tempered pearlitic malleable irons ... **A1:** 83
Unpaired electrons
ESR analysis of ... **A10:** 254
Unpigmented drawing compounds
removal of ... **M5:** 5-6, 8-9
Unreactive rubber
as toughener ... **EM3:** 185
Unresolved doublets
Kα aluminum and magnesium lines as ... **A10:** 570
UNS *See* Unified Numbering System
UNS designations ... **M1:** 140, 143
constructional steels for elevated temperature use ... **M1:** 648, 649
Unsaturated anhydrides
for polyesters ... **EM1:** 132
Unsaturated compounds ... **EM3:** 30
defined ... **EM1:** 25 **EM2:** 44
Unsaturated esters
suppliers of ... **EM1:** 133
Unsaturated polyester resins *See also* Polyester resins; Resins
for filament winding ... **EM1:** 137-138
for pultrusion ... **EM1:** 538
Unsaturated polyesters *See also* Polyesters; Thermosetting polyesters; Thermosetting resins ... **EM3:** 30
applications ... **EM2:** 246-247
characteristics ... **EM2:** 247-251
chemistry ... **EM2:** 65
compression molding ... **EM2:** 249-250
costs and production volume ... **EM2:** 246
filament winding ... **EM2:** 251
filled/unfilled ... **EM2:** 250-251
forms of ... **EM2:** 246
hand lay-up/spray-up ... **EM2:** 249
mechanical properties ... **EM2:** 247
for porosity sealing in castings ... **EM3:** 51
processing ... **EM2:** 249-251
pultrusion ... **EM2:** 251, 394
resin transfer molding ... **EM2:** 250-251
as structural plastic ... **EM2:** 65
Unsaturated polymers *See also* Polymers; Unsaturated polyesters
defined ... **EM2:** 44
Unsaturation
determined ... **A10:** 219
EFG determination in polymers ... **A10:** 212
Unsharpness
film, defined ... **A17:** 300
geometric, in shadow formation ... **A17:** 311-314
image, from stem radiation ... **A17:** 305
in-motion radiography ... **A17:** 338

Unsharpness (continued)
screen, radiography **A17:** 300
total, radiography **A17:** 300
Unstable emulsion cleaners
use of **M5:** 33-35
Unstable equilibrium .... **A3:** 1 • 1
Unstable fast fracture
as fatigue stage **A12:** 175
Unsupported hole
defined **EL1:** 1160
Unsymmetric laminate *See also* Laminate(s) **EM3:** 30
defined **EM2:** 44
Unsymmetrical bending
of straight beams **A8:** 119
Unsymmetrical holding
in tests with hold period in tension **A8:** 349
Unsymmetrical laminate *See also* Laminate(s)
defined **EM1:** 25
Untapered bonded joints **EM3:** 475, 476
Untempered martensite (UTM) **A16:** 25-29
"Unzipping" **A6:** 240
Up stroke *See* Stroke
Up-and-down procedure *See* Staircase method
Updraw process **EM4:** 400, 1038
Upgraded cast iron *See* Compacted graphite iron
Upgrading
of castings **A6:** 495
Upholstery construction wire
description **M1:** 266, 268
tensile strength requirements **M1:** 268
Upholstery spring construction wire **A1:** 285
Upper bainite **A9:** 663-664
defined **A9:** 179
Upper bound method
analytical modeling **A14:** 394-395, 425
Upper control limit (UCL)
in control chart method **A17:** 726
Upper limit, design
and distortion **A11:** 136-138
Upper punch **A7:** 12
Upper ram **A7:** 12
Upper shelf energies
inclusion shape effects **A15:** 91
Upper shelf impact energy, HSLA steel
effect of cerium content **M1:** 418
Upper yield point
defined **A8:** 21-22
Upright bench microscope **A9:** 72
Upright incident-light microscopes **A9:** 72
light path **A9:** 71
Upright research-quality optical microscopes **A9:** 73
UPS *See* Ultraviolet photoelectron spectroscopy
Upset
cavities, location of **A14:** 87-88
defined **A14:** 14
definition **A6:** 1215 **M6:** 18
test, for forgeability **A14:** 215
Upset butt weld
failure origins **A11:** 443-444
steel wire, burrs on **A11:** 443
wire-end preparation for **A11:** 443
Upset butt welding
definition **A6:** 1215
Upset cylinder grids for strain measurement on **A8:** 579
tensile and axial compressive stresses at equator **A8:** 578
Upset cylinders **A7:** 410
Upset distance
definition **A6:** 1215
Upset force **A6:** 888
Upset forging
of aluminum alloys **A14:** 244
and cold working **A7:** 690
of copper and copper alloys **A14:** 255
defined **A14:** 14
forging modes and stress conditions on pores **A7:** 414
of heat-resistant alloys **A14:** 231
load vs displacement curve **A14:** 37
of stainless steels **A14:** 222
of titanium alloys **A14:** 272
Upset forging, hot *See* Hot upset forging
Upset forgings
inspection techniques **A17:** 495
Upset reduction
effect on forging load, heat-resistant alloys **A14:** 232
vs forging pressure, stainless steels **A14:** 223-224
Upset resistance butt welding
failure origins **A11:** 443-444
Upset test **A7:** 410
fracture limits **A8:** 580-581
specimens, fracture loci in **A8:** 580-581
and tension test fracture strains compared **A8:** 581
Upset test, standard
for material defects **A11:** 307
Upset welding *See also* Forge welding **A6:** 249-251
advantages **A6:** 249-250
alternative energy supply **A6:** 250-251
applications **A6:** 250
definition **A6:** 249, 1215 **M6:** 19
deformation **A6:** 249
difficult-to-weld materials joined **A6:** 249
equipment **A6:** 249
fixturing arrangement for cylindrical parts **A6:** 250
fixturing of parts **A6:** 250
homopolar generator **A6:** 250-251
limitations **A6:** 250
plug welds **A6:** 250
safety precautions **M6:** 59
types of welds **A6:** 250
Upsetter heading tools and dies
types **A14:** 85-86
Upsetters *See also* Forging machines; Headers; Horizontal forging machines; Upset forging; Upsetting; Upsetting and piercing
for cold extrusion **A14:** 303
defined **A14:** 14
die and die materials for **A14:** 43
rated sizes for **A14:** 83, 85
Upsetting *See also* Cold heading; Forging; Heading; Upset forging; Upsetters
center **A14:** 295
defined **A14:** 14
double-end **A14:** 90
effect on flash weld strength **M6:** 578
fracture in **A7:** 411
in header **A14:** 311
lubricant films used **A18:** 146
mechanisms for flash welding **M6:** 561-562
offset **A14:** 90
in open-die forging **A14:** 64
and piercing **A14:** 88-89
pipe and tubing **A14:** 91-93
plane-strain, and rolling, compared **A14:** 344-345
powder forging modes **A14:** 190
sequence in flash welding **M6:** 569-571
severity, stainless steels **A14:** 223
simple **A14:** 87-88
with sliding dies **A14:** 90-91
strain distribution in cube after **A14:** 437
and trimming, tooling setup for **A14:** 88
wrought aluminum alloy **A2:** 34
Upsetting and piercing **A14:** 88-89
double **A14:** 89
production examples of **A14:** 88-89
vs cold extrusion **A14:** 95
Upsetting force
definition **M6:** 19
Upsetting in fabricating steel wire products **M1:** 589, 590
Upsetting stresses
toot steel fracture by **A12:** 377
Upsetting time
definition **M6:** 19
Upslope time
definition **M6:** 19
Uptake delay
in concentric nebulizers **A10:** 35
Upturned-fiber flaws
defined **A17:** 562
Upward inflection
ferritic steels **A8:** 332
Urania-beryllia **EM4:** 191
Uranium *See also* Depleted uranium; Depleted uranium, heat treating; Optically anisotropic metals; Pure uranium; Uranium alloys; Uranium alloys, specific types
as actinide metal, properties **A2:** 1197
Uranium (continued)
AFS analysis of **A10:** 46
airborne particle standard for **A9:** 477
alloy concentration, effects on structure and properties of quenched alloys **A9:** 476
alloys, relative hydrogen susceptibility **A8:** 542
assay by delayed-neutron counting **A10:** 238
beta-ray dose rate of depleted uranium **A9:** 477
chemical toxicity of **A9:** 477
corrosion of **A13:** 813-822
cutting induced deformation in **A9:** 478
density of **A2:** 670
determined by controlled-potential coulometry **A10:** 209, 211
effect of heat during sectioning on hardness and microstructures **A9:** 477-478
effect of working temperature on phases **A9:** 481
electroplating **A13:** 819-820
electropolishing solutions for **A9:** 478
environmental considerations
epithermal neutron activation analysis **A10:** 239
etchants for **A9:** 480
etching of **A9:** 479-480
fissionable isotope U-235 and U-238 .... **A2:** 670
flow diagram for quantification of **A10:** 80
galvanic reactions during polishing **A9:** 478
gamma-radiation dose rate of depleted uranium **A9:** 477
grinding of **A9:** 478
health and safety in handling **A9:** 477
hot and cold fabrication techniques **A2:** 671-672
identification of inclusions **A9:** 479
ignition temperature of **A9:** 477
induction melting and molding **A2:** 671
ion plating **A13:** 821
isotopes of **A9:** 476
Jones reductor for **A10:** 176
low-temperature solid-state welding **A6:** 300
M lines used for **A10:** 86
macroetching procedures for **A9:** 479
macroexamination **A9:** 479
microalloyed **A2:** 677
microetching of **A9:** 479-480
microexamination **A9:** 479-480
microstructures **A9:** 480-481
as minor toxic metal, biologic effects **A2:** 1261-1262
in molten salts **A13:** 52-53
mounting of samples **A9:** 478
nickel-plated, moist nitrogen corrosion **A13:** 820
nuclear fuel application **A2:** 1261
oxidation **A13:** 813-815
phases **A9:** 476-487
pitting during polishing **A9:** 478
plate and sheet, cam plastometer **A8:** 194
polishing of **A9:** 478
polymorphism in **A9:** 476
preparation for bright-field microexamination **A9:** 480
preparation for polarized light microexamination **A9:** 480
preparation of metallographic samples **A9:** 477-480
processing and properties **A2:** 670-672
protective coatings for **A13:** 818-821
pure **M2:** 821-822, 832-833
pyrophoricity of **A2:** 670-671
radioactivity of **A2:** 670
recrystallization of **A9:** 476, 481, 483
roll welding **A6:** 314
safety and health considerations **A2:** 670-671
sectioning of **A9:** 477-478
solid-state-welded interlayers **A6:** 169, 171
solubilities of alloying elements in **A9:** 476
species weighed in gravimetry **A10:** 172
suitability for cladding combinations **M6:** 691
thermal diffusivity from 20 to 100 °C **A6:** 4
TNAA detection limits **A10:** 237
toxicity of **A2:** 670
ultratrace, determination by laser-induced fluorescence spectroscopy **A10:** 80-81
unalloyed **A2:** 671-672, 674
and uranium alloys **A2:** 670-682

Uranium (continued)
volumetric procedures for **A10:** 175
x-ray absorption curve as function of wavelength **A10:** 85
Uranium alloys *See also* Uranium; Uranium alloys, specific types
age hardening **A2:** 674-675
alloying **A2:** 672
alloying and hydrogen generation **A13:** 815
alloying effects on oxidation **A13:** 814-815
alloys, classes of **A2:** 677-681
annealed, mechanical properties **A2:** 673
anodizing **A9:** 142
bainitic-like microstructures **A9:** 665-666
cooling **A2:** 672-674
corrosion behavior **A13:** 815-816
corrosion of **A13:** 813-822
couples, galvanic corrosion **A13:** 818
delayed cracking **A2:** 675-676
density of **A2:** 670
diffusion welding **A6:** 886
effects of impurities on **A9:** 477
effects of quenching on **A9:** 477
electroplating **A13:** 820-821
environmental considerations **A2:** 670-671
galvanic behavior **A13:** 816-817
health and safety in handling **A9:** 477
heat tinting **A9:** 136
heat treatment **A2:** 672-674
image analysis of cellular decomposition of martensite in **A10:** 316
ion plating **A13:** 821
kinetics of cellular decomposition of martensite in **A10:** 316-318
microalloyed **A2:** 677
microstructure effect on corrosion properties **A13:** 816
microstructures, of quenched 'alloys **A2:** 674
phases **A9:** 476
polynary, SCC thresholds **A13:** 819
preparation of metallographic samples **A9:** 477-480
properties, of quenched alloys **A2:** 674
protective coatings for **A13:** 818-821
pyrophoricity **A2:** 670-671
quenched and aged, applications **A2:** 673
quenching **A2:** 673-674
radioactivity of **A2:** 670
residual stresses and stress relief **A2:** 675
rest potential vs. alloy content **A13:** 815
safety and health considerations **A2:** 670-671
solid-state phase transformations in welded joints **A9:** 581
solution heat treatment **A2:** 673-674
specific alloys **A2:** 677-681
stress-corrosion cracking **A13:** 817-818
tantalum corrosion in **A13:** 735
ternary, quaternary, higher-order alloys **A2:** 680-681
toxicity of **A2:** 670
Unicast process for **A15:** 251
and uranium **A2:** 670-682
uranium-molybdenum alloys **A2:** 680
uranium-niobium alloys **A2:** 679-680
uranium-titanium alloys **A2:** 677
welding **A2:** 676-677
Uranium alloys, specific types
U-0.3Mo **A9:** 481, 485
U-0.75Ti **A9:** 478, 481, 485-486 **A16:** 874, 876-877
aging
applications **M3:** 774
heat treatment **M3:** 775, 776-777
tensile properties **M3:** 775
U-0.75Ti aged, microstructure **A10:** 318
U-0.75Ti, aging temperature and time effects on hardness **A2:** 676
U-0.75Ti, aging temperature effect on tensile properties **A2:** 676
Uranium alloys, specific types (continued)
U-0.75Ti, annealed, mechanical properties **A2:** 673
U-0.75Ti, bainitic-like microstructure **A9:** 665-666
U-0.75Ti, cellular decomposition kinetics **A10:** 316
U-0.75Ti, coating protection **A13:** 820
U-0.75Ti, corrosion rate as function of porosity **A13:** 821
U-0.75Ti, effect of temperature on decomposition time **A10:** 319
U-0.75Ti, effect of time on cellular decomposition **A10:** 319
U-0.75Ti, fracture toughness **A2:** 678
U-0.75Ti, heat-treated, properties and applications **A2:** 674
U-0.75Ti, hydrogen content and strain rate effects on ductility **A2:** 678
U-0.75Ti, microstructure effect on corrosion properties **A13:** 816
U-0.75Ti, oxidation rate vs. water vapor pressure **A13:** 814
U-0.75Ti, plane-strain threshold for SCC **A13:** 818
U-0.75Ti, porosity in electroplated nickel on **A13:** 821
U-0.75Ti, quench rate effects on properties **A2:** 679
U-0.75Ti, SCC thresholds, air/aqueous environments **A13:** 818
U-0.75Ti, surface morphology after chemical etching **A13:** 820
U-2.0Mo **A9:** 481-482, 486-487
U-2.0Mo, annealed, mechanical properties **A2:** 673
U-2.0Mo, heat-treated, properties and applications **A2:** 674
U-2.3Nb, annealed, mechanical properties **A2:** 673
U-2.3Nb heat-treated, properties and applications **A2:** 674
U-2Mo
aging **M3:** 777
applications **M3:** 774
heat treatment **M3:** 775, 776-777
tensile properties **M3:** 775
U-4.5Nb, crack velocity v s SCC threshold, oxygen pressure **A13:** 818
U-4.5Nb, heat-treated, properties and applications **A2:** 674
U-6.0Nb, cooling rate effect on tensile properties **A2:** 675
U-6.0Nb, heat-treated, properties and applications **A2:** 674
U-6.3Nb, aging temperature effects on tensile properties **A2:** 680
U-6Nb **A9:** 478, 482, 487 **A16:** 874, 876-877
U-6Nb, microstructure effect on corrosion properties **A13:** 816-817
U-7.5Nb-2.5Zr **A9:** 482, 487
U-7.5Nb-2.5Zr (mulberry), SCC threshold stress **A13:** 818
U-7.5Nb-2.5Zr, heat-treated, properties and applications **A2:** 674
U-10Mo heat-treated, properties and applications **A2:** 674
U-14Zr, heat tinting **A9:** 136
U-33Al-25Co, two-phase structure **A9:** 8d
U-075Ti, corrosion kinetics **A13:** 821
U_{235} (enriched uranium) **A16:** 877
Uranium and uranium alloys **A16:** 874-878
biological effects, health hazard **A16:** 877
chip formation **A16:** 874, 875, 876
common uses **A16:** 874
cutting fluids **A16:** 874, 875, 876, 877
drilling **A16:** 875
electrochemical grinding **A16:** 543
external radiation exposure, health hazard **A16:** 877
Uranium and uranium alloys (continued)
facing **A16:** 875
fire hazard **A16:** 874-875, 876, 877
grinding **A16:** 874
health effects and required precautions **A16:** 877-878
heat treating **A16:** 874
internal radiation exposure health hazard **A16:** 877
machines used **A16:** 875
machining parameters **A16:** 874-875
metallurgical considerations **A16:** 874
milling **A16:** 875
problems associated with machining **A16:** 874
pyrophoric nature **A16:** 877
surface finish **A16:** 874, 875, 876
tapping **A16:** 875
threading **A16:** 875
tool life **A16:** 875-876
tools **A16:** 875-876, 877
turning **A16:** 874, 875
unique characteristics **A16:** 874
Uranium carbide
nuclear applications **A7:** 664
Uranium carbide cermets
application and properties **A2:** 978, 1002
Uranium carbides
heat tinting **A9:** 136
Uranium, depleted **M3:** 773-780
applications **M3:** 773-774
availability **M3:** 775
casting **M3:** 775-776
corrosion **M3:** 778
extrusions **M3:** 776
forging **M3:** 776
health hazards **M3:** 778
heat treating **M3:** 775, 776, 777
licensing **M3:** 779
machining **M3:** 777-778
mechanical properties **M3:** 774-775
melting **M3:** 775-776
physical properties **M3:** 774-775
pyrophoricity **M3:** 778
as radiographic screen **A17:** 316
rolling **M3:** 776
swaging **M3:** 776
Uranium dioxide
analysis of impurities **A10:** 149-150
fuel, hydrogen content **A7:** 664
fuel rod **A7:** 664, 665
nuclear applications **A7:** 664
nuclear fuel pellets **A7:** 664
stoichiometric **A7:** 665
Uranium dioxide cermets
application and properties **A2:** 978, 994
Uranium dioxide dust
biologic effects **A2:** 1261-1262
Uranium dioxide-aluminum fuel
blending of **A7:** 664
Uranium hexafluoride
gaseous lubricant for pumping equipment bearings **A18:** 522
Uranium nitride
nuclear applications **A7:** 664
Uranium oxide cermets
applications and properties **A2:** 993-994
Uranium oxide powders
packed density **A7:** 297
Uranium oxide-containing cermets **A7:** 804
Uranium powders
nuclear applications **A7:** 664
powder metallurgy **A7:** 18
production with magnesium **A7:** 131
pyrophoricity **A7:** 199
scrap recovery and reprocessing **A7:** 666
vacuum sintering atmospheres for **A7:** 345
virgin, for pellet fabrication **A7:** 665
Uranium, vapor pressure
relation to temperature **M4:** 309, 310

Uranium-molybdenum alloys
properties and processing **A2:** 680
stress-corrosion cracking in **A13:** 817
Uranium-niobium alloys
galvanic behavior **A13:** 816-817
properties and processing **A2:** 679-680
stress-corrosion cracking **A13:** 817-818
Uranium-niobium-zirconium alloys
corrosion resistance **A13:** 815
Uranium-titanium alloys
equilibrium phase diagrams **A2:** 673
properties and processing **A2:** 677-679
stress-corrosion cracking in **A13:** 817
Uranium-vanadium alloys, cast
tensile properties **A2:** 677
Uranium-zirconium alloys
heat tinting **A9:** 136
Uranyl
defined **A10:** 683
Uranyl ion
toxicity effects **A2:** 1261
Urban atmospheres
galvanized coatings in **A13:** 440
lead corrosion **A13:** 82
magnesium/magnesium alloys in **A13:** 743
simulated service testing **A13:** 204
Urban environment
effect on tin-bronze alloys **A12:** 403
Urban environments *See* Environment; Urban atmospheres
Urea
-based patterns, investment casting **A15:** 255
cast iron sensitivity to **A15:** 238
formaldehyde, as amino **EM2:** 230
formaldehyde resins, as injection moldable **EM2:** 321
formation of **EM3:** 203, 204
molding compounds, applications **EM2:** 230
as organic binder **A15:** 35
Urea-formaldehyde adhesive **EM3:** 30, 104
surface preparation **EM3:** 278
Urea-free resins
as organic binders **A15:** 35
Urethane
waterjet machinery **A16:** 522
Urethane acrylic polymers **EM2:** 268-271
Urethane coatings *See also* Coatings; Conformal coatings; Polyurethanes; Urethanes **A13:** 408-410
acrylic, as UV-curable **EL1:** 785
and acrylic, silicone, epoxy, compared **EL1:** 775
advantages/disadvantages **EL1:** 775-777
application methods **EL1:** 775, 778-779
brush coating **EL1:** 779
curing reaction **A13:** 410
dip coating **EL1:** 778-779
equipment **EL1:** 779-780
flow coating **EL1:** 779
isocyanates **A13:** 409
markets **EL1:** 775
polyesters **A13:** 410
polyethers **A13:** 410
properties/characteristics **EL1:** 776
removal **EL1:** 780
repair, of coated components **EL1:** 780
safety precautions **EL1:** 779-780
spray coating **EL1:** 778
substrate cleaning **EL1:** 776-778
Urethane hybrids *See also* Hybrid; Thermosetting resins **EM3:** 30
applications **EM2:** 268
characteristics **EM2:** 268-271
costs **EM2:** 268
defined **EM2:** 44
physical properties **EM2:** 269
prepreg, forming **EM2:** 271
processing **EM2:** 268-271
properties **EM2:** 268
RIM processing **EM2:** 269
SRIM processing **EM2:** 270
suppliers **EM2:** 271
Urethane plastics *See also* Isocyanate plastics; Plastics; Polyurethane; Polyurethanes; Urethane hybrids **EM3:** 30
defined **EM1:** 25 **EM2:** 44
Urethane resins and coatings **M5:** 474-475, 498, 502-503, 505
Urethane rubber **EM3:** 203
Urethane-acrylates
compared with epoxy-acrylates **EM3:** 92
cure mechanism **EM3:** 567
for motor magnet bonding **EM3:** 574, 575
for surface-mount technology bonding **EM3:** 570
Urethane-aliphatic diacrylate
properties **EM3:** 92
Urethane-aliphatic triacrylate
properties **EM3:** 92
Urethane-aromatic multiacrylate
properties **EM3:** 92
Urethanes *See also* Coatings; Conformal (coatings; Polyurethanes; Urethane coatings **EM3:** 44, 108-112, 594
acrylate, structure **EL1:** 858
acrylated, for solder masking **EL1:** 555
additives
for adhesion promoter in water-base coatings
advantages **EM3:** 600
aerospace industry applications **EM3:** 558, 559
aliphatic compared to aromatic **EM3:** 111
applications **EM3:** 205
aromatic compared to aliphatic
for automotive circuit devices **EM3:** 610
automotive electronic tacking and sealing **EM3:** 610
for automotive electronics bonding **EM3:** 553
for automotive filter element bonding **EM3:** 110
for automotive structural bonding **EM3:** 111, 554
for automotive windshield bonding **EM3:** 723-724
as body assembly sealants **EM3:** 554, 608, 609
for caulking **EM3:** 607-608
chemical structure **EL1:** 244
chemistry **EM3:** 108-110
for coating and encapsulation **EM3:** 580
for coating/encapsulation **EL1:** 242
colorants **EM3:** 111
compared to acrylics **EM3:** 119, 120, 121, 124
compared to polyether silicones **EM3:** 231
competing adhesives **EM3:** 111
component terminal sealant **EM3:** 612
for construction sealing **EM3:** 606, 607
for contour pattern flocking **EM3:** 110
cost factors **EM3:** 111
cross-linking **EM3:** 108-110
cure mechanism **EM3:** 112, 323, 554, 600
dielectric sealants **EM3:** 611
electrical contact assemblies **EM3:** 611
for electrically conductive bonding **EM3:** 572
for electronic insulation **EM3:** 110
for electronic packaging applications **EM3:** 600
for electrostatic flocking of plastics, textiles, or rubbers **EM3:** 110, 111
engineering adhesives family **EM3:** 567
fillers **EM3:** 111, 181
for flexible packaging **EM3:** 110
for foam and fabric laminations **EM3:** 110
for foam sponge bonding **EM3:** 110
formation of **EM3:** 203
functional types **EM3:** 110
gas tank seam sealant **EM3:** 610
handling **EM3:** 108
health hazard **EM3:** 108
heating and air conditioning duct sealing and bonding **EM3:** 610
for highway construction joints **EM3:** 57
for interior seals in window systems **EM3:** 57
for laminated film packaging **EM3:** 110
for laminating furniture during construction **EM3:** 110
for lamination of building panels **EM3:** 110
lap shear strength **EM3:** 294
for leather goods bonding **EM3:** 110
for lighting subcomponent bonding **EM3:** 552
market area **EM3:** 110
market size **EM3:** 111
methacrylate-capped and anaerobics **EM3:** 113
modifiers **EM3:** 111
for packaging film **EM3:** 110
for plastic film lamination **EM3:** 110
polymers used **EM3:** 112
polyurethane foam, bond breaker **EM3:** 549
polyurethane varnish, coating for aramid fibers **EM3:** 285
Urethanes (continued)
polyurethanes (PUR) **EM3:** 22, 74, 75, 594
advantages and disadvantages **EM3:** 675
advantages and limitations **EM3:** 77, 78
antimony oxide as flame retardant filler **EM3:** 179
applications **EM3:** 50, 53, 56
automotive market applications **EM3:** 608
for back light (rear window) installation **EM3:** 554
for body sealing and glazing material **EM3:** 57
for bonding metal to metal **EM3:** 46
for cabin pressure sealing **EM3:** 58, 59
characteristics **EM3:** 53, 675-676
chemical properties **EM3:** 52
chemical resistance properties **EM3:** 2
chemistry **EM3:** 50, 78
compared to epoxies **EM3:** 98
compared to polysulfides **EM3:** 195
as conductive adhesive **EM3:** 76
conformal overcoat **EM3:** 592
cure properties **EM3:** 51
curing method **EM3:** 78
for curtain wall construction **EM3:** 549
degradation **EM3:** 679
electrical/electronics applications, polyurethanes **EM3:** 44
epoxidized, modifier for epoxies **EM3:** 99
for expansion joint sealing **EM3:** 204
for faying surfaces **EM3:** 604
for fillets **EM3:** 604
flexibility **EM3:** 98
glass adherend showing interference separation **EM3:** 452
for granite construction **EM3:** 607
for handle assemblies bonding **EM3:** 45
for housing assemblies bonding **EM3:** 45
for insulated double-pane window construction **EM3:** 46
for insulated glass construction **EM3:** 58
IPN polymers **EM3:** 602
for lens bonding **EM3:** 575
modified hot-melt **EM3:** 50
no adverse effect by water-displacing corrosion inhibitors **EM3:** 641
for paving joint sealing **EM3:** 204
performance **EM3:** 674
for potting **EM3:** 585
predicted 1992 sales **EM3:** 77
properties **EM3:** 50, 78, 82, 677
reacting with acrylonitrile-butadiene rubber **EM3:** 148
resistant to many aggressive materials **EM3:** 637
for rivets **EM3:** 604
sealants **EM3:** 675-676
for sealing construction **EM3:** 606
for sheet molding compound (SMC) bonding **EM3:** 46
silane coupling agents **EM3:** 182
steel-to-polycarbonate shear strength **EM3:** 663, 664
suppliers **EM3:** 58, 78
for taillight and headlight bonding **EM3:** 46
thermal properties **EM3:** 52
thermoplastic **EM3:** 22
to fill unitized steel shells **EM3:** 577
as top coat **EM3:** 640, 641
for truck trailer joints **EM3:** 58
typical properties **EM3:** 83
viscosity and degree of cure **EM3:** 323
volume resistivity and conductivity **EM3:** 45
for windshield glazing **EM3:** 57
for windshield installation **EM3:** 554
for windshield sealing **EM3:** 608
zinc oxide as filler **EM3:** 179
pot life **EM3:** 111-112
primers **EM3:** 112
product design considerations **EM3:** 111-112
properties **EM3:** 109, 111, 600
for protecting electronic components **EM3:** 59
recreational vehicle modifications **EM3:** 610, 611
recycling of solvent emissions **EM3:** 112
for rubber athletic flooring **EM3:** 110

Urethanes (continued)
sealants EM3: 57, 188, 190-191, 203-207
additives and modifiers EM3: 205-206
application methods EM3: 205
application parameters EM3: 205
for automobile repair EM3: 204
for automobile windshield sealing EM3: 204
blocking agents EM3: 205
catalysts EM3: 204
characteristics EM3: 205
chemistry EM3: 203-204
commercial forms EM3: 204
competing with silicones and polysulfides EM3: 205
cost factors EM3: 204-205
cure rates EM3: 204
curing mechanism EM3: 203, 207
for drip rail steel molding sealing EM3: 204
flammability EM3: 205
functional types EM3: 204
for gutter repair EM3: 204
for horizontal joints EM3: 204
isocyanates, use in manufacture EM3: 203
joint failure EM3: 205
for liquid membranes or sheet goods between concrete slabs EM3: 204
markets EM3: 204
for opera window sealing EM3: 204
for parking deck sealing EM3: 204, 206
plasticizers EM3: 205
pot life EM3: 205, 206-207
primers EM3: 206, 207
processing parameters EM3: 206-207
product design considerations EM3: 206
properties EM3: 204
sag control materials EM3: 206
service life EM3: 204
shelf life EM3: 206-207
for sidewalk repair EM3: 204
silane adhesion promoters EM3: 206
suppliers EM3: 204-205
for T-top roof sealing EM3: 204
for taillight sealing EM3: 204
for tub perimeter repair EM3: 204
UV stabilizers and antioxidants EM3: 206
for vertical joints EM3: 204
for window repair EM3: 204
secondary seal for polyisobutylene EM3: 190
for shoe/boot manufacturing EM3: 110
for sporting goods manufacturing EM3: 576-577
storage life EM3: 111
storage requirements EM3: 112
suppliers EM3: 111, 204-205
surface preparation EM3: 112
tackifiers for EM3: 182-183
for trim bonding EM3: 110
UV-curing EM3: 568
for vinyl repair EM3: 110
for wall construction EM3: 56
for windshield bonding EM3: 53, 554
for windshield glazing EM3: 57
for windshield sealing EM3: 46
Urinary cadmium concentration
and renal dysfunction A2: 1241
Usable intrinsic device speed
maximum defined EL1: 2
Use failures, integrated circuits *See also* Service failures
Useful life phase
reliability life cycle EL1: 897
User acceptance
of corrosion test results A13: 316
User-requested functions
rules for EL1: 128
UTA8DV *See* Titanium alloys, specific types, Ti-8Al-1Mo-1V
Utility boiler drum, failure during
hydrotesting A11: 647
Utilization
of cooling curves (thermal analysis) A15: 182-185
of induction furnaces A15: 374
Utilization, corrosion test data
and computers A13: 317
UTW *See* Ultrathin window
UV *See* Ultraviolet
UV technology *See* Ultraviolet (UV) technology
UV-curable silicone coatings EL1: 824
UV/VIS *See* Ultraviolet/visible absorption spectroscopy
UV/VIS spectrophotometry
instrumentation of A10: 61
UVSOR
as synchrotron radiation source A10: 413

V

v *See* Electron velocity; Poisson's ratio
***V(z)* curve** A18: 407, 408
V-36
composition A4: 795 A6: 929
V-50Mo, refractory metal brazing
filler metal A6: 942
V-57
composition A4: 794 A16: 736
machining A16: 738, 741-743, 746-747, 749-757
V-bend die *See also* Three-point bending; V-dies
defined A14: 14
V-block and knife-edge support
in constant-stress testing A8: 320-321
V-block bending A8: 125-127
V-cone blender A7: 12
V-dies
for hot forging A14: 43
for open-die forging A14: 61
for press bending A14: 525
press brake forming by A11: 308
V-grip
for axial fatigue testing A8: 369
V-groove joints
radiographic inspection A17: 334
V-groove weld
definition A6: 1215
V-groove welds
arc welding of heat-resistant alloys M6: 356-357
cobalt-based alloys M6: 367-368
nickel-based alloys M6: 360
arc welding of nickel alloys M6: 437, 442-443
definition, illustration M6: 60-61
double, applications of M6: 61
electron beam welding M6: 614
flux cored arc welding M6: 105-106
gas tungsten arc welding M6: 201
of heat-resistant alloys M6: 356-357, 360
oxyacetylene braze welding M6: 597
oxyfuel gas welding M6: 590-591
preparation M6: 67-68
single, applications of M6: 61
V-guides A16: 111
V-mixer A7: 12
for metal-filled plastics A7: 606
V-notch bend bars
fracture stress measured in A8: 466
V-pores *See* Interparticle porosity
V-process *See also* Vacuum molding
defined A15: 11
V-ring seal
defined A18: 20
V-sections
suited to planing A16: 186
V-shape work guides
for rotary swaging A14: 134
V-W (Phase Diagram) ..... A3: 2•381
V-X diagram
defined A9: 19
V-Zr (Phase Diagram) ..... A3: 2•381
Vacancies *See also* Interstitials; Point defects
cation and anion A13: 65-66
as crack initiation sites A17: 216
diffusion A13: 68
effect, resistance-ratio test A2: 1096
formed by irradiation of crystals A9: 116
in iridium A10: 588
as point defects A10: 588
point-defect agglomerates A9: 116
supersaturation, and zone formation, wrought aluminum alloy A2: 39
Vacancy
defined A9: 19
Vacancy coalescence
alloy steels A12: 349
Vacancy precipitation
in germanium A9: 608
in pure metals A9: 608
Vacuum *See also* Vacuum leak testing systems
absence of stress-corrosion cracking in A8: 499
and air, fatigue fractures compared A12: 48, 55
atmospheres, for brazing A11: 450
brazing atmosphere sources A6: 628
chamber A8: 196, 412-415
chamber, scanning Auger microscope A11: 35
creep testing in A8: 303
effect, aluminum alloys A12: 46
effect, Astroloy A12: 48
effect, fatigue A12: 36, 46-49, 120
effect, Inconel X-750 ..... A12: 48
effect, stainless steel A12: 48
effect, titanium alloys A12: 47-48
electromagnetic radiation in A10: 83
electron beam welds A11: 444
environment, at ambient and elevated temperatures A8: 410-412
fatigue crack growth rate of nickel-base superalloy in A8: 412-413
as fatigue environment A12: 35
fatigue test system A8: 412-413
gas mass spectroscopy in A10: 151
and gaseous fatigue testing A8: 410-415
hard, for testing fracture surface A8: 37
induction melting, irons A12: 219
levels, as absolute pressure A17: 59
melted, abbreviation for A10: 691
-melted iron, continuous heating and cooling effect on flow stress A8: 177
-metal interfaces, SERS for A10: 136
nickel alloy cracking in A12: 396
and oxidizing gases, at elevated temperatures A8: 412-415
pumping system, SEM microscope A10: 491
requirement for XPS instruments A10: 571
stress to 1.0% strain, tantalum alloys in A8: 306
stressing, for optical holographic interferometry A17: 410
system, SEM imaging A12: 171
systems, LEISS analysis A10: 607
technology, unit of pressure A17: 59
test chamber, Auger spectrometer A10: 554
ultrahigh, abbreviation for A10: 691
ultrahigh, EXAFS electron detection for surface structure in A10: 418
ultrahigh, for atom probe microanalysis A10: 591
use in microanalytical techniques A11: 36
valve, bakeable, for environmental test chamber A8: 411
vapor deposition, for replica shadowing A12: 180
x-ray spectrometer detectors in A10: 91
Vacuum (inert) gas fusion
as trace element analysis A2: 1095
Vacuum aluminizing
zinc alloys A2: 530
Vacuum and plasma carburizing A18: 873

Vacuum annealing
water-atomized tool steel powders ... A7: 104
Vacuum arc degassing
ladle furnace and ... A15: 435-438
Vacuum arc double electrode remelting ... A15: 408
Vacuum arc melting, and electron beam melting
compared ... A15: 410
Vacuum arc remelted (VAR) steels
bearing steels ... A18: 726, 727, 731
factors influencing wear and failure in die-casting dies ... A18: 632
Vacuum arc remelting *See also* Consumable-electrode remelting
advantages ... A15: 406-407
atmosphere ... A15: 407
defined ... A15: 11
ingot defects ... A15: 407
melt rate ... A15: 407
of stainless steels ... A14: 222
vacuum arc double electrode remelting ... A15: 408
Vacuum arc remelting (VAR) ... A1: 930, 968, 970
of bearing steels ... A11: 490
effect on inclusions ... A11: 340
filler metals for ultrahigh-strength low-alloy steels ... A6: 673
ultrahigh-strength steels ... A4: 207, 217
Vacuum arc remelting (VAR) consumable electrode
NbTi superconductors ... A2: 1044
Vacuum arc skull melting
and casting ... A15: 409-410
of nickel and nickel alloys ... A2: 429
of zirconium alloys ... A15: 837
Vacuum atmospheres ... A7: 341, 345
in heat treating ... A7: 453
Vacuum atomization ... A7: 25, 26, 37-39
equipment ... A7: 43-45
powder applications ... A7: 45
powder properties ... A7: 44-45
Vacuum bag *See also* Pressure bags ... EM3: 30, 799
for cure processing ... EM1: 644
for epoxy composites ... EM1: 71
lay-up sequence ... EM1: 703
for polyimide resin curing ... EM1: 662
Vacuum bag molding *See also* Pressure bag molding ... EM1: 25, 549
defined ... EM2: 44
process ... EM2: 338, 340
size and shape effects ... EM2: 291
Vacuum bell jar
for carbon thin films ... A12: 173
Vacuum blasting ... A13: 414
Vacuum brazing
definition ... A6: 1215 M6: 19
dispersion-strengthened aluminum alloys ... A6: 543
Vacuum brazing of
aluminum alloys ... M6: 1029
heat-resistant steels ... M6: 1017
stainless steels ... M6: 1009-1010
steels ... M6: 935
tungsten sheet ... M6: 1060
Vacuum capacitance ... EM3: 428, 429
Vacuum capillary infiltration ... A7: 552
Vacuum carburized 8620H steel ... A9: 225
Vacuum carburizing ... A4: 262, 348-351 M4: 270-274
advantages ... A4: 351 M4: 273
carbon gradient control ... A4: 349-350 M4: 272, 273, 274
carbon gradients ... A4: 348 M4: 271, 272
carbon potential, control of ... A4: 348
carbon profile, effect on ... A4: 350 M4: 273
carbon uniformity ... A4: 350-351 M4: 274
case depth prediction ... A4: 349-350
costs ... M4: 272-273
diffusion time, effect on surface carbon ... A4: 348, 350 M4: 271
furnace design ... A4: 349
high-temperature ... A4: 351
operating cost comparison ... A4: 357
process cycles ... A4: 348-349 M4: 270, 271-272
pulse/pump method ... A4: 350-351
single cycle ... M4: 270
time versus diffusion time ... A4: 350 M4: 272
Vacuum casting
defined ... A15: 11
as permanent mold method ... A15: 276
Vacuum cathodic etching
for optical metallography samples ... A10: 301
Vacuum coating ... M5: 387-411
adhesion ... M5: 410-411
aluminum coatings ... M5: 388, 390-395, 399-400, 408
applications ... M5: 392-393
batch process ... M5: 397-398
carriage mechanism ... M5: 406
chromium coatings ... M5: 389-390, 392
conductor films, deposition of ... M5: 408
continuous processes ... M5: 392, 397, 404-405
corrosion protection ... M5: 395, 397, 402
cost factors ... M5: 410-411
crucible sources ... M5: 389-392
decorative coatings ... M5: 392-394, 396-403, 407-408
deposition rate ... M5: 388
dielectric films, deposition of ... M5: 408-409
electrical coatings ... M5: 394-399, 402-403, 408-409
electron beam heating coatings ... M5: 390, 404-405, 410
electronic coatings ... M5: 395, 398-399
electroplating in conjunction with ... M5: 394
encapsulation process ... M5: 397
equipment ... M5: 403-407, 410
maintenance ... M5: 405
evaporation of compounds mixtures, and alloys ... M5: 390-391
evaporation process ... M5: 387-393
flash ... M5: 391-392
multiple-source ... M5: 391, 393
reactive ... M5: 391
evaporation rates ... M5: 387
evaporation sources ... M5: 388-391
functional coatings ... M5: 392
gas analysis ... M5: 410
gas evolution in ... M5: 396
glass ... M5: 394, 396-397, 401-402
gold coatings ... M5: 388, 395, 400
heat treatment process ... M5: 409
high-temperature protection, thick films for ... M5: 395, 397, 399-400, 402-403, 409
ion plating process ... M5: 387
iron-nickel alloy films ... M5: 395
lacquering ... M5: 400
masking processes ... M5: 407-408
material compatability ... M5: 397
metal and metal compound coatings ... M5: 399-401
microprocessor-controlled system ... M5: 410
monitoring ... M5: 409-410
nickel-chromium alloy coatings ... M5: 402-403
noble metal coatings ... M5: 388, 395, 400
optical coatings ... M5: 395-396, 399-403
plastic ... M5: 394, 400-401
platinum alloy coatings ... M5: 388, 390-391, 394
post-treatment processes ... M5: 402-403, 408-409
powder feeder ... M5: 391, 393
pretreatment processes ... M5: 400-402, 408-409
process ... M5: 387-392, 397-400, 403-411
control ... M5: 409
examples of ... M5: 407-411
protective coatings applied over ... M5: 402, 408-409
pumping equipment ... M5: 404-406
quality control ... M5: 409-411
racks ... M5: 406-407
refractory oxide and metal coatings ... M5: 388-392, 400-401
resistance sources ... M5: 389-391
resistor films, deposition of ... M5: 408
rotating shafts ... M5: 405
semicontinuous process ... M5: 397-398, 403-404
silicon oxide coatings ... M5: 390-391, 394-395, 401
sputtering process ... M5: 387
sublimation sources ... M5: 389-392
substrate requirements ... M5: 396-397
temperature ... M5: 388, 390-391, 400
tests and testing ... M5: 402, 410
thermal stability ... M5: 396
thick films for high-temperature protection ... M5: 395, 397, 399-400, 402-403, 409
thickness ... M5: 394, 400-401, 410
traps and baffles ... M5: 404
vacuum seals ... M5: 404-405
vapor sources ... M5: 399-400
wear resistance ... M5: 395
Vacuum consumable electrode melting
titanium and titanium alloy castings ... A2: 642
Vacuum control ... EM3: 741
Vacuum deaeration
gas/oil production ... A13: 1244
Vacuum degassed steels, applications
rolling-element bearings ... A18: 260
Vacuum degassing *See also* Degassing ... A1: 228, 930 A7: 434-435 M1: 181
aluminum P/M alloys ... A2: 203
copper alloys ... A15: 467
cycle ... A7: 435
defined ... A15: 1084
for neutron embrittlement ... A11: 100
thermally conductive adhesives ... EM3: 571, 572
Vacuum deoxidation
of water-atomized tool steel powders ... A7: 104
Vacuum deposition ... A12: 180, 484-487
defined ... A13: 13 EL1: 1160
Vacuum deposition, carbon
for extraction replicas ... A17: 54
Vacuum deposition of interference films ... A9: 147-148
Vacuum desiccation ... EM3: 711
Vacuum diffusion bonding
mechanically alloyed oxide dispersion-strengthened (MA ODS) alloys ... A2: 949
Vacuum evaporation
plating waste recovery process ... M5: 316
for thin-film hybrids ... EL1: 313
Vacuum extractor
for component removal ... EL1: 726
Vacuum fluxing
for hydrogen removal ... A15: 460
Vacuum forming *See also* Forming ... A14: 847, 857
defined ... EM2: 44
Vacuum furnace
furnace brazing ... A6: 121
Vacuum furnaces *See* Vacuum heat treating, furnaces ... A7: 356-359, 381-382, 515
Vacuum fusion
capabilities ... A10: 226
of inorganic solids, types of information from ... A10: 4-6
Vacuum gages
Phillips/Penning types ... A17: 64
Vacuum heat treating ... A16: 55-56, 871
atmospheric, compared to ... A4: 492-493 M4: 308
measurements ... M4: 307-308
gas ballasting ... A4: 504
measurements ... A4: 492, 506-509
pressure control ... A4: 492-494, 497, 501-502, 507-509 M4: 322, 323-324
process control instrumentation ... A4: 507-509
quenching, gas ... A4: 497, 498, 501-502 M4: 317-318
quenching, liquid ... A4: 498, 499, 500, 502 M4: 318
temperature control ... A4: 499, 506-507 M4: 322-323
uses ... A4: 492, 494 M4: 307
vapor pressure ... A4: 493-495 M4: 308-311
Vacuum heat treating, furnaces ... A4: 492, 494-501
advantages ... A4: 492, 493, 497
aluminum brazing, fluxless ... A4: 497
applications ... A4: 492
backfilling, gases for ... A4: 493, 497, 501-502, 504 M4: 318-319
carbon-bonded carbon fiber ... A4: 501
cold wall ... A4: 494, 496-498, 501 M4: 312-313
cold wall, types ... A4: 496-498 M4: 313-315
disadvantages ... A4: 497, 501
graphite cloth heaters ... A4: 499, 500 M4: 316
heat insulation ... A4: 496, 500-501 M4: 316-317
heating elements ... A4: 494, 496, 498-499, 500 M4: 315-316
hot wall ... A4: 494, 495-496, 501 M4: 311
hot wall, types ... A4: 495-496 M4: 311-312
insulation maintenance ... A4: 496, 501 M4: 317
metallic shielding ... A4: 498, 500-501 M4: 316-317
multilayer graphite ... A4: 497, 501 M4: 317
power supplies ... A4: 499, 500
refractory metals ... A4: 499-500 M4: 316
sandwich construction ... A4: 501
solid graphite heaters ... A4: 500 M4: 316
vacuum chambers ... A4: 492, 494, 497, 499, 500, 503-504 M4: 319-320
workload support ... A4: 492, 496, 497, 498, 499, 502-503 M4: 319

Vacuum heat treating, pumping systems ... **A4:** 503-506
cryogenic ... **A4:** 505 **M4:** 322
diffusion ... **A4:** 504-505 **M4:** 321-322
liquid ring ... **A4:** 505
mechanical ... **A4:** 504, 505 **M4:** 320-321
oil booster ... **A4:** 505 **M4:** 322
steam ejector ... **A4:** 505 **M4:** 322
turbomolecular ... **A4:** 505
valves ... **A4:** 505-506 **M4:** 322
Vacuum hot pressing ... **A7:** 507-509
adapted for diffusion bonding ... **A7:** 517
Be ... **A16:** 870
beryllium powders ... **A7:** 171, 172, 758
of prealloyed P/M aluminum alloys ... **A14:** 250-251
sintering furnace ... **A7:** 515
system, all purpose ... **A7:** 515
temperatures ... **A7:** 172
to high density ... **A7:** 522
uniaxial ... **A7:** 170
Vacuum hot pressing (VHP) ... **EM1:** 25
beryllium powder ... **A2:** 685
defined ... **EM2:** 44
production scale ... **A2:** 987
Vacuum impregnation
for incorporating lead-based alloys ... **A7:** 559
of powder metallurgy specimens ... **A9:** 505
of wrought stainless steel specimens with surface cracks ... **A9:** 279
Vacuum impregnation as a mounting method ... **A9:** 31
for aluminum alloys ... **A9:** 352
Vacuum impregnation process
sealant application method ... **EM3:** 609
Vacuum impregnation sealing method ... **M5:** 369
Vacuum induction degassing
ladles ... **A15:** 438-440
Vacuum induction degassing and pouring
as furnace design ... **A15:** 396
Vacuum induction furnace
crucible materials ... **A15:** 394
degassing ... **A15:** 395
design ... **A15:** 396-397
melts, cleanliness ... **A15:** 396
metallurgy of ... **A15:** 393-396
trace element removal ... **A15:** 394-395
Vacuum induction melted/vacuum arc remelted (VIM/VAR) steels
applications, rolling-element bearings ... **A18:** 260
bearing steels ... **A18:** 726, 727, 731
Vacuum induction melting ... **A7:** 25 **M1:** 110-111
applications ... **A15:** 396
casting technologies ... **A15:** 396-397
cobalt and cobalt alloy powders ... **A7:** 146
commercially pure iron, as magnetically soft materials ... **A2:** 764
defined ... **A15:** 11
effect on inclusions ... **A11:** 340
furnace design ... **A15:** 396-397
furnace, metallurgy of ... **A15:** 393-396
as metal ultrapurification technique ... **A2:** 1094
of nickel and nickel alloys ... **A2:** 429
of nickel-titanium shape memory effect (SME) alloys ... **A2:** 899
nonferrous materials, production ... **A15:** 396
process control system ... **A15:** 398
process technology and automation ... **A15:** 397-399
of zirconium alloys ... **A15:** 837
Vacuum induction melting (VIM) ... **A1:** 970, 981
ferritic stainless steels ... **A6:** 443, 444
filler metals for ultrahigh-strength low-alloy steels ... **A6:** 673
of superalloys ... **A1:** 986-988
alternative melt techniques ... **A1:** 988
filters ... **A1:** 988
melt process ... **A1:** 988
primary purification reaction ... **A1:** 986-987
Vacuum induction melting (VIM) (continued)
refractory materials ... **A1:** 987-988
ultrahigh-strength steels ... **A4:** 207, 217
Vacuum induction precision casting furnaces ... **A15:** 399-401
Vacuum induction remelting
automation ... **A15:** 399-400
DS and SC furnaces ... **A15:** 400-401
furnaces ... **A15:** 399
processes ... **A15:** 399-400
products ... **A15:** 399
Vacuum induction shape casting ... **A15:** 399-401
DS and SC furnaces ... **A15:** 400-401
furnaces ... **A15:** 399
process automation ... **A15:** 399-400
products ... **A15:** 399
Vacuum infiltration ... **A7:** 554
of composites ... **A15:** 847
Vacuum injection molding *See also* Injection molding
defined ... **EM2:** 44
Vacuum ion plating ... **EM4:** 218
Vacuum ladle degassing
argon injection, effect on ... **A15:** 432
procedures ... **A15:** 433
Vacuum leak testing systems
of brazed assemblies ... **A17:** 604
maintenance ... **A17:** 67
methods ... **A17:** 66-68
of soldered joints ... **A17:** 606
Vacuum melted
abbreviation for ... **A10:** 691
Vacuum melting
defined ... **A15:** 11
nickel alloys ... **A15:** 820
Vacuum melting and remelting processes
electron beam melting and casting ... **A15:** 410-419
electroslag remelting ... **A15:** 401-406
plasma melting and casting ... **A15:** 419-425
vacuum arc remelting ... **A15:** 406-408
vacuum arc skull melting and casting ... **A15:** 409-410
vacuum induction melting ... **A15:** 393-399
vacuum induction remelting and shape casting ... **A15:** 399-401
Vacuum melting for purifying metals
Vacuum metallizing
defined ... **EM2:** 44
shielding alternatives ... **A7:** 612
of zinc alloy castings ... **A15:** 797
Vacuum metallizing application
refractory metals and alloys ... **A2:** 559-560
Vacuum metallizing coatings
powders used ... **A7:** 573
Vacuum molding *See also* V process
defined ... **A15:** 235-236
plastic film characteristics ... **A15:** 236
sequence of operations ... **A15:** 236
as special process ... **A15:** 37
Vacuum nitrocarburizing ... **M4:** 215, 218
Vacuum out-gassing and impregnation as a mounting technique for resin-matrix composites ... **A9:** 588
Vacuum oxygen decarburization
in ladles ... **A15:** 429-431, 434-435
Vacuum oxygen decarburization (VOD)
ferritic stainless steels ... **A6:** 443, 444
Vacuum pack oxidation-resistant coating ... **M5:** 664-666
Vacuum plasma spraying (VPS)
molten particle deposition ... **EM4:** 204, 205, 206, 207, 208
Vacuum plasma structural deposition
aluminum P/M alloys ... **A2:** 204
Vacuum precision casting furnaces ... **A15:** 399-401
Vacuum processing
titanium and tantalum carbides ... **A7:** 158
Vacuum pulse method
of component removal ... **EL1:** 716-717
Vacuum pump
for explosive forming ... **A14:** 637
Vacuum pumping
leakage measurement by ... **A17:** 61
Vacuum pumping system ... **A8:** 413-414
Vacuum refining *See also* Refining
defined ... **A15:** 11
Vacuum residue
defined ... **A18:** 20
Vacuum sintering ... **A7:** 12, 172 **A16:** 65, 72
of aluminum ... **A7:** 384, 743
atmosphere ... **A7:** 341, 345, 368-369
of cemented carbides ... **A7:** 386
furnace ... **A7:** 12
micrograin high-speed steels ... **A18:** 616
of stainless steels ... **A7:** 368-369
time-temperature cycle, cemented carbides ... **A7:** 386
of titanium powder and compacts ... **A7:** 393-394
to full density ... **A7:** 373-37
Vacuum snapback thermoforming ... **EM2:** 401
Vacuum spectrometers
application of ... **A10:** 41
polychromator ... **A10:** 37, 38
for vacuum ultraviolet ... **A10:** 29
Vacuum system
LEISS analysis ... **A10:** 607
Vacuum systems
autoclave ... **EM1:** 646-647
Vacuum thermal agglomeration
tantalum powder ... **A7:** 162
Vacuum tubes ... **A13:** 1120
corrosion failure analysis ... **EL1:** 1110
microwave ... **A17:** 209
Vacuum tumble dryer
for lacquer coating ... **A7:** 588
Vacuum, ultrahigh
contamination prevention by ... **A7:** 251
Vacuum ultraviolet
as spectral region ... **A10:** 61
Vacuum vapor deposition
oxidation- resistant coatings ... **M5:** 664-666
Vacuum vapor deposition of interference contrast layers ... **A9:** 60
Vacuum-and-helium testing
of brazed assemblies ... **A17:** 604
Vacuum-arc remelting ... **M1:** 111, 113
ultrahigh-strength steels ... **M1:** 422, 426-429, 437, 439, 441
Vacuum-assisted resin injection ... **EM1:** 564
Vacuum-carbon deoxidation ... **M1:** 112, 114
Vacuum-coating methods
to apply interlayers for solid-state welding ... **A6:** 165
Vacuum-deoxidized steels ... **M1:** 124
Vacuum-deposited aluminum
as electronic material defects ... **A12:** 484, 486-487
Vacuum-melted alloys
counter-gravity low-pressure casting ... **A15:** 317, 318
Vacuum-tube power units
induction brazing ... **M6:** 967
VADER *See* Vacuum arc double electrode remelting
Valence
defined ... **A13:** 13
Valence states
of transition-element ions ... **A10:** 253
Valence-site symmetry, effect on XANES spectrum ... **A10:** 415
Valentine measurement method
shot peen coverage ... **M5:** 139
Valley Forge Company
as early foundry ... **A15:** 25-26
Valley Forge Foundry
crucible steel from ... **A15:** 31
Value engineering (VE)
for assembly and manufacture design ... **EL1:** 120-121

SUBJECTS OF THE INDEXED VOLUMES: ASM Handbook (designated by the letter "A"): **A1:** Properties and Selection: Irons, Steels, and High-Performance Alloys (1990); **A2:** Properties and Selection: Nonferrous Alloys and Special-Purpose Materials (1990); **A3:** Alloy Phase Diagrams (1992); **A4:** Heat Treating (1991); **A6:** Welding, Brazing, and Soldering (1993); **A7:** Powder Metallurgy (1984); **A8:** Mechanical Testing (1985); **A9:** Metallography and Microstructures (1985); **A10:** Materials Characterization (1986); **A11:** Failure Analysis and Prevention (1986); **A12:** Fractography (1987); **A13:** Corrosion (1987); **A14:** Forming and Forging (1988); **A15:** Casting (1988); **A16:** Machining (1989); **A17:** Nondestructive Testing and Quality Control (1989); **A18:** Friction, Lubrication, and Wear Technology (1992). **Metals Handbook, 9th Edition** (designated by the letter "M"): **M1:** Properties and Selection: Irons and Steels (1978); **M2:** Properties and Selection: Nonferrous Alloys and Pure Metals (1979); **M3:** Properties and Selection: Stainless Steels, Tool Materials and Special-Purpose Materials (1980); **M4:** Heat Treating (1981); **M5:** Surface Cleaning, Finishing, and Coating (1982); **M6:** Welding, Brazing, and Soldering (1983). **Engineered Materials Handbook** (designated by the letters "EM"): **EM1:** Composites (1987); **EM2:** Engineering Plastics (1988); **EM3:** Adhesives and Sealants (1990); **EM4:** Ceramics and Glasses (1991); **Electronic Materials Handbook** (designated by the letters "EL"): **EL1:** Packaging (1989).

Value, expected
statistical **A8:** 624
Valve
servo **A8:** 216
solenoid **A8:** 216
Valve alloys, specific types
21-4N, aluminum coating process **M5:** 345-346
Silcrome, aluminum coating process **M5:** 345-346
Valve bodies
cadmium plating **M5:** 260-261
magnetizing **A17:** 94
Valve bronze *See also* Copper casting alloys; Leaded tin bronzes; Valve metal
nominal composition **A2:** 347
Valve caps, threaded
economy in manufacture **M3:** 851
Valve guttering
as thermal fatigue mode **A11:** 289
"Valve hammer" **A18:** 600
Valve inserts
powders used **A7:** 572
Valve metal *See also* Cast copper alloys; Copper alloys, specific types, C84400; Valve bronze
properties and applications **A2:** 365
Valve seats
cemented carbide **A2:** 973
Valve spring retainer
by precision forging **A14:** 163
Valve springs *See also* Valves
distortion failure in **A11:** 138-139
failure due to residual shrinkage pipe **A11:** 533
failure from seam **A12:** 64
fatigue failure **A11:** 553
transverse failure origin **A11:** 554
Valve stems
cemented carbide **A2:** 973
Valve-seat retainer spring
designs for **A11:** 561
Valve-spring quality (VSQ) wire
characteristics of **A1:** 307
Valve-spring wire *See also* Steels, ASTM specific types, A230, A232
carbon steel for **M1:** 255
cost **M1:** 305
load-loss curves for springs **M1:** 300
seams in **M1:** 290
stress limit for springs **M1:** 296
stress relieving **M1:** 291
temperature limit for springs **M1:** 296
testing of **A1:** 309
Valves *See also* Valve springs
cast iron, coatings for **M1:** 103
in internal-combustion engines, elevated- temperature failures in **A11:** 288-289
poppet, thermal fatigue failure **A11:** 289
rotary, expansion and distortion failure in **A11:** 374-376
safety, pressure vessels , failures of **A11:** 644
spool-type hydraulic, seizing in **A11:** 141
steel castings **M1:** 388
stem failure, fracture surface of **A11:** 288
Valves, steel
aluminum coating **M5:** 335, 339-341, 345-346
Van de Graaff accelerators
as neutron sources **A17:** 388-389
Van de Graaff principle
and neutron radiography **A17:** 389
Van der Waals bond
as chemical bonding **EL1:** 93
definition **M6:** 19
van der Waals forces
agglomeration by **A7:** 62
Vanadates
as tube corrosive **A11:** 618
Vanadium
as a beta stabilizer in titanium alloys **A9:** 358
addition to low-alloy steels for pressure vessels and piping **A6:** 667
addition to strengthen chromium equivalent **A6:** 100
additions to martensitic stainless steels **M6:** 348
in alloy cast irons **A1:** 89-90
alloying effect in titanium alloys **A6:** 508
alloying, in cast irons **A13:** 567
alloying, in microalloyed uranium **A2:** 677
alloying, magnetic property effect **A2:** 762
alloying, wrought aluminum alloy **A2:** 55

Vanadium (continued)
as an addition to austenitic manganese steel castings **A9:** 239
at elevated-temperature service **A1:** 640-641
in austenitic manganese steel **A1:** 825
in cast iron **A1:** 6, 28
in cobalt-base alloys **A1:** 985
in commercial CPM tool steel compositions **A16:** 63
in compacted graphite iron **A1:** 59
in composition, effect on ductile iron **A4:** 686
in composition, effect on gray irons **A4:** 671
in compounds, determined by coulometric titration **A10:** 206
content in heat-treatable low-alloy (HTLA) steels **A6:** 670
content in high-strength low-alloy (HSLA) steels and postweld heat treatments **A6:** 664
content in stainless steels **A16:** 682-683
content in tool and die steels **A6:** 674
content in ultrahigh-strength low-alloy steels **A6:** 673
content of weld deposits **A6:** 675
and crack growth **A8:** 487
determined by controlled-potential coulometry **A10:** 209
diffusion bonding with **A2:** 564
as dopant for tungsten carbide **A18:** 795
in ductile iron **A15:** 649
early lamp filaments **A7:** 16
effect of, on hardenability **A1:** 395, 413, 419
effect of, on notch toughness **A1:** 741-742
effect of, on steel composition and formability **A1:** 577
effect on borided steels **A4:** 441
effect on cast iron microstructure **A18:** 701
effect on equilibrium temperature, cast irons **A15:** 65
effect on maraging steels **A4:** 222
effect on tool steel grindability **A16:** 726-727, 728, 729, 731, 732
effects of thermoreactive deposition/diffusion process **A4:** 449, 451
electron beam drip melted **A15:** 413
electron channeling pattern **A10:** 504, 508
electroslag welding, reactions **A6:** 274
evaporation fields for **A10:** 587
in ferrite **A1:** 408
ferritic stainless steels without **A1:** 936-937
formation of intermetallic phases in austenitic stainless steels **A9:** 284
friction welding **A6:** 154
for grain refinement, inclusion-forming **A15:** 95
as gray iron alloying element **A15:** 639
in hardfacing alloys **A18:** 759
in high-speed tool steels **A16:** 52, 54, 59
hydrogen damage **A13:** 171
interlayer material **M6:** 681
intermetallic compounds, for A15 superconductors **A2:** 1060
ion removal from **A10:** 200
in iron-base alloys, flame AAS analysis **A10:** 56
K-edge XANES spectra **A10:** 415
lubricant indicators and range of sensitivities **A18:** 301
in microalloy steel **A1:** 358
microalloying of **A14:** 220
as minor toxic metal, biologic effects **A2:** 1262
in multipoint cutting tools **A16:** 59
neutron and x-ray scattering, and absorption compared **A10:** 421
in nickel-base superalloys **A1:** 984
as nitride-forming **A15:** 93
nitride-forming element **A18:** 878
oil-ash corrosion with **A11:** 618
oxygen cutting, effect on **M6:** 898
in P/M high-speed tool steels **A16:** 61
in tool steels **A16:** 53
permanganate titration for **A10:** 176
precipitate stability and grain-boundary pinning **A6:** 73
pure **M2:** 822-823
pure, properties **A2:** 1172

Vanadium (continued)
recovery from selected electrode coverings **A6:** 60
redox titrations **A10:** 175
resistance spot welding **M6:** 478
spectrometric metals analysis **A18:** 300
in steel weldments **A6:** 417, 418, 420
submerged arc welding **A6:** 206
effect on cracking **M6:** 127
effect on microstructure toughness **M6:** 119
thermal diffusivity from 20 to 100 °C **A6:** 4
thermal expansion coefficient **A6:** 907
TNAA detection limits **A10:** 237
to aid milling of carbon and alloy steels **A16:** 676
to form simple and complex carbides **A16:** 667
to promote hardness **A4:** 124, 128-129
in tool steels **A18:** 734, 735-736, 739
toxicity **A6:** 1195, 1196
trace amounts affecting induction hardening **A4:** 185
ultrapure, by chemical vapor deposition **A2:** 1094
ultrapure, by zone-refining technique **A2:** 1094
use in flux cored electrodes **M6:** 103
vapor pressure, relation to temperature **A4:** 495
volumetric procedures for **A10:** 175
and WC powder preparation **A16:** 71
wear resistance of die material **A18:** 635-636
x-ray characterization of surface wear results for various microstructures **A18:** 469
Vanadium alloys
diffusion welding **A6:** 886
electron-beam welding **A6:** 581
Vanadium alloys, Ti-6Al-4V
aerospace applications **EM3:** 264
anodization procedures **EM3:** 265, 266
bonded with polyphenylquinoxaline **EM3:** 166, 167
chromic acid anodization **EM3:** 266, 267, 269, 270
wedge-crack propagation test **EM3:** 268, 269
Vanadium borides in wrought heat-resistant alloys **A9:** 312
Vanadium carbide **A16:** 72-74
coating applied to die-casting dies **A18:** 632
properties **A18:** 795, 801
tap density **A7:** 277
Toyota Diffusion process **A18:** 645
Vanadium carbides in austenitic manganese steel castings **A9:** 239
Vanadium in cast iron
carbide stabilizer **M1:** 80
gray iron **M1:** 28
Vanadium in steel **M1:** 115, 183, 188-189, 411, 417
castings **M1:** 388
nitriding, effect **M1:** 540-541
notch toughness **M1:** 694
steel sheet, effect on formability **M1:** 554
Vanadium nitride in plate steels
examination for **A9:** 203
Vanadium oxide
catalysts, Raman analysis **A10:** 133
vanadium K-edge XANES spectra of **A10:** 415
Vanadium pentoxide
applications/biologic effects **A2:** 1262
effect on fatigue strength **A11:** 131
as tube corrosive **A11:** 618
Vanadium, vapor pressure
relation to temperature **M4:** 309, 310
Vanadium, zone refined
impurity concentration **M2:** 713
Vanadium-niobium alloy
overheating **A12:** 145
Vanadium-niobium microalloyed steels **A1:** 403
Vanadium-nitrogen microalloyed steels **A1:** 403
Vanadizing
comparison of coatings for cold upsetting **A18:** 645
tool steels **A18:** 645
Vanadylyanadates
as tube corrosives **A11:** 618
Vander Lugt filter
holographic **A17:** 228
Vane
fixed **A8:** 216

Vane (continued)
rotary ... **A8:** 216
Vaneaxial fans
pressure-air flow characteristics ... **EL1:** 55
Vanes
hydraulic dynamometer stator, liquid erosion on ... **A11:** 169-170
turbine, failure of ... **A11:** 285
Vanes, turbine
hot corrosion of ... **A13:** 1000-1001
Vant'Hoff relation
liquid-solid equilibrium ... **A15:** 102
Vapogels
alkoxide-derived gels ... **EM4:** 210, 211
Vapometallurgy
of nickel and nickel alloys ... **A2:** 429
Vapor ... **EM3:** 30
application, of parylene coatings ... **EL1:** 762
bubbles ... **A12:** 180-181
defined ... **EM2:** 44
deposition, aluminum ... **A12:** 484-487
pressures, embrittler ... **A13:** 186
released through epoxy ... **EL1:** 961
water *See* Steam ... **A12:** 40, 52
Vapor blasting
electrochemical machining ... **A16:** 539
Vapor degrease ... **EM3:** 34, 42
Vapor degreasing *See also* Grease, removal of, Solvent cleaning ... **M5:** 40, 44-57
1,1,1-trichloroethane process ... **M5:** 44-48, 57
aluminum ... **M5:** 45-46, 53-55
applications ... **M5:** 44-45, 53-57
baskets and racks ... **M5:** 50
boiling liquid-warm liquid- vapor system ... **M5:** 46-47, 51, 54-55
brass ... **M5:** 45, 53-54
cast iron ... **M5:** 54
castings ... **M5:** 54, 56
chips and cutting fluids removed by ... **M5:** 10
as cleaning ... **EL1:** 777
cleanness, degree obtainable ... **M5:** 55-56
conveyor systems ... **M5:** 49, 55
copper and copper alloys ... **M5:** 45, 617-618
distillation, solvent ... **M5:** 48
effect on fatigue strength ... **A11:** 126
equipment ... **M5:** 47-53
installation ... **M5:** 49-50
maintenance of ... **M5:** 53
operation-startup to shutdown ... **M5:** 50-53
fume emission, solvent, ozone formation by ... **M5:** 57
iron ... **M5:** 45, 54
limitations of ... **M5:** 55-56
for liquid penetrant inspection ... **A17:** 81-82
magnesium and magnesium alloys ... **M5:** 45-46, 55, 629
magnetic particles removed by ... **M5:** 53-54
methylene chloride process ... **M5:** 44-49, 57
operation of systems-startup to shutdown ... **M5:** 50-53
perchloroethylene process ... **M5:** 44-45, 47-48
pigmented drawing compounds removed by ... **M5:** 8
process ... **M5:** 5, 8, 10
radioactive soils removed by ... **M5:** 54-55
rustproofing step ... **M5:** 47
safety precautions ... **M5:** 21, 45-46, 49, 53, 57
time-weighted average (TWA) exposure standards ... **M5:** 57
small, medium, and large units capacities and operating requirements ... **M5:** 49-51
soils, difficult, removal of ... **M5:** 56-57
solvent compositions and operating conditions ... **M5:** 44-46
solvent conservation, reclamation and waste disposal ... **M5:** 48-49, 57
solvent contamination control of ... **M5:** 47-48
Vapor degreasing (continued)
mineral oil in, percentage ... **M5:** 47-48
stainless steel ... **M5:** 54-55
steel ... **M5:** 45, 53-55
still, solvent ... **M5:** 48-49
systems and procedures *See also* specific systems by name ... **M5:** 46-47, 50-51
titanium and titanium alloys ... **M5:** 8, 656
trichloroethylene process ... **M5:** 44-48
ultrasonic system ... **M5:** - 47
unpigmented oils and greases removed by ... **M5:** 8
vapor phase only system ... **M5:** 46-47, 50, 56
vapor-spray-vapor system ... **M5:** 46-48, 50, 53, 55-56
warm liquid-vapor system ... **M5:** 46-47, 51, 54-55
water separator ... **M5:** 48-49
workpiece size, shape, placement and quantity, effects of ... **M5:** 47, 50, 52, 55-56
zinc and zinc alloys ... **M5:** 45-46, 54, 676-677
Vapor degreasing of specimens before mounting ... **A9:** 28
Vapor degreasing techniques ... **A7:** 458
Vapor deposition *See also* Chemical vapor deposition; Physical vapor deposition; Sputtering; Vacuum deposition; Vapor-deposited coatings ... **EM4:** 124, 215-220
adherence ... **EM4:** 219
chemical vapor deposition ... **EM4:** 215-218
advantages/disadvantages ... **EM4:** 216
applications ... **EM4:** 215, 216-217
ceramic materials produced by ... **EM4:** 216
conventional process ... **EM4:** 215
plasma-assisted ... **EM4:** 217-218
fiber-reinforced composites ... **EM4:** 219-220
of interference films ... **A9:** 137-138
physical vapor deposition ... **EM4:** 215, 218-219
applications ... **EM4:** 219
evaporation ... **EM4:** 218
ion plating ... **EM4:** 218
sputtering ... **EM4:** 218-219
Raman microprobe, TEM analysis, in graphites ... **A10:** 133
for thin-film semiconductors ... **A10:** 601, 602
to make ferrite films ... **EM4:** 1163
to make gamet films ... **EM4:** 1163
Vapor deposition, chemical
for thin-film hybrids ... **EL1:** 313
Vapor deposition coating
aluminum coatings ... **M5:** 346-347
chemical *See* Chemical vapor deposition
physical, oxidation protective coatings ... **M5:** 379
silicide ceramics ... **M5:** 535
vacuum *See* Vacuum vapor deposition
Vapor deposition processing
high-temperature superconductors ... **A2:** 1087
Vapor phase cleaning
as vapor degreasing technique ... **A7:** 458
Vapor phase corrosion
in pulp bleach plants ... **A13:** 1193
Vapor phase only vapor degreasing system ... **M5:** 46-47, 50, 56
Vapor phase organic cleaning ... **A13:** 1139
Vapor phase processes ... **EM4:** 22
Vapor phase soldering
of interconnections ... **EL1:** 117
as mass soldering technique ... **EL1:** 694
outer lead bonding ... **EL1:** 286
of passive components ... **EL1:** 180
as surface-mount soldering ... **EL1:** 702-704
Vapor plating *See also* Vacuum deposition
defined ... **A13:** 13-14
Vapor pressure
for AES samples ... **A10:** 556
aluminum ... **A15:** 80
calculated, in aluminum melts ... **A15:** 79
of rare earths ... **A2:** 720
as selection criterion, electrical contact materials ... **A2:** 840
Vapor pressure (continued)
in silicon modification ... **A15:** 163
Vapor quenching *See also* Quenching
of amorphous materials/metallic glasses ... **A2:** 806-807
Vapor recompression process
plating waste recovery ... **M5:** 316
Vapor transport mechanism
in tungsten powder production ... **A7:** 154
Vapor(s)
heat transfer to ... **A11:** 628
pressures, embrittler ... **A11:** 243
water, corrosive effects ... **A11:** 618, 630, 634-635
Vapor-deposited coatings
for carbon steel ... **A13:** 523
chemical vapor deposition ... **A13:** 457
deposition processes ... **A13:** 456-457
evaporation ... **A13:** 456-457
ion plating ... **A13:** 457
materials/applications ... **A13:** 457-458
sputtering ... **A13:** 456
types ... **A13:** 456-458
Vapor-deposited interference contrast layers ... **A9:** 59-60
Vapor-deposited replica, defined ... **A9:** 19
Vapor-deposited specimen
test interpretation ... **A8:** 237
Vapor-grown carbon fibers
for polymer matrix composites ... **EL1:** 1117-1118
Vapor-liquid-solid (VLS) process ... **EM1:** 25
Vapor-phase deposition
chromium codeposited in electroplating ... **A18:** 838
Vapor-phase deposition techniques
in active metal process ... **EM3:** 305
Vapor-phase epitaxy (PV)
for gallium arsenide (GaAs) ... **A2:** 745
Vapor-phase lubrication
defined ... **A18:** 20
Vapor-phase purification process
for ultrapure metals ... **A2:** 1094
Vapor-phase reflow
defined ... **EL1:** 1160
Vapor-phase soldering ... **A6:** 369-370
applications ... **A6:** 369
definition ... **A6:** 369
equipment ... **A6:** 369, 370
parameters ... **A6:** 370
pelletized vapor-phase batch system ... **A6:** 369
personnel ... **A6:** 369
Vapor-spray-vapor cleaning techniques ... **A7:** 459
Vapor-spray-vapor degreasing system ... **M5:** 46-48, 50, 53, 55-56
Vapor-streaming cementation process
ceramic coating ... **M5:** 545
Vaporization
-atomization interferences ... **A10:** 33, 34
interferences, in flame spectroscopy ... **A10:** 29, 47
liquid-erosion bubbles formation by ... **A11:** 163
selective ... **A10:** 25
Vaporization curves ... **A3:** 1 • 2
Vaporization processes
ion plating ... **M5:** 418
Vaporization reduction deposition process
tungsten powder production ... **A7:** 153
Vapormetallurgy processing
carbonyl ... **A7:** 92-93
Vaporproofing
pharmaceutical production equipment ... **A13:** 1231
Vapors from castable resins ... **A9:** 30-31
VAR *See* Consumable electrode vacuum arc remelting; Vacuum arc remelting; Vacuum-arc remelting
Varestraint test ... **A1:** 612 **A6:** 89, 90, 497
Variability
defined ... **A9:** 19
increase with increasing mean life ... **A8:** 699
as material assumption ... **EM1:** 309
of material properties ... **A8:** 623

Variability (continued)
measures of ... **A8:** 625
of population characteristics, in sampling ... **A10:** 13
relative ... **A8:** 625
sample, and measurement ... **A10:** 12
Variability (statistical)
sources of ... **EM2:** 606
Variable
background ... **A8:** 639
experimental, defined ... **A8:** 639
independent, in fatigue testing ... **A8:** 698
random ... **A8:** 624
Variable 2θ geometry
in RDF analysis ... **A10:** 396
Variable effects (statistical)
average main effects, calculated ... **A17:** 744
variable interactions, meaning ... **A17:** 745
Variable load amplitude fatigue tests ... **A1:** 370, 372
Variable mesh simulator (VMS) ... **EL1:** 129
Variable penetration ... **A6:** 19
Variable polarity ... **A6:** 39
Variable polarity plasma arc (VPPA) welding ... **A6:** 195, 197, 199
of aluminum-lithium alloys ... **A2:** 184
Variable resistors
contamination ... **EL1:** 1001-1002
defined ... **EL1:** 1160
interconnect defects ... **EL1:** 1002
mechanical defects ... **EL1:** 1002
Variable takeoff angle method
for thin-film sample preparation ... **A10:** 95
Variable wavelength geometry
RDF analysis ... **A10:** 396
Variable-amplitude loading
predicted fatigue resistance in ... **A8:** 712
Variable-frequency drive (VFD)
pump motor ... **A18:** 594-595
Variable-temperature ESR investigations ... **A10:** 257
Variable-throw crank
for crank and lever testing machine ... **A8:** 369-370
Variable-volume method
thermal expansion molding ... **EM1:** 590
Variables
continuous and discrete ... **A17:** 746
Varian 9-Ghz cavity
insert for FMR high- temperature studies ... **A10:** 271
Variance ... **A18:** 481
binomial distribution ... **A8:** 636
defined ... **A8:** 14
equality of ... **EM1:** 305
exponential distribution ... **A8:** 635
normal distribution ... **A8:** 630, 631
Poisson distribution ... **A8:** 637
as second moment of population ... **A8:** 628
of statistical distributions ... **A8:** 629
Weibull distribution ... **A8:** 633
Variation (statistical)
countermeasures ... **A17:** 722-730
factors producing ... **A17:** 693
of NDE process ... **A17:** 689
process behavior, over time ... **A17:** 723-724
R control charts, construction interpretation ... **A17:** 725
rational sampling, importance ... **A17:** 728-730
reduction, and quality loss function ... **A17:** 721
sample means control chart construction/interpretation ... **A17:** 725
Shewhart control chart model ... **A17:** 725-728
sources of ... **A17:** 722-730
system, experimental study ... **A17:** 742
Variational noise
defined ... **A17:** 723
Varistors ... **EM4:** 1150-1154
application of zinc oxide varistors, critical parameters ... **EM4:** 1152
applications ... **EM4:** 1150
electrical characteristics of zinc oxide varistors ... **EM4:** 1150, 1151
fabrication of zinc oxide varistors ... **EM4:** 1153-1154
microstructures of zinc oxide varistors ... **EM4:** 1150-1152
properties ... **EM4:** 1150
reliability of zinc oxide varistors ... **EM4:** 1152-1153
Varnish ... **M5:** 497-499, 503
defined ... **A18:** 20
Varnish overprint
coating for ... **EL1:** 863
Varnishing
zinc alloys ... **A2:** 530
Vasco X2-M
nominal compositions ... **A18:** 726
Vat, galvanizing
failure of ... **A11:** 273-274
Vaulting
elimination in isostatic pressing ... **EM4:** 129
VAW electrolytic brightening
aluminum and aluminum alloys ... **M5:** 582
VCA *See* Titanium alloys, specific types, Ti-13V-11Cr-3Al
VCD *See* Vibrational circular dichroism
Vector network analysis
application ... **A17:** 216
FMR eddy current probes ... **A17:** 221
microwave inspection ... **A17:** 206, 222
schematic diagram ... **A17:** 223
Vegard's law ... **A6:** 545
Vegetable oils, unsaturated
epoxidization of ... **EL1:** 818
Vegetable oils, unsaturation
determined by electrometric titration ... **A10:** 205
Vegetables
nickel toxins in ... **A7:** 203
Vehicle
defined ... **EL1:** 1160
organic , for substrate screen printing ... **EL1:** 249
Vehicle structural design
full-scale tests for ... **EM1:** 346-351
Vehicle suspension
leaf springs for ... **A1:** 322, 325
Vehicles
military ... **A7:** 687-688
Veil *See also* Surfacing veil ... **EM3:** 30
defined ... **EM1:** 25 **EM2:** 44
Veining
as casting defect ... **A11:** 381
coating for ... **A15:** 240
Veining in grains ... **A9:** 604
defined ... **A9:** 19
Vello process ... **EM4:** 400, 1033, 1038
Velocity *See also* Angular velocity; Magnetically induced velocity changes (MIVC); Velocity measurement
abrasive ... **A15:** 511
-affected corrosion, in water ... **A11:** 188-190
analyzers ... **A10:** 570-571
angular, SI derived unit and symbol for ... **A10:** 685
in blast cleaning ... **A15:** 506
clastic wave ... **A8:** 209
conversion factors ... **A8:** 723 **A10:** 686
crack, as function of stress intensity ... **A13:** 170
critical, single-phase alloys ... **A15:** 114-119
dependence of propagating crack toughness ... **A8:** 284
distribution, as ALPID result ... **A14:** 427
effect on cellular and dendrite spacing ... **A15:** 117
effect on corrosion in seawater ... **M1:** 740-742
effect on corrosion inhibitors ... **A11:** 198
effect, produced fluids ... **A13:** 479
effects, copper/copper alloys ... **A13:** 623, 625
electron, abbreviation for ... **A10:** 691
in erosion/cavitation testing ... **A13:** 312-313
flow, effect on erosion damage ... **A11:** 165-166
flow path ... **A13:** 966
fluid, aqueous corrosion ... **A13:** 40
free surface, during spalling ... **A8:** 212
gradients, effect on powder segregation ... **A7:** 187-189
growth, partition coefficient change with hackle ... **A11:** 745
high, impact, of drop of liquid ... **A11:** 164
high, impact response curves for testing ... **A8:** 271
interface, single-phase alloys ... **A15:** 114-119
as interference, microwave inspection ... **A17:** 204
interferometer ... **A8:** 211, 231
intergranular crack ... **A13:** 160
of leaks ... **A17:** 58
maximum signal propagation ... **EL1:** 5
Velocity (continued)
maximum tubular, copper alloys ... **A13:** 624
normal stress-particle ... **A8:** 232
plastic wave ... **A8:** 209
plateau, for stress-corrosion cracking ... **A8:** 497
plots, disk-forging simulation ... **A14:** 430
porosity effects ... **A17:** 212
of propagation, defined ... **EL1:** 1160
of propagation, wave theory of ... **A10:** 83
of ram, in high-energy-rate forging ... **A14:** 100
relation to electromagnetic radiation ... **A10:** 83
SCC plateau crack ... **A13:** 268
seawater, corrosion effect ... **A13:** 333, 516, 623, 625
shear stress/transverse particle ... **A8:** 232
SI derived unit and symbol for ... **A10:** 685
SI unit for ... **A8:** 721
slide, drawing presses ... **A14:** 578
in slider-crank mechanism ... **A14:** 39
of solution, total immersion tests ... **A13:** 222
speed of light, abbreviation ... **A10:** 689
stress-corrosion cracking ... **A13:** 277-278
symbol for ... **A8:** 721, 726
transducers, for strain measurement ... **A8:** 193
ultrasonic ... **A17:** 162, 232-233, 597
ultrasonic, and powder mechanical properties ... **A7:** 484
vane tip, abrasive wheel ... **A15:** 512
Velocity hackle ... **EM4:** 636
Velocity interferometer ... **A8:** 211, 231
Velocity measurement *See also* Velocity
bulk-sound ... **A17:** 274-275
by laser inspection ... **A17:** 16-17
ultrasonic inspection ... **A17:** 274-275
as ultrasonic test technique ... **EM1:** 776
Venn diagram, for stress corrosion
corrosion fatigue, and hydrogen embrittlement ... **A13:** 291
Vent
broken casting at ... **A11:** 386
defined ... **A14:** 14 **A15:** 11 **EM1:** 25 **EM2:** 44
in permanent molds ... **A15:** 278
Vent cloth ... **EM3:** 30
defined ... **EM1:** 25 **EM2:** 44
Vent forming
analysis of ... **A14:** 924
Vent mark
defined ... **A14:** 14
Venting *See also* Air venting ... **EM1:** 5, 166-167 **EM3:** 30
air, die casting ... **A15:** 291
blow molding ... **EM2:** 354-355
and carbon monoxide poisoning ... **A7:** 349
coating effect ... **A15:** 281
of core ... **A15:** 241
defined ... **EM2:** 44
die ... **A15:** 289
gas displacement ... **A15:** 291
of molds ... **EM2:** 614
surface finish effects ... **A15:** 285
systems, for explosion suppression ... **A7:** 198
Venturi-type air cooler
for ultrasonic testing ... **A8:** 247
VEPP-2M
as synchrotron radiation source ... **A10:** 413
VEPP-3
as synchrotron radiation source ... **A10:** 413
VEPP-4
as synchrotron radiation source ... **A10:** 413
Verde antique copper coloring solution ... **M5:** 625
Verdet coefficient ... **EM4:** 854-855
Verification ... **A8:** 14, 88, 611-612
of alloys ... **A10:** 118
by nondestructive evaluation ... **A17:** 671-672
engineering ... **EL1:** 992-993
of fracture control policy ... **A17:** 669-673
of heat treatment ... **A14:** 249, 282
of logic design ... **EL1:** 129
mechanical ... **EL1:** 941
physical ... **EL1:** 941
procedure, fatigue testing ... **EL1:** 742
Verification, heat treatment
wrought titanium alloys ... **A2:** 620-622
Verification, of pattern/casting dimensions
automated ... **A15:** 199
Verified loading range
defined ... **A8:** 14

Verifilm EM3: 736-737, 767
and visual inspection nondestructive testing EM3: 751
Vermicular cast iron *See* Compacted graphite cast iron
Vermicular graphite *See also* Compacted graphite irons
eutectic compacted, growth of A15: 178-179
irons, welding metallurgy A15: 521-522
Vermicular graphite cast iron *See* Compacted graphite iron
Vermicular iron *See* Compacted graphite iron
Vermicularity, of graphite
fatigue fracture from A11: 360-361
Vermiculite
defined EM1: 25 EM2: 44
Vernier *See also* Least count
defined A8: 14
Versailles Project on Advanced Materials and Standards (VAMAS)
round-robin sliding wear tests A18: 486
Versatility
of rigid epoxies EL1: 810
Verson hydroform process
lubricants A14: 612-613
presses A14: 612
procedure A14: 612
single-draw operation A14: 613-614
surface finish A14: 613
tools A14: 612
Verson-Wheelon process *See also* Guerin process; Rubber-pad forming
presses A14: 609
procedure A14: 609-610
as rubber-pad forming A14: 609
secondary operations A14: 610
tools A14: 609
Vertical boring mills A16: 160, 164-165
Vertical centrifugal casting
defects in A15: 306-307
equipment A15: 307
mold design A15: 300-304
as permanent mold process A15: 34
process details A15: 304-306
processes A15: 300
Vertical centrifugal casting machines A15: 306-307
Vertical coordinate measuring machines A17: 20-21
Vertical core knockout machine A15: 504
Vertical counterblow hammer A14: 25, 28
Vertical direct-chill casting A15: 313-314
Vertical drill presses
trepanning A16: 176
Vertical drilling
with cemented carbide tools A2: 974
Vertical illumination
defined A9: 19
Vertical lighting
and oblique lighting, compared A12: 87
photomacrographic A12: 83
Vertical load train
in modified test machines A8: 159
Vertical multiple-spindle automatic chucking machines A16: 378-379
Vertical or short take-off and landing (V/STOL) aircraft
fatigue EM3: 501
Vertical position
definition A6: 1215 M6: 19
Vertical position (pipe welding)
definition A6: 1215
Vertical ring rolling machines A14: 109, 112-113
Vertical roughness parameter
defined A12: 200
Vertical sectioning
for true fracture surface area A12: 198-199, 211-212
Vertical sections of a ternary diagram .... A3: 1•5
Vertical semicontinuous casting
wrought copper and copper alloys A2: 243
Vertical shaft furnace
for sand reclamation A15: 227-228
Vertical stress distribution A7: 301
Vertical turbine pumps (VTPS) A18: 595
Vertical turret lathes
boring A16: 160
Vertical welding
arc welding of coppers M6: 402
flux cored arc welding M6: 107
gas tungsten arc welding of aluminum alloys M6: 391-392
indication by electrode classification M6: 84
oxyfuel, gas welding M6: 589
of pipe M6: 591-592
shielded metal arc welding M6: 76, 85, 441
of nickel alloys M6: 441
Vertical-plane machines
for press forming A14: 559
Vertically parted molding machines
green sand molding A15: 344
Verwey transition EM4: 751
Very high strain rates
in compression testing A8: 190-191
Very large scale integrated circuits (VLSIC) EM3: 579
Very large scale integration (microcircuit) products
gas mass spectroscopy in A10: 156-157
Very large scale integration (VLSI) generation EM3: 581, 584, 586
Very-high-impact polystyrenes *See also* High-impact polystyrenes (PS, HIPS)
grades of EM2: 199
Very-high-speed integrated chip
defined EL1: 1160
Very-high-speed integrated circuits (VHSICs) *See also* High-frequency digital systems
clock frequencies EL1: 76
cross talk noise EL1: 76
failure mechanisms EL1: 982-983
interconnects, line impedance EL1: 388
interconnects, signal line density EL1: 389
modeling/simulation requirements EL1: 77-81
test and maintenance EL1: 376
Very-large-scale integration (VLSI)
accelerated testing EL1: 887
capability EL1: 13
chip, assumptions EL1: 270
CMOS, as future trend EL1: 390
defined EL1: 1160
development EL1: 160
direct chip interconnect EL1: 231-232
effect, passive component fabrication EL1: 178
instrumentation and testing EL1: 365
manufacturing test coverage EL1: 374
mechanisms, failure kinetics for EL1: 889-893
-optimized architectures EL1: 2
packaging approaches EL1: 269-270
packaging, development EL1: 961
as primary technology for electronic packaging EL1: 12
processing yield, WSI vs VLSI EL1: 266-268
silicon -metal oxide semiconductors, effect on EL1: 2
tape automated bonding (TAB) with EL1: 274
testing phases EL1: 377-378
and wafer-scale integration, compared EL1: 263, 269-270
VESPEL (DuPont)
polyimide resin for thrust washer material A18: 567
Vessels *See* Pressure vessels
argon oxygen decarburization, schematic A15: 427
ladle furnace and vacuum arc degassing A15: 436
pouring, direct heating A15: 498-499
for sample dissolution treatments A10: 165-167
Vessels (continued)
for sinters/fusions A10: 166-167
for sodium peroxide fusions A10: 166
vacuum, ladle degassing A15: 432
VF *See* Vacuum fusion
VHP *See* Vacuum hot pressing; Vacuum hot pressing (VHP)
VHSICs *See* Very-high-speed integrated circuits (VHSICs)
VI improver
defined A18: 20
Via hole
defined EL1: 1160
formation EL1: 326-328
plating, thin-film hybrids EL1: 329
Via(s) *See also* Buried vias; Semiburied vias; Thermal vias; Via hole
blind, selection EL1: 112
buried inner layer EL1: 543
defined EL1: 13
density EL1: 613, 620-621
filling, and conductor resistor printing EL1: 463-464
formation, ceramic packages EL1: 463
interlayer, electrical effects EL1: 76
materials and processes selection EL1: 113-115
nested EL1: 304
placement EL1: 517
registration EL1: 464
signal, parallel with option for EL1: 112
thermal, effect of multichip module EL1: 53-54
Vias A6: 992, 994
Vibrated vacuum gas test
for hydrogen measurement A15: 459
Vibrating beams
theory of EM1: 209
Vibrating conveyor
as shakeout equipment A15: 503
Vibrating drum shakeout equipment A15: 348, 503
Vibrating media core knockout system A15: 506
Vibrating powders *See also* Tap density
Vibrating sample magnetometer A10: 268
Vibrating string technique
for stress-relaxation tension test A8: 325
tensile stress-relaxation curve A8: 325
Vibrating tub
for lacquer coating A7: 588
Vibrating-reed relay
defined EL1: 1160
Vibration *See also* Damping properties analysis
affecting microhardness testing A8: 96
analysis, by optical holographic interferometry A17: 424-425
applied A7: 297
-caused fatigue fracture, stainless steel lever A11: 114
and cavitation damage A11: 378
as centrifugal casting defect A15: 307
of composites EM1: 35
damping EM1: 190
with dirt and moisture, bearing damage by A11: 494, 496
distortion from A11: 138
effect, crack growth A12: 109
effect in creep and stress-rupture testing A8: 312
effect, low-carbon steel A12: 245
effect on density A7: 296
effect on mean coordination number A7: 296
effects, coordinate measuring machines A17: 26-27
effects, optical holographic interferometry A17: 412, 413
failures EL1: 64-65
fatigue failures by A11: 308-309, 518, 621
flexible printed boards EL1: 589
fuel pump failed by A11: 465

SUBJECTS OF THE INDEXED VOLUMES: ASM Handbook (designated by the letter "A"): **A1:** Properties and Selection: Irons, Steels, and High-Performance Alloys (1990); **A2:** Properties and Selection: Nonferrous Alloys and Special-Purpose Materials (1990); **A3:** Alloy Phase Diagrams (1992); **A4:** Heat Treating (1991); **A6:** Welding, Brazing, and Soldering (1993); **A7:** Powder Metallurgy (1984); **A8:** Mechanical Testing (1985); **A9:** Metallography and Microstructures (1985); **A10:** Materials Characterization (1986); **A11:** Failure Analysis and Prevention (1986); **A12:** Fractography (1987); **A13:** Corrosion (1987); **A14:** Forming and Forging (1988); **A15:** Casting (1988); **A16:** Machining (1989); **A17:** Nondestructive Testing and Quality Control (1989); **A18:** Friction, Lubrication, and Wear Technology (1992). **Metals Handbook, 9th Edition** (designated by the letter "M"): **M1:** Properties and Selection: Irons and Steels (1978); **M2:** Properties and Selection: Nonferrous Alloys and Pure Metals (1979); **M3:** Properties and Selection: Stainless Steels, Tool Materials and Special-Purpose Materials (1980); **M4:** Heat Treating (1981); **M5:** Surface Cleaning, Finishing, and Coating (1982); **M6:** Welding, Brazing, and Soldering (1983). **Engineered Materials Handbook** (designated by the letters "EM"): **EM1:** Composites (1987); **EM2:** Engineering Plastics (1988); **EM3:** Adhesives and Sealants (1990); **EM4:** Ceramics and Glasses (1991); **Electronic Materials Handbook** (designated by the letters "EL"): **EL1:** Packaging (1989).

Vibration (continued)
high-frequency, fatigue-fracture surface from ... **A11:** 112
holographic measurement ... **A17:** 405
package-level testing ... **EL1:** 937-938
pattern, holographic detection ... **A17:** 16
properties, defined ... **EM1:** 206
as screening environment ... **EL1:** 875
shaft fatigue fracture from ... **A11:** 476
and shock durability ... **EL1:** 62-65
as solder joint attachment failure mechanism ... **EL1:** 740
testing, as failure verification ... **EL1:** 944, 1061
to fill flexible envelopes for isostatic pressing ... **A7:** 297
torsional ... **A11:** 561-562
ultrasonic ... **A17:** 232-234
in ultrasonic hardness tester ... **A8:** 101
Vibration analysis ... **A18:** 293-297
analysis methods and problems ... **A18:** 294-295
averaging ... **A18:** 295-296
leakage and windowing ... **A18:** 295
relationship between friction and vibration ... **A18:** 296-297
contact between two flat plates ... **A18:** 297
free vibration frequency ... **A18:** 297
true area of contact ... **A18:** 297
wear rate ... **A18:** 297
statistical analysis approach ... **A18:** 296
surfaces in contact ... **A18:** 293-294
instrumentation ... **A18:** 293-294
Vibration as a method to control grain size in aluminum alloy ingots ... **A9:** 630
Vibration compaction
defined ... **A7:** 12
Vibration damping function ... **EM3:** 33, 36
Vibration density
defined ... **A7:** 12
Vibration, free *See* Free vibration
Vibration radiusing
refractory metals and alloys ... **A2:** 562
Vibration welding ... **EM2:** 724-725
Vibration-induced heating
for thermal inspection ... **A17:** 398
Vibrational analysis
Fourier-transform infrared spectroscopy ... **A10:** 126
high-resolution electron energy loss spectroscopy ... **A10:** 126
Raman and infrared as ... **A10:** 126, 127
Vibrational behavior
of atoms, and determination of crystal structure ... **A10:** 352
pyridine, model environments for ... **A10:** 134
Raman, in intercalated graphite species ... **A10:** 133
ring-breathing, of pyridine ... **A10:** 134
surface, surface-enhanced Raman scattering for ... **A10:** 136
of surfaces, SERS analysis ... **A10:** 136
Vibrational false brinelling *See* Fretting
Vibrational frequencies, infrared
calculation of ... **A10:** 110
Vibrations
information, by Raman analysis ... **A10:** 126-138
metal-ligand, Raman spectroscopy for ... **A10:** 126
molecular, effect in infrared spectroscopy ... **A10:** 109
molecular, in infrared spectroscopy ... **A10:** 111
molecular, in Raman spectroscopy ... **A10:** 127
weld microstructures and ... **A6:** 53-54
Vibrators *See also* Air vibrators
early practice ... **A15:** 28
recommended contact materials ... **A2:** 863
Vibratory ball milling ... **A7:** 23
Vibratory ball mills ... **A7:** 66-68
Vibratory cavitation
defined ... **A18:** 20
Vibratory compaction
defined ... **A7:** 12, 306
Vibratory deburring ... **A7:** 458, 669
Vibratory devices
erosion/cavitation testing ... **A13:** 313
Vibratory energy
ultrasonic ... **EM1:** 615
Vibratory finishing
aluminum and aluminum alloys ... **M5:** 573

Vibratory finishing (continued)
bowl process ... **M5:** 130-131
copper and copper alloys ... **M5:** 615-616
magnesium alloys ... **M5:** 631
tub process ... **M5:** 130
zinc alloys ... **M5:** 676
Vibratory mill ... **A7:** 12
Vibratory milling
silver powders ... **A7:** 148
temperature versus milling time curves ... **A7:** 61, 62
time, and x-ray line broadening ... **A7:** 61
time, effect on apparent density and Hall flowability ... **A7:** 59
Vibratory polishing
of aluminum-coated sheet steel ... **A9:** 197
copper and copper alloys ... **A9:** 400
defined ... **A9:** 19
of iron-nickel and iron-cobalt alloys ... **A9:** 532
powder metallurgy materials ... **A9:** 507
tin and tin alloys ... **A9:** 449
of tin plate ... **A9:** 198
titanium and titanium alloys ... **A9:** 459
wrought heat-resistant alloys ... **A9:** 307
Vibratory polishing methods ... **A9:** 42
for alpha-beta brass ... **A9:** 44
effect of load on ... **A9:** 44
effect of suspending liquid on low carbon steel ... **A9:** 43
Vibratory polishing of
beryllium ... **A9:** 389
electrical contact materials ... **A9:** 550
refractory metals ... **A9:** 439
uranium and uranium alloys ... **A9:** 478
Vibratory powder dispensers
for flame cutting ... **A7:** 843
Vibratory rotary machine finishing ... **M5:** 133
Vibratory stress relieving ... **M6:** 892
Vibratory tube mill ... **A7:** 66-68
Vibrothermography
as nondestructive test technique ... **EM1:** 777
Vicalloy *See* Iron-base alloys; Permanent magnet materials, specific types
Vicalloy alloys *See also* Magnetic materials ... **A9:** 538-539
Vicat softening point ... **EM3:** 30
defined ... **EM2:** 44
Vickers (microindentation) hardness number
defined ... **A18:** 20
Vickers hardness
abbreviation for ... **A11:** 797
carbides ... **A16:** 81
ceramics ... **A16:** 101
cermets ... **A16:** 91
number (HV), defined ... **A11:** 11
test, defined ... **A11:** 11
Vickers hardness number, (HV) *See also* Diamond pyramid hardness numbers
Brinell hardness conversions ... **A8:** 111
defined ... **A8:** 15, 90-91
equivalent hardness numbers ... **A8:** 112-113
equivalent Rockwell B numbers ... **A8:** 109-110
for indentations with same test load ... **A8:** 91, 97-99
Rockwell C hardness conversions ... **A8:** 110
vs. load ... **A8:** 94-96
Vickers hardness test ... **A8:** 71, 102, 725
Vickers hardness testing
of castings ... **A17:** 521
Vickers indentation
compared with Knoop ... **A8:** 90
elastic recovery ... **A8:** 95
filar units for measuring ... **A8:** 91
surface preparation ... **A8:** 93
Vickers indentation hardness ... **A18:** 433
Vickers indenter ... **A8:** 90-91, 95-96, 100, 102
Vickers microhardness testing *See also* Knoop mirohardness testing
applications ... **A8:** 96-98
determining hardness number ... **A8:** 91
diamond pyramid indenter ... **A8:** 91
hardness number vs. load ... **A8:** 94-96
hardness value determined ... **A8:** 90
indentation measuring ... **A8:** 91
indentation spacing ... **A8:** 94
indentations compared with Knoop ... **A8:** 90
indenter selection ... **A8:** 90-91

Vickers microhardness testing (continued)
and Knoop ... **A8:** 90-98
load vs. indentation size ... **A8:** 94
and Rockwell C scale test ... **A8:** 91
surface preparation ... **A8:** 93
test considerations ... **A8:** 94-96
testers ... **A8:** 91-93
Vickers microindentation tests ... **A18:** 415, 416, 417, 418
VID *See* Vacuum induction degassing
Video cassette recorders
thick-film hybrid application ... **EL1:** 385
Video crack growth measuring systems ... **A8:** 246
Videorecording, used in conjunction with
scanning electron microscopy ... **A9:** 97
Videoscopes *See also* Cameras; Television cameras
with CCD probes, as borescopes ... **A17:** 5-8
images ... **A17:** 5-6
working length ... **A17:** 7
Videotape/videodisk
digital image enhancement ... **A17:** 463
Vidicon
defined ... **A10:** 683
and diode array detectors, use in Raman spectroscopy ... **A10:** 128-129
Vidicon camera
dynamic range, radiography ... **A17:** 318
as image sensors, optical ... **A17:** 10
secondary electron-coupled (SEC) ... **A17:** 10
vision machine ... **A17:** 31-32
VIDP *See* Vacuum induction degassing and pouring
Vienna lime buffing compounds ... **M5:** 117
View
defined ... **A17:** 385
identification ... **A17:** 338-341
radiographic, selection of ... **A17:** 330-338
of radiographs ... **A17:** 347
selection, radiographic inspection ... **A17:** 330-338
View aliasing
computed tomography (CT) ... **A17:** 376
View camera
fractographic ... **A12:** 78-79, 85
with scanning light photomacrography system ... **A12:** 82
Vilella's reagent
composition of ... **A9:** 211
for etching stainless steel ... **A10:** 311
Vilella's reagent as an etchant for
alloy steel ... **A9:** 211
austenitic manganese steel casting specimens ... **A9:** 239
iron-chromium-nickel heat-resistant casting alloys ... **A9:** 330-333
wrought stainless steels ... **A9:** 281-282
Villard-circuit equipment
radiography ... **A17:** 305
VIM *See* Vacuum induction melting
Vinyl acetate plastics ... **EM3:** 30
defined ... **EM2:** 44
Vinyl acrylic
chemistry ... **EM3:** 50
Vinyl adhesives ... **EM3:** 75
film, surface parameter ... **EM3:** 41
for hemmed flange bonding ... **EM3:** 553-554
latex characteristics ... **EM3:** 53
preformed, for window glazing ... **EM3:** 56
silane coupling agents ... **EM3:** 182
Vinyl chloride
chemisorption and solid friction ... **A18:** 29
Vinyl chloride, in vinyl chloride and vinylidene chloride copolymer
Raman analysis ... **A10:** 132
Vinyl chloride plastics ... **EM3:** 30
defined ... **EM2:** 44
environmental effects ... **EM2:** 427
Vinyl coating
steel sheet ... **M1:** 176
Vinyl degradation
polyvinyl chlorides (PVC) ... **EM2:** 212
Vinyl ester resins *See also* Polyester resins
clear casting mechanical properties ... **EM1:** 91
for commercial application ... **EM1:** 31-32
defined ... **EM1:** 25
delayed gel times ... **EM1:** 133
in fiberglass-polyester resin composites ... **EM1:** 91
for filament winding ... **EM1:** 137-138

Vinyl ester resins (continued)
formulation **EM1:** 133, 137, 141
glass content effect **EM1:** 91
preparation/application **EM1:** 90
for pultrusion **EM1:** 538-539
for resin transfer molding **EM1:** 169
for sheet molding compounds **EM1:** 141-142
suppliers **EM1:** 133
for wet lay-up **EM1:** 133-134
Vinyl esters *See also* Thermosetting resins **EM3:** 30
applications **EM2:** 272-273
characteristics **EM2:** 273-275
commercial forms **EM2:** 272
costs and production volume **EM2:** 272
defined **EM2:** 44
mechanical properties **EM2:** 273-274
polymerization **EM2:** 275
processing **EM2:** 274-275
properties **EM2:** 273-275
for pultrusion **EM2:** 394
suppliers **EM2:** 275
Vinyl esters, room-temperature curing
in composites **A11:** 731
Vinyl films
identification of polymer and plasticizer materials in **A10:** 123-124
Vinyl groups in silicone
Raman analysis **A10:** 132
Vinyl plastisols **EM3:** 48
for automotive bonding and sealing **EM3:** 609
for body seam sealing **EM3:** 51
chemistry **EM3:** 50, 51
properties **EM3:** 50
Vinyl polymers
thermal degradation **EM2:** 423
thermal properties **EM2:** 447
Vinyl resins and coatings **M5:** 474-475, 497-498, 505
Vinyl toluene
as polyester diluent **EM1:** 132
Vinyl topcoats
as marine corrosion coating **A13:** 914
Vinyl urethanes **A13:** 410
Vinyl-base enamels
stripping of **M5:** 648
Vinyl-phenolics **EM3:** 76, 105
advantages and limitations **EM3:** 79
compared to nylon-epoxies **EM3:** 78
properties **EM3:** 106
typical film adhesive properties **EM3:** 78
Vinylic polysilane
used in silicon-base ceramics **EM4:** 223
Vinylidene chloride plastics **EM3:** 30
defined **EM2:** 44
Vinylidene fluoride, group
and polymer naming **EM2:** 57
Virgin filament
defined **EM1:** 25 **EM2:** 45
Virgin material
defined **EM2:** 45
Virgin materials **EM3:** 30
Virginia
early American foundries **A15:** 25
Virtual leaks
defined **A17:** 57
Viscoelastic structural adhesives
for auto tire tread cracking **EM3:** 513
Viscoelasticity *See also* Elasticity **EM2:** 412-422 **EM3:** 30, 382
analyses of **EM2:** 533
analysis, of fiber composites **EM1:** 190-191
dashpot model **EM2:** 414
defined **A18:** 20 **EM1:** 25 **EM2:** 45
experimental analysis **EM2:** 417-419
material parameters **EM2:** 419-422
of plastics **EM2:** 659
polyethylene terephthalates (PET) **EM2:** 174
of polymers **A11:** 758
Viscoelasticity (continued)
as relaxation modulus and creep compliance with time **EM1:** 190-191
spring model **EM2:** 414
of thermoplastic resins **EM2:** 625
viscoelastic behavior **EM2:** 412-417
Viscometers **EM3:** 322-323
Viscosimeter
defined **EL1:** 1160-1161
Viscosity *See also* Absolute viscosity; Intrinsic viscosity; Relative viscosity; Specific viscosity; Viscoelasticity **A18:** 60, 65, 67 **EM3:** 30, 426
absolute, symbol and units **A18:** 544
aluminum casting alloys **A2:** 145
Brookfield **EM2:** 533-534
coefficient, defined **EM2:** 45
complex **EM1:** 761
conversion factors **A8:** 723 **A10:** 686
defined **A18:** 20 **EL1:** 1161 **EM1:** 25 **EM2:** 45
during cure **EM1:** 649, 702
dynamic, as function of temperature **A15:** 110
dynamic, SI derived unit and symbol for **A10:** 685
dynamic, SI unit/symbol for **A8:** 721
effect, magabsorption powder measurement **A17:** 153
effect of temperature control **EM3:** 712, 713
effect on dispensing system **EM3:** 696-700, 701
effect on friction torque loss of internal combustion engine parts **A18:** 560
effect, solder paste print resolution **EL1:** 732
of epoxy resin **EM1:** 77, 736
for filament winding **EM1:** 135
fillers **EM3:** 177, 178
of flexible epoxies **EL1:** 821
and fluidity **A15:** 766
of fluxes **EL1:** 644
of fossil fuels, ESR determination of **A10:** 253
gear lubricants **A18:** 542
of glass fibers **EM1:** 47
grades, comparison of classifications **A18:** 85
heat effect **EM3:** 729
hold steps, effect **EM1:** 141
-index improvers **A11:** 154
iron-carbon alloys **A15:** 168
kinematic, SI derived unit and symbol for **A10:** 685
kinematic, SI unit/symbol for **A8:** 721
loss, permanent or temporary **A18:** 84
lubricant failure from **A11:** 153-154
lubricant indicators and range of sensitivities **A18:** 301
of lubricants **A14:** 516
of lubricants for ball and roller bearings **A11:** 511
lubricants for rolling-element bearings **A18:** 134, 135
and mean free path **A17:** 59
measurement of **EM3:** 322-323
melt, ionomers **EM2:** 122-123
melt, of thermoplastics **EM1:** 102
melt, polyvinyl chlorides (PVC) **EM2:** 210
of mixed resin systems, testing **EM1:** 737
modifiers, in nonengine lubricant formulations **A18:** 111
nomenclature for hydrostatic bearings with orifice or capillary restrictor **A18:** 92
oil, for magnetic particles **A17:** 101
plastisol sealants to effect **EM3:** 720-721
resin, during cure **EM1:** 655-656
rigid epoxies **EL1:** 810
of semisolid metals **A15:** 327
and shear stress, semisolid alloy **A15:** 328
of slag/dross, inclusion-forming effects **A15:** 91
solution **EM2:** 533
of solution, in flame spectroscopy **A10:** 29
and spin coating **EL1:** 326
Viscosity (continued)
steady shear, and normal stresses **EL1:** 842-843
symbol and units (at 40 °C, or 105 °F) **A18:** 544
of thermoplastics **EM1:** 294
units of **A18:** 20, 140
urethane coatings **EL1:** 775
of urethanes for windshield bonding **EM3:** 724
Viscosity coefficient **EM3:** 30
Viscosity improvers **A18:** 99, 108-110
applications **A18:** 110
dispersants **A18:** 109
in engine lubricant formulations **A18:** 111
formation **A18:** 109
functions **A18:** 108
multifunctional nature **A18:** 111
in nonengine lubricant formulations **A18:** 111
permanent viscosity loss **A18:** 110
shear stability **A18:** 109
temporary viscosity loss **A18:** 109-110
thickening efficiency **A18:** 109
Viscosity index (VI) ..... **A18:** 83 , 108, 134, 135, 140
defined **A18:** 20
engine oils **A18:** 168
Viscosity index character **A18:** 107
Viscosity index improvers (VIIS) **A18:** 84
grease additives **A18:** 125
Viscosity, solution
effect on stress-corrosion cracking **A13:** 147
Viscous
defined **A18:** 20
Viscous composite sintering (VCS) **EM4:** 285, 287
Viscous deformation *See also* Anelastic deformation; Elastic deformation **EM3:** 30
defined **EM2:** 45
Viscous dislocation glide
creep by **A8:** 308
Viscous flow
GMS analysis **A10:** 152
in leaks **A17:** 58
Viscous friction *See* Fluid friction
Viscous friction torque **A18:** 511
Viscous glass sintering (VGS) **EM4:** 285, 287
Viscous oils
for room-temperature compression testing **A8:** 195
Visibility
limited, magnetic rubber inspection methods **A17:** 123
of magnetic particles **A17:** 100
Visible **A10:** 683, 691
Visible emission
defined **EL1:** 1161
Visible light
emitting diode, defined **EL1:** 161
for polymerization of compounds **EL1:** 854
spectroscopy **EL1:** 1103
Visible light waves
inspection advantages **A17:** 10
in optical holography **A17:** 405
Visible penetrants
sensitivity level **A17:** 75, 77
Visible radiation
defined **A10:** 683
Visible spectra
DRS and FT-IR analysis **A10:** 114
Vision
human vs. machine capabilities **A17:** 30
Vision pouring system
automatic **A15:** 500-501
Vision probes
coordinate measuring machines **A17:** 25
Vision systems
classifications, machine vision process **A17:** 33
with component assembly equipment **EL1:** 732
computers for **A17:** 44
hard-wired **A17:** 44
Vision/laser inspection systems
robotic **EM2:** 845

SUBJECTS OF THE INDEXED VOLUMES: ASM Handbook (designated by the letter "A"): **A1:** Properties and Selection: Irons, Steels, and High-Performance Alloys (1990); **A2:** Properties and Selection: Nonferrous Alloys and Special-Purpose Materials (1990); **A3:** Alloy Phase Diagrams (1992); **A4:** Heat Treating (1991); **A6:** Welding, Brazing, and Soldering (1993); **A7:** Powder Metallurgy (1984); **A8:** Mechanical Testing (1985); **A9:** Metallography and Microstructures (1985); **A10:** Materials Characterization (1986); **A11:** Failure Analysis and Prevention (1986); **A12:** Fractography (1987); **A13:** Corrosion (1987); **A14:** Forming and Forging (1988); **A15:** Casting (1988); **A16:** Machining (1989); **A17:** Nondestructive Testing and Quality Control (1989); **A18:** Friction, Lubrication, and Wear Technology (1992). **Metals Handbook, 9th Edition** (designated by the letter "M"): **M1:** Properties and Selection: Irons and Steels (1978); **M2:** Properties and Selection: Nonferrous Alloys and Pure Metals (1979); **M3:** Properties and Selection: Stainless Steels, Tool Materials and Special-Purpose Materials (1980); **M4:** Heat Treating (1981); **M5:** Surface Cleaning, Finishing, and Coating (1982); **M6:** Welding, Brazing, and Soldering (1983). **Engineered Materials Handbook** (designated by the letters "EM"): **EM1:** Composites (1987); **EM2:** Engineering Plastics (1988); **EM3:** Adhesives and Sealants (1990); **EM4:** Ceramics and Glasses (1991); **Electronic Materials Handbook** (designated by the letters "EL"): **EL1:** Packaging (1989).

Visual examination *See also* Nondestructive evaluation (NDE)
defined **EL1:** 1161
embrittlement phenomena **A12:** 123-137
of failed parts **A11:** 16, 173
in failure verification/fault isolation **EL1:** 1058
fractographic **A12:** 78
fracture identification chart for **A11:** 80
of heat-exchanger failed parts **A11:** 628-629
interpretation of fractures **A12:** 96-123
and light microscopy **A12:** 91-165
preliminary, of specimen **A12:** 72-73
quality control applications **A12:** 140-143
sequence for fractured components **A12:** 92
techniques **A12:** 91-93
weld cracking **A12:** 137-140
Visual examination/inspection
of anodized coatings **A13:** 397
of exfoliation corrosion **A13:** 244
of marine corrosion **A13:** 920
pitting corrosion **A13:** 231
Visual fit
model for *da/dN* vs. stress intensity **A8:** 681
Visual image analysis
stages **EL1:** 366
Visual inspection *See also* Inspection; Machine vision; NDE reliability; Optical inspection; Testing; Vision systems **A6:** 1081-1082, 1085 **A17:** 3-11 **M6:** 847
of adhesive-bonded joints **A17:** 616-617
and automated equipment **A17:** 116
automatic **EL1:** 941-942
borescopes **A17:** 3-10
of brazed assemblies **A17:** 603-604
brazed joints **A6:** 1118-1119
of casting defects **A15:** 545 **A17:** 520
of casting surfaces **A17:** 512
computer aided **A15:** 572
as defect verification **EL1:** 872-873
defined **A17:** 3
electron-beam welding **A6:** 866
equipment **A17:** 3
fitness for service evaluation **A6:** 376
flexible borescopes **A17:** 5-10
of forgings **A17:** 498-499
of investment castings **A15:** 264
with machine vision **A17:** 38-40
magnifying systems **A17:** 10-11
methods **EL1:** 365-368
optical sensors **A17:** 10
plasma-MIG welding **A6:** 224
of powder metallurgy parts **A17:** 545-547
of pressure vessels **A17:** 646-647
printed board coupons **EL1:** 574
rigid borescopes **A17:** 4-5
of solder joints **EL1:** 735
of soldered joints **A6:** 981 **M6:** 1089-1090
to evaluate gas turbine ceramic components **EM4:** 718
of weldments **A17:** 590-591
Visual leak testing
of pressure systems **A17:** 66
Visual reference gaging
magnifying systems for **A17:** 10-11
Visualization, of airport runways
by microwave holography **A17:** 226-227
Vitallium (Co-Cr-Mo alloy)
use in interposition arthroplasty **A18:** 656
use in metal-on-metal total hip replacements **A18:** 657
Vitamins
powder used **A7:** 574
Vitreous
defined **EL1:** 1161
Vitreous bond systems
bonded-abrasive grains **A2:** 1014-1015
Vitreous bonded grinding wheels
mixing operations **EM4:** 98
ordered mixture formed **EM4:** 98
Vitreous carbon
Raman analysis **A10:** 132
Vitreous china **EM4:** 4
absorption **EM4:** 4
applications **EM4:** 4
characterization **EM4:** 6
composition **EM4:** 5
products **EM4:** 4, 5
Vitreous coatings
aluminum and aluminum alloys **M5:** 609-610
Vitreous dielectric compositions
applications **EL1:** 109
Vitreous floor tile
composition **EM4:** 5
Vitreous fractures
studies **A12:** 2
Vitreous sanitaryware
characterization **EM4:** 6
composition **EM4:** 5
properties **EM4:** 6
Vitreous silica, properties
non-CRT applications **EM4:** 1048-1049
Vitreous solder glass, properties
non-CRT applications **EM4:** 1048-1049
Vitrification **EM4:** 3, 35
defined
definition **EM4:** 3
properties **EM4:** 3
purpose **EM4:** 3
of structural ceramics **A2:** 1020-1021
Vitrified bond
thread grinding wheels **A16:** 271-272, 273
Vitrified silica fibers
as thermocouple wire insulation **A2:** 882
VLS process *See* Vapor-liquid-solid process
VM-1
composition **A18:** 821
mechanical properties **A18:** 823
wear and friction properties **A18:** 822
VM-2
composition **A18:** 821
mechanical properties **A18:** 823
wear and friction properties **A18:** 822
Vocabulary *See* Glossary of terms
VOD *See* Vacuum oxygen decarburization
Void *See also* Pores; Porosity **A7:** 12 **EM3:** 30
behavior during welding **A7:** 456
coalescence **EM3:** 507
defined **EM2:** 45
growth, ductile fracture **A8:** 571-572
growth, fracture mechanics of **A8:** 439
initiation, ductile fracture **A8:** 571
linking, ductile fracture **A8:** 571-572
nucleation, dimpled rupture **A8:** 479
nucleation rate, average **A8:** 288
particle nucleated ductile intergranular **A8:** 487
sheet **A8:** 479
Void content **EM3:** 30
defined **EM2:** 45
Void growth (McClintock) model
of fracture **A14:** 393
Void ratio **A6:** 147
Void(s) *See also* Defect(s); Voiding
adhesive **EL1:** 1046-1047
in cured epoxy systems **EL1:** 818
defined **EL1:** 1161
distribution, bonding layer **EL1:** 214
effect, eutectic die attach **EL1:** 214-215
effect, polymer die attach **EL1:** 218-219
electromigration **EL1:** 1014
from dewetting **EL1:** 676
from fluxes **EL1:** 684
Kirkendall **EL1:** 680, 1148
in leaded and leadless surface-mount joints **EL1:** 732
printed board coupons **EL1:** 575
solder, inspection of **EL1:** 942
Voiding
adhesive **EL1:** 1046-1047
electromigration-induced **EL1:** 890
in integrated circuits **A11:** 774
semiconductor chips **EL1:** 964
Voids *See also* Cavities; Inclusion; Microvoid coalescence; Porosity; Shrinkage
air-bubble **EM1:** 3-4
alloy steels **A12:** 339
along grain boundaries, steam-generator tubes **A11:** 606-607
alpha-stabilized, titanium forgings **A17:** 497-498
black, at grain boundaries **A12:** 346
by incomplete fusion **A11:** 93
cast aluminum alloys **A12:** 408
computed tomography (CT) **A17:** 361
content, defined **EM1:** 25
creep **A17:** 55
Voids (continued)
defined **A13:** 14 **A15:** 11 **EM1:** 25
dendritic **A15:** 119
determined, microwave inspection **A17:** 202
dimple formation along **A12:** 338
eddy current inspection of **A17:** 164
effect on bond testing **A17:** 638
elimination, by curing **EM1:** 655
in fatigue fractures **A11:** 128, 454
in filament winding **EM1:** 135
in foam adhesive joints **A17:** 614
formation, as quality-control variable **EM1:** 730
formation, during bonding **EM1:** 687
formation, in BMI curing **EM1:** 660-661
initiation, titanium alloys **A12:** 445
in integrated circuits **A11:** 774
as internal defect **A10:** 587
irradiated materials **A12:** 365
Kirkendall **A11:** 776
linked **A11:** 667
magnetic field testing detection **A17:** 129
maraging steels **A12:** 383
metal-to-metal, in adhesive-bonded joints **A17:** 610
microwave inspection **A17:** 202, 212
nodule-nucleated, ductile irons **A12:** 236, 237
noninterconnecting, sulfur concrete **A12:** 472
-nucleating sites, sulfide inclusions as **A12:** 14
with particles, titanium alloys **A12:** 455
in polymers **A11:** 759, 761
precipitation-hardening stainless steels **A12:** 370
quantitative metallography of **A9:** 129
r-type cavities as **A12:** 122, 140
radiographic methods **A17:** 296
in radomes **A17:** 202
rounded **A12:** 445
in semiconductors, radiographic appearance **A17:** 350
shrinkage **A12:** 67
in solidification shrinkage **A15:** 109
as source, nonrelevant indications **A17:** 106
as squeeze casting defect **A15:** 325
thermal inspection **A17:** 396, 402-403
time-to-failure in gold-aluminum wire bonds effects of **A11:** 776
titanium alloys **A12:** 445, 452, 455
toot steels **A12:** 379
ultrasonic inspection of **A17:** 232
as volumetric flaw **A17:** 50
welding arc **A11:** 93
Voids in
aluminum alloys **A9:** 358
arc welds of heat-resistant alloys **M6:** 364
copper-base powder metallurgy materials **A9:** 554
silver
flash welds **M6:** 580
si -base powder metallurgy materials **A9:** 558-560
tungsten-base powder metallurgy materials **A9:** 561
in uranium alloys **A9:** 477, 480, 483, 485-486
Voids in fiber composites
mounting to prevent **A9:** 588
resulting from manufacture **A9:** 591
Voigt elastic constants combination **A6:** 146
Voigt element
as viscoelasticity model **EM2:** 414
Voigt-Kelvin element
as model **EM2:** 661
Volatile compounds **A7:** 154
complex mixtures analysis in **A10:** 639
Volatile content **EM3:** 30, 37
defined **EM1:** 25 **EM2:** 45
Volatile inhibitors
carbon steels **A13:** 525
Volatile liquids
analytical methods for **A10:** 7, 10
Volatile materials
removal for XPS analysis **A10:** 575
Volatiles *See also* Fire, Flammability **EM3:** 30
defined **EM1:** 25 **EM2:** 45
effect in curing **EL1:** 733
effect in sands **A15:** 208
Volatility **A18:** 84
of flux **EL1:** 644
of nitric acid reactions **A10:** 166

Volatilization
and atomization, in graphite furnace atomizers **A10:** 53
and changes in mass through sintering **A7:** 309
of nitric acid reactions **A10:** 166
of residues, gravimetric analysis **A10:** 163
spark, for solid-sample analysis **A10:** 36
Volatilize
defined **A7:** 13
Volatilized silica
applications **EM4:** 47
composition **EM4:** 47
supply source **EM4:** 47
Volhard titration
for arsenic **A10:** 173
defined **A10:** 164
indirect, for zinc **A10:** 173
of silver **A10:** 173
Volt
defined **EL1:** 1161
Volt-ampere curves
submerged arc welding **M6:** 131-132
Voltage *See also* Amperage; Circuit voltage
applied, and electrolysis **A10:** 197
applied, in field ion microscope **A10:** 588
bipolar technology **EL1:** 156-157
bucking, in thermocouple thermometers **A2:** 869-870
capacitance variation with **EL1:** 158
constant increases, in electrogravimetry **A10:** 198
dendritic growth under application of **EL1:** 660
design effect **EL1:** 521
distribution, WSI **EL1:** 355-356
drop, in thermocouple thermometers **A2:** 869
effect on weld attributes **A6:** 182
in electrical testing **EM2:** 584
increases, effect in constant current methods electrogravimetry **A10:** 198
instability, twin-on, MOSFET **EL1:** 159
-pulsed and laser-pulsed atom probes compared **A10:** 597
regulators, recommended contact materials **A2:** 863
transfer, WSI **EL1:** 358
value, as decomposition potential **A10:** 199
Voltage (sensing) relay
defined **EL1:** 1161
Voltage, accelerating
in SEM **A12:** 167
Voltage alignment
defined **A9:** 19
Voltage breakdown test **EL1:** 561
Voltage contrast
defined **EL1:** 1100
electron beam **A11:** 768
and electron beam induced current (EBIC) **EL1:** 1094, 1100-1101
mechanism, schematic **EL1:** 372
phase-dependent **A11:** 768
scanning electron microscopy **A9:** 90
SEM micrograph **A11:** 768 **EL1:** 1101
as SEM special technique **A10:** 506-507, 683
Voltage diagram, inspection/reference coils
eddy current inspection **A17:** 177
Voltage regulation
defined **EL1:** 1161
Voltage regulator
definition **M6:** 19
Voltage standing-wave ratio
defined **EL1:** 1161
Voltage stressing
zero-risk groups **EL1:** 139
Voltage-specific corrosion
of aluminum **A11:** 771
Voltages
for in situ electropolishing **A9:** 55
Voltammetry *See also* Classical electrochemical and radiochemical analysis **A10:** 188-196
applications **A10:** 188, 194-195
capabilities **A10:** 181
cyclic **A10:** 192
defined **A10:** 683
electrometric titration and, compared **A10:** 202
estimated analysis time **A10:** 188
general uses **A10:** 188
improvements and developments **A10:** 193
information obtainable from **A10:** 188, 189, 193-194
introduction and principles **A10:** 189-191
limitations **A10:** 188
linear sweep **A10:** 191-192
mass transfer processes in **A10:** 189
with other electrodes **A10:** 191-193
polarography with DME as **A10:** 189
principle of differential pulse stripping **A10:** 193
related techniques **A10:** 188
samples **A10:** 188
solid-electrode, capabilities **A10:** 207
stripping **A10:** 192
Voltammogram, single-sweep peaked
with carbon-base electrode **A10:** 192
Voltmeter
for electrogravimetry **A10:** 200
Voltmeter circuit
thermocouple in **A2:** 869
Volume *See also* Volumetric
casting, defined *See* Casting volume
CBED patterns to analyze **A10:** 439
changes, quench cracks from **A11:** 122
conservation, in creep rupture testing **A8:** 305
constant, as material behavior **A8:** 343
control, precision forging **A14:** 160
conversion factors **A8:** 723 **A10:** 686
diffusion, compared to grain-boundary diffusion **A10:** 478
and electrical testing **EM2:** 585-586
electron scattering **A10:** 434
expansion, cyanates **EM2:** 234
expansion, hydrogen-caused **A11:** 46
feed metal, in riser design **A15:** 577-578
fraction, effect, tensile ductility, steel **A14:** 364
fraction measurement, image analysis **A10:** 313, 314
loss, from erosion **A11:** 155
loss measurement, of wear **A8:** 606
of metal casting shipments **A15:** 41-42
partial effect **A17:** 374
per unit time, conversion factors **A10:** 686
relation to x-ray spatial resolution **A10:** 448
resistivity **EM2:** 45, 227, 460
sample, production of inelastically backscattered electrons by **A10:** 499
SI derived unit and symbol for **A10:** 685
SI unit/symbol for **A8:** 721
of signals produced by electron beam **A10:** 498-500
small, AEM-EDS microanalytical techniques for **A10:** 446
of solution, total immersion tests **A13:** 222
specific, SI derived unit and symbol for **A10:** 685
system, leak detection effects **A17:** 70
water, effects, gas/oil wells **A13:** 480
Volume change on freezing
cast copper alloys **A2:** 356-391
Volume diffusion **A7:** 13, 313
Volume discount, of parts
defined **EM2:** 84
Volume filling
defined **A7:** 13
Volume fraction *See also* Fiber volume fraction **A3:** 1 • 29 **EM2:** 45, 506 **EM3:** 30
of carbide, in cermets **A2:** 991
cermets and metal-matrix composites **A2:** 978
Volume fraction (continued)
defined **EM1:** 25
and ductility, relationship of dispersions in copper **A9:** 125
equality to areal ratio, linear ratio and point ratio **A9:** 125
of particles, at interface **A15:** 144
phase, peritectics **A15:** 125
Volume fraction measurements
image analysis **A10:** 313, 314
Volume per unit time
conversion factors **A8:** 723
Volume resistance *See also* Volume resistivity **EM2:** 45, 227, 460 **EM3:** 31
defined **EM1:** 25
Volume resistivity **EM3:** 31
carbon fiber/fabric reinforced epoxy resin **EM1:** 412
defined **EL1:** 1161 **EM1:** 359
epoxy resin system composites **EM1:** 403, 405, 407, 409, 414
glass fabric reinforced epoxy resin **EM1:** 405
glass fiber reinforced epoxy resin **EM1:** 407
graphite fiber reinforced epoxy resin **EM1:** 414
high-temperature thermoset matrix composites **EM1:** 377
Kevlar 49 fiber/fabric reinforced epoxy resin **EM1:** 409
low-temperature thermoset matrix composites **EM1:** 398
and material selection **EM1:** 38
medium-temperature thermoset matrix composites **EM1:** 385
thermoplastic matrix composites **EM1:** 367, 370, 372
Volume, surface
as roughness parameter **A12:** 201
Volume wear
thermoplastic composites **A18:** 822
Volume(s)
flaws, particle **A7:** 59
fraction **A7:** 13
measuring displaced **A7:** 267-268
pore **A7:** 265
ratio **A7:** 13
shrinkage **A7:** 13, 322
specific **A7:** 265
specific surface, particle shape **A7:** 239
Volume-fraction measurements, application to establish phase boundaries in a two-phase field **A9:** 125
Volume-to-surface area ratios
bars/plates, gray iron **A15:** 635
in design **A15:** 611
Volume/area ratio **M1:** 15-16
Volume/area ratios
of gray iron **A1:** 15-16
Volumeter
Scott, for determining apparent density **A7:** 274, 275
Tap-Pak **A7:** 276
Volumetric
accuracy, coordinate measuring machines **A17:** 26
flaws, defined **A17:** 50
scanning, ultrasonic inspection **A17:** 231-232
Volumetric analysis
commonly used **A10:** 175
defined **A10:** 683
elastomeric tooling **EM1:** 591-595
electrometric titration as **A10:** 202
equilibrium in **A10:** 163
molarity and normality in **A10:** 162
oxidation-reduction **A10:** 163-164
precipitation titrations as **A10:** 164
redox reaction **A10:** 163
as redox titrations **A10:** 174-176
vs gravimetric analysis **A10:** 172

SUBJECTS OF THE INDEXED VOLUMES: ASM Handbook (designated by the letter "A"): **A1:** Properties and Selection: Irons, Steels, and High-Performance Alloys (1990); **A2:** Properties and Selection: Nonferrous Alloys and Special-Purpose Materials (1990); **A3:** Alloy Phase Diagrams (1992); **A4:** Heat Treating (1991); **A6:** Welding, Brazing, and Soldering (1993); **A7:** Powder Metallurgy (1984); **A8:** Mechanical Testing (1985); **A9:** Metallography and Microstructures (1985); **A10:** Materials Characterization (1986); **A11:** Failure Analysis and Prevention (1986); **A12:** Fractography (1987); **A13:** Corrosion (1987); **A14:** Forming and Forging (1988); **A15:** Casting (1988); **A16:** Machining (1989); **A17:** Nondestructive Testing and Quality Control (1989); **A18:** Friction, Lubrication, and Wear Technology (1992). **Metals Handbook, 9th Edition** (designated by the letter "M"): **M1:** Properties and Selection: Irons and Steels (1978); **M2:** Properties and Selection: Nonferrous Alloys and Pure Metals (1979); **M3:** Properties and Selection: Stainless Steels, Tool Materials and Special-Purpose Materials (1980); **M4:** Heat Treating (1981); **M5:** Surface Cleaning, Finishing, and Coating (1982); **M6:** Welding, Brazing, and Soldering (1983). **Engineered Materials Handbook** (designated by the letters "EM"): **EM1:** Composites (1987); **EM2:** Engineering Plastics (1988); **EM3:** Adhesives and Sealants (1990); **EM4:** Ceramics and Glasses (1991); **Electronic Materials Handbook** (designated by the letters "EL"): **EL1:** Packaging (1989).

Volumetric changes
in steel castings **A1:** 374, 376
Volumetric coefficient of thermal expansion *See* Thermal properties
wrought aluminum and aluminum alloys **A2:** 62-122
Volumetric modulus of elasticity *See* Bulk modulus of elasticity; Bulk modulus of elasticity (*K*)
Volumetric-displacement meter
for flow detection **A17:** 60-61
Volute springs **A1:** 302
von Mises
criterion, in multiaxial creep theories **A8:** 343
effective strain, triaxiality factor effect on **A8:** 344-345
effective stress-strain, torsion flow stress and compared **A8:** 162, 164
compression-tension data
relation, multiaxial and uniaxial stress states **A8:** 343
strain rate **A8:** 158
yield criterion **A8:** 576
von Mises stress **EM3:** 491-492, 493, 494, 495, 496, 497
Von Mises stress and strain criteria
solid-state-welded interlayers **A6:** 166, 169, 170
Vortex stabilization technique
Reed's **A10:** 32
Voxel
defined **A17:** 385
reconstruction, color images **A17:** 486
***Voyager* aircraft**
graphite fibers in **EM1:** 29, 30
VSA-11
properties **A18:** 803
VSMF Data Control Services
source of standards **EM3:** 64
Vulcanization *See also* Room-temperature vulcanizing (R7-V) **EM3:** 31
defined **EM1:** 25 **EM2:** 45
microwave inspection **A17:** 202
Vulcanization of polymers
Raman analysis **A10:** 132
Vulcanize **EM3:** 31
Vycor **EM4:** 427
applications
laboratory and process **EM4:** 1089
laboratory glassware **EM4:** 1088
lighting **EM4:** 1032, 1034
composition when used in lamps **EM4:** 1033
maximum operating temperatures **EM4:** 1035
properties **EM4:** 1033
laboratory glassware **EM4:** 1088
vessels for ion-exchange done in molten nitrate melts **EM4:** 461
Vycor crucibles
for fusions with acidic fluxes **A10:** 167
Vycor glass
encapsulation in HIP conforming to body configuration **EM4:** 197

W

W *See* Cycle
W temper, solution heat-treated
defined **A2:** 21
W-545
composition **A16:** 736
machining **A16:** 738, 741-743, 746-747, 749-757
W-type cracks *See* Triple-point cracking; Wedge cracks
W-Zr (Phase Diagram) .... **A3:** 2 • 382
W.W. Sly company (Cleveland)
tumbling mill of **A15:** 33
W1 52
composition **M4:** 653
W1, W2, W5 *See* Tool steels, specific types
W545
composition **A4:** 794
WAD 7823A
broaching **A16:** 209
Wadell shape factors **A7:** 239
Waelz kiln process
zinc recycling **A2:** 1224-1225
Wafer
defined **EL1:** 1161 **EM1:** 25
fabrication, failure mechanisms **EL1:** 978
initial probing **EL1:** 192
map **EL1:** 948
preparation **EL1:** 191- 92
-surface cleanliness **EL1:** 192
windows in **EL1:** 198
Wafer bumping **EL1:** 276-277, 477
Wafer fabrication
of polyimides **EL1:** 769-779
Wafer probe
defined **EL1:** 1161
Wafer processing
GaAs crystal fabrication **A2:** 744-745
Wafer-level physical test methods
analytical methods **EL1:** 918-926
Auger electron microscopy **EL1:** 922-924
materials analysis **EL1:** 917-918
Rutherford backscattering spectroscopy **EL1:** 926
scanning electron microscopy **EL1:** 918-920
secondary ion mass spectrometry **EL1:** 924-926
transmission electron microscopy **EL1:** 920-922
x-ray fluorescence spectroscopy **EL1:** 920
x-ray microprobe spectroscopy **EL1:** 920
Wafer-scale assemblies **EL1:** 250, 258
Wafer-scale integration (WSI) *See also* Hybrid wafer-scale integration; Silicon circuit boards (SCB); Wafer; Wafer-scale assemblies; Wafer-scale integration technology **EL1:** 354-364
advantages/limitations **EL1:** 258
of arrays **EL1:** 8
circuits, monolith **EL1:** 8 9
complete modules testing **EL1:** 363
cooling **EL1:** 363-364
disadvantages **EL1:** 7
as hybrids **EL1:** 250
instrumentation/testing **EL1:** 365
limitations **EL1:** 297-298
major problems **EL1:** 354
monolithic **EL1:** 8
as new technology **EL1:** 249
optical clocks **EL1:** 9-10
optical interconnections **EL1:** 10
power distribution **EL1:** 355-357
present status, summary **EL1:** 264-266
problem solving **EL1:** 354
self-test **EL1:** 376-377
signal transmission lines **EL1:** 357-362
system test interface **EL1:** 376
test procedures **EL1:** 373
test/restructure test process **EL1:** 377
testing and programming **EL1:** 362-363
testing phases **EL1:** 377-378
viability of **EL1:** 258
Wafer-scale integration technology
and alternatives, compared **EL1:** 269-270
monolithic like WSI approaches **EL1:** 271-272
multichip hybrids, advantages **EL1:** 270-271
overview of **EL1:** 263-273
packaging **EL1:** 272-273
present status, summary **EL1:** 264-266
processing yield, WSI vs VLSI **EL1:** 266-268
redundancy in **EL1:** 268
rerouting technologies **EL1:** 268-269
technology review **EL1:** 263-264
Wafering **A18:** 685-686
Wafering blades **A9:** 25
solution-annealed and aged, different illuminations compared **A9:** 81
Wafers, silicon
IR determination of oxygen and carbon determined in **A10:** 122-123
Waffles, master alloy
forms of **A15:** 164
Wagner, E.B
as inventor **A15:** 35
Wagner theory
of oxidation **A13:** 69-70
Wah Chang WC-1Zr *See* Niobium alloys, specific types
Wahl correction factor
steel springs **M1:** 303-304
Wahl corrections
for steel springs **A1:** 318-319, 320, 321
Wake hackle
definition **EM4:** 633
Walker equation **A11:** 53
Walker equivalent stress parameter
Goodman diagram plot for **A8:** 713
Walker model
crack growth rate **A8:** 681
Walking
estimation of load cycles for one leg during **A11:** 672
Walking-beam furnaces **A7:** 354-358
defined **A7:** 13
for sintering uranium dioxide pellets **A7:** 665
Wall
-cleaning fluxes, aluminum alloys **A15:** 446
mold, inspection of **A15:** 555-556
Wall effects
corrosion testing **A13:** 207
Wall neutrality number (N) **A6:** 204
Wall neutrality number method **A6:** 658
Wall plastics
powders used **A7:** 572
Wall thickness *See also* Casting section thickness; Mold walls; Section thickness; Thickness; Through-wall examination
defined **EL1:** 1161
design guidelines for **EM2:** 709
as error source, remote-field eddy current inspection **A17:** 200
as function of mold inside diameter **A15:** 303
metal molds, vertical centrifugal casting **A15:** 302-303
and mold temperature, permanent molds **A15:** 282
in parts design **EM2:** 615
polyamide-imides (PAI) **EM2:** 130
and riser location **A15:** 580-581
spindle, flaw detection through **A17:** 139
straightening effect **A14:** 690
testing of **A15:** 264
tube, for tube spinning **A14:** 677
of tubes, and bending **A14:** 665
Wall transition
polyamide-imides (PAI) **EM2:** 130
Wall-thinning
of tubes **A11:** 606
Wall/floor tiles
composition **EM4:** 45
properties of fired ware **EM4:** 45
Wallner line
definition **EM4:** 633-634
Wallner lines
cemented carbides **A12:** 470
in ceramic fracture **A11:** 747
defined **A8:** 15 **A11:** 11
and fatigue striations, compared **A11:** 26
formation **A12:** 13-14
as fracture surface feature **EM2:** 810
on WC-CO fracture surface **A12:** 18
and stress distribution at time of fracture **A11:** 746
and striations, compared **A12:** 119
tool steels **A12:** 378
Walnut shells
blasting with **M5:** 84, 93-94
Walsh transforms **A18:** 347
War of 1812
foundry history in **A15:** 27
Warburg impedance **EM3:** 435
Warm box
binders **A15:** 218
catalysts **A15:** 218
Warm components
gas-turbine **A11:** 284
Warm extrusion *See also* Extrusion
of cermet powder mixtures **A2:** 981-983
Warm forging
by HERF processing **A14:** 105
precision **A14:** 168, 171-172
steel phase transformation at **A14:** 162
Warm forming, temperatures
for steel **A14:** 620
Warm heading **A14:** 297-298
applications **A14:** 297
machines and heating devices **A14:** 297
tools **A14:** 297
Warm heading machines **A14:** 297

Warm liquid-vapor cleaning techniques **A7:** 459
Warm liquid-vapor degreasing system **M5:** 46-47, 51, 54-55
Warm rolling
in processing solid steel **A1:** 120, 124
Warm rotary forging **A14:** 177-179
Warm workability *See also* Warm forging; Warm heating; Warm working; Workability
of alloy and carbon steels **A14:** 174
Warm working *See also* Cold working; Hot working; Warm forging; Warm heading
defined **A14:** 14
Warm working temperature
ductile fracture **A8:** 572-574
Warm-runner molding
as thermoset **EM2:** 323
Warm-setting adhesive **EM3:** 31
Warp **EM3:** 31
defined **EM1:** 25 **EM2:** 45
Warp yarns *See also* Yarns
fabric direction **EM1:** 148
in fabric pattern **EM1:** 125
fiberglass fabric
heavy, in unidirectional fabrics **EM1:** 148
in rapier looms **EM1:** 127-128
Warpage **EM3:** 31
defined **A7:** 13 **A15:** 11 **EM2:** 45
part, liquid crystal polymers (LCP) **EM2:** 181
and pouring temperature **A15:** 283
thermoplastic injection molding **EM2:** 313
of wood patterns **A15:** 194
Warped casting
as casting defect **A11:** 387
Warping
in aluminum alloy forming **A14:** 797
as distortion **A11:** 141
silicon wafer **EL1:** 1053
Warranty requirements
damage analyses for **A8:** 683
Warwick Furnace, Pennsylvania
as early foundry **A15:** 25
Wash
defined **A15:** 11
mold, horizontal centrifugal casting **A15:** 296
permanent molds **A15:** 304
Wash primer
defined **A13:** 14
Wash primers **M5:** 477
Wash scab
as casting defect **A11:** 385
Washburn equation
mercury porosimetry and equation of forces **EM4:** 71
to determine open porosity **EM4:** 580, 581
Washer and screw assemblies
economy in manufacture **M3:** 847-848
Washers
interpretation by windowing **A17:** 36
mechanical fastening of **EM2:** 711
Washing
of radiographic film **A17:** 353
Washing machine P/M parts **A7:** 623
Washington, Augustine
as early founder **A15:** 25
Waspaloy *See also* Nickel alloys, specific types; Superalloys; Superalloys, specific types **A14:** 152, 236, 265, 374
aging **A4:** 796
aging cycle **A4:** 812 **M4:** 656
aging cycles **A6:** 574
annealing **M4:** 655
applications **A2:** 441 **A4:** 807
arc welding **M6:** 354
broaching **A16:** 203
chemical milling **A16:** 584
cobalt in **A1:** 1014-1016
composition **A4:** 794, 795 **A6:** 564 **A16:** 736 **M4:** 651-652 **M6:** 354
different heat treatments compared **A9:** 325-326
Waspaloy (continued)
electrochemical grinding **A16:** 542, 547
electrochemical machining **A16:** 539, 541
electron-beam welding **A6:** 865, 869
end milling **A16:** 539
fatigue fracture by embrittlement **A11:** 454-455
flash welding **M6:** 557
forging, grain-boundary carbide films **A11:** 267
grinding **A16:** 547
for hot-forging dies **A18:** 626
machining **A16:** 738, 741-743, 746-747, 749-758
mechanical properties **A4:** 807
microstructure **A9:** 309-311
milling **A16:** 547, 752, 755
postweld heat treatment **M6:** 357
sawing **A16:** 360
solution treating **M4:** 656
solution treatment **A6:** 574
solution-treating **A4:** 796, 807
stress relieving **M4:** 651-652
surface characteristics produced by electrochemical machining **A16:** 32
tensile properties **A4:** 808
thread grinding **A16:** 275
turning **A16:** 740
Waspaloy alloy
composition **A6:** 573
Waste *See* Material waste
blow molding, costs **EM2:** 299
coal, Miller numbers **A18:** 235
nickel, Miller numbers **A18:** 235
containers, nuclear, ultrasonic inspection **A17:** 275-276
incineration **A13:** 1368-1369
industrial chemical **A13:** 1368
in injection molding, and cost **EM2:** 294
municipal solid **A13:** 997-998, 1368
nuclear, acoustic emission inspection **A17:** 281
sewage sludge **A13:** 1368-1369
thermoforming, costs **EM2:** 301
treatment **A13:** 382, 395, 455
Waste disposal
metalworking lubricants **A18:** 142, 144, 145-146
Waste products
chemical, assay for toxic elements **A10:** 233
industrial, sampling of **A10:** 12-18
Waste recovery and treatment
acid cleaning processes **M5:** 65
alkaline etching process **M5:** 583-584
aluminum and aluminum alloy processing wastes **M5:** 583-584
cleaning processes **M5:** 20
copper plating process **M5:** 166-167
emulsion cleaning processes **M5:** 39
gold plating processes **M5:** 284
hard chromium plating process **M5:** 184, 186-187
mass finishing processes **M5:** 136-137
mechanical coating process **M5:** 302
phosphate coating processes **M5:** 448, 454-456
pickling processes **M5:** 81-82
plating wastes *See* Plating waste disposal and recovery
salt bath descaling, sludge removal **M5:** 98
solvent cleaners **M5:** 41
vapor degreasing solvents **M5:** 48-49, 57
Wastewater streams
UV/VIS analysis **A10:** 60
Wastewater treatment systems
with monolithic linings **A13:** 455
Watanabe number (J) **A6:** 278
Watchmaker's lathes **A16:** 153
Water *See also* Moisture; Rainwater; Rinse waters; Seawater; Water absorption; Water atomization; Water cooling; Water spraying; Water vapor
as a lubricant for diamond- impregnated wire sawing **A9:** 26
absorbed, and permittivity **EL1:** 600-601
adsorption, molding materials **A15:** 208-210
analysis **A10:** 7, 31, 41, 43, 60, 102, 141, 152, 212
Water (continued)
analysis by biamperometric titration **A10:** 204
applications, nickel-base alloys **A13:** 653-654
as aqueous cleaner **EL1:** 663-664
assay for toxic elements **A10:** 141, 148-149, 233
atomization process, electrical composite contacts **A2:** 857
batching, and microbiological corrosion **A13:** 314
behavior, in clay-water bonds **A15:** 212
bellows, effect on foundries **A15:** 27
cast steel corrosion in **A13:** 575
cationic, anionic, gaseous concentrations determined **A10:** 181
characteristics, prediction **A13:** 490
chemistry, LWR corrosion **A13:** 946
cleaning **A13:** 1143
cleaning, high pressure **A17:** 81, 82
-containing fuels, corrosion in **A11:** 191
content, as function of dewpoint in sintering atmospheres **A7:** 341
content, ideal, in silica-base bonds **A15:** 212
coolant, for ultrasonic testing **A8:** 247-248
-cooled aluminum combustion chamber cavitation damage **A11:** 168
-cooled grips **A8:** 158-160
-cooled locomotive diesel engine, corrosion- fatigue cracking in **A11:** 371-372
cooling **A13:** 1134-1135
-cooling system, blowout of connector tubes from **A11:** 312
in core cells, adhesive-bonded joints **A17:** 613
corrosion, aluminum/aluminum alloys **A13:** 597-598
corrosion, cast irons **A13:** 570
corrosion, copper/copper alloys **A13:** 621-627, 636
corrosion, precipitation-hardening stainless steels **A12:** 373
corrosion testing in **A13:** 207-208
corrosion, titanium/titanium alloys **A13:** 676-677
as crude oil contaminant **A13:** 1266
degree of hardness, symbol for **A11:** 798
dissolved gases in **A13:** 489
dissolved salts in **A13:** 490
distilled **A13:** 760-761, 689
distilled, as corrosive **A12:** 24, 430
distilled, titanium/titanium alloy SCC in **A13:** 689
domestic supply, effect on dezincification of cartridge brass water pipe **A11:** 178
E-pH diagram **A13:** 27
effect, chemical processing corrosion **A13:** 1136-1137
effect, epoxy resin matrices **EM1:** 76
effects on powder metallurgy material specimens **A9:** 503
effects on silicon-iron electrical steels **A9:** 531
elemental analysis **A10:** 102
evaluation, for microbiological corrosion **A13:** 314
explosions, aluminum-lithium alloys **A2:** 182
explosive reactivity of powders in **A7:** 194-198
exposure, of high-temperature epoxy resins **EM1:** 141
filtration of particles from **A10:** 94
flush, effects on tubes **A11:** 630
fresh, magnesium/magnesium alloys in **A13:** 742
and gas atomization **A7:** 25-34
glass-forming ability **EM4:** 494
ground, SSMS analysis **A10:** 148-149
heat transfer from **A11:** 628
heat-exchanger crevice corrosion in **A11:** 631-632
high-purity, aluminum/aluminum alloys in **A13:** 597
high-purity deaerated **A12:** 438
high-purity, stainless steel SCC in **A11:** 218

Water (continued)
high-temperature, light water reactor corrosion in ... **A13:** 946
high-temperature, nickel-base alloy SCC in ... **A13:** 649-650
hot, resistance, of porcelain enamels ... **A13:** 451-452
as inappropriate fire extinguisher ... **A7:** 200
industrial waste ... **A13:** 570
as injection molding binder ... **A7:** 498
jets, high pressure, for descaling ... **A14:** 87
Karl Fischer method to determine ... **A10:** 219
lead/lead alloys in ... **A13:** 784-787
lubricant indicators and range of sensitivities ... **A18:** 301
in lubricants ... **A14:** 514
mean free path ... **A17:** 59
melting point ... **EM4:** 494
mine ... **A13:** 1293-1294
molecules, in electrode process ... **A13:** 18-19
natural ... **A13:** 433, 435, 597
natural, analyses of ... **A10:** 41, 141
neutral, galvanic series in ... **A13:** 1288
nickel SCC testing in ... **A8:** 531
niobium corrosion in ... **A13:** 723
oxygen plus, loss on reduction (LOR) ... **A7:** 155
oxygenated, pitting corrosion by ... **A11:** 615
pH effect, steel corrosion rate in ... **A13:** 991
as plasticizer ... **EM1:** 141
pressure, effect, in liquid penetrant inspection ... **A17:** 71-74
pressure-temperature diagram ... **EL1:** 244
pressurized reactor ... **A8:** 423
pure tin in ... **A13:** 771
purest, conductance of ... **A10:** 203
purification ... **A8:** 421-424
quality, effects, water-recirculating systems ... **A13:** 489
quench, low-alloy steel ... **A11:** 393
rain, water drop impingement corrosion ... **A13:** 142
Raman scattering of ... **A10:** 133
resistance, of porcelain enamels ... **A13:** 449
rinsing, in cold extrusion ... **A14:** 304
as sample in gas analysis by mass spectroscopy ... **A10:** 152
SCC testing ... **A13:** 274
SCC testing of titanium alloys in ... **A8:** 531
-side corrosion, in boiler and steam equipment ... **A11:** 614-616
-side scale, removal of ... **A11:** 616
simulation in pressurized water reactor ... **A8:** 423-424
soft ... **A13:** 207
as solvent used in ceramics processing ... **EM4:** 117
sour ... **A13:** 1267
stainless steel corrosion in ... **A13:** 555-556
Standard Reference Materials for environmental/industrial ... **A10:** 95
structural changes in surfactant molecules determined ... **A10:** 118
surface tension ... **EM3:** 181
suspended solids in ... **A13:** 489
suspending liquid, for magnetic particles
swift-moving, corrosion in ... **A11:** 188
tank, for explosive forming ... **A14:** 636-637
tantalum resistance to ... **A13:** 725
tap ... **A13:** 761
tap, as corrosive ... **A12:** 24, 320
temper, defined ... **A15:** 212
temperature/pH effects ... **A13:** 489-490
testing methods ... **A13:** 208
trace analyses of ... **A10:** 57-58, 60, 204
treated vs. natural ... **A13:** 207
treatment, in steam power plants ... **A11:** 603
ultrapure, trace metal analysis for ... **A10:** 57-58
and ultrasonic waves ... **A17:** 232-233
uranium alloys in ... **A13:** 814-815
vapor ... **A13:** 143
vapor, condensed, as corrosive in boiler tubes ... **A11:** 618
vapor content, and dew point of sintering atmospheres ... **A7:** 361
vapor, effect of pH on copper alloy tubing ... **A11:** 634-635
vapor, effect on fatigue fracture appearance ... **A12:** 40, 52
vapor solubility, in copper alloys ... **A15:** 465

Water (continued)
variables ... **A13:** 207
velocity-affected corrosion in ... **A11:** 188-190
very pure ... **A13:** 207
viscosity at melting point ... **EM4:** 494
voltammetric monitoring of pollutant metals and nonmetals in ... **A10:** 188
volume, effects gas/oil wells ... **A13:** 480
walls, heat-transfer factors ... **A11:** 604
and wastewater streams , UV/VIS analysis of ... **A10:** 60
well, as ion chromatography solution ... **A10:** 658
well, graphitic corrosion from ... **A11:** 372-373
white, from paper- machines ... **A13:** 1188-1189
zinc corrosion in ... **A13:** 443, 759-762
zirconium/zirconium alloy corrosion in ... **A13:** 708
Water absorption *See also* Absorption; Moisture; Moisture absorption ... **EM3:** 31
defined ... **EM1:** 25 **EM2:** 45
electrical effects ... **EM2:** 466
epoxy resin matrices ... **EM1:** 74
in low-temperature thermoset matrix composites ... **EM1:** 393
on supplier data sheets ... **EM2:** 642
polyamide-imides (PAI) ... **EM2:** 133
polyaryl sulfones (PAS) ... **EM2:** 146
of selected polymers ... **EM2:** 761
thermoplastic resins ... **EM2:** 619
Water absorption soldering
flexible printed boards ... **EL1:** 590
Water absorption test ... **EL1:** 536-537, 1161
Water atomization *See also* Atomization; Explosivity; Moisture; Pyrophoricity; specific water-atomized powders; Water
of copier powders ... **A7:** 587
and gas atomization ... **A7:** 25-34
for hardfacing powders ... **A7:** 823
of iron ... **A7:** 84, 85
of low-carbon iron ... **A7:** 83-86
for nickel powder production ... **A7:** 134
of nonspherical copier powders ... **A7:** 587
production, schematic ... **A7:** 26
silicon effects in ... **A7:** 256
Water baths
for magnetic particle inspection ... **A7:** 578
Water break test ... **EM3:** 31, 743, 840-841, 843-844
for presence of oily contaminants ... **EM3:** 40
Water chemistry ... **A8:** 416, 422-423
Water cleaning ... **A13:** 1143
Water content of alcohols ... **A9:** 67
Water cooling
of chills, DS/SC furnaces ... **A15:** 400
cupolas ... **A15:** 384, 386
electric arc furnace ... **A15:** 359
of hot semiconductors ... **EL1:** 310
induction furnaces ... **A15:** 371-372
permanent steel molds ... **A15:** 304
Water corrosion ... **M5:** 430, 457-458
Water, corrosion in
nickel alloys ... **M3:** 172
zirconium ... **M3:** 786-787
Water, distilled
surface tension ... **EM3:** 181
Water drop impingement
in carbon steels ... **A13:** 519
and cavitation erosion, compared ... **A13:** 142
and design ... **A13:** 339
rain ... **A13:** 142
Water elutriation ... **A7:** 179
diagram ... **A7:** 180
René 95, superalloy ... **A7:** 180
Water filter housing, PVC
failure of ... **A11:** 763
Water for etchants ... **A9:** 67
Water injection plasma cutting ... **A14:** 730
Water injection systems
oil/gas wells ... **A13:** 479
Water insolubles test
to isolate impurities in soda ash ... **EM4:** 380
Water jet
defined ... **EM1:** 25 **EM2:** 45
Water jet configurations
atomization ... **A7:** 29
Water line attack
copper/copper alloys ... **A13:** 613
Water main pipe ... **A1:** 331-332

Water of hydration
in solid salts and acids ... **A9:** 68
Water pipe
graphitic corrosion in ... **A11:** 372-374
Water plasma spraying
ceramic coatings for adiabatic diesel engines ... **EM4:** 992
Water pump impeller
cavitation damage ... **A11:** 167-168
Water quenching
agitation ... **M4:** 40-41
contamination ... **M4:** 41
and infiltration ... **A7:** 558
system ... **M4:** 58, 63
Water resistance
water-base versus organic-solvent-base adhesives properties ... **EM3:** 86
Water rolling process
copper and copper alloys ... **M5:** 615
Water shield plasma cutting ... **A14:** 730
Water softener
ion exchange in ... **A10:** 658-659
Water spraying
in continuous casting ... **A15:** 311-312
conventional vs air-water mist spray ... **A15:** 312-213
systems, compared ... **A15:** 312
Water vapor *See also* Steam ... **A13:** 143, 813-814
and dewetting ... **EL1:** 676
and hydrogen content, forgings ... **A17:** 491-492
internal, content measurement ... **EL1:** 1064
sigma values for ionization ... **A17:** 68
Water walls
steam/water-side boilers ... **A13:** 991
Water wash
definition ... **M6:** 19
Water-atomized copper powders *See also* Copper powders ... **A7:** 106, 107
lithium-containing ... **A7:** 118
particle size measurement ... **A7:** 223, 224
properties ... **A7:** 119
Water-atomized high-carbon iron powders
annealing ... **A7:** 183
effect of residual carbon ... **A7:** 183
Water-atomized high-speed tool steel powders
annealing ... **A7:** 185
Water-atomized iron powders ... **A7:** 23
Water-atomized low-alloy steel powders ... **A7:** 101-103
Water-atomized low-carbon iron
annealing ... **A7:** 182
Water-atomized metal powders
green strengths of ... **A7:** 303
microstructural characteristics ... **A7:** 33-34
oxygen contents ... **A7:** 37
Water-atomized nickel
transverse-rupture strength ... **A7:** 397
Water-atomized particles
silicon surface ... **A7:** 252
Water-atomized powders ... **A13:** 833
Water-atomized steel powders
composition- depth profiles ... **A7:** 256
Water-atomized tin-based powders
membraneous shape formation ... **A7:** 28, 32
Water-atomized tool steel powders
composition and properties ... **A7:** 103, 104
Water-atomized welding powders ... **A7:** 818
Water-base adhesives ... **EM3:** 74, 86-90
advantages ... **EM3:** 75, 86
animal glues ... **EM3:** 89
for appliance market ... **EM3:** 87
bonding applications ... **EM3:** 87-90
for bonding electronics ... **EM3:** 87
bonding factors ... **EM3:** 87
for bonding machinery ... **EM3:** 87
casein ... **EM3:** 88
cellulosics ... **EM3:** 88
characteristics ... **EM3:** 74, 86-87
for construction applications ... **EM3:** 87, 88
for consumer goods ... **EM3:** 88
dextrin ... **EM3:** 88-89
foundry industry applications ... **EM3:** 47
for housewares applications ... **EM3:** 87
limitations ... **EM3:** 75, 86
markets ... **EM3:** 87
melamine formaldehyde ... **EM3:** 89
natural rubber latex ... **EM3:** 88

Water-base adhesives (continued)
neoprene (polychloroprene) **EM3:** 89
for nonrigid bonding **EM3:** 87, 88
phenol formaldehyde **EM3:** 89
polyvinyl acetate **EM3:** 89
polyvinyl alcohol **EM3:** 89
polyvinyl methyl ether **EM3:** 90
properties **EM3:** 87
reclaimed rubber **EM3:** 89
for rigid bonding **EM3:** 88
rosin **EM3:** 11, 188
sodium silicate water solution **EM3:** 89
starch adhesives **EM3:** 88
styrene-butadiene rubber **EM3:** 89
suppliers **EM3:** 90
synthetics **EM3:** 90
for tapes **EM3:** 88
for transportation applications **EM3:** 88
urea formaldehyde **EM3:** 89
wood bonding applications **EM3:** 89
for woodworking **EM3:** 46
Water-base contact cements
automotive applications **EM3:** 46
Water-base detergent cleaning
of fractures **A12:** 74
Water-base miscible oils
lubricant for tool steels **A18:** 738
Water-based rust preventives
safety precautions **M5:** 470
Water-blocked cable
defined **EL1:** 1161
Water-borne paints **M5:** 472, 500-502
Water-break test
cleaning process effectiveness **M5:** 37, 576
Water-column
designs, immersion-type ultrasonic search unit **A17:** 258-259
techniques, ultrasonic inspection **A17:** 248
Water-cooled tools **A18:** 627
Water-dispersible aluminum flake pastes **A7:** 594
Water-displacing polar rust-preventive compounds **M5:** 459-464, 467
applying, methods of **M5:** 467
Water-extended polyester **EM3:** 31
defined **EM2:** 45
Water-hardening steels **A1:** 768-771
Water-hardening tool steels *See* Tool steels, water-hardening
composition limits **A18:** 736
Water-hardening tools steels
forging temperatures **A14:** 81
Water-jacketed cooling sections **A7:** 354
Water-jacketed milling chambers **A7:** 62
Water-jet cutting **EM1:** 673-675
applications **EM1:** 674-675
benefits/problems **EM1:** 673-674
equipment/tools **EM1:** 675
future trends **EM1:** 675
materials cut by **EM1:** 675
Water-main pipe **M1:** 319
Water-recirculating systems
cooling system **A13:** 487
corrosion control **A13:** 494-497
corrosion processes **A13:** 487-489
environmental variables control in **A13:** 487-497
water quality influences **A13:** 489-494
Water-resistant adhesives **EM3:** 263
Water-resistant coatings
cast irons **M1:** 105, 106
Water-resistant porcelain enamel
composition of **M5:** 510
Water-side boilers
corrosion of **A13:** 990-993
Water-soluble alkali silicate coatings **M5:** 533-534
Water-soluble corrosion inhibitors **M5:** 432
Water-soluble developers (form B)
for liquid penetrant inspection **A17:** 77, 83
Water-soluble flux
corrosion effects **EL1:** 1110
Water-soluble oils
phosphate coatings supplemented with **M5:** 453, 455
Water-soluble organic fluxes **EL1:** 646-647
Water-soluble paper
leak detection with **A17:** 66
Water-suspendible developers (form C)
liquid penetrant inspection **A17:** 77, 83
Water-washable (method A) penetrants **A17:** 75, 77-78, 84
Water-washable visible penetrant method
of liquid penetrant inspection **A17:** 78
Water-well pipe **M1:** 315, 320-321
Water-wicking adhesives **EM3:** 263
Waterblasting **A13:** 414, 912
Waterbury's reagent
as etchant for copper ingot **A10:** 302
Waterjet cutting *See* Abrasive waterjet cutting
Waterjet/abrasive waterjet machining **EM4:** 313
Waterjet/abrasive waterjet machining (WJM) **A16:** 509, 520-527
abrasive waterjet applications **A16:** 527
advantages and disadvantages **A16:** 523-524
equipment **A16:** 520-522
flow rates **A16:** 521, 522-523
fluid additives **A16:** 522
MMCs **A16:** 894
pressure **A16:** 522, 523, 524
process characteristics **A16:** 522-523
waterjet applications **A16:** 524-526
Watt
defined **EL1:** 1161
Watts nickel plating bath **M5:** 199, 208-209, 212, 217
Watts solution **A18:** 836
Waukesha **A18:** 88
galling threshold load **A18:** 595
properties **A18:** 716
Wave
function, EXAFS analysis **A10:** 409
polarographic, defined **A10:** 190
speed **A8:** 197, 445
theory, as applied to electromagnetic radiation **A10:** 83
train *See* Acoustic wave train
Wave diagram
for stored-torque Kolsky bar **A8:** 220
Wave fluxers
through-hole soldering **EL1:** 682
Wave fluxing **EL1:** 648
Wave form
defined **A12:** 59
effect on fatigue **A12:** 58-63
and frequency **A12:** 58-63
strain, in fatigue testing **A12:** 62
Wave machines
specialized **EL1:** 632
Wave path
ultrasonic inspection **A17:** 597
Wave propagation
acoustic emission **A17:** 279-280
analysis, for measuring strain **A8:** 208
computer codes **A8:** 283
during split Hopkinson pressure bar test **A8:** 199
effects, and tensile test validity **A8:** 209
effects, impact bar tests with, strain rate ranges **A8:** 40
effects on high strain rate tension tests **A8:** 208-209
in elastic pressure bars **A8:** 199-200
for high strain rate generation **A8:** 190
in high strain rate testing **A8:** 188
importance in strain rate regimes in compression testing **A8:** 190-191
ring-up time **A8:** 209
straight-line, of x-rays/gamma-rays **A17:** 311
Wave propagation (continued)
stress, effect on strain rate **A8:** 40-41
ultrasonic **A17:** 233
Wave solder *See also* Solder waves; Wave soldering
for surface-mount joints **EL1:** 732
Wave soldering *See also* Dual-wave soldering **A6:** 366-368 **M6:** 19, 1087-1089
atmospheres **A6:** 367
defects **A6:** 366, 367, 368
defined **EL1:** 1161
definition **A6:** 366, 1215
and design **EL1:** 520
design considerations **A6:** 366-367
dross formation **A6:** 367
dual-in-line package (DIP) **A6:** 366
dwell times **A6:** 367, 368
equipment **A6:** 367
equipment, methods **EL1:** 681-682
equipment modifications to produce different wave geometries **A6:** 367
flexible printed boards **EL1:** 590
flux application methods **A6:** 366, 367
fluxes **A6:** 366, 367
fluxing operation **A6:** 367
as mass soldering system **EL1:** 681-696
problem analysis **EL1:** 1029-1030
process parameters **A6:** 366-367, 368
process schedule **A6:** 367
schematic of process **A6:** 366
sequence of processes **A6:** 366
small-out-line integrated circuit (SOIC) (surface-mount) **A6:** 366-367
as surface-mount soldering **EL1:** 701-702
Wave velocity, as measurement
microwave inspection **A17:** 205
Wave(s) *See also* specific wave types; Standing wave; Ultrasonic waves; Wave propagation; Wave velocity; Waveform; Waveguides Wavelengths
acoustic emission **A17:** 279-280
analysis package, acoustic emission inspection **A17:** 286
for optical holographic interferometry **A17:** 408-410
Wave-length-dispersive spectrometry
used to analyze x-ray line spectrum **A9:** 92
Waveform, electrical
effect in radiography **A17:** 304-305
Waveform, pressure
electromagnetic forming **A14:** 652
Waveform strain
and hold period effect on stainless steel **A8:** 348
Waveforms
distortion, in signal transmission **EL1:** 170-173
plating **EL1:** 511
RLC line **EL1:** 358
Waveguides *See* Electrical waveguides
microwave inspection **A17:** 202
pattern, transverse mode **A17:** 207
Wavelength
absorbance (UV/VIS) as function of **A10:** 63
calibration, for MFS analysis **A10:** 77
characteristic x-ray, and atomic number Mosely's relationship between **A10:** 433
Compton, defined **A10:** 84
conversion factors **A8:** 723 **A10:** 686
defined **EL1:** 1161
defined and symbol for **A10:** 683, 692
in electromagnetic radiation **A10:** 83
infrared spectral **A10:** 110
selectors, MFS **A10:** 76
sorters **A10:** 23
and ultrasonic fatigue testing specimens **A8:** 249
variable, geometry of **A10:** 396
x-ray absorption curve for uranium as function of **A10:** 85

Wavelength (x-rays)
defined **A9:** 19
Wavelength dispersion XRFS (WXRFS)
for chemical analysis **EM4:** 550-551
Wavelength dispersive analysis of x-rays (WDS)
for microstructural analysis **EM4:** 570
Wavelength dispersive detectors
electron microscopy of wrought stainless steels **A9:** 282
Wavelength dispersive spectroscopy (WDS) **EM3:** 237
depth profiling **EM3:** 247
Wavelength dispersive x-ray spectroscopy (WDS) **EL1:** 1043, 1094, 1098-1100
Wavelength x-ray analysis
compositional analysis of welds **A6:** 100
Wavelength-dispersive spectrometers
artifacts **A10:** 521
direct defocusing map **A10:** 527
for x-ray spectrometry **A10:** 83, 87, 89-93
Wavelength-dispersive spectrometry **A10:** 520-524
analysis of cartridge brass **A10:** 530
defined **A10:** 683-684
detection limits **A10:** 522
dot map for minor constituent **A10:** 527
dot mapping **A10:** 525-529
effect of bremsstrahlung sensitivity **A10:** 528
and energy-dispersive spectrometry compared **A10:** 521-522
instrument selection **A10:** 522
light-element analysis **A10:** 522
qualitative **A10:** 522-525
spectral resolution **A10:** 521-522
stainless steel **A10:** 87-88
Wavelength-dispersive x-ray spectrometer
development and schematic of **A11:** 32-33
Wavelength-dispersive x-ray spectroscopy **A12:** 168
Wavelengths
electromagnetic spectrum **A17:** 202
as interference, microwave inspection **A17:** 204
ultrasonic **A17:** 232-233
Wavelengths of light
role in color metallography **A9:** 136, 143, 151
Wavenumber
defined **A10:** 684
and depth of wave penetration, infrared spectroscopy **A10:** 113
infrared spectra, as frequency **A10:** 110
SI derived unit and symbol for **A10:** 685
SI unit/symbol for **A8:** 721
Wavy slip
in a single crystal of aluminum **A9:** 689
defined **A9:** 687
Wax
additives, investment casting **A15:** 254
components, in pattern assembly investment casting **A15:** 257
as expendable pattern material **A15:** 196
fillers, investment casting **A15:** 254
formulating materials **A15:** 197
as investment casting pattern materials **A15:** 253-255
pattern, defined **A15:** 11
phosphate coatings supplemented with **M5:** 435
as physical modeling material **A14:** 432
selection, investment casting **A15:** 254-255
selective cadmium plating using **M5:** 267, 268
stop-off medium, chrome plating **M5:** 187
stripping of **M5:** 19
for tool steel lubrication **A18:** 738
Wax binders
in rigid tool compaction **A7:** 322
Wax impregnation of powder metallurgy material specimens **A9:** 504
Wax pattern
definition **M6:** 19
Wax pattern (thermit welding)
definition **A6:** 1215
Wax technique
for TEM replicas **A12:** 185
WAX-20 (DS)
composition **A4:** 795
Wax-stearate binders
in rigid tool compaction **A7:** 322
Waxes *See also* Lubricant(s); Lubrication **A7:** 13, 190-193
Waywind
glass roving winder as **EM1:** 109
WC-1Zr *See* Niobium alloys, specific types, Nb-1Zr
WC-103 *See* Niobium alloys, specific types, C-103
WC-129Y *See* Niobium alloys, specific types, C-129Y
Weak beam imaging **A18:** 388
"Weak knee" **A6:** 994
Weak link superconducting
high-temperature superconductors **A2:** 1088
Weak-beam images **A9:** 111
compared to strong-beam images **A9:** 114
Weak-beam microscopy
bright-field image of dislocation tangle aluminum alloy **A10:** 467
for defect analysis **A10:** 466-467
for high-resolution diffraction contrast images **A10:** 446
showing dislocations in molybdenum-implanted aluminum **A10:** 484
Weak-beam technique
used for dislocations **A9:** 113
Weakest-link failure **EM1:** 193-194
Wear *See also* Abrasive wear; Adhesive wear; Fatigue; Lubricated wear; Oxidation; Wear failures; Wear resistance; Wear testing; Wear-resistant materials **A8:** 15, 601-608
abrasive **A11:** 1, 146-148, 158-159, 494-495, 595 **M1:** 599-605
adhesive **A11:** 1, 145-146, 158, 466, 596
AES analysis of **A10:** 566
analysis, procedures **A11:** 156
of antifriction bearing **A11:** 764-765
applications, cemented carbides for **A7:** 777-780
of cast iron pump parts **A11:** 365-367
catastrophic, defined **A11:** 1
checking, of solid graphite molds **A15:** 285
classification **M1:** 597, 599
of cobalt-base alloys **A2:** 447-448 **A13:** 663-664
coefficients **A11:** 146
combined, mechanisms of **A11:** 159
of continuous aluminum oxide fiber MMCs **EM1:** 875-876
of copper casting alloys **A2:** 352, 354-355
corrosive, defined **A11:** 2
data, cobalt-base alloys **A2:** 449-451
data for cobalt- and nickel-based hardfacing alloys **A7:** 828
debris, SEM analysis **A10:** 490
defined **A11:** 11 **A18:** 20
delamination theory of **A11:** 148
diffusion, coated carbide tools **A2:** 960
dry **M1:** 605-606
effect of material properties on **A11:** 158-159
effects, produced fluids **A13:** 479
electrical, on thrust bearing **A11:** 487
elements, determined in petroleum products by XRS **A10:** 82
embeddability of soft metals **M1:** 609-610
and erosion, electrical contact materials, life tests **A2:** 860
as failure mode in gears **A11:** 590
fatigue, defined **A11:** 4
in forging **A11:** 340
fractures, identification chart for **A11:** 80
from retained austenite in iron casting **A11:** 367-368
in gear teeth **A11:** 595
in gray iron **A1:** 24, 25-26
hardfacing to reduce **A7:** 823
of implants **A11:** 672
introduction **A18:** 175
method(s) for removing material from a solid surface **A18:** 175
iron castings failure from **A11:** 375
laboratory tests for **A11:** 161-162
of labyrinth seals, bearing failure by **A11:** 511-512
lubricated .... **A11:** 150-154, 361 **M1:** – 604-606
maximum admissible level **A18:** 493
metals, OES analysis in oils **A10:** 21
of nickel aluminide alloys **A18:** 772-777
nonlubricated **A11:** 154-155
on titanium bone screw head **A11:** 689, 692
oxidative, defined **A11:** 7
in polymers, due to friction **A11:** 764
Wear (continued)
of press forming dies **A14:** 505-507
properties, of carbon fiber reinforced polymers **EM1:** 36
rapid, from severe abrasion in shell liner **A11:** 375-377
rate, defined **A11:** 11
rate, Keller's equation for **A13:** 968-969
rate transitions **M1:** 598
rates, effect on dimensional accuracy **A15:** 284
of refiner plates, pulping systems **A13:** 1215-1217
resistance, of die materials **A14:** 45-46
rolling-element bearings failure by **A11:** 493-496
in shafts **A11:** 465
simulative testing **A18:** 175
in sliding bearings **A11:** 486
specific types **M6:** 805-806
of steel pinions **A11:** 597
surface, AES analysis for **A10:** 549
surface chemical analysis for **A7:** 250
surface property effect **A18:** 342
of telephone plugs, as life test example **A2:** 858
testing **M1:** 598, 599-604
tests, laboratory **A11:** 161-162
tip and notch, in turbine shroud **A11:** 283
of tooling, precision forging **A14:** 159
types of **A11:** 145-148
volume loss measurement **A8:** 606
weight-loss measurement **A8:** 606
weld overlays to overcome **M6:** 805-806
wire wooling **M1:** 606
Wear applications
copper alloy castings **M2:** 392-393
Wear coefficient **A18:** 478, 491
defined **A18:** 20
sliding and adhesive wear **A18:** 237
Wear constant
defined **A18:** 20- 21
Wear corrosion *See* Fretting
Wear damage **A18:** 274
Wear debris
defined **A18:** 21
dispersion in milling **A7:** 63
Wear depth **A18:** 267
Wear factor *See also* Specific wear rate
defined **A18:** 21
synovial joints, major load-bearing natural **A18:** 656
thermoplastic composites **A18:** 821, 822, 823-824
Wear failures *See also* Wear **A11:** 145-162
analysis **A11:** 156
by erosion **A11:** 155-156
by fretting **A11:** 148
by friction **A11:** 148-150
by hydrogen-assisted cracking **A11:** 399
combined mechanisms of **A11:** 159
delamination theory **A11:** 148
effect of material properties **A11:** 158-159
environmental effects **A11:** 159-160
laboratory tests for **A11:** 161-162
lubricated wear **A11:** 150-154
microstructure effects **A11:** 160-161
nonlubricated wear **A11:** 154-155
of parts, laboratory examination **A11:** 156-158
service history and **A11:** 158
surface configuration and **A11:** 159-160
wear, types of **A11:** 145-148
Wear fractures
identification chart for **A11:** 80
Wear index **A18:** 637
Wear lands
replacing **A7:** 462
Wear measurement **A18:** 362-369
applications **A18:** 363
area measures **A18:** 364-366
geometric wear volume calculation **A18:** 369
linear measures **A18:** 363-364
and lubrication **A18:** 365, 367, 369
mass loss measures of wear **A18:** 362-363
abrasive wear **A18:** 362
material transfer **A18:** 369
subsurface damage effect **A18:** 368-369
surface deformation effect **A18:** 367-369
volume measures **A18:** 366-367, 368
Wear of ceramics **EM4:** 605-608, 973-977
applications **EM4:** 605, 973-977
example **EM4:** 974-977

Wear of ceramics (continued)
parameters affecting wear ... **EM4:** 973-974
conventional laboratory tests ... **EM4:** 605
critical parameters for wear testing ... **EM4:** 606
dynamic behavior of the test apparatus ... **EM4:** 607
guidelines ... **EM4:** 608
Hertzian contact stresses ... **EM4:** 606
laboratory test/field test correlation ... **EM4:** 606
material/specimen history ... **EM4:** 607
postmeasurement handling ... **EM4:** 608
presentation of wear data ... **EM4:** 608
rolling contact fatigue test ... **EM4:** 605-606
simulated field tests ... **EM4:** 605
sliding motion ... **EM4:** 607
sliding wear test geometry ... **EM4:** 606-607
specimen fabrication costs ... **EM4:** 607
standard reference samples ... **EM4:** 607-608
step loading versus constant applied load testing ... **EM4:** 607
surface cleanliness ... **EM4:** 607
tests ... **EM4:** 605-606
wear maps ... **EM4:** 607
Wear of jet engine components *See* Jet engine components, wear of
Wear of pumps *See* Pumps, wear of
Wear or failure models
friction during metal forming ... **A18:** 60, 67
Wear particle concentration (WPC) ... **A18:** 302, 308, 309
Wear parts
refractory metals applications ... **A7:** 765
Wear plates, fifth-wheel mechanism
economy in manufacture ... **M3:** 854
Wear rate
defined ... **A11:** 11 **A18:** 21
Wear rate (of seals)
defined ... **A18:** 21
Wear rate per unit distance ... **A18:** 579
Wear resistance *See also* Wear ... **A18:** 185, 708 **M1:** 597-638
abrasive cloth wear tests ... **M1:** 603
aluminum-silicon alloys ... **A15:** 167
of anodized coatings ... **A13:** 397
ASP steels ... **A7:** 784
ASTM test methods ... **EM2:** 334
of austenitic manganese steel ... **A1:** 834
austenitic manganese steels ... **M1:** 618, 620
by ion implantation ... **A13:** 500
of cast steels ... **A1:** 376 **M1:** 400, 618
ceramic coatings ... **M5:** 547
chemical conversion coatings ... **M5:** 657-658
chemical vapor deposition coatings ... **M5:** 382-383
of cobalt-base alloys ... **A15:** 811
composite powders for ... **A7:** 175
correlation with hardness ... **M1:** 603
correlation with toughness ... **M1:** 607
defined ... **A18:** 21
electroplated chromium for ... **A13:** 875
flank, of cermets ... **A2:** 979
of gray iron ... **A1:** 25-26 **M1:** 24-26, 30
hard chromium plating ... **M5:** 170, 184
ion implantation coatings ... **M5:** 425
ion implantation to improve ... **M1:** 629
laser heat treatment to improve ... **M1:** 628-629
of magnesium alloys ... **A2:** 461
malleable iron ... **M1:** 71
microstructure, effect of ... **M1:** 611-615, 618, 620-621
of nickel alloys ... **A2:** 429
nickel plating ... **M3:** 181
nickel plating, electroless ... **M5:** 225-228, 237-240
of pearlitic and martensitic malleable iron ... **A1:** 83-84
pearlitic/martensitic malleable iron ... **A15:** 696
phosphate coatings ... **M5:** 436-437
of polyamide-imides (PAI) ... **EM2:** 137
rubber wheel wear tests ... **M1:** 600-601
seizure-resistance tests ... **M1:** 611-612

Wear resistance (continued)
selection of wear-resistant metals ... **M1:** 603-604, 606-611
steels, against hard and soft abrasives ... **M1:** 601-603
structural ceramics ... **A2:** 1019, 1021-1024
surface heat treatments to develop ... **M1:** 527, 533, 540, 625-631
thermal spray coating ... **M5:** 377
tool steels ... **M1:** 618
type of precipitate, effect of ... **M1:** 613
vacuum coatings ... **M5:** 395
versus hardness ... **A18:** 707-708
white cast irons ... **M1:** 619-622
zinc alloys ... **A15:** 786
Wear scar
defined ... **A18:** 21
Wear severity factor (*C*) ... **A18:** 268
Wear simulation testing ... **A8:** 604-606
Wear surfaces
scanning electron microscopy used to study ... **A9:** 99
Wear testing ... **A8:** 601-608
acceleration ... **A8:** 604, 606
ball-plane, stress and strokes combined ... **A8:** 603
case histories ... **A8:** 607
configurations ... **A8:** 601-602
control ... **A8:** 603-604, 606
dry sand low-stress rubber wheel abrasive ... **A8:** 605
elements ... **A8:** 604-607
equipment ... **A8:** 601
and friction ... **A8:** 604
lubrication ... **A8:** 604
measurement ... **A8:** 604, 606
metal band/metal platen, for high-speed printer ... **A8:** 607
reporting ... **A8:** 604, 606-607
scatter reduction in ... **A8:** 606
simulation ... **A8:** 604-606
sliding ... **A8:** 601-602
sliding contact ... **A8:** 605-606
specimen preparation ... **A8:** 604, 606
and tribology ... **A8:** 601
wear curve ... **A8:** 606
wear-in cycle ... **A8:** 604
Wear tracks ... **A18:** 476
Wear transition
defined ... **A18:** 21
Wear volume ... **A18:** 185
Wear-in ... **A8:** 604
Wear-resistant alloys, nonferrous
aluminum bronzes ... **M3:** 591-592
beryllium copper ... **M3:** 592-594
cobalt-base ... **M3:** 589-591
Wear-resistant manganese steels
weldability ... **A15:** 535
Wear-resistant materials, and steel-bonded titanium carbide cermets
compared ... **A2:** 997
Wear-resistant steel
submerged arc welding ... **A6:** 203
Wear-resistant tool steels
die materials for sheet metal forming ... **A18:** 628
Wearite 4-13 *See* Copper alloys, specific types, C62500
Wearout *See also* Failure rate
coating/passivation ... **EL1:** 245
failure rate, and attachment rigidity ... **EL1:** 740
mechanisms, leadless packaging ... **EL1:** 987-988
modes, temperature cycling ... **EL1:** 287
phase, life cycle ... **EL1:** 897
as reliability curve phase ... **EL1:** 740
Weather aging *See also* Aging; Weather resistance
degradation factors ... **EM2:** 575-576
and radiation susceptibility ... **EM2:** 575-580
test methods ... **EM2:** 576-580
weatherometers ... **EM2:** 577-578

Weather resistance
effect, electrical testing ... **EM2:** 584-585
polyarylates (PAR) ... **EM2:** 139-140
polyether sulfones (PES, PESV) ... **EM2:** 161
polyvinyl chlorides (PVC) ... **EM2:** 210
styrene-acrylonitriles (SAN, OSA ASA) ... **EM2:** 215
thermoplastic polyurethanes (TPUR) ... **EM2:** 206
thermoplastic resins ... **EM2:** 619
Weatherability *See* Aging; Weather aging; Weather resistance-e
of acrylics ... **EM2:** 105
of composites ... **EM1:** 36
Weathering ... **EM3:** 31
artificial ... **EM3:** 31
of building materials, simulated service testing ... **A13:** 204
defined ... **EM1:** 25
resistance, of porcelain enamels ... **A13:** 446, 451
steels ... **A13:** 515-521
Weathering and aging effect on adhesive joints ... **EM3:** 656-662
comparison of short- and long-term testing ... **EM3:** 661-662
environment, defined ... **EM3:** 656-657
long-term testing ... **EM3:** 658-661
short-term testing ... **EM3:** 657-658
Weathering, artificial *See* Artificial weathering
Weathering, defined *See also* Artificial weathering ... **EM2:** 45
Weathering of Polymers (Davis/Sims) ... **EM2:** 94
Weathering steel ... **A6:** 376
flux-cored arc welding, designator ... **A6:** 189
Weathering steels *See also* Low-alloy steets ... **A1:** 148, 399, 400 **A13:** 515-521
for art casting ... **A15:** 22
buried, corrosion protection ... **A13:** 519
copper-bearing steel ... **A13:** 516
corrosion behavior, different exposures ... **A13:** 516-517
corrosion resistance of ... **A1:** 400
design considerations ... **A13:** 518-520
examples ... **A13:** 518-520
and fire-retardant wood panels ... **A13:** 519
galvanic corrosion ... **A13:** 518-519
high-strength low-alloy steels ... **A13:** 516
low-alloy steels as ... **A13:** 82
packout rust formation, bolting and sealing ... **A13:** 519
painted ... **A13:** 519-520
protective oxide film, characteristics ... **A13:** 517-518
specifications of ... **A1:** 399
stadium, corrosion in ... **A13:** 1308-1309
storage and stacking of ... **A13:** 518
for structural corrosion protection ... **A13:** 1305-1306
tower legs and lighting standards, protection of ... **A13:** 520
Weatherometers
types ... **EM2:** 577-578
Weatherproofing
of pharmaceutical insulation systems ... **A13:** 1231
Weave *See also* specific weave patterns; Weaves
defined ... **EM1:** 25 **EM2:** 45
Weave bead
definition ... **A6:** 1215 **M6:** 19
Weave exposure
defined ... **EL1:** 1161
Weave pattern ... **EM1:** 148-150
Weaves *See also* Angle interlock; Basket weave; Fabric(s); Orthogonal weave; Plain weave; Polar weave; Rovings; Satin (crowfoot) weave; Twill weave
fiberglass fabric ... **EM1:** 110-111
geometry, for multidirectionally reinforced fabrics/prefon-ns ... **EM1:** 129-130
locking leno ... **EM1:** 125-127
low-temperature thermoset matrix composites ... **EM1:** 393
physical properties measured ... **EM1:** 286
radial and circumferential ... **EM1:** 131

SUBJECTS OF THE INDEXED VOLUMES: ASM Handbook (designated by the letter "A"): **A1:** Properties and Selection: Irons, Steels, and High-Performance Alloys (1990); **A2:** Properties and Selection: Nonferrous Alloys and Special-Purpose Materials (1990); **A3:** Alloy Phase Diagrams (1992); **A4:** Heat Treating (1991); **A6:** Welding, Brazing, and Soldering (1993); **A7:** Powder Metallurgy (1984); **A8:** Mechanical Testing (1985); **A9:** Metallography and Microstructures (1985); **A10:** Materials Characterization (1986); **A11:** Failure Analysis and Prevention (1986); **A12:** Fractography (1987); **A13:** Corrosion (1987); **A14:** Forming and Forging (1988); **A15:** Casting (1988); **A16:** Machining (1989); **A17:** Nondestructive Testing and Quality Control (1989); **A18:** Friction, Lubrication, and Wear Technology (1992). **Metals Handbook, 9th Edition** (designated by the letter "M"): **M1:** Properties and Selection: Irons and Steels (1978); **M2:** Properties and Selection: Nonferrous Alloys and Pure Metals (1979); **M3:** Properties and Selection: Stainless Steels, Tool Materials and Special-Purpose Materials (1980); **M4:** Heat Treating (1981); **M5:** Surface Cleaning, Finishing, and Coating (1982); **M6:** Welding, Brazing, and Soldering (1983). **Engineered Materials Handbook** (designated by the letters "EM"): **EM1:** Composites (1987); **EM2:** Engineering Plastics (1988); **EM3:** Adhesives and Sealants (1990); **EM4:** Ceramics and Glasses (1991); **Electronic Materials Handbook** (designated by the letters "EL"): **EL1:** Packaging (1989).

Weaves (continued)
styles of, as fabric parameter **EM1:** 125
three-dimensional **EM1:** 127
typical **EM1:** 355, 356
unidirectional/two-directional fabrics **EM1:** 125
for woven fabric prepregs **EM1:** 148
for woven roving **EM1:** 109
Weaving machines **EM1:** 129-131
Weaving wire **A1:** 851
Web
defined **A14:** 14 **EM2:** 45
Web materials
waterjet machining **A16:** 525, 526
Web plate, bridge
fracture surface **A11:** 710-711
Web thickness
steel forgings **M1:** 363-364
Webbing **EM3:** 31
Webbing, solder *See* Icicles
Weber total hip prosthesis **A11:** 670
Wedge bond
defined **EL1:** 1161
Wedge bonding *See* Thermocompression welding
Wedge cracking **A8:** 154, 572
Wedge cracks *See also* Triple-point cracking
austenitic stainless steels **A12:** 364
formation, in decohesive rupture **A12:** 19, 25-26
as intergranular **A12:** 121-123
Wedge deformation, three-dimensional
simulation of **A14:** 429
Wedge drive mechanism
in mechanical presses **A14:** 31
Wedge effect *See also* Wedge formation
defined **A18:** 21
Wedge formation *See also* Wedge effect
defined **A18:** 21
gold electroplating **A18:** 838
Wedge half-angle **A18:** 34
Wedge presses
load **A14:** 40
Wedge tensile test of bolts **A1:** 296
Wedge test
threaded fasteners **M1:** 277, 279
Wedge testing **EM3:** 386, 389, 657, 662, 804, 808
for chemical processing quality **EM3:** 738
durability assessment **EM3:** 666-668, 669
to evaluate service durability of surface preparation/adhesive combination **EM3:** 802, 803
Wedge tests
application **A12:** 141, 161
Wedge wire sectors
screening **A7:** 176
Wedge-forging test **A1:** 583, 584 **A8:** 587
for forgeability **A14:** 215, 382-384
Wedge-loaded compact specimen
for crack arrest testing **A8:** 453
R-curve measurement method with **A8:** 452
Wedge-opening load
abbreviated **A8:** 726
Wedge-opening load test ... **A8:** 538-539 **A13:** 284-285
and cantilever beam test compared **A8:** 538, 539
and contoured double-cantilever beam test compared **A8:** 538
Wedge-opening loading (WOL) **A11:** 64, 798
Wedge-type grips **A8:** 51, 371
Wedge-wedge bonding *See* Ultrasonic bonding
as component attachment **EL1:** 350
cycle **EL1:** 227
Wedgecut
for plating thickness inspection **EL1:** 943
Wedgelock fasteners **EM1:** 711
Wedging, mechanical
for sputtering problems **A10:** 556
Weepage
defined **A18:** 21
Weeping **EM3:** 31
defined **EM1:** 26 **EM2:** 45
Weft *See also* Fill; Filling; Filling yarn; Yarns
defined **EM1:** 26 **EM2:** 45
in unidirectional/two-directional fabrics **EM1:** 125
Weibull distribution **A8:** 628-629, 632-634, 714-717
estimated failure factors **EL1:** 902-903
failure density **EL1:** 897
scaling, failure plot **EL1:** 889
Weibull failure probability function **EM4:** 513
Weibull function
strength of glass relationship **EM4:** 567
Weibull normal stress averaging method **EM4:** 700, 701, 703, 704, 705, 706, 707
Weibull plots **A14:** 206
Weibull probability distribution (density function) **A16:** 45, 46, 47
Weibull statistical distribution
in fiber properties analysis **EM1:** 733-734
for fiber strength **EM1:** 193, 194
for mechanical properties testing **EM1:** 302-304
and parameters estimation **EM1:** 441
Weibull weakest link theory (WLT)
prediction of fast-fracture reliability of ceramics **EM4:** 700, 703
Weighing errors
and density errors **A7:** 483
Weighing, feature *See* Feature weighing
Weight *See also* Areal weight; Average molecular weight; Molecular weight
aluminum **A2:** 3
areal, fabric **EM1:** 125
casting, mold life effect **A15:** 281
control, in coining **A14:** 186
distribution, polymer **EM2:** 58, 62
ladle pouring by **A15:** 501
of magnesium castings **A15:** 808
of match plate patterns **A15:** 245
molecular, and melt properties **EM2:** 62
molecular, and polymers **EM2:** 58
of open-die forgings **A14:** 61
per length, test **EM1:** 286
of permanent mold castings **A15:** 275
of plaster mold castings **A15:** 242
precision, electric arc furnaces **A15:** 363
of precision forgings **A14:** 158
reduction, in magnesium and magnesium alloy parts **A2:** 476-479
reduction, thermoplastic structural foam parts **EM2:** 509
of selected structural metals **A2:** 478
as selection parameter **EM1:** 38-39
testing, for casting defects **A15:** 545
workpiece, in hot upset forging **A14:** 83
Weight, equivalent
defined **A10:** 162
Weight fraction of crystalline phases
XRPD determined **A10:** 333
Weight loss *See also* Coatings; Weight
by intergranular corrosion, stainless steel **A13:** 123
of cemented carbides in acids **A13:** 856
corrosion, in sour gas environments **A11:** 300
as corrosion test **A13:** 195
data, as nonlinear **A13:** 510
from molten lithium corrosion **A13:** 52-54
of galvanized coatings **A13:** 441
and pitting depth, aluminum alloy plate **A13:** 603
platinum-group metals, in air **A13:** 804
steel, from atmospheric corrosion **A11:** 193
Weight loss tests
pickling process **M5:** 70, 73-77
Weight, of coatings *See also* Coatings; Weight
loss **A13:** 385, 391, 397
Weight pan knife edge
constant-stress testing **A8:** 321-322
Weight percent
abbreviated **A8:** 726
Weight reduction
in aluminum P/M parts **A7:** 741
magnesium **M2:** 551-552
Weight testing
of castings **A17:** 521
Weight-and-lever system
for verification of Brinell test **A8:** 88
Weight-loss measurement
limitations **A8:** 606
Weighting of fiber composites to aid mounting **A9:** 588
Weights
of substrates **EL1:** 105
Weiner's expression
porosity and dielectric properties of ceramic-matrix composites **EM4:** 859-860
Weissenberg cameras
for diffraction patterns **A10:** 346
Weissenberg pattern
single-crystal diffraction **A10:** 330
Weld *See* Weld(s)
defined **A9:** 577
definition **A6:** 1215 **M6:** 19
Weld axis
definition **A6:** 1215
Weld backing rings
associated corrosion **A13:** 350-351
Weld bead
definition **A6:** 1215 **M6:** 19
stringer bead **M6:** 313
weave bead **M6:** 313
Weld bead morphology **A9:** 578
Weld bead terminology **A9:** 578
Weld beads
short weld sequence **A15:** 526
Weld beads, out-of-line
in steel pipe **A17:** 565
Weld bonding **A6:** 326
definition **A6:** 1215 **M6:** 19
Weld brazing
definition **A6:** 1215 **M6:** 19
Weld button spot testing **A6:** 453
Weld cladding **A6:** 789, 810, 811, 816-822
application considerations **A6:** 817
applications **A6:** 814, 816, 818, 822
bulkwelding process using metal powder joint fill **A6:** 816
carbon percentage and dilution factors **A6:** 810
composition control of stainless steel weld overlays **A6:** 817-820
definition **A6:** 816
dilution calculation **A6:** 812
dilution control **A6:** 819-820
dilution, penetration, and bead thickness **A6:** 815
electrodes **A6:** 813, 816, 819, 821
electroslag welding **A6:** 816, 819, 822
filler metals **A6:** 809, 816, 817, 818, 819, 820, 821
flux-cored arc welding **A6:** 816
overlays other than stainless steels **A6:** 822
plasma arc hot wire cladding process **A6:** 818
plasma arc welding **A6:** 816, 818, 819
series-arc deposits, chemical composition variations **A6:** 816
shielded metal arc welding **A6:** 819
stainless steel filler metals **A6:** 809
stainless steel, welding parameters **A6:** 817
stainless steels **A6:** 814
stainless steels by submerged arc welding **A6:** 813
stainless steels, procedures **A6:** 816, 820-822
alloying elements **A6:** 820
bulk welding **A6:** 820
electroslag overlays **A6:** 822
fluxes **A6:** 820
plasma arc hot wire process **A6:** 818, 819, 821
self-shielded flux-cored wire **A6:** 821
shingling **A6:** 820
submerged arc welding **A6:** 820-821
submerged arc welding **A6:** 816, 817, 820-821, 822
three-wire welding head for applications **A6:** 817
Weld contours *See also* Weld(s)
of arc welds **A11:** 413
of electrogas welds **A11:** 440
of electron beam welds **A11:** 446
of electroslag welds **A11:** 440
of flash welds **A11:** 442-443
Weld crack
definition **A6:** 1215 **M6:** 19
Weld cracking *See also* Cold cracking; Hot cracking; Lamellar tearing; Stress-relief cracking; Weld(s) **A1:** 606-608
cold cracking **A12:** 137-138
defined **A13:** 14
examination/interpretation **A12:** 72, 137-140
hot **A1:** 608
hot cracking **A12:** 138
hydrogen-induced **A1:** 606-607
inclusion **A1:** 608
lamellar **A1:** 607-608
lamellar tearing **A12:** 138-139
stress-relief **A1:** 607
stress-relief cracking **A12:** 139-140

Weld decay *See also* Sensitization **A6:** 1066 **A13:** 14, 125, 342, 349-351, 929
Weld defects *See also* Defects; Weld discontinuities and specific types; Welds
in bridge components **A11:** 708
brittle fracture from **A11:** 93-94
chain link failure from **A11:** 521-522
fatigue failure from **A11:** 127-128
from HAZ hot-cracking **A11:** 401
from stress concentrations **A11:** 400
in heat-exchanger tubing **A11:** 639-640
with incomplete fusion **A11:** 603
Weld defects in
arc welds of magnesium alloys **M6:** 435
arc welds of nickel alloys **M6:** 442-443
electroslag welds **M6:** 233
resistance welds of aluminum alloys **M6:** 543-544
resistance welds of stainless steels **M6:** 533-534
shielded metal arc welds **M6:** 92-94
solid-state welds **M6:** 681-682
Weld deposit quality
as electrode coating function **A7:** 816
Weld deposits
to improve abrasive wear resistance **A18:** 639
Weld discontinuities **M6:** 829-855
classification and definitions **M6:** 829-830
crack or crack-like **M6:** 830-835
causes and prevention **M6:** 834-835
fisheyes **M6:** 831-832, 834-835
graphitization **M6:** 834
hot cracking **M6:** 832-833
hydrogen-induced cold cracking **M6:** 830-832
lamellar tearing **M6:** 832
microfissures **M6:** 832
stress corrosion cracking **M6:** 833
definition **M6:** 829
effect of residual stresses **M6:** 840-841
brittle fracture **M6:** 840-841
fatigue failure **M6:** 841
fatigue and fracture control **M6:** 852
corrective actions **M6:** 852
examples of application **M6:** 852-853
geometric discontinuities **M6:** 835-837
melt-through **M6:** 837
overlapping **M6:** 837
undercut **M6:** 836-837
underfill **M6:** 837
lack of fusion and penetration **M6:** 837, 841-842
modes of failure **M6:** 842-843
brittle fracture **M6:** 842-843
comparison of modes **M6:** 843
fatigue failure **M6:** 843
nondestructive examination **M6:** 846-851
accuracy **M6:** 851
inspection methods **M6:** 847-850
reliability **M6:** 851
selection of technique **M6:** 850-851
sensitivity **M6:** 851
occurrence **M6:** 830
oxide inclusions **M6:** 838-839
porosity **M6:** 839-840, 845-846
repair **M6:** 840
significance **M6:** 841-846
slag inclusions **M6:** 837-838
tungsten inclusions **M6:** 839
types **M6:** 843-846
cracks **M6:** 843-844
lack of fusion and penetration **M6:** 844
porosity **M6:** 846, 850-851
slag inclusions **M6:** 845-846
weld undercut **M6:** 844
weld repair **M6:** 840
Weld discontinuities, overview of **A6:** 1073-1080
associated with specialized welding processes **A6:** 1076
classification of weld discontinuities **A6:** 1073-1076
cracks **A6:** 1075-1076
gas porosity **A6:** 1073-1074
Weld discontinuities, overview of (continued)
geometric weld discontinuities **A6:** 1075
lack of fusion and lack of penetration **A6:** 1075
metallurgical discontinuities **A6:** 1073
process-related discontinuities **A6:** 1073
slag inclusions **A6:** 1074
tungsten inclusions **A6:** 1074-1075
Weld face
definition **A6:** 1215
Weld failure origins *See also* Weld(s)
in arc welds **A11:** 412-415
in electrogas welds **A11:** 440
electron beam welds **A11:** 444-447
in electroslag welds **A11:** 439-440
in flash welds **A11:** 442-443
friction welds **A11:** 444
in high-frequency induction welds **A11:** 449
in laser beam welds **A11:** 447-449
in upset butt welds **A11:** 443-444
Weld filler metals
corrosivity **A13:** 344
effect, critical pitting temperature **A13:** 348
electrochemical properties **A13:** 49
high-alloy **A13:** 361
Weld gage
definition **M6:** 19
Weld interface
definition **A6:** 1215 **M6:** 20
Weld interval
definition **M6:** 20
Weld interval timer
definition **M6:** 20
Weld joint
definition **M6:** 60
design **M6:** 61
Weld leakage *See also* Welding **A7:** 431
Weld line *See also* Flow line
defined **EM1:** 26 **EM2:** 45
definition **A6:** 1215
integrity, thermoplastic injection molding **EM2:** 311
and meld lines **EM2:** 616
strength, liquid crystal polymers (LCP) **EM2:** 181
Weld mark *See* Flow line; Weld line
Weld metal
definition **A6:** 1215 **M6:** 20
microstructure **A10:** 478-480
stainless steel, measurement of 8-ferrite in **A10:** 287
Weld metal area
definition **M6:** 20
Weld metal microstructural analysis
by AEM **A10:** 478-481
elemental compositions examined **A10:** 479
experimental method **A10:** 479-480
Weld metal microstructure, effect of
on weldability **A1:** 604-605, 607
Weld microstructures
effect of temperature on **M1:** 561
Weld nugget *See* Fusion zone
Weld overlay **A13:** 652, 931
Weld overlays **M6:** 804-819
applicability **M6:** 805
application considerations **M6:** 808-809
composition control of
dilution **M6:** 807-808
parameters affecting dilution **M6:** 810-811
stainless steel **M6:** 809-811
stainless steel weld overlays **M6:** 811-814
use of Schaeffler diagram **M6:** 810
economics **M6:** 818-819
inspection **M6:** 817-818
in-service performance **M6:** 818
inspection during fabrication **M6:** 818
qualification tests **M6:** 817-818
types of discontinuities **M6:** 817
non-stainless steel materials **M6:** 816-817
Weld overlays (continued)
procedures for
electroslag overlays **M6:** 815-816
plasma arc processes **M6:** 814-815
self-shielded flux cored wire **M6:** 813-814
stainless steel overlays **M6:** 811-816
submerged arc welding **M6:** 811-814
tubular wire for filler metal **M6:** 813
use of agglomerated fluxes **M6:** 812
use of powdered alloys **M6:** 812-813
processes **M6:** 807-809
electroslag welding **M6:** 807-808
flux cored arc welding **M6:** 807
gas metal arc welding **M6:** 807
gas tungsten arc welding **M6:** 807
plasma arc welding **M6:** 807-808
shielded metal arc welding **M6:** 807
submerged arc welding **M6:** 807
thermal spray **M6:** 808
processes for composite structures **M6:** 804-805
processes for special applications **M6:** 816
purposes **M6:** 805-806
corrosion **M6:** 805
high-temperature service **M6:** 806
wear **M6:** 805-806
types of materials **M6:** 806-807
alloy steels **M6:** 807
ceramics **M6:** 807
copper alloys **M6:** 806
irons **M6:** 806-807
manganese alloys **M6:** 807
nickel and cobalt alloys **M6:** 806
stainless steels **M6:** 806
Weld pass
definition **A6:** 1215 **M6:** 20
Weld pass sequence
definition **A6:** 1215
Weld penetration
definition **A6:** 1215
Weld pool
definition **A6:** 1215
Weld procedure qualification **A6:** 1089-1093
codes, standards, and specifications **A6:** 1089
documentation **A6:** 1093
limitations on procedure qualification **A6:** 1092-1093
purpose of qualification **A6:** 1089
responsibility for the task **A6:** 1089
types of tests **A6:** 1089-1092
bend tests **A6:** 1090, 1092
butt-joint pipe test **A6:** 1090
fillet weld tests **A6:** 1091-1092
fracture tests **A6:** 1092
for full-penetration groove welds **A6:** 1090
hardness tests **A6:** 1092
notch toughness tests **A6:** 1090-1091
partial-penetration groove weld tests **A6:** 1091-1092
peel tests **A6:** 1092
procedure qualification tests **A6:** 1092
shear tests **A6:** 1092
spot and plug weld tests **A6:** 1091-1092
standard **A6:** 1089-1092
stud weld tests **A6:** 1092
surfacing tests **A6:** 1092
tension tests **A6:** 1090
tension-shear tests **A6:** 1091, 1092
torsion tests **A6:** 1092
toughness tests **A6:** 1090
transverse tension tests **A6:** 1090
weld cladding tests **A6:** 1092
weld-break tests **A6:** 1091, 1092
Weld puddle
definition **A6:** 1215
Weld reinforcement
definition **A6:** 1215
Weld repair *See also* Repair
titanium and titanium alloy castings **A2:** 643
of titanium castings **A15:** 833

Weld repair (continued)
of zirconium castings **A15:** 838
Weld root
definition **A6:** 1215
Weld set number (WSN) **A6:** 1061
Weld size
definition **A6:** 1215
Weld solidification fundamentals **A6:** 45-54
comparison of casting and welding solidification **A6:** 46
constitutional supercooling **A6:** 46-48, 49, 51, 52, 53
nonequilibrium effects: high-rate weld solidification and composition banding **A6:** 51-54
solidification of alloy welds (constitutional supercooling) **A6:** 46-48, 49, 51
weld microstructures, development of **A6:** 48-49
welding rate effect on weld pool shape and microstructure **A6:** 49-51
Weld spatter *See* Spatter
liquid penetrant inspection **A17:** 86
spring failure from **A11:** 559
Weld structure
defined **A9:** 19
Weld tab
definition **A6:** 1215 **M6:** 20
Weld tension test **A1:** 610
Weld throat
definition **A6:** 1215
Weld time
definition **M6:** 20
Weld timer
definition **M6:** 20
Weld toe
bridge, cracked **A11:** 707-708, 712
definition **A6:** 1215
hydrogen-induced cracking (HIC) in **A11:** 251
Weld twist *See also* Weld(s)
eddy current inspection **A17:** 563
flux leakage inspection **A17:** 564
ultrasonic inspection **A17:** 564-565
Weld wire
nonrelevant indications from **A17:** 106
Weld wires
chemical compositions **A9:** 577-578
Weld zone **A7:** 456
Weld(s) *See also* Arc welding; Arc welds; Boilers; Pressure vessels; specific weld types; Weld contours, Weld defects; Weld failure origins; Welding; Weldments; Weldments, failures of
acoustic emission inspection **A17:** 289-290
arc, failure origins **A11:** 412-415
in boiler tubes, fatigue fractures at **A11:** 620-621
carbon steel discharge line failed at **A11:** 639
in carbon steel pipe, inspection **A17:** 112
carbon-molybdenum desulfurizer, cracking in **A11:** 663
in chain links, magnetic particle inspection **A17:** 116
cleanup, poor **A11:** 444
corrosion failure of **A11:** 400-401
cracking, in arc-welded aluminum alloys **A11:** 435
cracking- tests, specimens for **A11:** 415
defect, effect in medium-carbon steel **A12:** 258
deposit, as crack origin **A12:** 254, 268
dissimilar-metal **A11:** 620-621, 796
double-submerged arc **A11:** 698
effect of martensite zone in **A11:** 426
electric-resistance, in pipe **A11:** 698
electrogas, failure origins **A11:** 440
electron beam **A11:** 444-447
electroslag, failure origins **A11:** 439-440
field girth, in pipe **A11:** 699
flange, fractured medium-carbon steel **A12:** 256
flash **A11:** 442-443, 698
fracture of brine-heater shell at **A11:** 637-638
fracture, titanium, from incomplete fusion **A12:** 65
friction **A11:** 444
girth **A11:** 422, 438, 648, 699
HAZ cracks, pressure vessel failure from **A11:** 650-652
HAZ, striations **A12:** 21
intergranular cracks and brittle fracture in **A11:** 653

Weld(s) (continued)
joint, grain-boundary embrittlement **A11:** 131-132
lap, in pipe **A11:** 698
laser beam **A11:** 447-449
laser beam, ductile fracture **A12:** 422
longitudinal, eddy current inspection **A17:** 186
longitudinal, fatigue cracking **A11:** 698
longitudinal, in pipe **A11:** 698-699, 704
magnetic paint inspected **A17:** 128
magnetizing **A17:** 94
metal, brittle fracture by corrosion products **A11:** 653
penetration, inadequate, as discontinuity **A12:** 65
penetration, incomplete, steam accumulator failure from **A11:** 647-648
photo effects of lighting on **A12:** 87-88
pipeline girth, inspection **A17:** 579-581
plug, cracking in **A11:** 655-656
properties, in resistance welds **A11:** 441
quality **A11:** 441, 448-449, 708
radiographic methods **A17:** 296
resistance, failures in **A11:** 440-442
resistance spot, fatigue fracture **A12:** 66, 67
robotic **A12:** 375
root, lamellar tearing at **A11:** 665
root, magnetic particle inspection **A17:** 111
root penetration, steam preheater shell failure from **A11:** 420
shape, in resistance welds **A11:** 441
shielded metal-arc, magnetic particle inspection **A17:** 114-115
spot **A11:** 310, 441, 495, 497
in stainless steel piping, SCC failure at **A11:** 216
stainless steel, shaft ductile fracture at **A11:** 481-482
submerged arc impact properties **A11:** 69
surfaces, in resistance welds **A11:** 441
termination, bridge, fatigue cracking at **A11:** 707
toe, hardness, fractured medium-carbon steel **A12:** 256
ultrasonic inspection **A17:** 272-273
under cyclic loading **A11:** 117
underbead crack **A12:** 155
upset butt **A11:** 443-444
videoscope monitoring **A17:** 8
in welded tubing and pipe, eddy current inspection **A17:** 186
Weld-area cracks
as flaw **A17:** 562, 565
Weld-delay time
definition **M6:** 19
Weld-deposited hardfacing
as shield against liquid impingement against erosion **A18:** 222
Weld-heat time
definition **M6:** 19
Weld-interface carbides
in friction welds **A11:** 444
Weld-metal quench effect **A6:** 178
Weldability *See also* Fabrication; Fabrication characteristics; Joining; specific welding techniques; Welding **A1:** 603-613
alloy steel **M1:** 561-564
aluminum alloys **A15:** 766
aluminum casting alloys **A2:** 150, 153-177
aluminum coated steel sheet **M1:** 173
of aluminum coatings **A1:** 220
aluminum-lithium alloys **A2:** 184, 190, 197
aluminum-silicon alloys **A15:** 159, 167
austenitic manganese steels **A15:** 735
carbon steel **M1:** 561-564
cast copper alloys **A2:** 357-391
cast iron **M1:** 563-564
cast steel **M1:** 400-401, 563
of cast steels **A15:** 532
of cast vs wrought steels **A15:** 535
of cobalt-base alloys **A13:** 662-665
defined **A15:** 532
definition **M1:** 561 **M6:** 19
heat treatable steels **M1:** 562-563
of heat-treatable low-alloy steels **A1:** 609
high-alloy steels **A15:** 730
of high-carbon steels **A1:** 609 **M1:** 561, 562-563
of high-strength low-alloy steel **A1:** 609
HSLA steels **M1:** 563
low-alloy steels **A15:** 719-720

Weldability (continued)
low-carbon low-alloy steels **M1:** 563
of low-carbon steels **A1:** 608 **M1:** 562, 563
of magnesium alloys **A2:** 474
maraging steels **M1:** 563
medium-carbon low-alloy steels **M1:** 561-563
of medium-carbon steels **A1:** 608-609 **M1:** 561, 562, 563
metallurgical factors affecting **A1:** 603-606, 607
chemical composition effect **A1:** 606, 609
hardenability and weldability **A1:** 603-604, 606
heat-affected zone microstructure **A1:** 605-606, 608
preweld and postweld heat treatments **A1:** 606
weld metal microstructure **A1:** 604-605, 607
and microstructure, cast irons **A15:** 522
of nickel-base alloys **A13:** 652
plain carbon steels **A15:** 705
plate **M1:** 197-198
platinum alloys **A2:** 712
of precoated steels **A1:** 609-610
of quenched and tempered steels **A1:** 609
rephosphorized steels **M1:** 563
resulfurized steels **M1:** 563
of stainless steels **A1:** 897-905
of steel castings **A1:** 378
of steel plate **A1:** 238-239
of steels **A1:** 603
terne coated sheet **M1:** 174
tool steels **M1:** 563
weld cracking **A1:** 606-608
hot cracking **A1:** 608
hydrogen-induced cracking **A1:** 606-607
inclusions **A1:** 608
lamellar cracking **A1:** 607-608, 609
stress-relief cracking **A1:** 607-608
wrought aluminum and aluminum alloys **A2:** 30-32, 104, 111
wrought steels **M1:** 563
of wrought tool steels **A1:** 775
Weldability testing **A6:** 603-612
Charpy V-notch impact toughness tests **A6:** 603
circular groove test **A6:** 604
circular patch test (cold and hot crack test) **A6:** 605, 606
cold cracking (hydrogen-induced cracking) **A6:** 604
controlled-thermal-severity test **A6:** 604, 605-606, 607
Cranfield test **A6:** 604
Cruciform test **A6:** 604, 606, 607
definition **A6:** 1215
drop weight notch toughness tests **A6:** 603
for evaluating cracking susceptibility **A6:** 603-608
for evaluating weld pool shape, fluid flow, and weld penetration **A6:** 608-609
externally loaded tests **A6:** 606-608
fillet weld break, shear, or fracture tests **A6:** 603
general characteristics of tests **A6:** 603
Gleeble testing **A6:** 603, 609-612
guided bend tests **A6:** 603
hot cracking **A6:** 603-604
alternate names **A6:** 604
hot ductility testing **A6:** 611-612
Houldcroft crack susceptibility test (hot crack test) **A6:** 605
implant test **A6:** 604, 606, 607
impulse decanting test **A6:** 609, 610
keyhole restraint cracking test (hot and cold cracks) **A6:** 604-605
keyhole slotted-plate restraint test (hot crack test) **A6:** 605
Lehigh cantilever test **A6:** 604
Lehigh restraint test **A6:** 604
liquid penetrant inspection **A6:** 603, 605
longitudinal, bead-on-plate test **A6:** 604
macroetch tests **A6:** 603
macroexamination tests **A6:** 603
magnetic particle inspection **A6:** 603
microexamination tests **A6:** 603
Navy circular patch test (cold and hot crack test) **A6:** 605, 606
nick bend test **A6:** 604
radiographic inspection **A6:** 603, 605

Weldability testing (continued)
rigid restraint (RRC) test **A6:** 604
self-restraint tests **A6:** 604-606
Sigmajig test (hot crack test) **A6:** 608, 609
slot test **A6:** 604
spot Varestraint test (hot crack test) **A6:** 607-608
stud weld/bend or torque tests **A6:** 603
subscale Varestraint test **A6:** 607, 608
Tekken test **A6:** 604, 605
tensile tests **A6:** 603
tension restraint cracking (TRC) test **A6:** 604
TIG-A-MA-JIG test **A6:** 607-608
trans Varestraint test **A6:** 607
Varestraint tests **A6:** 603, 604, 606-607, 608, 611
visual inspection **A6:** 603, 605
weld penetration tests **A6:** 608-609, 610
weld pool shape tests **A6:** 609
Weldability tests **A1:** 610-613
bend test **A1:** 610
Charpy V-notch test **A1:** 610-611
crack tip opening displacement test **A1:** 611
drop-weight test **A1:** 610
fabrication **A1:** 611-613
controlled thermal severity test **A1:** 612-613
Lehigh restraint test **A1:** 612
Varestraint test **A1:** 612
stress-corrosion cracking test **A1:** 611
weld tension test **A1:** 610
Weldalite 049 *See* Aluminum-lithium alloys, specific types
precipitation heat treatment **A4:** 843
WELDBEAD neural network system **A6:** 1060
Weldbonding
aluminum **M2:** 202
Welded
$2^1/_2$Cr-1Mo steel, strength of **M1:** 658-659
ductile iron **M1:** 56
HSLA steels **M1:** 409, 415, 419-420
HSLA steels, suitability of submerged-arc welding **M1:** 419-420
hydrogen embrittlement of steel **M1:** 687
malleable iron **M1:** 66-67, 71
maraging steels **M1:** 445, 446, 449
steel tubular products, production by **M1:** 316
steel wire coils **M1:** 261
steel wire fabrication **M1:** 588-589
steel wire rod coils **M1:** 253
ultrahigh-strength steels **M1:** 422, 424, 427, 429, 431, 432, 434, 438, 441
wire, steel wire rod for **M1:** 256
Welded blanks **A14:** 450-451
Welded chains
alloy steel wire for **A1:** 287 **M1:** 270
Welded coupons
corrosion testing **A13:** 198
Welded cover plates, bridge
fatigue cracks forming at **A11:** 707
Welded joint regions **A9:** 577
Welded joint sections **A9:** 579
Welded joints *See also* Joints; Weld(s); Welding; Weldments
bead morphology **A9:** 578
defects **A9:** 581
dissimilar metal **A9:** 582
etchants for **A9:** 580
in plate steels, examination **A9:** 202-203
sample preparation **A9:** 578
sections used in metallographic examination **A9:** 578
solid-state transformation structures **A9:** 580-581
solidification structures **A9:** 578-580
ultrasonic inspection **A17:** 272
Welded joints in steel tubulars
examination of **A9:** 211
Welded mechanical tubing **A1:** 334-335 **M1:** 324
Welded power take-off (PTO) handle **A7:** 677
Welded sheet metal preforms
explosive forming of **A14:** 642-643
Welded structures
fracture toughness of **A1:** 667-669, 670, 671, 672
Welded tubing
eddy current weld inspection **A17:** 186
Welder
definition **A6:** 1189, 1215 **M6:** 19
Welder certification
definition **M6:** 19
Welder performance qualification
definition **A6:** 1215 **M6:** 19
Welder registration
definition **M6:** 19
WELDEXCELL system **A6:** 1057-1064
backward chaining **A6:** 1059
blackboard **A6:** 1060
data base systems **A6:** 1058-1059
expert systems **A6:** 1059
FERRITEPREDICTOR expert system **A6:** 1059
forward chaining **A6:** 1059
frames **A6:** 1059
HEATFLOW neural network system **A6:** 1060
knowledge sources **A6:** 1058
neural network system (NNS) **A6:** 1059-1060, 1063
path planner **A6:** 1057, 1058
production rules **A6:** 1059
system integrator **A6:** 1057, 1060-1061
WELDBEAD neural network system **A6:** 1060
WELDHEAT expert system **A6:** 1059, 1060
welding schedule developer **A6:** 1057, 1058-1060
WELDPROSPEC expert system **A6:** 1059
WELDSELECTOR expert system **A6:** 1059
WELDHEAT expert system **A6:** 1059, 1060
Welding *See also* Arc welding; Fabrication characteristics; Joining; specific processes; specific welding techniques; Weld(s); Weldability
(atomic bonding) as milling process **A7:** 62
abrasive blasting of weldments **M5:** 91
aluminum **M2:** 191-199
aluminum alloy oil-line elbow, fractures by **A11:** 437
aluminum alloys **A15:** 763
aluminum-lithium alloys **A2:** 197
arc, of cast irons **A15:** 520-529
of austenitic manganese steel **A1:** 837-838 **M3:** 586
of beryllium **A2:** 683
beryllium-copper alloys **A2:** 414
cable, for input leads **A8:** 389
of cast carbon/low-alloy steels **A15:** 532-534
of cast irons **A15:** 520-531
of cast steels **A15:** 520, 531-537
of classical sculpture **A15:** 20
cold, and nonlubricated wear **A11:** 154-155
of composite materials, pressure vessels **A11:** 654-656
conditions, GMAW/FCAW of cast iron **A15:** 524
of containers, in containerized hot isostatic pressing **A7:** 431
of containers, in encapsulation techniques **A7:** 431
control of parameters by electrode coatings **A7:** 816
copper metals **M2:** 453-456
corrosion caused by weld metal
of corrosion-resistant composites **A13:** 652
of corrosion-resistant steel castings **A1:** 919, 928, 929
cost, in magnesium and magnesium alloys **A2:** 473-474
cracks, as planar flaws **A17:** 50
current ranges, cast irons **A15:** 524
defined **A8:** 15 **A18:** 21 **EL1:** 1161
definition **A6:** 1215 **M6:** 19
design/materials selection in **A13:** 321, 344
of discontinuous ceramic fiber MMCs **EM1:** 909
of dissimilar materials, neutron radiography **A17:** 393
Welding (continued)
ductile and brittle fractures from **A11:** 92-94
of ductile iron **A1:** 53-55
ductile iron castings **A15:** 664
ductile iron, joint properties **A15:** 528
effect, corrosion resistance **A13:** 49
effect of composition **A11:** 315
effects of surface-active agents and lubricants in milling **A7:** 63
of electrical contact materials **A2:** 841
electrical resistance alloys **A2:** 822
electrodes, improving production of **A7:** 818-820
electrodes, oxide dispersion-strengthened copper **A2:** 401
electrogas **A11:** 440
electromagnetic **A14:** 649 **EM2:** 724
electron beam **A11:** 444-447
electroslag **A11:** 439-440
equipment, contour roll forming **A14:** 627-628
failure mechanisms **EL1:** 1041-1046
of ferritic malleable iron **A1:** 76 **A15:** 693
and fit-up, inspection techniques **A17:** 645-646
flash **A11:** 442-443
friction **A11:** 444
of galvanized members **A13:** 441
-grade iron powders **A7:** 89, 818
of gray iron **A15:** 530-531
hard facing, processes used in **M3:** 567
heat, and sealing **EM2:** 724-725
heat, effect on corrosion potential **A13:** 345
of heat exchangers,effects of **A11:** 639-640
heat sink **A13:** 931
high-frequency **A11:** 448
high-frequency induction **A11:** 449
high-purity uranium **A2:** 672
of HSLA steels **A1:** 414-415
imperfections **A11:** 92-93
inertia, bending-fatigue fracture after **A11:** 469
of Invar **A2:** 892-893
of investment castings **A15:** 264
iron powder for **A7:** 817-818
iron powder shipments for **A7:** 23
last pass heat sink **A13:** 932
leakages in **A7:** 431
LME by copper during **A11:** 721
local cold, from fretting **A13:** 138
machine **A7:** 506
magnesium **M2:** 546-547
of magnesium alloys **A2:** 471-472
of maraging steels **A1:** 798
metallurgy, cast irons **A15:** 520-522
methods, for blanks **A14:** 450-451
monitoring, acoustic emission inspection **A17:** 289
neutron radiography of **A17:** 393
of nickel alloys **A15:** 822
nickel alloys, cleaning for **M5:** 673
of niobium-titanium superconducting materials **A2:** 1049
nonrelevant indications from **A17:** 106-108
in outer lead bonding **EL1:** 285
oxyacetylene, of cast irons **A15:** 529-531
particle effects in milling **A7:** 61
of pearlitic-martensitic malleable irons **A1:** 83, 84
pearlitic/martensitic malleable iron **A15:** 696-697
of PH martensitic stainless steels **A1:** 946
of PH semiaustenitic stainless steels **A1:** 943-944
post-, ductile and brittle fractures from **A11:** 94
postforging defects from **A11:** 333
postweld heat treatment, cast steels **A15:** 534-535
practice/sequence, and corrosion **A13:** 344
practices, effect on fatigue strength **A11:** 127
of pressure vessels **A11:** 644-645, 655
process, discontinuities from **A17:** 582
processes, for electronic fabrication **EL1:** 1041
of quenched and tempered martensitic stainless steels **A1:** 942
of railroad rail, longitudinal residual stress distribution in **A10:** 391-392

Welding (continued)
refractory metals and alloys **A2:** 563-564
relay weld integrity, gas mass spectroscopy for **A10:** 156
repair, distortion from **A11:** 141-142
residual stress measurement, by Barkhausen noise **A17:** 160
residual stress, SCC testing **A13:** 256
resistance **A11:** 440-442
resistance alloys **A2:** 831
resistance, electrodes for **A7:** 624-629
as secondary operation **A7:** 456-457
selective attack during pickling of welds **M5:** 565
selective plating compared to **M5:** 292-293
of shafts **A11:** 459, 480-481
in sintering process **A7:** 340
soundness, analyzed **A10:** 478-481
specifications, steel and alloy castings **A15:** 532
specimens **A8:** 510, 513, 520-521
spin **EM2:** 725
spot **A7:** 506, 624
spot ,ultrasonic **EM2:** 722
of stainless steels **A13:** 551-552
steel, aluminum coating of welds **M5:** 334-335
steel casting failure from **A11:** 399-401
for steel tubular products **A1:** 327-328
stress-relief, cast steels **A15:** 534-535
stresses, residual **A11:** 97-98
techniques, cast irons **A15:** 526
techniques, processes, applications **EL1:** 238
as tension source for stress-corrosion cracking **A8:** 502
tests **A7:** 625
of thermocouple junctions **A2:** 871
to seal metal packages **EL1:** 237-238
in tube/pipe rolling **A14:** 631
ultrasonic **EM2:** 721
upset resistance butt **A11:** 443-444
uranium alloys **A2:** 676-677
variable polarity plasma arc (VPPA) **A2:** 184
vibration **EM2:** 724-725
of wear-resistant manganese steels **A15:** 535
weld beads, stainless steel, grinding of **M5:** 555
with wire **A14:** 694
wrought titanium alloys **A2:** 617-618
zinc alloys **A15:** 795
zone, chromium depletion in **A10:** 179
zones, cast iron **A15:** 522
Welding alloys
nickel-base **A2:** 444-445
Welding arc **A6:** 64
Welding arc, gas metal
copper deposition with **A11:** 721
Welding blowpipe
definition **A6:** 1215
Welding current
definition **A6:** 1215 **M6:** 19
Welding cycle
definition **A6:** 1215 **M6:** 19
Welding dies
projection welding **M6:** 509-510
Welding electrode
definition **A6:** 1215
Welding factor
symbol and units **A18:** 544
Welding force **A6:** 888
Welding generator
definition **M6:** 19
Welding ground
definition **A6:** 1215
Welding head
definition **M6:** 19
Welding heat efficiency **A6:** 9
Welding in space and low-gravity environments **A6:** 1020-1025
adhesive bonding **A6:** 1023
automation need **A6:** 1025
brazing **A6:** 1023
categories of space welding applications **A6:** 1020
electron-beam welding **A6:** 1021-1022, 1023
examples of application **A6:** 1020-1021
extra-vehicular activity (EVA) welding **A6:** 1025
future of space welding **A6:** 1025
gas-tungsten arc welding **A6:** 1021, 1022
Welding in space and low-gravity environments (continued)
intelligent engineering/planning systems **A6:** 1025
laser-beam welding **A6:** 1021, 1022
mechanical fastening **A6:** 1023
metallurgy of low-gravity welds **A6:** 1023-1025
microgravity influences **A6:** 1023
Russian welding program **A6:** 1023
solar heat welding **A6:** 1022-1023
solid-state bonding **A6:** 1023
space welding environment, challenges and problems **A6:** 1021
United States M512 experiments **A6:** 1023-1025
welding processes **A6:** 1021-1023
Welding leads
definition **A6:** 1215 **M6:** 19
Welding machine *See also* specific processes
definition **A6:** 1215 **M6:** 19
Welding operator
definition **A6:** 1215 **M6:** 19
Welding position
definition **A6:** 1215
Welding pressure
definition **M6:** 19
Welding procedure *See also* specific process
definition **A6:** 1215 **M6:** 19
Welding procedure specification
definition **M6:** 19
Welding procedure specification (WPS)
definition **A6:** 1215
Welding process *See also* specific processes
definition **M6:** 19
Welding processes
aluminum **M2:** 196-199
Welding quality steel wire rod **M1:** 255, 256
Welding rectifier *See also* Rectifiers
definition **M6:** 19
Welding rod
definition **A6:** 1215
Welding rod grade powders
Domfer process **A7:** 91
Welding rods
for cast irons **A15:** 530
definition **M6:** 19
low-hydrogen **A11:** 251
for oxyfuel gas welding **M6:** 588-589
Welding rods, specific types
class RG45, oxyfuel gas welding **M6:** 588-589
class RG60, oxyfuel gas welding **M6:** 588-589
class RG65, oxyfuel gas welding **M6:** 588-589
Welding sequence *See also* specific processes
definition **A6:** 1215 **M6:** 19
Welding speed **A6:** 25
Welding technique
definition **M6:** 19
Welding tip
definition **A6:** 1215 **M6:** 19
types for oxyfuel gas welding **M6:** 586
Welding torch *See also* Torches
definition **M6:** 19-20
Welding torch (arc)
definition **A6:** 1215
Welding torch (oxyfuel gas)
definition **A6:** 1215
Welding transformer *See also* Transformers
definition **M6:** 20
Welding, use of phase diagrams in technique development ..... **A3:** 1 • 26-27
Welding wheel
definition **A6:** 1215
Welding wire
definition **A6:** 1215
Welding-quality rod **A1:** 273-274, 275
Weldment
definition **A6:** 1215 **M6:** 20
Weldment corrosion *See also* Weldments; Welds
aluminum alloy **A13:** 344-345
austenitic stainless steel **A13:** 347-355
carbon steel **A13:** 362-367
duplex stainless steel **A13:** 358-361
factors affecting **A13:** 344
ferritic stainless steel **A13:** 355-358
material selection for **A13:** 323-324
metallurgical factors **A13:** 344
nickel/high-nickel alloy **A13:** 361-362
tantalum/tantalum alloy **A13:** 345-347
Weldment(s) *See also* Joints; specific welding methods; Weld(s); Welding
acoustic emission inspection **A17:** 598-602
acoustical holography inspection **A17:** 445-446
arc welds, discontinuities in **A17:** 582-585
destructive evaluation **A17:** 582
diffusion bonding **A17:** 588-589
discontinuity signals **A17:** 596-598
eddy current inspection **A17:** 602
electric current perturbation inspection **A17:** 602
electron beam welding **A17:** 585-587
electroslag welding **A17:** 587-588
girder, magnetic particle inspection **A17:** 115
half-wave current, magnetic particle inspection **A17:** 91
in-service monitoring **A17:** 600-601
leak testing **A17:** 602
magnetic particle inspection **A17:** 114-115, 591-592
nondestructive inspection functions **A17:** 589-590
plasma arc welding **A17:** 587
postweld inspection **A17:** 599
radiographic appearance **A17:** 349-350
radiographic inspection **A17:** 334-335, 592-594
resistance welding **A17:** 588
specialized processes, discontinuities **A17:** 585
ultrasonic inspection **A17:** 232, 594-598
visual inspection **A17:** 590-591
Weldments *See also* Weldment corrosion; Weldments, failures of Weld(s); Welds
bend testing **A8:** 117
clapper, brittle fracture of **A11:** 645-646
corrosion of **A11:** 400-401 **A13:** 344-368
design, and corrosion **A13:** 344
hydrogen embrittlement in **A8:** 539
hydrogen-induced cracking in **A11:** 92, 93
preferential pitting **A13:** 345-346
radiographic inspection of **A11:** 17
SCC testing **A13:** 275
suction roll **A13:** 1206
testing of **A8:** 510, 513, 520-521
to pipe, failures of **A11:** 704
unmixed zones **A13:** 344
Weldments, failures of *See also* Arc welding; Weld(s) **A11:** 411-449
analysis procedures **A11:** 411-412
arc welds, failure origins in **A11:** 412-415
arc-welded alloy steel **A11:** 423-426
arc-welded aluminum alloys **A11:** 434-437
arc-welded hardenable carbon steel **A11:** 422-423
arc-welded heat-resisting alloys **A11:** 433-434
arc-welded low-carbon steel **A11:** 415 422
arc-welded stainless steel **A11:** 426-433
arc-welded titanium and titanium alloys **A11:** 437-439
electrogas welds, origins **A11:** 440
electron beam welds, origins **A11:** 444-447
electroslag welds, origins **A11:** 439-440
flash welds, origins **A11:** 442-443
friction welds, origins **A11:** 444
high-frequency induction welds, origins **A11:** 449
laser beam welds, origins **A11:** 447-449
resistance welds **A11:** 440-442
upset butt welds, origins **A11:** 443-444
Weldments in carbon and alloy steels
macroetching to reveal **A9:** 175-176
Weldnuts
measured by coordinate measuring machines **A17:** 18
WELDPROSPEC expert system **A6:** 1059
Welds *See also* Weldment corrosion; Weldments
corrosion in pulp bleach plants **A13:** 1194-1195
corrosion of **A13:** 652-653
corrosion potential **A13:** 344
definition **M6:** 26-28
deposit, chemical analysis **A13:** 366
discontinuities **M6:** 829-855
joint properties **M6:** 57
microstructure **M6:** 35-37
overlay repair **A13:** 931
penetration, incomplete **A13:** 344
pressure vessel **A13:** 1089
repair and cladding, nuclear reactors **A13:** 970-971
slag/spatter, corrosivity **A13:** 344
solidification **M6:** 28-31

Welds (continued)
specimens, SCC testing **A13:** 253
stress in, galvanizing effects **A13:** 438
Welds, characteristic features of **A1:** 603, 604, 605
multipass weldments **A1:** 603, 605
single-pass weldments **A1:** 603, 604
Welds, characterization of **A6:** 97-106
compositional analysis **A6:** 99, 100-105
defects found by macrostructural characterization **A6:** 97-98
description of ferrite/carbide microconstituents in low-carbon steel welds **A6:** 101
examples of weld characterization **A6:** 102-106
goals **A6:** 97
guidelines for selecting techniques **A6:** 100
internal characterization requiring destructive procedures **A6:** 99-100
composition **A6:** 99-100
macrostructure **A6:** 99
microstructure **A6:** 99
mechanical testing **A6:** 100-102, 103-104, 105, 106
nomenclature for fillet, lap, butt, and groove welds **A6:** 98
nondestructive characterization techniques **A6:** 97-99, 100, 102
Welds made with backing plates
preparation for examination **A9:** 578
WELDSELECTOR expert system **A6:** 1059
Well water
graphitic corrosion of gray iron pump bowl from **A11:** 372-373
Wellheads
oil/gas production **A13:** 1237
Wells, oil/gas production
corrosion of **A13:** 478-480, 1247-1258
Wells, protection
for thermocouples **A2:** 884
Wertime pyrotechnology
in casting history **A15:** 15
Western bentonite
as molding clay **A15:** 210, 341
Western Europe
telecommunication industry structure **EL1:** 384
Wet
defined **A7:** 13
Wet abrasive blasting *See also* Abrasive blasting **M5:** 93-96
abrasives used **M5:** 93-95
liquid carriers **M5:** 94-95
aluminum and aluminum alloys **M5:** 572
applications **M5:** 93
copper and copper alloys **M5:** 614
equipment **M5:** 95-96
maintenance **M5:** 96
nozzles, design and operation **M5:** 95-96
heat-resistant alloys **M5:** 563, 565
magnesium alloys **M5:** 629
precleaning process **M5:** 93-94
refractory and reactive metals **M5:** 653
stainless steel **M5:** 553
titanium and titanium alloys **M5:** 652-653
Wet analytical chemistry
for inorganic liquids and solutions **A10:** 7
for inorganic solids **A10:** 4-6
for organic liquids and solids **A10:** 10
Wet bag isostatic pressing **EM4:** 126, 127
Wet ball milling
aluminum pigments by **A7:** 593
Wet barrel finishing
aluminum and aluminum alloys **M5:** 572-574
magnesium alloys **M5:** 631
Wet baths
for wet-method magnetic particle inspection **A7:** 578
Wet blasting
for liquid penetrant inspection **A17:** 81-82
Wet blending
of uranium dioxide pellets **A7:** 665
Wet bottom ash systems
corrosion in **A13:** 1007-1008
Wet chemical etching
polyimides **EL1:** 326-327
Wet chemical methods
to analyze limestone **EM4:** 378
to analyze silica sand **EM4:** 378
to analyze soda ash **EM4:** 380
Wet chemistry
as advanced failure analysis technique **EL1:** 1106
Wet chemistry analysis **A13:** 1116-1117
Wet chlorine gas
titanium/titanium alloy resistance **A13:** 677
Wet cleaning
rust prevention after **A15:** 561
Wet corrosion **A13:** 80-82, 499-500
Wet cutting
for optical metallography specimen preparation **A10:** 300
Wet cutting abrasive wheels **A9:** 24
Wet developers
stations for **A17:** 79
Wet drawing *See also* Drawing
lubrication for **A14:** 338
for steel wire **A1:** 279
Wet etching
defined **A9:** 19
stainless steel **M5:** 560-561
Wet filament winding
cost **EM2:** 368
epoxies for **EM2:** 371
of epoxy composites **EM1:** 71
Wet fly ash systems **A13:** 1007
Wet friction applications **A7:** 702, 703
Wet glass bead shot peening **M5:** 144-145
Wet glass transition temperature (T_g) **EM1:** 32
Wet horizontal magnetizing equipment
head and tailstock type **A7:** 576
Wet horizontal-type magnetic particle inspection unit **A7:** 577
Wet hot dip galvanized coating **M5:** 326-328
Wet installation **EM3:** 31
defined **EM1:** 26 **EM2:** 45
Wet, laminar
defined **EM1:** 226
Wet lay-up *See also* Lay-up
defined **EM1:** 26 **EM2:** 45
of heat-exchanger tubing **A11:** 628
laminating, of epoxy composites **EM1:** 71-73, 75
resins **EM1:** 132-134
Wet lay-up method
of lamination **EL1:** 832
Wet lay-up resins **EM1:** 132-134
epoxies **EM1:** 134
future trends **EM1:** 134
polyesters **EM1:** 132-133
vinyl esters **EM1:** 133-134
Wet magnetic compaction **A7:** 327-328
Wet nitrogen
in zoned sintering atmospheres **A7:** 347
Wet particles *See also* Particles; Powder metallurgy parts
magnetic **A17:** 100
Wet peel test **EM3:** 657, 662
durability assessment **EM3:** 668-669
Wet power brush cleaning **M5:** 152
Wet process
defined **EL1:** 1161
Wet process acids
corrosion rates **A13:** 645, 647
Wet reclamation systems
for sands **A15:** 213, 351
Wet scrubbers **A7:** 73
Wet sieving **A7:** 216
Wet storage stain inhibitors
hot dip galvanized products **M5:** 331
Wet strength **EM3:** 31
defined **EM1:** 26 **EM2:** 45
Wet tumbling
heat-resistant alloys **M5:** 563, 565-566
Wet washing
liquid fire method **A10:** 166
for sand reclamation **A15:** 227
Wet welding
underwater welding **A6:** 1010, 1011
Wet winding *See also* Dry winding; Winding
defined **EM1:** 26 **EM2:** 45
Wet-assembly technique **EM3:** 36
Wet-bag method
for cermets **A2:** 982
Wet-film thickness
measurement **A13:** 417
Wet-method particles **A7:** 578
fluorescent **A7:** 579
of magnetic particle inspection **A7:** 577, 578
Wet-out
defined **EM1:** 26 **EM2:** 45
Wet-powder magnetic particles
suspending liquids for **A7:** 578
Wet-slug tantalum capacitors
failure mechanisms **EL1:** 997-998
Wetbag isostatic pressing **A7:** 444-445, 447-448
Wetbag tooling
defined **A7:** 13
Wettability
defined **A7:** 13
fluoropolymer coatings **EL1:** 782-783
of graphite **A10:** 543
of MMC fiber metals **A15:** 840-842
Wettability index (WI) **A6:** 116
Wetting *See also* Impregnation **EM3:** 31
agent, defined **A13:** 14
agents, and emulsifiers, for lubricant failure **A11:** 154
agents, ceramic shell molds, investment molding **A15:** 259
brazing **A6:** 114-115
of carbon fibers **EM1:** 52
cermet **A7:** 801
of cermets **A2:** 991
defects, from fluxes **EL1:** 684
defined **A13:** 14 **EL1:** 1161 **EM1:** 26 **EM2:** 45
definition **A6:** 115, 1215 **M6:** 20
in heterogeneous nucleation **A15:** 104
in inoculation **A15:** 105
of locomotive axle surface **A11:** 716
metal, by glass **EL1:** 455
of polymers **A11:** 761
rate of **A6:** 128
of sands **A15:** 208
and sizing **EM1:** 123
solder, SIMS analysis **EL1:** 1085-1086
as solderability mechanism **EL1:** 675-676, 1032-1034
of thermoplastics **EM1:** 101-103
of tin solders **A2:** 520
vs dewetting **EL1:** 990
in wet lay-up techniques **EM1:** 132
Wetting agents *See* Surfactants **A7:** 13
for slurry in spray drying **A7:** 75
torch brazing **M6:** 962
use with precious metal powders **A7:** 149
Wetting balance test **A6:** 136 **EL1:** 643, 677
test standards used to evaluate solderability **A6:** 136
Wetting characteristics
liquid penetrant inspection **A17:** 71-72
Wetting, surface energies, adhesion, and interface reaction thermodynamics **EM4:** 482-492
basic factors in bonding **EM4:** 482-483
chemical reactions leading to equilibrium **EM4:** 485-489
glass/metal seals **EM4:** 496
metal brazing **EM4:** 489-490
Mo-Mn metallizing process **EM4:** 490-491
non-oxide ceramics joining **EM4:** 491-492

Wetting, surface energies, adhesion, and interface reaction thermodynamics (continued)
nonwetting and spreading **EM4:** 483-485
Wetting time tests
for solderability **EL1:** 677
WF-11
composition **A6:** 929
WF-31
composition **A6:** 929
Wheatstone bridge
circuit, for torque calibration **A8:** 158
in quasi-static testing **A8:** 223
strain gage instruments **A17:** 450
in torsional Kolsky bar dynamic tests **A8:** 228
Wheel abrasive blasting systems **M5:** 86-87, 90, 92-93
Wheel buffing
copper and copper alloys **M5:** 616
process **M5:** 108-109, 118-119, 124-125
stainless steel **M5:** 552, 557
titanium and titanium alloys **M5:** 656
wheel speeds **M5:** 108-109, 119
wheel types **M5:** 118-119
wheels, problems with **M5:** 124-125
Wheel grinding
molybdenum **M5:** 659-660
stainless steel **M5:** 555-556, 559
tungsten **M5:** 659-660
Wheel loading
minimized in grinding **A7:** 462
Wheel polishing
adhesives used **M5:** 108
aluminum and aluminum alloys **M5:** 573
grit sizes **M5:** 108-109
lubricants used **M5:** 115
magnesium alloys **M5:** 632
process **M5:** 108, 111-114
stainless steel **M5:** 557-559
wheel speeds **M5:** 108-109
wheel types used **M5:** 109, 111-114
zinc alloys **M5:** 676
Wheel spindles, front-end loaders
economy in manufacture **M3:** 853
Wheel studs, steel
fatigue fracture **A11:** 531-537
Wheel test
of corrosion inhibitors **A13:** 483
Wheel(s)
ultrasonic inspection **A17:** 232
wire **A17:** 52
Wheel-and-band machines
continuous casting **A15:** 314-315
Wheel-type immersion ultrasonic search units **A17:** 259
Wheelabrating
defined **EM2:** 45
Wheelabrator-Frye Company (Mishawaka IN) **A15:** 33
Wheels, centrifugal *See* Centrifugal wheels
Wheels, coke-oven car
fatigue fracture in **A11:** 130
Whirl
in carbon aircraft brakes **A18:** 586
Whirl (oil)
defined **A18:** 21
Whirl gates
for inclusion control **A15:** 91
Whisker **EM3:** 31
defined **EM1:** 26, 64
definition **EM4:** 634
discontinuous silicon carbide **EM1:** 64
properties **EM1:** 62-63, 119
silicon carbide (SiC) **EM1:** 889-902, 941-944
silicon nitride **EM1:** 64
Whisker formation in an electronic circuit **A9:** 101
Whisker reinforcement
SiC **A16:** 99, 101-103
Whisker reinforcements **EM4:** 19
Whisker-lance mist hackle **EM4:** 639-640
Whisker-reinforced alumina
applications **A18:** 812
Whisker-reinforced ceramic-matrix composites
erosion of ceramics **A18:** 206
Whisker-reinforced ceramics **EM1:** 941-944
cutting speed and work material relationship **A18:** 616
Whisker-reinforced composites **EM4:** 1
Whisker-reinforced metal matrix composites **EM1:** 889-902
applications **EM1:** 901-902
engineering properties **EM1:** 900-901
manufacturing methods **EM1:** 897-898
materials **EM1:** 897, 901
matrix metals **EM1:** 896-897
secondary processing **EM1:** 898-900
surface treatments **EM1:** 900
whisker reinforcement **EM1:** 896
Whisker-toughened ceramic components
design practices **EM4:** 733-740
Whiskers *See also* Electromigration; Tin whiskers
defined **A7:** 13 **A11:** 11 **EL1:** 1161 **EM2:** 45
in discontinuous aluminum metal-matrix composites **A2:** 7, 906
as reinforcement **EL1:** 1119-1121
silicon carbide **A15:** 88, 840
silicon carbide whisker-reinforced alumina **A2:** 1023-1024
zone 2, package interior **EL1:** 1009-1010
Whiskers, tin *See* Tin whiskers
White bronze *See* Cast zinc
White cast iron *See also* Abrasion-resistant castirons **M1:** 3-5
abrasion resistance **M1:** 619-622
abrasion-resistant, compositions **M1:** 616-617
carbon content, effect of **M1:** 77, 78
continuous cooling transformation diagram **M1:** 84
conversion to malleable cast iron **M1:** 57-59
damping capacity **M1:** 32
density **M1:** 31
effect of alloying elements on structure **M1:** 4-5
hardenability **M1:** 80
hardness conversions **M1:** 87
hardness of microconstituents **M1:** 77, 89
magnetic properties **M1:** 31
martensitic
chemical composition **M1:** 82
mechanical properties **M1:** 86
transverse strength and relative toughness **M1:** 87
mechanical properties **M1:** 87-88, 619-620
microstructure **M1:** 4-5
patternmakers' rules for **M1:** 31, 33
pearlitic, transverse strengths and relative toughness **M1:** 87
physical properties **M1:** 87-88
rubber wheel abrasion tests **M1:** 622
structure **M1:** 84, 85
wear vs. carbon content **M1:** 608
White cast irons *See* Austenitic nodular irons; Nickel-chromium white irons **A1:** 107-108
abrasive wear materials **A18:** 186, 187, 189-190
alloy compositions and abrasion resistance **A18:** 189
corrosive wear **A18:** 276
erosion **A18:** 200, 206
erosion resistance for pump components **A18:** 598
microfracture **A18:** 186
superplasticity **A14:** 869-871
unalloyed **A13:** 567
White ceramics
replicas useful in failure analysis **EM4:** 630
White copper
as early nickel alloy **A2:** 428
White crown optical lens grade of glass
ceramic machining guidelines **EM4:** 333
White finish buffing compound **M5:** 117
White flare material **A7:** 602
White glass
iron content **EM4:** 378
White gold *See* Gold-nickel-copper alloys
White hexachloroethane smoke **A7:** 602
White iron *See also* Cast iron; High-alloy white irons **A1:** 3 **A9:** 245
abrasion-resistant, as-cast against a chill **A9:** 254
arc welding **M6:** 316
arc welding of **A15:** 528-529
as-cast **A9:** 255
bells cast in **A15:** 19
bells, Chinese **A15:** 19
White iron (continued)
composition limits **M6:** 309
as compound **A15:** 29
constitutional liquation **A6:** 75
defined **A15:** 11
high-alloy **A15:** 678-685
metal-to-earth abrasion alloys **A6:** 790, 791
microstructure , lamellar spacing **A15:** 120
oxyacetylene welding **M6:** 604-605
oxyacetylene welding of **A15:** 531
polished, minimum structural relief **A9:** 244
shrinkage allowances **A15:** 303
welded microstructure **A15:** 521
welding metallurgy **A15:** 520-521 **M6:** 307-308
White iron cast iron briquet roll
grinding cracks in **A11:** 362
White iron shell liners
mechanical abuse and wear of **A11:** 375-377
White irons **A18:** 649, 651, 652, 653
abrasive wear **A18:** 188
applications **A18:** 695, 701
chromium-molybdenum **A18:** 651
composition **A18:** 698, 806
fractographs **A12:** 238-239
fracture/failure causes illustrated **A12:** 216
hardness specification **A18:** 696, 698
high-chromium **A18:** 651, 654
abrasion resistance **A18:** 758
high-chromium, shear/tensile stress cracking **A12:** 239
ledeburite formation **A18:** 697
martensitic grinding media composition and hardness **A18:** 654
microstructure **A18:** 695, 697, 698, 700-701
mounting materials for **A9:** 243
nickel-chromium **A18:** 651
pearlitic **A18:** 651, 654
wear resistance relation to toughness **A18:** 707
White layer *See also* Beilby layer; Highly deformed layer
defined **A18:** 21
in gaseous nitrocarburizing **A7:** 455
nitrided surfaces **M1:** 540, 630-631
nitriding for **A11:** 121
worn surfaces **M1:** 598-599
White layer of nitrided steels **A9:** 217
preservation of, for examination **A9:** 218
White light
in liquid penetrant inspection **A17:** 73
method, speckle metrology **A17:** 434-435
White liquor
defined **A13:** 14
White manganese brass *See also* Cast copper alloys
properties and applications **A2:** 390
White manganese bronze *See also* Copper casting alloys
melt treatment **A15:** 775
nominal composition **A2:** 347
White martensite
metallographic sectioning **A11:** 24
White metal *See also* Pewter; Tin; Tin alloys
creep-rupture characteristics **A2:** 525
jewelry **A2:** 525
mechanical properties **A2:** 525
White metal (whitemetal)
defined **A18:** 21
White metals
for soft metal bearings **A11:** 483
White pine
as pattern material **A15:** 194
White radiation *See also* Continuum **EM4:** 558
defined **A10:** 325-326
sources, UV/VIS **A10:** 66
White rust *See* Wet storage stain
White spots, and shrinkage cavity
iron **A12:** 220
White spots, as defect
vacuum arc remelting **A15:** 407
White Tombasil *See* Copper alloys, specific types, C99700
properties and applications **A2:** 390
White water
from paper machines **A13:** 1188-1189
White wires
defined **EL1:** 7
White zone
high-carbon steel **A12:** 288

White-etching
brittle martensite as **A11:** 867-868
nitride surface layer, tool steel fracture from **A11:** 573, 576
White-etching layer
defined **A9:** 19
Whiteheart malleable cast iron
Whiteheart malleable iron **A1:** 74
Whitening, stress *See* Stress whitening
Whitewares **EM4:** 930-936
absorption **EM4:** 4
annual sales **EM4:** 936
applications **EM4:** 1, 4, 935
batching **EM4:** 95
body materials **EM4:** 930-931
competitive materials **EM4:** 936
composition **EM4:** 5, 45
development of the industry **EM4:** 936
fired properties **EM4:** 932
firing process **EM4:** 258
glaze materials **EM4:** 931
glazes **EM4:** 1061
manufacturing **EM4:** 931-932
market in U.S **EM4:** 1061
mixing operations **EM4:** 98
physical properties **EM4:** 45, 934, 935-936
production flowchart **EM4:** 933
products **EM4:** 4
raw materials **EM4:** 44
sanitaryware **EM4:** 932-934
strength measurement test method **EM4:** 567
tableware **EM4:** 934-935
testing **EM4:** 547
types **EM4:** 4
uniaxial strength **EM4:** 591
Whiting
in typical ceramic body compositions **EM4:** 5
Whiting, John H
as inventor **A15:** 30
Whitney, Asa
as inventor **A15:** 30
Whole-body motion, effect
optical holography **A17:** 412-413
WI-52
composition **A4:** 795 **A6:** 929 **A16:** 737
machining **A16:** 738, 741-743, 746-758
Wick lubrication
defined **A18:** 21
Wick test
for SCC evaluation **A13:** 273
Wicking **A6:** 369 **EM3:** 31
in carbon steels **A13:** 519, 521
condensation (vapor phase) soldering **EL1:** 704
defined **EL1:** 1161
method, of component removal **EL1:** 716
solder **EL1:** 694
Wicking (capillary action) **EM4:** 135, 136, 138
Wicking process
defined **A8:** 529
Wide face brushes **M5:** 152, 153
Wide-beam Auger electron spectroscopy spectrum **A7:** 254
Wideband transformer
defined **EL1:** 1161
Widmanstatten patterns in wrought heat- resistant alloys **A9:** 311-312
Widmanstatten precipitates in aluminum alloys **A9:** 359
Widmanstätten structure
defined **A8:** 15
micro, breakup and flow softening process **A8:** 172
in Ti alloy isothermal hot compression test specimen **A8:** 172
zirconium **A2:** 663
Widmanstatten structures
defined **A9:** 19
formation **A9:** 647
in titanium and titanium alloys **A9:** 460
Widmanstiitten structures
titanium alloys **A12:** 23, 443
Widmanst‡$$‡Adatten ferrite **A6:** 76, 77, 78, 1011
HSLA steels **A6:** 418
Widmanst‡$$‡Adatten structure
austenitic stainless steels **A6:** 459, 461, 462, 463
titanium alloys **A6:** 85, 86, 514, 517, 518, 519, 520
Width **EM3:** 31
defined **EM2:** 45
measurement, uniaxial tensile testing **A8:** 554-555
strip, contour roll forming **A14:** 629
symbol for **A8:** 724
of three-roll forming **A14:** 616
Width constraint method
plane-strain tensile testing **A8:** 557-558
Width, of dimples
measurement **A12:** 207
Width reduction
as near defect **EL1:** 568
Width to hole diameter ratio (*W*/*D*)
in pin bearing testing **A8:** 59
Wiederhorn theory
erosion rate of ceramics **A18:** 205
Wigner-Seitz radius **A6:** 144
Wilkinson cupola
development of **A15:** 29
Wilks' ATR attachment
effects in analysis **A10:** 120-121
Willemite **EM4:** 10
chemical system **EM4:** 870-871
island structure **EM4:** 758
primary phase Zn_2SiO_4, and other zinc-containing phases of non-commercial glass-ceramics **EM4:** 873
William Butcher Steel Works *See* Midvale Company
Williams-Landel-Ferry (WLF) equation **EM3:** 40, 354, 422
for rubbery material relationship between time and temperature **EM3:** 383
Wilson and Walowit's formula
inlet film thickness **A18:** 94
Wilson's disease
as copper toxicity **A2:** 1251
Winch drum
transverse cracking in **A11:** 395
Winches
ladle movement by **A15:** 27
Wind
effect in marine atmospheres **A13:** 905-906
Wind angle
defined **EM1:** 26 **EM2:** 45
Windage loss
Charpy impact tests **A8:** 262
Winder **EM1:** 45, 109
Winding *See also* Axial winding; Biaxial winding; Circumferential winding; Dry winding; Filament winding; Helical winding; Multicircuit winding; Planar helix winding; Planar winding; Reverse helical winding; Single-circuit winding; Spooling; Wetwinding
filament **EM1:** 135-138, 503-518
and fusion **EM1:** 138
machine, six-axis **EM1:** 152
machines, filament winding **EM2:** 375
pattern, defined **EM2:** 45-46
patterns **EM1:** 26, 508-509
preparation **EM1:** 507
resins for **EM1:** 135-138
tension, defined **EM1:** 26 **EM2:** 46
of towpregs **EM1:** 152
Winding defects
power inductors **EL1:** 1004
Windings
of A15 conductors **A2:** 1069-1070
compact, heat removal from **A2:** 1028
rotor generator, superconducting materials for **A2:** 1057
Window *See also* Fish-eye
beryllium **A17:** 306
Window (continued)
defined **EM2:** 46
display, defined **A17:** 385
function, EXAFS data analysis **A10:** 413, 414
spectrometer, ultrathin beryllium **A10:** 519
Window fracture
from hydrogen damage in boiler tubes **A11:** 612
Window functions **A18:** 295
Window, process
for microstructure control **A14:** 412-413
Window, processing *See* Processing window
Window technique
for ATEM sample preparation **A10:** 451
Window technique of preparing transmission electron microscopy specimens **A9:** 105-107
Windowbelt panels, aircraft
eddy current inspection **A17:** 193
Windowing **A18:** 295
in machine vision process **A17:** 33
Windowless detector
EPMA spectra from **A11:** 38, 39
Windows
in wafer **EL1:** 198-199
Winer coefficients **A10:** 290
Wing dies
for press bending **A14:** 527
Wing nut, aircraft
intergranular fracture **A11:** 29
Wing slat track, aircraft
bending distortion in **A11:** 140-141
Wing-attachment bolt, aircraft
seam cracking **A11:** 530, 532
Wipe acid cleaning **M5:** 60-65
Wipe bending test devices **A8:** 125
Wipe etching *See* Swabbing
Wipe tinning **M5:** 354
Wipe-in *See* Fill-and-wipe
Wipe-on paint stripping method **M5:** 19
Wiped joint
definition **A6:** 1215 **M6:** 20
Wiper
defined **A18:** 21 **EL1:** 1161
Wiper dies
for bending **A14:** 666
Wiper forming *See also* Tangent bending
defined **A14:** 14
Wiper spring
failure of **A11:** 558-559
Wiping *See* Wiper forming
defined **A18:** 21
Wiping action
defined **EL1:** 1161
Wiping dies
for press bending **A14:** 525-527
Wiping test method
cleaning process efficiency **M5:** 20
Wire *See also* High-temperature superconductors for wires and tape; Wire drawing; Wire rod; Wire stranding; Wireforming
aluminum, EPMA analysis of connection failure **A10:** 531-532
annealed, in open resistance heaters **A2:** 830
beryllium-copper alloys **A2:** 403, 411
biological corrosion of **A13:** 116
boron fibers as **EM1:** 31
brazing filler metals available in this form **A6:** 119
broken **EL1:** 1012
and cable, wrought copper and copper alloys **A2:** 250-260
cerclage, of sensitized stainless steel intercrystalline corrosion on **A11:** 676, 681
chip and, assembly **EL1:** 110
classifications, wrought copper and copper alloys **A2:** 251-253
coated, as samples, x-ray spectrometry **A10:** 95
coating, wrought copper and copper alloys **A2:** 256-257

SUBJECTS OF THE INDEXED VOLUMES: ASM Handbook (designated by the letter "A"): **A1:** Properties and Selection: Irons, Steels, and High-Performance Alloys (1990); **A2:** Properties and Selection: Nonferrous Alloys and Special-Purpose Materials (1990); **A3:** Alloy Phase Diagrams (1992); **A4:** Heat Treating (1991); **A6:** Welding, Brazing, and Soldering (1993); **A7:** Powder Metallurgy (1984); **A8:** Mechanical Testing (1985); **A9:** Metallography and Microstructures (1985); **A10:** Materials Characterization (1986); **A11:** Failure Analysis and Prevention (1986); **A12:** Fractography (1987); **A13:** Corrosion (1987); **A14:** Forming and Forging (1988); **A15:** Casting (1988); **A16:** Machining (1989); **A17:** Nondestructive Testing and Quality Control (1989); **A18:** Friction, Lubrication, and Wear Technology (1992). **Metals Handbook, 9th Edition** (designated by the letter "M"): **M1:** Properties and Selection: Irons and Steels (1978); **M2:** Properties and Selection: Nonferrous Alloys and Pure Metals (1979); **M3:** Properties and Selection: Stainless Steels, Tool Materials and Special-Purpose Materials (1980); **M4:** Heat Treating (1981); **M5:** Surface Cleaning, Finishing, and Coating (1982); **M6:** Welding, Brazing, and Soldering (1983). **Engineered Materials Handbook** (designated by the letters "EM"): **EM1:** Composites (1987); **EM2:** Engineering Plastics (1988); **EM3:** Adhesives and Sealants (1990); **EM4:** Ceramics and Glasses (1991); **Electronic Materials Handbook** (designated by the letters "EL"): **EL1:** Packaging (1989).

Wire (continued)
coiled ... **A14:** 567
copper ... **M2:** 265-274
copper and copper alloys ... **A2:** 239, 250-260
defect, spring failure from ... **A11:** 554
defect, spring, fracture at ... **A11:** 554
defined ... **A14:** 14
dental, wrought alloy ... **A13:** 1356-1357
drawing of ... **A14:** 333-334
drawn high-carbon steel, distortion in ... **A11:** 139
electrical, amino molding compounds ... **EM2:** 230
electrical resistance alloys ... **A2:** 822
Elgiloy ... **A13:** 1356
extruded ... **EM2:** 385
fiber texture in ... **A9:** 701
fiber textures in ... **A10:** 245
filaments, ternary molybdenum chalcogenides (chevrel phases) ... **A2:** 1079
fine, mechanically alloyed oxide alloys ... **A2:** 949
dispersion-strengthened (MA ODS)
flat or rectangular, production ... **A2:** 256
forming of ... **A14:** 694-697
friction welding ... **M6:** 720-721
fully reversed loading in ultrasonic testing ... **A8:** 242
gage, as strain gage ... **A8:** 618
graphite-reinforced, properties ... **EM1:** 869
heavily drawn body-centered cubic, curly grain structure in ... **A9:** 687
high-temperature superconductors for ... **A2:** 1085-1089
iron, 98% reduction, cell structure ... **A9:** 688
iron, curly grain structure in ... **A9:** 688
laser beam welding ... **M6:** 664-665
lead, attached to strain gages ... **A8:** 202
lead-tin solder, to reduce load cell ringing ... **A8:** 193
manufacturing, structural ceramic ... **A2:** 1019
mesh heaters, for elevated/low temperature tension testing ... **A8:** 36
metal, as reinforcement ... **EL1:** 1119-1121
metallic ... **EM1:** 118
microhardness testing for ... **A8:** 96
microstripline properties ... **EL1:** 602
mounting ... **A9:** 31
mounting with thermosetting epoxy ... **A9:** 167
multiple-slide forming of ... **A14:** 567
platinum, temperature effect on tensile strength ... **A13:** 800
preferred orientation in ... **A10:** 359
property limits, wrought aluminum and aluminum alloys ... **A2:** 106
refractory metal, as reinforcement, fiber-reinforced composites ... **A2:** 582
refractory metals and alloys ... **A2:** 558, 582-584
-related failures, package interior ... **EL1:** 1011-1013
rod, defined ... **A14:** 14
rolling, Turk's head machine ... **A14:** 694-695
rope ... **A13:** 141, 1294
special commodities ... **A1:** 851-852
specimen, for ultrasonic fatigue testing ... **A8:** 250
split, spring failure from ... **A11:** 555
stainless steel *See* Stainless steel, wire ... **A1:** 849-852 **M3:** 13-15
steel *See* Steel, wire
steel, curly lamellar structure ... **A9:** 688
steel rope, failures of ... **A11:** 515-521
steel, with coating ... **A14:** 697
superconducting multifilamentary ... **A14:** 341
as tantalum mill product ... **A7:** 770-771
tempers of ... **A1:** 850
testing grips for ... **A8:** 50
thermocouple, insulation ... **A2:** 882-883
thermoplastic polyurethanes (TPUR) ... **EM2:** 205
titanium *See* Titanium, wire
tungsten and non-sag, comparison ... **A7:** 767
wrought aluminum alloy ... **A2:** 33
wrought beryllium-copper alloys ... **A2:** 409
wrought copper ... **A2:** 250-260
wrought orthodontic ... **A13:** 1362
zinc-coated, atmospheric corrosion rates ... **A13:** 527
zirconium ... **A2:** 663
Wire bond
defined ... **EL1:** 1161
Wire bond (continued)
degradation, as environmental failure mechanism ... **EL1:** 494
failures, electron spectroscopy applications ... **EL1:** 1081-1083
lead frame assembly ... **EL1:** 487
positioning, as vitreous dielectric application ... **EL1:** 109
Wire bond strength tests
package- level ... **EL1:** 934-935
Wire bonding
at level 1 ... **EL1:** 76
automated ... **EL1:** 226
as chip interconnection method ... **EL1:** 224-236
as component attachment ... **EL1:** 349-350
cyanoacrylates ... **EM3:** 131
defined ... **EL1:** 1161
density limitation factors ... **EL1:** 228
destructive tests ... **EL1:** 229
failure mechanisms ... **EL1:** 977-978
failures, plastic packages ... **EL1:** 480
interconnects, molded plastic packages ... **EL1:** 472
methods ... **EL1:** 1042
pull strength ... **EL1:** 231
schematic ... **EL1:** 230
shear strength ... **EL1:** 231
thermal failures ... **EL1:** 62
thermocompression bonding ... **EL1:** 224-225
ultrasonic ... **EL1:** 225
Wire brushing
brushes, types used ... **M5:** 151-156
heat-resisting alloys ... **M5:** 566
for liquid penetrant inspection ... **A17:** 81, 82
magnesium alloys ... **M5:** 629, 649
nickel and nickel alloys ... **M5:** 674-675
painting process, abrasive blasting compared to ... **M5:** 476
rust and scale removal by ... **M5:** 12
satin finishing, aluminum and aluminum alloys ... **M5:** 574-575
stainless steel ... **M5:** 559
of stainless steel forgings ... **A14:** 230
titanium and titanium alloys ... **M5:** 656
Wire cloth screens ... **A7:** 176
Wire, copper
temper designations ... **M2:** 248
Wire diameter
effect on weld attributes ... **A6:** 182
Wire drawing *See also* Drawing; Wire ... **A18:** 223
accumulating-type continuous machine for ... **A14:** 333, 334
central burst prediction ... **A14:** 395
defined ... **A14:** 14
dies ... **A14:** 331, 336-337
of niobium-titanium superconducting materials ... **A2:** 1050
products, zinc and zinc alloy ... **A2:** 531
schematic ... **A18:** 59, 61, 62
as shape memory effect (SME) alloy application ... **A2:** 899
steel, from rods ... **A14:** 332
steel wire ... **M1:** 260-261
surface shear strain rate in torsion ... **A8:** 158
torsion testing ... **A14:** 373
torsional rotation rates ... **A8:** 158
von Mises effective strain rate ... **A8:** 158
of wrought copper and copper alloys ... **A2:** 255-258
Wire electrical discharge machining
in conjunction with EDM ... **A16:** 560
MMCs ... **A16:** 895-896, 897
refractory metals ... **A16:** 859
Wire flame guns
thermal spray coating ... **M5:** 365-366
Wire flame spraying *See* Oxyfuel wire spray process ... **A13:** 45
definition
Wire flattening rolls
cemented carbide ... **A2:** 970
Wire forming *See also* Wire ... **A14:** 694-697
accuracy ... **A14:** 695
lubrication ... **A14:** 696-697
manual and power bending ... **A14:** 695-696
material condition, effects ... **A14:** 694
in multiple-slide machines ... **A14:** 696
operations ... **A14:** 694
production problems/solutions ... **A14:** 696
rolling, in Turk's head machine ... **A14:** 694-695
Wire forming (continued)
speed ... **A14:** 694
spring coiling ... **A14:** 695
Wire forms ... **A1:** 302
applications ... **A14:** 694
steels for ... **M1:** 284
Wire gage systems ... **M1:** 259-260
Wire injection
plain carbon steels ... **A15:** 709-710
Wire jacketing
polyamide (PA) ... **EM2:** 125
Wire lead
defined ... **EL1:** 1161
Wire lines
scanning laser gages for ... **A17:** 12
Wire, nonferrous
spring materials ... **M1:** 283, 284, 285, 286
Wire products *See* Fasteners; Fence; Rope; Springs
Wire rod
continuous casting ... **A2:** 254
copper ... **M2:** 266-271
fabrication ... **A2:** 253-255
GE dip-form process ... **A2:** 255
Hazelett process ... **A2:** 255
Outokumpu process ... **A2:** 255
Properzi system ... **A2:** 255
rolling ... **A2:** 253-254
southwire continuous rod system ... **A2:** 254-255
Wire rod, steel *See also* Alloy steel wire rod; Carbon steel wire rod 253-257; Steel wire rod
Wire saws for sectioning ... **A9:** 25-26
Wire spring relay
defined ... **EL1:** 1161
Wire springs
failures in ... **A11:** 555
Wire, steel
annealing ... **M4:** 4, 7, 25
cleaning ... **M1:** 262
coatings ... **M1:** 262-265
container applications ... **M1:** 264
fasteners ... **M1:** 265-266
fence ... **M1:** 269
fine wire ... **M1:** 269
finishes ... **M1:** 261-262
gage systems ... **M1:** 259-260
heat treatment ... **M1:** 262
metal-coated wire ... **M1:** 263
packaging applications ... **M1:** 264
prestressed concrete applications ... **M1:** 264
quality descriptors and commodities ... **M1:** 263-264
rope ... **M1:** 265
shapes of wire ... **M1:** 259
sizes of wire ... **M1:** 259-260
specification wire ... **M1:** 262-263
spring materials ... **M1:** 266-268, 270, 283-285, 287-290, 296, 297, 301, 303, 305
standard size tolerances ... **M1:** 261
structural applications ... **M1:** 263, 264
tire beads ... **M1:** 269
upholstery construction ... **M1:** 266, 268
Wire straightener
definition ... **M6:** 20
Wire stranding
of wrought copper and copper alloys ... **A2:** 255-258
Wire stripping
waterjet machining ... **A16:** 525
Wire sweep
from encapsulation ... **EL1:** 809
Wire wheels
for surface preparation ... **A17:** 52
Wire wooling ... **M1:** 606
shaft wear by ... **A11:** 466
Wire wound hot isostatic pressure vessels ... **A7:** 420
Wire(s) *See also* Bar; Steel bar; Steel wire
eddy current inspection ... **A17:** 553-555
electromagnetic inspection methods ... **A17:** 552-555
flaw detection ... **A17:** 557
flaws, types of ... **A17:** 549-550
inspection methods ... **A17:** 550-555
inspection of ... **A17:** 549-556
liquid penetrant inspection ... **A17:** 550-551
magabsorption measurements on ... **A17:** 154-155
magnetic particle inspection ... **A17:** 550
magnetic permeability systems ... **A17:** 555-557
NDE equipment requirements ... **A17:** 557
quality control ... **A17:** 736-737

Wire(s) (continued)
reversible permeability curves **A17:** 146
sorting procedures **A17:** 557
thin, measured by diffraction pattern technique **A17:** 13
ultrasonic inspection **A17:** 551-552
Wire-drawing dies **A14:** 331, 336-337
Wire-feed speed
definition **M6:** 20
Wire-feed systems
automatic torch brazing **M6:** 959
electrogas welding **M6:** 240
electroslag welding, drive systems **M6:** 228
flux cored arc welding **M6:** 97-99
gas metal arc welding **M6:** 159-161
stoppages **M6:** 172
submerged arc welding **M6:** 132
Wire-on-bolt test
galvanic corrosion **A13:** 238
Wire-tacking adhesives **EM3:** 569, 572
Wire-type penetrameters
radiographic inspection **A17:** 340-341
Wire-wound chip inductor **EL1:** 179, 187
Wire-wound metallic resistors *See also* Resistors
construction **EL1:** 178
failure mechanisms **EL1:** 971
Wire-wrapped circuit boards
custom **EL1:** 7
Wireability, effect
interconnections **EL1:** 18-20
Wirebar
copper **M2:** 266
Wirebars, copper alloy
grain structures **A9:** 641-642
Wiredrawing **A1:** 277-279
copper **M2:** 271-273
Wiredrawing dies
abrasion **M3:** 521, 522
cemented tungsten carbide **M3:** 522-523, 524
diamond **M3:** 521-522, 524
die breakage **M3:** 523
die life **M3:** 521, 522, 523, 524
tool steel **M3:** 523
Wireframe
as geometric modeler **A15:** 858
Wiremaking practices **M1:** 260-262
Wiring *See* Wire
demand, performance modeling **EL1:** 13-14
density, and thermal expansion **EL1:** 613
density, metal cores **EL1:** 620-621
design, flexible printed boards **EL1:** 581
failures, zone 2, package interior **EL1:** 1011-1013
in metal-processing equipment **A13:** 1315-1316
sources, for surfaces **EL1:** 115
surface, materials and processes selection **EL1:** 113-115
yield, WSI **EL1:** 354
Wiring boards, printed
as cyanate application **EM2:** 232, 237
Withdrawal force
defined **EL1:** 1161
Withdrawal press cycle
cermet forming **A2:** 982
Withdrawal tooling systems **A7:** 334
Witness marks
in copper alloy ingots **A9:** 642
Wobbulator
for FMR measurement **A17:** 222
Wolfram
evaporation fields for **A10:** 587
Wolframite
as tungsten-bearing ore **A7:** 152
Wolfs ear
planes of weakness **A8:** 155
Wollaston prism
in fatigue study **A12:** 121
role in differential interference contrast **A9:** 150-151
Wollaston process **A7:** 15-16
Wollastonite **A18:** 569 **EM4:** 1010
in ceramic tiles **EM4:** 926
chain structure **EM4:** 759
chemical system **EM4:** 872, 873
composition **EM4:** 932
crystal structure **EM4:** 881
Wollastonite (calcium metasilicate) **EM3:** 175
as filler **EM3:** 177
Wood
brewery use **A13:** 1221
as pattern material **A15:** 194, 243
substrate cure rate and bond strength for cyanoacrylates **EM3:** 129
surface parameter **EM3:** 41
thermal expansion rate **A7:** 611
versus tile whiteware **EM4:** 929
Wood (clean)
friction coefficient data **A18:** 75
Wood Adhesives—Chemistry and Technology **EM3:** 69
Wood and Wood Products Redbook **EM3:** 71
Wood and wood-base products
PCD tooling **A16:** 110
Wood, compressed
carbides for machining **A16:** 75
Wood failure **EM3:** 31
Wood flour **EM3:** 175
as extender **EM3:** 176
as sand addition **A15:** 211
Wood laminate **EM3:** 31
Wood laminates
as pattern material **A15:** 194
Wood, particle board
PCD tooling **A16:** 110
Wood products
analytic methods for **A10:** 9
Wood screw quality carbon steel wire rod **M1:** 254
Wood screw quality rod **A1:** 273
Wood veneer **EM3:** 31
Wood's Nickel
strikes as plating for **EL1:** 679
Wood/acrylonitrile-butadiene-styrene
acrylic properties **EM3:** 122
Wood/wood
acrylic properties **EM3:** 122
Woods
drilling **A16:** 226, 229
grinding (sanding) **A16:** 435
Woodworking
market **EM3:** 46
Woody fractures
alloy steels **A12:** 319, 333
appearance, in tensile fractures **A12:** 104
high-carbon steels **A12:** 281
in iron **A12:** 224
studies **A12:** 1-3
surface, tool steels **A12:** 375
in wrought iron **A12:** 224
Woof *See* Weft
Wool fiberglas
defect inclusion levels **EM4:** 392
Wool wire **A1:** 852
Worcra process **EM4:** 903
Word processors
P/M parts for **A7:** 667
Work
SI unit/symbol for **A8:** 721
Work angle
definition **M6:** 20
Work coil
definition **M6:** 20
Work connection
definition **A6:** 1215 **M6:** 20
Work envelope
coordinate measuring machines **A17:** 19
Work factor
defined **A18:** 21
Work flow
in computer-aided design (CAD) **EL1:** 528-529
Work hardening *See also* Strain hardening **EM3:** 31
in austenitic manganese steel **A1:** 831-832
by abrasive waterjet cutting **A14:** 752
in cold-formed parts **A11:** 308
and creep **A8:** 305, 309
defined **A8:** 301 **A13:** 14 **EM1:** 26 **EM2:** 46
in dual-phase steels **A1:** 424, 426
in early casting **A15:** 16
effect on fatigue strength **A11:** 119
effect on microstructure during creep **A8:** 305
for embrittlement, in loose powder compaction **A7:** 298
and flow stress **A7:** 300
and green strength **A7:** 302
in hardness testing **A8:** 71
of heat-resistant alloys **A14:** 779
material, stress distribution in torsion testing of **A8:** 140
of metals **A14:** 299-300
of platinum dispersion-strengthened alloys **A7:** 722
rate, and fatigue **A11:** 102
and recovery, in creep **A8:** 301-302
in three-roll forming **A14:** 620
vs extrusion ratio **A14:** 300
Work hardening coefficient *See also* *n* value; Strain hardening coefficient
of sheet metals **A8:** 555, 556
Work hardening exponent *See* n value; Work hardening coefficient
Work lead
definition **A6:** 1215 **M6:** 20
Work materials *See* Work metal
Work metal *See also* Die materials; Tool materials; Workpiece(s)
aluminum alloy, for drop hammer forming **A14:** 656
for blanking **A14:** 449
composition, effect, contour roll forming **A14:** 624
contour roll forming **A14:** 624-635
and die life **A14:** 57
for fine-edge blanking and piercing **A14:** 472-473
finish, and press-brake forming **A14:** 540
hardness, piercing effects **A14:** 462
HERF processing **A14:** 105
for hot upset forging **A14:** 83
magnesium alloy, for drop hammer forming **A14:** 656-657
preparation, cold heading **A14:** 293
for press forming **A14:** 504
press-brake forming **A14:** 533
properties, spinning effects **A14:** 604
thickness, blanking effects **A14:** 456
thickness, press forming **A14:** 548-549
three-roll forming **A14:** 616
variables, in tube spinning **A14:** 678
Work of adhesion **A18:** 435 **EM3:** 623-624
Work of cohesion **A18:** 399, 400, 403
Work of fracture **EM4:** 36
Work rolls
for ring rolling **A14:** 124-125
in roll compacting **A7:** 406
Work-hardened aluminum alloys
identification of temper **A9:** 358
Work-hardening coefficient in the shear stress/shear strain flow equation **A18:** 34
Workability *See also* Bulk formability of steels; Bulk workability; Bulk workability testing; Ductility; Forgeability; Formability; Intrinsic workability Workability tests; Plastic deformation; Workability theory
of alloy systems, various **A8:** 575
alloys containing elements forming insoluble compounds **A8:** 575

SUBJECTS OF THE INDEXED VOLUMES: ASM Handbook (designated by the letter "A"): **A1:** Properties and Selection: Irons, Steels, and High-Performance Alloys (1990); **A2:** Properties and Selection: Nonferrous Alloys and Special-Purpose Materials (1990); **A3:** Alloy Phase Diagrams (1992); **A4:** Heat Treating (1991); **A6:** Welding, Brazing, and Soldering (1993); **A7:** Powder Metallurgy (1984); **A8:** Mechanical Testing (1985); **A9:** Metallography and Microstructures (1985); **A10:** Materials Characterization (1986); **A11:** Failure Analysis and Prevention (1986); **A12:** Fractography (1987); **A13:** Corrosion (1987); **A14:** Forming and Forging (1988); **A15:** Casting (1988); **A16:** Machining (1989); **A17:** Nondestructive Testing and Quality Control (1989); **A18:** Friction, Lubrication, and Wear Technology (1992). **Metals Handbook, 9th Edition** (designated by the letter "M"): **M1:** Properties and Selection: Irons and Steels (1978); **M2:** Properties and Selection: Nonferrous Alloys and Pure Metals (1979); **M3:** Properties and Selection: Stainless Steels, Tool Materials and Special-Purpose Materials (1980); **M4:** Heat Treating (1981); **M5:** Surface Cleaning, Finishing, and Coating (1982); **M6:** Welding, Brazing, and Soldering (1983). **Engineered Materials Handbook** (designated by the letters "EM"): **EM1:** Composites (1987); **EM2:** Engineering Plastics (1988); **EM3:** Adhesives and Sealants (1990); **EM4:** Ceramics and Glasses (1991); **Electronic Materials Handbook** (designated by the letters "EL"): **EL1:** Packaging (1989).

Workability (continued)
of alloys forming brittle second phase on cooling ... **A8:** 165, 575
alloys forming ductile second phase on cooling ... **A8:** 165, 575
of alloys forming ductile second phase on heating ... **A8:** 165, 575
alloys forming low-melting second phase on heating ... **A8:** 165, 575
alloys with elements forming soluble compounds ... **A8:** 575
application, bulk forming processes ... **A14:** 388-404
application of torsion test to determine ... **A8:** 154-184
assessment of bulk ... **A8:** 577-578
behaviors of alloy systems ... **A8:** 165
bulk ... **A8:** 571-597
of cast and wrought metals at varied temperatures ... **A8:** 574
in closed-die forging ... **A8:** 590-591
common specimen shapes for ... **A8:** 156
and compressive stress ... **A8:** 576
computer-aided finite element analysis of plastic deformation and heat flow ... **A8:** 577
criteria for centerbursting in aluminum alloy ... **A8:** 577-578
data, for forging process design ... **A14:** 439-441
defined ... **A8:** 154, 571 **A14:** 19, 159, 363
diagram, for free surface fracture ... **A14:** 19L -20
die chilling simulation effects, multiphase alloys ... **A8:** 180
in drawing ... **A8:** 592-593
dynamic material modeling of ... **A14:** 370-371
evaluating ... **A8:** 571-578
in extrusion and drawing ... **A8:** 591-593
and failure ... **A8:** 154
flow localization analyses ... **A8:** 169-173, 573
and flow stress as function of temperature ... **A14:** 166-170
and fracture ... **A8:** 154, 571-573, 577
fracture criteria ... **A14:** 370
fracture mechanisms ... **A14:** 363-364
friction ... **A8:** 575-576
gage length and failure ... **A8:** 156
grain size and structure ... **A8:** 573-574
of heat-resistant alloys ... **A14:** 232
insoluble compound alloys ... **A8:** 165
introduction ... **A14:** 363-372
limits, cold rolling ... **A8:** 594
material factors affecting ... **A8:** 155, 571-574 **A14:** 363-367
measurement and prediction of deformation limits before fracture ... **A8:** 571
measuring flow-localization-controlled ... **A8:** 169
metallurgical considerations ... **A8:** 573-574
microstructure development during deformation ... **A8:** 173-178
nonisothermal upset test ... **A8:** 588
platinum ... **A2:** 709
platinum-rhodium alloys ... **A2:** 710
of porous preforms ... **A14:** 193
prediction *See* Workability tests
process variables controlling ... **A8:** 574-577 **A14:** 367-370
pure metal ... **A8:** 165, 574-575
and reduction in area rating scale ... **A8:** 586
in rolling ... **A8:** 593-596
single-phase alloy ... **A8:** 165, 474-475
soluble compound alloys ... **A8:** 165
stainless steel ... **A8:** 166
and strain ... **A8:** 166-168, 572, 574-575
and stress state ... **A8:** 572, 576
and temperature ... **A8:** 165, 178, 572, 574-575
temperature dependence for alloys ... **A8:** 165
tests ... **A14:** 373-387
tests for forging ... **A8:** 587-591
theory, bulk forming processes ... **A14:** 388-404
thin-walled tubular specimen ... **A8:** 156
of titanium alloys ... **A14:** 838
and torsion testing ... **A8:** 155-160
warm, carbon content effects ... **A14:** 174
wrought aluminum and aluminum alloys ... **A2:** 30-32, 111
yielding ... **A8:** 576-577
ZGS platinum ... **A2:** 714

Workability test
Lee-Kuhn ... **A7:** 410, 411
Workability tests
bend test ... **A14:** 377
compression process analysis ... **A14:** 376-377
compression test ... **A14:** 374-375
ductility testing ... **A14:** 376
for flow localization ... **A14:** 384-385
forgeability ... **A14:** 382-384
forging defects ... **A14:** 385-386
hot tension testing ... **A14:** 381-382
plain-strain compression test ... **A14:** 377-379
plastic instability in compression ... **A14:** 376-377
primary ... **A14:** 373-377
ring compression test ... **A14:** 379-381
secondary-tension test ... **A14:** 379
specialized ... **A14:** 377-382
tension test ... **A14:** 373
torsion test ... **A14:** 373-374
Workability theory
and applications ... **A14:** 396-403
in bulk forming processes ... **A14:** 388
empirical criterion of fracture ... **A14:** 389-393
fracture models and criteria ... **A14:** 393-396
stress and strain states ... **A14:** 388-389
Workhardening *See also* Strain hardening
in plastic deformation ... **A9:** 685
Working
fluids, as corrosive ... **A11:** 209
mechanical, banding from ... **A11:** 315
of platinum group metals ... **A14:** 849-851
superficial, brittle fracture from ... **A11:** 327
Working channels
in borescopes/fiberscopes ... **A17:** 8
Working curves
low- and elevated- temperature design properties ... **A8:** 670-671
Working distance ... **A18:** 378
Working distance (WD)
secondary electron imaging (SEI) ... **EL1:** 1096-1097
Working distance of objective lenses ... **A9:** 73
defined ... **A9:** 19
Working electrode *See also* Electrodes; Reference electrodes
defined ... **A13:** 14
Working hardness
for coining ... **A14:** 182
Working length
of borescopes ... **A17:** 9
of videoscopes ... **A17:** 7
Working life *See also* Gelation time; Pot; Pot life ... **EM3:** 31
defined ... **EM1:** 26 **EM2:** 46
Working stress
polyaryl sulfones (PAS) ... **EM2:** 145
Workpiece *See also* Specimen
for Brinell testing ... **A8:** 84
with curved surfaces, Rockwell hardness testing of ... **A8:** 81, 83
cylindrical, Rockwell correction factors for ... **A8:** 82
mounting, Rockwell hardness testing ... **A8:** 80
primary and secondary deformation processes ... **A7:** 522
for Scleroscope hardness testing ... **A8:** 105
surface finish, for Scleroscope hardness testing ... **A8:** 105
thermocouple installation for hot isostatic pressing cycles ... **A7:** 438
thickness, in Brinell test ... **A8:** 85, 88
Workpiece How strength
nomenclature for lubrication regimes ... **A18:** 90
Workpiece lead
definition ... **A6:** 1215
Workpiece(s) *See also* Die materials; Shapes; Tool materials; Work metal
asymmetrical, drawing of ... **A14:** 586
complex, for cold heading ... **A14:** 294
configuration, radial forging ... **A14:** 17
configuration, rotary forging ... **A14:** 176-177
constitutive response, data base ... **A14:** 412
deformation, schematic ... **A14:** 178
design, and die life ... **A14:** 57
detail, in coining ... **A14:** 180
drawn, expanding of ... **A14:** 586-587
ejection, deep drawing ... **A14:** 587
electromagnetic forming ... **A14:** 645

Workpiece(s) (continued)
feeding, in rotary swaging ... **A14:** 134
with flanges, drawing of ... **A14:** 585-586
handling equipment for ... **A14:** 63
hardness, electric current effects ... **A17:** 110
heated, for beryllium forming ... **A14:** 806
large, hot upset forging of ... **A14:** 93
long, rotary swaging tools for ... **A14:** 134
magnesium alloy, heating of ... **A14:** 826
multicolored, for physical modeling ... **A14:** 437
for multiple-slide forming ... **A14:** 567
precleaning, for liquid penetrant inspection ... **A17:** 80-82
preparation, for coining ... **A14:** 180
rotation, rotary swaging ... **A14:** 129
shape, for press forming ... **A14:** 549
size, for coining ... **A14:** 180
swaged ... **A14:** 128, 141
tapered aluminum, swaged ... **A14:** 141
temperature, as critical forging factor ... **A14:** 231
temperature, precision forging ... **A14:** 161-162
thermophysical properties, data base ... **A14:** 412
tolerance, and die life ... **A14:** 57
unacceptable, by liquid penetrant inspection ... **A17:** 86
workability, in precision forging ... **A14:** 159
World Materials Congress ... **EM3:** 71
Worm gears
and worm gear sets ... **A11:** 586-589
Worm holes ... **A6:** 408
Wormhole porosity
in weldments ... **A17:** 583
Worn parts
laboratory examination ... **A11:** 156-158
Woven broad goods ... **EM1:** 125-127
Woven fabric
damping in ... **EM1:** 213
defined ... **EM1:** 26 **EM2:** 46
prepregs ... **EM1:** 148-150
tests for ... **EM1:** 291
Woven fabric prepregs ... **EM1:** 148-150
fabric construction ... **EM1:** 148-149
fabric mechanical properties ... **EM1:** 150
fabric prepreg forms ... **EM1:** 149
fabrication techniques ... **EM1:** 150
hybrids ... **EM1:** 149-150
resin application ... **EM1:** 149
Woven fabric properties
aramid fibers ... **EL1:** 615
Woven fibrous composites
damping analysis of ... **EM1:** 213
Woven preforms
multidirectional carbon/carbon composite ... **EM1:** 915-917
Woven roving *See also* Roving; Rovings ... **EM1:** 114
aramid fiber
defined ... **EM1:** 26 **EM2:** 46
and fiberglass mat ... **EM1:** 109
fiberglass, production process ... **EM1:** 109
Woven-wire fence
steel ... **M1:** 271
Wrap bending test devices ... **A8:** 125
Wrap dies
for precision forging ... **A14:** 52
Wrap forming *See* Stretch forming
Wrap seam
defined ... **EM2:** 46
Wrap-around bend
defined ... **EM2:** 46
Wrapped bush (bearing)
defined ... **A18:** 21
Wrapping
for composite tube ... **EM1:** 569-572
stretch ... **A14:** 594
WRC-1988 diagram ... **A6:** 457, 501, 678, 680, 686, 687, 688, 693, 703, 818-819, 825
WRC-1992 diagram ... **A6:** 82, 83, 457, 459, 460-461, 462, 463, 471, 473, 501, 678, 679, 680, 681, 685, 811, 812, 819, 823, 825
Wrinkle
defined ... **EM1:** 26 **EM2:** 46
Wrinkle depression
defined ... **EM2:** 46
Wrinkles
in adhesive-bonded joints ... **A17:** 613
in pipe ... **A11:** 704

Wrinkling
defined **A8:** 15 **A14:** 14
effect of *n* value in **A8:** 551
as formability problem **A8:** 548
and fracture limits, conical cup drawing **A8:** 564
and material properties **A8:** 552 **A14:** 881
sheet metal **A14:** 878
in sheet metal forming **A8:** 548
strains, on forming limit diagram **A8:** 564
tests **A14:** 892-893
tests for **A8:** 563-564
and true compressive hoop strain **A8:** 564
Wrist pin bearing
defined **A18:** 21
Wrought
aluminum, torsional stress vs. life in high-cycle regime **A8:** 150
grain structure, at varied temperatures **A8:** 574
iron, ultimate shear stress for **A8:** 148
metal, and cast metal workability at varied temperatures **A8:** 574
nonferrous metals, Brinell test application **A8:** 89
steel, strain-life curve for **A8:** 573-574
Wrought alloys
aluminum **M2:** 44-62
anisotropy of **A14:** 367
corrosion testing **A13:** 193-194
grain refining of **A15:** 479-480
intergranular corrosion evaluation **A13:** 240
prehistoric **A15:** 15
weldability **A15:** 535
for wires **A13:** 1356
zinc **M2:** 635-637
Wrought aluminum *See also* Aluminum; Aluminum alloys; Aluminum alloys, specific types; Wrought aluminum alloys; Wrought aluminum alloys, specific types
designation system **A2:** 15
properties **A2:** 62-122
Wrought aluminum alloy series, 1*xxx* through 7*xxx*
characteristics **A2:** 29, 32-33
Wrought aluminum alloys *See also* Aluminum; Aluminum alloys; Aluminum alloys, specific types; Cast aluminum; Cast aluminum alloys; Wrought aluminum; Wrought aluminum alloys, specific types **A7:** 525-527
alloy designation series **A2:** 29, 32-33
alloying, general effects **A2:** 44-46
antimony alloying **A2:** 46
applications **A2:** 29-32
arsenic alloying **A2:** 46
atmospheric corrosion **A13:** 596, 600
bend properties **A2:** 58
beryllium alloying **A2:** 46
bismuth alloying **A2:** 46
boron alloying **A2:** 46-47
cadmium alloying **A2:** 47
calcium alloying **A2:** 47
carbon alloying **A2:** 47
chromium alloying **A2:** 47
cliffs **A12:** 418
cobalt alloying **A2:** 47
composition and microstructure corrosion effects **A13:** 585-587
compositions **A2:** 17-21
compositions of **A7:** 526 **A9:** 359
copper alloying **A2:** 47-51
copper-magnesium alloying **A2:** 48
corrosion characteristics **A2:** 30-32
design of shapes **A2:** 34-36
designation system **A2:** 15
electrical and thermal conductivity **A7:** 742
elevated-temperature properties **A2:** 59
extrusions, fibrous interior structure **A12:** 416
fabrication characteristics **A2:** 30-32
fatigue behavior **A2:** 59
fatigue crack growth **A2:** 59
Wrought aluminum alloys (continued)
flowchart of processing techniques **A7:** 527
formability **A2:** 41-42
fractographs **A12:** 414-439
fracture toughness **A2:** 42-44, 58-60
fracture/failure causes illustrated **A12:** 217
gallium alloying **A2:** 51
general corrosion resistance ratings **A13:** 586
heat-treatable commercial, solution potentials **A13:** 584
heat-treatable, strengthening **A2:** 39-41
hydrogen alloying **A2:** 51
indium alloying **A2:** 51-52
iron alloying **A2:** 52
lead alloying **A2:** 52
lithium alloying **A2:** 52
low-temperature properties **A2:** 59-60
magnesium alloying **A2:** 52-53
magnesium-manganese alloying **A2:** 52
magnesium-silicide alloying **A2:** 52-53
manganese alloying **A2:** 53-54
mechanical property limits **A2:** 57
mercury alloying **A2:** 54
mill products, types **A2:** 33-34
molybdenum alloying **A2:** 54
niobium alloying **A2:** 54
nomenclatures **A2:** 4
non-heat-treatable, strengthening **A2:** 37-39
nonheat-treatable commercial, solution potentials **A13:** 584
phases in aluminum alloys **A2:** 36-37
phosphorus alloying **A2:** 54
physical metallurgy **A2:** 36-57
physical properties **A2:** 45-46
processing of **A7:** 525-527
properties of **A2:** 57-60, 62-122
SCC ratings **A13:** 593
silicon alloying **A2:** 54-55
silver alloying **A2:** 55
specific alloying elements and impurities **A2:** 46-57
strengthening mechanisms **A2:** 37-41
strontium alloying **A2:** 55
sulfur alloying **A2:** 55
superelastic eutectic, tensile-overload fracture **A12:** 437
tin alloying **A2:** 55
values, typical **A2:** 57
vanadium alloying **A2:** 55
zinc-magnesium alloying **A2:** 55-56
zinc-magnesium-copper alloying **A2:** 56
zirconium alloying **A2:** 56-57
Wrought aluminum alloys, specific types *See also* Aluminum; Aluminum alloys; Wrought aluminum; Wrought aluminum alloys; Wrought aluminum, specific types
67Al-33Cu, tensile overload fracture **A12:** 437
1100, fractured by gas explosion **A12:** 414
2011, applications and properties **A2:** 66-67
2014 Alclad, applications and properties **A2:** 67-68
2014, applications and properties **A2:** 67-68
2014-T6, fatigue fracture **A12:** 415-416
2014-T6 heat-treated forging, fatigue fracture **A12:** 417
2017, applications and properties **A2:** 68, 70
2024 Alclad, applications and properties **A2:** 70-71
2024, applications and properties **A2:** 70-71
2024-T3, dimple rupture **A12:** 417
2024-T3, fatigue fracture **A12:** 418
2025-T6, corrosion pit and toot mark fracture **A12:** 419
2025-T6, fatigue failure by inadequate shot peening **A12:** 420
2036, applications and properties **A2:** 71-72
2048, applications and properties **A2:** 74
2124, applications and properties **A2:** 74-75
2218, applications and properties **A2:** 75, 77-78
Wrought aluminum alloys, specific types (continued)
2219 Alclad, applications and properties **A2:** 79-80
2219, applications and properties **A2:** 79-80
2319, applications and properties **A2:** 80-81
2618, applications and properties **A2:** 81-82
3003 Alclad, applications and properties **A2:** 82-84
3003, applications and properties **A2:** 82-84
3105, applications and properties **A2:** 87
4032, applications and properties **A2:** 87-88
4043, applications and properties **A2:** 88-89
5005, applications and properties **A2:** 89
5050, applications and properties **A2:** 89-90
5052, applications and properties **A2:** 90-91
5056 Alclad, applications and properties **A2:** 91-92
5056, applications and properties **A2:** 91-92
5083, applications and properties **A2:** 92-93
5086 Alclad, applications and properties **A2:** 93-94
5086, applications and properties **A2:** 93-94
5154, applications and properties **A2:** 94-95
5182, applications and properties **A2:** 95
5252, applications and properties **A2:** 95-96
5254, applications and properties **A2:** 96-97
5356, applications and properties **A2:** 97
5454, applications and properties **A2:** 97-98
5456, applications and properties **A2:** 98-99
5456, laser beam weld, ductile fracture **A12:** 422
5457, applications and properties **A2:** 99-100
5652, applications and properties **A2:** 100
5657, applications and properties **A2:** 100
6005, applications and properties **A2:** 100-101
6009, applications and properties **A2:** 101
6010, applications and properties **A2:** 101-102
6061 Alclad, applications and properties **A2:** 102-103
6061, applications and properties **A2:** 102-103
6063, applications and properties **A2:** 103-104
6066, applications and properties **A2:** 104-105
6080, applications and properties **A2:** 105
6101, applications and properties **A2:** 105-106
6151, applications and properties **A2:** 106
6201, applications and properties **A2:** 106-107
6205, applications and properties **A2:** 107
6262, applications and properties **A2:** 107
6351, applications and properties **A2:** 107-108
6463, applications and properties **A2:** 108
7005, applications and properties **A2:** 108-109
7039, applications and properties **A2:** 109-111
7049, applications and properties **A2:** 111-113
7050, applications and properties **A2:** 113-114
7050-T7, fatigue fracture from cyclic stress **A12:** 421
7072, applications and properties **A2:** 114-115
7075 Alclad, applications and properties **A2:** 115-116
7075, applications and properties **A2:** 115-116
7075-T6, brittle fracture **A12:** 424
7075-T6, cleavage fracture **A12:** 424, 430
7075-T6, cold shut fracture **A12:** 435
7075-T6, cone-shaped fracture surface **A12:** 426
7075-T6, corrosion and leaves **A12:** 433
7075-T6, corrosion fatigue fracture, ductile striations **A12:** 431, 432
7075-T6, corrosion fatigue with brittle striations **A12:** 430, 432
7075-T6, corrosion-fatigue fracture **A12:** 430-433
7075-T6, fatigue fracture **A12:** 427, 429
7075-T6, fatigue fracture by cyclic stress **A12:** 430
7075-T6, high-cycle fatigue fracture **A12:** 426
7075-T6, intergranular and stress-corrosion cracking **A12:** 434
7075-T6, intergranular fracture **A12:** 431
7075-T6, solidification porosity **A12:** 431
7075-T6, stress-corrosion cracking **A12:** 433-436

Wrought aluminum alloys, specific types (continued)
7075-T6, stretching and serpentine glide ... A12: 434
7075-T6, tension-overload fracture ... A12: 423-425
7075-T6, tension-overload plane-strain fracture ... A12: 423
7075-T736 forging, effect of peening on dross inclusion ... A12: 422
7076, applications and properties ... A2: 116-118
7090, composition ... A7: 526
7091, composition ... A7: 526
7175, applications and properties ... A2: 118
7175-T736 forging, fatigue fracture by dross inclusion ... A12: 422
7178 Alclad, applications and properties ... A2: 119
7178, applications and properties ... A2: 119
7475, applications and properties ... A2: 119, 121-122
7475-T6, corrosion fatigue fracture ... A12: 432
9052, composition ... A7: 526
Al-5.6Zn-1.9Mg, corrosion-fatigue fracture ... A12: 438
Al-5.6Zn-1.9Mg, fatigue crack propagation ... A12: 439
Al-5.6Zn-1.9Mg, fatigue crack tip deformation ... A12: 438
Al-5.6Zn-1.9Mg, transgranular corrosion fatigue crack propagation ... A12: 438
Al-C, composition ... A7: 526
Al-Cu-C, composition ... A7: 526
Al-Fe-Ce, composition ... A7: 526
Al-Fe-Co, composition ... A7: 526
Al-Fe-Cr, composition ... A7: 526
Al-Li, composition ... A7: 526
Al-Mg-C, composition ... A7: 526
IN-9021, composition ... A7: 526
IN-9051, composition ... A7: 526

Wrought aluminum and aluminum alloys
cleaning and finishing ... M5: 574, 576, 583, 588, 597, 603-604, 606

Wrought aluminum, specific types
1050, applications and properties ... A2: 62
1060, applications and properties ... A2: 62-63
1100, applications and properties ... A2: 63-64
1145, applications and properties ... A2: 64
1199, applications and properties ... A2: 64–65
1350, applications and properties ... A2: 65-66

Wrought beryllium-copper alloys *See also* Beryllium-copper alloys
age hardening ... A2: 236
composition ... A2: 403
high-conductivity, age hardening ... A2: 406-407
high-strength, age hardening ... A2: 406
high-strength, composition ... A2: 403-404
high-strength, mechanical properties ... A2: 409
mechanical properties ... A2: 409
stamped, cold formed ... A2: 403

Wrought beryllium-nickel alloys *See also* Beryllium-nickel alloys
fabrication characteristics ... A2: 424
mechanical and physical properties ... A2: 423-424

Wrought carbon steels
stress-corrosion cracking in ... A11: 214-215

Wrought cobalt-base alloys
as implant materials ... A11: 672

Wrought cobalt-base superalloys
alloying elements, effect of ... A1: 951
applications for ... A1: 950, 967-968
compositions of ... A1: 965
mechanical properties
stress-rupture properties ... A1: 957, 962, 967
tensile properties ... A1: 958-959, 960-961
melting (incipient) temperatures ... A1: 956
microstructure ... A1: 965-967
oxidation and hot corrosion ... A1: 968
physical properties ... A1: 963-964

Wrought commercial purity titanium
specific types ... A7: 475

Wrought copper alloys ... A13: 617-618, 629

Wrought copper products *See also* Copper; Copper alloys; Copper alloys, specific types; Wrought copper alloys; Wrought copper alloys, specific types; Wrought coppers; Wrought coppers, specific types
applications ... A2: 241

Wrought copper products (continued)
sheet and strip ... A2: 241-248
stress-relaxation characteristics ... A2: 260-263
tubular products ... A2: 248-250
wire and cable ... A2: 250-260

Wrought copper sheet and strip
annealing ... A2: 245-247
casting in book molds ... A2: 242
cleaning ... A2: 247
cold rolling to final thickness ... A2: 244-245
horizontal continuous casting ... A2: 243
hot rolling ... A2: 243
melting ... A2: 242
milling or scalping ... A2: 243-244
raw materials ... A2: 241-242
semicontinuous and continuous casting ... A2: 243
slitting, cutting and leveling ... A2: 247-248
stress relief ... A2: 247
vertical direct-chill (DC) semicontinuous casting ... A2: 243
vertical semicontinuous casting ... A2: 243

Wrought copper tubular products
applications ... A2: 248
cold drawing ... A2: 250
extrusion ... A2: 249
joints in ... A2: 249
product specifications ... A2: 250
production of finished tubes ... A2: 250
production of tube shells ... A2: 249-250
tube properties ... A2: 249
tube reducing ... A2: 250

Wrought copper wire and cable
copper classifications, for conductors ... A2: 251
history ... A2: 250
insulation and jacketing ... A2: 258-260
round wire ... A2: 251-252
square and rectangular wire ... A2: 252
stranded wire ... A2: 252-253
tin-coated wire ... A2: 253
wire and cable classifications ... A2: 251-253
wire rod, fabrication of ... A2: 253-255
wiredrawing and wire stranding ... A2: 255-258

Wrought coppers
applications ... A2: 220-223
availability ... A2: 216
properties ... A2: 217-219

Wrought coppers and copper alloys
alloy systems ... A2: 238
aluminum bronze, applications and properties ... A2: 325-333
availability ... A2: 216
beryllium-copper, applications and properties ... A2: 284-290
brasses, applications and properties ... A2: 298-321
bronzes, applications and properties ... A2: 296-298, 312-321
color-controlled ... A2: 234
copper-nickel, applications and properties ... A2: 338-341
corrosion ratings ... A2: 229-230
free-machining, applications and properties ... A2: 277-280, 291-292
nickel-sliver alloys, applications and properties ... A2: 342-345
oxygen free, applications and properties ... A2: 265-269
phosphor bronze, applications and properties ... A2: 322-325
properties ... A2: 217-219, 265-345
silicon bronze, applications and properties ... A2: 334-336
temper designations ... A2: 234
tough pitch, applications and properties ... A2: 269-276

Wrought coppers and copper alloys, specific types
See also Wrought coppers and copper alloys
C10100, applications and properties ... A2: 265
C10200, applications and properties ... A2: 265
C10300, applications and properties ... A2: 265, 267
C10400, applications and properties ... A2: 267-268
C10500, applications and properties ... A2: 267-268
C10700, applications and properties ... A2: 267-268
C10800, applications and properties ... A2: 268-269
C11000, applications and properties ... A2: 269-272
C11100, applications and properties ... A2: 272-274
C11300, applications and properties ... A2: 274-275

Wrought coppers and copper alloys, specific types (continued)
C11400, applications and properties ... A2: 274-275
C11600, applications and properties ... A2: 274-275
C12500, applications and properties ... A2: 275-277
C12700, applications and properties ... A2: 275-277
C12800, applications and properties ... A2: 275-277
C12900, applications and properties ... A2: 275-277
C14300, applications and properties ... A2: 277
C14500, applications and properties ... A2: 277-278
C14700, applications and properties ... A2: 278-280
C15000, applications and properties ... A2: 280-281
C15100, applications and properties ... A2: 281
C15500, applications and properties ... A2: 281-282
C15710, applications and properties ... A2: 282
C15720, applications and properties ... A2: 282-283
C15735, applications and properties ... A2: 283
C16200, applications and properties ... A2: 283-284
C17000, applications and properties ... A2: 284-285
C17200, applications and properties ... A2: 285-287
C17300, applications and properties ... A2: 285-287
C17410, applications and properties ... A2: 287-288
C17500, applications and properties ... A2: 288-289
C17600, applications and properties ... A2: 289-290
C18100, applications and properties ... A2: 290
C18200, applications and properties ... A2: 290-291
C18400, applications and properties ... A2: 290-291
C18500, applications and properties ... A2: 290-291
C18700, applications and properties ... A2: 291-292
C19200, applications and properties ... A2: 292
C19210, applications and properties ... A2: 292-293
C19400, applications and properties ... A2: 293-294
C19500, applications and properties ... A2: 294
C19520, applications and properties ... A2: 294-295
C19700, applications and properties ... A2: 295
C21000, applications and properties ... A2: 295-296
C22000, applications and properties ... A2: 296-297
C22600, applications and properties ... A2: 297-298
C23000, applications and properties ... A2: 298-299
C24000, applications and properties ... A2: 299-300
C26000, applications and properties ... A2: 300-302
C26800, applications and properties ... A2: 302-304
C27000, applications and properties ... A2: 302-304
C28000, applications and properties ... A2: 302-305
C31400, applications and properties ... A2: 305
C31600, applications and properties ... A2: 305-306
C33000, applications and properties ... A2: 306
C33200, applications and properties ... A2: 306-307
C33500, applications and properties ... A2: 307
C34000, applications and properties ... A2: 307-308
C34200, applications and properties ... A2: 308-309
C34900, applications and properties ... A2: 309
C35000, applications and properties ... A2: 309
C35300, applications and properties ... A2: 308-309
C35600, applications and properties ... A2: 310
C36000, applications and properties ... A2: 310-311
C36500, applications and properties ... A2: 311
C36600, applications and properties ... A2: 311
C36700, applications and properties ... A2: 311
C36800, applications and properties ... A2: 311
C37000, applications and properties ... A2: 311
C37700, applications and properties ... A2: 312
C38500, applications and properties ... A2: 312-313
C40500, applications and properties ... A2: 313
C40800, applications and properties ... A2: 313-314
C41100, applications and properties ... A2: 314-315
C41500, applications and properties ... A2: 315
C41900, applications and properties ... A2: 315
C42200, applications and properties ... A2: 315-316
C42500, applications and properties ... A2: 316
C43000, applications and properties ... A2: 316-317
C43400, applications and properties ... A2: 317
C43500, applications and properties ... A2: 317-318
C44300, applications and properties ... A2: 318-319
C44400, applications and properties ... A2: 318-319
C44500, applications and properties ... A2: 318-319
C46400, applications and properties ... A2: 319-320
C46600, applications and properties ... A2: 319-320
C46700, applications and properties ... A2: 319-320
C48200, applications and properties ... A2: 320-321
C48500, applications and properties ... A2: 321
C50500, applications and properties ... A2: 321-322
C50710, applications and properties ... A2: 322
C51000, applications and properties ... A2: 322-323
C51100, applications and properties ... A2: 323
C52100, applications and properties ... A2: 324
C52400, applications and properties ... A2: 324-325

Wrought coppers and copper alloys, specific types (continued)
C54400, applications and properties A2: 325
C60600, applications and properties A2: 325
C60800, applications and properties A2: 325-326
C61000, applications and properties A2: 326
C61300, applications and properties A2: 326-327
C61400, applications and properties A2: 327-329
C61500, applications and properties A2: 329
C62300, applications and properties A2: 329-330
C62400, applications and properties A2: 330-331
C62500, applications and properties A2: 331
C63000, applications and properties A2: 331-332
C63200, applications and properties A2: 332-333
C63600, applications and properties A2: 333
C63800, applications and properties A2: 333-334
C65100, applications and properties A2: 334
C65400, applications and properties A2: 334-335
C65500, applications and properties A2: 335
C66400, applications and properties A2: 335-336
C68800, applications and properties A2: 336-337
C69000, applications and properties A2: 337
C69400, applications and properties A2: 337
C70250, applications and properties A2: 337-338
C70400, applications and properties A2: 338
C70600, applications and properties A2: 338-339
C71000, applications and properties A2: 339
C71500, applications and properties A2: 339-340
C71900, applications and properties A2: 340-341
C72200, applications and properties A2: 341
C72500, applications and properties A2: 342
C74500, applications and properties A2: 342-343
C75200, applications and properties A2: 343
C75400, applications and properties A2: 343-344
C75700, applications and properties A2: 344
C77000, applications and properties A2: 344-345
C78200, applications and properties A2: 345
Wrought duplex stainless steels *See also* Duplex stainless steels
intergranular corrosion A13: 127
Wrought extra-low interstitial titanium alloys for orthopedic implants A7: 658
Wrought heat-resistant alloys A9: 305-329 A14: 236, 779-784
compositions A9: 306
etchants A9: 308
grinding A9: 305-306
macroetchants A9: 306
macroetching A9: 305
microexamination A9: 307-309
microstructures A9: 308-312
mounting A9: 305
phases A9: 309-312
role of alloying elements A9: 309
sectioning A9: 305
specimen preparation A9: 305-307
Wrought heat-resistant alloys, specific types
16-25-6, after forging and stress relieving A9: 316-317
50Cr-50Ni, hot rolled, annealed A9: 324
50Cr-50Ni, sheet, annealed A9: 324
A-286, eta phase A9: 315
A-286, microstructure A9: 309
A-286, solution annealed and aged A9: 315
A-286, solution annealed and oil quenched A9: 314
AISI 650, after forging and stress relieving A9: 316-317
AISI 660, eta phase A9: 315
AISI 660, solution annealed and aged A9: 315
AISI 660, solution annealed and oil quenched A9: 314
AISI 661, different heat treatments compared A9: 316
AISI 662, solution annealed, effects of different aging processes A9: 315
AISI 680, different heat treatments compared A9: 317
Alloy 600, as-forged A9: 321
Wrought heat-resistant alloys, specific types (continued)
Alloy 625, solution annealed and air cooled A9: 321
Alloy 718, different heat treatments compared A9: 318-320
Astroloy, forgings, different heat treatments compared A9: 320-321
Discaloy, solution annealed, effects of different aging processes A9: 315
Elgiloy, cold drawn, and annealed A9: 326
Fe-18Cr-12Ni-1Nb, Z phase in A9: 312
Hastelloy X, different heat treatments compared A9: 317
Hastelloy X, microstructure A9: 309
Haynes 25, different heat treatments compared A9: 327-328
Haynes 25, heat treated and cold worked A9: 327-328
Haynes 188, cold rolled and heat treated A9: 328
Haynes 188, solution annealed and aged at different temperatures A9: 328
Incoloy 800, strip, mill-annealed condition A9: 316
Incoloy 901, heat treated, and creep tested to rupture A9: 316
Inconel 706, gamma double prime in A9: 311
Inconel 718, gamma double prime in A9: 311
Inconel 718, Laves phase in A9: 312
Inconel 718, microstructure A9: 309
Inconel X-750, different etchants compared A9: 322
Inconel X-750, solution annealed and aged A9: 321-322
MP35N, solution annealed, air cooled, cold worked A9: 329
MP35N, solution annealed and aged A9: 329
N-155, different heat treatments compared A9: 316
Nimonic 80, different heat treatments compared A9: 322
Nimonic 80A, carbides in A9: 311
Nimonic 80A, gamma prime in A9: 310-311
Pyromet 31, heat treated to form eta A9: 317-318
Pyromet, solution annealed and aged A9: 318
René 41, different heat treatments compared A9: 322-323
René 95, hot isostatically pressed, different illumination modes compared A9: 323
S-816, as-forged bar A9: 326
S-816, creep-rupture tested A9: 327
S-816, different heat treatments compared A9: 326
Stellite 6B, solution annealed and aged at different temperatures A9: 328
U-700, as-forged A9: 323
U-700, different heat treatments compared A9: 323-324
U-710, bar, different heat treatments compared A9: 324
Udimet 500, gamma prime in A9: 311
Udimet 630, gamma double prime in A9: 311
Udimet 700, borides in A9: 312
Udimet 700, gamma prime in A9: 311
Wrought heat-resisting alloys
intergranular corrosion and chromium depletion in A11: 277
Wrought ingot metallurgy A7: 717
Wrought iron
chemical analysis and sampling A7: 249
nonmetallic inclusion in A9: 39, 42
ocean corrosion rates for A13: 898
phosphorus segregation, macroetching to reveal A9: 176
wear rates for test plates in drag conveyor bottoms A18: 720
woody impact fracture A12: 224
Wrought irons
thermal expansion coefficient A6: 907
Wrought magnesium alloys *See also* Magnesium; Magnesium alloys; Wrought magnesium alloys, specific types
cleaning and finishing M5: 629, 646
extruded bars and shapes A2: 459
forgings A2: 459
metal-matrix composites A2: 460
properties of A2: 480-491
sheet and plate A2: 459-460
Wrought magnesium alloys, specific types *See also* Wrought magnesium alloys
AS41A, properties A2: 492-493
AZ10A, properties A2: 480
AZ21X1, properties A2: 480
AZ31B, properties A2: 480-481
AZ31C, properties A2: 480-481
AZ61A, properties A2: 481-482
AZ80A, properties A2: 482-483
AZ91B, properties A2: 496-497
HK31A, properties A2: 483
HM21A, properties A2: 483-484
HM31A, properties A2: 484-485
MIA, properties A2: 485-487
PE, properties A2: 487-488
ZC71, properties A2: 488-489
ZK21A, properties A2: 489-490
ZK40A, properties A2: 490
ZK60A, properties A2: 490-491
Wrought martensitic stainless steel selection A6: 432-441
all-weld-metal chemical compositions for filler metals A6: 439
carbon content A6: 432, 433, 437, 438, 441
chemical compositions A6: 432
chromium content A6: 432, 433
electron-beam welding A6: 441
filler-metal selection A6: 438-439
flash welding A6: 441
friction welding A6: 441
general welding characteristics A6: 433
hardness A6: 433, 435
heat treatment A6: 438
heat-affected zone ... A6: 433, 435, 436, 438, 440, 441
hydrogen-induced cold cracking A6: 436-437, 438-439
laser-beam welding A6: 440-441
material composition and selection of preheat temperature A6: 437-438
mechanical properties in various conditions A6: 433, 434
microstructure A6: 432
molybdenum content A6: 432
nickel content A6: 432
non-arc welding processes A6: 440-441
physical properties in the annealed condition A6: 433, 435
resistance welding A6: 441
specific welding recommendations A6: 439-440
weld microstructure A6: 433-436
weldability A6: 436-438
Wrought molybdenum *See also* Molybdenum; Refractory metals and alloys
joining A2: 563
Wrought nickel
electrical resistivity A7: 403
Wrought nickel-base superalloys
alloying elements, effect of A1: 951
alloying for corrosion resistance A1: 956-957
applications for A1: 950
coatings for oxidation and corrosion resistance A1: 957, 959
compositions of A1: 950-951
forming of A1: 971
hot forming temperatures A1: 969
heat treatment A1: 956
low-temperature corrosion A1: 956

SUBJECTS OF THE INDEXED VOLUMES: ASM Handbook (designated by the letter "A"): **A1:** Properties and Selection: Irons, Steels, and High-Performance Alloys (1990); **A2:** Properties and Selection: Nonferrous Alloys and Special-Purpose Materials (1990); **A3:** Alloy Phase Diagrams (1992); **A4:** Heat Treating (1991); **A6:** Welding, Brazing, and Soldering (1993); **A7:** Powder Metallurgy (1984); **A8:** Mechanical Testing (1985); **A9:** Metallography and Microstructures (1985); **A10:** Materials Characterization (1986); **A11:** Failure Analysis and Prevention (1986); **A12:** Fractography (1987); **A13:** Corrosion (1987); **A14:** Forming and Forging (1988); **A15:** Casting (1988); **A16:** Machining (1989); **A17:** Nondestructive Testing and Quality Control (1989); **A18:** Friction, Lubrication, and Wear Technology (1992). **Metals Handbook, 9th Edition** (designated by the letter "M"): **M1:** Properties and Selection: Irons and Steels (1978); **M2:** Properties and Selection: Nonferrous Alloys and Pure Metals (1979); **M3:** Properties and Selection: Stainless Steels, Tool Materials and Special-Purpose Materials (1980); **M4:** Heat Treating (1981); **M5:** Surface Cleaning, Finishing, and Coating (1982); **M6:** Welding, Brazing, and Soldering (1983). **Engineered Materials Handbook** (designated by the letters "EM"): **EM1:** Composites (1987); **EM2:** Engineering Plastics (1988); **EM3:** Adhesives and Sealants (1990); **EM4:** Ceramics and Glasses (1991); **Electronic Materials Handbook** (designated by the letters "EL"): **EL1:** Packaging (1989).

Wrought nickel-base superalloys (continued)
mechanical properties
stress-rupture properties **A1:** 954, 955, 956, 957, 959, 962, 966
tensile properties **A1:** 958-959, 960-961
melting (incipient) temperatures **A1:** 856
melting processes
alloy type versus melting process **A1:** 971
electroslag remelting **A1:** 970
vacuum arc remelting **A1:** 970
vacuum double electrode remelting **A1:** 970
vacuum induction melting **A1:** 970
microstructure **A1:** 951-956
borides **A1:** 955-956
carbides **A1:** 954-955
gamma double prime **A1:** 953-954
gamma matrix **A1:** 952-953
gamma prime **A1:** 952-953
grain-boundary chemistry **A1:** 954
topologically close-packed phases **A1:** 956
physical properties **A1:** 963-964
thermomechanical processing **A1:** 963-964
Wrought orthodontic wires
of precious metals **A2:** 696
Wrought oxide-dispersion strengthened P/M superalloys **A7:** 527-528
Wrought P/M processing **A7:** 522-529
Wrought P/M stainless steels **A13:** 826
Wrought permalloy
from powder **A7:** 641
Wrought products *See also* Aluminum mill and engineered wrought products; Cast products; Wrought copper products **A7:** 522-529
aluminum **A2:** 29-61
beryllium **A2:** 687 **A7:** 758-759
liquid penetrant inspection of **A17:** 71
modern developments **A7:** 18
in P/M history **A7:** 18
quality control tests **A12:** 141-142
tubular **A17:** 561-581
Wrought recrystallized metals
workabilities **A14:** 366
Wrought semisolid metalworking processes **A15:** 332-333
Wrought stainless steels **A1:** 841-907 **A9:** 279-296
classification of **A1:** 841-842
nonstandard types **A1:** 842, 847-848
standard types **A1:** 842, 843, 844, 845, 846
compositions **A9:** 280
compositions of **A1:** 843, 847-848
corrosion in specific environments **A1:** 873
architectural **A1:** 882
atmospheric corrosion **A1:** 873-874
corrosion in chemical environments **A1:** 875-878
corrosion in various applications **A1:** 878-882
corrosion in water **A1:** 874-875, 876
food and beverage industries **A1:** 878
oil and gas industry **A1:** 879-880
pharmaceutical industry **A1:** 879
power industry **A1:** 879-880
pulp and paper industry **A1:** 880-881
transportation industry **A1:** 881-882
corrosion properties **A1:** 869-884
corrosion testing **A1:** 882-883
pitting and crevice corrosion **A1:** 883-884
creep rupture of 304 stainless steel **A1:** 622
effect of composition on corrosion **A1:** 871-872
carbon content **A1:** 872
chromium content **A1:** 871
manganese content **A1:** 872
molybdenum content **A1:** 872
nickel content **A1:** 871-872
nitrogen content **A1:** 872
elevated-temperature properties **A1:** 861-863, 930-949
of austenitic stainless steels **A1:** 944-949
corrosion **A1:** 935-936
creep and stress rupture **A1:** 932-933
creep-fatigue interaction **A1:** 934-935
of ferritic stainless steels **A1:** 936-937
of martensitic stainless steels **A1:** 939-942
of precipitation-hardening alloys **A1:** 942-944
rupture ductility **A1:** 933-934
embrittlement (475 °C) **A1:** 708
etching **A9:** 281-282
fabrication characteristics **A1:** 887-895
Wrought stainless steels (continued)
factors in selection **A1:** 842
corrosion resistance **A1:** 842, 844-845, 849, 850
fabrication and cleaning **A1:** 845
mechanical properties **A1:** 845
surface finish **A1:** 845
fatigue strength **A1:** 861, 863
forgeability of **A1:** 889-894
austenitic stainless steels **A1:** 891-893
closed-die forgeability **A1:** 890-891
duplex stainless steels **A1:** 894
ferritic stainless steels **A1:** 893
ingot breakdown **A1:** 890
martensitic stainless steels **A1:** 892, 893
precipitation-hardening stainless steels **A1:** 893-894
formability of **A1:** 888-889
austenitic stainless steels **A1:** 888-889
duplex stainless steels **A1:** 889
ferritic stainless steels **A1:** 889
forms of corrosion **A1:** 869-873
crevice **A1:** 873
erosion-corrosion **A1:** 873
galvanic **A1:** 872
general **A1:** 872
intergranular **A1:** 873
oxidation **A1:** 873
pitting **A1:** 872-873
stress-corrosion cracking **A1:** 873
grinding **A9:** 279
hydrogen damage **A1:** 715
in-service care **A1:** 887
influence of product form on
cast structures **A1:** 867-868
cold reduced products **A1:** 868, 869
hot processing **A1:** 868
properties **A1:** 865, 867
interim surface protection **A1:** 886-887
machinability of **A1:** 894-897
austenitic stainless steels **A1:** 894-896
duplex stainless steels **A1:** 896
ferritic and martensitic stainless steels **A1:** 894
precipitation-hardening alloys **A1:** 896-897
macroetching **A9:** 279
macroexamination **A9:** 279
mechanism of corrosion resistance **A1:** 870-871
microexamination **A9:** 279-283
microstructures **A9:** 283-286
mill finishes **A1:** 885-886
plate finishes **A1:** 886
sheet finishes **A1:** 885-886
strip finishes **A1:** 886
wire finishes **A1:** 886
mounting **A9:** 279
notch toughness and transition temperature **A1:** 859-461, 865
for orthopedic implants **A7:** 658
physical properties **A1:** 868-869, 871
polishing **A9:** 279-281
product forms **A1:** 845-853, 930, 932
bar **A1:** 848-849
foil **A1:** 848
pipe, tubes, and tubing **A1:** 852-853
plate **A1:** 843, 845-846
semifinished products **A1:** 852
sheet **A1:** 843, 846
strip **A1:** 846, 848
wire **A1:** 849-852
production tonnage **A1:** 841
recycling **A1:** 1027-1028
blending **A1:** 1032
by industry **A1:** 1028
collection **A1:** 1029
degreasing **A1:** 1032
demand **A1:** 1028
processing **A1:** 1028-1032
secondary nickel refining **A1:** 1032
separation and sorting **A1:** 1028, 1029-1032
size reduction and compaction **A1:** 1032
recycling of metallurgical wastes **A1:** 1032
sectioning **A9:** 279
sensitization **A1:** 706-708
of austenitic stainless steels **A1:** 706-707
of duplex stainless steels **A1:** 707-708
sheet formability **A1:** 888-889
power requirements **A1:** 889
Wrought stainless steels (continued)
stress-strain relationships **A1:** 889
shipments of **A1:** 841, 842
sigma-phase embrittlement **A1:** 708-711
of austenitic stainless steels **A1:** 708-711
of duplex stainless steels **A1:** 711
of ferritic stainless steels **A1:** 709-711
specimen preparation **A9:** 279-283
stress-corrosion cracking **A1:** 725-728, 873
of austenitic stainless steels **A1:** 725-726
of duplex stainless steels **A1:** 727
subzero-temperature properties **A1:** 847-848, 863-865
fatigue strength **A1:** 865
fracture crack growth rates **A1:** 865, 869, 870
fracture toughness **A1:** 865, 867
tensile properties **A1:** 865, 866, 867, 868, 869
surface finishing of **A1:** 884-885
electropolishing **A1:** 885
grinding, polishing, and buffing **A1:** 884-885
tensile properties **A1:** 853
austenitic types **A1:** 853-858
duplex (austenite/ferrite) types **A1:** 856, 858
ferritic types **A1:** 856, 860
martensitic types **A1:** 858, 862-863
precipitation-hardening types **A1:** 858-859, 864-865
weldability of **A1:** 897-905
austenitic stainless steels **A1:** 898-901
duplex stainless steels **A1:** 904-905
ferritic stainless steels **A1:** 901-902
martensitic stainless steels **A1:** 902
precipitation-hardening steels **A1:** 902-904
Wrought stainless steels, specific types
7-Mo PLUS **A9:** 286
15-5PH, martensitic **A9:** 285
15-5PH, solution annealed and aged **A9:** 295
17-4PH, different etchants compared **A9:** 295-296
17-4PH, martensitic **A9:** 285
17-4PH, solution annealed and aged **A9:** 294
17-7PH, semiaustenitic **A9:** 285
18Cr-12Ni-1Nb, Z phase found in **A9:** 284
20Cb-3, solution annealed, different illumination modes compared **A9:** 290
22-13-5, solution annealed and cold drawn **A9:** 290
26Ni-15Cr, G phase found in **A9:** 284
44LN **A9:** 286
182-FM, resulfurized **A9:** 292
440A, different heat treatments compared **A9:** 293
A-286, austenitic **A9:** 285
A-286, G phase found in **A9:** 284
AISI 201, strip, annealed and rapidly cooled **A9:** 287
AISI 301, mill annealed and cold worked **A9:** 287
AISI 301, sheet, cold rolled, 10% and 40% reduction **A9:** 287
AISI 302, microstructure **A9:** 283
AISI 302, strip, annealed and rapidly cooled **A9:** 287
AISI 302B, microstructure **A9:** 283
AISI 303, manganese sulfide and manganese selenide inclusions **A9:** 291
AISI 303, resulfurized **A9:** 291
AISI 304, microstructure **A9:** 283
AISI 304, strip, annealed and air cooled **A9:** 287
AISI 304L, microstructure **A9:** 283
AISI 308, solution annealed and cold worked **A9:** 290
AISI 310, microstructure **A9:** 283
AISI 310, plate, hot rolled and annealed **A9:** 288
AISI 312, solution annealed and aged **A9:** 296
AISI 316, annealed, exposed to high temperature **A9:** 288
AISI 316, microstructure **A9:** 283-284
AISI 316, solution annealed and water quenched, different etches compared **A9:** 288
AISI 316, tubing, packed with boron nitride powder **A9:** 288
AISI 316L, cold drawn, different illumination modes compared **A9:** 290
AISI 316L, microstructure **A9:** 283

Wrought stainless steels, specific types (continued)
AISI 317, microstructure ... **A9:** 283
AISI 317L, microstructure ... **A9:** 283
AISI 321, annealed furnace part ... **A9:** 288
AISI 321, microstructure ... **A9:** 283-284
AISI 329 ... **A9:** 286
AISI 330, microstructure ... **A9:** 283
AISI 347, microstructure ... **A9:** 283-284
AISI 403, quenched and tempered ... **A9:** 292
AISI 403, sulfur added ... **A9:** 292
AISI 409, strip, annealed and air cooled ... **A9:** 291
AISI 410, tempering ... **A9:** 285
AISI 416, free-machining, annealed ... **A9:** 292-293
AISI 420, quenched and tempered ... **A9:** 292
AISI 420, sulfur added ... **A9:** 292
AISI 430, strip, annealed and air cooled ... **A9:** 291
AISI 430F, manganese sulfide stringers ... **A9:** 292
AISI 431, quenched and tempered ... **A9:** 293
AISI 434, modified ... **A9:** 292
AISI 440B, effects of polishing ... **A9:** 293
AISI 440C, different heat treatments compared ... **A9:** 293
AISI 440C, difficulties in preparation ... **A9:** 279
Alloy 2205 ... **A9:** 286
AM-350 ... **A9:** 285, 293
AM-355 ... **A9:** 285
AM-355, heat treated, different illumination modes compared ... **A9:** 294
Custom 450 ... **A9:** 285
Custom 450, solution annealed and aged ... **A9:** 294
Custom 455 ... **A9:** 285
Custom 455, solution annealed and aged ... **A9:** 294
E-Brite, plate ... **A9:** 292
Fe-12Cr-Co-Mo, R phase found in ... **A9:** 284
Ferralium Alloy 255 ... **A9:** 286
IN-744, sigma formation in ... **A9:** 286
PH13-8Mo ... **A9:** 285
PH13-8Mo, solution annealed and aged ... **A9:** 294
PH14-8Mo ... **A9:** 285
PH15-7Mo ... **A9:** 285
Stainless W ... **A9:** 285
Stainless W, solution annealed and aged ... **A9:** 296
Wrought steel
nonmetallic inclusions in ... **A11:** 317
Wrought steels *See also* Steels; Steels, wrought
anisotropy ... **A14:** 367
decarburization ... **A1:** 745, 746
dendritic structure revealed by Macroetching ... **A9:** 173
deoxidation practice ... **A1:** 741, 742-743
effects of surface condition on notch toughness in ... **A1:** 745
electroplating ... **A1:** 746
hot deformation temperature ... **A1:** 743-744
mechanical properties ... **A14:** 196
and powder forgings, compared ... **A14:** 196
section and part size ... **A1:** 745
toughness anisotropy ... **A1:** 744-745
weldability ... **M1:** 563
Wrought structure and ductility
in closed-die forgings ... **A1:** 340
Wrought superalloys *See* Wrought cobalt-base superalloys; Wrought heat- resistant alloys; Wrought nickel-base superalloys
Wrought titanium *See also* Titanium; Titanium alloys; Wrought titanium alloys; Wrought titanium alloys, specific types
alloys ... **A2:** 597-608
commercially pure, applications ... **A2:** 603
commercially pure titanium ... **A2:** 592-597
comparative properties of P/M and ... **A7:** 395
mechanical properties ... **A7:** 475
specific types ... **A7:** 475
Wrought titanium (continued)
specifications ... **A2:** 593
tensile properties ... **A2:** 630
Wrought titanium alloys *See also* Wrought titanium alloys, specific types
aged microstructures ... **A2:** 608
aging and overaging ... **A2:** 619
alloy classes ... **A2:** 600-605
alpha alloys ... **A2:** 600-601
alpha alloys, applications ... **A2:** 603
alpha double prime (orthorhombic martensite) ... **A2:** 606
alpha prime (hexagonal martensite) ... **A2:** 606
alpha stabilizers ... **A2:** 599
alpha structures ... **A2:** 606
alpha-beta alloys ... **A2:** 602
alpha-beta alloys, applications ... **A2:** 604
annealing ... **A2:** 619
applications ... **A2:** 603-604
beta alloys ... **A2:** 602, 605
beta alloys, applications ... **A2:** 604
beta microstructure ... **A2:** 607-608
beta transus temperatures ... **A2:** 623
compositions, alpha-beta alloys ... **A2:** 601
compositions, beta alloys ... **A2:** 602
effects, of alloy elements ... **A2:** 598-600
elevated-temperature mechanical properties ... **A2:** 624-627
eutectoid-forming group ... **A2:** 599
extrusion ... **A2:** 614
fatigue crack-growth rates ... **A2:** 631
fatigue properties ... **A2:** 623-624
forging ... **A2:** 611-614
forming ... **A2:** 614-616
fracture toughness ... **A2:** 623
heat treatment ... **A2:** 618-622
hydrogen contamination ... **A2:** 620
isomorphous group ... **A2:** 599
joining ... **A2:** 616-618
low-temperature properties ... **A2:** 628-631
mechanical properties ... **A2:** 621-622 **A7:** 475
microstructural elements ... **A2:** 606
near-alpha alloys, applications ... **A2:** 603-604
omega phase ... **A2:** 606-607
primary fabrication ... **A2:** 609-611
properties ... **A2:** 622-631
reduction to billet ... **A2:** 610
rolling (bar, plate, sheet) ... **A2:** 610-611
solution treating ... **A2:** 619
specifications ... **A2:** 593
stress relieving ... **A2:** 618
tensile properties ... **A2:** 630
tensile properties, room temperature ... **A2:** 621-623
thermal stability ... **A2:** 627-628
welding ... **A2:** 617-618
widely used types ... **A2:** 597-598
wrought alloy processing ... **A2:** 608-622
Wrought titanium alloys, specific types *See also* Wrought titanium; Wrought titanium alloys
IMI 829, for creep resistance ... **A2:** 598
Ti-5Al-2.5Sn, for weldability ... **A2:** 598
Ti-5Al-2Sn-2Zr-4Mo-4Cr *See* Ti-17
Ti-6Al-2Nb-1Ta-1Mo, characteristics ... **A2:** 598
Ti-6Al-2Sn-4Zr-2Mo *See* Ti-6242
Ti-6Al-2Sn-4Zr-2Mo, beta annealing ... **A2:** 620
Ti-6Al-2Sn-4Zr-6Mo, applications ... **A2:** 598
Ti-6Al-4V, as most widely used ... **A2:** 598
Ti-6Al-4V, beta annealing ... **A2:** 620
Ti-6Al-4V, characteristics ... **A2:** 598
Ti-6Al-4V, recrystallization annealing ... **A2:** 620
Ti-6Al-4V, solution treating and overaging ... **A2:** 619-620
Ti-6Al-4V-ELI , beta annealing ... **A2:** 620
Ti-6Al-4V-ELI, characteristics ... **A2:** 598
Ti-6Al-4V-ELI, recrystallization annealing ... **A2:** 620
Ti-6Al-6V-2Sn, characteristics ... **A2:** 598
TI-10V-2Fe-3Al, characteristics ... **A2:** 598
TI-17, applications ... **A2:** 598
Wrought titanium alloys, specific types (continued)
Ti-6242, for creep resistance ... **A2:** 598
Ti-6242S, for creep resistance ... **A2:** 598
Wrought tool steels *See also* specific types ... **A1:** 757-779
applications, selection guide ... **A1:** 763
classification and characteristics ... **A1:** 757-771
cold-work steels ... **A1:** 763-766
cross-reference of United States and foreign designations ... **A1:** 760
high-speed steels ... **A1:** 759-762
hot-work steels ... **A1:** 762-763
low-alloy special-purpose steels ... **A1:** 767
mold steels ... **A1:** 767-768
shock-resisting steels ... **A1:** 766-767
water-hardening steels ... **A1:** 768-771
cold-work steels ... **A1:** 763-766
air-hardening, medium alloy ... **A1:** 763-765
high-carbon, high-chromium ... **A1:** 765
oil-hardening ... **A1:** 765-766
composition of nonstandard steels ... **A1:** 762
distortion and safety in hardening ... **A1:** 777
fabrication of ... **A1:** 774-778
grindability ... **A1:** 775
hardenability ... **A1:** 775-777
machinability ... **A1:** 774-775
resistance to decarburization ... **A1:** 778
of standard steels ... **A1:** 758-759
weldability ... **A1:** 775
general properties factors in tool selection ... **A1:** 776-777
characteristics ... **A1:** 772-773
processing and service
heat treating hardening and tempering ... **A1:** 770
normalizing and annealing temperatures ... **A1:** 769
high-speed steels ... **A1:** 759-762
molybdenum ... **A1:** 759
tungsten ... **A1:** 759-762
hot-work steels ... **A1:** 762-763
chromium ... **A1:** 762
molybdenum ... **A1:** 763
tungsten ... **A1:** 762-763
low-alloy special-purpose steels ... **A1:** 767
machinability
of annealed tool steels ... **A1:** 777
grindability index ... **A1:** 778
machining allowances ... **A1:** 778-779
hot-rolled square and flat bars ... **A1:** 778
mechanical properties
hot hardness of die steels ... **A1:** 777
hot hardness of high-speed steels ... **A1:** 787
of low-alloy special-purpose steels at room temperature ... **A1:** 767
of shock-resisting steels at room temperature ... **A1:** 767
mold steels ... **A1:** 767-768
precision cast hot-work ... **A1:** 779
shock-resisting steels ... **A1:** 766-767
surface treatments ... **A1:** 779
carburizing ... **A1:** 779
nitriding ... **A1:** 779
oxide coatings ... **A1:** 779
plating ... **A1:** 779
sulfide treatment ... **A1:** 779
titanium nitride ... **A1:** 779
testing of ... **A1:** 771-778
performance in service ... **A1:** 771-774
thermal properties
resistance to softening of hot-work steels at elevated temperatures ... **A1:** 765
thermal conductivity ... **A1:** 775
thermal expansion ... **A1:** 774
typical heat treatments and properties ... **A1:** 771
water-hardening steels ... **A1:** 768-771
Wrought tungsten *See also* Refractory metals and alloys; Tungsten; Tungsten alloys
joining ... **A2:** 563

Wrought tungsten bar
microstructure **A7:** 627
Wrought unalloyed aluminum
compositions **A2:** 17-21
Wrought zinc and zinc alloy products
types **A2:** 531-532
WSI *See* Wafer-scale integration; Wafer-scale integration technology
Wullaert-Server Charpy/fracture toughness correlation **A8:** 265
Wunsch-Bell model
integrated circuits **A11:** 786
Wurtzite structure
silicon carbide **EM4:** 806

X

x control charts *See also* Control charts
and moving-range control charts **A17:** 732-733
X factor **A6:** 278
***x, y, z* part coordinate system**
coordinate measuring machines **A17:** 19
X zeolites
Raman analysis of pyridine on **A10:** 134
X-40
broaching **A16:** 744
composition **A4:** 795 **A6:** 1215 **A16:** 737
X-45
composition **A6:** 1215
machining **A16:** 757, 758
X-80 Arctic pipeline steel
laser-beam welding **A6:** 263
***x*-axis** *See also y*-axis; *z*-axis
defined **EM1:** 26 **EM2:** 46
fill yarns as **EM1:** 125
X-band microwave frequency
use in ESR **A10:** 255
X-radiation
defined **A9:** 19
X-radiography
detection of subsurface cracking **A18:** 369
X-ray absorption *See also* Absorption; X-ray absorption near-edge structure
effect of absorption-edge energy **A10:** 85
and fluorescence, effect in **A10:** 448
mass absorption **A10:** 84
photoelectric effect **A10:** 84
and scatter **A10:** 84
to analyze ceramic powder particle sizes **EM4:** 67
in x-ray spectrometry **A10:** 84
X-ray absorption near-edge spectroscopy (XANES) **EM3:** 237
X-ray absorption near-edge structure
in EXAFS analysis **A10:** 415-416
multiple scattering effects **A10:** 410
X-ray analysis, characteristic *See* Characteristic-x-ray analysis
X-ray analytical electron microscopy capabilities
and FIM/AP **A10:** 583
X-ray and nuclear applications
beryllium powders in **A7:** 761
X-ray and optical spectroscopy *See* Optical and x-ray spectroscopy; X-ray spectrometry
X-ray anomalous scattering
capabilities **A10:** 407
X-ray area scans
across pure-element standard, map of defocusing **A10:** 527
of dental alloy **A10:** 525-526
limit of spatial resolution **A10:** 527
X-ray "browning" **EM4:** 339
X-ray characterization of surface wear **A18:** 463-470
analysis of debris particles **A18:** 468-469, 470
future trends and conclusions **A18:** 469-470
near-surface gradients **A18:** 465-468
differential gradient **A18:** 467-468
integral gradients **A18:** 465-467
x-ray diffraction (XRD) **A18:** 463-465, 466, 468, 469-470
x-ray diffraction and fluorescence from flat surfaces **A18:** 463-464
x-ray diffraction and fluorescence from rough surfaces **A18:** 464-465
x-ray fluorescence (XRF) **A18:** 463-465, 469
X-ray compositional line broadening (XCLB) **A7:** 61, 316
X-ray computed tomography
to evaluate gas turbine ceramic components **EM4:** 718
X-ray detectable pigment
in tracer yarns **EM1:** 149
X-ray detectors *See also* Detectors; X-ray(s)
beryllium windows **A2:** 684
computed tomography (CT) **A17:** 369-371
detector linearity **A17:** 369
dynamic range **A17:** 369
effect with SEM **A10:** 498
response time **A17:** 369
scintillation **A17:** 370-371
X-ray diffraction *See also* Diffraction **A7:** 316 **A10:** 325-332 **A13:** 1115 **EM4:** 52, 53, 56, 59
as advanced failure analysis technique **EL1:** 1104
analysis, types of **A10:** 325
Bragg's law **A10:** 327, 329
capabilities **A10:** 277, 287, 402, 407, 420, 429, 490
for chemical analysis of polymer fibers **EM4:** 223
and crystallographic texture measurement and analysis, compared **A10:** 358
defined **A10:** 684
detection methods, XRPD **A10:** 331
diffraction experiments, types of **A10:** 329-332
and extended x-ray absorption fine structure compared **A10:** 417
geometry of powder diffraction **A10:** 331
of inorganic solids, types of information from **A10:** 4-6
introduction **A10:** 325, 332
line profile, factors controlling **A10:** 331, 332
microstructure and magnetic properties relationship between **A10:** 599
of organic solids, information from **A10:** 9
pattern, for silica glass **A10:** 398-399
for residual stress measurement **A17:** 51
residual stresses evaluated by **A11:** 125-126, 134
as residue analysis **A10:** 177
samples **A10:** 325
single crystal or polycrystalline **A10:** 325
spectral peaks **A10:** 520
stress measurement, principles of **A10:** 381-382
studies **A12:** 4
studies, of rocket-motor case fracture **A11:** 96
theory **A10:** 325-329
to analyze JK method coatings from molten particle deposition **EM4:** 204
to analyze Tyranno fiber **EM4:** 223
used to identify deformation twins **A9:** 688
used to identify sigma phase in heat-resistant casting alloys **A9:** 332
used to investigate crystallographic texture **A9:** 701-702
volume percent of theoretical density of ceramic powders **EM4:** 71
X-ray diffraction (XRD) **A18:** 463-465, 466, 468, 469-470
X-ray diffraction (XRD) analysis **EM2:** 179, 835
X-ray diffraction analysis
Si_3N_4 product surfaces **EM4:** 238
weld characterization **A6:** 104- 105, 106
X-ray diffraction examination
mounts for **A9:** 28
X-ray diffraction profile analysis **A18:** 468
X-ray diffraction residual stress techniques *See also* Diffraction methods **A10:** 380-392
applications **A10:** 380, 389-392
basic procedure **A10:** 384-387
as confined to surfaces **A10:** 382
defined **A10:** 684
effect of Bragg's law **A10:** 381
estimated analysis time **A10:** 380, 385
general uses **A10:** 380
instrumental and positioning errors **A10:** 387
introduction and principles **A10:** 381-382
limitations **A10:** 380
macrostress measurement by **A10:** 380
microstress measurement by **A10:** 380
plane-stress elastic model of **A10:** 382-384
and proton microprobe **A10:** 107
related techniques **A10:** 380
samples **A10:** 380, 384-385
sources of error **A10:** 387
X-ray diffraction residual stress techniques (continued)
stress measurement, principles of **A10:** 381-382
subsurface measurement and required corrections **A10:** 388-389
X-ray diffraction stress measurement
See also X-ray diffraction residual stress techniques **A10:** 381-382
X-ray diffraction studies
of wrought heat- resistant alloys **A9:** 308
X-ray diffractometer **A7:** 316
X-ray diffractometer technique **A6:** 1095, 1096
X-ray dot map of silver diffusion in an electronic circuit **A9:** 101
X-ray dot maps
used to study powder metallurgy materials **A9:** 508
X-ray elastic constants
determination for Inconel **A10:** 718, 388
recommended for ferrous and nonferrous alloys **A10:** 382
in X-ray diffraction residual stress techniques **A10:** 387
X-ray element analysis **A9:** 153
X-ray emission
characteristic **A10:** 84
continuum **A10:** 83-84
as radioactive decay mode **A10:** 245
ratio to Compton scatter peak, in x-ray spectrometry **A10:** 99
in x-ray spectrometry **A10:** 83-84
X-ray emission spectroscopy
defined **A10:** 684
X-ray emission spectroscopy (XES) **EM3:** 237
X-ray energy
abbreviation for **A10:** 690
X-ray energy dispersive diffractometry used to determine crystallographic texture **A9:** 703
X-ray film *See also* Radiographic film; X-ray(s)
characteristic curves **A17:** 324-325
density **A17:** 323-324
development, effects on film characteristics **A17:** 327
exposure **A17:** 324
film gradient **A17:** 325-326
film radiography **A17:** 323-327
graininess **A17:** 326
methods, modeling of **A17:** 711-713
as recording media **A17:** 314
spectral sensitivity **A17:** 326-327
speed **A17:** 325
X-ray film technique **A6:** 1095, 1096
X-ray fluorescence *See also* X-ray spectrometry **A7:** 250 **EL1:** 942-943
capabilities **A10:** 102, 197, 233
for coating weight measurement **A13:** 391-392
defined **A10:** 684
neutron activation analysis and compared **A10:** 233
and optical emission spectroscopy compared **A10:** 21
and PIXE, detection limits **A10:** 105-106
X-ray fluorescence (XRF) **A18:** 463-465, 469
spectrometric metals analysis **A18:** 300
X-ray fluorescence spectrometry (XRFS) **EM4:** 550-552, 553, 554, 555
for microstructural analysis **EM4:** 26
to analyze the bulk chemical composition of starting powders **EM4:** 72
X-ray fluorescence spectroscopy (XRF) **EL1:** 920
X-ray fluorescence techniques
compositional analysis of welds **A6:** 100
X-ray fluorescent analysis
low-carbon steel **A12:** 249
X-ray fluorescent analysis (XRS)
for microchemical analysis **EM4:** 25
for phase analysis **EM4:** 25
X-ray fractography
determined failure modes in refractory materials **A10:** 376
of molybdenum crystal, fracture surface **A10:** 376-377
X-ray gages
for chemical surface studies **A10:** 177
X-ray induced Auger electron spectroscopy (XAES) **EM3:** 237

X-ray industry
refractory metals in ... **A7:** 17
X-ray inspection
electron-beam welding ... **A6:** 866
real-time
of solder joints ... **EL1:** 738, 942
for weld characterization ... **A6:** 97, 98, 100
X-ray inspection, real time system
schematic ... **A15:** 554
X-ray interferometry
as x-ray topographical ... **A10:** 371
X-ray laminography ... **A6:** 1126
X-ray line broadening
changes due to recovery ... **A9:** 693
X-ray line broadening applications ... **A10:** 374
X-ray maps *See also* Dot mapping; Elemental mapping; Mapping
analog and digitally filtered, of iron in aluminum matrix ... **A10:** 448
by EMPA analog mapping ... **A10:** 525-529
of copper and tin penetration of bearing elements ... **A11:** 720
of copper penetration in axles ... **A11:** 725
defined ... **A10:** 684
or scan, of dental alloy ... **A10:** 526
section of failed melting pot ... **A11:** 39
and superimposed quantitative data, results of ... **A10:** 529
X-ray maps of iron-base alloys ... **A9:** 162
X-ray microanalysis
in analytical electron microscopy ... **A10:** 446-449
by energy-dispersive spectrometry ... **A10:** 461-464
of diffusion-induced grain-boundary migration ... **A10:** 462
X-ray microbeam method
for polycrystalline microstructure ... **A10:** 374
X-ray microfocus techniques
to detect weld discontinuities in diffusion welding ... **A6:** 1080
X-ray microprobe spectroscopy ... **EL1:** 920
X-ray microradiography
to examine crack paths in polycrystalline materials ... **EM4:** 656
X-ray monochromators
germanium single crystals as ... **A2:** 743
X-ray photoelectron microscopy
failure analysis ... **EM3:** 248-249
X-ray photoelectron spectroscopy *See also* Electron or x-ray spectroscopic methods ... **A7:** 250, 251, 255-257 **A10:** 568-580
applications ... **A10:** 568, 576-579
Auger parameter ... **A10:** 572
capabilities ... **A10:** 549, 603
capabilities, and FIM/AP ... **A10:** 583
capabilities, compared with classical wet analytical chemistry ... **A10:** 161
capabilities, compared with infrared spectroscopy ... **A10:** 109
chemical shifts measured ... **A10:** 572
of chromate conversion coatings ... **A13:** 391-392
as compared with x-ray analysis and AES ... **A10:** 569
defined ... **A10:** 684
depth analysis ... **A10:** 573-574
electron spectroscopy for chemical analysis ... **A10:** 568
electronic transitions ... **A10:** 569
general uses ... **A10:** 568
of inorganic solids, types of information from ... **A10:** 4-6
instrumentation ... **A10:** 570-571
introduction and principles ... **A10:** 568-569
kinetic energy measurement as basis of ... **A10:** 85
multiplet splitting ... **A10:** 572
nomenclature ... **A10:** 569
of organic solids, information from ... **A10:** 9
preparing and mounting samples ... **A10:** 574-576
qualitative analysis ... **A10:** 571-572
X-ray photoelectron spectroscopy (continued)
quantitative analysis ... **A10:** 572-573
related techniques ... **A10:** 568
samples ... **A10:** 568, 574-576
shake-up satellites ... **A10:** 572
spectrum, carbon 1s lines in ethyl trifluoracetate ... **A10:** 572
surface sensitivity ... **A10:** 569-570
use with Auger electron spectroscopy ... **A10:** 554
X-ray photoelectron spectroscopy (XPS) ... **A18:** 445-448, 449, 452, 456 **EL1:** 1075, 1107 **EM1:** 285 **EM3:** 237-238
advantages and limitations ... **EM3:** 239
of bimetal foil laminates ... **A11:** 44
chemical state information ... **EM3:** 243-244
compared to secondary ion mass spectrometry (SIMS) ... **EM3:** 237
data analysis ... **A18:** 447-448
development ... **A18:** 445
equipment ... **A18:** 446-447
fundamentals ... **A18:** 445-446
depth distribution effect on background of spectra ... **EM3:** 247
depth profiling by inelastic scattering ratio ... **EM3:** 247
depth profiling by multiple anodes/different transitions ... **EM3:** 247
development of ... **A11:** 35
for examining adsorption of NTMP ... **EM3:** 250
of integrated circuit leadframe ... **A11:** 45
ion oxide sputtering of titanium oxide ... **EM3:** 246
phosphoric acid anodized surfaces ... **EM3:** 249
of plating peeling ... **A11:** 43-45
process ... **EM3:** 237-238
quantification ... **EM3:** 242-243
small-area, combined with ion sputtering ... **EM3:** 244
spectrum ... **A18:** 447, 448
surface analysis ... **EM3:** 259 **EM4:** 25
surface behavior diagrams ... **EM3:** 247
survey spectrum ... **EM3:** 237, 238
to analyze the surface composition of ceramic powders ... **EM4:** 73
verifying CuO resulting from copper etching ... **EM3:** 269
versus SIMS ... **A18:** 458
X-ray photoelectron spectroscopy (XPS/ESCA) ... **EM2:** 811, 812-816
X-ray photon emission
and fluorescent yield ... **A10:** 86
inner-shell ionization and de-excitation by ... **A10:** 433
and secondary electron ejection ... **A10:** 86
X-ray photons *See* Photons
for Auger electron emission ... **A7:** 250
X-ray powder diffraction *See also* Diffraction methods ... **A10:** 333-343
applications ... **A10:** 333, 341-342
automated ... **A10:** 338
Bragg's law ... **A10:** 337
Debye-Scherrer camera ... **A10:** 335
direct comparison method ... **A10:** 340
estimated analysis time ... **A10:** 333
Gandolfi camera ... **A10:** 335
general uses ... **A10:** 333
geometry of ... **A10:** 331
$I/I_{corundum}$ method, XRPD analysis ... **A10:** 340
instrumentation ... **A10:** 334-338
internal/external standard methods ... **A10:** 340
introduction ... **A10:** 333-334
lattice-parameter method ... **A10:** 339
Laue camera ... **A10:** 334-335
limitations ... **A10:** 333
qualitative analysis ... **A10:** 338-339
quantitative analysis ... **A10:** 339-340
related techniques ... **A10:** 333
Rietveld refinement, capabilities of ... **A10:** 344
samples ... **A10:** 333
sources of error in ... **A10:** 340-341
X-ray powder diffraction (continued)
spiking method ... **A10:** 340
standardless method ... **A10:** 340
X-ray powder diffraction (XRD)
for phase analysis ... **EM4:** 25
to analyze the phase composition of ceramic powders ... **EM4:** 73
X-ray powder diffractometry ... **EM4:** 558-559
X-ray radiography *See also* Radiography; X-rays(s) ... **EM1:** 775-776
for adhesive-bonded joints ... **A17:** 624
defect detection ... **EM3:** 748, 749, 750, 751
for instrumentation/testing ... **EL1:** 369
of investment castings ... **A15:** 264
microstructure effect ... **A17:** 51
and microwave inspection, compared ... **A17:** 202
of powder metallurgy parts ... **A17:** 537-538
of soldered joints ... **A17:** 607-608
X-ray radiography, of compacts
for density distribution ... **A7:** 301
X-ray radiography of fiber composites ... **A9:** 591-592
X-ray satellites ... **A18:** 447
X-ray sedimentation
to analyze the ceramic powder particle size ... **EM4:** 68
X-ray spectrometers
EDS/WDS ... **A10:** 518-522
types of ... **A11:** 32
X-ray spectrometry *See also* Energy-dispersive x-ray spectrometry; Optical and x-ray spectroscopy; X-rays ... **A10:** 82-101
absorption edges ... **A10:** 85
advantages ... **A10:** 82
analysis time, qualitative ... **A10:** 82
applications ... **A10:** 82, 94, 99-101
basis for qualitative analysis ... **A10:** 96
boron and fluorine determined in borosilicate glass ... **A10:** 179
Bragg's law ... **A10:** 87
calculation of LLD ... **A10:** 96
capabilities ... **A10:** 197, 212, 243, 333
capabilities, compared with classical wet analytical chemistry ... **A10:** 161
capabilities, compared with infrared spectroscopy ... **A10:** 109
characteristic x-rays ... **A10:** 82, 83-84
of coal ... **A10:** 100
continuum ... **A10:** 83
defined ... **A10:** 684
electromagnetic radiation ... **A10:** 83
elements and x-rays, relationships between ... **A10:** 84-85
emission ... **A10:** 85-87
energy-dispersive x-ray spectrometers ... **A10:** 89-93
general use ... **A10:** 82
of inorganic liquids and solutions, information from ... **A10:** 7
of inorganic solids, types of information from ... **A10:** 4-6
instrumentation ... **A10:** 87-93
interelement effects ... **A10:** 87
introduction ... **A10:** 82-83
limitations ... **A10:** 82
mass absorption coefficient ... **A10:** 84
neutron activation analysis and compared ... **A10:** 233
operation ... **A10:** 89
of organic liquids and solutions, information from ... **A10:** 10
of organic solids, information from ... **A10:** 9
photoelectric effect ... **A10:** 84
Planck's constant ... **A10:** 83
qualitative analysis ... **A10:** 95-96
quantitative analysis ... **A10:** 96-99
related techniques ... **A10:** 82
sample preparation ... **A10:** 93-95
samples ... **A10:** 82, 88, 93-95, 99-101
scatter of x-rays ... **A10:** 84
selectivity of ... **A10:** 96

X-ray spectrometry (continued)
of stainless alloy ... **A10:** 178
wavelength-dispersive x-ray spectrometers ... **A10:** 87
x-ray emission and absorption ... **A10:** 84
X-ray spectroscopy *See* X-ray spectrometry ... **A1:** 1030-1031, 1032 **A9:** 91-93
applied to fracture studies ... **A9:** 99
for bulk chemistry analyses ... **EM4:** 24
conductive coatings used in ... **A9:** 98
spectra ... **A9:** 92
X-ray spectrum *See also* Radiography
defined ... **A17:** 303-304
R-output ... **A17:** 304
X-ray target
composite ... **A2:** 559
X-ray topography *See also* Diffraction methods; Reflection topography ... **A10:** 365-379
applications ... **A10:** 365, 375-379
estimated analysis time ... **A10:** 365
general uses ... **A10:** 365
introduction ... **A10:** 366-368
limitations ... **A10:** 365
methods and instrumentation ... **A10:** 368-375
related techniques ... **A10:** 365
samples ... **A10:** 365
types of ... **A10:** 368-375
X-ray transparency
of lithium-oxide-based glass ... **EM1:** 107
X-ray tube
defined ... **A9:** 19
X-ray tubes *See also* X-ray(s)
anode materials ... **A10:** 89-90
beryllium windows ... **A2:** 684
classified ... **A17:** 302
conventional ... **A17:** 302
Coolidge ... **A10:** 88
design and materials ... **A17:** 302-303
electrical characteristics ... **A17:** 303
electrical waveform, effects ... **A17:** 304-305
excitation vs secondary-target excitation, x-ray spectrometers ... **A10:** 89
heel effect ... **A17:** 305
inherent filtration ... **A17:** 306-307
microfocus ... **A17:** 302-303
minifocus ... **A17:** 302
molybdenum, in wavelength-dispersive x-ray spectrometer ... **A10:** 88
operating characteristics ... **A17:** 303-307
R-output ... **A17:** 304
radiographic inspection ... **A17:** 302-307
reciprocity law ... **A17:** 304
stem radiation ... **A17:** 305
tube rating ... **A17:** 305-306
voltage/current, effects ... **A17:** 304
for x-ray spectrometry ... **A10:** 87-90
x-ray spectrum ... **A17:** 303-304
X-ray turbidimetry ... **A7:** 219
X-ray wavelength and atomic number
Mosely's relationship between ... **A10:** 433
X-ray window assembly
brazed beryllium ... **A7:** 761
X-ray(s) *See also* X-ray detectors; X-ray film; X-ray radiography; X-ray tubes
collimators, computed tomography (CT) ... **A17:** 368
as computed tomography (CT) radiation source ... **A17:** 368-371
diffraction ... **A17:** 51
effective absorption ... **A17:** 310-311
energies, in radiography ... **A17:** 303-304
and gamma-rays, in neutron radiography ... **A17:** 387-388
high-energy, as neutron sources ... **A17:** 388-389
image, real-time filmless ... **EL1:** 369
and neutrons, attenuation compared ... **A17:** 387
production of ... **A17:** 298
radiographs ... **A17:** 393
showing electron beam/sample interaction ... **EL1:** 1095
sources, radiographic inspection ... **A17:** 297
sources, real-time radiography ... **A17:** 321-322
spectrum ... **A17:** 303-304
units for ... **A17:** 301
X-rays *See also* Electron or x-ray spectroscopic methods; Optical and x-ray spectroscopy
absorption ... **A10:** 84-85
absorption curve for uranium, as function of wavelength ... **A10:** 85
absorption edges ... **A10:** 85
beams, in x-ray spectrometry ... **A10:** 82
characteristic ... **A10:** 82-84, 326, 435
characteristic, defined ... **A12:** 168
collimating, basic methods ... **A10:** 403
continuum, as inelastic scattering process ... **A10:** 433
as continuum radiation ... **A10:** 83
defined ... **A9:** 19
detection of ... **A10:** 326
diffraction of ... **A10:** 326-327
and elements, relationship in x-ray spectrometry ... **A10:** 84
emission, as radioactive decay mode ... **A10:** 245
energy, vs mass absorption, in copper ... **A10:** 85, 87
family lines of ... **A10:** 522-523
fluorescent yield and, defined ... **A10:** 86, 87
hard ... **A10:** 83
history of development ... **A10:** 82-83
maps ... **A12:** 168, 473
mass absorption coefficients ... **A10:** 85
nature and generation of ... **A10:** 325-326
origin, in Bohr atom model ... **A12:** 168
photographic film detection ... **A10:** 334
primary, defined ... **A10:** 679
production, vs mass, PIXE analysis ... **A10:** 104
Rayleigh and Compton scatter of ... **A10:** 85
secondary, defined ... **A10:** 681
sequential or simultaneous detection ... **A10:** 87
signals, SEM ... **A12:** 168
soft ... **A10:** 83
sources ... **A10:** 395, 570
spatial resolution, effect in microanalysis ... **A10:** 448
spectra, measuring ... **A10:** 518-522
spectrograph, defined ... **A10:** 684
spectrometers, EDS/WDS ... **A10:** 518-522
studied by scanning transmission electron microscopy ... **A9:** 104
X-sections
as basic casting shape ... **A15:** 599, 604, 606-607, 609-610
solidification sequence ... **A15:** 604
***x-t* diagram**
of shear and longitudinal impact waves ... **A8:** 232
***x-y* plotter**
for fatigue testing machines ... **A8:** 368
for torsional tests ... **A8:** 147
***X-y* plotters**
as eddy current inspection readout ... **A17:** 179, 190
***x-y* recorder**
for quasi-static testing ... **A8:** 223
traces during fatigue crack growth test ... **A8:** 383-384
***X-Y* stages**
error characterization by interferometer ... **A17:** 15
***x-y* storage oscillators**
as eddy current inspection readout ... **A17:** 179
XANES *See* X-ray absorption near-edge structure
XCLB *See* X-ray compositional line broadening
XD intermetallic-matrix composites
development ... **A2:** 911
Xenon
ionization potentials and imaging fields for ... **A10:** 586
Xenon arc
effects on polyether-imides (PEI) ... **EM2:** 157
Xenon arc lamp
for weatherometers ... **EM2:** 578-579
Xenon as an ion bombardment etchant for fiber composites ... **A9:** 591
Xenon flash tube
for symmetric rod impact test ... **A8:** 204
Xenon ionization detectors
computed tomography (CT) ... **A17:** 370
Xenon lamps
arc AAS/AFS ... **A10:** 52
sources, spectral output ... **A10:** 76
Xenon lamps, pulsed
for radiation curing ... **EL1:** 864
Xenon-arc light source for microscopes ... **A9:** 72
Xerogels
alkoxide-derived gels ... **EM4:** 210, 211, 212, 213
Xerography ... **A7:** 580
Xeroradiography
as image conversion medium ... **A17:** 315
powder loud development ... **A7:** 582
XLPE *See* Cross-linked polyethylene
XMC compression molded sheet
composition/properties ... **EM1:** 560
XPS *See* X-ray photoelectron spectroscopy
XRD *See* X-ray diffraction
XRPD *See* X-ray powder diffraction
XRS *See* X-ray spectrometry
***xy*-plane** *See also* *z*-axis
defined ... **EM1:** 26 **EM2:** 46
Xydar *See* Liquid crystal, resins
Xylene
as organic cleaning solvent ... **A12:** 74
solvent resistance/repairability ... **EL1:** 822
surface tension ... **EM3:** 181
Xylene (solvent)
batch weight of formulation when used in oxidizing sintering atmospheres ... **EM4:** 163
Xylenol orange
as metallochromic indicator ... **A10:** 174
Xylenols ... **EM3:** 103
2,3-Xylenol, physical properties ... **EM3:** 104
2,4-Xylenol, physical properties ... **EM3:** 104
2,5-Xylenol, physical properties ... **EM3:** 104
2,6-Xylenol, physical properties ... **EM3:** 104
3,4-Xylenol, physical properties ... **EM3:** 104
3,5-Xylenol, physical properties ... **EM3:** 104
XYZ transition
as x-ray notation ... **A10:** 569

Y

Y lens
gas mass spectrometer ... **A10:** 153
Y zeolites
Raman analysis of pyridine on ... **A10:** 134
***y*-axis** *See also* *x*-axis ... **EM1:** 26, 125
defined ... **EM2:** 46
Y-block *See* Keel block
Y-blocks *See* Keel block
Y-branch fittings
pipe welding ... **M6:** 591
Y-joint *See* Knuckle area
y-modulation
roused for image modification in scanning electron microscopy ... **A9:** 95
Y-shaped parts
magnetic particle inspection ... **A17:** 113-114
Y-Zn (Phase Diagram) ... **A3:** 2 • 382
Y-Zr (Phase Diagram) ... **A3:** 2 • 382
Yachts
composite structures for ... **EM1:** 840-842
Yarn
defined ... **EM2:** 46
Yarn bundle *See* Bundle; Plied yarn; Yarn
Yarn count ... **EM1:** 125, 149
Yarn eyelet, steel
matrix hardness of ... **A11:** 160
Yarns *See also* Bundle; Continuous filament yarn; Filling yarn; Multifilament yarn; Plied yarn; Weft; Yarn bundle
aramid fiber ... **EM1:** 114
braiding ... **EM1:** 519
carbon fiber ... **EM1:** 51
defined ... **EM1:** 26
glass fibers ... **EM1:** 110-111
glass textile, production ... **EM1:** 110-111
graphite, modulus, and dry bundle tensile strength ... **EM1:** 362
Kevlar, sizes ... **EM1:** 114
in lamina cross section ... **EM1:** 176
spacing, in multidirectionally reinforced fabrics preforms ... **EM1:** 130
stuffer, angle-interlock fabric with ... **EM1:** 130
tracer ... **EM1:** 149
x, *y* coordinate system for ... **EM1:** 125-127

Yates method of analysis
applied **A8:** 654, 657
for factorial experiments **A8:** 653-654
for fractional factorial plans **A8:** 644
Yaw
defined **A17:** 18
Yb-Zn (Phase Diagram) .... **A3:** 2 • 383
YBCO superconducting materials *See also* High-temperature superconductors
properties **A2:** 1082
Yellow (K2) photographic lens filter
with ultraviolet illumination **A12:** 84
Yellow brass *See* Copper alloys, specific types
brazing **A6:** 629
gas-metal arc welding **A6:** 763
resistance spot welding **A6:** 850
weldability **A6:** 753
Yellow brass, 65%, microstructure of .... **A3:** 1 • 22
Yellow brass plating **M5:** 285
Yellow brasses *See also* Brasses; Copper casting alloys; Leaded yellow brasses; Wrought coppers and copper alloys
Antioch process for **A15:** 247
applications and properties **A2:** 225, 302-304, 306, 348
as electrical contact materials **A2:** 843
foundry properties for sand casting **A2:** 348
high strength, intermetallic inclusions in **A15:** 96
melt treatment **A15:** 774
nominal compositions **A2:** 347
recycling **A2:** 1214
shrinkage allowances **A15:** 303
Yellow chromate conversion coating
cadmium plate **M5:** 269
Yellow golds *See* Gold-sliver-copper alloys
Yellow Mill Pond Bridge
fatigue cracking in **A11:** 707-708
Yellow-green light filters **A9:** 72
Yellowness
optical testing for **EM2:** 594-595
Yield *See also* Creep; Flow; Flow, Ultimate strength; Plastic deformation; Plastic, flow; Plastic flow; Yield point; Yield point elongation; Yield strength; Yield stress; Yielding
casting, defined *See* Casting yield
casting, gating design and **A15:** 589
Charpy V-notch as screening test for **A8:** 263
chip, in WSI **EL1:** 354
compressive, in polystyrene **A11:** 759
condition for **A11:** 49
considerations, active-component fabrication **EL1:** 198-199
criteria **A8:** 576-577 **A14:** 369-370
curves, experimental, with interaction **A8:** 641-642
defined **A8:** 15 **A11:** 11 **A13:** 14 **A14:** 14 **A15:** 11 **EL1:** 1161
in dynamic notched round bar testing **A8:** 282
effect on crack growth and toughness **A11:** 54
experimental **A8:** 640, 650
failure analysis of **EM2:** 730-731
fracture toughness **A8:** 268
high strain rate effect on **A8:** 208
measures of **A8:** 21
point, defined **A11:** 11 **A13:** 14 **A14:** 14
process, surface-mount soldering **EL1:** 708
processing, WSI vs VLSI **EL1:** 266-268
of secondary and backscattered electrons **A10:** 502
silicon defect density effects **EL1:** 88
size of plastic zone **A8:** 281
solder joint **EL1:** 691
stress, defined **A13:** 14
stress, dependence on grain size **A11:** 68
tensile load vs. elongation **A9:** 684
types **A9:** 684
in wafer-scale integration **EL1:** 9
wiring, WSI **EL1:** 354
Yield (continued)
in workability **A8:** 576-577
Yield elongation
plastic deformation **A9:** 684
Yield point **EM3:** 31
defined **A8:** 15, 21-22 **EM1:** 26 **EM2:** 46, 433
diffusion processes and **A8:** 35
effect of low deformation rate on **A8:** 38, 39
elongation, defined **A8:** 15, 21-22
and Lüders bands **A8:** 22
point elongation, defined **EM2:** 46
tensile impact and **EM2:** 556
in torsional testing **A8:** 148
Yield point elongation **A1:** 574-575 **EM3:** 31
defined **A8:** 15, 21-22
Lüders bands in **A8:** 548
Yield point, stainless steels
and magabsorption measurement **A17:** 152
Yield shear strength **A8:** 139
Yield shear stress **A8:** 141
Yield strength *See also* compressive yield strength; Compressive yield strength; Mechanical properties; Microyield strength; Offset; Shear yield strength; Tensile properties; Tensile strength; Yield; Yield stretch; Yielding **A6:** 389 **A7:** 312 **EM3:** 31
of aluminum alloys **A7:** 474
aluminum and aluminum alloys **A2:** 8
amorphous materials and metallic glasses **A2:** 813-814
basis for classifying high-strength steels **M1:** 403, 409-410
in bending, for copper alloy strip **A8:** 134
beryllium-copper alloys **A2:** 418
by constant crosshead speed tests **A8:** 43
in cantilever beam bend test **A8:** 132, 134
compressive **A8:** 662, 667
copper alloy, lead frame materials **EL1:** 488-490
copper and copper alloys **A2:** 219, 223
copper casting alloys **A2:** 348-350, 355
of copper-based P/M materials **A7:** 470
and corrosion fatigue crack propagation **A8:** 405, 408
defined **A8:** 15, 21 **A11:** 11 **A13:** 14 **A14:** 14 **EL1:** 1161 **EM1:** 26 **EM2:** 46, 433-434
deformation rate effect, in niobium **A8:** 38-39
effect, contour roll forming **A14:** 624
effect, critical stress and sulfide fracture toughness **A13:** 534
effect, hydrogen damage **A13:** 169
effects of sintering temperature **A7:** 369
effects of sintering time **A7:** 370
effects of strain rate and temperature on **A8:** 38, 40
ferrite effect, high-alloy steels **A15:** 725
of ferrous P/M materials **A7:** 465, 466
from uniaxial tensile testing **A8:** 555
as function of die temperature **A14:** 153
gas content effect, aluminum alloy **A15:** 457
in heat-exchanger tubes **A11:** 628
hold time effect on **A8:** 36-37
of injection molded P/M materials **A7:** 471
of molybdenum and molybdenum alloys **A7:** 476
offset *See* Offset yield strength **A8:** 21, 132
ordered intermetallics **A2:** 913-914
overloading and **A11:** 136-138
of P/M and ingot metallurgy tool steels **A7:** 471, 472
of P/M and wrought titanium and alloys **A7:** 475
of P/M forged low-alloy steel powders **A7:** 470
in press-brake forming **A14:** 541
of rhenium and rhenium-containing alloys **A7:** 477
shear **A8:** 139
sintering temperature effects **A13:** 825
solid-state-welded interlayers **A6:** 169
Yield strength (continued)
in spinning **A14:** 604
of stainless steels **A7:** 468
static, and fatigue **A11:** 102
static design based on **A8:** 20
of superalloys **A7:** 473
for superalloys, high-pressure hydrogen gas **A8:** 408
of tantalum wire **A7:** 478
temperature dependence, nickel aluminides **A2:** 915-916
testing machines for **A8:** 547, 565
thermoplastic fluoropolymers **EM2:** 117
titanium PA P/M alloy compacts **A2:** 654
vs. tempering temperature, low-alloy steels **A13:** 954
Young and Dupré equation.... **A6:** 114-115, 128, 129, 619
Yield strength, aluminum-killed steels
inert gas fusion analysis **A10:** 231
Yield strength, of dual-phase steels **A1:** 426-427
of steel castings **A1:** 365
Yield strength of steels as a function of the mean free distance between cementite
particles **A9:** 130
Yield strength plotted against complexity index **A9:** 130
for beryllium-aluminum alloys **A9:** 131
Yield stress *See also* Yield
decreasing, in swaging **A14:** 143
defined **A8:** 15 **A11:** 11 **A14:** 14
in double-notch shear testing **A8:** 229
effect of environment on **A11:** 225
effect on ferrite grain diameter, steels **A11:** 68
elastic bending below **A14:** 881
elastic bending below, as springback **A8:** 552
fluid models with **EL1:** 848
models, of rheological behavior **EL1:** 848
on flow curve **A8:** 23
and residual stress **A11:** 57
shear **A8:** 141
and strain rate .... **A8:** 38 -39, 41, 229
in torsional Kolsky bar dynamic test **A8:** 228
Yield stretch *See also* Mechanical properties; Yield
wrought aluminum and aluminum alloys **A2:** 101
Yield value **EM3:** 31
Yielding *See also* Yield
cross, as distortion **A11:** 138
discontinuous **A11:** 25
localized, from residual stress **A11:** 97
overloading and **A11:** 136
in polymers **A11:** 760-761
reverse **A11:** 57
studied by x-ray topography and synchrotron radiation **A10:** 365
tolerance to, in limit analysis **A11:** 136-138
Yielding, anomalous
ordered intermetallics **A2:** 913-914
Yielding dislocations
as acoustic emission source **A17:** 287
Yields
with copier powders **A7:** 586
YIG *See* Yttrium-iron garnet
Yin-yang cells
of superconducting materials **A2:** 1057
Yoke frame closures
hot isostatic pressure vessels **A7:** 420, 421
Yoke, main landing gear deflection
SCC failure in **A11:** 23
Yokes
applications **A17:** 94
for demagnetization **A17:** 121-122
electromagnetic **A17:** 93-95
as magnetizers **A17:** 93-95, 130
permanent-magnet **A17:** 93
for subsurface discontinuities **A17:** 114
Yoloy (Ni-Cu alloy steel), applications
protection tubes and wells **A4:** 533

Yoshida buckling test
for sheet metals **A8:** 563-564
Youden square experimental plan **A8:** 647, 650, 651
for two sources of inhomogeneity **A8:** 643
Young's creep modulus **EM1:** 191
Young's equation **A6:** 128, 129
non-reacting steady-state sessile drop **EM4:** 484, 485
Young's modulus *See also* Coefficient of elasticity; Complex Young's modulus; Elasticity; Initial modulus; Modulus of elasticity; Modulus of Elasticity (E) **EM3:** 31 **EM4:** 424
cemented carbide properties testing **A16:** 77
cemented carbides **A16:** 79
ceramics **A16:** 101
defined **A8:** 15 **A14:** 14 **EM1:** 26 **EM2:** 46
and density of low-alloy ferrous parts **A7:** 463
effect, shape memory alloys **A2:** 898
of elasticity, fatigue striation spacing and stress-intensity with **A8:** 482, 484
of glass fibers **EM1:** 46
longitudinal, vs. fiber volume fraction **EM1:** 208
of molybdenum and molybdenum alloys **A7:** 476
rule-of-mixture, metal-matrix composites **A2:** 903
in uniaxial tensile testing **A8:** 555
Young's modulus, electrical contacts
friction and wear of **A18:** 683
Young's relaxation modulus **EM1:** 191
Ytterbium *See also* Rare earth metals
as divalent **A2:** 720
pure **M2:** 823
as rare earth metal, properties **A2:** 720, 1188
TNAA detection limits **A10:** 237
Ytterbium in garnets **A9:** 538
Ytterbium-169
for radiographic inspection **A17:** 308
Yttria **A16:** 98, 100, 101
in mechanically alloyed oxide dispersion-strengthened (MA ODS) alloys **A2:** 943-944
Yttria (Y_2O_3)
heat-treatable ceramic **A18:** 812
Yttria-alumina doped silicon nitride
cladless HIP process **EM4:** 196, 197
Yttria-partially stabilized zirconia (Y-PSZ) **EM4:** 1136, 1138
chemical integrity of seals **EM4:** 540
fracture toughness **EM4:** 602
Yttria-stabilized transformation-toughened zirconia *See also* Ceramics; Structural ceramics; Toughened ceramics
as structural ceramic, applications and properties **A2:** 1023
Yttria-stabilized zirconia
abradable seal material **A18:** 589
ceramic coatings for dies **A18:** 644
friction coefficient **A18:** 815
properties **A18:** 814
Yttria-stabilized ZrO_2
automotive electrical/electronic applications **EM4:** 1106
Yttrium *See also* Rare earth metals
fluoride separation **A10:** 169
heat-affected zone cracks **A6:** 91
in high-temperature YBCO system superconductors **A2:** 1085
ion implantation and oxidation resistance **A18:** 856
oxide, tool materials used for cast iron **A16:** 656
pure **M2:** 823-824
as rare earth metal, properties **A2:** 720, 1189
segregation, at grain boundaries **A10:** 483
SiAlYON **A16:** 100
solidification cracking **A6:** 90
species weighed in gravimetry **A10:** 172
ultrapure, by external gettering **A2:** 1094
weighed as the fluoride **A10:** 171
YAG (yttrium-aluminum-garnet) **A16:** 101
Yttrium (Y)
in heat-resistant alloys **A4:** 512
vapor pressure, relation to temperature **A4:** 495
Yttrium aluminum garnet laser
characteristics and advantages **M6:** 656
Yttrium aluminum garnet laser (continued)
characteristics of chemical composition **M6:** 651
high-production welding **M6:** 647-648
mounting **M6:** 667
Yttrium in garnets **A9:** 538
Yttrium iron garnet (YIG)
ferromagnetic resonance **A17:** 220
Yttrium oxide
Vickers and Knoop microindentation hardness numbers **A18:** 416
Yttrium oxide (Y_2O_3) **EM4:** 18
as additive for pressure densification **EM4:** 298-299, 300
additive to Si_3N_4 ...... **EM4:** 226
effect on glass hardness **EM4:** 851
grain-growth inhibitor **EM4:** 188
hot pressing **EM4:** 191
sintering aid **EM4:** 188
Yttrium oxide-stabilized tetragonal polycrystal ceramic (Y-TZP)
ceramic ceramic joints **EM4:** 516
Yttrium, vapor pressure
relation to temperature **M4:** 310
Yttrium-aluminum-garnet (YAG) **EM4:** 18
laser transformation hardening **A4:** 287, 290, 291
solidification cracking **A6:** 88
Yttrium-aluminum-garnet (YAG) laser pulse
for solder joints **EL1:** 942
Yttrium-indium-garnet (YIG) **EM4:** 18
Yttrium-iron garnet **A9:** 53
Yttrium-iron-garnet **EM4:** 60, 61
permeability **EM4:** 1162
resistivity **EM4:** 1162
saturation flux density **EM4:** 1162

Z

Z *See* Atomic number
as standard normal distribution (statistical) **A17:** 738
Z elements
x-ray tubes for **A10:** 101
Z lens
gas mass spectrometer **A10:** 153
Z mills
for wrought copper and copper alloys **A2:** 244
Z phase
in austenitic stainless steels **A9:** 284
in wrought heat-resistant alloys **A9:** 312
***z*-axis** *See also xy*-plane
defined **EM1:** 26 **EM2:** 46
***z*-axis strains**
epoxy-aramid/epoxy-glass **EL1:** 987
ZA-8 zinc alloy
die castings **A2:** 529
gravity castings **A2:** 530
properties **A2:** 535-536
ZA-12 *See* Zinc alloys, specific types, zinc foundry alloy ZA-12
ZA-12 zinc alloy
die castings **A2:** 529, 536-537
gravity casting **A2:** 530
ZA-27 zinc alloy
die castings **A2:** 529, 537-538
gravity castings **A2:** 530
Zachariasen model **EM4:** 493
ZAF corrections
defined
in EDS mapping **A10:** 528
Zamak 3
properties **A2:** 532-533
Zamak die casting alloy
development of **A15:** 35
Zamak-3 (die casting) *See* Zinc alloys, specific types, AG40A
Zamak-5 (die casting and sand casting) *See* Zinc alloys, specific types, AC41A
Zeeman effect
background correction for **A10:** 51-52
defined **A10:** 264, 684
Zener criterion
grain boundaries pinned according to **EM4:** 242
Zener-Hollomon parameter
for flow stress **A8:** 162-163
for grain size **A8:** 175
Zener-Hollomon parameter (continued)
mean free path between spheroidite particles **A8:** 176, 178
Zeolites **EM3:** 111
crystal structure **EM4:** 882
transition metals as source of fluorescence on **A10:** 130
X and Y, Raman spectroscopy of pyridine on **A10:** 134
Zephiran chloride
description **A9:** 68
Zephiran chloride as an etchant additive **A9:** 169
Zero bleed
defined **EM1:** 26 **EM2:** 46
Zero convention
of electrode potentials **A13:** 21
Zero gravity casting
effect on dendritic structures in copper alloy ingots **A9:** 638
Zero mean stress **A6:** 967, 968
Zero strain
fracture mechanics of **A11:** 57
Zero time **EM3:** 31
defined **A8:** 15 **EM2:** 46
Zero-CTE metal matrix composites
carbon fiber **EM1:** 52
Zero-dispersion double spectrometer
with Triplemate device **A10:** 129
Zero-draft forgings
from copper alloys **A14:** 258
Zero-field splitting
in ESR spectra **A10:** 261
Zero-insertion-force connector
defined **EL1:** 1161
Zero-loss peak **A18:** 390
Zero-order Laue zone
abbreviation for **A10:** 691
-CBED patterns **A10:** 439, 441
Zero-resistance ammeter
for galvanic corrosion **A13:** 236, 238
Zero-risk analysis
for life cycle testing **EL1:** 135-136, 139
Zero-risk stress testing
life cycle **EL1:** 136
Zero-wear limit (N_o) **A18:** 265-266, 268
Zerodur
composition **EM4:** 872
elastic constant **EM4:** 875
hardness **EM4:** 876
maximum use temperature **EM4:** 875
polishing **EM4:** 469
properties **EM4:** 872
seals to metals **EM4:** 499
thermal properties **EM4:** 876
Zeroing adjustments
for tension testing force measurement **A8:** 48
Zerol bevel gears
described **A11:** 587
Zeta potential *See* Electrokinetic potential **EM4:** 74
ZGS platinum *See* Zirconia grain-stabilized platinum and platinum alloys
ZHC copper
applications and properties **A2:** 281
Zhurkov's equation **EM3:** 354, 357-358
ZIA IRRS process
for zinc recycling **A2:** 1225
Ziegler-Natta catalysts *See also* Coordination catalysis
defined **EM2:** 46
Zielgler-Natta catalysts **EM3:** 31
Zig-zag in-line packages (ZIP) **EL1:** 443
Zilloy-15 *See* Zinc alloys, specific types, rolled zinc alloy
Zilloy-40 *See* Zinc alloys, specific types, copper-hardened rolled zinc
Zinc *See also* Pure metals; Pure zinc; Zinc alloy castings; Zinc alloys; Zinc alloys, specific types; Zinc coatings; Zinc recycling; Zinc-alloys
abrasion artifacts in **A9:** 34, 37-38
as addition to aluminum alloys **A4:** 842
addition to aluminum-base bearing alloys **A18:** 752
as addition to brazing filler metals **A6:** 904
adhesion layer for polyimides **EM3:** 158
alkaline cleaning of **M5:** 33, 35-36

Zinc (continued)
alloy castings ... **A2:** 528-532
alloying addition to heat-treatable aluminum alloys ... **A6:** 528, 530
alloying effect on copper alloys ... **M6:** 400
alloying effects ... **A13:** 759
alloying in aluminum alloys ... **M6:** 373
in alloys ... **A6:** 1165
alloys, suitability for journal bearings ... **M1:** 610-611
as aluminum alloying element ... **A2:** 16, 132-133
in aluminum alloys ... **A15:** 746
anodes ... **A13:** 470, 764-765, 921
anodes showing electrochemical and corrosion effects on adhesives ... **EM3:** 629-631, 632
anodized ... **EM3:** 417
applications ... **M2:** 629-637
aqueous corrosion, as function of pH ... **A13:** 1304
Arrhenius plots of delayed steel failure in ... **A11:** 240, 244
atmospheric corrosion ... **A13:** 82, 205, 756-758
austenitic stainless steel embrittlement by ... **A11:** 236-237
austenitic steels embrittlement by ... **A13:** 184
-bearing paints ... **A13:** 768-769
biologic effects and toxicity ... **A2:** 1255-1256
bracelet anode ... **A13:** 470
brass plating bath content ... **M5:** 285-286
buffing *See* Zinc, polishing and buffing of
capacitor discharge stud welding ... **M6:** 738
cast, historical use ... **A15:** 22
cast product applications ... **A2:** 530
cathodic protection of iron ... **A13:** 467
cavitation erosion ... **A18:** 216
chemical corrosion of ... **A13:** 762-763
chemical resistance ... **M5:** 4, 7, 10
chromating process sequence ... **A13:** 3')O
chromium plating of ... **M5:** 189-191, 194-195
removal of ... **M5:** 173
coated onto aluminum alloy sleeve bearing liners ... **A9:** 567
-coated steels, spangles formation ... **A8:** 548
coating for resistance seam welding ... **M6:** 502
coating, of metal fasteners ... **A11:** 542
coatings ... **A2:** 527-528 **A13:** 426, 460-461, 526-527, 756, 765-768, 911-912 **EM3:** 53
coatings, hydrogen embrittlement testing for ... **A8:** 542
concentration, in phosphate baths ... **A13:** 384
constant-current electrolysis ... **A10:** 200
-containing copper, digital composition map ... **A10:** 528
contamination ... **M6:** 321
content effect, electron-beam welding ... **A6:** 872
content, effect in copper alloys ... **A15:** 468
content in magnesium alloys ... **M6:** 427
in copper alloys ... **A6:** 752
and copper specification ... **A2:** 1057-1058
corrosion fatigue ... **A13:** 763-764
corrosion, in different waters ... **A13:** 443
corrosion, in neutral chloride solution ... **A13:** 378
corrosion of ... **A13:** 755-769
corrosion protection abilities ... **A13:** 755
corrosion resistance, in soils ... **A13:** 762
corrosion service ... **M2:** 646-655
critical relative humidity ... **A13:** 82
cyanide-to-zinc ratio, brass plating ... **M5:** 285-286
in dental amalgam ... **A18:** 669
deoxidizing, copper and copper alloys ... **A2:** 236
determined by controlled-potential coulometry ... **A10:** 209
dietary, and lead toxicity ... **A2:** 1246
diffusion ... **EL1:** 679
in dissolved salts, acids, and bases ... **A13:** 763
dot map, diffusion-induced grain-boundary migration in ... **A10:** 527
double kink band produced by axial compression ... **A9:** 689

Zinc (continued)
EDTA titration ... **A10:** 173
effect, in commercial bronze ... **A2:** 296-297
effect, in red brass ... **A2:** 298
effect on adhesive wear of brass versus tool steel ... **A18:** 237
effect on glass substrate bonding ... **EM3:** 283
effect on tin-base alloys ... **A18:** 748
effects, cartridge brass ... **A2:** 296, 300
effects on SCC in copper ... **A11:** 221
electrodeposited coatings ... **A13:** 426, 911-912
-electroplated steel fastener, hydrogen embrittlement failure ... **A11:** 548-549
electropolishing with alkali hydroxides ... **A9:** 54
embrittlement ... **A13:** 178
embrittlement of ferritic steels ... **A11:** 237-238
as embrittlement source ... **A11:** 234, 237-238
emulsion cleaning of ... **M5:** 7
in enamel cover coats ... **EM3:** 304
erosive attack of melt on die surface ... **A18:** 630
as essential metal ... **A2:** 1250, 1255-1256
etching by polarized light ... **A9:** 59
evaporation fields for ... **A10:** 587
ferritic steels embrittlement by ... **A13:** 184
friction coefficient data ... **A18:** 71
galvanic corrosion of ... **A13:** 85
galvanic corrosion with magnesium ... **M2:** 607
galvanized coatings *See* Hot dip galvanized coating
galvanizing ... **A2:** 527-528
gas tungsten arc welding ... **M6:** 183
general biological corrosion of ... **A13:** 87
as gold alloy ... **A2:** 690
gravimetric finishes ... **A10:** 171
heat-affected zone fissuring in nickel-base alloys ... **A6:** 588
high-purity, and SCC ... **A11:** 539
hot dip galvanizing of ... **A13:** 765-767
hot extrusion of ... **A14:** 322
hydrochloric acid corrosion ... **A13:** 467
ICP-determined in plant tissues ... **A10:** 41
ICP-determined in silver scrap metal ... **A10:** 41
impurity in solders ... **M6:** 1072
as inclusion-forming, aluminum alloys ... **A15:** 95
indirect Volhard titration ... **A10:** 173
intergranular corrosion, prevention ... **A13:** 765
interlayer for aluminum-base alloys being diffusion welded ... **A6:** 885
intermetallic inclusions in copper ... **A15:** 96
ions, as cathodic inhibitors ... **A13:** 495
lap welding ... **M6:** 673
lead, Rockwell scale for ... **A8:** 76
liquid, as embrittler of steels ... **A11:** 237
liquid-metal embrittlement, and material selection ... **A13:** 334
liquid-metal embrittlement of ... **A11:** 28, 233
lubricant indicators and range of sensitivities ... **A18:** 301
in magnesium alloys ... **A9:** 426-428
molten ... **A11:** 273
nickel alloy surface, removal from ... **M5:** 672
nonaqueous corrosion ... **A13:** 763
ores, indium occurrence in ... **A2:** 750
oxyfuel gas welding ... **M6:** 583
phosphate coating solution contaminated by ... **M5:** 455-456
plain carbon steel resistance to ... **A13:** 515
plating ... **A13:** 767
polishing and buffing ... **M5:** 108, 112-114
products ... **A2:** 527-532
PSG as diffusion barrier ... **EM3:** 583
pure ... **M2:** 824-826
pure, properties ... **A2:** 1174
pure, thermal properties ... **A18:** 42
radiographic absorption ... **A17:** 311
recommended impurity limits of solders ... **A6:** 986
recrystallization in ... **A9:** 34
recycling ... **A2:** 1223-1226

Zinc (continued)
relative solderability as a function of flux type ... **A6:** 129
relative weldability ratings, resistance spot welding ... **A6:** 834
removal, from lead alloys ... **A15:** 476
resistance welding ... **A6:** 833
-rich coatings, for corrosion control ... **A11:** 194
rolled, compositions ... **A13:** 760
sacrificial anodes ... **A13:** 469
safety precautions ... **A6:** 1196
safety standards for soldering ... **M6:** 1098
shielded metal arc welding ... **A6:** 179 **M6:** 75
shrinkage allowance ... **A15:** 303
slab, grades and compositions ... **A2:** 527
slip planes ... **A9:** 684
soil corrosion ... **A13:** 762
as solder impurity ... **EL1:** 639
solderability ... **A6:** 978
solubility in aluminum ... **A2:** 36-37
solubility in magnesium ... **M2:** 525
solution potential ... **M2:** 207
species weighed in gravimetry ... **A10:** 172
spectrometric metals analysis ... **A18:** 300
spray material for oxyfuel wire spray process ... **A18:** 829
steel embrittled by ... **M1:** 686, 688
stress-corrosion cracking of ... **A13:** 763-764
superplasticity of ... **A8:** 553
tantalum corrosion in ... **A13:** 735
temperature effect, corrosion rate ... **A13:** 910
thermal diffusivity from 20 to 100 °C ... **A6:** 4
thermal expansion coefficient ... **A6:** 907
as thermal spray coating ... **A13:** 460-461
in thermal spray coatings ... **A6:** 1004-1009
as tin solder impurity ... **A2:** 521
TNAA detection limits ... **A10:** 238
as trace element, cupolas ... **A15:** 388
TWA limits for particulates ... **A6:** 984
ultrapure, by zone-refining technique ... **A2:** 1094
untreated, material compatibility ... **A13:** 764
vapor degreasing of ... **M5:** 45-46, 54
vapor pressure ... **A4:** 493 **A6:** 621
vapor pressure, relation to temperature ... **A4:** 495
Vickers and Knoop microindentation hardness numbers ... **A18:** 416
volatilization losses in melting ... **EM4:** 389
volumetric procedures for ... **A10:** 175
in water ... **A13:** 443, 759-762
weighed as the phosphate ... **A10:** 171
weighed as the sulfide ... **A10:** 171
weight loss ... **A13:** 441
wetness, and corrosion rate ... **A13:** 908
and zinc alloys ... **A15:** 786-797
and zinc primers, for filiform corrosion ... **A13:** 108

Zinc acid chloride plating ... **M5:** 245, 250-253
advantages and limitations ... **M5:** 250-251
agitation ... **M5:** 252
anodes ... **M5:** 252
bleedout ... **M5:** 251
cathode current efficiency ... **M5:** 250, 252
iron contamination ... **M5:** 253
pH control ... **M5:** 252-253
solution composition and operating conditions ... **M5:** 250-253
temperature control ... **M5:** 252

Zinc alkaline noncyanide plating ... **M5:** 246-248
anodes ... **M5:** 250
bright plating range ... **M5:** 250
cathode current densities ... **M5:** 250
cathode current efficiency ... **M5:** 246-248, 250
efficiency ... **M5:** 246, 248
filtration ... **M5:** 250
metal content, effects of ... **M5:** 246-248
solution composition and operating conditions ... **M5:** 246-248, 250

SUBJECTS OF THE INDEXED VOLUMES: ASM Handbook (designated by the letter "A"): **A1:** Properties and Selection: Irons, Steels, and High-Performance Alloys (1990); **A2:** Properties and Selection: Nonferrous Alloys and Special-Purpose Materials (1990); **A3:** Alloy Phase Diagrams (1992); **A4:** Heat Treating (1991); **A6:** Welding, Brazing, and Soldering (1993); **A7:** Powder Metallurgy (1984); **A8:** Mechanical Testing (1985); **A9:** Metallography and Microstructures (1985); **A10:** Materials Characterization (1986); **A11:** Failure Analysis and Prevention (1986); **A12:** Fractography (1987); **A13:** Corrosion (1987); **A14:** Forming and Forging (1988); **A15:** Casting (1988); **A16:** Machining (1989); **A17:** Nondestructive Testing and Quality Control (1989); **A18:** Friction, Lubrication, and Wear Technology (1992). **Metals Handbook, 9th Edition** (designated by the letter "M"): **M1:** Properties and Selection: Irons and Steels (1978); **M2:** Properties and Selection: Nonferrous Alloys and Pure Metals (1979); **M3:** Properties and Selection: Stainless Steels, Tool Materials and Special-Purpose Materials (1980); **M4:** Heat Treating (1981); **M5:** Surface Cleaning, Finishing, and Coating (1982); **M6:** Welding, Brazing, and Soldering (1983). **Engineered Materials Handbook** (designated by the letters "EM"): **EM1:** Composites (1987); **EM2:** Engineering Plastics (1988); **EM3:** Adhesives and Sealants (1990); **EM4:** Ceramics and Glasses (1991); **Electronic Materials Handbook** (designated by the letters "EL"): **EL1:** Packaging (1989).

Zinc alkaline noncyanide plating (continued)
sodium hydroxide content **M5:** 246, 248
temperature control **M5:** 250
voltages **M5:** 250
Zinc alloy castings
gravity castings **A2:** 529-530
pressure die castings **A2:** 528-529
properties **A2:** 531
temperature effects **A2:** 533
Zinc alloy coated steels **A1:** 217-218
Zinc alloys *See also* High-zinc alloys; Zinc; Zinc alloy castings, specific types Zinc alloys, specific types **A16:** 831-834
55% aluminum-zinc alloy coating, steel sheet and wire **M5:** 348-350
acid dipping **M5:** 677
alkaline soak and electrocleaning **M5:** 677
applications **A15:** 797 **M2:** 630-637
blanking and piercing dies, use for **M3:** 485, 487
boring **A16:** 831-832
broaching **A16:** 204
as cast in plaster molds **A15:** 243
casting removal **A15:** 790-792
castings **A2:** 528-532
chip formation **A16:** 831, 832, 834
chromating process sequence **A13:** 390
chrome plating, hard, removal of **M5:** 185
cleaning and finishing processes **M5:** 676-677
climb milling **A16:** 834
coatings **A2:** 527-528
cold form tapping **A16:** 266
composition control **A15:** 787-788
conveyors **A15:** 792
copper strike **M5:** 677
cutting fluids **A16:** 831, 832, 833, 834
deep drawing dies, materials for **M3:** 499
deep-hole drilling **A16:** 832
degreasing **M5:** 676-677
design advantages **A15:** 792-797
dichromated, thermal energy method of deburring **A16:** 577-578
die casting advantages **A15:** 792-794
as die casting alloy **A15:** 35, 286, 786
die casting, intergranular corrosion evaluation **A13:** 241
die casting machines **A15:** 789
die cutting speeds **A16:** 301
die lubrication system **A15:** 790
die temperature **A15:** 790
die threading **A16:** 833-834
die(s) **A15:** 780-790
diphase cleaning of **M5:** 677
as draw tooling **A14:** 511
drilling **A16:** 832, 833
effect on Cu alloy machinability **A16:** 808
electrochemical grinding **A16:** 543
electron-beam welding **A6:** 872
end milling **A16:** 834
finishing **A15:** 795-796
finishing and secondary operations **A2:** 530-531
fluxing of **A15:** 451-452
form tapping **A16:** 833
furnaces **A15:** 788
galvanizing **A2:** 527-528
gravity castings **A2:** 529-530
grinding **A16:** 833, 834
hard spots from intermetallics **A16:** 831
hot extrusion of **A14:** 322
inclusions in **A15:** 488-489
inoculant for **A15:** 105
launder system **A15:** 788
machinability **A16:** 831
machining **A15:** 794
markets for **A15:** 42
melting heat for **A15:** 376
milling **A16:** 312, 313, 314, 834
nickel plating of **M5:** 144-200, 203-204, 210, 215-216, 218
contamination effects **M5:** 210
nuts, SCC of **A11:** 538-539
oxyfuel gas welding **A6:** 281
permanent mold casting **A15:** 275
photochemical machining **A16:** 588, 590
plating, preparation for **M5:** 676-677
polishing and buffing **M5:** 676
press forming dies, use for **M3:** 490, 491, 493
pressure die castings **A2:** 528-529

Zinc alloys (continued)
properties of **A2:** 532-542
reaming **A16:** 832-833
resistance welding **A6:** 833, 847
sawing **A16:** 834
scrap return **A15:** 788-789
semisolid forging/casting of **A15:** 327
slush casting, properties **A2:** 538-539
spade drilling **A16:** 225, 230
spotfacing **A16:** 833
strengths of **A15:** 786
stripping of **M5:** 218
surface finish of die castings **A16:** 831, 832, 834
tapping **A16:** 258, 833
temperature control **A15:** 790
temperature effect **A2:** 533
thermal expansion coefficient **A6:** 907
for thermal spray coatings **A13:** 460-461
tool life **A16:** 831, 832, 833
tools **A16:** 831, 832
turning **A16:** 94, 831-832
Unicast process for **A15:** 251
use of high-purity zinc in **A11:** 539
Zinc alloys, gravity casting
applications **M2:** 635
Zinc alloys, specific types *See also* Zinc; Zinc alloys
12%-Al applications **M2:** 635-636
27%-Al applications **M2:** 635-636
98Zn-2Al **A6:** 351
AC40A, as die casting, composition **A15:** 286
AC41A alloy **M2:** 639
composition **M2:** 630-631
creep data **M2:** 634-635
designations **M2:** 630-632
mechanical properties **M2:** 633
physical properties **M2:** 633
AG40A alloy **M2:** 638
composition **M2:** 631
creep data **M2:** 634-635
designations **M2:** 630-631
mechanical properties **M2:** 633
AG41A, as die casting, composition **A15:** 286
alloy 3 .... **A16:** 831
alloy 3, composition and use **A15:** 786
alloy 5 .... **A16:** 831
alloy 5, for tensile strength **A15:** 786
alloy 7 .... **A16:** 831
composition **M2:** 630-631
designations **M2:** 630-631
mechanical properties **M2:** 633
physical properties **M2:** 633
alloy 7, as die casting, composition **A15:** 286
alloy 7, composition and use **A15:** 786
Alloy No. 2, die castings **A2:** 529, 535
Alloy No. 3, die castings **A2:** 529, 532-533
Alloy No. 5, die castings **A2:** 529, 533-534
Alloy No. 7, die castings **A2:** 529, 534-535
Alloy ZA-8, die castings **A2:** 529, 535-536
Alloy ZA-12, die castings **A2:** 529, 536-537
commercial rolled zincs **M2:** 641-643
copper-hardened rolled zinc **M2:** 643
Cu-5Zn
annealing time and temperature effect on hardness **A4:** 828
annealing time and temperature effect on microstructure **A4:** 829
annealing time effect on annealing process **A4:** 830
microstructure showing deformation and bent annealing twins **A4:** 828
recrystallization **A4:** 829
Cu-40Zn
heat-treatment effect on hardness **A4:** 837, 838
microstructure after quenching **A4:** 838
two-phase structure development **A4:** 837, 838
Cu-42Zn, two-phase structure development **A4:** 838, 839
Cu-43Zn, two-phase structure development **A4:** 838, 839
Cu-Zn, plastic deformation **A4:** 827
ILZRO 16 .... **M2:** 640-641
composition **M2:** 632
designations **M2:** 631-632
mechanical properties **M2:** 633
physical properties **M2:** 633
ILZRO 16, as die casting, composition **A15:** 286
ILZRO 16, die castings **A2:** 529

Zinc alloys, specific types (continued)
Kirksite alloy, gravity casting **A2:** 530
rolled zinc alloy **M2:** 643-644
superplastic zinc alloy **M2:** 644-645
Z35631
casting method **A18:** 754
composition **A18:** 753
designations **A18:** 753
mechanical properties **A18:** 754
Z35831
composition **A18:** 753
designations **A18:** 753
Z35840
casting method **A18:** 754
mechanical properties **A18:** 754
ZA-8 .... **A16:** 831
ZA-8, composition **A15:** 786
ZA-8, for permanent mold casting **A15:** 786
ZA-8, gravity castings **A2:** 530
ZA-12 .... **A16:** 831
ZA-12, as general-purpose alloy **A15:** 786
ZA-12, composition **A15:** 786
ZA-12, for sand casting **A15:** 797
ZA-12, gravity castings **A2:** 530
ZA-27 .... **A16:** 831, 833
ZA-27, composition **A15:** 786
ZA-27, die castings **A2:** 529, 537-538
ZA-27, for sand casting **A15:** 797
ZA-27, gravity castings **A2:** 530
ZA-27, ultrahigh performance **A15:** 786
zinc foundry alloy ZA-12 .... **M2:** 640
zinc-base slush-casting alloys **M2:** 639-640
Zn-0.3Pb-0.03Cd (commercial rolled zinc) properties **A2:** 540
Zn-0.06Pb-0.06Cd (commercial rolled zinc) properties **A2:** 539-540
Zn-0.8Cu-0.15Ti, properties **A2:** 541-542
Zn-0.08Pb (commercial rolled zinc) properties **A2:** 539
Zn-1.0Cu, properties **A2:** 540-541
Zn-1.0Cu-0.010Mg (rolled-zinc alloy) properties **A2:** 541
Zn-1.25Cu-0.2Ti 0.15Cr (ILZRO 16) properties **A2:** 538
Zn-4Al, erosive attack of zinc melt on the die surface **A18:** 630
Zn-4Al-0.04Mg (AG40A), properties **A2:** 532-533
Zn-4Al-0.015Mg (AG40B), properties **A2:** 534-535
Zn-4Al-1Cu-0.05Mg (AC41A), properties **A2:** 533-534
Zn-4Al-2.5Cu-0.04Mg (AC43A), properties **A2:** 535
Zn-5Al-MM, hot dip coatings **A2:** 528
Zn-8Al-1Cu-0.02Mg (ZA-8) properties **A2:** 535-536
Zn-11Al-1Cu-0.025 Mg (ZA-12) properties **A2:** 536-537
Zn-22Al (superplastic zinc), properties **A2:** 542
Zn-27Al **M2:** 641
Zn-27Al-2Cu-0.015Mg (ZA-27) properties **A2:** 537-538
Zn-Cu-Ti alloy **M2:** 644
Zinc alloys, wrought
characteristics **M2:** 636
classification **M2:** 636
Zinc ammonium chloride
boiling point **M5:** 327
Zinc and zinc alloy die castings
chromium plating, decorative **M5:** 189, 191, 194-195
cleaning and finishing processes **M5:** 33, 35-36
copper plating of **M5:** 160-163, 168
electrolytic cleaning of **M5:** 33, 35-36, 677
emulsion cleaning of **M5:** 35, 676-677
nickel plating of **M5:** 203, 205, 207-208
polishing and buffing of **M5:** 112-114, 123, 676
compounds, removal of **M5:** 10-11
properties of **M5:** 676
vapor degreasing of **M5:** 54, 676-677
Zinc and zinc alloys **A9:** 488-496
etchants for **A9:** 488
grinding **A9:** 488
macroetching **A9:** 488
microetching **A9:** 488-489
microstructures **A9:** 489-490
mounting **A9:** 488
polishing **A9:** 488

Zinc and zinc alloys (continued)
sectioning ... **A9:** 488
specimen preparation ... **A9:** 488-489
Zinc and zinc alloys, specific types
1% Cu, hot rolled ... **A9:** 495
Alloy 3, as die cast, and aged ... **A9:** 493
Alloy 3, exposed to wet steam ... **A9:** 495
Alloy 3, fracture surface of tension test bar ... **A9:** 495
Alloy 3, gravity cast in a permanent mold ... **A9:** 494
Alloy 5, as die cast ... **A9:** 493
Alloy 5, die cast and aged ... **A9:** 494
Alloy 5, different casting methods compared ... **A9:** 494
ASTM AC41A, as die cast ... **A9:** 493
ASTM AC41A, die cast and aged ... **A9:** 494
ASTM AC41A, different casting methods compared ... **A9:** 494
ASTM AG40A, as die cast, and aged ... **A9:** 493
ASTM AG40A, exposed to wet steam ... **A9:** 495
ASTM AG40A, fracture sufrace of tension test bar ... **A9:** 495
ASTM AG40A, gravity cast in a permanent molds ... **A9:** 494
brass special zinc, cold rolled ... **A9:** 495
brass special zinc, hot-rolled ... **A9:** 495
cast zinc, 0.6% Cu and 0.14% Ti ... **A9:** 493
Galfan coated on steel ... **A9:** 496
prime western Zinc, as-cast ... **A9:** 493
special high-grade zinc, as-cast ... **A9:** 493
ZA-8, different casting methods compared ... **A9:** 490
ZA-8, eta phase in ... **A9:** 489
ZA-8, magnesium additions ... **A9:** 490
ZA-12, different casting methods compared ... **A9:** 491
ZA-12, eta phase in ... **A9:** 489
ZA-12, magnesium additions ... **A9:** 490
ZA-27, different casting methods compared ... **A9:** 491-492
ZA-27, eta and alpha phases in ... **A9:** 489
ZA-27, magnesium additions ... **A9:** 490
ZA-27 with 0.05% Fe, as sand cast ... **A9:** 493
ZA-27 with 0.13% Fe, as sand cast ... **A9:** 492
Zn-0.6Cu-0.14Ti, titanium-zinc stringers ... **A9:** 495
Zn-0.55Cu-0.12Ti, as die cast in a cold chamber ... **A9:** 494
Zn-1Cu, cold-rolled ... **A9:** 495
Zn-7Ni, peritectic structures ... **A9:** 676
Zn-12Al-0.75Cu-0.02Mg, as die cast in a cold chamber ... **A9:** 494
Zn-12Al-0.75Cu-0.02Mg, gravity cast in a permanent mold ... **A9:** 494
Zn-22Al, superplastic structure ... **A9:** 496
Zinc anodes ... **A13:** 470, 764-765, 921
cathodic protection ... **M2:** 654
Zinc borate
for flame retardance ... **EM3:** 179
Zinc chromate
as cathodic inhibitor ... **A13:** 496
as multicomponent cathodic inhibitor ... **A13:** 495-496
paints ... **A13:** 769
Zinc chromate primers ... **A1:** 222 **M1:** 175
Zinc coating
chromate passivation ... **A1:** 214-215
coating tests and designations ... **A1:** 212-214, 215
corrosion resistance ... **A1:** 581, 584
electrogalvanizing ... **A1:** 217
hot dip galvanizing ... **A1:** 216-217, 218
inorganic ... **M5:** 501-505
mechanical coating ... **M5:** 300-302
organic ... **M5:** 502, 505
oxidation-resistant types ... **M5:** 665-666
packaging and storage ... **A1:** 215-216
painting ... **A1:** 215
for threaded steel fasteners ... **A1:** 295
Zinc coating (continued)
zinc spraying ... **A1:** 218
Zincrometal ... **A1:** 217
Zinc coating(s)
atmospheric exposure tests ... **A13:** 757-758
effect, corrosion of structures ... **A13:** 1304
paints ... **A13:** 768-769
processes ... **A13:** 765-768
protection ... **A13:** 755-756
of steels ... **A13:** 432-434, 1011-1014
Zinc coatings *See also* Electrogalvanizing; Hot dip galvanizing; Zinc-rich primers
atmospheric corrosion resistance ... **M1:** 722
bridge wire ... **M1:** 272
cast irons ... **M1:** 102-104
corrosion protection afforded by ... **M1:** 752-754
electrogalvanizing ... **A2:** 528
hot dip galvanizing ... **A2:** 527-528
mechanical galvanizing ... **A2:** 528
seawater corrosion resistance ... **M1:** 745
sheet ... **M1:** 167-171
anodizing of ... **M1:** 169
chromate passivation of ... **M1:** 168-169
coating tests ... **M1:** 168, 169
corrosion resistance ... **M1:** 167-168, 170
designations ... **M1:** 168
electrogalvanizing ... **M1:** 170-171
heat reflection ... **M1:** 172-173
hot dip galvanizing ... **M1:** 169-170
packaging and storage ... **M1:** 169
painting of ... **M1:** 169, 170
service life ... **M1:** 168
spraying ... **M1:** 171
temper rolling of coated sheet ... **M1:** 170
springs ... **M1:** 291
threaded fasteners ... **M1:** 279
wire ... **M1:** 264
wire fence ... **M1:** 271
zinc-iron alloy in ... **M1:** 170
Zinc, corrosion environments
acids ... **M2:** 648
aqueous ... **M2:** 647-648
atmospheric ... **M2:** 649-650
gases ... **M2:** 649
indoor exposure ... **M2:** 649-650
inhibitors, use of ... **M2:** 648-649
nonaqueous liquids ... **M2:** 649
oxygen in water, effect of ... **M2:** 648-649
salts ... **M2:** 648
seacoast ... **M2:** 649-650
soils ... **M2:** 649
Zinc, corrosion resistance
anodic coatings, effect of ... **M2:** 646-647
composition, effect of ... **M2:** 646-647
electrochemical corrosion ... **M2:** 650, 652
fatigue ... **M2:** 650-652
rate, compared to iron ... **M2:** 646
Zinc cyanide plating ... **M5:** 244-250, 252-254
advantages and limitations ... **M5:** 244-245
agitation ... **M5:** 249, 252
anodes ... **M5:** 246-247
bright throwing and covering power ... **M5:** 244, 249-250
brighteners, use of ... **M5:** 246, 250
current densities ... **M5:** 248-249
current efficiency ... **M5:** 248-249
cyanide-to-zinc ratio ... **M5:** 248-249
efficiency ... **M5:** 245-246, 249-250
filtration ... **M5:** 249
grades of zinc used for ... **M5:** 247
hydrogen, embrittlement caused by ... **M5:** 253
low-cyanide system ... **M5:** 244-246, 249-250
microcyanide system ... **M5:** 245
midcyanide system ... **M5:** 244-249
solution compositions and operating conditions ... **M5:** 244-250
sodium carbonate content ... **M5:** 248
sodium cyanide concentration ... **M5:** 249-250
Zinc cyanide plating (continued)
zinc content ... **M5:** 247-249
solution preparation ... **M5:** 246
standard system ... **M5:** 244-249
temperature effects ... **M5:** 247-250
thickness, control of ... **M5:** 253-254
Zinc deficiencies
biologic effects ... **A2:** 1255
Zinc di-*n*-octyldithiophosphate ... **A18:** 253
Zinc dialkyl dithiophosphates (ZDDP) ... **A18:** 141
Zinc, die castings
aging ... **M2:** 631
alloys ... **M2:** 630-633
applications ... **M2:** 632-633
assembly ... **M2:** 631
finishing ... **M2:** 631-632
heat treatment ... **M2:** 631-632
relative solderability as a function of flux type ... **A6:** 129
Zinc, die castings engineering
casting design ... **M2:** 634
die design ... **M2:** 633-634
Zinc die-casting alloys *See* Zinc and zinc alloys
Zinc dithiophosphates
as antiwear agents ... **A18:** 111
as corrosion inhibitors ... **A18:** 111
lubricant analysis ... **A18:** 300, 301
as oxidation inhibitors ... **A18:** 111
Zinc dust
applications ... **M2:** 629-630
Zinc dust/zinc oxide paints and coatings ... **A13:** 443, 768-769
Zinc evaporation
in shape memory effect (SME) alloys ... **A2:** 900
Zinc ferrites ... **A9:** 538
Zinc flake powders ... **A7:** 596
Zinc flaring
in copper-zinc alloys ... **A15:** 466
Zinc fluoborate plating systems ... **M5:** 252-253
Zinc foundry alloy ZA-12 *See* Zinc alloys, specific types, zinc foundry alloy ZA-12
Zinc foundry alloy ZA-27 *See* Zinc alloys, specific types, Zn-27Al
Zinc foundry alloys
properties ... **A2:** 535-538
Zinc furnaces, historic
in Africa ... **A15:** 19
Zinc, galvanized
alloys layer ... **M2:** 652
coating thickness ... **M2:** 652-653
coating uniformity ... **M2:** 652-653
life of coating ... **M2:** 653
sheet ... **M2:** 651, 653
steel wire ... **M2:** 652-654
Zinc in copper ... **M2:** 242-243
Zinc in fusible alloys ... **M3:** 799
Zinc ores
as gallium source ... **A2:** 742-743
Zinc oxide
applications ... **M2:** 630
in binary phosphate glasses ... **A10:** 131
for curing butyl rubber ... **EM3:** 146
filler for polyurethanes ... **EM3:** 179
lattice image of ... **A10:** 446
LEISS spectra ... **A10:** 604
metal-to-metal oxide equilibria ... **A7:** 340
quantitative XRPD analysis in calcite ... **A10:** 342
Zinc oxide (ZnO) *See also* Engineering properties of single oxides
powders, gas phase reactions ... **EM4:** 62
role in glazes ... **EM4:** 1062
Zinc oxide fumes
metal fume fever from ... **A2:** 1255
Zinc oxide solubility
for cleaning ... **A13:** 381
Zinc oxide thin films
characteristics ... **EM4:** 1119
used for surface acoustic wave devices ... **EM4:** 1119

SUBJECTS OF THE INDEXED VOLUMES: ASM Handbook (designated by the letter "A"): **A1:** Properties and Selection: Irons, Steels, and High-Performance Alloys (1990); **A2:** Properties and Selection: Nonferrous Alloys and Special-Purpose Materials (1990); **A3:** Alloy Phase Diagrams (1992); **A4:** Heat Treating (1991); **A6:** Welding, Brazing, and Soldering (1993); **A7:** Powder Metallurgy (1984); **A8:** Mechanical Testing (1985); **A9:** Metallography and Microstructures (1985); **A10:** Materials Characterization (1986); **A11:** Failure Analysis and Prevention (1986); **A12:** Fractography (1987); **A13:** Corrosion (1987); **A14:** Forming and Forging (1988); **A15:** Casting (1988); **A16:** Machining (1989); **A17:** Nondestructive Testing and Quality Control (1989); **A18:** Friction, Lubrication, and Wear Technology (1992). **Metals Handbook, 9th Edition** (designated by the letter "M"): **M1:** Properties and Selection: Irons and Steels (1978); **M2:** Properties and Selection: Nonferrous Alloys and Pure Metals (1979); **M3:** Properties and Selection: Stainless Steels, Tool Materials and Special-Purpose Materials (1980); **M4:** Heat Treating (1981); **M5:** Surface Cleaning, Finishing, and Coating (1982); **M6:** Welding, Brazing, and Soldering (1983). **Engineered Materials Handbook** (designated by the letters "EM"): **EM1:** Composites (1987); **EM2:** Engineering Plastics (1988); **EM3:** Adhesives and Sealants (1990); **EM4:** Ceramics and Glasses (1991); **Electronic Materials Handbook** (designated by the letters "EL"): **EL1:** Packaging (1989).

Zinc oxide-eugenol
for dental ceramics ... **A18:** 666, 673
Zinc phosphate
application ... **A13:** 386-387
coating process ... **A13:** 383
as conversion coating ... **A13:** 387
for dental cements ... **A18:** 666, 673
as shaft conversion coating ... **A11:** 482
types ... **A13:** 386-387
for wire coating ... **A14:** 697
Zinc phosphate coating ... **M5:** 434-438, 441-444, 448-451, 454-455
characteristics of ... **M5:** 434
crystal structure ... **M5:** 449-451, 455
equipment ... **M5:** 445, 449-450, 452, 454
immersion systems ... **M5:** 434-437, 440-442, 450
iron concentration ... **M5:** 442
solution composition and operating conditions ... **M5:** 435, 444-445, 449-450, 452, 454
spray system ... **M5:** 435-438, 440-442, 449, 452, 454
weight, coating ... **M5:** 434, 436-437, 442, 448-449, 452, 454
Zinc phosphate coatings ... **A1:** 222 **M1:** 174
atmospheric corrosion resistance enhanced by ... **M1:** 722
corrosion protection afforded by ... **M1:** 754
hot rolled bars ... **M1:** 200
wire ... **M1:** 266
Zinc phosphate prepaint treatment ... **M5:** 476-478
Zinc phosphate treatment ... **EM3:** 42
Zinc phosphonates
as cathodic inhibitors ... **A13:** 496
Zinc plating ... **A13:** 767 **M5:** 244-255
acid *See* Zinc acid chloride plating
alkaline *See* Zinc alkaline noncyanide plating
aluminum and aluminum alloys ... **M5:** 601-606
ammonium chloride system ... **M5:** 251-252
anodes ... **M5:** 246-247, 250, 252
appearance ... **M5:** 255
applications ... **M5:** 254-255
cadmium plating compared to ... **M5:** 253-254, 264, 266
chloride process *See* Zinc acid chloride plating
chromate conversion coating of ... **M5:** 254-255
corrosion protection ... **M5:** 253-255
current densities ... **M5:** 248-250
current efficiency ... **M5:** 246-250, 252
cyanide *See* Zinc cyanide plating
equipment ... **M5:** 253
filtration process ... **M5:** 249-250
fluoborate process ... **M5:** 252-253
hydrogen embrittlement caused by ... **M5:** 251, 253
lacquering ... **M5:** 255
limitations of ... **M5:** 244-245, 250-251, 253-255
magnesium alloys ... **M5:** 638-639, 642, 644-647
solution composition and operating conditions ... **M5:** 644-645, 647
stripping of ... **M5:** 668
noncyanide *See* Zinc alkaline noncyanide plating
potassium chloride process ... **M5:** 251-252
processing steps ... **M5:** 253-254
rinsewater recovery ... **M5:** 318
sodium-ammonium chloride system ... **M5:** 251-252
steel ... **M5:** 254-255
stripping of ... **M5:** 646
sulfate process ... **M5:** 252-253
supplementary coating ... **M5:** 254-255
thickness ... **M5:** 253-255
types, general discussion ... **M5:** 244
Zinc plating of specimens for edge retention ... **A9:** 32
Zinc polyacrylate
for dental cements ... **A18:** 666, 673
Zinc polyphosphate
as cathodic inhibitor ... **A13:** 496
Zinc powder(s)
alloys ... **A7:** 249
chemical analysis and sampling ... **A7:** 249
contact angle with mercury ... **A7:** 269
electrolytic ... **A7:** 72
explosive reaction with moisture ... **A7:** 194-195
metal-to-metal oxide equilibria ... **A7:** 340
moderate explosivity class ... **A7:** 196
as plating material ... **A7:** 459
pressure-density relationships ... **A7:** 299
Zinc primers
for filiform corrosion ... **A13:** 108
Zinc recycling
from electric arc furnace (EAF) dust ... **A2:** 1224-1225
scrap sources ... **A2:** 1223-1224
technology ... **A2:** 1224
Zinc, resistance of
to liquid-metal corrosion ... **A1:** 635
Zinc, rolled
atmospheric corrosion, tests ... **M2:** 647, 650
Zinc selenide
as internal reflection element ... **A10:** 113
as internal reflection element, in surfactant study ... **A10:** 118
Zinc selenide as an evaporated interference layer material ... **A9:** 60
Zinc silicate
as filler for solder glass ... **EM4:** 1072
Zinc, slab
composition ... **M2:** 629
grades ... **M2:** 629
production ... **M2:** 629-630
superplastic ... **M2:** 629
Zinc, specific types
Prime Western, use in hot dip galvanized coating ... **M5:** 324
Zinc spraying ... **A1:** 212, 218
Zinc stearate
mixing and mixer selection for spray-dried powders ... **EM4:** 98
as mold release agent ... **EM1:** 158
Zinc stearate lubricant ... **A7:** 190-193
in Ancor MH-100 ... **A7:** 192
burn-off ... **A7:** 351, 352
effect of mixing time and compacting pressure ... **A7:** 191
effect on green strength ... **A7:** 302
mix flow and bulk density ... **A7:** 190
stripping pressure and bulk density ... **A7:** 190, 191
Zinc stearate soaps
for tool steel lubrication ... **A18:** 737
Zinc sulfate plating systems ... **M5:** 252-253
Zinc sulfide
applications ... **M2:** 629
crystals, for neutron radiography ... **A17:** 390
liquid impingement erosion ... **A18:** 226
rain erosion effects on coated surface ... **A18:** 222
screens, radiography ... **A17:** 319
sol-gel processing ... **EM4:** 447
Zinc sulfide as an evaporated interference layer material ... **A9:** 60
Zinc sulfide as an interference film ... **A9:** 147
Zinc telluride as an evaporated interference layer material ... **A9:** 60
Zinc thermal spraying ... **A13:** 767-768
Zinc titania cover glass, laboratory glassware
composition and properties ... **EM4:** 1088
Zinc, vapor pressure
relation to temperature ... **M4:** 309, 310
Zinc, wrought
classification ... **M2:** 636-637
extrusions ... **M2:** 636
fabrication ... **M2:** 636-637
finishing ... **M2:** 637
machining ... **M2:** 637
mechanical properties ... **M2:** 636-637
rolled products ... **M2:** 636
soldering ... **M2:** 637
welding ... **M2:** 637
wire drawing ... **M2:** 636
Zinc-alloy coated steels
automotive industry ... **A13:** 1014
Zinc-aluminum
filler metal for torch soldering ... **A6:** 352
Zinc-aluminum alloys ... **A9:** 489-490
for thermal spray coatings ... **A13:** 460-461
Zinc-aluminum solder
dip soldering ... **A6:** 356
Zinc-base alloys ... **A18:** 693
as bearing alloys ... **A18:** 748, 753-754
applications ... **A18:** 753, 754
casting method ... **A18:** 754
composition ... **A18:** 753
designations ... **A18:** 753
mechanical properties ... **A18:** 754
Zinc-base alloys (continued)
microstructures ... **A18:** 754
bearing material systems ... **A18:** 745, 746
applications ... **A18:** 746
bearing performance characteristics ... **A18:** 746
load capacity rating ... **A18:** 746
die material for sheet metal forming ... **A18:** 628
erosive attack on die surfaces ... **A18:** 630
heat and temperature effects on strength retention ... **A18:** 745
life-limiting factors for die-casting dies ... **A18:** 629
Zinc-base die cast alloys
atmospheric corrosion ... **A13:** 760
Zinc-bearing paints ... **A13:** 768-769
Zinc-coated fence wire ... **A1:** 285
Zinc-coated sheet steel
preparation of samples ... **A9:** 197
Zinc-coated steel
soldering ... **A6:** 631
Zinc-coated steels ... **A13:** 432-434, 1011-1014
resistance spot welding ... **M6:** 479-480, 491-492
Zinc-coated strand wire ... **A1:** 282
Zinc-copper alloys
electrolytic etching to reveal phases ... **A9:** 489
Zinc-copper-titanium alloy
properties ... **A2:** 541-542
Zinc-magnesium alloying
wrought aluminum alloy ... **A2:** 55-56
Zinc-magnesium-copper alloying
wrought aluminum alloy ... **A2:** 56
Zinc-phosphate coating
Auger imaging of ... **A10:** 556, 558
Zinc-plated steel
resistance spot welding ... **M6:** 480
Zinc-rich coatings ... **A13:** 410-412
for corrosion control ... **A11:** 194
inorganic ... **A13:** 411-412
organic ... **A13:** 410
Zinc-rich primers ... **A1:** 222-223 **A13:** 913
corrosion prevention with ... **M5:** 432-433
Zincate
precoating ... **A6:** 131
Zincating
aluminum and aluminum alloys ... **M5:** 601-606
double-immersion process ... **M5:** 603-604, 606
solution composition and operating conditions ... **M5:** 603-604
Zincrometal ... **A1:** 217, 223
for automobiles ... **A13:** 1014
press forming ... **A14:** 564
Zincrometal priming system ... **M1:** 175
Zincrometal, with black paint
mounting of ... **A9:** 167-168
Zircaloy
claddings, embrittled by liquid cesium ... **A11:** 230
effect of strain rate on ductility in ... **A8:** 38, 42
fuel cladding, cesium-cadmium LME cleavage fracture ... **A11:** 235
occurrence of SMIE in ... **A11:** 24
Zircaloy-2
contour band sawing ... **A16:** 363
Zircaloy-2 (R60802)
chemical composition per ASTM specification B 351-92 ... **A6:** 787
trace element impurity effect on GTA weld penetration ... **A6:** 20
ultrasonic welding ... **A6:** 326
Zircaloy-2 plate
eddy current inspection ... **A17:** 186-187
Zircaloy-4 (R60804)
chemical composition per ASTM Specification B 351-92 ... **A6:** 787
Zircaloy-boron carbide
nuclear applications ... **A7:** 666
Zircaloy-clad LWR fuel rods
corrosion of ... **A13:** 945-948
Zircar fiber ... **EM1:** 63
Zircon *See also* Mold(s); Sand(s)
applications ... **EM4:** 46
refractory ... **EM4:** 901, 902, 905, 906
in ceramic tiles ... **EM4:** 926
composition ... **EM4:** 46
defined ... **A15:** 11
fusion flux for ... **A10:** 167
island structure ... **EM4:** 758
as mold refractory, investment casting ... **A15:** 258
as molding sand, characteristics ... **A15:** 209

Zircon (continued)
refractory composition ... **EM4:** 896
refractory physical properties ... **EM4:** 897, 898, 899
sand molds, in Cosworth process ... **A15:** 38
slurries, formulations and properties ... **A15:** 260
supply sources ... **EM4:** 46
thermal expansion coefficient ... **A6:** 907
typical oxide compositions of raw materials ... **EM4:** 550
Zirconates ... **EM4:** 60
Zirconia ... **A16:** 98
abrasive wear ... **A18:** 186, 189
alumina abrasive ... **A16:** 432, 434, 436, 440
applications ... **A18:** 812
ceramic coatings of dies ... **A18:** 643, 644
chemical composition ... **A6:** 60
crystal structures ... **A18:** 814
fretting wear ... **A18:** 248, 250
friction surfacing inclusion ... **A6:** 323
ground by diamond wheels ... **A16:** 462
properties ... **A18:** 192, 813, 814
scanning acoustic microscopy wear studies ... **A18:** 409, 410
as structural ceramic, applications and properties ... **A2:** 1022
thermal expansion coefficient ... **A6:** 907
thermionic emission production ... **A6:** 30
toughened, as structural ceramics ... **A2:** 1022-1023
ultrasonic machining ... **A16:** 530
Zirconia (ZrO_2) *See also* Zirconium oxide
applications
aerospace ... **EM4:** 1005
wear ... **EM4:** 975, 977
as basis for solid-electrolyte sensors ... **EM4:** 1136-1137
ceramic coatings for adiabatic diesel engines ... **EM4:** 992
effect on chemical properties of glass ... **EM4:** 857
effects of adding a second phase in ceramic- matrix composites ... **EM4:** 862-863
engineering properties *See* Engineering properties of zirconia
flexural strength ... **EM4:** 974
fracture toughness ... **EM4:** 983, 974
heated sensors ... **EM4:** 1137, 1138
matrix material for ceramic-matrix composites ... **EM4:** 840
plasma-sprayed, thermal properties when used as engine wall insulator lining ... **EM4:** 992
properties, adiabatic engine use ... **EM4:** 990
property data of composite components ... **EM4:** 863
refractory materials ... **EM4:** 907
role in glazes ... **EM4:** 1062
scuffing temperatures and coefficients of friction between ring and cylinder liner materials ... **EM4:** 991
stabilized, for potentiometric sensors ... **EM4:** 1131
thermal conductivity ... **EM4:** 974
thermal expansion ... **EM4:** 974
Vickers hardness ... **EM4:** 974
Zirconia alumina
applications ... **EM4:** 331
bond type ... **EM4:** 331
Zirconia ceramic coatings ... **M5:** 534-536, 540-542
Zirconia ceramics
chemical etching ... **EM4:** 575
solid-state sintering ... **EM4:** 277
Zirconia electrode
to monitor pH ... **A8:** 422
Zirconia grain-stabilized platinum and platinum alloys ... **A2:** 713-714
Zirconia mullite
applications ... **EM4:** 46
composition ... **EM4:** 46
supply sources ... **EM4:** 46
Zirconia powder
as nucleating agent in brazing with glasses ... **EM4:** 520
Zirconia with 3% MgO
partially stabilized ... **A9:** 94
Zirconia-silica fibers ... **EM1:** 61
Zirconia-toughened alumina (ZTA)
applications ... **EM4:** 976
fracture toughness ... **EM4:** 973
mineral processing ... **EM4:** 961
property, comparison, mineral processing ... **EM4:** 962
as structural ceramic ... **A2:** 1022-1023
Zirconia/aluminosilicate
melting/fining ... **EM4:** 391
Zirconium *See also* Hafnium; Optically anisotropic metals; Reactive metals; Zirconium alloys; Zirconium alloys, specific types ... **A13:** 707-721 **M3:** 781-783
absorptivity ... **A6:** 265
acid attack from polishing slurry ... **A9:** 498
in active metal process ... **EM3:** 305
as addition to aluminum alloys ... **A4:** 843
addition to fluxes affecting ionization process ... **A6:** 57
addition to high-temperature alloys ... **A6:** 563
addition to molybdenum for electron-beam welding ... **A6:** 870-871
additions, for magnesium alloy grain refinement ... **A15:** 480
AFS analysis of ... **A10:** 46
air-carbon arc cutting ... **A6:** 1176
allotropic transformation ... **A2:** 665-666 **M3:** 781-782
alloying additions ... **A9:** 497
alloying effect in titanium alloys ... **A6:** 508, 509
alloying effect on nickel-base alloys ... **A6:** 590
alloying, in wrought titanium alloys ... **A2:** 599
alloying, wrought aluminum alloys ... **A2:** 56-57
alloys, effect of strain rate on ductility ... **A8:** 38, 42
alloys, hydrogen damage in ... **A11:** 338
aluminum pretreatment process ... **M5:** 458
as an addition to austenitic manganese steel castings ... **A9:** 239
as an addition to cobalt-base heat-resistant casting alloys ... **A9:** 334
as an addition to niobium alloys ... **A9:** 441
as an addition to tantalum alloys ... **A9:** 442
as an alloying element in titanium alloys ... **A9:** 458
anisotropy and preferred orientation ... **A2:** 667
annealing ... **A2:** 663
anodizing procedure for ... **A9:** 498-499
applications ... **A2:** 667-669 **A13:** 718-719 **M6:** 1051
applications, sheet metals ... **A6:** 400
arc welding *See* Arc welding of zirconium and hafnium
as-cast, with niobium-base and tantalum- base inclusions ... **A9:** 157
atomic interaction descriptions ... **A6:** 144
basis for organometer coupling agents ... **EM3:** 182
brazing ... **M6:** 1051-1052
applications ... **M6:** 1053-1054
filler metals ... **M6:** 1051-1052
preparation ... **M6:** 1053
procedure ... **M6:** 1051
cadmium-induced LME in ... **A11:** 234
in cast iron ... **A1:** 8
casting ... **A2:** 663-664
characteristics ... **M6:** 1051
chemical compositions ... **M6:** 457
chemical corrosion attack in ... **A9:** 499
chemical resistance ... **M5:** 4
chemical-mechanical polishing of ... **A9:** 497-498
classification in tungsten alloy electrodes for GTAW ... **A6:** 191
in cobalt-base alloys ... **A1:** 985
coextrusion welding ... **A6:** 311
cold rolling ... **A2:** 663
Zirconium (continued)
cold work and recrystallization ... **A2:** 666
cold working ... **M3:** 782
compositions of nuclear-grade alloys ... **M6:** 1052
contained in platelets in ZrC matrix for directed metal oxidation ... **EM4:** 234-235
corrosion ... **M3:** 784-791
corrosion forms ... **A13:** 717-718
corrosion protection ... **A13:** 718
corrosion resistance ... **A13:** 707-717
and crack growth ... **A8:** 487
crystal bar liner, cleavage fracture ... **A11:** 235
crystal structure ... **M3:** 781
deoxidation, low-alloy steels ... **A15:** 715
diffusion bonding ... **A6:** 156
diffusion welding ... **M6:** 677
distillation separation process ... **A2:** 661-662
drawing and spinning ... **A2:** 665
effect, hydrogen solubility in magnesium ... **A15:** 462
effect of, on hardenability of steel ... **A1:** 413, 470
effect on maraging steels ... **A4:** 222
effects of, on notch toughness ... **A1:** 742
electron-beam welding ... **A6:** 854
in enamel cover coats ... **EM3:** 304
in enameling ground coat ... **EM3:** 304
epithermal neutron activation analysis ... **A10:** 239
etchants for ... **A9:** 498
evaporation fields for ... **A10:** 587
explosion welding ... **A6:** 304, 896 **M6:** 710
explosive-bonded to carbon steel plate ... **A9:** 155
explosive-bonded to titanium ... **A9:** 157
extrusion of tubing ... **A2:** 663
extrusion welding ... **M6:** 677
in ferrite ... **A1:** 408
in filler metals for active metal brazing ... **EM4:** 523
in filler metals for direct brazing ... **EM4:** 518-519
forging ... **A2:** 662
friction coefficient data ... **A18:** 71
friction welding ... **M6:** 722
friction-welded to titanium ... **A9:** 157
galvanic corrosion of ... **A13:** 717
gas tungsten arc welding ... **M6:** 182
as grain refiner, copper alloys ... **A15:** 96
grain-refining effect on magnesium-zirconium alloys ... **A4:** 899
gravimetric finishes ... **A10:** 171
grinding ... **A2:** 664
grinding of ... **A9:** 497
and hafnium ... **A2:** 661-669
heat tinting ... **A9:** 136
heat-affected zone fissuring in nickel-base alloys ... **A6:** 588
high frequency resistance welding ... **M6:** 760
high-energy neutron irradiation of ... **A10:** 234
history ... **A2:** 661
hot rolling ... **A2:** 663
hydride platelets in ... **A9:** 499-500
hydrochloric acid corrosion of ... **A13:** 1163
as inoculant ... **A15:** 105
inoculant for aluminum ... **A6:** 53
intergranular attack in ... **A9:** 499
interlayer material ... **M6:** 681
in limestone ... **EM4:** 379
liquid-liquid separation process ... **A2:** 661
machining ... **A2:** 664
macroexamination of ... **A9:** 498
in magnesium alloys ... **A9:** 427-428
in malleable iron ... **A1:** 10
melting ... **A2:** 662
in metal powder-glass frit method ... **EM3:** 305
metal processing ... **A2:** 661-662
metallurgy ... **A2:** 665-667
microexamination of ... **A9:** 498-499
mounting of ... **A9:** 497
in nickel-base superalloys ... **A1:** 984
nitric acid corrosion of ... **A13:** 1156
as nitride-forming ... **A15:** 93

Zirconium (continued)
nuclear grades, compositions and tensile properties ... **A2:** 665
organic precipitant for ... **A10:** 169
oxidation potentials and bonding ... **EM4:** 482
oxidized, single crystal sphere ... **A9:** 137
oxygen, effect of ... **M3:** 783
oxygen, role of ... **A2:** 667
in pharmaceutical production facilities ... **A13:** 1227
photometric analysis methods ... **A10:** 64
physical properties ... **A6:** 941
physical/mechanical properties ... **A13:** 707
plasma arc welding ... **A6:** 197
polycrystalline, different illuminations compared ... **A9:** 80
preferred orientation ... **M3:** 783
primary fabrication ... **A2:** 662-664
principal ASTM specifications for weldable nonferrous sheet metals ... **A6:** 400
properties ... **A6:** 629
pure ... **M2:** 826-831
pure, properties ... **A2:** 1175
recrystallization ... **M3:** 782
recrystallization in ... **A9:** 501
rod and wire ... **A2:** 663
SCC resistance ... **A13:** 328
secondary fabrication ... **A2:** 664-665
sectioning of ... **A9:** 497
shielding gas purity ... **A6:** 65
solid-state bonding in joining non-oxide ceramics ... **EM4:** 525, 528
solvent extractant for ... **A10:** 170
species weighed in gravimetry ... **A10:** 172
in steel ... **A1:** 147
suitability for cladding combinations ... **M6:** 691
sulfuric acid corrosion of ... **A13:** 1152-1153
in superalloys ... **A1:** 954, 989
thermal diffusivity from 20 to 100 °C ... **A6:** 4
thermal expansion coefficient ... **A6:** 907
TNAA analysis for ... **A10:** 234
as trace element, cupolas ... **A15:** 388
tube bending ... **A2:** 664-665
ultrapure, by electrotransport purification ... **A2:** 1094-1095
ultrapure, by external gettering ... **A2:** 1094
ultrapure, by iodide/chemical vapor deposition ... **A2:** 1094
ultrapure, by zone-refining technique ... **A2:** 1094
ultrasonic machining ... **EM4:** 359
ultrasonic welding ... **M6:** 746
strengths of welds ... **M6:** 752
use in flux cored electrodes ... **M6:** 103
vapor pressure, relation to temperature ... **A4:** 495
in vapor-phase metallizing ... **EM3:** 306
weighed as the phosphate ... **A10:** 171
welding ... **A2:** 665
in wrought heat-resistant alloys ... **A9:** 311
and zirconium alloys ... **A15:** 836-839

Zirconium alloys *See also* Reactive metals; Zirconium; Zirconium alloys, specific types ... **A6:** 787-788 **A9:** 497-501 **A13:** 707-721 **A16:** 844-857 **M3:** 783
abrasive grit/shot blasting ... **A15:** 838
analysis for manganese by periodate method ... **A10:** 69
anodizing ... **A9:** 142-143
applications ... **A2:** 667-669 **A6:** 787
brazed, corrosion testing ... **A13:** 885
casting technology development ... **A15:** 836
categories ... **A6:** 787
commercial grade ... **A6:** 787
reactor grade ... **A6:** 787
chemical compositions ... **A13:** 709
cleaning processes ... **M5:** 667
composition ... **A6:** 787
compositions and mechanical properties ... **A2:** 666
containing tin ... **A2:** 526
contour band sawing ... **A16:** 363
corrosion resistance, in various media ... **A13:** 711-715
delayed hydride cracking ... **A6:** 788
densities ... **A13:** 709
diffusion welding ... **A6:** 885, 886
electrochemical machining ... **A16:** 535
electrodes ... **A6:** 788

Zirconium alloys (continued)
electrolyte for anodizing ... **A9:** 143
electron-beam welding ... **A6:** 581, 787
electroplating of ... **M5:** 668
elements implanted to improve wear and friction properties ... **A18:** 858
filler metals ... **A6:** 788
finishing processes ... **M5:** 667-668
fretting wear ... **A18:** 248
grades ... **A2:** 667-669
heat-affected zones ... **A6:** 787
hot isostatic pressing ... **A15:** 838
hydride platelets in ... **A9:** 499-500
hydrogen damage ... **A13:** 171
hydrogen entry ... **A13:** 329
iron-chromium phase in ... **A9:** 500
machining ... **A15:** 838
mechanical properties ... **A13:** 709
melting ... **A15:** 837
melting point ... **A16:** 601
metallurgy of ... **A2:** 665-667
molds ... **A15:** 836-837
nuclear applications ... **A2:** 667-668
nuclear grades, compositions and tensile properties ... **A2:** 665
patterns ... **A15:** 836
photochemical machining ... **A16:** 588, 590
plasma arc welding ... **A6:** 198
postweld heat treatment ... **A6:** 788
process procedures ... **A6:** 788
cleaning before welding ... **A6:** 788
heat treatment ... **A6:** 788
preheating ... **A6:** 788
process selection ... **A6:** 787
processing ... **A15:** 837-838
rammed graphite molding ... **A15:** 273
recrystallization in ... **A9:** 500
resistance seam welding ... **A6:** 787
roll welding ... **A6:** 312
room temperature properties ... **A15:** 838
sand systems ... **A15:** 836-837
shell molds ... **A15:** 837
shielding gases ... **A6:** 787-788
shrink cavity in ... **A9:** 501
solid-state phase transformations in welded joints ... **A9:** 581
specifications ... **A2:** 669
structure/properties ... **A15:** 838-839
thermal expansion coefficient ... **A6:** 907
tin addition ... **M2:** 615-616
vacuum arc skull melting ... **A15:** 837
vacuum induction melting ... **A15:** 837
weld repair ... **A15:** 838
zirconium phosphide intermetallic particles in ... **A9:** 500

Zirconium alloys, specific types *See also* zirconium; Zirconium alloys
alloy 702, chemical analysis/mechanical properties ... **A15:** 838
alloy 702, microstructure ... **A15:** 837
alloy 702, tensile properties ... **A15:** 837
Grade 702, commercial pure zirconium ... **A2:** 668
Grade 704, characteristics ... **A2:** 668
Grade 705, characteristics ... **A2:** 668
Grade 706, characteristics ... **A2:** 668-669
reactor grade ... **A2:** 667-668
zinc-niobium, pickling ... **M5:** 667
Zircaloy 2 ... **A9:** 500
Zircaloy 4 ... **A9:** 500-501
Zircaloy 4, as-cast ingot, heat tinted ... **A9:** 155
Zircaloy alloys ... **A9:** 498
Zircaloy, as-cast, differential interference contrast ... **A9:** 159
Zircaloy, cleaning and finishing ... **M5:** 667-668
Zircaloy, corrosion resistance ... **A13:** 946
Zircaloy, embrittlement ... **A13:** 179
Zircaloy, forging, differential interference contrast ... **A9:** 159
Zircaloy-2 ... **M3:** 782, 783
hydrogen pickup in aqueous environments ... **M3:** 788, 789
organic coolants, corrosion in ... **M3:** 785, 790
steam, corrosion in ... **M3:** 786
water, corrosion in ... **M3:** 786
Zircaloy-2, hydrogen pickup vs. hydrogen overpressure ... **A13:** 708
Zircaloy-2, metallurgy ... **A2:** 667

Zirconium alloys, specific types (continued)
Zircaloy-2, pressurized water and steam corrosion ... **A13:** 708
Zircaloy-2 sheet, brazing corrosion data ... **A13:** 881-883
Zircaloy-4 ... **M3:** 783
hydrogen pickup in aqueous environments ... **M3:** 789
organic coolants, corrosion in ... **M3:** 790
water, corrosion in ... **M3:** 786-789
Zircaloy-4, hydrogen pickup vs. hydrogen overpressure ... **A13:** 708
Zircaloy-4, metallurgy ... **A2:** 667
Zircaloy-4, pressurized water/steam corrosion ... **A13:** 708
Zr-1.5Sn, as grade 704 ... **A2:** 668
Zr-1.5Sn, brazing ... **A6:** 945
Zr-1.5Sn-0.2Fe-0.1Cr, physical properties ... **A6:** 941
Zr-2.5Nb
heat treatment, effect on corrosion ... **M3:** 789-790
hydrogen pickup in aqueous environments ... **M3:** 789-790
organic coolants, corrosion in ... **M3:** 790
Zr-2.5Nb, as grade 705 ... **A2:** 668
Zr-2.5Nb, corrosion resistance ... **A2:** 668
Zr-2.5Nb, plate, element contamination ... **A9:** 157
Zr-5Be
in argon and vacuum atmospheres ... **A6:** 116
brazing ... **A6:** 945
contact angles on beryllium at various test temperatures
reactive metal brazing, filler metal ... **A6:** 945
test conditions effect on interfacial energies ... **A6:** 117
wetting of beryllium ... **A6:** 115
Zr-8Cr-8Ni, reactive metal brazing, filler metal ... **A6:** 945
Zr-29Mn, brazing ... **A6:** 945
Zr-34Ti-33V, brazing ... **A6:** 943
Zr-50Ag, brazing ... **A6:** 945
Zr7O2, corrosion rates ... **A13:** 716
Zr7O2, hydrochloric acid corrosion ... **A13:** 719
Zr7O2, mechanical requirements, pressure vessels ... **A13:** 710
Zr7O2, welded, in hydrochloric acid solution ... **A13:** 719
Zr7O5, mechanical requirements, pressure vessels ... **A13:** 710
Zr702 ... **A9:** 499
Zr705 ... **A9:** 499

Zirconium alloys, welding of
friction welding ... **A6:** 787
gas-metal arc welding ... **A6:** 787
gas-tungsten arc welding ... **A6:** 787
laser-beam welding ... **A6:** 787
plasma arc welding ... **A6:** 787
resistance spot welding ... **A6:** 787
resistance welding ... **A6:** 787

Zirconium boride (ZrB_2)
adiabatic temperatures ... **EM4:** 229
binary phase diagram ... **EM4:** 792
hardness ... **EM4:** 799
properties ... **EM4:** 793, 796, 797, 799
strength ... **EM4:** 800
synthesized by SHS process ... **EM4:** 229

Zirconium boride cermets
application and properties ... **A2:** 978, 1003

Zirconium boride-based cermets ... **A7:** 811, 812

Zirconium carbide
in niobium alloys ... **A9:** 441
properties ... **A18:** 795
thermal expansion coefficient ... **A6:** 907

Zirconium carbide cermets ... **A7:** 810-811
applications and properties ... **A2:** 1001-1002

Zirconium carbonitrides
in austenitic manganese steel castings ... **A9:** 239

Zirconium, commercially pure
corrosion resistance ... **A6:** 585

Zirconium copper *See also* Copper alloys, specific types, C15000; Copper and copper alloys
examination of oxides ... **A9:** 400
heat treating ... **M2:** 258-259

Zirconium dioxide (ZrO_2)
properties ... **A18:** 801

Zirconium dioxide (ZrO_2) (continued)
x-ray characterization of surface wear results for various microstructures ... **A18:** 469
Zirconium dioxide (ZrO_2)(Y_2O_3)
properties ... **A6:** 629
Zirconium dioxide as an interference film ... **A9:** 147-148
Zirconium dioxide/zirconium silicon oxide (ZrO_2/ $ZrSiO_4$)
methods used for synthesis ... **A18:** 802
Zirconium in steel ... **M1:** 115, 411, 417
notch toughness, effect on ... **M1:** 694
Zirconium nitride
synthesized by SHS process ... **EM4:** 229
Zirconium nitride in plate steels
examination for ... **A9:** 203
Zirconium oxide
as core coating refractory ... **A15:** 240
diffusion welding ... **A6:** 886
in niobium alloys ... **A9:** 441
vacuum heat-treating support fixture material ... **A4:** 503
Zirconium oxide (ZrO_2) *See also* Zirconia ... **EM4:** 32
abrasive machining ... **EM4:** 318, 320, 321, 322, 326
as additive for pressure densification ... **EM4:** 298-299
additive to improve chemical durability and enhance translucency ... **EM4:** 1085
additives ... **EM4:** 50-51
in alkali-resistant enamels ... **EM4:** 953
alloyed with aluminum oxide for tough abrasives ... **EM4:** 332
applications ... **EM4:** 20, 46, 47, 48, 50, 51, 586
ceramic/metal joints ... **EM4:** 515
ceramic/metal seals ... **EM4:** 502
chloride removed after successive washes of powder ... **EM4:** 93
coating formation in molten particle deposition ... **EM4:** 206
commercially spray-dried granules ... **EM4:** 103
component in photochromic ophthalmic and flat glass composition ... **EM4:** 442
composition ... **EM4:** 46
corrosion resistance of refractories ... **EM4:** 391
fused ... **EM4:** 50, 51
hardness ... **EM4:** 351
high-purity stabilized ... **EM4:** 47
hot isostatic pressing ... **EM4:** 197
key products ... **EM4:** 48
Knoop hardness ... **EM4:** 30
mechanical properties ... **EM4:** 316
melting/fining ... **EM4:** 391, 392
metastable phase ... **EM4:** 25
non-oxide ceramic joining ... **EM4:** 480
partially stabilized ... **EM4:** 758
phase analysis ... **EM4:** 25
physical properties ... **EM4:** 316
Poisson's ratio ... **EM4:** 30
properties ... **EM4:** 30
radioactivity ... **EM4:** 50
raw materials ... **EM4:** 48
sintering ... **EM4:** 266
specific properties impared in CTV tubes ... **EM4:** 1039
spherical media composition for wet-milling ... **EM4:** 78
strength and fracture toughness ... **EM4:** 586
superplasticity ... **EM4:** 301
supply sources ... **EM4:** 46
ultrasonic machining ... **EM4:** 360
world production ... **EM4:** 50
Zirconium oxide cermets
applications and properties ... **A2:** 993
Zirconium oxide coating
molybdenum ... **M5:** 662
Zirconium oxide, partially stabilized ... **EM4:** 19
material selection for structural ceramics ... **EM4:** 29
Zirconium oxide-based cermets ... **A7:** 803
Zirconium powder
uses for ... **A9:** 497
Zirconium powders and powder alloys
chemical analysis and sampling ... **A7:** 249
early lamp filaments ... **A7:** 16
flammability in dust clouds ... **A7:** 195
high explosivity ... **A7:** 196
particle size requirements ... **A7:** 603
pellets, fragmentation device using ... **A7:** 691
powder metallurgy ... **A7:** 18
pyrophoricity ... **A7:** 199
pyrotechnic chemical requirements ... **A7:** 603
pyrotechnic requirements ... **A7:** 601-602
as reactive material, properties ... **A7:** 597
tap density ... **A7:** 277
toxicity and exposure limits ... **A7:** 207
in uranium dioxide fuel rods ... **A7:** 664
vacuum sintering atmospheres for ... **A7:** 345
Zirconium silicate
abrasive in commercial prophylactic paste ... **A18:** 666, 669
as filler ... **EM3:** 179
as filler for solder glass ... **EM4:** 1072
melting/fining ... **EM4:** 391, 392
as molding sand, characteristics ... **A15:** 209
as refractory, core coatings ... **A15:** 240
as refractory filler ... **EM4:** 1072
Zirconium silicon oxide ($ZrSiO_4$)
properties ... **A18:** 801
Zirconium titanium stannate (ZTS)
applications ... **EM4:** 48
key product properties ... **EM4:** 48
raw materials ... **EM4:** 48
Zirconium, vapor pressure
relation to temperature ... **M4:** 309, 310
Zirconium, zirconium alloys, heat treating
annealing ... **M4:** 787
processing ... **M4:** 787
Zirconium, zone refined
impurity concentration ... **M2:** 713
Zirconium-aluminum alloys
peritectic and peritectoid reactions ... **A9:** 678
Zirconium-arc light sources for microscopes ... **A9:** 72
Zirconium-barium lanthanum-aluminum fluoride (ZBLA) glasses
optical properties ... **EM4:** 854
Zirconium-base corrosion-resistant alloys, selection of ... **A6:** 598, 599
applications ... **A6:** 599
microstructure ... **A6:** 599
thermal expansion coefficient ... **A6:** 599
welding characteristics ... **A6:** 599
Zirconium-copper alloys
age hardenable ... **A2:** 236
applications and properties ... **A2:** 280-281
work hardening ... **A2:** 230
Zirconium-hafnium alloys
corrosion resistance ... **A13:** 720
Zirconium-niobium alloys
anodizing of ... **A9:** 498
weldment in ... **A9:** 499
Zirconium-opacified porcelain enamel
composition of ... **M5:** 510
Zirconium-oxide, commercial
powder washing effect on sintered microstructure ... **EM4:** 92
Zirconium-oxide materials
decomposition control ... **EM4:** 194
Zirconium-oxide-based ceramics
for oxygen sensors ... **EM4:** 17
Zirconium-titanium alloy
anodizing ... **A9:** 143
ZMC
as thermosetting process ... **EM2:** 287
ZOLZ *See* Zero-order Laue zone
ZOLZ-CBEDPs *See* Zero-order Laue zone
Zone ... **EL1:** 2
cupola ... **A15:** 389
defined ... **A9:** 19
melting, defined ... **A15:** 12
package exterior, failure mechanisms ... **EL1:** 1006-1008
package interior, failure mechanisms ... **EL1:** 1008-1013
of strain localization, medium-carbon steel ... **A12:** 42
transformed, medium-carbon steel ... **A12:** 42
welding, cast irons ... **A15:** 522
Zone annealing
mechanically alloyed oxide dispersion-strengthened (MA ODS) alloys ... **A2:** 947
Zone axes of crystals ... **A9:** 710
Zone axis patterns
use in identifying unknown phases/ particles ... **A10:** 456-457
Zone axis patterns (ZAPS) ... **A18:** 387
Zone definition
zero-risk analysis ... **EL1:** 139
Zone diagram, coupled
aluminum-silicon alloys ... **A15:** 162, 164
Zone formation
wrought aluminum alloy ... **A2:** 39
Zone, heat-affected
abbreviation ... **A8:** 724
Zone heaters
directional solidification/single- crystal furnaces ... **A15:** 400
Zone location, vs. point location
acoustic emission inspection ... **A17:** 292
Zone plates
microwave holography ... **A17:** 224
Zone refining
as ultrapurification technique ... **A2:** 1093-1094
Zone refining for purifying metals ... **M2:** 710
Zone rules *See also* Control charts
for control chart analysis ... **A17:** 730-732
Zone sintering ... **A7:** 13, 346-348
furnace ... **A7:** 346-348
Zone temperatures
wave soldering preheaters ... **EL1:** 686-688
Zone theory *See also* Band theory
applications ... **EL1:** 98
and band theory ... **EL1:** 97
Zone(s)
brittle reaction, composites, acoustic emission inspection ... **A17:** 288
dead, calibration for ultrasonic inspection ... **A17:** 267
fatigue-fracture ... **A11:** 109-110
final-fracture ... **A11:** 104-105
heat-affected (HAZ), defined ... **A11:** 5
of high deformation ... **A11:** 102
plastic, at crack tip ... **A17:** 287
remote-field vs. exciter coil/coupling ... **A17:** 195
Zone-refined
abbreviation for ... **A10:** 691
nickel rod, sulfur segregation in ... **A10:** 562
Zone-refined lead
composition ... **A2:** 544
Zoned atmospheres ... **A7:** 346-348
nitrogen-based ... **A7:** 346, 347
Zoomscope sight
for leaded brass reticle mount ... **A7:** 738-739
Zr_3Al trialuminides
properties ... **A2:** 932
ZRBSC-D
erosion test results ... **A18:** 200
ZRBSC-M
erosion test results ... **A18:** 200
ZTA *See* Zirconia-toughened alumina